EDITORS FOR THE CHEMICAL RUBBER CO.

Handbook
OF
Chemistry and Physics

A Ready-Reference Book of Chemical and Physical Data

FORTY-NINTH EDITION

EDITOR

ROBERT C. WEAST, Ph.D.

Vice President, Research, Consolidated Natural Gas Service Company, Inc.
Formerly Professor of Chemistry at Case Institute of Technology

In collaboration with a large number of professional chemists and physicists whose assistance is acknowledged in the list of general collaborators and in connection with the particular tables or sections involved.

Published by

THE CHEMICAL RUBBER CO.

18901 Cranwood Parkway, Cleveland, Ohio, 44128

Library of Congress Card No. 13-11056

PRINTED IN U. S. A.

Preface to the Forty-Ninth Edition

Revision of a single volume reference book which has been previously revised forty-seven times since the year 1913 is a challenge. Valuable information is not going out of date as rapidly as new, useful information and data are being developed. Thus, maintaining a book within a reasonable physical size presents a problem. The book must be maintained within certain bounds even though scientists demand, and deservedly so, more extensive tables of the highest accuracy in order that the previously existing frontiers of scientific knowledge can be exceeded.

Computers and calculators are commonplace today. This gives the editor serious concern as to the advisability of continuing the rather extensive section of mathematical tables in the Handbook of Chemistry and Physics. On the other hand, presence of computors and calculators does not dictate that all of the programs and data necessary for such programs are available in all of the laboratories. For that reason the decision was made not to reduce the mathematical content of the Handbook for the forty-ninth edition.

The Handbook of Chemistry and Physics cannot list all types of data that are useful to all areas of the physical sciences; space does not permit this. However, in this present edition there is a list of all laboratories, data evaluation centers, data compilation centers, etc. which are cooperating with the Office of Critical Tables. This rather lengthy list is located at the back of the Handbook immediately preceding the index. This list presents locations where critical data are being generated, evaluated or compiled. The availability of such data is also indicated in most instances.

This forty-ninth edition contains approximately two hundred pages of new information. Information on these pages includes: X-Ray Crystallographic Data of Minerals, X-Ray Atomic Energy Levels, X-Ray Wavelengths of the Elements, Tables of the Properties of Semiconductors, Heat Capacity of Some Common Organic Compounds, Physical Properties of Pigments, Properties of Some Commercial Plastics, Sources of Critical Data, Selected Values of Dipole Moments for Molecules in the Gas Phase, Volume Properties of Ordinary Water, Density of Deuterium Oxide, Viscosity of Water from 0 to 100C, certain tables from the ASME 1967 Steam Tables, and some new Definitions and Formulas.

In keeping with past practice, certain of the tables have been revised by use of more recent data and information. Among these tables are: Critical Temperatures and Pressures, Properties of Elementary Particles, Nuclear Spins, Moments and Magnetic Resonance Frequencies, Specific Gravity of Aqueous Invert Sugar Solutions, Derived Units from Fundamental Units, Definitions and Formulas, and Far Infrared Correlation Charts.

In addition to the revisions and additions, corrections have been made of all known errors which were in the 48th and earlier editions. Sincere gratitude is expressed to all who have written regarding these errors or who have made suggestions for further improvement of the book.

Robert C. Weast

May 1968

Physics Editorial Board

Collaborators and Contributors

ALBERTS, L., PH.D.
Department of Physics
Rand Afrikaans University
Johannesburg, South Africa

ALBRIGHT, J. G., PH.D.
Department of Physics
Westminster College
New Wilmington, Pennsylvania

ALLEN, GABRIEL, PH.D.
National Aeronautics and Space Administration
Cleveland, Ohio

ANDERSON, WESTON A., PH.D.
Varian Associates
Palo Alto, California

APFELBAUM, PERCY M., PH.D.
College of the City of New York

†ARNOLD, ERIC. A., PH.D.
Professor of Chemistry (Emeritus)
Case Western Reserve University
Cleveland, Ohio

ASKILL, JOHN, PH.D.
Chairman, Physics Dept.
Millikin University
Decatur, Illinois

BACON, CHARLES S., JR., PH.D.
Case Western Reserve University
Cleveland, Ohio

BAILAR, JOHN C., JR., PH.D.
Department of Chemistry
University of Illinois
Urbana, Illinois

BAXTER, ROBERT A., M.S.
Professor of Chemistry
Colorado School of Mines
Golden, Colorado

BEARDEN, J. A., PH.D.
The Johns Hopkins University
Baltimore, Maryland

BENADE, ARTHUR H., PH.D.
Department of Physics
Case Western Reserve University
Cleveland, Ohio

BENNETT, GEORGE W., PH.D.
Grove City College
Grove City, Pennsylvania

†Retired.

BENTLEY, FREEMAN F.
Air Force Materials Laboratory
Wright-Patterson Air Force Base, Ohio

BILLMAN, JOHN H., PH.D.
Indiana University
Bloomington, Indiana

BLOOM, ARNOLD L., PH.D.
Spectra-Physics, Inc.
Mountain View, California

BRADFORD, JOHN R., PH.D.
Dean
College of Engineering
Lubbock, Texas

BRADLEY, JAMES A., A.M.
Newark College of Engineering
Newark, New Jersey

BROWN, BANCROFT H., PH.D.
Dartmouth College
Hanover, New Hampshire

BROWN, MORDEN G., PH.D.
Research Supervisor
American Optical Company
Southbridge, Mass.

BURG, ANTON B., PH.D.
University of Southern California
Los Angeles, California

BURNS, G. PRESTON, PH.D.
Mary Washington College
Fredericksburg, Virginia

BURR, A. F., PH.D.
The Johns Hopkins University
Baltimore, Maryland

CARLETON, RALPH K., PH.D.
Boston College
Chestnut Hill, Mass.

CATCHPOLE, J. P.
Parlin Research Laboratory
E. I. DuPont de Nemours & Co.
Parlin, New Jersey

COOKE, GILES B., PH.D.
Chairman Science Dept.
Essex Community College
Essex, Maryland

COTTLE, D. L., PH.D.
Esso Research and Engineering Company
Linden, New Jersey

COLLABORATORS AND CONTRIBUTORS

Cunningham, B. B., Ph.D.
Lawrence Radiation Laboratory
University of California
Berkeley, California

Currie, Lauchlin, M., Ph.D.
Consultant
New York City

Davies, Mansel, Ph.D.
Edward Davies Laboratories
University of Wales
Aberystwyth, Wales

Degering, E. F., Ph.D.
U.S. Army Natick Laboratories
Pioneering Research Division
Radiation Chemistry Laboratory
Natick, Mass.

De Ment, Jack, D.Sc.
Dement Laboratories
Portland, Oregon

†Deming, H. G., Ph.D.
Sarasota, Florida

DeVries, Thomas, Ph.D.
Professor of Analytical Chemistry
Purdue University
Lafayette, Indiana

DiCyan, Erwin, Ph.D.
Consulting Chemist
New York, N.Y.

Donnay, J. D. H., Ph.D.
The Johns Hopkins University
Baltimore, Md.

Dunlop, A. P., Ph.D.
John Stuart Research Laboratories
The Quaker Oats Co.
Barrington, Illinois

Dunn, Max S., Ph.D.
Professor of Biochemistry
University of California
Los Angeles, California

Foster, L. M., Ph.D.
Thomas J. Watson Research Center
IBM Corporation
Yorktown Heights, New York

Foster, Laurence S., Ph.D.
U.S. Army Materials Research Agency
Watertown, Mass.

Fulford, G.
Parlin Research Laboratory
E. I. DuPont de Nemours & Co.
Parlin, New Jersey

Furtsch, E. F., Ph.D.
Virginia Polytechnic Institute
Blacksburg, Virginia

Gettens, R. J., M.A.
Freer Gallery of Art
Smithsonian Institution
Washington, D.C.

Gilbert, Paul T., Jr.
Beckman Instruments
Fullerton, California

Gillette, L. A., Ph.D.
Pennsalt Chemicals Corporation
Wyandotte, Michigan

Gilman, Henry, Ph.D.
Iowa State University
Ames, Iowa

Girling, Brian, M.Sc., F.I.M.A.
The City University
London, England

Gregg, E. C., Ph.D.
Case Western Reserve University
Cleveland, Ohio

†Hale, Harrison, Ph.D.
Marietta, Georgia

Hammond, C. R.
West Hartford, Conn.

Harris, W. E., Ph.D.
University of Alberta
Edmonton, Alberta, Canada

Harwood, H. J., Ph.D.
Durkee Famous Foods
Chicago, Illinois

Heath, R. L., Ph.D.
Atomic Energy Division
Phillips Petroleum Co.
Idaho Falls, Idaho

Hoffman, R. W., Ph.D.
Case Western Reserve University
Cleveland, Ohio

Hurd, C. D., Ph. D.
Northwestern University
Evanston, Illinois

Kassner, J. L., Ph.D.
School of Science
Department of Physics
The University of Missouri at Rolla
Rolla, Missouri

Kennard, Olga, Ph.D.
University Chemical Laboratory
Cambridge, England

Kennard, T. G., Ph.D.
Glendora, California

Kerr, J. A., Ph.D.
University College of Wales
Aberystwyth, Wales

King, Allen L., Ph.D.
Dartmouth College
Hanover, New Hampshire

Kinney, C. R., Ph.D.
Des Moines, Iowa

Kretz, Ralph, Ph.D.
Department of Chemistry
Imperial College of Science
London, England

†Retired

COLLABORATORS AND CONTRIBUTORS

Lang, Gernot
Austria

Lawden, D. F., Sc.D.
University of Aston
Birmingham, England

Lee, Kenneth, Ph.D.
Varian Associates
Palo Alto, California

Levitt, Albert P.
U.S. Army Materials Research Agency
Watertown, Mass.

MacMasters, M. M., Ph.D.
Department of Grain Science and Industry
Kansas State University
Manhattan, Kansas

Magill, Mary A., Ph.D.
Chemical Abstract Service
Ohio State University
Columbus, Ohio

Mahlig, Walter
W. S. Tyler Co.
Cleveland, Ohio

Mathers, F. C., Ph.D.
Bloomington, Indiana

Meites, Louis, Ph.D.
Polytechnic Institute of Brooklyn, N.Y.

Mellon, M. G., Ph.D.
Purdue University
Lafayette, Indiana

Michaelson, H. B.
International Business Machines Corp.
Poughkeepsie, New York

Morgan, Wm. M., Ph.D.
Mount Union College
Alliance, Ohio

McGowan, John C., D.Sc.
Plastics Division
Imperial Chemical Industries, Ltd.
Hertfordshire, England

Navias, Louis, Ph. D.
Schenectady, New York

Nimmo, R. R., Ph.D.
University of Otago
New Zealand

Norton, Francis J., Ph.D.
General Electric Company
Schenectady, New York

Pamplin, B. R., M.A., Ph.D.
Bath University of Technology
Bristol, England

Parks, W. G., Ph.D.
University of Rhode Island
Kingston, Rhode Island

Patai, Saul, Ph. D.
Hebrew University of Jerusalem
Jerusalem, Israel

Pearl, Irwin A., Ph.D.
Institute of Paper Chemistry
Appleton, Wisconsin

Peart, R. F., Ph.D.
University of Illinois
Urbana, Illinois

Peed, Allie C., Jr.
Eastman Kodak Co.
Rochester, N.Y.

Pepinsky, Raymond, Ph.D.

Peters, F. N., Ph.D.
Quaker Oats Co.
Barrington, Illinois

Potts, Wm. M., Ph.D.
A and M College of Texas
College Station, Texas

Prebluda, H. J., Ph.D.
U.S. Industrial Chemicals Co.
New York City

Prettyman, Irven B., M.S.
The Firestone Tire and Rubber Co.
Akron, Ohio

Prutton, Carl F., Ph.D.
Food Machinery Corp.
New York City

Rappoport, Zvi, Ph.D.
Hebrew University of Jerusalem
Jerusalem, Israel

Ratcliffe, E. H.
National Physical Laboratory
Teddington, Middlesex, England

Ray, Francis E., Ph.D.
University of Florida
Gainesville, Florida

Reck, R. A., Ph.D.
Armour Industrial Chemical Co.
Chicago, Illinois

Reed, Marion C., Ph.D.
Mountainside, New Jersey

Reines, Frederick, Ph.D.
Department of Physics
Case Western Reserve University
Cleveland, Ohio

Roberts, B. W., Ph.D.
General Electric Research Laboratory
Schenectady, New York

Roberts, R. Chester, Ph.D.
Professor Emeritus of Chemistry
Colgate University
Hamilton, New York

Robinson, R. A., Ph.D.
National Bureau of Standards
Washington, D.C.

Rosen, R. J.
Los Angeles, California

COLLABORATORS AND CONTRIBUTORS

ROSENDAHL, EDWARD
Bennett-Rosendahl Co., Inc.
New York, N.Y.

ROZEK, ADELE L.
Velsicol Chemical Corporation
Chicago, Illinois

SCOTT, JANET D., M.S.
South Kent, Conn.

SEABORG, GLENN T., PH.D.
U.S. Atomic Energy Commission
Washington, D.C.

SHANKLAND, ROBERT S., PH.D.
Department of Physics
Case Western Reserve University
Cleveland, Ohio

SHELTON, J. R., PH.D.
Department of Chemistry
Case Western Reserve University
Cleveland, Ohio

SMITH, GEORGE W., PH.D.
General Motors Corp.
Warren, Michigan

SMITH, J. M., B.S.E.E.
General Electric Co.
Cleveland, Ohio

SMITHSON, LEE D.
Air Force Materials Laboratory
Wright-Patterson Air Force Base, Ohio

SPEDDING, F. H., PH.D.
Iowa State University
Ames, Iowa

SPENCER, C. C., PH.D.
Syracuse, New York

STOKES, R. H., PH.D.
University of New England
Armidale, Australia

TARPINIAN, ARAM
Lee Sons-Moos Laboratories
Great Neck, L.I., New York

TOMLIN, D. H., PH.D.
University of Reading
Reading, England

TROTMAN-DICKENSON, A. F., PH.D.
University College of Wales
Aberystwyth, Wales

VAN DER MEULEN, P. A., PH.D.
Rutgers University
New Brunswick, N.J.

VICKERY, HUBERT B., PH.D.
Connecticut Agricultural Exp. Station
New Haven, Conn.

WENDLANDT, WESLEY W., PH.D.
Texas Technological College
Lubbock, Texas

WHETTEN, N. R., PH.D.
General Electric Research Laboratories
Schenectady, New York

WOLF, A. V., PH.D.
University of Illinois
College of Medicine
Chicago, Illinois

WOOSTER, CHARLES B., PH.D.
Ciba Products Corporation
Kimberton, Pennsylvania

YOE, JOHN H., PH.D.
Professor Emeritus of Chemistry
University of Virginia
Charlottesville, Virginia

YOHE, G. R., PH.D.
Illinois State Geological Survey
University of Illinois
Urbana, Illinois

YOUNG, THOMAS FRASER, PH.D.
Division of Chemical Engineering
Argonne National Laboratory
Argonne, Illinois

ZABICKY, J., PH.D.
Hebrew University of Jerusalem
Jerusalem, Israel

ZUFFANTI, SAVERIO, A.M.
Northeastern University
Boston, Mass.

The Publishers and Editors will be grateful to readers of this Handbook who will call their attention to errors which may be discovered. Suggestions for improvement are also welcome.

Contents

CONTENTS

MATHEMATICAL TABLES

SECTION A (Continued)

CONTENTS

MATHEMATICAL TABLES
SECTION A (Continued)

CONTENTS

ELEMENTS AND INORGANIC COMPOUNDS
SECTION B

CONTENTS

ORGANIC COMPOUNDS
SECTION C

CONTENTS

GENERAL CHEMICAL
SECTION D

CONTENTS

GENERAL CHEMICAL
SECTION D (Continued)

CONTENTS

GENERAL PHYSICAL CONSTANTS
SECTION E

CONTENTS

GENERAL PHYSICAL CONSTANTS

SECTION E (Continued)

CONTENTS

GENERAL PHYSICAL CONSTANTS

SECTION E (Continued)

CONTENTS

MISCELLANEOUS
SECTION F

CONTENTS

MISCELLANEOUS
SECTION F (Continued)

CONTENTS

MISCELLANEOUS

SECTION F (Continued)

Fire Precautions and Chemical Hazards

Acetone.—Dilute with a spray of water to avoid spread of burning liquid. Use suitable gas mask.

Alcohol.—See under acetone.

Ammonia.—Use water and dilute acid. Use suitable gas mask.

Benzol or Benzene.—Use water to cool containers which are endangered; extinguish flame with sand, earth, fire-foam or carbon tetrachloride fire extinguishers. Use suitable gas mask.

Calcium Carbide.—Do not use water as this generates acetylene, an inflammable and explosive gas; cut off electric current to avoid ignition of gas. Remove containers to a dry place. Use gas mask.

Carbon Disulfide.—Use water to cool containers which are endangered; extinguish blaze with sand, earth, fire-foam or carbon tetrachloride fire extinguishers. Use suitable gas mask.

Carbon Tetrachloride.—When a carbon tetrachloride type extinguisher is used on a fire in a confined space, the fire should be attacked from outside the enclosure, if possible, or the area should be vacated immediately after the fire is out. No one should return to the enclosure until the air is cleared of smoke and fumes. These precautions should be observed regardless of the means by which the fire is extinguished, however, since fire in a confined space rapidly produces a toxic atmosphere.

Do not put carbon tetrachloride on a sodium fire, violent explosions may be caused.

Celluloid.—Use large volumes of water and sand. The smoke contains oxides of nitrogen which are injurious. Use suitable gas mask.

Chlorine.—Spray with water. The pungent nature of the gas makes the use of a gas mask imperative.

Collodion.—See under carbon disulfide.

Ether.—See under carbon disulfide.

Gasoline.—See under carbon disulfide.

Hydrochloric Acid.—Use large volumes of water also chalk or soda. Use gas mask.

Hydrocyanic or Prussic Acid.—Suitable gas mask is essential because of the extremely poisonous nature of the vapors. Provide ventilation.

Lacquer Solvents.—See under carbon disulfide.

Magnesium.—Do not use water. Use sand or earth to extinguish flames. Remove containers to a dry place.

Nitric Acid and Oxides of Nitrogen.—Use large volumes of water. Do not use sand or earth. Use gas mask.

Potassium.—Do not use water. Remove containers to a dry place. Extinguish flames with sand or earth. For storage, potassium is kept immersed in petroleum.

Potassium Hydroxide.—Use large volumes of water or dilute acids.

Phosphorus.—Use water and wet sand. Use gas mask. For storage, white phosphorus must be kept immersed in water. Red phosphorus is less dangerous.

Sodium.—See under potassium.

Sodium Hydroxide.—See under potassium hydroxide.

Sulfur.—Extinguish with water or sand. Use gas mask.

Sulfuric Acid.—See under hydrochloric acid.

Turpentine.—See under acetone.

USE OF MATHEMATICAL TABLES

For a complete discussion of the principles and use of mathematical tables, textbooks on the subject should be consulted. The following brief statements are intended to give only sufficient information to make possible the intelligent use of the tables, omitting for the most part any attempt at treating the theory and principles.

Exponential Method of Expressing Numbers—For convenience in writing and manipulation, numbers are often expressed as factors of appropriate powers of 10. The following examples will illustrate:

2,380,000,000.	may be written	2.38×10^9
238.	may be written	2.38×10^2
.238	may be written	2.38×10^{-1}
.000000238	may be written	2.38×10^{-7}

Logarithms—The logarithm of a number is the exponent of that power to which another number, the base, must be raised to give the number first named. Any positive number greater than 1 might serve as a base. Two have been selected, yielding two systems of logarithms. One base, 2.718 . . . . usually indicated by the letter e, gives rise to a system of logarithms convenient in higher mathematics. These are called natural, Napierian, or hyperbolic logarithms. Reference will be made to their use in a subsequent paragraph.

The other base used is 10, giving logarithms particularly adapted to use in computation, called common or Briggsian logarithms. Tables of logarithms given without designation are invariably of this latter type.

Since most numbers are irrational powers of ten, a common logarithm, in general, consists of an integer which is called the characteristic and an endless decimal, the mantissa.

It is to be observed that the common logarithms of all numbers expressed by the same figures in the same order with the decimal point in different positions have different characteristics but the same mantissa. To illustrate:—if the decimal point stands after the first figure of a number, counting from the left, the characteristic is 0; if after two figures, it is 1; if after three figures, it is $\bar{2}$, and so forth. If the decimal point stands before the first significant figure the characteristic is -1, usually written $\bar{1}$; if there is one zero between the decimal point and the first significant figure it is 2 and so on. For example: $\log 256 = 2.40824$, $\log 2.56 = 0.40824$, $\log 0.256 = \bar{1}.40824$, $\log 0.00256 = \bar{3}.40824$. The two latter are often written $\log 0.256 = 9.40824 - 10$, $\log 0.00256 = 7.40824 - 10$.

A method of determining characteristics of logarithms is to write the number with one figure to the left of the decimal point multiplied by the appropriate power of 10. The characteristic is then the exponent used. For example:

$$256{,}000{,}000 = 2.56 \times 10^8 \qquad \log = 8.40824$$
$$0.000000256 = 2.56 \times 10^{-7} \qquad \log = \bar{7}.40824 \text{ or } 3.40824 - 10$$

Inasmuch as the characteristic may be determined by inspection the mantissas only are given in tables of common logarithms.

To find the logarithm of a number:

For a number of four figures, take out the tabular mantissa on a line with the first three figures of the number and under its fourth figure. The characteristic is determined as previously explained.

For a number of less than four figures, supply zeros to make a four figure number and take the value of the mantissa from the tables as before. For example: log 2 = log 2.000 = 0.30103.

For a number of more than four figures, take the tabular value of the mantissa for the first four figures; find the difference between this mantissa and the next greater tabular mantissa and multiply the difference so found by the remaining figures of the number as a decimal and add the product to the mantissa of the first four figures. For example: to find log 46.762.

$$\log 46.76 = 1.66987$$

Tabular difference between this mantissa and that for 4677 is .00010.

$$\begin{aligned} \therefore \log 46.762 &= 1.66987 + .2 \times .00010 \\ &= 1.66987 + .00002 \\ &= 1.66989 \end{aligned}$$

To find the number corresponding to a given logarithm:

If the mantissa is found exactly in the table, join the figure at the top which is directly above the given mantissa to the three figures on the line at the left and place the decimal point according to the characteristic of the logarithm. For example, $\log^{-1}$ (antilogarithm) 3.39967 = 2510.

If the mantissa is not found exactly in the table it is necessary to interpolate. For example, $\log^{-1} 3.40028 = 2513. + \frac{9}{18} = 2513.5$.

The column of proportional parts at the right of each page of the table shows, under the heading of the various tabular differences, the parts of these differences which correspond to the digits from 1 to 9 in the fifth place. This makes it possible to take out a logarithm for a five figure number or to find an antilogarithm of the same number of significant figures with increased facility, usually by inspection.

The following formulae express the relations on which the use of logarithms is based:

$$\log ab = \log a + \log b$$

$$\log \frac{a}{b} = \log a - \log b$$

$$\log a^n = n \times \log a$$

$$\log \sqrt[n]{a} = \frac{\log a}{n}.$$

The following examples will serve as illustrations:

1. $52600 \times 0.00381 \times 2.74 = 549.1$

$$\begin{aligned} \log 52600 &= 4.72099 \\ \log 0.00381 &= \bar{3}.58092 \\ \log 2.74 &= 0.43775 \\ \hline \text{Sum:} &= 2.73966 \\ \text{Antilogarithm} &= 549.1 \end{aligned}$$

The sum is the logarithm of the product, the mantissa of which is 73966. On looking up this mantissa in the logarithm tables we see that it corresponds to the digits 5491. The characteristic is 2, hence there are three figures before the decimal point. The number corresponding to the logarithm, called the antilogarithm, is 549.1.

2. $0.00123 \div 52.7 = 0.00002334$

$$\begin{array}{ll} \log 0.00123 = \bar{3}.08991 \\ \log \quad 52.7 = 1.72181 \\ \hline \text{Subtracting} \quad \bar{5}.36810 \\ \text{Antilog} \quad 0.00002334 \end{array}$$

An Alternative method:

$$\begin{array}{l} \log 0.00123 = 7.08991 - 10 \\ \log \quad 52.7 = 1.72181 \\ \hline 5.36810 - 10 \end{array}$$

The characteristic $\bar{5}$ (5. -10) shows four zeros after the decimal point before the first significant figure.

3. $\dfrac{273 \times 780}{292 \times 760} \times 15 \times 0.09 = 1.295$

$$\begin{array}{l} \log 273 = 2.43616 \\ \log 780 = 2.89209 \\ \log \quad 15 = 1.17609 \\ \log 0.09 = \bar{2}.95424 \\ \hline \log \text{sum} = 5.45858 \end{array}$$

$$\begin{array}{l} \log 292 = 2.46538 \\ \log 760 = 2.88081 \\ \hline \log \text{denominator} = 5.34619 \end{array}$$

$$\begin{array}{ll} \text{log numerator} & = 5.45858 \\ \text{log denominator} & = 5.34619 \\ \hline \text{subtracting} & = 0.11239 \\ \text{antilogarithm} & = 1.295 \end{array}$$

As division may be accomplished by multiplying by the reciprocal of a number, the above may be considerably simplified. The logarithm of the reciprocal of a number, called the cologarithm, is readily obtained from the table by subtracting the logarithm of the number from zero. This may readily be read off from the table of mantissas. Change the sign of the characteristic algebraically adding to it -1, then mentally subtract each figure of the mantissa from 9 proceeding from left to right, the last figure being subtracted from 10. The example then is:

$$\begin{array}{ll} \log \quad 273 & = 2.43616 \\ \log \quad 780 & = 2.89209 \\ \log \quad 15 & = 1.17609 \\ \log \quad 0.09 & = \bar{2}.95424 \\ \text{colog } 292 & = \bar{3}.53462 \\ \text{colog } 760 & = \bar{3}.11919 \\ \hline & \quad 0.11239 \end{array}$$

4. $(0.00098)^4 = 9.224 \times 10^{-13}$

$$\begin{array}{lll} \log 0.00098 & = \bar{4}.99123 & \\ & \quad\quad\quad 4 & \\ \hline & \quad 3.96492 & (a) \\ \bar{4} \times 4 & \quad \overline{16}. & (b) \\ \hline \log (0.00098)^4 & = \overline{13}.96492 & (c) \\ \text{antilog} = 9.224 \times 10^{-13} & & \end{array}$$

An alternative method:

$$\begin{array}{r} \log 0.00098 = 6.99123 - 10 \\ 4 \\ \hline 27.96492 - 40 \\ \text{or } 7.96492 - 20 \\ \text{or } \overline{13}.96492 \\ \text{antilog} = 9.224 \times 10^{-13} \end{array}$$

In the above it will be noted that the mantissa is always positive hence the multiplication of the mantissa shown at (a) while (b) shows the multiplication of the characteristic. (c) is the algebraic sum.

5. $\sqrt[5]{492} = 3.455$
$\log 492 = 2.69197$

Dividing the logarithm by 5 gives as the logarithm of the root 0.53839 the antilogarithm of which is 3.455 both characteristic and mantissa being positive. When the characteristic is negative and not evenly divisible by the root to be taken a modification of the logarithm is necessary.

6. $\sqrt[3]{0.000372} =$

$$\log 3.72 \times 10^{-4} = \bar{4}.57054 \qquad \text{(a)}$$
$$= 26.57054 - 30 \quad \text{(b)}$$

dividing (b) by 3 gives 8.85685 − 10 which may be written $\bar{2}.85685$ and is the logarithm of the root sought, the antilogarithm of which is 0.07192.

7. $0.000372^{1.2} = 0.000076674$
$\log 0.000372 = \bar{4}.57054$
or $6.57054 - 10$
1.2
$7.88465 - 12$
antilogarithm 0.000076674

8. $(0.000372)^{-1.32} = 33642$
colog 0.000372 = 3.42946
1.32
4.52689
antilogarithm 33642

Four-Place Logarithms—This short table on two facing pages makes possible logarithmic computation precise to four significant figures, (three without interpolation). The mantissa is given complete and the proportional parts indicated for each line.

Four-Place Antilogarithms—Some computers prefer to use separate tables for determining antilogarithms; the table being entered from the margins with the logarithm and the number being found in the body of the table. Such a table is given to accompany the four-place logarithms.

Five-Place Logarithms—For computation involving five significant figures, (four without interpolation) the five-place table will be adequate. Since the first two figures will be the same for several lines of the table they are given in the first line only. The point at which these first two figures change is indicated by an asterisk. While space does not permit the proportional parts for each line, tables will be found for each tabular difference.

The supplementary table following the five-place logarithms, giving seven-place logarithms for numbers of five significant figures from 10,000 to 12,000 will be found convenient to increase precision and avoid the inconvenience of interpolation where the differences are large.

Logarithms of the Trigonometric Functions—Logarithms of the functions are given for each minute from 0–360°.

The quantity −10 is to be appended to all logarithms of the sine and cosine, to logarithms of the tangent from 0–45° and of the cotangent from 45–90°.

With degrees indicated at either side of the top of the page use the column headings at the top. With degrees stated at the bottom of the page use the column designations at the bottom.

With degrees at the left (top or bottom) use the minute column at the left, and with degrees on the right side of the page use the minute column at the right.

To illustrate the proper employment of headings for angles in the four quadrants—

$\log \sin 6° 24' = 9.04715 - 10$ $\quad$ $\log \sin 186° 24' = 9.04715 - 10$
$\log \sin 83° 15' = 9.99698 - 10$ $\quad$ $\log \sin 263° 15' = 9.99698 - 10$
$\log \cos 96° 41' = 9.06589 - 10$ $\quad$ $\log \cos 276° 41' = 9.06589 - 10$
$\log \cos 173° 49' = 9.99747 - 10$ $\quad$ $\log \cos 353° 49' = 9.99747 - 10$

For the accurate determination of values where the tabular differences are large, the values of CS and CT are given. The following equations indicate their use.

To find the logarithm of the functions of an angle:

For angles 0–3°	For angles 87–90°
$\log \sin \theta = \log \theta'' - \text{CS}$	$\log \cos \theta = \log (90° - \theta)'' - \text{CS}$
$\log \tan \theta = \log \theta'' - \text{CT}$	$\log \cot \theta = \log (90° - \theta)'' - \text{CT}$
$\log \cot \theta = \text{colog} \tan \theta$	$\log \tan \theta = \text{colog} \cot \theta$

To find the angle:

For angles 0–3°	For angles 87–90°
$\log \theta'' = \log \sin \theta + \text{CS}$	$\log (90° - \theta)'' = \log \cos \theta + \text{CS}$
$\log \theta'' = \log \tan \theta + \text{CT}$	$\log (90° - \theta)'' = \log \cot \theta + \text{CT}$

In the above expressions, θ'' and $(90° - \theta)''$ are used to indicate the value of the angles expressed in seconds. The values in the body of the table are the cologarithms and should be used as indicated above.

The values of the logarithms S and T are also given in a separate table. For these the following relations hold:

To find the functions of an angle.

$$\log \sin \theta = \log \theta'' + \text{S} \qquad \log \cos \theta = \log (90° - \theta)'' + \text{S}$$
$$\log \tan \theta = \log \theta'' + \text{T} \qquad \log \cot \theta = \log (90° - \theta)'' + \text{T}$$

To find the angle.

$$\log \theta'' = \log \sin \theta - \text{S} \qquad \log (90° - \theta)'' = \log \cos \theta - \text{S}$$
$$\log \theta'' = \log \tan \theta - \text{T} \qquad \log (90° - \theta)'' = \log \cot \theta - \text{T}$$

Where the values of CS and CT are given, the angles expressed in seconds are given in the supplementary column at the left.

The tabular differences are given under the headings "d" and "c.d.," the latter referring to the common difference for the tangent and cotangent. Tables of proportional parts ("P.P.") facilitate interpolation. At the bottom of each column will be found special proportional parts between the tabular differences for the tangent or cotangent and those for the sine or cosine. These are useful when one function is to be obtained directly from the other without determining the angle.

For example, suppose log tan θ is given as 9.67644 and log cos θ is required. The difference between the given logarithm and that given in the table, 9.67622 (opposite 25° 23′), is 22. The tabular differences of the two logarithmic functions at this place are 32 and 6. In the proportional table for $\frac{6}{32}$, 22 corresponds to 4; this, subtracted from the tabular logarithmic cosine 9.95591, gives the required log cos θ = 9.95587.

The symbols $\bar{5}$ and $\dot{5}$ are used to indicate how the terminal 5 has been derived. For example, the logarithm 8.8307$\bar{5}$ is more fully given as 8.8307495 while the value 9.4082$\dot{5}$ is derived from 9.4082539. Thus, in rounding off to four places, a number ending in $\bar{5}$ should be decreased but a number ending in $\dot{5}$ should be increased.

Natural Trigonometric Functions—Values of the natural trigonometric functions of angles are given for each minute from 0–360°.

For degrees indicated at the top of the page use the column headings at the top. For degrees indicated at the bottom use the column indications at the bottom.

With degrees at the left of each block (top or bottom), use the minute column at the left and with degrees at the right of each block use the minute column at the right.

Natural Functions and their Logarithms are given for angles in degrees and tenths from 0 to 90 degrees.

Natural Functions and their Logarithms are given for angles in radians and hundredths, from 0 to 2 radians.

Haversines—Values of $(1 - \cos\theta)/2$ for angles between 0 and 180° are given to four significant figures. The four-place mantissas of the logarithms of the haversines are also given. The correct characteristic must be provided in each case.

The listed values of the haversines were derived from values which were computed to seven significant figures. The logarithms were independently derived from the more exact values of the haversines and are, therefore, in many cases not the exact value of the logarithm of the haversine as listed. This is notably true at the beginning of the table where the logarithm can be given with more exactness than the function.

Natural Logarithms—The natural logarithms of numbers from 0.000 to 999. are given in a group of four tables. The method of finding logarithms of numbers not included in the tables is indicated at the beginning of the third page. A convenient table of constants occurs at the top of the fourth page.

The first page gives the natural logarithms of numbers from 0.000 to 0.499. Since the characteristics change rapidly for the smaller numbers, they are indicated *above* the mantissa in the first line. In the second and following lines the characteristics are given at the left only. For example, $\log_e 0.004 = -5.52146$; $\log_e 0.014 = -4.26870$.

The succeeding pages give the natural logarithms of numbers up to 999.

Exponential Functions—Values of e^x, $\log e^x$ and e^{-x} where e is the base of the natural system of logarithms 2.71828 . . . and x has values from 0 to 10. Facilitating the solution of exponential equations, these tables also serve as a table of natural or Napierian antilogarithms. For instance if the logarithm or exponent $x = 3.26$ the corresponding number or value of e^x is 26.050. Its reciprocal e^{-x} is .038388.

Hyperbolic Functions—The table gives the values and logarithms of the hyperbolic sine x, cosine x, tangent x and cotangent x for values of x from 0 to 10.

Degrees-Radians—This table gives the value in radians to five significant figures; for each 10 minutes from 0° 0′ to 90° 0′; for each degree from 90 to 180; for each 10 degrees from 180 to 480. Values are also given for each minute from 0–60′ and for each second from 0–60″.

Tables are also provided to facilitate changing from degrees and decimal fractions to radians, from decimal fractions of a degree to minutes and seconds and the reverse operations.

Numerical Tables—The first section gives the reciprocals of numbers from 0 to 1000 and circumferences and areas of circles with diameters having these values. Reciprocals and circumferences for values not listed can be obtained by an appropriate shift of the decimal point.

The second section is devoted to squares, cubes and roots. The squares and cubes from 1 to 1000 are given exactly. The roots are given to seven significant figures. Since the square roots of $10n$ are given, values of the square roots from 1 to 10,000 may be found directly. For the square roots of numbers below and above this range, use may be made of the following relations: $\sqrt{100n} = 10\sqrt{n}$; $\sqrt{1000n} = 10\sqrt{10n}$; $\sqrt{\frac{1}{10}n} = \frac{1}{10}\sqrt{10n}$; $\sqrt{\frac{1}{100}n} = \frac{1}{10}\sqrt{n}$; $\sqrt{\frac{1}{1000}n} = \frac{1}{100}\sqrt{10n}$. For example, the square root of 0.268 may be found by using the form, $\sqrt{0.268} = \frac{1}{100}\sqrt{10 \times 268}$. The tabular value for the square root of $10n$ for 268 is 51.76872. Hence, the desired root is 0.5176872.

Values of cube roots for all numbers from 1 to 100,000 will be found directly in the table. Cube roots for numbers above or below this range will be found from the following relations: $\sqrt[3]{1000n} = 10\sqrt[3]{n}$; $\sqrt[3]{10{,}000n} = 10\sqrt[3]{10n}$; $\sqrt[3]{100{,}000n} = 10\sqrt[3]{100n}$; $\sqrt[3]{\frac{1}{10}n} = \frac{1}{10}\sqrt[3]{100n}$; $\sqrt[3]{\frac{1}{100}n} = \frac{1}{10}\sqrt[3]{10n}$; $\sqrt[3]{\frac{1}{1000}n} = \frac{1}{10}\sqrt[3]{n}$. For example, the cube root of 731,000 may be found by using the form, $\sqrt[3]{731{,}000} = 10\sqrt[3]{731}$. The tabular value of the root for 731 is 9.008223. The desired root is, therefore, 90.08223.

Powers of Numbers—This table is given to supplement the values of squares and cubes of numbers found in the preceding numerical table. The larger numbers are expressed exponentially to at least seven significant figures. The approximate value written as a whole number may be obtained by shifting the decimal point to the right by the number of places indicated in the exponent of 10 shown at the head of each group of values. For example: the approximate value of 33^8 is found in the table as 14.064086×10^{11}. Written as a whole number it is 1,406,408,600,000.

Factorials and their Logarithms—The product $n \times (n-1) \times (n-2) \times \cdots \times 1$ is called factorial n, expressed as $n!$ or $\underline{|n}$. For example: factorial $5 = 5 \times 4 \times 3 \times 2 \times 1 = 120$. Factorials are very often met with in series. For purposes of computation in such cases the table giving the values of the factorials and of their logarithms for numbers from 1 to 100 is provided. The values of the factorials are expressed exponentially to 5 significant figures.

A brief table of exact values and reciprocals of factorials is to be found on page A–91.

Factors for Computing Probable Errors—The probable error of a series of n measures $a_1, a_2, a_3 \ldots a_n$, the mean of which is m, is given by the expression,

$$e = \frac{0.6745}{\sqrt{n-1}} \sqrt{(m-a_1)^2 + (m-a_2)^2 + \cdots (m-a_n)^2}$$

The probable error of the mean is,

$$E = \frac{0.6745}{\sqrt{n(n-1)}} \sqrt{(m-a_1)^2 + (m-a_2)^2 + \cdots (m-a_n)^2}$$

The following approximate equations are convenient forms for computation,

$$e = 0.8453 \frac{\Sigma d}{\sqrt{n(n-1)}}$$

$$E = 0.8453 \frac{\Sigma d}{n\sqrt{n-1}}$$

The symbol Σd represents the arithmetical sum of the deviations.

For convenience in computing the probable error the value of several of the factors involved is given for values of n from 2 to 100.

Probability of Occurrence of Deviations—The significance of deviations is indicated by this table. The probability of occurrence of deviations as great as or greater than any specific value is given for various ratios of deviation to probable error and also with respect to the standard deviation σ. The probability of occurrence is stated in per cent or chances in 100. The odds against occurrence are also stated. The probable error is $0.6745 \times (\sigma)$.

Normal Curve of Error—If, for a large number of observations, the frequency y, of the occurrence of an error of magnitude t be plotted, a curve results whose equation may be written,

$$y = \frac{1}{\sqrt{2\pi}} e^{-t^2/2}$$

The area, ordinates and derivatives for this curve given in the table are useful in the treatment of observational data. A text on statistical methods should be consulted for a complete explanation.

Factors and Primes—The table presents the prime factors of *all* factorable numbers and the logarithms of all prime numbers from 1 to 2000.

CONVERSION TABLES

Inches	Centimeters	Centimeters	Inches
1	2.54	1	0.393701
2	5.08	2	0.787402
3	7.62	3	1.181103
4	10.16	4	1.574804
5	12.70	5	1.968505
6	15.24	6	2.362206
7	17.78	7	2.755907
8	20.32	8	3.149608
9	22.86	9	3.543309
Feet	**Meters**	**Meters**	**Feet**
1	0.3048	1	3.280840
2	0.6096	2	6.561680
3	0.9144	3	9.842520
4	1.2192	4	13.123360
5	1.5240	5	16.404200
6	1.8288	6	19.685040
7	2.1336	7	22.965880
8	2.4384	8	26.246720
9	2.7432	9	29.527560
Yards	**Meters**	**Meters**	**Yards**
1	0.9144	1	1.0936133
2	1.8288	2	2.1872266
3	2.7432	3	3.2808399
4	3.6576	4	4.3744532
5	4.5720	5	5.4680665
6	5.4864	6	6.5616798
7	6.4008	7	7.6552931
8	7.3152	8	8.7489064
9	8.2296	9	9.8425197
Miles (Statute)	**Kilometers**	**Kilometers**	**Miles**
1	1.609344	1	0.6213712
2	3.218688	2	1.2427424
3	4.828032	3	1.8641136
4	6.437376	4	2.4854848
5	8.046720	5	3.1068560
6	9.656064	6	3.7282272
7	11.265408	7	4.3495984
8	12.874752	8	4.9709696
9	14.484096	9	5.5923408
Pounds (av)	**Kilograms**	**Kilograms**	**Pounds (av)**
1	0.45359237	1	2.2046226
2	0.90718474	2	4.4092452
3	1.36077711	3	6.6138678
4	1.81436948	4	8.8184904
5	2.26796185	5	11.0231130
6	2.72155422	6	13.2277356
7	3.17514659	7	15.4323582
8	3.62873896	8	17.6369808
9	4.08233133	9	19.8416034

Conversion Factors

U. S. AND METRIC UNITS

Each unit in bold face type is followed by its equivalent in other units of the same quantity.

Acre—0.0015625 square mile (statute); 4.3560 $\times$ 10^4 square feet; 0.40468564 hectare.

Bushel—(U. S.)—1.244456 cubic feet; 2150.42 cubic inches; 0.035239 cubic meter; 35.23808 liters.

Centimeter—0.0328084 foot; 0.393701 inch.

Circular Mil—7.853982 $\times$ 10^{-7} square inches; 5.067075 $\times$ 10^{-6} square centimeters.

Cubic Centimeter—0.061024 cubic inch; 0.270512 dram (U. S. fluid); 16.230664 minims (U. S.); 0.999972 milliliter.

Cubic Foot—0.803564 bushel (U. S.); 7.480520 gallons (U. S. liquid); 0.028317 cubic meter; 28.31605 liters.

Cubic Inch—16.387064 cubic centimeters.

Cubic Meter—35.314667 cubic feet; 264.17205 gallons (U. S. liquid).

Foot—0.3048 meter.

Gallon (U. S. liquid)—0.1336816 cubic foot; 0.832675 gallon (British); 231 cubic inches; 0.0037854 cubic meter; 3.785306 liters.

Grain—0.06479891 gram.

Gram—0.00220462 pound (avoirdupois); 0.035274 ounce (avoirdupois); 15.432358 grains.

Hectare—2.471054 acres; 1.07639 $\times$ 10^5 square feet.

Inch—2.54 centimeters.

Kilogram—2.204623 pounds (avoirdupois).

Kilometer—0.621371 mile (statute).

Liter—0.264179 gallon (U. S. liquid); 0.0353157 cubic foot; 1.056718 quarts (U. S. liquid).

Meter—1.093613 yards; 3.280840 feet; 39.37008 inches.

Mile (statute)—1.609344 kilometers.

Ounce (U. S. fluid)—1.804688 cubic inches; 29.573730 cubic centimeters.

Ounce (avoirdupois)—28.349523 grams.

Ounce (apothecary or troy)—31.103486 grams.

Pint (U. S. liquid)—0.473163 liter; 473.17647 cubic centimeters.

Pound (avoirdupois)—0.453592 kilogram; 453.-59237 grams.

Pound (apothecary or troy)—0.3732417 kilo-gram, 373.24172 grams.

Quart (U. S. dry)—1.10119 liters.

Quart (liquid)—0.946326 liter.

Radian—57.295779 degrees.

Rod—5.0292 meters.

Square Centimeter—0.155000 square inch.

Square Foot—0.09290304 square meter.

Square Inch—645.16 square millimeters.

Square Meter—10.763910 square feet.

Square Yard—0.836127 square meter.

Ton (short)—907.18474 kilograms.

Yard—0.9144 meter.

NUMERICAL CONSTANTS

π Constants

π = 3.14159 26535 89793 23846 26433 83279 50288 41971 69399 37510
$1/\pi$ = 0.31830 98861 83790 67153 77675 26745 02872 40689 19291 48091
π^2 = 9.86960 44010 89358 61883 44909 99876 15113 53136 99407 24079
$\log_e \pi$ = 1.14472 98858 49400 17414 34273 51353 05871 16472 94812 91531
$\log_{10} \pi$ = 0.49714 98726 94133 85435 12682 88290 89887 36516 78324 38044
$\log_{10} \sqrt{2\pi}$ = 0.39908 99341 79057 52478 25035 91507 69595 02099 34102 92127

Logarithmic Constants

e = 2.71828 18284 59045 23536 02874 71352 66249 77572 47093 69995
$1/e$ = 0.36787 94411 71442 32159 55237 70161 46086 74458 11131 03176
e^2 = 7.38905 60989 30650 22723 04274 60575 00781 31803 15570 55184
$M = \log_{10} e$ = 0.43429 44819 03251 82765 11289 18916 60508 22943 97005 80366
$1/M = \log_e 10$ = 2.30258 50929 94045 68401 79914 54684 36420 76011 01488 62877
$\log_{10} M$ = 9.63778 43113 00536 78912 29674 98565 − 10

Miscellaneous π and e Constants

π^e = 22.45915 77183 61045 47342 71522
e^π = 23.14069 26327 79269 00572 90864
$e^{-\pi}$ = 0.04321 39182 63772 24977 44177
$e^{\frac{1}{2}\pi}$ = 4.81047 73809 65351 65547 30357
$i^i = e^{-\frac{1}{2}\pi}$ = 0.20787 95763 50761 90854 69556

Numerical Constants

$\sqrt{2}$ = 1.41421 35623 73095 04880 16887 24209 69807 85696 71875 37694
$\sqrt[3]{2}$ = 1.25992 10498 94873 16476 72106 07278 22835 05702 51464 70150
$\log_e 2$ = 0.69314 71805 59945 30941 72321 21458 17656 80755 00134 36025
$\log_{10} 2$ = 0.30102 99956 63981 19521 37388 94724 49302 67681 89881 46210
$\sqrt{3}$ = 1.73205 08075 68877 29352 74463 41505 87236 69428 05253 81038
$\sqrt[3]{3}$ = 1.44224 95703 07408 38232 16383 10780 10958 83918 69253 49935
$\log_e 3$ = 1.09861 22886 68109 69139 52452 36922 52570 46474 90557 82274
$\log_{10} 3$ = 0.47712 12547 19662 43729 50279 03255 11530 92001 28864 19069

Miscellaneous

Euler's Constant γ = 0.57721 56649 01532 86061
$\log_e \gamma$ = −0.54953 93129 81644 82234
Golden Ratio ϕ = 1.61803 39887 49894 84820 45868 34365 63811 77203 09180

Numbers Containing π

π = 3.14159 26536 $\quad$ $\log_{10} \pi$ = 0.49714 98727 $\quad$ $\log_e \pi$ = 1.14472 98858

	Number	Logarithm		Number	Logarithm
π	3.1415 927	0.4971 499	π^2	9.8696 044	0.9942 997
2π	6.2831 853	0.7981 799	$2\pi^2$	19.7392 088	1.2953 297
3π	9.4247 780	0.9742 711	$4\pi^2$	39.4784 176	1.5963 597
4π	12.5663 706	1.0992 099	$1/\pi^2$	0.1013 212	9.0057 003 − 10
8π	25.1327 412	1.4002 399	$1/(2\pi^2)$	0.0506 606	8.7046 703 − 10
$\pi/2$	1.5707 963	0.1961 199	$1/(4\pi^2)$	0.0253 303	8.4036 403 − 10
$\pi/3$	1.0471 976	0.0200 286	$\sqrt{\pi}$	1.7724 539	0.2485 749
$\pi/4$	0.7853 982	9.8950 899 − 10	$\sqrt{\pi/4}$ or	0.8862 269	9.9475 449 − 10
$\pi/6$	0.5235 988	9.7189 986 − 10	$\sqrt{\pi}/2$		
$\pi/8$	0.3926 991	9.5940 599 − 10	$\sqrt{\pi}/4$	0.4431 135	9.6465 149 − 10
$2\pi/3$	2.0943 951	0.3210 586	$\sqrt{\pi/2}$	1.2533 141	0.0980 599
$4\pi/3$	4.1887 902	0.6220 886	$\sqrt{2/\pi}$	0.7978 846	9.9019 401 − 10
$1/\pi$	0.3183 099	9.5028 501 − 10	π^3	31.0062 767	1.4914 496
$2/\pi$	0.6366 198	9.8038 801 − 10	$\sqrt[3]{\pi}$	1.4645 919	0.1657 166
$4/\pi$	1.2732 395	0.1049 101	$1/\sqrt[3]{\pi}$	0.6827 841	9.8342 834 − 10
$1/(2\pi)$	0.1591 549	9.2018 201 − 10	$\sqrt[3]{\pi^2}$	2.1450 294	0.3314 332
$1/(4\pi)$	0.0795 775	8.9007 901 − 10	$1/\sqrt{\pi}$	0.5641 896	9.7514 251 − 10
$1/(6\pi)$	0.0530 516	8.7246 989 − 10	$1/\sqrt{2\pi}$	0.3989 423	9.6009 101 − 10
$1/(8\pi)$	0.0397 887	8.5997 601 − 10	$2/\sqrt{\pi}$	1.1283 792	0.0524 551
$\pi/180$	0.0174 533	8.2418 774 − 10			
$180/\pi$	57.2957 795	1.7581 226			

Change of Base

$$\log_a x = \log_b x / \log_b a$$

$$\log_{10} x = \log_e x / \log_e 10 \qquad \log_e x = \log_{10} x / \log_{10} e$$

$$\log_e x = 1/M \log_{10} x = 2.30258\ 50930 \log_{10} x$$

$$\log_{10} x = M \log_e x = 0.43429\ 44819 \log_e x$$

MISCELLANEOUS CONSTANTS

Equatorial radius of the earth = 6378.388 km = 3963.34 miles (statute).
Polar radius of the earth, 6356.912 km = 3949.99 miles (statute).
1 degree of latitude at 40° = 69 miles.
1 international nautical mile = 1.15078 miles (statute) = 1852 m = 6076.115 ft.
Mean density of the earth = 5.522 g/cm^3 = 344.7 lb/ft^3.
Constant of gravitation, 6.673 ± 0.003 × 10^{-8} $cm^3/g \times sec^2$.
Acceleration due to gravity at sea level, latitude 45° = 980.621 cm/sec^2 = 32.1725 ft/sec^2.
Length of seconds pendulum at sea level, latitude 45° = 99.3574 cm = 39.1171 in.
1 knot (international) = 101.269 ft/min = 1.6878 ft/sec = 1.1508 miles (statute)/hr.
1 micron = 10^{-4} cm.
1 ångstrom 10^{-8} cm.
Mass of hydrogen atom = (1.67339 ± 0.00031) × 10^{-24} g.
Avogadro's number = (6.02257 ± 0.00009) × 10^{-23}/g mole.
Planck's constant = (6.62554 ± 0.00015) × 10^{-27} erg-sec.
Density of mercury at 0° C = 13.5955 g ml.
Density of water at 3.98° C = 1.000000 g/ml.
Density, maximum, of water, at 3.98° C = 0.999973 g/cm^3.
Density of dry air at 0° C, 760 mm = 1.2929 g/liter.
Velocity of sound in dry air at 0° C = 331.36 m/sec = 1087.1 ft/sec.
Velocity of light in vacuum = 2.997925 ± 0.000002) × 10^{10} cm/sec.
Heat of fusion of water 0° C = 79.71 cal/g.
Heat of vaporization of water 100° C = 539.55 cal/g.
Electrochemical equivalent of silver 0.001118 g/sec international amp.
Absolute wave length of red cadmium light in air at 15° C, 760 mm pressure = 6438.4696 A
Wave length of orange-red line of krypton 86 = 6057.802 A.
Gas Constant = 8.31432 ± 0.00034 × 10^7 erg (g mole)$^{-1}$ deg^{-1}.

DECIMAL EQUIVALENTS OF COMMON FRACTIONS

1/64 =	0.015 625	11/32 22/64 =	0.343 75	43/64 =	0.671 875
1/32 2/64 =	.031 25	23/64 =	.359 375	11/16 22/32 44/64 =	.687 5
3/64 =	.046 875	3/8 12/32 24/64 =	.375	45/64 =	.703 125
1/16 2/32 4/64 =	.062 5	25/64 =	.390 625	23/32 46/64 =	.718 75
5/64 =	.078 125	13/32 26/64 =	.406 25	47/64 =	.734 375
3/32 6/64 =	.093 75	27/64 =	.421 875	3/4 24/32 48/64 =	.75
7/64 =	.109 375	7/16 14/32 28/64 =	.437 5	49/64 =	.765 625
1/8 4/32 8/64 =	.125	29/64 =	.453 125	25/32 50/64 =	.781 25
9/64 =	.140 625	15/32 30/64 =	.468 75	51/64 =	.796 875
5/32 10/64 =	.156 25	31/64 =	.484 375	13/16 26/32 52/64 =	.812 5
11/64 =	.171 875	1/2 16/32 32/64 =	.50	53/64 =	.828 125
3/16 6/32 12/64 =	.187 5	33/64 =	.515 625	27/32 54/64 =	.843 75
13/64 =	.203 125	17/32 34/64 =	.531 25	55/64 =	.859 375
7/32 14/64 =	.218 75	35/64 =	.546 875	7/8 28/32 56/64 =	.875
15/64 =	.234 375	9/16 18/32 36/64 =	.562 5	57/64 =	.890 625
1/4 8/32 16/64 =	.25	37/64 =	.578 125	29/32 58/64 =	.906 25
17/64 =	.265 625	19/32 38/64 =	.593 75	59/64 =	.921 875
9/32 18/64 =	.281 25	39/64 =	.609 375	15/16 30/32 60/64 =	.937 5
19/64 =	.296 875	5/8 20/32 40/64 =	.625	61/64 =	.953 125
5/16 10/32 20/64 =	.312 5	41/64 =	.640 625	31/32 62/64 =	.968 75
21/64 =	.328 125	21/32 42/64 =	.656 25	63/64 =	.984 375

MULTIPLES OF $\frac{\pi}{2}$

$$\frac{\pi}{2} = \prod_{n=1}^{\infty} \frac{4n^2}{4n^2-1} = \prod_{n=1}^{\infty} \frac{(2n)^2}{(2n-1)(2n+1)} = \left(\frac{2\cdot 2}{1\cdot 3}\right)\cdot\left(\frac{4\cdot 4}{3\cdot 5}\right)\cdot\left(\frac{6\cdot 6}{5\cdot 7}\right)\cdots$$

n	$n\frac{\pi}{2}$	n	$n\frac{\pi}{2}$	n	$n\frac{\pi}{2}$	n	$n\frac{\pi}{2}$
1	1.57079 63268	26	40.84070 44967	51	80.11061 26665	76	119.38052 08364
2	3.14159 26536	27	42.41150 08235	52	81.68140 89933	77	120.95131 71632
3	4.71238 89804	28	43.98229 71503	53	83.25220 53201	78	122.52211 34900
4	6.28318 53072	29	45.55309 34771	54	84.82300 16469	79	124.09290 98168
5	7.85398 16340	30	47.12388 98038	55	86.39379 79737	80	125.66370 61436
6	9.42477 79608	31	48.69468 61306	56	87.96459 43005	81	127.23450 24704
7	10.99557 42876	32	50.26548 24574	57	89.53539 06273	82	128.80529 87972
8	12.56637 06144	33	51.83627 87842	58	91.10618 69541	83	130.37609 51240
9	14.13716 69412	34	53.40707 51110	59	92.67698 32809	84	131.94689 14508
10	15.70796 32679	35	54.97787 14378	60	94.24777 96077	85	133.51768 77776
11	17.27875 95947	36	56.54866 77646	61	95.81857 59345	86	135.08848 41044
12	18.84955 59215	37	58.11946 40914	62	97.38937 22613	87	136.65928 04312
13	20.42035 22483	38	59.69026 04182	63	98.96016 85881	88	138.23007 67580
14	21.99114 85751	39	61.26105 67450	64	100.53096 49149	89	139.80087 30847
15	23.56194 49019	40	62.83185 30718	65	102.10176 12417	90	141.37166 94115
16	25.13274 12287	41	64.40264 93986	66	103.67255 75685	91	142.94246 57383
17	26.70353 75555	42	65.97344 57254	67	105.24335 38953	92	144.51326 20651
18	28.27433 38823	43	67.54424 20522	68	106.81415 02221	93	146.08405 83919
19	29.84513 02091	44	69.11503 83790	69	108.38494 65488	94	147.65485 47187
20	31.41592 65359	45	70.68583 47058	70	109.95574 28756	95	149.22565 10455
21	32.98672 28627	46	72.25663 10326	71	111.52653 92024	96	150.79644 73723
22	34.55751 91895	47	73.82742 73594	72	113.09733 55292	97	152.36724 36991
23	36.12831 55163	48	75.39822 36862	73	114.66813 18560	98	153.93804 00259
24	37.69911 18431	49	76.96902 00129	74	116.23892 81828	99	155.50883 63527
25	39.26990 81699	50	78.53981 63397	75	117.80972 45096	100	157.07963 26795

FOUR-PLACE LOGARITHMS

N	0	1	2	3	4	5	6	7	8	9	Proportional Parts 1	2	3	4	5	6	7	8	9
10	0000	0043	0086	0128	0170	0212	0253	0294	0334	0374	*4	8	12	17	21	25	29	33	37
11	0414	0453	0492	0531	0569	0607	0645	0682	0719	0755	4	8	11	15	19	23	26	30	34
12	0792	0828	0864	0899	0934	0969	1004	1038	1072	1106	3	7	10	14	17	21	24	28	31
13	1139	1173	1206	1239	1271	1303	1335	1367	1399	1430	3	6	10	13	16	19	23	26	29
14	1461	1492	1523	1553	1584	1614	1644	1673	1703	1732	3	6	9	12	15	18	21	24	27
15	1761	1790	1818	1847	1875	1903	1931	1959	1987	2014	*3	6	8	11	14	17	20	22	25
16	2041	2068	2095	2122	2148	2175	2201	2227	2253	2279	3	5	8	11	13	16	18	21	24
17	2304	2330	2355	2380	2405	2430	2455	2480	2504	2529	2	5	7	10	12	15	17	20	22
18	2553	2577	2601	2625	2648	2672	2695	2718	2742	2765	2	5	7	9	12	14	16	19	21
19	2788	2810	2833	2856	2878	2900	2923	2945	2967	2989	2	4	7	9	11	13	16	18	20
20	3010	3032	3054	3075	3096	3118	3139	3160	3181	3201	2	4	6	8	11	13	15	17	19
21	3222	3243	3263	3284	3304	3324	3345	3365	3385	3404	2	4	6	8	10	12	14	16	18
22	3424	3444	3464	3483	3502	3522	3541	3560	3579	3598	2	4	6	8	10	12	14	15	17
23	3617	3636	3655	3674	3692	3711	3729	3747	3766	3784	2	4	6	7	9	11	13	15	17
24	3802	3820	3838	3856	3874	3892	3909	3927	3945	3962	2	4	5	7	9	11	12	14	16
25	3979	3997	4014	4031	4048	4065	4082	4099	4116	4133	2	3	5	7	9	10	12	14	15
26	4150	4166	4183	4200	4216	4232	4249	4265	4281	4298	2	3	5	7	8	10	11	13	15
27	4314	4330	4346	4362	4378	4393	4409	4425	4440	4456	2	3	5	6	8	9	11	13	14
28	4472	4487	4502	4518	4533	4548	4564	4579	4594	4609	2	3	5	6	8	9	11	12	14
29	4624	4639	4654	4669	4683	4698	4713	4728	4742	4757	1	3	4	6	7	9	10	12	13
30	4771	4786	4800	4814	4829	4843	4857	4871	4886	4900	1	3	4	6	7	9	10	11	13
31	4914	4928	4942	4955	4969	4983	4997	5011	5024	5038	1	3	4	6	7	8	10	11	12
32	5051	5065	5079	5092	5105	5119	5132	5145	5159	5172	1	3	4	5	7	8	9	11	12
33	5185	5198	5211	5224	5237	5250	5263	5276	5289	5302	1	3	4	5	6	8	9	10	12
34	5315	5328	5340	5353	5366	5378	5391	5403	5416	5428	1	3	4	5	6	8	9	10	11
35	5441	5453	5465	5478	5490	5502	5514	5527	5539	5551	1	2	4	5	6	7	9	10	11
36	5563	5575	5587	5599	5611	5623	5635	5647	5658	5670	1	2	4	5	6	7	8	10	11
37	5682	5694	5705	5717	5729	5740	5752	5763	5775	5786	1	2	3	5	6	7	8	9	10
38	5798	5809	5821	5832	5843	5855	5866	5877	5888	5899	1	2	3	5	6	7	8	9	10
39	5911	5922	5933	5944	5955	5966	5977	5988	5999	6010	1	2	3	4	5	7	8	9	10
40	6021	6031	6042	6053	6064	6075	6085	6096	6107	6117	1	2	3	4	5	6	8	9	10
41	6128	6138	6149	6160	6170	6180	6191	6201	6212	6222	1	2	3	4	5	6	7	8	9
42	6232	6243	6253	6263	6274	6284	6294	6304	6314	6325	1	2	3	4	5	6	7	8	9
43	6335	6345	6355	6365	6375	6385	6395	6405	6415	6425	1	2	3	4	5	6	7	8	9
44	6435	6444	6454	6464	6474	6484	6493	6503	6513	6522	1	2	3	4	5	6	7	8	9
45	6532	6542	6551	6561	6571	6580	6590	6599	6609	6618	1	2	3	4	5	6	7	8	9
46	6628	6637	6646	6656	6665	6675	6684	6693	6702	6712	1	2	3	4	5	6	7	7	8
47	6721	6730	6739	6749	6758	6767	6776	6785	6794	6803	1	2	3	4	5	5	6	7	8
48	6812	6821	6830	6839	6848	6857	6866	6875	6884	6893	1	2	3	4	4	5	6	7	8
49	6902	6911	6920	6928	6937	6946	6955	6964	6972	6981	1	2	3	4	4	5	6	7	8
50	6990	6998	7007	7016	7024	7033	7042	7050	7059	7067	1	2	3	3	4	5	6	7	8
51	7076	7084	7093	7101	7110	7118	7126	7135	7143	7152	1	2	3	3	4	5	6	7	8
52	7160	7168	7177	7185	7193	7202	7210	7218	7226	7235	1	2	2	3	4	5	6	7	7
53	7243	7251	7259	7267	7275	7284	7292	7300	7308	7316	1	2	2	3	4	5	6	6	7
54	7324	7332	7340	7348	7356	7364	7372	7380	7388	7396	1	2	2	3	4	5	6	6	7
N	0	1	2	3	4	5	6	7	8	9	1	2	3	4	5	6	7	8	9

*** Interpolation in this section of the table is inaccurate.**

FOUR-PLACE LOGARITHMS (Continued)

N	0	1	2	3	4	5	6	7	8	9	Proportional Parts								
											1	2	3	4	5	6	7	8	9
55	7404	7412	7419	7427	7435	7443	7451	7459	7466	7474	1	2	2	3	4	5	5	6	7
56	7482	7490	7497	7505	7513	7520	7528	7536	7543	7551	1	2	2	3	4	5	5	6	7
57	7559	7566	7574	7582	7589	7597	7604	7612	7619	7627	1	2	2	3	4	5	5	6	7
58	7634	7642	7649	7657	7664	7672	7679	7686	7694	7701	1	1	2	3	4	4	5	6	7
59	7709	7716	7723	7731	7738	7745	7752	7760	7767	7774	1	1	2	3	4	4	5	6	7
60	7782	7789	7796	7803	7810	7818	7825	7832	7839	7846	1	1	2	3	4	4	5	6	6
61	7853	7860	7868	7875	7882	7889	7896	7903	7910	7917	1	1	2	3	4	4	5	6	6
62	7924	7931	7938	7945	7952	7959	7966	7973	7980	7987	1	1	2	3	3	4	5	6	6
63	7993	8000	8007	8014	8021	8028	8035	8041	8048	8055	1	1	2	3	3	4	5	5	6
64	8062	8069	8075	8082	8089	8096	8102	8109	8116	8122	1	1	2	3	3	4	5	5	6
65	8129	8136	8142	8149	8156	8162	8169	8176	8182	8189	1	1	2	3	3	4	5	5	6
66	8195	8202	8209	8215	8222	8228	8235	8241	8248	8254	1	1	2	3	3	4	5	5	6
67	8261	8267	8274	8280	8287	8293	8299	8306	8312	8319	1	1	2	3	3	4	5	5	6
68	8325	8331	8338	8344	8351	8357	8363	8370	8376	8382	1	1	2	3	3	4	4	5	6
69	8388	8395	8401	8407	8414	8420	8426	8432	8439	8445	1	1	2	2	3	4	4	5	6
70	8451	8457	8463	8470	8476	8482	8488	8494	8500	8506	1	1	2	2	3	4	4	5	6
71	8513	8519	8525	8531	8537	8543	8549	8555	8561	8567	1	1	2	2	3	4	4	5	5
72	8573	8579	8585	8591	8597	8603	8609	8615	8621	8627	1	1	2	2	3	4	4	5	5
73	8633	8639	8645	8651	8657	8663	8669	8675	8681	8686	1	1	2	2	3	4	4	5	5
74	8692	8698	8704	8710	8716	8722	8727	8733	8739	8745	1	1	2	2	3	4	4	5	5
75	8751	8756	8762	8768	8774	8779	8785	8791	8797	8802	1	1	2	2	3	3	4	5	5
76	8808	8814	8820	8825	8831	8837	8842	8848	8854	8859	1	1	2	2	3	3	4	5	5
77	8865	8871	8876	8882	8887	8893	8899	8904	8910	8915	1	1	2	2	3	3	4	4	5
78	8921	8927	8932	8938	8943	8949	8954	8960	8965	8971	1	1	2	2	3	3	4	4	5
79	8976	8982	8987	8993	8998	9004	9009	9015	9020	9025	1	1	2	2	3	3	4	4	5
80	9031	9036	9042	9047	9053	9058	9063	9069	9074	9079	1	1	2	2	3	3	4	4	5
81	9085	9090	9096	9101	9106	9112	9117	9122	9128	9133	1	1	2	2	3	3	4	4	5
82	9138	9143	9149	9154	9159	9165	9170	9175	9180	9186	1	1	2	2	3	3	4	4	5
83	9191	9196	9201	9206	9212	9217	9222	9227	9232	9238	1	1	2	2	3	3	4	4	5
84	9243	9248	9253	9258	9263	9269	9274	9279	9284	9289	1	1	2	2	3	3	4	4	5
85	9294	9299	9304	9309	9315	9320	9325	9330	9335	9340	1	1	2	2	3	3	4	4	5
86	9345	9350	9355	9360	9365	9370	9375	9380	9385	9390	1	1	2	2	3	3	4	4	5
87	9395	9400	9405	9410	9415	9420	9425	9430	9435	9440	0	1	1	2	2	3	3	4	4
88	9445	9450	9455	9460	9465	9469	9474	9479	9484	9489	0	1	1	2	2	3	3	4	4
89	9494	9499	9504	9509	9513	9518	9523	9528	9533	9538	0	1	1	2	2	3	3	4	4
90	9542	9547	9552	9557	9562	9566	9571	9576	9581	9586	0	1	1	2	2	3	3	4	4
91	9590	9595	9600	9605	9609	9614	9619	9624	9628	9633	0	1	1	2	2	3	3	4	4
92	9638	9643	9647	9652	9657	9661	9666	9671	9675	9680	0	1	1	2	2	3	3	4	4
93	9685	9689	9694	9699	9703	9708	9713	9717	9722	9727	0	1	1	2	2	3	3	4	4
94	9731	9736	9741	9745	9750	9754	9759	9763	9768	9773	0	1	1	2	2	3	3	4	4
95	9777	9782	9786	9791	9795	9800	9805	9809	9814	9818	0	1	1	2	2	3	3	4	4
96	9823	9827	9832	9836	9841	9845	9850	9854	9859	9863	0	1	1	2	2	3	3	4	4
97	9868	9872	9877	9881	9886	9890	9894	9899	9903	9908	0	1	1	2	2	3	3	4	4
98	9912	9917	9921	9926	9930	9934	9939	9943	9948	9952	0	1	1	2	2	3	3	4	4
99	9956	9961	9965	9969	9974	9978	9983	9987	9991	9996	0	1	1	2	2	3	3	3	4
N	0	1	2	3	4	5	6	7	8	9	1	2	3	4	5	6	7	8	9

*FOUR-PLACE COMMON LOGARITHMS OF DECIMAL FRACTIONS

N	0	1	2	3	4	5	6	7	8	9
.10	−1.000	−.9957	−.9914	−.9872	−.9830	−.9788	−.9747	−.9706	−.9666	−.9626
.11	−.9586	−.9547	−.9508	−.9469	−.9431	−.9393	−.9355	−.9318	−.9281	−.9245
.12	−.9208	−.9172	−.9136	−.9101	−.9066	−.9031	−.8996	−.8962	−.8928	−.8894
.13	−.8861	−.8827	−.8794	−.8761	−.8729	−.8697	−.8665	−.8633	−.8601	−.8570
.14	−.8539	−.8508	−.8477	−.8447	−.8416	−.8386	−.8356	−.8327	−.8297	−.8268
.15	−.8239	−.8210	−.8182	−.8153	−.8125	−.8097	−.8069	−.8041	−.8013	−.7986
.16	−.7959	−.7932	−.7905	−.7878	−.7852	−.7825	−.7799	−.7773	−.7747	−.7721
.17	−.7696	−.7670	−.7645	−.7620	−.7595	−.7570	−.7545	−.7520	−.7496	−.7471
.18	−.7447	−.7423	−.7399	−.7375	−.7352	−.7328	−.7305	−.7282	−.7258	−.7235
.19	−.7212	−.7190	−.7167	−.7144	−.7122	−.7100	−.7077	−.7055	−.7033	−.7011
.20	−.6990	−.6968	−.6946	−.6925	−.6904	−.6882	−.6861	−.6840	−.6819	−.6799
.21	−.6778	−.6757	−.6737	−.6716	−.6696	−.6676	−.6655	−.6635	−.6615	−.6596
.22	−.6576	−.6556	−.6536	−.6517	−.6498	−.6478	−.6459	−.6440	−.6421	−.6402
.23	−.6383	−.6364	−.6345	−.6326	−.6308	−.6289	−.6271	−.6253	−.6234	−.6216
.24	−.6198	−.6180	−.6162	−.6144	−.6126	−.6108	−.6091	−.6073	−.6055	−.6038
.25	−.6021	−.6003	−.5986	−.5969	−.5952	−.5935	−.5918	−.5901	−.5884	−.5867
.26	−.5850	−.5834	−.5817	−.5800	−.5784	−.5768	−.5751	−.5735	−.5719	−.5702
.27	−.5686	−.5670	−.5654	−.5638	−.5622	−.5607	−.5591	−.5575	−.5560	−.5544
.28	−.5528	−.5513	−.5498	−.5482	−.5467	−.5452	−.5436	−.5421	−.5406	−.5391
.29	−.5376	−.5361	−.5346	−.5331	−.5317	−.5302	−.5287	−.5272	−.5258	−.5243
.30	−.5229	−.5214	−.5200	−.5186	−.5171	−.5157	−.5143	−.5129	−.5114	−.5100
.31	−.5086	−.5072	−.5058	−.5045	−.5031	−.5017	−.5003	−.4989	−.4976	−.4962
.32	−.4949	−.4935	−.4921	−.4908	−.4895	−.4881	−.4868	−.4855	−.4841	−.4828
.33	−.4815	−.4802	−.4789	−.4776	−.4763	−.4750	−.4737	−.4724	−.4711	−.4698
.34	−.4685	−.4672	−.4660	−.4647	−.4634	−.4622	−.4609	−.4597	−.4584	−.4572
.35	−.4559	−.4547	−.4535	−.4522	−.4510	−.4498	−.4486	−.4473	−.4461	−.4449
.36	−.4437	−.4425	−.4413	−.4401	−.4389	−.4377	−.4365	−.4353	−.4342	−.4330
.37	−.4318	−.4306	−.4295	−.4283	−.4271	−.4260	−.4248	−.4237	−.4225	−.4214
.38	−.4202	−.4191	−.4179	−.4168	−.4157	−.4145	−.4134	−.4123	−.4112	−.4101
.39	−.4089	−.4078	−.4067	−.4056	−.4045	−.4034	−.4023	−.4012	−.4001	−.3990
.40	−.3979	−.3969	−.3958	−.3947	−.3936	−.3925	−.3915	−.3904	−.3893	−.3883
.41	−.3872	−.3862	−.3851	−.3840	−.3830	−.3820	−.3809	−.3799	−.3788	−.3778
.42	−.3768	−.3757	−.3747	−.3737	−.3726	−.3716	−.3706	−.3696	−.3686	−.3675
.43	−.3665	−.3655	−.3645	−.3635	−.3625	−.3615	−.3605	−.3595	−.3585	−.3575
.44	−.3565	−.3556	−.3546	−.3536	−.3526	−.3516	−.3507	−.3497	−.3487	−.3478
.45	−.3468	−.3458	−.3449	−.3439	−.3429	−.3420	−.3410	−.3401	−.3391	−.3382
.46	−.3372	−.3363	−.3354	−.3344	−.3335	−.3325	−.3316	−.3307	−.3298	−.3288
.47	−.3279	−.3270	−.3261	−.3251	−.3242	−.3233	−.3224	−.3215	−.3206	−.3197
.48	−.3188	−.3179	−.3170	−.3161	−.3152	−.3143	−.3134	−.3125	−.3116	−.3107
.49	−.3098	−.3089	−.3080	−.3072	−.3063	−.3054	−.3045	−.3036	−.3028	−.3019
.50	−.3010	−.3002	−.2993	−.2984	−.2976	−.2967	−.2958	−.2950	−.2941	−.2933
.51	−.2924	−.2916	−.2907	−.2899	−.2890	−.2882	−.2874	−.2865	−.2857	−.2848
.52	−.2840	−.2832	−.2823	−.2815	−.2807	−.2798	−.2790	−.2782	−.2774	−.2765
.53	−.2757	−.2749	−.2741	−.2733	−.2725	−.2716	−.2708	−.2700	−.2692	−.2684
.54	−.2676	−.2668	−.2660	−.2652	−.2644	−.2636	−.2628	−.2620	−.2612	−.2604

* This table gives the logarithms of the decimal fractions which are negative numbers.

For example log 0.61 = −0.2147 = 9.7853 − 10. It should be noted that the entries as given can be used conveniently to find cologarithms of positive numbers. Every positive number $N = P \cdot (10)^k$, where $0 < P \leqq 1$. Since colog N = − log N, it follows that colog N = − log P − k.

For example colog 0.61 = − log 0.61 = 0.2147; colog 61 = 0.2147 − 2; and colog 0.00061 = 3.2147.

*FOUR-PLACE COMMON LOGARITHMS OF DECIMAL FRACTIONS (Continued)

N	0	1	2	3	4	5	6	7	8	9
.55	−.2596	−.2588	−.2581	−.2573	−.2565	−.2557	−.2549	−.2541	−.2534	−.2526
.56	−.2518	−.2510	−.2503	−.2495	−.2487	−.2480	−.2472	−.2464	−.2457	−.2449
.57	−.2441	−.2434	−.2426	−.2418	−.2411	−.2403	−.2396	−.2388	−.2381	−.2373
.58	−.2366	−.2358	−.2351	−.2343	−.2336	−.2328	−.2321	−.2314	−.2306	−.2299
.59	−.2291	−.2284	−.2277	−.2269	−.2262	−.2255	−.2248	−.2240	−.2233	−.2226
.60	−.2218	−.2211	−.2204	−.2197	−.2190	−.2182	−.2175	−.2168	−.2161	−.2154
.61	−.2147	−.2140	−.2132	−.2125	−.2118	−.2111	−.2104	−.2097	−.2090	−.2083
.62	−.2076	−.2069	−.2062	−.2055	−.2048	−.2041	−.2034	−.2027	−.2020	−.2013
.63	−.2007	−.2000	−.1993	−.1986	−.1979	−.1972	−.1965	−.1959	−.1952	−.1945
.64	−.1938	−.1931	−.1925	−.1918	−.1911	−.1904	−.1898	−.1891	−.1884	−.1878
.65	−.1871	−.1864	−.1858	−.1851	−.1844	−.1838	−.1831	−.1824	−.1818	−.1811
.66	−.1805	−.1798	−.1791	−.1785	−.1778	−.1772	−.1765	−.1759	−.1752	−.1746
.67	−.1739	−.1733	−.1726	−.1720	−.1713	−.1707	−.1701	−.1694	−.1688	−.1681
.68	−.1675	−.1669	−.1662	−.1656	−.1649	−.1643	−.1637	−.1630	−.1624	−.1618
.69	−.1612	−.1605	−.1599	−.1593	−.1586	−.1580	−.1574	−.1568	−.1561	−.1555
.70	−.1549	−.1543	−.1537	−.1530	−.1524	−.1518	−.1512	−.1506	−.1500	−.1494
.71	−.1487	−.1481	−.1475	−.1469	−.1463	−.1457	−.1451	−.1445	−.1439	−.1433
.72	−.1427	−.1421	−.1415	−.1409	−.1403	−.1397	−.1391	−.1385	−.1379	−.1373
.73	−.1367	−.1361	−.1355	−.1349	−.1343	−.1337	−.1331	−.1325	−.1319	−.1314
.74	−.1308	−.1302	−.1296	−.1290	−.1284	−.1278	−.1273	−.1267	−.1261	−.1255
.75	−.1249	−.1244	−.1238	−.1232	−.1226	−.1221	−.1215	−.1209	−.1203	−.1198
.76	−.1192	−.1186	−.1180	−.1175	−.1169	−.1163	−.1158	−.1152	−.1146	−.1141
.77	−.1135	−.1129	−.1124	−.1118	−.1113	−.1107	−.1101	−.1096	−.1090	−.1085
.78	−.1079	−.1073	−.1068	−.1062	−.1057	−.1051	−.1046	−.1040	−.1035	−.1029
.79	−.1024	−.1018	−.1013	−.1007	−.1002	−.0996	−.0991	−.0985	−.0980	−.0975
.80	−.0969	−.0964	−.0958	−.0953	−.0947	−.0942	−.0937	−.0931	−.0926	−.0921
.81	−.0915	−.0910	−.0904	−.0899	−.0894	−.0888	−.0883	−.0878	−.0872	−.0867
.82	−.0862	−.0857	−.0851	−.0846	−.0841	−.0835	−.0830	−.0825	−.0820	−.0814
.83	−.0809	−.0804	−.0799	−.0794	−.0788	−.0783	−.0778	−.0773	−.0768	−.0762
.84	−.0757	−.0752	−.0747	−.0742	−.0737	−.0731	−.0726	−.0721	−.0716	−.0711
.85	−.0706	−.0701	−.0696	−.0691	−.0685	−.0680	−.0675	−.0670	−.0665	−.0660
.86	−.0655	−.0650	−.0645	−.0640	−.0635	−.0630	−.0625	−.0620	−.0615	−.0610
.87	−.0605	−.0600	−.0595	−.0590	−.0585	−.0580	−.0575	−.0570	−.0565	−.0560
.88	−.0555	−.0550	−.0545	−.0540	−.0535	−.0531	−.0526	−.0521	−.0516	−.0511
.89	−.0506	−.0501	−.0496	−.0491	−.0487	−.0482	−.0477	−.0472	−.0467	−.0462
.90	−.0458	−.0453	−.0448	−.0443	−.0438	−.0434	−.0429	−.0424	−.0419	−.0414
.91	−.0410	−.0405	−.0400	−.0395	−.0391	−.0386	−.0381	−.0376	−.0372	−.0367
.92	−.0362	−.0357	−.0353	−.0348	−.0343	−.0339	−.0334	−.0329	−.0325	−.0320
.93	−.0315	−.0311	−.0306	−.0301	−.0297	−.0292	−.0287	−.0283	−.0278	−.0273
.94	−.0269	−.0264	−.0259	−.0255	−.0250	−.0246	−.0241	−.0237	−.0232	−.0227
.95	−.0223	−.0218	−.0214	−.0209	−.0205	−.0200	−.0195	−.0191	−.0186	−.0182
.96	−.0177	−.0173	−.0168	−.0164	−.0159	−.0155	−.0150	−.0146	−.0141	−.0137
.97	−.0132	−.0128	−.0123	−.0119	−.0114	−.0110	−.0106	−.0101	−.0097	−.0092
.98	−.0088	−.0083	−.0079	−.0074	−.0070	−.0066	−.0061	−.0057	−.0052	−.0048
.99	−.0044	−.0039	−.0035	−.0031	−.0026	−.0022	−.0017	−.0013	−.0009	−.0004

* See footnote page A–14.

ANTILOGARITHMS

	0	1	2	3	4	5	6	7	8	9	PP 1	PP 2	PP 3	PP 4	PP 5	PP 6	PP 7	PP 8	PP 9
.00	1000	1002	1005	1007	1009	1012	1014	1016	1019	1021	0	0	1	1	1	1	2	2	2
.01	1023	1026	1028	1030	1033	1035	1038	1040	1042	1045	0	0	1	1	1	1	2	2	2
.02	1047	1050	1052	1054	1057	1059	1062	1064	1067	1069	0	0	1	1	1	1	2	2	2
.03	1072	1074	1076	1079	1081	1084	1086	1089	1091	1094	0	0	1	1	1	1	2	2	2
.04	1096	1099	1102	1104	1107	1109	1112	1114	1117	1119	0	1	1	1	1	2	2	2	2
.05	1122	1125	1127	1130	1132	1135	1138	1140	1143	1146	0	1	1	1	1	2	2	2	2
.06	1148	1151	1153	1156	1159	1161	1164	1167	1169	1172	0	1	1	1	1	2	2	2	2
.07	1175	1178	1180	1183	1186	1189	1191	1194	1197	1199	0	1	1	1	1	2	2	2	2
.08	1202	1205	1208	1211	1213	1216	1219	1222	1225	1227	0	1	1	1	1	2	2	2	3
.09	1230	1233	1236	1239	1242	1245	1247	1250	1253	1256	0	1	1	1	1	2	2	2	3
.10	1259	1262	1265	1268	1271	1274	1276	1279	1282	1285	0	1	1	1	1	2	2	2	3
.11	1288	1291	1294	1297	1300	1303	1306	1309	1312	1315	0	1	1	1	2	2	2	3	3
.12	1318	1321	1324	1327	1330	1334	1337	1340	1343	1346	0	1	1	1	2	2	2	2	3
.13	1349	1352	1355	1358	1361	1365	1368	1371	1374	1377	0	1	1	1	2	2	2	3	3
.14	1380	1384	1387	1390	1393	1396	1400	1403	1406	1409	0	1	1	1	2	2	2	3	3
.15	1413	1416	1419	1422	1426	1429	1432	1435	1439	1442	0	1	1	1	2	2	2	3	3
.16	1445	1449	1452	1455	1459	1462	1466	1469	1472	1476	0	1	1	1	2	2	2	3	3
.17	1479	1483	1486	1489	1493	1496	1500	1503	1507	1510	0	1	1	1	2	2	2	3	3
.18	1514	1517	1521	1524	1528	1531	1535	1538	1542	1545	0	1	1	1	2	2	2	3	3
.19	1549	1552	1556	1560	1563	1567	1570	1574	1578	1581	0	1	1	1	2	2	3	3	3
.20	1585	1589	1592	1596	1600	1603	1607	1611	1614	1618	0	1	1	1	2	2	3	3	3
.21	1622	1626	1629	1633	1637	1641	1644	1648	1652	1656	0	1	1	2	2	2	3	3	3
.22	1660	1663	1667	1671	1675	1679	1683	1687	1690	1694	0	1	1	2	2	2	3	3	3
.23	1698	1702	1706	1710	1714	1718	1722	1726	1730	1734	0	1	1	2	2	2	3	3	4
.24	1738	1742	1746	1750	1754	1758	1762	1766	1770	1774	0	1	1	2	2	2	3	3	4
.25	1778	1782	1786	1791	1795	1799	1803	1807	1811	1816	0	1	1	2	2	2	3	3	4
.26	1820	1824	1828	1832	1837	1841	1845	1849	1854	1858	0	1	1	2	2	3	3	3	4
.27	1862	1866	1871	1875	1879	1884	1888	1892	1897	1901	0	1	1	2	2	3	3	3	4
.28	1905	1910	1914	1919	1923	1928	1932	1936	1941	1945	0	1	1	2	2	3	3	4	4
.29	1950	1954	1959	1963	1968	1972	1977	1982	1986	1991	0	1	1	2	2	3	3	4	4
.30	1995	2000	2004	2009	2014	2018	2023	2028	2032	2037	0	1	1	2	2	3	3	4	4
.31	2042	2046	2051	2056	2061	2065	2070	2075	2080	2084	0	1	1	2	2	3	3	4	4
.32	2089	2094	2099	2104	2109	2113	2118	2123	2128	2133	0	1	1	2	2	3	3	4	4
.33	2138	2143	2148	2153	2158	2163	2168	2173	2178	2183	0	1	1	2	2	3	3	4	4
.34	2188	2193	2198	2203	2208	2213	2218	2223	2228	2234	1	1	2	2	3	3	4	4	5
.35	2239	2244	2249	2254	2259	2265	2270	2275	2280	2286	1	1	2	2	3	3	4	4	5
.36	2291	2296	2301	2307	2312	2317	2323	2328	2333	2339	1	1	2	2	3	3	4	4	5
.37	2344	2350	2355	2360	2366	2371	2377	2382	2388	2393	1	1	2	2	3	3	4	4	5
.38	2399	2404	2410	2415	2421	2427	2432	2438	2443	2449	1	1	2	2	3	3	4	4	5
.39	2455	2460	2466	2472	2477	2483	2489	2495	2500	2506	1	1	2	2	3	3	4	5	5
.40	2512	2518	2523	2529	2535	2541	2547	2553	2559	2564	1	1	2	2	3	4	4	5	5
.41	2570	2576	2582	2588	2594	2600	2606	2612	2618	2624	1	1	2	2	3	4	4	5	5
.42	2630	2636	2642	2649	2655	2661	2667	2673	2679	2685	1	1	2	2	3	4	4	5	6
.43	2692	2698	2704	2710	2716	2723	2729	2735	2742	2748	1	1	2	3	3	4	4	5	6
.44	2754	2761	2767	2773	2780	2786	2793	2799	2805	2812	1	1	2	3	3	4	4	5	6
.45	2818	2825	2831	2838	2844	2851	2858	2864	2871	2877	1	1	2	3	3	4	5	5	6
.46	2884	2891	2897	2904	2911	2917	2924	2931	2938	2944	1	1	2	3	3	4	5	5	6
.47	2951	2958	2965	2972	2979	2985	2992	2999	3006	3013	1	1	2	3	3	4	5	5	6
.48	3020	3027	3034	3041	3048	3055	3062	3069	3076	3083	1	1	2	3	4	4	5	6	6
.49	3090	3097	3105	3112	3119	3126	3133	3141	3148	3155	1	1	2	3	4	4	5	6	6
	0	1	2	3	4	5	6	7	8	9	1	2	3	4	5	6	7	8	9

(PP = Proportional Parts)

ANTILOGARITHMS (Continued)

	0	1	2	3	4	5	6	7	8	9	Proportional Parts 1	2	3	4	5	6	7	8	9
.50	3162	3170	3177	3184	3192	3199	3206	3214	3221	3228	1	1	2	3	4	4	5	6	7
.51	3236	3243	3251	3258	3266	3273	3281	3289	3296	3304	1	2	2	3	4	5	5	6	7
.52	3311	3319	3327	3334	3342	3350	3357	3365	3373	3381	1	2	2	3	4	5	5	6	7
.53	3388	3396	3404	3412	3420	3428	3436	3443	3451	3459	1	2	2	3	4	5	6	6	7
.54	3467	3475	3483	3491	3499	3508	3516	3524	3532	3540	1	2	2	3	4	5	6	6	7
.55	3548	3556	3565	3573	3581	3589	3597	3606	3614	3622	1	2	2	3	4	5	6	7	7
.56	3631	3639	3648	3656	3664	3673	3681	3690	3698	3707	1	2	3	3	4	5	6	7	8
.57	3715	3724	3733	3741	3750	3758	3767	3776	3784	3793	1	2	3	3	4	5	6	7	8
.58	3802	3811	3819	3828	3837	3846	3855	3864	3873	3882	1	2	3	4	4	5	6	7	8
.59	3890	3899	3908	3917	3926	3936	3945	3954	3963	3972	1	2	3	4	5	5	6	7	8
.60	3981	3990	3999	4009	4018	4027	4036	4046	4055	4064	1	2	3	4	5	6	6	7	8
.61	4074	4083	4093	4102	4111	4121	4130	4140	4150	4159	1	2	3	4	5	6	7	8	9
.62	4169	4178	4188	4198	4207	4217	4227	4236	4246	4256	1	2	3	4	5	6	7	8	9
.63	4266	4276	4285	4295	4305	4315	4325	4335	4345	4355	1	2	3	4	5	6	7	8	9
.64	4365	4375	4385	4395	4406	4416	4426	4436	4446	4457	1	2	3	4	5	6	7	8	9
.65	4467	4477	4487	4498	4508	4519	4529	4539	4550	4560	1	2	3	4	5	6	7	8	9
.66	4571	4581	4592	4603	4613	4624	4634	4645	4656	4667	1	2	3	4	5	6	7	9	10
.67	4677	4688	4699	4710	4721	4732	4742	4753	4764	4775	1	2	3	4	5	7	8	9	10
.68	4786	4797	4808	4819	4831	4842	4853	4864	4875	4887	1	2	3	4	6	7	8	9	10
.69	4898	4909	4920	4932	4943	4955	4966	4977	4989	5000	1	2	3	5	6	7	8	9	10
.70	5012	5023	5035	5047	5058	5070	5082	5093	5105	5117	1	2	4	5	6	7	8	9	11
.71	5129	5140	5152	5164	5176	5188	5200	5212	5224	5236	1	2	4	5	6	7	8	10	11
.72	5248	5260	5272	5284	5297	5309	5321	5333	5346	5358	1	2	4	5	6	7	9	10	11
.73	5370	5383	5395	5408	5420	5433	5445	5458	5470	5483	1	3	4	5	6	8	9	10	11
.74	5495	5508	5521	5534	5546	5559	5572	5585	5598	5610	1	3	4	5	6	8	9	10	12
.75	5623	5636	5649	5662	5675	5689	5702	5715	5728	5741	1	3	4	5	7	8	9	10	12
.76	5754	5768	5781	5794	5808	5821	5834	5848	5861	5875	1	3	4	5	7	8	9	11	12
.77	5888	5902	5916	5929	5943	5957	5970	5984	5998	6012	1	3	4	5	7	8	10	11	12
.78	6026	6039	6053	6067	6081	6095	6109	6124	6138	6152	1	3	4	6	7	8	10	11	13
.79	6166	6180	6194	6209	6223	6237	6252	6266	6281	6295	1	3	4	6	7	9	10	11	13
.80	6310	6324	6339	6353	6368	6383	6397	6412	6427	6442	1	3	4	6	7	9	10	12	13
.81	6457	6471	6486	6501	6516	6531	6546	6561	6577	6592	2	3	5	6	8	9	11	12	14
.82	6607	6622	6637	6653	6668	6683	6699	6714	6730	6745	2	3	5	6	8	9	11	12	14
.83	6761	6776	6792	6808	6823	6839	6855	6871	6887	6902	2	3	5	6	8	9	11	13	14
.84	6918	6934	6950	6966	6982	6998	7015	7031	7047	7063	2	3	5	6	8	10	11	13	15
.85	7079	7096	7112	7129	7145	7161	7178	7194	7211	7228	2	3	5	7	8	10	12	13	15
.86	7244	7261	7278	7295	7311	7328	7345	7362	7379	7396	2	3	5	7	8	10	12	13	15
.87	7413	7430	7447	7464	7482	7499	7516	7534	7551	7568	2	3	5	7	9	10	12	14	16
.88	7586	7603	7621	7638	7656	7674	7691	7709	7727	7745	2	4	5	7	9	11	12	14	16
.89	7762	7780	7798	7816	7834	7852	7870	7889	7907	7925	2	4	5	7	9	11	13	14	16
.90	7943	7962	7980	7998	8017	8035	8054	8072	8091	8110	2	4	6	7	9	11	13	15	17
.91	8128	8147	8166	8185	8204	8222	8241	8260	8279	8299	2	4	6	8	9	11	13	15	17
.92	8318	8337	8356	8375	8395	8414	8433	8453	8472	8492	2	4	6	8	10	12	14	15	17
.93	8511	8531	8551	8570	8590	8610	8630	8650	8670	8690	2	4	6	8	10	12	14	16	18
.94	8710	8730	8750	8770	8790	8810	8831	8851	8872	8892	2	4	6	8	10	12	14	16	18
.95	8913	8933	8954	8974	8995	9016	9036	9057	9078	9099	2	4	6	8	10	12	15	17	19
.96	9120	9141	9162	9183	9204	9226	9247	9268	9290	9311	2	4	6	8	11	13	15	17	19
.97	9333	9354	9376	9397	9419	9441	9462	9484	9506	9528	2	4	7	9	11	13	15	17	20
.98	9550	9572	9594	9616	9638	9661	9683	9705	9727	9750	2	4	7	9	11	13	16	18	20
.99	9772	9795	9817	9840	9863	9886	9908	9931	9954	9977	2	5	7	9	11	14	16	18	20
	0	1	2	3	4	5	6	7	8	9	1	2	3	4	5	6	7	8	9

N.	0	1	2	3	4	5	6	7	8	9	Proportional parts			
100	00 000	043	087	130	173	217	260	303	346	389		**44**	**43**	**42**
101	432	475	518	561	604	647	689	732	775	817	1	4.4	4.3	4.2
102	860	903	945	988	*030	*072	*115	*157	*199	*242	2	8.8	8.6	8.4
103	01 284	326	368	410	452	494	536	578	620	662	3	13.2	12.9	12.6
104	703	745	787	828	870	912	953	995	*036	*078	4	17.6	17.2	16.8
105	02 119	160	202	243	284	325	366	407	449	490	5	22.0	21.5	21.0
106	531	572	612	653	694	735	776	816	857	898	6	26.4	25.8	25.2
107	938	979	*019	*060	*100	*141	*181	*222	*262	*302	7	30.8	30.1	29.4
108	03 342	383	423	463	503	543	583	623	663	703	8	35.2	34.4	33.6
109	743	782	822	862	902	941	981	*021	*060	*100	9	39.6	38.7	37.8
110	04 139	179	218	258	297	336	376	415	454	493		**41**	**40**	**39**
111	532	571	610	650	689	727	766	805	844	883	1	4.1	4.0	3.9
112	922	961	999	*038	*077	*115	*154	*192	*231	*269	2	8.2	8.0	7.8
113	05 308	346	385	423	461	500	538	576	614	652	3	12.3	12.0	11.7
114	690	729	767	805	843	881	918	956	994	*032	4	16.4	16.0	15.6
115	06 070	108	145	183	221	258	296	333	371	408	5	20.5	20.0	19.5
116	446	483	521	558	595	633	670	707	744	781	6	24.6	24.0	23.4
117	819	856	893	930	967	*004	*041	*078	*115	*151	7	28.7	28.0	27.3
118	07 188	225	262	298	335	372	408	445	482	518	8	32.8	32.0	31.2
119	555	591	628	664	700	737	773	809	846	882	9	36.9	36.0	35.1
120	918	954	990	*027	*063	*099	*135	*171	*207	*243		**38**	**37**	**36**
121	08 279	314	350	386	422	458	493	529	565	600	1	3.8	3.7	3.6
122	636	672	707	743	778	814	849	884	920	955	2	7.6	7.4	7.2
123	991	*026	*061	*096	*132	*167	*202	*237	*272	*307	3	11.4	11.1	10.8
124	09 342	377	412	447	482	517	552	587	621	656	4	15.2	14.8	14.4
125	691	726	760	795	830	864	899	934	968	*003	5	19.0	18.5	18.0
126	10 037	072	106	140	175	209	243	278	312	346	6	22.8	22.2	21.6
127	380	415	449	483	517	551	585	619	653	687	7	26.6	25.9	25.2
128	721	755	789	823	857	890	924	958	992	*025	8	30.4	29.6	28.8
129	11 059	093	126	160	193	227	261	294	327	361	9	34.2	33.3	32.4
130	394	428	461	494	528	561	594	628	661	694		**35**	**34**	**33**
131	727	760	793	826	860	893	926	959	992	*024	1	3.5	3.4	3.3
132	12 057	090	123	156	189	222	254	287	320	352	2	7.0	6.8	6.6
133	385	418	450	483	516	548	581	613	646	678	3	10.5	10.2	9.9
134	710	743	775	808	840	872	905	937	969	*001	4	14.0	13.6	13.2
135	13 033	066	098	130	162	194	226	258	290	322	5	17.5	17.0	16.5
136	354	386	418	450	481	513	545	577	609	640	6	21.0	20.4	19.8
137	672	704	735	767	799	830	862	893	925	956	7	24.5	23.8	23.1
138	988	*019	*051	*082	*114	*145	*176	*208	*239	*270	8	28.0	27.2	26.4
139	14 301	333	364	395	426	457	489	520	551	582	9	31.5	30.6	29.7
140	613	644	675	706	737	768	799	829	860	891		**32**	**31**	**30**
141	922	953	983	*014	*045	*076	*106	*137	*168	*198	1	3.2	3.1	3.0
142	15 229	259	290	320	351	381	412	442	473	503	2	6.4	6.2	6.0
143	534	564	594	625	655	685	715	746	776	806	3	9.6	9.3	9.0
144	836	866	897	927	957	987	*017	*047	*077	*107	4	12.8	12.4	12.0
145	16 137	167	197	227	256	286	316	346	376	406	5	16.0	15.5	15.0
146	435	465	495	524	554	584	613	643	673	702	6	19.2	18.6	18.0
147	732	761	791	820	850	879	909	938	967	997	7	22.4	21.7	21.0
148	17 026	056	085	114	143	173	202	231	260	289	8	25.6	24.8	24.0
149	319	348	377	406	435	464	493	522	551	580	9	28.8	27.9	27.0
150	609	638	667	696	725	754	782	811	840	869				
N.	0	1	2	3	4	5	6	7	8	9	Proportional parts			

FIVE-PLACE LOGARITHMS (Continued)

N.	0	1	2	3	4	5	6	7	8	9	Proportional parts		
150	17 609	638	667	696	725	754	782	811	840	869		**29**	**28**
151	898	926	955	984	*013	*041	*070	*099	*127	*156	1	2.9	2.8
152	18 184	213	241	270	298	327	355	384	412	441	2	5.8	5.6
153	469	498	526	554	583	611	639	667	696	724	3	8.7	8.4
154	752	780	808	837	865	893	921	949	977	*005	4	11.6	11.2
155	19 033	061	089	117	145	173	201	229	257	285	5	14.5	14.0
156	312	340	368	396	424	451	479	507	535	562	6	17.4	16.8
157	590	618	645	673	700	728	756	783	811	838	7	20.3	19.6
158	866	893	921	948	976	*003	*030	*058	*085	*112	8	23.2	22.4
159	20 140	167	194	222	249	276	303	330	358	385	9	26.1	25.2
160	412	439	466	493	520	548	575	602	629	656		**27**	**26**
161	683	710	737	763	790	817	844	871	898	925	1	2.7	2.6
162	952	978	*005	*032	*059	*085	*112	*139	*165	*192	2	5.4	5.2
163	21 219	245	272	299	325	352	378	405	431	458	3	8.1	7.8
164	484	511	537	564	590	617	643	669	696	722	4	10.8	10.4
165	748	775	801	827	854	880	906	932	958	985	5	13.5	13.0
166	22 011	037	063	089	115	141	167	194	220	246	6	16.2	15.6
167	272	298	324	350	376	401	427	453	479	505	7	18.9	18.2
168	531	557	583	608	634	660	686	712	737	763	8	21.6	20.8
169	789	814	840	866	891	917	943	968	994	*019	9	24.3	23.4
170	23 045	070	096	121	147	172	198	223	249	274		**25**	
171	300	325	350	376	401	426	452	477	502	528	1	2.5	
172	553	578	603	629	654	679	704	729	754	779	2	5.0	
173	805	830	855	880	905	930	955	980	*005	*030	3	7.5	
174	24 055	080	105	130	155	180	204	229	254	279	4	10.0	
175	304	329	353	378	403	428	452	477	502	527	5	12.5	
176	551	576	601	625	650	674	699	724	748	773	6	15.0	
177	797	822	846	871	895	920	944	969	993	*018	7	17.5	
178	25 042	066	091	115	139	164	188	212	237	261	8	20.0	
179	285	310	334	358	382	406	431	455	479	503	9	22.5	
180	527	551	575	600	624	648	672	696	720	744		**24**	**23**
181	768	792	816	840	864	888	912	935	959	983	1	2.4	2.3
182	26 007	031	055	079	102	126	150	174	198	221	2	4.8	4.6
183	245	269	293	316	340	364	387	411	435	458	3	7.2	6.9
184	482	505	529	553	576	600	623	647	670	694	4	9.6	9.2
185	717	741	764	788	811	834	858	881	905	928	5	12.0	11.5
186	951	975	998	*021	*045	*068	*091	*114	*138	*161	6	14.4	13.8
187	27 184	207	231	254	277	300	323	346	370	393	7	16.8	16.1
188	416	439	462	485	508	531	554	577	600	623	8	19.2	18.4
189	646	669	692	715	738	761	784	807	830	852	9	21.6	20.7
190	875	898	921	944	967	989	*012	*035	*058	*081		**22**	**21**
191	28 103	126	149	171	194	217	240	262	285	307	1	2.2	2.1
192	330	353	375	398	421	443	466	488	511	533	2	4.4	4.2
193	556	578	601	623	646	668	691	713	735	758	3	6.6	6.3
194	780	803	825	847	870	892	914	937	959	981	4	8.8	8.4
195	29 003	026	048	070	092	115	137	159	181	203	5	11.0	10.5
196	226	248	270	292	314	336	358	380	403	425	6	13.2	12.6
197	447	469	491	513	535	557	579	601	623	645	7	15.4	14.7
198	667	688	710	732	754	776	798	820	842	863	8	17.6	16.8
199	885	907	929	951	973	994	*016	*038	*060	*081	9	19.8	18.9
200	30 103	125	146	168	190	211	233	255	276	298			
N.	0	1	2	3	4	5	6	7	8	9	Proportional parts		

FIVE-PLACE LOGARITHMS (Continued)

N.	0	1	2	3	4	5	6	7	8	9	Proportional parts		
200	30 103	125	146	168	190	211	233	255	276	298		**22**	**21**
201	320	341	363	384	406	428	449	471	492	514	1	2.2	2.1
202	535	557	578	600	621	643	664	685	707	728	2	4.4	4.2
203	750	771	792	814	835	856	878	899	920	942	3	6.6	6.3
204	963	984	*006	*027	*048	*069	*091	*112	*133	*154	4	8.8	8.4
205	31 175	197	218	239	260	281	302	323	345	366	5	11.0	10.5
206	387	408	429	450	471	492	513	534	555	576	6	13.2	12.6
207	597	618	639	660	681	702	723	744	765	785	7	15.4	14.7
208	806	827	848	869	890	911	931	952	973	994	8	17.6	16.8
209	32 015	035	056	077	098	118	139	160	181	201	9	19.8	18.9
210	222	243	263	284	305	325	346	366	387	408		**20**	
211	428	449	469	490	510	531	552	572	593	613	1	2.0	
212	634	654	675	695	715	736	756	777	797	818	2	4.0	
213	838	858	879	899	919	940	960	980	*001	*021	3	6.0	
214	33 041	062	082	102	122	143	163	183	203	224	4	8.0	
215	244	264	284	304	325	345	365	385	405	425	5	10.0	
216	445	465	486	506	526	546	566	586	606	626	6	12.0	
217	646	666	686	706	726	746	766	786	806	826	7	14.0	
218	846	866	885	905	925	945	965	985	*005	*025	8	16.0	
219	34 044	064	084	104	124	143	163	183	203	223	9	18.0	
220	242	262	282	301	321	341	361	380	400	420		**19**	
221	439	459	479	498	518	537	557	577	596	616	1	1.9	
222	635	655	674	694	713	733	753	772	792	811	2	3.8	
223	830	850	869	889	908	928	947	967	986	*005	3	5.7	
224	35 025	044	064	083	102	122	141	160	180	199	4	7.6	
225	218	238	257	276	295	315	334	353	372	392	5	9.5	
226	411	430	449	468	488	507	526	545	564	583	6	11.4	
227	603	622	641	660	679	698	717	736	755	774	7	13.3	
228	793	813	832	851	870	889	908	927	946	965	8	15.2	
229	984	*003	*021	*040	*059	*078	*097	*116	*135	*154	9	17.1	
230	36 173	192	211	229	248	267	286	305	324	342		**18**	
231	361	380	399	418	436	455	474	493	511	530	1	1.8	
232	549	568	586	605	624	642	661	680	698	717	2	3.6	
233	736	754	773	791	810	829	847	866	884	903	3	5.4	
234	922	940	959	977	996	*014	*033	*051	*070	*088	4	7.2	
235	37 107	125	144	162	181	199	218	236	254	273	5	9.0	
236	291	310	328	346	365	383	401	420	438	457	6	10.8	
237	475	493	511	530	548	566	585	603	621	639	7	12.6	
238	658	676	694	712	731	749	767	785	803	822	8	14.4	
239	840	858	876	894	912	931	949	967	985	*003	9	16.2	
240	38 021	039	057	075	093	112	130	148	166	184		**17**	
241	202	220	238	256	274	292	310	328	346	364	1	1.7	
242	382	399	417	435	453	471	489	507	525	543	2	3.4	
243	561	578	596	614	632	650	668	686	703	721	3	5.1	
244	739	757	775	792	810	828	846	863	881	899	4	6.8	
245	917	934	952	970	987	*005	*023	*041	*058	*076	5	8.5	
246	39 094	111	129	146	164	182	199	217	235	252	6	10.2	
247	270	287	305	322	340	358	375	393	410	428	7	11.9	
248	445	463	480	498	515	533	550	568	585	602	8	13.6	
249	620	637	655	672	690	707	724	742	759	777	9	15.3	
250	794	811	829	846	863	881	898	915	933	950			
N.	0	1	2	3	4	5	6	7	8	9	Proportional parts		

N.	0	1	2	3	4	5	6	7	8	9	Proportional parts	
250	39 794	811	829	846	863	881	898	915	933	950		**18**
251	967	985	*002	*019	*037	*054	*071	*088	*106	*123	1	1.8
252	40 140	157	175	192	209	226	243	261	278	295	2	3.6
253	312	329	346	364	381	398	415	432	449	466	3	5.4
254	483	500	518	535	552	569	586	603	620	637	4	7.2
255	654	671	688	705	722	739	756	773	790	807	5	9.0
256	824	841	858	875	892	909	926	943	960	976	6	10.8
257	993	*010	*027	*044	*061	*078	*095	*111	*128	*145	7	12.6
258	41 162	179	196	212	229	246	263	280	296	313	8	14.4
259	330	347	363	380	397	414	430	447	464	481	9	16.2
260	497	514	531	547	564	581	597	614	631	647		**17**
261	664	681	697	714	731	747	764	780	797	814	1	1.7
262	830	847	863	880	896	913	929	946	963	979	2	3.4
263	996	*012	*029	*045	*062	*078	*095	*111	*127	*144	3	5.1
264	42 160	177	193	210	226	243	259	275	292	308	4	6.8
265	325	341	357	374	390	406	423	439	455	472	5	8.5
266	488	504	521	537	553	570	586	602	619	635	6	10.2
267	651	667	684	700	716	732	749	765	781	797	7	11.9
268	813	830	846	862	878	894	911	927	943	959	8	13.6
269	975	991	*008	*024	*040	*056	*072	*088	*104	*120	9	15.3
270	43 136	152	169	185	201	217	233	249	265	281		**16**
271	297	313	329	345	361	377	393	409	425	441	1	1.6
272	457	473	489	505	521	537	553	569	584	600	2	3.2
273	616	632	648	664	680	696	712	727	743	759	3	4.8
274	775	791	807	823	838	854	870	886	902	917	4	6.4
275	933	949	965	981	996	*012	*028	*044	*059	*075	5	8.0
276	44 091	107	122	138	154	170	185	201	217	232	6	9.6
277	248	264	279	295	311	326	342	358	373	389	7	11.2
278	404	420	436	451	467	483	498	514	529	545	8	12.8
279	560	576	592	607	623	638	654	669	685	700	9	14.4
280	716	731	747	762	778	793	809	824	840	855		**15**
281	871	886	902	917	932	948	963	979	994	*010	1	1.5
282	45 025	040	056	071	086	102	117	133	148	163	2	3.0
283	179	194	209	225	240	255	271	286	301	317	3	4.5
284	332	347	362	378	393	408	423	439	454	469	4	6.0
285	484	500	515	530	545	561	576	591	606	621	5	7.5
286	637	652	667	682	697	712	728	743	758	773	6	9.0
287	788	803	818	834	849	864	879	894	909	924	7	10.5
288	939	954	969	984	*000	*015	*030	*045	*060	*075	8	12.0
289	46 090	105	120	135	150	165	180	195	210	225	9	13.5
290	240	255	270	285	300	315	330	345	359	374		**14**
291	389	404	419	434	449	464	479	494	509	523	1	1.4
292	538	553	568	583	598	613	627	642	657	672	2	2.8
293	687	702	716	731	746	761	776	790	805	820	3	4.2
294	835	850	864	879	894	909	923	938	953	967	4	5.6
295	982	997	*012	*026	*041	*056	*070	*085	*100	*114	5	7.0
296	47 129	144	159	173	188	202	217	232	246	261	6	8.4
297	276	290	305	319	334	349	363	378	392	407	7	9.8
298	422	436	451	465	480	494	509	524	538	553	8	11.2
299	567	582	596	611	625	640	654	669	683	698	9	12.6
300	712	727	741	756	770	784	799	813	828	842		
N.	0	1	2	3	4	5	6	7	8	9	Proportional parts	

FIVE-PLACE LOGARITHMS (Continued)

N.		0	1	2	3	4	5	6	7	8	9	Proportional parts	
300	47	712	727	741	756	770	784	799	813	828	842		**15**
301		857	871	885	900	914	929	943	958	972	986	1	1.5
302	48	001	015	029	044	058	073	087	101	116	130	2	3.0
303		144	159	173	187	202	216	230	244	259	273	3	4.5
304		287	302	316	330	344	359	373	387	401	416	4	6.0
305		430	444	458	473	487	501	515	530	544	558	5	7.5
306		572	586	601	615	629	643	657	671	686	700	6	9.0
307		714	728	742	756	770	785	799	813	827	841	7	10.5
308		855	869	883	897	911	926	940	954	968	982	8	12.0
309		996	*010	*024	*038	*052	*066	*080	*094	*108	*122	9	13.5
310	49	136	150	164	178	192	206	220	234	248	262		
311		276	290	304	318	332	346	360	374	388	402		
312		415	429	443	457	471	485	499	513	527	541		
313		554	568	582	596	610	624	638	651	665	679		**14**
314		693	707	721	734	748	762	776	790	803	817	1	1.4
315		831	845	859	872	886	900	914	927	941	955	2	2.8
316		969	982	996	*010	*024	*037	*051	*065	*079	*092	3	4.2
317	50	106	120	133	147	161	174	188	202	215	229	4	5.6
318		243	256	270	284	297	311	325	338	352	365	5	7.0
319		379	393	406	420	433	447	461	474	488	501	6	8.4
												7	9.8
320		515	529	542	556	569	583	596	610	623	637	8	11.2
321		651	664	678	691	705	718	732	745	759	772	9	12.6
322		786	799	813	826	840	853	866	880	893	907		
323		920	934	947	961	974	987	*001	*014	*028	*041		
324	51	055	068	081	095	108	121	135	148	162	175		
325		188	202	215	228	242	255	268	282	295	308		
326		322	335	348	362	375	388	402	415	428	441		**13**
327		455	468	481	495	508	521	534	548	561	574	1	1.3
328		587	601	614	627	640	654	667	680	693	706	2	2.6
329		720	733	746	759	772	786	799	812	825	838	3	3.9
												4	5.2
330		851	865	878	891	904	917	930	943	957	970	5	6.5
331		983	996	*009	*022	*035	*048	*061	*075	*088	*101	6	7.8
332	52	114	127	140	153	166	179	192	205	218	231	7	9.1
333		244	257	270	284	297	310	323	336	349	362	8	10.4
334		375	388	401	414	427	440	453	466	479	492	9	11.7
335		504	517	530	543	556	569	582	595	608	621		
336		634	647	660	673	686	699	711	724	737	750		
337		763	776	789	802	815	827	840	853	866	879		
338		892	905	917	930	943	956	969	982	994	*007		**12**
339	53	020	033	046	058	071	084	097	110	122	135	1	1.2
												2	2.4
340		148	161	173	186	199	212	224	237	250	263	3	3.6
341		275	288	301	314	326	339	352	364	377	390	4	4.8
342		403	415	428	441	453	466	479	491	504	517	5	6.0
343		529	542	555	567	580	593	605	618	631	643	6	7.2
344		656	668	681	694	706	719	732	744	757	769	7	8.4
345		782	794	807	820	832	845	857	870	882	895	8	9.6
346		908	920	933	945	958	970	983	995	*008	*020	9	10.8
347	54	033	045	058	070	083	095	108	120	133	145		
348		158	170	183	195	208	220	233	245	258	270		
349		283	295	307	320	332	345	357	370	382	394		
350		407	419	432	444	456	469	481	494	506	518		
N.		0	1	2	3	4	5	6	7	8	9	Proportional parts	

N.	0	1	2	3	4	5	6	7	8	9
350	54 407	419	432	444	456	469	481	494	506	518
351	531	543	555	568	580	593	605	617	630	642
352	654	667	679	691	704	716	728	741	753	765
353	777	790	802	814	827	839	851	864	876	888
354	900	913	925	937	949	962	974	986	998	*011
355	55 023	035	047	060	072	084	096	108	121	133
356	145	157	169	182	194	206	218	230	242	255
357	267	279	291	303	315	328	340	352	364	376
358	388	400	413	425	437	449	461	473	485	497
359	509	522	534	546	558	570	582	594	606	618
360	630	642	654	666	678	691	703	715	727	739
361	751	763	775	787	799	811	823	835	847	859
362	871	883	895	907	919	931	943	955	967	979
363	991	*003	*015	*027	*038	*050	*062	*074	*086	*098
364	56 110	122	134	146	158	170	182	194	205	217
365	229	241	253	265	277	289	301	312	324	336
366	348	360	372	384	396	407	419	431	443	455
367	467	478	490	502	514	526	538	549	561	573
368	585	597	608	620	632	644	656	667	679	691
369	703	714	726	738	750	761	773	785	797	808
370	820	832	844	855	867	879	891	902	914	926
371	937	949	961	972	984	996	*008	*019	*031	*043
372	57 054	066	078	089	101	113	124	136	148	159
373	171	183	194	206	217	229	241	252	264	276
374	287	299	310	322	334	345	357	368	380	392
375	403	415	426	438	449	461	473	484	496	507
376	519	530	542	553	565	576	588	600	611	623
377	634	646	657	669	680	692	703	715	726	738
378	749	761	772	784	795	807	818	830	841	852
379	864	875	887	898	910	921	933	944	955	967
380	978	990	*001	*013	*024	*035	*047	*058	*070	*081
381	58 092	104	115	127	138	149	161	172	184	195
382	206	218	229	240	252	263	274	286	297	309
383	320	331	343	354	365	377	388	399	410	422
384	433	444	456	467	478	490	501	512	524	535
385	546	557	569	580	591	602	614	625	636	647
386	659	670	681	692	704	715	726	737	749	760
387	771	782	794	805	816	827	838	850	861	872
388	883	894	906	917	928	939	950	961	973	984
389	995	*006	*017	*028	*040	*051	*062	*073	*084	*095
390	59 106	118	129	140	151	162	173	184	195	207
391	218	229	240	251	262	273	284	295	306	318
392	329	340	351	362	373	384	395	406	417	428
393	439	450	461	472	483	494	506	517	528	539
394	550	561	572	583	594	605	616	627	638	649
395	660	671	682	693	704	715	726	737	748	759
396	770	780	791	802	813	824	835	846	857	868
397	879	890	901	912	923	934	945	956	966	977
398	988	999	*010	*021	*032	*043	*054	*065	*076	*086
399	60 097	108	119	130	141	152	163	173	184	195
400	206	217	228	239	249	260	271	282	293	304
N.	0	1	2	3	4	5	6	7	8	9

Proportional parts

	13
1	1.3
2	2.6
3	3.9
4	5.2
5	6.5
6	7.8
7	9.1
8	10.4
9	11.7

	12
1	1.2
2	2.4
3	3.6
4	4.8
5	6.0
6	7.2
7	8.4
8	9.6
9	10.8

	11
1	1.1
2	2.2
3	3.3
4	4.4
5	5.5
6	6.6
7	7.7
8	8.8
9	9.9

	10
1	1.0
2	2.0
3	3.0
4	4.0
5	5.0
6	6.0
7	7.0
8	8.0
9	9.0

Proportional parts

N.		0	1	2	3	4	5	6	7	8	9	Proportional parts	
400	60	206	217	228	239	249	260	271	282	293	304		
401		314	325	336	347	358	369	379	390	401	412		
402		423	433	444	455	466	477	487	498	509	520		
403		531	541	552	563	574	584	595	606	617	627		
404		638	649	660	670	681	692	703	713	724	735		
405		746	756	767	778	788	799	810	821	831	842		
406		853	863	874	885	895	906	917	927	938	949		
407		959	970	981	991	*002	*013	*023	*034	*045	*055		**11**
408	61	066	077	087	098	109	119	130	140	151	162	1	1.1
409		172	183	194	204	215	225	236	247	257	268	2	2.2
												3	3.3
410		278	289	300	310	321	331	342	352	363	374	4	4.4
411		384	395	405	416	426	437	448	458	469	479	5	5.5
412		490	500	511	521	532	542	553	563	574	584	6	6.6
413		595	606	616	627	637	648	658	669	679	690	7	7.7
414		700	711	721	731	742	752	763	773	784	794	8	8.8
415		805	815	826	836	847	857	868	878	888	899	9	9.9
416		909	920	930	941	951	962	972	982	993	*003		
417	62	014	024	034	045	055	066	076	086	097	107		
418		118	128	138	149	159	170	180	190	201	211		
419		221	232	242	252	263	273	284	294	304	315		
420		325	335	346	356	366	377	387	397	408	418		**10**
421		428	439	449	459	469	480	490	500	511	521	1	1.0
422		531	542	552	562	572	583	593	603	613	624	2	2.0
423		634	644	655	665	675	685	696	706	716	726	3	3.0
424		737	747	757	767	778	788	798	808	818	829	4	4.0
425		839	849	859	870	880	890	900	910	921	931	5	5.0
426		941	951	961	972	982	992	*002	*012	*022	*033	6	6.0
427	63	043	053	063	073	083	094	104	114	124	134	7	7.0
428		144	155	165	175	185	195	205	215	225	236	8	8.0
429		246	256	266	276	286	296	306	317	327	337	9	9.0
430		347	357	367	377	387	397	407	417	428	438		
431		448	458	468	478	488	498	508	518	528	538		
432		548	558	568	579	589	599	609	619	629	639		
433		649	659	669	679	689	699	709	719	729	739		
434		749	759	769	779	789	799	809	819	829	839		
435		849	859	869	879	889	899	909	919	929	939		
436		949	959	969	979	988	998	*008	*018	*028	*038		
437	64	048	058	068	078	088	098	108	118	128	137		
438		147	157	167	177	187	197	207	217	227	237		**9**
439		246	256	266	276	286	296	306	316	326	335	1	0.9
												2	1.8
440		345	355	365	375	385	395	404	414	424	434	3	2.7
441		444	454	464	473	483	493	503	513	523	532	4	3.6
442		542	552	562	572	582	591	601	611	621	631	5	4.5
443		640	650	660	670	680	689	699	709	719	729	6	5.4
444		738	748	758	768	777	787	797	807	816	826	7	6.3
445		836	846	856	865	875	885	895	904	914	924	8	7.2
446		933	943	953	963	972	982	992	*002	*011	*021	9	8.1
447	65	031	040	050	060	070	079	089	099	108	118		
448		128	137	147	157	167	176	186	196	205	215		
449		225	234	244	254	263	273	283	292	302	312		
450		321	331	341	350	360	369	379	389	398	408		
N.		0	1	2	3	4	5	6	7	8	9	Proportional parts	

N.	0	1	2	3	4	5	6	7	8	9
450	65 321	331	341	350	360	369	379	389	398	408
451	418	427	437	447	456	466	475	485	495	504
452	514	523	533	543	552	562	571	581	591	600
453	610	619	629	639	648	658	667	677	686	696
454	706	715	725	734	744	753	763	772	782	792
455	801	811	820	830	839	849	858	868	877	887
456	896	906	916	925	935	944	954	963	973	982
457	992	*001	*011	*020	*030	*039	*049	*058	*068	*077
458	66 087	096	106	115	124	134	143	153	162	172
459	181	191	200	210	219	229	238	247	257	266
460	276	285	295	304	314	323	332	342	351	361
461	370	380	389	398	408	417	427	436	445	455
462	464	474	483	492	502	511	521	530	539	549
463	558	567	577	586	596	605	614	624	633	642
464	652	661	671	680	689	699	708	717	727	736
465	745	755	764	773	783	792	801	811	820	829
466	839	848	857	867	876	885	894	904	913	922
467	932	941	950	960	969	978	987	997	*006	*015
468	67 025	034	043	052	062	071	080	089	099	108
469	117	127	136	145	154	164	173	182	191	201
470	210	219	228	237	247	256	265	274	284	293
471	302	311	321	330	339	348	357	367	376	385
472	394	403	413	422	431	440	449	459	468	477
473	486	495	504	514	523	532	541	550	560	569
474	578	587	596	605	614	624	633	642	651	660
475	669	679	688	697	706	715	724	733	742	752
476	761	770	779	788	797	806	815	825	834	843
477	852	861	870	879	888	897	906	916	925	934
478	943	952	961	970	979	988	997	*006	*015	*024
479	68 034	043	052	061	070	079	088	097	106	115
480	124	133	142	151	160	169	178	187	196	205
481	215	224	233	242	251	260	269	278	287	296
482	305	314	323	332	341	350	359	368	377	386
483	395	404	413	422	431	440	449	458	467	476
484	485	494	502	511	520	529	538	547	556	565
485	574	583	592	601	610	619	628	637	646	655
486	664	673	681	690	699	708	717	726	735	744
487	753	762	771	780	789	797	806	815	824	833
488	842	851	860	869	878	886	895	904	913	922
489	931	940	949	958	966	975	984	993	*002	*011
490	69 020	028	037	046	055	064	073	082	090	099
491	108	117	126	135	144	152	161	170	179	188
492	197	205	214	223	232	241	249	258	267	276
493	285	294	302	311	320	329	338	346	355	364
494	373	381	390	399	408	417	425	434	443	452
495	461	469	478	487	496	504	513	522	531	539
496	548	557	566	574	583	592	601	609	618	627
497	636	644	653	662	671	679	688	697	705	714
498	723	732	740	749	758	767	775	784	793	801
499	810	819	827	836	845	854	862	871	880	888
500	897	906	914	923	932	940	949	958	966	975
N.	0	1	2	3	4	5	6	7	8	9

Proportional parts

	10
1	1.0
2	2.0
3	3.0
4	4.0
5	5.0
6	6.0
7	7.0
8	8.0
9	9.0

	9
1	0.9
2	1.8
3	2.7
4	3.6
5	4.5
6	5.4
7	6.3
8	7.2
9	8.1

	8
1	0.8
2	1.6
3	2.4
4	3.2
5	4.0
6	4.8
7	5.6
8	6.4
9	7.2

Proportional parts

N.	0	1	2	3	4	5	6	7	8	9	Proportional parts	
500	69 897	906	914	923	932	940	949	958	966	975		
501	984	992	*001	*010	*018	*027	*036	*044	*053	*062		
502	70 070	079	088	096	105	114	122	131	140	148		
503	157	165	174	183	191	200	209	217	226	234		
504	243	252	260	269	278	286	295	303	312	321		
505	329	338	346	355	364	372	381	389	398	406		
506	415	424	432	441	449	458	467	475	484	492		
507	501	509	518	526	535	544	552	561	569	578		**9**
508	586	595	603	612	621	629	638	646	655	663	1	0.9
509	672	680	689	697	706	714	723	731	740	749	2	1.8
											3	2.7
510	757	766	774	783	791	800	808	817	825	834	4	3.6
511	842	851	859	868	876	885	893	902	910	919	5	4.5
512	927	935	944	952	961	969	978	986	995	*003	6	5.4
513	71 012	020	029	037	046	054	063	071	079	088	7	6.3
514	096	105	113	122	130	139	147	155	164	172	8	7.2
515	181	189	198	206	214	223	231	240	248	257	9	8.1
516	265	273	282	290	299	307	315	324	332	341		
517	349	357	366	374	383	391	399	408	416	425		
518	433	441	450	458	466	475	483	492	500	508		
519	517	525	533	542	550	559	567	575	584	592		
520	600	609	617	625	634	642	650	659	667	675		
521	684	692	700	709	717	725	734	742	750	759		**8**
522	767	775	784	792	800	809	817	825	834	842	1	0.8
523	850	858	867	875	883	892	900	908	917	925	2	1.6
524	933	941	950	958	966	975	983	991	999	*008	3	2.4
525	72 016	024	032	041	049	057	066	074	082	090	4	3.2
526	099	107	115	123	132	140	148	156	165	173	5	4.0
527	181	189	198	206	214	222	230	239	247	255	6	4.8
528	263	272	280	288	296	304	313	321	329	337	7	5.6
529	346	354	362	370	378	387	395	403	411	419	8	6.4
											9	7.2
530	428	436	444	452	460	469	477	485	493	501		
531	509	518	526	534	542	550	558	567	575	583		
532	591	599	607	616	624	632	640	648	656	665		
533	673	681	689	697	705	713	722	730	738	746		
534	754	762	770	779	787	795	803	811	819	827		
535	835	843	852	860	868	876	884	892	900	908		
536	916	925	933	941	949	957	965	973	981	989		
537	997	*006	*014	*022	*030	*038	*046	*054	*062	*070		
538	73 078	086	094	102	111	119	127	135	143	151		
539	159	167	175	183	191	199	207	215	223	231		**7**
											1	0.7
540	239	247	255	263	272	280	288	296	304	312	2	1.4
541	320	328	336	344	352	360	368	376	384	392	3	2.1
542	400	408	416	424	432	440	448	456	464	472	4	2.8
543	480	488	496	504	512	520	528	536	544	552	5	3.5
544	560	568	576	584	592	600	608	616	624	632	6	4.2
545	640	648	656	664	672	679	687	695	703	711	7	4.9
546	719	727	735	743	751	759	767	775	783	791	8	5.6
547	799	807	815	823	830	838	846	854	862	870	9	6.3
548	878	886	894	902	910	918	926	933	941	949		
549	957	965	973	981	989	997	*005	*013	*020	*028		
550	74 036	044	052	060	068	076	084	092	099	107		
N.	0	1	2	3	4	5	6	7	8	9	Proportional parts	

FIVE-PLACE LOGARITHMS (Continued)

N.	0	1	2	3	4	5	6	7	8	9	Proportional parts	
550	74 036	044	052	060	068	076	084	092	099	107		
551	115	123	131	139	147	155	162	170	178	186		
552	194	202	210	218	225	233	241	249	257	265		
553	273	280	288	296	304	312	320	327	335	343		
554	351	359	367	374	382	390	398	406	414	421		
555	429	437	445	453	461	468	476	484	492	500		
556	507	515	523	531	539	547	554	562	570	578		
557	586	593	601	609	617	624	632	640	648	656		
558	663	671	679	687	695	702	710	718	726	733		
559	741	749	757	764	772	780	788	796	803	811		
560	819	827	834	842	850	858	865	873	881	889		**8**
561	896	904	912	920	927	935	943	950	958	966	1	0.8
562	974	981	989	997	*005	*012	*020	*028	*035	*043	2	1.6
563	75 051	059	066	074	082	089	097	105	113	120	3	2.4
564	128	136	143	151	159	166	174	182	189	197	4	3.2
565	205	213	220	228	236	243	251	259	266	274	5	4.0
566	282	289	297	305	312	320	328	335	343	351	6	4.8
567	358	366	374	381	389	397	404	412	420	427	7	5.6
568	435	442	450	458	465	473	481	488	496	504	8	6.4
569	511	519	526	534	542	549	557	565	572	580	9	7.2
570	587	595	603	610	618	626	633	641	648	656		
571	664	671	679	686	694	702	709	717	724	732		
572	740	747	755	762	770	778	785	793	800	808		
573	815	823	831	838	846	853	861	868	876	884		
574	891	899	906	914	921	929	937	944	952	959		
575	967	974	982	989	997	*005	*012	*020	*027	*035		
576	76 042	050	057	065	072	080	087	095	103	110		
577	118	125	133	140	148	155	163	170	178	185		
578	193	200	208	215	223	230	238	245	253	260		
579	268	275	283	290	298	305	313	320	328	335		
580	343	350	358	365	373	380	388	395	403	410		**7**
581	418	425	433	440	448	455	462	470	477	485	1	0.7
582	492	500	507	515	522	530	537	545	552	559	2	1.4
583	567	574	582	589	597	604	612	619	626	634	3	2.1
584	641	649	656	664	671	678	686	693	701	708	4	2.8
585	716	723	730	738	745	753	760	768	775	782	5	3.5
586	790	797	805	812	819	827	834	842	849	856	6	4.2
587	864	871	879	886	893	901	908	916	923	930	7	4.9
588	938	945	953	960	967	975	982	989	997	*004	8	5.6
589	77 012	019	026	034	041	048	056	063	070	078	9	6.3
590	085	093	100	107	115	122	129	137	144	151		
591	159	166	173	181	188	195	203	210	217	225		
592	232	240	247	254	262	269	276	283	291	298		
593	305	313	320	327	335	342	349	357	364	371		
594	379	386	393	401	408	415	422	430	437	444		
595	452	459	466	474	481	488	495	503	510	517		
596	525	532	539	546	554	561	568	576	583	590		
597	597	605	612	619	627	634	641	648	656	663		
598	670	677	685	692	699	706	714	721	728	735		
599	743	750	757	764	772	779	786	793	801	808		
600	815	822	830	837	844	851	859	866	873	880		
N.	0	1	2	3	4	5	6	7	8	9	Proportional parts	

N.		0	1	2	3	4	5	6	7	8	9
600	77	815	822	830	837	844	851	859	866	873	880
601		887	895	902	909	916	924	931	938	945	952
602		960	967	974	981	988	996	*003	*010	*017	*025
603	78	032	039	046	053	061	068	075	082	089	097
604		104	111	118	125	132	140	147	154	161	168
605		176	183	190	197	204	211	219	226	233	240
606		247	254	262	269	276	283	290	297	305	312
607		319	326	333	340	347	355	362	369	376	383
608		390	398	405	412	419	426	433	440	447	455
609		462	469	476	483	490	497	504	512	519	526
610		533	540	547	554	561	569	576	583	590	597
611		604	611	618	625	633	640	647	654	661	668
612		675	682	689	696	704	711	718	725	732	739
613		746	753	760	767	774	781	789	796	803	810
614		817	824	831	838	845	852	859	866	873	880
615		888	895	902	909	916	923	930	937	944	951
616		958	965	972	979	986	993	*000	*007	*014	*021
617	79	029	036	043	050	057	064	071	078	085	092
618		099	106	113	120	127	134	141	148	155	162
619		169	176	183	190	197	204	211	218	225	232
620		239	246	253	260	267	274	281	288	295	302
621		309	316	323	330	337	344	351	358	365	372
622		379	386	393	400	407	414	421	428	435	442
623		449	456	463	470	477	484	491	498	505	511
624		518	525	532	539	546	553	560	567	574	581
625		588	595	602	609	616	623	630	637	644	650
626		657	664	671	678	685	692	699	706	713	720
627		727	734	741	748	754	761	768	775	782	789
628		796	803	810	817	824	831	837	844	851	858
629		865	872	879	886	893	900	906	913	920	927
630		934	941	948	955	962	969	975	982	989	996
631	80	003	010	017	024	030	037	044	051	058	065
632		072	079	085	092	099	106	113	120	127	134
633		140	147	154	161	168	175	182	188	195	202
634		209	216	223	229	236	243	250	257	264	271
635		277	284	291	298	305	312	318	325	332	339
636		346	353	359	366	373	380	387	393	400	407
637		414	421	428	434	441	448	455	462	468	475
638		482	489	496	502	509	516	523	530	536	543
639		550	557	564	570	577	584	591	598	604	611
640		618	625	632	638	645	652	659	665	672	679
641		686	693	699	706	713	720	726	733	740	747
642		754	760	767	774	781	787	794	801	808	814
643		821	828	835	841	848	855	862	868	875	882
644		889	895	902	909	916	922	929	936	943	949
645		956	963	969	976	983	990	996	*003	*010	*017
646	81	023	030	037	043	050	057	064	070	077	084
647		090	097	104	111	117	124	131	137	144	151
648		158	164	171	178	184	191	198	204	211	218
649		224	231	238	245	251	258	265	271	278	285
650		291	298	305	311	318	325	331	338	345	351
N.		0	1	2	3	4	5	6	7	8	9

Proportional parts

	8
1	0.8
2	1.6
3	2.4
4	3.2
5	4.0
6	4.8
7	5.6
8	6.4
9	7.2

	7
1	0.7
2	1.4
3	2.1
4	2.8
5	3.5
6	4.2
7	4.9
8	5.6
9	6.3

	6
1	0.6
2	1.2
3	1.8
4	2.4
5	3.0
6	3.6
7	4.2
8	4.8
9	5.4

FIVE-PLACE LOGARITHMS (Continued)

N.	0	1	2	3	4	5	6	7	8	9	Proportional parts	
650	81 291	298	305	311	318	325	331	338	345	351		
651	358	365	371	378	385	391	398	405	411	418		
652	425	431	438	445	451	458	465	471	478	485		
653	491	498	505	511	518	525	531	538	544	551		
654	558	564	571	578	584	591	598	604	611	617		
655	624	631	637	644	651	657	664	671	677	684		
656	690	697	704	710	717	723	730	737	743	750		
657	757	763	770	776	783	790	796	803	809	816		
658	823	829	836	842	849	856	862	869	875	882		
659	889	895	902	908	915	921	928	935	941	948		
660	954	961	968	974	981	987	994	*000	*007	*014		**7**
661	82 020	027	033	040	046	053	060	066	073	079	1	0.7
662	086	092	099	105	112	119	125	132	138	145	2	1.4
663	151	158	164	171	178	184	191	197	204	210	3	2.1
664	217	223	230	236	243	249	256	263	269	276	4	2.8
665	282	289	295	302	308	315	321	328	334	341	5	3.5
666	347	354	360	367	373	380	387	393	400	406	6	4.2
667	413	419	426	432	439	445	452	458	465	471	7	4.9
668	478	484	491	497	504	510	517	523	530	536	8	5.6
669	543	549	556	562	569	575	582	588	595	601	9	6.3
670	607	614	620	627	633	640	646	653	659	666		
671	672	679	685	692	698	705	711	718	724	730		
672	737	743	750	756	763	769	776	782	789	795		
673	802	808	814	821	827	834	840	847	853	860		
674	866	872	879	885	892	898	905	911	918	924		
675	930	937	943	950	956	963	969	975	982	988		
676	995	*001	*008	*014	*020	*027	*033	*040	*046	*052		
677	83 059	065	072	078	085	091	097	104	110	117		
678	123	129	136	142	149	155	161	168	174	181		
679	187	193	200	206	213	219	225	232	238	245		
680	251	257	264	270	276	283	289	296	302	308		**6**
681	315	321	327	334	340	347	353	359	366	372	1	0.6
682	378	385	391	398	404	410	417	423	429	436	2	1.2
683	442	448	455	461	467	474	480	487	493	499	3	1.8
684	506	512	518	525	531	537	544	550	556	563	4	2.4
685	569	575	582	588	594	601	607	613	620	626	5	3.0
686	632	639	645	651	658	664	670	677	683	689	6	3.6
687	696	702	708	715	721	727	734	740	746	753	7	4.2
688	759	765	771	778	784	790	797	803	809	816	8	4.8
689	822	828	835	841	847	853	860	866	872	879	9	5.4
690	885	891	897	904	910	916	923	929	935	942		
691	948	954	960	967	973	979	985	992	998	*004		
692	84 011	017	023	029	036	042	048	055	061	067		
693	073	080	086	092	098	105	111	117	123	130		
694	136	142	148	155	161	167	173	180	186	192		
695	198	205	211	217	223	230	236	242	248	255		
696	261	267	273	280	286	292	298	305	311	317		
697	323	330	336	342	348	354	361	367	373	379		
698	386	392	398	404	410	417	423	429	435	442		
699	448	454	460	466	473	479	485	491	497	504		
700	510	516	522	528	535	541	547	553	559	566		
N.	0	1	2	3	4	5	6	7	8	9	Proportional parts	

N.	0	1	2	3	4	5	6	7	8	9	Proportional parts	
700	84 510	516	522	528	535	541	547	553	559	566		
701	572	578	584	590	597	603	609	615	621	628		
702	634	640	646	652	658	665	671	677	683	689		
703	696	702	708	714	720	726	733	739	745	751		
704	757	763	770	776	782	788	794	800	807	813		
705	819	825	831	837	844	850	856	862	868	874		
706	880	887	893	899	905	911	917	924	930	936		
707	942	948	954	960	967	973	979	985	991	997		**7**
708	85 003	009	016	022	028	034	040	046	052	058	1	0.7
709	065	071	077	083	089	095	101	107	114	120	2	1.4
											3	2.1
710	126	132	138	144	150	156	163	169	175	181	4	2.8
711	187	193	199	205	211	217	224	230	236	242	5	3.5
712	248	254	260	266	272	278	285	291	297	303	6	4.2
713	309	315	321	327	333	339	345	352	358	364	7	4.9
714	370	376	382	388	394	400	406	412	418	425	8	5.6
715	431	437	443	449	455	461	467	473	479	485	9	6.3
716	491	497	503	509	516	522	528	534	540	546		
717	552	558	564	570	576	582	588	594	600	606		
718	612	618	625	631	637	643	649	655	661	667		
719	673	679	685	691	697	703	709	715	721	727		**6**
											1	0.6
720	733	739	745	751	757	763	769	775	781	788	2	1.2
721	794	800	806	812	818	824	830	836	842	848	3	1.8
722	854	860	866	872	878	884	890	896	902	908	4	2.4
723	914	920	926	932	938	944	950	956	962	968	5	3.0
724	974	980	986	992	998	*004	*010	*016	*022	*028	6	3.6
725	86 034	040	046	052	058	064	070	076	082	088	7	4.2
726	094	100	106	112	118	124	130	136	141	147	8	4.8
727	153	159	165	171	177	183	189	195	201	207	9	5.4
728	213	219	225	231	237	243	249	255	261	267		
729	273	279	285	291	297	303	308	314	320	326		
730	332	338	344	350	356	362	368	374	380	386		**5**
731	392	398	404	410	415	421	427	433	439	445	1	0.5
732	451	457	463	469	475	481	487	493	499	504	2	1.0
733	510	516	522	528	534	540	546	552	558	564	3	1.5
734	570	576	581	587	593	599	605	611	617	623	4	2.0
735	629	635	641	646	652	658	664	670	676	682	5	2.5
736	688	694	700	705	711	717	723	729	735	741	6	3.0
737	747	753	759	764	770	776	782	788	794	800	7	3.5
738	806	812	817	823	829	835	841	847	853	859	8	4.0
739	864	870	876	882	888	894	900	906	911	917	9	4.5
740	923	929	935	941	947	953	958	964	970	976		
741	982	988	994	999	*005	*011	*017	*023	*029	*035		
742	87 040	046	052	058	064	070	075	081	087	093		
743	099	105	111	116	122	128	134	140	146	151		
744	157	163	169	175	181	186	192	198	204	210		
745	216	221	227	233	239	245	251	256	262	268		
746	274	280	286	291	297	303	309	315	320	326		
747	332	338	344	349	355	361	367	373	379	384		
748	390	396	402	408	413	419	425	431	437	442		
749	448	454	460	466	471	477	483	489	495	500		
750	506	512	518	523	529	535	541	547	552	558		
N.	0	1	2	3	4	5	6	7	8	9	Proportional parts	

N.	0	1	2	3	4	5	6	7	8	9	Proportional parts	
750	87 506	512	518	523	529	535	541	547	552	558		
751	564	570	576	581	587	593	599	604	610	616		
752	622	628	633	639	645	651	656	662	668	674		
753	679	685	691	697	703	708	714	720	726	731		
754	737	743	749	754	760	766	772	777	783	789		
755	795	800	806	812	818	823	829	835	841	846		
756	852	858	864	869	875	881	887	892	898	904		
757	910	915	921	927	933	938	944	950	955	961		
758	967	973	978	984	990	996	*001	*007	*013	*018		
759	88 024	030	036	041	047	053	058	064	070	076		
760	081	087	093	098	104	110	116	121	127	133		**6**
761	138	144	150	156	161	167	173	178	184	190	1	0.6
762	195	201	207	213	218	224	230	235	241	247	2	1.2
763	252	258	264	270	275	281	287	292	298	304	3	1.8
764	309	315	321	326	332	338	343	349	355	360	4	2.4
765	366	372	377	383	389	395	400	406	412	417	5	3.0
766	423	429	434	440	446	451	457	463	468	474	6	3.6
767	480	485	491	497	502	508	513	519	525	530	7	4.2
768	536	542	547	553	559	564	570	576	581	587	8	4.8
769	593	598	604	610	615	621	627	632	638	643	9	5.4
770	649	655	660	666	672	677	683	689	694	700		
771	705	711	717	722	728	734	739	745	750	756		
772	762	767	773	779	784	790	795	801	807	812		
773	818	824	829	835	840	846	852	857	863	868		
774	874	880	885	891	897	902	908	913	919	925		
775	930	936	941	947	953	958	964	969	975	981		
776	986	992	997	*003	*009	*014	*020	*025	*031	*037		
777	89 042	048	053	059	064	070	076	081	087	092		
778	098	104	109	115	120	126	131	137	143	148		
779	154	159	165	170	176	182	187	193	198	204		
780	209	215	221	226	232	237	243	248	254	260		**5**
781	265	271	276	282	287	293	298	304	310	315	1	0.5
782	321	326	332	337	343	348	354	360	365	371	2	1.0
783	376	382	387	393	398	404	409	415	421	426	3	1.5
784	432	437	443	448	454	459	465	470	476	481	4	2.0
785	487	492	498	504	509	515	520	526	531	537	5	2.5
786	542	548	553	559	564	570	575	581	586	592	6	3.0
787	597	603	609	614	620	625	631	636	642	647	7	3.5
788	653	658	664	669	675	680	686	691	697	702	8	4.0
789	708	713	719	724	730	735	741	746	752	757	9	4.5
790	763	768	774	779	785	790	796	801	807	812		
791	818	823	829	834	840	845	851	856	862	867		
792	873	878	883	889	894	900	905	911	916	922		
793	927	933	938	944	949	955	960	966	971	977		
794	982	988	993	998	*004	*009	*015	*020	*026	*031		
795	90 037	042	048	053	059	064	069	075	080	086		
796	091	097	102	108	113	119	124	129	135	140		
797	146	151	157	162	168	173	179	184	189	195		
798	200	206	211	217	222	227	233	238	244	249		
799	255	260	266	271	276	282	287	293	298	304		
800	309	314	320	325	331	336	342	347	352	358		
N.	0	1	2	3	4	5	6	7	8	9	Proportional parts	

N.		0	1	2	3	4	5	6	7	8	9	Proportional parts	
800	90	309	314	320	325	331	336	342	347	352	358		
801		363	369	374	380	385	390	396	401	407	412		
802		417	423	428	434	439	445	450	455	461	466		
803		472	477	482	488	493	499	504	509	515	520		
804		526	531	536	542	547	553	558	563	569	574		
805		580	585	590	596	601	607	612	617	623	628		
806		634	639	644	650	655	660	666	671	677	682		
807		687	693	698	703	709	714	720	725	730	736		
808		741	747	752	757	763	768	773	779	784	789		
809		795	800	806	811	816	822	827	832	838	843		
810		849	854	859	865	870	875	881	886	891	897		**6**
811		902	907	913	918	924	929	934	940	945	950	1	0.6
812		956	961	966	972	977	982	988	993	998	*004	2	1.2
813	91	009	014	020	025	030	036	041	046	052	057	3	1.8
814		062	068	073	078	084	089	094	100	105	110	4	2.4
815		116	121	126	132	137	142	148	153	158	164	5	3.0
816		169	174	180	185	190	196	201	206	212	217	6	3.6
817		222	228	233	238	243	249	254	259	265	270	7	4.2
818		275	281	286	291	297	302	307	312	318	323	8	4.8
819		328	334	339	344	350	355	360	365	371	376	9	5.4
820		381	387	392	397	403	408	413	418	424	429		
821		434	440	445	450	455	461	466	471	477	482		
822		487	492	498	503	508	514	519	524	529	535		
823		540	545	551	556	561	566	572	577	582	587		
824		593	598	603	609	614	619	624	630	635	640		
825		645	651	656	661	666	672	677	682	687	693		
826		698	703	709	714	719	724	730	735	740	745		
827		751	756	761	766	772	777	782	787	793	798		
828		803	808	814	819	824	829	834	840	845	850		
829		855	861	866	871	876	882	887	892	897	903		
830		908	913	918	924	929	934	939	944	950	955		**5**
831		960	965	971	976	981	986	991	997	*002	*007	1	0.5
832	92	012	018	023	028	033	038	044	049	054	059	2	1.0
833		065	070	075	080	085	091	096	101	106	111	3	1.5
834		117	122	127	132	137	143	148	153	158	163	4	2.0
835		169	174	179	184	189	195	200	205	210	215	5	2.5
836		221	226	231	236	241	247	252	257	262	267	6	3.0
837		273	278	283	288	293	298	304	309	314	319	7	3.5
838		324	330	335	340	345	350	355	361	366	371	8	4.0
839		376	381	387	392	397	402	407	412	418	423	9	4.5
840		428	433	438	443	449	454	459	464	469	474		
841		480	485	490	495	500	505	511	516	521	526		
842		531	536	542	547	552	557	562	567	572	578		
843		583	588	593	598	603	609	614	619	624	629		
844		634	639	645	650	655	660	665	670	675	681		
845		686	691	696	701	706	711	716	722	727	732		
846		737	742	747	752	758	763	768	773	778	783		
847		788	793	799	804	809	814	819	824	829	834		
848		840	845	850	855	860	865	870	875	881	886		
849		891	896	901	906	911	916	921	927	932	937		
850		942	947	952	957	962	967	973	978	983	988		
N.		0	1	2	3	4	5	6	7	8	9	Proportional parts	

N.	0	1	2	3	4	5	6	7	8	9	Proportional parts
850	92 942	947	952	957	962	967	973	978	983	988	
851	993	998	*003	*008	*013	*018	*024	*029	*034	*039	
852	93 044	049	054	059	064	069	075	080	085	090	
853	095	100	105	110	115	120	125	131	136	141	
854	146	151	156	161	166	171	176	181	186	192	
855	197	202	207	212	217	222	227	232	237	242	
856	247	252	258	263	268	273	278	283	288	293	
857	298	303	308	313	318	323	328	334	339	344	
858	349	354	359	364	369	374	379	384	389	394	**6**
859	399	404	409	414	420	425	430	435	440	445	1 | 0.6
											2 | 1.2
860	450	455	460	465	470	475	480	485	490	495	3 | 1.8
861	500	505	510	515	520	526	531	536	541	546	4 | 2.4
862	551	556	561	566	571	576	581	586	591	596	5 | 3.0
863	601	606	611	616	621	626	631	636	641	646	6 | 3.6
864	651	656	661	666	671	676	682	687	692	697	7 | 4.2
865	702	707	712	717	722	727	732	737	742	747	8 | 4.8
866	752	757	762	767	772	777	782	787	792	797	9 | 5.4
867	802	807	812	817	822	827	832	837	842	847	
868	852	857	862	867	872	877	882	887	892	897	
869	902	907	912	917	922	927	932	937	942	947	
870	952	957	962	967	972	977	982	987	992	997	**5**
871	94 002	007	012	017	022	027	032	037	042	047	1 | 0.5
872	052	057	062	067	072	077	082	086	091	096	2 | 1.0
873	101	106	111	116	121	126	131	136	141	146	3 | 1.5
874	151	156	161	166	171	176	181	186	191	196	4 | 2.0
875	201	206	211	216	221	226	231	236	240	245	5 | 2.5
876	250	255	260	265	270	275	280	285	290	295	6 | 3.0
877	300	305	310	315	320	325	330	335	340	345	7 | 3.5
878	349	354	359	364	369	374	379	384	389	394	8 | 4.0
879	399	404	409	414	419	424	429	433	438	443	9 | 4.5
880	448	453	458	463	468	473	478	483	488	493	
881	498	503	507	512	517	522	527	532	537	542	
882	547	552	557	562	567	571	576	581	586	591	
883	596	601	606	611	616	621	626	630	635	640	
884	645	650	655	660	665	670	675	680	685	689	
885	694	699	704	709	714	719	724	729	734	738	
886	743	748	753	758	763	768	773	778	783	787	**4**
887	792	797	802	807	812	817	822	827	832	836	1 | 0.4
888	841	846	851	856	861	866	871	876	880	885	2 | 0.8
889	890	895	900	905	910	915	919	924	929	934	3 | 1.2
											4 | 1.6
890	939	944	949	954	959	963	968	973	978	983	5 | 2.0
891	988	993	998	*002	*007	*012	*017	*022	*027	*032	6 | 2.4
892	95 036	041	046	051	056	061	066	071	075	080	7 | 2.8
893	085	090	095	100	105	109	114	119	124	129	8 | 3.2
894	134	139	143	148	153	158	163	168	173	177	9 | 3.6
895	182	187	192	197	202	207	211	216	221	226	
896	231	236	240	245	250	255	260	265	270	274	
897	279	284	289	294	299	303	308	313	318	323	
898	328	332	337	342	347	352	357	361	366	371	
899	376	381	386	390	395	400	405	410	415	419	
900	424	429	434	439	444	448	453	458	463	468	
N.	0	1	2	3	4	5	6	7	8	9	Proportional parts

FIVE-PLACE LOGARITHMS (Continued)

N.	0	1	2	3	4	5	6	7	8	9	Proportional parts
900	95 424	429	434	439	444	448	453	458	463	468	
901	472	477	482	487	492	497	501	506	511	516	
902	521	525	530	535	540	545	550	554	559	564	
903	569	574	578	583	588	593	598	602	607	612	
904	617	622	626	631	636	641	646	650	655	660	
905	665	670	674	679	684	689	694	698	703	708	
906	713	718	722	727	732	737	742	746	751	756	
907	761	766	770	775	780	785	789	794	799	804	
908	809	813	818	823	828	832	837	842	847	852	
909	856	861	866	871	875	880	885	890	895	899	
910	904	909	914	918	923	928	933	938	942	947	**5**
911	952	957	961	966	971	976	980	985	990	995	1 | 0.5
912	999	*004	*009	*014	*019	*023	*028	*033	*038	*042	2 | 1.0
913	96 047	052	057	061	066	071	076	080	085	090	3 | 1.5
914	095	099	104	109	114	118	123	128	133	137	4 | 2.0
915	142	147	152	156	161	166	171	175	180	185	5 | 2.5
916	190	194	199	204	209	213	218	223	227	232	6 | 3.0
917	237	242	246	251	256	261	265	270	275	280	7 | 3.5
918	284	289	294	298	303	308	313	317	322	327	8 | 4.0
919	332	336	341	346	350	355	360	365	369	374	9 | 4.5
920	379	384	388	393	398	402	407	412	417	421	
921	426	431	435	440	445	450	454	459	464	468	
922	473	478	483	487	492	497	501	506	511	515	
923	520	525	530	534	539	544	548	553	558	562	
924	567	572	577	581	586	591	595	600	605	609	
925	614	619	624	628	633	638	642	647	652	656	
926	661	666	670	675	680	685	689	694	699	703	
927	708	713	717	722	727	731	736	741	745	750	
928	755	759	764	769	774	778	783	788	792	797	
929	802	806	811	816	820	825	830	834	839	844	
930	848	853	858	862	867	872	876	881	886	890	**4**
931	895	900	904	909	914	918	923	928	932	937	1 | 0.4
932	942	946	951	956	960	965	970	974	979	984	2 | 0.8
933	988	993	997	*002	*007	*011	*016	*021	*025	*030	3 | 1.2
934	97 035	039	044	049	053	058	063	067	072	077	4 | 1.6
935	081	086	090	095	100	104	109	114	118	123	5 | 2.0
936	128	132	137	142	146	151	155	160	165	169	6 | 2.4
937	174	179	183	188	192	197	202	206	211	216	7 | 2.8
938	220	225	230	234	239	243	248	253	257	262	8 | 3.2
939	267	271	276	280	285	290	294	299	304	308	9 | 3.6
940	313	317	322	327	331	336	340	345	350	354	
941	359	364	368	373	377	382	387	391	396	400	
942	405	410	414	419	424	428	433	437	442	447	
943	451	456	460	465	470	474	479	483	488	493	
944	497	502	506	511	516	520	525	529	534	539	
945	543	548	552	557	562	566	571	575	580	585	
946	589	594	598	603	607	612	617	621	626	630	
947	635	640	644	649	653	658	663	667	672	676	
948	681	685	690	695	699	704	708	713	717	722	
949	727	731	736	740	745	749	754	759	763	768	
950	772	777	782	786	791	795	800	804	809	813	
N.	0	1	2	3	4	5	6	7	8	9	Proportional parts

N.	0	1	2	3	4	5	6	7	8	9	Proportional parts
950	97 772	777	782	786	791	795	800	804	809	813	
951	818	823	827	832	836	841	845	850	855	859	
952	864	868	873	877	882	886	891	896	900	905	
953	909	914	918	923	928	932	937	941	946	950	
954	955	959	964	968	973	978	982	987	991	996	
955	98 000	005	009	014	019	023	028	032	037	041	
956	046	050	055	059	064	068	073	078	082	087	
957	091	096	100	105	109	114	118	123	127	132	
958	137	141	146	150	155	159	164	168	173	177	
959	182	186	191	195	200	204	209	214	218	223	
960	227	232	236	241	245	250	254	259	263	268	**5**
961	272	277	281	286	290	295	299	304	308	313	1 | 0.5
962	318	322	327	331	336	340	345	349	354	358	2 | 1.0
963	363	367	372	376	381	385	390	394	399	403	3 | 1.5
964	408	412	417	421	426	430	435	439	444	448	4 | 2.0
965	453	457	462	466	471	475	480	484	489	493	5 | 2.5
966	498	502	507	511	516	520	525	529	534	538	6 | 3.0
967	543	547	552	556	561	565	570	574	579	583	7 | 3.5
968	588	592	597	601	605	610	614	619	623	628	8 | 4.0
969	632	637	641	646	650	655	659	664	668	673	9 | 4.5
970	677	682	686	691	695	700	704	709	713	717	
971	722	726	731	735	740	744	749	753	758	762	
972	767	771	776	780	784	789	793	798	802	807	
973	811	816	820	825	829	834	838	843	847	851	
974	856	860	865	869	874	878	883	887	892	896	
975	900	905	909	914	918	923	927	932	936	941	
976	945	949	954	958	963	967	972	976	981	985	
977	989	994	998	*003	*007	*012	*016	*021	*025	*029	
978	99 034	038	043	047	052	056	061	065	069	074	
979	078	083	087	092	096	100	105	109	114	118	
980	123	127	131	136	140	145	149	154	158	162	**4**
981	167	171	176	180	185	189	193	198	202	207	1 | 0.4
982	211	216	220	224	229	233	238	242	247	251	2 | 0.8
983	255	260	264	269	273	277	282	286	291	295	3 | 1.2
984	300	304	308	313	317	322	326	330	335	339	4 | 1.6
985	344	348	352	357	361	366	370	374	379	383	5 | 2.0
986	388	392	396	401	405	410	414	419	423	427	6 | 2.4
987	432	436	441	445	449	454	458	463	467	471	7 | 2.8
988	476	480	484	489	493	498	502	506	511	515	8 | 3.2
989	520	524	528	533	537	542	546	550	555	559	9 | 3.6
990	564	568	572	577	581	585	590	594	599	603	
991	607	612	616	621	625	629	634	638	642	647	
992	651	656	660	664	669	673	677	682	686	691	
993	695	699	704	708	712	717	721	726	730	734	
994	739	743	747	752	756	760	765	769	774	778	
995	782	787	791	795	800	804	808	813	817	822	
996	826	830	835	839	843	848	852	856	861	865	
997	870	874	878	883	887	891	896	900	904	909	
998	913	917	922	926	930	935	939	944	948	952	
999	957	961	965	970	974	978	983	987	991	996	
1000	00 000	004	009	013	017	022	026	030	035	039	
N.	0	1	2	3	4	5	6	7	8	9	Proportional parts

N.		0	1	2	3	4	5	6	7	8	9	d.
1000	000	0000	0434	0869	1303	1737	2171	2605	3039	3473	3907	434
1001		4341	4775	5208	5642	6076	6510	6943	7377	7810	8244	434
1002		8677	9111	9544	9977	*0411	*0844	*1277	*1710	*2143	*2576	433
1003	001	3009	3442	3875	4308	4741	5174	5607	6039	6472	6905	433
1004		7337	7770	8202	8635	9067	9499	9932	*0364	*0796	*1228	432
1005	002	1661	2093	2525	2957	3389	3821	4253	4685	5116	5548	432
1006		5980	6411	6843	7275	7706	8138	8569	9001	9432	9863	431
1007	003	0295	0726	1157	1588	2019	2451	2882	3313	3744	4174	431
1008		4605	5036	5467	5898	6328	6759	7190	7620	8051	8481	431
1009		8912	9342	9772	*0203	*0633	*1063	*1493	*1924	*2354	*2784	430
1010	004	3214	3644	4074	4504	4933	5363	5793	6223	6652	7082	430
1011		7512	7941	8371	8800	9229	9659	*0088	*0517	*0947	*1376	429
1012	005	1805	2234	2663	3092	3521	3950	4379	4808	5237	5666	429
1013		6094	6523	6952	7380	7809	8238	8666	9094	9523	9951	429
1014	006	0380	0808	1236	1664	2092	2521	2949	3377	3805	4233	428
1015		4660	5088	5516	5944	6372	6799	7227	7655	8082	8510	428
1016		8937	9365	9792	*0219	*0647	*1074	*1501	*1928	*2355	*2782	427
1017	007	3210	3637	4064	4490	4917	5344	5771	6198	6624	7051	427
1018		7478	7904	8331	8757	9184	9610	*0037	*0463	*0889	*1316	426
1019	008	1742	2168	2594	3020	3446	3872	4298	4724	5150	5576	426
1020		6002	6427	6853	7279	7704	8130	8556	8981	9407	9832	426
1021	009	0257	0683	1108	1533	1959	2384	2809	3234	3659	4084	425
1022		4509	4934	5359	5784	6208	6633	7058	7483	7907	8332	425
1023		8756	9181	9605	*0030	*0454	*0878	*1303	*1727	*2151	*2575	424
1024	010	3000	3424	3848	4272	4696	5120	5544	5967	6391	6815	424
1025		7239	7662	8086	8510	8933	9357	9780	*0204	*0627	*1050	424
1026	011	1474	1897	2320	2743	3166	3590	4013	4436	4859	5282	423
1027		5704	6127	6550	6973	7396	7818	8241	8664	9086	9509	423
1028		9931	*0354	*0776	*1198	*1621	*2043	*2465	*2887	*3310	*3732	422
1029	012	4154	4576	4998	5420	5842	6264	6685	7107	7529	7951	422
1030		8372	8794	9215	9637	*0059	*0480	*0901	*1323	*1744	*2165	422
1031	013	2587	3008	3429	3850	4271	4692	5113	5534	5955	6376	421
1032		6797	7218	7639	8059	8480	8901	9321	9742	*0162	*0583	421
1033	014	1003	1424	1844	2264	2685	3105	3525	3945	4365	4785	420
1034		5205	5625	6045	6465	6885	7305	7725	8144	8564	8984	420
1035		9403	9823	*0243	*0662	*1082	*1501	*1920	*2340	*2759	*3178	420
1036	015	3598	4017	4436	4855	5274	5693	6112	6531	6950	7369	419
1037		7788	8206	8625	9044	9462	9881	*0300	*0718	*1137	*1555	419
1038	016	1974	2392	2810	3229	3647	4065	4483	4901	5319	5737	418
1039		6155	6573	6991	7409	7827	8245	8663	9080	9498	9916	418
1040	017	0333	0751	1168	1586	2003	2421	2838	3256	3673	4090	417
1041		4507	4924	5342	5759	6176	6593	7010	7427	7844	8260	417
1042		8677	9094	9511	9927	*0344	*0761	*1177	*1594	*2010	*2427	417
1043	018	2843	3259	3676	4092	4508	4925	5341	5757	6173	6589	416
1044		7005	7421	7837	8253	8669	9084	9500	9916	*0332	*0747	416
1045	019	1163	1578	1994	2410	2825	3240	3656	4071	4486	4902	415
1046		5317	5732	6147	6562	6977	7392	7807	8222	8637	9052	415
1047		9467	9882	*0296	*0711	*1126	*1540	*1955	*2369	*2784	*3198	415
1048	020	3613	4027	4442	4856	5270	5684	6099	6513	6927	7341	414
1049		7755	8169	8583	8997	9411	9824	*0238	*0652	*1066	*1479	414
1050	021	1893	2307	2720	3134	3547	3961	4374	4787	5201	5614	413
N.		0	1	2	3	4	5	6	7	8	9	d.

N.		0	1	2	3	4	5	6	7	8	9	d.
1050	021	1893	2307	2720	3134	3547	3961	4374	4787	5201	5614	413
1051		6027	6440	6854	7267	7680	8093	8506	8919	9332	9745	413
1052	022	0157	0570	0983	1396	1808	2221	2634	3046	3459	3871	413
1053		4284	4696	5109	5521	5933	6345	6758	7170	7582	7994	412
1054		8406	8818	9230	9642	*0054	*0466	*0878	*1289	*1701	*2113	412
1055	023	2525	2936	3348	3759	4171	4582	4994	5405	5817	6228	411
1056		6639	7050	7462	7873	8284	8695	9106	9517	9928	*0339	411
1057	024	0750	1161	1572	1982	2393	2804	3214	3625	4036	4446	411
1058		4857	5267	5678	6088	6498	6909	7319	7729	8139	8549	410
1059		8960	9370	9780	*0190	*0600	*1010	*1419	*1829	*2239	*2649	410
1060	025	3059	3468	3878	4288	4697	5107	5516	5926	6335	6744	410
1061		7154	7563	7972	8382	8791	9200	9609	*0018	*0427	*0836	409
1062	026	1245	1654	2063	2472	2881	3289	3698	4107	4515	4924	409
1063		5333	5741	6150	6558	6967	7375	7783	8192	8600	9008	408
1064		9416	9824	*0233	*0641	*1049	*1457	*1865	*2273	*2680	*3088	408
1065	027	3496	3904	4312	4719	5127	5535	5942	6350	6757	7165	408
1066		7572	7979	8387	8794	9201	9609	*0016	*0423	*0830	*1237	407
1067	028	1644	2051	2458	2865	3272	3679	4086	4492	4899	5306	407
1068		5713	6119	6526	6932	7339	7745	8152	8558	8964	9371	406
1069		9777	*0183	*0590	*0996	*1402	*1808	*2214	*2620	*3026	*3432	406
1070	029	3833	4244	4649	5055	5461	5867	6272	6678	7084	7489	406
1071		7895	8300	8706	9111	9516	9922	*0327	*0732	*1138	*1543	405
1072	030	1948	2353	2758	3163	3568	3973	4378	4783	5188	5592	405
1073		5997	6402	6807	7211	7616	8020	8425	8830	9234	9638	405
1074	031	0043	0447	0851	1256	1660	2064	2468	2872	3277	3681	404
1075		4085	4489	4893	5296	5700	6104	6508	6912	7315	7719	404
1076		8123	8526	8930	9333	9737	*0140	*0544	*0947	*1350	*1754	403
1077	032	2157	2560	2963	3367	3770	4173	4576	4979	5382	5785	403
1078		6188	6590	6993	7396	7799	8201	8604	9007	9409	9812	403
1079	033	0214	0617	1019	1422	1824	2226	2629	3031	3433	3835	402
1080		4238	4640	5042	5444	5846	6248	6650	7052	7453	7855	402
1081		8257	8659	9060	9462	9864	*0265	*0667	*1068	*1470	*1871	402
1082	034	2273	2674	3075	3477	3878	4279	4680	5081	5482	5884	401
1083		6285	6686	7087	7487	7888	8289	8690	9091	9491	9892	401
1084	035	0293	0693	1094	1495	1895	2296	2696	3096	3497	3897	400
1085		4297	4698	5098	5498	5898	6298	6698	7098	7498	7898	400
1086		8298	8698	9098	9498	9898	*0297	*0697	*1097	*1496	*1896	400
1087	036	2295	2695	3094	3494	3893	4293	4692	5091	5491	5890	399
1088		6289	6688	7087	7486	7885	8284	8683	9082	9481	9880	399
1089	037	0279	0678	1076	1475	1874	2272	2671	3070	3468	3867	399
1090		4265	4663	5062	5460	5858	6257	6655	7053	7451	7849	398
1091		8248	8646	9044	9442	9839	*0237	*0635	*1033	*1431	*1829	398
1092	038	2226	2624	3022	3419	3817	4214	4612	5009	5407	5804	398
1093		6202	6599	6996	7393	7791	8188	8585	8982	9379	9776	397
1094	039	0173	0570	0967	1364	1761	2158	2554	2951	3348	3745	397
1095		4141	4538	4934	5331	5727	6124	6520	6917	7313	7709	397
1096		8106	8502	8898	9294	9690	*0086	*0482	*0878	*1274	*1670	396
1097	040	2066	2462	2858	3254	3650	4045	4441	4837	5232	5628	396
1098		6023	6419	6814	7210	7605	8001	8396	8791	9187	9582	395
1099		9977	*0372	*0767	*1162	*1557	*1952	*2347	*2742	*3137	*3532	395
1100	041	3927	4322	4716	5111	5506	5900	6295	6690	7084	7479	395
N.		0	1	2	3	4	5	6	7	8	9	d.

N.		0	1	2	3	4	5	6	7	8	9	d.
1100	041	3927	4322	4716	5111	5506	5900	6295	6690	7084	7479	395
1101		7873	8268	8662	9056	9451	9845	*0239	*0633	*1028	*1422	394
1102	042	1816	2210	2604	2998	3392	3786	4180	4574	4968	5361	394
1103		5755	6149	6543	6936	7330	7723	8117	8510	8904	9297	394
1104		9691	*0084	*0477	*0871	*1264	*1657	*2050	*2444	*2837	*3230	393
1105	043	3623	4016	4409	4802	5195	5587	5980	6373	6766	7159	393
1106		7551	7944	8337	8729	9122	9514	9907	*0299	*0692	*1084	393
1107	044	1476	1869	2261	2653	3045	3437	3829	4222	4614	5006	392
1108		5398	5790	6181	6573	6965	7357	7749	8140	8532	8924	392
1109		9315	9707	*0099	*0490	*0882	*1273	*1664	*2056	*2447	*2839	392
1110	045	3230	3621	4012	4403	4795	5186	5577	5968	6359	6750	391
1111		7141	7531	7922	8313	8704	9095	9485	9876	*0267	*0657	391
1112	046	1048	1438	1829	2219	2610	3000	3391	3781	4171	4561	390
1113		4952	5342	5732	6122	6512	6902	7292	7682	8072	8462	390
1114		8852	9242	9632	*0021	*0411	*0801	*1190	*1580	*1970	*2359	390
1115	047	2749	3138	3528	3917	4306	4696	5085	5474	5864	6253	389
1116		6642	7031	7420	7809	8198	8587	8976	9365	9754	*0143	389
1117	048	0532	0921	1309	1698	2087	2475	2864	3253	3641	4030	389
1118		4418	4806	5195	5583	5972	6360	6748	7136	7525	7913	388
1119		8301	8689	9077	9465	9853	*0241	*0629	*1017	*1405	*1792	388
1120	049	2180	2568	2956	3343	3731	4119	4506	4894	5281	5669	388
1121		6056	6444	6831	7218	7606	7993	8380	8767	9154	9541	387
1122		9929	*0316	*0703	*1090	*1477	*1863	*2250	*2637	*3024	*3411	387
1123	050	3798	4184	4571	4958	5344	5731	6117	6504	6890	7277	387
1124		7663	8049	8436	8822	9208	9595	9981	*0367	*0753	*1139	386
1125	051	1525	1911	2297	2683	3069	3455	3841	4227	4612	4998	386
1126		5384	5770	6155	6541	6926	7312	7697	8083	8468	8854	386
1127		9239	9624	*0010	*0395	*0780	*1166	*1551	*1936	*2321	*2706	385
1128	052	3091	3476	3861	4246	4631	5016	5400	5785	6170	6555	385
1129		6939	7324	7709	8093	8478	8862	9247	9631	*0016	*0400	385
1130	053	0784	1169	1553	1937	2321	2706	3090	3474	3858	4242	384
1131		4626	5010	5394	5778	6162	6546	6929	7313	7697	8081	384
1132		8464	8848	9232	9615	9999	*0382	*0766	*1149	*1532	*1916	384
1133	054	2299	2682	3066	3449	3832	4215	4598	4981	5365	5748	383
1134		6131	6514	6896	7279	7662	8045	8428	8811	9193	9576	383
1135		9959	*0341	*0724	*1106	*1489	*1871	*2254	*2636	*3019	*3401	382
1136	055	3783	4166	4548	4930	5312	5694	6077	6459	6841	7223	382
1137		7605	7987	8369	8750	9132	9514	9896	*0278	*0659	*1041	382
1138	056	1423	1804	2186	2567	2949	3330	3712	4093	4475	4856	381
1139		5237	5619	6000	6381	6762	7143	7524	7905	8287	8668	381
1140		9049	9429	9810	*0191	*0572	*0953	*1334	*1714	*2095	*2476	381
1141	057	2856	3237	3618	3998	4379	4759	5140	5520	5900	6281	381
1142		6661	7041	7422	7802	8182	8562	8942	9322	9702	*0082	380
1143	058	0462	0842	1222	1602	1982	2362	2741	3121	3501	3881	380
1144		4260	4640	5019	5399	5778	6158	6537	6917	7296	7676	380
1145		8055	8434	8813	9193	9572	9951	*0330	*0709	*1088	*1467	379
1146	059	1846	2225	2604	2983	3362	3741	4119	4498	4877	5256	379
1147		5634	6013	6391	6770	7148	7527	7905	8284	8662	9041	379
1148		9419	9797	*0175	*0554	*0932	*1310	*1688	*2066	*2444	*2822	378
1149	060	3200	3578	3956	4334	4712	5090	5468	5845	6223	6601	378
1150		6978	7356	7734	8111	8489	8866	9244	9621	9999	*0376	378
N.		0	1	2	3	4	5	6	7	8	9	d.

N.		0	1	2	3	4	5	6	7	8	9	d.
1150	060	6978	7356	7734	8111	8489	8866	9244	9621	9999	*0376	378
1151	061	0753	1131	1508	1885	2262	2639	3017	3394	3771	4148	377
1152		4525	4902	5279	5656	6032	6409	6786	7163	7540	7916	377
1153		8293	8670	9046	9423	9799	*0176	*0552	*0929	*1305	*1682	377
1154	062	2058	2434	2811	3187	3563	3939	4316	4692	5068	5444	376
1155		5820	6196	6572	6948	7324	7699	8075	8451	8827	9203	376
1156		9578	9954	*0330	*0705	*1081	*1456	*1832	*2207	*2583	*2958	376
1157	063	3334	3709	4084	4460	4835	5210	5585	5960	6335	6711	375
1158		7086	7461	7836	8211	8585	8960	9335	9710	*0085	*0460	375
1159	064	0834	1209	1584	1958	2333	2708	3082	3457	3831	4205	375
1160		4580	4954	5329	5703	6077	6451	6826	7200	7574	7948	374
1161		8322	8696	9070	9444	9818	*0192	*0566	*0940	*1314	*1688	374
1162	065	2061	2435	2809	3182	3556	3930	4303	4677	5050	5424	374
1163		5797	6171	6544	6917	7291	7664	8037	8410	8784	9157	373
1164		9530	9903	*0276	*0649	*1022	*1395	*1768	*2141	*2514	*2886	373
1165	066	3259	3632	4005	4377	4750	5123	5495	5868	6241	6613	373
1166		6986	7358	7730	8103	8475	8847	9220	9592	9964	*0336	372
1167	067	0709	1081	1453	1825	2197	2569	2941	3313	3685	4057	372
1168		4428	4800	5172	5544	5915	6287	6659	7030	7402	7774	372
1169		8145	8517	8888	9259	9631	*0002	*0374	*0745	*1116	*1487	371
1170	068	1859	2230	2601	2972	3343	3714	4085	4456	4827	5198	371
1171		5569	5940	6311	6681	7052	7423	7794	8164	8535	8906	371
1172		9276	9647	*0017	*0388	*0758	*1129	*1499	*1869	*2240	*2610	370
1173	069	2980	3350	3721	4091	4461	4831	5201	5571	5941	6311	370
1174		6681	7051	7421	7791	8160	8530	8900	9270	9639	*0009	370
1175	070	0379	0748	1118	1487	1857	2226	2596	2965	3335	3704	369
1176		4073	4442	4812	5181	5550	5919	6288	6658	7027	7396	369
1177		7765	8134	8503	8871	9240	9609	9978	*0347	*0715	*1084	369
1178	071	1453	1822	2190	2559	2927	3296	3664	4033	4401	4770	369
1179		5138	5506	5875	6243	6611	6979	7348	7716	8084	8452	368
1180		8820	9188	9556	9924	*0292	*0660	*1028	*1396	*1763	*2131	368
1181	072	2499	2867	3234	3602	3970	4337	4705	5072	5440	5807	368
1182		6175	6542	6910	7277	7644	8011	8379	8746	9113	9480	367
1183		9847	*0215	*0582	*0949	*1316	*1683	*2050	*2416	*2783	*3150	367
1184	073	3517	3884	4251	4617	4984	5351	5717	6084	6450	6817	367
1185		7184	7550	7916	8283	8649	9016	9382	9748	*0114	*0481	366
1186	074	0847	1213	1579	1945	2311	2677	3043	3409	3775	4141	366
1187		4507	4873	5239	5605	5970	6336	6702	7068	7433	7799	366
1188		8164	8530	8895	9261	9626	9992	*0357	*0723	*1088	*1453	365
1189	075	1819	2184	2549	2914	3279	3644	4010	4375	4740	5105	365
1190		5470	5835	6199	6564	6929	7294	7659	8024	8388	8753	365
1191		9118	9482	9847	*0211	*0576	*0940	*1305	*1669	*2034	*2398	364
1192	076	2763	3127	3491	3855	4220	4584	4948	5312	5676	6040	364
1193		6404	6768	7132	7496	7860	8224	8588	8952	9316	9680	364
1194	077	0043	0407	0771	1134	1498	1862	2225	2589	2952	3316	364
1195		3679	4042	4406	4769	5133	5496	5859	6222	6585	6949	363
1196		7312	7675	8038	8401	8764	9127	9490	9853	*0216	*0579	363
1197	078	0942	1304	1667	2030	2393	2755	3118	3480	3843	4206	363
1198		4568	4931	5293	5656	6018	6380	6743	7105	7467	7830	362
1199		8192	8554	8916	9278	9640	*0003	*0365	*0727	*1089	*1451	362
1200	079	1812	2174	2536	2898	3260	3622	3983	4345	4707	5068	362
N.		0	1	2	3	4	5	6	7	8	9	d.

Min.	Values of S, −10 to be appended 0°	1°	2°	3°	4°	Values of T, −10 to be appended 0°	1°	2°	3°	4°	Sec.
0′	4.68 557	555	549	538	522	4.68 557	562	575	597	628	0′
1	557	555	549	537	522	557	562	575	598	629	60
2	557	555	548	537	522	557	562	576	598	629	120
3	557	555	548	537	521	557	562	576	599	630	180
4	557	555	548	537	521	558	563	576	599	631	240
5	557	555	548	537	521	558	563	577	599	631	300
6	557	555	548	536	520	558	563	577	600	632	360
7	557	555	548	536	520	558	563	577	600	632	420
8	557	555	548	536	520	558	563	578	601	633	480
9	557	555	547	536	520	558	563	578	601	634	540
10	4.68 557	555	547	535	519	4.68 558	564	578	602	634	600
11	557	554	547	535	519	558	564	579	602	635	660
12	557	554	547	535	519	558	564	579	603	635	720
13	557	554	547	535	518	558	564	579	603	636	780
14	557	554	547	534	518	558	564	580	604	637	840
15	557	554	546	534	518	558	564	580	604	637	900
16	557	554	546	534	517	558	565	580	605	638	960
17	557	554	546	534	517	558	565	581	605	639	1020
18	557	554	546	534	517	558	565	581	606	639	1080
19	557	554	546	533	516	558	565	581	606	640	1140
20	4.68 557	554	546	533	516	4.68 558	565	582	607	640	1200
21	557	554	545	533	516	558	566	582	607	641	1260
22	557	553	545	533	515	558	566	582	608	642	1320
23	557	553	545	532	515	558	566	583	608	642	1380
24	557	553	545	532	515	558	566	583	609	643	1440
25	557	553	545	532	515	558	566	583	609	644	1500
26	557	553	544	532	514	558	567	584	610	644	1560
27	557	553	544	531	514	558	567	584	610	645	1620
28	557	553	544	531	514	558	567	584	611	646	1680
29	557	553	544	531	513	559	567	585	611	646	1740
30	4.68 557	553	544	531	513	4.68 559	567	585	612	647	1800
31	557	552	544	530	513	559	568	585	612	648	1860
32	557	552	543	530	512	559	568	586	613	648	1920
33	557	552	543	530	512	559	568	586	613	649	1980
34	557	552	543	529	512	559	568	587	614	650	2040
35	557	552	543	529	511	559	569	587	614	650	2100
36	557	552	543	529	511	559	569	587	615	651	2160
37	557	552	542	529	511	559	569	588	615	652	2220
38	557	552	542	528	510	559	569	588	616	652	2280
39	557	552	542	528	510	559	570	589	616	653	2340
40	4.68 557	551	542	528	510	4.68 559	570	589	617	654	2400
41	556	551	542	528	509	560	570	589	617	654	2460
42	556	551	541	527	509	560	570	590	618	655	2520
43	556	551	541	527	508	560	571	590	619	656	2580
44	556	551	541	527	508	560	571	591	619	656	2640
45	556	551	541	527	508	560	571	591	620	657	2700
46	556	551	541	526	507	560	571	591	620	658	2760
47	556	551	540	526	507	560	572	592	621	659	2820
48	556	550	540	526	507	560	572	592	621	659	2880
49	556	550	540	525	506	560	572	593	622	660	2940
50	4.68 556	550	540	525	506	4.68 561	572	593	622	661	3000
51	556	550	540	525	506	561	573	593	623	661	3060
52	556	550	539	525	505	561	573	594	624	662	3120
53	556	550	539	524	505	561	573	594	624	663	3180
54	556	550	539	524	505	561	573	595	625	664	3240
55	556	549	539	524	504	561	574	595	625	664	3300
56	556	549	539	523	504	561	574	596	626	665	3360
57	556	549	538	523	503	562	574	596	626	666	3420
58	555	549	538	523	503	562	575	596	627	667	3480
59	555	549	538	523	503	562	575	597	628	667	3540
60	4.68 555	549	538	522	502	4.68 562	575	597	628	668	3600

LOGARITHMS OF THE FUNCTIONS

0° (180°) (359°) **179°**

″	′	L. Sin.	*d.	C. S.	C. T.	L. Tan.	*c.d.	L. Cot.	L. Cos.	′
0	**0**	—		—	—	—		—	0.00 000	**60**
60	1	6.46 373	30103	5.31 443	5.31 443	6.46 373	30103	3.53 627	0.00 000	59
120	2	6.76 476	17609	5.31 443	5.31 443	6.76 476	17609	3.23 524	0.00 000	58
180	3	6.94 085	12494	5.31 443	5.31 443	6.94 085	12494	3.05 915	0.00 000	57
240	4	7.06 579	9691	5.31 443	5.31 442	7.06 579	9691	2.93 421	0.00 000	56
300	5	7.16 270	7918	5.31 443	5.31 442	7.16 270	7918	2.83 730	0.00 000	55
360	6	7.24 188	6694	5.31 443	5.31 442	7.24 188	6694	2.75 812	0.00 000	54
420	7	7.30 882	5800	5.31 443	5.31 442	7.30 882	5800	2.69 118	0.00 000	53
480	8	7.36 682	5115	5.31 443	5.31 442	7.36 682	5115	2.63 318	0.00 000	52
540	9	7.41 797	4576	5.31 443	5.31 442	7.41 797	4576	2.58 203	0.00 000	51
600	**10**	7.46 373	4139	5.31 443	5.31 442	7.46 373	4139	2.53 627	0.00 000	**50**
660	11	7.50 512	3779	5.31 443	5.31 442	7.50 512	3779	2.49 488	0.00 000	49
720	12	7.54 291	3476	5.31 443	5.31 442	7.54 291	3476	2.45 709	0.00 000	48
780	13	7.57 767	3218	5.31 443	5.31 442	7.57 767	3219	2.42 233	0.00 000	47
840	14	7.60 985	2997	5.31 443	5.31 442	7.60 986	2996	2.39 014	0.00 000	46
900	15	7.63 982	2802	5.31 443	5.31 442	7.63 982	2803	2.36 018	0.00 000	45
960	16	7.66 784	2633	5.31 443	5.31 442	7.66 785	2633	2.33 215	0.00 000	44
1020	17	7.69 417	2483	5.31 443	5.31 442	7.69 418	2482	2.30 582	9.99 999	43
1080	18	7.71 900	2348	5.31 443	5.31 442	7.71 900	2348	2.28 100	9.99 999	42
1140	19	7.74 248	2227	5.31 443	5.31 442	7.74 248	2228	2.25 752	9.99 999	41
1200	**20**	7.76 475	2119	5.31 443	5.31 442	7.76 476	2119	2.23 524	9.99 999	**40**
1260	21	7.78 594	2021	5.31 443	5.31 442	7.78 595	2020	2.21 405	9.99 999	39
1320	22	7.80 615	1930	5.31 443	5.31 442	7.80 615	1931	2.19 385	9.99 999	38
1380	23	7.82 545	1848	5.31 443	5.31 442	7.82 546	1848	2.17 454	9.99 999	37
1440	24	7.84 393	1773	5.31 443	5.31 442	7.84 394	1773	2.15 606	9.99 999	36
1500	25	7.86 166	1704	5.31 443	5.31 442	7.86 167	1704	2.13 833	9.99 999	35
1560	26	7.87 870	1639	5.31 443	5.31 442	7.87 871	1639	2.12 129	9.99 999	34
1620	27	7.89 509	1579	5.31 443	5.31 442	7.89 510	1579	2.10 490	9.99 999	33
1680	28	7.91 088	1524	5.31 443	5.31 442	7.91 089	1524	2.08 911	9.99 999	32
1740	29	7.92 612	1472	5.31 443	5.31 441	7.92 613	1473	2.07 387	9.99 998	31
1800	**30**	7.94 084	1424	5.31 443	5.31 441	7.94 086	1424	2.05 914	9.99 998	**30**
1860	31	7.95 508	1379	5.31 443	5.31 441	7.95 510	1379	2.04 490	9.99 998	29
1920	32	7.96 887	1336	5.31 443	5.31 441	7.96 889	1336	2.03 111	9.99 998	28
1980	33	7.98 223	1297	5.31 443	5.31 441	7.98 225	1297	2.01 775	9.99 998	27
2040	34	7.99 520	1259	5.31 443	5.31 441	7.99 522	1259	2.00 478	9.99 998	26
2100	35	8.00 779	1223	5.31 443	5.31 441	8.00 781	1223	1.99 219	9.99 998	25
2160	36	8.02 002	1190	5.31 443	5.31 441	8.02 004	1190	1.97 996	9.99 998	24
2220	37	8.03 192	1158	5.31 443	5.31 441	8.03 194	1159	1.96 806	9.99 997	23
2280	38	8.04 350	1128	5.31 443	5.31 441	8.04 353	1128	1.95 647	9.99 997	22
2340	39	8.05 478	1100	5.31 443	5.31 441	8.05 481	1100	1.94 519	9.99 997	21
2400	**40**	8.06 578	1072	5.31 443	5.31 441	8.06 581	1072	1.93 419	9.99 997	**20**
2460	41	8.07 650	1046	5.31 444	5.31 440	8.07 653	1047	1.92 347	9.99 997	19
2520	42	8.08 696	1022	5.31 444	5.31 440	8.08 700	1022	1.91 300	9.99 997	18
2580	43	8.09 718	999	5.31 444	5.31 440	8.09 722	998	1.90 278	9.99 997	17
2640	44	8.10 717	976	5.31 444	5.31 440	8.10 720	976	1.89 280	9.99 996	16
2700	45	8.11 693	954	5.31 444	5.31 440	8.11 696	955	1.88 304	9.99 996	15
2760	46	8.12 647	934	5.31 444	5.31 440	8.12 651	934	1.87 349	9.99 996	14
2820	47	8.13 581	914	5.31 444	5.31 440	8.13 585	915	1.86 415	9.99 996	13
2880	48	8.14 495	896	5.31 444	5.31 440	8.14 500	895	1.85 500	9.99 996	12
2940	49	8.15 391	877	5.31 444	5.31 440	8.15 395	878	1.84 605	9.99 996	11
3000	**50**	8.16 268	860	5.31 444	5.31 439	8.16 273	860	1.83 727	9.99 995	**10**
3060	51	8.17 128	843	5.31 444	5.31 439	8.17 133	843	1.82 867	9.99 995	9
3120	52	8.17 971	827	5.31 444	5.31 439	8.17 976	828	1.82 024	9.99 995	8
3180	53	8.18 798	812	5.31 444	5.31 439	8.18 804	812	1.81 196	9.99 995	7
3240	54	8.19 610	797	5.31 444	5.31 439	8.19 616	797	1.80 384	9.99 995	6
3300	55	8.20 407	782	5.31 444	5.31 439	8.20 413	782	1.79 587	9.99 994	5
3360	56	8.21 189	769	5.31 444	5.31 439	8.21 195	769	1.78 805	9.99 994	4
3420	57	8.21 958	755	5.31 445	5.31 439	8.21 964	756	1.78 036	9.99 994	3
3480	58	8.22 713	743	5.31 445	5.31 438	8.22 720	742	1.77 280	9.99 994	2
3540	59	8.23 456	730	5.31 445	5.31 438	8.23 462	730	1.76 538	9.99 994	1
3600	**60**	8.24 186		5.31 445	5.31 438	8.24 192		1.75 808	9.99 993	**0**
	′	L. Cos.	d.			L. Cot.	c.d.	L. Tan.	L. Sin.	′

90° (270°) (269°) **89°**

* Interpolation in this section of the table is not accurate.

1° (181°) — (358°) 178°

"	'	L. Sin.	*d.	C. S.	C. T.	L. Tan.	*c.d.	L. Cot.	L. Cos.	'
3600	**0**	8.24 186	717	5.31 445	5.31 438	8.24 192	718	1.75 808	9.99 993	**60**
3660	1	8.24 903	706	5.31 445	5.31 438	8.24 910	706	1.75 090	9.99 993	59
3720	2	8.25 609	695	5.31 445	5.31 438	8.25 616	696	1.74 384	9.99 993	58
3780	3	8.26 304	684	5.31 445	5.31 438	8.26 312	684	1.73 688	9.99 993	57
3840	4	8.26 988	673	5.31 445	5.31 437	8.26 996	673	1.73 004	9.99 992	56
3900	5	8.27 661	663	5.31 445	5.31 437	8.27 669	663	1.72 331	9.99 992	55
3960	6	8.28 324	653	5.31 445	5.31 437	8.28 332	654	1.71 668	9.99 992	54
4020	7	8.28 977	644	5.31 445	5.31 437	8.28 986	643	1.71 014	9.99 992	53
4080	8	8.29 621	634	5.31 445	5.31 437	8.29 629	634	1.70 371	9.99 992	52
4140	9	8.30 255	624	5.31 445	5.31 437	8.30 263	625	1.69 737	9.99 991	51
4200	**10**	8.30 879	616	5.31 446	5.31 437	8.30 888	617	1.69 112	9.99 991	**50**
4260	11	8.31 495	608	5.31 446	5.31 436	8.31 505	607	1.68 495	9.99 991	49
4320	12	8.32 103	599	5.31 446	5.31 436	8.32 112	599	1.67 888	9.99 990	48
4380	13	8.32 702	590	5.31 446	5.31 436	8.32 711	591	1.67 289	9.99 990	47
4440	14	8.33 292	583	5.31 446	5.31 436	8.33 302	584	1.66 698	9.99 990	46
4500	15	8.33 875	575	5.31 446	5.31 436	8.33 886	575	1.66 114	9.99 990	45
4560	16	8.34 450	568	5.31 446	5.31 435	8.34 461	568	1.65 539	9.99 989	44
4620	17	8.35 018	560	5.31 446	5.31 435	8.35 029	561	1.64 971	9.99 989	43
4680	18	8.35 578	553	5.31 446	5.31 435	8.35 590	553	1.64 410	9.99 989	42
4740	19	8.36 131	547	5.31 446	5.31 435	8.36 143	546	1.63 857	9.99 989	41
4800	**20**	8.36 678	539	5.31 446	5.31 435	8.36 689	540	1.63 311	9.99 988	**40**
4860	21	8.37 217	533	5.31 447	5.31 434	8.37 229	533	1.62 771	9.99 988	39
4920	22	8.37 750	526	5.31 447	5.31 434	8.37 762	527	1.62 238	9.99 988	38
4980	23	8.38 276	520	5.31 447	5.31 434	8.38 289	520	1.61 711	9.99 987	37
5040	24	8.38 796	514	5.31 447	5.31 434	8.38 809	514	1.61 191	9.99 987	36
5100	25	8.39 310	508	5.31 447	5.31 434	8.39 323	509	1.60 677	9.99 987	35
5160	26	8.39 818	502	5.31 447	5.31 433	8.39 832	502	1.60 168	9.99 986	34
5220	27	8.40 320	496	5.31 447	5.31 433	8.40 334	496	1.59 666	9.99 986	33
5280	28	8.40 816	491	5.31 447	5.31 433	8.40 830	491	1.59 170	9.99 986	32
5340	29	8.41 307	485	5.31 447	5.31 433	8.41 321	486	1.58 679	9.99 985	31
5400	**30**	8.41 792	480	5.31 447	5.31 433	8.41 807	480	1.58 193	9.99 985	**30**
5460	31	8.42 272	474	5.31 448	5.31 432	8.42 287	475	1.57 713	9.99 985	29
5520	32	8.42 746	470	5.31 448	5.31 432	8.42 762	470	1.57 238	9.99 984	28
5580	33	8.43 216	464	5.31 448	5.31 432	8.43 232	464	1.56 768	9.99 984	27
5640	34	8.43 680	459	5.31 448	5.31 432	8.43 696	460	1.56 304	9.99 984	26
5700	35	8.44 139	455	5.31 448	5.31 431	8.44 156	455	1.55 844	9.99 983	25
5760	36	8.44 594	450	5.31 448	5.31 431	8.44 611	450	1.55 389	9.99 983	24
5820	37	8.45 044	445	5.31 448	5.31 431	8.45 061	446	1.54 939	9.99 983	23
5880	38	8.45 489	441	5.31 448	5.31 431	8.45 507	441	1.54 493	9.99 982	22
5940	39	8.45 930	436	5.31 449	5.31 431	8.45 948	437	1.54 052	9.99 982	21
6000	**40**	8.46 366	433	5.31 449	5.31 430	8.46 385	432	1.53 615	9.99 982	**20**
6060	41	8.46 799	427	5.31 449	5.31 430	8.46 817	428	1.53 183	9.99 981	19
6120	42	8.47 226	424	5.31 449	5.31 430	8.47 245	424	1.52 755	9.99 981	18
6180	43	8.47 650	419	5.31 449	5.31 430	8.47 669	420	1.52 331	9.99 981	17
6240	44	8.48 069	416	5.31 449	5.31 429	8.48 089	416	1.51 911	9.99 980	16
6300	45	8.48 485	411	5.31 449	5.31 429	8.48 505	412	1.51 495	9.99 980	15
6360	46	8.48 896	408	5.31 449	5.31 429	8.48 917	408	1.51 083	9.99 979	14
6420	47	8.49 304	404	5.31 450	5.31 428	8.49 325	404	1.50 675	9.99 979	13
6480	48	8.49 708	400	5.31 450	5.31 428	8.49 729	401	1.50 271	9.99 979	12
6540	49	8.50 108	396	5.31 450	5.31 428	8.50 130	397	1.49 870	9.99 978	11
6600	**50**	8.50 504	393	5.31 450	5.31 428	8.50 527	393	1.49 473	9.99 978	**10**
6660	51	8.50 897	390	5.31 450	5.31 427	8.50 920	390	1.49 080	9.99 977	9
6720	52	8.51 287	386	5.31 450	5.31 427	8.51 310	386	1.48 690	9.99 977	8
6780	53	8.51 673	382	5.31 450	5.31 427	8.51 696	383	1.48 304	9.99 977	7
6840	54	8.52 055	379	5.31 450	5.31 427	8.52 079	380	1.47 921	9.99 976	6
6900	55	8.52 434	376	5.31 451	5.31 426	8.52 459	376	1.47 541	9.99 976	5
6960	56	8.52 810	373	5.31 451	5.31 426	8.52 835	373	1.47 165	9.99 975	4
7020	57	8.53 183	369	5.31 451	5.31 426	8.53 208	370	1.46 792	9.99 975	3
7080	58	8.53 552	367	5.31 451	5.31 425	8.53 578	367	1.46 422	9.99 974	2
7140	59	8.53 919	363	5.31 451	5.31 425	8.53 945	363	1.46 055	9.99 974	1
7200	**60**	8.54 282		5.31 451	5.31 425	8.54 308		1.45 692	9.99 974	**0**
	'	L. Cos.	d.			L. Cot.	c.d.	L. Tan.	L. Sin.	'

91° (271°) — (268°) 88°

* Interpolation in this section of the table is not accurate.

2° (182°) — (357°) 177°

"	'	L. Sin.	*d.	C. S.	C. T.	L. Tan.	*c.d.	L. Cot.	L. Cos.	'
7200	**0**	8.54 282	360	5.31 451	5.31 425	8.54 308	361	1.45 692	9.99 974	**60**
7260	1	8.54 642	357	5.31 451	5.31 425	8.54 669	358	1.45 331	9.99 973	59
7320	2	8.54 999	355	5.31 452	5.31 424	8.55 027	355	1.44 973	9.99 973	58
7380	3	8.55 354	351	5.31 452	5.31 424	8.55 382	352	1.44 618	9.99 972	57
7440	4	8.55 705	349	5.31 452	5.31 424	8.55 734	349	1.44 266	9.99 972	56
7500	5	8.56 054	346	5.31 452	5.31 423	8.56 083	346	1.43 917	9.99 971	55
7560	6	8.56 400	343	5.31 452	5.31 423	8.56 429	344	1.43 571	9.99 971	54
7620	7	8.56 743	341	5.31 452	5.31 423	8.56 773	341	1.43 227	9.99 970	53
7680	8	8.57 084	337	5.31 453	5.31 422	8.57 114	338	1.42 886	9.99 970	52
7740	9	8.57 421	336	5.31 453	5.31 422	8.57 452	336	1.42 548	9.99 969	51
7800	**10**	8.57 757	332	5.31 453	5.31 422	8.57 788	333	1.42 212	9.99 969	**50**
7860	11	8.58 089	330	5.31 453	5.31 421	8.58 121	330	1.41 879	9.99 968	49
7920	12	8.58 419	328	5.31 453	5.31 421	8.58 451	328	1.41 549	9.99 968	48
7980	13	8.58 747	325	5.31 453	5.31 421	8.58 779	326	1.41 221	9.99 967	47
8040	14	8.59 072	323	5.31 454	5.31 421	8.59 105	323	1.40 895	9.99 967	46
8100	15	8.59 395	320	5.31 454	5.31 420	8.59 428	321	1.40 572	9.99 967	45
8160	16	8.59 715	318	5.31 454	5.31 420	8.59 749	319	1.40 251	9.99 966	44
8220	17	8.60 033	316	5.31 454	5.31 420	8.60 068	316	1.39 932	9.99 966	43
8280	18	8.60 349	313	5.31 454	5.31 419	8.60 384	314	1.39 616	9.99 965	42
8340	19	8.60 662	311	5.31 454	5.31 419	8.60 698	311	1.39 302	9.99 964	41
8400	**20**	8.60 973	309	5.31 455	5.31 418	8.61 009	310	1.38 991	9.99 964	**40**
8460	21	8.61 282	307	5.31 455	5.31 418	8.61 319	307	1.38 681	9.99 963	39
8520	22	8.61 589	305	5.31 455	5.31 418	8.61 626	305	1.38 374	9.99 963	38
8580	23	8.61 894	302	5.31 455	5.31 417	8.61 931	303	1.38 069	9.99 962	37
8640	24	8.62 196	301	5.31 455	5.31 417	8.62 234	301	1.37 766	9.99 962	36
8700	25	8.62 497	298	5.31 455	5.31 417	8.62 535	299	1.37 465	9.99 961	35
8760	26	8.62 795	296	5.31 456	5.31 416	8.62 834	297	1.37 166	9.99 961	34
8820	27	8.63 091	294	5.31 456	5.31 416	8.63 131	295	1.36 869	9.99 960	33
8880	28	8.63 385	293	5.31 456	5.31 416	8.63 426	292	1.36 574	9.99 960	32
8940	29	8.63 678	290	5.31 456	5.31 415	8.63 718	291	1.36 282	9.99 959	31
9000	**30**	8.63 968	288	5.31 456	5.31 415	8.64 009	289	1.35 991	9.99 959	**30**
9060	31	8.64 256	287	5.31 456	5.31 415	8.64 298	287	1.35 702	9.99 958	29
9120	32	8.64 543	284	5.31 457	5.31 414	8.64 585	285	1.35 415	9.99 958	28
9180	33	8.64 827	283	5.31 457	5.31 414	8.64 870	284	1.35 130	9.99 957	27
9240	34	8.65 110	281	5.31 457	5.31 413	8.65 154	281	1.34 846	9.99 956	26
9300	35	8.65 391	279	5.31 457	5.31 413	8.65 435	280	1.34 565	9.99 956	25
9360	36	8.65 670	277	5.31 457	5.31 413	8.65 715	278	1.34 285	9.99 955	24
9420	37	8.65 947	276	5.31 458	5.31 412	8.65 993	276	1.34 007	9.99 955	23
9480	38	8.66 223	274	5.31 458	5.31 412	8.66 269	274	1.33 731	9.99 954	22
9540	39	8.66 497	272	5.31 458	5.31 412	8.66 543	273	1.33 457	9.99 954	21
9600	**40**	8.66 769	270	5.31 458	5.31 411	8.66 816	271	1.33 184	9.99 953	**20**
9660	41	8.67 039	269	5.31 458	5.31 411	8.67 087	269	1.32 913	9.99 952	19
9720	42	8.67 308	267	5.31 459	5.31 410	8.67 356	268	1.32 644	9.99 952	18
9780	43	8.67 575	266	5.31 459	5.31 410	8.67 624	266	1.32 376	9.99 951	17
9840	44	8.67 841	263	5.31 459	5.31 410	8.67 890	264	1.32 110	9.99 951	16
9900	45	8.68 104	263	5.31 459	5.31 409	8.68 154	263	1.31 846	9.99 950	15
9960	46	8.68 367	260	5.31 459	5.31 409	8.68 417	261	1.31 583	9.99 949	14
10020	47	8.68 627	259	5.31 460	5.31 408	8.68 678	260	1.31 322	9.99 949	13
10080	48	8.68 886	258	5.31 460	5.31 408	8.68 938	258	1.31 062	9.99 948	12
10140	49	8.69 144	256	5.31 460	5.31 408	8.69 196	257	1.30 804	9.99 948	11
10200	**50**	8.69 400	254	5.31 460	5.31 407	8.69 453	255	1.30 547	9.99 947	**10**
10260	51	8.69 654	253	5.31 460	5.31 407	8.69 708	254	1.30 292	9.99 946	9
10320	52	8.69 907	252	5.31 461	5.31 406	8.69 962	252	1.30 038	9.99 946	8
10380	53	8.70 159	250	5.31 461	5.31 406	8.70 214	251	1.29 786	9.99 945	7
10440	54	8.70 409	249	5.31 461	5.31 405	8.70 465	249	1.29 535	9.99 944	6
10500	55	8.70 658	247	5.31 461	5.31 405	8.70 714	248	1.29 286	9.99 944	5
10560	56	8.70 905	246	5.31 461	5.31 405	8.70 962	246	1.29 038	9.99 943	4
10620	57	8.71 151	244	5.31 462	5.31 404	8.71 208	245	1.28 792	9.99 942	3
10680	58	8.71 395	243	5.31 462	5.31 404	8.71 453	244	1.28 547	9.99 942	2
10740	59	8.71 638	242	5.31 462	5.31 403	8.71 697	243	1.28 303	9.99 941	1
10800	**60**	8.71 880		5.31 462	5.31 403	8.71 940		1.28 060	9.99 940	**0**
	'	L. Cos.	d.			L. Cot.	c.d.	L. Tan.	L. Sin.	'

92° (272°) — (267°) 87°

* Interpolation in this section of the table is not accurate.

3° (183°)					(356°) **176°**								
′	L. Sin.	*d.	L. Tan.	*c.d.	L. Cot.	L. Cos.	′	P. P.					
0	8.71 880	240	8.71 940	241	1.28 060	9.99 940	**60**	″	**241**	**239**	**237**	**235**	**234**
1	8.72 120	239	8.72 181	239	1.27 819	9.99 940	59	1	4.0	4.0	4.0	3.9	3.9
2	8.72 359	238	8.72 420	239	1.27 580	9.99 939	58	2	8.0	8.0	7.9	7.8	7.8
3	8.72 597	237	8.72 659	237	1.27 341	9.99 938	57	3	12.0	12.0	11.8	11.8	11.7
4	8.72 834	235	8.72 896	236	1.27 104	9.99 938	56	4	16.1	15.9	15.8	15.7	15.6
5	8.73 069	234	8.73 132	234	1.26 868	9.99 937	55	5	20.1	19.9	19.8	19.6	19.5
6	8.73 303	232	8.73 366	234	1.26 634	9.99 936	54	6	24.1	23.9	23.7	23.5	23.4
7	8.73 535	232	8.73 600	232	1.26 400	9.99 936	53	7	28.1	27.9	27.6	27.4	27.3
8	8.73 767	230	8.73 832	231	1.26 168	9.99 935	52	8	32.1	31.9	31.6	31.3	31.2
9	8.73 997	229	8.74 063	229	1.25 937	9.99 934	51	9	36.2	35.8	35.6	35.2	35.1
10	8.74 226	228	8.74 292	229	1.25 708	9.99 934	**50**	″	**232**	**229**	**227**	**225**	**223**
11	8.74 454	226	8.74 521	227	1.25 479	9.99 933	49	1	3.9	3.8	3.8	3.8	3.7
12	8.74 680	226	8.74 748	226	1.25 252	9.99 932	48	2	7.7	7.6	7.6	7.5	7.4
13	8.74 906	224	8.74 974	225	1.25 026	9.99 932	47	3	11.6	11.4	11.4	11.2	11.2
14	8.75 130	223	8.75 199	224	1.24 801	9.99 931	46	4	15.5	15.3	15.1	15.0	14.9
15	8.75 353	222	8.75 423	222	1.24 577	9.99 930	45	5	19.3	19.1	18.9	18.8	18.6
16	8.75 575	220	8.75 645	222	1.24 355	9.99 929	44	6	23.2	22.9	22.7	22.5	22.3
17	8.75 795	220	8.75 867	220	1.24 133	9.99 929	43	7	27.1	26.7	26.5	26.2	26.0
18	8.76 015	219	8.76 087	219	1.23 913	9.99 928	42	8	30.9	30.5	30.3	30.0	29.7
19	8.76 234	217	8.76 306	219	1.23 694	9.99 927	41	9	34.8	34.4	34.0	33.8	33.4
20	8.76 451	216	8.76 525	217	1.23 475	9.99 926	**40**	″	**222**	**220**	**217**	**215**	**213**
21	8.76 667	216	8.76 742	216	1.23 258	9.99 926	39	1	3.7	3.7	3.6	3.6	3.6
22	8.76 883	214	8.76 958	215	1.23 042	9.99 925	38	2	7.4	7.3	7.2	7.2	7.1
23	8.77 097	213	8.77 173	214	1.22 827	9.99 924	37	3	11.1	11.0	10.8	10.8	10.6
24	8.77 310	212	8.77 387	213	1.22 613	9.99 923	36	4	14.8	14.7	14.5	14.3	14.2
25	8.77 522	211	8.77 600	211	1.22 400	9.99 923	35	5	18.5	18.3	18.1	17.9	17.8
26	8.77 733	210	8.77 811	211	1.22 189	9.99 922	34	6	22.2	22.0	21.7	21.5	21.3
27	8.77 943	209	8.78 022	210	1.21 978	9.99 921	33	7	25.9	25.7	25.3	25.1	24.8
28	8.78 152	208	8.78 232	209	1.21 768	9.99 920	32	8	29.6	29.3	28.9	28.7	28.4
29	8.78 360	208	8.78 441	208	1.21 559	9.99 920	31	9	33.3	33.0	32.6	32.2	32.0
30	8.78 568	206	8.78 649	206	1.21 351	9.99 919	**30**	″	**211**	**208**	**206**	**203**	**201**
31	8.78 774	205	8.78 855	206	1.21 145	9.99 918	29	1	3.5	3.5	3.4	3.4	3.4
32	8.78 979	204	8.79 061	205	1.20 939	9.99 917	28	2	7.0	6.9	6.9	6.8	6.7
33	8.79 183	203	8.79 266	204	1.20 734	9.99 917	27	3	10.6	10.4	10.3	10.2	10.0
34	8.79 386	202	8.79 470	203	1.20 530	9.99 916	26	4	14.1	13.9	13.7	13.5	13.4
35	8.79 588	201	8.79 673	202	1.20 327	9.99 915	25	5	17.6	17.3	17.2	16.9	16.8
36	8.79 789	201	8.79 875	201	1.20 125	9.99 914	24	6	21.1	20.8	20.6	20.3	20.1
37	8.79 990	199	8.80 076	201	1.19 924	9.99 913	23	7	24.6	24.3	24.0	23.7	23.4
38	8.80 189	199	8.80 277	199	1.19 723	9.99 913	22	8	28.1	27.7	27.5	27.1	26.8
39	8.80 388	197	8.80 476	198	1.19 524	9.99 912	21	9	31.6	31.2	30.9	30.4	30.2
40	8.80 585	197	8.80 674	198	1.19 326	9.99 911	**20**	″	**199**	**197**	**195**	**193**	**192**
41	8.80 782	196	8.80 872	196	1.19 128	9.99 910	19	1	3.3	3.3	3.2	3.2	3.2
42	8.80 978	195	8.81 068	196	1.18 932	9.99 909	18	2	6.6	6.6	6.5	6.4	6.4
43	8.81 173	194	8.81 264	195	1.18 736	9.99 909	17	3	10.0	9.8	9.8	9.6	9.6
44	8.81 367	193	8.81 459	194	1.18 541	9.99 908	16	4	13.3	13.1	13.0	12.9	12.8
45	8.81 560	192	8.81 653	193	1.18 347	9.99 907	15	5	16.6	16.4	16.2	16.1	16.0
46	8.81 752	192	8.81 846	192	1.18 154	9.99 906	14	6	19.9	19.7	19.5	19.3	19.2
47	8.81 944	190	8.82 038	192	1.17 962	9.99 905	13	7	23.2	23.0	22.8	22.5	22.4
48	8.82 134	190	8.82 230	190	1.17 770	9.99 904	12	8	26.5	26.3	26.0	25.7	25.6
49	8.82 324	189	8.82 420	190	1.17 580	9.99 904	11	9	29.8	29.6	29.2	29.0	28.8
50	8.82 513	188	8.82 610	189	1.17 390	9.99 903	**10**	″	**189**	**187**	**185**	**183**	**181**
51	8.82 701	187	8.82 799	188	1.17 201	9.99 902	9	1	3.2	3.1	3.1	3.0	3.0
52	8.82 888	187	8.82 987	188	1.17 013	9.99 901	8	2	6.3	6.2	6.2	6.1	6.0
53	8.83 075	186	8.83 175	186	1.16 825	9.99 900	7	3	9.4	9.4	9.2	9.2	9.0
54	8.83 261	185	8.83 361	186	1.16 639	9.99 899	6	4	12.6	12.5	12.3	12.2	12.1
55	8.83 446	184	8.83 547	185	1.16 453	9.99 898	5	5	15.8	15.6	15.4	15.2	15.1
56	8.83 630	183	8.83 732	184	1.16 268	9.99 898	4	6	18.9	18.7	18.5	18.3	18.1
57	8.83 813	183	8.83 916	184	1.16 084	9.99 897	3	7	22.0	21.8	21.6	21.4	21.1
58	8.83 996	181	8.84 100	182	1.15 900	9.99 896	2	8	25.2	24.9	24.7	24.4	24.1
59	8.84 177	181	8.84 282	182	1.15 718	9.99 895	1	9	28.4	28.0	27.8	27.4	27.2
60	8.84 358		8.84 464		1.15 536	9.99 894	**0**	10	31.5	31.2	30.8	30.5	30.2
′	L. Cos.	d.	L. Cot.	c.d.	L. Tan.	L. Sin.	′	P. P.					
93° (273°)					(266°) **86°**								

* Interpolation in this section of the table is not accurate.

4° (184°)					(355°) **175°**								
′	L. Sin.	d.	L. Tan.	*c.d.	L. Cot.	L. Cos.	′	P. P.					
0	8.84 358	181	8.84 464	182	1.15 536	9.99 894	**60**	″	**182**	**181**	**179**	**178**	**177**
1	8.84 539	179	8.84 646	180	1.15 354	9.99 893	59	1	3.0	3.0	3.0	3.0	3.0
2	8.84 718	179	8.84 826	180	1.15 174	9.99 892	58	2	6.1	6.0	6.0	5.9	5.9
3	8.84 897	178	8.85 006	179	1.14 994	9.99 891	57	3	9.1	9.0	9.0	8.9	8.8
4	8.85 075	177	8.85 185	178	1.14 815	9.99 891	56	4	12.1	12.1	11.9	11.9	11.8
5	8.85 252	177	8.85 363	177	1.14 637	9.99 890	55	5	15.2	15.1	14.9	14.8	14.8
6	8.85 429	176	8.85 540	177	1.14 460	9.99 889	54	6	18.2	18.1	17.9	17.8	17.7
7	8.85 605	175	8.85 717	176	1.14 283	9.99 888	53	7	21.2	21.1	20.9	20.8	20.6
8	8.85 780	175	8.85 893	176	1.14 107	9.99 887	52	8	24.3	24.1	23.9	23.7	23.6
9	8.85 955	173	8.86 069	174	1.13 931	9.99 886	51	9	27.3	27.2	26.8	26.7	26.6
10	8.86 128	173	8.86 243	174	1.13 757	9.99 885	**50**	″	**176**	**175**	**174**	**173**	**172**
11	8.86 301	173	8.86 417	174	1.13 583	9.99 884	49	1	2.9	2.9	2.9	2.9	2.9
12	8.86 474	171	8.86 591	172	1.13 409	9.99 883	48	2	5.9	5.8	5.8	5.8	5.7
13	8.86 645	171	8.86 763	172	1.13 237	9.99 882	47	3	8.8	8.8	8.7	8.6	8.6
14	8.86 816	171	8.86 935	171	1.13 065	9.99 881	46	4	11.7	11.7	11.6	11.5	11.5
15	8.86 987	169	8.87 106	171	1.12 894	9.99 880	45	5	14.7	14.6	14.5	14.4	14.3
16	8.87 156	169	8.87 277	170	1.12 723	9.99 879	44	6	17.6	17.5	17.4	17.3	17.2
17	8.87 325	169	8.87 447	169	1.12 553	9.99 879	43	7	20.5	20.4	20.3	20.2	20.1
18	8.87 494	167	8.87 616	169	1.12 384	9.99 878	42	8	23.5	23.3	23.2	23.1	22.9
19	8.87 661	168	8.87 785	168	1.12 215	9.99 877	41	9	26.4	26.2	26.1	26.0	25.8
20	8.87 829	166	8.87 953	167	1.12 047	9.99 876	**40**	″	**171**	**170**	**169**	**168**	**167**
21	8.87 995	166	8.88 120	167	1.11 880	9.99 875	39	1	2.8	2.8	2.8	2.8	2.8
22	8.88 161	165	8.88 287	166	1.11 713	9.99 874	38	2	5.7	5.7	5.6	5.6	5.6
23	8.88 326	164	8.88 453	165	1.11 547	9.99 873	37	3	8.6	8.5	8.4	8.4	8.4
24	8.88 490	164	8.88 618	165	1.11 382	9.99 872	36	4	11.4	11.3	11.3	11.2	11.1
25	8.88 654	163	8.88 783	165	1.11 217	9.99 871	35	5	14.2	14.2	14.1	14.0	13.9
26	8.88 817	163	8.88 948	163	1.11 052	9.99 870	34	6	17.1	17.0	16.9	16.8	16.7
27	8.88 980	162	8.89 111	163	1.10 889	9.99 869	33	7	20.0	19.8	19.7	19.6	19.5
28	8.89 142	162	8.89 274	163	1.10 726	9.99 868	32	8	22.8	22.7	22.5	22.4	22.3
29	8.89 304	160	8.89 437	161	1.10 563	9.99 867	31	9	25.6	25.5	25.4	25.2	25.0
30	8.89 464	161	8.89 598	162	1.10 402	9.99 866	**30**	″	**166**	**165**	**164**	**163**	**162**
31	8.89 625	159	8.89 760	160	1.10 240	9.99 865	29	1	2.8	2.8	2.7	2.7	2.7
32	8.89 784	159	8.89 920	160	1.10 080	9.99 864	28	2	5.5	5.5	5.5	5.4	5.4
33	8.89 943	159	8.90 080	160	1.09 920	9.99 863	27	3	8.3	8.2	8.2	8.2	8.1
34	8.90 102	158	8.90 240	159	1.09 760	9.99 862	26	4	11.1	11.0	10.9	10.9	10.8
35	8.90 260	157	8.90 399	158	1.09 601	9.99 861	25	5	13.8	13.8	13.7	13.6	13.5
36	8.90 417	157	8.90 557	158	1.09 443	9.99 860	24	6	16.6	16.5	16.4	16.3	16.2
37	8.90 574	156	8.90 715	157	1.09 285	9.99 859	23	7	19.4	19.2	19.1	19.0	18.9
38	8.90 730	155	8.90 872	157	1.09 128	9.99 858	22	8	22.1	22.0	21.9	21.7	21.6
39	8.90 885	155	8.91 029	156	1.08 971	9.99 857	21	9	24.9	24.8	24.6	24.4	24.3
40	8.91 040	155	8.91 185	155	1.08 815	9.99 856	**20**	″	**161**	**160**	**159**	**158**	**157**
41	8.91 195	154	8.91 340	155	1.08 660	9.99 855	19	1	2.7	2.7	2.6	2.6	2.6
42	8.91 349	153	8.91 495	155	1.08 505	9.99 854	18	2	5.4	5.3	5.3	5.3	5.2
43	8.91 502	153	8.91 650	153	1.08 350	9.99 853	17	3	8.0	8.0	8.0	7.9	7.8
44	8.91 655	152	8.91 803	154	1.08 197	9.99 852	16	4	10.7	10.7	10.6	10.5	10.5
45	8.91 807	152	8.91 957	153	1.08 043	9.99 851	15	5	13.4	13.3	13.2	13.2	13.1
46	8.91 959	151	8.92 110	152	1.07 890	9.99 850	14	6	16.1	16.0	15.9	15.8	15.7
47	8.92 110	151	8.92 262	152	1.07 738	9.99 848	13	7	18.8	18.7	18.6	18.4	18.3
48	8.92 261	150	8.92 414	151	1.07 586	9.99 847	12	8	21.5	21.3	21.2	21.1	20.9
49	8.92 411	150	8.92 565	151	1.07 435	9.99 846	11	9	24.2	24.0	23.8	23.7	23.6
50	8.92 561	149	8.92 716	150	1.07 284	9.99 845	**10**	″	**156**	**155**	**154**	**153**	**152**
51	8.92 710	149	8.92 866	150	1.07 134	9.99 844	9	1	2.6	2.6	2.6	2.6	2.5
52	8.92 859	148	8.93 016	149	1.06 984	9.99 843	8	2	5.2	5.2	5.1	5.1	5.1
53	8.93 007	147	8.93 165	148	1.06 835	9.99 842	7	3	7.8	7.8	7.7	7.6	7.6
54	8.93 154	147	8.93 313	149	1.06 687	9.99 841	6	4	10.4	10.3	10.3	10.2	10.1
55	8.93 301	147	8.93 462	147	1.06 538	9.99 840	5	5	13.0	12.9	12.8	12.8	12.7
56	8.93 448	146	8.93 609	147	1.06 391	9.99 839	4	6	15.6	15.5	15.4	15.3	15.2
57	8.93 594	146	8.93 756	147	1.06 244	9.99 838	3	7	18.2	18.1	18.0	17.8	17.7
58	8.93 740	145	8.93 903	146	1.06 097	9.99 837	2	8	20.8	20.7	20.5	20.4	20.3
59	8.93 885	145	8.94 049	146	1.05 951	9.99 836	1	9	23.4	23.2	23.1	23.0	22.8
60	8.94 030		8.94 195		1.05 805	9.99 834	**0**	10	26.0	25.8	25.7	25.5	25.3
′	L. Cos.	d.	L. Cot.	c.d.	L. Tan.	L. Sin.	′	P. P.					
94° (274°)					(265°) **85°**								

* Interpolation in this section of the table is not accurate.

5° (185°) (354°) **174°**

′	L. Sin.	d.	L. Tan.	c.d.	L. Cot.	L. Cos.	′	P.P.					
0	8.94 030	144	8.94 195	145	1.05 805	9.99 834	**60**	″	**151**	**149**	**148**	**147**	**146**
1	8.94 174	143	8.94 340	145	1.05 660	9.99 833	59	1	2.5	2.5	2.5	2.4	2.4
2	8.94 317	144	8.94 485	145	1.05 515	9.99 832	58	2	5.0	5.0	4.9	4.9	4.9
3	8.94 461	142	8.94 630	143	1.05 370	9.99 831	57	3	7.6	7.4	7.4	7.4	7.3
4	8.94 603	143	8.94 773	144	1.05 227	9.99 830	56	4	10.1	9.9	9.9	9.8	9.7
5	8.94 746	141	8.94 917	143	1.05 083	9.99 829	55	5	12.6	12.4	12.3	12.2	12.2
6	8.94 887	142	8.95 060	142	1.04 940	9.99 828	54	6	15.1	14.9	14.8	14.7	14.6
7	8.95 029	141	8.95 202	142	1.04 798	9.99 827	53	7	17.6	17.4	17.3	17.2	17.0
8	8.95 170	140	8.95 344	142	1.04 656	9.99 825	52	8	20.1	19.9	19.7	19.6	19.5
9	8.95 310	140	8.95 486	141	1.04 514	9.99 824	51	9	22.6	22.4	22.2	22.0	21.9
10	8.95 450	139	8.95 627	140	1.04 373	9.99 823	**50**	″	**145**	**144**	**143**	**142**	**141**
11	8.95 589	139	8.95 767	141	1.04 233	9.99 822	49	1	2.4	2.4	2.4	2.4	2.4
12	8.95 728	139	8.95 908	139	1.04 092	9.99 821	48	2	4.8	4.8	4.8	4.7	4.7
13	8.95 867	138	8.96 047	140	1.03 953	9.99 820	47	3	7.2	7.2	7.2	7.1	7.0
14	8.96 005	138	8.96 187	138	1.03 813	9.99 819	46	4	9.7	9.6	9.5	9.5	9.4
15	8.96 143	137	8.96 325	139	1.03 675	9.99 817	45	5	12.1	12.0	11.9	11.8	11.8
16	8.96 280	137	8.96 464	138	1.03 536	9.99 816	44	6	14.5	14.4	14.3	14.2	14.1
17	8.96 417	136	8.96 602	137	1.03 398	9.99 815	43	7	16.9	16.8	16.7	16.6	16.4
18	8.96 553	136	8.96 739	138	1.03 261	9.99 814	42	8	19.3	19.2	19.1	18.9	18.8
19	8.96 689	136	8.96 877	136	1.03 123	9.99 813	41	9	21.8	21.6	21.4	21.3	21.2
20	8.96 825	135	8.97 013	137	1.02 987	9.99 812	**40**	″	**140**	**139**	**138**	**137**	**136**
21	8.96 960	135	8.97 150	135	1.02 850	9.99 810	39	1	2.3	2.3	2.3	2.3	2.3
22	8.97 095	134	8.97 285	136	1.02 715	9.99 809	38	2	4.7	4.6	4.6	4.6	4.5
23	8.97 229	134	8.97 421	135	1.02 579	9.99 808	37	3	7.0	7.0	6.9	6.8	6.8
24	8.97 363	133	8.97 556	135	1.02 444	9.99 807	36	4	9.3	9.3	9.2	9.1	9.1
25	8.97 496	133	8.97 691	134	1.02 309	9.99 806	35	5	11.7	11.6	11.5	11.4	11.3
26	8.97 629	133	8.97 825	134	1.02 175	9.99 804	34	6	14.0	13.9	13.8	13.7	13.6
27	8.97 762	132	8.97 959	133	1.02 041	9.99 803	33	7	16.3	16.2	16.1	16.0	15.9
28	8.97 894	132	8.98 092	133	1.01 908	9.99 802	32	8	18.7	18.5	18.4	18.3	18.1
29	8.98 026	131	8.98 225	133	1.01 775	9.99 801	31	9	21.0	20.8	20.7	20.6	20.4
30	8.98 157	131	8.98 358	132	1.01 642	9.99 800	**30**	″	**135**	**134**	**133**	**132**	**131**
31	8.98 288	131	8.98 490	132	1.01 510	9.99 798	29	1	2.2	2.2	2.2	2.2	2.2
32	8.98 419	130	8.98 622	131	1.01 378	9.99 797	28	2	4.5	4.5	4.4	4.4	4.4
33	8.98 549	130	8.98 753	131	1.01 247	9.99 796	27	3	6.8	6.7	6.6	6.6	6.6
34	8.98 679	129	8.98 884	131	1.01 116	9.99 795	26	4	9.0	8.9	8.9	8.8	8.7
35	8.98 808	129	8.99 015	130	1.00 985	9.99 793	25	5	11.2	11.2	11.1	11.0	10.9
36	8.98 937	129	8.99 145	130	1.00 855	9.99 792	24	6	13.5	13.4	13.3	13.2	13.1
37	8.99 066	128	8.99 275	130	1.00 725	9.99 791	23	7	15.8	15.6	15.5	15.4	15.3
38	8.99 194	128	8.99 405	129	1.00 595	9.99 790	22	8	18.0	17.9	17.7	17.6	17.5
39	8.99 322	128	8.99 534	128	1.00 466	9.99 788	21	9	20.2	20.1	20.0	19.8	19.6
40	8.99 450	127	8.99 662	129	1.00 338	9.99 787	**20**	″	**130**	**129**	**128**	**127**	**126**
41	8.99 577	127	8.99 791	128	1.00 209	9.99 786	19	1	2.2	2.2	2.1	2.1	2.1
42	8.99 704	126	8.99 919	127	1.00 081	9.99 785	18	2	4.3	4.3	4.3	4.2	4.2
43	8.99 830	126	9.00 046	128	0.99 954	9.99 783	17	3	6.5	6.4	6.4	6.4	6.3
44	8.99 956	126	9.00 174	127	0.99 826	9.99 782	16	4	8.7	8.6	8.5	8.5	8.4
45	9.00 082	125	9.00 301	126	0.99 699	9.99 781	15	5	10.8	10.8	10.7	10.6	10.5
46	9.00 207	125	9.00 427	126	0.99 573	9.99 780	14	6	13.0	12.9	12.8	12.7	12.6
47	9.00 332	124	9.00 553	126	0.99 447	9.99 778	13	7	15.2	15.0	14.9	14.8	14.7
48	9.00 456	125	9.00 679	126	0.99 321	9.99 777	12	8	17.3	17.2	17.1	16.9	16.8
49	9.00 581	123	9.00 805	125	0.99 195	9.99 776	11	9	19.5	19.4	19.2	19.0	18.9
50	9.00 704	124	9.00 930	125	0.99 070	9.99 775	**10**	″	**125**	**124**	**123**	**122**	**121**
51	9.00 828	123	9.01 055	124	0.98 945	9.99 773	9	1	2.1	2.1	2.0	2.0	2.0
52	9.00 951	123	9.01 179	124	0.98 821	9.99 772	8	2	4.2	4.1	4.1	4.1	4.0
53	9.01 074	122	9.01 303	124	0.98 697	9.99 771	7	3	6.2	6.2	6.2	6.1	6.0
54	9.01 196	122	9.01 427	123	0.98 573	9.99 769	6	4	8.3	8.3	8.2	8.1	8.1
55	9.01 318	122	9.01 550	123	0.98 450	9.99 768	5	5	10.4	10.3	10.2	10.2	10.1
56	9.01 440	121	9.01 673	123	0.98 327	9.99 767	4	6	12.5	12.4	12.3	12.2	12.1
57	9.01 561	121	9.01 796	122	0.98 204	9.99 765	3	7	14.6	14.5	14.4	14.2	14.1
58	9.01 682	121	9.01 918	122	0.98 082	9.99 764	2	8	16.7	16.5	16.4	16.3	16.1
59	9.01 803	120	9.02 040	122	0.97 960	9.99 763	1	9	18.8	18.6	18.4	18.3	18.2
60	9.01 923		9.02 162		0.97 838	9.99 761	**0**	10	20.8	20.7	20.5	20.3	20.2
′	L. Cos.	d.	L. Cot.	c.d.	L. Tan.	L. Sin.	′	P.P.					

95° (275°) (264°) **84°**

6° (186°) (353°) **173°**

′	L. Sin.	d.	L. Tan.	c.d.	L. Cot.	L. Cos.	′	P. P.				
0	9.01 923	120	9.02 162	121	0.97 838	9.99 761	**60**	″	**121**	**120**	**119**	**118**
1	9.02 043	120	9.02 283	121	0.97 717	9.99 760	59	1	2.0	2.0	2.0	2.0
2	9.02 163	120	9.02 404	121	0.97 596	9.99 759	58	2	4.0	4.0	4.0	3.9
3	9.02 283	119	9.02 525	120	0.97 475	9.99 757	57	3	6.0	6.0	6.0	5.9
4	9.02 402	118	9.02 645	121	0.97 355	9.99 756	56	4	8.1	8.0	7.9	7.9
5	9.02 520	119	9.02 766	119	0.97 234	9.99 755	55	5	10.1	10.0	9.9	9.8
6	9.02 639	118	9.02 885	120	0.97 115	9.99 753	54	6	12.1	12.0	11.9	11.8
7	9.02 757	117	9.03 005	119	0.96 995	9.99 752	53	7	14.1	14.0	13.9	13.8
8	9.02 874	118	9.03 124	118	0.96 876	9.99 751	52	8	16.1	16.0	15.9	15.7
9	9.02 992	117	9.03 242	119	0.96 758	9.99 749	51	9	18.2	18.0	17.8	17.7
10	9.03 109	117	9.03 361	118	0.96 639	9.99 748	**50**	10	20.2	20.0	19.8	19.7
11	9.03 226	116	9.03 479	118	0.96 521	9.99 747	49	20	40.3	40.0	39.7	39.3
12	9.03 342	116	9.03 597	117	0.96 403	9.99 745	48	30	60.5	60.0	59.5	59.0
13	9.03 458	116	9.03 714	118	0.96 286	9.99 744	47	40	80.7	80.0	79.3	78.7
14	9.03 574	116	9.03 832	116	0.96 168	9.99 742	46	50	100.8	100.0	99.2	98.3
15	9.03 690	115	9.03 948	117	0.96 052	9.99 741	45	″	**117**	**116**	**115**	**114**
16	9.03 805	115	9.04 065	116	0.95 935	9.99 740	44	1	2.0	1.9	1.9	1.9
17	9.03 920	114	9.04 181	116	0.95 819	9.99 738	43	2	3.9	3.9	3.8	3.8
18	9.04 034	115	9.04 297	116	0.95 703	9.99 737	42	3	5.8	5.8	5.8	5.7
19	9.04 149	113	9.04 413	115	0.95 587	9.99 736	41	4	7.8	7.7	7.7	7.6
20	9.04 262	114	9.04 528	115	0.95 472	9.99 734	**40**	5	9.8	9.7	9.6	9.5
21	9.04 376	114	9.04 643	115	0.95 357	9.99 733	39	6	11.7	11.6	11.5	11.4
22	9.04 490	113	9.04 758	115	0.95 242	9.99 731	38	7	13.6	13.5	13.4	13.3
23	9.04 603	112	9.04 873	114	0.95 127	9.99 730	37	8	15.6	15.5	15.3	15.2
24	9.04 715	113	9.04 987	114	0.95 013	9.99 728	36	9	17.6	17.4	17.2	17.1
25	9.04 828	112	9.05 101	113	0.94 899	9.99 727	35	10	19.5	19.3	19.2	19.0
26	9.04 940	112	9.05 214	114	0.94 786	9.99 726	34	20	39.0	38.7	38.3	38.0
27	9.05 052	112	9.05 328	113	0.94 672	9.99 724	33	30	58.5	58.0	57.5	57.0
28	9.05 164	111	9.05 441	112	0.94 559	9.99 723	32	40	78.0	77.3	76.7	76.0
29	9.05 275	111	9.05 553	113	0.94 447	9.99 721	31	50	97.5	96.7	95.8	95 0
30	9.05 386	111	9.05 666	112	0.94 334	9.99 720	**30**	″	**113**	**112**	**111**	**110**
31	9.05 497	110	9.05 778	112	0.94 222	9.99 718	29	1	1.9	1.9	1.8	1.8
32	9.05 607	110	9.05 890	112	0.94 110	9.99 717	28	2	3.8	3.7	3.7	3.7
33	9.05 717	110	9.06 002	111	0.93 998	9.99 716	27	3	5.6	5.6	5.6	5.5
34	9.05 827	110	9.06 113	111	0.93 887	9.99 714	26	4	7.5	7.5	7.4	7.3
35	9.05 937	109	9.06 224	111	0.93 776	9.99 713	25	5	9.4	9.3	9.2	9.2
36	9.06 046	109	9.06 335	110	0.93 665	9.99 711	24	6	11.3	11.2	11.1	11.0
37	9.06 155	109	9.06 445	111	0.93 555	9.99 710	23	7	13.2	13.1	13.0	12.8
38	9.06 264	108	9.06 556	110	0.93 444	9.99 708	22	8	15.1	14.9	14.8	14.7
39	9.06 372	109	9.06 666	109	0.93 334	9.99 707	21	9	17.0	16.8	16.6	16.5
40	9.06 481	108	9.06 775	110	0.93 225	9.99 705	**20**	10	18.8	18.7	18.5	18.3
41	9.06 589	107	9.06 885	109	0.93 115	9.99 704	19	20	37.7	37.3	37.0	36.7
42	9.06 696	108	9.06 994	109	0.93 006	9.99 702	18	30	56.5	56.0	55.5	55.0
43	9.06 804	107	9.07 103	108	0.92 897	9.99 701	17	40	75.3	74.7	74.0	73.3
44	9.06 911	107	9.07 211	109	0.92 789	9.99 699	16	50	94.2	93.3	92.5	91.7
45	9.07 018	106	9.07 320	108	0.92 680	9.99 698	15	″	**109**	**108**	**107**	**106**
46	9.07 124	107	9.07 428	108	0.92 572	9.99 696	14	1	1.8	1.8	1.8	1.8
47	9.07 231	106	9.07 536	107	0.92 464	9.99 695	13	2	3.6	3.6	3.6	3.5
48	9.07 337	105	9.07 643	108	0.92 357	9.99 693	12	3	5.4	5.4	5.4	5.3
49	9.07 442	106	9.07 751	107	0.92 249	9.99 692	11	4	7.3	7.2	7.1	7.1
50	9.07 548	105	9.07 858	106	0.92 142	9.99 690	**10**	5	9.1	9.0	8.9	8.8
51	9.07 653	105	9.07 964	107	0.92 036	9.99 689	9	6	10.9	10.8	10.7	10.6
52	9.07 758	105	9.08 071	106	0.91 929	9.99 687	8	7	12.7	12.6	12.5	12.4
53	9.07 863	105	9.08 177	106	0.91 823	9.99 686	7	8	14.5	14.4	14.3	14.1
54	9.07 968	104	9.08 283	106	0.91 717	9.99 684	6	9	16.4	16.2	16.0	15.9
55	9.08 072	104	9.08 389	106	0.91 611	9.99 683	5	10	18.2	18.0	17.8	17.7
56	9.08 176	104	9.08 495	105	0.91 505	9.99 681	4	20	36.3	36.0	35.7	35.3
57	9.08 280	103	9.08 600	105	0.91 400	9.99 680	3	30	54.5	54.0	53.5	53.0
58	9.08 383	103	9.08 705	105	0.91 295	9.99 678	2	40	72.7	72.0	71.3	70.7
59	9.08 486	103	9.08 810	104	0.91 190	9.99 677	1	50	90.8	90.0	89.2	88.3
60	9.08 589		9.08 914		0.91 086	9.99 675	**0**					
′	L. Cos.	d.	L. Cot.	c.d.	L. Tan.	L. Sin.	′	P. P.				

96° (276°) (263°) **83°**

LOGARITHMS OF THE FUNCTIONS (Continued)

7° (187°) — (352°) **172°**

′	L. Sin.	d.	L. Tan.	c.d.	L. Cot.	L. Cos.	′
0	9.08 589	103	9.08 914	105	0.91 086	9.99 675	**60**
1	9.08 692	103	9.09 019	104	0.90 981	9.99 674	59
2	9.08 795	102	9.09 123	104	0.90 877	9.99 672	58
3	9.08 897	102	9.09 227	103	0.90 773	9.99 670	57
4	9.08 999	102	9.09 330	104	0.90 670	9.99 669	56
5	9.09 101	101	9.09 434	103	0.90 566	9.99 667	55
6	9.09 202	102	9.09 537	103	0.90 463	9.99 666	54
7	9.09 304	101	9.09 640	102	0.90 360	9.99 664	53
8	9.09 405	101	9.09 742	103	0.90 258	9.99 663	52
9	9.09 506	100	9.09 845	102	0.90 155	9.99 661	51
10	9.09 606	101	9.09 947	102	0.90 053	9.99 659	**50**
11	9.09 707	100	9.10 049	101	0.89 951	9.99 658	49
12	9.09 807	100	9.10 150	102	0.89 850	9.99 656	48
13	9.09 907	99	9.10 252	101	0.89 748	9.99 655	47
14	9.10 006	100	9.10 353	101	0.89 647	9.99 653	46
15	9.10 106	99	9.10 454	101	0.89 546	9.99 651	45
16	9.10 205	99	9.10 555	101	0.89 445	9.99 650	44
17	9.10 304	98	9.10 656	100	0.89 344	9.99 648	43
18	9.10 402	99	9.10 756	100	0.89 244	9.99 647	42
19	9.10 501	98	9.10 856	100	0.89 144	9.99 645	41
20	9.10 599	98	9.10 956	100	0.89 044	9.99 643	**40**
21	9.10 697	98	9.11 056	99	0.88 944	9.99 642	39
22	9.10 795	98	9.11 155	99	0.88 845	9.99 640	38
23	9.10 893	97	9.11 254	99	0.88 746	9.99 638	37
24	9.10 990	97	9.11 353	99	0.88 647	9.99 637	36
25	9.11 087	97	9.11 452	99	0.88 548	9.99 635	35
26	9.11 184	97	9.11 551	98	0.88 449	9.99 633	34
27	9.11 281	96	9.11 649	98	0.88 351	9.99 632	33
28	9.11 377	97	9.11 747	98	0.88 253	9.99 630	32
29	9.11 474	96	9.11 845	98	0.88 155	9.99 629	31
30	9.11 570	96	9.11 943	97	0.88 057	9.99 627	**30**
31	9.11 666	95	9.12 040	98	0.87 960	9.99 625	29
32	9.11 761	96	9.12 138	97	0.87 862	9.99 624	28
33	9.11 857	95	9.12 235	97	0.87 765	9.99 622	27
34	9.11 952	95	9.12 332	96	0.87 668	9.99 620	26
35	9.12 047	95	9.12 428	97	0.87 572	9.99 618	25
36	9.12 142	94	9.12 525	96	0.87 475	9.99 617	24
37	9.12 236	95	9.12 621	96	0.87 379	9.99 615	23
38	9.12 331	94	9.12 717	96	0.87 283	9.99 613	22
39	9.12 425	94	9.12 813	96	0.87 187	9.99 612	21
40	9.12 519	93	9.12 909	95	0.87 091	9.99 610	**20**
41	9.12 612	94	9.13 004	95	0.86 996	9.99 608	19
42	9.12 706	93	9.13 099	95	0.86 901	9.99 607	18
43	9.12 799	93	9.13 194	95	0.86 806	9.99 605	17
44	9.12 892	93	9.13 289	95	0.86 711	9.99 603	16
45	9.12 985	93	9.13 384	94	0.86 616	9.99 601	15
46	9.13 078	93	9.13 478	95	0.86 522	9.99 600	14
47	9.13 171	92	9.13 573	94	0.86 427	9.99 598	13
48	9.13 263	92	9.13 667	94	0.86 333	9.99 596	12
49	9.13 355	92	9.13 761	93	0.86 239	9.99 595	11
50	9.13 447	92	9.13 854	94	0.86 146	9.99 593	**10**
51	9.13 539	91	9.13 948	93	0.86 052	9.99 591	9
52	9.13 630	92	9.14 041	93	0.85 959	9.99 589	8
53	9.13 722	91	9.14 134	93	0.85 866	9.99 588	7
54	9.13 813	91	9.14 227	93	0.85 773	9.99 586	6
55	9.13 904	90	9.14 320	92	0.85 680	9.99 584	5
56	9.13 994	91	9.14 412	92	0.85 588	9.99 582	4
57	9.14 085	90	9.14 504	93	0.85 496	9.99 581	3
58	9.14 175	91	9.14 597	91	0.85 403	9.99 579	2
59	9.14 266	90	9.14 688	92	0.85 312	9.99 577	1
60	9.14 356		9.14 780		0.85 220	9.99 575	**0**
′	L. Cos.	d.	L. Cot.	c.d.	L. Tan.	L. Sin.	′

P. P.

″	**105**	**104**	**103**	**102**
1	1.8	1.7	1.7	1.7
2	3.5	3.5	3.4	3.4
3	5.2	5.2	5.2	5.1
4	7.0	6.9	6.9	6.8
5	8.8	8.7	8.6	8.5
6	10.5	10.4	10.3	10.2
7	12.2	12.1	12.0	11.9
8	14.0	13.9	13.7	13.6
9	15.8	15.6	15.4	15.3
10	17.5	17.3	17.2	17.0
20	35.0	34.7	34.3	34.0
30	52.5	52.0	51.5	51.0
40	70.0	69.3	68.7	68.0
50	87.5	86.7	85.8	85.0
″	**101**	**100**	**99**	**98**
1	1.7	1.7	1.6	1.6
2	3.4	3.3	3.3	3.3
3	5.0	5.0	5.0	4.9
4	6.7	6.7	6.6	6.5
5	8.4	8.3	8.2	8.2
6	10.1	10.0	9.9	9.8
7	11.8	11.7	11.6	11.4
8	13.5	13.3	13.2	13.1
9	15.2	15.0	14.8	14.7
10	16.8	16.7	16.5	16.3
20	33.7	33.3	33.0	32.7
30	50.5	50.0	49.5	49.0
40	67.3	66.7	66.0	65.3
50	84.2	83.3	82.5	81.7
″	**97**	**96**	**95**	**94**
1	1.6	1.6	1.6	1.6
2	3.2	3.2	3.2	3.1
3	4.8	4.8	4.8	4.7
4	6.5	6.4	6.3	6.3
5	8.1	8.0	7.9	7.8
6	9.7	9.6	9.5	9.4
7	11.3	11.2	11.1	11.0
8	12.9	12.8	12.7	12.5
9	14.6	14.4	14.2	14.1
10	16.2	16.0	15.8	15.7
20	32.3	32.0	31.7	31.3
30	48.5	48.0	47.5	47.0
40	64.7	64.0	63.3	62.7
50	80.8	80.0	79.2	78.3
″	**93**	**92**	**91**	**90**
1	1.6	1.5	1.5	1.5
2	3.1	3.1	3.0	3.0
3	4.6	4.6	4.6	4.5
4	6.2	6.1	6.1	6.0
5	7.8	7.7	7.6	7.5
6	9.3	9.2	9.1	9.0
7	10.8	10.7	10.6	10.5
8	12.4	12.3	12.1	12.0
9	14.0	13.8	13.6	13.5
10	15.5	15.3	15.2	15.0
20	31.0	30.7	30.3	30.0
30	46.5	46.0	45.5	45.0
40	62.0	61.3	60.7	60.0
50	77.5	76.7	75.8	75.0

97° (277°) — (262°) **82°**

8° (188°) — (351°) **171°**

′	L. Sin.	d.	L. Tan.	c. d.	L. Cot.	L. Cos.	′
0	9.14 356	89	9.14 780	92	0.85 220	9.99 575	**60**
1	9.14 445	90	9.14 872	91	0.85 128	9.99 574	59
2	9.14 535	89	9.14 963	91	0.85 037	9.99 572	58
3	9.14 624	90	9.15 054	91	0.84 946	9.99 570	57
4	9.14 714	89	9.15 145	91	0.84 855	9.99 568	56
5	9.14 803	88	9.15 236	91	0.84 764	9.99 566	55
6	9.14 891	89	9.15 327	90	0.84 673	9.99 565	54
7	9.14 980	89	9.15 417	91	0.84 583	9.99 563	53
8	9.15 069	88	9.15 508	90	0.84 492	9.99 561	52
9	9.15 157	88	9.15 598	90	0.84 402	9.99 559	51
10	9.15 245	88	9.15 688	89	0.84 312	9.99 557	**50**
11	9.15 333	88	9.15 777	90	0.84 223	9.99 556	49
12	9.15 421	87	9.15 867	89	0.84 133	9.99 554	48
13	9.15 508	88	9.15 956	90	0.84 044	9.99 552	47
14	9.15 596	87	9.16 046	89	0.83 954	9.99 550	46
15	9.15 683	87	9.16 135	89	0.83 865	9.99 548	45
16	9.15 770	87	9.16 224	88	0.83 776	9.99 546	44
17	9.15 857	87	9.16 312	89	0.83 688	9.99 545	43
18	9.15 944	86	9.16 401	88	0.83 599	9.99 543	42
19	9.16 030	86	9.16 489	88	0.83 511	9.99 541	41
20	9.16 116	87	9.16 577	88	0.83 423	9.99 539	**40**
21	9.16 203	86	9.16 665	88	0.83 335	9.99 537	39
22	9.16 289	85	9.16 753	88	0.83 247	9.99 535	38
23	9.16 374	86	9.16 841	87	0.83 159	9.99 533	37
24	9.16 460	85	9.16 928	88	0.83 072	9.99 532	36
25	9.16 545	86	9.17 016	87	0.82 984	9.99 530	35
26	9.16 631	85	9.17 103	87	0.82 897	9.99 528	34
27	9.16 716	85	9 17 190	87	0.82 810	9.99 526	33
28	9.16 801	85	9.17 277	86	0.82 723	9.99 524	32
29	9.16 886	84	9.17 363	87	0.82 637	9.99 522	31
30	9.16 970	85	9.17 450	86	0.82 550	9.99 520	**30**
31	9.17 055	84	9.17 536	86	0.82 464	9.99 518	29
32	9.17 139	84	9.17 622	86	0 82 378	9.99 517	28
33	9.17 223	84	9.17 708	86	0.82 292	9.99 515	27
34	9.17 307	84	9.17 794	86	0.82 206	9.99 513	26
35	9.17 391	83	9.17 880	85	0.82 120	9.99 511	25
36	9.17 474	84	9.17 965	86	0.82 035	9.99 509	24
37	9.17 558	83	9.18 051	85	0.81 949	9.99 507	23
38	9.17 641	83	9.18 136	85	0.81 864	9.99 505	22
39	9.17 724	83	9.18 221	85	0.81 779	9.99 503	21
40	9.17 807	83	9.18 306	85	0.81 694	9.99 501	**20**
41	9.17 890	83	9.18 391	84	0.81 609	9.99 499	19
42	9.17 973	82	9.18 475	85	0.81 525	9.99 497	18
43	9.18 055	82	9.18 560	84	0.81 440	9.99 495	17
44	9.18 137	83	9.18 644	84	0.81 356	9.99 494	16
45	9.18 220	82	9.18 728	84	0.81 272	9.99 492	15
46	9.18 302	81	9.18 812	84	0.81 188	9.99 490	14
47	9.18 383	82	9.18 896	83	0.81 104	9.99 488	13
48	9.18 465	82	9.18 979	84	0.81 021	9.99 486	12
49	9.18 547	81	9.19 063	83	0.80 937	9.99 484	11
50	9.18 628	81	9.19 146	83	0.80 854	9.99 482	**10**
51	9.18 709	81	9.19 229	83	0.80 771	9.99 480	9
52	9.18 790	81	9.19 312	83	0.80 688	9.99 478	8
53	9.18 871	81	9.19 395	83	0.80 605	9.99 476	7
54	9.18 952	81	9.19 478	83	0.80 522	9.99 474	6
55	9.19 033	80	9.19 561	82	0.80 439	9.99 472	5
56	9.19 113	80	9.19 643	82	0.80 357	9.99 470	4
57	9.19 193	80	9.19 725	82	0.80 275	9.99 468	3
58	9.19 273	80	9.19 807	82	0.80 193	9.99 466	2
59	9.19 353	80	9.19 889	82	0.80 111	9.99 464	1
60	9.19 433		9.19 971		0.80 029	9.99 462	**0**
′	L. Cos.	d.	L. Cot.	c. d.	L. Tan.	L. Sin.	′

P. P.

″	**92**	**91**	**90**
1	1.5	1.5	1.5
2	3.1	3.0	3.0
3	4.6	4.6	4.5
4	6.1	6.1	6.0
5	7.7	7.6	7.5
6	9.2	9.1	9.0
7	10.7	10.6	10.5
8	12.3	12.1	12.0
9	13.8	13.6	13.5
10	15.3	15.2	15.0
20	30.7	30.3	30.0
30	46.0	45.5	45.0
40	61.3	60.7	60.0
50	76.7	75.8	75.0
″	**89**	**88**	**87**
1	1.5	1.5	1.4
2	3.0	2.9	2.9
3	4.4	4.4	4.4
4	5.9	5.9	5.8
5	7.4	7.3	7.2
6	8.9	8.8	8.7
7	10.4	10.3	10.2
8	11.9	11.7	11.6
9	13.4	13.2	13.0
10	14.8	14.7	14.5
20	29.7	29.3	29.0
30	44.5	44.0	43.5
40	59.3	58.7	58.0
50	74.2	73.3	72.5
″	**86**	**85**	**84**
1	1.4	1.4	1.4
2	2.9	2.8	2.8
3	4.3	4.2	4.2
4	5.7	5.7	5.6
5	7.2	7.1	7.0
6	8.6	8.5	8.4
7	10.0	9.9	9.8
8	11.5	11.3	11.2
9	12.9	12.8	12.6
10	14.3	14.2	14.0
20	28.7	28.3	28.0
30	43.0	42.5	42.0
40	57.3	56.7	56.0
50	71.7	70.8	70.0
″	**83**	**82**	**81**
1	1.4	1.4	1.4
2	2.8	2.7	2.7
3	4.2	4.1	4.0
4	5.5	5.5	5.4
5	6.9	6.8	6.8
6	8.3	8.2	8.1
7	9.7	9.6	9.4
8	11.1	10.9	10.8
9	12.4	12.3	12.2
10	13.8	13.7	13.5
20	27.7	27.3	27.0
30	41.5	41.0	40.5
30	55.3	54.7	54.0
50	69.2	68.3	67.5

98° (278°) — (261°) **81°**

9° (189°) (350°) **170°**

′	L. Sin.	d.	L. Tan.	c.d.	L. Cot.	L. Cos.	′
0	9.19 433	80	9.19 971	82	0.80 029	9.99 462	**60**
1	9.19 513	79	9.20 053	81	0.79 947	9.99 460	59
2	9.19 592	80	9.20 134	82	0.79 866	9.99 458	58
3	9.19 672	79	9.20 216	81	0.79 784	9.99 456	57
4	9.19 751	79	9.20 297	81	0.79 703	9.99 454	56
5	9.19 830	79	9.20 378	81	0.79 622	9.99 452	55
6	9.19 909	79	9.20 459	81	0.79 541	9.99 450	54
7	9.19 988	79	9.20 540	81	0.79 460	9.99 448	53
8	9.20 067	78	9.20 621	80	0.79 379	9.99 446	52
9	9.20 145	78	9.20 701	81	0.79 299	9.99 444	51
10	9.20 223	79	9.20 782	80	0.79 218	9.99 442	**50**
11	9.20 302	78	9.20 862	80	0.79 138	9.99 440	49
12	9.20 380	78	9.20 942	80	0.79 058	9.99 438	48
13	9.20 458	77	9.21 022	80	0.78 978	9.99 436	47
14	9.20 535	78	9.21 102	80	0.78 898	9.99 434	46
15	9.20 613	78	9.21 182	79	0.78 818	9.99 432	45
16	9.20 691	77	9.21 261	80	0.78 739	9.99 429	44
17	9.20 768	77	9.21 341	79	0.78 659	9.99 427	43
18	9.20 845	77	9.21 420	79	0.78 580	9.99 425	42
19	9.20 922	77	9.21 499	79	0.78 501	9.99 423	41
20	9.20 999	77	9.21 578	79	0.78 422	9.99 421	**40**
21	9.21 076	77	9.21 657	79	0.78 343	9.99 419	39
22	9.21 153	76	9.21 736	78	0.78 264	9.99 417	38
23	9.21 229	77	9.21 814	79	0.78 186	9.99 415	37
24	9.21 306	76	9.21 893	78	0.78 107	9.99 413	36
25	9.21 382	76	9.21 971	78	0.78 029	9.99 411	35
26	9.21 458	76	9.22 049	78	0.77 951	9.99 409	34
27	9.21 534	76	9.22 127	78	0.77 873	9.99 407	33
28	9.21 610	75	9.22 205	78	0.77 795	9.99 404	32
29	9.21 685	76	9.22 283	78	0.77 717	9.99 402	31
30	9.21 761	75	9.22 361	77	0.77 639	9.99 400	**30**
31	9.21 836	76	9.22 438	78	0.77 562	9.99 398	29
32	9.21 912	75	9.22 516	77	0.77 484	9.99 396	28
33	9.21 987	75	9.22 593	77	0.77 407	9.99 394	27
34	9.22 062	75	9.22 670	77	0.77 330	9.99 392	26
35	9.22 137	74	9.22 747	77	0.77 253	9.99 390	25
36	9.22 211	75	9.22 824	77	0.77 176	9.99 388	24
37	9.22 286	75	9.22 901	76	0.77 099	9.99 385	23
38	9.22 361	74	9.22 977	77	0.77 023	9.99 383	22
39	9.22 435	74	9.23 054	76	0.76 946	9.99 381	21
40	9.22 509	74	9.23 130	76	0.76 870	9.99 379	**20**
41	9.22 583	74	9.23 206	77	0.76 794	9.99 377	19
42	9.22 657	74	9.23 283	76	0.76 717	9.99 375	18
43	9.22 731	74	9.23 359	76	0.76 641	9.99 372	17
44	9.22 805	73	9.23 435	75	0.76 565	9.99 370	16
45	9.22 878	74	9.23 510	76	0.76 490	9.99 368	15
46	9.22 952	73	9.23 586	75	0.76 414	9.99 366	14
47	9.23 025	73	9.23 661	76	0.76 339	9.99 364	13
48	9.23 098	73	9.23 737	75	0.76 263	9.99 362	12
49	9.23 171	73	9.23 812	75	0.76 188	9.99 359	11
50	9.23 244	73	9.23 887	75	0.76 113	9.99 357	**10**
51	9.23 317	73	9.23 962	75	0.76 038	9.99 355	9
52	9.23 390	72	9.24 037	75	0.75 963	9.99 353	8
53	9.23 462	73	9.24 112	74	0.75 888	9.99 351	7
54	9.23 535	72	9.24 186	75	0.75 814	9.99 348	6
55	9.23 607	72	9.24 261	74	0.75 739	9.99 346	5
56	9.23 679	73	9.24 335	75	0.75 665	9.99 344	4
57	9.23 752	71	9.24 410	74	0.75 590	9.99 342	3
58	9.23 823	72	9.24 484	74	0.75 516	9.99 340	2
59	9.23 895	72	9.24 558	74	0.75 442	9.99 337	1
60	9.23 967		9.24 632		0.75 368	9.99 335	**0**
′	L. Cos.	d.	L. Cot.	c.d.	L. Tan.	L. Sin.	′

99° (279°) (260°) **80°**

P. P.

″	**80**	**79**	**78**	**77**
1	1.3	1.3	1.3	1.3
2	2.7	2.6	2.6	2.6
3	4.0	4.0	3.9	3.8
4	5.3	5.3	5.2	5.1
5	6.7	6.6	6.5	6.4
6	8.0	7.9	7.8	7.7
7	9.3	9.2	9.1	9.0
8	10.7	10.5	10.4	10.3
9	12.0	11.8	11.7	11.6
10	13.3	13.2	13.0	12.8
20	26.7	26.3	26.0	25.7
30	40.0	39.5	39.0	38.5
40	53.3	52.7	52.0	51.3
50	66.7	65.8	65.0	64.2

″	**76**	**75**	**74**	**73**
1	1.3	1.2	1.2	1.2
2	2.5	2.5	2.5	2.4
3	3.8	3.8	3.7	3.6
4	5.1	5.0	4.9	4.9
5	6.3	6.2	6.2	6.1
6	7.6	7.5	7.4	7.3
7	8.9	8.8	8.6	8.5
8	10.1	10.0	9.9	9.7
9	11.4	11.2	11.1	11.0
10	12.7	12.5	12.3	12.2
20	25.3	25.0	24.7	24.3
30	38.0	37.5	37.0	36.5
40	50.7	50.0	49.3	48.7
50	63.3	62.5	61.7	60.8

″	**72**	**71**	**3**	**2**
1	1.2	1.2	0.0	0.0
2	2.4	2.4	0.1	0.1
3	3.6	3.6	0.2	0.1
4	4.8	4.7	0.2	0.1
5	6.0	5.9	0.2	0.2
6	7.2	7.1	0.3	0.2
7	8.4	8.3	0.4	0.2
8	9.6	9.5	0.4	0.3
9	10.8	10.6	0.4	0.3
10	12.0	11.8	0.5	0.3
20	24.0	23.7	1.0	0.7
30	36.0	35.5	1.5	1.0
40	48.0	47.3	2.0	1.3
50	60.0	59.2	2.5	1.7

	3/79	**3/78**	**3/77**
0	13.2	13.0	12.8
1	39.5	39.0	38.5
2	65.8	65.0	64.2
3			

	3/76	**3/75**	**3/74**
0	12.7	12.5	12.3
1	38.0	37.5	37.0
2	63.3	62.5	61.7
3			

10° (190°) (349°) **169°**

′	L. Sin.	d.	L. Tan.	c.d.	L. Cot.	L. Cos.	d.	′
0	9.23 967	72	9.24 632	74	0.75 368	9.99 335	2	**60**
1	9.24 039	71	9.24 706	73	0.75 294	9.99 333	2	59
2	9.24 110	71	9.24 779	74	0.75 221	9.99 331	3	58
3	9.24 181	72	9.24 853	73	0.75 147	9.99 328	2	57
4	9.24 253	71	9.24 926	74	0.75 074	9.99 326	2	56
5	9.24 324	71	9.25 000	73	0.75 000	9.99 324	2	55
6	9.24 395	71	9.25 073	73	0.74 927	9.99 322	3	54
7	9.24 466	70	9.25 146	73	0.74 854	9.99 319	2	53
8	9.24 536	71	9.25 219	73	0.74 781	9.99 317	2	52
9	9.24 607	70	9.25 292	73	0.74 708	9.99 315	2	51
10	9.24 677	71	9.25 365	72	0.74 635	9.99 313	3	**50**
11	9.24 748	70	9.25 437	73	0.74 563	9.99 310	2	49
12	9.24 818	70	9.25 510	72	0.74 490	9.99 308	2	48
13	9.24 888	70	9.25 582	73	0.74 418	9.99 306	2	47
14	9.24 958	70	9.25 655	72	0.74 345	9.99 304	3	46
15	9.25 028	70	9.25 727	72	0.74 273	9.99 301	2	45
16	9.25 098	70	9.25 799	72	0.74 201	9.99 299	2	44
17	9.25 168	69	9.25 871	72	0.74 129	9.99 297	3	43
18	9.25 237	70	9.25 943	72	0.74 057	9.99 294	2	42
19	9.25 307	69	9.26 015	71	0.73 985	9.99 292	2	41
20	9.25 376	69	9.26 086	72	0.73 914	9.99 290	2	**40**
21	9.25 445	69	9.26 158	71	0.73 842	9.99 288	3	39
22	9.25 514	69	9.26 229	72	0.73 771	9.99 285	2	38
23	9.25 583	69	9.26 301	71	0.73 699	9.99 283	2	37
24	9.25 652	69	9.26 372	71	0.73 628	9.99 281	3	36
25	9.25 721	69	9.26 443	71	0.73 557	9.99 278	2	35
26	9.25 790	68	9.26 514	71	0.73 486	9.99 276	2	34
27	9.25 858	69	9.26 585	70	0.73 415	9.99 274	3	33
28	9.25 927	68	9.26 655	71	0.73 345	9.99 271	2	32
29	9.25 995	68	9.26 726	71	0.73 274	9.99 269	2	31
30	9.26 063	68	9.26 797	70	0.73 203	9.99 267	3	**30**
31	9.26 131	68	9.26 867	70	0.73 133	9.99 264	2	29
32	9.26 199	68	9.26 937	71	0.73 063	9.99 262	2	28
33	9.26 267	68	9.27 008	70	0.72 992	9.99 260	3	27
34	9.26 335	68	9.27 078	70	0.72 922	9.99 257	2	26
35	9.26 403	67	9.27 148	70	0.72 852	9.99 255	3	25
36	9.26 470	68	9.27 218	70	0.72 782	9.99 252	2	24
37	9.26 538	67	9.27 288	69	0.72 712	9.99 250	2	23
38	9.26 605	67	9.27 357	70	0.72 643	9.99 248	3	22
39	9.26 672	67	9.27 427	69	0.72 573	9.99 245	2	21
40	9.26 739	67	9.27 496	70	0.72 504	9.99 243	2	**20**
41	9.26 806	67	9.27 566	69	0.72 434	9.99 241	3	19
42	9.26 873	67	9.27 635	69	0.72 365	9.99 238	2	18
43	9.26 940	67	9.27 704	69	0.72 296	9.99 236	3	17
44	9.27 007	66	9.27 773	69	0.72 227	9.99 233	2	16
45	9.27 073	67	9.27 842	69	0.72 158	9.99 231	2	15
46	9.27 140	66	9.27 911	69	0.72 089	9.99 229	3	14
47	9.27 206	67	9.27 980	69	0.72 020	9.99 226	2	13
48	9.27 273	66	9.28 049	68	0.71 951	9.99 224	3	12
49	9.27 339	66	9.28 117	69	0.71 883	9.99 221	2	11
50	9.27 405	66	9.28 186	68	0.71 814	9.99 219	2	**10**
51	9.27 471	66	9.28 254	69	0.71 746	9.99 217	3	9
52	9.27 537	65	9.28 323	68	0.71 677	9.99 214	2	8
53	9.27 602	66	9.28 391	68	0.71 609	9.99 212	3	7
54	9.27 668	66	9.28 459	68	0.71 541	9.99 209	2	6
55	9.27 734	65	9.28 527	68	0.71 473	9.99 207	3	5
56	9.27 799	65	9.28 595	67	0.71 405	9.99 204	2	4
57	9.27 864	66	9.28 662	68	0.71 338	9.99 202	2	3
58	9.27 930	65	9.28 730	68	0.71 270	9.99 200	3	2
59	9.27 995	65	9.28 798	67	0.71 202	9.99 197	2	1
60	9.28 060		9.28 865		0.71 135	9.99 195		**0**
′	L. Cos.	d.	L. Cot.	c.d.	L. Tan.	L. Sin.	d.	′

100° (280°) (259°) **79°**

P. P.

″	**74**	**73**	**72**
1	1.2	1.2	1.2
2	2.5	2.4	2.4
3	3.7	3.6	3.6
4	4.9	4.9	4.8
5	6.2	6.1	6.0
6	7.4	7.3	7.2
7	8.6	8.5	8.4
8	9.9	9.7	9.6
9	11.1	11.0	10.8
10	12.3	12.2	12.0
20	24.7	24.3	24.0
30	37.0	36.5	36.0
40	49.3	48.7	48.0
50	61.7	60.8	60.0

″	**71**	**70**	**69**
1	1.2	1.2	1.2
2	2.4	2.3	2.3
3	3.6	3.5	3.4
4	4.7	4.7	4.6
5	5.9	5.8	5.8
6	7.1	7.0	6.9
7	8.3	8.2	8.0
8	9.5	9.3	9.2
9	10.6	10.5	10.4
10	11.8	11.7	11.5
20	23.7	23.3	23.0
30	35.5	35.0	34.5
40	47.3	46.7	46.0
50	59.2	58.3	57.5

″	**68**	**67**	**66**
1	1.1	1.1	1.1
2	2.3	2.2	2.2
3	3.4	3.4	3.3
4	4.5	4.5	4.4
5	5.7	5.6	5.5
6	6.8	6.7	6.6
7	7.9	7.8	7.7
8	9.1	8.9	8.8
9	10.2	10.0	9.9
10	11.3	11.2	11.0
20	22.7	22.3	22.0
30	34.0	33.5	33.0
40	45.3	44.7	44.0
50	56.7	55.8	55.0

	3/74	**3/73**	**3/72**
0	12.3	12.2	12.0
1	37.0	36.5	36.0
2	61.7	60.8	60.0
3			

	3/71	**3/70**	**3/69**	**3/68**
0	11.8	11.7	11.5	11.3
1	35.5	35.0	34.5	34.0
2	59.2	58.3	57.5	56.7
3				

11° (191°) (348°) **168°**

′	L. Sin.	d.	L. Tan.	c.d.	L. Cot.	L. Cos.	d.	′
0	9.28 060	65	9.28 865	68	0.71 135	9.99 195	3	**60**
1	9.28 125	65	9.28 933	67	0.71 067	9.99 192	2	59
2	9.28 190	64	9.29 000	67	0.71 000	9.99 190	3	58
3	9.28 254	65	9.29 067	67	0.70 933	9.99 187	2	57
4	9.28 319	65	9.29 134	67	0.70 866	9.99 185	3	56
5	9.28 384	64	9.29 201	67	0.70 799	9.99 182	2	55
6	9.28 448	64	9.29 268	67	0.70 732	9.99 180	3	54
7	9.28 512	65	9.29 335	67	0.70 665	9.99 177	2	53
8	9.28 577	64	9.29 402	66	0.70 598	9.99 175	3	52
9	9.28 641	64	9 29 468	67	0.70 532	9.99 172	2	51
10	9.28 705	64	9.29 535	66	0.70 465	9.99 170	3	**50**
11	9.28 769	64	9.29 601	67	0.70 399	9.99 167	2	49
12	9.28 833	63	9.29 668	66	0.70 332	9.99 165	3	48
13	9.28 896	64	9.29 734	66	0.70 266	9.99 162	2	47
14	9.28 960	64	9.29 800	66	0.70 200	9.99 160	3	46
15	9.29 024	63	9.29 866	66	0.70 134	9.99 157	2	45
16	9.29 087	63	9.29 932	66	0.70 068	9.99 155	3	44
17	9.29 150	64	9.29 998	66	0.70 002	9.99 152	2	43
18	9.29 214	63	9.30 064	66	0.69 936	9.99 150	3	42
19	9.29 277	63	9.30 130	65	0.69 870	9.99 147	2	41
20	9.29 340	63	9.30 195	66	0.69 805	9.99 145	3	**40**
21	9.29 403	63	9.30 261	65	0.69 739	9.99 142	2	39
22	9.29 466	63	9.30 326	65	0.69 674	9.99 140	3	38
23	9.29 529	62	9.30 391	66	0.69 609	9.99 137	2	37
24	9.29 591	63	9.30 457	65	0.69 543	9.99 135	3	36
25	9.29 654	62	9.30 522	65	0.69 478	9.99 132	2	35
26	9.29 716	63	9.30 587	65	0.69 413	9.99 130	3	34
27	9.29 779	62	9.30 652	65	0.69 348	9.99 127	3	33
28	9.29 841	62	9.30 717	65	0.69 283	9.99 124	2	32
29	9.29 903	63	9.30 782	64	0.69 218	9.99 122	3	31
30	9.29 966	62	9.30 846	65	0.69 154	9.99 119	2	**30**
31	9.30 028	62	9.30 911	64	0.69 089	9.99 117	3	29
32	9.30 090	61	9.30 975	65	0.69 025	9.99 114	2	28
33	9.30 151	62	9.31 040	64	0.68 960	9.99 112	3	27
34	9.30 213	62	9.31 104	64	0.68 896	9.99 109	3	26
35	9.30 275	61	9.31 168	65	0.68 832	9.99 106	2	25
36	9.30 336	62	9.31 233	64	0.68 767	9.99 104	3	24
37	9.30 398	61	9.31 297	64	0.68 703	9.99 101	2	23
38	9.30 459	62	9.31 361	64	0.68 639	9.99 099	3	22
39	9.30 521	61	9.31 425	64	0.68 575	9.99 096	3	21
40	9.30 582	61	9.31 489	63	0.68 511	9.99 093	2	**20**
41	9.30 643	61	9.31 552	64	0.68 448	9.99 091	3	19
42	9.30 704	61	9.31 616	63	0.68 384	9.99 088	2	18
43	9.30 765	61	9.31 679	64	0.68 321	9.99 086	3	17
44	9.30 826	61	9.31 743	63	0.68 257	9.99 083	3	16
45	9.30 887	60	9.31 806	64	0.68 194	9.99 080	2	15
46	9.30 947	61	9.31 870	63	0.68 130	9.99 078	3	14
47	9.31 008	60	9.31 933	63	0.68 067	9.99 075	3	13
48	9.31 068	61	9.31 996	63	0.68 004	9.99 072	2	12
49	9.31 129	60	9.32 059	63	0.67 941	9.99 070	3	11
50	9.31 189	61	9.32 122	63	0.67 878	9.99 067	3	**10**
51	9.31 250	60	9.32 185	63	0.67 815	9.99 064	2	9
52	9.31 310	60	9.32 248	63	0.67 752	9.99 062	3	8
53	9.31 370	60	9.32 311	62	0.67 689	9.99 059	3	7
54	9.31 430	60	9.32 373	63	0.67 627	9.99 056	2	6
55	9.31 490	59	9.32 436	62	0.67 564	9.99 054	3	5
56	9.31 549	60	9.32 498	63	0.67 502	9.99 051	3	4
57	9.31 609	60	9.32 561	62	0.67 439	9.99 048	2	3
58	9.31 669	59	9.32 623	62	0.67 377	9.99 046	3	2
59	9.31 728	60	9.32 685	62	0.67 315	9.99 043	3	1
60	9.31 788		9.32 747		0.67 253	9.99 040		**0**
′	L. Cos.	d.	L. Cot.	c.d.	L. Tan.	L. Sin.	d.	′

P. P.

″	**65**	**64**	**63**
1	1.1	1.1	1.0
2	2.2	2.1	2.1
3	3.2	3.2	3.2
4	4.3	4.3	4.2
5	5.4	5.3	5.2
6	6.5	6.4	6.3
7	7.6	7.5	7.4
8	8.7	8.5	8.4
9	9.8	9.6	9.4
10	10.8	10.7	10.5
20	21.7	21.3	21.0
30	32.5	32.0	31.5
40	43.3	42.7	42.0
50	54.2	53.3	52.5
″	**62**	**61**	**60**
1	1.0	1.0	1.0
2	2.1	2.0	2.0
3	3.1	3.0	3.0
4	4.1	4.1	4.0
5	5.2	5.1	5.0
6	6.2	6.1	6.0
7	7.2	7.1	7.0
8	8.3	8.1	8.0
9	9.3	9.2	9.0
10	10.3	10.2	10.0
20	20.7	20.3	20.0
30	31.0	30.5	30.0
40	41.3	40.7	40.0
50	51.7	50.8	50.0
″	**59**	**3**	**2**
1	1.0	0.0	0.0
2	2.0	0.1	0.1
3	3.0	0.2	0.1
4	3.9	0.2	0.1
5	4.9	0.2	0.2
6	5.9	0.3	0.2
7	6.9	0.4	0.2
8	7.9	0.4	0.3
9	8.8	0.4	0.3
10	9.8	0.5	0.3
20	19.7	1.0	0.7
30	29.5	1.5	1.0
40	39.3	2.0	1.3
50	49.2	2.5	1.7

	3/67	**3/66**	**3/65**
0	11.2	11.0	10.8
1	33.5	33.0	32.5
2	55.8	55.0	54.2
3			
	3/64	**3/63**	**3/62**
0	10.7	10.5	10.3
1	32.0	31.5	31.0
2	53.3	52.5	51.7
3			

101° (281°) (258°) **78°**

12° (192°) (347°) **167°**

′	L. Sin.	d.	L. Tan.	c.d.	L. Cot.	L. Cos.	d.	′
0	9.31 788	59	9.32 747	63	0.67 253	9.99 040	2	**60**
1	9.31 847	60	9.32 810	62	0.67 190	9.99 038	3	59
2	9.31 907	59	9.32 872	61	0.67 128	9.99 035	3	58
3	9.31 966	59	9.32 933	62	0.67 067	9.99 032	2	57
4	9.32 025	59	9.32 995	62	0.67 005	9.99 030	3	56
5	9.32 084	59	9.33 057	62	0.66 943	9.99 027	3	55
6	9.32 143	59	9.33 119	61	0.66 881	9.99 024	2	54
7	9.32 202	59	9.33 180	62	0.66 820	9.99 022	3	53
8	9.32 261	58	9.33 242	61	0.66 758	9.99 019	3	52
9	9.32 319	59	9.33 303	62	0.66 697	9.99 016	3	51
10	9.32 378	59	9.33 365	61	0.66 635	9.99 013	2	**50**
11	9.32 437	58	9.33 426	61	0.66 574	9.99 011	3	49
12	9.32 495	58	9.33 487	61	0.66 513	9.99 008	3	48
13	9.32 553	59	9.33 548	61	0.66 452	9.99 005	3	47
14	9.32 612	58	9.33 609	61	0.66 391	9.99 002	2	46
15	9.32 670	58	9.33 670	61	0.66 330	9.99 000	3	45
16	9.32 728	58	9.33 731	61	0.66 269	9.98 997	3	44
17	9.32 786	58	9.33 792	61	0.66 208	9.98 994	3	43
18	9.32 844	58	9.33 853	60	0.66 147	9.98 991	2	42
19	9.32 902	58	9.33 913	61	0.66 087	9.98 989	3	41
20	9.32 960	58	9.33 974	60	0.66 026	9.98 986	3	**40**
21	9.33 018	57	9.34 034	61	0.65 966	9.98 983	3	39
22	9.33 075	58	9.34 095	60	0.65 905	9.98 980	2	38
23	9.33 133	57	9.34 155	60	0.65 845	9.98 978	3	37
24	9.33 190	58	9.34 215	61	0.65 785	9.98 975	3	36
25	9.33 248	57	9.34 276	60	0.65 724	9.98 972	3	35
26	9.33 305	57	9.34 336	60	0.65 664	9.98 969	2	34
27	9.33 362	58	9.34 396	60	0.65 604	9.98 967	3	33
28	9.33 420	57	9.34 456	60	0.65 544	9.98 964	3	32
29	9.33 477	57	9.34 516	60	0.65 484	9.98 961	3	31
30	9.33 534	57	9.34 576	59	0.65 424	9.98 958	3	**30**
31	9.33 591	56	9.34 635	60	0.65 365	9.98 955	2	29
32	9.33 647	57	9.34 695	60	0.65 305	9.98 953	3	28
33	9.33 704	57	9.34 755	59	0.65 245	9.98 950	3	27
34	9.33 761	57	9.34 814	60	0.65 186	9.98 947	3	26
35	9.33 818	56	9.34 874	59	0.65 126	9.98 944	3	25
36	9.33 874	57	9.34 933	59	0.65 067	9.98 941	3	24
37	9.33 931	56	9.34 992	59	0.65 008	9.98 938	2	23
38	9.33 987	56	9.35 051	60	0.64 949	9.98 936	3	22
39	9.34 043	57	9.35 111	59	0.64 889	9.98 933	3	21
40	9.34 100	56	9.35 170	59	0.64 830	9.98 930	3	**20**
41	9.34 156	56	9.35 229	59	0.64 771	9.98 927	3	19
42	9.34 212	56	9.35 288	59	0.64 712	9.98 924	3	18
43	9.34 268	56	9.35 347	58	0.64 653	9.98 921	2	17
44	9.34 324	56	9.35 405	59	0.64 595	9.98 919	3	16
45	9.34 380	56	9.35 464	59	0.64 536	9.98 916	3	15
46	9.34 436	55	9.35 523	58	0.64 477	9.98 913	3	14
47	9.34 491	56	9.35 581	59	0.64 419	9.98 910	3	13
48	9.34 547	55	9.35 640	58	0.64 360	9.98 907	3	12
49	9.34 602	56	9.35 698	59	0.64 302	9.98 904	3	11
50	9.34 658	55	9.35 757	58	0.64 243	9.98 901	3	**10**
51	9.34 713	56	9.35 815	58	0.64 185	9.98 898	2	9
52	9.34 769	55	9.35 873	58	0.64 127	9.98 896	3	8
53	9.34 824	55	9.35 931	58	0.64 069	9.98 893	3	7
54	9.34 879	55	9.35 989	58	0.64 011	9.98 890	3	6
55	9.34 934	55	9.36 047	58	0.63 953	9.98 887	3	5
56	9.34 989	55	9.36 105	58	0.63 895	9.98 884	3	4
57	9.35 044	55	9.36 163	58	0.63 837	9.98 881	3	3
58	9.35 099	55	9.36 221	58	0.63 779	9.98 878	3	2
59	9.35 154	55	9.36 279	57	0.63 721	9.98 875	3	1
60	9.35 209		9.36 336		0.63 664	9.98 872		**0**
′	L. Cos.	d.	L. Cot.	c.d.	L. Tan.	L. Sin.	d.	′

P. P.

″	**63**	**62**	**61**
1	1.0	1.0	1.0
2	2.1	2.1	2.0
3	3.2	3.1	3.0
4	4.2	4.1	4.1
5	5.2	5.2	5.1
6	6.3	6.2	6.1
7	7.4	7.2	7.1
8	8.4	8.3	8.1
9	9.4	9.3	9.2
10	10.5	10.3	10.2
20	21.0	20.7	20.3
30	31.5	31.0	30.5
40	42.0	41.3	40.7
50	52.5	51.7	50.8
″	**60**	**59**	**58**
1	1.0	1.0	1.0
2	2.0	2.0	1.9
3	3.0	3.0	2.9
4	4.0	3.9	3.9
5	5.0	4.9	4.8
6	6.0	5.9	5.8
7	7.0	6.9	6.8
8	8.0	7.9	7.7
9	9.0	8.8	8.7
10	10.0	9.8	9.7
20	20.0	19.7	19.3
30	30.0	29.5	29.0
40	40.0	39.3	38.7
50	50.0	49.2	48.3
″	**57**	**56**	**55**
1	1.0	0.9	0.9
2	1.9	1.9	1.8
3	2.8	2.8	2.8
4	3.8	3.7	3.7
5	4.8	4.7	4.6
6	5.7	5.6	5.5
7	6.6	6.5	6.4
8	7.6	7.5	7.3
9	8.6	8.4	8.2
10	9.5	9.3	9.2
20	19.0	18.7	18.3
30	28.5	28.0	27.5
40	38.0	37.3	36.7
50	47.5	46.7	45.8

	3/62	**3/61**	**3/60**
0	10.3	10.2	10.0
1	31.0	30.5	30.0
2	51.7	50.8	50.0
3			
	3/59	**3/58**	**3/57**
0	9.8	9.7	9.5
1	29.5	29.0	28.5
2	49.2	48.3	47.5
3			

102° (282°) (257°) **77°**

LOGARITHMS OF THE FUNCTIONS (Continued)

LOGARITHMS OF THE FUNCTIONS (Continued)

13° (193°) (346°) **166°**

′	L. Sin.	d.	L. Tan.	c.d.	L. Cot.	L. Cos.	d.	′
0	9.35 209	54	9.36 336	58	0.63 664	9.98 872	3	**60**
1	9.35 263	55	9.36 394	58	0.63 606	9.98 869	2	59
2	9.35 318	55	9.36 452	57	0.63 548	9.98 867	3	58
3	9.35 373	54	9.36 509	57	0.63 491	9.98 864	3	57
4	9.35 427	54	9.36 566	58	0.63 434	9.98 861	3	56
5	9.35 481	55	9.36 624	57	0.63 376	9.98 858	3	55
6	9.35 536	54	9.36 681	57	0.63 319	9.98 855	3	54
7	9.35 590	54	9.36 738	57	0.63 262	9.98 852	3	53
8	9.35 644	54	9.36 795	57	0.63 205	9.98 849	3	52
9	9.35 698	54	9.36 852	57	0.63 148	9.98 846	3	51
10	9.35 752	54	9.36 909	57	0.63 091	9.98 843	3	**50**
11	9.35 806	54	9.36 966	57	0.63 034	9.98 840	3	49
12	9.35 860	54	9.37 023	57	0.62 977	9.98 837	3	48
13	9.35 914	54	9.37 080	57	0.62 920	9.98 834	3	47
14	9.35 968	54	9.37 137	56	0.62 863	9.98 831	3	46
15	9.36 022	53	9.37 193	57	0.62 807	9.98 828	3	45
16	9.36 075	54	9.37 250	56	0.62 750	9.98 825	3	44
17	9.36 129	53	9.37 306	57	0.62 694	9.98 822	3	43
18	9.36 182	54	9.37 363	56	0.62 637	9.98 819	3	42
19	9.36 236	53	9.37 419	57	0.62 581	9.98 816	3	41
20	9.36 289	53	9.37 476	56	0.62 524	9.98 813	3	**40**
21	9.36 342	53	9.37 532	56	0.62 468	9.98 810	3	39
22	9.36 395	54	9.37 588	56	0.62 412	9.98 807	3	38
23	9.36 449	53	9.37 644	56	0.62 356	9.98 804	3	37
24	9.36 502	53	9.37 700	56	0.62 300	9.98 801	3	36
25	9.36 555	53	9.37 756	56	0.62 244	9.98 798	3	35
26	9.36 608	52	9.37 812	56	0.62 188	9.98 795	3	34
27	9.36 660	53	9.37 868	56	0.62 132	9.98 792	3	33
28	9.36 713	53	9.37 924	56	0.62 076	9.98 789	3	32
29	9.36 766	53	9.37 980	55	0.62 020	9.98 786	3	31
30	9.36 819	52	9.38 035	56	0.61 965	9.98 783	3	**30**
31	9.36 871	53	9.38 091	56	0.61 909	9.98 780	3	29
32	9.36 924	52	9.38 147	55	0.61 853	9.98 777	3	28
33	9.36 976	52	9.38 202	55	0.61 798	9.98 774	3	27
34	9.37 028	53	9.38 257	56	0.61 743	9.98 771	3	26
35	9.37 081	52	9.38 313	55	0.61 687	9.98 768	3	25
36	9.37 133	52	9.38 368	55	0.61 632	9.98 765	3	24
37	9.37 185	52	9.38 423	56	0.61 577	9.98 762	3	23
38	9.37 237	52	9.38 479	55	0.61 521	9.98 759	3	22
39	9.37 289	52	9.38 534	55	0.61 466	9.98 756	3	21
40	9.37 341	52	9.38 589	55	0.61 411	9.98 753	3	**20**
41	9.37 393	52	9.38 644	55	0.61 356	9.98 750	4	19
42	9.37 445	52	9.38 699	55	0.61 301	9.98 746	3	18
43	9.37 497	52	9.38 754	54	0.61 246	9.98 743	3	17
44	9.37 549	51	9.38 808	55	0.61 192	9.98 740	3	16
45	9.37 600	52	9.38 863	55	0.61 137	9.98 737	3	15
46	9.37 652	51	9.38 918	54	0.61 082	9.98 734	3	14
47	9.37 703	52	9.38 972	55	0.61 028	9.98 731	3	13
48	9.37 755	51	9.39 027	55	0.60 973	9.98 728	3	12
49	9.37 806	52	9.39 082	54	0.60 918	9.98 725	3	11
50	9.37 858	51	9.39 136	54	0.60 864	9.98 722	3	**10**
51	9.37 909	51	9.39 190	55	0.60 810	9.98 719	4	9
52	9.37 960	51	9.39 245	54	0.60 755	9.98 715	3	8
53	9.38 011	51	9.39 299	54	0.60 701	9.98 712	3	7
54	9.38 062	51	9.39 353	54	0.60 647	9.98 709	3	6
55	9.38 113	51	9.39 407	54	0.60 593	9.98 706	3	5
56	9.38 164	51	9.39 461	54	0.60 539	9.98 703	3	4
57	9.38 215	51	9.39 515	54	0.60 485	9.98 700	3	3
58	9.38 266	51	9.39 569	54	0.60 431	9.98 697	3	2
59	9.38 317	51	9.39 623	54	0.60 377	9.98 694	4	1
60	9.38 368		9.39 677		0.60 323	9.98 690		**0**
′	L. Cos.	d.	L. Cot.	c.d.	L. Tan.	L. Sin.	d.	′

P. P.

″	57	56	55
1	1.0	0.9	0.9
2	1.9	1.9	1.8
3	2.8	2.8	2.8
4	3.8	3.7	3.7
5	4.8	4.7	4.6
6	5.7	5.6	5.5
7	6.6	6.5	6.4
8	7.6	7.5	7.3
9	8.6	8.4	8.2
10	9.5	9.3	9.2
20	19.0	18.7	18.3
30	28.5	28.0	27.5
40	38.0	37.3	36.7
50	47.5	46.7	45.8

″	54	53	52
1	0.9	0.9	0.9
2	1.8	1.8	1.7
3	2.7	2.6	2.6
4	3.6	3.5	3.5
5	4.5	4.4	4.3
6	5.4	5.3	5.2
7	6.3	6.2	6.1
8	7.2	7.1	6.9
9	8.1	8.0	7.8
10	9.0	8.8	8.7
20	18.0	17.7	17.3
30	27.0	26.5	26.0
40	36.0	35.3	34.7
50	45.0	44.2	43.3

″	51	4	3	2
1	0.8	0.1	0.0	0.0
2	1.7	0.1	0.1	0.1
3	2.6	0.2	0.2	0.1
4	3.4	0.3	0.2	0.1
5	4.2	0.3	0.2	0.2
6	5.1	0.4	0.3	0.2
7	6.0	0.5	0.4	0.2
8	6.8	0.5	0.4	0.3
9	7.6	0.6	0.4	0.3
10	8.5	0.7	0.5	0.3
20	17.0	1.3	1.0	0.7
30	25.5	2.0	1.5	1.0
40	34.0	2.7	2.0	1.3
50	42.5	3.3	2.5	1.7

	4/55	4/54	3/58	3/57
0	6.9	6.8	9.7	9.5
1	20.6	20.2	29.0	28.5
2	34.4	33.8	48.3	47.5
3	48.1	47.2	—	—
4				

	3/56	3/55	3/54
0	9.3	9.2	9.0
1	28.0	27.5	27.0
2	46.7	45.8	45.0
3			

103° (283°) (256°) **76°**

14° (194°) (345°) **165°**

′	L. Sin.	d.	L. Tan.	c.d.	L. Cot.	L. Cos.	d.	′
0	9.38 368	50	9.39 677	54	0.60 323	9.98 690	3	**60**
1	9.38 418	51	9.39 731	54	0.60 269	9.98 687	3	59
2	9.38 469	50	9.39 785	53	0.60 215	9.98 684	3	58
3	9.38 519	51	9.39 838	54	0.60 162	9.98 681	3	57
4	9.38 570	50	9.39 892	53	0.60 108	9.98 678	3	56
5	9.38 620	50	9.39 945	54	0.60 055	9.98 675	4	55
6	9.38 670	51	9.39 999	53	0.60 001	9.98 671	3	54
7	9.38 721	50	9.40 052	54	0.59 948	9.98 668	3	53
8	9.38 771	50	9.40 106	53	0.59 894	9.98 665	3	52
9	9.38 821	50	9.40 159	53	0.59 841	9.98 662	3	51
10	9.38 871	50	9.40 212	54	0.59 788	9.98 659	3	**50**
11	9.38 921	50	9.40 266	53	0.59 734	9.98 656	4	49
12	9.38 971	50	9.40 319	53	0.59 681	9.98 652	3	48
13	9.39 021	50	9.40 372	53	0.59 628	9.98 649	3	47
14	9.39 071	50	9.40 425	53	0.59 575	9.98 646	3	46
15	9.39 121	49	9.40 478	53	0.59 522	9.98 643	3	45
16	9.39 170	50	9.40 531	53	0.59 469	9.98 640	4	44
17	9.39 220	50	9.40 584	52	0.59 416	9.98 636	3	43
18	9.39 270	49	9.40 636	53	0.59 364	9.98 633	3	42
19	9.39 319	50	9.40 689	53	0.59 311	9.98 630	3	41
20	9.39 369	49	9.40 742	53	0.59 258	9.98 627	4	**40**
21	9.39 418	49	9.40 795	52	0.59 205	9.98 623	3	39
22	9.39 467	50	9.40 847	53	0.59 153	9.98 620	3	38
23	9.39 517	49	9.40 900	52	0.59 100	9.98 617	3	37
24	9.39 566	49	9.40 952	53	0.59 048	9.98 614	4	36
25	9.39 615	49	9.41 005	52	0.58 995	9.98 610	3	35
26	9.39 664	49	9.41 057	52	0.58 943	9.98 607	3	34
27	9.39 713	49	9.41 109	52	0.58 891	9.98 604	3	33
28	9.39 762	49	9.41 161	53	0.58 839	9.98 601	4	32
29	9.39 811	49	9.41 214	52	0.58 786	9.98 597	3	31
30	9.39 860	49	9.41 266	52	0.58 734	9.98 594	3	**30**
31	9.39 909	49	9.41 318	52	0.58 682	9.98 591	3	29
32	9.39 958	48	9.41 370	52	0.58 630	9.98 588	4	28
33	9.40 006	49	9.41 422	52	0.58 578	9.98 584	3	27
34	9.40 055	48	9.41 474	52	0.58 526	9.98 581	3	26
35	9.40 103	49	9.41 526	52	0.58 474	9.98 578	4	25
36	9.40 152	48	9.41 578	51	0.58 422	9.98 574	3	24
37	9.40 200	49	9.41 629	52	0.58 371	9.98 571	3	23
38	9.40 249	48	9.41 681	52	0.58 319	9.98 568	3	22
39	9.40 297	49	9.41 733	51	0.58 267	9.98 565	4	21
40	9.40 346	48	9.41 784	52	0.58 216	9.98 561	3	**20**
41	9.40 394	48	9.41 836	51	0.58 164	9.98 558	3	19
42	9.40 442	48	9.41 887	52	0.58 113	9.98 555	4	18
43	9.40 490	48	9.41 939	51	0.58 061	9.98 551	3	17
44	9.40 538	48	9.41 990	51	0.58 010	9.98 548	3	16
45	9.40 586	48	9.42 041	52	0.57 959	9.98 545	4	15
46	9.40 634	48	9.42 093	51	0.57 907	9.98 541	3	14
47	9.40 682	48	9.42 144	51	0.57 856	9.98 538	3	13
48	9.40 730	48	9.42 195	51	0.57 805	9.98 535	4	12
49	9.40 778	47	9.42 246	51	0.57 754	9.98 531	3	11
50	9.40 825	48	9.42 297	51	0.57 703	9.98 528	3	**10**
51	9.40 873	48	9.42 348	51	0.57 652	9.98 525	4	9
52	9.40 921	47	9.42 399	51	0.57 601	9.98 521	3	8
53	9.40 968	48	9.42 450	51	0.57 550	9.98 518	3	7
54	9.41 016	47	9.42 501	51	0.57 499	9.98 515	4	6
55	9.41 063	48	9.42 552	51	0.57 448	9.98 511	3	5
56	9.41 111	47	9.42 603	50	0.57 397	9.98 508	3	4
57	9.41 158	47	9.42 653	51	0.57 347	9.98 505	4	3
58	9.41 205	47	9.42 704	51	0.57 296	9.98 501	3	2
59	9.41 252	48	9.42 755	50	0.57 245	9.98 498	4	1
60	9.41 300		9.42 805		0.57 195	9.98 494		**0**
′	L. Cos.	d.	L. Cot.	c.d.	L. Tan.	L. Sin.	d.	′

P. P.

″	54	53	52
1	0.9	0.9	0.9
2	1.8	1.8	1.7
3	2.7	2.6	2.6
4	3.6	3.5	3.5
5	4.5	4.4	4.3
6	5.4	5.3	5.2
7	6.3	6.2	6.1
8	7.2	7.1	6.9
9	8.1	8.0	7.8
10	9.0	8.8	8.7
20	18.0	17.7	17.3
30	27.0	26.5	26.0
40	36.0	35.3	34.7
50	45.0	44.2	43.3

″	51	50	49
1	0.8	0.8	0.8
2	1.7	1.7	1.6
3	2.6	2.5	2.4
4	3.4	3.3	3.3
5	4.2	4.2	4.1
6	5.1	5.0	4.9
7	6.0	5.8	5.7
8	6.8	6.7	6.5
9	7.6	7.5	7.4
10	8.5	8.3	8.2
20	17.0	16.7	16.3
30	25.5	25.0	24.5
40	34.0	33.3	32.7
50	42.5	41.7	40.8

″	48	47	4	3
1	0.8	0.8	0.1	0.0
2	1.6	1.6	0.1	0.1
3	2.4	2.4	0.2	0.2
4	3.2	3.1	0.3	0.2
5	4.0	3.9	0.3	0.2
6	4.8	4.7	0.4	0.3
7	5.6	5.5	0.5	0.4
8	6.4	6.3	0.5	0.4
9	7.2	7.0	0.6	0.4
10	8.0	7.8	0.7	0.5
20	16.0	15.7	1.3	1.0
30	24.0	23.5	2.0	1.5
40	32.0	31.3	2.7	2.0
50	40.0	39.2	3.3	2.5

	4/54	4/53	4/52	4/51
0	6.8	6.6	6.5	6.4
1	20.2	19.9	19.5	19.1
2	33.8	33.1	32.5	31.9
3	47.2	46.4	45.5	44.6
4				

	3/54	3/53	3/52	3/51
0	9.0	8.8	8.7	8.5
1	27.0	26.5	26.0	25.5
2	45.0	44.2	43.3	42.5
3				

104° (284°) (255°) **75°**

LOGARITHMS OF THE FUNCTIONS (Continued)

15° (195°) — (344°) **164°**

′	L. Sin.	d.	L. Tan.	c.d.	L. Cot.	L. Cos.	d.	′
0	9.41 300	47	9.42 805	51	0.57 195	9.98 494	3	**60**
1	9.41 347	47	9.42 856	50	0.57 144	9.98 491	3	59
2	9.41 394	47	9.42 906	51	0.57 094	9.98 488	4	58
3	9.41 441	47	9.42 957	50	0.57 043	9.98 484	3	57
4	9.41 488	47	9.43 007	50	0.56 993	9.98 481	4	56
5	9.41 535	47	9.43 057	51	0.56 943	9.98 477	3	55
6	9.41 582	46	9.43 108	50	0.56 892	9.98 474	3	54
7	9.41 628	47	9.43 158	50	0.56 842	9.98 471	4	53
8	9.41 675	47	9.43 208	50	0.56 792	9.98 467	3	52
9	9.41 722	46	9.43 258	50	0.56 742	9.98 464	4	51
10	9.41 768	47	9.43 308	50	0.56 692	9.98 460	3	**50**
11	9.41 815	46	9.43 358	50	0.56 642	9.98 457	4	49
12	9.41 861	47	9.43 408	50	0.56 592	9.98 453	3	48
13	9.41 908	46	9.43 458	50	0.56 542	9.98 450	3	47
14	9.41 954	47	9.43 508	50	0.56 492	9.98 447	4	46
15	9.42 001	46	9.43 558	49	0.56 442	9.98 443	3	45
16	9.42 047	46	9.43 607	50	0.56 393	9.98 440	4	44
17	9.42 093	47	9.43 657	50	0.56 343	9.98 436	3	43
18	9.42 140	46	9.43 707	49	0.56 293	9.98 433	4	42
19	9.42 186	46	9.43 756	50	0.56 244	9.98 429	3	41
20	9.42 232	46	9.43 806	49	0.56 194	9.98 426	4	**40**
21	9.42 278	46	9.43 855	50	0.56 145	9.98 422	3	39
22	9.42 324	46	9.43 905	49	0.56 095	9.98 419	4	38
23	9.42 370	46	9.43 954	50	0.56 046	9.98 415	3	37
24	9.42 416	45	9.44 004	49	0.55 996	9.98 412	3	36
25	9.42 461	46	9.44 053	49	0.55 947	9.98 409	4	35
26	9.42 507	46	9.44 102	49	0.55 898	9.98 405	3	34
27	9.42 553	46	9.44 151	50	0.55 849	9.98 402	4	33
28	9.42 599	45	9.44 201	49	0.55 799	9.98 398	3	32
29	9.42 644	46	9.44 250	49	0.55 750	9.98 395	4	31
30	9.42 690	45	9.44 299	49	0.55 701	9.98 391	3	**30**
31	9.42 735	46	9.44 348	49	0.55 652	9.98 388	4	29
32	9.42 781	45	9.44 397	49	0.55 603	9.98 384	3	28
33	9.42 826	46	9.44 446	49	0.55 554	9.98 381	4	27
34	9.42 872	45	9.44 495	49	0.55 505	9.98 377	4	26
35	9.42 917	45	9.44 544	48	0.55 456	9.98 373	3	25
36	9.42 962	46	9.44 592	49	0.55 408	9.98 370	4	24
37	9.43 008	45	9.44 641	49	0.55 359	9.98 366	3	23
38	9.43 053	45	9.44 690	48	0.55 310	9.98 363	4	22
39	9.43 098	45	9.44 738	49	0.55 262	9.98 359	3	21
40	9.43 143	45	9.44 787	49	0.55 213	9.98 356	4	**20**
41	9.43 188	45	9.44 836	48	0.55 164	9.98 352	3	19
42	9.43 233	45	9.44 884	49	0.55 116	9.98 349	4	18
43	9.43 278	45	9.44 933	48	0.55 067	9.98 345	3	17
44	9.43 323	44	9.44 981	48	0.55 019	9.98 342	4	16
45	9.43 367	45	9.45 029	49	0.54 971	9.98 338	4	15
46	9.43 412	45	9.45 078	48	0.54 922	9.98 334	3	14
47	9.43 457	45	9.45 126	48	0.54 874	9.98 331	4	13
48	9.43 502	44	9.45 174	48	0.54 826	9.98 327	3	12
49	9.43 546	45	9.45 222	49	0.54 778	9.98 324	4	11
50	9.43 591	44	9.45 271	48	0.54 729	9.98 320	3	**10**
51	9.43 635	45	9.45 319	48	0.54 681	9.98 317	4	9
52	9.43 680	44	9.45 367	48	0.54 633	9.98 313	4	8
53	9.43 724	45	9.45 415	48	0.54 585	9.98 309	3	7
54	9.43 769	44	9.45 463	48	0.54 537	9.98 306	4	6
55	9.43 813	44	9.45 511	48	0.54 489	9.98 302	3	5
56	9.43 857	44	9.45 559	47	0.54 441	9.98 299	4	4
57	9.43 901	45	9.45 606	48	0.54 394	9.98 295	4	3
58	9.43 946	44	9.45 654	48	0.54 346	9.98 291	3	2
59	9.43 990	44	9.45 702	48	0.54 298	9.98 288	4	1
60	9.44 034		9.45 750		0.54 250	9.98 284		**0**
′	L. Cos.	d.	L. Cot.	c.d.	L. Tan.	L. Sin.	d.	′

P. P.

″	51	50	49
1	0.8	0.8	0.8
2	1.7	1.7	1.6
3	2.6	2.5	2.4
4	3.4	3.3	3.3
5	4.2	4.2	4.1
6	5.1	5.0	4.9
7	6.0	5.8	5.7
8	6.8	6.7	6.5
9	7.6	7.5	7.4
10	8.5	8.3	8.2
20	17.0	16.7	16.3
30	25.5	25.0	24.5
40	34.0	33.3	32.7
50	42.5	41.7	40.8

″	48	47	46
1	0.8	0.8	0.8
2	1.6	1.6	1.5
3	2.4	2.4	2.3
4	3.2	3.1	3.1
5	4.0	3.9	3.8
6	4.8	4.7	4.6
7	5.6	5.5	5.4
8	6.4	6.3	6.1
9	7.2	7.0	6.9
10	8.0	7.8	7.7
20	16.0	15.7	15.3
30	24.0	23.5	23.0
40	32.0	31.3	30.7
50	40.0	39.2	38.3

″	45	44	4	3
1	0.8	0.7	0.1	0.0
2	1.5	1.5	0.1	0.1
3	2.2	2.2	0.2	0.2
4	3.0	2.9	0.3	0.2
5	3.8	3.7	0.3	0.2
6	4.5	4.4	0.4	0.3
7	5.2	5.1	0.5	0.4
8	6.0	5.9	0.5	0.4
9	6.8	6.6	0.6	0.4
10	7.5	7.3	0.7	0.5
20	15.0	14.7	1.3	1.0
30	22.5	22.0	2.0	1.5
40	30.0	29.3	2.7	2.0
50	37.5	36.7	3.3	2.5

	4/50	4/49	4/48	4/47
0	6.2	6.1	6.0	5.9
1	18.8	18.4	18.0	17.6
2	31.2	30.6	30.0	29.4
3	43.8	42.9	42.0	41.1

	3/51	3/50	3/49	3/48
0	8.5	8.3	8.2	8.0
1	25.5	25.0	24.5	24.0
2	42.5	41.7	40.8	40.0

P. P.

105° (285°) — (254°) **74°**

16° (196°) — (343°) **163°**

′	L. Sin.	d.	L. Tan.	c.d.	L. Cot.	L. Cos.	d.	′
0	9.44 034	44	9.45 750	47	0.54 250	9.98 284	3	**60**
1	9.44 078	44	9.45 797	48	0.54 203	9.98 281	4	59
2	9.44 122	44	9.45 845	47	0.54 155	9.98 277	4	58
3	9.44 166	44	9.45 892	48	0.54 108	9.98 273	3	57
4	9.44 210	43	9.45 940	47	0.54 060	9.98 270	4	56
5	9.44 253	44	9.45 987	48	0.54 013	9.98 266	4	55
6	9.44 297	44	9.46 035	47	0.53 965	9.98 262	3	54
7	9.44 341	44	9.46 082	48	0.53 918	9.98 259	4	53
8	9.44 385	43	9.46 130	47	0.53 870	9.98 255	4	52
9	9.44 428	44	9.46 177	47	0.53 823	9.98 251	3	51
10	9.44 472	44	9.46 224	47	0.53 776	9.98 248	4	**50**
11	9.44 516	43	9.46 271	48	0.53 729	9.98 244	4	49
12	9.44 559	43	9.46 319	47	0.53 681	9.98 240	3	48
13	9.44 602	44	9.46 366	47	0.53 634	9.98 237	4	47
14	9.44 646	43	9.46 413	47	0.53 587	9.98 233	4	46
15	9.44 689	44	9.46 460	47	0.53 540	9.98 229	3	45
16	9.44 733	43	9.46 507	47	0.53 493	9.98 226	4	44
17	9.44 776	43	9.46 554	47	0.53 446	9.98 222	4	43
18	9.44 819	43	9.46 601	47	0.53 399	9.98 218	3	42
19	9.44 862	43	9.46 648	46	0.53 352	9.98 215	4	41
20	9.44 905	43	9.46 694	47	0.53 306	9.98 211	4	**40**
21	9.44 948	44	9.46 741	47	0.53 259	9.98 207	3	39
22	9.44 992	43	9.46 788	47	0.53 212	9.98 204	4	38
23	9.45 035	42	9.46 835	46	0.53 165	9.98 200	4	37
24	9.45 077	43	9.46 881	47	0.53 119	9.98 196	4	36
25	9.45 120	43	9.46 928	47	0.53 072	9.98 192	3	35
26	9.45 163	43	9.46 975	46	0.53 025	9.98 189	4	34
27	9.45 206	43	9.47 021	47	0.52 979	9.98 185	4	33
28	9.45 249	43	9.47 068	46	0.52 932	9.98 181	4	32
29	9.45 292	42	9.47 114	46	0.52 886	9.98 177	3	31
30	9.45 334	43	9.47 160	47	0.52 840	9.98 174	4	**30**
31	9.45 377	42	9.47 207	46	0.52 793	9.98 170	4	29
32	9.45 419	43	9.47 253	46	0.52 747	9.98 166	4	28
33	9.45 462	42	9.47 299	47	0.52 701	9.98 162	3	27
34	9.45 504	43	9.47 346	46	0.52 654	9.98 159	4	26
35	9.45 547	42	9.47 392	46	0.52 608	9.98 155	4	25
36	9.45 589	43	9.47 438	46	0.52 562	9.98 151	4	24
37	9.45 632	42	9.47 484	46	0.52 516	9.98 147	3	23
38	9.45 674	42	9.47 530	46	0.52 470	9.98 144	4	22
39	9.45 716	42	9.47 576	46	0.52 424	9.98 140	4	21
40	9.45 758	43	9.47 622	46	0.52 378	9.98 136	4	**20**
41	9.45 801	42	9.47 668	46	0.52 332	9.98 132	3	19
42	9.45 843	42	9.47 714	46	0.52 286	9.98 129	4	18
43	9.45 885	42	9.47 760	46	0.52 240	9.98 125	4	17
44	9.45 927	42	9.47 806	46	0.52 194	9.98 121	4	16
45	9.45 969	42	9.47 852	45	0.52 148	9.98 117	4	15
46	9.46 011	42	9.47 897	46	0.52 103	9.98 113	3	14
47	9.46 053	42	9.47 943	46	0.52 057	9.98 110	4	13
48	9.46 095	41	9.47 989	46	0.52 011	9.98 106	4	12
49	9.46 136	42	9.48 035	45	0.51 965	9.98 102	4	11
50	9.46 178	42	9.48 080	46	0.51 920	9.98 098	4	**10**
51	9.46 220	42	9.48 126	45	0.51 874	9.98 094	4	9
52	9.46 262	41	9.48 171	46	0.51 829	9.98 090	3	8
53	9.46 303	42	9.48 217	45	0.51 783	9.98 087	4	7
54	9.46 345	41	9.48 262	45	0.51 738	9.98 083	4	6
55	9.46 386	42	9.48 307	46	0.51 693	9.98 079	4	5
56	9.46 428	41	9.48 353	45	0.51 647	9.98 075	4	4
57	9.46 469	42	9.48 398	45	0.51 602	9.98 071	4	3
58	9.46 511	41	9.48 443	46	0.51 557	9.98 067	4	2
59	9.46 552	42	9.48 489	45	0.51 511	9.98 063	3	1
60	9.46 594		9.48 534		0.51 466	9.98 060		**0**
′	L. Cos.	d.	L. Cot.	c.d.	L. Tan.	L. Sin.	d.	′

P.P.

″	48	47	46
1	0.8	0.8	0.8
2	1.6	1.6	1.5
3	2.4	2.4	2.3
4	3.2	3.1	3.1
5	4.0	3.9	3.8
6	4.8	4.7	4.6
7	5.6	5.5	5.4
8	6.4	6.3	6.1
9	7.2	7.0	6.9
10	8.0	7.8	7.7
20	16.0	15.7	15.3
30	24.0	23.5	23.0
40	32.0	31.3	30.7
50	40.0	39.2	38.3

″	45	44	43
1	0.8	0.7	0.7
2	1.5	1.5	1.4
3	2.2	2.2	2.2
4	3.0	2.9	2.9
5	3.8	3.7	3.6
6	4.5	4.4	4.3
7	5.2	5.1	5.0
8	6.0	5.9	5.7
9	6.8	6.6	6.4
10	7.5	7.3	7.2
20	15.0	14.7	14.3
30	22.5	22.0	21.5
40	30.0	29.3	28.7
50	37.5	36.7	35.8

″	42	41	4	3
1	0.7	0.7	0.1	0.0
2	1.4	1.4	0.1	0.1
3	2.1	2.0	0.2	0.2
4	2.8	2.7	0.3	0.2
5	3.5	3.4	0.3	0.2
6	4.2	4.1	0.4	0.3
7	4.9	4.8	0.5	0.4
8	5.6	5.5	0.5	0.4
9	6.3	6.2	0.6	0.4
10	7.0	6.8	0.7	0.5
20	14.0	13.7	1.3	1.0
30	21.0	20.5	2.0	1.5
40	28.0	27.3	2.7	2.0
50	35.0	34.2	3.3	2.5

	4/48	4/47	4/46	4/45
0	6.0	5.9	5.8	5.6
1	18.0	17.6	17.2	16.9
2	30.0	29.4	28.8	28.1
3	42.0	41.1	40.2	39.4

	3/48	3/47	3/46	3/45
0	8.0	7.8	7.7	7.5
1	24.0	23.5	23.0	22.5
2	40.0	39.2	38.3	37.5

P.P.

106° (286°) — (253°) **73°**

17° (197°) (342°) **162°**

′	L. Sin.	d.	L. Tan.	c.d.	L. Cot.	L. Cos.	d.	′
0	9.46 594	41	9.48 534	45	0.51 466	9.98 060	4	**60**
1	9.46 635	41	9.48 579	45	0.51 421	9.98 056	4	59
2	9.46 676	41	9.48 624	45	0.51 376	9.98 052	4	58
3	9.46 717	41	9.48 669	45	0.51 331	9.98 048	4	57
4	9.46 758	42	9.48 714	45	0.51 286	9.98 044	4	56
5	9.46 800	41	9.48 759	45	0.51 241	9.98 040	4	55
6	9.46 841	41	9.48 804	45	0.51 196	9.98 036	4	54
7	9.46 882	41	9.48 849	45	0.51 151	9.98 032	3	53
8	9.46 923	41	9.48 894	45	0.51 106	9.98 029	4	52
9	9.46 964	41	9.48 939	45	0.51 061	9.98 025	4	51
10	9.47 005	40	9.48 984	45	0.51 016	9.98 021	4	**50**
11	9.47 045	41	9.49 029	44	0.50 971	9.98 017	4	49
12	9.47 086	41	9.49 073	45	0.50 927	9.98 013	4	48
13	9.47 127	41	9.49 118	45	0.50 882	9.98 009	4	47
14	9.47 168	41	9.49 163	44	0.50 837	9.98 005	4	46
15	9.47 209	40	9.49 207	45	0.50 793	9.98 001	4	45
16	9.47 249	41	9.49 252	44	0.50 748	9.97 997	4	44
17	9.47 290	40	9.49 296	45	0.50 704	9.97 993	4	43
18	9.47 330	41	9.49 341	44	0.50 659	9.97 989	3	42
19	9.47 371	40	9.49 385	45	0.50 615	9.97 986	4	41
20	9.47 411	41	9.49 430	44	0.50 570	9.97 982	4	**40**
21	9.47 452	40	9.49 474	45	0.50 526	9.97 978	4	39
22	9.47 492	41	9.49 519	44	0.50 481	9.97 974	4	38
23	9.47 533	40	9.49 563	44	0.50 437	9.97 970	4	37
24	9.47 573	40	9.49 607	45	0.50 393	9.97 966	4	36
25	9.47 613	41	9.49 652	44	0.50 348	9.97 962	4	35
26	9.47 654	40	9.49 696	44	0.50 304	9.97 958	4	34
27	9.47 694	40	9.49 740	44	0.50 260	9.97 954	4	33
28	9.47 734	40	9.49 784	44	0.50 216	9.97 950	4	32
29	9.47 774	40	9.49 828	44	0.50 172	9.97 946	4	31
30	9.47 814	40	9.49 872	44	0.50 128	9.97 942	4	**30**
31	9.47 854	40	9.49 916	44	0.50 084	9.97 938	4	29
32	9.47 894	40	9.49 960	44	0.50 040	9.97 934	4	28
33	9.47 934	40	9.50 004	44	0.49 996	9.97 930	4	27
34	9.47 974	40	9.50 048	44	0.49 952	9.97 926	4	26
35	9.48 014	40	9.50 092	44	0.49 908	9.97 922	4	25
36	9.48 054	40	9.50 136	44	0.49 864	9.97 918	4	24
37	9.48 094	39	9.50 180	43	0.49 820	9.97 914	4	23
38	9.48 133	40	9.50 223	44	0.49 777	9.97 910	4	22
39	9.48 173	40	9.50 267	44	0.49 733	9.97 906	4	21
40	9.48 213	39	9.50 311	44	0.49 689	9.97 902	4	**20**
41	9.48 252	40	9.50 355	43	0.49 645	9.97 898	4	19
42	9.48 292	40	9.50 398	44	0.49 602	9.97 894	4	18
43	9.48 332	39	9.50 442	43	0.49 558	9.97 890	4	17
44	9.48 371	40	9.50 485	44	0.49 515	9.97 886	4	16
45	9.48 411	39	9.50 529	43	0.49 471	9.97 882	4	15
46	9.48 450	40	9.50 572	44	0.49 428	9.97 878	4	14
47	9.48 490	39	9.50 616	43	0.49 384	9.97 874	4	13
48	9.48 529	39	9.50 659	44	0.49 341	9.97 870	4	12
49	9.48 568	39	9.50 703	43	0.49 297	9.97 866	5	11
50	9.48 607	40	9.50 746	43	0.49 254	9.97 861	4	**10**
51	9.48 647	39	9.50 789	44	0.49 211	9.97 857	4	9
52	9.48 686	39	9.50 833	43	0.49 167	9.97 853	4	8
53	9.48 725	39	9.50 876	43	0.49 124	9.97 849	4	7
54	9.48 764	39	9.50 919	43	0.49 081	9.97 845	4	6
55	9.48 803	39	9.50 962	43	0.49 038	9.97 841	4	5
56	9.48 842	39	9.51 005	43	0.48 995	9.97 837	4	4
57	9.48 881	39	9.51 048	44	0.48 952	9.97 833	4	3
58	9.48 920	39	9.51 092	43	0.48 908	9.97 829	4	2
59	9.48 959	39	9.51 135	43	0.48 865	9.97 825	4	1
60	9.48 998		9.51 178		0.48 822	9.97 821		**0**
′	L. Cos.	d.	L. Cot.	c.d.	L. Tan.	L. Sin.	d.	′

P.P.

″	**45**	**44**	**43**
1	0.8	0.7	0.7
2	1.5	1.5	1.4
3	2.2	2.2	2.2
4	3.0	2.9	2.9
5	3.8	3.7	3.6
6	4.5	4.4	4.3
7	5.2	5.1	5.0
8	6.0	5.9	5.7
9	6.8	6.6	6.4
10	7.5	7.3	7.2
20	15.0	14.7	14.3
30	22.5	22.0	21.5
40	30.0	29.3	28.7
50	37.5	36.7	35.8

″	**42**	**41**	**40**
1	0.7	0.7	0.7
2	1.4	1.4	1.3
3	2.1	2.0	2.0
4	2.8	2.7	2.7
5	3.5	3.4	3.3
6	4.2	4.1	4.0
7	4.9	4.8	4.7
8	5.6	5.5	5.3
9	6.3	6.2	6.0
10	7.0	6.8	6.7
20	14.0	13.7	13.3
30	21.0	20.5	20.0
40	28.0	27.3	26.7
50	35.0	34.2	33.3

″	**39**	**5**	**4**	**3**
1	0.6	0.1	0.1	0.0
2	1.3	0.2	0.1	0.1
3	2.0	0.2	0.2	0.2
4	2.6	0.3	0.3	0.2
5	3.2	0.4	0.3	0.2
6	3.9	0.5	0.4	0.3
7	4.6	0.6	0.5	0.4
8	5.2	0.7	0.5	0.4
9	5.8	0.8	0.6	0.4
10	6.5	0.8	0.7	0.5
20	13.0	1.7	1.3	1.0
30	19.5	2.5	2.0	1.5
40	26.0	3.3	2.7	2.0
50	32.5	4.2	3.3	2.5

	5/43	**4/45**	**4/44**
0	4.3	5.6	5.5
1	12.	16.9	16.5
2	21	28.1	27.5
3	30.1	39.4	38.5
4	38.7	—	—
5			

	4/43	**3/45**	**3/44**
0	5.4	7.5	7.3
1	16.1	22.5	22.0
2	26.9	37.5	36.7
3	37.6	—	—
4			

107° (287°) (252°) **72°**

18° (198°) (341°) **161°**

′	L. Sin.	d.	L. Tan.	c.d.	L. Cot.	L. Cos.	d	′
0	9.48 998	39	9.51 178	43	0.48 822	9.97 821	4	**60**
1	9.49 037	39	9.51 221	43	0.48 779	9.97 817	5	59
2	9.49 076	39	9.51 264	42	0.48 736	9.97 812	4	58
3	9.49 115	38	9.51 306	43	0.48 694	9.97 808	4	57
4	9.49 153	39	9.51 349	43	0.48 651	9.97 804	4	56
5	9.49 192	39	9.51 392	43	0.48 608	9.97 800	4	55
6	9.49 231	38	9.51 435	43	0.48 565	9.97 796	4	54
7	9.49 269	39	9.51 478	42	0.48 522	9.97 792	4	53
8	9.49 308	39	9.51 520	43	0.48 480	9.97 788	4	52
9	9.49 347	38	9.51 563	43	0.48 437	9.97 784	5	51
10	9.49 385	39	9.51 606	42	0.48 394	9.97 779	4	**50**
11	9.49 424	38	9.51 648	43	0.48 352	9.97 775	4	49
12	9.49 462	38	9.51 691	43	0.48 309	9.97 771	4	48
13	9.49 500	39	9.51 734	42	0.48 266	9.97 767	4	47
14	9.49 539	38	9.51 776	43	0.48 224	9.97 763	4	46
15	9.49 577	38	9.51 819	42	0.48 181	9.97 759	5	45
16	9.49 615	39	9.51 861	42	0.48 139	9.97 754	4	44
17	9.49 654	38	9.51 903	43	0.48 097	9.97 750	4	43
18	9.49 692	38	9.51 946	42	0.48 054	9.97 746	4	42
19	9.49 730	38	9.51 988	43	0.48 012	9.97 742	4	41
20	9.49 768	38	9.52 031	42	0.47 969	9.97 738	4	**40**
21	9.49 806	38	9.52 073	42	0.47 927	9.97 734	5	39
22	9.49 844	38	9.52 115	42	0.47 885	9.97 729	4	38
23	9.49 882	38	9.52 157	43	0.47 843	9.97 725	4	37
24	9.49 920	38	9.52 200	42	0.47 800	9.97 721	4	36
25	9.49 958	38	9.52 242	42	0.47 758	9.97 717	4	35
26	9.49 996	38	9.52 284	42	0.47 716	9.97 713	5	34
27	9.50 034	38	9.52 326	42	0.47 674	9.97 708	4	33
28	9.50 072	38	9.52 368	42	0.47 632	9.97 704	4	32
29	9.50 110	38	9.52 410	42	0.47 590	9.97 700	4	31
30	9.50 148	37	9.52 452	42	0.47 548	9.97 696	5	**30**
31	9.50 185	38	9.52 494	42	0.47 506	9.97 691	4	29
32	9.50 223	38	9.52 536	42	0.47 464	9.97 687	4	28
33	9.50 261	37	9.52 578	42	0.47 422	9.97 683	4	27
34	9.50 298	38	9.52 620	41	0.47 380	9.97 679	5	26
35	9.50 336	38	9.52 661	42	0.47 339	9.97 674	4	25
36	9.50 374	37	9.52 703	42	0.47 297	9.97 670	4	24
37	9.50 411	38	9.52 745	42	0.47 255	9.97 666	4	23
38	9.50 449	37	9.52 787	42	0.47 213	9.97 662	5	22
39	9.50 486	37	9.52 829	41	0.47 171	9.97 657	4	21
40	9.50 523	38	9.52 870	42	0.47 130	9.97 653	4	**20**
41	9.50 561	37	9.52 912	41	0.47 088	9.97 649	4	19
42	9.50 598	37	9.52 953	42	0.47 047	9.97 645	5	18
43	9.50 635	38	9.52 995	42	0.47 005	9.97 640	4	17
44	9.50 673	37	9.53 037	41	0.46 963	9.97 636	4	16
45	9.50 710	37	9.53 078	42	0.46 922	9.97 632	4	15
46	9.50 747	37	9.53 120	41	0.46 880	9.97 628	5	14
47	9.50 784	37	9.53 161	41	0.46 839	9.97 623	4	13
48	9.50 821	37	9.53 202	42	0.46 798	9.97 619	4	12
49	9.50 858	38	9.53 244	41	0.46 756	9.97 615	5	11
50	9.50 896	37	9.53 285	42	0.46 715	9.97 610	4	**10**
51	9.50 933	37	9.53 327	41	0.46 673	9.97 606	4	9
52	9.50 970	37	9.53 368	41	0.46 632	9.97 602	5	8
53	9.51 007	36	9.53 409	41	0.46 591	9.97 597	4	7
54	9.51 043	37	9.53 450	42	0.46 550	9.97 593	4	6
55	9.51 080	37	9.53 492	41	0.46 508	9.97 589	5	5
56	9.51 117	37	9.53 533	41	0.46 467	9.97 584	4	4
57	9.51 154	37	9.53 574	41	0.46 426	9.97 580	4	3
58	9.51 191	36	9.53 615	41	0.46 385	9.97 576	5	2
59	9.51 227	37	9.53 656	41	0.46 344	9.97 571	4	1
60	9.51 264		9.53 697		0.46 303	9.97 567		**0**
′	L. Cos.	d.	L. Cot.	c.d.	L. Tan.	L. Sin.	d	′

P. P.

″	**43**	**42**	**41**
1	0.7	0.7	0.7
2	1.4	1.4	1.4
3	2.2	2.1	2.0
4	2.9	2.8	2.7
5	3.6	3.5	3.4
6	4.3	4.2	4.1
7	5.0	4.9	4.8
8	5.7	5.6	5.5
9	6.4	6.3	6.2
10	7.2	7.0	6.8
20	14.3	14.0	13.7
30	21.5	21.0	20.5
40	28.7	28.0	27.3
50	35.8	35.0	34.2

″	**39**	**38**	**37**
1	0.6	0.6	0.6
2	1.3	1.3	1.2
3	2.0	1.9	1.8
4	2.6	2.5	2.5
5	3.2	3.2	3.1
6	3.9	3.8	3.7
7	4.6	4.4	4.3
8	5.2	5.1	4.9
9	5.8	5.7	5.6
10	6.5	6.3	6.2
20	13.0	12.7	12.3
30	19.5	19.0	18.5
40	26.0	25.3	24.7
50	32.5	31.7	30.8

″	**36**	**5**	**4**
1	0.6	0.1	0.1
2	1.2	0.2	0.1
3	1.8	0.2	0.2
4	2.4	0.3	0.3
5	3.0	0.4	0.3
6	3.6	0.5	0.4
7	4.2	0.6	0.5
8	4.8	0.7	0.5
9	5.4	0.8	0.6
10	6.0	0.8	0.7
20	12.0	1.7	1.3
30	18.0	2.5	2.0
40	24.0	3.3	2.7
50	30.0	4.2	3.3

	5/43	**5/42**	**5/41**
0	4.3	4.2	4.1
1	12.9	12.6	12.3
2	21.5	21.0	20.5
3	30.1	29.4	28.7
4	38.7	37.8	36.9
5			

	4/43	**4/42**	**4/41**
0	5.4	5.2	5.1
1	16.1	15.8	15.4
2	26.9	26.2	25.6
3	37.6	36.8	35.9
4			

108° (288°) (251°) **71°**

LOGARITHMS OF THE FUNCTIONS (Continued)

19° (199°) (340°) **160°**

′	L. Sin.	d.	L. Tan.	c.d.	L. Cot.	L. Cos.	d.	′	″	P.P.		
0	9.51 264		9.53 697		0.46 303	9.97 567		**60**	″	**41**	**40**	**39**
1	9.51 301	37	9.53 738	41	0.46 262	9.97 563	4	59	1	0.7	0.7	0.6
2	9.51 338	37	9.53 779	41	0.46 221	9.97 558	5	58	2	1.4	1.3	1.3
3	9.51 374	36	9.53 820	41	0.46 180	9.97 554	4	57	3	2.0	2.0	2.0
4	9.51 411	37	9.53 861	41	0.46 139	9.97 550	4	56	4	2.7	2.7	2.6
5	9.51 447	36	9.53 902	41	0.46 098	9.97 545	5	55	5	3.4	3.3	3.2
6	9.51 484	37	9.53 943	41	0.46 057	9.97 541	4	54	6	4.1	4.0	3.9
7	9.51 520	36	9.53 984	41	0.46 016	9.97 536	5	53	7	4.8	4.7	4.6
8	9.51 557	37	9.54 025	41	0.45 975	9.97 532	4	52	8	5.5	5.3	5.2
9	9.51 593	36	9.54 065	40	0.45 935	9.97 528	4	51	9	6.2	6.0	5.8
10	9.51 629	36	9.54 106	41	0.45 894	9.97 523	5	**50**	10	6.8	6.7	6.5
11	9.51 666	37	9.54 147	41	0.45 853	9.97 519	4	49	20	13.7	13.3	13.0
12	9.51 702	36	9.54 187	40	0.45 813	9.97 515	4	48	30	20.5	20.0	19.5
13	9.51 738	36	9.54 228	41	0.45 772	9.97 510	5	47	40	27.3	26.7	26.0
14	9.51 774	36	9.54 269	41	0.45 731	9.97 506	4	46	50	34.2	33.3	32.5
15	9.51 811	37	9.54 309	40	0.45 691	9.97 501	5	45	″	**37**	**36**	**35**
16	9.51 847	36	9.54 350	41	0.45 650	9.97 497	4	44	1	0.6	0.6	0.6
17	9.51 883	36	9.54 390	40	0.45 610	9.97 492	5	43	2	1.2	1.2	1.2
18	9.51 919	36	9.54 431	41	0.45 569	9.97 488	4	42	3	1.8	1.8	1.8
19	9.51 955	36	9.54 471	40	0.45 529	9.97 484	4	41	4	2.5	2.4	2.3
20	9.51 991	36	9.54 512	41	0.45 488	9.97 479	5	**40**	5	3.1	3.0	2.9
21	9.52 027	36	9.54 552	40	0.45 448	9.97 475	4	39	6	3.7	3.6	3.5
22	9.52 063	36	9.54 593	41	0.45 407	9.97 470	5	38	7	4.3	4.2	4.1
23	9.52 099	36	9.54 633	40	0.45 367	9.97 466	4	37	8	4.9	4.8	4.7
24	9.52 135	36	9.54 673	40	0.45 327	9.97 461	5	36	9	5.6	5.4	5.2
25	9.52 171	36	9.54 714	41	0.45 286	9.97 457	4	35	10	6.2	6.0	5.8
26	9.52 207	36	9.54 754	40	0.45 246	9.97 453	4	34	20	12.3	12.0	11.7
27	9.52 242	35	9.54 794	40	0.45 206	9.97 448	5	33	30	18.5	18.0	17.5
28	9.52 278	36	9.54 835	41	0.45 165	9.97 444	4	32	40	24.7	24.0	23.3
29	9.52 314	36	9.54 875	40	0.45 125	9.97 439	5	31	50	30.8	30.0	29.2
30	9.52 350	36	9.54 915	40	0.45 085	9.97 435	4	**30**	″	**34**	**5**	**4**
31	9.52 385	35	9.54 955	40	0.45 045	9.97 430	5	29	1	0.6	0.1	0.1
32	9.52 421	36	9.54 995	40	0.45 005	9.97 426	4	28	2	1.1	0.2	0.1
33	9.52 456	35	9.55 035	40	0.44 965	9.97 421	5	27	3	1.7	0.2	0.2
34	9.52 492	36	9.55 075	40	0.44 925	9.97 417	4	26	4	2.3	0.3	0.3
35	9.52 527	35	9.55 115	40	0.44 885	9.97 412	5	25	5	2.8	0.4	0.3
36	9.52 563	36	9.55 155	40	0.44 845	9.97 408	4	24	6	3.4	0.5	0.4
37	9.52 598	35	9.55 195	40	0.44 805	9.97 403	5	23	7	4.0	0.6	0.5
38	9.52 634	36	9.55 235	40	0.44 765	9.97 399	4	22	8	4.5	0.7	0.5
39	9.52 669	35	9.55 275	40	0.44 725	9.97 394	5	21	9	5.1	0.8	0.6
40	9.52 705	36	9.55 315	40	0.44 685	9.97 390	4	**20**	10	5.7	0.8	0.7
41	9.52 740	35	9.55 355	40	0.44 645	9.97 385	5	19	20	11.3	1.7	1.3
42	9.52 775	35	9.55 395	40	0.44 605	9.97 381	4	18	30	17.0	2.5	2.0
43	9.52 811	36	9.55 434	39	0.44 566	9.97 376	5	17	40	22.7	3.3	2.7
44	9.52 846	35	9.55 474	40	0.44 526	9.97 372	4	16	50	28.3	4.2	3.3
45	9.52 881	35	9.55 514	40	0.44 486	9.97 367	5	15				
46	9.52 916	35	9.55 554	40	0.44 446	9.97 363	4	14		**5/41**	**5/40**	**5/39**
47	9.52 951	35	9.55 593	39	0.44 407	9.97 358	5	13				
48	9.52 986	35	9.55 633	40	0.44 367	9.97 353	5	12	0	4.1	4 0	3.9
49	9.53 021	35	9.55 673	40	0.44 327	9.97 349	4	11	1	12.3	12.0	11.7
50	9.53 056	35	9.55 712	39	0.44 288	9.97 344	5	**10**	2	20.5	20.0	19.5
51	9.53 092	36	9.55 752	40	0.44 248	9.97 340	4	9	3	28.7	28.0	27.3
52	9.53 126	34	9.55 791	39	0.44 209	9.97 335	5	8	4	36.9	36.0	35.1
53	9.53 161	35	9.55 831	40	0.44 169	9.97 331	4	7	5			
54	9.53 196	35	9.55 870	39	0.44 130	9.97 326	5	6		**4/41**	**4/40**	**4/39**
55	9.53 231	35	9.55 910	40	0.44 090	9.97 322	4	5				
56	9.53 266	35	9.55 949	39	0.44 051	9.97 317	5	4	0	5.1	5.0	4.9
57	9.53 301	35	9.55 989	40	0.44 011	9.97 312	5	3	1	15.4	15.0	14.6
58	9.53 336	35	9.56 028	39	0.43 972	9.97 308	4	2	2	25.6	25.0	24.4
59	9.53 370	34	9.56 067	39	0.43 933	9.97 303	5	1	3	35.9	35.0	34.1
60	9.53 405	35	9.56 107	40	0.43 893	9.97 299	4	**0**	4			
′	L. Cos.	d.	L. Cot.	c.d.	L. Tan.	L. Sin.	d.	′		P.P.		

109° (289°) (250°) **70°**

20° (200°) (339°) **159°**

′	L. Sin.	d.	L. Tan.	c.d.	L. Cot.	L. Cos.	d.	′	″	P.P.		
0	9.53 405		9.56 107		0.43 893	9.97 299		**60**	″	**40**	**39**	**38**
1	9.53 440	35	9.56 146	39	0.43 854	9.97 294	5	59	1	0.7	0.6	0.6
2	9.53 475	35	9.56 185	39	0.43 815	9.97 289	5	58	2	1.3	1.3	1.3
3	9.53 509	34	9.56 224	39	0.43 776	9.97 285	4	57	3	2.0	2.0	1.9
4	9.53 544	35	9.56 264	40	0.43 736	9.97 280	5	56	4	2.7	2.6	2.5
5	9.53 578	34	9.56 303	39	0.43 697	9.97 276	4	55	5	3.3	3.2	3.2
6	9.53 613	35	9.56 342	39	0.43 658	9.97 271	5	54	6	4.0	3.9	3.8
7	9.53 647	34	9.56 381	39	0.43 619	9.97 266	5	53	7	4.7	4.6	4.4
8	9.53 682	35	9.56 420	39	0.43 580	9.97 262	4	52	8	5.3	5.2	5.1
9	9.53 716	34	9.56 459	39	0.43 541	9.97 257	5	51	9	6.0	5.8	5.7
10	9.53 751	35	9.56 498	39	0.43 502	9.97 252	5	**50**	10	6.7	6.5	6.3
11	9.53 785	34	9.56 537	39	0.43 463	9.97 248	4	49	20	13.3	13.0	12.7
12	9.53 819	34	9.56 576	39	0.43 424	9.97 243	5	48	30	20.0	19.5	19.0
13	9.53 854	35	9.56 615	39	0.43 385	9.97 238	5	47	40	26.7	26.0	25.3
14	9.53 888	34	9.56 654	39	0.43 346	9.97 234	4	46	50	33.3	32.5	31.7
15	9.53 922	34	9.56 693	39	0.43 307	9.97 229	5	45	″	**37**	**35**	**34**
16	9.53 957	35	9.56 732	39	0.43 268	9.97 224	5	44	1	0.6	0.6	0.6
17	9.53 991	34	9.56 771	39	0.43 229	9.97 220	4	43	2	1.2	1.2	1.1
18	9.54 025	34	9.56 810	39	0.43 190	9.97 215	5	42	3	1.8	1.8	1.7
19	9.54 059	34	9.56 849	39	0.43 151	9.97 210	5	41	4	2.5	2.3	2.3
20	9.54 093	34	9.56 887	38	0.43 113	9.97 206	4	**40**	5	3.1	2.9	2.8
21	9.54 127	34	9.56 926	39	0.43 074	9.97 201	5	39	6	3.7	3.5	3.4
22	9.54 161	34	9.56 965	39	0.43 035	9.97 196	5	38	7	4.3	4.1	4.0
23	9.54 195	34	9.57 004	39	0.42 996	9.97 192	4	37	8	4.9	4.7	4.5
24	9.54 229	34	9.57 042	38	0.42 958	9.97 187	5	36	9	5.6	5.2	5.1
25	9.54 263	34	9.57 081	39	0.42 919	9.97 182	5	35	10	6.2	5.8	5.7
26	9.54 297	34	9.57 120	39	0.42 880	9.97 178	4	34	20	12.3	11.7	11.3
27	9.54 331	34	9.57 158	38	0.42 842	9.97 173	5	33	30	18.5	17.5	17.0
28	9.54 365	34	9.57 197	39	0.42 803	9.97 168	5	32	40	24.7	23.3	22.7
29	9.54 399	34	9.57 235	38	0.42 765	9.97 163	5	31	50	30.8	29.2	28.3
30	9.54 433	34	9.57 274	39	0.42 726	9.97 159	4	**30**	″	**33**	**5**	**4**
31	9.54 466	33	9.57 312	38	0.42 688	9.97 154	5	29	1	0.6	0.1	0.1
32	9.54 500	34	9.57 351	39	0.42 649	9.97 149	5	28	2	1.1	0.2	0.1
33	9.54 534	34	9.57 389	38	0.42 611	9.97 145	4	27	3	1.6	0.2	0.2
34	9.54 567	33	9.57 428	39	0.42 572	9.97 140	5	26	4	2.2	0.3	0.3
35	9.54 601	34	9.57 466	38	0.42 534	9.97 135	5	25	5	2.8	0.4	0.3
36	9.54 635	34	9.57 504	38	0.42 496	9.97 130	5	24	6	3.3	0.5	0.4
37	9.54 668	33	9.57 543	39	0.42 457	9.97 126	4	23	7	3.8	0.6	0.5
38	9.54 702	34	9.57 581	38	0.42 419	9.97 121	5	22	8	4.4	0.7	0.5
39	9.54 735	33	9.57 619	38	0.42 381	9.97 116	5	21	9	5.0	0.8	0.6
40	9.54 769	34	9.57 658	39	0.42 342	9.97 111	5	**20**	10	5.5	0.8	0.7
41	9.54 802	33	9.57 696	38	0.42 304	9.97 107	4	19	20	11.0	1.7	1.3
42	9.54 836	34	9.57 734	38	0.42 266	9.97 102	5	18	30	16.5	2.5	2.0
43	9.54 869	33	9.57 772	38	0.42 228	9.97 097	5	17	40	22.0	3.3	2.7
44	9.54 903	34	9.57 810	38	0.42 190	9.97 092	5	16	50	27.5	4.2	3.3
45	9.54 936	33	9.57 849	39	0.42 151	9.97 087	5	15				
46	9.54 969	33	9.57 887	38	0.42 113	9.97 083	4	14		**5/40**	**5/39**	**5/38**
47	9.55 003	34	9.57 925	38	0.42 075	9.97 078	5	13				
48	9.55 036	33	9.57 963	38	0.42 037	9.97 073	5	12	0	4.0	3.9	3.8
49	9.55 069	33	9.58 001	38	0.41 999	9.97 068	5	11	1	12.0	11.7	11.4
50	9.55 102	33	9.58 039	38	0.41 961	9.97 063	5	**10**	2	20.0	19.5	19.0
51	9.55 136	34	9.58 077	38	0.41 923	9.97 059	4	9	3	28.0	27.3	26.6
52	9.55 169	33	9.58 115	38	0.41 885	9.97 054	5	8	4	36.0	35.1	34.2
53	9.55 202	33	9.58 153	38	0.41 847	9.97 049	5	7	5			
54	9.55 235	33	9.58 191	38	0.41 809	9.97 044	5	6		**5/37**	**4/39**	**4/38**
55	9.55 268	33	9.58 229	38	0.41 771	9.97 039	5	5				
56	9.55 301	33	9.58 267	38	0.41 733	9.97 035	4	4	0	3.7	4.9	4.8
57	9.55 334	33	9.58 304	37	0.41 696	9.97 030	5	3	1	11.1	14.6	14.2
58	9.55 367	33	9.58 342	38	0.41 658	9.97 025	5	2	2	18.5	24.4	23.8
59	9.55 400	33	9.58 380	38	0.41 620	9.97 020	5	1	3	25.9	34.1	33.2
60	9.55 433	33	9.58 418	38	0.41 582	9.97 015	5	**0**	4	33.3	—	—
									5			
′	L. Cos.	d.	L. Cot.	c.d.	L. Tan.	L. Sin.	d.	′		P.P.		

110° (290°) (249°) **69°**

LOGARITHMS OF THE FUNCTIONS (Continued)

21° (201°) (338°) **158°**

′	L. Sin.	d.	L. Tan.	c.d.	L. Cot.	L. Cos.	d.	′
0	9.55 433	33	9.58 418	37	0.41 582	9.97 015	5	**60**
1	9.55 466	33	9.58 455	38	0.41 545	9.97 010	5	59
2	9.55 499	33	9.58 493	38	0.41 507	9.97 005	4	58
3	9.55 532	32	9.58 531	38	0.41 469	9.97 001	5	57
4	9.55 564	33	9.58 569	37	0.41 431	9.96 996	5	56
5	9.55 597	33	9.58 606	38	0.41 394	9.96 991	5	55
6	9.55 630	33	9.58 644	37	0.41 356	9.96 986	5	54
7	9.55 663	32	9.58 681	38	0.41 319	9.96 981	5	53
8	9.55 695	33	9.58 719	38	0.41 281	9.96 976	5	52
9	9.55 728	33	9.58 757	37	0.41 243	9.96 971	5	51
10	9.55 761	32	9.58 794	38	0.41 206	9.96 966	4	**50**
11	9.55 793	33	9.58 832	37	0.41 168	9.96 962	5	49
12	9.55 826	32	9.58 869	38	0.41 131	9.96 957	5	48
13	9.55 858	33	9.58 907	37	0.41 093	9.96 952	5	47
14	9.55 891	32	9.58 944	37	0.41 056	9.96 947	5	46
15	9.55 923	33	9.58 981	38	0.41 019	9.96 942	5	45
16	9.55 956	32	9.59 019	37	0.40 981	9.96 937	5	44
17	9.55 988	33	9.59 056	38	0.40 944	9.96 932	5	43
18	9.56 021	32	9.59 094	37	0.40 906	9.96 927	5	42
19	9.56 053	32	9.59 131	37	0.40 869	9.96 922	5	41
20	9.56 085	33	9.59 168	37	0.40 832	9.96 917	5	**40**
21	9.56 118	32	9.59 205	38	0.40 795	9.96 912	5	39
22	9.56 150	32	9.59 243	37	0.40 757	9.96 907	4	38
23	9.56 182	33	9.59 280	37	0.40 720	9.96 903	5	37
24	9.56 215	32	9.59 317	37	0.40 683	9.96 898	5	36
25	9.56 247	32	9.59 354	37	0.40 646	9.96 893	5	35
26	9.56 279	32	9.59 391	38	0.40 609	9.96 888	5	34
27	9.56 311	32	9.59 429	37	0.40 571	9.96 883	5	33
28	9.56 343	32	9.59 466	37	0.40 534	9.96 878	5	32
29	9.56 375	33	9.59 503	37	0.40 497	9.96 873	5	31
30	9.56 408	32	9.59 540	37	0.40 460	9.96 868	5	**30**
31	9.56 440	32	9.59 577	37	0.40 423	9.96 863	5	29
32	9.56 472	32	9.59 614	37	0.40 386	9.96 858	5	28
33	9.56 504	32	9.59 651	37	0.40 349	9.96 853	5	27
34	9.56 536	32	9.59 688	37	0.40 312	9.96 848	5	26
35	9.56 568	31	9.59 725	37	0.40 275	9.96 843	5	25
36	9.56 599	32	9.59 762	37	0.40 238	9.96 838	5	24
37	9.56 631	32	9.59 799	36	0.40 201	9.96 833	5	23
38	9.56 663	32	9.59 835	37	0.40 165	9.96 828	5	22
39	9.56 695	32	9.59 872	37	0.40 128	9.96 823	5	21
40	9.56 727	32	9.59 909	37	0.40 091	9.96 818	5	**20**
41	9.56 759	31	9.59 946	37	0.40 054	9.96 813	5	19
42	9.56 790	32	9.59 983	36	0.40 017	9.96 808	5	18
43	9.56 822	32	9.60 019	37	0.39 981	9.96 803	5	17
44	9.56 854	32	9.60 056	37	0.39 944	9.96 798	5	16
45	9.56 886	31	9.60 093	37	0.39 907	9.96 793	5	15
46	9.56 917	32	9.60 130	36	0.39 870	9.96 788	5	14
47	9.56 949	31	9.60 166	37	0.39 834	9.96 783	5	13
48	9.56 980	32	9.60 203	37	0.39 797	9.96 778	6	12
49	9.57 012	32	9.60 240	36	0.39 760	9.96 772	5	11
50	9.57 044	31	9.60 276	37	0.39 724	9.96 767	5	**10**
51	9.57 075	32	9.60 313	36	0.39 687	9.96 762	5	9
52	9.57 107	31	9.60 349	37	0.39 651	9.96 757	5	8
53	9.57 138	31	9.60 386	36	0.39 614	9.96 752	5	7
54	9.57 169	32	9.60 422	37	0.39 578	9.96 747	5	6
55	9.57 201	31	9.60 459	36	0.39 541	9.96 742	5	5
56	9.57 232	32	9.60 495	37	0.39 505	9.96 737	5	4
57	9.57 264	31	9.60 532	36	0.39 468	9.96 732	5	3
58	9.57 295	31	9.60 568	37	0.39 432	9.96 727	5	2
59	9.57 326	32	9.60 605	36	0.39 395	9.96 722	5	1
60	9.57 358		9.60 641		0.39 359	9.96 717		**0**
′	L. Cos.	d.	L. Cot.	c.d.	L. Tan.	L. Sin.	d.	′

P.P.

″	38	37	36
1	0.6	0.6	0.6
2	1.3	1.2	1.2
3	1.9	1.8	1.8
4	2.5	2.5	2.4
5	3.2	3.1	3.0
6	3.8	3.7	3.6
7	4.4	4.3	4.2
8	5.1	4.9	4.8
9	5.7	5.6	5.4
10	6.3	6.2	6.0
20	12.7	12.3	12.0
30	19.0	18.5	18.0
40	25.3	24.7	24.0
50	31.7	30.8	30.0

″	33	32	31
1	0.6	0.5	0.5
2	1.1	1.1	1.0
3	1.6	1.6	1.6
4	2.2	2.1	2.1
5	2.8	2.7	2.6
6	3.3	3.2	3.1
7	3.8	3.7	3.6
8	4.4	4.3	4.1
9	5.0	4.8	4.6
10	5.5	5.3	5.2
20	11.0	10.7	10.3
30	16.5	16.0	15.5
40	22.0	21.3	20.7
50	27.5	26.7	25.8

″	6	5	4
1	0.1	0.1	0.1
2	0.2	0.2	0.1
3	0.3	0.2	0.2
4	0.4	0.3	0.3
5	0.5	0.4	0.3
6	0.6	0.5	0.4
7	0.7	0.6	0.5
8	0.8	0.7	0.5
9	0.9	0.8	0.6
10	1.0	0.8	0.7
20	2.0	1.7	1.3
30	3.0	2.5	2.0
40	4.0	3.3	2.7
50	5.0	4.2	3.3

	6/37	5/38	5/37
0	3.1	3.8	3.7
1	9.2	11.4	11.1
2	15.4	19.0	18.5
3	21.6	26.6	25.9
4	27.8	34.2	33.3
5	33.9	—	—
6			

	5/36	4/38	4/37
0	3.6	4.8	4.6
1	10.8	14.2	13.9
2	18.0	23.8	23.1
3	25.2	33.2	32.4
4	32.4	—	—
5			

111° (291°) (248°) **68°**

22° (202°) (337°) **157°**

′	L. Sin.	d.	L. Tan.	c.d.	L. Cot.	L. Cos.	d.	′
0	9.57 358	31	9.60 641	36	0.39 359	9.96 717	6	**60**
1	9.57 389	31	9.60 677	37	0.39 323	9.96 711	5	59
2	9.57 420	31	9.60 714	36	0.39 286	9.96 706	5	58
3	9.57 451	31	9.60 750	36	0.39 250	9.96 701	5	57
4	9.57 482	32	9.60 786	37	0.39 214	9.96 696	5	56
5	9.57 514	31	9.60 823	36	0.39 177	9.96 691	5	55
6	9.57 545	31	9.60 859	36	0.39 141	9.96 686	5	54
7	9.57 576	31	9.60 895	36	0.39 105	9.96 681	5	53
8	9.57 607	31	9.60 931	36	0.39 069	9.96 676	6	52
9	9.57 638	31	9.60 967	37	0.39 033	9.96 670	5	51
10	9.57 669	31	9.61 004	36	0.38 996	9.96 665	5	**50**
11	9.57 700	31	9.61 040	36	0.38 960	9.96 660	5	49
12	9.57 731	31	9.61 076	36	0.38 924	9.96 655	5	48
13	9.57 762	31	9.61 112	36	0.38 888	9.96 650	5	47
14	9.57 793	31	9.61 148	36	0.38 852	9.96 645	5	46
15	9.57 824	31	9.61 184	36	0.38 816	9.96 640	6	45
16	9.57 855	30	9.61 220	36	0.38 780	9.96 634	5	44
17	9.57 885	31	9.61 256	36	0.38 744	9.96 629	5	43
18	9.57 916	31	9.61 292	36	0.38 708	9.96 624	5	42
19	9.57 947	31	9.61 328	36	0.38 672	9.96 619	5	41
20	9.57 978	30	9.61 364	36	0.38 636	9.96 614	6	**40**
21	9.58 008	31	9.61 400	36	0.38 600	9.96 608	5	39
22	9.58 039	31	9.61 436	36	0.38 564	9.96 603	5	38
23	9.58 070	31	9.61 472	36	0.38 528	9.96 598	5	37
24	9.58 101	30	9.61 508	36	0.38 492	9.96 593	5	36
25	9.58 131	31	9.61 544	35	0.38 456	9.96 588	6	35
26	9.58 162	30	9.61 579	36	0.38 421	9.96 582	5	34
27	9.58 192	31	9.61 615	36	0.38 385	9.96 577	5	33
28	9.58 223	30	9.61 651	36	0.38 349	9.96 572	5	32
29	9.58 253	31	9.61 687	35	0.38 313	9.96 567	5	31
30	9.58 284	30	9.61 722	36	0.38 278	9.96 562	6	**30**
31	9.58 314	31	9.61 758	36	0.38 242	9.96 556	5	29
32	9.58 345	30	9.61 794	36	0.38 206	9.96 551	5	28
33	9.58 375	31	9.61 830	35	0.38 170	9.96 546	5	27
34	9.58 406	30	9.61 865	36	0.38 135	9.96 541	6	26
35	9.58 436	31	9.61 901	35	0.38 099	9.96 535	5	25
36	9.58 467	30	9.61 936	36	0.38 064	9.96 530	5	24
37	9.58 497	30	9.61 972	36	0.38 028	9.96 525	5	23
38	9.58 527	30	9.62 008	35	0.37 992	9.96 520	6	22
39	9.58 557	31	9.62 043	36	0.37 957	9.96 514	5	21
40	9.58 588	30	9.62 079	35	0.37 921	9.96 509	5	**20**
41	9.58 618	30	9.62 114	36	0.37 886	9.96 504	6	19
42	9.58 648	30	9.62 150	35	0.37 850	9.96 498	5	18
43	9.58 678	31	9.62 185	36	0.37 815	9.96 493	5	17
44	9.58 709	30	9.62 221	35	0.37 779	9.96 488	5	16
45	9.58 739	30	9.62 256	36	0.37 744	9.96 483	6	15
46	9.58 769	30	9.62 292	35	0.37 708	9.96 477	5	14
47	9.58 799	30	9.62 327	35	0.37 673	9.96 472	5	13
48	9.58 829	30	9.62 362	36	0.37 638	9.96 467	6	12
49	9.58 859	30	9.62 398	35	0.37 602	9.96 461	5	11
50	9.58 889	30	9.62 433	35	0.37 567	9.96 456	5	**10**
51	9.58 919	30	9.62 468	36	0.37 532	9.96 451	6	9
52	9.58 949	30	9.62 504	35	0.37 496	9.96 445	5	8
53	9.58 979	30	9.62 539	35	0.37 461	9.96 440	5	7
54	9.59 009	30	9.62 574	35	0.37 426	9.96 435	6	6
55	9.59 039	30	9.62 609	36	0.37 391	9.96 429	5	5
56	9.59 069	29	9.62 645	35	0.37 355	9.96 424	5	4
57	9.59 098	30	9.62 680	35	0.37 320	9.96 419	6	3
58	9.59 128	30	9.62 715	35	0.37 285	9.96 413	5	2
59	9.59 158	30	9.62 750	35	0.37 250	9.96 408	5	1
60	9.59 188		9.62 785		0.37 215	9.96 403		**0**
′	L. Cos.	d.	L. Cot.	c.d.	L. Tan.	L. Sin.	d.	′

P. P.

″	37	36	35
1	0.6	0.6	0.6
2	1.2	1.2	1.2
3	1.8	1.8	1.8
4	2.5	2.4	2.3
5	3.1	3.0	2.9
6	3.7	3.6	3.5
7	4.3	4.2	4.1
8	4.9	4.8	4.7
9	5.6	5.4	5.2
10	6.2	6.0	5.8
20	12.3	12.0	11.7
30	18.5	18.0	17.5
40	24.7	24.0	23.3
50	30.8	30.0	29.2

″	32	31	30
1	0.5	0.5	0.5
2	1.1	1.0	1.0
3	1.6	1.6	1.5
4	2.1	2.1	2.0
5	2.7	2.6	2.5
6	3.2	3.1	3.0
7	3.7	3.6	3.5
8	4.3	4.1	4.0
9	4.8	4.6	4.5
10	5.3	5.2	5.0
20	10.7	10.3	10.0
30	16.0	15.5	15.0
40	21.3	20.7	20.0
50	26.7	25.8	25.0

″	29	6	5
1	0.5	0.1	0.1
2	1.0	0.2	0.2
3	1.4	0.3	0.2
4	1.9	0.4	0.3
5	2.4	0.5	0.4
6	2.9	0.6	0.5
7	3.4	0.7	0.6
8	3.9	0.8	0.7
9	4.4	0.9	0.8
10	4.8	1.0	0.8
20	9.7	2.0	1.7
30	14.5	3.0	2.5
40	19.3	4.0	3.3
50	24.2	5.0	4.2

	6/36	6/35
0	3.0	2.9
1	9.0	8.8
2	15.0	14.6
3	21.0	20.4
4	27.0	26.2
5	33.0	32.1
6		

	5/37	5/36	5/35
0	3.7	3.6	3.5
1	11.1	10.8	10.5
2	18.5	18.0	17.5
3	25.9	25.2	24.5
4	33.3	32.4	31.5
5			

112° (292°) (247°) **67**

LOGARITHMS OF THE FUNCTIONS (Continued)

23° (203°) — (336°) **156°**

'	L. Sin.	d.	L. Tan.	c.d.	L. Cot.	L. Cos.	d.	'
0	9.59 188		9.62 785		0.37 215	9.96 403		**60**
1	9.59 218	30	9.62 820	35	0.37 180	9.96 397	6	59
2	9.59 247	29	9.62 855	35	0.37 145	9.96 392	5	58
3	9.59 277	30	9.62 890	35	0.37 110	9.96 387	5	57
4	9.59 307	30	9.62 926	36	0.37 074	9.96 381	6	56
5	9.59 336	29	9.62 961	35	0.37 039	9.96 376	5	55
6	9.59 366	30	9.62 996	35	0.37 004	9.96 370	6	54
7	9.59 396	30	9.63 031	35	0.36 969	9.96 365	5	53
8	9.59 425	29	9.63 066	35	0.36 934	9.96 360	5	52
9	9.59 455	30	9.63 101	35	0.36 899	9.96 354	6	51
10	9.59 484	29	9.63 135	34	0.36 865	9.96 349	5	**50**
11	9.59 514	30	9.63 170	35	0.36 830	9.96 343	6	49
12	9.59 543	29	9.63 205	35	0.36 795	9.96 338	5	48
13	9.59 573	30	9.63 240	35	0.36 760	9.96 333	5	47
14	9.59 602	29	9.63 275	35	0.36 725	9.96 327	6	46
15	9.59 632	30	9.63 310	35	0.36 690	9.96 322	5	45
16	9.59 661	29	9.63 345	35	0.36 655	9.96 316	6	44
17	9.59 690	29	9.63 379	34	0.36 621	9.96 311	5	43
18	9.59 720	30	9.63 414	35	0.36 586	9.96 305	6	42
19	9.59 749	29	9.63 449	35	0.36 551	9.96 300	5	41
20	9.59 778	29	9.63 484	35	0.36 516	9.96 294	6	**40**
21	9.59 808	30	9.63 519	35	0.36 481	9.96 289	5	39
22	9.59 837	29	9.63 553	34	0.36 447	9.96 284	5	38
23	9.59 866	29	9.63 588	35	0.36 412	9.96 278	6	37
24	9.59 895	29	9.63 623	35	0.36 377	9.96 273	5	36
25	9.59 924	29	9.63 657	34	0.36 343	9.96 267	6	35
26	9.59 954	30	9.63 692	35	0.36 308	9.96 262	5	34
27	9.59 983	29	9.63 726	34	0.36 274	9.96 256	6	33
28	9.60 012	29	9.63 761	35	0.36 239	9.96 251	5	32
29	9.60 041	29	9.63 796	35	0.36 204	9.96 245	6	31
30	9.60 070	29	9.63 830	34	0.36 170	9.96 240	5	**30**
31	9.60 099	29	9.63 865	35	0.36 135	9.96 234	6	29
32	9.60 128	29	9.63 899	34	0.36 101	9.96 229	5	28
33	9.60 157	29	9.63 934	35	0.36 066	9.96 223	6	27
34	9.60 186	29	9.63 968	34	0.36 032	9.96 218	5	26
35	9.60 215	29	9.64 003	35	0.35 997	9.96 212	6	25
36	9.60 244	29	9.64 037	34	0.35 963	9.96 207	5	24
37	9.60 273	29	9.64 072	35	0.35 928	9.96 201	6	23
38	9.60 302	29	9.64 106	34	0.35 894	9.96 196	5	22
39	9.60 331	29	9.64 140	34	0.35 860	9.96 190	6	21
40	9.60 359	28	9.64 175	35	0.35 825	9.96 185	5	**20**
41	9.60 388	29	9.64 209	34	0.35 791	9.96 179	6	19
42	9.60 417	29	9.64 243	34	0.35 757	9.96 174	5	18
43	9.60 446	29	9.64 278	35	0.35 722	9.96 168	6	17
44	9.60 474	28	9.64 312	34	0.35 688	9.96 162	6	16
45	9.60 503	29	9.64 346	34	0.35 654	9.96 157	5	15
46	9.60 532	29	9.64 381	35	0.35 619	9.96 151	6	14
47	9.60 561	29	9.64 415	34	0.35 585	9.96 146	5	13
48	9.60 589	28	9.64 449	34	0.35 551	9.96 140	6	12
49	9.60 618	29	9.64 483	34	0.35 517	9.96 135	5	11
50	9.60 646	28	9.64 517	34	0.35 483	9.96 129	6	**10**
51	9.60 675	29	9.64 552	35	0.35 448	9.96 123	6	9
52	9.60 704	29	9.64 586	34	0.35 414	9.96 118	5	8
53	9.60 732	28	9.64 620	34	0.35 380	9.96 112	6	7
54	9.60 761	29	9.64 654	34	0.35 346	9.96 107	5	6
55	9.60 789	28	9.64 688	34	0.35 312	9.96 101	6	5
56	9.60 818	29	9.64 722	34	0.35 278	9.96 095	6	4
57	9.60 846	28	9.64 756	34	0.35 244	9.96 090	5	3
58	9.60 875	29	9.64 790	34	0.35 210	9.96 084	6	2
59	9.60 903	28	9.64 824	34	0.35 176	9.96 079	5	1
60	9.60 931	28	9.64 858	34	0.35 142	9.96 073	6	**0**
'	L. Cos.	d.	L. Cot.	c.d.	L. Tan.	L. Sin.	d.	'

113° (293°) — (246°) **66°**

P. P.

"	36	35	34
1	0.6	0.6	0.6
2	1.2	1.2	1.1
3	1.8	1.8	1.7
4	2.4	2.3	2.3
5	3.0	2.9	2.8
6	3.6	3.5	3.4
7	4.2	4.1	4.0
8	4.8	4.7	4.5
9	5.4	5.2	5.1
10	6.0	5.8	5.7
20	12.0	11.7	11.3
30	18.0	17.5	17.0
40	24.0	23.3	22.7
50	30.0	29.2	28.3

"	30	29	28
1	0.5	0.5	0.5
2	1.0	1.0	0.9
3	1.5	1.4	1.4
4	2.0	1.9	1.9
5	2.5	2.4	2.3
6	3.0	2.9	2.8
7	3.5	3.4	3.3
8	4.0	3.9	3.7
9	4.5	4.4	4.2
10	5.0	4.8	4.7
20	10.0	9.7	9.3
30	15.0	14.5	14.0
40	20.0	19.3	18.7
50	25.0	24.2	23.3

"	6	5
1	0.1	0.1
2	0.2	0.2
3	0.3	0.2
4	0.4	0.3
5	0.5	0.4
6	0.6	0.5
7	0.7	0.6
8	0.8	0.7
9	0.9	0.8
10	1.0	0.8
20	2.0	1.7
30	3.0	2.5
40	4.0	3.3
50	5.0	4.2

	6/36	6/35	6/34
0			
1	3.0	2.9	2.8
2	9.0	8.8	8.5
3	15.0	14.6	14.2
4	21.0	20.4	19.8
5	27.0	26.2	25.5
6	33.0	32.1	31.2

	5/35	5/34
0		
1	3.5	3.4
2	10.5	10.2
3	17.5	17.0
4	24.5	23.8
5	31.5	30.6

24° (204°) — (335°) **155°**

'	L. Sin.	d.	L. Tan.	c.d.	L. Cot.	L. Cos.	d.	'
0	9.60 931		9.64 858		0.35 142	9.96 073		**60**
1	9.60 960	29	9.64 892	34	0.35 108	9.96 067	6	59
2	9.60 988	28	9.64 926	34	0.35 074	9.96 062	5	58
3	9.61 016	28	9.64 960	34	0.35 040	9.96 056	6	57
4	9.61 045	29	9.64 994	34	0.35 006	9.96 050	6	56
5	9.61 073	28	9.65 028	34	0.34 972	9.96 045	5	55
6	9.61 101	28	9.65 062	34	0.34 938	9.96 039	6	54
7	9.61 129	28	9.65 096	34	0.34 904	9.96 034	5	53
8	9.61 158	29	9.65 130	34	0.34 870	9.96 028	6	52
9	9.61 186	28	9.65 164	34	0.34 836	9.96 022	6	51
10	9.61 214	28	9.65 197	33	0.34 803	9.96 017	5	**50**
11	9.61 242	28	9.65 231	34	0.34 769	9.96 011	6	49
12	9.61 270	28	9.65 265	34	0.34 735	9.96 005	6	48
13	9.61 298	28	9.65 299	34	0.34 701	9.96 000	5	47
14	9.61 326	28	9.65 333	34	0.34 667	9.95 994	6	46
15	9.61 354	28	9.65 366	33	0.34 634	9.95 988	6	45
16	9.61 382	28	9.65 400	34	0.34 600	9.95 982	6	44
17	9.61 411	29	9.65 434	34	0.34 566	9.95 977	5	43
18	9.61 438	27	9.65 467	33	0.34 533	9.95 971	6	42
19	9.61 466	28	9.65 501	34	0.34 499	9.95 965	6	41
20	9.61 494	28	9.65 535	34	0.34 465	9.95 960	5	**40**
21	9.61 522	28	9.65 568	33	0.34 432	9.95 954	6	39
22	9.61 550	28	9.65 602	34	0.34 398	9.95 948	6	38
23	9.61 578	28	9.65 636	34	0.34 364	9.95 942	6	37
24	9.61 606	28	9.65 669	33	0.34 331	9.95 937	5	36
25	9.61 634	28	9.65 703	34	0.34 297	9.95 931	6	35
26	9.61 662	28	9.65 736	33	0.34 264	9.95 925	6	34
27	9.61 689	27	9.65 770	34	0.34 230	9.95 920	5	33
28	9.61 717	28	9.65 803	33	0.34 197	9.95 914	6	32
29	9.61 745	28	9.65 837	34	0.34 163	9.95 908	6	31
30	9.61 773	28	9.65 870	33	0.34 130	9.95 902	6	**30**
31	9.61 800	27	9.65 904	34	0.34 096	9.95 897	5	29
32	9.61 828	28	9.65 937	33	0.34 063	9.95 891	6	28
33	9.61 856	28	9.65 971	34	0.34 029	9.95 885	6	27
34	9.61 883	27	9.66 004	33	0.33 996	9.95 879	6	26
35	9.61 911	28	9.66 038	34	0.33 962	9.95 873	6	25
36	9.61 939	28	9.66 071	33	0.33 929	9.95 868	5	24
37	9.61 966	27	9.66 104	33	0.33 896	9.95 862	6	23
38	9.61 994	28	9.66 138	34	0.33 862	9.95 856	6	22
39	9.62 021	27	9.66 171	33	0.33 829	9.95 850	6	21
40	9.62 049	28	9.66 204	33	0.33 796	9.95 844	6	**20**
41	9.62 076	27	9.66 238	34	0.33 762	9.95 839	5	19
42	9.62 104	28	9.66 271	33	0.33 729	9.95 833	6	18
43	9.62 131	27	9.66 304	33	0.33 696	9.95 827	6	17
44	9.62 159	28	9.66 337	33	0.33 663	9.95 821	6	16
45	9.62 186	27	9.66 371	34	0.33 629	9.95 815	6	15
46	9.62 214	28	9.66 404	33	0.33 596	9.95 810	5	14
47	9.62 241	27	9.66 437	33	0.33 563	9.95 804	6	13
48	9.62 268	27	9.66 470	33	0.33 530	9.95 798	6	12
49	9.62 296	28	9.66 503	33	0.33 497	9.95 792	6	11
50	9.62 323	27	9.66 537	34	0.33 463	9.95 786	6	**10**
51	9.62 350	27	9.66 570	33	0.33 430	9.95 780	6	9
52	9.62 377	27	9.66 603	33	0.33 397	9.95 775	5	8
53	9.62 405	28	9.66 636	33	0.33 364	9.95 769	6	7
54	9.62 432	27	9.66 669	33	0.33 331	9.95 763	6	6
55	9.62 459	27	9.66 702	33	0.33 298	9.95 757	6	5
56	9.62 486	27	9.66 735	33	0.33 265	9.95 751	6	4
57	9.62 513	27	9.66 768	33	0.33 232	9.95 745	6	3
58	9.62 541	28	9.66 801	33	0.33 199	9.95 739	6	2
59	9.62 568	27	9.66 834	33	0.33 166	9.95 733	6	1
60	9.62 595	27	9.66 867	33	0.33 133	9.95 728	5	**0**
'	L. Cos.	d.	L. Cot.	c.d.	L. Tan.	L. Sin.	d.	'

114° (294°) — (245°) **65°**

P. P.

"	34	33
1	0.6	0.6
2	1.1	1.1
3	1.7	1.6
4	2.3	2.2
5	2.8	2.8
6	3.4	3.3
7	4.0	3.8
8	4.5	4.4
9	5.1	5.0
10	5.7	5.5
20	11.3	11.0
30	17.0	16.5
40	22.7	22.0
50	28.3	27.5

"	29	28	27
1	0.5	0.5	0.4
2	1.0	0.9	0.9
3	1.4	1.4	1.4
4	1.9	1.9	1.8
5	2.4	2.3	2.2
6	2.9	2.8	2.7
7	3.4	3.3	3.2
8	3.9	3.7	3.6
9	4.4	4.2	4.0
10	4.8	4.7	4.5
20	9.7	9.3	9.0
30	14.5	14.0	13.5
40	19.3	18.7	18.0
50	24.2	23.3	22.5

"	6	5
1	0.1	0.1
2	0.2	0.2
3	0.3	0.2
4	0.4	0.3
5	0.5	0.4
6	0.6	0.5
7	0.7	0.6
8	0.8	0.7
9	0.9	0.8
10	1.0	0.8
20	2.0	1.7
30	3.0	2.5
40	4.0	3.3
50	5.0	4.2

	6/34	6/33	5/34
0			
1	2.8	2.8	3.4
2	8.5	8.2	10.2
3	14.2	13.8	17.0
4	19.8	19.2	23.8
5	25.5	24.8	30.6
6	31.2	30.2	—

25° (205°) (334°) **154°**

′	L. Sin.	d.	L. Tan.	c.d.	L. Cot.	L. Cos.	d.	′
0	9.62 595	27	9.66 867	33	0.33 133	9.95 728	6	**60**
1	9.62 622	27	9.66 900	33	0.33 100	9.95 722	6	59
2	9.62 649	27	9.66 933	33	0.33 067	9.95 716	6	58
3	9.62 676	27	9.66 966	33	0.33 034	9.95 710	6	57
4	9.62 703	27	9.66 999	33	0.33 001	9.95 704	6	56
5	9.62 730	27	9.67 032	33	0.32 968	9.95 698	6	55
6	9.62 757	27	9.67 065	33	0.32 935	9.95 692	6	54
7	9.62 784	27	9.67 098	33	0.32 902	9.95 686	6	53
8	9.62 811	27	9.67 131	32	0.32 869	9.95 680	6	52
9	9.62 838	27	9.67 163	33	0.32 837	9.95 674	6	51
10	9.62 865	27	9.67 196	33	0.32 804	9.95 668	5	**50**
11	9.62 892	26	9.67 229	33	0.32 771	9.95 663	6	49
12	9.62 918	27	9.67 262	33	0.32 738	9.95 657	6	48
13	9.62 945	27	9.67 295	32	0.32 705	9.95 651	6	47
14	9.62 972	27	9.67 327	33	0.32 673	9.95 645	6	46
15	9.62 999	27	9.67 360	33	0.32 640	9.95 639	6	45
16	9.63 026	26	9.67 393	33	0.32 607	9.95 633	6	44
17	9.63 052	27	9.67 426	32	0.32 574	9.95 627	6	43
18	9.63 079	27	9.67 458	33	0.32 542	9.95 621	6	42
19	9.63 106	27	9.67 491	33	0.32 509	9.95 615	6	41
20	9.63 133	26	9.67 524	32	0.32 476	9.95 609	6	**40**
21	9.63 159	27	9.67 556	33	0.32 444	9.95 603	6	39
22	9.63 186	27	9.67 589	33	0.32 411	9.95 597	6	38
23	9.63 213	26	9.67 622	32	0.32 378	9.95 591	6	37
24	9.63 239	27	9.67 654	33	0.32 346	9.95 585	6	36
25	9.63 266	26	9.67 687	32	0.32 313	9.95 579	6	35
26	9.63 292	27	9.67 719	33	0.32 281	9.95 573	6	34
27	9.63 319	26	9.67 752	33	0.32 248	9.95 567	6	33
28	9.63 345	27	9.67 785	32	0.32 215	9.95 561	6	32
29	9.63 372	26	9.67 817	33	0.32 183	9.95 555	6	31
30	9.63 398	27	9.67 850	32	0.32 150	9.95 549	6	**30**
31	9.63 425	26	9.67 882	33	0.32 118	9.95 543	6	29
32	9.63 451	27	9.67 915	32	0.32 085	9.95 537	6	28
33	9.63 478	26	9.67 947	33	0.32 053	9.95 531	6	27
34	9.63 504	27	9.67 980	32	0.32 020	9.95 525	6	26
35	9.63 531	26	9.68 012	32	0.31 988	9.95 519	6	25
36	9.63 557	26	9.68 044	33	0.31 956	9.95 513	6	24
37	9.63 583	27	9.68 077	32	0.31 923	9.95 507	7	23
38	9.63 610	26	9.68 109	33	0.31 891	9.95 500	6	22
39	9.63 636	26	9.68 142	32	0.31 858	9.95 494	6	21
40	9.63 662	27	9.68 174	32	0.31 826	9.95 488	6	**20**
41	9.63 689	26	9.68 206	33	0.31 794	9.95 482	6	19
42	9.63 715	26	9.68 239	32	0.31 761	9.95 476	6	18
43	9.63 741	26	9.68 271	32	0.31 729	9.95 470	6	17
44	9.63 767	27	9.68 303	33	0.31 697	9.95 464	6	16
45	9.63 794	26	9.68 336	32	0.31 664	9.95 458	6	15
46	9.63 820	26	9.68 368	32	0.31 632	9.95 452	6	14
47	9.63 846	26	9.68 400	32	0.31 600	9.95 446	6	13
48	9.63 872	26	9.68 432	33	0.31 568	9.95 440	6	12
49	9.63 898	26	9.68 465	32	0.31 535	9.95 434	7	11
50	9.63 924	26	9.68 497	32	0.31 503	9.95 427	6	**10**
51	9.63 950	26	9.68 529	32	0.31 471	9.95 421	6	9
52	9.63 976	26	9.68 561	32	0.31 439	9.95 415	6	8
53	9.64 002	26	9.68 593	33	0.31 407	9.95 409	6	7
54	9.64 028	26	9.68 626	32	0.31 374	9.95 403	6	6
55	9.64 054	26	9.68 658	32	0.31 342	9.95 397	6	5
56	9.64 080	26	9.68 690	32	0.31 310	9.95 391	7	4
57	9.64 106	26	9.68 722	32	0.31 278	9.95 384	6	3
58	9.64 132	26	9.68 754	32	0.31 246	9.95 378	6	2
59	9.64 158	26	9.68 786	32	0.31 214	9.95 372	6	1
60	9.64 184		9.68 818		0.31 182	9.95 366		**0**
′	L. Cos.	d.	L. Cot.	c.d.	L. Tan.	L. Sin.	d.	′

115° (295°) (244°) **64°**

P.P.

″	33	32
1	0.6	0.5
2	1.1	1.1
3	1.6	1.6
4	2.2	2.1
5	2.8	2.7
6	3.3	3.2
7	3.8	3.7
8	4.4	4.3
9	5.0	4.8
10	5.5	5.3
20	11.0	10.7
30	16.5	16.0
40	22.0	21.3
50	27.5	26.7

″	27	26
1	0.4	0.4
2	0.9	0.9
3	1.4	1.3
4	1.8	1.7
5	2.2	2.2
6	2.7	2.6
7	3.2	3.0
8	3.6	3.5
9	4.0	3.9
10	4.5	4.3
20	9.0	8.7
30	13.5	13.0
40	18.0	17.3
50	22.5	21.7

″	7	6	5
1	0.1	0.1	0.1
2	0.2	0.2	0.2
3	0.4	0.3	0.2
4	0.5	0.4	0.3
5	0.6	0.5	0.4
6	0.7	0.6	0.5
7	0.8	0.7	0.6
8	0.9	0.8	0.7
9	1.0	0.9	0.8
10	1.2	1.0	0.8
20	2.3	2.0	1.7
30	3.5	3.0	2.5
40	4.7	4.0	3.3
50	5.8	5.0	4.2

	7/32	6/32	5/33
0			
1	2.3	2.7	3.3
2	6.9	8.0	9.9
3	11.4	13.3	16.5
4	16.0	18.7	23.1
5	20.6	24.0	29.7
6	25.1	29.3	—
7	29.7	—	—

26° (206°) (333°) **153°**

′	L. Sin.	d.	L. Tan.	c.d.	L. Cot.	L. Cos.	d.	′
0	9.64 184	26	9.68 818	32	0.31 182	9.95 366	6	**60**
1	9.64 210	26	9.68 850	32	0.31 150	9.95 360	6	59
2	9.64 236	26	9.68 882	32	0.31 118	9.95 354	6	58
3	9.64 262	26	9.68 914	32	0.31 086	9.95 348	7	57
4	9.64 288	25	9.68 946	32	0.31 054	9.95 341	6	56
5	9.64 313	26	9.68 978	32	0.31 022	9.95 335	6	55
6	9.64 339	26	9.69 010	32	0.30 990	9.95 329	6	54
7	9.64 365	26	9.69 042	32	0.30 958	9.95 323	6	53
8	9.64 391	26	9.69 074	32	0.30 926	9.95 317	7	52
9	9.64 417	25	9.69 106	32	0.30 894	9.95 310	6	51
10	9.64 442	26	9.69 138	32	0.30 862	9.95 304	6	**50**
11	9.64 468	26	9.69 170	32	0.30 830	9.95 298	6	49
12	9.64 494	25	9.69 202	32	0.30 798	9.95 292	6	48
13	9.64 519	26	9.69 234	32	0.30 766	9.95 286	7	47
14	9.64 545	26	9.69 266	32	0.30 734	9.95 279	6	46
15	9.64 571	25	9.69 298	31	0.30 702	9.95 273	6	45
16	9.64 596	26	9.69 329	32	0.30 671	9.95 267	6	44
17	9.64 622	25	9.69 361	32	0.30 639	9.95 261	7	43
18	9.64 647	26	9.69 393	32	0.30 607	9.95 254	6	42
19	9.64 673	25	9.69 425	32	0.30 575	9.95 248	6	41
20	9.64 698	26	9.69 457	31	0.30 543	9.95 242	6	**40**
21	9.64 724	25	9.69 488	32	0.30 512	9.95 236	7	39
22	9.64 749	26	9.69 520	32	0.30 480	9.95 229	6	38
23	9.64 775	25	9.69 552	32	0.30 448	9.95 223	6	37
24	9.64 800	26	9.69 584	31	0.30 416	9.95 217	6	36
25	9.64 826	25	9.69 615	32	0.30 385	9.95 211	7	35
26	9.64 851	26	9.69 647	32	0.30 353	9.95 204	6	34
27	9.64 877	25	9.69 679	31	0.30 321	9.95 198	6	33
28	9.64 902	25	9.69 710	32	0.30 290	9.95 192	7	32
29	9.64 927	26	9.69 742	32	0.30 258	9.95 185	6	31
30	9.64 953	25	9.69 774	31	0.30 226	9.95 179	6	**30**
31	9.64 978	25	9.69 805	32	0.30 195	9.95 173	6	29
32	9.65 003	26	9.69 837	31	0.30 163	9.95 167	7	28
33	9.65 029	25	9.69 868	32	0.30 132	9.95 160	6	27
34	9.65 054	25	9.69 900	32	0.30 100	9.95 154	6	26
35	9.65 079	25	9.69 932	31	0.30 068	9.95 148	7	25
36	9.65 104	26	9.69 963	32	0.30 037	9.95 141	6	24
37	9.65 130	25	9.69 995	31	0.30 005	9.95 135	6	23
38	9.65 155	25	9.70 026	32	0.29 974	9.95 129	7	22
39	9.65 180	25	9.70 058	31	0.29 942	9.95 122	6	21
40	9.65 205	25	9.70 089	32	0.29 911	9.95 116	6	**20**
41	9.65 230	25	9.70 121	31	0.29 879	9.95 110	7	19
42	9.65 255	26	9.70 152	32	0.29 848	9.95 103	6	18
43	9.65 281	25	9.70 184	31	0.29 816	9.95 097	7	17
44	9.65 306	25	9.70 215	32	0.29 785	9.95 090	6	16
45	9.65 331	25	9.70 247	31	0.29 753	9.95 084	6	15
46	9.65 356	25	9.70 278	31	0.29 722	9.95 078	7	14
47	9.65 381	25	9.70 309	32	0.29 691	9.95 071	6	13
48	9.65 406	25	9.70 341	31	0.29 659	9.95 065	6	12
49	9.65 431	25	9.70 372	32	0.29 628	9.95 059	7	11
50	9.65 456	25	9.70 404	31	0.29 596	9.95 052	6	**10**
51	9.65 481	25	9.70 435	31	0.29 565	9.95 046	7	9
52	9.65 506	25	9.70 466	32	0.29 534	9.95 039	6	8
53	9.65 531	25	9.70 498	31	0.29 502	9.95 033	6	7
54	9.65 556	24	9.70 529	31	0.29 471	9.95 027	7	6
55	9.65 580	25	9.70 560	32	0.29 440	9.95 020	6	5
56	9.65 605	25	9.70 592	31	0.29 408	9.95 014	7	4
57	9.65 630	25	9.70 623	31	0.29 377	9.95 007	6	3
58	9.65 655	25	9.70 654	31	0.29 346	9.95 001	6	2
59	9.65 680	25	9.70 685	32	0.29 315	9.94 995	7	1
60	9.65 705		9.70 717		0.29 283	9.94 988		**0**
′	L. Cos.	d.	L. Cot.	c.d.	L. Tan.	L. Sin.	d.	′

116° (296°) (243°) **63°**

P.P.

″	32	31
1	0.5	0.5
2	1.1	1.0
3	1.6	1.6
4	2.1	2.1
5	2.7	2.6
6	3.2	3.1
7	3.7	3.6
8	4.3	4.1
9	4.8	4.6
10	5.3	5.2
20	10.7	10.3
30	16.0	15.5
40	21.3	20.7
50	26.7	25.8

″	26	25	24
1	0.4	0.4	0.4
2	0.9	0.8	0.8
3	1.3	1.2	1.2
4	1.7	1.7	1.6
5	2.2	2.1	2.0
6	2.6	2.5	2.4
7	3.0	2.9	2.8
8	3.5	3.3	3.2
9	3.9	3.8	3.6
10	4.3	4.2	4.0
20	8.7	8.3	8.0
30	13.0	12.5	12.0
40	17.3	16.7	16.0
50	21.7	20.8	20.0

″	7	6
1	0.1	0.1
2	0.2	0.2
3	0.4	0.3
4	0.5	0.4
5	0.6	0.5
6	0.7	0.6
7	0.8	0.7
8	0.9	0.8
9	1.0	0.9
10	1.2	1.0
20	2.3	2.0
30	3.5	3.0
40	4.7	4.0
50	5.8	5.0

	7/32	7/31	6/32
0			
1	2.3	2.2	2.7
2	6.9	6.6	8.0
3	11.4	11.1	13.3
4	16.0	15.5	18.7
5	20.6	19.9	24.0
6	25.1	24.4	29.3
7	29.7	28.8	—

27° (207°) (332°) **152°**

′	L. Sin.	d.	L. Tan.	c.d.	L. Cot.	L. Cos.	d.	′
0	9.65 705	24	9.70 717	31	0.29 283	9.94 988	6	**60**
1	9.65 729	25	9.70 748	31	0.29 252	9.94 982	7	59
2	9.65 754	25	9.70 779	31	0.29 221	9.94 975	6	58
3	9.65 779	25	9.70 810	31	0.29 190	9.94 969	7	57
4	9.65 804	24	9.70 841	32	0.29 159	9.94 962	6	56
5	9.65 828	25	9.70 873	31	0.29 127	9.94 956	7	55
6	9.65 853	25	9.70 904	31	0.29 096	9.94 949	6	54
7	9.65 878	24	9.70 935	31	0.29 065	9.94 943	7	53
8	9.65 902	25	9.70 966	31	0.29 034	9.94 936	6	52
9	9.65 927	25	9.70 997	31	0.29 003	9.94 930	7	51
10	9.65 952	24	9.71 028	31	0.28 972	9.94 923	6	**50**
11	9.65 976	25	9.71 059	31	0.28 941	9.94 917	6	49
12	9.66 001	24	9.71 090	31	0.28 910	9.94 911	7	48
13	9.66 025	25	9.71 121	32	0.28 879	9.94 904	6	47
14	9.66 050	25	9.71 153	31	0.28 847	9.94 898	7	46
15	9.66 075	24	9.71 184	31	0.28 816	9.94 891	6	45
16	9.66 099	25	9.71 215	31	0.28 785	9.94 885	7	44
17	9.66 124	24	9.71 246	31	0.28 754	9.94 878	7	43
18	9.66 148	25	9.71 277	31	0.28 723	9.94 871	6	42
19	9.66 173	24	9.71 308	31	0.28 692	9.94 865	7	41
20	9.66 197	24	9.71 339	31	0.28 661	9.94 858	6	**40**
21	9.66 221	25	9.71 370	31	0.28 630	9.94 852	7	39
22	9.66 246	24	9.71 401	30	0.28 599	9.94 845	6	38
23	9.66 270	25	9.71 431	31	0.28 569	9.94 839	7	37
24	9.66 295	24	9.71 462	31	0.28 538	9.94 832	6	36
25	9.66 319	24	9.71 493	31	0.28 507	9.94 826	7	35
26	9.66 343	25	9.71 524	31	0.28 476	9.94 819	6	34
27	9.66 368	24	9.71 555	31	0.28 445	9.94 813	7	33
28	9.66 392	24	9.71 586	31	0.28 414	9.94 806	7	32
29	9.66 416	25	9.71 617	31	0.28 383	9.94 799	6	31
30	9.66 441	24	9.71 648	31	0.28 352	9.94 793	7	**30**
31	9.66 465	24	9.71 679	30	0.28 321	9.94 786	6	29
32	9.66 489	24	9.71 709	31	0.28 291	9.94 780	7	28
33	9.66 513	24	9.71 740	31	0.28 260	9.94 773	6	27
34	9.66 537	25	9.71 771	31	0.28 229	9.94 767	7	26
35	9.66 562	24	9.71 802	31	0.28 198	9.94 760	7	25
36	9.66 586	24	9.71 833	30	0.28 167	9.94 753	6	24
37	9.66 610	24	9.71 863	31	0.28 137	9.94 747	7	23
38	9.66 634	24	9.71 894	31	0.28 106	9.94 740	6	22
39	9.66 658	24	9.71 925	30	0.28 075	9.94 734	7	21
40	9.66 682	24	9.71 955	31	0.28 045	9.94 727	7	**20**
41	9.66 706	25	9.71 986	31	0.28 014	9.94 720	6	19
42	9.66 731	24	9.72 017	31	0.27 983	9.94 714	7	18
43	9.66 755	24	9.72 048	30	0.27 952	9.94 707	7	17
44	9.66 779	24	9.72 078	31	0.27 922	9.94 700	6	16
45	9.66 803	24	9.72 109	31	0.27 891	9.94 694	7	15
46	9.66 827	24	9.72 140	30	0.27 860	9.94 687	7	14
47	9.66 851	24	9.72 170	31	0.27 830	9.94 680	6	13
48	9.66 875	24	9.72 201	30	0.27 799	9.94 674	7	12
49	9.66 899	23	9.72 231	31	0.27 769	9.94 667	7	11
50	9.66 922	24	9.72 262	31	0.27 738	9.94 660	6	**10**
51	9.66 946	24	9.72 293	30	0.27 707	9.94 654	7	9
52	9.66 970	24	9.72 323	31	0.27 677	9.94 647	7	8
53	9.66 994	24	9.72 354	30	0.27 646	9.94 640	6	7
54	9.67 018	24	9.72 384	31	0.27 616	9.94 634	7	6
55	9.67 042	24	9.72 415	30	0.27 585	9.94 627	7	5
56	9.67 066	24	9.72 445	31	0.27 555	9.94 620	6	4
57	9.67 090	23	9.72 476	30	0.27 524	9.94 614	7	3
58	9.67 113	24	9.72 506	31	0.27 494	9.94 607	7	2
59	9.67 137	24	9.72 537	30	0.27 463	9.94 600	7	1
60	9.67 161		9.72 567		0.27 433	9.94 593		**0**
′	L. Cos.	d.	L. Cot.	c.d.	L. Tan.	L. Sin.	d.	′

P. P.

″	32	31	30
1	0.5	0.5	0.5
2	1.1	1.0	1.0
3	1.6	1.6	1.5
4	2.1	2.1	2.0
5	2.7	2.6	2.5
6	3.2	3.1	3.0
7	3.7	3.6	3.5
8	4.3	4.1	4.0
9	4.8	4.6	4.5
10	5.3	5.2	5.0
20	10.7	10.3	10.0
30	16.0	15.5	15.0
40	21.3	20.7	20.0
50	26.7	25.8	25.0

″	25	24	23
1	0.4	0.4	0.4
2	0.8	0.8	0.8
3	1.2	1.2	1.2
4	1.7	1.6	1.5
5	2.1	2.0	1.9
6	2.5	2.4	2.3
7	2.9	2.8	2.7
8	3.3	3.2	3.1
9	3.8	3.6	3.4
10	4.2	4.0	3.8
20	8.3	8.0	7.7
30	12.5	12.0	11.5
40	16.7	16.0	15.3
50	20.8	20.0	19.2

″	7	6
1	0.1	0.1
2	0.2	0.2
3	0.4	0.3
4	0.5	0.4
5	0.6	0.5
6	0.7	0.6
7	0.8	0.7
8	0.9	0.8
9	1.0	0.9
10	1.2	1.0
20	2.3	2.0
30	3.5	3.0
40	4.7	4.0
50	5.8	5.0

	7/30	6/31	6/30
0	2.1	2.6	2.5
1	6.4	7.8	7.5
2	10.7	12.9	12.5
3	15.0	18.1	17.5
4	19.3	23.2	22.5
5	23.6	28.4	27.5
6	27.9	—	—
7			

P. P.

117° (297°) (242°) **62°**

28° (208°) (331°) **151°**

′	L. Sin.	d.	L. Tan.	c.d.	L. Cot.	L. Cos.	d.	′
0	9.67 161	24	9.72 567	31	0.27 433	9.94 593	6	**60**
1	9.67 185	23	9.72 598	30	0.27 402	9.94 587	7	59
2	9.67 208	24	9.72 628	31	0.27 372	9.94 580	7	58
3	9.67 232	24	9.72 659	30	0.27 341	9.94 573	6	57
4	9.67 256	24	9.72 689	31	0.27 311	9.94 567	7	56
5	9.67 280	23	9.72 720	30	0.27 280	9.94 560	7	55
6	9.67 303	24	9.72 750	30	0.27 250	9.94 553	7	54
7	9.67 327	23	9.72 780	31	0.27 220	9.94 546	6	53
8	9.67 350	24	9.72 811	30	0.27 189	9.94 540	7	52
9	9.67 374	24	9.72 841	31	0.27 159	9.94 533	7	51
10	9.67 398	23	9.72 872	30	0.27 128	9.94 526	7	**50**
11	9.67 421	24	9.72 902	30	0.27 098	9.94 519	6	49
12	9.67 445	23	9.72 932	31	0.27 068	9.94 513	7	48
13	9.67 468	24	9.72 963	30	0.27 037	9.94 506	7	47
14	9.67 492	23	9.72 993	30	0.27 007	9.94 499	7	46
15	9.67 515	24	9.73 023	31	0.26 977	9.94 492	7	45
16	9.67 539	23	9.73 054	30	0.26 946	9.94 485	6	44
17	9.67 562	24	9.73 084	30	0.26 916	9.94 479	7	43
18	9.67 586	23	9.73 114	30	0.26 886	9.94 472	7	42
19	9.67 609	24	9.73 144	31	0.26 856	9.94 465	7	41
20	9.67 633	23	9.73 175	30	0.26 825	9.94 458	7	**40**
21	9.67 656	24	9.73 205	30	0.26 795	9.94 451	6	39
22	9.67 680	23	9.73 235	30	0.26 765	9.94 445	7	38
23	9.67 703	23	9.73 265	30	0.26 735	9.94 438	7	37
24	9.67 726	24	9.73 295	31	0.26 705	9.94 431	7	36
25	9.67 750	23	9.73 326	30	0.26 674	9.94 424	7	35
26	9.67 773	23	9.73 356	30	0.26 644	9.94 417	7	34
27	9.67 796	24	9.73 386	30	0.26 614	9.94 410	6	33
28	9.67 820	23	9.73 416	30	0.26 584	9.94 404	7	32
29	9.67 843	23	9.73 446	30	0.26 554	9.94 397	7	31
30	9.67 866	24	9.73 476	31	0.26 524	9.94 390	7	**30**
31	9.67 890	23	9.73 507	30	0.26 493	9.94 383	7	29
32	9.67 913	23	9.73 537	30	0.26 463	9.94 376	7	28
33	9.67 936	23	9.73 567	30	0.26 433	9.94 369	7	27
34	9.67 959	23	9.73 597	30	0.26 403	9.94 362	7	26
35	9.67 982	24	9.73 627	30	0.26 373	9.94 355	6	25
36	9.68 006	23	9.73 657	30	0.26 343	9.94 349	7	24
37	9.68 029	23	9.73 687	30	0.26 313	9.94 342	7	23
38	9.68 052	23	9.73 717	30	0.26 283	9.94 335	7	22
39	9.68 075	23	9.73 747	30	0.26 253	9.94 328	7	21
40	9.68 098	23	9.73 777	30	0.26 223	9.94 321	7	**20**
41	9.68 121	23	9.73 807	30	0.26 193	9.94 314	7	19
42	9.68 144	23	9.73 837	30	0.26 163	9.94 307	7	18
43	9.68 167	23	9.73 867	30	0.26 133	9.94 300	7	17
44	9.68 190	23	9.73 897	30	0.26 103	9.94 293	7	16
45	9.68 213	24	9.73 927	30	0.26 073	9.94 286	7	15
46	9.68 237	23	9.73 957	30	0.26 043	9.94 279	6	14
47	9.68 260	23	9.73 987	30	0.26 013	9.94 273	7	13
48	9.68 283	22	9.74 017	30	0.25 983	9.94 266	7	12
49	9.68 305	23	9.74 047	30	0.25 953	9.94 259	7	11
50	9.68 328	23	9.74 077	30	0.25 923	9.94 252	7	**10**
51	9.68 351	23	9.74 107	30	0.25 893	9.94 245	7	9
52	9.68 374	23	9.74 137	29	0.25 863	9.94 238	7	8
53	9.68 397	23	9.74 166	30	0.25 834	9.94 231	7	7
54	9.68 420	23	9.74 196	30	0.25 804	9.94 224	7	6
55	9.68 443	23	9.74 226	30	0.25 774	9.94 217	7	5
56	9.68 466	23	9.74 256	30	0.25 744	9.94 210	7	4
57	9.68 489	23	9.74 286	30	0.25 714	9.94 203	7	3
58	9.68 512	22	9.74 316	29	0.25 684	9.94 196	7	2
59	9.68 534	23	9.74 345	30	0.25 655	9.94 189	7	1
60	9.68 557		9.74 375		0.25 625	9.94 182		**0**
′	L. Cos.	d.	L. Cot.	c.d.	L. Tan.	L. Sin.	d.	′

P. P.

″	31	30	29
1	0.5	0.5	0.5
2	1.0	1.0	1.0
3	1.6	1.5	1.4
4	2.1	2.0	1.9
5	2.6	2.5	2.4
6	3.1	3.0	2.9
7	3.6	3.5	3.4
8	4.1	4.0	3.9
9	4.6	4.5	4.4
10	5.2	5.0	4.8
20	10.3	10.0	9.7
30	15.5	15.0	14.5
40	20.7	20.0	19.3
50	25.8	25.0	24.2

″	24	23	22
1	0.4	0.4	0.4
2	0.8	0.8	0.7
3	1.2	1.2	1.1
4	1.6	1.5	1.5
5	2.0	1.9	1.8
6	2.4	2.3	2.2
7	2.8	2.7	2.6
8	3.2	3.1	2.9
9	3.6	3.4	3.3
10	4.0	3.8	3.7
20	8.0	7.7	7.3
30	12.0	11.5	11.0
40	16.0	15.3	14.7
50	20.0	19.2	18.3

″	7	6
1	0.1	0.1
2	0.2	0.2
3	0.4	0.3
4	0.5	0.4
5	0.6	0.5
6	0.7	0.6
7	0.8	0.7
8	0.9	0.8
9	1.0	0.9
10	1.2	1.0
20	2.3	2.0
30	3.5	3.0
40	4.7	4.0
50	5.8	5.0

	7/31	6/31	6/30
0	2.2	2.6	2.5
1	6.6	7.8	7.5
2	11.1	12.9	12.5
3	15.5	18.1	17.5
4	19.9	23.2	22.5
5	24.4	28.4	27.5
6	28.8	—	—
7			

P. P.

118° (298°) (241°) **61°**

LOGARITHMS OF THE FUNCTIONS (Continued)

29° (209°) | (330°) 150°

′	L. Sin.	d.	L. Tan.	c.d.	L. Cot.	L. Cos.	d.	′
0	9.68 557	23	9.74 375	30	0.25 625	9.94 182	7	60
1	9.68 580	23	9.74 405	30	0.25 595	9.94 175	7	59
2	9.68 603	22	9.74 435	30	0.25 565	9.94 168	7	58
3	9.68 625	23	9.74 465	29	0.25 535	9.94 161	7	57
4	9.68 648	23	9.74 494	30	0.25 506	9.94 154	7	56
5	9.68 671	23	9.74 524	30	0.25 476	9.94 147	7	55
6	9.68 694	22	9.74 554	29	0.25 446	9.94 140	7	54
7	9.68 716	23	9.74 583	30	0.25 417	9.94 133	7	53
8	9.68 739	23	9.74 613	30	0.25 387	9.94 126	7	52
9	9.68 762	22	9.74 643	30	0.25 357	9.94 119	7	51
10	9.68 784	23	9.74 673	29	0.25 327	9.94 112	7	50
11	9.68 807	22	9.74 702	30	0.25 298	9.94 105	7	49
12	9.68 829	23	9.74 732	30	0.25 268	9.94 098	8	48
13	9.68 852	23	9.74 762	29	0.25 238	9.94 090	7	47
14	9.68 875	22	9.74 791	30	0.25 209	9.94 083	7	46
15	9.68 897	23	9.74 821	30	0.25 179	9.94 076	7	45
16	9.68 920	22	9.74 851	29	0.25 149	9.94 069	7	44
17	9.68 942	23	9.74 880	30	0.25 120	9.94 062	7	43
18	9.68 965	22	9.74 910	29	0.25 090	9.94 055	7	42
19	9.68 987	23	9.74 939	30	0.25 061	9.94 048	7	41
20	9.69 010	22	9.74 969	29	0.25 031	9.94 041	7	40
21	9.69 032	23	9.74 998	30	0.25 002	9.94 034	7	39
22	9.69 055	22	9.75 028	30	0.24 972	9.94 027	7	38
23	9.69 077	23	9.75 058	29	0.24 942	9.94 020	8	37
24	9.69 100	22	9.75 087	30	0.24 913	9.94 012	7	36
25	9.69 122	22	9.75 117	29	0.24 883	9.94 005	7	35
26	9.69 144	23	9.75 146	30	0.24 854	9.93 998	7	34
27	9.69 167	22	9.75 176	29	0.24 824	9.93 991	7	33
28	9.69 189	23	9.75 205	30	0.24 795	9.93 984	7	32
29	9.69 212	22	9.75 235	29	0.24 765	9.93 977	7	31
30	9.69 234	22	9.75 264	30	0.24 736	9.93 970	7	30
31	9.69 256	23	9.75 294	29	0.24 706	9.93 963	8	29
32	9.69 279	22	9.75 323	30	0.24 677	9.93 955	7	28
33	9.69 301	22	9.75 353	29	0.24 647	9.93 948	7	27
34	9.69 323	22	9.75 382	29	0.24 618	9.93 941	7	26
35	9.69 345	23	9.75 411	30	0.24 589	9.93 934	7	25
36	9.69 368	22	9.75 441	29	0.24 559	9.93 927	7	24
37	9.69 390	22	9.75 470	30	0.24 530	9.93 920	8	23
38	9.69 412	22	9.75 500	29	0.24 500	9.93 912	7	22
39	9.69 434	22	9.75 529	29	0.24 471	9.93 905	7	21
40	9.69 456	23	9.75 558	30	0.24 442	9.93 898	7	20
41	9.69 479	22	9.75 588	29	0.24 412	9.93 891	7	19
42	9.69 501	22	9.75 617	30	0.24 383	9.93 884	8	18
43	9.69 523	22	9.75 647	29	0.24 353	9.93 876	7	17
44	9.69 545	22	9.75 676	29	0.24 324	9.93 869	7	16
45	9.69 567	22	9.75 705	30	0.24 295	9.93 862	7	15
46	9.69 589	22	9.75 735	29	0.24 265	9.93 855	8	14
47	9.69 611	22	9.75 764	29	0.24 236	9.93 847	7	13
48	9.69 633	22	9.75 793	29	0.24 207	9.93 840	7	12
49	9.69 655	22	9.75 822	30	0.24 178	9.93 833	7	11
50	9.69 677	22	9.75 852	29	0.24 148	9.93 826	7	10
51	9.69 699	22	9.75 881	29	0.24 119	9.93 819	8	9
52	9.69 721	22	9.75 910	29	0.24 090	9.93 811	7	8
53	9.69 743	22	9.75 939	30	0.24 061	9.93 804	7	7
54	9.69 765	22	9.75 969	29	0.24 031	9.93 797	8	6
55	9.69 787	22	9.75 998	29	0.24 002	9.93 789	7	5
56	9.69 809	22	9.76 027	29	0.23 973	9.93 782	7	4
57	9.69 831	22	9.76 056	30	0.23 944	9.93 775	7	3
58	9.69 853	22	9.76 086	29	0.23 914	9.93 768	8	2
59	9.69 875	22	9.76 115	29	0.23 885	9.93 760	7	1
60	9.69 897		9.76 144		0.23 856	9.93 753		0
′	L. Cos.	d.	L. Cot.	c.d.	L. Tan.	L. Sin.	d.	′

119° (299°) | (240°) 60°

P. P.

″	30	29	23
1	0.5	0.5	0.4
2	1.0	1.0	0.8
3	1.5	1.4	1.2
4	2.0	1.9	1.5
5	2.5	2.4	1.9
6	3.0	2.9	2.3
7	3.5	3.4	2.7
8	4.0	3.9	3.1
9	4.5	4.4	3.4
10	5.0	4.8	3.8
20	10.0	9.7	7.7
30	15.0	14.5	11.5
40	20.0	19.3	15.3
50	25.0	24.2	19.2

″	22	8	7
1	0.4	0.1	0.1
2	0.7	0.3	0.2
3	1.1	0.4	0.4
4	1.5	0.5	0.5
5	1.8	0.7	0.6
6	2.2	0.8	0.7
7	2.6	0.9	0.8
8	2.9	1.1	0.9
9	3.3	1.2	1.0
10	3.7	1.3	1.2
20	7.3	2.7	2.3
30	11.0	4.0	3.5
40	14.7	5.3	4.7
50	18.3	6.7	5.8

	8/30	8/29
0	1.9	1.8
1	5.6	5.4
2	9.4	9.1
3	13.1	12.7
4	16.9	16.3
5	20.6	19.9
6	24.4	23.6
7	28.1	27.2
8		

	7/30	7/29
0	2.1	2.1
1	6.4	6.2
2	10.7	10.4
3	15.0	14.5
4	19.3	18.6
5	23.6	22.8
6	27.9	26.9
7		

P. P.

30° (210°) | (329°) 149°

′	L. Sin.	d.	L. Tan.	c.d.	L. Cot.	L. Cos.	d.	′
0	9.69 897	22	9.76 144	29	0.23 856	9.93 753	7	60
1	9.69 919	22	9.76 173	29	0.23 827	9.93 746	8	59
2	9.69 941	22	9.76 202	29	0.23 798	9.93 738	7	58
3	9.69 963	21	9.76 231	30	0.23 769	9.93 731	7	57
4	9.69 984	22	9.76 261	29	0.23 739	9.93 724	7	56
5	9.70 006	22	9.76 290	29	0.23 710	9.93 717	8	55
6	9.70 028	22	9.76 319	29	0.23 681	9.93 709	7	54
7	9.70 050	22	9.76 348	29	0.23 652	9.93 702	7	53
8	9.70 072	21	9.76 377	29	0.23 623	9.93 695	8	52
9	9.70 093	22	9.76 406	29	0.23 594	9.93 687	7	51
10	9.70 115	22	9.76 435	29	0.23 565	9.93 680	7	50
11	9.70 137	22	9.76 464	29	0.23 536	9.93 673	8	49
12	9.70 159	21	9.76 493	29	0.23 507	9.93 665	7	48
13	9.70 180	22	9.76 522	29	0.23 478	9.93 658	8	47
14	9.70 202	22	9.76 551	29	0.23 449	9.93 650	7	46
15	9.70 224	21	9.76 580	29	0.23 420	9.93 643	7	45
16	9.70 245	22	9.76 609	30	0.23 391	9.93 636	8	44
17	9.70 267	21	9.76 639	29	0.23 361	9.93 628	7	43
18	9.70 288	22	9.76 668	29	0.23 332	9.93 621	7	42
19	9.70 310	22	9.76 697	28	0.23 303	9.93 614	8	41
20	9.70 332	21	9.76 725	29	0.23 275	9.93 606	7	40
21	9.70 353	22	9.76 754	29	0.23 246	9.93 599	8	39
22	9.70 375	21	9.76 783	29	0.23 217	9.93 591	7	38
23	9.70 396	22	9.76 812	29	0.23 188	9.93 584	7	37
24	9.70 418	21	9.76 841	29	0.23 159	9.93 577	8	36
25	9.70 439	22	9.76 870	29	0.23 130	9.93 569	7	35
26	9.70 461	21	9.76 899	29	0.23 101	9.93 562	8	34
27	9.70 482	22	9.76 928	29	0.23 072	9.93 554	7	33
28	9.70 504	21	9.76 957	29	0.23 043	9.93 547	8	32
29	9.70 525	22	9.76 986	29	0.23 014	9.93 539	7	31
30	9.70 547	21	9.77 015	29	0.22 985	9.93 532	7	30
31	9.70 568	22	9.77 044	29	0.22 956	9.93 525	8	29
32	9.70 590	21	9.77 073	28	0.22 927	9.93 517	7	28
33	9.70 611	22	9.77 101	29	0.22 899	9.93 510	8	27
34	9.70 633	21	9.77 130	29	0.22 870	9.93 502	7	26
35	9.70 654	21	9.77 159	29	0.22 841	9.93 495	8	25
36	9.70 675	22	9.77 188	29	0.22 812	9.93 487	7	24
37	9.70 697	21	9.77 217	29	0.22 783	9.93 480	8	23
38	9.70 718	21	9.77 246	28	0.22 754	9.93 472	7	22
39	9.70 739	22	9.77 274	29	0.22 726	9.93 465	8	21
40	9.70 761	21	9.77 303	29	0.22 697	9.93 457	7	20
41	9.70 782	21	9.77 332	29	0.22 668	9.93 450	8	19
42	9.70 803	21	9.77 361	29	0.22 639	9.93 442	7	18
43	9.70 824	22	9.77 390	28	0.22 610	9.93 435	8	17
44	9.70 846	21	9.77 418	29	0.22 582	9.93 427	7	16
45	9.70 867	21	9.77 447	29	0.22 553	9.93 420	8	15
46	9.70 888	21	9.77 476	29	0.22 524	9.93 412	7	14
47	9.70 909	22	9.77 505	28	0.22 495	9.93 405	8	13
48	9.70 931	21	9.77 533	29	0.22 467	9.93 397	7	12
49	9.70 952	21	9.77 562	29	0.22 438	9.93 390	8	11
50	9.70 973	21	9.77 591	28	0.22 409	9.93 382	7	10
51	9.70 994	21	9.77 619	29	0.22 381	9.93 375	8	9
52	9.71 015	21	9.77 648	29	0.22 352	9.93 367	7	8
53	9.71 036	22	9.77 677	29	0.22 323	9.93 360	8	7
54	9.71 058	21	9.77 706	28	0.22 294	9.93 352	8	6
55	9.71 079	21	9.77 734	29	0.22 266	9.93 344	7	5
56	9.71 100	21	9.77 763	28	0.22 237	9.93 337	8	4
57	9.71 121	21	9.77 791	29	0.22 209	9.93 329	7	3
58	9.71 142	21	9.77 820	29	0.22 180	9.93 322	8	2
59	9.71 163	21	9.77 849	28	0.22 151	9.93 314	7	1
60	9.71 184		9.77 877		0.22 123	9.93 307		0
′	L. Cos.	d.	L. Cot.	c.d.	L. Tan.	L. Sin.	d.	′

120° (300°) | (239°) 59°

P.P.

″	30	29	28
1	0.5	0.5	0.5
2	1.0	1.0	0.9
3	1.5	1.4	1.4
4	2.0	1.9	1.9
5	2.5	2.4	2.3
6	3.0	2.9	2.8
7	3.5	3.4	3.3
8	4.0	3.9	3.7
9	4.5	4.4	4.2
10	5.0	4.8	4.7
20	10.0	9.7	9.3
30	15.0	14.5	14.0
40	20.0	19.3	18.7
50	25.0	24.2	23.3

″	22	21
1	0.4	0.4
2	0.7	0.7
3	1.1	1.0
4	1.5	1.4
5	1.8	1.8
6	2.2	2.1
7	2.6	2.4
8	2.9	2.8
9	3.3	3.2
10	3.7	3.5
20	7.3	7.0
30	11.0	10.5
40	14.7	14.0
50	18.3	17.5

″	8	7
1	0.1	0.1
2	0.3	0.2
3	0.4	0.4
4	0.5	0.5
5	0.7	0.6
6	0.8	0.7
7	0.9	0.8
8	1.1	0.9
9	1.2	1.0
10	1.3	1.2
20	2.7	2.3
30	4.0	3.5
40	5.3	4.7
50	6.7	5.8

	7/30	7/29	7/28
0	2.1	2.1	2.0
1	6.4	6.2	6.0
2	10.7	10.4	10.0
3	15.0	14.5	14.0
4	19.3	18.6	18.0
5	23.6	22.8	22.0
6	27.9	26.9	26.0
7			

P.P.

LOGARITHMS OF THE FUNCTIONS (Continued)

31° (211°) — (328°) **148°**

′	L. Sin.	d.	L. Tan.	c.d.	L. Cot.	L. Cos.	d.	′
0	9.71 184	21	9.77 877	29	0.22 123	9.93 307	8	**60**
1	9.71 205	21	9.77 906	29	0.22 094	9.93 299	8	59
2	0.71 226	21	9.77 935	28	0.22 065	9.93 291	7	58
3	9.71 247	21	9.77 963	29	0.22 037	9.93 284	8	57
4	9.71 268	21	9.77 992	28	0.22 008	9.93 276	7	56
5	9.71 289	21	9.78 020	29	0.21 980	9.93 269	8	55
6	9.71 310	21	9.78 049	28	0.21 951	9.93 261	8	54
7	9.71 331	21	9.78 077	29	0.21 923	9.93 253	7	53
8	9.71 352	21	9.78 106	29	0.21 894	9.93 246	8	52
9	9.71 373	20	9.78 135	28	0.21 865	9.93 238	8	51
10	9.71 393	21	9.78 163	29	0.21 837	9.93 230	7	**50**
11	9.71 414	21	9.78 192	28	0.21 808	9.93 223	8	49
12	9.71 435	21	9.78 220	29	0.21 780	9.93 215	8	48
13	9.71 456	21	9.78 249	28	0.21 751	9.93 207	7	47
14	9.71 477	21	9.78 277	29	0.21 723	9.93 200	8	46
15	9.71 498	21	9.78 306	28	0.21 694	9.93 192	8	45
16	9.71 519	20	9.78 334	29	0.21 666	9.93 184	7	44
17	9.71 539	21	9.78 363	28	0.21 637	9.93 177	8	43
18	9.71 560	21	9.78 391	28	0.21 609	9.93 169	8	42
19	9.71 581	21	9.78 419	29	0.21 581	9.93 161	7	41
20	9.71 602	20	9.78 448	28	0.21 552	9.93 154	8	**40**
21	9.71 622	21	9.78 476	29	0.21 524	9.93 146	8	39
22	9.71 643	21	9.78 505	28	0.21 495	9.93 138	7	38
23	9.71 664	21	9.78 533	29	0.21 467	9.93 131	8	37
24	9.71 685	20	9.78 562	28	0.21 438	9.93 123	8	36
25	9.71 705	21	9.78 590	28	0.21 410	9.93 115	7	35
26	9.71 726	21	9.78 618	29	0.21 382	9.93 108	8	34
27	9.71 747	20	9.78 647	28	0.21 353	9.93 100	8	33
28	9.71 767	21	9.78 675	29	0.21 325	9.93 092	8	32
29	9.71 788	21	9.78 704	28	0.21 296	9.93 084	7	31
30	9.71 809	20	9.78 732	28	0.21 268	9.93 077	8	**30**
31	9.71 829	21	9.78 760	29	0.21 240	9.93 069	8	29
32	9.71 850	20	9.78 789	28	0.21 211	9.93 061	8	28
33	9.71 870	21	9.78 817	28	0.21 183	9.93 053	7	27
34	9.71 891	20	9.78 845	29	0.21 155	9.93 046	8	26
35	9.71 911	21	9.78 874	28	0.21 126	9.93 038	8	25
36	9.71 932	20	9.78 902	28	0.21 098	9.93 030	8	24
37	9.71 952	21	9.78 930	29	0.21 070	9.93 022	8	23
38	9.71 973	21	9.78 959	28	0.21 041	9.93 014	7	22
39	9.71 994	20	9.78 987	28	0.21 013	9.93 007	8	21
40	9.72 014	20	9.79 015	28	0.20 985	9.92 999	8	**20**
41	9.72 034	21	9.79 043	29	0.20 957	9.92 991	8	19
42	9.72 055	20	9.79 072	28	0.20 928	9.92 983	7	18
43	9.72 075	21	9.79 100	28	0.20 900	9.92 976	8	17
44	9.72 096	20	9.79 128	28	0.20 872	9.92 968	8	16
45	9.72 116	21	9.79 156	29	0.20 844	9.92 960	8	15
46	9.72 137	20	9.79 185	28	0.20 815	9.92 952	8	14
47	9.72 157	20	9.79 213	28	0.20 787	9.92 944	8	13
48	9.72 177	21	9.79 241	28	0.20 759	9.92 936	7	12
49	9.72 198	20	9.79 269	28	0.20 731	9.92 929	8	11
50	9.72 218	20	9.79 297	29	0.20 703	9.92 921	8	**10**
51	9.72 238	21	9.79 326	28	0.20 674	9.92 913	8	9
52	9.72 259	20	9.79 354	28	0.20 646	9.92 905	8	8
53	9.72 279	20	9.79 382	28	0.20 618	9.92 897	8	7
54	9.72 299	21	9.79 410	28	0.20 590	9.92 889	8	6
55	9.72 320	20	9.79 438	28	0.20 562	9.92 881	7	5
56	9.72 340	20	9.79 466	29	0.20 534	9.92 874	8	4
57	9.72 360	21	9.79 495	28	0.20 505	9.92 866	8	3
58	9.72 381	20	9.79 523	28	0.20 477	9.92 858	8	2
59	9.72 401	20	9.79 551	28	0.20 449	9.92 850	8	1
60	9.72 421		9.79 579		0.20 421	9.92 842		**0**
′	L. Cos.	d.	L. Cot.	c.d.	L. Tan.	L. Sin.	d.	′

121° (301°) — (238°) **58°**

P.P.

″	29	28
1	0.5	0.5
2	1.0	0.9
3	1.4	1.4
4	1.9	1.9
5	2.4	2.3
6	2.9	2.8
7	3.4	3.3
8	3.9	3.7
9	4.4	4.2
10	4.8	4.7
20	9.7	9.3
30	14.5	14.0
40	19.3	18.7
50	24.2	23.3

″	21	20
1	0.4	0.3
2	0.7	0.7
3	1.0	1.0
4	1.4	1.3
5	1.8	1.7
6	2.1	2.0
7	2.4	2.3
8	2.8	2.7
9	3.2	3.0
10	3.5	3.3
20	7.0	6.7
30	10.5	10.0
40	14.0	13.3
50	17.5	16.7

″	8	7
1	0.1	0.1
2	0.3	0.2
3	0.4	0.4
4	0.5	0.5
5	0.7	0.6
6	0.8	0.7
7	0.9	0.8
8	1.1	0.9
9	1.2	1.0
10	1.3	1.2
20	2.7	2.3
30	4.0	3.5
40	5.3	4.7
50	6.7	5.8

	8/30	8/29	8/28
0			
1	1.9	1.8	1.8
2	5.6	5.4	5.2
3	9.4	9.1	8.8
4	13.1	12.7	12.2
5	16.9	16.3	15.8
6	20.6	19.9	19.2
7	24.4	23.6	22.8
8	28.1	27.2	26.2

32° (212°) — (327°) **147°**

′	L. Sin.	d.	L. Tan.	c.d.	L. Cot.	L. Cos.	d.	′
0	9.72 421	20	9.79 579	28	0.20 421	9.92 842	8	**60**
1	9.72 441	20	9.79 607	28	0.20 393	9.92 834	8	59
2	9.72 461	21	9.79 635	28	0.20 365	9.92 826	8	58
3	9.72 482	20	9.79 663	28	0.20 337	9.92 818	8	57
4	9.72 502	20	9.79 691	28	0.20 309	9.92 810	7	56
5	9.72 522	20	9.79 719	28	0.20 281	9.92 803	8	55
6	9.72 542	20	9.79 747	29	0.20 253	9.92 795	8	54
7	9.72 562	20	9.79 776	28	0.20 224	9.92 787	8	53
8	9.72 582	20	9.79 804	28	0.20 196	9.92 779	8	52
9	9.72 602	20	9.79 832	28	0.20 168	9.92 771	8	51
10	9.72 622	21	9.79 860	28	0.20 140	9.92 763	8	**50**
11	9.72 643	20	9.79 888	28	0.20 112	9.92 755	8	49
12	9.72 663	20	9.79 916	28	0.20 084	9.92 747	8	48
13	9.72 683	20	9.79 944	28	0.20 056	9.92 739	8	47
14	9.72 703	20	9.79 972	28	0.20 028	9.92 731	8	46
15	9.72 723	20	9.80 000	28	0.20 000	9.92 723	8	45
16	9.72 743	20	9.80 028	28	0.19 972	9.92 715	8	44
17	9.72 763	20	9.80 056	28	0.19 944	9.92 707	8	43
18	9.72 783	20	9.80 084	28	0.19 916	9.92 699	8	42
19	9.72 803	20	9.80 112	28	0.19 888	9.92 691	8	41
20	9.72 823	20	9.80 140	28	0.19 860	9.92 683	8	**40**
21	9.72 843	20	9.80 168	27	0.19 832	9.92 675	8	39
22	9.72 863	20	9.80 195	28	0.19 805	9.92 667	8	38
23	9.72 883	19	9.80 223	28	0.19 777	9.92 659	8	37
24	9.72 902	20	9.80 251	28	0.19 749	9.92 651	8	36
25	9.72 922	20	9.80 279	28	0.19 721	9.92 643	8	35
26	9.72 942	20	9.80 307	28	0.19 693	9.92 635	8	34
27	9.72 962	20	9.80 335	28	0.19 665	9.92 627	8	33
28	9.72 982	20	9.80 363	28	0.19 637	9.92 619	8	32
29	9.73 002	20	9.80 391	28	0.19 609	9.92 611	8	31
30	9.73 022	19	9.80 419	28	0.19 581	9.92 603	8	**30**
31	9.73 041	20	9.80 447	27	0.19 553	9.92 595	8	29
32	9.73 061	20	9.80 474	28	0.19 526	9.92 587	8	28
33	9.73 081	20	9.80 502	28	0.19 498	9.92 579	8	27
34	9.73 101	20	9.80 530	28	0.19 470	9.92 571	8	26
35	9.73 121	19	9.80 558	28	0.19 442	9.92 563	8	25
36	9.73 140	20	9.80 586	28	0.19 414	9.92 555	9	24
37	9.73 160	20	9.80 614	28	0.19 386	9.92 546	8	23
38	9.73 180	20	9.80 642	27	0.19 358	9.92 538	8	22
39	9.73 200	19	9.80 669	28	0.19 331	9.92 530	8	21
40	9.73 219	20	9.80 697	28	0.19 303	9.92 522	8	**20**
41	9.73 239	20	9.80 725	28	0.19 275	9.92 514	8	19
42	9.73 259	19	9.80 753	28	0.19 247	9.92 506	8	18
43	9.73 278	20	9.80 781	27	0.19 219	9.92 498	8	17
44	9.73 298	20	9.80 808	28	0.19 192	9.92 490	8	16
45	9.73 318	19	9.80 836	28	0.19 164	9.92 482	9	15
46	9.73 337	20	9.80 864	28	0.19 136	9.92 473	8	14
47	9.73 357	20	9.80 892	27	0.19 108	9.92 465	8	13
48	9.73 377	19	9.80 919	28	0.19 081	9.92 457	8	12
49	9.73 396	20	9.80 947	28	0.19 053	9.92 449	8	11
50	9.73 416	19	9.80 975	28	0.19 025	9.92 441	8	**10**
51	9.73 435	20	9.81 003	27	0.18 997	9.92 433	8	9
52	9.73 455	19	9.81 030	28	0.18 970	9.92 425	9	8
53	9.73 474	20	9.81 058	28	0.18 942	9.92 416	8	7
54	9.73 494	19	9.81 086	27	0.18 914	9.92 408	8	6
55	9.73 513	20	9.81 113	28	0.18 887	9.92 400	8	5
56	9.73 533	19	9.81 141	28	0.18 859	9.92 392	8	4
57	9.73 552	20	9.81 169	27	0.18 831	9.92 384	8	3
58	9.73 572	19	9.81 196	28	0.18 804	9.92 376	9	2
59	9.73 591	20	9.81 224	28	0.18 776	9.92 367	8	1
60	9.73 611		9.81 252		0.18 748	9.92 359		**0**
′	L. Cos.	d.	L. Cot.	c.d.	L. Tan.	L. Sin.	d.	′

122° (302°) — (237°) **57°**

P. P.

″	29	28	27
1	0.5	0.5	0.4
2	1.0	0.9	0.9
3	1.4	1.4	1.4
4	1.9	1.9	1.8
5	2.4	2.3	2.2
6	2.9	2.8	2.7
7	3.4	3.3	3.2
8	3.9	3.7	3.6
9	4.4	4.2	4.0
10	4.8	4.7	4.5
20	9.7	9.3	9.0
30	14.5	14.0	13.5
40	19.3	18.7	18.0
50	24.2	23.3	22.5

″	21	20	19
1	0.4	0.3	0.3
2	0.7	0.7	0.6
3	1.0	1.0	1.0
4	1.4	1.3	1.3
5	1.8	1.7	1.6
6	2.1	2.0	1.9
7	2.4	2.3	2.2
8	2.8	2.7	2.5
9	3.2	3.0	2.8
10	3.5	3.3	3.2
20	7.0	6.7	6.3
30	10.5	10.0	9.5
40	14.0	13.3	12.7
50	17.5	16.7	15.8

″	9	8	7
1	0.2	0.1	0.1
2	0.3	0.3	0.2
3	0.4	0.4	0.4
4	0.6	0.5	0.5
5	0.8	0.7	0.6
6	0.9	0.8	0.7
7	1.0	0.9	0.8
8	1.2	1.1	0.9
9	1.4	1.2	1.0
10	1.5	1.3	1.2
20	3.0	2.7	2.3
30	4.5	4.0	3.5
40	6.0	5.3	4.7
50	7.5	6.7	5.8

	8/29	8/28	7/28
0			
1	1.8	1.8	2.0
2	5.4	5.2	6.0
3	9.1	8.8	10.0
4	12.7	12.2	14.0
5	16.3	15.8	18.0
6	19.9	19.2	22.0
7	23.6	22.8	26.0
8	27.2	26.2	—

LOGARITHMS OF THE FUNCTIONS (Continued)

33° (213°) — (326°) **146°**

′	L. Sin.	d.	L. Tan.	c.d.	L. Cot.	L. Cos.	d.	′
0	9.73 611	19	9.81 252	27	0.18 748	9.92 359	8	**60**
1	9.73 630	20	9.81 279	28	0.18 721	9.92 351	8	59
2	9.73 650	19	9.81 307	28	0.18 693	9.92 343	8	58
3	9.73 669	20	9.81 335	27	0.18 665	9.92 335	9	57
4	9.73 689	19	9.81 362	28	0.18 638	9.92 326	8	56
5	9.73 708	19	9.81 390	28	0.18 610	9.92 318	8	55
6	9.73 727	20	9.81 418	27	0.18 582	9.92 310	8	54
7	9.73 747	19	9.81 445	28	0.18 555	9.92 302	9	53
8	9.73 766	19	9.81 473	27	0.18 527	9.92 293	8	52
9	9.73 785	20	9.81 500	28	0.18 500	9.92 285	8	51
10	9.73 805	19	9.81 528	28	0.18 472	9.92 277	8	**50**
11	9.73 824	19	9.81 556	27	0.18 444	9.92 269	9	49
12	9.73 843	20	9.81 583	28	0.18 417	9.92 260	8	48
13	9.73 863	19	9.81 611	27	0.18 389	9.92 252	8	47
14	9.73 882	19	9.81 638	28	0.18 362	9.92 244	9	46
15	9.73 901	20	9.81 666	27	0.18 334	9.92 235	8	45
16	9.73 921	19	9.81 693	28	0.18 307	9.92 227	8	44
17	9.73 940	19	9.81 721	27	0.18 279	9.92 219	8	43
18	9.73 959	19	9.81 748	28	0.18 252	9.92 211	9	42
19	9.73 978	19	9.81 776	27	0.18 224	9.92 202	8	41
20	9.73 997	20	9.81 803	28	0.18 197	9.92 194	8	**40**
21	9.74 017	19	9.81 831	27	0.18 169	9.92 186	9	39
22	9.74 036	19	9.81 858	28	0.18 142	9.92 177	8	38
23	9.74 055	19	9.81 886	27	0.18 114	9.92 169	8	37
24	9.74 074	19	9.81 913	28	0.18 087	9.92 161	9	36
25	9.74 093	20	9.81 941	27	0.18 059	9.92 152	8	35
26	9.74 113	19	9.81 968	28	0.18 032	9.92 144	8	34
27	9.74 132	19	9.81 996	27	0.18 004	9.92 136	9	33
28	9.74 151	19	9.82 023	28	0.17 977	9.92 127	8	32
29	9.74 170	19	9.82 051	27	0.17 949	9.92 119	8	31
30	9.74 189	19	9.82 078	28	0.17 922	9.92 111	9	**30**
31	9.74 208	19	9.82 106	27	0.17 894	9.92 102	8	29
32	9.74 227	19	9.82 133	28	0.17 867	9.92 094	8	28
33	9.74 246	19	9.82 161	27	0.17 839	9.92 086	9	27
34	9.74 265	19	9.82 188	27	0.17 812	9.92 077	8	26
35	9.74 284	19	9.82 215	28	0.17 785	9.92 069	9	25
36	9.74 303	19	9.82 243	27	0.17 757	9.92 060	8	24
37	9.74 322	19	9.82 270	28	0.17 730	9.92 052	8	23
38	9.74 341	19	9.82 298	27	0.17 702	9.92 044	9	22
39	9.74 360	19	9.82 325	27	0.17 675	9.92 035	8	21
40	9.74 379	19	9.82 352	28	0.17 648	9.92 027	9	**20**
41	9.74 398	19	9.82 380	27	0.17 620	9.92 018	8	19
42	9.74 417	19	9.82 407	28	0.17 593	9.92 010	8	18
43	9.74 436	19	9.82 435	27	0.17 565	9.92 002	9	17
44	9.74 455	19	9.82 462	27	0.17 538	9.91 993	8	16
45	9.74 474	19	9.82 489	28	0.17 511	9.91 985	9	15
46	9.74 493	19	9.82 517	27	0.17 483	9.91 976	8	14
47	9.74 512	19	9.82 544	27	0.17 456	9.91 968	9	13
48	9.74 531	18	9.82 571	28	0.17 429	9.91 959	8	12
49	9.74 549	19	9.82 599	27	0.17 401	9.91 951	9	11
50	9.74 568	19	9.82 626	27	0.17 374	9.91 942	8	**10**
51	9.74 587	19	9.82 653	28	0.17 347	9.91 934	9	9
52	9.74 606	19	9.82 681	27	0.17 319	9.91 925	8	8
53	9.74 625	19	9.82 708	27	0.17 292	9.91 917	9	7
54	9.74 644	18	9.82 735	27	0.17 265	9.91 908	8	6
55	9.74 662	19	9.82 762	28	0.17 238	9.91 900	9	5
56	9.74 681	19	9.82 790	27	0.17 210	9.91 891	8	4
57	9.74 700	19	9.82 817	27	0.17 183	9.91 883	9	3
58	9.74 719	18	9.82 844	27	0.17 156	9.91 874	8	2
59	9.74 737	19	9.82 871	28	0.17 129	9.91 866	9	1
60	9.74 756		9.82 899		0.17 101	9.91 857		**0**
′	L. Cos.	d.	L. Cot.	c.d.	L. Tan	L. Sin.	d.	′

123° (303°) — (236°) **56°**

P. P.

″	28	27
1	0.5	0.4
2	0.9	0.9
3	1.4	1.4
4	1.9	1.8
5	2.3	2.2
6	2.8	2.7
7	3.3	3.2
8	3.7	3.6
9	4.2	4.0
10	4.7	4.5
20	9.3	9.0
30	14.0	13.5
40	18.7	18.0
50	23.3	22.5

″	20	19	18
1	0.3	0.3	0.3
2	0.7	0.6	0.6
3	1.0	1.0	0.9
4	1.3	1.3	1.2
5	1.7	1.6	1.5
6	2.0	1.9	1.8
7	2.3	2.2	2.1
8	2.7	2.5	2.4
9	3.0	2.8	2.7
10	3.3	3.2	3.0
20	6.7	6.3	6.0
30	10.0	9.5	9.0
40	13.3	12.7	12.0
50	16.7	15.8	15.0

″	9	8
1	0.2	0.1
2	0.3	0.3
3	0.4	0.4
4	0.6	0.5
5	0.8	0.7
6	0.9	0.8
7	1.0	0.9
8	1.2	1.1
9	1.4	1.2
10	1.5	1.3
20	3.0	2.7
30	4.5	4.0
40	6.0	5.3
50	7.5	6.7

	9/28	9/27	8/27
0			
1	1.6	1.5	1.7
2	4.7	4.5	5.1
3	7.8	7.5	8.4
4	10.9	10.5	11.8
5	14.0	13.5	15.2
6	17.1	16.5	18.6
7	20.2	19.5	21.9
8	23.3	22.5	25.3
9	26.4	25.5	—

34° (214°) — (325°) **145°**

′	L. Sin.	d.	L. Tan.	c.d.	L. Cot.	L. Cos.	d.	′
0	9.74 756	19	9.82 899	27	0.17 101	9.91 857	8	**60**
1	9.74 775	19	9.82 926	27	0.17 074	9.91 849	9	59
2	9.74 794	18	9.82 953	27	0.17 047	9.91 840	8	58
3	9.74 812	19	9.82 980	28	0.17 020	9.91 832	9	57
4	9.74 831	19	9.83 008	27	0.16 992	9.91 823	8	56
5	9.74 850	18	9.83 035	27	0.16 965	9.91 815	9	55
6	9.74 868	19	9.83 062	27	0.16 938	9.91 806	8	54
7	9.74 887	19	9.83 089	28	0.16 911	9.91 798	9	53
8	9.74 906	18	9.83 117	27	0.16 883	9.91 789	8	52
9	9.74 924	19	9.83 144	27	0.16 856	9.91 781	9	51
10	9.74 943	18	9.83 171	27	0.16 829	9.91 772	9	**50**
11	9.74 961	19	9.83 198	27	0.16 802	9.91 763	8	49
12	9.74 980	19	9.83 225	27	0.16 775	9.91 755	9	48
13	9.74 999	18	9.83 252	28	0.16 748	9.91 746	8	47
14	9.75 017	19	9.83 280	27	0.16 720	9.91 738	9	46
15	9.75 036	18	9.83 307	27	0.16 693	9.91 729	9	45
16	9.75 054	19	9.83 334	27	0.16 666	9.91 720	8	44
17	9.75 073	18	9.83 361	27	0.16 639	9.91 712	9	43
18	9.75 091	19	9.83 388	27	0.16 612	9.91 703	8	42
19	9.75 110	18	9.83 415	27	0.16 585	9.91 695	9	41
20	9.75 128	19	9.83 442	28	0.16 558	9.91 686	9	**40**
21	9.75 147	18	9.83 470	27	0.16 530	9.91 677	8	39
22	9.75 165	19	9.83 497	27	0.16 503	9.91 669	9	38
23	9.75 184	18	9.83 524	27	0.16 476	9.91 660	9	37
24	9.75 202	19	9.83 551	27	0.16 449	9.91 651	8	36
25	9.75 221	18	9.83 578	27	0.16 422	9.91 643	9	35
26	9.75 239	19	9.83 605	27	0.16 395	9.91 634	9	34
27	9.75 258	18	9.83 632	27	0.16 368	9.91 625	8	33
28	9.75 276	18	9.83 659	27	0.16 341	9.91 617	9	32
29	9.75 294	19	9.83 686	27	0.16 314	9.91 608	9	31
30	9.75 313	18	9.83 713	27	0.16 287	9.91 599	8	**30**
31	9.75 331	19	9.83 740	28	0.16 260	9.91 591	9	29
32	9.75 350	18	9.83 768	27	0.16 232	9.91 582	9	28
33	9.75 368	18	9.83 795	27	0.16 205	9.91 573	8	27
34	9.75 386	19	9.83 822	27	0.16 178	9.91 565	9	26
35	9.75 405	18	9.83 849	27	0.16 151	9.91 556	9	25
36	9.75 423	18	9.83 876	27	0.16 124	9.91 547	9	24
37	9.75 441	18	9.83 903	27	0.16 097	9.91 538	8	23
38	9.75 459	19	9.83 930	27	0.16 070	9.91 530	9	22
39	9.75 478	18	9.83 957	27	0.16 043	9.91 521	9	21
40	9.75 496	18	9.83 984	27	0.16 016	9.91 512	8	**20**
41	9.75 514	19	9.84 011	27	0.15 989	9.91 504	9	19
42	9.75 533	18	9.84 038	27	0.15 962	9.91 495	9	18
43	9.75 551	18	9.84 065	27	0.15 935	9.91 486	9	17
44	9.75 569	18	9.84 092	27	0.15 908	9.91 477	8	16
45	9.75 587	18	9.84 119	27	0.15 881	9.91 469	9	15
46	9.75 605	19	9.84 146	27	0.15 854	9.91 460	9	14
47	9.75 624	18	9.84 173	27	0.15 827	9.91 451	9	13
48	9.75 642	18	9.84 200	27	0.15 800	9.91 442	9	12
49	9.75 660	18	9.84 227	27	0.15 773	9.91 433	8	11
50	9.75 678	18	9.84 254	26	0.15 746	9.91 425	9	**10**
51	9.75 696	18	9.84 280	27	0.15 720	9.91 416	9	9
52	9.75 714	19	9.84 307	27	0.15 693	9.91 407	9	8
53	9.75 733	18	9.84 334	27	0.15 666	9.91 398	9	7
54	9.75 751	18	9.84 361	27	0.15 639	9.91 389	8	6
55	9.75 769	18	9.84 388	27	0.15 612	9.91 381	9	5
56	9.75 787	18	9.84 415	27	0.15 585	9.91 372	9	4
57	9.75 805	18	9.84 442	27	0.15 558	9.91 363	9	3
58	9.75 823	18	9.84 469	27	0.15 531	9.91 354	9	2
59	9.75 841	18	9.84 496	27	0.15 504	9.91 345	9	1
60	9.75 859		9.84 523		0.15 477	9.91 336		**0**
′	L. Cos.	d.	L. Cot.	c.d.	L. Tan.	L. Sin.	d.	′

124° (304°) — (235°) **55°**

P.P.

″	28	27	26
1	0.5	0.4	0.4
2	0.9	0.9	0.9
3	1.4	1.4	1.3
4	1.9	1.8	1.7
5	2.3	2.2	2.2
6	2.8	2.7	2.6
7	3.3	3.2	3.0
8	3.7	3.6	3.5
9	4.2	4.0	3.9
10	4.7	4.5	4.3
20	9.3	9.0	8.7
30	14.0	13.5	13.0
40	18.7	18.0	17.3
50	23.3	22.5	21.7

″	19	18
1	0.3	0.3
2	0.6	0.6
3	1.0	0.9
4	1.3	1.2
5	1.6	1.5
6	1.9	1.8
7	2.2	2.1
8	2.5	2.4
9	2.8	2.7
10	3.2	3.0
20	6.3	6.0
30	9.5	9.0
40	12.7	12.0
50	15.8	15.0

″	9	8
1	0.2	0.1
2	0.3	0.3
3	0.4	0.4
4	0.6	0.5
5	0.8	0.7
6	0.9	0.8
7	1.0	0.9
8	1.2	1.1
9	1.4	1.2
10	1.5	1.3
20	3.0	2.7
30	4.5	4.0
40	6.0	5.3
50	7.5	6.7

	9/28	8/28	8/27
0			
1	1.6	1.8	1.7
2	4.7	5.2	5.1
3	7.8	8.8	8.4
4	10.9	12.2	11.8
5	14.0	15.8	15.2
6	17.1	19.2	18.6
7	20.2	22.8	21.9
8	23.3	26.2	25.3
9	26.4	—	—

LOGARITHMS OF THE FUNCTIONS (Continued)

35° (215°) (324°) **144°**

′	L. Sin.	d.	L. Tan.	c.d.	L. Cot.	L. Cos.	d.	′
0	9.75 859	18	9.84 523	27	0.15 477	9.91 336	8	**60**
1	9.75 877	18	9.84 550	26	0.15 450	9.91 328	9	59
2	9.75 895	18	9.84 576	27	0.15 424	9.91 319	9	58
3	9.75 913	18	9.84 603	27	0.15 397	9.91 310	9	57
4	9.75 931	18	9.84 630	27	0.15 370	9.91 301	9	56
5	9.75 949	18	9.84 657	27	0.15 343	9.91 292	9	55
6	9.75 967	18	9.84 684	27	0.15 316	9.91 283	9	54
7	9.75 985	18	9.84 711	27	0.15 289	9.91 274	8	53
8	9.76 003	18	9.84 738	26	0.15 262	9.91 266	9	52
9	9.76 021	18	9.84 764	27	0.15 236	9.91 257	9	51
10	9.76 039	18	9.84 791	27	0.15 209	9.91 248	9	**50**
11	9.76 057	18	9.84 818	27	0.15 182	9.91 239	9	49
12	9.76 075	18	9.84 845	27	0.15 155	9.91 230	9	48
13	9.76 093	18	9.84 872	27	0.15 128	9.91 221	9	47
14	9.76 111	18	9.84 899	26	0.15 101	9.91 212	9	46
15	9.76 129	17	9.84 925	27	0.15 075	9.91 203	9	45
16	9.76 146	18	9.84 952	27	0.15 048	9.91 194	9	44
17	9.76 164	18	9.84 979	27	0.15 021	9.91 185	9	43
18	9.76 182	18	9.85 006	27	0.14 994	9.91 176	9	42
19	9.76 200	18	9.85 033	26	0.14 967	9.91 167	9	41
20	9.76 218	18	9.85 059	27	0.14 941	9.91 158	9	**40**
21	9.76 236	17	9.85 086	27	0.14 914	9.91 149	8	39
22	9.76 253	18	9.85 113	27	0.14 887	9.91 141	9	38
23	9.76 271	18	9.85 140	26	0.14 860	9.91 132	9	37
24	9.76 289	18	9.85 166	27	0.14 834	9.91 123	9	36
25	9.76 307	17	9.85 193	27	0.14 807	9.91 114	9	35
26	9.76 324	18	9.85 220	27	0.14 780	9.91 105	9	34
27	9.76 342	18	9.85 247	26	0.14 753	9.91 096	9	33
28	9.76 360	18	9.85 273	27	0.14 727	9.91 087	9	32
29	9.76 378	17	9.85 300	27	0.14 700	9.91 078	9	31
30	9.76 395	18	9.85 327	27	0.14 673	9.91 069	9	**30**
31	9.76 413	18	9.85 354	26	0.14 646	9.91 060	9	29
32	9.76 431	17	9.85 380	27	0.14 620	9.91 051	9	28
33	9.76 448	18	9.85 407	27	0.14 593	9.91 042	9	27
34	9.76 466	18	9.85 434	26	0.14 566	9.91 033	10	26
35	9.76 484	17	9.85 460	27	0.14 540	9.91 023	9	25
36	9.76 501	18	9.85 487	27	0.14 513	9.91 014	9	24
37	9.76 519	18	9.85 514	26	0.14 486	9.91 005	9	23
38	9.76 537	17	9.85 540	27	0.14 460	9.90 996	9	22
39	9.76 554	18	9.85 567	27	0.14 433	9.90 987	9	21
40	9.76 572	18	9.85 594	26	0.14 406	9.90 978	9	**20**
41	9.76 590	17	9.85 620	27	0.14 380	9.90 969	9	19
42	9.76 607	18	9.85 647	27	0.14 353	9.90 960	9	18
43	9.76 625	17	9.85 674	26	0.14 326	9.90 951	9	17
44	9.76 642	18	9.85 700	27	0.14 300	9.90 942	9	16
45	9.76 660	17	9.85 727	27	0.14 273	9.90 933	9	15
46	9.76 677	18	9.85 754	26	0.14 246	9.90 924	9	14
47	9.76 695	17	9.85 780	27	0.14 220	9.90 915	9	13
48	9.76 712	18	9.85 807	27	0.14 193	9.90 906	10	12
49	9.76 730	17	9.85 834	26	0.14 166	9.90 896	9	11
50	9.76 747	18	9.85 860	27	0.14 140	9.90 887	9	**10**
51	9.76 765	17	9.85 887	26	0.14 113	9.90 878	9	9
52	9.76 782	18	9.85 913	27	0.14 087	9.90 869	9	8
53	9.76 800	17	9.85 940	27	0.14 060	9.90 860	9	7
54	9.76 817	18	9.85 967	26	0.14 033	9.90 851	9	6
55	9.76 835	17	9.85 993	27	0.14 007	9.90 842	10	5
56	9.76 852	18	9.86 020	26	0.13 980	9.90 832	9	4
57	9.76 870	17	9.86 046	27	0.13 954	9.90 823	9	3
58	9.76 887	17	9.86 073	27	0.13 927	9.90 814	9	2
59	9.76 904	18	9.86 100	26	0.13 900	9.90 805	9	1
60	9.76 922		9.86 126		0.13 874	9.90 796		**0**
′	L. Cos.	d.	L. Cot.	c.d.	L. Tan.	L. Sin.	d.	′

125° (305°) (234°) **54°**

P.P.

″	**27**	**26**	**18**
1	0.4	0.4	0.3
2	0.9	0.9	0.6
3	1.4	1.3	0.9
4	1.8	1.7	1.2
5	2.2	2.2	1.5
6	2.7	2.6	1.8
7	3.2	3.0	2.1
8	3.6	3.5	2.4
9	4.0	3.9	2.7
10	4.5	4.3	3.0
20	9.0	8.7	6.0
30	13.5	13.0	9.0
40	18.0	17.3	12.0
50	22.5	21.7	15.0

″	**17**	**10**	**9**	**8**
1	0.3	0.2	0.2	0.1
2	0.6	0.3	0.3	0.3
3	0.8	0.5	0.4	0.4
4	1.1	0.7	0.6	0.5
5	1.4	0.8	0.8	0.7
6	1.7	1.0	0.9	0.8
7	2.0	1.2	1.0	0 9
8	2.3	1.3	1.2	1.1
9	2.6	1.5	1.4	1.2
10	2.8	1.7	1.5	1.3
20	5.7	3.3	3.0	2.7
30	8.5	5.0	4.5	4.0
40	11.3	6.7	6.0	5.3
50	14.2	8.3	7.5	6.7

	10/27	**10/26**
0	1.4	1.3
1	4.1	3.9
2	6.8	6.5
3	9.4	9.1
4	12.2	11.7
5	14.8	14.3
6	17.6	16.9
7	20.2	19.5
8	22.9	22.1
9	25.6	24.7
10		

	9/27	**9/26**
0	1.5	1.4
1	4.5	4.3
2	7.5	7.2
3	10.5	10.1
4	13.5	13.0
5	16.5	15.9
6	19.5	18.8
7	22.5	21.7
8	25.5	24.6
9		

36° (216°) (323°) **143°**

′	L. Sin.	d.	L. Tan.	c.d.	L. Cot.	L. Cos.	d.	′
0	9.76 922	17	9.86 126	27	0.13 874	9.90 796	9	**60**
1	9.76 939	18	9.86 153	26	0.13 847	9.90 787	10	59
2	9.76 957	17	9.86 179	27	0.13 821	9.90 777	9	58
3	9.76 974	17	9.86 206	26	0.13 794	9.90 768	9	57
4	9.76 991	18	9.86 232	27	0.13 768	9.90 759	9	56
5	9.77 009	17	9.86 259	26	0.13 741	9.90 750	9	55
6	9.77 026	17	9.86 285	27	0.13 715	9.90 741	10	54
7	9.77 043	18	9.86 312	26	0.13 688	9.90 731	9	53
8	9.77 061	17	9.86 338	27	0.13 662	9.90 722	9	52
9	9.77 078	17	9.86 365	27	0.13 635	9.90 713	9	51
10	9.77 095	17	9.86 392	26	0.13 608	9.90 704	10	**50**
11	9.77 112	18	9.86 418	27	0.13 582	9.90 694	9	49
12	9.77 130	17	9.86 445	26	0.13 555	9.90 685	9	48
13	9.77 147	17	9.86 471	27	0.13 529	9.90 676	9	47
14	9.77 164	17	9.86 498	26	0.13 502	9.90 667	10	46
15	9.77 181	18	9.86 524	27	0.13 476	9.90 657	9	45
16	9.77 199	17	9.86 551	26	0.13 449	9.90 648	9	44
17	9.77 216	17	9.86 577	26	0.13 423	9.90 639	9	43
18	9.77 233	17	9.86 603	27	0.13 397	9.90 630	10	42
19	9.77 250	18	9.86 630	26	0.13 370	9.90 620	9	41
20	9.77 268	17	9.86 656	27	0.13 344	9.90 611	9	**40**
21	9.77 285	17	9.86 683	26	0.13 317	9.90 602	10	39
22	9.77 302	17	9.86 709	27	0.13 291	9.90 592	9	38
23	9.77 319	17	9.86 736	26	0.13 264	9.90 583	9	37
24	9.77 336	17	9.86 762	27	0.13 238	9.90 574	9	36
25	9.77 353	17	9.86 789	26	0.13 211	9.90 565	10	35
26	9.77 370	17	9.86 815	27	0.13 185	9.90 555	9	34
27	9.77 387	18	9.86 842	26	0.13 158	9.90 546	9	33
28	9.77 405	17	9.86 868	26	0.13 132	9.90 537	10	32
29	9.77 422	17	9.86 894	27	0.13 106	9.90 527	9	31
30	9.77 439	17	9.86 921	26	0.13 079	9.90 518	9	**30**
31	9.77 456	17	9.86 947	27	0.13 053	9.90 509	10	29
32	9.77 473	17	9.86 974	26	0.13 026	9.90 499	9	28
33	9.77 490	17	9.87 000	27	0.13 000	9.90 490	10	27
34	9.77 507	17	9.87 027	26	0.12 973	9.90 480	9	26
35	9.77 524	17	9.87 053	26	0.12 947	9.90 471	9	25
36	9.77 541	17	9.87 079	27	0.12 921	9.90 462	10	24
37	9.77 558	17	9.87 106	26	0.12 894	9.90 452	9	23
38	9.77 575	17	9.87 132	26	0.12 868	9.90 443	9	22
39	9.77 592	17	9.87 158	27	0.12 842	9.90 434	10	21
40	9.77 609	17	9.87 185	26	0.12 815	9.90 424	9	**20**
41	9.77 626	17	9.87 211	27	0.12 789	9.90 415	10	19
42	9.77 643	17	9.87 238	26	0.12 762	9.90 405	9	18
43	9.77 660	17	9.87 264	26	0.12 736	9.90 396	10	17
44	9.77 677	17	9.87 290	27	0.12 710	9.90 386	9	16
45	9.77 694	17	9.87 317	26	0.12 683	9.90 377	9	15
46	9.77 711	17	9.87 343	26	0.12 657	9.90 368	10	14
47	9.77 728	16	9.87 369	27	0.12 631	9.90 358	9	13
48	9.77 744	17	9.87 396	26	0.12 604	9.90 349	10	12
49	9.77 761	17	9.87 422	26	0.12 578	9.90 339	9	11
50	9.77 778	17	9.87 448	27	0.12 552	9.90 330	10	**10**
51	9.77 795	17	9.87 475	26	0.12 525	9.90 320	9	9
52	9.77 812	17	9.87 501	26	0.12 499	9.90 311	10	8
53	9.77 829	17	9.87 527	27	0.12 473	9.90 301	9	7
54	9.77 846	16	9.87 554	26	0.12 446	9.90 292	10	6
55	9.77 862	17	9.87 580	26	0.12 420	9.90 282	9	5
56	9.77 879	17	9.87 606	27	0.12 394	9.90 273	10	4
57	9.77 896	17	9.87 633	26	0.12 367	9.90 263	9	3
58	9.77 913	17	9.87 659	26	0.12 341	9.90 254	10	2
59	9.77 930	16	9.87 685	26	0.12 315	9.90 244	9	1
60	9.77 946		9.87 711		0.12 289	9.90 235		**0**
′	L. Cos.	d.	L. Cot.	c.d.	L. Tan.	L. Sin.	d.	′

126° (306°) (233°) **53°**

P. P.

″	**27**	**26**
1	0.4	0.4
2	0.9	0.9
3	1.4	1.3
4	1.8	1.7
5	2.2	2.2
6	2.7	2.6
7	3.2	3.0
8	3.6	3.5
9	4.0	3.9
10	4.5	4.3
20	9.0	8.7
30	13.5	13.0
40	18.0	17.3
50	22.5	21.7

″	**18**	**17**	**16**
1	0.3	0.3	0.3
2	0.6	0.6	0.5
3	0.9	0.8	0.8
4	1.2	1.1	1.1
5	1.5	1.4	1.3
6	1.8	1.7	1.6
7	2.1	2.0	1.9
8	2.4	2.3	2.1
9	2.7	2.6	2.4
10	3.0	2.8	2.7
20	6.0	5.7	5.3
30	9.0	8.5	8.0
40	12.0	11.3	10.7
50	15.0	14.2	13.3

″	**10**	**9**
1	0.2	0.2
2	0.3	0.3
3	0.5	0.4
4	0.7	0.6
5	0.8	0.8
6	1.0	0.9
7	1.2	1.0
8	1.3	1.2
9	1.5	1.4
10	1.7	1.5
20	3.3	3.0
30	5.0	4.5
40	6.7	6.0
50	8.3	7.5

	9/27	**9/26**
0	1.5	1.4
1	4.5	4.3
2	7.5	7.2
3	10.5	10.1
4	13.5	13.0
5	16.5	15.9
6	19.5	18.8
7	22.5	21.7
8	25.5	24.6
9		

LOGARITHMS OF THE FUNCTIONS (Continued)

37° (217°) — (322°) **142°**

′	L. Sin	d.	L. Tan.	c.d.	L. Cot.	L. Cos.	d.	′
0	9.77 946	17	9.87 711	27	0.12 289	9.90 235	10	**60**
1	9.77 963	17	9.87 738	26	0.12 262	9.90 225	9	59
2	9.77 980	17	9.87 764	26	0.12 236	9.90 216	10	58
3	9.77 997	16	9.87 790	27	0.12 210	9.90 206	9	57
4	9.78 013	17	9.87 817	26	0.12 183	9.90 197	10	56
5	9.78 030	17	9.87 843	26	0.12 157	9.90 187	9	55
6	9.78 047	16	9.87 869	26	0.12 131	9.90 178	10	54
7	9.78 063	17	9.87 895	27	0.12 105	9.90 168	9	53
8	9.78 080	17	9.87 922	26	0.12 078	9.90 159	10	52
9	9.78 097	16	9.87 948	26	0.12 052	9.90 149	10	51
10	9.78 113	17	9.87 974	26	0.12 026	9.90 139	9	**50**
11	9.78 130	17	9.88 000	27	0.12 000	9.90 130	10	49
12	9.78 147	16	9.88 027	26	0.11 973	9.90 120	9	48
13	9.78 163	17	9.88 053	26	0.11 947	9.90 111	10	47
14	9.78 180	17	9.88 079	26	0.11 921	9.90 101	10	46
15	9.78 197	16	9.88 105	26	0.11 895	9.90 091	9	45
16	9.78 213	17	9.88 131	27	0.11 869	9.90 082	10	44
17	0.78 230	16	9.88 158	26	0.11 842	9.90 072	9	43
18	9.78 246	17	9.88 184	26	0.11 816	9.90 063	10	42
19	9.78 263	17	9.88 210	26	0.11 790	9.90 053	10	41
20	9.78 280	16	9.88 236	26	0.11 764	9.90 043	9	**40**
21	9.78 296	17	9.88 262	27	0.11 738	9.90 034	10	39
22	9.78 313	16	9.88 289	26	0.11 711	9.90 024	10	38
13	9.78 329	17	9.88 315	26	0.11 685	9.90 014	9	37
24	9.78 346	16	9.88 341	26	0.11 659	9.90 005	10	36
25	9.78 362	17	9.88 367	26	0.11 633	9.89 995	10	35
26	9.78 379	16	9.88 393	27	0.11 607	9.89 985	9	34
27	9.78 395	17	9.88 420	26	0.11 580	9.89 976	10	33
28	9.78 412	16	9.88 446	26	0.11 554	9.89 966	10	32
29	9.78 428	17	9.88 472	26	0.11 528	9.89 956	9	31
30	9.78 445	16	9.88 498	26	0.11 502	9.89 947	10	**30**
31	9.78 461	17	9.88 524	26	0.11 476	9.89 937	10	29
32	9.78 478	16	9.88 550	27	0.11 450	9.89 927	9	28
33	9.78 494	16	9.88 577	26	0.11 423	9.89 918	10	27
34	9.78 510	17	9.88 603	26	0.11 397	9.89 908	10	26
35	9.78 527	16	9.88 629	26	0.11 371	9.89 898	10	25
36	9.78 543	17	9.88 655	26	0.11 345	9.89 888	9	24
37	9.78 560	16	9.88 681	26	0.11 319	9.89 879	10	23
38	9.78 576	16	9.88 707	26	0.11 293	9.89 869	10	22
39	9.78 592	17	9.88 733	26	0.11 267	9.89 859	10	21
40	9.78 609	16	9.88 759	27	0.11 241	9.89 849	9	**20**
41	9.78 625	17	9.88 786	26	0.11 214	9.89 840	10	19
42	9.78 642	16	9.88 812	26	0.11 188	9.89 830	10	18
43	9.78 658	16	9.88 838	26	0.11 162	9.89 820	10	17
44	9.78 674	17	9.88 864	26	0.11 136	9.80 810	9	16
45	9.78 691	16	9.88 890	26	0.11 110	9.89 801	10	15
46	9.78 707	16	9.88 916	26	0.11 084	9.89 791	10	14
47	9.78 723	16	9.88 942	26	0.11 058	9.89 781	10	13
48	9.78 739	17	9.88 968	26	0.11 032	9.89 771	10	12
49	9.78 756	16	9.88 994	26	0.11 006	9.89 761	9	11
50	9.78 772	16	9.89 020	26	0.10 980	9.89 752	10	**10**
51	9.78 788	17	9.89 046	27	0.10 954	9.89 742	10	9
52	9.78 805	16	9.89 073	26	0.10 927	9.89 732	10	8
53	9.78 821	16	9.89 099	26	0.10 901	9.89 722	10	7
54	9.78 837	16	9.89 125	26	0.10 875	9.89 712	10	6
55	9.78 853	16	9.89 151	26	0.10 849	9.89 702	9	5
56	9.78 869	17	9.89 177	26	0.10 823	9.89 693	10	4
57	9.78 886	16	9.89 203	26	0.10 797	9.89 683	10	3
58	9.78 902	16	9.89 229	26	0.10 771	9.89 673	10	2
59	9.78 918	16	9.89 255	26	0.10 745	9.89 663	10	1
60	9.78 934		9.89 281		0.10 719	9.89 653		**0**
′	L. Cos.	d.	L. Cot.	c.d.	L. Tan.	L. Sin.	d.	′

127° (307°) — (232°) **52°**

P. P.

″	**27**	**26**
1	0.4	0.4
2	0.9	0.9
3	1.4	1.3
4	1.8	1.7
5	2.2	2.2
6	2.7	2.6
7	3.2	3.0
8	3.6	3.5
9	4.0	3.9
10	4.5	4.3
20	9.0	8.7
30	13.5	13.0
40	18.0	17.3
50	22.5	21.7

″	**17**	**16**
1	0.3	0.3
1	0.6	0.5
3	0.8	0.8
4	1.1	1.1
5	1.4	1.3
6	1.7	1.6
7	2.0	1.9
8	2.3	2.1
9	2.6	2.4
10	2.8	2.7
20	5.7	5.3
30	8.5	8.0
40	11.3	10.7
50	14.2	13.3

″	**10**	**9**
1	0.2	0.2
2	0.3	0.3
3	0.5	0.4
4	0.7	0.6
5	0.8	0.8
6	1.0	0.9
7	1.2	1.0
8	1.3	1.2
9	1.5	1.4
10	1.7	1.5
20	3.3	3.0
30	5.0	4.5
40	6.7	6.0
50	8.3	7.5

	10/27	**10/26**
0	1.4	1.3
1	4.1	3.9
2	6.8	6.5
3	9.4	9.1
4	12.2	11.7
5	14.8	14.3
6	17.6	16.9
7	20.2	19.5
8	22.9	22.1
9	25.6	24.7
10		

38° (218°) — (321°) **141°**

′	L. Sin.	d.	L. Tan.	c.d.	L. Cot.	L. Cos.	d.	′
0	9.78 934	16	9.89 281	26	0.10 719	9.89 653	10	**60**
1	9.78 950	17	9.89 307	26	0.10 693	9.89 643	10	59
2	9.78 967	16	9.89 333	26	0.10 667	9.89 633	9	58
3	9.78 983	16	9.89 359	26	0.10 641	9.89 624	10	57
4	9.78 999	16	9.89 385	26	0.10 615	9.89 614	10	56
5	9.79 015	16	9.89 411	26	0.10 589	9.89 604	10	55
6	9.79 031	16	9.89 437	26	0.10 563	9.89 594	10	54
7	9.79 047	16	9.89 463	26	0.10 537	9.89 584	10	53
8	9.79 063	16	9.89 489	26	0.10 511	9.89 574	10	52
9	9.79 079	16	9.89 515	26	0.10 485	9.89 564	10	51
10	9.79 095	16	9.89 541	26	0.10 459	9.89 554	10	**50**
11	9.79 111	17	9.89 567	26	0.10 433	9.89 544	10	49
12	9.79 128	16	9.89 593	26	0.10 407	9.89 534	10	48
13	9.79 144	16	9.89 619	26	0.10 381	9.89 524	10	47
14	9.79 160	16	9.89 645	26	0.10 355	9.89 514	10	46
15	9.79 176	16	9.89 671	26	0.10 329	9.89 504	9	45
16	9.79 192	16	9.89 697	26	0.10 303	9.89 495	10	44
17	9.79 208	16	9.89 723	26	0.10 277	9.89 485	10	43
18	9.79 224	16	9.89 749	26	0.10 251	9.89 475	10	42
19	9.79 240	16	9.89 775	26	0.10 225	9.89 465	10	41
20	9.79 256	16	9.89 801	26	0.10 199	9.89 455	10	**40**
21	9.79 272	16	9.89 827	26	0.10 173	9.89 445	10	39
22	9.79 288	16	9.89 853	26	0.10 147	9.89 435	10	38
23	9.79 304	15	9.89 879	26	0.10 121	9.89 425	10	37
24	9.79 319	16	9.89 905	26	0.10 095	9.89 415	10	36
25	9.79 335	16	9.89 931	26	0.10 069	9.89 405	10	35
26	9.79 351	16	9.89 957	26	0.10 043	9.89 395	10	34
27	9.79 367	16	9.89 983	26	0.10 017	9.89 385	10	33
28	9.79 383	16	9.90 009	26	0.09 991	9.89 375	11	32
29	9.79 399	16	9.90 035	26	0.09 965	9.89 364	10	31
30	9.79 415	16	9.90 061	25	0.09 939	9.89 354	10	**30**
31	9.79 431	16	9.90 086	26	0.09 914	9.89 344	10	29
32	9.79 447	16	9.90 112	26	0.09 888	9.89 334	10	28
33	9.79 463	15	9.90 138	26	0.09 862	9.89 324	10	27
34	9.79 478	16	9.90 164	26	0.09 836	9.89 314	10	26
35	9.79 494	16	9.90 190	26	0.09 810	9.89 304	10	25
36	9.79 510	16	9.90 216	26	0.09 784	9.89 294	10	24
37	9.79 526	16	9.90 242	26	0.09 758	9.89 284	10	23
38	9.79 542	16	9.90 268	26	0.09 732	9.89 274	10	22
39	9.79 558	15	9.90 294	26	0.09 706	9.89 264	10	21
40	9.79 573	16	9.90 320	26	0.09 680	9.89 254	10	**20**
41	9.79 589	16	9.90 346	25	0.09 654	9.89 244	11	19
42	9.79 605	16	9.90 371	26	0.09 629	9.89 233	10	18
43	9.79 621	15	9.90 397	26	0.09 603	9.89 223	10	17
44	9.79 636	16	9.90 423	26	0.09 577	9.89 213	10	16
45	9.79 652	16	9.90 449	26	0.09 551	9.89 203	10	15
46	9.79 668	16	9.90 475	26	0.09 525	9.89 193	10	14
47	9.79 684	15	9.90 501	26	0.09 499	9.89 183	10	13
48	9.79 699	16	9.90 527	26	0.09 473	9.89 173	11	12
49	9.79 715	16	9.90 553	25	0.09 447	9.89 162	10	11
50	9.79 731	15	9.90 578	26	0.09 422	9.89 152	10	**10**
51	9.79 746	16	9.90 604	26	0.09 396	9.89 142	10	9
52	9.79 762	16	9.90 630	26	0.09 370	9.89 132	10	8
53	9.79 778	15	9.90 656	26	0.09 344	9.89 122	10	7
54	9.79 793	16	9.90 682	26	0.09 318	9.89 112	11	6
55	9.79 809	16	9.90 708	26	0.09 292	9.89 101	10	5
56	9.79 825	15	9.90 734	25	0.09 266	9.89 091	10	4
57	9.79 840	16	9.90 759	26	0.09 241	9.89 081	10	3
58	9.79 856	16	9.90 785	26	0.09 215	9.89 071	11	2
59	9.79 872	15	9.90 811	26	0.09 189	9.89 060	10	1
60	9.79 887		9.90 837		0.09 163	9.89 050		**0**
′	L. Cos.	d.	L. Cot.	c.d.	L. Tan.	L. Sin.	d.	′

128° (308°) — (231°) **51°**

P. P.

″	**26**	**25**
1	0.4	0.4
2	0.9	0.8
3	1.3	1.2
4	1.7	1.7
5	2.2	2.1
6	2.6	2.5
7	3.0	2.9
8	3.5	3.3
9	3.9	3.8
10	4.3	4.2
20	8.7	8.3
30	13.0	12.5
40	17.3	16.7
50	21.7	20.8

″	**17**	**16**	**15**
1	0.3	0.3	0.2
2	0.6	0.5	0.5
3	0.8	0.8	0.8
4	1.1	1.1	1.0
5	1.4	1.3	1.2
6	1.7	1.6	1.5
7	2.0	1.9	1.8
8	2.3	2.1	2.0
9	2.6	2.4	2.2
10	2.8	2.7	2.5
20	5.7	5.3	5.0
30	8.5	8.0	7.5
40	11.3	10.7	10.0
50	14.2	13.3	12.5

″	**11**	**10**	**9**
1	0.2	0.2	0.2
2	0.4	0.3	0.3
3	0.6	0.5	0.4
4	0.7	0.7	0.6
5	0.9	0.8	0.8
6	1.1	1.0	0.9
7	1.3	1.2	1.0
8	1.5	1.3	1.2
9	1.6	1.5	1.4
10	1.8	1.7	1.5
20	3.7	3.3	3.0
30	5.5	5.0	4.5
40	7.3	6.7	6.0
50	9.2	8.3	7.5

	10/26	**10/25**	**9/26**
0	1.3	1.2	1.4
1	3.9	3.8	4.3
2	6.5	6.2	7.2
3	9.1	8.8	10.1
4	11.7	11.2	13.0
5	14.3	13.8	15.9
6	16.9	16.2	18.8
7	19.5	18.8	21.7
8	22.1	21.2	24.6
9	24.7	23.8	—
10			

LOGARITHMS OF THE FUNCTIONS (Continued)

39° (219°) (320°) 140°

′	L. Sin.	d.	L. Tan.	c.d.	L. Cot.	L. Cos.	d.	′
0	9.79 887	16	9.90 837	26	0.09 163	9.89 050	10	**60**
1	9.79 903	15	9.90 863	26	0.09 137	9.89 040	10	59
2	9.79 918	16	9.90 889	25	0.09 111	9.89 030	10	58
3	9.79 934	16	9.90 914	26	0.09 086	9.89 020	11	57
4	9.79 950	15	9.90 940	26	0.09 060	9.89 009	10	56
5	9.79 965	16	9.90 966	26	0.09 034	9.88 999	10	55
6	9.79 981	15	9.90 992	26	0.09 008	9.88 989	11	54
7	9.79 996	16	9.91 018	25	0.08 982	9.88 978	10	53
8	9.80 012	15	9.91 043	26	0.08 957	9.88 968	10	52
9	9.80 027	16	9.91 069	26	0.08 931	9.88 958	10	51
10	9.80 043	15	9.91 095	26	0.08 905	9.88 948	11	**50**
11	9.80 058	16	9.91 121	26	0.08 879	9.88 937	10	49
12	9.80 074	15	9.91 147	25	0.08 853	9.88 927	10	48
13	9.80 089	16	9.91 172	26	0.08 828	9.88 917	11	47
14	9.80 105	15	9.91 198	26	0.08 802	9.88 906	10	46
15	9.80 120	16	9.91 224	26	0.08 776	9.88 896	10	45
16	9.80 136	15	9.91 250	26	0.08 750	9.88 886	11	44
17	9.80 151	15	9.91 276	25	0.08 724	9.88 875	10	43
18	9.80 166	16	9.91 301	26	0.08 699	9.88 865	10	42
19	9.80 182	15	9.91 327	26	0.08 673	9.88 855	11	41
20	9.80 197	16	9.91 353	26	0.08 647	9.88 844	10	**40**
21	9.80 213	15	9.91 379	25	0.08 621	9.88 834	10	39
22	9.80 228	16	9.91 404	26	0.08 596	9.88 824	11	38
23	9.80 244	15	9.91 430	26	0.08 570	9.88 813	10	37
24	9.80 259	15	9.91 456	26	0.08 544	9.88 803	10	36
25	9.80 274	16	9.91 482	25	0.08 518	9.88 793	11	35
26	9.80 290	15	9.91 507	26	0.08 493	9.88 782	10	34
27	9.80 305	15	9.91 533	26	0.08 467	9.88 772	11	33
28	9.80 320	16	9.91 559	26	0.08 441	9.88 761	10	32
29	9.80 336	15	9.91 585	25	0.08 415	9.88 751	10	31
30	9.80 351	15	9.91 610	26	0.08 390	9.88 741	11	**30**
31	9.80 366	16	9.91 636	26	0.08 364	9.88 730	10	29
32	9.80 382	15	9.91 662	26	0.08 338	9.88 720	11	28
33	9.80 397	15	9.91 688	25	0.08 312	9.88 709	10	27
34	9.80 412	16	9.91 713	26	0.08 287	9.88 699	11	26
35	9.80 428	15	9.91 739	26	0.08 261	9.88 688	10	25
36	9.80 443	15	9.91 765	26	0.08 235	9.88 678	10	24
37	9.80 458	15	9.91 791	25	0.08 209	9.88 668	11	23
38	9.80 473	16	9.91 816	26	0.08 184	9.88 657	10	22
39	9.80 489	15	9.91 842	26	0.08 158	9.88 647	11	21
40	9.80 504	15	9.91 868	25	0.08 132	9.88 636	10	**20**
41	9.80 519	15	9.91 893	26	0.08 107	9.88 626	11	19
42	9.80 534	16	9.91 919	26	0.08 081	9.88 615	10	18
43	9.80 550	15	9.91 945	26	0.08 055	9.88 605	11	17
44	9.80 565	15	9.91 971	25	0.08 029	9.88 594	10	16
45	9.80 580	15	9.91 996	26	0.08 004	9.88 584	11	15
46	9.80 595	15	9.92 022	26	0.07 978	9.88 573	10	14
47	9.80 610	15	9.92 048	25	0.07 952	9.88 563	11	13
48	9.80 625	16	9.92 073	26	0.07 927	9.88 552	10	12
49	9.80 641	15	9.92 099	26	0.07 901	9.88 542	11	11
50	9.80 656	15	9.92 125	25	0.07 875	9.88 531	10	**10**
51	9.80 671	15	9.92 150	26	0.07 850	9.88 521	11	9
52	9.80 686	15	9.92 176	26	0.07 824	9.88 510	11	8
53	9.80 701	15	9.92 202	25	0.07 798	9.88 499	10	7
54	9.80 716	15	9.92 227	26	0.07 773	9.88 489	11	6
55	9.80 731	15	9.92 253	26	0.07 747	9.88 478	10	5
56	9.80 746	16	9.92 279	25	0.07 721	9.88 468	11	4
57	9.80 762	15	9.92 304	26	0.07 696	9.88 457	10	3
58	9.80 777	15	9.92 330	26	0.07 670	9.88 447	11	2
59	9.80 792	15	9.92 356	25	0.07 644	9.88 436	11	1
60	9.80 807		9.92 381		0.07 619	9.88 425		**0**
′	L. Cos.	d.	L. Cot.	c.d.	L. Tan.	L. Sin.	d.	′

P. P.

″	**26**	**25**
1	0.4	0.4
2	0.9	0.8
3	1.3	1.2
4	1.7	1.7
5	2.2	2.1
6	2.6	2.5
7	3.0	2.9
8	3.5	3.3
9	3.9	3.8
10	4.3	4.2
20	8.7	8.3
30	13.0	12.5
40	17.3	16.7
50	21.7	20.8

″	**16**	**15**
1	0.3	0.2
2	0.5	0.5
3	0.8	0.8
4	1.1	1.0
5	1.3	1.2
6	1.6	1.5
7	1.9	1.8
8	2.1	2.0
9	2.4	2.2
10	2.7	2.5
20	5.3	5.0
30	8.0	7.5
40	10.7	10.0
50	13.3	12.5

″	**11**	**10**
1	0.2	0.2
2	0.4	0.3
3	0.6	0.5
4	0.7	0.7
5	0.9	0.8
6	1.1	1.0
7	1.3	1.2
8	1.5	1.3
9	1.6	1.5
10	1.8	1.7
20	3.7	3.3
30	5.5	5.0
40	7.3	6.7
50	9.2	8.3

	11/26	**11/25**
0	1.2	1.1
1	3.5	3.4
2	5.9	5.7
3	8.3	7.9
4	10.6	10.2
5	13.0	12.5
6	15.4	14.8
7	17.7	17.1
8	20.1	19.3
9	22.5	21.6
10	24.8	23.9
11		

129° (309°) (230°) 50°

40° (220°) (319°) 139°

′	L. Sin.	d.	L. Tan.	c.d.	L. Cot.	L. Cos.	d.	′
0	9.80 807	15	9.92 381	26	0.07 619	9.88 425	10	**60**
1	9.80 822	15	9.92 407	26	0.07 593	9.88 415	11	59
2	9.80 837	15	9.92 433	25	0.07 567	9.88 404	10	58
3	9.80 852	15	9.92 458	26	0.07 542	9.88 394	11	57
4	9.80 867	15	9.92 484	26	0.07 516	9.88 383	11	56
5	9.80 882	15	9.92 510	25	0.07 490	9.88 372	10	55
6	9.80 897	15	9.92 535	26	0.07 465	9.88 362	11	54
7	9.80 912	15	9.92 561	26	0.07 439	9.88 351	11	53
8	9.80 927	15	9.92 587	25	0.07 413	9.88 340	10	52
9	9.80 942	15	9.92 612	26	0.07 388	9.88 330	11	51
10	9.80 957	15	9.92 638	25	0.07 362	9.88 319	11	**50**
11	9.80 972	15	9.92 663	26	0.07 337	9.88 308	10	49
12	9.80 987	15	9.92 689	26	0.07 311	9.88 298	11	48
13	9.81 002	15	9.92 715	25	0.07 285	9.88 287	11	47
14	9.81 017	15	9.92 740	26	0.07 260	9.88 276	10	46
15	9.81 032	15	9.92 766	26	0.07 234	9.88 266	11	45
16	9.81 047	14	9.92 792	25	0.07 208	9.88 255	11	44
17	9.81 061	15	9.92 817	26	0.07 183	9.88 244	10	43
18	9.81 076	15	9.92 843	25	0.07 157	9.88 234	11	42
19	9.81 091	15	9.92 868	26	0.07 132	9.88 223	11	41
20	9.81 106	15	9.92 894	26	0.07 106	9.88 212	11	**40**
21	9.81 121	15	9.92 920	25	0.07 080	9.88 201	10	39
22	9.81 136	15	9.92 945	26	0.07 055	9.88 191	11	38
23	9.81 151	15	9.92 971	25	0.07 029	9.88 180	11	37
24	9.81 166	14	9.92 996	26	0.07 004	9.88 169	11	36
25	9.81 180	15	9.93 022	26	0.06 978	9.88 158	10	35
26	9.81 195	15	9.93 048	25	0.06 952	9.88 148	11	34
27	9.81 210	15	9.93 073	26	0.06 927	9.88 137	11	33
28	9.81 225	15	9.93 099	25	0.06 901	9.88 126	11	32
29	9.81 240	14	9.93 124	26	0.06 876	9.88 115	10	31
30	9.81 254	15	9.93 150	25	0.06 850	9.88 105	11	**30**
31	9.81 269	15	9.93 175	26	0.06 825	9.88 094	11	29
32	9.81 284	15	9.93 201	26	0.06 799	9.88 083	11	28
33	9.81 299	15	9.93 227	25	0.06 773	9.88 072	11	27
34	9.81 314	14	9.93 252	26	0.06 748	9.88 061	10	26
35	9.81 328	15	9.93 278	25	0.06 722	9.88 051	11	25
36	9.81 343	15	9.93 303	26	0.06 697	9.88 040	11	24
37	9.81 358	14	9.93 329	25	0.06 671	9.88 029	11	23
38	9.81 372	15	9.93 354	26	0.06 646	9.88 018	11	22
39	9.81 387	15	9.93 380	26	0.06 620	9.88 007	11	21
40	9.81 402	15	9.93 406	25	0.06 594	9.87 996	11	**20**
41	9.81 417	14	9.93 431	26	0.06 569	9.87 985	10	19
42	9.81 431	15	9.93 457	25	0.06 543	9.87 975	11	18
43	9.81 446	15	9.93 482	26	0.06 518	9.87 964	11	17
44	9.81 461	14	9.93 508	25	0.06 492	9.87 953	11	16
45	9.81 475	15	9.93 533	26	0.06 467	9.87 942	11	15
46	9.81 490	15	9.93 559	25	0.06 441	9.87 931	11	14
47	9.81 505	14	9.93 584	26	0.06 416	9.87 920	11	13
48	9.81 519	15	9.93 610	26	0.06 390	9.87 909	11	12
49	9.81 534	15	9.93 636	25	0.06 364	9.87 898	11	11
50	9.81 549	14	9.93 661	26	0.06 339	9.87 887	10	**10**
51	9.81 563	15	9.93 687	25	0.06 313	9.87 877	11	9
52	9.81 578	14	9.93 712	26	0.06 288	9.87 866	11	8
53	9.81 592	15	9.93 738	25	0.06 262	9.87 855	11	7
54	9.81 607	15	9.93 763	26	0.06 237	9.87 844	11	6
55	9.81 622	14	9.93 789	25	0.06 211	9.87 833	11	5
56	9.81 636	15	9.93 814	26	0.06 186	9.87 822	11	4
57	9.81 651	14	9.93 840	25	0.06 160	9.87 811	11	3
58	9.81 665	15	9.93 865	26	0.06 135	9.87 800	11	2
59	9.81 680	14	9.93 891	25	0.06 109	9.87 789	11	1
60	9.81 694		9.93 916		0.06 084	9.87 778		**0**
′	L. Cos.	d.	L. Cot.	c.d.	L. Tan.	L. Sin.	d.	′

P. P.

″	**26**	**25**
1	0.4	0.4
2	0.9	0.8
3	1.3	1.2
4	1.7	1.7
5	2.2	2.1
6	2.6	2.5
7	3.0	2.9
8	3.5	3.3
9	3.9	3.8
10	4.3	4.2
20	8.7	8.3
30	13.0	12.5
40	17.3	16.7
50	21.7	20.8

″	**15**	**14**
1	0.2	0.2
2	0.5	0.5
3	0.8	0.7
4	1.0	0.9
5	1.2	1.2
6	1.5	1.4
7	1.8	1.6
8	2.0	1.9
9	2.2	2.1
10	2.5	2.3
20	5.0	4.7
30	7.5	7.0
40	10.0	9.3
50	12.5	11.7

″	**11**	**10**
1	0.2	0.2
2	0.4	0.3
3	0.6	0.5
4	0.7	0.7
5	0.9	0.8
6	1.1	1.0
7	1.3	1.2
8	1.5	1.3
9	1.6	1.5
10	1.8	1.7
20	3.7	3.3
30	5.5	5.0
40	7.3	6.7
50	9.2	8.3

		11/26	**10/26**	**10/25**
0	10	1.2	1.3	1.2
1	9	3.5	3.9	3.8
2	8	5.9	6.5	6.2
3	7	8.3	9.1	8.8
4	6	10.6	11.7	11.2
5	5	13.0	14.3	13.8
6	4	15.4	16.9	16.2
7	3	17.7	19.5	18.8
8	2	20.1	22.1	21.2
9	1	22.5	24.7	23.8
10	0	24.8	—	—
11				

130° (310°) (229°) 49°

LOGARITHMS OF THE FUNCTIONS (Continued)

41° (221°) | (318°) **138°**

′	L. Sin.	d.	L. Tan.	c.d.	L. Cot.	L. Cos.	d.	′
0	9.81 694	15	9.93 916	26	0.06 084	9.87 778	11	**60**
1	9.81 709	14	9.93 942	25	0.06 058	9.87 767	11	59
2	9.81 723	15	9.93 967	26	0.06 033	9.87 756	11	58
3	9.81 738	14	9.93 993	25	0.06 007	9.87 745	11	57
4	9.81 752	15	9.94 018	26	0.05 982	9.87 734	11	56
5	9.81 767	14	9.94 044	25	0.05 956	9.87 723	11	55
6	9.81 781	15	9.94 069	26	0.05 931	9.87 712	11	54
7	9.81 796	14	9.94 095	25	0.05 905	9.87 701	11	53
8	9.81 810	15	9.94 120	26	0.05 880	9.87 690	11	52
9	9.81 825	14	9.94 146	25	0.05 854	9.87 679	11	51
10	9.81 839	15	9.94 171	26	0.05 829	9.87 668	11	**50**
11	9.81 854	14	9.94 197	25	0.05 803	9.87 657	11	49
12	9.81 868	14	9.94 222	26	0.05 778	9.87 646	11	48
13	9.81 882	15	9.94 248	25	0.05 752	9.87 635	11	47
14	9.81 897	14	9.94 273	26	0.05 727	9.87 624	11	46
15	9.81 911	15	9.94 299	25	0.05 701	9.87 613	12	45
16	9.81 926	14	9.94 324	26	0.05 676	9.87 601	11	44
17	9.81 940	15	9.94 350	25	0.05 650	9.87 590	11	43
18	9.81 955	14	9.94 375	26	0.05 625	9.87 579	11	42
19	9.81 969	14	9.94 401	25	0.05 599	9.87 568	11	41
20	9.81 983	15	9.94 426	26	0.05 574	9.87 557	11	**40**
21	9.81 998	14	9.94 452	25	0.05 548	9.87 546	11	39
22	9.82 012	14	9.94 477	26	0.05 523	9.87 535	11	38
23	9.82 026	15	9.94 503	25	0.05 497	9.87 524	11	37
24	9.82 041	14	9.94 528	26	0.05 472	9.87 513	12	36
25	9.82 055	14	9.94 554	25	0.05 446	9.87 501	11	35
26	9.82 069	15	9.94 579	25	0.05 421	9.87 490	11	34
27	9.82 084	14	9.94 604	26	0.05 396	9.87 479	11	33
28	9.82 098	14	9.94 630	25	0.05 370	9.87 468	11	32
29	9.82 112	14	9.94 655	26	0.05 345	9.87 457	11	31
30	9.82 126	15	9.94 681	25	0.05 319	9.87 446	12	**30**
31	9.82 141	14	9.94 706	26	0.05 294	9.87 434	11	29
32	9.82 155	14	9.94 732	25	0.05 268	9.87 423	11	28
33	9.82 169	15	9.94 757	26	0.05 243	9.87 412	11	27
34	9.82 184	14	9.94 783	25	0.05 217	9.87 401	11	26
35	9.82 198	14	9.94 808	26	0.05 192	9.87 390	12	25
36	9.82 212	14	9.94 834	25	0.05 166	9.87 378	11	24
37	9.82 226	14	9.94 859	25	0.05 141	9.87 367	11	23
38	9.82 240	15	9.94 884	26	0.05 116	9.87 356	11	22
39	9.82 255	14	9.94 910	25	0.05 090	9.87 345	11	21
40	9.82 269	14	9.94 935	26	0.05 065	9.87 334	12	**20**
41	9.82 283	14	9.94 961	25	0.05 039	9.87 322	11	19
42	9.82 297	14	9.94 986	26	0.05 014	9.87 311	11	18
43	9.82 311	15	9.95 012	25	0.04 988	9.87 300	12	17
44	9.82 326	14	9.95 037	25	0.04 963	9.87 288	11	16
45	9.82 340	14	9.95 062	26	0.04 938	9.87 277	11	15
46	9.82 354	14	9.95 088	25	0.04 912	9.87 266	11	14
47	9.82 368	14	9.95 113	26	0.04 887	9.87 255	12	13
48	9.82 382	14	9.95 139	25	0.04 861	9.87 243	11	12
49	9.82 396	14	9.95 164	26	0.04 836	9.87 232	11	11
50	9.82 410	14	9.95 190	25	0.04 810	9.87 221	12	**10**
51	9.82 424	15	9.95 215	25	0.04 785	9.87 209	11	9
52	9.82 439	14	9.95 240	26	0.04 760	9.87 198	11	8
53	9.82 453	14	9.95 266	25	0.04 734	9.87 187	12	7
54	9.82 467	14	9.95 291	26	0.04 709	9.87 175	11	6
55	9.82 481	14	9.95 317	25	0.04 683	9.87 164	11	5
56	9.82 495	14	9.95 342	26	0.04 658	9.87 153	12	4
57	9.82 509	14	9.95 368	25	0.04 632	9.87 141	11	3
58	9.82 523	14	9.95 393	25	0.04 607	9.87 130	11	2
59	9.82 537	14	9.95 418	26	0.04 582	9.87 119	12	1
60	9.82 551		9.95 444		0.04 556	9.87 107		**0**
′	L. Cos.	d.	L. Cot.	c.d.	L. Tan.	L. Sin.	d.	′

P. P.

″	**26**	**25**
1	0.4	0.4
2	0.9	0.8
3	1.3	1.2
4	1.7	1.7
5	2.2	2.1
6	2.6	2.5
7	3.0	2.9
8	3.5	3.3
9	3.9	3.8
10	4.3	4.2
20	8.7	8.3
30	13.0	12.5
40	17.3	16.7
50	21.7	20.8

″	**15**	**14**
1	0.2	0.2
2	0.5	0.5
3	0.8	0.7
4	1.0	0.9
5	1.2	1.2
6	1.5	1.4
7	1.8	1.6
8	2.0	1.9
9	2.2	2.1
10	2.5	2.3
20	5.0	4.7
30	7.5	7.0
40	10.0	9.3
50	12.5	11.7

″	**12**	**11**
1	0.2	0.2
2	0.4	0.4
3	0.6	0.6
4	0.8	0.7
5	1.0	0.9
6	1.2	1.1
7	1.4	1.3
8	1.6	1.5
9	1.8	1.6
10	2.0	1.8
20	4.0	3.7
30	6.0	5.5
40	8.0	7.3
50	10.0	9.2

	12/26	**12/25**	**11/25**
0	1.1	1.1	1.1
1	3.2	3.1	3.4
2	5.4	5.2	5.7
3	7.6	7.3	7.9
4	9.8	9.4	10.2
5	11.9	11.5	12.5
6	14.1	13.6	14.8
7	16.2	15.6	17.1
8	18.4	17.7	19.3
9	20.6	19.8	21.6
10	22.8	21.9	23.9
11	24.9	23.9	—
12			

131° (311°) | (228°) **48°**

42° (222°) | (317°) **137°**

′	L. Sin.	d.	L. Tan.	c.d.	L. Cot.	L. Cos.	d.	′
0	9.82 551	14	9.95 444	25	0.04 556	9.87 107	11	**60**
1	9.82 565	14	9.95 469	26	0.04 531	9.87 096	11	59
2	9.82 579	14	9.95 495	25	0.04 505	9.87 085	12	58
3	9.82 593	14	9.95 520	25	0.04 480	9.87 073	11	57
4	9.82 607	14	9.95 545	26	0.04 455	9.87 062	12	56
5	9.82 621	14	9.95 571	25	0.04 429	9.87 050	11	55
6	9.82 635	14	9.95 596	26	0.04 404	9.87 039	11	54
7	9.82 649	14	9.95 622	25	0.04 378	9.87 028	12	53
8	9.82 663	14	9.95 647	25	0.04 353	9.87 016	11	52
9	9.82 677	14	9.95 672	26	0.04 328	9.87 005	12	51
10	9.82 691	14	9.95 698	25	0.04 302	9.86 993	11	**50**
11	9.82 705	14	9.95 723	25	0.04 277	9.86 982	12	49
12	9.82 719	14	9.95 748	26	0.04 252	9.86 970	11	48
13	9.82 733	14	9.95 774	25	0.04 226	9.86 959	12	47
14	9.82 747	14	9.95 799	26	0.04 201	9.86 947	11	46
15	9.82 761	14	9.95 825	25	0.04 175	9.86 936	12	45
16	9.82 775	13	9.95 850	25	0.04 150	9.86 924	11	44
17	9.82 788	14	9.95 875	26	0.04 125	9.86 913	11	43
18	9.82 802	14	9.95 901	25	0.04 099	9.86 902	12	42
19	9.82 816	14	9.95 926	26	0.04 074	9.86 890	11	41
20	9.82 830	14	9.95 952	25	0.04 048	9.86 879	12	**40**
21	9.82 844	14	9.95 977	25	0.04 023	9.86 867	12	39
22	9.82 858	14	9.96 002	26	0.03 998	9.86 855	11	38
23	9.82 872	13	9.96 028	25	0.03 972	9.86 844	12	37
24	9.82 885	14	9.96 053	25	0.03 947	9.86 832	11	36
25	9.82 899	14	9.96 078	26	0.03 922	9.86 821	12	35
26	9.82 913	14	9.96 104	25	0.03 896	9.86 809	11	34
27	9.82 927	14	9.96 129	26	0.03 871	9.86 798	12	33
28	9.82 941	14	9.96 155	25	0.03 845	9.86 786	11	32
29	9.82 955	13	9.96 180	25	0.03 820	9.86 775	12	31
30	9.82 968	14	9.96 205	26	0.03 795	9.86 763	11	**30**
31	9.82 982	14	9.96 231	25	0.03 769	9.86 752	12	29
32	9.82 996	14	9.96 256	25	0.03 744	9.86 740	12	28
33	9.83 010	13	9.96 281	26	0.03 719	9.86 728	11	27
34	9.83 023	14	9.96 307	25	0.03 693	9.86 717	12	26
35	9.83 037	14	9.96 332	25	0.03 668	9.86 705	11	25
36	9.83 051	14	9.96 357	26	0.03 643	9.86 694	12	24
37	9.83 065	13	9.96 383	25	0.03 617	9.86 682	12	23
38	9.83 078	14	9.96 408	25	0.03 592	9.86 670	11	22
39	9.83 092	14	9.96 433	26	0.03 567	9.86 659	12	21
40	9.83 106	14	9.96 459	25	0.03 541	9.86 647	12	**20**
41	9.83 120	13	9.96 484	26	0.03 516	9.86 635	11	19
42	9.83 133	14	9.96 510	25	0.03 490	9.86 624	12	18
43	9.83 147	14	9.96 535	25	0.03 465	9.86 612	12	17
44	9.83 161	13	9.96 560	26	0.03 440	9.86 600	11	16
45	9.83 174	14	9.96 586	25	0.03 414	9.86 589	12	15
46	9.83 188	14	9.96 611	25	0.03 389	9.86 577	12	14
47	9.83 202	13	9.96 636	26	0.03 364	9.86 565	11	13
48	9.83 215	14	9.96 662	25	0.03 338	9.86 554	12	12
49	9.83 229	13	9.96 687	25	0.03 313	9.86 542	12	11
50	9.83 242	14	9.96 712	26	0.03 288	9.86 530	12	**10**
51	9.83 256	14	9.96 738	25	0.03 262	9.86 518	11	9
52	9.83 270	13	9.96 763	25	0.03 237	9.86 507	12	8
53	9.83 283	14	9.96 788	26	0.03 212	9.86 495	12	7
54	9.83 297	13	9.96 814	25	0.03 186	9.86 483	11	6
55	9.83 310	14	9.96 839	25	0.03 161	9.86 472	12	5
56	9.83 324	14	9.96 864	26	0.03 136	9.86 460	12	4
57	9.83 338	13	9.96 890	25	0.03 110	9.86 448	12	3
58	9.83 351	14	9.96 915	25	0.03 085	9.86 436	11	2
59	9.83 365	13	9.96 940	26	0.03 060	9.86 425	12	1
60	9.83 378		9.96 966		0.03 034	9.86 413		**0**
′	L. Cos.	d.	L. Cot.	c.d.	L. Tan.	L. Sin.	d.	′

P. P.

″	**26**	**25**
1	0.4	0.4
2	0.9	0.8
3	1.3	1.2
4	1.7	1.7
5	2.2	2.1
6	2.6	2.5
7	3.0	2.9
8	3.5	3.3
9	3.9	3.8
10	4.3	4.2
20	8.7	8.3
30	13.0	12.5
40	17.3	16.7
50	21.7	20.8

″	**14**	**13**
1	0.2	0.2
2	0.5	0.4
3	0.7	0.6
4	0.9	0.9
5	1.2	1.1
6	1.4	1.3
7	1.6	1.5
8	1.9	1.7
9	2.1	2.0
10	2.3	2.2
20	4.7	4.3
30	7.0	6.5
40	9.3	8.7
50	11.7	10.8

″	**12**	**11**
1	0.2	0.2
2	0.4	0.4
3	0.6	0.6
4	0.8	0.7
5	1.0	0.9
6	1.2	1.1
7	1.4	1.3
8	1.6	1.5
9	1.8	1.6
10	2.0	1.8
20	4.0	3.7
30	6.0	5.5
40	8.0	7.3
50	10.0	9.2

	12/26	**11/26**	**11/25**
0	1.1	1.2	1.1
1	3.2	3.5	3.4
2	5.4	5.9	5.7
3	7.6	8.3	7.9
4	9.8	10.6	10.2
5	11.9	13.0	12.5
6	14.1	15.4	14.8
7	16.2	17.7	17.1
8	18.4	20.1	19.3
9	20.6	22.5	21.6
10	22.8	24.8	23.9
11	24.9	—	—
12			

132° (312°) | (227°) **47°**

LOGARITHMS OF THE FUNCTIONS (Continued)

43° (223°) (316°) **136°**

′	L. Sin.	d.	L. Tan.	c.d.	L. Cot.	L. Cos.	d.	′
0	9.83 378	14	9.96 966	25	0.03 034	9.86 413	12	**60**
1	9.83 392	13	9.96 991	25	0.03 009	9.86 401	12	59
2	9.83 405	14	9.97 016	26	0.02 984	9.86 389	12	58
3	9.83 419	13	9.97 042	25	0.02 958	9.86 377	11	57
4	9.83 432	14	9.97 067	25	0.02 933	9.86 366	12	56
5	9.83 446	13	9.97 092	26	0.02 908	9.86 354	12	55
6	9.83 459	14	9.97 118	25	0.02 882	9.86 342	12	54
7	9.83 473	13	9.97 143	25	0.02 857	9.86 330	12	53
8	9.83 486	14	9.97 168	25	0.02 832	9.86 318	12	52
9	9.83 500	13	9.97 193	26	0.02 807	9.86 306	11	51
10	9.83 513	14	9.97 219	25	0.02 781	9.86 295	12	**50**
11	9.83 527	13	9.97 244	25	0.02 756	9.86 283	12	49
12	9.83 540	14	9.97 269	26	0.02 731	9.86 271	12	48
13	9.83 554	13	9.97 295	25	0.02 705	9.86 259	12	47
14	9.83 567	14	9.97 320	25	0.02 680	9.86 247	12	46
15	9.83 581	13	9.97 345	26	0.02 655	9.86 235	12	45
16	9.83 594	14	9.97 371	25	0.02 629	9.86 223	12	44
17	9.83 608	13	9.97 396	25	0.02 604	9.86 211	11	43
18	9.83 621	13	9.97 421	26	0.02 579	9.86 200	12	42
19	9.83 634	14	9.97 447	25	0.02 553	9.86 188	12	41
20	9.83 648	13	9.97 472	25	0.02 528	9.86 176	12	**40**
21	9.83 661	13	9.97 497	26	0.02 503	9.86 164	12	39
22	9.83 674	14	9.97 523	25	0.02 477	9.86 152	12	38
23	9.83 688	13	9.97 548	25	0.02 452	9.86 140	12	37
24	9.83 701	14	9.97 573	25	0.02 427	9.86 128	12	36
25	9.83 715	13	9.97 598	26	0.02 402	9.86 116	12	35
26	9.83 728	13	9.97 624	25	0.02 376	9.86 104	12	34
27	9.83 741	14	9.97 649	25	0.02 351	9.86 092	12	33
28	9.83 755	13	9.97 674	26	0.02 326	9.86 080	12	32
29	9.83 768	13	9.97 700	25	0.02 300	9.86 068	12	31
30	9.83 781	14	9.97 725	25	0.02 275	9.86 056	12	**30**
31	9.83 795	13	9.97 750	26	0.02 250	9.86 044	12	29
32	9.83 808	13	9.97 776	25	0.02 224	9.86 032	12	28
33	9.83 821	13	9.97 801	25	0.02 199	9.86 020	12	27
34	9.83 834	14	9.97 826	25	0.02 174	9.86 008	12	26
35	9.83 848	13	9.97 851	26	0.02 149	9.85 996	12	25
36	9.83 861	13	9.97 877	25	0.02 123	9.85 984	12	24
37	9.83 874	13	9.97 902	25	0.02 098	9.85 972	12	23
38	9.83 887	14	9.97 927	26	0.02 073	9.85 960	12	22
39	9.83 901	13	9.97 953	25	0.02 047	9.85 948	12	21
40	9.83 914	13	9.97 978	25	0.02 022	9.85 936	12	**20**
41	9.83 927	13	9.98 003	26	0.01 997	9.85 924	12	19
42	9.83 940	14	9.98 029	25	0.01 971	9.85 912	12	18
43	9.83 954	13	9.98 054	25	0.01 946	9.85 900	12	17
44	9.83 967	13	9.98 079	25	0.01 921	9.85 888	12	16
45	9.83 980	13	9.98 104	26	0.01 896	9.85 876	12	15
46	9.83 993	13	9.98 130	25	0.01 870	9.85 864	13	14
47	9.84 006	14	9.98 155	25	0.01 845	9.85 851	12	13
48	9.84 020	13	9.98 180	26	0.01 820	9.85 839	12	12
49	9.84 033	13	9.98 206	25	0.01 794	9.85 827	12	11
50	9.84 046	13	9.98 231	25	0.01 769	9.85 815	12	**10**
51	9.84 059	13	9.98 256	25	0.01 744	9.85 803	12	9
52	9.84 072	13	9.98 281	26	0.01 719	9.85 791	12	8
53	9.84 085	13	9.98 307	25	0.01 693	9.85 779	13	7
54	9.84 098	14	9.98 332	25	0.01 668	9.85 766	12	6
55	9.84 112	13	9.98 357	26	0.01 643	9.85 754	12	5
56	9.84 125	13	9.98 383	25	0.01 617	9.85 742	12	4
57	9.84 138	13	9.98 408	25	0.01 592	9.85 730	12	3
58	9.84 151	13	9.98 433	25	0.01 567	9.85 718	12	2
59	9.84 164	13	9.98 458	26	0.01 542	9.85 706	13	1
60	9.84 177		9.98 484		0.01 516	9.85 693		**0**
′	L. Cos.	d.	L. Cot.	c.d.	L. Tan.	L. Sin.	d.	′

133° (313°) (226°) **46°**

P. P.

″	**26**	**25**
1	0.4	0.4
2	0.9	0.8
3	1.3	1.2
4	1.7	1.7
5	2.2	2.1
6	2.6	2.5
7	3.0	2.9
8	3.5	3.3
9	3.9	3.8
10	4.3	4.2
20	8.7	8.3
30	13.0	12.5
40	17.3	16.7
50	21.7	20.8

″	**14**	**13**
1	0.2	0.2
2	0.5	0.4
3	0.7	0.6
4	0.9	0.9
5	1.2	1.1
6	1.4	1.3
7	1.6	1.5
8	1.9	1.7
9	2.1	2.0
10	2.3	2.2
20	4.7	4.3
30	7.0	6.5
40	9.3	8.7
50	11.7	10.8

″	**12**	**11**
1	0.2	0.2
2	0.4	0.4
3	0.6	0.6
4	0.8	0.7
5	1.0	0.9
6	1.2	1.1
7	1.4	1.3
8	1.6	1.5
9	1.8	1.6
10	2.0	1.8
20	4.0	3.7
30	6.0	5.5
40	8.0	7.3
50	10.0	9.2

	13/26	**13/25**	**12/25**
0			
1	1.0	0.9	1.1
2	3.0	2.9	3.1
3	5.0	4.8	5.2
4	7.0	6.7	7.3
5	9.0	8.7	9.4
6	11.0	10.6	11.5
7	13.0	12.5	13.5
8	15.0	14.4	15.6
9	17.0	16.3	17.7
10	19.0	18.3	19.8
11	21.0	20.2	21.9
12	23.0	22.1	23.9
13	25.0	24.1	—

LOGARITHMS OF THE FUNCTIONS (Continued)

44° (224°) (315°) **135°**

′	L. Sin.	d.	L. Tan.	c.d.	L. Cot.	L. Cos.	d.	′
0	9.84 177	13	9.98 484	25	0.01 516	9.85 693	12	**60**
1	9.84 190	13	9.98 509	25	0.01 491	9.85 681	12	59
2	9.84 203	13	9.98 534	26	0.01 466	9.85 669	12	58
3	9.84 216	13	9.98 560	25	0.01 440	9.85 657	12	57
4	9.84 229	13	9.98 585	25	0.01 415	9.85 645	13	56
5	9.84 242	13	9.98 610	25	0.01 390	9.85 632	12	55
6	9.84 255	14	9.98 635	26	0.01 365	9.85 620	12	54
7	9.84 269	13	9.98 661	25	0.01 339	9.85 608	12	53
8	9.84 282	13	9.98 686	25	0.01 314	9.85 596	13	52
9	9.84 295	13	9.98 711	26	0.01 289	9.85 583	12	51
10	9.84 308	13	9.98 737	25	0.01 263	9.85 571	12	**50**
11	9.84 321	13	9.98 762	25	0.01 238	9.85 559	12	49
12	9.84 334	13	9.98 787	25	0.01 213	9.85 547	13	48
13	9.84 347	13	9.98 812	26	0.01 188	9.85 534	12	47
14	9.84 360	13	9.98 838	25	0.01 162	9.85 522	12	46
15	9.84 373	12	9.98 863	25	0.01 137	9.85 510	13	45
16	9.84 385	13	9.98 888	25	0.01 112	9.85 497	12	44
17	9.84 398	13	9.98 913	26	0.01 087	9.85 485	12	43
18	9.84 411	13	9.98 939	25	0.01 061	9.85 473	13	42
19	9.84 424	13	9.98 964	25	0.01 036	9.85 460	12	41
20	9.84 437	13	9.98 989	26	0.01 011	9.85 448	12	**40**
21	9.84 450	13	9.99 015	25	0.00 985	9.85 436	13	39
22	9.84 463	13	9.99 040	25	0.00 960	9.85 423	12	38
23	9.84 476	13	9.99 065	25	0.00 935	9.85 411	12	37
24	9.84 489	13	9.99 090	26	0.00 910	9.85 399	13	36
25	9.84 502	13	9.99 116	25	0.00 884	9.85 386	12	35
26	9.84 515	13	9.99 141	25	0.00 859	9.85 374	13	34
27	9.84 528	12	9.99 166	25	0.00 834	9.85 361	12	33
28	9.84 540	13	9.99 191	26	0.00 809	9.85 349	12	32
29	9.84 553	13	9.99 217	25	0.00 783	9.85 337	13	31
30	9.84 566	13	9.99 242	25	0.00 758	9.85 324	12	**30**
31	9.84 579	13	9.99 267	26	0.00 733	9.85 312	13	29
32	9.84 592	13	9.99 293	25	0.00 707	9.85 299	12	28
33	9.84 605	13	9.99 318	25	0.00 682	9.85 287	13	27
34	9.84 618	12	9.99 343	25	0.00 657	9.85 274	12	26
35	9.84 630	13	9.99 368	26	0.00 632	9.85 262	12	25
36	9.84 643	13	9.99 394	25	0.00 606	9.85 250	13	24
37	9.84 656	13	9.99 419	25	0.00 581	9.85 237	12	23
38	9.84 669	13	9.99 444	25	0.00 556	9.85 225	13	22
39	9.84 682	12	9.99 469	26	0.00 531	9.85 212	12	21
40	9.84 694	13	9.99 495	25	0.00 505	9.85 200	13	**20**
41	9.84 707	13	9.99 520	25	0.00 480	9.85 187	12	19
42	9.84 720	13	9.99 545	25	0.00 455	9.85 175	13	18
43	9.84 733	12	9.99 570	26	0.00 430	9.85 162	12	17
44	9.84 745	13	9.99 596	25	0.00 404	9.85 150	13	16
45	9.84 758	13	9.99 621	25	0.00 379	9.85 137	12	15
46	9.84 771	13	9.99 646	26	0.00 354	9.85 125	13	14
47	9.84 784	12	9.99 672	25	0.00 328	9.85 112	12	13
48	9.84 796	13	9.99 697	25	0.00 303	9.85 100	13	12
49	9.84 809	13	9.99 722	25	0.00 278	9.85 087	13	11
50	9.84 822	13	9.99 747	26	0.00 253	9.85 074	12	**10**
51	9.84 835	12	9.99 773	25	0.00 227	9.85 062	13	9
52	9.84 847	13	9.99 798	25	0.00 202	9.85 049	12	8
53	9.84 860	13	9.99 823	25	0.00 177	9.85 037	13	7
54	9.84 873	12	9.99 848	26	0.00 152	9.85 024	12	6
55	9.84 885	13	9.99 874	25	0.00 126	9.85 012	13	5
56	9.84 898	13	9.99 899	25	0.00 101	9.84 999	13	4
57	9.84 911	12	9.99 924	25	0.00 076	9.84 986	12	3
58	9.84 923	13	9.99 949	26	0.00 051	9.84 974	13	2
59	9.84 936	13	9.99 975	25	0.00 025	9.84 961	12	1
60	9.84 949		0.00 000		0.00 000	9.84 949		**0**
′	L. Cos.	d.	L. Cot.	c.d.	L. Tan.	L. Sin.	d.	′

134° (314°) (225°) **45°**

P. P.

″	**26**	**25**
1	0.4	0.4
2	0.9	0.8
3	1.3	1.2
4	1.7	1.7
5	2.2	2.1
6	2.6	2.5
7	3.0	2.9
8	3.5	3.3
9	3.9	3.8
10	4.3	4.2
20	8.7	8.3
30	13.0	12.5
40	17.3	16.7
50	21.7	20.8

″	**14**	**13**	**12**
1	0.2	0.2	0.2
2	0.5	0.4	0.4
3	0.7	0.6	0.6
4	0.9	0.9	0.8
5	1.2	1.1	1.0
6	1.4	1.3	1.2
7	1.6	1.5	1.4
8	1.9	1.7	1.6
9	2.1	2.0	1.8
10	2.3	2.2	2.0
20	4.7	4.3	4.0
30	7.0	6.5	6.0
40	9.3	8.7	8.0
50	11.7	10.8	10.0

	13/26	**13/25**
0		
1	1.0	0.9
2	3.0	2.9
3	5.0	4.8
4	7.0	6.7
5	9.0	8.7
6	11.0	10.6
7	13.0	12.5
8	15.0	14.4
9	17.0	16.3
10	19.0	18.3
11	21.0	20.2
12	23.0	22.1
13	25.0	24.1

	12/26	**12/25**
0		
1	1.1	1.1
2	3.2	3.1
3	5.4	5.2
4	7.6	7.3
5	9.8	9.4
6	11.9	11.5
7	14.1	13.5
8	16.2	15.6
9	18.4	17.7
10	20.6	19.8
11	22.8	21.9
12	24.9	23.9

Natural Trigonometric Functions

Values of the trigonometric functions of angles for each minute from 0–360°.

For degrees indicated at the top of the page use the column headings at the top. For degrees indicated at the bottom use the column indications at the bottom.

With degrees at the left of each block (top or bottom), use the minute column at the left and with degrees at the right of each block use the minute column at the right.

If natural trigonometric function tables are used for angle measures greater than 90° and less than 360°, appropriate signs for the functions must be supplied in accordance with the quadrant in which the angle measure belongs.

NATURAL TRIGONOMETRIC FUNCTIONS

0° (180°) — (359°) **179°**

′	Sin	Tan	Cot	Cos	Sec	Csc	′
0	.00000	.00000	———	1.0000	1.0000	———	**60**
1	.00029	.00029	3437.7	1.0000	1.0000	3437.7	59
2	.00058	.00058	1718.9	1.0000	1.0000	1718.9	58
3	.00087	.00087	1145.9	1.0000	1.0000	1145.9	57
4	.00116	.00116	859.44	1.0000	1.0000	859.44	56
5	.00145	.00145	687.55	1.0000	1.0000	687.55	**55**
6	.00175	.00175	572.96	1.0000	1.0000	572.96	54
7	.00204	.00204	491.11	1.0000	1.0000	491.11	53
8	.00233	.00233	429.72	1.0000	1.0000	429.72	52
9	.00262	.00262	381.97	1.0000	1.0000	381.97	51
10	.00291	.00291	343.77	1.0000	1.0000	343.78	**50**
11	.00320	.00320	312.52	.99999	1.0000	312.52	49
12	.00349	.00349	286.48	.99999	1.0000	286.48	48
13	.00378	.00378	264.44	.99999	1.0000	264.44	47
14	.00407	.00407	245.55	.99999	1.0000	245.55	46
15	.00436	.00436	229.18	.99999	1.0000	229.18	**45**
16	.00465	.00465	214.86	.99999	1.0000	214.86	44
17	.00495	.00495	202.22	.99999	1.0000	202.22	43
18	.00524	.00524	190.98	.99999	1.0000	190.99	42
19	.00553	.00553	180.93	.99998	1.0000	180.93	41
20	.00582	.00582	171.89	.99998	1.0000	171.89	**40**
21	.00611	.00611	163.70	.99998	1.0000	163.70	39
22	.00640	.00640	156.26	.99998	1.0000	156.26	38
23	.00669	.00669	149.47	.99998	1.0000	149.47	37
24	.00698	.00698	143.24	.99998	1.0000	143.24	36
25	.00727	.00727	137.51	.99997	1.0000	137.51	**35**
26	.00756	.00756	132.22	.99997	1.0000	132.22	34
27	.00785	.00785	127.32	.99997	1.0000	127.33	33
28	.00814	.00815	122.77	.99997	1.0000	122.78	32
29	.00844	.00844	118.54	.99996	1.0000	118.54	31
30	.00873	.00873	114.59	.99996	1.0000	114.59	**30**
31	.00902	.00902	110.89	.99996	1.0000	110.90	29
32	.00931	.00931	107.43	.99996	1.0000	107.43	28
33	.00960	.00960	104.17	.99995	1.0000	104.18	27
34	.00989	.00989	101.11	.99995	1.0000	101.11	26
35	.01018	.01018	98.218	.99995	1.0001	98.223	**25**
36	.01047	.01047	95.489	.99995	1.0001	95.495	24
37	.01076	.01076	92.908	.99994	1.0001	92.914	23
38	.01105	.01105	90.463	.99994	1.0001	90.469	22
39	.01134	.01135	88.144	.99994	1.0001	88.149	21
40	.01164	.01164	85.940	.99993	1.0001	85.946	**20**
41	.01193	.01193	83.844	.99993	1.0001	83.849	19
42	.01222	.01222	81.847	.99993	1.0001	81.853	18
43	.01251	.01251	79.943	.99992	1.0001	79.950	17
44	.01280	.01280	78.126	.99992	1.0001	78.133	16
45	.01309	.01309	76.390	.99991	1.0001	76.397	**15**
46	.01338	.01338	74.729	.99991	1.0001	74.736	14
47	.01367	.01367	73.139	.99991	1.0001	73.146	13
48	.01396	.01396	71.615	.99990	1.0001	71.622	12
49	.01425	.01425	70.153	.99990	1.0001	70.160	11
50	.01454	.01455	68.750	.99989	1.0001	68.757	**10**
51	.01483	.01484	67.402	.99989	1.0001	67.409	9
52	.01513	.01513	66.105	.99989	1.0001	66.113	8
53	.01542	.01542	64.858	.99988	1.0001	64.866	7
54	.01571	.01571	63.657	.99988	1.0001	63.665	6
55	.01600	.01600	62.499	.99987	1.0001	62.507	**5**
56	.01629	.01629	61.383	.99987	1.0001	61.391	4
57	.01658	.01658	60.306	.99986	1.0001	60.314	3
58	.01687	.01687	59.266	.99986	1.0001	59.274	2
59	.01716	.01716	58.261	.99985	1.0001	58.270	1
60	.01745	.01746	57.290	.99985	1.0002	57.299	**0**
′	Cos	Cot	Tan	Sin	Csc	Sec	′

90° (270°) — (269°) **89°**

1° (181°) — (358°) **178°**

′	Sin	Tan	Cot	Cos	Sec	Csc	′
0	.01745	.01746	57.290	.99985	1.0002	57.299	**60**
1	.01774	.01775	56.351	.99984	1.0002	56.359	59
2	.01803	.01804	55.442	.99984	1.0002	55.451	58
3	.01832	.01833	54.561	.99983	1.0002	54.570	57
4	.01862	.01862	53.709	.99983	1.0002	53.718	56
5	.01891	.01891	52.882	.99982	1.0002	52.892	**55**
6	.01920	.01920	52.081	.99982	1.0002	52.090	54
7	.01949	.01949	51.303	.99981	1.0002	51.313	53
8	.01978	.01978	50.549	.99980	1.0002	50.558	52
9	.02007	.02007	49.816	.99980	1.0002	49.826	51
10	.02036	.02036	49.104	.99979	1.0002	49.114	**50**
11	.02065	.02066	48.412	.99979	1.0002	48.422	49
12	.02094	.02095	47.740	.99978	1.0002	47.750	48
13	.02123	.02124	47.085	.99977	1.0002	47.096	47
14	.02152	.02153	46.449	.99977	1.0002	46.460	46
15	.02181	.02182	45.829	.99976	1.0002	45.840	**45**
16	.02211	.02211	45.226	.99976	1.0002	45.237	44
17	.02240	.02240	44.639	.99975	1.0003	44.650	43
18	.02269	.02269	44.066	.99974	1.0003	44.077	42
19	.02298	.02298	43.508	.99974	1.0003	43.520	41
20	.02327	.02328	42.964	.99973	1.0003	42.976	**40**
21	.02356	.02357	42.433	.99972	1.0003	42.445	39
22	.02385	.02386	41.916	.99972	1.0003	41.928	38
23	.02414	.02415	41.411	.99971	1.0003	41.423	37
24	.02443	.02444	40.917	.99970	1.0003	40.930	36
25	.02472	.02473	40.436	.99969	1.0003	40.448	**35**
26	.02501	.02502	39.965	.99969	1.0003	39.978	34
27	.02530	.02531	39.506	.99968	1.0003	39.519	33
28	.02560	.02560	39.057	.99967	1.0003	39.070	32
29	.02589	.02589	38.618	.99966	1.0003	38.631	31
30	.02618	.02619	38.188	.99966	1.0003	38.202	**30**
31	.02647	.02648	37.769	.99965	1.0004	37.782	29
32	.02676	.02677	37.358	.99964	1.0004	37.371	28
33	.02705	.02706	36.956	.99963	1.0004	36.970	27
34	.02734	.02735	36.563	.99963	1.0004	36.576	26
35	.02763	.02764	36.178	.99962	1.0004	36.191	**25**
36	.02792	.02793	35.801	.99961	1.0004	35.815	24
37	.02821	.02822	35.431	.99960	1.0004	35.445	23
38	.02850	.02851	35.070	.99959	1.0004	35.084	22
39	.02879	.02881	34.715	.99959	1.0004	34.730	21
40	.02908	.02910	34.368	.99958	1.0004	34.382	**20**
41	.02938	.02939	34.027	.99957	1.0004	34.042	19
42	.02967	.02968	33.694	.99956	1.0004	33.708	18
43	.02996	.02997	33.366	.99955	1.0004	33.381	17
44	.03025	.03026	33.045	.99954	1.0005	33.060	16
45	.03054	.03055	32.730	.99953	1.0005	32.746	**15**
46	.03083	.03084	32.421	.99952	1.0005	32.437	14
47	.03112	.03114	32.118	.99952	1.0005	32.134	13
48	.03141	.03143	31.821	.99951	1.0005	31.836	12
49	.03170	.03172	31.528	.99950	1.0005	31.544	11
50	.03199	.03201	31.242	.99949	1.0005	31.258	**10**
51	.03228	.03230	30.960	.99948	1.0005	30.976	9
52	.03257	.03259	30.683	.99947	1.0005	30.700	8
53	.03286	.03288	30.412	.99946	1.0005	30.428	7
54	.03316	.03317	30.145	.99945	1.0006	30.161	6
55	.03345	.03346	29.882	.99944	1.0006	29.899	**5**
56	.03374	.03376	29.624	.99943	1.0006	29.641	4
57	.03403	.03405	29.371	.99942	1.0006	29.388	3
58	.03432	.03434	29.122	.99941	1.0006	29.139	2
59	.03461	.03463	28.877	.99940	1.0006	28.894	1
60	.03490	.03492	28.636	.99939	1.0006	28.654	**0**
′	Cos	Cot	Tan	Sin	Csc	Sec	′

91° (271°) — (268°) **88°**

2° (182°) — (357°) **177°**

′	Sin	Tan	Cot	Cos	Sec	Csc	′
0	.03490	.03492	28.636	.99939	1.0006	28.654	**60**
1	.03519	.03521	28.399	.99938	1.0006	28.417	59
2	.03548	.03550	28.166	.99937	1.0006	28.184	58
3	.03577	.03579	27.937	.99936	1.0006	27.955	57
4	.03606	.03609	27.712	.99935	1.0007	27.730	56
5	.03635	.03638	27.490	.99934	1.0007	27.508	**55**
6	.03664	.03667	27.271	.99933	1.0007	27.290	54
7	.03693	.03696	27.057	.99932	1.0007	27.075	53
8	.03723	.03725	26.845	.99931	1.0007	26.864	52
9	.03752	.03754	26.637	.99930	1.0007	26.655	51
10	.03781	.03783	26.432	.99929	1.0007	26.451	**50**
11	.03810	.03812	26.230	.99927	1.0007	26.249	49
12	.03839	.03842	26.031	.09926	1.0007	26.050	48
13	.03868	.03871	25.835	.99925	1.0007	25.854	47
14	.03897	.03900	25.642	.99924	1.0008	25.661	46
15	.03926	.03929	25.452	.99923	1.0008	25.471	**45**
16	.03955	.03958	25.264	.99922	1.0008	25.284	44
17	.03984	.03987	25.080	.99921	1.0008	25.100	43
18	.04013	.04016	24.898	.99919	1.0008	24.918	42
19	.04042	.04046	24.719	.99918	1.0008	24.739	41
20	.04071	.04075	24.542	.99917	1.0008	24.562	**40**
21	.04100	.04104	24.368	.99916	1.0008	24.388	39
22	.04129	.04133	24.196	.99915	1.0009	24.216	38
23	.04159	.04162	24.026	.99913	1.0009	24.047	37
24	.04188	.04191	23.859	.99912	1.0009	23.880	36
25	.04217	.04220	23.695	.99911	1.0009	23.716	**35**
26	.04246	.04250	23.532	.99910	1.0009	23.553	34
27	.04275	.04279	23.372	.99909	1.0009	23.393	33
28	.04304	.04308	23.214	.99907	1.0009	23.235	32
29	.04333	.04337	23.058	.99906	1.0009	23.079	31
30	.04362	.04366	22.904	.99905	1.0010	22.926	**30**
31	.04391	.04395	22.752	.99904	1.0010	22.774	29
32	.04420	.04424	22.602	.99902	1.0010	22.624	28
33	.04449	.04454	22.454	.99901	1.0010	22.476	27
34	.04478	.04483	22.308	.99900	1.0010	22.330	26
35	.04507	.04512	22.164	.99898	1.0010	22.187	**25**
36	.04536	.04541	22.022	.99897	1.0010	22.044	24
37	.04565	.04570	21.881	.99896	1.0010	21.904	23
38	.04594	.04599	21.743	.99894	1.0011	21.766	22
39	.04623	.04628	21.606	.99893	1.0011	21.629	21
40	.04653	.04658	21.470	.99892	1.0011	21.494	**20**
41	.04682	.04687	21.337	.99890	1.0011	21.360	19
42	.04711	.04716	21.205	.99889	1.0011	21.229	18
43	.04740	.04745	21.075	.99888	1.0011	21.098	17
44	.04769	.04774	20.946	.99886	1.0011	20.970	16
45	.04798	.04803	20.819	.99885	1.0012	20.843	**15**
46	.04827	.04833	20.693	.99883	1.0012	20.717	14
47	.04856	.04862	20.569	.99882	1.0012	20.593	13
48	.04885	.04891	20.446	.99881	1.0012	20.471	12
49	.04914	.04920	20.325	.99879	1.0012	20.350	11
50	.04943	.04949	20.206	.99878	1.0012	20.230	**10**
51	.04972	.04978	20.087	.99876	1.0012	20.112	9
52	.05001	.05007	19.970	.99875	1.0013	19.995	8
53	.05030	.05037	19.855	.99873	1.0013	19.880	7
54	.05059	.05066	19.740	.99872	1.0013	19.766	6
55	.05088	.05095	19.627	.99870	1.0013	19.653	**5**
56	.05117	.05124	19.516	.99869	1.0013	19.541	4
57	.05146	.05153	19.405	.99867	1.0013	19.431	3
58	.05175	.05182	19.296	.99866	1.0013	19.322	2
59	.05205	.05212	19.188	.99864	1.0014	19.214	1
60	.05204	.05241	19.081	.99863	1.0014	19.107	**0**
′	Cos	Cot	Tan	Sin	Csc	Sec	′

92° (272°) — (267°) **87°**

3° (183°) — (356°) **176°**

′	Sin	Tan	Cot	Cos	Sec	Csc	′
0	.05234	.05241	19.081	.99863	1.0014	19.107	**60**
1	.05263	.05270	18.976	.99861	1.0014	19.002	59
2	.05292	.05299	18.871	.99860	1.0014	18.898	58
3	.05321	.05328	18.768	.99858	1.0014	18.794	57
4	.05350	.05357	18.666	.99857	1.0014	18.692	56
5	.05379	.05387	18.564	.99855	1.0014	18.591	**55**
6	.05408	.05416	18.464	.99854	1.0015	18.492	54
7	.05437	.05445	18.366	.99852	1.0015	18.393	53
8	.05466	.05474	18.268	.99851	1.0015	18.295	52
9	.05495	.05503	18.171	.99849	1.0015	18.198	51
10	.05524	.05533	18.075	.99847	1.0015	18.103	**50**
11	.05553	.05562	17.980	.99846	1.0015	18.008	49
12	.05582	.05591	17.886	.99844	1.0016	17.914	48
13	.05611	.05620	17.793	.99842	1.0016	17.822	47
14	.05640	.05649	17.702	.99841	1.0016	17.730	46
15	.05669	.05678	17.611	.99839	1.0016	17.639	**45**
16	.05698	.05708	17.521	.99838	1.0016	17.549	44
17	.05727	.05737	17.431	.99836	1.0016	17.460	43
18	.05756	.05766	17.343	.99834	1.0017	17.372	42
19	.05785	.05795	17.256	.99833	1.0017	17.285	41
20	.05814	.05824	17.169	.99831	1.0017	17.198	**40**
21	.05844	.05854	17.084	.99829	1.0017	17.113	39
22	.05873	.05883	16.999	.99827	1.0017	17.028	38
23	.05902	.05912	16.915	.99826	1.0017	16.945	37
24	.05931	.05941	16.832	.99824	1.0018	16.862	36
25	.05960	.05970	16.750	.99822	1.0018	16.779	**35**
26	.05989	.05999	16.668	.99821	1.0018	16.698	34
27	.06018	.06029	16.587	.99819	1.0018	16.618	33
28	.06047	.06058	16.507	.99817	1.0018	16.538	32
29	.06076	.06087	16.428	.99815	1.0019	16.459	31
30	.06105	.06116	16.350	.99813	1.0019	16.380	**30**
31	.06134	.06145	16.272	.99812	1.0019	16.303	29
32	.06163	.06175	16.195	.99810	1.0019	16.226	28
33	.06192	.06204	16.119	.99808	1.0019	16.150	27
34	.06221	.06233	16.043	.99806	1.0019	16.075	26
35	.06250	.06262	15.969	.99804	1.0020	16.000	**25**
36	.06279	.06291	15.895	.99803	1.0020	15.926	24
37	.06308	.06321	15.821	.99801	1.0020	15.853	23
38	.06337	.06350	15.748	.99799	1.0020	15.780	22
39	.06366	.06379	15.676	.99797	1.0020	15.708	21
40	.06395	.06408	15.605	.99795	1.0021	15.637	**20**
41	.06424	.06438	15.534	.99793	1.0021	15.566	19
42	.06453	.06467	15.464	.99792	1.0021	15.496	18
43	.06482	.06496	15.394	.99790	1.0021	15.427	17
44	.06511	.06525	15.325	.99788	1.0021	15.358	16
45	.06540	.06554	15.257	.99786	1.0021	15.290	**15**
46	.06569	.06584	15.189	.99784	1.0022	15.222	14
47	.06598	.06613	15.122	.99782	1.0022	15.155	13
48	.06627	.06642	15.056	.99780	1.0022	15.089	12
49	.06656	.06671	14.990	.99778	1.0022	15.023	11
50	.06685	.06700	14.924	.99776	1.0022	14.958	**10**
51	.06714	.06730	14.860	.99774	1.0023	14.893	9
52	.06743	.06759	14.795	.99772	1.0023	14.829	8
53	.06773	.06788	14.732	.99770	1.0023	14.766	7
54	.06802	.06817	14.669	.99768	1.0023	14.703	6
55	.06831	.06847	14.606	.99766	1.0023	14.640	**5**
56	.06860	.06876	14.544	.99764	1.0024	14.578	4
57	.06889	.06905	14.482	.99762	1.0024	14.517	3
58	.06918	.06934	14.421	.99760	1.0024	14.456	2
59	.06947	.06963	14.361	.99758	1.0024	14.395	1
60	.06976	.06993	14.301	.99756	1.0024	14.336	**0**
′	Cos	Cot	Tan	Sin	Csc	Sec	′

93° (273°) — (266°) **86°**

4° (184°) (355°) **175°**

′	Sin	Tan	Cot	Cos	Sec	Csc	′
0	.06976	.06993	14.301	.99756	1.0024	14.336	**60**
1	.07005	.07022	14.241	.99754	1.0025	14.276	59
2	.07034	.07051	14.182	.99752	1.0025	14.217	58
3	.07063	.07080	14.124	.99750	1.0025	14.159	57
4	.07092	.07110	14.065	.99748	1.0025	14.101	56
5	.07121	.07139	14.008	.99746	1.0025	14.044	**55**
6	.07150	.07168	13.951	.99744	1.0026	13.987	54
7	.07179	.07197	13.894	.99742	1.0026	13.930	53
8	.07208	.07227	13.838	.99740	1.0026	13.874	52
9	.07237	.07256	13.782	.99738	1.0026	13.818	51
10	.07266	.07285	13.727	.99736	1.0027	13.763	**50**
11	.07295	.07314	13.672	.99734	1.0027	13.708	49
12	.07324	.07344	13.617	.99731	1.0027	13.654	48
13	.07353	.07373	13.563	.99729	1.0027	13.600	47
14	.07382	.07402	13.510	.99727	1.0027	13.547	46
15	.07411	.07431	13.457	.99725	1.0028	13.494	**45**
16	.07440	.07461	13.404	.99723	1.0028	13.441	44
17	.07469	.07490	13.352	.99721	1.0028	13.389	43
18	.07498	.07519	13.300	.99719	1.0028	13.337	42
19	.07527	.07548	13.248	.99716	1.0028	13.286	41
20	.07556	.07578	13.197	.99714	1.0029	13.235	**40**
21	.07585	.07607	13.146	.99712	1.0029	13.184	39
22	.07614	.07636	13.096	.99710	1.0029	13.134	38
23	.07643	.07665	13.046	.99708	1.0029	13.084	37
24	.07672	.07695	12.996	.99705	1.0030	13.035	36
25	.07701	.07724	12.947	.99703	1.0030	12.985	**35**
26	.07730	.07753	12.898	.99701	1.0030	12.937	34
27	.07759	.07782	12.850	.99699	1.0030	12.888	33
28	.07788	.07812	12.801	.99696	1.0030	12.840	32
29	.07817	.07841	12.754	.99694	1.0031	12.793	31
30	.07846	.07870	12.706	.99692	1.0031	12.745	**30**
31	.07875	.07899	12.659	.99689	1.0031	12.699	29
32	.07904	.07929	12.612	.99687	1.0031	12.652	28
33	.07933	.07958	12.566	.99685	1.0032	12.606	27
34	.07962	.07987	12.520	.99683	1.0032	12.560	26
35	.07991	.08017	12.474	.99680	1.0032	12.514	**25**
36	.08020	.08046	12.429	.99678	1.0032	12.469	24
37	.08049	.08075	12.384	.99676	1.0033	12.424	23
38	.08078	.08104	12.339	.99673	1.0033	12.379	22
39	.08107	.08134	12.295	.99671	1.0033	12.335	21
40	.08136	.08163	12.251	.99668	1.0033	12.291	**20**
41	.08165	.08192	12.207	.99666	1.0034	12.248	19
42	.08194	.08221	12.163	.99664	1.0034	12.204	18
43	.08223	.08251	12.120	.99661	1.0034	12.161	17
44	.08252	.08280	12.077	.99659	1.0034	12.119	16
45	.08281	.08309	12.035	.99657	1.0034	12.076	**15**
46	.08310	.08339	11.992	.99654	1.0035	12.034	14
47	.08339	.08368	11.950	.99652	1.0035	11.992	13
48	.08368	.08397	11.909	.99649	1.0035	11.951	12
49	.08397	.08427	11.867	.99647	1.0035	11.909	11
50	.08426	.08456	11.826	.99644	1.0036	11.868	**10**
51	.08455	.08485	11.785	.99642	1.0036	11.828	9
52	.08484	.08514	11.745	.99639	1.0036	11.787	8
53	.08513	.08544	11.705	.99637	1.0036	11.747	7
54	.08542	.08573	11.664	.99635	1.0037	11.707	6
55	.08571	.08602	11.625	.99632	1.0037	11.668	**5**
56	.08600	.08632	11.585	.99630	1.0037	11.628	4
57	.08629	.08661	11.546	.99627	1.0037	11.589	3
58	.08658	.08690	11.507	.99625	1.0038	11.551	2
59	.08687	.08720	11.468	.99622	1.0038	11.512	1
60	.08716	.08749	11.430	.99619	1.0038	11.474	**0**
′	Cos	Cot	Tan	Sin	Csc	Sec	′

94° (274°) (265°) **85°**

5° (185°) (354°) **174°**

′	Sin	Tan	Cot	Cos	Sec	Csc	′
0	.08716	.08749	11.430	.99619	1.0038	11.474	**60**
1	.08745	.08778	11.392	.99617	1.0038	11.436	59
2	.08774	.08807	11.354	.99614	1.0039	11.398	58
3	.08803	.08837	11.316	.99612	1.0039	11.360	57
4	.08831	.08866	11.279	.99609	1.0039	11.323	56
5	.08860	.08895	11.242	.99607	1.0039	11.286	**55**
6	.08889	.08925	11.205	.99604	1.0040	11.249	54
7	.08918	.08954	11.168	.99602	1.0040	11.213	53
8	.08947	.08983	11.132	.99599	1.0040	11.176	52
9	.08976	.09013	11.095	.99596	1.0041	11.140	51
10	.09005	.09042	11.059	.99594	1.0041	11.105	**50**
11	.09034	.09071	11.024	.99591	1.0041	11.069	49
12	.09063	.09101	10.988	.99588	1.0041	11.034	48
13	.09092	.09130	10.953	.99586	1.0042	10.998	47
14	.09121	.09159	10.918	.99583	1.0042	10.963	46
15	.09150	.09189	10.883	.99580	1.0042	10.929	**45**
16	.09179	.09218	10.848	.99578	1.0042	10.894	44
17	.09208	.09247	10.814	.99575	1.0043	10.860	43
18	.09237	.09277	10.780	.99572	1.0043	10.826	42
19	.09266	.09306	10.746	.99570	1.0043	10.792	41
20	.09295	.09335	10.712	.99567	1.0043	10.758	**40**
21	.09324	.09365	10.678	.99564	1.0044	10.725	39
22	.09353	.09394	10.645	.99562	1.0044	10.692	38
23	.09382	.09423	10.612	.99559	1.0044	10.659	37
24	.09411	.09453	10.579	.99556	1.0045	10.626	36
25	.09440	.09482	10.546	.99553	1.0045	10.593	**35**
26	.09469	.09511	10.514	.99551	1.0045	10.561	34
27	.09498	.09541	10.481	.99548	1.0045	10.529	33
28	.09527	.09570	10.449	.99545	1.0046	10.497	32
29	.09556	.09600	10.417	.99542	1.0046	10.465	31
30	.09585	.09629	10.385	.99540	1.0046	10.433	**30**
31	.09614	.09658	10.354	.99537	1.0047	10.402	29
32	.09642	.09688	10.322	.99534	1.0047	10.371	28
33	.09671	.09717	10.291	.99531	1.0047	10.340	27
34	.09700	.09746	10.260	.99528	1.0047	10.309	26
35	.09729	.09776	10.229	.99526	1.0048	10.278	**25**
36	.09758	.09805	10.199	.99523	1.0048	10.248	24
37	.09787	.09834	10.168	.99520	1.0048	10.217	23
38	.09816	.09864	10.138	.99517	1.0049	10.187	22
39	.09845	.09893	10.108	.99514	1.0049	10.157	21
40	.09874	.09923	10.078	.99511	1.0049	10.128	**20**
41	.09903	.09952	10.048	.99508	1.0049	10.098	19
42	.09932	.09981	10.019	.99506	1.0050	10.068	18
43	.09961	.10011	9.9893	.99503	1.0050	10.039	17
44	.09990	.10040	9.9601	.99500	1.0050	10.010	16
45	.10019	.10069	9.9310	.99497	1.0051	9.9812	**15**
46	.10048	.10099	9.9021	.99494	1.0051	9.9525	14
47	.10077	.10128	9.8734	.99491	1.0051	9.9239	13
48	.10106	.10158	9.8448	.99488	1.0051	9.8955	12
49	.10135	.10187	9.8164	.99485	1.0052	9.8672	11
50	.10164	.10216	9.7882	.99482	1.0052	9.8391	**10**
51	.10192	.10246	9.7601	.99479	1.0052	9.8112	9
52	.10221	.10275	9.7322	.99476	1.0053	9.7834	8
53	.10250	.10305	9.7044	.99473	1.0053	9.7558	7
54	.10279	.10334	9.6768	.99470	1.0053	9.7283	6
55	.10308	.10363	9.6493	.99467	1.0054	9.7010	**5**
56	.10337	.10393	9.6220	.99464	1.0054	9.6739	4
57	.10366	.10422	9.5949	.99461	1.0054	9.6469	3
58	.10395	.10452	9.5679	.99458	1.0054	9.6200	2
59	.10424	.10481	9.5411	.99455	1.0055	9.5933	1
60	.10453	.10510	9.5144	.99452	1.0055	9.5668	**0**
′	Cos	Cot	Tan	Sin	Csc	Sec	′

95° (275°) (264°) **84°**

NATURAL TRIGONOMETRIC FUNCTIONS (*Continued*)

6° (186°) — (353°) **173°**

′	Sin	Tan	Cot	Cos	Sec	Csc	′
0	.10453	.10510	9.5144	.99452	1.0055	9.5668	**60**
1	.10482	.10540	9.4878	.99449	1.0055	9.5404	59
2	.10511	.10569	9.4614	.99446	1.0056	9.5141	58
3	.10540	.10599	9.4352	.99443	1.0056	9.4880	57
4	.10569	.10628	9.4090	.99440	1.0056	9.4620	56
5	.10597	.10657	9.3831	.99437	1.0057	9.4362	**55**
6	.10626	.10687	9.3572	.99434	1.0057	9.4105	54
7	.10655	.10716	9.3315	.99431	1.0057	9.3850	53
8	.10684	.10746	9.3060	.99428	1.0058	9.3596	52
9	.10713	.10775	9.2806	.99424	1.0058	9.3343	51
10	.10742	.10805	9.2553	.99421	1.0058	9.3092	**50**
11	.10771	.10834	9.2302	.99418	1.0059	9.2842	49
12	.10800	.10863	9.2052	.99415	1.0059	9.2593	48
13	.10829	.10893	9.1803	.99412	1.0059	9.2346	47
14	.10858	.10922	9.1555	.99409	1.0059	9.2100	46
15	.10887	.10952	9.1309	.99406	1.0060	9.1855	**45**
16	.10916	.10981	9.1065	.99402	1.0060	9.1612	44
17	.10945	.11011	9.0821	.99399	1.0060	9.1370	43
18	.10973	.11040	9.0579	.99396	1.0061	9.1129	42
19	.11002	.11070	9.0338	.99393	1.0061	9.0890	41
20	.11031	.11099	9.0098	.99390	1.0061	9.0652	**40**
21	.11060	.11128	8.9860	.99386	1.0062	9.0415	39
22	.11089	.11158	8.9623	.99383	1.0062	9.0179	38
23	.11118	.11187	8.9387	.99380	1.0062	8.9944	37
24	.11147	.11217	8.9152	.99377	1.0063	8.9711	36
25	.11176	.11246	8.8919	.99374	1.0063	8.9479	**35**
26	.11205	.11276	8.8686	.99370	1.0063	8.9248	34
27	.11234	.11305	8.8455	.99367	1.0064	8.9019	33
28	.11263	.11335	8.8225	.99364	1.0064	8.8790	32
29	.11291	.11364	8.7996	.99360	1.0064	8.8563	31
30	.11320	.11394	8.7769	.99357	1.0065	8.8337	**30**
31	.11349	.11423	8.7542	.99354	1.0065	8.8112	29
32	.11378	.11452	8.7317	.99351	1.0065	8.7888	28
33	.11407	.11482	8.7093	.99347	1.0066	8.7665	27
34	.11436	.11511	8.6870	.99344	1.0066	8.7444	26
35	.11465	.11541	8.6648	.99341	1.0066	8.7223	**25**
36	.11494	.11570	8.6427	.99337	1.0067	8.7004	24
37	.11523	.11600	8.6208	.99334	1.0067	8.6786	23
38	.11552	.11629	8.5989	.99331	1.0067	8.6569	22
39	.11580	.11659	8.5772	.99327	1.0068	8.6353	21
40	.11609	.11688	8.5555	.99324	1.0068	8.6138	**20**
41	.11638	.11718	8.5340	.99320	1.0068	8.5924	19
42	.11667	.11747	8.5126	.99317	1.0069	8.5711	18
43	.11696	.11777	8.4913	.99314	1.0069	8.5500	17
44	.11725	.11806	8.4701	.99310	1.0069	8.5289	16
45	.11754	.11836	8.4490	.99307	1.0070	8.5079	**15**
46	.11783	.11865	8.4280	.99303	1.0070	8.4871	14
47	.11812	.11895	8.4071	.99300	1.0070	8.4663	13
48	.11840	.11924	8.3863	.99297	1.0071	8.4457	12
49	.11869	.11954	8.3656	.99293	1.0071	8.4251	11
50	.11898	.11983	8.3450	.99290	1.0072	8.4047	**10**
51	.11927	.12013	8.3245	.99286	1.0072	8.3843	9
52	.11956	.12042	8.3041	.99283	1.0072	8.3641	8
53	.11985	.12072	8.2838	.99279	1.0073	8.3439	7
54	.12014	.12101	8.2636	.99276	1.0073	8.3238	6
55	.12043	.12131	8.2434	.99272	1.0073	8.3039	**5**
56	.12071	.12160	8.2234	.99269	1.0074	8.2840	4
57	.12100	.12190	8.2035	.99265	1.0074	8.2642	3
58	.12129	.12219	8.1837	.99262	1.0074	8.2446	2
59	.12158	.12249	8.1640	.99258	1.0075	8.2250	1
60	.12187	.12278	8.1443	.99255	1.0075	8.2055	**0**
′	Cos	Cot	Tan	Sin	Csc	Sec	′

96° (276°) — (263°) **83°**

7° (187°) — (352°) **172°**

′	Sin	Tan	Cot	Cos	Sec	Csc	′
0	.12187	.12278	8.1443	.99255	1.0075	8.2055	**60**
1	.12216	.12308	8.1248	.99251	1.0075	8.1861	59
2	.12245	.12338	8.1054	.99248	1.0076	8.1668	58
3	.12274	.12367	8.0860	.99244	1.0076	8.1476	57
4	.12302	.12397	8.0667	.99240	1.0077	8.1285	56
5	.12331	.12426	8.0476	.99237	1.0077	8.1095	**55**
6	.12360	.12456	8.0285	.99233	1.0077	8.0905	54
7	.12389	.12485	8.0095	.99230	1.0078	8.0717	53
8	.12418	.12515	7.9906	.99226	1.0078	8.0529	52
9	.12447	.12544	7.9718	.99222	1.0078	8.0342	51
10	.12476	.12574	7.9530	.99219	1.0079	8.0156	**50**
11	.12504	.12603	7.9344	.99215	1.0079	7.9971	49
12	.12533	.12633	7.9158	.99211	1.0079	7.9787	48
13	.12562	.12662	7.8973	.99208	1.0080	7.9604	47
14	.12591	.12692	7.8789	.99204	1.0080	7.9422	46
15	.12620	.12722	7.8606	.99200	1.0081	7.9240	**45**
16	.12649	.12751	7.8424	.99197	1.0081	7.9059	44
17	.12678	.12781	7.8243	.99193	1.0081	7.8879	43
18	.12706	.12810	7.8062	.99189	1.0082	7.8700	42
19	.12735	.12840	7.7882	.99186	1.0082	7.8522	41
20	.12764	.12869	7.7704	.99182	1.0082	7.8344	**40**
21	.12793	.12899	7.7525	.99178	1.0083	7.8168	39
22	.12822	.12929	7.7348	.99175	1.0083	7.7992	38
23	.12851	.12958	7.7171	.99171	1.0084	7.7817	37
24	.12880	.12988	7.6996	.99167	1.0084	7.7642	36
25	.12908	.13017	7.6821	.99163	1.0084	7.7469	**35**
26	.12937	.13047	7.6647	.99160	1.0085	7.7296	34
27	.12966	.13076	7.6473	.99156	1.0085	7.7124	33
28	.12995	.13106	7.6301	.99152	1.0086	7.6953	32
29	.13024	.13136	7.6129	.99148	1.0086	7.6783	31
30	.13053	.13165	7.5958	.99144	1.0086	7.6613	**30**
31	.13081	.13195	7.5787	.99141	1.0087	7.6444	29
32	.13110	.13224	7.5618	.99137	1.0087	7.6276	28
33	.13139	.13254	7.5449	.99133	1.0087	7.6109	27
34	.13168	.13284	7.5281	.99129	1.0088	7.5942	26
35	.13197	.13313	7.5113	.99125	1.0088	7.5776	**25**
36	.13226	.13343	7.4947	.99122	1.0089	7.5611	24
37	.13254	.13372	7.4781	.99118	1.0089	7.5446	23
38	.13283	.13402	7.4615	.99114	1.0089	7.5282	22
39	.13312	.13432	7.4451	.99110	1.0090	7.5119	21
40	.13341	.13461	7.4287	.99106	1.0090	7.4957	**20**
41	.13370	.13491	7.4124	.99102	1.0091	7.4795	19
42	.13399	.13521	7.3962	.99098	1.0091	7.4635	18
43	.13427	.13550	7.3800	.99094	1.0091	7.4474	17
44	.13456	.13580	7.3639	.99091	1.0092	7.4315	16
45	.13485	.13609	7.3479	.99087	1.0092	7.4156	**15**
46	.13514	.13639	7.3319	.99083	1.0093	7.3998	14
47	.13543	.13669	7.3160	.99079	1.0093	7.3840	13
48	.13572	.13698	7.3002	.99075	1.0093	7.3684	12
49	.13600	.13728	7.2844	.99071	1.0094	7.3527	11
50	.13629	.13758	7.2687	.99067	1.0094	7.3372	**10**
51	.13658	.13787	7.2531	.99063	1.0095	7.3217	9
52	.13687	.13817	7.2375	.99059	1.0095	7.3063	8
53	.13716	.13846	7.2220	.99055	1.0095	7.2909	7
54	.13744	.13876	7.2066	.99051	1.0096	7.2757	6
55	.13773	.13906	7.1912	.99047	1.0096	7.2604	**5**
56	.13802	.13935	7.1759	.99043	1.0097	7.2453	4
57	.13831	.13965	7.1607	.99039	1.0097	7.2302	3
58	.13860	.13995	7.1455	.99035	1.0097	7.2152	2
59	.13889	.14024	7.1304	.99031	1.0098	7.2002	1
60	.13917	.14054	7.1154	.99027	1.0098	7.1853	**0**
′	Cos	Cot	Tan	Sin	Csc	Sec	′

97° (277°) — (262°) **82°**

8° (188°) (351°) **171°**

′	Sin	Tan	Cot	Cos	Sec	Csc	′
0	.13917	.14054	7.1154	.99027	1.0098	7.1853	**60**
1	.13946	.14084	7.1004	.99023	1.0099	7.1705	59
2	.13975	.14113	7.0855	.99019	1.0099	7.1557	58
3	.14004	.14143	7.0706	.99015	1.0100	7.1410	57
4	.14033	.14173	7.0558	.99011	1.0100	7.1263	56
5	.14061	.14202	7.0410	.99006	1.0100	7.1117	**55**
6	.14090	.14232	7.0264	.99002	1.0101	7.0972	54
7	.14119	.14262	7.0117	.98998	1.0101	7.0827	53
8	.14148	.14291	6.9972	.98994	1.0102	7.0683	52
9	.14177	.14321	6.9827	.98990	1.0102	7.0539	51
10	.14205	.14351	6.9682	.98986	1.0102	7.0396	**50**
11	.14234	.14381	6.9538	.98982	1.0103	7.0254	49
12	.14263	.14410	6.9395	.98978	1.0103	7.0112	48
13	.14292	.14440	6.9252	.98973	1.0104	6.9971	47
14	.14320	.14470	6.9110	.98969	1.0104	6.9830	46
15	.14349	.14499	6.8969	.98965	1.0105	6.9690	**45**
16	.14378	.14529	6.8828	.98961	1.0105	6.9550	44
17	.14407	.14559	6.8687	.98957	1.0105	6.9411	43
18	.14436	.14588	6.8548	.98953	1.0106	6.9273	42
19	.14464	.14618	6.8408	.98948	1.0106	6.9135	41
20	.14493	.14648	6.8269	.98944	1.0107	6.8998	**40**
21	.14522	.14678	6.8131	.98940	1.0107	6.8861	39
22	.14551	.14707	6.7994	.98936	1.0108	6.8725	38
23	.14580	.14737	6.7856	.98931	1.0108	6.8589	37
24	.14608	.14767	6.7720	.98927	1.0108	6.8454	36
25	.14637	.14796	6.7584	.98923	1.0109	6.8320	**35**
26	.14666	.14826	6.7448	.98919	1.0109	6.8186	34
27	.14695	.14856	6.7313	.98914	1.0110	6.8052	33
28	.14723	.14886	6.7179	.98910	1.0110	6.7919	32
29	.14752	.14915	6.7045	.98906	1.0111	6.7787	31
30	.14781	.14945	6.6912	.98902	1.0111	6.7655	**30**
31	.14810	.14975	6.6779	.98897	1.0112	6.7523	29
32	.14838	.15005	6.6646	.98893	1.0112	6.7392	28
33	.14867	.15034	6.6514	.98889	1.0112	6.7262	27
34	.14896	.15064	6.6383	.98884	1.0113	6.7132	26
35	.14925	.15094	6.6252	.98880	1.0113	6.7003	**25**
36	.14954	.15124	6.6122	.98876	1.0114	6.6874	24
37	.14982	.15153	6.5992	.98871	1.0114	6.6745	23
38	.15011	.15183	6.5863	.98867	1.0115	6.6618	22
39	.15040	.15213	6.5734	.98863	1.0115	6.6490	21
40	.15069	.15243	6.5606	.98858	1.0116	6.6363	**20**
41	.15097	.15272	6.5478	.98854	1.0116	6.6237	19
42	.15126	.15302	6.5350	.98849	1.0116	6.6111	18
43	.15155	.15332	6.5223	.98845	1.0117	6.5986	17
44	.15184	.15362	6.5097	.98841	1.0117	6.5861	16
45	.15212	.15391	6.4971	.98836	1.0118	6.5736	**15**
46	.15241	.15421	6.4846	.98832	1.0118	6.5612	14
47	.15270	.15451	6.4721	.98827	1.0119	6.5489	13
48	.15299	.15481	6.4596	.98823	1.0119	6.5366	12
49	.15327	.15511	6.4472	.98818	1.0120	6.5243	11
50	.15356	.15540	6.4348	.98814	1.0120	6.5121	**10**
51	.15385	.15570	6.4225	.98809	1.0120	6.4999	9
52	.15414	.15600	6.4103	.98805	1.0121	6.4878	8
53	.15442	.15630	6.3980	.98800	1.0121	6.4757	7
54	.15471	.15660	6.3859	.98796	1.0122	6.4637	6
55	.15500	.15689	6.3737	.98791	1.0122	6.4517	**5**
56	.15529	.15719	6.3617	.98787	1.0123	6.4398	4
57	.15557	.15749	6.3496	.98782	1.0123	6.4279	3
58	.15586	.15779	6.3376	.98778	1.0124	6.4160	2
59	.15615	.15809	6.3257	.98773	1.0124	6.4042	1
60	.15643	.15838	6.3138	.98769	1.0125	6.3925	**0**
′	Cos	Cot	Tan	Sin	Csc	Sec	′

98° (278°) (261°) **81°**

9° (189°) (350°) **170°**

′	Sin	Tan	Cot	Cos	Sec	Csc	′
0	.15643	.15838	6.3138	.98769	1.0125	6.3925	**60**
1	.15672	.15868	6.3019	.98764	1.0125	6.3807	59
2	.15701	.15898	6.2901	.98760	1.0126	6.3691	58
3	.15730	.15928	6.2783	.98755	1.0126	6.3574	57
4	.15758	.15958	6.2666	.98751	1.0127	6.3458	56
5	.15787	.15988	6.2549	.98746	1.0127	6.3343	**55**
6	.15816	.16017	6.2432	.98741	1.0127	6.3228	54
7	.15845	.16047	6.2316	.98737	1.0128	6.3113	53
8	.15873	.16077	6.2200	.98732	1.0128	6.2999	52
9	.15902	.16107	6.2085	.98728	1.0129	6.2885	51
10	.15931	.16137	6.1970	.98723	1.0129	6.2772	**50**
11	.15959	.16167	6.1856	.98718	1.0130	6.2659	49
12	.15988	.16196	6.1742	.98714	1.0130	6.2546	48
13	.16017	.16226	6.1628	.98709	1.0131	6.2434	47
14	.16046	.16256	6.1515	.98704	1.0131	6.2323	46
15	.16074	.16286	6.1402	.98700	1.0132	6.2211	**45**
16	.16103	.16316	6.1290	.98695	1.0132	6.2100	44
17	.16132	.16346	6.1178	.98690	1.0133	6.1990	43
18	.16160	.16376	6.1066	.98686	1.0133	6.1880	42
19	.16189	.16405	6.0955	.98681	1.0134	6.1770	41
20	.16218	.16435	6.0844	.98676	1.0134	6.1661	**40**
21	.16246	.16465	6.0734	.98671	1.0135	6.1552	39
22	.16275	.16495	6.0624	.98667	1.0135	6.1443	38
23	.16304	.16525	6.0514	.98662	1.0136	6.1335	37
24	.16333	.16555	6.0405	.98657	1.0136	6.1227	36
25	.16361	.16585	6.0296	.98652	1.0137	6.1120	**35**
26	.16390	.16615	6.0188	.98648	1.0137	6.1013	34
27	.16419	.16645	6.0080	.98643	1.0138	6.0906	33
28	.16447	.16674	5.9972	.98638	1.0138	6.0800	32
29	.16476	.16704	5.9865	.98633	1.0139	6.0694	31
30	.16505	.16734	5.9758	.98629	1.0139	6.0589	**30**
31	.16533	.16764	5.9651	.98624	1.0140	6.0483	29
32	.16562	.16794	5.9545	.98619	1.0140	6.0379	28
33	.16591	.16824	5.9439	.98614	1.0141	6.0274	27
34	.16620	.16854	5.9333	.98609	1.0141	6.0170	26
35	.16648	.16884	5.9228	.98604	1.0142	6.0067	**25**
36	.16677	.16914	5.9124	.98600	1.0142	5.9963	24
37	.16706	.16944	5.9019	.98595	1.0143	5.9860	23
38	.16734	.16974	5.8915	.98590	1.0143	5.9758	22
39	.16763	.17004	5.8811	.98585	1.0144	5.9656	21
40	.16792	.17033	5.8708	.98580	1.0144	5.9554	**20**
41	.16820	.17063	5.8605	.98575	1.0145	5.9452	19
42	.16849	.17093	5.8502	.98570	1.0145	5.9351	18
43	.16878	.17123	5.8400	.98565	1.0146	5.9250	17
44	.16906	.17153	5.8298	.98561	1.0146	5.9150	16
45	.16935	.17183	5.8197	.98556	1.0147	5.9049	**15**
46	.16964	.17213	5.8095	.98551	1.0147	5.8950	14
47	.16992	.17243	5.7994	.98546	1.0148	5.8850	13
48	.17021	.17273	5.7894	.98541	1.0148	5.8751	12
49	.17050	.17303	5.7794	.98536	1.0149	5.8652	11
50	.17078	.17333	5.7694	.98531	1.0149	5.8554	**10**
51	.17107	.17363	5.7594	.98526	1.0150	5.8456	9
52	.17136	.17393	5.7495	.98521	1.0150	5.8358	8
53	.17164	.17423	5.7396	.98516	1.0151	5.8261	7
54	.17193	.17453	5.7297	.98511	1.0151	5.8164	6
55	.17222	.17483	5.7199	.98506	1.0152	5.8067	**5**
56	.17250	.17513	5.7101	.98501	1.0152	5.7970	4
57	.17279	.17543	5.7004	.98496	1.0153	5.7874	3
58	.17308	.17573	5.6906	.98491	1.0153	5.7778	2
59	.17336	.17603	5.6809	.98486	1.0154	5.7683	1
60	.17365	.17633	5.6713	.98481	1.0154	5.7588	**0**
′	Cos	Cot	Tan	Sin	Csc	Sec	′

99° (279°) (260°) **80°**

NATURAL TRIGONOMETRIC FUNCTIONS (*Continued*)

10° (190°) (349°) **169°**

′	Sin	Tan	Cot	Cos	Sec	Csc	′
0	.17365	.17633	5.6713	.98481	1.0154	5.7588	**60**
1	.17393	.17663	5.6617	.98476	1.0155	5.7493	59
2	.17422	.17693	5.6521	.98471	1.0155	5.7398	58
3	.17451	.17723	5.6425	.98466	1.0156	5.7304	57
4	.17479	.17753	5.6329	.98461	1.0156	5.7210	56
5	.17508	.17783	5.6234	.98455	1.0157	5.7117	**55**
6	.17537	.17813	5.6140	.98450	1.0157	5.7023	54
7	.17565	.17843	5.6045	.98445	1.0158	5.6930	53
8	.17594	.17873	5.5951	.98440	1.0158	5.6838	52
9	.17623	.17903	5.5857	.98435	1.0159	5.6745	51
10	.17651	.17933	5.5764	.98430	1.0160	5.6653	**50**
11	.17680	.17963	5.5671	.98425	1.0160	5.6562	49
12	.17708	.17993	5.5578	.98420	1.0161	5.6470	48
13	.17737	.18023	5.5485	.98414	1.0161	5.6379	47
14	.17766	.18053	5.5393	.98409	1.0162	5.6288	46
15	.17794	.18083	5.5301	.98404	1.0162	5.6198	**45**
16	.17823	.18113	5.5209	.98399	1.0163	5.6107	44
17	.17852	.18143	5.5118	.98394	1.0163	5.6017	43
18	.17880	.18173	5.5026	.98389	1.0164	5.5928	42
19	.17909	.18203	5.4936	.98383	1.0164	5.5838	41
20	.17937	.18233	5.4845	.98378	1.0165	5.5749	**40**
21	.17966	.18263	5.4755	.98373	1.0165	5.5660	39
22	.17995	.18293	5.4665	.98368	1.0166	5.5572	38
23	.18023	.18323	5.4575	.98362	1.0166	5.5484	37
24	.18052	.18353	5.4486	.98357	1.0167	5.5396	36
25	.18081	.18384	5.4397	.98352	1.0168	5.5308	**35**
26	.18109	.18414	5.4308	.98347	1.0168	5.5221	34
27	.18138	.18444	5.4219	.98341	1.0169	5.5134	33
28	.18166	.18474	5.4131	.98336	1.0169	5.5047	32
29	.18195	.18504	5.4043	.98331	1.0170	5.4960	31
30	.18224	.18534	5.3955	.98325	1.0170	5.4874	**30**
31	.18252	.18564	5.3868	.98320	1.0171	5.4788	29
32	.18281	.18594	5.3781	.98315	1.0171	5.4702	28
33	.18309	.18624	5.3694	.98310	1.0172	5.4617	27
34	.18338	.18654	5.3607	.98304	1.0173	5.4532	26
35	.18367	.18684	5.3521	.98299	1.0173	5.4447	**25**
36	.18395	.18714	5.3435	.98294	1.0174	5.4362	24
37	.18424	.18745	5.3349	.98288	1.0174	5.4278	23
38	.18452	.18775	5.3263	.98283	1.0175	5.4194	22
39	.18481	.18805	5.3178	.98277	1.0175	5.4110	21
40	.18509	.18835	5.3093	.98272	1.0176	5.4026	**20**
41	.18538	.18865	5.3008	.98267	1.0176	5.3943	19
42	.18567	.18895	5.2924	.98261	1.0177	5.3860	18
43	.18595	.18925	5.2839	.98256	1.0178	5.3777	17
44	.18624	.18955	5.2755	.98250	1.0178	5.3695	16
45	.18652	.18986	5.2672	.98245	1.0179	5.3612	**15**
46	.18681	.19016	5.2588	.98240	1.0179	5.3530	14
47	.18710	.19046	5.2505	.98234	1.0180	5.3449	13
48	.18738	.19076	5.2422	.98229	1.0180	5.3367	12
49	.18767	.19106	5.2339	.98223	1.0181	5.3286	11
50	.18795	.19136	5.2257	.98218	1.0181	5.3205	**10**
51	.18824	.19166	5.2174	.98212	1.0182	5.3124	9
52	.18852	.19197	5.2092	.98207	1.0183	5.3044	8
53	.18881	.19227	5.2011	.98201	1.0183	5.2963	7
54	.18910	.19257	5.1929	.98196	1.0184	5.2883	6
55	.18938	.19287	5.1848	.98190	1.0184	5.2804	**5**
56	.18967	.19317	5.1767	.98185	1.0185	5.2724	4
57	.18995	.19347	5.1686	.98179	1.0185	5.2645	3
58	.19024	.19378	5.1606	.98174	1.0186	5.2566	2
59	.19052	.19408	5.1526	.98168	1.0187	5.2487	1
60	.19081	.19438	5.1446	.98163	1.0187	5.2408	**0**
′	Cos	Cot	Tan	Sin	Csc	Sec	′

100° (280°) (259°) **79°**

11° (191°) (348°) **168°**

′	Sin	Tan	Cot	Cos	Sec	Csc	′
0	.19081	.19438	5.1446	.98163	1.0187	5.2408	**60**
1	.19109	.19468	5.1366	.98157	1.0188	5.2330	59
2	.19138	.19498	5.1286	.98152	1.0188	5.2252	58
3	.19167	.19529	5.1207	.98146	1.0189	5.2174	57
4	.19195	.19559	5.1128	.98140	1.0189	5.2097	56
5	.19224	.19589	5.1049	.98135	1.0190	5.2019	**55**
6	.19252	.19619	5.0970	.98129	1.0191	5.1942	54
7	.19281	.19649	5.0892	.98124	1.0191	5.1865	53
8	.19309	.19680	5.0814	.98118	1.0192	5.1789	52
9	.19338	.19710	5.0736	.98112	1.0192	5.1712	51
10	.19366	.19740	5.0658	.98107	1.0193	5.1636	**50**
11	.19395	.19770	5.0581	.98101	1.0194	5.1560	49
12	.19423	.19801	5.0504	.98096	1.0194	5.1484	48
13	.19452	.19831	5.0427	.98090	1.0195	5.1409	47
14	.19481	.19861	5.0350	.98084	1.0195	5.1333	46
15	.19509	.19891	5.0273	.98079	1.0196	5.1258	**45**
16	.19538	.19921	5.0197	.98073	1.0197	5.1183	44
17	.19566	.19952	5.0121	.98067	1.0197	5.1109	43
18	.19595	.19982	5.0045	.98061	1.0198	5.1034	42
19	.19623	.20012	4.9969	.98056	1.0198	5.0960	41
20	.19652	.20042	4.9894	.98050	1.0199	5.0886	**40**
21	.19680	.20073	4.9819	.98044	1.0199	5.0813	39
22	.19709	.20103	4.9744	.98039	1.0200	5.0739	38
23	.19737	.20133	4.9669	.98033	1.0201	5.0666	37
24	.19766	.20164	4.9594	.98027	1.0201	5.0593	36
25	.19794	.20194	4.9520	.98021	1.0202	5.0520	**35**
26	.19823	.20224	4.9446	.98016	1.0202	5.0447	34
27	.19851	.20254	4.9372	.98010	1.0203	5.0375	33
28	.19880	.20285	4.9298	.98004	1.0204	5.0302	32
29	.19908	.20315	4.9225	.97998	1.0204	5.0230	31
30	.19937	.20345	4.9152	.97992	1.0205	5.0159	**30**
31	.19965	.20376	4.9078	.97987	1.0205	5.0087	29
32	.19994	.20406	4.9006	.97981	1.0206	5.0016	28
33	.20022	.20436	4.8933	.97975	1.0207	4.9944	27
34	.20051	.20466	4.8860	.97969	1.0207	4.9873	26
35	.20079	.20497	4.8788	.97963	1.0208	4.9803	**25**
36	.20108	.20527	4.8716	.97958	1.0209	4.9732	24
37	.20136	.20557	4.8644	.97952	1.0209	4.9662	23
38	.20165	.20588	4.8573	.97946	1.0210	4.9591	22
39	.20193	.20618	4.8501	.97940	1.0210	4.9521	21
40	.20222	.20648	4.8430	.97934	1.0211	4.9452	**20**
41	.20250	.20679	4.8359	.97928	1.0212	4.9382	19
42	.20279	.20709	4.8288	.97922	1.0212	4.9313	18
43	.20307	.20739	4.8218	.97916	1.0213	4.9244	17
44	.20336	.20770	4.8147	.97910	1.0213	4.9175	16
45	.20364	.20800	4.8077	.97905	1.0214	4.9106	**15**
46	.20393	.20830	4.8007	.97899	1.0215	4.9037	14
47	.20421	.20861	4.7937	.97893	1.0215	4.8969	13
48	.20450	.20891	4.7867	.97887	1.0216	4.8901	12
49	.20478	.20921	4.7798	.97881	1.0217	4.8833	11
50	.20507	.20952	4.7729	.97875	1.0217	4.8765	**10**
51	.20535	.20982	4.7659	.97869	1.0218	4.8697	9
52	.20563	.21013	4.7591	.97863	1.0218	4.8630	8
53	.20592	.21043	4.7522	.97857	1.0219	4.8563	7
54	.20620	.21073	4.7453	.97851	1.0220	4.8496	6
55	.20649	.21104	4.7385	.97845	1.0220	4.8429	**5**
56	.20677	.21134	4.7317	.97839	1.0221	4.8362	4
57	.20706	.21164	4.7249	.97833	1.0222	4.8296	3
58	.20734	.21195	4.7181	.97827	1.0222	4.8229	2
59	.20763	.21225	4.7114	.97821	1.0223	4.8163	1
60	.20791	.21256	4.7046	.97815	1.0223	4.8097	**0**
′	Cos	Cot	Tan	Sin	Csc	Sec	′

101° (281°) (258°) **78°**

12° (192°) (347°) **167°**

′	Sin	Tan	Cot	Cos	Sec	Csc	′
0	.20791	.21256	4.7046	.97815	1.0223	4.8097	**60**
1	.20820	.21286	4.6979	.97809	1.0224	4.8032	59
2	.20848	.21316	4.6912	.97803	1.0225	4.7966	58
3	.20877	.21347	4.6845	.97797	1.0225	4.7901	57
4	.20905	.21377	4.6779	.97791	1.0226	4.7836	56
5	.20933	.21408	4.6712	.97784	1.0227	4.7771	**55**
6	.20962	.21438	4.6646	.97778	1.0227	4.7706	54
7	.20990	.21469	4.6580	.97772	1.0228	4.7641	53
8	.21019	.21499	4.6514	.97766	1.0228	4.7577	52
9	.21047	.21529	4.6448	.97760	1.0229	4.7512	51
10	.21076	.21560	4.6382	.97754	1.0230	4.7448	**50**
11	.21104	.21590	4.6317	.97748	1.0230	4.7384	49
12	.21132	.21621	4.6252	.97742	1.0231	4.7321	48
13	.21161	.21651	4.6187	.97735	1.0232	4.7257	47
14	.21189	.21682	4.6122	.97729	1.0232	4.7194	46
15	.21218	.21712	4.6057	.97723	1.0233	4.7130	**45**
16	.21246	.21743	4.5993	.97717	1.0234	4.7067	44
17	.21275	.21773	4.5928	.97711	1.0234	4.7004	43
18	.21303	.21804	4.5864	.97705	1.0235	4.6942	42
19	.21331	.21834	4.5800	.97698	1.0236	4.6879	41
20	.21360	.21864	4.5736	.97692	1.0236	4.6817	**40**
21	.21388	.21895	4.5673	.97686	1.0237	4.6755	39
22	.21417	.21925	4.5609	.97680	1.0238	4.6693	38
23	.21445	.21956	4.5546	.97673	1.0238	4.6631	37
24	.21474	.21986	4.5483	.97667	1.0239	4.6569	36
25	.21502	.22017	4.5420	.97661	1.0240	4.6507	**35**
26	.21530	.22047	4.5357	.97655	1.0240	4.6446	34
27	.21559	.22078	4.5294	.97648	1.0241	4.6385	33
28	.21587	.22108	4.5232	.97642	1.0241	4.6324	32
29	.21616	.22139	4.5169	.97636	1.0242	4.6263	31
30	.21644	.22169	4.5107	.97630	1.0243	4.6202	**30**
31	.21672	.22200	4.5045	.97623	1.0243	4.6142	29
32	.21701	.22231	4.4983	.97617	1.0244	4.6081	28
33	.21729	.22261	4.4922	.97611	1.0245	4.6021	27
34	.21758	.22292	4.4860	.97604	1.0245	4.5961	26
35	.21786	.22322	4.4799	.97598	1.0246	4.5901	**25**
36	.21814	.22353	4.4737	.97592	1.0247	4.5841	24
37	.21843	.22383	4.4676	.97585	1.0247	4.5782	23
38	.21871	.22414	4.4615	.97579	1.0248	4.5722	22
39	.21899	.22444	4.4555	.97573	1.0249	4.5663	21
40	.21928	.22475	4.4494	.97566	1.0249	4.5604	**20**
41	.21956	.22505	4.4434	.97560	1.0250	4.5545	19
42	.21985	.22536	4.4373	.97553	1.0251	4.5486	18
43	.22013	.22567	4.4313	.95747	1.0251	4.5428	17
44	.22041	.22597	4.4253	.97541	1.0252	4.5369	16
45	.22070	.22628	4.4194	.97534	1.0253	4.5311	**15**
46	.22098	.22658	4.4134	.97528	1.0253	4.5253	14
47	.22126	.22689	4.4075	.97521	1.0254	4.5195	13
48	.22155	.22719	4.4015	.97515	1.0255	4.5137	12
49	.22183	.22750	4.3956	.97508	1.0256	4.5079	11
50	.22212	.22781	4.3897	.97502	1.0256	4.5022	**10**
51	.22240	.22811	4.3838	.97496	1.0257	4.4964	9
52	.22268	.22842	4.3779	.97489	1.0258	4.4907	8
53	.22297	.22872	4.3721	.97483	1.0258	4.4850	7
54	.22325	.22903	4.3662	.97476	1.0259	4.4793	6
55	.22353	.22934	4.3604	.97470	1.0260	4.4736	**5**
56	.22382	.22964	4.3546	.97463	1.0260	4.4679	4
57	.22410	.22995	4.3488	.97457	1.0261	4.4623	3
58	.22438	.23026	4.3430	.97450	1.0262	4.4566	2
59	.22467	.23056	4.3372	.97444	1.0262	4.4510	1
60	.22495	.23087	4.3315	.97437	1.0263	4.4454	**0**
′	Cos	Cot	Tan	Sin	Csc	Sec	′

102° (282°) (257°) **77°**

13° (193°) (346°) **166°**

′	Sin	Tan	Cot	Cos	Sec	Csc	′
0	.22495	.23087	4.3315	.97437	1.0263	4.4454	**60**
1	.22523	.23117	4.3257	.97430	1.0264	4.4398	59
2	.22552	.23148	4.3200	.97424	1.0264	4.4342	58
3	.22580	.23179	4.3143	.97417	1.0265	4.4287	57
4	.22608	.23209	4.3086	.97411	1.0266	4.4231	56
5	.22637	.23240	4.3029	.97404	1.0266	4.4176	**55**
6	.22665	.23271	4.2972	.97398	1.0267	4.4121	54
7	.22693	.23301	4.2916	.97391	1.0268	4.4066	53
8	.22722	.23332	4.2859	.97384	1.0269	4.4011	52
9	.22750	.23363	4.2803	.97378	1.0269	4.3956	51
10	.22778	.23393	4.2747	.97371	1.0270	4.3901	**50**
11	.22807	.23424	4.2691	.97365	1.0271	4.3847	49
12	.22835	.23455	4.2635	.97358	1.0271	4.3792	48
13	.22863	.23485	4.2580	.97351	1.0272	4.3738	47
14	.22892	.23516	4.2524	.97345	1.0273	4.3684	46
15	.22920	.23547	4.2468	.97338	1.0273	4.3630	**45**
16	.22948	.23578	4.2413	.97331	1.0274	4.3576	44
17	.22977	.23608	4.2358	.97325	1.0275	4.3522	43
18	.23005	.23639	4.2303	.97318	1.0276	4.3469	42
19	.23033	.23670	4.2248	.97311	1.0276	4.3415	41
20	.23062	.23700	4.2193	.97304	1.0277	4.3362	**40**
21	.23090	.23731	4.2139	.97298	1.0278	4.3309	39
22	.23118	.23762	4.2084	.97291	1.0278	4.3256	38
23	.23146	.23793	4.2030	.97284	1.0279	4.3203	37
24	.23175	.23823	4.1976	.97278	1.0280	4.3150	36
25	.23203	.23854	4.1922	.97271	1.0281	4.3098	**35**
26	.23231	.23885	4.1868	.97264	1.0281	4.3045	34
27	.23260	.23916	4.1814	.97257	1.0282	4.2993	33
28	.23288	.23946	4.1760	.97251	1.0283	4.2941	32
29	.23316	.23977	4.1706	.97244	1.0283	4.2889	31
30	.23345	.24008	4.1653	.97237	1.0284	4.2837	**30**
31	.23373	.24039	4.1600	.97230	1.0285	4.2785	29
32	.23401	.24069	4.1547	.97223	1.0286	4.2733	28
33	.23429	.24100	4.1493	.97217	1.0286	4.2681	27
34	.23458	.24131	4.1441	.97210	1.0287	4.2630	26
35	.23486	.24162	4.1388	.97203	1.0288	4.2579	**25**
36	.23514	.24193	4.1335	.97196	1.0288	4.2527	24
37	.23542	.24223	4.1282	.97189	1.0289	4.2476	23
38	.23571	.24254	4.1230	.97182	1.0290	4.2425	22
39	.23599	.24285	4.1178	.97176	1.0291	4.2375	21
40	.23627	.24316	4.1126	.97169	1.0291	4.2324	**20**
41	.23656	.24347	4.1074	.97162	1.0292	4.2273	19
42	.23684	.24377	4.1022	.97155	1.0293	4.2223	18
43	.23712	.24408	4.0970	.97148	1.0294	4.2173	17
44	.23740	.24439	4.0918	.97141	1.0294	4.2122	16
45	.23769	.24470	4.0867	.97134	1.0295	4.2072	**15**
46	.23797	.24501	4.0815	.97127	1.0296	4.2022	14
47	.23825	.24532	4.0764	.97120	1.0297	4.1973	13
48	.23853	.24562	4.0713	.97113	1.0297	4.1923	12
49	.23882	.24593	4.0662	.97106	1.0298	4.1873	11
50	.23910	.24624	4.0611	.97100	1.0299	4.1824	**10**
51	.23938	.24655	4.0560	.97093	1.0299	4.1774	9
52	.23966	.24686	4.0509	.97086	1.0300	4.1725	8
53	.23995	.24717	4.0459	.97079	1.0301	4.1676	7
54	.24023	.24747	4.0408	.97072	1.0302	4.1627	6
55	.24051	.24778	4.0358	.97065	1.0302	4.1578	**5**
56	.24079	.24809	4.0308	.97058	1.0303	4.1529	4
57	.24108	.24840	4.0257	.97051	1.0304	4.1481	3
58	.24136	.24871	4.0207	.97044	1.0305	4.1432	2
59	.24164	.24902	4.0158	.97037	1.0305	4.1384	1
60	.24192	.24933	4.0108	.97030	1.0306	4.1336	**0**
′	Cos	Cot	Tan	Sin	Csc	Sec	′

103° (283°) (256°) **76°**

14° (194°) (345°) **165°**

′	Sin	Tan	Cot	Cos	Sec	Csc	′
0	.24192	.24933	4.0108	.97030	1.0306	4.1336	**60**
1	.24220	.24964	4.0058	.97023	1.0307	4.1287	59
2	.24249	.24995	4.0009	.97015	1.0308	4.1239	58
3	.24277	.25026	3.9959	.97008	1.0308	4.1191	57
4	.24305	.25056	3.9910	.97001	1.0309	4.1144	56
5	.24333	.25087	3.9861	.96994	1.0310	4.1096	**55**
6	.24362	.25118	3.9812	.96987	1.0311	4.1048	54
7	.24390	.25149	3.9763	.96980	1.0311	4.1001	53
8	.24418	.25180	3.9714	.96973	1.0312	4.0954	52
9	.24446	.25211	3.9665	.96966	1.0313	4.0906	51
10	.24474	.25242	3.9617	.96959	1.0314	4.0859	**50**
11	.24503	.25273	3.9568	.96952	1.0314	4.0812	49
12	.24531	.25304	3.9520	.96945	1.0315	4.0765	48
13	.24559	.25335	3.9471	.96937	1.0316	4.0718	47
14	.24587	.25366	3.9423	.96930	1.0317	4.0672	46
15	.24615	.25397	3.9375	.96923	1.0317	4.0625	**45**
16	.24644	.25428	3.9327	.96916	1.0318	4.0579	44
17	.24672	.25459	3.9279	.96909	1.0319	4.0532	43
18	.24700	.25490	3.9232	.96902	1.0320	4.0486	42
19	.24728	.25521	3.9184	.96894	1.0321	4.0440	41
20	.24756	.25552	3.9136	.96887	1.0321	4.0394	**40**
21	.24784	.25583	3.9089	.96880	1.0322	4.0348	39
22	.24813	.25614	3.9042	.96873	1.0323	4.0302	38
23	.24841	.25645	3.8995	.96866	1.0324	4.0256	37
24	.24869	.25676	3.8947	.96858	1.0324	4.0211	36
25	.24897	.25707	3.8900	.96851	1.0325	4.0165	**35**
26	.24925	.25738	3.8854	.96844	1.0326	4.0120	34
27	.24954	.25769	3.8807	.96837	1.0327	4.0075	33
28	.24982	.25800	3.8760	.96829	1.0327	4.0029	32
29	.25010	.25831	3.8714	.96822	1.0328	3.9984	31
30	.25038	.25862	3.8667	.96815	1.0329	3.9939	**30**
31	.25066	.25893	3.8621	.96807	1.0330	3.9894	29
32	.25094	.25924	3.8575	.96800	1.0331	3.9850	28
33	.25122	.25955	3.8528	.96793	1.0331	3.9805	27
34	.25151	.25986	3.8482	.96786	1.0332	3.9760	26
35	.25179	.26017	3.8436	.96778	1.0333	3.9716	**25**
36	.25207	.26048	3.8391	.96771	1.0334	3.9672	24
37	.25235	.26079	3.8345	.96764	1.0334	3.9627	23
38	.25263	.26110	3.8299	.96756	1.0335	3.9583	22
39	.25291	.26141	3.8254	.96749	1.0336	3.9539	21
40	.25320	.26172	3.8208	.96742	1.0337	3.9495	**20**
41	.25348	.26203	3.8163	.96734	1.0338	3.9451	19
42	.25376	.26235	3.8118	.96727	1.0338	3.9408	18
43	.25404	.26266	3.8073	.96719	1.0339	3.9364	17
44	.25432	.26297	3.8028	.96712	1.0340	3.9320	16
45	.25460	.26328	3.7983	.96705	1.0341	3.9277	**15**
46	.25488	.26359	3.7938	.96697	1.0342	3.9234	14
47	.25516	.26390	3.7893	.96690	1.0342	3.9190	13
48	.25545	.26421	3.7848	.96682	1.0343	3.9147	12
49	.25573	.26452	3.7804	.96675	1.0344	3.9104	11
50	.25601	.26483	3.7760	.96667	1.0345	3.9061	**10**
51	.25629	.26515	3.7715	.96660	1.0346	3.9018	9
52	.25657	.26546	3.7671	.96653	1.0346	3.8976	8
53	.25685	.26577	3.7627	.96645	1.0347	3.8933	7
54	.25713	.26608	3.7583	.96638	1.0348	3.8890	6
55	.25741	.26639	3.7539	.96630	1.0349	3.8848	**5**
56	.25769	.26670	3.7495	.96623	1.0350	3.8806	4
57	.25798	.26701	3.7451	.96615	1.0350	3.8763	3
58	.25826	.26733	3.7408	.96608	1.0351	3.8721	2
59	.25854	.26764	3.7364	.96600	1.0352	3.8679	1
60	.25882	.26795	3.7321	.96593	1.0353	3.8637	**0**
′	Cos	Cot	Tan	Sin	Csc	Sec	′

104° (284°) (255°) **75°**

15° (195°) (344°) **164°**

′	Sin	Tan	Cot	Cos	Sec	Csc	′
0	.25882	.26795	3.7321	.96593	1.0353	3.8637	**60**
1	.25910	.26826	3.7277	.96585	1.0354	3.8595	59
2	.25938	.26857	3.7234	.96578	1.0354	3.8553	58
3	.25966	.26888	3.7191	.96570	1.0355	3.8512	57
4	.25994	.26920	3.7148	.96562	1.0356	3.8470	56
5	.26022	.26951	3.7105	.96555	1.0357	3.8428	**55**
6	.26050	.26982	3.7062	.96547	1.0358	3.8387	54
7	.26079	.27013	3.7019	.96540	1.0358	3.8346	53
8	.26107	.27044	3.6976	.96532	1.0359	3.8304	52
9	.26135	.27076	3.6933	.96524	1.0360	3.8263	51
10	.26163	.27107	3.6891	.96517	1.0361	3.8222	**50**
11	.26191	.27138	3.6848	.96509	1.0362	3.8181	49
12	.26219	.27169	3.6806	.96502	1.0363	3.8140	48
13	.26247	.27201	3.6764	.96494	1.0363	3.8100	47
14	.26275	.27232	3.6722	.96486	1.0364	3.8059	46
15	.26303	.27263	3.6680	.96479	1.0365	3.8018	**45**
16	.26331	.27294	3.6638	.96471	1.0366	3.7978	44
17	.26359	.27326	3.6596	.96463	1.0367	3.7937	43
18	.26387	.27357	3.6554	.96456	1.0367	3.7897	42
19	.26415	.27388	3.6512	.96448	1.0368	3.7857	41
20	.26443	.27419	3.6470	.96440	1.0369	3.7817	**40**
21	.26471	.27451	3.6429	.96433	1.0370	3.7777	39
22	.26500	.27482	3.6387	.96425	1.0371	3.7737	38
23	.26528	.27513	3.6346	.96417	1.0372	3.7697	37
24	.26556	.27545	3.6305	.96410	1.0372	3.7657	36
25	.26584	.27576	3.6264	.94602	1.0373	3.7617	**35**
26	.26612	.27607	3.6222	.96394	1.0374	3.7577	34
27	.26640	.27638	3.6181	.96386	1.0375	3.7538	33
28	.26668	.27670	3.6140	.96379	1.0376	3.7498	32
29	.26696	.27701	3.6100	.96371	1.0377	3.7459	31
30	.26724	.27732	3.6059	.96363	1.0377	3.7420	**30**
31	.26752	.27764	3.6018	.96355	1.0378	3.7381	29
32	.26780	.27795	3.5978	.96347	1.0379	3.7341	28
33	.26808	.27826	3.5937	.96340	1.0380	3.7302	27
34	.26836	.27858	3.5897	.96332	1.0381	3.7263	26
35	.26864	.27889	3.5856	.96324	1.0382	3.7225	**25**
36	.26892	.27921	3.5816	.96316	1.0382	3.7186	24
37	.26920	.27952	3.5776	.96308	1.0383	3.7147	23
38	.26948	.27983	3.5736	.96301	1.0384	3.7108	22
39	.26976	.28015	3.5696	.96293	1.0385	3.7070	21
40	.27004	.28046	3.5656	.96285	1.0386	3.7032	**20**
41	.27032	.28077	3.5616	.96277	1.0387	3.6993	19
42	.27060	.28109	3.5576	.96269	1.0388	3.6955	18
43	.27088	.28140	3.5536	.96261	1.0388	3.6917	17
44	.27116	.28172	3.5497	.96253	1.0389	3.6879	16
45	.27144	.28203	3.5457	.96246	1.0390	3.6840	**15**
46	.27172	.28234	3.5418	.96238	1.0391	3.6803	14
47	.27200	.28266	3.5379	.96230	1.0392	3.6765	13
48	.27228	.28297	3.5339	.96222	1.0393	3.6727	12
49	.27256	.28329	3.5300	.96214	1.0394	3.6689	11
50	.27284	.28360	3.5261	.96206	1.0394	3.6652	**10**
51	.27312	.28391	3.5222	.96198	1.0395	3.6614	9
52	.27340	.28423	3.5183	.96190	1.0396	3.6575	8
53	.27368	.28454	3.5144	.96182	1.0397	3.6539	7
54	.27396	.28486	3.5105	.96174	1.0398	3.6502	6
55	.27424	.28517	3.5067	.96166	1.0399	3.6465	**5**
56	.27452	.28549	3.5028	.96158	1.0400	3.6427	4
57	.27480	.28580	3.4989	.96150	1.0400	3.6390	3
58	.27508	.28612	3.4951	.96142	1.0401	3.6353	2
59	.27536	.28643	3.4912	.96134	1.0402	3.6316	1
60	.27564	.28675	3.4874	.96126	1.0403	3.6280	**0**
′	Cos	Cot	Tan	Sin	Csc	Sec	′

105° (285°) (254°) **74°**

16° (196°) (343°) **163°**

′	Sin	Tan	Cot	Cos	Sec	Csc	′
0	.27564	.28675	3.4874	.96126	1.0403	3.6280	**60**
1	.27592	.28706	3.4836	.96118	1.0404	3.6243	59
2	.27620	.28738	3.4798	.96110	1.0405	3.6206	58
3	.27648	.28769	3.4760	.96102	1.0406	3.6169	57
4	.27676	.28801	3.4722	.96094	1.0406	3.6133	56
5	.27704	.28832	3.4684	.96086	1.0407	3.6097	**55**
6	.27731	.28864	3.4646	.96078	1.0408	3.6060	54
7	.27759	.28895	3.4608	.96070	1.0409	3.6024	53
8	.27787	.28927	3.4570	.96062	1.0410	3.5988	52
9	.27815	.28958	3.4533	.96054	1.0411	3.5951	51
10	.27843	.28990	3.4495	.96046	1.0412	3.5915	**50**
11	.27871	.29021	3.4458	.96037	1.0413	3.5879	49
12	.27899	.29053	3.4420	.96029	1.0413	3.5843	48
13	.27927	.29084	3.4383	.96021	1.0414	3.5808	47
14	.27955	.29116	3.4346	.96013	1.0415	3.5772	46
15	.27983	.29147	3.4308	.96005	1.0416	3.5736	**45**
16	.28011	.29179	3.4271	.95997	1.0417	3.5700	44
17	.28039	.29210	3.4234	.95989	1.0418	3.5665	43
18	.28067	.29242	3.4197	.95981	1.0419	3.5629	42
19	.28095	.29274	3.4160	.95972	1.0420	3.5594	41
20	.28123	.29305	3.4124	.95964	1.0421	3.5559	**40**
21	.28150	.29337	3.4087	.95956	1.0421	3.5523	39
22	.28178	.29368	3.4050	.95948	1.0422	3.5488	38
23	.28206	.29400	3.4014	.95940	1.0423	3.5453	37
24	.28234	.29432	3.3977	.95931	1.0424	3.5418	36
25	.28262	.29463	3.3941	.95923	1.0425	3.5383	**35**
26	.28290	.29495	3.3904	.95915	1.0426	3.5348	34
27	.28318	.29526	3.3868	.95907	1.0427	3.5313	33
28	.28346	.29558	3.3832	.95898	1.0428	3.5279	32
29	.28374	.29590	3.3796	.95890	1.0429	3.5244	31
30	.28402	.29621	3.3759	.95882	1.0429	3.5209	**30**
31	.28429	.29653	3.3723	.95874	1.0430	3.5175	29
32	.28457	.29685	3.3687	.95865	1.0431	3.5140	28
33	.28485	.29716	3.3652	.95857	1.0432	3.5106	27
34	.28513	.29748	3.3616	.95849	1.0433	3.5072	26
35	.28541	.29780	3.3580	.95841	1.0434	3.5037	**25**
36	.28569	.29811	3.3544	.95832	1.0435	3.5003	24
37	.28597	.29843	3.3509	.95824	1.0436	3.4969	23
38	.28625	.29875	3.3473	.95816	1.0437	3.4935	22
39	.28652	.29906	3.3438	.95807	1.0438	3.4901	21
40	.28680	.29938	3.3402	.95799	1.0439	3.4867	**20**
41	.28708	.29970	3.3367	.95791	1.0439	3.4833	19
42	.28736	.30001	3.3332	.95782	1.0440	3.4799	18
43	.28764	.30033	3.3297	.95774	1.0441	3.4766	17
44	.28792	.30065	3.3261	.95766	1.0442	3.4732	16
45	.28820	.30097	3.3226	.95757	1.0443	3.4699	**15**
46	.28847	.30128	3.3191	.95749	1.0444	3.4665	14
47	.28875	.30160	3.3156	.95740	1.0445	3.4632	13
48	.28903	.30192	3.3122	.95732	1.0446	3.4598	12
49	.28931	.30224	3.3087	.95724	1.0447	3.4565	11
50	.28959	.30255	3.3052	.95715	1.0448	3.4532	**10**
51	.28987	.30287	3.3017	.95707	1.0449	3.4499	9
52	.29015	.30319	3.2983	.95698	1.0450	3.4465	8
53	.29042	.30351	3.2948	.95690	1.0450	3.4432	7
54	.29070	.30382	3.2914	.95681	1.0451	3.4399	6
55	.29098	.30414	3.2879	.95673	1.0452	3.4367	**5**
56	.29126	.30446	3.2845	.95664	1.0453	3.4334	4
57	.29154	.30478	3.2811	.95656	1.0454	3.4301	3
58	.29182	.30509	3.2777	.95647	1.0455	3.4268	2
59	.29209	.30541	3.2743	.95639	1.0456	3.4236	1
60	.29237	.30573	3.2709	.95630	1.0457	3.4203	**0**
′	Cos	Cot	Tan	Sin	Csc	Sec	′

106° (286°) (253°) 73°

17° (197°) (342°) **162°**

′	Sin	Tan	Cot	Cos	Sec	Csc	′
0	.29237	.30573	3.2709	.95630	1.0457	3.4203	**60**
1	.29265	.30605	3.2675	.95622	1.0458	3.4171	59
2	.29293	.30637	3.2641	.95613	1.0459	3.4138	58
3	.29321	.30669	3.2607	.95605	1.0460	3.4106	57
4	.29348	.30700	3.2573	.95596	1.0461	3.4073	56
5	.29376	.30732	3.2539	.95588	1.0462	3.4041	**55**
6	.29404	.30764	3.2506	.95579	1.0463	3.4009	54
7	.29432	.30796	3.2472	.95571	1.0463	3.3977	53
8	.29460	.30828	3.2438	.95562	1.0464	3.3945	52
9	.29487	.30860	3.2405	.95554	1.0465	3.3913	51
10	.29515	.30891	3.2371	.95545	1.0466	3.3881	**50**
11	.29543	.30923	3.2338	.95536	1.0467	3.3849	49
12	.29571	.30955	3.2305	.95528	1.0468	3.3817	48
13	.29599	.30987	3.2272	.95519	1.0469	3.3785	47
14	.29626	.31019	3.2238	.95511	1.0470	3.3754	46
15	.29654	.31051	3.2205	.95502	1.0471	3.3722	**45**
16	.29682	.31083	3.2172	.95493	1.0472	3.3691	44
17	.29710	.31115	3.2139	.95485	1.0473	3.3659	43
18	.29737	.31147	3.2106	.95476	1.0474	3.3628	42
19	.29765	.31178	3.2073	.95467	1.0475	3.3596	41
20	.29793	.31210	3.2041	.95459	1.0476	3.3565	**40**
21	.29821	.31242	3.2008	.95450	1.0477	3.3534	39
22	.29849	.31274	3.1975	.95441	1.0478	3.3502	38
23	.29876	.31306	3.1943	.95433	1.0479	3.3471	37
24	.29904	.31338	3.1910	.95424	1.0480	3.3440	36
25	.29932	.31370	3.1878	.95415	1.0480	3.3409	**35**
26	.29960	.31402	3.1845	.95407	1.0481	3.3378	34
27	.29987	.31434	3.1813	.95398	1.0482	3.3347	33
28	.30015	.31466	3.1780	.95389	1.0483	3.3317	32
29	.30043	.31498	3.1748	.95380	1.0484	3.3286	31
30	.30071	.31530	3.1716	.95372	1.0485	3.3255	**30**
31	.30098	.31562	3.1684	.95363	1.0486	3.3224	29
32	.30126	.31594	3.1652	.95354	1.0487	3.3194	28
33	.30154	.31626	3.1620	.95345	1.0488	3.3163	27
34	.30182	.31658	3.1588	.95337	1.0489	3.3133	26
35	.30209	.31690	3.1556	.95328	1.0490	3.3102	**25**
36	.30237	.31722	3.1524	.95319	1.0491	3.3072	24
37	.30265	.31754	3.1492	.95310	1.0492	3.3042	23
38	.30292	.31786	3.1460	.95301	1.0493	3.3012	22
39	.30320	.31818	3.1429	.95293	1.0494	3.2981	21
40	.30348	.31850	3.1397	.95284	1.0495	3.2951	**20**
41	.30376	.31882	3.1366	.95275	1.0496	3.2921	19
42	.30403	.31914	3.1334	.95266	1.0497	3.2891	18
43	.30431	.31946	3.1303	.95257	1.0498	3.2861	17
44	.30459	.31978	3.1271	.95248	1.0499	3.2831	16
45	.30486	.32010	3.1240	.95240	1.0500	3.2801	**15**
46	.30514	.32042	3.1209	.95231	1.0501	3.2772	14
47	.30542	.32074	3.1178	.95222	1.0502	3.2742	13
48	.30570	.32106	3.1146	.95213	1.0503	3.2712	12
49	.30597	.32139	3.1115	.95204	1.0504	3.2683	11
50	.30625	.32171	3.1084	.95195	1.0505	3.2653	**10**
51	.30653	.32203	3.1053	.95186	1.0506	3.2624	9
52	.30680	.32235	3.1022	.95177	1.0507	3.2594	8
53	.30708	.32267	3.0991	.95168	1.0508	3.2565	7
54	.30736	.32299	3.0961	.95159	1.0509	3.2535	6
55	.30763	.32331	3.0930	.95150	1.0510	3.2506	**5**
56	.30791	.32363	3.0899	.95142	1.0511	3.2477	4
57	.30819	.32396	3.0868	.95133	1.0512	3.2448	3
58	.30846	.32428	3.0838	.95124	1.0513	3.2419	2
59	.30874	.32460	3.0807	.95115	1.0514	3.2390	1
60	.30902	.32492	3.0777	.95106	1.0515	3.2361	**0**
′	Cos	Cot	Tan	Sin	Csc	Sec	′

107° (287°) (252°) **72°**

18° (198°) (341°) **161°**

′	Sin	Tan	Cot	Cos	Sec	Csc	′
0	.30902	.32492	3.0777	.95106	1.0515	3.2361	**60**
1	.30929	.32524	3.0746	.95097	1.0516	3.2332	59
2	.30957	.32556	3.0716	.95088	1.0517	3.2303	58
3	.30985	.32588	3.0686	.95079	1.0518	3.2274	57
4	.31012	.32621	3.0655	.95070	1.0519	3.2245	56
5	.31040	.32653	3.0625	.95061	1.0520	3.2217	**55**
6	.31068	.32685	3.0595	.95052	1.0521	3.2188	54
7	.31095	.32717	3.0565	.95043	1.0522	3.2159	53
8	.31123	.32749	3.0535	.95033	1.0523	3.2131	52
9	.31151	.32782	3.0505	.95024	1.0524	3.2102	51
10	.31178	.32814	3.0475	.95015	1.0525	3.2074	**50**
11	.31206	.32846	3.0445	.95006	1.0526	3.2045	49
12	.31233	.32878	3.0415	.94997	1.0527	3.2017	48
13	.31261	.32911	3.0385	.94988	1.0528	3.1989	47
14	.31289	.32943	3.0356	.94979	1.0529	3.1960	46
15	.31316	.32975	3.0326	.94970	1.0530	3.1932	**45**
16	.31344	.33007	3.0296	.94961	1.0531	3.1904	44
17	.31372	.33040	3.0267	.94952	1.0532	3.1876	43
18	.31399	.33072	3.0237	.94943	1.0533	3.1848	42
19	.31427	.33104	3.0208	.94933	1.0534	3.1820	41
20	.31454	.33136	3.0178	.94924	1.0535	3.1792	**40**
21	.31482	.33169	3.0149	.94915	1.0536	3.1764	39
22	.31510	.33201	3.0120	.94906	1.0537	3.1736	38
23	.31537	.33233	3.0090	.94897	1.0538	3.1708	37
24	.31565	.33266	3.0061	.94888	1.0539	3.1681	36
25	.31593	.33298	3.0032	.94878	1.0540	3.1653	**35**
26	.31620	.33330	3.0003	.94869	1.0541	3.1625	34
27	.31648	.33363	2.9974	.94860	1.0542	3.1598	33
28	.31675	.33395	2.9945	.94851	1.0543	3.1570	32
29	.31703	.33427	2.9916	.94842	1.0544	3.1543	31
30	.31730	.33460	2.9887	.94832	1.0545	3.1515	**30**
31	.31758	.33492	2.9858	.94823	1.0546	3.1488	29
32	.31786	.33524	2.9829	.94814	1.0547	3.1461	28
33	.31813	.33557	2.9800	.94805	1.0548	3.1433	27
34	.31841	.33589	2.9772	.94795	1.0549	3.1406	26
35	.31868	.33621	2.9743	.94786	1.0550	3.1379	**25**
36	.31896	.33654	2.9714	.94777	1.0551	3.1352	24
37	.31923	.33686	2.9686	.94768	1.0552	3.1325	23
38	.31951	.33718	2.9657	.94758	1.0553	3.1298	22
39	.31979	.33751	2.9629	.94749	1.0554	3.1271	21
40	.32006	.33783	2.9600	.94740	1.0555	3.1244	**20**
41	.32034	.33816	2.9572	.94730	1.0556	3.1217	19
42	.32061	.33848	2.9544	.94721	1.0557	3.1190	18
43	.32089	.33881	2.9515	.94712	1.0558	3.1163	17
44	.32116	.33913	2.9487	.94702	1.0559	3.1137	16
45	.32144	.33945	2.9459	.94693	1.0560	3.1110	**15**
46	.32171	.33978	2.9431	.94684	1.0561	3.1083	14
47	.32199	.34010	2.9403	.94674	1.0563	3.1057	13
48	.32227	.34043	2.9375	.94665	1.0564	3.1030	12
49	.32254	.34075	2.9347	.94656	1.0565	3.1004	11
50	.32282	.34108	2.9319	.94646	1.0566	3.0977	**10**
51	.32309	.34140	2.9291	.94637	1.0567	3.0951	9
52	.32337	.34173	2.9263	.94627	1.0568	3.0925	8
53	.32364	.34205	2.9235	.94618	1.0569	3.0898	7
54	.32392	.34238	2.9208	.94609	1.0570	3.0872	6
55	.32419	.34270	2.9180	.94599	1.0571	3.0846	**5**
56	.32447	.34303	2.9152	.94590	1.0572	3.0820	4
57	.32474	.34335	2.9125	.94580	1.0573	3.0794	3
58	.32502	.34368	2.9097	.94571	1.0574	3.0768	2
59	.32529	.34400	2.9070	.94561	1.0575	3.0742	1
60	.32557	.34433	2.9042	.94552	1.0576	3.0716	**0**
′	Cos	Cot	Tan	Sin	Csc	Sec	′

108° (288°) (251°) **71°**

19° (199°) (340°) **160°**

′	Sin	Tan	Cot	Cos	Sec	Csc	′
0	.32557	.34433	2.9042	.94552	1.0576	3.0716	**60**
1	.32584	.34465	2.9015	.94542	1.0577	3.0690	59
2	.32612	.34498	2.8987	.94533	1.0578	3.0664	58
3	.32639	.34530	2.8960	.94523	1.0579	3.0638	57
4	.32667	.34563	2.8933	.94514	1.0580	3.0612	56
5	.32694	.34596	2.8905	.94504	1.0582	3.0586	**55**
6	.32722	.34628	2.8878	.94495	1.0583	3.0561	54
7	.32749	.34661	2.8851	.94485	1.0584	3.0535	53
8	.32777	.34693	2.8824	.94476	1.0585	3.0509	52
9	.32804	.34726	2.8797	.94466	1.0586	3.0484	51
10	.32832	.34758	2.8770	.94457	1.0587	3.0458	**50**
11	.32859	.34791	2.8743	.94447	1.0588	3.0433	49
12	.32887	.34824	2.8716	.94438	1.0589	3.0407	48
13	.32914	.34856	2.8689	.94428	1.0590	3.0382	47
14	.32942	.34889	2.8662	.94418	1.0591	3.0357	46
15	.32969	.34922	2.8636	.94409	1.0592	3.0331	**45**
16	.32997	.34954	2.8609	.94399	1.0593	3.0306	44
17	.33024	.34987	2.8582	.94390	1.0594	3.0281	43
18	.33051	.35020	2.8556	.94380	1.0595	3.0256	42
19	.33079	.35052	2.8529	.94370	1.0597	3.0231	41
20	.33106	.35085	2.8502	.94361	1.0598	3.0206	**40**
21	.33134	.35118	2.8476	.94351	1.0599	3.0181	39
22	.33161	.35150	2.8449	.94342	1.0600	3.0156	38
23	.33189	.35183	2.8423	.94332	1.0601	3.0131	37
24	.33216	.35216	2.8397	.94322	1.0602	3.0106	36
25	.33244	.35248	2.8370	.94313	1.0603	3.0081	**35**
26	.33271	.35281	2.8344	.94303	1.0604	3.0056	34
27	.33298	.35314	2.8318	.94293	1.0605	3.0031	33
28	.33326	.35346	2.8291	.94284	1.0606	3.0007	32
29	.33353	.35379	2.8265	.94274	1.0607	2.9982	31
30	.33381	.35412	2.8239	.94264	1.0608	2.9957	**30**
31	.33408	.35445	2.8213	.94254	1.0610	2.9933	29
32	.33436	.35477	2.8187	.94245	1.0611	2.9908	28
33	.33463	.35510	2.8161	.94235	1.0612	2.9884	27
34	.33490	.35543	2.8135	.94225	1.0613	2.9859	26
35	.33518	.35576	2.8109	.94215	1.0614	2.9835	**25**
36	.33545	.35608	2.8083	.94206	1.0615	2.9811	24
37	.33573	.35641	2.8057	.94196	1.0616	2.9786	23
38	.33600	.35674	2.8032	.94186	1.0617	2.9762	22
39	.33627	.35707	2.8006	.94176	1.0618	2.9738	21
40	.33655	.35740	2.7980	.94167	1.0619	2.9713	**20**
41	.33682	.35772	2.7955	.94157	1.0621	2.9689	19
42	.33710	.35805	2.7929	.94147	1.0622	2.9665	18
43	.33737	.35838	2.7903	.94137	1.0623	2.9641	17
44	.33764	.35871	2.7878	.94127	1.0624	2.9617	16
45	.33792	.35904	2.7852	.94118	1.0625	2.9593	**15**
46	.33819	.35937	2.7827	.94108	1.0626	2.9569	14
47	.33846	.35969	2.7801	.94098	1.0627	2.9545	13
48	.33874	.36002	2.7776	.94088	1.0628	2.9521	12
49	.33901	.36035	2.7751	.94078	1.0629	2.9498	11
50	.33929	.36068	2.7725	.94068	1.0631	2.9474	**10**
51	.33956	.36101	2.7700	.94058	1.0632	2.9450	9
52	.33983	.36134	2.7675	.94049	1.0633	2.9426	8
53	.34011	.36167	2.7650	.94039	1.0634	2.9403	7
54	.34038	.36199	2.7625	.94029	1.0635	2.9379	6
55	.34065	.36232	2.7600	.94019	1.0636	2.9355	**5**
56	.34093	.36265	2.7575	.94009	1.0637	2.9332	4
57	.34120	.36298	2.7550	.93999	1.0638	2.9308	3
58	.34147	.36331	2.7525	.93989	1.0640	2.9285	2
59	.34175	.36364	2.7500	.93979	1.0641	2.9261	1
60	.34202	.36397	2.7475	.93969	1.0642	2.9238	**0**
′	Cos	Cot	Tan	Sin	Csc	Sec	′

109° (289°) (250°) **70°**

20° (200°) (339°) **159°**

′	Sin	Tan	Cot	Cos	Sec	Csc	′
0	.34202	.36397	2.7475	.93969	1.0642	2.9238	**60**
1	.34229	.36430	2.7450	.93959	1.0643	2.9215	59
2	.34257	.36463	2.7425	.93949	1.0644	2.9191	58
3	.34284	.36496	2.7400	.93939	1.0645	2.9168	57
4	.34311	.36529	2.7376	.93929	1.0646	2.9145	56
5	.34339	.36562	2.7351	.93919	1.0647	2.9122	**55**
6	.34366	.36595	2.7326	.93909	1.0649	2.9099	54
7	.34393	.36628	2.7302	.93899	1.0650	2.9075	53
8	.34421	.36661	2.7277	.93889	1.0651	2.9052	52
9	.34448	.36694	2.7253	.93879	1.0652	2.9029	51
10	.34475	.36727	2.7228	.93869	1.0653	2.9006	**50**
11	.34503	.36760	2.7204	.93859	1.0654	2.8983	49
12	.34530	.36793	2.7179	.93849	1.0655	2.8960	48
13	.34557	.36826	2.7155	.93839	1.0657	2.8938	47
14	.34584	.36859	2.7130	.93829	1.0658	2.8915	46
15	.34612	.36892	2.7106	.93819	1.0659	2.8892	**45**
16	.34639	.36925	2.7082	.93809	1.0660	2.8869	44
17	.34666	.36958	2.7058	.93799	1.0661	2.8846	43
18	.34694	.36991	2.7034	.93789	1.0662	2.8824	42
19	.34721	.37024	2.7009	.93779	1.0663	2.8801	41
20	.34748	.37057	2.6985	.93769	1.0665	2.8779	**40**
21	.34775	.37090	2.6961	.93759	1.0666	2.8756	39
22	.34803	.37123	2.6937	.93748	1.0667	2.8733	38
23	.34830	.37157	2.6913	.93738	1.0668	2.8711	37
24	.34857	.37190	2.6889	.93728	1.0669	2.8688	36
25	.34884	.37223	2.6865	.93718	1.0670	2.8666	**35**
26	.34912	.37256	2.6841	.93708	1.0671	2.8644	34
27	.34939	.37289	2.6818	.93698	1.0673	2.8621	33
28	.34966	.37322	2.6794	.93688	1.0674	2.8599	32
29	.34993	.37355	2.6770	.93677	1.0675	2.8577	31
30	.35021	.37388	2.6746	.93667	1.0676	2.8555	**30**
31	.35048	.37422	2.6723	.93657	1.0677	2.8532	29
32	.35075	.37455	2.6699	.93647	1.0678	2.8510	28
33	.35102	.37488	2.6675	.93637	1.0680	2.8488	27
34	.35130	.37521	2.6652	.93626	1.0681	2.8466	26
35	.35157	.37554	2.6628	.93616	1.0682	2.8444	**25**
36	.35184	.37588	2.6605	.93606	1.0683	2.8422	24
37	.35211	.37621	2.6581	.93596	1.0684	2.8400	23
38	.35239	.37654	2.6558	.93585	1.0685	2.8378	22
39	.35266	.37687	2.6534	.93575	1.0687	2.8356	21
40	.35293	.37720	2.6511	.93565	1.0688	2.8334	**20**
41	.35320	.37754	2.6488	.93555	1.0689	2.8312	19
42	.35347	.37787	2.6464	.93544	1.0690	2.8291	18
43	.35375	.37820	2.6441	.93534	1.0691	2.8269	17
44	.35402	.37853	2.6418	.93524	1.0692	2.8247	16
45	.35429	.37887	2.6395	.93514	1.0694	2.8225	**15**
46	.35456	.37920	2.6371	.93503	1.0695	2.8204	14
47	.35484	.37953	2.6348	.93493	1.0696	2.8182	13
48	.35511	.37986	2.6325	.93483	1.0697	2.8161	12
49	.35538	.38020	2.6302	.93472	1.0698	2.8139	11
50	.35565	.38053	2.6279	.93462	1.0700	2.8117	**10**
51	.35592	.38086	2.6256	.93452	1.0701	2.8096	9
52	.35619	.38120	2.6233	.93441	1.0702	2.8075	8
53	.35647	.38153	2.6210	.93431	1.0703	2.8053	7
54	.35674	.38186	2.6187	.93420	1.0704	2.8032	6
55	.35701	.38220	2.6165	.93410	1.0705	2.8010	**5**
56	.35728	.38253	2.6142	.93400	1.0707	2.7989	4
57	.35755	.38286	2.6119	.93389	1.0708	2.7968	3
58	.35782	.38320	2.6096	.93379	1.0709	2.7947	2
59	.35810	.38353	2.6074	.93368	1.0710	2.7925	1
60	.35837	.38386	2.6051	.93358	1.0711	2.7904	**0**
′	Cos	Cot	Tan	Sin	Csc	Sec	′

110° (290°) (249°) **69°**

21° (201°) (338°) **158°**

′	Sin	Tan	Cot	Cos	Sec	Csc	′
0	.35837	.38386	2.6051	.93358	1.0711	2.7904	**60**
1	.35864	.38420	2.6028	.93348	1.0713	2.7883	59
2	.35891	.38453	2.6006	.93337	1.0714	2.7862	58
3	.35918	.38487	2.5983	.93327	1.0715	2.7841	57
4	.35945	.38520	2.5961	.93316	1.0716	2.7820	56
5	.35973	.38553	2.5938	.93306	1.0717	2.7799	**55**
6	.36000	.38587	2.5916	.93295	1.0719	2.7778	54
7	.36027	.38620	2.5893	.93285	1.0720	2.7757	53
8	.36054	.38654	2.5871	.93274	1.0721	2.7736	52
9	.36081	.38687	2.5848	.93264	1.0722	2.7715	51
10	.36108	.38721	2.5826	.93253	1.0723	2.7695	**50**
11	.36135	.38754	2.5804	.93243	1.0725	2.7674	49
12	.36162	.38787	2.5782	.93232	1.0726	2.7653	48
13	.36190	.38821	2.5759	.93222	1.0727	2.7632	47
14	.36217	.38854	2.5737	.93211	1.0728	2.7612	46
15	.36244	.38888	2.5715	.93201	1.0730	2.7591	**45**
16	.36271	.38921	2.5693	.93190	1.0731	2.7570	44
17	.36298	.38955	2.5671	.93180	1.0732	2.7550	43
18	.36325	.38988	2.5649	.93169	1.0733	2.7529	42
19	.36352	.39022	2.5627	.93159	1.0734	2.7509	41
20	.36379	.39055	2.5605	.93148	1.0736	2.7488	**40**
21	.36406	.39089	2.5583	.93137	1.0737	2.7468	39
22	.36434	.39122	2.5561	.93127	1.0738	2.7447	38
23	.36461	.39156	2.5539	.93116	1.0739	2.7427	37
24	.36488	.39190	2.5517	.93106	1.0740	2.7407	36
25	.36515	.39223	2.5495	.93095	1.0742	2.7386	**35**
26	.36542	.39257	2.5473	.93084	1.0743	2.7366	34
27	.36569	.39290	2.5452	.93074	1.0744	2.7346	33
28	.36596	.39324	2.5430	.93063	1.0745	2.7325	32
29	.36623	.39357	2.5408	.93052	1.0747	2.7305	31
30	.36650	.39391	2.5386	.93042	1.0748	2.7285	**30**
31	.36677	.39425	2.5365	.93031	1.0749	2.7265	29
32	.36704	.39458	2.5343	.93020	1.0750	2.7245	28
33	.36731	.39492	2.5322	.93010	1.0752	2.7225	27
34	.36758	.39526	2.5300	.92999	1.0753	2.7205	26
35	.36785	.39559	2.5279	.92988	1.0754	2.7185	**25**
36	.36812	.39593	2.5257	.92978	1.0755	2.7165	24
37	.36839	.39626	2.5236	.92967	1.0757	2.7145	23
38	.36867	.39660	2.5214	.92956	1.0758	2.7125	22
39	.36894	.39694	2.5193	.92945	1.0759	2.7105	21
40	.36921	.39727	2.5172	.92935	1.0760	2.7085	**20**
41	.36948	.39761	2.5150	.92924	1.0761	2.7065	19
42	.36975	.39795	2.5129	.92913	1.0763	2.7046	18
43	.37002	.39829	2.5108	.92902	1.0764	2.7026	17
44	.37029	.39862	2.5086	.92892	1.0765	2.7006	16
45	.37056	.39896	2.5065	.92881	1.0766	2.6986	**15**
46	.37083	.39930	2.5044	.92870	1.0768	2.6967	14
47	.37110	.39963	2.5023	.92859	1.0769	2.6947	13
48	.37137	.39997	2.5002	.92849	1.0770	2.6927	12
49	.37164	.40031	2.4981	.92838	1.0771	2.6908	11
50	.37191	.40065	2.4960	.92827	1.0773	2.6888	**10**
51	.37218	.40098	2.4939	.92816	1.0774	2.6869	9
52	.37245	.40132	2.4918	.92805	1.0775	2.6849	8
53	.37272	.40166	2.4897	.92794	1.0777	2.6830	7
54	.37299	.40200	2.4876	.92784	1.0778	2.6811	6
55	.37326	.40234	2.4855	.92773	1.0779	2.6791	**5**
56	.37353	.40267	2.4834	.92762	1.0780	2.6772	4
57	.37380	.40301	2.4813	.92751	1.0782	2.6752	3
58	.37407	.40335	2.4792	.92740	1.0783	2.6733	2
59	.37434	.40369	2.4772	.92729	1.0784	2.6714	1
60	.37461	.40403	2.4751	.92718	1.0785	2.6695	**0**
′	Cos	Cot	Tan	Sin	Csc	Sec	′

111° (291°) (248°) **68°**

NATURAL TRIGONOMETRIC FUNCTIONS (*Continued*)

22° (202°) (337°) **157°**

′	Sin	Tan	Cot	Cos	Sec	Csc	′
0	.37461	.40403	2.4751	.92718	1.0785	2.6695	**60**
1	.37488	.40436	2.4730	.92707	1.0787	2.6675	59
2	.37515	.40470	2.4709	.92697	1.0788	2.6656	58
3	.37542	.40504	2.4689	.92686	1.0789	2.6637	57
4	.37569	.40538	2.4668	.92675	1.0790	2.6618	56
5	.37595	.40572	2.4648	.92664	1.0792	2.6599	**55**
6	.37622	.40606	2.4627	.92653	1.0793	2.6580	54
7	.37649	.40640	2.4606	.92642	1.0794	2.6561	53
8	.37676	.40674	2.4586	.92631	1.0796	2.6542	52
9	.37703	.40707	2.4566	.92620	1.0797	2.6523	51
10	.37730	.40741	2.4545	.92609	1.0798	2.6504	**50**
11	.37757	.40775	2.4525	.92598	1.0799	2.6485	49
12	.37784	.40809	2.4504	.92587	1.0801	2.6466	48
13	.37811	.40843	2.4484	.92576	1.0802	2.6447	47
14	.37838	.40877	2.4464	.92565	1.0803	2.6429	46
15	.37865	.40911	2.4443	.92554	1.0804	2.6410	**45**
16	.37892	.40945	2.4423	.92543	1.0806	2.6391	44
17	.37919	.40979	2.4403	.92532	1.0807	2.6372	43
18	.37946	.41013	2.4383	.92521	1.0808	2.6354	42
19	.37973	.41047	2.4362	.92510	1.0810	2.6335	41
20	.37999	.41081	2.4342	.92499	1.0811	2.6316	**40**
21	.38026	.41115	2.4322	.92488	1.0812	2.6298	39
22	.38053	.41149	2.4302	.92477	1.0814	2.6279	38
23	.38080	.41183	2.4282	.92466	1.0815	2.6260	37
24	.38107	.41217	2.4262	.92455	1.0816	2.6242	36
25	.38134	.41251	2.4242	.92444	1.0817	2.6223	**35**
26	.38161	.41285	2.4222	.92432	1.0819	2.6205	34
27	.38188	.41319	2.4202	.92421	1.0820	2.6186	33
28	.38215	.41353	2.4182	.92410	1.0821	2.6168	32
29	.38241	.41387	2.4162	.92399	1.0823	2.6150	31
30	.38268	.41421	2.4142	.92388	1.0824	2.6131	**30**
31	.38295	.41455	2.4122	.92377	1.0825	2.6113	29
32	.38322	.41490	2.4102	.92366	1.0827	2.6095	28
33	.38349	.41524	2.4083	.92355	1.0828	2.6076	27
34	.38376	.41558	2.4063	.92343	1.0829	2.6058	26
35	.38403	.41592	2.4043	.92332	1.0830	2.6040	**25**
36	.38430	.41626	2.4023	.92321	1.0832	2.6022	24
37	.38456	.41660	2.4004	.92310	1.0833	2.6003	23
38	.38483	.41694	2.3984	.92299	1.0834	2.5985	22
39	.38510	.41728	2.3964	.92287	1.0836	2.5967	21
40	.38537	.41763	2.3945	.92276	1.0837	2.5949	**20**
41	.38564	.41797	2.3925	.92265	1.0838	2.5931	19
42	.38591	.41831	2.3906	.92254	1.0840	2.5913	18
43	.38617	.41865	2.3886	.92243	1.0841	2.5895	17
44	.38644	.41899	2.3867	.92231	1.0842	2.5877	16
45	.38671	.41933	2.3847	.92220	1.0844	2.5859	**15**
46	.38698	.41968	2.3828	.92209	1.0845	2.5841	14
47	.38725	.42002	2.3808	.92198	1.0846	2.5823	13
48	.38752	.42036	2.3789	.92186	1.0848	2.5805	12
49	.38778	.42070	2.3770	.92175	1.0849	2.5788	11
50	.38805	.42105	2.3750	.92164	1.0850	2.5770	**10**
51	.38832	.42139	2.3731	.92152	1.0852	2.5752	9
52	.38859	.42173	2.3712	.92141	1.0853	2.5734	8
53	.38886	.42207	2.3693	.92130	1.0854	2.5716	7
54	.38912	.42242	2.3673	.92119	1.0856	2.5699	6
55	.38939	.42276	2.3654	92107	1.0857	2.5681	**5**
56	.38966	.42310	2.3635	92096	1.0858	2.5663	4
57	.38993	.42345	2.3616	.92085	1.0860	2.5646	3
58	.39020	.42379	2.3597	.92073	1.0861	2.5628	2
59	.39046	.42413	2.3578	.92062	1.0862	2.5611	1
60	.39073	.42447	2.3559	.92050	1.0864	2.5593	**0**
′	Cos	Cot	Tan	Sin	Csc	Sec	′

112° (292°) (247°) **67°**

23° (203°) (336°) **156°**

′	Sin	Tan	Cot	Cos	Sec	Csc	′
0	.39073	.42447	2.3559	.92050	1.0864	2.5593	**60**
1	.39100	.42482	2.3539	.92039	1.0865	2.5576	59
2	.39127	.42516	2.3520	.92028	1.0866	2.5558	58
3	.39153	.42551	2.3501	.92016	1.0868	2.5541	57
4	.39180	.42585	2 3483	92005	1.0869	2.5523	56
5	.39207	.42619	2.3464	.91994	1.0870	2.5506	**55**
6	.39234	.42654	2.3445	.91982	1.0872	2.5488	54
7	.39260	.42688	2.3426	.91971	1.0873	2.5471	53
8	.39287	.42722	2.3407	.91959	1.0874	2.5454	52
9	.39314	.42757	2.3388	.91948	1.0876	2.5436	51
10	.39341	.42791	2.3369	.91936	1.0877	2.5419	**50**
11	.39367	.42826	2.3351	.91925	1.0878	2.5402	49
12	.39394	.42860	2.3332	.91914	1.0880	2.5384	48
13	.39421	.42894	2.3313	.91902	1.0881	2.5367	47
14	.39448	.42929	2.3294	.91891	1.0883	2.5350	46
15	.39474	.42963	2.3276	.91879	1.0884	2.5333	**45**
16	.39501	.42998	2.3257	.91868	1.0885	2.5316	44
17	.39528	.43032	2.3238	.91856	1.0887	2.5299	43
18	.39555	.43067	2.3220	.91845	1.0888	2.5282	42
19	.39581	.43101	2.3201	.91833	1.0889	2.5264	41
20	.39608	.43136	2.3183	.91822	1.0891	2.5247	**40**
21	.39635	.43170	2.3164	.91810	1.0892	2.5230	39
22	.39661	.43205	2.3146	.91799	1.0893	2.5213	38
23	.39688	.43239	2.3127	.91787	1.0895	2.5196	37
24	.39715	.43274	2.3109	.91775	1.0896	2.5180	36
25	.39741	.43308	2.3090	.91764	1.0898	2.5163	**35**
26	.39768	.43343	2.3072	.91752	1.0899	2.5146	34
27	.39795	.43378	2.3053	.91741	1.0900	2.5129	33
28	.39822	.43412	2.3035	.91729	1.0902	2.5112	32
29	.39848	.43447	2.3017	.91718	1.0903	2.5095	31
30	.39875	.43481	2.2998	.91706	1.0904	2.5078	**30**
31	.39902	.43516	2.2980	.91694	1.0906	2.5062	29
32	.39928	.43550	2.2962	.91683	1.0907	2.5045	28
33	.39955	.43585	2.2944	.91671	1.0909	2.5028	27
34	.39982	.43620	2.2925	.91660	1.0910	2.5012	26
35	.40008	.43654	2.2907	.91648	1.0911	2.4995	**25**
36	.40035	.43689	2.2889	.91636	1.0913	2.4978	24
37	.40062	.43724	2.2871	.91625	1.0914	2.4962	23
38	.40088	.43758	2.2853	.91613	1.0915	2.4945	22
39	.40115	.43793	2.2835	.91601	1.0917	2.4928	21
40	.40141	.43828	2.2817	.91590	1.0918	2.4912	**20**
41	.40168	.43862	2.2799	.91578	1.0920	2.4895	19
42	.40195	.43897	2.2781	.91566	1.0921	2.4879	18
43	.40221	.43932	2.2763	.91555	1.0922	2.4862	17
44	.40248	.43966	2.2745	.91543	1.0924	2.4846	16
45	.40275	.44001	2.2727	.91531	1.0925	2.4830	**15**
46	.40301	.44036	2.2709	.91519	1.0927	2.4813	14
47	.40328	.44071	2.2691	.91508	1.0928	2.4797	13
48	.40355	.44105	2.2673	.91496	1.0929	2.4780	12
49	.40381	.44140	2.2655	.91484	1.0931	2.4764	11
50	.40408	.44175	2.2637	.91472	1.0932	2.4748	**10**
51	.40434	.44210	2.2620	.91461	1.0934	2.4731	9
52	.40461	.44244	2.2602	.91449	1.0935	2.4715	8
53	.40488	.44279	2.2584	.91437	1.0936	2.4699	7
54	.40514	.44314	2.2566	.91425	1.0938	2.4683	6
55	.40541	.44349	2.2549	.91414	1.0939	2.4667	**5**
56	.40567	.44384	2.2531	.91402	1.0941	2.4650	4
57	.40594	.44418	2.2513	.91390	1.0942	2.4634	3
58	.40621	.44453	2.2496	.91378	1.0944	2.4618	2
59	.40647	.44488	2.2478	.91366	1.0945	2.4602	1
60	.40674	.44523	2.2460	.91355	1.0946	2.4586	**0**
′	Cos	Cot	Tan	Sin	Csc	Sec	′

113° (293°) (246°) **66°**

24° (204°) (335°) **155°**

′	Sin	Tan	Cot	Cos	Sec	Csc	′
0	.40674	.44523	2.2460	.91355	1.0946	2.4586	**60**
1	.40700	.44558	2.2443	.91343	1.0948	2.4570	59
2	.40727	.44593	2.2425	.91331	1.0949	2.4554	58
3	.40753	.44627	2.2408	.91319	1.0951	2.4538	57
4	.40780	.44662	2.2390	.91307	1.0952	2.4522	56
5	.40806	.44697	2.2373	.91295	1.0953	2.4506	**55**
6	.40833	.44732	2.2355	.91283	1.0955	2.4490	54
7	.40860	.44767	2.2338	.91272	1.0956	2.4474	53
8	.40886	.44802	2.2320	.91260	1.0958	2.4458	52
9	.40913	.44837	2.2303	.91248	1.0959	2.4442	51
10	.40939	.44872	2.2286	.91236	1.0961	2.4426	**50**
11	.40966	.44907	2.2268	.91224	1.0962	2.4411	49
12	.40992	.44942	2.2251	.91212	1.0963	2.4395	48
13	.41019	.44977	2.2234	.91200	1.0965	2.4379	47
14	.41045	.45012	2.2216	.91188	1.0966	2.4363	46
15	.41072	.45047	2.2199	.91176	1.0968	2.4348	**45**
16	.41098	.45082	2.2182	.91164	1.0969	2.4332	44
17	.41125	.45117	2.2165	.91152	1.0971	2.4316	43
18	.41151	.45152	2.2148	.91140	1.0972	2.4300	42
19	.41178	.45187	2.2130	.91128	1.0974	2.4285	41
20	.41204	.45222	2.2113	.91116	1.0975	2.4269	**40**
21	.41231	.45257	2.2096	.91104	1.0976	2.4254	39
22	.41257	.45292	2.2079	.91092	1.0978	2.4238	38
23	.41284	.45327	2.2062	.91080	1.0979	2.4222	37
24	.41310	.45362	2.2045	.91068	1.0981	2.4207	36
25	.41337	.45397	2.2028	.91056	1.0982	2.4191	**35**
26	.41363	.45432	2.2011	.91044	1.0984	2.4176	34
27	.41390	.45467	2.1994	.91032	1.0985	2.4160	33
28	.41416	.45502	2.1977	.91020	1.0987	2.4145	32
29	.41443	.45538	2.1960	.91008	1.0988	2.4130	31
30	.41469	.45573	2.1943	.90996	1.0989	2.4114	**30**
31	.41496	.45608	2.1926	.90984	1.0991	2.4099	29
32	.41522	.45643	2.1909	.90972	1.0992	2.4083	28
33	.41549	.45678	2.1892	.90960	1.0994	2.4068	27
34	.41575	.45713	2.1876	.90948	1.0995	2.4053	26
35	.41602	.45748	2.1859	.90936	1.0997	2.4038	**25**
36	.41628	.45784	2.1842	.90924	1.0998	2.4022	24
37	.41655	.45819	2.1825	.90911	1.1000	2.4007	23
38	.41681	.45854	2.1808	.90899	1.1001	2.3992	22
39	.41707	.45889	2.1792	.90887	1.1003	2.3977	21
40	.41734	.45924	2.1775	.90875	1.1004	2.3961	**20**
41	.41760	.45960	2.1758	.90863	1.1006	2.3946	19
42	.41787	.45995	2.1742	.90851	1.1007	2.3931	18
43	.41813	.46030	2.1725	.90839	1.1009	2.3916	17
44	.41840	.46065	2.1708	.90826	1.1010	2.3901	16
45	.41866	.46101	2.1692	.90814	1.1011	2.3886	**15**
46	.41892	.46136	2.1675	.90802	1.1013	2.3871	14
47	.41919	.46171	2.1659	.90790	1.1014	2.3856	13
48	.41945	.46206	2.1642	.90778	1.1016	2.3841	12
49	.41972	.46242	2.1625	.90766	1.1017	2.3826	11
50	.41998	.46277	2.1609	.90753	1.1019	2.3811	**10**
51	.42024	.46312	2.1592	.90741	1.1020	2.3796	9
52	.42051	.46348	2.1576	.90729	1.1022	2.3781	8
53	.42077	.46383	2.1560	.90717	1.1023	2.3766	7
54	.42104	.46418	2.1543	.90704	1.1025	2.3751	6
55	.42130	.46454	2.1527	.90692	1.1026	2.3736	**5**
56	.42156	.46489	2.1510	.90680	1.1028	2.3721	4
57	.42183	.46525	2.1494	.90668	1.1029	2.3706	3
58	.42209	.46560	2.1478	.90655	1.1031	2.3692	2
59	.42235	.46595	2.1461	.90643	1.1032	2.3677	1
60	.42262	.46631	2.1445	.90631	1.1034	2.3662	**0**
′	Cos	Cot	Tan	Sin	Csc	Sec	′

114° (294°) (245°) **65°**

25° (205°) (334°) **154°**

′	Sin	Tan	Cot	Cos	Sec	Csc	′
0	.42262	.46631	2.1445	.90631	1.1034	2.3662	**60**
1	.42288	.46666	2.1429	.90618	1.1035	2.3647	59
2	.42315	.46702	2.1413	.90606	1.1037	2.3633	58
3	.42341	.46737	2.1396	.90594	1.1038	2.3618	57
4	.42367	.46772	2.1380	.90582	1.1040	2.3603	56
5	.42394	.46808	2.1364	.90569	1.1041	2.3588	**55**
6	.42420	.46843	2.1348	.90557	1.1043	2.3574	54
7	.42446	.46879	2.1332	.90545	1.1044	2.3559	53
8	.42473	.46914	2.1315	.90532	1.1046	2.3545	52
9	.42499	.46950	2.1299	.90520	1.1047	2.3530	51
10	.42525	.46985	2.1283	.90507	1.1049	2.3515	**50**
11	.42552	.47021	2.1267	.90495	1.1050	2.3501	49
12	.42578	.47056	2.1251	.90483	1.1052	2.3486	48
13	.42604	.47092	2.1235	.90470	1.1053	2.3472	47
14	.42631	.47128	2.1219	.90458	1.1055	2.3457	46
15	.42657	.47163	2.1203	.90446	1.1056	2.3443	**45**
16	.42683	.47199	2.1187	.90433	1.1058	2.3428	44
17	.42709	.47234	2.1171	.90421	1.1059	2.3414	43
18	.42736	.47270	2.1155	.90408	1.1061	2.3400	42
19	.42762	.47305	2.1139	.90396	1.1062	2.3385	41
20	.42788	.47341	2.1123	.90383	1.1064	2.3371	**40**
21	.42815	.47377	2.1107	.90371	1.1066	2.3356	39
22	.42841	.47412	2.1092	.90358	1.1067	2.3342	38
23	.42867	.47448	2.1076	.90346	1.1069	2.3328	37
24	.42894	.47483	2.1060	.90334	1.1070	2.3314	36
25	.42920	.47519	2.1044	.90321	1.1072	2.3299	**35**
26	.42946	.47555	2.1028	.90309	1.1073	2.3285	34
27	.42972	.47590	2.1013	.90296	1.1075	2.3271	33
28	.42999	.47626	2.0997	.90284	1.1076	2.3257	32
29	.43025	.47662	2.0981	.90271	1.1078	2.3242	31
30	.43051	.47698	2.0965	.90259	1.1079	2.3228	**30**
31	.43077	.47733	2.0950	.90246	1.1081	2.3214	29
32	.43104	.47769	2.0934	.90233	1.1082	2.3200	28
33	.43130	.47805	2.0918	.90221	1.1084	2.3186	27
34	.43156	.47840	2.0903	.90208	1.1085	2.3172	26
35	.43182	.47876	2.0887	.90196	1.1087	2.3158	**25**
36	.43209	.47912	2.0872	.90183	1.1089	2.3144	24
37	.43235	.47948	2.0856	.90171	1.1090	2.3130	23
38	.43261	.47984	2.0840	.90158	1.1092	2.3115	22
39	.43287	.48019	2.0825	.90146	1.1093	2.3101	21
40	.43313	.48055	2.0809	.90133	1.1095	2.3088	**20**
41	.43340	.48091	2.0794	.90120	1.1096	2.3074	19
42	.43366	.48127	2.0778	.90108	1.1098	2.3060	18
43	.43392	.48163	2.0763	.90095	1.1099	2.3046	17
44	.43418	.48198	2.0748	.90082	1.1101	2.3032	16
45	.43445	.48234	2.0732	.90070	1.1102	2.3018	**15**
46	.43471	.48270	2.0717	.90057	1.1104	2.3004	14
47	.43497	.48306	2.0701	.90045	1.1106	2.2990	13
48	.43523	.48342	2.0686	.90032	1.1107	2.2976	12
49	.43549	.48378	2.0671	.90019	1.1109	2.2962	11
50	.43575	.48414	2.0655	.90007	1.1110	2.2949	**10**
51	.43602	.48450	2.0640	.89994	1.1112	2.2935	9
52	.43628	.48486	2.0625	.89981	1.1113	2.2921	8
53	.43654	.48521	2.0609	.89968	1.1115	2.2907	7
54	.43680	.48557	2.0594	.89956	1.1117	2.2894	6
55	.43706	.48593	2.0579	.89943	1.1118	2.2880	**5**
56	.43733	.48629	2.0564	.89930	1.1120	2.2866	4
57	.43759	.48665	2.0549	.89918	1.1121	2.2853	3
58	.43785	.48701	2.0533	.89905	1.1123	2.2839	2
59	.43811	.48737	2.0518	.89892	1.1124	2.2825	1
60	.43837	.48773	2.0503	.89879	1.1126	2.2812	**0**
′	Cos	Cot	Tan	Sin	Csc	Sce	′

115° (295°) (244°) **64°**

26° (206°) (333°) **153°**

′	Sin	Tan	Cot	Cos	Sec	Csc	′
0	.43837	.48773	2.0503	.89879	1.1126	2.2812	**60**
1	.43863	.48809	2.0488	.89867	1.1128	2.2798	59
2	.43889	.48845	2.0473	.89854	1.1129	2.2785	58
3	.43916	.48881	2.0458	.89841	1.1131	2.2771	57
4	.43942	.48917	2.0443	.89828	1.1132	2.2757	56
5	.43968	.48953	2.0428	.89816	1.1134	2.2744	**55**
6	.43994	.48989	2.0413	.89803	1.1136	2.2730	54
7	.44020	.49026	2.0398	.89790	1.1137	2.2717	53
8	.44046	.49062	2.0383	.89777	1.1139	2.2703	52
9	.44072	.49098	2.0368	.89764	1.1140	2.2690	51
10	.44098	.49134	2.0353	.89752	1.1142	2.2677	**50**
11	.44124	.49170	2.0338	.89739	1.1143	2.2663	49
12	.44151	.49206	2.0323	.89726	1.1145	2.2650	48
13	.44177	.49242	2.0308	.89713	1.1147	2.2636	47
14	.44203	.49278	2.0293	.89700	1.1148	2.2623	46
15	.44229	.49315	2.0278	.89687	1.1150	2.2610	**45**
16	.44255	.49351	2.0263	.89674	1.1151	2.2596	44
17	.44281	.49387	2.0248	.89662	1.1153	2.2583	43
18	.44307	.49423	2.0233	.89649	1.1155	2.2570	42
19	.44333	.49459	2.0219	.89636	1.1156	2.2556	41
20	.44359	.49495	2.0204	.89623	1.1158	2.2543	**40**
21	.44385	.49532	2.0189	.89610	1.1159	2.2530	39
22	.44411	.49568	2.0174	.89597	1.1161	2.2517	38
23	.44437	.49604	2.0160	.89584	1.1163	2.2504	37
24	.44464	.49640	2.0145	.89571	1.1164	2.2490	36
25	.44490	.49677	2.0130	.89558	1.1166	2.2477	**35**
26	.44516	.49713	2.0115	.89545	1.1168	2.2464	34
27	.44542	.49749	2.0101	.89532	1.1169	2.2451	33
28	.44568	.49786	2.0086	.89519	1.1171	2.2438	32
29	.44594	.49822	2.0072	.89506	1.1172	2.2425	31
30	.44620	.49858	2.0057	.89493	1.1174	2.2412	**30**
31	.44646	.49894	2.0042	.89480	1.1176	2.2399	29
32	.44672	.49931	2.0028	.89467	1.1177	2.2385	28
33	.44698	.49967	2.0013	.89454	1.1179	2.2372	27
34	.44724	.50004	1.9999	.89441	1.1180	2.2359	26
35	.44750	.50040	1.9984	.89428	1.1182	2.2346	**25**
36	.44776	.50076	1.9970	.89415	1.1184	2.2333	24
37	.44802	.50113	1.9955	.89402	1.1185	2.2320	23
38	.44828	.50149	1.9941	.89389	1.1187	2.2308	22
39	.44854	.50185	1.9926	.89376	1.1189	2.2295	21
40	.44880	.50222	1.9912	.89363	1.1190	2.2282	**20**
41	.44906	.50258	1.9897	.89350	1.1192	2.2269	19
42	.44932	.50295	1.9883	.89337	1.1194	2.2256	18
43	.44958	.50331	1.9868	.89324	1.1195	2.2243	17
44	.44984	.50368	1.9854	.89311	1.1197	2.2230	16
45	.45010	.50404	1.9840	.89298	1.1198	2.2217	**15**
46	.45036	.50441	1.9825	.89285	1.1200	2.2205	14
47	.45062	.50477	1.9811	.89272	1.1202	2.2192	13
48	.45088	.50514	1.9797	.89259	1.1203	2.2179	12
49	.45114	.50550	1.9782	.89245	1.1205	2.2166	11
50	.45140	.50587	1.9768	.89232	1.1207	2.2153	**10**
51	.45166	.50623	1.9754	.89219	1.1208	2.2141	9
52	.45192	.50660	1.9740	.89206	1.1210	2.2128	8
53	.45218	.50696	1.9725	.89193	1.1212	2.2115	7
54	.45243	.50733	1.9711	.89180	1.1213	2.2103	6
55	.45269	.50769	1.9697	.89167	1.1215	2.2090	**5**
56	.45295	.50806	1.9683	.89153	1.1217	2.2077	4
57	.45321	.50843	1.9669	.89140	1.1218	2.2065	3
58	.45347	.50879	1.9654	.89127	1.1220	2.2052	2
59	.45373	.50916	1.9640	.89114	1.1222	2.2039	1
60	.45399	.50953	1.9626	.89101	1.1223	2.2027	**0**
′	Cos	Cot	Tan	Sin	Csc	Sec	′

116° (296°) (243°) **63°**

27° (207°) (332°) **152°**

′	Sin	Tan	Cot	Cos	Sec	Csc	′
0	.45399	.50953	1.9626	.89101	1.1223	2.2027	**60**
1	.45425	.50989	1.9612	.89087	1.1225	2.2014	59
2	.45451	.51026	1.9598	.89074	1.1227	2.2002	58
3	.45477	.51063	1.9584	.89061	1.1228	2.1989	57
4	.45503	.51099	1.9570	.89048	1.1230	2.1977	56
5	.45529	.51136	1.9556	.89035	1.1232	2.1964	**55**
6	.45554	.51173	1.9542	.89021	1.1233	2.1952	54
7	.45580	.51209	1.9528	.89008	1.1235	2.1939	53
8	.45606	.51246	1.9514	.88995	1.1237	2.1927	52
9	.45632	.51283	1.9500	.88981	1.1238	2.1914	51
10	.45658	.51319	1.9486	.88968	1.1240	2.1902	**50**
11	.45684	.51356	1.9472	.88955	1.1242	2.1890	49
12	.45710	.51393	1.9458	.88942	1.1243	2.1877	48
13	.45736	.51430	1.9444	.88928	1.1245	2.1865	47
14	.45762	.51467	1.9430	.88915	1.1247	2.1852	46
15	.45787	.51503	1.9416	.88902	1.1248	2.1840	**45**
16	.45813	.51540	1.9402	.88888	1.1250	2.1828	44
17	.45839	.51577	1.9388	.88875	1.1252	2.1815	43
18	.45865	.51614	1.9375	.88862	1.1253	2.1803	42
19	.45891	.51651	1.9361	.88848	1.1255	2.1791	41
20	.45917	.51688	1.9347	.88835	1.1257	2.1779	**40**
21	.45942	.51724	1.9333	.88822	1.1259	2.1766	39
22	.45968	.51761	1.9319	.88808	1.1260	2.1754	38
23	.45994	.51798	1.9306	.88795	1.1262	2.1742	37
24	.46020	.51835	1.9292	.88782	1.1264	2.1730	36
25	.46046	.51872	1.9278	.88768	1.1265	2.1718	**35**
26	.46072	.51909	1.9265	.88755	1.1267	2.1705	34
27	.46097	.51946	1.9251	.88741	1.1269	2.1693	33
28	.46123	.51983	1.9237	.88728	1.1270	2.1681	32
29	.46149	.52020	1.9223	.88715	1.1272	2.1669	31
30	.46175	.52057	1.9210	.88701	1.1274	2.1657	**30**
31	.46201	.52094	1.9196	.88688	1.1276	2.1645	29
32	.46226	.52131	1.9183	.88674	1.1277	2.1633	28
33	.46252	.52168	1.9169	.88661	1.1279	2.1621	27
34	.46278	.52205	1.9155	.88647	1.1281	2.1609	26
35	.46304	.52242	1.9142	.88634	1.1282	2.1596	**25**
36	.46330	.52279	1.9128	.88620	1.1284	2.1584	24
37	.46355	.52316	1.9115	.88607	1.1286	2.1572	23
38	.46381	.52353	1.9101	.88593	1.1288	2.1560	22
39	.46407	.52390	1.9088	.88580	1.1289	2.1549	21
40	.46433	.52427	1.9074	.88566	1.1291	2.1537	**20**
41	.46458	.52464	1.9061	.88553	1.1293	2.1525	19
42	.46484	.52501	1.9047	.88539	1.1294	2.1513	18
43	.46510	.52538	1.9034	.88526	1.1296	2.1501	17
44	.46536	.52575	1.9020	.88512	1.1298	2.1489	16
45	.46561	.52613	1.9007	.88499	1.1300	2.1477	**15**
46	.46587	.52650	1.8993	.88485	1.1301	2.1465	14
47	.46613	.52687	1.8980	.88472	1.1303	2.1453	13
48	.46639	.52724	1.8967	.88458	1.1305	2.1441	12
49	.46664	.52761	1.8953	.88445	1.1307	2.1430	11
50	.46690	.52798	1.8940	.88431	1.1308	2.1418	**10**
51	.46716	.52836	1.8927	.88417	1.1310	2.1406	9
52	.46742	.52873	1.8913	.88404	1.1312	2.1394	8
53	.46767	.52910	1.8900	.88390	1.1313	2.1382	7
54	.46793	.52947	1.8887	.88377	1.1315	2.1371	6
55	.46819	.52985	1.8873	.88363	1.1317	2.1359	**5**
56	.46844	.53022	1.8860	.88349	1.1319	2.1347	4
57	.46870	.53059	1.8847	.88336	1.1320	2.1336	3
58	.46896	.53096	1.8834	.88322	1.1322	2.1324	2
59	.46921	.53134	1.8820	.88308	1.1324	2.1312	1
60	.46947	.53171	1.8807	.88295	1.1326	2.1301	**0**
′	Cos	Cot	Tan	Sin	Csc	Sec	′

117° (297°) (242°) **62°**

28° (208°) | (331°) **151°**

′	Sin	Tan	Cot	Cos	Sec	Csc	′
0	.46947	.53171	1.8807	.88295	1.1326	2.1301	**60**
1	.46973	.53208	1.8794	.88281	1.1327	2.1289	59
2	.46999	.53246	1.8781	.88267	1.1329	2.1277	58
3	.47024	.53283	1.8768	.88254	1.1331	2.1266	57
4	.47050	.53320	1.8755	.88240	1.1333	2.1254	56
5	.47076	.53358	1.8741	.88226	1.1334	2.1242	**55**
6	.47101	.53395	1.8728	.88213	1.1336	2.1231	54
7	.47127	.53432	1.8715	.88199	1.1338	2.1219	53
8	.47153	.53470	1.8702	.88185	1.1340	2.1208	52
9	.47178	.53507	1.8689	.88172	1.1342	2.1196	51
10	.47204	.53545	1.8676	.88158	1.1343	2.1185	**50**
11	.47229	.53582	1.8663	.88144	1.1345	2.1173	49
12	.47255	.53620	1.8650	.88130	1.1347	2.1162	48
13	.47281	.53657	1.8637	.88117	1.1349	2.1150	47
14	.47306	.53694	1.8624	.88103	1.1350	2.1139	46
15	.47332	.53732	1.8611	.88089	1.1352	2.1127	**45**
16	.47358	.53769	1.8598	.88075	1.1354	2.1116	44
17	.47383	.53807	1.8585	.88062	1.1356	2.1105	43
18	.47409	.53844	1.8572	.88048	1.1357	2.1093	42
19	.47434	.53882	1.8559	.88034	1.1359	2.1082	41
20	.47460	.53920	1.8546	.88020	1.1361	2.1070	**40**
21	.47486	.53957	1.8533	.88006	1.1363	2.1059	39
22	.47511	.53995	1.8520	.87993	1.1365	2.1048	38
23	.47537	.54032	1.8507	.87979	1.1366	2.1036	37
24	.47562	.54070	1.8495	.87965	1.1368	2.1025	36
25	.47588	.54107	1.8482	.87951	1.1370	2.1014	**35**
26	.47614	.54145	1.8469	.87937	1.1372	2.1002	34
27	.47639	.54183	1.8456	.87923	1.1374	2.0991	33
28	.47665	.54220	1.8443	.87909	1.1375	2.0980	32
29	.47690	.54258	1.8430	.87896	1.1377	2.0969	31
30	.47716	.54296	1.8418	.87882	1.1379	2.0957	**30**
31	.47741	.54333	1.8405	.87868	1.1381	2.0946	29
32	.47767	.54371	1.8392	.87854	1.1383	2.0935	28
33	.47793	.54409	1.8379	.87840	1.1384	2.0924	27
34	.47818	.54446	1.8367	.87826	1.1386	2.0913	26
35	.47844	.54484	1.8354	.87812	1.1388	2.0901	**25**
36	.47869	.54522	1.8341	.87798	1.1390	2.0890	24
37	.47895	.54560	1.8329	.87784	1.1392	2.0879	23
38	.47920	.54597	1.8316	.87770	1.1393	2.0868	22
39	.47946	.54635	1.8303	.87756	1.1395	2.0857	21
40	.47971	.54673	1.8291	.87743	1.1397	2.0846	**20**
41	.47997	.54711	1.8278	.87729	1.1399	2.0835	19
42	.48022	.54748	1.8265	.87715	1.1401	2.0824	18
43	.48048	.54786	1.8253	.87701	1.1402	2.0813	17
44	.48073	.54824	1.8240	.87687	1.1404	2.0802	16
45	.48099	.54862	1.8228	.87673	1.1406	2.0791	**15**
46	.48124	.54900	1.8215	.87659	1.1408	2.0779	14
47	.48150	.54938	1.8202	.87645	1.1410	2.0768	13
48	.48175	.54975	1.8190	.87631	1.1412	2.0757	12
49	.48201	.55013	1.8177	.87617	1.1413	2.0747	11
50	.48226	.55051	1.8165	.87603	1.1415	2.0736	**10**
51	.48252	.55089	1.8152	.87589	1.1417	2.0725	9
52	.48277	.55127	1.8140	.87575	1.1419	2.0714	8
53	.48303	.55165	1.8127	.87561	1.1421	2.0703	7
54	.48328	.55203	1.8115	.87546	1.1423	2.0692	6
55	.48354	.55241	1.8103	.87532	1.1424	2.0681	**5**
56	.48379	.55279	1.8090	.87518	1.1426	2.0670	4
57	.48405	.55317	1.8078	.87504	1.1428	2.0659	3
58	.48430	.55355	1.8065	.87490	1.1430	2.0648	2
59	.48456	.55393	1.8053	.87476	1.1432	2.0637	1
60	.48481	.55431	1.8040	.87462	1.1434	2.0627	**0**
′	Cos	Cot	Tan	Sin	Csc	Sec	′

118° (298°) | (241°) **61°**

29° (209°) | (330)° **150°**

′	Sin	Tan	Cot	Cos	Sec	Csc	′
0	.48481	.55431	1.8040	.87462	1.1434	2.0627	**60**
1	.48506	.55469	1.8028	.87448	1.1435	2.0616	59
2	.48532	.55507	1.8016	.87434	1.1437	2.0605	58
3	.48557	.55545	1.8003	.87420	1.1439	2.0594	57
4	.48583	.55583	1.7991	.87406	1.1441	2.0583	56
5	.48608	.55621	1.7979	.87391	1.1443	2.0573	**55**
6	.48634	.55659	1.7966	.87377	1.1445	2.0562	54
7	.48659	.55697	1.7954	.87363	1.1446	2.0551	53
8	.48684	.55736	1.7942	.87349	1.1448	2.0540	52
9	.48710	.55774	1.7930	.87335	1.1450	2.0530	51
10	.48735	.55812	1.7917	.87321	1.1452	2.0519	**50**
11	.48761	.55850	1.7905	.87306	1.1454	2.0508	49
12	.48786	.55888	1.7893	.87292	1.1456	2.0498	48
13	.48811	.55926	1.7881	.87278	1.1458	2.0487	47
14	.48837	.55964	1.7868	.87264	1.1460	2.0476	46
15	.48862	.56003	1.7856	.87250	1.1461	2.0466	**45**
16	.48888	.56041	1.7844	.87235	1.1463	2.0455	44
17	.48913	.56079	1.7832	.87221	1.1465	2.0445	43
18	.48938	.56117	1.7820	.87207	1.1467	2.0434	42
19	.48964	.56156	1.7808	.87193	1.1469	2.0423	41
20	.48989	.56194	1.7796	.87178	1.1471	2.0413	**40**
21	.49014	.56232	1.7783	.87164	1.1473	2.0402	39
22	.49040	.56270	1.7771	.87150	1.1474	2.0392	38
23	.49065	.56309	1.7759	.87136	1.1476	2.0381	37
24	.49090	.56347	1.7747	.87121	1.1478	2.0371	36
25	.49116	.56385	1.7735	.87107	1.1480	2.0360	**35**
26	.49141	.56424	1.7723	.87093	1.1482	2.0350	34
27	.49166	.56462	1.7711	.87079	1.1484	2.0339	33
28	.49192	.56501	1.7699	.87064	1.1486	2.0329	32
29	.49217	.56539	1.7687	.87050	1.1488	2.0318	31
30	.49242	.56577	1.7675	.87036	1.1490	2.0308	**30**
31	.49268	.56616	1.7663	.87021	1.1491	2.0297	29
32	.49293	.56654	1.7651	.87007	1.1493	2.0287	28
33	.49318	.56693	1.7639	.86993	1.1495	2.0276	27
34	.49344	.56731	1.7627	.86978	1.1497	2.0266	26
35	.49369	.56769	1.7615	.86964	1.1499	2.0256	**25**
36	.49394	.56808	1.7603	.86949	1.1501	2.0245	24
37	.49419	.56846	1.7591	.86935	1.1503	2.0235	23
38	.49445	.56885	1.7579	.86921	1.1505	2.0225	22
39	.49470	.56923	1.7567	.86906	1.1507	2.0214	21
40	.49495	.56962	1.7556	.86892	1.1509	2.0204	**20**
41	.49521	.57000	1.7544	.86878	1.1510	2.0194	19
42	.49546	.57039	1.7532	.86863	1.1512	2.0183	18
43	.49571	.57078	1.7520	.86849	1.1514	2.0173	17
44	.49596	.57116	1.7508	.86834	1.1516	2.0163	16
45	.49622	.57155	1.7496	.86820	1.1518	2.0152	**15**
46	.49647	.57193	1.7485	.86805	1.1520	2.0142	14
47	.49672	.57232	1.7473	.86791	1.1522	2.0132	13
48	.49697	.57271	1.7461	.86777	1.1524	2.0122	12
49	.49723	.57309	1.7449	.86762	1.1526	2.0112	11
50	.49748	.57348	1.7437	.86748	1.1528	2.0101	**10**
51	.49773	.57386	1.7426	.86733	1.1530	2.0091	9
52	.49798	.57425	1.7414	.86719	1.1532	2.0081	8
53	.49824	.57464	1.7402	.86704	1.1533	2.0071	7
54	.49849	.57503	1.7391	.86690	1.1535	2.0061	6
55	.49874	.57541	1.7379	.86675	1.1537	2.0051	**5**
56	.49899	.57580	1.7367	.86661	1.1539	2.0040	4
57	.49924	.57619	1.7355	.86646	1.1541	2.0030	3
58	.49950	.57657	1.7344	.86632	1.1543	2.0020	2
59	.49975	.57696	1.7332	.86617	1.1545	2.0010	1
60	.50000	.57735	1.7321	.86603	1.1547	2.0000	**0**
′	Cos	Cot	Tan	Sin	Csc	Sec	′

119° (299°) | (240°) **60°**

30° (210°) (329°) **149°**

′	Sin	Tan	Cot	Cos	Sec	Csc	′
0	.50000	.57735	1.7321	.86603	1.1547	2.0000	**60**
1	.50025	.57774	1.7309	.86588	1.1549	1.9990	59
2	.50050	.57813	1.7297	.86573	1.1551	1.9980	58
3	.50076	.57851	1.7286	.86559	1.1553	1.9970	57
4	.50101	.57890	1.7274	.86544	1.1555	1.9960	56
5	.50126	.57929	1.7262	.86530	1.1557	1.9950	**55**
6	.50151	.57968	1.7251	.86515	1.1559	1.9940	54
7	.50176	.58007	1.7239	.86501	1.1561	1.9930	53
8	.50201	.58046	1.7228	.86486	1.1563	1.9920	52
9	.50227	.58085	1.7216	.86471	1.1565	1.9910	51
10	.50252	.58124	1.7205	.86457	1.1566	1.9900	**50**
11	.50277	.58162	1.7193	.86442	1.1568	1.9890	49
12	.50302	.58201	1.7182	.86427	1.1570	1.9880	48
13	.50327	.58240	1.7170	.86413	1.1572	1.9870	47
14	.50352	.58279	1.7159	.86398	1.1574	1.9860	46
15	.50377	.58318	1.7147	.86384	1.1576	1.9850	**45**
16	.50403	.58357	1.7136	.86369	1.1578	1.9840	44
17	.50428	.58396	1.7124	.86354	1.1580	1.9830	43
18	.50453	.58435	1.7113	.86340	1.1582	1.9821	42
19	.50478	.58474	1.7102	.86325	1.1584	1.9811	41
20	.50503	.58513	1.7090	.86310	1.1586	1.9801	**40**
21	.50528	.58552	1.7079	.86295	1.1588	1.9791	39
22	.50553	.58591	1.7067	.86281	1.1590	1.9781	38
23	.50578	.58631	1.7056	.86266	1.1592	1.9771	37
24	.50603	.58670	1.7045	.86251	1.1594	1.9762	36
25	.50628	.58709	1.7033	.86237	1.1596	1.9752	**35**
26	.50654	.58748	1.7022	.86222	1.1598	1.9742	34
27	.50679	.58787	1.7011	.86207	1.1600	1.9732	33
28	.50704	.58826	1.6999	.86192	1.1602	1.9722	32
29	.50729	.58865	1.6988	.86178	1.1604	1.9713	31
30	.50754	.58905	1.6977	.86163	1.1606	1.9703	**30**
31	.50779	.58944	1.6965	.86148	1.1608	1.9693	29
32	.50804	.58983	1.6954	.86133	1.1610	1.9684	28
33	.50829	.59022	1.6943	.86119	1.1612	1.9674	27
34	.50854	.59061	1.6932	.86104	1.1614	1.9664	26
35	.50879	.59101	1.6920	.86089	1.1616	1.9654	**25**
36	.50904	.59140	1.6909	.86074	1.1618	1.9645	24
37	.50929	.59179	1.6898	.86059	1.1620	1.9635	23
38	.50954	.59218	1.6887	.86045	1.1622	1.9625	22
39	.50979	.59258	1.6875	.86030	1.1624	1.9616	21
40	.51004	.59297	1.6864	.86015	1.1626	1.9606	**20**
41	.51029	.59336	1.6853	.86000	1.1628	1.9597	19
42	.51054	.59376	1.6842	.85985	1.1630	1.9587	18
43	.51079	.59415	1.6831	.85970	1.1632	1.9577	17
44	.51104	.59454	1.6820	.85956	1.1634	1.9568	16
45	.51129	.59494	1.6808	.85941	1.1636	1.9558	**15**
46	.51154	.59533	1.6797	.85926	1.1638	1.9549	14
47	.51179	.59573	1.6786	.85911	1.1640	1.9539	13
48	.51204	.59612	1.6775	.85896	1.1642	1.9530	12
49	.51229	.59651	1.6764	.85881	1.1644	1.9520	11
50	.51254	.59691	1.6753	.85866	1.1646	1.9511	**10**
51	.51279	.59730	1.6742	.85851	1.1648	1.9501	9
52	.51304	.59770	1.6731	.85836	1.1650	1.9492	8
53	.51329	.59809	1.6720	.85821	1.1652	1.9482	7
54	.51354	.59849	1.6709	.85806	1.1654	1.9473	6
55	.51379	.59888	1.6698	.85792	1.1656	1.9463	**5**
56	.51404	.59928	1.6687	.85777	1.1658	1.9454	4
57	.51429	.59967	1.6676	.85762	1.1660	1.9444	3
58	.51454	.60007	1.6665	.85747	1.1662	1.9435	2
59	.51479	.60046	1.6654	.85732	1.1664	1.9425	1
60	.51504	.60086	1.6643	.85717	1.1666	1.9416	**0**
′	Cos	Cot	Tan	Sin	Csc	Sec	′

120° (300°) (239°) **59°**

31° (211°) (328°) **148°**

′	Sin	Tan	Cot	Cos	Sec	Csc	′
0	.51504	.60086	1.6643	.85717	1.1666	1.9416	**60**
1	.51529	.60126	1.6632	.85702	1.1668	1.9407	59
2	.51554	.60165	1.6621	.85687	1.1670	1.9397	58
3	.51579	.60205	1.6610	.85672	1.1672	1.9388	57
4	.51604	.60245	1.6599	.85657	1.1675	1.9379	56
5	.51628	.60284	1.6588	.85642	1.1677	1.9369	**55**
6	.51653	.60324	1.6577	.85627	1.1679	1.9360	54
7	.51678	.60364	1.6566	.85612	1.1681	1.9351	53
8	.51703	.60403	1.6555	.85597	1.1683	1.9341	52
9	.51728	.60443	1.6545	.85582	1.1685	1.9332	51
10	.51753	.60483	1.6534	.85567	1.1687	1.9323	**50**
11	.51778	.60522	1.6523	.85551	1.1689	1.9313	49
12	.51803	.60562	1.6512	.85536	1.1691	1.9304	48
13	.51828	.60602	1.6501	.85521	1.1693	1.9295	47
14	.51852	.60642	1.6490	.85506	1.1695	1.9285	46
15	.51877	.60681	1.6479	.85491	1.1697	1.9276	**45**
16	.51902	.60721	1.6469	.85476	1.1699	1.9267	44
17	.51927	.60761	1.6458	.85461	1.1701	1.9258	43
18	.51952	.60801	1.6447	.85446	1.1703	1.9249	42
19	.51977	.60841	1.6436	.85431	1.1705	1.9239	41
20	.52002	.60881	1.6426	.85416	1.1707	1.9230	**40**
21	.52026	.60921	1.6415	.85401	1.1710	1.9221	39
22	.52051	.60960	1.6404	.85385	1.1712	1.9212	38
23	.52076	.61000	1.6393	.85370	1.1714	1.9203	37
24	.52101	.61040	1.6383	.85355	1.1716	1.9194	36
25	.52126	.61080	1.6372	.85340	1.1718	1.9184	**35**
26	.52151	.61120	1.6361	.85325	1.1720	1.9175	34
27	.52175	.61160	1.6351	.85310	1.1722	1.9166	33
28	.52200	.61200	1.6340	.85294	1.1724	1.9157	32
29	.52225	.61240	1.6329	.85279	1.1726	1.9148	31
30	.52250	.61280	1.6319	.85264	1.1728	1.9139	**30**
31	.52275	.61320	1.6308	.85249	1.1730	1.9130	29
32	.52299	.61360	1.6297	.85234	1.1732	1.9121	28
33	.52324	.61400	1.6287	.85218	1.1735	1.9112	27
34	.52349	.61440	1.6276	.85203	1.1737	1.9103	26
35	.52374	.61480	1.6265	.85188	1.1739	1.9094	**25**
36	.52399	.61520	1.6255	.85173	1.1741	1.9084	24
37	.52423	.61561	1.6244	.85157	1.1743	1.9075	23
38	.52448	.61601	1.6234	.85142	1.1745	1.9066	22
39	.52473	.61641	1.6223	.85127	1.1747	1.9057	21
40	.52498	.61681	1.6212	.85112	1.1749	1.9048	**20**
41	.52522	.61721	1.6202	.85096	1.1751	1.9039	19
42	.52547	.61761	1.6191	.85081	1.1753	1.9031	18
43	.52572	.61801	1.6181	.85066	1.1756	1.9022	17
44	.52597	.61842	1.6170	.85051	1.1758	1.9013	16
45	.52621	.61882	1.6160	.85035	1.1760	1.9004	**15**
46	.52646	.61922	1.6149	.85020	1.1762	1.8995	14
47	.52671	.61962	1.6139	.85005	1.1764	1.8986	13
48	.52696	.62003	1.6128	.84989	1.1766	1.8977	12
49	.52720	.62043	1.6118	.84974	1.1768	1.8968	11
50	.52745	.62083	1.6107	.84959	1.1770	1.8959	**10**
51	.52770	.62124	1.6097	.84943	1.1773	1.8950	9
52	.52794	.62164	1.6087	.84928	1.1775	1.8941	8
53	.52819	.62204	1.6076	.84913	1.1777	1.8933	7
54	.52844	.62245	1.6066	.84897	1.1779	1.8924	6
55	.52869	.62285	1.6055	.84882	1.1781	1.8915	**5**
56	.52893	.62325	1.6045	.84866	1.1783	1.8906	4
57	.52918	.62366	1.6034	.84851	1.1785	1.8897	3
58	.52943	.62406	1.6024	.84836	1.1788	1.8888	2
59	.52967	.62446	1.6014	.84820	1.1790	1.8880	1
60	.52992	.62487	1.6003	.84805	1.1792	1.8871	**0**
′	Cos	Cot	Tan	Sin	Csc	Sec	′

121° (301°) (238°) **58°**

NATURAL TRIGONOMETRIC FUNCTIONS (*Continued*)

32° (212°) — (327°) **147°**

′	Sin	Tan	Cot	Cos	Sec	Csc	′
0	.52992	.62487	1.6003	.84805	1.1792	1.8871	**60**
1	.53017	.62527	1.5993	.84789	1.1794	1.8862	59
2	.53041	.62568	1.5983	.84774	1.1796	1.8853	58
3	.53066	.62608	1.5972	.84759	1.1798	1.8844	57
4	.53091	.62649	1.5962	.84743	1.1800	1.8836	56
5	.53115	.62689	1.5952	.84728	1.1803	1.8827	**55**
6	.53140	.62730	1.5941	.84712	1.1805	1.8818	54
7	.53164	.62770	1.5931	.84697	1.1807	1.8810	53
8	.53189	.62811	1.5921	.84681	1.1809	1.8801	52
9	.53214	.62852	1.5911	.84666	1.1811	1.8792	51
10	.53238	.62892	1.5900	.84650	1.1813	1.8783	**50**
11	.53263	.62933	1.5890	.84635	1.1815	1.8775	49
12	.53288	.62973	1.5880	.84619	1.1818	1.8766	48
13	.53312	.63014	1.5869	.84604	1.1820	1.8757	47
14	.53337	.63055	1.5859	.84588	1.1822	1.8749	46
15	.53361	.63095	1.5849	.84573	1.1824	1.8740	**45**
16	.53386	.63136	1.5839	.84557	1.1826	1.8731	44
17	.53411	.63177	1.5829	.84542	1.1828	1.8723	43
18	.53435	.63217	1.5818	.84526	1.1831	1.8714	42
19	.53460	.63258	1.5808	.84511	1.1833	1.8706	41
20	.53484	.63299	1.5798	.84495	1.1835	1.8697	**40**
21	.53509	.63340	1.5788	.84480	1.1837	1.8688	39
22	.53534	.63380	1.5778	.84464	1.1830	1.8680	38
23	.53558	.63421	1.5768	.84448	1.1842	1.8671	37
24	.53583	.63462	1.5757	.84433	1.1844	1.8663	36
25	.53607	.63503	1.5747	.84417	1.1846	1.8654	**35**
26	.53632	.63544	1.5737	.84402	1.1848	1.8646	34
27	.53656	.63584	1.5727	.84386	1.1850	1.8637	33
28	.53681	.63625	1.5717	.84370	1.1852	1.8629	32
29	.53705	.63666	1.5707	.84355	1.1855	1.8620	31
30	.53730	.63707	1.5697	.84339	1.1857	1.8612	**30**
31	.53754	.63748	1.5687	.84324	1.1859	1.8603	29
32	.53779	.63789	1.5677	.84308	1.1861	1.8595	28
33	.53804	.63830	1.5667	.84292	1.1863	1.8586	27
34	.53828	.63871	1.5657	.84277	1.1866	1.8578	26
35	.53853	.63912	1.5647	.84261	1.1868	1.8569	**25**
36	.53877	.63953	1.5637	.84245	1.1870	1.8561	24
37	.53902	.63994	1.5627	.84230	1.1872	1.8552	23
38	.53926	.64035	1.5617	.84214	1.1875	1.8544	22
39	.53951	.64076	1.5607	.84198	1.1877	1.8535	21
40	.53975	.64117	1.5597	.84182	1.1879	1.8527	**20**
41	.54000	.64158	1.5587	.84167	1.1881	1.8519	19
42	.54024	.64199	1.5577	.84151	1.1883	1.8510	18
43	.54049	.64240	1.5567	.84135	1.1886	1.8502	17
44	.54073	.64281	1.5557	.84120	1.1888	1.8494	16
45	.54097	.64322	1.5547	.84104	1.1890	1.8485	**15**
46	.54122	.64363	1.5537	.84088	1.1892	1.8477	14
47	.54146	.64404	1.5527	.84072	1.1895	1.8468	13
48	.54171	.64446	1.5517	.84057	1.1897	1.8460	12
49	.54195	.64487	1.5507	.84041	1.1899	1.8452	11
50	.54220	.64528	1.5497	.84025	1.1901	1.8443	**10**
51	.54244	.64569	1.5487	.84009	1.1903	1.8435	9
52	.54269	.64610	1.5477	.83994	1.1906	1.8427	8
53	.54293	.64652	1.5468	.83978	1.1908	1.8419	7
54	.54317	.64693	1.5458	.83962	1.1910	1.8410	6
55	.54342	.64734	1.5448	.83946	1.1912	1.8402	**5**
56	.54366	.64775	1.5438	.83930	1.1915	1.8394	4
57	.54391	.64817	1.5428	.83915	1.1917	1.8385	3
58	.54415	.64858	1.5418	.83899	1.1919	1.8377	2
59	.54440	.64899	1.5408	.83883	1.1921	1.8369	1
60	.54464	.64941	1.5399	.83867	1.1924	1.8361	**0**
′	Cos	Cot	Tan	Sin	Csc	Sec	′

122° (302°) — (237°) **57°**

33° (213°) — (326°) **146°**

′	Sin	Tan	Cot	Cos	Sec	Csc	′
0	.54464	.64941	1.5399	.83867	1.1924	1.8361	**60**
1	.54488	.64982	1.5389	.83851	1.1926	1.8353	59
2	.54513	.65024	1.5379	.83835	1.1928	1.8344	58
3	.54537	.65065	1.5369	.83819	1.1930	1.8336	57
4	.54561	.65106	1.5359	.83804	1.1933	1.8328	56
5	.54586	.65148	1.5350	.83788	1.1935	1.8320	**55**
6	.54610	.65189	1.5340	.83772	1.9137	1.8312	54
7	.54635	.65231	1.5330	.83756	1.1939	1.8303	53
8	.54659	.65272	1.5320	.83740	1.1942	1.8295	52
9	.54683	.65314	1.5311	.83724	1.1944	1.8287	51
10	.54708	.65355	1.5301	.83708	1.1946	1.8279	**50**
11	.54732	.65397	1.5291	.83692	1.1949	1.8271	49
12	.54756	.65438	1.5282	.83676	1.1951	1.8263	48
13	.54781	.65480	1.5272	.83660	1.1953	1.8255	47
14	.54805	.65521	1.5262	.83645	1.1955	1.8247	46
15	.54829	.65563	1.5253	.83629	1.1958	1.8238	**45**
16	.54854	.65604	1.5243	.83613	1.1960	1.8230	44
17	.54878	.65646	1.5233	.83597	1.1962	1.8222	43
18	.54902	.65688	1.5224	.83581	1.1964	1.8214	42
19	.54927	.65729	1.5214	.83565	1.1967	1.8206	41
20	.54951	.65771	1.5204	.83549	1.1969	1.8198	**40**
21	.54975	.65813	1.5195	.83533	1.1971	1.8190	39
22	.54999	.65854	1.5185	.83517	1.1974	1.8182	38
23	.55024	.65896	1.5175	.83501	1.1976	1.8174	37
24	.55048	.65938	1.5166	.83485	1.1978	1.8166	36
25	.55072	.65980	1.5156	.83469	1.1981	1.8158	**35**
26	.55097	.66021	1.5147	.83453	1.1983	1.8150	34
27	.55121	.66063	1.5137	.83437	1.1985	1.8142	33
28	.55145	.66105	1.5127	.83421	1.1987	1.8134	32
29	.55169	.66147	1.5118	.83405	1.1990	1.8126	31
30	.55194	.66189	1.5108	.83389	1.1992	1.8118	**30**
31	.55218	.66230	1.5099	.83373	1.1994	1.8110	29
32	.55242	.66272	1.5089	.83356	1.1997	1.8102	28
33	.55266	.66314	1.5080	.83340	1.1999	1.8094	27
34	.55291	.66356	1.5070	.83324	1.2001	1.8086	26
35	.55315	.66398	1.5061	.83308	1.2004	1.8078	**25**
36	.55339	.66440	1.5051	.83292	1.2006	1.8070	24
37	.55363	.66482	1.5042	.83276	1.2008	1.8062	23
38	.55388	.66524	1.5032	.83260	1.2011	1.8055	22
39	.55412	.66566	1.5023	.83244	1.2013	1.8047	21
40	.55436	.66608	1.5013	.83228	1.2015	1.8039	**20**
41	.55460	.66650	1.5004	.83212	1.2018	1.8031	19
42	.55484	.66692	1.4994	.83195	1.2020	1.8023	18
43	.55509	.66734	1.4985	.83179	1.2022	1.8015	17
44	.55533	.66776	1.4975	.83163	1.2025	1.8007	16
45	.55557	.66818	1.4966	.83147	1.2027	1.8000	**15**
46	.55581	.66860	1.4957	.83131	1.2029	1.7992	14
47	.55605	.66902	1.4947	.83115	1.2032	1.7984	13
48	.55630	.66944	1.4938	.83098	1.2034	1.7976	12
49	.55654	.66986	1.4928	.83082	1.2036	1.7968	11
50	.55678	.67028	1.4919	.83066	1.2039	1.7960	**10**
51	.55702	.67071	1.4910	.83050	1.2041	1.7953	9
52	.55726	.67113	1.4900	.83034	1.2043	1.7945	8
53	.55750	.67155	1.4891	.83017	1.2046	1.7937	7
54	.55775	.67197	1.4882	.83001	1.2048	1.7929	6
55	.55799	.67239	1.4872	.82985	1.2050	1.7922	**5**
56	.55823	.67282	1.4863	.82969	1.2053	1.7914	4
57	.55847	.67324	1.4854	.82953	1.2055	1.7906	3
58	.55871	.67366	1.4844	.82936	1.2057	1.7898	2
59	.55895	.67409	1.4835	.82920	1.2060	1.7891	1
60	.55919	.67451	1.4826	.82904	1.2062	1.7883	**0**
′	Cos	Cot	Tan	Sin	Csc	Sec	′

123° (303°) — (236°) 56°

NATURAL TRIGONOMETRIC FUNCTIONS (*Continued*)

34° (214°) (325°) **145°**

′	Sin	Tan	Cot	Cos	Sec	Csc	′
0	.55919	.67451	1.4826	.82904	1.2062	1.7883	**60**
1	.55943	.67493	1.4816	.82887	1.2065	1.7875	59
2	.55968	.67536	1.4807	.82871	1.2067	1.7868	58
3	.55992	.67578	1.4798	.82855	1.2069	1.7860	57
4	.56016	.67620	1.4788	.82839	1.2072	1.7852	56
5	.56040	.67663	1.4779	.82822	1.2074	1.7844	**55**
6	.56064	.67705	1.4770	.82806	1.2076	1.7837	54
7	.56088	.67748	1.4761	.82790	1.2079	1.7829	53
8	.56112	.67790	1.4751	.82773	1.2081	1.7821	52
9	.56136	.67832	1.4742	.82757	1.2084	1.7814	51
10	.56160	.67875	1.4733	.82741	1.2086	1.7806	**50**
11	.56184	.67917	1.4724	.82724	1.2088	1.7799	49
12	.56208	.67960	1.4715	.82708	1.2091	1.7791	48
13	.56232	.68002	1.4705	.82692	1.2093	1.7783	47
14	.56256	.68045	1.4696	.82675	1.2096	1.7776	46
15	.56280	.68088	1.4687	.82659	1.2098	1.7768	**45**
16	.56305	.68130	1.4678	.82643	1.2100	1.7761	44
17	.56329	.68173	1.4669	.82626	1.2103	1.7753	43
18	.56353	.68215	1.4659	.82610	1.2105	1.7745	42
19	.56377	.68258	1.4650	.82593	1.2108	1.7738	41
20	.56401	.68301	1.4641	.82577	1.2110	1.7730	**40**
21	.56425	.68343	1.4632	.82561	1.2112	1.7723	39
22	.56449	.68386	1.4623	.82544	1.2115	1.7715	38
23	.56473	.68429	1.4614	.82528	1.2117	1.7708	37
24	.56497	.68471	1.4605	.82511	1.2120	1.7700	36
25	.56521	.68514	1.4596	.82495	1.2122	1.7693	**35**
26	.56545	.68557	1.4586	.82478	1.2124	1.7685	34
27	.56569	.68600	1.4577	.82462	1.2127	1.7678	33
28	.56593	.68642	1.4568	.82446	1.2129	1.7670	32
29	.56617	.68685	1.4559	.82429	1.2132	1.7663	31
30	.56641	.68728	1.4550	.82413	1.2134	1.7655	**30**
31	.56665	.68771	1.4541	.82396	1.2136	1.7648	29
32	.56689	.68814	1.4532	.82380	1.2139	1.7640	28
33	.56713	.68857	1.4523	.82363	1.2141	1.7633	27
34	.56736	.68900	1.4514	.82347	1.2144	1.7625	26
35	.56760	.68942	1.4505	.82330	1.2146	1.7618	**25**
36	.56784	.68985	1.4496	.82314	1.2149	1.7610	24
37	.56808	.69028	1.4487	.82297	1.2151	1.7603	23
38	.56832	.69071	1.4478	.82281	1.2154	1.7596	22
39	.56856	.69114	1.4469	.82264	1.2156	1.7588	21
40	.56880	.69157	1.4460	.82248	1.2158	1.7581	**20**
41	.56904	.69200	1.4451	.82231	1.2161	1.7573	19
42	.56928	.69243	1.4442	.82214	1.2163	1.7566	18
43	.56952	.69286	1.4433	.82198	1.2166	1.7559	17
44	.56976	.69329	1.4424	.82181	1.2168	1.7551	16
45	.57000	.69372	1.4415	.82165	1.2171	1.7544	**15**
46	.57024	.69416	1.4406	.82148	1.2173	1.7537	14
47	.57047	.69459	1.4397	.82132	1.2176	1.7529	13
48	.57071	.69502	1.4388	.82115	1.2178	1.7522	12
49	.57095	.69545	1.4379	.82098	1.2181	1.7515	11
50	.57119	.69588	1.4370	.82082	1.2183	1.7507	**10**
51	.57143	.69631	1.4361	.82065	1.2185	1.7500	9
52	.57167	.69675	1.4352	.82048	1.2188	1.7493	8
53	.57191	.69718	1.4344	.82032	1.2190	1.7485	7
54	.57215	.69761	1.4335	.82015	1.2193	1.7478	6
55	.57238	.69804	1.4326	.81999	1.2195	1.7471	**5**
56	.57262	.69847	1.4317	.81982	1.2198	1.7463	4
57	.57286	.69891	1.4308	.81965	1.2200	1.7456	3
58	.57310	.69934	1.4299	.81949	1.2203	1.7449	2
59	.57334	.69977	1.4290	.81932	1.2205	1.7442	1
60	.57358	.70021	1.4281	.81915	1.2208	1.7434	**0**
′	Cos	Cot	Tan	Sin	Csc	Sec	′

124° (304°) (235°) **55°**

35° (215°) (324°) **144°**

′	Sin	Tan	Cot	Cos	Sec	Csc	′
0	.57358	.70021	1.4281	.81915	1.2208	1.7434	**60**
1	.57381	.70064	1.4273	.81899	1.2210	1.7427	59
2	.57405	.70107	1.4264	.81882	1.2213	1.7420	58
3	.57429	.70151	1.4255	.81865	1.2215	1.7413	57
4	.57453	.70194	1.4246	.81848	1.2218	1.7406	56
5	.57477	.70238	1.4237	.81832	1.2220	1.7398	**55**
6	.57501	.70281	1.4229	.81815	1.2223	1.7391	54
7	.57524	.70325	1.4220	.81798	1.2225	1.7384	53
8	.57548	.70368	1.4211	.81782	1.2228	1.7377	52
9	.57572	.70412	1.4202	.81765	1.2230	1.7370	51
10	.57596	.70455	1.4193	.81748	1.2233	1.7362	**50**
11	.57619	.70499	1.4185	.81731	1.2235	1.7355	49
12	.57643	.70542	1.4176	.81714	1.2238	1.7348	48
13	.57667	.70586	1.4167	.81698	1.2240	1.7341	47
14	.57691	.70629	1.4158	.81681	1.2243	1.7334	46
15	.57715	.70673	1.4150	.81664	1.2245	1.7327	**45**
16	.57738	.70717	1.4141	.81647	1.2248	1.7320	44
17	.57762	.70760	1.4132	.81631	1.2250	1.7312	43
18	.57786	.70804	1.4124	.81614	1.2253	1.7305	42
19	.57810	.70848	1.4115	.81597	1.2255	1.7298	41
20	.57833	.70891	1.4106	.81580	1.2258	1.7291	**40**
21	.57857	.70935	1.4097	.81563	1.2260	1.7284	39
22	.57881	.70979	1.4089	.81546	1.2263	1.7277	38
23	.57904	.71023	1.4080	.81530	1.2265	1.7270	37
24	.57928	.71066	1.4071	.81513	1.2268	1.7263	36
25	.57952	.71110	1.4063	.81496	1.2271	1.7256	**35**
26	.57976	.71154	1.4054	.81479	1.2273	1.7249	34
27	.57999	.71198	1.4045	.81462	1.2276	1.7242	33
28	.58023	.71242	1.4037	.81445	1.2278	1.7235	32
29	.58047	.71285	1.4028	.81428	1.2281	1.7228	31
30	.58070	.71329	1.4019	.81412	1.2283	1.7221	**30**
31	.58094	.71373	1.4011	.81395	1.2286	1.7213	29
32	.58118	.71417	1.4002	.81378	1.2288	1.7206	28
33	.58141	.71461	1.3994	.81361	1.2291	1.7199	27
34	.58165	.71505	1.3985	.81344	1.2293	1.7192	26
35	.58189	.71549	1.3976	.81327	1.2296	1.7185	**25**
36	.58212	.71593	1.3968	.81310	1.2299	1.7179	24
37	.58236	.71637	1.3959	.81293	1.2301	1.7172	23
38	.58260	.71681	1.3951	.81276	1.2304	1.7165	22
39	.58283	.71725	1.3942	.81259	1.2306	1.7158	21
40	.58307	.71769	1.3934	.81242	1.2309	1.7151	**20**
41	.58330	.71813	1.3925	.81225	1.2311	1.7144	19
42	.58354	.71857	1.3916	.81208	1.2314	1.7137	18
43	.58378	.71901	1.3908	.81191	1.2317	1.7130	17
44	.58401	.71946	1.3899	.81174	1.2319	1.7123	16
45	.58425	.71990	1.3891	.81157	1.2322	1.7116	**15**
46	.58449	.72034	1.3882	.81140	1.2324	1.7109	14
47	.58472	.72078	1.3874	.81123	1.2327	1.7102	13
48	.58496	.72122	1.3865	.81106	1.2329	1.7095	12
49	.58519	.72167	1.3857	.81089	1.2332	1.7088	11
50	.58543	.72211	1.3848	.81072	1.2335	1.7081	**10**
51	.58567	.72255	1.3840	.81055	1.2337	1.7075	9
52	.58590	.72299	1.3831	.81038	1.2340	1.7068	8
53	.58614	.72344	1.3823	.81021	1.2342	1.7061	7
54	.58637	.72388	1.3814	.81004	1.2345	1.7054	6
55	.58661	.72432	1.3806	.80987	1.2348	1.7047	5
56	.58684	.72477	1.3798	.80970	1.2350	1.7040	4
57	.58708	.72521	1.3789	.80953	1.2353	1.7033	3
58	.58731	.72565	1.3781	.80936	1.2355	1.7027	2
59	.58755	.72610	1.3772	.80919	1.2358	1.7020	1
60	.58779	.72654	1.3764	.80902	1.2361	1.7013	**0**
′	Cos	Cot	Tan	Sin	Csc	Sec	′

125° (305°) (234°) **54°**

36° (216°) (323°) **143°**

′	Sin	Tan	Cot	Cos	Sec	Csc	′
0	.58779	.72654	1.3764	.80902	1.2361	1.7013	**60**
1	.58802	.72699	1.3755	.80885	1.2363	1.7006	59
2	.58826	.72743	1.3747	.80867	1.2366	1.6999	58
3	.58849	.72788	1.3739	.80850	1.2369	1.6993	57
4	.58873	.72832	1.3730	.80833	1.2371	1.6986	56
5	.58896	.72877	1.3722	.80816	1.2374	1.6979	**55**
6	.58920	.72921	1.3713	.80799	1.2376	1.6972	54
7	.58943	.72966	1.3705	.80782	1.2379	1.6966	53
8	.58967	.73010	1.3697	.80765	1.2382	1.6959	52
9	.58990	.73055	1.3688	.80748	1.2384	1.6952	51
10	.59014	.73100	1.3680	.80730	1.2387	1.6945	**50**
11	.59037	.73144	1.3672	.80713	1.2390	1.6939	49
12	.59061	.73189	1.3663	.80696	1.2392	1.6932	48
13	.59084	.73234	1.3655	.80679	1.2395	1.6925	47
14	.59108	.73278	1.3647	.80662	1.2397	1.6918	46
15	.59131	.73323	1.3638	.80644	1.2400	1.6912	**45**
16	.59154	.73368	1.3630	.80627	1.2403	1.6905	44
17	.59178	.73413	1.3622	.80610	1.2405	1.6898	43
18	.59201	.73457	1.3613	.80593	1.2408	1.6892	42
19	.59225	.73502	1.3605	.80576	1.2411	1.6885	41
20	.59248	.73547	1.3597	.80558	1.2413	1.6878	**40**
21	.59272	.73592	1.3588	.80541	1.2416	1.6871	39
22	.59295	.73637	1.3580	.80524	1.2419	1.6865	38
23	.59318	.73681	1.3572	.80507	1.2421	1.6858	37
24	.59342	.73726	1.3564	.80489	1.2424	1.6852	36
25	.59365	.73771	1.3555	.80472	1.2427	1.6845	**35**
26	.59389	.73816	1.3547	.80455	1.2429	1.6838	34
27	.59412	.73861	1.3539	.80438	1.2432	1.6832	33
28	.59436	.73906	1.3531	.80420	1.2435	1.6825	32
29	.59459	.73951	1.3522	.80403	1.2437	1.6818	31
30	.59482	.73996	1.3514	.80386	1.2440	1.6812	**30**
31	.59506	.74041	1.3506	.80368	1.2443	1.6805	29
32	.59529	.74086	1.3498	.80351	1.2445	1.6799	28
33	.59552	.74131	1.3490	.80334	1.2448	1.6792	27
34	.59576	.74176	1.3481	.80316	1.2451	1.6785	26
35	.59599	.74221	1.3473	.80299	1.2453	1.6779	**25**
36	.59622	.74267	1.3465	.80282	1.2456	1.6772	24
37	.59646	.74312	1.3457	.80264	1.2459	1.6766	23
38	.59669	.74357	1.3449	.80247	1.2462	1.6759	22
39	.59693	.74402	1.3440	.80230	1.2464	1.6753	21
40	.59716	.74447	1.3432	.80212	1.2467	1.6746	**20**
41	.59739	.74492	1.3424	.80195	1.2470	1.6739	19
42	.59763	.74538	1.3416	.80178	1.2472	1.6733	18
43	.59786	.74583	1.3408	.80160	1.2475	1.6726	17
44	.59809	.74628	1.3400	.80143	1.2478	1.6720	16
45	.59832	.74674	1.3392	.80125	1.2480	1.6713	**15**
46	.59856	.74719	1.3384	.80108	1.2483	1.6707	14
47	.59879	.74764	1.3375	.80091	1.2486	1.6700	13
48	.59902	.74810	1.3367	.80073	1.2489	1.6694	12
49	.59926	.74855	1.3359	.80056	1.2491	1.6687	11
50	.59949	.74900	1.3351	.80038	1.2494	1.6681	**10**
51	.59972	.74946	1.3343	.80021	1.2497	1.6674	9
52	.59995	.74991	1.3335	.80003	1.2499	1.6668	8
53	.60019	.75037	1.3327	.79986	1.2502	1.6661	7
54	.60042	.75082	1.3319	.79968	1.2505	1.6655	6
55	.60065	.75128	1.3311	.79951	1.2508	1.6649	**5**
56	.60089	.75173	1.3303	.79934	1.2510	1.6642	4
57	.60112	.75219	1.3295	.79916	1.2513	1.6636	3
58	.60135	.75264	1.3287	.79899	1.2516	1.6629	2
59	.60158	.75310	1.3278	.79881	1.2519	1.6623	1
60	.60182	.75355	1.3270	.79864	1.2521	1.6616	**0**
′	Cos	Cot	Tan	Sin	Csc	Sec	′

126° (306°) (233°) **53°**

37° (217°) (322°) **142°**

′	Sin	Tan	Cot	Cos	Sec	Csc	′
0	.60182	.75355	1.3270	.79864	1.2521	1.6616	**60**
1	.60205	.75401	1.3262	.79846	1.2524	1.6610	59
2	.60228	.75447	1.3254	.79829	1.2527	1.6604	58
3	.60251	.75492	1.3246	.79811	1.2530	1.6597	57
4	.60274	.75538	1.3238	.79793	1.2532	1.6591	56
5	.60298	.75584	1.3230	.79776	1.2535	1.6584	**55**
6	.60321	.75629	1.3222	.79758	1.2538	1.6578	54
7	.60344	.75675	1.3214	.79741	1.2541	1.6572	53
8	.60367	.75721	1.3206	.79723	1.2543	1.6565	52
9	.60390	.75767	1.3198	.79706	1.2546	1.6559	51
10	.60414	.75812	1.3190	.79688	1.2549	1.6553	**50**
11	.60437	.75858	1.3182	.79671	1.2552	1.6546	49
12	.60460	.75904	1.3175	.79653	1.2554	1.6540	48
13	.60483	.75950	1.3167	.79635	1.2557	1.6534	47
14	.60506	.75996	1.3159	.79618	1.2560	1.6527	46
15	.60529	.76042	1.3151	.79600	1.2563	1.6521	**45**
16	.60553	.76088	1.3143	.79583	1.2566	1.6515	44
17	.60576	.76134	1.3135	.79565	1.2568	1.6508	43
18	.60599	.76180	1.3127	.79547	1.2571	1.6502	42
19	.60622	.76226	1.3119	.79530	1.2574	1.6496	41
20	.60645	.76272	1.3111	.79512	1.2577	1.6489	**40**
21	.60668	.76318	1.3103	.79494	1.2579	1.6483	39
22	.60691	.76364	1.3095	.79477	1.2582	1.6477	38
23	.60714	.76410	1.3087	.79459	1.2585	1.6471	37
24	.60738	.76456	1.3079	.79441	1.2588	1.6464	36
25	.60761	.76502	1.3072	.79424	1.2591	1.6458	**35**
26	.60784	.76548	1.3064	.79406	1.2593	1.6452	34
27	.60807	.76594	1.3056	.79388	1.2596	1.6446	33
28	.60830	.76640	1.3048	.79371	1.2599	1.6439	32
29	.60853	.76686	1.3040	.79353	1.2602	1.6433	31
30	.60876	.76733	1.3032	.79335	1.2605	1.6427	**30**
31	.60899	.76779	1.3024	.79318	1.2608	1.6421	29
32	.60922	.76825	1.3017	.79300	1.2610	1.6414	28
33	.60945	.76871	1.3009	.79282	1.2613	1.6408	27
34	.60968	.76918	1.3001	.79264	1.2616	1.6402	26
35	.60991	.76964	1.2993	.79247	1.2619	1.6396	**25**
36	.61015	.77010	1.2985	.79229	1.2622	1.6390	24
37	.61038	.77057	1.2977	.79211	1.2624	1.6383	23
38	.61061	.77103	1.2970	.79193	1.2627	1.6377	22
39	.61084	.77149	1.2962	.79176	1.2630	1.6371	21
40	.61107	.77196	1.2954	.79158	1.2633	1.6365	**20**
41	.61130	.77242	1.2946	.79140	1.2636	1.6359	19
42	.61153	.77289	1.2938	.79122	1.2639	1.6353	18
43	.61176	.77335	1.2931	.79105	1.2641	1.6346	17
44	.61199	.77382	1.2923	.79087	1.2644	1.6340	16
45	.61222	.77428	1.2915	.79069	1.2647	1.6334	**15**
46	.61245	.77475	1.2907	.79051	1.2650	1.6328	14
47	.61268	.77521	1.2900	.79033	1.2653	1.6322	13
48	.61291	.77568	1.2892	.79016	1.2656	1.6316	12
49	.61314	.77615	1.2884	.78998	1.2659	1.6310	11
50	.61337	.77661	1.2876	.78980	1.2661	1.6303	**10**
51	.61360	.77708	1.2869	.78962	1.2664	1.6297	9
52	.61383	.77754	1.2861	.78944	1.2667	1.6291	8
53	.61406	.77801	1.2853	.78926	1.2670	1.6285	7
54	.61429	.77848	1.2846	.78908	1.2673	1.6279	6
55	.61451	.77895	1.2838	.78891	1.2676	1.6273	**5**
56	.61474	.77941	1.2830	.78873	1.2679	1.6267	4
57	.61497	.77988	1.2822	.78855	1.2682	1.6261	3
58	.61520	.78035	1.2815	.78837	1.2684	1.6255	2
59	.61543	.78082	1.2807	.78819	1.2687	1.6249	1
60	.61566	.78129	1.2799	.78801	1.2690	1.6243	**0**
′	Cos	Cot	Tan	Sin	Csc	Sec	′

127° (307°) (232°) **52°**

38° (218°) (321°) **141°**

′	Sin	Tan	Cot	Cos	Sec	Csc	′
0	.61566	.78129	1.2799	.78801	1.2690	1.6243	**60**
1	.61589	.78175	1.2792	.78783	1.2693	1.6237	59
2	.61612	.78222	1.2784	.78765	1.2696	1.6231	58
3	.61635	.78269	1.2776	.78747	1.2699	1.6225	57
4	.61658	.78316	1.2769	.78729	1.2702	1.6219	56
5	.61681	.78363	1.2761	.78711	1.2705	1.6213	**55**
6	.61704	.78410	1.2753	.78694	1.2708	1.6207	54
7	.61726	.78457	1.2746	.78676	1.2710	1.6201	53
8	.61749	.78504	1.2738	.78658	1.2713	1.6195	52
9	.61772	.78551	1.2731	.78640	1.2716	1.6189	51
10	.61795	.78598	1.2723	.78622	1.1719	1.6183	**50**
11	.61818	.78645	1.2715	.78604	1.2722	1.6177	49
12	.61841	.78692	1.2708	.78586	1.2725	1.6171	48
13	.61864	.78739	1.2700	.78568	1.2728	1.6165	47
14	.61887	.78786	1.2693	.78550	1.2731	1.6159	46
15	.61909	.78834	1.2685	.78532	1.2734	1.6153	**45**
16	.61932	.78881	1.2677	.78514	1.2737	1.6147	44
17	.61955	.78928	1.2670	.78496	1.2740	1.6141	43
18	.61978	.78975	1.2662	.78478	1.2742	1.6135	42
19	.62001	.79022	1.2655	.78460	1.2745	1.6129	41
20	.62024	.79070	1.2647	.78442	1.2748	1.6123	**40**
21	.62046	.79117	1.2640	.78424	1.2751	1.6117	39
22	.62069	.79164	1.2632	.78405	1.2754	1.6111	38
23	.62092	.79212	1.2624	.78387	1.2757	1.6105	37
24	.62115	.79259	1.2617	.78369	1.2760	1.6099	36
25	.62138	.79306	1.2609	.78351	1.2763	1.6093	**35**
26	.62160	.79354	1.2602	.78333	1.2766	1.6087	34
27	.62183	.79401	1.2594	.78315	1.2769	1.6082	33
28	.62206	.79449	1.2587	.78297	1.2772	1.6076	32
29	.62229	.79496	1.2579	.78279	1.2775	1.6070	31
30	.62251	.79544	1.2572	.78261	1.2778	1.6064	**30**
31	.62274	.79591	1.2564	.78243	1.2781	1.6058	29
32	.62297	.79639	1.2557	.78225	1.2784	1.6052	28
33	.62320	.79686	1.2549	.78206	1.2787	1.6046	27
34	.62342	.79734	1.2542	.78188	1.2790	1.6040	26
35	.62365	.79781	1.2534	.78170	1.2793	1.6035	**25**
36	.62388	.79829	1.2527	.78152	1.2796	1.6029	24
37	.62411	.79877	1.2519	.78134	1.2799	1.6023	23
38	.62433	.79924	1.2512	.78116	1.2802	1.6017	22
39	.62456	.79972	1.2504	.78098	1.2804	1.6011	21
40	.62479	.80020	1.2497	.78079	1.2807	1.6005	**20**
41	.62502	.80067	1.2489	.78061	1.2810	1.6000	19
42	.62524	.80115	1.2482	.78043	1.2813	1.5994	18
43	.62547	.80163	1.2475	.78025	1.2816	1.5988	17
44	.62570	.80211	1.2467	.78007	1.2819	1.5982	16
45	.62592	.80258	1.2460	.77988	1.2822	1.5976	**15**
46	.62615	.80306	1.2452	.77970	1.2825	1.5971	14
47	.62638	.80354	1.2445	.77952	1.2828	1.5965	13
48	.62660	.80402	1.2437	.77934	1.2831	1.5959	12
49	.62683	.80450	1.2430	.77916	1.2834	1.5953	11
50	.62706	.80498	1.2423	.77897	1.2837	1.5948	**10**
51	.62728	.80546	1.2415	.77879	1.2840	1.5942	9
52	.62751	.80594	1.2408	.77861	1.2843	1.5936	8
53	.62774	.80642	1.2401	.77843	1.2846	1.5930	7
54	.62796	.80690	1.2393	.77824	1.2849	1.5925	6
55	.62819	.80738	1.2386	.77806	1.2852	1.5919	**5**
56	.62842	.80786	1.2378	.77788	1.2855	1.5913	4
57	.62864	.80834	1.2371	.77769	1.2859	1.5907	3
58	.62887	.80882	1.2364	.77751	1.2862	1.5902	2
59	.62909	.80930	1.2356	.77733	1.2865	1.5896	1
60	.62932	.80978	1.2349	.77715	1.2868	1.5890	**0**
′	Cos	Cot	Tan	Sin	Csc	Sec	′

128° (308°) (231°) **51°**

39° (219°) (320°) **140°**

′	Sin	Tan	Cot	Cos	Sec	Csc	′
0	.62932	.80978	1.2349	.77715	1.2868	1.5890	**60**
1	.62955	.81027	1.2342	.77696	1.2871	1.5884	59
2	.62977	.81075	1.2334	.77678	1.2874	1.5879	58
3	.63000	.81123	1.2327	.77660	1.2877	1.5873	57
4	.63022	.81171	1.2320	.77641	1.2880	1.5867	56
5	.63045	.81220	1.2312	.77623	1.2883	1.5862	**55**
6	.63068	.81268	1.2305	.77605	1.2886	1.5856	54
7	.63090	.81316	1.2298	.77586	1.2889	1.5850	53
8	.63113	.81364	1.2290	.77568	1.2892	1.5845	52
9	.63135	.81413	1.2283	.77550	1.2895	1.5839	51
10	.63158	.81461	1.2276	.77531	1.2898	1.5833	**50**
11	.63180	.81510	1.2268	.77513	1.2901	1.5828	49
12	.63203	.81558	1.2261	.77494	1.2904	1.5822	48
13	.63225	.81606	1.2254	.77476	1.2907	1.5816	47
14	.63248	.81655	1.2247	.77458	1.2910	1.5811	46
15	.63271	.81703	1.2239	.77439	1.2913	1.5805	**45**
16	.63293	.81752	1.2232	.77421	1.2916	1.5800	44
17	.63316	.81800	1.2225	.77402	1.2919	1.5794	43
18	.63338	.81849	1.2218	.77384	1.2923	1.5788	42
19	.63361	.81898	1.2210	.77366	1.2926	1.5783	41
20	.63383	.81946	1.2203	.77347	1.2929	1.5777	**40**
21	.63406	.81995	1.2196	.77329	1.2932	1.5771	39
22	.63428	.82044	1.2189	.77310	1.2935	1.5766	38
23	.63451	.82092	1.2181	.77292	1.2938	1.5760	37
24	.63473	.82141	1.2174	.77273	1.2941	1.5755	36
25	.63496	.82190	1.2167	.77255	1.2944	1.5749	**35**
26	.63518	.82238	1.2160	.77236	1.2947	1.5744	34
27	.63540	.82287	1.2153	.77218	1.2950	1.5738	33
28	.63563	.82336	1.2145	.77199	1.2953	1.5732	32
29	.63585	.82385	1.2138	.77181	1.2957	1.5727	31
30	.63608	.82434	1.2131	.77162	1.2960	1.5721	**30**
31	.63630	.82483	1.2124	.77144	1.2963	1.5716	29
32	.63653	.82531	1.2117	.77125	1.2966	1.5710	28
33	.63675	.82580	1.2109	.77107	1.2969	1.5705	27
34	.63698	.82629	1.2102	.77088	1.2972	1.5699	26
35	.63720	.82678	1.2095	.77070	1.2975	1.5694	**25**
36	.63742	.82727	1.2088	.77051	1.2978	1.5688	24
37	.63765	.82776	1.2081	.77033	1.2981	1.5683	23
38	.63787	.82825	1.2074	.77014	1.2985	1.5677	22
39	.63810	.82874	1.2066	.76996	1.2988	1.5672	21
40	.63832	.82923	1.2059	.76977	1.2991	1.5666	**20**
41	.63854	.82972	1.2052	.76959	1.2994	1.5661	19
42	.63877	.83022	1.2045	.76940	1.2997	1.5655	18
43	.63899	.83071	1.2038	.76921	1.3000	1.5650	17
44	.63922	.83120	1.2031	.76903	1.3003	1.5644	16
45	.63944	.83169	1.2024	.76884	1.3007	1.5639	**15**
46	.63966	.83218	1.2017	.76866	1.3010	1.5633	14
47	.63989	.83268	1.2009	.76847	1.3013	1.5628	13
48	.64011	.83317	1.2002	.76828	1.3016	1.5622	12
49	.64033	.83366	1.1995	.76810	1.3019	1.5617	11
50	.64056	.83415	1.1988	.76791	1.3022	1.5611	**10**
51	.64078	.83465	1.1981	.76772	1.3026	1.5606	9
52	.64100	.83514	1.1974	.76754	1.3029	1.5601	8
53	.64123	.83564	1.1967	.76735	1.3032	1.5595	7
54	.64145	.83613	1.1960	.76717	1.3035	1.5590	6
55	.64167	.83662	1.1953	.76698	1.3038	1.5584	**5**
56	.64190	.83712	1.1946	.76679	1.3041	1.5579	4
57	.64212	.83761	1.1939	.76661	1.3045	1.5573	3
58	.64234	.83811	1.1932	.76642	1.3048	1.5568	2
59	.64256	.83860	1.1925	.76623	1.3051	1.5563	1
60	.64279	.83910	1.1918	.76604	1.3054	1.5557	**0**
′	Cos	Cot	Tan	Sin	Csc	Sec	′

129° (309°) (230°) **50°**

NATURAL TRIGONOMETRIC FUNCTIONS (*Continued*)

40° (220°) — (319°) **139°**

′	Sin	Tan	Cot	Cos	Sec	Csc	′
0	.64279	.83910	1.1918	.76604	1.3054	1.5557	**60**
1	.64301	.83960	1.1910	.76586	1.3057	1.5552	59
2	.64323	.84009	1.1903	.76567	1.3060	1.5546	58
3	.64346	.84059	1.1896	.76548	1.3064	1.5541	57
4	.64368	.84108	1.1889	.76530	1.3067	1.5536	56
5	.64390	.84158	1.1882	.76511	1.3070	1.5530	**55**
6	.64412	.84208	1.1875	.76492	1.3073	1.5525	54
7	.64435	.84258	1.1868	.76473	1.3076	1.5520	53
8	.64457	.84307	1.1861	.76455	1.3080	1.5514	52
9	.64479	.84357	1.1854	.76436	1.3083	1.5509	51
10	.64501	.84407	1.1847	.76417	1.3086	1.5504	**50**
11	.64524	.84457	1.1840	.76398	1.3089	1.5498	49
12	.64546	.84507	1.1833	.76380	1.3093	1.5493	48
13	.64568	.84556	1.1826	.76361	1.3096	1.5488	47
14	.64590	.84606	1.1819	.76342	1.3099	1.5482	46
15	.64612	.84656	1.1812	.76323	1.3102	1.5477	**45**
16	.64635	.84706	1.1806	.76304	1.3105	1.5472	44
17	.64657	.84756	1.1799	.76286	1.3109	1.5466	43
18	.64679	.84806	1.1792	.76267	1.3112	1.5461	42
19	.64701	.84856	1.1785	.76248	1.3115	1.5456	41
20	.64723	.84906	1.1778	.76229	1.3118	1.5450	**40**
21	.64746	.84956	1.1771	.76210	1.3122	1.5445	39
22	.64768	.85006	1.1764	.76192	1.3125	1.5440	38
23	.64790	.85057	1.1757	.76173	1.3128	1.5435	37
24	.64812	.85107	1.1750	.76154	1.3131	1.5429	36
25	.64834	.85157	1.1743	.76135	1.3135	1.5424	**35**
26	.64856	.85207	1.1736	.76116	1.3138	1.5419	34
27	.64878	.85257	1.1729	.76097	1.3141	1.5413	33
28	.64901	.85308	1.1722	.76078	1.3144	1.5408	32
29	.64923	.85358	1.1715	.76059	1.3148	1.5403	31
30	.64945	.85408	1.1708	.76041	1.3151	1.5398	**30**
31	.64967	.85458	1.1702	.76022	1.3154	1.5392	29
32	.64989	.85509	1.1695	.76003	1.3157	1.5387	28
33	.65011	.85559	1.1688	.75984	1.3161	1.5382	27
34	.65033	.85609	1.1681	.75965	1.3164	1.5377	26
35	.65055	.85660	1.1674	.75946	1.3167	1.5372	**25**
36	.65077	.85710	1.1667	.75927	1.3171	1.5366	24
37	.65100	.85761	1.1660	.75908	1.3174	1.5361	23
38	.65122	.85811	1.1653	.75889	1.3177	1.5356	22
39	.65144	.85862	1.1647	.75870	1.3180	1.5351	21
40	.65166	.85912	1.1640	.75851	1.3184	1.5345	**20**
41	.65188	.85963	1.1633	.75832	1.3187	1.5340	19
42	.65210	.86014	1.1626	.75813	1.3190	1.5335	18
43	.65232	.86064	1.1619	.75794	1.3194	1.5330	17
44	.65254	.86115	1.1612	.75775	1.3197	1.5325	16
45	.65276	.86166	1.1606	.75756	1.3200	1.5320	**15**
46	.65298	.86216	1.1599	.75738	1.3203	1.5314	14
47	.65320	.86267	1.1592	.75719	1.3207	1.5309	13
48	.65342	.86318	1.1585	.75700	1.3210	1.5304	12
49	.65364	.86368	1.1578	.75680	1.3213	1.5299	11
50	.65386	.86419	1.1571	.75661	1.3217	1.5294	**10**
51	.65408	.86470	1.1565	.75642	1.3220	1.5289	9
52	.65430	.86521	1.1558	.75623	1.3223	1.5283	8
53	.65452	.86572	1.1551	.75604	1.3227	1.5278	7
54	.65474	.86623	1.1544	.75585	1.3230	1.5273	6
55	.65496	.86674	1.1538	.75566	1.3233	1.5268	**5**
56	.65518	.86725	1.1531	.75547	1.3237	1.5263	4
57	.65540	.86776	1.1524	.75528	1.3240	1.5258	3
58	.65562	.86827	1.1517	.75509	1.3243	1.5253	2
59	.65584	.86878	1.1510	.75490	1.3247	1.5248	1
60	.65606	.86929	1.1504	.75471	1.3250	1.5243	**0**
′	Cos	Cot	Tan	Sin	Csc	Sec	′

130° (310°) — (229°) **49°**

41° (221°) — (318°) **138°**

′	Sin	Tan	Cot	Cos	Sec	Csc	′
0	.65606	.86929	1.1504	.75471	1.3250	1.5243	**60**
1	.65628	.86980	1.1497	.75452	1.3253	1.5237	59
2	.65650	.87031	1.1490	.75433	1.3257	1.5232	58
3	.65672	.87082	1.1483	.75414	1.3260	1.5227	57
4	.65694	.87133	1.1477	.75395	1.3264	1.5222	56
5	.65716	.87184	1.1470	.75375	1.3267	1.5217	**55**
6	.65738	.87236	1.1463	.75356	1.3270	1.5212	54
7	.65759	.87287	1.1456	.75337	1.3274	1.5207	53
8	.65781	.87338	1.1450	.75318	1.3277	1.5202	52
9	.65803	.87389	1.1443	.75299	1.3280	1.5197	51
10	.65825	.87441	1.1436	.75280	1.3284	1.5192	**50**
11	.65847	.87492	1.1430	.75261	1.3287	1.5187	49
12	.65869	.87543	1.1423	.75241	1.3291	1.5182	48
13	.65891	.87595	1.1416	.75222	1.3294	1.5177	47
14	.65913	.87646	1.1410	.75203	1.3297	1.5172	46
15	.65935	.87698	1.1403	.75184	1.3301	1.5167	**45**
16	.65956	.87749	1.1396	.75165	1.3304	1.5162	44
17	.65978	.87801	1.1389	.75146	1.3307	1.5156	43
18	.66000	.87852	1.1383	.75126	1.3311	1.5151	42
19	.66022	.87904	1.1376	.75107	1.3314	1.5146	41
20	.66044	.87955	1.1369	.75088	1.3318	1.5141	**40**
21	.66066	.88007	1.1363	.75069	1.3321	1.5136	39
22	.66088	.88059	1.1356	.75050	1.3325	1.5131	38
23	.66109	.88110	1.1349	.75030	1.3328	1.5126	37
24	.66131	.88162	1.1343	.75011	1.3331	1.5121	36
25	.66153	.88214	1.1336	.74992	1.3335	1.5116	**35**
26	.66175	.88265	1.1329	.74973	1.3338	1.5111	34
27	.66197	.88317	1.1323	.74953	1.3342	1.5107	33
28	.66218	.88369	1.1316	.74934	1.3345	1.5102	32
29	.66240	.88421	1.1310	.74915	1.3348	1.5097	31
30	.66262	.88473	1.1303	.74896	1.3352	1.5092	**30**
31	.66284	.88524	1.1296	.74876	1.3355	1.5087	29
32	.66306	.88576	1.1290	.74857	1.3359	1.5082	28
33	.66327	.88628	1.1283	.74838	1.3362	1.5077	27
34	.66349	.88680	1.1276	.74818	1.3366	1.5072	26
35	.66371	.88732	1.1270	.74799	1.3369	1.5067	**25**
36	.66393	.88784	1.1263	.74780	1.3373	1.5062	24
37	.66414	.88836	1.1257	.74760	1.3376	1.5057	23
38	.66436	.88888	1.1250	.74741	1.3380	1.5052	22
39	.66458	.88940	1.1243	.74722	1.3383	1.5047	21
40	.66480	.88992	1.1237	.74703	1.3386	1.5042	**20**
41	.66501	.89045	1.1230	.74683	1.3390	1.5037	19
42	.66523	.89097	1.1224	.74664	1.3393	1.5032	18
43	.66545	.89149	1.1217	.74644	1.3397	1.5027	17
44	.66566	.89201	1.1211	.74625	1.3400	1.5023	16
45	.66588	.89253	1.1204	.74606	1.3404	1.5018	**15**
46	.66610	.89306	1.1197	.74586	1.3407	1.5013	14
47	.66632	.89358	1.1191	.74567	1.3411	1.5008	13
48	.66653	.89410	1.1184	.74548	1.3414	1.5003	12
49	.66675	.89463	1.1178	.74528	1.3418	1.4998	11
50	.66697	.89515	1.1171	.74509	1.3421	1.4993	**10**
51	.66718	.89567	1.1165	.74489	1.3425	1.4988	9
52	.66740	.89620	1.1158	.74470	1.3428	1.4984	8
53	.66762	.89672	1.1152	.74451	1.3432	1.4979	7
54	.66783	.89725	1.1145	.74431	1.3435	1.4974	6
55	.66805	.89777	1.1139	.74412	1.3439	1.4969	**5**
56	.66827	.89830	1.1132	.74392	1.3442	1.4964	4
57	.66848	.89883	1.1126	.74373	1.3446	1.4959	3
58	.66870	.89935	1.1119	.74353	1.3449	1.4954	2
59	.66891	.89988	1.1113	.74334	1.3453	1.4950	1
60	.66913	.90040	1.1106	.74314	1.3456	1.4945	**0**
′	Cos	Cot	Tan	Sin	Csc	Sec	′

131° (311°) — (228°) **48°**

42° (222°) (317°) **137°**

′	Sin	Tan	Cot	Cos	Sec	Csc	′
0	.66913	.90040	1.1106	.74314	1.3456	1.4945	**60**
1	.66935	.90093	1.1100	.74295	1.3460	1.4940	59
2	.66956	.90146	1.1093	.74276	1.3463	1.4935	58
3	.66978	.90199	1.1087	.74256	1.3467	1.4930	57
4	.66999	.90251	1.1080	.74237	1.3470	1.4925	56
5	.67021	.90304	1.1074	.74217	1.3474	1.4921	**55**
6	.67043	.90357	1.1067	.74198	1.3478	1.4916	54
7	.67064	.90410	1.1061	.74178	1.3481	1.4911	53
8	.67086	.90463	1.1054	.74159	1.3485	1.4906	52
9	.67107	.90516	1.1048	.74139	1.3488	1.4901	51
10	.67129	.90569	1.1041	.74120	1.3492	1.4897	**50**
11	.67151	.90621	1.1035	.74100	1.3495	1.4892	49
12	.67172	.90674	1.1028	.74080	1.3499	1.4887	48
13	.67194	.90727	1.1022	.74061	1.3502	1.4882	47
14	.67215	.90781	1.1016	.74041	1.3506	1.4878	46
15	.67237	.90834	1.1009	.74022	1.3510	1.4873	**45**
16	.67258	.90887	1.1003	.74002	1.3513	1.4868	44
17	.67280	.90940	1.0996	.73983	1.3517	1.4863	43
18	.67301	.90993	1.0990	.73963	1.3520	1.4859	42
19	.67323	.91046	1.0983	.73944	1.3524	1.4854	41
20	.67344	.91099	1.0977	.73924	1.3527	1.4849	**40**
21	.67366	.91153	1.0971	.73904	1.3531	1.4844	39
22	.67387	.91206	1.0964	.73885	1.3535	1.4840	38
23	.67409	.91259	1.0958	.73865	1.3538	1.4835	37
24	.67430	.91313	1.0951	.73846	1.3542	1.4830	36
25	.67452	.91366	1.0945	.73826	1.3545	1.4825	**35**
26	.67473	.91419	1.0939	.73806	1.3549	1.4821	34
27	.67495	.91473	1.0932	.73787	1.3553	1.4816	33
28	.67516	.91526	1.0926	.73767	1.3556	1.4811	32
29	.67538	.91580	1.0919	.73747	1.3560	1.4807	31
30	.67559	.91633	1.0913	.73728	1.3563	1.4802	**30**
31	.67580	.91687	1.0907	.73708	1.3567	1.4797	29
32	.67602	.91740	1.0900	.73688	1.3571	1.4792	28
33	.67623	.91794	1.0894	.73669	1.3574	1.4788	27
34	.67645	.91847	1.0888	.73649	1.3578	1.4783	26
35	.67666	.91901	1.0881	.73629	1.3582	1.4778	**25**
36	.67688	.91955	1.0875	.73610	1.3585	1.4774	24
37	.67709	.92008	1.0869	.73590	1.3589	1.4769	23
38	.67730	.92062	1.0862	.73570	1.3592	1.4764	22
39	.67752	.92116	1.0856	.73551	1.3596	1.4760	21
40	.67773	.92170	1.0850	.73531	1.3600	1.4755	**20**
41	.67795	.92224	1.0843	.73511	1.3603	1.4750	19
42	.67816	.92277	1.0837	.73491	1.3607	1.4746	18
43	.67837	.92331	1.0831	.73472	1.3611	1.4741	17
44	.67859	.92385	1.0824	.73452	1.3614	1.4737	16
45	.67880	.92439	1.0818	.73432	1.3618	1.4732	**15**
46	.67901	.92493	1.0812	.73413	1.3622	1.4727	14
47	.67923	.92547	1.0805	.73393	1.3625	1.4723	13
48	.67944	.92601	1.0799	.73373	1.3629	1.4718	12
49	.67965	.92655	1.0793	.73353	1.3633	1.4713	11
50	.67987	.92709	1.0786	.73333	1.3636	1.4709	**10**
51	.68008	.92763	1.0780	.73314	1.3640	1.4704	9
52	.68029	.92817	1.0774	.73294	1.3644	1.4700	8
53	.68051	.92872	1.0768	.73274	1.3647	1.4695	7
54	.68072	.92926	1.0761	.73254	1.3651	1.4690	6
55	.68093	.92980	1.0755	.73234	1.3655	1.4686	**5**
56	.68115	.93034	1.0749	.73215	1.3658	1.4681	4
57	.68136	.93088	1.0742	.73195	1.3662	1.4677	3
58	.68157	.93143	1.0736	.73175	1.3666	1.4672	2
59	.68179	.93197	1.0730	.73155	1.3670	1.4667	1
60	.68200	.93252	1.0724	.73135	1.3673	1.4663	**0**
′	Cos	Cot	Tan	Sin	Csc	Sec	′

132° (312°) (227°) **47°**

43° (223°) (316°) **136°**

′	Sin	Tan	Cot	Cos	Sec	Csc	′
0	.68200	.93252	1.0724	.73135	1.3673	1.4663	**60**
1	.68221	.93306	1.0717	.73116	1.3677	1.4658	59
2	.68242	.93360	1.0711	.73096	1.3681	1.4654	58
3	.68264	.93415	1.0705	.73076	1.3684	1.4649	57
4	.68285	.93469	1.0699	.73056	1.3688	1.4645	56
5	.68306	.93524	1.0692	.73036	1.3692	1.4640	**55**
6	.68327	.93578	1.0686	.73016	1.3696	1.4635	54
7	.68349	.93633	1.0680	.72996	1.3699	1.4631	53
8	.68370	.93688	1.0674	.72976	1.3703	1.4626	52
9	.68391	.93742	1.0668	.72957	1.3707	1.4622	51
10	.68412	.93797	1.0661	.72937	1.3711	1.4617	**50**
11	.68434	.93852	1.0655	.72917	1.3714	1.4613	49
12	.68455	.93906	1.0649	.72897	1.3718	1.4608	48
13	.68476	.93961	1.0643	.72877	1.3722	1.4604	47
14	.68497	.94016	1.0637	.72857	1.3726	1.4599	46
15	.68518	.94071	1.0630	.72837	1.3729	1.4595	**45**
16	.68539	.94125	1.0624	.72817	1.3733	1.4590	44
17	.68561	.94180	1.0618	.72797	1.3737	1.4586	43
18	.68582	.94235	1.0612	.72777	1.3741	1.4581	42
19	.68603	.94290	1.0606	.72757	1.3744	1.4577	41
20	.68624	.94345	1.0599	.72737	1.3748	1.4572	**40**
21	.68645	.94400	1.0593	.72717	1.3752	1.4568	39
22	.68666	.94455	1.0587	.72697	1.3756	1.4563	38
23	.68688	.94510	1.0581	.72677	1.3759	1.4559	37
24	.68709	.94565	1.0575	.72657	1.3763	1.4554	36
25	.68730	.94620	1.0569	.72637	1.3767	1.4550	**35**
26	.68751	.94676	1.0562	.72617	1.3771	1.4545	34
27	.68772	.94731	1.0556	.72597	1.3775	1.4541	33
28	.68793	.94786	1.0550	.72577	1.3778	1.4536	32
29	.68814	.94841	1.0544	.72557	1.3782	1.4532	31
30	.68835	.94896	1.0538	.72537	1.3786	1.4527	**30**
31	.68857	.94952	1.0532	.72517	1.3790	1.4523	29
32	.68878	.95007	1.0526	.72497	1.3794	1.4518	28
33	.68899	.95062	1.0519	.72477	1.3797	1.4514	27
34	.68920	.95118	1.0513	.72457	1.3801	1.4510	26
35	.68941	.95173	1.0507	.72437	1.3805	1.4505	**25**
36	.68962	.95229	1.0501	.72417	1.3809	1.4501	24
37	.68983	.95284	1.0495	.72397	1.3813	1.4496	23
38	.69004	.95340	1.0489	.72377	1.3817	1.4492	22
39	.69025	.95395	1.0483	.72357	1.3820	1.4487	21
40	.69046	.95451	1.0477	.72337	1.3824	1.4483	**20**
41	.69067	.95506	1.0470	.72317	1.3828	1.4479	19
42	.69088	.95562	1.0464	.72297	1.3832	1.4474	18
43	.69109	.95618	1.0458	.72277	1.3836	1.4470	17
44	.69130	.95673	1.0452	.72257	1.3840	1.4465	16
45	.69151	.95729	1.0446	.72236	1.3843	1.4461	**15**
46	.69172	.95785	1.0440	.72216	1.3847	1.4457	14
47	.69193	.95841	1.0434	.72196	1.3851	1.4452	13
48	.69214	.95897	1.0428	.72176	1.3855	1.4448	12
49	.69235	.95952	1.0422	.72156	1.3859	1.4443	11
50	.69256	.96008	1.0416	.72136	1.3863	1.4439	**10**
51	.69277	.96064	1.0410	.72116	1.3867	1.4435	9
52	.69298	.96120	1.0404	.72095	1.3871	1.4430	8
53	.69319	.96176	1.0398	.72075	1.3874	1.4426	7
54	.69340	.96232	1.0392	.72055	1.3878	1.4422	6
55	.69361	.96288	1.0385	.72035	1.3882	1.4417	**5**
56	.69382	.96344	1.0379	.72015	1.3886	1.4413	4
57	.69403	.96400	1.0373	.71995	1.3890	1.4409	3
58	.69424	.96457	1.0367	.71974	1.3894	1.4404	2
59	.69445	.96513	1.0361	.71954	1.3898	1.4400	1
60	.69466	.96569	1.0355	.71934	1.3902	1.4396	**0**
′	Cos	Cot	Tan	Sin	Csc	Sec	′

133° (313°) (226°) **46°**

44° (224°) (315°) 135°

′	Sin	Tan	Cot	Cos	Sec	Csc	′
0	.69466	.96569	1.0355	.71934	1.3902	1.4396	**60**
1	.69487	.96625	1.0349	.71914	1.3906	1.4391	59
2	.69508	.96681	1.0343	.71894	1.3909	1.4387	58
3	.69529	.96738	1.0337	.71873	1.3913	1.4383	57
4	.69549	.96794	1.0331	.71853	1.3917	1.4378	56
5	.69570	.96850	1.0325	.71833	1.3921	1.4374	**55**
6	.69591	.96907	1.0319	.71813	1.3925	1.4370	54
7	.69612	.96963	1.0313	.71792	1.3929	1.4365	53
8	.69633	.97020	1.0307	.71772	1.3933	1.4361	52
9	.69654	.97076	1.0301	.71752	1.3937	1.4357	51
10	.69675	.97133	1.0295	.71732	1.3941	1.4352	**50**
11	.69696	.97189	1.0289	.71711	1.3945	1.4348	49
12	.69717	.97246	1.0283	.71691	1.3949	1.4344	48
13	.69737	.97302	1.0277	.71671	1.3953	1.4340	47
14	.69758	.97359	1.0271	.71650	1.3957	1.4335	46
15	.69779	.97416	1.0265	.71630	1.3961	1.4331	**45**
16	.69800	.97472	1.0259	.71610	1.3965	1.4327	44
17	.69821	.97529	1.0253	.71590	1.3969	1.4322	43
18	.69842	.97586	1.0247	.71569	1.3972	1.4318	42
19	.69862	.97643	1.0241	.71549	1.3976	1.4314	41
20	.69883	.97700	1.0235	.71529	1.3980	1.4310	**40**
21	.69904	.97756	1.0230	.71508	1.3984	1.4305	39
22	.69925	.97813	1.0224	.71488	1.3988	1.4301	38
23	.69946	.97870	1.0218	.71468	1.3992	1.4297	37
24	.69966	.97927	1.0212	.71447	1.3996	1.4293	36
25	.69987	.97984	1.0206	.71427	1.4000	1.4288	**35**
26	.70008	.98041	1.0200	.71407	1.4004	1.4284	34
27	.70029	.98098	1.0194	.71386	1.4008	1.4280	33
28	.70049	.98155	1.0188	.71366	1.4012	1.4276	32
29	.70070	.98213	1.0182	.71345	1.4016	1.4271	31
30	.70091	.98270	1.0176	.71325	1.4020	1.4267	**30**
31	.70112	.98327	1.0170	.71305	1.4024	1.4263	29
32	.70132	.98384	1.0164	.71284	1.4028	1.4259	28
33	.70153	.98441	1.0158	.71264	1.4032	1.4255	27
34	.70174	.98499	1.0152	.71243	1.4036	1.4250	26
35	.70195	.98556	1.0147	.71223	1.4040	1.4246	**25**
36	.70215	.98613	1.0141	.71203	1.4044	1.4242	24
37	.70236	.98671	1.0135	.71182	1.4048	1.4238	23
38	.70257	.98728	1.0129	.71162	1.4052	1.4234	22
39	.70277	.98786	1.0123	.71141	1.4057	1.4229	21
40	.70298	.98843	1.0117	.71121	1.4061	1.4225	**20**
41	.70319	.98901	1.0111	.71100	1.4065	1.4221	19
42	.70339	.98958	1.0105	.71080	1.4069	1.4217	18
43	.70360	.99016	1.0099	.71059	1.4073	1.4213	17
44	.70381	.99073	1.0094	.71039	1.4077	1.4208	16
45	.70401	.99131	1.0088	.71019	1.4081	1.4204	**15**
46	.70422	.99189	1.0082	.70998	1.4085	1.4200	14
47	.70443	.99247	1.0076	.70978	1.4089	1.4196	13
48	.70463	.99304	1.0070	.70957	1.4093	1.4192	12
49	.70484	.99362	1.0064	.70937	1.4097	1.4188	11
50	.70505	.99420	1.0058	.70916	1.4101	1.4183	**10**
51	.70525	.99478	1.0052	.70896	1.4105	1.4179	9
52	.70546	.99536	1.0047	.70875	1.4109	1.4175	8
53	.70567	.99594	1.0041	.70855	1.4113	1.4171	7
54	.70587	.99652	1.0035	.70834	1.4118	1.4167	6
55	.70608	.99710	1.0029	.70813	1.4122	1.4163	**5**
56	.70628	.99768	1.0023	.70793	1.4126	1.4159	4
57	.70649	.99826	1.0017	.70772	1.4130	1.4154	3
58	.70670	.99884	1.0012	.70752	1.4134	1.4150	2
59	.70690	.99942	1.0006	.70731	1.4138	1.4146	1
60	.70711	1.0000	1.0000	.70711	1.4142	1.4142	**0**
′	Cos	Cot	Tan	Sin	Csc	Sec	′

134° (314°) (225°) 45°

NATURAL TRIGONOMETRIC FUNCTIONS FOR ANGLES IN DEGREES AND DECIMALS

Deg.	Sin	*Tan	*Cot	Cos	Deg.
0.0	.00000	.00000	∞	1.0000	**90.0**
.1	.00175	.00175	573.0	1.0000	89.9
.2	.00349	.00349	286.5	1.0000	.8
.3	.00524	.00524	191.0	1.0000	.7
.4	.00698	.00698	143.24	1.0000	.6
.5	.00873	.00873	114.59	1.0000	.5
.6	.01047	.01047	95.49	0.9999	.4
.7	.01222	.01222	81.85	.9999	.3
.8	.01396	.01396	71.62	.9999	.2
.9	.01571	.01571	63.66	.9999	89.1
1.0	.01745	.01746	57.29	0.9998	**89.0**
.1	.01920	.01920	52.08	.9998	88.9
.2	.02094	.02095	47.74	.9998	.8
.3	.02269	.02269	44.07	.9997	.7
.4	.02443	.02444	40.92	.9997	.6
.5	.02618	.02619	38.19	.9997	.5
.6	.02792	.02793	35.80	.9996	.4
.7	.02967	.02968	33.69	.9996	.3
.8	.03141	.03143	31.82	.9995	.2
.9	.03316	.03317	30.14	.9995	88.1
2.0	.03490	.03492	28.64	0.9994	**88.0**
.1	.03664	.03667	27.27	.9993	87.9
.2	.03839	.03842	26.03	.9993	.8
.3	.04013	.04016	24.90	.9992	.7
.4	.04188	.04191	23.86	.9991	.6
.5	.04362	.04366	22.90	.9990	.5
.6	.04536	.04541	22.02	.9990	.4
.7	.04711	.04716	21.20	.9989	.3
.8	.04885	.04891	20.45	.9988	.2
.9	.05059	.05066	19.74	.9987	87.1
3.0	.05234	.05241	19.081	0.9986	**87.0**
.1	.05408	.05416	18.464	.9985	86.9
.2	.05582	.05591	17.886	.9984	.8
.3	.05756	.05766	17.343	.9983	.7
.4	.05931	.05941	16.832	.9982	.6
.5	.06105	.06116	16.350	.9981	.5
.6	.06279	.06291	15.895	.9980	.4
.7	.06453	.06467	15.464	.9979	.3
.8	.06627	.06642	15.056	.9978	.2
.9	.06802	.06817	14.669	.9977	86.1
4.0	.06976	.06993	14.301	0.9976	**86.0**
.1	.07150	.07168	13.951	.9974	85.9
.2	.07324	.07344	13.617	.9973	.8
.3	.07498	.07519	13.300	.9972	.7
.4	.07672	.07695	12.996	.9971	.6
.5	.07846	.07870	12.706	.9969	.5
.6	.08020	.08046	12.429	.9968	.4
.7	.08194	.08221	12.163	.9966	.3
.8	.08368	.08397	11.909	.9965	.2
.9	.08542	.08573	11.664	.9963	85.1
5.0	.08716	.08749	11.430	0.9962	**85.0**
.1	.08889	.08925	11.205	.9960	84.9
.2	.09063	.09101	10.988	.9959	.8
.3	.09237	.09277	10.780	.9957	.7
.4	.09411	.09453	10.579	.9956	.6
.5	.09585	.09629	10.385	.9954	.5
.6	.09758	.09805	10.199	.9952	.4
.7	.09932	.09981	10.019	.9951	.3
.8	.10106	.10158	9.845	.9949	.2
.9	.10279	.10334	9.677	.9947	84.1
6.0	.10453	.10510	9.514	0.9945	**84.0**
Deg.	**Cos**	***Cot**	***Tan**	**Sin**	**Deg.**

* Interpolation in this section of the table is inaccurate.

Deg.	Sin	Tan	Cot	Cos	Deg
6.0	.10453	.10510	9.514	0.9945	**84.0**
.1	.10626	.10687	9.357	.9943	83.9
.2	.1080	.10863	9.205	.9942	.8
.3	.10973	.11040	9.058	.9940	.7
.4	.11147	.11217	8.915	.9938	.6
.5	.11320	.11394	8.777	.9936	.5
.6	.11494	.11570	8.643	.9934	.4
.7	.11667	.11747	8.513	.9932	.3
.8	.11840	.11924	8.386	.9930	.2
.9	.12014	.12101	8.264	.9928	83.1
7.0	.12187	.12278	8.144	0.9925	**83.0**
.1	.12360	.12456	8.028	.9923	82.9
.2	.12533	.12633	7.916	.9921	.8
.3	.12706	.12810	7.806	.9919	.7
.4	.12880	.12988	7.700	.9917	.6
.5	.13053	.13165	7.596	.9914	.5
.6	.13226	.13343	7.495	.9912	.4
.7	.13399	.13521	7.396	.9910	.3
.8	.13572	.13698	7.300	.9907	.2
.9	.13744	.13876	7.207	.9905	82.1
8.0	.13917	.14054	7.115	0.9903	**82.0**
.1	.14090	.14232	7.026	.9900	81.9
.2	.14263	.14410	6.940	.9898	.8
.3	.14436	.14588	6.855	.9895	.7
.4	.14608	.14767	6.772	.9893	.6
.5	.14781	.14945	6.691	.9890	.5
.6	.14954	.15124	6.612	.9888	.4
.7	.15126	.15302	6.535	.9885	.3
.8	.15299	.15481	6.460	.9882	.2
.9	.15471	.15660	6.386	.9880	81.1
9.0	.15643	.15838	6.314	0.9877	**81.0**
.1	.15816	.16017	6.243	.9874	80.9
.2	.15988	.16196	6.174	.9871	.8
.3	.16160	.16376	6.107	.9869	.7
.4	.16333	.16555	6.041	.9866	.6
.5	.16505	.16734	5.976	.9863	.5
.6	.16677	.16914	5.912	.9860	.4
.7	.16849	.17093	5.850	.9857	.3
.8	.17021	.17273	5.789	.9854	.2
.9	.17193	.17453	5.730	.9851	80.1
10.0	.1736	.1763	5.671	0.9848	**80.0**
.1	.1754	.1781	5.614	.9845	79.9
.2	.1771	.1799	5.558	.9842	.8
.3	.1788	.1817	5.503	.9839	.7
.4	.1805	.1835	5.449	.9836	.6
.5	.1822	.1853	5.396	.9833	.5
.6	.1840	.1871	5.343	.9829	.4
.7	.1857	.1890	5.292	.9826	.3
.8	.1874	.1908	5.242	.9823	.2
.9	.1891	.1926	5.193	.9820	79.1
11.0	.1908	.1944	5.145	0.9816	**79.0**
.1	.1925	.1962	5.097	.9813	78.9
.2	.1942	.1980	5.050	.9810	.8
.3	.1959	.1998	5.005	.9806	.7
.4	.1977	.2016	4.959	.9803	.6
.5	.1994	.2035	4.915	.9799	.5
.6	.2011	.2053	4.872	.9796	.4
.7	.2028	.2071	4.829	.9792	.3
.8	.2045	.2089	4.787	.9789	.2
.9	.2062	.2107	4.745	.9785	78.1
12.0	.2079	.2126	4.705	0.9781	**78.0**
Deg.	**Cos**	**Cot**	**Tan**	**Sin**	**Deg.**

Deg.	Sin	Tan	Cot	Cos	Deg
12.0	0.2079	0.2126	4.705	0.9781	**78.0**
.1	.2096	.2144	4.665	.9778	77.9
.2	.2113	.2162	4.625	.9774	.8
.3	.2130	.2180	4.586	.9770	.7
.4	.2147	.2199	4.548	.9767	.6
.5	.2164	.2217	4.511	.9763	.5
.6	.2181	.2235	4.474	.9759	.4
7	.2198	.2254	4.437	.9755	.3
.8	.2215	.2272	4.402	.9751	.2
.9	.2233	.2290	4.366	.9748	77.1
13.0	0.2250	0.2309	4.331	0.9744	**77.0**
.1	.2267	.2327	4.297	.9740	76.9
.2	.2284	.2345	4.264	.9736	.8
.3	.2300	.2364	4.230	.9732	.7
.4	.2317	.2382	4.198	.9728	.6
.5	.2334	.2401	4.165	.9724	.5
.6	.2351	.2419	4.134	.9720	.4
.7	.2368	.2438	4.102	.9715	.3
.8	.2385	.2456	4.071	.9711	.2
.9	.2402	.2475	4.041	.9707	76.1
14.0	0.2419	0.2493	4.011	0.9703	**76.0**
.1	.2436	.2512	3.981	.9699	75.9
.2	.2453	.2530	3.952	.9694	.8
.3	.2470	.2549	3.923	.9690	.7
.4	.2487	.2568	3.895	.9686	.6
.5	.2504	.2586	3.867	.9681	.5
.6	.2521	.2605	3.839	.9677	.4
.7	.2538	.2623	3.812	.9673	.3
.8	.2554	.2642	3.785	.9668	.2
.9	.2571	.2661	3.758	.9664	75.1
15.0	0.2588	0.2679	3.732	0.9659	**75.0**
.1	.2605	.2698	3.706	.9655	74.9
.2	.2622	.2717	3.681	.9650	.8
.3	.2639	.2736	3.655	.9646	.7
.4	.2656	.2754	3.630	.9641	.6
.5	.2672	.2773	3.606	.9636	.5
.6	.2689	.2792	3.582	.9632	.4
.7	.2706	.2811	3.558	.9627	.3
.8	.2723	.2830	3.534	.9622	.2
.9	.2740	.2849	3.511	.9617	74.1
16.0	0.2756	0.2867	3.487	0.9613	**74.0**
.1	.2773	.2886	3.465	.9608	73.9
.2	.2790	.2905	3.442	.9603	.8
.3	.2807	.2924	3.420	.9598	.7
.4	.2823	.2943	3.398	.9593	.6
.5	.2840	.2962	3.376	.9588	.5
.6	.2857	.2981	3.354	.9583	.4
.7	.2874	.3000	3.333	.9578	.3
.8	.2890	.3019	3.312	.9573	.2
.9	.2907	.3038	3.291	.9568	73.1
17.0	0.2924	0.3057	3.271	0.9563	**73.0**
.1	.2940	.3076	3.251	.9558	72.9
.2	.2957	.3096	3.230	.9553	.8
.3	.2974	.3115	3.211	.9548	.7
.4	.2990	.3134	3.191	.9542	.6
.5	.3007	.3153	3.172	.9537	.5
.6	.3024	.3172	3.152	.9532	.4
.7	.3040	.3191	3.133	.9527	.3
.8	.3057	.3211	3.115	.9521	.2
.9	.3074	.3230	3.096	.9516	72.1
18.0	0.3090	0.3249	3.078	0.9511	**72.0**
Deg.	**Cos**	**Cot**	**Tan**	**Sin**	**Deg.**

Deg.	Sin	Tan	Cot	Cos	Deg
18.0	0.3090	0.3249	3.078	0.9511	**72.0**
.1	.3107	.3269	3.060	.9505	71.9
.2	.3123	.3288	3.042	.9500	.8
.3	.3140	.3307	3.024	.9494	.7
.4	.3156	.3327	3.006	.9489	.6
.5	.3173	.3346	2.989	.9483	.5
.6	.3190	.3365	2.971	.9478	.4
.7	.3206	.3385	2.954	.9472	.3
.8	.3223	.3404	2.937	.9466	.2
.9	.3239	.3424	2.921	.9461	71.1
19.0	0.3256	0.3443	2.904	0.9455	**71.0**
.1	.3272	.3463	2.888	.9449	70.9
.2	.3289	.3482	2.872	.9444	.8
.3	.3305	.3502	2.856	.9438	.7
.4	.3322	.3522	2.840	.9432	.6
.5	.3338	.3541	2.824	.9426	.5
.6	.3355	.3561	2.808	.9421	.4
.7	.3371	.3581	2.793	.9415	.3
.8	.3387	.3600	2.778	.9409	.2
.9	.3404	.3620	2.762	.9403	70.1
20.0	0.3420	0.3640	2.747	0.9397	**70.0**
.1	.3437	.3659	2.733	.9391	69.9
.2	.3453	.3679	2.718	.9385	.8
.3	.3469	.3699	2.703	.9379	.7
.4	.3486	.3719	2.689	.9373	.6
.5	.3502	.3739	2.675	.9367	.5
.6	.3518	.3759	2.660	.9361	.4
.7	.3535	.3779	2.646	.9354	.3
.8	.3551	.3799	2.633	.9348	.2
.9	.3567	.3819	2.619	.9342	69.1
21.0	0.3584	0.3839	2.605	0.9336	**69.0**
.1	.3600	.3859	2.592	.9330	68.9
.2	.3616	.3879	2.578	.9323	.8
.3	.3633	.3899	2.565	.9317	.7
.4	.3649	.3919	2.552	.9311	.6
.5	.3665	.3939	2.539	.9304	.5
.6	.3681	.3959	2.526	.9298	.4
.7	.3697	.3979	2.513	.9291	.3
.8	.3714	.4000	2.500	.9285	.2
.9	.3730	.4020	2.488	.9278	68.1
22.0	0.3746	0.4040	2.475	0.9272	**68.0**
.1	.3762	.4061	2.463	.9265	67.9
.2	.3778	.4081	2.450	.9259	.8
.3	.3795	.4101	2.438	.9252	.7
.4	.3811	.4122	2.426	.9245	.6
.5	.3827	.4142	2.414	.9239	.5
.6	.3843	.4163	2.402	.9232	.4
.7	.3859	.4183	2.391	.9225	.3
.8	.3875	.4204	2.379	.9219	.2
.9	.3891	.4224	2.367	.9212	67.1
23.0	0.3907	0.4245	2.356	0.9205	**67.0**
.1	.3923	.4265	2.344	.9198	66.9
.2	.3939	.4286	2.333	.9191	.8
.3	.3955	.4307	2.322	.9184	.7
.4	.3971	.4327	2.311	.9178	.6
.5	.3987	.4348	2.300	.9171	.5
.6	.4003	.4369	2.289	.9164	.4
.7	.4019	.4390	2.278	.9157	.3
.8	.4035	.4411	2.267	.9150	.2
.9	.4051	.4431	2.257	.9143	66.1
24.0	0.4067	0.4452	2.246	0.9135	**66.0**
Deg.	**Cos**	**Cot**	**Tan**	**Sin**	**Deg.**

NATURAL TRIGONOMETRIC FUNCTIONS FOR ANGLES IN DEGREES AND DECIMALS

NATURAL FUNCTIONS FOR DEGREES AND DECIMALS (Continued)

Deg.	Sin	Tan	Cot	Cos	Deg.
24.0	0.4067	0.4452	2.246	0.9135	**66.0**
.1	.4083	.4473	2.236	.9128	65.9
.2	.4099	.4494	2.225	.9121	.8
.3	.4115	.4515	2.215	.9114	.7
.4	.4131	.4536	2.204	.9107	.6
.5	.4147	.4557	2.194	.9100	.5
.6	.4163	.4578	2.184	.9092	.4
.7	.4179	.4599	2.174	.9085	.3
.8	.4195	.4621	2.164	.9078	.2
.9	.4210	.4642	2.154	.9070	65.1
25.0	0.4226	0.4663	2.145	0.9063	**65.0**
.1	.4242	.4684	2.135	.9056	64.9
.2	.4258	.4706	2.125	.9048	.8
.3	.4274	.4727	2.116	.9041	.7
.4	.4289	.4748	2.106	.9033	.6
.5	.4305	.4770	2.097	.9026	.5
.6	.4321	.4791	2.087	.9018	.4
.7	.4337	.4813	2.078	.9011	.3
.8	.4352	.4834	2.069	.9003	.2
.9	.4368	.4856	2.059	.8996	64.1
26.0	0.4384	0.4877	2.050	0.8988	**64.0**
.1	.4399	.4899	2.041	.8980	63.9
.2	.4415	.4921	2.032	.8973	.8
.3	.4431	.4942	2.023	.8965	.7
.4	.4446	.4964	2.014	.8957	.6
.5	.4462	.4986	2.006	.8949	.5
.6	.4478	.5008	1.997	.8942	.4
.7	.4493	.5029	1.988	.8934	.3
.8	.4509	.5051	1.980	.8926	.2
.9	.4524	.5073	1.971	.8918	63.1
27.0	0.4540	0.5095	1.963	0.8910	**63.0**
.1	.4555	.5117	1.954	.8902	62.9
.2	.4571	.5139	1.946	.8894	.8
.3	.4586	.5161	1.937	.8886	.7
.4	.4602	.5184	1.929	.8878	.6
.5	.4617	.5206	1.921	.8870	.5
.6	.4633	.5228	1.913	.8862	.4
.7	.4648	.5250	1.905	.8854	.3
.8	.4664	.5272	1.897	.8846	.2
.9	.4679	.5295	1.889	.8838	62.1
28.0	0.4695	0.5317	1.881	0.8829	**62.0**
.1	.4710	.5340	1.873	.8821	61.9
.2	.4726	.5362	1.865	.8813	.8
.3	.4741	.5384	1.857	.8805	.7
.4	.4756	.5407	1.849	.8796	.6
.5	.4772	.5430	1.842	.8788	.5
.6	.4787	.5452	1.834	.8780	.4
.7	.4802	.5475	1.827	.8771	.3
.8	.4818	.5498	1.819	.8763	.2
.9	.4833	.5520	1.811	.8755	61.1
29.0	0.4848	0.5543	1.804	0.8746	**61.0**
.1	.4863	.5566	1.797	.8738	60.9
.2	.4879	.5589	1.789	.8729	.8
.3	.4894	.5612	1.782	.8721	.7
.4	.4909	.5635	1.775	.8712	.6
.5	.4924	.5658	1.767	.8704	.5
.6	.4939	.5681	1.760	.8695	.4
.7	.4955	.5704	1.753	.8686	.3
.8	.4970	.5727	1.746	.8678	.2
.9	.4985	.5750	1.739	.8669	60.1
30.0	0.5000	0.5774	1.732	0.8660	**60.0**
Deg.	Cos	Cot	Tan	Sin	Deg.

Deg.	Sin	Tan	Cot	Cos	Deg.
30.0	0.5000	0.5774	1.7321	0.8660	**60.0**
.1	.5015	.5797	1.7251	.8652	59.9
.2	.5030	.5820	1.7182	.8643	.8
.3	.5045	.5844	1.7113	.8634	.7
.4	.5060	.5867	1.7045	.8625	.6
.5	.5075	.5890	1.6977	.8616	.5
.6	.5090	.5914	1.6909	.8607	.4
.7	.5105	.5938	1.6842	.8599	.3
.8	.5120	.5961	1.6775	.8590	.2
.9	.5135	.5985	1.6709	.8581	59.1
31.0	0.5150	0.6009	1.6643	0.8572	**59.0**
.1	.5165	.6032	1.6577	.8563	58.9
.2	.5180	.6056	1.6512	.8554	.8
.3	.5195	.6080	1.6447	.8545	.7
.4	.5210	.6104	1.6383	.8536	.6
.5	.5225	.6128	1.6319	.8526	.5
.6	.5240	.6152	1.6255	.8517	.4
.7	.5255	.6176	1.6191	.8508	.3
.8	.5270	.6200	1.6128	.8499	.2
.9	.5284	.6224	1.6066	.8490	58.1
32.0	0.5299	0.6249	1.6003	0.8480	**58.0**
.1	.5314	.6273	1.5941	.8471	57.9
.2	.5329	.6297	1.5880	.8462	.8
.3	.5344	.6322	1.5818	.8453	.7
.4	.5358	.6346	1.5757	.8443	.6
.5	.5373	.6371	1.5697	.8434	.5
.6	.5388	.6395	1.5637	.8425	.4
.7	.5402	.6420	1.5577	.8415	.3
.8	.5417	.6445	1.5517	.8406	.2
.9	.5432	.6469	1.5458	.8396	57.1
33.0	0.5446	0.6494	1.5399	0.8387	**57.0**
.1	.5461	.6519	1.5340	.8377	56.9
.2	.5476	.6544	1.5282	.8368	.8
.3	.5490	.6569	1.5224	.8358	.7
.4	.5505	.6594	1.5166	.8348	.6
.5	.5519	.6619	1.5108	.8339	.5
.6	.5534	.6644	1.5051	.8329	.4
.7	.5548	.6669	1.4994	.8320	.3
.8	.5563	.6694	1.4938	.8310	.2
.9	.5577	.6720	1.4882	.8300	56.1
34.0	0.5592	0.6745	1.4826	0.8290	**56.0**
.1	.5606	.6771	1.4770	.8281	55.9
.2	.5621	.6796	1.4715	.8271	.8
.3	.5635	.6822	1.4659	.8261	.7
.4	.5650	.6847	1.4605	.8251	.6
.5	.5664	.6873	1.4550	.8241	.5
.6	.5678	.6899	1.4496	.8231	.4
.7	.5693	.6924	1.4442	.8221	.3
.8	.5707	.6950	1.4388	.8211	.2
.9	.5721	.6976	1.4335	.8202	55.1
35.0	0.5736	0.7002	1.4281	0.8192	**55.0**
.1	.5750	.7028	1.4229	.8181	54.9
.2	.5764	.7054	1.4176	.8171	.8
.3	.5779	.7080	1.4124	.8161	.7
.4	.5793	.7107	1.4071	.8151	.6
.5	.5807	.7133	1.4019	.8141	.5
.6	.5821	.7159	1.3968	.8131	.4
.7	.5835	.7186	1.3916	.8121	.3
.8	.5850	.7212	1.3865	.8111	.2
.9	.5864	.7239	1.3814	.8100	54.1
36.0	0.5878	0.7265	1.3764	0.8090	**54.0**
Deg.	Cos	Cot	Tan	Sin	Deg.

Deg.	Sin	Tan	Cot	Cos	Deg.
36.0	0.5878	0.7265	1.3764	0.8090	**54.0**
.1	.5892	.7292	1.3713	.8080	53.9
.2	.5906	.7319	1.3663	.8070	.8
.3	.5920	.7346	1.3613	.8059	.7
.4	.5934	.7373	1.3564	.8049	.6
.5	.5948	.7400	1.3514	.8039	.5
.6	.5962	.7427	1.3465	.8028	.4
.7	.5976	.7454	1.3416	.8018	.3
.8	.5990	.7481	1.3367	.8007	.2
.9	.6004	.7508	1.3319	.7997	53.1
37.0	0.6018	0.7536	1.3270	0.7986	**53.0**
.1	.6032	.7563	1.3222	.7976	52.9
.2	.6046	.7590	1.3175	.7965	.8
.3	.6060	.7618	1.3127	.7955	.7
.4	.6074	.7646	1.3079	.7944	.6
.5	.6088	.7673	1.3032	.7934	.5
.6	.6101	.7701	1.2985	.7923	.4
.7	.6115	.7729	1.2938	.7912	.3
.8	.6129	.7757	1.2892	.7902	.2
.9	.6143	.7785	1.2846	.7891	52.1
38.0	0.6157	0.7813	1.2799	0.7880	**52.0**
.1	.6170	.7841	1.2753	.7869	51.9
.2	.6184	.7869	1.2708	.7859	.8
.3	.6198	.7898	1.2662	.7848	.7
.4	.6211	.7926	1.2617	.7837	.6
.5	.6225	.7954	1.2572	.7826	.5
.6	.6239	.7983	1.2527	.7815	.4
.7	.6252	.8012	1.2482	.7804	.3
.8	.6266	.8040	1.2437	.7793	.2
.9	.6280	.8069	1.2393	.7782	51.1
39.0	0.6293	0.8098	1.2349	0.7771	**51.0**
.1	.6307	.8127	1.2305	.7760	50.9
.2	.6320	.8156	1.2261	.7749	.8
.3	.6334	.8185	1.2218	.7738	.7
.4	.6347	.8214	1.2174	.7727	.6
.5	.6361	.8243	1.2131	.7716	.5
.6	.6374	.8273	1.2088	.7705	.4
.7	.6388	.8302	1.2045	.7694	.3
.8	.6401	.8332	1.2002	.7683	.2
.9	.6414	.8361	1.1960	.7672	50.1
40.0	0.6428	0.8391	1.1918	0.7660	**50.0**
.1	.6441	.8421	1.1875	.7649	49.9
.2	.6455	.8451	1.1833	.7638	.8
.3	.6468	.8481	1.1792	.7627	.7
.4	.6481	.8511	1.1750	.7615	.6
40.5	0.6494	0.8541	1.1708	0.7604	**49.5**
Deg.	Cos	Cot	Tan	Sin	Deg.

Deg.	Sin	Tan	Cot	Cos	Deg.
40.5	0.6494	0.8541	1.1708	0.7604	**49.5**
.6	.6508	.8571	1.1667	.7593	.4
.7	.6521	.8601	1.1626	.7581	.3
.8	.6534	.8632	1.1585	.7570	.2
.9	.6547	.8662	1.1544	.7559	49.1
41.0	0.6561	0.8693	1.1504	0.7547	**49.0**
.1	.6574	.8724	1.1463	.7536	48.9
.2	.6587	.8754	1.1423	.7524	.8
.3	.6600	.8785	1.1383	.7513	.7
.4	.6613	.8816	1.1343	.7501	.6
.5	.6626	.8847	1.1303	.7490	.5
.6	.6639	.8878	1.1263	.7478	.4
.7	.6652	.8910	1.1224	.7466	.3
.8	.6665	.8941	1.1184	.7455	.2
.9	.6678	.8972	1.1145	.7443	48.1
42.0	0.6691	0.9004	1.1106	0.7431	**48.0**
.1	.6704	.9036	1.1067	.7420	47.9
.2	.6717	.9067	1.1028	.7408	.8
.3	.6730	.9099	1.0990	.7396	.7
.4	.6743	.9131	1.0951	.7385	.6
.5	.6756	.9163	1.0913	.7373	.5
.6	.6769	.9195	1.0875	.7361	.4
.7	.6782	.9228	1.0837	.7349	.3
.8	.6794	.9260	1.0799	.7337	.2
.9	.6807	.9293	1.0761	.7325	47.1
43.0	0.6820	0.9325	1.0724	0.7314	**47.0**
.1	.6833	.9358	1.0686	.7302	46.9
.2	.6845	.9391	1.0649	.7290	.8
.3	.6858	.9424	1.0612	.7278	.7
.4	.6871	.9457	1.0575	.7266	.6
.5	.6884	.9490	1.0538	.7254	.5
.6	.6896	.9523	1.0501	.7242	.4
.7	.6909	.9556	1.0464	.7230	.3
.8	.6921	.9590	1.0428	.7218	.2
.9	.6934	.9623	1.0392	.7206	46.1
44.0	0.6947	0.9657	1.0355	0.7193	**46.0**
.1	.6959	.9691	1.0319	.7181	45.9
.2	.6972	.9725	1.0283	.7169	.8
.3	.6984	.9759	1.0247	.7157	.7
.4	.6997	.9793	1.0212	.7145	.6
.5	.7009	.9827	1.0176	.7133	.5
.6	.7022	.9861	1.0141	.7120	.4
.7	.7034	.9896	1.0105	.7108	.3
.8	.7046	.9930	1.0070	.7096	.2
.9	.7059	.9965	1.0035	.7083	45.1
45.0	0.7071	1.0000	1.0000	0.7071	**45.0**
Deg.	Cos	Cot	Tan	Sin	Deg.

NATURAL FUNCTIONS FOR DEGREES AND DECIMALS (Continued)

LOGARITHMS OF TRIGONOMETRIC FUNCTIONS FOR ANGLES IN DEGREES AND DECIMALS

Deg.	L. Sin	*L. Tan	*L. Cot	L. Cos	Deg.
0.0	— ∞	— ∞	∞	0.0000	**90.0**
.1	7.2419	7.2419	2.7581	0.0000	89.9
.2	7.5429	7.5429	2.4571	0.0000	.8
.3	7.7190	7.7190	2.2810	0.0000	.7
.4	7.8439	7.8439	2.1561	0.0000	.6
.5	7.9408	7.9409	2.0591	0.0000	.5
.6	8.0200	8.0200	1.9800	0.0000	.4
.7	8.0870	8.0870	1.9130	0.0000	.3
.8	8.1450	8.1450	1.8550	0.0000	.2
.9	8.1961	8.1962	1.8038	9.9999	89.1
1.0	8.2419	8.2419	1.7581	9.9999	**89.0**
.1	8.2832	8.2833	1.7167	9.9999	88.9
.2	8.3210	8.3211	1.6789	9.9999	.8
.3	8.3558	8.3559	1.6441	9.9999	.7
.4	8.3880	8.3881	1.6119	9.9999	.6
.5	8.4179	8.4181	1.5819	9.9999	.5
.6	8.4459	8.4461	1.5539	9.9998	.4
.7	8.4723	8.4725	1.5275	9.9998	.3
.8	8.4971	8.4973	1.5027	9.9998	.2
.9	8.5206	8.5208	1.4792	9.9998	88.1
2.0	8.5428	8.5431	1.4569	9.9997	**88.0**
.1	8.5640	8.5643	1.4357	9.9997	87.9
.2	8.5842	8.5845	1.4155	9.9997	.8
.3	8.6035	8.6038	1.3962	9.9996	.7
.4	8.6220	8.6223	1.3777	9.9996	.6
.5	8.6397	8.6401	1.3599	9.9996	.5
.6	8.6567	8.6571	1.3429	9.9996	.4
.7	8.6731	8.6736	1.3264	9.9995	.3
.8	8.6889	8.6894	1.3106	9.9995	.2
.9	8.7041	8.7046	1.2954	9.9994	87.1
3.0	8.7188	8.7194	1.2806	9.9994	**87.0**
.1	8.7330	8.7337	1.2663	9.9994	86.9
.2	8.7468	8.7475	1.2525	9.9993	.8
.3	8.7602	8.7609	1.2391	9.9993	.7
.4	8.7731	8.7739	1.2261	9.9992	.6
.5	8.7857	8.7865	1.2135	9.9992	.5
.6	8.7979	8.7988	1.2012	9.9991	.4
.7	8.8098	8.8107	1.1893	9.9991	.3
.8	8.8213	8.8223	1.1777	9.9990	.2
.9	8.8326	8.8336	1.1664	9.9990	86.1
4.0	8.8436	8.8446	1.1554	9.9989	**86.0**
.1	8.8543	8.8554	1.1446	9.9989	85.9
.2	8.8647	8.8659	1.1341	9.9988	.8
.3	8.8749	8.8762	1.1238	9.9988	.7
.4	8.8849	8.8862	1.1138	9.9987	.6
.5	8.8946	8.8960	1.1040	9.9987	.5
.6	8.9042	8.9056	1.0944	9.9986	.4
.7	8.9135	8.9150	1.0850	9.9985	.3
.8	8.9226	8.9241	1.0759	9.9985	.2
.9	8.9315	8.9331	1.0669	9.9984	85.1
5.0	8.9403	8.9420	1.0580	9.9983	**85.0**
.1	8.9489	8.9506	1.0494	9.9983	84.9
.2	8.9573	8.9591	1.0409	9.9982	.8
.3	8.9655	8.9674	1.0326	9.9981	.7
.4	8.9736	8.9756	1.0244	9.9981	.6
.5	8.9816	8.9836	1.0164	9.9980	.5
.6	8.9894	8.9915	1.0085	9.9979	.4
.7	8.9970	8.9992	1.0008	9.9978	.3
.8	9.0046	9.0068	0.9932	9.9978	.2
.9	9.0120	9.0143	0.9857	9.9977	84.1
6.0	9.0192	9.0216	0.9784	9.9976	**84.0**
Deg.	L. Cos	*L. Cot	*L. Tan	L. Sin	Deg.

* Interpolation in this section of the table is inaccurate.

Deg.	L. Sin	L. Tan	L. Cot	L. Cos	Deg.
6.0	9.0192	9.0216	0.9784	9.9976	**84.0**
.1	9.0264	9.0289	0.9711	9.9975	83.9
.2	9.0334	9.0360	0.9640	9.9975	.8
.3	9.0403	9.0430	0.9570	9.9974	.7
.4	9.0472	9.0499	0.9501	9.9973	.6
.5	9.0539	9.0567	0.9433	9.9972	.5
.6	9.0605	9.0633	0.9367	9.9971	.4
.7	9.0670	9.0699	0.9301	9.9970	.3
.8	9.0734	9.0764	0.9236	9.9969	.2
.9	9.0797	9.0828	0.9172	9.9968	83.1
7.0	9.0859	9.0891	0.9109	9.9968	**83.0**
.1	9.0920	9.0954	0.9046	9.9967	82.9
.2	9.0981	9.1015	0.8985	9.9966	.8
.3	9.1040	9.1076	0.8924	9.9965	.7
.4	9.1099	9.1135	0.8865	9.9964	.6
.5	9.1157	9.1194	0.8806	9.9963	.5
.6	9.1214	9.1252	0.8748	9.9962	.4
.7	9.1271	9.1310	0.8690	9.9961	.3
.8	9.1326	9.1367	0.8633	9.9960	.2
.9	9.1381	9.1423	0.8577	9.9959	82.1
8.0	9.1436	9.1478	0.8522	9.9958	**82.0**
.1	9.1489	9.1533	0.8467	9.9956	81.9
.2	9.1542	9.1587	0.8413	9.9955	.8
.3	9.1594	9.1640	0.8360	9.9954	.7
.4	9.1646	9.1693	0.8307	9.9953	.6
.5	9.1697	9.1745	0.8255	9.9952	.5
.6	9.1747	9.1797	0.8203	9.9951	.4
.7	9.1797	9.1848	0.8152	9.9950	.3
.8	9.1847	9.1898	0.8102	9.9949	.2
.9	9.1895	9.1948	0.8052	9.9947	81.1
9.0	9.1943	9.1997	0.8003	9.9946	**81.0**
.1	9.1991	9.2046	0.7954	9.9945	80.9
.2	9.2038	9.2094	0.7906	9.9944	.8
.3	9.2085	9.2142	0.7858	9.9943	.7
.4	9.2131	9.2189	0.7811	9.9941	.6
.5	9.2176	9.2236	0.7764	9.9940	.5
.6	9.2221	9.2282	0.7718	9.9939	.4
.7	9.2266	9.2328	0.7672	9.9937	.3
.8	9.2310	9.2374	0.7626	9.9936	.2
.9	9.2353	9.2419	0.7581	9.9935	80.1
10.0	9.2397	9.2463	0.7537	9.9934	**80.0**
.1	9.2439	9.2507	0.7493	9.9932	79.9
.2	9.2482	9.2551	0.7449	9.9931	.8
.3	9.2524	9.2594	0.7406	9.9929	.7
.4	9.2565	9.2637	0.7363	9.9928	.6
.5	9.2606	9.2680	0.7320	9.9927	.5
.6	9.2647	9.2722	0.7278	9.9925	.4
.7	9.2687	9.2764	0.7236	9.9924	.3
.8	9.2727	9.2805	0.7195	9.9922	.2
.9	9.2767	9.2846	0.7154	9.9921	79.1
11.0	9.2806	9.2887	0.7113	9.9919	**79.0**
.1	9.2845	9.2927	0.7073	9.9918	78.9
.2	9.2883	9.2967	0.7033	9.9916	.8
.3	9.2921	9.3006	0.6994	9.9915	.7
.4	9.2959	9.3046	0.6954	9.9913	.6
.5	9.2997	9.3085	0.6915	9.9912	.5
.6	9.3034	9.3123	0.6877	9.9910	.4
.7	9.3070	9.3162	0.6838	9.9909	.3
.8	9.3107	9.3200	0.6800	9.9907	.2
.9	9.3143	9.3237	0.6763	9.9906	78.1
12.0	9.3179	9.3275	0.6725	9.9904	**78.0**
Deg.	L. Cos	L. Cot	L. Tan	L. Sin	Deg.

Deg	L. Sin	L. Tan	L. Cot	L. Cos	Deg
12.0	9.3179	9.3275	0.6725	9.9904	**78.0**
.1	9.3214	9.3312	0.6688	9.9902	77.9
.2	9.3250	9.3349	0.6651	9.9901	.8
.3	9.3284	9.3385	0.6615	9.9899	.7
.4	9.3319	9.3422	0.6578	9.9897	.6
.5	9.3353	9.3458	0.6542	9.9896	.5
.6	9.3387	9.3493	0.6507	9.9894	.4
.7	9.3421	9.3529	0.6471	9.9892	.3
.8	9.3455	9.3564	0.6436	9.9891	.2
.9	9.3488	9.3599	0.6401	9.9889	77.1
13.0	9.3521	9.3634	0.6366	9.9887	**77.0**
.1	9.3554	9.3668	0.6332	9.9885	76.9
.2	9.3586	9.3702	0.6298	9.9884	.8
.3	9.3618	9.3736	0.6264	9.9882	.7
.4	9.3650	9.3770	0.6230	9.9880	.6
.5	9.3682	9.3804	0.6196	9.9878	.5
.6	9.3713	9.3837	0.6163	9.9876	.4
.7	9.3745	9.3870	0.6130	9.9875	.3
.8	9.3775	9.3903	0.6097	9.9873	.2
.9	9.3806	9.3935	0.6065	9.9871	76.1
14.0	9.3837	9.3968	0.6032	9.9869	**76.0**
.1	9.3867	9.4000	0.6000	9.9867	75.9
.2	9.3897	9.4032	0.5968	9.9865	.8
.3	9.3927	9.4064	0.5936	9.9863	.7
.4	9.3957	9.4095	0.5905	9.9861	.6
.5	9.3986	9.4127	0.5873	9.9859	.5
.6	9.4015	9.4158	0.5842	9.9857	.4
.7	9.4044	9.4189	0.5811	9.9855	.3
.8	9.4073	9.4220	0.5780	9.9853	.2
.9	9.4102	9.4250	0.5750	9.9851	75.1
15.0	9.4130	9.4281	0.5719	9.9849	**75.0**
.1	9.4158	9.4311	0.5689	9.9847	74.9
.2	9.4186	9.4341	0.5659	9.9845	.8
.3	9.4214	9.4371	0.5629	9.9843	.7
.4	9.4242	9.4400	0.5600	9.9841	.6
.5	9.4269	9.4430	0.5570	9.9839	.5
.6	9.4296	9.4459	0.5541	9.9837	.4
.7	9.4323	9.4488	0.5512	9.9835	.3
.8	9.4350	9.4517	0.5483	9.9833	.2
.9	9.4377	9.4546	0.5454	9.9831	74.1
16.0	9.4403	9.4575	0.5425	9.9828	**74.0**
.1	9.4430	9.4603	0.5397	9.9826	73.9
.2	9.4456	9.4632	0.5368	9.9824	.8
.3	9.4482	9.4660	0.5340	9.9822	.7
.4	9.4508	9.4688	0.5312	9.9820	.6
.5	9.4533	9.4716	0.5284	9.9817	.5
.6	9.4559	9.4744	0.5256	9.9815	.4
.7	9.4584	9.4771	0.5229	9.9813	.3
.8	9.4609	9.4799	0.5201	9.9811	.2
.9	9.4634	9.4826	0.5174	9.9808	73.1
17.0	9.4659	9.4853	0.5147	9.9806	**73.0**
.1	9.4684	9.4880	0.5120	9.9804	72.9
.2	9.4709	9.4907	0.5093	9.9801	.8
.3	9.4733	9.4934	0.5066	9.9799	.7
.4	9.4757	9.4961	0.5039	9.9797	.6
.5	9.4781	9.4987	0.5013	9.9794	.5
.6	9.4805	9.5014	0.4986	9.9792	.4
.7	9.4829	9.5040	0.4960	9.9789	.3
.8	9.4853	9.5066	0.4934	9.9787	.2
.9	9.4876	9.5092	0.4908	9.9785	72.1
18.0	9.4900	9.5118	0.4882	9.9782	**72.0**
Deg.	L. Cos	L. Cot	L. Tan	L. Sin	Deg.

Deg.	L. Sin	L. Tan	L. Cot	L. Cos	Deg
18.0	9.4900	9.5118	0.4882	9.9782	**72.0**
.1	9.4923	9.5143	0.4857	9.9780	71.9
.2	9.4946	9.5169	0.4831	9.9777	.8
.3	9.4969	9.5195	0.4805	9.9775	.7
.4	9.4992	9.5220	0.4780	9.9772	.6
.5	9.5015	9.5245	0.4755	9.9770	.5
.6	9.5037	9.5270	0.4730	9.9767	.4
.7	9.5060	9.5295	0.4705	9.9764	.3
.8	9.5082	9.5320	0.4680	9.9762	.2
.9	9.5104	9.5345	0.4655	9.9759	71.1
19.0	9.5126	9.5370	0.4630	9.9757	**71.0**
.1	9.5148	9.5394	0.4606	9.9754	70.9
.2	9.5170	9.5419	0.4581	9.9751	.8
.3	9.5192	9.5443	0.4557	9.9749	.7
.4	9.5213	9.5467	0.4533	9.9746	.6
.5	9.5235	9.5491	0.4509	**9.9743**	.5
.6	9.5256	9.5516	0.4484	**9.9741**	.4
.7	9.5278	9.5539	0.4461	**9.9738**	.3
.8	9.5299	9.5563	0.4437	**9.9735**	.2
.9	9.5320	9.5587	0.4413	**9.9733**	70.1
20.0	9.5341	9.5611	0.4389	9.9730	**70.0**
.1	9.5361	9.5634	0.4366	9.9727	69.9
.2	9.5382	9.5658	0.4342	9.9724	.8
.3	9.5402	9.5681	0.4319	9.9722	.7
.4	9.5423	9.5704	0.4296	9.9719	.6
.5	9.5443	9.5727	0.4273	9.9716	.5
.6	9.5463	9.5750	0.4250	9.9713	.4
.7	9.5484	9.5773	0.4227	9.9710	.3
.8	9.5504	9.5796	0.4204	9.9707	.2
.9	9.5523	9.5819	0.4181	9.9704	69.1
21.0	9.5543	9.5842	0.4158	9.9702	**69.0**
.1	9.5563	9.5864	0.4136	9.9699	68.9
.2	9.5583	9.5887	0.4113	9.9696	.8
.3	9.5602	9.5909	0.4091	9.9693	.7
.4	9.5621	9.5932	0.4068	9.9690	.6
.5	9.5641	9.5954	0.4046	9.9687	.5
.6	9.5660	9.5976	0.4024	9.9684	.4
.7	9.5679	9.5998	0.4002	9.9681	.3
.8	9.5698	9.6020	0.3980	9.9678	.2
.9	9.5717	9.6042	0.3958	9.9675	68.1
22.0	9.5736	9.6064	0.3936	9.9672	**68.0**
.1	9.5754	9.6086	0.3914	9.9669	67.9
.2	9.5773	9.6108	0.3892	9.9666	.8
.3	9.5792	9.6129	0.3871	9.9662	.7
.4	9.5810	9.6151	0.3849	9.9659	.6
.5	9.5828	9.6172	0.3828	9.9656	.5
.6	9.5847	9.6194	0.3806	9.9653	.4
.7	9.5865	9.6215	0.3785	9.9650	.3
.8	9.5883	9.6236	0.3764	9.9647	.2
.9	9.5901	9.6257	0.3743	9.9643	67.1
23.0	9.5919	9.6279	0.3721	9.9640	**67.0**
.1	9.5937	9.6300	0.3700	9.9637	66.9
.2	9.5954	9.6321	0.3679	9.9634	.8
.3	9.5972	9.6341	0.3659	9.9631	.7
.4	9.5990	9.6362	0.3638	9.9627	.6
.5	9.6007	9.6383	0.3617	9.9624	.5
.6	9.6024	9.6404	0.3596	9.9621	.4
.7	9.6042	9.6424	0.3576	9.9617	.3
.8	9.6059	9.6445	0.3555	9.9614	.2
.9	9.6076	9.6465	0.3535	9.9611	66.1
24.0	9.6093	9.6486	0.3514	9.9607	**66.0**
Deg.	L. Cos	L. Cot	L. Tan	L. Sin	Deg.

LOGARITHMS OF FUNCTIONS FOR DEGREES AND DECIMALS (Continued)

Deg.	L. Sin	L. Tan	L. Cot	L. Cos	Deg.
24.0	9.6093	9.6486	0.3514	9.9607	**66.0**
.1	9.6110	9.6506	0.3494	9.9604	65.9
.2	9.6127	9.6527	0.3473	9.9601	.8
.3	9.6144	9.6547	0.3453	9.9597	.7
.4	9.6161	9.6567	0.3433	9.9594	.6
.5	9.6177	9.6587	0.3413	9.9590	.5
.6	9.6194	9.6607	0.3393	9.9587	.4
.7	9.6210	9.6627	0.3373	9.9583	.3
.8	9.6227	9.6647	0.3353	9.9580	.2
.9	9.6243	9.6667	0.3333	9.9576	65.1
25.0	9.6259	9.6687	0.3313	9.9573	**65.0**
.1	9.6276	9.6706	0.3294	9.9569	64.9
.2	9.6292	9.6726	0.3274	9.9566	.8
.3	9.6308	9.6746	0.3254	9.9562	.7
.4	9.6324	9.6765	0.3235	9.9558	.6
.5	9.6340	9.6785	0.3215	9.9555	.5
.6	9.6356	9.6804	0.3196	9.9551	.4
.7	9.6371	9.6824	0.3176	9.9548	.3
.8	9.6387	9.6843	0.3157	9.9544	.2
.9	9.6403	9.6863	0.3137	9.9540	64.1
26.0	9.6418	9.6882	0.3118	9.9537	**64.0**
.1	9.6434	9.6901	0.3099	9.9533	63.9
.2	9.6449	9.6920	0.3080	9.9529	.8
.3	9.6465	9.6939	0.3061	9.9525	.7
.4	9.6480	9.6958	0.3042	9.9522	.6
.5	9.6495	9.6977	0.3023	9.9518	.5
.6	9.6510	9.6996	0.3004	9.9514	.4
.7	9.6526	9.7015	0.2985	9.9510	.3
.8	9.6541	9.7034	0.2966	9.9506	.2
.9	9.6556	9.7053	0.2947	9.9503	63.1
27.0	9.6570	9.7072	0.2928	9.9499	**63.0**
.1	9.6585	9.7090	0.2910	9.9495	62.9
.2	9.6600	9.7109	0.2891	9.9491	.8
.3	9.6615	9.7128	0.2872	9.9487	.7
.4	9.6629	9.7146	0.2854	9.9483	.6
.5	9.6644	9.7165	0.2835	9.9479	.5
.6	9.6659	9.7183	0.2817	9.9475	.4
.7	9.6673	9.7202	0.2798	9.9471	.3
.8	9.6687	9.7220	0.2780	9.9467	.2
.9	9.6702	9.7238	0.2762	9.9463	62.1
28.0	9.6716	9.7257	0.2743	.9459	**62.0**
.1	9.6730	9.7275	0.2725	9.9455	61.9
.2	9.6744	9.7293	0.2707	9.9451	.8
.3	9.6759	9.7311	0.2689	9.9447	.7
.4	9.6773	9.7330	0.2670	9.9443	.6
.5	9.6787	9.7348	0.2652	9.9439	.5
.6	9.6801	9.7366	0.2634	9.9435	.4
.7	9.6814	9.7384	0.2616	9.9431	.3
.8	9.6828	9.7402	0.2598	9.9427	.2
.9	9.6842	9.7420	0.2580	9.9422	61.1
29.0	9.6856	9.7438	0.2562	9.9418	**61.0**
.1	9.6869	9.7455	0.2545	9.9414	60.9
.2	9.6883	9.7473	0.2527	9.9410	.8
.3	9.6896	9.7491	0.2509	9.9406	.7
.4	9.6910	9.7509	0.2491	9.9401	.6
.5	9.6923	9.7526	0.2474	9.9397	.5
.6	9.6937	9.7544	0.2456	9.9393	.4
.7	9.6950	9.7562	0.2438	9.9388	.3
.8	9.6963	9.7579	0.2421	9.9384	.2
.9	9.6977	9.7597	0.2403	9.9380	60.1
30.0	9.6990	9.7614	0.2386	9.9375	**60.0**
Deg.	L. Cos	L. Cot	L. Tan	L. Sin	Deg.

Deg.	L. Sin	L. Tan	L. Cot	L. Cos	Deg.
30.0	9.6990	9.7614	0.2386	9.9375	**60.0**
.1	9.7003	9.7632	0.2368	9.9371	59.9
.2	9.7016	9.7649	0.2351	9.9367	.8
.3	9.7029	9.7667	0.2333	9.9362	.7
.4	9.7042	9.7684	0.2316	9.9358	.6
.5	9.7055	9.7701	0.2299	9.9353	.5
.6	9.7068	9.7719	0.2281	9.9349	.4
.7	9.7080	9.7736	0.2264	9.9344	.3
.8	9.7093	9.7753	0.2247	9.9340	.2
.9	9.7106	9.7771	0.2229	9.9335	59.1
31.0	9.7118	9.7788	0.2212	9.9331	**59.0**
.1	9.7131	9.7805	0.2195	9.9326	58.9
.2	9.7144	9.7822	0.2178	9.9322	.8
.3	9.7156	9.7839	0.2161	9.9317	.7
.4	9.7168	9.7856	0.2144	9.9312	.6
.5	9.7181	9.7873	0.2127	9.9308	.5
.6	9.7193	9.7890	0.2110	9.9303	.4
.7	9.7205	9.7907	0.2093	9.9298	.3
.8	9.7218	9.7924	0.2076	9.9294	.2
.9	9.7230	9.7941	0.2059	9.9289	58.1
32.0	9.7242	9.7958	0.2042	9.9284	**58.0**
.1	9.7254	9.7975	0.2025	9.9279	57.9
.2	9.7266	9.7992	0.2008	9.9275	.8
.3	9.7278	9.8008	0.1992	9.9270	.7
.4	9.7290	9.8025	0.1975	9.9265	.6
.5	9.7302	9.8042	0.1958	9.9260	.5
.6	9.7314	9.8059	0.1941	9.9255	.4
.7	9.7326	9.8075	0.1925	9.9251	.3
.8	9.7338	9.8092	0.1908	9.9246	.2
.9	9.7349	9.8109	0.1891	9.9241	57.1
33.0	9.7361	9.8125	0.1875	9.9236	**57.0**
.1	9.7373	9.8142	0.1858	9.9231	56.9
.2	9.7384	9.8158	0.1842	9.9226	.8
.3	9.7396	9.8175	0.1825	9.9221	.7
.4	9.7407	9.8191	0.1809	9.9216	.6
.5	9.7419	9.8208	0.1792	9.9211	.5
.6	9.7430	9.8224	0.1776	9.9206	.4
.7	9.7442	9.8241	0.1759	9.9201	.3
.8	9.7453	9.8257	0.1743	9.9196	.2
.9	9.7464	9.8274	0.1726	9.9191	56.1
34.0	9.7476	9.8290	0.1710	9.9186	**56.0**
.1	9.7487	9.8306	0.1694	9.9181	55.9
.2	9.7498	9.8323	0.1677	9.9175	.8
.3	9.7509	9.8339	0.1661	9.9170	.7
.4	9.7520	9.8355	0.1645	9.9165	.6
.5	9.7531	9.8371	0.1629	9.9160	.5
.6	9.7542	9.8388	0.1612	9.9155	.4
.7	9.7553	9.8404	0.1596	9.9149	.3
.8	9.7564	9.8420	0.1580	9.9144	.2
.9	9.7575	9.8436	0.1564	9.9139	55.1
35.0	9.7586	9.8452	0.1548	9.9134	**55.0**
.1	9.7597	9.8468	0.1532	9.9128	54.9
.2	9.7607	9.8484	0.1516	9.9123	.8
.3	9.7618	9.8501	0.1499	9.9118	.7
.4	9.7629	9.8517	0.1483	9.9112	.6
.5	9.7640	9.8533	0.1467	9.9107	.5
.6	9.7650	9.8549	0.1451	9.9101	.4
.7	9.7661	9.8565	0.1435	9.9096	.3
.8	9.7671	9.8581	0.1419	9.9091	.2
.9	9.7682	9.8597	0.1403	9.9085	54.1
36.0	9.7692	9.8613	0.1387	9.9080	**54.0**
Deg.	L. Cos	L. Cot	L. Tan	L. Sin	Deg.

Deg.	L. Sin	L. Tan	L. Cot	L. Cos	Deg.
36.0	9.7692	9.8613	0.1387	9.9080	**54.0**
.1	9.7703	9.8629	0.1371	9.9074	53.9
.2	9.7713	9.8644	0.1356	9.9069	.8
.3	9.7723	9.8660	0.1340	9.9063	.7
.4	9.7734	9.8676	0.1324	9.9057	.6
.5	9.7744	9.8692	0.1308	9.9052	.5
.6	9.7754	9.8708	0.1292	9.9046	.4
.7	9.7764	9.8724	0.1276	9.9041	.3
.8	9.7774	9.8740	0.1260	9.9035	.2
.9	9.7785	9.8755	0.1245	9.9029	53.1
37.0	9.7795	9.8771	0.1229	9.9023	**53.0**
.1	9.7805	9.8787	0.1213	9.9018	52.9
.2	9.7815	9.8803	0.1197	9.9012	.8
.3	9.7825	9.8818	0.1182	9.9006	.7
.4	9.7835	9.8834	0.1166	9.9000	.6
.5	9.7844	9.8850	0.1150	9.8995	.5
.6	9.7854	9.8865	0.1135	9.8989	.4
.7	9.7864	9.8881	0.1119	9.8983	.3
.8	9.7874	9.8897	0.1103	9.8977	.2
.9	9.7884	9.8912	0.1088	9.8971	52.1
38.0	9.7893	9.8928	0.1072	9.8965	**52.0**
.1	9.7903	9.8944	0.1056	9.8959	51.9
.2	9.7913	9.8959	0.1041	9.8953	.8
.3	9.7922	9.8975	0.1025	9.8947	.7
.4	9.7932	9.8990	0.1010	9.8941	.6
.5	9.7941	9.9006	0.0994	9.8935	.5
.6	9.7951	9.9022	0.0978	9.8929	.4
.7	9.7960	9.9037	0.0963	9.8923	.3
.8	9.7970	9.9053	0.0947	9.8917	.2
.9	9.7979	9.9068	0.0932	9.8911	51.1
39.0	9.7989	9.9084	0.0916	9.8905	**51.0**
.1	9.7998	9.9099	0.0901	9.8899	50.9
.2	9.8007	9.9115	0.0885	9.8893	.8
.3	9.8017	9.9130	0.0870	9.8887	.7
.4	9.8026	9.9146	0.0854	9.8880	.6
.5	9.8035	9.9161	0.0839	9.8874	.5
.6	9.8044	9.9176	0.0824	9.8868	.4
.7	9.8053	9.9192	0.0808	9.8862	.3
.8	9.8063	9.9207	0.0793	9.8855	.2
.9	9.8072	9.9223	0.0777	9.8849	50.1
40.0	9.8081	9.9238	0.0762	9.8843	**50.0**
.1	9.8090	9.9254	0.0746	9.8836	49.9
.2	9.8099	9.9269	0.0731	9.8830	.8
.3	9.8108	9.9284	0.0716	9.8823	.7
.4	9.8117	9.9300	0.0700	9.8817	.6
.5	9.8125	9.9315	0.0685	9.8810	.5
.6	9.8134	9.9330	0.0670	9.8804	.4
.7	9.8143	9.9346	0.0654	9.8797	.3
.8	9.8152	9.9361	0.0639	9.8791	.2
.9	9.8161	9.9376	0.0624	9.8784	49.1
41.0	9.8169	9.9392	0.0608	9.8778	**49.0**
.1	9.8178	9.9407	0.0593	9.8771	48.9
.2	9.8187	9.9422	0.0578	9.8765	.8
.3	9.8195	9.9438	0.0562	9.8758	.7
.4	9.8204	9.9453	0.0547	9.8751	.6
.5	9.8213	9.9468	0.0532	9.8745	.5
.6	9.8221	9.9483	0.0517	9.8738	.4
.7	9.8230	9.9499	0.0501	9.8731	.3
.8	9.8238	9.9514	0.0486	9.8724	.2
.9	9.8247	9.9529	0.0471	9.8718	48.1
Deg.	L. Cos	L. Cot	L. Tan	L. Sin	Deg.

LOGARITHMS OF FUNCTIONS FOR DEGREES AND DECIMALS (Continued)

Deg.	L. Sin	L. Tan	L. Cot	L. Cos	Deg.
42.0	9.8255	9.9544	0.0456	9.8711	**48.0**
.1	9.8264	9.9560	0.0440	9.8704	47.9
.2	9.8272	9.9575	0.0425	9.8697	.8
.3	9.8280	9.9590	0.0410	9.8690	.7
.4	9.8289	9.9605	0.0395	9.8683	.6
.5	9.8297	9.9621	0.0379	9.8676	.5
.6	9.8305	9.9636	0.0364	9.8669	.4
.7	9.8313	9.9651	0.0349	9.8662	.3
.8	9.8322	9.9666	0.0334	9.8655	.2
.9	9.8330	9.9681	0.0319	9.8648	47.1
43.0	9.8338	9.9697	0.0303	9.8641	**47.0**
.1	9.8346	9.9712	0.0288	9.8634	46.9
.2	9.8354	9.9727	0.0273	9.8627	.8
.3	9.8362	9.9742	0.0258	9.8620	.7
.4	9.8370	9.9757	0.0243	9.8613	.6
.5	9.8378	9.9772	0.0228	9.8606	.5
.6	9.8386	9.9788	0.0212	9.8598	.4
.7	9.8394	9.9803	0.0197	9.8591	.3
.8	9.8402	9.9818	0.0182	9.8584	.2
.9	9.8410	9.9833	0.0167	9.8577	46.1
44.0	9.8418	9.9848	0.0152	9.8569	**46.0**
.1	9.8426	9.9864	0.0136	9.8562	.9
.2	9.8433	9.9879	0.0121	9.8555	.8
.3	9.8441	9.9894	0.0106	9.8547	.7
.4	9.8449	9.9909	0.0091	9.8540	.6
.5	9.8457	9.9924	0.0076	9.8532	.5
.6	9.8464	9.9939	0.0061	9.8525	.4
.7	9.8472	9.9955	0.0045	9.8517	.3
.8	9.8480	9.9970	0.0030	9.8510	.2
.9	9.8487	9.9985	0.0015	9.8502	.1
45.0	9.8495	0.0000	0.0000	9.8495	**45.0**
Deg.	L. Cos	L. Cot	L. Tan	L. Sin	Deg.

Rad.	Sin	Tan	Cot	Cos
.00	.00000	.00000	∞	1.0000
.01	.01000	.01000	99.997	.99995
.02	.02000	.02000	49.993	.99980
.03	.03000	.03001	33.323	.99955
.04	.03999	.04002	24.987	.99920
.05	.04998	.05004	19.983	.99875
.06	.05996	.06007	16.647	.99820
.07	.06994	.07011	14.262	.99755
.08	.07991	.08017	12.473	.99680
.09	.08988	.09024	11.081	.99595
.10	.09983	.10033	9.9666	.99500
.11	.10978	.11045	9.0542	.99396
.12	.11971	.12058	8.2933	.99281
.13	.12963	.13074	7.6489	.99156
.14	.13954	.14092	7.0961	.99022
.15	.14944	.15114	6.6166	.98877
.16	.15932	.16138	6.1966	.98723
.17	.16918	.17166	5.8256	.98558
.18	.17903	.18197	5.4954	.98384
.19	.18886	.19232	5.1997	.98200
.20	.19867	.20271	4.9332	.98007
.21	.20846	.21314	4.6917	.97803
.22	.21823	.22362	4.4719	.97590
.23	.22798	.23414	4.2709	.97367
.24	.23770	.24472	4.0864	.97134
.25	.24740	.25534	3.9163	.96891
.26	.25708	.26602	3.7591	.96639
.27	.26673	.27676	3.6133	.96377
.28	.27636	.28755	3.4776	.96106
.29	.28595	.29841	3.3511	.95824
.30	.29552	.30934	3.2327	.95534
.31	.30506	.32033	3.1218	.95233
.32	.31457	.33139	3.0176	.94924
.33	.32404	.34252	2.9195	.94604
.34	.33349	.35374	2.8270	.94275
.35	.34290	.36503	2.7395	.93937
.36	.35227	.37640	2.6567	.93590
.37	.36162	.38786	2.5782	.93233
.38	.37092	.39941	2.5037	.92866
.39	.38019	.41105	2.4328	.92491
.40	.38942	.42279	2.3652	.92106
.41	.39861	.43463	2.3008	.91712
.42	.40776	.44657	2.2393	.91309
.43	.41687	.45862	2.1804	.90897
.44	.42594	.47078	2.1241	.90475
.45	.43497	.48306	2.0702	.90045
.46	.44395	.49545	2.0184	.89605
.47	.45289	.50797	1.9686	.89157
.48	.46178	.52061	1.9208	.88699
.49	.47063	.53339	1.8748	.88233
.50	.47943	.54630	1.8305	.87758
Rad.	Sin	Tan	Cot	Cos

Rad.	Sin	Tan	Cot	Cos
.50	.47943	.54630	1.8305	.87758
.51	.48818	.55936	1.7878	.87274
.52	.49688	.57256	1.7465	.86782
.53	.50553	.58592	1.7067	.86281
.54	.51414	.59943	1.6683	.85771
.55	.52269	.61311	1.6310	.85252
.56	.53119	.62695	1.5950	.84726
.57	.53963	.64097	1.5601	.84190
.58	.54802	.65517	1.5263	.83646
.59	.55636	.66956	1.4935	.83094
.60	.56464	.68414	1.4617	.82534
.61	.57287	.69892	1.4308	.81965
.62	.58104	.71391	1.4007	.81388
.63	.58914	.72911	1.3715	.80803
.64	.59720	.74454	1.3431	.80210
.65	.60519	.76020	1.3154	.79608
.66	.61312	.77610	1.2885	.78999
.67	.62099	.79225	1.2622	.78382
.68	.62879	.80866	1.2366	.77757
.69	.63654	.82534	1.2116	.77125
.70	.64422	.84229	1.1872	.76484
.71	.65183	.85953	1.1634	.75836
.72	.65938	.87707	1.1402	.75181
.73	.66687	.89492	1.1174	.74517
.74	.67429	.91309	1.0952	.73847
.75	.68164	.93160	1.0734	.73169
.76	.68892	.95045	1.0521	.72484
.77	.69614	.96967	1.0313	.71791
.78	.70328	.98926	1.0109	.71091
.79	.71035	1.0092	.99084	.70385
.80	.71736	1.0296	.97121	.69671
.81	.72429	1.0505	.95197	.68950
.82	.73115	1.0717	.93309	.68222
.83	.73793	1.0934	.91455	.67488
.84	.74464	1.1156	.89635	.66746
.85	.75128	1.1383	.87848	.65998
.86	.75784	1.1616	.86091	.65244
.87	.76433	1.1853	.84365	.64483
.88	.77074	1.2097	.82668	.63715
.89	.77707	1.2346	.80998	.62941
.90	.78333	1.2602	.79355	.62161
.91	.78950	1.2864	.77738	.61375
.92	.79560	1.3133	.76146	.60582
.93	.80162	1.3409	.74578	.59783
.94	.80756	1.3692	.73034	.58979
.95	.81342	1.3984	.71511	.58168
.96	.81919	1.4284	.70010	.57352
.97	.82489	1.4592	.68531	.56530
.98	.83050	1.4910	.67071	.55702
.99	.83603	1.5237	.65631	.54869
1.00	.84147	1.5574	.64209	.54030
Rad.	Sin	Tan	Cot	Cos

Rad.	Sin	Tan	Cot	Cos
1.00	.84147	1.5574	.64209	.54030
1.01	.84683	1.5922	.62806	.53186
1.02	.85211	1.6281	.61420	.52337
1.03	.85730	1.6652	.60051	.51482
1.04	.86240	1.7036	.58699	.50622
1.05	.86742	1.7433	.57362	.49757
1.06	.87236	1.7844	.56040	.48887
1.07	.87720	1.8270	.54734	.48012
1.08	.88196	1.8712	.53441	.47133
1.09	.88663	1.9171	.52162	.46249
1.10	.89121	1.9648	.50897	.45360
1.11	.89570	2.0143	.49644	.44466
1.12	.90010	2.0660	.48404	.43568
1.13	.90441	2.1198	.47175	.42666
1.14	.90863	2.1759	.45959	.41759
1.15	.91276	2.2345	.44753	.40849
1.16	.91680	2.2958	.43558	.39934
1.17	.92075	2.3600	.42373	.39015
1.18	.92461	2.4273	.41199	.38092
1.19	.92837	2.4979	.40034	.37166
1.20	.93204	2.5722	.38878	.36236
1.21	.93562	2.6503	.37731	.35302
1.22	.93910	2.7328	.36593	.34365
1.23	.94249	2.8198	.35463	.33424
1.24	.94578	2.9119	.34341	.32480
1.25	.94898	3.0096	.33227	.31532
1.26	.95209	3.1133	.32121	.30582
1.27	.95510	3.2236	.31021	.29628
1.28	.95802	3.3413	.29928	.28672
1.29	.96084	3.4672	.28842	.27712
1.30	.96356	3.6021	.27762	.26750
1.31	.96618	3.7471	.26687	.25785
1.32	.96872	3.9033	.25619	.24818
1.33	.97115	4.0723	.24556	.23848
1.34	.97348	4.2556	.23498	.22875
1.35	.97572	4.4552	.22446	.21901
1.36	.97786	4.6734	.21398	.20924
1.37	.97991	4.9131	.20354	.19945
1.38	.98185	5.1774	.19315	.18964
1.39	.98370	5.4707	.18279	.17981
1.40	.98545	5.7979	.17248	.16997
1.41	.98710	6.1654	.16220	.16010
1.42	.98865	6.5811	.15195	.15023
1.43	.99010	7.0555	.14173	.14033
1.44	.99146	7.6018	.13155	.13042
1.45	.99271	8.2381	.12139	.12050
1.46	.99387	8.9886	.11125	.11057
1.47	.99492	9.8874	.10114	.10063
1.48	.99588	10.983	.09105	.09067
1.49	.99674	12.350	.08097	.08071
1.50	.99749	14.101	.07091	.07074
Rad.	Sin	Tan	Cot	Cos

Rad.	Sin	Tan	Cot	Cos
1.50	.99749	14.101	.07091	.07074
1.51	.99815	16.428	.06087	.06076
1.52	.99871	19.670	.05084	.05077
1.53	.99917	24.498	.04082	.04079
1.54	.99953	32.461	.03081	.03079
1.55	.99978	48.078	.02080	.02079
1.56	.99994	92.621	.01080	.01080
1.57	1.0000	1255.8	.00080	.00080
1.58	.99996	−108.65	−.00920	−.00920
1.59	.99982	−52.067	−.01921	−.01920
1.60	.99957	−34.233	−.02921	−.02920
1.61	.99923	−25.495	−.03922	−.03919
1.62	.99879	−20.307	−.04924	−.04918
1.63	.99825	−16.871	−.05927	−.05917
1.64	.99761	−14.427	−.06931	−.06915
1.65	.99687	−12.599	−.07937	−.07912
1.66	.99602	−11.181	−.08944	−.08909
1.67	.99508	−10.047	−.09953	−.09904
1.68	.99404	−9.1208	−.10964	−.10899
1.69	.99290	−8.3492	−.11977	−.11892
1.70	.99166	−7.6966	−.12993	−.12884
1.71	.99033	−7.1373	−.14011	−.13875
1.72	.98889	−6.6524	−.15032	−.14865
1.73	.98735	−6.2281	−.16056	−.15853
1.74	.98572	−5.8535	−.17084	−.16840
1.75	.98399	−5.5204	−.18115	−.17825
1.76	.98215	−5.2221	−.19149	−.18808
1.77	.98022	−4.9534	−.20188	−.19789
1.78	.97820	−4.7101	−.21231	−.20768
1.79	.97607	−4.4887	−.22278	−.21745
1.80	.97385	−4.2863	−.23330	−.22720
1.81	.97153	−4.1005	−.24387	−.23693
1.82	.96911	−3.9294	−.25449	−.24663
1.83	.96659	−3.7712	−.26517	−.25631
1.84	.96398	−3.6245	−.27590	−.26596
1.85	.96128	−3.4881	−.28669	−.27559
1.86	.95847	−3.3608	−.29755	−.28519
1.87	.95557	−3.2419	−.30846	−.29476
1.88	.95258	−3.1304	−.31945	−.30430
1.89	.94949	−3.0257	−.33051	−.31381
1.90	.94630	−2.9271	−.34164	−.32329
1.91	.94302	−2.8341	−.35284	−.33274
1.92	.93965	−2.7463	−.36413	−.34215
1.93	.93618	−2.6632	−.37549	−.35153
1.94	.93262	−2.5843	−.38695	−.36087
1.95	.92896	−2.5095	−.39849	−.37018
1.96	.92521	−2.4383	−.41012	−.37945
1.97	.92137	−2.3705	−.42185	−.38868
1.98	.91744	−2.3058	−.43368	−.39788
1.99	.91341	−2.2441	−.44562	−.40703
2.00	.90930	−2.1850	−.45766	−.41615
Rad.	Sin	Tan	Cot	Cos

NATURAL FUNCTIONS FOR ANGLES IN RADIANS

LOGARITHMS OF THE FUNCTIONS FOR ANGLES IN RADIANS

Rad	Sin	Tan	Cot	Cos
.00	− ∞	− ∞	∞	0.00000
.01	7.99999	8.00001	1.99999	9.99998
.02	8.30100	8.30109	1.69891	9.99991
.03	8.47706	8.47725	1.52275	9.99980
.04	8.60194	8.60229	1.39771	9.99965
.05	8.69879	8.69933	1.30067	9.99946
.06	8.77789	8.77867	1.22133	9.99922
.07	8.84474	8.84581	1.15419	9.99894
.08	8.90263	8.90402	1.09598	9.99861
.09	8.95366	8.95542	1.04458	9.99824
.10	8.99928	9.00145	0.99855	9.99782
.11	9.04052	9.04315	0.95685	9.99737
.12	9.07814	9.08127	0.91873	9.99687
.13	9.11272	9.11640	0.88360	9.99632
.14	9.14471	9.14898	0.85102	9.99573
.15	9.17446	9.17937	0.82063	9.99510
.16	9.20227	9.20785	0.79215	9.99442
.17	9.22836	9.23466	0.76534	9.99369
.18	9.25292	9.26000	0.74000	9.99293
.19	9.27614	9.28402	0.71598	9.99211
.20	9.29813	9.30688	0.69312	9.99126
.21	9.31902	9.32867	0.67133	9.99035
.22	9.33891	9.34951	0.65049	9.98940
.23	9.35789	9.36948	0.63052	9.98841
.24	9.37603	9.38866	0.61134	9.98737
.25	9.39341	9.40712	0.59288	9.98628
.26	9.41007	9.42492	0.57508	9.98515
.27	9.42607	9.44210	0.55790	9.98397
.28	9.44147	9.45872	0.54128	9.98275
.29	9.45629	9.47482	0.52518	9.98148
.30	9.47059	9.49043	0.50957	9.98016
.31	9.48438	9.50559	0.49441	9.97879
.32	9.49771	9.52034	0.47966	9.97737
.33	9.51060	9.53469	0.46531	9.97591
.34	9.52308	9.54868	0.45132	9.97440
.35	9 53516	9.56233	0.43767	9.97284
.36	9.54688	9.57565	0.42435	9.97123
.37	9.55825	9.58868	0.41132	9.96957
.38	9.56928	9.60142	0.39858	9.96786
.39	9.58000	9.61390	0.38610	9.96610
.40	9.59042	9.62613	0.37387	9.96429
.41	9.60055	9.63812	0.36188	9.96243
.42	9.61041	9.64989	0.35011	9.96051
.43	9.62000	9.66145	0.33855	9.95855
.44	9.62935	9.67282	0.32718	9.95653
.45	9.63845	9.68400	0.31600	9.95446
.46	9.64733	9.69500	0 30500	9.95233
.47	9.65599	9.70583	0.29417	9.95015
.48	9.66443	9.71651	0.28349	9 94792
.49	9.67268	9.72704	0.27296	9.94563
.50	9.68072	9.73743	0.26257	9.94329
Rad	L. Sin	L. Tan	L. Cot	L. Cos

Rad.	L. Sin	L. Tan	L. Cot	L. Cos
.50	9.68072	9.73743	0.26257	9.94329
.51	9.68858	9.74769	0.25231	9.94089
.52	9.69625	9.75782	0.24218	9.93843
.53	9.70375	9.76784	0.23216	9.93591
.54	9.71108	9.77774	0.22226	9.93334
.55	9.71824	9.78754	0.21246	9.93071
.56	9.72525	9.79723	0.20277	9.92801
.57	9.73210	9.80684	0.19316	9.92526
.58	9.73880	9.81635	0.18365	9.92245
.59	9.74536	9.82579	0.17421	9.91957
.60	9.75177	9.83514	0.16486	9.91663
.61	9.75805	9.84443	0.15557	9.91363
.62	9.76420	9.85364	0.14636	9.91056
.63	9.77022	9.86280	0.13720	9.90743
.64	9.77612	9.87189	0.12811	9 90423
.65	9.78189	9.88093	0.11907	9.90096
.66	9.78754	9.88992	0.11008	9.89762
.67	9.79308	9.89886	0.10114	9.89422
.68	9.79851	9.90777	0.09223	9.89074
.69	9.80382	9.91663	0.08337	9.88719
70	9.80903	9.92546	0.07454	9.88357
.71	9.81414	9.93426	0.06574	9.87988
.72	9.81914	9.94303	0.05697	9.87611
.73	9.82404	9.95178	0.04822	9.87226
.74	9.82885	9.96051	0.03949	9.86833
.75	9.83355	9 96923	0.03077	9.86433
.76	9.83817	9.97793	0.02207	9.86024
.77	9.84269	9.98662	0.01338	9.85607
.78	9.84713	9.99531	0.00469	9.85182
.79	9.85147	0.00400	9.99600	9.84748
.80	9.85573	0.01268	9.98732	9.84305
.81	9.85991	0.02138	9.97862	9.83853
.82	9.86400	0.03008	9.96992	9.83393
.83	9.86802	0.03879	9.96121	9.82922
.84	9.87195	0.04752	9.95248	9.82443
.85	9.87580	0.05627	9.94373	9.81953
.86	9.87958	0.06504	9.93496	9.81454
.87	9.88328	0.07384	9.92616	9.80944
.88	9.88691	0.08266	9.91734	9.80424
.89	9.89046	0.09153	9.90847	9.79894
.90	9.89394	0.10043	9.89957	9.79352
.91	9.89735	0.10937	9.89063	9.78799
.92	9.90070	0.11835	9.88165	9.78234
.93	9.90397	0.12739	9.87261	9.77658
.94	9.90717	0.13648	9.86352	9.77070
95	9.91031	0.14563	9.85437	9.76469
.96	9.91339	0.15484	9.84516	9.75855
.97	9.91639	0.16412	9.83588	9.75228
.98	9.91934	0.17347	9.82653	9.74587
.99	9.92222	0.18289	9.81711	9.73933
1.00	9.92504	0.19240	9.80760	9.73264
Rad.	L. Sin	L. Tan	L. Cot	L. Cos

Rad.	L. Sin	L. Tan	L. Cot	L. Cos
1.00	9.92504	0.19240	9.80760	9.73264
1.01	9.92780	0.20200	9.79800	9.72580
1.02	9.93049	0.21169	9.78831	9.71881
1.03	9.93313	0.22148	9.77852	9.71165
1.04	9.93571	0.23137	9.76863	9.70434
1.05	9.93823	0.24138	9.75862	9.69686
1.06	9.94069	0.25150	9.74850	9.68920
1.07	9.94310	0.26175	9.73825	9.68135
1.08	9.94545	0.27212	9.72788	9.67332
1.09	9.94774	0.28264	9.71736	9.66510
1.10	9.94998	0.29331	9.70669	9.65667
1.11	9.95216	0.30413	9.69587	9.64803
1.12	9.95429	0.31512	9.68488	9.63917
1.13	9.95637	0.32628	9.67372	9 63008
1.14	9.95839	0.33763	9.66237	9.62075
1.15	9.96036	0 34918	9.65082	9.61118
1.16	9.96228	0.36093	9.63907	9.60134
1.17	9.96414	0.37291	9.62709	9 59123
1.18	9.96596	0.38512	9.61488	9.58084
1.19	9.96772	0.39757	9.60243	9.57015
1.20	9.96943	0.41030	9.58970	9.55914
1.21	9.97110	0.42330	9.57670	9.54780
1.22	9.97271	0.43660	9.56340	9.53611
1.23	9.97428	0.45022	9.54978	9.52406
1.24	9.97579	0.46418	9.53582	9.51161
1.25	9 97726	0.47850	9.52150	9 49875
1.26	9.97868	0.49322	9.50678	9.48546
1.27	9.98005	0.50835	9.49165	9.47170
1.28	9.98137	0.52392	9.47608	9.45745
1.29	9.98265	0.53998	9.46002	9.44267
1.30	9.98388	0.55656	9.44344	9.42732
1.31	9.98506	0.57369	9.42631	9.41137
1.32	9.98620	0.59144	9.40856	9.39476
1.33	9.98729	0.60984	9.39016	9.37744
1.34	9.98833	0.62896	9.37104	9.35937
1.35	9.98933	0.64887	9.35113	9.34046
1.36	9.99028	0.66964	9.33036	9.32064
1.37	9.99119	0.69135	9.30865	9.29983
1.38	9.99205	0.71411	9.28589	9.27793
1.39	9.99286	0.73804	9.26196	9.25482
1.40	9.99363	0.76327	9.23673	9.23036
1.41	9.99436	0.78996	9.21004	9.20440
1.42	9.99504	0.81830	9.18170	9.17674
1.43	9.99568	0.84853	9.15147	9.14716
1.44	9.99627	0.88092	9.11908	9.11536
1.45	9.99682	0.91583	9.08417	9.08100
1.46	9.99733	0.95369	9.04631	9.04364
1.47	9.99779	0.99508	9.00492	9.00271
1.48	9.99821	1.04074	8.95926	8.95747
1.49	9.99858	1.09166	8.90834	8.90692
1.50	9.99891	1.14926	8.85074	8.84965
Rad.	L. Sin	L. Tan	L. Cot	L. Cos

Rad	L. Sin	L. Tan	L. Cot	L. Cos
1.50	9.99891	1.14926	8.85074	8 84965
1.51	9.99920	1.21559	8.78441	8.78361
1.52	9.99944	1.29379	8.70621	8.70565
1.53	9.99964	1.38914	8.61086	8.61050
1.54	9.99979	1.51136	8.48864	8 48843
1.55	9.99991	1.68195	8.31805	8.31796
1.56	9.99997	1.96671	8.03329	8.03327
1.57	0.00000	3.09891	6.90109	6.90109
1.58	9.99998	2.03603*	7.96397*	7 96396*
1.59	9.99992	1.71656	8.28344	8.28336
1.60	9.99981	1.53444	8 46556	8.46538
1.61	9.99967	1.40645	8.59355	8.59323
1.62	9.99947	1.30765	8.69235	8.69182
1.63	9.99924	1 22714	8.77286	8.77209
1.64	9.99896	1.15918	8.84082	8.83978
1.65	9.99864	1.10035	8.89965	8.89829
1.66	9.99827	1.04847	8.95154	8.94981
1.67	9.99786	1.00204	8.99796	8.99582
1.68	9.99741	0.96003	9.03997	9.03737
1.69	9.99691	0.92165	9.07835	9.07526
1.70	9.99636	0 88630	9.11370	9.11007
1.71	9.99578	0.85353	9.14647	9.14225
1.72	9.99515	0.82298	9.17702	9.17217
1.73	9.99447	0.79436	9.20564	9.20012
1.74	9.99375	0.76742	9.23258	9.22634
1.75	9.99299	0.74197	9.25803	9.25102
1.76	9.99218	0.71784	9.28216	9.27434
1.77	9.99133	0.69490	9.30510	9.29642
1.78	9.99043	0.67303	9.32697	9.31740
1.79	9 98948	0.65212	9.34788	9.33736
1.80	9.98849	0.63208	9.36792	9.35641
1.81	9.98745	0.61284	9.38716	9.37462
1.82	9.98637	0.59432	9.40568	9.39205
1.83	9.98524	0.57648	9.42352	9.40877
1.84	9.98407	0.55925	9.44075	9.42482
1.85	9.98285	0.54258	9.45742	9.44026
1.86	9.98158	0.52645	9.47355	9.45513
1.87	9.98026	0.51080	9.48920	9.46947
1.88	9.97890	0.49560	9.50440	9.48330
1.89	9.97749	0.48082	9.51918	9.49667
1.90	9.97603	0.46644	9.53356	9.50959
1.91	9.97452	0.45242	9.54758	9.52210
1.92	9.97296	0.43875	9.56125	9.53422
1.93	9.97136	0.42540	9.57460	9.54597
1.94	9.96970	0.41235	9.58765	9.55735
1.95	9.96800	0.39958	9.60042	9.56841
1.96	9.96624	0.38708	9.61292	9.57916
1.97	9.96443	0.37484	9.62516	9.58960
1.98	9.96258	0.36283	9.63717	9.59975
1.99	9.96067	0.35104	9.64896	9.60963
2.00	9.95871	0.33946	9.66054	9.61925
Rad.	L. Sin	L. Tan	L. Cot	L. Cos

*Values of the cosine, tangent and cotangent for angles in the table, 1.58 radians and above, are negative. As a consequence the logarithms of the functions involved are for their absolute values. For example log |cos 1.90 = 9.50959 − 10.

LOGARITHMS OF THE FUNCTIONS FOR ANGLES IN RADIANS

NATURAL FUNCTIONS FOR ANGLES IN π RADIANS

$x\pi$ Radians		Sin	Tan	Cot	Cos
x =					
.00 or	1.00	.00000	.00000	inf	1.00000
.01	.99	.03141	.03143	31.821	.99951
.02	.98	.06279	.06291	15.895	.99803
.03	.97	.09411	.09453	10.579	.99556
.04	.96	.12533	.12633	7.9158	.99211
.05	.95	.15643	.15838	6.3138	.98769
.06	.94	.18738	.19076	5.2422	.98229
.07	.93	.21814	.22353	4.4737	.97592
.08	.92	.24869	.25676	3.8947	.96858
.09	.91	.27899	.29053	3.4420	.96029
.10	.90	.30902	.32492	3.0777	.95106
.11	.89	.33874	.36002	2.7776	.94088
.12	.88	.36812	.39593	2.5257	.92978
.13	.87	.39715	.43274	2.3109	.91775
.14	.86	.42578	.47056	2.1251	.90483
.15	.85	.45399	.50953	1.9626	.89101
.16	.84	.48175	.54975	1.8190	.87631
.17	.83	.50904	.59140	1.6909	.86074
.18	.82	.53583	.63462	1.5757	.84433
.19	.81	.56208	.67960	1.4715	.82708
.20	.80	.58779	.72654	1.3764	.80902
.21	.79	.61291	.77568	1.2892	.79016
.22	.78	.63742	.82727	1.2088	.77051
.23	.77	.66131	.88162	1.1343	.75011
.24	.76	.68455	.93906	1.0649	.72897
.25	.75	.70711	1.0000	1.0000	.70711
.26	.74	.72897	1.0649	.93906	.68455
.27	.73	.75011	1.1343	.88162	.66131
.28	.72	.77051	1.2088	.82727	.63742
.29	.71	.79016	1.2892	.77568	.61291
.30	.70	.80902	1.3764	.72654	.58779
.31	.69	.82708	1.4715	.67960	.56208
.32	.68	.84433	1.5757	.63462	.53583
.33	.67	.86074	1.6909	.59140	.50904
.34	.66	.87631	1.8190	.54975	.48175
.35	.65	.89101	1.9626	.50953	.45399
.36	.64	.90483	2.1251	.47056	.42578
.37	.63	.91775	2.3109	.43274	.39715
.38	.62	.92978	2.5257	.39593	.36812
.39	.61	.94088	2.7776	.36002	.33874
.40	.60	.95106	3.0777	.32492	.30902
.41	.59	.96029	3.4420	.29053	.27899
.42	.58	.96858	3.8947	.25676	.24869
.43	.57	.97592	4.4737	.22353	.21814
.44	.56	.98229	5.2422	.19076	.18738
.45	.55	.98769	6.3138	.15838	.15643
.46	.54	.99211	7.9158	.12633	.12533
.47	.53	.99556	10.579	.09453	.09411
.48	.52	.99803	15.895	.06291	.06279
.49	.51	.99951	31.821	.03143	.03141
.50	.50	1.0000	inf	.00000	.00000

These functions are useful in the solution of wave equations such as the displacement equation of a sound wave in the form:

$$y = A \sin 2\pi n x$$

without the necessity of reducing the angular rotation either to radians or to degrees in order to find the value of the function.

The algebraic sign of the function follows the familiar Quadrant Law for the particular function desired. Thus a numerical value of 9.13π radians becomes (by the subtraction of the greatest multiple of 2π radians) 1.13π radians. This is the same as $.13\pi$ radians in the 3rd Quadrant, which would give, from the tables above, a value of the sine function of −.39715.

Submitted by J. A. Blythe Jr.

HAVERSINES

$\text{hav}\ \theta = \frac{1}{2}\ \text{vers}\ \theta = \frac{1}{2}(1 - \cos\theta) = \sin^2 \frac{1}{2}\theta$
$\text{hav}\ (-\theta) = \text{hav}\ \theta$
$\text{hav}\ (180° - \theta) = \text{hav}\ (180° + \theta) = 1 - \text{hav}\ \theta$
Characteristics of the logarithms are omitted.

θ°	0′ Value	0′ Log	10′ Value	10′ Log	20′ Value	20′ Log	30′ Value	30′ Log	40′ Value	40′ Log	50′ Value	50′ Log
0	.0000	—	.0000	$\bar{6}$.3254	.0000	$\bar{6}$.9275	.0000	$\bar{5}$.2796	.0000	$\bar{5}$.5295	.0001	$\bar{5}$.7233
1	.0001	$\bar{5}$.8817	.0001	.0156	.0001	.1316	.0002	.2339	.0002	.3254	.0003	.4081
2	.0003	.4837	.0004	.5532	.0004	.6176	.0005	.6775	.0005	.7336	.0006	.7862
3	.0007	.8358	.0008	.8828	.0008	.9273	.0009	.9697	.0010	.0101	.0011	.0487
4	.0012	.0856	.0013	.1211	.0014	.1551	.0015	.1879	.0017	.2195	.0018	.2499
5	.0019	.2794	.0020	.3078	.0022	.3354	.0023	.3621	.0024	.3880	.0026	.4132
6	.0027	.4376	.0029	.4614	.0031	.4845	.0032	.5071	.0034	.5290	.0036	.5504
7	.0037	.5714	.0039	.5918	.0041	.6117	.0043	.6312	.0045	.6503	.0047	.6689
8	.0049	.6872	.0051	.7051	.0053	.7226	.0055	.7397	.0057	.7566	.0059	.7731
9	.0062	.7893	.0064	.8052	.0066	.8208	.0069	.8361	.0071	.8512	.0073	.8660
10	.0076	.8806	.0079	.8949	.0081	.9090	.0084	.9229	.0086	.9365	.0089	.9499
11	.0092	.9631	.0095	.9762	.0097	.9890	.0100	.0016	.0103	.0141	.0106	.0264
12	.0109	.0385	.0112	.0504	.0115	.0622	.0119	.0738	.0122	.0852	.0125	.0966
13	.0128	.1077	.0131	.1187	.0135	.1296	.0138	.1404	.0142	.1510	.0145	.1614
14	.0149	.1718	.0152	.1820	.0156	.1921	.0159	.2021	.0163	.2120	.0167	.2217
15	.0170	.2314	.0174	.2409	.0178	.2504	.0182	.2597	.0186	.2689	.0190	.2781
16	.0194	.2871	.0198	.2961	.0202	.3049	.0206	.3137	.0210	.3223	.0214	.3309
17	.0218	.3394	.0223	.3478	.0227	.3561	.0231	.3644	.0236	.3726	.0240	.3807
18	.0245	.3887	.0249	.3966	.0254	.4045	.0258	.4123	.0263	.4200	.0268	.4276
19	.0272	.4352	.0277	.4427	.0282	.4502	.0287	.4576	.0292	.4649	.0297	.4721
20	.0302	.4793	.0307	.4865	.0312	.4935	.0317	.5006	.0322	.5075	.0327	.5144
21	.0332	.5213	.0337	.5281	.0343	.5348	.0348	.5415	.0353	.5481	.0359	.5547
22	.0364	.5612	.0370	.5677	.0375	.5741	.0381	.5805	.0386	.5868	.0392	.5931
23	.0397	.5993	.0403	.6055	.0409	.6116	.0415	.6177	.0421	.6238	.0426	.6298
24	.0432	.6358	.0438	.6417	.0444	.6476	.0450	.6534	.0456	.6592	.0462	.6650
25	.0468	.6707	.0475	.6764	.0481	.6820	.0487	.6876	.0493	.6932	.0500	.6987
26	.0506	.7042	.0512	.7096	.0519	.7150	.0525	.7204	.0532	.7258	.0538	.7311
27	.0545	.7364	.0552	.7416	.0558	.7468	.0565	.7520	.0572	.7572	.0578	.7623
28	.0585	.7674	.0592	.7724	.0599	.7774	.0606	.7824	.0613	.7874	.0620	.7923
29	.0627	.7972	.0634	.8021	.0641	.8069	.0648	.8117	.0655	.8165	.0663	.8213
30	.0670	.8260	.0677	.8307	.0684	.8354	.0692	.8400	.0699	.8446	.0707	.8492
31	.0714	.8538	.0722	.8583	.0729	.8629	.0737	.8673	.0744	.8718	.0752	.8763
32	.0760	.8807	.0767	.8851	.0775	.8894	.0783	.8938	.0791	.8981	.0799	.9024
33	.0807	.9067	.0815	.9109	.0823	.9152	.0831	.9194	.0839	.9236	.0847	.9277
34	.0855	.9319	.0863	.9360	.0871	.9401	.0879	.9442	.0888	.9482	.0896	.9523
35	.0904	.9563	.0913	.9603	.0921	.9643	.0929	.9682	.0938	.9721	.0946	.9761
36	.0955	.9800	.0963	.9838	.0972	.9877	.0981	.9915	.0989	.9954	.0998	.9992
37	.1007	.0030	.1016	.0067	.1024	.0105	.1033	.0142	.1042	.0179	.1051	.0216
38	.1060	.0253	.1069	.0289	.1078	.0326	.1087	.0362	.1096	.0398	.1105	.0434
39	.1114	.0470	.1123	.0505	.1133	.0541	.1142	.0576	.1151	.0611	.1160	.0646
40	.1170	.0681	.1179	.0716	.1189	.0750	.1198	.0784	.1207	.0819	.1217	.0853
41	.1226	.0887	.1236	.0920	.1246	.0954	.1255	.0987	.1265	.1020	.1275	.1054
42	.1284	.1087	.1294	.1119	.1304	.1152	.1314	.1185	.1323	.1217	.1333	.1249
43	.1343	.1282	.1353	.1314	.1363	.1345	.1373	.1377	.1383	.1409	.1393	.1440
44	.1403	.1472	.1413	.1503	.1424	.1534	.1434	.1565	.1444	.1596	.1454	.1626
45	.1464	.1657	.1475	.1687	.1485	.1718	.1495	.1748	.1506	.1778	.1516	.1808
46	.1527	.1838	.1537	.1867	.1548	.1897	.1558	.1926	.1569	.1956	.1579	.1985
47	.1590	.2014	.1601	.2043	.1611	.2072	.1622	.2101	.1633	.2129	.1644	.2158
48	.1654	.2186	.1665	.2215	.1676	.2243	.1687	.2271	.1698	.2299	.1709	.2327
49	.1720	.2355	.1731	.2382	.1742	.2410	.1753	.2437	.1764	.2465	.1775	.2492
50	.1786	.2519	.1797	.2546	.1808	.2573	.1820	.2600	.1831	.2627	.1842	.2653
51	.1853	.2680	.1865	.2706	.1876	.2732	.1887	.2759	.1899	.2785	.1910	.2811
52	.1922	.2837	.1933	.2863	.1945	.2888	.1956	.2914	.1968	.2940	.1979	.2965
53	.1991	.2991	.2003	.3016	.2014	.3041	.2026	.3066	.2038	.3091	.2049	.3116
54	.2061	.3141	.2073	.3166	.2085	.3190	.2096	.3215	.2108	.3239	.2120	.3264
55	.2132	.3288	.2144	.3312	.2156	.3336	.2168	.3361	.2180	.3384	.2192	.3408
56	.2204	.3432	.2216	.3456	.2228	.3480	.2240	.3503	.2252	.3527	.2265	.3550
57	.2277	.3573	.2289	.3596	.2301	.3620	.2314	.3643	.2326	.3666	.2338	.3689
58	.2350	.3711	.2363	.3734	.2375	.3757	.2388	.3779	.2400	.3802	.2412	.3824
59	.2425	.3847	.2437	.3869	.2450	.3891	.2462	.3913	.2475	.3935	.2487	.3957

HAVERSINES (Continued)

Characteristics of the logarithms are omitted.

θ°	0′ Value	0′ Log	10′ Value	10′ Log	20′ Value	20′ Log	30′ Value	30′ Log	40′ Value	40′ Log	50′ Value	50′ Log
60	.2500	.3979	.2513	.4001	.2525	.4023	.2538	.4045	.2551	.4066	.2563	.4088
61	.2576	.4109	.2589	.4131	.2601	.4152	.2614	.4173	.2627	.4195	.2640	.4216
62	.2653	.4237	.2665	.4258	.2678	.4279	.2691	.4300	.2704	.4320	2717	.4341
63	.2730	.4362	.2743	.4382	.2756	.4403	.2769	.4423	.2782	.4444	.2795	.4464
64	.2808	.4484	.2821	.4504	.2834	.4524	.2847	.4545	.2861	.4565	.2874	.4584
65	.2887	.4604	.2900	.4624	.2913	.4644	.2927	.4664	.2940	.4683	.2953	.4703
66	.2966	.4722	.2980	.4742	.2993	.4761	.3006	.4780	.3020	.4799	.3033	.4819
67	.3046	.4838	.3060	.4857	.3073	.4876	.3087	.4895	.3100	.4914	.3113	.4932
68	.3127	.4951	.3140	.4970	.3154	.4989	.3167	.5007	.3181	.5026	.3195	.5044
69	.3208	.5063	.3222	.5081	.3235	.5099	.3249	.5117	.3263	.5136	.3276	.5154
70	.3290	.5172	.3304	.5190	.3317	.5208	.3331	.5226	.3345	.5244	.3358	.5261
71	.3372	.5279	.3386	.5297	.3400	.5314	.3413	.5332	.3427	.5349	.3441	.5367
72	.3455	.5384	.3469	.5402	.3483	.5419	.3496	.5436	.3510	.5454	.3524	.5471
73	.3538	.5488	.3552	.5505	.3566	.5522	.3580	.5539	.3594	.5556	.3608	.5572
74	.3622	.5589	.3636	.5606	.3650	.5623	.3664	.5639	.3678	.5656	.3692	.5672
75	.3706	.5689	.3720	.5705	.3734	.5722	.3748	.5738	.3762	.5754	.3776	.5771
76	.3790	.5787	.3805	.5803	.3819	.5819	.3833	.5835	.3847	.5851	.3861	.5867
77	.3875	.5883	.3889	.5899	.3904	.5915	.3918	.5930	.3932	.5946	.3946	.5962
78	.3960	.5977	.3975	.5993	.3989	.6009	.4003	.6024	.4017	.6039	.4032	.6055
79	.4046	.6070	.4060	.6086	.4075	.6101	.4089	.6116	.4103	.6131	.4117	.6146
80	.4132	.6161	.4146	.6176	.4160	.6191	.4175	.6206	.4189	.6221	.4203	.6236
81	.4218	.6251	.4232	.6266	.4247	.6280	.4261	.6295	.4275	.6310	.4290	.6324
82	.4304	.6339	.4319	.6353	.4333	.6368	.4347	.6382	.4362	.6397	.4376	.6411
83	.4391	.6425	.4405	.6440	.4420	.6454	.4434	.6468	.4448	.6482	.4463	.6496
84	.4477	.6510	.4492	.6524	.4506	.6538	.4521	.6552	.4535	.6566	.4550	.6580
85	.4564	.6594	.4579	.6607	.4593	.6621	.4608	.6635	.4622	.6648	.4637	.6662
86	.4651	.6676	.4666	.6689	.4680	.6703	.4695	.6716	.4709	.6730	.4724	.6743
87	.4738	.6756	.4753	.6770	.4767	.6783	.4782	.6796	.4796	.6809	.4811	.6822
88	.4826	.6835	.4840	.6848	.4855	.6862	.4869	.6875	.4884	.6887	.4898	.6900
89	.4913	.6913	.4927	.6926	.4942	.6939	.4956	.6952	.4971	.6964	.4985	.6977
90	.5000	.6990	.5015	.7002	.5029	.7015	.5044	.7027	.5058	.7040	.5073	.7052
91	.5087	.7065	.5102	.7077	.5116	.7090	.5131	.7102	.5145	.7114	.5160	.7126
92	.5174	.7139	.5189	.7151	.5204	.7163	.5218	.7175	.5233	.7187	.5247	.7199
93	.5262	.7211	.5276	.7223	.5291	.7235	.5305	.7247	.5320	.7259	.5334	.7271
94	.5349	.7283	.5363	.7294	.5378	.7306	.5392	.7318	.5407	.7329	.5421	.7341
95	.5436	.7353	.5450	.7364	.5465	.7376	.5479	.7387	.5494	.7399	.5508	.7410
96	.5523	.7421	.5537	.7433	.5552	.7444	.5566	.7455	.5580	.7467	.5595	.7478
97	.5609	.7489	.5624	.7500	.5638	.7511	.5653	.7523	.5667	.7534	.5681	.7545
98	.5696	.7556	.5710	.7567	.5725	.7577	.5739	.7588	.5753	.7599	.5768	.7610
99	.5782	.7621	.5797	.7632	.5811	.7642	.5825	.7653	.5840	.7664	.5854	.7674
100	.5868	.7685	.5883	.7696	.5897	.7706	.5911	.7717	.5925	.7727	.5940	.7738
101	.5954	.7748	.5968	.7759	.5983	.7769	.5997	.7779	.6011	.7790	.6025	.7800
102	.6040	.7810	.6054	.7820	.6068	.7830	.6082	.7841	.6096	.7851	.6111	.7861
103	.6125	.7871	.6139	.7881	.6153	.7891	.6167	.7901	.6181	.7911	.6195	.7921
104	.6210	.7931	.6224	.7940	.6238	.7950	.6252	.7960	.6266	.7970	.6280	.7980
105	.6294	.7989	.6308	.7999	.6322	.8009	.6336	.8018	.6350	.8028	.6364	.8037
106	.6378	.8047	.6392	.8056	.6406	.8066	.6420	.8075	.6434	.8085	.6448	.8094
107	.6462	.8104	.6476	.8113	.6490	.8122	.6504	.8131	.6517	.8141	.6531	.8150
108	.6545	.8159	.6559	.8168	.6573	.8177	.6587	.8187	.6600	.8196	.6614	.8205
109	.6628	.8214	.6642	.8223	.6655	.8232	.6669	.8241	.6683	.8250	.6696	.8258
110	.6710	.8267	.6724	.8276	.6737	.8285	.6751	.8294	.6765	.8302	.6778	.8311
111	.6792	.8320	.6805	.8329	.6819	.8337	.6833	.8346	.6846	.8354	.6860	.8363
112	.6873	.8371	.6887	.8380	.6900	.8388	.6913	.8397	.6927	.8405	.6940	.8414
113	.6954	.8422	.6967	.8430	.6980	.8439	.6994	.8447	.7007	.8455	.7020	.8464
114	.7034	.8472	.7047	.8480	.7060	.8488	.7073	.8496	.7087	.8504	.7100	.8513
115	.7113	.8521	.7126	.8529	.7139	.8537	.7153	.8545	.7166	.8553	.7179	.8561
116	.7192	.8568	.7205	.8576	.7218	.8584	.7231	.8592	.7244	.8600	.7257	.8608
117	.7270	.8615	.7283	.8623	.7296	.8631	.7309	.8638	.7322	.8646	.7335	.8654
118	.7347	.8661	.7360	.8669	.7373	.8676	.7386	.8684	.7399	.8691	.7411	.8699
119	.7424	.8706	.7437	.8714	.7449	.8721	.7462	.8729	.7475	.8736	.7487	.8743

θ°	0′ Value	0′ Log	10′ Value	10′ Log	20′ Value	20′ Log	30′ Value	30′ Log	40′ Value	40′ Log	50′ Value	50′ Log
120	.7500	.8751	.7513	.8758	.7525	.8765	.7538	.8772	.7550	.8780	.7563	.8787
121	.7575	.8794	.7588	.8801	.7600	.8808	.7612	.8815	.7625	.8822	.7637	.8829
122	.7650	.8836	.7662	.8843	.7674	.8850	.7686	.8857	.7699	.8864	.7711	.8871
123	.7723	.8878	.7735	.8885	.7748	.8892	.7760	.8898	.7772	.8905	.7784	.8912
124	.7796	.8919	.7808	.8925	.7820	.8932	.7832	.8939	.7844	.8945	.7856	.8952
125	.7868	.8959	.7880	.8965	.7892	.8972	.7904	.8978	.7915	.8985	.7927	.8991
126	.7939	.8998	.7951	.9004	.7962	.9010	.7974	.9017	.7986	.9023	.7997	.9030
127	.8009	.9036	.8021	.9042	.8032	.9048	.8044	.9055	.8055	.9061	.8067	.9067
128	.8078	.9073	.8090	.9079	.8101	.9085	.8113	.9092	.8124	.9098	.8135	.9104
129	.8147	.9110	.8158	.9116	.8169	.9122	.8180	.9128	.8192	.9134	.8203	.9140
130	.8214	.9146	.8225	.9151	.8236	.9157	.8247	.9163	.8258	.9169	.8269	.9175
131	.8280	.9180	.8291	.9186	.8302	.9192	.8313	.9198	.8324	.9203	.8335	.9209
132	.8346	.9215	.8356	.9220	.8367	.9226	.8378	.9231	.8389	.9237	.8399	.9242
133	.8410	.9248	.8421	.9253	.8431	.9259	.8442	.9264	.8452	.9270	.8463	.9275
134	.8473	.9281	.8484	.9286	.8494	.9291	.8505	.9297	.8515	.9302	.8525	.9307
135	.8536	.9312	.8546	.9318	.8556	.9323	.8566	.9328	.8576	.9333	.8587	.9338
136	.8597	.9343	.8607	.9348	.8617	.9353	.8627	.9359	.8637	.9364	.8647	.9369
137	.8657	.9374	.8667	.9379	.8677	.9383	.8686	.9388	.8696	.9393	.8706	.9398
138	.8716	.9403	.8725	.9408	.8735	.9413	.8745	.9417	.8754	.9422	.8764	.9427
139	.8774	.9432	.8783	.9436	.8793	.9441	.8802	.9446	.8811	.9450	.8821	.9455
140	.8830	.9460	.8840	.9464	.8849	.9469	.8858	.9473	.8867	.9478	.8877	.9482
141	.8886	.9487	.8895	.9491	.8904	.9496	.8913	.9500	.8922	.9505	.8931	.9509
142	.8940	.9513	.8949	.9518	.8958	.9522	.8967	.9526	.8976	.9531	.8984	.9535
143	.8993	.9539	.9002	.9543	.9011	.9548	.9019	.9552	.9028	.9556	.9037	.9560
144	.9045	.9564	.9054	.9568	.9062	.9572	.9071	.9576	.9079	.9580	.9087	.9584
145	.9096	.9588	.9104	.9592	.9112	.9596	.9121	.9600	.9129	.9604	.9137	.9608
146	.9145	.9612	.9153	.9616	.9161	.9620	.9169	.9623	.9177	.9627	.9185	.9631
147	.9193	.9635	.9201	.9638	.9209	.9642	.9217	.9646	.9225	.9650	.9233	.9653
148	.9240	.9657	.9248	.9660	.9256	.9664	.9263	.9668	.9271	.9671	.9278	.9675
149	.9286	.9678	.9293	.9682	.9301	.9685	.9308	.9689	.9316	.9692	.9323	.9695
150	.9330	.9699	.9337	.9702	.9345	.9706	.9352	.9709	.9359	.9712	.9366	.9716
151	.9373	.9719	.9380	.9722	.9387	.9725	.9394	.9729	.9401	.9732	.9408	.9735
152	.9415	.9738	.9422	.9741	.9428	.9744	.9435	.9747	.9442	.9751	.9448	.9754
153	.9455	.9757	.9462	.9760	.9468	.9763	.9475	.9766	.9481	.9769	.9488	.9772
154	.9494	.9774	.9500	.9777	.9507	.9780	.9513	.9783	.9519	.9786	.9525	.9789
155	.9532	.9792	.9538	.9794	.9544	.9797	.9550	.9800	.9556	.9803	.9562	.9805
156	.9568	.9808	.9574	.9811	.9579	.9813	.9585	.9816	.9591	.9819	.9597	.9821
157	.9603	.9824	.9608	.9826	.9614	.9829	.9619	.9831	.9625	.9834	.9630	.9836
158	.9636	.9839	.9641	.9841	.9647	.9844	.9652	.9846	.9657	.9849	.9663	.9851
159	.9668	.9853	.9673	.9856	.9678	.9858	.9683	.9860	.9688	.9863	.9693	.9865
160	.9698	.9867	.9703	.9869	.9708	.9871	.9713	.9874	.9718	.9876	.9723	.9878
161	.9728	.9880	.9732	.9882	.9737	.9884	.9742	.9886	.9746	.9888	.9751	.9890
162	.9755	.9892	.9760	.9894	.9764	.9896	.9769	.9898	.9773	.9900	.9777	.9902
163	.9782	.9904	.9786	.9906	.9790	.9908	.9794	.9910	.9798	.9911	.9802	.9913
164	.9806	.9915	.9810	.9917	.9814	.9919	.9818	.9920	.9822	.9922	.9826	.9924
165	.9830	.9925	.9833	.9927	.9837	.9929	.9841	.9930	.9844	.9932	.9848	.9933
166	.9851	.9935	.9855	.9937	.9858	.9938	.9862	.9940	.9865	.9941	.9869	.9943
167	.9872	.9944	.9875	.9945	.9878	.9947	.9881	.9948	.9885	.9950	.9888	.9951
168	.9891	.9952	.9894	.9954	.9897	.9955	.9900	.9956	.9903	.9957	.9905	.9959
169	.9908	.9960	.9911	.9961	.9914	.9962	.9916	.9963	.9919	.9965	.9921	.9966
170	.9924	.9967	.9927	.9968	.9929	.9969	.9931	.9970	.9934	.9971	.9936	.9972
171	.9938	.9973	.9941	.9974	.9943	.9975	.9945	.9976	.9947	.9977	.9949	.9978
172	.9951	.9979	.9953	.9980	.9955	.9981	.9957	.9981	.9959	.9982	.9961	.9983
173	.9963	.9984	.9964	.9985	.9966	.9985	.9968	.9986	.9969	.9987	.9971	.9987
174	.9973	.9988	.9974	.9989	.9976	.9989	.9977	.9990	.9978	.9991	.9980	.9991
175	.9981	.9992	.9982	.9992	.9983	.9993	.9985	.9993	.9986	.9994	.9987	.9994
176	.9988	.9995	.9989	.9995	.9990	.9996	.9991	.9996	.9992	.9996	.9992	.9997
177	.9993	.9997	.9994	.9997	.9995	.9998	.9995	.9998	.9996	.9998	.9996	.9998
178	.9997	.9999	.9997	.9999	.9998	.9999	.9998	.9999	.9999	.9999	.9999	.9999
179	.9999	.9999	.9999	.9999	1.0000	.0000	1.0000	.0000	1.0000	.0000	1.0000	.0000
180	1.0000	.0000										

NATURAL OR NAPERIAN LOGARITHMS

0.000–0.499

N	0	1	2	3	4	5	6	7	8	9
0.00	− ∞	−6‡	−6	−5	−5	−5	−5	−4	−4	−4
		.90776	.21461	.80914	.52146	.29832	.11000	.96185	.82831	.71053
.01	−4.60517	.50986	.42285	.34281	.26870	.19971	.13517	.07454	.01738	*.96332
.02	−3.91202	.86323	.81671	.77226	.72970	.68888	.64966	.61192	.57555	.54046
.03	.50656	.47377	.44202	.41125	.38139	.35241	.32424	.29684	.27017	.24419
.04	.21888	.19418	.17009	.14656	.12357	.10109	.07911	.05761	.03655	.01593
.05	−2.99573	.97593	.95651	.93746	.91877	.90042	.88240	.86470	.84731	.83022
.06	.81341	.79688	.78062	.76462	.74887	.73337	.71810	.70306	.68825	.67365
.07	.65926	.64508	.63109	.61730	.60369	.59027	.57702	.56395	.55105	.53831
.08	.52573	.51331	.50104	.48891	.47694	.46510	.45341	.44185	.43042	.41912
.09	.40795	.39690	.38597	.37516	.36446	.35388	.34341	.33304	.32279	.31264
0.10	−2.30259	.29263	.28278	.27303	.26336	.25379	.24432	.23493	.22562	.21641
.11	.20727	.19823	.18926	.18037	.17156	.16282	.15417	.14558	.13707	.12863
.12	.12026	.11196	.10373	.09557	.08747	.07944	.07147	.06357	.05573	.04794
.13	.04022	.03256	.02495	.01741	.00992	.00248	*.99510	*.98777	*.98050	*.97328
.14	−1.96611	.95900	.95193	.94491	.93794	.93102	.92415	.91732	.91054	.90381
.15	.89712	.89048	.88387	.87732	.87080	.86433	.85790	.85151	.84516	.83885
.16	.83258	.82635	.82016	.81401	.80789	.80181	.79577	.78976	.78379	.77786
.17	.77196	.76609	.76026	.75446	.74870	.74297	.73727	.73161	.72597	.72037
.18	.71480	.70926	.70375	.69827	.69282	.68740	.68201	.67665	.67131	.66601
.19	.66073	.65548	.65026	.64507	.63990	.63476	.62964	.62455	.61949	.61445
0.20	−1.60944	.60445	.59949	.59455	.58964	.58475	.57988	.57504	.57022	.56542
.21	.56065	.55590	.55117	.54646	.54178	.53712	.53248	.52786	.52326	.51868
.22	.51413	.50959	.50508	.50058	.49611	.49165	.48722	.48281	.47841	.47403
.23	.46968	.46534	.46102	.45672	.45243	.44817	.44392	.43970	.43548	.43129
.24	.42712	.42296	.41882	.41469	.41059	.40650	.40242	.39837	.39433	.39030
.25	.38629	.38230	.37833	.37437	.37042	.36649	.36258	.35868	.35480	.35093
.26	.34707	.34323	.33941	.33560	.33181	.32803	.32426	.32051	.31677	.31304
.27	.30933	.30564	.30195	.29828	.29463	.29098	.28735	.28374	.28013	.27654
.28	.27297	.26940	.26585	.26231	.25878	.25527	.25176	.24827	.24479	.24133
.29	.23787	.23443	.23100	.22758	.22418	.22078	.21740	.21402	.21066	.20731
0.30	−1.20397	.20065	.19733	.19402	.19073	.18744	.18417	.18091	.17766	.17441
.31	.17118	.16796	.16475	.16155	.15836	.15518	.15201	.14885	.14570	.14256
.32	.13943	.13631	.13320	.13010	.12701	.12393	.12086	.11780	.11474	.11170
.33	.10866	.10564	.10262	.09961	.09661	.09362	.09064	.08767	.08471	.08176
.34	.07881	.07587	.07294	.07002	.06711	.06421	.06132	.05843	.05555	.05268
.35	−1.04982	.04697	.04412	.04129	.03846	.03564	.03282	.03002	.02722	.02443
.36	.02165	.01888	.01611	.01335	.01060	.00786	.00512	.00239	*.99967	*.99696
.37	−0.99425	.99155	.98886	.98618	.98350	.98083	.97817	.97551	.97286	.97022
.38	.96758	.96496	.96233	.95972	.95711	.95451	.95192	.94933	.94675	.94418
.39	.94161	.93905	.93649	.93395	.93140	.92887	.92634	.92382	.92130	.91879
0.40	−0.91629	.91379	.91130	.90882	.90634	.90387	.90140	.89894	.89649	.89404
.41	.89160	.88916	.88673	.88431	.88189	.87948	.87707	.87467	.87227	.86988
.42	.86750	.86512	.86275	.86038	.85802	.85567	.85332	.85097	.84863	.84630
.43	.84397	.84165	.83933	.83702	.83471	.83241	.83011	.82782	.82554	.82326
.44	.82098	.81871	.81645	.81419	.81193	.80968	.80744	.80520	.80296	.80073
.45	.79851	.79629	.79407	.79186	.78966	.78746	.78526	.78307	.78089	.77871
.46	.77653	.77436	.77219	.77003	.76787	.76572	.76357	.76143	.75929	.75715
.47	.75502	.75290	.75078	.74866	.74655	.74444	.74234	.74024	.73814	.73605
.48	.73397	.73189	.72981	.72774	.72567	.72361	.72155	.71949	.71744	.71539
.49	.71335	.71131	.70928	.70725	.70522	.70320	.70118	.69917	.69716	.69515

‡ Note that the whole number values are given above the decimal values for the first line. In the second and following lines they are given at the left. All decimal values are negative on this page.

NATURAL OR NAPERIAN LOGARITHMS (Continued)
0.500–0.999

N	0	1	2	3	4	5	6	7	8	9
0.50	−0.69315	.69115	.68916	.68717	.68518	.68320	.68122	.67924	.67727	.67531
.51	.67334	.67139	.66934	.66748	.66553	.66359	.66165	.65971	.65778	.65585
.52	.65393	.65201	.65009	.64817	.64626	.64436	.64245	.64055	.63866	.63677
.53	.63488	.63299	.63111	.62923	.62736	.62549	.62362	.62176	.61990	.61804
.54	.61619	.61434	.61249	.61065	.60881	.60697	.60514	.60331	.60148	.59966
.55	.59784	.59602	.59421	.59240	.59059	.58879	.58699	.58519	.58340	.58161
.56	.57982	.57803	.57625	.57448	.57270	.57093	.56916	.56740	.56563	.56387
.57	.56212	.56037	.55862	.55687	.55513	.55339	.55165	.54991	.54818	.54645
.58	.54473	.54300	.54128	.53957	.53785	.53614	.53444	.53273	.53103	.52933
.59	.52763	.52594	.52425	.52256	.52088	.51919	.51751	.51584	.51416	.51249
0.60	−0.51083	.50916	.50750	.50584	.50418	.50253	.50088	.49923	.49758	.49594
.61	.49430	.49266	.49102	.48939	.48776	.48613	.48451	.48289	.48127	.47965
.62	.47804	.47642	.47482	.47321	.47160	.47000	.46840	.46681	.46522	.46362
.63	.46204	.46045	.45887	.45728	.45571	.45413	.45256	.45099	.44942	.44785
.64	.44629	.44473	.44317	.44161	.44006	.43850	.43696	.43541	.43386	.43232
.65	.43078	.42925	.42771	.42618	.42465	.42312	.42159	.42007	.41855	.41703
.66	.41552	.41400	.41249	.41098	.40947	.40797	.40647	.40497	.40347	.40197
.67	.40048	.39899	.39750	.39601	.39453	.39304	.39156	.39008	.38861	.38713
.68	.38566	.38419	.38273	.38126	.37980	.37834	.37688	.37542	.37397	.37251
.69	.37106	.36962	.36817	.36673	.36528	.36384	.36241	.36097	.35954	.35810
0.70	−0.35667	.35525	.35382	.35240	.35098	.34956	.34814	.34672	.34531	.34390
.71	.34249	.34108	.33968	.33827	.33687	.33547	.33408	.33268	.33129	.32989
.72	.32850	.32712	.32573	.32435	.32296	.32158	.32021	.31883	.31745	.31608
.73	.31471	.31334	.31197	.31061	.30925	.30788	.30653	.30517	.30381	.30246
.74	.30111	.29975	.29841	.29706	.29571	.29437	.29303	.29169	.29035	.28902
.75	.28768	.28635	.28502	.28369	.28236	.28104	.27971	.27839	.27707	.27575
.76	.27444	.27312	.27181	.27050	.26919	.26788	.26657	.26527	.26397	.26266
.77	.26136	.26007	.25877	.25748	.25618	.25489	.25360	.25231	.25103	.24974
.78	.24846	.24718	.24590	.24462	.24335	.24207	.24080	.23953	.23826	.23699
.79	.23572	.23446	.23319	.23193	.23067	.22941	.22816	.22690	.22565	.22439
0.80	−0.22314	.22189	.22065	.21940	.21816	.21691	.21567	.21433	.21319	.21196
.81	.21072	.20949	.20825	.20702	.20579	.20457	.20334	.20212	.20089	.19967
.82	.19845	.19723	.19601	.19480	.19358	.19237	.19116	.18995	.18874	.18754
.83	.18633	.18513	.18392	.18272	.18152	.18032	.17913	.17793	.17674	.17554
.84	.17435	.17316	.17198	.17079	.16960	.16842	.16724	.16605	.16487	.16370
.85	−0.16252	.16134	.16017	.15900	.15782	.15665	.15548	.15432	.15315	.15199
.86	.15032	.14966	.14850	.14734	.14618	.14503	.14387	.14272	.14156	.14041
.87	.13926	.13811	.13697	.13582	.13467	.13353	.13239	.13125	.13011	.12897
.88	.12783	.12670	.12556	.12443	.12330	.12217	.12104	.11991	.11878	.11766
.89	.11653	.11541	.11429	.11317	.11205	.11093	.10981	.10870	.10759	.10647
0.90	−0.10536	.10425	.10314	.10203	.10093	.09982	.09872	.09761	.09651	.09541
.91	.09431	.09321	.09212	.09102	.08992	.08883	.08744	.08665	.08556	.08447
.92	.08338	.08230	.08121	.08013	.07904	.07796	.07688	.07580	.07472	.07365
.93	.07257	.07150	.07042	.06935	.06828	.06721	.06614	.06507	.06401	.06294
.94	.06188	.06081	.05975	.05869	.05763	.05657	.05551	.05446	.05340	.05235
.95	.05129	.05024	.04919	.04814	.04709	.04604	.04500	.04395	.04291	.04186
.96	.04082	.03978	.03874	.03770	.03666	.03563	.03459	.03356	.03252	.03149
.97	.03046	.02943	.02840	.02737	.02634	.02532	.02429	.02327	.02225	.02122
.98	.02020	.01918	.01816	.01715	.01613	.01511	.01410	.01309	.01207	.01106
.99	.01005	.00904	.00803	.00702	.00602	.00501	.00401	.00300	.00200	.00100

NATURAL OR NAPERIAN LOGARITHMS (Continued)

To find the natural logarithm of a number which is $\frac{1}{10}$, $\frac{1}{100}$, $\frac{1}{1000}$, etc. of a number whose logarithm is given, subtract from the given logarithm $\log_e 10$, $2 \log_e 10$, $3 \log_e 10$, etc.

To find the natural logarithm of a number which is 10, 100, 1000, etc. times a number whose logarithm is given, add to the given logarithm $\log_e 10$, $2 \log_e 10$, $3 \log_e 10$, etc.

$\log_e 10 = 2.30258\ 50930$
$2 \log_e 10 = 4.60517\ 01860$
$3 \log_e 10 = 6.90775\ 52790$
$4 \log_e 10 = 9.21034\ 03720$
$5 \log_e 10 = 11.51292\ 54650$
$6 \log_e 10 = 13.81551\ 05580$
$7 \log_e 10 = 16.11809\ 56510$
$8 \log_e 10 = 18.42068\ 07440$
$9 \log_e 10 = 20.72326\ 58369$
$10 \log_e 10 = 23.02585\ 09299$

See preceding table for logarithms for numbers between 0.000 and 0.999.

1.00–4.99

N	0	1	2	3	4	5	6	7	8	9
1.0	0.00000	.00995	.01980	.02956	.03922	.04879	.05827	.06766	.07696	.08618
.1	.09531	.10436	.11333	.12222	.13103	.13976	.14842	.15700	.16551	.17395
.2	.18232	.19062	.19885	.20701	.21511	.22314	.23111	.23902	.24686	.25464
.3	.26236	.27003	.27763	.28518	.29267	.30010	.30748	.31481	.32208	.32930
.4	.33647	.34359	.35066	.35767	.36464	.37156	.37844	.38526	.39204	.39878
.5	.40547	.41211	.41871	.42527	.43178	.43825	.44469	.45108	.45742	.46373
.6	.47000	.47623	.48243	.48858	.49470	.50078	.50682	.51282	.51879	.52473
.7	.53063	.53649	.54232	.54812	.55389	.55962	.56531	.57098	.57661	.58222
.8	.58779	.59333	.59884	.60432	.60977	.61519	.62058	.62594	.63127	.63658
.9	.64185	.64710	.65233	.65752	.66269	.66783	.67294	.67803	.68310	.68813
2.0	0.69315	.69813	.70310	.70804	.71295	.71784	.72271	.72755	.73237	.73716
.1	.74194	.74669	.75142	.75612	.76081	.76547	.77011	.77473	.77932	.78390
.2	.78846	.79299	.79751	.80200	.80648	.81093	.81536	.81978	.82418	.82855
.3	.83291	.83725	.84157	.84587	.85015	.85442	.85866	.86289	.86710	.87129
.4	.87547	.87963	.88377	.88789	.89200	.89609	.90016	.90422	.90826	.91228
.5	.91629	.92028	.92426	.92822	.93216	.93609	.94001	.94391	.94779	.95166
.6	.95551	.95935	.96317	.96698	.97078	.97456	.97833	.98208	.98582	.98954
.7	.99325	.99695	*.00063	*.00430	*.00796	*.01160	*.01523	*.01885	*.02245	*.02604
.8	1.02962	.03318	.03674	.04028	.04380	.04732	.05082	.05431	.05779	.06126
.9	.06471	.06815	.07158	.07500	.07841	.08181	.08519	.08856	.09192	.09527
3.0	1.09861	.10194	.10526	.10856	.11186	.11514	.11841	.12168	.12493	.12817
.1	.13140	.13462	.13783	.14103	.14422	.14740	.15057	.15373	.15688	.16002
.2	.16315	.16627	.16938	.17248	.17557	.17865	.18173	.18479	.18784	.19089
.3	.19392	.19695	.19996	.20297	.20597	.20896	.21194	.21491	.21788	.22083
.4	.22378	.22671	.22964	.23256	.23547	.23837	.24127	.24415	.24703	.24990
.5	.25276	.25562	.25846	.26130	.26413	.26695	.26976	.27257	.27536	.27815
.6	.28093	.28371	.28647	.28923	.29198	.29473	.29746	.30019	.30291	.30563
.7	.30833	.31103	.31372	.31641	.31909	.32176	.32442	.32708	.32972	.33237
.8	.33500	.33763	.34025	.34286	.34547	.34807	.35067	.35325	.35584	.35841
.9	.36098	.36354	.36609	.36864	.37118	.37372	.37624	.37877	.38128	.38379
4.0	1.38629	.38879	.39128	.39377	.39624	.39872	.40118	.40364	.40610	.40854
.1	.41099	.41342	.41585	.41828	.42070	.42311	.42552	.42792	.43031	.43270
.2	.43508	.43746	.43984	.44220	.44456	.44692	.44927	.45161	.45395	.45629
.3	.45862	.46094	.46326	.46557	.46787	.47018	.47247	.47476	.47705	.47933
.4	.48160	.48387	.48614	.48840	.49065	.49290	.49515	.49739	.49962	.50185
.5	.50408	.50630	.50851	.51072	.51293	.51513	.51732	.51951	.52170	.52388
.6	.52606	.52823	.53039	.53256	.53471	.53687	.53902	.54116	.54330	.54543
.7	.54756	.54969	.55181	.55393	.55604	.55814	.56025	.56235	.56444	.56653
.8	.56862	.57070	.57277	.57485	.57691	.57898	.58104	.58309	.58515	.58719
.9	.58924	.59127	.59331	.59534	.59737	.59939	.60141	.60342	.60543	.60744

NATURAL OR NAPERIAN LOGARITHMS (Continued)

5.00–9.99

N	0	1	2	3	4	5	6	7	8	9
5.0	1.60944	.61144	.61343	.61542	.61741	.61939	.62137	.62334	.62531	.62728
.1	.62924	.63120	.63315	.63511	.63705	.63900	.64094	.64287	.64481	.64673
.2	.64866	.65058	.65250	.65441	.65632	.65823	.66013	.66203	.66393	.66582
.3	.66771	.66959	.67147	.67335	.67523	.67710	.67896	.68083	.68269	.68455
.4	.68640	.68825	.69010	.69194	.69378	.69562	.69745	.69928	.70111	.70293
.5	.70475	.70656	.70838	.71019	.71199	.71380	.71560	.71740	.71919	.72098
.6	.72277	.72455	.72633	.72811	.72988	.73166	.73342	.73519	.73695	.73871
.7	.74047	.74222	.74397	.74572	.74746	.74920	.75094	.75267	.75440	.75613
.8	.75786	.75958	.76130	.76302	.76473	.76644	.76815	.76985	.77156	.77326
.9	.77495	.77665	.77834	.78002	.78171	.78339	.78507	.78675	.78842	.79009
6.0	1.79176	.79342	.79509	.79675	.79840	.80006	.80171	.80336	.80500	.80665
.1	.80829	.80993	.81156	.81319	.81482	.81645	.81808	.81970	.82132	.82294
.2	.82455	.82616	.82777	.82938	.83098	.83258	.83418	.83578	.83737	.83896
.3	.84055	.84214	.84372	.84530	.84688	.84845	.85003	.85160	.85317	.85473
.4	.85630	.85786	.85942	.86097	.86253	.86408	.86563	.86718	.86872	.87026
.5	.87180	.87334	.87487	.87641	.87794	.87947	.88099	.88251	.88403	.88555
.6	.88707	.88858	.89010	.89160	.89311	.89462	.89612	.89762	.89912	.90061
.7	.90211	.90360	.90509	.90658	.90806	.90954	.91102	.91250	.91398	.91545
.8	.91692	.91839	.91986	.92132	.92279	.92425	.92571	.92716	.92862	.93007
.9	.93152	.93297	.93442	.93586	.93730	.93874	.94018	.94162	.94305	.94448
7.0	1.94591	.94734	.94876	.95019	.95161	.95303	.95445	.95586	.95727	.95869
.1	.96009	.96150	.96291	.96431	.96571	.96711	.96851	.96991	.97130	.97269
.2	.97408	.97547	.97685	.97824	.97962	.98100	.98238	.98376	.98513	.98650
.3	.98787	.98924	.99061	.99198	.99334	.99470	.99606	.99742	.99877	*.00013
.4	2.00148	.00283	.00418	.00553	.00687	.00821	.00956	.01089	.01223	.01357
.5	.01490	.01624	.01757	.01890	.02022	.02155	.02287	.02419	.02551	.02683
.6	.02815	.02946	.03078	.03209	.03340	.03471	.03601	.03732	.03862	.03992
.7	.04122	.04252	.04381	.04511	.04640	.04769	.04898	.05027	.05156	.05284
.8	.05412	.05540	.05668	.05796	.05924	.06051	.06179	.06306	.06433	.06560
.9	.06686	.06813	.06939	.07065	.07191	.07317	.07443	.07568	.07694	.07819
8.0	2.07944	.08069	.08194	.08318	.08443	.08567	.08691	.08815	.08939	.09063
.1	.09186	.09310	.09433	.09556	.09679	.09802	.09924	.10047	.10169	.10291
.2	.10413	.10535	.10657	.10779	.10900	.11021	.11142	.11263	.11384	.11505
.3	.11626	.11746	.11866	.11986	.12106	.12226	.12346	.12465	.12585	.12704
.4	.12823	.12942	.13061	.13180	.13298	.13417	.13535	.13653	.13771	.13889
.5	.14007	.14124	.14242	.14359	.14476	.14593	.14710	.14827	.14943	.15060
.6	.15176	.15292	.15409	.15524	.15640	.15756	.15871	.15987	.16102	.16217
.7	.16332	.16447	.16562	.16677	.16791	.16905	.17020	.17134	.17248	.17361
.8	.17475	.17589	.17702	.17816	.17929	.18042	.18155	.18267	.18380	.18493
.9	.18605	.18717	.18830	.18942	.19054	.19165	.19277	.19389	.19500	.19611
9.0	2.19722	.19834	.19944	.20055	.20166	.20276	.20387	.20497	.20607	.20717
.1	.20827	.20937	.21047	.21157	.21266	.21375	.21485	.21594	.21703	.21812
.2	.21920	.22029	.22138	.22246	.22354	.22462	.22570	.22678	.22786	.22894
.3	.23001	.23109	.23216	.23324	.23431	.23538	.23645	.23751	.23858	.23965
.4	.24071	.24177	.24284	.24390	.24496	.24601	.24707	.24813	.24918	.25024
.5	.25129	.25234	.25339	.25444	.25549	.25654	.25759	.25863	.25968	.26072
.6	.26176	.26280	.26384	.26488	.26592	.26696	.26799	.26903	.27006	.27109
.7	.27213	.27316	.27419	.27521	.27624	.27727	.27829	.27932	.28034	.28136
.8	.28238	.28340	.28442	.28544	.28646	.28747	.28849	.28950	.29051	.29152
.9	.29253	.29354	.29455	.29556	.29657	.29757	.29858	.29958	.30058	.30158

NATURAL OR NAPERIAN LOGARITHMS (Continued)

Constants

$\log_e 10 = 2.30258\ 50930$	$6 \log_e 10 = 13.81551\ 05580$
$2 \log_e 10 = 4.60517\ 01860$	$7 \log_e 10 = 16.11809\ 56510$
$3 \log_e 10 = 6.90775\ 52790$	$8 \log_e 10 = 18.42068\ 07440$
$4 \log_e 10 = 9.21034\ 03720$	$9 \log_e 10 = 20.72326\ 58369$
$5 \log_e 10 = 11.51292\ 54650$	$10 \log_e 10 = 23.02585\ 09299$

10.0–49.9

N	0	1	2	3	4	5	6	7	8	9
10.	2.30259	.31254	.32239	.33214	.34181	.35138	.36085	.37024	.37955	.38876
11.	.39790	.40695	.41591	.42480	.43361	.44235	.45101	.45959	.46810	.47654
12.	.48491	.49321	.50144	.50960	.51770	.52573	.53370	.54160	.54945	.55723
13.	.56495	.57261	.58022	.58776	.59525	.60269	.61007	.61740	.62467	.63189
14.	.63906	.64617	.65324	.66026	.66723	.67415	.68102	.68785	.69463	.70136
15.	.70805	.71469	.72130	.72785	.73437	.74084	.74727	.75366	.76001	.76632
16.	.77259	.77882	.78501	.79117	.79728	.80336	.80940	.81541	.82138	.82731
17.	.83321	.83908	.84491	.85071	.85647	.86220	.86790	.87356	.87920	.88480
18.	.89037	.89591	.90142	.90690	.91235	.91777	.92316	.92852	.93386	.93916
19.	.94444	.94969	.95491	.96011	.96527	.97041	.97553	.98062	.98568	.99072
20.	2.99573	*.00072	*.00568	*.01062	*.01553	*.02042	*.02529	*.03013	*.03495	*.03975
21.	3.04452	.04927	.05400	.05871	.06339	.06805	.07269	.07731	.08191	.08649
22.	.09104	.09558	.10009	.10459	.10906	.11352	.11795	.12236	.12676	.13114
23.	.13549	.13983	.14415	.14845	.15274	.15700	.16125	.16548	.16969	.17388
24.	.17805	.18221	.18635	.19048	.19458	.19867	.20275	.20680	.21084	.21487
25.	.21888	.22287	.22684	.23080	.23475	.23868	.24259	.24649	.25037	.25424
26.	.25810	.26194	.26576	.26957	.27336	.27714	.28091	.28466	.28840	.29213
27.	.29584	.29953	.30322	.30689	.31054	.31419	.31782	.32143	.32504	.32863
28.	.33220	.33577	.33932	.34286	.34639	.34990	.35341	.35690	.36038	.36384
29.	.36730	.37074	.37417	.37759	.38099	.38439	.38777	.39115	.39451	.39786
30.	3.40120	.40453	.40784	.41115	.41444	.41773	.42100	.42426	.42751	.43076
31.	.43399	.43721	.44042	.44362	.44681	.44999	.45316	.45632	.45947	.46261
32.	.46574	.46886	.47197	.47507	.47816	.48124	.48431	.48738	.49043	.49347
33.	.49651	.49953	.50255	.50556	.50856	.51155	.51453	.51750	.52046	.52342
34.	.52636	.52930	.53223	.53515	.53806	.54096	.54385	.54674	.54962	.55249
35.	.55535	.55820	.56105	.56388	.56671	.56953	.57235	.57515	.57795	.58074
36.	.58352	.58629	.58906	.59182	.59457	.59731	.60005	.60278	.60550	.60821
37.	.61092	.61362	.61631	.61899	.62167	.62434	.62700	.62966	.63231	.63495
38.	.63759	.64021	.64284	.64545	.64806	.65066	.65325	.65584	.65842	.66099
39.	.66356	.66612	.66868	.67122	.67377	.67630	.67883	.68135	.68387	.68638
40.	3.68888	.69138	.69387	.69635	.69883	.70130	.70377	.70623	.70868	.71113
41.	.71357	.71601	.71844	.72086	.72328	.72569	.72810	.73050	.73290	.73529
42.	.73767	.74005	.74242	.74479	.74715	.74950	.75185	.75420	.75654	.75887
43.	.76120	.76352	.76584	.76815	.77046	.77276	.77506	.77735	.77963	.78191
44.	.78419	.78646	.78872	.79098	.79324	.79549	.79773	.79997	.80221	.80444
45.	.80666	.80888	.81110	.81331	.81551	.81771	.81991	.82210	.82428	.82647
46.	.82864	.83081	.83298	.83514	.83730	.83945	.84160	.84374	.84588	.84802
47.	.85015	.85227	.85439	.85651	.85862	.86073	.86283	.86493	.86703	.86912
48.	.87120	.87328	.87536	.87743	.87950	.88156	.88362	.88568	.88773	.88978
49.	.89182	.89386	.89589	.89792	.89995	.90197	.90399	.90600	.90801	.91002

NATURAL OR NAPERIAN LOGARITHMS (Continued)

50.0–99.9

N	0	1	2	3	4	5	6	7	8	9
50.	3.91202	.91402	.91602	.91801	.91999	.92197	.92395	.92593	.92790	.92986
51.	.93183	.93378	.93574	.93769	.93964	.94158	.94352	.94546	.94739	.94932
52.	.95124	.95316	.95508	.95700	.95891	.96081	.96272	.96462	.96651	.96840
53.	.97029	.97218	.97406	.97594	.97781	.97968	.98155	.98341	.98527	.98713
54.	.98898	.99083	.99268	.99452	.99636	.99820	*.00003	*.00186	*.00369	*.00551
55.	4.00733	.00915	.01096	.01277	.01458	.01638	.01818	.01998	.02177	.02356
56.	.02535	.02714	.02892	.03069	.03247	.03424	.03601	.03777	.03954	.04130
57.	.04305	.04480	.04655	.04830	.05004	.05178	.05352	.05526	.05699	.05872
58.	.06044	.06217	.06389	.06560	.06732	.06903	.07073	.07244	.07414	.07584
59.	.07754	.07923	.08092	.08261	.08429	.08598	.08766	.08933	.09101	.09268
60.	4.09434	.09601	.09767	.09933	.10099	.10264	.10429	.10594	.10759	.10923
61.	.11087	.11251	.11415	.11578	.11741	.11904	.12066	.12228	.12390	.12552
62.	.12713	.12875	.13036	.13196	.13357	.13517	.13677	.13836	.13996	.14155
63.	.14313	.14472	.14630	.14789	.14946	.15104	.15261	.15418	.15575	.15732
64.	.15888	.16044	.16200	.16356	.16511	.16667	.16821	.16976	.17131	.17285
65.	.17439	.17592	.17746	.17899	.18052	.18205	.18358	.18510	.18662	.18814
66.	.18965	.19117	.19268	.19419	.19570	.19720	.19870	.20020	.20170	.20320
67.	.20469	.20618	.20767	.20916	.21065	.21213	.21361	.21509	.21656	.21804
68.	.21951	.22098	.22244	.22391	.22537	.22683	.22829	.22975	.23120	.23266
69.	.23411	.23555	.23700	.23844	.23989	.24133	.24276	.24420	.24563	.24707
70.	4.24850	.24992	.25135	.25277	.25419	.25561	.25703	.25845	.25986	.26127
71.	.26268	.26409	.26549	.26690	.26830	.26970	.27110	.27249	.27388	.27528
72.	.27667	.27805	.27944	.28082	.28221	.28359	.28496	.28634	.28772	.28909
73.	.29046	.29183	.29320	.29456	.29592	.29729	.29865	.30000	.30136	.30271
74.	.30407	.30542	.30676	.30811	.30946	.31080	.31214	.31348	.31482	.31615
75.	.31749	.31882	.32015	.32149	.32281	.32413	.32546	.32678	.32810	.32942
76.	.33073	.33205	.33336	.33467	.33598	.33729	.33860	.33990	.34120	.34251
77.	.34381	.34510	.34640	.34769	.34899	.35028	.35157	.35286	.35414	.35543
78.	.35671	.35800	.35927	.36055	.36182	.36310	.36437	.36564	.36691	.36818
79.	.36945	.37071	.37198	.37324	.37450	.37576	.37701	.37827	.37952	.38078
80.	4.38203	.38328	.38452	.38577	.38701	.38826	.38950	.39074	.39198	.39321
81.	.39445	.39568	.39692	.39815	.39938	.40060	.40183	.40305	.40428	.40550
82.	.40672	.40794	.40916	.41037	.41159	.41280	.41401	.41522	.41643	.41764
83.	.41884	.42004	.42125	.42245	.42365	.42485	.42604	.42724	.42843	.42963
84.	.43082	.43201	.43319	.43438	.43557	.43675	.43793	.43912	.44030	.44147
85.	.44265	.44383	.44500	.44617	.44735	.44852	.44969	.45085	.45202	.45318
86.	.45435	.45551	.45667	.45783	.45899	.46014	.46130	.46245	.46361	.46476
87.	.46591	.46706	.46820	.46935	.47050	.47164	.47278	.47392	.47506	.47620
88.	.47734	.47847	.47961	.48074	.48187	.48300	.48413	.48526	.48639	.48751
89.	.48864	.48976	.49088	.49200	.49312	.49424	.49536	.49647	.49758	.49870
90.	4.49981	.50092	.50203	.50314	.50424	.50535	.50645	.50756	.50866	.50976
91.	.51086	.51196	.51305	.51415	.51525	.51634	.51743	.51852	.51961	.52070
92.	.52179	.52287	.52396	.52504	.52613	.52721	.52829	.52937	.53045	.53152
93.	.53260	.53367	.53475	.53582	.53689	.53796	.53903	.54010	.54116	.54223
94.	.54329	.54436	.54542	.54648	.54754	.54860	.54966	.55071	.55177	.55282
95.	.55388	.55493	.55598	.55703	.55808	.55913	.56017	.56122	.56226	.56331
96.	.56435	.56539	.56643	.56747	.56851	.56954	.57058	.57161	.57265	.57368
97.	.57471	.57574	.57677	.57780	.57883	.57985	.58088	.58190	.58292	.58395
98.	.58497	.58599	.58701	.58802	.58904	.59006	.59107	.59208	.59310	.59411
99.	.59512	.59613	.59714	.59815	.59915	.60016	.60116	.60217	.60317	.60417

NATURAL OR NAPERIAN LOGARITHMS (Continued)

0–499

N	0	1	2	3	4	5	6	7	8	9
0	−∞	0.00000	0.69315	1.09861	.38629	.60944	.79176	.94591	*.07944	*.19722
1	2.30259	.39790	.48491	.56495	.63906	.70805	.77259	.83321	.89037	.94444
2	.99573	*.04452	*.09104	*.13549	*.17805	*.21888	*.25810	*.29584	*.33220	*.36730
3	3.40120	.43399	.46574	.49651	.52636	.55535	.58352	.61092	.63759	.66356
4	.68888	.71357	.73767	.76120	.78419	.80666	.82864	.85015	.87120	.89182
5	.91202	.93183	.95124	.97029	.98898	*.00733	.*02535	*.04305	*0.6044	*.07754
6	4.09434	.11087	.12713	.14313	.15888	.17439	.18965	.20469	.21951	.23411
7	.24850	.26268	.27667	.29046	.30407	.31749	.33073	.34381	.35671	.36945
8	.38203	.39445	.40672	.41884	.43082	.44265	.45435	.46591	.47734	.48864
9	.49981	.51086	.52179	.53260	.54329	.55388	.56435	.57471	.58497	.59512
10	4.60517	.61512	.62497	.63473	.64439	.65396	.66344	.67283	.68213	.69135
11	.70048	.70953	.71850	.72739	.73620	.74493	.75359	.76217	.77068	.77912
12	.78749	.79579	.80402	.81218	.82028	.82831	.83628	.84419	.85203	.85981
13	.86753	.87520	.88280	.89035	.89784	.90527	.91265	.91998	.92725	.93447
14	.94164	.94876	.95583	.96284	.96981	.97673	.98361	.99043	.99721	*.00395
15	5.01064	.01728	.02388	.03044	.03695	.04343	.04986	.05625	.06260	.06890
16	.07517	.08140	.08760	.09375	.09987	.10595	.11199	.11799	.12396	.12990
17	.13580	.14166	.14749	.15329	.15906	.16479	.17048	.17615	.18178	.18739
18	.19296	.19850	.20401	.20949	.21494	.22036	.22575	.23111	.23644	.24175
19	.24702	.25227	.25750	.26269	.26786	.27300	.27811	.28320	.28827	.29330
20	5.29832	.30330	.30827	.31321	.31812	.32301	.32788	.33272	.33754	.34233
21	.34711	.35186	.35659	.36129	.36598	.37064	.37528	.37990	.38450	.38907
22	.39363	.39816	.40268	.40717	.41165	.41610	.42053	.42495	.42935	.43372
23	.43808	.44242	.44674	.45104	.45532	.45959	.46383	.46806	.47227	.47646
24	.48064	.48480	.48894	.49306	.49717	.50126	.50533	.50939	.51343	.51745
25	.52146	.52545	.52943	.53339	.53733	.54126	.54518	.54908	.55296	.55683
26	.56068	.56452	.56834	.57215	.57595	.57973	.58350	.58725	.59099	.59471
27	.59842	.60212	.60580	.60947	.61313	.61677	.62040	.62402	.62762	.63121
28	.63479	.63835	.64191	.64545	.64897	.65249	.65599	.65948	.66296	.66643
29	.66988	.67332	.67675	.68017	.68358	.68698	.69036	.69373	.69709	.70044
30	5.70378	.70711	.71043	.71373	.71703	.72031	.72359	.72685	.73010	.73334
31	.73657	.73979	.74300	.74620	.74939	.75257	.75574	.75890	.76205	.76519
32	.76832	.77144	.77455	.77765	.78074	.78383	.78690	.78996	.79301	.79606
33	.79909	.80212	.80513	.80814	.81114	.81413	.81711	.82008	.82305	.82600
34	.82895	.83188	.83481	.83773	.84064	.84354	.84644	.84932	.85220	.85507
35	.85793	.86079	.86363	.86647	.86930	.87212	.87493	.87774	.88053	.88332
36	.88610	.88888	.89164	.89440	.89715	.89990	.90263	.90536	.90808	.91080
37	.91350	.91620	.91889	.92158	.92426	.92693	.92959	.93225	.93489	.93754
38	.94017	.94280	.94542	.94803	.95064	.95324	.95584	.95842	.96101	.96358
39	.96615	.96871	.97126	.97381	.97635	.97889	.98141	.98394	.98645	.98896
40	5.99146	.99396	.99645	.99894	*.00141	*.00389	*.00635	*.00881	*.01127	*.01372
41	6.01616	.01859	.02102	.02345	.02587	.02828	.03069	.03309	.03548	.03787
42	.04025	.04263	.04501	.04737	.04973	.05209	.05444	.05678	.05912	.06146
43	.06379	.06611	.06843	.07074	.07304	.07535	.07764	.07993	.08222	.08450
44	.08677	.08904	.09131	.09357	.09582	.09807	.10032	.10256	.10479	.10702
45	.10925	.11147	.11368	.11589	.11810	.12030	.12249	.12468	.12687	.12905
46	.13123	.13340	.13556	.13773	.13988	.14204	.14419	.14633	.14847	.15060
47	.15273	.15486	.15698	.15910	.16121	.16331	.16542	.16752	.16961	.17170
48	.17379	.17587	.17794	.18002	.18208	.18415	.18621	.18826	.19032	.19236
49	.19441	.19644	.19848	.20051	.20254	.20456	.20658	.20859	.21060	.21261

NATURAL OR NAPERIAN LOGARITHMS (Continued)

500–999

N	0	1	2	3	4	5	6	7	8	9
50	6.21461	.21661	.21860	.22059	.22258	.22456	.22654	.22851	.23048	.23245
51	.23441	.23637	.23832	.24028	.24222	.24417	.24611	.24804	.24998	.25190
52	.25383	.25575	.25767	.25958	.26149	.26340	.26530	.26720	.26910	.27099
53	.27288	.27476	.27664	.27852	.28040	.28227	.28413	.28600	.28786	.28972
54	.29157	.29342	.29527	.29711	.29895	.30079	.30262	.30445	.30628	.30810
55	.30992	.31173	.31355	.31536	.31716	.31897	.32077	.32257	.32436	.32615
56	.32794	.32972	.33150	.33328	.33505	.33683	.33859	.34036	.34212	.34388
57	.34564	.34739	.34914	.35089	.35263	.35437	.35611	.35784	.35957	.36130
58	.36303	.36475	.36647	.36819	.36990	.37161	.37332	.37502	.37673	.37843
59	.38012	.38182	.38351	.38519	.38688	.38856	.39024	.39192	.39359	.39526
60	6.30693	.39859	.40026	.40192	.40357	.40523	.40688	.40853	.41017	.41182
61	.41346	.41510	.41673	.41836	.41999	.42162	.42325	.42487	.42649	.42811
62	.42972	.43133	.43294	.43455	.43615	.43775	.43935	.44095	.44254	.44413
63	.44572	.44731	.44889	.45047	.45205	.45362	.45520	.45677	.45834	.45990
64	.46147	.46303	.46459	.46614	.46770	.46925	.47080	.47235	.47389	.47543
65	.47697	.47851	.48004	.48158	.48311	.48464	.48616	.48768	.48920	.49072
66	.49224	.49375	.49527	.49677	.49828	.49979	.50129	.50279	.50429	.50578
67	.50728	.50877	.51026	.51175	.51323	.51471	.51619	.51767	.51915	.52062
68	.52209	.52356	.52503	.52649	.52796	.52942	.53088	.53233	.53379	.53524
69	.53669	.53814	.53959	.54103	.54247	.54391	.54535	.54679	.54822	.54965
70	6.55108	.55251	.55393	.55536	.55678	.55820	.55962	.56103	.56244	.56386
71	.56526	.56667	.56808	.56948	.57088	.57228	.57368	.57508	.57647	.57786
72	.57925	.58064	.58203	.58341	.58479	.58617	.58755	.58893	.59030	.59167
73	.59304	.59441	.59578	.59715	.59851	.59987	.60123	.60259	.60394	.60530
74	.60665	.60800	.60935	.61070	.61204	.61338	.61473	.61607	.61740	.61874
75	.62007	.62141	.62274	.62407	.62539	.62672	.62804	.62936	.63068	.63200
76	.63332	.63463	.63595	.63726	.63857	.63988	.64118	.64249	.64379	.64509
77	.64639	.64769	.64898	.65028	.65157	.65286	.65415	.65544	.65673	.65801
78	.65929	.66058	.66185	.66313	.66441	.66568	.66696	.66823	.66950	.67077
79	.67203	.67330	.67456	.67582	.67708	.67834	.67960	.68085	.68211	.68336
80	6.68461	.68586	.68711	.68835	.68960	.69084	.69208	.69332	.69456	.69580
81	.69703	.69827	.69950	.70073	.70196	.70319	.70441	.70564	.70686	.70808
82	.70930	.71052	.71174	.71296	.71417	.71538	.71659	.71780	.71901	.72022
83	.72143	.72263	.72383	.72503	.72623	.72743	.72863	.72982	.73102	.73221
84	.73340	.73459	.73578	.73697	.73815	.73934	.74052	.74170	.74288	.74406
85	.74524	.74641	.74759	.74876	.74993	.75110	.75227	.75344	.75460	.75577
86	.75693	.75809	.75926	.76041	.76157	.76273	.76388	.76504	.76619	.76734
87	.76849	.76964	.77079	.77194	.77308	.77422	.77537	.77651	.77765	.77878
88	.77992	.78106	.78219	.78333	.78446	.78559	.78672	.78784	.78897	.79010
89	.79122	.79234	.79347	.79459	.79571	.79682	.79794	.79906	.80017	.80128
90	6.80239	.80351	.80461	.80572	.80683	.80793	.80904	.81014	.81124	.81235
91	.81344	.81454	.81564	.81674	.81783	.81892	.82002	.82111	.82220	.82329
92	.82437	.82546	.82655	.82763	.82871	.82979	.83087	.83195	.83303	.83411
93	.83518	.83626	.83733	.83841	.83948	.84055	.84162	.84268	.84375	.84482
94	.84588	.84694	.84801	.84907	.85013	.85118	.85224	.85330	.85435	.85541
95	.85646	.85751	.85857	.85961	.86066	.86171	.86276	.86380	.86485	.86589
96	.86693	.86797	.86901	.87005	.87109	.87213	.87316	.87420	.87523	.87626
97	.87730	.87833	.87936	.88038	.88141	.88244	.88346	.88449	.88551	.88653
98	.88755	.88857	.88959	.89061	.89163	.89264	.89366	.89467	.89568	.89669
99	.89770	.89871	.89972	.90073	.90174	.90274	.90375	.90475	.90575	.90675

EXPONENTIAL FUNCTIONS

x	e^x	$\text{Log}_{10}(e^x)$	e^{-x}	x	e^x	$\text{Log}_{10}(e^x)$	e^{-x}
0.00	1.0000	0.00000	1.000000	**0.50**	1.6487	0.21715	0.606531
0.01	1.0101	.00434	0.990050	0.51	1.6653	.22149	.600496
0.02	1.0202	.00869	.980199	0.52	1.6820	.22583	.594521
0.03	1.0305	.01303	.970446	0.53	1.6989	.23018	.588605
0.04	1.0408	.01737	.960789	0.54	1.7160	.23452	.582748
0.05	1.0513	0.02171	0.951229	**0.55**	1.7333	0.23886	0.576950
0.06	1.0618	.02606	.941765	0.56	1.7507	.24320	.571209
0.07	1.0725	.03040	.932394	0.57	1.7683	.24755	.565525
0.08	1.0833	.03474	.923116	0.58	1.7860	.25189	.559898
0.09	1.0942	.03909	.913931	0.59	1.8040	.25623	.554327
0.10	1.1052	0.04343	0.904837	**0.60**	1.8221	0.26058	0.548812
0.11	1.1163	.04777	.895834	0.61	1.8404	.26492	.543351
0.12	1.1275	.05212	.886920	0.62	1.8589	.26926	.537944
0.13	1.1388	.05646	.878095	0.63	1.8776	.27361	.532592
0.14	1.1503	.06080	.869358	0.64	1.8965	.27795	.527292
0.15	1.1618	0.06514	0.860708	**0.65**	1.9155	0.28229	0.522046
0.16	1.1735	.06949	.852144	0.66	1.9348	.28663	.516851
0.17	1.1853	.07383	.843665	0.67	1.9542	.29098	.511709
0.18	1.1972	.07817	.835270	0.68	1.9739	.29532	.506617
0.19	1.2092	.08252	.826959	0.69	1.9937	.29966	.501576
0.20	1.2214	0.08686	0.818731	**0.70**	2.0138	0.30401	0.496585
0.21	1.2337	.09120	.810584	0.71	2.0340	.30835	.491644
0.22	1.2461	.09554	.802519	0.72	2.0544	.31269	.486752
0.23	1.2586	.09989	.794534	0.73	2.0751	.31703	.481909
0.24	1.2712	.10423	.786628	0.74	2.0959	.32138	.477114
0.25	1.2840	0.10857	0.778801	**0.75**	2.1170	0.32572	0.472367
0.26	1.2969	.11292	.771052	0.76	2.1383	.33006	.467666
0.27	1.3100	.11726	.763379	0.77	2.1598	.33441	.463013
0.28	1.3231	.12160	.755784	0.78	2.1815	.33875	.458406
0.29	1.3364	.12595	.748264	0.79	2.2034	.34309	.453845
0.30	1.3499	0.13029	0.740818	**0.80**	2.2255	0.34744	0.449329
0.31	1.3634	.13463	.733447	0.81	2.2479	.35178	.444858
0.32	1.3771	.13897	.726149	0.82	2.2705	.35612	.440432
0.33	1.3910	.14332	.718924	0.83	2.2933	.36046	.436049
0.34	1.4049	.14766	.711770	0.84	2.3164	.36481	.431711
0.35	1.4191	0.15200	0.704688	**0.85**	2.3396	0.36915	0.427415
0.36	1.4333	.15635	.697676	0.86	2.3632	.37349	.423162
0.37	1.4477	.16069	.690734	0.87	2.3869	.37784	.418952
0.38	1.4623	.16503	.683861	0.88	2.4109	.38218	.414783
0.39	1.4770	.16937	.677057	0.89	2.4351	.38652	.410656
0.40	1.4918	0.17372	0.670320	**0.90**	2.4596	0.39087	0.406570
0.41	1.5068	.17806	.663650	0.91	2.4843	.39521	.402524
0.42	1.5220	.18240	.657047	0.92	2.5093	.39955	.398519
0.43	1.5373	.18675	.650509	0.93	2.5345	.40389	.394554
0.44	1.5527	.19109	.644036	0.94	2.5600	.40824	.390628
0.45	1.5683	0.19543	0.637628	**0.95**	2.5857	0.41258	0.386741
0.46	1.5841	.19978	.631284	0.96	2.6117	.41692	.382893
0.47	1.6000	.20412	.625002	0.97	2.6379	.42127	.379083
0.48	1.6161	.20846	.618783	0.98	2.6645	.42561	.375311
0.49	1.6323	.21280	.612626	0.99	2.6912	.42995	.371577
0.50	1.6487	0.21715	0.606531	**1.00**	2.7183	0.43429	0.367879

EXPONENTIAL FUNCTIONS (Continued)

x	e^x	$\text{Log}_{10}(e^x)$	e^{-x}
1.00	2.7183	0.43429	0.367879
1.01	2.7456	.43864	.364219
1.02	2.7732	.44298	.360595
1.03	2.8011	.44732	.357007
1.04	2.8292	.45167	.353455
1.05	2.8577	0.45601	0.349938
1.06	2.8864	.46035	.346456
1.07	2.9154	.46470	.343009
1.08	2.9447	.46904	.339596
1.09	2.9743	.47338	.336216
1.10	3.0042	0.47772	0.332871
1.11	3.0344	.48207	.329559
1.12	3.0649	.48641	.326280
1.13	3.0957	.49075	.323033
1.14	3.1268	.49510	.319819
1.15	3.1582	0.49944	0.316637
1.16	3.1899	.50378	.313486
1.17	3.2220	.50812	.310367
1.18	3.2544	.51247	.307279
1.19	3.2871	.51681	.304221
1.20	3.3201	0.52115	0.301194
1.21	3.3535	.52550	.298197
1.22	3.3872	.52984	.295230
1.23	3.4212	.53418	.292293
1.24	3.4556	.53853	.289384
1.25	3.4903	0.54287	0.286505
1.26	3.5254	.54721	.283654
1.27	3.5609	.55155	.280832
1.28	3.5966	.55590	.278037
1.29	3.6328	.56024	.275271
1.30	3.6693	0.56458	0.272532
1.31	3.7062	.56893	.269820
1.32	3.7434	.57327	.267135
1.33	3.7810	.57761	.264477
1.34	3.8190	.58195	.261846
1.35	3.8574	0.58630	0.259240
1.36	3.8962	.59064	.256661
1.37	3.9354	.59498	.254107
1.38	3.9749	.59933	.251579
1.39	4.0149	.60367	.249075
1.40	4.0552	0.60801	0.246597
1.41	4.0960	.61236	.244143
1.42	4.1371	.61670	.241714
1.43	4.1787	.62104	.239309
1.44	4.2207	.62538	.236928
1.45	4.2631	0.62973	0.234570
1.46	4.3060	.63407	.232236
1.47	4.3492	.63841	.229925
1.48	4.3929	.64276	.227638
1.49	4.4371	.64710	.225373
1.50	4.4817	0.65144	0.223130

x	e^x	$\text{Log}_{10}(e^x)$	e^{-x}
1.50	4.4817	0.65144	0.223130
1.51	4.5267	.65578	.220910
1.52	4.5722	.66013	.218712
1.53	4.6182	.66447	.216536
1.54	4.6646	.66881	.214381
1.55	4.7115	0.67316	0.212248
1.56	4.7588	.67750	.210136
1.57	4.8066	.68184	.208045
1.58	4.8550	.68619	.205975
1.59	4.9037	.69053	.203926
1.60	4.9530	0.69487	0.201897
1.61	5.0028	.69921	.199888
1.62	5.0531	.70356	.197899
1.63	5.1039	.70790	.195930
1.64	5.1552	.71224	.193980
1.65	5.2070	0.71659	0.192050
1.66	5.2593	.72093	.190139
1.67	5.3122	.72527	.188247
1.68	5.3656	.72961	.186374
1.69	5.4195	.73396	.184520
1.70	5.4739	0.73830	0.182684
1.71	5.5290	.74264	.180866
1.72	5.5845	.74699	.179066
1.73	5.6407	.75133	.177284
1.74	5.6973	.75567	.175520
1.75	5.7546	0.76002	0.173774
1.76	5.8124	.76436	.172045
1.77	5.8709	.76870	.170333
1.78	5.9299	.77304	.168638
1.79	5.9895	.77739	.166960
1.80	6.0496	0.78173	0.165299
1.81	6.1104	.78607	.163654
1.82	6.1719	.79042	.162026
1.83	6.2339	.79476	.160414
1.84	6.2965	.79910	.158817
1.85	6.3598	0.80344	0.157237
1.86	6.4237	.80779	.155673
1.87	6.4883	.81213	.154124
1.88	6.5535	.81647	.152590
1.89	6.6194	.82082	.151072
1.90	6.6859	0.82516	0.149569
1.91	6.7531	.82950	.148080
1.92	6.8210	.83385	.146607
1.93	6.8895	.83819	.145148
1.94	6.9588	.84253	.143704
1.95	7.0287	0.84687	0.142274
1.96	7.0993	.85122	.140858
1.97	7.1707	.85556	.139457
1.98	7.2427	.85990	.138069
1.99	7.3155	.86425	.136695
2.00	7.3891	0.86859	0.135335

EXPONENTIAL FUNCTIONS (Continued)

x	e^x	$\text{Log}_{10}(e^x)$	e^{-x}
2.00	7.3891	0.86859	0.135335
2.01	7.4633	.87293	.133989
2.02	7.5383	.87727	.132655
2.03	7.6141	.88162	.131336
2.04	7.6906	.88596	.130029
2.05	7.7679	0.89030	0.128735
2.06	7.8460	.89465	.127454
2.07	7.9248	.89899	.126186
2.08	8.0045	.90333	.124930
2.09	8.0849	.90768	.123687
2.10	8.1662	0.91202	0.122456
2.11	8.2482	.91636	.121238
2.12	8.3311	.92070	.120032
2.13	8.4149	.92505	.118837
2.14	8.4994	.92939	.117655
2.15	8.5849	0.93373	0.116484
2.16	8.6711	.93808	.115325
2.17	8.7583	.94242	.114178
2.18	8.8463	.94676	.113042
2.19	8.9352	.95110	.111917
2.20	9.0250	0.95545	0.110803
2.21	9.1157	.95979	.109701
2.22	9.2073	.96413	.108609
2.23	9.2999	.96848	.107528
2.24	9.3933	.97282	.106459
2.25	9.4877	0.97716	0.105399
2.26	9.5831	.98151	.104350
2.27	9.6794	.98585	.103312
2.28	9.7767	.99019	.102284
2.29	9.8749	.99453	.101266
2.30	9.9742	0.99888	0.100259
2.31	10.074	1.00322	.099261
2.32	10.176	1.00756	.098274
2.33	10.278	1.01191	.097296
2.34	10.381	1.01625	.096328
2.35	10.486	1.02059	0.095369
2.36	10.591	1.02493	.094420
2.37	10.697	1.02928	.093481
2.38	10.805	1.03362	.092551
2.39	10.913	1.03796	.091630
2.40	11.023	1.04231	0.090718
2.41	11.134	1.04665	.089815
2.42	11.246	1.05099	.088922
2.43	11.359	1.05534	.088037
2.44	11.473	1.05968	.087161
2.45	11.588	1.06402	0.086294
2.46	11.705	1.06836	.085435
2.47	11.822	1.07271	.084585
2.48	11.941	1.07705	.083743
2.49	12.061	1.08139	.082910
2.50	12.182	1.08574	0.082085

x	e^x	$\text{Log}_{10}(e^x)$	e^{-x}
2.50	12.182	1.08574	0.082085
2.51	12.305	1.09008	.081268
2.52	12.429	1.09442	.080460
2.53	12.554	1.09877	.079659
2.54	12.680	1.10311	.078866
2.55	12.807	1.10745	0.078082
2.56	12.936	1.11179	.077305
2.57	13.066	1.11614	.076536
2.58	13.197	1.12048	.075774
2.59	13.330	1.12482	.075020
2.60	13.464	1.12917	0.074274
2.61	13.599	1.13351	.073535
2.62	13.736	1.13785	.072803
2.63	13.874	1.14219	.072078
2.64	14.013	1.14654	.071361
2.65	14.154	1.15088	0.070651
2.66	14.296	1.15522	.069948
2.67	14.440	1.15957	.069252
2.68	14.585	1.16391	.068563
2.69	14.732	1.16825	.067881
2.70	14.880	1.17260	0.067206
2.71	15.029	1.17694	.066537
2.72	15.180	1.18128	.065875
2.73	15.333	1.18562	.065219
2.74	15.487	1.18997	.064570
2.75	15.643	1.19431	0.063928
2.76	15.800	1.19865	.063292
2.77	15.959	1.20300	.062662
2.78	16.119	1.20734	.062039
2.79	16.281	1.21168	.061421
2.80	16.445	1.21602	0.060810
2.81	16.610	1.22037	.060205
2.82	16.777	1.22471	.059606
2.83	16.945	1.22905	.059013
2.84	17.116	1.23340	.058426
2.85	17.288	1.23774	0.057844
2.86	17.462	1.24208	.057269
2.87	17.637	1.24643	.056699
2.88	17.814	1.25077	.056135
2.89	17.993	1.25511	.055576
2.90	18.174	1.25945	0.055023
2.91	18.357	1.26380	.054476
2.92	18.541	1.26814	.053934
2.93	18.728	1.27248	.053397
2.94	18.916	1.27683	.052866
2.95	19.106	1.28117	0.052340
2.96	19.298	1.28551	.051819
2.97	19.492	1.28985	.051303
2.98	19.688	1.29420	.050793
2.99	19.886	1.29854	.050287
3.00	20.086	1.30288	0.049787

EXPONENTIAL FUNCTIONS (Continued)

x	e^x	$\text{Log}_{10}(e^x)$	e^{-x}	x	e^x	$\text{Log}_{10}(e^x)$	e^{-x}
3.00	20.086	1.30288	0.049787	**3.50**	33.115	1.52003	0.030197
3.01	20.287	1.30723	.049292	3.51	33.448	1.52437	.029897
3.02	20.491	1.31157	.048801	3.52	33.784	1.52872	.029599
3.03	20.697	1.31591	.048316	3.53	34.124	1.53306	.029305
3.04	20.905	1.32026	.047835	3.54	34.467	1.53740	.029013
3.05	21.115	1.32460	0.047359	**3.55**	34.813	1.54175	0.028725
3.06	21.328	1.32894	.046888	3.56	35.163	1.54609	.028439
3.07	21.542	1.33328	.046421	3.57	35.517	1.55043	.028156
3.08	21.758	1.33763	.045959	3.58	35.874	1.55477	.027876
3.09	21.977	1.34197	.045502	3.59	36.234	1.55912	.027598
3.10	22.198	1.34631	0.045049	**3.60**	36.598	1.56346	0.027324
3.11	22.421	1.35066	.044601	3.61	36.966	1.56780	.027052
3.12	22.646	1.35500	.044157	3.62	37.338	1.57215	.026783
3.13	22.874	1.35934	.043718	3.63	37.713	1.57649	.026516
3.14	23.104	1.36368	.043283	3.64	38.092	1.58083	.026252
3.15	23.336	1.36803	0.042852	**3.65**	38.475	1.58517	0.025991
3.16	23.571	1.37237	.042426	3.66	38.861	1.58952	.025733
3.17	23.807	1.37671	.042004	3.67	39.252	1.59386	.025476
3.18	24.047	1.38106	.041586	3.68	39.646	1.59820	.025223
3.19	24.288	1.38540	.041172	3.69	40.045	1.60255	.024972
3.20	24.533	1.38974	0.040762	**3.70**	40.447	1.60689	0.024724
3.21	24.779	1.39409	.040357	3.71	40.854	1.61123	.024478
3.22	25.028	1.39843	.039955	3.72	41.264	1.61558	.024234
3.23	25.280	1.40277	.039557	3.73	41.679	1.61992	.023993
3.24	25.534	1.40711	.039164	3.74	42.098	1.62426	.023754
3.25	25.790	1.41146	0.038774	**3.75**	42.521	1.62860	0.023518
3.26	26.050	1.41580	.038388	3.76	42.948	1.63295	.023284
3.27	26.311	1.42014	.038006	3.77	43.380	1.63729	.023052
3.28	26.576	1.42449	.037628	3.78	43.816	1.64163	.022823
3.29	26.843	1.42883	.037254	3.79	44.256	1.64598	.022596
3.30	27.113	1.43317	0.036883	**3.80**	44.701	1.65032	0.022371
3.31	27.385	1.43751	.036516	3.81	45.150	1.65466	.022148
3.32	27.660	1.44186	.036153	3.82	45.604	1.65900	.021928
3.33	27.938	1.44620	.035793	3.83	46.063	1.66335	.021710
3.34	28.219	1.45054	.035437	3.84	46.525	1.66769	.021494
3.35	28.503	1.45489	0.035084	**3.85**	46.993	1.67203	0.021280
3.36	28.789	1.45923	.034735	3.86	47.465	1.67638	.021068
3.37	29.079	1.46357	.034390	3.87	47.942	1.68072	.020858
3.38	29.371	1.46792	.034047	3.88	48.424	1.68506	.020651
3.39	29.666	1.47226	.033709	3.89	48.911	1.68941	.020445
3.40	29.964	1.47660	0.033373	**3.90**	49.402	1.69375	0.020242
3.41	30.265	1.48094	.033041	3.91	49.899	1.69809	.020041
3.42	30.569	1.48529	.032712	3.92	50.400	1.70243	.019841
3.43	30.877	1.48963	.032387	3.93	50.907	1.70678	.019644
3.44	31.187	1.49397	.032065	3.94	51.419	1.71112	.019448
3.45	31.500	1.49832	0.031746	**3.95**	51.935	1.71546	0.019255
3.46	31.817	1.50266	.031430	3.96	52.457	1.71981	.019063
3.47	32.137	1.50700	.031117	3.97	52.985	1.72415	.018873
3.48	32.460	1.51134	.030807	3.98	53.517	1.72849	.018686
3.49	32.786	1.51569	.030501	3.99	54.055	1.73283	.018500
3.50	33.115	1.52003	0.030197	**4.00**	54.598	1.73718	0.018316

EXPONENTIAL FUNCTIONS (Continued)

x	e^x	$\text{Log}_{10}(e^x)$	e^{-x}
4.00	54.598	1.73718	0.018316
4.01	55.147	1.74152	.018133
4.02	55.701	1.74586	.017953
4.03	56.261	1.75021	.017774
4.04	56.826	1.75455	.017597
4.05	57.397	1.75889	0.017422
4.06	57.974	1.76324	.017249
4.07	58.557	1.76758	.017077
4.08	59.145	1.77192	.016907
4.09	59.740	1.77626	.016739
4.10	60.340	1.78061	0.016573
4.11	60.947	1.78495	.016408
4.12	61.559	1.78929	.016245
4.13	62.178	1.79364	.016083
4.14	62.803	1.79798	.015923
4.15	63.434	1.80232	0.015764
4.16	64.072	1.80667	.015608
4.17	64.715	1.81101	.015452
4.18	65.366	1.81535	.015299
4.19	66.023	1.81969	.015146
4.20	66.686	1.82404	0.014996
4.21	67.357	1.82838	.014846
4.22	68.033	1.83272	.014699
4.23	68.717	1.83707	.014552
4.24	69.408	1.84141	.014408
4.25	70.105	1.84575	0.014264
4.26	70.810	1.85009	.014122
4.27	71.522	1.85444	.013982
4.28	72.240	1.85878	.013843
4.29	72.966	1.86312	.013705
4.30	73.700	1.86747	0.013569
4.31	74.440	1.87181	.013434
4.32	75.189	1.87615	.013300
4.33	75.944	1.88050	.013168
4.34	76.708	1.88484	.013037
4.35	77.478	1.88918	0.012907
4.36	78.257	1.89352	.012778
4.37	79.044	1.89787	.012651
4.38	79.838	1.90221	.012525
4.39	80.640	1.90655	.012401
4.40	81.451	1.91090	0.012277
4.41	82.269	1.91524	.012155
4.42	83.096	1.91958	.012034
4.43	83.931	1.92392	.011914
4.44	84.775	1.92827	.011796
4.45	85.627	1.93261	0.011679
4.46	86.488	1.93695	.011562
4.47	87.357	1.94130	.011447
4.48	88.235	1.94564	.011333
4.49	89.121	1.94998	.011221
4.50	90.017	1.95433	0.011109

x	e^x	$\text{Log}_{10}(e^x)$	e^{-x}
4.50	90.017	1.95433	0.011109
4.51	90.922	1.95867	.010998
4.52	91.836	1.96301	.010889
4.53	92.759	1.96735	.010781
4.54	93.691	1.97170	.010673
4.55	94.632	1.97604	0.010567
4.56	95.583	1.98038	.010462
4.57	96.544	1.98473	.010358
4.58	97.514	1.98907	.010255
4.59	98.494	1.99341	.010153
4.60	99.484	1.99775	0.010052
4.61	100.48	2.00210	.009952
4.62	101.49	2.00644	.009853
4.63	102.51	2.01078	.009755
4.64	103.54	2.01513	.009658
4.65	104.58	2.01947	0.009562
4.66	105.64	2.02381	.009466
4.67	106.70	2.02816	.009372
4.68	107.77	2.03250	.009279
4.69	108.85	2.03684	.009187
4.70	109.95	2.04118	0.009095
4.71	111.05	2.04553	.009005
4.72	112.17	2.04987	.008915
4.73	113.30	2.05421	.008826
4.74	114.43	2.05856	.008739
4.75	115.58	2.06290	0.008652
4.76	116.75	2.06724	.008566
4.77	117.92	2.07158	.008480
4.78	119.10	2.07593	.008396
4.79	120.30	2.08027	.008312
4.80	121.51	2.08461	0.008230
4.81	122.73	2.08896	.008148
4.82	123.97	2.09330	.008067
4.83	125.21	2.09764	.007987
4.84	126.47	2.10199	.007907
4.85	127.74	2.10633	0.007828
4.86	129.02	2.11067	.007750
4.87	130.32	2.11501	.007673
4.88	131.63	2.11936	.007597
4.89	132.95	2.12370	.007521
4.90	134.29	2.12804	0.007447
4.91	135.64	2.13239	.007372
4.92	137.00	2.13673	.007299
4.93	138.38	2.14107	.007227
4.94	139.77	2.14541	.007155
4.95	141.17	2.14976	0.007083
4.96	142.59	2.15410	.007013
4.97	144.03	2.15844	.006943
4.98	145.47	2.16279	.006874
4.99	146.94	2.16713	.006806
5.00	148.41	2.17147	0.006738

EXPONENTIAL FUNCTIONS (Continued)

x	e^x	$\text{Log}_{10}(e^x)$	e^{-x}	x	e^x	$\text{Log}_{10}(e^x)$	e^{-x}
5.00	148.41	2.17147	0.006738	**5.50**	244.69	2.38862	0.0040868
5.01	149.90	2.17582	.006671	5.55	257.24	2.41033	.0038875
5.02	151.41	2.18016	.006605	5.60	270.43	2.43205	.0036979
5.03	152.93	2.18450	.006539	5.65	284.29	2.45376	.0035175
5.04	154.47	2.18884	.006474	5.70	298.87	2.47548	.0033460
5.05	156.02	2.19319	0.006409	**5.75**	314.19	2.49719	0.0031828
5.06	157.59	2.19753	.006346	5.80	330.30	2.51891	.0030276
5.07	159.17	2.20187	.006282	5.85	347.23	2.54062	.0028799
5.08	160.77	2.20622	.006220	5.90	365.04	2.56234	.0027394
5.09	162.39	2.21056	.006158	5.95	383.75	2.58405	.0026058
5.10	164.02	2.21490	0.006097	**6.00**	403.43	2.60577	0.0024788
5.11	165.67	2.21924	.006036	6.05	424.11	2.62748	.0023579
5.12	167.34	2.22359	.005976	6.10	445.86	2.64920	.0022429
5.13	169.02	2.22793	.005917	6.15	468.72	2.67091	.0021335
5.14	170.72	2.23227	.005858	6.20	492.75	2.69263	.0020294
5.15	172.43	2.23662	0.005799	**6.25**	518.01	2.71434	0.0019305
5.16	174.16	2.24096	.005742	6.30	544.57	2.73606	.0018363
5.17	175.91	2.24530	.005685	6.35	572.49	2.75777	.0017467
5.18	177.68	2.24965	.005628	6.40	601.85	2.77948	.0016616
5.19	179.47	2.25399	.005572	6.45	632.70	2.80120	.0015805
5.20	181.27	2.25833	0.005517	**6.50**	665.14	2.82291	0.0015034
5.21	183.09	2.26267	.005462	6.55	699.24	2.84463	.0014301
5.22	184.93	2.26702	.005407	6.60	735.10	2.86634	.0013604
5.23	186.79	2.27136	.005354	6.65	772.78	2.88806	.0012940
5.24	188.67	2.27570	.005300	6.70	812.41	2.90977	.0012309
5.25	190.57	2.28005	0.005248	**6.75**	854.06	2.93149	0.0011709
5.26	192.48	2.28439	.005195	6.80	897.85	2.95320	.0011138
5.27	194.42	2.28873	.005144	6.85	943.88	2.97492	.0010595
5.28	196.37	2.29307	.005092	6.90	992.27	2.99663	.0010078
5.29	198.34	2.29742	.005042	6.95	1043.1	3.01835	.0009586
5.30	200.34	2.30176	0.004992	**7.00**	1096.6	3.04006	0.0009119
5.31	202.35	2.30610	.004942	7.05	1152.9	3.06178	.0008674
5.32	204.38	2.31045	.004893	7.10	1212.0	3.08349	.0008251
5.33	206.44	2.31479	.004844	7.15	1274.1	3.10521	.0007849
5.34	208.51	2.31913	.004796	7.20	1339.4	3.12692	.0007466
5.35	210.61	2.32348	0.004748	**7.25**	1408.1	3.14863	0.0007102
5.36	212.72	2.32782	.004701	7.30	1480.3	3.17035	.0006755
5.37	214.86	2.33216	.004654	7.35	1556.2	3.19206	.0006426
5.38	217.02	2.33650	.004608	7.40	1636.0	3.21378	.0006113
5.39	219.20	2.34085	.004562	7.45	1719.9	3.23549	.0005814
5.40	221.41	2.34519	0.004517	**7.50**	1808.0	3.25721	0.0005531
5.41	223.63	2.34953	.004472	7.55	1900.7	3.27892	.0005261
5.42	225.88	2.35388	.004427	7.60	1998.2	3.30064	.0005005
5.43	228.15	2.35822	.004383	7.65	2100.6	3.32235	.0004760
5.44	230.44	2.36256	.004339	7.70	2208.3	3.34407	.0004528
5.45	232.76	2.36690	0.004296	**7.75**	2321.6	3.36578	0.0004307
5.46	235.10	2.37125	.004254	7.80	2440.6	3.38750	.0004097
5.47	237.46	2.37559	.004211	7.85	2565.7	3.40921	.0003898
5.48	239.85	2.37993	.004169	7.90	2697.3	3.43093	.0003707
5.49	242.26	2.38428	.004128	7.95	2835.6	3.45264	.0003527
5.50	244.69	2.38862	0.004087	**8.00**	2981.0	3.47436	0.0003355

EXPONENTIAL FUNCTIONS (Continued)

x	e^x	$\text{Log}_{10}(e^x)$	e^{-x}
8.00	2981.0	3.47436	0.0003355
8.05	3133.8	3.49607	.0003191
8.10	3294.5	3.51779	.0003035
8.15	3463.4	3.53950	.0002887
8.20	3641.0	3.56121	.0002747
8.25	3827.6	3.58293	0.0002613
8.30	4023.9	3.60464	.0002485
8.35	4230.2	3.62636	.0002364
8.40	4447.1	3.64807	.0002249
8.45	4675.1	3.66979	.0002139
8.50	4914.8	3.69150	0.0002035
8.55	5166.8	3.71322	.0001935
8.60	5431.7	3.73493	.0001841
8.65	5710.1	3.75665	.0001751
8.70	6002.9	3.77836	.0001666
8.75	6310.7	3.80008	0.0001585
8.80	6634.2	3.82179	.0001507
8.85	6974.4	3.84351	.0001434
8.90	7332.0	3.86522	.0001364
8.95	7707.9	3.88694	.0001297
9.00	8103.1	3.90865	0.0001234
9.05	8518.5	3.93037	.0001174
9.10	8955.3	3.95208	.0001117
9.15	9414.4	3.97379	.0001062
9.20	9897.1	3.99551	.0001010
9.25	10405	4.01722	0.0000961
9.30	10938	4.03894	.0000914
9.35	11499	4.06065	.0000870
9.40	12088	4.08237	.0000827
9.45	12708	4.10408	.0000787
9.50	13360	4.12580	0.0000749
9.55	14045	4.14751	.0000712
9.60	14765	4.16923	.0000677
9.65	15522	4.19094	.0000644
9.70	16318	4.21266	.0000613
9.75	17154	4.23437	0.0000583
9.80	18034	4.25609	.0000555
9.85	18958	4.27780	.0000527
9.90	19930	4.29952	.0000502
9.95	20952	4.32123	0.0000477
10.00	22026	4.34294	0.0000454

HYPERBOLIC FUNCTIONS

The logarithms given below show the mantissa only. The proper characteristic must be added.

x	Sinh x		Cosh x		Tanh x		Coth x	
	Value	$\log_{10}$	Value	$\log_{10}$	Value	$\log_{10}$	Value	$\log_{10}$
0.00	0.00000	$-\infty$	1.00000	.00000	0.00000	$-\infty$	∞	∞
0.01	.01000	.00001	1.00005	.00002	.01000	.99999	100.003	.00001
0.02	.02000	.30106	1.00020	.00009	.02000	.30097	50.007	.69903
0.03	.03000	.47719	1.00045	.00020	.02999	.47699	33.343	.52301
0.04	.04001	.60218	1.00080	.00035	.03998	.60183	25.013	.39817
0.05	0.05002	.69915	1.00125	.00054	0.04996	.69861	20.017	.30139
0.06	.06004	.77841	1.00180	.00078	.05993	.77763	16.687	.22237
0.07	.07006	.84545	1.00245	.00106	.06989	.84439	14.309	.15561
0.08	.08009	.90355	1.00320	.00139	.07983	.90216	12.527	.09784
0.09	.09012	.95483	1.00405	.00176	.08976	.95307	11.141	.04693
0.10	0.10017	.00072	1.00500	.00217	0.09967	.99856	10.0333	.00144
0.11	.11022	.04227	1.00606	.00262	.10956	.03965	9.1275	.96035
0.12	.12029	.08022	1.00721	.00312	.11943	.07710	8.3733	.92290
0.13	.13037	.11517	1.00846	.00366	.12927	.11151	7.7356	.88849
0.14	.14046	.14755	1.00982	.00424	.13909	.14330	7.1895	.85670
0.15	0.15056	.17772	1.01127	.00487	0.14889	.17285	6.7166	.82715
0.16	.16068	.20597	1.01283	.00554	.15865	.20044	6.3032	.79956
0.17	.17082	.23254	1.01448	.00625	.16838	.22629	5.9389	.77371
0.18	.18097	.25762	1.01624	.00700	.17808	.25062	5.6154	.74938
0.19	.19115	.28136	1.01810	.00779	.18775	.27357	5.3263	.72643
0.20	0.20134	.30392	1.02007	.00863	0.19738	.29529	5.0665	.70471
0.21	.21155	.32541	1.02213	.00951	.20697	.31590	4.8317	.68410
0.22	.22178	.34592	1.02430	.01043	.21652	.33549	4.6186	.66451
0.23	.23203	.36555	1.02657	.01139	.22603	.35416	4.4242	.64584
0.24	.24231	.38437	1.02894	.01239	.23550	.37198	4.2464	.62802
0.25	0.25261	.40245	1.03141	.01343	0.24492	.38902	4.0830	.61098
0.26	.26294	.41986	1.03399	.01452	.25430	.40534	3.9324	.59466
0.27	.27329	.43663	1.03667	.01564	.26362	.42099	3.7933	.57901
0.28	.28367	.45282	1.03946	.01681	.27291	.43601	3.6643	.56399
0.29	.29408	.46847	1.04235	.01801	.28213	.45046	3.5444	.54954
0.30	0.30452	.48362	1.04534	.01926	0.29131	.46436	3.4327	.53564
0.31	.31499	.49830	1.04844	.02054	.30044	.47775	3.3285	.52225
0.32	.32549	.51254	1.05164	.02187	.30951	.49067	3.2309	.50933
0.33	.33602	.52637	1.05495	.02323	.31852	.50314	3.1395	.49686
0.34	.34659	.53981	1.05836	.02463	.32748	.51518	3.0536	.48482
0.35	0.35719	.55290	1.06188	.02607	0.33638	.52682	2.9729	.47318
0.36	.36783	.56564	1.06550	.02755	.34521	.53809	2.8968	.46191
0.37	.37850	.57807	1.06923	.02907	.35399	.54899	2.8249	.45101
0.38	.38921	.59019	1.07307	.03063	.36271	.55956	2.7570	.44044
0.39	.39996	.60202	1.07702	.03222	.37136	.56980	2.6928	.43020
0.40	0.41075	.61358	1.08107	.03385	0.37995	.57973	2.6319	.42027
0.41	.42158	.62488	1.08523	.03552	.38847	.58936	2.5742	.41064
0.42	.43246	.63594	1.08950	.03723	.39693	.59871	2.5193	.40129
0.43	.44337	.64677	1.09388	.03897	.40532	.60780	2.4672	.39220
0.44	.45434	.65738	1.09837	.04075	.41364	.61663	2.4175	.38337
0.45	0.46534	.66777	1.10297	.04256	0.42190	.62521	2.3702	.37479
0.46	.47640	.67797	1.10768	.04441	.43008	.63355	2.3251	.36645
0.47	.48750	.68797	1.11250	.04630	.43820	.64167	2.2821	.35833
0.48	.49865	.69779	1.11743	.04822	.44624	.64957	2.2409	.35043
0.49	.50984	.70744	1.12247	.05018	.45422	.65726	2.2016	.34274

HYPERBOLIC FUNCTIONS (Continued)

x	Sinh x		Cosh x		Tanh x		Coth x	
	Value	$\log_{10}$	Value	$\log_{10}$	Value	$\log_{10}$	Value	$\log_{10}$
0.50	0.52110	.71692	1.12763	.05217	0.46212	.66475	2.1640	.33525
0.51	.53240	.72624	1.13289	.05419	.46995	.67205	2.1279	.32795
0.52	.54375	.73540	1.13827	.05625	.47770	.67916	2.0934	.32084
0.53	.55516	.74442	1.14377	.05834	.48538	.68608	2.0602	.31392
0.54	.56663	.75330	1.14938	.06046	.49299	.69284	2.0284	.30716
0.55	0.57815	.76204	1.15510	.06262	0.50052	.69942	1.9979	.30058
0.56	.58973	.77065	1.16094	.06481	.50798	.70584	1.9686	.29416
0.57	.60137	.77914	1.16690	.06703	.51536	.71211	1.9404	.28789
0.58	.61307	.78751	1.17297	.06929	.52267	.71822	1.9133	.28178
0.59	.62483	.79576	1.17916	.07157	.52990	.72419	1.8872	.27581
0.60	0.63665	.80390	1.18547	.07389	0.53705	.73001	1.8620	.26999
0.61	.64854	.81194	1.19189	.07624	.54413	.73570	1.8378	.26430
0.62	.66049	.81987	1.19844	.07861	.55113	.74125	1.8145	.25875
0.63	.67251	.82770	1.20510	.08102	.55805	.74667	1.7919	.25333
0.64	.68459	.83543	1.21189	.08346	.56490	.75197	1.7702	.24803
0.65	0.69675	.84308	1.21879	.08593	0.57167	.75715	1.7493	.24285
0.66	.70897	.85063	1.22582	.08843	.57836	.76220	1.7290	.23780
0.67	.72126	.85809	1.23297	.09095	.58498	.76714	1.7095	.23286
0.68	.73363	.86548	1.24025	.09351	.59152	.77197	1.6906	.22803
0.69	.74607	.87278	1.24765	.09609	.59798	.77669	1.6723	.22331
0.70	0.75858	.88000	1.25517	.09870	0.60437	.78130	1.6546	.21870
0.71	.77117	.88715	1.26282	.10134	.61068	.78581	1.6375	.21419
0.72	.78384	.89423	1.27059	.10401	.61691	.79022	1.6210	.20978
0.73	.79659	.90123	1.27849	.10670	.62307	.79453	1.6050	.20547
0.74	.80941	.90817	1.28652	.10942	.62915	.79875	1.5895	.20125
0.75	0.82232	.91504	1.29468	.11216	0.63515	.80288	1.5744	.19712
0.76	.83530	.92185	1.30297	.11493	.64108	.80691	1.5599	.19309
0.77	.84838	.92859	1.31139	.11773	.64693	.81086	1.5458	.18914
0.78	.86153	.93527	1.31994	.12055	.65271	.81472	1.5321	.18528
0.79	.87478	.94190	1.32862	.12340	.65841	.81850	1.5188	.18150
0.80	0.88811	.94846	1.33743	.12627	0.66404	.82219	1.5059	.17781
0.81	.90152	.95498	1.34638	.12917	.66959	.82581	1.4935	.17419
0.82	.91503	.96144	1.35547	.13209	.67507	.82935	1.4813	.17065
0.83	.92863	.96784	1.36468	.13503	.68048	.83281	1.4696	.16719
0.84	.94233	.97420	1.37404	.13800	.68581	.83620	1.4581	.16380
0.85	0.95612	.98051	1.38353	.14099	0.69107	.83952	1.4470	.16048
0.86	.97000	.98677	1.39316	.14400	.69626	.84277	1.4362	.15723
0.87	.98398	.99299	1.40293	.14704	.70137	.84595	1.4258	.15405
0.88	.99806	.99916	1.41284	.15009	.70642	.84906	1.4156	.15094
0.89	1.01224	.00528	1.42289	.15317	.71139	.85211	1.4057	.14789
0.90	1.02652	.01137	1.43309	.15627	0.71630	.85509	1.3961	.14491
0.91	1.04090	.01741	1.44342	.15939	.72113	.85801	1.3867	.14199
0.92	1.05539	.02341	1.45390	.16254	.72590	.86088	1.3776	.13912
0.93	1.06998	.02937	1.46453	.16570	.73059	.86368	1.3687	.13632
0.94	1.08468	.03530	1.47530	.16888	.73522	.86642	1.3601	.13358
0.95	1.09948	.04119	1.48623	.17208	0.73978	.86910	1.3517	.13090
0.96	1.11440	.04704	1.49729	.17531	.74428	.87173	1.3436	.12827
0.97	1.12943	.05286	1.50851	.17855	.74870	.87431	1.3356	.12569
0.98	1.14457	.05864	1.51988	.18181	.75307	.87683	1.3279	.12317
0.99	1.15983	.06439	1.53141	.18509	.75736	.87930	1.3204	.12070

HYPERBOLIC FUNCTIONS (Continued)

x	Sinh x		Cosh x		Tanh x		Coth x	
	Value	$\log_{10}$	Value	$\log_{10}$	Value	$\log_{10}$	Value	$\log_{10}$
1.00	1.17520	.07011	1.54308	.18839	0.76159	.88172	1.3130	.11828
1.01	1.19069	.07580	1.55491	.19171	.76576	.88409	1.3059	.11591
1.02	1.20630	.08146	1.56689	.19504	.76987	.88642	1.2989	.11358
1.03	1.22203	.08708	1.57904	.19839	.77391	.88869	1.2921	.11131
1.04	1.23788	.09268	1.59134	.20176	.77789	.89092	1.2855	.10908
1.05	1.25386	.09825	1.60379	.20515	0.78181	.89310	1.2791	.10690
1.06	1.26996	.10379	1.61641	.20855	.78566	.89524	1.2728	.10476
1.07	1.28619	.10930	1.62919	.21197	.78946	.89733	1.2667	.10267
1.08	1.30254	.11479	1.64214	.21541	.79320	.89938	1.2607	.10062
1.09	1.31903	.12025	1.65525	.21886	.79688	.90139	1.2549	.09861
1.10	1.33565	.12569	1.66852	.22233	0.80050	.90336	1.2492	.09664
1.11	1.35240	.13111	1.68196	.22582	.80406	.90529	1.2437	.09471
1.12	1.36929	.13649	1.69557	.22931	.80757	.90718	1.2383	.09282
1.13	1.38631	.14186	1.70934	.23283	.81102	.90903	1.2330	.09097
1.14	1.40347	.14720	1.72329	.23636	.81441	.91085	1.2279	.08915
1.15	1.42078	.15253	1.73741	.23990	0.81775	.91262	1.2229	.08738
1.16	1.43822	.15783	1.75171	.24346	.82104	.91436	1.2180	.08564
1.17	1.45581	.16311	1.76618	.24703	.82427	.91607	1.2132	.08393
1.18	1.47355	.16836	1.78083	.25062	.82745	.91774	1.2085	.08226
1.19	1.49143	.17360	1.79565	.25422	.83058	.91938	1.2040	.08062
1.20	1.50946	.17882	1.81066	.25784	0.83365	.92099	1.1995	.07901
1.21	1.52764	.18402	1.82584	.26146	.83668	.92256	1.1952	.07744
1.22	1.54598	.18920	1.84121	.26510	.83965	.92410	1.1910	.07590
1.23	1.56447	.19437	1.85676	.26876	.84258	.92561	1.1868	.07439
1.24	1.58311	.19951	1.87250	.27242	.84546	.92709	1.1828	.07291
1.25	1.60192	.20464	1.88842	.27610	0.84828	.92854	1.1789	.07146
1.26	1.62088	.20975	1.90454	.27979	.85106	.92996	1.1750	.07004
1.27	1.64001	.21485	1.92084	.28349	.85380	.93135	1.1712	.06865
1.28	1.65930	.21993	1.93734	.28721	.85648	.93272	1.1676	.06728
1.29	1.67876	.22499	1.95403	.29093	.85913	.93406	1.1640	.06594
1.30	1.69838	.23004	1.97091	.29467	0.86172	.93537	1.1605	.06463
1.31	1.71818	.23507	1.98800	.29842	.86428	.93665	1.1570	.06335
1.32	1.73814	.24009	2.00528	.30217	.86678	.93791	1.1537	.06209
1.33	1.75828	.24509	2.02276	.30594	.86925	.93914	1.1504	.06086
1.34	1.77860	.25008	2.04044	.30972	.87167	.94035	1.1472	.05965
1.35	1.79909	.25505	2.05833	.31352	0.87405	.94154	1.1441	.05846
1.36	1.81977	.26002	2.07643	.31732	.87639	.94270	1.1410	.05730
1.37	1.84062	.26496	2.09473	.32113	.87869	.94384	1.1381	.05616
1.38	1.86166	.26990	2.11324	.32495	.88095	.94495	1.1351	.05505
1.39	1.88289	.27482	2.13196	.32878	.88317	.94604	1.1323	.05396
1.40	1.90430	.27974	2.15090	.33262	0.88535	.94712	1.1295	.05288
1.41	1.92591	.28464	2.17005	.33647	.88749	.94817	1.1268	.05183
1.42	1.94770	.28952	2.18942	.34033	.88960	.94919	1.1241	.05081
1.43	1.96970	.29440	2.20900	.34420	.89167	.95020	1.1215	.04980
1.44	1.99188	.29926	2.22881	.34807	.89370	.95119	1.1189	.04881
1.45	2.01427	.30412	2.24884	.35196	0.89569	.95216	1.1165	.04784
1.46	2.03686	.30896	2.26910	.35585	.89765	.95311	1.1140	.04689
1.47	2.05965	.31379	2.28958	.35976	.89958	.95404	1.1116	.04596
1.48	2.08265	.31862	2.31029	.36367	.90147	.95495	1.1093	.04505
1.49	2.10586	.32343	2.33123	.36759	.90332	.95584	1.1070	.04416

x	Sinh x		Cosh x		Tanh x		Coth x	
	Value	$\log_{10}$	Value	$\log_{10}$	Value	$\log_{10}$	Value	$\log_{10}$
1.50	2.12928	.32823	2.35241	.37151	0.90515	.95672	1.1048	.04328
1.51	2.15291	.33303	2.37382	.37545	.90694	.95758	1.1026	.04242
1.52	2.17676	.33781	2.39547	.37939	.90870	.95842	1.1005	.04158
1.53	2.20082	.34258	2.41736	.38334	.91042	.95924	1.0984	.04076
1.54	2.22510	.34735	2.43949	.38730	.91212	.96005	1.0963	.03995
1.55	2.24961	.35211	2.46186	.39126	0.91379	.96084	1.0943	.03916
1.56	2.27434	.35686	2.48448	.39524	.91542	.96162	1.0924	.03838
1.57	2.29930	.36160	2.50735	.39921	.91703	.96238	1.0905	.03762
1.58	2.32449	.36633	2.53047	.40320	.91860	.96313	1.0886	.03687
1.59	2.34991	.37105	2.55384	.40719	.92015	.96386	1.0868	.03614
1.60	2.37557	.37577	2.57746	.41119	0.92167	.96457	1.0850	.03543
1.61	2.40146	.38048	2.60135	.41520	.92316	.96528	1.0832	.03472
1.62	2.42760	.38518	2.62549	.41921	.92462	.96597	1.0815	.03403
1.63	2.45397	.38987	2.64990	.42323	.92606	.96664	1.0798	.03336
1.64	2.48059	.39456	2.67457	.42725	.92747	.96730	1.0782	.03270
1.65	2.50746	.39923	2.69951	.43129	0.92886	.96795	1.0766	.03205
1.66	2.53459	.40391	2.72472	.43532	.93022	.96858	1.0750	.03142
1.67	2.56196	.40857	2.75021	.43937	.93155	.96921	1.0735	.03079
1.68	2.58959	.41323	2.77596	.44341	.93286	.96982	1.0720	.03018
1.69	2.61748	.41788	2.80200	.44747	.93415	.97042	1.0705	.02958
1.70	2.64563	.42253	2.82832	.45153	.93541	.97100	1.0691	.02900
1.71	2.67405	.42717	2.85491	.45559	.93665	.97158	1.0676	.02842
1.72	2.70273	.43180	2.88180	.45966	.93786	.97214	1.0663	.02786
1.73	2.73168	.43643	2.90897	.46374	.93906	.97269	1.0649	.02731
1.74	2.76091	.44105	2.93643	.46782	.94023	.97323	1.0636	.02677
1.75	2.79041	.44567	2.96419	.47191	0.94138	.97376	1.0623	.02624
1.76	2.82020	.45028	2.99224	.47600	.94250	.97428	1.0610	.02572
1.77	2.85026	.45488	3.02059	.48009	.94361	.97479	1.0598	.02521
1.78	2.88061	.45948	3.04925	.48419	.94470	.97529	1.0585	.02471
1.79	2.91125	.46408	3.07821	.48830	.94576	.97578	1.0574	.02422
1.80	2.94217	.46867	3.10747	.49241	0.94681	.97626	1.0562	.02374
1.81	2.97340	.47325	3.13705	.49652	.94783	.97673	1.0550	.02327
1.82	3.00492	.47783	3.16694	.50064	.94884	.97719	1.0539	.02281
1.83	3.03674	.48241	3.19715	.50476	.94983	.97764	1.0528	.02236
1.84	3.06886	.48698	3.22768	.50889	.95080	.97809	1.0518	.02191
1.85	3.10129	.49154	3.25853	.51302	0.95175	.97852	1.0507	.02148
1.86	3.13403	.49610	3.28970	.51716	.95268	.97895	1.0497	.02105
1.87	3.16709	.50066	3.32121	.52130	.95359	.97936	1.0487	.02064
1.88	3.20046	.50521	3.35305	.52544	.95449	.97977	1.0477	.02023
1.89	3.23415	.50976	3.38522	.52959	.95537	.98017	1.0467	.01983
1.90	3.26816	.51430	3.41773	.53374	0.95624	.98057	1.0458	.01943
1.91	3.30250	.51884	3.45058	.53789	.95709	.98095	1.0448	.01905
1.92	3.33718	.52338	3.48378	.54205	.95792	.98133	1.0439	.01867
1.93	3.37218	.52791	3.51733	.54621	.95873	.98170	1.0430	.01830
1.94	3.40752	.53244	3.55123	.55038	.95953	.98206	1.0422	.01794
1.95	3.44321	.53696	3.58548	.55455	0.96032	.98242	1.0413	.01758
1.96	3.47923	.54148	3.62009	.55872	.96109	.98276	1.0405	.01724
1.97	3.51561	.54600	3.65507	.56290	.96185	.98311	1.0397	.01689
1.98	3.55234	.55051	3.69041	.56707	.96259	.98344	1.0389	.01656
1.99	3.58942	.55502	3.72611	.57126	.96331	.98377	1.0381	.01623

HYPERBOLIC FUNCTIONS (Continued)

x	Sinh x Value	Sinh x $\log_{10}$	Cosh x Value	Cosh x $\log_{10}$	Tanh x Value	Tanh x $\log_{10}$	Coth x Value	Coth x $\log_{10}$
2.00	3.62686	.55953	3.76220	.57544	0.96403	.98409	1.0373	.01591
2.01	3.66466	.56403	3.79865	.57963	.96473	.98440	1.0366	.01560
2.02	3.70283	.56853	3.83549	.58382	.96541	.98471	1.0358	.01529
2.03	3.74138	.57303	3.87271	.58802	.96609	.98502	1.0351	.01498
2.04	3.78029	.57753	3.91032	.59221	.96675	.98531	1.0344	.01469
2.05	3.81958	.58202	3.94832	.59641	0.96740	.98560	1.0337	.01440
2.06	3.85926	.58650	3.98671	.60061	.96803	.98589	1.0330	.01411
2.07	3.89932	.59099	4.02550	.60482	.96865	.98617	1.0324	.01383
2.08	3.93977	.59547	4.06470	.60903	.96926	.98644	1.0317	.01356
2.09	3.98061	.59995	4.10430	.61324	.96986	.98671	1.0311	.01329
2.10	4.02186	.60443	4.14431	.61745	0.97045	.98697	1.0304	.01303
2.11	4.06350	.60890	4.18474	.62167	.97103	.98723	1.0298	.01277
2.12	4.10555	.61337	4.22558	.62589	.97159	.98748	1.0292	.01252
2.13	4.14801	.61784	4.26685	.63011	.97215	.98773	1.0286	.01227
2.14	4.19089	.62231	4.30855	.63433	.97269	.98798	1.0281	.01202
2.15	4.23419	.62677	4.35067	.63856	0.97323	.98821	1.0275	.01179
2.16	4.27791	.63123	4.39323	.64278	.97375	.98845	1.0270	.01155
2.17	4.32205	.63569	4.43623	.64701	.97426	.98868	1.0264	.01132
2.18	4.36663	.64015	4.47967	.65125	.97477	.98890	1.0259	.01110
2.19	4.41165	.64460	4.52356	.65548	.97526	.98912	1.0254	.01088
2.20	4.45711	.64905	4.56791	.65972	0.97574	.98934	1.0249	.01066
2.21	4.50301	.65350	4.61271	.66396	.97622	.98955	1.0244	.01045
2.22	4.54936	.65795	4.65797	.66820	.97668	.98975	1.0239	.01025
2.23	4.59617	.66240	4.70370	.67244	.97714	.98996	1.0234	.01004
2.24	4.64344	.66684	4.74989	.67668	.97759	.99016	1.0229	.00984
2.25	4.69117	.67128	4.79657	.68093	0.97803	.99035	1.0225	.00965
2.26	4.73937	.67572	4.84372	.68518	.97846	.99054	1.0220	.00946
2.27	4.78804	.68016	4.89136	.68943	.97888	.99073	1.0216	.00927
2.28	4.83720	.68459	4.93948	.69368	.97929	.99091	1.0211	.00909
2.29	4.88684	.68903	4.98810	.69794	.97970	.99109	1.0207	.00891
2.30	4.93696	.69346	5.03722	.70219	0.98010	.99127	1.0203	.00873
2.31	4.98758	.69789	5.08684	.70645	.98049	.99144	1.0199	.00856
2.32	5.03870	.70232	5.13697	.71071	.98087	.99161	1.0195	00839
2.33	5.09032	.70675	5.18762	.71497	.98124	.99178	1.0191	.00822
2.34	5.14245	.71117	5.23878	.71923	.98161	.99194	1.0187	.00806
2.35	5.19510	.71559	5.29047	.72349	0.98197	.99210	1.0184	.00790
2.36	5.24827	.72002	5.34269	.72776	.98233	.99226	1.0180	.00774
2.37	5.30196	.72444	5.39544	.73203	.98267	.99241	1.0176	.00759
2.38	5.35618	.72885	5.44873	.73630	.98301	.99256	1.0173	.00744
2.39	5.41093	.73327	5.50256	.74056	.98335	.99271	1.0169	.00729
2.40	5.46623	.73769	5.55695	.74484	0.98367	.99285	1.0166	.00715
2.41	5.52207	.74210	5.61189	.74911	.98400	.99299	1.0163	.00701
2.42	5.57847	.74652	5.66739	.75338	.98431	.99313	1.0159	.00687
2.43	5.63542	.75093	5.72346	.75766	.98462	.99327	1.0156	.00673
2.44	5.69294	.75534	5.78010	.76194	.98492	.99340	1.0153	.00660
2.45	5.75103	.75975	5.83732	.76621	0.98522	.99353	1.0150	.00647
2.46	5.80969	.75415	5.89512	.77049	.98551	.99366	1.0147	.00634
2.47	5.86893	.76856	5.95352	.77477	.98579	.99379	1.0144	.00621
2.48	5.92876	.77296	6.01250	.77906	.98607	.99391	1.0141	.00609
2.49	5.98918	.77737	6.07209	.78334	.98635	.99403	1.0138	.00597

HYPERBOLIC FUNCTIONS (Continued)

x	Sinh x		Cosh x		Tanh x		Coth x	
	Value	$\log_{10}$	Value	$\log_{10}$	Value	$\log_{10}$	Value	$\log_{10}$
2.50	6.05020	.78177	6.13229	.78762	0.98661	.99415	1.0136	.00585
2.51	6.11183	.78617	6.19310	.79191	.98688	.99426	1.0133	.00574
2.52	6.17407	.79057	6.25453	.79619	.98714	.99438	1.0130	.00562
2.53	6.23692	.79497	6.31658	.80048	.98739	.99449	1.0128	.00551
2.54	6.30040	.79937	6.37927	.80477	.98764	.99460	1.0125	.00540
2.55	6.36451	.80377	6.44259	.80906	0.98788	.99470	1.0123	.00530
2.56	6.42926	.80816	6.50656	.81335	.98812	.99481	1.0120	.00519
2.57	6.49464	.81256	6.57118	.81764	.98835	.99491	1.0118	.00509
2.58	6.56068	.81695	6.63646	.82194	.98858	.99501	1.0115	.00499
2.59	6.62738	.82134	6.70240	.82623	.98881	.99511	1.0113	.00489
2.60	6.69473	.82573	6.76901	.83052	0.98903	.99521	1.0111	.00479
2.61	6.76276	.83012	6.83629	.83482	.98924	.99530	1.0109	.00470
2.62	6.83146	.83451	6.90426	.83912	.98946	.99540	1.0107	.00460
2.63	6.90085	.83890	6.97292	.84341	.98966	.99549	1.0104	.00451
2.64	6.97092	.84329	7.04228	.84771	.98987	.99558	1.0102	.00442
2.65	7.04169	.84768	7.11234	.85201	0.99007	.99566	1.0100	.00434
2.66	7.11317	.85206	7.18312	.85631	.99026	.99575	1.0098	.00425
2.67	7.18536	.85645	7.25461	.86061	.99045	.99583	1.0096	.00417
2.68	7.25827	.86083	7.32683	.86492	.99064	.99592	1.0094	.00408
2.69	7.33190	.86522	7.39978	.86922	.99083	.99600	1.0093	.00400
2.70	7.40626	.86960	7.47347	.87352	0.99101	.99608	1.0091	.00392
2.71	7.48137	.87398	7.54791	.87783	.99118	.99615	1.0089	.00385
2.72	7.55722	.87836	7.62310	.88213	.99136	.99623	1.0087	.00377
2.73	7.63383	.88274	7.69905	.88644	.99153	.99631	1.0085	.00369
2.74	7.71121	.88712	7.77578	.89074	.99170	.99638	1.0084	.00362
2.75	7.78935	.89150	7.85328	.89505	0.99186	.99645	1.0082	.00355
2.76	7.86828	.89588	7.93157	.89936	.99202	.99652	1.0080	.00348
2.77	7.94799	.90026	8.01065	.90367	.99218	.99659	1.0079	.00341
2.78	8.02849	.90463	8.09053	.90798	.99233	.99666	1.0077	.00334
2.79	8.10980	.90901	8.17122	.91229	.99248	.99672	1.0076	.00328
2.80	8.19192	.91339	8.25273	.91660	0.99263	.99679	1.0074	.00321
2.81	8.27486	.91776	8.33506	.92091	.99278	.99685	1.0073	.00315
2.82	8.35862	.92213	8.41823	.92522	.99292	.99691	1.0071	.00309
2.83	8.44322	.92651	8.50224	.92953	.99306	.99698	1.0070	.00302
2.84	8.52867	.93088	8.58710	.93385	.99320	.99704	1.0069	.00296
2.85	8.61497	.93525	8.67281	.93816	0.99333	.99709	1.0067	.00291
2.86	8.70213	.93963	8.75940	.94247	.99346	.99715	1.0066	.00285
2.87	8.79016	.94400	8.84686	.94679	.99359	.99721	1.0065	.00279
2.88	8.87907	.94837	8.93520	.95110	.99372	.99726	1.0063	.00274
2.89	8.96887	.95274	9.02444	.95542	.99384	.99732	1.0062	.00268
2.90	9.05956	.95711	9.11458	.95974	0.99396	.99737	1.0061	.00263
2.91	9.15116	.96148	9.20564	.96405	.99408	.99742	1.0060	.00258
2.92	9.24368	.96584	9.29761	.96837	.99420	.99747	1.0058	.00253
2.93	9.33712	.97021	9.39051	.97269	.99431	.99752	1.0057	.00248
2.94	9.43149	.97458	9.48436	.97701	.99443	.99757	1.0056	.00243
2.95	9.52681	.97895	9.57915	.98133	0.99454	.99762	1.0055	.00238
2.96	9.62308	.98331	9.67490	.98565	.99464	.99767	1.0054	.00233
2.97	9.72031	.98768	9.77161	.98997	.99475	.99771	1.0053	.00229
2.98	9.81851	.99205	9.86930	.99429	.99485	.99776	1.0052	.00224
2.99	9.91770	.99641	9.96798	.99861	.99496	.99780	1.0051	.00220

HYPERBOLIC FUNCTIONS (Continued)

x	Sinh x		Cosh x		Tanh x		Coth x	
	Value	$\log_{10}$	Value	$\log_{10}$	Value	$\log_{10}$	Value	$\log_{10}$
3.0	10.0179	.00078	10.0677	.00293	0.99505	.99785	1.0050	.00215
3.1	11.0765	.04440	11.1215	.04616	.99595	.99824	1.0041	.00176
3.2	12.2459	.08799	12.2866	.08943	.99668	.99856	1.0033	.00144
3.3	13.5379	.13155	13.5748	.13273	.99728	.99882	1.0027	.00118
3.4	14.9654	.17509	14.9987	.17605	.99777	.99903	1.0022	.00097
3.5	16.5426	.21860	16.5728	.21940	0.99818	.99921	1.0018	.00079
3.6	18.2855	.26211	18.3128	.26275	.99851	.99935	1.0015	.00065
3.7	20.2113	.30559	20.2360	.30612	.99878	.99947	1.0012	.00053
3.8	22.3394	.34907	22.3618	.34951	.99900	.99957	1.0010	.00043
3.9	24.6911	.39254	24.7113	.39290	.99918	.99964	1.0008	.00036
4.0	27.2899	.43600	27.3082	.43629	0.99933	.99971	1.0007	.00029
4.1	30.1619	.47946	30.1784	.47970	.99945	.99976	1.0005	.00024
4.2	33.3357	.52291	33.3507	.52310	.99955	.99980	1.0004	.00020
4.3	36.8431	.56636	36.8567	.56652	.99963	.99984	1.0004	.00016
4.4	40.7193	.60980	40.7316	.60993	.99970	.99987	1.0003	.00013
4.5	45.0030	.65324	45.0141	.65335	0.99975	.99989	1.0002	.00011
4.6	49.7371	.69668	49.7472	.69677	.99980	.99991	1.0002	.00009
4.7	54.9690	.74012	54.9781	.74019	.99983	.99993	1.0002	.00007
4.8	60.7511	.78355	60.7593	.78361	.99986	.99994	1.0001	.00006
4.9	67.1412	.82699	67.1486	.82704	.99989	.99995	1.0001	.00005
5.0	74.2032	.87042	74.2099	.87046	0.99991	.99996	1.0001	.00004
5.1	82.008	.91386	82.014	.91389	.99993	.99997	1.0001	.00003
5.2	90.633	.95729	90.639	.95731	.99994	.99997	1.0001	.00003
5.3	100.17	.00072	100.17	.00074	.99995	.99998	1.0000	.00002
5.4	110.70	.04415	110.71	.04417	.99996	.99998	1.0000	.00002
5.5	122.34	.08758	122.35	.08760	0.99997	.99999	1.0000	.00001
5.6	135.21	.13101	135.22	.13103	.99997	.99999	1.0000	.00001
5.7	149.43	.17444	149.44	.17445	.99998	.99999	1.0000	.00001
5.8	165.15	.21787	165.15	.21788	.99998	.99999	1.0000	.00001
5.9	182.52	.26130	182.52	.26131	.99998	.99999	1.0000	.00001
6.0	201.71	.30473	201.72	.30474	0.99999	.00000	1.0000	.00000
6.1	222.93	.34817	222.93	.34817	.99999	.00000	1.0000	.00000
6.2	246.37	.39159	246.38	.39161	.99999	.00000	1.0000	.00000
6.3	272.29	.43503	272.29	.43503	.99999	.00000	1.0000	.00000
6.4	300.92	.47845	300.92	.47845	.99999	.00000	1.0000	.00000
6.5	332.57	.52188	332.57	.52188	1.0000	.00000	1.0000	.00000
6.6	367.55	.56532	367.55	.56532	1.0000	.00000	1.0000	.00000
6.7	406.20	.60874	406.20	.60874	1.0000	.00000	1.0000	.00000
6.8	448.92	.65217	448.92	.65217	1.0000	.00000	1.0000	.00000
6.9	496.14,	.69560	496.14	.69560	1.0000	.00000	1.0000	.00000
7.0	548.32	.73903	548.32	.73903	1.0000	.00000	1.0000	.00000
7.1	605.98	.78246	605.98	.78246	1.0000	.00000	1.0000	.00000
7.2	669.72	.82589	669.72	.82589	1.0000	.00000	1.0000	.00000
7.3	740.15	.86932	740.15	.86932	1.0000	.00000	1.0000	.00000
7.4	817.99	.91275	817.99	.91275	1.0000	.00000	1.0000	.00000
7.5	904.02	.95618	904.02	.95618	1.0000	.00000	1.0000	.00000
7.6	999.10	.99961	999.10	.99961	1.0000	.00000	1.0000	.00000
7.7	1104.2	.04305	1104.2	.04305	1.0000	.00000	1.0000	.00000
7.8	1220.3	.08647	1220.3	.08647	1.0000	.00000	1.0000	.00000
7.9	1348.6	.12988	1348.6	.12988	1.0000	.00000	1.0000	.00000

HYPERBOLIC FUNCTIONS (Continued)

x	Sinh x		Cosh x		Tanh x		Coth x	
	Value	$\log_{10}$	Value	$\log_{10}$	Value	$\log_{10}$	Value	$\log_{10}$
8.0	1490.5	.17333	1490.5	.17333	1.0000	.00000	1.0000	.00000
8.1	1647.2	.21675	1647.2	.21675	1.0000	.00000	1.0000	.00000
8.2	1820.5	.26019	1820.5	.26019	1.0000	.00000	1.0000	.00000
8.3	2011.9	.30360	2011.9	.30360	1.0000	.00000	1.0000	.00000
8.4	2223.5	.34704	2223.5	.34704	1.0000	.00000	1.0000	.00000
8.5	2457.4	.39048	2457.4	.39048	1.0000	.00000	1.0000	.00000
8.6	2715.8	.43390	2715.8	.43390	1.0000	.00000	1.0000	.00000
8.7	3001.5	.47734	3001.5	.47734	1.0000	.00000	1.0000	.00000
8.8	3317.1	.52076	3317.1	.52076	1.0000	.00000	1.0000	.00000
8.9	3666.0	.56419	3666.0	.56419	1.0000	.00000	1.0000	.00000
9.0	4051.5	.60762	4051.5	.60762	1.0000	.00000	1.0000	.00000
9.1	4477.6	.65105	4477.6	.65105	1.0000	.00000	1.0000	.00000
9.2	4948.6	.69448	4948.6	.69448	1.0000	.00000	1.0000	.00000
9.3	5469.0	.73791	5469.0	.73791	1.0000	.00000	1.0000	.00000
9.4	6044.2	.78134	6044.2	.78134	1.0000	.00000	1.0000	.00000
9.5	6679.9	.82477	6679.9	.82477	1.0000	.00000	1.0000	.00000
9.6	7382.4	.86820	7382.4	.86820	1.0000	.00000	1.0000	.00000
9.7	8158.8	.91163	8158.8	.91163	1.0000	.00000	1.0000	.00000
9.8	9016.9	.95506	9016.9	.95506	1.0000	.00000	1.0000	.00000
9.9	9965.2	.99849	9965.2	.99849	1.0000	.00000	1.0000	.00000
10.0	11013.2	.04191	11013.2	.04191	1.0000	.00000	1.0000	.00000

MILS—RADIANS—DEGREES

1 mil = 0.00098175 radians = 0.05625° = 3.375′ = 202.5″
1000 mils = 0.98175 radians = 56.25°
6400 mils = 360° = 2π radians
1 radian = 1018.6 mils
1° = 17.777778 mils
1′ = 0.296296 mils
1″ = 0.0049382 mils

DEGREES—RADIANS

1 radian = 57° 17′ 44″ .80625

	log
1 radian = 57.29577 95131 degrees	1.75812 26324
1 radian = 3437.74677 07849 minutes	3.53627 38828
1 radian = 206264.80625 seconds	5.31442 51332
1 degree = 0.01745 32925 19943 radians	8.24187 73676–10
1 minute = 0.00029 08882 08666 radians	6.46372 61172–10
1 second = 0.00000 48481 36811 radians	4.68557 48668–10

DEGREES—RADIANS

The table gives in radians the angle which is expressed in degrees and minutes at the side and top. Angles expressed to the nearest minute and second can readily be converted to radians by adding to the equivalent of the whole number of degrees the equivalents of the minutes and seconds found on the third page of this table.

°	00′	10	20	30	40	50
0	0.00000	0.00291	0.00582	0.00873	0.01164	0.01454
1	0.01745	0.02036	0.02327	0.02618	0.02909	0.03200
2	0.03491	0.03782	0.04072	0.04363	0.04654	0.04945
3	0.05236	0.05527	0.05818	0.06109	0.06400	0.06690
4	0.06981	0.07272	0.07563	0.07854	0.08145	0.08436
5	0.08727	0.09018	0.09308	0.09599	0.09890	0.10181
6	0.10472	0.10763	0.11054	0.11345	0.11636	0.11926
7	0.12217	0.12508	0.12799	0.13090	0.13381	0.13672
8	0.13963	0.14254	0.14544	0.14835	0.15126	0.15417
9	0.15708	0.15999	0.16290	0.16581	0.16872	0.17162
10	0.17453	0.17744	0.18035	0.18326	0.18617	0.18908
11	0.19199	0.19490	0.19780	0.20071	0.20362	0.20653
12	0.20944	0.21235	0.21526	0.21817	0.22108	0.22398
13	0.22689	0.22980	0.23271	0.23562	0.23853	0.24144
14	0.24435	0.24725	0.25016	0.25307	0.25598	0.25889
15	0.26180	0.26471	0.26762	0.27053	0.27343	0.27634
16	0.27925	0.28216	0.28507	0.28798	0.29089	0.29380
17	0.29671	0.29961	0.30252	0.30543	0.30834	0.31125
18	0.31416	0.31707	0.31998	0.32289	0.32579	0.32870
19	0.33161	0.33452	0.33743	0.34034	0.34325	0.34616
20	0.34907	0.35197	0.35488	0.35779	0.36070	0.36361
21	0.36652	0.36943	0.37234	0.37525	0.37815	0.38106
22	0.38397	0.38688	0.38979	0.39270	0.39561	0.39852
23	0.40143	0.40433	0.40724	0.41015	0.41306	0.41597
24	0.41888	0.42179	0.42470	0.42761	0.43051	0.43342
25	0.43633	0.43924	0.44215	0.44506	0.44797	0.45088
26	0.45379	0.45669	0.45960	0.46251	0.46542	0.46833
27	0.47124	0.47415	0.47706	0.47997	0.48287	0.48578
28	0.48869	0.49160	0.49451	0.49742	0.50033	0.50324
29	0.50615	0.50905	0.51196	0.51487	0.51778	0.52069
30	0.52360	0.52651	0.52942	0.53233	0.53523	0.53814
31	0.54105	0.54396	0.54687	0.54978	0.55269	0.55560
32	0.55851	0.56141	0.56432	0.56723	0.57014	0.57305
33	0.57596	0.57887	0.58178	0.58469	0.58759	0.59050
34	0.59341	0.59632	0.59923	0.60214	0.60505	0.60796
35	0.61087	0.61377	0.61668	0.61959	0.62250	0.62541
36	0.62832	0.63123	0.63414	0.63705	0.63995	0.64286
37	0.64577	0.64868	0.65159	0.65450	0.65741	0.66032
38	0.66323	0.66613	0.66904	0.67195	0.67486	0.67777
39	0.68068	0.68359	0.68650	0.68941	0.69231	0.69522
40	0.69813	0.70104	0.70395	0.70686	0.70977	0.71268
41	0.71558	0.71849	0.72140	0.72431	0.72722	0.73013
42	0.73304	0.73595	0.73886	0.74176	0.74467	0.74758
43	0.75049	0.75340	0.75631	0.75922	0.76213	0.76504
44	0.76794	0.77085	0.77376	0.77667	0.77958	0.78249
45	0.78540	0.78831	0.79122	0.79412	0.79703	0.79994
46	0.80285	0.80576	0.80867	0.81158	0.81449	0.81740
47	0.82030	0.82321	0.82612	0.82903	0.83194	0.83485
48	0.83776	0.84067	0.84358	0.84648	0.84939	0.85230
49	0.85521	0.85812	0.86103	0.86394	0.86685	0.86976

DEGREES—RADIANS (Continued)

°	00′	10	20	30	40	50
50	0.87266	0.87557	0.87848	0.88139	0.88430	0.88721
51	0.89012	0.89303	0.89594	0.89884	0.90175	0.90466
52	0.90757	0.91048	0.91339	0.91630	0.91921	0.92212
53	0.92502	0.92793	0.93084	0.93375	0.93666	0.93957
54	0.94248	0.94539	0.94830	0.95120	0.95411	0.95702
55	0.95993	0.96284	0.96575	0.96866	0.97157	0.97448
56	0.97738	0.98029	0.98320	0.98611	0.98902	0.99193
57	0.99484	0.99775	1.00066	1.00356	1.00647	1.00938
58	1.01229	1.01520	1.01811	1.02102	1.02393	1.02684
59	1.02974	1.03265	1.03556	1.03847	1.04138	1.04429
60	1.04720	1.05011	1.05302	1.05592	1.05883	1.06174
61	1.06465	1.06756	1.07047	1.07338	1.07629	1.07920
62	1.08210	1.08501	1.08792	1.09083	1.09374	1.09665
63	1.09956	1.10247	1.10538	1.10828	1.11119	1.11410
64	1.11701	1.11992	1.12283	1.12574	1.12865	1.13156
65	1.13446	1.13737	1.14028	1.14319	1.14610	1.14901
66	1.15192	1.15483	1.15774	1.16064	1.16355	1.16646
67	1.16937	1.17228	1.17519	1.17810	1.18101	1.18392
68	1.18682	1.18973	1.19264	1.19555	1.19846	1.20137
69	1.20428	1.20719	1.21009	1.21300	1.21591	1.21882
70	1.22173	1.22464	1.22755	1.23046	1.23337	1.23627
71	1.23918	1.24209	1.24500	1.24791	1.25082	1.25373
72	1.25664	1.25955	1.26245	1.26536	1.26827	1.27118
73	1.27409	1.27700	1.27991	1.28282	1.28573	1.28863
74	1.29154	1.29445	1.29736	1.30027	1.30318	1.30609
75	1.30900	1.31191	1.31481	1.31772	1.32063	1.32354
76	1.32645	1.32936	1.33227	1.33518	1.33809	1.34099
77	1.34390	1.34681	1.34972	1.35263	1.35554	1.35845
78	1.36136	1.36427	1.36717	1.37008	1.37299	1.37590
79	1.37881	1.38172	1.38463	1.38754	1.39045	1.39335
80	1.39626	1.39917	1.40208	1.40499	1.40790	1.41081
81	1.41372	1.41663	1.41953	1.42244	1.42535	1.42826
82	1.43117	1.43408	1.43699	1.43990	1.44281	1.44571
83	1.44862	1.45153	1.45444	1.45735	1.46026	1.46317
84	1.46608	1.46899	1.47189	1.47480	1.47771	1.48062
85	1.48353	1.48644	1.48935	1.49226	1.49517	1.49807
86	1.50098	1.50389	1.50680	1.50971	1.51262	1.51553
87	1.51844	1.52135	1.52425	1.52716	1.53007	1.53298
88	1.53589	1.53880	1.54171	1.54462	1.54753	1.55043
89	1.55334	1.55625	1.55916	1.56207	1.56498	1.56789
90	1.57080	1.57371	1.57661	1.57952	1.58243	1.58534
91	1.58825	1.59116	1.59407	1.59698	1.59989	1.60279
92	1.60570	1.60861	1.61152	1.61443	1.61734	1.62025
93	1.62316	1.62607	1.62897	1.63188	1.63479	1.63770
94	1.64061	1.64352	1.64643	1.64934	1.65225	1.65515
95	1.65806	1.66097	1.66388	1.66679	1.66970	1.67261
96	1.67552	1.67842	1.68133	1.68424	1.68715	1.69006
97	1.69297	1.69588	1.69879	1.70170	1.70460	1.70751
98	1.71042	1.71333	1.71624	1.71915	1.72206	1.72497
99	1.72788	1.73078	1.73369	1.73660	1.73951	1.74242
100	1.74533	1.74824	1.75115	1.75406	1.75696	1.75987
101	1.76278	1.76569	1.76860	1.77151	1.77442	1.77733
102	1.78024	1.78314	1.78605	1.78896	1.79187	1.79478
103	1.79769	1.80060	1.80351	1.80642	1.80932	1.81223
104	1.81514	1.81805	1.82096	1.82387	1.82678	1.82969
105	1.83260	1.83550	1.83841	1.84132	1.84423	1.84714
106	1.85004	1.85296	1.85587	1.85878	1.86168	1.86459
107	1.86750	1.87041	1.87332	1.87623	1.87914	1.88205
108	1.88496	1.88786	1.89077	1.89368	1.89659	1.89950
109	1.90241	1.90532	1.90823	1.91114	1.91404	1.91695
110	1.91986	1.92277	1.92568	1.92859	1.93150	1.93441

DEGREES—RADIANS (Continued)

Deg.	Radians	Deg.	Radians	Min.	Radians	Sec.	Radians
90	1.57080	**150**	2.61799	**0**	0.00000	**0**	0.00000
91	1.58825	151	2.63545	1	0.00029	1	0.00000
92	1.60570	152	2.65290	2	0.00058	2	0.00001
93	1.62316	153	2.67035	3	0.00087	3	0.00001
94	1.64061	154	2.68781	4	0.00116	4	0.00002
95	1.65806	**155**	2.70526	**5**	0.00145	**5**	0.00002
96	1.67552	156	2.72271	6	0.00175	6	0.00003
97	1.69297	157	2.74017	7	0.00204	7	0.00003
98	1.71042	158	2.75762	8	0.00233	8	0.00004
99	1.72788	159	2.77507	9	0.00262	9	0.00004
100	1.74533	**160**	2.79253	**10**	0.00291	**10**	0.00005
101	1.76278	161	2.80998	11	0.00320	11	0.00005
102	1.78024	162	2.82743	12	0.00349	12	0.00006
103	1.79769	163	2.84489	13	0.00378	13	0.00006
104	1.81514	164	2.86234	14	0.00407	14	0.00007
105	1.83260	**165**	2.87979	**15**	0.00436	**15**	0.00007
106	1.85005	166	2.89725	16	0.00465	16	0.00008
107	1.86750	167	2.91470	17	0.00495	17	0.00008
108	1.88496	168	2.93215	18	0.00524	18	0.00009
109	1.90241	169	2.94961	19	0.00553	19	0.00009
110	1.91986	**170**	2.96706	**20**	0.00582	**20**	0.00010
111	1.93732	171	2.98451	21	0.00611	21	0.00010
112	1.95477	172	3.00197	22	0.00640	22	0.00011
113	1.97222	173	3.01942	23	0.00669	23	0.00011
114	1.98968	174	3.03687	24	0.00698	24	0.00012
115	2.00713	**175**	3.05433	**25**	0.00727	**25**	0.00012
116	2.02458	176	3.07178	26	0.00756	26	0.00013
117	2.04204	177	3.08923	27	0.00785	27	0.00013
118	2.05949	178	3.10669	28	0.00814	28	0.00014
119	2.07694	179	3.12414	29	0.00844	29	0.00014
120	2.09440	**180**	3.14159	**30**	0.00873	**30**	0.00015
121	2.11185	190	3.31613	31	0.00902	31	0.00015
122	2.12930	200	3.49066	32	0.00931	32	0.00016
123	2.14676	210	3.66519	33	0.00960	33	0.00016
124	2.16421	220	3.83972	34	0.00989	34	0.00016
125	2.18166	**230**	4.01426	**35**	0.01018	**35**	0.00017
126	2.19911	240	4.18879	36	0.01047	36	0.00017
127	2.21657	250	4.36332	37	0.01076	37	0.00018
128	2.23402	260	4.53786	38	0.01105	38	0.00018
129	2.25147	270	4.71239	39	0.01134	39	0.00019
130	2.26893	**280**	4.88692	**40**	0.01164	**40**	0.00019
131	2.28638	290	5.06145	41	0.01193	41	0.00020
132	2.30383	300	5.23599	42	0.01222	42	0.00020
133	2.32129	310	5.41052	43	0.01251	43	0.00021
134	2.33874	320	5.58505	44	0.01280	44	0.00021
135	2.35619	**330**	5.75959	**45**	0.01309	**45**	0.00022
136	2.37365	340	5.93412	46	0.01338	46	0.00022
137	2.39110	350	6.10865	47	0.01367	47	0.00023
138	2.40855	360	6.28319	48	0.01396	48	0.00023
139	2.42601	370	6.45772	49	0.01425	49	0.00024
140	2.44346	**380**	6.63225	**50**	0.01454	**50**	0.00024
141	2.46091	390	6.80678	51	0.01484	51	0.00025
142	2.47837	400	6.98132	52	0.01513	52	0.00025
143	2.49582	410	7.15585	53	0.01542	53	0.00026
144	2.51327	420	7.33038	54	0.01571	54	0.00026
145	2.53073	**430**	7.50492	**55**	0.01600	**55**	0.00027
146	2.54818	440	7.67945	56	0.01629	56	0.00027
147	2.56563	450	7.85398	57	0.01658	57	0.00028
148	2.58309	460	8.02851	58	0.01687	58	0.00028
149	2.60054	470	8.20305	59	0.01716	59	0.00029
150	2.61799	**480**	8.37758	**60**	0.01745	**60**	0.00029

DEGREES, MINUTES, AND SECONDS TO RADIANS

Units in degrees, minutes *or* seconds	Degrees to Radians	Minutes to Radians	Seconds to Radians
10	0.174 5329	0.002 9089	0.000 0485
20	0.349 0659	0.005 8178	0.000 0970
30	0.523 5988	0.008 7266	0.000 1454
40	0.698 1317	0.011 6355	0.000 1939
50	0.872 6646	0.014 5444	0.000 2424
60	1.047 1976	0.017 4533	0.000 2909
70	1.221 7305	(0.020 3622)	(0.000 3394)
80	1.396 2634	(0.023 2711)	(0.000 3879)
90	1.570 7963	(0.026 1800)	(0.000 4364)
100	1.745 3293	....	....
200	3.490 6585	....	....
300	5.235 9878	....	....

where n = 1, 2, 3, 4, etc. n(100°) = n(1.745 3293)

*RADIANS TO DEGREES, MINUTES, AND SECONDS

Radians	1.0	0.1	0.01	0.001	0.0001
1	57° 17′ 44.8″	5° 43′ 46.5″	0° 34′ 22.6″	0° 03′ 26.3″	0° 00′ 20.6″
2	114° 35′ 29.6″	11° 27′ 33.0″	1° 08′ 45.3″	0° 06′ 52.5″	0° 00′ 41.3″
3	171° 53′ 14.4″	17° 11′ 19.4″	1° 43′ 07.9′	0° 10′ 18.8″	0° 01′ 01.9″
4	229° 10′ 59.2″	22° 55′ 05.9″	2° 17′ 30.6″	0° 13′ 45.1′	0° 01′ 22.5″
5	286° 28′ 44.0″	28° 38′ 52.4″	2° 51′ 53.2″	0° 17′ 11.3″	0° 01′ 43.1″
6	343° 46′ 28.8″	34° 22′ 38.9″	3° 26′ 15.9″	0° 20′ 37.6″	0° 02′ 03.8″
7	401° 04′ 13.6″	40° 06′ 25.4″	4° 00′ 38.5″	0° 24′ 03.9″	0° 02′ 24.4″
8	458° 21′ 58.4″	45° 50′ 11.8″	4° 35′ 01.2″	0° 27′ 30.1″	0° 02′ 45.0″
9	515° 39′ 43.3″	51° 33′ 58.3″	5° 09′ 23.8″	0° 30′ 56.4″	0° 03′ 05.6″

* If 3.214 is desired in degrees, minutes and seconds it is obtained as follows:

3 = 171° 53′ 14.4″
.2 = 11° 27′ 33.0″
.01 = 0° 34′ 22.6″
.004 = 0° 13′ 45.1″

3.214 = 184° 8′ 55.1″

DEGREES AND DECIMAL FRACTIONS TO RADIANS

The table below facilitates conversion of an angle expressed in degrees and decimal fractions into radians. To convert 25.78 into radians, find the equivalents, successively, of 20°, 5°, 0°.7, 0°.08 and add.

Deg.	Radians	Deg.	Radians	Deg.	Radians	Deg.	Radians	Deg.	Radians
10	0.174533	1	0.017453	0.1	0.001745	0.01	0.000175	0.001	0.000017
20	0.349066	2	.034907	.2	.003491	.02	.000349	.002	.000035
30	0.523599	3	.052360	.3	.005236	.03	.000524	.003	.000052
40	0.698132	4	.069813	.4	.006981	.04	.000698	.004	.000070
50	0.872665	5	.087266	.5	.008727	.05	.000873	.005	.000087
60	1.047198	6	.104720	.6	.010472	.06	.001047	.006	.000105
70	1.221730	7	.122173	.7	.012217	.07	.001222	.007	.000122
80	1.396263	8	.139626	.8	.013963	.08	.001396	.008	.000140
90	1.570796	9	.157080	.9	.015708	.09	.001571	.009	.000157

RADIANS TO DEGREES AND DECIMALS

Radians	Degrees	Radians	Degrees	Radians	Degrees	Radians	Degrees
1	57.2958	0.1	5.7296	0.01	0.5730	0.001	0.0573
2	114.5916	.2	11.4592	.02	1.1459	.002	.1146
3	171.8873	.3	17.1887	.03	1.7189	.003	.1719
4	229.1831	.4	22.9183	.04	2.2918	.004	.2292
5	286.4789	.5	28.6479	.05	2.8648	.005	.2865
6	343.7747	.6	34.3775	.06	3.4377	.006	.3438
7	401.0705	.7	40.1070	.07	4.0107	.007	.4011
8	458.3662	.8	45.8366	.08	4.5837	.008	.4584
9	515.6620	.9	51.5662	.09	5.1566	.009	.5157
10	572.9578	1.0	57.2958	.10	5.7296	.010	.5730

RADIANS—DEGREES

Multiples and Fractions of π Radians in Degrees

Radians	Radians	Deg.	Radians	Radians	Deg.	Radians	Radians	Deg.
π	3.1416	180	$\pi/2$	1.5708	90	$2\pi/3$	2.0944	120
2π	6.2832	360	$\pi/3$	1.0472	60	$3\pi/4$	2.3562	135
3π	9.4248	540	$\pi/4$	0.7854	45	$5\pi/6$	2.6180	150
4π	12.5664	720	$\pi/5$	0.6283	36	$7\pi/6$	3.6652	210
5π	15.7080	900	$\pi/6$	0.5236	30	$5\pi/4$	3.9270	225
6π	18.8496	1080	$\pi/7$	0.4488	25.714	$4\pi/3$	4.1888	240
7π	21.9911	1260	$\pi/8$	0.3927	22.5	$3\pi/2$	4.7124	270
8π	25.1327	1440	$\pi/9$	0.3491	20	$5\pi/3$	5.2360	300
9π	28.2743	1620	$\pi/10$	0.3142	18	$7\pi/4$	5.4978	315
10π	31.4159	1800	$\pi/12$	0.2618	15	$11\pi/6$	5.7596	330

CONVERSION OF ANGLES FROM ARC TO TIME

Arc	Time	Arc	Time	Arc	Time	Arc	Time
°	h m	°	h m	″	s	″	s
′	m s	′	m s				
0	0 00	20	1 20	0	0.00	8	0.53
1	0 04	30	2 00	1	0.07	9	0.60
2	0 08	40	2 40	2	0.13	10	0.67
3	0 12	50	3 20	3	0.20	20	1.33
4	0 16	60	4 00	4	0.27	30	2.00
5	0 20	70	4 40	5	0.33	40	2.67
6	0 24	80	5 20	6	0.40	50	3.33
7	0 28	90	6 00	7	0.47	60	4.00
8	0 32	100	6 40				
9	0 36	200	13 20				
10	0 40	300	20 00				

MINUTES AND SECONDS TO DECIMAL PARTS OF A DEGREE

MINUTES AND SECONDS TO DECIMAL PARTS OF A DEG.

Min.	Degrees	Sec.	Degrees
0	0.00000	**0**	0.00000
1	.01667	1	.00028
2	.03333	2	.00056
3	.05	3	.00083
4	.06667	4	.00111
5	.08333	5	.00139
6	.10	6	.00167
7	.11667	7	.00194
8	.13333	8	.00222
9	.15	9	.0025
10	0.16667	**10**	0.00278
11	.18333	11	.00306
12	.20	12	.00333
13	.21667	13	.00361
14	.23333	14	.00389
15	.25	15	.00417
16	.26667	16	.00444
17	.28333	17	.00472
18	.30	18	.005
19	.31667	19	.00528
20	0.33333	**20**	0.00556
21	.35	21	.00583
22	.36667	22	.00611
23	.38333	23	.00639
24	.40	24	.00667
25	.41667	25	.00694
26	.43333	26	.00722
27	.45	27	.0075
28	.46667	28	.00778
29	.48333	29	.00806
30	0.50	**30**	0.00833
31	.51667	31	.00861
32	.53333	32	.00889
33	.55	33	.00917
34	.56667	34	.00944
35	.58333	35	.00972
36	.60	36	.01
37	.61667	37	.01028
38	.63333	38	.01056
39	.65	39	.01083
40	0.66667	**40**	0.01111
41	.68333	41	.01139
42	.70	42	.01167
43	.71667	43	.01194
44	.73333	44	.01222
45	.75	45	.0125
46	.76667	46	.01278
47	.78333	47	.01306
48	.80	48	.01333
49	.81667	49	.01361
50	0.83333	**50**	0.01389
51	.85	51	.01417
52	.86667	52	.01444
53	.88333	53	.01472
54	.90	54	.015
55	.91667	55	.01528
56	.93333	56	.01556
57	.95	57	.01583
58	.96667	58	.01611
59	.98333	59	.01639
60	1.00	**60**	0.01667

DECIMAL PARTS OF A DEGREE TO MINUTES AND SECONDS

Deg.	′	″	Deg.	′	″
0.00	0	00	**0.60**	36	
.01	0	36	.61	36	36
.02	1	12	.62	37	12
.03	1	48	.63	37	48
.04	2	24	.64	38	24
.05	3		.65	39	
.06	3	36	.66	39	36
.07	4	12	.67	40	12
.08	4	48	.68	40	48
.09	5	24	.69	41	24
0.10	6		**0.70**	42	
.11	6	36	.71	42	36
.12	7	12	.72	43	12
.13	7	48	.73	43	48
.14	8	24	.74	44	24
.15	9		.75	45	
.16	9	36	.76	45	36
.17	10	12	.77	46	12
.18	10	48	.78	46	48
.19	11	24	.79	47	24
0.20	12		**0.80**	48	
.21	12	36	.81	48	36
.22	13	12	.82	49	12
.23	13	48	.83	49	48
.24	14	24	.84	50	24
.25	15		.85	51	
.26	15	36	.86	51	36
.27	16	12	.87	52	12
.28	16	48	.88	52	48
.29	17	24	.89	53	24
0.30	18		**0.90**	54	
.31	18	36	.91	54	36
.32	19	12	.92	55	12
.33	19	48	.93	55	48
.34	20	24	.94	56	24
.35	21		.95	57	
.36	21	36	.96	57	36
.37	22	12	.97	58	12
.38	22	48	.98	58	48
.39	23	24	.99	59	24
0.40	24		**1.00**	60	
.41	24	36			
.42	25	12			
.43	25	48			
.44	26	24			
.45	27				
.46	27	36			
.47	28	12			
.48	28	48			
.49	29	24			
0.50	30				
.51	30	36			
.52	31	12			
.53	31	48			
.54	32	24			
.55	33				
.56	33	36			
.57	34	12			
.58	34	48			
.59	35	24			
0.60	36				

Deg.	Sec.
0.000	0.0
.001	3.6
.002	7.2
.003	10.8
.004	14.4
.005	18.
.006	21.6
.007	25.2
.008	28.8
.009	32.4
0.010	36.

RECIPROCALS, CIRCUMFERENCE AND AREA OF CIRCLES

As a matter of convenience, the values of 1000 × (1/*n*) are given in the table. To obtain the actual value of the reciprocal, shift the decimal point three places to the left.

Circumferences and areas of circles are given for the values of *n* as the diameter.

n	$1000\frac{1}{n}$	Circumference πn	Area $\frac{\pi n^2}{4}$	n	$1000\frac{1}{n}$	Circumference πn	Area $\frac{\pi n^2}{4}$
0	∞	0.000000	.0000000	**50**	20.00000	157.0796	1963.495
1	1000.000	3.141593	.7853982	51	19.60784	160.2212	2042.821
2	500.0000	6.283185	3.141593	52	19.23077	163.3628	2123.717
3	333.3333	9.424778	7.068583	53	18.86792	166.5044	2206.183
4	250.0000	12.56637	12.56637	54	18.51852	169.6460	2290.221
5	200.0000	15.70796	19.63495	55	18.18182	172.7876	2375.829
6	166.6667	18.84956	28.27433	56	17.85714	175.9292	2463.009
7	142.8571	21.99115	38.48451	57	17.54386	179.0708	2551.759
8	125.0000	25.13274	50.26548	58	17.24138	182.2124	2642.079
9	111.1111	28.27433	63.61725	59	16.94915	185.3540	2733.971
10	100.0000	31.41593	78.53982	**60**	16.66667	188.4956	2827.433
11	90.90909	34.55752	95.03318	61	61.39344	191.6372	2922.467
12	83.33333	37.69911	113.0973	62	16.12903	194.7787	3019.071
13	76.92308	40.84070	132.7323	63	15.87302	197.9203	3117.245
14	71.42857	43.98230	153.9380	64	15.62500	201.0619	3216.991
15	66.66667	47.12389	176.7146	65	15.38462	204.2035	3318.307
16	62.50000	50.26548	201.0619	66	15.15152	207.3451	3421.194
17	58.82353	53.40708	226.9801	67	14.92537	210.4867	3525.652
18	55.55556	56.54867	254.4690	68	14.70588	213.6283	3631.681
19	52.63158	59.69026	283.5287	69	14.49275	216.7699	3739.281
20	50.00000	62.83185	314.1593	**70**	14.28571	219.9115	3848.451
21	47.61905	65.97345	346.3606	71	14.08451	223.0531	3959.192
22	45.45455	69.11504	380.1327	72	13.88889	226.1947	4071.504
23	43.47826	72.25663	415.4756	73	13.69863	229.3363	4185.387
24	41.66667	75.39822	452.3893	74	13.51351	232.4779	4300.840
25	40.00000	78.53982	490.8739	75	13.33333	235.6194	4417.865
26	38.46154	81.68141	530.9292	76	13.15789	238.7610	4536.460
27	37.03704	84.82300	572.5553	77	12.98701	241.9026	4656.626
28	35.71429	87.96459	615.7522	78	12.82051	245.0442	4778.362
29	34.48276	91.10619	660.5199	79	12.65823	248.1858	4901.670
30	33.33333	94.24778	706.8583	**80**	12.50000	251.3274	5026.548
31	32.25806	97.38937	754.7676	81	12.34568	254.4690	5152.997
32	31.25000	100.5310	804.2477	82	12.19512	257.6106	5281.017
33	30.30303	103.6726	855.2986	83	12.04819	260.7522	5410.608
34	29.41176	106.8142	907.9203	84	11.90476	263.8938	5541.769
35	28.57143	109.9557	962.1128	85	11.76471	267.0354	5674.502
36	27.77778	113.0973	1017.876	86	11.62791	270.1770	5808.805
37	27.02703	116.2389	1075.210	87	11.49425	273.3186	5944.679
38	26.31579	119.3805	1134.115	88	11.36364	276.4602	6082.123
39	25.64103	122.5221	1194.591	89	11.23596	279.6017	6221.139
40	25.00000	125.6637	1256.637	**90**	11.11111	282.7433	6361.725
41	24.39024	128.8053	1320.254	91	10.98901	285.8849	6503.882
42	23.80952	131.9469	1385.442	92	10.86957	289.0265	6647.610
43	23.25581	135.0885	1452.201	93	10.75269	292.1681	6792.909
44	22.72727	138.2301	1520.531	94	10.63830	295.3097	6939.778
45	22.22222	141.3717	1590.431	95	10.52632	298.4513	7088.218
46	21.73913	144.5133	1661.903	96	10.41667	301.5929	7238.229
47	21.27660	147.6549	1734.945	97	10.30928	304.7345	7389.811
48	20.83333	150.7964	1809.557	98	10.20408	307.8761	7542.964
49	20.40816	153.9380	1885.741	99	10.10101	311.0177	7697.687
50	20.00000	157.0796	1963.495	**100**	10.00000	314.1593	7853.982

n	$1000\frac{1}{n}$	Circumference πn	Area $\frac{\pi n^2}{4}$	n	$1000\frac{1}{n}$	Circumference πn	Area $\frac{\pi n^2}{4}$
100	10.00000	314.1593	7853.982	**150**	6.666 667	471.2389	17671.46
101	9.900 990	317.3009	8011.847	151	6.622 517	474.3805	17907.86
102	9.803 922	320.4425	8171.282	152	6.578 947	477.5221	18145.84
103	9.708 738	323.5840	8332.289	153	6.535 948	480.6637	18385.39
104	9.615 385	326.7256	8494.867	154	6.493 506	483.8053	18626.50
105	9.523 810	329.8672	8659.015	155	6.451 613	486.9469	18869.19
106	9.433 962	333.0088	8824.734	156	6.410 256	490.0885	19113.45
107	9.345 794	336.1504	8992.024	157	6.369 427	493.2300	19359.28
108	9.259 259	339.2920	9160.884	158	6.329 114	496.3716	19606.68
109	9.174 312	342.4336	9331.316	159	6.289 308	499.5132	19855.65
110	9.090 909	345.5752	9503.318	**160**	6.250 000	502.6548	20106.19
111	9.009 009	348.7168	9676.891	161	6.211 180	505.7964	20358.31
112	8.928 571	351.8584	9852.035	162	6.172 840	508.9380	20611.99
113	8.849 558	355.0000	10028.75	163	6.134 969	512.0796	20867.24
114	8.771 930	358.1416	10207.03	164	6.097 561	515.2212	21124.07
115	8.695 652	361.2832	10386.89	165	6.060 606	518.3628	21382.46
116	8.620 690	364.4247	10568.32	166	6.024 096	521.5044	21642.43
117	8.547 009	367.5663	10751.32	167	5.988 024	524.6460	21903.97
118	8.474 576	370.7079	10935.88	168	5.952 381	527.7876	22167.08
119	8.403 361	373.8495	11122.02	169	5.917 160	530.9292	22431.76
120	8.333 333	376.9911	11309.73	**170**	5.882 353	534.0708	22698.01
121	8.264 463	380.1327	11499.01	171	5.847 953	537.2123	22965.83
122	8.196 721	383.2743	11689.87	172	5.813 953	540.3539	23235.22
123	8.130 081	386.4159	11882.29	173	5.780 347	543.4955	23506.18
124	8.064 516	389.5575	12076.28	174	5.747 126	546.6371	23778.71
125	8.000 000	392.6991	12271.85	175	5.714 286	549.7787	24052.82
126	7.936 508	395.8407	12468.98	176	5.681 818	552.9203	24328.49
127	7.874 016	398.9823	12667.69	177	5.649 718	556.0619	24605.74
128	7.812 500	402.1239	12867.96	178	5.617 978	559.2035	24884.56
129	7.751 938	405.2655	13069.81	179	5.586 592	562.3451	25164.94
130	7.692 308	408.4070	13273.23	**180**	5.555 556	565.4867	25446.90
131	7.633 588	411.5486	13478.22	181	5.524 862	568.6283	25730.43
132	7.575 758	414.6902	13684.78	182	5.494 505	571.7699	26015.53
133	7.518 797	417.8318	13892.91	183	5.464 481	574.9115	26302.20
134	7.462 687	420.9734	14102.61	184	5.434 783	578.0530	26590.44
135	7.407 407	424.1150	14313.88	185	5.405 405	581.1946	26880.25
136	7.352 941	427.2566	14526.72	186	5.376 344	584.3362	27171.63
137	7.299 270	430.3982	14741.14	187	5.347 594	587.4778	27464.59
138	7.246 377	433.5398	14957.12	188	5.319 149	590.6194	27759.11
139	7.194 245	436.6814	15174.68	189	5.291 005	593.7610	28055.21
140	7.142 857	439.8230	15393.80	**190**	5.263 158	596.9026	28352.87
141	7.092 199	442.9646	15614.50	191	5.235 602	600.0442	28652.11
142	7.042 254	446.1062	15836.77	192	5.208 333	603.1858	28952.92
143	6.993 007	449.2477	16060.61	193	5.181 347	606.3274	29255.30
144	6.944 444	452.3893	16286.02	194	5.154 639	609.4690	29559.25
145	6.896 552	455.5309	16513.00	195	5.128 205	612.6106	29864.77
146	6.849 315	458.6725	16741.55	196	5.102 041	615.7522	30171.86
147	6.802 721	461.8141	16971.67	197	5.076 142	618.8938	30480.52
148	6.756 757	464.9557	17203.36	198	5.050 505	622.0353	30790.75
149	6.711 409	468.0973	17436.62	199	5.025 126	625.1769	31102.55
150	6.666 667	471.2389	17671.46	**200**	5.000 000	628.3185	31415.93

n	$1000\frac{1}{n}$	Circum-ference πn	Area $\frac{\pi n^2}{4}$	n	$1000\frac{1}{n}$	Circum-ference πn	Area $\frac{\pi n^2}{4}$
200	5.000 000	628.3185	31415.93	**250**	4.000 000	785.3982	49087.39
201	4.975 124	631.4601	31730.87	251	3.984 066	788.5398	49480.87
202	4.950 495	634.6017	32047.39	252	3.968 254	791.6813	49875.92
203	4.926 108	637.7433	32365.47	253	3.952 569	794.8229	50272.55
204	4.901 961	640.8849	32685.13	254	3.937 008	797.9645	50670.75
205	4.878 049	644.0265	33006.36	255	3.921 569	801.1061	51070.52
206	4.854 369	647.1681	33329.16	256	3.906 250	804.2477	51471.85
207	4.830 918	650.3097	33653.53	257	3.891 051	807.3893	51874.76
208	4.807 692	653.4513	33979.47	258	3.875 969	810.5309	52279.24
209	4.784 689	656.5929	34306.98	259	3.861 004	813.6725	52685.29
210	4.761 905	659.7345	34636.06	**260**	3.846 154	816.8141	53092.92
211	4.739 336	662.8760	34966.71	261	3.831 418	819.9557	53502.11
212	4.716 981	666.0176	35298.94	262	3.816 794	823.0973	53912.87
213	4.694 836	669.1592	35632.73	263	3.802 281	826.2389	54325.21
214	4.672 897	672.3008	35968.09	264	3.787 879	829.3805	54739.11
215	4.651 163	675.4424	36305.03	265	3.773 585	832.5221	55154.59
216	4.629 630	678.5840	36643.54	266	3.759 398	835.6636	55571.63
217	4.608 295	681.7256	36983.61	267	3.745 318	838.8052	55990.25
218	4.587 156	684.8672	37325.26	268	3.731 343	841.9468	56410.44
219	4.566 210	688.0088	37668.48	269	3.717 472	845.0884	56832.20
220	4.545 455	691.1504	38013.27	**270**	3.703 704	848.2300	57255.53
221	4.524 887	694.2920	38359.63	271	3.690 037	851.3716	57680.43
222	4.504 505	697.4336	38707.56	272	3.676 471	854.5132	58106.90
223	4.484 305	700.5752	39057.07	273	3.663 004	857.6548	58534.94
224	4.464 286	703.7168	39408.14	274	3.649 635	860.7964	58964.55
225	4.444 444	706.8583	39760.78	275	3.636 364	863.9380	59395.74
226	4.424 779	709.9999	40115.00	276	3.623 188	867.0796	59828.49
227	4.405 286	713.1415	40470.78	277	3.610 108	870.2212	60262.82
228	4.385 965	716.2831	40828.14	278	3.597 122	873.3628	60698.71
229	4.366 812	719.4247	41187.07	279	3.584 229	876.5044	61136.18
230	4.347 826	722.5663	41547.56	**280**	3.571 429	879.6459	61575.22
231	4.329 004	725.7079	41909.63	281	3.558 719	882.7875	62015.82
232	4.310 345	728.8495	42273.27	282	3.546 099	885.9291	62458.00
233	4.291 845	731.9911	42638.48	283	3.533 569	889.0707	62901.75
234	4.273 504	735.1327	43005.26	284	3.521 127	892.2123	63347.07
235	4.255 319	738.2743	43373.61	285	3.508 772	895.3539	63793.97
236	4.237 288	741.4159	43743.54	286	3.496 503	898.4955	64242.43
237	4.219 409	744.5575	44115.03	287	3.484 321	901.6371	64692.46
238	4.201 681	747.6991	44488.09	288	3.472 222	904.7787	65144.07
239	4.184 100	750.8406	44862.73	289	3.460 208	907.9203	65597.24
240	4.166 667	753.9822	45238.93	**290**	3.448 276	911.0619	66051.99
241	4.149 378	757.1238	45616.71	291	3.436 426	914.2035	66508.30
242	4.132 231	760.2654	45996.06	292	3.424 658	917.3451	66966.19
243	4.115 226	763.4070	46376.98	293	3.412 969	920.4866	67425.65
244	4.098 361	766.5486	46759.47	294	3.401 361	923.6282	67886.68
245	4.081 633	769.6902	47143.52	295	3.389 831	926.7698	68349.28
246	4.065 041	772.8318	47529.16	296	3.378 378	929.9114	68813.45
247	4.048 583	775.9734	47916.36	297	3.367 003	933.0530	69279.19
248	4.032 258	779.1150	48305.13	298	3.355 705	936.1946	69746.50
249	4.016 064	782.2566	48695.47	299	3.344 482	939.3362	70215.38
250	4.000 000	785.3982	49087.39	**300**	3.333 333	942.4778	70685.83

RECIPROCALS, CIRCUMFERENCE AND AREA OF CIRCLES (Continued)

n	$1000\frac{1}{n}$	Circum-ference πn	Area $\frac{\pi n^2}{4}$	n	$1000\frac{1}{n}$	Circum-ference πn	Area $\frac{\pi n^2}{4}$
300	3.333 333	942.4778	70685.83	**350**	2.857 143	1099.557	96211.28
301	3.322 259	945.6194	71157.86	351	2.849 003	1102.699	96761.84
302	3.311 258	948.7610	71631.45	352	2.840 909	1105.841	97313.97
303	3.300 330	951.9026	72106.62	353	2.832 861	1108.982	97867.68
304	3.289 474	955.0442	72583.36	354	2.824 859	1112.124	98422.96
305	3.278 689	958.1858	73061.66	355	2.816 901	1115.265	98979.80
306	3.267 974	961.3274	73541.54	356	2.808 989	1118.407	99538.22
307	3.257 329	964.4689	74022.99	357	2.801 120	1121.549	100 098.2
308	3.246 753	967.6105	74506.01	358	2.793 296	1124.690	100 659.8
309	3.236 246	970.7521	74990.60	359	2.785 515	1127.832	101 222.9
310	3.225 806	973.8937	75476.76	**360**	2.777 778	1130.973	101 787.6
311	3.215 434	977.0353	75964.50	361	2.770 083	1134.155	102 353.9
312	3.205 128	980.1769	76453.80	362	2.762 431	1137.257	102 921.7
313	3.194 888	983.3185	76944.67	363	2.754 821	1140.398	103 491.1
314	3.184 713	986.4601	77437.12	364	2.747 253	1143.540	104 062.1
315	3.174 603	989.6017	77931.13	365	2.739 726	1146.681	104 634.7
316	3.164 557	992.7433	78426.72	366	2.732 240	1149.823	105 208.8
317	3.154 574	995.8849	78923.88	367	2.724 796	1152.965	105 784.5
318	3.144 654	999.0265	79422.60	368	2.717 391	1156.106	106 361.8
319	3.134 796	1002.168	79922.90	369	2.710 027	1159.248	106 940.6
320	3.125 000	1005.310	80424.77	**370**	2.702 703	1162.389	107 521.0
321	3.115 265	1008.451	80928.21	371	2.695 418	1165.531	108 103.0
322	3.105 590	1011.593	81433.22	372	2.688 172	1168.672	108 686.5
323	3.095 975	1014.734	81939.80	373	2.680 965	1171.814	109 271.7
324	3.086 420	1017.876	82447.96	374	2.673 797	1174.956	109 858.4
325	3.076 923	1021.018	82957.68	375	2.666 667	1178.097	110 446.6
326	3.067 485	1024.159	83468.98	376	2.659 574	1181.239	111 036.5
327	3.058 104	1027.301	83981.84	377	2.652 520	1184.380	111 627.9
328	3.048 780	1030.442	84496.28	378	2.645 503	1187.522	112 220.8
329	3.039 514	1033.584	85012.28	379	2.638 522	1190.664	112 815.4
330	3.030 303	1036.726	85529.86	**380**	2.631 579	1193.805	113 411.5
331	3.021 148	1039.867	86049.01	381	2.624 672	1196.947	114 009.2
332	3.012 048	1043.009	86569.73	382	2.617 801	1200.088	114 608.4
333	3.003 003	1046.150	87092.02	383	2.610 966	1203.230	115 209.3
334	2.994 012	1049.292	87615.88	384	2.604 167	1206.372	115 811.7
335	2.985 075	1052.434	88141.31	385	2.597 403	1209.513	116 415.6
336	2.976 190	1055.575	88668.31	386	2.590 674	1212.655	117 021.2
337	2.967 359	1058.717	89196.88	387	2.583 979	1215.796	117 628.3
338	2.958 580	1061.858	89727.03	388	2.577 320	1218.938	118 237.0
339	2.949 853	1065.000	90258.74	389	2.570 694	1222.080	118 847.2
340	2.941 176	1068.142	90792.03	**390**	2.564 103	1225.221	119 459.1
341	2.932 551	1071.283	91326.88	391	2.557 545	1228.363	120 072.5
342	2.923 977	1074.425	91863.31	392	2.551 020	1231.504	120 687.4
343	2.915 452	1077.566	92401.31	393	2.544 529	1234.646	121 304.0
344	2.906 977	1080.708	92940.88	394	2.538 071	1237.788	121 922.1
345	2.898 551	1083.849	93482.02	395	2.531 646	1240.929	122 541.7
346	2.890 173	1086.991	94024.73	396	2.525 253	1244.071	123 163.0
347	2.881 844	1090.133	94569.01	397	2.518 892	1247.212	123 785.8
348	2.873 563	1093.274	95114.86	398	2.512 563	1250.354	124 410.2
349	2.865 330	1096.416	95662.28	399	2.506 266	1253.495	125 036.2
350	2.857 143	1099.557	96211.28	**400**	2.500 000	1256.637	125 663.7

RECIPROCALS, CIRCUMFERENCE AND AREA OF CIRCLES (Continued)

n	$1000\frac{1}{n}$	Circum-ference πn	Area $\frac{\pi n^2}{4}$	n	$1000\frac{1}{n}$	Circum-ference πn	Area $\frac{\pi n^2}{4}$
400	2.500 000	1256.637	125 663.7	**450**	2.222 222	1413.717	159 043.1
401	2.493 766	1259.779	126 292.8	451	2.217 295	1416.858	159 750.8
402	2.487 562	1262.920	126 923.5	452	2.212 389	1420.000	160 460.0
403	2.481 390	1266.062	127 555.7	453	2.207 506	1423.141	161 170.8
404	2.475 248	1269.203	128 189.5	454	2.202 643	1426.283	161 883.1
405	2.469 136	1272.345	128 824.9	455	2.197 802	1429.425	162 597.1
406	2.463 054	1275.487	129 461.9	456	2.192 982	1432.566	163 312.6
407	2.457 002	1278.628	130 100.4	457	2.188 184	1435.708	164 029.6
408	2.450 980	1281.770	130 740.5	458	2.183 406	1438.849	164 748.3
409	2.444 988	1284.911	131 382.2	459	2.178 649	1441.991	165 468.5
410	2.439 024	1288.053	132 025.4	**460**	2.173 913	1445.133	166 190.3
411	2.433 090	1291.195	132 670.2	461	2.169 197	1448.274	166 913.6
412	2.427 184	1294.336	133 316.6	462	2.164 502	1451.416	167 638.5
413	2.421 308	1297.478	133 964.6	463	2.159 827	1454.557	168 365.0
414	2.415 459	1300.619	134 614.1	464	2.155 172	1457.699	169 093.1
415	2.409 639	1303.761	135 265.2	465	2.150 538	1460.841	169 822.7
416	2.403 846	1306.903	135 917.9	466	2.145 923	1463.982	170 553.9
417	2.398 082	1310.044	136 572.1	467	2.141 328	1467.124	171 286.7
418	2.392 344	1313.186	137 227.9	468	2.136 752	1470.265	172 021.0
419	2.386 635	1316.327	137 885.3	469	2.132 196	1473.407	172 757.0
420	2.380 952	1319.469	138 544.2	**470**	2.127 660	1476.549	173 494.5
421	2.375 297	1322.611	139 204.8	471	2.123 142	1479.690	174 233.5
422	2.369 668	1325.752	139 866.8	472	2.118 644	1482.832	174 974.1
423	2.364 066	1328.894	140 530.5	473	2.114 165	1485.973	175 716.3
424	2.358 491	1332.035	141 195.7	474	2.109 705	1489.115	176 460.1
425	2.352 941	1335.177	141 862.5	475	2.105 263	1492.257	177 205.5
426	2.347 418	1338.318	142 530.9	476	2.100 840	1495.398	177 952.4
427	2.341 920	1341.460	143 200.9	477	2.096 436	1498.540	178 700.9
428	2.336 449	1344.602	143 872.4	478	2.092 050	1501.681	179 450.9
429	2.331 002	1347.743	144 545.5	479	2.087 683	1504.823	180 202.5
430	2.325 581	1350.885	145 220.1	**480**	2.083 333	1507.964	180 955.7
431	2.320 186	1354.026	145 896.3	481	2.079 002	1511.106	181 710.5
432	2.314 815	1357.168	146 574.1	482	2.074 689	1514.248	182 466.8
433	2.309 469	1360.310	147 253.5	483	2.070 393	1517.389	183 224.8
434	2.304 147	1363.451	147 934.5	484	2.066 116	1520.531	183 984.2
435	2.298 851	1366.593	148 617.0	485	2.061 856	1523.672	184 745.3
436	2.293 578	1369.734	149 301.0	486	2.057 613	1526.814	185 507.9
437	2.288 330	1372.876	149 986.7	487	2.053 388	1529.956	186 272.1
438	2.283 105	1376.018	150 673.9	488	2.049 180	1533.097	187 037.9
439	2.277 904	1379.159	151 362.7	489	2.044 990	1536.239	187 805.2
440	2.272 727	1382.301	152 053.1	**490**	2.040 816	1539.380	188 574.1
441	2.267 574	1385.442	152 745.0	491	2.036 660	1542.522	189 344.6
442	2.262 443	1388.584	153 438.5	492	2.032 520	1545.664	190 116.6
443	2.257 336	1391.726	154 133.6	493	2.028 398	1548.805	190 890.2
444	2.252 252	1394.867	154 830.3	494	2.024 291	1551.947	191 665.4
445	2.247 191	1398.009	155 528.5	495	2.020 202	1555.088	192 442.2
446	2.242 152	1401.150	156 228.3	496	2.016 129	1558.230	193 220.5
447	2.237 136	1404.292	156 929.6	497	2.012 072	1561.372	194 000.4
448	2.232 143	1407.434	157 632.6	498	2.008 032	1564.513	194 781.9
449	2.227 171	1410.575	158 337.1	499	2.004 008	1567.655	195 564.9
450	2.222 222	1413.717	159 043.1	**500**	2.000 000	1570.796	196 349.5

n	$1000\frac{1}{n}$	Circum-ference πn	Area $\frac{\pi n^2}{4}$	n	$1000\frac{1}{n}$	Circum-ference πn	Area $\frac{\pi n^2}{4}$
500	2.000 000	1570.796	196 349.5	**550**	1.818 182	1727.876	237 582.9
501	1.996 008	1573.938	197 135.7	551	1.814 882	1731.018	238 447.7
502	1.992 032	1577.080	197 923.5	552	1.811 594	1734.159	239 314.0
503	1.988 072	1580.221	198 712.8	553	1.808 318	1737.301	240 181.8
504	1.984 127	1583.363	199 503.7	554	1.805 054	1740.442	241 051.3
505	1.980 198	1586.504	200 296.2	555	1.801 802	1743.584	241 922.3
506	1.976 285	1589.646	201 090.2	556	1.798 561	1746.726	242 794.8
507	1.972 387	1592.787	201 885.8	557	1.795 332	1749.867	243 669.0
508	1.968 504	1595.929	202 683.0	558	1.792 115	1753.009	244 544.7
509	1.964 637	1599.071	203 481.7	559	1.788 909	1756.150	245 422.0
510	1.960 784	1602.212	204 282.1	**560**	1.785 714	1759.292	246 300.9
511	1.956 947	1605.354	205 084.0	561	1.782 531	1762.433	247 181.3
512	1.953 125	1608.495	205 887.4	562	1.779 359	1765.575	248 063.3
513	1.949 318	1611.637	206 692.4	563	1.776 199	1768.717	248 946.9
514	1.945 525	1614.779	207 499.1	564	1.773 050	1771.858	249 832.0
515	1.941 748	1617.920	208 307.2	565	1.769 912	1775.000	250 718.7
516	1.937 984	1621.062	209 117.0	566	1.766 784	1778.141	251 607.0
517	1.934 236	1624.203	209 928.3	567	1.763 668	1781.283	252 496.9
518	1.930 502	1627.345	210 741.2	568	1.760 563	1784.425	253 388.3
519	1.926 782	1630.487	211 555.6	569	1.757 469	1787.566	254 281.3
520	1.923 077	1633.628	212 371.7	**570**	1.754 386	1790.708	255 175.9
521	1.919 386	1636.770	213 189.3	571	1.751 313	1793.849	256 072.0
522	1.915 709	1639.911	214 008.4	572	1.748 252	1796.991	256 969.7
523	1.912 046	1643.053	214 829.2	573	1.745 201	1800.133	257 869.0
524	1.908 397	1646.195	215 651.5	574	1.742 160	1803.274	258 769.8
525	1.904 762	1649.336	216 475.4	575	1.739 130	1806.416	259 672.3
526	1.901 141	1652.478	217 300.8	576	1.736 111	1809.557	260 576.3
527	1.897 533	1655.619	218 127.8	577	1.733 102	1812.699	261 481.8
528	1.893 939	1658.761	218 956.4	578	1.730 104	1815.841	262 389.0
529	1.890 359	1661.903	219 786.6	579	1.727 116	1818.982	263 297.7
530	1.886 792	1665.044	220 618.3	**580**	1.724 138	1822.124	264 207.9
531	1.883 239	1668.186	221 451.7	581	1.721 170	1825.265	265 119.8
532	1.879 699	1671.327	222 286.5	582	1.718 213	1828.407	266 033.2
533	1.876 173	1674.469	223 123.0	583	1.715 266	1831.549	266 948.2
534	1.872 659	1677.610	223 961.0	584	1.712 329	1834.690	267 864.8
535	1.869 159	1680.752	224 800.6	585	1.709 402	1837.832	268 782.9
536	1.865 672	1683.894	225 641.8	586	1.706 485	1840.973	269 702.6
537	1.862 197	1687.035	226 484.5	587	1.703 578	1844.115	270 623.9
538	1.858 736	1690.177	227 328.8	588	1.700 680	1847.256	271 546.7
539	1.855 288	1693.318	228 174.7	589	1.697 793	1850.398	272 471.1
540	1.851 852	1696.460	229 022.1	**590**	1.694 915	1853.540	273 397.1
541	1.848 429	1699.602	229 871.1	591	1.692 047	1856.681	274 324.7
542	1.845 018	1702.743	230 721.7	592	1.689 189	1859.823	275 253.8
543	1.841 621	1705.885	231 573.9	593	1.686 341	1862.964	276 184.5
544	1.838 235	1709.026	232 427.6	594	1.683 502	1866.106	277 116.7
545	1.834 862	1712.168	233 282.9	595	1.680 672	1869.248	278 050.6
546	1.831 502	1715.310	234 139.8	596	1.677 852	1872.389	278 986.0
547	1.828 154	1718.451	234 998.2	597	1.675 042	1875.531	279 923.0
548	1.824 818	1721.593	235 858.2	598	1.672 241	1878.672	280 861.5
549	1.821 494	1724.734	236 719.8	599	1.669 449	1881.814	281 801.6
550	1.818 182	1727.876	237 582.9	**600**	1.666 667	1884.956	282 743.3

RECIPROCALS, CIRCUMFERENCE AND AREA OF CIRCLES (Continued)

n	$1000\frac{1}{n}$	Circum-ference πn	Area $\frac{\pi n^2}{4}$	n	$1000\frac{1}{n}$	Circum-ference πn	Area $\frac{\pi n^2}{4}$
600	1.666 667	1884.956	282 743.3	**650**	1.538 462	2042.035	331 830.7
601	1.663 894	1888.097	283 686.6	651	1.536 098	2045.177	332 852.5
602	1.661 130	1891.239	284 631.4	652	1.533 742	2048.318	333 875.9
603	1.658 375	1894.380	285 577.8	653	1.531 394	2051.460	334 900.8
604	1.655 629	1897.522	286 525.8	654	1.529 052	2054.602	335 927.4
605	1.652 893	1900.664	287 475.4	655	1.526 718	2057.743	336 955.4
606	1.650 165	1903.805	288 426.5	656	1.524 390	2060.885	337 985.1
607	1.647 446	1906.947	289 379.2	657	1.522 070	2064.026	339 016.3
608	1.644 737	1910.088	290 333.4	658	1.519 757	2067.168	340 049.1
609	1.642 036	1913.230	291 289.3	659	1.517 451	2070.310	341 083.5
610	1.639 344	1916.372	292 246.7	**660**	1.515 152	2073.451	342 119.4
611	1.636 661	1919.513	293 205.6	661	1.512 859	2076.593	343 157.0
612	1.633 987	1922.655	294 166.2	662	1.510 574	2079.734	344 196.0
613	1.631 321	1925.796	295 128.3	663	1.508 296	2082.876	345 236.7
614	1.628 664	1928.938	296 092.0	664	1.506 024	2086.018	346 278.9
615	1.626 016	1932.079	297 057.2	665	1.503 759	2089.159	347 322.7
616	1.623 377	1935.221	298 024.0	666	1.501 502	2092.301	348 368.1
617	1.620 746	1938.363	298 992.4	667	1.499 250	2095.442	349 415.0
618	1.618 123	1941.504	299 962.4	668	1.497 006	2098.584	350 463.5
619	1.615 509	1944.646	300 933.9	669	1.494 768	2101.725	351 513.6
620	1.612 903	1947.787	301 907.1	**670**	1.492 537	2104.867	352 565.2
621	1.610 306	1950.929	302 881.7	671	1.490 313	2108.009	353 618.5
622	1.607 717	1954.071	303 858.0	672	1.488 095	2111.150	354 673.2
623	1.605 136	1957.212	304 835.8	673	1.485 884	2114.292	355 729.6
624	1.602 564	1960.354	305 815.2	674	1.483 680	2117.433	356 787.5
625	1.600 000	1963.495	306 796.2	675	1.481 481	2120.575	357 847.0
626	1.597 444	1966.637	307 778.7	676	1.479 290	2123.717	358 908.1
627	1.594 896	1969.779	308 762.8	677	1.477 105	2126.858	359 970.8
628	1.592 357	1972.920	309 748.5	678	1.474 926	2130.000	361 035.0
629	1.589 825	1976.062	310 735.7	679	1.472 754	2133.141	362 100.8
630	1.587 302	1979.203	311 724.5	**680**	1.470 588	2136.283	363 168.1
631	1.584 786	1982.345	312 714.9	681	1.468 429	2139.425	364 237.0
632	1.582 278	1985.487	313 706.9	682	1.466 276	2142.566	365 307.5
633	1.579 779	1988.628	314 700.4	683	1.464 129	2145.708	366 379.6
634	1.577 287	1991.770	315 695.5	684	1.461 988	2148.849	367 453.2
635	1.574 803	1994.911	316 692.2	685	1.459 854	2151.991	368 528.5
636	1.572 327	1998.053	317 690.4	686	1.457 726	2155.133	369 605.2
637	1.569 859	2001.195	318 690.2	687	1.455 604	2158.274	370 683.6
638	1.567 398	2004.336	319 691.6	688	1.453 488	2161.416	371 763.5
639	1.564 945	2007.478	320 694.6	689	1.451 379	2164.557	372 845.0
640	1.562 500	2010.619	321 699.1	**690**	1.449 275	2167.699	373 928.1
641	1.560 062	2013.761	322 705.2	691	1.447 178	2170.841	375 012.7
642	1.557 632	2016.902	323 712.8	692	1.445 087	2173.982	376 098.9
643	1.555 210	2020.044	324 722.1	693	1.443 001	2177.124	377 186.7
644	1.552 795	2023.186	325 732.9	694	1.440 922	2180.265	378 276.0
645	1.550 388	2026.327	326 745.3	695	1.438 849	2183.407	379 366.9
646	1.547 988	2029.469	327 759.2	696	1.436 782	2186.548	380 459.4
647	1.545 595	2032.610	328 774.7	697	1.434 720	2189.690	381 553.5
648	1.543 210	2035.752	329 791.8	698	1.432 665	2192.832	382 649.1
649	1.540 832	2038.894	330 810.5	699	1.430 615	2195.973	383 746.3
650	1.538 462	2042.035	331 830.7	**700**	1.428 571	2199.115	384 845.1

RECIPROCALS, CIRCUMFERENCE AND AREA OF CIRCLES (Continued)

n	$1000\frac{1}{n}$	Circum-ference πn	Area $\frac{\pi n^2}{4}$	n	$1000\frac{1}{n}$	Circum-ference πn	Area $\frac{\pi n^2}{4}$
700	1.428 571	2199.115	384 845.1	**750**	1.333 333	2356.194	441 786.5
701	1.426 534	2202.256	385 945.4	751	1.331 558	2359.336	442 965.3
702	1.424 501	2205.398	387 047.4	752	1.329 787	2362.478	444 145.8
703	1.422 475	2208 540	388 150.8	753	1.328 021	2365.619	445 327.8
704	1.420 455	2211.681	389 255.9	754	1.326 260	2368.761	446 511.4
705	1.418 440	2214.823	390 362.5	755	1.324 503	2371.902	447 696.6
706	1.416 431	2217.964	391 470.7	756	1.322 751	2375.044	448 883.3
707	1.414 427	2221.106	392 580.5	757	1.321 004	2378.186	450 071.6
708	1.412 429	2224.248	393 691.8	758	1.319 261	2381.327	451 261.5
709	1.410 437	2227.389	394 804.7	759	1.317 523	2384.469	452 453.0
710	1.408 451	2230.531	395 919.2	**760**	1.315 789	2387.610	453 646.0
711	1.406 470	2233.672	397 035.3	761	1.314 060	2390.752	454 840.6
712	1.404 494	2236.814	398 152.9	762	1.312 336	2393.894	456 036.7
713	1.402 525	2239.956	399 272.1	763	1.310 616	2397.035	457 234.5
714	1.400 560	2243.097	400 392.8	764	1.308 901	2400.177	458 433.8
715	1.398 601	2246.239	401 515.2	765	1.307 190	2403.318	459 634.6
716	1.396 648	2249.380	402 639.1	766	1.305 483	2406.460	460 837.1
717	1.394 700	2252.522	403 764.6	767	1.303 781	2409.602	462 041.1
718	1.392 758	2255.664	404 891.6	768	1.302 083	2412.743	463 246.7
719	1.390 821	2258.805	406 020.2	769	1.300 390	2415.885	464 453.8
720	1.388 889	2261.947	407 150.4	**770**	1.298 701	2419.026	465 662.6
721	1.386 963	2265.088	408 282.2	771	1.297 017	2422.168	466 872.9
722	1.385 042	2268.230	409 415.5	772	1.295 337	2425.310	468 084.7
723	1.383 126	2271.371	410 550.4	773	1.293 661	2428.451	469 298.2
724	1.381 215	2274.513	411 686.9	774	1.291 990	2431.593	470 513.2
725	1.379 310	2277.655	412 824.9	775	1.290 323	2434.734	471 729.8
726	1.377 410	2280.796	413 964.5	776	1.288 660	2437.876	472 947.9
727	1.375 516	2283.938	415 105.7	777	1.287 001	2441.017	474 167.6
728	1.373 626	2287.079	416 248.5	778	1.285 347	2444.159	475 388.9
729	1.371 742	2290.221	417 392.8	779	1.283 697	2447.301	476 611.8
730	1.369 863	2293.363	418 538.7	**780**	1.282 051	2450.442	477 836.2
731	1.367 989	2296.504	419 686.1	781	1.280 410	2453.584	479 062.2
732	1.366 120	2299.646	420 835.2	782	1.278 772	2456.725	480 289.8
733	1.364 256	2302.787	421 985.8	783	1.277 139	2459.867	481 519.0
734	1.362 398	2305.929	423 138.0	784	1.275 510	2463.009	482 749.7
735	1.360 544	2309.071	424 291.7	785	1.273 885	2466.150	483 982.0
736	1.358 696	2312.212	425 447.0	786	1.272 265	2469.292	485 215.8
737	1.356 852	2315.354	426 603.9	787	1.270 648	2472.433	486 451.3
738	1.355 014	2318.495	427 762.4	788	1.269 036	2475.575	487 688.3
739	1.353 180	2321.637	428 922.4	789	1.267 427	2478.717	488 926.9
740	1.351 351	2324.779	430 084.0	**790**	1.265 823	2481.858	490 167.0
741	1.349 528	2327.920	431 247.2	791	1.264 223	2485.000	491 408.7
742	1.347 709	2331.062	432 412.0	792	1.262 626	2488.141	492 652.0
743	1.345 895	2334.203	433 578.3	793	1.261 034	2491.283	493 896.8
744	1.344 086	2337.345	434 746.2	794	1.259 446	2494.425	495 143.3
745	1.342 282	2340.487	435 915.6	795	1.257 862	2497.566	496 391.3
746	1.340 483	2343.628	437 086.6	796	1.256 281	2500.708	497 640.8
747	1.338 688	2346.770	438 259.2	797	1.254 705	2503.849	498 892.0
748	1.336 898	2349.911	439 433.4	798	1.253 133	2506.991	500 144.7
749	1.335 113	2353.053	440 609.2	799	1.251 564	2510.133	501 399.0
750	1.333 333	2356.194	441 786.5	**800**	1.250 000	2513.274	502 654.8

n	$1000\frac{1}{n}$	Circumference πn	Area $\frac{\pi n^2}{4}$	n	$1000\frac{1}{n}$	Circumference πn	Area $\frac{\pi n^2}{4}$
800	1.250 000	2513.274	502 654.8	**850**	1.176 471	2670.354	567 450.2
801	1.248 439	2516.416	503 912.2	851	1.175 088	2673.495	568 786.1
802	1.246 883	2519.557	505 171.2	852	1.173 709	2676.637	570 123.7
803	1.245 330	2522.699	506 431.8	853	1.172 333	2679.779	571 462.8
804	1.243 781	2525.840	507 693.9	854	1.170 960	2682.920	572 803.4
805	1.242 236	2528.982	508 957.6	855	1.169 591	2686.062	574 145.7
806	1.240 695	2532.124	510 222.9	856	1.168 224	2689.203	575 489.5
807	1.239 157	2535.265	511 489.8	857	1.166 861	2692.345	576 834.9
808	1.237 624	2538.407	512 758.2	858	1.165 501	2695.486	578 181.9
809	1.236 094	2541.548	514 028.2	859	1.164 144	2698.628	579 530.4
810	1.234 568	2544.690	515 299.7	**860**	1.162 791	2701.770	580 880.5
811	1.233 046	2547.832	516 572.9	861	1.161 440	2704.911	582 232.2
812	1.231 527	2550.973	517 847.6	862	1.160 093	2708.053	583 585.4
813	1.230 012	2554.115	519 123.8	863	1.158 749	2711.194	584 940.2
814	1.228 501	2557.256	520 401.7	864	1.157 407	2714.336	586 296.6
815	1.226 994	2560.398	521 681.1	865	1.156 069	2717.478	587 654.5
816	1.225 490	2563.540	522 962.1	866	1.154 734	2720.619	589 014.1
817	1.223 990	2566.681	524 244.6	867	1.153 403	2723.761	590 375.2
818	1.222 494	2569.823	525 528.8	868	1.152 074	2726.902	591 737.8
819	1.221 001	2572.964	526 814.5	869	1.150 748	2730.044	593 102.1
820	1.219 512	2576.106	528 101.7	**870**	1.149 425	2733.186	594 467.9
821	1.218 027	2579.248	529 390.6	871	1.148 106	2736.327	595 835.2
822	1.216 545	2582.389	530 681.0	872	1.146 789	2739.469	597 204.2
823	1.215 067	2585.531	531 973.0	873	1.145 475	2742.610	598 574.7
824	1.213 592	2588.672	533 266.5	874	1.144 165	2745.752	599 946.8
825	1.212 121	2591.814	534 561.6	875	1.142 857	2748.894	601 320.5
826	1.210 654	2594.956	535 858.3	876	1.141 553	2752.035	602 695.7
827	1.209 190	2598.097	537 156.6	877	1.140 251	2755.177	604 072.5
828	1.207 729	2601.239	538 456.4	878	1.138 952	2758.318	605 450.9
829	1.206 273	2604.380	539 757.8	879	1.137 656	2761.460	606 830.8
830	1.204 819	2607.522	541 060.8	**880**	1.136 364	2764.602	608 212.3
831	1.203 369	2610.663	542 365.3	881	1.135 074	2767.743	609 595.4
832	1.201 923	2613.805	543 671.5	882	1.133 787	2770.885	610 980.1
833	1.200 480	2616.947	544 979.1	883	1.132 503	2774.026	612 366.3
834	1.199 041	2620.088	546 288.4	884	1.131 222	2777.168	613 754.1
835	1.197 605	2623.230	547 599.2	885	1.129 944	2780.309	615 143.5
836	1.196 172	2626.371	548 911.6	886	1.128 668	2783.451	616 534.4
837	1.194 743	2629.513	550 225.6	887	1.127 396	2786.593	617 926.9
838	1.193 317	2632.655	551 541.1	888	1.126 126	2789.734	619 321.0
839	1.191 895	2635.796	552 858.3	889	1.124 859	2792.876	620 716.7
840	1.190 476	2638.938	554 176.9	**890**	1.123 596	2796.017	622 113.9
841	1.189 061	2642.079	555 497.2	891	1.122 334	2799.159	623 512.7
842	1.187 648	2645.221	556 819.0	892	1.121 076	2802.301	624 913.0
843	1.186 240	2648.363	558 142.4	893	1.119 821	2805.442	626 315.0
844	1.184 834	2651.504	559 467.4	894	1.118 568	2808.584	627 718.5
845	1.183 432	2654.646	560 793.9	895	1.117 318	2811.725	629 123.6
846	1.182 033	2657.787	562 122.0	896	1.116 071	2814.867	630 530.2
847	1.180 638	2660.929	563 451.7	897	1.114 827	2818.009	631 938.4
848	1.179 245	2664.071	564 783.0	898	1.113 586	2821.150	633 348.2
849	1.177 856	2667.212	566 115.8	899	1.112 347	2824.292	634 759.6
850	1.176 471	2670.354	567 450.2	**900**	1.111 111	2827.433	636 172.5

n	$1000\frac{1}{n}$	Circum-ference πn	Area $\frac{\pi n^2}{4}$	n	$1000\frac{1}{n}$	Circum-ference πn	Area $\frac{\pi n^2}{4}$
900	1.111 111	2827.433	636 172.5	**950**	1.052 632	2984.513	708 821.8
901	1.109 878	2830.575	637 587.0	951	1.051 525	2987.655	710 314.9
902	1.108 647	2833.717	639 003.1	952	1.050 420	2990.796	711 809.5
903	1.107 420	2836.858	640 420.7	953	1.049 318	2993.938	713 305.7
904	1.106 195	2840.000	641 839.9	954	1.048 218	2997.079	714 803.4
905	1.104 972	2843.141	643 260.7	955	1.047 120	3000.221	716 302.8
906	1.103 753	2846.283	644 683.1	956	1.046 025	3003.363	717 803.7
907	1.102 536	2849.425	646 107.0	957	1.044 932	3006.504	719 306.1
908	1.101 322	2852.566	647 532.5	958	1.043 841	3009.646	720 810.2
909	1.100 110	2855.708	648 959.6	959	1.042 753	3012.787	722 315.8
910	1.098 901	2858.849	650 388.2	**960**	1.041 667	3015.929	723 822.9
911	1.097 695	2861.991	651 818.4	961	1.040 583	3019.071	725 331.7
912	1.096 491	2865.133	653 250.2	962	1.039 501	3022.212	726 842.0
913	1.095 290	2868.274	654 683.6	963	1.038 422	3025.354	728 353.9
914	1.094 092	2871.416	656 118.5	964	1.037 344	3028.495	729 867.4
915	1.092 896	2874.557	657 555.0	965	1.036 269	3031.637	731 382.4
916	1.091 703	2877.699	658 993.0	966	1.035 197	3034.779	732 899.0
917	1.090 513	2880.840	660 432.7	967	1.034 126	3037.920	734 417.2
918	1.089 325	2883.982	661 873.9	968	1.033 058	3041.062	735 936.9
919	1.088 139	2887.124	663 316.7	969	1.031 992	3044.203	737 458.2
920	1.086 957	2890.265	664 761.0	**970**	1.030 928	3047.345	738 981.1
921	1.085 776	2893.407	666 206.9	971	1.029 866	3050.486	740 505.6
922	1.084 599	2896.548	667 654.4	972	1.028 807	3053.628	742 031.6
923	1.083 424	2899.690	669 103.5	973	1.027 749	3056.770	743 559.2
924	1.082 251	2902.832	670 554.1	974	1.026 694	3059.911	745 088.4
925	1.081 081	2905.973	672 006.3	975	1.025 641	3063.053	746 619.1
926	1.079 914	2909.115	673 460.1	976	1.024 590	3066.194	748 151.4
927	1.078 749	2912.256	674 915.4	977	1.023 541	3069.336	749 685.3
928	1.077 586	2915.398	676 372.3	978	1.022 495	3072.478	751 220.8
929	1.076 426	2918.540	677 830.8	979	1.021 450	3075.619	752 757.8
930	1.075 269	2921.681	679 290.9	**980**	1.020 408	3078.761	754 296.4
931	1.074 114	2924.823	680 752.5	981	1.019 368	3081.902	755 836.6
932	1.072 961	2927.964	682 215.7	982	1.018 330	3085.044	757 378.3
933	1.071 811	2931.106	683 680.5	983	1.017 294	3088.186	758 921.6
934	1.070 664	2934.248	685 146.8	984	1.016 260	3091.327	760 466.5
935	1.069 519	2937.389	686 614.7	985	1.015 228	3094.469	762 012.9
936	1.068 376	2940.531	688 084.2	986	1.014 199	3097.610	763 561.0
937	1.067 236	2943.672	689 555.2	987	1.013 171	3100.752	765 110.5
938	1.066 098	2946.814	691 027.9	988	1.012 146	3103.894	766 661.7
939	1.064 963	2949.956	692 502.1	989	1.011 122	3107.035	768 214.4
940	1.063 830	2953.097	693 977.8	**990**	1.010 101	3110.177	769 768.7
941	1.062 699	2956.239	695 455.2	991	1.009 082	3113.318	771 324.6
942	1.061 571	2959.380	696 934.1	992	1.008 065	3116.460	772 882.1
943	1.060 445	2962.522	698 414.5	993	1.007 049	3119.602	774 441.1
944	1.059 322	2965.663	699 896.6	994	1.006 036	3122.743	776 001.7
945	1.058 201	2968.805	701 380.2	995	1.005 025	3125.885	777 563.8
946	1.057 082	2971.947	702 865.4	996	1.004 016	3129.026	779 127.5
947	1.055 966	2975.088	704 352.1	997	1.003 009	3132.168	780 692.8
948	1.054 852	2978.230	705 840.5	998	1.002 004	3135.309	782 259.7
949	1.053 741	2981.371	707 330.4	999	1.001 001	3138.451	783 828.2
950	1.052 632	2984.513	708 821.8	**1000**	1.000 000	3141.593	785 398.2

SQUARES, CUBES AND ROOTS

Roots of numbers other than those given directly may be found by the following relations:

$\sqrt{100n} = 10\sqrt{n}$; $\sqrt{1000n} = 10\sqrt{10n}$; $\sqrt{\frac{1}{10}n} = \frac{1}{10}\sqrt{10n}$; $\sqrt{\frac{1}{100}n} = \frac{1}{10}\sqrt{n}$, $\sqrt{\frac{1}{1000}n} = \frac{1}{100}\sqrt{10n}$; $\sqrt[3]{1000n} = 10\sqrt[3]{n}$; $\sqrt[3]{10{,}000n} = 10\sqrt[3]{10n}$; $\sqrt[3]{100{,}000n} = 10\sqrt[3]{100n}$; $\sqrt[3]{\frac{1}{10}n} = \frac{1}{10}\sqrt[3]{100n}$; $\sqrt[3]{\frac{1}{100}n} = \frac{1}{10}\sqrt[3]{10n}$; $\sqrt[3]{\frac{1}{1000}n} = \frac{1}{10}\sqrt[3]{n}$.

n	n^2	$\sqrt{n}$	$\sqrt{10n}$	n^3	$\sqrt[3]{n}$	$\sqrt[3]{10n}$	$\sqrt[3]{100n}$
1	1	1.000 000	3.162 278	1	1.000 000	2.154 435	4.641 589
2	4	1.414 214	4.472 136	8	1.259 921	2.714 418	5.848 035
3	9	1.732 051	5.477 226	27	1.442 250	3.107 233	6.694 330
4	16	2.000 000	6.324 555	64	1.587 401	3.419 952	7.368 063
5	25	2.236 068	7.071 068	125	1.709 976	3.684 031	7.937 005
6	36	2.449 490	7.745 967	216	1.817 121	3.914 868	8.434 327
7	49	2.645 751	8.366 600	343	1.912 931	4.121 285	8.879 040
8	64	2.828 427	8.944 272	512	2.000 000	4.308 869	9.283 178
9	81	3.000 000	9.486 833	729	2.080 084	4.481 405	9.654 894
10	100	3.162 278	10.00000	1 000	2.154 435	4.641 589	10.00000
11	121	3.316 625	10.48809	1 331	2.223 980	4.791 420	10.32280
12	144	3.464 102	10.95445	1 728	2.289 428	4.932 424	10.62659
13	169	3.605 551	11.40175	2 197	2.351 335	5.065 797	10.91393
14	196	3.741 657	11.83216	2 744	2.410 142	5.192 494	11.18689
15	225	3.872 983	12.24745	3 375	2.466 212	5.313 293	11.44714
16	256	4.000 000	12.64911	4 096	2.519 842	5.428 835	11.69607
17	289	4.123 106	13.03840	4 913	2.571 282	5.539 658	11.93483
18	324	4.242 641	13.41641	5 832	2.620 741	5.646 216	12.16440
19	361	4.358 899	13.78405	6 859	2.668 402	5.748 897	12.38562
20	400	4.472 136	14.14214	8 000	2.714 418	5.848 035	12.59921
21	441	4.582 576	14.49138	9 261	2.758 924	5.943 922	12.80579
22	484	4.690 416	14.83240	10 648	2.802 039	6.036 811	13.00591
23	529	4.795 832	15.16575	12 167	2.843 867	6.126 926	13.20006
24	576	4.898 979	15.49193	13 824	2.884 499	6.214 465	13.38866
25	625	5.000 000	15.81139	15 625	2.924 018	6.299 605	13.57209
26	676	5.099 020	16.12452	17 576	2.962 496	6.382 504	13.75069
27	729	5.196 152	16.43168	19 683	3.000 000	6.463 304	13.92477
28	784	5.291 503	16.73320	21 952	3.036 589	6.542 133	14.09460
29	841	5.385 165	17.02939	24 389	3.072 317	6.619 106	14.26043
30	900	5.477 226	17.32051	27 000	3.107 233	6.694 330	14.42250
31	961	5.567 764	17.60682	29 791	3.141 381	6.767 899	14.58100
32	1 024	5.656 854	17.88854	32 768	3.174 802	6.839 904	14.73613
33	1 089	5.744 563	18.16590	35 937	3.207 534	6.910 423	14.88806
34	1 156	5.830 952	18.43909	39 304	3.239 612	6.979 532	15.03695
35	1 225	5.916 080	18.70829	42 875	3.271 066	7.047 299	15.18294
36	1 296	6.000 000	18.97367	46 656	3.301 927	7.113 787	15.32619
37	1 369	6.082 763	19.23538	50 653	3.332 222	7.179 054	15.46680
38	1 444	6.164 414	19.49359	54 872	3.361 975	7.243 156	15.60491
39	1 521	6.244 998	19.74842	59 319	3.391 211	7.306 144	15.74061
40	1 600	6.324 555	20.00000	64 000	3.419 952	7.368 063	15.87401
41	1 681	6.403 124	20.24846	68 921	3.448 217	7.428 959	16.00521
42	1 764	6.480 741	20.49390	74 088	3.476 027	7.488 872	16.13429
43	1 849	6.557 439	20.73644	79 507	3.503 398	7.547 842	16.26133
44	1 936	6.633 250	20.97618	85 184	3.530 348	7.605 905	16.38643
45	2 025	6.708 204	21.21320	91 125	3.556 893	7.663 094	16.50964
46	2 116	6.782 330	21.44761	97 336	3.583 048	7.719 443	16.63103
47	2 209	6.855 655	21.67948	103 823	3.608 826	7.774 980	16.75069
48	2 304	6.928 203	21.90890	110 592	3.634 241	7.829 735	16.86865
49	2 401	7.000 000	22.13594	117 649	3.659 306	7.883 735	16.98499
50	2 500	7.071 068	22.36068	125 000	3.684 031	7.937 005	17.09976

SQUARES, CUBES AND ROOTS (Continued)

n	n^2	$\sqrt{n}$	$\sqrt{10n}$	n^3	$\sqrt[3]{n}$	$\sqrt[3]{10n}$	$\sqrt[3]{100n}$
50	2 500	7.071 068	22.36068	125 000	3.684 031	7.937 005	17.09976
51	2 601	7.141 428	22.58318	132 651	3.708 430	7.989 570	17.21301
52	2 704	7.211 103	22.80351	140 608	3.732 511	8.041 452	17.32478
53	2 809	7.280 110	23.02173	148 877	3.756 286	8.092 672	17.43513
54	2 916	7.348 469	23.23790	157 464	3.779 763	8.143 253	17.54411
55	3 025	7.416 198	23.45208	166 375	3.802 952	8.193 213	17.65174
56	3 136	7.483 315	23.66432	175 616	3.825 862	8.242 571	17.75808
57	3 249	7.549 834	23.87467	185 193	3.848 501	8.291 344	17.86316
58	3 364	7.615 773	24.08319	195 112	3.870 877	8.339 551	17.96702
59	3 481	7.681 146	24.28992	205 379	3.892 996	8.387 207	18.06969
60	3 600	7.745 967	24.49490	216 000	3.914 868	8.434 327	18.17121
61	3 721	7.810 250	24.69818	226 981	3.936 497	8.480 926	18.27160
62	3 844	7.874 008	24.89980	238 328	3.957 892	8.527 019	18.37091
63	3 969	7.937 254	25.09980	250 047	3.979 057	8.572 619	18.46915
64	4 096	8.000 000	25.29822	262 144	4.000 000	8.617 739	18.56636
65	4 225	8.062 258	25.49510	274 625	4.020 726	8.662 391	18.66256
66	4 356	8.124 038	25.69047	287 496	4.041 240	8.706 588	18.75777
67	4 489	8.185 353	25.88436	300 763	4.061 548	8.750 340	18.85204
68	4 624	8.246 211	26.07681	314 432	4.081 655	8.793 659	18.94536
69	4 761	8.306 624	26.26785	328 509	4.101 566	8.836 556	19.03778
70	4 900	8.366 600	26.45751	343 000	4.121 285	8.879 040	19.12931
71	5 041	8.426 150	26.64583	357 911	4.140 818	8.921 121	19.21997
72	5 184	8.485 281	26.83282	373 248	4.160 168	8.962 809	19.30979
73	5 329	8.544 004	27.01851	389 017	4.179 339	9.004 113	19.39877
74	5 476	8.602 325	27.20294	405 224	4.198 336	9.045 042	19.48695
75	5 625	8.660 254	27.38613	421 875	4.217 163	9.085 603	19.57434
76	5 776	8.717 798	27.56810	438 976	4.235 824	9.125 805	19.66095
77	5 929	8.774 964	27.74887	456 533	4.254 321	9.165 656	19.74681
78	6 084	8.831 761	27.92848	474 552	4.272 659	9.205 164	19.83192
79	6 241	8.888 194	28.10694	493 039	4.290 840	9.244 335	19.91632
80	6 400	8.944 272	28.28427	512 000	4.308 869	9.283 178	20.00000
81	6 561	9.000 000	28.46050	531 441	4.326 749	9.321 698	20.08299
82	6 724	9.055 385	28.63564	551 368	4.344 481	9.359 902	20.16530
83	6 889	9.110 434	28.80972	571 787	4.362 071	9.397 796	20.24694
84	7 056	9.165 151	28.98275	592 704	4.379 519	9.435 388	20.32793
85	7 225	9.219 544	29.15476	614 125	4.396 830	9.472 682	20.40828
86	7 396	9.273 618	29.32576	636 056	4.414 005	9.509 685	20.48800
87	7 569	9.327 379	29.49576	658 503	4.431 048	9.546 403	20.56710
88	7 744	9.380 832	29.66479	681 472	4.447 960	9.582 840	20.64560
89	7 921	9.433 981	29.83287	704 969	4.464 745	9.619 002	20.72351
90	8 100	9.486 833	30.00000	729 000	4.481 405	9.654 894	20.80084
91	8 281	9.539 392	30.16621	753 571	4.497 941	9.690 521	20.87759
92	8 464	9.591 663	30.33150	778 688	4.514 357	9.725 888	20.95379
93	8 649	9.643 651	30.49590	804 357	4.530 655	9.761 000	21.02944
94	8 836	9.695 360	30.65942	830 584	4.546 836	9.795 861	21.10454
95	9 025	9.746 794	30.82207	857 375	4.562 903	9.830 476	21.17912
96	9 216	9.797 959	30.98387	884 736	4.578 857	9.864 848	21.25317
97	9 409	9.848 858	31.14482	912 673	4.594 701	9.898 983	21.32671
98	9 604	9.899 495	31.30495	941 192	4.610 436	9.932 884	21.39975
99	9 801	9.949 874	31.46427	970 299	4.626 065	9.966 555	21.47229
100	10 000	10.00000	31.62278	1 000 000	4.641 589	10.00000	21.54435

SQUARES, CUBES AND ROOTS (Continued)

n	n^2	$\sqrt{n}$	$\sqrt{10n}$	n^3	$\sqrt[3]{n}$	$\sqrt[3]{10n}$	$\sqrt[3]{100n}$
100	10 000	10.00000	31.62278	1 000 000	4.641 589	10.00000	21.54435
101	10 201	10.04988	31.78050	1 030 301	4.657 010	10.03322	21.61592
102	10 404	10.09950	31.93744	1 061 208	4.672 329	10.06623	21.68703
103	10 609	10.14889	32.09361	1 092 727	4.687 548	10.09902	21.75767
104	10 816	10.19804	32.24903	1 124 864	4.702 669	10.13159	21.82786
105	11 025	10.24695	32.40370	1 157 625	4.717 694	10.16396	21.89760
106	11 236	10.29563	32.55764	1 191 016	4.732 623	10.19613	21.96689
107	11 449	10.34408	32.71085	1 225 043	4.747 459	10.22809	22.03575
108	11 664	10.39230	32.86335	1 259 712	4.762 203	10.25986	22.10419
109	11 881	10.44031	33.01515	1 295 029	4.776 856	10.29142	22.17220
110	12 100	10.48809	33.16625	1 331 000	4.791 420	10.32280	22.23980
111	12 321	10.53565	33.31666	1 367 631	4.805 896	10.35399	22.30699
112	12 544	10.58301	33.46640	1 404 928	4.820 285	10.38499	22.37378
113	12 769	10.63015	33.61547	1 442 897	4.834 588	10.41580	22.44017
114	12 996	10.67708	33.76389	1 481 544	4.848 808	10.44644	22.50617
115	13 225	10.72381	33.91165	1 520 875	4.862 944	10.47690	22.57179
116	13 456	10.77033	34.05877	1 560 896	4.876 999	10.50718	22.63702
117	13 689	10.81665	34.20526	1 601 613	4.890 973	10.53728	22.70189
118	13 924	10.86278	34.35113	1 643 032	4.904 868	10.56722	22.76638
119	14 161	10.90871	34.49638	1 685 159	4.918 685	10.59699	22.83051
120	14 400	10.95445	34.64102	1 728 000	4.932 424	10.62659	22.89428
121	14 641	11.00000	34.78505	1 771 561	4.946 087	10.65602	22.95770
122	14 884	11.04536	34.92850	1 815 848	4.959 676	10.68530	23.02078
123	15 129	11.09054	35.07136	1 860 867	4.973 190	10.71441	23.08350
124	15 376	11.13553	35.21363	1 906 624	4.986 631	10.74337	23.14589
125	15 625	11.18034	35.35534	1 953 125	5.000 000	10.77217	23.20794
126	15 876	11.22497	35.49648	2 000 376	5.013 298	10.80082	23.26967
127	16 129	11.26943	35.63706	2 048 383	5.026 526	10.82932	23.33107
128	16 384	11.31371	35.77709	2 097 152	5.039 684	10.85767	23.39214
129	16 641	11.35782	35.91657	2 146 689	5.052 774	10.88587	23.45290
130	16 900	11.40175	36.05551	2 197 000	5.065 797	10.91393	23.51335
131	17 161	11.44552	36.19392	2 248 091	5.078 753	10.94184	23.57348
132	17 424	11.48913	36.33180	2 299 968	5.091 643	10.96961	23.63332
133	17 689	11.53256	36.46917	2 352 637	5.104 469	10.99724	23.69285
134	17 956	11.57584	36.60601	2 406 104	5.117 230	11.02474	23.75208
135	18 225	11.61895	36.74235	2 460 375	5.129 928	11.05209	23.81102
136	18 496	11.66190	36.87818	2 515 456	5.142 563	11.07932	23.86966
137	18 769	11.70470	37.01351	2 571 353	5.155 137	11.10641	23.92803
138	19 044	11.74734	37.14835	2 628 072	5.167 649	11.13336	23.98610
139	19 321	11.78983	37.28270	2 685 619	5.180 101	11.16019	24.04390
140	19 600	11.83216	37.41657	2 744 000	5.192 494	11.18689	24.10142
141	19 881	11.87434	37.54997	2 803 221	5.204 828	11.21346	24.15867
142	20 164	11.91638	37.68289	2 863 288	5.217 103	11.23991	24.21565
143	20 449	11.95826	37.81534	2 924 207	5.229 322	11.26623	24.27236
144	20 736	12.00000	37.94733	2 985 984	5.241 483	11.29243	24.32881
145	21 025	12.04159	38.07887	3 048 625	5.253 588	11.31851	24.38499
146	21 316	12.08305	38.20995	3 112 136	5.265 637	11.34447	24.44092
147	21 609	12.12436	38.34058	3 176 523	5.277 632	11.37031	24.49660
148	21 904	12.16553	38.47077	3 241 792	5.289 572	11.39604	24.55202
149	22 201	12.20656	38.60052	3 307 949	5.301 459	11.42165	24.60719
150	22.500	12.24745	38.72983	3 375 000	5.313 293	11.44714	24.66212

SQUARES, CUBES AND ROOTS (Continued)

n	n^2	$\sqrt{n}$	$\sqrt{10n}$	n^3	$\sqrt[3]{n}$	$\sqrt[3]{10n}$	$\sqrt[3]{100n}$
150	22 500	12.24745	38.72983	3 375 000	5.313 293	11.44714	24.66212
151	22 801	12.28821	38.85872	3 442 951	5.325 074	11.47252	24.71680
152	23 104	12.32883	38.98718	3 511 808	5.336 803	11.49779	24.77125
153	23 409	12.36932	39.11521	3 581 577	5.348 481	11.52295	24.82545
154	23 716	12.40967	39.24283	3 652 264	5.360 108	11.54800	24.87942
155	24 025	12.44990	39.37004	3 723 875	5.371 685	11.57295	24.93315
156	24 336	12.49000	39.49684	3 796 416	5.383 213	11.59778	24.98666
157	24 649	12.52996	39.62323	3 869 893	5.394 691	11.62251	25.03994
158	24 964	12.56981	39.74921	3 944 312	5.406 120	11.64713	25.09299
159	25 281	12.60952	39.87480	4 019 679	5.417 502	11.67165	25.14581
160	25 600	12.64911	40.00000	4 096 000	5.428 835	11.69607	25.19842
161	25 921	12.68858	40.12481	4 173 281	5.440 122	11.72039	25.25081
162	26 244	12.72792	40.24922	4 251 528	5.451 362	11.74460	25.30298
163	26 569	12.76715	40.37326	4 330 747	5.462 556	11.76872	25.35494
164	26 896	12.80625	40.49691	4 410 944	5.473 704	11.79274	25.40668
165	27 225	12.84523	40.62019	4 492 125	5.484 807	11.81666	25.45822
166	27 556	12.88410	40.74310	4 574 296	5.495 865	11.84048	25.50954
167	27 889	12.92285	40.86563	4 657 463	5.506 878	11.86421	25.56067
168	28 224	12.96148	40.98780	4 741 632	5.517 848	11.88784	25.61158
169	28 561	13.00000	41.10961	4 826 809	5.528 775	11.91138	25.66230
170	28 900	13.03840	41.23106	4 913 000	5.539 658	11.93483	25.71282
171	29 241	13.07670	41.35215	5 000 211	5.550 499	11.95819	25.76313
172	29 584	13.11488	41.47288	5 088 448	5.561 298	11.98145	25.81326
173	29 929	13.15295	41.59327	5 177 717	5.572 055	12.00463	25.86319
174	30 276	13.19091	41.71331	5 268 024	5.582 770	12.02771	25.91292
175	30 625	13.22876	41.83300	5 359 375	5.593 445	12.05071	25.96247
176	30 976	13.26650	41.95235	5 451 776	5.604 079	12.07362	26.01183
177	31 329	13.30413	42.07137	5 545 233	5.614 672	12.09645	26.06100
178	31 684	13.34166	42.19005	5 639 752	5.625 226	12.11918	26.10999
179	32 041	13.37909	42.30839	5 735 339	5.635 741	12.14184	26.15879
180	32 400	13.41641	42.42641	5 832 000	5.646 216	12.16440	26.20741
181	32 761	13.45362	42.54409	5 929 741	5.656 653	12.18689	26.25586
182	33 124	13.49074	42.66146	6 028 568	5.667 051	12.20929	26.30412
183	33 489	13.52775	42.77850	6 128 487	5.677 411	12.23161	26.35221
184	33 856	13.56466	42.89522	6 229 504	5.687 734	12.25385	26.40012
185	34 225	13.60147	43.01163	6 331 625	5.698 019	12.27601	26.44786
186	34 596	13.63818	43.12772	6 434 856	5.708 267	12.29809	26.49543
187	34 969	13.67479	43.24350	6 539 203	5.718 479	12.32009	26.54283
188	35 344	13.71131	43.35897	6 644 672	5.728 654	12.34201	26.59006
189	35 721	13.74773	43.47413	6 751 269	5.738 794	12.36386	26.63712
190	36 100	13.78405	43.58899	6 859 000	5.748 897	12.38562	26.68402
191	36 481	13.82027	43.70355	6 967 871	5.758 965	12.40731	26.73075
192	36 864	13.85641	43.81780	7 077 888	5.768 998	12.42893	26.77732
193	37 249	13.89244	43.93177	7 189 057	5.778 997	12.45047	26.82373
194	37 636	13.92839	44.04543	7 301 384	5.788 960	12.47194	26.86997
195	38 025	13.96424	44.15880	7 414 875	5.798 890	12.49333	26.91606
196	38 416	14.00000	44.27189	7 529 536	5.808 786	12.51465	26.96199
197	38 809	14.03567	44.38468	7 645 373	5.818 648	12.53590	27.00777
198	39 204	14.07125	44.49719	7 762 392	5.828 477	12.55707	27.05339
199	39 601	14.10674	44.60942	7 880 599	5.838 272	12.57818	27.09886
200	40 000	14.14214	44.72136	8 000 000	5.848 035	12.59921	27.14418

SQUARES, CUBES AND ROOTS (Continued)

n	n^2	$\sqrt{n}$	$\sqrt{10n}$	n^3	$\sqrt[3]{n}$	$\sqrt[3]{10n}$	$\sqrt[3]{100n}$
200	40 000	14.14214	44.72136	8 000 000	5.848 035	12.59921	27.14418
201	40 401	14.17745	44.83302	8 120 601	5.857 766	12.62017	27.18934
202	40 804	14.21267	44.94441	8 242 408	5.867 464	12.64107	27.23436
203	41 209	14.24781	45.05552	8 365 427	5.877 131	12.66189	27.27922
204	41 616	14.28286	45.16636	8 489 664	5.886 765	12.68265	27.32394
205	42 025	14.31782	45.27693	8 615 125	5.896 369	12.70334	27.36852
206	42 436	14.35270	45.38722	8 741 816	5.905 941	12.72396	27.41295
207	42 849	14.38749	45.49725	8 869 743	5.915 482	12.74452	27.45723
208	43 264	14.42221	45.60702	8 998 912	5.924 992	12.76501	27.50138
209	43 681	14.45683	45.71652	9 129 329	5.934 472	12.78543	27.54538
210	44 100	14.49138	45.82576	9 261 000	5.943 922	12.80579	27.58924
211	44 521	14.52584	45.93474	9 393 931	5.953 342	12.82609	27.63296
212	44 944	14.56022	46.04346	9 528 128	5.962 732	12.84632	27.67655
213	45 369	14.59452	46.15192	9 663 597	5.972 093	12.86648	27.72000
214	45 796	14.62874	46.26013	9 800 344	5.981 424	12.88659	27.76331
215	46 225	14.66288	46.36809	9 938 375	5.990 726	12.90663	27.80649
216	46 656	14.69694	46.47580	10 077 696	6.000 000	12.92661	27.84953
217	47 089	14.73092	46.58326	10 218 313	6.009 245	12.94653	27.89244
218	47 524	14.76482	46.69047	10 360 232	6.018 462	12.96638	27.93522
219	47 961	14.79865	46.79744	10 503 459	6.027 650	12.98618	27.97787
220	48 400	14.83240	46.90416	10 648 000	6.036 811	13.00591	28.02039
221	48 841	14.86607	47.01064	10 793 861	6.045 944	13.02559	28.06278
222	49 284	14.89966	47.11688	10 941 048	6.055 049	13.04521	28.10505
223	49 729	14.93318	47.22288	11 089 567	6.064 127	13.06477	28.14718
224	50 176	14.96663	47.32864	11 239 424	6.073 178	13.08427	28.18919
225	50 625	15.00000	47.43416	11 390 625	6.082 202	13.10371	28.23108
226	51 076	15.03330	47.53946	11 543 176	6.091 199	13.12309	28.27284
227	51 529	15.06652	47.64452	11 697 083	6.100 170	13.14242	28.31448
228	51 984	15.09967	47.74935	11 852 352	6.109 115	13.16169	28.35600
229	52 441	15.13275	47.85394	12 008 989	6.118 033	13.18090	28.39739
230	52 900	15.16575	47.95832	12 167 000	6.126 926	13.20006	28.43867
231	53 361	15.19868	48.06246	12 326 391	6.135 792	13.21916	28.47983
232	53 824	15.23155	48.16638	12 487 168	6.144 634	13.23821	28.52086
233	54 289	15.26434	48.27007	12 649 337	6.153 449	13.25721	28.56178
234	54 756	15.29706	48.37355	12 812 904	6.162 240	13.27614	28.60259
235	55 225	15.32971	48.47680	12 977 875	6.171 006	13.29503	28.64327
236	55 696	15.36229	48.57983	13 144 256	6.179 747	13.31386	28.68384
237	56 169	15.39480	48.68265	13 312 053	6.188 463	13.33264	28.72430
238	56 644	15.42725	48.78524	13 481 272	6.197 154	13.35136	28.76464
239	57 121	15.45962	48.88763	13 651 919	6.205 822	13.37004	28.80487
240	57 600	15.49193	48.98979	13 824 000	6.214 465	13.38866	28.84499
241	58 081	15.52417	49.09175	13 997 521	6.223 084	13.40723	28.88500
242	58 564	15.55635	49.19350	14 172 488	6.231 680	13.42575	28.92489
243	59 049	15.58846	49.29503	14 348 907	6.240 251	13.44421	28.96468
244	59 536	15.62050	49.39636	14 526 784	6.248 800	13.46263	29.00436
245	60 025	15.65248	49.49747	14 706 125	6.257 325	13.48100	29.04393
246	60 516	15.68439	49.59839	14 886 936	6.265 827	13.49931	29.08339
247	61 009	15.71623	49.69909	15 069 223	6.274 305	13.51758	29.12275
248	61 504	15.74802	49.79960	15 252 992	6.282 761	13.53580	29.16199
249	62 001	15.77973	49.89990	15 438 249	6.291 195	13.55397	29.20114
250	62 500	15.81139	50.00000	15 625 000	6.299 605	13.57209	29.24018

SQUARES, CUBES AND ROOTS (Continued)

n	n^2	$\sqrt{n}$	$\sqrt{10n}$	n^3	$\sqrt[3]{n}$	$\sqrt[3]{10n}$	$\sqrt[3]{100n}$
250	62 500	15.81139	50.00000	15 625 000	6.299 605	13.57209	29.24018
251	63 001	15.84298	50.09990	15 813 251	6.307 994	13.59016	29.27911
252	63 504	15.87451	50.19960	16 003 008	6.316 360	13.60818	29.31794
253	64 009	15.90597	50.29911	16 194 277	6.324 704	13.62616	29.35667
254	64 516	15.93738	50.39841	16 387 064	6.333 026	13.64409	29.39530
255	65 025	15.96872	50.49752	16 581 375	6.341 326	13.66197	29.43383
256	65 536	16.00000	50.59644	16 777 216	6.349 604	13.67981	29.47225
257	66 049	16.03122	50.69517	16 974 593	6.357 861	13.69760	29.51058
258	66 564	16.06238	50.79370	17 173 512	6.366 097	13.71534	29.54880
259	67 081	16.09348	50.89204	17 373 979	6.374 311	13.73304	29.58693
260	67 600	16.12452	50.99020	17 576 000	6.382 504	13.75069	29.62496
261	68 121	16.15549	51.08816	17 779 581	6.390 677	13.76830	29.66289
262	68 644	16.18641	51.18594	17 984 728	6.398 828	13.78586	29.70073
263	69 169	16.21727	51.28353	18 191 447	6.406 959	13.80337	29.73847
264	69 696	16.24808	51.38093	18 399 744	6.415 069	13.82085	29.77611
265	70 225	16.27882	51.47815	18 609 625	6.423 158	13.83828	29.81366
266	70 756	16.30951	51.57519	18 821 096	6.431 228	13.85566	29.85111
267	71 289	16.34013	51.67204	19 034 163	6.439 277	13.87300	29.88847
268	71 824	16.37071	51.76872	19 248 832	6.447 306	13.89030	29.92574
269	72 361	16.40122	51.86521	19 465 109	6.455 315	13.90755	29.96292
270	72 900	16.43168	51.96152	19 683 000	6.463 304	13.92477	30.00000
271	73 441	16.46208	52.05766	19 902 511	6.471 274	13.94194	30.03699
272	73 984	16.49242	52.15362	20 123 648	6.479 224	13.95906	30.07389
273	74 529	16.52271	52.24940	20 346 417	6.487 154	13.97615	30.11070
274	75 076	16.55295	52.34501	20 570 824	6.495 065	13.99319	30.14742
275	75 625	16.58312	52.44044	20 796 875	6.502 957	14.01020	30.18405
276	76 176	16.61325	52.53570	21 024 576	6.510 830	14.02716	30.22060
277	76 729	16.64332	52.63079	21 253 933	6.518 684	14.04408	30.25705
278	77 284	16.67333	52.72571	21 484 952	6.526 519	14.06096	30.29342
279	77 841	16.70329	52.82045	21 717 639	6.534 335	14.07780	30.32970
280	78 400	16.73320	52.91503	21 952 000	6.542 133	14.09460	30.36589
281	78 961	16.76305	53.00943	22 188 041	6.549 912	14.11136	30.40200
282	79 524	16.79286	53.10367	22 425 768	6.557 672	14.12808	30.43802
283	80 089	16.82260	53.19774	22 665 187	6.565 414	14.14476	30.47395
284	80 656	16.85230	53.29165	22 906 304	6.573 138	14.16140	30.50981
285	81 225	16.88194	53.38539	23 149 125	6.580 844	14.17800	30.54557
286	81 796	16.91153	53.47897	23 393 656	6.588 532	14.19456	30.58126
287	82 369	16.94107	53.57238	23 639 903	6.596 202	14.21109	30.61686
288	82 944	16.97056	53.66563	23 887 872	6.603 854	14.22757	30.65238
289	83 521	17.00000	53.75872	24 137 569	6.611 489	14.24402	30.68781
290	84 100	17.02939	53.85165	24 389 000	6.619 106	14.26043	30.72317
291	84 681	17.05872	53.94442	24 642 171	6.626 705	14.27680	30.75844
292	85 264	17.08801	54.03702	24 897 088	6.634 287	14.29314	30.79363
293	85 849	17.11724	54.12947	25 153 757	6.641 852	14.30944	30.82875
294	86 436	17.14643	54.22177	25 412 184	6.649 400	14.32570	30.86378
295	87 025	17.17556	54.31390	25 672 375	6.656 930	14.34192	30.89873
296	87 616	17.20465	54.40588	25 934 336	6.664 444	14.35811	30.93361
297	88 209	17.23369	54.49771	26 198 073	6.671 940	14.37426	30.96840
298	88 804	17.26268	54.58938	26 463 592	6.679 420	14.39037	31.00312
299	89 401	17.29162	54.68089	26 730 899	6.686 883	14.40645	31.03776
300	90 000	17.32051	54.77226	27 000 000	6.694 330	14.42250	31.07233

SQUARES, CUBES AND ROOTS (Continued)

n	n^2	$\sqrt{n}$	$\sqrt{10n}$	n^3	$\sqrt[3]{n}$	$\sqrt[3]{10n}$	$\sqrt[3]{100n}$
300	90 000	17.32051	54.77226	27 000 000	6.694 330	14.42250	31.07233
301	90 601	17.34935	54.86347	27 270 901	6.701 759	14.43850	31.10681
302	91 204	17.37815	54.95453	27 543 608	6.709 173	14.45447	31.14122
303	91 809	17.40690	55.04544	27 818 127	6.716 570	14.47041	31.17556
304	92 416	17.43560	55.13620	28 094 464	6.723 951	14.48631	31.20982
305	93 025	17.46425	55.22681	28 372 625	6.731 315	14.50218	31.24400
306	93 636	17.49286	55.31727	28 652 616	6.738 664	14.51801	31.27811
307	94 249	17.52142	55.40758	28 934 443	6.745 997	14.53381	31.31214
308	94 864	17.54993	55.49775	29 218 112	6.753 313	14.54957	31.34610
309	95 481	17.57840	55.58777	29 503 629	6.760 614	14.56530	31.37999
310	96 100	17.60682	55.67764	29 791 000	6.767 899	14.58100	31.41381
311	96 721	17.63519	55.76737	30 080 231	6.775 169	14.59666	31.44755
312	97 344	17.66352	55.85696	30 371 328	6.782 423	14.61229	31.48122
313	97 969	17.69181	55.94640	30 664 297	6.789 661	14.62788	31.51482
314	98 596	17.72005	56.03570	30 959 144	6.796 884	14.64344	31.54834
315	99 225	17.74824	56.12486	31 255 875	6.804 092	14.65897	31.58180
316	99 856	17.77639	56.21388	31 554 496	6.811 285	14.67447	31.61518
317	100 489	17.80449	56.30275	31 855 013	6.818 462	14.68993	31.64850
318	101 124	17.83255	56.39149	32 157 432	6.825 624	14.70536	31.68174
319	101 761	17.86057	56.48008	32 461 759	6.832 771	14.72076	31.71492
320	102 400	17.88854	56.56854	32 768 000	6.839 904	14.73613	31.74802
321	103 041	17.91647	56.65686	33 076 161	6.847 021	14.75146	31.78106
322	103 684	17.94436	56.74504	33 386 248	6.854 124	14.76676	31.81403
323	104 329	17.97220	56.83309	33 698 267	6.861 212	14.78203	31.84693
324	104 976	18.00000	56.92100	34 012 224	6.868 285	14.79727	31.87976
325	105 625	18.02776	57.00877	34 328 125	6.875 344	14.81248	31.91252
326	106 276	18.05547	57.09641	34 645 976	6.882 389	14.82766	31.94522
327	106 929	18.08314	57.18391	34 965 783	6.889 419	14.84280	31.97785
328	107 584	18.11077	57.27128	35 287 552	6.896 434	14.85792	32.01041
329	108 241	18.13836	57.35852	35 611 289	6.903 436	14.87300	32.04291
330	108 900	18.16590	57.44563	35 937 000	6.910 423	14.88806	32.07534
331	109 561	18.19341	57.53260	36 264 691	6.917 396	14.90308	32.10771
332	110 224	18.22087	57.61944	36 594 368	6.924 356	14.91807	32.14001
333	110 889	18.24829	57.70615	36 926 037	6.931 301	14.93303	32.17225
334	111 556	18.27567	57.79273	37 259 704	6.938 232	14.94797	32.20442
335	112 225	18.30301	57.87918	37 595 375	6.945 150	14.96287	32.23653
336	112 896	18.33030	57.96551	37 933 056	6.952 053	14.97774	32.26857
337	113 569	18.35756	58.05170	38 272 753	6.958 943	14.99259	32.30055
338	114 244	18.38478	58.13777	38 614 472	6.965 820	15.00740	32.33247
339	114 921	18.41195	58.22371	38 958 219	6.972 683	15.02219	32.36433
340	115 600	18.43909	58.30952	39 304 000	6.979 532	15.03695	32.39612
341	116 281	18.46619	58.39521	39 651 821	6.986 368	15.05167	32.42785
342	116 964	18.49324	58.48077	40 001 688	6.993 191	15.06637	32.45952
343	117 649	18.52026	58.56620	40 353 607	7.000 000	15.08104	32.49112
344	118 336	18.54724	58.65151	40 707 584	7.006 796	15.09568	32.52267
345	119 025	18.57418	58.73670	41 063 625	7.013 579	15.11030	32.55415
346	119 716	18.60108	58.82176	41 421 736	7.020 349	15.12488	32.58557
347	120 409	18.62794	58.90671	41 781 923	7.027 106	15.13944	32.61694
348	121 104	18.65476	58.99152	42 144 192	7.033 850	15.15397	32.64824
349	121 801	18.68154	59.07622	42 508 549	7.040 581	15.16847	32.67948
350	122 500	18.70829	59.16080	42 875 000	7.047 299	15.18294	32.71066

SQUARES, CUBES AND ROOTS (Continued)

n	n^2	$\sqrt{n}$	$\sqrt{10n}$	n^3	$\sqrt[3]{n}$	$\sqrt[3]{10n}$	$\sqrt[3]{100n}$
350	122 500	18.70829	59.16080	42 875 000	7.047 299	15.18294	32.71066
351	123 201	18.73499	59.24525	43 243 551	7.054 004	15.19739	32.74179
352	123 904	18.76166	59.32959	43 614 208	7.060 697	15.21181	32.77285
353	124 609	18.78829	59.41380	43 986 977	7.067 377	15.22620	32.80386
354	125 316	18.81489	59.49790	44 361 864	7.074 044	15.24057	32.83480
355	126 025	18.84144	59.58188	44 738 875	7.080 699	15.25490	32.86569
356	126 736	18.86796	59.66574	45 118 016	7.087 341	15.26921	32.89652
357	127 449	18.89444	59.74948	45 499 293	7.093 971	15.28350	32.92730
358	128 164	18.92089	59.83310	45 882 712	7.100 588	15.29775	32.95801
359	128 881	18.94730	59.91661	46 268 279	7.107 194	15.31198	32.98867
360	129 600	18.97367	60.00000	46 656 000	7.113 787	15.32619	33.01927
361	130 321	19.00000	60.08328	47 045 881	7.120 367	15.34037	33.04982
362	131 044	19.02630	60.16644	47 437 928	7.126 936	15.35452	33.08031
363	131 769	19.05256	60.24948	47 832 147	7.133 492	15.36864	33.11074
364	132 496	19.07878	60.33241	48 228 544	7.140 037	15.38274	33.14112
365	133 225	19.10497	60.41523	48 627 125	7.146 569	15.39682	33.17144
366	133 956	19.13113	60.49793	49 027 896	7.153 090	15.41087	33.20170
367	134 689	19.15724	60.58052	49 430 863	7.159 599	15.42489	33.23191
368	135 424	19.18333	60.66300	49 836 032	7.166 096	15.43889	33.26207
369	136 161	19.20937	60.74537	50 243 409	7.172 581	15.45286	33.29217
370	136 900	19.23538	60.82763	50 653 000	7.179 054	15.46680	33.32222
371	137 641	19.26136	60.90977	51 064 811	7.185 516	15.48073	33.35221
372	138 384	19.28730	60.99180	51 478 848	7.191 966	15.49462	33.38215
373	139 129	19.31321	61.07373	51 895 117	7.198 405	15.50849	33.41204
374	139 876	19.33908	61.15554	52 313 624	7.204 832	15.52234	33.44187
375	140 625	19.36492	61.23724	52 734 375	7.211 248	15.53616	33.47165
376	141 376	19.39072	61.31884	53 157 376	7.217 652	15.54996	33.50137
377	142 129	19.41649	61.40033	53 582 633	7.224 045	15.56373	33.53105
378	142 884	19.44222	61.48170	54 010 152	7.230 427	15.57748	33.56067
379	143 641	19.46792	61.56298	54 439 939	7.236 797	15.59121	33.59024
380	144 400	19.49359	61.64414	54 872 000	7.243 156	15.60491	33.61975
381	145 161	19.51922	61.72520	55 306 341	7.249 505	15.61858	33.64922
382	145 924	19.54482	61.80615	55 742 968	7.255 842	15.63224	33.67863
383	146 689	19.57039	61.88699	56 181 887	7.262 167	15.64587	33.70800
384	147 456	19.59592	61.96773	56 623 104	7.268 482	15.65947	33.73731
385	148 225	19.62142	62.04837	57 066 625	7.274 786	15.67305	33.76657
386	148 996	19.64688	62.12890	57 512 456	7.281 079	15.68661	33.79578
387	149 769	19.67232	62.20932	57 960 603	7.287 362	15.70014	33.82494
388	150 544	19.69772	62.28965	58 411 072	7.293 633	15.71366	33.85405
389	151 321	19.72308	62.36986	58 863 869	7.299 894	15.72714	33.88310
390	152 100	19.74842	62.44998	59 319 000	7.306 144	15.74061	33.91211
391	152 881	19.77372	62.52999	59 776 471	7.312 383	15.75405	33.94107
392	153 664	19.79899	62.60990	60 236 288	7.318 611	15.76747	33.96999
393	154 449	19.82423	62.68971	60 698 457	7.324 829	15.78087	33.99885
394	155 236	19.84943	62.76942	61 162 984	7.331 037	15.79424	34.02766
395	156 025	19.87461	62.84903	61 629 875	7.337 234	15.80759	34.05642
396	156 816	19.89975	62.92853	62 099 136	7.343 420	15.82092	34.08514
397	157 609	19.92486	63.00794	62 570 773	7.349 597	15.83423	34.11381
398	158 404	19.94994	63.08724	63 044 792	7.355 762	15.84751	34.14242
399	159 201	19.97498	63.16645	63 521 199	7.361 918	15.86077	34.17100
400	160 000	20.00000	63.24555	64 000 000	7.368 063	15.87401	34.19952

n	n^2	$\sqrt{n}$	$\sqrt{10n}$	n^3	$\sqrt[3]{n}$	$\sqrt[3]{10n}$	$\sqrt[3]{100n}$
400	160 000	20.00000	63.24555	64 000 000	7.368 063	15.87401	34.19952
401	160 801	20.02498	63.32456	64 481 201	7.374 198	15.88723	34.22799
402	161 604	20.04994	63.40347	64 964 808	7.380 323	15.90042	34.25642
403	162 409	20.07486	63.48228	65 450 827	7.386 437	15.91360	34.28480
404	163 216	20.09975	63.56099	65 939 264	7.392 542	15.92675	34.31314
405	164 025	20.12461	63.63961	66 430 125	7.398 636	15.93988	34.34143
406	164 836	20.14944	63.71813	66 923 416	7.404 721	15.95299	34.36967
407	165 649	20.17424	63.79655	67 419 143	7.410 795	15.96607	34.39786
408	166 464	20.19901	63.87488	67 917 312	7.416 860	15.97914	34.42601
409	167 281	20.22375	63.95311	68 417 929	7.422 914	15.99218	34.45412
410	168 100	20.24846	64.03124	68 921 000	7.428 959	16.00521	34.48217
411	168 921	20.27313	64.10928	69 426 531	7.434 994	16.01821	34.51018
412	169 744	20.29778	64.18723	69 934 528	7.441 019	16.03119	34.53815
413	170 569	20.32240	64.26508	70 444 997	7.447 034	16.04415	34.56607
414	171 396	20.34699	64.34283	70 957 944	7.453 040	16.05709	34.59395
415	172 225	20.37155	64.42049	71 473 375	7.459 036	16.07001	34.62178
416	173 056	20.39608	64.49806	71 991 296	7.465 022	16.08290	34.64956
417	173 889	20.42058	64.57554	72 511 713	7.470 999	16.09578	34.67731
418	174 724	20.44505	64.65292	73 034 632	7.476 966	16.10864	34.70500
419	175 561	20.46949	64.73021	73 560 059	7.482 924	16.12147	34.73266
420	176 400	20.49390	64.80741	74 088 000	7.488 872	16.13429	34.76027
421	177 241	20.51828	64.88451	74 618 461	7.494 811	16.14708	34.78783
422	178 084	20.54264	64.96153	75 151 448	7.500 741	16.15986	34.81535
423	178 929	20.56696	65.03845	75 686 967	7.506 661	16.17261	34.84283
424	179 776	20.59126	65.11528	76 225 024	7.512 572	16.18534	34.87027
425	180 625	20.61553	65.19202	76 765 625	7.518 473	16.19806	34.89766
426	181 476	20.63977	65.26868	77 308 776	7.524 365	16.21075	34.92501
427	182 329	20.66398	65.34524	77 854 483	7.530 248	16.22343	34.95232
428	183 184	20.68816	65.42171	78 402 752	7.536 122	16.23608	34.97958
429	184 041	20.71232	65.49809	78 953 589	7.541 987	16.24872	35.00680
430	184 900	20.73644	65.57439	79 507 000	7.547 842	16.26133	35.03398
431	185 761	20.76054	65.65059	80 062 991	7.553 689	16.27393	35.06112
432	186 624	20.78461	65.72671	80 621 568	7.559 526	16.28651	35.08821
433	187 489	20.80865	65.80274	81 182 737	7.565 355	16.29906	35.11527
434	188 356	20.83267	65.87868	81 746 504	7.571 174	16.31160	35.14228
435	189 225	20.85665	65.95453	82 312 875	7.576 985	16.32412	35.16925
436	190 096	20.88061	66.03030	82 881 856	7.582 787	16.33662	35.19618
437	190 969	20.90454	66.10598	83 453 453	7.588 579	16.34910	35.22307
438	191 844	20.92845	66.18157	84 027 672	7.594 363	16.36156	35.24991
439	192 721	20.95233	66.25708	84 604 519	7.600 139	16.37400	35.27672
440	193 600	20.97618	66.33250	85 184 000	7.605 905	16.38643	35.30348
441	194 481	21.00000	66.40783	85 766 121	7.611 663	16.39883	35.33021
442	195 364	21.02380	66.48308	86 350 888	7.617 412	16.41122	35.35689
443	196 249	21.04757	66.55825	86 938 307	7.623 152	16.42358	35.38354
444	197 136	21.07131	66.63332	87 528 384	7.628 884	16.43593	35.41014
445	198 025	21.09502	66.70832	88 121 125	7.634 607	16.44826	35.43671
446	198 916	21.11871	66.78323	88 716 536	7.640 321	16.46057	35.46323
447	199 809	21.14237	66.85806	89 314 623	7.646 027	16.47287	35.48971
448	200 704	21.16601	66.93280	89 915 392	7.651 725	16.48514	35.51616
449	201 601	21.18962	67.00746	90 518 849	7.657 414	16.49740	35.54257
450	202 500	21.21320	67.08204	91 125 000	7.663 094	16.50964	35.56893

SQUARES, CUBES AND ROOTS (Continued)

n	n^2	$\sqrt{n}$	$\sqrt{10n}$	n^3	$\sqrt[3]{n}$	$\sqrt[3]{10n}$	$\sqrt[3]{100n}$
450	202 500	21.21320	67.08204	91 125 000	7.663 094	16.50964	35.56893
451	203 401	21.23676	67.15653	91 733 851	7.668 766	16.52186	35.59526
452	204 304	21.26029	67.23095	92 345 408	7.674 430	16.53406	35.62155
453	205 209	21.28380	67.30527	92 959 677	7.680 086	16.54624	35.64780
454	206 116	21.30728	67.37952	93 576 664	7.685 733	16.55841	35.67401
455	207 025	21.33073	67.45369	94 196 375	7.691 372	16.57056	35.70018
456	207 936	21.35416	67.52777	94 818 816	7.697 002	16.58269	35.72632
457	208 849	21.37756	67.60178	95 443 993	7.702 625	16.59480	35.75242
458	209 764	21.40093	67.67570	96 071 912	7.708 239	16.60690	35.77848
459	210 681	21.42429	67.74954	96 702 579	7.713 845	16.61897	35.80450
460	211 600	21.44761	67.82330	97 336 000	7.719 443	16.63103	35.83048
461	212 521	21.47091	67.89698	97 972 181	7.725 032	16.64308	35.85642
462	213 444	21.49419	67.97058	98 611 128	7.730 614	16.65510	35.88233
463	214 369	21.51743	68.04410	99 252 847	7.736 188	16.66711	35.90820
464	215 296	21.54066	68.11755	99 897 344	7.741 753	16.67910	35.93404
465	216 225	21.56386	68.19091	100 544 625	7.747 311	16.69108	35.95983
466	217 156	21.58703	68.26419	101 194 696	7.752 861	16.70303	35.98559
467	218 089	21.61018	68.33740	101 847 563	7.758 402	16.71497	36.01131
468	219 024	21.63331	68.41053	102 503 232	7.763 936	16.72689	36.03700
469	219 961	21.65641	68.48357	103 161 709	7.769 462	16.73880	36.06265
470	220 900	21.67948	68.55655	103 823 000	7.774 980	16.75069	36.08826
471	221 841	21.70253	68.62944	104 487 111	7.780 490	16.76256	36.11384
472	222 784	21.72556	68.70226	105 154 048	7.785 993	16.77441	36.13938
473	223 729	21.74856	68.77500	105 823 817	7.791 488	16.78625	36.16488
474	224 676	21.77154	68.84766	106 496 424	7.796 975	16.79807	36.19035
475	225 625	21.79449	68.92024	107 171 875	7.802 454	16.80988	36.21578
476	226 576	21.81742	68.99275	107 850 176	7.807 925	16.82167	36.24118
477	227 529	21.84033	69.06519	108 531 333	7.813 389	16.83344	36.26654
478	228 484	21.86321	69 13754	109 215 352	7.818 846	16.84519	36.29187
479	229 441	21.88607	69.20983	109 902 239	7.824 294	16.85693	36.31716
480	230 400	21.90890	69 28203	110 592 000	7.829 735	16.86865	36.34241
481	231 361	21.93171	69.35416	111 284 641	7.835 169	16.88036	36.36763
482	232 324	21.95450	69.42622	111 980 168	7.840 595	16.89205	36.39282
483	233 289	21.97726	69 49820	112 678 587	7.846 013	16.90372	36.41797
484	234 256	22.00000	69.57011	113 379 904	7.851 424	16.91538	36.44308
485	235 225	22.02272	69.64194	114 084 125	7.856 828	16.92702	36.46817
486	236 196	22.04541	69.71370	114 791 256	7.862 224	16.93865	36.49321
487	237 169	22.06808	69.78539	115 501 303	7.867 613	16.95026	36.51822
488	238 144	22.09072	69.85700	116 214 272	7.872 994	16.96185	36.54320
489	239 121	22.11334	69.92853	116 930 169	7.878 368	16.97343	36.56815
490	240 100	22.13594	70.00000	117 649 000	7.883 735	16.98499	36.59306
491	241 081	22.15852	70.07139	118 370 771	7.889 095	16.99654	36.61793
492	242 064	22.18107	70.14271	119 095 488	7.894 447	17.00807	36.64278
493	243 049	22.20360	70.21396	119 823 157	7.899 792	17.01959	36.66758
494	244 036	22.22611	70.28513	120 553 784	7.905 129	17.03108	36.69236
495	245 025	22.24860	70.35624	121 287 375	7.910 460	17.04257	36.71710
496	246 016	22.27106	70.42727	122 023 936	7.915 783	17.05404	36.74181
497	247 009	22.29350	70.49823	122 763 473	7.921 099	17.06549	36.76649
498	248 004	22.31591	70.56912	123 505 992	7.926 408	17.07693	36.79113
499	249 001	22.33831	70.63993	124 251 499	7.931 710	17.08835	36.81574
500	250 000	22.36068	70.71068	125 000 000	7.937 005	17.09976	36.84031

SQUARES, CUBES AND ROOTS (Continued)

n	n^2	$\sqrt{n}$	$\sqrt{10n}$	n^3	$\sqrt[3]{n}$	$\sqrt[3]{10n}$	$\sqrt[3]{100n}$
500	250 000	22.36068	70.71068	125 000 000	7.937 005	17.09976	36.84031
501	251 001	22.38303	70.78135	125 751 501	7.942 293	17.11115	36.86486
502	252 004	22.40536	70.85196	126 506 008	7.947 574	17.12253	36.88937
503	253 009	22.42766	70.92249	127 263 527	7.952 848	17.13389	36.91385
504	254 016	22.44994	70.99296	128 024 064	7.958 114	17.14524	36.93830
505	255 025	22.47221	71.06335	128 787 625	7.963 374	17.15657	36.96271
506	256 036	22.49444	71.13368	129 554 216	7.968 627	17.16789	36.98709
507	257 049	22.51666	71.20393	130 323 843	7.973 873	17.17919	37.01144
508	258 064	22.53886	71.27412	131 096 512	7.979 112	17.19048	37.03576
509	259 081	22.56103	71.34424	131 872 229	7.984 344	17.20175	37.06004
510	260 100	22.58318	71.41428	132 651 000	7.989 570	17.21301	37.08430
511	261 121	22.60531	71.48426	133 432 831	7.994 788	17.22425	37.10852
512	262 144	22.62742	71.55418	134 217 728	8.000 000	17.23548	37.13271
513	263 169	22.64950	71.62402	135 005 697	8.005 205	17.24669	37.15687
514	264 196	22.67157	71.69379	135 796 744	8.010 403	17.25789	37.18100
515	265 225	22.69361	71.76350	136 590 875	8.015 595	17.26908	37.20509
516	266 256	22.71563	71.83314	137 388 096	8.020 779	17.28025	37.22916
517	267 289	22.73763	71.90271	138 188 413	8.025 957	17.29140	37.25319
518	268 324	22.75961	71.97222	138 991 832	8.031 129	17.30254	37.27720
519	269 361	22.78157	72.04165	139 798 359	8.036 293	17.31367	37.30117
520	270 400	22.80351	72.11103	140 608 000	8.041 452	17.32478	37.32511
521	271 441	22.82542	72.18033	141 420 761	8.046 603	17.33588	37.34902
522	272 484	22.84732	72.24957	142 236 648	8.051 748	17.34696	37.37290
523	273 529	22.86919	72.31874	143 055 667	8.056 886	17.35804	37.39675
524	274 576	22.89105	72.38784	143 877 824	8.062 018	17.36909	37.42057
525	275 625	22.91288	72.45688	144 703 125	8.067 143	17.38013	37.44436
526	276 676	22.93469	72.52586	145 531 576	8.072 262	17.39116	37.46812
527	277 729	22.95648	72.59477	146 363 183	8.077 374	17.40218	37.49185
528	278 784	22.97825	72.66361	147 197 952	8.082 480	17.41318	37.51555
529	279 841	23.00000	72.73239	148 035 889	8.087 579	17.42416	37.53922
530	280 900	23.02173	72.80110	148 877 000	8.092 672	17.43513	37.56286
531	281 961	23.04344	72.86975	149 721 291	8.097 759	17.44609	37.58647
532	283 024	23.06513	72.93833	150 568 768	8.102 839	17.45704	37.61005
533	284 089	23.08679	73.00685	151 419 437	8.107 913	17.46797	37.63360
534	285 156	23.10844	73.07530	152 273 304	8.112 980	17.47889	37.65712
535	286 225	23.13007	73.14369	153 130 375	8.118 041	17.48979	37.68061
536	287 296	23.15167	73.21202	153 990 656	8.123 096	17.50068	37.70407
537	288 369	23.17326	73.28028	154 854 153	8.128 145	17.51156	37.72751
538	289 444	23.19483	73.34848	155 720 872	8.133 187	17.52242	37.75091
539	290 521	23.21637	73.41662	156 590 819	8.138 223	17.53327	37.77429
540	291 600	23.23790	73.48469	157 464 000	8.143 253	17.54411	37.79763
541	292 681	23.25941	73.55270	158 340 421	8.148 276	17.55493	37.82095
542	293 764	23.28089	73.62065	159 220 088	8.153 294	17.56574	37.84424
543	294 849	23.30236	73.68853	160 103 007	8.158 305	17.57654	37.86750
544	295 936	23.32381	73.75636	160 989 184	8.163 310	17.58732	37.89073
545	297 025	23.34524	73.82412	161 878 625	8.168 309	17.59809	37.91393
546	298 116	23.36664	73.89181	162 771 336	8.173 302	17.60885	37.93711
547	299 209	23.38803	73.95945	163 667 323	8.178 289	17.61959	37.96025
548	300 304	23.40940	74.02702	164 566 592	8.183 269	17.63032	37.98337
549	301 401	23.43075	74.09453	165 469 149	8.188 244	17.64104	38.00646
550	302 500	23.45208	74.16198	166 375 000	8.193 213	17.65174	38.02952

SQUARES, CUBES AND ROOTS (Continued)

n	n^2	$\sqrt{n}$	$\sqrt{10n}$	n^3	$\sqrt[3]{n}$	$\sqrt[3]{10n}$	$\sqrt[3]{100n}$
550	302 500	23.45208	74.16198	166 375 000	8.193 213	17.65174	38.02952
551	303 601	23.47339	74.22937	167 284 151	8.198 175	17.66243	38.05256
552	304 704	23.49468	74.29670	168 196 608	8.203 132	17.67311	38.07557
553	305 809	23.51595	74.36397	169 112 377	8.208 082	17.68378	38.09854
554	306 916	23.53720	74.43118	170 031 464	8.213 027	17.69443	38.12149
555	308 025	23.55844	74.49832	170 953 875	8.217 966	17.70507	38.14442
556	309 136	23.57965	74.56541	171 879 616	8.222 899	17.71570	38.16731
557	310 249	23.60085	74.63243	172 808 693	8.227 825	17.72631	38.19018
558	311 364	23.62202	74.69940	173 741 112	8.232 746	17.73691	38.21302
559	312 481	23.64318	74.76630	174 676 879	8.237 661	17.74750	38.23584
560	313 600	23.66432	74.83315	175 616 000	8.242 571	17.75808	38.25862
561	314 721	23.68544	74.89993	176 558 481	8.247 474	17.76864	38.28138
562	315 844	23.70654	74.96666	177 504 328	8.252 372	17.77920	38.30412
563	316 969	23.72762	75.03333	178 453 547	8.257 263	17.78973	38.32682
564	318 096	23.74868	75.09993	179 406 144	8.262 149	17.80026	38.34950
565	319 225	23.76973	75.16648	180 362 029	8.267 029	17.81077	38.37215
566	320 356	23.79075	75.23297	181 321 496	8.271 904	17.82128	38.39478
567	321 489	23.81176	75.29940	182 284 263	8.276 773	17.83177	38.41737
568	322 624	23.83275	75.36577	183 250 432	8.281 635	17.84224	38.43995
569	323 761	23.85372	75.43209	184 220 009	8.286 493	17.85271	38.46249
570	324 900	23.87467	75.49834	185 193 000	8.291 344	17.86316	38.48501
571	326 041	23.89561	75.56454	186 169 411	8.296 190	17.87360	38.50750
572	327 184	23.91652	75.63068	187 149 248	8.301 031	17.88403	38.52997
573	328 329	23.93742	75.69676	188 132 517	8.305 865	17.89444	38.55241
574	329 476	23.95830	75.76279	189 119 224	8.310 694	17.90485	38.57482
575	330 625	23.97916	75.82875	190 109 375	8.315 517	17.91524	38.59721
576	331 776	24.00000	75.89466	191 102 976	8.320 335	17.92562	38.61958
577	332 929	24.02082	75.96052	192 100 033	8.325 148	17.93599	38.64191
578	334 084	24.04163	76.02631	193 100 552	8.329 954	17.94634	38.66422
579	335 241	24.06242	76.09205	194 104 539	8.334 755	17.95669	38.68651
580	336 400	24.08319	76.15773	195 112 000	8.339 551	17.96702	38.70877
581	337 561	24.10394	76.22336	196 122 941	8.344 341	17.97734	38.73100
582	338 724	24.12468	76.28892	197 137 368	8.349 126	17.98765	38.75321
583	339 889	24.14539	76.35444	198 155 287	8.353 905	17.99794	38.77539
584	341 056	24.16609	76.41989	199 176 704	8.358 678	18.00823	38.79755
585	342 225	24.18677	76.48529	200 201 625	8.363 447	18.01850	38.81968
586	343 396	24.20744	76.55064	201 230 056	8.368 209	18.02876	38.84179
587	344 569	24.22808	76.61593	202 262 003	8.372 967	18.03901	38.86387
588	345 744	24.24871	76.68116	203 297 472	8.377 719	18.04925	38.88593
589	346 921	24.26932	76.74634	204 336 469	8.382 465	18.05947	38.90796
590	348 100	24.28992	76.81146	205 379 000	8.387 207	18.06969	38.92996
591	349 281	24.31049	76.87652	206 425 071	8.391 942	18.07989	38.95195
592	350 464	24.33105	76.94154	207 474 688	8.396 673	18.09008	38.97390
593	351 649	24.35159	77.00649	208 527 857	8.401 398	18.10026	38.99584
594	352 836	24.37212	77.07140	209 584 584	8.406 118	18.11043	39.01774
595	354 025	24.39262	77.13624	210 644 875	8.410 833	18.12059	39.03963
596	355 216	24.41311	77.20104	211 708 736	8.415 542	18.13074	39.06149
597	356 409	24.43358	77.26578	212 776 173	8.420 246	18.14087	39.08332
598	357 604	24.45404	77.33046	213 847 192	8.424 945	18.15099	39.10513
599	358 801	24.47448	77.39509	214 921 799	8.429 638	18.16111	39.12692
600	360 000	24.49490	77.45967	216 000 000	8.434 327	18.17121	39.14868

SQUARES, CUBES AND ROOTS (Continued)

n	n^2	$\sqrt{n}$	$\sqrt{10n}$	n^3	$\sqrt[3]{n}$	$\sqrt[3]{10n}$	$\sqrt[3]{100n}$
600	360 000	24.49490	77.45967	216 000 000	8.434 327	18.17121	39.14868
601	361 201	24.51530	77.52419	217 081 801	8.439 010	18.18130	39.17041
602	362 404	24.53569	77.58866	218 167 208	8.443 688	18.19137	39.19213
603	363 609	24.55606	77.65307	219 256 227	8.448 361	18.20144	39.21382
604	364 816	24.57641	77.71744	220 348 864	8.453 028	18.21150	39.23548
605	366 025	24.59675	77.78175	221 445 125	8.457 691	18.22154	39.25712
606	367 236	24.61707	77.84600	222 545 016	8.462 348	18.23158	39.27874
607	368 449	24.63737	77.91020	223 648 543	8.467 000	18.24160	39.30033
608	369 664	24.65766	77.97435	224 755 712	8.471 647	18.25161	39.32190
609	370 881	24.67793	78.03845	225 866 529	8.476 289	18.26161	39.34345
610	372 100	24.69818	78.10250	226 981 000	8.480 926	18.27160	39.36497
611	373 321	24.71841	78.16649	228 099 131	8.485 558	18.28158	39.38647
612	374 544	24.73863	78.23043	229 220 928	8.490 185	18.29155	39.40795
613	375 769	24.75884	78.29432	230 346 397	8.494 807	18.30151	39.42940
614	376 996	24.77902	78.35815	231 475 544	8.499 423	18.31145	39.45083
615	378 225	24.79919	78.42194	232 608 375	8.504 035	18.32139	39.47223
616	379 456	24.81935	78.48567	233 744 896	8.508 642	18.33131	39.49362
617	380 689	24.83948	78.54935	234 885 113	8.513 243	18.34123	39.51498
618	381 924	24.85961	78.61298	236 029 032	8.517 840	18.35113	39.53631
619	383 161	24.87971	78.67655	237 176 659	8.522 432	18.36102	39.55763
620	384 400	24.89980	78.74008	238 328 000	8.527 019	18.37091	39.57892
621	385 641	24.91987	78.80355	239 483 061	8.531 601	18.38078	39.60018
622	386 884	24.93993	78.86698	240 641 848	8.536 178	18.39064	39.62143
623	388 129	24.95997	78.93035	241 804 367	8.540 750	18.40049	39.64265
624	389 376	24.97999	78.99367	242 970 624	8.545 317	18.41033	39.66385
625	390 625	25.00000	79.05694	244 140 625	8.549 880	18.42016	39.68503
626	391 876	25.01999	79.12016	245 314 376	8.554 437	18.42998	39.70618
627	393 129	25.03997	79.18333	246 491 883	8.558 990	18.43978	39.72731
628	394 384	25.05993	79.24645	247 673 152	8.563 538	18.44958	39.74842
629	395 641	25.07987	79.30952	248 858 189	8.568 081	18.45937	39.76951
630	396 900	25.09980	79.37254	250 047 000	8.572 619	18.46915	39.79057
631	398 161	25.11971	79.43551	251 239 591	8.577 152	18.47891	39.81161
632	399 424	25.13961	79.49843	252 435 968	8.581 681	18.48867	39.83263
633	400 689	25.15949	79.56130	253 636 137	8.586 205	18.49842	39.85363
634	401 956	25.17936	79.62412	254 840 104	8.590 724	18.50815	39.87461
635	403 225	25.19921	79.68689	256 047 875	8.595 238	18.51788	39.89556
636	404 496	25.21904	79.74961	257 259 456	8.599 748	18.52759	39.91649
637	405 769	25.23886	79.81228	258 474 853	8.604 252	18.53730	39.93740
638	407 044	25.25866	79.87490	259 694 072	8.608 753	18.54700	39.95829
639	408 321	25.27845	79.93748	260 917 119	8.613 248	18.55668	39.97916
640	409 600	25.29822	80.00000	262 144 000	8.617 739	18.56636	40.00000
641	410 881	25.31798	80.06248	263 374 721	8.622 225	18.57602	40.02082
642	412 164	25.33772	80.12490	264 609 288	8.626 706	18.58568	40.04162
643	413 449	25.35744	80.18728	265 847 707	8.631 183	18.59532	40.06240
644	414 736	25 37716	80.24961	267 089 984	8.635 655	18.60495	40.08316
645	416 025	25.39685	80.31189	268 336 125	8.640 123	18.61458	40.10390
646	417 316	25.41653	80.37413	269 586 136	8.644 585	18.62419	40.12461
647	418 609	25.43619	80.43631	270 840 023	8.649 044	18.63380	40.14530
648	419 904	25.45584	80.49845	272 097 792	8.653 497	18.64340	40.16598
649	421 201	25.47548	80.56054	273 359 449	8.657 947	18.65298	40.18663
650	422 500	25.49510	80.62258	274 625 000	8.662 391	18.66256	40.20726

n	n^2	$\sqrt{n}$	$\sqrt{10n}$	n^3	$\sqrt[3]{n}$	$\sqrt[3]{10n}$	$\sqrt[3]{100n}$
650	422 500	25 49510	80.62258	274 625 000	8.662 391	18.66256	40.20726
651	423 801	25.51470	80.68457	275 894 451	8.666 831	18.67212	40.22787
652	425 104	25.53429	80.74652	277 167 808	8.671 266	18.68168	40.24845
653	426 409	25.55386	80.80842	278 445 077	8.675 697	18.69122	40.26902
654	427 716	25.57342	80.87027	279 726 264	8.680 124	18.70076	40.28957
655	429 025	25 59297	80.93207	281 011 375	8.684 546	18.71029	40.31009
656	430 336	25.61250	80.99383	282 300 416	8.688 963	18.71980	40.33059
657	431 649	25.63201	81.05554	283 593 393	8.693 376	18.72931	40.35108
658	432 964	25.65151	81.11720	284 890 312	8.697 784	18.73881	40.37154
659	434.281	25.67100	81.17881	286 191 179	8.702 188	18.74830	40.39198
660	435 600	25.69047	81.24038	287 496 000	8.706 588	18.75777	40.41240
661	436 921	25.70992	81.30191	288 804 781	8.710 983	18.76724	40.43280
662	438 244	25.72936	81.36338	290 117 528	8.715 373	18.77670	40.45318
663	439 569	25.74879	81.42481	291 434 247	8.719 760	18.78615	40.47354
664	440 896	25.76820	81.48620	292 754 944	8.724 141	18.79559	40.49388
665	442 225	25.78759	81.54753	294 079 625	8.728 519	18.80502	40.51420
666	443 556	25.80698	81.60882	295 408 296	8.732 892	18.81444	40.53449
667	444 889	25.82634	81.67007	296 740 963	8.737 260	18.82386	40.55477
668	446 224	25.84570	81.73127	298 077 632	8.741 625	18.83326	40.57503
669	447 561	25.86503	81.79242	299 418 309	8.745 985	18.84265	40.59526
670	448 900	25.88436	81.85353	300 763 000	8.750 340	18.85204	40.61548
671	450 241	25.90367	81.91459	302 111 711	8.754 691	18.86141	40.63568
672	451 584	25.92296	81.97561	303 464 448	8.759 038	18.87078	40.65585
673	452 929	25.94224	82.03658	304 821 217	8.763 381	18.88013	40.67601
674	454 276	25.96151	82.09750	306 182 024	8.767 719	18.88948	40.69615
675	455 625	25.98076	82.15838	307 546 875	8.772 053	18.89882	40.71626
676	456 976	26.00000	82.21922	308 915 776	8.776 383	18.90814	40.73636
677	458 329	26.01922	82.28001	310 288 733	8.780 708	18.91746	40.75644
678	459 684	26.03843	82.34076	311 665 752	8.785 030	18.92677	40.77650
679	461 041	26.05763	82.40146	313 046 839	8.789 347	18.93607	40.79653
680	462 400	26.07681	82.46211	314 432 000	8.793 659	18.94536	40.81655
681	463 761	26.09598	82.52272	315 821 241	8.797 968	18.95465	40.83655
682	465 124	26.11513	82.58329	317 214 568	8.802 272	18.96392	40.85653
683	466 489	26.13427	82.64381	318 611 987	8.806 572	18.97318	40.87649
684	467 856	26.15339	82.70429	320 013 504	8.810 868	18.98244	40.89643
685	469 225	26.17250	82.76473	321 419 125	8.815 160	18.99169	40.91635
686	470 596	26.19160	82.82512	322 828 856	8.819 447	19.00092	40.93625
687	471 969	26.21068	82.88546	324 242 703	8.823 731	19.01015	40.95613
688	473 344	26.22975	82.94577	325 660 672	8.828 010	19.01937	40.97599
689	474 721	26.24881	83.00602	327 082 769	8.832 285	19.02858	40.99584
690	476 100	26.26785	83.06624	328 509 000	8.836 556	19.03778	41.01566
691	477 481	26.28688	83.12641	329 939 371	8.840 823	19.04698	41.03546
692	478 864	26.30589	83.18654	331 373 888	8.845 085	19.05616	41.05525
693	480 249	26.32489	83.24662	332 812 557	8.849 344	19.06533	41.07502
694	481 636	26.34388	83.30666	334 255 384	8.853 599	19.07450	41.09476
695	483 025	26.36285	83.36666	335 702 375	8.857 849	19.08366	41.11449
696	484 416	26.38181	83.42661	337 153 536	8.862 095	19.09281	41.13420
697	485 809	26.40076	83.48653	338 608 873	8.866 338	19.10195	41.15389
698	487 204	26.41969	83.54639	340 068 392	8.870 576	19.11108	41.17357
699	488 601	26.43861	83.60622	341 532 099	8.874 810	19.12020	41.19322
700	490 000	26.45751	83.66600	343 000 000	8.879 040	19.12931	41.21285

n	n^2	$\sqrt{n}$	$\sqrt{10n}$	n^3	$\sqrt[3]{n}$	$\sqrt[3]{10n}$	$\sqrt[3]{100n}$
700	490 000	26.45751	83.66600	343 000 000	8.879 040	19.12931	41.21285
701	491 401	26.47640	83.72574	344 472 101	8.883 266	19.13842	41.23247
702	492 804	26.49528	83.78544	345 948 408	8.887 488	19.14751	41.25207
703	494 209	26.51415	83.84510	347 428 927	8.891 706	19.15660	41.27164
704	495 616	26.53300	83.90471	348 913 664	8.895 920	19.16568	41.29120
705	497 025	26.55184	83.96428	350 402 625	8.900 130	19.17475	41.31075
706	498 436	26.57066	84.02381	351 895 816	8.904 337	19.18381	41.33027
707	499 849	26.58947	84.08329	353 393 243	8.908 539	19.19286	41.34977
708	501 264	26.60827	84.14274	354 894 912	8.912 737	19.20191	41.36926
709	502 681	26.62705	84.20214	356 400 829	8.916 931	19.21095	41.38873
710	504 100	26.64583	84.26150	357 911 000	8.921 121	19.21997	41.40818
711	505 521	26.66458	84.32082	359 425 431	8.925 308	19.22899	41.42761
712	506 944	26.68333	84.38009	360 944 128	8.929 490	19.23800	41.44702
713	508 369	26.70206	84.43933	362 467 097	8.933 669	19.24701	41.46642
714	509 796	26.72078	84.49852	363 994 344	8.937 843	19.25600	41.48579
715	511 225	26.73948	84.55767	365 525 875	8.942 014	19.26499	41.50515
716	512 656	26.75818	84.61678	367 061 696	8.946 181	19.27396	41.52449
717	514 089	26.77686	84.67585	368 601 813	8.950 344	19.28293	41.54382
718	515 524	26.79552	84.73488	370 146 232	8.954 503	19.29189	41.56312
719	516 961	26.81418	84.79387	371 694 959	8.958 658	19.30084	41.58241
720	518 400	26.83282	84.85281	373 248 000	8.962 809	19.30979	41.60168
721	519 841	26.85144	84.91172	374 805 361	8.966 957	19.31872	41.62093
722	521 284	26.87006	84.97058	376 367 048	8.971 101	19.32765	41.64016
723	522 729	26.88866	85.02941	377 933 067	8.975 241	19.33657	41.65938
724	524 176	26.90725	85.08819	379 503 424	8.979 377	19.34548	41.67857
725	525 625	26.92582	85.14693	381 078 125	8.983 509	19.35438	41.69775
726	527 076	26.94439	85.20563	382 657 176	8.987 637	19.36328	41.71692
727	528 529	26.96294	85.26429	384 240 583	8.991 762	19.37216	41.73606
728	529 984	26.98148	85.32292	385 828 352	8.995 883	19.38104	41.75519
729	531 441	27.00000	85.38150	387 420 489	9.000 000	19.38991	41.77430
730	532 900	27.01851	85.44004	389 017 000	9.004 113	19.39877	41.79339
731	534 361	27.03701	85.49854	390 617 891	9.008 223	19.40763	41.81247
732	535 824	27.05550	85.55700	392 223 168	9.012 329	19.41647	41.83152
733	537 289	27.07397	85.61542	393 832 837	9.016 431	19.42531	41.85056
734	538 756	27.09243	85.67380	395 446 904	9.020 529	19.43414	41.86959
735	540 225	27.11088	85.73214	397 065 375	9.024 624	19.44296	41.88859
736	541 696	27.12932	85.79044	398 688 256	9.028 715	19.45178	41.90758
737	543 169	27.14774	85.84870	400 315 553	9.032 802	19.46058	41.92655
738	544 644	27.16616	85.90693	401 947 272	9.036 886	19.46938	41.94551
739	546 121	27.18455	85.96511	403 583 419	9.040 966	19.47817	41.96444
740	547 600	27.20294	86.02325	405 224 000	9.045 042	19.48695	41.98336
741	549 081	27.22132	86.08136	406 869 021	9.049 114	19.49573	42.00227
742	550 564	27.23968	86.13942	408 518 488	9.053 183	19.50449	42.02115
743	552 049	27.25803	86.19745	410 172 407	9.057 248	19.51325	42.04002
744	553 536	27.27636	86.25543	411 830 784	9.061 310	19.52200	42.05887
745	555 025	27.29469	86.31338	413 493 625	9.065 368	19.53074	42.07771
746	556 516	27.31300	86.37129	415 160 936	9.069 422	19.53948	42.09653
747	558 009	27.33130	86.42916	416 832 723	9.073 473	19.54820	42.11533
748	559 504	27.34959	86.48699	418 508 992	9.077 520	19.55692	42.13411
749	561 001	27.36786	86.54479	420 189 749	9.081 563	19.56563	42.15288
750	562 500	27.38613	86.60254	421 875 000	9.085 603	19.57434	42.17163

SQUARES, CUBES AND ROOTS (Continued)

n	n^2	$\sqrt{n}$	$\sqrt{10n}$	n^3	$\sqrt[3]{n}$	$\sqrt[3]{10n}$	$\sqrt[3]{100n}$
750	562 500	27.38613	86.60254	421 875 000	9.085 603	19.57434	42.17163
751	564 001	27.40438	86.66026	423 564 751	9.089 639	19.58303	42.19037
752	565 504	27.42262	86.71793	425 259 008	9.093 672	19.59172	42.20909
753	567 009	27.44085	86.77557	426 957 777	9.097 701	19.60040	42.22779
754	568 516	27.45906	86.83317	428 661 064	9.101 727	19.60908	42.24647
755	570 025	27.47726	86.89074	430 368 875	9.105 748	19.61774	42.26514
756	571 536	27.49545	86.94826	432 081 216	9.109 767	19.62640	42.28379
757	573 049	27.51363	87.00575	433 798 093	9.113 782	19.63505	42.30243
758	574 564	27.53180	87.06320	435 519 512	9.117 793	19.64369	42.32105
759	576 081	27.54995	87.12061	437 245 479	9.121 801	19.65232	42.33965
760	577 600	27.56810	87.17798	438 976 000	9.125 805	19.66095	42.35824
761	579 121	27.58623	87.23531	440 711 081	9.129 806	19.66957	42.37681
762	580 644	27.60435	87.29261	442 450 728	9.133 803	19.67818	42.39536
763	582 169	27.62245	87.34987	444 194 947	9.137 797	19.68679	42.41390
764	583 696	27.64055	87.40709	445 943 744	9.141 787	19.69538	42.43242
765	585 225	27.65863	87.46428	447 697 125	9.145 774	19.70397	42.45092
766	586 756	27.67671	87.52143	449 455 096	9.149 758	19.71256	42.46941
767	588 289	27.69476	87.57854	451 217 663	9.153 738	19.72113	42.48789
768	589 824	27.71281	87.63561	452 984 832	9.157 714	19.72970	42.50634
769	591 361	27.73085	87.69265	454 756 609	9.161 687	19.73826	42.52478
770	592 900	27.74887	87.74964	456 533 000	9.165 656	19.74681	42.54321
771	594 441	27.76689	87.80661	458 314 011	9.169 623	19.75535	42.56162
772	595 984	27.78489	87.86353	460 099 648	9.173 585	19.76389	42.58001
773	597 529	27.80288	87.92042	461 889 917	9.177 544	19.77242	42.59839
774	599 076	27.82086	87.97727	463 684 824	9.181 500	19.78094	42.61675
775	600 625	27.83882	88.03408	465 484 375	9.185 453	19.78946	42.63509
776	602 176	27.85678	88.09086	467 288 576	9.189 402	19.79797	42.65342
777	603 729	27.87472	88.14760	469 097 433	9.193 347	19.80647	42.67174
778	605 284	27.89265	88.20431	470 910 952	9.197 290	19.81496	42.69004
779	606 841	27.91057	88.26098	472 729 139	9.201 229	19.82345	42.70832
780	608 400	27.92848	88.31761	474 552 000	9.205 164	19.83192	42.72659
781	609 961	27.94638	88.37420	476 379 541	9.209 096	19.84040	42.74484
782	611 524	27.96426	88.43076	478 211 768	9.213 025	19.84886	42.76307
783	613 089	27.98214	88.48729	480 048 687	9.216 950	19.85732	42.78129
784	614 656	28.00000	88.54377	481 890 304	9.220 873	19.86577	42.79950
785	616 225	28.01785	88.60023	483 736 625	9.224 791	19.87421	42.81769
786	617 796	28.03569	88.65664	485 587 656	9.228 707	19.88265	42.83586
787	619 369	28.05352	88.71302	487 443 403	9.232 619	19.89107	42.85402
788	620 944	28.07134	88.76936	489 303 872	9.236 528	19.89950	42.87216
789	622 521	28.08914	88.82567	491 169 069	9.240 433	19.90791	42.89029
790	624 100	28.10694	88.88194	493 039 000	9.244 335	19.91632	42.90840
791	625 681	28.12472	88.93818	494 913 671	9.248 234	19.92472	42.92650
792	627 264	28.14249	88.99438	496 793 088	9.252 130	19.93311	42.94458
793	628 849	28.16026	89.05055	498 677 257	9.256 022	19.94150	42.96265
794	630 436	28.17801	89.10668	500 566 184	9.259 911	19.94987	42.98070
795	632 025	28.19574	89.16277	502 459 875	9.263 797	19.95825	42.99874
796	633 616	28.21347	89.21883	504 358 336	9.267 680	19.96661	43.01676
797	635 209	28.23119	89.27486	506 261 573	9.271 559	19.97497	43.03477
798	636 804	28.24889	89.33085	508 169 592	9.275 435	19.98332	43.05276
799	638 401	28.26659	89.38680	510 082 399	9.279 308	19.99166	43.07073
800	640 000	28.28427	89.44272	512 000 000	9.283 178	20.00000	43.08869
n	n^2	$\sqrt{n}$	$\sqrt{10n}$	n^3	$\sqrt[3]{n}$	$\sqrt[3]{10n}$	$\sqrt[3]{100n}$

n	n^2	$\sqrt{n}$	$\sqrt{10n}$	n^3	$\sqrt[3]{n}$	$\sqrt[3]{10n}$	$\sqrt[3]{100n}$
800	640 000	28.28427	89.44272	512 000 000	9.283 178	20.00000	43.08869
801	641 601	28.30194	89.49860	513 922 401	9.287 044	20.00833	43.10664
802	643 204	28.31960	89.55445	515 849 608	9.290 907	20.01665	43.12457
803	644 809	28.33725	89.61027	517 781 627	9.294 767	20.02497	43.14249
804	646 416	28.35489	89.66605	519 718 464	9.298 624	20.03328	43.16039
805	648 025	28.37252	89.72179	521 660 125	9.302 477	20.04158	43.17828
806	649 636	28.39014	89.77750	523 606 616	9.306 328	20.04988	43.19615
807	651 249	28.40775	89.83318	525 557 943	9.310 175	20.05816	43.21400
808	652 864	28.42534	89.88882	527 514 112	9.314 019	20.06645	43.23185
809	654 481	28.44293	89.94443	529 475 129	9.317 860	20.07472	43.24967
810	656 100	28.46050	90.00000	531 441 000	9.321 698	20.08299	43.26749
811	657 721	28.47806	90.05554	533 411 731	9.325 532	20.09125	43.28529
812	659 344	28.49561	90.11104	535 387 328	9.329 363	20.09950	43.30307
813	660 969	28.51315	90.16651	537 367 797	9.333 192	20.10775	43.32084
814	662 596	28.53069	90.22195	539 353 144	9.337 017	20.11599	43.33859
815	664 225	28.54820	90.27735	541 343 375	9.340 839	20.12423	43.35633
816	665 856	28.56571	90.33272	543 338 496	9.344 657	20.13245	43.37406
817	667 489	28.58321	90.38805	545 338 513	9.348 473	20.14067	43.39177
818	669 124	28.60070	90.44335	547 343 432	9.352 286	20.14889	43.40947
819	670 761	28.61818	90.49862	549 353 259	9.356 095	20.15710	43.42715
820	672 400	28.63564	90.55385	551 368 000	9.359 902	20.16530	43.44481
821	674 041	28.65310	90.60905	553 387 661	9.363 705	20.17349	43.46247
822	675 684	28.67054	90.66422	555 412 248	9.367 505	20.18168	43.48011
823	677 329	28.68798	90.71935	557 441 767	9.371 302	20.18986	43.49773
824	678 976	28.70540	90.77445	559 476 224	9.375 096	20.19803	43.51534
825	680 625	28.72281	90.82951	561 515 625	9.378 887	20.20620	43.53294
826	682 276	28.74022	90.88454	563 559 976	9.382 675	20.21436	43.55052
827	683 929	28.75761	90.93954	565 609 283	9.386 460	20.22252	43.56809
828	685 584	28.77499	90.99451	567 663 552	9.390 242	20.23066	43.58564
829	687 241	28.79236	91.04944	569 722 789	9.394 021	20.23880	43.60318
830	688 900	28.80972	91.10434	571 787 000	9.397 796	20.24694	43.62071
831	690 561	28.82707	91.15920	573 856 191	9.401 569	20.25507	43.63822
832	692 224	28.84441	91.21403	575 930 368	9.405 339	20.26319	43.65572
833	693 889	28.86174	91.26883	578 009 537	9.409 105	20.27130	43.67320
834	695 556	28.87906	91.32360	580 093 704	9.412 869	20.27941	43.69067
835	697 225	28.89637	91.37833	582 182 875	9.416 630	20.28751	43.70812
836	698 896	28.91366	91.43304	584 277 056	9.420 387	20.29561	43.72556
837	700 569	28.93095	91.48770	586 376 253	9.424 142	20.30370	43.74299
838	702 244	28.94823	91.54234	588 480 472	9.427 894	20.31178	43.76041
839	703 921	28 96550	91.59694	590 589 719	9.431 642	20.31986	43.77781
840	705 600	28.98275	91.65151	592 704 000	9.435 388	20.32793	43.79519
841	707 281	29.00000	91.70605	594 823 321	9.439 131	29.33599	43.81256
842	708 964	29.01724	91.76056	596 947 688	9.442 870	20.34405	43.82992
843	710 649	29.03446	91.81503	599 077 107	9.446 607	20.35210	43.84727
844	712 336	29.05168	91.86947	601 211 584	9.450 341	20.36014	43.86460
845	714 025	29.06888	91.92388	603 351 125	9.454 072	20.36818	43.88191
846	715 716	29.08608	91.97826	605 495 736	9.457 800	20.37621	43.89922
847	717 409	29.10326	92.03260	607 645 423	9.461 525	20.38424	43.91651
848	719 104	29.12044	92.08692	609 800 192	9.465 247	20.39226	43.93378
849	720 801	29.13760	92.14120	611 960 049	9.468 966	20.40027	43.95105
850	722 500	29.15476	92.19544	614 125 000	9.472 682	20.40828	43.96830

n	n^2	$\sqrt{n}$	$\sqrt{10n}$	n^3	$\sqrt[3]{n}$	$\sqrt[3]{10n}$	$\sqrt[3]{100n}$
850	722 500	29.15476	92.19544	614 125 000	9.472 682	20.40828	43.96830
851	724 201	29.17190	92.24966	616 295 051	9.476 396	20.41628	43.98553
852	725 904	29.18904	92.30385	618 470 208	9.480 106	20.42427	44.00275
853	727 609	29.20616	92.35800	620 650 477	9.483 814	20.43226	44.01996
854	729 316	29.22328	92.41212	622 835 864	9.487 518	20.44024	44.03716
855	731 025	29.24038	92.46621	625 026 375	9.491 220	20.44821	44.05434
856	732 736	29.25748	92.52027	627 222 016	9.494 919	20.45618	44.07151
857	734 449	29.27456	92.57429	629 422 793	9.498 615	20.46415	44.08866
858	736 164	29.29164	92.62829	631 628 712	9.502 308	20.47210	44.10581
859	737 881	29.30870	92.68225	633 839 779	9.505 998	20.48005	44.12293
860	739 600	29.32576	92.73618	636 056 000	9.509 685	20.48800	44.14005
861	741 321	29.34280	92.79009	638 277 381	9.513 370	20.49593	44.15715
862	743 044	29.35984	92.84396	640 503 928	9.517 052	20.50387	44.17424
863	744 769	29.37686	92.89779	642 735 647	9.520 730	20.51179	44.19132
864	746 496	29.39388	92.95160	644 972 544	9.524 406	20.51971	44.20838
865	748 225	29.41088	93.00538	647 214 625	9.528 079	20.52762	44.22543
866	749 956	29.42788	93.05912	649 461 896	9.531 750	20.53553	44.24246
867	751 689	29.44486	93.11283	651 714 363	9.535 417	20.54343	44.25949
868	753 424	29.46184	93.16652	653 972 032	9.539 082	20.55133	44.27650
869	755 161	29.47881	93.22017	656 234 909	9.542 744	20.55922	44.29349
870	756 900	29.49576	93.27379	658 503 000	9.546 403	20.56710	44.31048
871	758 641	29.51271	93.32738	660 776 311	9.550 059	20.57498	44.32745
872	760 384	29.52965	93.38094	663 054 848	9.553 712	20.58285	44.34440
873	762 129	29.54657	93.43447	665 338 617	9.557 363	20.59071	44.36135
874	763 876	29.56349	93.48797	667 627 624	9.561 011	20.59857	44.37828
875	765 625	29.58040	93.54143	669 921 875	9.564 656	20.60643	44.39520
876	767 376	29.59730	93.59487	672 221 376	9.568 298	20.61427	44.41211
877	769 129	29.61419	93.64828	674 526 133	9.571 938	20.62211	44.42900
878	770 884	29.63106	93.70165	676 836 152	9.575 574	20.62995	44.44588
879	772 641	29.64793	93.75500	679 151 439	9.579 208	20.63778	44.46275
880	774 400	29.66479	93.80832	681 472 000	9.582 840	20.64560	44.47960
881	776 161	29.68164	93.86160	683 797 841	9.586 468	20.65342	44.49644
882	777 924	29.69848	93.91486	686 128 968	9.590 094	20.66123	44.51327
883	779 689	29.71532	93.96808	688 465 387	9.593 717	20.66904	44.53009
884	781 456	29.73214	94.02127	690 807 104	9.597 337	20.67684	44.54689
885	783 225	29.74895	94.07444	693 154 125	9.600 955	20.68463	44.56368
886	784 996	29.76575	94.12757	695 506 456	9.604 570	20.69242	44.58046
887	786 769	29.78255	94.18068	697 864 103	9.608 182	20.70020	44.59723
888	788 544	29.79933	94.23375	700 227 072	9.611 791	20.70798	44.61398
889	790 321	29.81610	94.28680	702 595 369	9.615 398	20.71575	44.63072
890	792 100	29.83287	94.33981	704 969 000	9.619 002	20.72351	44.64745
891	793 881	29.84962	94.39280	707 347 971	9.622 603	20.73127	44.66417
892	795 664	29.86637	94.44575	709 732 288	9.626 202	20.73902	44.68087
893	797 449	29.88311	94.49868	712 121 957	9.629 797	20.74677	44.69756
894	799 236	29.89983	94.55157	714 516 984	9.633 391	20.75451	44.71424
895	801 025	29.91655	94.60444	716 917 375	9.636 981	20.76225	44.73090
896	802 816	29.93326	94.65728	719 323 136	9.640 569	20.76998	44.74756
897	804 609	29.94996	94.71008	721 734 273	9.644 154	20.77770	44.76420
898	806 404	29.96665	94.76286	724 150 792	9.647 737	20.78542	44.78083
899	808 201	29.98333	94.81561	726 572 699	9.651 317	20.79313	44.79744
900	810 000	30.00000	94.86833	729 000 000	9.654 894	20.80084	44.81405

n	n^2	$\sqrt{n}$	$\sqrt{10n}$	n^3	$\sqrt[3]{n}$	$\sqrt[3]{10n}$	$\sqrt[3]{100n}$
900	810 000	30.00000	94.86833	729 000 000	9.654 894	20.80084	44.81405
901	811 801	30.01666	94.92102	731 432 701	9.658 468	20.80854	44.83064
902	813 604	30.03331	94.97368	733 870 808	9.662 040	20.81623	44.84722
903	815 409	30.04996	95.02631	736 314 327	9.665 610	20.82392	44.86379
904	817 216	30.06659	95.07891	738 763 264	9.669 176	20.83161	44.88034
905	819 025	30.08322	95.13149	741 217 625	9.672 740	20.83929	44.89688
906	820 836	30.09983	95.18403	743 677 416	9.676 302	20.84696	44.91341
907	822 649	30.11644	95.23655	746 142 643	9.679 860	20.85463	44.92993
908	824 464	30.13304	95.28903	748 613 312	9.683 417	20.86229	44.94644
909	826 281	30.14963	95.34149	751 089 429	9.686 970	20.86994	44.96293
910	828 100	30.16621	95.39392	753 571 000	9.690 521	20.87759	44.97941
911	829 921	30.18278	95.44632	756 058 031	9.694 069	20.88524	44.99588
912	831 744	30.19934	95.49869	758 550 528	9.697 615	20.89288	45.01234
913	833 569	30.21589	95.55103	761 048 497	9.701 158	20.90051	45.02879
914	835 396	30.23243	95.60335	763 551 944	9.704 699	20.90814	45.04522
915	837 225	30.24897	95.65563	766 060 875	9.708 237	20.91576	45.06164
916	839 056	30.26549	95.70789	768 575 296	9.711 772	20.92338	45.07805
917	840 889	30.28201	95.76012	771 095 213	9.715 305	20.93099	45.09445
918	842 724	30.29851	95.81232	773 620 632	9.718 835	20.93860	45.11084
919	844 561	30.31501	95.86449	776 151 559	9.722 363	20.94620	45.12721
920	846 400	30.33150	95.91663	778 688 000	9.725 888	20.95379	45.14357
921	848 241	30.34798	95.96874	781 229 961	9.729 411	20.96138	45.15992
922	850 084	30.36445	96.02083	783 777 448	9.732 931	20.96896	45.17626
923	851 929	30.38092	96.07289	786 330 467	9.736 448	20.97654	45.19259
924	853 776	30.39737	96.12492	788 889 024	9.739 963	20.98411	45.20891
925	855 625	30.41381	96.17692	791 453 125	9.743 476	20.99168	45.22521
926	857 476	30.43025	96.22889	794 022 776	9.746 986	20.99924	45.24150
927	859 329	30.44667	96.28084	796 597 983	9.750 493	21.00680	45.25778
928	861 184	30.46309	96.33276	799 178 752	9.753 998	21.01435	45.27405
929	863 041	30.47950	96.38465	801 765 089	9.757 500	21.02190	45.29030
930	864 900	30.49590	96.43651	804 357 000	9.761 000	21.02944	45.30655
931	866 761	30.51229	96.48834	806 954 491	9.764 497	21.03697	45.32278
932	868 624	30.52868	96.54015	809 557 568	9.767 992	21.04450	45.33900
933	870 489	30.54505	96.59193	812 166 237	9.771 485	21.05203	45.35521
934	872 356	30.56141	96.64368	814 780 504	9.774 974	21.05954	45.37141
935	874 225	30.57777	96.69540	817 400 375	9.778 462	21.06706	45.38760
936	876 096	30.59412	96.74709	820 025 856	9.781 946	21.07456	45.40377
937	877 969	30.61046	96.79876	822 656 953	9.785 429	21.08207	45.41994
938	879 844	30.62679	96.85040	825 293 672	9.788 909	21.08956	45.43609
939	881 721	30.64311	96.90201	827 936 019	9.792 386	21.09706	45.45223
940	883 600	30.65942	96.95360	830 584 000	9.795 861	21.10454	45.46836
941	885 481	30.67572	97.00515	833 237 621	9.799 334	21.11202	45.48448
942	887 364	30.69202	97.05668	835 896 888	9.802 804	21.11950	45.50058
943	889 249	30.70831	97.10819	838 561 807	9.806 271	21.12697	45.51668
944	891 136	30.72458	97.15966	841 232 384	9.809 736	21.13444	45.53276
945	893 025	30.74085	97.21111	843 908 625	9.813 199	21.14190	45.54883
946	894 916	30.75711	97.26253	846 590 536	9.816 659	21.14935	45.56490
947	896 809	30.77337	97.31393	849 278 123	9.820 117	21.15680	45.58095
948	898.704	30.78961	97.36529	851 971 392	9.823 572	21.16424	45.59698
949	900 601	30.80584	97.41663	854 670 349	9.827 025	21.17168	45.61301
950	902 500	30.82207	97.46794	857 375 000	9.830 476	21.17912	45.62903

SQUARES, CUBES AND ROOTS (Continued)

n	n^2	$\sqrt{n}$	$\sqrt{10n}$	n^3	$\sqrt[3]{n}$	$\sqrt[3]{10n}$	$\sqrt[3]{100n}$
950	902 500	30.82207	97.46794	857 375 000	9.830 476	21.17912	45.62903
951	904 401	30.83829	97.51923	860 085 351	9.833 924	21.18655	45.64503
952	906 304	30.85450	97.57049	862 801 408	9.837 369	21.19397	45.66102
953	908 209	30.87070	97.62172	865 523 177	9.840 813	21.20139	45.67701
954	910 116	30.88689	97.67292	868 250 664	9.844 254	21.20880	45.69298
955	912 025	30.90307	97.72410	870 983 875	9.847 692	21.21621	45.70894
956	913 936	30.91925	97.77525	873 722 816	9.851 128	21.22361	45.72489
957	915 849	30.93542	97.82638	876 467 493	9.854 562	21.23101	45.74082
958	917 764	30.95158	97.87747	879 217 912	9.857 993	21.23840	45.75675
959	919 681	30.96773	97.92855	881 974 079	9.861 422	21.24579	45.77267
960	921 600	30.98387	97.97959	884 736 000	9.864 848	21.25317	45.78857
961	923 521	31.00000	98.03061	887 503 681	9.868 272	21.26055	45.80446
962	925 444	31.01612	98.08160	890 277 128	9.871 694	21.26792	45.82035
963	927 369	31.03224	98.13256	893 056 347	9.875 113	21.27529	45.83622
964	929 296	31.04835	98.18350	895 841 344	9.878 530	21.28265	45.85208
965	931 225	31.06445	98.23441	898 632 125	9.881 945	21.29001	45.86793
966	933 156	31.08054	98.28530	901 428 696	9.885 357	21.29736	45.88376
967	935 089	31.09662	98.33616	904 231 063	9.888 767	21.30470	45.89959
968	937 024	31.11270	98.38699	907 039 232	9.892 175	21.31204	45.91541
969	938 961	31.12876	98.43780	909 853 209	9.895 580	21.31938	45.93121
970	940 900	31.14482	98.48858	912 673 000	9.898 983	21.32671	45.94701
971	942 841	31.16087	98.53933	915 498 611	9.902 384	21.33404	45.96279
972	944 784	31.17691	98.59006	918 330 048	9.905 782	21.34136	45.97857
973	946 729	31.19295	98.64076	921 167 317	9.909 178	21.34868	45.99433
974	948 676	31.20897	98.69144	924 010 424	9.912 571	21.35599	46.01008
975	950 625	31.22499	98.74209	926 859 375	9.915 962	21.36329	46.02582
976	952 576	31.24100	98.79271	929 714 176	9.919 351	21.37059	46.04155
977	954 529	31.25700	98.84331	932 574 833	9.922 738	21.37789	46.05727
978	956 484	31.27299	98.89388	935 441 352	9.926 122	21.38518	46.07298
979	958 441	31.28898	98.94443	938 313 739	9.929 504	21.39247	46.08868
980	960 400	31.30495	98.99495	941 192 000	9.932 884	21.39975	46.10436
981	962 361	31.32092	99.04544	944 076 141	9.936 261	21.40703	46.12004
982	964 324	31.33688	99.09591	946 966 168	9.939 636	21.41430	46.13571
983	966 289	31.35283	99.14636	949 862 087	9.943 009	21.42156	46.15136
984	968 256	31.36877	99.19677	952 763 904	9.946 380	21.42883	46.16700
985	970 225	31.38471	99.24717	955 671 625	9.949 748	21.43608	46.18264
986	972 196	31.40064	99.29753	958 585 256	9.953 114	21.44333	46.19826
987	974 169	31.41656	99.34787	961 504 803	9.956 478	21.45058	46.21387
988	976 144	31.43247	99.39819	964 430 272	9.959 839	21.45782	46.22948
989	978 121	31.44837	99.44848	967 361 669	9.963 198	21.46506	46.24507
990	980 100	31.46427	99.49874	970 299 000	9.966 555	21.47229	46.26065
991	982 081	31.48015	99.54898	973 242 271	9.969 910	21.47952	46.27622
992	984 064	31.49603	99.59920	976 191 488	9.973 262	21.48674	46.29178
993	986 049	31.51190	99.64939	979 146 657	9.976 612	21.49396	46.30733
994	988 036	31.52777	99.69955	982 107 784	9.979 960	21.50117	46.32287
995	990 025	31.54362	99.74969	985 074 875	9.983 305	21.50838	46.33840
996	992 016	31.55947	99.79980	988 047 936	9.986 649	21.51558	46.35392
997	994 009	31.57531	99.84989	991 026 973	9.989 990	21.52278	46.36943
998	996 004	31.59114	99.89995	994 011 992	9.993 329	21.52997	46.38492
999	998 001	31.60696	99.94999	997 002 999	9.996 666	21.53716	46.40041
1000	1 000 000	31.62278	100.00000	1 000 000 000	10.000 000	21.54435	46.41589

n	n^4	n^5	n^6	n^7	n^8
1	1	1	1	1	1
2	16	32	64	128	256
3	81	243	729	2187	6561
4	256	1024	4096	16384	65536
5	625	3125	15625	78125	390625
6	1296	7776	46656	279936	1679616
7	2401	16807	117649	823543	5764801
8	4096	32768	262144	2097152	16777216
9	6561	59049	531441	4782969	43046721
					$\times 10^8$
10	10000	100000	1000000	10000000	1.000000
11	14641	161051	1771561	19487171	2.143589
12	20736	248832	2985984	35831808	4.299817
13	28561	371293	4826809	62748517	8.157307
14	38416	537824	7529536	105413504	14.757891
15	50625	759375	11390625	170859375	25.628906
16	65536	1048576	16777216	268435456	42.949673
17	83521	1419857	24137569	410338673	69.757574
18	104976	1889568	34012224	612220032	110.199606
19	130321	2476099	47045881	893871739	169.835630
				$\times 10^9$	$\times 10^{10}$
20	160000	3200000	64000000	1.280000	2.560000
21	194481	4084101	85766121	1.801089	3.782286
22	234256	5153632	113379904	2.494358	5.487587
23	279841	6436343	148035889	3.404825	7.831099
24	331776	7962624	191102976	4.586471	11.007531
25	390625	9765625	244140625	6.103516	15.258789
26	456976	11881376	308915776	8.031810	20.882706
27	531441	14348907	387420489	10.460353	28.242954
28	614656	17210368	481890304	13.492929	37.780200
29	707281	20511149	594823321	17.249876	50.024641
			$\times 10^8$	$\times 10^{10}$	$\times 10^{11}$
30	810000	24300000	7.290000	2.187000	6.561000
31	923521	28629151	8.875037	2.751261	8.528910
32	1048576	33554432	10.737418	3.435974	10.995116
33	1185921	39135393	12.914680	4.261844	14.064086
34	1336336	45435424	15.448044	5.252335	17.857939
35	1500625	52521875	18.382656	6.433930	22.518754
36	1679616	60466176	21.767823	7.836416	28.211099
37	1874161	69343957	25.657264	9.493188	35.124795
38	2085136	79235168	30.109364	11.441558	43.477921
39	2313441	90224199	35.187438	13.723101	53.520093
			$\times 10^9$	$\times 10^{10}$	$\times 10^{12}$
40	2560000	102400000	4.096000	16.384000	6.553600
41	2825761	115856201	4.750104	19.475427	7.984925
42	3111696	130691232	5.489032	23.053933	9.682652
43	3418801	147008443	6.321363	27.181861	11.688200
44	3748096	164916224	7.256314	31.927781	14.048224
45	4100625	184528125	8.303766	37.366945	16.815125
46	4477456	205962976	9.474297	43.581766	20.047612
47	4879681	229345007	10.779215	50.662312	23.811287
48	5308416	254803968	12.230590	58.706834	28.179280
49	5764801	282475249	13.841287	67.822307	33.232931
50	6250000	312500000	15.625000	78.125000	39.062500

n	n^4	n^5	n^6	n^7	n^8
			$\times 10^9$	$\times 10^{11}$	$\times 10^{13}$
50	6250000	312500000	15.625000	7.812500	3.906250
51	6765201	345025251	17.596288	8.974107	4.576794
52	7311616	380204032	19.770610	10.280717	5.345973
53	7890481	418195493	22.164361	11.747111	6.225969
54	8503056	459165024	24.794911	13.389252	7.230196
55	9150625	503284375	27.680641	15.224352	8.373394
56	9834496	550731776	30.840979	17.270948	9.671731
57	10556001	601692057	34.296447	19.548975	11.142916
58	11316496	656356768	38.068693	22.079842	12.806308
59	12117361	714924299	42.180534	24.886515	14.683044
		$\times 10^8$	$\times 10^{10}$	$\times 10^{11}$	$\times 10^{13}$
60	12960000	7.776000	4.665600	27.993600	16.796160
61	13845841	8.445963	5.152037	31.427428	19.170731
62	14776336	9.161328	5.680024	35.216146	21.834011
63	15752961	9.924365	6.252350	39.389806	24.815578
64	16777216	10.737418	6.871948	43.980465	28.147498
65	17850625	11.602906	7.541889	49.022279	31.864481
66	18974736	12.523326	8.265395	54.551607	36.004061
67	20151121	13.501251	9.045838	60.607116	40.606768
68	21381376	14.539336	9.886748	67.229888	45.716324
69	22667121	15.640313	10.791816	74.463533	51.379837
		$\times 10^8$	$\times 10^{10}$	$\times 10^{12}$	$\times 10^{14}$
70	24010000	16.807000	11.764900	8.235430	5.764801
71	25411681	18.042294	12.810028	9.095120	6.457535
72	26873856	19.349176	13.931407	10.030613	7.222041
73	28398241	20.730716	15.133423	11.047399	8.064601
74	29986576	22.190066	16.420649	12.151280	8.991947
75	31640625	23.730469	17.797852	13.348389	10.011292
76	33362176	25.355254	19.269993	14.645195	11.130348
77	35153041	27.067842	20.842238	16.048523	12.357363
78	37015056	28.871744	22.519960	17.565569	13.701144
79	38950081	30.770564	24.308746	19.203909	15.171088
		$\times 10^8$	$\times 10^{10}$	$\times 10^{12}$	$\times 10^{14}$
80	40960000	32.768000	26.214400	20.971520	16.777216
81	43046721	34.867844	28.242954	22.876792	18.530202
82	45212176	37.073984	30.400667	24.928547	20.441409
83	47458321	39.390406	32.694037	27.136051	22.522922
84	49787136	41.821194	35.129803	29.509035	24.787589
85	52200625	44.370531	37.714952	32.057709	27.249053
86	54700816	47.042702	40.456724	34.792782	29.921793
87	57289761	49.842092	43.362620	37.725479	32.821167
88	59969536	52.773192	46.440409	40.867560	35.963452
89	62742241	55.840594	49.698129	44.231335	39.365888
		$\times 10^9$	$\times 10^{11}$	$\times 10^{13}$	$\times 10^{15}$
90	65610000	5.904900	5.314410	4.782969	4.304672
91	68574961	6.240321	5.678693	5.167610	4.702525
92	71639296	6.590815	6.063550	5.578466	5.132189
93	74805201	6.956884	6.469902	6.017009	5.595818
94	78074896	7.339040	6.898698	6.484776	6.095689
95	81450625	7.737809	7.350919	6.983373	6.634204
96	84934656	8.153727	7.827578	7.514475	7.213896
97	88529281	8.587340	8.329720	8.079828	7.837434
98	92236816	9.039208	8.858424	8.681255	8.507630
99	96059601	9.509900	9.414801	9.320653	9.227447
100	100000000	10.000000	10.000000	10.000000	10.000000

POWERS OF NUMBERS

POWERS OF TWO

n	2^n	n	2^n
1	2	41	21990 23255 552
2	4	42	43980 46511 104
3	8	43	87960 93022 208
4	16	44	17592 18604 4416
5	32	45	35184 37208 8832
6	64	46	70368 74417 7664
7	128	47	14073 74883 55328
8	256	48	28147 49767 10656
9	512	49	56294 99534 21312
10	1024	50	11258 99906 84262 4
11	2048	51	22517 99813 68524 8
12	4096	52	45035 99627 37049 6
13	8192	53	90071 99254 74099 2
14	16384	54	18014 39850 94819 84
15	32768	55	36028 79701 89639 68
16	65536	56	72057 59403 79279 36
17	13107 2	57	14411 51880 75855 872
18	26214 4	58	28823 03761 51711 744
19	52428 8	59	57646 07523 03423 488
20	10485 76	60	11529 21504 60684 6976
21	20971 52	61	23058 43009 21369 3952
22	41943 04	62	46116 86018 42738 7904
23	83886 08	63	92233 72036 85477 5808
24	16777 216	64	18446 74407 37095 51616
25	33554 432	65	36893 48814 74191 03232
26	67108 864	66	73786 97629 48382 06464
27	13421 7728	67	14757 39525 89676 41292 8
28	26843 5456	68	29514 79051 79352 82585 6
29	53687 0912	69	59029 58103 58705 65171 2
30	10737 41824	70	11805 91620 71741 13034 24
31	21474 83648	71	23611 83241 43482 26068 48
32	42949 67296	72	47223 66482 86964 52136 96
33	85899 34592	73	94447 32965 73929 04273 92
34	17179 86918 4	74	18889 46593 14785 80854 784
35	34359 73836 8	75	37778 93186 29571 61709 568
36	68719 47673 6	76	75557 86372 59143 23419 136
37	13743 89534 72	77	15111 57274 51828 64683 8272
38	27487 79069 44	78	30223 14549 03657 29367 6544
39	54975 58138 88	79	60446 29098 07314 58735 3088
40	10995 11627 776	80	12089 25819 61462 91747 06176

n	2^n
81	24178 51639 22925 83494 12352
82	48357 03278 45851 66988 24704
83	96714 06556 91703 33976 49408
84	19342 81311 38340 66795 29881 6
85	38685 62622 76681 33590 59763 2
86	77371 25245 53362 67181 19526 4
87	15474 25049 10672 53436 23905 28
88	30948 50098 21345 06872 47810 56
89	61897 00196 42690 13744 95621 12
90	12379 40039 28538 02748 99124 224
91	24758 80078 57076 05497 98248 448
92	49517 60157 14152 10995 96496 896
93	99035 20314 28304 21991 92993 792
94	19807 04062 85660 84398 38598 7584
95	39614 08125 71321 68796 77197 5168
96	79228 16251 42643 37593 54395 0336
97	15845 63250 28528 67518 70879 00672
98	31691 26500 57057 35037 41758 01344
99	63382 53001 14114 70074 83516 02688
100	12676 50600 22822 94014 96703 20537 6
101	25353 01200 45645 88029 93406 41075 2

FACTORIALS AND THEIR LOGARITHMS

n	$n!$	$\log n!$	n	$n!$	$\log n!$
			50	3.0414×10^{64}	64.48307
1	1.0000	0.00000	51	1.5511×10^{66}	66.19065
2	2.0000	0.30103	52	8.0658×10^{67}	67.90665
3	6.0000	0.77815	53	4.2749×10^{69}	69.63092
4	2.4000×10	1.38021	54	2.3084×10^{71}	71.36332
5	1.2000×10^{2}	2.07918	**55**	1.2696×10^{73}	73.10368
6	7.2000×10^{2}	2.85733	56	7.1100×10^{74}	74.85187
7	5.0400×10^{3}	3.70243	57	4.0527×10^{76}	76.60774
8	4.0320×10^{4}	4.60552	58	2.3506×10^{78}	78.37117
9	3.6288×10^{5}	5.55976	59	1.3868×10^{80}	80.14202
10	3.6288×10^{6}	6.55976	**60**	8.3210×10^{81}	81.92017
11	3.9917×10^{7}	7.60116	61	5.0758×10^{83}	83.70550
12	4.7900×10^{8}	8.68034	62	3.1470×10^{85}	85.49790
13	6.2270×10^{9}	9.79428	63	1.9826×10^{87}	87.29724
14	8.7178×10^{10}	10.94041	64	1.2689×10^{89}	89.10342
15	1.3077×10^{12}	12.11650	**65**	8.2477×10^{90}	90.91633
16	2.0923×10^{13}	13.32062	66	5.4435×10^{92}	92.73587
17	3.5569×10^{14}	14.55107	67	3.6471×10^{94}	94.56195
18	6.4024×10^{15}	15.80634	68	2.4800×10^{96}	96.39446
19	1.2165×10^{17}	17.08509	69	1.7112×10^{98}	98.23331
20	2.4329×10^{18}	18.38612	**70**	1.1979×10^{100}	100.07841
21	5.1091×10^{19}	19.70834	71	8.5048×10^{101}	101.92966
22	1.1240×10^{21}	21.05077	72	6.1234×10^{103}	103.78700
23	2.5852×10^{22}	22.41249	73	4.4701×10^{105}	105.65032
24	6.2045×10^{23}	23.79271	74	3.3079×10^{107}	107.51955
25	1.5511×10^{25}	25.19065	**75**	2.4809×10^{109}	109.39461
26	4.0329×10^{26}	26.60562	76	1.8855×10^{111}	111.27543
27	1.0889×10^{28}	28.03698	77	1.4518×10^{113}	113.16192
28	3.0489×10^{29}	29.48414	78	1.1324×10^{115}	115.05401
29	8.8418×10^{30}	30.94654	79	8.9462×10^{116}	116.95164
30	2.6525×10^{32}	32.42366	**80**	7.1569×10^{118}	118.85473
31	8.2228×10^{33}	33.91502	81	5.7971×10^{120}	120.76321
32	2.6313×10^{35}	35.42017	82	4.7536×10^{122}	122.67703
33	8.6833×10^{36}	36.93869	83	3.9455×10^{124}	124.59610
34	2.9523×10^{38}	38.47016	84	3.3142×10^{126}	126.52038
35	1.0333×10^{40}	40.01423	**85**	2.8171×10^{128}	128.44980
36	3.7199×10^{41}	41.57054	86	2.4227×10^{130}	130.38430
37	1.3764×10^{43}	43.13874	87	2.1078×10^{132}	132.32382
38	5.2302×10^{44}	44.71852	88	1.8548×10^{134}	134.26830
39	2.0398×10^{46}	46.30959	89	1.6508×10^{136}	136.21769
40	8.1592×10^{47}	47.91165	**90**	1.4857×10^{138}	138.17194
41	3.3453×10^{49}	49.52443	91	1.3520×10^{140}	140.13098
42	1.4050×10^{51}	51.14768	92	1.2438×10^{142}	142.09477
43	6.0415×10^{52}	52.78115	93	1.1568×10^{144}	144.06325
44	2.6583×10^{54}	54.42460	94	1.0874×10^{146}	146.03638
45	1.1962×10^{56}	56.07781	**95**	1.0330×10^{148}	148.01410
46	5.5026×10^{57}	57.74057	96	9.9168×10^{149}	149.99637
47	2.5862×10^{59}	59.41267	97	9.6193×10^{151}	151.98314
48	1.2414×10^{61}	61.09391	98	9.4269×10^{153}	153.97437
49	6.0828×10^{62}	62.78410	99	9.3326×10^{155}	155.97000
50	3.0414×10^{64}	64.48307	**100**	9.3326×10^{157}	157.97000

FACTORS FOR COMPUTING PROBABLE ERRORS

n	$\frac{1}{\sqrt{n}}$	$\frac{1}{\sqrt{n(n-1)}}$	$\frac{.6745}{\sqrt{n-1}}$	$\frac{.6745}{\sqrt{n(n-1)}}$	$\frac{.8453}{n\sqrt{n-1}}$	$\frac{.8453}{\sqrt{n(n-1)}}$
2	.707107	.707107	.6745	.4769	.4227	.5978
3	.577350	.408248	.4769	.2754	.1993	.3451
4	.500000	.288675	.3894	.1947	.1220	.2440
5	.447214	.223607	.3372	.1508	.0845	.1890
6	.408248	.182574	.3016	.1231	.0630	.1543
7	.377964	.154303	.2754	.1041	.0493	.1304
8	.353553	.133631	.2549	.0901	.0399	.1130
9	.333333	.117851	.2385	.0795	.0332	.0996
10	.316228	.105409	.2248	.0711	.0282	.0891
11	.301511	.095346	.2133	.0643	.0243	.0806
12	.288675	.087039	.2034	.0587	.0212	.0736
13	.277350	.080064	.1947	.0540	.0188	.0677
14	.267261	.074125	.1871	.0500	.0167	.0627
15	.258199	.069007	.1803	.0465	.0151	.0583
16	.250000	.064550	.1742	.0435	.0136	.0546
17	.242536	.060634	.1686	.0409	.0124	.0513
18	.235702	.057166	.1636	.0386	.0114	.0483
19	.229416	.054074	.1590	.0365	.0105	.0457
20	.223607	.051299	.1547	.0346	.0097	.0434
21	.218218	.048795	.1508	.0329	.0090	.0412
22	.213201	.046524	.1472	.0314	.0084	.0393
23	.208514	.044455	.1438	.0300	.0078	.0376
24	.204124	.042563	.1406	.0287	.0073	.0360
25	.200000	.040825	.1377	.0275	.0069	.0345
26	.196116	.039223	.1349	.0265	.0065	.0332
27	.192450	.037743	.1323	.0255	.0061	.0319
28	.188982	.036370	.1298	.0245	.0058	.0307
29	.185695	.035093	.1275	.0237	.0055	.0297
30	.182574	.033903	.1252	.0229	.0052	.0287
31	.179605	.032791	.1231	.0221	.0050	.0277
32	.176777	.031750	.1211	.0214	.0047	.0268
33	.174078	.030773	.1192	.0208	.0045	.0260
34	.171499	.029854	.1174	.0201	.0043	.0252
35	.169031	.028989	.1157	.0196	.0041	.0245
36	.166667	.028172	.1140	.0190	.0040	.0238
37	.164399	.027400	.1124	.0185	.0038	.0232
38	.162221	.026669	.1109	.0180	.0037	.0225
39	.160128	.025976	.1094	.0175	.0035	.0220
40	.158114	.025318	.1080	.0171	.0034	.0214
41	.156174	.024693	.1066	.0167	.0033	.0209
42	.154303	.024098	.1053	.0163	.0031	.0204
43	.152499	.023531	.1041	.0159	.0030	.0199
44	.150756	.022990	.1029	.0155	.0029	.0194
45	.149071	.022473	.1017	.0152	.0028	.0190
46	.147442	.021979	.1005	.0148	.0027	.0186
47	.145865	.021507	.0994	.0145	.0027	.0182
48	.144338	.021054	.0984	.0142	.0026	.0178
49	.142857	.020620	.0974	.0139	.0025	.0174
50	.141421	.020203	.0964	.0136	.0024	.0171
50	.141421	.020203	.0964	.0136	.0024	.0171
51	.140028	.019803	.0954	.0134	.0023	.0167
52	.138675	.019418	.0945	.0131	.0023	.0164
53	.137361	.019048	.0935	.0129	.0022	.0161
54	.136083	.018692	.0927	.0126	.0022	.0158
55	.134840	.018349	.0918	.0124	.0021	.0155
56	.133631	.018019	.0910	.0122	.0020	.0152
57	.132453	.017700	.0901	.0119	.0020	.0150
58	.131306	.017392	.0893	.0117	.0019	.0147
59	.130189	.017095	.0886	.0115	.0019	.0145
60	.129099	.016807	.0878	.0113	.0018	.0142
61	.128037	.016529	.0871	.0112	.0018	.0140
62	.127000	.016261	.0864	.0110	.0018	.0138
63	.125988	.016001	.0857	.0108	.0017	.0135
64	.125000	.015749	.0850	.0106	.0017	.0133
65	.124035	.015504	.0843	.0105	.0016	.0131
66	.123091	.015268	.0837	.0103	.0016	.0129
67	.122169	.015038	.0830	.0101	.0016	.0127
68	.121268	.014815	.0824	.0100	.0015	.0125
69	.120386	.014599	.0818	.0099	.0015	.0123
70	.119523	.014389	.0812	.0097	.0015	.0122
71	.118678	.014185	.0806	.0096	.0014	.0120
72	.117851	.013986	.0801	.0094	.0014	.0118
73	.117041	.013793	.0795	.0093	.0014	.0117
74	.116248	.013606	.0789	.0092	.0013	.0115
75	.115470	.013423	.0784	.0091	.0013	.0113
76	.114708	.013245	.0779	.0089	.0013	.0112
77	.113961	.013072	.0773	.0088	.0013	.0111
78	.113228	.012904	.0769	.0087	.0012	.0109
79	.112509	.012739	.0764	.0086	.0012	.0108
80	.111803	.012579	.0759	.0085	.0012	.0106
81	.111111	.012423	.0754	.0084	.0012	.0105
82	.110432	.012270	.0749	.0083	.0012	.0104
83	.109764	.012121	.0745	.0082	.0011	.0103
84	.109109	.011976	.0740	.0081	.0011	.0101
85	.108465	.011835	.0736	.0080	.0011	.0100
86	.107833	.011696	.0732	.0079	.0011	.0099
87	.107211	.011561	.0727	.0078	.0011	.0098
88	.106600	.011429	.0723	.0077	.0010	.0097
89	.106000	.011300	.0719	.0076	.0010	.0096
90	.105409	.011173	.0715	.0075	.0010	.0094
91	.104828	.011050	.0711	.0075	.0010	.0093
92	.104257	.010929	.0707	.0074	.0010	.0092
93	.103695	.010811	.0703	.0073	.0010	.0091
94	.103142	.010695	.0699	.0072	.0009	.0090
95	.102598	.010582	.0696	.0071	.0009	.0089
96	.102062	.010471	.0692	.0071	.0009	.0089
97	.101535	.010363	.0688	.0070	.0009	.0088
98	.101015	.010257	.0685	.0069	.0009	.0087
99	.100504	.010152	.0681	.0069	.0009	.0086
100	.100000	.010050	.0678	.0068	.0008	.0085

FACTORS FOR COMPUTING PROBABLE ERRORS

PROBABILITY OF OCCURRENCE OF DEVIATIONS

Valid for thirty or more samples.

Probability of occurrence, expressed as per cent, and odds against a deviation as great or greater than that designated is given for various ratios of the deviation to the probable error and to the standard deviation.

(From Pearl, Medical Biometry and Statistics, W. B. Saunders Company, publishers, by permission.)

Ratio dev. to P.E.	Probable occurrence %	Odds against, to 1	Ratio dev. to std. dev.	Probable occurrence %	Odds against, to 1
1.0	50.00	1.00	0.67449	50.00	1.00
1.1	45.81	1.18	0.7	48.39	1.07
1.2	41.83	1.39	0.8	42.37	1.36
1.3	38.06	1.63	0.9	36.81	1.72
1.4	34.50	1.90	1.0	31.73	2.15
1.5	31.17	2.21	1.1	27.13	2.69
1.6	28.05	2.57	1.2	23.01	3.35
1.7	25.15	2.98	1.3	19.36	4.17
1.8	22.47	3.45	1.4	16.15	5.19
1.9	20.00	4.00	1.5	13.36	6.48
2.0	17.73	4.64	1.6	10.96	8.12
2.1	15.67	5.38	1.7	8.91	10.22
2.2	13.78	6.25	1.8	7.19	12.92
2.3	12.08	7.28	1.9	5.74	16.41
2.4	10.55	8.48	2.0	4.55	20.98
2.5	9.18	9.90	2.1	3.57	26.99
2.6	7.95	11.58	2.2	2.78	34.96
2.7	6.86	13.58	2.3	2.14	45.62
2.8	5.89	15.96	2.4	1.64	59.99
2.9	5.05	18.82	2.5	1.24	79.52
3.0	4.30	22.24	2.6	.932	106.3
3.1	3.65	26.37	2.7	.693	143.2
3.2	3.09	31.36	2.8	.511	194.7
3.3	2.60	37.42	2.9	.373	267.0
3.4	2.18	44.80	3.0	.270	369.4
3.5	1.82	53.82	3.1	.194	515.7
3.6	1.52	64.89	3.2	.137	726.7
3.7	1.26	78.53	3.3	.0967	1,033.
3.8	1.04	95.38	3.4	.0674	1,483.
3.9	.853	116.3	3.5	.0465	2,149.
4.0	.698	142.3	3.6	.0318	3,142.
4.1	.569	174.9	3.7	.0216	4,637.
4.2	.461	215.8	3.8	.0145	6,915.
4.3	.373	267.2	3.9	.00962	10,394.
4.4	.300	332.4	4.0	.00634	15,772.
4.5	.240	415.0	5.0	5.73×10^{-5}	1.744×10^{6}
4.6	.192	520.4	6.0	2.0×10^{-7}	5.0×10^{8}
4.7	.152	655.3	7.0	2.6×10^{-10}	3.9×10^{11}
4.8	.121	828.3			
4.9	.0950	1,052.			
5.0	.0745	1,341.			
6.0	.0052	19,300.			
7.0	.00023	4.27×10^{5}			
8.0	6.8×10^{-6}	1.47×10^{7}			
9.0	1.3×10^{-7}	7.30×10^{8}			
10.0	1.5×10^{-9}	$6\ 5\times10^{10}$			

FACTORIALS, EXACT VALUES AND RECIPROCALS

n	$n!$	n	$n!$	n	$1/n!$	n	$1/n!$
1	1	11	39916800	1	1.	11	$.25052 \times 10^{-7}$
2	2	12	479001600	2	0.5	12	$.20877 \times 10^{-8}$
3	6	13	6227020800	3	.16667	13	$.16059 \times 10^{-9}$
4	24	14	87178291200	4	$.41667 \times 10^{-1}$	14	$.11471 \times 10^{-10}$
5	120	15	1307674368000	5	$.83333 \times 10^{-2}$	15	$.76472 \times 10^{-12}$
6	720	16	20922789888000	6	$.13889 \times 10^{-2}$	16	$.47795 \times 10^{-13}$
7	5040	17	355687428096000	7	$.19841 \times 10^{-3}$	17	$.28115 \times 10^{-14}$
8	40320	18	6402373705728000	8	$.24802 \times 10^{-4}$	18	$.15619 \times 10^{-15}$
9	362880	19	121645100408832000	9	$.27557 \times 10^{-5}$	19	$.82206 \times 10^{-17}$
10	3628800	20	2432902008176640000	10	$.27557 \times 10^{-6}$	20	$.41103 \times 10^{-18}$

NORMAL CURVE OF ERROR

If x is measured in "σ-units" from the mean* the following tables give the values for the area integral

$$\int_0^t \phi(x)dx,\ \phi(x),\ \phi''(x) = (x^2 - 1)\phi(x),\ \phi'''(x)$$
$$= (3x - x^3)\phi(x), \text{ and } \phi^{iv}(x) = (x^4 - 6x^2 + 3)\phi(x)$$

for value $x = t$.

It should be noted that other probability integrals can be calculated from these tables. For example:

$$\int_0^t \phi(x)dx = \frac{1}{2}\operatorname{erf}\left(\frac{t}{\sqrt{2}}\right), \text{ where } \operatorname{erf}\left(\frac{t}{\sqrt{2}}\right),$$

represents the error function associated with the normal curve.

To evaluate erf (2.3) proceed as follows:

Since $\frac{t}{\sqrt{2}} = 2.3$, one finds $t = (2.3)\sqrt{2} = (2.3)(1.414) = 3.25$.

In entry for area opposite $t = 3.25$, the value 0.4994 is given. Thus erf $(2.3) = 2\ (0.4994) = 0.9988$.

* See index for Probability, formulae.

NORMAL CURVE OF ERROR

$$\text{Normal Curve } \phi(x) = \frac{1}{\sqrt{2\pi}} e^{-\frac{x^2}{2}}$$

t	Area	Ordinate $\phi(x)$	Second derivative $\phi''(x)$	Third derivative $\phi'''(x)$	Fourth derivative $\phi^{iv}(x)$
.00	.0000	.3989	−.3989	.0000	1.1968
.01	.0040	.3989	−.3989	.0120	1.1965
.02	.0080	.3989	−.3987	.0239	1.1956
.03	.0120	.3988	−.3984	.0359	1.1941
.04	.0160	.3986	−.3980	.0478	1.1920
.05	.0199	.3984	−.3975	.0597	1.1894
.06	.0239	.3982	−.3968	.0716	1.1861
.07	.0279	.3980	−.3960	.0834	1.1822
.08	.0319	.3977	−.3951	.0952	1.1778
.09	.0359	.3973	−.3941	.1070	1.1727
.10	.0398	.3970	−.3930	.1187	1.1671
.11	.0438	.3965	−.3917	.1303	1.1609
.12	.0478	.3961	−.3904	.1419	1.1541
.13	.0517	.3956	−.3889	.1534	1.1468
.14	.0557	.3951	−.3873	.1648	1.1389
.15	.0596	.3945	−.3856	.1762	1.1304
.16	.0636	.3939	−.3838	.1874	1.1214
.17	.0675	.3932	−.3819	.1986	1.1118
.18	.0714	.3925	−.3798	.2097	1.1017
.19	.0754	.3918	−.3777	.2206	1.0911
.20	.0793	.3910	−.3754	.2315	1.0799
.21	.0832	.3902	−.3730	.2422	1.0682
.22	.0871	.3894	−.3706	.2529	1.0560
.23	.0910	.3885	−.3680	.2634	1.0434
.24	.0948	.3876	−.3653	.2737	1.0302
.25	.0987	.3867	−.3625	.2840	1.0165
.26	.1026	.3857	−.3596	.2941	1.0024
.27	.1064	.3847	−.3566	.3040	0.9878
.28	.1103	.3836	−.3535	.3138	0.9727
.29	.1141	.3825	−.3504	.3235	0.9572
.30	.1179	.3814	−.3471	.3330	0.9413
.31	.1217	.3802	−.3437	.3423	0.9250
.32	.1255	.3790	−.3402	.3515	0.9082
.33	.1293	.3778	−.3367	.3605	0.8910
.34	.1331	.3765	−.3330	.3693	0.8735
.35	.1368	.3752	−.3293	.3779	0.8556
.36	.1406	.3739	−.3255	.3864	0.8373
.37	.1443	.3726	−.3216	.3947	0.8186
.38	.1480	.3712	−.3176	.4028	0.7996
.39	.1517	.3697	−.3135	.4107	0.7803
.40	.1554	.3683	−.3094	.4184	0.7607
.41	.1591	.3668	−.3051	.4259	0.7408
.42	.1628	.3653	−.3008	.4332	0.7206
.43	.1664	.3637	−.2965	.4403	0.7001
.44	.1700	.3621	−.2920	.4472	0.6793
.45	.1736	.3605	−.2875	.4539	0.6583
.46	.1772	.3589	−.2830	.4603	0.6371
.47	.1808	.3572	−.2783	.4666	0.6156
.48	.1844	.3555	−.2736	.4727	0.5940
.49	.1879	.3538	−.2689	.4785	0.5721
.50	.1915	.3521	−.2641	.4841	0.5501

t	Area	Ordinate $\phi(x)$	Second derivative $\phi''(x)$	Third derivative $\phi'''(x)$	Fourth derivative $\phi^{iv}(x)$
.50	.1915	.3521	−.2641	.4841	.5501
.51	.1950	.3503	−.2592	.4895	.5279
.52	.1985	.3485	−.2543	.4947	.5056
.53	.2019	.3467	−.2493	.4996	.4831
.54	.2054	.3448	−.2443	.5043	.4605
.55	.2088	.3429	−.2392	.5088	.4378
.56	.2123	.3411	−.2341	.5131	.4150
.57	.2157	.3391	−.2289	.5171	.3921
.58	.2190	.3372	−.2238	.5209	.3691
.59	.2224	.3352	−.2185	.5245	.3461
.60	.2258	.3332	−.2133	.5278	.3231
.61	.2291	.3312	−.2080	.5309	.3000
.62	.2324	.3292	−.2027	.5338	.2770
.63	.2357	.3271	−.1973	.5365	.2539
.64	.2389	.3251	−.1919	.5389	.2309
.65	.2422	.3230	−.1865	.5411	.2078
.66	.2454	.3209	−.1811	.5431	.1849
.67	.2486	.3187	−.1757	.5448	.1620
.68	.2518	.3166	−.1702	.5463	.1391
.69	.2549	.3144	−.1647	.5476	.1164
.70	.2580	.3123	−.1593	.5486	.0937
.71	.2612	.3101	−.1538	.5495	.0712
.72	.2642	.3079	−.1483	.5501	.0487
.73	.2673	.3056	−.1428	.5504	.0265
.74	.2704	.3034	−.1373	.5506	.0043
.75	.2734	.3011	−.1318	.5505	−.0176
.76	.2764	.2989	−.1262	.5502	−.0394
.77	.2794	.2966	−.1207	.5497	−.0611
.78	.2823	.2943	−.1153	.5490	−.0825
.79	.2852	.2920	−.1098	.5481	−.1037
.80	.2881	.2897	−.1043	.5469	−.1247
.81	.2910	.2874	−.0988	.5456	−.1455
.82	.2939	.2850	−.0934	.5440	−.1660
.83	.2967	.2827	−.0880	.5423	−.1862
.84	.2996	.2803	−.0825	.5403	−.2063
.85	.3023	.2780	−.0771	.5381	−.2260
.86	.3051	.2756	−.0718	.5358	−.2455
.87	.3079	.2732	−.0664	.5332	−.2646
.88	.3106	.2709	−.0611	.5305	−.2835
.89	.3133	.2685	−.0558	.5276	−.3021
.90	.3159	.2661	−.0506	.5245	−.3203
.91	.3186	.2637	−.0453	.5212	−.3383
.92	.3212	.2613	−.0401	.5177	−.3559
.93	.3238	.2589	−.0350	.5140	−.3731
.94	.3264	.2565	−.0299	.5102	−.3901
.95	.3289	.2541	−.0248	.5062	−.4066
.96	.3315	.2516	−.0197	.5021	−.4228
.97	.3340	.2492	−.0147	.4978	−.4387
.98	.3365	.2468	−.0098	.4933	−.4541
.99	.3389	.2444	−.0049	.4887	−.4692
1.00	.3413	.2420	.0000	.4839	−.4839

t	Area	Ordinate $\phi(x)$	Second derivative $\phi''(x)$	Third derivative $\phi'''(x)$	Fourth derivative $\phi^{iv}(x)$
1.00	.3413	.2420	.0000	.4839	−.4839
1.01	.3438	.2396	.0048	.4790	−.4983
1.02	.3461	.2371	.0096	.4740	−.5122
1.03	.3485	.2347	.0143	.4688	−.5257
1.04	.3508	.2323	.0190	.4635	−.5389
1.05	.3531	.2299	.0236	.4580	−.5516
1.06	.3554	.2275	.0281	.4524	−.5639
1.07	.3577	.2251	.0326	.4467	−.5758
1.08	.3599	.2227	.0371	.4409	−.5873
1.09	.3621	.2203	.0414	.4350	−.5984
1.10	.3643	.2179	.0458	.4290	−.6091
1.11	.3665	.2155	.0500	.4228	−.6193
1.12	.3686	.2131	.0542	.4166	−.6292
1.13	.3708	.2107	.0583	.4102	−.6386
1.14	.3729	.2083	.0624	.4038	−.6476
1.15	.3749	.2059	.0664	.3973	−.6561
1.16	.3770	.2036	.0704	.3907	−.6643
1.17	.3790	.2012	.0742	.3840	−.6720
1.18	.3810	.1989	.0780	.3772	−.6792
1.19	.3830	.1965	.0818	.3704	−.6861
1.20	.3849	.1942	.0854	.3635	−.6926
1.21	.3869	.1919	.0890	.3566	−.6986
1.22	.3888	.1895	.0926	.3496	−.7042
1.23	.3907	.1872	.0960	.3425	−.7094
1.24	.3925	.1849	.0994	.3354	−.7141
1.25	.3944	.1827	.1027	.3282	−.7185
1.26	.3962	.1804	.1060	.3210	−.7224
1.27	.3980	.1781	.1092	.3138	−.7259
1.28	.3997	.1759	.1123	.3065	−.7291
1.29	.4015	.1736	.1153	.2992	−.7318
1.30	.4032	.1714	.1182	.2918	−.7341
1.31	.4049	.1692	.1211	.2845	−.7361
1.32	.4066	.1669	.1239	.2771	−.7376
1.33	.4082	.1647	.1267	.2697	−.7388
1.34	.4099	.1626	.1293	.2624	−.7395
1.35	.4115	.1604	.1319	.2550	−.7399
1.36	.4131	.1582	.1344	.2476	−.7400
1.37	.4147	.1561	.1369	.2402	−.7396
1.38	.4162	.1540	.1392	.2328	−.7389
1.39	.4177	.1518	.1415	.2254	−.7378
1.40	.4192	.1497	.1437	.2180	−.7364
1.41	.4207	.1476	.1459	.2107	−.7347
1.42	.4222	.1456	.1480	.2033	−.7326
1.43	.4236	.1435	.1500	.1960	−.7301
1.44	.4251	.1415	.1519	.1887	−.7274
1.45	.4265	.1394	.1537	.1815	−.7243
1.46	.4279	.1374	.1555	.1742	−.7209
1.47	.4292	.1354	.1572	.1670	−.7172
1.48	.4306	.1334	.1588	.1599	−.7132
1.49	.4319	.1315	.1604	.1528	−.7089
1.50	.4332	.1295	.1619	.1457	−.7043

t	Area	Ordinate $\phi(x)$	Second derivative $\phi''(x)$	Third derivative $\phi'''(x)$	Fourth derivative $\phi^{iv}(x)$
1.50	.4332	.1295	.1619	.1457	−.7043
1.51	.4345	.1276	.1633	.1387	−.6994
1.52	.4357	.1257	.1647	.1317	−.6942
1.53	.4370	.1238	.1660	.1248	−.6888
1.54	.4382	.1219	.1672	.1180	−.6831
1.55	.4394	.1200	.1683	.1111	−.6772
1.56	.4406	.1182	.1694	.1044	−.6710
1.57	.4418	.1163	.1704	.0977	−.6646
1.58	.4430	.1145	.1714	.0911	−.6580
1.59	.4441	.1127	.1722	.0846	−.6511
1.60	.4452	.1109	.1730	.0781	−.6441
1.61	.4463	.1092	.1738	.0717	−.6368
1.62	.4474	.1074	.1745	.0654	−.6293
1.63	.4485	.1057	.1751	.0591	−.6216
1.64	.4495	.1040	.1757	.0529	−.6138
1.65	.4505	.1023	.1762	.0468	−.6057
1.66	.4515	.1006	.1766	.0408	−.5975
1.67	.4525	.0989	.1770	.0349	−.5891
1.68	.4535	.0973	.1773	.0290	−.5806
1.69	.4545	.0957	.1776	.0233	−.5720
1.70	.4554	.0941	.1778	.0176	−.5632
1.71	.4564	.0925	.1779	.0120	−.5542
1.72	.4573	.0909	.1780	.0065	−.5452
1.73	.4582	.0893	.1780	.0011	−.5360
1.74	.4591	.0878	.1780	−.0042	−.5267
1.75	.4599	.0863	.1780	−.0094	−.5173
1.76	.4608	.0848	.1778	−.0146	−.5079
1.77	.4616	.0833	.1777	−.0196	−.4983
1.78	.4625	.0818	.1774	−.0245	−.4887
1.79	.4633	.0804	.1772	−.0294	−.4789
1.80	.4641	.0790	.1769	−.0341	−.4692
1.81	.4649	.0775	.1765	−.0388	−.4593
1.82	.4656	.0761	.1761	−.0433	−.4494
1.83	.4664	.0748	.1756	−.0477	−.4395
1.84	.4671	.0734	.1751	−.0521	−.4295
1.85	.4678	.0721	.1746	−.0563	−.4195
1.86	.4686	.0707	.1740	−.0605	−.4095
1.87	.4693	.0694	.1734	−.0645	−.3995
1.88	.4700	.0681	.1727	−.0685	−.3894
1.89	.4706	.0669	.1720	−.0723	−.3793
1.90	.4713	.0656	.1713	−.0761	−.3693
1.91	.4719	.0644	.1705	−.0797	−.3592
1.92	.4726	.0632	.1697	−.0832	−.3492
1.93	.4732	.0620	.1688	−.0867	−.3392
1.94	.4738	.0608	.1679	−.0900	−.3292
1.95	.4744	.0596	.1670	−.0933	−.3192
1.96	.4750	.0584	.1661	−.0964	−.3093
1.97	.4756	.0573	.1651	−.0994	−.2994
1.98	.4762	.0562	.1641	−.1024	−.2895
1.99	.4767	.0551	.1630	−.1052	−.2797
2.00	.4773	.0540	.1620	−.1080	−.2700

NORMAL CURVE OF ERROR

Normal Curve $\phi(x) = \frac{1}{\sqrt{2\pi}} e^{-\frac{x^2}{2}}$

NORMAL CURVE OF ERROR

Normal Curve $\phi(x) = \frac{1}{\sqrt{2\pi}} e^{-\frac{x^2}{2}}$ **(Continued)**

t	Area	Ordinate $\phi(x)$	Second derivative $\phi''(x)$	Third derivative $\phi'''(x)$	Fourth derivative $\phi^{iv}(x)$
2.00	.4773	.0540	.1620	−.1080	−.2700
2.01	.4778	.0529	.1609	−.1106	−.2603
2.02	.4783	.0519	.1598	−.1132	−.2506
2.03	.4788	.0508	.1586	−.1157	−.2411
2.04	.4793	.0498	.1575	−.1180	−.2316
2.05	.4798	.0488	.1563	−.1203	−.2222
2.06	.4803	.0478	.1550	−.1225	−.2129
2.07	.4808	.0468	.1538	−.1245	−.2036
2.08	.4812	.0459	.1526	−.1265	−.1945
2.09	.4817	.0449	1513	−.1284	−.1854
2.10	.4821	.0440	.1500	−.1302	−.1765
2.11	.4826	.0431	.1487	−.1320	−.1676
2.12	.4830	.0422	.1474	−.1336	−.1588
2.13	.4834	.0413	.1460	−.1351	−.1502
2.14	.4838	.0404	.1446	−.1366	−.1416
2.15	.4842	.0396	.1433	−.1380	−.1332
2.16	.4846	.0387	.1419	−.1393	−.1249
2.17	.4850	.0379	.1405	−.1405	−.1167
2.18	.4854	.0371	.1391	−.1416	−.1086
2.19	.4857	.0363	.1377	−.1426	−.1006
2.20	.4861	.0355	.1362	−.1436	−.0927
2.21	.4865	.0347	.1348	−.1445	−.0850
2.22	.4868	.0339	.1333	−.1453	−.0774
2.23	.4871	.0332	.1319	−.1460	−.0700
2.24	.4875	.0325	.1304	−.1467	−.0626
2.25	.4878	.0317	.1289	−.1473	−.0554
2.26	.4881	.0310	.1275	−.1478	−.0484
2.27	.4884	.0303	.1260	−.1483	−.0414
2.28	.4887	.0297	.1245	−.1486	−.0346
2.29	.4890	.0290	.1230	−.1490	−.0279
2.30	.4893	.0283	.1215	−.1492	−.0214
2.31	.4896	.0277	.1200	−.1494	−.0150
2.32	.4898	.0271	.1185	−.1495	−.0088
2.33	.4901	.0264	.1170	−.1496	−.0027
2.34	.4904	.0258	.1155	−.1496	.0033
2.35	.4906	.0252	.1141	−.1495	.0092
2.36	.4909	.0246	.1126	−.1494	.0149
2.37	.4911	.0241	.1111	−.1492	.0204
2.38	.4913	.0235	.1096	−.1490	.0258
2.39	.4916	.0229	.1081	−.1487	.0311
2.40	.4918	.0224	.1066	−.1483	.0362
2.41	.4920	.0219	.1051	−.1480	.0412
2.42	.4922	.0213	.1036	−.1475	.0461
2.43	.4925	.0208	.1022	−.1470	.0508
2.44	.4927	.0203	.1007	−.1465	.0554
2.45	.4929	.0198	.0992	−.1459	.0598
2.46	.4931	.0194	.0978	−.1453	.0641
2.47	.4932	.0189	.0963	−.1446	.0683
2.48	.4934	.0184	.0949	−.1439	.0723
2.49	.4936	.0180	.0935	−.1432	.0762
2.50	.4938	.0175	.0920	−.1424	.0800

t	Area	Ordinate $\phi(x)$	Second derivative $\phi''(x)$	Third derivative $\phi'''(x)$	Fourth derivative $\phi^{iv}(x)$
2.50	.4938	.0175	.0920	−.1424	.0800
2.51	.4940	.0171	.0906	−.1416	.0836
2.52	.4941	.0167	.0892	−.1408	.0871
2.53	.4943	.0163	.0878	−.1399	.0905
2.54	.4945	.0159	.0864	−.1389	.0937
2.55	.4946	.0155	.0850	−.1380	.0968
2.56	.4948	.0151	.0836	−.1370	.0998
2.57	.4949	.0147	.0823	−.1360	.1027
2.58	.4951	.0143	.0809	−.1350	.1054
2.59	.4952	.0139	.0796	−.1339	.1080
2.60	.4953	.0136	.0782	−.1328	.1105
2.61	.4955	.0132	.0769	−.1317	.1129
2.62	.4956	.0129	.0756	−.1305	.1152
2.63	.4957	.0126	.0743	−.1294	.1173
2.64	.4959	.0122	.0730	−.1282	.1194
2.65	.4960	.0119	.0717	−.1270	.1213
2.66	.4961	.0116	.0705	−.1258	.1231
2.67	.4962	.0113	.0692	−.1245	.1248
2.68	.4963	.0110	.0680	−.1233	.1264
2.69	.4964	.0107	.0668	−.1220	.1279
2.70	.4965	.0104	.0656	−.1207	.1293
2.71	.4966	.0101	.0644	−.1194	.1306
2.72	.4967	.0099	.0632	−.1181	.1317
2.73	.4968	.0096	.0620	−.1168	.1328
2.74	.4969	.0094	.0608	−.1154	.1338
2.75	.4970	.0091	.0597	−.1141	.1347
2.76	.4971	.0089	.0585	−.1127	.1356
2.77	.4972	.0086	.0574	−.1114	.1363
2.78	.4973	.0084	.0563	−.1100	.1369
2.79	.4974	.0081	.0552	−.1087	.1375
2.80	.4974	.0079	.0541	−.1073	.1379
2.81	.4975	.0077	.0531	−.1059	.1383
2.82	.4976	.0075	.0520	−.1045	.1386
2.83	.4977	.0073	.0510	−.1031	.1389
2.84	.4977	.0071	.0500	−.1017	.1390
2.85	.4978	.0069	.0490	−.1003	.1391
2.86	.4979	.0067	.0480	−.0990	.1391
2.87	.4980	.0065	.0470	−.0976	.1391
2.88	.4980	.0063	.0460	−.0962	.1389
2.89	.4981	.0061	.0451	−.0948	.1388
2.90	.4981	.0060	.0441	−.0934	.1385
2.91	.4982	.0058	.0432	−.0920	.1382
2.92	.4983	.0056	.0423	−.0906	.1378
2.93	.4983	.0055	.0414	−.0893	.1374
2.94	.4984	.0053	.0405	−.0879	.1369
2.95	.4984	.0051	.0396	−.0865	.1364
2.96	.4985	.0050	.0388	−.0852	.1358
2.97	.4985	.0049	.0379	−.0838	.1352
2.98	.4986	.0047	.0371	−.0825	.1345
2.99	.4986	.0046	.0363	−.0811	.1337
3.00	.4987	.0044	.0355	−.0798	.1330

t	Area	Ordinate $\phi(x)$	Second derivative $\phi''(x)$	Third derivative $\phi'''(x)$	Fourth derivative $\phi^{iv}(x)$
3.00	.4987	.0044	.0355	−.0798	.1330
3.01	.4987	.0043	.0347	−.0785	.1321
3.02	.4987	.0042	.0339	−.0771	.1313
3.03	.4988	.0041	.0331	−.0758	.1304
3.04	.4988	.0039	.0324	−.0745	.1294
3.05	.4989	.0038	.0316	−.0732	.1285
3.06	.4989	.0037	.0309	−.0720	.1275
3.07	.4989	.0036	.0302	−.0707	.1264
3.08	.4990	.0035	.0295	−.0694	.1254
3.09	.4990	.0034	.0288	−.0682	.1243
3.10	.4990	.0033	.0281	−.0669	.1231
3.11	.4991	.0032	.0275	−.0657	.1220
3.12	.4991	.0031	.0268	−.0645	.1208
3.13	.4991	.0030	.0262	−.0633	.1196
3.14	.4992	.0029	.0256	−.0621	.1184
3.15	.4992	.0028	.0249	−.0609	.1171
3.16	.4992	.0027	.0243	−.0598	.1159
3.17	.4992	.0026	.0237	−.0586	.1146
3.18	.4993	.0025	.0232	−.0575	.1133
3.19	.4993	.0025	.0226	−.0564	.1120
3.20	.4993	.0024	.0220	−.0552	.1107
3.21	.4993	.0023	.0215	−.0541	.1093
3.22	.4994	.0022	.0210	−.0531	.1080
3.23	.4994	.0022	.0204	−.0520	.1066
3.24	.4994	.0021	.0199	−.0509	.1053
3.25	.4994	.0020	.0194	−.0499	.1039
3.26	.4994	.0020	.0189	−.0488	.1025
3.27	.4995	.0019	.0184	−.0478	.1011
3.28	.4995	.0018	.0180	−.0468	.0997
3.29	.4995	.0018	.0175	−.0458	.0983
3.30	.4995	.0017	.0170	−.0449	.0969
3.31	.4995	.0017	.0166	−.0439	.0955
3.32	.4996	.0016	.0162	−.0429	.0941
3.33	.4996	.0016	.0157	−.0420	.0927
3.34	.4996	.0015	.0153	−.0411	.0913
3.35	.4996	.0015	.0149	−.0402	.0899
3.36	.4996	.0014	.0145	−.0393	.0885
3.37	.4996	.0014	.0141	−.0384	.0871
3.38	.4996	.0013	.0138	−.0376	.0857
3.39	.4997	.0013	.0134	−.0367	.0843
3.40	.4997	.0012	.0130	−.0359	.0829
3.41	.4997	.0012	.0127	−.0350	.0815
3.42	.4997	.0012	.0123	−.0342	.0801
3.43	.4997	.0011	.0120	−.0334	.0788
3.44	.4997	.0011	.0116	−.0327	.0774
3.45	.4997	.0010	.0113	−.0319	.0761
3.46	.4997	.0010	.0110	−.0311	.0747
3.47	.4997	.0010	.0107	−.0304	.0734
3.48	.4998	.0009	.0104	−.0297	.0721
3.49	.4998	.0009	.0101	−.0290	.0707
3.50	.4998	.0009	.0098	−.0283	.0694

NORMAL CURVE OF ERROR

Normal Curve $\phi(x) = \frac{1}{\sqrt{2\pi}} e^{-\frac{x^2}{2}}$ (Continued)

t	Area	Ordinate $\phi(x)$	Second derivative $\phi''(x)$	Third derivative $\phi'''(x)$	Fourth derivative $\phi^{iv}(x)$
3.50	.4998	.0009	.0098	−.0283	.0694
3.51	.4998	.0008	.0095	−.0276	.0681
3.52	.4998	.0008	.0093	−.0269	.0669
3.53	.4998	.0008	.0090	−.0262	.0656
3.54	.4998	.0008	.0087	−.0256	.0643
3.55	.4998	.0007	.0085	−.0249	.0631
3.56	.4998	.0007	.0082	−.0243	.0618
3.57	.4998	.0007	.0080	−.0237	.0606
3.58	.4998	.0007	.0078	−.0231	.0594
3.59	.4998	.0006	.0075	−.0225	.0582
3.60	.4998	.0006	.0073	−.0219	.0570
3.61	.4999	.0006	.0071	−.0214	.0559
3.62	.4999	.0006	.0069	−.0208	.0547
3.63	.4999	.0006	.0067	−.0203	.0536
3.64	.4999	.0005	.0065	−.0198	.0524
3.65	.4999	.0005	.0063	−.0192	.0513
3.66	.4999	.0005	.0061	−.0187	.0502
3.67	.4999	.0005	.0059	−.0182	.0492
3.68	.4999	.0005	.0057	−.0177	.0481
3.69	.4999	.0004	.0056	−.0173	.0470
3.70	.4999	.0004	.0054	−.0168	.0460
3.71	.4999	.0004	.0052	−.0164	.0450
3.72	.4999	.0004	.0051	−.0159	.0440
3.73	.4999	.0004	.0049	−.0155	.0430
3.74	.4999	.0004	.0048	−.0150	.0420
3.75	.4999	.0004	.0046	−.0146	.0410
3.76	.4999	.0003	.0045	−.0142	.0401
3.77	.4999	.0003	.0043	−.0138	.0392
3.78	.4999	.0003	.0042	−.0134	.0382
3.79	.4999	.0003	.0041	−.0131	.0373
3.80	.4999	.0003	.0039	−.0127	.0365
3.81	.4999	.0003	.0038	−.0123	.0356
3.82	.4999	.0003	.0037	−.0120	.0347
3.83	.4999	.0003	.0036	−.0116	.0339
3.84	.4999	.0003	.0034	−.0113	.0331
3.85	.4999	.0002	.0033	−.0110	.0323
3.86	.4999	.0002	.0032	−.0107	.0315
3.87	.5000	.0002	.0031	−.0104	.0307
3.88	.5000	.0002	.0030	−.0100	.0299
3.89	.5000	.0002	.0029	−.0098	.0292
3.90	.5000	.0002	.0028	−.0095	.0284
3.91	.5000	.0002	.0027	−.0092	.0277
3.92	.5000	.0002	.0026	−.0089	.0270
3.93	.5000	.0002	.0026	−.0086	.0263
3.94	.5000	.0002	.0025	−.0084	.0256
3.95	.5000	.0002	.0024	−.0081	.0250
3.96	.5000	.0002	.0023	−.0079	.0243
3.97	.5000	.0002	.0022	−.0076	.0237
3.98	.5000	.0001	.0022	−.0074	.0230
3.99	.5000	.0001	.0021	−.0072	.0224
4.00	.5000	.0001	.0020	−.0070	.0218

NORMAL CURVE OF ERROR

Normal Curve $\phi(x) = \frac{1}{\sqrt{2\pi}} e^{-\frac{x^2}{2}}$ (Continued)

t	Area	Ordinate $\phi(x)$	Second derivative $\phi''(x)$	Third derivative $\phi'''(x)$	Fourth derivative $\phi^{iv}(x)$
4.00	.5000	.0001	.0020	−.0070	.0218
4.01	.5000	.0001	.0019	−.0067	.0212
4.02	.5000	.0001	.0019	−.0065	.0207
4.03	.5000	.0001	.0018	−.0063	.0201
4.04	.5000	.0001	.0018	−.0061	.0195
4.05	.5000	.0001	.0017	−.0059	.0190
4.06	.5000	.0001	.0016	−.0058	.0185
4.07	.5000	.0001	.0016	−.0056	.0180
4.08	.5000	.0001	.0015	−.0054	.0175
4.09	.5000	.0001	.0015	−.0052	.0170
4.10	.5000	.0001	.0014	−.0051	.0165
4.11	.5000	.0001	.0014	−.0049	.0160
4.12	.5000	.0001	.0013	−.0047	.0156
4.13	.5000	.0001	.0013	−.0046	.0151
4.14	.5000	.0001	.0012	−.0044	.0147
4.15	.5000	.0001	.0012	−.0043	.0143
4.16	.5000	.0001	.0011	−.0042	.0138
4.17	.5000	.0001	.0011	−.0040	.0134
4.18	.5000	.0001	.0011	−.0039	.0130
4.19	.5000	.0001	.0010	−.0038	.0127
4.20	.5000	.0001	.0010	−.0036	.0123
4.21	.5000	.0001	.0009	−.0035	.0119
4.22	.5000	.0001	.0009	−.0034	.0116
4.23	.5000	.0001	.0009	−.0033	.0112
4.24	.5000	.0001	.0009	−.0032	.0109
4.25	.5000	.0001	.0008	−.0031	.0105
4.26	.5000	.0001	.0008	−.0030	.0102
4.27	.5000	.0000	.0008	−.0029	.0099
4.28	.5000	.0000	.0007	−.0028	.0096
4.29	.5000	.0000	.0007	−.0027	.0093
4.30	.5000	.0000	.0007	−.0026	.0090
4.31	.5000	.0000	.0007	−.0025	.0087
4.32	.5000	.0000	.0006	−.0024	.0085
4.33	.5000	.0000	.0006	−.0023	.0082
4.34	.5000	.0000	.0006	−.0022	.0079
4.35	.5000	.0000	.0006	−.0022	.0077
4.36	.5000	.0000	.0005	−.0021	.0074
4.37	.5000	.0000	.0005	−.0020	.0072
4.38	.5000	.0000	.0005	−.0019	.0070
4.39	.5000	.0000	.0005	−.0019	.0067
4.40	.5000	.0000	.0005	−.0018	.0065
4.41	.5000	.0000	.0004	−.0017	.0063
4.42	.5000	.0000	.0004	−.0017	.0061
4.43	.5000	.0000	.0004	−.0016	.0059
4.44	.5000	.0000	.0004	−.0016	.0057
4.45	.5000	.0000	.0004	−.0015	.0055
4.46	.5000	.0000	.0004	−.0014	.0053
4.47	.5000	.0000	.0004	−.0014	.0052
4.48	.5000	.0000	.0003	−.0013	.0050
4.49	.5000	.0000	.0003	−.0013	.0048
4.50	.5000	.0000	.0003	−.0012	.0047

t	Area	Ordinate $\phi(x)$	Second derivative $\phi''(x)$	Third derivative $\phi'''(x)$	Fourth derivative $\phi^{iv}(x)$
4.50	.5000	.0000	.0003	−.0012	.0047
4.51	.5000	.0000	.0003	−.0012	.0045
4.52	.5000	.0000	.0003	−.0012	.0044
4.53	.5000	.0000	.0003	−.0011	.0042
4.54	.5000	.0000	.0003	−.0011	.0041
4.55	.5000	.0000	.0003	−.0010	.0039
4.56	.5000	.0000	.0002	−.0010	.0038
4.57	.5000	.0000	.0002	−.0010	.0037
4.58	.5000	.0000	.0002	−.0009	.0035
4.59	.5000	.0000	.0002	−.0009	.0034
4.60	.5000	.0000	.0002	−.0009	.0033
4.61	.5000	.0000	.0002	−.0008	.0032
4.62	.5000	.0000	.0002	−.0008	.0031
4.63	.5000	.0000	.0002	−.0008	.0030
4.64	.5000	.0000	.0002	−.0007	.0028
4.65	.5000	.0000	.0002	−.0007	.0027
4.66	.5000	.0000	.0002	−.0007	.0026
4.67	.5000	.0000	.0002	−.0006	.0026
4.68	.5000	.0000	.0002	−.0006	.0025
4.69	.5000	.0000	.0001	−.0006	.0024
4.70	.5000	.0000	.0001	−.0006	.0023
4.71	.5000	.0000	.0001	−.0006	.0022
4.72	.5000	.0000	.0001	−.0005	.0021
4.73	.5000	.0000	.0001	−.0005	.0020
4.74	.5000	.0000	.0001	−.0005	.0020
4.75	.5000	.0000	.0001	−.0005	.0019
4.76	.5000	.0000	.0001	−.0005	.0018
4.77	.5000	.0000	.0001	−.0004	.0018
4.78	.5000	.0000	.0001	−.0004	.0017
4.79	.5000	.0000	.0001	−.0004	.0016
4.80	.5000	.0000	.0001	−.0004	.0016
4.81	.5000	.0000	.0001	−.0004	.0015
4.82	.5000	.0000	.0001	−.0004	.0015
4.83	.5000	.0000	.0001	−.0003	.0014
4.84	.5000	.0000	.0001	−.0003	.0013
4.85	.5000	.0000	.0001	−.0003	.0013
4.86	.5000	.0000	.0001	−.0003	.0012
4.87	.5000	.0000	.0001	−.0003	.0012
4.88	.5000	.0000	.0001	−.0003	.0012
4.89	.5000	.0000	.0001	−.0003	.0011
4.90	.5000	.0000	.0001	−.0003	.0011
4.91	.5000	.0000	.0001	−.0002	.0010
4.92	.5000	.0000	.0001	−.0002	.0010
4.93	.5000	.0000	.0001	−.0002	.0009
4.94	.5000	.0000	.0001	−.0002	.0009
4.95	.5000	.0000	.0000	−.0002	.0009
4.96	.5000	.0000	.0000	−.0002	.0008
4.97	.5000	.0000	.0000	−.0002	.0008
4.98	.5000	.0000	.0000	−.0002	.0008
4.99	.5000	.0000	.0000	−.0002	.0007

TESTS OF SIGNIFICANCE

"t" Test of Significance Between Two Sample Means ($\bar{x}_1$ and $\bar{x}_2$).

(Use Fisher's t distribution)

Paired variates:

$$t = \frac{\bar{d}}{\sqrt{\frac{\Sigma(d_i - \bar{d})^2}{N(N-1)}}} \text{ with } N - 1 \text{ degrees of freedom}$$

where $\bar{d} = \bar{x}_1 - \bar{x}_2$
$d_1 = x_{11} - x_{21}$
$d_2 = x_{12} - x_{22}$ etc.
N = sample size

Unpaired variates:

$$t = \frac{\bar{x}_1 - \bar{x}_2}{\sqrt{\frac{\Sigma_i(x_{1i} - \bar{x}_1)^2 + \Sigma_i(x_{2i} - \bar{x}_2)^2}{N_1 + N_2 - 2}\left(\frac{1}{N_1} + \frac{1}{N_2}\right)}}$$

with $N_1 + N_2 - 2$ degrees of freedom

where N_1 = size of sample 1
N_2 = size of sample 2

Chi Square Table

The column headings "P" are probabilities, while the body of the table gives values of χ^2; the left column the degrees of freedom. The probability of a value $\chi^2 \geqq$ than the value specified is thus obtained. For $n = 20$, $P(\chi^2 \geqq 10.851)$ the table gives .95, and accordingly by subtraction $P(\chi^2 \leqq 10.851)$ is .05.

F Test for Equality of Variances

$$F = \frac{\sigma_1^2}{\sigma_2^2}$$

where σ_1^2 = variance of sample with size N_1
σ_2^2 = variance of sample with size N_2
with $N_1 - 1$ = degrees of freedom for numerator
$N_2 - 1$ = degrees of freedom for denominator

(Use Snedecor's F distribution)

TABLE FOR t TEST OF SIGNIFICANCEBETWEEN TWO SAMPLE MEANS ($\bar{x}_1$ AND $\bar{x}_2$)

TABLE FOR t TEST OF SIGNIFICANCE BETWEEN TWO SAMPLE MEANS ($\bar{x}_1$ AND $\bar{x}_2$)

−t 0 +t

Degrees of freedom	*P = 0.9	0.8	0.7	0.6	0.5	0.4	0.3	0.2	0.1	0.05	0.02	0.01
1	0.158	0.325	0.510	0.727	1.000	1.376	1.963	3.078	6.314	12.706	31.821	63.557
2	0.142	0.289	0.445	0.617	0.816	1.061	1.386	1.886	2.920	4.303	6.965	9.925
3	0.137	0.277	0.424	0.584	0.765	0.978	1.250	1.638	2.353	3.182	4.541	5.841
4	0.134	0.271	0.414	0.569	0.741	0.941	1.190	1.533	2.132	2.776	3.747	4.604
5	0.132	0.267	0.408	0.559	0.727	0.920	1.156	1.476	2.015	2.571	3.365	4.032
6	0.131	0.265	0.404	0.553	0.718	0.906	1.134	1.440	1.943	2.447	3.143	3.707
7	0.130	0.263	0.402	0.549	0.711	0.896	1.119	1.415	1.895	2.365	2.998	3.499
8	0.130	0.262	0.399	0.546	0.706	0.889	1.108	1.397	1.860	2.306	2.896	3.355
9	0.129	0.261	0.398	0.543	0.703	0.883	1.100	1.383	1.833	2.262	2.821	3.250
10	0.129	0.260	0.397	0.542	0.700	0.879	1.093	1.372	1.812	2.228	2.764	3.169
11	0.129	0.260	0.396	0.540	0.697	0.876	1.088	1.363	1.796	2.201	2.718	3.106
12	0.128	0.259	0.395	0.539	0.695	0.873	1.083	1.356	1.782	2.179	2.681	3.055
13	0.128	0.259	0.394	0.538	0.694	0.870	1.079	1.350	1.771	2.160	2.650	3.012
14	0.128	0.258	0.393	0.537	0.692	0.868	1.076	1.345	1.761	2.145	2.624	2.977
15	0.128	0.258	0.393	0.536	0.691	0.866	1.074	1.341	1.753	2.131	2.602	2.947
16	0.128	0.258	0.392	0.535	0.690	0.865	1.071	1.337	1.746	2.120	2.583	2.921
17	0.128	0.257	0.392	0.534	0.689	0.863	1.069	1.333	1.740	2.110	2.567	2.898
18	0.127	0.257	0.392	0.534	0.688	0.862	1.067	1.330	1.734	2.101	2.552	2.878
19	0.127	0.257	0.391	0.533	0.688	0.861	1.066	1.328	1.729	2.093	2.539	2.861
20	0.127	0.257	0.391	0.533	0.687	0.860	1.064	1.325	1.725	2.086	2.528	2.845
21	0.127	0.257	0.391	0.532	0.686	0.859	1.063	1.323	1.721	2.080	2.518	2.831
22	0.127	0.256	0.390	0.532	0.686	0.858	1.061	1.321	1.717	2.074	2.508	2.819
23	0.127	0.256	0.390	0.532	0.685	0.858	1.060	1.319	1.714	2.069	2.500	2.807
24	0.127	0.256	0.390	0.531	0.685	0.857	1.059	1.318	1.711	2.064	2.492	2.797
25	0.127	0.256	0.390	0.531	0.684	0.856	1.058	1.316	1.708	2.060	2.485	2.787
26	0.127	0.256	0.390	0.531	0.684	0.856	1.058	1.315	1.706	2.056	2.479	2.779
27	0.127	0.256	0.389	0.531	0.684	0.855	1.057	1.314	1.703	2.052	2.473	2.771
28	0.127	0.256	0.389	0.530	0.683	0.855	1.056	1.313	1.701	2.048	2.467	2.763
29	0.127	0.256	0.389	0.530	0.683	0.854	1.055	1.311	1.699	2.045	2.462	2.756
30	0.127	0.256	0.389	0.530	0.683	0.854	1.055	1.310	1.697	2.042	2.457	2.750
∞	0.12566	0.25335	0.38532	0.52440	0.67449	0.84162	1.03643	1.28155	1.64485	1.95996	2.32634	2.57582

Reproduced from *Statistical Methods for Research Workers*, 6th ed., with the permission of the author, R. A. Fisher, and his publisher, Oliver and Boyd, Edinburgh.

* P is the probability of having t this large or larger in size by chance.

χ^2 TABLE

Degrees of freedom	P = 0.99	0.98	0.95	0.90	0.80	0.70
1	0.000157	0.000628	0.00393	0.0158	0.0642	0.148
2	0.0201	0.0404	0.103	0.211	0.446	0.713
3	0.115	0.185	0.352	0.584	1.005	1.424
4	0.297	0.429	0.711	1.064	1.649	2.195
5	0.554	0.752	1.145	1.610	2.343	3.000
6	0.872	1.134	1.635	2.204	3.070	3.828
7	1.239	1.564	2.167	2.833	3.822	4.671
8	1.646	2.032	2.733	3.490	4.594	5.527
9	2.088	2.532	3.325	4.168	5.380	6.393
10	2.558	3.059	3.940	4.865	6.179	7.267
11	3.053	3.609	4.575	5.578	6.989	8.148
12	3.571	4.178	5.226	6.304	7.807	9.034
13	4.107	4.765	5.892	7.042	8.634	9.926
14	4.660	5.368	6.571	7.790	9.467	10.821
15	5.229	5.985	7.261	8.547	10.307	11.721
16	5.812	6.614	7.962	9.312	11.152	12.624
17	6.408	7.255	8.672	10.085	12.002	13.531
18	7.015	7.906	9.390	10.865	12.857	14.440
19	7.633	8.567	10.117	11.651	13.716	15.352
20	8.260	9.237	10.851	12.443	14.578	16.266
21	8.897	9.915	11.591	13.240	15.445	17.182
22	9.542	10.600	12.338	14.041	16.314	18.101
23	10.196	11.293	13.091	14.848	17.187	19.021
24	10.856	11.992	13.848	15.659	18.062	19.943
25	11.524	12.697	14.611	16.473	18.940	20.867
26	12.198	13.409	15.379	17.292	19.820	21.792
27	12.879	14.125	16.151	18.114	20.703	22.719
28	13.565	14.847	16.928	18.939	21.588	23.647
29	14.256	15.574	17.708	19.768	22.475	24.577
30	14.953	16.306	18.493	20.599	23.364	25.508

For degrees of freedom greater than 30, the expression $\sqrt{2\chi^2} - \sqrt{2n' - 1}$ may be used as a normal deviate with unit variance, where n' is the number of degrees of freedom.

Reproduced from *Statistical Methods for Research Workers*, 6th ed., with the permission of the author, R. A. Fisher, and his publisher, Oliver and Boyd, Edinburgh.

χ^2 TABLE

Degrees of freedom	0.50	0.30	0.20	0.10	0.05	0.02	0.01
1	0.455	1.074	1.642	2.706	3.841	5.412	6.635
2	1.386	2.408	3.219	4.605	5.991	7.824	9.210
3	2.366	3.665	4.642	6.251	7.815	9.837	11.341
4	3.357	4.878	5.989	7.779	9.488	11.668	13.277
5	4.351	6.064	7.289	9.236	11.070	13.388	15.086
6	5.348	7.231	8.558	10.645	12.592	15.033	16.812
7	6.346	8.383	9.803	12.017	14.067	16.622	18.475
8	7.344	9.524	11.030	13.362	15.507	18.168	20.090
9	8.343	10.656	12.242	14.684	16.919	19.679	21.666
10	9.342	11.781	13.442	15.987	18.307	21.161	23.209
11	10.341	12.899	14.631	17.275	19.675	22.618	24.725
12	11.340	14.011	15.812	18.549	21.026	24.054	26.217
13	12.340	15.119	16.985	19.812	22.362	25.472	27.688
14	13.339	16.222	18.151	21.064	23.685	26.873	29.141
15	14.339	17.322	19.311	22.307	24.996	28.259	30.578
16	15.338	18.418	20.465	23.542	26.296	29.633	32.000
17	16.338	19.511	21.615	24.769	27.587	30.995	33.409
18	17.338	20.601	22.760	25.989	28.869	32.346	34.805
19	18.338	21.689	23.900	27.204	30.144	33.687	36.191
20	19.337	22.775	25.038	28.412	31.410	35.020	37.566
21	20.337	23.858	26.171	29.615	32.671	36.343	38.932
22	21.337	24.939	27.301	30.813	33.924	37.659	40.289
23	22.337	26.018	28.429	32.007	35.172	38.968	41.638
24	23.337	27.096	29.553	33.196	36.415	40.270	42.980
25	24.337	28.172	30.675	34.382	37.652	41.566	44.314
26	25.336	29.246	31.795	35.563	38.885	42.856	45.642
27	26.336	30.319	32.912	36.741	40.113	44.140	46.963
28	27.336	31.391	34.027	37.916	41.337	45.419	48.278
29	28.336	32.461	35.139	39.087	42.557	46.693	49.588
30	29.336	33.530	36.250	40.256	43.773	47.962	50.892

TABLE FOR F TEST FOR EQUALITY OF VARIANCES

*5 % (Roman Type) and 1 % (Bold-Face Type)

Points for the Distribution of F

Degrees of freedom for lesser mean square	Degrees of freedom for greater mean square											
	1	2	3	4	5	6	7	8	9	10	11	12
1	161 **4052**	200 **4999**	216 **5403**	225 **5625**	230 **5764**	234 **5859**	237 **5928**	239 **5981**	241 **6022**	242 **6056**	243 **6082**	244 **6106**
2	18.51 **98.49**	19.00 **99.01**	19.16 **99.17**	19.25 **99.25**	19.30 **99.30**	19.33 **99.33**	19.36 **99.34**	19.37 **99.36**	19.38 **99.38**	19.39 **99.40**	19.40 **99.41**	19.41 **99.42**
3	10.13 **34.12**	9.55 **30.81**	9.28 **29.46**	9.12 **28.71**	9.01 **28.24**	8.94 **27.91**	8.88 **27.67**	8.84 **27.49**	8.81 **27.34**	8.78 **27.23**	8.76 **27.13**	8.74 **27.05**
4	7.71 **21.20**	6.94 **18.00**	6.59 **16.69**	6.39 **15.98**	6.26 **15.52**	6.16 **15.21**	6.09 **14.98**	6.04 **14.80**	6.00 **14.66**	5.96 **14.54**	5.93 **14.45**	5.91 **14.37**
5	6.61 **16.26**	5.79 **13.27**	5.41 **12.06**	5.19 **11.39**	5.05 **10.97**	4.95 **10.67**	4.88 **10.45**	4.82 **10.27**	4.78 **10.15**	4.74 **10.05**	4.70 **9.96**	4.68 **9.89**
6	5.99 **13.74**	5.14 **10.92**	4.76 **9.78**	4.53 **9.15**	4.39 **8.75**	4.28 **8.47**	4.21 **8.26**	4.15 **8.10**	4.10 **7.98**	4.06 **7.87**	4.03 **7.79**	4.00 **7.72**
7	5.59 **12.25**	4.74 **9.55**	4.35 **8.45**	4.12 **7.85**	3.97 **7.46**	3.87 **7.19**	3.79 **7.00**	3.73 **6.84**	3.68 **6.71**	3.63 **6.62**	3.60 **6.54**	3.57 **6.47**
8	5.32 **11.26**	4.46 **8.65**	4.07 **7.59**	3.84 **7.01**	3.69 **6.63**	3.58 **6.37**	3.50 **6.19**	3.44 **6.03**	3.39 **5.91**	3.34 **5.82**	3.31 **5.74**	3.28 **5.67**
9	5.12 **10.56**	4.26 **8.02**	3.86 **6.99**	3.63 **6.42**	3.48 **6.06**	3.37 **5.80**	3.29 **5.62**	3.23 **5.47**	3.18 **5.35**	3.13 **5.26**	3.10 **5.18**	3.07 **5.11**
10	4.96 **10.04**	4.10 **7.56**	3.71 **6.55**	3.48 **5.99**	3.33 **5.64**	3.22 **5.39**	3.14 **5.21**	3.07 **5.06**	3.02 **4.95**	2.97 **4.85**	2.94 **4.78**	2.91 **4.71**
11	4.84 **9.65**	3.98 **7.20**	3.59 **6.22**	3.36 **5.67**	3.20 **5.32**	3.09 **5.07**	3.01 **4.88**	2.95 **4.74**	2.90 **4.63**	2.86 **4.54**	2.82 **4.46**	2.79 **4.40**
12	4.75 **9.33**	3.88 **6.93**	3.49 **5.95**	3.26 **5.41**	3.11 **5.06**	3.00 **4.82**	2.92 **4.65**	2.85 **4.50**	2.80 **4.39**	2.76 **4.30**	2.72 **4.22**	2.69 **4.16**
13	4.67 **9.07**	3.80 **6.70**	3.41 **5.74**	3.18 **5.20**	3.02 **4.86**	2.92 **4.62**	2.84 **4.44**	2.77 **4.30**	2.72 **4.19**	2.67 **4.10**	2.63 **4.02**	2.60 **3.96**
14	4.60 **8.86**	3.74 **6.51**	3.34 **5.56**	3.11 **5.03**	2.96 **4.69**	2.85 **4.46**	2.77 **4.28**	2.70 **4.14**	2.65 **4.03**	2.60 **3.94**	2.56 **3.86**	2.53 **3.80**
15	4.54 **8.68**	3.68 **6.36**	3.29 **5.42**	3.06 **4.89**	2.90 **4.56**	2.79 **4.32**	2.70 **4.14**	2.64 **4.00**	2.59 **3.89**	2.55 **3.80**	2.51 **3.73**	2.48 **3.67**
16	4.49 **8.53**	3.63 **6.23**	3.24 **5.29**	3.01 **4.77**	2.85 **4.44**	2.74 **4.20**	2.66 **4.03**	2.59 **3.89**	2.54 **3.78**	2.49 **3.69**	2.45 **3.61**	2.42 **3.55**

Degrees of freedom for lesser mean square	Degrees of freedom for greater mean square											
	14	16	20	24	30	40	50	75	100	200	500	∞
1	245 **6142**	246 **6169**	248 **6208**	249 **6234**	250 **6258**	251 **6286**	252 **6302**	253 **6323**	253 **6334**	254 **6352**	254 **6361**	254 **6366**
2	19.42 **99.43**	19.43 **99.44**	19.44 **99.45**	19.45 **99.46**	19.46 **99.47**	19.47 **99.48**	19.47 **99.48**	19.48 **99.49**	19.49 **99.49**	19.49 **99.49**	19.50 **99.50**	19.50 **99.50**
3	8.71 **26.92**	8.69 **26.83**	8.66 **26.69**	8.64 **26.60**	8.62 **26.50**	8.60 **26.41**	8.58 **26.30**	8.57 **26.27**	8.56 **26.23**	8.54 **26.18**	8.54 **26.14**	8.53 **26.12**
4	5.87 **14.24**	5.84 **14.15**	5.80 **14.02**	5.77 **13.93**	5.74 **13.83**	5.71 **13.74**	5.70 **13.69**	5.68 **13.61**	5.66 **13.57**	5.65 **13.52**	5.64 **13.48**	5.63 **13.46**
5	4.64 **9.77**	4.60 **9.68**	4.56 **9.55**	4.53 **9.47**	4.50 **9.38**	4.46 **9.29**	4.44 **9.24**	4.42 **9.17**	4.40 **9.13**	4.38 **9.07**	4.37 **9.04**	4.36 **9.02**
6	3.96 **7.60**	3.92 **7.52**	3.87 **7.39**	3.84 **7.31**	3.81 **7.23**	3.77 **7.14**	3.75 **7.09**	3.72 **7.02**	3.71 **6.99**	3.69 **6.94**	3.68 **6.90**	3.67 **6.88**
7	3.52 **6.35**	3.49 **6.27**	3.44 **6.15**	3.41 **6.07**	3.38 **5.98**	3.34 **5.90**	3.32 **5.85**	3.29 **5.78**	3.28 **5.75**	3.25 **5.70**	3.24 **5.67**	3.23 **5.65**
8	3.23 **5.56**	3.20 **5.48**	3.15 **5.36**	3.12 **5.28**	3.08 **5.20**	3.05 **5.11**	3.03 **5.06**	3.00 **5.00**	2.98 **4.96**	2.96 **4.91**	2.94 **4.88**	2.93 **4.86**
9	3.02 **5.00**	2.98 **4.92**	2.93 **4.80**	2.90 **4.73**	2.86 **4.64**	2.82 **4.56**	2.80 **4.51**	2.77 **4.45**	2.76 **4.41**	2.73 **4.36**	2.72 **4.33**	2.71 **4.31**
10	2.86 **4.60**	2.82 **4.52**	2.77 **4.41**	2.74 **4.33**	2.70 **4.25**	2.67 **4.17**	2.64 **4.12**	2.61 **4.05**	2.59 **4.01**	2.56 **3.96**	2.55 **3.93**	2.54 **3.91**
11	2.74 **4.29**	2.70 **4.21**	2.65 **4.10**	2.61 **4.02**	2.57 **3.94**	2.53 **3.86**	2.50 **3.80**	2.47 **3.74**	2.45 **3.70**	2.42 **3.66**	2.41 **3.62**	2.40 **3.60**
12	2.64 **4.05**	2.60 **3.98**	2.54 **3.86**	2.50 **3.78**	2.46 **3.70**	2.42 **3.61**	2.40 **3.56**	2.36 **3.49**	2.35 **3.46**	2.32 **3.41**	2.31 **3.38**	2.30 **3.36**
13	2.55 **3.85**	2.51 **3.78**	2.46 **3.67**	2.42 **3.59**	2.38 **3.51**	2.34 **3.42**	2.32 **3.37**	2.28 **3.30**	2.26 **3.27**	2.24 **3.21**	2.22 **3.18**	2.21 **3.16**
14	2.48 **3.70**	2.44 **3.62**	2.39 **3.51**	2.35 **3.43**	2.31 **3.34**	2.27 **3.26**	2.24 **3.21**	2.21 **3.14**	2.19 **3.11**	2.16 **3.06**	2.14 **3.02**	2.13 **3.00**
15	2.43 **3.56**	2.39 **3.48**	2.33 **3.36**	2.29 **3.29**	2.25 **3.20**	2.21 **3.12**	2.18 **3.07**	2.15 **3.00**	2.12 **2.97**	2.10 **2.92**	2.08 **2.89**	2.07 **2.87**
16	2.37 **3.45**	2.33 **3.37**	2.28 **3.25**	2.24 **3.18**	2.20 **3.10**	2.16 **3.01**	2.13 **2.96**	2.09 **2.89**	2.07 **2.86**	2.04 **2.80**	2.02 **2.77**	2.01 **2.75**

Reprinted, by permission, from Snedecor, *Statistical Methods*, Collegiate Press, Iowa State College, Ames.

* This table gives values of F which one would expect to exceed by chance alone 5% and 1% of the time.

TABLE FOR F TEST FOR EQUALITY OF VARIANCES

TABLE FOR *F* TEST FOR EQUALITY OF VARIANCES (Continued)

5% (Roman Type) and 1% (Bold-Face Type) Points for the Distribution of *F* (Continued)

Degrees of freedom for lesser mean square	Degrees of freedom for greater mean square											
	1	2	3	4	5	6	7	8	9	10	11	12
17	4.45 **8.40**	3.59 **6.11**	3.20 **5.18**	2.96 **4.67**	2.81 **4.34**	2.70 **4.10**	2.62 **3.93**	2.55 **3.79**	2.50 **3.68**	2.45 **3.59**	2.41 **3.52**	2.38 **3.45**
18	4.41 **8.28**	3.55 **6.01**	3.16 **5.09**	2.93 **4.58**	2.77 **4.25**	2.66 **4.01**	2.58 **3.85**	2.51 **3.71**	2.46 **3.60**	2.41 **3.51**	2.37 **3.44**	2.34 **3.37**
19	4.38 **8.18**	3.52 **5.93**	3.13 **5.01**	2.90 **4.50**	2.74 **4.17**	2.63 **3.94**	2.55 **3.77**	2.48 **3.63**	2.43 **3.52**	2.38 **3.43**	2.34 **3.36**	2.31 **3.30**
20	4.35 **8.10**	3.49 **5.85**	3.10 **4.94**	2.87 **4.43**	2.71 **4.10**	2.60 **3.87**	2.52 **3.71**	2.45 **3.56**	2.40 **3.45**	2.35 **3.37**	2.31 **3.30**	2.28 **3.23**
21	4.32 **8.02**	3.47 **5.78**	3.07 **4.87**	2.84 **4.37**	2.68 **4.04**	2.57 **3.81**	2.49 **3.65**	2.42 **3.51**	2.37 **3.40**	2.32 **3.31**	2.28 **3.24**	2.25 **3.17**
22	4.30 **7.94**	3.44 **5.72**	3.05 **4.82**	2.82 **4.31**	2.66 **3.99**	2.55 **3.76**	2.47 **3.59**	2.40 **3.45**	2.35 **3.35**	2.30 **3.26**	2.26 **3.18**	2.23 **3.12**
23	4.28 **7.88**	3.42 **5.66**	3.03 **4.76**	2.80 **4.26**	2.64 **3.94**	2.53 **3.71**	2.45 **3.54**	2.38 **3.41**	2.32 **3.30**	2.28 **3.21**	2.24 **3.14**	2.20 **3.07**
24	4.26 **7.82**	3.40 **5.61**	3.01 **4.72**	2.78 **4.22**	2.62 **3.90**	2.51 **3.67**	2.43 **3.50**	2.36 **3.36**	2.30 **3.25**	2.26 **3.17**	2.22 **3.09**	2.18 **3.03**
25	4.24 **7.77**	3.38 **5.57**	2.99 **4.68**	2.76 **4.18**	2.60 **3.86**	2.49 **3.63**	2.41 **3.46**	2.34 **3.32**	2.28 **3.21**	2.24 **3.13**	2.20 **3.05**	2.16 **2.99**
26	4.22 **7.72**	3.37 **5.53**	2.89 **4.64**	2.74 **4.14**	2.59 **3.82**	2.47 **3.59**	2.39 **3.42**	2.32 **3.29**	2.27 **3.17**	2.22 **3.09**	2.18 **3.02**	2.15 **2.96**
27	4.21 **7.68**	3.35 **5.49**	2.96 **4.60**	2.73 **4.11**	2.57 **3.79**	2.46 **3.56**	2.37 **3.39**	2.30 **3.26**	2.25 **3.14**	2.20 **3.06**	2.16 **2.98**	2.13 **2.93**
28	4.20 **7.64**	3.34 **5.45**	2.95 **4.57**	2.71 **4.07**	2.56 **3.76**	2.44 **3.53**	2.36 **3.36**	2.29 **3.23**	3.24 **3.11**	2.19 **3.03**	2.15 **2.95**	2.12 **2.90**
29	4.18 **7.60**	3.33 **5.52**	2.93 **4.54**	2.70 **4.04**	2.54 **3.73**	2.43 **3.50**	2.35 **3.33**	2.28 **3.20**	2.22 **3.08**	2.18 **3.00**	2.14 **2.92**	2.10 **2.87**
30	4.17 **7.56**	3.32 **5.39**	2.92 **4.51**	2.69 **4.02**	2.53 **3.70**	2.42 **3.47**	2.34 **3.30**	2.27 **3.17**	2.21 **3.06**	2.16 **2.98**	2.12 **2.90**	2.09 **2.84**
32	4.15 **7.50**	3.30 **5.34**	2.90 **4.46**	2.67 **3.97**	2.51 **3.66**	2.40 **3.42**	2.32 **3.25**	2.25 **3.12**	2.19 **3.01**	2.14 **2.94**	2.10 **2.86**	2.07 **2.80**
34	4.13 **7.44**	3.28 **5.29**	2.88 **4.42**	2.65 **3.93**	2.49 **3.61**	2.38 **3.38**	2.30 **3.21**	2.23 **3.08**	2.17 **2.97**	2.12 **2.89**	2.08 **2.82**	2.05 **2.76**
36	4.11 **7.39**	3.26 **5.25**	2.86 **4.38**	2.63 **3.89**	2.48 **3.58**	2.36 **3.35**	2.28 **3.18**	2.21 **3.04**	2.15 **2.94**	2.10 **2.86**	2.06 **2.78**	2.03 **2.72**
38	4.10 **7.35**	3.25 **5.21**	2.85 **4.34**	2.62 **3.86**	2.46 **3.54**	2.35 **3.32**	2.26 **3.15**	2.19 **3.02**	2.14 **2.91**	2.09 **2.82**	2.05 **2.75**	2.02 **2.69**
40	4.08 **7.31**	3.23 **5.18**	2.84 **4.31**	2.61 **3.83**	2.45 **3.51**	2.34 **3.29**	2.25 **3.12**	2.18 **2.99**	2.12 **2.88**	2.07 **2.80**	2.04 **2.73**	2.00 **2.66**
42	4.07 **7.27**	3.22 **5.15**	2.83 **4.29**	2.59 **3.80**	2.44 **3.49**	2.32 **3.26**	2.24 **3.10**	2.17 **2.96**	2.11 **2.86**	2.06 **2.77**	2.02 **2.70**	1.99 **2.64**
44	4.06 **7.24**	3.21 **5.12**	2.82 **4.26**	2.58 **3.78**	2.43 **3.46**	2.31 **3.24**	2.23 **3.07**	2.16 **2.94**	2.10 **2.84**	2.05 **2.75**	2.01 **2.68**	1.98 **2.62**

Degrees of freedom for lesser mean square	Degrees of freedom for greater mean square											
	14	16	20	24	30	40	50	75	100	200	500	∞
17	2.33 **3.35**	2.29 **3.27**	2.23 **3.16**	2.19 **3.08**	2.15 **3.00**	2.11 **2.92**	2.08 **2.86**	2.04 **2.79**	2.02 **2.76**	1.99 **2.70**	1.97 **2.67**	1.96 **2.65**
18	2.29 **3.27**	2.25 **3.19**	2.19 **3.07**	2.15 **3.00**	2.11 **2.91**	2.07 **2.83**	2.04 **2.78**	2.00 **2.71**	1.98 **2.68**	1.95 **2.62**	1.93 **2.59**	1.92 **2.57**
19	2.26 **3.19**	2.21 **3.12**	2.15 **3.00**	2.11 **2.92**	2.07 **2.84**	2.02 **2.76**	2.00 **2.70**	1.96 **2.63**	1.94 **2.60**	1.91 **2.54**	1.90 **2.51**	1.88 **2.49**
20	2.23 **3.13**	2.18 **3.05**	2.12 **2.94**	2.08 **2.86**	2.04 **2.77**	1.99 **2.69**	1.96 **2.63**	1.92 **2.56**	1.90 **2.53**	1.87 **2.47**	1.85 **2.44**	1.84 **2.42**
21	2.20 **3.07**	2.15 **2.99**	2.09 **2.88**	2.05 **2.80**	2.00 **2.72**	1.96 **2.63**	1.93 **2.58**	1.89 **2.51**	1.87 **2.47**	1.84 **2.42**	1.82 **2.38**	1.81 **2.36**
22	2.18 **3.02**	2.13 **2.94**	2.07 **2.83**	2.03 **2.75**	1.98 **2.67**	1.93 **2.58**	1.91 **2.53**	1.87 **2.46**	1.84 **2.42**	1.81 **2.37**	1.80 **2.33**	1.78 **2.31**
23	2.14 **2.97**	2.10 **2.89**	2.04 **2.78**	2.00 **2.70**	1.96 **2.62**	1.91 **2.53**	1.88 **2.48**	1.84 **2.41**	1.82 **2.37**	1.79 **2.32**	1.77 **2.28**	1.76 **2.26**
24	2.13 **2.93**	2.09 **2.85**	2.02 **2.74**	1.98 **2.66**	1.94 **2.58**	1.89 **2.49**	1.86 **2.44**	1.82 **2.36**	1.80 **2.33**	1.76 **2.27**	1.74 **2.23**	1.73 **2.21**
25	2.11 **2.89**	2.06 **2.81**	2.00 **2.70**	1.96 **2.62**	1.92 **2.54**	1.87 **2.45**	1.84 **2.40**	1.80 **2.32**	1.77 **2.29**	1.74 **2.23**	1.72 **2.19**	1.71 **2.17**
26	2.10 **2.86**	2.05 **2.77**	1.99 **2.66**	1.95 **2.58**	1.90 **2.50**	1.85 **2.41**	1.82 **2.36**	1.78 **2.28**	1.76 **2.25**	1.72 **2.19**	1.70 **2.15**	1.69 **2.13**
27	2.08 **2.83**	2.03 **2.74**	1.97 **2.63**	1.93 **2.55**	1.88 **2.47**	1.84 **2.38**	1.80 **2.33**	1.76 **2.25**	1.74 **2.21**	1.71 **2.16**	1.68 **2.12**	1.67 **2.10**
28	2.06 **2.80**	2.02 **2.71**	1.96 **2.60**	1.91 **2.52**	1.87 **2.44**	1.81 **2.35**	1.78 **2.30**	1.75 **2.22**	1.72 **2.18**	1.69 **2.13**	1.67 **2.09**	1.65 **2.06**
29	2.05 **2.77**	2.00 **2.68**	1.94 **2.57**	1.90 **2.49**	1.85 **2.41**	1.80 **2.32**	1.77 **2.27**	1.73 **2.19**	1.71 **2.15**	1.68 **2.10**	1.65 **2.06**	1.64 **2.03**
30	2.04 **2.74**	1.99 **2.66**	1.93 **2.55**	1.89 **2.47**	1.84 **2.38**	1.79 **2.29**	1.76 **2.24**	1.72 **2.16**	1.69 **2.13**	1.66 **2.07**	1.64 **2.03**	1.62 **2.01**
32	2.02 **2.70**	1.97 **2.62**	1.91 **2.51**	1.86 **2.42**	1.82 **2.34**	1.76 **2.25**	1.74 **2.20**	1.69 **2.12**	1.67 **2.08**	1.64 **2.02**	1.61 **1.98**	1.59 **1.96**
34	2.00 **2.66**	1.95 **2.58**	1.89 **2.47**	1.84 **2.38**	1.80 **2.30**	1.74 **2.21**	1.71 **2.15**	1.67 **2.08**	1.64 **2.04**	1.61 **1.98**	1.59 **1.94**	1.57 **1.91**
36	1.89 **2.62**	1.93 **2.54**	1.87 **2.43**	1.82 **2.35**	1.78 **2.26**	1.72 **2.17**	1.69 **2.12**	1.65 **2.04**	1.62 **2.00**	1.59 **1.94**	1.56 **1.90**	1.55 **1.87**
38	1.96 **2.59**	1.92 **2.51**	1.85 **2.40**	1.80 **2.32**	1.76 **2.22**	1.71 **2.14**	1.67 **2.08**	1.63 **2.00**	1.60 **1.97**	1.57 **1.90**	1.54 **1.86**	1.53 **1.84**
40	1.95 **2.56**	1.90 **2.49**	1.84 **2.37**	1.79 **2.29**	1.74 **2.20**	1.69 **2.11**	1.66 **2.05**	1.61 **1.97**	1.59 **1.94**	1.55 **1.88**	1.53 **1.84**	1.51 **1.81**
42	1.94 **2.54**	1.89 **2.46**	1.82 **2.35**	1.78 **2.26**	1.73 **2.17**	1.68 **2.08**	1.64 **2.02**	1.60 **1.94**	1.57 **1.91**	1.54 **1.85**	1.51 **1.80**	1.49 **1.78**
44	1.92 **2.52**	1.88 **2.44**	1.81 **2.32**	1.76 **2.24**	1.72 **2.15**	1.66 **2.06**	1.63 **2.00**	1.58 **1.92**	1.56 **1.88**	1.52 **1.82**	1.50 **1.78**	1.48 **1.75**

TABLE FOR *F* TEST FOR EQUALITY OF VARIANCES (Continued)

5% (Roman Type) and 1% (Bold-Face Type) Points for the Distribution of *F* (Continued)

TABLE FOR *F* TEST FOR EQUALITY OF VARIANCES (Continued)

5% (Roman Type) and 1% (**Bold-Face Type**) Points for the Distribution of F (Continued)

Degrees of freedom for lesser mean square	Degrees of freedom for greater mean square											
	1	2	3	4	5	6	7	8	9	10	11	12
46	4.05 **7.21**	3.20 **5.10**	2.81 **4.24**	2.57 **3.76**	2.42 **3.44**	2.30 **3.22**	2.22 **3.05**	2.14 **2.92**	2.09 **2.82**	2.00 **2.66**	2.04 **2.73**	1.97 **2.60**
48	4.04 **7.19**	3.19 **5.08**	2.80 **4.22**	2.56 **3.74**	2.41 **3.42**	2.30 **3.20**	2.21 **3.04**	2.14 **2.90**	2.08 **2.80**	1.99 **2.64**	2.03 **2.71**	1.96 **2.58**
50	4.03 **7.17**	3.18 **5.06**	2.79 **4.20**	2.56 **3.72**	2.40 **3.41**	2.29 **3.18**	2.20 **3.02**	2.13 **2.88**	2.07 **2.78**	1.98 **2.62**	2.02 **2.70**	1.95 **2.56**
55	4.02 **7.12**	3.17 **5.01**	2.78 **4.16**	2.54 **3.68**	2.38 **3.37**	2.27 **3.15**	2.18 **2.98**	2.11 **2.85**	2.05 **2.75**	1.97 **2.59**	2.00 **2.66**	1.93 **2.53**
60	4.00 **7.08**	3.15 **4.98**	2.76 **4.13**	2.52 **3.65**	2.37 **3.34**	2.25 **3.12**	2.17 **2.95**	2.10 **2.82**	2.04 **2.72**	1.95 **2.56**	1.99 **2.63**	1.92 **2.50**
65	3.99 **7.04**	3.14 **4.95**	2.75 **4.10**	2.51 **3.62**	2.36 **3.31**	2.24 **3.09**	2.15 **2.93**	2.08 **2.79**	2.02 **2.70**	1.94 **2.54**	1.98 **2.61**	1.90 **2.47**
70	3.98 **7.01**	3.13 **4.92**	2.74 **4.08**	2.50 **3.60**	2.35 **3.29**	2.32 **3.07**	2.14 **2.91**	2.07 **2.77**	2.01 **2.67**	1.93 **2.51**	1.97 **2.59**	1.89 **2.45**
80	3.96 **6.96**	3.11 **4.88**	2.72 **4.04**	2.48 **3.56**	2.33 **3.25**	2.21 **3.04**	2.12 **2.87**	2.05 **2.74**	1.99 **2.64**	1.91 **2.48**	1.95 **2.55**	1.88 **2.41**
100	3.94 **6.90**	3.09 **4.82**	2.70 **3.98**	2.46 **3.51**	2.30 **3.20**	2.19 **2.99**	2.10 **2.82**	2.03 **2.69**	1.97 **2.59**	1.88 **2.43**	1.92 **2.51**	1.85 **2.36**
125	3.92 **6.84**	3.07 **4.78**	2.68 **3.94**	2.44 **3.47**	2.29 **3.17**	2.17 **2.95**	2.08 **2.79**	2.01 **2.65**	1.95 **2.56**	1.86 **2.40**	1.90 **2.47**	1.83 **2.33**
150	3.91 **6.81**	3.06 **4.75**	2.67 **3.91**	2.43 **3.44**	2.27 **3.13**	2.16 **2.92**	2.07 **2.76**	2.00 **2.62**	1.94 **2.53**	1.85 **2.37**	1.89 **2.44**	1.82 **2.30**
200	3.89 **6.76**	3.04 **4.71**	2.65 **3.88**	2.41 **3.41**	2.26 **3.11**	2.14 **2.90**	2.05 **2.73**	1.98 **2.60**	1.92 **2.50**	1.83 **2.34**	1.87 **2.41**	1.80 **2.28**
400	3.86 **6.70**	3.02 **4.66**	2.62 **3.83**	2.39 **3.36**	2.23 **3.06**	2.12 **2.85**	2.03 **2.69**	1.96 **2.55**	1.90 **2.46**	1.81 **2.29**	1.85 **2.37**	1.78 **2.23**
1000	3.85 **6.66**	3.00 **4.62**	2.61 **3.80**	2.38 **3.34**	2.22 **3.04**	2.10 **2.82**	2.02 **2.66**	1.95 **2.53**	1.89 **2.43**	1.80 **2.26**	1.84 **2.34**	1.76 **2.20**
∞	3.84 **6.64**	2.99 **4.60**	2.60 **3.78**	2.37 **3.32**	2.21 **3.02**	2.09 **2.80**	2.01 **2.64**	1.94 **2.51**	1.88 **2.41**	1.79 **2.24**	1.83 **2.32**	1.75 **2.13**

Degrees of freedom for lesser mean square	Degrees of freedom for greater mean square											
	14	16	20	24	30	40	50	75	100	200	500	∞
46	1.91 **2.50**	1.87 **2.42**	1.80 **2.30**	1.75 **2.22**	1.71 **2.13**	1.65 **2.04**	1.62 **1.98**	1.57 **1.90**	1.54 **1.86**	1.51 **1.80**	1.48 **1.76**	1.46 **1.72**
48	1.90 **2.48**	1.86 **2.40**	1.79 **2.28**	1.74 **2.20**	1.70 **2.11**	1.64 **2.02**	1.61 **1.96**	1.56 **1.88**	1.53 **1.84**	1.50 **1.78**	1.47 **1.73**	1.45 **1.70**
50	1.90 **2.46**	1.85 **2.39**	1.78 **2.26**	1.74 **2.18**	1.69 **2.10**	1.63 **2.00**	1.60 **1.94**	1.55 **1.86**	1.52 **1.82**	1.48 **1.76**	1.46 **1.71**	1.44 **1.68**
55	1.88 **2.43**	1.83 **2.35**	1.76 **2.23**	1.72 **2.15**	1.67 **2.06**	1.61 **1.96**	1.58 **1.90**	1.52 **1.82**	1.50 **1.78**	1.46 **1.71**	1.43 **1.66**	1.41 **1.64**
60	1.86 **2.40**	1.81 **2.32**	1.75 **2.20**	1.70 **2.12**	1.65 **2.03**	1.59 **1.93**	1.56 **1.87**	1.50 **1.79**	1.48 **1.74**	1.44 **1.68**	1.41 **1.63**	1.39 **1.60**
65	1.85 **2.37**	1.80 **2.30**	1.73 **2.18**	1.68 **2.09**	1.63 **2.00**	1.57 **1.90**	1.54 **1.84**	1.49 **1.76**	1.46 **1.71**	1.42 **1.64**	1.39 **1.60**	1.37 **1.56**
70	1.84 **2.35**	1.79 **2.28**	1.72 **2.15**	1.67 **2.07**	1.62 **1.98**	1.56 **1.88**	1.53 **1.82**	1.47 **1.74**	1.45 **1.69**	1.40 **1.62**	1.37 **1.56**	1.35 **1.53**
80	1.82 **2.32**	1.77 **2.24**	1.70 **2.11**	1.65 **2.03**	1.60 **1.94**	1.54 **1.84**	1.51 **1.78**	1.45 **1.70**	1.42 **1.65**	1.38 **1.57**	1.35 **1.52**	1.32 **1.49**
100	1.79 **2.26**	1.75 **2.19**	1.68 **2.06**	1.63 **1.98**	1.57 **1.89**	1.51 **1.79**	1.48 **1.73**	1.42 **1.64**	1.39 **1.59**	1.34 **1.51**	1.30 **1.46**	1.28 **1.43**
125	1.77 **2.23**	1.72 **2.15**	1.65 **2.03**	1.60 **1.94**	1.55 **1.85**	1.49 **1.75**	1.45 **1.68**	1.39 **1.59**	1.36 **1.54**	1.31 **1.46**	1.27 **1.40**	1.25 **1.37**
150	1.76 **2.20**	1.71 **2.12**	1.64 **2.00**	1.59 **1.91**	1.54 **1.83**	1.47 **1.72**	1.44 **1.66**	1.37 **1.56**	1.34 **1.51**	1.29 **1.43**	1.25 **1.37**	1.22 **1.33**
200	1.74 **1.17**	1.69 **2.09**	1.62 **1.97**	1.57 **1.88**	1.52 **1.79**	1.45 **1.69**	1.42 **1.62**	1.35 **1.53**	1.32 **1.48**	1.26 **1.39**	1.22 **1.33**	1.19 **1.28**
400	1.72 **2.12**	1.67 **2.04**	1.60 **1.92**	1.54 **1.84**	1.49 **1.74**	1.42 **1.64**	1.38 **1.57**	1.32 **1.47**	1.28 **1.42**	1.22 **1.32**	1.16 **1.24**	1.13 **1.19**
1000	1.70 **2.09**	1.65 **2.01**	1.58 **1.89**	1.53 **1.81**	1.47 **1.71**	1.41 **1.61**	1.36 **1.54**	1.30 **1.44**	1.26 **1.38**	1.19 **1.28**	1.13 **1.19**	1.08 **1.11**
∞	1.69 **2.07**	1.64 **1.99**	1.57 **1.87**	1.52 **1.79**	1.46 **1.69**	1.40 **1.59**	1.35 **1.52**	1.28 **1.41**	1.24 **1.36**	1.17 **1.25**	1.11 **1.15**	1.00 **1.00**

COMPLETE ELLIPTIC INTEGRALS

$$K = \int_0^{\pi/2} \frac{d\Phi}{\sqrt{1 - k^2 \sin^2 \Phi}}.$$

$$E = \int_0^{\pi/2} \sqrt{1 - k^2 \sin^2 \Phi} \cdot d\Phi = E\left(k, \frac{\pi}{2}\right)$$

$\sin^{-1} k$	K	$\log K$	$\sin^{-1} k$	K	$\log K$
0°	1.5708	0.196120	45	1.8541	0.268127
1	1.5709	0.196153	46	1.8691	0.271644
2	1.5713	0.196252	47	1.8848	0.275267
3	1.5719	0.196418	48	1.9011	0.279001
4	1.5727	0.196649	49	1.9180	0.282848
5	1.5738	0.196947	50	1.9356	0.286811
6	1.5751	0.197312	51	1.9539	0.290895
7	1.5767	0.197743	52	1.9729	0.295101
8	1.5785	0.198241	53	1.9927	0.299435
9	1.5805	0.198806	54	2.0133	0.303901
10	1.5828	0.199438	55	2.0347	0.308504
11	1.5854	0.200137	56	2.0571	0.313247
12	1.5882	0.200904	57	2.0804	0.318138
13	1.5913	0.201740	58	2.1047	0.323182
14	1.5946	0.202643	59	2.1300	0.328384
15	1.5981	0.203615	60	2.1565	0.333753
16	1.6020	0.204657	61	2.1842	0.339295
17	1.6061	0.205768	62	2.2132	0.345020
18	1.6105	0.206948	63	2.2435	0.350936
19	1.6151	0.208200	64	2.2754	0.357053
20	1.6200	0.209522	65	2.3088	0.363384
21	1.6252	0.210916	66	2.3439	0.369940
22	1.6307	0.212382	67	2.3809	0.376736
23	1.6365	0.213921	68	2.4198	0.383787
24	1.6426	0.215533	69	2.4610	0.391112
25	1.6490	0.217219	70	2.5046	0.398730
26	1.6557	0.218981	71	2.5507	0.406665
27	1.6627	0.220818	72	2.5998	0.414943
28	1.6701	0.222732	73	2.6521	0.423596
29	1.6777	0.224723	74	2.7081	0.432660
30	1.6858	0.226793	75	2.7681	0.442176
31	1.6941	0.228943	76	2.8327	0.452196
32	1.7028	0.231173	77	2.9026	0.462782
33	1.7119	0.233485	78	2.9786	0.474008
34	1.7214	0.235880	79	3.0617	0.485967
35	1.7312	0.238359	80°	3.1534	0.498777
36	1.7415	0.240923	81	3.2553	0.512591
37	1.7522	0.243575	82	3.3699	0.527613
38	1.7633	0.246315	83	3.5004	0.544120
39	1.7748	0.249146	84	3.6519	0.562514
40°	1.7868	0.252068	85°	3.8317	0.583396
41	1.7992	0.255085	86	4.0528	0.607751
42	1.8122	0.258197	87	4.3387	0.637355
43	1.8256	0.261406	88	4.7427	0.676027
44	1.8396	0.264716	89	5.4349	0.735192
			90	∞	∞

Values of K for $\sin^{-1} k$ = 85° to 89° by 0.1° and 89° to 90° by minutes

$\sin^{-1} k$	K	$\log K$	$\sin^{-1} k$	K	$\log K$
85.0°	3.832	0.58343	89° 0′	5.435	0.73520
85.1	3.852	0.58569	89 2	5.469	0.73791
85.2	3.872	0.58794	89 4	5.504	0.74068
85.3	3.893	0.59028	89 6	5.540	0.74351
85.4	3.914	0.59262	89 8	5.578	0.74648
85.5	3.936	0.59506	89 10	5.617	0.74950
85.6	3.958	0.59748	89 12	5.658	0.75266
85.7	3.981	0.59999	89 14	5.700	0.75587
85.8	4.004	0.60249	89 16	5.745	0.75929
85.9	4.028	0.60509	89 18	5.791	0.76275
86.0	4.053	0.60778	89 20	5.840	0.76641
86.1	4.078	0.61045	89 22	5.891	0.77019
86.2	4.104	0.61321	89 24	5.946	0.77422
86.3	4.130	0.61595	89 26	6.003	0.77837
86.4	4.157	0.61878	89 28	6.063	0.78269
86.5	4.185	0.62170	89 30	6.128	0.78732
86.6	4.214	0.62469	89 32	6.197	0.79218
86.7	4.244	0.62778	89 34	6.271	0.79734
86.8	4.274	0.63083	89 36	6.351	0.80284
86.9	4.306	0.63407	89 38	6.438	0.80875
87.0	4.339	0.63739	89 40	6.533	0.81511
87.1	4.372	0.64068	89 41	6.584	0.81849
87.2	4.407	0.64414	89 42	6.639	0.82210
87.3	4.444	0.64777	89 43	6.696	0.82582
87.4	4.481	0.65137	89 44	6.756	0.82969
87.5	4.520	0.65514	89 45	6.821	0.83385
87.6	4.562	0.65916	89 46	6.890	0.83822
87.7	4.603	0.66304	89 47	6.964	0.84286
87.8	4.648	0.66727	89 48	7.044	0.84782
87.9	4.694	0.67154	89 49	7.131	0.85315
88.0	4.743	0.67605	89 50	7.226	0.85890
88.1	4.794	0.68070	89 51	7.332	0.86522
88.2	4.848	0.68556	89 52	7.449	0.87210
88.3	4.905	0.69064	89 53	7.583	0.87984
88.4	4.965	0.69592	89 54	7.737	0.88857
88.5	5.030	0.70157	89 55	7.919	0.89867
88.6	5.099	0.70749	89 56	8.143	0.91078
88.7	5.173	0.71374	89 57	8.430	0.92583
88.8	5.253	0.72041	89 58	8.836	0.94626
88.9	5.340	0.72754	89 59	9.529	0.97905
89.0	5.435	0.73520	90 0	∞	∞

$\sin^{-1} k$	E	$\log E$	$\sin^{-1} k$	E	$\log E$	$\sin^{-1} k$	E	$\log E$	$\sin^{-1} k$	E	$\log E$
0°	1.5708	0.196120	25	1.4981	0.175545	50	1.3055	0.115790	75	1.0764	0.031976
1	1.5707	0.196087	26	1.4924	0.173876	51	1.2963	0.112698	76	1.0686	0.028819
2	1.5703	0.195988	27	1.4864	0.172144	52	1.2870	0.109563	77	1.0611	0.025740
3	1.5697	0.195822	28	1.4803	0.170348	53	1.2776	0.106386	78	1.0538	0.022749
4	1.5689	0.195591	29	1.4740	0.168489	54	1.2681	0.103169	79	1.0468	0.019858
5	1.5678	0.195293	30	1.4675	0.166567	55	1.2587	0.099915	80	1.0401	0.017081
6	1.5665	0.194930	31	1.4608	0.164583	56	1.2492	0.096626	81	1.0338	0.014432
7	1.5649	0.194500	32	1.4539	0.162537	57	1.2397	0.093303	82	1.0278	0.011927
8	1.5632	0.194004	33	1.4469	0.160429	58	1.2301	0.089950	83	1.0223	0.009584
9	1.5611	0.193442	34	1.4397	0.158261	59	1.2206	0.086569	84	1.0172	0.007422
10	1.5589	0.192815	35	1.4323	0.156031	60	1.2111	0.083164	85	1.0127	0.005465
11	1.5564	0.192121	36	1.4248	0.153742	61	1.2015	0.079738	86	1.0086	0.003740
12	1.5537	0.191362	37	1.4171	0.151393	62	1.1920	0.076293	87	1.0053	0.002278
13	1.5507	0.190537	38	1.4092	0.148985	63	1.1826	0.072834	88	1.0026	0.001121
14	1.5476	0.189646	39	1.4013	0.146519	64	1.1732	0.069364	89	1.0008	0.000326
15	1.5442	0.188690	40	1.3931	0.143995	65	1.1638	0.065889	90	1.0000	0.000000
16	1.5405	0.187668	41	1.3849	0.141414	66	1.1545	0.062412			
17	1.5367	0.186581	42	1.3765	0.138778	67	1.1453	0.058937			
18	1.5326	0.185428	43	1.3680	0.136086	68	1.1362	0.055472			
19	1.5283	0.184210	44	1.3594	0.133340	69	1.1272	0.052020			
20	1.5238	0.182928	45°	1.3506	0.130541	70	1.1184	0.048589			
21	1.5191	0.181580	46	1.3418	0.127690	71	1.1096	0.045183			
22	1.5141	0.180168	47	1.3329	0.124788	72	1.1011	0.041812			
23	1.5090	0.178691	48	1.3238	0.121836	73	1.0927	0.038481			
24	1.5037	0.177150	49	1.3147	0.118836	74	1.0844	0.035200			

ELLIPTIC INTEGRALS OF THE FIRST AND SECOND KIND

An elliptic integral has the form $\int R(x, \sqrt{f(x)})\, dx$, where R represents a rational function and $f(x) = a + bx + cx^2 + dx^3 + ex^4$, an algebraic function of the third or fourth degree.

1. Elliptic integrals of the *first kind* are represented by

$$F(k, \phi) = \int_0^\phi \frac{d\Phi}{\sqrt{1 - k^2 \sin^2 \Phi}} = \int_0^x \frac{d\xi}{\sqrt{(1 - \xi^2)(1 - k^2\xi^2)}}, \quad x = \sin \phi,\ k^2 < 1.$$

2. Elliptic integrals of the second kind are represented by

$$E(k, \phi) = \int_0^\phi \sqrt{1 - k^2 \sin^2 \Phi}\, d\Phi = \int_0^x \frac{\sqrt{1 - k^2\xi^2}}{\sqrt{1 - \xi^2}}\, d\xi, \quad x = \sin \phi,\ k^2 < 1.$$

3. Elliptic integrals of the third kind are represented as

$$\pi(k, n, \phi) = \int_0^\phi \frac{d\Phi}{(1 + n \sin^2 \Phi)\sqrt{1 - k^2 \sin^2 \Phi}}, \quad k^2 < 1,\ n \text{ an integer.}$$

Elliptic integrals of the third kind are also represented as

$$\pi_1(k, n, x) = \int_0^x \frac{d\xi}{(1 + n\xi^2)\sqrt{(1 - \xi^2)(1 - k^2\xi^2)}}, \quad x = \sin \phi,\ k^2 < 1,\ n \text{ an integer.}$$

4. The complete integrals are

$$K = F\left(k, \frac{\pi}{2}\right) = \frac{\pi}{2}\left[1 + \left(\frac{1}{2}\right)^2 k^2 + \left(\frac{3}{2 \cdot 4}\right)^2 k^4 + \left(\frac{3 \cdot 5}{2 \cdot 4 \cdot 6}\right)^2 k^6 + \cdots\right]$$

$$E = E\left(k, \frac{\pi}{2}\right) = \frac{\pi}{2}\left[1 - \left(\frac{1}{2^2}\right) k^2 - \left(\frac{3^2}{2^2 \cdot 4^2}\right)\frac{k^4}{3} - \left(\frac{3^2 \cdot 5^2}{2^2 \cdot 4^2 \cdot 6^2}\right)\frac{k^6}{5} - \left(\frac{3^2 \cdot 5^2 \cdot 7^2}{2^2 \cdot 4^2 \cdot 6^2 \cdot 8^2}\right)\frac{k^8}{7} - \cdots\right].$$

$$K' = F\left(\sqrt{1 - k^2}, \frac{\pi}{2}\right),\quad E' = E\left(\sqrt{1 - k^2}, \frac{\pi}{2}\right)$$

5. The following relation holds between K, K', E, E', namely

$$KE' + EK' - KK' = \frac{\pi}{2}$$

6. To evaluate elliptic integrals for values outside the range contained in the following tables, the following relations are useful

$$F(k, \pi) = 2K;\ E(k, \pi) = 2E$$

$$F(k, \phi + m\pi) = mF(k, \pi) + F(k, \phi) = 2mK + F(k, \phi) \qquad m = 0, 1, 2, 3, \ldots$$

$$E(k, \phi + m\pi) = mE(k, \pi) + E(k, \phi) = 2mE + E(k, \phi) \qquad m = 0, 1, 2, 3, \ldots$$

7. If $u = F(k, \phi) = \int_0^\phi \frac{d\Phi}{\sqrt{1 - k^2 \sin^2 \Phi}}$ $(k^2 < 1)$,
= elliptic integral of the first kind.

$$u = \int_0^x \frac{dx}{\sqrt{(1 - \xi^2)(1 - k^2\xi^2)}}, \quad \text{where } x = \sin \phi.$$

ϕ is called the amplitude of u or am u.
k is called the modulus.

$$k' = \sqrt{1 - k^2} = \text{the complementary modulus.}$$

$$\sin \phi = \text{sn}\, u = x \qquad \tan \phi = \text{tn}\, u = \frac{x}{\sqrt{1 - x^2}}.$$

$\cos \phi = \text{cn}\, u = \sqrt{1 - x^2}$. $\quad \Delta\phi = \text{dn}\, u = \sqrt{1 - k^2x^2}$.

am $0 = 0$. $\quad$ sn $0 = 0$.
cn $0 = 1$. $\quad$ dn $0 = 1$.
am $(-u) = -$am u. $\quad$ sn $(-u) = -$sn u.
cn $(-u) =$ cn u. $\quad$ dn $(-u) =$ dn u.

tn $(-u) = -$tn u.
$\text{sn}^2 u + \text{cn}^2 u = 1$.
$\text{dn}^2 u + k^2 \text{sn}^2 u = 1$.
$\text{dn}^2 u - k^2 \text{cn}^2 u = 1 - k^2 = k'^2$.

ELLIPTIC INTEGRALS OF THE FIRST KIND: $F(k, \phi)$

$$F(k, \phi) = \int_0^{\phi} \frac{d\Phi}{\sqrt{1 - k^2 \sin^2 \Phi}}, \qquad \theta = \sin^{-1} k$$

ϕ \ θ	5°	10°	15°	20°	25°	30°	35°	40°	45°
1°	0.0175	0.0175	0.0175	0.0175	0.0175	0.0175	0.0175	0.0175	0.0175
2°	0.0349	0.0349	0.0349	0.0349	0.0349	0.0349	0.0349	0.0349	0.0349
3°	0.0524	0.0524	0.0524	0.0524	0.0524	0.0524	0.0524	0.0524	0.0524
4°	0.0698	0.0698	0.0698	0.0698	0.0698	0.0698	0.0698	0.0698	0.0698
5°	0.0873	0.0873	0.0873	0.0873	0.0873	0.0873	0.0873	0.0873	0.0873
6°	0.1047	0.1047	0.1047	0.1047	0.1048	0.1048	0.1048	0.1048	0.1048
7°	0.1222	0.1222	0.1222	0.1222	0.1222	0.1222	0.1223	0.1223	0.1223
8°	0.1396	0.1396	0.1397	0.1397	0.1397	0.1397	0.1398	0.1398	0.1399
9°	0.1571	0.1571	0.1571	0.1572	0.1572	0.1572	0.1573	0.1573	0.1574
10°	0.1745	0.1746	0.1746	0.1746	0.1747	0.1748	0.1748	0.1749	0.1750
11°	0.1920	0.1920	0.1921	0.1921	0.1922	0.1923	0.1924	0.1925	0.1926
12°	0.2095	0.2095	0.2095	0.2096	0.2097	0.2098	0.2099	0.2101	0.2102
13°	0.2269	0.2270	0.2270	0.2271	0.2272	0.2274	0.2275	0.2277	0.2279
14°	0.2444	0.2444	0.2445	0.2446	0.2448	0.2450	0.2451	0.2453	0.2456
15°	0.2618	0.2619	0.2620	0.2621	0.2623	0.2625	0.2628	0.2630	0.2633
16°	0.2793	0.2794	0.2795	0.2797	0.2799	0.2802	0.2804	0.2808	0.2811
17°	0.2967	0.2968	0.2970	0.2972	0.2975	0.2978	0.2981	0.2985	0.2989
18°	0.3142	0.3143	0.3145	0.3148	0.3151	0.3154	0.3159	0.3163	0.3167
19°	0.3317	0.3318	0.3320	0.3323	0.3327	0.3331	0.3336	0.3341	0.3347
20°	0.3491	0.3493	0.3495	0.3499	0.3503	0.3508	0.3514	0.3520	0.3526
21°	0.3666	0.3668	0.3671	0.3675	0.3680	0.3685	0.3692	0.3699	0.3706
22°	0.3840	0.3842	0.3846	0.3851	0.3856	0.3863	0.3871	0.3879	0.3887
23°	0.4015	0.4017	0.4021	0.4027	0.4033	0.4041	0.4049	0.4059	0.4068
24°	0.4190	0.4192	0.4197	0.4203	0.4210	0.4219	0.4229	0.4239	0.4250
25°	0.4364	0.4367	0.4372	0.4379	0.4387	0.4397	0.4408	0.4420	0.4433
26°	0.4539	0.4542	0.4548	0.4556	0.4565	0.4576	0.4588	0.4602	0.4616
27°	0.4714	0.4717	0.4724	0.4732	0.4743	0.4755	0.4769	0.4784	0.4800
28°	0.4888	0.4893	0.4899	0.4909	0.4921	0.4934	0.4950	0.4967	0.4985
29°	0.5063	0.5068	0.5075	0.5086	0.5099	0.5114	0.5132	0.5150	0.5170
30°	0.5238	0.5243	0.5251	0.5263	0.5277	0.5294	0.5313	0.5334	0.5356
31°	0.5412	0.5418	0.5427	0.5440	0.5456	0.5475	0.5496	0.5519	0.5543
32°	0.5587	0.5593	0.5603	0.5617	0.5635	0.5656	0.5679	0.5704	0.5731
33°	0.5762	0.5769	0.5780	0.5795	0.5814	0.5837	0.5862	0.5890	0.5920
34°	0.5937	0.5944	0.5956	0.5973	0.5994	0.6018	0.6046	0.6077	0.6109
35°	0.6111	0.6119	0.6133	0.6151	0.6173	0.6200	0.6231	0.6264	0.6300
36°	0.6286	0.6295	0.6309	0.6329	0.6353	0.6383	0.6416	0.6452	0.6491
37°	0.6461	0.6470	0.6486	0.6507	0.6534	0.6565	0.6602	0.6641	0.6684
38°	0.6636	0.6646	0.6662	0.6685	0.6714	0.6749	0.6788	0.6831	0.6877
39°	0.6810	0.6821	0.6839	0.6864	0.6895	0.6932	0.6975	0.7021	0.7071
40°	0.6985	0.6997	0.7016	0.7043	0.7076	0.7116	0.7162	0.7213	0.7267
41°	0.7160	0.7173	0.7193	0.7222	0.7258	0.7301	0.7350	0.7405	0.7463
42°	0.7335	0.7348	0.7370	0.7401	0.7440	0.7486	0.7539	0.7598	0.7661
43°	0.7510	0.7524	0.7548	0.7580	0.7622	0.7671	0.7728	0.7791	0.7859
44°	0.7685	0.7700	0.7725	0.7760	0.7804	0.7857	0.7918	0.7986	0.8059
45°	0.7859	0.7876	0.7903	0.7940	0.7987	0.8044	0.8109	0.8181	0.8260

ϕ \ θ	50°	55°	60°	65°	70°	75°	80°	85°	90°
1°	0.0175	0.0175	0.0175	0.0175	0.0175	0.0175	0.0175	0.0175	0.0175
2°	0.0349	0.0349	0.0349	0.0349	0.0349	0.0349	0.0349	0.0349	0.0349
3°	0.0524	0.0524	0.0524	0.0524	0.0524	0.0524	0.0524	0.0524	0.0524
4°	0.0698	0.0699	0.0699	0.0699	0.0699	0.0699	0.0699	0.0699	0.0699
5°	0.0873	0.0873	0.0873	0.0874	0.0874	0.0874	0.0874	0.0874	0.0874
6°	0.1048	0.1048	0.1049	0.1049	0.1049	0.1049	0.1049	0.1049	0.1049
7°	0.1224	0.1224	0.1224	0.1224	0.1224	0.1225	0.1225	0.1225	0.1225
8°	0.1399	0.1399	0.1400	0.1400	0.1400	0.1401	0.1401	0.1401	0.1401
9°	0.1575	0.1575	0.1576	0.1576	0.1577	0.1577	0.1577	0.1577	0.1577
10°	0.1751	0.1751	0.1752	0.1753	0.1753	0.1754	0.1754	0.1754	0.1754
11°	0.1927	0.1928	0.1929	0.1930	0.1930	0.1931	0.1931	0.1932	0.1932
12°	0.2103	0.2105	0.2106	0.2107	0.2108	0.2109	0.2109	0.2110	0.2110
13°	0.2280	0.2282	0.2284	0.2285	0.2286	0.2287	0.2288	0.2288	0.2289
14°	0.2458	0.2460	0.2462	0.2464	0.2465	0.2466	0.2467	0.2468	0.2468
15°	0.2636	0.2638	0.2641	0.2643	0.2645	0.2646	0.2647	0.2648	0.2648
16°	0.2814	0.2817	0.2820	0.2823	0.2825	0.2827	0.2828	0.2829	0.2830
17°	0.2993	0.2997	0.3000	0.3003	0.3006	0.3008	0.3010	0.3011	0.3012
18°	0.3172	0.3177	0.3181	0.3185	0.3188	0.3191	0.3193	0.3194	0.3195
19°	0.3352	0.3357	0.3362	0.3367	0.3371	0.3374	0.3377	0.3378	0.3379
20°	0.3533	0.3539	0.3545	0.3550	0.3555	0.3559	0.3561	0.3563	0.3564
21°	0.3714	0.3721	0.3728	0.3734	0.3740	0.3744	0.3747	0.3749	0.3750
22°	0.3896	0.3904	0.3912	0.3919	0.3926	0.3931	0.3935	0.3937	0.3938
23°	0.4078	0.4088	0.4097	0.4105	0.4113	0.4119	0.4123	0.4126	0.4127
24°	0.4261	0.4272	0.4283	0.4292	0.4301	0.4308	0.4313	0.4316	0.4317
25°	0.4446	0.4458	0.4470	0.4481	0.4490	0.4498	0.4504	0.4508	0.4509
26°	0.4630	0.4645	0.4658	0.4670	0.4681	0.4690	0.4697	0.4701	0.4702
27°	0.4816	0.4832	0.4847	0.4861	0.4873	0.4884	0.4891	0.4896	0.4897
28°	0.5003	0.5021	0.5038	0.5053	0.5067	0.5079	0.5087	0.5092	0.5094
29°	0.5190	0.5210	0.5229	0.5247	0.5262	0.5275	0.5285	0.5291	0.5293
30°	0.5379	0.5401	0.5422	0.5442	0.5459	0.5474	0.5484	0.5491	0.5493
31°	0.5568	0.5593	0.5617	0.5639	0.5658	0.5674	0.5686	0.5693	0.5696
32°	0.5759	0.5786	0.5812	0.5837	0.5858	0.5876	0.5889	0.5898	0.5900
33°	0.5950	0.5980	0.6010	0.6037	0.6060	0.6080	0.6095	0.6104	0.6107
34°	0.6143	0.6176	0.6208	0.6238	0.6265	0.6287	0.6303	0.6313	0.6317
35°	0.6336	0.6373	0.6408	0.6441	0.6471	0.6495	0.6513	0.6525	0.6528
36°	0.6531	0.6571	0.6610	0.6647	0.6679	0.6706	0.6726	0.6739	0.6743
37°	0.6727	0.6771	0.6814	0.6854	0.6890	0.6919	0.6941	0.6955	0.6960
38°	0.6925	0.6973	0.7019	0.7063	0.7102	0.7135	0.7159	0.7175	0.7180
39°	0.7123	0.7176	0.7227	0.7275	0.7318	0.7353	0.7380	0.7397	0.7403
40°	0.7323	0.7380	0.7436	0.7488	0.7535	0.7575	0.7604	0.7623	0.7629
41°	0.7524	0.7586	0.7647	0.7704	0.7756	0.7799	0.7831	0.7852	0.7859
42°	0.7727	0.7794	0.7860	0.7922	0.7979	0.8026	0.8062	0.8084	0.8092
43°	0.7931	0.8004	0.8075	0.8143	0.8204	0.8256	0.8295	0.8320	0.8328
44°	0.8136	0.8215	0.8293	0.8367	0.8433	0.8490	0.8533	0.8560	0.8569
45°	0.8343	0.8428	0.8512	0.8592	0.8665	0.8727	0.8774	0.8804	0.8814

ELLIPTIC INTEGRALS OF THE FIRST KIND: $F(k, \phi)$

ELLIPTIC INTEGRALS OF THE FIRST KIND: $F(k, \phi)$ (Continued)

$$F(k, \phi) = \int_0^{\phi} \frac{d\Phi}{\sqrt{1 - k^2 \sin^2 \Phi}}, \qquad \theta = \sin^{-1} k$$

φ \ θ	5°	10°	15°	20°	25°	30°	35°	40°	45°
46°	0.8034	0.8052	0.8080	0.8120	0.8170	0.8230	0.8300	0.8378	0.8462
47°	0.8209	0.8227	0.8258	0.8300	0.8353	0.8418	0.8492	0.8575	0.8666
48°	0.8384	0.8403	0.8436	0.8480	0.8537	0.8606	0.8685	0.8773	0.8870
49°	0.8559	0.8579	0.8614	0.8661	0.8721	0.8794	0.8878	0.8972	0.9076
50°	0.8734	0.8756	0.8792	0.8842	0.8905	0.8982	0.9072	0.9173	0.9283
51°	0.8909	0.8932	0.8970	0.9023	0.9090	0.9172	0.9267	0.9374	0.9491
52°	0.9084	0.9108	0.9148	0.9204	0.9275	0.9361	0.9462	0.9575	0.9701
53°	0.9259	0.9284	0.9326	0.9385	0.9460	0.9551	0.9658	0.9778	0.9912
54°	0.9434	0.9460	0.9505	0.9567	0.9646	0.9742	0.9855	0.9982	1.0124
55°	0.9609	0.9637	0.9683	0.9748	0.9832	0.9933	1.0052	1.0187	1.0337
56°	0.9784	0.9813	0.9862	0.9930	1.0018	1.0125	1.0250	1.0393	1.0552
57°	0.9959	0.9989	1.0041	1.0112	1.0204	1.0317	1.0449	1.0600	1.0768
58°	1.0134	1.0166	1.0219	1.0295	1.0391	1.0509	1.0648	1.0807	1.0985
59°	1.0309	1.0342	1.0398	1.0477	1.0578	1.0702	1.0848	1.1016	1.1204
60°	1.0484	1.0519	1.0577	1.0660	1.0766	1.0896	1.1049	1.1226	1.1424
61°	1.0659	1.0695	1.0757	1.0843	1.0953	1.1089	1.1250	1.1436	1.1646
62°	1.0834	1.0872	1.0936	1.1026	1.1141	1.1284	1.1452	1.1648	1.1868
63°	1.1009	1.1049	1.1115	1.1209	1.1330	1.1478	1.1655	1.1860	1.2093
64°	1.1184	1.1225	1.1295	1.1392	1.1518	1.1674	1.1859	1.2073	1.2318
65°	1.1359	1.1402	1.1474	1.1575	1.1707	1.1869	1.2063	1.2288	1.2545
66°	1.1534	1.1579	1.1654	1.1759	1.1896	1.2065	1.2267	1.2503	1.2773
67°	1.1709	1.1756	1.1833	1.1943	1.2085	1.2262	1.2472	1.2719	1.3002
68°	1.1884	1.1932	1.2013	1.2127	1.2275	1.2458	1.2678	1.2936	1.3232
69°	1.2059	1.2109	1.2193	1.2311	1.2465	1.2655	1.2885	1.3154	1.3464
70°	1.2234	1.2286	1.2373	1.2495	1.2655	1.2853	1.3092	1.3372	1.3697
71°	1.2410	1.2463	1.2553	1.2680	1.2845	1.3051	1.3209	1.3592	1.3931
72°	1.2585	1.2640	1.2733	1.2864	1.3036	1.3249	1.3507	1.3812	1.4167
73°	1.2760	1.2817	1.2913	1.3049	1.3226	1.3448	1.3715	1.4033	1.4403
74°	1.2935	1.2994	1.3093	1.3234	1.3417	1.3647	1.3924	1.4254	1.4640
75°	1.3110	1.3171	1.3273	1.3418	1.3608	1.3846	1.4134	1.4477	1.4879
76°	1.3285	1.3348	1.3454	1.3603	1.3800	1.4045	1.4344	1.4700	1.5118
77°	1.3460	1.3525	1.3634	1.3788	1.3991	1.4245	1.4554	1.4923	1.5359
78°	1.3636	1.3702	1.3814	1.3974	1.4183	1.4445	1.4765	1.5147	1.5600
79°	1.3811	1.3879	1.3995	1.4159	1.4374	1.4645	1.4976	1.5372	1.5842
80°	1.3986	1.4056	1.4175	1.4344	1.4566	1.4846	1.5187	1.5597	1.6085
81°	1.4161	1.4234	1.4356	1.4530	1.4758	1.5046	1.5399	1.5823	1.6328
82°	1.4336	1.4411	1.4536	1.4715	1.4950	1.5247	1.5611	1.6049	1.6572
83°	1.4512	1.4588	1.4717	1.4901	1.5143	1.5448	1.5823	1.6276	1.6817
84°	1.4687	1.4765	1.4897	1.5086	1.5335	1.5649	1.6035	1.6502	1.7062
85°	1.4862	1.4942	1.5078	1.5272	1.5527	1.5850	1.6248	1.6730	1.7308
86°	1.5037	1.5120	1.5259	1.5457	1.5720	1.6052	1.6461	1.6957	1.7554
87°	1.5212	1.5297	1.5439	1.5643	1.5912	1.6253	1.6673	1.7184	1.7801
88°	1.5388	1.5474	1.5620	1.5829	1.6105	1.6454	1.6886	1.7412	1.8047
89°	1.5563	1.5651	1.5801	1.6015	1.6297	1.6656	1.7099	1.7640	1.8294
90°	1.5738	1.5828	1.5981	1.6200	1.6490	1.6858	1.7312	1.7868	1.8541

φ \ θ	50°	55°	60°	65°	70°	75°	80°	85°	90°
46°	0.8552	0.8643	0.8734	0.8821	0.8900	0.8968	0.9019	0.9052	0.9063
47°	0.8761	0.8860	0.8958	0.9053	0.9139	0.9212	0.9269	0.9304	0.9316
48°	0.8973	0.9079	0.9185	0.9287	0.9381	0.9461	0.9523	0.9561	0.9575
49°	0.9186	0.9300	0.9415	0.9525	0.9627	0.9714	0.9781	0.9824	0.9838
50°	0.9401	0.9523	0.9647	0.9766	0.9876	0.9971	1.0044	1.0091	1.0107
51°	0.9617	0.9748	0.9881	1.0010	1.0130	1.0233	1.0313	1.0364	1.0381
52°	0.9835	0.9976	1.0118	1.0258	1.0387	1.0499	1.0587	1.0642	1.0662
53°	1.0055	1.0205	1.0359	1.0509	1.0649	1.0771	1.0866	1.0927	1.0948
54°	1.0277	1.0437	1.0602	1.0764	1.0915	1.1048	1.1152	1.1219	1.1242
55°	1.0500	1.0672	1.0848	1.1022	1.1186	1.1331	1.1444	1.1517	1.1542
56°	1.0725	1.0908	1.1097	1.1285	1.1462	1.1619	1.1743	1.1823	1.1851
57°	1.0952	1.1147	1.1349	1.1551	1.1743	1.1914	1.2049	1.2136	1.2167
58°	1.1180	1.1389	1.1605	1.1822	1.2030	1.2215	1.2362	1.2458	1.2492
59°	1.1411	1.1632	1.1864	1.2097	1.2321	1.2522	1.2684	1.2789	1.2826
60°	1.1643	1.1879	1.2125	1.2376	1.2619	1.2837	1.3014	1.3129	1.3170
61°	1.1877	1.2128	1.2392	1.2660	1.2922	1.3159	1.3352	1.3480	1.3524
62°	1.2113	1.2379	1.2661	1.2949	1.3231	1.3490	1.3701	1.3841	1.3890
63°	1.2351	1.2633	1.2933	1.3242	1.3547	1.3828	1.4059	1.4214	1.4268
64°	1.2591	1.2890	1.3209	1.3541	1.3870	1.4175	1.4429	1.4599	1.4659
65°	1.2833	1.3149	1.3489	1.3844	1.4199	1.4532	1.4810	1.4998	1.5065
66°	1.3076	1.3411	1.3773	1.4153	1.4536	1.4898	1.5203	1.5411	1.5485
67°	1.3321	1.3675	1.4060	1.4467	1.4880	1.5274	1.5610	1.5840	1.5923
68°	1.3568	1.3942	1.4351	1.4786	1.5232	1.5661	1.6030	1.6287	1.6379
69°	1.3817	1.4212	1.4646	1.5111	1.5591	1.6059	1.6466	1.6752	1.6856
70°	1.4068	1.4484	1.4944	1.5441	1.5959	1.6468	1.6918	1.7237	1.7354
71°	1.4320	1.4759	1.5246	1.5777	1.6335	1.6891	1.7388	1.7745	1.7877
72°	1.4574	1.5036	1.5552	1.6118	1.6720	1.7326	1.7876	1.8277	1.8427
73°	1.4830	1.5315	1.5862	1.6465	1.7113	1.7774	1.8384	1.8837	1.9008
74°	1.5087	1.5597	1.6175	1.6818	1.7516	1.8237	1.8915	1.9427	1.9623
75°	1.5345	1.5882	1.6492	1.7176	1.7927	1.8715	1.9468	2.0050	2.0276
76°	1.5606	1.6168	1.6812	1.7540	1.8347	1.9207	2.0047	2.0711	2.0973
77°	1.5867	1.6457	1.7136	1.7909	1.8777	1.9716	2.0653	2.1414	2.1721
78°	1.6130	1.6748	1.7462	1.8284	1.9215	2.0240	2.1288	2.2164	2.2528
79°	1.6394	1.7040	1.7792	1.8664	1.9663	2.0781	2.1954	2.2969	2.3404
80°	1.6660	1.7335	1.8125	1.9048	2.0119	2.1339	2.2653	2.3836	2.4362
81°	1.6926	1.7631	1.8461	1.9438	2.0584	2.1913	2.3387	2.4775	2.5421
82°	1.7193	1.7929	1.8799	1.9831	2.1057	2.2504	2.4157	2.5795	2.6603
83°	1.7462	1.8228	1.9140	2.0229	2.1537	2.3110	2.4965	2.6911	2.7942
84°	1.7731	1.8528	1.9482	2.0630	2.2024	2.3731	2.5811	2.8136	2.9487
85°	1.8001	1.8830	1.9826	2.1035	2.2518	2.4366	2.6694	2.9487	3.1313
86°	1.8271	1.9132	2.0172	2.1442	2.3017	2.5013	2.7612	3.0978	3.3547
87°	1.8542	1.9435	2.0519	2.1852	2.3520	2.5670	2.8561	3.2620	3.6425
88°	1.8813	1.9739	2.0867	2.2263	2.4026	2.6336	2.9537	3.4412	4.0481
89°	1.9084	2.0043	2.1216	2.2675	2.4535	2.7007	3.0530	3.6328	4.7413
90°	1.9356	2.0347	2.1565	2.3088	2.5046	2.7681	3.1534	3.8317	—

ELLIPTIC INTEGRALS OF THE SECOND KIND: $E(k, \phi)$

$$E(k, \phi) = \int_0^\phi \sqrt{(1 - k^2 \sin^2 \Phi)}\, d\Phi, \qquad \theta = \sin^{-1} k$$

ϕ \ θ	5°	10°	15°	20°	25°	30°	35°	40°	45°
1°	0.0175	0.0175	0.0175	0.0175	0.0175	0.0175	0.0175	0.0175	0.0175
2°	0.0349	0.0349	0.0349	0.0349	0.0349	0.0349	0.0349	0.0349	0.0349
3°	0.0524	0.0524	0.0524	0.0524	0.0524	0.0524	0.0524	0.0523	0.0523
4°	0.0698	0.0698	0.0698	0.0698	0.0698	0.0698	0.0698	0.0698	0.0698
5°	0.0873	0.0873	0.0873	0.0873	0.0872	0.0872	0.0872	0.0872	0.0872
6°	0.1047	0.1047	0.1047	0.1047	0.1047	0.1047	0.1047	0.1046	0.1046
7°	0.1222	0.1222	0.1222	0.1221	0.1221	0.1221	0.1221	0.1220	0.1220
8°	0.1396	0.1396	0.1396	0.1396	0.1395	0.1395	0.1395	0.1394	0.1394
9°	0.1571	0.1571	0.1570	0.1570	0.1570	0.1569	0.1569	0.1568	0.1568
10°	0.1745	0.1745	0.1745	0.1744	0.1744	0.1743	0.1742	0.1742	0.1741
11°	0.1920	0.1920	0.1919	0.1918	0.1918	0.1917	0.1916	0.1915	0.1914
12°	0.2094	0.2094	0.2093	0.2093	0.2092	0.2091	0.2089	0.2088	0.2087
13°	0.2269	0.2268	0.2268	0.2267	0.2265	0.2264	0.2263	0.2261	0.2259
14°	0.2443	0.2443	0.2442	0.2441	0.2439	0.2437	0.2436	0.2433	0.2431
15°	0.2618	0.2617	0.2616	0.2615	0.2613	0.2611	0.2608	0.2606	0.2603
16°	0.2792	0.2791	0.2790	0.2788	0.2786	0.2784	0.2781	0.2778	0.2775
17°	0.2967	0.2966	0.2964	0.2962	0.2959	0.2956	0.2953	0.2949	0.2946
18°	0.3141	0.3140	0.3138	0.3136	0.3133	0.3129	0.3125	0.3121	0.3116
19°	0.3316	0.3314	0.3312	0.3309	0.3305	0.3301	0.3296	0.3291	0.3286
20°	0.3490	0.3489	0.3486	0.3483	0.3478	0.3473	0.3468	0.3462	0.3456
21°	0.3665	0.3663	0.3660	0.3656	0.3651	0.3645	0.3639	0.3632	0.3625
22°	0.3839	0.3837	0.3834	0.3829	0.3823	0.3817	0.3809	0.3802	0.3793
23°	0.4013	0.4011	0.4007	0.4002	0.3996	0.3988	0.3980	0.3971	0.3961
24°	0.4188	0.4185	0.4181	0.4175	0.4168	0.4159	0.4150	0.4139	0.4129
25°	0.4362	0.4359	0.4354	0.4348	0.4339	0.4330	0.4319	0.4308	0.4296
26°	0.4537	0.4533	0.4528	0.4520	0.4511	0.4500	0.4488	0.4475	0.4462
27°	0.4711	0.4707	0.4701	0.4693	0.4682	0.4670	0.4657	0.4643	0.4628
28°	0.4886	0.4881	0.4874	0.4865	0.4854	0.4840	0.4825	0.4809	0.4793
29°	0.5060	0.5055	0.5048	0.5037	0.5025	0.5010	0.4993	0.4975	0.4957
30°	0.5234	0.5229	0.5221	0.5209	0.5195	0.5179	0.5161	0.5141	0.5120
31°	0.5409	0.5403	0.5394	0.5381	0.5366	0.5348	0.5327	0.5306	0.5283
32°	0.5583	0.5577	0.5567	0.5553	0.5536	0.5516	0.5494	0.5470	0.5446
33°	0.5757	0.5751	0.5740	0.5725	0.5706	0.5684	0.5660	0.5634	0.5607
34°	0.5932	0.5924	0.5912	0.5896	0.5876	0.5852	0.5826	0.5797	0.5768
35°	0.6106	0.6098	0.6085	0.6067	0.6045	0.6019	0.5991	0.5960	0.5928
36°	0.6280	0.6272	0.6258	0.6238	0.6214	0.6186	0.6155	0.6122	0.6087
37°	0.6455	0.6445	0.6430	0.6409	0.6383	0.6353	0.6319	0.6283	0.6245
38°	0.6629	0.6619	0.6602	0.6580	0.6552	0.6519	0.6483	0.6444	0.6403
39°	0.6803	0.6792	0.6775	0.6750	0.6720	0.6685	0.6646	0.6604	0.6559
40°	0.6977	0.6966	0.6947	0.6921	0.6888	0.6851	0.6808	0.6763	0.6715
41°	0.7152	0.7139	0.7119	0.7091	0.7056	0.7016	0.6970	0.6921	0.6870
42°	0.7326	0.7313	0.7291	0.7261	0.7224	0.7180	0.7132	0.7079	0.7024
43°	0.7500	0.7486	0.7463	0.7431	0.7391	0.7345	0.7293	0.7237	0.7178
44°	0.7674	0.7659	0.7634	0.7600	0.7558	0.7508	0.7453	0.7393	0.7330
45°	0.7849	0.7832	0.7806	0.7770	0.7725	0.7672	0.7613	0.7549	0.7482

ϕ \ θ	50°	55°	60°	65°	70°	75°	80°	85°	90°
1°	0.0175	0.0175	0.0175	0.0175	0.0175	0.0175	0.0175	0.0175	0.0175
2°	0.0349	0.0349	0.0349	0.0349	0.0349	0.0349	0.0349	0.0349	0.0349
3°	0.0523	0.0523	0.0523	0.0523	0.0523	0.0523	0.0523	0.0523	0.0523
4°	0.0698	0.0698	0.0698	0.0698	0.0698	0.0698	0.0698	0.0698	0.0698
5°	0.0872	0.0872	0.0872	0.0872	0.0872	0.0872	0.0872	0.0872	0.0872
6°	0.1046	0.1046	0.1046	0.1046	0.1046	0.1045	0.1045	0.1045	0.1045
7°	0.1220	0.1220	0.1219	0.1219	0.1219	0.1219	0.1219	0.1219	0.1219
8°	0.1394	0.1393	0.1393	0.1393	0.1392	0.1392	0.1392	0.1392	0.1392
9°	0.1567	0.1566	0.1566	0.1566	0.1565	0.1565	0.1565	0.1564	0.1564
10°	0.1740	0.1739	0.1739	0.1738	0.1738	0.1737	0.1737	0.1737	0.1736
11°	0.1913	0.1912	0.1911	0.1910	0.1909	0.1909	0.1908	0.1908	0.1908
12°	0.2085	0.2084	0.2083	0.2082	0.2081	0.2080	0.2080	0.2079	0.2079
13°	0.2258	0.2256	0.2254	0.2253	0.2252	0.2251	0.2250	0.2250	0.2250
14°	0.2429	0.2427	0.2425	0.2424	0.2422	0.2421	0.2420	0.2419	0.2419
15°	0.2601	0.2598	0.2596	0.2594	0.2592	0.2590	0.2589	0.2588	0.2588
16°	0.2771	0.2768	0.2765	0.2763	0.2761	0.2759	0.2757	0.2757	0.2756
17°	0.2942	0.2938	0.2935	0.2932	0.2929	0.2927	0.2925	0.2924	0.2924
18°	0.3112	0.3107	0.3103	0.3099	0.3096	0.3094	0.3092	0.3091	0.3090
19°	0.3281	0.3276	0.3271	0.3267	0.3263	0.3260	0.3258	0.3256	0.3256
20°	0.3450	0.3444	0.3438	0.3433	0.3429	0.3425	0.3422	0.3421	0.3420
21°	0.3618	0.3611	0.3604	0.3598	0.3593	0.3589	0.3586	0.3584	0.3584
22°	0.3785	0.3777	0.3770	0.3763	0.3757	0.3752	0.3749	0.3747	0.3746
23°	0.3952	0.3943	0.3935	0.3927	0.3920	0.3915	0.3911	0.3908	0.3907
24°	0.4118	0.4108	0.4098	0.4090	0.4082	0.4076	0.4071	0.4068	0.4067
25°	0.4284	0.4272	0.4261	0.4251	0.4243	0.4236	0.4230	0.4227	0.4226
26°	0.4449	0.4436	0.4423	0.4412	0.4402	0.4394	0.4389	0.4385	0.4384
27°	0.4613	0.4598	0.4584	0.4572	0.4561	0.4552	0.4545	0.4541	0.4540
28°	0.4776	0.4760	0.4744	0.4730	0.4718	0.4708	0.4701	0.4696	0.4695
29°	0.4938	0.4920	0.4903	0.4887	0.4874	0.4863	0.4855	0.4850	0.4848
30°	0.5100	0.5080	0.5061	0.5044	0.5029	0.5016	0.5007	0.5002	0.5000
31°	0.5261	0.5239	0.5218	0.5199	0.5182	0.5169	0.5159	0.5152	0.5150
32°	0.5421	0.5396	0.5373	0.5352	0.5334	0.5319	0.5308	0.5301	0.5299
33°	0.5580	0.5553	0.5528	0.5505	0.5485	0.5468	0.5456	0.5449	0.5446
34°	0.5738	0.5709	0.5681	0.5656	0.5634	0.5616	0.5603	0.5595	0.5592
35°	0.5895	0.5863	0.5833	0.5806	0.5782	0.5762	0.5748	0.5739	0.5736
36°	0.6051	0.6017	0.5984	0.5954	0.5928	0.5907	0.5891	0.5881	0.5878
37°	0.6207	0.6169	0.6134	0.6101	0.6073	0.6050	0.6032	0.6022	0.6018
38°	0.6361	0.6321	0.6282	0.6247	0.6216	0.6191	0.6172	0.6160	0.6157
39°	0.6515	0.6471	0.6429	0.6391	0.6357	0.6330	0.6310	0.6297	0.6293
40°	0.6667	0.6620	0.6575	0.6533	0.6497	0.6468	0.6446	0.6432	0.6428
41°	0.6818	0.6767	0.6719	0.6674	0.6636	0.6604	0.6580	0.6566	0.6561
42°	0.6969	0.6914	0.6862	0.6814	0.6772	0.6738	0.6712	0.6697	0.6691
43°	0.7118	0.7059	0.7003	0.6952	0.6907	0.6870	0.6843	0.6826	0.6820
44°	0.7266	0.7204	0.7144	0.7088	0.7040	0.7000	0.6971	0.6953	0.6947
45°	0.7414	0.7346	0.7282	0.7223	0.7171	0.7129	0.7097	0.7078	0.7071

ELLIPTIC INTEGRALS OF THE SECOND KIND: $E(k, \phi)$
(Continued)

$$E(k, \phi) = \int_0^{\phi} \sqrt{(1 - k^2 \sin^2 \Phi)}\, d\Phi, \qquad \theta = \sin^{-1} k$$

ϕ \ θ	5°	10°	15°	20°	25°	30°	35°	40°	45°
46°	0.8023	0.8006	0.7977	0.7939	0.7891	0.7835	0.7772	0.7704	0.7633
47°	0.8197	0.8179	0.8149	0.8108	0.8057	0.7998	0.7931	0.7858	0.7782
48°	0.8371	0.8352	0.8320	0.8277	0.8223	0.8160	0.8089	0.8012	0.7931
49°	0.8545	0.8525	0.8491	0.8446	0.8389	0.8322	0.8247	0.8165	0.8079
50°	0.8719	0.8698	0.8663	0.8614	0.8554	0.8483	0.8404	0.8317	0.8227
51°	0.8894	0.8871	0.8834	0.8783	0.8719	0.8644	0.8560	0.8469	0.8373
52°	0.9068	0.9044	0.9004	0.8951	0.8884	0.8805	0.8716	0.8620	0.8518
53°	0.9242	0.9217	0.9175	0.9119	0.9048	0.8965	0.8872	0.8770	0.8663
54°	0.9416	0.9389	0.9345	0.9287	0.9212	0.9125	0.9026	0.8919	0.8806
55°	0.9590	0.9562	0.9517	0.9454	0.9376	0.9284	0.9181	0.9068	0.8949
56°	0.9764	0.9735	0.9687	0.9622	0.9540	0.9443	0.9335	0.9216	0.9091
57°	0.9938	0.9908	0.9858	0.9789	0.9703	0.9602	0.9488	0.9363	0.9232
58°	1.0112	1.0080	1.0028	0.9956	0.9866	0.9760	0.9641	0.9510	0.9372
59°	1.0286	1.0253	1.0198	1.0123	1.0029	0.9918	0.9793	0.9656	0.9511
60°	1.0460	1.0426	1.0368	1.0290	1.0191	1.0076	0.9945	0.9801	0.9650
61°	1.0634	1.0598	1.0538	1.0456	1.0354	1.0233	1.0096	0.9946	0.9787
62°	1.0808	1.0771	1.0708	1.0623	1.0516	1.0389	1.0246	1.0090	0.9924
63°	1.0982	1.0943	1.0878	1.0789	1.0678	1.0546	1.0397	1.0233	1.0060
64°	1.1156	1.1115	1.1048	1.0955	1.0839	1.0702	1.0547	1.0376	1.0195
65°	1.1330	1.1288	1.1218	1.1121	1.1001	1.0858	1.0696	1.0518	1.0329
66°	1.1504	1.1460	1.1387	1.1287	1.1162	1.1013	1.0845	1.0660	1.0463
67°	1.1678	1.1632	1.1557	1.1453	1.1323	1.1168	1.0993	1.0801	1.0596
68°	1.1852	1.1805	1.1726	1.1618	1.1483	1.1323	1.1141	1.0941	1.0728
69°	1.2026	1.1977	1.1896	1.1784	1.1644	1.1478	1.1289	1.1081	1.0859
70°	1.2200	1.2149	1.2065	1.1949	1.1804	1.1632	1.1436	1.1221	1.0990
71°	1.2374	1.2321	1.2234	1.2114	1.1964	1.1786	1.1583	1.1359	1.1120
72°	1.2548	1.2493	1.2403	1.2280	1.2124	1.1939	1.1729	1.1498	1.1250
73°	1.2722	1.2666	1.2573	1.2445	1.2284	1.2093	1.1875	1.1636	1.1379
74°	1.2896	1.2838	1.2742	1.2609	1.2443	1.2246	1.2021	1.1773	1.1507
75°	1.3070	1.3010	1.2911	1.2774	1.2603	1.2399	1.2167	1.1910	1.1635
76°	1.3244	1.3182	1.3080	1.2939	1.2762	1.2552	1.2312	1.2047	1.1762
77°	1.3418	1.3354	1.3249	1.3104	1.2921	1.2704	1.2457	1.2183	1.1889
78°	1.3592	1.3526	1.3417	1.3268	1.3080	1.2856	1.2601	1.2319	1.2015
79°	1.3765	1.3698	1.3586	1.3432	1.3239	1.3009	1.2746	1.2454	1.2141
80°	1.3939	1.3870	1.3755	1.3597	1.3398	1.3161	1.2890	1.2590	1.2266
81°	1.4113	1.4042	1.3924	1.3761	1.3556	1.3312	1.3034	1.2725	1.2391
82°	1.4287	1.4214	1.4093	1.3925	1.3715	1.3464	1.3177	1.2859	1.2516
83°	1.4461	1.4386	1.4261	1.4090	1.3873	1.3616	1.3321	1.2994	1.2640
84°	1.4635	1.4558	1.4430	1.4254	1.4032	1.3767	1.3464	1.3128	1.2765
85°	1.4809	1.4729	1.4598	1.4418	1.4190	1.3919	1.3608	1.3262	1.2889
86°	1.4983	1.4901	1.4767	1.4582	1.4348	1.4070	1.3751	1.3396	1.3012
87°	1.5156	1.5073	1.4936	1.4746	1.4507	1.4221	1.3894	1.3530	1.3136
88°	1.5330	1.5245	1.5104	1.4910	1.4665	1.4372	1.4037	1.3664	1.3260
89°	1.5504	1.5417	1.5273	1.5074	1.4823	1.4523	1.4180	1.3798	1.3383
90°	1.5678	1.5589	1.5442	1.5238	1.4981	1.4675	1.4323	1.3931	1.3506

ϕ \ θ	50°	55°	60°	65°	70°	75°	80°	85°	90°
46°	0.7560	0.7488	0.7419	0.7356	0.7301	0.7255	0.7221	0.7200	0.7193
47°	0.7705	0.7628	0.7555	0.7488	0.7429	0.7380	0.7344	0.7321	0.7314
48°	0.7849	0.7768	0.7690	0.7618	0.7555	0.7502	0.7464	0.7440	0.7431
49°	0.7992	0.7905	0.7822	0.7746	0.7679	0.7623	0.7581	0.7556	0.7547
50°	0.8134	0.8042	0.7954	0.7872	0.7801	0.7741	0.7697	0.7670	0.7660
51°	0.8275	0.8177	0.8084	0.7997	0.7921	0.7858	0.7811	0.7781	0.7771
52°	0.8414	0.8311	0.8212	0.8120	0.8039	0.7972	0.7922	0.7891	0.7880
53°	0.8553	0.8444	0.8339	0.8241	0.8155	0.8084	0.8031	0.7998	0.7986
54°	0.8690	0.8575	0.8464	0.8361	0.8270	0.8194	0.8137	0.8102	0.8090
55°	0.8827	0.8705	0.8588	0.8479	0.8382	0.8302	0.8242	0.8204	0.8192
56°	0.8962	0.8834	0.8710	0.8595	0.8493	0.8408	0.8344	0.8304	0.8290
57°	0.9096	0.8961	0.8831	0.8709	0.8601	0.8511	0.8443	0.8401	0.8387
58°	0.9230	0.9088	0.8950	0.8822	0.8707	0.8612	0.8540	0.8496	0.8480
59°	0.9362	0.9213	0.9068	0.8932	0.8812	0.8711	0.8635	0.8588	0.8572
60°	0.9493	0.9336	0.9184	0.9042	0.8914	0.8808	0.8728	0.8677	0.8660
61°	0.9623	0.9459	0.9299	0.9149	0.9015	0.8903	0.8817	0.8764	0.8746
62°	0.9752	0.9580	0.9412	0.9254	0.9113	0.8995	0.8905	0.8849	0.8829
63°	0.9880	0.9700	0.9524	0.9358	0.9210	0.9085	0.8990	0.8930	0.8910
64°	1.0007	0.9818	0.9634	0.9460	0.9304	0.9173	0.9072	0.9009	0.8988
65°	1.0133	0.9936	0.9743	0.9561	0.9397	0.9258	0.9152	0.9086	0.9063
66°	1.0258	1.0052	0.9850	0.9659	0.9487	0.9341	0.9230	0.9159	0.9135
67°	1.0383	1.0167	0.9956	0.9756	0.9576	0.9422	0.9305	0.9230	0.9205
68°	1.0506	1.0281	1.0061	0.9852	0.9662	0.9501	0.9377	0.9299	0.9272
69°	1.0628	1.0394	1.0164	0.9946	0.9747	0.9578	0.9447	0.9364	0.9336
70°	1.0750	1.0506	1.0266	1.0038	0.9830	0.9652	0.9514	0.9427	0.9397
71°	1.0871	1.0617	1.0367	1.0129	0.9911	0.9724	0.9579	0.9487	0.9455
72°	1.0991	1.0727	1.0467	1.0218	0.9990	0.9794	0.9642	0.9544	0.9511
73°	1.1110	1.0836	1.0565	1.0306	1.0067	0.9862	0.9702	0.9599	0.9563
74°	1.1228	1.0944	1.0662	1.0392	1.0143	0.9928	0.9759	0.9650	0.9613
75°	1.1346	1.1051	1.0759	1.0477	1.0217	0.9992	0.9814	0.9699	0.9659
76°	1.1463	1.1158	1.0854	1.0561	1.0290	1.0053	0.9867	0.9745	0.9703
77°	1.1580	1.1263	1.0948	1.0643	1.0361	1.0113	0.9917	0.9789	0.9744
78°	1.1695	1.1368	1.1041	1.0724	1.0430	1.0171	0.9965	0.9829	0.9781
79°	1.1811	1.1472	1.1133	1.0805	1.0498	1.0228	1.0011	0.9867	0.9816
80°	1.1926	1.1576	1.1225	1.0884	1.0565	1.0282	1.0054	0.9902	0.9848
81°	1.2040	1.1678	1.1316	1.0962	1.0630	1.0335	1.0096	0.9935	0.9877
82°	1.2154	1.1781	1.1406	1.1040	1.0695	1.0387	1.0135	0.9965	0.9903
83°	1.2267	1.1883	1.1495	1.1116	1.0758	1.0437	1.0173	0.9992	0.9925
84°	1.2381	1.1984	1.1584	1.1192	1.0821	1.0486	1.0209	1.0017	0.9945
85°	1.2493	1.2085	1.1673	1.1267	1.0882	1.0534	1.0244	1.0039	0.9962
86°	1.2606	1.2186	1.1761	1.1342	1.0944	1.0581	1.0277	1.0060	0.9976
87°	1.2719	1.2286	1.1848	1.1417	1.1004	1.0628	1.0309	1.0078	0.9986
88°	1.2831	1.2386	1.1936	1.1491	1.1064	1.0673	1.0340	1.0095	0.9994
89°	1.2943	1.2487	1.2023	1.1565	1.1124	1.0719	1.0371	1.0111	0.9998
90°	1.3055	1.2587	1.2111	1.1638	1.1184	1.0764	1.0401	1.0127	1.0000

FACTORS AND PRIMES

If n is prime the mantissa of its logarithm is given. If n is not prime its prime factors are given.

n	0	1	2	3	4
0	....		**3010300**	**4771213**	2^2
1	$2 \cdot 5$	**0413927**	$2^2 \cdot 3$	**1139434**	$2 \cdot 7$
2	$2^2 \cdot 5$	$3 \cdot 7$	$2 \cdot 11$	**3617278**	$2^3 \cdot 3$
3	$2 \cdot 3 \cdot 5$	**4913617**	2^5	$3 \cdot 11$	$2 \cdot 17$
4	$2^3 \cdot 5$	**6127839**	$2 \cdot 3 \cdot 7$	**6334685**	$2^2 \cdot 11$
5	$2 \cdot 5^2$	$3 \cdot 17$	$2^2 \cdot 13$	**7242759**	$2 \cdot 3^3$
6	$2^2 \cdot 3 \cdot 5$	**7853298**	$2 \cdot 31$	$3^2 \cdot 7$	2^6
7	$2 \cdot 5 \cdot 7$	**8512583**	$2^3 \cdot 3^2$	**8633229**	$2 \cdot 37$
8	$2^4 \cdot 5$	3^4	$2 \cdot 41$	**9190781**	$2^2 \cdot 3 \cdot 7$
9	$2 \cdot 3^2 \cdot 5$	$7 \cdot 13$	$2^2 \cdot 23$	$3 \cdot 31$	$2 \cdot 47$
10	$2^2 \cdot 5^2$	**0043214**	$2 \cdot 3 \cdot 17$	**0128372**	$2^3 \cdot 13$
11	$2 \cdot 5 \cdot 11$	$3 \cdot 37$	$2^4 \cdot 7$	**0530784**	$2 \cdot 3 \cdot 19$
12	$2^3 \cdot 3 \cdot 5$	11^2	$2 \cdot 61$	$3 \cdot 41$	$2^2 \cdot 31$
13	$2 \cdot 5 \cdot 13$	**1172713**	$2^2 \cdot 3 \cdot 11$	$7 \cdot 19$	$2 \cdot 67$
14	$2^2 \cdot 5 \cdot 7$	$3 \cdot 47$	$2 \cdot 71$	$11 \cdot 13$	$2^4 \cdot 3^2$
15	$2 \cdot 3 \cdot 5^2$	**1789769**	$2^3 \cdot 19$	$3^2 \cdot 17$	$2 \cdot 7 \cdot 11$
16	$2^5 \cdot 5$	$7 \cdot 23$	$2 \cdot 3^3$	**2121876**	$2^2 \cdot 41$
17	$2 \cdot 5 \cdot 17$	$3^2 \cdot 19$	$2^2 \cdot 43$	**2380461**	$2 \cdot 3 \cdot 29$
18	$2^2 \cdot 3^2 \cdot 5$	**2576786**	$2 \cdot 7 \cdot 13$	$3 \cdot 61$	$2^3 \cdot 23$
19	$2 \cdot 5 \cdot 19$	**2810334**	$2^6 \cdot 3$	**2855573**	$2 \cdot 97$
20	$2^3 \cdot 5^2$	$3 \cdot 67$	$2 \cdot 101$	$7 \cdot 29$	$2^2 \cdot 3 \cdot 17$
21	$2 \cdot 3 \cdot 5 \cdot 7$	**3242825**	$2^2 \cdot 53$	$3 \cdot 71$	$2 \cdot 107$
22	$2^2 \cdot 5 \cdot 11$	$13 \cdot 17$	$2 \cdot 3 \cdot 37$	**3483049**	$2^5 \cdot 7$
23	$2 \cdot 5 \cdot 23$	$3 \cdot 7 \cdot 11$	$2^3 \cdot 29$	**3673559**	$2 \cdot 3^2 \cdot 13$
24	$2^4 \cdot 3 \cdot 5$	**3820170**	$2 \cdot 11^2$	3^5	$2^2 \cdot 61$
25	$2 \cdot 5^3$	**3996737**	$2^2 \cdot 3^2 \cdot 7$	$11 \cdot 23$	$2 \cdot 127$
26	$2^2 \cdot 5 \cdot 13$	$3^2 \cdot 29$	$2 \cdot 131$	**4199557**	$2^3 \cdot 3 \cdot 11$
27	$2 \cdot 3^3 \cdot 5$	**4329693**	$2^4 \cdot 17$	$3 \cdot 7 \cdot 13$	$2 \cdot 137$
28	$2^3 \cdot 5 \cdot 7$	**4487063**	$2 \cdot 3 \cdot 47$	**4517864**	$2^2 \cdot 71$
29	$2 \cdot 5 \cdot 29$	$3 \cdot 97$	$2^2 \cdot 73$	**4668676**	$2 \cdot 3 \cdot 7^2$
30	$2^2 \cdot 3 \cdot 5^2$	$7 \cdot 43$	$2 \cdot 151$	$3 \cdot 101$	$2^4 \cdot 19$
31	$2 \cdot 5 \cdot 31$	**4927604**	$2^3 \cdot 3 \cdot 13$	**4955443**	$2 \cdot 157$
32	$2^6 \cdot 5$	$3 \cdot 107$	$2 \cdot 7 \cdot 23$	$17 \cdot 19$	$2^2 \cdot 3^4$
33	$2 \cdot 3 \cdot 5 \cdot 11$	**5198280**	$2^2 \cdot 83$	$3^2 \cdot 37$	$2 \cdot 167$
34	$2^2 \cdot 5 \cdot 17$	$11 \cdot 31$	$2 \cdot 3^2 \cdot 19$	7^3	$2^3 \cdot 43$
35	$2 \cdot 5^2 \cdot 7$	$3^3 \cdot 13$	$2^5 \cdot 11$	**5477747**	$2 \cdot 3 \cdot 59$
36	$2^3 \cdot 3^2 \cdot 5$	19^2	$2 \cdot 181$	$3 \cdot 11^2$	$2^2 \cdot 7 \cdot 13$
37	$2 \cdot 5 \cdot 37$	$7 \cdot 53$	$2^2 \cdot 3 \cdot 31$	**5717088**	$2 \cdot 11 \cdot 17$
38	$2^2 \cdot 5 \cdot 19$	$3 \cdot 127$	$2 \cdot 191$	**5831988**	$2^7 \cdot 3$
39	$2 \cdot 3 \cdot 5 \cdot 13$	$17 \cdot 23$	$2^3 \cdot 7^2$	$3 \cdot 131$	$2 \cdot 197$
40	$2^4 \cdot 5^2$	**6031444**	$2 \cdot 3 \cdot 67$	$13 \cdot 31$	$2^2 \cdot 101$
41	$2 \cdot 5 \cdot 41$	$3 \cdot 137$	$2^2 \cdot 103$	$7 \cdot 59$	$2 \cdot 3^2 \cdot 23$
42	$2^2 \cdot 3 \cdot 5 \cdot 7$	**6242821**	$2 \cdot 211$	$3^2 \cdot 47$	$2^3 \cdot 53$
43	$2 \cdot 5 \cdot 43$	**6344773**	$2^4 \cdot 3^3$	**6364879**	$2 \cdot 7 \cdot 31$
44	$2^3 \cdot 5 \cdot 11$	$3^2 \cdot 7^2$	$2 \cdot 13 \cdot 17$	**6464037**	$2^2 \cdot 3 \cdot 37$
45	$2 \cdot 3^2 \cdot 5^2$	$11 \cdot 41$	$2^2 \cdot 113$	$3 \cdot 151$	$2 \cdot 227$
46	$2^2 \cdot 5 \cdot 23$	**6637009**	$2 \cdot 3 \cdot 7 \cdot 11$	**6655810**	$2^4 \cdot 29$
47	$2 \cdot 5 \cdot 47$	$3 \cdot 157$	$2^3 \cdot 59$	$11 \cdot 43$	$2 \cdot 3 \cdot 79$
48	$2^5 \cdot 3 \cdot 5$	$13 \cdot 37$	$2 \cdot 241$	$3 \cdot 7 \cdot 23$	$2^2 \cdot 11^2$
49	$2 \cdot 5 \cdot 7^2$	**6910815**	$2^2 \cdot 3 \cdot 41$	$17 \cdot 29$	$2 \cdot 13 \cdot 19$
50	$2^2 \cdot 5^3$	$3 \cdot 167$	$2 \cdot 251$	**7015680**	$2^3 \cdot 3^2 \cdot 7$

FACTORS AND PRIMES (Continued)

n	5	6	7	8	9
0	**6989700**	$2 \cdot 3$	**8450980**	2^3	3^2
1	$3 \cdot 5$	2^4	**2304489**	$2 \cdot 3^2$	**2787536**
2	5^2	$2 \cdot 13$	3^3	$2^2 \cdot 7$	**4623980**
3	$5 \cdot 7$	$2^2 \cdot 3^2$	**5682017**	$2 \cdot 19$	$3 \cdot 13$
4	$3^2 \cdot 5$	$2 \cdot 23$	**6720979**	$2^4 \cdot 3$	7^2
5	$5 \cdot 11$	$2^3 \cdot 7$	$3 \cdot 19$	$2 \cdot 29$	**7708520**
6	$5 \cdot 13$	$2 \cdot 3 \cdot 11$	**8260748**	$2^2 \cdot 17$	$3 \cdot 23$
7	$3 \cdot 5^2$	$2^2 \cdot 19$	$7 \cdot 11$	$2 \cdot 3 \cdot 13$	**8976271**
8	$5 \cdot 17$	$2 \cdot 43$	$3 \cdot 29$	$2^3 \cdot 11$	**9493900**
9	$5 \cdot 19$	$2^5 \cdot 3$	**9867717**	$2 \cdot 7^2$	$3^2 \cdot 11$
10	$3 \cdot 5 \cdot 7$	$2 \cdot 53$	**0293838**	$2^2 \cdot 3^3$	**0374265**
11	$5 \cdot 23$	$2^2 \cdot 29$	$3^2 \cdot 13$	$2 \cdot 59$	$7 \cdot 17$
12	5^3	$2 \cdot 3^2 \cdot 7$	**1038037**	2^7	$3 \cdot 43$
13	$3^3 \cdot 5$	$2^3 \cdot 17$	**1367206**	$2 \cdot 3 \cdot 23$	**1430148**
14	$5 \cdot 29$	$2 \cdot 73$	$3 \cdot 7^2$	$2^2 \cdot 37$	**1731863**
15	$5 \cdot 31$	$2^2 \cdot 3 \cdot 13$	**1958997**	$2 \cdot 79$	$3 \cdot 53$
16	$3 \cdot 5 \cdot 11$	$2 \cdot 83$	**2227165**	$2^3 \cdot 3 \cdot 7$	13^2
17	$5^2 \cdot 7$	$2^4 \cdot 11$	$3 \cdot 59$	$2 \cdot 89$	**2528530**
18	$5 \cdot 37$	$2 \cdot 3 \cdot 31$	$11 \cdot 17$	$2^2 \cdot 47$	$3^3 \cdot 7$
19	$3 \cdot 5 \cdot 13$	$2^2 \cdot 7^2$	**2944662**	$2 \cdot 3^2 \cdot 11$	**2988531**
20	$5 \cdot 41$	$2 \cdot 103$	$3^2 \cdot 23$	$2^4 \cdot 13$	$11 \cdot 19$
21	$5 \cdot 43$	$2^3 \cdot 3^3$	$7 \cdot 31$	$2 \cdot 109$	$3 \cdot 73$
22	$3^2 \cdot 5^2$	$2 \cdot 113$	**3560259**	$2^2 \cdot 3 \cdot 19$	**3598355**
23	$5 \cdot 47$	$2^2 \cdot 59$	$3 \cdot 79$	$2 \cdot 7 \cdot 17$	**3783979**
24	$5 \cdot 7^2$	$2 \cdot 3 \cdot 41$	$13 \cdot 19$	$2^3 \cdot 31$	$3 \cdot 83$
25	$3 \cdot 5 \cdot 17$	2^8	**4099331**	$2 \cdot 3 \cdot 43$	$7 \cdot 37$
26	$5 \cdot 53$	$2 \cdot 7 \cdot 19$	$3 \cdot 89$	$2^2 \cdot 67$	**4297523**
27	$5^2 \cdot 11$	$2^2 \cdot 3 \cdot 23$	**4424798**	$2 \cdot 139$	$3^2 \cdot 31$
28	$3 \cdot 5 \cdot 19$	$2 \cdot 11 \cdot 13$	$7 \cdot 41$	$2^5 \cdot 3^2$	17^2
29	$5 \cdot 59$	$2^3 \cdot 37$	$3^3 \cdot 11$	$2 \cdot 149$	$13 \cdot 23$
30	$5 \cdot 61$	$2 \cdot 3^2 \cdot 17$	**4871384**	$2^2 \cdot 7 \cdot 11$	$3 \cdot 103$
31	$3^2 \cdot 5 \cdot 7$	$2^2 \cdot 79$	**5010593**	$2 \cdot 3 \cdot 53$	$11 \cdot 29$
32	$5^2 \cdot 13$	$2 \cdot 163$	$3 \cdot 109$	$2^3 \cdot 41$	$7 \cdot 47$
33	$5 \cdot 67$	$2^4 \cdot 3 \cdot 7$	**5276299**	$2 \cdot 13^2$	$3 \cdot 113$
34	$3 \cdot 5 \cdot 23$	$2 \cdot 173$	**5403295**	$2^2 \cdot 3 \cdot 29$	**5428254**
35	$5 \cdot 71$	$2^2 \cdot 89$	$3 \cdot 7 \cdot 17$	$2 \cdot 179$	**5550944**
36	$5 \cdot 73$	$2 \cdot 3 \cdot 61$	**5646661**	$2^4 \cdot 23$	$3^2 \cdot 41$
37	$3 \cdot 5^3$	$2^3 \cdot 47$	$13 \cdot 29$	$2 \cdot 3^3 \cdot 7$	**5786392**
38	$5 \cdot 7 \cdot 11$	$2 \cdot 193$	$3^2 \cdot 43$	$2^2 \cdot 97$	**5899496**
39	$5 \cdot 79$	$2^2 \cdot 3^2 \cdot 11$	**5987905**	$2 \cdot 199$	$3 \cdot 7 \cdot 19$
40	$3^4 \cdot 5$	$2 \cdot 7 \cdot 29$	$11 \cdot 37$	$2^3 \cdot 3 \cdot 17$	**6117233**
41	$5 \cdot 83$	$2^5 \cdot 13$	$3 \cdot 139$	$2 \cdot 11 \cdot 19$	**6222140**
42	$5^2 \cdot 17$	$2 \cdot 3 \cdot 71$	$7 \cdot 61$	$2^2 \cdot 107$	$3 \cdot 11 \cdot 13$
43	$3 \cdot 5 \cdot 29$	$2^2 \cdot 109$	$19 \cdot 23$	$2 \cdot 3 \cdot 73$	**6424645**
44	$5 \cdot 89$	$2 \cdot 223$	$3 \cdot 149$	$2^6 \cdot 7$	**6522463**
45	$5 \cdot 7 \cdot 13$	$2^3 \cdot 3 \cdot 19$	**6599162**	$2 \cdot 229$	$3^3 \cdot 17$
46	$3 \cdot 5 \cdot 31$	$2 \cdot 233$	**6693169**	$2^2 \cdot 3^2 \cdot 13$	$7 \cdot 67$
47	$5^2 \cdot 19$	$2^2 \cdot 7 \cdot 17$	$3^2 \cdot 53$	$2 \cdot 239$	**6803355**
48	$5 \cdot 97$	$2 \cdot 3^5$	**6875290**	$2^3 \cdot 61$	$3 \cdot 163$
49	$3^2 \cdot 5 \cdot 11$	$2^4 \cdot 31$	$7 \cdot 71$	$2 \cdot 3 \cdot 83$	**6981005**
50	$5 \cdot 101$	$2 \cdot 11 \cdot 23$	$3 \cdot 13^2$	$2^2 \cdot 127$	**7067178**

FACTORS AND PRIMES (Continued)

n	0	1	2	3	4
50	$2^2 \cdot 5^3$	$3 \cdot 167$	$2 \cdot 251$	**7015680**	$2^3 \cdot 3^2 \cdot 7$
51	$2 \cdot 3 \cdot 5 \cdot 17$	$7 \cdot 73$	2^9	$3^3 \cdot 19$	$2 \cdot 257$
52	$2^3 \cdot 5 \cdot 13$	**7168377**	$2 \cdot 3^2 \cdot 29$	**7185017**	$2^2 \cdot 131$
53	$2 \cdot 5 \cdot 53$	$3^2 \cdot 59$	$2^2 \cdot 7 \cdot 19$	$13 \cdot 41$	$2 \cdot 3 \cdot 89$
54	$2^2 \cdot 3^3 \cdot 5$	**7331973**	$2 \cdot 271$	$3 \cdot 181$	$2^5 \cdot 17$
55	$2 \cdot 5^2 \cdot 11$	$19 \cdot 29$	$2^3 \cdot 3 \cdot 23$	$7 \cdot 79$	$2 \cdot 277$
56	$2^4 \cdot 5 \cdot 7$	$3 \cdot 11 \cdot 17$	$2 \cdot 281$	**7505084**	$2^2 \cdot 3 \cdot 47$
57	$2 \cdot 3 \cdot 5 \cdot 19$	**7566361**	$2^2 \cdot 11 \cdot 13$	$3 \cdot 191$	$2 \cdot 7 \cdot 41$
58	$2^2 \cdot 5 \cdot 29$	$7 \cdot 83$	$2 \cdot 3 \cdot 97$	$11 \cdot 53$	$2^3 \cdot 73$
59	$2 \cdot 5 \cdot 59$	$3 \cdot 197$	$2^4 \cdot 37$	**7730547**	$2 \cdot 3^3 \cdot 11$
60	$2^3 \cdot 3 \cdot 5^2$	**7788745**	$2 \cdot 7 \cdot 43$	$3^2 \cdot 67$	$2^2 \cdot 151$
61	$2 \cdot 5 \cdot 61$	$13 \cdot 47$	$2^2 \cdot 3^2 \cdot 17$	**7874605**	$2 \cdot 307$
62	$2^2 \cdot 5 \cdot 31$	$3^3 \cdot 23$	$2 \cdot 311$	$7 \cdot 89$	$2^4 \cdot 3 \cdot 13$
63	$2 \cdot 3^2 \cdot 5 \cdot 7$	**8000294**	$2^3 \cdot 79$	$3 \cdot 211$	$2 \cdot 317$
64	$2^7 \cdot 5$	**8068580**	$2 \cdot 3 \cdot 107$	**8082110**	$2^2 \cdot 7 \cdot 23$
65	$2 \cdot 5^2 \cdot 13$	$3 \cdot 7 \cdot 31$	$2^2 \cdot 163$	**8149132**	$2 \cdot 3 \cdot 109$
66	$2^2 \cdot 3 \cdot 5 \cdot 11$	**8202015**	$2 \cdot 331$	$3 \cdot 13 \cdot 17$	$2^3 \cdot 83$
67	$2 \cdot 5 \cdot 67$	$11 \cdot 61$	$2^5 \cdot 3 \cdot 7$	**8280151**	$2 \cdot 337$
68	$2^3 \cdot 5 \cdot 17$	$3 \cdot 227$	$2 \cdot 11 \cdot 31$	**8344207**	$2^2 \cdot 3^2 \cdot 19$
69	$2 \cdot 3 \cdot 5 \cdot 23$	**8394780**	$2^2 \cdot 173$	$3^2 \cdot 7 \cdot 11$	$2 \cdot 347$
70	$2^2 \cdot 5^2 \cdot 7$	**8457180**	$2 \cdot 3^3 \cdot 13$	$19 \cdot 37$	$2^6 \cdot 11$
71	$2 \cdot 5 \cdot 71$	$3^2 \cdot 79$	$2^3 \cdot 89$	$23 \cdot 31$	$2 \cdot 3 \cdot 7 \cdot 17$
72	$2^4 \cdot 3^2 \cdot 5$	$7 \cdot 103$	$2 \cdot 19^2$	$3 \cdot 241$	$2^2 \cdot 181$
73	$2 \cdot 5 \cdot 73$	$17 \cdot 43$	$2^2 \cdot 3 \cdot 61$	**8651040**	$2 \cdot 367$
74	$2^2 \cdot 5 \cdot 37$	$3 \cdot 13 \cdot 19$	$2 \cdot 7 \cdot 53$	**8709888**	$2^3 \cdot 3 \cdot 31$
75	$2 \cdot 3 \cdot 5^3$	**8756399**	$2^4 \cdot 47$	$3 \cdot 251$	$2 \cdot 13 \cdot 29$
76	$2^3 \cdot 5 \cdot 19$	**8813847**	$2 \cdot 3 \cdot 127$	$7 \cdot 109$	$2^2 \cdot 191$
77	$2 \cdot 5 \cdot 7 \cdot 11$	$3 \cdot 257$	$2^2 \cdot 193$	**8881795**	$2 \cdot 3^2 \cdot 43$
78	$2^2 \cdot 3 \cdot 5 \cdot 13$	$11 \cdot 71$	$2 \cdot 17 \cdot 23$	$3^3 \cdot 29$	$2^4 \cdot 7^2$
79	$2 \cdot 5 \cdot 79$	$7 \cdot 113$	$2^3 \cdot 3^2 \cdot 11$	$13 \cdot 61$	$2 \cdot 397$
80	$2^5 \cdot 5^2$	$3^2 \cdot 89$	$2 \cdot 401$	$11 \cdot 73$	$2^2 \cdot 3 \cdot 67$
81	$2 \cdot 3^4 \cdot 5$	**9090209**	$2^2 \cdot 7 \cdot 29$	$3 \cdot 271$	$2 \cdot 11 \cdot 37$
82	$2^2 \cdot 5 \cdot 41$	**9143432**	$2 \cdot 3 \cdot 137$	**9153998**	$2^3 \cdot 103$
83	$2 \cdot 5 \cdot 83$	$3 \cdot 277$	$2^6 \cdot 13$	$7^2 \cdot 17$	$2 \cdot 3 \cdot 139$
84	$2^3 \cdot 3 \cdot 5 \cdot 7$	29^2	$2 \cdot 421$	$3 \cdot 281$	$2^2 \cdot 211$
85	$2 \cdot 5^2 \cdot 17$	$23 \cdot 37$	$2^2 \cdot 3 \cdot 71$	**9309490**	$2 \cdot 7 \cdot 61$
86	$2^2 \cdot 5 \cdot 43$	$3 \cdot 7 \cdot 41$	$2 \cdot 431$	**9360108**	$2^5 \cdot 3^3$
87	$2 \cdot 3 \cdot 5 \cdot 29$	$13 \cdot 67$	$2^3 \cdot 109$	$3^2 \cdot 97$	$2 \cdot 19 \cdot 23$
88	$2^4 \cdot 5 \cdot 11$	**9449759**	$2 \cdot 3^2 \cdot 7^2$	**9459607**	$2^2 \cdot 13 \cdot 17$
89	$2 \cdot 5 \cdot 89$	$3^4 \cdot 11$	$2^2 \cdot 223$	$19 \cdot 47$	$2 \cdot 3 \cdot 149$
90	$2^2 \cdot 3^2 \cdot 5^2$	$17 \cdot 53$	$2 \cdot 11 \cdot 41$	$3 \cdot 7 \cdot 43$	$2^3 \cdot 113$
91	$2 \cdot 5 \cdot 7 \cdot 13$	**9595184**	$2^4 \cdot 3 \cdot 19$	$11 \cdot 83$	$2 \cdot 457$
92	$2^3 \cdot 5 \cdot 23$	$3 \cdot 307$	$2 \cdot 461$	$13 \cdot 71$	$2^2 \cdot 3 \cdot 7 \cdot 11$
93	$2 \cdot 3 \cdot 5 \cdot 31$	$7^2 \cdot 19$	$2^2 \cdot 233$	$3 \cdot 311$	$2 \cdot 467$
94	$2^2 \cdot 5 \cdot 47$	**9735896**	$2 \cdot 3 \cdot 157$	$23 \cdot 41$	$2^4 \cdot 59$
95	$2 \cdot 5^2 \cdot 19$	$3 \cdot 317$	$2^3 \cdot 7 \cdot 17$	**9790929**	$2 \cdot 3^2 \cdot 53$
96	$2^6 \cdot 3 \cdot 5$	31^2	$2 \cdot 13 \cdot 37$	$3^2 \cdot 107$	$2^2 \cdot 241$
97	$2 \cdot 5 \cdot 97$	**9872192**	$2^2 \cdot 3^5$	$7 \cdot 130$	$2 \cdot 487$
98	$2^2 \cdot 5 \cdot 7^2$	$3^2 \cdot 109$	$2 \cdot 491$	**9925535**	$2^3 \cdot 3 \cdot 41$
99	$2 \cdot 3^2 \cdot 5 \cdot 11$	**9960737**	$2^5 \cdot 31$	$3 \cdot 331$	$2 \cdot 7 \cdot 71$
100	$2^3 \cdot 5^3$	$7 \cdot 11 \cdot 13$	$2 \cdot 3 \cdot 167$	$17 \cdot 59$	$2^2 \cdot 251$

FACTORS AND PRIMES (Continued)

n	5	6	7	8	9
50	$5 \cdot 101$	$2 \cdot 11 \cdot 23$	$3 \cdot 13^2$	$2^2 \cdot 127$	**7067178**
51	$5 \cdot 103$	$2^2 \cdot 3 \cdot 43$	$11 \cdot 47$	$2 \cdot 7 \cdot 37$	$3 \cdot 173$
52	$3 \cdot 5^2 \cdot 7$	$2 \cdot 263$	$17 \cdot 31$	$2^4 \cdot 3 \cdot 11$	23^2
53	$5 \cdot 107$	$2^3 \cdot 67$	$3 \cdot 179$	$2 \cdot 269$	$7^2 \cdot 11$
54	$5 \cdot 109$	$2 \cdot 3 \cdot 7 \cdot 13$	**7379873**	$2^2 \cdot 137$	$3^2 \cdot 61$
55	$3 \cdot 5 \cdot 37$	$2^2 \cdot 139$	**7458552**	$2 \cdot 3^2 \cdot 31$	$13 \cdot 43$
56	$5 \cdot 113$	$2 \cdot 283$	$3^4 \cdot 7$	$2^3 \cdot 71$	**7551123**
57	$5^2 \cdot 23$	$2^6 \cdot 3^2$	**7611758**	$2 \cdot 17^2$	$3 \cdot 193$
58	$3^2 \cdot 5 \cdot 13$	$2 \cdot 293$	**7686381**	$2^2 \cdot 3 \cdot 7^2$	$19 \cdot 31$
59	$5 \cdot 7 \cdot 17$	$2^2 \cdot 149$	$3 \cdot 199$	$2 \cdot 13 \cdot 23$	**7774268**
60	$5 \cdot 11^2$	$2 \cdot 3 \cdot 101$	**7831887**	$2^5 \cdot 19$	$3 \cdot 7 \cdot 29$
61	$3 \cdot 5 \cdot 41$	$2^3 \cdot 7 \cdot 11$	**7902852**	$2 \cdot 3 \cdot 103$	**7916906**
62	5^4	$2 \cdot 313$	$3 \cdot 11 \cdot 19$	$2^2 \cdot 157$	$17 \cdot 37$
63	$5 \cdot 127$	$2^2 \cdot 3 \cdot 53$	$7^2 \cdot 13$	$2 \cdot 11 \cdot 29$	$3^2 \cdot 71$
64	$3 \cdot 5 \cdot 43$	$2 \cdot 17 \cdot 19$	**8109043**	$2^3 \cdot 3^4$	$11 \cdot 59$
65	$5 \cdot 131$	$2^4 \cdot 41$	$3^2 \cdot 73$	$2 \cdot 7 \cdot 47$	**8188854**
66	$5 \cdot 7 \cdot 19$	$2 \cdot 3^2 \cdot 37$	$23 \cdot 29$	$2^2 \cdot 167$	$3 \cdot 223$
67	$3^3 \cdot 5^2$	$2^2 \cdot 13^2$	**8305887**	$2 \cdot 3 \cdot 113$	$7 \cdot 97$
68	$5 \cdot 137$	$2 \cdot 7^3$	$3 \cdot 229$	$2^4 \cdot 43$	$13 \cdot 53$
69	$5 \cdot 139$	$2^3 \cdot 3 \cdot 29$	$17 \cdot 41$	$2 \cdot 349$	$3 \cdot 233$
70	$3 \cdot 5 \cdot 47$	$2 \cdot 353$	$7 \cdot 101$	$2^2 \cdot 3 \cdot 59$	**8506462**
71	$5 \cdot 11 \cdot 13$	$2^2 \cdot 179$	$3 \cdot 239$	$2 \cdot 359$	**8567289**
72	$5^2 \cdot 29$	$2 \cdot 3 \cdot 11^2$	**8615344**	$2^3 \cdot 7 \cdot 13$	3^6
73	$3 \cdot 5 \cdot 7^2$	$2^5 \cdot 23$	$11 \cdot 67$	$2 \cdot 3^2 \cdot 41$	**8686444**
74	$5 \cdot 149$	$2 \cdot 373$	$3^2 \cdot 83$	$2^2 \cdot 11 \cdot 17$	$7 \cdot 107$
75	$5 \cdot 151$	$2^2 \cdot 3^3 \cdot 7$	**8790959**	$2 \cdot 379$	$3 \cdot 11 \cdot 23$
76	$3^2 \cdot 5 \cdot 17$	$2 \cdot 383$	$13 \cdot 59$	$2^8 \cdot 3$	**8859263**
77	$5^2 \cdot 31$	$2^3 \cdot 97$	$3 \cdot 7 \cdot 37$	$2 \cdot 389$	$19 \cdot 41$
78	$5 \cdot 157$	$2 \cdot 3 \cdot 131$	**8959747**	$2^2 \cdot 197$	$3 \cdot 263$
79	$3 \cdot 5 \cdot 53$	$2^2 \cdot 199$	**9014583**	$2 \cdot 3 \cdot 7 \cdot 19$	$17 \cdot 47$
80	$5 \cdot 7 \cdot 23$	$2 \cdot 13 \cdot 31$	$3 \cdot 269$	$2^3 \cdot 101$	**9079485**
81	$5 \cdot 163$	$2^4 \cdot 3 \cdot 17$	$19 \cdot 43$	$2 \cdot 409$	$3^2 \cdot 7 \cdot 13$
82	$3 \cdot 5^2 \cdot 11$	$2 \cdot 7 \cdot 59$	**9175055**	$2^2 \cdot 3^2 \cdot 23$	**9185545**
83	$5 \cdot 167$	$2^2 \cdot 11 \cdot 19$	$3^3 \cdot 31$	$2 \cdot 419$	**9237620**
84	$5 \cdot 13^2$	$2 \cdot 3^2 \cdot 47$	$7 \cdot 11^2$	$2^4 \cdot 53$	$3 \cdot 283$
85	$3^2 \cdot 5 \cdot 19$	$2^3 \cdot 107$	**9329808**	$2 \cdot 3 \cdot 11 \cdot 13$	**9339932**
86	$5 \cdot 173$	$2 \cdot 433$	$3 \cdot 17^2$	$2^2 \cdot 7 \cdot 31$	$11 \cdot 79$
87	$5^3 \cdot 7$	$2^2 \cdot 3 \cdot 73$	**9429996**	$2 \cdot 439$	$3 \cdot 293$
88	$3 \cdot 5 \cdot 59$	$2 \cdot 443$	**9479236**	$2^3 \cdot 3 \cdot 37$	$7 \cdot 127$
89	$5 \cdot 179$	$2^7 \cdot 7$	$3 \cdot 13 \cdot 23$	$2 \cdot 449$	$29 \cdot 31$
90	$5 \cdot 181$	$2 \cdot 3 \cdot 151$	**9576073**	$2^2 \cdot 227$	$3^2 \cdot 101$
91	$3 \cdot 5 \cdot 61$	$2^2 \cdot 229$	$7 \cdot 131$	$2 \cdot 3^3 \cdot 17$	**9633155**
92	$5^2 \cdot 37$	$2 \cdot 463$	$3^2 \cdot 103$	$2^5 \cdot 29$	**9680157**
93	$5 \cdot 11 \cdot 17$	$2^3 \cdot 3^2 \cdot 13$	**9717396**	$2 \cdot 7 \cdot 67$	$3 \cdot 313$
94	$3^3 \cdot 5 \cdot 7$	$2 \cdot 11 \cdot 43$	**9763500**	$2^2 \cdot 3 \cdot 79$	$13 \cdot 73$
95	$5 \cdot 191$	$2^2 \cdot 239$	$3 \cdot 11 \cdot 29$	$2 \cdot 479$	$7 \cdot 137$
96	$5 \cdot 193$	$2 \cdot 3 \cdot 7 \cdot 23$	**9854265**	$2^3 \cdot 11^2$	$3 \cdot 17 \cdot 19$
97	$3 \cdot 5^2 \cdot 13$	$2^4 \cdot 61$	**9898946**	$2 \cdot 3 \cdot 163$	$11 \cdot 89$
98	$5 \cdot 197$	$2 \cdot 17 \cdot 29$	$3 \cdot 7 \cdot 47$	$2^2 \cdot 13 \cdot 19$	$23 \cdot 43$
99	$5 \cdot 199$	$2^2 \cdot 3 \cdot 83$	**9986952**	$2 \cdot 499$	$3^3 \cdot 37$
100	$3 \cdot 5 \cdot 67$	$2 \cdot 503$	$19 \cdot 53$	$2^4 \cdot 3^2 \cdot 7$	**0038912**

FACTORS AND PRIMES (Continued)

n	0	1	2	3	4
100	$2^3 \cdot 5^3$	$7 \cdot 11 \cdot 13$	$2 \cdot 3 \cdot 167$	$17 \cdot 59$	$2^2 \cdot 251$
101	$2 \cdot 5 \cdot 101$	$3 \cdot 337$	$2^2 \cdot 11 \cdot 23$	**0056094**	$2 \cdot 3 \cdot 13^2$
102	$2^2 \cdot 3 \cdot 5 \cdot 17$	**0090257**	$2 \cdot 7 \cdot 73$	$3 \cdot 11 \cdot 31$	2^{10}
103	$2 \cdot 5 \cdot 103$	**0132587**	$2^3 \cdot 3 \cdot 43$	**0141003**	$2 \cdot 11 \cdot 47$
104	$2^4 \cdot 5 \cdot 13$	$3 \cdot 347$	$2 \cdot 521$	$7 \cdot 149$	$2^2 \cdot 3^2 \cdot 29$
105	$2 \cdot 3 \cdot 5^2 \cdot 7$	**0216027**	$2^2 \cdot 263$	$3^4 \cdot 13$	$2 \cdot 17 \cdot 31$
106	$2^2 \cdot 5 \cdot 53$	**0257154**	$2 \cdot 3^2 \cdot 59$	**0265333**	$2^3 \cdot 7 \cdot 19$
107	$2 \cdot 5 \cdot 107$	$3^2 \cdot 7 \cdot 17$	$2^4 \cdot 67$	$29 \cdot 37$	$2 \cdot 3 \cdot 179$
108	$2^3 \cdot 3^3 \cdot 5$	$23 \cdot 47$	$2 \cdot 541$	$3 \cdot 19^2$	$2^2 \cdot 271$
109	$2 \cdot 5 \cdot 109$	**0378248**	$2^2 \cdot 3 \cdot 7 \cdot 13$	**0386202**	$2 \cdot 547$
110	$2^2 \cdot 5^2 \cdot 11$	$3 \cdot 367$	$2 \cdot 19 \cdot 29$	**0425755**	$2^4 \cdot 3 \cdot 23$
111	$2 \cdot 3 \cdot 5 \cdot 37$	$11 \cdot 101$	$2^3 \cdot 139$	$3 \cdot 7 \cdot 53$	$2 \cdot 557$
112	$2^5 \cdot 5 \cdot 7$	$19 \cdot 59$	$2 \cdot 3 \cdot 11 \cdot 17$	**0503798**	$2^2 \cdot 281$
113	$2 \cdot 5 \cdot 113$	$3 \cdot 13 \cdot 29$	$2^2 \cdot 283$	$11 \cdot 103$	$2 \cdot 3^4 \cdot 7$
114	$2^2 \cdot 3 \cdot 5 \cdot 19$	$7 \cdot 163$	$2 \cdot 571$	$3^2 \cdot 127$	$2^3 \cdot 11 \cdot 13$
115	$2 \cdot 5^2 \cdot 23$	**0610753**	$2^7 \cdot 3^2$	**0618293**	$2 \cdot 577$
116	$2^3 \cdot 5 \cdot 29$	$3^3 \cdot 43$	$2 \cdot 7 \cdot 83$	**0655797**	$2^2 \cdot 3 \cdot 97$
117	$2 \cdot 3^2 \cdot 5 \cdot 13$	**0685569**	$2^2 \cdot 293$	$3 \cdot 17 \cdot 23$	$2 \cdot 587$
118	$2^2 \cdot 5 \cdot 59$	**0722499**	$2 \cdot 3 \cdot 197$	$7 \cdot 13^2$	$2^5 \cdot 37$
119	$2 \cdot 5 \cdot 7 \cdot 17$	$3 \cdot 397$	$2^3 \cdot 149$	**0766404**	$2 \cdot 3 \cdot 199$
120	$2^4 \cdot 3 \cdot 5^2$	**0795430**	$2 \cdot 601$	$3 \cdot 401$	$2^2 \cdot 7 \cdot 43$
121	$2 \cdot 5 \cdot 11^2$	$7 \cdot 173$	$2^2 \cdot 3 \cdot 101$	**0838608**	$2 \cdot 607$
122	$2^2 \cdot 5 \cdot 61$	$3 \cdot 11 \cdot 37$	$2 \cdot 13 \cdot 47$	**0874265**	$2^3 \cdot 3^2 \cdot 17$
123	$2 \cdot 3 \cdot 5 \cdot 41$	**0902581**	$2^4 \cdot 7 \cdot 11$	$3^2 \cdot 137$	$2 \cdot 617$
124	$2^3 \cdot 5 \cdot 31$	$17 \cdot 73$	$2 \cdot 3^3 \cdot 23$	$11 \cdot 113$	$2^2 \cdot 311$
125	$2 \cdot 5^4$	$3^2 \cdot 139$	$2^2 \cdot 313$	$7 \cdot 179$	$2 \cdot 3 \cdot 11 \cdot 19$
126	$2^2 \cdot 3^2 \cdot 5 \cdot 7$	$13 \cdot 97$	$2 \cdot 631$	$3 \cdot 421$	$2^4 \cdot 79$
127	$2 \cdot 5 \cdot 127$	$31 \cdot 41$	$2^3 \cdot 3 \cdot 53$	$19 \cdot 67$	$2 \cdot 7^2 \cdot 13$
128	$2^8 \cdot 5$	$3 \cdot 7 \cdot 61$	$2 \cdot 641$	**1082267**	$2^2 \cdot 3 \cdot 107$
129	$2 \cdot 3 \cdot 5 \cdot 43$	**1109262**	$2^2 \cdot 17 \cdot 19$	$3 \cdot 431$	$2 \cdot 647$
130	$2^2 \cdot 5^2 \cdot 13$	**1142773**	$2 \cdot 3 \cdot 7 \cdot 31$	**1149444**	$2^3 \cdot 163$
131	$2 \cdot 5 \cdot 131$	$3 \cdot 19 \cdot 23$	$2^5 \cdot 41$	$13 \cdot 101$	$2 \cdot 3^2 \cdot 73$
132	$2^3 \cdot 3 \cdot 5 \cdot 11$	**1209028**	$2 \cdot 661$	$3^3 \cdot 7^2$	$2^2 \cdot 331$
133	$2 \cdot 5 \cdot 7 \cdot 19$	11^3	$2^2 \cdot 3^2 \cdot 37$	$31 \cdot 43$	$2 \cdot 23 \cdot 29$
134	$2^2 \cdot 5 \cdot 67$	$3^2 \cdot 149$	$2 \cdot 11 \cdot 61$	$17 \cdot 79$	$2^6 \cdot 3 \cdot 7$
135	$2 \cdot 3^3 \cdot 5^2$	$7 \cdot 193$	$2^3 \cdot 13^2$	$3 \cdot 11 \cdot 41$	$2 \cdot 677$
136	$2^4 \cdot 5 \cdot 17$	**1338581**	$2 \cdot 3 \cdot 227$	$29 \cdot 47$	$2^2 \cdot 11 \cdot 31$
137	$2 \cdot 5 \cdot 137$	$3 \cdot 457$	$2^2 \cdot 7^3$	**1376705**	$2 \cdot 3 \cdot 229$
138	$2^2 \cdot 3 \cdot 5 \cdot 23$	**1401937**	$2 \cdot 691$	$3 \cdot 461$	$2^3 \cdot 173$
139	$2 \cdot 5 \cdot 139$	$13 \cdot 107$	$2^4 \cdot 3 \cdot 29$	$7 \cdot 199$	$2 \cdot 17 \cdot 41$
140	$2^3 \cdot 5^2 \cdot 7$	$3 \cdot 467$	$2 \cdot 701$	$23 \cdot 61$	$2^2 \cdot 3^3 \cdot 13$
141	$2 \cdot 3 \cdot 5 \cdot 47$	$17 \cdot 83$	$2^2 \cdot 353$	$3^2 \cdot 157$	$2 \cdot 7 \cdot 101$
142	$2^2 \cdot 5 \cdot 71$	$7^2 \cdot 29$	$2 \cdot 3^2 \cdot 79$	**1532049**	$2^4 \cdot 89$
143	$2 \cdot 5 \cdot 11 \cdot 13$	$3^3 \cdot 53$	$2^3 \cdot 179$	**1562462**	$2 \cdot 3 \cdot 239$
144	$2^5 \cdot 3^2 \cdot 5$	$11 \cdot 131$	$2 \cdot 7 \cdot 103$	$3 \cdot 13 \cdot 37$	$2^2 \cdot 19^2$
145	$2 \cdot 5^2 \cdot 29$	**1616674**	$2^2 \cdot 3 \cdot 11^2$	**1622656**	$2 \cdot 727$
146	$2^2 \cdot 5 \cdot 73$	$3 \cdot 487$	$2 \cdot 17 \cdot 43$	$7 \cdot 11 \cdot 19$	$2^3 \cdot 3 \cdot 61$
147	$2 \cdot 3 \cdot 5 \cdot 7^2$	**1676127**	$2^6 \cdot 23$	$3 \cdot 491$	$2 \cdot 11 \cdot 67$
148	$2^3 \cdot 5 \cdot 37$	**1705551**	$2 \cdot 3 \cdot 13 \cdot 19$	**1711412**	$2^2 \cdot 7 \cdot 53$
149	$2 \cdot 5 \cdot 149$	$3 \cdot 7 \cdot 71$	$2^2 \cdot 373$	**1740598**	$2 \cdot 3^2 \cdot 83$
150	$2^2 \cdot 3 \cdot 5^3$	$19 \cdot 79$	$2 \cdot 751$	$3^2 \cdot 167$	$2^5 \cdot 47$

FACTORS AND PRIMES (Continued)

n	5	6	7	8	9
100	$3 \cdot 5 \cdot 67$	$2 \cdot 503$	$19 \cdot 53$	$2^4 \cdot 3^2 \cdot 7$	**0038912**
101	$5 \cdot 7 \cdot 29$	$2^3 \cdot 127$	$3^2 \cdot 113$	$2 \cdot 509$	**0081742**
102	$5^2 \cdot 41$	$2 \cdot 3^3 \cdot 19$	$13 \cdot 79$	$2^2 \cdot 257$	$3 \cdot 7^3$
103	$3^2 \cdot 5 \cdot 23$	$2^2 \cdot 7 \cdot 37$	$17 \cdot 61$	$2 \cdot 3 \cdot 173$	**0166155**
104	$5 \cdot 11 \cdot 19$	$2 \cdot 523$	$3 \cdot 349$	$2^3 \cdot 131$	**0207755**
105	$5 \cdot 211$	$2^5 \cdot 3 \cdot 11$	$7 \cdot 151$	$2 \cdot 23^2$	$3 \cdot 353$
106	$3 \cdot 5 \cdot 71$	$2 \cdot 13 \cdot 41$	$11 \cdot 97$	$2^2 \cdot 3 \cdot 89$	**0289777**
107	$5^2 \cdot 43$	$2^2 \cdot 269$	$3 \cdot 359$	$2 \cdot 7^2 \cdot 11$	$13 \cdot 83$
108	$5 \cdot 7 \cdot 31$	$2 \cdot 3 \cdot 181$	**0362295**	$2^6 \cdot 17$	$3^2 \cdot 11^2$
109	$3 \cdot 5 \cdot 73$	$2^3 \cdot 137$	**0402066**	$2 \cdot 3^2 \cdot 61$	$7 \cdot 157$
110	$5 \cdot 13 \cdot 17$	$2 \cdot 7 \cdot 79$	$3^3 \cdot 41$	$2^2 \cdot 277$	**0449315**
111	$5 \cdot 223$	$2^2 \cdot 3^2 \cdot 31$	**0480532**	$2 \cdot 13 \cdot 43$	$3 \cdot 373$
112	$3^2 \cdot 5^3$	$2 \cdot 563$	$7^2 \cdot 23$	$2^3 \cdot 3 \cdot 47$	**0526939**
113	$5 \cdot 227$	$2^4 \cdot 71$	$3 \cdot 379$	$2 \cdot 569$	$17 \cdot 67$
114	$5 \cdot 229$	$2 \cdot 3 \cdot 191$	$31 \cdot 37$	$2^2 \cdot 7 \cdot 41$	$3 \cdot 383$
115	$3 \cdot 5 \cdot 7 \cdot 11$	$2^2 \cdot 17^2$	$13 \cdot 89$	$2 \cdot 3 \cdot 193$	$19 \cdot 61$
116	$5 \cdot 233$	$2 \cdot 11 \cdot 53$	$3 \cdot 389$	$2^4 \cdot 73$	$7 \cdot 167$
117	$5^2 \cdot 47$	$2^3 \cdot 3 \cdot 7^2$	$11 \cdot 107$	$2 \cdot 19 \cdot 31$	$3^2 \cdot 131$
118	$3 \cdot 5 \cdot 79$	$2 \cdot 593$	**0744507**	$2^2 \cdot 3^3 \cdot 11$	$29 \cdot 41$
119	$5 \cdot 239$	$2^3 \cdot 13 \cdot 23$	$3^2 \cdot 7 \cdot 19$	$2 \cdot 599$	$11 \cdot 109$
120	$5 \cdot 241$	$2 \cdot 3^2 \cdot 67$	$17 \cdot 71$	$2^3 \cdot 151$	$3 \cdot 13 \cdot 31$
121	$3^5 \cdot 5$	$2^6 \cdot 19$	**0852906**	$2 \cdot 3 \cdot 7 \cdot 29$	$23 \cdot 53$
122	$5^2 \cdot 7^2$	$2 \cdot 613$	$3 \cdot 409$	$2^2 \cdot 307$	**0895519**
123	$5 \cdot 13 \cdot 19$	$2^2 \cdot 3 \cdot 103$	**0923697**	$2 \cdot 619$	$3 \cdot 7 \cdot 59$
124	$3 \cdot 5 \cdot 83$	$2 \cdot 7 \cdot 89$	$29 \cdot 43$	$2^5 \cdot 3 \cdot 13$	**0965624**
125	$5 \cdot 251$	$2^3 \cdot 157$	$3 \cdot 419$	$2 \cdot 17 \cdot 37$	**1000257**
126	$5 \cdot 11 \cdot 23$	$2 \cdot 3 \cdot 211$	$7 \cdot 181$	$2^2 \cdot 317$	$3^3 \cdot 47$
127	$3 \cdot 5^2 \cdot 17$	$2^2 \cdot 11 \cdot 29$	**1061909**	$2 \cdot 3^2 \cdot 71$	**1068705**
128	$5 \cdot 257$	$2 \cdot 643$	$3^2 \cdot 11 \cdot 13$	$2^3 \cdot 7 \cdot 23$	**1102529**
129	$5 \cdot 7 \cdot 37$	$2^4 \cdot 3^4$	**1129400**	$2 \cdot 11 \cdot 59$	$3 \cdot 433$
130	$3^2 \cdot 5 \cdot 29$	$2 \cdot 653$	**1162756**	$2^2 \cdot 3 \cdot 109$	$7 \cdot 11 \cdot 17$
131	$5 \cdot 263$	$2^2 \cdot 7 \cdot 47$	$3 \cdot 439$	$2 \cdot 659$	**1202448**
132	$5^2 \cdot 53$	$2 \cdot 3 \cdot 13 \cdot 17$	**1228709**	$2^4 \cdot 83$	$3 \cdot 443$
133	$3 \cdot 5 \cdot 89$	$2^3 \cdot 167$	$7 \cdot 191$	$2 \cdot 3 \cdot 223$	$13 \cdot 103$
134	$5 \cdot 269$	$2 \cdot 673$	$3 \cdot 449$	$2^2 \cdot 337$	$19 \cdot 71$
135	$5 \cdot 271$	$2^2 \cdot 3 \cdot 113$	$23 \cdot 59$	$2 \cdot 7 \cdot 97$	$3^2 \cdot 151$
136	$3 \cdot 5 \cdot 7 \cdot 13$	$2 \cdot 683$	**1357685**	$2^3 \cdot 3^2 \cdot 19$	37^2
137	$5^3 \cdot 11$	$2^5 \cdot 43$	$3^4 \cdot 17$	$2 \cdot 13 \cdot 53$	$7 \cdot 197$
138	$5 \cdot 277$	$2 \cdot 3^2 \cdot 7 \cdot 11$	$19 \cdot 73$	$2^2 \cdot 347$	$3 \cdot 463$
139	$3^2 \cdot 5 \cdot 31$	$2^2 \cdot 349$	$11 \cdot 127$	$2 \cdot 3 \cdot 233$	**1458177**
140	$5 \cdot 281$	$2 \cdot 19 \cdot 37$	$3 \cdot 7 \cdot 67$	$2^7 \cdot 11$	**1489110**
141	$5 \cdot 283$	$2^3 \cdot 3 \cdot 59$	$13 \cdot 109$	$2 \cdot 709$	$3 \cdot 11 \cdot 43$
142	$3 \cdot 5^2 \cdot 19$	$2 \cdot 23 \cdot 31$	**1544240**	$2^2 \cdot 3 \cdot 7 \cdot 17$	**1550322**
143	$5 \cdot 7 \cdot 41$	$2^2 \cdot 359$	$3 \cdot 479$	$2 \cdot 719$	**1580608**
144	$5 \cdot 17^2$	$2 \cdot 3 \cdot 241$	**1604685**	$2^3 \cdot 181$	$3^2 \cdot 7 \cdot 23$
145	$3 \cdot 5 \cdot 97$	$2^4 \cdot 7 \cdot 13$	$31 \cdot 47$	$2 \cdot 3^6$	**1640553**
146	$5 \cdot 293$	$2 \cdot 733$	$3^2 \cdot 163$	$2^2 \cdot 367$	$13 \cdot 113$
147	$5^2 \cdot 59$	$2^2 \cdot 3^2 \cdot 41$	$7 \cdot 211$	$2 \cdot 739$	$3 \cdot 17 \cdot 29$
148	$3^3 \cdot 5 \cdot 11$	$2 \cdot 743$	**1723110**	$2^4 \cdot 3 \cdot 31$	**1728947**
149	$5 \cdot 13 \cdot 23$	$2^3 \cdot 11 \cdot 17$	$3 \cdot 499$	$2 \cdot 7 \cdot 107$	**1758016**
150	$5 \cdot 7 \cdot 43$	$2 \cdot 3 \cdot 251$	$11 \cdot 137$	$2^2 \cdot 13 \cdot 29$	$3 \cdot 503$

FACTORS AND PRIMES (Continued)

n	0	1	2	3	4
150	$2^2 \cdot 3 \cdot 5^3$	$19 \cdot 79$	$2 \cdot 751$	$3^2 \cdot 167$	$2^5 \cdot 47$
151	$2 \cdot 5 \cdot 151$	**1792645**	$2^3 \cdot 3^3 \cdot 7$	$17 \cdot 89$	$2 \cdot 757$
152	$2^4 \cdot 5 \cdot 19$	$3^2 \cdot 13^2$	$2 \cdot 761$	**1826999**	$2^2 \cdot 3 \cdot 127$
153	$2 \cdot 3^2 \cdot 5 \cdot 17$	**1849752**	$2^2 \cdot 383$	$3 \cdot 7 \cdot 73$	$2 \cdot 13 \cdot 59$
154	$2^2 \cdot 5 \cdot 7 \cdot 11$	$23 \cdot 67$	$2 \cdot 3 \cdot 257$	**1883659**	$2^3 \cdot 193$
155	$2 \cdot 5^2 \cdot 31$	$3 \cdot 11 \cdot 47$	$2^4 \cdot 97$	**1911715**	$2 \cdot 3 \cdot 7 \cdot 37$
156	$2^3 \cdot 3 \cdot 5 \cdot 13$	$7 \cdot 223$	$2 \cdot 11 \cdot 71$	$3 \cdot 521$	$2^2 \cdot 17 \cdot 23$
157	$2 \cdot 5 \cdot 157$	**1961762**	$2^2 \cdot 3 \cdot 131$	$11^2 \cdot 13$	$2 \cdot 787$
158	$2^2 \cdot 5 \cdot 79$	$3 \cdot 17 \cdot 31$	$2 \cdot 7 \cdot 113$	**1994809**	$2^4 \cdot 3^2 \cdot 11$
159	$2 \cdot 3 \cdot 5 \cdot 53$	$37 \cdot 43$	$2^3 \cdot 199$	$3^3 \cdot 59$	$2 \cdot 797$
160	$2^6 \cdot 5^2$	**2043913**	$2 \cdot 3^2 \cdot 89$	$7 \cdot 229$	$2^2 \cdot 401$
161	$2 \cdot 5 \cdot 7 \cdot 23$	$3^2 \cdot 179$	$2^2 \cdot 13 \cdot 31$	**2076344**	$2 \cdot 3 \cdot 269$
162	$2^2 \cdot 3^4 \cdot 5$	**2097830**	$2 \cdot 811$	$3 \cdot 541$	$2^3 \cdot 7 \cdot 29$
163	$2 \cdot 5 \cdot 163$	$7 \cdot 233$	$2^5 \cdot 3 \cdot 17$	$23 \cdot 71$	$2 \cdot 19 \cdot 43$
164	$2^3 \cdot 5 \cdot 41$	$3 \cdot 547$	$2 \cdot 821$	$31 \cdot 53$	$2^2 \cdot 3 \cdot 137$
165	$2 \cdot 3 \cdot 5^2 \cdot 11$	$13 \cdot 127$	$2^2 \cdot 7 \cdot 59$	$3 \cdot 19 \cdot 29$	$2 \cdot 827$
166	$2^2 \cdot 5 \cdot 83$	$11 \cdot 151$	$2 \cdot 3 \cdot 277$	**2208922**	$2^7 \cdot 13$
167	$2 \cdot 5 \cdot 167$	$3 \cdot 557$	$2^3 \cdot 11 \cdot 19$	$7 \cdot 239$	$2 \cdot 3^3 \cdot 31$
168	$2^4 \cdot 3 \cdot 5 \cdot 7$	41^2	$2 \cdot 29^2$	$3^2 \cdot 11 \cdot 17$	$2^2 \cdot 421$
169	$2 \cdot 5 \cdot 13^2$	$19 \cdot 89$	$2^2 \cdot 3^2 \cdot 47$	**2286570**	$2 \cdot 7 \cdot 11^2$
170	$2^2 \cdot 5^2 \cdot 17$	$3^5 \cdot 7$	$2 \cdot 23 \cdot 37$	$13 \cdot 131$	$2^3 \cdot 3 \cdot 71$
171	$2 \cdot 3^2 \cdot 5 \cdot 19$	$29 \cdot 59$	$2^4 \cdot 107$	$3 \cdot 571$	$2 \cdot 857$
172	$2^3 \cdot 5 \cdot 43$	**2357809**	$2 \cdot 3 \cdot 7 \cdot 41$	**2362853**	$2^2 \cdot 431$
173	$2 \cdot 5 \cdot 173$	$3 \cdot 577$	$2^2 \cdot 433$	**2387986**	$2 \cdot 3 \cdot 17^2$
174	$2^2 \cdot 3 \cdot 5 \cdot 29$	**2407988**	$2 \cdot 13 \cdot 67$	$3 \cdot 7 \cdot 83$	$2^4 \cdot 109$
175	$2 \cdot 5^3 \cdot 7$	$17 \cdot 103$	$2^3 \cdot 3 \cdot 73$	**2437819**	$2 \cdot 877$
176	$2^5 \cdot 5 \cdot 11$	$3 \cdot 587$	$2 \cdot 881$	$41 \cdot 43$	$2^2 \cdot 3^2 \cdot 7^2$
177	$2 \cdot 3 \cdot 5 \cdot 59$	$7 \cdot 11 \cdot 23$	$2^2 \cdot 443$	$3^2 \cdot 197$	$2 \cdot 887$
178	$2^2 \cdot 5 \cdot 89$	$13 \cdot 137$	$2 \cdot 3^4 \cdot 11$	**2511513**	$2^3 \cdot 223$
179	$2 \cdot 5 \cdot 179$	$3^2 \cdot 199$	$2^8 \cdot 7$	$11 \cdot 163$	$2 \cdot 3 \cdot 13 \cdot 23$
180	$2^3 \cdot 3^2 \cdot 5^2$	**2555137**	$2 \cdot 17 \cdot 53$	$3 \cdot 601$	$2^2 \cdot 11 \cdot 41$
181	$2 \cdot 5 \cdot 181$	**2579185**	$2^2 \cdot 3 \cdot 151$	$7^2 \cdot 37$	$2 \cdot 907$
182	$2^2 \cdot 5 \cdot 7 \cdot 13$	$3 \cdot 607$	$2 \cdot 911$	**2607867**	$2^5 \cdot 3 \cdot 19$
183	$2 \cdot 3 \cdot 5 \cdot 61$	**2626883**	$2^3 \cdot 229$	$3 \cdot 13 \cdot 47$	$2 \cdot 7 \cdot 131$
184	$2^4 \cdot 5 \cdot 23$	$7 \cdot 263$	$2 \cdot 3 \cdot 307$	$19 \cdot 97$	$2^2 \cdot 461$
185	$2 \cdot 5^2 \cdot 37$	$3 \cdot 617$	$2^2 \cdot 463$	$17 \cdot 109$	$2 \cdot 3^2 \cdot 103$
186	$2^2 \cdot 3 \cdot 5 \cdot 31$	**2697464**	$2 \cdot 7^2 \cdot 19$	$3^4 \cdot 23$	$2^3 \cdot 233$
187	$2 \cdot 5 \cdot 11 \cdot 17$	**2720738**	$2^4 \cdot 3^2 \cdot 13$	**2725378**	$2 \cdot 937$
188	$2^3 \cdot 5 \cdot 47$	$3^2 \cdot 11 \cdot 19$	$2 \cdot 941$	$7 \cdot 269$	$2^2 \cdot 3 \cdot 157$
189	$2 \cdot 3^3 \cdot 5 \cdot 7$	$31 \cdot 61$	$2^2 \cdot 11 \cdot 43$	$3 \cdot 631$	$2 \cdot 947$
190	$2^2 \cdot 5^2 \cdot 19$	**2789821**	$2 \cdot 3 \cdot 317$	$11 \cdot 173$	$2^4 \cdot 7 \cdot 17$
191	$2 \cdot 5 \cdot 191$	$3 \cdot 7^2 \cdot 13$	$2^3 \cdot 239$	**2817150**	$2 \cdot 3 \cdot 11 \cdot 29$
192	$2^7 \cdot 3 \cdot 5$	$17 \cdot 113$	$2 \cdot 31^2$	$3 \cdot 641$	$2^2 \cdot 13 \cdot 37$
193	$2 \cdot 5 \cdot 193$	**2857823**	$2^2 \cdot 3 \cdot 7 \cdot 23$	**2862319**	$2 \cdot 967$
194	$2^2 \cdot 5 \cdot 97$	$3 \cdot 647$	$2 \cdot 971$	$29 \cdot 67$	$2^3 \cdot 3^5$
195	$2 \cdot 3 \cdot 5^2 \cdot 13$	**2902573**	$2^5 \cdot 61$	$3^2 \cdot 7 \cdot 31$	$2 \cdot 977$
196	$2^3 \cdot 5 \cdot 7^2$	$37 \cdot 53$	$2 \cdot 3^2 \cdot 109$	$13 \cdot 151$	$2^2 \cdot 491$
197	$2 \cdot 5 \cdot 197$	$3^3 \cdot 73$	$2^2 \cdot 17 \cdot 29$	**2951271**	$2 \cdot 3 \cdot 7 \cdot 47$
198	$2^2 \cdot 3^2 \cdot 5 \cdot 11$	$7 \cdot 283$	$2 \cdot 991$	$3 \cdot 661$	$2^6 \cdot 31$
199	$2 \cdot 5 \cdot 199$	$11 \cdot 181$	$2^3 \cdot 3 \cdot 83$	**2995073**	$2 \cdot 997$
200	$2^4 \cdot 5^3$	$3 \cdot 23 \cdot 29$	$2 \cdot 7 \cdot 11 \cdot 13$	**3016809**	$2^2 \cdot 3 \cdot 167$

FACTORS AND PRIMES (Continued)

n	5	6	7	8	9
150	$5 \cdot 7 \cdot 43$	$2 \cdot 3 \cdot 251$	$11 \cdot 137$	$2^2 \cdot 13 \cdot 29$	$3 \cdot 503$
151	$3 \cdot 5 \cdot 101$	$2^2 \cdot 379$	$37 \cdot 41$	$2 \cdot 3 \cdot 11 \cdot 23$	$7^2 \cdot 31$
152	$5^2 \cdot 61$	$2 \cdot 7 \cdot 109$	$3 \cdot 509$	$2^3 \cdot 191$	$11 \cdot 139$
153	$5 \cdot 307$	$2^9 \cdot 3$	$29 \cdot 53$	$2 \cdot 769$	$3^4 \cdot 19$
154	$3 \cdot 5 \cdot 103$	$2 \cdot 773$	$7 \cdot 13 \cdot 17$	$2^2 \cdot 3^2 \cdot 43$	**1900514**
155	$5 \cdot 311$	$2^2 \cdot 389$	$3^2 \cdot 173$	$2 \cdot 19 \cdot 41$	**1928461**
156	$5 \cdot 313$	$2 \cdot 3^3 \cdot 29$	**1950690**	$2^5 \cdot 7^2$	$3 \cdot 523$
157	$3^2 \cdot 5^2 \cdot 7$	$2^3 \cdot 197$	$19 \cdot 83$	$2 \cdot 3 \cdot 263$	**1983821**
158	$5 \cdot 317$	$2 \cdot 13 \cdot 61$	$3 \cdot 23^2$	$2^2 \cdot 397$	$7 \cdot 227$
159	$5 \cdot 11 \cdot 29$	$2^2 \cdot 3 \cdot 7 \cdot 19$	**2033049**	$2 \cdot 17 \cdot 47$	$3 \cdot 13 \cdot 41$
160	$3 \cdot 5 \cdot 107$	$2 \cdot 11 \cdot 73$	**2060159**	$2^3 \cdot 3 \cdot 67$	**2065560**
161	$5 \cdot 17 \cdot 19$	$2^4 \cdot 101$	$3 \cdot 7^2 \cdot 11$	$2 \cdot 809$	**2092468**
162	$5^3 \cdot 13$	$2 \cdot 3 \cdot 271$	**2113876**	$2^2 \cdot 11 \cdot 37$	$3^2 \cdot 181$
163	$3 \cdot 5 \cdot 109$	$2^2 \cdot 409$	**2140487**	$2 \cdot 3^2 \cdot 7 \cdot 13$	$11 \cdot 149$
164	$5 \cdot 7 \cdot 47$	$2 \cdot 823$	$3^3 \cdot 61$	$2^4 \cdot 103$	$17 \cdot 97$
165	$5 \cdot 331$	$2^3 \cdot 3^2 \cdot 23$	**2193225**	$2 \cdot 829$	$3 \cdot 7 \cdot 79$
166	$3^2 \cdot 5 \cdot 37$	$2 \cdot 7^2 \cdot 17$	**2219356**	$2^2 \cdot 3 \cdot 139$	**2224563**
167	$5^2 \cdot 67$	$2^2 \cdot 419$	$3 \cdot 13 \cdot 43$	$2 \cdot 839$	$23 \cdot 73$
168	$5 \cdot 337$	$2 \cdot 3 \cdot 281$	$7 \cdot 241$	$2^3 \cdot 211$	$3 \cdot 563$
169	$3 \cdot 5 \cdot 113$	$2^5 \cdot 53$	**2296818**	$2 \cdot 3 \cdot 283$	**2301934**
170	$5 \cdot 11 \cdot 31$	$2 \cdot 853$	$3 \cdot 569$	$2^2 \cdot 7 \cdot 61$	**2327421**
171	$5 \cdot 7^3$	$2^2 \cdot 3 \cdot 11 \cdot 13$	$17 \cdot 101$	$2 \cdot 859$	$3^2 \cdot 191$
172	$3 \cdot 5^2 \cdot 23$	$2 \cdot 863$	$11 \cdot 157$	$2^6 \cdot 3^3$	$7 \cdot 13 \cdot 19$
173	$5 \cdot 347$	$2^3 \cdot 7 \cdot 31$	$3^2 \cdot 193$	$2 \cdot 11 \cdot 79$	$37 \cdot 47$
174	$5 \cdot 349$	$2 \cdot 3^2 \cdot 97$	**2422929**	$2^2 \cdot 19 \cdot 23$	$3 \cdot 11 \cdot 53$
175	$3^3 \cdot 5 \cdot 13$	$2^2 \cdot 439$	$7 \cdot 251$	$2 \cdot 3 \cdot 293$	**2452658**
176	$5 \cdot 353$	$2 \cdot 883$	$3 \cdot 19 \cdot 31$	$2^3 \cdot 13 \cdot 17$	$29 \cdot 61$
177	$5^2 \cdot 71$	$2^4 \cdot 3 \cdot 37$	**2496874**	$2 \cdot 7 \cdot 127$	$3 \cdot 593$
178	$3 \cdot 5 \cdot 7 \cdot 17$	$2 \cdot 19 \cdot 47$	**2521246**	$2^2 \cdot 3 \cdot 149$	**2526103**
179	$5 \cdot 359$	$2^2 \cdot 449$	$3 \cdot 599$	$2 \cdot 29 \cdot 31$	$7 \cdot 257$
180	$5 \cdot 19^2$	$2 \cdot 3 \cdot 7 \cdot 43$	$13 \cdot 139$	$2^4 \cdot 113$	$3^3 \cdot 67$
181	$3 \cdot 5 \cdot 11^2$	$2^3 \cdot 227$	$23 \cdot 79$	$2 \cdot 3^2 \cdot 101$	$17 \cdot 107$
182	$5^2 \cdot 73$	$2 \cdot 11 \cdot 83$	$3^2 \cdot 7 \cdot 29$	$2^2 \cdot 457$	$31 \cdot 59$
183	$5 \cdot 367$	$2^2 \cdot 3^3 \cdot 17$	$11 \cdot 167$	$2 \cdot 919$	$3 \cdot 613$
184	$3^2 \cdot 5 \cdot 41$	$2 \cdot 13 \cdot 71$	**2664669**	$2^3 \cdot 3 \cdot 7 \cdot 11$	43^2
185	$5 \cdot 7 \cdot 53$	$2^6 \cdot 29$	$3 \cdot 619$	$2 \cdot 929$	$11 \cdot 13^2$
186	$5 \cdot 373$	$2 \cdot 3 \cdot 311$	**2711443**	$2^2 \cdot 467$	$3 \cdot 7 \cdot 89$
187	$3 \cdot 5^4$	$2^2 \cdot 7 \cdot 67$	**2734643**	$2 \cdot 3 \cdot 313$	**2739268**
188	$5 \cdot 13 \cdot 29$	$2 \cdot 23 \cdot 41$	$3 \cdot 17 \cdot 37$	$2^5 \cdot 59$	**2762320**
189	$5 \cdot 379$	$2^3 \cdot 3 \cdot 79$	$7 \cdot 271$	$2 \cdot 13 \cdot 73$	$3^2 \cdot 211$
190	$3 \cdot 5 \cdot 127$	$2 \cdot 953$	**2803507**	$2^2 \cdot 3^2 \cdot 53$	$23 \cdot 83$
191	$5 \cdot 383$	$2^2 \cdot 479$	$3^3 \cdot 71$	$2 \cdot 7 \cdot 137$	$19 \cdot 101$
192	$5^2 \cdot 7 \cdot 11$	$2 \cdot 3^2 \cdot 107$	$41 \cdot 47$	$2^3 \cdot 241$	$3 \cdot 643$
193	$3^2 \cdot 5 \cdot 43$	$2^4 \cdot 11^2$	$13 \cdot 149$	$2 \cdot 3 \cdot 17 \cdot 19$	$7 \cdot 277$
194	$5 \cdot 389$	$2 \cdot 7 \cdot 139$	$3 \cdot 11 \cdot 59$	$2^2 \cdot 487$	**2898118**
195	$5 \cdot 17 \cdot 23$	$2^2 \cdot 3 \cdot 163$	$19 \cdot 103$	$2 \cdot 11 \cdot 89$	$3 \cdot 653$
196	$3 \cdot 5 \cdot 131$	$2 \cdot 983$	$7 \cdot 281$	$2^4 \cdot 3 \cdot 41$	$11 \cdot 179$
197	$5^2 \cdot 79$	$2^3 \cdot 13 \cdot 19$	$3 \cdot 659$	$2 \cdot 23 \cdot 43$	**2964458**
198	$5 \cdot 397$	$2 \cdot 3 \cdot 331$	**2981979**	$2^2 \cdot 7 \cdot 71$	$3^2 \cdot 13 \cdot 17$
199	$3 \cdot 5 \cdot 7 \cdot 19$	$2^2 \cdot 499$	**3003781**	$2 \cdot 3^3 \cdot 37$	**3008128**
200	$5 \cdot 401$	$2 \cdot 17 \cdot 59$	$3^2 \cdot 223$	$2^3 \cdot 251$	$7^2 \cdot 41$

EXTENDED TABLES OF FACTORS AND PRIMES

The following procedure makes possible the determination of a number between 2009 and 19,949 as being either prime, or if not, what its factors will be. It is herewith included with the kind permission of its author, Professor Leonard Caners.

The following symbols will be used:

N is the number whose factors, if any, are to be determined.
A is N with the last digit dropped. Thus if $N = 17{,}873$, $A = 1787$.
R is the range and is the integral part of the square root of N with its last two digits dropped. Thus R of $17{,}873 = 13$.
K is the key number and is found in the table below.
P_1 is a possible factor corresponding to K.
The twin series are given in the table for the sake of completeness but are not written down in actual practice.
There are two steps as outlined below.

Step I

N	K	P_1	Series	Procedure
Any number ending in 1, 3, 7, or 9, within limits stated above	$A - R$	$10R$ + last digit	$K + n;\ P_1 - 10n$ $n = 0, 1, \ldots, 2R + 1$	Beginning with K read table of Factors and Primes from left to right and look for corresponding possible factors

Step II

N ending in	K	P_1	Series	Procedure
1	$(A - 2) - 3R$	$10R + 7$	$K + 3n;\ P_1 - 10n$ $n = 0, 1, \ldots, 2R + 1$	Same as in Step I except that only every *third* entry in table of Factors and Primes is examined for corresponding possible factors
3	$A - 3R$	$10R + 1$	$n = 0, 1, \ldots, 2R + 1$	
7	$(A - 2) - 3R$	$10R + 9$	$n = 0, 1, \ldots, 2R + 1$	
9	$A - 3R$	$10R + 3$	$n = 0, 1, \ldots, 2R + 1$	

As an example consider 7519. Using tables on preceding pages proceed as follows:

Step I

From Table $N = 7519$; $A = 751$; $R = 8$, $K = 743$; $P_1 = 89$ and the series obtained:

743	89
744	79
745	69
746	59
—	—
751	9
752	1
753	11
—	—
760	81

Examining the Table of Factors and Primes beginning with 743 note whether 743 has 89 as a factor; whether 744 has 79 as a factor, etc. Then continue to read from left to right keeping in mind the possible factors 69, 59, . . . 81.

Since none are found one concludes that 7519 has no factors within the given range ending in 9 or 1.

Step II

From Table $K = 727$ and $P_1 = 83$; the series obtained is:

727	83
730	73
733	63
—	—
751	3
754	7
—	—
778	87

Proceed exactly as in Step I. Begin with 727 and note whether it has 83 as a factor. Thereafter examine every *third* entry for the remaining corresponding possible factors 73, 63 . . . 3, 7, . . . 87. Since 730 yields the factor 73 one concludes that 73 is the factor of 7519. By division $7519 = 73 \times 103$. Had no factor been found in either Step I or Step II the conclusion would be that the number under consideration was prime.

Proof for Step I

Let N end in 1. Then, if N has a factor ending in 1, write

$$(10a + 1)(10b + 1) = N = 10A + 1$$
$$10ab + a + b = A$$

or

$$b(10a + 1) = (A - a)$$

This proves that any factor $10a + 1$ of N is also a factor of $A - a$. Let $a = (R - n)$, then $(10a + 1)$ becomes $(10R + 1) - 10n$ and $(A - a)$ becomes $(A - R) + n$, i.e., $P_1 - 10n$ and $K + n$ respectively as given in the table. Since $P_1 - 10n$ becomes a number ending in 9 when $P_1 < 10n$, all possible factors ending in 9 are also provided for.

The proof for Step II is similar. For possible factors ending in 3 or 7 write

$$(10a + 7)(10b + 3) = N = 10A + 1.$$

By identical reasoning this results in $P_1 - 10n$ and $K + 3n$ respectively given in the table. This completes the proof for numbers ending in 1.

Identical reasoning establishes the key numbers and series for ending in 3, 7, or 9.

*Derivatives

In the following formulas u, v, w represent functions of x, while a, c, n represent fixed real numbers. All arguments in the trigonometric functions are measured in radians, and all inverse trigonometric and hyperbolic functions represent principal values.

1. $\frac{d}{dx}(a) = 0$

2. $\frac{d}{dx}(x) = 1$

3. $\frac{d}{dx}(au) = a\frac{du}{dx}$

4. $\frac{d}{dx}(u + v - w) = \frac{du}{dx} + \frac{dv}{dx} - \frac{dw}{dx}$

5. $\frac{d}{dx}(uv) = u\frac{dv}{dx} + v\frac{du}{dx}$

6. $\frac{d}{dx}(uvw) = uv\frac{dw}{dx} + vw\frac{du}{dx} + uw\frac{dv}{dx}$

7. $\frac{d}{dx}\left(\frac{u}{v}\right) = \frac{v\frac{du}{dx} - u\frac{dv}{dx}}{v^2} = \frac{1}{v}\frac{du}{dx} - \frac{u}{v^2}\frac{dv}{dx}$

8. $\frac{d}{dx}(u^n) = nu^{n-1}\frac{du}{dx}$

9. $\frac{d}{dx}(\sqrt{u}) = \frac{1}{2\sqrt{u}}\frac{du}{dx}$

10. $\frac{d}{dx}\left(\frac{1}{u}\right) = -\frac{1}{u^2}\frac{du}{dx}$

11. $\frac{d}{dx}[f(u)] = \frac{d}{du}[f(u)] \cdot \frac{du}{dx}$

12. $\frac{d^2}{dx^2}[f(u)] = \frac{df(u)}{du} \cdot \frac{d^2u}{dx^2} + \frac{d^2f(u)}{du^2} \cdot \left(\frac{du}{dx}\right)^2$

13. $\frac{d^n}{dx^n}[uv] = \binom{n}{0} v \frac{d^nu}{dx^n} + \binom{n}{1} \frac{dv}{dx}\frac{d^{n-1}u}{dx^{n-1}} + \binom{n}{2} \frac{d^2v}{dx^2}\frac{d^{n-2}u}{dx^{n-2}} + \cdots + \binom{n}{k} \frac{d^kv}{dx^k}\frac{d^{n-k}u}{dx^{n-k}} + \cdots + \binom{n}{n} u \frac{d^nv}{dx^n}$

    where $\binom{n}{r} = \frac{n!}{r!(n-r)!}$ the binomial coefficient, n non-negative integer and $\binom{n}{0} = 1$.

14. $\frac{du}{dx} = \frac{1}{\frac{dx}{du}}$ if $\frac{dx}{du} \neq 0$

* Let $y = f(x)$ and $\frac{dy}{dx} = \frac{d[f(x)]}{dx} = f'(x)$ define respectively a function and its derivative for any value x in their common domain. The differential for the function at such a value x is accordingly defined as

$$dy = d[f(x)] = \frac{dy}{dx}dx = \frac{d[f(x)]}{dx}dx = f'(x)\,dx$$

Each derivative formula has an associated differential formula. For example, formula 6 above has the differential formula

$$d(uvw) = uv\,dw + vw\,du + uw\,dv$$

15. $\frac{d}{dx}(\log_a u) = (\log_a e)\frac{1}{u}\frac{du}{dx}$

16. $\frac{d}{dx}(\log_e u) = \frac{1}{u}\frac{du}{dx}$

17. $\frac{d}{dx}(a^u) = a^u(\log_e a)\frac{du}{dx}$

18. $\frac{d}{dx}(e^u) = e^u\frac{du}{dx}$

19. $\frac{d}{dx}(u^v) = vu^{v-1}\frac{du}{dx} + (\log_e u)u^v\frac{dv}{dx}$

20. $\frac{d}{dx}(\sin u) = \frac{du}{dx}(\cos u)$

21. $\frac{d}{dx}(\cos u) = -\frac{du}{dx}(\sin u)$

22. $\frac{d}{dx}(\tan u) = \frac{du}{dx}(\sec^2 u)$

23. $\frac{d}{dx}(\cot u) = -\frac{du}{dx}(\csc^2 u)$

24. $\frac{d}{dx}(\sec u) = \frac{du}{dx}\sec u \cdot \tan u$

25. $\frac{d}{dx}(\csc u) = -\frac{du}{dx}\csc u \cdot \cot u$

26. $\frac{d}{dx}(\text{vers } u) = \frac{du}{dx}\sin u$

27. $\frac{d}{dx}(\text{arc sin } u) = \frac{1}{\sqrt{1-u^2}} \cdot \frac{du}{dx}, \left(-\frac{\pi}{2} \leq \text{arc sin } u \leq \frac{\pi}{2}\right)$

28. $\frac{d}{dx}(\text{arc cos } u) = -\frac{1}{\sqrt{1-u^2}}\frac{du}{dx}, (0 \leq \text{arc cos } u \leq \pi)$

29. $\frac{d}{dx}(\text{arc tan } u) = \frac{1}{1+u^2}\frac{du}{dx}, \left(-\frac{\pi}{2} < \text{arc tan } u < \frac{\pi}{2}\right)$

30. $\frac{d}{dx}(\text{arc cot } u) = -\frac{1}{1+u^2}\frac{du}{dx}, (0 \leq \text{arc cot } u \leq \pi)$

31. $\frac{d}{dx}(\text{arc sec } u) = \frac{1}{u\sqrt{u^2-1}}\frac{du}{dx}, \left(0 \leq \text{arc sec } u < \frac{\pi}{2}, -\pi \leq \text{arc sec } u < -\frac{\pi}{2}\right)$

32. $\frac{d}{dx}(\text{arc csc } u) = -\frac{1}{u\sqrt{u^2-1}}\frac{du}{dx}, \left(0 < \text{arc csc } u \leq \frac{\pi}{2}, -\pi < \text{arc csc } u \leq -\frac{\pi}{2}\right)$

33. $\frac{d}{dx}(\text{arc vers } u) = \frac{1}{\sqrt{2u-u^2}}\frac{du}{dx}, (0 \leq \text{arc vers } u \leq \pi)$

34. $\frac{d}{dx}(\sinh u) = \frac{du}{dx}(\cosh u)$

35. $\frac{d}{dx}(\cosh u) = \frac{du}{dx}(\sinh u)$

36. $\frac{d}{dx}(\tanh u) = \frac{du}{dx}(\text{sech}^2 u)$

37. $\frac{d}{dx}(\coth u) = -\frac{du}{dx}(\text{csch}^2 u)$

38. $\frac{d}{dx}(\text{sech } u) = -\frac{du}{dx}(\text{sech } u \cdot \tanh u)$

39. $\frac{d}{dx}(\text{csch } u) = -\frac{du}{dx}(\text{csch } u \cdot \coth u)$

40. $\frac{d}{dx}(\sinh^{-1} u) = \frac{d}{dx}[\log(u + \sqrt{u^2+1})] = \frac{1}{\sqrt{u^2+1}}\frac{du}{dx}$

41. $\frac{d}{dx}(\cosh^{-1} u) = \frac{d}{dx}[\log(u + \sqrt{u^2-1})] = \frac{1}{\sqrt{u^2-1}}\frac{du}{dx}, \ (u > 1, \cosh^{-1} u > 0)$

42. $\frac{d}{dx}(\tanh^{-1} u) = \frac{d}{dx}\left[\frac{1}{2}\log\frac{1+u}{1-u}\right] = \frac{1}{1-u^2}\frac{du}{dx}, \ (u^2 < 1)$

43. $\frac{d}{dx}(\coth^{-1} u) = \frac{d}{dx}\left[\frac{1}{2}\log\frac{u+1}{u-1}\right] = \frac{1}{1-u^2}\frac{du}{dx}, \ (u^2 > 1)$

44. $\frac{d}{dx}(\operatorname{sech}^{-1} u) = \frac{d}{dx}\left[\log\frac{1+\sqrt{1-u^2}}{u}\right] = -\frac{1}{u\sqrt{1-u^2}}\frac{du}{dx}, \ (0 < u < 1)$

45. $\frac{d}{dx}(\operatorname{csch}^{-1} u) = \frac{d}{dx}\left[\log\frac{1+\sqrt{1+u^2}}{u}\right] = -\frac{1}{|u|\sqrt{1+u^2}}\frac{du}{dx}$

46. $\frac{d}{dq}\int_p^q f(x)\,dx = f(q), \ [p \text{ constant}]$

47. $\frac{d}{dp}\int_p^q f(x)\,dx = -f(p), \ [q \text{ constant}]$

48. $\frac{d}{da}\int_p^q f(x, a)\,dx = \int_p^q \frac{\partial}{\partial a}[f(x, a)]\,dx + f(q, a)\frac{dq}{da} - f(p, a)\frac{dp}{da}$

INTEGRATION

The following is a brief discussion of some integration techniques. A more complete discussion can be found in a number of good text books. However, the purpose of this introduction is simply to discuss a few of the important techniques which may be used, in conjunction with the integral table which follows, to integrate particular functions.

No matter how extensive the integral table, it is a fairly uncommon occurrence to find in the table the exact integral desired. Usually some form of transformation will have to be made. The simplest type of transformation, and yet the most general, is substitution. Simple forms of substitution, such as $y = ax$, are employed almost unconsciously by experienced users of integral tables. Other substitutions may require more thought. In some sections of the tables, appropriate substitutions are suggested for integrals which are similar to, but not exactly like, integrals in the table. Finding the right substitution is largely a matter of intuition and experience.

Several precautions must be observed when using substitutions:

1. Be sure to make the substitution in the dx term, as well as everywhere else in the integral.
2. Be sure that the function substituted is one-to-one and continuous. If this is not the case, the integral must be restricted in such a way as to make it true. See the example following.
3. With definite integrals, the limits should also be expressed in terms of the new dependent variable. With indefinite integrals, it is necessary to perform the reverse substitutions to obtain the answer in terms of the original independent variable. This may also be done for definite integrals, but it is usually easier to change the limits.

Example:

$$\int \frac{x^4}{\sqrt{a^2 - x^2}}\, dx$$

Here we make the substitution $x = |a| \sin \theta$. Then $dx = |a| \cos \theta d\theta$, and

$$\sqrt{a^2 - x^2} = \sqrt{a^2 - a^2 \sin^2 \theta} = |a| \sqrt{1 - \sin^2 \theta} = |a \cos \theta|$$

Notice the absolute value signs. It is very important to keep in mind that a square root radical always denotes the positive square root, and to assure the sign is always kept positive. Thus $\sqrt{x^2} = |x|$. Failure to observe this is a common cause of errors in integration.

Notice also that the indicated substitution is not a one-to-one function, that is, it does not have a unique inverse. Thus we must restrict the range of θ in such a way as to make the function one-to-one. Fortunately, this is easily done by solving for θ

$$\theta = \sin^{-1} \frac{x}{|a|}$$

and restricting the inverse sine to the principal values, $-\frac{\pi}{2} \leq \theta \leq \frac{\pi}{2}$.

Thus the integral becomes

$$\int \frac{a^4 \sin^4 \theta \, |a| \cos \theta d\theta}{|a| \; |\cos \theta|}$$

Now, however, in the range of values chosen for θ, $\cos\theta$ is always positive. Thus we may remove the absolute value signs from $\cos\theta$ in the denominator. (This is one of the reasons that the principal values of the inverse trigonometric functions are defined as they are.)

Then the $\cos\theta$ terms cancel, and the integral becomes

$$a^4 \int \sin^4\theta \, d\theta$$

By application of integral formulas 244 and 242, we integrate this to

$$-a^4 \frac{\sin^3\theta\cos\theta}{4} - \frac{3a^4}{8}\cos\theta\sin\theta + \frac{3a^4}{8}\theta + C$$

We now must perform the inverse substitution to get the result in terms of x. We have

$$\theta = \sin^{-1}\frac{x}{|a|}$$

$$\sin\theta = \frac{x}{|a|}$$

Then

$$\cos\theta = \pm\sqrt{1 - \sin^2\theta} = \pm\sqrt{1 - \frac{x^2}{a^2}} = \pm\frac{\sqrt{a^2 - x^2}}{|a|}.$$

Because of the previously mentioned fact that $\cos\theta$ is positive, we may omit the $\pm$ sign. The reverse substitution then produces the final answer

$$\int \frac{x^4}{\sqrt{a^2 - x^2}} dx = -\tfrac{1}{4}x^3\sqrt{a^2 - x^2} - \tfrac{3}{8}a^2 x\sqrt{a^2 - x^2} + \frac{3a^4}{8}\sin^{-1}\frac{x}{|a|} + C.$$

Any rational function of x may be integrated, if the denominator is factored into linear and irreducible quadratic factors. The function may then be broken into partial fractions, and the individual partial fractions integrated by use of the appropriate formula from the integral table. See the section on partial fractions for further information.

Many integrals may be reduced to rational functions by proper substitutions. For example,

$$z = \tan\frac{x}{2}$$

will reduce any rational function of the six trigonometric functions of x to a rational function of z. (Frequently there are other substitutions which are simpler to use, but this one will always work. See integral formula number 374.)

Any rational function of x and $\sqrt{ax + b}$ may be reduced to a rational function of z by making the substitution

$$z = \sqrt{ax + b}.$$

Other likely substitutions will be suggested by looking at the form of the integrand.

The other main method of transforming integrals is integration by parts. This involves applying formula number 5 or 6 in the accompanying integral table. The critical factor in this method is the choice of the functions u and v. In order for the method to be successful, $v = \int dv$ and $\int v\,du$ must be easier to integrate than the original integral. Again, this choice is largely a matter of intuition and experience.

Example:

$$\int x \sin x \, dx$$

Two obvious choices are $u = x$, $dv = \sin x \, dx$, or $u = \sin x$, $dv = x \, dx$. Since a preliminary mental calculation indicates that $\int v\,du$ in the second choice would be more, rather than less, complicated than the original integral (it would contain x^2), we use the first choice.

$$u = x \qquad du = dx$$

$$dv = \sin x\,dx \qquad v = -\cos x$$

$$\int x \sin x\,dx = \int u\,dv = uv - \int v\,du = -x\cos x + \int \cos x\,dx$$

$$= \sin x - x\cos x$$

Of course, this result could have been obtained directly from the integral table, but it provides a simple example of the method. In more complicated examples the choice of u and v may not be so obvious, and several different choices may have to be tried. Of course, there is no guarantee that any of them will work.

Integration by parts may be applied more than once, or combined with substitution. A fairly common case is illustrated by the following example.

Example:

$$\int e^x \sin x\,dx$$

Let $\quad u = e^x \quad$ Then $\quad du = e^x\,dx$

$$dv = \sin x\,dx \qquad v = -\cos x$$

$$\int e^x \sin x\,dx = \int u\,dv = uv - \int v\,du = -e^x\cos x + \int e^x \cos x\,dx$$

In this latter integral,

let $\quad u = e^x \quad$ Then $\quad du = e^x\,dx$

$$dv = \cos x\,dx \qquad v = \sin x$$

$$\int e^x \sin x\,dx = -e^x\cos x + \int e^x\cos x\,dx = -e^x\cos x + \int u\,dv$$

$$= -e^x\cos x + uv - \int v\,du$$

$$= -e^x\cos x + e^x\sin x - \int e^x \sin x\,dx$$

This looks as if a circular transformation has taken place, since we are back at the same integral we started from. However, the above equation can be solved algebraically for the required integral:

$$\int e^x \sin x\,dx = \tfrac{1}{2}(e^x \sin x - e^x \cos x)$$

In the second integration by parts, if the parts had been chosen as $u = \cos x$, $dv = e^x\,dx$, we would indeed have made a circular transformation, and returned to the starting place. In general, when doing repeated integration by parts, one should never choose the function u at any stage to be the same as the function v at the previous stage, or a constant times the previous v.

The following rule is called the extended rule for integration by parts. It is the result of $n + 1$ successive applications of integration by parts.

If

$$g_1(x) = \int g(x)\,dx, \qquad g_2(x) = \int g_1(x)\,dx,$$

$$g_3(x) = \int g_2(x)\,dx, \ldots, g_m(x) = \int g_{m-1}(x)\,dx, \ldots,$$

then

$$\int f(x)\cdot g(x)\,dx = f(x)\cdot g_1(x) - f'(x)\cdot g_2(x) + f''(x)\cdot g_3(x) - + \cdots$$

$$+ (-1)^n f^{(n)}(x) g_{n+1}(x) + (-1)^{n+1} \int f^{(n+1)}(x) g_{n+1}(x)\,dx.$$

A useful special case of the above rule is when $f(x)$ is a polynomial of degree n. Then $f^{(n+1)}(x) = 0$, and

$$\int f(x) \cdot g(x)\,dx = f(x) \cdot g_1(x) - f'(x) \cdot g_2(x) + f''(x) \cdot g_3(x) - + \cdots + (-1)^n f^{(n)}(x) g_{n+1}(x) + C$$

Example:

If $f(x) = x^2$, $g(x) = \sin x$

$$\int x^2 \sin x\,dx = -x^2 \cos x + 2x \sin x + 2 \cos x + C$$

Another application of this formula occurs if

$$f''(x) = af(x) \quad \text{and} \quad g''(x) = bg(x),$$

where a and b are unequal constants. In this case, by a process similar to that used in the above example for $\int e^x \sin x\,dx$, we get the formula

$$f(x) g(x)\,dx = \frac{f(x) \cdot g'(x) - f'(x) \cdot g(x)}{b - a} + C$$

This formula could have been used in the example mentioned. Here is another example.

Example:

If $f(x) = e^{2x}$, $g(x) = \sin 3x$, then $a = 4$, $b = -9$, and

$$\int e^{2x} \sin 3x\,dx = \frac{3e^{2x} \cos 3x - 2e^{2x} \sin 3x}{-9 - 4} + C = \frac{e^{2x}}{13}(2 \sin 3x - 3 \cos 3x) + C$$

The following additional points should be observed when using this table.

1. A constant of integration is to be supplied with the answers for indefinite integrals.
2. Logarithmic expressions are to base $e = 2.71828\ldots$, unless otherwise specified, and are to be evaluated for the absolute value of the arguments involved therein.
3. All angles are measured in radians, and inverse trigonometric and hyperbolic functions represent principal values, unless otherwise indicated.
4. If the application of a formula produces either a zero denominator or the square root of a negative number in the result, there is always available another form of the answer which avoids this difficulty. In many of the results, the excluded values are specified, but when such are omitted it is presumed that one can tell what these should be, especially when difficulties of the type herein mentioned are obtained.
5. When inverse trigonometric functions occur in the integrals, be sure that any replacements made for them are strictly in accordance with the rules for such functions. This causes little difficulty when the argument of the inverse trigonometric function is positive, since then all angles involved are in the first quadrant. However, if the argument is negative, special care must be used. Thus if $u > 0$,

   $$\sin^{-1} u = \cos^{-1} \sqrt{1 - u^2} = \csc^{-1} \frac{1}{u}, \text{ etc.}$$

   However, if $u < 0$,

   $$\sin^{-1} u = -\cos^{-1} \sqrt{1 - u^2} = -\pi - \csc^{-1} \frac{1}{u}, \text{ etc.}$$

   See the section on inverse trigonometric functions for a full treatment of the allowable substitutions.
6. In integrals 259–263, the right side includes expressions of the form

   $$A \tan^{-1} [B + C \tan f(x)].$$

In these formulas, the $\tan^{-1}$ does not necessarily represent the principal value. Instead of always employing the principal branch of the inverse tangent function, one must instead use that branch of the inverse tangent function upon which $f(x)$ lies for any particular choice of x.

Example:

Using Integral Formula 259.

$$\int_0^{4\pi} \frac{dx}{2 + \sin x} = \frac{2}{\sqrt{3}} \tan^{-1} \frac{2 \tan \frac{x}{2} + 1}{\sqrt{3}} \Bigg]_0^{4\pi}$$

$$= \frac{2}{\sqrt{3}} \left[\tan^{-1} \frac{2 \tan 2\pi + 1}{\sqrt{3}} - \tan^{-1} \frac{2 \tan 0 + 1}{\sqrt{3}} \right]$$

$$= \frac{2}{\sqrt{3}} \left[\frac{13\pi}{6} - \frac{\pi}{6} \right] = \frac{4\pi}{\sqrt{3}} = \frac{4\sqrt{3}\pi}{3}$$

Here

$$\tan^{-1} \frac{2 \tan 2\pi + 1}{\sqrt{3}} = \tan^{-1} \frac{1}{\sqrt{3}} = \frac{13\pi}{6},$$

since $f(x) = 2\pi$; and

$$\tan^{-1} \frac{2 \tan 0 + 1}{\sqrt{3}} = \tan^{-1} \frac{1}{\sqrt{3}} = \frac{\pi}{6},$$

since $f(x) = 0$.

ELEMENTARY FORMS

1. $\int a\,dx = ax$

2. $\int a \cdot f(x)dx = a \int f(x)dx$

3. $\int \phi(y)\,dx = \int \frac{\phi(y)}{y'}\,dy,\qquad \text{where } y' = \frac{dy}{dx}$

4. $\int (u + v)\,dx = \int u\,dx + \int v\,dx$, where u and v are any functions of x

5. $\int u\,dv = u\int dv - \int v\,du = uv - \int v\,du$

6. $\int u\frac{dv}{dx}\,dx = uv - \int v\frac{du}{dx}\,dx$

7. $\int x^n\,dx = \frac{x^{n+1}}{n+1},\qquad \text{except } n = -1$

8. $\int \frac{f'(x)\,dx}{f(x)} = \log f(x),\qquad (df(x) = f'(x)\,dx)$

9. $\int \frac{dx}{x} = \log x$

10. $\int \frac{f'(x)\,dx}{2\sqrt{f(x)}} = \sqrt{f(x)},\qquad (df(x) = f'(x)\,dx)$

11. $\int e^x\,dx = e^x$

12. $\int e^{ax}\,dx = e^{ax}/a$

13. $\int b^{ax}\,dx = \frac{b^{ax}}{a\log b},\ (b > 0)$

14. $\int \log x\,dx = x\log x - x$

15. $\int a^x \log a\,dx = a^x,\ (a > 0)$

16. $\int \frac{dx}{a^2 + x^2} = \frac{1}{a}\tan^{-1}\frac{x}{a}$

17. $$\int \frac{dx}{a^2 - x^2} = \begin{cases} \frac{1}{a}\tanh^{-1}\frac{x}{a} \\ \text{or} \\ \frac{1}{2a}\log\frac{a+x}{a-x}, \qquad (a^2 > x^2) \end{cases}$$

18. $$\int \frac{dx}{x^2 - a^2} = \begin{cases} -\frac{1}{a}\coth^{-1}\frac{x}{a} \\ \text{or} \\ \frac{1}{2a}\log\frac{x-a}{x+a}, \qquad (x^2 > a^2) \end{cases}$$

19. $$\int \frac{dx}{\sqrt{a^2 - x^2}} = \begin{cases} \sin^{-1}\frac{x}{|a|} \\ \text{or} \\ -\cos^{-1}\frac{x}{|a|} \end{cases}$$

20. $$\int \frac{dx}{\sqrt{x^2 \pm a^2}} = \log\left(x + \sqrt{x^2 \pm a^2}\right)$$

21. $$\int \frac{dx}{x\sqrt{x^2 - a^2}} = \frac{1}{|a|}\sec^{-1}\frac{x}{a}$$

22. $$\int \frac{dx}{x\sqrt{a^2 \pm x^2}} = -\frac{1}{a}\log\left(\frac{a + \sqrt{a^2 \pm x^2}}{x}\right)$$

23. $$\int \frac{dx}{x\sqrt{a + bx}} = \begin{cases} \frac{2}{\sqrt{-a}}\tan^{-1}\sqrt{\frac{a+bx}{-a}}, \qquad (a < 0) \\ \text{or} \\ \frac{-2}{\sqrt{a}}\tanh^{-1}\sqrt{\frac{a+bx}{a}} \\ \text{or} \\ \frac{1}{\sqrt{a}}\log\frac{\sqrt{a+bx} - \sqrt{a}}{\sqrt{a+bx} + \sqrt{a}} \end{cases}$$

FORMS CONTAINING $(a + bx)$

For forms containing $a + bx$, but not listed in the table, the substitution $u = \frac{a + bx}{x}$ may prove helpful.

24. $$\int (a + bx)^n\,dx = \frac{(a + bx)^{n+1}}{(n+1)b}, \qquad (n \neq -1)$$

25. $$\int x(a + bx)^n\,dx$$
$$= \frac{1}{b^2(n+2)}(a + bx)^{n+2} - \frac{a}{b^2(n+1)}(a + bx)^{n+1}, \quad (n \neq -1, -2)$$

26. $$\int x^2(a + bx)^n\,dx = \frac{1}{b^3}\left[\frac{(a+bx)^{n+3}}{n+3} - 2a\,\frac{(a+bx)^{n+2}}{n+2} + a^2\,\frac{(a+bx)^{n+1}}{n+1}\right]$$

27. $$\int x^m(a + bx)^n\,dx = \frac{x^{m+1}(a+bx)^n}{m+n+1} + \frac{an}{m+n+1}\int x^m(a+bx)^{n-1}\,dx$$

28. $$\int x^m(a+bx)^n\,dx = \begin{cases} \dfrac{1}{a(n+1)}\left[-x^{m+1}(a+bx)^{n+1} + (m+n+2)\displaystyle\int x^m(a+bx)^{n+1}\,dx\right] \\ \text{or} \\ \dfrac{1}{b(m+n+1)}\left[x^m(a+bx)^{n+1} - ma\displaystyle\int x^{m-1}(a+bx)^n\,dx\right] \end{cases}$$

29. $$\int \frac{dx}{a+bx} = \frac{1}{b}\log(a+bx)$$

30. $$\int \frac{dx}{(a+bx)^2} = -\frac{1}{b(a+bx)}$$

31. $$\int \frac{dx}{(a+bx)^3} = -\frac{1}{2b(a+bx)^2}$$

32. $$\int \frac{xdx}{a+bx} = \begin{cases} \dfrac{1}{b^2}[a+bx-a\log(a+bx)] \\ \text{or} \\ \dfrac{x}{b} - \dfrac{a}{b^2}\log(a+bx) \end{cases}$$

33. $$\int \frac{xdx}{(a+bx)^2} = \frac{1}{b^2}\left[\log(a+bx) + \frac{a}{a+bx}\right]$$

34. $$\int \frac{xdx}{(a+bx)^3} = \frac{1}{b^2}\left[-\frac{1}{a+bx} + \frac{a}{2(a+bx)^2}\right]$$

35. $$\int \frac{xdx}{(a+bx)^n} = \frac{1}{b^2}\left[\frac{-1}{(n-2)(a+bx)^{n-2}} + \frac{a}{(n-1)(a+bx)^{n-1}}\right], \quad n \neq 1, 2$$

36. $$\int \frac{x^2dx}{a+bx} = \frac{1}{b^3}\left[\frac{1}{2}(a+bx)^2 - 2a(a+bx) + a^2\log(a+bx)\right]$$

37. $$\int \frac{x^2\,dx}{(a+bx)^2} = \frac{1}{b^3}\left[a+bx-2a\log(a+bx) - \frac{a^2}{a+bx}\right]$$

38. $$\int \frac{x^2\,dx}{(a+bx)^3} = \frac{1}{b^3}\left[\log(a+bx) + \frac{2a}{a+bx} - \frac{a^2}{2(a+bx)^2}\right]$$

39. $$\int \frac{x^2\,dx}{(a+bx)^n} = \frac{1}{b^3}\left[\frac{-1}{(n-3)(a+bx)^{n-3}} + \frac{2a}{(n-2)(a+bx)^{n-2}} - \frac{a^2}{(n-1)(a+bx)^{n-1}}\right], \quad n \neq 1, 2, 3$$

40. $$\int \frac{dx}{x(a+bx)} = -\frac{1}{a}\log\frac{a+bx}{x}$$

41. $$\int \frac{dx}{x(a+bx)^2} = \frac{1}{a(a+bx)} - \frac{1}{a^2}\log\frac{a+bx}{x}$$

42. $$\int \frac{dx}{x(a+bx)^3} = \frac{1}{a^3}\left[\frac{1}{2}\left(\frac{2a+bx}{a+bx}\right)^2 + \log\frac{x}{a+bx}\right]$$

43. $$\int \frac{dx}{x^2(a+bx)} = -\frac{1}{ax} + \frac{b}{a^2}\log\frac{a+bx}{x}$$

44. $$\int \frac{dx}{x^3(a+bx)} = \frac{2bx-a}{2a^2x^2} + \frac{b^2}{a^3}\log\frac{x}{a+bx}$$

45. $$\int \frac{dx}{x^2(a+bx)^2} = -\frac{a+2bx}{a^2x(a+bx)} + \frac{2b}{a^3}\log\frac{a+bx}{x}$$

FORMS CONTAINING $c^2 \pm x^2$, $x^2 - c^2$

46. $$\int \frac{dx}{c^2+x^2} = \frac{1}{c}\tan^{-1}\frac{x}{c}$$

47. $$\int \frac{dx}{ax^2+c} = \frac{1}{\sqrt{ac}}\tan^{-1}\left(x\sqrt{\frac{a}{c}}\right), \qquad (a,\ c > 0)$$

48. $$\int \frac{dx}{c^2-x^2} = \frac{1}{2c}\log\frac{c+x}{c-x}, \qquad (c^2 > x^2)$$

49. $$\int \frac{dx}{ax^2+c} = \begin{cases} \dfrac{1}{2\sqrt{-ac}}\log\dfrac{x\sqrt{a}-\sqrt{-c}}{x\sqrt{a}+\sqrt{-c}}, & (a > 0,\ c < 0) \\ \text{or} & \\ \dfrac{1}{2\sqrt{-ac}}\log\dfrac{\sqrt{c}+x\sqrt{-a}}{\sqrt{c}-x\sqrt{-a}}, & (a < 0,\ c > 0) \end{cases}$$

50. $$\int \frac{dx}{x^2-c^2} = \frac{1}{2c}\log\frac{x-c}{x+c}, \qquad (x^2 > c^2)$$

FORMS CONTAINING $a + bx$ AND $a' + b'x$

51. $$\int \frac{dx}{(a+bx)(a'+b'x)} = \frac{1}{ab'-a'b}\cdot\log\left(\frac{a'+b'x}{a+bx}\right)$$

52. $$\int \frac{x\,dx}{(a+bx)(a'+b'x)} = \frac{1}{ab'-a'b}\left[\frac{a}{b}\log(a+bx) - \frac{a'}{b'}\log(a'+b'x)\right]$$

53. $$\int \frac{dx}{(a+bx)^2(a'+b'x)} = \frac{1}{ab'-a'b}\left(\frac{1}{a+bx} + \frac{b'}{ab'-a'b}\log\frac{a'+b'x}{a+bx}\right)$$

54. $$\int \frac{x\,dx}{(a+bx)^2(a'+b'x)} = \frac{-a}{b(ab'-a'b)(a+bx)} - \frac{a'}{(ab'-a'b)^2}\log\frac{a'+b'x}{a+bx}$$

55. $$\int \frac{x^2\,dx}{(a+bx)^2(a'+b'x)} = \frac{a^2}{b^2(ab'-a'b)(a+bx)} + \frac{1}{(ab'-a'b)^2}\left[\frac{a'^2}{b'}\log(a'+b'x) + \frac{a(ab'-2a'b)}{b^2}\log(a+bx)\right]$$

56. $$\int \frac{dx}{(a+bx)^n(a'+b'x)^m} = \frac{1}{(m-1)(ab'-a'b)}\left[\frac{-1}{(a+bx)^{n-1}(a'+b'x)^{m-1}} - (m+n-2)b\int \frac{dx}{(a+bx)^n(a'+b'x)^{m-1}}\right]$$

57. $$\int \frac{a+bx}{a'+b'x}\,dx = \frac{bx}{b'} + \frac{ab'-a'b}{b'^2}\log(a'+b'x)$$

58. $$\int \frac{(a+bx)^m dx}{(a'+b'x)^n} = \begin{cases} -\dfrac{1}{(n-1)(ab'-a'b)}\left[\dfrac{(a+bx)^{m+1}}{(a'+b'x)^{n-1}} + b(n-m-2)\displaystyle\int \dfrac{(a+bx)^m\,dx}{(a'+b'x)^{n-1}}\right] \\ \text{or} \\ -\dfrac{1}{b'(n-m-1)}\left[\dfrac{(a+bx)^m}{(a'+b'x)^{n-1}} + m(ab'-a'b)\displaystyle\int \dfrac{(a+bx)^{m-1}\,dx}{(a'+b'x)^n}\right] \\ \text{or} \\ \dfrac{-1}{(n-1)b'}\left[\dfrac{(a+bx)^m}{(a'+b'x)^{n-1}} - mb\displaystyle\int \dfrac{(a+bx)^{m-1}\,dx}{(a'+b'x)^{n-1}}\right] \end{cases}$$

FORMS CONTAINING $\sqrt{a+bx} = \sqrt{u}$ AND $\sqrt{a'+b'x} = \sqrt{v}$ WITH $k = ab' - a'b$

In integrals 59–66, if $k = 0$, then $v = \frac{a'}{a}u$, and formulas starting with 112 should be used in place of these.

59. $$\int \sqrt{uv}\,dx = \frac{k+2bv}{4bb'}\sqrt{uv} - \frac{k^2}{8bb'}\int \frac{dx}{\sqrt{uv}}$$

60. $$\int \frac{dx}{v\sqrt{u}} = \begin{cases} \dfrac{1}{\sqrt{kb'}}\log\dfrac{b'\sqrt{u}-\sqrt{kb'}}{b'\sqrt{u}+\sqrt{kb'}} \\ \text{or} \\ \dfrac{2}{\sqrt{-kb'}}\tan^{-1}\dfrac{b'\sqrt{u}}{\sqrt{-kb'}} \end{cases}$$

61. $$\int \frac{dx}{\sqrt{uv}} = \begin{cases} \dfrac{2}{\sqrt{bb'}}\tanh^{-1}\dfrac{\sqrt{bb'uv}}{bv}, & (bb' > 0) \\ \text{or} \\ \dfrac{1}{\sqrt{bb'}}\log\dfrac{bv+\sqrt{bb'uv}}{bv-\sqrt{bb'uv}}, & (bb' > 0) \\ \text{or} \\ \dfrac{1}{\sqrt{bb'}}\log\dfrac{(bv+\sqrt{bb'uv})^2}{|v|}, & (bb' > 0) \\ \text{or} \\ \dfrac{2}{\sqrt{-bb'}}\tan^{-1}\dfrac{\sqrt{-bb'uv}}{bv}, & (bb' < 0) \end{cases}$$

62. $$\int \frac{x\,dx}{\sqrt{uv}} = \frac{\sqrt{uv}}{bb'} - \frac{ab'+a'b}{2bb'}\int \frac{dx}{\sqrt{uv}}$$

63. $\int \frac{dx}{v\sqrt{uv}} = \frac{-2\sqrt{uv}}{kv}$ or $= \frac{-2\sqrt{u}}{k\sqrt{v}}, \quad (v > 0)$

64. $\int \frac{\sqrt{v}}{\sqrt{u}}\, dx = \int \frac{v\, dx}{\sqrt{uv}}$ or $= \frac{\sqrt{uv}}{b} - \frac{k}{2b}\int \frac{dx}{\sqrt{uv}}, \quad (v > 0)$

65. $\int v^m \sqrt{u}\, dx = \frac{1}{(2m+3)b'}\left(2v^{m+1}\sqrt{u} + k\int \frac{v^m\, dx}{\sqrt{u}}\right)$

66. $\int \frac{dx}{v^m\sqrt{u}} = -\frac{1}{(m-1)k}\left(\frac{\sqrt{u}}{v^{m-1}} + \left(m - \frac{3}{2}\right) b \int \frac{dx}{v^{m-1}\sqrt{u}}\right)$

FORMS CONTAINING $(a + bx^n)$

67. $\int \frac{dx}{a + bx^2} = \frac{1}{\sqrt{ab}} \tan^{-1} \frac{x\sqrt{ab}}{a}$

68. $\int \frac{dx}{a + bx^2} = \begin{cases} \frac{1}{2\sqrt{-ab}} \log \frac{a + x\sqrt{-ab}}{a - x\sqrt{-ab}}, \\ \text{or} \\ \frac{1}{\sqrt{-ab}} \tanh^{-1} \frac{x\sqrt{-ab}}{a}, \quad (ab < 0) \end{cases}$

69. $\int \frac{dx}{a^2 + b^2x^2} = \frac{1}{ab} \tan^{-1} \frac{bx}{a}$

70. $\int \frac{xdx}{a + bx^2} = \frac{1}{2b} \log (a + bx^2)$

71. $\int \frac{xdx}{a^2 + b^2x^2} = \frac{1}{2b^2} \log (a^2 + b^2x^2)$

72. $\int \frac{x^2\, dx}{a + bx^2} = \frac{x}{b} - \frac{a}{b}\int \frac{dx}{a + bx^2}$

73. $\int \frac{dx}{(a + bx^2)^2} = \frac{x}{2a(a + bx^2)} + \frac{1}{2a}\int \frac{dx}{a + bx^2}$

74. $\int \frac{dx}{(x^2 + a^2)^2} = \frac{1}{2a^3} \tan^{-1} \frac{x}{a} + \frac{x}{2a^2(x^2 + a^2)}$

75. $\int \frac{dx}{a^2 - b^2x^2} = \frac{1}{2ab} \log \frac{a + bx}{a - bx}$

76. $\int \frac{dx}{(x^2 - a^2)^2} = -\frac{x}{2a^2(x^2 - a^2)} + \frac{1}{4a^3} \log \frac{a + x}{a - x}$

77. $\int \frac{dx}{(a + bx^2)^{m+1}} = \frac{1}{2ma} \frac{x}{(a + bx^2)^m} + \frac{2m - 1}{2ma}\int \frac{dx}{(a + bx^2)^m}$

78. $\int \frac{xdx}{(a + bx^2)^{m+1}} = -\frac{1}{2bm(a + bx^2)^m}$

79. $\int \frac{x^2\, dx}{(a + bx^2)^{m+1}} = \frac{-x}{2mb(a + bx^2)^m} + \frac{1}{2mb}\int \frac{dx}{(a + bx^2)^m}$

80. $\int \frac{dx}{x(a+bx^2)} = \frac{1}{2a}\log\frac{x^2}{a+bx^2}$

81. $\int \frac{dx}{x^2(a+bx^2)} = -\frac{1}{ax} - \frac{b}{a}\int\frac{dx}{a+bx^2}$

82. $\int \frac{dx}{x(a+bx^2)^{m+1}} = \frac{1}{2am(a+bx^2)^m} + \frac{1}{a}\int\frac{dx}{x(a+bx^2)^m}, \qquad (m \neq 0)$

83. $\int \frac{dx}{x^2(a+bx^2)^{m+1}} = \frac{1}{a}\int\frac{dx}{x^2(a+bx^2)^m} - \frac{b}{a}\int\frac{dx}{(a+bx^2)^{m+1}}$

84. $\int \frac{dx}{a+bx^3} = \frac{k}{3a}\left[\frac{1}{2}\log\frac{(k+x)^2}{k^2-kx+x^2} + \sqrt{3}\tan^{-1}\frac{2x-k}{k\sqrt{3}}\right], \qquad (bk^3 = a)$

85. $\int \frac{xdx}{a+bx^3} = \frac{1}{3bk}\left[\frac{1}{2}\log\frac{k^2-kx+x^2}{(k+x)^2} + \sqrt{3}\tan^{-1}\frac{2x-k}{k\sqrt{3}}\right], \qquad (bk^3 = a)$

86. $\int \frac{dx}{x(a+bx^n)} = \frac{1}{an}\log\frac{x^n}{a+bx^n}.$

87. $\int \frac{dx}{(a+bx^n)^{m+1}} = \frac{1}{a}\int\frac{dx}{(a+bx^n)^m} - \frac{b}{a}\int\frac{x^n\,dx}{(a+bx^n)^{m+1}}$

88. $\int \frac{x^m\,dx}{(a+bx^n)^{p+1}} = \frac{1}{b}\int\frac{x^{m-n}\,dx}{(a+bx^n)^p} - \frac{a}{b}\int\frac{x^{m-n}\,dx}{(a+bx^n)^{p+1}}$

89. $\int \frac{dx}{x^m(a+bx^n)^{p+1}} = \frac{1}{a}\int\frac{dx}{x^m(a+bx^n)^p} - \frac{b}{a}\int\frac{dx}{x^{m-n}(a+bx^n)^{p+1}}$

90. $\int x^m(a+bx^n)^p\,dx$

$$= \frac{x^{m-n+1}(a+bx^n)^{p+1}}{b(np+m+1)} - \frac{a(m-n+1)}{b(np+m+1)}\int x^{m-n}(a+bx^n)^p\,dx$$

91. $\int x^m(a+bx^n)^p\,dx = \frac{x^{m+1}(a+bx^n)^p}{np+m+1} + \frac{anp}{np+m+1}\int x^m(a+bx^n)^{p-1}\,dx$

92. $\int x^{m-1}(a+bx^n)^p\,dx$

$$= \frac{1}{b(m+np)}\left[x^{m-n}(a+bx^n)^{p+1} - (m-n)a\int x^{m-n-1}(a+bx^n)^p\,dx\right]$$

93. $\int x^{m-1}(a+bx^n)^p\,dx$

$$= \frac{1}{m+np}\left[x^m(a+bx^n)^p + npa\int x^{m-1}(a+bx^n)^{p-1}\,dx\right]$$

94. $\int x^{m-1}(a+bx^n)^p\,dx$

$$= \frac{1}{ma}\left[x^m(a+bx^n)^{p+1} - (m+np+n)b\int x^{m+n-1}(a+bx^n)^p\,dx\right]$$

95. $\int x^{m-1}(a+bx^n)^p\,dx$

$$= \frac{1}{an(p+1)}\left[-x^m(a+bx^n)^{p+1} + (m+np+n)\int x^{m-1}(a+bx^n)^{p+1}\,dx\right]$$

FORMS CONTAINING $(a + bx + cx^2)$

$X = a + bx + cx^2$ and $q = 4\,ac - b^2$

For integrals 96–111, if $q = 0$, then $X = c\left(x + \frac{b}{2c}\right)^2$, and formulas starting with 24 should be used in place of these.

96. $\int \frac{dx}{X} = \frac{2}{\sqrt{q}}\tan^{-1}\frac{2cx+b}{\sqrt{q}}$

97. $\int \frac{dx}{X} = \frac{-2}{\sqrt{-q}}\tanh^{-1}\frac{2cx+b}{\sqrt{-q}}$

98. $\int \frac{dx}{X} = \frac{1}{\sqrt{-q}}\log\frac{2cx+b-\sqrt{-q}}{2cx+b+\sqrt{-q}}$

99. $\int \frac{dx}{X^2} = \frac{2cx+b}{qX} + \frac{2c}{q}\int\frac{dx}{X}$

100. $\int \frac{dx}{X^3} = \frac{2cx+b}{q}\left(\frac{1}{2X^2} + \frac{3c}{qX}\right) + \frac{6c^2}{q^2}\int\frac{dx}{X}$

101. $$\int \frac{dx}{X^{n+1}} = \begin{cases} \dfrac{2cx+b}{nqX^n} + \dfrac{2(2n-1)c}{qn}\displaystyle\int\frac{dx}{X^n} \\ \text{or} \\ \dfrac{(2n)!}{(n!)^2}\left(\dfrac{c}{q}\right)^n\left[\dfrac{2cx+b}{q}\displaystyle\sum_{r=1}^{n}\left(\dfrac{q}{cX}\right)^r\left(\dfrac{(r-1)!r!}{(2r)!}\right) + \displaystyle\int\frac{dx}{X}\right] \end{cases}$$

102. $\int \frac{x\,dx}{X} = \frac{1}{2c}\log X - \frac{b}{2c}\int\frac{dx}{X}$

103. $\int \frac{x\,dx}{X^2} = -\frac{bx+2a}{qX} - \frac{b}{q}\int\frac{dx}{X}$

104. $\int \frac{x\,dx}{X^{n+1}} = -\frac{2a+bx}{nqX^n} - \frac{b(2n-1)}{nq}\int\frac{dx}{X^n}$

105. $\int \frac{x^2}{X}\,dx = \frac{x}{c} - \frac{b}{2c^2}\log X + \frac{b^2-2ac}{2c^2}\int\frac{dx}{X}$

106. $\int \frac{x^2}{X^2}\,dx = \frac{(b^2-2ac)x+ab}{cqX} + \frac{2a}{q}\int\frac{dx}{X}$

107. $$\int \frac{x^m\,dx}{X^{n+1}} = -\frac{x^{m-1}}{(2n-m+1)cX^n} - \frac{n-m+1}{2n-m+1}\cdot\frac{b}{c}\int\frac{x^{m-1}\,dx}{X^{n+1}} + \frac{m-1}{2n-m+1}\cdot\frac{a}{c}\int\frac{x^{m-2}\,dx}{X^{n+1}}$$

108. $\int \frac{dx}{xX} = \frac{1}{2a} \log \frac{x^2}{X} - \frac{b}{2a} \int \frac{dx}{X}$

109. $\int \frac{dx}{x^2X} = \frac{b}{2a^2} \log \frac{X}{x^2} - \frac{1}{ax} + \left(\frac{b^2}{2a^2} - \frac{c}{a}\right) \int \frac{dx}{X}$

110. $\int \frac{dx}{xX^n} = \frac{1}{2a(n-1)X^{n-1}} - \frac{b}{2a} \int \frac{dx}{X^n} + \frac{1}{a} \int \frac{dx}{xX^{n-1}}$

111. $$\int \frac{dx}{x^m X^{n+1}} = -\frac{1}{(m-1)ax^{m-1}X^n} - \frac{n+m-1}{m-1} \frac{b}{a} \int \frac{dx}{x^{m-1}X^{n+1}} - \frac{2n+m-1}{m-1} \cdot \frac{c}{a} \int \frac{dx}{x^{m-2}X^{n+1}}$$

FORMS CONTAINING $\sqrt{a + bx}$

112. $\int \sqrt{a+bx}\, dx = \frac{2}{3b} \sqrt{(a+bx)^3}$

113. $\int x\sqrt{a+bx}\, dx = -\frac{2(2a-3bx)\sqrt{(a+bx)^3}}{15b^2}$

114. $\int x^2\sqrt{a+bx}\, dx = \frac{2(8a^2 - 12abx + 15b^2x^2)\sqrt{(a+bx)^3}}{105b^3}$

115. $$\int x^m \sqrt{a+bx}\, dx = \frac{2}{b(2m+3)}\left[x^m\sqrt{(a+bx)^3} - ma\int x^{m-1}\sqrt{a+bx}\, dx\right]$$

116. $\int \frac{\sqrt{a+bx}}{x} dx = 2\sqrt{a+bx} + a\int \frac{dx}{x\sqrt{a+bx}}$

(see No. 123 and No. 124)

117. $\int \frac{\sqrt{a+bx}}{x^2} dx = -\frac{\sqrt{a+bx}}{x} + \frac{b}{2}\int \frac{dx}{x\sqrt{a+bx}}$

(see No. 123 and No. 124)

118. $$\int \frac{\sqrt{a+bx}}{x^m} = -\frac{1}{(m-1)a}\left[\frac{\sqrt{(a+bx)^3}}{x^{m-1}} + \frac{(2m-5)b}{2}\int \frac{\sqrt{a+bx}\, dx}{x^{m-1}}\right], \quad (m \neq 1)$$

119. $\int \frac{dx}{\sqrt{a+bx}} = \frac{2\sqrt{a+bx}}{b}$

120. $\int \frac{xdx}{\sqrt{a+bx}} = -\frac{2(2a-bx)}{3b^2}\sqrt{a+bx}$

121. $\int \frac{x^2\, dx}{\sqrt{a+bx}} = \frac{2(8a^2 - 4abx + 3b^2x^2)}{15b^3}\sqrt{a+bx}$

122. $$\int \frac{x^m\,dx}{\sqrt{a+bx}} = \frac{2x^m\sqrt{a+bx}}{(2m+1)b} - \frac{2ma}{(2m+1)b}\int \frac{x^{m-1}\,dx}{\sqrt{a+bx}}$$

123. $$\int \frac{dx}{x\sqrt{a+bx}} = \frac{1}{\sqrt{a}}\log\left(\frac{\sqrt{a+bx}-\sqrt{a}}{\sqrt{a+bx}+\sqrt{a}}\right), \qquad (a>0)$$

124. $$\int \frac{dx}{x\sqrt{a+bx}} = \frac{2}{\sqrt{-a}}\tan^{-1}\sqrt{\frac{a+bx}{-a}}, \qquad (a<0)$$

125. $$\int \frac{dx}{x^2\sqrt{a+bx}} = -\frac{\sqrt{a+bx}}{ax} - \frac{b}{2a}\int \frac{dx}{x\sqrt{a+bx}}$$

126. $$\int \frac{dx}{x^n\sqrt{a+bx}} = \begin{cases} -\dfrac{\sqrt{a+bx}}{(n-1)ax^{n-1}} - \dfrac{(2n-3)b}{(2n-2)a}\displaystyle\int \frac{dx}{x^{n-1}\sqrt{a+bx}} \\ \text{or} \\ \dfrac{(2n-2)!}{[(n-1)!]^2}\left[\dfrac{4\sqrt{a+bx}}{b}\displaystyle\sum_{r=1}^{n}\frac{r!(r-1)!}{x^r(2r)!} \right. \\ \left. \qquad + \left(\dfrac{-b}{4a}\right)^{n-1}\displaystyle\int \frac{dx}{x\sqrt{a+bx}}\right] \end{cases}$$

127. $$\int (a+bx)^{\pm\frac{n}{2}}\,dx = \frac{2(a+bx)^{\frac{2\pm n}{2}}}{b(2\pm n)}$$

128. $$\int x(a+bx)^{\pm\frac{n}{2}}\,dx = \frac{2}{b^2}\left[\frac{(a+bx)^{\frac{4\pm n}{2}}}{4\pm n} - \frac{a(a+bx)^{\frac{2\pm n}{2}}}{2\pm n}\right]$$

129. $$\int \frac{dx}{x(a+bx)^{\frac{m}{2}}} = \frac{1}{a}\int \frac{dx}{x(a+bx)^{\frac{m-2}{2}}} - \frac{b}{a}\int \frac{dx}{(a+bx)^{\frac{m}{2}}}$$

130. $$\int \frac{(a+bx)^{\frac{n}{2}}\,dx}{x} = b\int (a+bx)^{\frac{n-2}{2}}\,dx + a\int \frac{(a+bx)^{\frac{n-2}{2}}}{x}\,dx$$

131. $$\int f(x,\sqrt{a+bx})\,dx = \frac{2}{b}\int f\left(\frac{z^2-a}{b}, z\right), zdz \qquad (z^2 = a+bx)$$

FORMS CONTAINING $\sqrt{x^2 \pm a^2}$

132. $$\int \sqrt{x^2 \pm a^2}\,dx = \tfrac{1}{2}[x\sqrt{x^2\pm a^2} \pm a^2\log(x+\sqrt{x^2\pm a^2})]$$

133. $$\int \frac{dx}{\sqrt{x^2\pm a^2}} = \log(x+\sqrt{x^2\pm a^2})$$

134. $$\int \frac{dx}{x\sqrt{x^2-a^2}} = \frac{1}{|a|}\sec^{-1}\frac{x}{a}$$

135. $$\int \frac{dx}{x\sqrt{x^2+a^2}} = -\frac{1}{a}\log\left(\frac{a+\sqrt{x^2+a^2}}{x}\right)$$

136. $$\int \frac{\sqrt{x^2+a^2}}{x}\,dx = \sqrt{x^2+a^2} - a\log\left(\frac{a+\sqrt{x^2+a^2}}{x}\right)$$

137. $$\int \frac{\sqrt{x^2-a^2}}{x}\,dx = \sqrt{x^2-a^2} - |a|\sec^{-1}\frac{x}{a}$$

138. $$\int \frac{x\,dx}{\sqrt{x^2\pm a^2}} = \sqrt{x^2\pm a^2}$$

139. $$\int x\sqrt{x^2\pm a^2}\,dx = \tfrac{1}{3}\sqrt{(x^2\pm a^2)^3}$$

140. $$\int \sqrt{(x^2\pm a^2)^3}\,dx$$
$$= \frac{1}{4}\left[x\sqrt{(x^2\pm a^2)^3} \pm \frac{3a^2x}{2}\sqrt{x^2\pm a^2} + \frac{3a^4}{2}\log\left(x+\sqrt{x^2\pm a^2}\right)\right]$$

141. $$\int \frac{dx}{\sqrt{(x^2\pm a^2)^3}} = \frac{\pm x}{a^2\sqrt{x^2\pm a^2}}$$

142. $$\int \frac{x\,dx}{\sqrt{(x^2\pm a^2)^3}} = \frac{-1}{\sqrt{x^2\pm a^2}}$$

143. $$\int x\sqrt{(x^2\pm a^2)^3}\,dx = \tfrac{1}{5}\sqrt{(x^2\pm a^2)^5}$$

144. $$\int x^2\sqrt{x^2\pm a^2}\,dx$$
$$= \frac{x}{4}\sqrt{(x^2\pm a^2)^3} \mp \frac{a^2}{8}x\sqrt{x^2\pm a^2} - \frac{a^4}{8}\log\left(x+\sqrt{x^2\pm a^2}\right)$$

145. $$\int x^3\sqrt{x^2+a^2}\,dx = \left(\tfrac{1}{5}x^2 - \tfrac{2}{15}a^2\right)\sqrt{(a^2+x^2)^3}$$

146. $$\int x^3\sqrt{x^2-a^2}\,dx = \frac{1}{5}\sqrt{(x^2-a^2)^5} + \frac{a^2}{3}\sqrt{(x^2-a^2)^3}$$

147. $$\int \frac{x^2dx}{\sqrt{x^2\pm a^2}} = \frac{x}{2}\sqrt{x^2\pm a^2} \mp \frac{a^2}{2}\log\left(x+\sqrt{x^2\pm a^2}\right)$$

148. $$\int \frac{x^3\,dx}{\sqrt{x^2\pm a^2}} = \frac{1}{3}\sqrt{(x^2\pm a^2)^3} \mp a^2\sqrt{x^2\pm a^2}$$

149. $$\int \frac{dx}{x^2\sqrt{x^2\pm a^2}} = \mp\frac{\sqrt{x^2\pm a^2}}{a^2x}$$

150. $$\int \frac{dx}{x^3\sqrt{x^2+a^2}} = -\frac{\sqrt{x^2+a^2}}{2a^2x^2} + \frac{1}{2a^3}\log\frac{a+\sqrt{x^2+a^2}}{x}$$

151. $$\int \frac{dx}{x^3\sqrt{x^2-a^2}} = \frac{\sqrt{x^2-a^2}}{2a^2x^2} + \frac{1}{2|a^3|}\sec^{-1}\frac{x}{a}$$

152. $$\int x^2\sqrt{(x^2 \pm a^2)^3}\,dx = \frac{x}{6}\sqrt{(x^2 \pm a^2)^5} \mp \frac{a^2x}{24}\sqrt{(x^2 \pm a^2)^3} - \frac{a^4x}{16}\sqrt{x^2 \pm a^2} \mp \frac{a^6}{16}\log\left(x + \sqrt{x^2 \pm a^2}\right)$$

153. $$\int x^3\sqrt{(x^2 \pm a^2)^3}\,dx = \frac{1}{7}\sqrt{(x^2 \pm a^2)^7} \mp \frac{a^2}{5}\sqrt{(x^2 \pm a^2)^5}$$

154. $$\int \frac{\sqrt{x^2 \pm a^2}\,dx}{x^2} = -\frac{\sqrt{x^2 \pm a^2}}{x} + \log\left(x + \sqrt{x^2 \pm a^2}\right)$$

155. $$\int \frac{\sqrt{x^2 + a^2}}{x^3}\,dx = -\frac{\sqrt{x^2 + a^2}}{2x^2} - \frac{1}{2a}\log\frac{a + \sqrt{x^2 + a^2}}{x}$$

156. $$\int \frac{\sqrt{x^2 - a^2}}{x^3}\,dx = -\frac{\sqrt{x^2 - a^2}}{2x^2} + \frac{1}{2|a|}\sec^{-1}\frac{x}{a}$$

157. $$\int \frac{x^2\,dx}{\sqrt{(x^2 \pm a^2)^3}} = \frac{-x}{\sqrt{x^2 \pm a^2}} + \log\left(x + \sqrt{x^2 \pm a^2}\right)$$

158. $$\int \frac{x^3\,dx}{\sqrt{(x^2 \pm a^2)^3}} = \sqrt{x^2 \pm a^2} \pm \frac{a^2}{\sqrt{x^2 \pm a^2}}$$

159. $$\int \frac{dx}{x\sqrt{(x^2 + a^2)^3}} = \frac{1}{a^2\sqrt{x^2 + a^2}} - \frac{1}{a^3}\log\frac{a + \sqrt{x^2 + a^2}}{x}$$

160. $$\int \frac{dx}{x\sqrt{(x^2 - a^2)^3}} = -\frac{1}{a^2\sqrt{x^2 - a^2}} - \frac{1}{|a^3|}\sec^{-1}\frac{x}{a}$$

161. $$\int \frac{dx}{x^2\sqrt{(x^2 \pm a^2)^3}} = -\frac{1}{a^4}\left[\frac{\sqrt{x^2 \pm a^2}}{x} + \frac{x}{\sqrt{x^2 \pm a^2}}\right]$$

162. $$\int \frac{dx}{x^3\sqrt{(x^2 + a^2)^3}} = -\frac{1}{2a^2x^2\sqrt{x^2 + a^2}} - \frac{3}{2a^4\sqrt{x^2 + a^2}} + \frac{3}{2a^5}\log\frac{a + \sqrt{x^2 + a^2}}{x}$$

163. $$\int \frac{dx}{x^3\sqrt{(x^2 - a^2)^3}} = \frac{1}{2a^2x^2\sqrt{x^2 - a^2}} - \frac{3}{2a^4\sqrt{x^2 - a^2}} - \frac{3}{2|a^5|}\sec^{-1}\frac{x}{a}$$

164. $$\int f\left(x, \sqrt{x^2 + a^2}\right)dx = a\int f(a\tan u, |a\sec u|)\sec^2 u\,du, \qquad (x = a\tan u)$$

165. $$\int f\left(x, \sqrt{x^2 - a^2}\right)dx = a\int f(a\sec u, |a\tan u|)\sec u\tan u\,du, \qquad (x = a\sec u)$$

FORMS CONTAINING $\sqrt{a^2 - x^2}$

166. $$\int \sqrt{a^2 - x^2}\, dx = \frac{1}{2}\left[x\sqrt{a^2 - x^2} + a^2 \sin^{-1}\frac{x}{|a|}\right]$$

167. $$\int \frac{dx}{\sqrt{a^2 - x^2}} = \begin{cases} \sin^{-1}\dfrac{x}{|a|} \\ \quad\text{or} \\ -\cos^{-1}\dfrac{x}{|a|} \end{cases}$$

168. $$\int \frac{dx}{x\sqrt{a^2 - x^2}} = -\frac{1}{a}\log\left(\frac{a + \sqrt{a^2 - x^2}}{x}\right)$$

169. $$\int \frac{\sqrt{a^2 - x^2}}{x}\, dx = \sqrt{a^2 - x^2} - a\log\left(\frac{a + \sqrt{a^2 - x^2}}{x}\right)$$

170. $$\int \frac{x\, dx}{\sqrt{a^2 - x^2}} = -\sqrt{a^2 - x^2}$$

171. $$\int x\sqrt{a^2 - x^2}\, dx = -\tfrac{1}{3}\sqrt{(a^2 - x^2)^3}$$

172. $$\int \sqrt{(a^2 - x^2)^3}\, dx$$
$$= \frac{1}{4}\left[x\sqrt{(a^2 - x^2)^3} + \frac{3a^2x}{2}\sqrt{a^2 - x^2} + \frac{3a^4}{2}\sin^{-1}\frac{x}{|a|}\right]$$

173. $$\int \frac{dx}{\sqrt{(a^2 - x^2)^3}} = \frac{x}{a^2\sqrt{a^2 - x^2}}$$

174. $$\int \frac{x\, dx}{\sqrt{(a^2 - x^2)^3}} = \frac{1}{\sqrt{a^2 - x^2}}$$

175. $$\int x\sqrt{(a^2 - x^2)^3}\, dx = -\tfrac{1}{5}\sqrt{(a^2 - x^2)^5}$$

176. $$\int x^2\sqrt{a^2 - x^2}\, dx = -\frac{x}{4}\sqrt{(a^2 - x^2)^3} + \frac{a^2}{8}\left(x\sqrt{a^2 - x^2} + a^2\sin^{-1}\frac{x}{|a|}\right)$$

177. $$\int x^3\sqrt{a^2 - x^2}\, dx = (-\tfrac{1}{5}x^2 - \tfrac{2}{15}a^2)\sqrt{(a^2 - x^2)^3}$$

178. $$\int x^2\sqrt{(a^2 - x^2)^3}\, dx = -\frac{1}{6}x\sqrt{(a^2 - x^2)^5} + \frac{a^2x}{24}\sqrt{(a^2 - x^2)^3}$$
$$+ \frac{a^4x}{16}\sqrt{a^2 - x^2} + \frac{a^6}{16}\sin^{-1}\frac{x}{|a|}$$

179. $$\int x^3\sqrt{(a^2 - x^2)^3}\, dx = \frac{1}{7}\sqrt{(a^2 - x^2)^7} - \frac{a^2}{5}\sqrt{(a^2 - x^2)^5}$$

180. $$\int \frac{x^2\, dx}{\sqrt{a^2 - x^2}} = -\frac{x}{2}\sqrt{a^2 - x^2} + \frac{a^2}{2}\sin^{-1}\frac{x}{|a|}$$

181. $$\int \frac{dx}{x^2\sqrt{a^2-x^2}} = -\frac{\sqrt{a^2-x^2}}{a^2x}$$

182. $$\int \frac{\sqrt{a^2-x^2}}{x^2}\,dx = -\frac{\sqrt{a^2-x^2}}{x} - \sin^{-1}\frac{x}{|a|}$$

183. $$\int \frac{\sqrt{a^2-x^2}}{x^3}\,dx = -\frac{\sqrt{a^2-x^2}}{2x^2} + \frac{1}{2a}\log\frac{a+\sqrt{a^2-x^2}}{x}$$

184. $$\int \frac{x^2\,dx}{\sqrt{(a^2-x^2)^3}} = \frac{x}{\sqrt{a^2-x^2}} - \sin^{-1}\frac{x}{|a|}$$

185. $$\int \frac{x^3\,dx}{\sqrt{a^2-x^2}} = -\frac{2}{3}(a^2-x^2)^{\frac{3}{2}} - x^2(a^2-x^2)^{\frac{1}{2}} = -\frac{1}{3}\sqrt{a^2-x^2}\,(x^2-2a^2)$$

186. $$\int \frac{x^3\,dx}{\sqrt{(a^2-x^2)^3}} = 2(a^2-x^2)^{\frac{1}{2}} + \frac{x^2}{(a^2-x^2)^{\frac{1}{2}}} = \frac{a^2}{\sqrt{a^2-x^2}} + \sqrt{a^2-x^2}$$

187. $$\int \frac{dx}{x^3\sqrt{a^2-x^2}} = -\frac{\sqrt{a^2-x^2}}{2a^2x^2} - \frac{1}{2a^3}\log\frac{a+\sqrt{a^2-x^2}}{x}$$

188. $$\int \frac{dx}{x\sqrt{(a^2-x^2)^3}} = \frac{1}{a^2\sqrt{a^2-x^2}} - \frac{1}{a^3}\log\frac{a+\sqrt{a^2-x^2}}{x}$$

189. $$\int \frac{dx}{x^2\sqrt{(a^2-x^2)^3}} = \frac{1}{a^4}\left[-\frac{\sqrt{a^2-x^2}}{x} + \frac{x}{\sqrt{a^2-x^2}}\right]$$

190. $$\int \frac{dx}{x^3\sqrt{(a^2-x^2)^3}} = -\frac{1}{2a^2x^2\sqrt{a^2-x^2}} + \frac{3}{2a^4\sqrt{a^2-x^2}} - \frac{3}{2a^5}\log\frac{a+\sqrt{a^2-x^2}}{x}$$

191. $$\int \frac{\sqrt{a^2-x^2}}{b^2+x^2}\,dx = \frac{\sqrt{a^2+b^2}}{|b|}\sin^{-1}\frac{x\sqrt{a^2+b^2}}{|a|\sqrt{x^2+b^2}} - \sin^{-1}\frac{x}{|a|}$$

192. $$\int f(x, \sqrt{a^2-x^2})\,dx = a\int f(a\sin u, |a\cos u|)\cos u\,du, \qquad (x = a\sin u)$$

FORMS CONTAINING $\sqrt{a+bx+cx^2}$

$$X = a + bx + cx^2,\ q = 4\,ac - b^2, \text{ and } k = \frac{4c}{q}$$

For integrals 193–219, when $q = 0$, then $X = c\left(x + \frac{b}{2c}\right)^2$ and formulas starting with 24 should be used.

193. $$\int \frac{dx}{\sqrt{X}} = \frac{1}{\sqrt{c}}\log\left(\sqrt{X} + x\sqrt{c} + \frac{b}{2\sqrt{c}}\right) \quad \text{if } c > 0$$

194. $$\int \frac{dx}{\sqrt{X}} = \frac{1}{\sqrt{c}}\sinh^{-1}\left(\frac{2cx+b}{\sqrt{4ac-b^2}}\right) \quad \text{if } c > 0$$

195. $$\int \frac{dx}{\sqrt{X}} = \frac{1}{\sqrt{-c}} \sin^{-1}\left(\frac{-2cx - b}{\sqrt{b^2 - 4ac}}\right) \qquad \text{if } c < 0$$

196. $$\int \frac{dx}{X\sqrt{X}} = \frac{2(2cx + b)}{q\sqrt{X}}$$

197. $$\int \frac{dx}{X^2\sqrt{X}} = \frac{2(2cx + b)}{3q\sqrt{X}}\left(\frac{1}{X} + 2k\right)$$

198. $$\int \frac{dx}{X^n\sqrt{X}} = \begin{cases} \dfrac{2(2cx + b)\sqrt{X}}{(2n - 1)qX^n} + \dfrac{2k(n - 1)}{2n - 1}\displaystyle\int \frac{dx}{X^{n-1}\sqrt{X}} \\ \quad \text{or} \\ \dfrac{(2cx + b)(n!)(n - 1)!4^n k^{n-1}}{q[(2n)!]\sqrt{X}} \displaystyle\sum_{r=0}^{n-1} \frac{(2r)!}{(4kX)^r(r!)^2} \end{cases}$$

199. $$\int \sqrt{X}\,dx = \frac{(2cx + b)\sqrt{X}}{4c} + \frac{1}{2k}\int \frac{dx}{\sqrt{X}}$$

200. $$\int X\sqrt{X}\,dx = \frac{(2cx + b)\sqrt{X}}{8c}\left(X + \frac{3}{2k}\right) + \frac{3}{8k^2}\int \frac{dx}{\sqrt{X}}$$

201. $$\int X^2\sqrt{X}\,dx = \frac{(2cx + b)\sqrt{X}}{12c}\left(X^2 + \frac{5X}{4k} + \frac{15}{8k^2}\right) + \frac{5}{16k^3}\int \frac{dx}{\sqrt{X}}$$

202. $$\int X^n\sqrt{X}\,dx = \begin{cases} \dfrac{(2cx + b)X^n\sqrt{X}}{4(n + 1)c} + \dfrac{2n + 1}{2(n + 1)k}\displaystyle\int \frac{X^n\,dx}{\sqrt{X}} \\ \quad \text{or} \\ \dfrac{(2n + 2)!}{[(n + 1)!]^2(4k)^{n+1}}\left[\dfrac{k(2cx + b)\sqrt{X}}{c}\displaystyle\sum_{r=0}^{n} \frac{r!(r + 1)!(4kX)^r}{(2r + 2)!} + \int \frac{dx}{\sqrt{X}}\right] \end{cases}$$

203. $$\int \frac{x\,dx}{\sqrt{X}} = \frac{\sqrt{X}}{c} - \frac{b}{2c}\int \frac{dx}{\sqrt{X}}$$

204. $$\int \frac{x\,dx}{X\sqrt{X}} = -\frac{2(bx + 2a)}{q\sqrt{X}}$$

205. $$\int \frac{x\,dx}{X^n\sqrt{X}} = -\frac{\sqrt{X}}{(2n - 1)cX^n} - \frac{b}{2c}\int \frac{dx}{X^n\sqrt{X}}$$

206. $$\int \frac{x^2\,dx}{\sqrt{X}} = \left(\frac{x}{2c} - \frac{3b}{4c^2}\right)\sqrt{X} + \frac{3b^2 - 4ac}{8c^2}\int \frac{dx}{\sqrt{X}}$$

207. $$\int \frac{x^2\,dx}{X\sqrt{X}} = \frac{(2b^2 - 4ac)x + 2ab}{cq\sqrt{X}} + \frac{1}{c}\int \frac{dx}{\sqrt{X}}$$

208. $$\int \frac{x^2\,dx}{X^n\sqrt{X}} = \frac{(2b^2 - 4ac)x + 2ab}{(2n - 1)cqX^{n-1}\sqrt{X}} + \frac{4ac + (2n - 3)b^2}{(2n - 1)cq}\int \frac{dx}{X^{n-1}\sqrt{X}}$$

209. $$\int \frac{x^3\,dx}{\sqrt{X}} = \left(\frac{x^2}{3c} - \frac{5bx}{12c^2} + \frac{5b^2}{8c^3} - \frac{2a}{3c^2}\right)\sqrt{X} + \left(\frac{3ab}{4c^2} - \frac{5b^3}{16c^3}\right)\int \frac{dx}{\sqrt{X}}$$

210. $$\int x\sqrt{X}\,dx = \frac{X\sqrt{X}}{3c} - \frac{b}{2c}\int \sqrt{X}\,dx$$

211. $$\int xX\sqrt{X}\,dx = \frac{X^2\sqrt{X}}{5c} - \frac{b}{2c}\int X\sqrt{X}\,dx$$

212. $$\int \frac{xX^n\,dx}{\sqrt{X}} = \frac{X^n\sqrt{X}}{(2n+1)c} - \frac{b}{2c}\int \frac{X^n\,dx}{\sqrt{X}}$$

213. $$\int x^2\sqrt{X}\,dx = \left(x - \frac{5b}{6c}\right)\frac{X\sqrt{X}}{4c} + \frac{5b^2 - 4ac}{16c^2}\int \sqrt{X}\,dx$$

214. $$\int \frac{dx}{x\sqrt{X}} = -\frac{1}{\sqrt{a}}\log\left(\frac{\sqrt{X}+\sqrt{a}}{x} + \frac{b}{2\sqrt{a}}\right), \qquad (a > 0)$$

215. $$\int \frac{dx}{x\sqrt{X}} = \frac{1}{\sqrt{-a}}\sin^{-1}\left(\frac{bx+2a}{|x|\sqrt{-q}}\right), \qquad (a < 0)$$

216. $$\int \frac{dx}{x\sqrt{X}} = -\frac{2\sqrt{X}}{bx}, \qquad (a = 0)$$

217. $$\int \frac{dx}{x^2\sqrt{X}} = -\frac{\sqrt{X}}{ax} - \frac{b}{2a}\int \frac{dx}{x\sqrt{X}}$$

218. $$\int \frac{\sqrt{X}\,dx}{x} = \sqrt{X} + \frac{b}{2}\int \frac{dx}{\sqrt{X}} + a\int \frac{dx}{x\sqrt{X}}$$

219. $$\int \frac{\sqrt{X}\,dx}{x^2} = -\frac{\sqrt{X}}{x} + \frac{b}{2}\int \frac{dx}{x\sqrt{X}} + c\int \frac{dx}{\sqrt{X}}$$

FORMS INVOLVING $\sqrt{2ax - x^2}$

220. $$\int \sqrt{2ax - x^2}\,dx = \frac{1}{2}\left[(x-a)\sqrt{2ax - x^2} + a^2\sin^{-1}\frac{x-a}{|a|}\right]$$

221. $$\int \frac{dx}{\sqrt{2ax - x^2}} = \begin{cases} \cos^{-1}\dfrac{a-x}{|a|} \text{ or } \sin^{-1}\dfrac{x-a}{|a|} \\ \text{or} \\ 2\sin^{-1}\sqrt{\dfrac{x}{2a}} \text{ or } 2\cos^{-1}\sqrt{\dfrac{x}{2a}}, \ (a > 0) \end{cases}$$

222. $$\int x^n\sqrt{2ax - x^2}\,dx$$
$$= -\frac{x^{n-1}(2ax - x^2)^{\frac{3}{2}}}{n+2} + \frac{(2n+1)a}{n+2}\int x^{n-1}\sqrt{2ax - x^2}\,dx, \qquad (n \neq -2)$$

223. $$\int \frac{\sqrt{2ax - x^2}}{x^n}\,dx = \frac{(2ax - x^2)^{\frac{3}{2}}}{(3-2n)ax^n} + \frac{n-3}{(2n-3)a}\int \frac{\sqrt{2ax - x^2}}{x^{n-1}}\,dx, \qquad \left(n \neq \frac{3}{2}\right)$$

224. $\int \frac{x^n\,dx}{\sqrt{2ax-x^2}} = \frac{-x^{n-1}\sqrt{2ax-x^2}}{n} + \frac{a(2n-1)}{n}\int \frac{x^{n-1}}{\sqrt{2ax-x^2}}\,dx, \ (n \neq 0)$

225. $\int \frac{dx}{x^n\sqrt{2ax-x^2}} = \frac{\sqrt{2ax-x^2}}{a(1-2n)x^n} + \frac{n-1}{(2n-1)a}\int \frac{dx}{x^{n-1}\sqrt{2ax-x^2}}, \ \left(n \neq \frac{1}{2}\right)$

226. $\int \frac{dx}{(2ax-x^2)^{\frac{3}{2}}} = \frac{x-a}{a^2\sqrt{2ax-x^2}}$

227. $\int \frac{x\,dx}{(2ax-x^2)^{\frac{3}{2}}} = \frac{x}{a\sqrt{2ax-x^2}}$

228. $\int \frac{dx}{\sqrt{2ax+x^2}} = \log\left(x+a+\sqrt{2ax+x^2}\right)$

MISCELLANEOUS ALGEBRAIC FORMS

229. $$\int \sqrt{ax^2+c}\,dx = \begin{cases} \frac{x}{2}\sqrt{ax^2+c} + \frac{c}{2\sqrt{a}}\log\left(x\sqrt{a}+\sqrt{ax^2+c}\right), & (a>0) \\ \text{or} & \\ \frac{x}{2}\sqrt{ax^2+c} + \frac{c}{2\sqrt{-a}}\sin^{-1}\left(x\sqrt{\frac{-a}{c}}\right), & (a<0) \end{cases}$$

230. $$\int \frac{dx}{\sqrt{a+bx}\cdot\sqrt{a'+b'x}} = \begin{cases} \frac{2}{\sqrt{bb'}}\tanh^{-1}\frac{\sqrt{bb'uv}}{bv}, & (bb'>0) \\ \text{or} & \\ \frac{1}{\sqrt{bb'}}\log\frac{bv+\sqrt{bb'uv}}{bv-\sqrt{bb'uv}}, & (bb'>0) \\ \text{or} & \\ \frac{1}{\sqrt{bb'}}\log\frac{(bv+\sqrt{bb'uv})^2}{|v|} & (bb'>0) \\ \text{or} & \\ \frac{2}{\sqrt{-bb'}}\tan^{-1}\frac{\sqrt{-bb'uv}}{bv}, & (bb'<0) \end{cases}$$

(see integral 61)

231. $\int \sqrt{\frac{1+x}{1-x}}\,dx = \sin^{-1}x - \sqrt{1-x^2}$

232. $\int \frac{dx}{\sqrt{a \pm 2bx + cx^2}}$

$$= \frac{1}{\sqrt{c}}\log\left(\pm b + cx + \sqrt{c}\sqrt{a \pm 2bx + cx^2}\right), \qquad (b^2 - ac \neq 0)$$

233. $\int \frac{dx}{\sqrt{a \pm 2bx - cx^2}} = \frac{1}{\sqrt{c}}\sin^{-1}\frac{cx \mp b}{\sqrt{b^2+ac}}$

234. $\int \frac{x\,dx}{\sqrt{a \pm 2bx + cx^2}} = \frac{1}{c}\sqrt{a \pm 2bx + cx^2}$

$$\mp \frac{b}{\sqrt{c^3}}\log\left(\pm b + cx + \sqrt{c}\sqrt{a \pm 2bx + cx^2}\right), \qquad (b^2 - ac \neq 0)$$

235. $$\int \frac{x\,dx}{\sqrt{a \pm 2bx - cx^2}} = -\frac{1}{c}\sqrt{a \pm 2bx - cx^2} \pm \frac{b}{\sqrt{c^3}} \sin^{-1} \frac{cx \mp b}{\sqrt{b^2 + ac}}$$

FORMS INVOLVING TRIGONOMETRIC FUNCTIONS

236. $$\int \sin x\,dx = -\cos x$$

237. $$\int \cos x\,dx = \sin x$$

238. $$\int \tan x\,dx = -\log \cos x = \log \sec x$$

239. $$\int \cot x\,dx = \log \sin x = -\log \csc x$$

240. $$\int \sec x\,dx = \log(\sec x + \tan x) = \log \tan\left(\frac{\pi}{4} + \frac{x}{2}\right)$$

241. $$\int \csc x\,dx = \log(\csc x - \cot x) = \log \tan \frac{x}{2}$$

242. $$\int \sin^2 x\,dx = -\tfrac{1}{2}\cos x \sin x + \tfrac{1}{2}x = \tfrac{1}{2}x - \tfrac{1}{4}\sin 2x$$

243. $$\int \sin^3 x\,dx = -\tfrac{1}{3}\cos x\,(\sin^2 x + 2)$$

244. $$\int \sin^n x\,dx = -\frac{\sin^{n-1} x \cos x}{n} + \frac{n-1}{n}\int \sin^{n-2} x\,dx$$

245. $$\int \cos^2 x\,dx = \tfrac{1}{2}\sin x \cos x + \tfrac{1}{2}x = \tfrac{1}{2}x + \tfrac{1}{4}\sin 2x$$

246. $$\int \cos^3 x\,dx = \tfrac{1}{3}\sin x\,(\cos^2 x + 2)$$

247. $$\int \cos^n x\,dx = \frac{1}{n}\cos^{n-1} x \sin x + \frac{n-1}{n}\int \cos^{n-2} x\,dx$$

248. $$\int \sin \frac{x}{a}\,dx = -a \cos \frac{x}{a}$$

249. $$\int \cos \frac{x}{a}\,dx = a \sin \frac{x}{a}$$

250. $\int \sin(a+bx)\,dx = -\frac{1}{b}\cos(a+bx)$

251. $\int \cos(a+bx)\,dx = \frac{1}{b}\sin(a+bx)$

252. $$\int \frac{dx}{\sin x} = \begin{cases} \int \csc x\,dx = \log(\csc x - \cot x) \\ \text{or} \\ -\frac{1}{2}\log\frac{1+\cos x}{1-\cos x} = \log\tan\frac{x}{2} \end{cases}$$

253. $$\int \frac{dx}{\cos x} = \begin{cases} \int \sec x\,dx = \log(\sec x + \tan x) \\ \text{or} \\ \frac{1}{2}\log\left(\frac{1+\sin x}{1-\sin x}\right) = \log\tan\left(\frac{\pi}{4}+\frac{x}{2}\right) \end{cases}$$

254. $\int \frac{dx}{\cos^2 x} = \int \sec^2 x\,dx = \tan x$

255. $\int \frac{dx}{\cos^n x} = \frac{1}{n-1}\cdot\frac{\sin x}{\cos^{n-1} x} + \frac{n-2}{n-1}\int\frac{dx}{\cos^{n-2} x}$

256. $\int \frac{dx}{1 \pm \sin x} = \mp\tan\left(\frac{\pi}{4}\mp\frac{x}{2}\right)$

257. $\int \frac{dx}{1+\cos x} = \tan\frac{x}{2}$

258. $\int \frac{dx}{1-\cos x} = -\cot\frac{x}{2}$

259. $$\int \frac{dx}{a+b\sin x} = \begin{cases} \frac{2}{\sqrt{a^2-b^2}}\tan^{-1}\frac{a\tan\frac{x}{2}+b}{\sqrt{a^2-b^2}} \\ \text{or} \\ \frac{1}{\sqrt{b^2-a^2}}\log\frac{a\tan\frac{x}{2}+b-\sqrt{b^2-a^2}}{a\tan\frac{x}{2}+b+\sqrt{b^2-a^2}} \end{cases}$$

260. $$\int \frac{dx}{a+b\cos x} = \begin{cases} \frac{2}{\sqrt{a^2-b^2}}\tan^{-1}\frac{\sqrt{a^2-b^2}\tan\frac{x}{2}}{a+b} \\ \text{or} \\ \frac{1}{\sqrt{b^2-a^2}}\log\left(\frac{\sqrt{b^2-a^2}\tan\frac{x}{2}+a+b}{\sqrt{b^2-a^2}\tan\frac{x}{2}-a-b}\right) \end{cases}$$

261. $$\int \frac{dx}{a + b\sin x + c\cos x}$$

$$= \begin{cases} \dfrac{1}{\sqrt{b^2 + c^2 - a^2}} \log \dfrac{b - \sqrt{b^2 + c^2 - a^2} + (a - c)\tan \frac{x}{2}}{b + \sqrt{b^2 + c^2 - a^2} + (a - c)\tan \frac{x}{2}}, & \text{if } a^2 < b^2 + c^2, \; a \neq c. \\ \text{or} & \\ \dfrac{2}{\sqrt{a^2 - b^2 - c^2}} \tan^{-1} \dfrac{b + (a - c)\tan \frac{x}{2}}{\sqrt{a^2 - b^2 - c^2}}, & \text{if } a^2 > b^2 + c^2 \\ \text{or} & \\ \dfrac{1}{a}\left[\dfrac{a - (b + c)\cos x - (b - c)\sin x}{a - (b - c)\cos x + (b + c)\sin x}\right], & \text{if } a^2 = b^2 + c^2, \; a \neq c. \end{cases}$$

262. $$\int \frac{\sin^2 x\, dx}{a + b\cos^2 x} = \frac{1}{b}\sqrt{\frac{a + b}{a}} \tan^{-1}\left(\sqrt{\frac{a}{a + b}} \tan x\right) - \frac{x}{b},$$
$[ab > 0, \text{ or } |a| > |b|]$

263. $$\int \frac{dx}{a^2\cos^2 x + b^2 \sin^2 x} = \frac{1}{ab}\tan^{-1}\left(\frac{b\tan x}{a}\right)$$

264. $$\int \sqrt{1 - \cos x}\, dx = \pm 2\sqrt{2}\cos \frac{x}{2},$$
[use + when $(4k - 2)\pi < x \leq 4k\pi$, otherwise −; k an integer]

265. $$\int \sqrt{1 + \cos x}\, dx = \pm 2\sqrt{2}\sin \frac{x}{2},$$
[use + when $(4k - 1)\pi < x \leq (4k + 1)\pi$, otherwise −; k an integer]

266. $$\int \sqrt{1 + \sin x}\, dx = \pm 2\left(\sin \frac{x}{2} - \cos \frac{x}{2}\right),$$
[use + if $(8k - 1)\frac{\pi}{2} < x \leq (8k + 3)\frac{\pi}{2}$, otherwise −; k an integer]

267. $$\int \sqrt{1 - \sin x}\, dx = \pm 2\left(\sin \frac{x}{2} + \cos \frac{x}{2}\right),$$
[use + if $(8k - 3)\frac{\pi}{2} < x \leq (8k + 1)\frac{\pi}{2}$, otherwise −; k an integer]

268. $$\int \frac{dx}{\sqrt{1 - \cos x}} = \pm \sqrt{2}\log \tan \frac{x}{4},$$
[use + if $4k\pi < x < (4k + 2)\pi$, otherwise −; k an integer]

269. $$\int \frac{dx}{\sqrt{1 + \cos x}} = \pm \sqrt{2}\log \tan\left(\frac{x + \pi}{4}\right),$$
[use + if $(4k - 1)\pi < x < (4k + 1)\pi$, otherwise −; k an integer]

270. $$\int \frac{dx}{\sqrt{1 - \sin x}} = \pm \sqrt{2}\log \tan\left(\frac{x}{4} - \frac{\pi}{8}\right)$$
[use + if $(8k + 1)\frac{\pi}{2} < x < (8k + 5)\frac{\pi}{2}$, otherwise −; k an integer]

271. $\int \frac{dx}{\sqrt{1+\sin x}} = \pm\sqrt{2}\log\tan\left(\frac{x}{4}+\frac{\pi}{8}\right),$

[use + if $(8k-1)\frac{\pi}{2} < x < (8k+3)\frac{\pi}{2}$, otherwise −; k an integer]

272. $\int \sin mx \sin nx\, dx = \frac{\sin(m-n)x}{2(m-n)} - \frac{\sin(m+n)x}{2(m+n)}, \quad [m^2 \neq n^2]$

273. $\int x\sin^2 x\, dx = \frac{x^2}{4} - \frac{x\sin 2x}{4} - \frac{\cos 2x}{8}$

274. $\int x^2\sin^2 x\, dx = \frac{x^3}{6} - \left(\frac{x^2}{4}-\frac{1}{8}\right)\sin 2x - \frac{x\cos 2x}{4}$

275. $\int x\sin^3 x\, dx = \frac{x\cos 3x}{12} - \frac{\sin 3x}{36} - \frac{3}{4}x\cos x + \frac{3}{4}\sin x$

276. $\int \sin^4 x\, dx = \frac{3x}{8} - \frac{\sin 2x}{4} + \frac{\sin 4x}{32}$

277. $\int \cos mx \cos nx\, dx = \frac{\sin(m-n)x}{2(m-n)} + \frac{\sin(m+n)x}{2(m+n)}, \quad [m^2 \neq n^2]$

278. $\int x\cos^2 x\, dx = \frac{x^2}{4} + \frac{x\sin 2x}{4} + \frac{\cos 2x}{8}$

279. $\int x^2\cos^2 x\, dx = \frac{x^3}{6} + \left(\frac{x^2}{4}-\frac{1}{8}\right)\sin 2x + \frac{x\cos 2x}{4}$

280. $\int x\cos^3 x\, dx = \frac{x\sin 3x}{12} + \frac{\cos 3x}{36} + \frac{3}{4}x\sin x + \frac{3}{4}\cos x$

281. $\int \cos^4 x\, dx = \frac{3x}{8} + \frac{\sin 2x}{4} + \frac{\sin 4x}{32}$

282. $\int \frac{\sin x\, dx}{x^m} = -\frac{\sin x}{(m-1)x^{m-1}} + \frac{1}{m-1}\int \frac{\cos x\, dx}{x^{m-1}}$

283. $\int \frac{\cos x\, dx}{x^m} = -\frac{\cos x}{(m-1)x^{m-1}} - \frac{1}{m-1}\int \frac{\sin x\, dx}{x^{m-1}}$

284. $\int \tan^3 x\, dx = \frac{1}{2}\tan^2 x + \log\cos x$

285. $\int \tan^n x\, dx = \frac{\tan^{n-1} x}{n-1} - \int \tan^{n-2} x\, dx$

286. $\int \cot^3 x\, dx = -\frac{1}{2}\cot^2 x - \log\sin x$

287. $\int \cot^4 x\, dx = -\frac{1}{3}\cot^3 x + \cot x + x$

288. $\int \cot^n x\, dx = -\frac{\cot^{n-1} x}{n-1} - \int \cot^{n-2} x\, dx, \quad [n \neq 1]$

289. $$\int \sin x \cos x \, dx = \tfrac{1}{2} \sin^2 x$$

290. $$\int \sin mx \cos nx \, dx = -\frac{\cos (m-n)x}{2(m-n)} - \frac{\cos (m+n)x}{2(m+n)}, \qquad (m^2 \neq n^2)$$

291. $$\int \sin^2 x \cos^2 x \, dx = -\tfrac{1}{8}(\tfrac{1}{4} \sin 4x - x)$$

292. $$\int \sin x \cos^m x \, dx = -\frac{\cos^{m+1} x}{m+1}$$

293. $$\int \sin^m x \cos x \, dx = \frac{\sin^{m+1} x}{m+1}$$

294. $$\int \cos^m x \sin^n x \, dx = \frac{\cos^{m-1} x \sin^{n+1} x}{m+n} + \frac{m-1}{m+n} \int \cos^{m-2} x \sin^n x \, dx, \qquad (m \neq -n)$$

295. $$\int \cos^m x \sin^n x \, dx = -\frac{\sin^{n-1} x \cos^{m+1} x}{m+n} + \frac{n-1}{m+n} \int \cos^m x \sin^{n-2} x \, dx, \qquad (m \neq -n)$$

296. $$\int \frac{\cos^m x \, dx}{\sin^n x} = -\frac{\cos^{m+1} x}{(n-1) \sin^{n-1} x} - \frac{m-n+2}{n-1} \int \frac{\cos^m x \, dx}{\sin^{n-2} x}$$

297. $$\int \frac{\cos^m x \, dx}{\sin^n x} = \frac{\cos^{m-1} x}{(m-n) \sin^{n-1} x} + \frac{m-1}{m-n} \int \frac{\cos^{m-2} x \, dx}{\sin^n x}, \qquad (m \neq n)$$

298. $$\int \frac{\sin^m x \, dx}{\cos^n x} = -\int \frac{\cos^m \left(\frac{\pi}{2} - x\right) d\left(\frac{\pi}{2} - x\right)}{\sin^n \left(\frac{\pi}{2} - x\right)}$$

299. $$\int \frac{\sin x \, dx}{\cos^2 x} = \frac{1}{\cos x} = \sec x$$

300. $$\int \frac{\sin^2 x \, dx}{\cos x} = -\sin x + \log \tan \left(\frac{\pi}{4} + \frac{x}{2}\right)$$

301. $$\int \frac{\cos x \, dx}{\sin^2 x} = \frac{-1}{\sin x} = -\operatorname{cosec} x$$

302. $$\int \frac{dx}{\sin x \cos x} = \log \tan x$$

303. $$\int \frac{dx}{\sin x \cos^2 x} = \frac{1}{\cos x} + \log \tan \frac{x}{2}$$

304. $$\int \frac{dx}{\sin x \cos^n x} = \frac{1}{(n-1) \cos^{n-1} x} + \int \frac{dx}{\sin x \cos^{n-2} x}, \qquad (n \neq 1)$$

305. $$\int \frac{dx}{\sin^2 x \cos x} = -\frac{1}{\sin x} + \log \tan \left(\frac{\pi}{4} + \frac{x}{2}\right)$$

306. $$\int \frac{dx}{\sin^2 x \cos^2 x} = -2 \cot 2x$$

307. $$\int \frac{dx}{\sin^m x \cos^n x} = -\frac{1}{m-1} \cdot \frac{1}{\sin^{m-1} x \cdot \cos^{n-1} x} + \frac{m+n-2}{m-1} \int \frac{dx}{\sin^{m-2} x \cdot \cos^n x}$$

308. $$\int \frac{dx}{\sin^m x} = -\frac{1}{m-1} \cdot \frac{\cos x}{\sin^{m-1} x} + \frac{m-2}{m-1} \int \frac{dx}{\sin^{m-2} x}$$

309. $$\int \frac{dx}{\sin^2 x} = -\cot x$$

310. $$\int \tan^2 x\, dx = \tan x - x$$

311. $$\int \tan^n x\, dx = \frac{\tan^{n-1} x}{n-1} - \int \tan^{n-2} x\, dx, \qquad (n \neq 1)$$

312. $$\int \cot^2 x\, dx = -\cot x - x$$

313. $$\int \cot^n x\, dx = -\frac{\cot^{n-1} x}{n-1} - \int \cot^{n-2} x\, dx$$

314. $$\int \sec^2 x\, dx = \tan x$$

315. $$\int \sec^n x\, dx = \int \frac{dx}{\cos^n x} = \frac{1}{n-1} \frac{\sin x}{\cos^{n-1} x} + \frac{n-2}{n-1} \int \frac{dx}{\cos^{n-2} x}$$

316. $$\int \csc^2 x\, dx = -\cot x$$

317. $$\int \csc^n x\, dx = \int \frac{dx}{\sin^n x} = -\frac{1}{n-1} \frac{\cos x}{\sin^{n-1} x} + \frac{n-2}{n-1} \int \frac{dx}{\sin^{n-2} x}$$

318. $$\int x \sin x\, dx = \sin x - x \cos x$$

319. $$\int x \sin (ax)\, dx = \frac{1}{a^2} \sin (ax) - \frac{x}{a} \cos (ax)$$

320. $$\int x^2 \sin x\, dx = 2x \sin x - (x^2 - 2) \cos x$$

321. $$\int x^2 \sin (ax)\, dx = \frac{2x}{a^2} \sin (ax) - \frac{a^2x^2 - 2}{a^3} \cos (ax)$$

322. $\int x^3 \sin x \, dx = (3x^2 - 6) \sin x - (x^3 - 6x) \cos x$

323. $\int x^3 \sin (ax) \, dx = \frac{3a^2x^2 - 6}{a^4} \sin (ax) - \frac{a^2x^3 - 6x}{a^3} \cos (ax)$

324. $\int x^m \sin x \, dx = -x^m \cos x + m \int x^{m-1} \cos x \, dx$

325. $\int x^m \sin (ax) \, dx = -\frac{1}{a} x^m \cos (ax) + \frac{m}{a} \int x^{m-1} \cos (ax) \, dx$

326. $\int x \cos x \, dx = \cos x + x \sin x$

327. $\int x \cos (ax) \, dx = \frac{1}{a^2} \cos (ax) + \frac{x}{a} \sin (ax)$

328. $\int x^2 \cos x \, dx = 2x \cos x + (x^2 - 2) \sin x$

329. $\int x^2 \cos (ax) \, dx = \frac{2x \cos (ax)}{a^2} + \frac{a^2x^2 - 2}{a^3} \sin (ax)$

330. $\int x^3 \cos x \, dx = (3x^2 - 6) \cos x + (x^3 - 6x) \sin x$

331. $\int x^3 \cos (ax) \, dx = \frac{(3a^2x^2 - 6)}{a^4} \cos (ax) + \frac{a^2x^3 - 6x}{a^3} \sin (ax)$

332. $\int x^m \cos x \, dx = x^m \sin x - m \int x^{m-1} \sin x \, dx$

333. $\int x^m \cos (ax) \, dx = \frac{1}{a} x^m \sin (ax) - \frac{m}{a} \int x^{m-1} \sin (ax) \, dx$

334. $\int \frac{\sin x}{x} \, dx = x - \frac{x^3}{3 \cdot 3!} + \frac{x^5}{5 \cdot 5!} - \frac{x^7}{7 \cdot 7!} + \frac{x^9}{9 \cdot 9!} \cdots$

335. $\int \frac{\sin (ax)}{x} \, dx = ax - \frac{a^3x^3}{3 \cdot 3!} + \frac{a^5x^5}{5 \cdot 5!} - \frac{a^7x^7}{7 \cdot 7!} + \frac{a^9x^9}{9 \cdot 9!} + - \cdots$

336. $\int \frac{\cos x}{x} \, dx = \log x - \frac{x^2}{2 \cdot 2!} + \frac{x^4}{4 \cdot 4!} - \frac{x^6}{6 \cdot 6!} + \frac{x^8}{8 \cdot 8!} \cdots$

337. $\int \frac{\cos (ax)}{x} \, dx = \log x - \frac{a^2x^2}{2 \cdot 2!} + \frac{a^4x^4}{4 \cdot 4!} - \frac{a^6x^6}{6 \cdot 6!} + \frac{a^8x^8}{8 \cdot 8!} - + \cdots$

FORMS INVOLVING INVERSE TRIGONOMETRIC FUNCTIONS

338. $\int \sin^{-1} x \, dx = x \sin^{-1} x + \sqrt{1 - x^2}$

339. $\int \cos^{-1} x \, dx = x \cos^{-1} x - \sqrt{1 - x^2}$

340. $\int \tan^{-1} x\, dx = x \tan^{-1} x - \frac{1}{2} \log (1 + x^2)$

341. $\int \cot^{-1} x\, dx = x \cot^{-1} x + \frac{1}{2} \log (1 + x^2)$

342. $\int \sec^{-1} x\, dx = x \sec^{-1} x - \log (x + \sqrt{x^2 - 1})$

343. $\int \csc^{-1} x\, dx = x \csc^{-1} x + \log (x + \sqrt{x^2 - 1})$

344. $\int \operatorname{vers}^{-1} x\, dx = (x - 1) \operatorname{vers}^{-1} x + \sqrt{2x - x^2}$

345. $\int \sin^{-1} \frac{x}{a}\, dx = x \sin^{-1} \frac{x}{a} + \sqrt{a^2 - x^2}, \qquad (a > 0)$

346. $\int \cos^{-1} \frac{x}{a}\, dx = x \cos^{-1} \frac{x}{a} - \sqrt{a^2 - x^2}, \qquad (a > 0)$

347. $\int \tan^{-1} \frac{x}{a}\, dx = x \tan^{-1} \frac{x}{a} - \frac{a}{2} \log (a^2 + x^2)$

348. $\int \cot^{-1} \frac{x}{a}\, dx = x \cot^{-1} \frac{x}{a} + \frac{a}{2} \log (a^2 + x^2)$

349. $\int (\sin^{-1} x)^2\, dx = x (\sin^{-1} x)^2 - 2x + 2 \sqrt{1 - x^2} (\sin^{-1} x)$

350. $\int (\cos^{-1} x)^2\, dx = x (\cos^{-1} x)^2 - 2x - 2 \sqrt{1 - x^2} (\cos^{-1} x)$

351. $\int x \sin^{-1} x\, dx = \frac{1}{4}[(2x^2 - 1) \sin^{-1} x + x \sqrt{1 - x^2}]$

352. $\int x \sin^{-1} (ax)\, dx = \frac{1}{4a^2} [(2a^2x^2 - 1) \sin^{-1} (ax) + ax \sqrt{1 - a^2x^2}]$

353. $\int x \cos^{-1} x\, dx = \frac{1}{4}[(2x^2 - 1) \cos^{-1} x - x \sqrt{1 - x^2}]$

354. $\int x \cos^{-1} (ax)\, dx = \frac{1}{4a^2} [(2a^2x^2 - 1) \cos^{-1} (ax) - ax \sqrt{1 - a^2x^2}]$

355. $\int x^n \sin^{-1} x\, dx = \frac{x^{n+1} \sin^{-1} x}{n + 1} - \frac{1}{n + 1} \int \frac{x^{n+1}\, dx}{\sqrt{1 - x^2}}$

356. $\int x^n \sin^{-1} (ax)\, dx = \frac{x^{n+1}}{n + 1} \sin^{-1} (ax) - \frac{a}{n + 1} \int \frac{x^{n+1}\, dx}{\sqrt{1 - a^2x^2}}, \qquad (n \neq -1)$

357. $\int x^n \cos^{-1} x\, dx = \frac{x^{n+1} \cos^{-1} x}{n + 1} + \frac{1}{n + 1} \int \frac{x^{n+1}\, dx}{\sqrt{1 - x^2}}$

358. $\int x^n \cos^{-1} (ax)\, dx = \frac{x^{n+1}}{n + 1} \cos^{-1} (ax) + \frac{a}{n + 1} \int \frac{x^{n+1}\, dx}{\sqrt{1 - a^2x^2}}, \qquad (n \neq -1)$

359. $\int x \tan^{-1} x\, dx = \frac{1}{2}(1 + x^2) \tan^{-1} x - \frac{x}{2}$

360. $\int x^n \tan^{-1}(ax)\, dx = \frac{x^{n+1}}{n+1} \tan^{-1}(ax) - \frac{a}{n+1} \int \frac{x^{n+1}\, dx}{1 + a^2x^2}, \qquad (n \neq -1)$

361. $\int x \cot^{-1} x\, dx = \frac{1}{2}(1 + x^2) \cot^{-1} x + \frac{x}{2}$

362. $\int x^n \cot^{-1} x\, dx = \frac{x^{n+1}}{n+1} \cot^{-1} x + \frac{1}{n+1} \int \frac{x^{n+1}}{1 + x^2}\, dx$

363. $\int \frac{\sin^{-1} x\, dx}{x^2} = \log\left(\frac{1 - \sqrt{1 - x^2}}{x}\right) - \frac{\sin^{-1} x}{x}$

364. $\int \frac{\sin^{-1}(ax)}{x^2}\, dx = a \log\left(\frac{1 - \sqrt{1 - a^2x^2}}{x}\right) - \frac{\sin^{-1}(ax)}{x}$

365. $\int \frac{\cos^{-1}(ax)}{x}\, dx = \frac{\pi}{2} \log x - ax - \frac{1}{2 \cdot 3 \cdot 3}(ax)^3$

$$- \frac{1 \cdot 3}{2 \cdot 4 \cdot 5 \cdot 5}(ax)^5 - \frac{1 \cdot 3 \cdot 5}{2 \cdot 4 \cdot 6 \cdot 7 \cdot 7}(ax)^7 - \cdots$$

366. $\int \frac{\cos^{-1}(ax)}{x^2}\, dx = -\frac{1}{x} \cos^{-1}(ax) + a \log \frac{1 + \sqrt{1 - a^2x^2}}{x}$

367. $\int \frac{\tan^{-1} x\, dx}{x^2} = \log x - \frac{1}{2} \log(1 + x^2) - \frac{\tan^{-1} x}{x}$

368. $\int \frac{\tan^{-1}(ax)}{x^2}\, dx = -\frac{1}{x} \tan^{-1}(ax) - \frac{a}{2} \log \frac{1 + a^2x^2}{x^2}$

FORMS INVOLVING TRIGONOMETRIC SUBSTITUTIONS

369. $\int f(\sin x)\, dx = 2 \int f\left(\frac{2z}{1 + z^2}\right) \cdot \frac{dz}{1 + z^2}, \qquad \left(z = \tan \frac{x}{2}\right)$

370. $\int f(\cos x)\, dx = 2 \int f\left(\frac{1 - z^2}{1 + z^2}\right) \frac{dz}{1 + z^2}, \qquad \left(z = \tan \frac{x}{2}\right)$

371.* $\int f(\sin x)\, dx = \int f(u) \frac{du}{\sqrt{1 - u^2}}, \qquad (u = \sin x)$

372.* $\int f(\cos x)\, dx = -\int f(u) \frac{du}{\sqrt{1 - u^2}}, \qquad (u = \cos x)$

373.* $\int f(\sin x, \cos x)\, dx = \int f(u, \sqrt{1 - u^2}) \frac{du}{\sqrt{1 - u^2}}, \qquad (u = \sin x)$

* The square roots appearing in these formulas may be plus or minus, depending on the quadrant of x. Care must be used to give them the proper sign.

374. $\int f(\sin x, \cos x)\, dx = 2 \int f\left(\frac{2z}{1 + z^2}, \frac{1 - z^2}{1 + z^2}\right) \frac{dz}{1 + z^2}, \qquad \left(z = \tan \frac{x}{2}\right)$

375. $\int \frac{dx}{a + b \tan x} = \frac{1}{a^2 + b^2}[ax + b \log(a \cos x + b \sin x)]$

376. $$\int \frac{dx}{a + b \cot x} = \frac{1}{a^2 + b^2}[ax - b \log (a \sin x + b \cos x)]$$

LOGARITHMIC FORMS

377. $$\int \log x \, dx = x \log x - x$$

378. $$\int x \log x \, dx = \frac{x^2}{2} \log x - \frac{x^2}{4}$$

379. $$\int x^2 \log x \, dx = \frac{x^3}{3} \log x - \frac{x^3}{9}$$

380. $$\int (\log X) \, dx = \begin{cases} \left(x + \frac{b}{2c}\right) \log X - 2x + \frac{\sqrt{4ac - b^2}}{c} \tan^{-1} \frac{2cx + b}{\sqrt{4ac - b^2}}, \\ \qquad (b^2 - 4ac < 0) \\ \text{or} \\ \left(x + \frac{b}{2c}\right) \log X - 2x + \frac{\sqrt{b^2 - 4ac}}{c} \tanh^{-1} \frac{2cx + b}{\sqrt{b^2 - 4ac}}, \\ \qquad (b^2 - 4ac > 0) \\ \text{where} \\ X = a + bx + cx^2 \end{cases}$$

381. $$\int x^p \log (ax) \, dx = \frac{x^{p+1}}{p + 1} \log (ax) - \frac{x^{p+1}}{(p + 1)^2}, \qquad (p \neq -1)$$

382. $$\int x^n \log X \, dx = \frac{x^{n+1}}{n + 1} \log X - \frac{2c}{n + 1} \int \frac{x^{n+2}}{X} dx - \frac{b}{n + 1} \int \frac{x^{n+1}}{X} dx$$
where $X = a + bx + cx^2$

383. $$\int (\log x)^2 \, dx = x (\log x)^2 - 2x \log x + 2x$$

384. $$\int (\log x)^n \, dx = x (\log x)^n - n \int (\log x)^{n-1} \, dx, \qquad (n \neq -1)$$

385. $$\int \frac{(\log x)^n}{x} dx = \frac{1}{n + 1} (\log x)^{n+1}$$

386. $$\int \frac{dx}{\log x} = \log (\log x) + \log x + \frac{(\log x)^2}{2 \cdot 2!} + \frac{(\log x)^3}{3 \cdot 3!} + \cdots$$

387. $$\int \frac{dx}{x \log x} = \log (\log x)$$

388. $$\int \frac{dx}{x(\log x)^n} = -\frac{1}{(n - 1)(\log x)^{n-1}}$$

389. $$\int \frac{x^m \, dx}{(\log x)^n} = -\frac{x^{m+1}}{(n - 1)(\log x)^{n-1}} + \frac{m + 1}{n - 1} \int \frac{x^m \, dx}{(\log x)^{n-1}}$$

390. $$\int x^m \log x \, dx = x^{m+1} \left[\frac{\log x}{m + 1} - \frac{1}{(m + 1)^2}\right]$$

391. $\int x^m(\log x)^n\,dx = \frac{x^{m+1}(\log x)^n}{m+1} - \frac{n}{m+1}\int x^m(\log x)^{n-1}\,dx, \qquad [m \neq -1]$

392. $\int \sin\log x\,dx = \tfrac{1}{2}x\sin\log x - \tfrac{1}{2}x\cos\log x$

393. $\int \cos\log x\,dx = \tfrac{1}{2}x\sin\log x + \tfrac{1}{2}x\cos\log x$

EXPONENTIAL FORMS

394. $\int e^x\,dx = e^x$

395. $\int e^{-x}\,dx = -e^{-x}$

396. $\int e^{ax}\,dx = \frac{e^{ax}}{a}$

397. $\int x\,e^{ax}\,dx = \frac{e^{ax}}{a^2}(ax-1)$

398. $$\int x^m\,e^{ax}\,dx = \begin{cases} \dfrac{x^m e^{ax}}{a} - \dfrac{m}{a}\displaystyle\int x^{m-1}e^{ax}\,dx \\ \text{or} \\ e^{ax}\displaystyle\sum_{r=0}^{m}(-1)^r\frac{m!x^{m-r}}{(m-r)!a^{r+1}} \end{cases}$$

399. $\int \frac{e^{ax}\,dx}{x} = \log x + \frac{ax}{1!} + \frac{a^2x^2}{2\cdot 2!} + \frac{a^3x^3}{3\cdot 3!} + \cdots$

400. $\int \frac{e^{ax}}{x^m}\,dx = -\frac{1}{m-1}\frac{e^{ax}}{x^{m-1}} + \frac{a}{m-1}\int \frac{e^{ax}}{x^{m-1}}\,dx$

401. $\int e^{ax}\log x\,dx = \frac{e^{ax}\log x}{a} - \frac{1}{a}\int\frac{e^{ax}}{x}\,dx$

402. $\int \frac{dx}{1+e^x} = x - \log(1+e^x) = \log\frac{e^x}{1+e^x}$

403. $\int \frac{dx}{a+be^{px}} = \frac{x}{a} - \frac{1}{ap}\log(a+be^{px})$

404. $\int \frac{dx}{ae^{mx}+be^{-mx}} = \frac{1}{m\sqrt{ab}}\tan^{-1}\left(e^{mx}\sqrt{\frac{a}{b}}\right), \qquad (a>0,\ b>0)$

405. $$\int \frac{dx}{ae^{mx}-be^{-mx}} = \begin{cases} \dfrac{1}{2m\sqrt{ab}}\log\dfrac{\sqrt{a}e^{mx}-\sqrt{b}}{\sqrt{a}e^{mx}+\sqrt{b}} \\ \text{or} \\ \dfrac{-1}{m\sqrt{ab}}\tanh^{-1}\left(\sqrt{\dfrac{a}{b}}\,e^{mx}\right) \\ \text{or} \\ -\dfrac{1}{m\sqrt{ab}}\coth^{-1}\left(\sqrt{\dfrac{a}{b}}\,e^{mx}\right), \qquad (a>0,\ b>0) \end{cases}$$

406. $\int (a^x - a^{-x}) (\log a) dx = a^x + a^{-x}$

407. $\int e^{ax} \sin (bx)\, dx = \frac{e^{ax}[a \sin (bx) - b \cos (bx)]}{a^2 + b^2}$

408. $\int e^{ax} \sin (bx) \sin (cx)\, dx = \frac{e^{ax}[(b - c) \sin (b - c)x + a \cos (b - c)x]}{2[a^2 + (b - c)^2]} - \frac{e^{ax}[(b + c) \sin (b + c)x + a \cos (b + c)x]}{2[a^2 + (b + c)^2]}$

409. $\int e^{ax} \sin (bx) \cos (cx)\, dx = \frac{e^{ax}[a \sin (b - c)x - (b - c) \cos (b - c)x]}{2[a^2 + (b - c)^2]} + \frac{e^{ax}[a \sin (b + c)x - (b + c) \cos (b + c)x]}{2[a^2 + (b + c)^2]}$

410. $\int e^{ax} \sin (bx) \sin (bx + c)\, dx$

$$= \frac{e^{ax} \cos c}{2a} - \frac{e^{ax}[a \cos (2bx + c) + 2b \sin (2bx + c)]}{2(a^2 + 4b^2)}$$

411. $\int e^{ax} \sin (bx) \cos (bx + c)\, dx$

$$= \frac{-e^{ax} \sin c}{2a} + \frac{e^{ax}[a \sin (2bx + c) - 2b \cos (2bx + c)]}{2(a^2 + 4b^2)}$$

412. $\int e^{ax} \cos (bx)\, dx = \frac{e^{ax}}{a^2 + b^2} [a \cos (bx) + b \sin (bx)]$

413. $\int e^{ax} \cos (bx) \cos (cx)\, dx = \frac{e^{ax}[(b - c) \sin (b - c)x + a \cos (b - c)x]}{2[a^2 + (b - c)^2]} + \frac{e^{ax}[(b + c) \sin (b + c)x + a \cos (b + c)x]}{2[a^2 + (b + c)^2]}$

414. $\int e^{ax} \cos (bx) \cos (bx + c)\, dx$

$$= \frac{e^{ax} \cos c}{2a} + \frac{e^{ax}[a \cos (2bx + c) + 2b \sin (2bx + c)]}{2(a^2 + 4b^2)}$$

415. $\int e^{ax} \cos (bx) \sin (bx + c)\, dx$

$$= \frac{e^{ax} \sin c}{2a} + \frac{e^{ax}[a \sin (2bx + c) - 2b \cos (2bx + c)]}{2(a^2 + 4b^2)}$$

416. $\int e^{ax} \sin^n bx\, dx = \frac{1}{a^2 + n^2b^2} \Big[(a \sin bx - nb \cos bx) e^{ax} \sin^{n-1} bx + n(n - 1)b^2 \int e^{ax} \sin^{n-2} bx \cdot dx \Big]$

417. $\int e^{ax} \cos^n bx\, dx = \frac{1}{a^2 + n^2b^2} \Big[(a \cos bx + nb \sin bx) e^{ax} \cos^{n-1} bx + n(n - 1)b^2 \int e^{ax} \cos^{n-2} bx\, dx \Big]$

418. $\int x^m e^x \sin x\, dx = \frac{1}{2} x^m e^x(\sin x - \cos x) - \frac{m}{2} \int x^{m-1} e^x \sin x\, dx + \frac{m}{2} \int x^{m-1} e^x \cos x\, dx$

419. $$\int x^m e^{ax} \sin bx\, dx = \begin{cases} = x^m e^{ax} \dfrac{a \sin bx - b \cos bx}{a^2 + b^2} - \dfrac{m}{a^2 + b^2} \displaystyle\int x^{m-1} e^{ax}(a \sin bx - b \cos bx)\, dx \\ \text{or} \\ = e^{ax} \left[\dfrac{1}{\rho} x^m \sin (bx - \alpha) - \dfrac{m}{\rho^2} x^{m-1} \sin (bx - 2\alpha) \pm \dfrac{m(m-1) \cdots 1}{\rho^{m+1}} \sin \{bx - (m+1)\alpha\}\right] \\ \text{where } a + b\sqrt{-1} = \rho(\cos \alpha + \sqrt{-1} \sin \alpha) \end{cases}$$

420. $\int x^m e^x \cos x\, dx = \frac{1}{2} x^m e^x(\sin x + \cos x) - \frac{m}{2} \int x^{m-1} e^x \sin x\, dx - \frac{m}{2} \int x^{m-1} e^x \cos x\, dx$

421. $$\int x^m e^{ax} \cos bx\, dx = \begin{cases} x^m e^{ax} \dfrac{a \cos bx + b \sin bx}{a^2 + b^2} - \dfrac{m}{a^2 + b^2} \displaystyle\int x^{m-1} e^{ax}(a \cos bx + b \sin bx)\, dx \\ \text{or} \\ e^{ax} \left[\dfrac{1}{\rho} x^m \cos (bx - \alpha) - \dfrac{m}{\rho^2} x^{m-1} \cos (bx - 2\alpha) + \cdots \pm \dfrac{m(m-1) \cdots 1}{\rho^{m+1}} \cos (bx - (m+1)\alpha)\right] \\ \text{where } a + b\sqrt{-1} = \rho(\cos \alpha + \sqrt{-1} \sin \alpha) \end{cases}$$

422. $\int e^{ax} \sin (bx) \cos (cx)\, dx =$

$$\frac{e^{ax}}{\rho} [(a \sin (bx) - b \cos (bx)) \cos (cx - \alpha) - c \sin (bx) \sin (cx - \alpha)]$$

where $\rho = \sqrt{(a^2 + b^2 - c^2)^2 + 4a^2c^2}$
$\rho \cos \alpha = a^2 + b^2 - c^2$
$\rho \sin \alpha = 2ac$

423. $$\int e^{ax} \cos^m x \sin^n x\, dx = \frac{e^{ax} \cos^{m-1} x \sin^n x\{a \cos x + (m+n) \sin x\}}{(m+n)^2 + a^2} - \frac{na}{(m+n)^2 + a^2} \int e^{ax} \cos^{m-1} x \sin^{n-1} x\, dx + \frac{(m-1)(m+n)}{(m+n)^2 + a^2} \int e^{ax} \cos^{m-2} x \sin^n x\, dx$$

423. (Continued)

or

$$= \frac{e^{ax}\cos^m x \sin^{n-1} x\{a \sin x - (m+n)\cos x\}}{(m+n)^2 + a^2} + \frac{ma}{(m+n)^2 + a^2}\int e^{ax}\cos^{m-1} x \sin^{n-1} x\, dx + \frac{(n-1)(m+n)}{(m+n)^2 + a^2}\int e^{ax}\cos^m x \sin^{n-2} x\, dx$$

or

$$= \frac{e^{ax}\cos^{m-1} x \sin^{n-1} x(a \sin x \cos x + m \sin^2 x - n \cos^2 x)}{(m+n)^2 + a^2} + \frac{m(m-1)}{(m+n)^2 + a^2}\int e^{ax}\cos^{m-2} x \sin^n x\, dx + \frac{n(n-1)}{(m+n)^2 + a^2}\int e^{ax}\cos^m x \sin^{n-2} x\, dx$$

or

$$= \frac{e^{ax}\cos^{m-1} x \sin^{n-1} x(a \cos x \sin x + m \sin^2 x - n \cos^2 x)}{(m+n)^2 + a^2} + \frac{m(m-1)}{(m+n)^2 + a^2}\int e^{ax}\cos^{m-2} x \sin^{n-2} x\, dx + \frac{(n-m)(n+m-1)}{(m+n)^2 + a^2}\int e^{ax}\cos^m x \sin^{n-2} x\, dx$$

424. $\int \frac{e^{ax}}{\sin^n x}\, dx = -\frac{e^{ax}\{a \sin x + (n-2)\cos x\}}{(n-1)(n-2)\sin^{n-1} x} + \frac{a^2 + (n-2)^2}{(n-1)(n-2)}\int \frac{e^{ax}}{\sin^{n-2} x}\, dx$

425. $\int \frac{e^{ax}}{\cos^n x}\, dx = -\frac{e^{ax}\{a \cos x - (n-2)\sin x\}}{(n-1)(n-2)\cos^{n-1} x} + \frac{a^2 + (n-2)^2}{(n-1)(n-2)}\int \frac{e^{ax}}{\cos^{n-2} x}\, dx$

426. $\int e^{ax}\tan^n x\, dx = e^{ax}\frac{\tan^{n-1} x}{n-1} - \frac{a}{n-1}\int e^{ax}\tan^{n-1} x\, dx - \int e^{ax}\tan^{n-2} x\, dx$

HYPERBOLIC FORMS

427. $\int \sinh x\, dx = \cosh x$

428. $\int \cosh x\, dx = \sinh x$

429. $\int \tanh x\, dx = \log \cosh x$

430. $\int \coth x\, dx = \log \sinh x$

431. $\int \operatorname{sech} x\, dx = \tan^{-1}(\sinh x)$

432. $\int \operatorname{csch} x\, dx = \log \tanh\left(\frac{x}{2}\right)$

433. $\int x \sinh x\, dx = x \cosh x - \sinh x$

434. $\int x^n \sinh x\, dx = x^n \cosh x - n \int x^{n-1} \cosh x\, dx$

435. $\int x \cosh x\, dx = x \sinh x - \cosh x$

436. $\int x^n \cosh x\, dx = x^n \sinh x - n \int x^{n-1} \sinh x\, dx$

437. $\int \operatorname{sech} x \tanh x\, dx = -\operatorname{sech} x$

438. $\int \operatorname{csch} x \coth x\, dx = -\operatorname{csch} x$

439. $\int \sinh^2 x\, dx = \frac{\sinh 2x}{4} - \frac{x}{2}$

440. $$\int \sinh^m x \cosh^n x\, dx = \begin{cases} \frac{1}{m+n} \sinh^{m+1} x \cosh^{n-1} x \\ \qquad + \frac{n-1}{m+n} \int \sinh^m x \cosh^{n-2} x\, dx \\ \text{or} \\ \frac{1}{m+n} \sinh^{m-1} x \cosh^{n+1} x \\ \qquad - \frac{m-1}{m+n} \int \sinh^{m-2} x \cosh^n x\, dx, \qquad (m+n \neq 0) \end{cases}$$

441. $$\int \frac{dx}{\sinh^m x \cosh^n x} = \begin{cases} -\frac{1}{(m-1)\sinh^{m-1} x \cosh^{n-1} x} \\ \qquad - \frac{m+n-2}{m-1} \int \frac{dx}{\sinh^{m-2} x \cosh^n x}, \qquad (m \neq 1) \\ \text{or} \\ \frac{1}{(n-1)\sinh^{m-1} x \cosh^{n-1} x} \\ \qquad + \frac{m+n-2}{n-1} \int \frac{dx}{\sinh^m x \cosh^{n-2} x}, \qquad (n \neq 1) \end{cases}$$

442. $\int \tanh^2 x\, dx = x - \tanh x$

443. $\int \tanh^n x\, dx = -\frac{\tanh^{n-1} x}{n-1} + \int \tanh^{n-2} x\, dx, \qquad (n \neq 1)$

444. $\int \operatorname{sech}^2 x\, dx = \tanh x$

445. $\int \cosh^2 x\, dx = \frac{\sinh 2x}{4} + \frac{x}{2}$

446. $\int \coth^2 x\, dx = x - \coth x$

447. $\int \coth^n x\,dx = -\frac{\coth^{n-1} x}{n-1} + \int \coth^{n-2} x\,dx, \qquad (n \neq 1)$

448. $\int \operatorname{csch}^2 x\,dx = -\operatorname{ctnh} x$

449. $\int \sinh mx \sinh nx\,dx = \frac{\sinh(m+n)x}{2(m+n)} - \frac{\sinh(m-n)x}{2(m-n)}, \qquad (m^2 \neq n^2)$

450. $\int \cosh mx \cosh nx\,dx = \frac{\sinh(m+n)x}{2(m+n)} + \frac{\sinh(m-n)x}{2(m-n)}, \qquad (m^2 \neq n^2)$

451. $\int \sinh mx \cosh nx\,dx = \frac{\cosh(m+n)x}{2(m+n)} + \frac{\cosh(m-n)x}{2(m-n)}, \qquad (m^2 \neq n^2)$

452. $\int \sinh^{-1}\frac{x}{a}\,dx = x\sinh^{-1}\frac{x}{a} - \sqrt{x^2+a^2}, \qquad (a > 0)$

453. $\int x\sinh^{-1}\frac{x}{a}\,dx = \left(\frac{x^2}{2}+\frac{a^2}{4}\right)\sinh^{-1}\frac{x}{a} - \frac{x}{4}\sqrt{x^2+a^2}, \qquad (a > 0)$

454. $\int x^n \sinh^{-1} x\,dx = \frac{x^{n+1}}{n+1}\sinh^{-1} x - \frac{1}{n+1}\int \frac{x^{n+1}}{(1+x^2)^{\frac{1}{2}}}\,dx, \qquad (n \neq -1)$

455. $\int \cosh^{-1}\frac{x}{a}\,dx = \begin{cases} x\cosh^{-1}\frac{x}{a} - \sqrt{x^2-a^2}, & \left(\cosh^{-1}\frac{x}{a} > 0\right) \\ \text{or} & \\ x\cosh^{-1}\frac{x}{a} + \sqrt{x^2-a^2}, & \left(\cosh^{-1}\frac{x}{a} < 0\right), \ (a > 0) \end{cases}$

456. $\int x\cosh^{-1}\frac{x}{a}\,dx = \frac{2x^2-a^2}{4}\cosh^{-1}\frac{x}{a} - \frac{x}{4}(x^2-a^2)^{\frac{1}{2}}$

457. $\int x^n \cosh^{-1} x\,dx = \frac{x^{n+1}}{n+1}\cosh^{-1} x - \frac{1}{n+1}\int \frac{x^{n+1}}{(x^2-1)^{\frac{1}{2}}}\,dx, \qquad (n \neq -1)$

458. $\int \tanh^{-1}\frac{x}{a}\,dx = x\tanh^{-1}\frac{x}{a} + \frac{a}{2}\log(a^2-x^2), \qquad \left(\left|\frac{x}{a}\right| < 1\right)$

459. $\int \coth^{-1}\frac{x}{a}\,dx = x\coth^{-1}\frac{x}{a} + \frac{a}{2}\log(x^2-a^2), \qquad \left(\left|\frac{x}{a}\right| > 1\right)$

460. $\int x\tanh^{-1}\frac{x}{a}\,dx = \frac{x^2-a^2}{2}\tanh^{-1}\frac{x}{a} + \frac{ax}{2}, \qquad \left(\left|\frac{x}{a}\right| < 1\right)$

461. $\int x^n \tanh^{-1} x\,dx = \frac{x^{n+1}}{n+1}\tanh^{-1} x - \frac{1}{n+1}\int \frac{x^{n+1}}{1-x^2}\,dx, \qquad (n \neq -1)$

462. $\int x\coth^{-1}\frac{x}{a}\,dx = \frac{x^2-a^2}{2}\coth^{-1}\frac{x}{a} + \frac{ax}{2}, \qquad \left(\left|\frac{x}{a}\right| > 1\right)$

463. $\int x^n \coth^{-1} x\,dx = \frac{x^{n+1}}{n+1}\coth^{-1} x + \frac{1}{n+1}\int \frac{x^{n+1}}{x^2-1}\,dx, \qquad (n \neq -1)$

464. $\int \operatorname{sech}^{-1} x\,dx = x\operatorname{sech}^{-1} x + \arcsin x$

465. $\int x \operatorname{sech}^{-1} x\, dx = \frac{x^2}{2} \operatorname{sech}^{-1} x - \frac{1}{2}(1 - x^2)$

466. $\int x^n \operatorname{sech}^{-1} x\, dx = \frac{x^{n+1}}{n+1} \operatorname{sech}^{-1} x + \frac{1}{n+1} \int \frac{x^n}{(1 - x^2)^{\frac{1}{2}}}\, dx, \qquad (n \neq -1)$

467. $\int \operatorname{csch}^{-1} x\, dx = x \operatorname{csch}^{-1} x + \frac{x}{|x|} \sinh^{-1} x$

468. $\int x \operatorname{csch}^{-1} x\, dx = \frac{x^2}{2} \operatorname{csch}^{-1} x + \frac{1}{2} \frac{x}{|x|} \sqrt{1 + x^2}$

469. $\int x^n \operatorname{csch}^{-1} x\, dx = \frac{x^{n+1}}{n+1} \operatorname{csch}^{-1} x + \frac{1}{n+1} \int \frac{x^n}{(x^2 + 1)^{\frac{1}{2}}}\, dx, \qquad (n \neq -1)$

DEFINITE INTEGRALS

470. $\int_0^\infty x^{n-1} e^{-x}\, dx = \int_0^1 \left(\log \frac{1}{x}\right)^{n-1} dx = \frac{1}{n} \prod_{m=1}^{\infty} \frac{\left(1 + \frac{1}{m}\right)^n}{1 + \frac{n}{m}}$

$= \Gamma(n),\ n \neq 0, -1, -2, -3, \ldots$ (Gamma Function)

471. $\int_0^\infty t^n p^{-t}\, dt = \frac{n!}{(\log p)^{n+1}}, \qquad (n = 0, 1, 2, 3, \ldots \text{ and } p > 0)$

472. $\int_0^\infty t^{n-1} e^{-(a+1)t}\, dt = \frac{\Gamma(n)}{(a+1)^n}, \qquad (n > 0, a > -1)$

473. $\int_0^1 x^m \left(\log \frac{1}{x}\right)^n dx = \frac{\Gamma(n+1)}{(m+1)^{n+1}}, \qquad (m > -1, n > -1)$

474. $\Gamma(n)$ is finite if $n > 0$, $\Gamma(n+1) = n\Gamma(n)$

475. $\Gamma(n) \cdot \Gamma(1 - n) = \frac{\pi}{\sin n\pi}$

476. $\Gamma(n) = (n - 1)!$ if $n =$ integer > 0

477. $\Gamma(\tfrac{1}{2}) = 2 \int_0^\infty e^{-t^2}\, dt = \sqrt{\pi} = 1.7724538509 \ldots = (-\tfrac{1}{2})!$

478. $\Gamma\left(n + \frac{1}{2}\right) = \frac{1 \cdot 3 \cdot 5 \cdot 7 \cdots (2n - 1)}{2^n} \sqrt{\pi}$, where n is an integer and > 0 (see values of $\Gamma(n)$ at end of integral table)

479. $\int_0^1 x^{m-1}(1 - x)^{n-1}\, dx = B(m, n)$, (Beta function)

480. $B(m, n) = B(n, m) = \frac{\Gamma(m)\Gamma(n)}{\Gamma(m + n)}$, where m and n are any positive real numbers

481. $\int_0^1 x^{m-1}(1 - x)^{n-1}\, dx = \int_0^\infty \frac{x^{m-1}\, dx}{(1 + x)^{m+n}} = \frac{\Gamma(m)\Gamma(n)}{\Gamma(m + n)}$

482. $$\int_a^b (x-a)^m(b-x)^n\,dx = (b-a)^{m+n+1}\frac{\Gamma(m+1)\cdot\Gamma(n+1)}{\Gamma(m+n+2)}, \qquad (m>-1,\ n>-1,\ b>a)$$

483. $$\int_1^\infty \frac{dx}{x^m} = \frac{1}{m-1}, \qquad [m>1]$$

484. $$\int_0^\infty \frac{dx}{(1+x)x^p} = \pi\csc p\pi, \qquad [p<1]$$

485. $$\int_0^\infty \frac{dx}{(1-x)x^p} = -\pi\cot p\pi, \qquad [p<1]$$

486. $$\int_0^\infty \frac{x^{p-1}\,dx}{1+x} = \frac{\pi}{\sin p\pi} = \mathrm{B}(p, 1-p) = \Gamma(p)\Gamma(1-p), \qquad [0<p<1]$$

487. $$\int_0^\infty \frac{x^{m-1}\,dx}{1+x^n} = \frac{\pi}{n\sin\frac{m\pi}{n}}, \qquad [0<m<n]$$

488. $$\int_0^\infty \frac{x^a\,dx}{(m+x^b)^c} = m^{\frac{a+1}{b}-c}\left[\frac{\Gamma\left(\frac{a+1}{b}\right)\Gamma\left(c-\frac{a+1}{b}\right)}{\Gamma(c)}\right], \qquad \left(a>-1,\ b>0,\ m>0,\ c>\frac{a+1}{b}\right)$$

489. $$\int_0^\infty \frac{dx}{(1+x)\sqrt{x}} = \pi$$

490. $$\int_0^\infty \frac{a\,dx}{a^2+x^2} = \frac{\pi}{2}, \text{ if } a>0;\ 0, \text{ if } a=0;\ -\frac{\pi}{2}, \text{ if } a<0$$

491. $$\int_{-a}^{a}(a^2-x^2)^{\frac{n}{2}}\,dx = \frac{1}{2}\int_0^a (a^2-x^2)^{\frac{n}{2}}\,dx = \frac{1\cdot 3\cdot 5\cdots n}{2\cdot 4\cdot 6\cdots (n+1)}\cdot\frac{\pi}{2}\cdot a^{n+1} \qquad (n \text{ odd})$$

492. $$\int_0^a x^m(a^2-x^2)^{\frac{n}{2}}\,dx = \begin{cases} \frac{1}{2}a^{m+n+1}B\left(\frac{m+1}{2}, \frac{n+2}{2}\right) \\ \text{or} \\ \frac{1}{2}a^{m+n+1}\frac{\Gamma\left(\frac{m+1}{2}\right)\Gamma\left(\frac{n+2}{2}\right)}{\Gamma\left(\frac{m+n+3}{2}\right)} \end{cases}$$

493. $$\int_0^{\pi/2} (\sin^n x)\, dx = \begin{cases} \displaystyle\int_0^{\pi/2} (\cos^n x)\, dx \\ \text{or} \\ \dfrac{1 \cdot 3 \cdot 5 \cdot 7 \cdots (n-1)}{2 \cdot 4 \cdot 6 \cdot 8 \cdots (n)} \dfrac{\pi}{2}, & (n \text{ an even integer, } n \neq 0) \\ \text{or} \\ \dfrac{2 \cdot 4 \cdot 6 \cdot 8 \cdots (n-1)}{1 \cdot 3 \cdot 5 \cdot 7 \cdots (n)}, & (n \text{ an odd integer, } n \neq 1) \\ \text{or} \\ \dfrac{\sqrt{\pi}}{2} \dfrac{\Gamma\left(\dfrac{n+1}{2}\right)}{\Gamma\left(\dfrac{n}{2}+1\right)}, & (n > -1) \end{cases}$$

494. $$\int_0^\infty \frac{\sin mx\, dx}{x} = \frac{\pi}{2}, \text{ if } m > 0;\ 0, \text{ if } m = 0;\ -\frac{\pi}{2}, \text{ if } m < 0$$

495. $$\int_0^\infty \frac{\cos x\, dx}{x} = \infty$$

496. $$\int_0^\infty \frac{\tan x\, dx}{x} = \frac{\pi}{2}$$

497. $$\int_0^\pi \sin ax \cdot \sin bx\, dx = \int_0^\pi \cos ax \cdot \cos bx\, dx = 0, \qquad (a \neq b;\ a, b \text{ integers})$$

498. $$\int_0^{\pi/a} [\sin(ax)][\cos(ax)]\, dx = \int_0^\pi [\sin(ax)][\cos(ax)]\, dx = 0$$

499. $$\int_0^\pi [\sin(ax)][\cos(bx)]\, dx = \frac{2a}{a^2 - b^2}, \text{ if } a - b \text{ is odd, or zero if } a - b \text{ is even}$$

500. $$\int_0^\infty \frac{\sin x \cos mx\, dx}{x} = 0, \text{ if } m < -1 \text{ or } m > 1,\ = \frac{\pi}{4}, \text{ if } m = \pm 1;\ = \frac{\pi}{2}, \text{ if } m^2 < 1$$

501. $$\int_0^\infty \frac{\sin ax \sin bx}{x^2}\, dx = \frac{\pi a}{2}, \qquad (a \leq b)$$

502. $$\int_0^\pi \sin^2 mx\, dx = \int_0^\pi \cos^2 mx\, dx = \frac{\pi}{2}$$

503. $$\int_0^\infty \frac{\sin^2 x\, dx}{x^2} = \frac{\pi}{2}$$

504. $$\int \frac{\cos mx}{1 + x^2}\, dx = \frac{\pi}{2} e^{-|m|}$$

505. $$\int_0^\infty \cos(x^2)\, dx = \int_0^\infty \sin(x^2)\, dx = \frac{1}{2}\sqrt{\frac{\pi}{2}}$$

506. $$\int_0^\infty \frac{\sin x\, dx}{\sqrt{x}} = \int_0^\infty \frac{\cos x\, dx}{\sqrt{x}} = \sqrt{\frac{\pi}{2}}$$

507. $\int_0^{\pi/2} \frac{dx}{1 + a\cos x} = \frac{\cos^{-1} a}{\sqrt{1 - a^2}}, \qquad (a < 1)$

508. $\int_0^{\infty} \frac{dx}{a + b\cos x} = \frac{\pi}{\sqrt{a^2 - b^2}}, \qquad (a > b \geq 0)$

509. $\int_0^{2\pi} \frac{dx}{1 + a\cos x} = \frac{2\pi}{\sqrt{1 - a^2}}, \qquad (a^2 < 1)$

510. $\int_0^{\infty} \frac{\cos ax - \cos bx}{x}\, dx = \log \frac{b}{a}$

511. $\int_0^{\pi/2} \frac{dx}{a^2 \sin^2 x + b^2 \cos^2 x} = \frac{\pi}{2ab}$

512. $\int_0^{\pi/2} \frac{dx}{(a^2 \sin^2 x + b^2 \cos^2 x)^2} = \frac{\pi(a^2 + b^2)}{4a^3 b^3}, \qquad (a, b > 0)$

513. $\int_0^{\pi/2} \sin^{n-1} x \cos^{m-1} x\, dx = \frac{1}{2} \mathrm{B}\left(\frac{n}{2}, \frac{m}{2}\right),$ m and n positive integers

514. $\int_0^{\pi/2} (\sin^{2n+1} \theta)\, d\theta = \frac{2 \cdot 4 \cdot 6 \cdots (2n)}{1 \cdot 3 \cdot 5 \cdots (2n + 1)}, \qquad (n = 1, 2, 3 \ldots)$

515. $\int_0^{\pi/2} (\sin^{2n} \theta)\, d\theta = \frac{1 \cdot 3 \cdot 5 \cdots (2n - 1)}{2 \cdot 4 \cdots (2n)} \left(\frac{\pi}{2}\right), \qquad (n = 1, 2, 3 \ldots)$

516. $\int_0^{\pi/2} \sqrt{\cos \theta}\, d\theta = \frac{(2\pi)^{\frac{3}{2}}}{[\Gamma(\frac{1}{4})]^2}$

517. $\int_0^{\pi/2} (\tan^h \theta)\, d\theta = \frac{\pi}{2 \cos\left(\frac{h\pi}{2}\right)}, \qquad (0 < h < 1)$

518. $\int_0^{\infty} \frac{\tan^{-1}(ax) - \tan^{-1}(bx)}{x}\, dx = \frac{\pi}{2} \log \frac{a}{b}, \qquad (a, b > 0)$

519. The area enclosed by a curve defined through the equation $x^{\frac{b}{c}} + y^{\frac{b}{c}} = a^{\frac{b}{c}}$ where $a > 0$, c a positive odd integer and b a positive even integer is given by

$$\frac{\left[\Gamma\left(\frac{c}{b}\right)\right]^2}{\Gamma\left(\frac{2c}{b}\right)} \left(\frac{2ca^2}{b}\right)$$

520. $I = \iiint_R x^{h-1} y^{m-1} z^{n-1}\, dv$, where R denotes the region of space bounded by the co-ordinate planes and that portion of the surface $\left(\frac{x}{a}\right)^p + \left(\frac{y}{b}\right)^q + \left(\frac{z}{c}\right)^k = 1$, which lies in the first octant and where $h, m, n, p, q, k, a, b, c,$ denote positive real numbers is given by

520. (Continued)

$$\int_0^a x^{h-1}\,dx \int_0^{b\left[1-\left(\frac{x}{a}\right)^p\right]^{\frac{1}{q}}} y^m\,dy \int_0^{c\left[1-\left(\frac{x}{a}\right)^p-\left(\frac{y}{b}\right)^q\right]^{\frac{1}{k}}} z^{n-1}\,dz = \frac{a^h b^m c^n}{pqk}\,\frac{\Gamma\left(\frac{h}{p}\right)\Gamma\left(\frac{m}{q}\right)\Gamma\left(\frac{n}{k}\right)}{\Gamma\left(\frac{h}{p}+\frac{m}{q}+\frac{n}{k}+1\right)}$$

521. $\int_0^\infty e^{-ax}\,dx = \frac{1}{a}, \qquad (a > 0)$

522. $\int_0^\infty \frac{e^{-ax}-e^{-bx}}{x}\,dx = \log\frac{b}{a}, \qquad (a, b > 0)$

523. $\int_0^\infty x^n e^{-ax^p}\,dx = \frac{\Gamma\left(\frac{n+1}{p}\right)}{pa^{\frac{n+1}{p}}}, \quad (n > -1, p > o, a > o)$

524. $\int_0^\infty e^{-a^2x^2}\,dx = \frac{1}{2a}\sqrt{\pi} = \frac{1}{2a}\,\Gamma\left(\frac{1}{2}\right), \qquad (a > 0)$

525. $\int_0^\infty xe^{-x^2}\,dx = \frac{1}{2}$

526. $\int_0^\infty x^2e^{-x^2}\,dx = \frac{\sqrt{\pi}}{4}$

527. $\int_0^\infty x^{2n}e^{-ax^2}\,dx = \frac{1\cdot3\cdot5\ \cdots\ (2n-1)}{2^{n+1}a^n}\sqrt{\frac{\pi}{a}}$

528. $\int_0^1 x^m e^{-ax}\,dx = \frac{m!}{a^{m+1}}\left[1 - e^{-a}\sum_{r=0}^{m}\frac{a^r}{r!}\right]$

529. $\int_0^\infty e^{\left(-x^2-\frac{a^2}{x^2}\right)}\,dx = \frac{e^{-2a}\sqrt{\pi}}{2}, \qquad (a \geq 0)$

530. $\int_0^\infty e^{-nx}\sqrt{x}\,dx = \frac{1}{2n}\sqrt{\frac{\pi}{n}}$

531. $\int_0^\infty \frac{e^{-nx}}{\sqrt{x}}\,dx = \sqrt{\frac{\pi}{n}}$

532. $\int_0^\infty e^{-ax}\cos mx\,dx = \frac{a}{a^2+m^2}, \qquad (a > 0)$

533. $\int_0^\infty e^{-ax}\sin mx\,dx = \frac{m}{a^2+m^2}, \qquad (a > 0)$

534. $\int_0^\infty xe^{-ax}[\sin(bx)]\,dx = \frac{2ab}{(a^2+b^2)^2}, \qquad (a > 0)$

535. $\int_0^\infty xe^{-ax}[\cos(bx)]\,dx = \frac{a^2-b^2}{(a^2+b^2)^2}, \qquad (a > 0)$

536. $\int_0^\infty x^n e^{-ax}[\sin(bx)]\,dx = \frac{n![(a-ib)^{n+1}-(a+ib)^{n+1}]}{2(a^2+b^2)^{n+1}}, \qquad (i^2=-1,\ a>0)$

537. $\int_0^\infty x^n e^{-ax}[\cos(bx)]\,dx = \frac{n![(a-ib)^{n+1}+(a+ib)^{n+1}]}{2(a^2+b^2)^{n+1}}, \qquad (i^2=-1,\ a>0)$

538. $\int_0^\infty \frac{e^{-ax}\sin x}{x}\,dx = \cot^{-1} a, \qquad (a>0)$

539. $\int_0^\infty e^{-a^2x^2}\cos bx\,dx = \frac{\sqrt{\pi}}{2a}\,e^{\frac{-b^2}{4a^2}}, \qquad (ab\neq 0)$

540. $\int_0^\infty e^{-t\cos\phi}t^{b-1}[\sin(t\sin\phi)\,dt = [\Gamma(b)]\sin(b\phi), \qquad \left(b>0,\ -\frac{\pi}{2}<\phi<\frac{\pi}{2}\right)$

541. $\int_0^\infty e^{-t\cos\phi}t^{b-1}[\cos(t\sin\phi)]\,dt = [\Gamma(b)]\cos(b\phi), \qquad \left(b>0,\ -\frac{\pi}{2}<\phi<\frac{\pi}{2}\right)$

542. $\int_0^\infty t^{b-1}\cos t\,dt = [\Gamma(b)]\cos\left(\frac{b\pi}{2}\right), \qquad (0<b<1)$

543. $\int_0^\infty t^{b-1}(\sin t)\,dt = [\Gamma(b)]\sin\left(\frac{b\pi}{2}\right), \qquad (0<b<1)$

544. $\int_0^1 (\log x)^n\,dx = (-1)^n\cdot n!$

545. $\int_0^1 \left(\log\frac{1}{x}\right)^{\frac{1}{2}}dx = \frac{\sqrt{\pi}}{2}$

546. $\int_0^1 \left(\log\frac{1}{x}\right)^{-\frac{1}{2}}dx = \sqrt{\pi}$

547. $\int_0^1 \left(\log\frac{1}{x}\right)^{n}dx = n!$

548. $\int_0^1 x\log(1-x)\,dx = -\frac{3}{4}$

549. $\int_0^1 x\log(1+x)\,dx = \frac{1}{4}$

550. $\int_0^1 \frac{\log x}{1+x}\,dx = -\frac{\pi^2}{12}$

551. $\int_0^1 \frac{\log x}{1-x}\,dx = -\frac{\pi^2}{6}$

552. $\int_0^1 \frac{\log x}{1-x^2}\,dx = -\frac{\pi^2}{8}$

553. $\int_0^1 \log\left(\frac{1+x}{1-x}\right)\cdot\frac{dx}{x} = \frac{\pi^2}{4}$

554. $\int_0^1 \frac{\log x\,dx}{\sqrt{1-x^2}} = -\frac{\pi}{2}\log 2$

555. $\int_0^1 x^m \left[\log\left(\frac{1}{x}\right)\right]^n dx = \frac{\Gamma(n+1)}{(m+1)^{n+1}}$, if $m+1>0$, $n+1>0$

556. $\int_0^1 \frac{(x^p - x^q)\,dx}{\log x} = \log\left(\frac{p+1}{q+1}\right)$, $\quad (p+1>0,\ q+1>0)$

557. $\int_0^1 \frac{dx}{\sqrt{\log\left(\frac{1}{x}\right)}} = \sqrt{\pi}$

558. $\int_0^\infty \log\left(\frac{e^x+1}{e^x-1}\right) dx = \frac{\pi^2}{4}$

559. $\int_0^{\pi/2} \log \sin x\,dx = \int_0^{\pi/2} \log \cos x\,dx = -\frac{\pi}{2}\log 2$

560. $\int_0^{\pi/2} \log \sec x\,dx = \int_0^{\pi/2} \log \csc x\,dx = \frac{\pi}{2}\log 2$

561. $\int_0^{\pi} x \log \sin x\,dx = -\frac{\pi^2}{2}\log 2$

562. $\int_0^{\pi/2} \sin x \log \sin x\,dx = \log 2 - 1$

563. $\int_0^{\pi/2} \log \tan x\,dx = 0$

564. $\int_0^{\pi} \log(a \pm b\cos x)\,dx = \pi \log\left(\frac{a+\sqrt{a^2-b^2}}{2}\right)$, $\quad (a \geqq b)$

565. $\int_0^\infty \frac{dx}{\cosh ax} = \frac{\pi}{2a}$

566. $\int_0^\infty \frac{x\,dx}{\sinh ax} = \frac{\pi^2}{4a^2}$

567. $\int_0^\infty e^{-ax}\cosh bx\,dx = \frac{a}{a^2-b^2}$, $\quad (0 \le |b| < a)$

568. $\int_0^\infty e^{-ax}\sinh bx\,dx = \frac{b}{a^2-b^2}$, $\quad (0 \le |b| < a)$

569. $\int_{+\infty}^1 \frac{e^{-xu}}{u}\,du = \gamma + \log x - x + \frac{x^2}{2\cdot 2!} - \frac{x^3}{3\cdot 3!} + \frac{x^4}{4\cdot 4!} - \cdots,$

$$\text{where } \gamma = \lim_{z\to\infty}\left(1 + \frac{1}{2} + \frac{1}{3} + \cdots + \frac{1}{z} - \log z\right)$$
$$= 0.5772157 \cdots, \qquad (0 < x < \infty)$$

570. $\int_0^{\pi/2} \frac{dx}{\sqrt{1-k^2\sin^2 x}} = \frac{\pi}{2}\left[1 + \left(\frac{1}{2}\right)^2 k^2 + \left(\frac{1\cdot 3}{2\cdot 4}\right)^2 k^4 + \left(\frac{1\cdot 3\cdot 5}{2\cdot 4\cdot 6}\right)^2 k^6 + \cdots\right]$, if $k^2 < 1$

571. $$\int_0^{\pi/2} \sqrt{1 - k^2 \sin^2 x}\, dx = \frac{\pi}{2}\left[1 - \left(\frac{1}{2}\right)^2 k^2 - \left(\frac{1 \cdot 3}{2 \cdot 4}\right)^2 \frac{k^4}{3} - \left(\frac{1 \cdot 3 \cdot 5}{2 \cdot 4 \cdot 6}\right)^2 \frac{k^6}{5} - \cdots\right], \text{ if } k^2 < 1$$

572. $$\int_0^{\infty} e^{-x} \log x\, dx = -\gamma = -0.5772157 \cdots$$

573. $$\int_0^{\infty} \left(\frac{1}{1 - e^{-x}} - \frac{1}{x}\right) e^{-x}\, dx = \gamma = 0.5772157 \cdots$$ [Euler's Constant]

574. $$\int_0^{\infty} \frac{1}{x}\left(\frac{1}{1 + x} - e^{-x}\right) dx = \gamma = 0.5772157 \cdots$$

*GAMMA FUNCTION

Values of $\Gamma(n) = \int_0^\infty e^{-x}x^{n-1}dx;\ \Gamma(n+1) = n\Gamma(n)$

n	$\Gamma(n)$	n	$\Gamma(n)$	n	$\Gamma(n)$	n	$\Gamma(n)$
1.00	1.00000	1.25	.90640	1.50	.88623	1.75	.91906
1.01	.99433	1.26	.90440	1.51	.88659	1.76	.92137
1.02	.98884	1.27	.90250	1.52	.88704	1.77	.92376
1.03	.98355	1.28	.90072	1.53	.88757	1.78	.92623
1.04	.97844	1.29	.89904	1.54	.88818	1.79	.92877
1.05	.97350	1.30	.89747	1.55	.88887	1.80	.93138
1.06	.96874	1.31	.89600	1.56	.88964	1.81	.93408
1.07	.96415	1.32	.89464	1.57	.89049	1.82	.93685
1.08	.95973	1.33	.89338	1.58	.89142	1.83	.93969
1.09	.95546	1.34	.89222	1.59	.89243	1.84	.94261
1.10	.95135	1.35	.89115	1.60	.89352	1.85	.94561
1.11	.94739	1.36	.89018	1.61	.89468	1.86	.94869
1.12	.94359	1.37	.88931	1.62	.89592	1.87	.95184
1.13	.93993	1.38	.88854	1.63	.89724	1.88	.95507
1.14	.93642	1.39	.88785	1.64	.89864	1.89	.95838
1.15	.93304	1.40	.88726	1.65	.90012	1.90	.96177
1.16	.92980	1.41	.88676	1.66	.90167	1.91	.96523
1.17	.92670	1.42	.88636	1.67	.90330	1.92	.96878
1.18	.92373	1.43	.88604	1.68	.90500	1.93	.97240
1.19	.92088	1.44	.88580	1.69	.90678	1.94	.97610
1.20	.91817	1.45	.88565	1.70	.90864	1.95	.97988
1.21	.91558	1.46	.88560	1.71	.91057	1.96	.98374
1.22	.91311	1.47	.88563	1.72	.91258	1.97	.98768
1.23	.91075	1.48	.88575	1.73	.91466	1.98	.99171
1.24	.90852	1.49	.88595	1.74	.91683	1.99	.99581
						2.00	1.00000

* For large positive values of x, $\Gamma(x)$ approximates the asymptotic series

$$x^x e^{-x}\sqrt{\frac{2\pi}{x}}\left[1+\frac{1}{12x}+\frac{1}{288x^2}-\frac{139}{51840x^3}-\frac{571}{2488320x^4}+\cdots\right].$$

BESSEL FUNCTIONS $J_0(x)$ AND $J_1(x)$

$$J_0(x) = 1 - \left(\frac{x}{2}\right)^2 + \frac{\left(\frac{x}{2}\right)^4}{1^2\cdot 2^2} - \frac{\left(\frac{x}{2}\right)^6}{1^2\cdot 2^2\cdot 3^2} + - \cdots$$

$$J_1(x) = -J_0{}^1(x) = \frac{x}{2} - \frac{\left(\frac{x}{2}\right)^3}{1^2\cdot 2} + \frac{\left(\frac{x}{2}\right)^5}{1^2\cdot 2^2\cdot 3} - + \cdots.$$

$$J_n(x) = \sum_{k=0}^{\infty}\frac{(-1)^k X^{n+2k}}{2^{n+2k}k!(n+k)!} \qquad (n = 0, 1, 2, 3, \ldots)$$

x	$J_0(x)$	$J_1(x)$	x	$J_0(x)$	$J_1(x)$	x	$J_0(x)$	$J_1(x)$
0.0	1.0000	.0000	**3.5**	−.3801	.1374	**7.0**	.3001	.0047
0.1	.9975	.0499	3.6	−.3918	.0955	7.1	.2991	.0252
0.2	.9900	.0995	3.7	−.3992	.0538	7.2	.2951	.0543
0.3	.9776	.1483	3.8	−.4026	.0128	7.3	.2882	.0826
0.4	.9604	.1960	3.9	−.4018	−.0272	7.4	.2786	.1096
0.5	.9385	.2423	**4.0**	−.3971	−.0660	**7.5**	.2663	.1352
0.6	.9120	.2867	4.1	−.3887	−.1033	7.6	.2516	.1592
0.7	.8812	.3290	4.2	−.3766	−.1386	7.7	.2346	.1813
0.8	.8463	.3688	4.3	−.3610	−.1719	7.8	.2154	.2014
0.9	.8075	.4059	4.4	−.3423	−.2028	7.9	.1944	.2192
1.0	.7652	.4401	**4.5**	−.3205	−.2311	**8.0**	.1717	.2346
1.1	.7196	.4709	4.6	−.2961	−.2566	8.1	.1475	.2476
1.2	.6711	.4983	4.7	−.2693	−.2791	8.2	.1222	.2580
1.3	.6201	.5220	4.8	−.2404	−.2985	8.3	.0960	.2657
1.4	.5669	.5419	4.9	−.2097	−.3147	8.4	.0692	.2708
1.5	.5118	.5579	**5.0**	−.1776	−.3276	**8.5**	.0419	.2731
1.6	.4554	.5699	5.1	−.1443	−.3371	8.6	.0146	.2728
1.7	.3980	.5778	5.2	−.1103	−.3432	8.7	−.0125	.2697
1.8	.3400	.5815	5.3	−.0758	−.3460	8.8	−.0392	.2641
1.9	.2818	.5812	5.4	−.0412	−.3453	8.9	−.0653	.2559
2.0	.2239	.5767	**5.5**	−.0068	−.3414	**9.0**	−.0903	.2453
2.1	.1666	.5683	5.6	.0270	−.3343	9.1	−.1142	.2324
2.2	.1104	.5560	5.7	.0599	−.3241	9.2	−.1367	.2174
2.3	.0555	.5399	5.8	.0917	−.3110	9.3	−.1577	.2004
2.4	.0025	.5202	5.9	.1220	−.2951	9.4	−.1768	.1816
2.5	−.0484	4971	**6.0**	.1506	−.2767	**9.5**	−.1939	.1613
2.6	−.0968	.4708	6.1	.1773	−.2559	9.6	−.2090	.1395
2.7	−.1424	.4416	6.2	.2017	−.2329	9.7	−.2218	.1166
2.8	−.1850	.4097	6.3	.2238	−.2081	9.8	−.2323	0928
2.9	−.2243	.3754	6.4	.2433	−.1816	9.9	−.2403	.0684
3.0	−.2601	.3391	**6.5**	.2601	−.1538	**10.0**	−.2459	.0435
3.1	−.2921	.3009	6.6	.2740	−.1250	10.1	−.2490	.0184
3.2	−.3202	.2613	6.7	.2851	−.0953	10.2	−.2496	−.0066
3.3	−.3443	.2207	6.8	.2931	−.0652	10.3	−.2477	−.0313
3.4	−.3643	.1792	6.9	.2981	−.0349	10.4	−.2434	−.0555

BESSEL FUNCTIONS $J_0(x)$ AND $J_1(x)$ (Continued)

x	$J_0(x)$	$J_1(x)$	x	$J_0(x)$	$J_1(x)$	x	$J_0(x)$	$J_1(x)$
10.5	−.2366	−.0789	**12.5**	.1469	−.1655	**14.5**	.0875	.193
10.6	−.2276	−.1012	12.6	.1626	−.1487	14.6	.0679	.199
10.7	−.2164	−.1224	12.7	.1766	−.1307	14.7	.0476	.204
10.8	−.2032	−.1422	12.8	.1887	−.1114	14.8	.0271	.206
10.9	−.1881	−.1603	12.9	.1988	−.0912	14.9	.0064	.206
11.0	−.1712	−.1768	**13.0**	.2069	−.0703			
11.1	−.1528	−.1913	13.1	.2129	−.0489			
11.2	−.1330	−.2039	13.2	.2167	−.0271			
11.3	−.1121	−.2143	13.3	.2183	−.0052			
11.4	−.0902	−.2225	13.4	.2177	.0166			
11.5	−.0677	−.2284	**13.5**	.2150	.0380			
11.6	−.0446	−.2320	13.6	.2101	.0590			
11.7	−.0213	−.2333	13.7	.2032	.0791			
11.8	.0020	−.2323	13.8	.1943	.0984			
11.9	.0250	−.2290	13.9	.1836	.1165			
12.0	.0477	−.2234	**14.0**	.1711	.1334			
12.1	.0697	−.2157	14.1	.1570	.1488			
12.2	.0908	−.2060	14.2	.1414	.1626			
12.3	.1108	−.1943	14.3	.1245	.1747			
12.4	.1296	−.1807	14.4	.1065	.1850			

$J_0(x) = 0$; $x = 2.405, 5.520, 8.654, 11.792, 14.931, 18.071, 21.212, 24.35$
$J_1(x) = 0$: $x = 3.832, 7.016, 10.173, 13.324, 16.471, 19.616.$

BESSEL FUNCTIONS FOR SPHERICAL COORDINATES

$$j_n(x) = \sqrt{\pi/2x}\,J_{n+\frac{1}{2}}(x), \quad n_n(x) = \sqrt{\pi/2x}\,N_{n+\frac{1}{2}}(x)$$

x	$j_0(x)$	$n_0(x)$	$j_1(x)$	$n_1(x)$	$j_2(x)$	$n_2(x)$
0.0	1.0000	$-\infty$	0.0000	$-\infty$	0.0000	$-\infty$
0.1	0.9983	−9.9500	0.0333	−100.50	0.0007	−3005.0
0.2	0.9933	−4.9003	0.0664	−25.495	0.0027	−377.52
0.4	0.9735	−2.3027	0.1312	−6.7302	0.0105	−48.174
0.6	0.9411	−1.3756	0.1929	−3.2337	0.0234	−14.793
0.8	0.8967	−0.8709	0.2500	−1.9853	0.0408	−6.5740
1.0	0.8415	−0.5403	0.3012	−1.3818	0.0620	−3.6050
1.2	0.7767	−0.3020	0.3453	−1.0283	0.0865	−2.2689
1.4	0.7039	−0.1214	0.3814	−0.7906	0.1133	−1.5728
1.6	0.6247	+0.0183	0.4087	−0.6133	0.1416	−1.1682
1.8	0.5410	0.1262	0.4268	--0.4709	0.1703	−0.9111
2.0	0.4546	0.2081	0.4354	−0.3506	0.1985	−0.7340
2.2	0.3675	0.2675	0.4346	−0.2459	0.2251	−0.6028
2.4	0.2814	0.3072	0.4245	−0.1534	0.2492	−0.4990
2.6	0.1983	0.3296	0.4058	−0.0715	0.2700	−0.4121
2.8	0.1196	0.3365	0.3792	+0.0005	0.2867	−0.3359
3.0	+0.0470	0.3300	0.3457	0.0630	0.2986	−0.2670
3.2	−0.0182	0.3120	0.3063	0.1157	0.3054	−0.2035
3.4	−0.0752	0.2844	0.2623	0.1588	0.3066	−0.1442
3.6	−0.1229	0.2491	0.2150	0.1921	0.3021	−0.0890
3.8	−0.1610	0.2082	0.1658	0.2158	0.2919	−0.0378
4.0	−0.1892	0.1634	0.1161	0.2300	0.2763	+0.0091
4.2	−0.2075	0.1167	0.0673	0.2353	0.2556	0.0514
4.4	−0.2163	0.0699	+0.0207	0.2321	0.2304	0.0884
4.6	−0.2160	+0.0244	−0.0226	0.2213	0.2013	0.1200
4.8	−0.2075	−0.0182	−0.0615	0.2037	0.1691	0.1456
5.0	−0.1918	−0.0567	−0.0951	0.1804	0.1347	0.1650
5.2	−0.1699	−0.0901	−0.1228	0.1526	0.0991	0.1781
5.4	−0.1431	−0.1175	−0.1440	0.1213	0.0631	0.1850
5.6	−0.1127	−0.1385	−0.1586	0.0880	+0.0278	0.1856
5.8	−0.0801	−0.1527	−0.1665	0.0538	−0.0060	0.1805
6.0	−0.0466	−0.1600	−0.1678	+0.0199	−0.0373	0.1700
6.2	−0.0134	−0.1607	−0.1629	−0.0125	−0.0654	0.1547
6.4	+0.0182	−0.1552	−0.1523	−0.0425	−0.0896	0.1353
6.6	0.0472	−0.1440	−0.1368	−0.0690	−0.1094	0.1126
6.8	0.0727	−0.1278	−0.1172	−0.0915	−0.1243	0.0875
7.0	0.0939	−0.1077	−0.0943	−0.1092	−0.1343	0.0609
7.2	0.1102	−0.0845	−0.0692	−0.1220	−0.1391	0.0337
7.4	0.1215	−0.0593	−0.0429	−0.1294	−0.1388	+0.0068
7.6	0.1274	−0.0331	−0.0163	−0.1317	−0.1338	−0.0189
7.8	0.1280	−0.0069	+0.0095	−0.1289	−0.1244	−0.0427
8.0	0.1237	+0.0182	0.0336	−0.1214	−0.1111	−0.0637

Taken from Vibration and Sound with the permission of Philip Mors author, and McGraw-Hill Book Company, Inc., publisher.

HYPERBOLIC BESSEL FUNCTIONS

$$I_m(z) = i^{-m}J_m(iz)$$

z	$I_0(z)$	$I_1(z)$	$I_2(z)$
0.0	1.0000	0.0000	0.0000
0.1	1.0025	0.0501	0.0012
0.2	1.0100	0.1005	0.0050
0.4	1.0404	0.2040	0.0203
0.6	1.0921	0.3137	0.0464
0.8	1.1665	0.4329	0.0843
1.0	1.2661	0.5652	0.1358
1.2	1.3937	0.7147	0.2026
1.4	1.5534	0.8861	0.2876
1.6	1.7500	1.0848	0.3940
1.8	1.9895	1.3172	0.5260
2.0	2.2796	1.5906	0.6890
2.2	2.6292	1.9141	0.8891
2.4	3.0492	2.2981	1.1111
2.6	3.5532	2.7554	1.4338
2.8	4.1574	3.3011	1.7994
3.0	4.8808	3.9534	2.2452
3.2	5.7472	4.7343	2.7884
3.4	6.7848	5.6701	3.4495
3.6	8.0278	6.7926	4.2538
3.8	9.5169	8.1405	5.2323
4.0	11.302	9.7594	6.4224
4.2	13.443	11.705	7.8683
4.4	16.010	14.046	9.6259
4.6	19.097	16.863	11.761
4.8	22.794	20.253	14.355
5.0	27.240	24.335	17.505
5.2	32.584	29.254	21.332
5.4	39.010	35.181	25.980
5.6	46.738	42.327	31.621
5.8	56.039	50.945	38.472
6.0	67.235	61.341	46.788
6.2	80.717	73.888	56.882
6.4	96.963	89.025	69.143
6.6	116.54	107.31	84.021
6.8	140.14	129.38	102.08
7.0	168.59	156.04	124.01
7.2	202.92	188.25	150.63
7.4	244.34	227.17	182.94
7.6	294.33	274.22	222.17
7.8	354.68	331.10	269.79
8.0	427.57	399.87	327.60

Taken from Vibration and Sound with the permission of Philip Morse, author, and McGraw-Hill Book Company, Inc., publisher.

SINE, COSINE, AND EXPONENTIAL INTEGRALS

$$Si(x) = \int_0^x \frac{\sin v}{v}\,dv; \qquad Ci(x) = \int_\infty^x \frac{\cos v}{v}\,dv;$$

$$Ei(x) = \int_{-\infty}^x \frac{e^v}{v}\,dv; \qquad -Ei(-x) = \int_x^\infty \frac{e^{-v}}{v}\,dv$$

x	$Si(x)$	$Ci(x)$	$Ei(x)$	$-Ei(-x)$
0.0	0.00000	$-\infty$	$-\infty$	$+\infty$
0.1	0.09994	−1.72787	−1.62281	1.82292
0.2	.19956	−1.04221	− .82176	1.22265
0.3	.29850	− .64917	− .30267	.90568
0.4	.39646	− .37881	.10477	.70238
0.5	.49311	− .17778	.45422	.55977
0.6	.58813	− .02227	.76988	.45438
0.7	.68122	.10051	1.06491	.37377
0.8	.77210	.19828	1.34740	.31060
0.9	.86047	.27607	1.62281	.26018
1.0	.94608	.33740	1.89512	.21938
1.1	1.02869	.38487	2.16738	.18599
1.2	1.10805	.42046	2.44209	.15841
1.3	1.18396	.44574	2.72140	.13545
1.4	1.25623	.46201	3.00721	.11622
1.5	1.32468	.47036	3.30129	.10002
1.6	1.38918	.47173	3.60532	.08631
1.7	1.44959	.46697	3.92096	.07465
1.8	1.50582	.45681	4.24987	.06471
1.9	1.55778	.44194	4.59371	.05620
2.0	1.60541	.42298	4.95423	.04890
2.1	1.64870	.40051	5.33324	.04261
2.2	1.68762	.37507	5.73261	.03719
2.3	1.72221	.34718	6.15438	.03250
2.4	1.75249	.31729	6.60067	.02844

SINE, COSINE, AND EXPONENTIAL INTEGRALS (Continued)

x	$Si(x)$	$Ci(x)$	$Ei(x)$	$-Ei(-x)$
2.5	1.77852	.28587	7.07377	.02491
2.6	1.80039	.25334	7.57611	.02185
2.7	1.81821	.22008	8.11035	.01918
2.8	1.83210	.18649	8.67930	.01686
2.9	1.84219	.15290	9.28602	.01482
3.0	1.84865	.11963	9.93383	.01305
3.1	1.85166	.08699	10.6263	.01149
3.2	1.85140	.05526	11.3673	.01013
3.3	1.84808	.02468	12.1610	.00894
3.4	1.84191	− .00452	13.0121	.00789
3.5	1.83313	− .03213	13.9254	.00697
3.6	1.82195	− .05797	14.9063	.00616
3.7	1.80862	− .08190	15.9606	.00545
3.8	1.79339	− .10378	17.0948	.00482
3.9	1.77650	− .12350	18.3157	.00427
4.0	1.75820	− .14098	19.6309	.00378
4.1	1.73874	− .15617	21.0485	.00335
4.2	1.71837	− .16901	22.5774	.00297
4.3	1.69732	− .17951	24.2274	.00263
4.4	1.67583	− .18766	26.0090	.00234
4.5	1.65414	− .19349	27.9337	.00207
4.6	1.63246	− .19705	30.0141	.00184
4.7	1.61101	− .19839	32.2639	.00164
4.8	1.58998	− .19760	34.6979	.00145
4.9	1.56956	− .19478	37.3325	.00129
5.0	1.54993	− .19003	40.1853	.00115
5.1	1.53125	− .18348	43.2757	.00102
5.2	1.51367	− .17525	46.6249	.00091
5.3	1.49731	− .16551	50.2557	.00081
5.4	1.48230	− .15439	54.1935	.00072
5.5	1.46872	− .14205	58.4655	.00064
5.6	1.45667	− .12867	63.1018	.00057
5.7	1.44620	− .11441	68.1350	.00051
5.8	1.43736	− .09944	73.6008	.00045
5.9	1.43018	− .08393	79.5382	.00040
6.0	1.42469	− .06806	85.9898	.00036
6.1	1.42087	− .05198	93.0020	.00032
6.2	1.41871	− .03587	100.626	.00029
6.3	1.41817	− .01988	108.916	.00026
6.4	1.41922	− .00418	117.935	.00023
6.5	1.42179	+ .01110	127.747	.00020
6.6	1.42582	+ .02582	138.426	.00018
6.7	1.43121	.03986	150.050	.00016
6.8	1.43787	.05308	162.707	.00014
6.9	1.44570	.06539	176.491	.00013
7.0	1.45460	.07670	191.505	.00012
7.1	1.46443	.08691	207.863	.00010
7.2	1.47509	.09596	225.688	.00009
7.3	1.48644	.10379	245.116	.00008
7.4	1.49834	.11036	266.296	.00007
7.5	1.51068	.11563	289.388	.00007
7.6	1.52331	.11960	314.572	.00006
7.7	1.53611	.12225	342.040	.00005
7.8	1.54894	.12359	372.006	.00005
7.9	1.56167	.12364	404.701	.00004
8.0	1.57419	.12243	440.380	.00004
8.1	1.58637	.12002	479.322	.00003
8.2	1.59810	.11644	521.831	.00003
8.3	1.60928	.11177	568.242	.00003
8.4	1.61981	.10607	618.919	00002
8.5	1.62960	.09943	674.264	.00002
8.6	1.63857	.09194	734.714	.00002
8.7	1.64665	.08368	800.749	.00002
8.8	1.65379	.07476	872.895	.00002
8.9	1.65993	.06528	951.728	.00001
9.0	1.66504	.05535	1037.88	.00001
9.1	1.66908	.04507	1132.04	.00001
9.2	1.67205	.03455	1234.96	.00001
9.3	1.67393	.02391	1347.48	.00001
9.4	1.67473	.01325	1470.51	.00001
9.5	1.67446	.00268	1605.03	.00001
9.6	1.67316	− .00771	1752.14	.00001
9.7	1.67084	− .01780	1913.05	.00001
9.8	1.66757	− .02752	2089.05	.00001
9.9	1.66338	− .03676	2281.58	.00000
10.0	1.65835	− .04546	2492.23	.00000
10.5	1.62294	− .07828	3883.74	.00000
11.0	1.57831	− .08956	6071.41	.00000
11.5	1.53572	− .07857	9518.20	.00000
12.0	1.50497	− .04978	14959.5	.00000
12.5	1.49234	− .01141	23565.1	.00000
13.0	1.49936	+ .02676	37197.7	.00000
13.5	1.52291	+ .05576	58827.0	.00000
14.0	1.55621	.06940	93193.0	.00000
14.5	1.59072	.06554	147866.	.00000
15.0	1.61819	.04628	234956.	.00000

LAPLACE TRANSFORMS

These tables of Operations and Transforms were taken from "Modern Operational Mathematics in Engineering" by permission from the author, R. V. Churchill, and the publisher, McGraw-Hill Book Company, Inc.

The operational method for solving differential equations makes use of the Laplace transformation which associates a given function $F(t)$ with a second function $f(s)$, so that

$$f(s) = L\{F(t)\} = \int_0^\infty e^{-st}F(t)\,dt,$$

where t = a real variable, $F(t)$ = a real function of t; $F(t) = 0$, $t < 0$; $f(s)$ = a function of s; s = a complex variable; and $e = 2.71828\ldots$

$f(s)$ is called the Laplace transform of $F(t)$.

$F(t)$ can also be expressed as follows:

$$F(t) = \frac{1}{2\pi i}\int_{a-i\infty}^{a+i\infty} e^{st}f(s)\,ds,$$

where a is chosen to the right of any singularity of $f(s)$.

FOURIER TRANSFORMS

For a sectionally continuous function $F(x)$ over a finite interval of the variable x, in which the origin and a unit of length have been so chosen that the end points of the interval for x become $x = 0$ and $x = \pi$, the Fourier finite sine and cosine transforms for $F(x)$ are defined and symbolized respectively as follows:

$$S\{F(x)\} = \int_0^\pi F(x)\sin nx\,dx = f_s(n) \qquad (n = 1, 2, \cdots)$$

$$C\{F(x)\} = \int_0^\pi F(x)\cos nx\,dx = f_c(n) \qquad (n = 1, 2, \cdots)$$

If the interval for $F(x)$ is defined so that the end points are $x = 0$ and $x = l$, the substitution $\alpha = \dfrac{\pi x}{l}$ makes possible the use of these tables with

$$\int_0^l F(x)\sin\frac{n\pi x}{l}\,dx = \frac{l}{\pi}S\left\{F\left(\frac{lx}{\pi}\right)\right\}$$

$$\int_0^l F(x)\cos\frac{n\pi x}{l}\,dx = \frac{l}{\pi}C\left\{F\left(\frac{lx}{\pi}\right)\right\}$$

Thus if $F(x) = x$ for interval $(0,l)$

$$\int_0^l x\sin\frac{n\pi x}{l}\,dx = \frac{l}{\pi}S\left\{\frac{l}{\pi}x\right\} = \frac{l^2}{\pi^2}S\{x\} = \frac{l^2}{\pi^2}\frac{(-1)^{n+1}}{n} \qquad (n = 1, 2, \cdots)$$

LAPLACE OPERATIONS

	$F(t)$	$f(s)$
1	$F(t)$	$\int_0^\infty e^{-st}F(t)\,dt$
2	$AF(t) + BG(t)$	$Af(s) + Bg(s)$
3	$F'(t)$	$sf(s) - F(+0)$
4	$F^{(n)}(t)$	$s^nf(s) - s^{n-1}F(+0) - s^{n-2}F'(+0) - \cdots - F^{(n-1)}(+0)$
5	$\int_0^t F(\tau)\,d\tau$	$\frac{1}{s}f(s)$
6	$\int_0^t\int_0^\tau F(\lambda)\,d\lambda\,d\tau$	$\frac{1}{s^2}f(s)$
7	$\int_0^t F_1(t-\tau)F_2(\tau)\,d\tau = F_1 * F_2$	$f_1(s)f_2(s)$
8	$tF(t)$	$-f'(s)$
9	$t^nF(t)$	$(-1)^nf^{(n)}(s)$
10	$\frac{1}{t}F(t)$	$\int_s^\infty f(x)\,dx$
11	$e^{at}F(t)$	$f(s-a)$
12	$F(t-b)$, where $F(t) = 0$ when $t < 0$	$e^{-bs}f(s)$

LAPLACE OPERATIONS (Continued)

	$F(t)$	$f(s)$
13	$\frac{1}{c}F\left(\frac{t}{c}\right)$	$f(cs)$
14	$\frac{1}{c}e^{\frac{bt}{c}}F\left(\frac{t}{c}\right)$	$f(cs-b)$
15	$F(t+a) = F(t)$	$\dfrac{\int_0^a e^{-st}F(t)\,dt}{1-e^{-as}}$
16	$F(t+a) = -F(t)$	$\dfrac{\int_0^a e^{-st}F(t)\,dt}{1+e^{-as}}$
17	$F_1(t)$, the half-wave rectification of $F(t)$ in No. 16	$\dfrac{f(s)}{1-e^{-as}}$
18	$F_2(t)$, the full-wave rectification of $F(t)$ in No. 16	$f(s)\coth\frac{as}{2}$
19	$\sum_1^m \frac{p(a_n)}{q'(a_n)}e^{a_nt}$	$\frac{p(s)}{q(s)}$, $q(s) = (s-a_1)(s-a_2)\cdots(s -$
20	$e^{at}\sum_{n=1}^r \frac{\phi^{(r-n)}(a)}{(r-n)!}\frac{t^{n-1}}{(n-1)!} + \cdots$	$\frac{p(s)}{q(s)} = \frac{\phi(s)}{(s-a)^r}$

LAPLACE TRANSFORMS

	$f(s)$	$F(t)$
1	$\frac{1}{s}$	1
2	$\frac{1}{s^2}$	t
3	$\frac{1}{s^n}$ $(n = 1, 2, \cdots)$	$\frac{t^{n-1}}{(n-1)!}$
4	$\frac{1}{\sqrt{s}}$	$\frac{1}{\sqrt{\pi t}}$
5	$s^{-\frac{3}{2}}$	$2\sqrt{\frac{t}{\pi}}$
6	$s^{-(n+\frac{1}{2})}$ $(n = 1, 2, \cdots)$	$\frac{2^nt^{n-\frac{1}{2}}}{1\cdot3\cdot5\cdots(2n-1)\sqrt{\pi}}$
7	$\frac{\Gamma(k)}{s^k}$ $(k > 0)$	t^{k-1}
8	$\frac{1}{s-a}$	e^{at}
9	$\frac{1}{(s-a)^2}$	te^{at}
10	$\frac{1}{(s-a)^n}$ $(n = 1, 2, \cdots)$	$\frac{1}{(n-1)!}t^{n-1}e^{at}$
11	$\frac{\Gamma(k)}{(s-a)^k}$ $(k > 0)$	$t^{k-1}e^{at}$
12*	$\frac{1}{(s-a)(s-b)}$	$\frac{1}{a-b}(e^{at} - e^{bt})$
13*	$\frac{s}{(s-a)(s-b)}$	$\frac{1}{a-b}(ae^{at} - be^{bt})$
14*	$\frac{1}{(s-a)(s-b)(s-c)}$	$-\frac{(b-c)e^{at} + (c-a)e^{bt} + (a -}{(a-b)(b-c)(c-a)}$
15	$\frac{1}{s^2+a^2}$	$\frac{1}{a}\sin at$
16	$\frac{s}{s^2+a^2}$	$\cos at$
17	$\frac{1}{s^2-a^2}$	$\frac{1}{a}\sinh at$
18	$\frac{s}{s^2-a^2}$	$\cosh at$
19	$\frac{1}{s(s^2+a^2)}$	$\frac{1}{a^2}(1-\cos at)$
20	$\frac{1}{s^2(s^2+a^2)}$	$\frac{1}{a^3}(at - \sin at)$
21	$\frac{1}{(s^2+a^2)^2}$	$\frac{1}{2a^3}(\sin at - at\cos at)$
22	$\frac{s}{(s^2+a^2)^2}$	$\frac{t}{2a}\sin at$
23	$\frac{s^2}{(s^2+a^2)^2}$	$\frac{1}{2a}(\sin at + at\cos at)$

* Here a, b, and (in 14) c represent distinct constants.

	$f(s)$	$F(t)$
24	$\frac{s^2 - a^2}{(s^2 + a^2)^2}$	$t \cos at$
25	$\frac{s}{(s^2 + a^2)(s^2 + b^2)}$ $(a^2 \neq b^2)$	$\frac{\cos at - \cos bt}{b^2 - a^2}$
26	$\frac{1}{(s - a)^2 + b^2}$	$\frac{1}{b} e^{at} \sin bt$
27	$\frac{s - a}{(s - a)^2 + b^2}$	$e^{at} \cos bt$
28	$\frac{3a^2}{s^3 + a^3}$	$e^{-at} - e^{\frac{at}{2}} \left(\cos \frac{at\sqrt{3}}{2} - \sqrt{3} \sin \frac{at\sqrt{3}}{2} \right)$
29	$\frac{4a^3}{s^4 + 4a^4}$	$\sin at \cosh at - \cos at \sinh at$
30	$\frac{s}{s^4 + 4a^4}$	$\frac{1}{2a^2} \sin at \sinh at$
31	$\frac{1}{s^4 - a^4}$	$\frac{1}{2a^3} (\sinh at - \sin at)$
32	$\frac{s}{s^4 - a^4}$	$\frac{1}{2a^2} (\cosh at - \cos at)$
33	$\frac{8a^3 s^2}{(s^2 + a^2)^3}$	$(1 + a^2 t^2) \sin at - at \cos at$
34*	$\frac{1}{s} \left(\frac{s - 1}{s} \right)^n$	$L_n(t) = \frac{e^t}{n!} \frac{d^n}{dt^n} (t^n e^{-t})$
35	$\frac{s}{(s - a)^{\frac{3}{2}}}$	$\frac{1}{\sqrt{\pi t}} e^{at} (1 + 2at)$
36	$\sqrt{s - a} - \sqrt{s - b}$	$\frac{1}{2\sqrt{\pi t^3}} (e^{bt} - e^{at})$
37	$\frac{1}{\sqrt{s} + a}$	$\frac{1}{\sqrt{\pi t}} - a e^{a^2 t} \operatorname{erfc} (a\sqrt{t})$
38	$\frac{\sqrt{s}}{s - a^2}$	$\frac{1}{\sqrt{\pi t}} + a e^{a^2 t} \operatorname{erf} (a\sqrt{t})$
39	$\frac{\sqrt{s}}{s + a^2}$	$\frac{1}{\sqrt{\pi t}} - \frac{2a}{\sqrt{\pi}} e^{-a^2 t} \int_0^{a\sqrt{t}} e^{\lambda^2} d\lambda$
40	$\frac{1}{\sqrt{s}(s - a^2)}$	$\frac{1}{a} e^{a^2 t} \operatorname{erf} (a\sqrt{t})$
41	$\frac{1}{\sqrt{s}(s + a^2)}$	$\frac{2}{a\sqrt{\pi}} e^{-a^2 t} \int_0^{a\sqrt{t}} e^{\lambda^2} d\lambda$
42	$\frac{b^2 - a^2}{(s - a^2)(b + \sqrt{s})}$	$e^{a^2 t}[b - a \operatorname{erf} (a\sqrt{t})] - b e^{b^2 t} \operatorname{erfc} (b\sqrt{t})$
43	$\frac{1}{\sqrt{s}(\sqrt{s} + a)}$	$e^{a^2 t} \operatorname{erfc} (a\sqrt{t})$
44	$\frac{1}{(s + a)\sqrt{s + b}}$	$\frac{1}{\sqrt{b - a}} e^{-at} \operatorname{erf} (\sqrt{b - a}\sqrt{t})$
45	$\frac{b^2 - a^2}{\sqrt{s}(s - a^2)(\sqrt{s} + b)}$	$e^{a^2 t} \left[\frac{b}{a} \operatorname{erf} (a\sqrt{t}) - 1 \right] + e^{b^2 t} \operatorname{erfc} (b\sqrt{t})$
46*	$\frac{(1 - s)^n}{s^{n + \frac{1}{2}}}$	$\frac{n!}{(2n)!\sqrt{\pi t}} H_{2n}(\sqrt{t})$
47	$\frac{(1 - s)^n}{s^{n + \frac{3}{2}}}$	$-\frac{n!}{\sqrt{\pi}(2n + 1)!} H_{2n+1}(\sqrt{t})$
48†	$\frac{\sqrt{s + 2a}}{\sqrt{s}} - 1$	$a e^{-at}[I_1(at) + I_0(at)]$
49	$\frac{1}{\sqrt{s + a}\sqrt{s + b}}$	$e^{-\frac{1}{2}(a+b)t} I_0 \left(\frac{a - b}{2} t \right)$
50	$\frac{\Gamma(k)}{(s + a)^k (s + b)^k}$ $(k > 0)$	$\sqrt{\pi} \left(\frac{t}{a - b} \right)^{k - \frac{1}{2}} e^{-\frac{1}{2}(a+b)t} I_{k-\frac{1}{2}} \left(\frac{a - b}{2} t \right)$
51	$\frac{1}{(s + a)^{\frac{1}{2}}(s + b)^{\frac{3}{2}}}$	$t e^{-\frac{1}{2}(a+b)t} \left[I_0 \left(\frac{a - b}{2} t \right) + I_1 \left(\frac{a - b}{2} t \right) \right]$
52	$\frac{\sqrt{s + 2a} - \sqrt{s}}{\sqrt{s + 2a} + \sqrt{s}}$	$\frac{1}{t} e^{-at} I_1(at)$
53	$\frac{(a - b)^k}{(\sqrt{s + a} + \sqrt{s + b})^{2k}}$ $(k > 0)$	$\frac{k}{t} e^{-\frac{1}{2}(a+b)t} I_k \left(\frac{a - b}{2} t \right)$

* $L_n(t)$ is the Laguerre polynomial of degree n.

LAPLACE TRANSFORMS (Continued)

	$f(s)$	$F(t)$
54	$\frac{(\sqrt{s + a} + \sqrt{s})^{-2\nu}}{\sqrt{s}\sqrt{s + a}}$ $(\nu > -1)$	$\frac{1}{a^\nu} e^{-\frac{1}{2}at} I_\nu \left(\frac{1}{2} at \right)$
55	$\frac{1}{\sqrt{s^2 + a^2}}$	$J_0(at)$
56	$\frac{(\sqrt{s^2 + a^2} - s)^\nu}{\sqrt{s^2 + a^2}}$ $(\nu > -1)$	$a^\nu J_\nu(at)$
57	$\frac{1}{(s^2 + a^2)^k}$ $(k > 0)$	$\frac{\sqrt{\pi}}{\Gamma(k)} \left(\frac{t}{2a} \right)^{k - \frac{1}{2}} J_{k-\frac{1}{2}}(at)$
58	$(\sqrt{s^2 + a^2} - s)^k (k > 0)$	$\frac{k a^k}{t} J_k(at)$
59	$\frac{(s - \sqrt{s^2 - a^2})^\nu}{\sqrt{s^2 - a^2}}$ $(\nu > -1)$	$a^\nu I_\nu(at)$
60	$\frac{1}{(s^2 - a^2)^k}$ $(k > 0)$	$\frac{\sqrt{\pi}}{\Gamma(k)} \left(\frac{t}{2a} \right)^{k - \frac{1}{2}} I_{k-\frac{1}{2}}(at)$
61	$\frac{e^{-ks}}{s}$	$S_k(t) = \begin{cases} 0 \text{ when } 0 < t < k \\ 1 \text{ when } t > k \end{cases}$
62	$\frac{e^{-ks}}{s^2}$	$\begin{cases} 0 & \text{when } 0 < t < k \\ t - k & \text{when } t > k \end{cases}$
63	$\frac{e^{-ks}}{s^\mu}$ $(\mu > 0)$	$\begin{cases} 0 & \text{when } 0 < t < k \\ \frac{(t - k)^{\mu - 1}}{\Gamma(\mu)} & \text{when } t > k \end{cases}$
64	$\frac{1 - e^{-ks}}{s}$	$\begin{cases} 1 \text{ when } 0 < t < k \\ 0 \text{ when } t > k \end{cases}$
65	$\frac{1}{s(1 - e^{-ks})} = \frac{1 + \coth \frac{1}{2}ks}{2s}$	$S(k, t) = n$ when $(n - 1)k < t < nk (n = 1, 2, \cdots)$
66	$\frac{1}{s(e^{ks} - a)}$	$\begin{cases} 0 \text{ when } 0 < t < k \\ 1 + a + a^2 + \cdots + a^{n-1} \\ \quad \text{when } nk < t < (n + 1)k \\ \quad (n = 1, 2, \cdots) \end{cases}$
67	$\frac{1}{s} \tanh ks$	$M(2k, t) = (-1)^{n-1}$ when $2k(n - 1) < t < 2kn$ $(n = 1, 2, \cdots)$
68	$\frac{1}{s(1 + e^{-ks})}$	$\frac{1}{2} M(k, t) + \frac{1}{2} = \frac{1 - (-1)^n}{2}$ when $(n - 1)k < t < nk$
69*	$\frac{1}{s^2} \tanh ks$	$H(2k, t)$
70	$\frac{1}{s \sinh ks}$	$2S(2k, t + k) - 2 = 2(n - 1)$ when $(2n - 3)k < t < (2n - 1)k$ $(t > 0)$
71	$\frac{1}{s \cosh ks}$	$M(2k, t + 3k) + 1 = 1 + (-1)^n$ when $(2n - 3)k < t < (2n - 1)k$ $(t > 0)$
72	$\frac{1}{s} \coth ks$	$2S(2k, t) - 1 = 2n - 1$ when $2k(n - 1) < t < 2kn$
73	$\frac{k}{s^2 + k^2} \coth \frac{\pi s}{2k}$	$\|\sin kt\|$
74	$\frac{1}{(s^2 + 1)(1 - e^{-\pi s})}$	$\begin{cases} \sin t & \text{when } (2n - 2)\pi < t < (2n - 1)\pi \\ 0 & \text{when } (2n - 1)\pi < t < 2n\pi \end{cases}$
75	$\frac{1}{s} e^{-\frac{k}{s}}$	$J_0(2\sqrt{kt})$
76	$\frac{1}{\sqrt{s}} e^{-\frac{k}{s}}$	$\frac{1}{\sqrt{\pi t}} \cos 2\sqrt{kt}$
77	$\frac{1}{\sqrt{s}} e^{\frac{k}{s}}$	$\frac{1}{\sqrt{\pi t}} \cosh 2\sqrt{kt}$
78	$\frac{1}{s^{\frac{3}{2}}} e^{-\frac{k}{s}}$	$\frac{1}{\sqrt{\pi k}} \sin 2\sqrt{kt}$
79	$\frac{1}{s^{\frac{3}{2}}} e^{\frac{k}{s}}$	$\frac{1}{\sqrt{\pi k}} \sinh 2\sqrt{kt}$
80	$\frac{1}{s^\mu} e^{-\frac{k}{s}}$ $(\mu > 0)$	$\left(\frac{t}{k} \right)^{\frac{\mu - 1}{2}} J_{\mu-1}(2\sqrt{kt})$

* $H_n(x)$ is the Hermite polynomial, $H_n(x) = e^{x^2} \frac{d^n}{dx^n} (e^{-x^2})$.

† $I_n(x) = i^{-n} J_n(ix)$, where J_n is Bessel's function of the first kind.

* $H(2k,t) = k + (r - k)(-1)^n$ where $t = 2kn + r$; $0 \leq r < 2k$; $n = 0, 1, 2, \ldots$.

LAPLACE TRANSFORMS (Continued)

	$f(s)$	$F(t)$
81	$\frac{1}{s^{\mu}} e^{\frac{k}{s}}\ (\mu > 0)$	$\left(\frac{t}{k}\right)^{\frac{\mu-1}{2}} I_{\mu-1}(2\sqrt{kt})$
82	$e^{-k\sqrt{s}}\ (k > 0)$	$\frac{k}{2\sqrt{\pi t^3}} \exp\left(-\frac{k^2}{4t}\right)$
83	$\frac{1}{s} e^{-k\sqrt{s}}\ (k \geqq 0)$	$\operatorname{erfc}\left(\frac{k}{2\sqrt{t}}\right)$
84	$\frac{1}{\sqrt{s}} e^{-k\sqrt{s}}\ (k \geqq 0)$	$\frac{1}{\sqrt{\pi t}} \exp\left(-\frac{k^2}{4t}\right)$
85	$s^{-\frac{3}{2}} e^{-k\sqrt{s}}\ (k \geqq 0)$	$2\sqrt{\frac{t}{\pi}} \exp\left(-\frac{k^2}{4t}\right) - k \operatorname{erfc}\left(\frac{k}{2\sqrt{t}}\right)$
86	$\frac{ae^{-k\sqrt{s}}}{s(a+\sqrt{s})}\ (k \geqq 0)$	$-e^{ak}e^{a^2t} \operatorname{erfc}\left(a\sqrt{t} + \frac{k}{2\sqrt{t}}\right) + \operatorname{erfc}\left(\frac{k}{2\sqrt{t}}\right)$
87	$\frac{e^{-k\sqrt{s}}}{\sqrt{s}(a+\sqrt{s})}\ (k \geqq 0)$	$e^{ak}e^{a^2t} \operatorname{erfc}\left(a\sqrt{t} + \frac{k}{2\sqrt{t}}\right)$
88	$\frac{e^{-k\sqrt{s(s+a)}}}{\sqrt{s(s+a)}}$	$\begin{cases} 0 & \text{when } 0 < t < k \\ e^{-\frac{1}{2}at} I_0(\frac{1}{2}a\sqrt{t^2-k^2}) & \text{when } t > k \end{cases}$
89	$\frac{e^{-k\sqrt{s^2+a^2}}}{\sqrt{s^2+a^2}}$	$\begin{cases} 0 & \text{when } 0 < t < k \\ J_0(a\sqrt{t^2-k^2}) & \text{when } t > k \end{cases}$
90	$\frac{e^{-k\sqrt{s^2-a^2}}}{\sqrt{s^2-a^2}}$	$\begin{cases} 0 & \text{when } 0 < t < k \\ I_0(a\sqrt{t^2-k^2}) & \text{when } t > k \end{cases}$
91	$\frac{e^{-k(\sqrt{s^2+a^2}-s)}}{\sqrt{s^2+a^2}}\ (k \geqq 0)$	$J_0(a\sqrt{t^2+2kt})$
92	$e^{-ks} - e^{-k\sqrt{s^2+a^2}}$	$\begin{cases} 0 & \text{when } 0 < t < k \\ \frac{ak}{\sqrt{t^2-k^2}} J_1(a\sqrt{t^2-k^2}) & \text{when } t > k \end{cases}$
93	$e^{-k\sqrt{s^2-a^2}} - e^{-ks}$	$\begin{cases} 0 & \text{when } 0 < t < k \\ \frac{ak}{\sqrt{t^2-k^2}} I_1(a\sqrt{t^2-k^2}) & \text{when } t > k \end{cases}$
94	$\frac{a^{\nu} e^{-k\sqrt{s^2+a^2}}}{\sqrt{s^2+a^2}(\sqrt{s^2+a^2}+s)^{\nu}}\ (\nu > -1)$	$\begin{cases} 0 & \text{when } 0 < t < k \\ \left(\frac{t-k}{t+k}\right)^{\frac{1}{2}\nu} J_{\nu}(a\sqrt{t^2-k^2}) & \text{when } t > k \end{cases}$
95	$\frac{1}{s}\log s$	$\Gamma'(1) - \log t\quad [\Gamma'(1) = -0.5772]$
96	$\frac{1}{s^k}\log s\ (k > 0)$	$t^{k-1}\left\{\frac{\Gamma'(k)}{[\Gamma(k)]^2} - \frac{\log t}{\Gamma(k)}\right\}$
97	$\frac{\log s}{s-a}\ (a > 0)$	$e^{at}[\log a - \operatorname{Ei}(-at)]$
98	$\frac{\log s}{s^2+1}$	$\cos t\, \operatorname{Si}(t) - \sin t\, \operatorname{Ci}(t)$
99	$\frac{s\log s}{s^2+1}$	$-\sin t\, \operatorname{Si}(t) - \cos t\, \operatorname{Ci}(t)$
100	$\frac{1}{s}\log(1+ks)\ (k > 0)$	$-\operatorname{Ei}\left(-\frac{t}{k}\right)$
101	$\log\frac{s-a}{s-b}$	$\frac{1}{t}(e^{bt} - e^{at})$
102	$\frac{1}{s}\log(1+k^2s^2)$	$-2\operatorname{Ci}\left(\frac{t}{k}\right)$
103	$\frac{1}{s}\log(s^2+a^2)\ (a > 0)$	$2\log a - 2\operatorname{Ci}(at)$
104	$\frac{1}{s^2}\log(s^2+a^2)\ (a > 0)$	$\frac{2}{a}[at\log a + \sin at - at\, \operatorname{Ci}(at)]$
105	$\log\frac{s^2+a^2}{s^2}$	$\frac{2}{t}(1 - \cos at)$
106	$\log\frac{s^2-a^2}{s^2}$	$\frac{2}{t}(1 - \cosh at)$
107	$\arctan\frac{k}{s}$	$\frac{1}{t}\sin kt$
108	$\frac{1}{s}\arctan\frac{k}{s}$	$\operatorname{Si}(kt)$

LAPLACE TRANSFORMS (Continued)

	$f(s)$	$F(t)$
109	$e^{k^2s^2}\operatorname{erfc}(ks)\ (k > 0)$	$\frac{1}{k\sqrt{\pi}}\exp\left(-\frac{t^2}{4k^2}\right)$
110	$\frac{1}{s}e^{k^2s^2}\operatorname{erfc}(ks)\ (k > 0)$	$\operatorname{erf}\left(\frac{t}{2k}\right)$
111	$e^{ks}\operatorname{erfc}(\sqrt{ks})(k > 0)$	$\frac{\sqrt{k}}{\pi\sqrt{t}(t+k)}$
112	$\frac{1}{\sqrt{s}}\operatorname{erfc}(\sqrt{ks})$	$\begin{cases} 0 & \text{when } 0 < t < k \\ (\pi t)^{-\frac{1}{2}} & \text{when } t > k \end{cases}$
113	$\frac{1}{\sqrt{s}}e^{ks}\operatorname{erfc}(\sqrt{ks})\ (k > 0)$	$\frac{1}{\sqrt{\pi(t+k)}}$
114	$\operatorname{erf}\left(\frac{k}{\sqrt{s}}\right)$	$\frac{1}{\pi t}\sin(2k\sqrt{t})$
115	$\frac{1}{\sqrt{s}}e^{\frac{k^2}{s}}\operatorname{erfc}\left(\frac{k}{\sqrt{s}}\right)$	$\frac{1}{\sqrt{\pi t}}e^{-2k\sqrt{t}}$
116*	$K_0(ks)$	$\begin{cases} 0 & \text{when } 0 < t < k \\ (t^2-k^2)^{-\frac{1}{2}} & \text{when } t > k \end{cases}$
117	$K_0(k\sqrt{s})$	$\frac{1}{2t}\exp\left(-\frac{k^2}{4t}\right)$
118	$\frac{1}{s}e^{ks}K_1(ks)$	$\frac{1}{k}\sqrt{t(t+2k)}$
119	$\frac{1}{\sqrt{s}}K_1(k\sqrt{s})$	$\frac{1}{k}\exp\left(-\frac{k^2}{4t}\right)$
120	$\frac{1}{\sqrt{s}}e^{\frac{k}{s}}K_0\left(\frac{k}{s}\right)$	$\frac{2}{\sqrt{\pi t}}K_0(2\sqrt{2kt})$
121	$\pi e^{-ks}I_0(ks)$	$\begin{cases} [t(2k-t)]^{-\frac{1}{2}} & \text{when } 0 < t < 2k \\ 0 & \text{when } t > 2k \end{cases}$
122*	$\pi e^{-ks}I_1(ks)$	$\begin{cases} \frac{k-t}{\pi k\sqrt{t(2k-t)}} & \text{when } 0 < t < 2k \\ 0 & \text{when } t > 2k \end{cases}$

* $K_n(x)$ is Bessel's function of the second kind for the imaginary argume

* Several additional transforms, especially those involving other Bes[…] functions, can be found in the tables by G. A. Campbell and R. M. Fost[…] "Fourier Integrals for Practical Applications," or N. W. McLachlan a[…] P. Humbert, "Formulaire pour le calcul symbolique." In the tables [of] Campbell and Foster, only those entries containing the condition $0 < g$ [$-$] $k < g$, where g is our t, are Laplace transforms.

FOURIER TRANSFORMS

Finite Sine Transforms

	$f_s(n)$	$F(x)$
1	$f_s(n) = \int_0^{\pi} F(x)\sin nx\, dx\quad (n = 1, 2, \cdots)$	$F(x)$
2	$(-1)^{n+1}f_s(n)$	$F(\pi - x)$
3	$\frac{1}{n}$	$\frac{\pi - x}{\pi}$
4	$\frac{(-1)^{n+1}}{n}$	$\frac{x}{\pi}$
5	$\frac{1-(-1)^n}{n}$	1
6	$\frac{2}{n^2}\sin\frac{n\pi}{2}$	$\begin{cases} x & \text{when } 0 < x < \pi/2 \\ \pi - x & \text{when } \pi/2 < x < \pi \end{cases}$
7	$\frac{(-1)^{n+1}}{n^3}$	$\frac{x(\pi^2 - x^2)}{6\pi}$
8	$\frac{1-(-1)^n}{n^3}$	$\frac{x(\pi - x)}{2}$
9	$\frac{\pi^2(-1)^{n-1}}{n} - \frac{2[1-(-1)^n]}{n^3}$	x^2
10	$\pi(-1)^n\left(\frac{6}{n^3} - \frac{\pi^2}{n}\right)$	x^3
11	$\frac{n}{n^2+c^2}[1 - (-1)^n e^{c\pi}]$	e^{cx}
12	$\frac{n}{n^2+c^2}$	$\frac{\sinh c(\pi - x)}{\sinh c\pi}$

FOURIER TRANSFORMS (Continued)

	$f_s(n)$	$F(x)$
13	$\frac{n}{n^2 - k^2}$ $(k \neq 0, 1, 2, \cdots)$	$\frac{\sin k(\pi - x)}{\sin k\pi}$
14	$\begin{cases} \frac{\pi}{2} \text{ when } n = m \\ 0 \text{ when } n \neq m \end{cases}$ $(m = 1, 2, \cdots)$	$\sin mx$
15	$\frac{n}{n^2 - k^2} [1 - (-1)^n \cos k\pi]$ $(k \neq 1, 2, \cdots)$	$\cos kx$
16	$\begin{cases} \frac{n}{n^2 - m^2} [1 - (-1)^{n+m}] & \text{when } n \neq m = 1, 2, \cdots \\ 0 & \text{when } n = m \end{cases}$	$\cos mx$
17	$\frac{n}{(n^2 - k^2)^2}$ $(k \neq 0, 1, 2, \cdots)$	$\frac{\pi \sin kx}{2k \sin^2 k\pi} - \frac{x \cos k(\pi - x)}{2k \sin k\pi}$
18	$\frac{b^n}{n}$ $(\|b\| \leq 1)$	$\frac{2}{\pi} \arctan \frac{b \sin x}{1 - b \cos x}$
19	$\frac{1 - (-1)^n}{n} b^n$ $(\|b\| \leq 1)$	$\frac{2}{\pi} \arctan \frac{2b \sin x}{1 - b^2}$

Finite Cosine Transforms

	$f_c(n)$	$F(x)$
1	$f_c(n) = \int_0^\pi F(x) \cos nx\, dx$ $(n = 0, 1, 2, \cdots)$	$F(x)$
2	$(-1)^n f_c(n)$	$F(\pi - x)$
3	0 when $n = 1, 2, \cdots$; $f_c(0) = \pi$	1
4	$\frac{2}{n} \sin \frac{n\pi}{2}$; $f_c(0) = 0$	$\begin{cases} 1 \text{ when } 0 < x < \pi/2 \\ -1 \text{ when } \pi/2 < x < \pi \end{cases}$
5	$-\frac{1 - (-1)^n}{n^2}$; $f_c(0) = \frac{\pi^2}{2}$	x
6	$\frac{(-1)^n}{n^2}$; $f_c(0) = \frac{\pi^2}{6}$	$\frac{x^2}{2\pi}$
7	$\frac{1}{n^2}$; $f_c(0) = 0$	$\frac{(\pi - x)^2}{2\pi} - \frac{\pi}{6}$
8	$3\pi^2 \frac{(-1)^n}{n^2} - 6 \frac{1 - (-1)^n}{n^4}$; $f_c(0) = \frac{\pi^4}{4}$	x^3
9	$\frac{(-1)^n e^{c\pi} - 1}{n^2 + c^2}$	$\frac{1}{c} e^{cx}$
10	$\frac{1}{n^2 + c^2}$	$\frac{\cosh c(\pi - x)}{c \sinh c\pi}$
11	$\frac{k}{n^2 - k^2} [(-1)^n \cos \pi k - 1]$ $(k \neq 0, 1, 2, \cdots)$	$\sin kx$
12	$\frac{(-1)^{n+m} - 1}{n^2 - m^2}$; $f_c(m) = 0$ $(m = 1, 2, \cdots)$	$\frac{1}{m} \sin mx$
13	$\frac{1}{n^2 - k^2}$ $(k \neq 0, 1, 2, \cdots)$	$-\frac{\cos k(\pi - x)}{k \sin k\pi}$
14	0 when $n = 1, 2, \cdots$; $f_c(m) = \frac{\pi}{2}$ $(m = 1, 2, \cdots)$	$\cos mx$

The following functions appear among the entries of the preceding tables on transforms.

Function	Definition	Name
$Ei(x)$	$\int_{-\infty}^{x} \frac{e^v}{v} dv$; or sometimes defined as $-Ei(-x) = \int_x^\infty \frac{e^{-v}}{v} dv$	See index for Sine, Cosine, and Exponential Integral tables
$Si(x)$	$\int_0^x \frac{\sin v}{v} dv$	See index for Sine, Cosine, and Exponential Integral tables
$Ci(x)$	$\int_\infty^x \frac{\cos v}{v} dv$; or sometimes defined as negative of this integral	See index for Sine, Cosine, and Exponential Integral tables
$erf(x)$	$\frac{2}{\sqrt{\pi}} \int_0^x e^{-v^2} dv$	Error function
$erfc(x)$	$1 - erf(x) = \frac{2}{\sqrt{\pi}} \int_x^\infty e^{-v^2} dv$	Complementary function to error function
$L_n(x)$	$\frac{e^x}{n!} \frac{d^n}{dx^n} (x^n e^{-x})$, $n = 0, 1, \cdots$	Laguerre polynomial of degree n

Special Formulas

Certain types of differential equations occur sufficiently often to justify the use of formulas for the corresponding particular solutions. The following set of tables I to XIV covers all first-second and n'th order ordinary linear differential equations with constant coefficients for which the right members are of the form $P(x)e^{rx}\sin sx$ or $P(x)e^{rx}\cos sx$ where r and s are constants and $P(x)$ is a polynomial of degree n.

When the right member of a reducible partial linear differential equation with constant coefficients is not zero, particular solutions for certain types of right members are contained in tables XV to XXI. In these tables both F and P are used to denote polynomials, and it is assumed that no denominator is zero. In any formula the roles of x and y may be reversed throughout, changing a formula in which x dominates to one in which y dominates. Tables XIX, XX, XXI are applicable whether the equations are reducible or not. The symbol $\binom{m}{n}$ stands for $\frac{m!}{(m-n)!n!}$ and is the $n+1$ st coefficient in the expansion of $(a+b)^m$. Also $0! = 1$ by definition.

The tables as herewith given are those contained in the text Differential Equations by Ginn and Company (1955) and are published with their kind permission and that of the author, Professor Frederick H. Steen.

TABLE I: $(D-a)y = R$

R	y_p
1. e^{rx}	$\frac{e^{rx}}{r-a}$
2. $\sin sx$*	$-\frac{a\sin sx + s\cos sx}{a^2+s^2} = -\frac{1}{\sqrt{a^2+s^2}}\sin\left(sx+\tan^{-1}\frac{s}{a}\right)$
3. $P(x)$	$-\frac{1}{a}\left[P(x)+\frac{P'(x)}{a}+\frac{P''(x)}{a^2}+\cdots+\frac{P^{(n)}(x)}{a^n}\right]$
4. $e^{rx}\sin sx$*	Replace a by $a-r$ in formula 2 and multiply by e^{rx}.
5. $P(x)e^{rx}$	Replace a by $a-r$ in formula 3 and multiply by e^{rx}.
6. $P(x)\sin sx$*	$-\sin sx\left[\frac{a}{a^2+s^2}P(x)+\frac{a^2-s^2}{(a^2+s^2)^2}P'(x)+\frac{a^3-3as^2}{(a^2+s^2)^3}P''(x)+\cdots+\frac{a^k-\binom{k}{2}a^{k-2}s^2+\binom{k}{4}a^{k-4}s^4-\cdots}{(a^2+s^2)^k}P^{(k-1)}(x)+\cdots\right]$ $-\cos sx\left[\frac{s}{a^2+s^2}P(x)+\frac{2as}{(a^2+s^2)^2}P'(x)+\frac{3a^2s-s^3}{(a^2+s^2)^3}P''(x)+\cdots+\frac{\binom{k}{1}a^{k-1}s-\binom{k}{3}a^{k-3}s^3+\cdots}{(a^2+s^2)^k}P^{(k-1)}(x)+\cdots\right]$
7. $P(x)e^{rx}\sin sx$*	Replace a by $a-r$ in formula 6 and multiply by e^{rx}.
8. e^{ax}	xe^{ax}
9. $e^{ax}\sin sx$*	$-\frac{e^{ax}\cos sx}{s}$
10. $P(x)e^{ax}$	$e^{ax}\int P(x)\,dx$
11. $P(x)e^{ax}\sin sx$*	$\frac{e^{ax}\sin sx}{s}\left[\frac{P'(x)}{s^3}-\frac{P'''(x)}{s^3}+\frac{P^{v}(x)}{s^5}-\cdots\right]-\frac{e^{ax}\cos sx}{s}\left[P(x)-\frac{P''(x)}{s^2}+\frac{P^{iv}(x)}{s^4}-\cdots\right]$

* For $\cos sx$ in R replace "sin" by "cos" and "cos" by "− sin" in y_p.

$$D^n = \frac{d^n}{dx^n} \qquad \binom{m}{n} = \frac{m!}{(m-n)!n!} \qquad 0! = 1$$

TABLE II: $(D - a)^2 y = R$

R	y_p
12. e^{rx}	$\dfrac{e^{rx}}{(r-a)^2}$
13. $\sin sx$*	$\dfrac{1}{(a^2+s^2)^2}[(a^2-s^2)\sin sx + 2as\cos sx] = \dfrac{1}{a^2+s^2}\sin\left(sx + \tan^{-1}\dfrac{2as}{a^2-s^2}\right)$
14. $P(x)$	$\dfrac{1}{a^2}\left[P(x) + \dfrac{2P'(x)}{a} + \dfrac{3P''(x)}{a^2} + \cdots + \dfrac{(n+1)P^{(n)}(x)}{a^n}\right]$
15. $e^{rx}\sin sx$*	Replace a by $a - r$ in formula 13 and multiply by e^{rx}.
16. $P(x)e^{rx}$	Replace a by $a - r$ in formula 14 and multiply by e^{rx}.
17. $P(x)\sin sx$*	$\sin sx\left[\dfrac{a^2-s^2}{(a^2+s^2)^2}P(x) + 2\dfrac{a^3-3as^2}{(a^2+s^2)^3}P'(x) + 3\dfrac{a^4-6a^2s^2+s^4}{(a^2+s^2)^4}P''(x) + \cdots + (k-1)\dfrac{a^k - \binom{k}{2}a^{k-2}s^2 + \binom{k}{4}a^{k-4}s^4 - \cdots}{(a^2+s^2)^k}P^{(k-2)}(x) + \cdots\right]$ $+ \cos sx\left[\dfrac{2as}{(a^2+s^2)^2}P(x) + 2\dfrac{3a^2s-s^3}{(a^2+s^2)^3}P'(x) + 3\dfrac{4a^3s-4as^3}{(a^2+s^2)^4}P''(x) + \cdots + (k-1)\dfrac{\binom{k}{1}a^{k-1}s - \binom{k}{3}a^{k-3}s^3 + \cdots}{(a^2+s^2)^k}P^{(k-2)}(x) + \cdots\right]$
18. $P(x)e^{rx}\sin sx$*	Replace a by $a - r$ in formula 17 and multiply by e^{rx}.
19. e^{ax}	$\frac{1}{2}x^2e^{ax}$
20. $e^{ax}\sin sx$*	$-\dfrac{e^{ax}\sin sx}{s^2}$
21. $P(x)e^{ax}$	$e^{ax}\displaystyle\int\int P(x)dx\,dx$
22. $P(x)e^{ax}\sin sx$*	$-\dfrac{e^{ax}\sin sx}{s^2}\left[P(x) - \dfrac{3P''(x)}{s^2} + \dfrac{5P^{iv}(x)}{s^4} - \dfrac{7P^{vi}(x)}{s^6} + \cdots\right] - \dfrac{e^{ax}\cos sx}{s^2}\left[\dfrac{2P'(x)}{s} - \dfrac{4P'''(x)}{s^3} + \dfrac{6P^{v}(x)}{s^5} - \cdots\right]$

TABLE III: $(D^2 + q)y = R$

R	y_p
23. e^{rx}	$\dfrac{e^{rx}}{r^2+q}$
24. $\sin sx$*	$\dfrac{\sin sx}{-s^2+q}$
25. $P(x)$	$\dfrac{1}{q}\left[P(x) - \dfrac{P''(x)}{q} + \dfrac{P^{iv}(x)}{q^2} - \cdots + (-1)^k\dfrac{P^{(2k)}(x)}{q^k}\cdots\right]$
26. $e^{rx}\sin sx$*	$\dfrac{(r^2-s^2+q)e^{rx}\sin sx - 2rse^{rx}\cos sx}{(r^2-s^2+q)^2+(2rs)^2} = \dfrac{e^{rx}}{\sqrt{(r^2-s^2+q)^2+(2rs)^2}}\sin\left[sx - \tan^{-1}\dfrac{2rs}{r^2-s^2+q}\right]$
27 $P(x)e^{rx}$	$\dfrac{e^{rx}}{r^2+q}\left[P(x) - \dfrac{2r}{r^2+q}P'(x) + \dfrac{3r^2-q}{(r^2+q)^2}P''(x) - \dfrac{4r^3-4qr}{(r^2+q)^3}P'''(x) + \cdots + (-1)^{k-1}\dfrac{\binom{k}{1}r^{k-1} - \binom{k}{3}r^{k-3}q + \binom{k}{5}r^{k-5}q^2 - \cdots}{(r^2+q)^{k-1}}P^{(k-1)}(x) + \cdots\right]$
28. $P(x)\sin sx$*	$\dfrac{\sin sx}{(-s^2+q)}\left[P(x) - \dfrac{3s^2+q}{(-s^2+q)^2}P''(x) + \dfrac{5s^4+10s^2q+q^2}{(-s^2+q)^4}P^{iv}(x) + \cdots + (-1)^k\dfrac{\binom{2k+1}{1}s^{2k} + \binom{2k+1}{3}s^{2k-2}q + \binom{2k+1}{5}s^{2k-4}q^2 + \cdots}{(-s^2+q)^{2k}}P^{(2k)}(x) + \cdots\right]$ $-\dfrac{s\cos sx}{(-s^2+q)}\left[\dfrac{2P'(x)}{(-s^2+q)} - \dfrac{4s^2+4q}{(-s^2+q)^3}P'''(x) + \cdots + (-1)^{k+1}\dfrac{\binom{2k}{1}s^{2k-2} + \binom{2k}{3}s^{2k-4}q + \cdots}{(-s^2+q)^{2k-1}}P^{(2k-1)}(x) + \cdots\right]$

TABLE IV: $(D^2 + b^2)y = R$

R	y_p
29. $\sin bx$*	$-\dfrac{x\cos bx}{2b}$
30. $P(x)\sin bx$*	$\dfrac{\sin bx}{(2b)^2}\left[P(x) - \dfrac{P''(x)}{(2b)^2} + \dfrac{P^{iv}(x)}{(2b)^4} - \cdots\right] - \dfrac{\cos bx}{2b}\displaystyle\int\left[P(x) - \dfrac{P''(x)}{(2b)^2} + \cdots\right]dx$

* For $\cos sx$ in R replace "sin" by "cos" and "cos" by "− sin" in y_p.

TABLE V: $(D^2 + pD + q)y = R$

R	y_p
31. e^{rx}	$\dfrac{e^{rx}}{r^2 + pr + q}$
32. $\sin sx$*	$\dfrac{(q - s^2)\sin sx - ps\cos sx}{(q - s^2)^2 + (ps)^2} = \dfrac{1}{\sqrt{(q - s^2)^2 + (ps)^2}} \sin\left(sx - \tan^{-1}\dfrac{ps}{q - s^2}\right)$
33. $P(x)$	$\dfrac{1}{q}\left[P(x) - \dfrac{p}{q}P'(x) + \dfrac{p^2 - q}{q^2}P''(x) - \dfrac{p^3 - 2pq}{q^3}P'''(x) + \cdots + (-1)^n \dfrac{p^n - \binom{n-1}{1}p^{n-2}q + \binom{n-2}{2}p^{n-4}q^2 - \cdots}{q^n} P^{(n)}(x)\right]$
34. $e^{rx}\sin sx$*	Replace p by $p + 2r$, q by $q + pr + r^2$ in formula 32 and multiply by e^{rx}.
35. $P(x)e^{rx}$	Replace p by $p + 2r$, q by $q + pr + r^2$ in formula 33 and multiply by e^{rx}.

TABLE VI: $(D - b)(D - a)y = R$

36. $P(x)\sin sx$*

$$\frac{\sin sx}{b - a}\left[\left(\frac{a}{a^2 + s^2} - \frac{b}{b^2 + s^2}\right)P(x) + \left(\frac{a^2 - s^2}{(a^2 + s^2)^2} - \frac{b^2 - s^2}{(b^2 + s^2)^2}\right)P'(x) + \left(\frac{a^3 - 3as^2}{(a^2 + s^2)^3} - \frac{b^3 - 3bs^2}{(b^2 + s^2)^3}\right)P''(x) + \cdots\right]$$
$$+ \frac{\cos sx}{b - a}\left[\left(\frac{s}{a^2 + s^2} - \frac{s}{b^2 + s^2}\right)P(x) + \left(\frac{2as}{(a^2 + s^2)^2} - \frac{2bs}{(b^2 + s^2)^2}\right)P'(x) + \left(\frac{3a^2s - s^3}{(a^2 + s^2)^3} - \frac{3b^2s - s^3}{(b^2 + s^2)^3}\right)P''(x) + \cdots\right]^{\dagger}$$

37. $P(x)e^{rx}\sin sx$* — Replace a by $a - r$, b by $b - r$ in formula 36 and multiply by e^{rx}.

38. $P(x)e^{ax}$ — $\dfrac{e^{ax}}{a - b}\left[\int P(x)dx + \dfrac{P(x)}{(b - a)} + \dfrac{P'(x)}{(b - a)^2} + \dfrac{P''(x)}{(b - a)^3} + \cdots + \dfrac{P^{(n)}(x)}{(b - a)^{n+1}}\right]$

TABLE VII: $(D^2 - 2aD + a^2 + b^2)y = R$

R — y_p

39. $P(x)\sin sx$*

$$\frac{\sin sx}{2b}\left[\left(\frac{s + b}{a^2 + (s + b)^2} - \frac{s - b}{a^2 + (s - b)^2}\right)P(x) + \left(\frac{2a(s + b)}{[a^2 + (s + b)^2]^2} - \frac{2a(s - b)}{[a^2 + (s - b)^2]^2}\right)P'(x) + \left(\frac{3a^2(s + b) - (s + b)^3}{[a^2 + (s + b)^2]^3} - \frac{3a^2(s - b) - (s - b)^3}{[a^2 + (s - b)^2]^3}\right)P''(x) + \cdots\right]$$
$$- \frac{\cos sx}{2b}\left[\left(\frac{a}{a^2 + (s + b)^2} - \frac{a}{a^2 + (s - b)^2}\right)P(x) + \left(\frac{a^2 - (s + b)^2}{[a^2 + (s + b)^2]^2} - \frac{a^2 - (s - b)^2}{[a^2 + (s - b)^2]^2}\right]P'(x) + \left(\frac{a^3 - 3a(s + b)^2}{[a^2 + (s + b)^2]^3} - \frac{a^3 - 3a(s - b)^2}{[a^2 + (s - b)^2]^3}\right)P''(x) + \cdots\right]^{\dagger}$$

40. $P(x)e^{rx}\sin sx$* — Replace a by $a - r$ in formula 39 and multiply by e^{rx}.

41. $P(x)e^{ax}$ — $\dfrac{e^{ax}}{b^2}\left[P(x) - \dfrac{P''(x)}{b^2} + \dfrac{P^{iv}(x)}{b^4} - \cdots\right]$

42. $e^{ax}\sin sx$* — $\dfrac{e^{ax}\sin sx}{-s^2 + b^2}$

43. $e^{ax}\sin bx$* — $-\dfrac{xe^{ax}\cos bx}{2b}$

44. $P(x)e^{ax}\sin bx$* — $\dfrac{e^{ax}\sin bx}{(2b)^2}\left[P(x) - \dfrac{P''(x)}{(2b)^2} + \dfrac{P^{iv}(x)}{(2b)^4} - \cdots\right] - \dfrac{e^{ax}\cos bx}{2b}\int\left[P(x) - \dfrac{P''(x)}{(2b)^2} + \dfrac{P^{iv}(x)}{(2b)^4} - \cdots\right]dx$

TABLE VIII: $f(D)y = [D^n + a_{n-1}D^{n-1} + \cdots + a_1D + a_0]y = R$

45. e^{rx} — $\dfrac{e^{rx}}{f(r)}$

46. $\sin sx$* — $\dfrac{[a_0 - a_2s^2 + a_4s^4 - \cdots]\sin sx - [a_1s - a_3s^3 + a_5s^5 + \cdots]\cos sx}{[a_0 - a_2s^2 + a_4s^4 - \cdots]^2 + [a_1s - a_3s^3 + a_5s^5 - \cdots]^2}$

TABLE IX: $f(D^2)y = R$

R	y_p
47. $\sin sx$*	$\dfrac{\sin sx}{f(-s^2)} = \dfrac{\sin sx}{a_0 - a_2s^2 + \cdots \pm s^{2n}}$

* For $\cos sx$ in R replace "sin" by "cos" and "cos" by "− sin" in y_p.
† For additional terms, compare with formula 6.

TABLE X: $(D - a)^n = R$

48. e^{rx} $\quad \dfrac{e^{rx}}{(r-a)^n}$

49. $\sin sx$* $\quad \dfrac{(-1)^n}{(a^2+s^2)^n}\left\{\left[a^n - \binom{n}{2}a^{n-2}s^2 + \binom{n}{4}a^{n-4}s^4 - \cdots\right]\sin sx + \left[\binom{n}{1}a^{n-1}s - \binom{n}{3}a^{n-3}s^3 + \cdots\right]\cos sx\right\}$

50. $P(x)$ $\quad \dfrac{(-1)^n}{a^n}\left[P(x) + \binom{n}{1}\dfrac{P'(x)}{a} + \binom{n+1}{2}\dfrac{P''(x)}{a^2} + \binom{n+2}{3}\dfrac{P'''(x)}{a^3} + \cdots\right]$

51. $e^{rx}\sin sx$* Replace a by $a - r$ in formula 49 and multiply by e^{rx}.

52. $e^{rx}P(x)$ Replace a by $a - r$ in formula 50 and multiply by e^{rx}.

53 $P(x)\sin sx$*

$$(-1)^n \sin sx\left[A_nP(x) + \binom{n}{1}A_{n+1}P'(x) + \binom{n+1}{2}A_{n+2}P''(x) + \binom{n+2}{3}A_{n+3}P'''(x) + \cdots\right]$$
$$+ (-1)^n \cos sx\left[B_nP(x) + \binom{n}{1}B_{n+1}P'(x) + \binom{n+1}{2}B_{n+2}P''(x) + \binom{n+2}{3}B_{n+3}P'''(x) + \cdots\right]$$
$$A_1 = \frac{a}{a^2+s^2},\ A_2 = \frac{a^2-s^2}{(a^2+s^2)^2},\ \cdots,\ A_k = \frac{a^k - \binom{k}{2}a^{k-2}s^2 + \binom{k}{4}a^{k-4}s^4 - \cdots}{(a^2+s^2)^k}.$$
$$B_1 = \frac{s}{a^2+s^2},\ B_2 = \frac{2as}{(a^2+s^2)^2},\ \cdots,\ B_k = \frac{\binom{k}{1}a^{k-1}s - \binom{k}{3}a^{k-3}s^3 + \cdots}{(a^2+s^2)^k}.$$

54. $P(x)e^{rx}\sin sx$* Replace a by $a - r$ in formula 53 and multiply by e^{rx}.

55. $e^{ax}P(x)$ $\quad e^{ax}\displaystyle\iint\cdots\int P(x)dx^n$

R $\quad y_p$

56. $P(x)e^{ax}\sin sx$*

$$\frac{(-1)^{\frac{n-1}{2}}e^{ax}\sin sx}{s^n}\left[\binom{n}{n-1}\frac{P'(x)}{s} - \binom{n+2}{n-1}\frac{P'''(x)}{s^3} + \binom{n+4}{n-1}\frac{P^{v}(x)}{s^5} - \cdots\right]$$
$$+ \frac{(-1)^{\frac{n+1}{2}}e^{ax}\cos sx}{s^n}\left[\binom{n-1}{n-1}P(x) - \binom{n+1}{n-1}\frac{P''(x)}{s^2} + \binom{n+3}{n-1}\frac{P^{iv}(x)}{s^4} - \cdots\right] \quad (n \text{ odd})$$

$$\frac{(-1)^{\frac{n}{2}}e^{ax}\sin sx}{s^n}\left[\binom{n-1}{n-1}P(x) - \binom{n+1}{n-1}\frac{P''(x)}{s^2} + \binom{n+3}{n-1}\frac{P^{iv}(x)}{s^4} - \cdots\right]$$
$$+ \frac{(-1)^{\frac{n}{2}}e^{ax}\cos sx}{s^n}\left[\binom{n}{n-1}\frac{P'(x)}{s} - \binom{n+2}{n-1}\frac{P'''(x)}{s^3} + \binom{n+4}{n-1}\frac{P^{v}(x)}{s^5} - \cdots\right] \quad (n \text{ even})$$

TABLE XI: $(D - a)^n f(D)y = R$

57. e^{ax} $\quad \dfrac{x^n}{n!}\cdot\dfrac{e^{ax}}{f(a)}$

TABLE XII: $(D^2 + q)^n y = R$

58. e^{rx} $\quad e^{rx}/(r^2+q)^n$

59. $\sin sx$* $\quad \sin sx/(q-s^2)^n$

60. $P(x)$ $\quad \dfrac{1}{q^n}\left[P(x) - \binom{n}{1}\dfrac{P''(x)}{q} + \binom{n+1}{2}\dfrac{P^{iv}(x)}{q^2} - \binom{n+2}{3}\dfrac{P^{vi}(x)}{q^3} + \cdots\right]$

61. $e^{rx}\sin sx$* $\quad \dfrac{e^{rx}}{(A^2+B^2)^n}\left\{\left[A^n - \binom{n}{2}A^{n-2}B^2 + \binom{n}{4}A^{n-4}B^4 - \cdots\right]\sin sx - \left[\binom{n}{1}A^{n-1}B - \binom{n}{3}A^{n-3}B^3 + \cdots\right]\cos sx\right\}$

$$A = r^2 - s^2 + q, \qquad B = 2rs$$

TABLE XIII: $(D^2 + b^2)^n y = R$

R $\quad y_p$

62. $\sin bx$* $\quad (-1)^{\frac{n+1}{2}}\dfrac{x^n\cos bx}{n!(2b)^n}$ (n odd), $\qquad (-1)^{\frac{n}{2}}\dfrac{x^n\sin bx}{n!(2b)^n}$ (n even)

* For $\cos sx$ in R replace "sin" by "cos" and "cos" by "$-$ sin" in y_p.

TABLE XIV: $(D^n - q)y = R$

63. e^{rx} $\quad e^{rx}/(r^n - q)$

64. $P(x)$ $\quad -\frac{1}{q}\left[P(x)\frac{P^{(n)}(x)}{q} + \frac{P^{(2n)}(x)}{q^2} + \cdots\right]$

65. $\sin sx$* $\quad -\frac{q \sin sx + (-1)^{\frac{n-1}{2}} s^n \cos sx}{q^2 + s^{2n}}$ (n odd), $\quad \frac{\sin sx}{(-s^2)^{n/2} - q}$ (n even)

66. $e^{rx} \sin sx$* $\quad \frac{Ae^{rx}\sin sr - Be^{rx}\cos sx}{A^2 + B^2} = \frac{e^{rx}}{\sqrt{A^2 + B^2}} \sin\left(sx - \tan^{-1}\frac{B}{A}\right)$

$$A = [r^n - \binom{n}{2}r^{n-2}s^2 + \binom{n}{4}r^{n-4}s^4 - \cdots] - q, \qquad B = [\binom{n}{1}r^{n-1}s - \binom{n}{3}r^{n-3}s^3 + \cdots]$$

TABLE XV: $(D_x + mD_y)z = R$

R	z_p
67. e^{ax+by}	$\frac{e^{ax+by}}{a + mb}$
68. $f(ax + by)$	$\frac{\int f(u)du}{a + mb}$, $u = ax + by$
69. $f(y - mx)$	$xf(y - mx)$
70. $\phi(x, y)f(y - mx)$	$f(y - mx)\int\phi(x, a + mx)dx$ ($a = y - mx$ after integration)

TABLE XVI: $(D_x + mD_y - k)z = R$

R	z_p
71. e^{ax+by}	$\frac{e^{ax+by}}{a + mb - k}$
72. $\sin(ax + by)$*	$-\frac{(a + bm)\cos(ax + by) + k\sin(ax + by)}{(a + bm)^2 + k^2}$
73. $e^{\alpha x+\beta y}\sin(ax + by)$*	Replace k in 72 by $k - \alpha - m\beta$ and multiply by $e^{\alpha x+\beta y}$
74. $e^{xk}f(ax + by)$	$\frac{e^{kx}\int f(u)du}{a + mb}$, $u = ax + by$
75. $f(y - mx)$	$-\frac{f(y - mx)}{k}$
76. $P(x)f(y - mx)$	$-\frac{1}{k}f(y - mx)\left[P(x) + \frac{P'(x)}{k} + \frac{P''(x)}{k^2} + \cdots + \frac{P^{(n)}(x)}{k^n}\right]$
77. $e^{kx}f(y - mx)$	$xe^{kx}f(y - mx)$

TABLE XVII: $(D_x + mD_y)^n z = R$

R	z_p
78. e^{ax+by}	$\frac{e^{ax+by}}{(a + mb)^n}$
79. $f(ax + by)$	$\frac{\int\int\cdots\int f(u)du^n}{(a + mb)^n}$, $u = ax + by$
80. $f(y - mx)$	$\frac{x^n}{n!}f(y - mx)$
81. $\phi(x, y)f(y + mx)$	$f(y - mx)\int\int\cdots\int\phi(x, a + mx)dx^n$ ($a = y - mx$ after integration)

TABLE XVIII: $(D_x + mD_y - k)^n z = R$

R	z_p
82. e^{ax+by}	$\frac{e^{ax+by}}{(a + mb - k)^n}$
83. $f(y - mx)$	$\frac{(-1)^n f(y - mx)}{k^n}$
84. $P(x)f(y - mx)$	$\frac{(-1)^n}{k^n}f(y - mx)\left[P(x) + \binom{n}{1}\frac{P'(x)}{k} + \binom{n+1}{2}\frac{P''(x)}{k^2} + \binom{n+2}{3}\frac{P'''(x)}{k^3} + \cdots\right]$
85. $e^{kx}f(ax + by)$	$\frac{e^{kx}\int\int\cdots\int f(u)du^n}{(a + mb)^n}$, $u = ax + by$
86. $e^{kx}f(y - mx)$	$\frac{x^n}{n!}e^{kx}f(y - mx)$

* For $\cos(ax + by)$ replace "sin" by "cos," and "cos" by "$-$ sin" in z_p.

$$D_x = \frac{\partial}{\partial x}; \quad D_y = \frac{\partial}{\partial y}; \quad D_x^k D_y^r = \frac{\partial^{k+r}}{\partial x^k \partial y^r}$$

TABLE XIX: $[D_x^n + a_1D_x^{n-1}D_y + a_2D_x^{n-2}D_y^2 + \cdots + a^nD_y^n]z = R$

87. e^{ax+by} $\qquad \dfrac{e^{ax+by}}{a^n + a_1a^{n-1}b + a_2a^{n-2}b^2 + \cdots + a_nb^n}$

88. $f(ax + by)$ $\qquad \dfrac{\int\int \cdots \int f(u)du^n}{a^n + a_1a^{n-1}b + a_2a^{n-2}b^2 + \cdots + a^nb^n}, \ u = ax + by$

TABLE XX: $F(D_x, D_y)z = R$

89. e^{ax+by} $\qquad \dfrac{e^{ax+by}}{F(a, b)}$

TABLE XXI: $F(D_x^2, D_xD_y, D_y^2)z = R$

90. $\sin (ax + by)$* $\qquad \dfrac{\sin (ax + by)}{F(-a^2, -ab, -b^2)}$

* For $\cos (ax + by)$ replace "sin" by "cos," and "cos" by "$-$ sin" in z_p.

The following examples illustrate the use of the tables.

EXAMPLE 1. Solve $(D^2 - 4)y = \sin 3x$.
Substitution of $q = -4$, $s = 3$ in formula 24 gives

$$y_p = \frac{\sin 3x}{-9 - 4},$$

wherefore the general solution is

$$y = c_1e^{2x} + c_2e^{-2x} - \frac{\sin 3x}{13}.$$

EXAMPLE 2. Obtain a particular solution of $(D^2 - 4D + 5)y = x^2e^{3x} \sin x$.

Applying formula 40 with $a = 2$, $b = 1$, $r = 3$, $s = 1$, $P(x) = x^2$, $s + b = 2$, $s - b = 0$, $a - r = -1$, $(a - r)^2 + (s + b)^2 = 5$, $(a - r)^2 + (s - b)^2 = 1$, we have

$$y_p = \frac{e^{3x} \sin x}{2}\left[\left(\frac{2}{5} - \frac{0}{1}\right)x^2 + \left(\frac{2 \cdot -1 \cdot 2}{25} - \frac{2 \cdot -1 \cdot 0}{1}\right)2x + \left(\frac{3 \cdot 1 \cdot 2 - 2^3}{125} - \frac{3 \cdot 1 \cdot 0 - 0}{1}\right)2\right]$$
$$- \frac{e^{3x} \cos x}{2}\left[\left(\frac{-1}{5} - \frac{-1}{1}\right)x^2 + \left(\frac{1 - 4}{25} - \frac{1 - 0}{1}\right)2x + \left(\frac{-1 - 3 \cdot -1 \cdot 4}{125} - \frac{-1 - 3 \cdot -1 \cdot 0}{1}\right)2\right]$$
$$= \left(\tfrac{1}{5}x^2 - \tfrac{4}{25}x - \tfrac{2}{125}\right)e^{3x} \sin x + \left(-\tfrac{2}{5}x^2 + \tfrac{28}{25}x - \tfrac{136}{125}\right)e^{3x} \cos x.$$

The special formulas effect a very considerable saving of time in problems of this type.

EXAMPLE 3. Obtain a particular solution of $(D^2 - 4D + 5)y = x^2e^{2x} \cos x$. (Compare with Example 2.)

Formula 40 is not applicable here since for this equation

$r = a$, $s = b$, wherefore the denominator $(a - r)^2 + (s - b)^2 = 0$. We turn instead to formula 44. Substituting $a = 2$, $b = 1$, $P(x) = x^2$ and replacing sin by cos, cos by $-$ sin, we obtain

$$y_p = \frac{e^{2x} \cos x}{4}(x^2 - \tfrac{2}{4}) + \frac{e^{2x} \sin x}{2}\int (x^2 - \tfrac{1}{2})dx$$
$$= \left(\frac{x^2}{4} - \frac{1}{8}\right)e^{2x} \cos x + \left(\frac{x^3}{6} - \frac{x}{4}\right)e^{2x} \sin x,$$

which is the required solution.

EXAMPLE 4. Find z_p for $(D_x - 3D_y)z = \ln (y + 3x)$.
Referring to Table XV we note that formula 69 (not 68) is applicable. This gives

$$z_p = x \ln (y + 3x).$$

It is easily seen that $-\dfrac{y}{3} \ln (y + 3x)$ would serve equally well.

EXAMPLE 5. Solve $(D_x + 2D_y - 4)z = y \cos (y - 2x)$.
Since R in formula 76 contains a polynomial in x, not y, we rewrite the given equation in the form

$$(D_y + \tfrac{1}{2}D_x - 2)z = \tfrac{1}{2}y \cos (y - 2x).$$

Then $\qquad z_c = e^{2y}F(x - \tfrac{1}{2}y) = e^{2y}f(2x - y),$

and by the formula

$$z_p = -\tfrac{1}{2} \cos (y - 2x) \cdot \left(\frac{y}{2} + \frac{\frac{1}{2}}{2}\right)$$
$$= -\tfrac{1}{8}(2y + 1) \cos (y - 2x).$$

EXAMPLE 6. Find z_p for $(D_x + 4D_y)^3z = (2x - y)^2$.
Using formula 79, we obtain

$$z_p = \frac{\int\int\int u^2du^3}{[2 + 4(-1)]^3} = \frac{u^5}{5 \cdot 4 \cdot 3 \cdot (-8)} = -\frac{(2x - y)^5}{480}.$$

EXAMPLE 7. Find z_p for $(D_x^3 + 5D_x^2D_y - 7D_x + 4)z = e^{2x+3y}$.
By formula 87

$$z_p = \frac{e^{2x+3y}}{2^3 + 5 \cdot 2^2 \cdot 3 - 7 \cdot 2 + 4} = \frac{e^{2x+3y}}{58}.$$

EXAMPLE 8. Find z_p for

$$(D_x^4 + 6D_x^3D_y + D_xD_y + D_y^2 + 9)z = \sin (3x + 4y).$$

Since every term in the left member is of *even* degree in the two operators D_x and D_y, formula 90 is applicable.
It gives

$$z_p = \frac{\sin (3x + 4y)}{(-9)^2 + 6(-9)(-12) + (-12) + (-16) + 9}$$
$$= \frac{\sin (3x + 4y)}{710}$$

Factors and Expansions

$$(a \pm b)^2 = a^2 \pm 2ab + b^2.$$
$$(a \pm b)^3 = a^3 \pm 3a^2b + 3ab^2 \pm b^3.$$
$$(a \pm b)^4 = a^4 \pm 4a^3b + 6a^2b^2 \pm 4ab^3 + b^4.$$
$$a^2 - b^2 = (a - b)(a + b).$$
$$a^2 + b^2 = (a + b\sqrt{-1})(a - b\sqrt{-1}).$$
$$a^3 - b^3 = (a - b)(a^2 + ab + b^2).$$
$$a^3 + b^3 = (a + b)(a^2 - ab + b^2).$$
$$a^4 + b^4 = (a^2 + ab\sqrt{2} + b^2)(a^2 - ab\sqrt{2} + b^2).$$
$$a^n - b^n = (a - b)(a^{n-1} + a^{n-2}b + \ldots + b^{n-1}).$$
$$a^n - b^n = (a + b)(a^{n-1} - a^{n-2}b + \ldots - b^{n-1}), \text{ for even values of } n.$$
$$a^n + b^n = (a + b)(a^{n-1} - a^{n-2}b + \ldots + b^{n-1}), \text{ for odd values of } n.$$
$$a^4 + a^2b^2 + b^4 = (a^2 + ab + b^2)(a^2 - ab + b^2).$$
$$(a + b + c)^2 = a^2 + b^2 + c^2 + 2ab + 2ac + 2bc.$$
$$(a + b + c)^3 = a^3 + b^3 + c^3 + 3a^2(b + c) + 3b^2(a + c) + 3c^2(a + b) + 6abc.$$
$$(a + b + c + d + \ldots)^2 = a^2 + b^2 + c^2 + d^2 + \ldots + 2a(b + c + d + \ldots) + 2b(c + d + \ldots) + 2c(d + \ldots) + \ldots$$

See also under Series

Powers and Roots

$a^x \times a^y = a^{(x+y)}$. $a^0 = 1$ [if $a \neq 0$]. $(ab)^x = a^x b^x$.

$\dfrac{a^x}{a^y} = a^{(x-y)}$. $a^{-x} = \dfrac{1}{a^x}$. $\left(\dfrac{a}{b}\right)^x = \dfrac{a^x}{b^x}$.

$(a^x)^y = a^{xy}$. $a^{\frac{1}{x}} = \sqrt[x]{a}$. $\sqrt[x]{ab} = \sqrt[x]{a}\sqrt[x]{b}$.

$\sqrt[x]{\sqrt[y]{a}} = \sqrt[xy]{a}$. $a^{\frac{x}{y}} = \sqrt[y]{a^x}$. $\sqrt[x]{\dfrac{a}{b}} = \dfrac{\sqrt[x]{a}}{\sqrt[x]{b}}$.

Proportion

If $\dfrac{a}{b} = \dfrac{c}{d}$, then $\dfrac{a+b}{b} = \dfrac{c+d}{d}$,

$\dfrac{a-b}{b} = \dfrac{c-d}{d}$, $\dfrac{a-b}{a+b} = \dfrac{c-d}{c+d}$.

SUM OF POWERS OF INTEGERS, $\sum_{k=1}^{n} k^m$

($m = 1, 2, 3, 4$); $1 \leq n \leq 40$

n	Σk	Σk^2	Σk^3	Σk^4
1	1	1	1	1
2	3	5	9	17
3	6	14	36	98
4	10	30	100	354
5	15	55	225	979
6	21	91	441	2275
7	28	140	784	4676
8	36	204	1296	8772
9	45	285	2025	15333
10	55	385	3025	25333
11	66	506	4356	39974
12	78	650	6084	60710
13	91	819	8281	89271
14	105	1015	11025	127687
15	120	1240	14400	178312
16	136	1496	18496	243848
17	153	1785	23409	327369
18	171	2109	29241	432345
19	190	2470	36100	562666
20	210	2870	44100	722666
21	231	3311	53361	917147
22	253	3795	64009	1151403
23	276	4324	76176	1431244
24	300	4900	90000	1763020
25	325	5525	105625	2153645
26	351	6201	123201	2610621
27	378	6930	142884	3142062
28	406	7714	164836	3756718
29	435	8555	189225	4463999
30	465	9455	216225	5273999
31	496	10416	246016	6197520
32	528	11440	278784	7246096
33	561	12529	314721	8432017
34	595	13685	354025	9768353
35	630	14910	396900	11268978
36	666	16206	443556	12948594
37	703	17575	494209	14822755
38	741	19019	549081	16907891
39	780	20540	608400	19221332
40	820	22140	672400	21781332

ALGEBRA

SUMS OF NUMBERS

The sum of the first n numbers,—

$$\Sigma(n) = 1+2+3+4+5 \ldots . + n = \frac{n(n+1)}{2}$$

The sum of the squares of the first n numbers,

$$\Sigma(n^2) = 1^2+2^2+3^2+4^2+5^2 \ldots + n^2 = \frac{n(n+1)(2n+1)}{6}$$

The sum of the cubes of the first n numbers,

$$\Sigma(n^3) = 1^3+2^3+3^3+4^3+5^3 \ldots . + n^3 = \frac{n^2(n+1)^2}{4}$$

ARITHMETICAL PROGRESSION

If a is the first term; l, the last term; d, the common difference; n, the number of terms and s, the sum of n terms,—

$$l = a + (n-1)d \qquad s = \frac{n}{2}(a+l)$$

$$s = \frac{n}{2}\left\{ 2a + (n-1)d \right\}$$

GEOMETRICAL PROGRESSION

If a is the first term; l, the last term; r, the common ratio; n, the number of terms and s, the sum of n terms,—

$$l = ar^{n-1} \qquad s = a\frac{(1-r^n)}{1-r}$$

$$s = a\frac{(r^n-1)}{r-1} \qquad s = \frac{lr-a}{r-1}$$

If n is infinity and r^2 less than unity,—

$$s = \frac{a}{1-r}$$

FACTORIALS

$\underline{n|} = n! = e^{-n}n^n\sqrt{2\pi n}$, approximately.

PERMUTATIONS

If M denote the number of permutations of n things taken p at a time,—

$$M = n(n-1)(n-2) \ldots . (n-p+1)$$

COMBINATIONS

If M denote the number of combinations of n things taken p at a time,—

$$M = \frac{n(n-1)(n-2) \ldots . (n-p+1)}{p!}$$

$$M = \frac{n!}{p!(n-p)!}$$

Quadratic Equations

Any quadratic equation may be reduced to the form, —

$$ax^2 + bx + c = 0$$

$$\text{Then} \quad x = \frac{-b \pm \sqrt{b^2 - 4ac}}{2a}.$$

If a, b, and c, are real, then
if $b^2 - 4ac$ is positive, the roots are real and unequal;
if $b^2 - 4ac$ is zero, the roots are real and equal;
if $b^2 - 4ac$ is negative, the roots are imaginary and unequal;
a, b, and c, are rational.

Cubic Equations

A cubic equation, $y^3 + py^2 + qy + r = 0$ may be reduced to the form, —

$$x^3 + ax + b = 0$$

by substituting for y the value, $\left(x - \frac{p}{3}\right)$. Here

$$a = \tfrac{1}{3}(3q - p^2) \text{ and } b = \tfrac{1}{27}(2p^3 - 9pq + 27r).$$

For solution let, —

$$A = \sqrt[3]{-\frac{b}{2} + \sqrt{\frac{b^2}{4} + \frac{a^3}{27}}}, \qquad B = \sqrt[3]{-\frac{b}{2} - \sqrt{\frac{b^2}{4} + \frac{a^3}{27}}},$$

then the values of x will be given by,

$$x = A + B, \quad -\frac{A+B}{2} + \frac{A-B}{2}\sqrt{-3}, \quad -\frac{A+B}{2} - \frac{A-B}{2}\sqrt{-3}$$

If $\frac{b^2}{4} + \frac{a^3}{27} > 0$, there will be one real root and two conjugate imaginary roots.

If $\frac{b^2}{4} + \frac{a^3}{27} = 0$, there will be three real roots of which two at least are equal

If $\frac{b^2}{4} + \frac{a^3}{27} < 0$, there will be three real and unequal roots.

In the last case a trigonometric solution is useful. Compute the value of the angle ϕ in the expression, —

$$\cos\phi = -\frac{b}{2} \div \sqrt{\left(-\frac{a^3}{27}\right)},$$

then x will have the following values:—

$$2\sqrt{-\frac{a}{3}}\cos\frac{\phi}{3}, \qquad 2\sqrt{-\frac{a}{3}}\cos\left(\frac{\phi}{3} + 120°\right),$$

$$2\sqrt{-\frac{a}{3}}\cos\left(\frac{\phi}{3} + 240°\right)$$

PARTIAL FRACTIONS

Compiled by Professor M. D. Prince

This section applies only to rational algebraic fractions with numerator of lower degree than the denominator. Improper fractions can be reduced to proper fractions by long division.

Every fraction may be expressed as the sum of component fractions whose denominators are factors of the denominator of the original fraction.

Let $N(x)$ = numerator, a polynomial of the form

$$n_0 + n_1x + n_2x^2 + \cdots + n_ix^i$$

I. *Non-repeated Linear Factors*

$$\frac{N(x)}{(x-a)G(x)} = \frac{A}{x-a} + \frac{F(x)}{G(x)}$$

$$A = \left[\frac{N(x)}{G(x)}\right]_{x=a}$$

$F(x)$ determined by methods discussed in the following sections. Example:

$$\frac{x^2+3}{x(x-2)(x^2+2x+4)} = \frac{A}{x} + \frac{B}{x-2} + \frac{F(x)}{x^2+2x+4}$$

$$A = \left[\frac{x^2+3}{(x-2)(x^2+2x+4)}\right]_{x=0} = -\frac{3}{8}$$

$$B = \left[\frac{x^2+3}{x(x^2+2x+4)}\right]_{x=2} = \frac{4+3}{2(4+4+4)} = \frac{7}{24}$$

II. *Repeated Linear Factors*

$$\frac{N(x)}{x^mG(x)} = \frac{A_0}{x^m} + \frac{A_1}{x^{m-1}} + \cdots + \frac{A_{m-1}}{x} + \frac{F(x)}{G(x)}$$

$$F(x) = f_0 + f_1x + f_2x^2 + \cdots, \quad G(x) = g_0 + g_1x + g_2x^2 + \cdots$$

$$A_0 = \frac{n_0}{g_0}, \qquad A_1 = \frac{n_1 - A_0g_1}{g_0}, \qquad A_2 = \frac{n_2 - A_0g_2 - A_1g_1}{g_0}$$

$$\text{General term: } A_k = \frac{1}{g_0}\left[n_k - \sum_{i=0}^{k-1} A_ig_{k-i}\right]$$

$${}^*m = 1 \begin{cases} f_0 = n_1 - A_0g_1 \\ f_1 = n_2 - A_0g_2 \\ f_j = n_{j+1} - A_0g_{j+1} \end{cases}$$

$$m = 2 \begin{cases} f_0 = n_2 - A_0g_2 - A_1g_1 \\ f_1 = n_3 - A_0g_3 - A_1g_2 \\ f_j = n_{j+2} - [A_0g_{j+2} + A_1g_{j+1}] \end{cases}$$

$$m = 3 \begin{cases} f_0 = n_3 - A_0g_3 - A_1g_2 - A_2g_1 \\ f_1 = n_3 - A_0g_4 - A_1g_3 - A_2g_2 \\ f_j = n_{j+3} - [A_0g_{j+3} + A_1g_{j+2} + A_2g_{j+1}] \end{cases}$$

* Note: If $G(x)$ contains linear factors, $F(x)$ may be determined by previous section I.

$$\text{any } m: \quad f_j = n_{m+j} - \sum_{i=0}^{m-1} A_ig_{m+j-i}$$

Example: $$\frac{x^2+1}{x^3(x^2-3x+6)} = \frac{A_0}{x^3} + \frac{A_1}{x^2} + \frac{A_2}{x} + \frac{f_1x+f_0}{x^2-3x+6}$$

$$A_0 = \frac{1}{6}, \qquad A_1 = \frac{0 - (\frac{1}{6})(-3)}{6} = \frac{1}{12},$$

$$A_2 = \frac{1 - (\frac{1}{6})(1) - (\frac{1}{12})(-3)}{6} = \frac{13}{72}$$

$$m = 3, \quad f_0 = 0 - \tfrac{1}{6}(0) - \tfrac{1}{12}(1) - \tfrac{13}{72}(-3) = \tfrac{11}{24}$$
$$f_1 = 0 - \tfrac{1}{6}(0) - \tfrac{1}{12}(0) - \tfrac{13}{72}(1) = -\tfrac{13}{72}$$

III. *Repeated Linear Factors*

$$\frac{N(x)}{(x-a)^mG(x)} = \frac{A_0}{(x-a)^m} + \frac{A_1}{(x-a)^{m-1}} + \cdots + \frac{A_{m-1}}{(x-a)} + \frac{F(x)}{G(x)}$$

Change to form $\frac{N'(y)}{y^mG'(y)}$ by substitution of $x = y + a$. Resolve into partial fractions in terms of y as described in Section II. Then express in terms of x by substitution $y = x - a$.

Example: $\frac{x-3}{(x-2)^2(x^2+x+1)}$. Let $x - 2 = y$, $x = y + 2$

$$\frac{(y+2)-3}{y^2[(y+2)^2+(y+2)+1]} = \frac{y-1}{y^2(y^2+5y+7)}$$
$$= \frac{A_0}{y^2} + \frac{A_1}{y} + \frac{f_1y+f_0}{y^2+5y+7}$$

$$A_0 = -\tfrac{1}{7} \qquad A_1 = \frac{1 - (-\frac{1}{7})(5)}{7} = \frac{12}{49}$$

$$(m = 2)$$
$$f_0 = 0 - (-\tfrac{1}{7})(1) - (\tfrac{12}{49})(5) = -\tfrac{53}{49}$$
$$f_1 = 0 - (-\tfrac{1}{7})(0) - (\tfrac{12}{49})(1) = -\tfrac{12}{49}$$

$$\therefore \frac{y-1}{y^2(y^2+5y+7)} = \frac{-\frac{1}{7}}{y^2} + \frac{\frac{12}{49}}{y} + \frac{-\frac{12}{49}y - \frac{53}{49}}{y^2+5y+7}$$

Let $y = x - 2$, then

$$\frac{x-3}{(x-2)^2(x^2+x+1)} = \frac{-\frac{1}{7}}{(x-2)^2} + \frac{\frac{12}{35}}{(x-2)}$$
$$+ \frac{-\frac{12}{49}(x-2) - \frac{53}{49}}{x^2+x+1} = -\frac{1}{7(x-2)^2} + \frac{12}{35(x-2)}$$
$$+ \frac{-12x-29}{49(x^2+x+1)}$$

IV. *Repeated Linear Factors*

Alternative method of determining coefficients:

$$\frac{N(x)}{(x-a)^m G(x)} = \frac{A_0}{(x-a)^m} + \cdots + \frac{A_k}{(x-a)^{m-k}} + \cdots + \frac{A_{m-1}}{x-a} + \frac{F(x)}{G(x)}$$

$$A_k = \frac{1}{k!}\left\{D_x^k\left[\frac{N(x)}{G(x)}\right]\right\}_{x=a}$$

Where D_x^k is the differentiating operator, and the derivative of zero order is defined as:

$$D_x^0 u = u.$$

V. *Factors of Higher Degree*

Factors of higher degree have the corresponding numerators indicated.

$$\frac{N(x)}{(x^2+h_1x+h_0)G(x)} = \frac{a_1x+a_0}{x^2+h_1x+h_0} + \frac{F(x)}{G(x)}$$

$$\frac{N(x)}{(x^2+h_1x+h_0)^2 G(x)} = \frac{a_1x+a_0}{(x^2+h_1x+h_0)^2} + \frac{b_1x+b_0}{(x^2+h_1x+h_0)} + \frac{F(x)}{G(x)}$$

$$\frac{N(x)}{(x^3+h_2x^2+h_1x+h_0)G(x)} = \frac{a_2x^2+a_1x+a_0}{x^3+h_2x^2+h_1x+h_0} + \frac{F(x)}{G(x)}$$

etc.

Problems of this type are determined first by solving for the coefficients due to linear factors as shown above, and then determining the remaining coefficients by the general methods given below.

VI. *General Methods for Evaluating Coefficients*

1. $$\frac{N(x)}{D(x)} = \frac{N(x)}{G(x)H(x)L(x)} = \frac{A(x)}{G(x)} + \frac{B(x)}{H(x)} + \frac{C(x)}{L(x)} + \cdots$$

Multiply both sides of equation by $D(x)$ to clear fractions. Then collect terms, equate like powers of x, and solve the resulting simultaneous equations for the unknown coefficients.

2. Clear fractions as above. Then let x assume certain convenient values ($x = 1, 0, -1, \ldots$). Solve the resulting equations for the unknown coefficients.

3. $$\frac{N(x)}{G(x)H(x)} = \frac{A(x)}{G(x)} + \frac{B(x)}{H(x)}$$

Then $$\frac{N(x)}{G(x)H(x)} - \frac{A(x)}{G(x)} = \frac{B(x)}{H(x)}$$

If $A(x)$ can be determined, such as by Method I, then $B(x)$ can be found as above.

APPROXIMATIONS

If a and b are small quantities, the following relations are approximately true,—

$$(1 \pm a)^m = 1 \pm ma,$$

$$(1 \pm a)^m (1 \pm b)^n = 1 \pm ma \pm nb.$$

If n is nearly equal to m,

$$\sqrt{mn} = \frac{n+m}{2}, \text{ approximately.}$$

If θ is a very small angle expressed in radians,—

$$\frac{\sin\theta}{\theta} = 1 \text{ and } \frac{\tan\theta}{\theta} = 1, \text{ approximately.}$$

SERIES

The expression in parentheses following certain of the series indicates the region of convergence. If not otherwise indicated it is to be understood that the series converges for all finite values of x.

Binomial

$$(x+y)^n = x^n + nx^{n-1}y + \frac{n(n-1)}{2!}x^{n-2}y^2 + \frac{n(n-1)(n-2)}{3!}x^{(n-3)}y^3 + \cdots \quad (y^2 < x^2)$$

$$(1 \pm x)^n = 1 \pm nx + \frac{n(n-1)x^2}{2!} \pm \frac{n(n-1)(n-2)x^3}{3!} + \ldots \text{etc.} \quad (x^2 < 1)$$

$$(1 \pm x)^{-n} = 1 \mp nx + \frac{n(n+1)x^2}{2!} \mp \frac{n(n+1)(n+2)x^3}{3!} + \ldots \text{etc.} \quad (x^2 < 1)$$

$$(1 \pm x)^{-1} = 1 \mp x + x^2 \mp x^3 + x^4 \mp x^5 + \ldots \quad (x^2 < 1)$$

$$(1 \pm x)^{-2} = 1 \mp 2x + 3x^2 \mp 4x^3 + 5x^4 \mp 6x^5 + \ldots \quad (x^2 < 1)$$

Reversion of Series

Let a series be represented by

$$y = a_1x + a_2x^2 + a_3x^3 + a_4x^4 + a_5x^5 + a_6x^6 + \cdots \quad (a_1 \neq 0)$$

to find the coefficients of the series

$$x = A_1y + A_2y^2 + A_3y^3 + A_4y^4 + \cdots$$

$$A_1 = \frac{1}{a_1} \qquad A_2 = -\frac{a_2}{a_1^3} \qquad A_3 = \frac{1}{a_1^5}(2a_2^2 - a_1a_3)$$

$$A_4 = \frac{1}{a_1^7}(5a_1a_2a_3 - a_1^2a_4 - 5a_2^3)$$

$$A_5 = \frac{1}{a_1^9}(6a_1^2a_2a_4 + 3a_1^2a_3^2 + 14a_2^4 - a_1^3a_5 - 21a_1a_2^2a_3)$$

$$A_6 = \frac{1}{a_1^{11}}(7a_1^3a_2a_5 + 7a_1^3a_3a_4 + 84a_1a_2^3a_3 - a_1^4a_6 - 28a_1^2a_2^2a_4 - 28a_1^2a_2a_3^2 - 42a_2^5)$$

$$A_7 = \frac{1}{a_1^{13}}(8a_1^4a_2a_6 + 8a_1^4a_3a_5 + 4a_1^4a_4^2 + 120a_1^2a_2^3a_4 + 180a_1^2a_2^2a_3^2 + 132a_2^6 - a_1^5a_7 - 36a_1^3a_2^2a_5 - 72a_1^3a_2a_3a_4 - 12a_1^3a_3^3 - 330a_1a_2^4a_3)$$

Taylor's Series

$$f(x+h) = f(x) + hf'(x) + \frac{h^2}{2!}f''(x) + \frac{h^3}{3!}f'''(x) + \cdots$$

$$= f(h) + xf'(h) + \frac{x^2}{2!}f''(h) + \frac{x^3}{3!}f'''(h) + \cdots$$

Maclaurin's Series

$$f(x) = f(o) + xf'(o) + \frac{x^2}{2!}f''(o) + \frac{x^3}{3!}f'''(o) + \cdots$$

Exponential

$$e = 1 + \frac{1}{1} + \frac{1}{2!} + \frac{1}{3!} + \frac{1}{4!} + \cdots$$

$$e^x = 1 + x + \frac{x^2}{2!} + \frac{x^3}{3!} + \frac{x^4}{4!} + \cdots \qquad \text{(all real values of } x\text{)}$$

$$a^x = 1 + x\log_e a + \frac{(x\log a)^2}{2!} + \frac{(x\log a)^3}{3!} + \cdots$$

Logarithmic

$$\log_e x = \frac{x-1}{x} + \frac{1}{2}\left(\frac{x-1}{x}\right)^2 + \frac{1}{3}\left(\frac{x-1}{x}\right)^3 + \cdots \qquad (x > \tfrac{1}{2})$$

$$\log_e x = (x-1) - \tfrac{1}{2}(x-1)^2 + \tfrac{1}{3}(x-1)^3 - \cdots \qquad (2 > x > 0)$$

$$\log_e x = 2\left[\frac{x-1}{x+1} + \frac{1}{3}\left(\frac{x-1}{x+1}\right)^3 + \frac{1}{5}\left(\frac{x-1}{x+1}\right)^5 + \cdots\right] \qquad (x > 0)$$

$$\log_e(1+x) = x - \tfrac{1}{2}x^2 + \tfrac{1}{3}x^3 - \tfrac{1}{4}x^4 + \cdots \qquad (-1 < x < 1)$$

$$\log_e(n+1) - \log_e(n-1) = 2\left[\frac{1}{n} + \frac{1}{3n^3} + \frac{1}{5n^5} + \cdots\right]$$

$$\log_e(a+x) = \log_e a + 2\left[\frac{x}{2a+x} + \frac{1}{3}\left(\frac{x}{2a+x}\right)^3 + \frac{1}{5}\left(\frac{x}{2a+x}\right)^5 + \cdots\right] \qquad (a > 0,\ -a < x < +\infty)$$

Trigonometric

$$\sin x = x - \frac{x^3}{3!} + \frac{x^5}{5!} - \frac{x^7}{7!} + \cdots \qquad \text{(all real values of } x\text{)}$$

$$\cos x = 1 - \frac{x^2}{2!} + \frac{x^4}{4!} - \frac{x^6}{6!} + \cdots \qquad \text{(all real values of } x\text{)}$$

$$\tan x = x + \frac{x^3}{3} + \frac{2x^5}{15} + \frac{17x^7}{315} + \frac{62x^9}{2835} + \cdots \qquad \left(x^2 < \frac{\pi^2}{4}\right)$$

$$\sin^{-1} x = x + \frac{x^3}{6} + \frac{1}{2}\cdot\frac{3}{4}\cdot\frac{x^5}{5} + \frac{1}{2}\cdot\frac{3}{4}\cdot\frac{5}{6}\cdot\frac{x^7}{7} + \cdots \qquad (x^2 < 1)$$

$$\tan^{-1} x = x - \tfrac{1}{3}x^3 + \tfrac{1}{5}x^5 - \tfrac{1}{7}x^7 + \cdots \qquad (x^2 < 1)$$

$$= \frac{\pi}{2} - \frac{1}{x} + \frac{1}{3x^3} - \frac{1}{5x^5} + \cdots \qquad (x^2 > 1)$$

$$\log_e \sin x = \log_e x - \frac{x^2}{6} - \frac{x^4}{180} - \frac{x^6}{2835} - \cdots \qquad (x^2 < \pi^2)$$

$$\log_e \cos x = -\frac{x^2}{2} - \frac{x^4}{12} - \frac{x^6}{45} - \frac{17x^8}{2520} - \cdots \qquad \left(x^2 < \frac{\pi^2}{4}\right)$$

$$\log_e \tan x = \log_e x + \frac{x^2}{3} + \frac{7x^4}{90} + \frac{62x^6}{2835} + \cdots \qquad \left(x^2 < \frac{\pi^2}{4}\right)$$

$$e^{\sin x} = 1 + x + \frac{x^2}{2!} - \frac{3x^4}{4!} - \frac{8x^5}{5!} - \frac{3x^6}{6!} + \frac{56x^7}{7!} + \cdots$$

$$e^{\cos x} = e\left(1 - \frac{x^2}{2!} + \frac{4x^4}{4!} - \frac{31x^6}{6!} + \cdots\right)$$

$$e^{\tan x} = 1 + x + \frac{x^2}{2!} + \frac{3x^3}{3!} + \frac{9x^4}{4!} + \frac{37x^5}{5!} + \cdots \qquad \left(x^2 < \frac{\pi^2}{4}\right)$$

FOURIER SERIES

$$f(x) = \frac{a_0}{2} + a_1\cos\frac{2\pi x}{T} + a_2\cos\frac{4\pi x}{T} + \cdots + a_m\cos\frac{2\pi m x}{T} + \cdots + b_1\sin\frac{2\pi x}{T} + b_2\sin\frac{4\pi x}{T} + b_3\sin\frac{6\pi x}{T} + \cdots + b_n\sin\frac{2\pi n x}{T} + \cdots$$

where

$a_0 = \frac{2}{T}\int_c^{c+T} f(x)dx$, $a_m = \frac{2}{T}\int_c^{c+T} f(x)\cos\frac{2\pi m x}{T}dx$, $b_n = \frac{2}{T}\int_c^{c+T} f(x)\sin\frac{2\pi n x}{T}dx$, and $f(x+T) = f(x)$, with c and T constants.

ALGEBRA (Continued)

(See index for Cosine and Sine Transforms)

Auxiliary Formulas for Fourier Series

$$1 = \frac{4}{\pi}\left[\sin\frac{\pi x}{k} + \frac{1}{3}\sin\frac{3\pi x}{k} + \frac{1}{5}\sin\frac{5\pi x}{k} + \cdots\right] \quad [0 < x < k]$$

$$x = \frac{2k}{\pi}\left[\sin\frac{\pi x}{k} - \frac{1}{2}\sin\frac{2\pi x}{k} + \frac{1}{3}\sin\frac{3\pi x}{k} - \cdots\right] \quad [-k < x < k]$$

$$x = \frac{k}{2} - \frac{4k}{\pi^2}\left[\cos\frac{\pi x}{k} + \frac{1}{3^2}\cos\frac{3\pi x}{k} + \frac{1}{5^2}\cos\frac{5\pi x}{k} + \cdot\right] \quad [0 < x < k]$$

$$x^2 = \frac{2k^2}{\pi^3}\left[\left(\frac{\pi^2}{1} - \frac{4}{1}\right)\sin\frac{\pi x}{k} - \frac{\pi^2}{2}\sin\frac{2\pi x}{k} + \left(\frac{\pi^2}{3} - \frac{4}{3^3}\right)\sin\frac{3\pi x}{k} - \frac{\pi^2}{4}\sin\frac{4\pi x}{k} + \left(\frac{\pi^2}{5} - \frac{4}{5^3}\right)\sin\frac{5\pi x}{k} + \cdots\right] \quad [0 < x < k]$$

$$x^2 = \frac{k^2}{3} - \frac{4k^2}{\pi^2}\left[\cos\frac{\pi x}{k} - \frac{1}{2^2}\cos\frac{2\pi x}{k} + \frac{1}{3^2}\cos\frac{3\pi x}{k} - \frac{1}{4^2}\cos\frac{4\pi x}{k} + \cdots\right] \quad [-k < x < k]$$

$$1 - \frac{1}{3} + \frac{1}{5} - \frac{1}{7} + \cdots = \frac{\pi}{4}$$

$$1 + \frac{1}{2^2} + \frac{1}{3^2} + \frac{1}{4^2} + \cdots = \frac{\pi^2}{6} = B_1\pi^2$$

$$1 - \frac{1}{2^2} + \frac{1}{3^2} - \frac{1}{4^2} + \cdots = \frac{\pi^2}{12} = \frac{B_1\pi^2}{2}$$

$$1 + \frac{1}{3^2} + \frac{1}{5^2} + \frac{1}{7^2} + \cdots = \frac{\pi^2}{8} = \frac{3B_1\pi^2}{4}$$

$$\frac{1}{2^2} + \frac{1}{4^2} + \frac{1}{6^2} + \frac{1}{8^2} + \cdots = \frac{\pi^2}{24} = \frac{B_2\pi^4}{90}$$

In the above $B_1, B_2 \ldots$ represent the Bernoulli Numbers.

Bernoulli Numbers and Euler Numbers

The sequence of numbers $E_1, E_2, E_3, \ldots$ obtained from

$$E_n = \frac{(2n)!}{(2n-2)!2!}E_{n-1} - \frac{(2n)!}{(2n-4)!4!}E_{n-2} + \cdots + (-1)^{n-1}$$

are called Euler Numbers. Thus $E_1 = 1$, $E_2 = 5$, $E_3 = 61$, $E_4 = 1385$, $E_5 = 50{,}521, \ldots$.

In the above $0! = 1$ and $E_0 = 1$ by definition.

The sequence of numbers $B_1, B_2, B_3, B_4, \ldots$ obtained from

$$B_n = \frac{2n}{2^{2n}(2^{2n}-1)}\left[\frac{(2n-1)!}{(2n-2)!1!}E_{n-1} - \frac{(2n-1)!}{(2n-4)!3!}E_{n-2} + \cdots + (-1)^{n-1}\right]$$

are called Bernoulli numbers. Thus

$B_1 = \frac{1}{6}$, $B_2 = \frac{1}{30}$, $B_3 = \frac{1}{42}$, $B_4 = \frac{1}{30}$, $B_5 = \frac{5}{66}, \ldots$

$$B_n = \frac{(2n)!}{\pi^{2n}2^{2n-1}}\left[1 + \frac{1}{2^{2n}} + \frac{1}{3^{2n}} + \frac{1}{4^{2n}} + \cdots\right]$$

$$B_n = \frac{(2n)!}{\pi^{2n}(2^{2n-1}-1)}\left[1 - \frac{1}{2^{2n}} + \frac{1}{3^{2n}} - \frac{1}{4^{2n}} + \cdots\right]$$

$$B_n = \frac{2(2n)!}{\pi^{2n}(2^{2n}-1)}\left[1 + \frac{1}{3^{2n}} + \frac{1}{5^{2n}} + \frac{1}{7^{2n}} + \cdots\right]$$

$$E_n = \frac{2^{2n+2}(2n)!}{\pi^{2n+1}}\left[1 - \frac{1}{3^{2n+1}} + \frac{1}{5^{2n+1}} + \frac{1}{7^{2n+1}} + \cdots\right]$$

VECTOR ANALYSIS

Definitions

Any quantity which is completely determined by its magni tude is called a *scalar*. Examples of such are mass, density, temperature, etc. Any quantity which is completely determined by its magnitude and direction is called a *vector*. Examples of such are velocity, acceleration, force, etc. A vector quantity is represented by a directed line segment, the length of which represents the magnitude of the vector. A vector quantity is usually represented by a bold-faced letter such as $\mathbf{V}$. Two vectors $\mathbf{V}_1$ and $\mathbf{V}_2$ are equal to one another if they have equal magnitudes and are acting in the same directions. A negative vector, written as $-\mathbf{V}$, is one which acts in the opposite direction to $\mathbf{V}$, but is of equal magnitude to it. If we represent the magnitude of $\mathbf{V}$ by v, we write $|\mathbf{V}| = v$. A vector parallel to $\mathbf{V}$, but equal to the reciprocal of its magnitude is written as $\mathbf{V}^{-1}$ or as $\frac{1}{\mathbf{V}}$.

The unit vector $\frac{\mathbf{V}}{|\mathbf{V}|}$ $(\mathbf{V} \neq 0)$ is the one which has the same direction as $\mathbf{V}$, but has a magnitude of 1. (Sometimes represented as $\mathbf{V}^0$.)

Vector Algebra

The vector sum of $\mathbf{V}_1$ and $\mathbf{V}_2$ is represented by $\mathbf{V}_1 + \mathbf{V}_2$. The vector sum of $\mathbf{V}_1$ and $-\mathbf{V}_2$, or the difference of the vector $\mathbf{V}_2$ from $\mathbf{V}_1$ is represented by $\mathbf{V}_1 - \mathbf{V}_2$.

If r is a scalar, then $r\mathbf{V} = \mathbf{V}r$, and represents a vector r times the magnitude of $\mathbf{V}$, in the same direction as $\mathbf{V}$ if r is positive, and in the opposite direction if r is negative. If r and s are scalars, $\mathbf{V}_1, \mathbf{V}_2, \mathbf{V}_3$, vectors, then the following rules of scalars and vectors hold:

$$\mathbf{V}_1 + \mathbf{V}_2 = \mathbf{V}_2 + \mathbf{V}_1$$
$$(r + s)\mathbf{V}_1 = r\mathbf{V}_1 + s\mathbf{V}_1; \qquad r(\mathbf{V}_1 + \mathbf{V}_2) = r\mathbf{V}_1 + r\mathbf{V}_2$$
$$\mathbf{V}_1 + (\mathbf{V}_2 + \mathbf{V}_3) = (\mathbf{V}_1 + \mathbf{V}_2) + \mathbf{V}_3 = \mathbf{V}_1 + \mathbf{V}_2 + \mathbf{V}_3$$

If $\mathbf{V}_1$ is a vector in space, and a_1, b_1, c_1, the respective magnitudes of the projections of the vector along the coordinate Ox, Oy, Oz axes, then

$\mathbf{V}_1 = a_1\mathbf{i} + b_1\mathbf{j} + c_1\mathbf{k}$, where $\mathbf{i}, \mathbf{j}, \mathbf{k}$ are respectively the unit vectors along Ox, Oy, and Oz. Its magnitude is

$$|\mathbf{V}_1| = \sqrt{a_1^2 + b_1^2 + c_1^2},$$

and its direction cosines satisfy the proportion

$$a_1 : b_1 : c_1 = \cos\alpha : \cos\beta : \cos\gamma.$$

Thus, if $\mathbf{V}_1 = a_1\mathbf{i} + b_1\mathbf{j} + c_1\mathbf{k}$ and $\mathbf{V}_2 = a_2\mathbf{i} + b_2\mathbf{j} + c_2\mathbf{k}$,
Then $\mathbf{V}_1 + \mathbf{V}_2 = (a_1 + a_2)\mathbf{i} + (b_1 + b_2)\mathbf{j} + (c_1 + c_2)\mathbf{k}$

The Scalar, Dot, or Inner Product of Two Vectors $\mathbf{V}_1$ and $\mathbf{V}_2$

This product is represented as $\mathbf{V}_1 \cdot \mathbf{V}_2$ and is defined to be equal to $|\mathbf{V}_1|\,|\mathbf{V}_2| \cos\theta$, where θ is the angle from $\mathbf{V}_1$ to $\mathbf{V}_2$, i.e.,

$$\mathbf{V}_1 \cdot \mathbf{V}_2 = |\mathbf{V}_1|\,|\mathbf{V}_2| \cos\theta$$

The following rules apply for this product:

$$\mathbf{V}_1 \cdot \mathbf{V}_2 = a_1a_2 + b_1b_2 + c_1c_2 = \mathbf{V}_2 \cdot \mathbf{V}_1$$

It should be noted that scalar multiplication is commutative.

$$\begin{aligned} &\text{If } \mathbf{V}_1 = \mathbf{V}_2, \text{ then } \mathbf{V}_1 \cdot \mathbf{V}_2 = |\mathbf{V}_1|^2 \\ &(\mathbf{V}_1 + \mathbf{V}_2) \cdot \mathbf{V}_3 = \mathbf{V}_1 \cdot \mathbf{V}_3 + \mathbf{V}_2 \cdot \mathbf{V}_3 \\ &\mathbf{V}_1 \cdot (\mathbf{V}_2 + \mathbf{V}_3) = \mathbf{V}_1 \cdot \mathbf{V}_2 + \mathbf{V}_1 \cdot \mathbf{V}_3 \\ &\mathbf{i} \cdot \mathbf{i} = \mathbf{j} \cdot \mathbf{j} = \mathbf{k} \cdot \mathbf{k} = 1; \qquad \mathbf{i} \cdot \mathbf{j} = \mathbf{j} \cdot \mathbf{k} = \mathbf{k} \cdot \mathbf{i} = 0 \end{aligned}$$

The Vector or Cross Product of Vectors $\mathbf{V}_1$ and $\mathbf{V}_2$

This product is represented as $\mathbf{V}_1 \times \mathbf{V}_2$ and is defined to be equal to $|\mathbf{V}_1|\,|\mathbf{V}_2| (\sin\theta)\mathbf{l}$, where θ is the angle from $\mathbf{V}_1$ to $\mathbf{V}_2$ and $\mathbf{l}$ is a unit vector perpendicular to the plane of $\mathbf{V}_1$ and $\mathbf{V}_2$ and so directed that a right-handed screw driven in the direction of $\mathbf{l}$ would carry $\mathbf{V}_1$ into $\mathbf{V}_2$, i.e.,

$$\mathbf{V}_1 \times \mathbf{V}_2 = |\mathbf{V}_1|\,|\mathbf{V}_2| (\sin\theta)\mathbf{l}$$

The following rules apply for vector products:

$$\begin{aligned} &\mathbf{V}_1 \times \mathbf{V}_2 = -\mathbf{V}_2 \times \mathbf{V}_1 \\ &\mathbf{V}_1 \times (\mathbf{V}_2 + \mathbf{V}_3) = \mathbf{V}_1 \times \mathbf{V}_2 + \mathbf{V}_1 \times \mathbf{V}_3 \\ &(\mathbf{V}_1 + \mathbf{V}_2) \times \mathbf{V}_3 = \mathbf{V}_1 \times \mathbf{V}_3 + \mathbf{V}_2 \times \mathbf{V}_3 \\ &\mathbf{V}_1 \times (\mathbf{V}_2 \times \mathbf{V}_3) = \mathbf{V}_2(\mathbf{V}_3 \cdot \mathbf{V}_1) - \mathbf{V}_3(\mathbf{V}_1 \cdot \mathbf{V}_2) \\ &\mathbf{i} \times \mathbf{i} = \mathbf{j} \times \mathbf{j} = \mathbf{k} \times \mathbf{k} = 0.\mathbf{1} \text{ (zero vector)} \\ &\qquad = 0 \\ &\mathbf{i} \times \mathbf{j} = \mathbf{k}, \qquad \mathbf{j} \times \mathbf{k} = \mathbf{i}, \qquad \mathbf{k} \times \mathbf{i} = \mathbf{j} \\ &\text{If } \mathbf{V}_1 = a_1\mathbf{i} + b_1\mathbf{j} + c_1\mathbf{k}, \qquad \mathbf{V}_2 = a_2\mathbf{i} + b_2\mathbf{j} + c_2\mathbf{k}, \\ &\quad \mathbf{V}_3 = a_3\mathbf{i} + b_3\mathbf{j} + c_3\mathbf{k}, \end{aligned}$$

then

$$\mathbf{V}_1 \times \mathbf{V}_2 = \begin{vmatrix} \mathbf{i} & \mathbf{j} & \mathbf{k} \\ a_1 & b_1 & c_1 \\ a_2 & b_2 & c_2 \end{vmatrix} = (b_1c_2 - b_2c_1)\mathbf{i} + (c_1a_2 - c_2a_1)\mathbf{j} + (a_1b_2 - a_2b_1)\mathbf{k}$$

It should be noted that, since $\mathbf{V}_1 \times \mathbf{V}_2 = -\mathbf{V}_2 \times \mathbf{V}_1$, the vector product is not commutative.

$$\mathbf{V}_1 \cdot (\mathbf{V}_2 \times \mathbf{V}_3) = (\mathbf{V}_1 \times \mathbf{V}_2) \cdot \mathbf{V}_3 = \mathbf{V}_2 \cdot (\mathbf{V}_3 \times \mathbf{V}_1)$$

$$= (\mathbf{V}_1\mathbf{V}_2\mathbf{V}_3) = \begin{vmatrix} a_1 & b_1 & c_1 \\ a_2 & b_2 & c_2 \\ a_3 & b_3 & c_3 \end{vmatrix}$$

which is equal to the volume of a parallelepiped whose three determining edges are $\mathbf{V}_1$, $\mathbf{V}_2$, $\mathbf{V}_3$.

Differentiation of Vectors

If $\mathbf{V}_1 = a_1\mathbf{i} + b_1\mathbf{j} + c_1\mathbf{k}$, and $\mathbf{V}_2 = a_2\mathbf{i} + b_2\mathbf{j} + c_2\mathbf{k}$, and if $\mathbf{V}_1$ and $\mathbf{V}_2$ are functions of the scalar t, then

$$\frac{d}{dt}(\mathbf{V}_1 + \mathbf{V}_2 + \cdots) = \frac{d\mathbf{V}_1}{dt} + \frac{d\mathbf{V}_2}{dt} + \cdots,$$

where $\dfrac{d\mathbf{V}_1}{dt} = \dfrac{da_1}{dt}\mathbf{i} + \dfrac{db_1}{dt}\mathbf{j} + \dfrac{dc_1}{dt}\mathbf{k}$, etc.

$$\frac{d}{dt}(\mathbf{V}_1 \cdot \mathbf{V}_2) = \frac{d\mathbf{V}_1}{dt} \cdot \mathbf{V}_2 + \mathbf{V}_1 \cdot \frac{d\mathbf{V}_2}{dt}$$

$$\frac{d}{dt}(\mathbf{V}_1 \times \mathbf{V}_2) = \frac{d\mathbf{V}_1}{dt} \times \mathbf{V}_2 + \mathbf{V}_1 \times \frac{d\mathbf{V}_2}{dt}$$

Differential Operators—Rectangular Coordinates

By definition

$$\nabla \equiv \text{del} \equiv \mathbf{i}\frac{\partial}{\partial x} + \mathbf{j}\frac{\partial}{\partial y} + \mathbf{k}\frac{\partial}{\partial z}$$

$$\nabla^2 \equiv \text{Laplacian} \equiv \frac{\partial^2}{\partial x^2} + \frac{\partial^2}{\partial y^2} + \frac{\partial^2}{\partial z^2}$$

If S is a scalar function, then

$$\nabla S \equiv \text{grad } S \equiv \frac{\partial S}{\partial x}\mathbf{i} + \frac{\partial S}{\partial y}\mathbf{j} + \frac{\partial S}{\partial z}\mathbf{k}$$

The *distributive* law holds for finding a gradient. Thus if S and T are scalar functions

$$\nabla(S + T) = \nabla S + \nabla T$$

The *associative* law becomes the rule for differentiating a product:

$$\nabla(ST) = S\nabla T + T\nabla S$$

If $\mathbf{V}$ is a vector function with the magnitudes of the components parallel to the three coordinate axes V_x, V_y, V_z, then

$$\nabla \cdot \mathbf{V} \equiv \text{div } \mathbf{V} \equiv \frac{\partial V_x}{\partial x} + \frac{\partial V_y}{\partial y} + \frac{\partial V_z}{\partial z}$$

ALGEBRA (Continued)
VECTOR ANALYSIS (Continued)

VECTOR ANALYSIS (Continued)

The divergence obeys the distributive law. Thus, if **V** and **U** are vector functions, then

$$\nabla \cdot (\mathbf{V} + \mathbf{U}) = \nabla \cdot \mathbf{V} + \nabla \cdot \mathbf{U}$$

$$\nabla \times \mathbf{V} \equiv \text{curl } \mathbf{V} \equiv \text{rot } \mathbf{V} \equiv \begin{vmatrix} \mathbf{i} & \mathbf{j} & \mathbf{k} \\ \frac{\partial}{\partial x} & \frac{\partial}{\partial y} & \frac{\partial}{\partial z} \\ V_x & V_y & V_z \end{vmatrix}$$

The operator ∇ can be used more than once. The number of possibilities where ∇ is used twice are

$$\begin{aligned} \nabla \cdot (\nabla \theta) &\equiv \text{div grad } \theta \\ \nabla \times (\nabla \theta) &\equiv \text{curl grad } \theta \\ \nabla(\nabla \cdot \mathbf{V}) &\equiv \text{grad div } \mathbf{V} \\ \nabla \cdot (\nabla \times \mathbf{V}) &\equiv \text{div curl } \mathbf{V} \\ \nabla \times (\nabla \times \mathbf{V}) &\equiv \text{curl curl } \mathbf{V} \end{aligned}$$

Thus: $\text{div grad } S \equiv \nabla \cdot (\nabla S) \equiv \text{Laplacian } S \equiv \nabla^2 S$

$$\equiv \frac{\partial^2 S}{\partial x^2} + \frac{\partial^2 S}{\partial y^2} + \frac{\partial^2 S}{\partial z^2}$$

$$\text{curl grad } S \equiv 0; \qquad \text{curl curl } \mathbf{V} \equiv \text{grad div } \mathbf{V} - \nabla^2 \mathbf{V}; \quad \text{div curl } \mathbf{V} \equiv 0$$

Green's Theorem

Let **F** be a vector function and v a volume bounded by a surface s, then

$$\iiint_{(v)} \text{div } \mathbf{F}\, dv = \iiint_{(v)} \nabla \cdot \mathbf{F}\, dv = \iint_{(s)} \mathbf{F} \cdot d\mathbf{s},$$

where the integrations are to be carried out over the volume v and the surface s.

Stokes Theorem

Let **F** be a vector function and s a surface bounded by a simple closed curve c, then

$$\int_{(c)} \mathbf{F} \cdot d\mathbf{l} = \iint_{(s)} (\nabla \times \mathbf{F}) \cdot d\mathbf{s} = \iint_{(s)} \text{curl } \mathbf{F} \cdot d\mathbf{s},$$

where the integrations are to be carried out over the curve c and the surface s. It should be noted that

$$d\mathbf{l} = dx\,\mathbf{i} + dy\,\mathbf{j} + dz\,\mathbf{k}$$

The theorem implies that the line integral of **F** over the contour c equals the surface integral of $\nabla \times \mathbf{F}$ over a surface s which is bounded by c.

MISCELLANEOUS

The Sum (Σ, = Sigma) and Product (Π, = Pi) Notations

Σ denotes the **sum**, and Π, the **product** of all quantities of a given collection. In particular,

$\sum_{i=m}^{m+n} x_i$ means $x_m + x_{m+1} + \ldots + x_{m+n}$, $(n + 1$ terms in all),

$\prod_{i=m}^{m+n} x_i$ means $x_m x_{m+1} \ldots x_{m+n}$, $(n + 1$ factors in all).

For indicated **range,** R, (such as $m \leqq i \leqq m + n$), one may write $\sum_R x_i$, $\prod_R x_i$, respectively. Where the range is clear from the context one writes Σx_i, Πx_i, or even Σx, Πx, respectively. For c a constant and for x_i and y_i with common range (say of n elements),

$$\Sigma c x_i = c \Sigma x_i,\ \Sigma(x_i + y_i) = \Sigma x_i + \Sigma y_i,\ \Sigma(x_i + c) = nc + \Sigma x_i$$

Special Numerical Relations

(i) For range, $i = 1, 2, \ldots, n$, with $x_i = i$.
$\Sigma x_i = n(n + 1)/2$, $\Sigma(2x_i - 1) = n^2$,
$\Sigma x_i^2 = n(n + 1)(2n + 1)/6$.
$\Sigma x_i^3 = (\Sigma x_i)^2$, $\Sigma x_i^4 = (\Sigma x_i^2)[6(\Sigma x_i) - 1]/5$.
$\Pi(c + 1 - x_i) = c^{(n)}$, $\Pi x_i = n^{(n)} = n!$ (**"factorial n"**).
Hence $n! = n \,.\, (n - 1)!$ $\;0!$ is defined to be 1.
Stirling's formula (used for n large),

$$\sqrt{2n\pi}(n/e)^n < n! < \sqrt{2n\pi}(n/e)^n\left(1 + \frac{1}{12n - 1}\right),$$

$$(\pi = 3.14159 \ldots,\ e = 2.71828 \ldots).$$

$n!/(n - m)!$ gives the number of **permutations of n distinct things taken m at a time.**

(ii) For range, $i = -\left(\frac{n-1}{2}\right), -\left(\frac{n-1}{2}\right) + 1, \ldots, \left(\frac{n-1}{2}\right) - 1, \left(\frac{n-1}{2}\right)$, with $x_i = i$ (whether n is odd or even),

$$\sum x_i = \sum x_i^3 = 0,\ \sum x_i^2 = \frac{n(n^2 - 1)}{12},\ \sum x_i^4 = \frac{3n^2 - 7}{20} \sum x_i^2.$$

Special Numerical Relations

(Continued)

(iii) The Binomial Coefficients, $\binom{n}{m}$.
$\binom{n}{m} = n!/[(n-m)!m!]$, for integers m, n, $0 \leq m \leq n$. $\binom{n}{0} = \binom{n}{n} = 1$.

$(x+c)^n = \sum_r \binom{n}{r} x^{n-r} c^r$, $(0 \leq r \leq n)$, the **binomial expansion.** $\binom{n}{m}$ gives also the number of **combinations of n distinct things taken m at a time.**

$\binom{n}{m} + \binom{n}{m+1} = \binom{n+1}{m+1}$, **recursion relation** for binomial coefficients.

$$\binom{n}{n-m} = \binom{n}{m},\quad \sum_r (-1)^r \binom{n}{r} = 0,\quad \sum_r \binom{n}{r}^2 = \binom{2n}{n},\quad \sum_{s=m}^{n} \binom{s}{m} = \binom{n+1}{m+1}.$$

Table of Binomial Coefficients

n	$\binom{n}{0}$	$\binom{n}{1}$	$\binom{n}{2}$	$\binom{n}{3}$	$\binom{n}{4}$	$\binom{n}{5}$	$\binom{n}{6}$	$\binom{n}{7}$	$\binom{n}{8}$	$\binom{n}{9}$	$\binom{n}{10}$
0	1										
1	1	1									
2	1	2	1								
3	1	3	3	1							
4	1	4	6	4	1						
5	1	5	10	10	5	1					
6	1	6	15	20	15	6	1				
7	1	7	21	35	35	21	7	1			
8	1	8	28	56	70	56	28	8	1		
9	1	9	36	84	126	126	84	36	9	1	
10	1	10	45	120	210	252	210	120	45	10	1
11	1	11	55	165	330	462	462	330	165	55	11
12	1	12	66	220	495	792	924	792	495	220	66
13	1	13	78	286	715	1287	1716	1716	1287	715	286
14	1	14	91	364	1001	2002	3003	3432	3003	2002	1001
15	1	15	105	455	1365	3003	5005	6435	6435	5005	3003
16	1	16	120	560	1820	4368	8008	11440	12870	11440	8008
17	1	17	136	680	2380	6188	12376	19448	24310	24310	19448
18	1	18	153	816	3060	8568	18564	31824	43758	48620	43758
19	1	19	171	969	3876	11628	27132	50388	75582	92378	92378
20	1	20	190	1140	4845	15504	38760	77520	125970	167960	184756

NOTE: $\binom{n}{m} = \frac{n(n-1)(n-2)\ldots(n-m+1)}{m(m-1)(m-2)\ldots3.2.1}$; $\binom{n}{0} = 1$; $\binom{n}{1} = n$.

For coefficients missing from the above table, use the relation

$\binom{n}{m} = \binom{n}{n-m}$, e.g. $\binom{20}{11} = \binom{20}{9} = 167960$.

Finite Differences

For equi-spaced arguments x_i, and associated y_i, the successive **advancing y-differences** are, $\Delta^0 y_i = y_i$, $\Delta y_i = y_{i+1} - y_i$, $\Delta^2 y_i = \Delta y_{i+1} - \Delta y_i = y_{i+2} - 2y_{i+1} + y_i, \ldots,$ $\Delta^m y_i = \Delta^{m-1} y_{i+1} - \Delta^{m-1} y_i = \sum_r (-1)^r \binom{m}{r} y_{i+m-r}$. With arbitrary origin A and class-interval length, $x_{i+1} - x_i = h$, using $u_i = (x_i - A)/h$, write $y(u_i)$ for y_i. Then if for some fixed m, for the portion of the table considered, the values of $\Delta^{m+1} y_i$ be zero (or approximately, if these be regarded as negligible) **Newton's formula** gives

$$y(u) = \sum \frac{u^{(r)}}{r!} \Delta^r y(0) = y(0) + u\Delta y(0) + \frac{u(u-1)}{1\cdot2}\Delta^2 y(0) + \cdots + \frac{u(u-1)\ldots(u-m+1)}{m!}\Delta^m y(0).$$

This formula reduces to an identity for $u = u_0, u_1, \ldots, u_n$, $(u_i = i)$, and may be used to interpolate for intermediate values.

Example. Given

x	-4,	-2,	0,	2,	4,	6,	8, …
y	10,	14,	30,	64,	122,	210,	334, …

to find a value for y when $x = 10$, and when $x = 1$. Suppose for some reason A has been taken at $x = 2$. The work may be arranged as follows:

u	x	y	Δ	Δ^2	Δ^3	Δ^4
-3	-4	10				
			4			
-2	-2	14		12		
			16		6	
-1	0	30		18		0
			34		6	
0	2	64		24		0
			58		6	
1	4	122		30		0
			88		6	
2	6	210		36		—
			124		—	
3	8	334		—		—
			—		—	
—	—	—		—		—

$$y(u) = 64 + 58u + 30\frac{u(u-1)}{1\cdot2} + 6\frac{u(u-1)(u-2)}{1\cdot2\cdot3}$$
$$= 64 + 58u + 15u(u-1) + u(u-1)(u-2).$$

At $x = 10$, $u = 4$. Substituting $u = 4$, one has $y|_{x=10} = 500$.
At $x = 1$, $u = -\frac{1}{2}$. Substituting $u = -\frac{1}{2}$, one has $y|_{x=1} = 44\frac{3}{8}$.

Central Measures

Here the range of i is from 1 to n. With each value x_i is associated a weighting factor $f_i \geq 0$ (such as the frequency the probability, the mass, the reliability, or other multiplier).

N, the **total weight,** $= \Sigma f_i$.

$\bar{x}$, the **arithmetic mean,** $= \Sigma f_i x_i / N = \Sigma f_i x_i / \Sigma f_i$.

GM, the **geometric mean** (available when each x_i is positive), $= \sqrt[N]{\Pi x_i^{f_i}}$. $\text{Log } GM = \Sigma f_i \log x_i / N$.

Mo, the **mode,** = value among $(x_1, \ldots, x_n)$ having maximum associated f_i (usually obtained by interpolating after the data are graduated). For unweighted items, x_i, a mode is a value about which the values of x_i cluster most densely.

RMS, the **root-mean-square,** $= \sqrt{\Sigma f_i x_i^2 / N}$.

Md, the **median** (see below). For unweighted items, the median is the value, equaled or exceeded by exactly half of the values x_i in the given list. In case of a central pair, the median is usually taken as the arithmetic mean of this pair.

Mm, the **mid-mean** (see below). For unweighted items, the mid-mean is the arithmetic mean of the half-list obtained upon dropping out the highest quarter and lowest quarter of the items.

Cum $f|_X$, the value of "cumulative f" at X, $= \sum_{x_i < X} f_i$ (interpolation being used for X if necessary).

The M-Tiles

For **ungrouped data,** X is called the **rth m-tile** (or rth **m-tile mark**) $(r = 0, 1, \ldots, m)$ if simultaneously, $\sum_{x_i < X} f_i/N \leq r/m$, and $\sum_{x_i > X} f_i/N \leq (m - r)/m$. In particular the zeroth m-tile is **min,** the minimal value among the list $(x_1, \ldots, x_n)$, and the mth m-tile is **max,** the maximal value among the list.

For **grouped data,** the rth m-tile mark, X, is such that

$$\text{Cum } f|_X = Nr/m, \quad (r = 0, 1, 2, \ldots, m).$$

$$\text{Cum } f|_{\min} = 0, \qquad \text{Cum } f|_{\max} = N.$$

In particular, certain intermediate $(0 < r < m)$ m-tile marks are named as follows:

m	$r = 1$	2	3
2	Md (median)		
3	T_1 (lower tertile)	T_2 (upper tertile)	
4	Q_1 (lower quartile)	Md	Q_3 (upper quartile)
10	D_1 (first decile)	D_2	D_3 etc
100	PC_1 (first percentile)	PC_2	PC_3 etc

The term "rth m-tile" $(r = 1, \ldots, m)$ is also used to denote the class interval extending from the $(r - 1)$st to rth m-tile mark as defined above.

Mm, the **mid-mean,** =

$$2 \sum_{Q_1 \leq x_i \leq Q_3} f_i x_i / N = \sum_{Q_1 \leq x_i \leq Q_3} f_i x_i \Big/ \sum_{Q_1 \leq x_i \leq Q_3} f_i.$$

When each x_i is positive, and not all are equal, one always has $0 < \min < GM < \bar{x} < RMS < \max$.

For moderately-skewed distributions, one has approximately $Mo - \bar{x} = 3(Md - \bar{x})$, or $3Md = Mo + 2\bar{x}$.

Measures of Dispersion and Skewness

Here A is an arbitrary reference value, usually a convenient integral measure near $\bar{x}$.

ν_k, **kth moment about** A, $= \Sigma f_i (x_i - A)^k / N$, $(k = 0, 1, \ldots)$.

$\nu_0 = 1$, $\nu_1 = \bar{x} - A$. ν_2 as function of A is minimum for $A = \bar{x}$.

μ_k, **kth moment about** $\bar{x}$, $= \Sigma f_i (x_i - \bar{x})^k / N$, $(k = 0, 1, \ldots)$.

$$\mu_0 = 1,$$
$$\mu_1 = 0,$$
$$\mu_2 = \nu_2 - \nu_1^2 \quad (\mu_2 = \text{variance}),$$
$$\mu_3 = \nu_3 - 3\nu_1\nu_2 + 2\nu_1^3,$$
$$\mu_4 = \nu_4 - 4\nu_1\nu_3 + 6\nu_1^2\nu_2 - 3\nu_1^4.$$
$$\beta_1 = \mu_3^2/\mu_2^3. \quad \beta_2 = \mu_4/\mu_2^2.$$

σ, **standard deviation,** $= \sqrt{\mu_2}$.

$\alpha_3/2$, **momental skewness;** $\alpha_3 = \sqrt{\beta_1} = \mu_3/\sigma^3$.

$(\alpha_4 - 3)/2$, **kurtosis;** $\alpha_4 = \beta_2$.

MD, **mean deviation** (from the mean), $= \Sigma f_i |x_i - \bar{x}| / N$

$$= 2\left[\bar{x} \sum_{x_i < \bar{x}} f_i - \sum_{x_i < \bar{x}} f_i x_i\right] / N.$$ (This latter form is convenient for computation.)

s, **quartile deviation,** $= |Q_3 - Q_1|/2$.

$P.E.$, **probable error,** $= 0.6745\sigma$.

V, **coefficient of variation,** $= 100\sigma/\bar{x}\,\%$.

Pearson's measure of skewness $= (\bar{x} - Mo)/\sigma$. (Usually approximately $\alpha_3/2$.)

Bowley's measure of skewness $= (Q_3 - 2Md + Q_1)/(2s)$. (Bowley's measure of skewness lies between -1 and $+1$.)

STATISTICS

The Class Interval

$$\Delta x_i = x_{i+1} - x_i.$$

For equi-spaced arguments, $\Delta x_i = h$, the **length of the class interval,** x_i **is the mid-value or class mark.** The interval from $x_i - (h/2)$ to $x_i + (h/2)$ is the **class interval** with these as given **initial and terminal end values.**

$$u_i = (x_i - A)/h.$$
$$\bar{u} = \Sigma f_i u_i/N,\ \bar{x} = h\bar{u} + A.$$
$$(\mu_k)_x = h^k(\mu_k)_u,\ (k = 0, 1, \ldots).$$
$$\sigma_u^2 = [\Sigma f_i u_i^2/N] - \bar{u}^2,\ \sigma_x = h\sigma_u.$$
$$(\beta_1)_x = (\beta_1)_u,\ (\beta_2)_x = (\beta_2)_u.$$

Sheppard's corrections (to correct approximately for the error due to treating all elements in a given class interval of length h as though concentrated at the class mark).

For μ_0, μ_1, μ_3, no corrections.

In x-units,

corrected $(\mu_2)_x$ = uncorrected $(\mu_2)_x - h^2/12$,

corrected $(\mu_4)_x$ = uncorrected $(\mu_4)_x - h^2$ uncorrected $(\mu_2)_x/2 + 7h^4/240$.

In u-units, replace h by 1 in the formulae given above.

Least Squares

The **normal equations** for finding coefficients, $a_0, a_1, \ldots, a_m$, in fitting a curve of the form $y = a_0 + a_1x + \ldots + a_mx^m$ to data (X_i, Y_i), $i = 1, \ldots, n$, $(n > m)$, are $m + 1$ in number as follows:

$$\Sigma Y_i = a_0 n + a_1\Sigma X_i + a_2\Sigma X_i^2 + \ldots + a_m\Sigma X_i^m,$$
$$\Sigma X_iY_i = a_0\Sigma X_i + a_1\Sigma X_i^2 + a_2\Sigma X_i^3 + \ldots + a_m\Sigma X_i^{m+1},$$
$$\cdots\cdots\cdots\cdots\cdots\cdots\cdots\cdots$$
$$\Sigma X_i^mY_i = a_0\Sigma X_i^m + a_1\Sigma X_i^{m+1} + a_2\Sigma X_i^{m+2} + \ldots + a_m\Sigma X_i^{2m}.$$

Deviation from fitted curve,

$$d_i = Y_i - (a_0 + a_1X_i + \ldots + a_mX_i^m).$$
$$\Sigma d_i^2 = \Sigma Y_i^2 - (a_0\Sigma Y_i + a_1\Sigma X_iY_i + \ldots + a_m\Sigma X_i^mY_i).$$

For $z = ab^x$, use $y = \log z$, $a_0 = \log a$, $a_1 = \log b$.

For $z = at^p$, use $y = \log z$, $a_0 = \log a$, $a_1 = p$, $x = \log t$.

S_y, **standard error of estimate,** = root-mean-square of the y-deviations about a fitted curve $= \sqrt{\Sigma d_i^2/n}$.

Simple Correlation

Product Moment Method

Given n equi-spaced measurements X_i, $i = 1, 2, \ldots, n$, with $h = X_{i+1} - X_i$, $x_i = X_i - \bar{X}$; and m equi-spaced measurements Y_j, $j = 1, 2, \ldots, m$, with $k = Y_{j+1} - Y_j$, $y_j = Y_j - \bar{Y}$; and a weight (frequency, probability, etc.) e_{ij} (≥ 0), associated with (X_i, Y_j). Here e_{ij} is an entry in the table.

$$f_i = \sum_j e_{ij},\ g_j = \sum_i e_{ij}.$$
$$N = \sum_{ij} e_{ij} = \sum_i f_i = \sum_j g_j. \quad \text{(Check)}$$
$$\bar{x} = \sum_{ij} e_{ij}X_i/N = \sum_i f_iX_i/N;\ \bar{y} = \sum_{ij} e_{ij}Y_j/N = \sum_j g_jY_j/N.$$

Let A and B be arbitrary reference values, usually convenient integral measures near $\bar{X}$ and $\bar{Y}$, respectively.

$$u_i = (X_i - A)/h,\ v_j = (Y_j - B)/k;$$
$$\bar{u} = \Sigma f_iu_i/N,\ \bar{X} = h\bar{u} + A,\ \bar{v} = \Sigma g_jv_j/N,\ \bar{Y} = k\bar{v} + B.$$
$$\left.\begin{array}{l}\sigma_u^2 = (\mu_2)_u = (\Sigma f_iu_i^2/N) - \bar{u}^2,\ \sigma_x = h\sigma_u.\\ \sigma_v^2 = (\mu_2)_v = (\Sigma g_jv_j^2)/N - \bar{v}^2,\ \sigma_y = k\sigma_v.\end{array}\right\}\text{Apply Sheppard's corrections.}$$
$$U_j = \sum_i e_{ij}u_i,\ V_i = \sum_j e_{ij}v_j,\ P = \sum u_iV_i = \sum v_jU_j, \quad \text{(Check)}$$
$$p_{uv} = \sum_{ij} e_{ij}(u_i - \bar{u})(v_j - \bar{v})/N$$
$$= (P/N) - \overline{uv}.$$
$$p_{xy} = hkp_{uv}.$$

$r = p_{uv}/(\sigma_u\sigma_v) = p_{xy}/(\sigma_x\sigma_y)$ **(product-moment) coefficient of correlation.** In every case $-1 \leq r \leq 1$.

$Y - \bar{Y} = r\dfrac{\sigma_y}{\sigma_x}(X - \bar{X})$, or $y = r\dfrac{\sigma_y}{\sigma_x}x$, **regression line of** y **on** x.

$X - \bar{X} = r\dfrac{\sigma_x}{\sigma_y}(Y - \bar{Y})$, or $x = r\dfrac{\sigma_x}{\sigma_y}y$, **regression line of** x **on** y.

Example of Computation for Product-Moment Coefficient of Correlation

v_j	y_j \ u_i, x_i	-3 (12)	-2 (16)	-1 (20)	0 (24)	1 (28)	2 (32)	g_j	g_jv_j	$g_jv_j^2$	U_j $\left(-\sum_i e_{ij}u_i\right)$	v_jU_j
2	21			1	5	7	1	14	28	56	8	16
1	18		1	3	7	5	2	18	18	18	4	4
0	15		2	3	4	1		10	0	0		0
−1	12		3	1	1			5	−5	5	−7	7
−2	9	2	1					3	−6	12	−8	16
	f_i	2	7	8	17	13	3	50	35	91		43
	f_iu_i	−6	−14	−8	0	13	6	−9				
	$f_iu_i^2$	18	28	8	0	13	12	79				
	$V_i,\left(-\sum_j e_{ij}v_j\right)$	−4	−4	4		19	4					
	u_iV_i	12	8	−4	0	19	8	43				

$A = 24$, $B = 15$,
$h = 4$, $k = 3$,
$N = \Sigma f_i = \Sigma g_j = 50$,
$\Sigma f_iu_i = -9$, $\Sigma g_jv_j = 35$,
$\Sigma f_iu_i^2 = 79$, $\Sigma g_jv_j^2 = 91$,
$P = \Sigma u_iV_i = \Sigma v_jU_j = 43$.

$\bar{u} = -9/50 = -.18 \qquad \bar{v} = 35/50 = .70$

$\sigma_u^2 = (79/50) - (-.18)^2 - .083 = 1.46, \qquad \sigma_u = 1.21$

$\sigma_v^2 = (91/50) - (.70)^2 - .083 = 1.247. \qquad \sigma_v = 1.117$

$p_{uv} = (43/50) - (-.18)(.70) = +0.986$

$r = +0.986/(1.21 \times 1.117) = +.730. \qquad$ Ans. $r = +.730$

STATISTICS (Continued)

RANK DIFFERENCE METHOD

Given n corresponding pairs of measured items (X_i, Y_i), $(i = 1, \ldots, n)$. Let (u_i, v_i) be the corresponding rank numbers. Here $u_i = 1$ for the largest X_i, 2 for the next largest X_i, etc., and similarly $v_i = 1$ for the largest Y_i, 2 for the next largest Y_i, etc. $\rho = 1 - \dfrac{6\Sigma(u_i - v_i)^2}{n(n^2-1)}$, (rank difference) coefficient of correlation. In every case $-1 \le \rho \le 1$. **Check:** $\Sigma(u_i - v_i) = 0$.

Example of Computation for Rank-Difference Coefficient of Correlation

X_i	Y_i	u_i	v_i	$u_i - v_i$	$(u_i - v_i)^2$	
76	52	3	1	+2	4	Check: $\Sigma(u_i - v_i) = 0$.
66	34	8	9	−1	1	
63	32	10	10	0	0	
74	45	4	4	0	0	
79	50	1	2	−1	1	$\rho = 1 - \dfrac{6 \times 62}{10(10^2 - 1)}$
69	37	7	7	0	0	
77	35	2	8	−6	36	$= +0.63$
65	42	9	5	+4	16	
71	40	6	6	0	0	Ans. $\rho = +.63$
73	48	5	3	+2	4	
$N = 10$				0	62	

Probability

If among $a + b$ equi-probable and mutually exclusive events, a are regarded as favorable and b unfavorable, then for a single trial

p, probability of favorable outcome, $= \dfrac{a}{a+b}$,

q, probability of unfavorable outcome, $= 1 - p = \dfrac{b}{a+b}$.

The successive terms in the binomial expansion $(p+q)^n = \sum_r \binom{n}{r} p^{n-r} q^r$ give the respective probabilities that in n trials, the event will be favorable exactly $n - r$ times, $r = 0, \ldots, n$.

The mean number of favorable events is np, of unfavorable, nq; the standard deviation is $\sigma = \sqrt{npq}$, $\alpha_3 = (p - q)/\sigma$ (the positive direction being that of increasing unfavorability).

Normal curve (x measured in σ-units from the mean, and with area = 1):

$$y = \frac{1}{\sqrt{2\pi}} e^{-x^2/2} = 0.3989 e^{-x^2/2}.$$

MD (mean deviation from the mean) $= \sigma\sqrt{2/\pi} = 0.7979\sigma$.
s (quartile deviation from the mean) $= 0.6745\sigma = 0.845\,MD$.

Percentage areas, under normal curve, for successive class intervals measured from the mean:

Multiples of σ: 34 %, 14 %, 2 %.
Multiples of s: 25 %, 16 %, 7 %, 2 %.

Normal surface (x measured in σ_x-units y in σ_y-units from their means),

$$z = \frac{1}{2\pi\sqrt{1 - r^2}} e^{-(x^2 - 2rxy + y^2)/[2(1-r^2)]}.$$

***Goodness of Fit.** For a universe of objects falling into n mutually exclusive classes with class marks, $x_i (i = 1, 2, \ldots, n)$, let p_i be the probability for the ith class. Given a sample of N items, with f_i items in the ith class $(\Sigma f_i = N)$, the probability that a random sample of N items gives no better fit, expressed in terms of n and χ^2 ("Chi square"), $= \Sigma(f_i - Np_i)^2/(Np_i)$, is given by a table, portions of which are as follows:

Probability that a Random Sample Gives no Better Fit

n \ χ^2	1	2	3	4	6	8	10	15	20
3	.607	.368	.223	.135	.050	.018	.007	.001	.000
4	.801	.572	.392	.261	.112	.046	.019	.002	.000
5	.910	.736	.558	.406	.199	.092	.040	.005	.000
6	.963	.849	.700	.549	.306	.156	.075	.010	.001
7	.986	.920	.809	.677	.423	.238	.125	.020	.003
8	.995	.960	.885	.780	.540	.333	.189	.036	.006
9	.998	.981	.934	.857	.647	.433	.265	.059	.010
10	.999	.991	.964	.911	.740	.534	.350	.091	.018
11	1.000	.996	.981	.947	.815	.629	.440	.132	.029
12	1.000	.998	.991	.970	.873	.713	.530	.182	.045

n \ χ^2	8	10	12	14	16	18	20	25	30
10	.534	.350	.213	.122	.067	.035	.018	.003	.000
11	.629	.440	.285	.173	.100	.055	.029	.005	.001
12	.713	.530	.363	.233	.141	.082	.045	.009	.002
13	.785	.616	.446	.301	.191	.116	.067	.015	.003
14	.844	.694	.528	.374	.249	.158	.095	.023	.005
15	.889	.762	.606	.450	.313	.207	.130	.035	.008
16	.924	.820	.679	.526	.382	.263	.172	.050	.012
17	.949	.867	.744	.599	.453	.324	.220	.070	.018
18	.967	.904	.800	.667	.524	.389	.274	.095	.026
19	.979	.932	.847	.729	.593	.456	.333	.125	.037
20	.987	.953	.886	.784	.657	.522	.395	.161	.052

* See index for t Test and F Test Tables.

Plane Figures Bounded by Straight Lines

The **area of a triangle** whose base is b and altitude h

$$= \frac{hb}{2}.$$

The area of a **triangle** with angles A, B, and C and sides opposite a, b, and c, respectively

$$= \tfrac{1}{2}ab \sin C.$$

or

$$= \sqrt{s(s-a)(s-b)(s-c)},$$

where $s = \tfrac{1}{2}(a + b + c)$.

A **rectangle** with sides a and b has an area $= ab$.

The area of a **parallelogram** with side b and the perpendicular distance to the parallel side h

$$= bh.$$

The area of a **parallelogram** with sides a and b and the included angle θ

$$= ab \sin \theta.$$

The area of a **rhombus** with diagonals c and d,

$$= \tfrac{1}{2}cd.$$

The area of a **trapezoid** whose parallel sides are a and b and altitude h

$$= \tfrac{1}{2}(a + b)h.$$

The area of any **quadrilateral** with diagonals a and b and the angle between them θ

$$= \tfrac{1}{2}ab \sin \theta.$$

The area of a **regular polygon** with n sides, each of length l,

$$= \tfrac{1}{4}nl^2 \cot \frac{180°}{n}.$$

For a regular polygon of n sides, each side of length l, the radius of the **inscribed circle,**

$$= \frac{l}{2} \cot \frac{180°}{n}.$$

The **radius of the circumscribed circle,**

$$= \frac{l}{2} \operatorname{cosec} \frac{180°}{n}.$$

Area, Radius of Inscribed and Circumscribed Circles for Regular Polygons

l = length of one side.

Name.	Number of sides.	Area.	Radius of inscribed circle.	Radius of circumscribed circle.
Triangle, equilateral........	3	$0.43301l^2$	$0.28867l$	$0.57735l$
Square........	4	$1.00000l^2$	$0.50000l$	$0.70710l$
Pentagon........	5	$1.72048l^2$	$0.68819l$	$0.85065l$
Hexagon........	6	$2.59808l^2$	$0.86602l$	$1.0000l$
Heptagon........	7	$3.63391l^2$	$1.0383l$	$1.1523l$
Octagon........	8	$4.82843l^2$	$1.2071l$	$1.3065l$
Nonagon........	9	$6.18182l^2$	$1.3737l$	$1.4619l$
Decagon........	10	$7.69421l^2$	$1.5388l$	$1.6180l$
Undecagon........	11	$9.36564l^2$	$1.7028l$	$1.7747l$
Dodecagon........	12	$11.19615l^2$	$1.8660l$	$1.9318l$

Radius of circle inscribed in any triangle, whose sides are a, b, and c, where $s = \tfrac{1}{2}(a + b + c)$ is given by r

$$= \frac{\sqrt{s(s-a)(s-b)(s-c)}}{s}.$$

The radius of the **circumscribed** circle is given by R

$$= \frac{abc}{4\sqrt{s(s-a)(s-b)(s-c)}}.$$

The **perimeter of a polygon inscribed in a circle** of radius where n is the number of sides,

$$= 2nr \sin \frac{\pi}{n}. \qquad (\pi \text{ radians} = 180°)$$

The area of the **inscribed** polygon,

$$= \tfrac{1}{2}nr^2 \sin \frac{2\pi}{n}.$$

The perimeter of a polygon circumscribed about a circle of radius r, number of sides n

$$= 2nr \tan \frac{\pi}{n}.$$

The area of the **circumscribed** polygon

$$= nr^2 \tan \frac{\pi}{n}.$$

MENSURATION FORMULÆ

Plane Figures Bounded by Curved Lines

The **circumference of a circle** whose radius is r and diameter $d (d = 2r)$

$= 2\pi r = \pi d.$ $(\pi = 3.14159)$

The area of a circle

$= \pi r^2 = \frac{1}{4}\pi d^2 = .7854d^2.$

The length of an arc of a circle for an arc of θ degrees

$= \frac{\pi r \theta}{180}.$

NOTE.—In this and following similar formulæ r denotes the radius of the circle, $(OC$, Fig. 1).

For an arc of θ radians the length s

$= r\theta.$

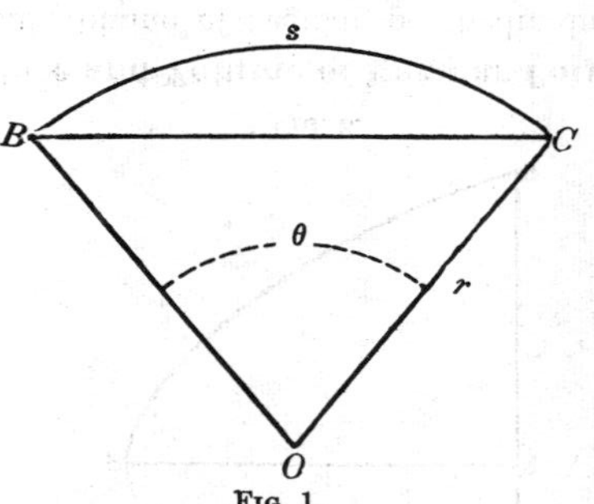

FIG. 1.

The length of a chord subtending an angle θ.

$= 2r \sin \frac{1}{2}\theta.$

The **area of a sector** where θ is the angle between the radii in degrees

$= \frac{\pi r^2 \theta}{360}.$

If s is the length of the arc, the area of the sector

$= \frac{sr}{2}.$

The **area of a segment** where θ is the angle between the two radii in degrees

$= \frac{\pi r^2 \theta}{360} - \frac{r^2 \sin \theta}{2}$ $\left[\begin{array}{l}\theta^\circ = 180^\circ - [2 \sin^{-1}(x/r)] \\ x = \perp \text{ dist. center to chord}\end{array}\right]$

If θ is in radians the area $= \frac{1}{2}r^2(\theta - \sin \theta).$

The area of the segment of a circle

$$= \frac{\pi r^2}{2} - \left[x\sqrt{r^2 - x^2} + r^2 \operatorname{Sin}^{-1}\left(\frac{x}{r}\right)\right]$$

where r is the radius of the circle and x the perpendicular distance of the chord from the center. The principal angle must be used in this formula.

The **area of the ring** between two circles of radius r_1 and r_2, one of which encloses the other,

$= \pi(r_1 + r_2)(r_1 - r_2).$

The two circles are not necessarily concentric.

Area of the sector of an annulus. (Fig. 2.)—If angle $GOH = \theta$ and the lines GO and $JO = r_1$ and r_2 respectively, the area $GHIJ = \frac{1}{2}\theta(r_1 + r_2)(r_1 - r_2).$

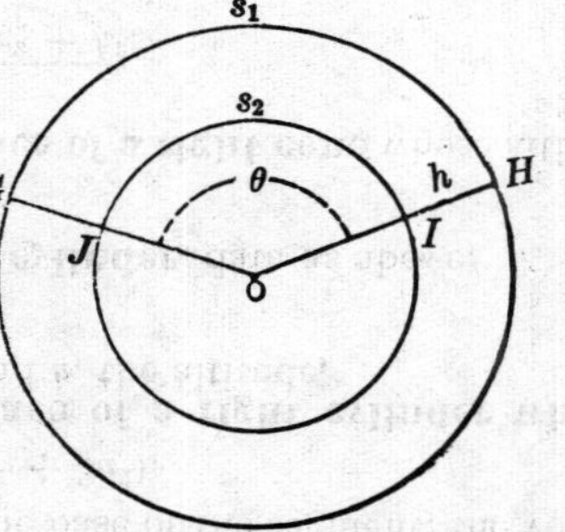

FIG. 2.

If s_1 = the length of the arc GH and s_2 = the arc JI and $h = HI = r_1 - r_2$, the area $GHIJ = \frac{1}{2}h(s_1 + s_2).$

The **circumference of an ellipse** whose semiaxes are a and b

$= 2\pi\sqrt{\frac{a^2 + b^2}{2}}$ (approx.) $= 4aE$ exactly. See tables of elliptic integrals for E, using $k = \frac{\sqrt{a^2 - b^2}}{a}$

The **area of an ellipse** $= \pi ab.$

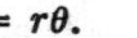

The length of the **arc of a parabola,** as arc SPQ in Fig. 3, where $x = PR$, and $y = QR$

$$\sqrt{4x^2 + y^2} + \frac{y^2}{2x} \log_e \frac{2x + \sqrt{4x^2 + y^2}}{y}$$

The area of the section of the parabola $PQRS$, $= \frac{4}{3}xy.$

Solids Bounded by Planes

The **lateral area of a regular prism** = perimeter of a right section × the length.

The **volume of a regular prism** = area of base × the altitude.

The **lateral area of a regular pyramid,** slant height l, length of one side of base a, and a number of sides n,

$$= \tfrac{1}{2}nal.$$

The **volume of a pyramid** = $\frac{1}{3}$ area of base × altitude.

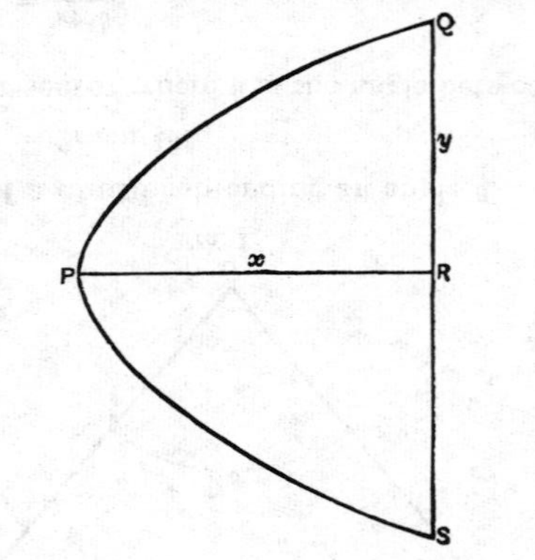

FIG. 3.

Surface and Volume of Regular Polyhedra

Surface and volume of regular polyhedra in terms of the length of one edge l.

Name.	Nature of surface.	Surface.	Volume.
Tetrahedron...	4 equilateral triangles	$1.73205l^2$	$0.11785l^3$
Hexahedron or cube........	6 squares..........	$6.00000l^2$	$1.00000l^3$
Octahedron....	8 equilateral triangles	$3.46410l^2$	$0.47140l^3$
Dodecahedron .	12 pentagons.........	$20.64573l^2$	$7.66312l^3$
Icosahedron...	20 equilateral triangles	$8.66025l^2$	$2.18170l^3$

Solids Bounded by Curved Surfaces

The **surface of a sphere** of radius r and diameter $d(= 2r)$

$$= 4\pi r^2 = \pi d^2 = 12.57r^2.$$

The **volume of a sphere**

$$= \tfrac{4}{3}\pi r^3 = \tfrac{1}{6}\pi d^3 = 4.189r^3.$$

The **area of a lune** on the surface of a sphere of radius r, included between two great circles whose inclination is θ radians.

$$= 2r^2\theta.$$

The **area of a spherical triangle** whose angles are A, B, and C (radians) on a sphere of radius r

$$= (A + B + C - \pi)r^2.$$

The **area of a spherical polygon** of n sides where θ is the sum of its angles in radians

$$= [\theta - (n - 2)\pi]r^2.$$

The area of the curved surface of a **spherical segment** of height h, radius of sphere r

$$= 2\pi rh.$$

The **volume of a spherical segment,** data as above

$$= \tfrac{1}{3}\pi h^2 (3r - h).$$

If a = radius of the base of the segment, the volume

$$= \tfrac{1}{6}\pi h(h^2 + 3a^2).$$

The **curved surface of a right cylinder where** r = the radius of the base and h, the altitude,

$$= 2\pi rh.$$

The **volume of a cylinder,** data as above,

$$= \pi r^2 h.$$

The **curved surface of a right cone** whose altitude is h and radius of base r

$$= \pi r\sqrt{r^2 + h^2}.$$

The **volume of a cone,** data as above,

$$= \frac{\pi}{3}r^2h = 1.047\, r^2h.$$

The **curved surface of the frustum of a right cone,** radius of base r_1, of top r_2 and altitude h,

$$= \pi(r_1 + r_2)\sqrt{h^2 + (r_1 - r_2)^2}.$$

The **volume of the frustum of a cone,** data as above,

$$= \pi\frac{h}{3}(r_1^2 + r_1r_2 + r_2^2).$$

The **oblate spheroid** is formed by the rotation of an ellipse about its minor axis. If a and b are the major and minor semi-axes respectively, and ϵ the eccentricity, the surface

$$= 2\pi a^2 + \pi\frac{b^2}{\epsilon}\log_e\frac{1+\epsilon}{1-\epsilon},$$

and volume $= \tfrac{4}{3}\pi a^2 b$.

The prolate spheroid is formed by the rotation of an ellipse about its major axis ($2a$), data as above.

Surface $\quad = 2\pi b^2 + 2\pi \frac{ab}{\epsilon} \sin^{-1} \epsilon,$

volume $\quad = \frac{4}{3}\pi ab^2.$

SIMPSON'S RULE FOR IRREGULAR AREAS

Divide the area into an even number ($2m$) of panels by means of $2m+1$ parallel lines, drawn at constant distance h apart; and denote the lengths of the intercepted segments by $y_0, y_1, \ldots, y_{2m-1}, y_{2m}$. The first and last of these may be zero. The area will then be

$$A = \tfrac{1}{3}h[(y_0 + y_{2m}) + 4(y_1 + y_3 + \ldots + y_{2m-1}) + 2(y_2 + y_4 + \ldots + y_{2m-2})]$$

While the formula is exact in many simple cases, ordinarily the formula provides only an approximation, for which the accuracy increases with an increase in the number of divisions. Simpson's Rule may be applied to finding volumes, if the measures $y_0, y_1, \ldots, y_{2m}$ be interpreted as the areas of parallel plane sections at constant distance h apart.

PRISMOIDAL FORMULA

As a special case where m=1, and H, (=2h) is the distance between two limiting parallel planes, one has for the volume of a solid figure,

$$V = \tfrac{1}{6}H(S_0 + 4S_1 + S_2).$$

Here S_0 and S_2 are the cross-sectional areas in these limiting planes (lower and upper bases, respectively), and S_1 is the cross section of the mid-section. The formula is exact for the cone, sphere, ellipsoid, and prismoid.

TRIGONOMETRIC FUNCTIONS IN A RIGHT-ANGLED TRIANGLE

If A, B, and C are the vertices (C the right angle), and a, b, and h the sides opposite respectively,

sine A = sin A = $\frac{a}{h}$, cosine A = cos A = $\frac{b}{h}$,

tangent A = tan A = $\frac{a}{b}$, cotangent A = cot A = ctn A = $\frac{b}{a}$,

secant A = sec A = $\frac{h}{b}$, cosecant A = csc A = $\frac{h}{a}$.

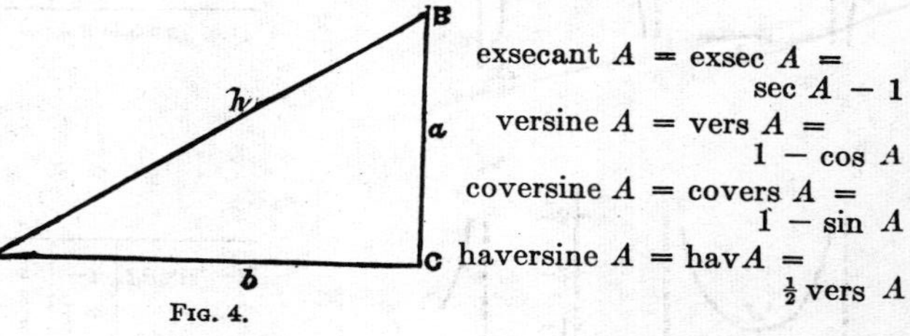

Fig. 4.

exsecant A = exsec A = sec $A - 1$

versine A = vers A = $1 - \cos A$

coversine A = covers A = $1 - \sin A$

haversine A = hav$A$ = $\frac{1}{2}$ vers A

RELATIONS BETWEEN DEGREE OF ACCURACY OF COMPUTED LENGTHS AND ANGLES

When solving a triangle for any of its parts the following should be observed:

Length to	(requires)	Angle to:
2 significant digits		nearest 30′ = 0.5°
3 significant digits		nearest 05′ = 0.083°
4 significant digits		nearest 01′ = 0.0167°
5 significant digits		nearest 0.1′ = 0.00167°

SIGNS AND LIMITS OF VALUE ASSUMED BY THE FUNCTIONS

Function	Quadrant I		Quadrant II		Quadrant III		Quadrant IV	
	Sign	Value	Sign	Value	Sign	Value	Sign	Value
sin.	+	0 to 1	+	1 to 0	−	0 to 1	−	1 to 0
cos.	+	1 to 0	−	0 to 1	−	1 to 0	+	0 to 1
tan.	+	0 to ∞	−	∞ to 0	+	0 to ∞	−	∞ to 0
cot.	+	∞ to 0	−	0 to ∞	+	∞ to 0	−	0 to ∞
sec.	+	1 to ∞	−	∞ to 1	−	1 to ∞	+	∞ to 1
cosec.	+	∞ to 1	+	1 to ∞	−	∞ to 1	−	1 to ∞

TRIGONOMETRIC FORMULAE (Continued)

GRAPHS OF THE TRIGONOMETRIC FUNCTIONS

$y = \sin x$

x	y	Point
0	0	$P_0(0, 0)$
$\frac{\pi}{6} = .52$	.50	$P_1(.52, .50)$
$\frac{\pi}{4} = .79$	.71	$P_2(.79, .71)$
$\frac{\pi}{3} = 1.05$	.87	$P_3(1.05, .87)$
$\frac{\pi}{2} = 1.57$	1	$P_4(1.57, 1)$
$\pi = 3.14$	0	$P_5(3.14, 0)$
$\frac{3\pi}{2} = 4.71$	−1	$P_6(4.71, -1)$
$2\pi = 6.28$	0	$P_7(6.28, 0)$

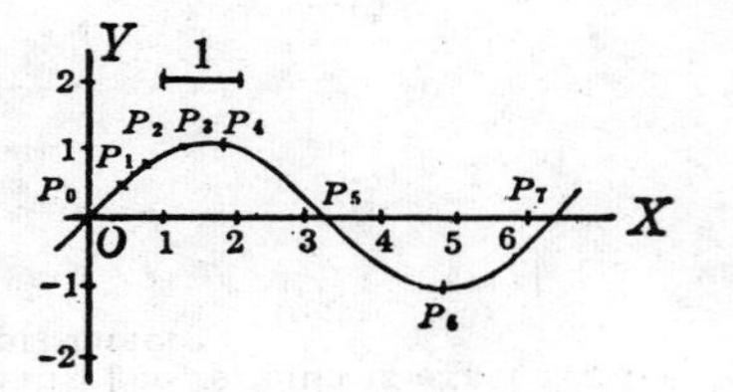

$y = \cos x$

x	y	Point
0	1	$P_0(0, 1)$
$\frac{\pi}{6} = .52$	.87	$P_1(.52, .87)$
$\frac{\pi}{4} = .79$	.71	$P_2(.79, .71)$
$\frac{\pi}{3} = 1.05$	.5	$P_3(1.05, .5)$
$\frac{\pi}{2} = 1.57$	0	$P_4(1.57, 0)$
$\pi = 3.14$	−1	$P_5(3.14, -1)$
$\frac{3\pi}{2} = 4.71$	0	$P_6(4.71, 0)$
$2\pi = 6.28$	1	$P_7(6.28, 1)$

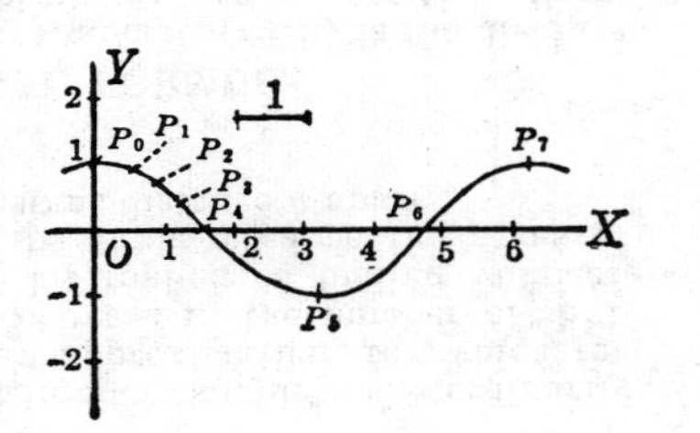

$y = \tan x$

x	y	Point
0	0	$P_0(0, 0)$
$\frac{\pi}{4}$	1	$P_1(.79, 1)$
$\frac{\pi}{2}$	∞	....
$\frac{3\pi}{4}$	−1	$P_3(2.36, -1)$
π	0	$P_4(3.14, 0)$
$\frac{5\pi}{4}$	1	$P_5(3.93, 1)$
$\frac{3\pi}{2}$	∞	....
2π	0	$P_6(6.28, 0)$

$y = \cot x$

x	y	Point
0	±∞	....
$\frac{\pi}{4}$	1	$P_1(.79, 1)$
$\frac{\pi}{2}$	0	$P_2(1.57, 0)$
$\frac{3\pi}{4}$	−1	$P_3(2.36, -1)$
π	±∞	....

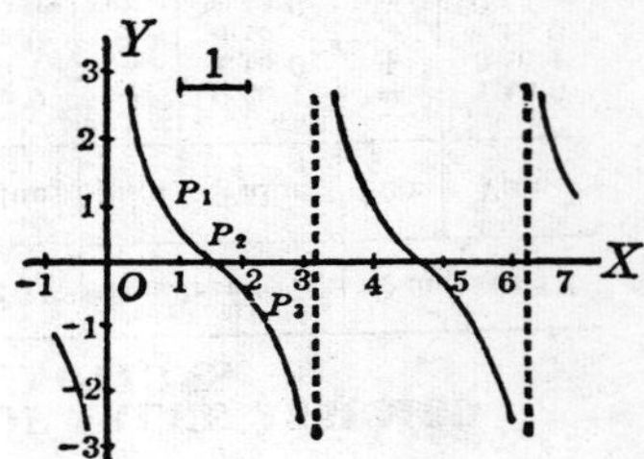

$y = \sec x$

x	y	Point
0	1	$P_0(0, 1)$
$\frac{\pi}{4}$	$\sqrt{2}$	$P_1(.79, 1.4)$
$\frac{\pi}{2}$	±∞	....
π	−1	$P_2(3.14, -1)$

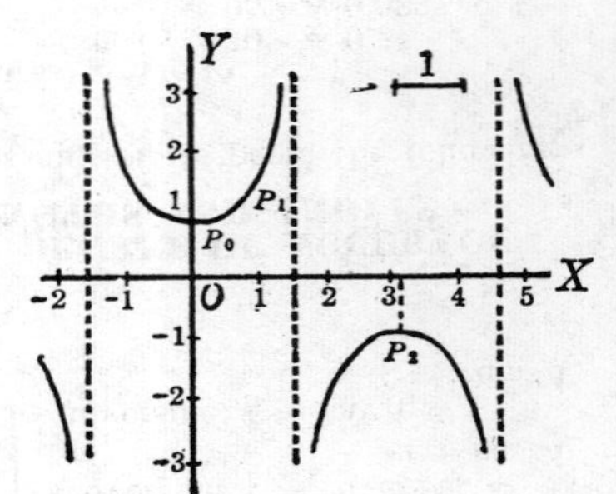

$y = \csc x$

x	y	Point
0	±∞	....
$\frac{\pi}{4}$	$\sqrt{2}$	$P_1(.79, 1.4)$
$\frac{\pi}{2}$	1	$P_2(1.57, 1)$
π	±∞	....

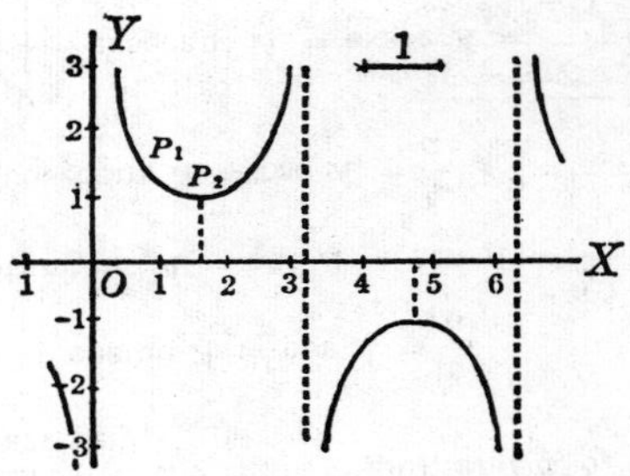

TRIGONOMETRIC FORMULAE (Continued)
GRAPHS OF THE TRIGONOMETRIC FUNCTIONS

GRAPHS OF THE TRIGONOMETRIC FUNCTIONS (Continued)

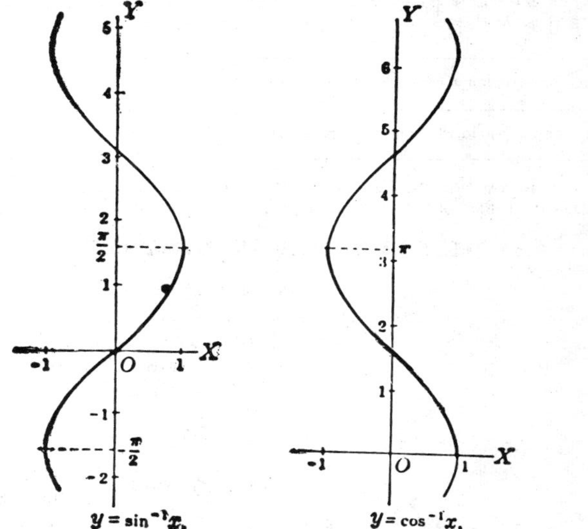

VALUE OF THE FUNCTIONS OF VARIOUS ANGLES

	0°	30°	45°	60°	90°	180°	270°
sin..........	0	$\frac{1}{2}$	$\frac{1}{2}\sqrt{2}$	$\frac{1}{2}\sqrt{3}$	1	0	−1
cos..........	1	$\frac{1}{2}\sqrt{3}$	$\frac{1}{2}\sqrt{2}$	$\frac{1}{2}$	0	−1	0
tan..........	0	$\frac{1}{3}\sqrt{3}$	1	$\sqrt{3}$	∞	0	∞
cot..........	∞	$\sqrt{3}$	1	$\frac{1}{3}\sqrt{3}$	0	∞	0
sec..........	1	$\frac{2\sqrt{3}}{3}$	$\sqrt{2}$	2	∞	−1	∞
cosec........	∞	2	$\sqrt{2}$	$\frac{2\sqrt{3}}{3}$	1	∞	−1

EXPONENTIAL DEFINITIONS OF CIRCULAR FUNCTIONS

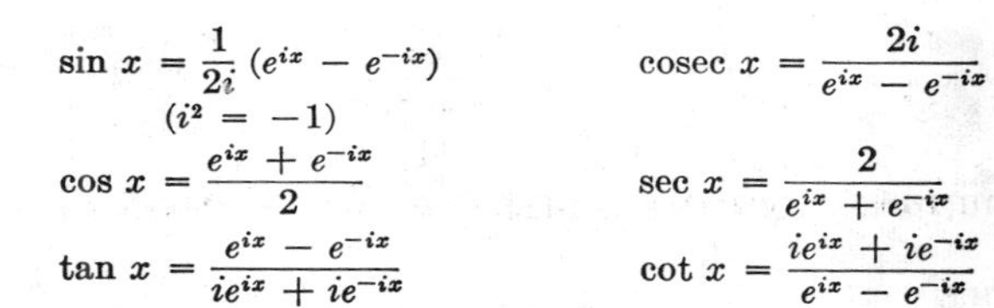

$$\sin x = \frac{1}{2i}(e^{ix} - e^{-ix}) \quad (i^2 = -1) \qquad \operatorname{cosec} x = \frac{2i}{e^{ix} - e^{-ix}}$$

$$\cos x = \frac{e^{ix} + e^{-ix}}{2} \qquad \sec x = \frac{2}{e^{ix} + e^{-ix}}$$

$$\tan x = \frac{e^{ix} - e^{-ix}}{ie^{ix} + ie^{-ix}} \qquad \cot x = \frac{ie^{ix} + ie^{-ix}}{e^{ix} - e^{-ix}}$$

RELATIONS OF THE FUNCTIONS

$$\sin x = \frac{1}{\operatorname{cosec} x}. \qquad \operatorname{cosec} x = \frac{1}{\sin x}.$$

$$\cos x = \frac{1}{\sec x}. \qquad \sec x = \frac{1}{\cos x}.$$

$$\tan x = \frac{1}{\cot x} = \frac{\sin x}{\cos x}. \qquad \sin^2 x + \cos^2 x = 1.$$

$$\cot x = \frac{1}{\tan x} = \frac{\cos x}{\sin x}. \qquad 1 + \tan^2 x = \sec^2 x. \quad 1 + \cot^2 x = \operatorname{cosec}^2 x.$$

$$^*\sin x = \pm\sqrt{1 - \cos^2 x}. \qquad \cos x = \pm\sqrt{1 - \sin^2 x}.$$

$$\tan x = \pm\sqrt{\sec^2 x - 1}. \qquad \sec x = \pm\sqrt{\tan^2 x + 1}.$$

$$\cot x = \pm\sqrt{\operatorname{cosec}^2 x - 1}. \qquad \operatorname{cosec} x = \pm\sqrt{\cot^2 x + 1}.$$

$$\sin x = \cos(90° - x) = \sin(180° - x).$$

$$\cos x = \sin(90° - x) = -\cos(180° - x).$$

$$\tan x = \cot(90° - x) = -\tan(180° - x).$$

$$\cot x = \tan(90° - x) = -\cot(180° - x).$$

$$\operatorname{cosec} x = \cot\frac{x}{2} - \cot x.$$

* The sign in front of radical depends on quadrant in which x falls.

FUNCTIONS OF SUMS OF ANGLES

$$\sin(x \pm y) = \sin x \cos y \pm \cos x \sin y.$$

$$\cos(x \pm y) = \cos x \cos y \mp \sin x \sin y.$$

$$\tan(x \pm y) = \frac{\tan x \pm \tan y}{1 \mp \tan x \tan y}.$$

FUNCTIONS OF MULTIPLE ANGLES

$$\sin 2x = 2 \sin x \cos x.$$

$$\cos 2x = \cos^2 x - \sin^2 x = 2\cos^2 x - 1 = 1 - 2\sin^2 x.$$

$$\sin 3x = 3 \sin x - 4 \sin^3 x.$$

$$\cos 3x = 4\cos^3 x - 3 \cos x.$$

$$\sin 4x = 8\cos^3 x \sin x - 4 \cos x \sin x.$$

$$\cos 4x = 8\cos^4 x - 8\cos^2 x + 1.$$

$$\sin 5x = 5 \sin x - 20\sin^3 x + 16 \sin^5 x.$$

$$\cos 5x = 16\cos^5 x - 20\cos^3 x + 5 \cos x.$$

$$\sin 6x = 32\cos^5 x \sin x - 32\cos^3 x \sin x + 6 \cos x \sin x.$$

$$\cos 6x = 32\cos^6 x - 48\cos^4 x + 18\cos^2 x - 1.$$

$$\tan 2x = \frac{2 \tan x}{1 - \tan^2 x}$$

$$\cot 2x = \frac{\cot^2 x - 1}{2 \cot x}.$$

FUNCTIONS OF MULTIPLE ANGLES (Continued)

$$\tan 3x = \frac{3\tan x - \tan^3 x}{1 - 3\tan^2 x}.$$

$$^*\sin \tfrac{1}{2}x = \pm\sqrt{\frac{1-\cos x}{2}}.$$

$$\cos \tfrac{1}{2}x = \pm\sqrt{\frac{1+\cos x}{2}}.$$

$$\tan \tfrac{1}{2}x = \pm\sqrt{\frac{1-\cos x}{1+\cos x}} = \frac{1-\cos x}{\sin x} = \frac{\sin x}{1+\cos x}.$$

* The sign in front of radical depends on quadrant in which $\frac{x}{2}$ falls.

MISCELLANEOUS RELATIONS

$$\sin x \pm \sin y = 2\sin \tfrac{1}{2}(x \pm y)\cdot\cos\tfrac{1}{2}(x \mp y).$$

$$\cos x + \cos y = 2\cos\tfrac{1}{2}(x+y)\cdot\cos\tfrac{1}{2}(x-y).$$

$$\cos x - \cos y = -2\sin\tfrac{1}{2}(x+y)\cdot\sin\tfrac{1}{2}(x-y).$$

$$\tan x \pm \tan y = \frac{\sin(x \pm y)}{\cos x\cdot\cos y}. \qquad \cot x \pm \cot y = \frac{\pm\sin(x \pm y)}{\sin x\cdot\sin y}.$$

$$\frac{1+\tan x}{1-\tan x} = \tan(45° + x). \qquad \frac{\cot x + 1}{\cot x - 1} = \cot(45° - x)$$

$$\frac{\sin x \pm \sin y}{\cos x + \cos y} = \tan\tfrac{1}{2}(x \pm y).$$

$$\frac{\sin x \pm \sin y}{\cos x - \cos y} = -\cot\tfrac{1}{2}(x \mp y).$$

$$\frac{\sin x + \sin y}{\sin x - \sin y} = \frac{\tan\frac{1}{2}(x+y)}{\tan\frac{1}{2}(x-y)}.$$

$$\sin^2 x - \sin^2 y = \sin(x+y)\cdot\sin(x-y).$$

$$\cos^2 x - \cos^2 y = -\sin(x+y)\sin(x-y).$$

$$\cos^2 x - \sin^2 y = \cos(x+y)\cos(x-y).$$

INVERSE TRIGONOMETRIC FUNCTIONS

The following table lists each of the six inverse trigonometric functions together with the interval of its principal value:

Function	Interval containing principal value	
	x positive or zero	x negative
$y = \sin^{-1} x$ and $\tan^{-1} x$..	$0 \leqq y \leqq \pi/2$	$-\pi/2 \leqq y < 0$
$y = \cos^{-1} x$ and $\cot^{-1} x$..	$0 \leqq y \leqq \pi/2$	$\pi/2 < y \leqq \pi$
$y = \sec^{-1} x$ and $\csc^{-1} x$..	$0 \leqq y \leqq \pi/2$	$-\pi \leqq y \leqq -\pi/2$

Usually the first letter in "arc" or the name of the inverse trigonometric functions is capitalized if the principal value is desired. Thus

$$\text{Arc}\sin\frac{1}{2} = \text{Sin}^{-1}\frac{1}{2} = \frac{\pi}{6},$$

while
$$\text{arc}\sin\frac{1}{2} = \frac{\pi}{6} + 2\pi n \text{ or } \frac{5\pi}{6} + 2\pi n$$

In the calculus both for differentiation and integral formulas, capitalization is not adhered to strictly, but principal values are always understood for inverse trigonometric functions when used unless specifically stated otherwise.

RELATIONS BETWEEN SIDES AND ANGLES OF ANY PLANE TRIANGLE

In a triangle with angles A, B, and C and sides opposite a, b, and c respectively,

$$\frac{a}{\sin A} = \frac{b}{\sin B} = \frac{c}{\sin C} = \text{diameter of the circumscribed circle.}$$

$$a^2 = b^2 + c^2 - 2bc\cos A.$$

$$a = b\cos C + c\cos B.$$

$$\cos A = \frac{b^2 + c^2 - a^2}{2bc}.$$

$$\tan\frac{A-B}{2} = \frac{a-b}{a+b}\cot\frac{C}{2}.$$

$$\sin A = \frac{2}{bc}\sqrt{s(s-a)(s-b)(s-c)},$$

where $s = \frac{1}{2}(a+b+c)$ and $r = \sqrt{\frac{(s-a)(s-b)(s-c)}{s}}$.

$$\sin\frac{A}{2} = \sqrt{\frac{(s-b)(s-c)}{bc}}.$$

$$\cos\frac{A}{2} = \sqrt{\frac{s(s-a)}{bc}}.$$

$$\tan\frac{A}{2} = \sqrt{\frac{(s-b)(s-c)}{s(s-a)}} = \frac{r}{s-a}.$$

$$\frac{a+b}{a-b} = \frac{\sin A + \sin B}{\sin A - \sin B} = \frac{\tan\frac{1}{2}(A+B)}{\tan\frac{1}{2}(A-B)} = \frac{\cot\frac{1}{2}C}{\tan\frac{1}{2}(A-B)}.$$

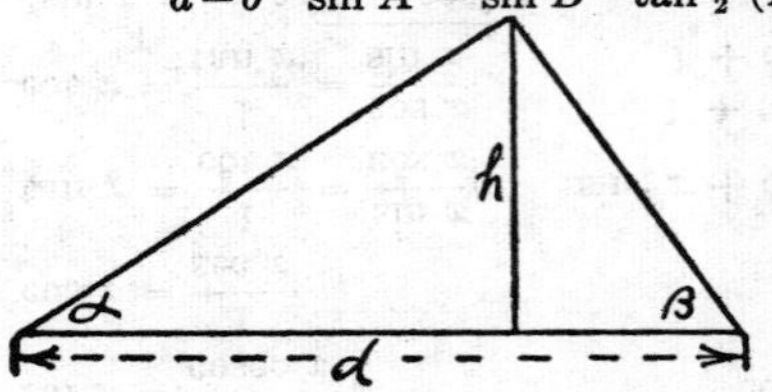

$$h = d\frac{\sin\alpha\sin\beta}{\sin(\alpha+\beta)} = \frac{d}{\cot\alpha + \cot\beta}$$

Similarly

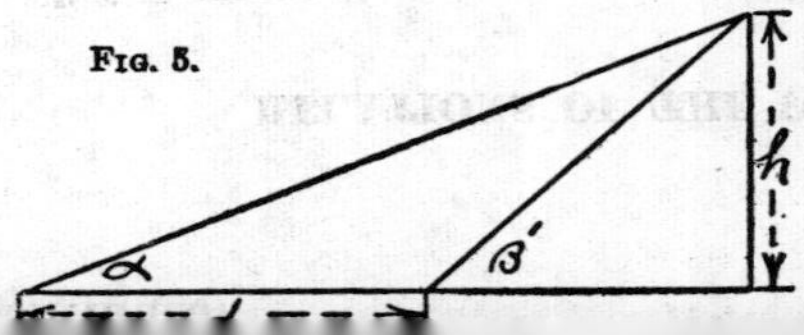

$$h = d\frac{\sin\alpha\sin\beta'}{\sin(\beta'-\alpha)} = \frac{d}{\cot\alpha - \cot\beta'}$$

Fig. 5.

RELATIONS IN ANY SPHERICAL TRIANGLE

If A, B and C be the three angles and a, b, and c the opposite sides,

$$\frac{\sin A}{\sin a}=\frac{\sin B}{\sin b}=\frac{\sin C}{\sin c}.$$

$$\cos a=\cos b\cos c+\sin b\sin c\cos A=\frac{\cos b\cos(c\pm\theta)}{\cos\theta}.$$

where $\tan\theta=\tan b\cos A$.

$$\cos A=-\cos B\cos C+\sin B\sin C\cos a.$$

$$\sin\tfrac{1}{2}A=\sqrt{\frac{\sin(s-b)\sin(s-c)}{\sin b\sin c}}.$$

where $s=\frac{1}{2}(a+b+c)$.

$$\cos\tfrac{1}{2}A=\sqrt{\frac{\sin s\sin(s-a)}{\sin b\sin c}}.$$

$$\tan\tfrac{1}{2}A=\frac{r}{\sin(s-a)}.$$

where $r=\sqrt{\dfrac{\sin(s-a)\sin(s-b)\sin(s-c)}{\sin s}}$.

$$\cos\tfrac{1}{2}a=\sqrt{\frac{\cos(S-B)\cos(S-C)}{\sin B\sin C}}.$$

where $S=\frac{1}{2}(A+B+C)$.

$$\sin\tfrac{1}{2}a=\sqrt{-\frac{\cos S\cos(S-A)}{\sin B\sin C}}.$$

$$\tan\tfrac{1}{2}a=R\cos(S-A)$$

where $R=\sqrt{\dfrac{-\cos S}{\cos(S-A)\cos(S-B)\cos(S-C)}}$.

$$\frac{\tan\frac{a+b}{2}}{\tan\frac{c}{2}}=\frac{\cos\frac{A-B}{2}}{\cos\frac{A+B}{2}},\quad\frac{\tan\frac{A+B}{2}}{\cot\frac{C}{2}}=\frac{\cos\frac{a-b}{2}}{\cos\frac{a+b}{2}}.$$

$$\frac{\tan\frac{a-b}{2}}{\tan\frac{c}{2}}=\frac{\sin\frac{A-B}{2}}{\sin\frac{A+B}{2}},\quad\frac{\tan\frac{A-B}{2}}{\cot\frac{C}{2}}=\frac{\sin\frac{a-b}{2}}{\sin\frac{a+b}{2}}$$

$$\text{hav } a=\text{hav }(b-c)+\sin b\sin c\text{ hav }A$$

$$\text{hav } A=\frac{\sqrt{\text{hav }[a+(b-c)]\text{ hav }[a-(b-c)]}}{\sin b\sin c}$$

NAPIER'S RULES FOR RIGHT SPHERICAL TRIANGLES

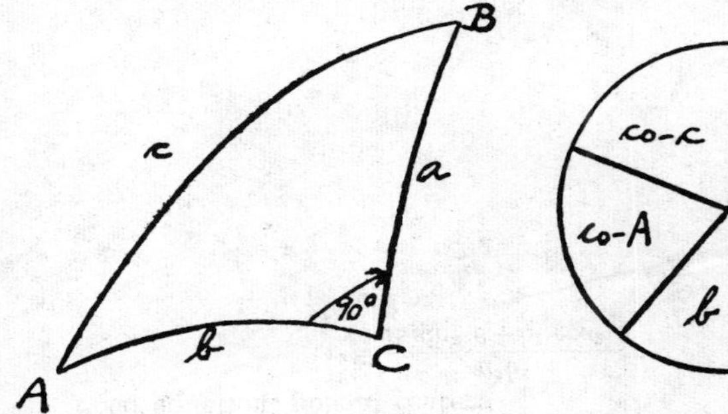

Take the five parts, *excluding the right angle* and consider them to be in a circular arrangement.

Attach a "Co" to the two angles and the hypotenuse, meaning "Complement of A" etc. Then:

1. The sine of the middle part equals the product of the tangents of the adjacent parts.
2. The sine of the middle part equals the product of the cosines of the opposite parts. The following rules for determining the quadrant of a calculated part apply:

1. A leg and the angle opposite it are always of the same quadrant.
2. If the hypotenuse is less than 90°, the legs are of the same quadrant. If the hypotenuse is greater than 90°, the legs are of opposite quadrants.

TRIGONOMETRIC FORMULAE (Continued)

The distance between two points x_1, y_1, and x_2, y_2,—rectangular coördinates:

$$d = \sqrt{(x_2 - x_1)^2 + (y_2 - y_1)^2}$$

For polar coördinates and points r_1, θ_1, and r_2, θ_2:

$$d = \sqrt{r_1^2 + r_2^2 - 2r_1r_2 \cos(\theta_1 - \theta_2)}$$

The area of a triangle whose vertices are x_1, y_1; x_2, y_2, and x_3, y_3:

$$A = \tfrac{1}{2}|x_1y_2 - x_2y_1 + x_2y_3 - x_3y_2 + x_3y_1 - x_1y_3|$$

For polar coördinates and vertices, r_1, θ_1; r_2, θ_2, and r_3, θ_3:

$$A = \tfrac{1}{2}\{r_1r_2 \sin(\theta_2 - \theta_1) + r_2r_3 \sin(\theta_3 - \theta_2) + r_3r_1 \sin(\theta_1 - \theta_3)\}$$

The equation of a straight line where m is the tangent of the angle of inclination and c, the distance of intersection with the Y axis from the origin:

$$y = mx + c$$

If a line of slope m passes through the point x_1, y_1 its equation is:

$$y - y_1 = m(x - x_1)$$

The equation of a line through the points x_1, y_1, and x_2, y_2 is:

$$\frac{y - y_1}{y_2 - y_1} = \frac{x - x_1}{x_2 - x_1}$$

If the intercepts on the X and Y axes are a and b respectively, the equation is:

$$\frac{x}{a} + \frac{y}{b} = 1$$

If the length of the perpendicular from the origin is p and its angle of inclination θ the equation is:

$$x \cos\theta + y \sin\theta = p$$

General equation of the straight line:

$Ax + By + C = 0$, where slope $m = \dfrac{-A}{B}$; x-intercept $a = \dfrac{-C}{A}$; y-intercept $b = \dfrac{-C}{B}$.

The equation of a circle whose center is at a, b, and whose radius is c:

$$(x - a)^2 + (y - b)^2 = c^2$$

If the origin is at the center:

$$x^2 + y^2 = c^2$$

The polar equation of a circle with the origin on the circumference and its center at point c, α:

$$r = 2c \cos(\theta - \alpha).$$

If the origin is not on the circumference, the radius a and the center at l, α, the equation becomes

$$a^2 = r^2 + l^2 - 2rl \cos(\theta - \alpha).$$

The equation of a parabola with the origin at the vertex, where f is the distance from the focus to the vertex:

$$y^2 = 4fx$$

If p is the semi-latus rectum $(=2f)$ the equation is:

$$y^2 = 2px$$

The polar equation where the pole is at the focus and p the semi-latus rectum is:

$$r = \frac{p}{1 - \cos\theta}$$

If the pole is at the vertex and p as above:

$$r = \frac{2p \cos\theta}{\sin^2\theta}$$

The equation of the ellipse with the origin at the center and semi-axes a and b:

$$\frac{x^2}{a^2} + \frac{y^2}{b^2} = 1$$

Polar equation where the pole is at the center:

$$r^2 = \frac{a^2b^2}{a^2 \sin^2\theta + b^2 \cos^2\theta}$$

The equation of the hyperbola with the origin at the center, semi-axes a and b:

$$\frac{x^2}{a^2} - \frac{y^2}{b^2} = 1$$

Polar equation, pole at center:

$$r^2 = \frac{a^2b^2}{a^2 \sin^2\theta - b^2 \cos^2\theta}$$

TRANSFORMATION OF COORDINATES

Rectangular System

(1) Translation only of axes parallel to themselves.
The coordinates of new origin with respect to old axes: $x = h$, $y = k$. Primed letters designate new coordinates.

$$x' = x - h \qquad x = x' + h$$
$$y' = y - k \qquad y = y' + k$$

(2) Rotation of axes with fixed origin.

Angle of rotation $= \theta$

$$x' = x \cos\theta + y \sin\theta \qquad x = x' \cos\theta - y' \sin\theta$$
$$y' = y \cos\theta - x \sin\theta \qquad y = y' \cos\theta + x' \sin\theta$$

(3) Origin translated and axes rotated.
Symbols same as above

$$x' = (x - h)\cos\theta + (y - k)\sin\theta$$
$$y' = (y - k)\cos\theta - (x - h)\sin\theta$$
$$x = x'\cos\theta - y'\sin\theta + h$$
$$y = y'\cos\theta + x'\sin\theta + k$$

Relation between Rectangular and Polar Coordinates

$$x = r\cos\theta \qquad r = \sqrt{x^2 + y^2} \qquad \sin\theta = \frac{y}{\sqrt{x^2 + y^2}}$$

$$y = r\sin\theta \qquad \theta = \tan^{-1}\frac{y}{x} \qquad \cos\theta = \frac{x}{\sqrt{x^2 + y^2}}$$

Rotation of Polar Axes through Angle θ

$$x' = r\cos\alpha$$
$$y' = r\sin\alpha$$
$$x = r\cos(\alpha + \theta)$$
$$y = r\sin(\alpha + \theta)$$

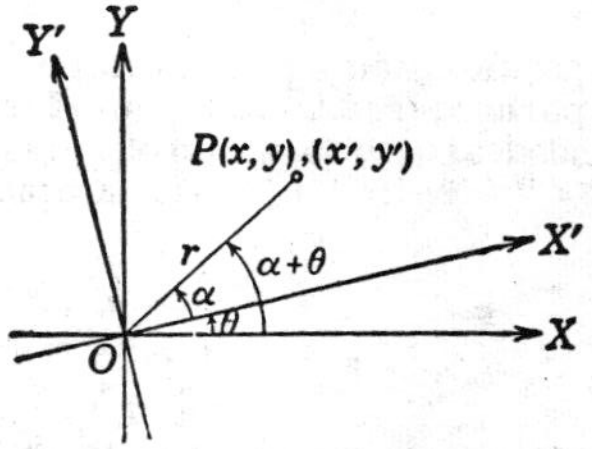

Y′ Y P(x, y), (x′, y′) r α+θ α θ X′ X O

TRANSFORMATION OF COORDINATES (Continued)

SOLID ANALYTICAL GEOMETRY

Space Coordinates

1. Rectangular System x, y, z.
2. Cylindrical System r, θ, z.
3. Spherical System ρ, θ, ϕ. In certain situations it is desirable to interchange the symbols ϕ for θ and vice versa.
4. Polar Space System ρ, α, β, γ.

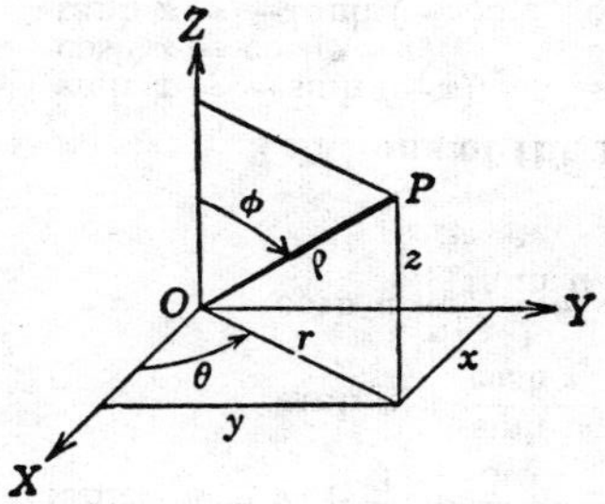

Z P φ ρ z O Y r x θ y X

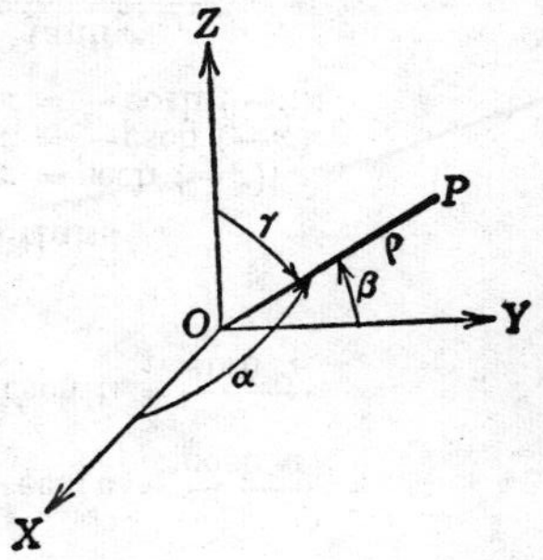

Z P γ ρ β O Y α X

Relations of Coordinates of Systems in Terms of x, y, z

Cylindrical	Spherical	Polar Space
$r = \sqrt{x^2 + y^2}$	$\rho = \sqrt{x^2 + y^2 + z^2}$	$\rho = \sqrt{x^2 + y^2 + z^2}$
$\theta = \tan^{-1}\frac{y}{x}$	$\theta = \tan^{-1}\frac{y}{x}$	$\alpha = \cos^{-1}\left(\frac{x}{\sqrt{x^2 + y^2 + z^2}}\right)$
$z = z$	$\phi = \cos^{-1}\left(\frac{z}{\sqrt{x^2 + y^2 + z^2}}\right)$	$\beta = \cos^{-1}\left(\frac{y}{\sqrt{x^2 + y^2 + z^2}}\right)$
		$\gamma = \cos^{-1}\left(\frac{z}{\sqrt{x^2 + y^2 + z^2}}\right)$

Relations of Rectangular Coordinates (x, y, z) in Terms of Cylindrical, Spherical and Polar Space Coordinates

Cylindrical	Spherical	Polar Space
$x = r\cos\theta$	$x = \rho\sin\phi\cos\theta$	$x = \rho\cos\alpha$
$y = r\sin\theta$	$y = \rho\sin\phi\sin\theta$	$y = \rho\cos\beta$
$z = z$	$z = \rho\cos\phi$	$z = \rho\cos\gamma$

ANALYTICAL GEOMETRY (Continued)

TRANSFORMATION OF COORDINATES (Continued)

TRANSFORMATION OF RECTANGULAR SPACE COORDINATES FOR TRANSLATION WITHOUT ROTATION

Coordinates of new origin are (h, k, l) in terms of the old coordinates. New coordinates are (x', y', z')

$$x' = x - h \qquad x = x' + h$$
$$y' = y - k \qquad y = y' + k$$
$$z' = z - l \qquad z = z' + l$$

For rotation only about the origin where $\lambda_1, \mu_1, \nu_1; \lambda_2, \mu_2, \nu_2$ and λ_3, μ_3, ν_3 are the direction cosines of the new axes with respect to the old axes. The coordinates of point P are (x, y, z) referred to the original axes and (x', y', z') referred to the new rotated axes.

$$x' = \lambda_1 x + \mu_1 y + \nu_1 z$$
$$y' = \lambda_2 x + \mu_2 y + \nu_2 z$$
$$z' = \lambda_3 x + \mu_3 y + \nu_3 z$$
$$x = \lambda_1 x' + \lambda_2 y' + \lambda_3 z'$$
$$y = \mu_1 x' + \mu_2 y' + \mu_3 z'$$
$$z = \nu_1 x' + \nu_2 y' + \nu_3 z'$$

HYPERBOLIC FUNCTIONS

Definitions

An hyperbolic function represents a relation between the coordinates of a given point on the arc of a rectangular hyperbola.

If O is the center, A the vertex, and P any point of the hyperbola APB,

$OM = x,$

$MP = y,$

$OA = a.$

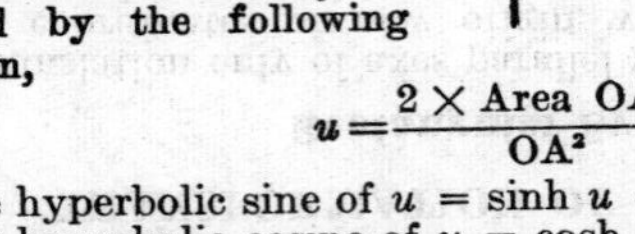

Fig. 6.

The function u may be defined by the following relation,

$$u = \frac{2 \times \text{Area OAP}}{OA^2}$$

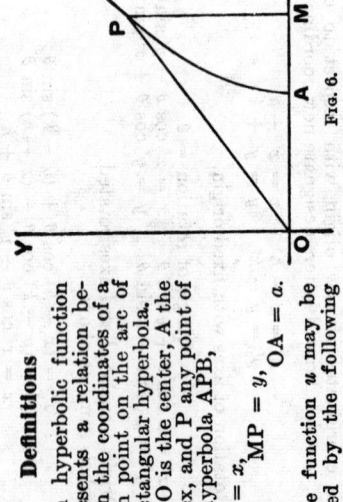

The hyperbolic sine of $u = \sinh u = y/a$.
The hyperbolic cosine of $u = \cosh u = x/a$.

Exponential Definitions of Hyperbolic Functions and Their Power Series

$$\sinh u = \frac{e^u - e^{-u}}{2} = \frac{1}{2}(e^u - e^{-u}) = u + \frac{u^3}{3!} + \frac{u^5}{5!} + \cdots$$

$$\cosh u = \frac{e^u + e^{-u}}{2} = \frac{1}{2}(e^u + e^{-u}) = 1 + \frac{u^2}{2!} + \frac{u^4}{4!} + \cdots$$

$$\tanh u = \frac{e^u - e^{-u}}{e^u + e^{-u}} = u - \frac{u^3}{3} + \frac{2u^5}{15} - \frac{17u^7}{315} + \cdots \qquad \left(u^2 < \frac{1}{4}\pi^2\right).$$

$$\sinh^{-1} u = u - \frac{1}{2}\cdot\frac{u^3}{3} + \frac{1\cdot 3}{2\cdot 4}\cdot\frac{u^5}{5} - \frac{1\cdot 3\cdot 5}{2\cdot 4\cdot 6}\cdot\frac{u^7}{7} + \cdots \qquad (u^2 < 1).$$

$$\sinh^{-1} u = \log 2u + \frac{1}{2}\cdot\frac{1}{2u^2} - \frac{1\cdot 3}{2\cdot 4}\cdot\frac{1}{4u^4} + \frac{1\cdot 3\cdot 5}{2\cdot 4\cdot 6}\cdot\frac{1}{6u^6} - \cdots \qquad (u^2 > 1).$$

$$\cosh^{-1} u = \log 2u - \frac{1}{2}\cdot\frac{1}{2u^2} - \frac{1\cdot 3}{2\cdot 4}\cdot\frac{1}{4u^4} - \frac{1\cdot 3\cdot 5}{2\cdot 4\cdot 6}\cdot\frac{1}{6u^6} - \cdots \qquad (u^2 > 1).$$

$$\tanh^{-1} u = u + \frac{u^3}{3} + \frac{u^5}{5} + \frac{u^7}{7} + \cdots \qquad (u^2 < 1).$$

$$\tanh u = \frac{\sinh u}{\cosh u}. \qquad \operatorname{sech} u = \frac{1}{\cosh u}.$$

$$\coth u = \frac{1}{\tanh u}. \qquad \operatorname{csch} u = \frac{1}{\sinh u}.$$

Relations of the Functions

$$\sinh x = -\sinh(-x). \qquad \operatorname{sech} x = \operatorname{sech}(-x).$$
$$\cosh x = \cosh(-x). \qquad \operatorname{csch} x = -\operatorname{csch}(-x).$$
$$\tanh x = -\tanh(-x). \qquad \coth x = -\coth(-x).$$

$$\sinh x = \frac{2\tanh\frac{1}{2}x}{1 - \tanh^2\frac{1}{2}x} = \frac{\tanh x}{\sqrt{1 - \tanh^2 x}}.$$

$$\cosh x = \frac{1 + \tanh^2\frac{1}{2}x}{1 - \tanh^2\frac{1}{2}x} = \frac{1}{\sqrt{1 - \tanh^2 x}}.$$

$$\cosh^2 x - \sinh^2 x = 1.$$

$$\tanh x = \sqrt{1 - \operatorname{sech}^2 x}. \qquad \operatorname{sech} x = \sqrt{1 - \tanh^2 x}.$$
$$\coth x = \sqrt{\operatorname{csch}^2 x + 1}. \qquad \operatorname{csch} x = \sqrt{\coth^2 x - 1}.$$

HYPERBOLIC FUNCTIONS (Continued)

Relations of the Functions (Continued)

$\sinh(\frac{1}{2}x) = \sqrt{\frac{1}{2}(\cosh x - 1)}.$
$\cosh(\frac{1}{2}x) = \sqrt{\frac{1}{2}(\cosh x + 1)}.$
$\tanh(\frac{1}{2}x) = (\cosh x - 1) \div \sinh x = \sinh x \div (\cosh x + 1).$
$\sinh(2x) = 2 \sinh x \cosh x.$
$\cosh(2x) = \cosh^2 x + \sinh^2 x = 2\cosh^2 x - 1 = 1 + 2\sinh^2 x$
$\tanh(2x) = 2 \tanh x \div (1 + \tanh^2 x).$
$\sinh 3x = 3 \sinh x + 4 \sinh^3 x.$
$\cosh 3x = 4 \cosh^3 x - 3 \cosh x.$
$\tanh 3x = (3 \tanh x + \tanh^3 x) \div (1 + 3 \tanh^2 x).$
$\sinh(x \pm y) = \sinh x \cdot \cosh y \pm \cosh x \cdot \sinh y.$
$\cosh(x \pm y) = \cosh x \cdot \cosh y \pm \sinh x \cdot \sinh y.$
$\tanh(x \pm y) = (\tanh x \pm \tanh y) \div (1 \pm \tanh x \cdot \tanh y).$
$\sinh x + \sinh y = 2 \sinh \frac{1}{2}(x + y) \cdot \cosh \frac{1}{2}(x - y).$
$\sinh x - \sinh y = 2 \cosh \frac{1}{2}(x + y) \cdot \sinh \frac{1}{2}(x - y).$
$\cosh x + \cosh y = 2 \cosh \frac{1}{2}(x + y) \cdot \cosh \frac{1}{2}(x - y).$
$\cosh x - \cosh y = 2 \sinh \frac{1}{2}(x + y) \cdot \sinh \frac{1}{2}(x - y).$

$$\sinh x + \cosh x = \frac{1 + \tanh \frac{1}{2}x}{1 - \tanh \frac{1}{2}x}$$

$$\tanh x \pm \tanh y = \frac{\sinh(x \pm y)}{\cosh x \cosh y}.$$

$$\coth x \pm \coth y = \pm \frac{\sinh(x \pm y)}{\sinh x \sinh y}.$$

Inverse Functions

$$\sinh^{-1} x = \log(x + \sqrt{x^2 + 1}) = \int \frac{dx}{\sqrt{x^2 + 1}} = \cosh^{-1} \sqrt{x^2 + 1}$$

$$\cosh^{-1} x = \log(x + \sqrt{x^2 - 1}) = \int \frac{dx}{\sqrt{x^2 - 1}} = \sinh^{-1} \sqrt{x^2 - 1}$$

$$\tanh^{-1} x = \frac{1}{2}\log(1 + x) - \frac{1}{2}\log(1 - x) = \int \frac{dx}{1 - x^2}.$$

$$\coth^{-1} x = \frac{1}{2}\log(1 + x) - \frac{1}{2}\log(x - 1) = \int \frac{dx}{1 - x^2}.$$

$$\operatorname{sech}^{-1} x = \log\left(\frac{1}{x} + \sqrt{\frac{1}{x^2} - 1}\right) = -\int \frac{dx}{x\sqrt{1 - x^2}}.$$

$$\operatorname{csch}^{-1} x = \log\left(\frac{1}{x} + \sqrt{\frac{1}{x^2} + 1}\right) = -\int \frac{dx}{x\sqrt{x^2 + 1}}.$$

Relations to Circular Functions

$\sinh x = -i \sin ix.$	$\sinh ix = i \sin x.$
$\cosh x = \cos ix.$	$\cosh ix = \cos x.$
$\tanh x = -i \tan ix.$	$\tanh ix = i \tan x.$

If $x = \log \tan\left(\frac{\pi}{4} + \frac{\theta}{2}\right) = \log(\sec\theta + \tan\theta),$

θ = the **gudermannian** of $x = \operatorname{gd} x.$

$\sinh x = \tan \operatorname{gd} x.$	$\tanh x = \sin \operatorname{gd} x.$
$\cosh x = \sec \operatorname{gd} x.$	$\tanh \frac{1}{2} x = \tan \frac{1}{2} \operatorname{gd} x.$

$$\frac{d \operatorname{gd} x}{dx} = \operatorname{sech} x.$$

COMPLETE ELLIPTIC INTEGRALS

$$K = \int_0^{\pi/2} \frac{d\Phi}{\sqrt{1 - k^2 \sin^2 \Phi}} = F\left(k, \frac{\pi}{2}\right)$$

$$E = \int_0^{\pi/2} \sqrt{1 - k^2 \sin^2 \Phi}\, d\Phi = E\left(k, \frac{\pi}{2}\right)$$

See index for other elliptic integral tables.

ATOMIC WEIGHTS

For the sake of completeness all known elements are included in the list. Several of those more recently discovered are represented only by the unstable isotopes. The value in parenthesis in the atomic weight column is, in each case, the mass number of the most stable isotope.**

Name	Symbol	At. No.	International atomic weight		Valence
			1966	1959	
Actinium	Ac	89	....	(227)	....
Aluminum	Al	13	26.9815	26.98	3
Americium	Am	95	....	(243)	3, 4, 5, 6
Antimony, stibium	Sb	51	121.75	121.76	3, 5
Argon	Ar	18	39.948	39.944	0
Arsenic	As	33	74.9216	74.92	3, 5
Astatine	At	85	....	(210)	1, 3, 5, 7
Barium	Ba	56	137.34	137.36	2
Berkelium	Bk	97	....	(247)	3, 4
Beryllium	Be	4	9.0122	9.013	2
Bismuth	Bi	83	208.980	208.99	3, 5
Boron	B	5	10.811	10.82	3
Bromine	Br	35	79.904[1]	79.916	1, 3, 5, 7
Cadmium	Cd	48	112.40	112.41	2
Calcium	Ca	20	40.08	40.08	2
Californium	Cf	98	....	(251)	....
Carbon	C	6	12.01115	12.011	2, 4
Cerium	Ce	58	140.12	140.13	3, 4
Cesium	Cs	55	132.905	132.91	1
Chlorine	Cl	17	35.453	35.457	1, 3, 5, 7
Chromium	Cr	24	51.996	52.01	2, 3, 6
Cobalt	Co	27	58.9332	58.94	2, 3
Columbium, see *Niobium*					
Copper	Cu	29	63.546[2]	63.54	1, 2
Curium	Cm	96	....	(247)	3
Dysprosium	Dy	66	162.50	162.51	3
Einsteinium	Es	99	....	(254)	....
Erbium	Er	68	167.26	167.27	3
Europium	Eu	63	151.96	152.0	2, 3
Fermium	Fm	100	....	(257)	....
Fluorine	F	9	18.9984	19.00	1
Francium	Fr	87	....	(223)	1
Gadolinium	Gd	64	157.25	157.26	3
Gallium	Ga	31	69.72	69.72	2, 3
Germanium	Ge	32	72.59	72.60	4
Gold, aurum	Au	79	196.967	197.0	1, 3
Hafnium	Hf	72	178.49	178.50	4
Helium	He	2	4.0026	4.003	0
Holmium	Ho	67	164.930	164.94	3
Hydrogen	H	1	1.00797	1.0080	1
Indium	In	49	114.82	114.82	3
Iodine	I	53	126.9044	126.91	1, 3, 5, 7
Iridium	Ir	77	192.2	192.2	3, 4
Iron, ferrum	Fe	26	55.847	55.85	2, 3
Krypton	Kr	36	83.80	83.80	0
Lanthanum	La	57	138.91	138.92	3
Lawrencium	Lr	103	(257)		
Lead, plumbum	Pb	82	207.19	207.21	2, 4
Lithium	Li	3	6.939	6.940	1
Lutetium	Lu	71	174.97	174.99	3
Magnesium	Mg	12	24.312	24.32	2
Manganese	Mn	25	54.9380	54.94	2, 3, 4, 6, 7
Mendelevium	Md	101	....	(256)	....
Mercury, hydrargyrum	Hg	80	200.59	200.61	1, 2
Molybdenum	Mo	42	95.94	95.95	3. 4. 6
Neodymium	Nd	60	144.24	144.27	3
Neon	Ne	10	20.183	20.183	0
Neptunium	Np	93	....	(237)	4, 5, 6
Nickel	Ni	28	58.71	58.71	2, 3
Niobium (columbium)	Nb	41	92.906	92.91	3, 5
Nitrogen	N	7	14.0067	14.008	3, 5
Nobelium	No	102	....	(254)	....
Osmium	Os	76	190.2	190.2	2, 3, 4, 8
Oxygen	O	8	15.9994	16.000	2
Palladium	Pd	46	106.4	106.4	2, 4, 6
Phosphorus	P	15	30.9738	30.975	3, 5
Platinum	Pt	78	195.09	195.09	2, 4
Plutonium	Pu	94	....	(244)	3, 4, 5, 6
Polonium	Po	84	....	(209)	....
Potassium, kalium	K	19	39.102	39.100	1
Praseodymium	Pr	59	140.907	140.92	3
Promethium	Pm	61	....	(145)	3
Protactinium	Pa	91	....	(231)	....
Radium	Ra	88	....	(226)	2
Radon	Rn	86	....	(222)	0
Rhenium	Re	75	186.2	186.22	....
Rhodium	Rh	45	102.905	102.91	3
Rubidium	Rb	37	85.47	85.48	1
Ruthenium	Ru	44	101.07	101.1	3, 4, 6, 8
Samarium	Sm	62	150.35	150.35	2, 3
Scandium	Sc	21	44.956	44.96	3
Selenium	Se	34	78.96	78.96	2, 4, 6
Silicon	Si	14	28.086	28.09	4
Silver, argentum	Ag	47	107.868[3]	107.873	1
Sodium, natrium	Na	11	22.9898	22.991	1
Strontium	Sr	38	87.62	87.63	2
Sulfur	S	16	32.064	32.066*	2, 4, 6
Tantalum	Ta	73	180.948	180.95	5
Technetium	Tc	43	....	(97)	6, 7
Tellurium	Te	52	127.60	127.61	2, 4, 6
Terbium	Tb	65	158.924	158.93	3
Thallium	Tl	81	204.37	204.39	1, 3
Thorium	Th	90	232.038	(232)	4
Thulium	Tm	69	168.934	168.94	3
Tin, stannum	Sn	50	118.69	118.70	2, 4
Titanium	Ti	22	47.90	47.90	3, 4
Tungsten (wolfram)	W	74	183.85	183.86	6
Uranium	U	92	238.03	238.07	4, 6
Vanadium	V	23	50.942	50.95	3, 5
Xenon	Xe	54	131.30	131.30	0
Ytterbium	Yb	70	173.04	173.04	2, 3
Yttrium	Y	39	88.905	88.91	3
Zinc	Zn	30	65.37	65.38	2
Zirconium	Zr	40	91.22	91.22	4

1. ±0.002
2. ±0.001
3. ±0.001

* Because of natural variations in the relative abundances of the isotopes of sulfur the atomic weight of this element has a range of ±0.003.

** The 1959 atomic weights are based on O = 16.000 whereas those of 1966 are based on the isotope C^{12}.

ELECTRONIC CONFIGURATION OF THE ELEMENTS

By Laurence S. Foster

References

Inorganic Chemistry by Therald Moeller, published by John Wiley and Sons, New York, 1952, p.p. 98–101.

The Chemistry of the Actinide Elements by Joseph J. Katz and Glenn T. Seaborg, published by Methuen and Company, Ltd., London, 1957; John Wiley and Sons, Inc. New York, 1957, p. 464.

Atomic No.	Element	K	L	M	N	O	P	Q
		1	2	3	4	5	6	7
		s	s p	s p d	s p d f	s p d f	s p d f	s p d f
1	H	1						
2	He	2						
3	Li	2	1					
4	Be	2	2					
5	B	2	2 1					
6	C	2	2 2					
7	N	2	2 3					
8	O	2	2 4					
9	F	2	2 5					
10	Ne	2	2 6					
11	Na	2	2 6	1				
12	Mg	2	2 6	2				
13	Al	2	2 6	2 1				
14	Si	2	2 6	2 2				
15	P	2	2 6	2 3				
16	S	2	2 6	2 4				
17	Cl	2	2 6	2 5				
18	Ar	2	2 6	2 6				
19	K	2	2 6	2 6 ..	1			
20	Ca	2	2 6	2 6 ..	2			
21	Sc	2	2 6	2 6 1	2			
22	Ti	2	2 6	2 6 2	2			
23	V	2	2 6	2 6 3	2			
24	Cr	2	2 6	2 6 5*	1			
25	Mn	2	2 6	2 6 5	2			
26	Fe	2	2 6	2 6 6	2			
27	Co	2	2 6	2 6 7	2			
28	Ni	2	2 6	2 6 8	2			
29	Cu	2	2 6	2 6 10*	1			
30	Zn	2	2 6	2 6 10	2			
31	Ga	2	2 6	2 6 10	2 1			
32	Ge	2	2 6	2 6 10	2 2			
33	As	2	2 6	2 6 10	2 3			
34	Se	2	2 6	2 6 10	2 4			
35	Br	2	2 6	2 6 10	2 5			
36	Kr	2	2 6	2 6 10	2 6			
37	Rb	2	2 6	2 6 10	2 6 ..	1		
38	Sr	2	2 6	2 6 10	2 6 ..	2		
39	Y	2	2 6	2 6 10	2 6 1	2		
40	Zr	2	2 6	2 6 10	2 6 2 .	2		
41	Cb	2	2 6	2 6 10	2 6 4* ..	1		
42	Mo	2	2 6	2 6 10	2 6 5 ..	1		
43	Tc	2	2 6	2 6 10	2 6 6 ..	1		
44	Ru	2	2 6	2 6 10	2 6 7 ..	1		
45	Rh	2	2 6	2 6 10	2 6 8 ..	1		
46	Pd	2	2 6	2 6 10	2 6 10* ..	.		
47	Ag	2	2 6	2 6 10	2 6 10 ..	1		
48	Cd	2	2 6	2 6 10	2 6 10 ..	2		
49	In	2	2 6	2 6 10	2 6 10 ..	2 1		
50	Sn	2	2 6	2 6 10	2 6 10 ..	2 2		
51	Sb	2	2 6	2 6 10	2 6 10 ..	2 3		
52	Te	2	2 6	2 6 10	2 6 10 ..	2 4		
53	I	2	2 6	2 6 10	2 6 10 ..	2 5		

Atomic No.	Element	K	L	M	N	O	P	Q
		1	2	3	4	5	6	7
		s	s p	s p d	s p d f	s p d f	s p d f	s p d
54	Xe	2	2 6	2 6 10	2 6 10	2 6		
55	Cs	2	2 6	2 6 10	2 6 10 ..	2 6	1	
56	Ba	2	2 6	2 6 10	2 6 10 ..	2 6	2	
57	La	2	2 6	2 6 10	2 6 10 ..	2 6 1 ..	2	
58	Ce	2	2 6	2 6 10	2 6 10 2*	2 6	2	
59	Pr	2	2 6	2 6 10	2 6 10 3	2 6	2	
60	Nd	2	2 6	2 6 10	2 6 10 4	2 6	2	
61	Pm	2	2 6	2 6 10	2 6 10 5	2 6	2	
62	Sm	2	2 6	2 6 10	2 6 10 6	2 6	2	
63	Eu	2	2 6	2 6 10	2 6 10 7	2 6	2	
64	Gd	2	2 6	2 6 10	2 6 10 7	2 6 1 ..	2	
65	Tb	2	2 6	2 6 10	2 6 10 9*	2 6	2	
66	Dy	2	2 6	2 6 10	2 6 10 10	2 6	2	
67	Ho	2	2 6	2 6 10	2 6 10 11	2 6	2	
68	Er	2	2 6	2 6 10	2 6 10 12	2 6	2	
69	Tm	2	2 6	2 6 10	2 6 10 13	2 6	2	
70	Yb	2	2 6	2 6 10	2 6 10 14	2 6	2	
71	Lu	2	2 6	2 6 10	2 6 10 14	2 6 1 ..	2	
72	Hf	2	2 6	2 6 10	2 6 10 14	2 6 2 ..	2	
73	Ta	2	2 6	2 6 10	2 6 10 14	2 6 3 ..	2	
74	W	2	2 6	2 6 10	2 6 10 14	2 6 4 ..	2	
75	Re	2	2 6	2 6 10	2 6 10 14	2 6 5 ..	2	
76	Os	2	2 6	2 6 10	2 6 10 14	2 6 6 ..	2	
77	Ir	2	2 6	2 6 10	2 6 10 14	2 6 9* ..	0	
78	Pt	2	2 6	2 6 10	2 6 10 14	2 6 9 ..	1	
79	Au	2	2 6	2 6 10	2 6 10 14	2 6 10 ..	1	
80	Hg	2	2 6	2 6 10	2 6 10 14	2 6 10 ..	2	
81	Tl	2	2 6	2 6 10	2 6 10 14	2 6 10 ..	2 1	
82	Pb	2	2 6	2 6 10	2 6 10 14	2 6 10 ..	2 2	
83	Bi	2	2 6	2 6 10	2 6 10 14	2 6 10 ..	2 3	
84	Po	2	2 6	2 6 10	2 6 10 14	2 6 10 ..	2 4	
85	At	2	2 6	2 6 10	2 6 10 14	2 6 10 ..	2 5	
86	Rn	2	2 6	2 6 10	2 6 10 14	2 6 10 ..	2 6	
87	Fr	2	2 6	2 6 10	2 6 10 14	2 6 10 ..	2 6	1
88	Ra	2	2 6	2 6 10	2 6 10 14	2 6 10 ..	2 6	2
89	Ac	2	2 6	2 6 10	2 6 10 14	2 6 10 ..	2 6 1 ..	2
90	Th	2	2 6	2 6 10	2 6 10 14	2 6 10 ..	2 6 2 .	2
91	Pa	2	2 6	2 6 10	2 6 10 14	2 6 10 2*	2 6 1 .	2
92	U	2	2 6	2 6 10	2 6 10 14	2 6 10 3	2 6 1 ..	2
93	Np	2	2 6	2 6 10	2 6 10 14	2 6 10 4	2 6 1 ..	2
94	Pu	2	2 6	2 6 10	2 6 10 14	2 6 10 6	2 6	2
95	Am	2	2 6	2 6 10	2 6 10 14	2 6 10 7	2 6	2
96	Cm	2	2 6	2 6 10	2 6 10 14	2 6 10 7	2 6 1 ..	2
97	Bk	2	2 6	2 6 10	2 6 10 14	2 6 10 8	2 6 1 ..	2
98	Cf	2	2 6	2 6 10	2 6 10 14	2 6 10 10	2 6	2
99	Es	2	2 6	2 6 10	2 6 10 14	2 6 10 11	2 6	2
100	Fm	2	2 6	2 6 10	2 6 10 14	2 6 10 12	2 6	2
101	Md	2	2 6	2 6 10	2 6 10 14	2 6 10 13	2 6	2

*** Note irregularity.**

PERIODIC TABLE OF THE ELEMENTS

PERIODIC TABLE OF THE ELEMENTS

1a	2a	3b	4b	5b	6b	7b	8			1b	2b	3a	4a	5a	6a	7a	0	Orbit
1 H +1 −1 1.00797 1		Atomic Number → **50** +2 +4 ← Oxidation States Symbol → Sn KEY TO CHART Atomic Weight → 118.69 −18−18−4 ← Electron Configuration															**2** He 0 4.0026 2	K
3 Li +1 6.939 2−1	**4** Be +2 9.0122 2−2											**5** B +3 10.811 2−3	**6** C +2 +4 −4 12.01115 2−4	**7** N +1 +2 +3 +4 +5 −1 −2 −3 14.0067 2−5	**8** O −2 15.9994 2−6	**9** F −1 18.9984 2−7	**10** Ne 0 20.183 2−8	K−L
11 Na +1 22.9898 2−8−1	**12** Mg +2 24.312 2−8−2	Transition Elements					Group 8			Transition Elements		**13** Al +3 26.9815 2−8−3	**14** Si +2 +4 −4 28.086 2−8−4	**15** P +3 +5 −3 30.9738 2−8−5	**16** S +4 +6 −2 32.064 2−8−6	**17** Cl +1 +5 +7 −1 35.453 2−8−7	**18** Ar 0 39.948 2−8−8	K−L−M
19 K +1 39.102 −8−8−1	**20** Ca +2 40.08 −8−8−2	**21** Sc +3 44.956 −8−9−2	**22** Ti +2 +3 +4 47.90 −8−10−2	**23** V +2 +3 +4 +5 50.942 −8−11−2	**24** Cr +2 +3 +6 51.996 −8−13−1	**25** Mn +2 +3 +4 +7 54.9380 −8−13−2	**26** Fe +2 +3 55.847 −8−14−2	**27** Co +2 +3 58.9332 −8−15−2	**28** Ni +2 +3 58.71 −8−16−2	**29** Cu +1 +2 63.546 −8−18−1	**30** Zn +2 65.37 −8−18−2	**31** Ga +3 69.72 −8−18−3	**32** Ge +2 +4 72.59 −8−18−4	**33** As +3 +5 −3 74.9216 −8−18−5	**34** Se +4 +6 −2 78.96 −8−18−6	**35** Br +1 +5 −1 79.904 −8−18−7	**36** Kr 0 83.80 −8−18−8	−L−M−N
37 Rb +1 85.47 −18−8−1	**38** Sr +2 87.62 −18−8−2	**39** Y +3 88.905 −18−9−2	**40** Zr +4 91.22 −18−10−2	**41** Nb +3 +5 92.906 −18−12−1	**42** Mo +6 95.94 −18−13−1	**43** Tc +4 +6 +7 (97) −18−13−2	**44** Ru +3 101.07 −18−15−1	**45** Rh +3 102.905 −18−16−1	**46** Pd +2 +4 106.4 −18−18−0	**47** Ag +1 107.868 −18−18−1	**48** Cd +2 112.40 −18−18−2	**49** In +3 114.82 −18−18−3	**50** Sn +2 +4 118.69 −18−18−4	**51** Sb +3 +5 −3 121.75 −18−18−5	**52** Te +4 +6 −2 127.60 −18−18−6	**53** I +1 +5 +7 −1 126.9044 −18−18−7	**54** Xe 0 131.30 −18−18−8	−M−N−O
55 Cs +1 132.905 −18−8−1	**56** Ba +2 137.34 −18−8−2	**57*** La +3 138.91 −18−9−2	**72** Hf +4 178.49 −32−10−2	**73** Ta +5 180.948 −32−11−2	**74** W +6 183.85 −32−12−2	**75** Re +4 +6 +7 186.2 −32−13−2	**76** Os +3 +4 190.2 −32−14−2	**77** Ir +3 +4 192.2 −32−15−2	**78** Pt +2 +4 195.09 −32−16−2	**79** Au +1 +3 196.967 −32−18−1	**80** Hg +1 +2 200.59 −32−18−2	**81** Tl +1 +3 204.37 −32−18−3	**82** Pb +2 +4 207.19 −32−18−4	**83** Bi +3 +5 208.980 −32−18−5	**84** Po +2 +4 (209) −32−18−6	**85** At (210) −32−18−7	**86** Rn 0 (222) −32−18−8	−N−O−P
87 Fr +1 (223) −18−8−1	**88** Ra +2 (226) −18−8−2	**89**** Ac +3 (227) −18−9−2																−O−P−Q

															Orbit
*Lanthanides	**58** Ce +3 +4 140.12 −19−9−2	**59** Pr +3 140.907 −20−9−2	**60** Nd +3 144.24 −22−8−2	**61** Pm +3 (145) −23−8−2	**62** Sm +2 +3 150.35 −24−8−2	**63** Eu +2 +3 151.96 −25−8−2	**64** Gd +3 157.25 −25−9−2	**65** Tb +3 158.924 −26−9−2	**66** Dy +3 162.50 −28−8−2	**67** Ho +3 164.930 −29−8−2	**68** Er +3 167.26 −30−8−2	**69** Tm +3 168.934 −31−8−2	**70** Yb +2 +3 173.04 −32−8−2	**71** Lu +3 174.97 −32−9−2	−N−O−P
**Actinides	**90** Th +4 (232) −19−9−2	**91** Pa +5 +4 (231) −20−9−2	**92** U +3 +4 +5 +6 (238) −21−9−2	**93** Np +3 +4 +5 +6 (237) −22−9−2	**94** Pu +3 +4 +5 +6 (244) −23−9−2	**95** Am +3 +4 +5 +6 (243) −24−9−2	**96** Cm +3 (247) −25−9−2	**97** Bk +3 +4 (247) −26−9−2	**98** Cf +3 (251) −28−8−2	**99** Es (254) −29−8−2	**100** Fm (257) −30−8−2	**101** Md (256) −31−8−2	**102** No (254) −32−8−2	**103** Lr	−O−P−Q

Numbers in parentheses are mass numbers of most stable isotope of that element.

Table of the Isotopes

compiled by

R. L. Heath

Atomic Energy Division, Phillips Petroleum Company
National Reactor Testing Station
Idaho Falls, Idaho

Data on all the currently known stable and radioactive isotopes are included in this compilation. Major sources of data for this compilation are the following:

1. "Nuclear Data Sheets" of the Nuclear Data Project of the National Academy of Sciences—National Research Council, including all data through May, 1963.
2. D. Strominger, J. M. Hollander, and G. T. Seaborg, "Table of Isotopes," *Revs. Modern Phys.*, **30,** 585 (1958).
3. Where available, current data from original published papers (through May, 1963).
4. Supplementary reference was made to R. L. Heath, "Scintillation Spectrometry Gamma Ray Spectrum Catalogue," U.S. AEC Report IDO-16408 (1958).
5. Values for thermal-neutron cross sections were obtained from BNL-325.

Z (Column 1): This column lists isotopes in order of atomic number.

Isotope (Column 2): Isotopes are listed by chemical designation and atomic number. Isomers are designated by the addition of the superscript *m*.

Atomic mass (Column 4): Values for the atomic masses of the isotopes are listed in this column. Values are given in the physical scale. The atomic-weight system used is based on the arbitrary assignment of the weight of the Carbon 12 isotope to be 12.00000 amu. This system has been recommended by the International Union of Pure and Applied Chemistry and the International Commission of Atomic Weights.

$t_{1/2}$ (Column 5): This column lists the half life of the radioactive isotopes. The notation used for these values is as follows: y = years, d = days, h = hours, m = minutes, s = seconds, ms = milliseconds, and μs = microseconds.

Modes of decay (Column 6): This column lists the observed modes of decay for all radioactive isotopes. In cases of branched decay between two or more modes, the branching ratios in percent are listed as a superscript for each mode of decay. Symbols used are: β^- = negative beta emission, β^+ = positron emission, α = alpha emission, EC = orbital electron capture, IT = isomeric transition from upper to lower isomeric state, n = neutron emission, SF = spontaneous fission.

Decay energy (Column 7): This lists the currently adopted values for the total disintegration energies (Q) in Mev. Where more than one mode of decay exists—separate values will be given if necessary.

Particle energies (Column 8): This column lists end-point energies of beta particles and measured energies of alpha particles in Mev.

Particle intensities (Column 9): In this column intensities of beta groups and alpha particles are listed. Values are either given in percent of decay or as relative intensities where the complete decay scheme is not known. Relative intensities are noted by the symbol†.

Gamma-ray energies (Column 10): Values for gamma-ray energies are listed in Mev. For those nuclides which emit large numbers of gamma rays, only the major gamma rays are listed. In general this includes only gamma rays with relative intensities in the first decade. For some nuclides which have short-lived daughters, the major gamma rays of the daughters may be included in parenthesis. Delayed gamma rays are noted by the symbol D. The abbreviation Ann. Rad. refers to the 0.511 Mev radiation resulting from the annihilation of positrons.

Gamma-ray intensities (Column 11): This column lists intensities corresponding to gamma rays given in Column 10. Values are given in percent of decays for a given transition, and include a correction for internal conversion. Relative intensities for gamma rays are indicated by the symbol †. In cases where the conversion coefficients are large, both relative gamma intensities and percent decay for the transition are given. The intensity of the conversion electrons for a given gamma ray are sometimes listed in percent of decays with the symbol CE.

Thermal-neutron capture cross sections (Column 12): This column lists cross sections for thermal-neutron capture in barns ($10^{-24}cm^2$). Where neutron capture for a given nucleus can produce two isomers the separate cross sections are listed as (A + B), where A is the cross section for production of the isomeric level and B is the cross section for production of the ground state. Fission cross sections are listed for some heavy elements together with the symbol σ_f.

TABLE OF THE ISOTOPES

z	Isotope	% nat. abundance	Atomic mass (A)	Lifetime $t_{\frac{1}{2}}$	Modes of decay	Decay energies (Mev)	Particle energies (Mev)	Particle intensities	Gamma energies (Mev)	Gamma intensities	Thermal neutron capture cross-sections (Barns) σ_c
	n^1		1.008665	12m	β^-	0.78	0.78	100%			
H			1.00797								0.33
1	$_1H$	99.985	1.007825								0.33
	$_1H^2$	0.015	2.01410								0.00057
	$_1H^3$			12.26y	β^-	0.0181	0.0181	100%			
He			4.0026								0.007
2	$_2He^3$	0.00013	3.01603								
	$_2He^4$	99.99987	4.00260								0
	$_2He^5$		5.0123	2×10^{-21}s	n,α						
	$_2He^6$			0.81s	β^-	3.50	3.50	100%			
	$_2He^7$			$\sim$50μs	β^-	$\sim$10			0.5		
Li			6.939								71
3	$_3Li^5$		5.0125	$\sim10^{-21}$s	P,α						
	$_3Li^6$	7.42	6.01513								0.028
	$_3Li^7$	92.58	7.01601								0.033
	$_3Li^8$			0.85s	β^- (2α3)	16.0	β^{-13}	100%			
	$_3Li^9$			0.17s	β^- (n,2α)	14.0					
Be			9.0122								0.010
4	$_4Be^6$		6.0198	$\gtrsim4\times10^{-21}$s	P, α Li^5						
	$_4Be^7$			53.6d	EC	0.86			0.4773	10.32%	
	$_4Be^8$		8.00531	$\sim3\times10^{-16}$s	2α						
	$_4Be^9$	100	9.01219								0.010
	$_4Be^{10}$			2.7×10^6y	β^-	0.56	0.56	100%			
	$_4Be^{11}$			13.6s	β^-	11.5	β^-11.5 9.3	61% 29%	2.12 6.8	32 4	
B			10.811								755
5	$_5B^8$			0.78s	β^+	18.0	14		Ann. Rad.		
	$_5B^9$		9.01333	$\geqq3\times10^{-19}$s	P, (2α)						
	$_5B^{10}$	19.6	10.01294								0.5
	$_5B^{11}$	80.4	11.00931								<0.05
	$_5B^{12}$			0.019s	β^-	13.4	β^- 13.4 9.1	97% 1%	4.4		
	$_5B^{13}$			0.035s	β^-	13					
C			12.01115								0.0037

TABLE OF THE ISOTOPES (Continued)

z	Isotope	% nat. abundance	Atomic mass (Å)	Lifetime $t_{\frac{1}{2}}$	Modes of decay	Decay energies (Mev)	Particle energies (Mev)	Particle intensities	Gamma energies (Mev)	Gamma intensities	Thermal neutron capture cross-sections (Barns) σ_c
6	$_6C^{10}$			19.0s	β^+	3.6	β^+ 1.9		0.72 1.04 Ann. Rad.	98 % 2 %	
	$_6C^{11}$			20.5m	β^+	1.98	β^+ 0.96	100 %	Ann. Rad.		
	$_6C^{12}$	98.89	12.00000								0.0037
	$_6C^{13}$	1.11	13.00335								0.0009
	$_6C^{14}$			5770y	β^-	0.156	β^-0.156	100 %			$<10^{-6}$
	$_6C^{15}$			2.25s	β^-	9.81	β^- 4.51 9.81	68 % 32 %	5.30	32 %	
	$_6C^{16}$			0.74s	β^-	8.0					
N			14.0067								1.9
7	$_7N^{12}$			0.011s	β^+	17.6	β^+ 16.4		Ann. Rad.		
	$_7N^{13}$			10.0m	β^+	2.22	β^+ 1.19	100 %	Ann. Rad.		
	$_7N^{14}$	99.63	14.00307								0.1
	$_7N^{15}$	0.37	15.00011								0.00002
	$_7N^{16}$			7.35s	β^-	10.4	β^- 4.3 10.4	68 % 26 %	6.13 7.11	72 % 5 %	
	$_7N^{17}$			4.14s	β^- n	8.7	β^- 3.7 1.1 0.43				
O			15.9994								<0.0002
8	$_8O^{14}$			73s	β^+	5.16	β^+ 1.83 4.14	99 % 1 %	2.31 Ann. Rad.	99 %	
	$_8O^{15}$			2.03m	β^+	2.75	β^+ 1.73	100 %	Ann. Rad.		
	$_8O^{16}$	99.759	15.99491								<0.0002
	$_8O^{17}$	0.037	16.99914								
	$_8O^{18}$	0.204	17.99916								0.0002
	$_8O^{19}$			29.4s	β^-	4.80	β^- 3.25 4.60	58 % 42 %	0.200 1.36 0.122	96 % 54 %	
	$_8O^{20}$			14s	β^-	3.75	β^- 2.69	100 %	1.06	100 %	
F			18.9984								0.009
9	$_9F^{16}$		16.01171	$\sim10^{-19}$s	β^+ 97 %						
	$_9F^{17}$			66s	β^+	2.77	1.75		Ann. Rad.		
	$_9F^{18}$			1.87h	β^+ 97 % EC 3 %	1.67	0.65	97 %	Ann. Rad.		
	$_9F^{19}$	100	18.99840								0.009
	$_9F^{20}$			11s	β^-	7.03	5.42	100 %	1.63	100 %	
	$_9F^{21}$			5s	β^-	5.7					

TABLE OF THE ISOTOPES (Continued)

z	Isotope	% nat. abundance	Atomic mass (A)	Lifetime $t_{\frac{1}{2}}$	Modes of decay	Decay energies (Mev)	Particle energies (Mev)	Particle intensities	Gamma energies (Mev)	Gamma intensities	Thermal neutron capture cross-sections (Barns) σ_c
Ne			20.183								0.035
10	$_{10}Ne^{18}$			1.46s	β^+	4.4	β^+ 3.4		1.04 Ann. Rad.		
	$_{10}Ne^{19}$			18s	β^+	3.25	β^+ 2.23		Ann. Rad.		
	$_{10}Ne^{20}$	90.92	19.99244								
	$_{10}Ne^{21}$	0.257	20.99395								
	$_{10}Ne^{22}$	8.82	21.99138								0.036
	$_{10}Ne^{23}$			38s	β^-	4.39	β^- 4.39 3.95	67 % 32 %	0.44 1.65	33 %	
	$_{10}Ne^{24}$			3.38m	β^-	2.45	β^- 1.98 1.10	92 % 8 %	0.472 0.878	100 % 8 %	
Na			22.9898								0.53
11	$_{11}Na^{20}$			0.4s	β^+	15.0			Ann. Rad.		
	$_{11}Na^{21}$			23s	β^+	3.52	2.50		0.35 Ann. Rad.	2 %	
	$_{11}Na^{22}$			2.58y	β^+ 89 % EC 11 %	2.84	β^+ 0.54	90 %	1.277 Ann. Rad.	99 %	
	$_{11}Na^{23}$	100	22.98977								0.53
	$_{11}Na^{24m}$			0.020 s	IT	0.47			0.47		
	$_{11}Na^{24}$			15.0h	β^-	5.52	β^- 1.39		2.753 1.37 3.8 5.3	100 % 100 %	
	$_{11}Na^{25}$			60s	β^-	3.8	β^- 3.8 2.8		0.39 0.583 0.980 1.61 1.96	15 % 14 % 15 % 6 % w	
	$_{11}Na^{26}$			1.0s	β^-	8.5	β^- 6.7		1.82		
Mg			24.312								0.063
12	$_{12}Mg^{23}$			12s	β^+	4.06	β^+ 3.0		Ann. Rad. 0.44	 9 %	
	$_{12}Mg^{24}$	78.70	23.98504								0.03
	$_{12}Mg^{25}$	10.13	24.98584								0.27
	$_{12}Mg^{26}$	11.17	25.98259								0.03
	$_{12}Mg^{27}$			9.5m	β^-	2.62	β^- 1.75 1.59	70 % 30 %	0.180 0.837 1.013	 70 % 30 %	<0.030
	$_{12}Mg^{28}$			21.3h	β^-	1.84	β^- 0.45	100 %	0.032 1.35 0.95 0.396	100 % 70 % 30 % 30 %	
Al			26.9815								0.23

TABLE OF THE ISOTOPES (Continued)

Z	Isotope	% nat. abundance	Atomic mass (A)	Lifetime $t_{\frac{1}{2}}$	Modes of decay	Decay energies (Mev)	Particle energies (Mev)	Particle intensities	Gamma energies (Mev)	Gamma intensities	Thermal neutron capture cross-sections (Barns) σ_c
13	$_{13}Al^{23}$			0.13s		5.1					
	$_{13}Al^{24}$			2.1s	β^+	14	β^+ 8.5		1.385 2.72 4.20 5.40 7.05 Ann. Rad.	† 40 † 32 † 15 † 6 † 7	
	$_{13}Al^{25}$			7.2s	β^+	4.26	β^+ 3.24		1.58 Ann. Rad.	ω	
	$_{13}Al^{26m}$			6.5s	β^+	4.23	β^+ 3.21		Ann. Rad.		
	$_{13}Al^{26}$			7.4×10^5y	β^+ EC	4.01	β^+ 1.16 EC	85 % 15 %	1.11 1.83 2.95 Ann. Rad.	3.7 % 99.3 % 0.3 %	
	$_{13}Al^{27}$	100	26.98153								0.23
	$_{13}Al^{28}$			2.30h	β^-	4.65	β^- 2.87	100 %	1.78	100 %	
	$_{13}Al^{29}$			6.6m	β^-	3.8	β^- 2.5 1.4	85 % 15 %	1.28 2.43 2.1	85 % 15 %	
	$_{13}Al^{30}$			3.3s	β^-	7.3	β^- 5.05		2.26 3.52		
Si			28.086								0.16
14	$_{14}Si^{26}$			2s	β^+	5.1	β^+ 3.8 2.9		0.82 Ann. Rad.		
	$_{14}Si^{27}$			42s	β^+	4.82	β^+ 3.85 1.5		0.84 1.01 Ann. Rad.	<0.2 % <0.2 %	
	$_{14}Si^{28}$	92.21	27.97693								0.08
	$_{14}Si^{29}$	4.70	28.97649								0.28
	$_{14}Si^{30}$	3.09	29.97376								0.11
	$_{14}Si^{31}$			2.62h	β^-	1.48	β^- 1.48		1.26	0.07 %	
	$_{14}Si^{32}$			≈700y	β^-	0.1	0.1	100 %			
P			30.9738								0.20
15	$_{15}P^{28}$			0.28s	β^+	14	β^+ 11 8	50 %	1.78 2.6–7.6 Ann. Rad.	75 %	
	$_{15}P^{29}$			4.4s	β^+	4.96	β^+ 3.94	99 %	1.28 2.43 Ann. Rad.	1 % 1 %	
	$_{15}P^{30}$			2.6m	β^+	4.3	β^+ 3.3		2.24 Ann. Rad.		
	$_{15}P^{31}$	100	30.97376								0.20
	$_{15}P^{32}$			14.3d	β^-	1.71	β^- 1.71				
	$_{15}P^{33}$			25d	β^-	0.25	β^- 0.25				

TABLE OF THE ISOTOPES (Continued)

z	Isotope	% nat. abundance	Atomic mass (A)	Lifetime $t_{\frac{1}{2}}$	Modes of decay	Decay energies (Mev)	Particle energies (Mev)	Particle intensities	Gamma energies (Mev)	Gamma intensities	Thermal neutron capture cross-sections (Barns) σ_c
	$_{15}P^{34}$			12.4s	β^-	5.1	β^- 5.1 3.0	75% 25%	2.1 4.0	25% 0.2%	
S			32.064								0.52
16	$_{16}S^{30}$			1.4s	β^+	6.0	β^+ 4.98 4.30		0.68 Ann. Rad.		
	$_{16}S^{31}$			2.6s	β^+	5.4	β^+ 4.42	99%	1.27 Ann. Rad.	1%	
	$_{16}S^{32}$	95.0	31.97207								
	$_{16}S^{33}$	0.76	32.97146								
	$_{16}S^{34}$	4.22	33.96786								0.26
	$_{16}S^{35}$			86.7d	β^-	0.168	β^- 0.167				
	$_{16}S^{36}$	0.014	35.96709								0.14
	$_{16}S^{37}$			5.1m	β^-	4.8	β^- 1.6 4.7	90% 10%	3.095	90%	
	$_{16}S^{38}$			2.87h	β^-	3.0	β^- 1.1 3.0	95% 5%	1.88	95%	
Cl			35.453								34
17	$_{17}Cl^{32}$			0.31s	β^+ α		β^+ 10 α 3	50% 2%	2.23 3.790 4.30 4.80 Ann. Rad.	70% 10% 7% 14%	
	$_{17}Cl^{33}$			2.5s	β^+	5.58	β^+ 4.5		2.8 Ann. Rad.	0.3%	
	$_{17}Cl^{34m}$			32.4m	β^+ IT	5.65 0.14	β^+ 2.54 1.32 IT	31% 30% 39%	0.14 1.15 2.27 3.22 4.0 Ann. Rad.	39% 18% 43% 7% 0.2%	
	$_{17}Cl^{34}$			1.5s	β^+	5.5	β^+ 4.5		Ann. Rad.		
	$_{17}Cl^{35}$	75.53	34.96885								45
	$_{17}Cl^{36}$			3×10^5y	β^- EC	0.71 1.14	β^- 0.71 EC	98% 2%			90
	$_{17}Cl^{37}$	24.47	36.96590								(0.005+0.56)
	$_{17}Cl^{38m}$			1s	IT	0.66			0.66		
	$_{17}Cl^{38}$			37.3m	β^-	4.8	β^- 4.8 1.1 2.8	53% 32% 16%	2.16 1.64	47% 31%	
	$_{17}Cl^{39}$			55.5m	β^-	3.45	β^- 1.91 2.18 3.45	85% 8% 7%	0.246 0.52 1.27	43% 42% 51%	
	$_{17}Cl^{40}$			1.4m	β^-	7.5	β^- 3.2 7.5		2.75 1.46 6.0	† 100 † 100	

TABLE OF THE ISOTOPES (Continued)

z	Isotope	% nat. abundance	Atomic mass (A)	Lifetime $t_{\frac{1}{2}}$	Modes of decay	Decay energies (Mev)	Particle energies (Mev)	Particle intensities	Gamma energies (Mev)	Gamma intensities	Thermal neutron capture cross-sections (Barns) σ_c
Ar			39.948								0.63
18	$_{18}Ar^{35}$			1.83s	β^+	5.98	β^+ 4.96	93 %	1.19 1.73 Ann. Rad.	5 % 2 %	
	$_{18}Ar^{36}$	0.337	35.96755								6
	$_{18}Ar^{37}$			34.3d	EC	0.82	EC	100 %			
	$_{18}Ar^{38}$	0.063	37.96272								0.8
	$_{18}Ar^{39}$			260y	β^-	0.57	β^- 0.57				
	$_{18}Ar^{40}$	99.60	39.96238								0.53
	$_{18}Ar^{41}$			1.83h	β^-	2.49	β^- 1.20 2.49	99 % 1 %	1.29	99 %	70.06
	$_{18}Ar^{42}$			>3.5y	β^-	0.6					
K			39.102								
19	$_{19}K^{37}$			1.2s	β^+	6.1	β^+ 5.1		Ann. Rad.		
	$_{19}K^{38m}$			0.97s	β^+	5.8	β^+ 4.8	100 %	Ann. Rad.		
	$_{19}K^{38}$			7.7m	β^+	5.8	β^+ 2.7	100 %	2.16 Ann. Rad.	100 %	
	$_{19}K^{39}$	93.10	38.96371								2.2
	$_{19}K^{40}$	0.0118		1.3×10^{9}y	β^- EC	1.32 1.51	β^- 1.32 EC	9 % 11 %	1.46	11 %	
	$_{19}K^{41}$	6.88	40.96184								1.1
	$_{19}K^{42}$			12.4h	β^-	3.53	β^- 3.53 2.03	82 % 18 %	0.315 1.516	0.15 % 18 %	
	$_{19}K^{43}$			22.4h	β^-	1.82	β^- 0.83 0.24–1.84	87 %	0.375 0.388 0.3935 0.591 0.617 1.015	85 % 11 % 13 % 17 % 81 % 3 %	
	$_{19}K^{44}$			22m	β^-	6.1	β^- 4.9 1.5	2.48	1.13 2.07 3.6		
	$_{19}K^{45}$			20m	β^-				0.18 1.7		
Ca			40.08								0.43
20	$_{20}Ca^{38}$			0.7s					3.5		
	$_{20}Ca^{39}$			0.9s	β^+	6.5	β^+ 5.5		Ann. Rad.		
	$_{20}Ca^{40}$	96.97	39.96259								0.2
	$_{20}Ca^{41}$			1.1×10^{5}y	EC	0.41	EC				
	$_{20}Ca^{42}$	0.64	41.95863								40
	$_{20}Ca^{43}$	0.145	42.95878								

TABLE OF THE ISOTOPES (Continued)

z	Isotope	% nat. abundance	Atomic mass (A)	Lifetime $t_{\frac{1}{2}}$	Modes of decay	Decay energies (Mev)	Particle energies (Mev)	Particle intensities	Gamma energies (Mev)	Gamma intensities	Thermal neutron capture cross-sections (Barns) σ_c
	$_{20}Ca^{44}$	2.06	43.95549								0.7
	$_{20}Ca^{45}$			165d	β^-	0.25	0.25	100 %			
	$_{20}Ca^{46}$	0.0033	45.9537								0.3
	$_{20}Ca^{47}$			4.7d	β^-	1.86	0.66 1.94	83 % 17 %	0.150 0.234 0.48 0.83 1.31	 6 % 6 % 77 %	
	$_{20}Ca^{48}$	0.18	47.9524	$>2\times10^{16}$y	β^-	0.12					1.1
	$_{20}Ca^{49}$			8.7m	β^-	5.1	2.0 1.0	89	3.10 4.05 4.68	89 % 8 % 3 %	
Sc			44.956								23
21	$_{21}Sc^{40}$			0.18s	β^+	14	β^+ 9.2		3.75 3.48 Ann. Rad.		
	$_{21}Sc^{41}$			0.87s	β^+	6	β^+ 5		Ann, Rad.		
	$_{21}Sc^{42m}$			62s	β^+	7.13	β^+ 2.82		1.52 0.43 Ann. Rad.		
	$_{21}Sc^{42}$			0.66s	β^+	6.55	β^+ 5.53		Ann. Rad.		
	$_{21}Sc^{43}$			3.9h	β^+	2.22	β^+ 1.19 0.82	72 % 16 %	0.371 0.627 0.25 Ann. Rad.	16 % 4 % 1 %	
	$_{21}Sc^{44m}$			2.4d	IT	0.27	IT		0.270		
	$_{21}Sc^{44}$			4.0h	β^+ EC	 3.65	β^+ 1.47 EC	93 % 7 %	1.16 2.54 Ann. Rad.	99 % 0.1 %	
	$_{21}Sc^{45}$	100	44.95592								(10+13)
	$_{21}Sc^{46m}$			20s	IT	0.14			0.14		
	$_{21}Sc^{46}$			84.0d	β^-	2.37	β^- 0.36		1.12 0.89	100 % 100 %	0.25
	$_{21}Sc^{47}$			3.4d	β^-	0.61	β^- 0.45 0.61	64 % 36 %	0.161	60 %	
	$_{21}Sc^{48}$			44h	β^-	3.99	β^- 0.65	100 %	1.31 1.04 0.99	100 % 100 % 100 %	
	$_{21}Sc^{49}$			57.5m	β^-	2.01	β^- 2.0	100 %			
	$_{21}Sc^{50m}$			23m	IT	0.49			0.49		
	$_{21}Sc^{50}$			1.8m	β^-	6.3	β^- 3.5	100 %	1.59 1.17	100 % 100 %	
Ti			47.90								5.8
22	$_{22}Ti^{43}$			0.6s	β^+	6.8	β^+		Ann. Rad.		

TABLE OF THE ISOTOPES (Continued)

z	Isotope	% nat. abundance	Atomic mass (A)	Lifetime $t_{\frac{1}{2}}$	Modes of decay	Decay energies (Mev)	Particle energies (Mev)	Particle intensities	Gamma energies (Mev)	Gamma intensities	Thermal neutron captue cross-sections (Barns) σ_c
	$_{22}Ti^{44}$			$\approx 10^3$y	[EC]	0.155	EC		0.079 0.070	100 % 100 %	
	$_{22}Ti^{45}$			3.08h	β^+ EC	2.06	β^+ 1.02 EC		0.45 0.80 Ann. Rad.	Weak <1.5 %	
	$_{22}Ti^{46}$	7.93	45.95263								0.6
	$_{22}Ti^{47}$	7.28	46.9518								1.7
	$_{22}Ti^{48}$	73.94	47.94795								8.3
	$_{22}Ti^{49}$	5.51	48.94787								1.9
	$_{22}Ti^{50}$	5.34	49.9448								0.14
	$_{22}Ti^{51}$			5.80m	β^-	2.46	β^- 2.13 1.52	95 % 5 %	0.323 0.608 0.93	96 % 1 % 4 %	
V			50.942								4.9
23	$_{23}V^{46}$			$\approx$1s	β^+		β^+		Ann. Rad.		
	$_{23}V^{46}$			0.4s	β^+	7.1	β^+ >6.05		Ann. Rad.		
	$_{23}V^{47}$			32m	β^+ EC	2.96	β^+ 1.94		1.54 1.87 Ann. Rad.		
	$_{23}V^{48}$			1.61d	β^+ EC	4.02	β^+ 0.70 EC	56 % 44 %	0.99 1.31 2.25 Ann. Rad.	100 % 100 % 2 %	
	$_{23}V^{49}$			330d	EC	0.62	EC				
	$_{23}V^{50}$	0.24	49.9472	$\approx 6 \times 10^{14}$y	β^- EC				0.71 1.59		~200
	$_{23}V^{51}$	99.76	50.9440								4.5
	$_{23}V^{52}$			3.77m	β^-	4.0	β^- 2.6		1.44	100 %	
	$_{23}V^{53}$			2.0m	β^-	2.53	β^- 2.50		1.00		
	$_{23}V^{54}$			55s	β^-		β^- 3.3		0.99 0.84 2.21	100 % 100 % w	
Cr			51.996								3.1
24	$_{24}Cr^{46}$			1.1s	β^+		β^+		Ann. Rad.		
	$_{24}Cr^{47}$			0.4s	β^+				Ann. Rad.		
	$_{24}Cr^{48}$			23h	EC	1.4	EC		0.310 0.117	100 % 95 %	
	$_{24}Cr^{49}$			42m	β^+	2.56	β^+ 1.54 1.35	50 % 25 %	0.089 0.152 0.063 Ann. Rad.	30 % 15 % 15 %	
	$_{24}Cr^{50}$	4.31	49.9461								17.0

TABLE OF THE ISOTOPES (Continued)

z	Isotope	% nat. abundance	Atomic mass (A)	Lifetime $t_{\frac{1}{2}}$	Modes of decay	Decay energies (Mev)	Particle energies (Mev)	Particle intensities	Gamma energies (Mev)	Gamma intensities	Thermal neutron capture cross-sections (Barns) σ_c
	$_{24}Cr^{51}$			27.8d	EC	0.75	EC		0.3200 0.324 0.624 0.650	10^{-3}% 9% 5×10^{-4}%	
	$_{24}Cr^{52}$	83.76	51.9405								0.8
	$_{24}Cr^{53}$	9.55	52.9407								18
	$_{24}Cr^{54}$	2.38	53.9389								0.38
	$_{24}Cr^{55}$			3.5m	β^-	2.8	β^- 2.8	100%			
	$_{24}Cr^{56}$			5.9m	β^-	1.6	β^- 1.5		0.083 0.026	100% 100%	
Mn			54.9380								13.3
25	$_{25}Mn^{50m}$			2m	β^+				0.66 Ann. Rad.		
	$_{25}Mn^{50}$			0.29s	β^+	7.6	β^+ 6.6		Ann. Rad.		
	$_{25}Mn^{51}$			45m	β^+	3.2	0.99 1.40 2.16		0.74 1.17 Ann. Rad.		
	$_{25}Mn^{52m}$			20m	β^+ IT	5.09 0.39	β^+ 2.63	100% 0.05%	1.46 0.39 Ann. Rad.	100% 0.05%	
	$_{25}Mn^{52}$			5.6d	EC β^+	4.7	EC β^+ 0.6	67% 33%	1.46 0.94 0.73 Ann. Rad.	100% 100% 100%	
	$_{25}Mn^{53}$			$\sim10^6$y	EC	0.60	EC	100%			
	$_{25}Mn^{54}$			291d	EC	1.38	EC	100%	0.835	100%	
	$_{25}Mn^{55}$	100	54.9381								13.3
	$_{25}Mn^{56}$			2.58h	β^-	3.71	β^- 2.86 1.05 0.75	60% 24% 15%	0.85 1.81 2.11 2.52 2.66 2.98	100% 23% 14%	
	$_{25}Mn^{57}$			1.7m	β^-	2.7	β^- 2.6		0.117 0.134 0.220 0.350 0.690	S S W W W	
	$_{25}Mn^{58}$			1.1m	β^-	6.5					
Fe			55.847								2.6
26	$_{26}Fe^{52}$			8h	β^+ EC	2.38	β^+ 0.80 EC	55% 45%	0.163 Ann. Rad.	100%	
	$_{26}Fe^{53}$			9m	β^+	4.0	β^+ 2.6		0.37 Ann. Rad. 0.88 1.27		
	$_{26}Fe^{54}$	5.82	53.9396								2.5

TABLE OF THE ISOTOPES (Continued)

z	Isotope	% nat. abundance	Atomic mass (A)	Lifetime $t_{\frac{1}{2}}$	Modes of decay	Decay energies (Mev)	Particle energies (Mev)	Particle intensities	Gamma energies (Mev)	Gamma intensities	Thermal neutron capture cross-sections (Barns) σ_c
	$_{26}Fe^{55}$			2.7y	EC	0.231	EC	100%			
	$_{26}Fe^{56}$	91.66	55.9349								2.7
	$_{26}Fe^{57}$	2.19	56.9354								2.5
	$_{26}Fe^{58}$	0.33	57.9333								1.0
	$_{26}Fe^{59}$			45d	β^-	1.56	0.46 0.27 1.56	53% 46% 0.3%	0.145 0.191 0.337 1.102 1.290	0.8% 2.5% 0.3% 56% 44%	
	$_{26}Fe^{60}$			$\approx 3 \times 10^5$y	β^-		$\beta^- \approx 0.24$				
	$_{26}Fe^{61}$			6.0m	β^-	3.9	β^- 2.8		0.29		
Co			58.9332								37
27	$_{27}Co^{54}$			0.18s	β^+	8.2	β^+ 7.2		Ann. Rad.		
	$_{27}Co^{55}$			18h	β^+ ~60% EC~40%	3.46	β^+ 1.51 1.04 EC	40% 36% 21%	0.247 0.476 0.937 1.41 Ann. Rad.	† 1.2 † 22.8 † 96.0 † 15.6	
	$_{27}Co^{56}$			77.3d	EC β^+	4.60	EC β^+ 1.46	82% 18%	0.845 1.24 1.0−3.47 Ann. Rad.	100% 70%	
	$_{27}Co^{57}$			270d	EC	0.84	EC	100%	0.01437 0.12205 0.13640	100%(γ+CE) 90% 10%	
	$_{27}Co^{58m}$			9.0h	IT	0.025	IT	100%	0.025		
	$_{27}CO^{58}$			71d	EC β^+	2.31	EC β^+ 0.48	85% 15%	0.81 1.64 ~0.81 Ann. Rad.	100% 0.5% 1.6%	
	$_{27}Co^{59}$	100	58.9332								(18+19)
	$_{27}Co^{60m}$			10.5m	IT β^-	0.059	IT β^- 1.54	99.7% 0.3%	0.058 1.332	99.7% 0.3%	
	$_{27}Co^{60}$			5.27y	β^-	2.82	β^- 0.31 1.48	100% 0.01%	1.332 1.1724 2.158	99+% 99+% 1.2×10^{-3}%	
	$_{27}Co^{61}$			9.90m	β^-	1.29	β^- 1.22	100%	0.072	100%(γ+CE)	
	$_{27}Co^{62m}$			1.9m	β^-		β^-		γ		
	$_{27}Co^{62}$			13.9m	β^-	5.22	β^- 2.88 0.88	75% 25%	1.17 1.47 1.73 2.03	†100 † 11 † 11 † 4	
	$_{27}Co^{63m}$			1.4h							
	$_{27}Co^{63}$			52s	β^-		β^- 3.6				
	$_{27}Co^{64m}$			2m	IT				γ		

TABLE OF THE ISOTOPES (Continued)

z	Isotope	% nat. abundance	Atomic mass (A)	Lifetime $t_{\frac{1}{2}}$	Modes of decay	Decay energies (Mev)	Particle energies (Mev)	Particle intensities	Gamma energies (Mev)	Gamma intensities	Thermal neutron capture cross-sections (Barns) σ_c
	$_{27}Co^{64}$			7.8m	β^-		β^-		γ		
Ni			58.71								4.6
28	$_{28}Ni^{56}$			6.4d	EC			100%	0.16 0.28 0.48 0.79 0.96 1.33 1.58 1.75	†100 † 30 † 40 †80 †10 † 5 †15 † 2	
	$_{28}Ni^{57}$			36h	EC β^+	3.23	EC β^+ 0.84 0.71	53% 40%	0.125 1.38 1.90 Ann. Rad.		
	$_{28}Ni^{58}$	67.88	57.9353								4.4
	$_{28}Ni^{59}$			8×10^4y	EC	1.07	EC	100%			
	$_{28}Ni^{60}$	26.23	59.9332								2.6
	$_{28}Ni^{61}$	1.19	60.9310								2
	$_{28}Ni^{62}$	3.66	61.9283								15
	$_{28}Ni^{63}$			92y	β^-	0.067	β^- 0.067	100%			
	$_{28}Ni^{64}$	1.08	63.9280								1.6
	$_{28}Ni^{65}$			2.56h	β^-	2.10	β^- 2.10 0.6 1.01	69% 23% 8%	0.368 1.114 1.480 0.51–1.725	4.5% 15.8% 25%	20
	$_{28}Ni^{66}$			55h	β^-	0.20	β^- 0.20	100%			
Cu			63.54								3.8
29	$_{29}Cu^{58}$			3.3s	β^+	8.50	β^+ 7.44 4.6		1.45 2.9 Ann. Rad.		
	$_{29}Cu^{59}$			81s	β^+	4.80	β^+ 3.75	67%	0.34 0.46 0.87 1.30 1.69 Ann. Rad.	5% 5% 9% 11% 1%	
	$_{29}Cu^{60}$			2.4m	β^+ EC	6.27	β^+ 2.00 3.00 3.92 EC	71% 18% 6% 5%	0.85 1.33 1.76 2.13 2.64 3.13–4.0 Ann. Rad.	15% 80% 52% 5.7% 5.5%	
	$_{29}Cu^{61}$			3.3h	β^+ EC	2.23	β^+ 1.21 EC	68% 32%	0.072 0.281 0.380 0.580 0.656 0.94–1.22 Ann. Rad.	4% 12% 2.5% 1.5% 11%	

TABLE OF THE ISOTOPES (Continued)

Z	Isotope	% nat. abundance	Atomic mass (A)	Lifetime $t_{\frac{1}{2}}$	Modes of decay	Decay energies (Mev)	Particle energies (Mev)	Particle intensities	Gamma energies (Mev)	Gamma intensities	Thermal neutron capture cross-sections (Barns) σ_c
	$_{29}Cu^{62}$			9.73m	β^+	3.93	β^+ 2.91 EC		0.67 0.87 1.18 1.35 1.46 1.46 1.98 2.24 Ann. Rad.	2%	
	$_{29}Cu^{63}$	69.09	62.9298								4.5
	$_{29}Cu^{64}$			12.8h	EC β^+ β^-	1.68 0.57	EC β^- 0.57 β^+ 0.66	43% 38% 19%	1.35 Ann. Rad.	1%	
	$_{29}Cu^{65}$	30.91	64.9278								2.3
	$_{29}Cu^{66}$			5.1m	β^-	2.63	β^- 2.63 1.59	91% 9%	1.04 0.83	9% 0.2%	
	$_{29}Cu^{67}$			61h	β^-	0.58	β^- 0.40 0.48 0.58	45% 35% 20%	0.186 0.094 0.300 0.391	†100 † 28 † ≈5 † ≈3	
	$_{29}Cu^{68}$			32s	β^-		β^- 3.0		1.08 1.24 1.53		
Zn			65.37								1.10
30	$_{30}Zn^{60}$			2.1m							
	$_{30}Zn^{61}$			89s	β^+	5.40	β^+ 4.38	80%	0.48 0.69 0.98 1.04 Ann. Rad.	11% 2% 3% 6%	
	$_{30}Zn^{62}$			9.3h	EC β^+	1.69	EC β^+ 0.67	80% 19%	0.0416 0.60 0.25 0.40 Ann. Rad.	36% 22%	
	$_{30}Zn^{63}$			38.3m	EC β^+	3.38	β^+ 2.36 EC	93% 7%	0.67 0.97 0.81–2.9 $>10\gamma$'s		
	$_{30}Zn^{64}$	48.89	63.9291								0.47
	$_{30}Zn^{65}$			245d	EC β^+	1.35	EC β^+ 0.33	98% 2%	1.114 Ann. Rad.	50%	
	$_{30}Zn^{66}$	27.81	65.9260								
	$_{30}Zn^{67}$	4.11	66.9271								
	$_{30}Zn^{68}$	18.57	67.9249								(0.1+1.0)
	$_{30}Zn^{69m}$			14h	IT	0.44	IT	100%	0.44		
	$_{30}Zn^{69}$			55m	β^-	0.91	β^- 0.90	100%	no γ		
	$_{30}Zn^{70}$	0.62	69.9253								0.09

TABLE OF THE ISOTOPES (Continued)

z	Isotope	% nat. abundance	Atomic mass (A)	Lifetime $t_{\frac{1}{2}}$	Modes of decay	Decay energies (Mev)	Particle energies (Mev)	Particle intensities	Gamma energies (Mev)	Gamma intensities	Thermal neutron capture cross-sections (Barns) σ_c
	$_{30}Zn^{71m}$			3.9h	β^-	3.0	β^- 1.5	100%	0.61 0.48 0.38	100% 100% 100%	
	$_{30}Zn^{71}$			2.2m	β^-	2.91	β^- 2.4	100%	0.12 0.51 0.90 1.09	100%	
	$_{30}Zn^{72}$			46.5h	β^-	1.6	β^- 0.3 1.6	95% 5%			
Ga			69.72								3.0
31	$_{31}Ga^{64}$			2.6m	β^+	7.0	β^+ 6.1 2.8	61% 38%	0.975 1.30 2.225 3.25 Ann. Rad.		
	$_{31}Ga^{65}$			15m	β^+ EC	3.26	β^+ 2.11 1.39 2.24 0.82	46%	0.12 0.054–1.86 >10γ's Ann. Rad.		
	$_{31}Ga^{66}$			9.5h	β^+ EC	5.17	β^+4.15 EC	49%	1.04 2.75 0.83–4.8 Ann. Rad.	37% 25%	
	$_{31}Ga^{67}$			78h	EC	1.00	EC	100%	0.0920 0.182 0.30 0.090–0.87	69% 24% 22%	
	$_{31}Ga^{68}$			68m	β^+ EC	2.91	1.89 EC	86% 14%	0.81 1.06 1.24 1.88 Ann. Rad.	† 6 †100 † 3 † 4	
	$_{31}Ga^{69}$	60.4	68.9257								1.9
	$_{31}Ga^{70}$			21m	β^-	1.65	β^- 1.65	99%	0.1735 1.039 1.215	0.44% 0.76% 0.003%	
	$_{31}Ga^{71}$	39.6	70.9249								5.0
	$_{31}Ga^{72}$			14.1h	β^-	3.99	β^- 0.96 0.66–3.16		0.84 0.69 0.11–2.82 >10γ's		
	$_{31}Ga^{73}$			4.8h	β^-	1.55	β^-1.19	91%	0.30 0.74	90% 7%	
	$_{31}Ga^{74}$			7.8m	β^-	3.4	β^- 2.65 2.0 1.1		0.60 2.3 0.30–2.8 >10γ's		
	$_{31}Ga^{75}$			2.0m	β^-		β^- 3.3		0.58		
	$_{31}Ga^{76}$			32s	β^-		β^- ≈6		0.57 0.96 1.12		

TABLE OF THE ISOTOPES (Continued)

z	Isotope	% nat. abundance	Atomic mass (A)	Lifetime $t_{\frac{1}{2}}$	Modes of decay	Decay energies (Mev)	Particle energies (Mev)	Particle intensities	Gamma energies (Mev)	Gamma intensities	Thermal neutron capture cross-sections (Barns) σ_c
Ge			72.59								2.4
32	$_{32}Ge^{65}$			1.5m	β^+	4.7	β^+ 3.7	94%	0.67 1.72 Ann. Rad.	3% 2%	
	$_{32}Ge^{66}$			2.4h	β^+ EC	3.0	β^+ 1.3 2.0 EC	69% 7% 24%	0.045 0.183 0.38 >10γ's Ann. Rad.	50% 30% 60%	
	$_{32}Ge^{67}$			19m	β^+ EC	4.3	β^+ 3.1 2.3 1.6 EC	60% 24% 8% 8%	0.170 0.92 0.34–3.4 >10γ's Ann. Rad.		
	$_{32}Ge^{68}$			280d	EC	0.7	EC		See Ga^{68}		
	$_{32}Ge^{69}$			40h	EC β^+	2.23	EC β^+ 1.21 0.61	67% 29% 3%	1.12 0.58 0.88 0.09–2.00 >10γ's Ann. Rad.		
	$_{32}Ge^{70}$	20.52	69.9243								3.5
	$_{32}Ge^{71m}$			20ms	IT	0.198	IT		0.175 0.023	100% 100%	
	$_{32}Ge^{71}$			11d	EC	0.24	EC		no γ		
	$_{32}Ge^{72}$	27.43	71.9217								1.0
	$_{32}Ge^{73m}$			0.53s	IT	0.068	IT		0.054 0.014	100% 100%	
	$_{32}Ge^{73}$	7.76	72.9234								14
	$_{32}Ge^{74}$	36.54	72.9212								(0.04+0.3)
	$_{32}Ge^{75m}$			49s	IT	0.139	IT		0.139	100%	
	$_{32}Ge^{75}$			82m	β^-	1.18	β^- 1.18 0.92	86% 11%	0.066 0.199 0.2645 >10γ's	† 2.2 † 12 †100	
	$_{32}Ge^{76}$	7.76	75.9214								(0.09+0.08)
	$_{32}Ge^{77m}$			54s	IT β^-	0.159 2.9	β^- 2.9 IT	78% 14%	0.159 0.215	14% 8%	
	$_{32}Ge^{77}$			11h	β^-	2.67	β^- 2.20 1.38 0.71	42% 35% 23%	0.215 0.27 0.04–2.3 >10γ's		
	$_{32}Ge^{78}$			2.1h	β^-		β^- 0.9				
As			74.9216								4.3
33	$_{33}As^{68}$			≈7m	β^+		β^+		Ann. Rad.		
	$_{33}As^{69}$			15m	β^+	3.9	β^+ 2.9		0.23 Ann. Rad.		

TABLE OF THE ISOTOPES (Continued)

z	Isotope	% nat. abundance	Atomic mass (A)	Lifetime $t_{\frac{1}{2}}$	Modes of decay	Decay energies (Mev)	Particle energies (Mev)	Particle intensities	Gamma energies (Mev)	Gamma intensities	Thermal neutron capture cross-sections (Barns) σ_c
	${}_{33}As^{70}$			50m	β^+ EC	 6.6	β^+ 1.4 2.4 EC	54% 26% 20%	1.04 2.0 0.18–1.7 $>10\gamma$'s Ann. Rad.	100%	
	${}_{33}As^{71}$			62h	EC β^+	2.01	EC β^+ 0.81	65% 35%	0.175 0.0232 Ann. Rad.	100% 6%	
	${}_{33}As^{72}$			26h	β^+ EC	4.36	β^+ 2.50 3.34 EC	56% 17% 23%	0.84 0.63 0.69−3.7 $>10\gamma$'s Ann. Rad.	77% 8%	
	${}_{33}As^{73}$			76d	EC	0.37	EC		See Ge^{73m}		
	${}_{33}As^{74m}$			8s	IT				0.28 IT		
	${}_{33}As^{74}$			18d	EC β^+ β^-	 2.53 1.36	EC β^+ .91 1.51 β^- 1.36	38% 26% 4% 18%	0.593 0.633 1.00–2.2 Ann. Rad.	60% 14%	
	${}_{33}As^{75m}$			17ms	IT	0.305			0.025 0.280 0.305	81% 81% 19%	
	${}_{33}As^{75}$	100	74.9216								4.3
	${}_{33}As^{76}$			26.5h	β^-	2.97	β^- 2.97 2.41	56% 31%	0.560 0.646 1.205 0.64–2.66	45% 6% 6%	
	${}_{33}As^{77}$			39h	β^-	0.69	β^- 0.69	98%	0.24 0.52 0.086 0.160 Ann. Rad.	2% 1%	
	${}_{33}As^{78m}$			6m	IT	0.50	IT		0.50		
	${}_{33}As^{78}$			91m	β^-	4.1	β^- 4.1 1.4	70%	0.615 0.700 1.32 0.08–2.68	42% 11% 16%	
	${}_{33}As^{79}$			9m	β^-	2.3	β^- 2.3		0.096 0.36 0.43 0.89		
	${}_{33}As^{80}$			15s	β^-	6.0	β^- 6.0 5.4 3.0–4.5		0.66 0.8–2.35		
	${}_{33}As^{81}$			33s	β^-	3.8	β^- 3.8		no γ		
	${}_{33}As^{85}$			0.43s	β^-						
Se			78.96								12
34	${}_{34}Se^{70}$			44m	β^+		β^+		Ann. Rad.		
	${}_{34}Se^{71}$			45m	β^+	4.4	β^+ 3.4		0.160 Ann. Rad.		

TABLE OF THE ISOTOPES (Continued)

z	Isotope	% nat. abundance	Atomic mass (A)	Lifetime $t_{\frac{1}{2}}$	Modes of decay	Decay energies (Mev)	Particle energies (Mev)	Particle intensities	Gamma energies (Mev)	Gamma intensities	Thermal neutron capture cross-sections (Barns) σ_c
	$_{34}Se^{72}$			8.4d	EC		EC		0.048	100 %	
	$_{34}Se^{73m}$			44m	β^+		β^+ 1.86		0.88 Ann. Rad.		
	$_{34}Se^{73}$			7.1h	β^+	2.75	β^+ 1.30 1.65	99 % 1 %	0.360 0.0665 0.860 1.310 Ann. Rad.	99 % 100 %	
	$_{34}Se^{74}$	0.87	73.9225								30
	$_{34}Se^{75}$			120d	EC	0.86	EC		0.265 0.136 0.280 0.024–0.58 >10γ's		
	$_{34}Se^{76}$	9.02	75.9192								(7+78)
	$_{34}Se^{77m}$			17.5s	IT	0.16			0.162		
	$_{34}Se^{77}$	7.58	76.9199								42
	$_{34}Se^{78}$	23.52	77.9173								0.4
	$_{34}Se^{79m}$			3.9m	IT	0.096	IT		0.0960		
	$_{34}Se^{79}$			7×10⁴y	β^-	0.16	β^- 0.16	100 %	No γ		
	$_{34}Se^{80}$	49.82	79.9165								(0.03+0.5)
	$_{34}Se^{81m}$			61m	IT	0.103	IT		0.103		
	$_{34}Se^{81}$			18m	β^-	1.6	β^- 1.6		0.28		
	$_{34}Se^{82}$	9.19	81.9167								(0.05+0.004)
	$_{34}Se^{83m}$			69s	β^-	3.7	β^- 3.4 1.5		0.350 0.650 1.01 2.02	† 16 † 20 †100 † 40	
	$_{34}Se^{83}$			25m	β^-	3.45	β^- 0.45 1.0 1.7		0.22 0.36 0.52 0.83 0.71–2.3 >10γ's		
	$_{34}Se^{84}$			3m	β^-						
	$_{34}Se^{85}$			39s	β^-						
	$_{34}Se^{87}$			16s	β^-						
Br			79.909								6.7
35	$_{35}Br^{74}$			42m	β^+	6.0	β^+ 4.7		0.64 Ann. Rad.		
	$_{35}Br^{75}$			1.6h	β^+ EC	2.72	β^+ 1.7 EC		0.28 Ann. Rad.		
	$_{35}Br^{76}$			16.5h	β^+ EC	4.6	β^+ 3.1 3.6 EC	30 % 11 % 34 %	0.25 0.33 0.42		

TABLE OF THE ISOTOPES (Continued)

z	Isotope	% nat. abundance	Atomic mass (A)	Lifetime $t_{\frac{1}{2}}$	Modes of decay	Decay energies (Mev)	Particle energies (Mev)	Particle intensities	Gamma energies (Mev)	Gamma intensities	Thermal neutron capture cross-sections (Barns) σ_c
									0.56 0.67 0.75 0.96 1.20 Ann. Rad.	70%	
	$_{35}Br^{77m}$			4.2m	IT		IT		0.107		
	$_{35}Br^{77}$			58h	EC β^+	1.36	EC β^+ 0.34	98% 2%	0.52 0.81 0.086–1.0 >10γ's		
	$_{35}Br^{78m}$			120μs	IT	0.149	IT		0.149		
	$_{35}Br^{78}$			6.5m	β^+	3.46	β^+ 2.46		0.62 Ann. Rad.		
	$_{35}Br^{79m}$			4.8s	IT		IT		0.21		
	$_{35}Br^{79}$	50.54	78.9183								(2.9+8.5)
	$_{35}Br^{80m}$			4.5h	IT	0.087	IT		0.049 0.037	100% 100%	
	$_{35}Br^{80}$			18m	EC β^+ β^-	2.00	β^- 2.00 1.38 EC β^+ 0.86	78% 14% 5% 3%	0.620 Ann. Rad.		
	$_{35}Br^{81m}$			37μs	IT	0.55	IT		0.27 0.28	100% 100%	
	$_{35}Br^{81}$	49.46	80.9163								3
	$_{35}Br^{82}$			35.7h	β^-	3.09	β^- 0.44	100%	0.55 0.611 0.69 0.76 0.822 1.030 1.30 1.46	† 80 † 50 † 33 †100 † 30 † 36 † 36 † 18	
	$_{35}Br^{83}$			24h	β^-	1.0	β^- 0.94		0.048 (See Kr^{83})	20%	
	$_{35}Br^{84m}$			6m	β^-		β^- 1.9 0.8 3.2	72% 20% 8%	0.88 1.46 0.44 1.89	70% 60% 60% 20%	
	$_{35}Br^{84}$			32m	β^-	4.7	β^- 4.7 2.8	32% 15%	0.88 2.1 1.9 0.27–3.9 >10γ's		
	$_{35}Br^{85}$			3.0m	β^-	2.5	β^- 2.5	100%	no γ		
	$_{35}Br^{87}$			5.5s	β^-	8.0	β^- 2.6 8.0 (n 0.3)	70% 30% 2%	3.2 5.4		
	$_{35}Br^{88}$			16s	β^- n						

z	Isotope	% nat. abundance	Atomic mass (A)	Lifetime $t_{\frac{1}{2}}$	Modes of decay	Decay energies (Mev)	Particle energies (Mev)	Particle intensities	Gamma energies (Mev)	Gamma intensities	Thermal neutron capture cross-sections (Barns) σ_c
	${}_{35}Br^{89}$			4.5s	β^- n		(n 0.5)				
	${}_{35}Br^{90}$			1.6s	β^- n						
Kr			83.80								31
36	${}_{36}Kr^{74}$			15m	β^+		β^+ 3.1		no γ Ann. Rad.		
	${}_{36}Kr^{75}$			5m							
	${}_{36}Kr^{76}$			10h	EC				0.028 0.093 0.267 0.316 0.40		
	${}_{36}Kr^{77}$			1.2h	β^+ EC	2.88	β^+ 1.86 1.67 EC	50 % 25 % 20 %	0.0242 0.1076 0.131 0.149 0.246 0.281 0.313 0.665 0.87 Ann. Rad.		
	${}_{36}Kr^{78}$	0.35	77.9204								(?+2)
	${}_{36}Kr^{79m}$			55s	IT	0.127			0.127		
	${}_{36}Kr^{79}$			34.5h	EC β^+	1.62	EC β^+ 0.60	92 % 7 %	0.044 0.261 0.08–0.83 >10γ's		
	${}_{36}Kr^{80}$	2.27	79.9164								95
	${}_{36}Kr^{81m}$			13s	IT	0.190	IT		0.190	100 %	
	${}_{36}Kr^{81}$			2×10^5y	EC	0.25	EC		0.012		
	${}_{36}Kr^{82}$	11.56	81.9135								45
	${}_{36}Kr^{83m}$			1.86h	IT	0.041	IT		0.0093 0.031	100 % 100 %	
	${}_{36}Kr^{83}$	11.55	82.9141								220
	${}_{36}Kr^{84}$	56.90	83.9115								(0.10+0.06)
	${}_{36}Kr^{85m}$			4.4h	IT β^-	31 0.98	β^- 0.83 IT	78 % 22 %	0.1495 0.305	78 % 22 %	
	${}_{36}Kr^{85}$			10.4y	β^-	0.68	β^- 0.67 0.15	99 % 1 %	0.52	1 %	<15
	${}_{36}Kr^{86}$	17.37	85.9106								0.6
	${}_{36}Kr^{87}$			78m	β^-	3.9	β^- 3.8 1.3 3.3		0.403 0.85 1.75 2.05 2.57	†100 † 19 † 42	<600

TABLE OF THE ISOTOPES (Continued)

z	Isotope	% nat. abundance	Atomic mass (A)	Lifetime $t_{\frac{1}{2}}$	Modes of decay	Decay energies (Mev)	Particle energies (Mev)	Particle intensities	Gamma energies (Mev)	Gamma intensities	Thermal neutron capture cross-sections (Barns) σ_c
	$_{36}Kr^{88}$			2.8h	β^-	2.9	β^- 0.52 2.7	68% 20%	0.028 0.166 0.191 0.36 0.85 1.55 2.19 2.40	† 20 †100 † 14 † 65 † 40 †100	
	$_{36}Kr^{89}$			3.2m	β^-		β^- 4.0		0.60 0.22 0.38–1.52		
	$_{36}Kr^{90}$			33s	β^-		β^- 1.8 3.2		0.125 0.54 1.10 1.54		
	$_{36}Kr^{91}$			10s	β^-		β^- 3.6				
	$_{36}Kr^{92}$			3s	β^-						
	$_{36}Kr^{93}$			6s	β^-						
	$_{36}Kr^{94}$			1s	β^-						
	$_{36}Kr^{95}$			Short	β^-						
Rb			85.47								0.73
37	$_{37}Rb^{79}$			21m	β^+		β^+		0.15 0.19 Ann. Rad.		
	$_{37}Rb^{80}$			34s	β^+	5.1	β^+ 4.1 EC	98% 2%	0.62 Ann. Rad.		
	$_{37}Rb^{81m}$			32m	IT β^+	0.085	β^+ 1.4 IT		0.085 Ann. Rad.		
	$_{37}Bb^{81}$			4.7h	EC β^+	2.05	EC β^+ 1.03	87% 7%	10.253 0.450 1.03 Ann. Rad.		
	$_{37}Rb^{82m}$			6.3h	EC β^+		EC β^+ 0.80	94% 6%	0.78 0.62 0.55 0.70–1.47 Ann. Rad.		
	$_{37}Rb^{82}$			75s	β^+	4.17	β^+ 3.17		0.78 1.4 Ann. Rad.		
	$_{37}Rb^{83}$			83d	EC		EC		0.53 (See Kr^{83})	100%	
	$_{37}Rb^{84m}$			23m	IT EC		IT EC	90% 10%	0.22 0.24 0.46 0.88	90%	
	$_{37}Rb^{84}$			33d	EC β^+ β^-	2.65 0.91	EC β^+ 0.8 1.63 β^- 0.91	76% 12% 9% 3%	0.88 1.01 1.90 Ann. Rad.		

TABLE OF THE ISOTOPES (Continued)

z	Isotope	% nat. abundance	Atomic mass (A)	Lifetime $t_{\frac{1}{2}}$	Modes of decay	Decay energies (Mev)	Particle energies (Mev)	Particle intensities	Gamma energies (Mev)	Gamma intensities	Thermal neutron capture cross-sections (Barns) σ_c
	$_{37}Rb^{85}$	72.15	84.9117								(0.007+1.0)
	$_{37}Rb^{86m}$			1.02m	IT	0.56	IT		0.56 0.78		
	$_{37}Rb^{86}$			18.77d	β^-	1.77	β^- 1.77 0.7	85%	0.527 1.080	9%	
	$_{37}Rb^{87}$	27.85		4.7×10^{10}y	β^-	0.27	β^- 0.27		no γ		0.12
	$_{37}Rb^{88}$			18m	β^-	5.2	β^- 5.2 3.3 2.0	70% 17% 13%	0.90 1.835 2.7 1.39–4.9 $>10\gamma$'s		
	$_{37}Rb^{89}$			15m	β^-	3.9	β^- 3.9 2.8	7%	0.663 1.05 1.26 1.55 2.20 2.59 2.75 3.52	† 22 †100 † 72 † 5 †19 †17 †3.7 †2.9	
	$_{37}Rb^{90}$			2.9m	β^-	6.6	β^- 6.6 5.8 2.2		0.84 0.53−5.23 $>10\gamma$'s		
	$_{37}Rb^{91}$			72s	β^-		β^- 4.6		0.095 0.35 (See Sr^{91})		
	$_{37}Rb^{92}$			5s	β^-		β^-				
	$_{37}Rb^{93}$			6s			β^-				
	$_{37}Rb^{94}$			3s			β^-				
	$_{37}Rb^{95}$			<2.5s			β^-				
Sr			87.62								1.3
38	$_{38}Sr^{80}$			1.7h	EC		EC		0.58		
	$_{38}Sr^{81}$			29m	β^+		β^+		Ann. Rad.		
	$_{38}Sr^{82}$			25.5d	EC		EC		0.15 0.40 0.95		
	$_{38}Sr^{83}$			33h	EC β^+		EC β^+ 1.2	75% 25%	0.040 0.074 0.101 0.151 0.165 Ann. Rad.		
	$_{38}Sr^{84}$	0.56	83.9134								1
	$_{38}Sr^{85m}$			70m	IT EC	0.233	IT EC	86% 14%	0.150 0.225 0.008 0.233	86% 13% 13% 1%	
	$_{38}Sr^{85}$			64d	EC	1.1	EC		0.5133	100%	
	$_{38}Sr^{86}$	9.86	85.9094								(1.3+?)

TABLE OF THE ISOTOPES (Continued)

z	Isotope	% nat. abundance	Atomic mass (A)	Lifetime $t_{\frac{1}{2}}$	Modes of decay	Decay energies (Mev)	Particle energies (Mev)	Particle intensities	Gamma energies (Mev)	Gamma intensities	Thermal neutron capture cross-sections (Barns) σ_c
	$_{38}Sr^{87m}$			2.8h	IT	0.39	IT		0.389	100%	
	$_{38}Sr^{87}$	7.02	86.9089								
	$_{38}Sr^{88}$	82.56	87.9056								0.005
	$_{38}Sr^{89}$			50.4d	β^-	1.47	β^- 1.46	100%			
	$_{38}Sr^{90}$			28y	β^-	0.54	β^- 0.54		no γ		
	$_{38}Sr^{91}$			9.7h	β^-	2.67	β^- 1.09 1.36 0.62–2.67	33% 29%	0.551 0.645 0.748 0.93 1.025 1.413	†59 †15 †27 † 3 †30 † 5	
	$_{38}Sr^{92}$			2.7h	β^-	1.93	β^- 0.54 1.5	90% 10%	0.23 0.44 1.37	3.6% 4% 90%	
	$_{38}Sr^{93}$			8m	β^-		β^- 2.0		0.18 0.25 0.60 0.72 0.88		
	$_{38}Sr^{94}$			1.3m	β^-		β^-				
	$_{38}Sr^{95}$			0.8m	β^-		β^-				
Y			88.905								1.3
39	$_{39}Y^{82}$			1.2h	β^+		β^+ 2		Ann. Rad.		
	$_{39}Y^{83}$			3.5h							
	$_{39}Y^{84}$			48m			β^+ 3.0		0.76 0.96 1.07		
	$_{39}Y^{85}$			5h	β^+	3.0	β^+ 2.0				
	$_{39}Y^{86m}$			49m	IT	0.218			0.010 0.208	100% 100%	
	$_{39}Y^{86}$			14.6h	β^+ EC	5.4	EC β^+ 0.76– 3.3		1.93 1.08 0.6 –2.6 Ann. Rad.		
	$_{39}Y^{87m}$			14h	IT	0.38	IT		0.384		
	$_{39}Y^{87}$			80h	EC β^+	1.7	EC β^+ 0.7	100% 0.3%	0.484 Ann. Rad. (See Sr^{87}) (0.388)		
	$_{39}Y^{88m_2}$			300μs	IT	0.39			0.39		
	$_{39}Y^{88m_1}$			0.014s	IT	0.24	IT		0.24		
	$_{39}Y^{88}$			108d	EC β^+	3.62	EC β^+ 0.57	100% 0.7%	0.90 1.835 2.76	96% 100% 0.05%	

TABLE OF THE ISOTOPES (Continued)

z	Isotope	% nat. abundance	Atomic mass (A)	Lifetime $t_{\frac{1}{2}}$	Modes of decay	Decay energies (Mev)	Particle energies (Mev)	Particle intensities	Gamma energies (Mev)	Gamma intensities	Thermal neutron capture cross-sections (Barns) σ_c
	$_{39}Y^{89m}$			16.1s	IT	0.91	IT		0.914		
	$_{39}Y^{89}$	100	88.9054								(0.001+1.3)
	$_{39}Y^{90m}$			3.2h 3	IT	0.685			0.203 0.482	100% 100% (10% CE)	
	$_{39}Y^{90}$			64.2h	β^-	2.26	β^- 2.27	100%	1.739	0.02%	
	$_{39}Y^{91m}$			50m	IT	0.55	IT		0.551	100%	
	$_{39}Y^{91}$			57.5d	β^-	1.55	β^- 1.55	100%	1.21	0.3%	
	$_{39}Y^{92}$			3.60h	β^-	3.64	β^- 3.64 1.32 1.59–2.71		0.21 0.48 0.94 1.45 1.9 2.4	† 10 † 11 † 19 †≈10 †≈1 †≈0.2	
	$_{39}Y^{93}$			10.4h	β^-	2.89	β^- 2.89	90%	0.27 0.94 0.38–2.4 >10γ's	6% 2%	
	$_{39}Y^{94}$			20m	β^-	5.0	β^- 5.0	40%	0.92 0.56 1.13 1.65–3.5 >10γ's	43% 6% 5%	
	$_{39}Y^{95}$			11m	β^-		β^-				
	$_{39}Y^{96}$			2.3m	β^-		β^- 3.5		0.7 1.0		
Zr			91.22								0.18
40	$_{40}Zr^{86}$			17h	EC				0.241 (See Y^{86})		
	$_{40}Zr^{87}$			1.6h	β^+ EC	3.51	β^+ 2.10		0.65 0.35 Ann. Rad. (See Y^{87})		
	$_{40}Zr^{88}$			85d	EC				0.39 (Y^{88m})	100%	
	$_{40}Zr^{89m}$			4.4m	IT β^+ EC	0.59 3.4	IT EC β^+ 0.9 2.4	93% 5%	0.59 1.53 Ann. Rad.	93% 7%	
	$_{40}Zr^{89}$			79h	EC β^+	2.84	EC β^+ 0.90	75% 25%	1.7 (See Y^{89})		
	$_{40}Zr^{90m}$			0.8s	IT	2.32	IT		2.32	100%	
	$_{40}Zr^{90}$	51.46	89.9043								0.1
	$_{40}Zr^{91}$	11.23	90.9053								1
	$_{40}Zr^{92}$	17.11	91.9046								0.2
	$_{40}Zr^{93}$			9.5×10^6y	β^-	0.063	β^- 0.063 0.034	75% 25%	(0.029) (See Nb^{93})	25%	

z	Isotope	% nat. abundance	Atomic mass (A)	Lifetime $t_{\frac{1}{2}}$	Modes of decay	Decay energies (Mev)	Particle energies (Mev)	Particle intensities	Gamma energies (Mev)	Gamma intensities	Thermal neutron capture cross-sections (Barns) σ_c
	$_{40}Zr^{94}$	17.40	93.9061								0.1
	$_{40}Zr^{95}$			65d	β^-	1.12	β^- 0.40 0.36 0.89	55 % 43 % 2 %	0.723 0.757 (See Nb^{95})	55 % 42 %	
	$_{40}Zr^{96}$	2.80	95.9082								0.1
	$_{40}Zr^{97}$			17h	β^-	2.66	β^- 1.91	90 %	0.5–2.6 (0.75 See Nb^{97})		
	$_{40}Zr^{99}$			30s	β^-						
Nb			92.906								1.1
41	$_{41}Nb^{89m}$			≈1h	β^+ [EC]		β^+		(See Zr^{89m} 0.59)		
	$_{41}Nb^{89}$			1.9h	β^+	3.9	β^+ 2.9				
	$_{41}Nb^{90m_2}$			24s	IT	0.12			0.120	100 %	
	$_{41}Nb^{90m_1}$			0.010s	IT	0.25	IT		0.25	100 %	
	$_{41}Nb^{90}$			14.6h	β^+	6.12	β^+ 1.50	97 %	1.14 2.32D 0.018–2.0		
	$_{41}Nb^{91m}$			64d	IT	0.105	IT EC	90 % 10 %	0.1043 1.21	90 % *10 10 % *61	
	$_{41}Nb^{91}$			long	EC	1.6					
	$41Nb^{92}$			10.1d	EC	2.07			0.931 0.90 1.83	97 % 2 % 1 %	
	$41Nb^{93m}$			3.7y	IT	0.029	IT		0.029		
	$_{41}Nb^{93}$	100	92.9060								(1+?)
	$_{41}Nb^{94m}$			6.6m	β^- IT	2.2 0.042	1.3	0.1 % 100 %	0.0415 0.87	100 % 0.1 %	
	$_{41}Nb^{94}$			2.0×10^4y	β^-	2.1	β^- 0.5		0.89 0.703	100 % 100 %	
	$_{41}Nb^{95m}$			90h	IT	0.23	IT		0.234	100 %	
	$_{41}Nb^{95}$			35d	β^-	0.93	β^- 0.16 0.93	99 % 1 %	0.768	99 %	
	$_{41}Nb^{96}$			23.35h	β^-	3.1	β^- 0.7 0.4	92 % 8 %	0.453 0.55 0.77 0.87 1.08 1.15	27 % 61 % 100 % 16 % 52 % 32 %	
	$_{41}Nb^{97m}$			1m	IT	0.75			0.75	100 %	
	$_{41}Nb^{97}$			72m	β^-	1.93	β^- 1.27	99 %	0.665 1.02	100 % 1 %	
	$_{41}Nb^{98}$			51.5m	β^-	4.6	β^- 3.1		0.78 0.72 0.33–2.7		

TABLE OF THE ISOTOPES (Continued)

z	Isotope	% nat. abundance	Atomic mass (A)	Lifetime $t_{\frac{1}{2}}$	Modes of decay	Decay energies (Mev)	Particle energies (Mev)	Particle intensities	Gamma energies (Mev)	Gamma intensities	Thermal neutron capture cross-sections (Barns) σ_c
	$_{41}Nb^{99}$			2.5m	β^-	3.2	β^- 3.2		0.10 0.26		
	$_{41}Nb^{100m}$			?11.5m	β^-		β^- 4.2		0.53		
	$_{41}Nb^{100}$			3m	β^-		β^- 3.2		0.53 0.36 0.45 0.14–2.9		
	$_{41}Nb^{101}$			1m			β^-				
Mo			95.94								2.7
42	$_{42}Mo^{90}$			5.7h	β^+ EC	2.5	EC β^+ 1.2		Ann. Rad. (0.21, 0.25 See Nb^{90m})		
	$_{42}Mo^{91m}$			65s	IT β^+	0.65 5.11	IT β^+ 2.5–3.99	60%	0.65 1.21 1.51 Ann. Rad.	60% 16% 22%	
	$_{42}Mo^{91}$			15.6m	β^+	4.44	β^+ 3.42		Ann. Rad.		
	$_{42}Mo^{92}$	15.84	91.9063								(<0.006+?)
	$_{42}Mo^{93m}$			6.9h	IT	2.43			0.263 0.6845 1.4790	100% 100% 100%	
	$_{42}Mo^{93}$			$\approx 10^4$y	EC	0.48	EC				
	$_{42}Mo^{94}$	9.04	93.9047								
	$_{42}Mo^{95}$	15.72	94.9046								14
	$_{42}Mo^{96}$	16.53	95.9046								1
	$_{42}Mo^{97}$	9.46	96.9058								2
	$_{42}Mo^{98}$	23.78	97.9055								0.15
	$_{42}Mo^{99}$			66h	β^-	1.38	β^- 1.23 0.45	85% 14%	0.74 0.041–0.78 0.140D 0.002D 0.142D	11% 80% 80% 9%	
	$_{42}Mo^{100}$	9.13	99.9076								0.2
	$_{42}Mo^{101}$			14.6m	β^-	2.82	β^- 2.23		1.02 0.59 2.98 0.08–1.66 (0.19D) >10γ's	25% 21% 16% 25%	
	$_{42}Mo^{102}$			11m			β^- 1.2				
Tc			99								22
43	$_{43}Tc^{92}$			4.3m	β^+ EC	7.9	β^+ 4.1		0.33 0.79 1.54 Ann. Rad.		

TABLE OF THE ISOTOPES (Continued)

Z	Isotope	% nat. abundance	Atomic mass (A)	Lifetime $t_{\frac{1}{2}}$	Modes of decay	Decay energies (Mev)	Particle energies (Mev)	Particle intensities	Gamma energies (Mev)	Gamma intensities	Thermal neutron capture cross-sections (Barns) σ_c
	$_{43}Tc^{93m}$			44m	IT EC	0.39 3.58	IT EC	80% 20%	0.3896 2.7	80% 20%	
	$_{43}Tc^{93}$			2.7h	EC β^+	3.19	EC β^+ 0.82 0.64	88% 10% 2%	1.35 1.48 2.0 Ann. Rad.	63% 32% 10%	
	$_{43}Tc^{94m}$			4.5h	EC β^+	4.46	β^+ 1.00		0.71 0.86 0.875 Ann. Rad.		
	$_{43}Tc^{94}$			52m	β^- EC	4.32	β^+ 2.41 EC	65% 25%	0.87 1.5 1.8 2.7 3.3 Ann. Rad.	98% 24% 12% 1% 1%	
	$_{43}Tc^{95m}$			60d	EC IT β^+	0.039 1.70	EC IT β^+ 0.68	96% 4% 0.1%	0.039 0.84 0.76–1.04 Ann. Rad.	4% 32%	
	$_{43}Tc^{95}$			20h	EC	1.66	EC		0.762 0.932 0.953 1.069	85% 6% 9%	
	$_{43}Tc^{96m}$			52m	IT	0.034	β^+ IT	0.01% 100%	0.034	100%	
	$_{43}Tc^{96}$			4.3d	EC	3.0	EC		0.77 0.80 0.84 1.12	100% 100% 100% 17%	
	$_{43}Tc^{97m}$			91d	IT	0.096	IT		0.096	100%	
	$_{43}Tc^{97}$			2.6×10^6y			EC				
	$_{43}Tc^{98}$			1.5×10^6y	β^-	1.7	β^- 0.3		0.655 0.745	100% 100%	(2+?)
	$_{43}Tc^{99m}$			6.0h	IT	0.142			0.140 0.020 0.142	90% 90% 10%	
	$_{43}Tc^{99}$			2.1×10^5y	β^-	0.29	β^- 0.29				22
	$_{42}Tc^{100}$			16s	β^-	3.4	β^- 3.4 2.9		0.54 0.60 $>10\gamma$'s		
	$_{42}Tc^{101}$			14m	β^-	1.63	β^- 1.32 1.07	87% 6%	0.31 0.54 0.13–0.94 $>10\gamma$'s	91% 5%	
	$_{43}Tc^{102m}$			4.5m			β^- 2.0		0.47–2.0		
	$_{43}Tc^{102}$			5s	β^-	4.1	β^- 4				
	$_{43}Tc^{103}$			1.2m			β^- 2.5				
	$_{43}Tc^{104}$			18m			β^- 2.4		0.310 0.360		

z	Isotope	% nat. abundance	Atomic mass (A)	Lifetime $t_{\frac{1}{2}}$	Modes of decay	Decay energies (Mev)	Particle energies (Mev)	Particle intensities	Gamma energies (Mev)	Gamma intensities	Thermal neutron capture cross-sections (Barns) σ_c
									0.490 0.680 0.870		
	$_{43}Tc^{105}$			10m			β^-				
Ru			101.07								
44	$_{44}Ru^{93}$			50s			β^+		Ann. Rad.		
	$_{44}Ru^{94}$			57m	EC β^+				Ann. Rad.		
	$_{44}Ru^{95}$			99m	EC β^+	2.2	EC β^+ 1.2		0.145 0.340 0.640 1.053 Ann. Rad.	† 1 † 4 † 0.5 † 1	
	$_{44}Ru^{96}$	5.51	95.9076								0.2
	$_{44}Ru^{97}$			2.9d			EC		0.1091 0.2180 0.325 0.567	 † 14 † 1	
	$_{44}Ru^{98}$	1.87	97.9055								
	$_{44}Ru^{99}$	12.72	98.9061								
	$_{44}Ru^{100}$	12.62	99.9030								
	$_{44}Ru^{101}$	17.07	100.9041								
	$_{44}Ru^{102}$	31.61	101.9037								1.4
	$_{44}Ru^{103}$			40d	β^-	0.75	β^- 0.21 0.13 0.71	90%	0.055 0.2956 0.35 0.497 0.560 0.6100	† 4 † ≈4 † ≈3 †1000 † 80	
	$_{44}Ru^{104}$	18.58	103.9055								0.7
	$_{44}Ru^{105}$			4.45h	β^-	1.91	β^- 1.88 1.15 1.08		0.130D 0.265 0.315 0.400 0.475 0.670 0.725 0.870 0.960	100%	
	$_{44}Ru^{106}$			1.0y	β^-	0.04	β^- 0.04	100%	(See Rh^{106})		
	$_{44}Ru^{107}$			4.2m	β^-	3.2	β^- 2.1–3.1		0.22		
	$_{44}Ru^{108}$			4.4m	β^-	1.3	β^- 1.1–1.3		0.17		
Rh			102.905								150
45	$_{45}Rh^{96}$			~11m							
	$_{45}Rh^{97}$			35m	β^+				Ann. Rad.		

TABLE OF THE ISOTOPES (Continued)

z	Isotope	% nat. abundance	Atomic mass (A)	Lifetime $t_{\frac{1}{2}}$	Modes of decay	Decay energies (Mev)	Particle energies (Mev)	Particle intensities	Gamma energies (Mev)	Gamma intensities	Thermal neutron capture cross-sections (Barns) σ_c
	$_{45}$Rh^{98}			8.7m	β^+ EC	4.2	β^+ 2.5		0.650 Ann. Rad.		
	$_{45}$Rh^{99m}			4.7h	EC β^+	2.11	EC β^+ 0.75		0.286 0.35 0.61 0.89 1.26 1.41 Ann. Rad.	 70 % 20 %	
	$_{45}$Rh^{99}			16d	EC β^+	2.1	EC β^+ 1.03		0.350 0.089 Ann. Rad.		
	$_{45}$Rh^{100}			21h	EC β^+	3.63	EC β^+ 2.61		0.54 0.44 0.30 − 2.4 >10γ's Ann. Rad.		
	$_{45}$Rh^{101m}			4.7d	IT	0.158	EC IT	96 % 4 %	0.158 0.31	4 % 96 %	
	$_{45}$Rh^{101}			5y		EC			0.195 0.127	100 % 100 %	
	$_{45}$Rh^{102}			206d	EC β^- β^+	 1.15 2.3	EC β^- 1.15 β^+ 1.30 0.82	65 % 20 % 10 %	0.48 0.51 1.42–2.06 >10γ's Ann. Rad.		
	$_{45}$Rh^{103m}			57m	IT	0.040	IT		0.0400	100 % (CE 98 %)	
	$_{45}$Rh^{103}	100	102.9048								(12+138)
	$_{45}$Rh^{104m}			4.4m	IT β^-	0.128	IT β^-	99 % 0.1 %	0.0514 0.0772 0.56–1.53	100 %, CE 100 %, CE	800
	$_{45}$Rh^{104}			42s	β^-	2.44	β^- 2.44 1.88	89 % 2 %	0.555 1.24	2 % 0.1 %	 40
	$_{45}$Rh^{105m}			40s	IT	0.130	IT		0.130	100 % (CE 75 %)	
	$_{45}$Rh^{105}			36h	β^-	0.56	β^- 0.56 0.24	80 % 20 %	0.0800 0.160 0.220	 20 %	15,000
	$_{45}$Rh^{105}								0.308 0.316 0.415 0.550	 20 %	
	$_{45}$Rh^{106m}			2.2h	β^-		β^- 0.79– 1.62		0.51 0.22–1.22 >10γ's		
	$_{45}$Rh^{106}			30s	β^-	3.53	β^- 3.53	72 %	0.51 0.62 0.7–3.4 >10γ's	20 % 11 %	
	$_{45}$Rh^{107}			21.7m	β^-	1.5	β^- 1.2 1.5		0.095 0.145		

z	Isotope	% nat. abundance	Atomic mass (A)	Lifetime $t_{\frac{1}{2}}$	Modes of decay	Decay energies (Mev)	Particle energies (Mev)	Particle intensities	Gamma energies (Mev)	Gamma intensities	Thermal neutron capture cross-sections (Barns) σ_c
	${}_{45}Rh^{107}$ (cont.)								0.305 0.388 0.475 0.575 0.680	†100 † 18 † 1.6 † 3	
	${}_{45}Rh^{108}$			17s	β^-		β^- 4		0.43 0.62 0.51 1.52		
	${}_{45}Rh^{109m}$			50s	IT	0.11			0.11	100%	
	${}_{45}Rh^{109}$			30s	β^-		β^-		0.49–0.31		
	${}_{45}Rh^{110}$			3.6s	β^-						
Pd			106.4								8
46	${}_{46}Pd^{98}$			17.5m	β^+				Ann. Rad.		
	${}_{46}Pd^{99}$			22m	β^+	3.8	β^+ 2.0		0.140 0.280 0.420 0.670 Ann. Rad.		
	${}_{46}Pd^{100}$			4.0d	EC				0.073 0.082		
	${}_{46}Pd^{101}$			8.5h	EC β^+	1.76	EC β^+ 0.58	96% 4%	0.288 0.590 0.720 1.190 1.280 Ann. Rad.	15% 15%	
	${}_{46}Pd^{102}$	0.96	101.9049								4.8
	${}_{46}Pd^{103}$			17d	EC	0.56			0.0402 0.301 0.324 0.364 0.499 (See Rh^{103m})	† 2.1 † 0.3 † 2.1 † 2.1 † 0.5	
	${}_{46}Pd^{104}$	10.97	103.9036								
	${}_{46}Pd^{105m}$			37μs	IT	0.50			0.32 0.18	100% 100%	
	${}_{46}Pd^{105}$	22.23	105.9046								
	${}_{46}Pd^{106}$	27.33	105.9032								
	${}_{46}Pd^{107m}$			21.3s	IT	0.21			0.213	100%	
	${}_{46}Pd^{107}$			7×10^6y	β^-	0.035	β^- 0.035	100%			
	${}_{46}Pd^{108}$	26.71	107.9030								(0.2+12)
	${}_{46}Pd^{109m}$			4.8m	IT	0.18	IT		0.18	100%	
	${}_{46}Pd^{109}$			13.6h	β^-	1.11	β^- 1.0		0.088D (See Ag^{109m})	100% (CE 92%)	
	${}_{46}Pd^{110}$	11.81	109.9045								(0.05+0.2)

TABLE OF THE ISOTOPES (Continued)

z	Isotope	% nat. abundance	Atomic mass (A)	Lifetime $t_{\frac{1}{2}}$	Modes of decay	Decay energies (Mev)	Particle energies (Mev)	Particle intensities	Gamma energies (Mev)	Gamma intensities	Thermal neutron capture cross-sections (Barns) σ_c
	$_{46}Pd^{111m}$			5.5h	IT	0.17			0.17	100 %	
	$_{46}Pd^{111}$			22m	β^-	2.2	β^- 2.13		0.070 (See Ag^{111m})	100 %	
	$_{46}Pd^{112}$			21.0h	β^-	0.30	β^- 0.28		0.018		
	$_{46}Pd^{113}$			1.5m	β^-						
	$_{46}Pd^{114}$			2.4m	β^-				no γ		
	$_{46}Pd^{115}$			45s	β^-						
Ag			107.870								63
47	$_{47}Ag^{102}$			15m	EC β^+				Ann. Rad.		
	$_{47}Ag^{103}$			1.0h	β^+	2.2	$\beta^+ \approx 1.3$		0.11 0.13 0.15 Ann. Rad.		
	$_{47}Ag^{104m}$			27m	β^+ IT	4.28 0.118	β^+ 2.70		0.1184 0.556 Ann. Rad.		
	$_{47}Ag^{104}$			69m	EC β^+	4.3	EC β^+ .99		0.56 0.77 0.94 0.17–1.81 Ann. Rad.		
	$_{47}Ag^{105}$			40d	EC		EC		0.064 0.28 0.35 0.65 0.16–1.1 >10γ's		
	$_{47}Ag^{106m}$			8.3d	EC		EC		0.22 0.41 0.51–2.63 >10γ's		
	$_{47}Ag^{106}$			24m	β^+ 1.95 EC	2.97	β^+ 1.95 1.45 EC	54 % 7 % 39 %	0.512 Ann. Rad.	18 %	
	$_{47}Ag^{107m}$			44s	IT	0.093			0.0934	100 % (CE 15 %)	
	$_{47}Ag^{107}$	51.82	106.9041								(?+40)
	$_{47}Ag^{108m}$			≥5y	IT EC	0.081	EC EC	90 % 10 %	0.72 0.62 0.43 0.081	90 % 90 % 90 % 10 %	
	$_{47}Ag^{108}$			2.4m	β^- β^+ EC	1.8	β^- 1.77 EC β^+ 0.8	97 % 2 % 0.1 %	0.63 0.42 Ann. Rad.	1 % 0.2 %	
	$_{47}Ag^{109m}$			40s	IT	0.088	IT		0.088	100 % (CE 92 %)	
	$_{47}Ag^{109}$	48.18	108.9047								(2+82)

TABLE OF THE ISOTOPES (Continued)

z	Isotope	% nat. abundance	Atomic mass (A)	Lifetime $t_{\frac{1}{2}}$	Modes of decay	Decay energies (Mev)	Particle energies (Mev)	Particle intensities	Gamma energies (Mev)	Gamma intensities	Thermal neutron capture cross-sections (Barns) σ_c
	$_{47}Ag^{110m}$			249d	IT β^-	0.116 2.99	β^- 0.086 0.54 0.43 IT	55% 2%	0.66 0.88 0.116 IT 0.44–2.0 >10γ's	100% 80% 2%	
	$_{47}Ag^{110}$			24s	β^-	2.87	β^- 2.87		0.66 0.72 0.81 0.88 0.95		
	$_{47}Ag^{111m}$			74s	IT	0.065			0.065	100%	
	$_{47}Ag^{111}$			7.5d	β^-	1.05	β^- 1.05 0.71	93% 6%	0.2470 0.337 (See Cd^{111m})	93% 6%	
	$_{47}Ag^{112}$			3.2h	β^-	4.05	β^- 4.05 3.44		0.618 1.10 1.39 1.62 1.83 2.11 2.51 2.79	†100 † 8 † 20 † 9 † 6 † 9 † 4 † 2	
	$_{47}Ag^{113m}$			1.2m	IT β^-		IT β^- <2	90% 10%	0.14 0.31 0.39 0.56 0.70		
	$_{47}Ag^{113}$			5.3h	β^-	2.00	β^- 2.00		0.31 0.12–1.18		
	$_{47}Ag^{114}$			5s	β^-	4.6	β^- 4.6		0.56		
	$_{47}Ag^{115m}$			20s	β^-						
	$_{47}Ag^{115}$			21.1m	β^-	2.9	β^- 2.9		0.138 0.227		
	$_{47}Ag^{116}$			2.5m	β^-		β 5.9		0.70 0.52		
	$_{47}Ag^{117}$			1.1m	β^-						
Cd			112.40								2500
48	$_{48}Cd^{103}$			10m	β^+		β^+		0.22 0.62 0.82 Ann. Rad.		
	$_{48}Cd^{104}$			59m	EC β^+		β^+		0.084 0.067 0.124 0.134		
	$_{48}Cd^{105}$			55m	EC β^+		EC β^+ 1.69 0.80		0.31 0.025–2.32 <10γ's Ann. Rad.		
	$_{48}Cd^{106}$	1.22	105.907								

TABLE OF THE ISOTOPES (Continued)

z	Isotope	% nat. abundance	Atomic mass (A)	Lifetime $t_{\frac{1}{2}}$	Modes of decay	Decay energies (Mev)	Particle energies (Mev)	Particle intensities	Gamma energies (Mev)	Gamma intensities	Thermal neutron capture cross-sections (Barns) σ_c
	$_{48}Cd^{107}$			6.7h	EC β^+ 0.32	1.44	EC β^+ 0.32	99+ % 0.3 %	0.848 0.094D Ann. Rad. (see Ag^{107m})	0.4 %	
	$_{48}Cd^{108}$	0.88	107.9050								
	$_{48}Cd^{109m}$			12μs	IT	0.058			0.058	100 %	
	$_{48}Cd^{109}$			470d	EC	0.16	EC		0.0875D (see Ag^{109m})	100 %	
	$_{48}Cd^{110}$	12.39	109.9030								(0.2+?)
	$_{48}Cd^{111m}$			49m	IT	0.347			0.1495 0.2465D	100 % 100 % (CE 60 %)	
	$_{48}Cd^{111}$	12.75	110.9042								
	$_{48}Cd^{112}$	24.07	111.9028								(0.03+?)
	$_{48}Cd^{113m}$			14y	IT β^-	0.27 0.58	β^- 0.58 IT	99 % 0.1 %	0.27	0.1 %	
	$_{48}Cd^{113}$	12.26	112.9046								27,000
	$_{48}Cd^{114}$	28.86	113.9036								(0.14+1.1)
	$_{48}Cd^{115m}$			43d	β^-	1.63	β^- 1.63 0.69	97 % 2 %	0.485 0.943 1.29	 2 % 1 %	
	$_{48}Cd^{115}$			2.3d	β^-	1.45	β^- 1.11 0.59	62 % 25 %	0.26 0.52 0.49 0.335D (see In^{115m})		
	$_{48}Cd^{116}$	7.58	115.905								1.4
	$_{48}Cd^{117m}$			3.2h	β^-		1.0				
	$_{48}Cd^{117}$			2.8h	β^-	2.5	β^- 1.8		0.27–2.25		
	$_{48}Cd^{118}$			50m	β^-	2.5					
	$_{48}Cd^{119m}$			2.7m	β^-						
	$_{48}Cd^{119}$			95m	β^-				0.3 0.82		
	$_{48}Cd^{120}$			<1m	β^-						
	$_{48}Cd^{121}$			3.5m	β^-				0.85		
In			114.82								194
49	$_{49}In^{106}$			5.3m			β^+ 3.1 4.85		1.65 1.85 Ann. Rad.		
	$_{49}In^{107}$			33m	β^+	3	β^+ 2		0.22 Ann. Rad.		
	$_{49}In^{108m}$			55m	IT EC β^+	0.25	EC B^+ 1.4 IT	61 % 37 % 2 %	Ann. Rad. 0.63 0.88		

z	Isotope	% nat. abundance	Atomic mass (A)	Lifetime $t_{\frac{1}{2}}$	Modes of decay	Decay energies (Mev)	Particle energies (Mev)	Particle intensities	Gamma energies (Mev)	Gamma intensities	Thermal neutron capture cross-sections (Barns) σ_c
									1.05 0.33 0.25	 2 %	
	$_{49}In^{108}$			40m	β^+ EC	5.1	β^+ 3.5 EC	90 % 10 %	0.637 Ann. Rad.		
	$_{49}In^{109m}$			1.3m	IT	0.66	IT		0.658(CE)	100 % (7 %)	
	$_{49}In^{109}$			4.3h	EC β^+	2.0	EC β^+ 0.80	94 % 6 %	0.20 0.63 0.23–1.15 0.058D (see Cd^{109m})		
	$_{49}In^{110m}$			5.0h	IT EC	0.120	EC IT	99+ % 0.6 %	0.120 0.659 0.8845 0.936 0.11–0.94	0.6 % 99 % 99 % 99 %	
	$_{49}In^{110}$			66m	β^+ EC	3.96	β^+ 2.25 EC	70 % 30 %	0.656 Ann. Rad.	100 %	
	$_{49}In^{111m}$			~10m	IT	0.53			0.53	100 %	
	$_{49}In^{111}$			2.81d	EC	1.24	EC		0.247 0.173	100 % 100 %	
	$_{49}In^{112m_2}$			0.042s	IT	0.31	IT		0.31	100 %	
	$_{49}In^{112m_1}$			21m	IT	0.155	IT		0.155	100 %	
	$_{49}In^{112}$			14m	β^- β^+ EC	0.66 2.6	β^- 0.66 EC β^+ 1.6	44 % 33 % 23 %	0.63 0.71 Ann. Rad.	7 %	
	$_{49}In^{113m}$			1.73h	IT	0.39			0.391(CE)	100 % (30 %)	
	$_{49}In^{113}$	4.28	112.9043								(61+2)
	$_{49}In^{114m_2}$			2.5s	IT	0.150			0.150	100 %	
	$_{49}In^{114m_1}$			50d	IT	0.191	IT EC	96 % 4 %	0.191(CE) 0.722 0.553 1.29	96.5 % (80 %) 3.5 % 3.6 % 0.2 %	
	$_{49}In^{114}$			72s	EC β^+ β^-	 1.42 1.98	β^- 1.98 β^+ 0.40 EC	99 % 0.003 % 1.0 %	0.311 1.29	 0.09 %	
	$_{49}In^{115m}$			4.4h	IT β^- 0.84	0.34 0.84	IT β^- 0.84	95 % 5 %	0.335(CE)	95 % (47 %)	
	$_{49}In^{115}$	95.72	114.9041	6×10^{14}y			β^- 0.6	100 %			(150+50)
	$_{49}In^{116m_2}$			2.2s	IT	0.16	IT		0.16(CE)	100 % (50 %)	
	$_{49}In^{116m_1}$			54m	β^-	3.35	β^- 1.00 0.85 0.60	51 % 28 % 21 %	0.1371 0.41 1.08 1.29 1.48 2.086	3 % 25 % 54 % 75 % 21 % 25 %	
	$_{49}In^{116}$			14s	β^-	3.29	β^- 3.29	100 %			

TABLE OF THE ISOTOPES (Continued)

z	Isotope	% nat. abundance	Atomic mass (A)	Lifetime $t_{\frac{1}{2}}$	Modes of decay	Decay energies (Mev)	Particle energies (Mev)	Particle intensities	Gamma energies (Mev)	Gamma intensities	Thermal neutron capture cross-sections (Barns) σ_c
	$_{49}In^{117m}$			1.9h	IT β^-	0.31 1.77	β^- 1.77 1.61 IT	55 % 23 % 22 %	0.1605 0.313IT(CE) 0.550 0.720	23 % 22 % (16 %)	
	$_{49}In^{117}$			45 min	β^-	1.47	β^- 0.74		0.1605 0.563	100 % 100 %	
	$_{49}In^{118m}$			4.5m	β^-		β^- 1.5		1.22 0.69		
	$_{49}In^{118}$			5.1s	β^-	4.2	4.2		1.22		
	$_{49}In^{119m}$			18m	IT β^-	0.3 2.7	β^- 2.7 1.8		0.3 IT 0.9		
	$_{49}In^{119}$			2.0m	β^-	2.4	β^- 1.6		0.82		
	$_{49}In^{120m}$			3s	IT						
	$_{49}In^{120}$			≈50s	β^-		β^- 2.2		0.73 0.85 1.02 1.18		
	$_{49}In^{121m}$			3.1m	β^-		β^- 3.7				
	$_{49}In^{121}$			30s	β^-				0.94		
	$_{49}In^{123m}$			36s	β^-		β^- 4.6				
	$_{49}In^{123}$			10s	β^-				1.10		
Sn			118.69								0.63
50	$_{50}Sn^{108}$			9m	EC		EC				
	$_{50}Sn^{109}$			18m	EC β^+		EC β^+ 1.6	80 % 20 %	0.112 0.34 0.52 0.89 Ann. Rad. 0.66 D (See In^{109m})		
	$_{50}Sn^{110}$			4.0h	EC				0.283 (See In^{110})	100 %	
	$_{50}Sn^{111}$			35m	EC β^+	2.52	EC β^+ 1.51	71 % 29 %	Ann. Rad.		
	$_{50}Sn^{112}$	0.96	111.9040								(0.4+0.9)
	$_{50}Sn^{113m}$			20m	IT	0.079			0.079		
	$_{50}Sn^{113}$			118d	EC	0.69			0.256 0.392D(CE) (See In^{113m})	2 % 100 % (30 %)	
	$_{50}Sn^{114}$	0.66	113.9030								
	$_{50}Sn^{115}$	0.35	114.9035								
	$_{50}Sn^{116}$	14.30	115.9021								(0.006+?)
	$_{50}Sn^{117m}$			14d	IT	0.320			0.159 0.161	100 % 100 %	

TABLE OF THE ISOTOPES (Continued)

z	Isotope	% nat. abundance	Atomic mass (A)	Lifetime $t_{\frac{1}{2}}$	Modes of decay	Decay energies (Mev)	Particle energies (Mev)	Particle intensities	Gamma energies (Mev)	Gamma intensities	Thermal neutron capture cross-sections (Barns) σ_c
	$_{50}Sn^{117}$	7.61	116.9031								
	$_{50}Sn^{118}$	24.03	117.9018								(0.01+?)
	$_{50}Sn^{119m}$			250d	IT	0.089	IT		0.065 0.0240	100 % 100 %	
	$_{50}Sn^{119}$	8.58	118.9034								
	$_{50}Sn^{120}$	32.85	119.9021								(~0.001+0.14)
	$_{50}Sn^{121m}$			>5y	β^-		β^- 0.42				
	$_{50}Sn^{121}$			≈25h	β^-	0.38	β^- 0.38	100 %			
	$_{50}Sn^{122}$	4.92	121.9034								(0.001+0.2)
	$_{50}Sn^{123m}$			125d	β^-	1.42	β^- 1.42	98 %	1.08	2 %	
	$_{50}Sn^{123}$			40m	β^-	1.42	β^- 1.26	100 %	0.159	100 %	
	$_{50}Sn^{124}$	5.94	123.9052								(0.2+0.004)
	$_{50}Sn^{125m}$			9.7m	β^-	2.37	β^- 2.04	98 %	0.3260 0.64 1.07 1.38 1.86	99.7 % 0.3 % 0.3 % 1.9 %	
	$_{50}Sn^{125}$			9.4d	β^-	2.34	β^- 2.34	95 %	1.07 1.97 0.23–1.41	4 % 1 %	
	$_{50}Sn^{126}$			≈10^5y	β^-				0.092 0.07 (See Sb^{126})		
	$_{50}Sn^{127m}$			≈3m							
	$_{50}Sn^{127}$			2.1h	β^-				0.12–2.8		
	$_{50}Sn^{128}$			5m			β^-		0.04–0.48 (See Sb^{128})		
	$_{50}Sn^{129}$			6.2m	β^-				1.1–3.1		
	$_{50}Sn^{130}$			2.6m	β^-						
	$_{50}Sn^{132}$			2.2m	β^-						
Sb			121.75								5
51	$_{51}Sb^{112}$			0.9m	β^+				1.27 Ann. Rad.		
	$_{51}Sb^{113}$			7m	β^+ EC		β^+ 2.42 1.85 EC		Ann. Rad.		
	$_{51}Sb^{114}$			3.4m	β^+	6.3	β^+ 4.0 2.7		0.9 1.3 Ann. Rad.		
	$_{51}Sb^{115}$			31m	EC β^+	3.03	β^+ 1.51		0.50 0.98 1.24 Ann. Rad.		

z	Isotope	% nat. abundance	Atomic mass (A)	Lifetime $t_{\frac{1}{2}}$	Modes of decay	Decay energies (Mev)	Particle energies (Mev)	Particle intensities	Gamma energies (Mev)	Gamma intensities	Thermal neutron capture cross-sections (Barns) σ_c
	$_{51}Sb^{116m}$			60m	β^+ EC		β^+ 1.45		0.107 0.140 0.400 0.920 1.30 Ann. Rad.		
	$_{51}Sb^{116}$			15m	β^+ EC		β^+ 2.4 1.5 EC		0.90 1.305 2.21 Ann. Rad.	† 31 †140 † 16	
	$_{51}Sb^{117}$			2.8h	EC β^+	1.82	EC β^+ 0.64	97 % 3 %	0.160		
	$_{51}Sb^{118m}$			3.5m	β^+ IT	0.108	β^+ 2.6		0.108 1.24		
	$_{51}Sb^{118}$			5.1h	EC β^+	3.9			0.040 0.257 1.03 1.22	100 % 100 % 100 %	
	$_{51}Sb^{119}$			38h	EC	0.58	EC		0.0238 (CE)	100 % (86 %)	
	$_{51}Sb^{120m}$			5.8d			EC		0.0895 0.1995 1.0400 1.175	100 % 100 % 100 % 100 %	
	$_{51}Sb^{120}$			16.4m	β^+ EC	2.72	β^+ 1.70		1.175		
	$_{51}Sb^{121}$	57.25	120.9038								(0.06+6)
	$_{51}Sb^{122m}$			3.5m	IT	0.136			0.061 0.075	100 % 100 %	
	$_{51}Sb^{122}$			2.8d	β^- β^+ EC	1.97 1.59	β^- 1.4 1.97 EC β^+ 0.57	63 % 30 % 3 % 0.01 %	0.095 0.5639 0.616 0.647 0.691 1.13 1.24 1.9	 66 % 3.4 % 0.7 %	
	$_{51}Sb^{123}$	42.75	122.9041								(0.03+0.03+3
	$_{51}Sb^{124m_2}$			21m	IT β^-	0.018 3	β^- 3		0.018		
	$_{51}Sb^{124m_1}$			1.3m	IT β^-	0.012 3	β^- 3		0.012		
	$_{51}Sb^{124}$			60.9d	β^-	2.92	β^- 0.61 2.31 0.22	51 % 23 % 11 %	0.605 1.69 0.72 2.09 0.63-2.3	98 % 46 % 10 % 7 %	
	$_{51}Sb^{125}$			2.0y	β^-	0.76	β^- 0.30 0.12 0.61	45 % 29 % 44 %	0.43 0.60 0.035–0.64 >10γ's (See Te^{125m})		
	$_{51}Sb^{126m}$			19m	β^-		β^- 2.0		0.42 0.66		

TABLE OF THE ISOTOPES (Continued)

z	Isotope	% nat. abundance	Atomic mass (A)	Lifetime $t_{\frac{1}{2}}$	Modes of decay	Decay energies (Mev)	Particle energies (Mev)	Particle intensities	Gamma energies (Mev)	Gamma intensities	Thermal neutron capture cross-sections (Barns) σ_c
	$_{51}Sb^{126}$			12d	β^-	3.7	β^- 2.0		0.42 0.66 0.3–1.0		
	$_{51}Sb^{127}$			3.7d	β^-	1.6	β^- 1.13 1.5	35% 20%	0.059 0.244 0.310 0.460 0.768	† 6 † 26 † 11 †100 † 45	
	$_{51}Sb^{128m}$			8.9h	β^-		β^- 1.0		0.32 0.4 0.75 0.90		
	$_{51}Sb^{128}$			10.3m	β^-		β^- 2.9		0.75 0.32		
	$_{51}Sb^{129}$			4.2h	β^-		β^- 1.87	20%	0.165 0.308 0.534 0.788 (See Te^{129})		
	$_{51}Sb^{130m}$			6m	β^-						
	$_{51}Sb^{130}$			37m	β^-				0.19–.094		
	$_{51}Sb^{131}$			23m	β^-						
	$_{51}Sb^{132}$			2.1m	β^-						
	$_{51}Sb^{133}$			4.1m	β^-						
	$_{51}Sb^{134}$			0.8m	β^-						
	$_{51}Sb^{135}$			<0.4m	β^- (n)						
Te			127.60								4.7
52	$_{52}Te^{114}$			16m	β^+				Ann. Rad. (See Sb^{114})		
	$_{52}Te^{115}$			6m							
	$_{52}Te^{116}$			2.5h	EC	1.6			0.094 (See Sb^{116})		
	$_{52}Te^{117}$			61m	EC β^+	3.5	β^+ 1.74		0.71		
	$_{52}Te^{118}$			60d	EC						
	$_{52}Te^{119m}$			4.5d	EC IT				0.15–2.1		
	$_{52}Te^{119}$			16h	EC β^+	2.29	β^+ 0.63		0.64 1.8		
	$_{52}Te^{120}$	0.089	119.9045								(2.4+<1)
	$_{52}Te^{121m}$			154d	IT	0.295			0.2133 0.0819	100% 100%	
	$_{52}Te^{121}$			17d	EC				0.070 0.506 0.574	2% 13% 87%	

TABLE OF THE ISOTOPES (Continued)

z	Isotope	% nat. abundance	Atomic mass (A)	Lifetime $t_{\frac{1}{2}}$	Modes of decay	Decay energies (Mev)	Particle energies (Mev)	Particle intensities	Gamma energies (Mev)	Gamma intensities	Thermal neutron capture cross-sections (Barns) σ_c
	$_{52}Te^{122}$	2.46	121.9030								(1+2)
	$_{52}Te^{123m}$			104d	IT	0.248	IT		0.0886 0.1591D	100 % 100 %	
	$_{52}Te^{123}$	0.87	122.9042								400
	$_{52}Te^{124}$	4.61	123.9028								(5+2)
	$_{52}Te^{125m}$			58d	IT	0.145			0.1097 0.0352D	100 % 100 %	
	$_{52}Te^{125}$	6.99	124.9044								1.5
	$_{52}Te^{126}$	18.71	125.9032								(0.09+0.8)
	$_{52}Te^{127m}$			105d	IT β^-	0.089	IT β^-	98 % 2 %	0.089 0.0585 0.635	100 % 2 %	
	$_{52}Te^{127}$			9.3h	β^-	0.69	β^- 0.69	99 %	0.0585 0.145 0.360 0.418	0.05 % 0.007 % 0.1 % 0.8 %	
	$_{52}Te^{128}$	31.79	127.9047								(0.016+0.14)
	$_{52}Te^{129m}$			33d	IT	0.106			0.106 (CE)	100 % (95 %)	
	$_{52}Te^{129}$			67.3m	β^-	1.48	β^- 1.45 1.00	71 % 15 %	0.027 0.47 0.75 1.10	86 % 9 % 0.7 %	
	$_{52}Te^{130}$	34.48	129.9067								(<0.01+0.2)
	$_{52}Te^{131m}$			1.2d	β^- IT	2.47 0.182	β^- 0.42 0.57–2.51 IT	52 % 22 %	0.147 0.77 1.12 0.182 IT 0.10–1.63	 22 %	
	$_{52}Te^{131}$			25m	β^-	2.28	β^- 2.14 1.69	60 % 25 %	0.147 0.448 0.595 0.74 0.950 1.140	94 % 23 % 6 % 4 % 8 %	
	$_{52}Te^{132}$			78h	β^-	0.51	β^- 0.22	60 %	0.225 0.053 (CE) (See I^{132})	100 % 100 % (90 %)	
	$_{52}Te^{133m}$			63m	IT β^-	0.334	β^- 1.3 2.4 IT	 13 %	0.3340 ≈0.4 0.6 1.0	13 %	
	$_{52}Te^{133}$			2m			β^- 1.3 2.4		0.6 1.0 ≈0.4		
	$_{52}Te^{134}$			42m			β^-		0.079–0.262		
I			126.9044								6.4
53	$_{53}I^{117}$			≈10m							

TABLE OF THE ISOTOPES (Continued)

z	Isotope	% nat. abundance	Atomic mass (A)	Lifetime $t_{\frac{1}{2}}$	Modes of decay	Decay energies (Mev)	Particle energies (Mev)	Particle intensities	Gamma energies (Mev)	Gamma intensities	Thermal neutron capture cross-sections (Barns) σ_c
	$_{53}I^{118}$			17m							
	$_{53}I^{119}$			18m			β^+		Ann. Rad.		
	$_{53}I^{120}$			1.4h	β^+	5.0	β^+ 4.0		Ann. Rad.		
	$_{53}I^{121}$			1.4h	EC β^+	2.36	EC β^+ 1.13	90 % 10 %	0.213 0.32 0.27 Ann. Rad.	100 % 7 % 3 %	
	$_{53}I^{122}$			3.5m	β^+	4.14	β^+ 3.12		Ann. Rad.		
	$_{53}I^{123}$			13h	EC		EC		0.1595 0.53 0.27	100 % 2 %	
	$_{53}I^{124}$			4.2d	EC β^+	 3.2	EC β^+ 1.55 2.15	72 % 13 % 13 %	0.603 0.73 1.72 1.95 2.24 Ann. Rad.	62 % 11 % 8 %	
	$_{53}I^{125}$			60.0d	EC	0.15			0.0352	100 %	
	$_{53}I^{126m}$			?2.6h							
	$_{53}I^{126}$			13.2d	EC β^- β^+	— 1.25 2.12	EC β^- 0.38 1.25 β^+ 1.11	55 % — 44 % 1 %	0.65 0.745 0.384 0.481 0.861 Ann. Rad.	33 % 4 % 36 % 5 % 1 %	
	$_{53}I^{127}$	100	126.9044								6.4
	$_{53}I^{128}$			25.0m	β^- EC	2.12 1.0	β^- 2.12 1.67 EC	76 % 16 % 6 %	0.753 0.44 0.53 0.98	0.3 % 17.0 % 1.8 % 0.3 %	
	$_{53}I^{129}$			1.72×10^7y	β^-	0.19	β^- 0.15	100 %	0.040 (CE)	100 % (90 %)	31
	$_{53}I^{130}$			12.5h	β^-	2.96	β^- 0.60 1.02	54 % 46 %	0.413 0.532 0.664 0.7440 1.15	24 % 100 % 100 % 70 % 30 %	1
	$_{53}I^{131}$			8.05d	β^-	0.97	β^- 0.61 0.25– 0.81	87 %	0.080164 0.17705 0.284307 0.364467 0.510 0.6384 0.7238	2.2 % 5.3 % 82 % 9 % 3 %	≈50
	$_{53}I^{132}$			2.3h	β^-	3.57	β^- 1.53 1.16 2.12	24 % 23 % 18 %	0.528 0.624 0.677 0.778 1.1 0.96 0.62–2.5	30 % 6 % 100 % 80 % 20 % 20 %	
	$_{53}I^{133}$			21h	β^-	1.78	β^- 1.4≈0.5	94 %	0.535 ≈0.87 ≈1.27	94 % 5 % 1 %	

TABLE OF THE ISOTOPES (Continued)

z	Isotope	% nat. abundance	Atomic mass (A)	Lifetime $t_{\frac{1}{2}}$	Modes of decay	Decay energies (Mev)	Particle energies (Mev)	Particle intensities	Gamma energies (Mev)	Gamma intensities	Thermal neutron capture cross-sections (Barns) σ_c
	$_{53}I^{134}$			53m	β^-	4.1	β^- 1.5 2.4 1.0–2.2	70% 30%	0.14 0.41 0.86 1.07 1.78 0.2–2.5		
	$_{53}I^{135}$			6.7h	β^-		β^- 1.0 0.5 1.4		0.42 0.86 1.04 1.14 1.275 1.46 1.72 1.80 (See Xe^{135}) (0.53D Xe^{133m})	6.9% 11% 9% 37% 34% 12% 19% 11%	
	$_{53}I^{136}$			86s	β^-	7.0	β^- 2.7 4.2 5.6 7.0		1.32 2.4 0.20–3.2		
	$_{53}I^{137}$			24s	β^- n		n 0.6				
	$_{53}I^{138}$			6.3s	β^- n						
	$_{53}I^{139}$			2.0s	β^- n						
Xe			131.30								3.5
54	$_{54}Xe^{121}$			40m	β^+ EC		β^+ 2.8		0.096 0.080 0.132 0.44 Ann. Rad.		
	$_{54}Xe^{122}$			19h	EC				0.148 0.090 0.187 0.238 (See I^{122})		
	$_{54}Xe^{123}$			1.8h	EC β^+		β^+ 1.5		0.149 0.178 0.328 Ann. Rad.		
	$_{54}Xe^{124}$	0.096	123.9061								74
	$_{54}Xe^{125m}$			55s	IT	0.186			0.075 0.110		
	$_{54}Xe^{125}$			18h	EC				0.055 0.096 0.106 0.1870 0.2430 0.460		
	$_{54}Xe^{126}$	0.090	125.9042								
	$_{54}Xe^{127m}$			75s	IT	0.300			0.175 0.125	100% 100%	

z	Isotope	% nat. abundance	Atomic mass (A)	Lifetime $t_{\frac{1}{2}}$	Modes of decay	Decay energies (Mev)	Particle energies (Mev)	Particle intensities	Gamma energies (Mev)	Gamma intensities	Thermal neutron capture cross-sections (Barns) σ_c
	$_{54}Xe^{127}$			36.4d	EC	0.7			0.057 0.145 0.170 0.203		
	$_{54}Xe^{128}$	1.92	127.9035								<5
	$_{54}Xe^{129m}$			8d	IT	0.236			0.196 0.0400	100% 100%	
	$_{54}Xe^{129}$	26.44	128.9048								45
	$_{54}Xe^{130}$	4.08	129.9035								<5
	$_{54}Xe^{131m}$			12d	IT	0.164			0.1640 (CE)	100% (96%)	
	$_{54}Xe^{131}$	21.18	130.9051								120
	$_{54}Xe^{132}$	26.89	131.9042								(<5+0.2)
	$_{54}Xe^{133m}$			2.3d	IT	0.233			0.233 (CE)	100% (90%)	
	$_{54}Xe^{133}$			5.27d	β^-	0.43	β^- 0.35	99%	0.0809 (CE) 0.160 0.30 0.38	100% (60%) 0.1%	190
	$_{54}Xe^{134}$	10.44	133.9054								(<5+0.2)
	$_{54}Xe^{135m}$			15m	IT	0.53			0.53 (CE)	100% (15%)	
	$_{54}Xe^{135}$			9.13h	β^-	1.16	β^- 0.91 0.55	97% 3%	0.249 0.36 0.607	†100 †≈0.1% † 3	2.7×10^6
	$_{54}Xe^{136}$	8.87	135.9072								0.15
	$_{54}Xe^{137}$			3.9m	β^-		β^- 3.5				
	$_{54}Xe^{138}$			17m	β^-		β^- 2.4		0.42 0.51 1.78 2.01	†100 † 20	
	$_{54}Xe^{139}$			41s	β^-		β^- ≈4.6		0.22 0.30 0.17 0.40		
	$_{54}Xe^{140}$			16s	β^-						
	$_{54}Xe^{141}$			2s	β^-						
	$_{54}Xe^{142}$			≈1.5s	β^-						
	$_{54}Xe^{143}$			1s	β^-						
Cs			132.905								31
55	$_{55}Cs^{123}$			6m	β^+				Ann. Rad.		
	$_{55}Cs^{125}$			45m	β^+ EC	3.07	EC β^+ 2.05	50% 50%	0.112 Ann. Rad.		
	$_{55}Cs^{126}$			1.6m	β^+ EC	4.8	β^+ 3.8 EC	82% 18%	0.385 Ann. Rad.	38%	
	$_{55}Cs^{127}$			6.2h	EC β^+	2.09	EC β^+ 0.68	94% 3%	0.1245 0.169	† 10	

TABLE OF THE ISOTOPES (Continued)

z	Isotope	% nat. abundance	Atomic mass (A)	Lifetime $t_{\frac{1}{2}}$	Modes of decay	Decay energies (Mev)	Particle energies (Mev)	Particle intensities	Gamma energies (Mev)	Gamma intensities	Thermal neutron capture cross-sections (Barns) σ_c
							1.06		0.196 0.2855 0.363 0.406 Ann. Rad.	 † 80	
	$_{55}Cs^{128}$			3.8m	β^+ EC	3.9	β^+ 2.9 2.5 1.5 EC	50 % 20 % 24 %	0.135 0.450 0.98 1.5 Ann. Rad.	60 % 20 %	
	$_{55}Cs^{129}$			0.7h	EC				0.040 0.092 0.174 0.283 0.315 0.373 0.415 0.55 0.585		
	$_{55}Cs^{130}$			30m	EC β^+ β^-	2.99 0.44	EC β^+ 1.97 β^- 0.44	52 % 46 % 2 %	Ann. Rad.		
	$_{55}Cs^{131}$			9.7h	EC	0.35			no γ		
	$_{55}Cs^{132}$			6.5d	EC	2.1			0.68 0.77 1.09 1.20 1.28	98 % 0.7 % 0.7 % 0.8 %	
	$_{55}Cs^{133}$	100	132.9041								(2.6+28)
	$_{55}Cs^{134m}$			2.90h	IT β^- $\approx$1 %	0.137	IT β^- 0.55	99 % 1 %	0.1374 0.126 0.0105		
	$_{55}Cs^{134}$			2.19y	β^-	2.06	β^- 0.65 0.09 0.21–1.45	75 % 30 %	0.605 0.794 0.565 0.20–1.37 >10γ's	97 % 94 % 13 %	134
	$_{55}Cs^{135}$			2.0×10^{6}y	β^-	0.21	β^- 0.21	100 %	no γ		
	$_{55}Cs^{136}$			13d	β^-	2.58	β^- 0.34 0.66	93 % 7 %	1.04 0.82 0.065–2.5 >10γ's		
	$_{55}Cs^{137}$			30y	β^-	1.18	β^- 0.52 1.18	92 % 8 %	0.6616D(CE) (See Ba^{137m})	92 % (11 %)	0.11
	$_{55}Cs^{138}$			32.2m	β^-	4.83	β^- 3.40	21 %	0.46 1.02 1.45 2.24 0.11–3.3 >10γ's	23 % 25 % 73 % 18 %	
	$_{55}Cs^{139}$			9.5m	β^-	4.3	β^- 4.3		0.41		
	$_{55}Cs^{140}$			66s	β^-				0.61		
	$_{55}Cs^{141}$			25s	β^-						

TABLE OF THE ISOTOPES (Continued)

z	Isotope	% nat. abundance	Atomic mass (A)	Lifetime $t_{\frac{1}{2}}$	Modes of decay	Decay energies (Mev)	Particle energies (Mev)	Particle intensities	Gamma energies (Mev)	Gamma intensities	Thermal neutron capture cross-sections (Barns) σ_c
	$_{55}Cs^{142}$			≦8s	β^-						
	$_{55}Cs^{143}$			Short	β^-						
	$_{55}Cs^{144}$			Short	β^-						
Ba			137.34								1.2
56	$_{56}Ba^{126}$			97m	EC				0.225 0.70 0.9 (See Cs^{126})		
	$_{56}Ba^{127}$			~12m	β^+				Ann. Rad.		
	$_{56}Ba^{128}$			2.4d	EC				0.135 0.28 (See Cs^{128})	 20 %	
	$_{56}Ba^{129}$			2.5h	EC β^+	2.9	EC β^+ 1.43 1.25 0.98		0.127 0.18 0.210 1.45 Ann. Rad.		
	$_{56}Ba^{130}$	0.101	129.9062								8.8
	$_{56}Ba^{131}$			11.6d			EC		0.50 0.126 0.22 0.055–1.04 >10γ's		
	$_{56}Ba^{132}$	0.097	131.9057								(4+3)
	$_{56}Ba^{133m}$			39h	IT	0.288			0.2757 0.0117	100 % 100 %	
	$_{56}Ba^{133}$			7.5y	EC	0.49			0.070 0.081 0.158 0.276 0.300 0.358	† 6 †32 †26 †74	
	$_{56}Ba^{134}$	2.42	133.9043								(<0.05+<4)
	$_{56}Ba^{135m}$			29h	IT	0.268			0.268	100 %	
	$_{56}Ba^{135}$	6.59	134.9056								5
	$_{56}Ba^{136}$	7.81	135.9044								(0.016+<1)
	$_{56}Ba^{137m}$			2.6m	IT	0.6616			0.6616 (CE)	100 % (11 %)	
	$_{56}Ba^{137}$	11.32	136.9056								4
	$_{56}Ba^{138}$	71.66	137.9050								0.5
	$_{56}Ba^{139}$			83m	β^-	2.38	β^- 2.38 2.23 0.95		0.166	30 %	4
	$_{56}Ba^{140}$			12.8d	β^-	1.05	β^- 1.02 0.48	60 %	0.0298 0.131 0.161 0.306 0.436 0.540 (See La^{140})	 25 %	<20

TABLE OF THE ISOTOPES (Continued)

z	Isotope	% nat. abundance	Atomic mass (A)	Lifetime $t_{\frac{1}{2}}$	Modes of decay	Decay energies (Mev)	Particle energies (Mev)	Particle intensities	Gamma energies (Mev)	Gamma intensities	Thermal neutron capture cross-sections (Barns) σ_c
	$_{56}Ba^{141}$			18m	β^-	2.9	β^- 2.9		0.12–0.74		
	$_{56}Ba^{142}$			11m	β^-		β^- 4		0.26 0.89 1.20–1.68		
	$_{56}Ba^{143}$			12s	β^-						
	$_{56}Ba^{144}$				β^-						
La			138.91								8.9
57	$_{57}La^{126}$			≈1m	β^+ EC				0.26 Ann. Rad.		
	$_{57}La^{128}$			6.5m	β^+ EC				0.28 Ann. Rad.		
	$_{57}La^{130}$			9m	β^+ EC				0.36 Ann. Rad.		
	$_{57}La^{131}$			61m	β^+ EC		EC β^+ 1.42 1.94	72 % 16 % 8 %	0.115 0.28 0.36 0.42 Ann. Rad.	23 % 17 % 20 % 20 %	
	$_{57}La^{132}$			4.5h	β^+	4.8	β^+ 3.8		1.0 Ann. Rad.		
	$_{57}La^{133}$			4h	EC β^+	2.2	EC β^+ 1.2		0.8 Ann. Rad.		
	$_{57}La^{134}$			6.5m	EC β^+	3.7	EC β^+ 2.7	56 % 44 %	0.60 Ann. Rad.		
	$_{57}La^{135}$			19.8h	EC		EC β^+	99 % 0.1 %	0.104 0.218 0.265 0.295 0.367 0.483 0.588 0.65 0.862		
	$_{57}La^{136}$			9.5m	EC β^+	2.9	EC β^+ 1.8	67 % 33 %	0.83 Ann. Rad.		
	$_{57}La^{137}$			6×10^4y	EC						
	$_{57}La^{138}$	0.089	137.9068	1.1×10^{11}y	EC β^+	1.67 1.02	EC β^- 0.205	70 % 30 %	0.54 0.810 1.43	15 % 30 % 70 %	
	$_{57}La^{139}$	99.9111	138.9061								8.9
	$_{57}La^{140}$			40.2h	β^-	3.8	β^- 1.34 1.10 0.42		>10γ's 0.33 0.49 0.82 0.92 1.60 2.53	35 % 44 % 40 % 88 %	
	$_{57}La^{141}$			3.9h	β^-	2.43	β^- 2.4		1.37	2 %	
	$_{57}La^{142}$			1.4h	β^-	4.4	β^- ≈4		0.630 0.870	†100 55 % † 30	

TABLE OF THE ISOTOPES (Continued)

Z	Isotope	% nat. abundance	Atomic mass (A)	Lifetime $t_{\frac{1}{2}}$	Modes of decay	Decay energies (Mev)	Particle energies (Mev)	Particle intensities	Gamma energies (Mev)	Gamma intensities	Thermal neutron capture cross-sections (Barns) σ_c
									1.0 1.8 2.0 2.4 2.9 3.4	† 20 † 15 † 30 † 55 † 15 † 5	
	$_{57}La^{143}$			14m	β^-	3.3					
	$_{57}La^{144}$			Short	β^-						
Ce			140.12								0.6
58	$_{58}Ce^{131}$			30m	β^+		β^+ 4.2		Ann. Rad.		
	$_{58}Ce^{132}$			4.2h	β^+				Ann. Rad.		
	$_{58}Ce^{133}$			6.3h	EC β^+		β^+ 1.3		1.8 Ann. Rad.		
	$_{58}Ce^{134}$			72h	EC				no γ		
	$_{58}Ce^{135}$			22h	EC β^+		β^+ 0.81		Ann. Rad.		
	$_{58}Ce^{136}$	0.193	135.907								(0.6+6.3)
	$_{58}Ce^{137m}$			34.5h	IT EC	0.26	IT EC	99+ % ≈0.1 %	0.255	100 %	
	$_{58}Ce^{137}$			9h	EC				0.010 0.443	100 % 3 %	
	$_{58}Ce^{138}$	0.250	137.9057								(~0.007+0.7)
	$_{58}Ce^{139m}$			55s	IT	0.74			0.740	100 %	
	$_{58}Ce^{139}$			140d	EC	0.27			0.167	100 %	
	$_{58}Ce^{140}$	88.48	139.9053								0.5
	$_{58}Ce^{141}$			32.5d	β^-	0.58	β^- 0.43 0.58	70 % 30 %	0.144	70 %	
	$_{58}Ce^{142}$	11.07	141.9090	5×10^{15}y	α		α 1.5				1
	$_{58}Ce^{143}$			33h	β^-	1.44	β^- 1.09 1.38	40 % 37 %	0.058 0.295 0.66 0.72 0.89 1.10	80 % 43 % 10 % 9 %	6
	$_{58}Ce^{144}$			285d	β^-	0.32	β^- 0.31 0.19	72 % 20 %	0.0337 0.0412 0.0535 0.0590 0.0806 0.098 0.1339 (See Pr^{144})		6
	$_{58}Ce^{145}$			3.0m	β^-	2.0	β^- 2.0				
	$_{58}Ce^{146}$			1.4m	β^- 0.7	1.0	β^- 0.7		0.05 0.110 0.142 0.22	weak † 20 † 42 † 50	

z	Isotope	% nat. abundance	Atomic mass (A)	Lifetime $t_{\frac{1}{2}}$	Modes of decay	Decay energies (Mev)	Particle energies (Mev)	Particle intensities	Gamma energies (Mev)	Gamma intensities	Thermal neutron capture cross-sections (Barns) σ_c
									0.25 0.27 0.32	weak † 12 †100	
	$_{58}Ce^{147}$			1.2m	β^-						
	$_{58}Ce^{148}$			1.2m	β^-						
Pr			140.907								11
59	$_{59}Pr^{134}$			1h					0.13 0.22 0.94		
	$_{59}Pr^{135}$			22m	β^+		β^+ 2.5		0.080 0.22 0.30 Ann. Rad.		
	$_{59}Pr^{136}$			7.0m	β^+		β^+ 2.0		0.17 0.8 1.1 Ann. Rad.		
	$_{59}Pr^{137}$			1.5h	EC β^+	2.8	EC β^+ 1.8	83 % 17 %			
	$_{59}Pr^{138}$			2.0h	EC β^+		EC β^+ 1.4	90 % 10 %	0.30 0.80 1.05 ≈1.4 ≈1.7 Ann. Rad.		
	$_{59}Pr^{139}$			4.5h	EC β^+	2.0	EC β^+ 1.0	94 % 6 %	1.0 1.3 1.6 Ann. Rad.		
	$_{59}Pr^{140}$			3.4m	β^+ EC	3.40	β^+ 2.23		Ann. Rad.		
	$_{59}Pr^{141}$	100	140.9074								11
	$_{59}Pr^{142}$			19.2h	β^-	2.15	β^- 2.15 0.58	96 % 4 %	1.57	4 %	20
	$_{59}Pr^{143}$			13.7d	β^-	0.93	β^- 0.933	100 %	no γ		
	$_{59}Pr^{144}$			17.3m	β^-	2.98	β^- 2.99 0.80 2.29	98 % 1 % 1.2 %	0.696 2.18 1.49	1.6 % 0.8 % 0.3 %	
	$_{59}Pr^{145}$			5.9h	β^-	1.80	β^- 1.80	95 %	0.022 0.34 0.92 0.97 1.04 1.15		
	$_{59}Pr^{146}$			24m	β^-	4.2	β^- 3.7 2.3	56 % 44 %	0.455 0.597 0.75 1.49	100 % 11 % 33 %	
	$_{59}Pr^{147}$			12m	β^-				0.32 ≈1.7		

TABLE OF THE ISOTOPES (Continued)

z	Isotope	% nat. abundance	Atomic mass (A)	Lifetime $t_{\frac{1}{2}}$	Modes of decay	Decay energies (Mev)	Particle energies (Mev)	Particle intensities	Gamma energies (Mev)	Gamma intensities	Thermal neutron capture cross-sections (Barns) σ_c
	$_{59}Pr^{148}$			12m	β^-						
Nd			144.24								50
60	$_{60}Nd^{138}$			22m	β^+	3.4	β^+	2.4%	Ann. Rad.		
	$_{60}Nd^{139}$			5.2h	EC β^+	4.1	EC β^+ 3.1	90% 10%	1.3 Ann. Rad.	9%	
	$_{60}Nd^{140}$			3.3d	EC				0.11 0.19 0.32 0.5 Ann. Rad.		
	$_{60}Nd^{141}$			2.4h	EC β^+	1.80	EC β^+ 0.78	95% 5%	0.42 0.88 1.15 1.30 Ann. Rad.	0.13% 0.26% 0.5% 0.25%	
	$_{60}Nd^{141m}$			64s	IT	0.76			0.76	100%	
	$_{60}Nd^{142}$	27.11	141.9075								17
	$_{60}Nd^{143}$	12.17	142.9096								330
	$_{60}Nd^{144}$	23.85	143.9099	$\approx 5\times10^{15}$y	α		α 1.8				5
	$_{60}Nd^{145}$	8.30	144.9122								50
	$_{60}Nd^{146}$	17.22	145.9127								2
	$_{60}Nd^{147}$			11.1d	β^-	0.91	β^- 0.82 0.38	77% 20%	0.091 0.53 0.12–0.69 >10γ's	29% 15%	
	$_{60}Nd^{148}$	5.73	147.9165								4
	$_{60}Nd^{149}$			1.8h	β^-	1.6	β^- 1.5 1.1 0.95		0.112 0.114 0.210 0.24–0.65 >10γ's		
	$_{60}Nd^{150}$	5.62	149.9207								1.5
	$_{60}Nd^{151}$			12m	β^-	2.4	β^- 1.93		0.12 0.21 0.030–2.17 >10γ's		
Pm											
61	$_{61}Pm^{141}$			22m	β^+	3.6	β^+ 2.6		Ann. Rad.		
	$_{61}Pm^{142}$			34s	β^+	4.82	β^+ 3.8 EC	69% 31%	1.57	0.2%	
	$_{61}Pm^{143}$			265d	EC	1.1	EC		0.742	45%	
	$_{61}Pm^{144}$			$\approx$400d	EC		EC		0.476 0.616 0.696	45% 100% 100%	
	$_{61}Pm^{145}$			18y	EC	0.14	EC		0.067 0.0725	12% 8%	

TABLE OF THE ISOTOPES (Continued)

z	Isotope	% nat. abundance	Atomic mass (A)	Lifetime $t_{\frac{1}{2}}$	Modes of decay	Decay energies (Mev)	Particle energies (Mev)	Particle intensities	Gamma energies (Mev)	Gamma intensities	Thermal neutron capture cross-sections (Barns) σ_c
	$_{61}Pm^{146}$			≈710d	EC β^-	1.72 1.53	EC β^- 0.78	 35 %	0.453 0.742 0.749	65 % 30 % 35 %	
	$_{61}Pm^{147}$			2.5y	β^-	0.225	β^- 0.225 0.10	100 % ≈10^{-2} %	0.121	very weak	(11+120)
	$_{61}Pm^{148m}$			40.6d	IT β^-	2.59	β^- 0.5 0.6 0.68 0.8 IT	37 % 19 % 8 % 24 % 6 %	0.286 0.412 0.548 0.627 0.723 0.913 1.011	10 % 18 % 91 % 94 % 31 % 18 % 20 %	<3000
	$_{61}Pm^{148}$			5.39d	β^-	2.46	β^- 1.1 2.0 2.6	41 % 14 % 45 %	0.550 0.910 1.46	31 % 17 % 24 %	27,000
	$_{61}Pm^{149}$			54.4h	β^-	1.06	β^- 1.064 0.784	89 % 10 %	0.285 0.256 0.548 0.582 0.850	10 %	
	$_{61}Pm^{150}$			2.7h	β^-	3.49	β^- 2.3 1.8 3.2 1.4	80 % 20 %	0.334 1.175 1.33 0.845 1.77		
	$_{61}Pm^{151}$			28.4h	β^-	≈1.8	β^- 1.3 0.49–1.1		0.0647 0.163 0.177 0.240 0.275 0.340 0.44 0.65 0.715 1.5		
	$_{61}Pm^{152}$			6m	β^-		β^- 2.2		0.12 0.24 1.0		
	$_{61}Pm^{154}$			2.5m	β^-		β^- 2.5				
Sm			150.35								5800
62	$_{62}Sm^{141}$			≈20d	EC β^+				Ann. Rad.		
	$_{62}Sm^{142}$			72m	EC β^+		EC β^+ 1.03	90 % 10 %	 Ann. Rad.		
	$_{62}Sm^{143m}$			2.33m	IT	0.68			0.68		
	$_{62}Sm^{143}$			1.0m	EC β^+	3.5	EC β^+ 2.5	57 % 43 %	Ann. Rad.		
	$_{62}Sm^{144}$	3.09	143.9117								~0.03
	$_{62}Sm^{145}$			340d	EC	0.64	EC		0.0613 (CE) 0.485	92 % (78 %)	
	$_{62}Sm^{146}$		146.9129	5×10^7y	α		2.55				

TABLE OF THE ISOTOPES (Continued)

z	Isotope	% nat. abundance	Atomic mass (A)	Lifetime $t_{\frac{1}{2}}$	Modes of decay	Decay energies (Mev)	Particle energies (Mev)	Particle intensities	Gamma energies (Mev)	Gamma intensities	Thermal neutron capture cross-sections (Barns) σ_c
	$_{62}Sm^{147}$	14.97	146.9146	1.06×10^{11}y	α		α 2.24				~90
	$_{62}Sm^{148}$	11.24	147.9146	1.2×10^{13}y	α		α 2.14				
	$_{62}Sm^{149}$	13.83	148.9169	$\sim4\times10^{14}$y	α		α 1.84				41,500
	$_{62}Sm^{150}$	7.44	149.9170								
	$_{62}Sm^{151}$			≈93y	β^-	0.076	β^- 0.076	98.3 %	0.021	1.7 %	12,000
	$_{62}Sm^{152}$	76.72	151.9195								220
	$_{62}Sm^{153}$			47.1h	β^-	0.803	β^- 0.698 0.640 0.803	40 % 38 % 22 %	0.06966 0.0834 0.099 0.10323 0.171 0.53 0.600	† 7 †100 † 0.07 † 0.2 † 0.04	
	$_{62}Sm^{154}$	22.71	153.9220								5
	$_{62}Sm^{155}$			22m	β^-	1.74	β^- 1.65 1.50	93 % 7 %	0.105 0.248 0.142	96 % 4 % 1 %	
	$_{62}Sm^{156}$			9h	β^-	0.9	β^- 0.9				
	$_{62}Sm^{157}$			0.5m	β^-				0.57		
Eu			151.96								43000
63	$_{63}Eu^{144}$			18m	β^+	3.4	β^+ 2.4		Ann. Rad.		
	$_{63}Eu^{145}$			5.6d	EC				0.63 0.53 0.73 0.89	24 % 3 % 60 %	
	$_{63}Eu^{146m}$			?38h	EC		EC		0.035 0.16 0.89		
	$_{63}Eu^{146}$			4.6d	EC β^+	3.89			0.635 0.749 0.67–2.42 Ann. Rad.		
	$_{63}Eu^{147}$			24d	EC α	1.8	EC α 2.9	100 % 10^{-3} %	0.076 0.121 0.198 0.602 0.678 0.800 0.88–1.08		
	$_{63}Eu^{148}$			54d	EC	≈3.4			0.56 0.90 1.46		
	$_{63}Eu^{149}$			120d	EC	0.80			0.285 0.330 0.256 0.582		
	$_{63}Eu^{150m}$			14h	β^- β^+ EC	1.01 2.27	β^- 1.01 β^+ 1.25		0.41 – 1.97 Ann. Rad.		

TABLE OF THE ISOTOPES (Continued)

z	Isotope	% nat. abundance	Atomic mass (A)	Lifetime $t_{\frac{1}{2}}$	Modes of decay	Decay energies (Mev)	Particle energies (Mev)	Particle intensities	Gamma energies (Mev)	Gamma intensities	Thermal neutron capture cross-sections (Barns) σ_c
	$_{63}Eu^{150}$			≈5y	EC				0.33 0.44 0.59–1.07		
	$_{63}Eu^{151m}$			58μs	IT	0.18	IT		0.18		
	$_{63}Eu^{151}$	47.82	150.9196								(1700+7000)
	$_{63}Eu^{152m}$			9.2h	EC β^+ β^-	1.92	β^- 1.87 EC β^+ 0.90	74 % 21 % 0.01 %	0.122 0.344 0.56 0.841 0.963 >10γ's	13 % 3 % 24 % 11 % 9 %	
	$_{63}Eu^{152}$			13y	EC β^+	1.86	EC β^- 0.22–1.47 β^+ 0.72	74 % 26 % 0.02 %	0.122 0.34 1.11 1.41	62 % 26 % 13 % 25 %	5000
	$_{63}Eu^{153}$	52.18	152.9209								320
	$_{63}Eu^{154}$			16y	β^-	1.97	β^- 0.15–1.85		0.123 0.248 0.72 0.875 0.998 1.006 1.28 >10γ's	88 % 7 % 23 % 15 % 12 % 18 % 36 %	1400
	$_{63}Eu^{155}$			1.7y	β^-	0.25	β^- 0.15–0.190 0.25	80 % 20 %	0.019 0.026 0.060 0.086 0.105 0.145 >10γ's		13,000
	$_{63}Eu^{156}$			15d	β^-	2.45	β^- 2.45 1.18 0.72 0.48	40 %	0.089 0.64 0.72 0.78 0.2–2.19 >10γ's		
	$_{63}Eu^{157}$			15h	β^-		β^- 1.0 1.7	75 % 25 %	0.6 0.2		
	$_{63}Eu^{158}$			60m	β^-		β^- 2.6				
	$_{63}Eu^{159}$			19m	β^-	2.2	β^- 2.2		0.045 0.176		
	$_{63}Eu^{160}$			≈2.5m	β^-		β^- 3.6				
Gd			157.25								46,000
64	$_{64}Gd^{145}$			25m	EC β^+	5.2	EC β^+ 2.4	55 % 45 %	1.75 1.04 0.80	90 % 9 % 8 %	
	$_{64}Gd^{146}$			48d	EC				0.1155 0.155 0.1148	100 % 100 % 100 %	
	$_{64}Gd^{147}$			35h	EC		EC		0.23 0.370 0.396 >10γ's 0.14–1.33		

z	Isotope	% nat. abundance	Atomic mass (A)	Lifetime $t_{\frac{1}{2}}$	Modes of decay	Decay energies (Mev)	Particle energies (Mev)	Particle intensities	Gamma energies (Mev)	Gamma intensities	Thermal neutron capture cross-sections (Barns) σ_c
	$_{64}Gd^{148}$		147.9177	≈130y	α 3.2		α 3.2				
	$_{64}Gd^{149}$		148.9189	9d	EC α		α 3.0	0.0007 %	0.159 0.298 0.346 0.535		
	$_{64}Gd^{150}$		149.9185	≈3×10⁵y	α		α 2.7	100 %			
	$_{64}Gd^{151}$			120d	EC				0.0215 0.080 0.140 0.1530 0.1753 0.242 0.305	†4.4 †6.0 †2.7 †4.5 †0.8	
	$_{64}Gd^{152}$	0.200	151.9195	1.1×10¹⁴y	α		α 2.15				<180
	$_{64}Gd^{153}$			200d	EC	0.28			0.069 0.0975 0.1032		
	$_{64}Gd^{154}$	2.15	153.9207								
	$_{64}Gd^{155}$	14.73	154.9226								58,000
	$_{64}Gd^{156}$	20.47	155.9221								
	$_{64}Gd^{157}$	15.68	156.9339								240,000
	$_{64}Gd^{158}$	24.87	157.9241								3.9
	$_{64}Gd^{159}$			18h	β^-	0.95	β^- 0.59 0.948 0.89	20 % 74 % 6 %	0.362 0.058	19 % 6 %	
	$_{64}Gd^{160}$	21.90	159.9071								0.8
	$_{64}Gd^{161}$			3.7m	β^-	2.02	β^- 0.6 1.54 1.44	91 % 5 % 4 %	0.102 0.316 0.363 0.485	22 % 24 % 60 % 2 %	
	$_{64}Gd^{162}$			>1y	β^-	≈1			0.041 0.117 0.34		
Tb			158.924								46
65	$_{65}Tb^{147}$			24m	β^+				0.305 0.14 Ann. Rad.		
	$_{65}Tb^{148}$			70m	β^+				0.780 1.120 Ann. Rad.		
	$_{65}Tb^{149m}$			4.0m	EC α		3.99				
	$_{65}Tb^{149}$			4.1h	EC α		EC α 3.95	90 % 10 %	0.170 0.230 0.360 ≈0.500 0.640		

TABLE OF THE ISOTOPES (Continued)

z	Isotope	% nat. abundance	Atomic mass (A)	Lifetime $t_{\frac{1}{2}}$	Modes of decay	Decay energies (Mev)	Particle energies (Mev)	Particle intensities	Gamma energies (Mev)	Gamma intensities	Thermal neutron capture cross-sections (Barns) σ_c
	$_{65}Tb^{150}$			3.1h					0.51 0.64 0.93		
	$_{65}Tb^{151}$		150.9230	19h	EC α		EC α 3.44	99+ % 0.0003 %	0.1082 0.1802 0.1921 0.252 0.2711 0.2875 0.3441 0.4321 0.6152		
	$_{65}Tb^{152m}$			4m	EC β^+ α		α	0.002 %	0.23 0.14 Ann. Rad.		
	$_{65}Tb^{152}$			18h	EC				0.344 0.271 0.411 1.32 >10γ's		
	$_{65}Tb^{153}$			62h	EC				0.0415 0.0517 0.0681 0.0875 0.1021 0.1098 0.1744 0.1952 0.2122 0.250		
	$_{65}Tb^{154m}$			8h	EC				0.123 0.248 0.180 0.35 0.30–0.65		
	$_{65}Tb^{154}$			21h	EC β^+		EC β^+ 2.7	99+ %	0.123 0.25 0.180 0.35 0.30–0.65 Ann. Rad.		
	$_{65}Tb^{155}$			5.6d	EC		EC		0.085 0.105 0.155 0.26 0.019–0.37		
	$_{65}Tb^{156m}$			5h	IT β^- β^+?	0.089			0.089		
	$_{65}Tb^{156}$			5.4d	EC	$\geqq$3			0.089 0.199 0.534 1.223 1.42 >10γ's	90 % 48 % 76 % 34 % 18 %	
	$_{65}Tb^{157}$			>30y	EC						
	$_{65}Tb^{158m}$			11s	IT	0.111			0.111	100 %	

z	Isotope	% nat. abundance	Atomic mass (A)	Lifetime $t_{\frac{1}{2}}$	Modes of decay	Decay energies (Mev)	Particle energies (Mev)	Particle intensities	Gamma energies (Mev)	Gamma intensities	Thermal neutron capture cross-sections (Barns) σ_c
	$_{65}Tb^{158}$			≈1000y	EC β^-	0.94	EC		0.08–1.19		
	$_{65}Tb^{159}$	100	158.9250								46
	$_{65}Tb^{160}$			73d	β^-	1.80	β^-0.57 0.87 0.45	40 %	0.087 0.196 0.215 0.298 0.875 0.962 0.966 1.18 1.27	75 % 6 % 4 % 30 % 33 % 11 % 25 % 17 % 8 %	
	$_{65}Tb^{161}$			7.1d	β^-	0.54	β^-0.54 0.49 0.46 0.41	60 % 12 % 11 % 17 %	0.0256 0.0277 0.04890 0.0571 0.0747 0.078 0.1062–0.175 >10γ's		
	$_{65}Tb^{162}$			2h	β^-						
	$_{65}Tb^{163m}$			?7m	β^-				0.18		
	$_{65}Tb^{163}$			7h	β^-						
	$_{65}Tb^{164}$			1d	β^-						
Dy			162.50								960
66	$_{66}Dy^{149}$			≈15m	EC						
	$_{66}Dy^{150}$			8m	α β^+		α 4.2 β^+		0.39 Ann. Rad.		
	$_{66}Dy^{151}$			19m	α		α 4.1				
	$_{66}Dy^{152}$		151.9244	2.5h	β^+ α		β^+ α 3.7	0.02 %	0.255 Ann. Rad.		
	$_{66}Dy^{153}$		152.9254	5h	EC α		α 3.5		γ		
	$_{66}Dy^{154m}$			?13h	α		α 3.35				
	$_{66}Dy^{154}$		153.9248	≈10^6y	α		α 2.85				
	$_{66}Dy^{155}$			10h	EC β^+		EC β^+ 0.85		0.06543 0.09038 0.1154 0.1558 0.1614 0.2057 0.2270 0.2714 Ann. Rad.		
	$_{66}Dy^{156}$	0.052	155.9238								
	$_{66}Dy^{157}$			8.2h	EC				0.0608 0.0830 0.1439 0.1725 0.2655 0.3266		

z	Isotope	% nat. abundance	Atomic mass (A)	Lifetime $t_{\frac{1}{2}}$	Modes of decay	Decay energies (Mev)	Particle energies (Mev)	Particle intensities	Gamma energies (Mev)	Gamma intensities	Thermal neutron capture cross-sections (Barns) σ_c
	$_{66}Dy^{158}$	0.090	157.9240								100
	$_{66}Dy^{159}$			144d	EC	0.38	EC		0.0580	25 %	
	$_{66}Dy^{160}$	2.29	159.9248								
	$_{66}Dy^{161}$	18.88	160.9266								380
	$_{66}Dy^{162}$	25.53	162.9265								140
	$_{66}Dy^{163}$	24.97	162.9284								120
	$_{66}Dy^{164}$	28.18	163.9288								(2000+800)
	$_{66}Dy^{165m}$			75s	IT β^-	0.108	IT β^- 0.84	97 % 3 %	0.108 0.156 0.361 0.515	97 % 3 %	
	$_{66}Dy^{165}$			2.3h	β^-	1.28	β^- 1.19 1.28 0.84 0.29	13 % 84 % 3 % 1.4 %	0.09479 0.284 0.361 0.560 0.62 0.71–1.08 $>10\gamma$'s		4700
	$_{66}Dy^{166}$			82h	β	0.48	β^- 0.402 0.481 0.06	92 % 5 % 3 %	0.028 0.054 0.082 0.288 0.344 0.375 0.428	12 % 16 % 77 % 0.1 % 0.3 % 1.1 % 1.4 %	
	$_{66}Dy^{167}$			4.4m	β^-						64
Ho			164.930								
67	$_{67}Ho^{151m}$			50s	α		α 4.58				
	$_{67}Ho^{151}$			37	α		α 4.48				
	$_{67}Ho^{152m}$			2.4m	α		α 4.34				
	$_{67}Ho^{152}$			52s	α		α 4.43				
	$_{67}Ho^{153m}$			6.5m	α		α 4.10				
	$_{67}Ho^{153}$			9m	α		α 3.98				
	$_{67}Ho^{155}$			50m	β^+		β^+ 2.1		0.138 Ann. Rad.		
	$_{67}Ho^{156}$			57m	β^+	$\approx$5	β^+ 1.8		0.14 0.27 0.37 0.69 Ann. Rad.		
	$_{67}Ho^{158}$			0.5h	β^+		β^+ 1.3		Ann. Rad.		
	$_{67}Ho^{159}$			33m	EC				0.125 0.180 0.250 0.305		

z	Isotope	% nat. abundance	Atomic mass (A)	Lifetime $t_{\frac{1}{2}}$	Modes of decay	Decay energies (Mev)	Particle energies (Mev)	Particle intensities	Gamma energies (Mev)	Gamma intensities	Thermal neutron capture cross-sections (Barns) σ_c
	$_{67}Ho^{160m}$			5.0h	IT	0.060	IT EC β^+	66 % 34 %	0.60 0.107–2.76 $>10\gamma$'s Ann. Rad.		
	$_{67}Ho^{160}$			28m	EC β^+		EC β^+ 0.57	99+ % 0.25 %	0.090 0.20 0.65 0.73 0.89 0.97 0.107–2.76 Ann. Rad.		
	$_{67}Ho^{161}$			2.5h	EC				0.02565 0.0775 0.1754 0.025–0.175 $>10\gamma$'s		
	$_{67}Ho^{162m}$			68m	IT [EC]				0.0382 0.0577 0.0808 0.1848 0.2828		
	$_{67}Ho^{162}$			12m	β^+	2.16	β^+ 1.14		0.76 Ann. Rad.		
	$_{67}Ho^{163m}$			0.8s	IT	0.305			0.305		
	$_{67}Ho^{163}$			$>10^3$y	EC	0.01					
	$_{67}Ho^{164}$			37m	β^- EC IT	0.99 1.4 0.046	β^- 0.99 0.90 EC IT	25 % 15 % 40 % 20 %	0.0373 0.046 0.091	12 % 20 % 15 %	
	$_{67}Ho^{165}$	100	164.9303								(?+?+64)
	$_{67}Ho^{166m_2}$			>30y			β^- 0.28 0.18 1.00 1.45	70 % 20 % 9 % 1 %	0.0804 0.185 0.282 0.415 0.706 0.817 $>10\gamma$'s	100 % 100 % 20 % 19 % 80 % 80 %	
	$_{67}Ho^{166m}$			200μs	IT						
	$_{67}Ho^{166}$			26.8h	β^-	1.85	β^- 1.84 1.76	52 % 47 %	0.0802 1.38 1.58 1.66	48 % 1 % 0.2 % 0.1 %	
	$_{67}Ho^{167}$			3.0h	β^-	1.0	β^- 0.28 1.00		0.35 0.70	≈18 % ≈ 6 %	
	$_{67}Ho^{168}$			3m	β^-		β^- ≈3.3		0.85		
	$_{67}Ho^{169}$			1.6h	β^-				0.13		
	$_{67}Ho^{170}$			45s	β^-		β ≈3.1				
Er			167.26								170
68	$_{68}Er^{158}$			2.5h	β^+		β^+ 0.8 0.3		Ann. Rad.		

z	Isotope	% nat. abundance	Atomic mass (A)	Lifetime $t_{\frac{1}{2}}$	Modes of decay	Decay energies (Mev)	Particle energies (Mev)	Particle intensities	Gamma energies (Mev)	Gamma intensities	Thermal neutron capture cross-sections (Barns) σ_c
	$_{68}Er^{160}$			29h					(See Ho^{160})		
	$_{68}Er^{161}$			3.1h	EC β^+		β^+ 1.21		0.065 0.195 0.82 1.12 Ann. Rad.		
	$_{68}Er^{162}$	0.136	161.9288								2
	$_{68}Er^{163}$			75m	EC				0.43 1.1		
	$_{68}Er^{164}$	1.56	163.9293								1.7
	$_{68}Er^{165}$			10h	EC	≥0.2					
	$_{68}Er^{166}$	33.41	165.9304								10
	$_{68}Er^{167m}$			2.5s	IT	0.208			0.208 (CE)	100% (61%)	
	$_{68}Er^{167}$	22.94	166.9320								700
	$_{68}Er^{168}$	27.07	167.9324								2
	$_{68}Er^{169}$			9.4d	β^-	0.34	β^- 0.332 0.340	42% 58%	0.0084 (CE)	42% (42%)	
	$_{68}Er^{170}$	14.88	169.9355								9
	$_{68}Er^{171}$			7.8h	β^-	1.48	β^- 1.05 0.5–1.48	93%	0.308 0.296 0.124 0.01–0.88 >10γ's		
	$_{68}Er^{172}$			50h	β^-	0.9			0.05–0.61		
Tm			168.934								125
69	$_{69}Tm^{161}$			30m	EC				0.084 0.147 0.172 0.144		
	$_{69}Tm^{162}$			77m	EC				0.102 0.24		
	$_{69}Tm^{163}$			2h	EC				0.104 0.242 0.069 0.240 0.022–0.66		
	$_{69}Tm^{164}$			2m	β^+	3.96	β^+ 2.9 1.3		0.091 0.21–0.93 Ann. Rad.		
	$_{69}Tm^{165}$			29h	EC				0.047 0.054 0.077 0.243 0.015–1.38 >10γ's		
	$_{69}Tm^{166}$			7.7h	EC β^+		EC β^+ 2.09	99+% 1%	0.0806 0.1546 0.182		

z	Isotope	% nat. abundance	Atomic mass (A)	Lifetime $t_{\frac{1}{2}}$	Modes of decay	Decay energies (Mev)	Particle energies (Mev)	Particle intensities	Gamma energies (Mev)	Gamma intensities	Thermal neutron capture cross-sections (Barns) σ_c
									0.706 0.788 0.596–2.06 >10γ's Ann. Rad.		
	$_{69}Tm^{167}$			9.6d	EC				0.208D 0.0570 0.53	99 % 33 % 1 %	
	$_{69}Tm^{168}$			85d	EC	≈2			0.0798 0.185 0.198 0.448 0.817 0.822 0.28–1.464 >10γ's	83 % 28 % 53 % 28 % 36 % 17 %	
	$_{69}Tm^{169}$	100	168.9344								125
	$_{69}Tm^{170}$			127d	β^-	0.96	β^-0.967 0.883	78 % 22 %	0.0842	22 %	
	$_{69}Tm^{171}$			1.9y	β^-	0.097	β^- 0.097 0.03	98 % 2 %	0.0667	2 %	
	$_{69}Tm^{172}$			64h	β^-	1.76	β^- 1.5		0.0787 0.181 1.09 1.49	100 %	
	$_{69}Tm^{173}$			7.3h	β^-		β^- 0.9		0.40 0.47		
	$_{69}Tm^{174}$			5.5m	β^-		β^- 2.5				
	$_{69}Tm^{175}$			20m	β^-		β^- 2.0		0.51		
	$_{69}Tm^{176}$			1.5m	β^-		β^- 4.2				
Yb			173.04								37
70	$_{70}Yb^{164}$			75m	EC β^+		β^+ 2.9		Ann. Rad.		
	$_{70}Yb^{165}$			82m	β^+		β^+ 29		Ann. Rad.		
	$_{70}Yb^{166}$			54h	EC				0.1120 0.1400 (See Tm^{166})		
	$_{70}Yb^{167}$			18m	EC				0.063 0.106 0.116 0.026–0.17		
	$_{70}Yb^{168}$	0.135	167.9339								11,000
	$_{70}Yb^{169m}$			50s	IT	0.0242			0.0242		
	$_{70}Yb^{169}$			32d	EC				>10γ's 0.0084 0.063 0.093 0.1098 0.241 0.261 0.308		

TABLE OF THE ISOTOPES (Continued)

z	Isotope	% nat. abundance	Atomic mass (A)	Lifetime $t_{\frac{1}{2}}$	Modes of decay	Decay energies (Mev)	Particle energies (Mev)	Particle intensities	Gamma energies (Mev)	Gamma intensities	Thermal neutron capture cross-sections (Barns) σ_c
	$_{70}Yb^{170}$	3.03	169.9349								
	$_{70}Yb^{171}$	14.31	170.9365								
	$_{70}Yb^{172}$	21.82	171.9366								
	$_{70}Yb^{173}$	16.13	172.9383								
	$_{70}Yb^{174}$	31.84	173.9390								
	$_{70}Yb^{175m}$			0.067s	IT	0.46			0.46		
	$_{70}Yb^{175}$			4.2d	β^-	0.47	β^-0.67	87%	0.11381 0.13765 0.14485 0.2513 0.2826 0.3961	† 31 † 2.2 † 5.9 † 3.8 † 62 †100	
	$_{70}Yb^{176}$	12.73	175.9427								
	$_{70}Yb^{177m}$			6.5s	IT				0.10 0.21		
	$_{70}Yb^{177}$			1.9h	β^-	1.38	β^- 1.38 1.23	87% 7%	0.1185 0.147 0.95 1.085 1.12 1.235	† 18 †100 † 6 † 30 † 4 † 27	
Lu			174.91								80
71	$_{71}Lu^{167}$			54m	EC		EC		0.030 0.24 0.05–.140 (see Yb^{167})		
	$_{71}Lu^{168m}$			7m	EC β^+	≈4.5	EC β^+ 1.2		0.087 0.987 0.900 1.4–2.13 Ann. Rad.	49% 11% 9%	
	$_{71}Lu^{168}$			2m	EC				0.089		
	$_{71}Lu^{169}$			1.5d	EC		EC		0.087 0.0242 0.0628 0.110 0.070–1.39		
	$_{71}Lu^{170}$			2.0d	EC β^+	3.5	EC β^+ 1.8		0.0842 0.1935 0.153–3.02 Ann. Rad.		
	$_{71}Lu^{171}$			8.3d			EC		0.0556 0.0667 0.0758		
	$_{71}Lu^{172}$			6.7d					0.0787 0.0906 0.1128 0.1815 0.2038 0.809 0.900	100% 24% 4% 35% 10% 20% 34%	

z	Isotope	% nat. abundance	Atomic mass (A)	Lifetime $t_{\frac{1}{2}}$	Modes of decay	Decay energies (Mev)	Particle energies (Mev)	Particle intensities	Gamma energies (Mev)	Gamma intensities	Thermal neutron capture cross-sections (Barns) σ_c
	$_{71}Lu^{173}$			≈1.3y	EC	0.7			0.0787 0.1007 0.272 0.637	91 % 31 % 19 % 2 %	
	$_{71}Lu^{174m_2}$			75μs	IT	0.133			0.133		
	$_{71}Lu^{174m_1}$			165d	EC	0.077	EC		0.0766 1.23 0.059 0.067 0.045–0.99	59 % 10 %	
	$_{71}Lu^{174}$			short	EC						
	$_{71}Lu^{175}$	97.41	174.9409								(18+5)
	$_{71}Lu^{176m}$			3.7h	β^-	1.2	β^- 1.2 1.1		0.0883		
	$_{71}Lu^{176}$	2.59		2.1×10^{10}y	β^-	1.02	β^- 0.42	100 %	0.0883 0.202 0.309	100 % 100 % 100 %	
	$_{71}Lu^{177}$			6.8d	β^-	0.50	β^- 0.50 0.18 0.38	90 % 7 % 3 %	0.07164 0.1130 0.2085 0.2500 0.3124	† 2 †100 †220 † 3 † 3	
	$_{71}Lu^{178m}$			16m	β^-		β^- 1.5		0.060 0.090 0.125 0.342 0.445	 †100 † 10	
	$_{71}Lu^{178}$			30m	β^-	2.3	β^- 2.3				
	$_{71}Lu^{179}$			7.5h	β^-		β^- 1.35		0.090 0.22		
	$_{71}Lu^{180}$			2.5m	β^-		β^- 3.3				
Hf			178.49								105
72	$_{72}Hf^{168}$			22m	EC β^+		β^+ 1.7		Ann. Rad.		
	$_{72}Hf^{169}$			1.9h	EC β^+		β^+ 2.9		Ann. Rad.		
	$_{72}Hf^{170}$			9h	EC β^+				Ann. Rad.		
	$_{72}Hf^{171}$			16h	EC		EC		0.18 0.29 0.63 1.02 1.4		
	$_{72}Hf^{172}$			≈5y	EC		EC		0.080 0.123	30 % 70 %	
	$_{72}Hf^{173}$			24h	EC				0.123 0.135 0.140 0.163 0.2985	†100 † 75	

TABLE OF THE ISOTOPES (Continued)

z	Isotope	% nat. abundance	Atomic mass (A)	Lifetime $t_{\frac{1}{2}}$	Modes of decay	Decay energies (Mev)	Particle energies (Mev)	Particle intensities	Gamma energies (Mev)	Gamma intensities	Thermal neutron capture cross-sections (Barns) σ_c
									0.308 0.312 0.542 0.63 1.02		
	$_{72}\mathrm{Hf}^{174}$	0.18	173.9403	4.3×10^{15}y	α		$\alpha\approx2.5$				400
	$_{72}\mathrm{Hf}^{175}$			70d	EC				0.08936 0.1613 0.2296 0.3189 0.3434 0.4330	† 4 † 1 †100 † 2	
	$_{72}\mathrm{Hf}^{176}$	5.20	175.9435								<30
	$_{72}\mathrm{Hf}^{177}$	18.50	176.9435								370
	$_{72}\mathrm{Hf}^{178m}$			5s	IT	1.148			0.0888 0.427 0.326 0.214 0.0931	100 % 100 % 100 % 100 % 100 %	
	$_{72}\mathrm{Hf}^{178}$	27.14	177.9439								80
	$_{72}\mathrm{Hf}^{179m}$			19s	IT	0.378			0.161 0.217	100 % 100 %	
	$_{72}\mathrm{Hf}^{179}$	13.75	178.9460								(.2+65)
	$_{72}\mathrm{Hf}^{180m}$			5.5h	IT	1.143			0.058 0.501 0.332 0.216 0.093 0.442	83 % 17 % 100 % 100 % 100 % 83 %	
	$_{72}\mathrm{Hf}^{180}$	35.24	179.9468								10
	$_{72}\mathrm{Hf}^{181}$			44.6d	β^-	1.023	β^-0.408 0.060–0.541	92 %	0.13302 0.13625 0.13686 0.3459 0.482 0.62	86 % 15 % 4 % 14 % 82 %	40
	$_{72}\mathrm{Hf}^{182}$			9×10^6y	β^-	~0.5	β^- ~0.5		0.27		
	$_{72}\mathrm{Hf}^{183}$			1.1h	β^-	2.2	β^- 1.4				
Ta			180.948								21
73	$_{73}\mathrm{Ta}^{172}$			30m	β^+				Ann. Rad.		
	$_{73}\mathrm{Ta}^{173}$			2.5h	EC				0.070 0.082 0.170 0.037–0.181		
	$_{73}\mathrm{Ta}^{174}$			1.3h	β^+				Ann. Rad. 0.0909 0.125 0.16 0.20 0.28 0.38		

z	Isotope	% nat. abundance	Atomic mass (A)	Lifetime $t_{\frac{1}{2}}$	Modes of decay	Decay energies (Mev)	Particle energies (Mev)	Particle intensities	Gamma energies (Mev)	Gamma intensities	Thermal neutron capture cross-sections (Barns) σ_c
	$_{73}$Ta^{175}			11h	EC				0.0816 0.104 0.207 0.267 0.18–1.15	18 % 11 % 17 % 13 %	
	$_{73}$Ta^{176}			8.0h	EC				0.0883 0.202 1.16 1.22 0.100–2.08	†42 †24	
	$_{73}$Ta^{177}			53h	EC	1.15			0.1730 0.2080 0.321 0.425 0.510 0.63 0.75 0.96 1.07	30 % 2 % 0.4 % 0.5 %	
	$_{73}$Ta^{178m}			9.4m	EC β^+	≈1.9	EC β^+ 0.88	98 % 2 %	0.09307 1.16 1.37 1.48 Ann. Rad.	42 %	
	$_{73}$Ta^{178}			2.2h	EC β^+	≈1.9	EC β^+ 1	98 % 2 %	0.0888D 0.0931 0.214 0.326 0.332 0.427 Ann. Rad.	100 % 100 % 100 % 100 % 40 % 100 % (see Hf^{178})^m	
	$_{73}$Ta^{179}			≈1.6y	EC	0.094	EC	100 %	no γ		
	$_{73}$Ta^{180m}			8.1h	EC β^-		EC β^- 0.60 0.71	83 % 9 % 8 %	0.093 0.102	41 % 9 %	
	$_{73}$Ta^{180}	0.0123	179.9475								
	$_{73}$Ta^{181m_2}			20μs	IT	0.615			0.1330 0.1361 0.3458 0.4820 0.6155		
	$_{73}$Ta^{181m}			7μs	IT	0.0063			0.0063		
	$_{73}$Ta^{181}	99.988	180.9480								(.07+21)
	$_{73}$Ta^{182m}			16.5m	IT	0.503			0.147 0.172 0.184 0.319 0.356		
	$_{73}$Ta^{182}			115d	β^-	1.74	β^- 0.18 0.36 0.44 0.51 0.25–0.48	38 % 20 % 23 % 8 %	>10γ's 0.100 1.0 1.231 1.122 1.222 1.189 0.033–1.61	 56 % 11 % 33 % 28 % 15 %	

TABLE OF THE ISOTOPES (Continued)

z	Isotope	% nat. abundance	Atomic mass (A)	Lifetime $t_{\frac{1}{2}}$	Modes of decay	Decay energies (Mev)	Particle energies (Mev)	Particle intensities	Gamma energies (Mev)	Gamma intensities	Thermal neutron capture cross-sections (Barns) σ_c
	$_{73}Ta^{183}$			5.0d	β^-	1.068	β^- 0.615 0.514–0.776	91% 9%	0.0465 0.0526 0.099 0.1079 0.1614 0.2443 0.2460 0.3540 >10γ's	59% 48% 29% 44% 18% 11% 35% 12%	
	$_{73}Ta^{184}$			8.7h	β^-	2.67	β^- 1.06 1.26	25% 72%	0.111 0.21 0.30 0.405 0.793 0.904 0.895 1.18	62% 6% 19% 60% 11% 13% 44% 30%	
	$_{73}Ta^{185}$			50m	β^-	1.9	β^- 1.72 1.48	95% 5%	0.075 0.10 0.175 0.245	35% 35% 65% 3%	
	$_{73}Ta^{186}$			10.5m	β^-	3.7	β^- 2.2	100%	0.122 0.20 0.30 0.41 0.51 0.608 0.730 0.94	† 25 †100 † 25 † 20 † 45 † 45 † 65 † 15	
W			183.85								19
74	$_{74}W^{176}$			1.3h	EC β^+		$\beta^+ \approx 2$		≈1.3 Ann. Rad.		
	$_{74}W^{177}$			2.2h	EC				0.5 1.2		
	$_{74}W^{178}$			22d	EC				≈0.3		
	$_{74}W^{179m}$			5m	IT	0.222			0.222		
	$_{74}W^{179}$			30m	EC	≈1.2			0.031		
	$_{74}W^{180m}$			0.005s	IT				0.102 0.35 0.5 0.22		
	$_{74}W^{180}$	0.14	179.9470								<20
	$_{74}W^{181m}$			14μs	IT				0.37		
	$_{74}W^{181}$			130d	EC	0.18			0.0063 0.1525 0.1361	35% 0.1% 0.1%	
	$_{74}W^{182}$	26.41	181.9483								(0.5+20)
	$_{74}W^{183m}$			5.5s	IT	0.39			0.060 0.105 0.155	† 25 † 10	
	$_{74}W^{183}$	14.40	182.9503								11

TABLE OF THE ISOTOPES (Continued)

Z	Isotope	% nat. abundance	Atomic mass (A)	Lifetime $t_{\frac{1}{2}}$	Modes of decay	Decay energies (Mev)	Particle energies (Mev)	Particle intensities	Gamma energies (Mev)	Gamma intensities	Thermal neutron capture cross-sections (Barns) σ_c
	$_{74}W^{184}$	30.64	183.9510								2.1
	$_{74}W^{185m}$			1.62m	IT				0.075 0.135 0.175 0.10	35% 100% 65% 35%	
	$_{74}W^{185}$			75.8d	β^-	0.432	β^- 0.432	99.98%	0.125	0.02%	
	$_{74}W^{186}$	28.41	185.9543								40
	$_{74}W^{187}$			24h	β^-	1.315	β^- 0.63 1.315 0.34	70% 20% 10%	0.134 0.480 0.619 0.686 0.774 0.866 $>10\gamma$'s		90
	$_{74}W^{188}$			65d	β^-	~0.8			(see Re^{188})		
Re			186.2								85
75	$_{75}Re^{177}$			17m	β^+		β^+ >0.4		Ann. Rad.		
	$_{75}Re^{178}$			15m	β^+ EC	≧4.1	β^+ 3.1		Ann. Rad.		
	$_{75}Re^{179}$			18m	EC				0.222D 0.28		
	$_{75}Re^{180m}$			2.4m	EC β^+		EC β^+ 1.1		0.88 0.106 Ann. Rad.		
	$_{75}Re^{180}$			20h	β^+	2.9	β^+ 1.9				
	$_{75}Re^{181}$			20h	EC				0.113 0.252 0.365 $>10\gamma$'s 0.020–1.54		
	$_{75}Re^{182m}$			13h	EC				0.100 0.229 1.1 1.2 0.020–1.44 $>10\gamma$'s		
	$_{75}Re^{182}$			64h					0.100 1.12 1.22 1.18 0.066–1.44 $>10\gamma$'s		
	$_{75}Re^{183}$			71d					$>10\gamma$'s 0.0465 0.108 0.246 0.085 0.041–0.41		
	$_{75}Re^{184m}$			2.2d	EC or IT				0.043 0.159		

TABLE OF THE ISOTOPES (Continued)

z	Isotope	% nat. abundance	Atomic mass (A)	Lifetime $t_{\frac{1}{2}}$	Modes of decay	Decay energies (Mev)	Particle energies (Mev)	Particle intensities	Gamma energies (Mev)	Gamma intensities	Thermal neutron capture cross-sections (Barns) σ_c
	$_{75}Re^{184}$			50d	EC				>10γ's 0.111 0.793 0.895 0.904	45% 40% 15% 4%	
	$_{75}Re^{185}$	37.07	184.9530								110
	$_{75}Re^{186}$			90h	β^- EC	1.07 0.70	β^- 1.07 0.93 EC	73% 23% 4%	0.1226 0.13722 0.628 0.767	2% 23%	
	$_{75}Re^{187}$	62.93	186.9560	7×10^{10}y	β^-	<0.01	β^- $\leqq$0.008				70
	$_{75}Re^{188m}$			18.7m	IT	0.169			0.064 0.105		
	$_{75}Re^{188}$			17h	β^-	2.116	β^- 2.12 2.0	79% 20%	0.155 0.633 0.48–1.96 >10γ's	18% 1%	
	$_{75}Re^{189}$			23h	β^-		β^- 1.0 0.8		0.219 0.150 0.070		
	$_{75}Re^{190}$			2.8m	β^-	3.3	β^- 1.6 1.7 1.9 1.4	21% 25% 29% 25%	0.187 0.361 1.02 1.33		
	$_{75}Re^{191}$			10m	β^-		β^- 1.8				
Os			190.2								15
76	$_{76}Os^{181}$			23m	EC				0.167 0.174		
	$_{76}Os^{182}$			22h	EC				0.0276 0.0555 0.1802 0.510		
	$_{76}Os^{183m}$			10h	IT EC	0.171	EC IT	54% 46%	0.0673 0.1707 1.036 1.103 1.1095	6% 46% 6% 26% 22%	
	$_{76}Os^{183}$			12h	EC		EC		0.3818 0.1144 0.1679 0.496 0.2362 >10γ's	87% 92% 17% 4% 6%	
	$_{76}Os^{184}$	0.018	183.9526								<200
	$_{76}Os^{185}$			94d	EC	0.982	EC		0.646 0.872 0.879 0.718	80% 7.4% 7% 4%	
	$_{76}Os^{186}$	1.59	185.9539								
	$_{76}Os^{187}$	1.64	186.9560								

z	Isotope	% nat. abundance	Atomic mass (A)	Lifetime $t_{\frac{1}{2}}$	Modes of decay	Decay energies (Mev)	Particle energies (Mev)	Particle intensities	Gamma energies (Mev)	Gamma intensities	Thermal neutron capture cross-sections (Barns) σ_c
	$_{76}Os^{188}$	13.3	187.9560								
	$_{76}Os^{189m}$			5.7h	IT	0.03			0.0304 (CE)	100% (100%)	
	$_{76}Os^{189}$	16.1	188.9586								(<.02+?)
	$_{76}Os^{190m}$			49.5m	IT	1.698			0.039 0.1865 0.359 0.505 0.617	100% 100% 100% 100% 100%	
	$_{76}Os^{190}$	26.4	189.9586								(32+8)
	$_{76}Os^{191m}$			14h	IT	0.0742	IT		0.07420	100%	
	$_{76}Os^{191}$			15d	β^-	0.314	β^- 0.143	99+%	0.042D 0.129D (see Ir^{191m})		
	$_{76}Os^{192}$	41.0	191.9612								1.6
	$_{76}Os^{193}$			32h	β^-	1.132	β^- 1.13 1.05 0.99 0.67	66% 12% 8% 8%	0.072 0.139 0.281 0.460 0.559 >10γ's	14% 10% 2% 4% 2%	200
	$_{76}Os^{194}$			~2y	β^-				(See Ir^{194})		
	$_{76}Os^{195}$			6m	β^-		β^- 2.0				
Ir			192.2								460
7	$_{77}Ir^{182}$			15m	EC β^+		EC β^+		Ann. Rad.		
	$_{77}Ir^{183}$			55m	EC				0.24		
	$_{77}Ir^{184}$			3.2h	EC β^+				0.13 0.27 0.29–4.3 Ann. Rad.		
	$_{77}Ir^{185}$			15h	EC				0.0374 0.0599 0.0973 0.1008 0.1044 0.2544		
	$_{77}Ir^{186}$			+5h	EC β^+	3.8			0.138 0.298 0.436 0.628 0.769 0.774 0.9230 Ann. Rad.	†100 †100 † 80	
	$_{77}Ir^{187}$			12h	EC				0.025–0.99 >10γ's		
	$_{77}Ir^{188}$			41h	EC β^+	2.8	EC β^+ 1.8	99+% 0.6%	0.155 0.633 0.32–2.2 >10γ's Ann. Rad.		

TABLE OF THE ISOTOPES (Continued)

z	Isotope	% nat. abundance	Atomic mass (A)	Lifetime $t_{\frac{1}{2}}$	Modes of decay	Decay energies (Mev)	Particle energies (Mev)	Particle intensities	Gamma energies (Mev)	Gamma intensities	Thermal neutron capture cross-sections (Barns) σ_c
	$_{77}Ir^{189}$			11d	EC				0.0695 0.245 0.033–.276 >10γ's		
	$_{77}Ir^{190m}$			3.2h	EC β^+		EC β^+ 2.0	88% 12%	0.186 0.361 ≈0.56 0.62 Ann. Rad.	100% 100% 100% 100%	
	$_{77}Ir^{190}$			11d	EC	2.1			0.1865 0.359 0.371 0.404 ≈0.52 0.5575 0.568 0.607 0.80 1.33	72% 27% 28% 32%	
	$_{77}Ir^{191m}$			4.9s	IT	0.042	IT		0.0416 0.129	100% 100%	
	$_{77}Ir^{191}$	37.3	190.9609								(250+750)
	$_{77}Ir^{192m}$			1.4m	IT β^-	0.048		99.9% 0.1%	0.058 0.316	99.9% 0.1%	
	$_{77}Ir^{192}$			74d	β^- EC	1.45 1.15	β^- 0.67 0.53 0.24 EC	41% 38% 15% 5%	0.3165 0.468 0.6045 >10γ's		
	$_{77}Ir^{193m}$			12d	IT	0.08			0.08019	100%	
	$_{77}Ir^{193}$	62.7	192.9633								1.20
	$_{77}Ir^{194}$			19h	β^-	2.24	β^- 2.24 1.90 0.98 0.19–0.74	68% 15% 9%	0.294 0.328 0.645 0.939 0.6–2.04 >10γ's	5.5% 29% 6% 3%	
	$_{77}Ir^{195}$			2.3h	β^-	1.1	β^- 1.1 0.6		0.070 0.42 0.66 0.88		
	$_{77}Ir^{197}$			7m	β^-	2.0	β^- 2.0 1.5				
	$_{77}Ir^{198}$			50s	β^-	4.4	β^- 3.6		0.78		
Pt			195.09								10
78	$_{78}Pt^{184}$			≈3h							
	$_{78}Pt^{186}$			2.5h	EC						
	$_{78}Pt^{187}$			≈2h							
	$_{78}Pt^{188}$			10d	EC	0.523			0.055 0.144 0.188 0.196		

TABLE OF THE ISOTOPES (Continued)

z	Isotope	% nat. abundance	Atomic mass (A)	Lifetime $t_{\frac{1}{2}}$	Modes of decay	Decay energies (Mev)	Particle energies (Mev)	Particle intensities	Gamma energies (Mev)	Gamma intensities	Thermal neutron capture cross-sections (Barns) σ_c
									0.382 0.410 0.470		
	$_{78}Pt^{189}$			11h	EC				0.14		
	$_{78}Pt^{190}$	0.0127	189.9600	7×10^{11}y	α		α 3.11				~90
	$_{78}Pt^{191}$			3.0d	EC				0.083 0.129 0.36 0.04–0.46 $>10\gamma$'s		
	$_{78}Pt^{192}$	0.78	191.9614	$\sim10^{15}$y	α		$\alpha\sim$2.6				<16
	$_{78}Pt^{193m}$			4.4d	IT	0.148			0.136 0.013	100 % 100 %	
	$_{78}Pt^{193}$			<500y	EC	0.045					
	$_{78}Pt^{194}$	32.9	193.9628								1.2
	$_{78}Pt^{195m}$			3.45d	IT	0.259			0.031 0.099 0.130	 100 %	
	$_{78}Pt^{195}$	33.8	194.9648								27
	$_{78}Pt^{196}$	25.3	195.9650								(~+1.0)
	$_{78}Pt^{197m}$			1.3h	IT	0.34			0.341 (CE)	100 % (100 %)	
	$_{78}Pt^{197}$			20h	β^-	0.75	β^- 0.67 0.48	94 % 5 %	0.0771 0.1910 0.269	99 % 5 % 1 %	
	$_{78}Pt^{198}$	7.21	197.9675								(0.03+4)
	$_{78}Pt^{199m}$			14s	IT	0.39	IT		0.39 0.032	100 % 100 %	
	$_{78}Pt^{199}$			30m	β^-	1.8	β^- 1.7		0.074 0.197 0.246 0.318 0.475 0.540 0.715 0.790 0.960		≈15
	$_{78}Pt^{200}$			11.5h	β^-		β^-		(see Au^{200})		
Au			196.967								99
79	$_{79}Au^{\leq187}$			4.3m	EC β^+ α		EC β^+ α 5.1	0.01 %			
	$_{79}Au^{186}$			15m	EC						
	$_{79}Au^{188}$			4.5m	EC	5.2					
	$_{79}Au^{189}$			42m	EC				0.135 0.290	†100 † 10	
	$_{79}Au^{190}$			39m	EC β^+				Ann. Rad.		

TABLE OF THE ISOTOPES (Continued)

Z	Isotope	% nat. abundance	Atomic mass (A)	Lifetime $t_{\frac{1}{2}}$	Modes of decay	Decay energies (Mev)	Particle energies (Mev)	Particle intensities	Gamma energies (Mev)	Gamma intensities	Thermal neutron capture cross-sections (Barns) σ_c
	$_{79}Au^{191}$			3.2h	EC				0.048 0.0910 0.135 0.1587 0.300 0.390 0.475 0.600	 † 10 † 60 † 5 † 4 † 10	
	$_{79}Au^{192}$			4.8h	EC β^+	3.24	EC β^+ 2.22	99 % 1 %	0.296 0.316 0.613 0.10–1.06 >10γ's		
	$_{79}Au^{193m}$			3.9s	IT EC	0.032	IT EC	99 % 0.03 %	0.0322 0.219 0.257 0.2906	99.5 % 7 % 92 % 0.5 %	
	$_{79}Au^{193}$			15.8h	EC				0.112 0.173 0.186 0.256 0.268 0.440 >10γ's	16 % 6 % 21 % 7 % 5 % 3 %	
	$_{79}Au^{194}$			39h	EC β^+	2.51	EC β^+ 1.50 1.50	97 % 1 % 1 %	0.293 0.328 0.10–2.04 >10γ's		
	$_{79}Au^{195m}$			30.6s	IT	0.318			0.057 0.061 0.200 0.261 0.318		
	$_{79}Au^{195}$			200d	EC	0.22			0.031 0.0987 0.129	† 1 †12 † 2	
	$_{79}Au^{196m}$			10h	IT	0.59	IT		0.084 0.148 0.175 0.186	100 % 100 % 100 % 100 %	
	$_{79}Au^{196}$			5.55d	EC β^-	1.12 0.68	EC β^- 0.27	88 % 12 %	0.331 0.356 0.427	24 % 88 % 12 %	
	$_{79}Au^{197m}$			7.2s	IT	0.409			0.1300 0.409 0.279	99 % 1 % 99 %	
	$_{79}Au^{197}$	100	196.9666								99
	$_{79}Au^{198}$			64.8h	β^-	1.37	β^- 0.96 0.28 1.37	96 % 1 % 0.2 %	0.411770 0.6758 1.086	99 % 1 % 0.2 %	26,000
	$_{79}Au^{199}$			3.15d	β^-	0.46	β^- 0.30 0.25 0.46	70 % 24 % 46 %	0.0500 0.1581 0.2081	8 % 78 % 16 %	≈30
	$_{79}Au^{200}$			48m	β^-	2.21	β^- 2.21 1.84 0.61	69 % 6 % 25 %	0.368 1.227 1.59	30 % 24 % 1 %	

TABLE OF THE ISOTOPES (Continued)

z	Isotope	% nat. abundance	Atomic mass (A)	Lifetime $t_{\frac{1}{2}}$	Modes of decay	Decay energies (Mev)	Particle energies (Mev)	Particle intensities	Gamma energies (Mev)	Gamma intensities	Thermal neutron capture cross-sections (Barns) σ_c
	$_{79}Au^{201}$			26m	β^-	1.5	β^- 1.5 0.9	95 % 2 %	0.55	5 %	
	$_{79}Au^{202}$			~25s	β^- or IT						
	$_{79}Au^{203}$			55s	β^-	1.9	β^- 1.9		0.69		
Hg			200.59								360
80	$_{80}Hg^{<195}$			0.7m	α		α 5.6				
	$_{80}Hg^{<191}$			~3h							
	$_{80}Hg^{<191}$			1.5h	EC						
	$_{80}Hg^{189}$			23m	EC				0.029 0.14 0.22		
	$_{80}Hg^{190}$			20m	EC		EC		0.12 0.38		
	$_{80}Hg^{191}$			57m	EC				0.253 0.274		
	$_{80}Hg^{192}$			5.7h	EC β^+		β^+ 1.2		0.0313 0.1143 0.1423 0.1460 0.1574 0.275 Ann. Rad.		
	$_{80}Hg^{193m}$			11h	EC β^+ IT		EC β^+ 1.17 IT	84 % 16 %	0.039 0.101 0.157–1.65 Ann. Rad. >10γ's	16 % 16 %	
	$_{80}Hg^{193}$			≈6h	EC				0.0379 0.1865 0.570 0.860 0.920		
	$_{80}Hg^{194m}$			0.4s	IT				0.048 0.134		
	$_{80}Hg^{194}$			130d	EC				(See Au^{194})		
	$_{80}Hg^{195m}$			40h	EC IT	0.123			>10γ's 0.016 0.037 0.123 0.56 0.173–1.24		
	$_{80}Hg^{195}$			9.5h	EC				0.061 0.180 0.78 0.20–1.17 >10γ's		
	$_{80}Hg^{196}$	0.146	195.9658								(400+900)

TABLE OF THE ISOTOPES (Continued)

z	Isotope	% nat. abundance	Atomic mass (A)	Lifetime $t_{\frac{1}{2}}$	Modes of decay	Decay energies (Mev)	Particle energies (Mev)	Particle intensities	Gamma energies (Mev)	Gamma intensities	Thermal neutron capture cross-sections (Barns) σ_c
	$_{80}Hg^{197m}$			24h	IT EC	0.165	IT EC	96% 4%	0.1301 0.1335 0.1646 0.279 0.407	96% 96%	
	$_{80}Hg^{197}$			65h	EC				0.0775 0.1916 0.269	100%	
	$_{80}Hg^{198}$	10.02	197.9668								(.02+?)
	$_{80}Hg^{199m}$			42m	IT	0.368	IT		0.368 0.158	100% 100%	
	$_{80}Hg^{199}$	16.84	198.9683								2000
	$_{80}Hg^{200}$	23.13	199.9683								<50
	$_{80}Hg^{201}$	13.22	200.9703								<50
	$_{80}Hg^{202}$	29.80	201.9706								3
	$_{80}Hg^{203}$			47d	β^-	0.49	β^- 0.21	100%	0.279	100%	
	$_{80}Hg^{204}$	6.85	203.9735								0.4
	$_{80}Hg^{205}$			5.2m	β^-	1.6	β^- 1.6 1.4		0.203		
	$_{80}Hg^{206}$			7.5m	β^-						
Tl			204.37								3.3
81	$_{81}Tl^{191}$			10m	EC						
	$_{81}Tl^{192}$			≈11m							
	$_{81}Tl^{193}$			23m	EC						
	$_{81}Tl^{194m}$			?33m	EC				0.097		
	$_{81}Tl^{194}$			33m	EC				0.43		
	$_{81}Tl^{195m}$			3.5s	IT	0.875	IT		0.0991 0.383 0.393	100% 100% 100%	
	$_{81}Tl^{195}$			1.2h	EC		EC		0.0370 0.23–0.88		
	$_{81}Tl^{196m}$			1.4h	IT	0.395	IT EC	4% 96%	0.084 0.120 0.275 0.241 0.034	96% 4% 3% 0.7% 0.7%	
	$_{81}Tl^{196}$			1.8h	EC				0.426		
	$_{81}Tl^{197m}$			0.54s	IT	0.994	IT		0.387 0.385 0.2222	100% 100% 100%	
	$_{81}Tl^{197}$			2.8h	EC				0.133 0.152 0.1731 0.269 0.425		

z	Isotope	% nat. abundance	Atomic mass (A)	Lifetime $t_{\frac{1}{2}}$	Modes of decay	Decay energies (Mev)	Particle energies (Mev)	Particle intensities	Gamma energies (Mev)	Gamma intensities	Thermal neutron capture cross-sections (Barns) σ_c
	$_{81}Tl^{198m}$			1.9h	IT EC	0.543			0.0478 0.194 0.261 0.282 0.411 0.588 0.638	 †13 †10 †10	
	$_{81}Tl^{198}$			5.3h	EC	3.5			0.412 0.676 1.089	100 % 10 %	
	$_{81}Tl^{199m}$			0.042s	IT	0.037	IT		0.037		
	$_{81}Tl^{199}$			7.4h	EC				0.0495 0.078 0.103 0.158 0.207 0.246 0.334 0.455 0.491		
	$_{81}Tl^{200}$			26h	EC β^+	2.45	EC β^+ 1.44	99 % 0.3 %	Ann. Rad. 0.368 0.579 1.207 1.227 1.36 >10γ's	 95 % 17 % 23 % 8 % 5 %	
	$_{81}Tl^{201m}$			?0.005s	IT						
	$_{81}Tl^{201}$			73h	EC	0.41			0.0306 0.0321 0.135 0.1672	10 % 10 % 10 % 23 %	
	$_{81}Tl^{202}$			12d	EC	1.1			0.440 0.523 0.965	96 % 0.4 % 0.05 %	
	$_{81}Tl^{203}$	29.50	202.9723								11
	$_{81}Tl^{204}$			3.9y	β^- EC	0.77 0.38	β^- 0.77 EC	98 % 2 %			
	$_{81}Tl^{205}$	70.50	204.9745								0.11
	$_{81}Tl^{206}$			4.20m	β^-	1.51	β^- 1.51		no γ		
	$_{81}Tl^{207}$			4.78m	β^-	1.44	β^- 1.44		0.890	≈0.2 %	
	$_{81}Tl^{208}$			3.1m	β^-	4.99	β^- 1.80 1.52 1.28	50 % 21 % 25 %	2.615 0.511 0.860 0.583 0.277	100 % 25 % 12 % 87 % 10 %	
	$_{81}Tl^{209}$			2.2m	β^-	3.9	β^- 1.8	100 %	1.56 0.45 0.12	100 % 100 % 100 %	
	$_{81}Tl^{210}$			1.3m	β^-	5.43	β^- 1.96	100 %	0.297 0.783 1.1	100 % 100 %	

z	Isotope	% nat. abundance	Atomic mass (A)	Lifetime $t_{\frac{1}{2}}$	Modes of decay	Decay energies (Mev)	Particle energies (Mev)	Particle intensities	Gamma energies (Mev)	Gamma intensities	Thermal neutron capture cross-sections (Barns) σ_c
	$_{81}Tl^{210}$ (cont.)								1.3 2.36	 100%	
Pb			207.19								0.7
82	$_{82}Pb^{194}$			11m	EC				0.204		
	$_{82}Pb^{195}$			17m	EC				0.099D 0.383D 0.393D (see Tl^{195m})		
	$_{82}Pb^{196}$			37m	EC				0.191 0.241 0.253		
	$_{82}Pb^{197m}$			42m	IT EC	0.234 ≈4.5	IT EC	20% 80%	0.2340 IT 0.222 0.385 0.387	20% 80% 80% 80%	
	$_{82}Pb^{198}$			2.4h	EC				0.1169 0.1732 0.2904 0.2593 0.3655 0.3820 0.3978 0.575 0.865	23% 41% 25% 12% 36% 5% 4% 4% 7%	
	$_{82}Pb^{199m}$			12m	IT	0.424			0.424		
	$_{82}Pb^{199}$			1.5h	EC β^+	4.2	β^+ 2.8	17%	0.3528 0.2670 0.7210 1.132 Ann. Rad.	19% 90% 10%	
	$_{82}Pb^{200}$			21h	EC				0.033 0.110 0.142 0.148 0.235 0.257 0.268		
	$_{82}Pb^{201m}$			61s	IT	0.629			0.629		
	$_{82}Pb^{201}$			9.4h	EC β^+				Ann. Rad. 0.330 0.361 0.406 0.692 0.766 1.099		
	$_{82}Pb^{202m}$			3.6h	IT EC	0.13	IT EC	90% 10%	>10γ's 0.787 0.129 0.961 0.422 0.240–0.66		
	$_{82}Pb^{202}$			~3×10⁵y	EC	0.05					
	$_{82}Pb^{203m}$			6.1s	IT	0.825			0.825		

TABLE OF THE ISOTOPES (Continued)

z	Isotope	% nat. abundance	Atomic mass (A)	Lifetime $t_{\frac{1}{2}}$	Modes of decay	Decay energies (Mev)	Particle energies (Mev)	Particle intensities	Gamma energies (Mev)	Gamma intensities	Thermal neutron capture cross-sections (Barns) σ_c
	$_{82}Pb^{203}$			52h	EC	1.00			0.279 0.401 0.678	†10 † 4.6 † 0.85	
	$_{82}Pb^{204m}$			67m	IT	2.91			0.2894 0.3746 0.6220 0.6633 0.897 0.910	 100 %	
	$_{82}Pb^{204}$	1.48	203.973	1.4×10^{19}y	α		α 2.6				0.7
	$_{82}Pb^{205m}$			0.004s	IT	0.026			0.026 0.99 0.703 0.284	100 % 91 % 9 % 9 %	
	$_{82}Pb^{205}$			3×10^{7}y	EC	0.05					
	$_{82}Pb^{206}$	23.6	205.9745								0.03
	$_{82}Pb^{207m}$			0.8s	IT	1.63			1.062 0.569	100 % 100 %	
	$_{82}Pb^{207}$	22.6	206.9759								0.73
	$_{82}Pb^{208}$	52.3	207.9766								0.0005
	$_{82}Pb^{209}$			3.3h	β^-	0.635	β^- 0.635		no γ		
	$_{82}Pb^{210}$			21y	β^-	0.061	β^- 0.015 0.061	85 % 15 %	0.047	85 %	
	$_{82}Pb^{211}$			36.1m	β^-	1.39	β^- 1.39 0.56	80 %	0.065 0.083 0.404 0.829	 5 % 6 % 13 %	
	$_{82}Pb^{212}$			10.64h	β^-	0.58	β^- 0.34 0.58	84 % 12 %	0.1151 0.1764 0.2383 0.3000 0.4152	 80 % 4 %	
	$_{82}Pb^{214}$			26.8m	β^-	1.0	β^- 0.7 1.03	6 %	0.352 0.053–0.259 $>10\gamma$'s		
Bi			208.980								0.034
83	$_{83}Bi^{196}$			7m	EC α		α 5.83	≈0.05 %			
	$_{83}Bi^{<198}$			1.7m	α		α 6.2				
	$_{83}Bi^{199}$			≈25m	EC α		EC α 5.47	99+ % 0.01 %			
	$_{83}Bi^{200}$			35m	EC				0.4623 1.027		
	$_{83}Bi^{201m}$			1.8h	EC α				0.629D		
	$_{83}Bi^{201}$			1.0h	EC α		EC α 5.2	100 % 0.003 %			

TABLE OF THE ISOTOPES (Continued)

Z	Isotope	% nat. abundance	Atomic mass (A)	Lifetime $t_{\frac{1}{2}}$	Modes of decay	Decay energies (Mev)	Particle energies (Mev)	Particle intensities	Gamma energies (Mev)	Gamma intensities	Thermal neutron capture cross-sections (Barns) σ_c
	$_{83}Bi^{202}$			1.6h	EC				0.422 0.961		
	$_{83}Bi^{203}$			12.3h	EC β^+	3.19	EC β^+ 1.35 0.73 α 4.85	 0.001 %	Ann. Rad. 0.825 0.264 0.847 0.820 1.864 >10γ's		
	$_{83}Bi^{204}$			11.6h	EC	4			>10γ's 0.22 0.899 0.28–1.023		
	$_{83}Bi^{205}$			15.3d	EC β^+	2.65	β^+ 0.95		0.026 0.550 0.571 0.703 0.988 1.766 1.864 >10γ's	11 % 3 % 5 % 30 % 18 % 29 % 6 %	
	$_{83}Bi^{206m}$			7μs	IT	0.060	IT		0.060		
	$_{83}Bi^{206}$			6.3d	EC β^+	3.6	β^+	0.004 %	0.184 0.516 0.343 0.538 0.880 0.803 1.099 1.72 >10γ's	54 % 43 % 35 % 31 % 72 % 100 % 16 % 35 %	
	$_{83}Bi^{207}$			30y	EC	2.4			0.570 1.064 1.46 1.766	†100 † 95 † 9	
	$_{83}Bi^{208m}$			.0027s	IT	1.43			0.51 0.92	† 90 †100	
	$_{83}Bi^{208}$			≈7.5×10^{5}y	EC	2.88			2.615		
	$_{83}Bi^{209}$	100	208.9804								(0.019+0.015)
	$_{83}Bi^{210m}$			2.6×10^{6}y	α		α 4.93 4.89 4.59	60 % 34 % 5 %		no γ	
	$_{83}Bi^{210}$			5.0d	β^-	1.16	β^- 1.16 α 4.7	 ≈10^{-5} %	no γ		
	$_{83}Bi^{211}$		210.9873	2.15m	α β^-	 0.60	α 6.617 6.273 β^- 0.060	83 % 17 % 0.3 %	0.351	17 %	
	$_{83}Bi^{212}$			60.6m	β^- α	2.25	β^- 0.085– 2.25 α 6.049 6.088	65 % 25 % 10 %	0.040 0.727 0.124–2.2 >10γ's	25 %	
	$_{83}Bi^{213}$			47m	β^- α	1.39	β^- 1.39 0.96 α 5.9	98 % 2 %	0.44		

TABLE OF THE ISOTOPES (Continued)

z	Isotope	% nat. abundance	Atomic mass (A)	Lifetime $t_{\frac{1}{2}}$	Modes of decay	Decay energies (Mev)	Particle energies (Mev)	Particle intensities	Gamma energies (Mev)	Gamma intensities	Thermal neutron capture cross-sections (Barns) σ_c
	$_{83}Bi^{214}$			19.7m	β^- α	3.2	β^- 0.4–3.2 α 5.5	99+ % 0.04 %	0.609 1.12 1.76 0.45–2.43 >10γ's		
	$_{83}Bi^{215}$			8m	β^-	2.26					
Po			210								
84	$_{84}Po^{192}$			0.5s	α		α 6.58				
	$_{84}Po^{193}$			4s	α		α 6.47				
	$_{84}Po^{194}$			13s	α		α 6.38				
	$_{84}Po^{195}$			30s	α		α 6.24				
	$_{84}Po^{196}$			1.9m	α		α 6.14				
	$_{84}Po^{197}$			4m	α		α 6.04				
	$_{84}Po^{198}$			7m	α		α 5.93				
	$_{84}Po^{199}$			12m	α		α 5.85				
	$_{84}Po^{200}$			10m	α		α 5.77				
	$_{84}Po^{201}$			18m	α EC		α 5.68 5.55 5.77 EC				
	$_{84}Po^{202}$			0.8h	EC α		EC α 5.58	98 % 2 %			
	$_{84}Po^{203}$			45m	EC α		EC α 5.48	93 % 7 %			
	$_{84}Po^{204}$			3.5h	EC α		EC α 5.37	99 % 1 %			
	$_{84}Po^{205}$			1.8h	EC α		EC α 5.2	100 % 0.07 %			
	$_{84}Po^{206}$		205.9805	8.8d	EC α	2.63	EC α 5.22	95 % 5 %	0.060 0.286 0.338 0.511 0.522 1.033 >10γ's	100 % 33 % 23 % 24 % 17 % 33 %	
	$_{84}Po^{207m}$			45μs	IT				0.31 0.81		
	$_{84}Po^{207}$		206.9816	5.7h	EC β^+ α	 2.90	EC β^+ 1.14 0.893 α 5.10	99+ % 0.5 % 0.01 %	0.74 0.99 0.41 0.25 0.10–2.06 >10γ's		
	$_{84}Po^{208}$		207.9813	2.9y	α EC		α 5.11 EC	100 % 0.003 %	0.600 0.288	0.006 % 0.003 %	
	$_{84}Po^{209}$		208.9825	103y	α EC	 1.90	α 4.88 EC	99+ % 0.5 %	0.91 0.26	0.5 % 0.4 %	

z	Isotope	% nat. abundance	Atomic mass (A)	Lifetime $t_{\frac{1}{2}}$	Modes of decay	Decay energies (Mev)	Particle energies (Mev)	Particle intensities	Gamma energies (Mev)	Gamma intensities	Thermal neutron capture cross-sections (Barns) σ_c
	$_{84}Po^{210}$		209.9829	138.40d	α		5.30		0.79	1.2×10^{-3}%	
	$_{84}Po^{211m}$			25s	α		7.14 8.70 7.85	90% 7% 3%	0.89 0.57D 1.064D (see Pb^{207})	3% 90% 9%	
	$_{84}Po^{211}$		210.9866	0.52s	α		7.44 6.90 6.57	99% 0.5% 0.5%	0.88 0.562	0.5% 0.5%	
	$_{84}Po^{212}$		211.9889	0.30μs	α		8.78				
	$_{84}Po^{213}$		212.9928	4μs	α		8.37				
	$_{84}Po^{214}$		213.9952	164μs	α		7.68				
	$_{84}Po^{215}$		214.9995	.0018s	α β^-		α 7.36 β^-	99+% 0.005%			
	$_{84}Po^{216}$		216.0019	0.16s	α		6.78				
	$_{84}Po^{217}$			<10s	α		6.54				
	$_{84}Po^{218}$		218.0089	3.05m	α β^-		α 6.00	99+% 0.02%			
At											
85	$_{85}At^{198}$?			43s	α		6.50				
	$_{85}At^{201}$?			1.5m	α		6.35				
	$_{85}At^{202}$			3m	EC α		α 6.13 6.23				
	$_{85}At^{203}$			7.2m	EC α		α 6.09				
	$_{85}At^{204}$			9m	EC α		α 5.95				
	$_{85}At^{205}$			26m	EC α		α 5.90				
	$_{85}At^{206}$			29m	EC α		EC α 5.70	95% 5%			
	$_{85}At^{207}$			1.6h	EC α	3.9	EC α 5.75	≈90% ≈10%			
	$_{85}At^{208m}$			6.2h	EC		EC	100%			
	$_{85}At^{208}$			1.6h	EC α	4.9	EC α 5.65	99+% 0.5%	0.120 0.175 0.250 0.660	 25% 99%	
	$_{85}At^{209}$			5.5h	EC α	3.4	EC α 5.64	95% 5%	0.838 0.0910 0.19500 0.547 0.782	 † 24 † 66 †100	
	$_{85}At^{210}$			8.3h	EC α	3.92	EC α 5.36 5.52 5.44	99+% 0.06% 0.05% 0.05%	0.047 0.116 1.18 0.245	 99% 99%	

z	Isotope	% nat. abundance	Atomic mass (A)	Lifetime $t_{\frac{1}{2}}$	Modes of decay	Decay energies (Mev)	Particle energies (Mev)	Particle intensities	Gamma energies (Mev)	Gamma intensities	Thermal neutron capture cross-sections (Barns) σ_c
	$_{85}At^{210}$ (cont.)								1.44 1.48 1.60		
	$_{85}At^{211}$		210.9875	7.2h	EC α	0.79	EC α 5.862	59% 41%	0.67	0.2%	
	$_{85}At^{212}$			0.22s	α		α 7.3				
	$_{85}At^{213}$		212.9963	<2s	α		α 9.2				
	$_{85}At^{214}$		213.9963	<5s	α		α 8.78				
	$_{85}At^{215}$		214.9987	≈100μs	α	μ	α 8.00				
	$_{85}At^{216}$		216.0024	≈300μs	α		α 7.79				
	$_{85}At^{217}$		217.0046	0.018s	α		α 7.02				
	$_{85}At^{218}$		218.0086	1.35	α		α 6.70 6.65				
	$_{85}At^{219}$		219.0114	54s	α β^-		α 6.27				
Rn			222								
86	$_{86}Rn^{204}$			3m	α		6.28				
	$_{86}Rn^{206}$			6.5m	α EC		α 6.25 EC	65% 35%			
	$_{86}Rn^{207}$			11m	EC α		EC α 6.12	96% 4%			
	$_{86}Rn^{208}$			23m	EC α		EC α 6.12	80% 20%			
	$_{86}Rn^{209}$			30m	EC α		EC α 6.04	83% 17%			
	$_{86}Rn^{210}$		209.9897	2.7h	EC α		α 6.04 EC	96% 4%			
	$_{86}Rn^{211}$		210.9906	16h	EC α	0.79	EC α 5.78 5.85 5.61	74% 17% 9%	0.68 0.44 0.069 1.37 >10γ's	74% 29% 17% 38%	
	$_{86}Rn^{212}$		211.9907	23m	α		6.26				
	$_{86}Rn^{215}$		214.9989	<1m	α		8.6				
	$_{86}Rn^{216}$		216.0002	45μs	α		8.04				
	$_{86}Rn^{217}$		217.0039	500μs	α		α 7.74				
	$_{86}Rn^{218}$		218.0056	0.030s	α		α 7.13 6.52	99+% 0.2%	0.61	0.2%	
	$_{86}Rn^{219}$		219.0095	4.0s	α		α 6.81 6.54 6.41	82% 13% 5%	 0.270 0.399	 13% 4.8%	
	$_{86}Rn^{220}$		220.0114	51.5s	α		6.282 5.747	99+% 0.3%	0.542	≈0.3%	

TABLE OF THE ISOTOPES (Continued)

z	Isotope	% nat. abundance	Atomic mass (A)	Lifetime $t_{\frac{1}{2}}$	Modes of decay	Decay energies (Mev)	Particle energies (Mev)	Particle intensities	Gamma energies (Mev)	Gamma intensities	Thermal neutron capture cross-sections (Barns) σ_c
	$_{86}Rn^{221}$			25m	β^- α		β^- α 6.0	80 % 20 %			
	$_{86}Rn^{222}$		222.0175	3.823d	α		α 4.586 4.98 4.83	99+ % 0.08 %	0.510	≈0.08 %	
	$_{86}Rn^{223}$			≈12m	β^-						
	$_{86}Rn^{224}$			≈5h	β^-						
Fr											
87	$_{87}Fr^{212}$		211.996	19m	EC α	5.0	EC α 6.39 6.41 6.34	56 % 17 % 16 % 11 %			
	$_{87}Fr^{217}$		217.0048	<2s	α		8.3				
	$_{87}Fr^{218}$		218.0075	<5s	α		7.85				
	$_{87}Fr^{219}$		219.0092	0.02s	α		7.30				
	$_{87}Fr^{220}$		220.0123	28s	α		6.69				
	$_{87}Fr^{221}$		221.0142	4.8m	α		6.30 6.11	84 % 16 %	0.22	14 %	
	$_{87}Fr^{222}$			15m	β^- α		β^- α	99+ % 0.1 %			
	$_{87}Fr^{223}$		223.0198	22m	β^- α	1.15	β^- 1.15 α 5.34	99+ % 0.005 %	0.049 0.080 0.215 0.31	40 % 24 % 3 % 0.8 %	
	$_{87}Fr^{224}$			≈2m	β^-						
Ra			226								20
88	$_{88}Ra^{213}$			2.7	α		6.90				
	$_{88}Ra^{219}$		219.0100	<1m	α		8.0				
	$_{88}Ra^{220}$		220.0100	0.025s	α		7.45 6.95	99 % 1 %	0.46	1 %	
	$_{88}Ra^{221}$		221.0139	30s	α		6.60 6.75 6.66 6.75	38 % 31 % 20 % 8 %	0.089 0.150 0.176	 12 % 3 %	
	$_{88}Ra^{222}$		222.0154	38s	α		6.55 6.23	95 % 5 %	0.3248 0.480 0.520 0.800	3.6 % 7×10^{-3} % 2×10^{-3} % 0.02 %	
	$_{88}Ra^{223}$		223.0186	11.7d	α		5.71 5.60 5.23 5.33–5.87	50 % 24 % 10 %	0.155 0.122 0.270 0.338 0.031–0.45 >10γ's		
	$_{88}Ra^{224}$		224.0202	3.64d	α		5.68 5.44	95 % 5 %	0.24098 0.290	3.7 % ≈8×10^{-3} %	

z	Isotope	% nat. abundance	Atomic mass (A)	Lifetime $t_{\frac{1}{2}}$	Modes of decay	Decay energies (Mev)	Particle energies (Mev)	Particle intensities	Gamma energies (Mev)	Gamma intensities	Thermal neutron capture cross-sections (Barns) σ_c
	$^{88}Ra^{224}$ (cont.)								0.410 0.650	≈4×10⁻³ % ≈9×10⁻³ %	12
	$_{88}Ra^{225}$			14.8d	β^-	0.32	β^- 0.32	1/m	0.0400	33 %	
	$_{88}Ra^{226}$		226.0254	1622y	α		4.78 4.59	95 % 4 %	0.187 0.260 0.420 0.64	4 % 0.001 %	20
	$_{88}Ra^{227}$			41m	β^-	1.31	1.31		0.0275 0.295 0.499	4 % 0.6 %	
	$_{88}Ra^{228}$			6.7y	β^-	0.055	0.055		≈0.03		36
	$_{88}Ra^{229}$			<5m	β^-						
	$_{88}Ra^{230}$			1h	β^-		1.2				
Ac			227								800
89	$^{89}Ac^{221}$		221.0157	<2s	α		7.6				
	$^{89}Ac^{222}$		222.0178	5s	α		6.96				
	$_{89}Ac^{223}$		223.0191	2.2m	α EC		α 6.64				
	$_{89}Ac^{224}$		224.0217	2.9h	EC α		EC α 6.17	90 % 10 %	0.217 0.133	†224 †100	
	$_{89}Ac^{225}$		225.0231	10.0d	α		5.818 5.782 5.721 5.713 5.627 5.54–5.67	54 % 28 % 9.5 % 2.6 % 3.8 %	0.0366 0.0384 0.0628 0.0873 0.0994 0.150 0.187		
	$_{89}Ac^{226}$			29h	β^- EC	1.4 0.77	β^- 1.2 EC	80 % 20 %	0.0676 0.0721 0.1581 0.185 0.2503 0.253	 †100 †100 †151 †132	
	$_{89}Ac^{227}$		227.0278	21.6y	β^- α	0.043	β^- 0.043 α 4.949 4.936 4.517–4.866	99 % 1 %	0.009–0.190 9γ's		800
	$_{89}Ac^{228}$			6.13h	β^-	2.25	β^- 1.11 0.45–2.18	53 %	0.057 0.10 0.91 0.08–0.966 >10γ's		
	$_{89}Ac^{229}$			66m	β^-	1.0	≈1.0	100 %			
	$^{89}Ac^{230}$			<1m	β^-	2.2	2.2	100 %			
	$_{89}Ac^{231}$			15m	β^-	2.1	2.1		0.18		
Th			232.038								7.5
90	$^{90}Th^{223}$		223.0209	0.9s	α		7.55	100 %			

TABLE OF THE ISOTOPES (Continued)

z	Isotope	% nat. abundance	Atomic mass (A)	Lifetime $t_{\frac{1}{2}}$	Modes of decay	Decay energies (Mev)	Particle energies (Mev)	Particle intensities	Gamma energies (Mev)	Gamma intensities	Thermal neutron capture cross-sections (Barns) σ_c
	$_{90}Th^{224}$		224.0214	~1s	α		7.17 6.90		0.18 0.09		
	$_{90}Th^{225}$		225.0237	8m	EC α	0.48	α 6.47 6.30 6.79 EC	42% 10%	0.245 0.32 0.36	5% 30% 7%	
	$_{90}Th^{226}$		226.0249	31m	α		6.33 6.22 6.095 6.029	79% 19% 1.7% 0.6%	0.1116 0.131 0.197 0.242	4.8% 0.4% 0.4% 1.2%	
	$_{90}Th^{227}$		227.0278	18.17d	α		5.976 6.036 5.755 5.667–6.007	24% 23% 2.1%	0.050 0.061 0.236 0.029–0.334 >10γ's	14% 9% 11%	$\sigma_f = 1500$
	$_{90}Th^{228}$		228.0287	1.91y	α		5.421 5.338 5.137– 5.208	71% 28%	0.085 0.134 0.169 0.205 0.214	1.6% 0.16% 0.13% 0.03% 0.27%	120
	$_{90}Th^{229}$		229.0316	7340y	α		4.85 4.94 5.02	70% 20% 10%	0.148 0.200 0.087	† 5 † 15 †100	$\sigma_f = 30$
	$_{90}Th^{230}$		230.0331	80.000y	α		4.682 4.615 4.240– 4.474	76% 24%	0.0677 0.110 0.144 0.19 0.203 0.235 0.255	0.59% 1×10^{-4}% 0.77% 1.4×10^{-2}% $\approx 5 \times 10^{-6}$% $\approx 5 \times 10^{-6}$% 1.7×10^{-2}%	
	$_{90}Th^{231}$			25.6h	β^-	0.386	0.30	78%	0.084 0.256 0.017–0.31 >10γ's	11% 13%	
	$_{90}Th^{232}$		232.0382	1.39×10^{10}	α SF		α 4.007 3.99 SF	76% 24%	0.059	24%	7.5
	$_{90}Th^{233}$			22.1m	β^-	1.23	1.23		0.029 0.086 0.056–0.89 >10γ's	2% 3%	
	$_{90}Th^{234}$			24.10d	β^-	0.192	0.192 0.10	65% 35%	0.092 0.063 0.029		1.8
	$_{90}Th^{235}$			<5m	β^-						
Pa			231								200
91	$_{91}Pa^{225}$			2.0s	α						
	$_{91}Pa^{226}$		226.0278	1.8m	α		6.81				
	$_{91}Pa^{227}$		227.0289	38.3m	α EC	1.0	α 6.46 EC	85% 15%			

TABLE OF THE ISOTOPES (Continued)

z	Isotope	% nat. abundance	Atomic mass (A)	Lifetime $t_{\frac{1}{2}}$	Modes of decay	Decay energies (Mev)	Particle energies (Mev)	Particle intensities	Gamma energies (Mev)	Gamma intensities	Thermal neutron capture cross-sections (Barns) σ_c
	$_{91}Pa^{228}$		228.0310	22h	EC α	2.09	EC α 5.795 6.074 6.101 6.11 5.707– 6.087	98 %	0.97 0.47 0.058–1.88 >10γ's		
	$_{91}Pa^{229}$			1.5d	EC	0.3	EC α 5.69	99 % 1 %	0.042		
	$_{91}Pa^{230}$			17.7d	β α EC	0.41 1.2	β^- 0.41 α EC	15 % 0.003 % 85 %	0.0528 0.121 0.255 0.445 0.535 0.636 0.712 1.013 >10γ's		
	$_{91}Pa^{231}$		231.0359	3.43×10^4y	α		α 5.001 5.017 5.046 4.938 4.666– 4.971	24 % 23 % 10 % 22 %	0.29 0.027– 0.356 >10γ's		200 $\sigma_f=0.01$
	$_{91}Pa^{232}$			1.32d	β^-	1.25	β^- 0.28 0.4–1.25		0.0475 0.38 0.820 0.895 0.971 0.035– 0.868 >10γ's	80 % 8 % 8 % 21 % 41 %	890 $\sigma_f=700$
	$_{91}Pa^{233}$			27.4d	β^-	0.568	β^- 0.26 0.15 0.57	58 % 37 % 5 %	0.086 0.300 0.3119 0.340 0.415 0.017– 0.415 >10γ's	17 % 13 % 70 % 7 % 1.7 %	(20+19)
	$_{91}Pa^{234m}$			1.18m	β^- IT	2.31	β^- 0.58 2.31 IT	99 % 1 %	0.043 0.23–1.83 >10γ's all very weak		
	$_{91}Pa^{234}$		234.043	6.66h	β^-	2.3	0.23–1.35		0.044 0.100 0.228 0.126– 1.85 >10γ's	92 % 71 % 33 %	
	$_{91}Pa^{235}$			24m	β^-	1.4	1.4	100 %			
	$_{91}Pa^{237}$			39m	β^-	2.30	β^- 2.30 1.35	60 % 30 %	0.090 0.91 0.87 0.46 0.145–1.42		
U			238.03								$\sigma=3.4$ $\sigma_f=4.2$

z	Isotope	% nat. abundance	Atomic mass (A)	Lifetime $t_{\frac{1}{2}}$	Modes of decay	Decay energies (Mev)	Particle energies (Mev)	Particle intensities	Gamma energies (Mev)	Gamma intensities	Thermal neutron capture cross-sections (Barns) σ_c
92	$_{92}U^{227}$		227.0309	1.3m	α EC		α 6.8				
	$_{92}U^{228}$		228.0313	9.3m	α EC		α 6.68 6.59 EC	80 % 20 %			
	$_{92}U^{229}$		229.0332	58m	EC α	1.16	EC α 6.36 6.33 6.30	80 % 13 % 5 % 2 %			
	$_{92}U^{230}$		230.0339	20.8d	α		α 5.88 5.81 5.66	67 % 32 % 0.7 %	0.07213 0.1543 0.158 0.232	0.75 % 0.16 % 0.17 % 0.24 %	$\sigma_f \approx 25$
	$_{92}U^{231}$			4.2d	EC α	0.37	EC α 5.45	99+ % 0.005 %	0.0181 0.07564 0.05854 0.0685 0.0813 0.0821 0.08418 0.1082		$\sigma_f \approx 400$
	$_{92}U^{232}$		232.0372	73.6y	α SF		α 5.318 5.261 5.134	68 % 32 % 0.3 %	0.057 0.1305 0.27 0.33	0.21 % 0.075 % 4×10^{-3} % 4×10^{-3} %	≈300 $\sigma_f = 80$
	$_{92}U^{233}$		233.0395	1.62×10^5y	α		α 4.816 4.773 4.717	83 % 15 % 1.6 %	0.0424 0.054 0.097 0.029	16 % 1 % 0.6 % 0.7 %	50 $\sigma_f = 530$
	$_{92}U^{234}$	0.0057	234.0409	2.48×10^5y	α SF		4.768 4.717	72 % 28 %	0.053 0.118		100
	$_{92}U^{235}$	0.72	235.0439	7.13×10^8y	α SF		4.559 4.370 4.354 4.333 4.318 4.117	6.7 % 25 % 35 % 14 % 8 % 5.8 %	0.074 0.094 0.1096 0.144 0.165 0.185 0.203 0.2890 0.367 0.385	 9 % 5 % 12 % >4 % 55 % >4 %	100 $\sigma_f = 580$
	$_{92}U^{235m}$			26.2m	IT	<0.001			0.000023 or 0.000075		
	$_{92}U^{236}$		236.0457	2.39×10^7y	α SF		α 4.499 4.45	73 % 27 %	≈0.050	27 %	6
	$_{92}U^{237}$			6.75d	β^-	0.51	β^- 0.25	95 %	0.060 0.208 0.026–0.37 >10γ's	81 % 89 %	
	$_{92}U^{238}$	99.27	238.0508	4.51×10^9y	α SF		α 4.195 4.14	77 % 23 %	0.048	23 %	2.7
	$_{92}U^{239}$			23.5m	β^-	1.28	1.21	100 %	0.074		22 $\sigma_f = 14$

z	Isotope	% nat. abundance	Atomic mass (A)	Lifetime $t_{\frac{1}{2}}$	Modes of decay	Decay energies (Mev)	Particle energies (Mev)	Particle intensities	Gamma energies (Mev)	Gamma intensities	Thermal neutron capture cross-sections (Barns) σ_c
	$_{92}U^{240}$			14.1h	β^-	0.49	β^- 0.36 0.32	75% 25%	0.044	25%	
Np			237								$\sigma = 170$ $\sigma_f = 0.019$
93	$_{93}Np^{231}$		231.0383	~50m	EC α	1.9	α 6.28	100%			
	$_{93}Np^{232}$			≈13m	EC	2.7					
	$_{93}Np^{233}$		233.0406	35m	EC α	1.03	EC α 5.53	99+% 0.001%			
	$_{93}Np^{234}$			4.4d	EC β^+	1.80	EC β^+ 0.8	99+% 0.05%	0.043–1.61 >10γ's		$\sigma_f \approx 900$
	$_{93}Np^{235}$			410d	α EC	0.13	EC α4.86–5.10	99+%	0.086 0.026		
	$_{93}Np^{236m}$			22h	β^- EC	0.52	β^- 0.52 EC	57% 43%	0.045 0.150 0.0443		
	$_{93}Np^{236}$		236.0466	>5000y	β^-						$\sigma_f = 2800$
	$_{93}Np^{237}$		237.0480	2.20×10^6y	α		4.787 4.767 4.52–4.87	53% 29%	0.020 0.087 0.0296 0.0568 0.143 0.175 0.200	 14% 14% 0.8% 0.1% 0.3%	$\sigma = 170$ $\sigma_f = 0.019$
	$_{93}Np^{238}$			2.10d	β^-	1.30	β^- 1.24 0.25	40% 30%	0.044 0.986 1.03 0.102 0.04–1.09 >10γ's	 20% 17%	$\sigma_f = 1600$
	$_{93}Np^{239}$			2.35d	β^-	0.72	0.07–0.72		0.045–0.33 >10γ's		$\sigma = (.35 + 25)$
	$_{93}Np^{240m}$			7.3m	β^-	2.18	0.65–2.18		0.043 0.55 0.60 0.04–1.62		
	$_{93}Np^{240}$			60m	β^-	2.05	β^- 0.89		0.085 0.155 0.20 0.25 0.44 0.575 0.60 0.92 1.00 1.16		
	$_{93}Np^{241}$			16m	β^-	1.36	1.36				
Pu			239								$\sigma = 270$ $\sigma f = 740$
94	$_{94}Pu^{232}$		232.0411	36m	EC α	1.0	EC α 6.58	98% 2%			

TABLE OF THE ISOTOPES (Continued)

z	Isotope	% nat. abundance	Atomic mass (A)	Lifetime $t_{\frac{1}{2}}$	Modes of decay	Decay energies (Mev)	Particle energies (Mev)	Particle intensities	Gamma energies (Mev)	Gamma intensities	Thermal neutron capture cross-sections (Barns) σ_c
	$_{94}Pu^{233}$			20m	EC α	1.95	EC α 6.3	99+ % 0.1 %			
	$_{94}Pu^{234}$		234.0433	9h	EC α	0.43	EC α 6.19	94 % 6 %	0.047		
	$_{94}Pu^{235}$			26m	EC α	1.18	EC α 5.85	99+ % 0.003 %			
	$_{94}Pu^{236}$		236.0461	2.85y	α SF		α 5.763 5.716 5.610	69 % 31 % 0.18 %	0.046 0.110 0.165	0.047 % 0.012 % 6.6×10^{-4} %	$\sigma_f = 270$
	$_{94}Pu^{237m}$			0.18s	IT SF	0.145			0.145	100 %	
	$_{94}Pu^{237}$		237.0483	45.6d	α EC SF	0.23	α 5.65 5.36 EC	99+ %	0.0264 0.0332 0.0435 0.0555 0.0596 0.064 0.0764		$\sigma_f = 2500$
	$_{94}Pu^{238}$		238.0495	89y	α SF		5.491 5.448	72 % 28 %	0.0436 0.0996 0.152 0.203 0.760 0.810 0.875	0.038 % 8×10^{-3} % 1×10^{-3} % 4×10^{-6} % 25×10^{-5} % $\approx 2 \times 10^{-5}$ %	$\sigma = 400$ $\sigma_f = 17$
	$_{94}Pu^{239}$		239.0522	24,360y	α SF		5.147 5.134 5.096	72 % 17 % 11 %	0.003 0.0125 0.038 0.052 0.121 0.207 0.340 0.3800 0.420	†320 † 70 † 20 † 30 † 60 † 40	270 $\sigma_f = 740$
	$_{94}Pu^{240}$		240.0540	6.58×10^3y	α SF		α 5.159 5.114 5.01	76 % 24 % 0.1 %	0.042 0.047	24 %	290
	$_{94}Pu^{241}$			13y	β^- α	0.021	β^- 0.02 α 4.893 4.848	99+ %	0.100 0.145	†100 † 20	$\sigma = 400$ $\sigma_f = 1000$
	$_{94}Pu^{242}$		242.0587	3.79×10^5y	α SF		α 4.898 4.854	76 % 24 %	0.045	24 %	$\sigma = 23$
	$_{94}Pu^{243}$			4.98h	β^-	0.57	β^- 0.57 0.49	62 % 38 %	0.0122 0.0297 0.0367 0.0422 0.052 0.082 0.0955 0.120 0.134 0.384	1 % 21 % 0.7 %	170
	$_{94}Pu^{244}$			7.6×10^7y	α SF		α				
	$_{94}Pu^{245}$			10.1h	β^-						260

TABLE OF THE ISOTOPES (Continued)

z	Isotope	% nat. abundance	Atomic mass (A)	Lifetime $t_{\frac{1}{2}}$	Modes of decay	Decay energies (Mev)	Particle energies (Mev)	Particle intensities	Gamma energies (Mev)	Gamma intensities	Thermal neutron capture cross-sections (Barns) σ_c
	${}_{94}Pu^{246}$			10.85d	β^-	0.38	β^- 0.15 0.33	73 % 27 %	0.047 0.027 0.08 0.175 0.215		
Am											
95	${}_{95}Am^{237}$			1.3h	EC	1.4	EC α 6.01	99+ % 0.005 %			
	${}_{95}Am^{238}$			1.86h	EC				0.98 0.58 0.37 0.95 1.35	 † 20 † 75 	
	${}_{95}Am^{239}$			12h	EC α	0.76	EC α 5.77	99+ % 0.004 %	0.04470 0.049 0.05731 0.06791 0.1818 0.2099 0.2265 0.2283 0.2776		
	${}_{95}Am^{240}$			51h	EC				0.04287 0.0989 0.92 1.02 1.40	 †14 †70 †15	
	${}_{95}Am^{241}$		241.0567	458y			α 5.482 5.439 5.386	85 % 13 % 1.6 %	0.060 0.027–0.37		$\sigma = (50+750)$ $\sigma_f = 32$
	${}_{95}Am^{242m}$			~152y	IT	0.048			0.048		
	${}_{95}Am^{242}$			16h	β^- EC	0.63	β^- 0.63		0.04220 0.04453		$\sigma = 1600$ $\sigma_f = 6400$
	${}_{95}Am^{243}$		243.0614	7.95×10^3y	α		α^- 5.267 5.224 5.17–5.34	87 % 12 %	0.075		74
	${}_{95}Am^{244m}$			11h	β^-		β^- 0.38				
	${}_{95}Am^{244}$			25m	β^- EC	1.5	β^- 1.5	99+ %			
	${}_{95}Am^{245}$			207h	β^-		β^-0.91		0.036 0.06 0.078 0.140 0.153 0.230 0.25		
	${}_{95}Am^{246}$			25m	β^-	2.29	β^- 1.31 1.60 2.10	79 % 14 % 7 %	0.035 0.106 0.245 0.78–1.06		
Cm											
96	${}_{96}Cm^{238}$		238.0530	2.5h	EC α		EC α 6.52	<90 % >10 %			

TABLE OF THE ISOTOPES (Continued)

z	Isotope	% nat. abundance	Atomic mass (A)	Lifetime $t_{\frac{1}{2}}$	Modes of decay	Decay energies (Mev)	Particle energies (Mev)	Particle intensities	Gamma energies (Mev)	Gamma intensities	Thermal neutron capture cross-sections (Barns) σ_c
	$_{96}Cm^{239}$			≈3h	EC		EC	100 %	0.188		
	$_{96}Cm^{240}$		240.0555	26.8d	α SF		α 6.26	70 %			
	$_{96}Cm^{241}$			35d	EC α	0.76	EC α 5.95	99 % 1 %	0.48 ≈0.60 0.145	97 % 1 %	
	$_{96}Cm^{242}$		242.0588	163d	α SF		α 6.110 6.066 5.967	74 % 26 % 10^{-3} %	0.04409	0.04 %	~20
	$_{96}Cm^{243}$		243.0614	35y	α EC		6.05 5.987 α 5.780 5.736 5.63–6.06 EC	5 % 6 % 73 % 12 % 0.3 %	0.106 0.2110 0.2280 0.278	 † 50 † 65 †100	σ = 250 σ_f = 700
	$_{96}Cm^{244}$		244.0629	17.6y	α		α 5.801 5.759 5.661 SF	76.7 % 23.3 % 0.017 %	0.0430 0.100 0.150	2.1×10^{-2} % 1.5×10^{-3} % 1.3×10^{-3} %	≈15
	$_{96}Cm^{245}$		245.0653	8×10^3y	α		α 5.45 5.36 5.30	15 % 77 % 8 %	0.173 0.130	14 % 5 %	σ = 200 σ_f = 1900
	$_{96}Cm^{246}$		246.0674	5480y	α		α 5.373 SF	100 %			~15
	$_{96}Cm^{247}$			≧4×10^7y	α						180
	$_{96}Cm^{248}$			4.7×10^5y	α		α 5.054 SF	89 %			≈6
	$_{96}Cm^{249}$			65m	β^-	0.86	β^- 0.86				
	$_{96}Cm^{250}$			2×10^4y			SF	100 %			
Bk											
97	$_{97}Bk^{243}$			4.5h	EC α	1.43	EC α 6.72 6.55 6.20	99+ % †30 †53 †17	0.042 0.146 0.187 0.54 0.74 0.84 0.96	 † 10 † 30 † 30	
	$_{97}Bk^{244}$			4.4h	EC α		EC α 6.67	99+ % 0.006 %	0.200 0.90 1.06 1.16 1.23 1.37 1.50 1.72	 †100 † 7 † 5 † 5 † 0.7 † 2 † 0.2	
	$_{97}Bk^{245}$			4.95d	EC α	0.84	EC α 6.37 6.17 5.89	99+ % †33 †41 †26	0.164 0.206 0.252 0.380 0.480	 31 % 5.1 %	
	$_{97}Bk^{246}$			1.8d	EC				0.82 1.09	≈40 %	

TABLE OF THE ISOTOPES (Continued)

z	Isotope	% nat. abundance	Atomic mass (A)	Lifetime $t_{\frac{1}{2}}$	Modes of decay	Decay energies (Mev)	Particle energies (Mev)	Particle intensities	Gamma energies (Mev)	Gamma intensities	Thermal neutron capture cross-sections (Barns) σ_c
	$_{97}Bk^{247}$		247.0702	~10^4y	α		α 5.67 5.51 5.30	37 % 58 % 5 %	0.084 0.265	40 % 30 %	
	$_{97}Bk^{248}$			23h	β^- EC	0.65	β^- 0.65 EC	70 % 30 %			
	$_{97}Bk^{249}$			314d	β^- α SF	0.135	β^- 0.125 α 5.417 5.03	99+ % † 4 †96	0.32	<1 %	
	$_{97}Bk^{250}$			3.2h	β^-	1.8	β^- 0.72 1.72 1.76	89 % 5 % 5 %	0.098 0.990 1.032	3 % 47 % 39 %	
Cf											
98	$_{98}Cf^{244}$		244.0659	25m	α		α 7.17				
	$_{98}Cf^{245}$			44m	EC α	1.54	EC α 7.11	70 % 30 %			
	$_{98}Cf^{246}$		246.0688	35.7h	α		α 6.753 6.711 SF	78 % 22 %	0.042 0.096 0.146	22 %	
	$_{98}Cf^{247}$			2.4h	EC				0.295 0.42 0.46	†20 †13 † 9	
	$_{98}Cf^{248}$		248.0724	350d	α SF		α 6.23	82 %	0.045		
	$_{98}Cf^{249}$		249.0748	360y	α SF		α 6.194 5.941 5.842 5.806 5.749 >10α's	2 % 3 % 3 % 84 % 4 %	0.255 0.340 0.394	3 % 15 % 72 %	$\sigma = 270$ $\sigma_f \approx 600$
	$_{98}Cf^{250}$		250.0766	10y	α SF		α 6.024 5.980	83 % 17 %	0.043	17 %	1500
	$_{98}Cf^{251}$			≈800y	α		5.841 5.675		0.180		≈3000
	$_{98}Cf^{252}$			2.55y	α SF		α 6.112 6.069 SF	87 % 10 % 3 %	0.043 0.100	1.4×10^{-2} % 1×10^{-2} %	30
	$_{98}Cf^{253}$			19d	β^-	0.27	0.27				
	$_{98}Cf^{254}$			56d	SF		SF	100 %			<2
Es											
99	$_{99}Es^{246}$			7.3m	α		α 7.35				
	$_{99}Es^{248}$			25m	EC α		EC α 6.87	99+ % 0.25 %			
	$_{99}Es^{249}$			2h	EC α	1.4	EC α 6.76	99+ % 0.13 %			
	$_{99}Es^{250}$			8h	EC		EC	100 %			
	$_{99}Es^{252}$		252.0829	~140d	α		α 6.64	100 %			

z	Isotope	% nat. abundance	Atomic mass (A)	Lifetime $t_{\frac{1}{2}}$	Modes of decay	Decay energies (Mev)	Particle energies (Mev)	Particle intensities	Gamma energies (Mev)	Gamma intensities	Thermal neutron capture cross-sections (Barns) σ_c
	$_{99}Es^{253}$		253.0847	20.0d	α SF		α 6.633 6.592 6.17–6.54	90 % 6.6 %	0.0419 0.051 0.389 0.427		300
	$_{99}Es^{254m}$			38h	β EC SF		β^- 1.04 EC	99+ % 0.1 %	0.660	40 %	
	$_{99}Es^{254}$		254.0881	480d	α		α 6.42		0.062		σ<40 σ_f=2.700
	$_{99}Es^{255}$			24d	β^-						≈40
	$_{99}Es^{256}$			short	β^-						
Fm											
100	$_{100}Fm^{248}$		248.0772	0.6m	α						
	$_{100}Fm^{249}$			2.5m	α		α 7.9				
	$_{100}Fm^{250}$		250.0795	30m	α		α 7.43	100 %			
	$_{100}Fm^{251}$			7h	EC α		EC α 6.89	99 % 1 %			
	$_{100}Fm^{252}$		252.0827	23h	α		α 7.05	100 %			
	$_{100}Fm^{253}$			~4.5d	EC α		EC α 6.94	89 % 11 %			
	$_{100}Fm^{254}$		254.0870	3.24h	SF	α	α 7.20 7.16 7.06 SF	82 % 17 % 1 % 5×10⁻² %	0.041 0.098	0.02 % 0.028 %	
	$_{100}Fm^{255}$			22h	α		α 7.03		≈0.055 ≈0.082	1 % 1 %	<100
	$_{100}Fm^{256}$			2.7h	SF		SF	100 %			
Md											
101	$_{101}Md^{255}$		255.0906	0.5h	EC α		α 7.34				
	$_{101}Md^{256}$			1.5h	EC		EC	100 %			
No											
102	$_{102}No^{253}$			~10m	α		8.5				
	$_{102}No^{254}$			~3s	α		8.8				
	$_{102}No^{255}$			~15s	α		8.2				
Lw											
103	$_{103}Lw^{257}$			8s	α		8.6				

THERMAL NEUTRON CROSS SECTIONS

The table of Thermal Neutron Cross Sections is taken from data published by the Neutron Cross Section Advisory Group of the Atomic Energy Commission, July 1, 1958, by Hughes and Schwartz.

All values given for an element refer to the natural mixture of isotopes, that is, they are atomic cross sections, while those given for specific isotopes are isotopic cross sections. All cross sections, unless marked mb (millibarns) are in barns (10^{-24} cm^2).

The reaction cross sections listed are those for a neutron velocity of 2200 meters per second. The reaction cross sections refer to all cases in which the neutron is not re-emitted, that is, to (n,γ), (n,p), and (n,α) reactions. Practically all the reaction cross sections are for (n,γ) reactions and the few that are not are so marked. The absorption cross sections, σ_{abs}, are the particular reaction cross sections that are measured by observing the reaction itself in which the neutron is absorbed. The activation cross sections, σ_{act}, are those determined from the radioactivity of the product nucleus, usually the result of an (n,γ) reaction, and in a few cases, which are specially marked, by (n,p) or (n,α) reactions. The activation cross sections always refer to particular isotopes and hence are isotopic cross sections; for mono-isotopic elements they are atomic cross sections as well. For a few cases in which the 2200 m/s value could not be determined, the cross sections are still included but are marked with an asterisk* indicating "pile neutrons."

The scattering cross sections are usually constant with energy in the thermal region, except for crystal effects, and are hence not quoted for 2200 m/s. The average scattering cross section, $\bar{\sigma}_s$, is that averaged over the Maxwell distribution and will depend on the crystalline form of the sample and even upon the size of the crystal grain, but it is listed here because of utility in certain practical applications.

Element	Isotope	Reaction Cross Sections (2200 m/s) σ_{abs}	σ_{act}		Scattering Cross Sections $\bar{\sigma}_s$
H		332 ± 2 mb			38 ± 4 (gas)
	H^{1}				
	H^{2}	0.46 ± 0.10 mb	12.4y	0.57 ± 0.01 mb	7 ± 1
He					0.8 ± 0.2
	He^{3}	5500 ± 300	np	5400 ± 200	1.0 ± 0.7
	He^{4}	0		0	
Li		71.0 ± 1.0			1.4 ± 0.3
	Li^{6}	(945)	nγ	28 ± 8 mb	
	Li^{7}		0.89s	36 ± 4 mb	
Be	Be^{7}		np	54,000 ± 8000	
			nα	<1	
	Be^{9}	10 ± 1 mb	2.7 × 10^{6}y	9 ± 3 mb	7 ± 1
B	B^{10}	4017 ± 32	nγ	0.5 ± 0.2	
			np	<0.2	
	B^{11}		0.03s	5 ± 3 mb	
C		3.73 ± 0.07 mb			4.8 ± 0.2
	C^{12}		nγ	3.3 ± 0.2 mb	
	C^{13}	0.5 ± 0.2 mb	5570y	0.9 ± 0.3 mb	
	C^{14}	<200	2.4s	<1 μb	
N		1.88 ± 0.05			10 ± 1
	N^{14}		np	1.75 ±.0.05	
			nγ	0.08 ± 0.02	
	N^{15}		7.4s	24 ± 8 μb	
O		<0.2 mb			4.2 ± 0.3
	O^{16}				
	O^{17}		5570y C^{14}	0.4 ± 0.1	
	O^{18}		29s	0.21 ± 0.04 mb	
F	F^{19}	<10 mb	11s	9 ± 2 mb	3.9 ± 0.2
Ne		<2.8			2.4 ± 0.3
	Ne^{20}				
	Ne^{21}				
	Ne^{22}		40s	36 ± 15 mb	
Na	Na^{23}	525 ± 10 mb	15.0h	536 ± 10 mb	4.0 ± 0.5
Mg		69 ± 2 mb			3.6 ± 0.4
	Mg^{24}	34 ± 10 mb			
	Mg^{25}	280 ± 90 mb			
	Mg^{26}	60 ± 60 mb	9.5m	27 ± 5 mb	
	Mg^{27}		21h	<30 mb*	
Al	Al^{27}	241 ± 3 mb	2.30m	0.21 ± 0.02	1.4 ± 0.1
Si		0.16 ± 0.02			1.7 ± 0.3
	Si^{28}	80 ± 30 mb			
	Si^{29}	0.28 ± 0.09			
	Si^{30}	0.4 ± 0.4	2.62h	110 ± 10 mb	
P	P^{31}	0.20 ± 0.02	14.3d	0.19 ± 0.01	5 ± 1
S		0.52 ± 0.02			1.1 ± 0.2
	S^{32}		nα	1.8 ± 1.0 mb	
	S^{33}		25.1d P^{33}	15 ± 10 mb	
			nα	<8 mb	
	S^{34}		87d	0.26 ± 0.05	
	S^{36}		5.0m	0.14 ± 0.04	

* Pile neutrons.

Element	Isotope	Reaction Cross Sections (2200 m/s) σ_{abs}	σ_{act}		Scattering Cross Sections $\bar{\sigma}_s$
$_{17}Cl$		33.8 ± 1.1			16 ± 3
	Cl^{35}		3.08 × 10^{5}y	30 ± 20	
			87d S^{35}	0.19 ± 0.05	
			14.3d P^{32}	<0.05 mb	
	Cl^{36}			90 ± 30	
	Cl^{37}		1.0s	5 ± 3* mb	
			37.5m	0.56 ± 0.12	
$_{18}A$		0.66 ± 0.04			1.5 ± 0.5
	A^{36}		35d	6 ± 2	
	A^{38}		265y	0.8 ± 0.2	
	A^{40}		109m	0.53 ± 0.02	
	A^{41}		>3.5y	>60 mb	
$_{19}K$		2.07 ± 0.07			1.5 ± 0.3
	K^{39}	1.94 ± 0.15	1.3 × 10^{9}y	3 ± 2*	
	K^{40}	70 ± 20	np	3.8 ± 0.7	
	K^{41}	1.24 ± 0.10	12.46h	1.30 ± 0.15	
$_{20}Ca$		0.44 ± 0.02			
	Ca^{40}	0.22 ± 0.04			
	Ca^{42}	42 ± 3			
	Ca^{43}				
	Ca^{44}		164d	0.72 ± 0.10	
	Ca^{46}		4.8d	0.25 ± 0.10	
	Ca^{48}		8.5m	1.1 ± 0.1	
$_{21}Sc$	Sc^{45}	24.0 ± 1.0	20s	10 ± 4	24 ± 2
			85d	22.3 ± 2.[illegible]	
			20s + 85d	22 ± 2	
$_{22}Ti$		5.8 ± 0.4			4 ± 1
	Ti^{46}	0.6 ± 0.2			2 ± 2
	Ti^{47}	1.7 ± 0.3			4 ± 1
	Ti^{48}	8.3 ± 0.6			4 ± 2
	Ti^{49}	1.9 ± 0.5			1 ± 1
	Ti^{50}	<0.2	5.8m	0.14 ± 0.03	3 ± 1
$_{23}V$		5.00 ± 0.01			5 ± 1
	V^{50}		nγ	250 ± 200	
	V^{51}		3.76m	4.5 ± 0.9	
$_{24}Cr$		3.1 ± 0.2			3.0 ± 0.5
	Cr^{50}	17.0 ± 1.4	27.7d	15.9 ± 1.6	
	Cr^{52}	0.76 ± 0.06			
	Cr^{53}	18.2 ± 1.5			
	Cr^{54}	<0.3	3.6m	0.38 ± 0.04	
$_{25}Mn$	Mn^{55}	13.2 ± 0.2	2.58h	13.3 ± 0.2	2.3 ± 0.3
$_{26}Fe$		2.62 ± 0.06	nα	<5 mb	11 ± 1
	Fe^{54}	2.3 ± 0.2	2.94y	2.8 ± 0.4	
	Fe^{56}	2.7 ± 0.2			
	Fe^{57}	2.5 ± 0.2			
	Fe^{58}	2.5 ± 2.0	44.3d	1.01 ± 0.10	
			3.5m Cr^{55}	<1.5 mb	
$_{27}Co$	Co^{59}	38.0 ± 0.7	10.4m	16 ± 3	7 ± 1
			5.28y	20 ± 3	
			10.4m + 5.28y	36.3 ± 1.5	
			(99.7% of 10.4m → 5.28y)		
	Co^{60m}		1.75h	100 ± 50	
	Co^{60}		1.75h	6 ± 2	

* Pile neutrons.

THERMAL NEUTRON CROSS SECTIONS (Continued)

Element	Isotope	Reaction Cross Sections (2200 m/s) σ_{abs}	σ_{act}		Scattering Cross Sections $\bar{\sigma}_s$
$_{28}$Ni		4.6 ± 0.1			17.5 ± 1.0
	Ni^{58}	4.4 ± 0.3			
	Ni^{60}	2.6 ± 0.2			
	Ni^{61}	2.0 ± 1.0			
	Ni^{62}	15 ± 2			
	Ni^{64}		2.56h	1.52 ± 0.14	
	Ni^{65}		56h	20 ± 2	
$_{29}$Cu		3.85 ± 0.03			7.2 ± 0.7
	Cu^{63}	4.5 ± 0.1	12.87h	4.51 ± 0.23	
	Cu^{65}	2.2 ± 0.2	5.15m	1.8 ± 0.4	
	Cu^{66}		59h	130 ± 30*	
$_{30}$Zn		1.10 ± 0.02			3.6 ± 0.4
	Zn^{64}		246.4d	0.47 ± 0.05	
			12.8h Cu^{64}	<10 μb	
			nα	15 ± 10 μb	
	Zn^{66}		80y Ni^{63}	<20 μb	
	Zn^{67}		nα	6 ± 4 μb	
	Zn^{68}		13.8h	99 ± 10 mb	
			52m	1.0 ± 0.2	
			nα	<20 μb	
	Zn^{70}		2.2m	85 ± 20 mb	
$_{31}$Ga		2.80 ± 0.15			4 ± 1
	Ga^{69}	2.1 ± 0.2	20.2m	1.4 ± 0.3	
	Ga^{71}	5.1 ± 0.4	14.2h	5.0 ± 0.5	
$_{32}$Ge		2.45 ± 0.20			3 ± 1
	Ge^{70}	3.4 ± 0.3	12d	3.42 ± 0.35	
	Ge^{72}	0.98 ± 0.09			
	Ge^{73}	14 ± 1			
	Ge^{74}	0.62 ± 0.06	48s	0.040 ± 0.008	
			82m	0.21 ± 0.08	
	Ge^{76}	0.36 ± 0.07	57s	80 ± 20 mb	
			12h	80 ± 20 mb	
			(~50% of 57s → 12h)		
$_{33}$As	As^{75}	4.3 ± 0.2	27h	5.4 ± 1.0	6 ± 1
$_{34}$Se		12.3 ± 0.4			11 ± 2
	Se^{74}	50 ± 7	123d	26 ± 6	
	Se^{76}	85 ± 7	18s	7 ± 3	
	Se^{77}	42 ± 4			
	Se^{78}	0.4 ± 0.4			
	Se^{80}	0.61 ± 0.06	57m	30 ± 10 mb	
			18m	0.5 ± 0.1	
	Se^{82}	2.1 ± 1.5	67s	50 ± 25 mb	
			25m	4 ± 2 mb	
			(None of 67s → 25m)		
$_{35}$Br		6.7 ± 0.3			6 ± 1
	Br^{79}		4.6h	2.9 ± 0.5	
			18m	8.5 ± 1.4	
			4.6h + 18m	10.4 ± 1.0	
	Br^{81}		35.9h	3.3 ± 0.4	
$_{36}$Kr		31 ± 2			7.2 ± 0.7
	Kr^{78}		34.5h	2.0 ± 0.5	
	Kr^{80}		13s + 2 × 10^{5}y	95 ± 15	
	Kr^{82}		nγ	45 ± 15	
	Kr^{83}		nγ	205 ± 30	
	Kr^{84}		4.4h	0.10 ± 0.03	
			9.4y	60 ± 20 mb	
			(23% of 4.4h → 9.4y)		
	Kr^{85}		nγ	<15	
	Kr^{86}		77m	60 ± 20 mb	
	Kr^{87}		2.8h	<600	
$_{37}$Rb		0.73 ± 0.07			
	Rb^{85}		18.7d	0.91 ± 0.08	
	Rb^{87}		17.8m	0.12 ± 0.03	
	Rb^{88}		15.4m	1.0 ± 0.2	
$_{38}$Sr		1.21 ± 0.06			10 ± 1
	Sr^{84}	<3	70m	<1	
			65d	1.4 ± 0.3	
			(80% of 70m → 65d)		
	Sr^{86}		2.8h	1.65 ± 0.16	
	Sr^{87}				
	Sr^{88}		53d	5 ± 1 mb	
	Sr^{89}		28y	0.5 ± 0.1*	
	Sr^{90}		64h	1.0 ± 0.6	
$_{39}$Y	Y^{89}	1.31 ± 0.08	63h	1.26 ± 0.08	
	Y^{90}		61d	<7	
	Y^{91}		3.5h	1.07 ± 0.09	

* Pile neutrons.

Element	Isotope	Reaction Cross Sections (2200 m/s) σ_{abs}	σ_{act}		Scattering Cross Sections $\bar{\sigma}_s$
$_{40}$Zr		185 ± 4 mb			8 ± 1
	Zr^{90}	0.10 ± 0.07			
	Zr^{91}	1.58 ± 0.12			
	Zr^{92}	0.25 ± 0.12			
	Zr^{93}	<4			
	Zr^{94}	0.08 ± 0.06	63d	0.076 ± 0.008	
	Zr^{96}	0.1 ± 0.1	17.0h	0.053 ± 0.005	
$_{41}$Nb	Nb^{93}	1.16 ± 0.02	6.6m	1.0 ± 0.5	5 ± 1
	Nb^{94}		36d	15 ± 4	
$_{42}$Mo		2.70 ± 0.04			7 ± 1
	Mo^{92}	<0.3	6.9h	<6 mb	
	Mo^{94}				
	Mo^{95}	13.9 ± 1.4			
	Mo^{96}	1.2 ± 0.6			
	Mo^{97}	2.2 ± 0.7			
	Mo^{98}	0.4 ± 0.4	67h	0.51 ± 0.06	
	Mo^{100}	0.5 ± 0.5	14.3m	0.20 ± 0.05	
$_{43}$Tc	Tc^{99}	22 ± 3	6h	2.6 ± 1.3	
$_{44}$Ru		2.56 ± 0.12			6 ± 1
	Ru^{96}		2.8d	0.21 ± 0.02	
	Ru^{98}				
	Ru^{99}				
	Ru^{100}				
	Ru^{101}				
	Ru^{102}		41d	1.44 ± 0.16	
	Ru^{104}		4.5h	0.7 ± 0.2	
	Ru^{105}		1.0y	0.2 ± 0.06	
$_{45}$Rh	Rh^{103}	149 ± 4	4.4m	12 ± 2	5 ± 1
			42s	140 ± 30	
			(99.9% of 4.4m → 42s)		
	Rh^{104}		36h	800 ± 100*	
	Rh^{104}		36h	40 ± 30*	
$_{46}$Pd		8.0 ± 1.5			3.6 ± 0.6
	Pd^{102}		17.0d	4.8 ± 1.5	
	Pd^{104}				
	Pd^{105}				
	Pd^{106}				
	Pd^{108}		13.6h	10.4 ± 0.8	
	Pd^{108}		4.8m	0.26 ± 0.04	
	Pd^{110}		23.6m	0.21 ± 0.03	
	Pd^{110}		5.5h	<0.05	
$_{47}$Ag		63 ± 1			6 ± 1
	Ag^{107}	31 ± 2	2.3m	45 ± 4	
	Ag^{109}	87 ± 7	270d	3.2 ± 0.4	
			24.2s	113 ± 13	
			(5% of 270d → 24.2s)		
$_{48}$Cd		2450 ± 50			7 ± 1
	Cd^{106}		6.7h	1.0 ± 0.5	
	Cd^{108}				
	Cd^{110}		49m	0.2 ± 0.1	
	Cd^{111}				
	Cd^{112}		5.1y	30 ± 15 mb	
	Cd^{113}	20,000 ± 300			
	Cd^{114}		43d	0.14 ± 0.03	
			53h	1.1 ± 0.3	
			(None of 43d → 53h)		
	Cd^{116}		2.9h	1.5 ± 0.3	
$_{49}$In		191 ± 3			2.2 ± 0.5
	In^{113}		49d	56 ± 12	
			72s	2.0 ± 0.6	
			(96.5% of 49d → 72s)		
	In^{115}		54.2m	155 ± 10	
			13s	52 ± 6	
			(None of 54.2m → 13s)		
$_{50}$Sn		625 ± 15 mb			4 ± 1
	Sn^{112}		112d	1.3 ± 0.3	
	Sn^{114}				
	Sn^{115}				
	Sn^{116}		14.5d	6 ± 2 mb	
	Sn^{117}				
	Sn^{118}		250d	10 ± 6 mb	
	Sn^{119}				
	Sn^{120}		>400d	1 ± 1 mb	
			27.5h	0.14 ± 0.03	
	Sn^{122}		130d	1.0 ± 0.5 mb	
			40m	0.16 ± 0.04	
	Sn^{124}		10m	0.2 ± 0.1	
			10d	4 ± 2 mb	
			(None of 10m → 10d)		

* Pile neutrons.

THERMAL NEUTRON CROSS SECTIONS (Continued)

Element	Isotope	Reaction Cross Sections (2200 m/s) σ_{abs}	σ_{act}		Scattering Cross Sections $\bar{\sigma}_s$
$_{51}$Sb		5.7 ± 1.0			4.3 ± 0.5
	Sb^{121}	5.9 ± 0.5	2.8d	6.8 ± 1.5	
			3.3m	0.19 ± 0.03	
	Sb^{123}	4.1 ± 0.3	21m	30 ± 15 mb	
			1.3m	30 ± 15 mb	
			60d	2.5 ± 0.5	
			(% of 21m & 1.3m → 60d unknown)		
$_{52}$Te		4.7 ± 0.1			5 ± 1
	Te^{120}	70 ± 70			
	Te^{122}	2.8 ± 0.9	110d	1.1 ± 0.5	
	Te^{123}	410 ± 30			
	Te^{124}	6.8 ± 1.3	58d	5 ± 3	
	Te^{125}	1.56 ± 0.16			
	Te^{126}	0.8 ± 0.2	110d	90 ± 20 mb	
			9.3h	0.8 ± 0.2	
			(98% of 110d → 9.3h)		
	Te^{128}	0.3 ± 0.3	33d	15 ± 5 mb	
			72m	0.13 ± 0.03	
	Te^{130}	0.5 ± 0.3	30h	<8 mb	
			25m	0.22 ± 0.05	
			(22% of 30h → 25m)		
$_{53}$I	I^{127}	7.0 ± 0.6	25.0m	5.6 ± 0.3	3.6 ± 0.5
	I^{129}	32 ± 5	12.6h	24 ± 3	
	I^{130}		8.05d	18 ± 3*	
	I^{131}		2.4h	50 ± 40	
$_{54}$Xe					4.3 ± 0.4
	Xe^{124}	74 ± 1			
	Xe^{126}				
	Xe^{128}		nγ	<5	
	Xe^{129}		nγ	45 ± 15	
	Xe^{130}		nγ	<5	
	Xe^{131}		nγ	120 ± 15	
	Xe^{132}		5.3d	0.2 ± 0.1	
	Xe^{133}		nγ	190 ± 90	
	Xe^{134}		9.13h	0.2 ± 0.1	
	Xe^{135}	2.72 ± 0.11 × 10^6			
	Xe^{136}		3.9m	0.15 ± 0.08	
$_{55}$Cs	Cs^{133}	28 ± 1	3.1h	3.0 ± 0.3	
			2.3y	30 ± 1	
			(~99% of 3.2h → 2.3y)		
	Cs^{134}		2.6 × 10^6y	134 ± 12*	
	Cs^{135}		13.7d	8.7 ± 0.5	
	Cs^{137}		33m	<2	
$_{56}$Ba		1.2 ± 0.1			8 ± 1
	Ba^{130}		12.0d	10 ± 1	
	Ba^{132}		7.2y	7 ± 2	
	Ba^{134}	2 ± 2			
	Ba^{135}	5.8 ± 0.9			
	Ba^{136}	0.4 ± 0.4			
	Ba^{137}	5.1 ± 0.4			
	Ba^{138}	0.7 ± 0.1	85m	0.5 ± 0.1	
	Ba^{139}		12.8d	4 ± 1	
	Ba^{140}		18m	<20*	
$_{57}$La		8.9 ± 0.2			
	La^{138}				
	La^{139}		40h	8.2 ± 0.8	
	La^{140}		3.7h	3.1 ± 1.0	
$_{58}$Ce		0.73 ± 0.08			
	Ce^{136}	25 ± 25	34.5h	0.6 ± 0.2	
			8.7h	6.3 ± 1.5	
	Ce^{138}	9 ± 6	55s	7 ± 5 mb	
			140d	0.6 ± 0.3	
	Ce^{140}	0.66 ± 0.06	32d	0.31 ± 0.10	
	Ce^{142}	1.0 ± 0.2	32h	0.94 ± 0.05	
	Ce^{143}		290d	6.0 ± 0.7*	
$_{59}$Pr	Pr^{141}	11.3 ± 0.2	19.3h	10.8 ± 1.0	
	Pr^{142}		13.7d	18 ± 3*	
	Pr^{143}		17m	89 ± 10	
$_{60}$Nd		46 ± 2			
	Nd^{142}	18 ± 2			
	Nd^{143}	324 ± 10	nγ	240 ± 50	
	Nd^{144}	5.0 ± 0.6			
	Nd^{145}	60 ± 6			
	Nd^{146}	10 ± 1	11.3d	1.8 ± 0.6	
	Nd^{118}	3.4 ± 1.0	1.8h	3.7 ± 1.2	
	Nd^{150}	3.0 ± 1.5	15m	1.5 ± 0.2	
$_{61}$Pm	Pm^{147}	150 ± 50*	5.3d	60 ± 20	
$_{62}$Sm		5600 ± 200			
	Sm^{144}		400d	<2	
	Sm^{147}	87 ± 60			
	Sm^{148}				
	Sm^{149}	40,800 ± 900			
	Sm^{150}				
	Sm^{151}	12,400 ± 400*			
	Sm^{152}	216 ± 6	47h	140 ± 40	
	Sm^{154}		24m	5.5 ± 1.1	
$_{63}$Eu		4300 ± 100			8 ± 1
	Eu^{151}	7800 ± 200	9.2h	1400 ± 300*	
	Eu^{152}		nγ	5500 ± 1500*	
	Eu^{153}	440 ± 25	16y	420 ± 100*	
	Eu^{154}		1.7y	1500 ± 400*	
	Eu^{155}		15.4d	14,000 ± 4000*	
$_{64}$Gd		46,000 ± 1000			
	Gd^{152}		230d	<125	
	Gd^{154}				
	Gd^{155}	56,200 ± 1000	nγ	70,000 ± 20,000*	
	Gd^{156}				
	Gd^{157}	242,000 ± 4000	nγ	160,000 ± 60,000*	
	Gd^{158}		18.0h	3.9 ± 0.4	
	Gd^{160}		3.6m	0.8 ± 0.3	
$_{65}$Tb	Tb^{159}	46 ± 4	73d	>22	
	Tb^{160}		7.0d	525 ± 100*	
$_{66}$Dy		950 ± 50			100 ± 20
	Dy^{156}				
	Dy^{158}		136d	96 ± 20	
	Dy^{160}				
	Dy^{161}				
	Dy^{162}				
	Dy^{163}				
	Dy^{164}		1.3m	2000 ± 200	
			140m	800 ± 100	
	Dy^{165}		82h	5000 ± 2000*	
$_{67}$Ho	Ho^{165}	65 ± 3	27.3h	60 ± 12	
$_{68}$Er		173 ± 17			
	Er^{162}		75m	2.03 ± 0.20	
	Er^{164}		10h	1.65 ± 0.17	
	Er^{166}				
	Er^{167}				
	Er^{168}		9.4d	2.0 ± 0.4	
	Er^{170}		2.5s + 7.5h	9 ± 2	
$_{69}$Tm	Tm^{169}	127 ± 4	129d	130 ± 30	7 ± 3
	Tm^{170}		1.9y	150 ± 20*	
$_{70}$Yb		37 ± 4			12 ± 5
	Yb^{168}		32d	11,000 ± 3000*	
	Yb^{170}				
	Yb^{171}				
	Yb^{172}				
	Yb^{173}				
	Yb^{174}		101h	60 ± 40	
	Yb^{176}		1.8h	5.5 ± 1.0	
$_{71}$Lu		112 ± 5			
	Lu^{175}		3.7h	35 ± 15	
	Lu^{176}		6.8d	4000 ± 800	
$_{72}$Hf		105 ± 5			8 ± 2
	Hf^{174}	1500 ± 1000			
	Hf^{176}	15 ± 15			
	Hf^{177}	380 ± 30			
	Hf^{178}	75 ± 10			
	Hf^{179}	65 ± 15			
	Hf^{180}	14 ± 5	46d	10 ± 3	
$_{73}$Ta	Ta^{181}	21 ± 1	16.4m	30 ± 10 mb	5 ± 1
			111d	19 ± 7	
			(~95% of 16.4m → 111d)		
	Ta^{182}		5.5d	17,000 ± 2000*	
$_{74}$W		19.2 ± 1.0			5 ± 1
	W^{180}	60 ± 60	140d	10 ± 10	
	W^{182}	20 ± 2			
	W^{183}	11 ± 1			

* Pile neutrons.

* Pile neutrons.

THERMAL NEUTRON CROSS SECTIONS (Continued)

Element	Isotope	Reaction Cross Sections (2200 m/s) σ_{abs}	σ_{act}		Scattering Cross Sections $\bar{\sigma}_s$
$_{74}$W (Contd.)	W^{184}	2.0 ± 0.3	74d	2.24 ± 0.22	
	W^{186}	35 ± 3	24h	34 ± 7	
	W^{187}		65d	90 ± 40	
$_{75}$Re		86 ± 4			14 ± 4
	Re^{185}	104 ± 8	91h	120 ± 12	
	Re^{187}	66 ± 5	17h	69 ± 7	
	Re^{188}		150d	<2	
$_{76}$Os		15.3 ± 0.7			
	Os^{184}		97d	<200	
	Os^{186}				
	Os^{187}				
	Os^{188}				
	Os^{189}				
	Os^{190}		16.0d	8 ± 3	
	Os^{192}		31h	1.6 ± 0.4	
	Os^{193}		700d	600 ± 200	
$_{77}$Ir		440 ± 20			
	Ir^{191}		1.4m	260 ± 100	
			74d	700 ± 200	
	Ir^{192}		nγ	700 ± 200	
	Ir^{193}		19.0h	130 ± 30	
$_{78}$Pt		8.8 ± 0.4			10 ± 1
	Pt^{190}	150 ± 150	18h	0.76 ± 0.10	
	Pt^{192}	8 ± 8	4.3d	90 ± 40	
	Pt^{194}	1.2 ± 0.9			
	Pt^{195}	27 ± 2			
	Pt^{196}	0.7 ± 0.7	18h	0.87 ± 0.09	
	Pt^{198}	4.0 ± 0.5	31m	3.9 ± 0.8	
	Pt^{199}		11.5h	15 ± 10	
$_{79}$Au	Au^{197}	98.8 ± 0.3	2.7d	96 ± 10	9.3 ± 1.0
	Au^{198}		3.15d	26,000 ± 1200	
	Au^{199}		48m	30 ± 15*	
$_{80}$Hg		380 ± 20			20 ± 5
	Hg^{196}		nγ	3100 ± 1000*	
			24h	420 ± 80	
			65h	880 ± 175	
	Hg^{198}		46.6m	0.018 ± 0.004	
	Hg^{199}		nγ	2500 ± 800*	
	Hg^{200}		nγ	<60*	
	Hg^{201}		nγ	<60*	
	Hg^{202}		47d	3.8 ± 0.8	
	Hg^{204}		5.5m	0.43 ± 0.10	
$_{81}$Tl		3.4 ± 0.5			14 ± 2
	Tl^{203}	11.4 ± 0.9	2.7y	8 ± 3	
	Tl^{205}	0.80 ± 0.08	4.2m	0.10 ± 0.03	
$_{82}$Pb		170 ± 2 mb			11 ± 1
	Pb^{204}	0.8 ± 0.6	5 × 10^{7}y	0.7 ± 0.2*	
	Pb^{206}	25 ± 5 mb			
	Pb^{207}	0.70 ± 0.03			
	Pb^{208}	<30 mb	3.2h	0.6 ± 0.2 mb	
$_{83}$Bi	Bi^{209}	34 ± 2 mb	5.0d	19 ± 2 mb	9 ± 1
$_{86}$Rn	Rn^{220}	25m	<0.2*		
	Rn^{222}	11.7d Ra^{223}	0.72 ± 0.07*		
$_{88}$Ra	Ra^{223}		3.64d	130 ± 20*	<100
	Ra^{224}		14.8d	12.0 ± 0.5*	
	Ra^{226}		41.2m	20 ± 3*	<0.1 mb
	Ra^{228}		<10m	36 ± 5*	<2

* Pile neutrons

Element	Isotope	Reaction Cross Sections (2200m/s) σ_{abs}	σ_{act}		Scattering Cross Sections $\bar{\sigma}_s$
$_{89}$Ac	Ac^{227}	510 ± 40	6.13h	795 ± 20	<2
$_{90}$Th	Th^{227}				1500 ± 1000
	Th^{228}		7.3 ± 10^{3}y	123 ± 15*	<0.3
	Th^{229}				45 ± 11
					ν 2.13 ± 0.03§
	Th^{230}	27 ± 2	25.6h	21.4 ± 0.3	<1 mb
	Th^{232}	7.56 ± 0.11	23.3m	7.33 ± 0.12	<0.2 mb
	Th^{233}		24.1d	1400 ± 200*	15 ± 2*
	Th^{234}		10m	1.8 ± 0.5*	<10 mb
$_{91}$Pa	Pa^{230}				1500 ± 250
	Pa^{231}	200 ± 20	1.31d	200 ± 15	10 ± 5 mb
	Pa^{232}		27.4d	760 ± 100*	700 ± 100
	Pa^{233}		1.18m	29 ± 5	<0.1
			6.7h	22 ± 3	
	Pa^{234}				<500
$_{92}$U		7.68 ± 0.07			4.18 ± 0.06
					n 1.34 ± 0.02‡
	U^{230}				25 ± 10
	U^{231}				400 ± 300
	U^{232}		1.62 × 10^{5}y	300 ± 200*	80 ± 20
	U^{233}	581 ± 7	2.52 × 10^{5}y	52 ± 2	527 ± 4
					α 0.102 ± 0.00
					n 2.28 ± 0.02‡
					ν 2.51 ± 0.03§
	U^{234}	105 ± 4	7.1 × 10^{8}y	90 ± 30	<0.65
	U^{235}	694 ± 8	2.40 × 10^{7}y	107 ± 5	582 ± 6
					α 0.19 ± 0.01
					n 2.07 ± 0.02‡
					ν 2.47 ± 0.03§
					$\bar{\sigma}_s$ 10 ± 2‖
	U^{236}	7 ± 2	6.7d	6 ± 1	
	U^{238}	2.71 ± 0.02	23.5m	2.74 ± 0.06	<0.5 mb
	U^{239}		17h	22 ± 5*	14 ± 3*
$_{93}$Np	Np^{234}				900 ± 300
	Np^{236}				2800 ± 800
	Np^{237}	170 ± 5	2.10d	169 ± 6	19 ± 3 mb
	Np^{238}				1600 ± 100
	Np^{239}		7.3m	35 ± 10*	<1
			60m	25 ± 15*	
			(<5% of 7.3m → 60m)		
$_{91}$Pu	Pu^{236}				170 ± 35
	Pu^{237}				2500 ± 500
	Pu^{238}		2.44 × 10^{4}y	403 ± 10	16.8 ± 0.3
	Pu^{239}	1026 ± 13	6.6 × 10^{3}y	315 ± 16	746 ± 8
					α 0.38 ± 0.02
					n 2.10 ± 0.02‡
					ν 2.90 ± 0.04§
					$\bar{\sigma}_s$ 9.6 ± 0.5‖
	Pu^{240}	286 ± 7	13.2y	250 ± 40	0.030 ± 0.045
	Pu^{241}	1400 ± 80	3.7 × 10^{5}y	390 ± 50*	1010 ± 13
					ν 3.06 ± 0.04§
	Pu^{242}	30 ± 2	4.98h	19 ± 1	<0.2
	Pu^{243}		7.5 × 10^{7}y	170 ± 90*	
	Pu^{244}		10h	1.8 ± 0.3*	
	Pu^{245}		11.2d	260 ± 150*	
$_{95}$Am	Am^{241}	630 ± 35	16.0h	750 ± 80*	3.1 ± 0.2
			100y	50 ± 40*	
			(<6% of 16h → 100y)		
	Am^{242m}				2500 ± 1000
	Am^{242}	8000 ± 1000*			6400 ± 500*
	Am^{243}		26m	74 ± 4	<75 mb
			19.2y	250 ± 50	
$_{96}$Cm	Cm^{242}		35y	20 ± 10*	<5*
	Cm^{243}	500 ± 300*	18y	250 ± 150*	700 ± 50
	Cm^{244}		2 × 10^{4}y	15 ± 10*	
	Cm^{245}		6.6 × 10^{3}y	200 ± 100*	1900 ± 200
	Cm^{246}		>10^{6}y	15 ± 10*	
	Cm^{248}		65m	6 ± 4*	
$_{97}$Bk	Bk^{249}		3.1h	500 ± 200*	
$_{98}$Cf	Cf^{249}	900 ± 400*	10y	270 ± 100*	600 ± 400*
	Cf^{250}		~700y	1500 ± 1000*	
	Cf^{251}		2.2y	3000 ± 2000*	
	Cf^{252}		18d	28 ± 7	
	Cf^{254}			<2*	
$_{99}$Es	E^{253}		38h	300 ± 150*	
	E^{254}	2700 ± 600*	24d	<40*	

* Pile neutrons.

† $\alpha = \dfrac{\sigma_{abs} - \sigma_F}{\sigma_F}$.

‡ n = Fission neutrons produced per thermal neutron absorbed.

§ $\nu = n(1 + \alpha)$.

‖ $\bar{\sigma}_s$ = Scattering cross section.

The Elements

C. R. Hammond

One of the most striking facts about the elements is their unequal distribution and occurrence in nature. Present knowledge of the chemical composition of the universe, obtained from the study of the spectra of stars and nebulae, indicates that hydrogen is by far the most abundant, and may comprise more than 90% of the atoms or about three quarters of the mass of the universe. Helium atoms account for most of the remainder. All of the other elements together probably contribute only slightly more than 1 per cent to the mass.

It is now thought that the chemical composition of the universe is undergoing continuous change. Hydrogen is being converted into helium, and helium is being changed into the heavier elements. As time goes on, the ratio of heavier elements increases relative to hydrogen. The process presumably is not reversible.

Studies of the solar spectrum have led to the identification of 67 elements in the sun's atmosphere; however all elements cannot be identified with the same degree of certainty. Other elements may be present in the sun although they have not yet been detected spectroscopically. The element helium was discovered on the sun before it was found on earth. Some elements, such as scandium, are relatively more plentiful in the sun and the stars than here on earth. No elements are found elsewhere in the universe that cannot now be accounted for on earth; however technetium, an unstable element which probably does not occur naturally on earth, has been identified in the spectra of certain late-type stars. This presents one of the most puzzling current problems of astrophysics.

F. W. Clarke and others have carefully studied the composition of igneous rocks making up the crust of the earth. It has been found that oxygen accounts for about 47% of the crust by weight, while silicon comprises about 28%, and aluminum about 8%. These elements, plus iron, calcium, sodium, potassium, and magnesium, account for about 99% of the earth's crust.

It is surprising that many elements, such as tin, copper, zinc, lead, mercury, silver, platinum, antimony, arsenic, and gold, that are essential to our needs and civilization, are among some of the rarest elements in the earth's crust. These are made available to us only by the processes of concentration in ore bodies. Some of the so-called "rare-earth" elements are now thought to be much more plentiful than originally thought, and are about as abundant as uranium, mercury, lead, or bismuth. The least abundant rare-earth, thulium, is now believed to be more plentiful than silver, gold, or platinum. It is also startling to find in light of recent knowledge that rubidium is the 16th most abundant element and is more plentiful than chlorine, although its compounds are so little known in chemistry and commerce.

Each element, from atomic number 1 to 100 has at least one radioactive isotope. About 1400 different nuclides (the name given to different kinds of nuclei whether they are of the same or different elements) are now recognized. Of these, about 260 are stable forms of natural elements. About 1130 are unstable, 65 of which occur in nature principally among the heaviest elements. More than 300 stable and radioactive isotopes are now produced and distributed by the Oak Ridge National Laboratory to customers licensed by the U. S. Atomic Energy Commission.

The available evidence leads to the conclusion that elements 89 (actinium) through 103 are so chemically similar to the *rare earths* or lanthanide elements (atomic numbers 57–71) that their electronic structures must also be similar. They have, therefore, been named *actinides*, after the first member of this series. An inner electron shell, consisting of fourteen 5f electrons is filled in progressing across the series. Those elements beyond uranium have been produced artificially on earth by nuclear reactions and synthesis and are known as the *transuranium* elements. The presently known transuranium elements have the following names and symbols: 93, neptunium (Np); 94, plutonium (Pu); 95, americium (Am); 96, curium (Cm); 97, berkelium (Bk); 98, californium (Cf); 99, einsteinium (E); 100, fermium (Fm); 101, mendelevium (Mv); element 102, (unnamed); and element 103, lawrencium (Lw).

Chemically the transuranium elements are very similar, although the observed differences are those expected and anticipated from their unique position in the periodic system as part of a second rare-earth series. All have trivalent ions, which form inorganic complex ions and organic chelates. Also in common are acid-insoluble trifluorides and oxalates, soluble sulfates, nitrates, chlorides, and perchlorates. Neptunium, plutonium, and americium have higher oxidation states in aqueous solution (similar to uranium), but the relative stability of these states to the common trivalent ion becomes progressively less as one proceeds to the higher atomic numbers. This is a direct consequence, indeed an identifying feature, of the actinide role as a second rare-earth type transition series.

One of the most important methods for study and elucidation of chemical behavior of the actinide elements has been ion-exchange chromatography. Adsorption on and elution from ion-exchange columns has made possible the identification and separation of trace quantities of all of the actinides and in particular the transuranium elements. The behavior of each actinide and transuranium element in this respect is very similar to its analogue rare-earth element. This has made it possible to detect as little as one or two atoms when this small a number has been made in some of the transmutation experiments.

At present, eleven transuranium elements have been created and discovered, with a total of about one hundred isotopes. All these new elements are unstable and, therefore, radioactive. The half-life of the various isotopes decreases, in general, with increasing atomic number, which means that as heavier and heavier elements are created, they exist for shorter and shorter periods, making their production, separation, and identification progressively more difficult.

It may still be possible to synthesize, separate, and identify a half-dozen or so more of the transuranium elements, but barring unknown experimental breakthroughs or unknown regions of stability, the end should come somewhere in the region of element 110. The elements up to and including einsteinium, element 99, have isotopes sufficiently long-lived to be isolated in macroscopic quantities, but this does not seem to be true beyond einsteinium. Unfortunately for the prospect of producing ever-higher elements, the longest-lived isotopes that can be made beyond elements 104 and 105 will probably not exist long enough for conventional chemical identification.

ACTINIUM (Gr. *aktis, aktinos*, beam or ray), Ac; at. wt. 227.02 (calc.) at. no. 89; m.p. 1050°C, b.p. 3200°C, ±300°C (est.); sp. g. 10.07 (calc.). Discovered by Andre Debierne in 1899 and independently by F. Giesel in 1902. Occurs naturally in association with uranium minerals. Actinium-227, a decay product of uranium-235, is a beta emitter with a 22-yr. half-life. Its principal decay products are thorium-227 (18.6-day half life), radium-223 (11.2-day half-life), and a number of short-lived products including radon, bismuth, polonium and lead isotopes. In equilibrium with its decay products, it is a powerful source of alpha rays. Actinium has been isolated, and commercial production of the element has been reported. Purified actinium comes into equilibrium with its decay products at the end of 185 days, and then decays according to its 22-yr. half-life. It is about 150 times as active as radium, making it of value in the production of neutrons.

ALUMINUM (L. *alumen, alum*), Al; at. wt. 26.9815; at. no. 13; m.p. 660.2°C, b.p. 2467°C; sp. gr. 2.6989 (20°C); valence 3. Wöhler is generally accredited with isolating the metal in 1827, although an impure form was prepared by Oersted two years earlier. The method of obtaining the metal by electrolysis of alumina dissolved in cryolite was discovered in 1886 by Hall in the U.S. and independently about the same time by Heroult in France. Although aluminum occurs in larger quantities in the earth's crust than any other metal, it does not appear free. It is found as the silicate in clays, feldspars, etc., while the commercial ore is *bauxite*, an impure hydrated oxide. Its production from clay is possible but not economically feasible at present. Pure aluminum, a silvery white metal, possesses many desirable characteristics. It is light, nontoxic, has a pleasing appearance, can easily be formed, machined, or cast, has a high thermal conductivity, and has excellent corrosion resistance. It is nonmagnetic and nonsparking, stands second among metals in the scale of malleability, and sixth in ductility. It is extensively used for kitchen utensils, outside building decoration, and in thousands of industrial applications where a strong, light, easily constructed material is needed. Although its electrical conductivity is only about 60% that of copper per area of cross section, it is used in electrical transmission lines because of its light weight. Pure aluminum is soft and lacks strength, but it can be alloyed with small amounts of copper, magnesium, silicon, manganese, and other elements to impart a variety of useful properties. These alloys are of vital importance in the construction of modern aircraft and rockets. Aluminum, evaporated in a vacuum, forms a highly reflective coating for both visible light and radiant heat. These coatings soon form a thin layer of the protective oxide and do not deteriorate as do silver coatings. They have found application in coatings for telescope mirrors, in making decorative paper, packages, toys, and in many other uses. The compounds of greatest importance are aluminum oxide, the sulfate, and the soluble sulfate with potassium (alum). The oxide, alumina, occurs naturally as ruby, sapphire, corundum, and emery and is used in glassmaking. Synthetic ruby and sapphire have found application in the construction of lasers for producing coherent light. In 1856, the price of aluminum was about 90 dollars a pound, and just before Hall's discovery in 1866, about 5 dollars. The price rapidly dropped to 30 cents and has been as low as 15 cents.

AMERICIUM (the Americas), Am; at. no. 95; m.p. > 850°C; b.p. . .; sp. gr. 11.7; valence 3, 4, 5, or 6. Americium was the fourth transuranium element to be discovered; the isotope Am^{241} was identified by Seaborg, James, Morgan and Ghiorso late in 1944 at the wartime Metallurgical Laboratory (now the Argonne National Laboratory) of the University of Chicago as the result of successive neutron capture reactions by plutonium isotopes in a nuclear reactor:

$$Pu^{239}(n,\gamma)Pu^{240}(n,\gamma)Pu^{241} \xrightarrow{\beta^-} Am^{241}.$$

Since the isotope Am^{241} can be prepared in relatively pure form by extraction as a decay product over a period of years from strongly neutron-bombarded plutonium Pu^{241}, this isotope is used for much of the chemical investigation of this element. Better suited is the isotope Am^{243} due to its longer half-life (8.8×10^3 years as compared to 470 years for Am^{241}). A mixture of the isotopes Am^{241}, Am^{242}, and Am^{243} can be prepared by intense neutron irradiation of Am^{241} according to the reactions $Am^{241}(n,\gamma)$ $Am^{242}(n,\gamma)$ Am^{243}. Nearly isotopically pure Am^{243} can be prepared by a sequence of neutron bombardments and chemical separations as follows: neutron-bombardment of Am^{241} yields Pu^{242} by the reactions $Am^{241}(n,\gamma)$ $Am^{242} \xrightarrow{EC} Pu^{242}$; after chemical separation the Pu^{242} can be transformed to Am^{243} via the reactions $Pu^{242}(n,\gamma)$ $Pu^{243} \xrightarrow{\beta^-} Am^{243}$, and the Am^{243} can be chemically separated. Fairly pure Pu^{242} can be prepared more simply by very intense neutron irradiation of Pu^{239} as the result of successive neutron-capture reactions.

Americium can be obtained by reduction of americium trifluoride with barium vapor at 1000°–1200°C. The luster of freshly prepared americium metal is whiter and more silvery than plutonium or neptunium prepared in the same manner. It appears to be more malleable than uranium or neptunium and tarnishes slowly in dry air at room temperature.

The element exists in three oxidation states in aqueous solution; Am^{+3} (light salmon), AmO_2^+ (color unknown), and AmO_2^{+2} (light tan). The trivalent state is highly stable and difficult to oxidize. AmO_2^+, like plutonium, is unstable with respect to disproportionation into Am^{+3} and AmO_2^{+2}. The ion Am^{+4} is so unstable in solution that it has not yet been detected, although tetravalent solid compounds are well known. There is some evidence that Am^{+2} has been prepared in tracer experiments at very low concentrations; this would be very similar to the analogous lanthanide, europium, which can be reduced to the divalent state.

Americium dioxide, AmO_2, is the important oxide; Am_2O_3 and, as with previous actinide elements, oxides of variable compositions between $AmO_{1.5}$ and AmO_2 are known. The halides AmF_3, AmF_4, $AmCl_3$, $AmBr_3$, and AmI_3 have also been prepared.

In 1962, the A.E.C. made an initial allotment of 200 gms. of Americium-241 available for sale from the Oak Ridge National Laboratory.

ANTIMONY (L. *antimonium*), Sb, (L. *stibium*, mark); at. wt. 121.75; at. no. 51; m.p. 630.5°C; b.p. 1380°C; sp. gr. 6.691 (20°C); valence 3 or 5. Recognized in compounds by the ancients; known as a metal at the beginning of the seventeenth century and possibly before that date. Antimony is a metallic element, common, but neither abundant nor widely diffused; sometimes found native, but more frequently as the sulfide, *stibnite* (Sb_2S_3); also as antimonides and sulfantimonides of the heavy metals, and as oxides. It is extracted from the sulfide by roasting to the oxide, which is reduced by salt and scrap iron; from its oxides it is also prepared by reduction with carbon. Four allotropic forms of antimony exist: yellow antimony, black antimony, explosive antimony, and metallic or ordinary antimony. Metallic antimony is an extremely brittle metal of a flaky, crystalline texture, blue-white color and metallic luster; hardness, 3 to 3.5; not acted on by air at room temperature, but burns brilliantly when heated with formation of white fumes of oxide Sb_2O_3; it is a poor conductor of heat and electricity. This form is available commercially with a purity of 99.999+%. The metal is widely used in alloys in a percentage varying from one to twenty. It greatly increases the hardness and mechanical strength of lead. Batteries, antifriction alloys, type metal, cable sheathing, and minor products use about half of the metal. Compounds taking up the other half are oxides, sulfide, sodium antimonate, and antimony trichloride. These are used in the manufacture of flame-proofing compounds, in paints, ceramic enamels, glass and pottery. Tartar emetic (hydrated potassium antimonyl tartrate) is used as a medicine. Antimony and some of its compounds are toxic. The maximum allowable concentration of antimony dust in the air is recommended to be 0.5 mg/cubic meter.

ARGON (Gr. *argon*, inactive, Ar; at. wt. 39.948; at. no. 18; freezing pt. − 189.2°C; b.p. −185.7°C; density 1.7837 g/1). Its presence in air was suspected by Cavendish in 1785; discovered by Lord Rayleigh and Sir William Ramsay in 1894. The gas is prepared by fractionation of liquid air, the atmosphere containing 0.94% argon. It is $2\frac{1}{2}$ times as soluble in water as nitrogen, having about the same solubility as oxygen; best recognized by the characteristic lines in the red end of the spectrum. It is used in electric light bulbs and in fluorescent tubes at a pressure of about 3 mm, and in filling photo tubes, glow tubes, etc. Argon is also used as an inert gas shield for arc welding and cutting, as a blanket for the production of titanium and other reactive elements, and as a protective atmosphere for growing silicon and germanium crystals. Argon is colorless and odorless, both as a gas and liquid. It is available in high-purity form. Commercial argon is available at a cost of about

10 cents per cu. ft. Although it is considered to be a very inert gas, a few compounds of argon have been reported in the literature.

ARSENIC (L. *arsenicum*, Gr. *arsenikon*, yellow orpiment—identified with *arsenikos*, male, from the belief that metals were different sexes—Arab. *az-zernikh*, the orpiment from Persian *zerni-zar*. gold–), As; at. wt. 74.9216; at. no. 33; valence +3 or +5. Elemental arsenic occurs in three solid modifications: yellow, black, and gray, with specific gravities of 1.97, 4.73, and 5.73, respectively. Gray arsenic, the ordinary stable form, has a m.p. of 817°C (28 atm.) and sublimes at 613°C. It is believed that Albertus Magnus obtained the element in 1250. In 1649 Schroeder published two methods of preparing it. Found native, in sufides *realgar* and *orpiment*, as arsenides and sulfarsenides of heavy metals and as oxide, and arsenates. *Mispickel* or arsenopyrite (FeSAs) is the most common mineral, from which on heating the arsenic sublimes leaving ferrous sulfide. The element is a steel gray, very brittle, crystalline, semi-metallic solid; it tarnishes in air and when heated is rapidly oxidized to arsenous oxide (As_2O_3) with the odor of garlic. Arsenic and its compounds are poisonous. The maximum allowable concentration of arsenic is recommended to be 0.5 mg/cu. meter of air. Arsenic is also used in bronzing, pyrotechny, and for hardening and improving the sphericity of shot. The most important compounds are white arsenic (As_2O_3), the sulfide, Paris green $3Cu(AsO_2)_2 \cdot Cu(C_2H_3O_2)_2$, calcium arsenic and lead arsenate, the last three being used as agricultural insecticides and poisons. Marsh's test makes use of the formation and ready decomposition of arsine (AsH_3). Arsenic is available in high-purity form. It is finding increasing uses as a doping agent in solid-state devices, such as transistors. Gallium arsenide is finding use as a laser material to convert electricity directly into coherent light.

ASTATINE (Gr. *astatos*, unstable) At; at. wt. 210—most stable isotope; at. no. 85; valence probably 1, 3, 5, or 7. Synthesized in 1940 by D. R. Corson, K. R. MacKenzie, and E. Segré at the University of California by bombarding bismuth with alpha-particles. The longest lived isotope, At-210, has a half-life of only 8.3 hrs. At-219, with a half-life of 0.9 min., is reported to be present in uranium ores. Astatine is the only member of the halogen family without stable isotopes. It is said to be more metallic than iodine, and like iodine, it accumulates in the thyroid gland.

BARIUM (Gr. *barys*, heavy), Ba; at. wt. 137.34; at. no. 56; m.p. 725°C; b. p. 1140°C; sp. gr. 3.5 (20°C); valence 2. Baryta was distinguished from lime by Scheele in 1774; the element was discovered by Sir Humphry Davy in 1808. It is found only in combination with other elements chiefly in *barite* or *heavy spar* (sulfate) and *witherite* (carbonate) and is prepared by electrolysis of the chloride. Barium is a metallic element, soft, and when pure is silvery-white like lead; it belongs to the alkaline earth group, resembling calcium chemically. The metal oxidizes very easily and should be kept under petroleum or other suitable oxygen-free liquids to exclude air. It is decomposed by water or alcohol. The metal is used as a "getter" in vacuum tubes. The most important compounds are the peroxide (BaO_2), chloride, sulfate, carbonate, nitrate and chlorate. Lithopone, a pigment containing barium sulfate and zinc sulfide, has good covering power, and does not darken in the presence of sulfides. The sulfate, as permanent white or *blanc fixé*, is also used in paint, in x-ray diagnostic work, and in glassmaking. The carbonate is used as a rat poison, while the nitrate and chlorate give colors in pyrotechny. The impure sulfide phosphoresces after exposure to the light. The compounds and the metal are not expensive. All barium compounds that are water or acid soluble are poisonous.

BERKELIUM (Berkeley, home of Univ. of Calif.) Bk; at. no. 97; valence 3 or 4. Berkelium, the eighth member of the actinide transition series, was discovered in December, 1949, by Thompson, Ghiorso, and Seaborg, and was the fifth transuranium element synthesized. It was produced by cyclotron bombardment of milligram amounts of Am^{241} with helium ions at Berkeley, California. The first isotope produced had a mass number of 243 and decayed with a half-life of 4.5 hrs. by electron capture.

The chemical properties of berkelium have been studied entirely to date through the use of tracer amounts. These studies have demonstrated that berkelium exists in aqueous solutions in two oxidation states, berkelium (III) and berkelium (IV). The chemistry and solubility of compounds appears to closely follow the other transuranium elements.

The existence of Bk^{249}, with a half-life of about 300 days, makes it feasible to isolate berkelium in weighable amounts so that its properties can be investigated with macroscopic quantities.

This isotope can be prepared by the intense neutron bombardment of Cm^{244} as the result of the capture of successive neutrons by the reactions

$$Cm^{244}(n,\gamma)Cm^{245}(n,\gamma)Cm^{246}(n,\gamma)Cm^{247}(n,\gamma)Cm^{248}(n,\gamma)Cm^{249} \xrightarrow{\beta^-} Bk^{249}.$$

One of the first visible amounts of a pure berkelium compound—berkelium chloride—was produced in 1962. It weighed about 3 billionths of a gram.

BERYLLIUM (Gr. *berryllos, beryl;* also called Glucinium or Glucinum, Gr. *glykys,* sweet;) Be; at. wt. 9.0122; at. no. 4; m.p. 1278 ± 5°C; b.p. 2970^{760}°C (5mm); sp. gr. 1.848 (20°C), valence 2. Discovered as the oxide by Vauquelin in beryl and in emeralds in 1798. The metal was isolated in 1828 by Wöhler and by Bussy independently by the action of potassium on beryllium chloride. Beryl (beryllium silicate) is the most important commercial source of the element and its compounds. Most of the metal is now prepared by reducing beryllium fluoride with magnesium metal. Beryllium metal did not become readily available to industry until 1957. The metal, steel gray in color, has many desirable properties. It is one of the lightest of all metals, and has one of the highest melting points of the light metals. Its modulus of elasticity is about one third greater than that of steel. It resists attack by concentrated nitric acid, has excellent thermal conductivity, and is nonmagnetic. It has a high permeability to x-rays, and when bombarded by alpha particles, as from radium or polonium, neutrons are produced in the ratio of about 30 neutrons/million alpha particles. At ordinary temperatures beryllium resists oxidation in air, although its ability to scratch glass is probably due to the formation of a thin layer of the oxide. Beryllium is used as an alloying agent in producing beryllium copper, which is extensively used for springs, electrical contacts, spot-welding electrodes, and nonsparking tools. It is finding application as a structural material for high-speed aircraft, missiles, and spacecraft. It is used in nuclear reactors as a reflector or moderator for it has a low thermal neutron absorption cross section. It is used in gyroscopes, computer parts, and inertial guidance instruments where lightness, stiffness, and dimensional stability are required. The oxide has a very high melting point and is also used in nuclear work and ceramic applications. Beryllium and its salts are toxic and should be handled with the greatest of care. Beryllium and its compounds should not be tasted to verify the sweetish nature of beryllium (as did early experimenters). The metal, its alloys, and its salts can be handled safely if certain work codes are observed, but no attempt should be made to work with beryllium before becoming familiar with proper safeguards. The maximum allowable concentration of beryllium dust is recommended to be about 1–2 μ gms/cu. meter in working areas or .01 μ gms/cu. meter in non-working areas. Beryllium metal in vacuum cast billet form is priced roughly at 50 dollars to 100 dollars per lb., depending on the quantity ordered. Fabricated forms are more expensive. High-purity beryllium is available commercially at about 2700 dollars per lb. As techniques and supplies become more plentiful, these costs will undoubtedly be much lower.

BISMUTH (Ger. *Weisse Masse,* white mass; later *Wismuth* and *Bisemutum*). Bi; at. wt. 208.980; at no. 83; m.p. 271.3°C; b.p. 1560 ± 5^{760}°C; sp. gr. 9.747; valence 3 or 5. In early times bismuth was confused with tin and lead. Claude Geoffroy the Younger showed it to be distinct from lead in 1753. It is a white, crystalline, brittle metal with a pinkish tinge. It occurs native. The most important ores are *bismuthinite* or *bismuth glance* (Bi_2S_3) and *bismite* (Bi_2O_3). Most of the bismuth produced in the U.S. is obtained as a by-product in refining lead, copper, tin, silver, and gold ores. Bismuth is the most diamagnetic of all metals, and the thermal conductivity is lower than any metal, except mercury. It has a high electrical resistance, and has the highest Hall effect of any metal (i.e., greatest increase in electrical resistance when placed in a magnetic field). Bismuth expands 3.32% on solidification. This property makes bismuth alloys particularly suited to the making of sharp castings of objects subject to damage by high temperatures. With other metals, such as tin, cadmium, etc., bismuth forms low-melting alloys which are extensively used for safety devices used in fire detection and extinguishing systems. When bismuth is heated in air it burns with a blue flame forming yellow fumes of the oxide. The metal is also used as a thermocouple material (has highest negativity known), and has found recent application as a carrier for U^{235} or U^{233} fuel in atomic reactors. Its soluble salts are characterized by forming insoluble basic salts on the addition of water—a property sometimes used in detection. Bismuth subnitrate and subcarbonate are used in medicine. High-purity bismuth metal is available for about 5 dollars per lb.

BORON (Ar. *Būraq*, Pers. *Būrah*), B; at. wt. 10.811; at. no. 5; m.p. 2300°C; b.p. sublimes 2550°C; sp. gr. of crystals 2.34, of amorphous variety 2.37; valence 3. Discovered in 1808 by Sir Humphry Davy and by Gay-Lussac and Thenard. The element is not found free in nature, but occurs as orthoboric acid usually in certain volcanic spring waters and as borates in *borax* and *colemanite. Ulexite,* another boron mineral, is interesting as it is nature's own version of "fiber optics." By far the most important source of boron is the mineral *rasorite,* also known as *kernite,* found in the Mojave desert of California. High-purity crystalline boron may be prepared by the vapor phase reduction of boron trichloride or tribromide with hydrogen on electrically heated filaments. The impure, or amorphous boron, a brownish-black powder, can be obtained by heating the trioxide with magnesium powder. Boron of 99.9999% purity has been produced and is available commercially. Elemental boron has an energy band gap of 1.50 to 1.56 electron volts, which is higher than that of either silicon or germanium. It has interesting optical characteristics, transmitting portions of the infra red, and is a poor conductor of electricity at room temperature, but a good conductor at high temperature. The most important compounds of boron are boric, or boracic acid widely used as a mild antiseptic, and borax ($Na_2B_4O_7{\cdot}10H_2O$), which serves as a cleansing flux in welding and as a water softener in washing powders. Boron compounds are used in production of enamels for covering steel of refrigerators, washing machines, and like products. Boron compounds are also extensively used in the manufacture of borosilicate glasses. The isotope boron 10 is used as a control for nuclear reactors, as a shield for nuclear radiation, and in instruments used for detecting neutrons. Boron nitride has remarkable properties and can be used to make a material as hard as diamond. It also has lubricating properties similar to graphite. The hydrides are easily oxidized with considerable energy liberation, and are being studied for use as rocket fuels.

BROMINE (Gr. *bromos,* stench), Br; at. wt. 79.909; at. no. 35; m.p. −7.2°C; b.p. 58.78°C; density of gas 7.59 g/l, liquid 3.12 (20°C); valence 1, 3, 5, or 7. Discovered by Balard in 1826, but not prepared in quantity until 1860. A member of the halogen group of elements, it is obtained from natural brines from wells in Michigan and West Virginia and from sea water by displacement with chlorine; electrolysis might be used. Bromine is the only liquid nonmetallic element. It is a heavy, mobile, reddish-brown liquid, volatilizing readily at room temperature to a red vapor with a strong disagreeable odor, resembling chlorine, and having a very irritating effect on the eyes and throat; it is readily soluble in water or carbon disulfide, forming a red solution; it is less active than chlorine but more so than iodine; it unites readily with many elements and has a bleaching action; when spilled on the skin it produces painful sores. About 80% of the bromine output in the U.S. is used in the production of ethylene dibromide, a lead scavenger used in making gasoline anti-knock compounds. Bromine is also used in making fumigants, flameproofing agents, water purification compounds, dyes, medicinals, sanitizers, inorganic bromides for photography, etc. Organic bromides are also important.

CADMIUM (L. *cadmia;* Gr. *kadmeia*—ancient name for calamine, zinc carbonate) Cd; at. wt. 112.40; at. no. 48; m.p. 320.9°C; b.p. 765°C; sp. gr. 8.65 (20°C); valence 2. Discovered by Stromeyer in 1817 from an impurity in zinc carbonate. Cadmium most often occurs in small quantities associated with zinc ores, such as *sphalerite* (ZnS). *Greenockite* (CdS) is the only mineral of any consequence bearing cadmium. Almost all cadmium is obtained as a by-product in the treatment of zinc, copper, and lead ores. It is a soft, bluish-white metal which is easily cut with a knife. It is similar in many respects to zinc. It is a component of some of the lowest melting alloys; it is used in bearing alloys with low coefficients of friction and great resistance to fatigue; it is used extensively in electroplating, which accounts for about 60% of its use. It is also used in many types of solder, for standard E.M.F. cells, for batteries, and as a barrier to control atomic fission. It forms a number of salts of which the sulfate is the most common; the sulfide is used as a yellow pigment. Cadmium and solutions of its compounds are toxic. Failure to appreciate the toxic properties of cadmium may cause workers to be unwittingly exposed to dangerous fumes. The recommended maximum allowable concentration of cadmium vapor in air is 0.1 mg/cu. meter. The current price of cadmium is about 2 dollars per pound. It is available in high-purity form.

CALCIUM (L. *calx,* lime), Ca; at. wt. 40.08; at. no. 20; m.p. 842–8°C; b.p. 1487°C; sp. gr. 1.55 (20°C); valence 2. Though lime was prepared by the Romans in the first century under the name calx, not until 1808 was the metal discovered. After learning that Berzelius and Pontin prepared calcium amalgam by electrolyzing lime in mercury, Davy was able to isolate the impure metal. Calcium is a metallic element, fifth in abundance in the earth's crust, of which it forms more than three per cent. It is an essential constituent of leaves, bones, teeth and shells. Never found in nature uncombined, it occurs abundantly as *limestone* ($CaCO_3$), *gypsum* ($CaSO_4{\cdot}2H_2O$) and *fluorite* (CaF_2); *apatite* is the fluophosphate or chlorophosphate of calcium. The metal is a silvery color, rather hard, and is prepared by electrolysis of the fused chloride to which calcium fluoride is added to lower the melting point. Chemically it is one of the alkaline earth elements; it readily forms a white coating of nitride in air, reacts with water, burns with a yellow red flame, forming largely the nitride. The metal is used as a reducing agent in preparing other metals, such as thorium, uranium, zirconium, etc., and is used as a deoxidizer, desulfurizer, or decarburizer for various ferrous and nonferrous alloys. It is also used as an alloying agent for aluminum, beryllium, copper, lead and magnesium alloys, and serves as a "getter" for residual gases in vacuum tubes, etc. Its natural and prepared compounds are widely used. Quicklime (CaO), made by heating limestone and changed into slaked lime by the careful addition of water, is the great cheap base of chemical industry with countless uses. Mixed with sand it hardens as mortar and plaster by taking up carbon dioxide from the air. Calcium from limestone is an important element in Portland cement. The solubility of the carbonate in water containing carbon dioxide causes the formation of caves with stalactites and stalagmites and hardness in water. Other important compounds are the carbide (CaC_2), chloride ($CaCl_2$), cyanamide ($Ca(CN_2)$), hypochlorite ($Ca(OCl)_2$), nitrate ($Ca(NO_3)_2$), and sulfide (CaS).

CALIFORNIUM (State and University of California; Cf; at. no. 98) Californium, the sixth transuranium element to be discovered, was produced by Thompson, Street, Ghiorso, and Seaborg in January, 1950 by bombarding microgram quantities of Cm^{242} with 35 MeV helium ions in the Berkeley 60-inch cyclotron. Like berkelium, its chemical properties have been studied solely with tracer amounts. Californium (III) is the only ion stable in aqueous solutions, all attempts to reduce or oxidize californium (III) having failed. The isotope Cf^{249} results from the beta decay of Bk^{249} while the heavier isotopes are produced by intense neutron irradiation by the reactions.

$$Bk^{249}(n,\gamma)\ Bk^{250} \xrightarrow{\beta^-} Cf^{250} \text{ and } Cf^{249}(n,\gamma)Cf^{250} \text{ followed by } Cf^{250}(n,\gamma)\ Cf^{251}(n,\gamma)Cf^{252}.$$

The existence of the isotopes Cf^{249}, Cf^{250}, Cf^{251}, and Cf^{252} makes it feasible to isolate californium in weighable amounts so that its properties can be investigated with macroscopic quantities. Californium 252 is a very strong neutron emitter. One microgram releases 170 million neutrons per minute which presents biological hazards. In 1960 a few tenths of a microgram of californium trichloride $CfCl_3$, californium oxychloride, CfOCl, and californium oxide, Cf_2O_3, were prepared. Pure californium metal has not yet been produced. Eventually, it is thought that gram quantities of californium may be prepared.

It has been postulated that californium may be produced in certain stellar explosions, called supernovae, for the radioactive decay of californium-254 (55-day half-life) agrees with the characteristics of the light curves of such explosions observed through telescopes.

CARBON (L. *carbo*, charcoal), C; at. wt. 12.01115; at. no. 6; m.p. 3550°C, sublimes above 3500°C; b.p. 4827°C; sp.g. amorphous 1.8–2.1, graphite 1.9–2.3, diamond 3.15–3.53 (depending on variety); gem diamond 3.513 (25°C); valence 2, 3, or 4. Carbon, an element of prehistoric discovery, is very widely distributed in nature. It is found in abundance in the sun, stars, comets, and atmospheres of most planets. Carbon in the form of microscopic diamonds is found in some meteorites. The energy of the sun and stars can be attributed at least in part to the well-known carbon-nitrogen cycle. Carbon is found free in nature in three allotropic forms: amorphous, graphite, and diamond. Graphite is one of the softest known materials while diamond is one of the hardest. In combination, carbon is found as carbon dioxide in the atmosphere of the earth and dissolved in all natural waters. It is a component of great rock masses in the form of carbonates of calcium (limestone), magnesium, and iron. Coal, petroleum, and natural gas are chiefly hydrocarbons. Carbon is unique among the elements in the vast number and variety of compounds it can form. With hydrogen, oxygen, and nitrogen, and other elements, it forms almost an infinite number of compounds, carbon atom often being linked to carbon atom. There are upwards of a million or more known carbon compounds, many thousands of which are vital to organic and life processes. Without carbon, the basis for life would be impossible. While it has been thought that silicon might take the place of carbon in forming a host of similar compounds, it is now not possible to form stable compounds with very long chains of silicon atoms. Some of the most important compounds of carbon are: carbon dioxide (CO_2), carbon monoxide (CO), carbon disulfide (CS_2), chloroform ($CHCl_3$), carbon tetrachloride (CCl_4), methane (CH_4), ethylene (C_2H_4), acetylene (C_2H_2), benzene (C_6H_6), ethyl alcohol (C_2H_5OH), acetic acid (CH_3COOH), and their derivatives. Carbon has five isotopes. In 1961 the International Union of Pure and Applied Chemistry adopted the isotope carbon-12 as the basis for atomic weights. Carbon-14, an isotope with a half-life of 5700 yrs, has been widely used to date such materials as wood, archeological specimens, etc.

CERIUM (named for the asteroid *Ceres*, which was discovered in 1801 only two years before the element), Ce; at. wt. 140.12; at. no. 58; m.p. 795°C; b.p. 3468°C; sp. gr. 6.67 to 8.23 (depending on allotropic form); valence 3 or 4. Discovered in 1803 by Klaproth and by Berzelius and Hisinger; metal prepared by Hillebrand and Norton in 1875. Cerium is the most abundant of the metals of the so-called rare earths; it is found in a number of minerals including *monazite, bastnaesite, cerite* and *samarskite*. Monazite and bastnaesite are presently the two most important sources of cerium. Large deposits of monazite found on the beaches of Travancore, India, in river sands in Brazil, and deposits of bastnaesite in Southern California will supply cerium, thorium and the other rare-earth metals for many years to come. The metal exists in four allotropic states. It is prepared by the electrolysis of the chloride. Cerium is an iron-gray lustrous metal. It is malleable and stable in dry air, but slowly decomposed by cold, and rapidly by hot water. The pure metal is likely to ignite if scratched with a knife. Ceric salts are orange-red or yellowish; cerous salts are usually white. Cerium is a component of misch metal, which is extensively used in the manufacture of pyrophoric alloys for cigarette lighters, etc. While cerium is not radioactive, the impure commercial grade may contain traces of thorium, which is radioactive. The oxide is an important constituent of incandescent gas mantles; as ceric sulfate it finds extensive use as a volumetric oxidizing agent in quantitative analysis. Cerium compounds are used in the manufacture of glass, both as a component and as a decolorizer. The oxide is finding increased use as a glass polishing agent instead of rouge, for it is much faster than rouge in polishing glass surfaces. Cerium, with other rare earths, is used in carbon-arc lighting, especially in the motion picture industry. It also finds use in metallurgical and nuclear applications. Commercial cerium metal costs about 10 dollars to 30 dollars/lb. depending on quantity and purity. In small lots, 99.9% cerium costs about 40 cents/gm.

CESIUM (L. *caesius*, sky blue), Cs; at. wt. 132.905; at. no. 55; m.p. 28.5°C; b.p. 690°C; sp. gr. 1.873 (20°C) valence 1. Cesium was discovered spectroscopically in 1860 in mineral water from Dürkheim, by Bunsen and Kirchhoff. Cesium, an alkali metal, occurs in *lepidolite, pollucite* (a hydrated silicate of aluminum and cesium) and other sources. It is isolated by electrolysis of the fused cyanide. The metal is characterized by a spectrum containing two bright lines in the blue along with several others in the red, yellow and green. It is silvery-white, soft, and ductile. It is the most electropositive and most alkaline element. Cesium, gallium and mercury are the only three metals that are liquid at room temperature. Cesium reacts explosively with cold water, and reacts with ice at temperatures above −116°C. Cesium hydroxide, the strongest base known, attacks glass. Because of its great affinity for oxygen the metal is used as a "getter" in radio tubes. It is also used in photo-electric cells, as well as a catalyst in the hydrogenation of certain organic compounds. The metal has recently found application in ion propulsion systems. Although these are not useable in the earth's atmosphere, 1 lb. of cesium in outer space theoretically will propel a vehicle 140 times as far as the burning of the same amount of any known liquid or solid. Its chief compounds are the chloride and the nitrate. The present price of cesium is about 100 dollars to 150 dollars/lb., depending on quantity and purity.

CHLORINE (Gr. *chloros*, greenish-yellow), Cl; at. wt. 35.453; at. no. 17; f.p. −100.98°C; b.p. −34.6°C; density 3.214 g/l, sp. gr. 1.56 (−33.6°C); valence 1, 3, 5, or 7. Discovered in 1774 by Scheele, who thought it contained oxygen; named in 1810 by Davy, who insisted it was an element. In nature it is found in the combined state only, chiefly with sodium as common salt (NaCl), *carnallite* ($KMgCl_3 \cdot 6H_2O$), and *sylvite* (KCl). It is a member of the halogen (salt forming) group of elements and is obtained from chlorides by the action of oxidizing agents and more often by electrolysis; it is a greenish-yellow gas, combining directly with nearly all elements. At 10°C one volume of water dissolves 3.10 volumes of chlorine, at 30° only 1.77 volumes. Chlorine is widely used in making many every-day products. It is used for producing safe drinking water the world over. Even the smallest water supplies are now usually chlorinated. It is also extensively used in the production of paper products, dyestuffs, textiles, petroleum products, medicines, antiseptics, insecticides, foodstuffs, solvents, paints, plastics, and many other consumer products. Most of the chlorine produced is used in the manufacture of chlorinated compounds for sanitation, pulp bleaching, disinfectants, and textile processing. Further use is in the manufacture of chlorates, chloroform, carbon tetrachloride and in the extraction of bromine. Organic chemistry demands much from chlorine, both as an oxidizing agent and in substitution, since it often brings desired properties in an organic compound when substituted for hydrogen, as in one form of synthetic rubber. Chlorine is a respiratory irritant. The gas irritates the mucous membranes and the liquid burns the skin. As little as 3.5 ppm. can be detected as an odor, and 1000 ppm. is likely to be fatal after a few deep breaths. It was used as a war gas in 1915. The recommended maximum allowable concentration in air is about 1 ppm. for prolonged exposure.

CHROMIUM (Gr. *chroma*, color), Cr; at. wt. 51.996; at. no. 24; m.p. 1890°C; b.p. 2482°C; sp. gr. 7.18–7.20 (20°C); valence, 2, 3, or 6. Discovered in 1797 by Vauquelin, who prepared the metal the next year. Chromium is a steel gray, lustrous, hard metal that takes a high polish. The principal ore is *chromite* ($FeO{\cdot}Cr_2O_3$), which is found in Southern Rhodesia, Russia, Republic of South Africa, Cuba, New Caledonia, and the Philippines. The metal is produced by reducing the oxide with aluminum. Chromium is used to harden steel, to manufacture stainless steel, and to form many very useful alloys. Much is used in plating to produce a hard, beautiful surface and to prevent corrosion. Chromium is used to give glass an emerald green color. It finds wide use as a catalyst. All compounds of chromium are colored; the most important are the chromates of sodium and potassium (K_2CrO_4) and the dichromates ($K_2Cr_2O_7$) and the potassium and ammonium chrome alums, as $KCr(SO_4)_2{\cdot}12H_2O$. The dichromates are used as oxidizing agents in quantitative analysis, also in tanning leather. Other compounds are of industrial value; lead chromate is chrome yellow, a valued pigment. Chromium compounds are used in the textile industry as mordants, and by the aircraft and other industries for anodizing aluminum. The refractory industry has found chromite useful for forming bricks and shapes, as it has a high melting point, moderate thermal expansion, and stability of crystalline structure. Hexavalent chromium compounds are toxic. The recommended maximum allowable concentration of dusts and mists in air, measured as CrO_3, is 0.1 mg./cu.m for daily 8-hr. exposure. Chromium is available in high-purity form.

COBALT (*Kobold*, from the German, goblin or evil spirit), Co; at. wt. 58.9332; at. no. 27; m.p. 1495°C; b.p. 2900°C; sp. gr. 8.9 (20°C); valence 2 or 3. Discovered by Brandt about 1735. Cobalt occurs in the minerals *cobaltite*, *smaltite*, and *erythrite*, and is often associated with nickel, silver, lead, copper, and iron ores, from which it is most frequently obtained as a by-product. It is also present in meteorites. Cobalt is a brittle, hard metal, closely resembling iron and nickel in appearance. It has a magnetic permeability of about two-thirds that of iron. It is alloyed with iron, nickel, and other metals to make Alnico, an alloy of unusual magnetic strength with many important uses. Stellite alloys, containing cobalt, chromium, and tungsten, are used for high-speed, heavy-duty, high-temperature cutting tools, and for dies. Cobalt is also used in other magnet steels and stainless steels, and in alloys used in jet turbines and turbo-superchargers. The metal is used in electroplating because of its appearance, hardness, and resistance to oxidation. The salts have been used for centuries for the production of brilliant and permanent blue colors in porcelain, glass, pottery, tiles and enamels. It is the principal ingredient in *Serves* and *Thenard's blue*. A solution of the chloride ($CoCl_2{\cdot}6H_2O$) is used as sympathetic ink. The cobalt ammines are of interest; the oxide and the nitrate are important. Cobalt carefully used in the form of the chloride, sulfate, acetate, or nitrate has been found effective in correcting a certain mineral deficiency disease in animals. Soils should contain 0.13 to 0.30 ppm. of cobalt for proper animal nutrition. Cobalt-60, an artificial isotope, is an important gamma ray source, and is extensively used as a radio-therapeutic agent. Single compact sources of Cobalt-60, are readily available and have an equivalent gamma ray output equal to thousands of grams of radium. The cost of Cobalt-60 is about 2 dollars/curie.

COLUMBIUM (see Niobium).

COPPER (L. *cuprum,* from the island of Cyprus), Cu; at. wt. 63.54; at. no. 29; m.p. 1083.0 ± 0.1°C; b.p. 2595°C; sp. gr. 8.96 (20°C); valence 1 or 2. The discovery of copper dates from prehistoric times; it is said to have been mined for more than 5000 years. It is one of man's most important metals. Copper is reddish-colored, takes on a bright metallic luster and is malleable, ductile, and a good conductor of heat and electricity (second only to silver in electrical conductivity). The electrical industry is one of the greatest users of copper. Copper occasionally occurs native, and is found in many minerals, such as *cuprite, malachite, azurite, chalcopyrite,* and *bornite;* the most important of these compounds are sulfides, oxides, and carbonates. From these it is obtained by smelting, leaching or electrolysis. Its alloys, brass and bronze, long used, are still very important; all American coins are copper alloys; monel and gun metals also contain copper. The most important compounds are the oxide and the sulfate, blue vitriol; the latter has wide use as an agricultural poison and as an algicide in water purification. Copper compounds are widely used in analytical chemistry, as Fehling's solution in tests for sugar. High-purity copper (99.999+ %) is available commercially.

CURIUM (Pierre and Marie Curie), Cm; at. no. 96; sp. g. about 7. Although curium follows americium in the periodic system, it was actually known before americium and was the third transuranium element to be discovered. It was identified by Seaborg, James, and Ghiorso in 1944 at the wartime Metallurgical Laboratory in Chicago as a result of helium-ion bombardment of Pu^{239} in the Berkeley, California, 60-inch cyclotron. The isotope Cm^{242} (half-life 162.5 days), produced from Am^{241} by the reactions $Am^{241}(n,\gamma)Am^{242} \xrightarrow{\beta^-} Cm^{242}$, has been used for much work with macroscopic quantities, although this is difficult due to the extremely high specific alpha activity. Cm^{244} is an excellent isotope for the investigation of curium in weighable amounts because it has a half-life of 19 years and can be prepared in fairly pure form by a sequence of neutron bombardments and chemical separations as follows: The neutron-bombardment of Am^{241} yields Pu^{242} by the reactions $Am^{241}(n,\gamma)$ $Am^{242} \xrightarrow{EC} Pu^{242}$; after chemical separation the Pu^{242} can be transformed to Am^{243} via the reactions $Pu^{242}(n,\gamma)$ $Pu^{243} \xrightarrow{\beta^-} Am^{243}$. Fairly pure Pu^{242} can be prepared more simply by very intense neutron irradiation of Pu^{239} as the result of successive neutron-capture reactions. The Am^{243} can be chemically separated and finally transmuted to Cm^{244} through neutron bombardment by the reactions $Am^{243}(n,\gamma)$ $Am^{244} \xrightarrow{\beta^-} Cm^{244}$. Further neutron bombardment of Cm^{244} produces higher-mass isotopes of longer half-life which are even better for use in the investigation of curium. Curium metal has been isolated, is silvery in appearance, and closely resembles the other actinide metals. The metal is about as malleable as plutonium prepared under the same conditions. It retains its bright appearance in a dry atmosphere of nitrogen for a few hours, but gradually tarnishes and badly corrodes after 24 hours. Also, in common with other metals, it can be prepared by heating curium trifluoride with barium vapor at 1275°C. The only oxidation state in aqueous solutions which has been definitely identified is Cm^{+3}, although a solid, black CmO_2 has been prepared. Both Cm_2O_3 and CmF_3 are white compounds. The general trend of decreasing stability of higher oxidation states with increasing atomic number indicates that oxidation states of curium higher than curium (III) probably is not stable in aqueous solutions. It has further been demonstrated that curium (III) cannot be reduced in aqueous solutions. The experimentally determined magnetic susceptibility of CmF_3 is in good agreement with the value expected for the electronic structure $5f^7$ in curium (III), assuming a Russell-Saunders coupling scheme. In 1960, it was reported that about 12kg. of plutonium was put into a reactor at the Savannah River facility for 2 to 3 yr. neutron bombardment. From this it was hoped to get 100 gms. or so of curium. Some of this would be bombarded to produce heavier elements. About 100 mg. of curium have already been distributed to various national laboratories for research purposes.

DEUTERIUM, an isotope of hydrogen—see Hydrogen.

DYSPROSIUM (Gr. *dysprositos,* hard to get at), Dy; at. wt. 162.50; at. no. 66; m.p. 1407°C; b.p.2600°C; sp. gr. 8.536; valence 3. Discovered but not isolated in 1886 by Lecoq de Boisbaudran. Dysprosium occurs along with other so-called rare-earth or "lanthanide" elements in a variety of minerals, such as *xenotime, fergusonite, gadolinite, euxenite, polycrase* and *blomstrandine.* The most important sources, however, are from *monazite and bastnaesite.* Dysprosium has only been isolated within recent years. Some very pure rare-earth metals, including dysprosium, can be prepared by reduction of the rare-earth salts with calcium followed by volatilization of the calcium from the resulting rare-earth calcium alloy. Dysprosium, as well as certain other rare-earth metals, can also be made by reducing the anhydrous fluoride or chloride with calcium, sodium, potassium, etc. Dysprosium, like the other rare-earth metals, has a metallic, bright silver luster. In moist air, the metal tarnishes with the formation of a loosely adhering oxide film which spalls off and exposes more metal surface to oxidation. The metal reacts slowly with water and is soluble in dilute acids. The metal is soft enough to be cut with a knife and can be machined without sparking if overheating is avoided. Small amounts of impurities can greatly affect the physical properties of the rare earth metals. While dysprosium has not yet found many applications, its thermal neutron absorption cross-section and high melting point suggest metallurgical uses in nuclear control applications for alloying with special stainless steels. Dysprosium oxide is a white powder. In small lots, dysprosium metal (99+% pure) costs less than one dollar per gram.

EINSTEINIUM (Albert Einstein), Es; at. no. 99. Einsteinium, the seventh trans-uranium element of the actinide series to be discovered, was identified by Ghiorso and co-workers at Berkeley in December 1952 in debris from the first large thermonuclear or "hydrogen" bomb explosion, which took place in the Pacific in November 1952. The isotope produced was the 20-day Es^{253}, originating from beta decay of U^{253} and daughters. Its chemical properties have been studied solely with tracer amounts and indicate that the (III) oxidation state may be the only one which exists in aqueous solution. The isotope Es^{253} and heavier isotopes have been produced by intense neutron bombardment of lower elements, such as plutonium, by a process of successive neutron capture interspersed with beta decays until these mass numbers and atomic numbers are reached. In 1961 a sufficient amount of einsteinium was produced by this method to permit separation of a macroscopic amount of Es^{253}. This sample weighed about .01 microgram. A special magnetic-type balance was used in making this determination. Es^{253} so produced was used to produce mendelevium (Element 101). Several isotopes of einsteinium exist. Es^{254} is relatively long-lived, and may be used to prepare larger weighable quantities of the element.

ELEMENT 102. Element 102 was unambiguously discovered and identified in April 1958 by A. Ghiorso, T. Sikkeland, J. R. Walton, and G. T. Seaborg, who prepared the three-second alpha-emitting isotope 102^{254} by bombarding Cm^{246} with C^{12} nuclei, accelerated in a heavy-ion linear accelerator (HILAC), and collecting the recoiling transmutation product atoms. The alpha decay daughter was shown by its ion exchange behavior to be an isotope of fermium (element 100); hence the parent was an isotope of element 102. The chemical properties of element 102 are not known. (In 1957, workers from the Argonne National Laboratory, The Atomic Energy Research Establishment of Harwell, England, and the Nobel Institute for Physics in Stockholm announced the discovery of an isotope of element 102, as a result of bombarding Cm^{244} with C^{13} nuclei in the Nobel Institute's cyclotron. This isotope was said to have a half-life of about 10 min., and to decay by emitting 8.5 m.e.v. alpha particles. The name Nobelium, and the symbol No was proposed, and the name was accepted by the Commission on Atomic Wts. of the International Union of Pure and Applied Chemistry. The acceptance was apparently premature, for all attempts to duplicate the Stockholm experiment in the U.S. and U.S.S.R. have failed, even though much more sensitive research tools were used than in the original experiment.

ERBIUM (*Ytterby*, a town in Sweden), Er; at. wt. 167.26; at. no. 68; m.p. 1497°C; b.p. 2900°C; sp. gr. 9.051; valence 3. Discovered in 1843 by Mosander, who reportedly isolated the element as a dark gray metallic powder. Erbium, one of the so-called rare-earth elements of the lanthanide series, is found in the minerals mentioned under dysprosium above. The element has been isolated in pure form only in recent years. The pure metal is soft and malleable, and has a bright, silvery, metallic luster. As with other rare-earth metals, its properties depend to a certain extent on the impurities present. The metal is fairly stable in air and does not oxidize as rapidly as some of the other rare-earth metals. Erbium can be produced by electrolysis of the fused anhydrous chloride. Recent production techniques, using ion-exchange reactions, have resulted in much lower prices of the rare-earth metals and their compounds in recent years. The cost of 99.9+% erbium metal is about one dollar per gram in small quantities. Erbium is finding nuclear and metallurgical uses. Added to vanadium, for example, erbium lowers the hardness and improves workability. Most of the rare earth oxides have sharp absorption bands in the visible, ultraviolet, and near infrared. This property, associated with the electronic structure, gives beautiful pastel colors to many of the rare-earth salts. Erbium oxide gives a pink color and is used as a colorant in glasses and porcelain enamel glazes.

EUROPIUM (Europe), Eu; at. wt. 151.96; at. no. 63; m.p. 826°C; b.p. 1439°C; sp. gr. 5.259; valence 2 or 3. Europium was discovered and prepared in the form of the oxide by Demarcay in 1896. It occurs in *monazite*, *bastnaesite*, and other minerals along with other rare-earth elements of the lanthanide series. It has been identified spectroscopically in the sun and certain stars. The pure metal was not isolated until recent years. Europium has a relatively low vapor pressure and its oxide can be reduced with lanthanum metal using special techniques. It is silvery-white in appearance and has a metallic luster. As with other rare-earth metals, except for lanthanum, europium ignites in air at about 150 to 180°C. Europium is about as hard as lead and is quite ductile. It is the most reactive of the rare-earth metals, quickly oxidizing in air; it resembles calcium in its reaction with water. Europium isotopes are good neutron absorbers and are being studied for use in nuclear control applications. Europium-doped plastic has been used as a laser material. With the development of ion-exchange techniques and special processes, the cost of the metal has been greatly reduced recently. Europium is one of the rarest and most costly of the rare-earth metals. It is priced at about 4 dollars to 8 dollars/gram, depending on purity and quantity.

FERMIUM (Enrico Fermi), Fm; at. no. 100. Fermium, the eighth transuranium element of the actinide series to be discovered, was identified by Ghiorso and co-workers early in 1953 in the debris from a thermonuclear explosion in the Pacific in work involving the University of California Radiation Laboratory, the Argonne National Laboratory, and the Los Alamos Scientific Laboratory. The isotope produced was the 16-hour Fm^{255}, originating from the beta decay of U^{255} and daughters. Its chemical properties have been studied solely with tracer amounts, and in normal aqueous media only the (III) oxidation state appears to exist. The isotope Fm^{254} and heavier isotopes can be produced by intense neutron irradiation of lower elements, such as plutonium, by a process of successive neutron capture interspersed with beta decays until these mass numbers and atomic numbers are reached. Nine isotopes of fermium are known to exist. Fm^{252}, with a half-life of about

23 hrs., is the longest lived. Fm^{250}, with a half-life of 30 min., has been shown to be a product of decay of Element 102^{254}. It was by chemical identification of Fm^{250} that it was certain that Element 102 had been produced.

FLUORINE (L. and F. *fluere*, flow, or flux), F; at. wt. 18.9984; at. no. 9; f.p. –219.62°C (1 atm.); b.p. – 188.14°C (1 atm.); density 1.696 g/1 (0°C, 1 atm.); sp. gr. of liquid 1.108 at b.p.; valence 1. In 1529, Georgius Agricola described the use of fluorspar as a flux, and as early as 1670 Schwandhard found that glass was etched when exposed to fluorspar treated with acid. Scheele and many later investigators, including Davy, Gay-Lussac, Lavoisier, and Thenard, experimented with hydrofluoric acid, some experiments ending in tragedy. The element was finally isolated in 1886 by Moisson after nearly 74 years of continuous effort. Fluorine occurs chiefly in *fluorspar* (CaF_2) and *cryolite* (Na_3AlF_6), but is rather widely distributed in other minerals. It is a member of the halogen family of elements, and obtained by electrolyzing a solution of potassium hydrogen fluoride in anhydrous hydrogen fluoride in a vessel of metal or transparent fluorspar. Modern commercial production methods are essentially variations on the procedures first used by Moisson. Fluorine is the most electronegative and reactive of all elements. It is a pale yellow, corrosive gas, which reacts with practically all organic and inorganic substances. Finely divided metals, glass, ceramics, carbon, and even water burns in fluorine with a bright flame. Until World War II, there was no commercial production of elemental fluorine. The atom-bomb project and nuclear energy applications, however, made it necessary to produce large quantities. Safe handling techniques have now been developed and it is possible at present to transport liquid fluorine by the ton. Fluorine and its compounds are used in producing uranium (from the hexafluoride) and more than 100 commercial fluorochemicals, including many well-known high-temperature plastics. Hydrofluoric acid is extensively used for etching the glass of light bulbs, etc. Fluorochloro hydrocarbons are extensively used in air conditioning and refrigeration. It has been suggested that fluorine can be substituted for hydrogen wherever it occurs in organic compounds, which could lead to an astronomical number of new fluorine compounds. The presence of fluorine as a soluble fluoride in drinking water to the extent of 2 ppm may cause mottled enamel in teeth, when used by children acquiring permanent teeth; in smaller amounts, however, fluorides are said to be beneficial and used in water supplies to prevent dental cavities. Elemental fluorine is being studied as a rocket propellant as it has an exceptionally high specific impulse value. Although compounds of fluorine with rare gases were reported many years ago, their existence now seems to be confirmed. Fluorides of xenon, radon, and krypton are among those recently reported. Elemental fluorine and the fluoride ion are highly toxic. The free element has a characteristic pungent odor, detectable in concentrations as low as 20 parts per billion, which is below the safe working level. The recommended maximum allowable concentration for a daily 8-hr. exposure is 0.1 ppm.

FRANCIUM (France), Fr; at. no. 87; at. wt. 223 (most stable isotope); valence 1. Discovered in 1939 by Mlle. Marguerite Perey of the Curie Institute, Paris. Francium, the last member of the alkali metal series, occurs as a result of an alpha disintegration of actinium. It can also be made artificially by bombarding thorium with protons. While it occurs naturally in uranium minerals, there is probably less than an ounce of francium at any time in the total crust of the earth. It has the highest equivalent weight of any element, and is the most unstable of the first 101 elements of the periodic system. There are possibly as many as 19 isotopes of francium; the longest lived, Fr^{223}(AcK) a daughter of Ac^{227}, has a half-life of 21 minutes. The chemistry of francium can only be studied on the tracer level.

GADOLINIUM (*gadolinite*—a mineral named for Gadolin, a Finnish chemist), Gd; at. wt. 157.25; at. no. 64; m.p. 1312°C; b.p. ~3000°C; sp. gr. 7.895 and 7.8 (depending on form); valence 3. Gadolinia, the oxide of gadolinium, was separated by Marignac in 1880 and by Lecoq de Boisbaudran in 1886. The element was named for the mineral from which this rare-earth was originally obtained. Gadolinium is found in several other minerals, including *monazite* and *bastnaesite,* which are of commercial importance. The element has been isolated only in recent years. With the development of ion-exchange techniques, the availability and price of gadolinium and the other rare-earth metals has greatly improved. The metal can be prepared in several ways, among which is the reduction of the anhydrous fluoride with metallic calcium. Like other related rare-earth metals, it is silvery-white, has a metallic luster, and is malleable and ductile. The metal is relatively stable in dry air, but in moist air, it tarnishes with the formation of a loosely adhering oxide film which spalls off and exposes more surface to oxidation. The metal reacts slowly with water and is soluble in dilute acid. Gadolinium has the highest neutron absorption cross-section of any known element (46,000 barns) and therefore may be useful for nuclear control rods and as a nuclear shield. It has been used in making gadolinium yttrium garnets, which have microwave applications. The metal has unusual superconductive properties. As little as 1% gadolinium has been found to improve the workability and resistance of iron, chromium, and related alloys to high temperatures and oxidation. Gadolinium ethyl sulfate has extremely low noise characteristics and may find use in duplicating the performance of h.f. amplifiers, such as the maser. The metal is ferromagnetic. The price of the metal is about 200 dollars/lb.

GALLIUM (L. *Gallia,* France), Ga; at. wt. 69.72; at. no. 31; m.p. 29.78°C; b.p. 2403°C; sp. gr. 5.907 (20°C); valence 2 or 3. Predicted and described by Mendeleev as ekaaluminum, and discovered spectroscopically by Lecoq de Boisbaudran in 1875, who in the same year obtained the free metal by electrolysis of a solution of the hydroxide in KOH. Gallium is often found as a trace element in *diaspore, sphalerite, germanite, bauxite,* and *coal.* Some flue dusts from burning coal have been shown to contain as much as 1.5% gallium. It is the only metal, except for mercury, cesium, and rubidium, which can be liquid near room temperatures; this makes possible its use in high-temperature thermometers. It has one of the longest liquid range of any metal and has a low vapor pressure even at high temperatures. Ultra-pure gallium has a beautiful, silvery appearance and the solid metal exhibits a conchoidal fracture similar to glass. The metal expands 3.1% on solidifying; therefore, it should not be stored in glass or metal containers as they may break as the metal solidifies. Gallium wets glass or porcelain, and forms a brilliant mirror when it is painted on glass. It has found recent use in doping semiconductors and producing solid-state devices, such as transistors. Gallium arsenide is capable of converting electricity directly into coherent light. Gallium readily alloys with most metals, and has been used as a component in low-melting alloys. Its toxicity appears to be of a low order, but should be handled with care until more data are forthcoming. The metal can be supplied in ultra-pure form (99.99999+%).

GERMANIUM (L. *Germania,* Germany), Ge; at. wt. 72.59; at. no. 32; m.p. 937.4°C; b.p. 2830°C; sp. gr. 5.323 (25°C); valence 2 and 4. Predicted by Mendeleev in 1871 as ekasilicon, and discovered by Winkler in 1886. The metal is found in *argyrodite,* a sulfide of germanium and silver; in *germanite,* which contains 8% of the element; in zinc ores; in coal; and in other minerals. The element is frequently obtained commercially from flue dusts of smelters processing zinc ores, and has been recovered from the by-products of combustion of certain coals. Its presence in coal insures a large reserve of the element in the years to come. Germanium can be separated from other metals by fractional distillation of its volatile tetrachloride. The tetrachloride may then be hydrolyzed to give GeO_2; then the dioxide can be reduced with hydrogen to give the metal. Recently developed zone-refining techniques permit the production of germanium of ultra-high purity. The element is a gray-white metalloid, and in its pure state is crystalline and brittle, retaining its luster in air at room temperature. It is a very important semiconductor material. Doped with arsenic or gallium, it is used as a transistor element, which is rapidly displacing vacuum tubes in electronics. It is used also in the construction of rectifiers, photodiodes, etc. Transistors now provide the largest use for the element, but germanium is also finding many other applications including use as an alloying agent, as a phosphor in fluorescent lamps, and as a catalyst. The element is transparent to the infrared and is used in infrared spectroscopes and other optical equipment, including extremely sensitive infrared detectors. Its high index of refraction and dispersion has made germanium useful as a component of glasses used in wide angle camera lenses and microscope objectives. The field of organogermanium chemistry is becoming increasingly important. Certain germanium compounds have a low mammalian toxicity, but a marked activity against certain bacteria, which makes them of interest as chemotherapeutic agents. The cost of germanium is about 50 cents/gm. for 99.9+% purity in small lots.

GOLD (Sanskrit *Jval*; anglo-Saxon *gold*), Au (L. *aurum*, shining dawn); at. wt. 196.967; at. no. 79; m.p. 1063.0°C: b.p. 2966°C; sp. gr. 19.32 (20°C); valence 1 or 3. Known and highly valued from earliest times. Gold is found in nature as the free metal and in tellurides; it is very widely distributed and is almost always associated with quartz or pyrite. It occurs in veins and alluvial deposits, and is often separated from rocks and other minerals by sluicing or panning operations. The metal is also obtained from its ores by cyaniding, amalgamating, and smelting processes. Refining is also frequently done by electrolysis. Gold occurs in sea water to the extent of 0.1 to 2 mg/ton, depending on the location where the sample is taken. As yet no method has been found for recovering gold from sea water profitably. It is estimated that all the gold in the world, so far refined, could be placed in a single cube fifty feet on a side. Of all the elements, gold in its pure state is undoubtedly the most beautiful. It is metallic, having a yellow color when in a mass, but when finely divided it may be black, ruby, or purple. The Purple of Cassius is a delicate test for auric gold. It is the most malleable and ductile metal; one ounce of gold can be beaten out to 300 sq. ft. It is a soft metal and is usually alloyed to give it more strength. It is a good conductor of heat and electricity, and is unaffected by air and most reagents. It is used in coinage and is a standard for monetary systems in many countries. It is also extensively used for jewelry, decoration, dental work, and for plating. It has recently been used for coating certain space satellites, as it is a good reflector of infrared and is inert. Gold, like other precious metals, is measured in troy weight; when alloyed with other metals, the term *carat* is used to express the amount of gold present—24 carats being pure gold. For many years the value of gold was set by the United States at $20.67/troy oz; since 1934 this value has been fixed by law and Presidential Order at 35 dollars/troy oz. The commonest compounds are auric chloride ($AuCl_3$) and chlorauric acid ($HAuCl_4$), the latter being used in photography for toning the silver image. Gold has many isotopes; Au^{198}, with a half-life of 2.7 days is used for treating cancer and other diseases. A mixture of one part nitric acid with three of hydrochloric acid is called *aqua regia*, because it dissolves Gold, the King of Metals. Gold is available commercially with a purity of 99.999+%.

HAFNIUM (*Hafnia*, Latin name for Copenhagen), Hf; at. wt. 178.49; at. no. 72; m.p. 2150°C; b.p. 5400°C; sp. gr. 13.29; valence 4. Was suspected as being present in various minerals and concentrations many years prior to its discovery in 1923 by D. Coster and G. von Hevesey. On the basis of the Bohr theory, the new element was expected to be associated with zirconium, and it was finally identified in *zircon* from Norway, by means of x-ray spectroscopic analysis. It was named in honor of the city in which the discovery was made. Most zirconium minerals contain hafnium to the extent of 1 to 5%. It was originally separated from zirconium by repeated recrystallization of the double ammonium or potassium fluorides by von Hevesey and Jantzen. Metallic hafnium was first prepared by van Arkel and deBoer by passing the vapor of the tetraiodide over a heated tungsten filament. Since the time hafnium was first separated from zirconium, many other methods of separation have been devised but only three have been found suitable for large-scale commercial production. The hot wire process, developed by van Arkel and deBoer, and the Kroll process are now commonly used to produce the metal. Hafnium has a brilliant silver luster. Its properties are considerably influenced by the impurities of zirconium present. Of all the elements, zirconium and hafnium are the two most difficult to separate. Their chemistry is almost identical; however the density of zirconium is about half that of hafnium. Very pure hafnium has been produced with zirconium being the major impurity. Because hafnium has a good absorption cross-section for thermal neutrons (almost 600 times that of zirconium), has excellent mechanical properties, and is extremely corrosion resistant, it is used for reactor control rods. Such rods are used in nuclear submarines, such as the *Nautilus.* Hafnium has been successfully alloyed with iron, titanium, niobium, tantalum, and other metals. Hafnium carbide is the most refractory binary composition known, and the nitride is the most refractory of all known metal nitrides (m.p. 3310°C). Hafnium is used in gas-filled and incandescent lamps, and is an efficient "getter" for scavenging oxygen and nitrogen. The price of the metal is in the broad range of 30 to 175 dollars/lb., depending on purity and quantity. The yearly production of hafnium in the U. S. is now in excess of 50,000 lb.

HELIUM (Gr. *helios,* the sun), He; at. wt. 4.0026; at. no. 2; m.p. below −272.2°C (26 atm.); b.p. −268.6°C; density 0.177 g/l; liquid density 7.62 lb./cu. ft. at. b.p.; valence 0. Evidence of the existence of helium was first obtained by Janssen during the solar eclipse of 1868 when he detected a new line in the solar spectrum; Lockyer and Frankland suggested the name *helium* for the new element; in 1895 Ramsay discovered helium in the uranium mineral *clevite,* and it was independently discovered in clevite by the Swedish chemists Cleve and Langlet about the same time. Except for hydrogen, helium is the most abundant element found throughout the universe. It has been detected spectroscopically in great abundance, especially in the hotter stars, and it is an important component in both the proton-proton reaction and the carbon cycle, which accounts for the energy of the sun and stars. The fusion of hydrogen into helium provides the energy of the hydrogen bomb. The helium content of the atmosphere is about 1 part in 200,000. Helium is chemically inert. While it is present in various radioactive minerals as a decay product, the bulk of the Free World's supply is obtained from wells in Texas, Oklahoma, and Kansas. The cost of helium fell from 2500 dollars/cu. ft. in 1915 to 1.5 cent/cu. ft. in 1940. The current price of 99.99% pure helium is about 13 cents/cu. ft. in small quantities. Helium has the lowest m.p. of any element and has found wide use in cryogenic research as its b.p. is close to absolute zero. Its use in the study of superconductivity is vital. Using liquid helium, Kurti and co-workers, and others, have succeeded in obtaining temperatures of a few microdegrees Kelvin, by the adiabatic demagnetization of copper nuclei, starting from about 0.01°K. Four isotopes of helium are known. Liquid helium exists in two forms: He I and He II, with a sharp transition point at 2.174°K (3.83 cm Hg). He I (above this temperature) is a normal liquid, but He II (below it) is unlike any other known substance. It expands on cooling; its conductivity for heat is enormous; and neither its heat conduction nor viscosity obey normal rules. It has other peculiar properties. Helium is the only liquid that cannot be solidified by lowering the temperature. It remains liquid down to absolute zero at ordinary pressures, but it can readily be solidified by increasing the pressure. Solid He^3 and He^4 are unusual in that both can readily be changed in volume by more than 30% by application of pressure. Helium is widely used as an inert gas shield for arc welding; as a lifting gas for balloons (being much safer than hydrogen); as a protective gas in growing silicon and germanium crystals, and in titanium and zirconium production; as a cooling medium for nuclear reactors; and as a gas for supersonic wind tunnels. A mixture of 80% helium and 20% oxygen is used as an artificial atmosphere for divers and others working under pressure.

HOLMIUM (L. *Holmia*, for Stockholm), Ho; at. wt. 164.930; at. no. 67; m.p. 1461°C; b.p. 2600°C; sp. gr. 8.803; valence 3. The spectral absorption bands of holmium were noticed in 1878 by the Swiss chemists Delafontaine and Soret, who announced the existence of an "Element X." Cleve, of Sweden, later independently discovered the element while working on erbia earth. The element is named after Cleve's native city. Pure holmia, the yellow oxide, was prepared by Homberg in 1911. Holmium occurs in *gadolinite*, *monazite*, and in other rare-earth minerals. It is commercially obtained from *monazite*, occurring in that mineral to the extent of about 0.05%. It has been isolated in pure form only in recent years. It can be separated from other rare earths by ion-exchange techniques, and isolated by the reduction of its anhydrous fluoride with calcium metal. Pure holmium has a metallic to bright silver luster. It is relatively soft and malleable, and stable in dry air at room temperature, but rapidly oxidizes in moist air and at elevated temperatures. Few uses have yet been found for the element. The price of 99+% holmium metal is about 80 cents/gm.

HYDROGEN (Gr. *hydro*, water, and *genes*, forming), H; at. wt. 1.00797; at. no. 1; m.p. −259.14°C; b.p. −252.5°C; density 0.08988 g/l; sp. gr. liquid 0.070 (−252°C); valence 1. Hydrogen was prepared many years before it was recognized as a distinct substance by Cavendish in 1766. It was named by Lavoisier. Hydrogen is the most abundant of all elements in the universe, and it is thought that the heavier elements were, and still are being, built from hydrogen and helium. It has been estimated that hydrogen makes up more than 90% of all the atoms or three-quarters of the mass of the universe. It is found in the sun and most stars, and plays an important part in the proton-proton reaction and carbon-nitrogen cycle, which accounts for the energy of the sun and stars. Hydrogen is abundant in some planets, such as Jupiter; the core of this planet may consist of solid hydrogen, at least in part, and liquid hydrogen may be present elsewhere on the planet. On earth, hydrogen occurs chiefly in combination with oxygen in water, but it is also present in organic matter, such as living plants, petroleum, coal, etc. It is present as the free element in the atmosphere, but only to the extent of less than 1 part/million, by volume. It is the lightest of all gases, and combines with other elements, sometimes explosively, to form compounds. Great quantities of hydrogen are required commercially for the fixation of nitrogen from the air in the Haber ammonia process and for the hydrogenation of fats and oils. It is also used in large quantities in methanol production, in hydrodealkylation, hydrocracking, and hydrodesulfurization. It is also used as a rocket fuel, for welding, for production of hydrochloric acid, for the reduction of metallic ores, and for filling balloons. Production in the U. S. alone now amounts to hundreds of millions, of cubic feet per day. It is prepared by the action of steam on heated carbon, by decomposition of certain hydrocarbons with heat, by the electrolysis of water, or by the displacement from acids by certain metals. It is also produced by the action of sodium or potassium hydroxide on aluminum. Liquid hydrogen is important in cryogenics and in the study of superconductivity as its m.p. is only a few degrees above absolute zero. In 1932 Urey announced the preparation of a stable isotope, deuterium, with an atomic weight of 2. Two years later an unstable isotope, tritium, with an atomic weight of 3 was discovered. Tritium has a half-life of about 12.5 years. One part deuterium is found to about 6000 ordinary hydrogen atoms. Tritium atoms are also present but in much smaller proportion. Tritium is readily produced in nuclear reactors, and is used in the production of the hydrogen bomb. It is also used as a radioactive agent in making luminous paints. The current price of tritium, to authorized personnel, is about 2 dollars/curie; deuterium gas is readily available, without permit, at about 1 dollar/liter. Heavy water, deuterium oxide, which is used as a moderator to slow down neutrons, is available without permit at a cost of 6 cents to 1 dollar/gm., depending on quantity and purity. Quite apart from isotopes, it has been shown that hydrogen gas under ordinary conditions is a mixture of two kinds of molecules, known as ortho- and para-hydrogen, which differ from one another by the spins of their electrons and nuclei.

INDIUM (from the brilliant indigo line in its spectrum), In; at. wt. 114.82; at. no. 49; m.p. 156.61°C; b.p. 2000 ± 10°C; sp. gr. 7.31 (20°C); valence 1,2(?), or 3. Discovered by Reich and Richter, who later isolated the metal. Indium is most frequently associated with zinc minerals, and it is from these that most commercial indium is now obtained; however, it is also found in iron, lead, and copper ores. Until 1924, a gram or so constituted the world's supply of this element in isolated form. It is now available in moderate quantities, but it must still be considered to be a rare element. It is probably about as abundant as silver. The present cost of indium is about $1.50 to $5.00/troy oz., depending on quantity and purity. It is available in ultrapure form. Indium is a very soft, silvery-white metal with a brilliant luster. The pure metal gives a high-pitched "cry" when bent. It wets glass, as does gallium. It has found application in making low-melting alloys; an alloy of 24% indium–76% gallium is liquid at room temperature. It is used in making bearing alloys, germanium transistors, rectifiers, thermistors, and photoconductors. It can be plated onto metal and evaporated onto glass forming a mirror as good as that made with silver, but with more resistance to atmospheric corrosion. There is evidence that indium is toxic and may be potentially hazardous to workers under certain conditions. Normal hygenic precautions should provide adequate protection.

IODINE (Gr. *iodes,* violet), I; at. wt. 126.9044; at. no. 53; m.p. 113.5°C; b.p. 184.35°; density of the gas 11.27 g/l; sp. gr. solid 4.93 (20°C); valence 1,3,5, or 7. Discovered by Courtois in 1811. Iodine, a halogen, occurs sparingly in the form of iodides in sea water from which it is assimilated by seaweeds, in Chilean saltpeter and nitrate-bearing earth, known as *caliche,* in brines from old sea deposits, and in brackish waters from oil and salt wells. Ultrapure iodine can be obtained from the reaction of potassium iodide with copper sulfate. Several other methods of isolating the element are known. Iodine is a grayish-black, lustrous solid, volatilizing at ordinary temperatures into a blue-violet gas with an irritating odor; it forms compounds with many elements, but is less active than the other halogens, which displace it from iodides. It dissolves readily in chloroform, carbon tetrachloride, or carbon disulfide to form beautiful purple solutions. It is only slightly soluble in water. Iodine compounds are important in organic chemistry and very useful in medicine. The artificial radioisotope I^{131}, with a half-life of 8 days, has been used in treating the thyroid gland. The most common compounds are the iodides of sodium and potassium (KI) and the iodates (KIO_3). Lack of iodine is the cause of goiter. The iodide and thyroxin, which contain iodine, are used internally in medicine, and a solution of KI and iodine in alcohol is used for external wounds. Potassium iodide finds use in photography. The deep blue color with starch solution is characteristic of the free element. Care should be taken in handling and using iodine as contact with the skin can cause lesions; iodine vapor is intensely irritating to the eyes and mucous membranes. The recommended maximum allowable concentration in air is 1 mg/cu. meter.

IRIDIUM (L. *iris,* rainbow), Ir; at. wt. 192.2; at. no. 77; m.p. 2410°C; b.p. 4527 ± 100°C; sp. gr. 22.42 (17°C); valence 3 or 4. Discovered in 1803 by Tennant in the residue left when crude platinum is dissolved by aqua regia. The name iridium is appropriate, for its salts are highly colored. Iridium, a metal of the platinum family, is white, similar to platinum, but with a slight yellowish cast. It is very hard and brittle, making it very hard to machine, form, or work. It is the most corrosion-resistant metal known, and was used in making the standard meter in Paris, which is a 90% platinum–10% iridium alloy. Iridium occurs uncombined in nature with platinum and other metals of this family in alluvial deposits. It has found use in making crucibles and apparatus for use at high temperatures, and for electrical contacts. Its principal use is as a hardening agent for platinum. With osmium, it forms an alloy which is used for tipping pens and compass bearings. The specific gravity of iridium is only very slightly lower than that of osmium, which is generally credited as being the heaviest known element. The density of iridium and osmium, calculated from the space lattice, which with these elements may be more reliable than actual physical measurements, gives a density of 22.65 compared to 22.61 for osmium. At present, therefore, we know that either iridium or osmium is the densest known element, but the data do not allow selection between the two.

IRON (Anglo-Saxon, *iron*) Fe, (L. *ferrum*), at. wt. 55.847; at. no. 26; m.p. 1535°C; b.p. 3000°C; sp. gr. 7.874 (20°C); valence 2,3,4, or 6. The use of iron is prehistoric. Genesis mentions that Tubal-Cain, seven generations from Adam, was "an instructer of every artificer in brass and iron." Iron is a relatively abundant element in the universe. It is found in the sun and many types of stars in considerable quantity. Its nuclei are very stable. Iron is found native as a principal component of a class of meteorites known as *siderites*, and is a minor constituent of the other two classes. The core of the earth, 2150 miles in radius, is thought to be largely composed of iron with about 10% occluded hydrogen. The metal is the fourth most abundant element, by weight, making up the crust of the earth. The most common ore is *hematite* (Fe_2O_3), from which the metal is obtained by reduction with carbon. Iron is found in other widely distributed minerals, such as *magnetite* which is frequently seen as *black sands* along beaches and banks of streams. *Taconite* is becoming increasingly important as a commercial ore. Iron is a vital constituent of plant and animal life, and appears in hemoglobin. The pure metal is not often encountered in commerce, but is usually alloyed with carbon or other metals. The pure metal is very reactive chemically, and rapidly corrodes especially in moist air or at elevated temperatures. It has four allotropic forms, or ferrites, known as α, β, γ, and δ, with transition points at 770°, 928°, and 1530°C. The α form is magnetic but when transformed into the β form, the magnetism disappears although the lattice remains unchanged. The relations of these forms are peculiar. Pig iron is an alloy containing about 3% carbon with varying amounts of S, Si, Mn, and P. It is hard, brittle, fairly fusible, and is used to produce other alloys, including steel. Wrought iron contains only a few tenths of per cent of carbon, is tough, malleable, less fusible, and has usually a "fibrous" structure. Carbon steel is an alloy of iron with carbon, with small amounts of Mn, S, P, and Si. Alloy steels are carbon steels with other additives, such as nickel, chromium, vanadium, etc. Iron is the cheapest and most abundant, useful, and important of all metals.

KRYPTON (Gr. *kryptos*, hidden), Kr; at. wt. 83.80; at. no. 36; m.p. −156.6°C; b.p. −152.30 ± 0.10°C; density 3.733 g/l (0°C); valence 0. Discovered in 1898 by Ramsay and Travers in the residue left after liquid air had nearly boiled away. Krypton is present in the air to the extent of about 1 part per million. One of the rare gasses, it is characterized by its whitish color, similar to helium, when excited in a discharge tube, and by its brilliant green and orange spectral lines. Krypton has been used as a gas to fill small bright lamps, such as used by miners; to fill bactericidal lamp-starter tubes and special electron tubes; and is used in electron etching and ion engines. Krypton 85, one of the 19 known isotopes of krypton and an isotope which holds its radioactivity longer than other gases, is used to stabilize the starting potential of gas-filled electronic tubes by providing free electrons. While krypton is generally thought of as inert gas, the compound KrF_4 was recently reported as having been made. There is yet some question whether true compounds of inert gases exist. Krypton gas is readily available from commercial sources at about 3 cents per cc in small quantities.

LANTHANUM (Gr. *lanthanein,* to lie hidden), La; at. wt. 138.91; at. no. 57; m.p. 920°C; b.p. 3469°C; sp. gr. 5.98–6.186 (depending on crystal structure); valence 3. Mosander in 1839 extracted a new earth, lanthana, from cerium nitrate, and recognized the new element. It is found with other rare-earth minerals, including *cerite, monazite,* and *bastnaesite.* It occurs in monazite to the extent of 22%; in bastnaesite 26%. These two minerals are the most important commercial sources. Misch metal, used in making lighter flints, contains about 25% lanthanum. The metal was isolated in relatively pure form in 1923. Ion-exchange reactions have led to much easier isolation of the rare-earth elements and have resulted in much lower prices and abundance of the rare-earth metals. The metal can be produced commercially by reducing the anhydrous fluoride with calcium. Lanthanum is silvery white, malleable, ductile, and soft enough to be cut with a knife. Next to europium it is the most reactive of the rare earth metals, rapidly oxidizing when exposed to air. The metal lights in air at 440°C. It is a superconductor at 6°K. It exhibits three allotropic forms. Little is known of the toxicity of the rare earth metals; therefore they should be handled carefully until more data are forthcoming. Rare-earth compounds, containing lanthanum, are extensively used in carbon lighting applications, especially by the motion picture industry for studio lighting and projection. This application consumes about 25% of the rare earth compounds produced. The price of 99.9+% La_2O_3 is about 6 dollars/lb. The cost of the metal is about 110 dollars/lb.

LAWRENCIUM (Ernest O. Lawrence, inventor of the Cyclotron), Lr; at. no. 103; at. mass no. 257. The last member of the 5f transition elements (actinide series) was discovered March 1961 by A. Ghiorso, T. Sikkeland, A. E. Larsh, and R. M. Latimer. A three microgram californium target, consisting of a mixture of isotopes of mass number 249, 250, 251, and 252 was bombarded with either B^{10} or B^{11}. The electrically charged transmutation nuclei recoiled with an atmosphere of helium and were collected on a thin copper conveyor tape which was then moved to place collected atoms in front of a series of solid-state detectors. The isotope of element 103 produced in this way decayed by emitting an 8.6 Mev alpha particle with a half-life of 8 seconds. The chemical properties of lawrencium are not known, and it is doubtful if it will ever be possible to make chemical or physical studies with macroscopic amounts of this element. As with element 102, it has not been possible to obtain chemical identification of lawrencium; the discovery is based solely on nuclear evidence.

LEAD (Anglo-Saxon *lead*) Pb (L. *plumbum*); at. wt. 207.19; at. no. 82; m.p. 327.5; b.p. 1744°C; sp. gr. 11.35 (20°C); valence 2 or 4. Long known; mentioned in Exodus. The alchemists believed lead to be the oldest metal and associated it with the planet Saturn. Lead is obtained chiefly from galena (PbS) by a roasting process. It is a bluish-white metal of bright luster, very soft, highly malleable, is ductile and a poor conductor of electricity. It is very resistant to corrosion; lead pipes bearing the insignia of Roman emperors, used as drains from the baths, are still in service. It is used as containers for corrosive liquids, such as in sulfuric acid chambers, and it may be toughened by the addition of a small percentage of antimony or other metals. Its alloys include solder, type metal, and various antifriction metals. Great quantities of lead, both as the metal and as the dioxide, are used in storage batteries. Much metal also goes into cable covering, plumbing, ammunition, and in the manufacture of lead tetraethyl, used as an anti-knock compound in gasoline. The metal is very effective as a sound absorber, is used as a radiation shield around x-ray equipment and nuclear reactors, and is used to absorb vibration. White lead, the basic lead carbonate, sublimed white lead ($PbSO_4$), chrome yellow ($PbCrO_4$), red lead (Pb_3O_4) and other lead compounds are extensively used in paints. Lead oxide is used in producing fine "crystal glass" and "flint glass" of a high index of refraction for achromatic lenses. The nitrate and the acetate are soluble salts. Lead salts are sometimes used in medicine, as antiseptics and astringents. Care must be used in both medicine and industry as lead is a cumulative poison.

LITHIUM (Gr. *lithos,* stone) Li; at. wt. 6.939; at. no. 3; m.p. 179°C; b.p. 1317°C; sp. gr. 0.534 (20°C); valence 1. Discovered by Arfvedson in 1817. Lithium is the lightest of all metals, with a density only about half that of water. It does not occur free in nature; combined it is found in small amounts in nearly all igneous rocks and in the waters of many mineral springs. *Lepidolite, spodumene, petalite,* and *amblygonite* are the more important minerals containing it. The metal is produced electrolytically from the fused chloride. Lithium is silvery in appearance much like Na and K, other members of the alkali metal series. It reacts with water, but not as vigorously as sodium. Lithium imparts a beautiful crimson color to a flame, but when the metal burns strongly the flame is a dazzlirg white. Since World War II, the production of lithium metal and its compounds has increased greatly. Because the metal has the highest specific heat of any solid element, it has found use in heat transfer applications; however it is corrosive and requires special handling. The metal has been used as an alloying agent, is of interest in synthesis of organic compounds, and it has nuclear applications. Lithium is used in special glasses and ceramics. The glass for the 200-inch telescope at Mt. Palomar contains lithium as a minor ingredient. Lithium chloride is one of the most hygroscopic materials known, and it, as well as lithium bromide, is used in air conditioning and industrial drying systems. Lithium stearate is used as an all-purpose and high-temperature lubricant. Other lithium compounds are used in dry cells and storage batteries. The metal is priced at about 10 dollars per lb.

LUTETIUM (*Lutetia,* ancient name for Paris—sometimes called *cassiopeium* by the Germans). Lu; at. wt. 174.97; at. no. 71; m.p. 1652°C; b.p. 3327°C; sp. gr. 9.872; valence 3. In 1907 Urbain described a process by which Marignac's ytterbium (1879) could be separated into the two elements, ytterbium (neoytterbium) and lutetium. These elements were identical with "aldebaranium" and "cassiopeium" independently discovered by von Welsbach about the same time. Charles James of the University of New Hampshire also independently prepared the very pure oxide, lutetia, at this time. The spelling of the element was changed from lutecium to lutetium very recently. Lutetium occurs in very small amounts in nearly all minerals containing yttrium, and is present in *monazite* to the extent of about 0.003%, which is a commercial source. The pure metal has been isolated only in recent years and is one of the most difficult to prepare. The metal is silvery white and relatively stable in air. While new techniques, including ion exchange reactions, have been developed to separate the various rare earth elements, lutetium is still the most costly of all naturally occurring rare earths, although it is slightly more abundant than thulium. It is now priced at about 10 dollars/gm. or 2100 dollars/lb. Lu^{176} which occurs naturally (2.5%) with Lu^{175} (97.5%), is radioactive with a half-life of about 7×10^{10} yrs. Virtually no commercial uses have been found yet for lutetium, as it is still one of the most costly natural elements.

MAGNESIUM (*Magnesia,* district in Thessaly), Mg; at. wt. 24.312; at. no. 12; m.p. 651°C; b.p. 1107°C; sp. gr. 1.738 (20°C); valence 2. Compounds of magnesium have long been known. Black recognized magnesium as an element in 1755. It was isolated by Davy in 1808, and prepared in coherent form by Bussy in 1831. Magnesium is the eighth most abundant element in the earth's crust. It does not occur uncombined, but is found in large deposits in the form of *magnesite, dolomite,* and other minerals. The metal is now principally obtained in the U.S. by electrolysis of fused magnesium chloride derived from brines, from wells and from sea water. Magnesium is a light, silvery white, and fairly tough metal. It tarnishes slightly in air, and finely divided magnesium readily ignites upon heating in air and burns with a dazzling white flame. It is used in flash-light photography, flares, and pyrotechnics, including incendiary bombs. It is one third lighter than aluminum, and in alloys is essential for airplane and missile construction. The metal improves the mechanical, fabrication, and welding characteristics of aluminum, when used as an alloying agent. Magnesium is used in producing nodular graphite in cast iron, and is used as an additive to conventional propellants. It is also used as a reducing agent in the production of pure uranium and other metals from their salts. The oxide, chloride, sulfate, and citrate are used in medicine. Dead-burned magnesite is employed for refractory purposes, such as brick and liners in furnaces and converters. Organic magnesium compounds (Grignard's reaction) are important. Great care should be taken in handling magnesium, especially in the finely divided state, as serious fires can occur. Water should not be used on burning magnesium or on magnesium fires.

MANGANESE (L. *magnes,* magnet—from magnetic properties of pyrolusite; It. *manganese,* corrupt form of *magnesia*), Mn; at. wt. 54.9380; at. no. 25; m.p. 1244 ± 3°C; b.p. 2097°C; sp. gr. 7.21 to 7.44, depending on allotropic form; valence 1, 2, 3, 4, 6, or 7. Recognized by Scheele, Bergman, and others as an element and isolated by Gahn in 1774 by reduction of the dioxide with carbon. Manganese minerals are widely distributed; oxides, silicates, and carbonates are the most common. The recent discovery of large quantities of manganese nodules on the floor of the oceans appears promising as a new source of manganese. These nodules contain about 24% manganese together with many other elements in lesser abundance. Pyrolusite (MnO_2) and rhodochrosite ($MnCO_3$) are common ores. The metal is obtained by reduction of the oxide with sodium, magnesium, aluminum, or by electrolysis. It is gray-white, resembling iron, but is harder and very brittle. The metal is reactive chemically, and decomposes cold water slowly. Manganese is used to form many important alloys. In steel, manganese improves the rolling and forging qualities, strength, toughness, stiffness, wear resistance, hardness, and hardenability. With aluminum and antimony, especially with small amounts of copper, it forms highly ferromagnetic alloys. Manganese metal is ferromagnetic only after special treatment. The pure metal exists in four allotropic forms. The alpha form is stable at ordinary temperature; gamma manganese, which changes to alpha at ordinary temperatures, is said to be flexible, soft, easily cut, and capable of being bent. The dioxide (pyrolusite) is used as a depolarizer in dry cells, and is used to "decolorize" glass that is colored green by impurities of iron. Manganese by itself colors glass an amethyst color, and is responsible for the color of true amethyst. The dioxide is also used in the preparation of oxygen, chlorine, and in drying black paints. The permanganate is a powerful oxidizing agent and is used in quantitative analysis and in medicine.

MENDELEVIUM (Dmitri Mendeleev), Md; at. no. 101. Mendelevium, the ninth transuranium element of the actinide series to the discovered, was first identified by Ghiorso, Harvey, Choppin, Thompson, and Seaborg early in 1955 as a result of the bombardment of the isotope Es^{253} with helium ions in the Berkeley 60-inch cyclotron. The isotope produced was presumably Md^{256} which has a half-life of about 1.5 hours and apparently decays by electron capture to Fm^{256}, which in turn decays predominantly by spontaneous fission with a half-life of about 3 hours. This first identification was notable in that only of the order of one to three atoms per experiment was produced. The extreme sensitivity for detection depended on the fact that its chemical properties could be accurately predicted as eka-thulium. There was a high sensitivity for detection because of the spontaneous fission decay. The chemical properties have been investigated solely by the tracer technique and seem to indicate that the predominant oxidation state in aqueous solution is the (III) state. There seem to be no isotopes of sufficiently long half-life to make it possible to isolate this element in weighable quantity.

MERCURY (Planet *Mercury*), Hg (*hydrargyrum,* liquid silver); at. wt. 200.59; at. no. 80; m.p. −38.87°C; b.p. 356.58°C; sp. gr. 13.546 (20°C); valence 1 or 2. Known to ancient Chinese and Hindus; found in Egyptian tombs of 1500 B.C. Mercury is the only common metal liquid at ordinary temperatures. It occurs free in nature, but the chief source is *cinnabar* (HgS). Spain and Italy produce about 50% of the world's supply of the metal. The commercial unit for handling mercury is the "flask," which weighs 76 lbs. The metal is obtained by heating cinnabar in a current of air and by condensing the vapor. It is a heavy, silvery white metal; a rather poor conductor of heat, as compared with other metals; and a fair conductor of electricity. The metal is widely used in laboratory work for making thermometers, barometers, diffusion pumps, and many other instruments. It is used in making mercury-vapor lamps and advertising signs, etc., and is used in mercury switches and other electrical apparatus. Other uses are for making pesticides, mercury cells for caustic-chlorine production, dental preparations, anti-fouling paint, batteries, and catalysts. The most important salts are mercuric chloride ($HgCl_2$, corrosive sublimate—a violent poison), mercurous chloride Hg_2Cl_2 (calomel—occasionally still used in medicine), mercury fulminate ($Hg(ONC)_2$, a detonator widely used in explosives), and mercuric sulfide (HgS, vermillion, a high-grade paint pigment). Organic mercury compounds are important. It is not generally appreciated that mercury is a virulent poison and is readily absorbed through the respiratory tract, the gastrointestinal tract, or through unbroken skin. It acts as a cumulative poison since only small amounts of the element can be eliminated at a time by the human organism. The maximum allowable concentration of mercury vapor in air has been set at 0.1 mg./cu. meter. Since mercury is a very volatile element, dangerous levels are readily attained in air. Air saturated with mercury vapor at 20°C contains a concentration which exceeds the toxic limit by more than 100 times. The danger increases at higher temperatures. *It is therefore important that mercury be handled with utmost care.* Containers of mercury should be securely covered and spillage should be avoided. If it is necessary to heat mercury or mercury compounds, it should be done in a well-ventilated hood.

MOLYBDENUM (Gr. *molybdos,* lead), Mo; at. wt. 95.94; at. no. 42; m.p. 2610°C; b.p. 5560°C; sp. gr. 10.22 (20°C); valence 2, 3, 4?, 5?, or 6. Before Scheele recognized molybdenite as a distinct ore of a new element in 1778, it was confused with graphite and lead ore. The metal was prepared in an impure form in 1782 by Hjelm. Molybdenum does not occur native, but is obtained from *molybdenite* (MoS_2) and from *wulfenite* ($PbMoO_4$). The metal is prepared from the powder made by the hydrogen reduction of purified molybdenum trioxide or ammonium molybdate. The metal is silvery white, very hard, but is softer and more ductile than tungsten. It has a high elastic modulus, and only tungsten and tantalum, of the more readily available metals, have higher melting points. It is a valuable alloying agent, as it contributes to the hardenability and toughness of quenched and tempered steels. It also improves the strength of steel at high temperatures. It is used in certain nickel-based alloys, such as the "Hastelloys," which are heat-resistant and corrosion-resistant to chemical solutions. Molybdenum oxidizes at elevated temperatures. The metal has found recent application as electrodes for electrically heated glass furnaces and forehearths. The metal is also used in nuclear energy applications and for missile and aircraft parts. Molybdenum wire is valuable for use as a filament material for metal evaporation work, and as a filament, grid, and screen material for electronic tubes. Molybdenum is an essential trace element in plant nutrition. Some lands are barren for lack of this element in the soil. Molybdenum sulfide is useful as a lubricant, especially at high temperatures where oils would decompose. Molybdenum powder is priced at about four dollars/lb., and bars rolled from arc-cast ingots cost about 15 dollars/lb.

NEODYMIUM (Gr. *neos,* new, and *didymos,* twin), Nd; at. wt. 144.24; at. no. 60; m.p. 1024°C; b.p. 3027°C; sp. gr. 6.80 and 7.004, depending on allotropic form; valence 3. In 1841 Mosander extracted from *cerite* a new rose-colored oxide, which he believed to contain a new element. He named the element *didymium,* as it was "an inseparable twin brother of lanthanum." In 1885 von Welsbach separated didymium into two new elemental components, *neodymia* and *praseodymia*, by repeated fractionation of ammonium didymium nitrate. While the free metal is in *misch metal*, long known and used as a pyrophoric alloy for lighter flints, the element was not isolated in relatively pure form until 1925. Neodymium is present in misch metal to the extent of about 18%. It is present in the minerals *monazite* and *bastnaesite*, which are the principal sources of rare earth metals. The element may be obtained by separating neodymium salts from other rare earths by ion exchange techniques, and by reducing the salts by calcium metal, followed by volatilization of the calcium from the resulting neodymium-calcium alloy. Other separation techniques are possible. The metal has a bright silvery metallic luster. Neodymium is one of the more reactive rare-earth metals and quickly tarnishes in air, forming an oxide that spalls off and exposes metal to oxidation. The metal, therefore, should be kept under light mineral oil or sealed in a plastic material. Neodymium exists in two allotropic forms. Didymium, of which neodymium is a component, is used for coloring glass to make welder's goggles. By itself, neodymium colors glass a delicate lavender shade. Light transmitted through such glass shows unusually sharp absorption bands. This glass is useful in astronomical work to produce sharp bands by which other spectral lines may be calibrated, and it can be used as a laser material in place of ruby to produce coherent light. Neodymium salts are also used as a colorant for enamels. The price of the metal is about 50 cents/gm or 120 dollars/lb.

NEON (Gr. *neos*, new), Ne; at. wt. 20.183; at. no. 10; m.p. −248.67°C; b.p. −245.92°C (1 atm.); density of gas 0.89990 gm/l (1 atm. 0°C); density of liquid at b.p. 1.207 gm/cm^3; valence 0. Discovered by Ramsay and Travers in 1898. Neon is a rare gaseous element present in the atmosphere to the extent of 1 part in 65,000 of air. It is obtained by liquefaction of air and separated from the other gases by fractional distillation. It is a very inert element; however it is said to form a compound with fluorine. It is still questionable if true compounds of the rare gases exist, but evidence is mounting in favor of their existence. In a vacuum discharge tube, neon glows reddish orange. Of all the rare gases, the discharge of neon is the most intense at ordinary voltages and currents. Neon is used in making the common neon advertising signs, which accounts for its largest use. It is also used to make high-voltage indicators, lightning arrestors, wave meter tubes, and TV tubes. Liquid neon is now commercially available and is finding important application as an economical cryogenic refrigerant. It has over 40 times more refrigerating capacity per unit volume than liquid helium and more than three times that of liquid hydrogen. It is compact, inert, and is less expensive than helium when it meets refrigeration requirements. The gas costs about 1 dollar and 50 cents/liter.

NEPTUNIUM (planet *Neptune*), Np; at. no. 93; m.p. 640 ± 1°C; sp. gr. 18.0 to 20.45, depending on allotropic form; valence 3, 4, 5, and 6. Neptunium was the first of the synthetic transuranium elements of the actinide series to be discovered; the isotope Np^{239} was produced by McMillan and Abelson in 1940 at Berkeley, California, as the result of bombarding uranium with cyclotron-produced neutrons. The isotope Np^{237} (half-life of 2.2 ± 10^6 years) is currently obtained in gram quantities as a by-product from nuclear reactors in the production of plutonium. Trace quantities of the element are actually found in nature due to transmutation reactions in uranium ores produced by the neutrons which are present. Neptunium metal has a silvery appearance, is chemically reactive, and exists in at least three structural modifications: α-neptunium, orthorhombic, density—20.45 g/cm^3; β-neptunium (above 278°C), tetragonal, density (313°C)—19.36 g/cm^3; γ-neptunium (above 500°C) cubic, density (600°C) 18.0 g/cm^3. Neptunium has four ionic oxidation states in solution: Np^{+3} (pale purple), analogous to the rare earth ion Pm^{+3}, Np^{+4} (yellow green), NpO_2^+ (green blue), and NpO_2^{++} (pale pink). These latter oxygenated species are in contrast to the rare earths which exhibit only simple ions of the (II),

(III), and (IV) oxidation states in aqueous solution. The element forms tri- and tetrahalides such as NpF_3, NpF_4, $NpCl_4$, $NpBr_3$, NpI_3, and oxides of various compositions such as are found in the uranium-oxygen system, including Np_3O_8 and NpO_2. In 1962, the A.E.C. made Np^{237} available for sale to its licensees and for export. This isotope can be used as a component in neutron detection instruments. An initial allotment of 200 gms was offered at a price of 500 dollars/gm or 50 cents/mg plus handling and shipping charges.

NICKEL (Ger. *Nickel,* Satan or "Old Nick," and from *kupfernickel,* Old Nick's copper), Ni; at. wt. 58.71; at. no. 28; m.p. 1453°C; b.p. 2732°C; sp. gr. 8.902 (25°C); valence 0, 1, 2, 3. Discovered by Cronstedt in 1751 in kupfernickel (*niccolite*). Nickel is found as a constituent in most meteorites and often serves as one of the criteria for distinguishing a meteorite from other minerals. Iron meteorites, or *siderites,* may contain iron alloyed with from 5% to nearly 20% nickel. Nickel is obtained commercially from *pentlandite* and *pyrrhotite* of the Sudbury region of Ontario. This district produces about 70% of the nickel for the free world. Small deposits are found in Norway, New Caledonia, Cuba, Japan, and elsewhere. Nickel is silvery-white and takes on a high polish. It is hard, malleable, ductile, somewhat ferromagnetic, and a fair conductor of heat and electricity. It belongs to the iron-cobalt group of metals and is chiefly valuable for the alloys it forms. It is extensively used for making stainless steel and other corrosion-resistant alloys, such as Invar, Monel, Inconel, and the Hastelloys. It is also used in coinage, in making nickel steel for armor plate and burglar-proof vaults, and is a component in Nichrome, Permalloy, and Constantan. Nickel added to glass gives a green color. Nickel plating is often used to provide a protective coating for other metals, and finely divided nickel is a catalyst for hydrogenating vegetable oils. It is also used in ceramics, in the manufacture of Alnico magnets, and in the Edison storage battery. The sulfate and the oxides are important compounds.

NIOBIUM (*Niobe,* daughter of Tantalus), Nb; or **COLUMBIUM** (*Columbia,* name for America); Cb; at. wt. 92.906; at. no. 41; m.p. 2468 ± 10°C; b.p. 4927°C; sp. gr. 8.57 (20°C); valence 2, 3, 4?, 5. Discovered in 1801 by Hatchett in an ore sent to England more than a century before by John Winthrop the Younger, first governor of Connecticut. The metal was first prepared in 1864 by Blomstrand, who reduced the chloride by heating it in a hydrogen atmosphere. The name "niobium" was adopted by the International Union of Pure and Applied Chemistry in 1950 after 100 years of controversy. Many leading chemical societies and government organizations refer to it by this name. Most metallurgists, leading metals societies, and all but one of the leading U.S. commercial producers, however, still refer to the metal as "columbium." The element is found in *columbite-tantalite, pyrochlore,* and *euxenite.* Extensive ore reserves are found in Canada, Norway, Africa, and the U.S. The metal can be isolated from tantalum, and prepared in several ways. It is a shiny, white, soft, and ductile metal, and takes on a bluish cast when exposed to air at room temperatures for a long time. The metal starts to oxidize in air at 200°C, and when processed at even moderate temperatures must be placed in a protective atmosphere. It is used as an alloying agent in carbon and alloy steels as well as with nonferrous metals. These alloys have improved strength and other desirable properties. The metal has a low capture cross-section for thermal neutrons. It is used in arc-welding rods for stabilized grades of stainless steel. The element has superconductive properties; a superconductive magnet has been made with Nb-Zr wire, which retains its superconductivity in strong magnetic fields. This type of application is said to offer hope of direct large-scale generation of electric power. In small quantities, columbium metal costs from 50 cents to 1 dollar/gm.

NITROGEN (L. *nitrum,* Gr. *nitron,* native soda; genes, *forming*), N; at. wt. 14.0067; at. no. 7; m.p. −209.86°C; b.p. −195.8°C; density 1.2506 g/l; sp. gr. liquid 0.808

(−195.8°C), solid 1.026 (−252°C); valence 3 or 5. Discovered by Daniel Rutherford in 1772, but Scheele, Cavendish, Priestley, and others about the same time studied "burnt or dephlogisticated air," as air without oxygen was then called. Nitrogen makes up 78% of the air, by volume. The estimated amount of this element in the atmosphere is more than 4000 billion tons. From this inexhaustible source it can be obtained by liquefaction and fractional distillation. The element is so inert that Lavoisier named it *azote*, meaning without life, yet its compounds are so active as to be most important in foods, poisons, fertilizers, and explosives. Nitrogen can be also easily prepared by heating a water solution of ammonium nitrite. Nitrogen, as a gas, is colorless, odorless, and a generally inert element. As a liquid it is also colorless and odorless, and is similar in appearance to water. When nitrogen is heated, it combines directly with magnesium, lithium, or calcium; when mixed with oxygen and subjected to electric sparks, it forms first nitric oxide (NO) and then the dioxide (NO_2); when heated under pressure with a catalyst with hydrogen, ammonia is formed (Haber process). The ammonia thus formed is of the utmost importance as it is used in fertilizers, and it can be oxidized to nitric acid (Ostwald process). The ammonia industry is the largest consumer of nitrogen. Large amounts of the gas are also used by the electronics industry, which uses the gas as a blanketing medium during production of such components as transistors, diodes, etc. The drug industry also uses large quantities. Nitrogen holds promise as a refrigerant both for the immersion freezing of food products and for transportation of foods. Liquid nitrogen is also used in missile work as a purge for components, insulators for space chambers, etc., and by the oil industry to build up great pressures in wells to force crude oil upward. Sodium and potassium nitrates are formed by the decomposition of organic matter with compounds of the metals present. In certain dry areas of the world these saltpeters are found in quantity. Ammonia, nitric acid, the nitrates, the five oxides (N_2O, NO, N_2O_3, NO_2, and N_2O_5), TNT, the cyanides, etc., are but a few of the important compounds. Nitrogen gas prices vary from 2 cents to $2.75 per 100 cu. ft., depending on purity, etc. Production of nitrogen in the U.S. is about 40 billion cubic feet per year or more.

NOBELIUM (See Element No. 102)

OSMIUM (Gr. *osme*, a smell) Os; at. wt. 190.2; at. no. 76; m.p. 3000 ± 10°C; b.p. 5000°C; sp. gr. 22.57; valence 2, 3, 4, or 8. Discovered in 1803 by Tennant in the residue left when crude platinum is dissolved by aqua regia. Osmium occurs in *iridosime* and in platinum-bearing river sands of the Urals, North America, and South America. It is also found in the nickel-bearing ores of the Sudbury, Ontario region along with other platinum metals. While the quantity of platinum metals in these ores is very small, the large tonnages of nickel ores processed make commercial recovery possible. The metal is lustrous, bluish-white, extremely hard, and brittle even at high temperatures. It has the highest melting point and lowest vapor pressure of the platinum group. The metal is very difficult to fabricate, but the powder can be sintered in a hydrogen atmosphere at a temperature of 2000°C. The solid metal is not affected by air at room temperature, but the powered or spongy metal slowly gives off osmium tetroxide, which is a powerful oxidizing agent and has a strong smell. The tetroxide is highly toxic, and boils at 130°C (760 mm). Concentrations in air as low as 10^{-7} gm/cu meter can cause lung congestion, skin damage, or eye damage. The tetroxide has been used to detect fingerprints and to stain fatty tissue for microscopic slides. The metal is almost entirely used to produce very hard alloys, with other metals of the platinum group, for fountain pen tips, instrument pivots, phonograph needles, and electrical contacts. The price of 99% pure osmium powder—the form usually supplied commercially—is about 3 dollars to 8 dollars/gm, depending on quantity and supplier. The measured density of iridium and osmium seem to indicate that osmium is slightly more dense than iridium, and osmium has generally been credited with being the heaviest known element. Calculations of the density from the space lattice, which may be more reliable for these elements than actual measurements, however, give a density of 22.65 for iridium compared to 22.61 for osmium. At present, therefore, we know either iridium or osmium is the heaviest element, but the data do not allow selection between the two.

OXYGEN (Gr. *oxys*, sharp, acid, and *genes*, forming; acid former), O; at. wt. 15.9994; at. no. 8; m.p. −218.4°C; b.p. −183.0°C; density 1.429 g/l (0°C); sp. gr. liquid 1.14 (−182.96°C); valence 2. For many centuries, workers from time-to-time realized air was composed of more than one component. The behavior of oxygen and nitrogen as components of air, led to the advancement of the phlogiston theory of combustion, which captured the minds of chemists for a century. Oxygen was prepared by several workers, including Bayen and Borch, but they did not know how to collect it, did not study its properties, nor did they recognize it as an elementary substance. Priestley is generally credited with its discovery, although Scheele also discovered it independently. Oxygen is the third most abundant element found in the universe, and it plays a part in the carbon-nitrogen cycle—one process thought to give the sun and stars their energy. Oxygen, as a gaseous element, forms 21% of the atmosphere by volume from which it can be obtained by liquefaction and fractional distillation. The element and its compounds make up 49.2%, by weight, of the earth's crust. About two thirds of the human body, and nine tenths of water is oxygen. In the laboratory it can be prepared by the electrolysis of water or by heating potassium chlorate with manganese dioxide as a catalyst. The gas is colorless, odorless, and tasteless. The liquid and solid forms are a pale blue color and are magnetic, but much less so than iron. Ozone (O_3), a highly active allotropic form of oxygen, is formed by the action of an electrical discharge or ultra-violet light on oxygen. Ozone's presence in the atmosphere (amounting to the equivalent of a layer 3 mm thick at ordinary pressures and temperatures) is of vital importance in preventing ultraviolet rays from the sun from reaching the earth's surface and destroying life on earth. Oxygen is very reactive and capable of combining with most elements. It is a component of hundreds of thousands of organic compounds. It is essential for respiration of all plants and animals and for practically all combustion. In hospitals it is frequently used to aid respiration of patients. Its atomic weight was used as a standard of comparison for each of the other elements until 1961 when the International Union of Pure and Applied Chemistry adopted carbon 12 as the new basis. Oxygen has six isotopes. Oxygen 18 occurs naturally, is stable, and is available commercially. Water (H_2O^{18} with 1.5% O^{18}) is also available. Commercial oxygen consumption in the U.S. is estimated at more than 140 billion cubic feet/year, and the demand is expected to double in the next few years. Oxygen enrichment for steel blast furnaces is the greatest user of the gas. Large quantities are also used in making synthesis gas for ammonia and methanol, ethylene oxide, and for oxy-acetylene welding. Air separation plants produce about 99% of the gas; electrolysis plants about 1%. The gas costs about 2.5 cents/cubic foot in relatively small quantities.

PALLADIUM (named after the asteroid *Pallas*, discovered about the same time; Gr. *Pallas*, goddess of wisdom), Pd; at. wt. 106.4; at. no. 46; m.p. 1552°C; b.p. 2927°C; sp. gr. 12.02 (20°C); valence 2, 3, or 4. Discovered in 1803 by Wollaston. Palladium is found along with platinum and other metals of the platinum group in placer deposits of the U.S.S.R., South and North America, Abyssinia, and Australia. It is also found associated with the nickel-copper deposits of South Africa and Ontario. Its separation from the platinum metals depends upon the type of ore in which it is found. It is a steel-white metal, does not tarnish in air, and is the least dense and lowest melting of the platinum group of metals. When annealed, it is soft and ductile; cold working greatly increases its strength and hardness. Palladium is attacked by nitric and sulfuric acid. At room temperatures the metal has the unusual property of absorbing up to 900 times its own volume of hydrogen, possibly forming Pd_2H. It is not yet clear if this is a true compound. Hydrogen readily diffuses through heated palladium and this provides a means of purifying the gas. Finely divided palladium is a good catalyst and is used for hydrogenation and dehydrogenation reactions. It is alloyed and used in jewelry trades. White gold is an alloy of gold decolorized by the addition of palladium. Like gold, palladium can be beaten into leaf as thin as 1/250,000 inch. The metal is used in dentistry, watchmaking, and in making surgical instruments and electrical contacts. The metal sells for about 25 dollars/troy ounce.

PHOSPHORUS (Gr. *phosphoros,* light-bearing; ancient name for the planet Venus when appearing before sunrise), P; at. wt. 30.9738; at. no. 15; m.p. (white) 44.1°C; b.p. (white) 280°C; sp. gr. (white) 1.82, (red) 2.20, (black) 2.25 to 2.69; valence 3 or 5. Discovered in 1669 by Brand, who prepared it from urine. Phosphorus exists in three allotropic forms: white (or yellow), red, and black (or violet). Never found free in nature, it is widely distributed in combination with minerals. *Phosphate* rock, which contains the mineral *apatite*—an impure tri-calcium phosphate, is an important source of the element. Large deposits are found in the U.S.S.R., in Morocco, and in Florida, Tennessee, Utah, Idaho, and elsewhere. Phosphorus is an essential ingredient of all cell protoplasm, nervous tissue, and bones. Ordinary phosphorus is a waxy white solid; when pure it is colorless and transparent. It is insoluble in water, but soluble in carbon disulfide. It takes fire spontaneously in air, burning to the pentoxide. It is very poisonous—50 mg constituting an approximate fatal dose. The maximum recommended allowable concentration in air is 0.1 mg/cubic meter. White phosphorus should be kept under water as it is dangerously reactive in air, and it should be handled with forceps, as contact with the skin may cause severe burns. When exposed to sunlight or when heated in its own vapor to 250°C, it is converted to the red variety, which does not phosphoresce in air as does the white variety. This form does not ignite spontaneously and it is not as dangerous as white phosphorous. It should, however, be handled with care as it does convert to the white form at some temperatures and it emits highly toxic fumes of the oxides of phosphorus when heated. The red modification is fairly stable, sublimes with a vapor pressure of 1 atm. at 417°C, and is used in the manufacture of safety matches, pyrotechnics, pesticides, incendiary shells, smoke bombs, tracer bullets, etc. White phosphorus may be made by several methods. By one process, tri-calcium phosphate, the essential ingredient of phosphate rock, is heated in the presence of carbon and silica in an electric furnace or fuel-fired blast furnace. Elementary phosphorus is liberated as vapor and may be collected under water. If desired, the phosphorus vapor and carbon monoxide produced by the reaction can be oxidized at once in the presence of moisture or water to produce phosphoric acid—an important compound in making super-phosphate fertilizers. In recent years, concentrated phosphoric acids, which may contain as much as 70 to 75% P_2O_5 content, have become of great importance to agriculture and farm production. World-wide demand for fertilizers has caused record phosphate production in recent years. Phosphates are used in the production of special glasses, such as those used for sodium lamps. Bone-ash, calcium phosphate, is also used to produce fine china-ware and to produce mono-calcium phosphate used in baking powder. Phosphorus is also important in the production of steels, phosphor bronze, and many other products. Tri-sodium phosphate is important as a cleaning agent, as a water-softener, and for preventing boiler scale and corrosion of pipes and boiler tubes. Organic compounds of phosphorus are important.

PLATINUM (Sp. *platina,* silver), Pt; at. wt. 195.09; at. no. 78; m.p. 1769°C; b.p. 3827 ± 100°C; sp. gr. 21.45 (20°C); valence 1?, 2, 3, or 4. Discovered in South America by Ulloa in 1735 and by Wood in 1741. The metal was used by pre-Colombian Indians. Platinum occurs native, accompanied by small quantities of iridium, osmium, palladium, ruthenium, and rhodium, all belonging to the same group of metals. These are found in the alluvial deposits of the Ural mountains, of Colombia, and of certain western American states. *Sperrylite* ($PtAs_2$) occurring with the nickel-bearing deposits of Sudbury, Ontario, is the source of a considerable amount of the metal. The large production of nickel offsets the fact that there is only one part of the platinum metals in two million parts of ore. Platinum is a beautiful silvery-white metal, when pure, and is malleable and ductile. It has a coefficient of expansion almost equal to that of soda-lime-silica glass, and is therefore used to make sealed electrodes in glass systems. The metal does not oxidize in air at any temperature, but is corroded by halogens, cyanides, sulfur, and caustic alkalis. It is insoluble in hydrochloric and nitric acid, but dissolves when they are mixed as *aqua regia*, forming chloroplatinic acid (H_2PtCl_6), an important compound. The metal is extensively used in jewlery, in wire and vessels for laboratory use, and in many valuable instruments, including thermocouple elements. It is also used for electrical contacts, corrosion-resistant apparatus, and in dentistry. Platinum-cobalt alloys have magnetic properties. One such alloy made of 76.7% Pt and 23.3% Co, by weight, is an extremely powerful magnet that offers a B-H (max) almost twice that of Alnico V. Platinum resistance wires are used for constructing high-temperature electric furnaces. The metal is used for coating missile nose cones, jet engine fuel nozzles, etc., which must perform reliably for long periods of time at high temperatures. The metal, like palladium, absorbs large volumes of hydrogen, retaining it at ordinary temperatures but giving it up at red heat. In the finely divided state platinum is an excellent catalyst, having long been used in the contact process for sulfuric acid, of which it does not now have a monopoly. Fine platinum wire will glow red hot when placed in the vapor of methyl alcohol. It acts here as a catalyst, converting the alcohol to formaldehyde. This phenomenon has been used commercially to produce cigarette lighters and hand warmers. Hydrogen and oxygen explode in the presence of platinum. The price of platinum has varied widely; more than a century ago it was used to adulterate gold. It was nearly eight times as valuable as gold in 1920; the present price is about 135 dollars/troy ounce.

PLUTONIUM (Planet *Pluto*), Pu; at. no. 94; at. mass ($\sim$239); sp. gr. (α modification) 19.84 (25°C); m.p. 639.5 $\pm$ 2°C; b.p. 3235 $\pm$ 19°C; valence 3, 4, 5, or 6. Plutonium was the second transuranium element of the actinide series to be discovered. The isotope Pu^{238} was produced in 1940 by Seaborg, McMillan, Kennedy, and Wahl by deuteron bombardment of uranium in the 60-inch cyclotron at Berkeley, California. Plutonium also exists in trace quantities in naturally occurring uranium ores. It is formed in much the same manner as neptunium, by irradiation of natural uranium with the neutrons which are present. By far of greatest importance is the isotope Pu^{239}, with a half-life of 24,360 years, produced in extensive quantities in nuclear reactors from natural uranium:

$$U^{238}\ (n,\gamma)\ U^{239} \xrightarrow{\beta^-} Np^{239} \xrightarrow{\beta^-} Pu^{239}$$

Fifteen isotopes of plutonium are known. Plutonium has assumed the position of dominant importance among the transuranium elements because of its successful use as an explosive ingredient in nuclear weapons and the place which it holds as a key material in the development of industrial use of nuclear power. One pound is equivalent to about ten million kilowatt hours of heat energy. Its importance depends on the nuclear property of being readily fissionable with neutrons and its availability in quantity. As with neptunium and uranium, plutonium metal can be prepared by reduction of the trifluoride with alkaline-earth metals. The metal has a silvery appearance and takes on a yellow tarnish when slightly oxidized. It is chemically reactive. A relatively large piece of plutonium is warm to the touch because of the energy given off in alpha decay. Larger pieces will produce enough heat to boil water. The metal readily dissolves in concentrated hydrochloric acid, hydroiodic acid, or perchloric acid with formation of the Pu^{+3} ion. The metal exhibits six allotropic modifications having various crystalline structures. The densities of these vary from 15.92 to 19.84. Plutonium also exhibits four ionic valence states in aqueous solutions: Pu^{+3} (blue lavender), Pu^{+4} (yellow brown), PuO_2^{+} (pink?) and PuO_2^{+2} (pink orange). The ion PuO_2^{+} is unstable in aqueous solutions, disproportionating into Pu^{+4} and PuO_2^{+2}; the

Pu^{+4} thus formed, however, oxidizes the PuO_2^+ into PuO_2^{+2}, itself being reduced to Pu^{+3}, giving finally Pu^{+3} and PuO_2^{+2}. Plutonium forms binary compounds with oxygen: PuO, PuO_2, and intermediate oxides of variable composition; with the halides: PuF_3, PuF_4, $PuCl_3$, $PuBr_3$, PuI_3; with carbon, nitrogen, and silicon: PuC, PuN, $PuSi_2$. Oxyhalides are also well known: PuOCl, PuOBr, PuOI. Because of the high rate of emission of alpha particles and the fact that the element is specifically absorbed by bone marrow, plutonium, as well as all of the other transuranium elements except neptunium, are radiological poisons and must be handled with special equipment and precautions. The body burden, or the amount that can be maintained indefinitely in an adult without producing significant body injury, for Pu^{239} is now recommended to be 0.008 microcuries, which is equivalent to 0.0005 micrograms. When quantities in excess of 300 g are handled, criticality must be considered. Plutonium standard units, consisting of about 1/2 g of plutonium metal sealed in a glass ampoule under a reduced argon atmosphere, are supplied by the National Bureau of Standards to A.E.C. licensees at about 34 dollars.

POLONIUM (Poland, native country of Mme. Curie), Po; at. mass (~210); at. no. 84; m.p. 254°C; b.p. 962°C; sp. gr. (alpha modification) 9.32; valence 2, 4 or 6. First element discovered by Mme. Curie, in 1898, while seeking the cause of radioactivity of pitchblende from Joachimsthal, Bohemia. The electroscope showed it separating with bismuth. Polonium is also called Radium F. Polonium is a very rare natural element. Uranium ores contain only about 100 micrograms of the element per ton. Its abundance is only about 0.2% of that of radium. In 1934 it was found that when natural bismuth (Bi^{209}) was bombarded by neutrons, Bi^{210}, the parent of polonium, was obtained. Milligram amounts of polonium may now be prepared this way, by using the high neutron fluxes of nuclear reactors. Polonium-210 is a low-melting, fairly volatile metal, 50% of which is vaporized in air in 45 hours at 55°C. It is an alpha-emitter with a half-life of 138.39 days. A milligram emits as many alpha particles as 5 grams of radium. The energy released by its decay is so large (27.5 calories per curie per day) that a capsule containing about half a gram reaches a temperature above 500°C. The capsule also presents a contact gamma-ray dose rate of 1.2 roentgens per hour. A few curies of polonium exhibit a blue glow, caused by excitation of the surrounding gas. Because almost all alpha radiation is stopped within the solid source and its container, giving up its energy, polonium has attracted attention for uses as a light-weight heat source for thermoelectric power in space satellites. Twenty-seven isotopes of polonium are known, with atomic masses ranging from 192 to 218. Polonuim-210 is the most readily available. Isotopes of mass 209 (half-life of 103 yrs.) and mass 208 (half-life of 2.9 yrs.) can be prepared by alpha, proton, or deuteron bombardment of lead or bismuth in a cyclotron, but these are expensive to produce. Metallic polonium has been prepared from polonium hydroxide and some other polonium compounds in the presence of concentrated aqueous or anhydrous liquid ammonia. Two allotropic modifications are known to exist. Polonium is readily dissolved in dilute acids, but is only slightly soluble in alkalis. Polonium salts of organic acids char rapidly; halide ammines are reduced to the metal. Polonium can be mixed or alloyed with beryllium to provide a source of neutrons. It has been used in devices for eliminating static charges in textile mills, etc; however, beta sources are more commonly used and are less dangerous. It is also used on brushes for removing dust from photographic films. The polonium for these is carefully sealed and controlled, minimizing hazards to the user. Polonium-210 is very dangerous to handle in even milligram or microgram amounts and special equipment and strict control is necessary. Damage arises from the complete absorption of the energy of the alpha particle into tissue. The maximum permissible body-burden for ingested polonium is only 0.03 microcuries, which represents a particle weighing only 6.8×10^{-12} grams. Weight-for-weight it is about 2.5×10^{11} times as toxic as hydrocyanic acid. The maximum allowable concentration for soluble polonium compounds, in air is about 2×10^{-11} microcuries/cc. Polonium chloride or nitrate is available commercially with an A.E.C. permit at a cost of about 6 dollars/millicurie, plus service charges.

POTASSIUM (English, *potash*—pot ashes; L. *kalium;* Arab. *qali,* alkali), K; at. wt. 39.102; at. no. 19; m.p. 63.65°C; b.p. 774°C; sp. gr. 0.862 (20°C); valence 1. Discovered in 1807 by Davy, who obtained it from caustic potash (KOH); this was the first metal isolated by electrolysis. The metal is the seventh most abundant and makes up about 2.4% by weight of the earth's crust. Most potassium minerals are insoluble and the metal is obtained from them only with great difficulty. Certain minerals, however, such as *sylvite, carnallite, langbeinite,* and *polyhalite* are found in ancient lake and sea beds and form rather extensive deposits from which potassium and its salts can readily be obtained. Potash is mined in Germany, New Mexico, California, Utah, and elsewhere. Large deposits of potash, found at a depth of some 3000 ft. in Saskatchewan, promise to be important in coming years. Potassium is also found in the ocean, but is present only in relatively small amounts, compared to sodium. The greatest demand for potash has been in its use for fertilizers. Potassium is never found free in nature, but is obtained by electrolysis of the hydroxide, much in the same manner as prepared by Davy. It is one of the most reactive and electropositive of metals; except for lithium, it is the lightest known metal. It is soft, easily cut with a knife, and is silvery in appearance immediately after a fresh surface is exposed. It rapidly oxidizes in air and must be preserved in a mineral oil, such as kerosene. As with other metals of the alkali group, it decomposes in water with the evolution of hydrogen. It catches fire spontaneously on water. Potassium and its salts impart a violet color to flames. Nine isotopes of potassium are known. Ordinary potassium is composed of 0.0119%K^{40}, a radioactive isotope with a half-life of 1.4×10^9 yrs. The radioactivity is only about 1/1000 that of uranium, and therefore presents no appreciable hazard. An alloy of sodium and potassium (NaK) is used as a heat-transfer medium. Many potassium salts are of utmost importance. These include the hydroxide, nitrate, carbonate, chloride, chlorate, bromide, iodide, cyanide, sulfate, chromate, and dichromate. Metallic potassium is available commercially for about 2 dollars/lb. in quantity.

PRASEODYMIUM (Gr. *prasios,* green, and *didymos,* twin), Pr; at. wt. 140.907; at. no. 59; m.p. 935°C; b.p. 3127°C; sp. gr. (α) 6.782, (β) 6.64; valence 3 or 4. In 1841 Mosander extracted the rare earth *didymia* from *lanthana*; in 1879 Lecoq de Boisbaudran isolated a new earth, *samaria,* from didymia obtained from the mineral *samarskite.* Six years later, in 1885, von Welsbach separated didymia into two other earths, *praseodymia* and *neodymia,* which gave salts of different colors. As with other rare-earths, compounds of these elements in solution have distinctive sharp spectral absorption bands or lines, some of which are only a few Ångstroms wide. The element occurs along with other rare-earth elements in a variety of minerals. *Monazite* and *bastnaesite* are the two principal commercial sources of the rare-earth metals. Ion-exchange techniques have led to much easier isolation of the rare earths and the cost has dropped greatly in the past few years. Praseodymium can be prepared by several methods, such as by calcium reduction of the anhydrous chloride or fluoride. Misch metal, used in making cigarette lighters, contains about 5% praseodymium metal. Praseodymium is soft, silvery, malleable, and ductile. It was prepared in relatively pure form in 1931. It is somewhat more resistant to corrosion in air than europium, lanthanum, cerium or neodymium, but it does develop a green oxide coating that spalls off, when exposed to air. As with other rare earth metals is should be kept under a light mineral oil, or sealed in plastic. Along with other rare earths, it is widely used as a core material for carbon arcs used by the motion picture industry for studio lighting and projection. Salts of praseodymium are used to color glasses and enamels; when mixed with certain other materials, praseodymium produces an intense and unusually clean yellow color in glass. Didymium glass, of which praseodymium is a component, is a colorant for welder's goggles. The metal (99+ % pure) is priced at about 60 cents per gram or 13 dollars/lb.

PROMETHIUM (*Prometheus,* who, according to mythology, stole fire from heaven), Pm; at. no. 61; m.p. 1035°C; b.p. 2730°C; sp. gr.....; valence 3. In 1902 Branner predicted the existence of an element between neodymium and samarium, and this was confirmed by Moseley in 1914. In 1941, workers at Ohio State University irradiated neodymium and praseodymium with neutrons, deuterons, and alpha particles, resp., and produced several new radioactivities, which most likely were those of Element 61. Wu and Segré, and Bethe, in 1942, confirmed the formation; however chemical proof of the production of Element 61 was lacking because of the difficulty in separating the rare earths from each other at that time. In 1945, Marinsky, Glendenin, and Coryell made the first chemical identification by use of ion-exchange chromatography. Their work was done by fission of uranium and by neutron bombardment of neodymium. Searches for the element in nature have been fruitless, and it now appears that promethium is completely missing from the earth's crust. Fourteen isotopes of promethium, with atomic masses from 141 to 154, are now known. Promethium–147, with a half-life of 2.64 yrs., is the most generally useful. Promethium has a specific activity of about 1 curie per milligram. It is a soft beta emitter; although no gamma rays are emitted, x-radiation can be generated when beta particles impinge on elements of a high atomic number, and care must be taken in handling it. Ion-exchange methods led to the preparation of about 10g of promethium from atomic reactor fuel processing wastes in early 1963. The element has applications as a beta source for thickness gages, and it can be absorbed by a phosphor to produce light. Light produced in this manner can be used for signs or signals that require dependable operation; it can be used as a nuclear-powered battery by capturing light in photocells which convert it into electric current. Such a battery, using Pm^{147}, would have a useful life of about 5 yrs. Promethium shows promise as a portable x-ray unit, and it may become useful as a heat source to provide auxiliary power for space probes and satellites. Promethium–147 is available to A.E.C. licensees at a cost of about $3.25 per curie in small quantities.

PROTACTINIUM (Gr. *protos,* first), Pa; at. wt. (~231); at. no. 91; m.p. ~1230? b.p. ...; sp. gr. 15.37 (calc.); valence 4 or 5. Soddy, Russell, and Fajans independently predicted the existence of eka-tantalum (also known as Uranium X_2), a new member of the uranium series which would occupy the vacant space just below tantalum in the periodic table. The element was discovered in 1917 independently by Hahn and Meitner, by Fajans, and by Soddy, Cranston, and Fleck. Hahn and Meitner treated *pitchblende* repeatedly with hot nitric acid, and from the insoluble residue, they separated a new radioactive substance, which they called *protoactinium.* The name has since been shortened to *protactinium.* In 1927 Grosse prepared Pa_2O_5, and in 1934 obtained the metal. Protactinium is rarer than radium; it occurs in pitchblende to the extent of about 1 part Pa^{231} to 10 million of ore. Belgian Congo ores have about 3 ppm. The metal has a bright metallic luster which it retains for some time in air. It has twelve known isotopes, the most common of which is Pr^{231} with a half-life of 32,480 years. Some isotopes have a very high neutron cross-section; others have low cross-sections. In 1959 and 1961, it was announced that the Great Britain Atomic Energy Authority extracted by a 12-stage process 125 gms of 99.9% protactinium, the world's only stock of the metal for many years to come. The extraction was made from 60 tons of waste material at a cost of about 500,000 dollars. Protactinium is one of the rarest and most expensive natural-occurring elements. It was reported that this stock was being distributed to laboratories around the world at a cost of about $2800/gm. The element is an alpha emitter (5.0 mev), and is a radiological hazard similar to polonium.

RADIUM (L. *radius*, ray) Ra; at. wt. (226); at. no. 88; m.p. 700°C; b.p. <1737°C; sp. gr. 5?; valence 2. Radium was discovered in 1898 by M. and Mme. Curie in the *pitchblende* or *uraninite* of North Bohemia, in which it occurs. There is about 1 gm. of radium in 7 tons of pitchblende. The element was isolated in 1911 by Mme. Curie and Debierne by the electrolysis of a solution of pure radium chloride, employing a mercury cathode; on distillation in an atmosphere of hydrogen this amalgam yielded the pure metal. Originally, radium was obtained from the rich pitchblende ore found at Joachimsthal, Bohemia. The *carnotite* sands of Colorado furnish some radium, but richer ores are found in the Belgian Congo and in the Great Bear Lake region of Canada. Radium is present in all uranium minerals, and could be extracted, if desired, from the extensive wastes of uranium processing. Large uranium deposits are located in Ontario, New Mexico, Utah, Australia, and elsewhere. Radium is obtained commercially as the bromide or chloride; it is doubtful if any appreciable stock of the isolated element now exists. The pure metal is brilliant white when freshly prepared, but blackens on exposure to air, probably due to formation of the nitride. It exhibits luminescence, as do its salts; it decomposes in water and is somewhat more volatile than barium. It is a member of the alkaline-earth group of metals. Radium imparts a carmine red color to a flame. Radium emits alpha, beta, and gamma rays and when mixed with beryllium produces neutrons. Thirteen isotopes are now known; radium–226, the common isotope, has a half-life of 1620 yrs. One gram of radium produces about 0.0001 milliliter (stp) of emanation, or radon gas, per day. This is pumped from the radium and sealed in minute tubes, which are used in the treatment of cancer and other diseases. One gram of radium yields about 1000 cal. of heat per year. Radium is used in producing self-luminous paints, neutron sources, and in medicine for the treatment of disease. Some of the more recently discovered radioisotopes, such as Co^{60}, are now being used in place of radium. Some of these sources are much more powerful, and others are safer to use. Radium loses about 1% of its activity in 25 yrs., being transformed into elements of lower atomic weight. Lead is a final product of disintegration. The study of radium has greatly altered our ideas of the structure of the atom. Radium is a radiological hazard. (Stored radium should be ventilated to prevent build-up of radon.) Inhalation, injection, or body exposure to radium can cause cancer and other body disorders. The recommended maximum allowable concentration for total body content is 0.1 microgram and exposure to 2 roentgens/mo. Radium in the form of the chloride or bromide is available without A.E.C. permit at a cost of about 25 dollars to 50 dollars/milligram, plus service charges.

RADON (from *radium;* called *niton* at first, L. *nitens*, shining), Rn; at. wt. (~222); at. no. 86; m.p. −71°C; b.p. −61.8°C; density of gas 9.73 gms/l; sp. gr. liquid 4.4 at −62°C, solid 4; valence 0. The element was discovered in 1900 by Dorn, who called it *radium emanation.* In 1908 Ramsay and Gray, who named it *niton*, isolated the element and determined its density, finding it to be the heaviest known gas. It is inert and occupies the last place in the zero group of gases in the Periodic Table. Since 1923, it has been called radon. Eighteen isotopes are known. Radon–222, coming from radium, has a half-life of 3.823 days and is an alpha emitter; radon–220, emanating naturally from thorium and called *thoron*, has a half-life of 54.5 seconds and is also an alpha emitter. Radon–219, emanates from actinium and is called *actinon*. It has a half-life of 3.92 seconds and is both an alpha and gamma emitter. It is estimated that every square mile of soil to a depth of 6 inches contains about 1 gm of radium, which releases radon in tiny amounts to the atmosphere. On the average, one part of radon is present to 1 sextillion parts of air. At ordinary temperatures radon is a colorless gas; when cooled below the freezing point, radon exhibits a brilliant phosphorescence which becomes yellow as the temperature is lowered and orange-red at the temperature of liquid air. It has been reported that fluorine reacts with radon, forming radon fluoride. Radon is still produced for therapeutic use by a few hospitals by pumping it from a radium source and sealing it in minute tubes, called seeds or needles, for application to patients. This practice is now largely discontinued as hospitals can order the seeds directly from suppliers, who make up the seeds with the desired activity for the day of use. Radon is available at a cost of about 4 dollars/millicurie. Care must be taken in handling radon, as with other radioactive materials. The main hazard is from inhalation of the element and its solid daughters, which are collected on dust in the air. The permissible level in air has been given at 10^{-8} microcuries/milliliter. Good ventilation should be provided where radium, thorium, or actinium is stored to prevent buildup of this element.

RHENIUM (L. *Rhenus*, Rhine), Re; at. wt. 186.2; at. no. 75; m.p. 3180°C; b.p. 5627°C (est.); sp. gr. 21.02 (20°C); valence −1, 2, 3, 4, 5, 6, 7. Discovery of rhenium is generally attributed to Noddack, Tacke, and Berg, who announced in 1925 they had detected the element in platinum ores and *columbite*. They also found the element in *gadolinite* and *molybdenite*. By working up 660 kg. of molybdenite they were able in 1928 to extract 1 gm of rhenium. The price in 1928 was 10,000 dollars/gm. Rhenium does not occur free in nature or as a compound in a distinct mineral species. It is, however, widely spread throughout the earth's crust to the extent of about 0.001 ppm. Commercial rhenium in the U.S. today is obtained from molybdenite roaster-flue dusts obtained from copper-sulfide ores mined in the vicinity of Miami, Arizona, and elsewhere in Arizona and Utah. Some molybdenites contain from 0.002 to 0.2% rhenium. Some 100 lbs. of rhenium are now being produced yearly in powder and fabricated forms by one U.S. company. The total estimated free-world reserve of rhenium metal is 100 tons. Rhenium metal is prepared by reducing ammonium perrhenate with hydrogen at elevated temperatures. The element is silvery white with a metallic luster; its density is exceeded only by that of platinum, iridium, and osmium; and its melting point is exceeded only by that of tungsten and carbon. It has other useful properties. The usual commercial form of the element is as a powder, but it can be consolidated by pressing and resistance-sintering in a vacuum or hydrogen atmosphere. This produces a compact shape in excess of 90% of the density of the metal. Annealed rhenium is very ductile, and can be bent, coiled, or rolled. Rhenium is used as an additive to tungsten and molybdenum-based alloys to impart useful properties. It is widely used for filaments for mass spectrographs and ion gages. Rhenium-molybdenum alloys are superconductive at 10°K. Rhenium is also used as an electrical contact material as it has good wear resistance and withstands arc corrosion. Thermocouples made of Re-W are used for measuring temperatures up to 2200°C, and rhenium wire is used in photoflash lamps for photography. Rhenium powder sells for about 2 dollars/gm or 600 dollars/lb. Fabricated rhenium strip sells for about 800 dollars to 1500 dollars/lb.

RHODIUM (Gr. *rhodon*, rose), Rh; at. wt. 102.905; at. no. 45; m.p. 1966 ± 3°C; b.p. 3727 ± 100°C; sp. gr. 12.41 (20°C); valence 2, 3, 4, or 5. Wollaston discovered rhodium in 1803–4 in crude platinum ore he presumably obtained from South America. Rhodium occurs native with other platinum metals in river sands of the Urals and in North and South America. It is also found with other platinum metals in the copper-nickel sulfide ores of the Sudbury, Ontario, region. Although the quantity occurring here is very small, the large tonnages of nickel processed makes the recovery commercially feasible. The metal is silvery white and at red heat slowly changes in air to the sesquioxide. At higher temperatures it converts back to the element. Rhodium has a higher melting point and lower density than platinum. Its major use is as an alloying agent to platinum. Such alloys are used for furnace windings, thermocouple elements, bushings for glass fiber production, electrodes for aircraft spark plugs, and laboratory crucibles. It is useful as an electrical contact material as it has a low electrical resistance, a low and stable contact resistance, and is highly resistant to corrosion. Plated rhodium, produced by electroplating or evaporation, is exceptionally hard and is used for optical instruments. It has a high reflectance and is hard and durable. Rhodium is also used for jewelry, for decoration, and as a catalyst. The present cost of rhodium in small quantities is about 7 dollars/gm or 140 dollars/troy ounce.

RUBIDIUM (L. *rubidius*, deepest red), Rb; at. wt. 85.47; at. no. 37; m.p. 38.89°C; b.p. 688°C; sp. gr. (solid) 1.532 (20°C), (liquid) 1.475 (39°C); valence 1, 2, 3, 4. Discovered in 1861 by Bunsen and Kirchoff in the mineral *lepidolite* by use of the spectroscope. The element is much more abundant than was thought several years ago. It is now considered to be the 16th most abundant element in the earth's crust. Rubidium occurs in *pollucite, carnallite, leucite.* and *zinnwaldite*, which contain traces up to 1%, in the form of the oxide. It is found in lepidolite to the extent of about 1.5%, and is recovered commercially from this source. Potassium minerals, such as those found at Searles Lake, California, and potassium chloride recovered from brines in Michigan also contain the element and are commercial sources. Rubidium can be liquid at room temperature. It is a soft, silvery-white metallic element of the alkali group and is the second most electropositive and alkaline element. It ignites spontaneously in air and reacts violently in water, setting fire to the liberated hydrogen. As with other alkali metals, it forms amalgams with mercury and it alloys with gold, cesium, sodium, and potassium. It colors a flame yellowish violet. It is prepared by electrolysis of the chloride or cyanide, and by other methods. It must be kept under a dry mineral oil or in a vacuum or inert atmosphere. Seventeen isotopes of rubidium are known. Natural-occurring rubidium is made of two isotopes Rb^{85} and Rb^{87}. Rubidium–87 is present to the extent of 27.85% in natural rubidium and is a beta emitter with a half-life of 6×10^{10} yrs. Ordinary rubidium is sufficiently radioactive to expose a photographic film in about 30 to 60 days. Rubidium forms four oxides: Rb_2O, Rb_2O_2, Rb_2O_3, Rb_2O_4. Because rubidium can be easily ionized, it is being considered for use in "ion engines" for space vehicles; however cesium is somewhat more efficient for this purpose. It is also proposed for use as a working fluid for vapor turbines and for use in a thermoelectric generator using the magnetohydrodynamic principle where rubidium ions are formed by heat at high temperature and passed through a magnetic field. These conduct electricity and act like an armature of a generator and cause electricity to be generated. Rubidium is used as a getter in vacuum tubes and as a photocell component. It has been used in making special glasses. The present cost in small quantities is about 7 dollars/gm (99.9%). In large quantities the cost might be less than 100 dollars/lb.

RUTHENIUM (L. *Ruthenia*, Russia), Ru; at. wt. 101.07; at. no. 44; m.p. 2250°C; b.p. 3900°C; sp. gr. 12.41 (20°C); valence 0,1,2,3,4,5,6,7,8. Berzelius and Osann in 1827 examined the residues left after dissolving crude platinum from the Ural Mts. in *aqua regia*. While Berzelius found no unusual metals, Osann thought he found three new metals, one of which he named ruthenium. In 1844 Klaus, generally recognized as the discoverer, showed that Osann's ruthenium oxide was very impure and that it contained a new metal. Klaus obtained 6 gms of ruthenium from the portion of crude platinum that is insoluble in aqua regia. A member of the platinum group, ruthenium occurs native with other members of the group in ores found in the Ural Mts. and in North and South America. It is also found along with other platinum metals in small but commercial quantities in *pentlandite* of the Sudbury, Ontario, nickel-mining region, and in *pyroxinite* deposits of South Africa. The metal is isolated commercially by a complex chemical process, the final stage of which is the hydrogen reduction of ammonium ruthenium chloride, which yields a powder. The powder is consolidated by powder metallurgy techniques or by argon-arc welding. Ruthenium is a hard, white metal and has four crystal modifications. It does not tarnish at room temperatures, but oxidizes in air at about 800°C. The metal is not attacked by hot or cold acids or aqua regia, but when potassium chlorate is added to the solution, it oxidizes explosively. It is attacked by halogens, hydroxides, etc. Ruthenium can be plated by electrodeposition or by thermal decomposition methods. The metal is one of the most effective hardeners for platinum and palladium, and is alloyed with these metals to make electrical contacts for severe wear resistance. A ruthenium-molybdenum alloy is said to be superconductive at 10.6°K. The corrosion resistance of titanium is improved a hundredfold by addition of 0.1% ruthenium. It is a versatile catalyst. The metal is priced at about 4 dollars/gm or 60 dollars/troy ounce.

SAMARIUM (*Samarskite*, a mineral), Sm; at. wt. 150.35; at. no. 62; m.p. 1072°C; b.p. 1900°C; sp. gr. (α) 7.536, (β) 7.40; valence 2 or 3. Discovered spectroscopically by its sharp absorption lines in 1879 by Lecoq de Boisbaudran in the mineral *samarskite*, named in honor of a Russian mine official, Col. Samarski. Samarium is found along with other members of the rare-earth elements in many minerals, including *monazite* and *bastnaesite*, which are commercial sources. It occurs in monazite to the extent of 2.8%. While *misch metal*, containing about 1% of samarium metal, has long been used, samarium has not been isolated in relatively pure form until recent years. Ion exchange techniques have recently simplified separation of the rare earths from one another; more recently, electrochemical deposition, using an electrolytic solution of lithium citrate and a mercury electrode, is said to be a simple, fast, and highly specific way to separate the rare earths. Samarium metal can be produced by reducing the oxide with barium or lanthanum. Samarium has a bright silver luster and is reasonably stable in air. Two crystal modifications of the metal exist, with a transformation pt. at 917°C. The metal ignites in air at about 150°C. Seventeen isotopes of samarium exist. Samarium, along with other rare earths, is used for carbon-arc lighting for the motion-picture industry. The sulfide has excellent high-temperature stability and good thermoelectric efficiencies up to 1100°C. Samarium is used to dope calcium fluoride crystals for use in optical masers or lasers. Compounds of the metal act as sensitizers for phosphors excited in the infrared; the oxide exhibits catalytic properties in the dehydration and dehydrogenation of ethyl alcohol. It is used in infrared absorbing glass and as a neutron absorber in nuclear reactors. The metal is priced at about 75 cents/gm or 150 dollars/lb.

SCANDIUM (L. *Scandia*, Scandinavia), Sc; at. wt. 44.956; at. no. 21; m.p. 1539°C; b.p. 2727°C; sp. gr. 2.992; valence 3. On the basis of the Periodic System, Mendeleev predicted the existence of *ekaboron*, which would have an atomic weight between 40 of calcium and 48 of titanium. The element was discovered by Nilson in 1879 in the minerals *euxenite* and *gadolinite*, which had not yet been found anywhere except in Scandanavia. By processing 10 kg of euxenite and other residues of rare-earth minerals, Nilson was able to prepare about 2 gms of scandium oxide of high purity. Cleve later pointed out that Nilson's scandium was identical with Mendeleev's ekaboron. Scandium is apparently a much more abundant element in the sun and certain stars than here on earth. It is about the 23rd most abundant element in the sun compared to the 50th most abundant on earth. It is widely distributed on earth, occurring in very minute quantities in over 800 mineral species. The blue color of beryl (aquamarine variety) is said to be due to scandium. It occurs as a principal component in the rare mineral *thortveitite*, found in Scandinavia and Malagazy. It is also found in the residues remaining after the extraction of tungsten from Zinnwald *wolframite*, and in *wiikite* and *bazzite*. Most scandium is presently being recovered as a by-product of the extraction of uranium from *davidite*, which contains about 0.02% Sc_2O_3. Metallic scandium was first prepared in 1938 by Fischer, Brunger, and Grieneisen, who electrolyzed a eutectic melt of potassium, lithium, and scandium chlorides at 700–800°C. Tungsten wire and a pool of molten zinc served as the electrodes in a graphite crucible. Methods of producing the metal are now somewhat more complicated. The production of the first pound of 99% pure scandium metal was announced in 1960 as having been made under a U.S. Air Force contract. Scandium is becoming more readily available and it is reported that many pounds of the metal have now been prepared.

Scandium is a silvery white metal which develops a slightly yellowish or pinkish cast upon exposure to air. It is relatively soft, and is reported to resemble yttrium and the rare-earth metals more than it resembles aluminum or titanium. It is a very light metal and has a higher melting point than aluminum, making it of interest to designers of space missiles. Scandium oxide is now being produced in quantity and is available at a cost of about 3 dollars to 100 dollars/gm, depending on purity, quantity, and supplier. The metal is still relatively expensive, costing about 60 to 150 dollars or more per gram, depending on the purity and supplier.

SELENIUM (Gr. *Selene*, moon), Se; at. wt. 78.96; at. no. 34; m.p. (gray) 217°C; b.p. (gray) 684.9 ± 1.0°C; sp. gr. (gray) 4.79, (vitreous) 4.28; valence 2, 4, or 6. Discovered by Berzelius in 1817, who found it associated with tellurium, named for the earth. Selenium is found in a few rare minerals, such as *crooksite* and *clausthalite*. In years past it has been obtained from flue dusts remaining from processing copper sulfide ores, but the anode muds from electrolytic copper refineries now provide the source of most of the world's selenium. Selenium is recovered by roasting the muds with soda or sulfuric acid, or by smelting them with soda and niter. Selenium exists in several allotropic forms. Three are generally recognized, but as many as six have been claimed. Selenium can be prepared with either an amorphous or crystalline structure. The color of amorphous selenium is either red, in powder form, or black, in vitreous form. Crystalline monoclinic selenium is a deep red; crystalline hexagonal selenium, the most stable variety, is a metallic gray. The element is a member of the sulfur family and resembles sulfur both in its various forms and in its compounds. Selenium exhibits both photovoltaic action, where light is converted directly into electricity, and photo-conductive action, where the electrical resistance decreases with increased illumination. These properties make selenium useful in the production of photocells and exposure meters for photographic use, as well as solar cells. Selenium is also able to convert a.c. electricity to d.c., and is extensively used in rectifiers. Below its melting point selenium is a p-type semiconductor, and is finding many uses in electronic and solid state applications. It is used in Xerography for reproducing and copying documents, letters, etc. It is used by the glass industry to decolorize glass and to make ruby-colored glasses and enamels. It is also used as a photographic toner, and as an additive to stainless steel. Elemental selenium has been said to be practically nontoxic; however, hydrogen selenide and other selenium compounds are extremely toxic, and resemble arsenic in its physiological reactions. Hydrogen selenide in a concentration of 1.5 ppm. is intolerable to man. Selenium occurs in some soils in amounts sufficient to produce serious effects on animals feeding on plants, such as locoweed, grown in such soils. The maximum allowable concentration of selenium compounds in air has been recommended to be 0.1 mg/cu meter. Selenium is priced at about 7 dollars/lb. It is also available in high-purity form at a somewhat higher cost.

SILICON (L. *silex, silicis*, flint), Si; at. wt. 28.086; at. no. 14; m.p. 1410°C; b.p. 2355°C; sp. gr. 2.33 (25°C); valence 4. Davy in 1800 thought silica to be a compound and not an element; later in 1811, Gay Lussac and Thenard probably prepared impure amorphous silicon by heating potassium with silicon tetrafluoride. Berzelius, generally credited with the discovery, in 1824 succeeded in preparing amorphous silicon by the same general method as used earlier, but he purified the product by removing the fluosilicates by repeated washings. Deville in 1854 first prepared crystalline silicon, the second allotropic form of the element. Silicon is present in the sun and stars and is a principal component of a class of meteorites known as *aerolites*. It is also a component of *tektites*, a natural glass of uncertain origin, but believed by many to be meteoritic. Silicon makes up 25.7% of the earth's crust, by weight, and is the second most abundant element, being exceeded only by oxygen. Silicon is not found free in nature, but occurs chiefly as the oxide, and as silicates. *Sand, quartz, rock crystal, amethyst, agate, flint, jasper*, and *opal* are some of the forms in which the oxide appears. *Granite, hornblende, asbestos, feldspar, clay, mica,* etc. are but a few of the numerous silicate minerals. Silicon is prepared commercially by heating silica and carbon in an electric furnace, using carbon electrodes. Several other methods can be used for preparing the element. Amorphous silicon can be prepared as a brown powder, which can be easily melted or vaporized. Crystalline silicon has a metallic luster and grayish color. Hyper-pure silicon can be prepared by the thermal decomposition of ultra-pure trichlorosilane in a hydrogen atmosphere, and by a vacuum float zone process. This product can be doped with boron, gallium, phosphorus, or arsenic, etc. to produce silicon for use in transistors, solar cells, rectifiers, and other solid-state devices which are used extensively in the electronics and space-age industries. Silicon is a relatively inert element, but it is attacked by halogens and dilute alkali. Most acids, except hydrofluoric, do not affect it. Silicones are important products of silicon. They may be prepared by hydrolyzing a silicon organic chloride, such as dimethyl silicon chloride. Hydrolysis and condensation of various substituted chlorosilanes can be used to produce a very great number of polymeric products, or silicones, ranging from liquids to hard, glass-like solids with many useful properties. Elemental silicon transmits more than 95% of all wavelengths of infra-red, from 1.3 to 6.7 microns. Silicon is one of man's most useful elements. In the form of sand and clay it is used to make concrete and brick; it is a useful refractory material for high-temperature work, and in the form of silicates it is used in making enamels, pottery, etc. Silica, as sand, is a principal ingredient of glass, one of the most inexpensive of materials with excellent mechanical, optical, thermal, and electrical properties. Glass can be made in a very great variety of shapes, and is used as containers, window glass, insulators, and thousands of other uses. Silicon is important in plant and animal life. Diatoms in both fresh and salt water extract silica from the water to build up their cell walls. Silica is present in ashes of plants and in the human skeleton. Silicon is an important ingredient in steel; silicon carbide is one of the most important abrasives. Regular grade silicon (97%) costs about 20 cents/lb. Silicon 99.7% pure costs about 7 dollars/lb.; hyper-pure silicon may cost as much as 100 dollars/lb.

SILVER (Anglo-Saxon, *Seolfor, siolfur*), Ag (L. argentum); at. wt. 107.870; at. no. 47; m.p. 960.8°C; b.p. 2212°C; sp. gr. 10.50 (20°C); valence 1, 2. Silver has been known since ancient times. It is mentioned in *Genesis.* Slag dumps in Asia Minor and on islands in the Aegean Sea indicate that man learned to separate silver from lead as early as 3000 B.C. Silver occurs native and in ores, such as *argentite* (Ag_2S) and *horn silver* ($AgCl$); lead, lead-zinc, copper, gold, and copper-nickel ores are principal sources. Silver is also recovered during electrolytic refining of copper. Commercial fine silver contains at least 99.90% silver. Purities of 99.999+% are available commercially. Pure silver has a brilliant white metallic luster. It is a little harder than gold and is very ductile and malleable, being exceeded only by gold, and perhaps palladium. Pure silver has the highest electrical and thermal conductivity of all metals, and possesses the lowest contact resistance. It is stable in pure air and water, but tarnishes when exposed to ozone, hydrogen sulfide, or air containing sulfur. The alloys of silver are important. Sterling silver is used for jewelry, silverware, etc. where appearance is paramount. This alloy contains 92.5% silver, the remainder being copper or some other metal. Silver is of utmost importance in photography—about 30% of theU.S. industrial consumption going into this application. It is used for dental alloys. Silver is used in making solder and brazing alloys, electrical contacts, and high capacity silver–zinc and silver–cadmium batteries. Silver paints are used for making printed circuits. It is used in mirror production and may be deposited on glass or metals by chemical deposition, electrodeposition, or by evaporation. When freshly deposited, it is the best reflector of visible light known, but it rapidly tarnishes and loses much of its reflectance. It is a poor reflector of ultraviolet. Silver chloride has interesting optical properties as it can be made transparent; it also is a cement for glass. Silver nitrate, or *lunar caustic*, the most important silver compound, is used in photography and medicine. The price of silver was fixed by the U.S. Treasury at 71 cents/tr. oz. in 1939 and at 90.5 cents in 1946. In November 1961 the U.S. Treasury suspended sales of nonmonetized silver. The return to a free market has caused price fluctuations up to $1.28/tr. oz. The free world consumption of this valuable element in recent years has greatly exceeded the output.

SODIUM (English, *soda;* Medieval Latin, *sodanum*, headache remedy), Na (L. *natrium*); at. wt. 22.9898; at. no. 11; m.p. 97.81 ± 0.03°C; b.p. 892°C; sp. gr. 0.971 (20°C); valence 1. Long recognized in compounds, first isolated by Davy in 1807 by electrolysis of caustic soda. Sodium is present in fair abundance in the sun and stars. The D lines of sodium are among the most prominent in the solar spectrum. Sodium is the sixth most abundant element on earth, comprising about 2.6% of the earth's crust; it is the most abundant of the alkali group of metals of which it is a member. The most common compound is sodium chloride, but it occurs in many other minerals, such as *soda niter, cryolite, amphibole, zeolite, sodalite*, etc. It is a very reactive element and is never found free in nature. It is now obtained commercially by the electrolysis of absolutely dry fused sodium chloride. This method is much cheaper than that of electrolyzing sodium hydroxide, as was used several years ago. Sodium is a soft, bright, silvery metal which floats on water, decomposing it with the evolution of hydrogen and the formation of the hydroxide. It may or may not ignite spontaneously on water, depending on the amount of oxide and metal exposed to the water. It normally does not ignite in air at temperatures below 115°C. Sodium should be handled with respect as it can be dangerous when improperly handled. Metallic sodium is vital in the manufacture of sodamide and sodium cyanide, sodium peroxide, and sodium hydride. It is used in preparing tetraethyl lead, in the reduction of organic esters, and in the preparation of organic compounds. The metal may be used to improve the structure of certain alloys, to descale metal, to purify molten metals, and as a heat transfer agent. An alloy of sodium with potassium, NaK, is also an important heat transfer agent. Sodium compounds are important to the paper, glass, soap, textile, petroleum, chemical and metal industries. Soap is generally a sodium salt of certain fatty acids. Among the many compounds that are of the greatest industrial importance are: common salt (NaCl), soda ash (Na_2CO_3), baking soda ($NaHCO_3$), caustic soda (NaOH), Chile saltpeter ($NaNO_3$), di- and trisodium phosphates, sodium thiosulfate (hypo, $Na_2S_2O_3 \cdot 5H_2O$), and borax ($Na_2B_4O_7 \cdot 10H_2O$). Metallic sodium is priced at about 15 to 20 cents/lb. in quantity. On a per cubic inch basis, it is the cheapest of all metals.

STRONTIUM (*Strontian*, town in Scotland), Sr; at. wt. 87.62; at. no. 38; m.p. 769°C; b.p. 1384°C; sp. gr. 2.54; valence 2. Discovered by Davy by electrolysis in 1808. Strontium is found chiefly as *celestite* ($SrSO_4$) and *strontianite* ($SrCO_3$). The metal can be prepared by electrolysis of the fused chloride mixed with potassium chloride, or is made by reducing strontium oxide with aluminum in a vacuum at a temperature at which strontium distills off. Three allotropic forms of the metal exist, with transition points at 235°C and 540°C. Strontium is softer than calcium and decomposes water more vigorously. It does not absorb nitrogen below 380°C. It should be kept under kerosene to prevent oxidation. Freshly cut strontium has a silvery appearance, but rapidly turns a yellowish color with the formation of the oxide. The finely divided metal ignites spontaneously in air. Volatile strontium salts impart a beautiful crimson color to flames, and these salts are used in pyrotechnics. Strontium has sixteen isotopic forms. Of greatest importance is Sr^{90} with a half-life of 28 yrs. It is a product of nuclear fallout and presents a health problem. This isotope is one of the best long-lived high-energy beta emitters known, and is used in SNAP devices (Systems for Nuclear Auxiliary Power). These devices hold promise for use in space vehicles, remote weather stations, navigational buoys, etc. where a light-weight, long-lived, nuclear-electric power source is needed. Strontium hydroxide, has been used in sugar refining; however lime is replacing its use as it is cheaper. Strontium titanate is an interesting optical material as it has an extremely high refractive index and an optical dispersion greater than that of diamond. It has been used as a gemstone, but it is very soft. It does not occur naturally. The applications of strontium are similar to those of barium and calcium, but there are few advantages and the cost is much higher. Strontium metal costs about 6 to 8 dollars/lb.

SULFUR (Sanskrit, *sulvere;* L. *sulphurium*), S; at. wt. 32.064; at. no. 16; m.p. (rhombic) 112.8°C, (monoclinic) 119.0°C; b.p. 444.6°C; sp. gr. (rhombic) 2.07, (monoclinic) 1.957 (20°C); valence 2, 4, or 6. Known to the ancients; referred to in Genesis as *brimstone.* Sulfur is found in meteorites. A dark area near the crater Aristarchus on the moon has been studied by R. W. Wood with ultraviolet light. This study suggests strongly that it is a sulfur deposit. Sulfur occurs native in the vicinity of volcanoes and hot springs. It is widely distributed in nature as *iron pyrites, galena, sphalerite, cinnabar, stibnite, gypsum, epsom salts, celestite, barite,* etc. Sulfur is commercially recovered from wells sunk into the salt domes along the Gulf Coast of the U.S. It is obtained from these wells by the Frasch process, which forces heated water into the wells to melt the sulfur, which is then brought to the surface. Sulfur also occurs in natural gas and petroleum crudes and must be removed from these products. Formerly this was done chemically, which wasted the sulfur. New processes now permit recovery, and these sources promise to be very important. Large amounts of sulfur are being recovered from Alberta gas fields. Sulfur is a pale yellow, odorless, brittle solid, which is insoluble in water, but soluble in carbon disulfide. In every state, whether gas, liquid, or solid, elemental sulfur occurs in more than one allotropic form or modification; these present a confusing multitude of forms whose relations are not yet fully understood. Amorphous or "plastic" sulfur is obtained by fast cooling of the crystalline form. Recent x-ray studies indicate that amorphous sulfur may have a helical structure with eight atoms per spiral. Crystalline sulfur seems to be made of rings, each containing eight sulfur atoms, which fit together to give a normal x-ray pattern. Nine isotopic forms of sulfur exist. Four occur in natural sulfur, none of which is radioactive. A finely divided form of sulfur, known as *flowers of sulfur,* is obtained by sublimation. Sulfur readily forms sulfides with many elements. Sulfur is a component of black gunpowder, is used in the vulcanization of natural rubber, and is used as a fungicide. A tremendous tonnage is used to produce sulfuric acid, the most important manufactured chemical. It is used in making sulfite paper and other papers, is used as a fumigant, and in the bleaching of dried fruits. The element is a good electrical insulator. Organic compounds containing sulfur are very important. Sulfur is available in purities of 99.999+ % at about 50 dollars/lb.

TANTALUM (Gr. *Tantalos,* mythological character—father of *Niobe*), Ta; at. wt. 180.948; at. no. 73; m.p. 2996°C; b.p. 5425 ± 100°C; sp. gr. 16.6; valence 2?, 3, 4?, or 5. Discovered in 1802 by Ekeberg, but many chemists thought niobium and tantalum were identical elements until Rose, in 1844, and Marignac, in 1866, indicated and showed that niobic and tantalic acids were two different acids. The early investigators only isolated the impure metal. The first relatively pure ductile tantalum was produced by von Bolton in 1903. Tantalum occurs principally in the mineral *columbite-tantalite* (Fe, Mn) (Nb, $Ta)_2O_6$. Separation of tantalum from niobium requires several complicated steps. Several methods are commercially used to produce the element including: electrolysis of molten potassium fluotantalate, reduction of potassium fluotantalate with sodium, or reacting tantalum carbide with tantalum oxide. Tantalum is a gray, heavy, and very hard metal. When pure, it is ductile and can be drawn into fine wire, which is used as a filament for evaporating metals, such as aluminum. Tantalum is almost completely immune to chemical attack at temperatures below 150°C, and is attacked only by hydrofluoric acid, acidic solutions containing the fluoride ion, and free sulfur trioxide. Alkalis attack it only slowly. At higher temperatures, tantalum becomes much more reactive. The element has a melting point exceeded only by tungsten and rhenium. Tantalum is used to make a variety of alloys with desirable properties, such as high-melting point, high strength, good ductility, etc. The metal has good "gettering" ability at high temperatures, and tantalum oxide films are stable, and have good rectifying and dielectric properties. Tantalum is used to make electrolytic capacitors, lightning arrestors, and surge suppressors. The metal is widely used to fabricate chemical process equipment, nuclear reactors, aircraft and missile parts, and surgical equipment. Tantalum oxide is used to make special glass with a high index of refraction for camera lenses. The metal has many other uses. The metal in powdered form costs about 35 dollars/lb. Sheet tantalum and fabricated forms are more expensive.

TECHNETIUM (Gr. *technetos*, artificial), Tc; at. wt. (~98); at. no. 43; m.p. 2200 ± 50°C; sp. gr. 11.50 (calc.); valence 3?, 4, 6, or 7. Element 43 was predicted on the basis of the periodic table, and was erroneously reported as having been discovered in 1925, at which time it was named *masurium*. The element was actually discovered by Perrier and Segrè in Italy in 1937. It was found in a sample of molybdenum, bombarded by deuterons in the Berkeley cyclotron, which E. Lawrence sent to these investigators. Technetium was the first element to be produced artificially. Since its discovery, searches for the element in terrestrial materials have been made without success. If it does exist, the concentration must be very small. Surprisingly, it has been found in the spectrum of S, M, and N type stars, and its presence in stellar matter is leading to new theories of the production of heavy elements in the stars. Fourteen isotopes of technetium, with atomic masses ranging from 92 to 105, are known. Tc^{97g} has a half-life of 2.6×10^6 yrs. Tc^{98} has a half-life of 1.5×10^6 yrs. The isotope Tc^{95m}, with a half-life of 60 days, is useful for tracer work, as it produces energetic gamma rays. Technetium has been produced in sufficient quantities to allow isolation of the metal, and production is now said to be in excess of 400 gms./month. The metal was first prepared by passing hydrogen gas at 1100°C over Tc_2S_7. It has since been prepared by heating ammonium pertechnetate and ammonium sulfate in hydrogen gas at 500–600°C. Technetium is a silvery-gray metal that tarnishes slowly in moist air. Until 1960, technetium was available only in small amounts and the price was as high as 2800 dollars per gm. In 1962 it was offered commercially to holders of A.E.C. permits at a price of 90 dollars/gm. Technetium dissolves in nitric acid, aqua regia, and conc. sulfuric acid, but is not soluble in hydrochloric acid of any strength. The element is a remarkable corrosion inhibitor for steel. It is reported that mild carbon steels may be effectively protected by as little as 5 ppm of $KTcO_4$ in aerated distilled water at temperatures up to 250°C. The metal is an excellent superconductor at 11°K and below.

TELLURIUM (L. *tellus*, earth), Te; at. wt. 127.60; at. no. 52; m.p. 449.5 ± 0.3°C; b.p. 989.8 ± 3.8°C; sp. gr. 6.24 (20°C); valence 2, 4, or 6. Discovered by Müller von Reichenstein in 1782; named by Klaproth in 1798. Tellurium is occasionally found native, but is more often found as the telluride of gold and other metals. It is recovered commercially from the anode muds produced during the electrolytic refining of blister copper. Crystalline tellurium has a silvery white appearance, and when pure exhibits a metallic luster. It is brittle and easily pulverized. Amorphous tellurium is formed by precipitating tellurium from a solution of telluric or tellurous acid. Whether this form is truly amorphous, or made of minute crystals, is open to question. Tellurium is a p-type semiconductor, and shows greater conductivity in certain directions, depending on alignment of the atoms. Its conductivity increases slightly with exposure to light. It can be doped with silver, copper, gold, tin, or other elements. In air, tellurium burns with a greenish-blue flame, forming the dioxide. Molten tellurium corrodes iron, copper, and stainless steel. Tellurium and its compounds are probably toxic and should be handled with care. Workmen exposed to as little as 0.01 mg/cu meter of air, or less, develop "tellurium breath," which has a garlic-like odor. Twenty-two isotopes of tellurium are known, with atomic masses ranging from 114 to 135. Tellurium improves the machinability of stainless steel, and its addition to lead decreases the corrosive action of sulfuric acid to lead and improves its strength and hardness. Tellurium is used in ceramics. Bismuth telluride has been used in thermoelectric devices. One such device, using two Bi-Te semiconductors, is reportedly capable of freezing or boiling water in seconds with the power from two flash-light batteries. The unit is said to be capable of bringing the temperature down to −75°C, using only two amperes of current. Tellurium with a purity of 99.5% costs about 4 dollars/lb. It is also available with purities of 99.999+%.

TERBIUM (*Ytterby*, village in Sweden), Tb; at. wt. 158.924; at. no. 65; m.p. 1356°C; b.p. 2800°C; sp. gr. 8.272; valence 3, 4. Discovered by Mosander in 1843. Terbium is a member of the lanthanide or "rare earth" group of elements. It is found in *cerite*, *gadolinite*, and other minerals along with other rare earths. It is recovered commercially from *monazite* in which it is present to the extent of 0.03% Terbium has been isolated only in recent years with the development of ion-exchange techniques for separating the rare-earth elements. As with other rare earths, it can be produced by reducing the anhydrous chloride or fluoride with calcium metal in a tantalum crucible. Calcium and tantalum impurities can be removed by vacuum remelting. Other methods of isolation are possible. Terbium is reasonably stable in air. It is a silvery gray metal, and is malleable, ductile and soft enough to be cut with a knife. Two crystal modifications exist; eighteen isotopes with atomic masses ranging from 147 to 164 are recognized. The oxide is a chocolate or dark maroon color. Sodium terbium borate is used as a laser material and emits coherent light at 5460Å. Terbium is used to dope calcium fluoride, calcium tungstate, and strontium molybdate, used in solid-state devices. There are few other uses yet found for terbium. The element is priced at about 4 dollars/gm or 750 dollars/lb.

THALLIUM (Gr. *thallos*, a green shoot or twig), Tl; at. wt. 204.37; at. no. 81; m.p. 303.5°C; b.p. 1457 ± 10°C; sp. gr. 11.85 (20°C); valence 1 or 3. Discovered spectroscopically in 1861 by Crookes. The element was named after the beautiful green spectral line, which identified the element. The metal was isolated both by Crookes and Lamy in 1862 about the same time. Thallium occurs in *crooksite*, *lorandite*, and *hutchinsonite*. It is also present in *pyrites* and is recovered from the roasting of this ore in connection with the production of sulfuric acid. It is also obtained from the smelting of lead and zinc ores. Extraction is somewhat complex and depends on the source of the thallium. When freshly exposed to air thallium exhibits a metallic luster, but soon develops a bluish gray tinge, resembling lead in appearance. A heavy oxide builds up on thallium if left in air, and in the presence of water the hydroxide is formed. The metal is very soft and malleable. It can be cut with a knife. Twenty isotopic forms of thallium, with atomic masses ranging from 191 to 210 are recognized. The element and its compounds are toxic and should be carefully handled. Contact of the metal with the skin is dangerous, and when melting the metal, adequate ventilation should be provided. The maximum allowable concentration of soluble thallium compounds in air is 0.1 mg/cu meter. Thallium sulfate is widely employed as a rodenticide and ant killer. It is odorless and tasteless, giving no warning of its presence. The electrical conductivity of thallium sulfide changes with exposure to infrared light and this compound is used in photocells. Thallium bromide-iodide crystals have been used as infrared detectors. Thallium has been used, with sulfur or selenium and arsenic, to produce low melting glasses which become fluid between 125 and 150°C. These glasses have properties at room temperatures similar to ordinary glasses and are said to be durable and insoluble in water. Thallium oxide has been used to produce glasses with a high index of refraction. A mercury–thallium alloy, which forms a eutectic at 8.5% thallium, is reported to freeze at −60°C, some 20°C below the freezing point of mercury. Commercial thallium metal costs about 8 dollars/lb. It is available also in high-purity form.

THORIUM (*Thor*, Scandinavian god of war), Th; at. wt. 232.038; at. no. 90; m.p. ~1700°C; b.p. ~4000°C; sp. gr. ~11.66; valence 4. Discovered by Berzelius in 1828. Thorium occurs in *thorite* ($ThSiO_4$) and in *thorianite* ($ThO_2 + UO_2$). Large deposits of thorium minerals have been reported in New England and elsewhere, but these have not yet been exploited. Thorium is now thought to be about three times as abundant as uranium and about as abundant as lead or molybdenum. The metal is fissionable and is a source of nuclear power. There is probably more energy available for use from thorium in the minerals of the earth's crust than from both uranium and fossil fuels. Thorium is recovered commercially from the mineral *monazite*, which contains from 3 to 9% ThO_2 along with most rare-earth minerals. Several methods are available for producing thorium metal; it can be obtained by reducing thorium oxide with calcium; by electrolysis of anhydrous thorium chloride in a fused mixture of sodium and potassium chlorides; by calcium reduction of thorium tetrachloride mixed with anhydrous zinc chloride; and by reduction of thorium tetrachloride with an alkali metal. When pure, thorium is a silvery-white metal which is air-stable and retains its luster for several months. When contaminated with the oxide, thorium slowly tarnishes in air, becoming gray and finally black. The physical properties of thorium are greatly influenced by the degree of contamination with the oxide. For this reason, values for the melting point and specific gravity are still in question. The purest specimens contain several tenths of a per cent of the oxide. Thorium oxide has a melting point of 3300°C and is the highest of all oxides. Only a few elements, such as tungsten, and a few compounds, such as tantalum carbide, have higher melting points. Thorium is slowly attacked by water, but does not dissolve readily in most common acids, except hydrochloric. Powdered thorium metal is often pyrophoric and should be carefully handled. When heated in air, thorium turnings ignite and burn brilliantly with a white light. The principal use of thorium has been in the preparation of the Welsbach mantle, used for portable gas lights. These mantles, consisting of thorium oxide with about 1% cerium oxide and other ingredients, glow with a dazzling light when heated in a gas flame. Thorium is an important alloying element in magnesium, imparting high strength and creep resistance at elevated temperatures. Because thorium has a low work-function and high electron emission, it is used to coat tungsten wire used in electronic equipment. The oxide is also used to control the grain size of tungsten used for electric lamps; it is also used for high-temperature laboratory crucibles. Thirteen isotopes of thorium are known, all of which are radioactive. Thorium–232 occurs naturally and has a half-life of 1.41×10^{10} yrs. It is an α, β and γ emitter. It is sufficiently radioactive to expose a photographic film in a few hours. Thorium can be used as a nuclear fuel, but has not received the attention of uranium, due to the demand for plutonium. Thorium disintegrates with the production of thoron ($radon^{220}$), which is an alpha emitter and presents a radiation hazard. Good ventilation of areas where thorium is stored or handled is therefore recommended. Thorium and its compounds are subject to licensing and control by the U.S. Atomic Energy Commission. In small lots, thorium metal costs about 2 dollars/gm.

THULIUM (*Thule*, the earliest name for Scandinavia), Tm; at. wt. 168.934; at. no. 69; m.p. 1545°C; b.p. 1727°C; sp. gr. 9.332; valence 2,3. Discovered in 1879 by Cleve. Thulium occurs in small quantities along with other rare earths in a number of minerals. It is obtained commercially from *monazite*, which contains about 0.007% of the element. Thulium is the least abundant of the rare-earth elements, but with new sources recently discovered, it is now considered to be about as rare as silver, gold, or cadmium. Ion-exchange techniques have recently permitted much easier separation of the rare earths, with much lower costs. Thulium metal, only a few years ago, was not obtainable at any cost; in 1950 the oxide sold for 450 dollars/gm. The oxide is now available, however, for 2 dollars/gm in lb. quantities. Thulium can be isolated by reduction of the oxide with lanthanum metal or by calcium reduction of the anhydrous fluoride. The pure metal has a bright, silvery luster. It is reasonably stable in air, is soft, malleable and ductile, and can be cut with a knife. Sixteen isotopes are known, with atomic masses ranging from 161 to 176. Because of the relatively high price of the metal, thulium has not yet found many practical applications. Thulium metal costs about 7 dollars/gm or 2000 dollars lb.

TIN (Anglo-Saxon, *tin*), Sn (L. *stannum*); at. wt. 118.69; at. no. 50; m.p. 231.89; b.p. 2270°C; sp. gr. (gray) 5.75, (white) 7.31; valence, 2, 4. Known to the ancients. Tin is found chiefly in *cassiterite* (SnO_2). Most of the world's supply comes from Malaya, Bolivia, Indonesia, the Congo, Thailand, and Nigeria. The U.S. produces almost none, although occurrences have been found in Alaska and California. Tin is obtained by reducing the ore with coal in a reverberatory furnace. Ordinary tin is a silvery white metal, is malleable, somewhat ductile, and has a highly crystalline structure. Due to the breaking of these crystals, a "tin cry" is heard when a bar is bent. The element has two or perhaps three allotropic forms. On warming, gray or α tin, with a cubic structure, changes at 13.2°C into white or β tin, the ordinary form of the metal. White tin has a tetragonal structure. Some authorities believe a γ form exists between 161°C and the melting point; however other authorities discount its existence. When tin is cooled below 13.2°C, it changes slowly from white to gray. This change is affected by impurities, such as aluminum and zinc, and can be prevented by small additions of antimony or bismuth. This change from the α to β form is called "the tin pest." There are few if any uses for gray tin. Tin takes a high polish and is used to coat other metals to prevent corrosion or other chemical action. Such tin plate over steel is used in the so-called tin can for preserving food. Alloys of tin are very important. Soft solder, type metal, fusible metal, pewter, bronze, bell metal, Babbitt metal, White metal, die casting alloy, and phosphor bronze are some of the important alloys using tin. Tin resists distilled, sea, and soft tap water, but is attacked by strong acids, alkalis, and acid salts. Oxygen in solution accelerates the attack. When heated in air, tin forms SnO_2, which is feebly acid, forming stannate salts with basic oxides. The most important salt is the chloride ($SnCl_2.H_2O$), which is used as a reducing agent and as a mordant in calico printing. Tin salts sprayed onto glass are used to produce electrically conductive coatings on the glass. These have been used for panel lighting and for frost-free windshields. Of recent interest is a crystalline tin-niobium alloy that is superconductive at very low temperatures. This promises to be important in the construction of superconductive magnets that generate enormous field strengths, but use practically no power. Such magnets, made of tin-niobium wire, weigh but a few pounds and will produce magnetic fields, when started with a small battery, that are comparable to that of a 100-ton electromagnet operated continuously with a large power supply.

TITANIUM (L. *Titans*, the first sons of the Earth, myth.), Ti; at. wt. 47.90; at. no. 22; m.p. 1675°C; b.p. 3260°C; sp. gr. 4.54; valence 2, 3, or 4. Discovered by Gregor in 1791; named by Klaproth in 1795. Impure titanium was prepared by Nilson and Pettersson in 1887; however the pure metal (99.9%) was not made until 1910 by Hunter by heating $TiCl_4$ with sodium in a steel bomb. Titanium is present in meteorites and in the sun. Titanium oxide bands are prominent in the spectra of M Type stars. The element is the ninth most abundant in the crust of the earth. Titanium is almost always present in igneous rocks and in the sediments derived from them. It occurs in the minerals *rutile*, *ilmenite*, and *sphene*, and is present in titanates and in many iron ores. Titanium is present in the ash of coal, in plants, and in the human body. The metal was a laboratory curiosity until Kroll, in 1946, showed that titanium could be produced commercially by reducing titanium tetrachloride with magnesium. This method is largely used for producing the metal today. The metal can be purified by decomposing the iodide. Titanium, when pure, is a lustrous, white metal. It has a low density, good strength, is easily fabricated, and has excellent corrosion resistance. It is ductile only when it is free of oxygen. The metal burns in air and is the only element that burns in nitrogen. Titanium is resistant to dilute sulfuric and hydrochloric acid, most organic acids, moist chlorine gas, and chloride solutions. The metal is dimorphic. The hexagonal α form changes to the cubic β form very slowly at about 880°C. The metal combines with oxygen at red heat, and with chlorine at 500°C. Titanium is important as an alloying agent with aluminum, molybdenum, manganese, iron, and other metals. Alloys of titanium are principally used for aircraft and missiles where light-weight, strength, and ability to withstand extremes of temperature are important. The metal has excellent resistance to sea water and is used for propeller shafts, rigging, and other parts of ships exposed to salt water. A titanium anode, coated with platinum has been used to provide cathodic protection from corrosion by salt water. Titanium metal is considered to be physiologically inert. When pure, titanium dioxide is relatively clear and has an extremely high index of refraction with an optical dispersion higher than diamond. It is produced artificially for use as a gemstone, but it is relatively soft. Star sapphires and rubies exhibit their asterism as a result of the presence of TiO_2. Titanium dioxide is extensively used for both house paint and artist's paint, as it is permanent and has good covering power. The price of commercial titanium metal is about 6 to 7 dollars/lb.

TUNGSTEN (Swedish, *tung sten,* heavy stone); also known as WOLFRAM (from *wolframite,* said to be named from *wolf rahm* or *spumi lupi,* because the ore interfered with the smelting of tin and was supposed to devour the tin), W; at. wt. 183.85; at. no. 74; m.p. 3410 ± 20°C; b.p. 5927°C; sp. gr. 19.3 (20°C); valence 2, 3, 4, 5, or 6. In 1779 Peter Woulfe examined the mineral now known as *wolframite* and concluded it must contain a new substance. Scheele, in 1781, found that a new acid could be made from *tung sten* (a name first applied about 1758 to a mineral now known as *scheelite*). Scheele and Bergman suggested the possibility of obtaining a new metal by reducing this acid. The de Elhuyar brothers found an acid in *wolframite* in 1783 that was identical to the acid of *tung sten* (tungstic acid) of Scheele, and in that year they succeeded in obtaining the element by reduction of this acid with charcoal. Tungsten occurs in *wolframite,* $(Fe, Mn)WO_4$; *scheelite,* $CaWO_4$; *huebnerite,* $MnWO_4$; and *ferberite,* $FeWO_4$. Important deposits of tungsten occur in Nevada, California, North Carolina, China, Korea, Bolivia, U.S.S.R., and Portugal. The metal is obtained by reducing tungstic oxide with hydrogen, carbon, carbonaceous gases, or calcium. Pure tungsten is a steel gray to tin-white metal. Very pure tungsten can be cut with a hacksaw, and can be forged, spun, drawn, and extruded. The impure metal is brittle and can be worked only with difficulty. Tungsten has the highest melting point and lowest vapor pressure of all metals, and at temperatures over 1650°C has the highest tensile strength. The metal oxidizes in air and must be protected at elevated temperatures. It has excellent corrosion resistance and is attacked only slightly by most mineral acids. The thermal expansion is about the same as boro-silicate glass, which makes the metal useful for glass-to-metal seals. Tungsten and its alloys are used extensively for filaments for electric lamps, electron and television tubes, and for metal evaporation work; for electrical contact points for automobile distributors; x-ray targets; windings and heating elements for electrical furnaces; and for numerous space missile and high-temperature applications. High-speed tool steels, Hastelloys, Stellite, and many other alloys contain tungsten. Tungsten carbide is of great importance to the metalworking, mining, and petroleum industries. Calcium and magnesium tungstates are widely used in fluorescent lighting; other salts of tungsten are used in the chemical, pigment, and tanning industries. Tungsten disulfide is a dry, high-temperature lubricant, stable to 500°C.

URANIUM (Planet *Uranus*), U; at. wt. 238.03; at. no. 92; m.p. 1132.3 ± 0.8°C; b.p. 3818°C; sp. gr. ~18.95; valence 3, 4, 5, or 6. Yellow-colored glass, containing more than 1% uranium oxide and dating back to 79 A.D., has been found near Naples, Italy. Klaproth recognized an unknown element in *pitchblende* and attempted to isolate the metal in 1789. The metal apparently was first isolated in 1841 by Peligot, who reduced the anhydrous chloride with potassium. Uranium is not as rare as it was once thought. It is now considered to be more plentiful than mercury, antimony, silver, or cadmium, and is about as abundant as molybdenum or arsenic. It occurs in numerous minerals, such as *pitchblende, uraninite, carnotite, autunite, uranophane, davidite,* and *tobernite.* It is also found in *phosphate rock, lignite, monazite sands,* and can be recovered commercially from these sources. The A.E.C purchases uranium in the form of acceptable concentrates at an established price of $8/lb. of U_3O_8. This incentive program has greatly increased the known uranium reserves. Uranium can be prepared by reducing a uranium halide, such as UF_4 with magnesium or calcium. The metal can also be produced by electrolysis of KUF_5 or UF_4, dissolved in a molten mixture of $CaCl_2$–NaCl. High-purity uranium can be prepared by the thermal decomposition of uranium halides on a hot filament. Uranium exhibits three crystallographic modifications as follows:

$$\alpha \xrightarrow{667°C} \beta \xrightarrow{775°C} \gamma$$

Uranium is a heavy, silvery white metal, which is pyrophoric when finely divided. It is a little softer than steel, and is attacked by cold water in a finely divided state. In air, the metal becomes coated with a layer of oxide. Acids dissolve the metal, but it is unaffected by alkalis. Uranium has fourteen isotopes, all of which are radioactive. Natural-occurring uranium contains 99.28% U^{238}, 0.71% U^{235}, and 0.0058% U^{234}. It is sufficiently radioactive to expose a photographic plate in an hour or so. U^{238}, with a half-life of 4.5×10^9 years, has been used to estimate the age of igneous rocks. The origin of uranium, the highest member of the natural-occurring elements is not clearly understood, although it may be presumed to be a decay product of elements of higher atomic weight, which are produced only by artificial means on earth. Uranium is of great importance as a basic nuclear fuel. U^{235}, the only natural-occurring fissionable material, can be separated from U^{238} by gaseous diffusion processes and by other methods. U^{238} can be converted to fissionable plutonium by the following reactions:

$$_{92}U^{238} + \text{neutron} \rightarrow {}_{92}U^{239} + \gamma$$
$$_{92}U^{239} \rightarrow {}_{93}Np^{239} + \text{electron}$$
$$_{93}Np^{239} \rightarrow {}_{94}Pu^{239} + \text{electron.}$$

One pound of completely fissioned uranium has the fuel value of over 1500 tons of coal. Uranium, to be used as a nuclear fuel, must be "enriched" by increasing the percentage of U^{235} present in a given sample. It is supplied for this purpose in the form of the unalloyed or alloyed metal, or it may be furnished as the oxide, as a compound dissolved in water or a fused salt, or as a liquid-metal solution. Uranium in the U.S.A. is under control of the Atomic Energy Commission. New uses are being explored for "depleted" uranium (i.e. uranium with the percentage of U^{235} lowered or removed); however, it has found relatively few applications. Uranium metal is used for x-ray targets for production of high-energy x-rays; the nitrate has been used as photographic toner; and the acetate is used in analytical chemistry. Uranium salts have also been used for producing yellow "vaseline" glass and glazes. Uranium and its compounds are highly toxic, both from a chemical and radiological standpoint. Finely divided uranium metal, being pyrophoric, presents a fire hazard. The maximum recommended allowable concentration of soluble uranium compounds in air (based on chemical toxicity) is 0.05 mg/cu meter; for insoluble compounds the concentration is set at 0.25 mg/cu meter of air. The permissible body level of natural uranium (based on radiotoxicity) is 0.2 microcurie for soluble compounds; for insoluble compounds the level is 0.009 microcurie, or in air 1.7×10^{-11} microcurie per milliliter.

VANADIUM (Scandinavian goddess, *Vanadis*), V; at. wt. 50.942; at. no. 23; m.p. 1890° ± 10°C; b.p. ~3000°C; sp. gr. 6.11 (18.7°C); valence 2, 3, 4, or 5. Vanadium was first discovered by del Rio in 1801. Unfortunately a French chemist incorrectly declared del Rio's new element was only impure chromium; del Rio thought himself to be mistaken and accepted the French chemist's statement. The element was rediscovered by Sefström in 1830, and isolated in nearly pure form by Roscoe, in 1867, who reduced the chloride with hydrogen. Vanadium of 99.3 to 99.8% purity was not produced until 1927. Vanadium is found in about 50 different minerals among which are *carnotite, roscoelite, vanadinite,* and *patronite*—important sources of the metal. Vanadium is also found in phosphate rock, certain iron ores, and is present in some crude oils in the form of organic complexes. Commercial production from petroleum ash holds promise as an important source of the element. High-purity vanadium can be obtained by reduction of the oxide with calcium or the trichloride with magnesium. It can also be produced by thermal decomposition methods. Pure vanadium is a bright white metal, and is soft and ductile. It has good corrosion resistance to alkalis, sulfuric and hydrochloric acid, and salt waters, but the metal oxidizes readily above 660°C. The metal has good structural strength and a low-fission neutron cross section, making it useful in nuclear applications. Vanadium is used in producing rust-resistant, spring, and high-speed tool steels. It is an important carbide stabilizer in making steels. Vanadium foil is used as a bonding agent in cladding titanium to steel. Vanadium pentoxide is used in ceramics and as a catalyst. It is also used as a mordant in dyeing and printing fabrics and in the manufacture of aniline black. Vanadium and its compounds should be handled with care. The maximum allowable concentration of V_2O_5 dust in air is about 0.5 mg/cu meter; V_2O_5 fumes should not exceed about 0.1 mg/cu meter of air. Ductile vanadium has recently become commercially available at a cost of about $40/lb. Commercial vanadium metal, of about 95% purity, costs about $5/lb.

WOLFRAM (see Tungsten)

XENON (Gr. *xenon*, stranger); Xe; at. wt. 131.30; at. no. 54; m.p. −111.9°C; b.p. −107.1 ± 3°C; density 5.887 ± 0.009 g/l, sp. gr. (liquid) 3.52 (−109°C); valence 0?. Discovered by Ramsay and Travers in 1898 in the residue left after evaporating liquid air components. Xenon is a member of the so-called rare or "inert" gases. It is present in the atmosphere to the extent of about one part in twenty million. The element is found in the gases evolved from certain mineral springs, and is commercially obtained by extraction from liquid air. Until recently, xenon has been considered inert and unable to form compounds with other elements. Those compounds that were occasionally reported in the literature were considered not to be true compounds. Evidence has been mounting in the past few years that xenon, as well as other members of the zero valence elements, do form compounds. Among the "compounds" of xenon now reported are xenon hydrate, deuterate, difluoride, tetrafluoride, hexafluoride, and $XePtF_6$ and $XeRhF_6$. More recently, xenon trioxide, which is highly explosive, has been prepared. The structure of these substances is still open to question. Xenon in a vacuum tube produces a beautiful blue glow when excited by an electrical discharge. The gas is used in making stroboscopic lamps, bactericidal lamps, and lamps used to excite ruby lasers for generating coherent light. Xenon is used in the atomic energy field in bubble chambers, probes, and other applications where its high molecular weight is of value. It is also potentially useful as a gas for ion engines. Twenty-four isotopes of xenon are recognized. Xe^{133} and Xe^{135} are produced by neutron irradiation in air-cooled nuclear reactors. Xe^{133} has useful applications as a radioisotope. Xenon gas has been used as an experimental surgical anesthetic on human beings. The element is available in sealed glass containers for about $6/100 cc. of gas at standard pressure.

YTTERBIUM (*Ytterby*, village in Sweden), Yb; at. wt. 173.04; at. no. 70; m.p. 824 ± 5°C; b.p. 1427°C; sp. gr. (α) 6.977, (β) 6.54; valence 2, 3. Marignac in 1878 discovered a new component, which he called *ytterbia*, in the earth then known as *erbia*. In 1907, Urbain separated ytterbia into two components, which he called *neoytterbia* and *lutecia*.

The elements in these earths are now known as *ytterbium* and *lutetium*, respectively. These elements are identical with *aldebaranium and cassiopeium* discovered independently and at about the same time by von Welsbach. Ytterbium occurs along with other rare earths in a number of rare minerals. It is commercially recovered principally from *monazite sand*, which contains about 0.03%. Ion-exchange techniques developed in recent years have greatly simplified the separation of the rare earths from one another. It is only in recent years that ytterbium has been isolated in relatively pure form. The element can be prepared by reducing the oxide with lanthanum or misch metal by special techniques. Ytterbium has a bright silvery luster, is soft, malleable, and quite ductile. It is stable in air and reacts only slowly with water. Ytterbium has two allotropic forms with a transformation point at 798°C. Fourteen isotopes are recognized. As yet, because of its relatively high price, very few uses have been found for the metal or its compounds. One isotope is reported to have been used as a radiation source as a substitute for a portable x-ray machine where electricity is unavailable. Ytterbium metal is commercially available with a purity of about 99+% for about $1.50/gm, or $200/lb.

YTTRIUM (*Ytterby*, village in Sweden), Y; at. wt. 88.905; at. no. 39; m.p. 1495° ± 5°C; b.p. 2927°C; sp. gr. 4.45; valence 3. *Yttria*, which is an earth containing yttrium, was discovered by Gadolin in 1794; in 1843 Mosander showed that yttria could be resolved into the earths of three elements. The name yttria was reserved for the most basic one; the others were named *erbia* and *terbia*. Yttrium occurs in nearly all of the rare-earth minerals. It is recovered commercially from *monazite sand*, which contains about 3%, and from *bastnaesite*, which contains about 0.2%. Wöhler obtained the impure element in 1828 by reduction of the anhydrous chloride with potassium. The metal is now produced commercially by reduction of the fluoride with calcium metal in a calcium fluoride-lined bomb, using zinc fluoride as an additive. The metal is produced as a Y-Zn alloy. The zinc is removed by vacuum distillation; the yttrium metal is arc-melted in helium. It can also be prepared by other techniques. Yttrium has a silvery metallic luster and is relatively stable in air. Turnings of the metal, however, ignite readily in air, and finely divided yttrium is very unstable in air. Small amounts of yttrium (0.1 to 0.2%) can be used to reduce the grain size in chromium, molybdenum, zirconium, and titanium, and to increase strength of aluminum and magnesium alloys. Alloys with other useful properties can be obtained by using yttrium as an additive. The metal can be used as a deoxidizer for vanadium and other nonferrous metals. The metal has a low cross section for nuclear capture. Y^{90}, one of the fifteen recognized isotopes of yttrium, exists in equilibrium with its parent Sr^{90}, a product of atomic explosions. Yttrium has been considered for use as a nodulizer for producing nodular cast iron, in which the graphite forms compact nodules instead of the usual flakes. Such iron has increased ductility. Yttrium iron, aluminum, and gadolinium garnets, with formulas such as $Y_3Fe_5O_{12}$ and $Y_3Al_5O_{12}$, have interesting magnetic properties. Such garnets are used in microwave and other electronic applications. Yttrium iron garnet is also exceptionally efficient as both a transmitter and transducer of acoustic energy. Yttrium metal of 99+% purity is commercially available at a cost of about $1/gm or $200/lb.

ZINC (Ger. *Zink*, of obscure origin), Zn; at. wt. 65.37; at. no. 30; m.p. 419.4°C; b.p, 907°C; sp. gr. 7.133 (25°C); valence 2. Zinc ores were used for making brass centuries before zinc was recognized as a distinct element. Tubal-Cain, seven generations from Adam, is mentioned as being an "instructor of every artificer in brass and iron." An alloy containing 87% zinc has been found in prehistoric ruins in Transylvania. Metallic zinc was produced in the 13th Century A.D. in India by reducing calamine with organic substances, such as wool. The metal was rediscovered in Europe by Marggraf in 1746, who showed that the metal could be obtained by reducing *calamine* with charcoal. The principal ores of zinc are *sphalerite* or *blende* (sulfide), *smithsonite* (carbonate), *calamine* (silicate), and *franklinite* (zinc, manganese, iron oxide). Zinc can be obtained by roasting its ores to form the oxide and by reduction of the oxide with coal or carbon, with subsequent distillation of the metal. Other methods of extraction are possible. Zinc is a bluish-white lustrous metal. It is brittle at ordinary temperatures but malleable at 100° to 150°C. It is a fair conductor of electricity, and burns in air at high red heat with evolution of white clouds of the oxide. The metal is employed to form numerous alloys with other metals. Brass, nickel silver, typewriter metal, commercial bronze, spring brass, German silver, soft solder and aluminum solder are some of the more important alloys. Large quantities of zinc are used to produce die castings, used extensively by the automotive, electrical, and hardware industries. Zinc is also extensively used to galvanize other metals, such as iron, to prevent corrosion. Neither zinc nor zirconium is ferromagnetic, but $ZrZn_2$ exhibits ferromagnetism at temperatures below 35°K. Zinc oxide is a unique and very useful material to modern civilization. It is widely used in the manufacture of paints, rubber products, cosmetics, pharmaceuticals, floor coverings, plastics, printing inks, soap, storage batteries, textiles, electrical equipment, and other products. It has unusual electrical, thermal, optical, and solid-state properties that have not yet been fully investigated. Lithopone, a mixture of zinc sulfide and barium sulfate, is an important pigment. Zinc sulfide is used in making luminous dials, X-ray and TV screens, and fluorescent lights. The chloride and chromate are also important compounds.

ZIRCONIUM (Arabic *zargun*, gold color), Zr; at. wt. 91.22; at. no. 40; m.p. 1852 ± 2°C; b.p. 3578°C; sp. gr. 6.53 ± 0.01 (calc.). The name *zircon* probably originated from the arabic word *zargun*, which describes the color of the gemstone now known as *zircon*, *jargon*, *hyacinth*, *jacinth*, or *ligure*. This mineral, or its variations, is mentioned in biblical writings. The mineral was not known to contain a new element until Klaproth, in 1789, analyzed a jargon from Ceylon and found a new earth, which Werner named zircon (*silex circonius*), and Klaproth called *Zirkonerde* (*zirconia*). The impure metal was first isolated by Berzelius in 1824 by heating a mixture of potassium and potassium zirconium fluoride in a small iron tube. Pure zirconium was first prepared in 1914. Zirconium is found in abundance in S-type stars, and has been identified in the sun and meteorites. Zircon, $ZrSiO_4$, the principal ore, is found in deposits in Florida, South Carolina, Australia, and Brazil. Zirconium also occurs in some 30 other recognized mineral species. Zirconium is produced commercially by reduction of the chloride with magnesium (the Kroll Process), and by other methods. It is a grayish-white lustrous metal. When finely divided, the metal will ignite spontaneously in air. The solid metal is much more difficult to ignite. The inherent toxicity of zirconium compounds is low. Hafnium is invariably found in zirconium ores, and the separation is difficult. Commercial-grade zirconium contains from 1 to 3% hafnium. Zirconium has a low absorption cross section for neutrons, and is therefore used for nuclear energy applications, such as for cladding fuel elements. Reactor-grade zirconium is essentially free of hafnium. Zircaloy is an important alloy developed specifically for nuclear applications. Zirconium is exceptionally resistant to corrosion by common acids and alkalis, by sea water, and by other agents. It is used extensively by the chemical industry where corrosive agents are employed. Zirconium is used as a getter in vacuum tubes, as an alloying agent in steel, in making surgical appliances, photoflash bulbs, explosive primers, rayon spinnerets, lamp filaments etc. It is used in poison ivy lotions as the carbonate as it combines with urushiol. With columbium, zirconium is superconductive at low temperatures and is used to make superconductive magnets, which offer hope of direct large-scale generation of electric power. Alloyed with zinc, zirconium becomes magnetic at temperatures below 35°K. Zirconium oxide (zircon) has a high index of refraction and is used as a gem material. The impure oxide, zirconia, is used for laboratory crucibles that will withstand heat shock, for linings of metallurgical furnaces, and by the glass and ceramic industries as a refractory material. Commercial zirconium metal sponge is priced at about $5/lb; reactor grade sponge at about $7/lb. Fabricated zirconium parts may be much higher in cost.

NOMENCLATURE OF INORGANIC CHEMISTRY

AMERICAN VERSION

By permission of the Committee on Publications of the International Union of Pure and Applied Chemistry.

INDEX TO IUPAC INORGANIC RULES

1. ELEMENTS

1.1. Names and Symbols of the Elements

1.11.—The elements should have the symbols given in the following table (Table I). It is desirable that the names should differ as little as possible among the different languages, but as complete uniformity is hard to achieve, separate lists have been drawn up in English and in French. The English list only is reproduced here.

TABLE I

ELEMENTS

Name	Symbol	Atomic number	Name	Symbol	Atomic number	Name	Symbol	Atomic number
Actinium	Ac	89	Gold (Aurum)	Au	79	Praseodymium	Pr	59
Aluminum	Al	13	Hafnium	Hf	72	Promethium	Pm	61
Americium	Am	95	Helium	He	2	Protactinium	Pa	91
Antimony	Sb	51	Holmium	Ho	67	Radium	Ra	88
Argon	Ar	18	Hydrogen	H	1	Radon	Rn	86
Arsenic	As	33	Indium	In	49	Rhenium	Re	75
Astatine	At	85	Iodine	I	53	Rhodium	Rh	45
Barium	Ba	56	Iridium	Ir	77	Rubidium	Rb	37
Berkelium	Bk	97	Iron (Ferrum)	Fe	26	Ruthenium	Ru	44
Beryllium	Be	4	Krypton	Kr	36	Samarium	Sm	62
Bismuth	Bi	83	Lanthanum	La	57	Scandium	Sc	21
Boron	B	5	Lead (Plumbum)	Pb	82	Selenium	Se	34
Bromine	Br	35	Lithium	Li	3	Silicon	Si	14
Cadmium	Cd	48	Lutetium	Lu	71	Silver (Argentum)	Ag	47
Calcium	Ca	20	Magnesium	Mg	12	Sodium	Na	11
Californium	Cf	98	Manganese	Mn	25	Strontium	Sr	38
Carbon	C	6	Mendelevium	Md	101	Sulfur	S	16
Cerium	Ce	58	Mercury	Hg	80	Tantalum	Ta	73
Cesium	Cs	55	Molybdenum	Mo	42	Technetium	Tc	43
Chlorine	Cl	17	Neodymium	Nd	60	Tellurium	Te	52
Chromium	Cr	24	Neon	Ne	10	Terbium	Tb	65
Cobalt	Co	27	Neptunium	Np	93	Thallium	Tl	81
Copper (Cuprum)	Cu	29	Nickel	Ni	28	Thorium	Th	90
Curium	Cm	96	Niobium	Nb	41	Thulium	Tm	69
Dysprosium	Dy	66	Nitrogen	N	7	Tin (Stannum)	Sn	50
Einsteinium	Es	99	Nobelium	No	102	Titanium	Ti	22
Erbium	Er	68	Osmium	Os	76	Tungsten (Wolfram)	W	74
Europium	Eu	63	Oxygen	O	8	Uranium	U	92
Fermium	Fm	100	Palladium	Pd	46	Vanadium	V	23
Fluorine	F	9	Phosphorus	P	15	Xenon	Xe	54
Francium	Fr	87	Platinum	Pt	78	Ytterbium	Yb	70
Gadolinium	Gd	64	Plutonium	Pu	94	Yttrium	Y	39
Gallium	Ga	31	Polonium	Po	84	Zinc	Zn	30
Germanium	Ge	32	Potassium	K	19	Zirconium	Zr	40

◆The committees reaffirm the name niobium for element 41 in spite of the fact that many in the United States, particularly outside of chemical circles, still retain the name columbium.

1.12.—The names placed in parentheses (after the trivial names) in the list in Table I shall always be used when forming names derived from those of the elements, *e.g.*, aurate, ferrate, wolframate and not goldate, ironate, tungstate.

For some compounds of sulfur, nitrogen, and antimony, derivatives of the Greek name θεῖον, the French name *azote*, and the Latin name *stibium*, respectively, are used.

Although the name nickel agrees with the chemical symbol, it is essentially a trivial name and is spelled so differently in various languages (niquel, nikkel, *etc.*) that it is recommended that names of derivatives be formed from the Latin name *niccolum, e.g.*, niccolate instead of nickelate. The name mercury should be used as the root name also in languages where the element has another name (mercurate, *not* hydrargyrate).

In the cases in which different names have been used, the Commission has selected one based upon considerations of prevailing usage and practicability. It should be emphasized that their selection carries no implication regarding priority of discovery.

◆Tungstate and nickelate are both so well established in American practice that the committees object to changing them to wolframate and niccolate.

1.13.—Any new metallic elements should be given names ending in -ium. Molybdenum and a few other elements have long been spelled without an "i" in most languages, and the Commission hesitates to insert it.

1.14.—All new elements shall have two-letter symbols.

1.15.—All isotopes of an element should have the same name. For hydrogen the isotope names protium, deuterium, and tritium may be retained, but it is undesirable to assign isotopic names instead of numbers to other elements. They should be designated by mass numbers as, for example, "oxygen-18."

◆The list in 3.21 implies that D is an acceptable symbol for deuterium, whereas ^{2}H is used in 1.32. It is recommended that D and T be allowed for deuterium and tritium, respectively. *Cf.* comments made at 1.31, 1.32.

1.2. Names for Groups of Elements and their Subdivisions

1.21.—The use of the collective names: halogens (F, Cl, Br, I, and At), chalcogens (O, S, Se, Te, and Po), and halides and chalcogenides for their compounds, alkali metals (Li to Fr), alkaline earth metals (Ca to Ra), and inert gases may be continued. The name rare earth metals may be used for the elements Sc, Y, and La to Lu inclusive; the name lanthanum series for the elements no. 57–71 (La to Lu inclusive), and the name lanthanides for the elements 58–71 (Ce to Lu inclusive) are recommended. Elements no. 89 (Ac) to 103 form the actinium series, and the name actinides is reserved for the elements in which the 5*f* shell is being filled. The name transuranium elements is also approved for the elements following uranium.

◆The collective term halogenides used in the Rules has been replaced in this version by halides, which is almost universally used in English and is unambiguous.

The inclusion of Sc with the rare earths is questioned by some. No need is seen for the terms lanthanum series and actinium series, particularly since the latter term is used for a radioactive series. The use of a collective term for elements 58–71 is approved, although it is suggested that lanthan*oid* is preferable to lanthan*ide* because of the use of -ide for binary compounds; similarly, actin*oid* is preferable to actin*ide*. Definition by means of atomic numbers is recommended in both cases rather than on the basis of interpretation (*e.g.*, filling of 5*f* shells).

1.22.—The word metalloid should not be used to denote nonmetals.

1.3. Indication of Mass, Charge, etc., on Atomic Symbols

1.31.—The mass number, atomic number, number of atoms, and ionic charge of an element may be indicated by means of four indices placed around the symbol. The positions are to be occupied thus

left upper index	right lower index	mass number	number of atoms
left lower index	right upper index	atomic number	ionic charge

Ionic charge should be indicated by A^{n+} rather than by A^{+n}.

Example: ${}^{32}_{16}S_2^{2+}$ represents a doubly ionized molecule containing two atoms of sulfur, each of which has the atomic number 16 and mass number 32.

The following is an example of an equation for a nuclear reaction

$${}^{26}_{12}Mg + {}^{4}_{2}He = {}^{29}_{13}Al + {}^{1}_{1}H$$

◆Although the practice of American chemists and physicists in general has been to put the mass number at the upper right of the symbol, the committees recognize the advantage of putting it at the upper left so that the upper right is available for the ionic charge as needed.

1.32.—Isotopically labeled compounds may be described by adding to the name of the compounds the symbol of the isotope in parentheses.

Examples:

${}^{32}PCl_3$ phosphorus(${}^{32}P$) trichloride (spoken: phosphorus-32 trichloride)

$H^{36}Cl$ hydrogen chloride^{36}(Cl) (spoken: hydrogen chloride-36)

${}^{15}NH_3$ ammonia(${}^{15}N$) (spoken: ammonia nitrogen-15)

The position of the labeled atom may be indicated by placing the isotope symbol immediately after the locant (name of the group concerned).

Example: ${}^{2}H_2{}^{35}SO_4$ sulfuric(${}^{35}S$) acid(${}^{2}H$)

If this method gives names which are ambiguous or difficult to pronounce, the whole group containing the labeled atom may be indicated.

Examples:

$HOSO_2{}^{35}SH$	thiosulfuric(${}^{35}SH$) acid
${}^{15}NO_2NH_2$	nitramide(${}^{15}NO_2$), not nitr(${}^{15}N$)amide
$NO_2{}^{15}NH_2$	nitramide(${}^{15}NH_2$)
$HO_3S^{18}O{-}^{18}OSO_3H$	peroxo(${}^{18}O_2$)disulfuric acid

1.4. Allotropes

If systematic names for gaseous and liquid modifications are required, they should be based on the size of the molecule, which can be indicated by Greek numerical prefixes (listed in 2.251). If the number of atoms is large and unknown, the prefix poly may be used. To indicate ring and chain structures the prefixes cyclo and catena may be used.

Examples:

Symbol	Trivial Name	Systematic Name
H	atomic hydrogen	monohydrogen
O_2	(common) oxygen	dioxygen
O_3	ozone	trioxygen
P_4	white phosphorus (yellow phosphorus)	tetraphosphorus
S_8	λ-sulfur	cycloöctasulfur or octasulfur
S_n	μ-sulfur	catenapolysulfur or polysulfur

For the nomenclature of solid allotropic forms the rules in Section 8 may be applied.

◆The use of prefixes to indicate ring and chain structures is favored by the committees, but it should be pointed out that ino (not catena) has been used by mineralogists for indicating chain structures in silicates, along with cyclo and other prefixes (neso, phyllo, tecto, soro) denoting structure, all used without italics or hyphens (*e.g.*, inosilicates). *Cf.* use of catena in 7.42 for chains of alternating, not self-linking, atoms.

2. FORMULAS AND NAMES OF COMPOUNDS IN GENERAL

Many chemical compounds are essentially binary in nature and can be regarded as combinations of ions or radicals; others may be treated as such for the purpose of nomenclature.

Some chemists have expressed the opinion that the name of a compound should indicate whether it is ionic or covalent. Such a distinction is made in some languages (*e.g.*, in German: Natriumchlorid *but* Chlorwasserstoff), but it has not been made consistently, and indeed it seems impossible to introduce this distinction into a consistent system of nomenclature, because the line of demarcation between these two categories is not sharp. In these rules a system of nomenclature has been built on the basis of the endings -ide and -ate, and it should be emphasized that these are intended to be applied both to ionic and covalent compounds. If it is desired to avoid such endings for neutral molecules, names can be given as coördination compounds in accordance with **2.24** and Section **7.**

2.1. Formulas

2.11.—Formulas provide the simplest and clearest method of designating inorganic compounds. They are of particular importance in chemical equations and in descriptions of chemical procedure. However, their general use in text is not recommended, although in some cases a formula, on account of its compactness, may be preferable to a cumbersome and awkward name.

2.12.—The *empirical formula* is formed by juxtaposition of the atomic symbols to give the *simplest possible* formula expressing the stoichiometric composition of the compound in question. The empirical formula may be supplemented by indication of the crystal structure—see Section **8.**

2.13.—For compounds consisting of discrete molecules the *molecular formula, i.e.*, a formula corresponding with the correct molecular weight of the compound, should be used, *e.g.*, S_2Cl_2 and $H_4P_2O_6$ and not SCl and H_2PO_3. When the molecular weight varies with temperature, *etc.*, the simplest possible formula generally may be chosen, *e.g.*, S, P, and NO_2 instead of S_8, P_4, and N_2O_4, unless it is desirable to indicate the molecular complexity.

2.14.—In the *structural formula* the sequence and spatial arrangement of the atoms in a molecule are indicated.

2.15.—In formulas the *electropositive constituent* (cation) should always be placed first, *e.g.*, KCl, $CaSO_4$.

This also applies in Romance languages even though the electropositive constituent is placed last in the *name, e.g.*, KCl, chlorure de potassium.

If the compound contains more than one electropositive or more than one electronegative constituent, their sequence is determined by Rules **6.32** and **6.33.**

2.16.—In the case of binary compounds between nonmetals that constituent should be placed first which appears earlier in the sequence: B, Si, C, Sb, As, P, N, H, Te, Se, S, At, I, Br, Cl, O, F.

Examples: NH_3, H_2S, N_4S_4, S_2Cl_2, Cl_2O, OF_2

◆Because N_4S_4 is definitely a nitride, not a sulfide, the committees prefer to write the formula S_4N_4, name it sulfur nitride, and cite it as an exception rather than as an example.

2.161.—For compounds containing three or more elements, however, the sequence should in general follow the order in which the atoms are actually bound in the molecule or ion, *e.g.*, NCS^-, not CNS^-, HOCN (cyanic acid), and HONC (fulminic acid).

Although formulas such as HNO_3, $HClO_4$, H_2SO_4, do not agree with this rule and HNO_3 does not even follow the main rule in **2.16**, the Commission does not at this time wish to break the old custom of putting the central atom immediately after the hydrogen atom in such cases (*cf*. Section **5**). The formula for hypochlorous acid may be written HOCl or HClO.

2.17.—In intermetallic compounds the constituents should be placed in the order

Fr, Cs, Rb, K, Na, Li
Ra, Ba, Sr, Ca, Mg, Be
103, No, Md, Fm, Es, Cf, Bk, Cm, Am, Pu, Np, U, Pa, Th, Ac, Lu–La, Y, Sc
Hf, Zr, Ti
Ta, Nb, V
W, Mo, Cr
Re, Tc, Mn
Pt, Ir, Os, Pd, Rh, Ru, Ni, Co, Fe
Au, Ag, Cu
Hg, Cd, Zn
Tl, In, Ga, Al
Pb, Sn, Ge
Bi, Sb
Po

Nonmetals (except Sb) in the order given in **2.16.**

Deviations from this order may be allowed, *e.g.*, when compounds with analogous structures are compared (AgZn and AgMg).

2.18.—The number of identical atoms or atomic groups in a formula is indicated by means of Arabic numerals, placed below and to the right of the symbol or symbols in parentheses ( ) or brackets [ ] to which they refer. Water of crystallization and similar loosely bound molecules, however, are designated by means of Arabic numerals before their formulas.

Examples: $CaCl_2$ not $CaCl^2$
$[Co(NH_3)_6]Cl_3$ not $[Co6NH_3]Cl_3$
$[Co(NH_3)_6]_2(SO_4)_3$
$Na_2SO_4.10H_2O$

2.19.—The prefixes *cis, trans, sym, asym* may be used in their usual senses. The prefixes may be connected with the formula by a hyphen and it is recommended that they be italicized.
Example: *cis*-$[PtCl_2(NH_3)_2]$

2.2. Systematic Names

Systematic names of compounds are formed by indicating the constituents and their proportions according to the following rules. (For the order of the constituents see also the later sections.)

2.21.—The name of the *electropositive constituent* (or that treated as such according to **2.16**) will not be modified (see, however, **2.2531**).

In Germanic languages the electropositive constituent is placed first, but in Romance languages it is customary to place the electronegative constituent first.

2.22.—If the *electronegative constituent* is monatomic its name is modified to end in -ide. For binary compounds of the nonmetals the name of the element standing later in the sequence in **2.16** is modified to end in -ide.

Examples: Sodium chloride, calcium sulfide, lithium nitride, arsenic selenide, calcium phosphides, nickel arsenide, aluminum borides, iron carbides, boron hydrides, phosphorus hydrides, hydrogen chloride, hydrogen sulfide, silicon carbide, carbon disulfide, sulfur hexafluoride, chlorine dioxide, oxygen difluoride.

Certain polyatomic groups are also given the ending -ide—see **3.22**.

In the Romance languages the endings -ure, -uro, and -eto are used instead of -ide. In some languages the word *oxyde* is used, whereas the ending -ide is used in the names of other binary compounds; it is recommended that the ending -ide be universally adopted in these languages.

◆Nitrogen sulfide has been taken out of the examples. *Cf.* comment at 2.16.

2.23.—If the electronegative constituent is polyatomic it should be designated by the termination -ate. In certain exceptional cases the terminations -ide and -ite are used—see **3.22**.

2.24.—In inorganic compounds it is generally possible in a polyatomic group to indicate a *characteristic atom* (as in ClO^-) or a *central atom* (as in ICl_4^-). Such a polyatomic group is designated a *complex*, and the atoms, radicals, or molecules bound to the characteristic or central atom are termed *ligands*.

In this case the name of a negatively charged complex should be formed from the name of the characteristic or central element (as indicated in **1.12**) modified to end in -ate.

Anionic ligands are indicated by the termination -o. Further details concerning the designation of ligands, the definition of "central atom," *etc.*, appear in Section 7.

Although the terms sulfate, phosphate, *etc.*, were originally the names of the anions of particular oxo acids, the names sulfate, phosphate, *etc.*, should now designate quite generally a negative group containing sulfur or phosphorus, respectively, as the central atom, irrespective of its oxidation state (the designation of the oxidation state is discussed in later rules) and the number and nature of the ligands. The complex is indicated by brackets [ ], but this is not always necessary.

Examples:

$Na_2[SO_4]$	sodium tetraoxosulfate	$Na_3[PS_4]$	sodium tetrathiophosphate
$Na_2[SO_3]$	sodium trioxosulfate	$Na[PCl_6]$	sodium hexachlorophosphate
$Na_2[S_2O_3]$	sodium trioxothiosulfate	$K[PO_2F_2]$	potassium dioxodifluorophosphate
$Na[SO_3F]$	sodium trioxofluorosulfate	$K[POCl_2(NH)]$	potassium oxodichloroimidophosphate
$Na_3[PO_4]$	sodium tetraoxophosphate		

In many cases these names may be abbreviated, *e.g.*, sodium sulfate, sodium thiosulfate (see **2.26**), and in other cases trivial names may be used (*cf.* **2.3**, **3.224**, and Section **5**). It should be pointed out, however, that the principle is quite generally applicable, to compounds containing organic ligands also, and its use is recommended in all cases where trivial names do not exist.

The coördination principle applied in this rule may also be applied to complexes which are positive or neutral (*cf.* **3.1** and Section **7**). However, neutral complexes which are as a rule considered as binary compounds are given names according to **2.22**, **2.16**. Thus, SO_3, sulfur trioxide, not trioxosulfur.

◆In the examples it would seem that full cöordination-type names should be given, *e.g.*, either sodium tetraoxosulfate (VI) or disodium tetraoxosulfate. *Cf.* 7.32 and comment at 7.312.

2.25.—Indication of the Proportions of the Constituents.

2.251.—The *stoichiometric proportions* may be denoted by means of Greek numerical prefixes (mono, di, tri, tetra, penta, hexa, hepta, octa, ennea, deca, hendeca, and dodeca) preceding without hyphen the names of the elements to which they refer. It may be necessary in some languages to supplement these numerals with hemi ($^1/_2$) and the Latin sesqui ($^3/_2$).

The prefix mono may generally be omitted. Beyond 12, Greek prefixes are replaced by Arabic numerals (with or without hyphen according to the custom of the language), because they are more readily understood.

This sytem is applicable to all types of compounds and is especially suitable for binary compounds of the nonmetals.

When it is required to indicate the number of entire groups of atoms, particularly when the name includes a numerical prefix with a different significance, the multiplicative numerals (Latin bis, Greek tris, tetrakis, *etc.*) are used and the whole group to which they refer may be placed in parentheses if necessary.

Examples:

N_2O	dinitrogen oxide	Fe_3O_4	triiron tetraoxide
NO_2	nitrogen dioxide	U_3O_8	triuranium octaoxide
N_2O_4	dinitrogen tetraoxide	MnO_2	manganese dioxide
N_2S_5	dinitrogen pentasulfide	$Ca_3[PO_4]_2$	tricalcium diorthophosphate
S_2Cl_2	disulfur dichloride	$Ca[PCl_6]_2$	calcium bis(hexachlorophosphate)

In indexes it may be convenient to italicize a numerical prefix at the beginning of the name and connect it to the rest of the name with a hyphen, but this is not desirable in text, *e.g.*, *tri*-Uranium octaoxide.

Since the degree of polymerization of many substances varies with temperature, state of aggregation, *etc.*, the name to be used should normally be based upon the simplest possible formula of the substance except when it is required specifically to draw attention to the degree of polymerization.

Example: The name nitrogen dioxide may be used for the equilibrium mixture of NO_2 and N_2O_4. Dinitrogen tetraoxide means specifically N_2O_4.

◆In accordance with the organic nomenclature rules and well-established practice, it is recommended that "or Latin" be inserted between "Greek" and "numerical prefixes" in the first sentence and that "nona" replace "ennea," and "undeca" replace "hendeca."

Extreme caution is advised in the omission of numerical prefixes, including mono (*cf.* second set of examples in 5.23), because of the frequent use of names such as chloroplatinate (*cf.* 2.26 and the last sentence in 5.24).

2.252.—The proportions of the constituents also may be indicated indirectly by *Stock's system*, that is, by Roman numerals representing the oxidation number or stoichiometric valence of the element, placed in parentheses immediately following the name. For zero the Arabic 0 will be used. When used in conjunction with symbols the Roman numeral may be placed above and to the right.

The Stock notation can be applied to both cations and anions, but preferably should *not* be applied to compounds between nonmetals.

In employing the Stock notation, use of the Latin names of the elements (or Latin roots) is considered advantageous.

Examples:

$FeCl_2$	iron(II) chloride or ferrum(II) chloride
$FeCl_3$	iron(III) chloride or ferrum(III) chloride
MnO_2	manganese(IV) oxide
BaO_2	barium(II) peroxide
$Pb^{II}_2Pb^{IV}O_4$	dilead(II) lead(IV) oxide or trilead tetraoxide
$K_4[Ni(CN)_4]$	potassium tetracyanonicollate(0)
$K_4[Fe(CN)_6]$	potassium hexacyanoferrate(II)
$Na_2[Fe(CO)_4]$	sodium tetracarbonylferrate(—II)

◆While the committees favor the extended use of the Stock notation, they suggest that in some cases the system of Ewens and Bassett (designation of the aggregate charge of a complex ion by an Arabic numeral in parentheses following the name, similar to the use as superior notations with formulas) is advantageous and should be allowed as an alternate (*cf.* 3.17 and comment at 7.323). Mixed use of the two systems, while not desirable in any one context, does not affect indexing and should not lead to confusion. See comment at 1.12.

2.253.—The following systems are in use but are not recommended:

2.2531.—The system of indicating valence by means of the suffixes *-ous* and *-ic* added to the root of the name of the cation may be retained for elements exhibiting not more than two valences.

2.2532.—*"Functional"* nomenclature (such as "nitric anhydride" for N_2O_5) is not recommended apart from the name *acid* to designate the acid function (Section **5**).

◆Apparently there is no objection to acid anhydride as a class name (*cf.* 5.32). Other functional derivatives of acids are named as such in the Rules (5.3).

2.26.—In systematic names it is not always necessary to indicate stoichiometric proportions. In many instances it is permissible to omit the numbers of atoms, oxidation numbers, *etc.*, when they are not required in the particular circumstances. For instance, these indications are not generally necessary with elements of essentially constant valence.

Examples:

sodium sulfate instead of sodium tetraoxosulfate
aluminum sulfate instead of aluminum(III) sulfate
potassium chloroplatinate(IV) instead of potassium hexachloroplatinate(IV)
potassium cyanoferrate(III) instead of potassium hexacyanoferrate(III)
phosphorus pentaoxide instead of diphosphorus pentaoxide

2.3. Trivial Names

Certain well-established trivial names for oxo acids (Section **5**) and for hydrogen compounds (water, ammonia, hydrazine) are still acceptable. For some other hydrogen compounds these names are approved:

B_2H_6	diboran	PH_3	phosphine	SbH_3	stibine	P_2H_4	diphosphine
SiH_4	silane	AsH_3	arsine	Si_2H_6	disilane, *etc.*	As_2H_4	diarsine

In some languages names of the type "Chlorwasserstoff" are in use and may be retained if national nomenclature committees so wish.

Purely trivial names, free from false scientific implications, such as soda, Chile saltpeter, quicklime, are harmless in industrial and popular literature; but old incorrect scientific names such as sulfate of magnesia, Natronhydrat, sodium muriate, carbonate of lime, should be avoided under all circumstances, and they should be eliminated from technical and patent literature.

◆For BH_3 (omitted in the Rules) borane rather than the previously used borine has been recommended by the Advisory Committee on the Nomenclature of Organic Boron Compounds of the ACS in a report not yet published.

Because soda is an ambiguous term, it is suggested that it be replaced by soda ash.

3. NAMES FOR IONS AND RADICALS

3.1. Cations

3.11.—Monatomic cations should be named like the corresponding element, without change or suffix, except as provided by **2.2531**.

Examples:

Cu^+ the copper(I) ion Cu^{2+} the copper(II) ion I^+ the iodine cation

◆For I^+, iodide(I) cation is more consistent with recommended practice. *Cf.* 3.21.

3.12.—The preceding principle should apply also to polyatomic cations corresponding to radicals for which special names are given in **3.32**, *i.e.*, these names should be used without change or suffix.

Examples: NO^+ the nitrosyl cation NO_2^+ the nitryl cation

◆Polyatomic here and in 3.13, 3.14, 3.223, 3.32, and 5.2 seems to be limited to more than one *kind* of atom, and hence heteratomic would be a more precise term. It is agreed that nitryl and not nitronium should be used in all cases (*cf.* 3.151).

3.13.—Polyatomic cations formed from monatomic cations by the addition of other ions or neutral atoms or molecules (ligands) will be regarded as complex and will be named according to the rules given in Section **7**.

Examples:

$[Al(H_2O)_6]^{3+}$ the hexaaquoaluminum ion $[CoCl(NH_3)_5]^{2+}$ the chloropentamminecobalt ion

For some important polyatomic cations which fall in this section, radical names given in **3.32** may be used alternatively, *e.g.*, for UO_2^{2+} the name uranyl(VI) ion in place of dioxouranium(VI) ion.

3.14.—Names for polyatomic cations derived by addition of protons to monatomic anions are formed by adding the ending -onium to the root of the name of the anion element.

Examples: phosphonium, arsonium, stibonium, oxonium, sulfonium, selenonium, telluronium, and iodonium ions.

Organic ions derived by substitution in these parent cations should be named as such, whether the parent itself is a known compound or not: for example $(CH_3)_4Sb^+$, the tetramethylstibonium ion.

The ion H_3O^+, which is in fact the monohydrated proton, is to be known as the oxonium ion when it is believed to have this constitution, as for example in $H_2O^+ClO_4^-$, oxonium perchlorate. The widely used term hydronium should be kept for the cases where it is wished to denote an indefinite degree of hydration of the proton, as, for example, in aqueous solution. If, however, the hydration is of no particular importance to the matter under consideration, the simpler term hydrogen ion may be used. The latter also may be used for the indefinitely solvated proton in nonaqueous solvents; but definite ions such as $CH_3OH_2^+$ and $(CH_3)_2OH^+$ should be named as derivatives of the oxonium ion, *i.e.*, as methyl- and dimethyloxonium ions, respectively.

◆The committees concur in oxonium for the ion H_3O^+, but see little reason for encouraging retention of the term hydronium ion because hydrogen ion adequately designates an indeterminate degree of hydration.

3.15.—Ions from Nitrogen Bases.

3.151.—The name ammonium for the ion NH_4^+ does not conform to **3.14**, but should be retained. This decision does *not* release the word nitronium for other uses: this would lead to inconsistencies when the rules are applied to other elements.

3.152.—Substituted ammonium ions derived from nitrogen bases with names ending in -amine will receive names formed by changing -amine to -ammonium. For example, $HONH_3^+$, the hydroxylammonium ion.

3.153.—When the nitrogen base is known by a name ending otherwise than in -amine, the cation name is to be formed by adding the ending -ium to the name of the base (if necessary omitting a final -e or other vowel).

Examples: hydrazinium, anilinium, glycinium, pyridinium, guanidinium, imidazolium.

The names uronium and thiouronium, though inconsistent with this rule, may be retained.

3.16.—Cations formed by adding protons to nonnitrogenous bases may also be given names formed by adding -ium to the name of the compound to which the proton is added.

Examples: dioxanium, acetonium.

In the case of cations formed by adding protons to acids, however, their names are to be formed by adding the word acidium to the name of the corresponding anion, and not that of the acid itself. For example, $H_2NO_3^+$, the nitrate acidium ion; $H_2NO_2^+$, the nitrite acidium ion; and $CH_3COOH_2^+$, the acetate acidium ion. Note, however, that when the anion of the acid is monatomic **3.14** will apply; for example, FH_2^+ is the fluoronium ion.

◆In accord with present practice, nitric acidium ion, *etc.*, are preferred to nitrate acidium ion, *etc.* $CH_3COOH_2^+$ is organic.

3.17.—Where more than one ion is derived from one base, as, for example, $N_2H_5^+$ and $N_2H_6^{2+}$, their charges may be indicated in their names as the hydrazinium(1+) and the hydrazinium(2+) ion, respectively.

◆*Cf.* comment at 2.252 and the use of Stock notation with ions or radicals in 3.13 and 3.32.

3.2. Anions

3.21.—The names for monatomic anions shall consist of the name (sometimes abbreviated) of the element, with the termination -ide. Thus

H^-	hydride ion	Br^-	bromide ion	Se^{2-}	selenide ion	As^{3-}	arsenide ion
D^-	deuteride ion	I^-	iodide ion	Te^{2-}	telluride ion	Sb^{3-}	antimonide ion
F^-	fluoride ion	O^{2-}	oxide ion	N^{3-}	nitride ion	C^{4-}	carbide ion
Cl^-	chloride ion	S^{2-}	sulfide ion	P^{3-}	phosphide ion	Si^{4-}	silicide ion
		B^{3-}	boride ion				

Expressions of the type "chlorine ion" are used particularly in connection with crystal structure work and spectroscopy; the Commission recommends that whenever the charge corresponds with that indicated above, the termination -ide should be used.

◆*Cf.* comments at 1.15 and 1.32 regarding ^2H and D.

3.22. Polyatomic Anions.

3.221.—Certain polyatomic anions have names ending in -ide. These are

OH^-	hydroxide ion	I_3^-	triiodide ion	$NHOH^-$	hydroxylamide ion
O_2^{2-}	peroxide ion	HF_2^-	hydrogen difluoride ion	$N_2H_3^-$	hydrazide ion
O_2^-	hyperoxide ion	N_3^-	azide ion	CN^-	cyanide ion
O_3^-	ozonide ion	NH^{2-}	imide ion	C_2^{2-}	acetylide ion
S_2^{2-}	disulfide ion	NH_2^-	amide ion		

Names for other polysulfide, polyhalide, *etc.*, ions may be formed in analogous manner. The OH^- ion should not be called the hydroxyl ion. The name hydroxyl is reserved for the OH group when neutral or positively charged, whether free or as a substituent (*cf.* **3.12** and **3.32**).

◆Superoxide is well established in English for O_2^- and no advantage is seen in changing to hyperoxide.

3.222.—Ions such as SH^- and O_2H^- will be called the hydrogen sulfide ion and the hydrogen peroxide ion, respectively. This agrees with **6.2**, and names such as hydrosulfide are not needed.

◆*Cf.* comment at 6.2. All "fused" names (as hydrogensulfide here and methylisocyanide in 5.33) in the original version have been written as two words in this version. For a rule on the written form of the names of compounds see *J. Chem. Education,* 8, 1336–8 (1931)

3.223.—The names for other polyatomic anions shall consist of the name of the central atom with the termination -ate, which is used quite generally for complex anions. Atoms and groups attached to the central atom shall generally be treated as ligands in a complex (*cf.* **2.24** and Section 7) as, for example, $[Sb(OH)_6]^-$, the hexahydroxoantimonate(V) ion.

This applies also when the exact composition of the anion is not known; *e.g.*, by solution of aluminum hydroxide or zinc hydroxide in sodium hydroxide, aluminate and zincate ions are formed.

3.224.—It is quite practicable to treat oxygen in the same manner as other ligands (**2.24**), but it has long been customary to ignore the name of this element altogether in anions and to indicate its presence and proportion by means of a series of prefixes (hypo-, per-, *etc.*, see Section **5**) and sometimes also by the suffix -ite in place of -ate.

The termination -ite has been used to denote a lower oxidation state and may be retained in trivial names in these cases

NO_2^-	nitrite	$PH_2O_2^-$	hypophosphite	$S_2O_2^{2-}$	thiosulfite
$N_2O_2^{2-}$	hyponitrite	AsO_3^{3-}	arsenite	SeO_3^{2-}	selenite
NOO_2^-	peroxonitrite	SO_3^{2-}	sulfite	ClO_2^-	chlorite
PHO_3^{2-}	phosphite	$S_2O_5^{2-}$	disulfite (pyrosulfite)	ClO^-	hypochlorite
$P_2H_2O_5^{2-}$	diphosphite (pyrophosphite)	$S_2O_4^{2-}$	dithionite		

(and correspondingly for the other halogens)

The Commission does not recommend the use of any such names other than those listed. A number of other names ending in -ite have been used, *e.g.*, antimonite, tellurite, stannite, plumbite, ferrite, manganite, but in many cases such compounds are known in the solid state to be double oxides and are to be treated as such (*cf.* **6.5**), *e.g.*, $Cu(CrO_2)_2$ copper(II) chromium(III) oxide, not copper chromite. Where there is reason to believe that a definite salt with a discrete anion exists, the name is formed in accordance with **2.24**. By dissolving, for example, Sb_2O_3, SnO, or PbO in sodium hydroxide an antimonate(III), a stannate(II), or a plumbate(II) is formed in the solution.

Concerning the use of prefixes hypo-, per-, *etc.*, see the list of acids (table in **5.214**). For all new compounds and even for the less common ones listed in the table in **3.224** or derived from the acids listed in the table in **5.214**, it is preferable to use the system given in **2.24** and in Sections **5** and **7**.

◆For phosphite names see comment at 6.2.

3.3. Radicals

3.31.—A radical is here regarded as a group of atoms which occurs repeatedly in a number of different compounds. Sometimes the same radical fulfils different functions in different cases, and accordingly different names often have been assigned to the same group. The Commission considers it desirable to reduce this diversity and recommends that formulas or systematic names be used to denote all new radicals, instead of introducing new trivial names. The list of names for ions and radicals on page B-169 gives an extensive selection of radical names at present in use in inorganic chemistry.

◆The list of names for ions and radicals following the Rules is very useful and can be made more so by additions (as of dithio, nitrilo, and azido in the last column) and by greater attempt at uniformity with organic usage. Some of the terms (as nitride and amide) listed as anions are used also of covalent compounds. A single atom may function like a radical as defined above and may be named similarly, as chloro and oxo.

3.32.—Certain radicals containing oxygen or other chalcogens have special names ending in -yl, and the Commission approves the provisional retention of

HO	hydroxyl	S_2O_5	pyrosulfuryl	PO	phosphoryl	ClO_2	chloryl
CO	carbonyl	SeO	seleninyl	VO	vanadyl	ClO_3	perchloryl
NO	nitrosyl	SeO_2	selenonyl	PuO_2	plutonyl		(and similarly
SO	sulfinyl	CrO_2	chromyl		(similarly for		for other
	(thionyl)	UO_2	uranyl		other actinide		halogens)
SO_2	sulfonyl	NpO_2	neptunyl		elements)		
	(sulfuryl)	NO_2	nitryl[1]	ClO	chlorosyl		

[1]The name nitroxyl should not be used for this group since the name nitroxylic acid has been used for H_2NO_2. Although the word nitryl is firmly established in English, nitroyl may be a better model for many other languages.

Names such as the above should be used only to designate compounds containing these discrete groups. The use of thionyl and sulfuryl should be restricted to the halides. Names such as bismuthyl and antimonyl are not approved because the compounds do not contain BiO and SbO groups, respectively; such compounds are to be designated as oxide halides (**6.4**).

Radicals analogous to the above containing other chalcogens in place of oxygen are named by adding the prefixes thio-, seleno-, *etc.*

Examples:

PS thiophosphoryl   CSe selenocarbonyl

In cases where radicals may have different valences, the oxidation number of the characteristic element should be indicated by means of the Stock notation. For example, the uranyl group UO_2 may refer either to the ion UO_2^{2+} or to the ion UO_2^{+}; these can be distinguished as uranyl(VI) and uranyl(V), respectively. In like manner, VO may be vanadyl(V), vanadyl(IV), and vanadyl(III).

These polyatomic radicals always are treated as forming the positive part of the compound.

Examples:

$COCl_2$	carbonyl chloride	$NO_2HS_2O_7$	nitryl hydrogen disulfate
NOS	nitrosyl sulfide	S_2O_5ClF	pyrosulfuryl chloride fluoride
PON	phosphoryl nitride	$SO_2(N_3)_2$	sulfonyl azide
$PSCl_3$	thiophosphoryl chloride	SO_2NH	sulfonyl imide
POCl	phosphoryl(III) chloride	IO_2F	iodyl fluoride

By using the same radical names regardless of unknown or controversial polarity relationships, names can be formed without entering into any controversy. Thus, for example, the compounds NOCl and $NOClO_4$ are quite unambiguously denoted by the names nitrosyl chloride and nitrosyl perchlorate, respectively.

◆Caution is urged in the use of some of these radical names: Vanadyl, for example, has been used for VO_2 as well as for VO (*cf.* also the naming of $VOSO_4$ in 6.42). Most of these radical names (except hydroxyl and thionyl) can be regarded as derived from the names of acids which have lost all of their hydroxyls (analogous to -yl or -oyl organic acid radical names) by the use of -yl and -osyl for radicals from -ic and -ous acids, respectively; this is implied in the footnote about nitroxyl. The use of the Stock notation in only one example (phosphoryl(III) chloride) seems confusing. It might be clearer to indicate stoichiometric proportions, *e.g.*, phosphoryl (mono)chloride, thiophosphoryl trichloride (*cf.* phosphoryl triamide in 5.34) or to use phosphorosyl for phosphoryl(III).

The restriction of the use of thionyl and sulfuryl to the halides was agreed upon at a joint meeting of the inorganic and organic nomenclature commissions of the IUPAC in 1951.

3.33.—It should be noted that the same radical may have different names in inorganic and organic chemistry. To draw attention to such differences the prefix names of radicals as substituents in organic compounds have been listed together with the inorganic names in the list of names printed at the end of the Rules. Names of purely organic compounds, of which many are important in the chemistry of coördination compounds (Section 7), should agree with the nomenclature of organic chemistry.

Organic chemical nomenclature is to a large extent based on the principle of substitution, *i.e.*, replacement of hydrogen atoms by other atoms or groups. Such "substitutive names" are extremely rare in inorganic chemistry; they are used, *e.g.*, in the following cases: NH_2Cl is called chloramine, and $NHCl_2$ dichloramine. These names may be retained in the absence of better terms. Other substitutive names (derived from "sulfonic acid" as a name for HSO_3H) are fluoro- and chlorosulfonic acid, aminosulfonic acid, iminodisulfonic acid, and nitrilotrisulfonic acid. These names should preferably be replaced by the following

FSO_3H	fluorosulfuric acid	$NH(SO_3H)_2$	imidodisulfuric acid
$ClSO_3H$	chlorosulfuric acid	$N(SO_3H)_3$	nitridotrisulfuric acid
NH_2SO_3H	amidosulfuric acid		

Names such as chlorosulfuric acid and amidosulfuric acid might be considered to be substitutive names derived by substitution of *hydroxyl* groups in sulfuric acid. From a more fundamental point of view, however (see **2.24**), such names are formed by adding hydroxyl, amide, imide, *etc.*, groups together with oxygen atoms to a sulfur atom, "sulfuric acid" in this connection standing as an abbreviation for "trioxosulfuric acid."

Another organic-chemical type of nomenclature, the formation of "conjunctive names," is also met in only a few cases in inorganic chemistry, *e.g.*, the hydrazine- and hydroxylaminesulfonic acids. According to the principles of inorganic chemical nomenclature these compounds should be called hydrazido- and hydroxylamidosulfuric acid.

◆These are not true "conjunctive names" since sulfonic acid is not a compound. For the naming of partial amides *cf.* also 5.34.

4. CRYSTALLINE PHASES OF VARIABLE COMPOSITION

Isomorphous replacement, interstitial solutions, intermetallic compounds, and other nonstoichiometric compounds (berthollides)

4.1.—If an intermediate crystalline phase occurs in a two-component (or more complex) system, it may obey the law of constant composition very closely, as in the case of sodium chloride, or it may be capable of varying in composition over an appreciable range, as occurs for example with FeS. A substance showing such a variation is called a *berthollide*.

In connection with the berthollides the concept of a characteristic or ideal composition is frequently used. A unique definition of this concept seems to be lacking. In one case it may be necessary to use a definition based upon lattice geometry and in another to base it on the ratio of valence electrons to atoms. Sometimes one can state several characteristic compositions, and at other times it is impossible to say whether a phase corresponds to a characteristic composition or not.

In spite of these difficulties it seems that the concept of a characteristic composition can be used in its present undefined form for establishing a system of notation for phases of variable composition. It also seems possible to use the concept even if the characteristic composition is not included in the known homogeneity range of the phase.

4.2.—For the present, mainly formulas should be used for berthollides and solid solutions, since strictly logical names tend to become inconveniently cumbersome. The latter should be used only when unavoidable (*e.g.*, for indexing), and may be written in the style of iron(II) sulfide (iron-deficient); molybdenum dicarbide (excess carbon), or the like. Mineralogical names should be used only to designate actual minerals and not to define chemical composition; thus the name calcite refers to a particular mineral (contrasted with other minerals of similar composition) and is not a term for the chemical compound whose composition is properly expressed by the name calcium carbonate. (The mineral name may, however, be used to indicate the structure type—see **6.52.**)

4.3.—A general notation for the berthollides, which can be used even when the mechanism of the variation in composition is unknown, is to put the sign ~ (read as *circa*) before the formula. (In special cases it may also be printed above the formula.)

Examples: ~FeS, $\overset{\sim}{\mathrm{CuZn}}$

The direction of the deviation may be indicated when required:

~FeS (iron-deficient); ~MoC_2 (excess carbon)

4.4.—For a phase where the variable composition is solely or partially caused by replacement, atoms or atomic groups which replace each other are separated by a comma and placed together between parentheses.

If possible the formula ought to be written so that the limits of the homogeneity range are represented when one or other of the two atoms or groups is lacking. For example the symbol (Ni,Cu) denotes the complete range from pure Ni to pure Cu; likewise K(Br,Cl) comprises the range from pure KBr to pure KCl. If only part of the homogeneity range is referred to, the major constituent should be placed first.

Substitution accompanied by the appearance of vacant positions (combination of substitutional and interstitial solution) receives an analogous notation. For example, $(Li_2,Mg)Cl_2$ denotes the homogeneous phase from LiCl to $MgCl_2$ where the anion lattice structure remains the same but one vacant cation position appears for every substitution of $2Li^+$ by Mg^{2+}.

The formula $(Mg_3,Al_2)Al_6O_{12}$ represents the homogeneous phase from the spinel $MgAl_2O_4$ (= $Mg_3Al_6O_{12}$) to the spinel form of Al_2O_3 (= $Al_2Al_6O_{12}$).

The solid solutions between CaF_2 and YF_3, where cation substitution is accompanied by interstitial addition of F^-, would be represented by the formula $(Ca,YF)F_2$. It is important to note that this formula is based purely on considerations of composition, and it does not imply that YF^{2+} takes over the actual physical position of Ca^{2+}. On the same basis a notation for the plagioclases would be $(NaSi,CaAl)Si_2AlO_8$.

4.5.—A still more complete notation, which should always be used in more complex cases, may be constructed by indicating in a formula the variables that define the composition. Thus, a phase involving simple substitution may be written A_xB_{1-x}.

Examples: Ni_xCu_{1-x} and KBr_xCl_{1-x}

This shows immediately that the total number of atoms in the lattice is constant. Combined substitutional and interstitial or subtractive solution can be shown in an analogous way. The commas and parentheses called for in **4.4** are not required in this case.

For example, the homogeneous phase between LiCl and $MgCl_2$ becomes $Li_{2x}Mg_{1-x}Cl_2$ and the phase between $MgAl_2O_4$ and Al_2O_3 can be written $Mg_{3x}Al_{2(1-x)}Al_6O_{12}$, which shows that it cannot contain more Mg than that corresponding to $MgAl_2O_4$ ($x = 1$). The other examples given in **4.4** will be given the formulas $Ca_xY_{1-x}F_{3-x}$ and $Na_xCa_{1-x}Si_{2+x}Al_{2-x}O_8$. In the case of the γ-phase of the Ag–Cd system, which has the characteristic formula Ag_5Cd_8, the Ag and Cd atoms can replace one another to some extent and the notation would be $Ag_{5\pm x}Cd_{8\mp x}$.

Further examples:

$Fe_{1-x}Sb$ $Fe_{1-x}O$ $Fe_{1-x}S$ $Cu_{2-x}O$ $Na_{1-x}WO_3$ (sodium tungsten bronzes)

For $x = 0$ each of these formulas corresponds to a characteristic composition. If it is desired to show that the variable denoted by x can attain only small values, this may be done by substituting ϵ for x.

Likewise a solid solution of hydrogen in palladium can be written as PdH_x, and a phase of the composition M which has dissolved a variable amount of water can be written $M(H_2O)_x$.

When this notation is used, a particular composition can be indicated by stating the actual value of the variable x. Probably the best way of doing this is to put the value in parentheses after the general formula. For example, $Li_{4-x}Fe_{3x}Ti_{2(1-x)}O_6$ ($x = 0.35$). If it is desired to introduce the value of x into the formula itself, the mechanism of solution is more clearly understood if one writes $Li_{4-0.35}Fe_{3\times 0.35}Ti_{2(1-0.35)}O_6$ instead of $Li_{3.65}Fe_{1.05}Ti_{1.30}O_6$.

5. ACIDS

Many of the compounds which now according to some definitions are called acids do not fall into the classical province of acids. In other parts of inorganic chemistry functional names are disappearing and it would have been most satisfactory to abolish them also for those compounds generally called acids. Names for these acids may be derived from the names of the anions as in Section 2, *e.g.*, hydrogen sulfate instead of sulfuric acid. The nomenclature of acids has, however, a long history of established custom, and it appears impossible to systematize acid names without drastic alteration of the accepted names of many important and well-known substances.

The present rules are aimed at preserving the more useful of the older names while attempting to guide further development along directions which should allow new compounds to be named in a more systematic manner.

5.1. Binary and Pseudobinary Acids

Acids giving rise to the -ide anions defined by **3.21** and **3.221** will be named as binary and pseudobinary compounds of hydrogen, *e.g.*, hydrogen chloride, hydrogen sulfide, hydrogen cyanide.

For the compound HN_3 the name hydrogen azide is recommended in preference to hydrazoic acid.

5.2. Acids Derived from Polyatomic Anions

Acids giving rise to anions bearing names ending in -ate or in -ite may also be treated as in **5.1**, but names more in accordance with custom are formed by using the terminations -ic acid and -ous acid corresponding with the anion terminations -ate and -ite, respectively. Thus chloric acid corresponds to chlorate, sulfuric acid to sulfate, and phosphorous acid to phosphite.

This nomenclature may also be used for less common acids, *e.g.*, hexacyanoferric acids correspond to hexacyanoferrate ions. In such cases, however, systematic names of the type hydrogen hexacyanoferrate are preferable.

Most of the common acids are oxo acids, *i.e.*, they contain only oxygen atoms bound to the characteristic atom. It is a long-established custom not to indicate these oxygen atoms. It is mainly for these acids that long-established names will have to be retained. Most other acids may be considered as coördination compounds and be named as such.

◆Polyatomic means heteratomic here, as opposed to pseudobinary. *Cf.* comment at 3.12.

5.21. Oxo Acids.—For the oxo acids the ous–ic notation to distinguish between different oxidation states is applied in many cases. The -ous acid names are restricted to acids corresponding to the -ite anions listed in the table in **3.224**.

Further distinction between different acids with the same characteristic element is in some cases effected by means of prefixes. This notation should not be extended beyond the cases listed below.

5.211.—The prefix hypo- is used to denote a lower oxidation state, and may be retained in these cases

$H_4B_2O_4$	hypoboric acid	HPH_2O_2	hypophosphorous acid
$H_2N_2O_2$	hyponitrous acid	HOCl	hypochlorous acid (and similarly for the other halogens)
$H_4P_2O_6$	hypophosphoric acid		

5.212.—The prefix per- is used to designate a higher oxidation state and may be retained for $HClO_4$, perchloric acid, and similarly for the other elements in Group VII.

The prefix per- should not be confused with the prefix peroxo- (see **5.22**).

5.213.—The prefixes ortho- and meta- have been used to distinguish acids differing in "water content." These names are approved

H_3BO_3	orthoboric acid	H_5IO_6	orthoperiodic acid	$(H_2SiO_3)_n$	metasilicic acids
H_4SiO_4	orthosilicic acid	H_6TeO_6	orthotelluric acid	$(HPO_3)_n$	metaphosphoric acids
H_3PO_4	orthophosphoric acid	$(HBO_2)_n$	metaboric acids		

For the acids derived by removing water from orthoperiodic or orthotelluric acid, the systematic names should be used, *e.g.*, HIO_4 tetraoxoiodic(VII) acid.

The prefix pyro- has been used to designate an acid formed from two molecules of an ortho acid minus one molecule of water. Such acids can now generally be regarded as the simplest cases of isopoly acids (*cf.* **7.5**). The prefix pyro- may be retained for pyrosulfurous and pyrosulfuric acids and for pyrophosphorous and pyrophosphoric acids, although in these cases also the prefix di- is preferable.

◆The use of orthoperiodic acid and orthotelluric acid is approved, but the question of names for HIO_4, *etc.*, needs further study because of confusion in the literature.

5.214.—The accompanying Table II contains the accepted names of the oxo acids (whether known in the free state or not) and some of their thio and peroxo derivatives (**5.22** and **5.23**).

For the less common of these acids systematic names would seem preferable, for example

H_2MnO_4 manganic(VI) acid, to distinguish it from H_3MnO_4, manganic(V) acid

$HReO_4$ tetraoxorhenic(VII) acid, to distinguish it from H_3ReO_5, pentaoxorhenic(VII) acid

H_2ReO_4 tetraoxorhenic(VI) acid, to distinguish it from $HReO_3$, trioxorhenic(V) acid; H_3ReO_4, tetraoxorhenic(V) acid; and $H_4Re_2O_7$, heptaoxodirhenic(V) acid

H_2NO_2 dioxonitric(II) acid instead of nitroxylic acid.

Trivial names should not be given to such acids as HNO, $H_2N_2O_3$, $H_2N_2O_4$, of which salts have been described. These salts are to be designated systematically as oxonitrates(I), trioxodinitrates(II), tetraoxodinitrates(III), respectively.

The names germanic acid, stannic acid, antimonic acid, bismuthic acid, vanadic acid, niobic acid, tantalic acid, telluric acid, molybdic acid, wolframic acid, and uranic acid may be used for substances with indefinite "water content" and degree of polymerization.

◆Unless trivial names clash with good nomenclature practices or are ambiguous, retention of well-established ones or the use of formulas is urged (especially for HNO, $H_2N_2O_3$, *etc.*) until structures are known. Systematic coordination-type names in the case of these nitrogen acids, for example, imply a structure that is ruled out by our present state of knowledge.

If hexahydroxoantimonic acid is considered a trivial name, hexahydrox*y*- might be preferable (*cf.* the systematic name hexahydroxoantimonate(V) ion in 3.223). For the analogous use of peroxo and peroxy, see comment at 5.22.

TABLE II

NAMES FOR OXO ACIDS

Formula	Name	Formula	Name
H_3BO_3	orthoboric acid or (mono)boric acid	H_2SO_4	sulfuric acid
$(HBO_2)_n$	metaboric acids	$H_2S_2O_7$	disulfuric or pyrosulfuric acid
$(HBO_2)_3$	trimetaboric acid	H_2SO_5	peroxo(mono)sulfuric acid
$H_4B_2O_4$	hypoboric acid	$H_2S_2O_8$	peroxodisulfuric acid
H_2CO_3	carbonic acid	$H_2S_2O_3$	thiosulfuric acid
HOCN	cyanic acid	$H_2S_2O_6$	dithionic acid
HNCO	isocyanic acid	H_2SO_3	sulfurous acid
HONC	fulminic acid	$H_2S_2O_5$	disulfurous or pyrosulfurous acid
H_4SiO_4	orthosilicic acid	$H_2S_2O_2$	thiosulfurous acid
$(H_2SiO_3)_n$	metasilicic acids	$H_2S_2O_4$	dithionous acid
HNO_3	nitric acid	H_2SO_2	sulfoxylic acid
HNO_4	peroxonitric acid	$H_2S_xO_6$ ($x = 3,4...$)	polythionic acids
HNO_2	nitrous acid	H_2SeO_4	selenic acid
HOONO	peroxonitrous acid	H_2SeO_3	selenious acid
H_2NO_2	nitroxylic acid	H_6TeO_6	(ortho)telluric acid
$H_2N_2O_2$	hyponitrous acid	H_2CrO_4	chromic acid
H_3PO_4	(ortho)phosphoric acid	$H_2Cr_2O_7$	dichromic acid
$H_4P_2O_7$	diphosphoric or pyrophosphoric acid	$HClO_4$	perchloric acid
$H_5P_3O_{10}$	triphosphoric acid	$HClO_3$	chloric acid
$H_{n+2}P_nO_{3n+1}$	polyphosphoric acids	$HClO_2$	chlorous acid
$(HPO_3)_n$	metaphosphoric acids	HClO	hypochlorous acid
$(HPO_3)_3$	trimetaphosphoric acid	$HBrO_3$	bromic acid
$(HPO_3)_4$	tetrametaphosphoric acid	$HBrO_2$	bromous acid
H_3PO_5	peroxo(mono)phosphoric acid	HBrO	hypobromous acid
$H_4P_2O_8$	peroxodiphosphoric acid	H_5IO_6	(ortho)periodic acid
$(HO)_2OP–PO(OH)_2$	hypophosphoric acid.	HIO_3	iodic acid
$(HO)_2P–O–PO(OH)_2$	diphosphoric(III,V) acid	HIO	hypoiodous acid
H_2PHO_3	phosphorous acid	$HMnO_4$	permanganic acid
$H_4P_2O_5$	diphosphorous or pyrophosphorous acid	H_2MnO_4	manganic acid
		$HTcO_4$	pertechnetic acid
HPH_2O_2	hypophosphorous acid	H_2TcO_4	technetic acid
H_3AsO_4	arsenic acid	$HReO_4$	perrhenic acid
H_3AsO_3	arsenious acid	H_2ReO_4	rhenic acid
$HSb(OH)_6$	hexahydroxoantimonic acid		

5.22. Peroxo Acids.—The prefix peroxo, when used in conjunction with the trivial names of acids, indicates substitution of –O– by –O–O– (*cf.* **7.312**).

Examples:				
	HNO_4	peroxonitric acid	H_2SO_5	peroxosulfuric acid
	H_3PO_5	peroxophosphoric acid	$H_2S_2O_8$	peroxodisulfuric acid
	$H_4P_2O_8$	peroxodiphosphoric acid		

◆Peroxy, as recommended in the 1940 Rules (inorganic), is more acceptable than peroxo to organic chemists. It is not necessary that the use with trivial names conform with the use of peroxo denoting a coördinated ligand; *e.g.*, peroxysulfuric acid or trioxoperoxosulfuric(VI) acid.

5.23. Thio Acids.—Acids derived from oxo acids by replacement of oxygen by sulfur are called *thio acids* (*cf.* **7.312**).

Examples: $H_2S_2O_2$ thiosulfurous acid $H_2S_2O_3$ thiosulfuric acid HSCN thiocyanic acid

When more than one oxygen atom can be replaced by sulfur the number of sulfur atoms generally should be indicated

H_3PO_3S	monothiophosphoric acid	H_3AsS_3	trithioarsenious acid
$H_3PO_2S_2$	dithiophosphoric acid	H_3AsS_4	tetrathioarsenic acid
H_2CS_3	trithiocarbonic acid		

The prefixes seleno- and telluro- may be used in a similar manner.

5.24. Chloro Acids, etc.—Acids containing ligands other than oxygen and sulfur are generally designated according to the rules in Section 7.

Examples:

$HAuCl_4$	hydrogen tetrachloraurate(III) or tetrachloroauric(III) acid
H_2PtCl_4	hydrogen tetrachloroplatinate(II) or tetrachloroplatinic(II) acid
H_2PtCl_6	hydrogen hexachloroplatinate(IV) or hexachloroplatinic(IV) acid
$H_4Fe(CN)_6$	hydrogen hexacyanoferrate(II) or hexacyanoferric(II) acid
$H[PHO_2F]$	hydrogen hydridodioxofluorophosphate or hydridodioxofluorophosphoric acid
HPF_6	hydrogen hexafluorophosphate or hexafluorophosphoric acid
H_2SiF_6	hydrogen hexafluorosilicate or hexafluorosilicic acid
H_2SnCl_6	hydrogen hexachlorostannate(IV) or hexachlorostannic(IV) acid
HBF_4	hydrogen tetrafluoroborate or tetrafluoroboric acid
$H[B(OH)_2F_2]$	hydrogen dihydroxodifluoroborate or dihydroxodifluoroboric acid
$H[B(C_6H_5)_4]$	hydrogen tetraphenylborate or tetraphenylboric acid

It is preferable to use names of the type hydrogen tetrachloroaurate(III).

For some of the more important acids of this type abbreviated names may be used, *e.g.*, chloroplatinic acid, fluorosilicic acid.

◆For the use of hydrido in the fifth example see comment at 7.312.

5.3. Functional Derivatives of Acids

Functional derivatives of acids are compounds formed from acids by substitution of OH and sometimes also O by other groups. In this borderline between organic and inorganic chemistry organic-chemical nomenclature principles prevail.

◆The intention of the statement "organic-chemical nomenclature principles prevail" is not clear, since most of the examples given in the sections immediately following are not named according to organic practice. *Cf.* 3.33.

5.31. Acid Halides.—The names of acid halides are formed from the name of the corresponding acid radical if this has a special name, *e. g.*, sulfuryl chloride, phosphoryl chloride.

In other cases these compounds are named as oxide halides according to rule **6.41**, *e.g.*, MoO_2Cl_2, molybdenum dioxide dichloride.

5.32. Acid Anhydrides.—Anhydrides of inorganic acids generally should be given names as oxides, *e.g.*, N_2O_5 dinitrogen pentaoxide, *not* nitric anhydride or nitric acid anhydride.

5.33. Esters.—Esters of inorganic acids are given names in the same way as the salts, *e.g.*, dimethyl sulfate, diethyl hydrogen phosphate.

If, however, it is desired to specify the constitution of the compound, a name based on the nomenclature for cöordination compounds should be used.

Example:

$(CH_3)_4[Fe(CN)_6]$	tetramethyl hexacyanoferrate(II)
or	or
$[Fe(CN)_2(CH_3NC)_4]$	dicyanotetrakis(methyl isocyanide)iron(II)

◆According to common organic practice for esters (ethers, sulfides, *etc.*) and to the naming of inorganic salts (*e.g.*, sodium sulfate, not disodium sulfate), methyl sulfate would be used instead of dimethyl sulfate. However, no objection is seen to the more specific name. Such names as methyl sulfate are better for alphabetic listing, as in indexes.

5.34. Amides.—The names for amides may be derived from the names of acids by replacing acid by amide, or from the names of the acid radicals.

Examples:

$SO_2(NH_2)_2$	sulfuric diamide or sulfonyl diamide
$PO(NH_2)_3$	phosphoric triamide or phosphoryl triamide

If not all hydroxyl groups of the acid have been replaced by NH_2 groups, names ending in -amidic acid may be used: this is an alternative to naming the compounds as complexes.

Examples:

NH_2SO_3H	amidosulfuric acid or sulfamidic acid
$NH_2PO(OH)_2$	amidophosphoric acid or phosphoramidic acid
$(NH_2)_2PO(OH)$	diamidophosphoric acid or phosphorodiamidic acid

Abbreviated names (sulfamide, phosphamide, sulfamic acid) are often used but are not recommended.

◆The use of adjectives from names of inorganic acids may lead to confusion because an -ic or -ous adjective (as chromic) may refer to a higher- or lower-valent form of the element as well as to the acid (a possibility that does not arise, of course, with adjectives from names of organic acids).

Names of the type phosphoramidic acid are recommended in the report of the Advisory Committee on the Nomenclature of Organic Phosphorus Compounds of the Division of Organic Chemistry of the ACS published in 1952, but are not as acceptable to the inorganic nomenclature committees as a whole as the amido- or cöordination-type names. *Cf.* 3.33.

The retention of sulfamic acid and sulfamide as trivial names is favored by the committees. An acceptable systematic name for NH_2SO_3H would be ammonia-sulfur trioxide, in keeping with its probable structure.

5.35. Nitriles.—The suffix -nitrile has been used in the names of a few inorganic compounds, *e.g.* $(PNCl_2)_3$, trimeric phosphonitrile chloride. According to **2.22** such compounds can be designated as nitrides, *e.g.*, phosphorus nitride dichloride. Accordingly there seems to be no reason for retention of the name nitrile (and nitrilo, *c.f.* **3.33**) in inorganic chemistry.

◆Nitrilo is used in organic chemistry, though not given in the list at the end of the Rules.

6. SALTS AND SALT-LIKE COMPOUNDS

Among salts particularly there persist many old names which are bad and misleading, and the Commission wishes to emphasize that any which do not conform to these Rules should be discarded.

6.1. Simple Salts

Simple salts fall under the broad definition of binary compounds given in Section **2**, and their names are formed from those of the constituent ions (given in Section **3**) in the manner set out in Section **2**.

6.2. Salts Containing Acid Hydrogen ("Acid" Salts[1])

Names are formed by adding the word hydrogen, to denote the replaceable hydrogen present, immediately in front of the name of the anion.

The nonacidic hydrogen present, *e.g.*, in the phosphite ion, is included in the name of the anion and is not explicitly cited (*e.g.*, Na_2PHO_3, sodium phosphite).

Examples:

$NaHCO_3$	sodium hydrogen carbonate
NaH_2PO_4	sodium dihydrogen phosphate
$NaH[PHO_3]$	sodium hydrogen phosphite

◆The use of "fused" hydrogen names (as hydrogencarbonate in the original version) is not acceptable in English; the present practice of running hydrogen as a separate word is preferred and has been followed throughout this version. *Cf.* 3.222, 6.324, 6.333. If necessary for clarity, parentheses can be used, as in naming ligands, *e.g.*, (hydrogen carbonato). The use of hydro (as in hydrocarbonato) is not acceptable because of conflicts with organic usage, where hydro denotes addition of hydrogen to unsaturated compounds.

It seems safer to cite even nonacidic hydrogen present in an anion like PHO_3^{2-} (unless the ion has a specific name), because of current usage. (*Cf.* triethyl phosphite in 7.412, seventh example).

6.3. Double Salts, Triple Salts, etc.

6.31.—In formulas all the cations shall precede the anions; in names the principles embodied in Section 2 shall be applied. In those languages where cation names are placed after anion names the adjectives double, triple, *etc.* (their equivalents in the language concerned) may be added immediately after the anion name. The number so implied concerns the number of *kinds* of cation present and *not* the total number of such ions.

[1]For "basic" salts see 6.4.

6.32.—Cations.

6.321.—Cations shall be arranged in order of increasing valence (except hydrogen, *cf.* **6.2** and **6.324**).

6.322. The cations of each valence group shall be arranged in order of decreasing atomic number, with the polyatomic radical ions (*e.g.*, ammonium) at the end of their appropriate group.

◆Alphabetical order would be simpler here and even for 6.321.

6.323. Hydration of Cations.—Owing to the prevalence of hydrated cations, many of which are in reality complex, it seems unnecessary to disturb the cation order in order to allow for this; but if it is necessary to draw attention specifically to the presence of a particular hydrated cation this may be done by writing, for example, "hexaaquo" or "tetraaquo" before the name of the simple ion. Apart from this exception, however, all complex ions should be placed after simple ones in the appropriate valence group.

◆*Cf.* comment at 7.322.

6.324. Acidic Hydrogen.—When hydrogen is considered to be present as a cation its name shall be cited last among the cations. Actually acidic hydrogen will in most cases be bound to an anion and shall be cited together with this (**6.2**). If the salt contains only one anion, acidic hydrogen shall be cited in the same place whichever view is taken of the function of this hydrogen. Nonacidic hydrogen shall be either not explicitly cited (*cf.* **6.2**) or designated hydrido (*cf.* **5.24** and **7.311**). For salts with more than one anion see **6.333.**

Examples:

$KMgF_3$	potassium magnesium fluoride
$TlNa(NO_3)_2$	thallium(I) sodium nitrate or thallium sodium dinitrate
$KNaCO_3$	potassium sodium carbonate
$NH_4MgPO_4.6H_2O$	ammonium magnesium phosphate hexahydrate
$NaZn(UO_2)_3(C_2H_3O_2)_9.6H_2O$	sodium zinc triuranyl acetate hexahydrate
$Na[Zn(H_2O)_6](UO_2)_3(C_2H_3O_2)_9$	sodium hexaaquozinc triuranyl acetate
$NaNH_4HPO_4.4H_2O$	sodium ammonium hydrogen phosphate tetrahydrate

◆*Cf.* comments at 6.2, 7.312. In the fifth and sixth examples, either triuranyl(VI) or nonaacetate should be specified. *Cf.* 3.32, 6.34.

6.33.—Anions.

6.331.—Anions are to be cited in this group order

1. H^-
2. O_2^- and OH^- (in that order)
3. Simple (*i.e.*, one element only) inorganic anions, other than H^- and O^{2-}
4. Inorganic anions containing two or more elements, other than OH^-
5. Anions of organic acids and organic substances exerting an acid function

◆The committees consider it preferable to cite H^- last in accordance with usage.

6.332.—Within group 3 the ions shall be cited in the order given in **2.16**, the inclusion of O in that list being taken as referring to all oxygen anions apart from O^{2-} (*i.e.*,O_2^{2-}, *etc.*).

Within group 4, anions containing the smallest number of atoms shall be cited first, and in the case of two ions containing the same number of atoms they shall be cited in order of decreasing atomic number of the central atoms. Thus CO_3^{2-} should precede CrO_4^{2-}, and the latter should precede SO_4^{2-}.

Within group 5 the anions shall be cited in alphabetical order.

◆Again alphabetical order would be simpler, within groups 3 and 4 as well as 5. *Cf.* comment at 7.251.

6.333.—Acidic hydrogen should be cited together with the anion to which it is attached. If it is not known to which anion the hydrogen is bound, it should be cited last among the cations.

◆*Cf.* comment at 6.2.

6.34.—The stoichiometric method is the most practicable for indicating the proportions of the constituents. It is not always essential to give the numbers of all the anions, provided the valences of all the cations are either known or indicated.

Examples:

$NaCl.NaF.2Na_2SO_4$ or $Na_6ClF(SO_4)_2$	(hexa)sodium chloride fluoride (bis)sulfate
$Ca_5F(PO_4)_3$	(penta)calcium fluoride (tris)phosphate

The parentheses in these cases mean that numerical prefixes may not be necessary. The multiplicative numerical prefixes bis, tris, *etc.*, should be used in connection with anions, because disulfate, triphosphate, *etc.*, designate isopoly anions.

6.4. Oxide and Hydroxide Salts ("Basic" salts formerly oxy and hydroxy salts)

6.41.—For the purposes of nomenclature, these should be regarded as double salts containing O^{2-} and OH^- anions, and Section **6.3** may be applied in its entirety.

6.42. Use of the Prefixes Oxy and Hydroxy.—In some languages the citation in full of all the separate anion names presents no trouble and is strongly recommended (*e.g.*, copper oxide chloride), to the exclusion of the oxy form wherever possible. In some other languages, however, such names as "oxyde et chlorure double de cuivre" are so far removed from current practice that the present system of using oxy- and hydroxy-, *e.g.*, oxychlorure de cuivre, may be retained in such cases.

Examples:

$Mg(OH)Cl$	magnesium hydroxide chloride
$BiOCl$	bismuth oxide chloride
$LaOF$	lanthanum oxide fluoride
$VOSO_4$	vanadium(IV) oxide sulfate
$CuCl_2.3Cu(OH)_2$ or $Cu_2(OH)_3Cl$	dicopper trihydroxide chloride
$ZrOCl_2.8H_2O$	zirconium oxide (di)chloride octahydrate

6.5. Double Oxides and Hydroxides

The terms "mixed oxides" and "mixed hydroxides" are not recommended. Such substances preferably should be named double, triple, *etc.*, oxides or hydroxides as the case may be.

Many double oxides and hydroxides belong to several distinct groups, each having its own characteristic structure type, which is sometimes named after some well-known mineral of the same group (*e.g.*, perovskite, ilmenite, spinel). Thus, $NaNbO_3$, $CaTiO_3$, $CaCrO_3$, $CuSnO_3$, $YAlO_3$, $LaAlO_3$, and $LaGaO_3$ all have the same structure as perovskite, $CaTiO_3$. Names such as calcium titanate may convey false implications and it is preferable to name such compounds as double oxides and double hydroxides unless there is clear and generally accepted evidence of cations and oxo or hydroxo anions in the structure. This does not mean that names such as titanates or aluminates should always be abandoned, because such substances may exist in solution and in the solid state (*cf.* **3.223**).

◆"Multiple" has been used in English as a class term including double, triple, etc. (oxides or the like).

6.51.—In the double oxides and hydroxides the metals shall be named in the same order as for double salts (**6.32**).

6.52.—When required the structure type may be added in parentheses and in italics after the name, except that when the type name is also the mineral name of the substance itself then the italics should not be used (*cf.* **4.2**).

Examples:

$NaNbO_3$	sodium niobium trioxide (*perovskite* type)
$MgTiO_3$	magnesium titanium trioxide (*ilmenite* type)
$FeTiO_3$	iron(II) titanium trioxide (ilmenite)
$4CaO.AL_2O_3.nH_2O$ or $Ca_2Al(OH)_7.nH_2O$	dicalcium aluminum hydroxide hydrate
but $Ca_3[Al(OH)_6]_2$	(tri)calcium (bis)-[hexahydroxoaluminate]
$LiAl(OH)_4.2MnO_2$ or $LiAlMn^{IV}{}_2O_4(OH)_4$	lithium aluminum dimanganese(IV) tetraoxide tetrahydroxide

7. COÖRDINATION COMPOUNDS

7.1. Definitions

In its oldest sense the term *coördination compound* is taken as referring to molecules or ions in which there is an atom (A) to which are attached other atoms (B) or groups (C) to a number in excess of that corresponding to the oxidation number of the atom A. However, the system of nomenclature originally evolved for the compounds within this narrow definition has proved useful for a much wider class of compounds, and for the purposes of nomenclature the restriction "in excess . . . oxidation number" is to be omitted. Any compound formed by addition of one or several ions and/or molecules to one or more ions or/and molecules may be named according to the same system as strict coördination compounds.

The effect of this is to bring many simple and well-known compounds under the same nomenclature rules as accepted coöordination compounds; the result is to reduce the diversity of names and avoid many controversial issues, because it should be understood that there is no intention of implying that any structural analogy necessarily exists between different compounds merely on account of a common system of nomenclature. The system extends also to many addition compounds.

In the rules which follow certain terms are used in the senses here indicated: the atom referred to above as (A) is known as the *central* or *nuclear* atom, and all other atoms which are directly attached to A are known as *coördinating atoms.* Atoms (B) and groups (C) are called *ligands.* A group containing more than one *potential* coördinating atom is termed a *multidentate* ligand, the number of potential coöordinating atoms being indicated by the terms *unidentate, bidentate, etc.* A *chelate* ligand is a ligand *attached* to *one* central atom through *two or more* coördinating atoms, while a *bridging group* is attached to *more than one* atom. The whole assembly of one or more central atoms with their attached ligands is referred to as a *complex*, which may be an uncharged molecule or an ion of either polarity. A *polynuclear complex* is a complex which contains *more than one* nuclear atom, their number being designated by the terms *mononuclear, dinuclear, etc.*

◆Some dissatisfaction with the definition of coördination compounds was expressed, though the broad definition (last sentence of first paragraph) was generally approved.

Central atom or *center of coördination* is to be preferred to *nuclear* atom because of other senses of nucleus, especially of an atom. Possible replacements for polynuclear, etc., are polycentric, *etc.*, or bridged complex since bridging group is in common use.

In the United States the Greek-Latin hybrid terms polydentate and monodentate seem to be used more than the all-Latin multidentate and unidentate.

7.2. Formulas and Names for Complex Compounds in General

7.21. Central Atoms.—In *formulas* the symbol for the central atom(s) should be placed *first* (except in formulas which are primarily structural), the anionic and neutral, *etc.*, ligands following as prescribed in **7.25**, and the formula for the whole complex entity (ion or molecule) should be placed in brackets [ ].

In *names* the central atom(s) should be placed *after* the ligands.

◆It is considered preferable to place the whole complex in brackets, especially when more than one central atom is present, but not essential with only one central atom or with complex ions or nonionic complexes, because brackets may be needed sometimes for indicating concentrations.

7.22. Indication of Valence and Proportion of Constituents.—The oxidation number of the central atom is indicated by means of the Stock notation (**2.252**). Alternatively the proportion of constituents may be given by means of stoichiometric prefixes (**2.251**).

7.23.—Formulas and names may be supplemented with the prefixes *cis, trans, etc.* (**2.19**).

7.24. Terminations.—Complex anions shall be given the termination -ate (*cf.* **2.23**, **2.24** and **3.223**). Complex cations and neutral molecules are given no distinguishing termination. For further details concerning the names of ligands see **7.3.**

7.25. Order of Citation of Ligands in Complexes.—

first: anionic ligands
next: neutral and cationic ligands

7.251.—The anionic ligands shall be cited in the order

1. H^-
2. O^{2-}, OH^- (in that sequence)
3. Other simple anions (*i.e.*, one element only)
4. Polyatomic anions
5. Organic anions in alphabetical order

The sequence within categories 3 and 4 should be that given in **6.332.**

◆H^- is preferably named last, not first (*cf.* comment at 6.331). Alphabetical order is strongly recommended for simplicity at least within 3 and 4, where only rare uses are involved. The intention seems to be to include under 3 monatomic ions (*i.e.*, one *atom*—rather than one *element*—only), in other words to exclude N_3, I_3, etc. Under 4 the insertion of "inorganic" between "Polyatomic" and "anions" is recommended.

7.252.—Neutral and cationic ligands shall be cited in the order given

first: water, ammonia (in that sequence)
then: other inorganic ligands in the sequence in which their coördinating elements appear in the list given in **2.16**
last: organic ligands in alphabetical order.

◆For the inorganic ligands alphabetical order again is recommended. The use of parentheses wherever there is any possibility of ambiguity should be stressed, as illustrated in examples under 7.321: potassium trichloro(ethylene)platinate(II), where parentheses are given in the Rules, and tetra(pyridine)platinum(II) tetrachloroplatinate(II), where they have been added in this version. Parentheses might also be helpful with "thiocyanato" preceded by a numerical prefix (see last example in 7.311) and are definitely required with two-word names for ligands recommended instead of the fused names of the original version (6.2), as in the seventh example in 7.412: di-μ-carbonyl-bis{carbonyl(triethyl phosphite)cobalt}. Since brackets denote complexes, braces can be used in formulas or names where needed in order to avoid the use of two sets of parentheses (braces were so used in some but not all such cases in the original version).

7.3. Names for Ligands

7.31.—Anionic Ligands.

7.311.—The names for anionic ligands, whether inorganic or organic, end in -o (see, however, **7.324**). In general, if the anion name ends in -ide, -ite, or -ate, the final -e is replaced by -o, giving -ido, -ito, and -ato, respectively.

Examples:

$Li[AlH_4]$	lithium tetrahydridoaluminate
$Na[BH_4]$	sodium tetrahydridoborate
$K_2[OsNCl_5]$	potassium nitridopentachloroösmate(VI)
$[Co(NH_2)_2(NH_3)_4]OC_2H_5$	diamidotetraamminecobalt(III) ethanolate
$[CoN_3(NH_3)_5]SO_4$	azidopentaamminecobalt(III) sulfate
$Na_3[Ag(S_2O_3)_2]$	sodium bis(thiosulfato)argentate(I)
$[Ru(HSO_3)_2(NH_3)_4]$	bis(hydrogen sulfito)tetraammineruthenium
(II) $NH_4[Cr(SCN)_4(NH_3)_2]$	ammonium tetrathiocyanatodiamminechromate (III)

◆For "ethanolate" in the fourth example "ethoxide" may be preferred.

7.312.—These anions do not follow exactly the above rule; modified forms have become established:

F^-	fluoride	fluoro (*not* fluo)	O_2^{2-}	peroxide	peroxo
Cl^-	chloride	chloro	HS^-	hydrogen sulfide	thiolo
Br^-	bromide	bromo	S^{2-}	sulfide	thio[1] (sulfido)
I^-	iodide	iodo	(but:S_2^{2-}	disulfide	disulfido)
O^{2-}	oxide	oxo	CN^-	cyanide	cyano
OH^-	hydroxide	hydroxo			

[1]The name thio has long been used to denote the ligand S^{2-} when it can be regarded as replacing O^{2-} in an oxo acid or its anion. The general use of this name will prevent confusion between the two interpretations of disulfido as S_2^{2-} or two S^{2-} ligands.

By analogy with hydroxo, CH_3O^-, *etc.*, are called methoxo, *etc.* For CH_3S^-, *etc.*, the systematic names methanethiolato, *etc.*, are used.

Examples:

$K[AgF_4]$	potassium tetrafluoroargentate(III)
$K_2[NiF_6]$	potassium hexafluoroniccolate(IV)
$Ba[BrF_4]_2$	barium tetrafluorobromate(III)
$Na[AlCl_4]$	sodium tetrachloroaluminate
$Cs[ICl_4]$	cesium tetrachloroiodate(III)
$K[Au(OH)_4]$	potassium tetrahydroxoaurate(III)
$K[CrOF_4]$	potassium oxotetrafluorochromate(V)
$K_2[Cr(O)_2O_2(CN)_2(NH_3)]$	potassium dioxoperoxodicyanoamminechromate(VI)
$Na[BH(OCH_3)_3]$	sodium hydridotrimethoxoborate
$K_2[Fe_2S_2(NO)_4]$	dipotassium dithiotetranitrosyldiferrate

◆It is strongly recommended on the basis of past and present usage, analogy with chloro, *etc.*, and euphony that hydro be added to the list of modified forms of names for anionic ligands and that hydrido be abandoned (*cf.* also examples in 7.311).

While the Subcommittee on Coördination Compounds recognizes the usefulness of the invariable -o ending for anionic ligands, some members of the general committees do not see a sharp enough distinction between such ligands and the same groups in organic compounds to justify a departure from organic usage by using hydroxo, methoxo, *etc.*, instead of hydroxy, methoxy, *etc.* These members therefore favor adding hydroxy, methoxy, *etc.*, to the hydrocarbon radicals excepted from the rule of -o endings for anions (7.324). For the use of peroxo, see comment at 5.22.

The -o of a negative ligand should not be elided before another vowel (chloroösmate, chloroiodo). This agrees with organic practice. *Cf.* comment at 7.322.

It should be pointed out here as well as in the list of names for ions and radicals at the end of the Rules that the approved organic name for unsubstituted HS is mercapto (not thiol, as also given in the list) and for the alkyl- and aryl-substituted radicals methylthio, *etc.*

7.313.—Ligands derived from organic compounds not normally called acids, but which function as such in complex formation by loss of a proton, should be treated as anionic and given the ending -ato. If, however, no proton is lost, the ligand must be treated as neutral—see **7.32.**

Examples:

$[Ni(C_4H_7N_2O_2)_2]$ bis(dimethylglyoximato)nickel(II)

$[Cu(C_5H_7O_2)_2]$ bis(acetylacetonato)copper(II)

bis(8-quinolinolato)silver(II)

bis(4-fluorosalicylaldehydato)copper(II)

N,N′-ethylenebis(salicylideneiminato)cobalt(II)

◆Although according to 3.33 the name for the ligand in the second example should be derived from the systematic name 2,4-pentanedione instead of from acetylacetone, acetylacetonato perhaps conveys better the idea that it is the enol form that is involved. However, the coördination subcommittee questions the use of the termination -ato rather than plain -o especially in cases where the -ate terms (as dimethylglyoximate) are not accepted organic practice. Such -ato terms are especially misleading in the last two examples, where the -ato belongs with the hydroxyl part of the name, not the aldehyde or imine part to which it is attached.

7.32.—Neutral and Cationic Ligands.

7.321.—The name of the coördinated molecule or cation is to be used without change, except in the special cases provided for in **7.322.**

Examples:

$[CoCl_2(C_4H_8N_2O_2)_2]$	dichlorobis(dimethylglyoxime)cobalt(II) (*cf.* nickel derivative given in **7.313**)
cis-$[PtCl_2(Et_3P)_2]$	*cis*-dichlorobis(triethylphosphine)platinum(II)
$[CuCl_2(CH_3NH_2)_2]$	dichlorobis(methylamine)copper(II)
$[Pt\ py_4]\ [PtCl_4]$	tetra(pyridine)platinum(II) tetrachloroplatinate(II)
$[Fe(dipy)_3]Cl_2$	tris(dipyridyl)iron(II) chloride
$[Co\ en_3]_2(SO_4)_3$	tris(ethylenediamine)cobalt(III) sulfate
$[Zn\{NH_2CH_2CH(NH_2)CH_2NH_2\}_2]I_2$	bis(1,2,3-triaminopropane)zinc iodide
$K[PtCl_3(C_2H_4)]$	potassium trichloro(ethylene)platinate(II) or potassium trichloromonoethyleneplatinate(II)
$[PtCl_2\{H_2NCH_2CH(NH_2)CH_2NH_3\}]Cl$	dichloro(2,3 - diaminopropylammonium)platinum(II) chloride
$[Cr(C_6H_5NC)_6]$	hexakis(phenyl isocyanide)chromium

◆In the fifth example, bipyridine is preferred to dipyridyl in organic practice, and in the seventh example 1,2,3-propanetriamine to 1,2,3-triaminopropane.

7.322.—Water and ammonia as neutral ligands in coördination complexes are called "aquo" and "ammine," respectively.

In the tentative rules it was proposed to change the old-established "aquo" to "aqua," thus keeping the -o termination consistently for anionic ligands alone. However, as the old form is so widely used, many regarded the change as too pedantic, and the Commission has decided to retain "aquo" as an exception.

Examples:

$[Cr(H_2O)_6Cl_3$	hexaaquochromium(III) chloride or hexaaquochromium trichloride
$[Al(OH)(H_2O)_5]^{++}$	the hydroxopentaaquoaluminum ion
$[Co(NH_3)_6]ClSO_4$	hexaamminecobalt(III) chloride sulfate
$[CoCl(NH_3)_5]Cl_2$	chloropentaamminecobalt(III) chloride
$[CoCl_3(NH_3)_2\{(CH_3)_2NH\}]$	trichlorodiammine(dimethylamine)cobalt(III)

◆Hexaquo, pentaquo, *etc.*, are used in the examples in the original version, but have been changed in this version to hexaaquo, *etc.*, for conformity with the latest approved organic practice. (*Cf.* hexaammonium, given with two a's separated by a hyphen in 7.6, second example, in the original version.)

7.323.—The groups NO, NS, CO, and CS, when linked directly to a metal atom, are to be called nitrosyl, thionitrosyl, carbonyl, and thiocarbonyl, respectively. In computing the oxidation number these radicals are treated as neutral.

Examples:

$Na_2[Fe(CN)_5NO]$	disodium pentacyanonitrosylferrate
$K_3[Fe(CN)_5CO]$	tripotassium pentacyanocarbonylferrate
$K[Co(CN)(CO)_2(NO)]$	potassium cyanodicarbonylnitrosylcobaltate(0)
$HCo(CO)_4$	hydrogen tetracarbonylcobaltate(−I)
$[Ni(CO)_2(Ph_3P)_2]$	dicarbonylbis(triphenylphosphine)nickel(0)
$[Fe\ en_3]\ [Fe(CO)_4]$	tris(ethylenediamine)iron(II) tetracarbonylferrate (−II)
$Mn_2(CO)_{10}$ or $[CO_5)Mn\text{–}Mn(CO)_5]$	decacarbonyldimanganese(0) or bis(pentacarbonylmanganese)

◆The necessity of arbitrarily considering these groups as always neutral can be avoided by not using the oxidation number but instead the stoichiometric proportions (as in some of these examples) or the Ewens-Bassett system (*cf.* comment at 2.252). Thus the anion in the last example in 7.312 (where NO is known to be positive) would by this system be named dithiotetranitrosyldiferrate (2–).

7.324.—Anions derived from hydrocarbons are given radical names without -o, but are counted as negative when computing the oxidation number.

The consistent introduction of the ending -o would in this case lead to unfamiliar names, *e.g.*, phenylato or phenido for $C_6H_5^-$. On the other hand, if the radicals were counted as neutral ligands, the central atom would have to be given an unusual oxidation number, *e.g.*, −I for boron in $K[B(C_6H_5)_4]$, instead of III.

Examples:

$K[B(C_6H_5)_4]$	potassium tetraphenylborate
$K[SbCl_5C_6H_5]$	potassium pentachloro(phenyl)antimonate(V)
$K_2[Cu(C_2H)_3]$	potassium triethynylcuprate(I)
$K_4[Ni(C_2C_6H_5)_4]$	potassium tetrakis(phenylethynyl)niccolate(0)
$[Fe(CO)_4(C_2C_6H_5)_2]$	tetracarbonylbis(phenylethynyl)iron(II)
$Fe(C_5H_5)_2$	bis(cyclopentadienyl)iron(II)
$[Fe(C_5H_5)_2]Cl$	bis(cyclopentadienyl)iron(III) chloride
$[Ni(NO)(C_5H_5)]$	nitrosylcyclopentadienylnickel

◆The use of radical names such as cyclopentadienyl in the examples does not seem consistent with the use of Stock notations; for the sixth example ("ferrocene"), iron(II) cyclopentadienide is preferred by some. This matter is presumably part of the whole organometallic problem to be dealt with by the new IUPAC joint organic-inorganic subcommittee.

For niccolate, see comment at 1.12.

7.33.—Alternative Modes of Linkage of Some Ligands.—Where ligands are capable of attachment by different atoms this may be denoted by adding the symbol for the atom by which attachment occurs at the end of the name of the ligand. Thus the dithioöxalato group may be attached through S or O, and these are distinguished as dithioöxalato-*S,S′* and dithioöxalato-*O,O′*, respectively.

In some cases different names are already in use for alternative modes of attachment, as, for example, thiocyanato (−SCN) and isothiocyanato (−NCS), nitro (−NO_2), and nitrito (−ONO). In these cases existing custom may conveniently be retained.

Examples:

$K_2\left[Ni\left(\begin{matrix}S-CO\\ |\\ S-CO\end{matrix}\right)_2\right]$

potassium bis(dithioöxalato-*S,S′*)niccolate(II)

$\left[\begin{matrix}Cl & & H_2N-CH_2\\ & Pt & |\\ Cl & & S-CH_2\end{matrix}\right]$ with S bearing $CH_2-CH_2N(CH_3)_2$

dichloro(2-*N,N*-dimethylaminoethyl 2-aminoethyl sulfide-*N′,S*)platinum(II)

$K_2[Pt(NO_2)_4]$	potassium tetranitroplatinate(II)
$Na_3[Co(NO_2)_6]$	sodium hexanitrocobaltate(III)
$[Co(NO_2)_3(NH_3)_3]$	trinitrotriamminecobalt(III)
$[Co(ONO)(NH_3)_5]SO_4$	nitritopentaamminecobalt(III) sulfate
$[Co[NCS)(NH_3)_5]Cl_2$	isothiocyanatopentaamminecobalt(III) chloride

◆Thioöxalato in the original version has been changed to dithioöxalato.

7.4. Di- and Polynuclear Compounds

7.41.—Bridging Groups.

7.411.—A bridging group shall be indicated by adding the Greek letter μ immediately before its name and separating this from the rest of the complex by a hyphen. Two or more bridging groups of the same kind are indicated by di-μ-, *etc.*

7.412.—If the number of central atoms bound by one bridging group exceeds two, the number shall be indicated by adding a subscript numeral to the μ.

This system of notation allows simply of distinction between, for example, μ-disulfido (one S_2 bridge) and di-μ-sulfido (two S bridges). It is also capable of extension to much more complex and unsymmetrical structures by use of the conventional prefixes *cis, trans, asym,* and *sym* where necessary.

Examples:

$[(NH_3)_5Cr\text{–}OH\text{–}Cr(NH_3)_5]Cl_5$

μ-hydroxo-bis{pentaamminechromium(III)}chloride

$\left[\begin{matrix}Cl & & Cl & & AsEt_3\\ & Pt & & Pt & \\ Et_3As & & Cl & & Cl\end{matrix}\right]$

di-μ-chloro-dichlorobis(triethylarsine)diplatinum (II) (three possible isomers: *asym, sym-cis,* and *sym-trans;* the last is shown)

$\left[\begin{matrix} & & CN & & \\ NCS & & S & & P(C_3H_7)_3\\ & Pt & & Pt & \\ (C_3H_7)_3P & & S & & SCN\\ & & CN & & \end{matrix}\right]$

di-μ-thiocyanato-dithiocyanatobis(tripropylphosphine)diplatinum(II)

$[(CO)_3Fe(CO)_3Fe(CO)_3]$	tri-μ-carbonyl-bis(tricarbonyliron)
$[(CO)_3Fe(SEt)_2Fe(CO)_3]$	di-μ-ethanethiolato-bis(tricarbonyliron)
$[(C_5H_5)(CO)Fe(CO)_2Fe(CO)(C_5H_5)]$	di-μ-carbonyl-bis(carbonylcyclopentadienyliron)
$[(CO)\{P(OEt)_3\}Co(CO)_2Co(CO)\{P(OEt)_3\}]$	di-μ-carbonyl-bis{carbonyl(triethyl phosphite)-cobalt}
$[Au(CN)(C_3H_7)_2]_4$	cyclo-tetra-μ-cyano-tetrakis(dipropylgold)
$[CuI(Et_3As)]_4$	tetra-μ_3-iodo-tetrakis{triethylarsinecopper(I)}
$[Be_4O(CH_3COO)_6]$	μ_4-oxo-hexa-μ-acetato-tetraberyllium

7.42. Extended Structures.—Where bridging causes an indefinite extension of the structure it is best to name compounds primarily on the basis of their over-all composition; thus the compound having the composition represented by the formula $CsCuCl_3$ has an anion with the structure:

$$\left[\begin{array}{c} \quad Cl \qquad\quad Cl \qquad\quad Cl \qquad\quad Cl \\ \ldots Cl-Cu-Cl-Cu-Cl-Cu-Cl-Cu\ldots \\ \quad Cl \qquad\quad Cl \qquad\quad Cl \qquad\quad Cl \end{array}\right]^{n-}$$

This may be expressed in the formula $(Cs^+)_n{}^-[(CuCl_3)_n]^{n-}$ which leads to the simple name cesium catena-μ-chloro-dichlorocuprate(II). If the structure were in doubt, however, the substance would be called cesium copper(II) chloride (as a double salt).

◆*Cf.* comment on *catena* at 1.4.

7.5. Isopoly Anions

The structure of many complicated isopoly anions has now been cleared up by X-ray work and it turns out that the indication of the several μ-oxo and oxo atoms in the name does not convey any clear picture of the structure and is therefore of little value.

For the time being it is sufficient to indicate the number of atoms by Greek prefixes, at least until isomers are found. When all atoms have their "normal" oxidation states (*e.g.*, W^{VI}), it is not necessary to give the numbers of the oxygen atoms, if all the others are indicated.

Examples:

$K_2S_2O_7$	dipotassium disulfate
$K_2S_3O_{10}$	dipotassium trisulfate
$Na_5P_3O_{10}$	pentasodium triphosphate
$K_2Cr_4O_{13}$	dipotassium tetrachromate
$Na_2B_4O_7$	disodium tetraborate
NaB_5O_8	sodium pentaborate
$Ca_3Mo_7O_{24}$	tricalcium heptamolybdate
$Na_7HNb_6O_{19}.15H_2O$	heptasodium monohydrogen hexaniobate-15-water
$K_2Mg_2V_{10}O_{28}.16H_2O$	dipotassium dimagnesium decavanadate-16-water

7.6. Heteropoly Anions

The central atom or atoms should be cited last in the name and first in the formula of the anion (*cf.* **7.21**), *e.g.*, wolframophosphate, *not* phosphowolframate.

If the oxidation number has to be given, it may be necessary to place it immediately after the atom referred to and not after the ending -ate, in order to avoid ambiguity.

The method formerly recommended for naming iso- and heteropoly anions by giving the number of atoms in parentheses is not practicable in more complicated cases.

Examples:

$(NH_4)_3PW_{12}O_{40}$	triammonium dodecawolframophosphate
$(NH_4)_6TeMo_6O_{24}.7H_2O$	hexaammonium hexamolybdotellurate heptahydrate
$Li_3HSiW_{12}O_{40}.24H_2O$	trilithium (mono)hydrogen dodecawolframosilicate-24-water
$K_6Mn^{IV}Mo_9O_{32}$	hexapotassium enneamolybdomanganate(IV)
$Na_6P^V{}_2Mo_{18}O_{62}$	hexasodium 18-molybdodiphosphate(V)
$Na_4P^{III}{}_2Mo_{12}O_{41}$	tetrasodium dodecamolybdodiphosphate(III)
$K_7Co^{II}Co^{III}W_{12}O_{42}.16H_2O$	heptapotassiumdodecawolframocobalt(II)-cobalt(III)ate-16-water
$K_3PV_2Mo_{10}O_{39}$	tripotassium decamolybdodivanadophosphate

◆The coördination subcommittee would prefer not to have sections 7.5 and 7.6 included under 7 and recommends that they should be studied by a special subcommittee. Some cyclic isomers are already known, and isopoly cations also are being investigated.

Cf. comment at 1.12 for stand on wolframate and wolframo, and 2.251 for nona instead of ennea (fourth example).

Isopoly and heteropoly are not separate words in the original version.

7.7. Addition Compounds

The ending -ate is now the accepted ending for *anions* and should generally not be used for addition compounds. Alcoholates are the *salts* of alcohols and this name should not be used to indicate alcohol of crystallization. Analogously addition compounds containing ether, ammonia, *etc.*, should *not* be termed etherates, ammoniates, *etc.*

However, one exception has to be recognized. According to the commonly accepted meaning of the ending -ate, "hydrate" would be, and was formerly regarded as, the name for a *salt* of water, *i.e.*, what is now known as a hydroxide; the name hydrate has now a very firm position as the name of a compound containing water of crystallization and is allowed also in these Rules to designate water bound in an unspecified way; it is considered to be preferable even in this case to avoid the ending -ate by using the name "water" (or its equivalent in other languages) when possible.

The names of addition compounds may be formed by connecting the names of individual compounds by hyphens (short dashes) and indicating the number of molecules by Arabic numerals. When the added molecules are organic, however, it is recommended to use multiplicative numeral prefixes (bis, tris, tetrakis, *etc.*) instead of Arabic figures to avoid confusion with the organic-chemical use of Arabic figures to indicate position of substituents.

Examples:

$CaCl_2.6H_2O$	calcium chloride–6-water (or calcium chloride hexahydrate)
$3CdSO_4.8H_2O$	3-cadmium sulfate–8-water
$Na_2CO_3.10H_2O$	sodium carbonate–10-water (or sodium carbonate decahydrate)
$AlCl_3.4C_2H_5OH$	aluminum chloride–4-ethanol or –tetrakisethanol
$BF_3.(C_2H_5)_2O$	boron trifluoride–diethyl ether
$BF_3.2CH_3OH$	boron trifluoride–bismethanol
$BF_3.H_3PO_4$	boron trifluoride–phosphoric acid
$BiCl_3.3PCl_5$	bismuth trichloride–3-(phosphorus pentachloride)
$TeCl_4.2PCl_5$	tellurium tetrachloride–2-(phosphorus pentachloride)
$(CH_3)_4NAsCl_4.2AsCl_3$	tetramethylammonium tetrachloroarsenate(III)–2-(arsenic trichloride)
$CaCl_2.8NH_3$	calcium chloride–8-ammonia
$8H_2S.46H_2O$	8-(hydrogen sulfide)–46-water
$8Kr.46H_2O$	8-krypton–46-water
$6Br_2.46H_2O$	6-dibromine–46-water
$8CHCl_3.16H_2S.136H_2O$	8-chloroform–16-(hydrogen sulfide)–136-water

These names are not very different from a pure verbal description which may in fact be used, *e.g.*, calcium chloride with 6 water, compound of aluminum chloride with 4 ethanol, *etc.*

If it needs to be shown that added molecules form part of a complex, the names are given according to **7.2** and **7.3.**

Examples:

$FeSO_4.7H_2O$ or $[Fe(H_2O)_6]SO_4.H_2O$	iron(II) sulfate heptahydrate or hexaaquoiron(II) sulfate monohydrate
$PtCl_2.2PCl_3$ or $[PtCl_2(PCl_3)_2]$	platinum(II) chloride–2-(phosphorus trichloride) or dichlorobis(phosphorus trichloride)-platinum(II)
$AlCl_3.NOCl$ or $NO[AlCl_4]$	aluminum chloride–nitrosyl chloride or nitrosyl tetrachloroaluminate
$BF_3.Et_3N$ or $[BF_3(Et_3N)]$	boron trifluoride–triethylamine or trifluoro(triethylamine)boron

◆Only some so-called "addition compounds" are known to be coordination compounds, and the formulas and names can show such structures. Those that are lattice compounds and those of unknown structure do not really belong in 7.

The committees like hydrate and ammoniate and see no particular advantage in dropping them (*cf.* the use of hydrate terms in examples in 6.324; 6-hydrate, *etc.*, as well as hexahydrate, *etc.*, are considered acceptable).

No reason can be seen for deviating here from usual organic practice by using multiplicative prefixes with simple names: tetraethanol is just as clear as tetrakisethanol, and dimethanol as bismethanol. Parentheses can always be used if there is danger of any ambiguity.

The use of short dashes ("en" dashes) instead of hyphens between the names (as in the examples here, but not in the original version) makes for greater ease in reading.

It is considered preferable by some to place the electron donor first in both the formulas and name: $(C_2H_5)_2O.BF_3$, diethyl ether–boron trifluoride (note that organic usage favors ethyl ether over diethyl ether).

8. POLYMORPHISM

Minerals occurring in nature with similar compositions have different names according to their crystal structures; thus, zinc blende, wurtzite; quartz, tridymite, and cristobalite. Chemists and metallographers have designated polymorphic modifications with Greek letters or with Roman numerals (α-iron, ice-I, *etc.*). The method is similar to the use of trivial names, and is likely to continue to be of use in the future in cases where the existence of polymorphism is established, but not the structures underlying it. Regrettably there has been no consistent system, and some investigators have designated as α the form stable at ordinary temperatures, while others have used α for the form stable immediately below the melting point, and some have even changed an already established usage and renamed α-quartz β-quartz, thereby causing confusion. If the α–β nomenclature is used for two substances A and B, difficulties are encountered when the binary system A–B is studied.

NOMENCLATURE OF INORGANIC CHEMISTRY (Continued)

LIST OF NAMES FOR IONS AND RADICALS

In inorganic chemistry substitutive names seldom are used, but the organic-chemical names are shown to draw attention to certain differences between organic and inorganic nomenclature.

	Name				
Atom or group	as neutral molecule	as cation or cationic radical[1]	as anion	as ligand	as prefix for substituent in organic compounds
H	monohydrogen	hydrogen	hydride	hydrido	
F	monofluorine		fluoride	fluoro	fluoro
Cl	monochlorine	chlorine	chloride	chloro	chloro
Br	monobromine	bromine	bromide	bromo	bromo
I	monoiodine	iodine	iodide	iodo	iodo
I_3			triiodide		
ClO		chlorosyl	hypochlorite	hypochlorito	
ClO_2	chlorine dioxide	chloryl	chlorite	chlorito	
ClO_3		perchloryl	chlorate	chlorato	
ClO_4			perchlorate		
IO		iodosyl	hypoiodite		iodoso
IO_2		iodyl			iodyl or iodoxy
O	monoöxygen		oxide	oxo	oxo or keto
O_2	dioxygen		O_2^{2-}: peroxide O_2^{-}: hyperoxide	peroxo	peroxy
HO	hydroxyl		hydroxide	hydroxo	hydroxy
HO_2	(perhydroxyl)		hydrogen peroxide	hydrogen peroxo	hydroperoxy
S	monosulfur		sulfide	thio (sulfido)	thio
HS	(sulfhydryl)		hydrogen sulfide	thiolo	thiol or mercapto
S_2	disulfur		disulfide	disulfido	
SO	sulfur monoxide	sulfinyl (thionyl)			sulfinyl
SO_2	sulfur dioxide	sulfonyl (sulfuryl)	sulfoxylate		sulfonyl
SO_3	sulfur trioxide		sulfite	sulfito	
HSO_3			hydrogen sulfite	hydrogen sulfito	
S_2O_3			thiosulfate	thiosulfato	
SO_4			sulfate	sulfato	
Se	selenium		selenide	seleno	seleno
SeO		seleninyl			seleninyl
SeO_2		selenonyl			selenonyl
SeO_3	selenium trioxide		selenite	selenito	
SeO_4			selenate	selenato	
Te	tellurium		telluride	telluro	telluro
CrO_2		chromyl			
UO_2		uranyl			
NpO_2		neptunyl			
PuO_2		plutonyl			
AmO_2		americyl			
N	mononitrogen		nitride	nitrido	
N_3			azide	azido	
NH			imide	imido	imino
NH_2			amide	amido	amino
NHOH			hydroxylamide	hydroxylamido	hydroxylamino
N_2H_3			hydrazide	hydrazido	hydrazino
NO	nitrogen oxide	nitrosyl		nitrosyl	nitroso
NO_2	nitrogen dioxide	nitryl		nitro	nitro
ONO			nitrite	nitrito	
NS		thionitrosyl			
$(NS)_n$		thiazyl (*e.g.*, trithiazyl)			
NO_3			nitrate	nitrato	
N_2O_3			hyponitrite	hyponitrito	
P	phosphorus		phosphide	phosphido	
PO		phosphoryl			phosphoroso
PS		thiophosphoryl			
PH_2O_3			hypophosphite	hypophosphito	
PHO_3			phosphite	phosphito	
PO_4			phosphate	phosphato	
AsO_4			arsenate	arsenato	
VO		vanadyl			
CO	carbon monoxide	carbonyl		carbonyl	carbonyl
CS		thiocarbonyl			
CH_3O	methoxyl		methanolate	methoxo	methoxy
C_2H_5O	ethoxyl		ethanolate	ethoxo	ethoxy
CH_3S			methanethiolate	methanethiolato	methylthio
C_2H_5S			ethanethiolate	ethanethiolato	ethylthio
CN		cyanogen	cyanide	cyano	cyano
OCN			cyanate	cyanato	cyanato
SCN			thiocyanate	thiocyanato and isothiocyanato	thiocyanato and isothiocyanato
SeCN			selenocyanate	selenocyanato	selenocyanato
TeCN			tellurocyanate	tellurocyanato	
CO_3			carbonate	carbonato	
HCO_3			hydrogen carbonate	hydrogen carbonato	
CH_3CO_2			acetate	acetato	acetoxy
CH_3CO	acetyl	acetyl			acetyl
C_2O_4			oxalate	oxalato	

[1]If necessary, oxidation state is to be given by Stock notation.

◆Although some additions might be made to this list, especially in the last column, and a few changes suggested, no attempt has been made to do so at this time. *Cf.* comments at such rules as 3.31, 3.32, 5.35, 6.2, and 7.312.

A rational system should be based upon crystal structure, and the designations α, β, γ, *etc.*, should be regarded as provisional, or as trivial names. The designations should be as short and understandable as possible, and convey a maximum of information to the reader. The rules suggested here have been framed as a basis for future work, and it is hoped that experience in their use may enable more specific rules to be formulated at a later date.

8.1.—For chemical purposes (*i.e.*, when particular mineral occurrences are not under consideration) polymorphs should be indicated by adding the crystal system after the name or formula. For example, zinc sulfide(cub.) or ZnS(cub.) corresponds to zinc blende or sphalerite, and ZnS(hex.) to wurtzite. The Commission considers that these abbreviations might with advantage be standardized internationally:

cub.	=	cubic	hex.	=	hexagonal
c.	=	body-centered	trig.	=	trigonal
f.	=	face-centered	mon.	=	monoclinic
tetr.	=	tetragonal	tric.	=	triclinic
o-rh.	=	orthorhombic			

Slightly distorted lattices may be indicated by use of the *circa* sign, ∼. Thus, for example, a slightly distorted face-centered cubic lattice would be expressed as ∼f.cub.

8.2.—Crystallographers may find it valuable to add the space-group; it is doubtful whether this system would commend itself to chemists where **8.1** is sufficient.

8.3.—Simple well-known structures may also be designated by giving the type-compound in italics in parentheses; but this system often breaks down because many structures are not referable to a type in this way. Thus, AuCd above 70° may be written as AuCd(cub.) or as AuCd(*CsCl-type*); but at low temperature only as AuCd(o-rh.), as its structure cannot be referred to a type.

◆*Cf.* 6.5 and 6.52.

ABBREVIATIONS USED IN TABLE OF PHYSICAL CONSTANTS OF INORGANIC COMPOUNDS

a	acid
abs	absolute
ac. a	acetic acid
acet	acetone
act	active
al	alcohol
alk	alkali
amm	ammonium
amor	amorphous
anh	anhydrous
appr	approximately
aq	aqua, water
aq. reg	aqua regia
asym	asymmetrical
atm	atmospheres
bipyr	bipyramidal
bl	blue
blk	black
boil	boiling
br., brn	brown
brnsh	brownish
bz	benzene
c	cold
calc	calculated
carb	carbon
caust	caustic
chl	chloroform
choc	chocolate
cit. a	citric acid
col	colorless
coll	colloidal
com'l	commercial
comp	compounds
compl	completely
conc	concentrated
const	constant
cont	contains
corros	corrosive
cr	crystalline
cub	cubic
d., dec	decomposes
deliq	deliquescent
deriv	derivative
dibas	dibasic
di-H	dihydrogen
dil	dilute
dimorph	dimorphous
disg	disagreeable
dk	dark
doubt	doubtful
duct	ductile
effl	efflorescent
em	emerald
eth	ether
ev	evolves
evln	evolution
ex	excess
exist	existence
exp	explodes
extr	extreme(ly)
f., fr	from
feath	feathery
fl	flakes
floc	flocculent
fluo, fluores	fluorescent
form	formic
fum	fuming
fus	fused
fxd	fixed
gel., gelat	gelatinous
gl	glass
glac	glacial
glit	glittering
glob	globular
glyc	glycerin
gran	granular
greas	greasy
grn	green
h	hot
hex	hexagonal
ht	heat
hyd	hydrolyzed
hydx	hydroxides
hyg	hygroscopic
i	insoluble
ign	ignites
ind	indigo
indef	indefinite
infl., inflam	inflammable
infus	infusible
irid	iridescent
leaf	leaflets
lem	lemon
lgr	ligroin
lng	long
lq., liq	liquid
lt	light
lum	luminous
lust	lustrous
me., meth	methyl
met	metal or metallic
micr	microscopic
min	mineral
misc	miscible
mixt	mixture
mod	modifications
monbas	monobasic
mon-H	monohydrogen
monocl	monoclinic
near	nearly
need	needles
nit	nitrate
oct	octahedral
odorl	odorless
offen	offensive
olv	olive
opt	optical or optically
or	orange
ord	ordinary
org	organic
oxal	oxalate or oxalic
pa	pale
pet	petroleum
pl	plates
pois	poisonous
polymorph	polymorphous
powd	powder
ppt	precipitate
pr	prisms
press	pressure
prob	probably
prop	properties
purp	purple
pyr	pyridine
quad	quadrilateral
quest	questioned
rect	rectangular
redsh	reddish
reg	regular
rhbdr	rhombohedral
rhomb	rhombic, orthorhombic
s	soluble
satd	saturated
sld	solid
sensit	sensitive
sc	scales
sec	secondary
silv	silver
sl	slightly
sly	slowly
sm	small
sod	sodium
soln	solution
solv	solvents
spont	spontaneous
st	steel
stab	stable
subl	sublimes
suffoc	suffocating
sulfd	sulfides
sulf	sulfur
sym	symmetrical
tabl	tablets
tart. a	tartaric acid
tetr	tetragonal
tetrah	tetrahedral
tol	toluene
trac	trace, traces
trans	transparent
translu	translucent
tri., trig	trigonal
tribas	tribasic
tricl	triclinic
trim	trimetric
tr	transition point
turp	turpentine
unpleas	unpleasant
unst	unstable
v	very
vac	vacuum
var	various
viol	violent, violence
visc	viscous
vitr	vitreous
vlt	violet
volt., volat	volatizes
wh	white
wh. lt	white light
yel	yellow
yelsh	yellowish
∞	soluble in all proportions
$>$	above
$<$	below

PHYSICAL CONSTANTS OF INORGANIC COMPOUNDS

No.	Name	Synonyms and Formulae	Mol. wt.	Crystalline form, properties and index of refraction	Density or spec. gravity	Melting point, °C	Boiling point, °C	Solubility, in grams per 100 cc: Cold water	Hot water	Other solvents
	Actinium									
a1	**Actinium**	Ac	227	silv wh met, cub		1050	3200 ± 300	d to $Ac(OH)_3$		
a2	bromide	$AcBr_3$	466.73	wh, hex	5.85	subl 800		s		
a3	chloride, tri-	$AcCl_3$	333.36	wh cr, hex	4.81	subl 960				
a4	fluoride, tri-	AcF_3	284	wh cr, hex	7.88			i	i	
a5	hydroxide	$Ac(OH)_3$	278.02	wh				i		
a6	iodide	AcI_3	607.71	wh		subl 700–800		s		
a7	oxalate	$Ac_2(C_2O_4)_3$	718.06					i		
a8	oxide, sesqui-	Ac_2O_3	502	wh cr, hex	9.19			i		
a9	sulfide, sesqui-	Ac_2S_3	550.19	dk, cub	6.75					
a10	**Aluminum**	Al	26.9815	silv wh duct met, cub	2.702	660.2	2467	i	i	s alk, HCl, H_2SO_4; i conc HNO_3, h ac a
a11	acetate, tri-	$Al(C_2H_3O_2)_3$	204.12	wh solid		d		v sl s	d	
a12	acetylacetonate	$Al(C_5H_7O_2)_3$	324.31	col, monocl	1.27	193, subl vac	314	i	i	v s al; s eth, bz
a13	*ortho*arsenate	$AlAsO_4.8H_2O$	310.02	wh powd	3.001	$-H_2O$		i	i	sl s a
a14	benzoate	$Al(C_7H_5O_2)_3$	390.33	wh cr powd				v sl s		
a15	benzyloxide	Aluminum benzylate. $Al(C_7H_7O)_3$	348.38			59–60	$283–284^{0.5}$			
a16	boride	AlB_{12}	156.71	dk red-blk, monocl	2.55_{4}^{18}			i		s hot HNO_3; i a, alk
a17	boride, di-	AlB_2	48.60	copper red, hex.	3.19					
a18	bromate	$Al(BrO_3)_3.9H_2O$	572.84	wh cr, hygr		62.3	d 100	s	s	sl s a
a19	bromide	$AlBr_3$ (or Al_2Br_6)	266.71	col, rhomb pl, deliq	2.64^{10} (fused)	97.5	263.3^{747}	s with viol	d	s al, acet, CS_2
a20	bromide, hexahydrate	$AlBr_3.6H_2O$	374.80	col-yelsh need, deliq	2.54	93	d 135	s	d	s al, amyl al; sl s CS_2
a21	bromide, pentadecylhydrate	$AlBr_3.15H_2O$	536.94	col need		−7.5	d 7	s	s	s al
a22	butoxide, tert-	$Al(C_4H_9O)_3$	246.33	wh cr	1.0251_{0}^{20}	subl 180, m.p >300, sealed tube.				v s org solv
a23	carbide	Al_4C_3	143.96	yel-grn, hex	2.36	stab to 1400	d 2200^{400}	d to CH_4		d dil a; i acet
a24	chlorate	$Al(ClO_3)_3.6H_2O$	385.43	col, rhbdr, deliq		d		vs	vs	s dil HCl
a25	*per*chlorate	$Al(ClO_4)_3.6H_2O$	433.43	col, hygr	2.020	82	$-6H_2O$, 178 d 262	s	s	
a26	chloride	$AlCl_3$ (or Al_2Cl_6)	133.34	wh to col, hex, odor HCl, v deliq	fus 2.44^{25} liq 1.31^{200}	$190^{2.5}$ atm	182.7^{752} subl. 177.8	69.9^{15} with viol	s d	$100^{12.5}$ abs al; 0.072^{25} chl; sCCl_4, eth sl s bz
a27	chloride, hexahydrate	$AlCl_3.6H_2O$	241.43	col, rhomb, deliq, 1.6	2.398	d 100		s	v s ev HCl	50 abs al; s eth; sl s HCl
a28	chloride, hexammine	$AlCl_3.6NH_3$	235.52	col cr, hygr	1.412_{4}^{25}	d		s		
a29	diethylmalonate deriv	$Al(C_7H_{11}O_4)_3$	504.47		1.084^{100}	98		i		s org solv
a30	ethoxide	$Al(C_2H_5O)_3$	162.14	wh cr	1.142_{0}^{20}	134	205^{14}	d	s	v sl s al, eth
a31	α-ethylacetoacetate deriv	$Al(C_6H_9O_3)_3$	414.39	wh cr	1.101^{80}	78–79	$190–200^{11}$	s		s lgr
a32	ferrocyanide	$Al_4[Fe(CN)_6]_3.17H_2O$	1050.05	br powd				sl s	sl s	s dil a
a33	fluoride	AlF_3	83.98	col, tricl	2.882_{4}^{25}	1291 $subl^{760}$		0.559^{25}	s	i a, alk, al, acet
a34	fluoride	$AlF_3.3\frac{1}{2}H_2O$	147.03	wh cr powd	1.914_{4}^{25}	$-2H_2O$ 100	anhydr 250	i	sl s	d ac a
a35	fluoride, monohydrate	Nat. fluellite. $AlF_3.H_2O$.	101.99	col, rhomb, 1.473, 1.490, 1.511	2.17			sl s	sl s	
a36	fluosilicate	Nat topaz. $2AlFO.SiO_2$	184.04	rhomb, 1.619, 1.620, 1.627	3.58			i	i	
a37	hydroxide	Nat. boehmite. $AlO(OH)$	59.99	wh, orthorhomb microcr	3.01	$-H_2O$, trans to γ-Al_2O_3		i	i	s h a, h alk
a38	hydroxide	Nat. diaspore. $AlO(OH)$	59.99	col, rhomb cr	3.3–3.5	$-H_2O$, trans to Al_2O_3		i	i	s h a, h alk
a39	hydroxide	$Al(OH)_3$	78.00	wh, monocl	2.42	$-H_2O$, 300		i	i	s a, alk; i al
a40	iodide	AlI_3 (or Al_2I_6)	407.69	br pl, cont free I_2, deliq	3.98^{25}	191	360	s d	s	s al, eth, CS_2, liq NH_3
a41	iodide, hexahydrate	$AlI_3.6H_2O$	515.79	wh-yel cr, hygr	2.63	d 185	d	v s	v s	s al, CS_2
a42	isopropoxide	$Al(C_3H_7O)_3$	204.25	wh cr	1.0346_{0}^{20}	118.5	140.5^{8}	d		s al, bz, chl
a43	lactate	$Al(C_3H_5O_3)_3$	294.20	wh-yelsh powd				v s		

No.	Name	Synonyms and Formulae	Mol. wt.	Crystalline form, properties and index of refraction	Density or spec. gravity	Melting point, °C	Boiling point, °C	Solubility, in grams per 100 cc		
								Cold water	Hot water	Other solvents
	Aluminum									
a44	nitrate	$Al(NO_3)_3.9H_2O$	375.13	col, rhomb, deliq, 1.54		73.5	d 150	63.7^{25}	v s d	100 al; s alk, acet, HNO_3. a
a45	nitride	AlN	40.99	wh cr, hex	3.26	>2200 (in N_2)	subl 2000	d (NH_3)	d	d a, alk
a46	oleate (com'l)	$Al(C_{18}H_{33}O_2)_3$(?)	871.37	wh powd, existence doubted except as basic salt				d	s	i al; v sl s bz
a47	oxalate	$Al_2(C_2O_4)_3.4H_2O$	390.08	wh powd				i	i	i al; s a
a48	oxide	Al_2O_3	101.96	col, hex, 1.768, 1.760	3.965^{25}	2045	2980	i		v sl s a, alk
a49	oxide	α-Alumina, nat. corundum. Al_2O_3	101.96	col, rhomb cr, 1.765	3.97	2015 ± 15	2980 ± 60	0.000098^{29}	i	v sl s a, alk
a50	oxide	γ-Alumina. Al_2O_3	101.96	wh micr cr, 1.7	3.5–3.9	tr to α		i	i	sl s a, alk
a51	oxide, monohydrate	$Al_2O_3.H_2O$	119.98	col, rhomb, 1.624 ± 0.003	3.014			i	i	
a52	oxide, trihydrate	Nat. gibbsite, hydrargilite. $Al_2O_3.3H_2O$	156.01	wh monocl cr, 1.577, 1.577, 1.595	2.42	tr to $Al_2O_3.H_2O$ (Boehmite)		i	i	s h a, alk
a53	oxide, trihydrate	Nat. bayerite. $Al_2O_3.3H_2O$	156.01	wh micr cr, 1.583	2.53	tr to $Al_2O_3H_2O$ (Boehmite)		i	i	s hot a, alk
a54	*meta*phosphate	$Al(PO_3)_3$	263.90	col, tetr	2.779			i	i	i a
a55	palmitate, mono- (com'l)	$Al(OH)_2C_{16}H_{31}O_2$	316.41	wh	1.095	200		i		s alk, hydrocarb
a56	1-phenol-4-sulfonate	$Al(C_6H_5O_4S)_3$	546.49	redsh-wh powd				s		s al, glyc
a57	phenoxide	$Al(C_6H_5O)_3$	306.27	grayish-wh cr mass	1.23	d 265		d		s al, eth, chl
a58	*ortho*phosphate	$AlPO_4$	121.95	wh rhomb pl, 1.546, 1.556, 1.578	2.566	>1500		i	i	s a, alk, al
a59	propoxide	$Al(C_3H_7O)_3$	204.25	wh cr	1.0578_0^{20}	106	248^{14}	d	d	s al
a60	salicylate	$Al(C_7H_5O_3)_3$	438.33	redsh-wh powd				i		i al; s alk
a61	selenide	Al_2Se_3	290.84	lt brn powd, unstable in air	3.437_4^{18}			d	d	d a
a62	silicate	Nat. sillimanite, andalusite, cyanite. $Al_2O_3.SiO_2$	162.04	wh, rhomb, 1.66	3.247	1545 tr to $Al_2O_3.2SiO_2$	>1545	i	i	d HF; i HCl; s fus alk
a63	silicate	Nat. mullite. $3Al_2O_3.2SiO_2$	426.05	col, rhomb, 1.638, 1.642, 1.653	3.156	1920		i	i	i a, HF
a64	stearate, tri-	$Al(C_{18}H_{35}O_2)_3$	877.42	wh powd	1.010	103		i		s al, bz, turp, alk
a65	sulphate	$Al_2(SO_4)_3$	342.15	wh powd, 1.47	2.71	d 770		31.3^0	98.1^{100}	s dil a; sl s al
a66	sulfate, hydrate	Nat. alunogenite. $Al_2(SO_4)_3.18H_2O$	666.43	col, monocl, 1.474, 1.467, 1.483	1.69^{17}	d 86.5		86.9^0	1104^{100}	i al
a67	sulfide	Al_2S_3	150.16	yel, hex, odor H_2S, d moist air	2.02^{13}	1100	subl 1500 (N_2)	d		s a; i acet
a68	thallium sulfate	Aluminum thallium alum. $AlTl(SO_4)_2.12H_2O$	639.66	col, oct, 1.50112	2.325_4^{20}	91		4.84^0	65.19^{60}	
a69	**Americium**	Am	243.13	silvery, hex		>850	2600 (extrap)			s dil a
a70	bromide	$AmBr_3$	482.86	wh, orthorhomb		subl		s		
a71	chloride	$AmCl_3$	349.49	pink, hex	5.78	subl 850		s		
a72	fluoride	AmF_3	300.12	pink, hex	9.53			i		
a73	iodide	AmI_3	623.84	yel, orthoromb	6.9			s		
a74	oxide	Am_2O_3	534.26	redsh-brn, cub or tan, or hex						s min a
a75	oxide, di-	AmO_2	275.13	blk, cub	11.68					s min a
a76	**Ammonia**	NH_3	17.03	col gas; liq, 0.817^{-79}, $1.325^{16.5}$	0.7710 g/ml; 760 mm	−77.7	−33.35	89.9	7.4^{100}	13.20^{20} al; s eth, org solv
a77	Ammonia-d_3	Trideuterio ammonia. ND_3	20.05			−74	−30.9			
	Ammonium									
v78	acetate	$NH_4C_2H_3O_2$	77.08	wh cr, hygr	1.17_4^{20}	114	d	148^4	d	7.89^{15} MeOH; s al; sl s acet
a79	acetate, hydrogen	$(NH_4)H(C_2H_3O_2)_2$	137.14	col need, deliq		66		s		s al
a80	aluminum chloride	$NH_4Cl.AlCl_3$	186.83	wh cr		304		s		
a81	aluminum sulfate	$NH_4Al(SO_4)_2$	237.14	col, hex	2.45^{20}			s		s glyc; i al

No.	Name	Synonyms and Formulae	Mol. wt.	Crystalline form, properties and index of refraction	Density or spec. gravity	Melting point, °C	Boiling point, °C	Solubility, in grams per 100 cc		
								Cold water	Hot water	Other solvents
	Ammonium									
a82	aluminum sulfate, hydrate	Nat. tschernigite. $NH_4Al(SO_4)_2.12H_2O$	453.33	col, cub, 1.459	1.64	93.5	$-10H_2O$, 120	15^{20}	v s	s dil a; i al
a83	*ortho*arsenate	$(NH_4)_3AsO_4.3H_2O$	247.08	rhomb cr		d, $-NH_3$		sl s		
a84	*ortho*arsenate, di-H	$NH_4H_2AsO_4$	158.98	col, tetr, 1.577, 1.522	2.311^{9}	d, $-NH_3^{300}$		33.74^{0}	122.4^{90}	
a85	*ortho*arsenate mono-H	$(NH_4)_2HAsO_4$	176.00	col, monocl, odor NH_3	1.989	d		s	d	i al
a86	*meta*arsenite	NH_4AsO_2	124.96	col, rhomb pr, hygr				v s	d	i al, acet; sl s NH_4OH
a87	azide	NH_4N_3	60.06	col pl	1.346	160	subl 134 expl	20.16^{30}	27.04^{40}	1.06^{20} 80 % al; i eth, bz
a88	benzene sulfonate	$NH_4C_6H_5SO_3$	175.21	rhomb	1.342	d 271–275		98	320	19 c al; i eth, bz
a89	benzoate	$NH_4C_7H_5O_2$	139.16	col, rhomb	1.260	d 198	subl 160	$19.6^{14.5}$	83.3^{100}	1.63^{25} al; s glyc; i eth
a90	*penta*borate	Ammonium decaborate. $NH_4B_5O_8.4H_2O$	272.15					7.03^{18}		
a91	*peroxy*borate	$NH_4BO_3.\frac{1}{2}H_2O$	85.86	wh cr		d		$1.55^{17.5}$		i al
a92	*tetra*borate	Ammonium biborate. $(NH_4)_2B_4O_7.4H_2O$	263.38	col, tetr		d		7.27^{18}	52.68^{90}	sl s acet; i al
a93	bromate	NH_4BrO_3	145.95	col, hex		expl		v s	vs	
a94	bromide	NH_4Br	97.95	cub, coll hygr, 1.712^{25}	2.429	subl 452	235 vac	97^{25}	145.6^{100}	10^{78} al; s acet, eth, NH_3
a95	*di*bromoiodide	NH_4IBr_2	304.75	met-grn pr, hygr		198		v s	s	s eth
a96	bromoplatinate	$(NH_4)_2PtBr_6$	710.62	red-brn cub	4.265^{24}	d 145		0.59^{20}	0.36^{100}	
a97	bromoselenate	$(NH_4)_2SeBr_6$	594.49	red, oct cr	3.326			d	d	sl s eth
a98	bromostannate	$(NH_4)_2SnBr_6$	634.22	col, cub	3.50	d		v s		
a99	cadmium chloride	$4NH_4Cl.CdCl_2$	397.27	col, rhomb, 1.6038	2.01			s		
a100	calcium arsenate	$NH_4CaAsO_4.6H_2O$	305.13	col, monocl	1.905^{15}	d 140		0.02	s	s NH_4Cl; i NH_4OH
a101	calcium phosphate	$NH_4CaPO_4.7H_2O$	279.20	1.561^{15}		d		i	d	s a
a102	carbamate	$NH_4NH_2CO_2$	78.07	col, rhomb		subl 60		v s	d	v s NH_4OH; sl s al; i acet
a103	carbamate acid carbonate	Sal volatile. $NH_4NH_2CO_2.NH_4HCO_3$	157.13	wh cr		subl		25^{15}	67^{65}	d al; s glyc; i acet
a104	carbonate	$(NH_4)_2CO_3.H_2O$	114.10	col, cub		d 58		100^{15}	d	i al, NH_3, CS_2; s dil MeOH
a105	carbonate, hydrogen	Ammonium bicarbonate. NH_4HCO_3	79.06	col, rhomb or monocl, 1.423, 1.536, 1.555	1.58	107.5 (d 36–60)	subl	11.9^{0}	d	i al, acet
a106	cerium nitrate(ic)	$(NH_4)_2Ce(NO_3)_6$	548.23	or, monocl				141^{25}	227^{80}	s HNO_3, al
a107	cerium nitrate(ous)	$2NH_4NO_3.Ce(NO_3)_3.4H_2O$	558.28	monocl		74		318.20	817.4^{65}	
a108	cerium sulfate(ous)	$(NH_4)_2SO_4.Ce_2(SO_4)_3.8H_2O$	844.69	monocl	2.523	$-6H_2O$, 100	$-8H_2O$,150	3.29^{49} (anhydr)		
a109	chlorate	NH_4ClO_3	101.49	col, monocl need.	1.80	102 expl		28.7^{0}	115^{75}	sl s al
a110	*per*chlorate	NH_4ClO_4	117.49	col, rhomb, 1.482	1.95	d		10.74^{0}	42.45^{85}	s acet; sl s al
a111	chloride	Sal ammoniac. NH_4Cl	53.49	col, cub, 1.642	1.527	subl 340	520	29.7^{0}	75.8^{100}	0.6^{19} al; s liq NH_3; i acet, eth
a112	chloroaurate	NH_4AuCl_4	356.82	yel, monocl or rhomb			520	s		sl s al
a113	chloroaurate, hydrate	$(NH_4AuCl_4)_4.5H_2O$	1517.35	yel, monocl		$-5H_2O$, 100		s		s al
a114	chlorogallate	NH_4GaCl_4	229.57	wh cr		275		v s	v s	s al; i pet eth
a115	chloroiridate	$(NH_4)_2IrCl_6$	441.00	red-blk, cub	2.856	d		0.69^{14}	4.38^{80}	i al; s HCl
a116	chloroiridite	$(NH_4)_3IrCl_6.1\frac{1}{2}H_2O$	486.06					s		
a117	chloroosmate	$(NH_4)_2OsCl_6$	439.00	cub	2.93					
a118	chloropalladate	$(NH_4)_2PdCl_6$	355.20	red-brn, cub	2.418	d		sl s		
a119	chloropalladite	$(NH_4)_2PdCl_4$	284.29	olive grn, tetr	2.17	d		s		i al
a120	*hexa*chloroplatinate	$(NH_4)_2PtCl_6$	443.89	yel, cub, 1.8	3.065	d		0.7^{15}	1.25^{100}	0.005 al; i eth, conc HCl
a121	chloroplatinite	$(NH_4)_2PtCl_4$	372.98	red, rhomb (tetr)	2.936	d 140–150		s	s	i al
a122	chloroplumbate	$(NH_4)_2PbCl_6$	455.98	yel, cub	2.925	d 120		sl s	d	s a
a123	chlorostannate	$(NH_4)_2SnCl_6$	367.49	wh, cub	2.4	d		$33.^{14.5}$	v s	
a124	*tetra*chlorozincate	$ZnCl_2.2NH_4Cl$	243.26	wh pl, rhomb, hygr	1.879	d 150		v s		
a125	chromate	$(NH_4)_2CrO_4$	152.08	yel, monocl	1.91^{12}	d 180		40.5^{30}	d	i al; sl s NH_3, acet
a126	*di*chromate	$(NH_4)_2Cr_2O_7$	252.06	or, monocl	2.15^{25}	d 170		30.8^{15}	89^{30}	s al; i acet
a127	*peroxy*chromate	$(NH_4)_3CrO_8$	234.11	red-brn, cub		d 40	expl 50	sl s	d	i al, eth, sl s NH_3; expl H_2SO_4
a129	chromium sulfate(ic)	$(NH_4)Cr(SO_4)_2.12H_2O$	478.34	grn or vlt, cub; 1.4842	1.72	94, $-9H_2O$, 100		21.2^{25}	32.8^{40}	s al, dil a
a130	citrate, di(*sec.*)	$(NH_4)_2HC_6H_5O_7$	226.19	wh gran or powd	1.48			100		sl s al
a131	citrate, tri-(*tert.*)	$(NH_4)_3C_6H_5O_7$	243.22	wh cr, deliq		d		v s		i al, eth, acet
a132	cobalt *ortho*phosphate(ous)	$(NH_4)CoPO_4.H_2O$	189.96	vlt cr powd				i	d	s a

No.	Name	Synonyms and Formulae	Mol. wt.	Crystalline form, properties and index of refraction	Density or spec. gravity	Melting point, °C	Boiling point, °C	Solubility, in grams per 100 cc: Cold water	Hot water	Other solvents
	Ammonium									
a133	cobalt (II)-selenate	$(NH_4)_2SeO_4.CoSO_4.6H_2O$	489.02	ruby-red, monocl, 1.526, 1.532, 1.541	2.228^{20}_{4}					
a134	cobalt sulfate(ous)	$(NH_4)_2SO_4.CoSO_4.6H_2O$.	395.23	ruby-red, monocl, 1.490, 1.495, 1.503	1.902			20.5^{20}	45.4^{80}	i al
a135	copper chloride(ic)	$2NH_4Cl.CuCl_2.2H_2O$	277.46	blue, tetrag, 1.744, 1.724	1.993	d 110		33.8^{0}	99.3^{80}	s a, al; sl s NH_3
a136	copper iodide(ous)	$NH_4I.CuI.H_2O$	353.40	rhomb pl				d	d	s NH_4I
a137	cyanate	NH_4OCN	60.06	wh cr	1.342^{20}_{4}	d 60		v s	d	sl s al; i eth
a138	cyanide	NH_4CN	44.06	col, cub	1.02^{100}	d 36	subl 40	v s	d	v s al
a139	cyanaurate	$NH_4Au(CN)_4.H_2O$	337.09	col pl		d 200		v s		v s al; i eth
a140	cyanaurite	$NH_4Au(CN)_2$	267.04	col, cub		d 100		v s	v s	s al: i eth
a141	cyanoplatinite	$(NH_4)_2Pt(CN)_4.H_2O$	553.25	yel cr				s		
a142	ethylsulfate	$NH_4C_2H_5SO_4$	143.16			99		s		
a143	ferricyanide	$(NH_4)_3Fe(CN)_6$	266.06	red cr, rhomb		d		v s		
a144	ferrocyanide	$(NH_4)_4Fe(CN)_6.3H_2O$	338.15	yel, monocl, turns bl in air		d		s	d	i al
a145	fluoantimonite	$(NH_4)_2SbF_5$	252.84	col, rhomb		d, subl		108		
a146	fluoborate	NH_4BF_4	104.84	wh, rhomb	1.871^{15}	subl		25^{16}	97^{100}	s NH_4OH
a147	fluogallate	$(NH_4)_3GaF_6$	237.83	wh oct cr		d >250–GaF_3			sl s	
a148	fluogermanate	$(NH_4)_2GeF_6$	222.66	col hex pr and bipyram, 1.428, 1.425	2.564^{25}_{25}			s		i al, MeOH
a149	fluophosphate, di-	$NH_4PO_2F_2$	119.01	col, rhomb		213		s	s	s al, acet
a150	fluophosphate, hexa.	NH_4PF_6	163.00	col, pl	2.180^{18}	d		s	s	s al, acet
a151	fluoride	NH_4F	37.04	col, hex, deliq	1.009^{25}	subl		100^{0}	d	s al; i NH_3
a152	fluoride, hydrogen	NH_4HF_2	57.04	rhomb or tetr, deliq, 1.390	1.50	125.6		v s	v s	sl s al
a153	fluorosulfonate	NH_4FSO_3	117.10	wh need		d 245		v s	v s	sl s al; s MeOH
a154	fluosilicate	α-Nat. cryptohalite. $(NH_4)_2SiF_6$	178.14	α oct, β hex, col α 1.3696	α 2.011 β 2.152	d		18.6^{17}	55.5^{100}	sl s al; i acet
a155	fluosulfonate	NH_4SO_3F	117.10	col, need		244.7		s	s	sl s al; s MeOH; d NH_4OH
a156	fluotitanate	$(NH_4)_2TiF_6$	197.97	hex pr		d		s	s	i al, eth
a157	fluozirconate	$(NH_4)_2ZrF_6$	241.29	rhomb, hex	1.154					
a158	fluozirconate	$(NH_4)_3ZrF_7$	278.32	col, cub	1.433			sl s		
a159	formate	NH_4CHO_2	63.06	wh, monocl, deliq.	1.280	116	d 180	102^{0}	531^{80}	s al, eth, NH_3
a160	gallium sulfate	Ammonium gallium alum. $Ga(NH_4)(SO_4)_2.12H_2O$	496.07	cub, 1.468	1.777			30.84^{25}		0.00875^{25} 70 % EtOH
a161	*tri*hydrogen para-*per*iodate	$(NH_4)_2H_3IO_6$	262.00	col, rhomb	2.85					
a162	hydroxide	NH_4OH	35.05	at ord temp in sol only		−77		s		
a163	iodate	NH_4IO_3	192.94	col, rhomb or monocl	3.309^{21}	d 150		2.06^{15}	14.5^{101}	
164	iodide	NH_4I	144.94	col, cub, hygr, 1.7031	2.514^{25}	subl 551	220 vac	154.2^{0}	250.3^{100}	v s al, acet, NH_3; sl s eth
165	*tri*iodide	NH_4I_3	398.75	dk br, rhomb	3.749	d 175		s d	d	
166	iodoplatinate	$(NH_4)_2PtI_6$	992.59	blk, cub	4.61					i al
167	iridium chloride (III)	$(NH_4)_3IrCl_6.H_2O$	477.05	grn-br-bl, cub		d 350		10.5^{19}		
168	iridium sulfate	$NH_4Ir(SO_4)_2.12H_2O$	618.54	yel-red cr			106	s		
169	iron (III) chloride	$2NH_4Cl.FeCl_3.H_2O$	287.20	ruby-red, rhomb, hygr, 1.78	1.99	234		v s	v s	
170	iron (III) fluoride	$3NH_4F.FeF_3$	223.95	col to lt yel, oct	1.96			sl s	sl s	
171	iron (II) selenate	$(NH_4)_2SeO_4.FeSO_4.6H_2O$	485.93	lt grn, monocl, 1.5226, 1.5260, 1.5334	2.191^{20}_{4}					
172	iron (III) sulfate	$(NH_4)_2SO_4.Fe_2(SO_4)_3$	532.02	wh, hex	2.49^{22}	d 420		44.15^{25}		
173	iron sulfate(ic)	$NH_4Fe(SO_4)_2.12H_2O$	482.19	vlt, cub oct, 1.4854	1.71	39–41	$-12H_2O$, 230	124.0^{25}	400^{100}	i al; s dil a
174	iron sulfate(ous)	$(NH_4)_2SO_4.FeSO_4.6H_2O$.	392.14	grn, monocl, 1.487, 1.492, 1.499	1.864^{20}_{4}	d 100–110		26.9^{20}	73.0^{80}	i al
175	lactate	$NH_4C_3H_5O_3$	107.11	col-yelsh liq	$1.19–1.21^{15}$			∞		∞ al
176	laurate, acid (mixt.)	$NH_4C_{12}H_{23}O_2.C_{12}H_{24}O_2$	417.68	wh		75	d	s	s	4.8^{7} al; sl s eth, acet
177	magnesium arsenate	$NH_4MgAsO_4.6H_2O$	289.36	col, tetrg, 1.608	1.932^{15}	d		0.038^{20}	0.024^{80}	s a; i al
178	magnesium carbonate	$(NH_4)_2CO_3.MgCO_3.4H_2O$	252.47	wh				s	v s	s a; i al

PHYSICAL CONSTANTS OF INORGANIC COMPOUNDS (Continued)

No.	Name	Synonyms and Formulae	Mol. wt.	Crystalline form, properties and index of refraction	Density or spec. gravity	Melting point, °C	Boiling point, °C	Solubility, in grams per 100 cc		
								Cold water	Hot water	Other solvents
	Ammonium									
a179	magnesium chloride	$NH_4Cl.MgCl_2.6H_2O$	256.80	col, rhomb, deliq	1.456	$-2H_2O$, 100	d	16.7		d al
a180	magnesium chromate	$(NH_4)_2CrO_4.MgCrO_4.6H_2O$	400.51	yel, monocl, 1.636, 1.637, 1.653	1.84	d		v s	v s	
a181	magnesium phosphate	Guanite, struvite $NH_4MgPO_4.6H_2O$	245.41	col, rhomb, 1.495, 1.496, 1.504	1.711	d		0.0231^{0}	0.0195^{80}	v s dil a; i al
a182	magnesium selenate	$(NH_4)_2SeO_4.MgSeO_4.6H_2O$	454.40	col monocl pr, 1.507, 1.509, 1.517	2.058^{20}_{4}			sl s		
a183	magnesium sulfate	Nat. boussingaulite. $(NH_4)_2SO_4.MgSO_4.6H_2O$	360.60	col, monocl, 1.472, 1.473, 1.479	1.723	>120	d 250	17.92^{0}	64.7^{100} (anhydr)	
a184	*l*-malate, hydrogen	$NH_4HC_4H_4O_5$	151.08	col, rhomb	1.5	161	d	$32.2^{15.7}$		
a185	*per*manganate	NH_4MnO_4	136.97	rhomb	2.208^{10}	d 110		7.9^{15}	d	
a186	manganese phosphate(ic)	$NH_4MnPO_4.H_2O$	185.96	wh cr				0.0031	0.05	i al, NH_4 salts
a187	manganese sulfate(ous)	$(NH_4)_2SO_4.MnSO_4.6H_2O$	391.23	pa red, monocl, 1.480, 1.484, 1.491	1.83			51.3^{25}	v s	
a188	molybdate	$(NH_4)_2MoO_4$	196.01	col, monocl pr	2.276^{25}_{4}	d		s, d	d	s a; i al, NH_3, ace
a189	*para*molybdate	"Molybdic acid" com'l. $(NH_4)_6Mo_7O_{24}.4H_2O$	1235.86	col-yelsh, monocl	2.498	$-H_2O$, 90	d 190	43	d	i al; s a, alk
a190	*per*molybdate	$3(NH_4)_2.O.5MoO_3.2MoO_4.6H_2O$	608.17	lt yel monocl pr	2.975	d 170		v s	v s	sl s al
a191	molybdotellurate	$(NH_4)_6TeMo_6O_{24}.7H_2O$	1321.56	col, rhomb	2.78	d 550	d	s	s	
a192	myristate, acid (mixt.)	$NH_4C_{14}H_{27}O_2.C_{14}H_{28}O_2$	473.78	wh solid		75–90	d	sl s	s	s al; i eth
a193	nickel chloride	$NH_4Cl.NiCl_2.6H_2O$	291.20	grn, monocl, deliq	1.654			v s	v s	
a194	nickel sulfate	Double nickel salt. $(NH_4)_2SO_4.NiSO_4.6H_2O$	395.00	dk bl-grn, monocl, 1.495, 1.501, 1.508	1.923			10.4^{20}	30^{80}	i al; s $(NH_4)_2SO_4$
a195	nitrate	NH_4NO_3	80.04	col, rhomb, (monocl $>32.1^{0}$)	1.725^{25}	169.6	210^{11}	118.3^{0}	871^{100}	3.8^{20} al; 17.1^{20} MeOH s acet, NH_3; i eth
a196	nitratocerate	$(NH_4)_2Ce(NO_3)_6$	548.23	yel-red, monocl				142.6^{25}	232^{90}	s al; sl s HNO_3
a197	nitrite	NH_4NO_2	64.04	wh-yelsh cr	1.69	60–70 expl	30 subl vac	v s	d	s dil al; i eth
a198	oleate, acid (mixt.)	$NH_4C_{18}H_{33}O_2.C_{18}H_{34}O_2$	581.97	wh powd		d 78		s	s	80^{50} al; 13.3^{15} eth
a199	osmium chloride	$(NH_4)_2OsCl_6$	439.00	blk, oct	2.93^{25}	subl 170		d	d	s HCl
a200	oxalate	$(NH_4)_2C_2O_4.H_2O$	142.11	col, rhomb, 1.439, 1.546, 1.594	1.50	d		2.54^{0}	11.8^{50}	i NH_3
a201	oxalate, acid	Ammonium binoxalate. $NH_4HC_2O_4.H_2O$	125.08	col, rhomb	1.556	$-H_2O$, 170		s		i eth, bz
a202	oxalatoferrate (III)	$(NH_4)_3Fe(C_2O_4)_3.3H_2O$	428.07	grn, monocl	1.78	d 165		42.7^{0}	345^{100}	
a203	palladium (II)-chloride	$(NH_4)_2PdCl_4$	284.29	grnsh-yel, tetr	2.17	d		v s	v s	s dil al; i abs al
a204	palladium (IV)-chloride	$(NH_4)_2PdCl_6$	355.20	red oct cr	2.418	d −Cl		sl s	sl s	
a205	palmitate, (acid)- (mixt.)	$NH_4C_{16}H_{31}O_2.C_{16}H_{32}O_2$	529.90	yelsh soapy mass or yel powd		>100	d	sl s	s	8.8^{50} al; 0.23^{13} eth
a206	*meta*periodate	NH_4IO_4	208.94	col, tetr	3.056^{18}	expl		2.7^{16}		
a207	*hypo*phosphate	$(NH_4)_2H_2P_2O_6$	196.04	col cr		170		7^{25}	25^{100}	
a208	*ortho*phosphate	$(NH_4)_3PO_4.3H_2O$	203.13	wh pr				26.1^{25}		sl s dil NH_4OH; i NH_3, acet
a209	*ortho*phosphate, di-H	$NH_4H_2PO_4$	115.03	col, tetr, 1.525, 1.479	1.803^{19}	190		22.7^{0}	173.2^{100}	i acet
a210	*ortho*phqsphate-mono-H	$(NH_4)_2HPO_4$	132.05	col, monocl, 1.52	1.619	d 155	d	57.5^{10}	106.0^{70}	i al, acet, NH_3
a211	*hypo*phosphite	$NH_4H_2PO_2$	83.03	rhomb tabl	1.634	200	d 240	s	s	s al, NH_3; i acet
a212	*ortho*phosphite, di-H	$NH_4H_2PO_3$	99.03	col, monocl pr		123	d 145	171^{0}	260^{31}	i al
a213	phosphofluoride, hexa-	NH_4PF_6	163.00	col, cub	2.180^{18}_{4}	d		74.8^{20}		s al, acet; d h a
a214	phosphomolybdate	Ammonium molybdo phosphate. $(NH_4)_3P(Mo_3O_{10})_4$	1876.35	yel powd		d		sl s	sl s	s alk; i al, HNO_3
a215	phosphotungstate	$(NH_4)_3P(W_3O_{10})_4$	2931.27	wh				sl s	sl s	
a216	picramate	$NH_4C_6H_4N_3O_5$	216.15	redsh-brn cr powd				s		
a217	picrate	$NH_4C_6H_2N_3O_7$	246.14	red or yel, rhomb	1.719	d	expl 423	1.1^{20}	s	sl s al
a219	praseodymium sulfate	$(NH_4)_2SO_4.Pr_2(SO_4)_3.8H_2O$	846.26	cr	$2.531^{16.5}$	$-8H_2O$, 170		sl s		
a220	propionate	$NH_4C_3H_5O_2$	91.11	pr, deliq	1.108^{25}	45		v s		s al, ac a
a221	*per*rhenate	NH_4ReO_4	268.24	wh, hex pl	3.97	d	d	6.1^{20}	32.34^{80}	

No.	Name	Synonyms and Formulae	Mol. wt.	Crystalline form, properties and index of refraction	Density or spec. gravity	Melting point, °C	Boiling point, °C	Solubility, in grams per 100 cc: Cold water	Solubility, in grams per 100 cc: Hot water	Solubility, in grams per 100 cc: Other solvents
	Ammonium									
222	rhodanid	NH_4NCS	76.12	monocl cr, to rhomb at 92, ~1.685	1.305	149.6	d 170	128°	347^{60}	v s al; s MeOH, acet; i $CHCl_3$
223	rhodium chloride	$(NH_4)_3RhCl_6.H_2O$	387.75	dk red rhomb need		$-H_2O$, 140		s	s	sl s al; s dil NH_4Cl
224	rhodium sulfate	Ammonium rhodium alum. $Rh(NH_4)(SO_4)_2.12H_2O$	529.25	orange, 1.5150		102				
225	*d*-saccharate, hydrogen	$NH_4HC_6H_8O_8$	227.17	wh need or monocl pr				1.22^{15}	24.35^{100}	i c al; s h al
226	salicylate	$NH_4C_7H_5O_3$	155.15	col, monocl			subl	111^{25}	v s	28.8^{25} al
227	selenate	$(NH_4)_2SeO_4$	179.03	col, monocl, 1.561, 1.563, 1.585	2.194^{20}_{4}	d		117^{7}	197^{100}	i al, acet, NH_3
228	selenate, hydrogen	NH_4HSeO_4	162.00	rhomb	2.162	d				
229	selenide	$(NH_4)_2Se$	115.04	col, or wh cr		d		s		
230	sodium phosphate, hydrate	Microsmic salt. $NH_4NaHPO_4.4H_2O$	209.07	col, monocl, 1.439, 1.442, 1.469	1.574	d 97				
231	stearate, acid (mixt.)	$NH_4C_{18}H_{35}O_2.C_{18}H_{36}O_2$	586.00	wh cr		d 110		v s		0.3^{25} al; 0.19^{25} eth; 0.08^{25} acet
232	succinate	$(NH_4)_2.C_4H_4O_4$	152.15	col cr	1.37			s		sl s al
233	sulfamate	$NH_4NH_2SO_3$	114.12	large, pl, deliq		125	d 160	166.6^{10}	357^{50}	i al
234	sulfate	Nat. mascagnite. $(NH_4)_2SO_4$	132.14	col, rhomb, 1.521, 1.523, 1.533	1.769^{50}	d 235		70.6^{0}	103.8^{100}	i al, acet, NH_3
235	sulfate, hydrogen	Ammonium bisulfate. NH_4HSO_4	115.11	col, rhomb, 1.473	1.78	146.9	d	100	v s	sl s al; i acet
236	*peroxy*disulfate	$(NH_4)_2S_2O_8$	228.18	col, monocl, 1.498, 1.502, 1.587	1.982	d 120		58.2^{0}	v s	
237	sulfide, hydro-	NH_4HS	51.11	wh rhomb, 1.74	1.17	118^{150atm}	88.4^{19atm}	128.1^{0}	d	s al; NH_3
238	sulfide, mono-	$(NH_4)_2S$	68.14	col yel cr (>−18), hygr		d		v s	d	s al; v s NH_3
239	sulfide, penta-	$(NH_4)_2S_5$	196.40	yel pr		d 115		v s		s al, i eth, CS_2
240	sulfite	$(NH_4)_2SO_3.H_2O$	134.15	col, monocl, 1.515	1.41^{25}	d 60–70	subl 150	32.4^{0}	60.4^{100}	sl s al; i acet
241	sulfite, hydrogen	Ammonium bisulfite. NH_4HSO_3	99.10	rhomb pr, deliq	2.03	subl 150 (in N_2)		71.8^{0}	84.7^{60}	
242	*dl*-tartrate	$(NH_4)_2C_4H_4O_6$	184.15	col, monocl, *d*, α 1.55, β 1.581	1.601	d		58.01^{15}	81.17^{60}	sl s al
243	*dl*-tartrate, hydrogen	$NH_4HC_4H_4O_6$	167.12	col, monocl pr, 1.561, 1.591	1.636	d		2.35^{15}	3.24^{25}	i al, s a, alk
244	tellurate	$(NH_4)_2TeO_4$	227.67	wh powd	$3.024^{24.5}$	d		s	s	i al; s dil a
245	thallium chloride	$3NH_4Cl.TlCl_3.2H_2O$	507.23	col	2.39			s		
246	thioantimonate	$(NH_4)_3SbS_4.4H_2O$	376.18	yel pr		d		71.2^{0}	d	i al, eth
247	thiocarbamate	$NH_4CS_2NH_2$	110.20	yel cr		d 50		v s		s al; sl s eth
248	thiocarbonate, tri-	$(NH_4)_2CS_3$	144.28	yel cr, hygr		subl		v s	d	sl s al, eth
249	thiocyanate	NH_4SCN	76.12	col, monocl, deliq	1.305	149.6	d 170	128^{0}	v s	s al, acet, NH_3
250	*di*thionate	$(NH_4)_2S_2O_6.\frac{1}{2}H_2O$	205.21	col, monocl	1.704	d 130		135^{0}	v s	i al
251	thiosulfate	$(NH_4)_2S_2O_3$	148.20	col, monocl, hygr	1.679	d 150		v s	103.3^{100}	i al; sl s acet
252	titanium oxalate, basic	$(NH_4)_2TiO(C_2O_4)_2.H_2O$	294.03					v s		
253	uranylcarbonate	$2(NH_4)_2CO_3.UO_2CO_3.2H_2O$	558.24	yel, monocl	2.773	d 100		5.8^{19}	d	s$(NH_4)_2CO_3$, aq.SO_2
254	uranylfluoride, penta-	$(NH_4)_3UO_2F_5$	419.14	tetr cr, 1.495	3.186	subl		s		
255	valerate	$NH_4C_5H_9O_2$	119.16	col or wh cr		d		s		s al, eth
256	*meta*vanadate	NH_4VO_3	116.98	wh-yelsh or col cr	2.326	d 200		0.52^{15}	6.95^{96}, d	
257	vanadium sulfate	$NH_4V(SO_4)_2.12H_2O$	477.28	red to bl	1.687	49		28.45^{20}		
258	zinc sulfate	$(NH_4)_2SO_4.ZnSO_4.6H_2O$	401.66	wh monocl, 1.489, 1.493, 1.499	1.931	d		7^{0} (anhydr)	42^{80} (anhydr)	
259	**Antimony**	Sb	121.75	silv wh met, hex	6.684^{25}	630.5	1380	i	i	s hot conc H_2SO_4, aq reg
260	bromide, tri-	$SbBr_3$	361.48	col rhomb, 1.74	4.148^{23}	96.6	280	d	d	s HCl, HBr, CS_2, NH_3, al, acet
261	chloride, penta-	$SbCl_5$	299.02	wh liq or monocl, 1.601^{14}	liq 2.336^{20}_{4}	2.8	79^{22}	d	d	s HCl, tarta, $CHCl_3$
262	chloride, tri-	Butter of antimony. $SbCl_3$	228.11	col, rhomb, deliq	3.140^{25}	73.4	283	601.6^{0}	∞^{80}	s abs al, HCl, tart a, $CHCl_3$, bz, acet
263	fluoride, penta-	SbF_5	216.74	col oily liq	2.99^{23} liq	7	149.5	s		s KF
264	fluoride, tri-	SbF_3	178.75	col, rhomb	$4.379^{20.9}$	292	subl 319	384.7^{0}	5.636^{30}	i NH_3
265	hydride	SbH_3	124.77	inflamm gas	gas 4.36^{15}; liq 2.26^{-25}	−88	$-17.1^{75'}$	0.41^{0}		1500 ml A; 2500 ml CS_2
266	iodide, penta-	SbI_5	756.27	br		79	400.6			
267	iodide, tri-	SbI_3	502.46	ruby-red, hex, 2.78_{Li}, 2.36_{Li}	4.917^{17}	170	401	d	d	i al; s CS_2, bz, HI, HCl
268	iodosulfide	SbSI	280.72	dk red		392	d	i	i	d conc HCl; i CS_2

No.	Name	Synonyms and Formulae	Mol. wt.	Crystalline form, properties and index of refraction	Density or spec. gravity	Melting point, °C	Boiling point, °C	Solubility, in grams per 100 cc		
								Cold water	Hot water	Other solvents
	Antimony									
a269	mercaptoacetamide	Antimony thioglycolamide. $Sb(C_2H_4NOS)_3$	392.12	wh cr		139		200		
a270	nitrate, basic	$2Sb_2O_3.N_2O_5$(?)	691.01	wh glossy cr		d		d		d a; v sl s conc HNO_3
a271	nitride	SbN	135.76	or powd		d		d		
a272	oxide, penta-	Sb_2O_5 (or Sb_4O_{10})	323.50	yel powd	3.80 (dep on temp)	−O, 380 −O_2, 930		v sl s	v sl s	v sl s KOH, HCl, HI
a273	oxide, tetra-	Nat. cervantite. Sb_2O_4 (or $Sb_2O_3.Sb_2O_5$)	307.50	wh powd, 2.00	5.82	−O, 930		v sl s	v sl s	v sl s KOH, HCl, HI
a274	oxide, tri-	Nat. senarmontite. Sb_2O_3 (or Sb_4O_6)	291.50	wh, cub, 2.087	5.2	656	1550 subl	v sl s	sl s	s KOH HCl(3%) 0.03^{20} tart a, ac
a275	oxide, tri-	Nat. valentinite. Sb_2O_3(or Sb_4O_6)	291.50	col, rhomb, 2.18, 2.35, 2.35	5.67	656	1550	v sl s	sl s	s KOH, HCl, tart, a, ac a
a276	III oxychloride(ous)	SbOCl	173.20	wh monocl		d 170		i	d	s acet, HCl, CS_2; i al, eth, $CHCl_3$
a277	III oxychloride(ous)	$Sb_4O_5Cl_2$	637.90	monocl	5.01	d 320		i		s HCl; i al, eth
a278	oxyhydrate	$H_4Sb_2O_7$	359.13	wh, amorph		−H_2O, 200		sl s	sl s	s alk
a279	III oxysulfate, di-(ous)	$Sb_2O_2SO_4$	371.56	wh	4.89			d	d	s, H_2SO_4
a280	potassium tartrate	Tartar emetic. $K(SbO)C_4H_4O_6.\frac{1}{2}H_2O$	333.93	col cr	2.6	−½H_2O, 100		8.3	33.3	i al; 6.7 glyc
a281	selenide	Sb_2Se_3	480.38	gray cr		611		v sl s		s conc HCl
a282	III sulfate	$Sb_2(SO_4)_3$	531.68	wh powd, deliq	3.625^{4}	d		i	d	s a
a283	sulfide, penta-	Sb_2S_5	403.82	yel powd, prism	4.120	d 75		i	i	i al; s HCl, alk, NH_4HS
a284	sulfide, tri-	Nat. stibnite. Sb_2S_3	339.69	blk, rhomb, 3.194, 4.064, 4.303	4.64	550	*ca* 1150	0.000175^{18}		s al, NH_4HS. K_2S, HCl; i ac
a285	sulfide, tri-	Sb_2S_3	339.69	yel-red, amorph	4.12	550	*ca* 1150	0.000175^{18}		s al, NH_4SH, K_2S, HCl; i ac a
a286	*d*-tartrate	$Sb_2(C_4H_4O_6)_3.6H_2O$	795.81	wh cr powd				s		
a287	telluride, tri-	Sb_2Te_3	626.30	gray	6.50^{13}	629				s HNO_3, aq reg
a288	**Argon**	Ar	39.948	col inert gas, cr: 1.65^{-233}, liq: 1.40^{-186}	1.7840^{0}g/l	−189.2	−185.7	5.6^{0} cm³	3.01^{50} cm³	
a289	**Arsenic**	As	74.9216	gray met, hex-rhomb	5.727^{14}	817 atm.	subl 613	i	i	s HNO_3
a290	**Arsenic**	As_4	299.69	yel, cub	2.026^{18}	d 358		i	i	s CS_2, bz
a291	**Arsenic acid**, meta-	$HAsO_3$	123.93	wh, hygr		d	forms ortho-arsenic acid			
a292	ortho-	$H_3AsO_4.\frac{1}{2}H_2O$	150.95	wh translu hygr cr	2.0–2.5	35.5	−H_2O 160	16.7	50	s al, alk, glyc
a293	pyro-	$H_4As_2O_7$	265.87	col pr		forms ortho-arsenic acid 206 d				
	Arsenic									
a294	bromide, tri-	$AsBr_3$	314.65	col-yel hygr pr	3.54^{25}	32.8	221	d	d	s HCl, HBr, CS_2
a295	chloride, tri-	$AsCl_3$	181.28	oily liq or need, 1.621^{14}_{F}	liq 2.163^{20}	−8.5	63^{752}	d	d	s HBr, HCl, PCl_3, al, eth
a296	fluoride, penta-	AsF_5	169.91	col gas	7.71 g/l	−80	−53	s		s alk, al, eth, bz
a297	fluoride, tri-	AsF_3	131.92	oily liq	liq 2.666	−8.5	-63^{752}	d	d	s al, eth, bz, NH_4OH
a298	hydride	Arsine. AsH_3	77.95	col gas	gas 2.695 liq $1.689^{84.9}$	−116.3	−55 (d 300)	20 ml		s $CHCl_3$, bz
a299	iodide, di-	AsI_2	328.73	red pr		d 136		d		s al, eth, $CHCl_3$ bz, CS_2
a300	iodide, penta-	AsI_5	709.44		3.93	76				
a301	iodide, tri-	AsI_3	455.64	red hex, *ca.* 2.59, *ca.* 2.23	4.39^{13}	146	403	6^{25}	30 d	5.2 CS_2; s al, eth bz, $CHCl_3$
a302	oxide, pent-	As_2O_5	229.84	wh amor, deliq	4.32	d 315		150^{16}	76.7^{100}	s al, a, alk
a303	oxide, tri-	As_2O_3	197.84	amor or vitreous	3.738	315		3.7^{20}	10.14^{100}	s alk, alk carb, HCl
a304	oxide, tri-	Nat. arsenolite. As_2O_3 (or As_4O_6)	197.84	col, cub or fibr, 1.755	3.865^{25}	subl 193		1.2^{2}	11.46^{100}	s al, alk, HCl
a305	oxide, tri-	Nat. claudetite. As_2O_3 (or As_4O_6)	197.84	col, monocl, 1.871, 1.92, 2.01	4.15	193, subl 315	457.2	1.2^{2}	11.46^{100}	s al, alk, HCl
a306	(III)oxychloride(ous)	AsOCl	126.37	brnsh		d		d	d	
a307	phosphide, mono-	AsP	105.90	br-red powd		subl, d		d	d	sl s CS_2; s H_2SO_4 HCl; i al, eth, $CHCl_3$
a308	selenide	As_2Se_3	386.72	br cr	4.75	*ca.* 360		i	d	s alk
a309	sulfide, di-	Nat. realgar. As_2S_2	213.97	red-br monocl, 2.46, 2.59, 2.61	α 3.506^{19} β 3.254^{19}	α tr 267 β 307	565	i	i	s K_2S, $NaHCO_3$
a310	sulfide, penta-	As_2S_5	310.16	yel		subl, d 500		0.000136^{0}	i	s alk, alk sulf, HNO_3

No.	Name	Synonyms and Formulae	Mol. wt.	Crystalline form, properties and index of refraction	Density or spec. gravity	Melting point, °C	Boiling point, °C	Solubility, in grams per 100 cc		
								Cold water	Hot water	Other solvents
	Arsenic									
a311	sulfide, tri-	Nat. orpiment. As_2S_3	246.04	yel or red, monocl, 2.4, 2.81, 3.02 (Li)	3.43	300	707	0.00005^{18}	sl s	s al, alk, alk carb
	Auric or Aurous	*See* **Gold**								
b1	**Barium**	Ba	137.34	yel-silv met	3.51^{20}	725	1140	d, ev H_2	d	s al; i bz
b2	acetate	$Ba(C_2H_3O_2)_2$	255.43	col cr	2.468			58.8^{0}	75^{100}	
b3	acetate, hydrate	$Ba(C_2H_3O_2)_2.H_2O$	273.45	col, tricl, 1.500, 1.517, 1.525	2.19	$-H_2O$, 150		76.4^{26}	75^{100}	sl s al
b4	amide	$Ba(NH_2)_2$	169.39	gray-wh cr		280		d	d	i liq NH_3
b5	*ortho*arsenate	$Ba_3(AsO_4)_2$	689.86		5.10	1605		0.055		s a, NH_4Cl
b6	*ortho*arsenate, mono-H	$BaHAsO_4.H_2O$	295.28	col, rhomb or monocl, 1.635	1.93^{15}	$-H_2O$, 150		sl s	d	s HCl
b7	arsenide	Ba_3As_2	561.86	br	4.1^{15}			d		d Cl_2, F_2, Br_2
b8	azide	$Ba(N_3)_2$	221.38	monocl pr	2.936	$-N_2$, 120	expl	17.3^{17}		abs al 0.017^{16}; i eth
b9	azide, hydrate	$Ba(N_3)_2.H_2O$	239.40	tricl, 1.7		expl		v s	v s	sl s al; i eth
b10	benzoate	$Ba(C_7H_5O_2)_2.2H_2O$	415.60	col, nacreous leaf		$-2H_2O$, 100		s		sl s al
b11	boride, hexa-	BaB_6	202.21	met-blk, cub	4.36^{16}	2270		i	i	s HNO_3; i HCl
b12	bromate	$Ba(BrO_3)_2.H_2O$	411.16	col, monocl	3.99^{18}	d 260		0.3^{0}	5.67^{100}	i al; s acet
b13	bromide	$BaBr_2$	297.16	col cr, 1.75	4.781^{24}	847	d	104.1^{20}	149^{100}	v s al, MeOH
b14	bromide, dihydrate	$BaBr_2.2H_2O$	333.19	col, monocl, 1.713, 1.727, 1.744	3.58^{24}	880, ($-H_2O$, 75)	$-2H_2O$, 120	151^{20}	204^{100}	v s MeOH, s al
b15	bromofluoride	$BaBr_2.BaF_2$	472.49	col pl	4.96^{18}			d	d	i al; s conc HCl, conc HNO_3
b16	bromoplatinate	$BaPtBr_6.10H_2O$	992.04	monocl	3.71					
b17	butyrate	$Ba(C_4H_7O_2)_2.2H_2O$	347.57	col				37.42^{0}	42.12^{80}	
b18	carbide	BaC_2	161.36	gray, tetr	3.75			d to C_2H_2		d a
b19	carbonate(α)	$BaCO_3$	197.35	wh, hex	4.43	1740^{90atm}	d	0.002^{20}	0.006^{100}	s a, NH_4Cl; i al
b20	carbonate(β)	$BaCO_3$	197.35			tr to α, 982		0.0022^{18}	0.0065^{100}	s a, NH_4Cl; i al
b21	carbonate(γ)	Nat. witherite. $BaCO_3$	197.35	wh rhomb, 1.529, 1.676, 1.677	4.43	to β, 811	d 1450	0.0022^{18}	0.0065^{100}	s a, NH_4Cl; i al
b22	chlorate	$Ba(ClO_3)_2.H_2O$	322.26	col monocl, 1.5622, 1.577, 1.635	3.18	414 ($-H_2O$, 120)	$-O$, 250	27.4^{15}	111.2^{100}	sl s al, acet, HCl
b23	*per*chlorate	$Ba(ClO_4)_2$	336.24	col, hex	3.2	505		198.5^{25}	562.3^{100}	v s al
b24	*per*chlorate, hydrate	$Ba(ClO_4)_2.3H_2O$	390.29	col, hex, 1.533	2.74	d 400		198^{33}	s	al 124^{62}
b25	chloride α	$BaCl_2$	208.25	col, monocl; 1.7303, 1.7367, 1.7420	3.856^{24}	tr to cub 925	1560	37.5^{26}	59^{100}	sl s HCl, HNO_3; v sl s al
b26	chloride β	$BaCl_2$	208.25	col, cub	3.917	963	1560			v sl s al
b27	chloride, hydrate	$BaCl_2.2H_2O$	244.28	col, monocl, 1.629, 1.642, 1.658 n_D^{25}	3.097^{24}_{4}	$-2H_2O$, 113	35.7^{20}	58.7^{100}		sl s HCl, HNO_3; v sl s al
b28	*hypo*chlorite	$Ba(ClO)_2.2H_2O$	276.28	col cr		d				
b29	chlorofluoride	$BaCl_2.BaF_2$	383.58	col, tetr, 1.640	4.51^{18}			d	d	i al; s conc HCl, HNO_3
b30	chloroplatinate	$BaPtCl_6.6H_2O$	653.24	or-yel, monocl	2.868	$-5H_2O$, 70	d	s		d a, al; i MeOH, eth
b31	chloroplatinite	$BaPtCl_4.3H_2O$	528.35	dk red pr	2.868	$-3H_2O$, 150		v s		s al
b32	chromate	$BaCrO_4$	253.33	yel, rhomb	4.498^{15}			0.00034^{16}	0.00044^{28}	s min a
b33	*di*chromate	$BaCr_2O_7$	353.33	red, monocl				sl s		s h conc H_2SO_4
b34	*di*chromate, hydrate	$BaCr_2O_7.2H_2O$	389.36	bright red-yel need		$-2H_2O$, 120		d		s conc CrO_3 soln
b35	chromite	$BaO.4Cr_2O_3$	761.30	grn-blk, hex	5.4^{15}			i	i	s a, fus carb
b26	citrate	$Ba_3(C_6H_5O_7)_2.7H_2O$	916.33	wh powd		$-7H_2O$, 150		0.0406^{18}	0.0572^{25}	sl s al; s HCl
b37	cyanide	$Ba(CN)_2$	189.38	wh cr powd				80^{14}		18^{14} 70 % al
b38	cyanoplatinite	$BaPt(CN)_4.4H_2O$	508.56	(a) monocl, yel, α 1.6704 (b) grn, rhomb	a) 2.076 b) 2.085	$-2H_2O$, 100		3^{16}	25^{100}	i al
b39	dithionate	$Ba(SO_3)_2.2H_2O$	333.50	col, rhomb or monocl, 1.586, 1.595, 1.607	$4.536^{13.5}$	d 120		24.75^{18}	90.9^{100}	sl s al
b40	ethylsulfate	$Ba(C_2H_5SO_4)_2.2H_2O$	423.62	wh lust leaf				s		sl s al
b41	ferrocyanide	$Ba_2Fe(CN)_6.6H_2O$	594.73	yel, monocl	2.666	$-H_2O$, 40		0.17^{15}	0.9^{100}	i al
b42	fluogallate	$Ba_3(GaF_6)_2.H_2O$	797.46	wh cr	4.06	$-\frac{1}{2}H_2O$, 110; $-\frac{1}{2}H_2O$, 230		i		s HF
b43	fluoride	BaF_2	175.34	col, cub, 1.4741 n_D^{25}	4.89	1280	2137	0.12^{25}	sl s	s a, NH_4Cl
b44	fluoroiodide	$BaF_2.BaI_2$	566.49	pl	5.21^{18}			d	d	i al; s conc HCl, conc NHO_3

No.	Name	Synonyms and Formulae	Mol. wt.	Crystalline form, properties and index of refraction	Density or spec. gravity	Melting point, °C	Boiling point, °C	Solubility, in grams per 100 cc: Cold water	Hot water	Other solvents
	Barium									
b45	fluosilicate	$BaSiF_6$	279.42	rhomb need	4.29_4^{21}	d 300		0.026^{17}	0.09^{100}	i al; sl s a, NH_4Cl
b46	formate	$Ba(CHO_2)_2$	227.38	col, rhomb, 1.573, 1.597, 1.636	3.21	d		27.76^{0}	39.71^{80}	i al, eth
b47	gluconate	$Ba(C_6H_{11}O_7)_2.3H_2O$	581.69	pr or rhomb leaf		$-3H_2O$, 100; d 120		$3.3^{15.5}$		i al
b48	hydride	BaH_2	139.36	gray cr	4.21^{0}	d 675	1400(?)	d to $Ba(OH)_2$ $+H_2$		d a
b49	hydroxide	$Ba(OH)_2.8H_2O$	315.48	col, monocl, 1.471, 1.502, 1.50	2.18^{16}	78	$-8H_2O$, 780	5.6^{15}	94.7^{78}	sl s al; i acet
b50	hyponitrite	$BaN_2O_2.4H_2O$	269.41	wh cr powd	2.742^{25}					
b51	iodate	$Ba(IO_3)_2$	487.15	monocl	4.998	d		0.008^{0}	197^{100}	s HNO_3, HCl
b52	iodate, hydrate	$Ba(IO_3)_2.H_2O$	505.17	col, monocl	4.657^{18}	$-H_2O$, 200		v sl s	sl s	s HNO_3, HCl; i al, acet, H_2SO_4
b53	iodide	BaI_2	391.15	col cr	5.15_4^{25}	740		170^{0}		al 77^{20}
b54	iodide, hydrate	$BaI_2.2H_2O$	427.18	col rhomb, deliq	5.15	$-H_2O$, 98.9; $-2H_2O$, 539; d 740	200^{15}	269^{100}		1.07^{15} al; s acet
b55	iodide, hydrate	$BaI_2.6H_2O$	499.24	col, hex		25.7		410^{0}	v s	v s al
b56	laurate	$Ba(C_{12}H_{23}O_2)_2$	535.97	wh leaf cr		260		$0.008^{15.3}$	0.011^{50}	0.008^{25} al; 0.006^{25} eth
b57	*l*-malate	$BaC_4H_4O_5$	269.41					0.883^{20}	1.044^{80}	
b58	malonate	$BaC_3H_2O_4.H_2O$	257.40	col				0.143^{0}	0.326^{80}	
b59	manganate	$BaMnO_4$	256.28	gray-grn, hex	4.85			v sl s		s a
b60	*per*-manganate	$Ba(MnO_4)_2$	375.21	br-vlt cr	3.77	d 200		62.5^{11}	75.4^{25}	d al
b61	methylsulfate	$Ba(CH_3SO_4)_2.2H_2O$	395.56	col effl cr				s		s al
b62	molybdate	$BaMoO_4$	297.28	wh powd	4.65	1480		0.0058^{23}		sl s a
b63	myristate	$Ba(C_{14}H_{27}O_2)_2$	592.08					0.007^{25}	0.010^{50}	0.009^{25} al; 0.003^{25} eth 0.046^{15} MeOH
b64	nitrate	Nitrobarite. $Ba(NO_3)_2$	261.35	col cub, 1.572	3.24^{23}	592	d	8.7^{20}	34.2^{100}	i al; sl s a
b65	nitride	Ba_3N_2	440.03	yel-br	4.783_4^{25}		1000 vac	d	d	
b66	nitrite	$Ba(NO_2)_2$	229.35	col, hex	3.23^{23}	d 217		67.5^{20}	300^{100}	sl s al
b67	nitrite, hydrate	$Ba(NO_2)_2.H_2O$	247.37	col-yelsh, hex	3.173^{20}	d 115		63^{20}	109.6^{80}	1.6 al; v s HCl; i acet
b68	oxalate	BaC_2O_4	225.36	cr	2.658	d 400		0.0093^{18}	0.0228^{100}	i al; s NH_4Cl, a
b69	oxide	BaO	153.34	col, cub, wh-yelsh powd, 1.98	5.72	1923	*ca* 2000	3.48^{20}	90.8^{100}	s dil a, al; i acet, NH_3
b70	oxide, per-	BaO_2	169.34	wh-gray powd	4.96	450	$-O$, 800	v sl s	d	s dil a; i acet
b71	oxide, per-, hydrate	BaO_28H_2O	313.46	col, hex	2.292	$-8H_2O$, 100		0.168	d	s dil a; i al, eth, acet
b72	palmitate	$Ba(C_{16}H_{31}O_2)_2$	648.19	wh cr powd		d		0.004^{15}	0.007^{50}	$0.008^{16.5}$ al; 0.001^{15} eth
b73	*hypo*phosphate	$BaPO_3$	216.31	need				sl s		s al; v sl s ac a
b74	*ortho*phosphate di-	$BaHPO_4$	233.32	wh, rhomb, 1.635, 1.617	4.165^{15}	d 410^{710}		0.01–0.02		s a, NH_4Cl
b75	*ortho*phosphate, mono-	$Ba(H_2PO_4)_2$	331.31	tricl	2.9^{4}			d	d	s a
b76	*ortho*phosphate, tri-	$Ba_3(PO_4)_2$	601.96	wh, cub	4.1^{16}			i	i	s a
b77	*pyro*phosphate	$Ba_2P_4O_7$	448.62	wh, rhomb	3.9^{20}			0.01	sl s	s a, NH_4 salts
b78	*hypo*phosphite	$Ba(H_2PO_2)_2.H_2O$	285.33	wh, monocl	2.90_4^{17}	d 100−150		30^{15}	33^{100}	i al
b79	propionate	$Ba(C_3H_5O_2)_2.H_2O$	301.50	rhomb, β 1.518		d 300		48^{0}	67.9^{80}	0.08 al
b80	salycylate	$Ba(C_7H_5O_3)_2.H_2O$	437.65	wh need				s		
b81	selenate	$BaSeO_4$	280.30	wh, rhomb	4.75	d		0.0118	0.138^{100}	s HCl; i HNO_3
b82	selenide	BaSe	216.30	wh cub disc, n_D, 2.268	5.02			d	d	d HCl
b83	*meta*silicate	$BaSiO_3$	213.42	col, rhomb, 1.673, 1.674, 1.678	4.399	1604		i	d	s HCl
b84	*meta*silicate, hydrate	$BaSiO_3.6H_2O$	351.52	rhomb, 1.542, 1.548, 1.548	2.59			0.17	d	
b85	stearate	$Ba(C_{18}H_{35}O_2)_2$	704.13	wh powd				0.004^{15}	0.006^{50}	$0.005^{16.5}$ al, 0.008^{25} al, 0.001^{25} eth
b86	succinate	$BaC_4H_4O_4$	253.37	wh powd				0.421^{0}	0.237^{80}	sl s al
b87	sulfate	Nat. barite, prec. blanc fixe. $BaSO_4$	233.40	wh, rhomb (monocl), 1.637, 1.638, 1.649	4.50^{15}	1580	tr 1149 monocl	0.000222^{18} 0.000246^{25}	0.000336^{50} 0.000413^{100}	0.006 s 3 % HCl; sl s H_2SO_4
b88	*peroxydi*sulfate	$BaS_2O_8.4H_2O$	401.52	wh, monocl		d		52.2^{0}	d	d al
b89	sulfide, hydro-	$Ba(HS)_2.4H_2O$	275.56	yel, rhomb		d 50		s		i al
b90	sulfide, mono-	BaS	169.40	col, cub, n_D 2.155	4.25^{15}	1200		d	d	i al
b91	sulfide, tetra-	$BaS_4.H_2O$	283.61	red or yel, rhomb	2.988	d 300		41^{15}	v s	i al, CS_2
b92	sulfide, tri-	BaS_3	233.53	yel-grn cr		d 554		s	s	

No.	Name	Synonyms and Formulae	Mol. wt.	Crystalline form, properties and index of refraction	Density or spec. gravity	Melting point, °C	Boiling point, °C	Solubility, in grams per 100 cc: Cold water	Hot water	Other solvents
	Barium									
b93	sulfite	$BaSO_3$	217.40	col, cub hex		d		0.02^{20}	0.002^{80}	v s HCl
b94	tartrate	$BaC_4H_4O_6.H_2O$	303.52	wh cr	$2.980^{20.8}$			0.026^{18}	0.058^{80}	0.032^{18} al
b95	tellurate	$BaTeO_4.3H_2O$	383.07	volum wh	4.2^{200}	d >200		sl s	sl s	s HCl, NHO_3
b96	*pyro*tellurate, hydrogen	$Ba(HTe_2O_7)_2.H_2O$	891.76	volum ppt; hot: yel; cold: wh				s	s	s a
b97	telluride	BaTe	264.94	yel-wh, cub, disc, n_D 2.440	5.13					d a
b98	thiocarbonate	$BaCS_3$	245.54	yel, hex		d		1.08^0	d	i al
b99	thiocyanate	$Ba(SCN)_2.2H_2O$	289.53	need, deliq	2.286^{18}	$-H_2O$, 160		4.3^{20}	s	35^{20} al
b100	thiosulfate	BaS_2O_3	249.47	wh, rhomb		d 220		0.2		
b101	thiosulfate, hydrate	$BaS_2O_3.H_2O$	267.48	wh, rhomb	3.5^{18}	d 100		0.208^{20}		i al
b102	titanate	$BaTiO_3$	233.24	tetr and hex, 2.40	tetr 6.017, hex 5.806					
b103	tungstate	$BaWO_4$	385.19	col, tetr	5.04			sl s	sl s	d a
b104	*pyro*vanadate	$Ba_2V_2O_7$	488.56	wh cr		863				
b105	**Beryllium**	Glucinum. Be	9.0122	grey met, hex	1.85^{20}	1278 ± 5	2970^{760}	i	sl s, d	s dil a, alk; i Hg
b106	acetate	$Be(C_2H_3O_2)_2$	127.10	col pl		d 300		i		i al, eth, CCl_4
b107	acetate, basic	$BeO(C_2H_3O_2)_6$	406.32	oct	1.36^4	284	331	sl d	d	s chl, ac a; sl s al, eth
b108	acetate propionate, basic	$Be_4O(C_2H_3O_2)_3.(C_3H_5O_2)_3$	448.40			127	330			
b109	aluminate	Nat. chrysoberyl. $BeAl_2O_4$	126.97	rhomb, 1.747, 1.748, 1.757	3.76	1870				i a
b110	aluminum silicate	Nat. beryl. $Be_3Al_2(SiO_3)_6$	537.51	transp, hex, col, 1.580, 1.547	2.66	1410 ± 100				i a
b111	aluminum silicate	Nat. euclase. $Be_2Al_2(SiO_4)_2.(OH)_2$	290.17	monocl, 1.652, 1.655, 1.671	3.1					
b112	benzenesulfonate	$Be(C_6H_5O_3S)_2$	323.35	monocl				v s	v s	v s al, acet, ac a; i eth, bz, CS_2, CCl_4
b113	*ortho*borate, basic	Nat. hambergite. $Be_2(OH)BO_3$	93.84	rhomb, 1.560, 1.591, 1.631	2.35					
b114	bromide	$BeBr_2$	168.83	wh need, deliq	3.465^{25}	490 ± 10 subl	520	s	v s	s al, eth; i bz; 18.56 pyr
b115	butyrate, basic	$Be_4O(C_4H_7O_2)_6$	574.64				239^{19}			
b116	carbide	Be_2C	30.04	yel, hex	1.90^{15}	>2100 d		d	d	s a; d alk
b117	carbonate, basic	$BeCO_3+Be(OH)_2$	112.05	wh powd				i	d	s a, alk
b118	chloride	$BeCl_2$	79.92	col need, deliq	1.899^{25}	405	520 (488)	v s	v s, d	v s al, eth, pyr; sl s bz, chl, CS_2; i NH_3, acet
b119	fluoride	BeF_2	47.01	col, amorph, <1.33	1.986^{25}	subl 800		∞	∞	sl s al; s H_2SO_4
b120	hydride	BeH_2	11.03	wh cr		d 125		d	d	i eth, tol
b121	iodide	BeI_2	262.82	col need	4.325^{25}	510 ± 10	590	d	d	s al, eth, CS_2
b122	nitrate	$Be(NO_2)_2.3H_2O$	187.07	wh-yel cr, deliq	1.557	60	142	v s	v s	v s al
b123	nitride	Be_3N_2	55.05	col, cub	2200 ± 100	d 2240		d	d	d a, conc alk; i al
b124	oxalate	$BeC_2O_4.3H_2O$	151.08	rhomb, β 1.487		$-H_2O$, 100; $-3H_2O$, 220	d 350	38.22^{25}		
b125	oxide	Nat. bromellite. BeO	25.01	wh hex, 1.719, 1.733	3.01	2530 ± 30	*ca* 3900	0.00002^{30}		s conc H_2SO_4, fus KOH
b126	oxide	$BeO.xH_2O$		wh amorph powd or gel		d		i	i	s a, alk, $(NH_4)_2CO_3$
b127	2,4-pentanedione deriv	Beryllium acetylacetonate $Be(C_5H_7O_2)_2$	207.23	wh, monocl	1.168^4	108	270	sl s	d	s al, eth, a
b128	*ortho*phosphate	$Be_3(PO_4)_2.3H_2O$	271.03			$-H_2O$, 100		s	s	s ac a
b129	propionate basic	$Be_4O(C_3H_5O_2)_6$	490.43			120				
b130	selenate	$BeSeO_4.4H_2O$	224.03	col, rhomb, 1.466, 1.501, 1.503	2.03	$-2H_2O$, 100; $-4H_2O$, 300		56.7^{25}	s	
b131	*di*-silicate	Nat. bertrandite. $Be_4Si_2O_7(OH)_2$	238.23	rhomb, 1.591, 1.605, 1.604	2.6					
b132	*ortho*silicate	Nat. phenazite. Be_2SiO_4	110.11	tricl, col, 1.654, 1.670	3.0					i a
b133	stearate (com'l)	$Be(C_{18}H_{35}O_2)_2$	575.97	wh, waxy		45		i	i	i al; s eth, CCl_4
b134	sulfate	$BeSO_4$	105.07		2.443	d 550–600		i	d to $BeSO_4$ $4H_2O$	
b135	sulfate, hydrate	$BeSO_4.4H_2O$	177.14	col, tetr, 1.472, 1.440	$1.713^{10.5}$	$-2H_2O$, 100	$-4H_2O$, 400	42.5^{25}	100^{100}	sl s conc H_2SO_4; i al
b136	sulfide	BeS	41.08	reg	2.36			d	d	
b137	**Bismuth**	Bi	208.98	rhomb silver-wh or redsh met	9.80	271.3	1560 ± 5^{760}	i	i	s h H_2SO_4, HNO, aq, reg; sl s h HCl
b138	acetate	$Bi(C_2H_3O_2)_3$	386.12	wh cr		d		i	i	s ac a

No.	Name	Synonyms and Formulae	Mol. wt.	Crystalline form, properties and index of refraction	Density or spec. gravity	Melting point, °C	Boiling point, °C	Solubility, in grams per 100 cc		
								Cold water	Hot water	Other solvents
	Bismuth									
b139	*ortho*arsenate	$BiAsO_4$	347.90	wh, monocl, 2.14, 2.15, 2.18	7.14			i	i	sl s h conc HNO_3
b140	benzoate	$Bi(C_7H_5O_2)_3$	572.33	wh powd						s a; i eth
b141	bromide, tri-	$BiBr_3$	448.71	yel cr powd, deliq	5.72^{25}_{4}	218	453	d to BiOBr	d	i al; s HCl, HBr, eth; v s liq NH_3
b142	carbonate, basic	Bismuth oxycarbonate. $Bi_2O_2CO_3$	509.97	wh powd	6.86		d	i	i	s a
b143	chloride, tetra-	$BiCl_4$	350.79	col cr		226				
b144	chloride, tri-	$BiCl_3$	315.34	wh cr, deliq	4.75^{25}_{4}	230–232	447	d to BiOCl	d	s a, al, eth, acet
b145	*di*chromate, basic	$(BiO)_2.Cr_2O_7$	665.94	yel-or-red cr				i	i	s a; i alk
b146	citrate	$BiC_6H_5O_7$	398.08	wh cr	3.458	d		sl s	sl s	sl s al; s NH_4OH
b147	fluoride, tri-	BiF_3	265.98	gray cr, cub, 1.74	5.32^{20}	727				s inorg a; i liq NH_3, al
b148	gallate, basic	Com'l dermatol. $Bi(OH_2)C_7H_5O_5$ (appr)	412.11	yel, amorph		d		i		i al, eth
b149	hydroxide	$Bi(OH)_3$ (appr)	260.00	wh amorph powd.	4.36	$-H_2O$, 100; d 415	$-1\frac{1}{2}H_2O$, 400	0.00014	d	s a; i or sl s conc alk
b150	iodate	$Bi(IO_3)_3$	733.69	wh				i		i HNO_3
b151	iodide, di-	BiI_2	462.79	red need, rhomb		d 400	subl vac	s		s al, MeOH
b152	iodide, tri-	BiI_3	589.69	redsh, hex	5.778^{15}	408	*ca* 500	i	d	3.5^{20} al; i a; sl s NH_3; s HCl, HI
b153	lactate, *dl*	$Bi(C_3H_5O_3).7H_2O$	512.22	pr, need				14.4^{25}		i al
b154	molybdate	$Bi(MoO_4)_3$	897.78	yel-wh tetr need	6.07	643				v s a
b155	nitrate	$Bi(NO_3)_3.5H_2O$	485.07	col tricl, sl hygr	2.83	d 30	$-5H_2O$, 80	d	d	v s HNO_3; s a, glyc; i al, 42^{19} acet
b156	nitrate, basic	$BiONO_3.H_2O$	305.00	hex leaf	4.928^{18}	$-H_2O$, 105	$-HNO_3$, 260	i	i	s a; i al
b157	oxalate	$Bi_2(C_2O_4)_3.7H_2O$	808.73	wh powd		$-6H_2O$, *ca* 130		d		s inorg a; i al, eth
b158	oxide, mono-	BiO	224.97	dk-gray powd	7.15^{19}	d *ca* 180 (Bi_2O_3)		sl d	d	d dil a; s dil KOH
b159	oxide, pent-	Bi_2O_5	497.96	dk red or br	5.10	−O, 150	−2O, 357	i	i	s KOH
b160	oxide, tetr-	$Bi_2O_4.2H_2O$	517.99	br powd	5.6	$-H_2O$, 110	$-2H_2O$, 180	i	i	s a
b161	oxide, tri-	Bi_2O_3	495.96	yel, rhomb	8.9	820	1890 (?)	i	i	s a
b162	oxide, tri-	Bi_2O_3	495.96	gray-blk, cub	8.20	tr 704		i	i	s a
b163	oxide, tri-	Bi_2O_3	495.96	wh-lt yel, rhomb, 1.91	8.55	860		i	i	sl s a
b164	oxybromide	BiOBr	304.89	col cr or wh powd	8.082^{15}	d red ht		i	i	i al; s a
b165	oxychloride	BiOCl	260.43	wh cr or powd, 2.15	7.72^{15}	red ht.		i	i	s a; i NH_3, tart a, acet
b166	oxyfluoride	BiOF	243.98	wh cr or powd	7.5^{20}_{20}	d red ht		i		s a
b167	oxyiodide	BiOI	351.88	red cr, tetr	7.922	d red ht		i	i	s a; i al, $CHCl_3$
b168	*ortho*phosphate	$BiPO_4$	303.95	wh, monocl	6.323^{15}	d		i	i	s HCl; i al, dil HNO_3
b169	propionate, basic	$BiOC_3H_5O_2$	297.97	wh powd				i		v s dil HCl; i al
b170	salicylate, basic	Bismuth subsalicylate. $Bi(C_7H_5O_3)_3.Bi_2O_3$ (appr)	1086.29	wh micr cr (variable comp)				i		s a, alk; i al, eth
b171	selenide, tri-	Nat. guanajuatite. Bi_2Se_3	654.84	blk, rhomb	6.82	710	d			i alk
b172	silicate	Nat. eulytite. $2Bi_2O_3.3SiO_2$	1112.17	yel, cub, 2.05	6.11			i	i	s, d HCl, HNO_3
b173	sulfate	$Bi_2(SO_4)_3$	706.14	wh need	5.08^{15}	d 405		d	d	s a
b174	mono-sulfide,	BiS	241.04	dk gray powd	7.6–7.8	680 (in CO_2)	d	v sl s		
b175	sulfide, tri-	Nat. bismuthinite, bismuthglance. Bi_2S_3	514.15	br-blk, rhomb, 1.340, 1.456, 1.459	7.39	d 685		0.000018^{18}		s HNO_3; i dil al
b176	tartrate	$Bi_2(C_4H_4O_6)_3.6H_2O$	970.27	wh powd	2.595^{25}_{5}	$-3H_2O$, 105		i	i	s a, alk; i al
b177	tellurate	Montanite. $Bi_2TeO_6.2H_2O$	677.59	biaxial, β: 2.09	3.79					
b178	telluride, tri-	Nat. tetradymite. Bi_2Te_3	800.76	gray, rhbdr	7.7^{20}_{4}	573				d HNO_3
b179	vandate	Nat. pucherite. $Bi_2O_3.V_2O_5$	647.84	red-grn, rhomb, 2.41, 2.50, 2.51 (Li)	6.25^{25}					
b180	**Bismuthic acid**	$HBiO_3$	257.99	red	5.75	$-H_2O$, 120	−2O, 357	i	i	s a, KOH
b181	**Borazole**	$B_3N_3H_6$	80.50	col liq	1^{-65}, 0.8614^{05}	−58	55	sl d	d	
b182	**Boric Acid,** meta-	HBO_2	43.82	wh cr, cub, 1.619	2.486	236±1		v sl s	sl s	

No.	Name	Synonyms and Formulae	Mol. wt.	Crystalline form, properties and index of refraction	Density or spec. gravity	Melting point, °C	Boiling point, °C	Solubility, in grams per 100 cc		
								Cold water	Hot water	Other solvents
	Boric acid									
b183	ortho-	Boracic acid. H_3BO_3	61.83	col, tricl, 1.337, 1.461, 1.462	1.435^{15}	169±1 tr to HBO_2	$-1\frac{1}{2}H_2O$, 300	6.35^{30}	27.6^{100}	28^{20} glyc; 0.0078 eth; 5.56 al; 20. 20^{25} MeOH; 1.92^{25} liq NH_3; sl s acet
b184	tetra- (pyro-)	$H_2B_4O_7$	157.26	vitr or wh powd						
b185	fluo-	HBF_4	87.81	col liq			d 130	s ∞	s ∞	s al
b186	**Borinoaminoborine**	B_2H_7N	42.68	col liq		−66.5	76.2			s triborine triamine
b187	**Boron**	B	10.811	yel monocl or br amorph powd	2.34, 2.37 amorph	2300	2550 (sub)	i	i	v sl s HNO_3
b188	arsenate	$BAsO_4$	149.73	wh cr tetrag, 1.681, 1.690	3.64	subl *ca* 700		v sl s	1.4^{100}	i al; s inorg a
b189	bromide, tri-	BBr_3	250.54	col fum liq, $n_D^{16.3}$, 1.5312	$2.6431_4^{18.4}$	−46	91.3±0.25	d		s al, CCl_4
b190	bromide, di-, iodide	BBr_2I	297.53	col liq			125	d	d	
b191	bromide, mono-, diiodide	$BBrI_2$	344.53	col liq			180	d	d	
b192	(di-) bromide, mono-, pentahydride	B_2H_5Br	106.67	col gas		−104	*ca* 10	hyd to HBO_2, HBr+H_2		
b193	(tetra-) carbide	B_4C	55.26	blk rhbdr	2.52	2350	>3500	i	i	i a; s fus alk
b194	chloride, tri-	BCl_3	117.17	col fum liq, $1.4195^{5.7}$ α line H_2	1.349_4^{11}	−107.3	12.5		d to HCl +H_3BO_3	d al
b195	fluoride, tri-	BF_3	67.81	col gas	2.99 g/1	−126.7	−99.9	106 (762 mm)	d	d al; s conc H_2SO_4
b196	fluoride dihydrate	$BF_3.2H_2O$	103.84	col liq; n_{HE}^{20} 1.31498	1.6316_4^{20}	6		d	d	s eth, dioxan
b197	hydride	Diborane, boroethane. B_2H_6	27.67	col gas	liq: 0.447^{-11} sol: 0.577^{-183}	−165.5	−92.5	sl s d to H_3BO_3 +H_2		d 1.6^0 al; s NH_4OH, conc H_2SO_4
b198	hydride	Dihydrotetraborane, borobutane. B_4H_{10}	53.32	col gas, pois	0.56^{-35}	−120.8	16	sl s, d		s bz; d al
199	hydride	Pentaborane. B_5H_9	63.13	col liq	0.66^0	−46.82	58.4	d		
200	hydride	Hexaborane. B_6H_{10}	74.95	col liq	0.69^0	−65	0^{72}	d		
201	hydride	Decaborane. $B_{10}H_{14}$	122.22	wh cr	0.94^{25}	99.5	213	sl s	d	v s CS_2; s al, eth bz
202	iodide, tri-	BI_3	391.52	col pl, hygr	3.35^{50}	49.9	210	d	d	d al; v s CS_2, bz, CCl_4
203	nitride	BN	24.82	wh, hex	2.25	subl *ca* 3000		i	sl d	sl s h a
204	oxide	B_2O_3	69.62	rhomb cr, 1,64, 1.61	2.46±0.01	460	*ca* 1860	sl s	s	
205	oxide glass	B_2O_3	69.62	col, vitr 1.485	1.812_4^{25}	*ca* 450		1.1^0	15.7^{100}	s al, a
206	phosphide	BP	41.78	maroon powd		ign 200		i	i	i all solv
207	*tri*selenide	B_2Se_3	258.50	yel-gray powd				d	d	
208	(hexa-) silicide	B_6Si	92.95	blk cr	2.47			i		s HNO_3; d H_2SO_4; i KOH
209	(tri-) silicide	B_3Si	60.52	blk rhomb	2.52			i		sl s HNO_3; d H_2SO_4, KOH
210	sulfide, penta-	B_2S_5	181.94	col, tetrag	1.85	390		d	d	d al
211	sulfide, tri-	B_2S_3	117.81	wh cr or vitr	1.55	310		d		sl s PCl_3, SCl_2; d al
212	**Borotungstic acid**	$H_5BW_{12}O_{40}.30H_2O$	3402.49	tetr, cr	3	45–51		s		s al, eth
213	**Bromic acid**	$HBrO_3$	128.92	known in sol only, col or yelsh		d 100		v s	d	
214	**Bromine**	Br_2	159.81	dk red liq, 1.661	2.928^{59}	−7.2	58.78	4.17^0, 3.58^{20}	3.52^{50}	v s al, eth, chl, CS_2
215	azide	Bromoazide. BrN_3	121.93	cr, red liq		*ca* 45	exp			s eth, KI; sl s bz, ligr
216	chloride	BrCl	115.36	red-col liq or gas		*ca* −66 d 10	*ca* 5	s d		s eth, CS_2
217	fluoride, mono-	BrF	98.91	red-br gas		d −33	−20			
218	fluoride, penta-	BrF_5	174.90	col liq	2.466^{25}	−61.3	40.5	d	d	
219	fluoride, tri-	BrF_3	136.90	col-gray-yel liq	2.49^{135}	(−2) 8.8	135	d viol. to O_2, HOBr, HF, $HBrO_3$		d alk
220	hydrate	$Br_2.10H_2O$	339.97	red oct	1.49	d 6.8		s		
221	oxide, di-	BrO_2	111.91	lt yel		d 0				
222	oxide, mon-	Br_2O	175.82	dk br		−17 to −18				s, d CCl_4
223	(tri-) oxide, oct-	Br_3O_8 (or Br_2O_8)$_n$	367.72	wh		stable at −40				

No.	Name	Synonyms and Formulae	Mol. wt.	Crystalline form, properties and index of refraction	Density or spec. gravity	Melting point, °C	Boiling point, °C	Solubility, in grams per 100 cc		
								Cold water	Hot water	Other solvents
	Bromauric acid									
b224	**Bromauric acid**	$HAuBr_4.5H_2O$	607.69	red-br cr		27		v s		s al
b225	**Bromous acid, hypo-**	$HBrO$	96.92	known only in soln, col-yel		40 (vac)		s	s d	s al, eth, chl
b226	**Bromoplatinic acid**	$H_2PtBr_6.9H_2O$	838.70	red monocl, deliq		d −100		v s	v s	
c1	**Cadmium**	Cd	112.40	hex silv-wh malleable met	8.642	320.9	765	i	i	s a, NH_4NO_3, h H_2SO_4
c2	acetate	$Cd(C_2H_3O_2)_2$	230.50	col	2.341	256	d	v s		s MeOH
c3	acetate, hydrate	$Cd(C_2H_3O_2)_2.2H_2O$	266.54	col monocl, odor ac a	2.01	$-H_2O$, 130		v s	v s	v s al
c4	amide	$Cd(NH_2)_2$	144.45		3.05^{25}	d 120				
c5	ammonium chloride	$CdCl_2.NH_4Cl$	236.79	col need, rhomb	2.93	289		33.45^{16}	$43.99^{63.8}$	s al, MeOH
c6	ammonium sulfate	$Cd(NH_4)_2(SO_4)_2.6H_2O$	448.69	col monocl pr	2.061^{20}_{4}	100 d $-H_2O$		s	s	
c7	arsenate, hydrogen	$CdHAsO_4.H_2O$	270.34		4.164^{15}_{4}	>120				
c8	arsenide	Cd_3As_2	487.04	dk gray cub	6.21^{15}_{4}	721		i	i	sl s HCl; s HNO_3; i aq reg
c9	benzoate	$Cd(C_7H_5O_2)_2.2H_2O$	390.66					3.34^{20}		sl s al
c10	borate	$Cd(BO_2)_2.H_2O$	248.03	wh rhomb	3.758	d		125^{17}		i al
c11	borotungstate	$Cd_5(BW_{12}O_{40}).18H_2O$	6600.25	yel tricl		75		1250^{19}	v s	
c12	bromide	$CdBr_2$	272.22	yel cr	5.192^{25}	567	863	57^{10}	162^{104}	26.6^{15} al; 0.4^{15} eth; s HCl; 1.6^{18} acet
c13	bromide, tetra-hydrate	$CdBr_2.4H_2O$	344.28	sm wh need, effl		tr 36		121^{10}	v s	25 al; s acet; sl s eth
c14	carbonate	$CdCO_3$	172.41	wh, trig	4.258^{4}	d <500		i	i	s a, KCN, NH_4 salts; i NH_3
c15	chlorate	$Cd(ClO_3)_2.2H_2O$	315.33	col pr, deliq	2.28^{18}	80		298^{0}	487^{65}	s al, a, acet
c16	chloride	$CdCl_2$	183.32	col, hex	4.047^{25}	568	960	140^{20}	150^{100}	1.52^{15} al; $1.7^{15.5}$ MeOH; i acet, eth
c17	chloride	$CdCl_2.2\frac{1}{2}H_2O$	228.35	col monocl, 1.6513	3.327	tr 34		168^{20}	180^{100}	2.05^{15} MeOH; sl s al
c18	chloroacetate, di-	$Cd(C_2HCl_2O_2).2H_2O$	386.29	need	2.132^{15}					
c19	chloroacetate mono-	$Cd(C_2H_2ClO_2)_2.6H_2O$	407.47		1.942^{25}					
c20	chloroacetate, tri-	$Cd(C_2Cl_3O_2)_2.1\frac{1}{2}H_2O$	464.18	rhomb	2.093^{25}					
c21	chloroplatinate	$CdPtCl_6.3H_2O$	574.25	yel trig need	2.882	$-H_2O$, 170, d		s	s	
c23	chromite	$CdCr_2O_4$	280.39	grn to blk, cub	5.79^{17}			i	i	i a
c24	cyanide	$Cd(CN)_2$	164.44	cr	dec >200			1.7^{15}	s	s a, KCN, NH_4OH; i al
c25	ferrocyanide	$Cd_2Fe(CN)_6.xH_2O$						i	i	s HCl
c26	fluogallate	$[Cd(H_2O)_6].[GaF_5H_2O]$	403.22	col cr, 1.45	2.79	$-5H_2O$, 110		v s		
c27	fluoride	CdF_2	150.40	wh cub, 1.56	6.64	1100	1758	4.35^{25}		s a, HF; i al, NH_3
c28	fluosilicate	$CdSiF_6.6H_2O$	362.57	col hex				s	s	s 50% al
c29	formate	$Cd(CHO_2)_2.2H_2O$	238.47	monocl	2.44	d		v s		
c30	fumarate	$Cd(C_4H_2O_4)$	226.48					0.9^{30}		
c31	hydroxide	$Cd(OH)_2$	146.41	wh, trig or amorph	4.79^{15}	d 300		0.00026^{25}		s a, NH_4 salts; i alk
c32	iodate	$Cd(IO_3)_2$	462.21	wh cr	6.43	d		s		s NHO_3; NH_4OH
c33	iodide	CdI_2	366.21	grn-yel powd	5.670^{30}_{4}	387	796	86.2^{25}	125^{100}	110.5^{20} al; 41^{25} acet; sl s NH_3; 206.7^{25} MeOH
c34	lactate	$Cd(C_3H_5O_3)_2$	290.54	need				10	12.5	i al
c35	maleate	$Cd(C_4H_2O_4).2H_2O$	262.49					0.66^{30}		
c36	*per*manganate	$Cd(MnO_4)_2.6H_2O$	458.36		2.81	d 95		v s	v s	
c37	molybdate	$CdMoO_4$	272.34	yel pl	5.347			sl s		s a, NH_4OH, KCN
c38	nitrate	$Cd(NO_3)_2$	236.41	col		350		109^{0}	326^{60} 682^{100}	v s a; s et ac
c39	nitrate, tetra-hydrate	$Cd(NO_3)_2.4H_2O$	308.47	wh pr, need, hygr	2.455^{17}_{4}	59.4	132	215		s al, NH_3; i HNO_3
c40	nitrocobaltate (III)	Cadium cobaltinitrite. $Cd[Co(NO_2)_6]_2$	1007.13	yel		d 175		sl s	v s	d a, alk, org solv
c41	oxalate	CdC_2O_4	200.42	col cr	3.32^{18}_{4}	d 340				s a; i al
c42	oxalate, trihydrate	$CdC_2O_4.3H_2O$	254.45	col cr		d		0.005^{18}	0.009	
c43	oxide	CdO	128.40	br, amorph	6.95	<1426	d 900–1000	i	i	s a, NH_4 salts; i alk
c44	oxide	CdO	128.40	br cub, 2.49 (Li)	8.15	d 900	subl 1559	i	i	s a, NH_4 salts; i alk
c45	*ortho*phosphate	$Cd_3(PO_4)_2$	527.14	col, amorph		1500		i		s a, NH_4 salts

No.	Name	Synonyms and Formulae	Mol. wt.	Crystalline form, properties and index of refraction	Density or spec. gravity	Melting point, °C	Boiling point, °C	Solubility, in grams per 100 cc		
								Cold water	Hot water	Other solvents
	Cadmium									
c46	*pyro*phosphate	$Cd_2P_2O_7$	398.74	wh cr leaf	4.965^{15}	above red heat		sl s	s	s a, NH_3
c47	phosphate, dihydrogen	$Cd(H_2PO_4)_2.2H_2O$	342.41	col, tricl	2.74_4^{15}	d 100				i al, eth; s HCl
c48	phosphide	Cd_3P_2	399.15	grn, tetr need	5.60	700				s d HCl; s exp conc HNO_3
c49	potassium cyanide	$Cd(CN)_2.2KCN$	294.68	col, glossy, oct	1.847			33.3	100	i al
c50	potassium sulfate	$CdK_2(SO_4).2H_2O$	322.69	tricl col tab	2.922^{16}			42.89^{16}	47.40^{40}	
c51	salycilate	$Cd(C_7H_5O_3)_2.H_2O$	404.65	wh need				sl s	s	s al, eth, glycerol a, NH_4OH
c52	selenate	$CdSeO_4.2H_2O$	291.39	rhomb	3.63	$-1H_2O$, 100; $-2H_2O$, 170		v s		
c53	selenide	CdSe	191.36	grn-br or red powd, hex	5.81_4^{15}	>1350		i		d a
c54	*meta*silicate	$CdSiO_3$	188.48	col, rhomb, 1.739	4.93	1242		v sl s		
c55	sulfate	$CdSO_4$	208.46	wh, rhomb	4.691_4^{20}	1000		75.5^0	60.8^{100}	i al, acet, NH_3
c56	sulfate, hydrate	$CdSO_4.H_2O$	226.48	monocl	3.79^{20}	tr 108		s	s	i al
c57	sulfate, hydrate	$CdSO_4.7H_2O$	334.57	col, monocl	2.48	tr 4		s		i al
c58	sulfate, hydrate	$3CdSO_4.8H_2O$	769.50	col, monocl, 1.565	3.09	tr 41.5		113^0	s	
c59	sulfide	Nat. greenockite. CdS	144.46	yel-or, hex, 2.506, 2.529	4.82	1750^{100atm}	subl in N_2, 980	0.00013^{18}	colloid	s a; v sl s NH_4OH
c60	sulfite	$CdSO_3$	192.46	cr		d		sl s		i al; s a, NH_4OH
c61	tartrate	$CdC_4H_4O_6$	260.47	wh cr powd				sl s		s a, NH_4OH
c62	telluride	CdTe	240.00	blk, cub	6.20^{15}	1041		i		i a; d HNO_3
c63	tungstate	$CdWO_4$	360.25	yel cr				0.05		s NH_4OH
	Cadmium complexes									
c64	tetramminecadmium *per*rhenate	$[Cd(NH_3)_4](ReO_4)_2$	681.16		3.714_4^{25}					0.037 conc NH_4OH
c65	tetrapyridine cadmiumfluosilicate	$[Cd(C_5H_5N_4)]SiF_6$	570.89	wh, tricl	2.282					
c66	**Calcium**	Ca	40.08	silv wh soft met, cub	1.54	842-8	1487	d to H_2+ $Ca(OH)_2$	d	s a, liq NH_3; sl s al; i bz
c67	acetate	$Ca(C_2H_3O_2)_2$	158.17	col cr; 1.55, 1.56, 1.57		d		37.4^0	29.7^{100}	sl s al
c68	acetate, dihydrate	$Ca(C_2H_3O_2)_2.2H_2O$	194.21	col need		$-1H_2O$, 84		34.7^{20}	33.5^{80}	
c69	acetate, monohydrate	$Ca(C_2H_3O_2)_2.H_2O$	176.19	col need		d		43.6^0	34.3^{100}	sl s al
c70	aluminate	$CaAl_2O_4$ (or $CaO.Al_2O_3$)	158.04	wh monocl, tricl or rhomb; 1.643, 1.665, 1.663	2.981^{25}	1600		d		s HCl; i HNO_3, H_2SO_4
c71	(tri-)aluminate	$Ca_3Al_2O_6$ (or $3CaO.Al_2O_3$)	270.20	wh, cub, 1.710	3.038^{25}	d 1535		i		s a
c72	(tri-)aluminate hexahydrate	$3CaO.Al_2O_3.6H_2O$	378.29	col, oct, 1.603	2.52^{20}	d 700–800		d		
c73	aluminosilicate	$2Ca.Al_2O_3.SiO_2$	274.20	col, tetr, 1.669, 1.658	3.048	1590 ± 2				d a
c74	aluminosilicate	Nat. anorthite. $CaAl_2.Si_2O_8$ (or $CaO.Al_2O_3.2SiO_2$)	278.21	wh, tricl, 1.5832	2.765	1551				
c75	*ortho*arsenate	$Ca(AsO_4)_2$	398.08	col amorph powd	3.620	1.455		0.013^{25}		
c76	arsenate, trihydrate	Nat. haidingerite. $2CaO.As_2O_5.3H_2O$	396.04	col, rhomb, 1.590, 1.602, 1.638	2.967					
c77	arsenide	Ca_3As_2	270.08	red cr	3.031^{25}	d		d	d	d a; s h HNO_3
c78	azide	$Ca(N_3)_2$	124.12	col, rhomb, hyg		$-3H_2O$, 110; exp 144–156		38.1^0	45^{15}	0.211^{16} al; i eth
c79	benzoate	$Ca(C_7H_5O_2)_2.3H_2O$	336.36	col, rhomb	1.436	$-3H_2O$, 110		2.7^0	8.3^{80}	
c80	*meta*borate	$Ca(BO_2)_2$	125.70	col, flat rhomb pr, 1.550, 1.660, 1.680		1154		sl s		s a, NH_4 salts; sl s ac a
c81	*meta*borate, hexahydrate	$Ca(BO_2)_2.6H_2O$	233.79	col, tetr, 1.520, 1.502	1.88			0.25^{20}		
c82	*tetra*borate	CaB_4O_7	195.32	readily vitrified		986				
c83	boride	CaB_6	104.95	blk, cub	2.3^{20}	2235		i	i	s HNO_3; sl s conc H_2SO_4
c84	bromide	$CaBr_2$	199.90	col, rhomb need, deliq	3.353^{25}	sl d 730	806–812	142^{20}	312^{105}	s al, acet, a; sl s NH_3, MeOH
c85	bromate	$Ca(BrO_3)_2.H_2O$	313.91	monocl cr	3.329	$-H_2O$, 180		v s	v s	
c86	bromide, hexahydrate	$CaBr_2.6H_2O$	307.99	col, hex cr	2.295	38.2	149	594^0	1360^{25}	s al, acet, a

No.	Name	Synonyms and Formulae	Mol. wt.	Crystalline form, properties and index of refraction	Density or spec. gravity	Melting point, °C	Boiling point, °C	Solubility, in grams per 100 cc		
								Cold water	Hot water	Other solvents
	Calcium									
c87	butyrate	$Ca(C_4H_7O_2)_2.3H_2O$	268.32	col cr				s	sl s	
c88	carbide	CaC_2	64.10	col, tetr, 1.75	2.22	stab 25–447	2300	d	d	
c89	carbonate	Nat. aragonite. $CaCO_3$	100.09	col, rhomb, 1.530, 1.681, 1.685	2.930	tr to calcite 520	d 825	0.00153^{25}	0.00190^{75}	s a, NH_4Cl
c90	carbonate	Nat. calcite. $CaCO_3$	100.09	col, rhomb or hex, 1.6583, 1.4864	2.710^{18}	1339^{1025}	d 898.6	0.0014^{25}	0.0018^{75}	s a, NH_4Cl
c91	carbonate, hexahydrate	$CaCO_3.6H_2O$	208.18	col, monocl, 1.460, 1.535, 1.545	1.771^{0}					
c92	chlorate	$Ca(ClO_3)_2$	206.99	wh cr, hyg		340 ± 10 (−some O)		s	s	s al, acet
c93	chlorate, dihydrate	$Ca(ClO_3)_2.2H_2O$	243.01	wh-yelsh, rhomb, or monocl, deliq	2.711	$-H_2O$, 100		177.7^{8}	v s	s al, acet
c94	*per*chlorate	$(CaClO_4)_2$	238.98	col cr	2.651	d 270		188.6^{25}	v s	166.2_{25}^{25} al; 237.4 MeOH
c95	chloride	$CaCl_2$	110.99	col, cub, deliq 1.52	2.15_4^{25}	772	>1600	74.5^{20}	159^{100}	s al, acet, ace a
c96	chloride aluminate	$3CaO.Al_2O_3.CaCl_2.10H_2O$	561.33	col, monocl or hex, hex, 1.550, 1.535	1.892^{14}	$-H_2O$, 105	$-8H_2O$, 350	sl s	d	s a
c97	chloride, dihydrate	$CaCl_2.2H_2O$	147.02	col cr	0.835			97.7^{0}	326^{60}	50^{80} al
c98	chloride, hexahydrate	$CaCl_2.6H_2O$	219.08	col, trig, deliq, 1.417, 1.393	1.71^{25}	29.92	$-4H_2O$, 30, $-6H_2O$, 200	279^{0}	536^{20}	s al
c99	chloride, monohydrate	$CaCl_2.H_2O$	129.00	col cr, deliq		260		76.8^{0}	249^{100}	s al; i acet
c100	chloride fluoride *ortho*phosphate	$3Ca_3(PO_4)_2.CaClF$	1025.08	col cr, 1.634, 1.631	3.14	1270		v sl s		
c101	chlorite	$Ca(ClO_2)_2$	174.98	wh, cub	2.71			d	d	i al
c102	*hypo*chlorite	$Ca(ClO)_2$	142.98	wh powd or flat pl, 1.545, 1.69	2.35	d 100		s		i al
c103	chlorite, basic	$Ca(ClO)_2.2Ca(OH)_2$	257.16	wh, hex, 1.51, 1.585	2.10			sl s solns with 5–6 % avail Cl	d	d a
c104	*hypo*chlorite, basic	Bleaching powder, chlorinated lime. $Ca(ClO)_2.CaCl_2.xCa(OH)_2.xH_2O$	comp varies	wh powd strong Cl odor		d		d evln Cl		d a
c105	*hypo*chlorite, trihydrate	$Ca(ClO)_2.3H_2O$	197.03	tetr pl, 1.535, 1.63	2.1	$-3H_2O$, 60				
c106	chromate	$CaCrO_4.2H_2O$	192.09	yel, monocl pr.		$-2H_2O$, 200		16.3^{20}	18.2^{45}	s a, al
c107	chromite	$CaCr_2O_4$	208.07	ol grn, cub need	4.8^{18}	2090		i	i	i a; s fus K_2CO_3
c108	cinnamate	$Ca(C_9H_7O_2)_2.3H_2O$	388.44	col cr				0.22^{2}	1.34^{100}	
c109	citrate	$Ca_3(C_6H_5O_7)_24H_2O$	570.51	wh need		$-4H_2O$, 120		0.85^{18}	0.96^{23}	0.0065^{18} al
c110	cyanamide	$CaCN_2$	80.10	col, hex, rhbdr		1300 subl >1150		d evl NH_3	d	
c111	cyanide	$Ca(CN)_2$	92.12	wh powd		d >350		d	d	
c112	cyanoplatinite	$CaPt(CN)_4.5H_2O$	429.31	yel-grn fluoresc, rhomb, 1.6226		$-5H_2O$, 100		s		
c113	ferricyanide	$Ca_3[Fe(CN)_6]_212H_2O$	760.42	red need, deliq				v s	v s	
c114	ferrite, mono-	$CaO.Fe_2O_3$	215.77	dk redsh r, rhomb, 2.58, 2.43 (Na)	5.08	1250		i	i	v sl s a
c115	ferrocyanide	$Ca_2Fe(CN)_611$ or $12H_2O$	490.28	yel tricl, 1.570, 1.582, 1.596	1.68	d		86.8^{25}	115^{65}	i al
c116	fluosilicate	$CaSiF_6$	182.16	col, tetr	2.66^{18}			sl s		s al, HF, HCl
c117	fluoride	Nat. fluorite. CaF_2	78.08	col, cub luminisc w heat, 1.434	3.180	1360	*ca* 2500	0.0016^{18}	0.0017^{26}	s NH_4 salts; sl s a i acet
c118	fluosilicate, dihydrate	$CaSiF_6.2H_2O$	218.19	col, tetrag	2.254			sl s d		s HCl, HF; i al
c119	formate	$Ca(CHO_2)_2$	130.12	col, rhomb, 1.510, 1.514, 1.578	2.015	d		16.2^{0}	18.4^{100}	i al
c120	fumarate	$CaC_4H_2O_4.3H_2O$	208.18	col, rhomb				2.11^{30}		
c121	*d*-gluconate	$Ca(C_6H_{11}O_7)_2.H_2O$	448.40	wh cr powd, need		$-H_2O$, 120		3.3^{15}		v sl s al
c122	glycerophosphate	$CaC_3H_5(OH)_2PO_4$	210.16	wh cr powd, hyg		d 170		2^{25}	sl s	i al
c123	hydride	CaH_2	42.10	wh, rhomb cr	1.9	816 (in H_2) d *ca* 600		d H_2 + $Ca(OH)_2$		d a
c124	hydroxide	$Ca(OH)_2$	74.09	col, hex, 1.574, 1.545	2.24	$-H_2O$, 580	d	0.185^{0}	0.077^{100}	s NH_4 salts, a; i al
c125	hyponitrite	$CaN_2O_2.4H_2O$	172.15	wh cr	1.834	d 320				d dil a
c126	iodate	Nat. lautarite. $Ca(IO_3)_2$	389.89	col, monocl	4.519^{15}	d 540		0.20^{15}	0.67^{90}	s HNO_3; i al
c127	iodate, hexahydrate	$Ca(IO_3)_2.6H_2O$	497.98	col, rhomb		d 35		0.13^{0}	1.22^{100}	s HNO_3
c128	iodide	CaI_2	293.89	yelsh-wh, hex, deliq	3.956_4^{25}	740	*ca* 1100	209^{20}	426^{100}	126^{20} MeOH; s al acet, a

No.	Name	Synonyms and Formulae	Mol. wt.	Crystalline form, properties and index of refraction	Density or spec. gravity	Melting point, °C	Boiling point, °C	Solubility, in grams per 100 cc: Cold water	Hot water	Other solvents
	Calcium									
c129	iodide, hexahydrate	$CaI_2.6H_2O$	401.98	yel, hex need	2.55	d 42	160	757^{0}	1680^{30}	s a, al, acet
c130	iron (III) aluminate	Calcium (tetra-) alumino-ferite, nat. celite. $4CaO.Fe_2O_3.Al_2O_3$	485.97	brn, rhomb, 1.98, 2.05, 2.08 all for λ	3.77	1418		3		
c131	isobutyrate	$Ca(C_4H_7O_2)_2.5H_2O$	304.35	col powd				20	sl s	
c132	lactate	$Ca(C_3H_5O_3)_2.5H_2O$	308.30	wh need, effl		$-3H_2O$, 100		3.1^{0}	7.9^{30}	sl s a; i al, eth
c133	laurate	$Ca(C_{12}H_{23}O_2)_2.H_2O$	456.73	wh need, effl		182–183		0.004^{15}	0.055^{100}	0.059^{15}, 1.72^{78} al
c134	linoleate	$Ca(C_{18}H_{31}O_2)_2$	598.97	wh amorph powd				i		s al, eth
c135	magnesium carbonate	Nat. dolomite. $CaCO_3.MgCO_3$	184.41	col, trig, 1.6817, 1.5026	2.872	d 730–760		0.032^{18}		
c136	magnesium *meta*silicate	Nat. diopside. $CaO.MgO.2SiO_2$	216.56	col, monocl, 1.665, 1.672, 1.695	3.275	1391		i	i	i HCl
c137	magnesium *ortho*silicate	Nat. mervinite. $3CaO.MgO.SiO_2$	328.72	col to pa grn, monocl, 1.708, 1.711, 1.718	3.150					
c138	dl-malate	$CaC_4H_4O_5.3H_2O$	226.20	col, rhomb, 1.545, 1.555, 1.575				0.321^{0}	$0.451^{37.5}$	i al
c139	*l*-malate	$CaC_4H_4O_5.2H_2O$	208.18	col				0.812^{0}	$1.224^{37.5}$	s al
c140	malate, dihydrogen	$Ca(HC_4H_4O_5)_2.6H_2O$	414.33	rhomb, or wh cr powd, 1.493, 1.507, 1.545				sl s		
c141	maleate	$CaC_4H_2O_4.H_2O$	172.15	col rhomb, 1.495, 1.575, 1.640				2.89^{25}	3.21^{40}	
c142	malonate	$CaC_3H_2O_4.4H_2O$	214.19					0.44^{0}	0.72^{100}	
c143	*per*manganate	$Ca(MnO_4)_2.5H_2O$	368.03	purp cr	2.4	d		331^{14}	338^{25}	s NH_4OH
c144	α-methylbutyrate	Calvium ethylmethylacetate. $Ca(C_5H_9O_2)_2$	242.34					24.24^{0}	25.65^{70}	
c145	molybdate	Nat. pawellite. $CaMoO_4$	200.01	col, tetr, 1.967, 1.978	4.38–4.53			i	d	s a i al, eth
c146	nitrate	$Ca(NO_3)_2$	164.09	col, cub, hyg	2.504^{18}	561		121.2^{18}	376^{100}	14^{15} al; s MeOH, liq NH_3, acet; i eth
c147	nitrate, tetrahydrate	$Ca(NO_3)_2.4H_2O$	236.15	col, monocl, deliq, 1.465, 1.498, 1.504	α 1.896, β 1.82	α 42.7, β 39.7	d 132	266^{0}	660^{30}	s al, acet
c148	nitrate, trihydrate	$Ca(NO_3)_2.3H_2O$	218.14	col, tricl		51.1				
c149	nitride	Ca_3N_2	148.25	brn cr, hex	2.63^{17}	1195		d	d	s dil a; d abs al
c150	nitrite	$Ca(NO_2)_2.H_2O$	150.11	col-yelsh, hex, deliq	2.23^{34}	$-H_2O$, 100		45.9^{0}	89.6^{91}	sl s al
c151	nitrite, tetrahydrate	$Ca(NO_2)_2.4H_2O$	204.15	col cr, tetr	1.674^{0}_{0}	$-2H_2O$, 44		74.9^{0}	106^{42}	s al
c152	oleate	$Ca(C_{18}H_{33}O_2)_2$	603.01	wh wax-like cr		83–84		0.04^{25}	0.03^{50}	sl s eth
c153	oxalate	CaC_2O_4	128.10	col, cub	2.2^{4}	d		0.00067^{13}	0.0014^{95}	s a; i ac a
c154	oxalate, hydrate	$CaC_2O_4.H_2O$	146.12	col	2.2	$-H_2O$, 200		i	i	s a; i ac a
c155	oxide	Lime, calcia. CaO	56.08	col, cub, 1.838	3.25–3.38	2580	2850	0.131^{10} d	0.07^{80} d	s a
c156	oxide, per-	CaO_2	72.08	wh, tetr, 1.895	2.92^{25}_{4}	d 275		sl s		s a
c157	oxide, per- octahydrate	$CaO_2.8H_2O$	216.20	wh, tetr, pearly	1.70	$-8H_2O$, 200	d 275 expl	sl s	d	s a, NH_4 salts; i al, eth
c158	palmitate	$Ca(C_{16}H_{31}O_2)_2$	550.93	wh or yelsh, wh fatty powd				0.003^{25}		v sl s al; 0.008^{25} eth
c159	1-phenol-4 sulfonate(p-)	$Ca[C_6H_4(OH)SO_3]_2.H_2O$	404.43	wh to pinkish powd				s		s al
c160	phenoxide	$Ca(OC_6H_5)_2$	226.29	redsh powd				sl s		sl s al
c161	*hypo*phosphate	$Ca_2P_2O_6.2H_2O$	274.13	gel				i		s HCl
c162	*meta*phosphate	$Ca(PO_3)_2$	198.02	col, 1.588, 1.595	2.82	975		i	i	i a
c163	*ortho*phosphate, di-(sec)	Nat. brushite. $CaHPO_4.2H_2O$	172.09	wh, tricl, 1.5576, 1.5457, 1.5392	2.306^{16}	$-H_2O$, 109		0.0316^{38}	0.075^{100}	i al, s a
c164	*ortho*phosphate, mono-(prim.)	$Ca(H_2PO_4)_2.H_2O$	252.07	col, tricl, deliq, 1.5292, 1.5176, 1.4392	2.220^{16}	$-H_2O$, 109	d 203	1.8^{30}	d	s a
c165	*ortho*phosphate, tri-(tert.)	Nat. whitlockite. $Ca_3(PO_4)_2$	310.18	wh amorph powd, 1.629, 1.626	3.14	1670		0.002	d	i al; s a
c166	*pyro*phosphate	$Ca_2P_2O_7$	254.10	col, biax, 1.585, 1.604	3.09	1230		i		s a
c167	*pyro*phosphate, pentahydrate	$Ca_2P_2O_7.5H_2O$	344.18	col, monocl, 1.539, 1.545, 1.551	2.25			sl s		s a; i NH_4Cl
c168	phosphide	Ca_3P_2	182.19	gray lumps	2.51	*ca* 1600		d ev PH_3		s a; i al, eth, bz
c169	*hypo*phosphite	$Ca(H_2PO_2)_2$	170.06	wh-gray, monocl		d		15.4^{25}	12.5^{100}	i al
c170	*ortho*phosphite, di-	$2CaHPO_3.3H_2O$	294.17					sl s	d	s NH_4Cl
c171	*ortho*plumbate	Ca_2PbO_4	351.35	red-br cr	5.71	d		i	d	s a
c172	propionate	$Ca(C_3H_5O_2)_2.H_2O$	204.24	col, monocl tabl				49^{0}	55.8^{100}	i al
c173	*l*-quinate	$Ca(C_7H_{11}O_6)_2.10H_2O$	602.56	rhomb leaf		50, $-10H_2O$ 120		16^{18}		i al

No.	Name	Synonyms and Formulae	Mol. wt.	Crystalline form, properties and index of refraction	Density or spec. gravity	Melting point, °C	Boiling point, °C	Solubility, in grams per 100 cc		
								Cold water	Hot water	Other solvents
	Calcium									
c174	salicylate	$Ca(C_7H_5O_3)_2.2H_2O$	350.34	wh, oct		$-2H_2O$, 120		4^{25}	s	s al
c175	selenate	$CaSeO_4$	183.04	col	2.88			7.9^{6}	5.4^{67}	
c176	selenate, dihydrate	$CaSeO_4.2H_2O$	219.07	col, monocl	2.68					
c177	selenide	CaSe	119.04	cub, 2.274	3.57					
c178	*meta*silicate(α)	Nat. pseudowollastonite. $CaSiO_3$	116.16	col, monocl, 1.610, 1.611, 1.664	2.905	1540		0.0095^{17}		s HCl
c179	*meta*silicate(β)	Nat. wollastonite. $CaSiO_3$	116.16	col, monocl, 1.616, 1.629, 1.631	2.5	tr 1200				
c180	di-*ortho*silicate (I)	Ca_2SiO_4	172.24	col, monocl, 1.717, 1.735	3.27	2130				
c181	di-*ortho*silicate (II)	Ca_2SiO_4	172.24	col, rhomb, 1.717, 1.735	3.28	tr to (I) 1420				
c182	di-*ortho*silicate (III)	Ca_2SiO_4	172.24	col, monocl, 1.642, 1.645, 1.654	2.97	tr to 675				
c183	(tri-)silicate	Nat. alite. Ca_3SiO_5 or $(3CaO.SiO_2)$	228.32	col, monocl, α 1.718, β 1.724	1900 (incogr)					
c184	silicide	$CaSi_2$	96.25		2.5			i	d	s a, alk
c185	stearate	$Ca(C_{18}H_{35}O_2)_2$	607.04	cr powd		179–180		0.004^{15}		i al, eth
c186	succinate	$CaC_4H_4O_4.3H_2O$	212.22	col, 1.460, 1.540, 1.610				0.193^{10}	0.89^{80}	
c187	sulfate	Nat. anhydrite. $CaSO_4$	136.14	col, rhomb, or monocl, 1.569, 1.575, 1.613	2.960	monocl 1450	rhomb tr to monocl 1193	0.209^{30}	0.1619^{100}	s a, NH_4 salts, $Na_2S_2O_3$, glyc
c188	sulfate	Soluble anhydrite. $CaSO_4$	136.14	col, hex or tricl, 1.505, 1.548	2.61	tr to rhomb >200				
c189	sulfate half-hydrate	Plaster of Paris. $CaSO_4.\frac{1}{2}H_2O$	145.15	wh powd		$-\frac{1}{2}H_2O$, 163		0.3^{20}	sl s	s a, NH_4 salts, $Na_2S_2O_3$, glyc
c190	sulfate dihydrate	Nat. gypsum. $CaSO_4.2H_2O$	172.17	col, monocl, 1.521, 1.523, 1.530	2.32	$-1\frac{1}{2}H_2O$, 128	$-2H_2O$, 163	0.241	0.222^{100}	s a, NH_4 salts, $Na_2S_2O_3$, glyc
c191	sulfide	Nat. oldhamite. CaS	72.14	col, cub, 2.137	2.5	d		0.021^{15} d	0.048^{60} d	d a
c192	sulfide, hydro-	$Ca(HS)_2.6H_2O$	214.32	col pr		d 15–18		v s		s al,
c193	sulfite	$CaSO_3.2H_2O$	156.17	col, hex		$-2H_2O$, 100		0.0043^{18}	0.0011^{100}	s H_2SO_4
c194	sulfite, dihydrogen	$Ca(HSO_3)_2$	202.22	yelsh liq, strong SO_2 odor				s		s a
c195	*d*-tartrate	$CaC_4H_4O_6.4H_2O$	260.21	col, rhomb, 1.525, 1.535, 1.550		d		$0\ 0266^{0}$	$0.0689^{37.5}$	sl s al
c196	*dl*-tartrate	$CaC_4H_4O_6.4H_2O$	260.21	tricl, powd or need		$-4H_2O$, 200		0.0032^{0}	$0.0078^{37.5}$	s HCl; i ac, a
c197	*meso*tartrate	$CaC_4H_4O_6.3H_2O$	242.20	wh, monocl or tricl pr		$-3H_2O$ <170		i	0.16^{100}	0.28^{18}, 0.85^{100} ac a
c198	telluride	CaTe	167.68	cub, 2.51, 2.58	4.873					
c199	tellurite	$CaTeO_3$	215.68	wh fl		>960		sl s	s	s a
c200	thiocarbonate, tri-	$CaCS_3$	148.28	yel cr				s		i al
c201	thiocyanate	$Ca(SCN)_2.3H_2O$	210.29	wh cr, deliq				v s	v s	v s al
c202	*di*-thionate	$Ca(SO_3)_2.4H_2O$	272.27	col, trig, 1.5496	2.176			16^{0}	30^{80}	
c203	thiosulfate	$CaS_2O_3.6H_2O$	260.30	tricl	1.872	d		100^{3}	d	s al
c204	*meta*titanate	Nat. perovskite. $CaTiO_3$	135.98	col, cub, rhomb, β 2.34	4.10	1975				
c205	tungstate	$CaWO_4$	287.93	wh, tetr, 1.9263, 1.9107	6.062^{20}			0.00064^{15}	0.00012^{100}	
c206	tungstate	Nat. scheelite. $CaWO_4$	287.93	col or w sc, tetr. 1.918, 1.934	6.06			0.2		i al, a; s NH_4Cl
c207	*meta*tungstate	$Ca_3H_4[H_2(W_2O_7)_6].27H_2O$	3500.96	col, tric		$-7H_2O$, 105	$-10H_2O$, d			d a
c208	valerate	$Ca(C_5H_9O_2)_2$	242.33					8.28^{0}	7.39^{100}	
c209	*meta*zirconate	$CaZrO_3$	179.30	col, monocl	4.78	2550				
c210	**Carbon**	Diamond. C	12.01	col, cub, 2.4173	3.51	>3550	4827	i	i	i a, alk
c211	carbon	Graphite. C	12.01	blk, hex	2.25^{20}	subl 3652–97	4200	i	i	s liq Fe; i a, alk
c212	carbon, amorphous	C	12.01	amorph, blk	1.8–2.1	subl 3652–97	4200	i	i	i a, alk
c213	(di)-bromide, hexa-	Hexabromomethane. C_2Br_6	503.48	rhomb pr, 1.740, 1.847, 1.863	3.823	148–149 d	210	i		s CS_2; v sl s al, eth
c214	bromide, tetra-	Tetrabromomethane. CBr_4	331.65	col, monocl or oct	3.42	tr to oct 48.4; m.p. 90.1	189.5	0.024^{30}		s al, eth, chl
c215	(di)-bromide, tetra-	Tetrabromethylene. C_2Br_4	343.66			57.5	227			
c216	(di)-chloride, hexa-	Hexachloro ethane. C_2Cl_6	236.74	col, rhmb, tricl or cub	2.091	subl 187		i		s al, eth, oils
c217	chloride, tetra-	Tetrachloromethane. CCl_4	153.81	col liq, 1.4601.	1.5867^{20}_{20}	−23	76.8		v sl s	s al, bz, chl, eth
c218	(di)-chloride, tetra,-	Tetrachloroethylene. C_2Cl_4	165.83	col liq, eth odor, 1.5055	1.6311^{15}_{4}	−22.4	120.8			s al, eth

No.	Name	Synonyms and Formulae	Mol. wt.	Crystalline form, properties and index of refraction	Density or spec. gravity	Melting point, °C	Boiling point, °C	Solubility, in grams per 100 cc		
								Cold water	Hot water	Other solvents
	Carbon									
c219	fluoride, tetra-	Tetrafluoromethane. CF_4	87.99	col gas	1.96^{-184}	−184	−128	sl s		
c220	iodide, tetra-	Tetraiodomethane. CI_4	519.63	dk red, cub	4.34^{20}	d 171		i	d	s al, CS_2, eth, MeOH, bz
c221	oxide, di-	CO_2	44.01	col gas or col liq	1.977^{0}g/l, liq 1.101^{-37}, solid 1.56^{-79}	$-56.6^{5.2atm}$	−78.5 subl	171.3^{0} cm³ 0.348^{0}g 0.145^{25}g	90.1^{20} cm³ 0.097^{40} g 0.058^{60}g	31^{15} cm³ al, s acet
c222	oxide, mon-	CO	28.01	col odorl pois gas	1.250^{0}g/l liq 0.793	−199	−191.5	3.5^{0} cm	2.32^{20} cm³	s al, bz, ac a, Cu_2Cl_2
c223	oxide, sub-	C_3O_2	68.03	col gas or liq, 1.4538	liq 1.114^{0}	−111.3	7	d		
c224	oxysulfide	COS	60.07	col gas, pois	gas 1.073 g/1^{0} liq 1.24^{-87}	−138.2	−50.2	54^{20} ml		s al; v s CS_2
c225	selenide, di-	CSe_2	169.93	golden yel liq, 1.845^{20}	2.6626^{25}_{4}	−45.5	125–126	i		d al; s CS_2, tol
c226	selenide, sulfide	$CSeS$	123.04	yel oily liq	1.9874	−85	84.5	i	i	sl s al; s CS_2
c227	sulfide, di-	CS_2	76.14	col liq, inflamm, 1.62950^{18}	1.261^{22}_{20}	−110.8	46.3	0.22^{22}	0.14^{50}	s al, eth
c228	sulfide, mono-	CS	44.08	red powd	1.66	d 200		i		i al; s eth, CS_2
c229	sulfide, sub-	C_3S_2	100.16	red liq	1.274	−0.5	d 90			
c230	sulfide telluride	$CSTe$	171.68	yel-red liq	2.9^{-50}	−54	d > −54			s CS_2, bz
c231	sulfochloride	Thiophosgene. $CSCl_2$	114.98	yel-red liq	1.509^{15}		73.5			
c232	**Carbonic acid**	H_2CO_3	62.03	exists in solution only				s		
c233	**Carbonyl bromide**	Carbon oxybromide. $COBr_2$	187.83	col liq			64.5			
c234	**Carbonyl chloride**	Phosgeve, carbon oxychloride. $COCl_2$	98.92	col gas, pois	1.392	−104	8.3	d		d al, a; v s bz, tol s ac a
c235	**Carbonyl fluoride, di-**	COF_2	66.01	col gas, hyg	sol 1.388^{-190} liq 1.139^{-114}	−114	−83.1	d		
c236	**Carbonyl selenide**	$COSe$	106.97	col gas, very pois	liq $1.812^{4.1}$	−124.4	−21.7	d		s $COCl_2$
c237	**Cerium**	Ce	140.12	gray met, cub or hex	6.78	795	3468	sl d	d	s dil min a, i alk
c238	(III) acetate	$Ce(C_2H_3O_2)_3$	317.26	col		d 308		20^{15}	12^{75}	
c239	(III) acetate hydrate	$Ce(C_2H_3O_2)_3.1\frac{1}{2}H_2O$	344.28	wh-redsh cr powd		$-1\frac{1}{2}$ H_2O, 115	d	26.5^{15}	16.2^{75}	
c240	boride, hexa-	CeB_6	204.98	blue met, cub		219^{0}	d	i	i	i HCl
c241	boride, tetra-	CeB_4	183.36	tetr	5.74					
c242	III bromate	$Ce(BrO_3)_3.9H_2O$	685.98	redsh-wh, hex		49		s		
c243	bromide	$CeBr_3.H_2O$	397.86	col need, deliq		d		v s	v s	v s al
c244	carbide	CeC_2	164.14	red, hex	5.23			d	d	s a
c245	carbonate	$Ce_2(CO_3)_3.5H_2O$	550.37	wh cr				i		s a; sl s $(NH_4)_2CO_3$
c246	carbonate fluoride	Nat. bastnaesite. $CeFCO_3$	219.13	hex, 1.717, 1.817	5					
c247	chloride	$CeCl_3$	246.48	col cr, deliq	3.92^{0}	848	1727	100	d	30 al; s acet
c248	citrate	$Ce(C_6H_5O_7).3\frac{1}{2}H_2O$	392.28	wh powd				i		s dil min a
c249	(III) cyanoplatinite	$Ce_2[Pt(CN)_4]_3.18H_2O$	1502.00	yel-bl lust, monocl	2.657	$-13\frac{1}{2}H_2O$ 100–110	d	s		
c250	(III) fluoride	CeF_3	197.12	wh, hex	6.16	1460	2300	i		
c251	(IV) fluoride	$CeF_4.H_2O$	234.13	col microcr, 1.614	4.77	*ca* 650	d	i		s a
c252	hydride	CeH_3	143.14	dk bl amorph powd		ign		d		
c253	(III) hydroxide	$Ce(OH)_3$	191.14	wh gelat ppt						s a, $(NH_4)_2CO_3$; i alk
c254	(IV) iodate	$Ce(IO_3)_4$	839.73	yel cr				0.015^{20}		
c255	(III) iodate	$Ce(IO_3)_3.2H_2O$	700.86	cr				0.16^{25}		s HNO_3
c256	(III) iodide	$CeI_3.9H_2O$	682.97	wh or redsh-wh cr		752	1397	v s		v s al
c257	(III) molybdate	$Ce_2(MoO_4)_3$	760.05	yel, tetr, 2.019, 2.007	4.83	973				
c258	(III) nitrate	$Ce(NO_3)_3.6H_2O$	434.23	col or redsh cr (trac La, Di), deliq		$-3H_2O$, 150	d 200	v s	v s	50 al; s acet
c259	(IV) nitrate, basic	$Ce(OH)(NO_3)_3.3H_2O$	397.07	long red need				s		
c260	(III) oxalate	$Ce_2(C_2O_4)_3.9H_2O$	706.44	yel-wh cr		d		v sl s		s H_2SO_4, HCl; i $H_2C_2O_4$, alk, eth, al
c261	(IV) oxide	$CeO_2.xH_2O$		yelsh gelat ppt						s a; sl s alk carb, i alk
c262	(III) oxide	C_2O_3	328.24	gray-grn, trig	6.86	1692, ign 200		i	i	s H_2SO_4, i HCl

No.	Name	Synonyms and Formulae	Mol. wt.	Crystalline form, properties and index of refraction	Density or spec. gravity	Melting point, °C	Boiling point, °C	Solubility, in grams per 100 cc		
								Cold water	Hot water	Other solvents
	Cerium									
c263	(IV) oxide(di-)	Ceria. CeO_2	172.12	brn-wh, cub	7.132^{23}	*ca* 2600		i	i	s H_2SO_4, HNO_3; i dil a
c264	oxychloride	$CeOCl$	191.57	purp leaf				i		s dil a
c265	(III) 2,4-pentanedione	Cerium acetylacetonate. $Ce(C_5H_7O_2)_3.3H_2O$	491.50	lt yel cr ppt		131–132		d		v s al
c266	(III) *meta*phosphate	$Ce(PO_3)_3$	377.04	micr need	3.272					i a
c267	(III) *ortho*phosphate	Nat. monazite. $CePO_4$	235.09	red, monocl or yel, rhomb, 1.795	5.22			i	i	s a; i al
c268	(III) salicylate	$Ce(C_7H_5O_3)_3$	551.47	wh-redsh wh powd				i		i al
c269	(III) selenate	$Ce_2(SeO_4)_3$	709.11	rhomb	4.456			39.55^{0}	2.513^{100}	
c270	silicide	$CeSi_2$	196.29		5.67^{17}			i		
c271	(IV) sulfate	$Ce(SO_4)_2$	332.24	deep yel cr	3.91^{18}	d 195		sl d, forms basic salts		
c272	(III) sulfate	$Ce_2(SO_4)_3$	568.42	col to grn, monocl or rhomb	3.912	d 920^{746}		10.1^{0}	2.25^{100}	
c273	(IV) sulfate, dihydrate	$Ce(SO_4)_2.4H_2O$	404.31	yel, rhomb				v s d		s dil H_2SO_4
c274	(III) sulfate, monohydrate	$Ce_2(SO_4)_3.9H_2O$	730.56	hex need	2.831			11.87^{15}	0.42^{30}	
c275	(III) sulfate, octahydrate	$Ce_2(SO_4)_3.8H_2O$	712.54	pink cr, tricl	2.886^{17}	$-8H_2O$, 630		12^{20}	6^{50}	
c276	(III) sulfate, pentahydrate	$Ce_2(SO_4)_3.5H_2O$	658.50	monocl	3.17			3.90^{50}	0.514^{100}	
c277	(III) sulfide	Ce_2S_3	376.43	red cr, br-dk powd, purp	5.020^{11}	d 2100 (vac)		i	d	s dil a
c278	(III) tungstate	$Ce_2(WO_4)_3$	1023.78	yel, tetr	6.77^{17}	1089				
	Cerium complexes									
c280	hexaantipyrinecerium perchlorate	$[Ce(C_{11}H_{12}N_2O)_6].(ClO_4)_3$	1567.86	col, hex cr		d 295–300		1.08^{20}		
c281	hexaantipyrinecerium iodide (III)	$[Ce(C_{11}H_{12}H_2O)_6].I_3$	1650.22	large yel cr		268–270		15.10^{20}		
c282	**Cesium**	Cs	132.905	silv met cr hex	1.8785^{15}	28.5	690	d		s liq NH_3
c283	acetate	$CsC_2H_3O_2$	191.95	deliq		194		$945.1^{-2.5}$	$1345.5^{88.5}$	
c284	aluminum sulfate	$CsAl(SO_4)_2.12H_2O$	568.19	col, cub, 1.4587	1.97	117		0.34^{0}	42.54^{100}	s dil al
c285	amide	$CsNH_2$	148.93	wh need	3.44^{25}_{4}	262 ± 1		d		s liq NH_3
c286	azide	CsN_3	174.93	col need, deliq		310		224.2^{0}		1.037^{16} al; i eth
c287	benzoate	$CsC_7H_5O_2$	254.02					294.5^{0}	398.5^{100}	
c288	borofluoride	$CsBF_4$	219.71	rhomb, 1.350	3.20	550 d		1.6^{17}	*ca* 30^{100}	s dil NH_3
c289	borohydride	$CsBH_4$	147.75	wh, cub, 1.498	2.404			v s		sl s al; i eth, bz
c290	bromate	$CsBrO_3$	260.81	hex, *ca* 2.15	4.109^{16}	*ca* 420 d		3.66^{25}	5.32^{35}	
c291	bromide, mono-	$CsBr$	212.81	col, cub, 1.6984	4.44, liq 3.04^{700}	636	1300	124.3^{25}	v s	s a
c292	bromide, tri-	$CsBr_3$	372.63	rhomb		180				
c293	*di*bromochloride	$CsBr_2Cl$	328.18	yel-red, rhomb		191	150, $-Br_2$	s		d al, acet
c294	bromochloride iodide	$CsIBrCl$	375.17	yel-red, rhomb		235	d 290	s		s al
c295	bromoiodide di-	$CsIBr_2$	419.63	rhomb	4.25	248	d 320	4.61^{20}		s al
c296	carbonate	Cs_2CO_3	325.82	col cr, deliq		d 610		260.5^{15}	v s	11^{19} al; s eth
c297	carbonate, hydrogen	$CsHCO_3$	193.92	rhomb		$175-\frac{1}{2}H_2O$		209.3^{15}	v s	s al
c298	chlorate	$CsClO_3$	216.36	sm cr	3.57			$6.28^{19.8}$	76.5^{90}	s al
c299	*per*chlorate	$CsClO_4$	232.36	rhomb, at 219 cub, 1.4752, 1.4788, 1.4804	3.327^{4}	d 250		2.00^{25}	28.57^{99}	0.093^{25} al; 0.7878^{25} al; 0.150^{25} acet
c300	chlorobromide	$CsBrCl_2$	283.72	glossy-yel, rhomb.		205		s		d al, eth
c301	chloride	$CsCl$	168.36	col, cub, deliq, 1.6418	3.988	646	1290	$162.22^{0.7}$	$259.56^{89.5}$	33.7^{25} MeOH; v s al; i aceton
c302	chloroiodide	$CsICl_2$	330.74	or, trig	3.86	230	d 290	s		s al
c303	chloroaurate	$CsAuCl_4$	471.68	yel, monocl.				0.5^{10}	27.5^{100}	s al
c304	chlorobromide, di-	$CsBrCl_2$	283.72			205				
c305	chloro*di*bromide	$CsBr_2Cl$	328.18	yel		191				
c306	chloroiodide, di-	$CsICl_2$	330.72	pa or, rhomb	3.68	230	d 290	s		s al
c307	chloroplatinate	Cs_2PtCl_6	673.62	yel, cub	4.197 ± 0.004	d 570		0.024^{0}	0.377^{100}	i al
c309	chlorostannate	$CsSnCl_6$	597.22	wh, cub	3.33					
c310	chromate	Cs_2CrO_4	381.80	yel pr, rhomb	4.237			71.4^{13}	95.5^{30}	
c311	chromium sulfate	Cesium chromium alum. $Cs[Cr(H_2O)_6](SO_4)_2.6H_2O$	593.21	vlt cr	2.064	116		9.4^{25}		
c312	cyanide	$CsCN$	158.92	very sm wh cr	2.93			v s	v s	
c313	fluoride	CsF	151.90	cub, deliq, 1.478 ± 0.005^{18}	4.115	682	1251	367^{18}		191^{15} MeOH; i Diox, Pyr
c314	fluoride	$CsF.\,1\frac{1}{2}H_2O$	178.93					366.5^{18}		
c315	fluorogermanate	Cs_2GeF_6	452.39	isotrop cr, reg oct	4.10			sl s	v s	sl s a
c316	fluosilicate	Cs_2SiF_6	407.89	wh, cub	3.372^{17}			60^{17}	sl s	i al
c317	fluotellurite	$CsTeF_5$	355.50	col need				d	d	s HF soln
c318	formate	$CsCHO_2$	172.92		1.0169^{21}_{4}				$2012^{95.4}$	

No.	Name	Synonyms and Formulae	Mol. wt.	Crystalline form, properties and index of refraction	Density or spec. gravity	Melting point, °C	Boiling point, °C	Solubility, in grams per 100 cc		
								Cold water	Hot water	Other solvents
	Cesium									
c319	formate	$CsCHO_2.H_2O$	195.94			41, $-H_2O$				
c320	gallium selenate	$CsGa(SeO_4)_2.2H_2O$	704.72	col cr				4.14^{25}		
c321	gallium sulfate	$CsGa(SO_4)_2.12H_2O$	610.93	col cub, 1.46495		2.113		1.21^{25}		0.0035^{25} 75 % al
c322	hydride	CsH	133.91	wh cr, cub	3.41	d		d	d	d a; i org solv
c323	hydrofluoride	CsF.HF	171.91	need, deliq		160		v s		v s a; i al
c324	hydrogencarbide	$CsHC_2$	157.94	trsp cr		300		d		
c325	hydroxide	CsOH	149.91	lt yel, deliq	3.675	272.3		395.5^{15}		s al
c326	iodate	$CsIO_3$	307.81	wh, monocl	4.85			2.6^{24}		
c327	*meta*periodate	$CsIO_4$	323.81	wh rhomb pl	4.259^{15}_{4}			2.15^{15}	s	
c328	iodide	CsI	259.81	rhomb, deliq, 1.7876	4.510^{25}_{4}	621	1280	44^{0}	160^{61}	s al
c329	iodide, penta-	CsI_5	767.43	bl, tricl		73				
c330	iodide, tri-	CsI_3	513.62	blk, rhomb	4.47	207.5		sl s	sl s	s al
c331	iodotetrachloride	$CsICl_4$	401.62	pale or needles	3.374^{-10}	228	d	sl s	sl s	
c332	iron (II) sulfate	$Cs_2SO_4.FeSO_4.6H_2O$	621.87	lt grn, monocl, 1.500, 1.504, 1.509	2.791^{20}_{4}	*ca* 70		101.1^{25} (anhyd)		
c333	iron (III) sulfate	$CsFe(SO_4)_2.12H_2O$	597.06	pa-vlt cr, 1.484	2.061^{20}	*ca* 90		s	s	
c334	magnesium sulfate	$Cs_2SO_4.MgSO_4.6H_2O$	590.34	col, monocl, 1.486, 1.452	2.676^{20}_{4}					
c335	*per*manganate	$CsMnO_4$	251.84		3.597	d 320		0.097^{1}	1.27^{59}	
c336	mercury bromide(ic)	$CsBr.2HgBr_2$	933.63	rhomb				0.807^{17}		sl s al
c337	mercury chloride(ic)	$CsCl.HgCl_2$	439.85	col, cub or rhomb, 1.792				1.44^{17}		i abs al
c338	nitrate	$CsNO_3$	194.91	col, hex or cub, 1.55, 1.56	3.685 liq 2.71^{500}	414	d	9.16^{0}	196.8^{109}	s acet, sl s al
c339	nitrate, hydrogen	$CsNO_3.NHO_3$	257.92	oct		100				
c340	nitrate, dihydrogen	$CsNO_3.2HNO_3$	320.92	col pl		32–36				
c341	nitrate	$CsNO_2$	178.91	yel cr				v s	v s	
c342	oxalate	$Cs_2C_2O_4$	353.82		3.230^{15}			282.9^{25}		
c343	oxide	Cs_2O	281.81	or need	4.25	d 400; m.p. 490 (in N_2)		v s	d	s a
c344	oxide, per	Cs_2O_2	297.81	pa yel need	4.25	400	650. $-O_2$	s	d	s a
c345	oxide, tri-	Cesium oxide, sesqui- Cs_2O_3	313.81	choc br cr, cub	4.25	400		d		s a
c346	phthalate, hydrogen	$CsHC_8H_4O_4$	298.03	rhomb	2.178					
c347	polonium chloride	Cs_2PoCl_6	688.53	cub, 1.86	3.82					
c348	rhodium sulfate	$CsRh(SO_4)_2.12H_2O$	644.12	yel, oct	2.238	110–111		sl s		
c349	rhodium sulfate	$CsRh(SO_4)_2.12H_2O$	644.12	or cr	2.22^{20}_{4}	111		sl s	s	
c350	salicylate	$CsC_7.H_5O_3$	270.02					196.2^{0}	1522^{100}	
c351	selenate	Cs_2SeO_4	408.77	rhomb, deliq, 1.5950, 1.5060, 1.5964	4.4528^{20}_{4}			244.8^{12}		
c352	sulfate	Cs_2SO_4	361.87	col rhomb, or hex, 1.560, 1.564, 1.566	4.243	1010	tr hex 600	167^{0}	220^{100}	i al, acet
c353	sulfate, hydrogen	$CsHSO_4$	229.97	col rhomb pr	3.352^{16}	d		s		
c354	sulfide	$Cs_2S.4H_2O$	369.94	wh cr, deliq				v s	v s	
c355	sulfide, di-	Cs_2S_2	329.94	dk red, amorph		460	>800	hgr		
c356	sulfide, di-	$Cs_2S_2.H_2O$	347.95	tetr				s		
c357	sulfide, hexa-	Cs_2S_6	458.19	br red		186				
c358	sulfide, penta-	Cs_2S_5	426.13		2.806^{15}	210				
c359	sulfide, tetra-	Cs_2S_4	394.06	yel		d 160				
c360	sulfide, tri-	Cs_2S_3	362.00	yel leaf		217	780			
c361	tartrate, hydrogen	$CsHC_4H_4O_6$	281.99	wh, rhomb cr				9.7^{25}	98^{100}	
c362	*l*-tartrate	$Cs_2C_4H_4O_6$	413.88	col, trig	3.03^{14}			v s d	v s	
c363	vanadium sulfate	Cesium vanadium alum· $VCs(SO_4)_2.12H_2O$	592.15	red, cub, 1.4780	2.033^{20}_{4}	82	$-12H_2O$, 230; d 300	0.464^{10}	sl s	
c364	**Chloramine, mono-**	NH_2Cl	51.48	yel liq		−66		s		s al, eth; v sl s CCl_4, bz
c365	**Chloric acid**	$HClO_3.7H_2O$	210.57	known only as col sol	$1.282^{14.2}$	<−20	d 40	v s		
c366	**Chloric acid,** per	$HClO_4$	100.46	col liq unstable	1.764^{22}	−112	39^{56}	∞		
c367	**Chloric acid,** per	Hydronium perchlorate. $HClO_4.H_2O$ or $(H_3O)^+(ClO_4)^-$	118.47	need, fairly stab	1.88, liq 1.776^{50}	50	exp 110	v s	v s	
c368	per, dihydrate	$HClO_4.2HC$	136.49	stab liq	1.65	−17.8	200	v s	v s	s al

No.	Name	Synonyms and Formulae	Mol. wt.	Crystalline form, properties and index of refraction	Density or spec. gravity	Melting point, °C	Boiling point, °C	Solubility, in grams per 100 cc: Cold water	Hot water	Other solvents
	Chlorine									
c369	**Chlorine**	Cl_2	70.906	grnsh-yel gas, or liq, or rhomb cr; gas 1.000768, liq 1.367	3.214^{0}	−100.98	−34.6	310^{10} cm³ 1.46^{0} g	177^{30} cm³ 0.57^{30} g	s alk
c370	azide	chlor(o)azide ClN_3	77.48	gas, expl				sl s		d alk
c371	fluoride, mono-	ClF	54.45	col gas	1.62^{-100}	−154±5	−100.8	d	d	
c372	fluoride, tri-	ClF_3	92.45	col gas	1.77^{13}	−83	11.3	d	d	
c373	hydrate	$Cl_2.8H_2O$	215.03	lt yel, rhomb	1.23	d 9.6		i		s alk
c374	oxide, di-	ClO_2	67.45	yel red gas, or red cr, expl	3.09^{11} g/l	−59.5	9.9^{731} exp	2000^{4} cm³	d to $HClO_3$, Cl_2, O_2	s alk, H_2SO_4
c375	oxide, hept-	Cl_2O_7	182.90	col oil		−91.5	82	s d		s bz
c376	oxide, mono-	Cl_2O	86.91	yel-red gas, or red-br liq	3.89^{0} g/1	−20	3.8^{766} exp	200 cm³	d to HOCl	s alk, H_2SO_4
c377	oxide, tetr-	ClO_4 or Cl_2O_8	99.45				d	s d		s bz
c378	chloroauric acid	$HAuCl_4.4H_2O$	411.85	brt yel need, deliq		d		s	v s	s al, eth
c379	chloroplatinic acid	$H_2PtCl_6.6H_2O$	517.92	red br pr, deliq	2.431	60		v s	v s	s al, eth
c380	chlorostannic acid	$H_2SnCl_6.6H_2O$	441.52	col leaf	1.93	9		s		
c381	**Chlorosulfonic acid**	$ClSO_3H$	116.52	col fum liq, 1.437^{14}	1.766^{18}	−80	158	d to H_2SO_4+HCl		d al, a; i CS_2
c382	**Chlorotetroxy fluoride**	ClO_4F	118.45	col gas, v exp		−167.3	−15.9			
c383	**Chloryl (per-)fluoride**	ClO_3F	102.45	gas	1.392^{25}	−146	−46.8			
c384	**Chromium**	Cr	51.996	steel gray, cub v hard	7.20^{28}	1890	2482	i	i	s dil H_2SO_4, HCl; i HNO_3, aq reg
c385	(II) acetate	$Cr(C_2H_3O_2)_2$	170.09	red cr				sl s	s	sl s al
c386	(III) acetate	$Cr(C_2H_3O_2)_3.H_2O$	247.15	gray-grn powd or blsh-grn pasty mass				s		i al
c387	arsenide, mon-	CrAs	126.92	gray, hex	6.35^{16}			i	i	i a
c388	boride, mono-	CrB	62.81	silv cr, orthorhomb	6.17	2760(?)		i	i	s fus Na_2O_2
c389	(II) bromide	$CrBr_2$	211.81	wh cr	4.356	842		s	s	s al
c390	(III) bromide	$CrBr_3$	291.72	olv gr, hex	4.250^{25}_{4}	subl		i	s	v s al; d alk
c391	bromide, hexahydrate	$[CrBr_2(H_2O)_4]Br.2H_2O$	399.81	grn cr, deliq				s	s tr to vlt	s al; i eth
c392	bromide, hexahydrate	$[Cr(H_2O)_6]Br_3$	399.81	blsh gray to vlt	5.4^{17}			v s	v s	i al
c393	(tri-)carbide, di-	Cr_3C_2	180.02	gray, rhomb	6.68	1890	3800	i	i	
c394	carbonyl	$Cr(CO)_6$	220.06	col, orthorhomb	1.77	d 110	210 exp	i	i	i al, eth, ac a; sl s CHI_3, CCl_4
c395	(II) chloride	$CrCl_2$	122.90	wh need, deliq	2.878^{25}	824		v s	v s	i al, eth
c396	(III) chloride	$CrCl_3$	158.35	vlt, trig	2.76^{15}	*ca* 1150	subl 1300	i	sl s	i al, acet, MeOH, eth
c397	chloride, hexahydrate	$[Cr(H_2O)_4Cl_2].2H_2O$	266.45	vlt, monocl	1.76	83		58.5^{25}	s	s al; i eth; sl s acet
c398	(II) fluoride	CrF_2	89.99	grn, cr, monocl	4.11	1100	>1300	sl s		i al; s h HCl
c399	(III) fluoride	CrF_3	108.99	grn, rhomb	3.8	>1000	subl 1100−1200	i		i al, NH_3; sl s a; s HF
c400	(II) hydroxide	$Cr(OH)_2$	86.01	yel-br		v		d		s a
c401	iodate, hydrate	$[Cr(H_2O)_6]I_3.3H_2O$	594.85	dk vlt cr, hygro	4.915^{25}_{4}	41−HI		v s	v s	s al, acet; i CHI
c402	(II) iodide	CrI_2	305.80	graysh powd	5.196	856	subl vac 800	s		
c403	(III) iodide	CrI_3	432.71	shiny blk cr	4.915^{25}_{4}	>600	$-I_2$, vac 350			
c404	(III) nitrate	$Cr(NO_3)_3.7\frac{1}{2}H_2O$	373.13	br, monocl		100	d	s	s	
c405	(III) nitrate	$Cr(NO_3)_3.9H_2O$	400.15	purple, monocl		60	d 100	s	s	s a, alk, al, acet
c406	nitride, mono-	CrN	66.00	cub or amorph	5.9	d 1700		i	i	sl s aq reg
c407	(II) oxalate	$CrC_2O_4.H_2O$	158.03	yel cr powd	2.468			sl s	s	s dil a
c408	(III) oxalate	$Cr_2(C_2O_4)_3.6H_2O$	476.14	red, amorph, hyg		120, $-H_2O$ tr to grn		s	s	v s (red) al, eth; i (grn) al
c409	oxide, di-	CrO_2	84.00	br-blk powd		300, −O		i		s HNO_3
c410	(II) oxide, mon-	CrO	68.00	blk powd				i	i	i dil HNO_3
c411	(III) oxide, sesqui-	Cr_2O_3	151.99	grn, hex, 2.551	5.21	2435	4000	i	i	i a, alk, al
c412	(III) oxide, sesqui-	$Cr_2O_3.xH_2O$	varies	vlt, amorph or bl-gray grn gel				i	i	s a, alk; sl s NH_4OH
c413	oxide, tri-	Chromic anhydride, "chromic acid", CrO_3	99.99	red, rhomb, deliq	2.70	196	d	61.7^{0}	67.45^{100}	s al, eth, H_2SO_4 HNO_3
c414	oxychloride	CrO_2Cl_2	154.90	dk red liq	1.911	−96.5	117	d	d	d al; s eth, ac a
c415	2,4-pentandione	Chromium acetylacetonate. $Cr(C_5H_7O_2)_3$	349.33			216	340	i		s org solv; i lgr

No.	Name	Synonyms and Formulae	Mol. wt.	Crystalline form, properties and index of refraction	Density or spec. gravity	Melting point, °C	Boiling point, °C	Solubility, in grams per 100 cc: Cold water	Hot water	Other solvents
	Chromium									
c416	(III) *ortho*phosphate	$CrPO_4.2H_2O$	183.00	vlt cr	$2.42^{32.5}$			sl s		s a, alk; i ac a
c417	(III) *ortho*phosphate	$CrPO_4.6H_2O$	255.06	vlt, tricl, 1.568, 1.591, 1.699	2.121^{14}	100				
c418	*pyro*phosphate	$Cr_4(P_2O_7)_3$	729.81	pa grn, monocl	3.2			i	i	s alk
c419	phosphide, mono-	CrP	82.97	gray-blk cr	5.7^{15}			i		s HNO_3, HF
c420	silicide	Cr_3Si_2	212.17	gray, tetr pr	5.5^{0}			i	i	s HCl, HF; i H_2SO_4, HNO_3
c421	(II) sulfate	$CrSO_4.7H_2O$	274.17	bl cr				12.35^{0}	d	s ls al; s NH_4OH
c422	(III) sulfate	$Cr_2(SO_4)_3$	392.18	vlt or red powd	3.012			i, s*		sl s al; i a
c423	(III) sulfate	$Cr_2(SO_4)_3.15H_2O$	662.41	vlt, amorph sc	1.867^{17}	100	$-10H_2O$, 100	s	d 67	i al
c424	(III) sulfate	$Cr_2(SO_4)_3.18H_2O$	716.45	bl vlt, cub oct, 1.564	1.7^{22}	$-12H_2O$, 100		120^{20}	s	s al
c425	(II) sulfide, mono-	CrS	84.06	blk powd, hex	4.85	1550		i		v s a
c426	(III) sulfide, sesqui-	Cr_2S_3	200.18	brn-blk powd	3.77^{19}	−S, 1350		i, d	i	s HNO_3; d al
c427	(III) sulfite	$Cr_2(SO_3)_3$	344.18	grnsh-wh	2.2	d				
c428	(II) tartrate	$CrC_4H_4O_6$	200.07	bl powd	2.33			i	i	sl s a; i ac a
c429	**Chromium complexes** *hex*ammine chromium-(III) chloride	$[Cr(NH_3)_6]Cl_3.H_2O$	278.55	yel cr	1.585			s		
c430	*hex*aureachromium-(III) fluosilicate	$[Cr(CON_2H_4)_6]_2.[SiF_6]_3.3H_2O$	1304.94	lt grn leaf				0.522^{20}		i al
c431	*hex*aureachromium-(III) *per*rhenate	$[C_2(CON_2H_4)_6].(ReO_4)_3$	1162.92	grn need	2.652^{25}_{4}			1.786		0.667 al
c432	chloropentammine chromium chloride	$[Cr(NH_3)_5Cl]Cl_2$	243.51	red, oct	1.696			0.65^{16}		i HCl
c433	**Cobalt**	Co	58.933	silv gray met, cub	8.9	1495	2900	i	i	s a
c434	(III) acetate	$Co(C_2H_3O_2)_3$	236.07	grn, oct		d 100		hydr readily		s a, glac ac a
c435	(II) acetate	$Co(C_2H_3O_2)_2.4H_2O$	249.08	red-vlt, monocl, deliq, 1.542	1.705^{19}	$-4H_2O$, 140		s	s	s a, al
c436	aluminate	(approx) Thenard's blue. $CoAl_2O_4$	176.89	bl, cub				i	i	
c437	(II) *ortho*arsenate	$Co_3(AsO_4)_2.8H_2O$	598.75	vlt-red, monocl, 1.626, 1.661, 1.669	3.178^{15}	d		i	i	s dil a, NH_4OH
c438	arsenic sulfide	Nat. cobaltite. CoAsS	156.92	gray-redsh	6.2–6.3	d				
c439	arsenide	Co_2As	192.79	cr powd	8.28	950		i	i	i HCl, H_2SO_4; s HNO_3, aq reg
c440	(II) benzoate	$Co(C_7H_5O_2)_2.4H_2O$	373.23	gray red leaf		$-4H_2O$, 115		v s		
c441	boride, mono-	CoB	69.74	pr	7.25^{18}			d	d	s HNO_3, aq reg
c442	(II) bromate	$Co(BrO_3)_2.6H_2O$	422.84	red, oct				45.5^{17}		s NH_4OH
c443	(II) bromide	$CoBr_2$	218.75	grn, hex, deliq	4.909^{25}_{4}	678 (in N_2)		66.7^{59}	68.1^{97}	77.1^{20} al; 58.6^{30} MeOH; s eth, acet
c444	(II) bromide hexahydrate	$CoBr_2.6H_2O$	326.84	red-vlt pr, deliq	2.46	47–48, $-4H_2O$ 100	$-6H_2O$, 130	s red color	153.2^{97}	s blk color, al, a, eth
c445	bromoplatinate	$CoPtBr_6.12H_2O$	949.66	trig	2.762					
c446	carbonate	Nat. spherocobaltite. $CoCO_3$	118.94	red, trig, 1.855, 1.60	4.13	d		i	i	s a; i NH_3
c447	(II) carbonate, basic	$2CoCO_3.Co(OH)_2.H_2O$	534.74	vlt-red pr				i	d	s a, $(NH_4)_2CO_3$
c448	carbonyl tetra-	Dicobalt octacarbonyl. $[Co(CO)_4]_2$ or $Co_2(CO)_8$	341.95	or cr or dk br, microcr	1.73^{18}	51	d 52	i	i	sl s al; s CS_2, eth
c449	carbonyl, tri-	Tetracobalt dodeca-carbonyl. $[Co(CO)_3]_4$ or $Co_4(CO)_{12}$	571.86	blk cr				sl s		s bz; d Br
c450	(II) chlorate	$Co(ClO_3)_2.6H_2O$	333.93	red, cub, deliq, 1.55	1.92	50	d 100	558.3^{0}	v s	s al
c451	(II) *per*chlorate	$Co(ClO_4)_2$	257.83	red need 1.510, 1.490	3.327			100^{0}	115^{45}	s al, acet
c452	(II) *per*chlorate	$Co(ClO_4)_2.5H_2O$	347.88	red, hex		143		100.13^{0}	115.10^{65}	v s al, acet i $CHCl_3$
c453	*per*chlorate	$Co(ClO_4)_2.6H_2O$	365.93	red pr		d 1534	d	259^{18}		s al, acet
c454	(II) *per*chlorate	$Co(ClO_4)_2.6H_2O$	365.93	red, oct, deliq, 1.55		d 182		255^{18}		s al, acet
c455	(II) chloride	$CoCl_2$	129.84	bl, hex, hygr	3.356^{25}_{4}	724 (in HCl gas)	1049	45^{7}	105^{96}	54.4 al; 8.6 acet; 38.5 MeOH; sl s eth
c456	(III) chloride	$CoCl_3$	165.29	red cr	2.94	subl		s		

* Several chromic salts exist in two forms, a soluble and insoluble modification.

No.	Name	Synonyms and Formulae	Mol. wt.	Crystalline form, properties and index of refraction	Density or spec. gravity	Melting point, °C	Boiling point, °C	Solubility, in grams per 100 cc		
								Cold water	Hot water	Other solvents
	Cobalt									
c457	(II) chloride, dihydrate	$CoCl_2.2H_2O$	165.87	red-vlt, monocl or tricl, 1.625, 1.671, 1.67	2.477^{25}_{25}			s	s	v sl s eth
c458	(II) chloride, hexahydrate	$CoCl_2.6H_2O$	237.93	red, monocl	1.924^{25}_{25}	86	$-6H_2O$, 110	76.7^{0}	190.7^{100}	v s (bl col) al; s acet; 0.29 eth
c459	chloroplatinate	$CoPtCl_6.6H_2O$	574.83	trig	2.699	d				
c460	chlorostannate	$CoSnCl_6.6H_2O$	498.43	rhomb or trig		d 100				
c461	(II) chromate	$CoCrO_4$	174.93	gray blk cr		d		i	d	s a, NH_4OH
c462	(II) citrate	$Co_3(C_6H_5O_7)_2.2H_2O$	591.04	rose-red		$-2H_2O$, 150		0.8		
c463	(II) cyanide dihydrate	$Co(CN)_2.2H_2O$	147.00	buff anhydr bl-vlt powd	anhydr 1.872^{25}	$-2H_2O$, 280	d 300	0.00418^{18}		s KCN, HCl, NH_4OH
c464	(II) cyanide, trihydrate	$Co(CN)_2.3H_2O$	165.01	red-gray powd, amorph		$-3H_2O$, 250		i		s KCN
c465	(II) ferricyanide	$Co_3[Fe(CN)_6]_2$	600.71	red need				i		s NH_4OH; i HCl
c466	(II) ferrocyanide	$Co_2Fe(CN)_6.xH_2O$		gray-grn				i		s KCN; i HCl
c467	(II) fluogallate	$[Co(H_2O)_6][GaF_5.H_2O]$	349.75	pink cr, monocl (?), 1.45	2.35	$-5H_2O$, 110		sl s		d a
c468	(II) fluoride	CoF_2	96.93	pink monocl	4.46^{25}_{4}	*ca* 1200	1400	1.5^{25}	s	sl s a; i al, eth, b
c469	(III) fluoride	CoF_3	115.93	br, hex	3.88			d to $Co(OH)_3$		i, al, eth, bz
c470	fluoride	$Co_2F_6.7H_2O$	357.96	grn powd	2.314^{25}			d		s H_2SO_4
c471	(II) fluoride, tetrahydrate	$CoF_2.4H_2O$	168.99	α: red, rhomb oct, β: rose cr powd	2.192^{25}_{4}	d 200		s	s	i al
c472	fluosilicate	$CoSiF_6.6H_2O$	309.10	pink trig, 1.382, 1.387	2.113^{19}			$118.1^{21.5}$		
c473	(II) formate	$Co(CHO_2)_2.2H_2O$	185.00	red cr	2.129^{22}	$-2H_2O$, 140	d 175	5.03^{20}		
c474	(II) hydroxide	$Co(OH)_2$	92.95	rose-red, rhomb	3.597^{15}	d		0.00032		s a, NH_4 salts; i alk
c475	(III) hydroxide	$Co_2O_3.3H_2O$	219.91	blk-brn powd	4.46	d	$-H_2O$, 100	0.00032		s a; i al
c476	(II) iodate	$Co(IO_3)_2$	408.74	bl-vlt need	5.008^{18}	d 200		0.45^{18}	1.33^{100}	s HCl, HNO_3. h H_2SO_4
c477	(II) iodate, hexahydrate	$Co(IO_3)_2.6H_2O$	516.83	red, oct	3.689^{21}	d 61	$-4H_2O$, 135	s		
c478[1]	(II) iodide (α) stable	CoI_2	312.74	blk hex, hyg	5.68	515(vac)	570 (vac)	159^{0}	420^{100}	v s al, acet
c478[2]	(II) iodide (β)	CoI_2	312.74	yel need, unstab	5.45^{25}	d 400		s		
c479	(II) iodide, dihydrate	$CoI_2.2H_2O$	348.77	grn, deliq		d 100		376.2^{45}	s	
c480	(II) iodide, hexahydrate	$CoI_2.6H_2O$	420.83	br-red hex, hygr	2.90	d 27, $-6H_2O$, 130		s	s	s al, eth, chl
c481	iodoplatinate	$CoPtI_6.9H_2O$	1177.59	trig	3.618					
c482	linoleate	$Co(C_{18}H_{31}O_2)_2$	617.83	br, amorph				i		s al, eth, acet
c483	(II) nitrate	$Co(NO_3)_2.6H_2O$	291.04	red, monocl, 1.52	1.87^{25}_{4}	55–56	$-3H_2O$, 55	133.8^{0}	217^{80}	$100.0^{12.5}$ al; s acet; sl s NH_3
c484	nitrosylcarbonyl	$Co(NO)(CO)_3$	172.97	cherryred liq		−1.05	48.6; d 55	i		s al, eth, acet, bz
c485	(II) oleate	$Co(C_{18}H_{33}O_2)_2$	621.86	br, amorph						s al, eth, oils, bz
c486	(II) oxalate	CoC_2O_4	146.95	wh or redsh	3.021^{25}	d 250		i		s a, NH_4OH
c487	oxalate, dihydrate	$CoC_2O_4.2H_2O$	182.98	pink cr		$-H_2O$, *ca* 190		v sl s	sl s	v sl s a; s NH_4OH
c488	(II) oxide	CoO	74.93	grn-brn cub	6.45	1935		i	i	s a; i al, NH_4OH
c489	(III) oxide	Co_2O_3	165.86	blk-gray, hex, or rhomb	5.18	d 895		i	i	s a; i al
c490	(II, III) oxide	Co_3O_4	240.80	blk, cub	6.07	tr to CoO 900–950		i	i	v sl s a; i aq reg
c491	palmitate	$Co(C_{16}H_{31}O_2)_2$	569.78			70.5				s pyr, hot CS_2, CCl_4; sl s eth; i MeOH, acet
c492	(II) *ortho*phosphate	$Co_3(PO_4)_2$	366.74	redsh cr	2.587^{25}			i	i	s H_3PO_4, NH_4OH
c493	(II) *ortho*phosphate, dihydrate	$Co_3(PO_4)_2.2H_2O$	402.77	pink powd				i		s H_3PO_4
c494	(II) *ortho*phosphate, octahydrate	$Co_3(PO_4)_2.8H_2O$	510.87	redsh powd	2.769^{25}	$-8H_2O$, 200		sl s		s min a, H_3PO_4; i al
c495	phosphide	Co_2P	148.84	gray need	6.4^{15}	1386		i	i	s HNO_3, aq reg
c496	(II) propionate	$Co(C_3H_5O_2).3H_2O$	259.12	dk-red cr		*ca* 250		anh 33.5^{11}		v s al
c497	(II) *per*rhenate	$Co(ReO_4)_2.5H_2O$	649.41	dk pink		d		d		
c498	(II) selenate, heptahydrate	$CoSeO_4.7H_2O$	328.00	monocl	2.135					
c499	selenate, hexahydrate	$CoSeO_4.6H_2O$	309.98	red, monocl, 1.5225	2.25^{17}			s	s	
c500	(II) selenate, pentahydrate	$CoSeO_4.5H_2O$	291.97	ruby red, tricl-	2.512	d		v s		

No.	Name	Synonyms and Formulae	Mol. wt.	Crystalline form, properties and index of refraction	Density or spec. gravity	Melting point, °C	Boiling point, °C	Solubility, in grams per 100 cc: Cold water	Hot water	Other solvents
	Cobalt									
c501	selenide, mono-	CoSe	137.89	yel, hex	7.65	red heat				s HNO_3, aq reg; i alk
c502	(II) *ortho*silicate	Co_2SiO_4	209.95	vlt cr, rhomb	4.63	1345		i	i	s dil HCl
c503	silicide	CoSi	87.03	rhomb		1395				s HCl; i HNO_3, H_2SO_4
c504	silicide, di-	$CoSi_2$	115.11	rhomb	5.3	1277				
c505	(di-)silicide	Co_2Si	145.95	gray cr	7.28^{0}	1327				d a
c506	(II) *ortho*stannate	Co_2SnO_4	300.55	grnsh-bl, cub	6.30^{18}					i H_2SO_4; s h HCl
c507	(II) sulfate	$CoSO_4$	155.00	dk blsh, cub	3.71^{25}_{25}	d 735		36.2^{20}	83^{100}	1.04^{18} MeOH; i NH_3
c508	(II) sulfate, hepta-hydrate	Nat. bieberite. $CoSO_4.7H_2O$	281.10	red-pink, monocl, 1.477, 1.483, 1.489	1.948^{25}_{25}	96.8	$-7H_2O$, 420	60.4^{3}	67^{70}	2.5^{3} al; 54.5^{18} MeOH
c509	(II) sulfate, hexa-hydrate	$CoSO_4.6H_2O$	263.09	red, monocl, 1.531 1.549, 1.552	2.019^{15}_{15}	$-2H_2O$, 95				
c510	(II) sulfate, mono-hydrate	$CoSO_4.H_2O$	173.01	red cr, 1.603, 1.639, 1.683	3.075^{25}	d		s	s	
c511	(III) sulfate	$Co_2(SO_4)_3.18H_2O$	730.33	bl-grn		d 35		s d		s H_2SO_4; i pyr
c512	sulfide, di-	CoS_2	123.06	blk, cub	4.269			i		s HNO_3, aq reg
c513	sulfide, mono-	Nat. sycoporite. CoS	91.00	redsh, silv-wh, oct	5.45^{18}	>1116		0.00038^{18}		sl s a
c514	(III) sulfide, sesqui	Co_2S_3	214.06	blk cr	4.8					d a, aq reg
c515	(tri-) sulfide	Cobalt sulfide, tetra-(ous, ic) Nat. linneite. Co_3S_4	305.06	dk gray, cub	4.86	d 480				
c516	(II) sulfite	$CoSO_3.5H_2O$	229.07	red				i		s H_2SO_3
c517	tartrate	$CoC_4H_4O_6$	207.01	redsh, monocl				sl s		s dil a
c518	thiocyanate	$Co(SCN)_2.3H_2O$	229.14	vlt, rhomb		$-3H_2O$, 105		s		s al, MeOH, eth
c519	*ortho*titanate	Co_2TiO_4	229.76	grnsh-blk, cub	5.07–5.12					s conc HCl; sl s dil HCl
c520	(II) tungstate	$CoWO_4$	306.78	bl-grn, monocl	8.42			i		s h conc a; sl s c dil a
	Cobalt complexes									
c521	hexammine cobalt (II) bromide	$CoBr_2.6NH_3$	320.93	dk pink cr	1.871^{25}_{4}	d 258		d		
c522	diamminecobalt (II) chloride [α]	$CoCl_2.2NH_3$	163.90	rose cr	2.097	273				
c523	diamminecobalt (II) chloride(β)	$CoCl_2.2NH_3$	163.90	bl-vlt	2.073	tr to α, 210 (in NH_3)				
c524	hexamminecobalt (II) chloride	$[Co(NH_3)_6]Cl_2$	232.02	rose red, oct	1.497	d		d		s NH_4OH; i abs al
c525	hexamminecobalt (III) chloride	$Co(NH_3)_6Cl_3$	267.46	wine-red, monocl	1.710^{25}_{4}	$-1NH_3$, 215		5.9^{10}	$12.74^{46.6}$	s conc HCl; i al, NH_4OH
526	hexamminecobalt (II) iodide	$CoI_2.6NH_3$	414.93	dk pink, cub	2.096^{25}_{4}	141^{100mm}				
527	hexamminecobalt (III) nitrate	$Co(NH_3)_6.(NO_3)_3$	347.13	yel, tetr	1.804^{25}_{4}			1.7^{25}	v s	v sl s dil a
528	hexamminecobalt (III) *per*rhenate	$[Co(NH_3)_6](ReO_4)_3.2H_2O$	947.74	or-yel pr	3.329^{25}			0.0469		
529	hexamminecobalt (II) sulfate	$CoSO_4.6NH_3$	257.18	pink powd	1.654^{25}_{4}	d 116^{760}		d		v s dil NH_3
530	hexamminecobalt (III) sulfate	$[Co(NH_3)_6]_2(SO_4)_3.5H_2O$	700.50	dk yel, monocl	1.797^{25} anhydr	$-4H_2O$, 100	$-5H_2O$, 150	$1.4^{17.4}$		
531	ammonium tetra-nitrodiammine (III) cobaltate	Erdmann's salt. $NH_4[Co(NH_3)_2(NO_2)_4]$	295.12	redsh-pa brn, rhomb, 1.78, 1.78. 1.74	1.876^{25}					
532	aquapentammine-cobalt (III) chloride (roseo)	$[Co(NH_3)_5.H_2O]Cl_3$	268.45	brick red cr	1.7^{25}	d 100		24.87^{25}		sl s HCl; i al
533	aquapentammine-cobalt (III)-sulfate (roseo)	$[Co(NH_3)_5H_2O]_2(SO_4)_3.2H_2O$	638.34	red, tetr	1.854^{20}	$-3H_2O$, 99	d 110	$1^{17.2}$	1.72^{27}	s H_2SO_4
534	*cis*-chloroaquo-tetramminecobalt (III) chloride	$[Co(NH_3)_4(H_2O)Cl]Cl_2$	251.42	vlt, rhomb	1.847	d		1.4^{0}		s a; i al
535	chloropentammine cobalt-(III) chloride (purpureo)	$[Co(NH_3)_5Cl]Cl_2$	250.45	dk red-vlt, rhomb	1.819^{25}_{25}	d		0.4^{25}	$1.031^{46.6}$	s conc H_2SO_4; i al

No.	Name	Synonyms and Formulae	Mol. wt.	Crystalline form, properties and index of refraction	Density or spec. gravity	Melting point, °C	Boiling point, °C	Solubility, in grams per 100 cc		
								Cold water	Hot water	Other solvents
	Cobalt complexes									
c536	triethylenediamminecobalt-(III) chloride	$Co[C_2H_4(NH_2)_2]_3Cl_3.3H_2O$	399.64	br pr	1.542^{17}	256; $-3H_2O$, 100		v s		
c537	trinitrotriamminecobalt	$Co(NH_3)_3(NO_2)_3$	248.04	yel, rhomb pl or leaf	1.992^{25}_{4}	d 158	exp 164	$0.177^{16.5}$	0.28^{25}	
c538	trinitrotetramminecobalt-(III) nitrate	$[Co(NH_3)_4(NO_2)_2]NO_2$	265.07	yel, rhomb	1.922^{17}			3^{20}		
c539	potassium tetranitrodiamminecobaltate (III)	$K[Co(NH_3)_2(NO_2)_4]$	316.12	yel, rhomb	2.076^{15}			$1.758^{16.5}$		
	Colombium	see Niobium.								
c540	**Copper**	Cu	63.546	redsh met, cub	8.92	1083 ± 0.1	2595	i	i	s HNO_3, h H_2SO_4; v sl s HCl, NH_4OH
c541	acetate, basic	Blue verdigris. $Cu(C_2H_3O_2)_2.CuO.6H_2O$	369.26	grnsh-bl powd				sl s		s dil a, NH_4OH; sl s al
c542	(II) acetate	Neutral verdigris. $Cu(C_2H_3O_2)_2.H_2O$	199.65	dk grn powd, 1.545, 1.550	1.882, anhydr 1.93	115	d 240	7.2	20	7.14 al; s eth
c543	(II) acetate metaarsenate	Paris green. $Cu(C_2H_3O_2)_2.3Cu(AsO_2)_2$ (approx)	1013.77	em grn powd				i		s a, NH_4OH; i al
c544	(III) acetylide	Cu_2C_2	151.10	red, amorph, expl.		exp		v sl s		s a, KCN
c545	amine azide	$Cu(NH_3)_2(N_3)_2$	181.64	dk grn cr, exp		d 100–105	exp 202	i	d	d a; i MeOH
c546	(II) *di*amminechloride, di-	$Cu(NH_3)_2Cl_2$	168.51	grn cr	2.32^{25}_{4}	260–270	d 300	i		s NH_4OH; i abs a
c547	(II) *hex*amminechloride, di-	$Cu(NH_3)_6Cl_2$	236.63	bl, cub	1.48^{25}_{4}			v s		
c548	*tet*rammine dithionate	$[Cu(NH_3)_4]S_2O_6$	291.79	vlt-bl cr		d 160		s	d	
c549	(II) *tet*rammine nitrate	$[Cu(NH_3)_4](NO_3)_2$	255.67	dk-bl, oct	1.91^{25}_{4}	d 210 exp		s		
c550	(II) amine nitrate	$[Cu(NH_3)_4](NO_2)_2$	223.61	vlt-bl, tetr		$-2NH_3$ 97		v s		
c551	*tet*rammine sulfate	Cuprum ammoniacale. $[Cu(NH_3)_4]SO_4.H_2O$	245.74	dk-bl, rhomb, unstab	1.79^{25}_{4}	$-NH_3.H_2O$, 30		$18.05^{21.5}$		
c552	(tri-)antimonide	Cu_3Sb	312.37	gray	8.51	687				
c553	(II) *ortho*arsenate	$Cu_3(AsO_4)_2.4H_2O$	540.52	blsh-grn				i	i	s a, NH_4OH
c554	(II) *ortho*arsenate, di-H	$Cu_5H_2(AsO_4)_4.2H_2O$	911.42	bl				i		s a, NH_4OH
c555	arsenide	Cu_5As_2	467.54	bl, oct	7.56			i	i	s a, NH_4OH
c556	tri-arsenide	Nat. domeykite. Cu_3As.	265.54	hex	8.0	830				
c557	(II) *ortho*arsenite, hydrogen(?)	Scheele's green. $CuHAsO_3$(?)	187.47	grn powd		d		i	i	s a, NH_4OH; i al
c558	(I) azide	CuN_3	105.56	col cr, v exp	3.26			0.00075^{20}		d conc H_2SO_4; s NH_4Cl
c559	(II) azide	$Cu(N_3)_2$	147.58	brn-red or brn-yel cr, exp	2.604	exp 215		0.008^{20}		v s dil a
c560	(II) benzoate	$Cu(C_7H_5O_2)_2.2H_2O$	341.80	lt bl cr powd		$-H_2O$, 110		sl s		s dil a; sl s al
c561	(II) *meta*borate	$Cu(BO_2)_2$	149.16	blsh grn cr powd	3.859			s		
c562	boride	Cu_3B_2	212.24	yel	8.116					
c563	(II) bromate	$Cu(BrO_3)_2.6H_2O$	427.45	bl-grn, cub	2.583	d 180	$-6H_2O$, 200	v s		s NH_4OH
c564	(I) bromide	CuBr (or Cu_2Br_2)	143.45	wh, cub, 2.116	4.98	492	1345	v sl s	d	s HBr, HCl, HNO_3, NH_4OH; i acet
c565	(II) bromide	$CuBr_2$	223.31	blk, monocl, deliq	4.77^{25}_{4}	498		v s		s al, acet, NH_3, pyr; i bz
c566	*trioxy*bromide	$CuBr_2.3Cu(OH)_2$	516.02	em grn, rhomb	4.00	$-H_2O$, 210–215	d 240–250	i	d	s dil min a, NH_4OH; v s a
c567	(II) butyrate	$Cu(C_4H_7O_2)_2.2H_2O$	273.77	dk grn cr				v sl s		s al, eth, NH_4O, dil a
c568	(I) carbonate	Cu_2CO_3	187.09	yel	4.40	d		i	i	s a, NH_4OH
c569	(II) carbonate, basic	Nat. malachite. $CuCO_3.Cu(OH)_2$	221.11	dk grn, monocl, 1.655, 1.875, 1.909	4.0	d 200		i	d	0.026 aq CO_2; s a, NH_4OH, KCN; i al
c570	(II) carbonate, basic	Nat. azurite, chessylite. $2CuCO_3.Cu(OH)_2$	344.65	bl, monocl, 1.730, 1.758, 1.838	3.88	d 220		i	d	s NH_4OH, h $NaHCO_3$
c571	(II) chlorate	$Cu(ClO_3)_2.6H_2O$	338.53	grn, cub, deliq		65	d 100	207^{0}	v s	s al, acet
c572	(II) chlorate, basic	$Cu(ClO_3)_2.3Cu(OH)_2$	523.11	grn cr or amorph	3.55	d		i	i	s dil a

No.	Name	Synonyms and Formulae	Mol. wt.	Crystalline form, properties and index of refraction	Density or spec. gravity	Melting point, °C	Boiling point, °C	Solubility, in grams per 100 cc		
								Cold water	Hot water	Other solvents
	Copper									
573	*per*chlorate	$Cu(ClO_4)_2$	262.43	monocl, 1.495, 1.505, 1.522	2.225^{23}	82.3		s	s	
574	*per*chlorate, hexahydrate	$Cu(ClO_4)_2.6H_2O$	370.53	lt bl, tricl, deliq, 1.505	2.225^{25}_{4}	82	d 120	v s		s al, eth
575	(I) chloride(ous)	Nat. nantokite. $CuCl$ (or Cu_2Cl_2)	98.99	wh, cub, 1.93	4.14	430	1490	0.0062		s HCl, NH_4OH, eth; i al
576	(II) chloride	$CuCl_2$	134.44	br, yel powd, hygr	3.386^{25}_{4}	620	993 d to CuCl	70.6^{0}	107.9^{100}	53^{15} al; 68^{15} MeOH; s h H_2SO_4, acet
577	(II) chloride, basic	$CuCl_2.Cu(OH)_2$	232.00	yel-grn, hex	3.78	$-H_2O$, 250	d red heat	d	d	
578	(II) chloride, dihydrate	Nat. eriochalcite. $CuCl_2.2H_2O$	170.47	bl-grn, rhomb, deliq. 1.644, 1.683, 1.731	2.54	$-2H_2O$, 100	d	110.4^{0}	192.4^{100}	s al, NH_4OH
579	chloride, thioureate	$CuCl.3[CS(NH_2)_2]$	327.35	col pr, 1.758, 1.17719	1.73	168		v s		
580	(II) chromate, basic	$CuCrO_4.2CuO.2H_2O$	374.64	yel br		$-2H_2O$, 260		i		s dil a, NH_4OH; i al
581	(II) *di*chromate	$CuCr_2O_7.2H_2O$	315.56	blk cr, deliq	2.283	$-2H_2O$, 100		v s	d	s a, NH_4OH, al
582	(I) chromite	$Cu_2Cr_2O_4$	295.07	gray blk cub pl	5.24^{20}			i	i	s HNO_3
583	(II) citrate	$2Cu_2C_6H_4O_7.5H_2O$	720.43	blsh grn powd		$-H_2O$, 100		i		s a, NH_4OH
584	(I) cyanide	$CuCN$	89.56	wh, monocl pr	2.92	473 (in N_2)	d	i	i	s HCl, KCN, NH_4OH; sl s liq NH_3
585	(II) cyanide	$Cu(CN)_2$	115.58	yel-grn powd		d		i		s a, alk, KCN, pyr,
586	ethylacetoacetate	$Cu(C_6H_9O_3)_2$	321.81	grn need		192–193	subl	i		v s al, eth; 10^{80} bz
587	(I) ferricyanide	$Cu_3Fe(CN)_6$	402.57	br red				i		i HCl; s NH_4OH
588	(II) ferricyanide	$Cu_3[Fe(CN)_6]_2.14H_2O$	866.74	yel-grn				i		i HCl; s NH_4OH
589	(II) ferrocyanide	Hatchett's brown. $Cu_2Fe(CN)_6.xH_2O$		red brn				i	i	i a, NH_3; s NH_4OH
590	(I) fluogallate	$[Cu(H_2O)_6][GaF_5.H_2O]$	354.36	pa bl, monocl(?), 1.45	2.20	$-5H_2O$, 110		sl s		s HF
591	(I) fluoride	CuF (or Cu_2F_2)	82.54	red cr, (exist?)		908	subl, 1100	i		s HCl, HF; d HNO_3; i al
592	(II) fluoride	CuF_2	101.54	wh, tricl	4.23	d 950		4.7^{20}	s	s dil min a; i al
593	(II) fluoride dihydrate	$CuF_2.2H_2O$	137.57	bl, monocl	2.93^{25}_{4}	d		4.7^{20}	d	s HCl, HF, HNO_3, al; i acet, NH_3
594	(I) fluosilicate	Cu_2SiF_6	269.16	red powd		d to SiF_4			d 100	
595	(II) fluosilicate	$CuSiF_6.4H_2O$	277.68	monocl pr	2.158			42.8		
596	(II) fluosilicate hexahydrate	$CuSiF_6.6H_2O$	313.71	bl, rhomb, deliq, 1.409, 1.408	2.207			232^{17}		0.16^{20} 92 % al
597	(II) formate	$Cu(CHO_2)_2$	153.55	bl, monocl	1.831			12.5	d	0.25 al
598	(II) formate tetrahydrate	$Cu(CHO_2)_2.4H_2O$	225.61	bl cr	1.81	$-H_2O$, 130		6.2		s alk; sl s al
599	(II) glycerine deriv	$Cu(C_2H_4NO_2)_2.H_2O$	229.67	bl need		$-H_2O$, 130		0.57^{15}		s alk
600	hydride	CuH (or Cu_2H_2)	64.55	red-brn, (exist?)	6.38	d sl 55–60		i	d 65	d HCl
601	(II) hydroxide	$Cu(OH)_2$	97.56	bl gel cr powd	3.368	$-H_2O$, d		i	d	s a, NH_4OH, KCN
602	(II) *tri*hydroxychloride	γ: Paratacamite δ: atacamite $CuCl_2.3Cu(OH)_2$	427.11	γ: grn, hex, δ: grn rhomb; γ: 1.743, 1.849, δ:1.861, 1.861, 1.880, grn lt	(γ)3.75	$-H_2O$, 250		i	i	v s a
603	(II) *tri*hydroxynitrate	$Cu(NO_3)_2.3Cu(OH)_2$	480.27	dk grn, rhomb or monocl	rhomb, 3.41 monocl, 3.378	$-H_2O$ ~400		i	d	v s a
604	(II) iodate	$Cu(IO_3)_2$	413.35	grn, monocl	5.241^{15}	d		0.1364^{15}	i	s dil HNO_3, dil H_2SO_4
605	(II) iodate, basic	$Cu(OH)IO_3$	255.45	grn, rhomb	4.873	d 290		i	i	s dil H_2SO_4
606	(II) iodate, monohydrate	Nat. bellingerite. $Cu(IO_3)_2.H_2O$	431.36	bl, tricl	4.872	$-H_2O$, 248	d 290	0.33^{15}	0.65^{100}	s dil H_2SO_4, NH_4OH; i al, dil HNO_3
607	*paraper*iodate	Cu_5HIO_6	350.00	grn cr powd		d 110		i	i	s HNO_3, NH_4OH
608	(I) iodide	Nat. marshite. CuI (or Cu_2I_2)	190.44	wh or brnsh-wh, cub, 2.346	5.62	605	1290	0.0008^{18}		s dil HCl, KI, KCN, conc H_2SO_4, liq NH_3
609	(II) lactate	$Cu(C_3H_5O_3)_2.2H_2O$	277.71	dk bl, monocl				16.7	45^{100}	s NH_4OH; sl s al
610	(II) laurate	$Cu(C_{12}H_{23}O_2)_2$	462.17	lt bl powd		111–113		sl s	sl s	
611	mercury iodide (α)	Cu_2HgI_4	835.29	red, tetr	6.116	tr *ca* 67		i		
612	mercury iodide (β)	Cu_2HgI_4	835.29	choc, cub	6.102			i		

No.	Name	Synonyms and Formulae	Mol. wt.	Crystalline form, properties and index of refraction	Density or spec. gravity	Melting point, °C	Boiling point, °C	Solubility, in grams per 100 cc		
								Cold water	Hot water	Other solvents
	Copper									
c613	(II) nitrate, hexahydrate	$Cu(NO_3)_2.6H_2O$	295.64	bl cr, deliq	2.074	$-3H_2O$, 26.4		243.7^{0}	∞	s al
c614	(II) nitrate, trihydrate	$Cu(NO_3)_2.3H_2O$	241.60	bl cr, deliq	2.32^{25}_{4}	114.5	$-HNO_3$, 170	137.8^{0}	1270^{100}	$100^{12.5}$ al; v s liq NH_3
c615	nitride	Cu_3N	204.63	dk grn powd	5.84^{25}_{4}	d 300		d		d a
c616	(II) nitrite, basic	$Cu(NO_2)_2.3Cu(OH)_2$	448.22	grn powd		d 120		i	d	v s dil a; sl s al; s NH_4OH
c617	(II) *hypo*nitrite, basic	$Cu(NO)_2.Cu(OH)_2$	221.11	pea grn amorph, hygr		d<100		i		s dil a; v s NH_4OH; d NaOH
c618	(II) nitroprusside	$CuFe(CN)_5NO.2H_2O$	315.51	wh-grnsh powd				i		s alk; i al
c619	(II) oleate	$Cu(C_{18}H_{33}O_2)_2$	626.47	br powd or grn-bl mass, pois				i	i	s eth
c620	(II) oxalate	$CuC_2O_4.\frac{1}{2}H_2O$	160.57	bl wh				0.00253^{25}		s NH_4OH; i ac a
c621	(I) oxide	Nat. cuprite. Cu_2O	143.08	red, oct cub, 2.705	6.0	1235	−O, 1800	i	i	s HCl, NH_4Cl, NH_4OH; sl s HNO_3; i al
c622	(II) oxide	Nat. tenorite. CuO	79.54	blk, monocl, β 2.63	6.3–6.49	1326		i	i	s a, NH_4Cl, KCN
c623	oxide, per-	$CuO_2.H_2O$	113.55	br or brnsh-blk cr		d 60		i		i al, s d a
c624	oxide, sub-	Cu_4O	270.16	olv grn, (exist?).		d		i		d a
c625	(II) oxychloride	Nat. atacamite. $Cu_2(OH)_3Cl$ (or $CuCl_2.3Cu(OH)_2$)	213.56	grn, orthorhomb.	3.76–3.78					
c626	(II) oxychloride	Brunswick green. $CuCl_2.3CuO.4H_2O$(?)	445.13	grn powd, or em grn to grnsh-blk, rhomb		$-3H_2O$, 140			d 100	s a, NH_4OH
c627	(II) palmitate	$Cu(C_{16}H_{31}O_2)_2$	574.39	grn-bl powd		120		i		s h bz, CS_2, CCl_4; sl s al, eth; i MeOH, acet
c628	2,4-pentanedione	Copper acetylacetonate. $Cu(C_5H_7O_2)_2$	261.76	bl cr		>230	subl	i		sl s al; s chl
c629	(I) phenyl	C_6H_5Cu	140.65	col powd		d 80		d	d	i al, CS_2; s pyr
c630	(II) *ortho*phosphate	$Cu_3(PO_4)_2.3H_2O$	434.61	bl, rhomb		d		i	sl s	s a, NH_4OH, H_3PO_4; i NH_3
c631	(tri-) phosphide	Cu_3P	221.59	gray-blk	6.4–6.8	d		i		s HNO_3; i HCl
c632	(di-) pyridine chloride(di)	$Cu(C_5H_5N)_2Cl_2$	316.67	grn-bluish, monocl, 1.60, 1.75	1.76	d 263		s	d	sl s c al, chl
c633	(II) salicylate	$Cu(C_7H_5O_3)_2.4H_2O$	409.83	bl-grn need				v s		v s al, NH_4OH
c634	(II) selenate	$CuSeO_4.5H_2O$	296.57	bl, tricl, 1.56	2.559	$-4H_2O$, 50–100	$-5H_2O$, 150	25.7^{15}	d	s a, NH_4OH; v s acet; i a
c635	(I) selenide	Cu_2Se	206.04	blk, cub	6.749^{30}_{4}	1113				d HCl
c636	(II) selenide	CuSe	142.50	grn-blk hex pl, unstab	5.99	d red heat		i	i	sl s HCl, NH_4O; s h HNO_3
c637	selenite	$CuSeO_3.2H_2O$	226.53	bl-grn, rhomb	3.31^{25}_{4}	$-H_2O$, 100		i	i	
c638	silicide	Cu_4Si	282.25	wh met	7.53	850				i HCl; d HNO_3
c639	(II) stearate	$Cu(C_{18}H_{35}O_2)_2$	630.50	lt grn-bl amorph powd		125		i		s eth, h bz, chl, turp; sl s pyr; i MeOH, acet
c640	(I) sulfate	Cu_2SO_4	223.14	gray powd, 1.724, 1.733, 1.739	3.605	+O, 200		d		s conc HCl, NH; glac ac a
c641	(II) sulfate	Nat. hydrocyanite. $CuSO_4$	159.60	grn, wh, rhomb, 1.733	3.603	sl d above 200	d 650 to CuO	14.3^{0}	75.4^{100}	1.04^{18} MeOH; i
c642	(II) sulfate, basic	Nat. brochantite. $CuSO_4.3Cu(OH)_2$	452.27	grn, monocl, 1.728, 1.771, 1.800	3.78	d 300		i	i	s a, NH_4OH
c643	(II) sulfate, pentahydrate	Blue vitriol, nat. chalcanthite. $CuSO_4.5H_2O$	249.68	bl, tricl, 1.514, 1.537, 1.543	2.284	$-4H_2O$, 110	$-5H_2O$, 150	31.6^{0}	203.3^{100}	15.6^{18} MeOH; i
c644	(I) sulfide	Nat. chalcocite. Cu_2S	159.14	blk, rhomb	5.6	1100		$\times 10^{-14}$		s HNO_3, NH_4O; i acet
c645	(II) sulfide	Nat. covellite. CuS	95.60	blk, monocl or hex, 1.45	4.6	tr 103	d 220	0.000033^{18}		s HNO_3, KCN, h HCl, H_2SO_4; i al, alk
c646	(I) sulfite, monohydrate	$Cu_2SO_3.H_2O$	225.16	red pr	4.46^{15}			i		d dil a; s NH_4O
c647	(I) sulfite, monohydrate	$Cu_2SO_3.H_2O$	225.16	wh, hex	3.83^{15}	d		sl s		s HCl, NH_4OH; i al, eth
c648	(I, II) sulfite, dihydrate	Chevreul's salt. $Cu_2SO_3.CuSO_3.2H_2O$	386.78	red cr	3.57	d 200		i	i	s HCl, NH_4OH; sl s HNO_3
c649	(II) tartrate	$CuC_4H_4O_6$	211.61	lt bl powd				v sl s		s a, alk
c650	(II) tartrate, trihydrate	$CuC_4H_4O_6.3H_2O$	265.66	lt gray-bl powd		d		0.02^{15}	0.14^{85}	s a, alk

No.	Name	Synonyms and Formulae	Mol. wt.	Crystalline form, properties and index of refraction	Density or spec. gravity	Melting point, °C	Boiling point, °C	Solubility, in grams per 100 cc: Cold water	Hot water	Other solvents
	Copper									
c651	telluride	Cu_2Te	254.68	bl-blk, oct	7.27					s $Br+H_2O$; i HCl, H_2SO_4
c652	telluride	Nat. rickardite. Cu_4Te_3	636.96	purp, tetr	7.54					
c653	tellurite	$CuTeO_3$	239.10	blk glass				i	i	s conc HCl
c654	(I) thiocyanate	CuSCN	121.62	wh	2.843	1084		0.0005^{18}		s NH_4OH; sl s ac a; i al; d min a
c655	(II) thiocyanate	$Cu(SCN)_2$	179.70	blk		d 100		d	d	s a, NH_4OH
c656	(II) tungstate	$CuWO_4.2H_2O$	347.42	lt-grn, oct		red heat		0.1^{15}		s NH_4OH; sl s ac a; i al; d min a
c657	xanthate	Copper ethylxanthogenate. $Cu(C_3H_5OS_2)_2$	305.94	yel ppt		d		i		s NH_4OH; v sl s al; i CS_2
	Copper complexes									
c658	diamminecopper (II) acetate	$Cu(C_2H_3O_2)_2.2NH_3$	215.69	vlt bl cr		d *ca* 175		s d		s ac a, NH_4OH; i al
c659	tetrammine copper (II) sulfate	$[Cu(NH_3)_4]SO_4.H_2O$	245.74	bl, rhomb	1.81	d 150		$18.5^{21.5}$	d	i al
c660	tetrapyridine copper-(II) fluosilicate	$[Cu(C_5H_5N)_4]SiF_6$	522.03	purp-bl, rhomb	2.108					
c661	tetrapyridine copper-(II) *per*rhenate	$[Cu(C_5H_5N)_4](ReO_4)_2$	880.34	bl cr, monocl	2.338			0.5555		
c662	**Cyanic acid** isocyanic acid	HOCN	43.03	liq	1.14^{20}_{4}			s d		s eth, bz, tol
c663	**Cyanoauric acid**	$HAu(CN)_4.3H_2O$	356.09	tab		50	d	s		s al, eth
c664	**Cobalticyanic acid**	$[H_3Co(CN)_6]_2.H_2O$	454.14	col need, deliq		d 100		s		s al, HCl, dil HNO_3, dil H_2SO_4
c665	**Cyanogen**	$(CN)_2$	52.04	col gas, pungent odor v pois	2.335 g/l, liq: $0.9577^{-21.17}$	−27.9	−20.7	450^{20} cm^3		230 cm^3 al, 500 cm^3 eth
c666	**Cyanogen compounds**	See organic tables								
d1	**Deuterium**	Heavy hydrogen. D_2	4.032	col gas	liq $0.169^{-250.9}$	−254.6	−249.7	sl s		
d2	deuterium chloride	DCl	37.47	gas		−114.8	−81.6			
d3	deuterium oxide	Heavy water. D_2O	20.031	col liq or hex cr, 1.33844^{20}	1.105^{20}_{4}	3.82	101.42^{760}	24.1 cm^3 $11.9^{25}cm^3$	8.4^{50} cm^3 $7.12^{50}cm^3$	
d4	**Dysprosium**	Dy	162.50	met, hex	8.556	1407	2600	i	i	s a
d5	acetate	$Dy(C_2H_3O_2)_3.4H_2O$	411.64	yel need		d 120		s		v sl s al
d6	bromate	$Dy(BrO_3)_3.9H_2O$	708.36	yel hex need		78	$-6H_2O$, 110	v s		sl s al
d7	bromide	$DyBr_3$	402.23	col cr		881	1480	s	s	
d8	carbonate	$Dy_2(CO_3)_3.4H_2O$	577.06			$-3H_2O$, 15		i		
d9	chloride	$DyCl_3$	268.85	shining yel pl	3.67^{0}_{4}	718	1500	s	s	
d10	chromate	$Dy_2(CrO_4)_3.10H_2O$	853.13	yel cr		$-3\frac{1}{2}H_2O$, 150	d		1.002^{25}	
d11	fluoride	DyF_3	219.50	col cr		1360	>2200	i	i	i dil a
d12	iodide	DyI_3	543.21	yelsh grn cr		955	1320	s	s	
d13	nitrate	$Dy(NO_3)_3.5H_2O$	438.58	yel cr		88.6		s		
d14	oxalate	$Dy_2(C_2O_4)_3.10H_2O$	769.21	wh pr		$-H_2O$, 40		i	i	s dil a
d15	oxide	Dysprosia. Dy_2O_3	373.00	wh powd	7.81^{27}	2340±10				grn soln a
d16	*ortho*phosphate	$DyPO_4.5H_2O$	347.55	yelsh-wh powd		$-5H_2O$> 200		i		s dil a, ac a
d17	selenate	$Dy_2(SeO_4)_3.8H_2O$	898.00	yel need		$-8H_2O$, 200		v s		i al
d18	sulfate	$Dy_2(SO_4)_3.8H_2O$	757.31	bril yel cr		stab 110	$-8H_2O$, 360	5.072^{20}	3.34^{40}	
e1	**Erbium**	Er	167.28	dk gray powd	9.164	1497	2900	i	i	s a
e2	acetate	$Er(C_2H_3O_2)_3.4H_2O$	416.48	wh cr, tricl	2.114					
e3	bromide	$ErBr_3.9H_2O$	569.15	vlt-rose cr		950	1460	s	s	
e4	chloride	$ErCl_3.6H_2O$	381.73	pink cr, deliq		774	1500	s	s	s al
e5	fluoride	ErF_3	224.28	rose cr		1350	2200	i	i	i dil a
e6	iodide	ErI_3	547.99	vlt-red cr		1020	1280	s	s	
e7	nitrate	$Er(NO_3)_3.5H_2O$	443.37	redsh cr		$-4H_2O$, 130		s		s al, eth, acet
e8	oxalate	$Er_2(C_2O_4)_3.10H_2O$	778.77	redsh micr powd	2.64(?)	d 575		i	i	i dil a
e9	oxide	Erbia. Er_2O_3	382.56	rose red powd, tr to cub at 1300	8.640	infus		0.00049^{24}		sl s min a
e10	sulfate	$Er_2(SO_4)_3$	622.74	wh powd, hygr	3.678	d 630		43^{0}		
e11	sulfate, octahydrate	$Er_2(SO_4)_3.8H_2O$	766.87	rose red, monocl	3.217	$-8H_2O$, 400		16^{20}	6.53^{40}	
e12	**Europium**	Eu	151.96	steel gray met, cub	5.244	826	1439	i	i	
e13	(II) bromide	$EuBr_2$	311.78			677	1880	s	s	
e14	(III) bromide	$EuBr_3$	391.69			702	d	s	s	
e15	(II) chloride	$EuCl_2$	222.87	wh, amorph		727	>2000	s	s	s a
e16	(III) chloride	$EuCl_3$	258.32	yel need	4.89^{20}	850				
e17	(II) fluoride	EuF_2	189.96	brt yel	6.495	1380	>2400	i	i	
e18	(III) fluoride	EuF_3	208.96	col		1390	2280	i	i	i dil a

No.	Name	Synonyms and Formulae	Mol. wt.	Crystalline form, properties and index of refraction	Density or spec. gravity	Melting point, °C	Boiling point, °C	Solubility, in grams per 100 cc: Cold water	Hot water	Other solvents
	Europium									
e19	(II) iodide	EuI_2	405.77	br to olv grn cr	5.50^{25}_{4}	527	1580	s	s	
e20	(III) iodide	EuI_3	532.68			877	d	s	s	
e21	(III) nitrate	$Eu(NO_3)_3.6H_2O$	446.07	col		85 (sealed tube)		v s	v s	
e22	oxide	Eu_2O_3	351.92	pa rose powd	7.42					
e23	(II) sulfate	$EuSO_4$	248.02	col, orthorhomb	4.989^{20}			i	i	i dil a
e24	(III) sulfate	$Eu_2(SO_4)_3.8H_2O$	736.23	pa rose cr	4.95 (anh)	$-8H_2O$, 375		2.563^{20}	1.93^{40}	
—	**Ferric or ferrous**	See *Iron*								
f1	**Ferricyanic acid**	$H_3Fe(CN)_6$	214.98	grn-brn need, deliq		d		s	s	s al
f2	**Ferrocyanic acid**	$H_4Fe(CN)_6$	215.99	wh need, bl in moist air		d		s	s	s al; i eth
f3	**Fluoboric acid**	HBF_4	87.81	col liq		d 130		∞	s	∞ al
f4	**Fluorophosphoric acid, di-**	HPO_2F_2	101.98	col fum liq	1.583^{25}_{4}	−75	116	s		
f5	**Fluorophosphoric acid, hexa-**	HPF_6	145.97	col fum liq	*ca* 1.65 (65%)	31 ($6H_2O$)				
f6	**Fluorophosphonic acid, mono-**	H_2PO_3F	99.99	col visc liq	1.818^{25}_{4}			∞		
f7	**Fluorine**	F	18.998	grn yel gas, pois, 1.000195	1.69^{15}g/l 1.51^{-188}	-219.62^{1atm}	-188.14^{1atm}	$HF+O_3$	d	
f8	(di-)oxide	F_2O	54.00	col gas or yel brn liq	liq 1.65^{-190}	−223.8	−144.8	sl s d	i	sl s alk, a
f9	oxide, di-	Dioxygen fluoride. F_2O_2	70.00	brn gas, red liq, orange solid	sol 1.912^{-165} liq 1.45^{-57}	−163.5	−57			
f10	**Fluosilicic acid**	$H_2SiF_6.xH_2O$	hydr (not known)	col fum coros liq, 1.3465^{25} (60.97% soln)	1.4634^{25} (60.97% soln)		d	s	s	sl s alk
f11	dihydrate	$H_2SiF_6.2H_2O$	180.12	wh cr, fum, deliq		d		s	s	s alk
f12	**Fluosulfonic acid**	HSO_3F	100.07	col liq	1.743^{15}	−87.3	165.5	s		
g1	**Gadolinium**	Gd	157.25	col or lt yel met, hex	7.948	1312	~3000	i	i	s a
g2	acetate tetrahydrate	$Gd(C_2H_3O_2)_3.4H_2O$	406.45	col, tricl	1.611			11.6^{25}		
g3	acetylacetonate, trihydrate	$Gd[CH(COCH_3)_2]_3.3H_2O$	508.63			143.5–145		i	i	
g4	bromide, hexahydrate	$GdBr_3.6H_2O$	505.07	rhomb pl	2.844^{16}			s	s	s HBr
g5	chloride	$GdCl_3$	263.61	col monocl, pr	4.52^{0}	609		s	s	
g6	chloride, hexahydrate	$GdCl_3.6H_2O$	371.70	wh pr, deliq	2.424^{0}			s	s	
g7	iodide	GdI_3	537.96	citr yel		926	1340	s	s	
g8	fluoride	GdF_3	214.25					i		sl s h HF
g9	*di*methylphosphate	$Gd[(CH_3)_2PO_4]_3$	532.37					23.0^{25}	6.7^{100}	s al
g10	nitrate, hexahydrate	$Gd(NO_3)_3.6H_2O$	451.36	tricl, deliq	2.332	91		v s	v s	s al
g11	nitrate, pentahydrate	$Gd(NO_3)_3.5H_2O$	433.34	pr	2.406^{15}	92		i	i	v sl s conc HNO_3
g12	oxalate	$Gd_2(C_2O_4)_3.10H_2O$	758.71	monocl		$-6H_2O$, 110		i		s HNO_3; v sl s H_2SO_4
g13	oxide	Gadolinia. Gd_2O_3	362.50	wh amorph powd, hygr	7.407^{15}			v sl s		s a
g14	selenate	$Gd_2(SeO_4)_3.8H_2O$	887.50	monocl, pearly	3.309	$-8H_2O$, 130		s	s	
g15	sulfate	$Gd_2(SO_4)_3$	602.68	col	$4.139^{14.5}$	d 500		3.98^{0}	$2.26^{34.4}$	
g16	sulfate, octahydrate	$Gd_2(SO_4)_3.8H_2O$	746.81	col, monocl	$3.010^{14.6}$			3.28^{20}	2.54^{40}	
g17	sulfide	Gd_2S_3	410.69	yel mass, hyg	3.8			d		d a
g18	**Gallium**	Ga	69.72	gray-blk orthorhomb, tendency to undercool	sol $5.904^{29.6}$ liq $6.095^{29.8}$	29.78	2403	i	i	s a, i alk
g19	acetate, basic	$4Ga(C_2H_3O_2)_3.2Ga_2O_3.5H_2O$	1452.37	wh micro cr		d 160		s d	d	i ac a
g20	acetylacetonate	2,4-Pentanedione deriv. $Ga(C_5H_7O_2)_3$	367.05	monocl or pl, rhomb or pyram, rhomb pyram	1.42, 1.41	194–195	subl 140^{10}	s	s	s acet
g21	arsenide	GaAs	144.64	dk-gray cub cr		1238				
g22	bromide, tri-	$GaBr_3$	309.45	wh cr	3.69^{25}_{4}	121.5±0.6	278.8	s	s	sl s NH_3
g23	bromide, tri-, hexammine	$GaBr_3.6NH_3$	411.63	wh powd				d	d	sl s NH_3
g24	bromide, tri-, monammine	$GaBr_3.NH_3$	326.48	wh powd	3.112^{25}	124		d	d	sl s NH_3
g25	*per*chlorate	$Ga(ClO_4)_3.6H_2O$	476.16			d 175		v s		v s al
g26	chloride, di-	$GaCl_2$	140.63	wh cr, deliq		164	535	d	d	s bz
g27	chloride, tri-	$GaCl_3$	176.03	wh need, deliq	2.47^{25}_{4} liq 2.36^{80}_{80}	77.9±02	201.3	v s	v s	s bz, CCl_4, CS_2

No.	Name	Synonyms and Formulae	Mol. wt.	Crystalline form, properties and index of refraction	Density or spec. gravity	Melting point, °C	Boiling point, °C	Solubility, in grams per 100 cc		
								Cold water	Hot water	Other solvents
	Gallium									
g28	chloride, tri-, hexammine	$GaCl_3.6NH_3$	278.26					d	d	s NH_3
g29	chloride, tri-, monoammine	$GaCl_3.NH_3$	193.11	wh powd	2.189^{25}	124		d	d	s NH_3
g30	ferrocyanide	$Ga_4[Fe(CN)_6]_3$	914.74			d		d	d	i conc HCl
g31	fluoride, tri-	GaF_3	126.72	wh powd	4.47±0.01	subl 800 (in N_2)	*ca* 1000	0.002	i	v sl s dil a; s HF
g32	fluoride, tri-, trihydrate	$GaF_3.3H_2O$	180.76	wh powd		−H_2O(vac) 140		i	sl s	sl s dil H_2F_2; v s dil HCl
g33	fluoride, tri-, triammine	$GaF_3.3NH_3$	177.81	wh powd		−NH_3, 100		d	d	
g34	hydride	Digallane, galloethane. Ga_2H_6	145.49	col liq		−21.4	139, d>130	d	d	d a, alk
g35	hydroxide	$Ga(OH)_3$	120.74	wh		d 440		i	i	s dil a
g36	hydroxyquinoline deriv.	$Ga(C_9H_6NO)_3$	502.18	grn-yel cr		>150	subl vac	0.0001	0.0012	s a, alk; sl s al
g37	iodide, tri-	GaI_3	450.43	lt yel cr	4.15^{25}_{4}	212±1	subl 345		d	
g38	iodide, tri-, hexammine	$GaI_3.6HN_3$	552.62	wh powd				d	d	
g39	iodide, tri-, monoammine	$GaI_3.NH_3$	467.46	wh powd	3.635^{25}_{4}			d	d	
g40	nitrate	$Ga(NO_3)_3.xH_2O$		wh cr, deliq		d 110	d to Ga_2O_3 200	v s	v s	s abs al; i eth
g41	nitride	GaN	83.73	dk gray powd	6.1	subl 800		i	i	i dil a; sl s h conc H_2SO_4, h conc NaOH
g42	oxide, sesqui-(α)	Ga_2O_3	187.44	wh, hex, rhomb, 1.92, 1.95	6.44	1900;tr to β 600		i	i	s alk; v sl s h a
g43	oxide, sesqui-(β)	Ga_2O_3	187.44	monocl, rhomb	5.88	1900(1740)		i	i	s alk; v sl s ha a
g44	oxide, sesqui-, monohydrate	$Ga_2O_3.H_2O$	205.45	wh micr cr, orthorhom, 1.84	5.2	−H_2O, 400 tr to Ga_2O_3		i	i	sl s a; s alk
g45	oxide, sub-	Ga_2O	155.44	blk brn powd	4.77^{25}_{4}	>660	subl >500	i	i	s a, alk
g46	oxalate	$Ga_2(C_2O_4)_3.4H_2O$	475.56	wh micro cr, hygr		−$4H_2O$, 180	d 200	0.4		
g47	oxychloride	$6GaOCl.14H_2O$	979.25	oct				i		v s KOH; i dil HNO_3; s acet
g48	selenate	$Ga_2(Se_4O)_3.16H_2O$	856.56	col, monocl or tricl cr				57.5^{25}	v s	
g49	selenide, mono-	GaSe	148.68	dk red-br greasy leaf	5.03^{25}_{4}	960±10				
g50	selenide, sesqui-	Ga_2Se_3	376.32	rdsh-bl brittle, hard	4.92^{25}_{4}	>1020±10				
g51	selenide, sub-	Ga_2Se	218.40	bl	5.02^{25}_{4}					
g52	sulfate	$Ga_2(SO_4)_3$	427.62	wh powd		diss 690^{760}		v s	v s	s al; i eth
g53	sulfate, hydrate	$Ga_2(SO_4)_3.18H_2O$	751.90	oct cr				v s	v s	s 60% al; i eth
g54	sulfide, mono-	GaS	101.78	yel cr	3.86^{25}_{4}	965±10		i	d	s a, alk
g55	sulfide, sesqui-	Ga_2S_3	235.63	yel cr, or wh amorph	3.65^{25}	1255±10		d	d	s a, alk
g56	sulfide, sub-	Ga_2S	171.50	dk gray	4.18^{25}_{4}	d vac 800		d	d	s a, alk
g57	telluride, mono-	GaTe	197.32	blk soft cr	5.44^{25}_{4}	824±2				
g58	telluride, sesqui-	Ga_2Te_3	522.24	blk brittle cr	5.57^{25}_{4}	790±2				
	Germane									
g59	bromo-	GeH_3Br	155.52	col liq	$2.34^{29.5}$	−32	52	d	d	i al; d alk
g60	chloro-	GeH_3Cl	111.07	col liq	1.75^{-25}	−52	28.0	d	d	i al; d alk
g61	chloro trifluoro-	GeF_3Cl	165.04	col gas		−66.2	−20.63	d	d	s abs al
g62	dibromo-	GeH_2Br_2	234.42		2.80^{0}	−15.0	89.0	d	d	i al; d alk
g63	dichloro-	GeH_2Cl_2	145.51	col liq	1.90^{-68}	−68.0	69.5	d	d	i al; d alk
g64	dichlorodifluoro-	$GeCl_2F_2$	181.49	col gas		−51.8	−2.8	d		s abs al
g65	tribromo-	Germanium bromoform $GeHBr_3$	313.33	col liq		−24	d	d	d	d alk
g66	trichloro-	Germanium chloroform. $GeHCl_3$	179.96	col liq	1.93^{0}	−71	75.2 d	d	d	d alk
g67	trichlorofluoro-	$GeCl_3F$	197.95	col liq		−49.8	37.5	d		s abs al
g68	**Germanium**	Ge	72.59	gray-wh met-cub	5.35^{20}_{20}	937.4	2830	i	i	s h H_2SO_4, aq reg; i alk

No.	Name	Synonyms and Formulae	Mol. wt.	Crystalline form, properties and index of refraction	Density or spec. gravtiy	Melting point, °C	Boiling point, °C	Solubility, in grams per 100 cc Cold water	Hot water	Other solvents
	Germanium									
g69	bromide, di-	$GeBr_2$	232.41	col need or pl		122	d	d	d	s a, $GeBr_4$, al; i bz
g70	bromide, tetra-	$GeBr_4$	392.23	gray-wh oct, 1.6269	3.132^{29}_{29}	26.1	186.5	d	d	s abs al, eth, bz; i conc H_2SO_4
g71	chloride, di-	$GeCl_2$	143.50	wh powd		d to Ge+ $GeCl_4$		d	d	s $GeCl_4$; i al, chl
g72	chloride, tetra-	$GeCl_4$	214.41	col liq, 1.464	1.8443^{30}	−49.5	84	d	d	s al, eth; v s dil HCl; i conc HCl conc H_2SO_4
g73	fluoride, di-	GeF_2	110.59	wh cr, hygr		d>350	subl	s	v s	
g74	fluoride, tetra-	GeF_4	148.58	col gas or liq, not liq at atm press	$2.46^{-36.5}$	subl −37		d to GeO+ H_2GeF_6		
g75	fluoride, tetra-	$GeF_4.3H_2O$	202.63	wh cr, deliq		d		s	s	
g76	hydride	Digermane. Ge_2H_6	151.23	liq	1.98^{-100}	−109	29; d 215	d		s liq NH_3
g77	hydride	Trigermane. Ge_3H_8	225.83	col liq	2.2^{20}	−105.6	110.5; d 195	i	i	s CCl_4
g78	hydride, tetra-	Germane. GeH_4	76.62	col gas	1.523^{-142}	−165	−88.5; d 350	i	i	s liq NH_3, NaOCl sl s h HCl
g79	imide	$Ge(NH)_2$	102.62	wh amorph powd		d 150			d to NH_3 +GeO_2	
g80	iodide, di-	GeI_2	326.40	or hex pl	5.37	d	subl vac 240	s	s d	s conc HI, dil a; sl s CCl_4, chl; i CS_2
g81	iodide, tetra-	GeI_4	580.21	red-or, cub	4.322^{26}_{26}	144	d 440	s d		d al, acet; s CS_2, CCl_4, bz, MeOH
g82	(tri-) nitride, di-	Ge_3N_2	245.78	blk cr			subl 650			
g83	(tri-) nitride, tetra-	Ge_3N_4	273.80	wh-lt brn powd	5.25^{25}_{4}	d 450		i	i	i a, alk
g84	oxide, d-(insoluble)	GeO_2	104.59	tetr	6.239	1086±5		i		sl s NaOH; i HCl
g85	oxide, di-(soluble)	GeO_2	104.59	col, hex, 1.650	4.228^{25}	1115.0±4		0.447^{25}	1.07^{100}	s a, alk; i HCl, HF; one form NaOH, NH_4OH; one form
g86	oxide, mono-	GeO	88.59	blk cr powd, 1.607		subl 710		i	i	s Cl_2 water, H_2O_2+NH_4OH; i a, alk
g87	oxychloride	$GeOCl_2$	159.50	col liq		−56	d>20	d	d	i all solv
g88	selenide	$GeSe_2$	230.51	orange, rhomb(?)	4.56^{25}	707±3	d	i	i	v sl s a; sl s alk
g89	sulfide, di-	GeS_2	136.72	wh powd, or wh, orthorhomb	2.94^{14}	*ca* 800	subl >600	0.45 d	d to GeO_2 +H_2S	s alk, alk sulf; i al, eth, a; 3.112 liq NH_3
g90	sulfide, mono-	GeS	104.65	yel-red amorph, or rhomb bipyram, blk	amorph: 3.31 rhomb: 4.01	530	subl 430	0.24	i	s HCl, alk or alk sulf; sl s NH_4OH 0.0473 liq NH_3
	Glucinum	See **Beryllium**								
g91	**Gold**	Au	196.967	yel duct met, cub, coll blue-viol	19.3 liq 17.0^{1063}	1063	2966	i	i	s aq reg, KCN, h H_2SO_4; i a
g92	(I) bromide	AuBr	276.88	yel-gray mass, or cr powd	7.9	d 115		i	i	d a; s NaCN
g93	(III) bromide	$AuBr_3$	436.69	gray powd, or br cr		97.5–Br, 160		sl s		s eth, al
g94	(I) chloride	AuCl	232.42	yel cr	7.4	170 d to $AuCl_3$	d 289.5	v sl s d	d	s HCl, HBr
g95	(III) chloride	$AuCl_3$ or Au_2Cl_6	303.33	claret-red cr pr	3.9	d 254	subl 265	68	v s	s al, eth; sl s NH_3 i CS_2
g96	(I) cyanide	AuCN	222.98	lt yel cr powd	7.12^{25}	d		v sl s	v sl s	s KCN, NH_4OH; i eth, alk
g97	(III) cyanide	Cyanoauric acid. $Au(CN)_3.3H_2O$ or $HAu(CN_4).3H_2O$	329.07	col pl, hygr		d 50		v s	d, v s	s al, eth
g98	(I) iodide	AuI	323.87	grnsh-yel powd	8.25	d 120		v sl s	sl s d	s KI
g99	(III) iodide	AuI_3	577.68	dk grn				i	d	s iodides
g100	(III) nitrate, hydrogen	Nitratoauric acid. $AuH(NO_3)_4.3H_2O$ or $H[Au(NO_3)_4].3H_2O$	500.04	yel, tricl, oct	2.84	d 72		s, d		s HNO_3
g101	(III) oxide	Au_2O_3	441.93			−O, 160	−3O, 250	i	i	s HCl, conc HNO_3 NaCN
g102	(III) oxide	$Au_2O_3.xH_2O$				−1½H_2O, 250		5.7×10^{-11} 25		s HCl, NaCN, conc HNO_3
g103	phosphide	Au_2P_3	486.86	gray	6.67		d			i HCl, dil HNO_3
g104	selenide	Au_2Se_3	630.81		4.65^{22}					
g105	(I) sulfide	Au_2S	426.00	br blk powd		d 240		i fresh sol	ppt coll	s aq reg, KCN; i s
g106	(III) sulfide	Au_2S_3	490.13	br blk powd	8.754	d 197		i		i al, eth, s Na_2S

No.	Name	Synonyms and Formulae	Mol. wt.	Crystalline form, properties and index of refraction	Density or spec. gravity	Melting point, °C	Boiling point, °C	Solubility, in grams per 100 cc		
								Cold water	Hot water	Other solvents
	Gold									
107	telluride, di-	Nat. krennerite. $AuTe_2$	452.16	1) rhomb, 2) monocl, 3) tricl yel earthy to massive	8.2–9.3	d 472		i	i	
h1	**Hafnium**	Hf	178.49	hex	13.31^{20}	2150	5400	i	i	s HF
h2	bromide	$HfBr_4$	498.13	wh		subl 420				
h3	carbide	HfC	190.54		12.20	*ca* 3890		i		
h4	chloride	$HfCl_4$	320.30	wh		subl 319		d		s MeOH, acet
h5	fluoride	HfF_4	254.48	monocl, 1.56						
h6	iodide	HfI_4	686.11				400 subl vac			
h7	nitride	HfN	192.50	yel-brn, cub		3305				
h8	oxide	Hafnia. HfO_2	210.49	wh, cub	9.68^{20}	2812	~5400(?)	i	i	
h9	oxychloride	$HfOCl_2.8H_2O$	409.52	col				s		
h10	**Helium**	He	4.0026	col gas, inert odorless	0.1785^{0} g/l liq $0.147^{-270.8}$	$-272.2^{26 atm}$	268.6	0.94^{0} cm³ 0.94^{25} cm³	1.05^{50} cm³ 1.21^{75} cm³	i al; absorbed by Pt
h11	**Holmium**	Ho	164.93	met, hex	8.803	1461	2600	i	i	
h12	bromide	$HoBr_3$	404.66	lt yel		914	1470	s	s	
h13	chloride	$HoCl_3$	271.29	lt yel		718	1500	s	s	
h14	iodide	HoI_3	545.64	lt yel		989	1300	s	s	
h15	fluoride	HoF_3	221.93	lt yel		1143	>2200	i	i	i dil a
h16	oxalate	$Ho_2(C_2O_4)_3.10H_2O$	774.10	pa tan		$-H_2O$, 40	d	i	i	i dil a
h17	oxide	Holmia. Ho_2O_3	377.86	tan				i	i	s a
h18	**Hydrazine**	NH_2NH_2	32.05	col liq or wh cr, 1.470^{22}	liq 1.011^{15}	1.4	113.5	v s		s al
h19	azide	$N_2H_4.HN_3$	75.07	wh pr, deliq, 1.53, 1.76		75.4		v s	v s	1.2^{23} al; 6.1^{23} MeOH; i CS_2 bz
h20	fluogermanate	$2N_2H_4.H_2GeF_6$	252.80	monocl pr, 1.452, 1.460, 1.464	2.406^{25}_{25}			s		
h21	fluosilicate	$N_2H_4.H_2SiF_6$	176.14	cr		d 186		v s		sl s al
h22	formate	$N_2H_4.2CH_2O_2$	124.10			128		s		
h23	hydrate	$N_2H_4.H_2O$	50.07	col fum liq or cub cr, 1.42842	1.03^{21}	−40	118.5^{740}	∞	∞	s al; i eth, chl
h24	hydrochloride, di-	$N_2H_4.2HCl$	104.97	col vitr, oct	1.42	198, −HCl	d 200	27.2^{22}	v s	sl s al
h25	hydrochloride, mono-	$N_2H_4.HCl$	68.51	wh need		89	d 240	v s		v sl s al; v s liq NH_3
h26	hydroiodide	Hydrazine monoiodide. $N_2H_4.HI$	159.98	col pr		124–126		s		
h27	nitrate, di-	$N_2H_4.2HNO_3$	158.07	col cr		104 (rapid heat); d 80 (slow heat)		v s	d	
h28	nitrate, mono-	$N_2H_4.HNO_3$	95.06	col dimorph need (α, β)		α70.71; β62.09	subl 140	174.9^{10}	2127^{60}	sl s al
h29	oxalate	$2N_2H_4.H_2C_2O_4$	154.14	wh need		148		200^{35}		0.0003^{22} al; i eth
h30	*per*chlorate	$N_2H_4.HClO_4.\frac{1}{2}H_2O$	141.51	exp	1.939	137	d 145	d		s al; i eth, bz, chl, CS_2
h31	*hypo*phosphate	$N_2H_4.2H_2PO_3$	194.03			152				
h32	*ortho*phosphate	$N_2H_4.H_3PO_4$	130.05	cr, hygr		82		v s		
h33	*ortho*phosphite	$N_2H_4.2H_3PO_3$	196.06			82				
h34	*ortho*phosphite	$N_2H_4.H_3PO_3$	114.05			36		v s		
h35	picrate	$N_2H_4.HC_6H_2N_3O_7.\frac{1}{2}H_2O$	270.16			201.3		s	s	
h36	selenate	$N_2H_4.H_2SeO_4$	177.02	col cr powd, unstab		exp		v sl s	v s	
h37	sulfate	$N_2H_4H_2SO_4$	130.13	col, rhomb	1.37	254	d	3.415^{25}	14.39^{80}	i al
h38	sulfate	$(N_2H_4)_2.H_2SO_4$	162.18	col cr, hygr		85		202.2^{25}	554.4^{60}	i al
h39	tartrate	$(N_2H_4)C_4H_6O_6$	182.13	col cr; $[\alpha]^{20}_{4}$ +22.5		182–183		6.0^{0}		
h40	**Hydroazoic acid**	Azoimide. HN_3	43.03	col liq	1.09^{25}_{4}	−80	37	∞	∞	s al, alk, eth
h41	**Hydrogen**	H_2	2.0159	col gas, cub sol	gas 0.0899 g/l liq 0.070	−259.14	−252.5	2.14^{0} cm³ 1.91^{25} cm³	0.85^{80} cm³ 1.89^{60} cm³	6.925^{0} cm³ al
h42	antimonide	Stibine H_3Sb	124.77	col gas, pois	gas 5.30^{0} g/l liq 2.26^{-25}	−88.5	−17	20 cm³	4 cm³	1500 cm³ al; 2500 cm³ CS_2
h43	arsenide	Arsine H_3As	77.95	col gas, pois	3.484 g/l	−113.5	−55, d 230	20 cm³	sl s	sl s al, alk
h44	arsenide (solid)	H_2As_2	151.86	br powd		d 200		i	d	i al, eth, CS_2, alk; s HNO_3
h45	bismuthide	Bismuthine. H_3Bi	212.00	liq, v unstab			22			
h46	bromide	Hydrobromic acid. HBr	80.92	col gas or pa yel liq, 1.325	gas 3.5^{0} g/l liq 2.77^{-67}	−88.5	−67.0	221^{0}	130^{100}	s al
h47	bromide (const. boiling)	HBr(47%)+H_2O		col liq	1.49	−11	126			
h48	bromide, dihydrate	$HBr.2H_2O$	116.95	wh cr, col liq	2.11^{-15}	−11		s	s	

No.	Name	Synonyms and Formulae	Mol. wt.	Crystalline form, properties and index of refraction	Density or spec. gravity	Melting point, °C	Boiling point, °C	Solubility, in grams per 100 cc Cold water	Hot water	Other solvents
	Hydrogen									
h49	bromide, monohydrate	$HBr.H_2O$	98.93	col liq	1.78	stab	−3.6 to −15.5 between 1–2.5 atm			
h50	chloride	Hydrochloric acid. HCl	36.46	col gas or col liq, pois	$1.187^{-84.9}$ gas 1.00045 g/l	−114.8	−84.9	82.3^{0}	56.1^{60}	327 cm^3 al; s eth bz
h51	chloride (const. boiling)	HCl(20.24 %) + H_2O		col liq	1.097		110			
h52	chloride, dihydrate	$HCl.2H_2O$	72.49	col liq	$1.46^{18.3}$	−17.7	d	d	∞	s al
h53	chloride, monohydrate	$HCl.H_2O$	54.48	col liq	1.48	−15.35		∞	∞	s al
h54	chloride, trihydrate	$HCl.3H_2O$	90.51	col liq		−24.4	d	∞		s al
h55	cyanide	Hydrocyanic acid. HCN	27.03	col liq or gas, pois, liq 1.2675^{10}	gas 0.901 g/l liq 0.699^{22}	−14	26	∞	∞	∞ al; s eth
h56	fluoride	Hydrofluoric acid. HF	20.01	col fum cor liq, or gas; gas 1.90	$0.991^{19.54}$ liq 0.987	−83.1	19.54	∞	v s	
h57	fluoride (const. boiling)	HF(35.35 %) + H_2O		col liq			120			
h58	iodide	Hydroiodic acid. HI	127.91	col gas, or pa yel liq, $n_D^{16.5}$ 1.466	gas 5.66^{0} g/l liq $2.85^{-4.7}$	−50.8	-35.38^{4} atm	42.5^{0} cm^3	v s	s al
h59	iodide (const. boiling)	HI(57 %) + H_2O		col or pa yel fum liq	1.70^{15}		127^{774}			
h60	iodide, dihydrate	$HI.2H_2O$	163.94	col liq		−43		∞		
h61	iodide, tetrahydrate	$HI.4H_2O$	199.97	col liq		−36.5		∞		
h62	iodide, trihydrate	$HI.3H_2O$	181.96	col liq		−48		∞		
h63	oxide	Water. H_2O	18.0153	col liq or hex cr, liq 1.333, sol 1.309, 1.313	1.000_4^4	0.000	100.000			∞ al
h64	oxide, per-	H_2O_2	34.01	col liq; 1.414^{22}	1.4422^{25}	−0.41	150.2^{760}	∞		s al, eth; i pet et
h65	phosphide	Phosphine. H_3P	34.00	col pois inflam gas or col liq, 1.317 liq	gas 1.529 g/l liq 0.746^{-90}	−133.5	−87.4	26^{17} cm^3	i	s al, eth, Cu_2Cl_2
h66	phosphide	H_4P_2	65.98	col liq	1.012	−90	57.5^{735}	i	i	s al, turp
h67	phosphide	$(H_2P_4)_3$	377.73	yel solid	1.83^{19}	ign 160	d	i	i	i al; s P, P_2H_4
h68	sulfide	H_2S	34.08	col gas, infl, liq 1.374	1.539^{0} g/l	−85.5	−60.7	437^{0} cm^3	186^{40} cm^3	9.54^{20} cm^3 al; s CS_2
h69	sulfide, di-	H_2S_2	66.14	yel oil, 1.885	1.334^{20}	−89.6	70.7^{760}	d		s bz, eth, CS_2; i
h70	sulfide, penta-	H_2S_5	162.34	clear yel oil	1.67^{16}	−50	50^{4}			s bz, eth, CS_2; i
h71	sulfide, tetra-	H_2S_4	130.27	lt yel liq	1.588^{15}	−85				
h72	sulfide, tri-	H_2S_3	98.21	brt yel liq, 1.705^{15}	1.496^{15}	−52	d 90			s bz, eth, CS_2; i
h73	telluride	H_2Te	129.62	col gas or yel need	gas 5.81 g/l liq 2.57^{-20}	−49	−2	s unstab		s al, alk
h74	**Hydroxylamine**	NH_2OH	33.03	wh need or col liq, deliq	1.204	33.05	56.5	s	d	s a, al, MeOH; v sl s eth, chl, bz, CS_2
h75	acetate	$NH_2OH.CH_3CO_2$	92.08	col cr		87	subl 90	v s		
h76	bromide	$NH_2OH.HBr$	113.95	wh, monocl	2.35_4^{22}			v s	v s	i eth
h77	fluogermanate	$(NH_2OH)_2.H_2GeF_6.2H_2O$	290.69	monocl pr, 1.418, 1.438, 1.443	2.229^{25}			s		s abs al
h78	fluosilicate	$(NH_2OH)_2H_2SiF_6.2H_2O$	246.18	scales				v s		i al
h79	formate	$NH_2OH.HCO_2$	78.05	col need		76	d 80	v s	d	s h al; i eth
h80	hydrochloride	$NH_2OH.HCl$	69.49	col, monocl	1.67^{17}	151	d	83^{17}	v s	4.43^{20} al; 16.4^{20} MeOH; s gluc; i eth
h81	iodide	$NH_2OH.HI$	160.94	col need, hygr		83–84 exp		v s	d	v s MeOH; sl s et
h82	nitrate	$NH_2OH.HNO_3$	96.04	wh		48	d <100	v s	d	v s al
h83	*ortho*phosphate	$(NH_2OH)_3.H_3PO_4$	197.09			148 exp		1.9^{20}	16.8^{90}	
h84	sulfate	$(NH_2OH)_2.H_2SO_4$	164.14	col, monocl		d 170	d	32.9^{0}	68.5^{20}	sl s al; s eth
i1	**Indium**	In	114.82	soft silv wh met, tetr	7.30^{20}	156.61	2000 ± 10	i	i	s a; v sl s NaOH
i2	antimonide	InSb	236.57	cr		535				
i3	arsenide	InAs	189.74	met cr		943				i a
i4	bromide, di-	$InBr_2$	274.64	pa yel solid	4.22^{25}	235	632 subl	d		s a
i5	bromide, mono-	InBr	194.73	red br solid	4.96^{25}	220	662 subl	d		s a
i6	bromide, tri-	$InBr_3$	354.55	wh to yel need, deliq	4.74^{25}	436 ± 2	subl	v s		
i7	*per*chlorate	$In(ClO_4)_3.8H_2O$	557.29	col cr, deliq		*ca* 800	d 200	v s	d	s abs al; sl s eth
i8	chloride, di-	$InCl_2$	185.73	wh rhomb, deliq	3.655^{25}	235	550–570	d	d	
i9	chloride, mono-	InCl	150.27	1) yel or 2) dk red, deliq	4.19^{25} yel 4.18^{25} red	225 ± 1	608	d	d	s a

No.	Name	Synonyms and Formulae	Mol. wt.	Crystalline form, properties and index of refraction	Density or spec. gravity	Melting point, °C	Boiling point, °C	Solubility, in grams per 100 cc		
								Cold water	Hot water	Other solvents
	Indium									
i10	chloride, tri-	$InCl_3$	221.18	wh pl, deliq	3.46_4^{25}	586 subl 300	volat 600	v s	v s	sl s al, eth
i11	cyanide	$In(CN)_3$	192.87	wh ppt				unstab		i dil a; s HCN; v sl s NaOH
i12	fluoride	InF_3	171.82	col	$4.39^{25}\pm1$	1170 ± 10	>1200	0.040^{25}		
i13	fluoride, trihydrate	$InF_3.3H_2O$	225.86	cr		$-3H_2O$, 100		8.49^{22}		s a; i al, eth
i14	fluoride, nonahydrate	$InF_3.9H_2O$	333.95	wh need		d		sl s	d	s HCl, HNO_3; i al, eth
i15	hydroxide	$In(OH)_3$	165.84	wh ppt		$-H_2O<150$		i		s a; v sl s NaOH; i NH_4OH
i16	iodate	$In(IO_3)_3$	639.53	wh cr			d	0.067^{20}		s dil HNO_3, dil H_2SO_4; s d HCl
i17	iodide, di-	InI_2	368.63		4.71^{25}	212				s a
i18	iodide, mono-	InI	241.72	br red solid	5.31	351	711–715		sl d	i al, eth, chl; s dil a
i19	iodide, tri-	InI_3	495.53	carmine red, or yel cr	4.69	210		v unstable	s	s a, chl, bz yxl
i20	methylate	$In(CH_3)_3$	159.93	col cr	1.568_{19}^{19}			d		d al, MeOH; s liq NH_3, eth; v s acet, bz
i21	nitrate	$In(NO_3)_3.3H_2O$	354.88	pl, deliq		$-2H_2O$, 100	d	v s		s al
i22	nitrate	$In(NO_3)_3.4\frac{1}{2}H_2O$	381.90	need, deliq		$-4\frac{1}{2}H_2O$	d	v s		s al
i23	oxide, mon-	InO	130.81	wh gray				i		s a
i24	oxide, sesqui-	In_2O_3	277.64	red brn, (h) pa yel (c) amorph and trig	7.179		volat 850	i		amorph s a; cr i a
i25	oxide, sub-	In_2O	245.64	blk cr	6.99_4^{25}	subl vac 565–700				s HCl
i26	phosphide	InP	145.79	brittle mass met		1070				v sl s min a
i27	selenate	$In_2(SeO_4)_3.10H_2O$	838.67	cr, deliq				v s		
i28	selenide, sesqui-	In_2Se_3	466.52	blk cr, or soft dk scales	5.67_4^{25}	890 ± 10				s, d conc a
i29	sulfate	$In_2(SO_4)_3$	517.83	wh gray powd, monocl pr, hygr	3.438			s	v s	
i30	sulfate	$In_2(SO_4)_3.9H_2O$	679.96	wh powd, hygr	3.44	d 250		v s		
i31	sulfate, dihydrate	$In_2(SO_4)_3.H_2SO_4.7H_2O$	742.01	rhomb cr		$-7H_2O$, $-H_2SO_4$, *ca* 250				
i32	sulfide, mono-	InS	146.88	dk	5.18^{25}	692 ± 5	subl vac 850			s HCl, HNO_3
i33	sulfide, sesqui-	In_2S_3	325.83	red cr or yel ppt	4.90	1050	subl *ca* 850 in high vac	i		s a; sl s Na_2S
i34	sulfide, sub-	In_2S	261.70	yel or blk need	5.87^{25}	653 ± 5				
i35	sulfite, basic	$2In_2O_3.3SO_2.8H_2O$	891.59	wh cr		$-3H_2O$, 100	$-8H_2O$, 260	i		
i36	telluride, sesqui-	InTe	242.42	dk met, shiny	6.29_4^{25}	696 ± 2				i HCl, s HNO_3
i37	telluride	In_2Te_3	612.44	bl brittle cr	5.78	667				
i38	**Iodic acid**	HIO_3	175.91	col or pa yel cr powd, rhomb	4.629^0	d 110		286^0 310^{16}	473^{80} 576^{101}	v s 87 % al; sl s HNO_3; i abs al, eth, chl
i39	metaper-	HIO_4	191.91	col		subl 110	d 138	v s, d		
i40	*ortho*paraper-	H_5IO_6 or $HIO_4.2H_2O$	227.94	wh monocl, deliq			d 140	113	v s	s al, eth
i41	**Iodine**	I_2	253.809	vlt blk met lust, rhomb, 3.34	4.93	113.5	184.35^{atm}	0.029^{20} 0.030^{25}	0.078^{50}	20.5^{15} al; 16.46^{25} bz 20.6^{17} eth; s chl glyc, KI; 24^{25} eth 23^{25} MeOH 20.15^{25} CS_2; 2.91^{25} CCl_4
i42	azide	Iod(o)azide. IN_3	168.92	yel, exp				s d		s $Na_2S_2O_3$
i43	bromide, mono-	IBr	206.81	dk gray cr	4.4157^0	(42) subl 50	d 116	s d		s al, eth, chl, CS_2
i44	bromide, tri-	IBr_3	366.63	br liq				s		s al
i45	chloride, mono-(α)	ICl	162.36	dk-red need, cub, red br oily liq	3.1822^0	27.2	97.4	d to HIO_3 +Cl		s al, eth, CS_2, HCl
i46	chloride, mono-(β)	ICl	162.36	brn red, rhombic 6 sided pl	liq 3.24^{34}	13.92	97.4; d 100	d		s al, eth, HCl
i47	chloride, tri-	ICl_3	233.26	yel brn, rhomb, red liq	3.117^{15}	101^{16atm}	d 77	s d		s al, eth, CCl_4, ac a, bz
i48	cyanide	ICN	152.92	wh cr				sl s	sl s	s al, eth
i49	fluoride, hepta-	IF_7	259.89	col cr or liq	liq 2.8^6	5.5	4.5 subl	v s, d	d	d a, alk
i50	fluoride, penta-	IF_5	221.90	col liq	3.75	9.6	98	d	d	d a, ak
i51	oxide, di- (or tetra-)	IO_2 or I_2O_4	158.90	lem-yel cr	4.2_{10}^{10}	d slow 75 rap 130		d to HIO_3+I_2		s H_2SO_4; sl s acet; i al, eth

No.	Name	Synonyms and Formulae	Mol. wt.	Crystalline form, properties and index of refraction	Density or spec. gravity	Melting point, °C	Boiling point, °C	Solubility, in grams per 100 cc		
								Cold water	Hot water	Other solvents
	Iodine									
i52	oxide, pent-	I_2O_5	333.81	wh trim	4.799^{25}_{4}	d 300–350		187.4^{13}	v s	i abs al, eth, chl, CS_2; sl s dil a
i53	(tetra-) oxide non-	I_4O_9	651.61	yel powd, hygr			d 75			
i54	**Iodo platinic acid**	$H_2PtI_6.9H_2O$	1120.67	blk, monocl, deliq		>100		v s, d	d	
i55	**Iodous acid, hypo-**	Iodine hydroxide. HOI	143.91	only in sol, yel to grayish				d	d	
i56	**Iridium**	Ir	192.20	silv wh met, cub	22.421	2410	4527 ± 100	i	i	sl s aq reg; i a, al
i57	bromide, tetra-	$IrBr_4$	511.84	blk, deliq		d		s d		s al
i58	bromide, tri-	$IrBr_3.4H_2O$	503.99	olv-grn cr		$-3H_2O$, 100		v s		i al
i59	carbonyl	$Ir_2(CO)_8$	608.48	yel cr		subl 160 (in CO_2)				s CCl_4
i60	carbonyl	$Ir_4(CO)_{12}$	1104.93	yel cr		subl 250 (in CO_2)				i CCl_4
i61	carbonyl chloride	$Ir(CO)_2Cl_2$	319.13	col need		d 140		d		d HCl, KOH
i62	chloride, di-	$IrCl_2$	263.11	blk gray cr (exist ?)		d 773		s		i a, alk
i63	chloride, tetra-	$IrCl_4$	334.01	dk-brn, amorph, hygr		d		s	d	s al, dil HCl
i64	chloride,tri-	$IrCl_3$	298.56	olv-grn hex, or trig	5.30	d 763		i	i	i a, alk
i65	fluoride, hexa-	IrF_6	306.19	yel glass or tetr	6.0	44.4^{30}	53	d	d	
i66	iodide, tetra-	IrI_4	699.82	blk		d 100		i	i	i al, s KI
i67	iodide, tri-	IrI_3	572.91	grn		d		sl s	s	sl s al
i68	oxalic acid	$H_3[Ir(C_2O_4)_3.xH_2O$		pa yel cr		d		v s	v s	sl s al; i eth
i69	oxide, di-	IrO_2	224.20	blk tetr or bl cr	3.15	d 1100		0.0002^{20}		i a, alk
i70	oxide, di- hydrate	$IrO_2.2H_2O$ or $Ir(OH)_4$	260.23	indigo bl cr		$-2H_2O$, 350		i	i	s HCl
i71	oxide, sesqui-	Ir_2O_3	432.40	bl-blk (exist ?)		−O, 400		i		s H_2SO_4, h HCl; i alk
i72	oxide, sesqui-	$Ir_2O_3.xH_2O$		olive green		d		i		s a, alk
i73	phosphor chloride	IrP_3Cl_{12} or $IrCl_3.3PCl_3$	710.56	yel pr		d 250		sl s	d 100	sl s, d al
i74	selenide	$IrSe_2$	350.12	dk gray cr powd		d 600–700 (in CO_2)				sl s aq reg; i a
i75	sulfate	$Ir_2(SO_4)_3.xH_2O$		yel pr		d		s		
i76	sulfide, di	IrS_2	256.33	br-blk	8.43^{25}_{4}	d 300		i		s aq reg; i a
i77	sulfide, hydro-	$Ir(HS)_3.2H_2O$	327.45	choc br		d		i		s HNO_3
i78	sulfide, mono-	IrS	224.26	bl-blk			d	i		s K_2S, i a
i79	sulfide, sesqui-	Ir_2S_3	480.59	br-blk	9.64^{25}_{4}	d		sl s		s K_2S, HNO_3
i80	telluride	$IrTe_3$	575.00	dk gray cr	9.5^{25}_{4}			i	i	i a; s h aq reg
	Iridium complexes									
i81	aquopentammine iridium-chloride	$[Ir(NH_3)_5H_5(H_2O)]Cl_3$	402.54	wh micro cr	2.474^{15}_{4}	$-H_2O$, 100		*ca* 75^{25}	d	i al, eth
i82	chloropentammine iridiumchloride	$[Ir(NH_3)_5Cl]Cl_2$	383.71	pa yel, rhomb	$2.681^{15.5}_{4}$	d		sl s	s	i HCl
i83	hexammine iridium chloride	$[Ir(NH_3)_6]Cl_3$	400.74	col, rhomb,	$2.434^{15.5}_{4}$			20		
i84	hexammine iridium nitrate	$[Ir(NH_3)_6](NO_3)_3$	408.40	col, tetr micro cr	2.395^{15}			1.7^{14}		
i85	nitratopentammine iridium nitrate	$[Ir(NH_3)_5(NO_3)](NO_3)_2$	463.37	wh micro cr	2.515^{18}_{4}	heat exp		0.28	2.5^{100}	
i86	**Iron**	Fe	55.847	silv met, cub	7.86	1535	3000	i	i	s a; i alk, al, eth
i87	(II) acetate	$Fe(C_2H_3O_2)_2.4H_2O$	246.00	lt grn need, monocl		d		v s		
i88	(III) acetate, basic	$FeOH(C_2H_3O_2)_2$	190.94	br-red powd				i		s a, al
i89	(III) acetylacetonate	$Fe(C_5O_2H_7)_3$	353.18	rubyred, rhomb	1.33	184		sl s	sl s	s al, acet, bz, chl
i90	(II) *ortho*arsenate	$Fe_3(AsO_4)_2.6H_2O$	553.47	grn amorph powd		d		i	i	s dil HCl; sl s NH_4OH
i91	(III) *ortho*arsenate	Nat. scorodite. $FeAsO_4.2H_2O$	230.80	grn, rhomb, 1.765, 1.774, 1.797	3.18	d		i	i	s HCl; i HNO_3
i92	(III) *ortho*arsenite, basic	$2FeAsO_3.Fe_2O_3.5H_2O$	607.30	br-yel powd		d		sl s		s a, alk
i93	(II) *pyro*arsenite	$Fe_2As_2O_5$	341.53	grn-wh				i		s NH_4OH
i94	arsenide	FeAs	130.77	wh	7.83	1020		v sl s		
i95	arsenide, di-	Arsenoferrite. $FeAs_2$	205.69	silv gray, cub	7.4	990		i		sl s HNO_3; i HCl
i96	(III) benzoate	$Fe(C_7H_5O_2)_3$	419.20	br powd				i		s h al, h eth
i97	boride	FeB	66.66	gray cr	7.15^{18}			i		s HNO_3, h conc H_2SO_4

No.	Name	Synonyms and Formulae	Mol. wt.	Crystalline form, properties and index of refraction	Density or spec. gravity	Melting point, °C	Boiling point, °C	Solubility, in grams per 100 cc		
								Cold water	Hot water	Other solvents
	Iron									
i98	(II) bromide	$FeBr_2$	215.67	grn-yel, hex	4.636^{25}	d 684 (?)		109^{10}	170^{95}	s al; sl s bz
i99	(III) bromide	$FeBr_3$ or Fe_2Br_6	295.57	dk red-brn, rhomb (?) deliq		subl d		s	s	s al, eth, sl s NH_3
i100	(III) bromide-, hexahydrate	$FeBr_3.6H_2O$	403.68	dk grn		27		v s	v s	s al, eth
i101	(III) cacodylate	$Fe[(CH_3)_2AsO_2]_3$	466.82	yelsh amorph powd				6.67		sl s al
i102	carbide	Fe_3C	179.55	gray, cub	7.694	1837		i	i	s a
i103	(II) carbonate	Nat. siderite. $FeCO_3$	115.85	gray, trig, 1.875, 1.633	3.8	d		0.0067^{25}		s CO_2 sol
i104	(II) carbonate, hydrate	$FeCO_3.H_2O$	133.86	amorph		d		sl s		s a, CO_2 sol
i105	carbonyl, ennea-	$Fe_2(CO)_9$	363.79	yel met cr, hex	2.085^{18}	d 80		i	i	v sl s al, MeOH; d HNO_3; i a
i106	carbonyl, penta-	$Fe(CO)_5$	195.90	visc yel liq	liq 1.457^{21}	−21	102.8^{749}	i		s al, eth, bz, alk, conc H_2SO_4
i107	carbonyl, tetra-	$Fe(CO)_4$	167.89	dk grn lust cr, tetr	1.996^{18}	d 140–150		i		s org solv, conc HNO_3 h H_2SO_4
i108	(II) *per*chlorate	$Fe(ClO_4)_2$	254.75	wh or grnsh-wh need, hygr		d>100		v s		
i109	(II) *per*chlorate hexahydrate	$Fe(ClO_4)_2.6H_2O$	362.84	grn, 1.493, 1.478		d>100		97.8^{0}	116.1^{60}	86.5^{20} al; s $HClO_4$
i110	*oxy*chloride	FeOCl	107.30	brn, rhomb	3.1	d 200				
i111	(II) chloride	Nat. lawrencite. $FeCl_2$	126.75	grn to yel, hex deliq, 1.567	3.16^{25}_{4}	670–674	subl	64.4^{10}	105.7^{100}	100 al; s acet; i eth
i112	(II) chloride, dihydrate	$FeCl_2.2H_2O$	162.78	grn, monocl	2.358					
i113	(II) chloride, tetrahydrate	$FeCl_2.4H_2O$	198.81	bl-grn, monocl, deliq	1.93			160.1^{10}	415.5^{100}	s al; sl s acet
i114	(III) chloride	Nat. molysite. $FeCl_3$ or Fe_2Cl_6	162.21	blk-brn hex	2.898^{25}_{4}	306	d 315	74.4^{0}	535.7^{100}	v s al, MeOH, eth; 63^{18} acet
i115	(III) chloride, hydrate	$FeCl_3.2\frac{1}{2}H_2O$	207.24	dk yel-red, rhomb, deliq		56		v s	v s	v s al, eth
i116	(III) chloride, hexahydrate	$FeCl_3.6H_2O$	270.30	br yel cr mass, v deliq		37	280–285	91.9^{20}	∞	s al, eth
i117	(II) chloride, hexammine	$FeCl_2.6NH_3$	228.94	wh powd	1.928^{25}_{4}					
i118	(II) chloroplatinate	$FePtCl_6.6H_2O$	571.75	yel, hex	2.714	d		v s	v s	
i119	(III) *di*chromate	$Fe_2(Cr_2O_7)_3$	759.66	red-brn gran				s		s a
i120	(II) chromite	$FeCr_2O_4$	223.84	brn-blk, cub	4.97^{20}			i	i	sl s a
i121	(II) citrate	$FeC_6H_6O_7.H_2O$	263.97	wh micr, rhomb		d 350 (in H_2)		sl s		s NH_4OH
i122	(III) citrate	$FeC_6H_5O_7.5H_2O$	335.03	red-brn scales				sl s	s	i al
i123	hydroxide	Nat. goethite. FeO(OH)	88.85	brn, blksh, rhomb, 2.260, 2.394, 2.400	4.28	$-\frac{1}{2}H_2O$, 136				s HCl
i124	(II) ferricyanide	$Fe_3[Fe(CN)_6]_2$(?)	591.45	deep bl		d		i		i al, dil a
i125	(III) ferricyanide	Berlin green. $Fe[Fe(CN)_6]$	267.80	cub						
i126	(II) ferrocyanide	$Fe_2[Fe(CN)_6]$	323.65	bl-wh, amorph	1.601^{25}_{4}	d 100	d 430 (vac)	i		
i127	(III) ferrocyanide	$Fe_4[Fe(CN)_6]_3$	859.25	dk bl cr		d		i	i	s HCl, H_2SO_4; i al, eth
i128	(II) fluoride	FeF_2	93.84	wh, rhomb	4.09^{25}_{4}	>1000 (?)		sl s		s a; i al, eth
i129	(II) fluoride, octahydrate	$FeF_2.8H_2O$	237.97	grn-bl	4.20 (anh)	$-8H_2O$, 100		sl s	s	s HF, a; i al, eth
i130	(II) fluoride, tetrahydrate	$FeF_2.4H_2O$	165.91	wh, rhomb	2.095	d		v sl s		s a; sl s al, eth
i131	(III) fluoride	FeF_3 or Fe_2F_6	112.84	grn, rhomb	3.52	>1000		sl s	s	s a; i al, eth
i132	(III) fluoride, tetrahydrate	$FeF_3.4\frac{1}{2}H_2O$	193.91	yel cr		$-3H_2O$, 100	d	sl s	s	i al
i133	(II) fluosilicate	$FeSiF_6.6H_2O$	306.01	col, trig, 1.361, 1.385	1.961			128.2		
i134	(III) fluosilicate	$Fe_2(SiF_6)_3$ (exist ?)	537.92	flesh col, gel				s	s, d	
i135	(III) formate	$Fe(CHO_2)_3.H_2O$	208.92	red cr or powd				s		v sl s al
i136	(III) glycerophosphate	$Fe_2[C_3H_5(OH)_2.PO_4]_3$	677.72	yelsh-grn sc or powd				50^{25}		i al
i137	(II) hydrogen cyanide	$H_4[Fe(CN)_6]$	215.99	wh, rhomb	1.536^{25}_{4}	d 190		s	s	v s al; s a; i acet
i138	(III) hydrogen cyanide	$H_3[Fe(CN)_6]$	214.98	brn-yel need		d 50–60		s	s	v s al; i eth

No.	Name	Synonyms and Formulae	Mol. wt.	Crystalline form, properties and index of refraction	Density or spec. gravity	Melting point, °C	Boiling point, °C	Solubility, in grams per 100 cc		
								Cold water	Hot water	Other solvents
	Iron									
i139	(II) hydroxide	$Fe(OH)_2$	89.86	pa grn hex, or wh amorph	3.4	d		0.00015[18]		s a, NH_4Cl; i alk
i140	(III) hydrosulfate	Iron(ic) tetrasulfate. Nat. rhomboklas. $Fe_2O_3.4SO_3.9H_2O$	642.08	wh to pink powd, 1.533, 1.550, 1.635	2.172	$-6H_2O$, 80		s		sl s abs al
i141	(III) iodate	$Fe(IO_3)_3$	580.55	grn yel powd	4.80^{20}_{4}	d 130		sl s		i dil HNO_3
i142	(II) iodide	FeI_2	309.66	gray hex, hygr	5.315	red heat		s	s	s al, acet
i143	(II) iodide, tetrahydrate	$FeI_2.4H_2O$	381.72	gray blk cr, deliq	2.873	d 90–98		v s	d	s al, eth
i144	(II) lactate	$Fe(C_3H_5O_3)_2.3H_2O$	287.96	grn-wh cr or powd		d		2.1[10]	8.5[100]	s alk citrate; v sl s al; i eth
i145	(III) lactate	$Fe(C_3H_5O_3)_3$	323.06	br, amorph, deliq				s	v s	i eth
i146	(III) malate	$Fe_2(C_4H_4O_5)_3$	507.91	br scales, hygr				s		s al
i147	methanoarsenate	$Fe(CH_4AsO_3)_3$	525.60	redsh br lustr scales				50		i al, eth
i148	(II) nitrate	$Fe(NO_3)_2.6H_2O$	287.95	grn, rhomb		60.5		83.5[20]	166.7[61]	
i149	(III) nitrate	$Fe(NO_3)_3.6H_2O$	349.95	cub		35	d	150[0]	∞	
i150	(III) nitrate	$Fe(NO_3)_3.9H_2O$	404.02	col-pa vlt, monocl, deliq	1.684	47.2	d 125	s		s al, acet; sl s HNO_3
i151	nitride	Fe_4N	237.39		6.57(?)					
i152	nitride	Fe_2N or Fe_4N_2	125.70	gray	6.35	d 200		i		s HCl, H_2SO_4
i153	nitrosyl carbonyl	$Fe(NO)_2(CO)_2$	171.88	dk red cr	1.56	18.5	d 50, 110	i		s org solv
i154	(III) oleate	$Fe(C_{18}H_{33}O_2)_3$	900.23	br-red fatty lumps				i		s a, al, eth
i155	(II) oxalate	$FeC_2O_4.2H_2O$	179.90	pa yel, rhomb	2.28	d 190		0.022	0.026	s a
i156	(III) oxalate	$Fe_2(C_2O_4)_3.5H_2O$	465.83	yel micro cr powd		d 100		v s	v s	s a; i al
i157	(II) oxide	Nat. wuestite. FeO	71.85	blk, cub, 2.32	5.7	1420		i	i	s a; i al, alk
i158	(III) oxide	Nat. hematite. Fe_2O_3	159.69	red-brn to blk, trig, 3.01, 2.94(Li)	5.24	1565		i	i	s HCl, H_2SO_4; sl s HNO_3
i159	oxide	Iron ferrosoferric, nat. magnetite. Fe_3O_4	231.54	blk, cub or red-blk powd, 2.42	5.18	d 1538		i	i	s conc a; i al, eth
i160	(III) oxide, hydrate	$Fe_2O_3.xH_2O$		red-brn amorph powd or gelat	2.44–3.60	all H_2O, 350–400		i	i	s a; i al, eth
i161	(II) *ortho*phosphate	Nat. vivianite. $Fe_3(PO_4)_2.8H_2O$	501.61	wh-bl, monocl, 1.579, 1.603, 1.633	2.58			i	i	s a; i ac a
i162	(III) *ortho*phosphate	$FePO_4.2H_2O$	186.85	pink, monocl	2.74	d		v sl s	0.67[100]	s HCl, H_2SO_4; i HNO_3
i163	(III) *pyro*phosphate	$Fe_4(P_2O_7)_3.9H_2O$	907.36	yel-wh powd				i		s a, alk citr
i164	phosphide, mono-	FeP	86.82	rhomb	6.07 (5.2[20])					
i165	(di-)phosphide	Fe_2P	142.67	bl-gray cr or powd	6.56	1290		i	i	s aq reg, HNO_3+HF; i dil a
i166	(tri-)phosphide	Fe_3P	198.51	gray	6.74	1100		i		
i167	(III) *hypo*phosphite	$Fe(H_2PO_2)_3$	250.81	wh or gray-wh powd		d		0.043[25]	0.083[100]	s alk citr
i168	(II) *meta*silicate	Nat. gruenerite. $FeSiO_3$	131.93	gray-grn, rhomb, 1.672, 1.697, 1.717	3.5	1146				
i169	*ortho*silicate	Nat. fayalite. Fe_2SiO_4	203.78	col, rhomb	4.34	1503 (?)		i		d HCl
i170	silicide	FeSi	83.93	yel-gray, oct	6.1			i		i aq reg
i171	(II) sulfate	Nat. szomolnikite. $FeSO_4.H_2O$	169.96	off-wh, monocl	2.970[25]			sl s	s	
i172	(III) sulfate	$Fe_2(SO_4)_3$	399.87	yel rhomb, hygr, 1.814	3.097[18]	d 480		sl s	d	i H_2SO_4, NH_3
i173	(II) sulfate, heptahydrate	Nat. melanterite. $FeSO_4.7H_2O$	278.05	bl-grn, monocl, 1.471, 1.478, 1.486	1.898	64 $-6H_2O$, 90	$-7H_2O$, 300	15.65	48.6[50]	sl s al; s abs MeOH
i174	(II) sulfate, pentahydrate	Nat siderotil. $FeSO_4.5H_2O$	242.02	wh, tricl, 1.526, 1.536, 1.542	2.2	$-5H_2O$, 300		s	s	i al
i175	(II) sulfate, tetrahydrate	$FeSO_4.4H_2O$	224.01	grn monocl pr, 1.533, 1.535	2.23–2.29					
i176	(III) sulfate, enneahydrate	Nat. coquimbite. $Fe_2(SO_4)_3.9H_2O$	562.01	rhomb, deliq, 1.552, 1.558	2.1	$-7H_2O$, 175		440	d	s abs al
i177	sulfide, di-	Nat. pyrite. FeS_2	119.98	yel, cub	5.0	1171		0.00049		d HNO_3. dil a
i178	sulfide, di-	Nat. marcasite. FeS_2	119.98	yel, rhomb	4.87	tr 450	d	0.00049		d HNO_3, i dil a
i179	(II) sulfide	Nat. troilite. FeS	87.91	blk-brn, hex	4.74	1193–1199	d	0.00062[18]	d	s d a; i NH_3
i180	(III) sulfide	Fe_2S_3	207.87	yel-grn	4.3	d		sl d	d FeS+S	d a
i181	(II) sulfite	$FeSO_3.3H_2O$	189.96	grnsh or wh cr		d 250		v sl s		s SO_2 sol; i al
i182	tantalate	Nat. tapiolite. $Fe(TaO_3)_2$	513.73	lt brn, tetr, 2.27, 2.42 (Li)	7.33					
i183	*d*-tartrate	$FeC_4H_4O_6$	203.92	wh cr				0.877[16]	v sl s	v s min a; s NH_4OH
i184	(II) thiocyanate	$Fe(SCN)_2.3H_2O$	226.06	grn, rhomb		d		v s		s al, eth, acet
i185	(III) thiocyanate	$Fe(SCN)_3$ or $Fe_2(SCN)_6$	230.09	blk-red, cub, deliq				v s	d	s al, eth, acet

No.	Name	Synonyms and Formulae	Mol. wt.	Crystalline form, properties and index of refraction	Density or spec. gravity	Melting point, °C	Boiling point, °C	Solubility, in grams per 100 cc		
								Cold water	Hot water	Other solvents
	Iron									
i186	(II) thiosulfate	$FeS_2O_3.5H_2O$	258.05	grn cr, deliq				v s	d	v s al
i187	tungstate	Nat. ferberite. $FeWO_4$	303.69	tetr, 2.40 (Li)	6.64					
i188	*meta*vanadate	$Fe(VO_3)_3$	352.67	grayish-brn powd.				i		s a; i al
k1	**Krypton**	Kr	83.80	inert gas	gas 3.736g/l liq 2.155 *at* −152.9	−156.6	−152.30±0.10	11.0^{0} cm³ 6.0^{25} cm³	4.67^{50} cm³	
l1	**Lanthanum**	La	138.91	wh met, tarnish in air, α: hex, β: cub above 350	α 6.194, β 6.17	920	3469	d	d	s min a; i conc H_2SO_4
l2	acetate	$La(C_2H_3O_2)_3.1\frac{1}{2}H_2O$	343.07					16.88^{25}		
l3	boride, hexa-	LaB_6	203.78	purp met cub	2.61	2210	d	i	i	i HCl
l4	bromate	$La(BrO_3)_3.9H_2O$	684.77	hex pr		37.5	−7H₂O, 100	28.5^{15}		i al
l5	bromide	$LaBr_3.7H_2O$	504.74	col cr	5.057^{25} anh	783±3 anh	1577	v s		v s al; i eth
l6	carbide	LaC_2	162.93	yel cr	5.02			d	d	s H_2SO_4; i conc HNO_3
l7	carbonate	$La_2(CO_3)_3.8H_2O$	601.97	wh	2.6–2.7			i	i	s dil a; sl s aq CO_2; i acet
l8	chloride	$LaCl_3$	245.27	wh cr, deliq	3.842^{25}	860	>1000	v s	d	v s al, pyr; i eth, bz
l9	chloride, hepta-hydrate	$LaCl_3.7H_2O$	371.38	wh, tricl, hygr		d 91		v s	v s	v s al
l10	hexaantipyrin *per*-chlorate	$[La(C_{11}H_{12}N_2O)].(ClO_4)_3$	1566.65	col, hex cr		d 290–295		1.50^{20}		
l11	hydroxide	$La(OH)_3$	189.93	wh powd		d		i		s a
l12	iodate	$La(IO_3)_3$	663.62	col, cr				1.7^{25}		
l13	iodide	LaI_3	519.62	gray-wh, rhomb, hygr	5.63	772		v s		s acet
l14	molybdate	$La_2(MoO_4)_3$	757.63	tetr	4.77^{16}	1181		0.00179^{25}	0.0033^{85}	
l15	nitrate	$La(NO_3)_3.6H_2O$	433.02	col cr, deliq, tricl		40	d 126	151.1^{25}	v s	v s al; s acet
l16	oxalate	$La_2(C_2O_4)_3.9H_2O$	704.02	wh cr		d		0.00008^{25}		s min a
l17	oxide	Lanthana. La_2O_3	325.82	wh rhomb, or amorph	6.51^{15}	2315	4200	0.0004^{29}	d	s a, NH_4Cl; i acet
l18	sulfate	$La_2(SO_4)_3$	566.00	wh powd, hygr	3.60^{15}	d 1150		3.0	0.69^{100}	sl s al; i acet
l19	sulfate, hydrate	$La_2(SO_4)_3.9H_2O$	728.14	col hex, 1.564	2.821	d white heat		3.8^{0}	0.87^{100}	sl s HCl; i al
l20	sulfide	La_2S_3	374.01	red-yel cr, hex	4.911^{11}	2100–2150 vac		d	d	s a
l21	**Lead**	Pb	207.19	silv-blsh wh soft met, cub, 2.01	11.3437^{16} Ra-Pb 11.288^{20}_{20} UPb 11.2960^{16}	327.5	1744	i	i	s HNO_3, h conc H_2SO_4
l22	abietate	$Pb(C_{20}H_{29}O_2)_2$	810.10	brn lumps or yelsh-wh powd				i		
l23	acetate	$Pb(C_2H_3O_2)_2$	325.28	wh cr	3.25^{20}_{4}	280		44.3^{20}	221^{50}	s glyc; v sl s al
l24	acetate, basic	$Pb_2(OH)(C_2H_3O_2)_3$	608.52	wh				v s		sl s al
l25	acetate, basic	$Pb(C_2H_3O_2)_2.3PbO.H_2O$	1012.86	wh powd						
l26	acetate, basic	$Pb(C_2H_3O_2)_2.Pb(OH)_2.H_2O$	584.50	wh, monocl				v s		v s al
l27	acetate, decahydrate	$Pb(C_2H_3O_2)_2.10H_2O$	505.44	wh, rhomb cr	1.69	22		s	s	i al
l28	acetate, trihydrate	Sugar of lead. $Pb(C_2H_3O_2)_2.3H_2O$	379.33	wh, monocl, β 1.567	2.55	−H₂O, 75	d 200	45.61^{15}	200^{100}	i al
l29	acetate, tetra-	$Pb(C_2H_3O_2)_4$	443.37	col, monocl	2.228^{17}	175		d		d al; s chl, h ac a
l30	*di*antimonate	$Pb_2Sb_2O_7$	769.88	dk yel powd	6.72			i	i	v sl s HCl
l31	*ortho*antimonate	$Pb_3(SbO_4)_2$	993.07	or-yel powd	6.58^{20}_{4}			i	i	v sl s HCl
l32	*ortho*antimonate	Nat. monimolite. $Pb_3(SbO_4)_2$	993.07	orange powd	6.58^{20}_{4}			i		i dil a
l33	*meta*arsenate	$Pb(AsO_3)_2$	453.03	hex tabl	6.42^{15}			d	d	s HNO_3
l34	*ortho*arsenate	$Pb_3(AsO_4)_2$	899.41	wh cr, v pois	7.80	1042, sl d 1000		v sl s		s HNO_3
l35	*ortho*arsenate, di-	Nat. schultenite. $PbHAsO_4$	347.12	monocl leaf, α 1.90, γ 1.97	5.79	d 720	−H₂O, 220	i	sl s	s HNO_3, caust alk
l36	*ortho*arsenate, mono-	$Pb(H_2AsO_4)_2$	489.06	tricl, 1.74, 1.82	4.46^{15}	d 140		d		s HNO_3
l37	*pyro*arsenate	$Pb_2As_2O_7$	676.22	rhomb, β 2.03	6.85^{15}_{15}	802		i	d	s HCl, HNO_3; i ac a
l38	*meta*arsenite	$Pb(AsO_2)_2$	421.03	wh powd	5.85			i		s HNO_3
l39	*ortho*arsenite	$Pb_3(AsO_3)_2.xH_2O$		wh powd	5.85			i		s alk, HNO_3
l40	azide	$Pb(N_3)_2$	291.23	col need, or powd.			expl 350	0.023^{18}	0.09^{70}	v s ac a; i NH_4OH
l41	*meta*borate	$Pb(BO_2)_2.H_2O$	310.82	wh cr powd	5.598, anhydr	−H₂O, 160		i	i	s a; i alk

No.	Name	Synonyms and Formulae	Mol. wt.	Crystalline form, properties and index of refraction	Density or spec. gravity	Melting point, °C	Boiling point, °C	Solubility, in grams per 100 cc: Cold water	Hot water	Other solvents
	Lead									
142	borofluoride	$Pb(BF_4)_2$	380.80	cr pr				d		d al
143	bromate	$Pb(BrO_3)_2.H_2O$	481.02	col, monocl	5.53	d 180		1.38^{20}	sl s	
144	bromide	$PbBr_2$	367.01	wh, rhomb	6.66	373	916	0.4554^{0} 0.8441^{20}	4.71^{100}	s a, KBr; sl s NH_3; i al
145	butyrate	$Pb(C_4H_7O_2)_2$	381.39	col scales, pois		90		i	i	s dil HNO_3
146	caprate	$Pb(C_{10}H_{19}O_2)_2$	549.71			103–104		i	i	0.0029^{20} eth
147	caproate	$Pb(C_6H_{11}O_2)_2$	437.50			73–74				1.09^{25} eth
148	caprylate	Lead octoate. $Pb(C_8H_{15}O_2)_2$	493.60	wh leaf		83.5–84.5		i	i	s al; 0.0938 eth
149	carbonate	Nat. cerussite. $PbCO_3$	267.20	col, rhomb, 1.804, 2.076, 2.078	6.6	d 315		0.00011^{20}	d	s a, alk; i NH_3, al
150	carbonate, basic	White lead, hydrocerussite. $2PbCO_3.Pb(OH)_2$	775.60	wh powd, or hex.	61.4	d 400		i	i	sl s aq CO_2; s HNO_3; i al
151	cerotate	$Pb(C_{26}H_{51}O_2)_2$	998.57	wh need		113		i		i al, eth; s bz
152	chlorate	$Pb(ClO_3)_2$	374.09	wh monocl, deliq	3.89	d 230		v s		s al
153	chlorate, hydrate	$Pb(ClO_3)_2.H_2O$	392.11	wh, monocl, deliq.	4.037	d 110		151.3^{18}	171^{80}	s al
154	*per*chlorate	$Pb(ClO_4)_2.3H_2O$	460.14	wh, rhomb	2.6	d 100		499.7^{25}		s al
155	chloride	Nat. cotunite. $PbCl_2$	278.10	wh, rhomb, 2.199, 2.217, 2.260	5.85	501	950^{760}	0.99^{20}	3.34^{100}	sl s dil HCl,NH_3; i al; s NH_4 salts
156	chloride, tetra-	$PbCl_4$	349.00	yel oily liq	3.18^{0}	−15	expl 105	d (Cl_2)	d	s conc HCl
157	chloride, sulfide	$PbCl_2.3PbS$	995.86	red				i	d	d a, alk; i dil a
158	chlorite	$Pb(ClO_2)_2$	342.09	yel, monocl		expl 126		0.095^{20}	0.42^{100}	s KOH
159	chromate	Nat. crocoite, chrome yellow. $PbCrO_4$	323.18	yel, monocl, 2.31, 2.37(Li), 2.66	6.12^{15}	844	d	0.000058^{25}	i	s a, alk; i ac a, NH_3
160	chromate, basic	Chrome red. $PbCrO_4.PbO$	546.37	red cr powd	6.63			i	i	s a, alk
161	chromate, basic	$Pb_2(OH)_2CrO_4$	564.39	red amorph or cr	6.63	920		i	i	s KOH
162	*di*chromate	$PbCr_2O_7$	423.18	red cr				d		s a, alk
163	citrate	$Pb_3(C_6H_5O_7)_2.3H_2O$	1053.82	wh cr powd				s		v sl s al
164	cyanate	$Pb(OCN)_2$	291.22	wh need		d		i	sl s	
165	cyanide	$Pb(CN)_2$	259.23	yelsh-wh powd, pois				sl s	s	s KCN
166	enanthate	$Pb(C_7H_{13}O_2)_2$	465.55	wh leaf		91.5		sl s		i al
167	ethylsulfate	$Pb(C_2H_5SO_4)_2.2H_2O$	493.57	col liq, pois				s		
168	ferricyanide	$Pb_3[Fe(CN)_6]_2.5$ (or 6) H_2O	1135.55	blk-brn to red, monocl pr		$-H_2O$, 110–120 d		sl s	s, d 100	s alk, HNO_3
169	ferrite	$PbFe_2O_4$	382.88	hex		1530 d, 725				
170	ferrocyanide	$Pb_2Fe(CN)_6$ $3H_2O$	680.38	yelsh-wh powd		$-H_2O$, 100		i		sl s H_2SO_4
171	fluoride	PbF_2	245.19	col, rhomb, pois	8.24	855	1290	0.064^{20}		s HNO_3; i acet, NH_3
172	fluorochloride	Nat. matlockite. PbFCl	261.64	wh, tetr, 2.145, 2.006	7.05	601		0.037^{25}	0.1081^{100}	
173	fluosilicate	$PbSiF_6.2H_2O$	385.30	col, monocl		d		s	v s	
174	fluosilicate, tetrahydrate	$PbSiF_6.4H_2O$	421.33	col, monocl		d<100				
175	formate	$Pb(CHO_2)_2$	297.23	wh, rhomb, lust, 1.789, 1.852, 1.877	4.63	d 190		1.6^{15}	20^{100}	i al
176	hydride, di-	PbH_2	209.21	gray powd		d				
177	hydroxide	$Pb(OH)_2$	241.20	wh, amorph		d 145		0.0155^{20}	sl s	s a, alk; i ac a
178	hydroxide	$Pb_2O(OH)_2$ or $2PbO.H_2O$	464.39	wh cub, or amorph powd, pois	7.592	d 145		0.014	sl s	s alk, ac a, HNO_3
179	iodate	$Pb(IO_3)_2$	557.00	wh	6.155^{20}	d 300		0.0012^{2}	0.003^{25}	sl s HNO_3; i NH_3
180	*paraper*iodate	$PbHIO_5$	415.10	wh cr		d 130		i	i	s dil HNO_3
181	*paraper*iodate, hydrate	$PbHIO_5.H_2O$	433.11	amorph		$-H_2O$, 110		i	i	sl s dil HNO_3
182	iodide, basic	$PbI_2.PbO.H_2O$	702.20	rhomb cr	6.83^{20}	d 100				
183	iodide,di-	PbI_2	461.00	yel hex powd, pois	6.16	402	954	0.044^{0} 0.063^{20}	0.41^{100}	s alk, KI; i al
184	iodide, mono-	PbI	334.09	pa yel		d 300		0.1		
185	isobutyrate	$Pb(C_4H_7O_2)_2$	381.39	wh pr		<100		9.1^{15}		
186	lactate	$Pb(C_3H_5O_3)_2$	385.33	wh cr powd				s		s h al
187	laurate	$Pb(C_{12}H_{23}O_2)_2$	605.82	chalky wh powd		104.7		0.009^{35}		0.008^{25} al; $0.007^{14.5}$ eth
188	lignocenate	$Pb(C_{24}H_{47}O_2)_2$	942.47	wh powd		117		i		v s h bz; sl s al; i eth
189	malate	$Pb(C_4H_4O_5).3H_2O$	393.31	wh powd				sl s		v sl s al
190	melissate	$Pb(C_{31}H_{61}O_2)_2$	1138.85	wh powd		115–116		i	i	s boil tol, ac a; sl s h bz, chl; i al, eth
191	molybdate	Nat. wulfenite. $PbMoO_4$	367.13	col-lt yel, tetr pl	6.92^{25}_{4}	1060–1070			i	d conc H_2SO_4; s a, KOH; i al
192	myristate	$Pb(C_{14}H_{27}O_2)_2$	661.93	wh powd		107		0.005^{35}	0.006^{50}	0.004^{25} al; $0.010^{14.5}$ eth

No.	Name	Synonyms and Formulae	Mol. wt.	Crystalline form, properties and index of refraction	Density or spec. gravity	Melting point, °C	Boiling point, °C	Solubility, in grams per 100 cc		
								Cold water	Hot water	Other solvents
	Lead									
193	2-naphthalene-sulfonate	$Pb(C_{10}H_7SO_3)_2$	621.65	wh cr powd, pois				i		s al
194	nitrate	$Pb(NO_3)_2$	331.20	col, cub or monocl, pois, 1.782	4.53^{20}	d 470		37.65^{0} 56.5°	127^{100}	8.77^{22} 43 % al; s alk, NH_3
195	nitrate, basic	$Pb(OH)NO_3$	286.20	wh rhomb cr	5.93	d 180		19.4^{19}	s	s a
196	nitrite	$3PbO.N_2O_3.H_2O$	763.60	lt yel powd				v s		s dil HNO_3
197	oleate	$Pb(C_{18}H_{33}O_2)_2$	770.12					i		6.46^{20} eth; s pet eth; sl s al
198	oxalate	PbC_2O_4	295.21	wh powd	5.28	d 300		0.00016^{18}		s HNO_3; i al
199	oxide-, di	Plattnerite. PbO_2	239.19	bz, tetr, ω 2.3(Li)	9.375	d 290		i	i	s dil HCl; sl s ac a; i al
100	oxide, mono-	Litharge. PbO	223.19	yel, tetr	9.53	888		0.0017^{20}		s HNO_3. alk, Pb acet, NH_4Cl, $CaCl_2$, $SrCl_2$
101	oxide, mono-	Massicot. PbO	223.19	yel, rhomb, 2.51, 2.61(Li), 2.71	8.0			0.0023^{22}	i	s alk
102	oxide, red	Minium. Pb_3O_4	685.57	red cr sc, or amorph powd	9.1	d 500		i	i	s HCl, acet a; i al
103	oxide, sesqui-	Pb_2O_3	462.38	or-yel powd, amorph		d 370		i	d	d a
104	oxide, sub	Pb_2O	430.38	blk, amorph	8.342	d		i	i	s a, alk
105	oxychloride	$PbCl_2.3PbO$	947.66	yel,				0.0056^{18}	0.07^{74}	
106	oxychloride	Cassel yellow. $PbCl_2.7PbO$	1840.42	yel cr, or powd				i		
107	oxychloride	Fiedlerite. $2PbCl_2.PbO.H_2O$	797.40	monocl, 1.816, 2.1023, 2.026	5.88^{20}	d 150				s HNO_3
108	oxychloride	Nat. laurionite. $PbCl_2.Pb(OH)_2$	519.29	rhomb	6.24	d 142				
109	oxychloride	Matlockite. $PbCl_2.PbO$	501.29						i	s alkalies, hot conc HCl
110	oxychloride	Nat. matlockite. $PbCl_2.Pb(OH)_2$	519.29	wh, tetrag, 2.04, 2.15, 2.15	7.21	d 524		0.0095^{18}		s alk
111	oxychloride	Nat. mendipite. $PbCl_2.2PbO$	724.47	yel, rhomb, 2.24, 2.27, 2.31	7.08	693		i	i	s alk
112	oxychloride	Nat. paralaurionite. $PbCl_2.PbO.H_2O$	519.29	col to wh, monocl pr, 2.146	6.05^{15}	d 150				
113	palmitate	$Pb(C_{16}H_{31}O_2)_2$	718.04	wh powd		112.3		0.005^{35}	0.007^{50}	s al; 0.148^{20} eth
114	phenolate	Lead phenate, lead carbolate. $Pb(OH)OC_6H_5$	317.30	yelsh-wh powd				i		
115	phenolsulfonate	Lead sulfocarbolate. $Pb[C_6H_4(OH)SO_3]_2.5H_2O$	643.60	wh lustr need				s		s al
116	*meta*phosphate	$Pb(PO_3)_2$	365.13	col cr		800(?)			v sl s	
117	*ortho*phosphate	$Pb_3(PO_4)_2$	811.51	col or wh powd, hex, 1.970, 1.936	6.9–7.3	1014		0.000014^{20}	i	s HNO_3, alk; i ac a, al
118	*ortho*phosphate, di-	$PbHPO_4$	303.17	rhomb	5.661^{15}	d				s HNO_3, alk, NH_4Cl
119	*ortho*phosphate, mono-	$Pb(H_2PO_4)_2$	401.16	need						s alk, dil HNO_3, h conc HCl; i acet a
120	phosphide	PbP_5	362.06	blk unstable, inflam		d 400 (vac)		d	d	d dil a
122	*ortho*phosphite	$PbHPO_3$	287.17	wh, powd		d		i		s HNO_3
123	picrate	$Pb(C_6H_2N_3O_7)_2.H_2O$	681.45	yel, need	2.831^{20}	$-H_2O$, 130	expl	0.88^{15}		
124	proprionate, tetra-	$Pb(C_3H_5O_2)_4$	499.48	solid		132				
125	*pyro*phosphate	$Pb_2P_2O_7$	588.32	wh, rhomb	5.8^{20}	824		i	i	s HNO_3, KOH
126	*pyro*phosphate, hydrate	$Pb_2P_2O_7.H_2O$	606.34	wh, rhomb		806 anhydr		i	d	s HNO_3, KOH, $Na_4P_2O_7$
127	selenate	$PbSeO_4$	350.17	wh, rhomb	6.37^{20}_{4}	d		i	i	s conc a
128	selenide	Nat. clausthalite. PbSe	286.15	gray, cub	8.10^{15}	1065		i		s HNO_3
129	metasilicate	Nat. alamosite. $PbSiO_3$	283.27	col or wh, monocl	6,49	766		i		d a
130	*ortho*silicate, di-	Nat. barysilite. $Pb_2Si_2O_7$	582.55	wh, trig, 2.070, 2.050	6.707			i	i	
131	stearate	$Pb(C_{18}H_{35}O_2)_2$	774.15	wh powd		115.7		0.05^{35}	0.06^{50}	$0.005^{14.5}$ eth; i al
132	sulfate	Nat. anglesite. $PbSO_4$	303.25	wh, monocl, or rhomb, 1.877, 1.822, 1.894	6.2	1170		0.00425^{25}	0.0056^{40}	s NH_4 salts; sl s conc H_2SO_4; i a
133	sulfate, basic	Nat. lanarkite. $PbSO_4.PbO$	526.44	wh, monocl, 1.93, 1.99, 2.02	6.92	977		0.0044^{0}	v sl s	sl s H_2SO_4
134	sulfate, hydrogen	$Pb(HSO_4)_2.H_2O$	419.34	wh cr		d		0.0001^{18} d		sl s H_2SO_4
135	*peroxydi*sulfate	$PbS_2O_8.3H_2O$	453.36	deliq				v s		
136	sulfide	Nat. galena. PbS	239.25	bl met cub, 3.921	7.5	1114		0.01244^{20}		s a; i al, KOH
137	sulfite	$PbSO_3$	287.25	wh powd		d		i	i	s HNO_3
138	tartrate, *dl*-	$PbC_4H_4O_6$	355.26	wh cr powd	2.53^{19}			0.0025^{20}	0.0074^{100}	s HNO_3, KOH; i al, ac a, NH_4 ac

No.	Name	Synonyms and Formulae	Mol. wt.	Crystalline form, properties and index of refraction	Density or spec. gravity	Melting point, °C	Boiling point, °C	Solubility, in grams per 100 cc: Cold water	Hot water	Other solvents
	Lead									
1139	*di*thionate	$PbS_2O_6.4H_2O$	439.38	trig, 1.635, 1.653	3.22	d		115.0^{20}		
1140	thiosulfate	PbS_2O_3	319.32	wh cr	5.18	d		0.03		s a, $Na_2S_2O_3$
1141	*meta*titanate	$PbTiO_3$	303.09	yel, rhomb-pyr	7.52			i	i	
1142	telluride	Nat. altaite. PbTe	334.79	wh, cub	8.164^{20}_{4}	917				i a
1143	thiocyanate	$Pb(SCN)_2$	323.35	wh, monocl	3.82	d 190		0.05^{20}	0.2^{100}	s KCNS, HNO_3
1144	tungstate	Nat. stolzite. $PbWO_4$	455.04	tetr, 2.269, 2.182	8.23			i		i HNO_3, s KOH
1145	tungstate	Nat. raspite. $PbWO_4$	455.04	col, monocl, 2.27, 2.27, 2.30	1123			0.03		d a; i al
1146	*meta*vanadate	$Pb(VO_3)_2$	405.07	yel powd				sl s		d HCl; s dil HNO
1147	**Lithium**	Li	6.939	silver white, soft	0.534^{20}	179	1317	d		
1148	acetate	$LiC_2H_3O_2.2H_2O$	102.01	wh, rhomb, α 1.40, β 1.50		70	d	300^{15}	v s	21.5 al
1149	acetylsalicylate	$LiC_9H_7O_4$	186.09	wh powd hygr, d in moist air				100		25 al
1150	*meta*aluminate	$LiAlO_2$ (or $Li_2Al_2O_4$)	65.92	wh, rhomb, 1.604, 1.614	2.55^{25}_{4}	1900–2000		i		
1151	aluminum hydride	$LiAlH_4$	37.95	wh cr powd	0.917	d 125		d		*ca* 30 eth
1152	amide	$LiNH_2$	22.96	col need, cub	$1.178^{17.5}$	380–400	d 750–200 subl	s	d	sl s liq NH_3, al; i eth, bz
1153	antimonide	Li_3Sb	142.57		3.2^{17}	>950		d	d	d a
1154	*ortho*arsenate	Li_3AsO_4	159.74	wh powd, rhomb	3.07^{15}			v sl s		s dil ac a; i pyr
1155	azide	LiN_3	48.96	col cr, hygr		d 115–298		66.41^{16}		20.26^{16} abs al; i eth
1156	benzoate	$LiC_7H_5O_2$	128.06	wh cr or powd				33^{25}	40^{100}	7.7^{25} al, 10^{78} al
1157	*meta*borate	$LiBO_2$	49.75	wh, tricl	$1.397^{41.7}$	845		2.57^{20}	11.83^{80}	
1158	*meta*borate	$LiBO_2.8H_2O$	193.87	col, trig	$1.38^{14.9}$	47				
1159	*penta*borate	$Li_2B_{10}O_{16}.8H_2O$	522.10	wh	1.72	300–350 $-8H_2O$		36.3^{45}	194^{100}	3.9^{30} al; 22^{33} glycerine; i bz
1160	*tetra*borate	$Li_2B_4O_7$	169.11	wh cr		930		2.89^{20}	5.45^{100}	i org solv
1161	borohydrate	$LiBH_4$	21.78	rhomb cr	0.66	d 279		s d		s eth
1162	borohydrate	$LiBH_4$	21.78	wh, orthorhomb	0.666	275 d		v sl s		d al; 2.5 eth
1163	bromide	LiBr	86.85	wh, cub, deliq, 1.784	3.464^{25}	547 deliq	1265	145^{4}	254^{90}	73^{40} al; 8 MeOH; s al, eth; sl s pyr
1164	bromide, dihydrate	$LiBr.2H_2O$	122.28	wh cr		$-1H_2O$; 44		246.0^{20}	v s	s al
1165	carbide	Li_2C_2	37.90	wh cr or powd	1.65^{18}			d	d	s a
1166	carbonate	Li_2CO_3	73.89	wh, monocl, 1.428, 1.567, 1.572	2.11	723	d 1310^{760}	1.54^{0}	0.72^{100}	i al; acet
1167	carbonate, acid	Lithium bicarbonate. $LiHCO_3$	67.96	wh				5.5^{13}		
1168	chlorate	$LiClO_3$	90.39	col, rhomb need, deliq, α 1.63, γ 1.64	1.1190^{18}_{4} (18 %Soln)	127.6	300 d	500^{27}		v s al; 0.142^{25} acetone
1169	chlorate	$LiClO_3.\frac{1}{2}H_2O$ (or $\frac{1}{4}H_2O$)	99.39	wh, tetr, deliq		65(?)	$-\frac{1}{2}H_2O$, 90 d 290	v s	v s	v s al
1170	perchlorate	$LiClO_4$	106.39	wh	2.428	236	430 d	60.0^{25}	150^{89}	152^{25} al; 182^{25} MeOH; 114^{25} eth; 137^{25} acetone
1171	perchlorate, trihydrate	$LiClO_4.3H_2O$	160.44	wh, hex	1.841	95 deliq 236 (anhydr)	d 100 $-2H_2O$	130^{25}		72.9^{25} al; 156^{25} MeOH; 96.2^{25} acetone; 0.096^{25} eth
1172	chloride	LiCl	42.39	wh, cub, 1.662	2.068^{25}	614	1325–1360	63.7^{0}	130^{95}	25.10^{30} al; 42.36^{25} MeOH; 4.11^{25} acetone; 0.538^{33} NH_4OH
1173	chloride, monohydrate	$LiCl.H_2O$	60.41	wh cr, hygr	1.78	$-H_2O$>98		86.2^{20}	s	s HCl
1174	chloroplatinate	$Li_2PtCl_6.6H_2O$	529.78	or prism		$-6H_2O$, 180		v s	v s	v s al; i eth
1175	bichromate, dihydrate	$Li_2Cr_2O_7.2H_2O$	265.90	orange-red cr, deliq	2.34^{30}	187 d	110 $-2H_2O$	187^{30}	278^{100}	s reacts al
1176	*di*chromate	$Li_2Cr_2O_7.2H_2O$	265.90	blk-brn cr, deliq		$-2H_2O$, 130	d	151^{30}		
1177	citrate	$Li_3C_6H_5O_7.4H_2O$	281.98	col cr or powd, deliq		$-4H_2O$, 105		74.5^{25}	66.7^{100}	sl s al, eth
1178	fluoride	LiF	25.94	wh, cub, 1.3915	2.635^{20}	842	1676	0.27^{18}		i al; s HF
1179	fluosilicate	$Li_2SiF_6.2H_2O$	191.99	wh, monocl, 1.300, 1.296	2.33^{12}	$-2H_2O$, 100	d	73^{17}		s al; i eth, acet
1180	fluosulfonate	$LiSO_3F$	106.00	wh powd		360		v s	s	v s al, eth, acet; i ligorin
1181	formate, monohydrate	$H.COOLi.H_2O$	69.97	wh, rhomb	1.46	$-H_2O$, 94	d 230	27.85^{18}	57.05^{98}	sl s al, acet; i bz
1182	gallium hydride	$LiGaH_4$	80.69	wh cr				d	d	s eth
1183	gallium nitride	$LiGaN_2$	118.55	lt gr powd	3.35	d 800		d	d	s a, alk
1184	*meta*germanate	Li_2GeO_3	134.47	monocl, 1.7	3.53^{21}	1239		0.85^{25}		s a

No.	Name	Synonyms and Formulae	Mol. wt.	Crystalline form, properties and index of refraction	Density or spec. gravity	Melting point, °C	Boiling point, °C	Solubility, in grams per 100 cc		
								Cold water	Hot water	Other solvents
	Lithium									
1185	hydride	LiH	7.95	wh cr	0.82	680		d		v sl s a
1186	hydroxide	LiOH	23.95	wh tetr, 1.464, 1.452	1.46	450	d 924	12.8^{20}	17.5^{100}	sl s al
1187	hydroxide, mono-hydrate	$LiOH.H_2O$	41.96	wh monocl, 1.460, 1.524	1.51			22.3^{10}	26.8^{80}	sl s al; i eth
1188	iodate	$LiIO_3$	181.84	wh, hex, hygr	4.502^{32}_{4}			80.3^{18}		i al
1189	iodine	LiI	133.84	wh, cub, 1.955 ± 0.003	3.494 ± 0.015	450	1180 ± 10	165^{20}	433^{80}	250.8^{20} al; 42.6^{18} acet 343.4^{20} MeOH; v s NH_4OH
1190	iodide, trihydrate	$LiI.3H_2O$	187.89	col-yelsh, hex, hygr	3.48	$73-H_2O$	$-2H_2O$, 80 $-H_2O$, 300	151^{0}	201.2^{60}	s abs al, acet
1191	laurate	$LiC_{12}H_{23}O_2$	206.25	wh powd		229.2–229.8		$0.154^{16.3}$	0.178^{25}	0.322^{25} al; $0.008^{15.3}$ eth; 0.240^{25} acet
1192	*per*manganate	$LiMnO_4.3H_2O$	179.92	cub	2.06	d 190		71.43^{16}		d alk
1193	molybdate	Li_2MoO_4	173.82	wh trig, hygr	2.66	705		v s		
1194	myristate	$LiC_{14}H_{27}O_2$	234.31			223.6–224.2		$0.027^{16.3}$ 0.036^{25}	0.062^{50}	$0.010^{15.3}$ eth; 0.331^{15} acet; 0.155^{20} al
1195	nitrate	$LiNO_3$	68.94	wh, trig, 1.735, 1.735	2.38	264	d 600	$89.8^{27.55}$	234^{100}	s NH_4OH, al; 37.15 pyridine
1196	nitrate, trihydrate	$LiNO_3.3H_2O$	122.99	col need		$-2\frac{1}{2}H_2O$, 29.9	$-3H_2O$, 61.1	34.8^{0}	$57.48^{29.6}$	s al, MeOH, acet
1197	nitride	Li_3N	34.82	red-brn amorph, or blk-gray cr, cub		tr 840–850 (in N_2)				
1198	nitrite	$LiNO_2.H_2O$	70.96	col flat need	1.615^{0}	>100	d	125^{0}	459^{80}	v s abs al
1199	oxalate	$Li_2C_2O_4$	101.90	col, rhomb, 1.465, 1.53, 1.696	$2.121^{17.5}$	d		$8^{19.5}$		i al, eth
1200	oxalate, acid	$LiHC_2O_4.H_2O$	113.99			d		8^{17}		
1201	oxide	Li_2O	29.88	wh cr, cub, n_D 1.644	$2.013^{25.2}$	>1700	1200^{600}	6.67^{0} d	10.02^{100}	
1202	palmitate	$LiC_{16}H_{31}O_2$	262.36	wh powd		224.5		0.01^{18}	0.015^{25}	0.347^{15} acet; 0.077^{20} al; $0.005^{15.8}$ eth
1203	*meta*phosphate	$LiPO_3$	85.91	col pl	2.461	red heat		i	i	s a
1204	*ortho*phosphate	Li_3PO_4	115.79	col, rhomb	$2.537^{17.5}$	837		0.039^{18}		s a, NH_4OH; i acet
1205	*ortho*phosphate	$Li_3PO_4.\frac{1}{2}H_2O$	124.80	wh cr powd	2.41	$-\frac{1}{2}H_2O$, 100		0.04^{25}		s a
1206	phosphate, di- H	LiH_2PO_4	103.93	col cr, hygr	2.461	>100				
1207	potassium sulfate	$LiKSO_4$	142.10	col, hex; n_D 1.4723, 1.4717	2.393^{20}			s	s	
1208	potassium *dl*-tartrate	$LiKC_4H_4O_6.H_2O$	212.13	col, monocl, β 1.523 (red)	1.610			s		
1209	salicylate	$LiC_7H_5O_3$	144.06	wh, powd, deliq		d		133.3		50 al
1210	selenide	$Li_2Se.9H_2O$	254.98	col, rhomb, deliq				d		
1211	*meta*silicate	Li_2SiO_3	89.96	col, rhomb; α 1.584, γ 1.604	2.52^{25}_{4}	1204		i	s d	s dil HCl
1212	*ortho*silicate	Li_4SiO_4	119.84	col, rhomb; α 1.594, γ 1.614	2.392^{25}_{4}	1256		i	d	d a
1213	silicide	Li_6Si_2	97.81	bl cr, hygr	*ca.* 1.12	d 600 (vac)		d	d	d a; i NH_3 turp
1214	sodium fluoaluminate	$Li_3Na_3(AlF_6)_2$	371.73	cub cr, 1.3395	2.774	710		0.074^{18}		
1215	stearate	$LiC_{18}H_{35}O_2$	290.41	wh cr	220.5–221.5			0.010^{18}		0.010^{25} al; 0.040^{18} eth; 0.457^{15} acet
1216	sulfate	Li_2SO_4	109.94	α monocl; β hex or rhomb, γ cub 500°C; β 1.465	2.221		860	26.1^{0}	23^{100}	i abs al, acet
1217	sulfate, hydrogen	$LiHSO_4$	104.01	col pr	2.123^{13}	120		d		
1218	sulfate, mono-hydrate	$Li_2SO_4.H_2O$	127.95	col cr, monocl, 1.465, 1.477, 1.488	880			34.9^{25}	29.2^{100}	11.5^{30} al+H_2O (23.9 % alco); i acet, pryidine
1219	sulfide	Li_2S	45.94	wh-yel, cub, deliq	1.66	900–975		v s	v s	v s al
1220	sulfide, hydro-	LiHS	40.01	wh powd, hygr				s	s	s al
1221	sulfite, monohydrate	$Li_2SO_3.H_2O$	111.96	wh need, α 1.53, γ 1.59		455 d	140 $-H_2O$	24.9^{80}	22^{80}	i org solv
1222	tartrate	$Li_2C_4H_4O_6.H_2O$	179.97	wh cr powd				s		
1223	thallium *dl*-tartrate	$LiTlC_4H_4O_6.2H_2O$	395.41	tricl	3.144					
1224	thiocyanate	LiSCN	65.02	wh cr, deliq, n_D 1.333				v s		s methylacet
1225	dithionate	$Li_2S_2O_6.2H_2O$	210.03	col, rhomb, 1.5602	2.158	d		v s		
1226	tungstate	Li_2WO_4	261.73	col, trig	3.71	742		v s	v s	d a; i al
1227	**Lutetium**	Cassiopeium. Lu	174.97	met, hex	9.842	1652	**3327**			

No.	Name	Synonyms and Formulae	Mol. wt.	Crystalline form, properties and index of refraction	Density or spec. gravity	Melting point, °C	Boiling point, °C	Solubility, in grams per 100 cc		
								Cold water	Hot water	Other solvents
	Lutetium									
l228	bromide	$LuBr_3$	414.70		1025	1400		s		
l229	chloride	$LuCl_3$	281.33	col cr	3.98	905	subl 750	s		
l230	fluoride	LuF_3	231.97			1182	2200	i	i	
l231	iodide	LuI_3	555.68			1050	1200	s	s	
l232	oxalate	$Lu_2(C_2O_4)_3.6H_2O$	722.09	wh cr		50 $(-H_2O)$		i	i	i dil a
l233	oxide	Lu_2O_3	397.94	cub cr	9.42					
l234	sulfate	$Lu_2(SO_4)_3.8H_2O$	782.25	col cr				42.27^{20}	16.93^{40}	
m1	**Magnesium**	Mg	24.312	silv wh met, hex	1.74^{5}	651	1107	i	d to $Mg(OH)_2$	s min a, conc. HF; NH_4 salts; i CrO_3, alk.
m2	acetate	$Mg(C_2H_3O_2)_2$	142.40	wh cr	1.42	323 d		v s	v s	5.25^{15} MeOH
m3	acetate, tetrahydrate	$Mg(C_2H_3O_2)_2.4H_2O$	214.46	col, monocl deliq, β 1.491	1.454	80		120^{15}	∞	v s al
m4	aluminate	Nat. spinel. $MgAl_2O_4$	142.27	col, cub, 1.723	3.6	2135				sl s H_2SO_4; v sl s dil HCl; i HNO_3
m5	amide	$Mg(NH_2)_2$	56.36	gray powd		d 350–400	d	d		v sl s liq NH_3; d al
m6	antimonate	$MgO.Sb_2O_5.12H_2O$	579.99	hex or tricl cr	2.60 (hex)	$-12H_2O$, 200		v sl s		d HCl
m7	antimonide	Mg_3Sb_2	316.44	met hex pl	4.088^{25}_{4}	961		i		
m8	*ortho*arsenate	Nat. hoernesite. $Mg_3(AsO_4)_2.8H_2O$	494.90	wh monocl	2.60–2.61					
m9	*ortho*arsenate	$Mg_3(AsO_4)_2.22H_2O$	747.11	wh cr	1.788	$-17H_2O$, 100	$-21H_2O$, 220	i	i	s a, NH_4Cl
m10	*ortho*arsenate, mono-H	Nat. roesslerite. $MgHAsO_4.7H_2O$	290.35	monocl	1.943^{15}	$-5H_2O$, 100		d		
m11	arsenide	Mg_3As_2	222.78	brn-red, cub	3.148^{25}_{4}	800		d	d	d dil a, al
m12	*ortho*arsenite	$Mg_3(AsO_3)_2$	318.78	wh				s	v s	s a, NH_4Cl; i NH_4OH
m13	benzoate	$Mg(C_7H_5O_2)_2.3H_2O$	320.59	wh powd		$-3H_2O$, 110	d 200	6.16^{15}	19.6^{100}	s al
m14	bismuthide	Mg_3Bi_2	490.90	met, hex	5.945^{25}_{4}	823				
m15	bismuth nitrate	$3Mg(NO_3)_2.2Bi(NO_3)_3.24H_2O$	1543.31	col cr, deliq	2.32^{16}_{16}	71		d	d	s HNO_3
m16	*di*borate	Nat. ascharite. $Mg_2B_2O_5.H_2O$	168.26	orthorhmb, 1.54	2.60–2.70					
m17	*meta*borate	Nat. pinnoite. $Mg(BO_2)_2.3H_2O$	163.98	yel, tetr, pyram, 1.565, 1.575	2.27–2.30					
m18	metaborate, octahydrate	$Mg(BO_2)_2.8H_2O$	254.05	col, tetr, 1.565, 1.575	2.30			i	v sl s	s a
m19	*ortho*borate	$Mg_3(BO_3)_2$	190.55	col, rhomb, 1.6527, 1.6537, 1.6548	2.99^{21}			i	i	s min a; i ac a
m20	boride	MgB_2	89.18	bl		d 1200 (vac)		d		sl s a
m21	bromate	$Mg(BrO_3)_2.6H_2O$	388.22	col, cub, 1.514	2.29	$-6H_2O$, 200	d	42^{18}	v s	i al
m22	bromide	$MgBr_2$	184.13	wh hex cr, deliq	3.72^{25}_{4}	700		101.50^{20}	125.6^{100}	6.9 al; 21.8^{20} MeOH
m23	bromide, hexahydrate	$MgBr_2.6H_2O$	292.22	col, hex pr or need, hygr, fluo in x-rays	2.00	172.4		316^{0}	v s	s al, acet; sl s NH
m24	bromoplatinate	$MgPtBr_6.12H_2O$	915.04	trig	2.802					
m25	carbonate	Nat. magnesite. $MgCO_3$	84.32	wh, trig, 1.717, 1.515	2.958	d 350	$-CO_2$, 900	0.0106		s a, aq+CO_2; i acet, NH_3
m26	carbonate, basic artinite	Nat. artinite. $MgCO_3.Mg(OH_2).3H_2O$	196.69	wh, rhomb, 1.489, 1.534, 1.557	2.02^{20}					
m27	carbonate, basic	Nat. hydromagnesite. $3MgCO_3.Mg(OH)_2.3H_2O$	365.34	wh, rhomb, 1.527, 1.530, 1.540	2.16	d		0.04	0.011	s a, NH_4 salts
m28	carbonate, pentahydrate	Nat. lansfordite. $MgCO_3.5H_2O$	174.40	wh, monocl, 1.456, 1.476, 1.502	1.73	d in air		0.176^{7}	0.375^{20}	s HCl, $MgSO_4$ sol
m29	carbonate, trihydrate	Nat. nesquehonite. $MgCO_3.3H_2O$	138.37	col, rhomb need, 1.495, 1.501, 1.526	1.850	165		0.179^{16}	d	s a; 1.4 aq+CO_2
m30	chlorate	$Mg(ClO_3)_2.6H_2O$	299.31	wh, rhomb need, deliq	1.80^{25}	35	d 120	128.6^{18}	v s	s al
m31	*per*chlorate	$Mg(ClO_4)_2$	223.21	deliq	2.21^{18}	d 251		49.90^{25}	v s	23.96^{25} al
m32	*per*chlorate, hexahydrate	$Mg(ClO_4)_2.6H_2O$	331.31	wh, rhomb cr, 1.482, 1.458	1.98	185–190		v s	v s	
m33	*per*chlorate, hexammine	$Mg(ClO_4)_2.6NH_3$	325.40	wh, cub	1.41^{20}_{4}					s liq NH_3; s d al
m34	chloride	$MgCl_2$	95.22	wh, lustr hex cr, 1.675, 1.59	2.316–2.33	708	1412	54.25^{20}	72.7^{100}	7.40^{30} al

No.	Name	Synonyms and Formulae	Mol. wt.	Crystalline form, properties and index of refraction	Density or spec. gravity	Melting point, °C	Boiling point, °C	Solubility, in grams per 100 cc		
								Cold water	Hot water	Other solvents
	Magnesium									
m35	chloride, hexahydrate	Nat. bischofite. $MgCl_2.6H_2O$	203.31	col, monocl, deliq, 1.495, 1.507 1.528	1.569	d 116–118	d	167	367	s al
m36	chloropalladate	$MgPdCl_6.6H_2O$	451.52	hex	2.12	d				
m37	chloroplatinate	$MgPtCl_6.6H_2O$	540.21	yel, trig	2.692	−H_2O, 180		s	s	
m38	chlorostannate	$MgSnCl_6.6H_2O$	463.81	tricl	2.08	d 100				
m39	chromate	$MgCrO_4.7H_2O$	266.41	yel, rhomb, 1.521, 1.550, 1.568	1.695		211.5^{18}	v s		
m40	chromite	$MgCr_2O_4$	192.30	dk-grn or red, cub	4.6^{20}			i	i	s conc H_2SO_4; i dl a, dil alk
m41	citrate, nono-H	$MgHC_6H_5O_7.5H_2O$	304.50	wh gran powd				20^{25}	s	s a; i al
m42	cyanide	$Mg(CN)_2$	76.35			d 300 to $MgCN_2$	d 600	s	d	
m43	cyanoplatinite	$MgPt(CN)_4.7H_2O$	449.58	red, pr	2.185^{16}	−H_2O, 45		v s	v s	i al, eth
m44	ferrite	$MgFeO_4$	200.00	blk, oct, 2.35	4.44–4.60	1750 ± 25				s conc HCl; i dil a, h HNO_3, al
m45	ferrocyanide	$Mg_2Fe(CN)_6.12H_2O$	476.76	pa yel cr		d *ca* 200		33		i al
m46	fluoride	Nat. sellaite. MgF_2	62.31	col, tetr, faint vlt, lumin, 1.378, 1.390; 3.14		1266	2239	0.0076^{18}	i	s HNO_3; sl s a; i al
m47	fluosilicate	$MgSiF_6$	166.39	wh, cr or powd				65		
m48	fluosilicate, hexahydrate	$MgSiF_6.6H_2O$	274.48	wh, trig	1.788	d 120		$64.8^{17.5}$		i al
m49	formate	$Mg(CHO_2)_2.2H_2O$	150.38	col, rhomb		−$2H_2O$, 100		14^0 (anh)	24^{100} (anh)	i al, eth
m50	*ortho*germanate	Mg_2GeO_4	185.21	wh ppt				0.0016^{25}		s a; i alk
m51	germanide	Mg_2Ge	121.21			1115				
m52	hydride	MgH_2	26.33	wh tetr cr or mass		d 280 (vac)		d viol		i eth
m53	hydroxide	Nat. brucite. $Mg(OH)_2$	58.33	col, hex pl, 1.559, 1.580	2.36	−H_2O, 350		0.0009^{18}	0.004^{100}	s a, NH_4 salts
m54	iodate	$Mg(IO_3)_2.4H_2O$	446.18	wh, monocl	$3.3^{13.5}$	−$4H_2O$, 210	d	10.2^{20}	19.3^{100}	
m55	iodide	MgI_2	278.12	wh, hex, deliq	4.43^{25}_{4}	d >700		148^{18}	164.9^{110}	s al, eth, NH_3
m56	iodide, octahydrate	$MgI_2.8H_2O$	422.24	wh powd, deliq		d 41		81^{20}	90.3^{80}	s al, eth
m57	lactate	$Mg(C_3H_5O_3).3H_2O$	256.50	wh cr powd, v bitter taste				3.3	16.7^{100}	i al, eth
m58	laurate	$Mg(C_{12}H_{23}O_2)_2.2H_2O$	458.97	wh lumps		150.4		0.007^{25}	0.041^{100}	0.415^{15} al; 0.012^{25} eth
m59	*per*manganate	$Mg(MnO_4)_2.6H_2O$	370.27	dk purp need, deliq	2.18(?)	d		v s	d	s MeOH, ac a
m60	molybdate	$MgMoO_4$	184.25	rhomb, tricl	2.208			13.7^{25}		
m61	myristate	$Mg(C_{14}H_{27}O_2)_2$	479.05	wh powd		131.6		$0,006^{15}$	0.014^{50}	0.189^{25} al; 0.007^{25} eth
m62	nitrate, dihydrate	$Mg(NO_3)_2.2H_2O$	184.35	col pr	2.0256^{25}	129		s	s	s al, liq NH_3; sl s conc HNO_3
m63	nitrate, hexahydrate	$Mg(NO_3)_2.6H_2O$	256.41	col, monocl, deliq	1.6363^{25}_{4}	89	d 330	125	v s	s al, liq NH_3
m64	nitride	Mg_3N_2	100.95	grn-yel, powd or mass	2.712^{25}_{4}	d 800	subl 700 (vac)	d	d	s a; i al
m65	nitrite, trihydrate	$Mg(NO_2)_2.3H_2O$	170.37	wh pr, hygr		d 100		s		s al
m66	oleate	$Mg(C_{18}H_{33}O_2)_2$	587.24	yel powd or mass				0.024^{25}		6.64^{20} al; s linseed oil; sl s eth
m67	oxalate	$MgC_2O_4.2H_2O$	148.36	wh powd	2.45	d 150		0.07^{16}	0.08^{100}	s alk, a, oxalate
m68	oxide	Nat. periclase. MgO	40.31	col, cub, 1.736	3.58^{25}	2800	3600	0.00062	0.0086^{30}	s a, NH_4 salts; i al
m69	oxide, per-	MgO_2	56.31	wh powd				i	i	s a
m70	palmitate	$Mg(C_{16}H_{31}O_2)_2$	535.16	wh cr need or lumps		121.5		0.008^{25}	0.009^{50}	0.047^{25} al; 0.003^{25} eth
m71	*ortho*phosphate	$Mg_3(PO_4)_2$	262.88	rhomb pl, iridisc		1184		i	i	s NH_4 salts, i liq NH_3
m72	*ortho*phosphate	$Mg_3(PO_4)_2.22H_2O$	659.22	col, monocl pr	1.640^{15}	−$18H_2O$, 100	d 200	v sl s		d a
m73	*ortho*phosphate, mono-H	Nat. newberyite. $MgHPO_4.3H_2O$	174.34	wh, rhomb, 1.514, 1.518, 1.533	2.123^{15}	−H_2O, 205	d 550–650	sl s		s a
m74	*ortho*phosphate, mono-H, heptahydrate	$MgHPO_4.7H_2O$	246.40	wh, monocl need	1.728^{15}	−$4H_2O$, 100	d 550–650	0.3	0.2	s a; i al
m75	*ortho*phosphate, octahydrate.	Nat. bobierite. $Mg_3(PO_4)_2.8H_2O$	407.00	wh, monocl pl, 1.510, 1.520, 1.543	2.195^{15}	−$5H_2O$, 150	−$8H_2O$, 400	v sl s		s NH_4 citrate
m76	*ortho*phosphate, tetrahydrate	$Mg_3(PO_4)_2.4H_2O$	334.97	monocl	1.64^{15}			0.0205		s a; i NH_4 salts
m77	phosphide	Mg_3P_2	134.88	yel-grn cub cr	2.055			d	d	d dil min a; sl d conc H_2SO_4

PHYSICAL CONSTANTS OF INORGANIC COMPOUNDS (Continued)

No.	Name	Synonyms and Formulae	Mol. wt.	Crystalline form, properties and index of refraction	Density or spec. gravity	Melting point, °C	Boiling point, °C	Solubility, in grams per 100 cc		
								Cold water	Hot water	Other solvents
	Magnesium									
m78	*hypo*phosphite	$Mg(H_2PO_2)_2.6H_2O$	262.38	wh, ditetrag	$1.59_4^{12.5}$	$-5H_2O$, 100	$-6H_2O$, 180	20^{25}	....	i al, eth
m79	*ortho*phosphite	$MgHPO_3.3H_2O$	158.34	....	....	....	....	....	0.25	s a
m80	*pyro*phosphate	$Mg_2P_2O_7$	222.57	col, tab monocl, 1.602, 1.604, 1.615	2.559, (3.06)	1383	....	i	i	s a; i al
m81	platinocyanide	$MgPt(CN)_4.7H_2O$	449.58	red cr, 1.561	2.185^{16}	$-2H_2O$, 45		s	s	s al; i eth
m82	salicylate	$Mg(C_7H_5O_3)_2.4H_2O$	370.61	col or sl redsh cr powd, effl	....			s	....	s al
m83	selenate	$MgSeO_4.6H_2O$	275.36	col, monocl, 1.468, 1.489, 1.491	1.928			v s	v s	....
m84	selenide	MgSe	103.27	lght gray powd or cr, 2.44	4.21			d	d	d a
m85	*meta*silicate	Nat. clinoenstatite. $MgSiO_3$	100.40	wh, monocl, α 1.651, γ 1.660	3.192_4^{25}	d 1557	....	i	i	v sl s HF
m86	*ortho*silicate	Nat. forstérite. Mg_2SiO_4	140.71	wh, orthorhmb, 1.65, 1.66, 1.67	3.21	1910		i	....	d h HCl
m87	(di-) silicide	Mg_2Si	76.71	blue cub	1.94	1102	....	i	d	s a, NH_4Cl, HCl
m88	silicofluoride	$MgSiF_6.6H_2O$	274.48	wh hex-rhomb, 1.3439, 1.3602	1.788	d 100	....	60^{25}	5	s dil a, v sl s HF i al
m89	stannide	Mg_2Sn	167.31	blsh-wh met	....	778	....	s	....	s dil HCl
m90	stearate	$Mg(C_{18}H_{35}O_2)_2$	591.27	wh powd or lumps	....	145	....	0.003^{15} 0.004^{25}	0.008^{50}	0.020^{25} al; 0.003^{25} eth
m91	sulfate	$MgSO_4$	120.37	col, rhomb cr, 1.56	2.66	d 1124	....	26^{0}	73.8^{100}	s al, glyc; 1.16^{18} eth; i acet
m92	sulfate, heptahydrate	Epsom salt, nat. epsomite. $MgSO_4.7H_2O$	246.48	col, rhomb or monocl, 1.433, 1.455, 1.461	1.68	$-6H_2O$, 150	$-7H_2O$, 200	71^{20}	91^{40}	sl s al, glyc
m93	sulfate, monohydrate	Nat. kieserite, $MgSO_4.H_2O$	138.39	col, monocl pr, 1.523, 1.535, 1.586	2.445	....	....	....	68.4^{100}	....
m94	sulfide	MgS	56.38	pa red-brn, cub, phosph, 2.271	2.84	d >2000	....	d	d	s a, PCl_3
m95	sulfite	$MgSO_3.6H_2O$	212.47	wh, rhomb or hex, 1.511, 1.464 (hex)	1.725	$-6H_2O$, 200	d	66^{25}	s	i al, NH_3
m96	*d*-tartrate	$MgC_4H_4O_6.5H_2O$	262.46	wh, rhomb	1.67	$-4H_2O$, 100	$-5H_2O$, 200	0.8^{18}	1.44^{90}	s min a; i al, Nl
m97	*d*-tartrate, hydrogen	$Mg(HC_4H_4O_6)_2.4H_2O$	394.54	wh, rhomb	1.72	....	....	....	1.893^{100}	....
m98	telluride	MgTe	151.91	wh, hex cr	3.86			d	....	d a
m99	thiosulfate	$MgS_2O_3.6H_2O$	244.53	col, rhomb pr	1.818^{24}	$-3H_2O$, 170	d	v s	v s	i al
m100	thiotellurite	Mg_3TeS_6	360.86	pa yel cr mass	....	....	....	s	s	s al
m101	tungstate	$MgWO_4$	272.16	col, monocl	5.66	....	....	i	....	d a; i al
m102	**Manganese**	Mn	54.938	gray-pink met, cub or tetr	7.20	1244±3	2097	d	d	s dil a
m103	(II) acetate	$Mn(C_2H_3O_2)_2$	173.02	brn cr	1.74	....	....	s, d	....	s al
m104	acetate, tetrahydrate	$Mn(C_2H_3O_2)_2.4H_2O$	245.08	pa red, monocl	1.589	....	....	s	....	s al
m105	arsenide, mono-	MnAs	129.86	blk, hex	6.17–6.20 (5.55)	d 400	....	i	i	s HCl, aq reg
m106	arsenide, di-	Mn_2As	184.80	....	....	1400	....	i	i	s aq reg
m107	arsenide, tri-	Mn_3As_2	314.66	magnetic, (exist ?)	....	....	....	i	i	s aq reg
m108	(II) benzoate	$Mn(C_7H_5O_2)_2.4H_2O$	369.23	pa red pr	....	....	....	6.55^{15}	....	....
m109	boride, di-	MnB_2	76.56	gray-vlt cr	6.9	....	....	d	d	s a
m110	boride, mono-	MnB	65.75	cr powd	6.2^{15}	....	....	....	....	....
m111	bromide, di-	$MnBr_2$	214.76	rose cr	4.385_4^{25}	d	....	127.3^{0}	228^{100}	i NH_3
m112	bromide, di-, tetrahydrate	$MnBr_2.4H_2O$	286.82	α stable, rose monocl, deliq β labile, col, rhomb	....	d 64.3	....	296.7^{0}	....	....
m113	carbide	Mn_3C	176.83	tetr	6.89^{17}	....	....	d	d	s a
m114	(II) carbonate	Nat rodochrosite. $MnCO_3$	114.95	rose, rhomb, lt brn in air	3.125	d	....	0.0065^{25}	....	s dil a, aq CO_2; i al, NH_3
m115	chloride, di-	Scacchite. $MnCl_2$	125.84	pink, cub cr, deliq	2.977_4^{25}	650	1190	72.3^{25}	123.8^{100}	s al; i eth, NH_3
m116	chloride, di-, tetrahydrate	$MnCl_2.4H_2O$	197.91	rose, monocl, deliq	2.01	58	$-H_2O$, 106; $-4H_2O$, 198	151^{8}	656^{100}	s al; i eth
m117	chloride, tri-	$MnCl_3$	161.30	brn cr or grnsh-blk	....	d sl	....	....	....	s abs al
m118	chloroplatinate	$MnPtCl_6.6H_2O$	570.84	trig	2.692	d	....	....	....	....
m119	chromite	$MnCr_2O_4$	222.93	gray-blk, cub	4.97^{20}	....	....	i	i	i a
m120	(II) citrate	$Mn_3(C_6H_5O_7)_2$	543.02	wh-redsh powd	....	....	....	v sl s	....	s Na-citr sol
m121	(II) ferrocyanide	$Mn_2Fe(CN)_6.7H_2O$	447.94	grnsh-wh powd	....	....	....	i	....	s HCl; i NH_4 sal
m122	fluogallate	$[Mn(H_2O)_6][GaF_5.H_2O]$	345.76	pink, orthorhomb, 1.45	2.22	d 230	....	v s	....	s HF
m123	fluosilicate	$MnSiF_6.6H_2O$	305.11	rose, hex pr, 1.357, 1.374	1.903	d	....	140^{18}	v s	s al

No.	Name	Synonyms and Formulae	Mol. wt.	Crystalline form, properties and index of refraction	Density or spec. gravity	Melting point, °C	Boiling point, °C	Solubility, in grams per 100 cc		
								Cold water	Hot water	Other solvents
	Manganese									
m124	fluoride, di-	MnF_2	92.93	red, tetr, or redsh powd	3.98	856		0.66^{40}	0.48^{100}	s a; i al, eth
m125	fluoride, tri-	MnF_3	111.93	red cr	3.54	d		d	d	s a
m126	formate	$Mn(CHO_2)_2.2H_2O$	181.00	rhomb	1.953	d		s	s	
m127	(II) glycerophosphate	$MnC_3H_7O_6P$	225.00	wh or sl red powd				sl s		s a, citr a; i al
m128	hydroxide	$MnO(OH)_2$	104.95	blk-brn, amorph (exist ?)	2.58			v sl s		
m129	(II) hydroxide	Nat. pyrochroite. $Mn(OH)_2$	88.95	wh-pink, trig 1.723, 1.681	3.258^{13}	d		0.0002^{18}		s a, NH_4 salts; i alk
m130	(III) hydroxide	Magnanite. MnO(OH)	87.94	br-blk, rhomb. 2.24, 2.24, 2.53 (Li)	4.2–4.4	d		i	i	s HCl, h H_2SO_4
m131	iodide, di-	MnI_2	308.75	pink, hex cr, deliq, br. in air	5.0^{1}	638 (vac) d *ca* 80	subl vac 500	s	s	0.02^{25} NH_3
m132	iodide, di-, tetrahydrate	$MnI_2.4H_2O$	380.81	rose, monocl, deliq		d		s	v s	
m133	*hexa*iodoplatinate	$MnPtI_6.9H_2O$	1173.59	trig	3.604_4^{20}	d				
m134	(II) nitrate	$Mn(NO_3)_2.4H_2O$	251.01	col, or rose, monocl	1.82	25.8	129.4	426.4^{0}	∞	v s al
m135	(II) lactate	$Mn(C_3H_5O_3)_2.3H_2O$	287.04	pa red, monocl		d		10	v s	s al
m136	(II) oxalate	MnC_2O_4	142.96	wh cr powd	$2.43^{21.7}$	d 150		i	i	s a, NH_4Cl
m137	(II) oxalate, dihydrate	$MnC_2O_4.2H_2O$	178.98	redsh-wh oct cr powd		$-2H_2O$, 100	d	0.0312^{25}	0.037^{36}	
m138	(II) oxalate, trihydrate	$MnC_2O_4.3H_2O$	197.00	pink, tricl		$-H_2O$, 25				
m139	(II, III) oxide	Nat. hausmannite. Mn_3O_4	228.81	blk, tetr (rhomb), 2.46, (Li) 2.15 (Li)	4.856	1705		i	i	s HCl
m140	oxide, di-	Nat. pyrolusite. MnO_2	86.94	blk, rhomb, or brn-blk powd	5.026	−O, 535		i	i	s HCl; i HNO_3, acet
m141	oxide, hept-	Mn_2O_7	221.87	dk red oil, hyg, exp	2.396_4^{20}	5.9	d 55, exp 95	v s	d	s H_2SO_4
m142	(II) oxide, mon-	Nat. manganosite, MnO	70.94	grn, cub, 2.16	5.43–5.46 (3.7–3.9)			i	i	s a, NH_4Cl
m143	(III) oxide, sesqui-	Nat. braunite. Mn_2O_3	157.87	blk, cub (tetr)	4.50	−O, 1080		i	i	s a; i ac a
m144	oxide, tri-	MnO_3	102.94	redsh, deliq (exist ?)		d		s	d	s alk, H_2SO_4
m145	(III) *meta*phosphate	$Mn_2(PO_3)_6.2H_2O$	619.74					sl s	s	
m146	(II) *ortho*phosphate	Nat. reddingite. $Mn_3(PO_4)_2.3H_2O$	408.80	rose or yelsh-wh rhomb, 1.651, 1.656, 1.683	3.102					
m147	(III) *ortho*phosphate	$MnPO_4.H_2O$	167.92	gray cr powd		$-H_2O$, 300	d	i		s h conc H_2SO_4, conc HCl, molten H_3PO_4
m148	(II) *ortho*phosphate, di-H	$Mn(H_2PO_4)_2.2H_2O$	284.94			$-H_2O$, >100		s		i al
m149	(II) *ortho*phosphate mono-H	$MnHPO_4.3H_2O$	204.97	red, rhomb or pink powd, 1.656				sl s	d	s a; i al
m150	(II) *pyro*phosphate	$Mn_2P_2O_7$	283.82	br-pink, monocl, 1.695, 1.704, 1.710	3.707^{25}	1196		i		s a
m151	(II) *pyro*phosphate, trihydrate	$Mn_2P_2O_7.3H_2O$	337.87	wh, amorph powd				i		s $K_2P_2O_7$ sol, H_2SO_3 i acet
m152	phosphide, mono-	MnP	85.91	dk gray	5.39^{21}	1190		i	i	sl s HNO_3
m153	(tri-)phosphide, di-	Mn_3P_2	226.76	dk gray	5.12^{18}	1095		i	i	sl s dil HNO_3
m154	(II) *hypo*phosphite	$Mn(H_2PO_2)_2.H_2O$	202.93	rose cr or powd		$-H_2O$, >150		12.5	16.7	i al
m155	(II) *ortho*phosphite	$MnHPO_3.H_2O$	152.93	redsh		$-H_2O$, 200		sl s		s $MnSO_4$, $MnCl_2$
m156	selenate, dihydrate	$MnSeO_4.2H_2O$	233.93	rhomb	2.95–3.01			s	s	
m157	selenate, pentahydrate	$MnSeO_4.5H_2O$	287.97	pa red, trig	2.33–2.39					
m158	selenide	MnSe	133.90	gray, cub	5.55^{15}			i		d dil a
m159	selenite	$MnSeO_3.2H_2O$	217.93	monocl cr				v sl s	v sl s	
m160	(II) *meta*slicate	Nat. rhodonite. $MnSiO_3$	131.02	red, tricl, 1.733, 1.740, 1.744	3.72^{25}	1323		i		i HCl
m161	silicide, di-	$MnSi_2$	111.11	gray, oct	5.24^{13}			i	i	s HF, alk; i HNO_3, H_2SO_3
m162	silicide, mono-	MnSi	83.02	tetrah	5.90^{15}	1280		i	i	s HF; v sl s a
m163	(di-)silicide,	Mn_2Si	137.96	quadr pr	6.20^{15}	1316		i	i	s HCl, NaOH; i HNO_3
m164	(II) sulfate	$MnSO_4$	151.00	redsh	3.25	700	d 850	52^{5}	70^{70}	s al; i eth
m165	(III) sulfate	$Mn_2(SO_4)_3$	398.06	grn cr, deliq, hex	3.24	d 160		d	d	s HCl, dil H_2SO_4; i conc. H_2SO_4, HNO_3

No.	Name	Synonyms and Formulae	Mol. wt.	Crystalline form, properties and index of refraction	Density or spec. gravity	Melting point, °C	Boiling point, °C	Solubility, in grams per 100 cc		
								Cold water	Hot water	Other solvents
	Manganese									
m166	(II) sulfate, dihydrate	$MnSO_4.2H_2O$	187.03	(exist ?)	2.526^{15}	stab 57–117		85.27^{35}	106.8^{55}	
m167	(II) sulfate, heptahydrate	$MnSO_4.7H_2O$	277.11	red monocl or rhomb	2.09	$-7H_2O$, 280; stab +9		172	118^{13}	i al
m168	(II) sulfate, hexahydrate	$MnSO_4.6H_2O$	259.09	(exist?)		stab +5 to +8		147.4	1.345^{38}	
m169	(II) sulfate, monohydrate	Nat. szmikite. $MnSO_4.H_2O$	169.01	pa pink monocl, 1.562, 1.595, 1.632	2.95	stab 57–117		98.47^{48}	79.8^{100}	
m170	(II) sulfate, pentahydrate	$MnSO_4.5H_2O$	241.08	rose, tricl, 1.495, 1.508, 1.514	2.103^{15}	stab 9–26		124^{0}	142^{54}	
m171	(II) sulfate, tetrahydrate	Common form. $MnSO_4.4H_2O$	223.06	pink, monocl or rhomb effl, 1.508, 1.522	2.107	stab 26–27		105.3^{0}	111.2^{54}	i al
m172	(II) sulfate, trihydrate	$MnSO_4.3H_2O$	205.05	(exist?)	2.356^{15}	stab 30–40		74.22^{5}	99.31^{57}	
m173	(II) sulfide	Nat. alabandite. MnS	87.00	grn cub or pink amorph, 2.70 (Li)	3.99	d		0.00047^{18}		s dil a, al; i $(NH_4)_2S$
m174	(II) sulfide	$3MnS.H_2O$	279.05	gray-pink		d		0.0006	i	s dil a; i $(NH_4)_2$
m175	(IV) sulfide	Nat. hauerite. MnS_2	119.07	blk cub, 2.69 (Li)	3.463	d		i	i	d HCl
m176	(II) tantalate	$Mn(TaO_3)_2$	512.83	blk, rhomb, 2.22, 2.25, 2.29	7.03					
m177	(II) tartrate	$MnC_4H_4O_6$	203.01	wh powd				v sl s		s dil a
m178	(II) thiocyanate	$Mn(SCN)_2.3H_2O$	225.16	deliq		$-3H_2O$, 160–170		s	v s	v s al
m179	(II) *di*thionate	$Mn(SO_3)_2$	215.06	tricl cr	1.757			s	v s	
m180	(II) titanate	Nat. pyrophanite. $MnTiO_3$	150.84	yel, trig, 2.481, 2.210	4.54	1360				
m181	valerate	$Mn(C_5H_9O_2)_2.2H_2O$	293.22	br powd				s		
m182	**Manganic acid, per-**	$HMnO_4$	119.94					v s	d	
m183	**Manganocyanic acid**	$H_4Mn(CN)_6$	215.08			d		i		v s al; i eth
m184	**Mercury**	Quicksilver. Hg	200.59	silv wh met, liq	13.5939^{20}_{4}	−38.87	356.58	i	i	s HNO_3; i dil HCl, HBr, HI cold H_2SO_4
m185	(I) acetate	$Hg_2(C_2H_3O_2)_2$	519.27	micaceous plates		d		0.75^{12}		s HNO_3, H_2SO_4; i al, eth
m186	(II) acetate	$Hg(C_2H_3O_2)_2$	318.76	wh, sc or powd	3.270	d		25^{10}	100^{100}	s al, ac a
m187	(II) acetylide	$3HgC_2.H_2O$	691.85	wh powd	5.3	expl		i	i	i al
m188	(II) *ortho*arsenate	$Hg_3(AsO_4)_2$	879.61	yel				v sl s		s HCl, HNO_3
m189	(I) *ortho*arsenate mono-H	Hg_2HAsO_4	541.11					i		s HNO_3; i ac a, NH_4OH
m190	(I) azide	$Hg_2(N_3)_2$	485.22	wh cr		expl, d by light		0.025		
m191	(II) benzoate	$Hg(C_7H_5O_2)_2.H_2O$	460.84	wh cr powd		165		1.2^{15}	2.5^{100}	s al, NaCl, NH_4Cl, bz
m192	(I) bromate	$Hg_2(BrO_3)_2$	656.99	cr		d		d		sl s HNO_3
m193	(II) bromate	$Hg(BrO_3)_2.2H_2O$	492.44	cr		d 130–140		0.15	1.6	s HCl, HNO_3, $Hg(NO_3)_2$
m194	(I) bromide	Hg_2Br_2	561.00	wh, yel, tetr	7.307	subl 345		0.000004^{25}		s a; i al, acet
m195	(II) bromide	$HgBr_2$	360.41	col rhomb	6.109^{25} 5.12^{240}	236	322	0.61^{25}	4.0^{100}	15^{0} al; s MeOH; v sl s eth
m196	bromide iodide	HgBrI	407.40	yel, rhomb		229	360			s al, eth
m197	(I) carbonate	Hg_2CO_3	461.19	yel br cr		d 130		0.0000045	d	s NH_4Cl; i al
m198	(II) carbonate, basic	$HgCO_3.2HgO$	693.78	br red				i		s NH_4Cl, aq CO_2
m199	(I) chlorate	$Hg_2(ClO_3)_2$	568.08	wh, rhomb	6.409	d 250		s	d	s al, ac a
m200	(II) chlorate	$Hg(ClO_3)_2$	367.49	wh need	4.998	d		25		
m201	(I) chloride	Calomel. Hg_2Cl_2	472.09	wh, tetr, 1.973, 2.656	7.150	subl 400		0.00020^{25}	0.001^{43}	s aq reg, $Hg(NO_3)_2$; sl s HCl, h HNO_3; i al, eth
m202	(II) chloride	Corrosive sublimate. $HgCl_2$	271.50	col, rhomb or wh powd pois, 1.859	5.44^{25}, liq 4.44^{280}	276	302	6.9^{20}	48^{100}	33^{25} al; 4 eth; s ac a, pyr
m203	(I) chromate	Hg_2CrO_4	517.17	red need, or powd		d		v sl s	sl s	s HCN, HNO_3; i al, ac
m204	(II) chromate	$HgCrO_4$	316.58	red, rhomb		d		sl s, d	d	s HN_4Cl; d a; i acet
m205	(II) cyanide	$Hg(CN)_2$	262.63	col, tetr, or wh powd pois, 1.645	3.996	d		9.3^{14}	33^{100}	$25^{19.5}$ MeOH; 8 al; s NH_3, glyc; i bz
m206	(I) fluoride	Hg_2F_2	439.18	yel, cub	8.73^{15}_{4}	570	d	d to Hg_2O		
m207	(II) fluoride	HgF_2	238.59	col, cub	8.95^{15}	d 645	650	d		s HF, dil HNO_3
m208	(I) fluosilicate	$Hg_2SiF_6.2H_2O$	579.29	col pr	2.134			sl s		i HCl

No.	Name	Synonyms and Formulae	Mol. wt.	Crystalline form, properties and index of refraction	Density or spec. gravity	Melting point, °C	Boiling point, °C	Solubility, in grams per 100 cc		
								Cold water	Hot water	Other solvents
	Mercury									
m209	(II) fluosilicate	$HgSiF_6.6H_2O$	450.76	col, rhomb, deliq		d easily				
m210	(I) formate	$Hg_2(CHO_2)_2$	491.22	glist scales		d		0.4^{17}	d	i al
m211	(II) fulminate	$Hg(CNO)_2$	284.62	wh, cub	4.42	expl		sl s	s	s al, NH_4OH
m212	(I) iodate	$Hg_2(IO_3)_2$	750.99	yelsh		d 250		i	i	s dil HCl, conc HNO_3
m213	(II) iodate	$Hg(IO_3)_2$	550.48	wh, amorph powd				i		s HCl, NH_4Cl, NaCl, KI; i HNO_3
m214	(I) iodide	Hg_2I_2	654.99	yel, tetr or amorph powd	7.70	subl 140	d 290	v sl s		s KI, NH_4OH; i al, eth
m215	(II) iodide (α)	HgI_2	454.90	red, tetr	6.36^{25}_{4}	tr 127		0.01^{25}		3.18^{25} acet; 2.23^{25} al; s chl; d NH_4OH
m216	(II) iodide (β)	HgI_2	454.90	yel, rhomb cr or powd	6.094^{127}_{4}	259	354	v sl s	sl s	s eth, KI, $Na_2S_2O_3$; v sl s al
m217	(I) nitrate	$Hg_2(NO_3)_2.2H_2O$	561.22	col, monocl, effl.	4.79^{4}	70		d	s, d	s dil HNO_3; i NH_4OH
m218	(II) nitrate	$Hg(NO_3)_2.\frac{1}{2}H_2O$	333.61	wh-yelsh cr or powd, deliq	4.39	79	d	v s	d	s acet, HNO_3, NH_3; i al
m219	(II) nitrate	$Hg(NO_3)_2.H_2O$	342.61	col cr or wh powd, deliq	4.3			s		s HNO_3; i al
m220	(I) nitrite	$Hg_2(NO_2)_2$	493.19	yel	7.33	d 100		d		
m221	nitride	Hg_3N_2	629.78	br powd		expl		d		s NH_4OH, NH_4 salts; d a
m222	(I) oxalate	$Hg_2C_2O_4$	489.20					i	i	sl s HNO_3
m223	(II) oxalate	HgC_2O_4	288.61			d		0.0107^{20}		s HCl; sl s HNO_3
m224	(I) oxide	Hg_2O	417.18	blk or brnish-blk powd	9.8	d 100		i	i	s HNO_3
m225	(II) oxide	Nat. montroydite. HgO	216.59	yel or red, rhomb, 2.37, 2.5, 2.65	11.1^{4}	d 500		0.0053^{25}	0.0395^{100}	s a; i al, eth, acet, alk, NH_3
m226	(II) oxybromide	$HgBr_2.3HgO$	1010.17	yel cr				i	sl s	v s al
m227	(II) oxychloride	$HgCl_2.2HgO$	704.67	red hex, or blk monocl	red 8.16–8.43 blk 8.53					
m228	(II) oxychloride	$HgCl_2.3HgO$	921.26	yel, hex	7.93	d 260		i	d	
m229	(II) oxycyanide	$Hg(CN)_2.HgO$	469.22	wh need or cr powd	4.437^{19}	expl		1.25	s	
m230	(II) oxyfluoride	$HgF_2.HgO.H_2O$	473.19	yel cr		d 100		d		s dil HNO_3
m231	(II) oxyiodide	$HgI_2.3HgO$	1104.17	yel br					d	s HI
m232	(II) selenide	Nat. tiemannite. HgSe	279.55	gray plates	8.266	vac subl		i		s aq reg
m233	(I) sulfate	Hg_2SO_4	497.24	col monocl, wh-yelsh powd	7.56	d	d	0.06^{25}	0.09^{100}	s HNO_3, H_2SO_4
m234	(II) sulfate	$HgSO_4$	296.65	col rhomb or wh powd	6.47	d		d		s a, NaCl; i al, acet, NH_3
m235	(II) sulfate, basic	$HgSO_4.2HgO$	729.83	lem yel powd	6.44		volat	0.003^{16}	sl s	s a; i al
m236	(I) sulfide	Hg_2S	433.24	blk		d		i		i al, $(NH_4)_2S$
m237	(II) sulfide (α)	Cinnabar, vermillion. HgS	232.65	red cr hex, or powd, 2.854, 3.201	8.10	subl 583.5		0.000001^{18}		s aq reg, Na_2S; i al, HNO_3
m238	(II) sulfide (β)	Metacinnabar. HgS	232.65	blk, cub or amorph powd	7.73	583.5		i		s aq reg, Na_2S, alk; i al, HNO_3
m239	(I) tartrate	$Hg_2C_4H_4O_6$	549.25	yelsh-wh cr powd.				i		i a
m240	(I) *ortho*tellurate	HgH_4TeO_6	428.22	trans, orthorhomb		d 20		slow d	rapid d	
m241	(II) *ortho*tellurate	Hg_3TeO_6	825.37	amber, cub		unalt at 140		i	i	s HCl, HNO_3
m242	(I) thiocyanate	$Hg_2(SCN)_2$	517.34			d		i		s HCl, KCNS
m243	(II) thiocyanate	$Hg(SCN)_2$	316.75	wh powd, pois.		d		0.07^{25}	s	s NH_4 salts, HCl, NH_3, KCN; sl s al, eth
m244	(I) tungstate	Hg_2WO_4	649.03	yel, amorph		d		i	i	d a; i al
m245	(II) tungstate	$HgWO_4$	448.44	yel		d		i	d	d a; i al
	Mercury nitrogen compounds									
m246	mercury (II) bromide, ammonobasic	$Hg(NH_2)Br$	296.52	wh powd		d		d		s NH_4OH; i al
m247	mercury (II) bromide, diammine	$Hg(NH_3)_2Br_2$	394.47	wh powd		180		d		s NH_4Cl, NH_4Br, NH_4I
m248	mercury (II) chloride ammonobasic	Infusible ppt. $Hg(NH_2)Cl$	252.07	wh powd or sm pr	5.70	infus		0.14	d 100	d a; i al
m249	mercury (II) chloride, aquobasic ammonobasic	Chloride of Millon's base. OHg_2NH_2Cl	468.66	pa yel or wh powd		d >120		sl s		s HCl, HNO_3
m250	mercury (II) chloride, diammine	Fusible white ppt. $Hg(NH_3)_2Cl_2$	305.56	rhombd		300		i	d	s a, KI

No.	Name	Synonyms and Formulae	Mol. wt.	Crystalline form, properties and index of refraction	Density or spec. gravity	Melting point, °C	Boiling point, °C	Solubility, in grams per 100 cc: Cold water	Hot water	Other solvents
	Mercury nitrogen compounds									
m251	mercury (II) iodide, ammonobasic	$Hg(NH_2)I$	353.52							i eth
m252	mercury (II) iodide aquobasic ammonobasic	Iodide of Millon's base. OHg_2NH_2I	560.11	yel to brn		>128	expl	i		s d, HCl, KI soln
m253	mercury (II) iodide, diammine	$Hg(NH_3)_2I_2$	488.48	col or pa yel powd or need				d		s NH_4OH
m254	**Millons's base**	$(HO)_2Hg_2NH_2OH$	468.22		4.083^{18}					
m255	**Molybdenum**	Mo	95.94	silv-wh met, or gray-blk powd, cub	10.2	2610	5560	i	i	s h conc HNO_3, h conc H_2SO_4, aq reg; sl s HCl i HF, NH_3
m256	boride, (di-)	MoB_2	117.56	rhomb	7.12					
m257	boride, (mono-)	MoB	106.75	tetr	8.65					
m258	(di-) boride	Mo_2B	202.69	tetr	9.26					
m259	bromide, di-	$MoBr_2$ (or Mo_3Br_6)	255.76	yel red, amorph	$4.88^{17.5}$			i	i	s alk; i a, aq reg
m260	bromide, tetra-	$MoBr_4$	415.58	blk need, deliq		d	volat	v s		d alk
m261	bromide, tri-	$MoBr_3$	335.67	dk-grn need		d		i	i	d alk, NH_3; i a
m262	carbide, mono-	MoC	107.95	gray, hex	8.20^{20}	2692		i	i	sl s HNO_3, HF, h H_2SO_4, HCl; i alk hydr
m263	carbide(di-)	Mo_2C	203.89	wh, hex pr	8.9	2687		i		sl s HNO_3, HF h H_2SO_4, aq reg HCl; i alk
m264	carbonyl	$Mo(CO)_6$	264.00	wh cr, rhomb diamagnet	1.96	d 150, without meltg	156.4^{766}	i	i	s bz; sl s eth
m265	carbonyl, tri-pyridine, tri-	$Mo(CO)_3(C_5H_5N)_3$	287.08	yel-brn cr		d				
m266	chloride, di-	$MoCl_2$ (or Mo_3Cl_6)	166.85	yel, amorph	3.714^{25}	d		i	i	s HCl, H_2SO_4, alk, NH_4OH, al acet
m267	chloride, penta-	$MoCl_5$	273.21	grn-blk cr, trig, deliq	2.928	194	268	d	d	s conc min a, liq NH_3, CCl_4, chl; s d al, eth
m268	chloride, tetra-	$MoCl_4$	237.75	brn powd or cr, deliq		d	vol	d	d	s conc min a; d al, eth
m269	chloride, tri-	$MoCl_3$	202.30	dk red need or powd	3.578^{25}_{4}	d		i	sl d	s conc H_2SO_4, conc HNO_3; v sl s al, eth; i HCl; d alk
m270	fluoride, hexa-	MoF_6	209.93	col cr	liq $2.55^{17.5}$	17.5^{406}	35^{760}	s d	d	s NH_4OH, alk
m271	hydro*tetra*chloro-hydroxide, di-	$[Mo_3Cl_4(H_2O)_2](OH)_2.6H_2O$	607.77	lt yel cr		$-H_2O$, 35–300		i	i	i a, al
m272	hydroxide	$Mo(OH)_3$ (or $Mo_2O_3.3H_2O$)	146.96	blk powd		d		0.2		sl s H_2SO_4, HCl; s 30 % H_2O_2
m273	hydroxide	$MoO(OH)_3$ (or $Mo_2O_5.3H_2O$)	162.96	br to blk powd				0.2 (coll)		s a, alk carb; i alk hydr
m274	(VI) hydroxide	$MoO_3.2H_2O$	179.97	lt yel, monocl pr	3.124^{15}			0.05^{18}		s dil H_2O_2, alk hydr; sl s a
m275	hydroxy*tetra*-bromide, di-	$Mo_3Br_4(OH)_2$	641.47	red powd						s alk
m276	hydroxy*tetra*-bromide, diocta-hydrate	$Mo_3Br_4(OH)_2.8H_2O$	785.59	golden yel cr		d	d			s HCl; d alk, HNO_3
m277	hydroxy*tetra*-chloride, di-	$Mo_3Cl_4(OH)_2.2H_2O$	499.68	pa yel, amorph				i		s conc a; i al
m278	iodide, di-	MoI_2	349.75	brn powd	$5.278^{25.4}$			i	d	v sl s a
m279	iodide, tetra-	MoI_4	603.56	blk cr		d 100				
m280	oxide, di-	MoO_2	127.94	lead gray, tetr or monocl	6.47			i	i	sl s h conc H_2SO_4; i alk, HCl, HF
m281	oxide, pent-	Mo_2O_5	271.88	vlt-bl powd (exist?)						s h H_2SO_4, h HCl
m282	oxide, pent-	"Molybd. blue" $Mo_2O_5.xH_2O$ (variations in Mo and O)		dk blue coll or powd	3.6^{18}	d		s		s a, MeOH; i acet, bz, chl
m283	oxide, sesqui-	Mo_2O_3	239.88	blk, opaque (exist?)				i	i	i a, alk, NH_4OH
m284	oxide, tri-	Molybdic anhydride. Nat. Molybdite. MoO_3	143.94	col or wh-yel, rhomb	4.692^{21}	795	subl 1155^{760}	0.1066^{18}	2.055^{70}	s a, alk sulf, NH_4OH
m285	oxy*di*bromide, di-	MoO_2Br_2	287.76	yel-red tabl, deliq				s		

No.	Name	Synonyms and Formulae	Mol. wt.	Crystalline form, properties and index of refraction	Density or spec. gravity	Melting point, °C	Boiling point, °C	Solubility, in grams per 100 cc: Cold water	Hot water	Other solvents
	Molybdenum									
m286	oxy*tetra*chloride	$MoOCl_4$	253.75	grn cr, deliq		subl		s		
m287	oxy*tri*chloride	$MoOCl_3$	218.30	grn cr		subl 100		d		
m288	oxy*di*chloride, di-	MoO_2Cl_2	198.84	yelsh wh scaly cr	3.31^{17}	subl		s	s	s al, eth
m289	oxy*di*chloride, dihydrate	$MoO_2Cl_2.H_2O$	216.86	pa yel cr		subl		s	s	sl s al, acet, eth
m290	oxy*hexa*chloride, tri-	$Mo_2O_3Cl_6$	452.60	rubyred or dk vlt cr		d		d		s eth
m291	oxy*penta*chloride, tri-	$Mo_2O_3Cl_5$	417.14	dk brn-blk cr, deliq		melts easily	subl	s	s	
m292	oxychloride acid	$MoO(OH)_2Cl_2$	216.63	wh need, deliq		d 160		v s		s al, eth, acet
m293	oxy*di*fluoride, di-	MoO_2F_2	165.94	wh cr, hygr	3.494^{25}	subl 270		v s	v s	s al, MeOH; i eth, chl, tol
m294	oxy*tetra*fluoride	$MoOF_4$	187.93	col-wh, deliq	3.001^{25}	98	180	s		s al, eth, CCl_4, s d H_2SO_4; v sl s bz
m295	*meta*phosphate	$Mo(PO_3)_6$	569.77	yel powd	3.28^{0}			i	i	sl s h aq reg; i HCL, HNO_3, H_2SO_4'
m296	phosphide	MoP (or Mo_2P_2)	126.93	gray-grn cr powd	6.167					s h HNO_3
m297	phosphide	MoP_2	157.89	blk powd	5.35^{25}					s HNO_3, h conc H_2SO_4, aq reg; i conc HCl
m298	silicide	$MoSi_2$	152.11	gray, met, tetr	$6.31^{20.5}$					i a, aq reg; v s HF+HNO_3
m299	sulfide, di-	Nat. molybdenite. MoS_2	160.07	blk luster, hex	4.80^{14}	1185	subl 450, d in air	i	i	s h H_2SO_4, aq reg, HNO_3; i dil a, conc H_2SO_4
m300	sulfide, penta	$Mo_2S_5.3H_2O$	406.25	dk br powd		$-H_2O$, 135	d	i	i	s NH_4OH, alk, sulfides
m301	sulfide, sesqui-	Mo_2S_3	288.07	steel gray need	5.91^{15}	d 1100	vol 1200			i conc HCl; d h HNO_3
m302	sulfide, tetra-	MoS_4	224.20	brn powd		d		i	i	i a; s h H_2SO_4, alk sulfide
m303	sulfide, tri-	MoS_3	192.13	blk pl		d	d	sl s	s	s alk sulf, conc KOH
m304	**Molybdic acid**	$H_2MoO_4.H_2O$ (or $MoO_3.2H_2O$)	179.97	yel, monocl	3.124^{15}	$-H_2O$, 70	d	0.133^{18}	2.568^{70}	s alk hydr, alk carb; sl s a
m305	anhydrous	H_2MoO_4 (or $MoO_3.H_2O$)	161.95	wh or sl yelsh, hex	3.112	$-H_2O$, 70		sl s	sl s	s alk, NH_4OH, H_2SO_4; i NH_3
m306	**Molybdic arsenic acid**	$As_2O_5.6MoO_3.18H_2O$	1475.75	col, trig	$2.493^{19.8}$	$-15H_2O$, 150		v s		s abs al; i chl, CS_2
m307	**Molybdic phosphoric acid**	$H_7[P(Mo_2O_7)_6].28H_2O$	2365.71	yel, oct	2.53	78		d		
m308	**Molybdic silicic acid**	$H_8[Si(Mo_2O_7)_6].28H_2O$	2363.83	yel, tetr		45	d 100	600^{14}		s dil a; i bz, chl, CS_2
n1	**Neodymium**	Nd	144.24	silv-wh to yelsh met, hex to m.p. 868, cub from 868	hex 7.004 cub 6.80	1024	3027	d		
n2	acetate	$Nd(C_2H_3O_2)_3.H_2O$	339.39					26.2		
n3	acetylacetonate	$Nd[CH(COCH_3)_2]_3$	441.57	pink cr	1.618	150–152				
n4	*hexa*antipyrin *per*chlorate	$[Nd(C_{11}H_{12}N_2O)_6].(ClO_4)_3$	1571.99	rose, hex cr		d 285–289		0.99^{20}		
n5	bromate	$Nd(BrO_3)_3.9H_2O$	690.10	red, hex		66.7	$-9H_2O$, 150	151^{25}		
n6	bromide	$NdBr_3$	383.97	grn cr		684	1540	sl s		
n7	carbide	NdC_2	168.26	yel, hex leaf	5.15	d		d	d	s dil a; i conc HNO_3
n8	chloride	$NdCl_3$	250.60	rose-vlt pr	4.134^{25}	784	1600	96.7^{13}	140^{100}	44.5 al; i eth, chl
n9	chloride, hexahydrate	$NdCl_3.6H_2O$	358.69	red, rhomb		124	$-6H_2O$, 160	246^{13}	511^{100}	v s al
n10	chromate	$Nd_2(CrO_4)_3.8H_2O$	780.58	yel cr				0.027		
n11	fluoride	NdF_3	201.24	pa lilac		1410	2300	i	i	
n12	iodide	NdI_3	524.95	blk cr powd		775±3	1370	s	s	
n13	kojate	$Nd(C_6H_5O_4)_3$	567.60	lt choc		d 275		i		
n14	manganous nitrate	$2Nd(NO_3)_3.3Mn(NO_3)_2.24H_2O$	1629.72	vlt-red	2.114	77		77.4^{30}		
n15	magnesium nitrate	$2Nd(NO_3)_3.3Mg(NO_3)_2.24H_2O$	1537.84	vlt-red	2.020	109		69.5^{30}		
n16	dimethylphosphate	$Nd[(CH_3)_2PO_4]_3$	519.35	pa lilac, hex pl				56.1^{25}	22.3^{100}	
n17	molybdate	$Nd_2(MoO_4)_3$	768.29	tetr, 2.005	5.14^{18}	1176				
n18	nickel nitrate	$2Nd(NO_3)_3.3Ni(NO_3)_2.24H_2O$	1641.04	blsh-grn	2.202	105.6		71.5^{30}		
n19	nitrate	$Nd(NO_3)_3.6H_2O$	438.35	tricl				152.9^{25}		s al, acet
n20	nitride	NdN	158.25	blk powd				d		
n21	oxalate	$Nd_2(C_2O_4)_3.10H_2O$	732.69	rose cr				0.000074^{25}		
n22	oxide	Neodymia. Nd_2O_3	336.48	lt bl powd, red fluores	7.24	~1900		0.00019^{29}	0.003^{75}	s a

No.	Name	Synonyms and Formulae	Mol. wt.	Crystalline form, properties and index of refraction	Density or spec. gravity	Melting point, °C	Boiling point, °C	Solubility, in grams per 100 cc		
								Cold water	Hot water	Other solvents
	Neodymium									
n23	sulfate	$Nd_2(SO_4)_3.8H_2O$	720.79	red, monocl, 1.41, 1.551, 1.562	2.85	1176		8^{20}	5.4^{40}	
n24	sulfide	Nd_2S_3	384.67	oliv grn powd	5.179^{11}	d		i	d	s dil a
n25	**Neon**	Ne	20.183	inert gas col sol, cub	gas: 0.9002^{0} g/1; liq: $1.204^{-245.9}$	−248.67	−245.92	1.47^{20} cm³		s liq O_2
n26	**Neptunium**	Np	237.00	α: orthorhomb silvery β: tetr (above 278) γ: cub (above 500)	α; 20.45 β: 19.36^{313} γ: 18.0^{600}	640 ± 1 278 ± 5 stab 278–570				s HCl
n27	bromide, tri-	$NpBr_3$	476.73	α: hex β: grn orthorhomb	α 6.92	subl *ca* 800		s		
n28	chloride, tetra-	$NpCl_4$	378.81	red-brn tetr	4.92	538		s		
n29	chloride, tri-	$NpCl_3$	343.36	wh, hex	5.38	*ca* 800		s		
n30	fluoride, hexa-	NpF_6	350.99	brn, orthorhomb.		53	d	d		
n31	fluoride, tetra-	NpF_4	312.99	grn, monocl	6.8			i		i conc HNO_3
n32	fluoride, tri-	NpF_3	294.00	purple, hex	9.12			i		
n33	iodide, tri-	NpI_3	617.71	brn, orthorhomb.	6.82			s		
n34	oxide, di-	NpO_2	269.00	apple grn, cub.	11.11			i		s conc a
n35	(tri-) oxide, octa-	Np_3O_8	839.00	brn, cub		d 500				s HNO_3
n36	**Nickel**	Ni	58.71	silv met, cub	8.90	1453	2732	i	i	s dil HNO_3; sl s HCl, H_2SO_4; i NH_3
n37	acetate	$Ni(C_2H_3O_2)_2$	176.80	grn pr	1.798	d	16.6			i al
n38	acetate, tetrahydrate	$Ni(C_2H_3O_2)_2.4H_2O$	248.86	grn pr	1.744	d	16			s dil al
n39	antimonide	Nat. breithauptite. NiSb	180.46	lt copper red, hex	7.54	1158	d 1400			
n40	*ortho*arsenate, octahydrate	$Ni_3(AsO_4)_2.8H_2O$	598.09	yelsh-grn powd	4.98			i		s a
n41	arsenide	Nat. niccolite. NiAs,	133.63	hex	7.57^{0}	968		i	i	s aq reg
n42	*ortho*arsenite, acid	$Ni_3H_6(AsO_3)_4.H_2O$	691.87	grn-wh		d		i		s a, alk
n43	benzenesulfonate	$Ni(C_6H_5SO_3)_2.6H_2O$	481.10	grn, monocl.	1.628^{25}	$-H_2O$	d	14.3^{18}	51.5^{82}	5.9 al; 4.5 eth
n44	boride	NiB	69.52	pr	7.39^{18}			d	d	s aq reg, HNO_3
n45	bromate	$Ni(BrO_3)_2.6H_2O$	422.62	monocl	2.575	d		28		
n46	bromide	$NiBr_2$	218.53	yel brn, deliq	5.098^{27}	963		112.8^{0}	155.1^{100}	s al, eth, NH_4OH
n47	bromide, trihydrate	$NiBr_2.3H_2O$	272.57	yelsh-grn need, deliq		$-3H_2O$, 300		199^{0}	315.7^{100}	s al, eth, NH_4OH
n48	bromoplatinate	$NiPtBr_6.6H_2O$	841.35	trig	3.715					
n49	di-*N*-butyldithiocarbamate	(NBC).$Ni[(C_4H_9)_2NCSS]_2$ 467.47		dk olive grn powd	1.29	89–90		i	i	sl s bz, pet comp; i al
n50	carbide	Ni_3C	188.14	dk gray powd	7.957^{25}					
n51	carbonate	$NiCO_3$	118.72	lt grn, rhomb		d		0.0093^{25}	i	s a
n52	carbonate, basic	$2NiCO_3.3Ni(OH)_2.4H_2O$	587.67	lt grn cr or brn powd		d		i	d	s a, NH_4 salts
n53	carbonate, basic	Zaratite. $NiCO_3.2Ni(OH)_2.4H_2O$(?)	376.23	emerald grn, cub, 1.56–1.61	2.6			i	i	s h dil HCl, NH_4OH
n54	carbonyl	$Ni(CO)_4$	170.75	col, volat, inflamm, liq, or need	1.32^{17}	−25	43	$0.018^{9.8}$		s aq reg, al, eth, bz, HNO_3; i dil a, dil alk
n55	chlorate	$Ni(ClO_3)_2.6H_2O$	333.70	dk red	2.07	d 80		0.9^{27}		
n56	*per*chlorate	$Ni(ClO_4)_2.6H_2O$	365.70	grn, hex need, 1.518, 1.498		140		222.5^{0}	273.7^{45}	s al, acet, chl
n57	chloride	$NiCl_2$	129.62	yel sc, deliq	3.55	1001	subl 973	64.2^{20}	87.6^{100}	s al, NH_4OH; i NH_3
n58	chloride, hexahydrate	$NiCl_2.6H_2O$	237.70	grn, monocl, deliq ~1.57				254^{20}	599^{100}	v s al
n59	chloropalladate	$NiPdCl_6.6H_2O$	485.92	hex	2.353					
n60	chloroplatinate	$NiPtCl_6.6H_2O$	574.61	trig	2.798					
n61	cyanide	$Ni(CN)_2$	110.75	yel-brn				i	i	s KCN
n62	cyanide, tetrahydrate	$Ni(CN)_2.4H_2O$	182.81	lt grn pl or powd pois		$-4H_2O$, 200	d	i	i	s KCN, NH_4OH, alk; sl s dil a
n63	ferrocyanide	$Ni_2Fe(CN)_6.xH_2O$		grn-wh	1.892(?)			i		s KCN, NH_4OH; i HCl
n64	fluogallate	$[Ni(H_2O)_6][GaF_5.H_2O]$	349.53	pa grn, monocl(?), 1.45	2.45	$-5H_2O$, 110		sl s		s HF
n65	fluoride	NiF_2	96.71	grn, tetr	4.63	subl 1000 (in HF)		4^{25}		s a, alk, eth, NH
n66	fluosilicate	$NiSiF_6.6H_2O$	308.88	grn, trig, 1.391, 1.407	2.134	d				
n67	formate, dihydrate	$Ni(CHO_2)_2.2H_2O$	184.78	grn cr	2.154	d		s		
n68	(II) hydroxide	$Ni(OH)_2$ (or $NiO.xH_2O$)	92.72	grn cr, or amorph	4.15(3.65)	d 230		0.013		s a, NH_4OH
n69	iodate	$Ni(IO_3)_2$	408.52	yel need	5.07			1.1^{30}	1.0^{90}	
n70	iodate, tetrahydrate	$Ni(IO_3)_2.4H_2O$	480.59	hex		d 100		1.4^{30}	1.1^{90}	
n71	iodide	NiI_2	312.52	blk cr, deliq	5.834	797		124.2^{0}	188.2^{100}	s al

No.	Name	Synonyms and Formulae	Mol. wt.	Crystalline form, properties and index of refraction	Density or spec. gravity	Melting point, °C	Boiling point, °C	Solubility, in grams per 100 cc		
								Cold water	Hot water	Other solvents
	Nickel									
n72	dimethylglyoxime	$Ni(HC_2H_6N_2O_2)_2$	288.94	scarlet red cr		subl 250		i	i	s a, abs al; i a ac, NH_4OH
n73	nitrate, hexahydrate	$Ni(NO_3)_2.6H_2O$	290.81,	grn, monocl, deliq	2.05	56.7	136.7	238.5^{0}	v s	s al, NH_4OH
n74	oleate	$Ni(C_{18}H_{33}O_2)_2$	621.64	grn oil		18–20				
n75	oxalate	$NiC_2O_4.2H_2O$	182.76	lt grn powd				i		s a, NH_4 salts; v sl s oxal a
n76	oxide, mono-	Nat. bunsenite. NiO	74.71	grn-blk, cub, 2.1818(red)	6.67	1990		i	i	s a, NH_4OH
n77	*ortho*phosphate	$Ni_3(PO_4)_2.8H_2O$	510.20	apple grn pl or emerald cr gran		d		i	i	s a, NH_4 salts; i me acet. et acet
n78	*pyro*phosphate	$Ni_2P_2O_7.xH_2O$		grn	3.93 (anhydr)					s a, NH_4OH
n79	(di-)phosphide	Ni_2P	148.39	gray cr	6.31^{15}	1112		i		s HNO_3+HF; i a
n80	(penta)phosphide, (di)-	Ni_5P_2	355.50	need or tabl cr		1185				
n81	(tri-)phosphide, (di-)	Ni_3P_2	238.08	dk grn-blk	5.99			i	i	s HNO_3; i HCl
n82	*hypo*phosphite	$Ni(H_2PO_2)_2.6H_2O$	296.78	grn	$1.82^{19.8}$	d 100		s		
n83	selenate	$NiSeO_4.6H_2O$	309.76	grn, tetr, 1.5393	2.314			s		
n84	selenide	NiSe	137.67	wh or gray, cub	8.46	red heat		i		s aq reg, HNO_3; i a, HCl
n85	silicide	Ni_2Si	145.51		7.2^{17}	1309		i		i a
n86	stearate	$Ni(C_{18}H_{35}O_2)_2$	625.67	grn powd		100		i		s CCl_4, pyr; sl s acet; i MeOH, eth
n87	sulfate	$NiSO_4$	154.78	yel, cub	3.68	d 848^{760}		29.3^{0}	83.7^{100}	i al, eth, acet
n88	sulfate, heptahydrate	Morenosite. $NiSO_4.7H_2O$	280.88	grn, rhomb, 1.467, 1.489, 1.492	1.948	99; $-H_2O$ 31.5	$-6H_2O$, 103	$75.6^{15.5}$	475.8^{100}	s al
n89	sulfate, hexahydrate	Single nickel salt. $NiSO_4.6H_2O$	262.86	α: bl, tetr β: grn, monocl, 1.511, 1.487	2.07	tr: 53.3	$-6H_2O$, 103	62.52^{0}	340.7^{100}	12.5 MeOH; v s al, NH_4OH
n90	sulfide, mono-	Nat. millerite. NiS	90.77	blk; trig or amorph	5.3–5.65	797		0.00036^{18}		s aq reg, HNO_3, KHS; sl s a
n91	sulfide, sub-	Heazlewoodite. Ni_3S_2	240.26	pa yelsh bronze met, lust	5.82	790		i		s HNO_3
n92	(II, III) sulfide	Poydomite. Ni_3S_4	304.39	gray-blk, cub	4.7			i		s HNO_3
n93	sulfite	$NiSO_3.6H_2O$	246.86	grn, tetrah				i		s HCl, H_2SO_4
n94	dithionate	$NiS_2O_6.6H_2O$	326.93	grn, tricl	1.908	d				
	Nickel Complexes									
n95	diaquotetrammine nickel (II) nitrate	$[Ni(NH_3)_4(H_2O)_2].(NO_3)_2$	286.87	grn cr				s		i al
n96	hexamminenickel (II) bromide	$[Ni(NH_3)_6]Br_2$	320.71	vlt powd	1.837			v s	d	
n97	hexamminenickel (II) chlorate	$[Ni(NH_3)_6](ClO_3)_2$	327.80		1.52	180		d to $Ni(NH_3)_4$		
n98	hexamminenickel (II) chloride	$[Ni(NH_3)_6]Cl_2$	231.80	blsh, cub	1.468^{25}			s	d	s NH_4OH; i al
n99	hexamminenickel (II) iodide	$[Ni(NH_3)_6]I_2$	414.70	pa bl, cub	2.101	d		d		s NH_4OH
n100	hexamminenickel (II) nitrate	$[Ni(NH_3)_6](NO_3)_2$	284.90	bl, oct or cub				4.46		
n101	tetrapyridinnickel (II) fluosilic	$[Ni(C_5H_5N)_4]SiF_6$	517.20	bl grn, rhomb	2.307					
n102	**Niobium**	Columbium. Nb	92.906	steel gray, lustr met cub, 1.80	8.57	2468±10	4927	i	i	s fus alk; i HCl, HNO_3, aq reg
n103	boride	NbB_2	114.53	hex	6.97	2900(?)				
n104	bromide, penta-	$NbBr_5$	492.46	purp red		265.2	361.6	d		s al, ethyl bromide
n105	carbide	NbC	104.92	blk, cub or lavender-gray powd	7.6	3500		i		s HNO_3, HF
n106	chloride, penta-	$NbCl_5$	270.17	yel-wh, deliq	2.75	204.7	254	d		s al, HCl, CCl_4
n107	fluoride, penta-	NbF_5	187.90	col, monocl pr, hygr	3.293	72–73	236	d		s al; sl s chl, CS_2, H_2SO_4
n108	hydride	NbH	93.91	gray powd	6.6	infus				s HF, conc H_2SO_4; i HCl, alk, HNO_3
n109	nitride	NbN	106.91	blk, cub	8.4	2573		i		s $HF+HNO_3$; i HNO_3
n110	oxalate, hydrogen	$Nb(HC_2O_4)_5$	538.05	col, monocl				d	d	s $H_2C_2O_4$; d al
n111	oxide, di-	NbO_2	124.90	blk	5.9			i	i	sl s alk; i a
n112	oxide, mon- (or di-)	NbO (or Nb_2O_2)	108.91	blk, cub	7.30			i	i	s a, alk; i al, NHO_3
n113	oxide, pent-	Nb_2O_5	265.81	wh, rhomb	4.47	1460		i	i	s HF, alk; i a
n114	oxide, pent-, hydrate	$Nb_2O_5.xH_2O$				d		i		s conc H_2SO_4, conc HCl, HF, alk; i NH_3
n115	oxide, tri-(sesqui)	Nb_2O_3	233.81	bl-blk		1780				

No.	Name	Synonyms and Formulae	Mol. wt.	Crystalline form, properties and index of refraction	Density or spec. gravity	Melting point, °C	Boiling point, °C	Solubility, in grams per 100 cc		
								Cold water	Hot water	Other solvents
	Niobium									
n116	oxybromide	$NbOBr_3$	348.63	yel cr		subl		d		s a
n117	oxychloride	$NbOCl_3$	215.26	col need		subl 400		s, d	d	s al, H_2SO_4; i HCl
n118	potassium fluoride	$NbOF_3.2KF.H_2O$	300.12	monocl leaf, lustr fatty				7.8	v s	
n119	**Nitric acid**	HNO_3	63.01	col liq, corr, pois, $1.397^{16.4}$	1.5027^{25}_{4}	−42	83	∞	∞	d al, viol; s eth
n120	const boil	68% HNO_3+32% H_2O		col liq	1.41^{20}_{4}		120.5	∞	∞	
n121	**Nitrogen**	N_2	28.0134	col gas, col liq, sol cub cr	gas 1.2506 g/l liq $0.8081^{-195.8}$ sol $1.026^{-252.5}$	−209.86	−195.8	2.33^{0} cm³	1.42^{40} cm³	sl s al
n122	chloride, tri-	NCl_3	120.37	yel oil or rhomb cr	1.653	<−40	<71, expl 95	i	d	s chl, bz, CCl_4, CS_2, PCl_3
n123	fluoride, tri-	NF_3	71.00	col gas	liq: 1.537^{-129}	−206.60	−128.8	v sl s		
n124	iodide, tri-	NI_3	394.72	blk, expl		expl	subl vac	i	d	s KCNS, $Na_2S_2O_3$
n125	iodide, tri-, monoammine	$NI_3.NH_3$	411.75	dk red, rhomb	3.5	d>20	expl	i	d	s HCl, KCNS, $Na_2S_2O_3$; i abs al
n126	oxide(ic)	NO	30.01	col gas, bl liq, sol liq 1.330^{-90}	gas 1.3402 g/l liq; $1.269^{-150.2}$	−163.6	−151.8	7.34^{0} cm³	2.37^{60} cm³	3.4 cm³ H_2SO_4; 26.6 cm³ al; s $FeSO_4$, CS_2
n127	oxide(ous)	N_2O	44.01	col gas or liq or cub cr, 1.0005^{20}_{760}	1.977^{0}_{760} g/l	−90.8	−88.5	130^{0} cm³	56.7^{25} cm³	s al, eth, H_2SO_4
n128	oxide, pent-	Nitric anhydride. N_2O_5	108.01	wh, rhomb or hex	1.642^{18}	30	d 47	s	d to HNO_3	s chl
n129	oxide, tri-	NO_3	62.00	blsh gas		d at ord temp				s eth
n130	peroxide	Nitrogen oxide, di-. NO_2	46.01	col sol, yel liq or brn gas, 1.40^{20}	1.4494^{20}_{20}	−11.20	21.2	s, d		s alk, CS_2, chl
n131	(di-) oxide(tri-)	Nitrous anhydride. N_2O_3	76.01	red-brn gas, bl sol or liq	1.447^{2}	−102	d 3.5	s	d	s eth, a, alk
n132	oxi (tri-) fluoride	NO_3F	81.06	col gas expl	liq; $1.507^{-45.9}$ sol: $1.951^{-193.2}$	−175	−45.9	d		s acet; expl al, eth
n133	sulfide, penta-	N_2S_5	188.33	gray cr		10–11	d	d		s CS_2, eth; i bz, al
n134	sulfide, tetra-	Tetranitrogen tetrasulfide sulfurnitride. N_4S_4	184.28	yel cr, 2.046	2.24^{18}	d 178				s al, bz, CS_2
n135	**Nitrosyl bromide**	NOBr	109.92	br gas or dk br liq	>1.0	−55.5	−2	d		s alk
n136	*per*chlorate	$NOClO_4.H_2O$	147.47	rhomb, deliq	2.169	d 100		d		expl al, eth
n137	chloride	NOCl	65.46	yel gas or yel-red liq or cr	gas: 2.99 g/l liq: 1.417^{-12}	−64.5	−5.5	d	d	s fum H_2SO_4
n138	fluoborate	$NOBF_4$	116.81	col, rhomb cr, hygr	2.185^{25}_{4}	subl $250^{0.01}$		d		
n139	fluoride	NOF	49.00	col gas	2.176 g/l	−134	−56	d		d to HNO_2+HF
n140	**Nitrosylsulfuric acid**	Chamber crystals. $NOHSO_4$	127.08	col, rhomb		d 73.5		d		s H_2SO_4
n141	**Nitrosylsulfuric anhydride**	$(NOSO_3)_2O$	236.14	tetr		217	360	d		s H_2SO_4
n142	**Nitrous acid**	HNO_2	47.01	only in sol (pa bl)				d		
n143	hypo-	$H_2N_2O_2$	62.03	wh, sol		expl		s		
n144	**Nitryl** chloride	NO_2Cl	81.46	pa yel-br gas	gas: 2.57 g/l liq: 1.32^{14}	<−31	5	d		
n145	fluoride	NO_2F	65.00	col gas, col sol	2.90 g/l	−139	−63.5	d		d al, eth, chl
o1	**Osmium**	Os	190.20	gray-blsh met, hex	22.48^{20}	3000±10	~5000	i	i	sl s aq reg, HNO_3; i NH_3
o2	carbonyl chloride	$Os(CO)_3Cl_2$	345.14	col pr	269–273	d 280		i	i	s NaOH; i a
o3	chloride, di-	$OsCl_2$	261.11	dk brn, deliq	d			i	sl d	s al, eth, HNO_3; sl s alk
o4	chloride, tetra-	$OsCl_4$	332.01	red br need		subl		sl s, d		i al
o5	chloride, tri-	$OsCl_3$	296.56	br, cub		d 500–600		v s		s alk, al, a; sl s eth
o6	chloride, tri-, trihydrate	$OsCl_3.3H_2O$	350.61	dk gr cr		d		v s		s al
o7	fluoride, hexa-	OsF_6	304.19	grn cr		32.1	45.9	d	d	
o8	fluoride, tetra-	OsF_4	266.19	br powd				d	d	
o9	iodide	OsI_4	697.82	vlt-blk, hygr met, lust				v s	d	s al

No.	Name	Synonyms and Formulae	Mol. wt.	Crystalline form, properties and index of refraction	Density or spec. gravity	Melting point, °C	Boiling point, °C	Solubility, in grams per 100 cc		
								Cold water	Hot water	Other solvents
	Osmium									
o10	oxide, di-, brown	OsO_2	222.20	brn cr	$11.37^{21.4}$	30% tr to OsO_4, 500		i	i	i a
o11	oxide, di-, black	OsO_2	222.20	blk powd	7.71^{21}	tr to br 350–400		i	i	s dil HCl
o12	oxide, mon-	OsO	206.20	blk				i	i	
o13	oxide, sesqui-	Os_2O_3	428.40	dk brn		d		i	i	i a
o14	oxide, tetra-	OsO_4	254.10	a) col, monocl. b) yel mass	4.906^{22}	a) 39.5 b) 41.0	130	5.70^{10}	6.23^{25}	250 ± 10^{20} CCl_4; s al, eth, NH_4OH, $POCl_3$
o15	sulfide, di-	OsS_2	254.33	blk, cub	9.47	d		i	i	s HNO_3; i alk
o16	sulfide, tetra-	OsS_4	318.46	br blk (exist?)		d		i		s dil HNO_3; i $(NH_4)_2S$
o17	sulfite	$OsSO_3$	270.26	bl blk		d		i		s dil HCl, alk
o18	telluride	$OsTe_2$	445.40	gray-blk cr		*ca* 600				i a; d dil HNO_3
o19	**Oxygen**	O_2	31.9988	col gas, sol hex cr	gas: 1.429^{0} g/l, liq: 1.149^{-183} sol: $1.426^{-252.5}$	−218.4	−183.0	4.89^{0} cm³ 3.16^{25} cm³	2.46^{50} cm³ 2.30^{100} cm³ 3	2.78^{25} al
o20	fluoride	OF_2	54.00	col gas, unst	liq: $1.90^{-223.8}$	−223.8	−144.8	sl s, d	i	sl s a, alk
o21	**Ozone**	O_3	47.9982	col gas, or dk bl liq, or bl-blk cr, liq: 1.2226	gas: 2.144^{0} g/l liq: $1.614^{-195.4}$ g/l	192.7 ± 2 1.0	−111.9	49^{0} cm³		s alk sol, oils
p1	**Palladium**	Pd	106.40	silv wh, met, cub.	11.97^{0} $11.40^{22.5}$	1552	2927	i	i	s aq reg, h HNO_3, H_2SO_4; sl s HCl
p2	bromide	$PbBr_2$	266.22	red br	5.173^{16}	d		i	i	s HBr
p3[1]	chloride	$PdCl_2$	177.31	dk red, cub need, deliq	4.0^{18}	d 500		s	s	s HBr; acet
p3[2]	chloride, dihydrate	$PdCl_2.2H_2O$	213.34	br pr, deliq		d		v s	v s	s HCl, acet
p4	cyanide	$Pd(CN)_2$	158.44	yelsh-wh		d		i	i	s KCN, NH_4OH; i dil a
p5	fluoride, di-	PdF_2	144.40	br, tetr	5.80	volat	d red heat	sl s, d		s HF
p6	fluoride, tri-	PdF_3	163.40	blk, rhomb	5.06	d	d	d	d	s HF
p7	hydride	Pd_2H (or Pd_4H_2)	213.81	silv met (exist?)	10.76	d				
p8	iodide	PdI_2	360.21	blk powd	6.003^{18}	d 350		i	i	s KI; i al, eth, dil HCl
p9	nitrate	$Pd(NO_3)_2$	230.41	br yel, rhomb, deliq		d		s, d		s HNO_3
p10	oxide, di-	$PdO_2.xH_2O$		dull red		d − H_2O, −O		i	i,d	s a, alk
p11	oxide, mon-	PdO	122.40	grnsh-bl or amber mass, or blk powd	8.70^{20}_{4}	870		i	i	i aq reg
p12	oxide, mon-, hydrate	$PdO.xH_2O$		yel to brn		d		i	i	s a, NH_3, NH_4Cl
p13	selenate	$PdSeO_4$	249.36	dk brn-red, rhomb, deliq	6.5	d red heat		v s	v s	i al, eth, alk; s NH_3
p14	selenide	PdSe	185.36	dk gray		<960		i		s aq reg
p15	selenide, di-	$PdSe_2$	264.32	olive gray, hex		<1000		i	i	v s aq reg; v sl s HNO_3; i alk
p16	silicide	PdSi	134.49	cr	7.31^{15}			i		
p17	sulfate	$PdSO_4.2H_2O$	238.50	red-br cr, deliq		d		v s	d	
p18	sulfide, di-	PdS_2	170.53	dk br cr	$4.7–4.8^{25}_{4}$	d		i	i	s aq reg, $(NH_4)_2S$
p19	sulfide, mono-	PdS	138.46	brn-blk tetr	6.6^{25}_{4}	d 950		i	i	sl s HNO_3, aq reg; i HCl, $(NH_4)_2S$
p20	sulfide, sub-	Pd_2S	244.86	grn gray (exist?)	7.303^{15}	d 800		i	i	sl s a, aq reg
p21	telluride, di-	$PdTe_2$	361.60	silvery cr, hex				i	i	v s aq reg; s HNO_3; i alk
	Palladium Complexes									
p22	diamminepalladium (II) hydroxide	$Pd(NH_3)_2(OH)_2$	174.48	yel micr cr		>105		v s	d	
p23	dichlorodiammine-palladium (II) *trans* (or α)	$Pd(NH_3)_2.Cl_2$	211.37	yel, tetr	2.5	d		0.304^{10}	s, d	s, d a; s NH_4OH; i chl, acet
p24	tetramminepalladium (II) chloride	$Pd(NH_3)_4Cl_2.H_2O$	263.44	col, tetr	1.91^{18}	d 120		v s		
p25	tetramminepalladium tetrachloropalladate	Vauquelin's salt. $Pd(NH_3)_4.PdCl_4$	422.73	pink powd or need	2.489^{21}	tr yel 184 d above 192		i	sl s	sl s dil HCl; s KOH
p26	**Phospham**	PN_2H	60.00	wh, amorph		infus		i	d	s conc H_2SO_4, alk; i a
p27	**Phosphomolybdic acid**	Molybdophosphoric acid $H_3[P(Mo_3O_{10})_4]$		yel, tetr		78–90		s	s	

No.	Name	Synonyms and Formulae	Mol. wt.	Crystalline form, properties and index of refraction	Density or spec. gravity	Melting point, °C	Boiling point, °C	Solubility, in grams per 100 cc: Cold water	Hot water	Other solvents
	Phosphonium									
	Phosphonium									
p28	bromide	PH_4Br	114.91	col, cub	gas: 2.464 g/l	subl *ca* 30	38.8[794]	d	d	
p29	chloride	PH_4Cl	70.46	col, cub		28[46 atm]	subl	d		
p30	iodide	PH_4I	161.91	col, tetr, deliq	2.86	18.5, subl 61.8	80	d		d, s a, alk
p31	sulfate	$(PH_4)_2SO_4$	166.07					d		
p32	**Phosphoramide**	Phosphorylamide. $PO(NH_2)_3$	95.04	wh, amorph		d		i	i	s al; i a
	Phosphoric acid									
p33	difluoro-	$H_2PO_2F_2$	102.99	col, fum liq	1.583^{25}_{4}	−96.5 ± 0.1	115.9 sl d			
p34	hypo-	$H_4P_2O_6.2H_2O$	198.01	col, rhomb deliq		70	d 100	d to $H_3PO_3+HPO_3$		
p35	meta-	HPO_3	79.98	col, vitrous, deliq	2.2–2.5	subl		d to H_3PO_4	d	s al; i liq CO_2
p36	monofluo-	H_2PO_3F	99.99	oily, col liq	1.818	−80				
p37	ortho-	H_3PO_4	98.00	col, liq, or rhomb cr, deliq	1.834^{18}	42.35	−½H_2O, 213	548	v s	s al
p38	ortho-	$2H_3PO_4.H_2O$	214.01	col, hex pr deliq		29.32	d	v s		
p39	pyro-	$H_4P_2O_7$	177.98	col, need or liq, hygr		61		709[23]	d to H_3PO_4	v s al, eth
p40	**Phosphorus**, black	P_4	123.8952	blk, incombust	2.70					i CS_2, conc H_2SO_4
p41	red	P_4	123.8952	redsh-brn, cub, or amorph powd, (mix of col and vlt?)	2.34	590[43 atm]	ign 200 280	v sl s	i	s abs al; i CS_2, eth, NH_3
p42	violet	P_4	123.8952	vlt, monocl	2.36	590				i org solv
p43	yellow	Phosphorus, white. P_4	123.8952	yel (or wh) cub or wax like solid, 2.144	1.82^{20}	44.1	280	0.0003[15]	sl s	0.3 al; 880[10]CS_2; s bz, NH_3, alk, eth, chl, tol
p44	bromide, penta-	PBr_5	430.52	yel, rhomb		d < 100	d 106	d		s CS_2, CCl_4, bz
p45	bromide, tri-	PBr_3	270.70	col, fum liq, $1.697^{26.6}$	2.852^{15}	−40	172.9	d		d al; s eth, chl, CS_2, CCl_4
p46	bromide(di-) chloride, tri-	PBr_2Cl_3	297.15	or cr		d 35		d		
p47	bromide(hepta-) chloride, di-	PBr_7Cl_2	661.24	pr				d		s PCl_3, PCl_5
p48	bromide(mono-) chloride, tetra-	$PBrCl_4$	252.69	yel cr				d		
p49	bromide(octa-) chloride, tri-	PBr_8Cl_3	776.60	brn need		25		d		
p50	bromide(di-) fluoride, tri-	PBr_2F_3	247.79	pa yel liq		−20	d 15	d		d glass
p51	bromide nitride	$(PNBr_2)_3$	614.40	col, rhomb		190	subl v 150	i		s eth; sl s chl, CS_2
p52	chloride, di-	PCl_2 (or P_2Cl_4 ?)	101.88	col liq		−28	180	hydr		
p53	chloride, penta-	PCl_5	208.24	yelsh-wh, tetr, fum	gas: 4.65^{296} g/l	d 166.8 (press)	subl 162	d		d a; s CS_2, CCl_4
p54	chloride, tri-	PCl_3	137.33	col, fum liq, 1.516^{14}	1.574^{21}	−112	75.5[749]	d	d	s eth, bz, chl, CS_2, CCl_4
p55	chloride(di-) fluoride, tri-	PCl_2F_3	158.88	col liq	5.4 g/l	−8	10			
p56	chloride(tri-)iodide, di-	PCl_3I_2	391.14	red, hex		d 259		d		s CS_2
p57	chloride(di-)nitride	$(PNCl_2)_3$	347.66	rhomb	1.98	114	256.5	i	d	s al, eth, bz, chl, a ac, CS_2
p58	chloride(di-)nitride	$(PNCl_2)_4$	463.55		2.18^{24}_{24}	123.5	328.5			
p59	chloride(di-)nitride	$(PNCl_2)_5$	579.43			41	224[13], polym >250			
p60	chloride(di-)nitride	$(PNCl_2)_6$	695.32			90	262[13], polym >250			
p61	cyanide	$P(CN)_3$	109.03	wh need		subl 130		d		v s eth; sl s h bz
p62	fluoride, penta-	PF_5	125.97	col gas	5.805 g/l	−83	−75	d		
p63	fluoride, tri-	PF_3	87.97	col gas	3.907 g/l	−151.5	−101.5	d		s al; d alk
p64	hydride, tri-	Phosphine. PH_3	34.00	col gas, pois		−133	−87.7	0.26 vol 20		
p65	iodide, di	P_2I_4	569.57	or, tricl		110		d	d	s CS_2
p66	iodide, tri-	PI_3	411.68	red, hex, deliq	4.18	61	d	d	d	v s CS_2
p67	oxide, pent-	Phosphoric anhydride. P_2O_5 (or P_4O_{10})	141.94	wh, monocl or powd, v deliq	2.39	580–585	subl 300	d to H_3PO_4	d	s H_2SO_4; i acet, NH_3
p68	oxide, sesqui-	Phosphorus trioxide. P_4O_6 (or P_2O_3)	219.89	col, or wh powd, monocl cr, deliq	2.135^{21}	23.8	175.4	d to H_3PO_3	d	s chl, bz, eth, CS_2
p69	oxide, tetra-	P_2O_4	125.95	col, rhomb, deliq	2.54^{23}	>100	180 vac	v s to H_3PO_3	d	
p70	oxide, tri-	P_2O_3 (or P_4O_6)	109.95	col, or wh powd, or monocl, deliq	2.135^{21}	23.8	173.8 (in N_2)	d to H_3PO_3	d	s CS_2, eth, chl, bz

No.	Name	Synonyms and Formulae	Mol. wt.	Crystalline form, properties and index of refraction	Density or spec. gravity	Melting point, °C	Boiling point, °C	Solubility, in grams per 100 cc		
								Cold water	Hot water	Other solvents
	Phosphorus									
p71	oxybromide	$POBr_3$	286.70	col pl	2.822	56	189.5	d		s H_2SO_4, CS_2, eth, bz, chl
p72	oxy*di*bromide chloride	$POBr_2Cl$	242.27	sol or liq	liq: 2.45^{60}	30	165	d		
p73	oxybromide chloride, di	$POBrCl_2$	197.79	tabl or liq	liq: 2.104^{14}	13	137.6	d		
p74	oxychloride	$POCl_3$	153.33	col, fum liq, $1.460^{25.1}$	1.675	2	105.3	d	d	d al, a
p75	oxychloride	$P_2O_3Cl_4$	251.76	col fum liq	liq: 1.58^{7}	<-50	212	d		
p76	oxyfluoride	POF_3	103.97	col gas	4.69 g/l	−68	−39.8	d		d al
p77	oxynitride	PON	60.98	wh, amorph		red heat		i	i	i a, alk
p78	oxysulfide	$P_4O_6S_4$	348.15	wh, tetr, deliq		102	295	d		50 CS_2
p79	selenide, penta-	P_2Se_5	456.75	dk red-blk need		d		d		s CCl_4; i CS_2
p80	(tetra-)selenide, tri-	P_4Se_3	360.78	or red cr	1.31	242	360–400			
p81	(tetra-)sulfide, hepta-	P_4S_7	348.34	lt yel cr	2.19^{17}	310	523			sl s CS_2
p82	sulfide, penta-	P_2S_5 (or P_4S_{10})	222.27	gray-yel cr, deliq	2.03	286–90	514	i	d	0.22 CS_2; s alk
p83	sesquisulfide	Tetraphosphorus trisulfide. P_4S_3	220.09	yel, rhomb	2.03^{17}	174	408	i	d	100^{17} CS_2; 11.1^{80} bz
p84	thiobromide	$PSBr_3$	302.76	yel, oct	2.85^{17}	38	d 212	s		s eth, CS_2, PCl_3
p85	thiochloride	$PSCl_3$	169.40	col, fum liq	1.668	−35	125	sl d	d	s CS_2
p86	thiocyanate	$P(SCN)_3$	205.22	liq	1.625^{18}	*ca* −4	265	d		s al, eth, bz, CS_2
	Phosphorous acid									
p87	hypo-	$H(H_2PO_2)$	66.00	col oily liq or deliq cr	1.493^{19}	26.5	d 130	s	v s	v s al, eth
p88	meta-	HPO_2	63.98	feather like cr				d		
p89	ortho-	$H_2(HPO_3)$	82.00	col-yel cr, deliq	$1.651^{21.2}$	73.6	d 200	309^{0}	694^{40}	s al
p90	pyro-	$H_4P_2O_5$	145.98	need		38	d 120	d		
p91	**Phosphotungstic acid**	Tungstophosphoric acid $H_3[P(W_3O_{10})_4].14H_2O$	3132.39	yel-grn cr, tricl		d		s		s al, eth
p92	phosphotungstic acid.	Dodecatungtophosphoric acid. $H_3[P(W_3O_{10})_4].24H_2O$	3312.54	trig		89		s		
p93	**Platinic acid,** *hexa*chloro	$H_2PtCl_6.6H_2O$	517.92	red, brn, deliq	2.431	60		v s	v s	s al, eth
p94	*tetra*cyano	$H_2Pt(CN)_4$	301.18			d 100		v s	v s	v s al, eth, chl
p95	*hexa*hydroxy-	$H_2Pt(OH)_6$	299.15	yel need		$-2H_2O$, 100	$-3H_2O$, 120	i	v sl s	s H_2SiF_6, dil a, alk
p96	*hexa*iodo-	$H_2PtI_6.9H_2O$	1120.67	blk-red, deliq				d		
p97	**Platinum**	Pt	195.09	silv met, cub	21.45^{20}	1769	3827 ± 100			s aq reg, fus alk
p98	arsenide	Nat. sperrylith. $PtAs_2$	344.93	gray, cub	11.8	d >800		sl d	sl d	v sl s a
p99	bromic acid	$H_2PtBr_6.9H_2O$	838.70	br-red, monocl, hygr				v s	v s	v s al, eth
p100	(II) bromide, di-	$PtBr_2$	354.91	br	6.65^{25}_{4}	d 250		i	i	i al; s HBr, KBr, aq Br
p101	(IV) bromide, tetra-	$PtBr_4$	514.73	dk cr	5.69^{25}_{4}	d 180		0.41^{20}	sl s	v s al, eth, HBr
p102	carbonyl bromide	$[Pt_2(CO)_2]Br_4$	765.84	lt red, need, hygr.	5.115^{25}_{4}	d 180		s d		s abs al, CCl_4, bz
p103	carbonyl chloride, di-	$Pt(CO)Cl_2$	294.00	yel need	4.2346^{25}_{4}	195, subl 240 in CO_2	d 300	d	d	s conc HCl, H_2SO_4, al
p104	*di*carbonyl chloride, di-	$Pt(CO)_2Cl_2$	322.02	lt yel need	3.4882^{25}_{4}	142	−CO, 210	d	d	d HCl; s CCl_4
p105	*di*platinum dicarbonyl tetrachloride	$Pt_2(CO)_2Cl_4$	588.01	or-yel need	4.235^{25}_{4}	195	subl 240 (in CO_2)	d		d HCl
p106	*di*platinum tricarbonyl tetrachloride	$Pt_2(CO)_3Cl_4$	616.02	or-yel need		130	d 250	d		s h CCl_4, d al
p107	carbonyl diiodide	PtI_2CO	476.91	red cr	5.257	d 140–150		d		s d, al; s bz
p108	carbonyl sulfide	$Pt(CO)S$	255.16	br-blk		d 300–400		d		d alk, al
p109	chloric acid	$H_2PtCl_6.6H_2O$	517.92	brn-red cr, hygr	2.431	60	d >115	s	s	s abs al, acet; v s eth
p110	(II) chloride, di-	$PtCl_2$	266.00	olive grn, hex	6.05	d 581 (in Cl_2)		v sl s		i al, eth; s HCl, NH_4OH
p111	(IV) chloride, tetra-	$PtCl_4$	336.90	br-red cr	4.303^{25}_{4}	d 370 (in Cl_2)		58.7^{25}	v s	sl s al, NH_3; s acet; i eth
p112	(IV) chloride, tetra-, hydrate	$PtCl_4.5H_2O$	426.98	red, monocl	2.43	$-H_2O$, 100		v s	v s	s al, eth
p113	chloride, tri-	$PtCl_3$	301.45	grnsh-blk	5.256^{25}	435		sl s	s	s h HCl; v sl s conc HCl

PHYSICAL CONSTANTS OF INORGANIC COMPOUNDS (Continued)

No.	Name	Synonyms and Formulae	Mol. wt.	Crystalline form, properties and index of refraction	Density or spec. gravity	Melting point, °C	Boiling point, °C	Solubility, in grams per 100 cc		
								Cold water	Hot water	Other solvents
	Platinum									
p114	*di*chlorocarbonyl, dichloride	$Pt(COCl_2)Cl_2$	364.91	yel cr		d		v s		sl s al; v sl s CCl_4
p115	(II) cyanide, di-	$Pt(CN)_2$	247.13	yel-br cr				i	i	i al, a, alk; s KCN
p116	(II) fluoride, di-	PtF_2	233.09	yelsh-grn				i	i	
p117	fluoride, hexa-	PtF_6	309.08	dk red solid, very unstable		57.6				
p118	(IV) fluoride, tetra-	PtF_4	271.08	deep red, fused, or yel-lt brn cr, deliq		d red heat		s d	v s	s a, alk
p119	(II) hydroxide	$Pt(OH)_2$	229.10	blk		d		i	i	s HCl, HBr, alk; i H_2SO_4, dil HNO_3
p120	(II) hydroxide, hydrate	$Pt(OH)_2.2H_2O$	265.14			$-2H_2O$, 100		i	i	s conc a
p121	*mono*hydroxy chloric acid	$H_2[PtCl_4(OH)].H_2O$	409.39	red-brn cr, hygr						
p122	(II) iodide, di-	PtI_2	448.90	blk powd	6.403^{25}_{4}	d 360		i	i	i eth, a; s HI; sl s Na_2SO_3
p123	(IV) iodide, tetra-	PtI_4	702.71	brn, amorph, or blk cr	6.064^{25}_{4}	d 130		s d		s al, acet, alk, HI KI, liq NH_3
p124	iodide, tri-	PtI_3	575.80	blk, like graphite	7.414^{25}_{4}	d 270		i	i	i al, eth; s KI
p125	(II) oxide, mon-	PtO	211.09	vlt-blk	14.9^{15}	d 550		i	i	s HCl; i a, aq reg
p126	(II) oxide, mon-, dihydrate	$PtO.2H_2O$	247.12			$-2H_2O$, 140–150		i	i	s conc HCl, conc H_2SO_4, conc HNO_3
p127	(IV) oxide, di-	PtO_2	227.03	blk	10.2	450		i	i	i a, aq reg
p128	(IV) oxide, di-dihydrate	Platinic hydroxide. $PtO_2.2H_2O$ or $Pt(OH)_4$	263.12			$-2H_2O$, 100		i	i	s HCl, aq reg, KOH
p129	(IV) oxide, di-, hydrate	$PtO_2.H_2O$	245.10					i	i	i aq reg, ac a, HCl; sl s NaOH
p130	(IV) oxide, di-trihydrate	$PtO_2.3H_2O$	281.13	ochre		d 300		i	i	i aq reg, HCl
p131	(IV) oxide, di-, tetrahydrate	Hydroxoplatinic acid. $PtO_2.4H_2O$ or $H_2Pt(OH)_6$	299.15	yel need		$-2H_2O$, 100	$-3H_2O$, 120	i	i	s a, dil alk
p132	(II, IV) oxide,	Pt_3O_4	649.27			d		i		i a, aq reg
p133	oxide, sesqui,	$Pt_2O_3.3H_2O$	492.22			$-H_2O$, 100		i	i	s conc H_2SO_4, caust alk
p134	oxide, tri-	PtO_3	243.09	redsh-brn powd						s HCl, H_2SO_3; sl s HNO_3, H_2SO_4
p135	*pyro*phosphate	PtP_2O_7	369.03	grn-yel	4.85	d 600		v sl s		
p136	phosphide	PtP_2	257.04	met shine	9.01^{25}_{4}	*ca* 1500		i	i	i a; v sl s aq reg
p137	selenide, di-	$PtSe_2$	353.01	blk or gray cr or amorph	7.65	d when dry				s aq reg; sl s HNO_3, H_2SO_4
p138	selenide, tri-	$PtSe_3$	431.97	bl flakes	7.15	d 140		i	i	i conc a, CS_2; s aq reg
p139	sulfate	$Pt(SO_4)_2.4H_2O$	459.27	yel pl				s	d	s al, eth, a
p140	(IV) sulfide, di-	PtS_2	259.22	blk-brn powd	7.66^{25}_{4}	d 225–250		i	i	s HCl, HNO_3; i $(NH_4)_2S$
p141	(II) sulfide, mono-	PtS	227.15	blk, tetr	10.04^{25}_{4}	d		i	i	s $(NH_4)_2$ S; i a, alk
p142	sulfide, sesqui-	Pt_2S_3	486.37	gray (exist ?)	5.52	d		i	i	sl s aq reg; i a
p143	(III) sulfuric acid	$H_2[Pt_2(SO_4)_4(H_2O)_2].9\frac{1}{2}H_2O$	983.62	or, red, tricl		d 150		s	s	d alk
p144	telluride	$PtTe_2$	450.29	gray, hex		1200–1300				sl s Na_2S, $(NH_4)_2S$
	Platinum complexes									
p145	tetramine platinum (II) chloride	$[Pt(NH_3)_4]Cl_2.H_2O$	352.13	col, tetr, 1.672, 1.667	2.737	250, $-H_2O$, 100				
p146	tetrammineplatinum (II) chloroplatinite	Magnus, salt. $[Pt(NH_3)_4]PtCl_4$	600.11	grn or red, tetr	<4.1	d		sl s	sl s	
p147	tetrachlorodiammine platinum (IV), *trans-*	$[Pt(NH_3)_2]Cl_4$	370.96		3.3	200–216				
p148	tetrachlorodiammine platinum (IV), *cis-*	$[Pt(NH_3)_2]Cl_4$	370.96	or-yel, rhomb or hexag pl or need		240				
p149	**Plumbous, plumbic**	see Lead								
p150	**Plutonium α**	Pu	242.00	sil wh, monocl	19.84	639.5 ± 2	3235 ± 19			s HCl; i HNO_3, conc H_2SO_4

No.	Name	Synonyms and Formulae	Mol. wt.	Crystalline form, properties and index of refraction	Density or spec. gravity	Melting point, °C	Boiling point, °C	Solubility, in grams per 100 cc		
								Cold water	Hot water	Other solvents
	Plutonium									
p151	β	Pu	239.05	monocl	17.70	stab 122 ± 2 to 206 ± 3				
p152	γ	Pu	239.05	orthorhomb	17.14	stab 206 ± 3 to 319 ± 5				
p153	δ	Pu	239.05	cub	15.92	stab 319 ± 5 to 451 ± 4				
p154	δ'	Pu	239.05	tetrag	16.00	stab 451 ± 4 to 476 ± 5				
p155	ε	Pu	239.05	cub	16.51	stab 476 ± 5 to 639.5 ± 2				
p156	bromide, tri-	$PuBr_3$	481.73	grn, orthorhomb	6.69	681		s		
p157	chloride, tri-	$PuCl_3$	348.36	emerald grn, hex	5.70	760		s		s dil a
p158	fluoride, hexa-	PuF_6	355.99	redsh-brn, orthorhomb		50.75	62.3	d		
p159	fluoride, tetra-	PuF_4	317.99	pa brn, monocl	7.0 ± 0.2	1037				
p160	fluoride, tri-	PuF_3	299.00	purple, hex	9.32	1425(±3)		i		
p161	iodide, tri-	PuI_3	622.71	bright grn, orthorhomb	6.92	777		s		
p162	nitride	PuN	256.01	blk, cub	14.25			hydrol		s HCl, H_2SO_2
p163	oxalate	$Pu(C_2O_4)_2.6H_2O$	526.13	yel-grn				i		
p164	oxide, di-	PuO_2	274.00	yelsh-grn, cub	11.46					sl s h conc H_2SO_4, HNO_3, HF
p165	sulfate	$Pu(SO_4)_2$	434.12	light pink						s dil min a
p166	sulfate, tetrahydrate	$Pu(SO_4)_2.4H_2O$	506.18	coral pink		d 280				s dil min a
p167	**Polonium**	Po	210.05	α-Po: simple cub; β-Po: rhbr	9.4 (for β-Po)	254	962	sl s		s dil min a; v sl s dil KOH
p168	ammonium chloride	$(NH_4)_2PoCl_6$	458.85		2.76					
p169	*tetra*bromide	$PoBr_4$	529.67	bright red, cub		330 (in Br atm)	360^{200}			s al, acet; i bz, CCl_4
p170	*di*chloride	$PoCl_2$	280.96	ruby red, orthorhomb	6.50	subl 190				s dil HNO_3
p171	*tetra*chloride	$PoCl_4$	351.86	yel, monocl or tric		300 (in Cl atm)	390	s, d		s HCl; sl s al, acet
p172	*tetra*iodide	PoI_4	717.67	blk cr		200 (in N atm subl)				sl s al, acet; i bz, CCl_4
p173	*di*oxide	PoO_2	242.05	red, tetr		d 500				
p174	selenate	$2PoO_2.SeO_3$	611.06	wh powd		>400				s dil HCl
p175	sulfate, basic	$2PoO_2.SO_3$	564.16	wh powd		>400, d 550				s dil HCl
p176	*di*sulfate	$Po(SO_4)_2$	402.17	purp		d 550				i al; v s dil HCl
p177	monosulfide	PoS	242.11	blk		d 275				i al; sl s dil HCl
p178	**Potassium**	Kalium. K	39.102	cub silv met	0.86^{20}	63.65	774	d to KOH	d	d al; s a, Hg, NH_3
p179	acetate	$KC_2H_3O_2$	98.15	wh, lust powd, deliq	1.57^{25}	292		253^{20}	492^{62}	33 al; 24.24^{15} MeOH; s liq NH_3; i eth, acet
p180	acetate, acid	$K_2C_2H_3O_2.HC_2H_3O_2$	158.20	col, need or pl, hygr		148	d 200	d	d	s al, acet
p181	acetyl salicylate	$KC_9H_7O_4.2H_2O$	254.29			65				
p182	*meta*aluminate	$K_2Al_2O_4.3H_2O$	250.22	col cr				v s, d	v s, d	s alk; i al
p183	aluminosilicate	Nat. orthoclase. $KAlSi_3O_8$ (or $K_2O.Al_2O_3.6SiO_2$)	278.34	wh, monocl, 1.518, 1.524, 1.526	2.56	*ca* 1200				
p184	aluminosilicate	Nat. microcline. $KAlSi_3O_8$ (or $K_2O.Al_2O_3.6SiO_2$)	278.34	wh, tricl, 1.522, 1.526, 1.530	2.54–2.57	1140–1300				
p185	aluminosilicate	Nat. muscovite, white mica. $KAl_3Si_3O_{10}.(OH)_2$ (or $K_2O.3Al_2O_3.6SiO_2.2H_2O$)	398.31	col, monocl, 1.551, 1.587, 1.581	2.76–2.80	d		i		
p186	aluminum *meta*silicate	Nat. leucite. $KAlSi_2O_6$	218.25	col cr, 1.508	2.47	1686 ± 5		i	i	d a
p187	aluminum *ortho*silicate	Nat. kaliophilite. $KAlSiO_4$	158.17	col, hex or rhomb (hex→rhomb 1540°) hex: 1.532, 1.572; rhomb: 1.528, 1.536	2.5	*ca* 1800 (rhomb)				
p188	aluminum sulfate	Nat. kalinite. $KAl(SO_4)_2.12H_2O$	474.39	col, cub, oct or monocl, cub: 1.454, 1.4564; hex: 1.456, 1.429	1.757^{20}_{14}	92.5 $-9H_4O$, 64.5	$-12H_2O$, 200	11.4^{20}	v s	i al, acet; s dil a
p189	amide	Potassamide. KNH_2	55.12	col-wh, or yel-grn, hygr		335	subl 400	d	d	d al; s liq NH_3
p190	*peroxy*lammine sulfonate	$(KSO_3)_2NO$	268.39	yel cr, expl				0.62^3, d	6.6^{29}, d	i al

No.	Name	Synonyms and Formulae	Mol. wt.	Crystalline form, properties and index of refraction	Density or spec. gravity	Melting point, °C	Boiling point, °C	Solubility, in grams per 100 cc: Cold water	Hot water	Other solvents
	Potassium									
p191	ammonium tartrate	$KNH_4C_4H_4O_6$	205.21	wh, cr powd				v s		
p192	antimonate, hydroxo-	"*Pyro*"-antimonate. $KSb(OH)_6.\frac{1}{2}H_2O$	271.90	wh gran or cr powd				2.82^{20}	s	
p193	antimonide	K_3Sb	239.06	yel-grn		812		d		d air
p194	antimony tartrate	$KSbC_4H_4O_7.\frac{1}{2}H_2O$	333.93	col, rhomb, 1.620, 1.636, 1.638	2.607	$-H_2O$, 100		$5.26^{8.7}$	35.7^{100}	i al; 6.67^{25} glyc
p195	*ortho*arsenate	K_3AsO_4	256.23	col need, deliq		1310		18.87	v s	4 al
p196	*ortho*arsenate, di-H	KH_2AsO_4	180.04	col, tetr, 1.567, 1.518	2.867	288		19^{6}	v s	i al; s NH_3, a; 52.5 glyc
p197	*ortho*arsenate, mono H	K_2HAsO_4	218.13	col monocl pr		d 300		18.86^{6}	s	i al
p198	*meta*arsenite	$KAsO_2$	146.02	wh powd, hygr				s	s	sl s al
p199	*ortho*arsenite	K_3AsO_3	240.23	col need				v s		s al
p200	*meta*arsenite, acid	$KH(AsO_2)_2.H_2O$	271.97					s		sl s al
p201	aurate	$KAuO_2.3H_2O$ (or $2H_2O$)	322.11	lt yel need		d		s	d	s al
p202	azide	KN_3	81.12	col, tetr	2.04	350 (vac)		49.6^{17}	105.7^{100}	s al; i eth
p203	benzoate	$KC_7H_5O_2.3H_2O$	214.26	wh cr powd		$-3H_2O$, 110	d	52^{25}	112^{100}	s al
p204	*di*borane	Diboranidex. $K_2B_2H_6$	105.87	wh, cub cr, 1.493	1.18	subl 400, vac		d		
p205	*penta*borate	Pentaboranidex. $K_2B_5H_9$	141.32	wh powd		d <180		s, d	d	
p206	*di*borane, dihydroxy	$K_2B_2H_6O_2$	137.87	col, cub cr	1.39	d→K, 400–500		s, d		s al; d a
p207	*meta*borate	KBO_2 (or $K_2B_2O_4$)	81.91	col, hex 1.526, 1.450		950		71^{30}	v s	i al, eth
p208	*penta*borate	$KB_5O_8.4H_2O$	293.21	col, rhomb		780			0.007°	
p209	*peroxy*borate	$KBO_3.\frac{1}{2}H_2O$	106.92	wh cr		$-O_2$, 100	d 150	1.22^{0}		i al, eth
p210	*tetra*borate	$K_2B_4O_7.8H_2O$	377.57	col, monocl	1.74 (anhydr)	d		26.7^{30}	v s	
p211	borohydride	KBH_4	53.94	wh, cub, 1.494	1.178	d 500		19.3^{20}	v s	0.25 al; 0.56 MeOH; i eth
p213	boroxalate	$KHC_2O_4.HBO_2.2H_2O$	207.98			$-H_2O$, 110		v s	v s	i, d al
p214	borotartrate	Solution: cream of tartar. $KC_4H_4BO_7$	213.99	wh cr powd	1.832			v s		i al, eth, chl
p215	bromate	$KBrO_3$	167.01	col, trig	$3.27^{17.5}$	434 d 370		13.3^{40}	49.75^{100}	sl s al; i acet
p216	bromide	KBr	119.01	col, cub, sl hygr, 1.559	2.75^{25}	730	1435	53.48^{0}	102^{100}	0.142^{25} al; sl s eth; s glyc
p217	bromoaurate	$K[AuBr_4]$	555.71	red-brn, rhomb		d 120		sl s		s al
p218	bromoaurate, dihydrate	$K[AuBr_4].2H_2O$	591.74	vlt, monocl cr	4.08			19.5^{15}	204^{67}	s al, KBr; d eth
p219	bromoiodide, di-	$KIBr_2$	325.82	red, rhomb		60	d 180	v s		
p220	*hexa*bromoplatinate	$K_2[PtBr_6]$	752.75	dk red-brn, cub	4.66^{24}	d>400		2.02^{20}	10^{100}	i al
p221	*tetra*bromoplatinite	$K_2[PtBr_4]$	592.93	br, rhomb				v s	v s	
p222	bromoplatinite, dihydrate	$K_2[PtBr_4].2H_2O$	628.96	blk, rhomb	3.747^{25}_{4}	$-H_2O$, vac		v s	v s	
p223	bromostannate	$K_2[SnBr_6]$	676.35	wh cr	3.783					
p224	cacodylate	$K[(CN_3)_2AsO_2].H_2O$	194.10	wh cr				s		sl s al; i eth
p225	cadmium cyanide	$K_2[Cd(CN)_4]$	294.68	col, oct	1.846	450		33.3	100^{100}	s al
p226	cadmium iodide	$2KI.CdI_2.2H_2O$	734.25	wh-yelsh cr powd, deliq	3.359			137^{15}		s a, al, eth
p227	calcium chloride	Chlorocalcite. $KCl.CaCl_2$	185.54	col cub, β 1.52	754			s		
p228	calcium magnesium sulfate	Polyhalite. $K_2Ca_2Mg(SO_4)_4.2H_2O$	602.95	wh, trig, 1.548, 1.562, 1.567	2.775					
p228	calcium magnesium sulfate	Krugite $K_2Ca_4Mg(SO_4)_6.2H_2O$	875.24	gray cr	2.801					
p229	calcium sulfate	Kaluszite, syngenite. $K_2Ca(SO_4)_2.H_2O$	328.42	col, monocl, 1.500, 1.517, 1.518	2.60	1004		0.25	d	i al; s a
p230	*d*-camphorate	$K_2C_{10}H_{14}O_4.5H_2O$	366.50	col, need cluster, hygr		$-5H_2C$, 110		260^{14}		s al
p231	carbide	KHC_2	64.13	col, rhomb cr	1.37					
p232	carbonate	K_2CO_3	138.21	col, monocl, hygr, 1.531	2.428^{19}	891	d	112^{20}	156^{100}	i al, acet
p233	*peroxy*carbonate	$K_2C_2O_6.H_2O$	216.22			200–300		s		
p234	carbonate, dihydrate	$K_2CO_3.2H_2O$	174.24	col, monocl, hygr, 1.380, 1.432, 1.573	2.043	$-H_2O$, 130		146.9	331^{100}	
p235	carbonate, hydrogen	$KHCO_3$	100.12	col, monocl, 1.482	2.17	d 100–200		22.4	60^{60}	i al
p236	carbonate, trihydrate	$2K_2CO_3.3H_2O$	330.47	col, monocl, 1.380, 1.482, 1.573	2.043			129.4	268.3^{100}	i al, conc HN_4OH
p237	carbonyl	$(KCO)_6$	402.68	gray-red		expl		expl		d al
p238	chlorate	$KClO_3$	122.55	col, monocl, 1.409, 1.517, 1.524	2.32	356	d 400	7.1^{20}	57^{100}	14.1^{100} 50 % al; sl s glyc, liq NH_3; i acet; s alk
p239	*per*chlorate	$KClO_4$	138.55	col, rhomb, 1.4717, 1.4724, 1.476	2.52^{10}	610 ± 10	d 400	0.75^{0}	21.8^{100}	v sl s al; i eth

No.	Name	Synonyms and Formulae	Mol. wt.	Crystalline form, properties and index of refraction	Density or spec. gravity	Melting point, °C	Boiling point, °C	Solubility, in grams per 100 cc		
								Cold water	Hot water	Other solvents
	Potassium									
p240	chloride	Nat. sylvite. KCl	74.56	cub, col 1.490	1.984	776	subl 1500	34.7^{20}	56.7^{100}	sl s al; s eth, glyc, alk
p241	*hypo*chlorite	KClO	90.55	in sol only		d		v s	v s	
p242	chloroaquoruthenate (III) penta-	$K_2[Ru(H_2O)Cl_5]$	374.55	rose pr		$-H_2O$, 200		s	s	sl s al
p243[1]	chloroaurate	K[AuCl]	377.88	yel, monocl	3.75	d 357		61.8^{20}	80.2^{60}	25 al; s a
p243[2]	chloroaurate, dihydrate	$K[AuCl_4].2H_2O$	413.91	yel , rhomb pl				s	s	s al, eth
p244	chlorochromate	Peligot's salt. $KCrO_3Cl$.	174.55	red, monocl	2.497	d		s, d		s a
p245	chlorohydroxo-ruthenate	$K_2[Ru(OH)Cl_5]$	373.55	brn-red cr		d		s, d	d	i al
p246	chloroiodate (III)	$KICl_4$	307.82	yel, rhomb	1.76^{45}	d		d		d eth
p247	chloroiodide, di-	$KICl_2$	236.91	col, monocl		60	d 215	d		
p248	chloroiridate	$K_4[IrCl_6]$	483.22	blk, cub	3.546	d		125^{19}	6.67	i al, KCl, HN_4OH
p249	chloronitrosyl-ruthenate (III) penta-	$K_2Ru(NO)Cl_5$	386.55	dk red, rhomb		d		12^{25}	80^{60}	i al
p250	chloroösmate (III)	$K_3[OsCl_6].3H_2O$	574.27	dk red cr		$-3H_2O$, 150		v s		s a; i eth
p251	chloroösmate (IV)	K_2OsCl_6	481.12	red, cub		d		sl s	s	i al; s dil HCl
p252	chloropalladate	$K_2[PdCl_6]$	397.32	red, cub	2.738	d		sl s, d	d	i al; sl s HCl
p253	chloropalladite	K_2PdCl_4	326.42	red-brn, tetr (yel cub)	2.67	d 105		s	v s	s KCl, HN_4OH; i al
p254	*hexa*chloroplatinate	$K_2[PtCl_6]$	486.01	yel, cub, ~ 1.825	3.499^{24}	d 250		0.481^{2}	5.22^{100}	i al, eth
p255	*tetra*chloroplatinite	$K_2[PtCl_4]$	415.11	red-brn tetr, 1.64, 1.67	3.38	d		0.93^{16}	5.3^{100}	i al
p256	chloroplumbate	K_2PbCl_6	498.11	lt yel, cub		d 190		d		s h HCl
p257	chlororhenate (IV)	K_2ReCl_6	477.12	yel-grn, oct	3.34			0.8	d	d alk; sl s HCl
p258	chlororhenate (V)	K_2ReOCl_5	457.67	grn hex pl				d	d	s a; i al, eth
p259	chlororhodate, hexa-	$K_3RhCl_6.3H_2O$	486.98	red, tricl	3.291	d		d		sl s al, KCl
p260	chlororhodite, penta-	K_2RhCl_5	358.37	red, rhomb		d		sl s	d	i al
p261	chlororuthenate (IV)	K_2RuCl_6	391.99	blk, cub		d		s, d		i al
p262	chlorostannate	K_2SnCl_6	409.63	col, cub, 1.657	2.71			s	s	
p263	chlorotellurate	K_2TeCl_6	418.52	pale yel, octahedral				d	d	s HCl
p264	chromate	Nat. tarapacaite. K_2CrO_4	194.20	yel, rhomb, β 1.74	2.732^{18}	968.3		62.9^{20}	79.2^{100}	i al
p265	*di*chromate	$K_2Cr_2O_7$	294.19	red, monocl or tricl, 1.738	2.676^{25}_{4}	tricl→ monocl 241.6 m.p. 398	d 500	4.9^{0}	102^{100}	i al
p266	*peroxy*chromate	K_3CrO_8	297.30	brn-red, cub		d 170		sl s		i a, al, eth
p267	chromium sulfate	Potassium chromium alum. $K[Cr(SO_4)_2].12H_2O$	499.41	vlt-ruby red, cub, oct, 1.4814	1.826^{25}	89	$-10H_2O$, 100 $-12H_2O$, 400	24.39^{25}	50	i al; s dil a
p268	chromium chromate, basic	$K_2CrO_4.2[Cr(OH).CrO_4]$	564.19	vlt brn amorph powd	2.28^{14}	300		i		i al, acet a
p269	citrate	$K_3C_6H_5O_7.H_2O$	324.42	wh cr	1.98	d 230		167^{15}	199.7^{31}	sl s al; s glyc
p270	citrate, monobasic	$KH_2C_6H_5O_7$	230.22	wh cr powd				s		
p271	cobalt carbonate, hydrogen(ous)	$KHCO_3.CoCO_3.4H_2O$	291.12	rose need				d		
p272	cobalt (II) cyanide	$K_4[Co(CN)_6]$	371.45	redsh-brn cr, deliq	2.039^{25}_{4}			v s	v s	d a; i al, eth, $CHCl_3$
p273	cobalt (III) cyanide	$K_3[CO(CN)_6]$	332.35	wh-yel, monocl pr	1.878^{25}_{4}			sl s	sl s	s dil HCl, dil HNO_3; sl s al
p274	cobaltinitrite	Fischer's salt. $K_3[Co(NO_2)_6]$	452.27	yel pr, cub				0.9^{17}	d	i al
p275	cobaltinitrite, hydrate	$K_3[Co(NO_2)_6].H_2O$	470.29	yel cr powd				i	s, d	s min a; sl s ac a; i al, eth
p276	cobaltinitrite, hydrate	$K_3[Co(NO_2)_6].1\frac{1}{2}H_2O$	479.30	yel, tetr		d 200		0.089^{17}	sl s	i al, meth
p277	cobaltmalonate (II)	$K_2[Co(C_3H_2O_4)_2]$	341.23		2.234					
p278	cobalt sulfate (II)	$K_2SO_4.CoSO_4.6H_2O$	437.36	red pr, monocl, 1.481, 1.487, 1.500	2.218			25.5^{0}	108.4^{49}	
p279	copperchloride	$KCl.CuCl_2$	209.00	red need	2.86					
p280	cyanate	KOCN	81.12	col, tetrag	2.056^{20}	d 700–900		75^{25}	s	i al
p281	cyanide	KCN	65.12	col, cub, wh gran, deliq, very pois, 1.410	1.52^{16}	634.5		50	100	$0.88^{19.5}$ al; $4.91^{19.5}$ MeOH; s glyc
p282	cyanoargentate (I)	Potassium argento-cyanide. $K[Ag(CN)_2]$	199.01	cub, 1.625	2.36			25^{20}	100	4.85 % al; i a
p283	cyanoaurate	$K[Au(CN)_2]$	288.10	col, rhomb	3.45			14.3	200	sl s al; i eth
p284	cyanoaurate (III)	$K[Au(CN)_4].1\frac{1}{2}H_2O$	367.16	col tabl		d 200		s	v s	s al

No.	Name	Synonyms and Formulae	Mol. wt.	Crystalline form, properties and index of refraction	Density or spec. gravity	Melting point, °C	Boiling point, °C	Solubility, in grams per 100 cc Cold water	Hot water	Other solvents
	Potassium									
p285	cyanocadmate	$K_2[Cd(CN)_4]$	294.68	col, cub	1.85			33	100^{100}	sl s al
p286	cyanochromate (III)	$K_3[Cr(CN)_6]$	325.41	yel, rhomb	1.71			30.9^{20}		i al
p287	cyanocobaltate (II)	$K_4[Co(CN)_6]$	371.42					s	s	i al, eth
p288	cyanocobaltate (III)	$K_3[Co(CN)_6]$	332.32	yel, monocl	1.906	d		s	s	i al; sl s NH_3
p289	cyanocuprate (I)	$K_3[Cu(CN)_4]$	284.92	col, rhbdr		d		v s		
p290	cyanomanganate (II)	$K_4[Mn(CN)_6].3H_2O$	421.50	deep bl, tetr				s	d	
p291	cyanomanganate (III)	$K_3[Mn(CN)_6]$	328.35	red, rhomb, 1.553, 1.555, 1.571 (Li)				s		
p292	cyanomercurate	$K_2[Hg(CN)_4]$	382.87	col, cr pois				s		s al
p293	cyanomolybdate	$K_4[Mo(CN)_8].2H_2O$	496.52	yel, rhomb	2.337^{25}_{4} (anhyd)	$-H_2O$, 105–110		v s	v s	i eth; 0.0017^{20} al
p294	cyanonickelate (II)	$K_2[Ni(CN)_4].H_2O$	259.00	red-yel, monocl cr or powd	1.875^{11}	$-H_2O$, 100		s		d a
p295	cyanoosmite	$K_4[Os(CN)_6].6H_2O$	556.76	col, yel, monocl, β 1.607		d		sl s	s	i al, eth
p296	cyanoplatinite	$K_2[Pt(CN)_4].3H_2O$	431.41	col, yel, rhomb, blue fluor, deliq	2.455^{16}	$-3H_2O$, 100	d 400–600	sl s	s	sl s al, eth, H_2SO_4
p297	cyanotungstate (IV)	$K_4[W(CN)_8].2H_2O$	584.43	lt yel- grn cr powd	1.989^{25}_{4} (anhydr)	$-2H_2O$, 115		130^{18}	s	i al, eth
p298	ethylsulfate	$KC_2H_5SO_4$	164.23	wh, monocl	1.843			s		s al
p299	ferricyanide	$K_3Fe(CN)_6$	329.26	red, monocl, 1.566, 1.569, 1.583	1.85^{25}	d		33^{4}	77.5^{100}	i al; s acet
p300	ferrocyanide	Yellow prussate of potash. $K_4Fe(CN)_6.3H_2O$	422.41	lem yel, monocl, β 1.577	1.85^{17}	$-3H_2O$, 70	d	27.8^{12} anhydr 14.5^{0}	$90.6^{96.3}$ anhydr 74^{98}	s acet; i al, eth, NH_3
p301	fluoberyllate	K_2BeF_4	163.21	col, rhomb		red heat		2^{20}	5.26^{100}	
p302	fluoborate	Nat. avogadrite. KBF_4	125.91	col, rhomb or cub, 1.324, 1.325, 1.325	2.498^{20}	d 350	d	0.44^{20}	6.27^{100}	sl s al, eth; i alk
p303	fluogermanate	K_2GeF_6	264.78	wh, hex		730	*ca* 835	0.542^{18}	2.58^{100}	
p304	fluomanganate (IV)	K_2MnF_6	247.13	yel, hex, tab		d		d	d	s conc HCl
p305	fluoniobate, penta-	Potassium oxyniobate. $K_2NbOF_5.H_2O$	300.12	col, monocl pl or leaf				7.69	s	s d conc HF
p306	fluorescein deriv	$K_2C_{20}H_{10}O_5$	408.50	yelsh-red powd				s		
p307	fluoride	KF	58.10	col, cub deliq, 1.363	2.48	846	1505	92.3^{18}	v s	s HF, NH_3; i al
p308	fluoride, acid	KHF_2	78.11	col, cub, deliq	2.37	d *ca* 225	d	41^{21}	v s	s $KC_2H_3O_2$; i al
p309	fluoride, dihydrate	$KF.2H_2O$	94.13	col, monocl pr, deliq, 1.352	2.454	41	156	349.3^{18}	v s	s HF; i al
p310	*hexa*fluorophosphate	KPF_6	184.07			*ca* 575	d	9.3^{25}	20.6^{80}	
p311	fluorotungstate	$2KF.WO_2F_2.H_2O$	388.06	monocl		$-H_2O$, red heat		6^{17}	s	
p312	fluosilicate	Nat, hieratite. K_2SiF_6	220.25	col, cub or hex 1.3991	hex 3.08 cub 2.665^{17}	d		$0.12^{17.5}$ 6.9^{19}	0.954^{100}	s HCl; i NH_3; v sl s al
p313	fluostannate	$K_2SnF_6.H_2O$	328.90	monocl pr	3.053			3.7^{18}	33.3^{100}	i al, NH_3
p314	fluosulfonate	$KFSO_3$	138.16	short, thick pr		311		6.9^{19}		
p315	fluotantalate	K_2TaF_7	392.14	col, rhomb	4.56; 5.24			sl s, d		sl s HF
p316	fluotellurate, di-	$K_2TeO_2F_2.3H_2O$	345.84	micros oct, monocl		d		sl s	sl s	s HF
p317	fluothorate	$K_2ThF_6.4H_2O$	496.29	col		d		2.15	6.6	d a; i al
p318	fluotitanate	$K_2TiF_6.H_2O$	258.11	col, monocl lust leaf		$-H_2O$, 32 m.p. 780	d	0.556^{0}	1.27^{21}	sl s min a; i NH_3
p319	fluozirconate	K_2ZrF_6	283.41	col, monocl, 1.466, 1.455	3.48			0.781^{2}	25^{100}	i NH_3
p320	formate	$KCHO_2$	84.12	col, rhomb deliq	1.91	167.5	d	331^{18}	657^{80}	s al; i eth
p321	gadolinium sulfate	$K_2SO_4.Gd_2(SO_4)_3.2H_2O$	812.98	cr	3.503^{16}			s	s	s K_2SO_4
p322	gallium sulfate	$KGa(SO_4)_2.12H_2O$	517.13	col cr	1.895			s		
p323	*di*germanate	$K_2Ge_2O_5$	303.38	wh cr	$4.31^{21.5}$	>83		s		s a
p324	*meta*germanate	K_2GeO_3	198.79	wh cr	$3.40^{21.5}$	823		s		s a
p325	*tetra*germanate	$K_2Ge_4O_9$	512.55	wh cr	$4.12^{21.5}$	1083		s		s a
p326	glycerophosphate	$K_2C_3H_7PO_6$	248.26	col-sl yelsh mass, hygr				v s	v s	s al
p327	hydride	KH	40.11	wh need 1.453	1.47	d		d	d	i CS_2, eth, bz
p328	hydroxide	KOH	56.11	wh, rhomb deliq	2.044	360.4±0.7	1320–1324	107^{15}	178^{100}	v s al; i eth, NH_3
p329	(tri-)hydroxylammine trisulfonate	$(KSO_3)_3.NO.1\frac{1}{2}H_2O$	414.52	col, monocl pr		$-H_2O$, 100–200		4^{18}	sl d	
p330	*hexa*hydroxyplatinate	$K_2[Pt(OH)_6]$	375.34	yel, rhomb	5.18	d 160		s	s	i al
p331	imidolsulfonate	$(KSO_3)_2NH$	253.34	col, monocl	2.515	d 170–180	d 360–440, vac	1.3^{23}	d	i HNO_3
p332	iodate	KIO_3	214.00	col, monocl	3.93^{32}_{4}	560	d>100	4.74^{0}	32.3^{100}	s KI; i al, NH_3
p333	iodate, acid	$KIO_3.HIO_3$	389.92	col, monocl				1.33^{15}		i al

No.	Name	Synonyms and Formulae	Mol. wt.	Crystalline form, properties and index of refraction	Density or spec. gravity	Melting point, °C	Boiling point, °C	Solubility, in grams per 100 cc		
								Cold water	Hot water	Other solvents
	Potassium									
p334	iodate, acid	$KIO_3.2HIO_3$	565.83	col, tricl				4.15		
p335	*meta*periodate	KIO_4	230.00	col, tetr, 1.6205	3.618^{15}_{4}	582	−O, 300	0.66^{13}	s	v sl s KOH
p336	iodide	KI	166.01	col or wh, cub or gran, 1.677	3.13	686	1330	127.5^{0}	208^{100}	1.88^{25} al; 1.31^{25} acet; sl s eth; s NH_3
p337	iodide, tri-	$KI_3.\frac{1}{2}H_2O$	428.82	dk bl, monocl, deliq	3.498	31	d 225	v s		s al, KI
p338	iodoaurate	$KAuI_4$	743.69	blk lust cr		d 150		s, d		s dil KI sol
p339	iodoiridite	K_3IrI_6	1070.93	gr cr		d		i	i	i al
p340	iodomercurate (II) tetra-	K_2HgI_4 (or $KI.HgI_2$)	786.41	yel cr, deliq				v s		i al
p341	iodomercurate (II) tri-	Potassium mercury iodide. $KHgI_3$ (or $KI.HgI_2$)	620.47	yel pr, deliq		105		v s		341^{34} al; s KI sol, ac a, eth
p342	iodoplatinate	K_2PtI_6	1034.72	blk, rect	4.96^{25}_{4}			s	s, d	i al
p343	iridium chloride	Potassium hexachloro iridate. K_2IrCl_6	483.12	redsh-blk, cub	3.549	d		1.12^{20}	s	i al
p344	iridium oxalate	$K_3[Ir(C_2O_4)_3].4H_2O$	645.63	orange, tricl cr	2.510^{19}	$-H_2O$, 120	d 160	s	v s	i al, eth
p345	iron chloride, (III)	Nat. erythrosiderite. $2KCl.FeCl_3.H_2O$	329.33	red, orthorhomb.	2.372					
p346	iron chloride (II)	Rinneite. $3KCl.NaCl.FeCl_2$	408.86	rhbdr, 1.589, 1.590	2.3					
p347	iron (III) oxalate	$K_3Fe(C_2O_4)_3.3H_2O$	491.26	emerald grn, monocl, 1.5019, 1.5558, 1.5960	2.133^{20}_{4}	$-3H_2O$, 100	d 230	4.7^{0}	118^{100}	i al
p348	iron sulfate (III)	$KFe(SO_4)_2.12H_2O$	599.32	vlt, cub oct, 1.452	1.83	33		$20^{12.5}$	v s	i al
p349	iron sulfate (III)	Nat. krausite. $K_2SO_4.Fe_2(SO_4)_3.24H_2O$	1006.51	pa yel-grn, monocl, 1.482	1.806	28	d 33			
p350	iron sulfate (II)	$K_2SO_4.FeSO_4.6H_2O$	434.27	grn pr, monocl, 1.476, 1.482, 1.497	2.169	d		s	s	
p351	iron sulfide	$KFeS_2$	159.08	purp, hex	2.563			d		
p352	lactate	$KC_3H_5O_3.xH_2O$						s		s al; i eth
p353	laurate	$KC_{12}H_{23}O_2$	238.42	amorph						4.5^{15} al
p354	laurate, acid (mixt)	$KC_{12}H_{23}O_2.C_{12}H_{24}O_2$	438.74	wh, wax like sol		160				$0.904^{13.5}$ al
p355	lead chloride	Nat. pseudocotunnite. $2KCl.PbCl_2$	427.21	yel		490		s		
p356	magnesium carbonate, hydrogen	$KHCO_3.MgCO_3.4H_2O$	256.50	col, tricl or rhomb	2.98			d		
p357	magnesium chloride	Nat. carnalite. $KCl.MgCl_2.6H_2O$	277.86	col, rhomb, deliq, 1.466, 1.475, 1.494	1.61	265		64.5^{19}	d	d al
p358	magnesium chloride sulfate	Nat. kainite. $KCl.MgSO_4.3H_2O$	248.98	col, monocl	2.131			79.56^{18}		i al, eth
p359	magnesium chromate	$K_2CrO_4.MgCrO_4.2H_2O$	370.53	tricl	2.59					
p360	magnesium phosphate, hexahydrate	$KMgPO_4.6H_2O$	266.48	wh, rhomb cr		$-5H_2O$, 110		d		
p361	magnesium selenate	$K_2SeO_4.MgSeO_4.6H_2O$	496.52	col, monocl, 1.497, 1.499, 1.514	2.3645^{20}_{4}	$-2H_2O$, 33		s	s	
p362	magnesium sulfate	Nat. langbelinite. $K_2SO_4.2MgSO_4$	415.01	tetrah	2.829	927				
p363	magnesium sulfate	Nat. leonite. $K_2SO_4.MgSO_4.4H_2O$	366.70	col, monocl, 1.483, 1.487, 1.490	2.201^{20}			v s		
p364	magnesium sulfate	$K_2SO_4.MgSO_4.6H_2O$	402.73	col, monocl, 1.461, 1.463, 1.476	2.15	d 72		19.26^{0} 25^{20}	59.8^{75}	
p365	malate	$K_2C_4H_4O_5$	210.28	col, viscid mass				s		
p366	manganate	K_2MnO_4	197.14	grn, rhomb		d 190		d	d	s KOH
367	*per*manganate	$KMnO_4$	158.04	purple rhomb, 1.59	2.703	d < 240		6.38^{20}	25^{65}	d al; v s MeOH, acet; s H_2SO_4
p368	manganese chloride(ous)	Chloromanganokalite. $4KCl.MnCl_2$	424.06	trig. 1.50	2.31			s	s	
p369	manganese sulfate(ic)	$KMn(SO_4)_2.12H_2O$	502.35	vlt, cub, (oct)				d		
p370	manganese sulfate(ous)	Manganolongbeinite. $K_2SO_4{\cdot}2MnSO_4$	476.20	rose, tetrah, 1.572	3.02	850				
p371	mercury tartrate(ous)	$KHgC_4O_6$	387.76	wh cr powd				i		i al
p372	methionate	Potassium methane disulfonate. $K_2CH_2(SO_3)_2$	252.36	monocl, β 1.539	2.376			s		
p373	methylsulfate	$2KCH_3SO_4.H_2O$	318.41	wh cr				s		s al

No.	Name	Synonyms and Formulae	Mol. wt.	Crystalline form, properties and index of refraction	Density or spec. gravity	Melting point, °C	Boiling point, °C	Solubility, in grams per 100 cc: Cold water	Hot water	Other solvents
	Potassium									
p374	molybdate	K_2MoO_4	238.14	wh powd or 4-sid pr, deliq	2.91^{18}	919	d 1400	184.6^{25}	v s	i al
p375	*per*molybdate	$K_2O.3MoO_4.3H_2O$	596.06	lt yel cr, monocl.		d 180		sl s	s	v sl s al
p376	*tri*molybdate	$K_2O.3MoO_3.3H_2O$	580.06	wh need		571 (anhydr)		0.22^{15}	s	
p377	molybdenum cyanate	$K_4[Mo(CN)_8].2H_2O$	496.52	yel, rhomb	2.337^{25}_{4} (anhydr)	$-2H_2O$, 105–110		v s	v s	0.0017^{20} abs al; i eth; d HCl, H_2SO_4
p378	myristate, acid (mixt)	$KC_{14}H_{27}O_2.C_{14}H_{28}O_2$	494.85	wh, waxlike sol		153				$0.453^{13.5}$ al
p379	naphthalene−1,5-disulfonate	$K_2C_{10}H_6(SO_3)_2.2H_2O$	400.53	monocl, 1.485, 1.669, 1.697	1.797			s		
p380	nickelsulfate	$K_2SO_4.NiSO_4.6H_2O$	437.13	bl, monocl, 1.484, 1.492, 1.505	2.124	d < 100		7^{0}	60.8^{75}	
p381	nitrate	Saltpeter. KNO_3	101.11	col, rhomb or trig, 1.335, 1.5056, 1.5064	2.109^{16}	tr-trig 129 m.p. 334	d 400	13.3^{0}	247^{100}	i dil al, eth; s liq NH_3, glyc
p382	nitride	K_3N	131.31	grnsh-blk		d		d		
p383	nitrite	KNO_2	85.11	wh-yelsh pr, deliq	1.915	440	d	281^{0}	413^{100}	s hot al; v s liq NH_3
p384	*m*-nitrophenoxide	$KOC_6H_4NO_2.2H_2O$	213.24	flat or need	1.691^{20}	$-H_2O$, 130	d	16.3^{15}		s al
p385	*p*-nitrophenoxide	$KOC_6H_4NO_2.2H_2O$	213.24	yel leaf	1.652^{20}	$-2H_2O$, 130	d	7.5^{15}		sl s al
p386	nitroplatinite	$K_2Pt(NO_2)_4$	457.32	col, monocl			d	3.8^{15}	s	
p387	nitroprusside	$K_2[Fe(NO)(CN)_5].2H_2O$	330.18	red, monocl, hygr.				100^{16}		s al
p388	nitrososulfate	$K_2SO_3(NO)_2$	218.28	col need		d 127, expl		$12^{14.5}$, d		i al
p389	oleate	$KC_{18}H_{33}O_2$	320.56	yelsh or brnsh soft mass or cr, 1.452				25	s	$4.315^{13.5}$ al; 100^{50} al; 3.5^{25} eth
p390	oleate, acid (mixt)	$KC_{18}H_{33}O_2.C_{18}H_{34}O_2$	603.03	wh, wax-like solid		95		s	s	$5.2^{13.5}$ al
p391	osmate	$K_2OsO_4.2H_2O$	368.43	vlt, cub, hygr		$-H_2O$, >100		sl s	s, d	i al, eth
p392	osmiumchloride	K_2OsCl_6	481.12	blk, oct	3.42^{16}	d 600		s	s	i al; s HCl
p393	osmylchloride	$K_2OsO_2Cl_4$	442.21	red, tetr	3.42	d 200 (in H atm)		s	s	
p394	osmyloxalate	$K_2[OsO_2(C_2O_4)_2].2H_2O$	512.47	brn need, tricl		$-H_2O$, 80	d 180	0.75^{15}	3.0^{15}	
p395	oxalate	$K_2C_2O_4.H_2O$	184.24	wh, monocl, 1.440, 1.485 1.550	$2.127^{3.9}$	$-H_2O$, 100			33^{16}	
p396	oxalate, hydrogen	KHC_2O_4	128.11	col, monocl, 1.382, 1.553, 1.573	2.044	d		2.5	16.7^{100}	i al, eth
p396	oxalate, hydrogen. monohydrate	$KHC_2O_4.H_2O$	146.13	rhomb	$2.044^{3.9}$					
p398	oxalate, tetra-	$KHC_2O_4.H_2C_2O_4.2H_2O$	254.20	col, tricl, 1.415, 1.536, 1.560	1.836	d		1.8^{13}		d al
p399	oxaloferrate (II)	$K_2[Fe(C_2O_4)_2].2H_2O$	346.12	gold need		d		s	s	
p400	oxaloferrate (II)	$K[Fe(C_2O_4)_2].2\frac{1}{2}H_2O$	316.03	br cr		d		92^{21}	d	i al
p401	oxaloferrate (III)	$K_3[Fe(C_2O_4)_3].3H_2O$	491.25	grn, monocl		$-3H_2O$, 100	d 230	4.7^{0}	117.7^{100}	i al, NH_3; s acet
p402	oxalatoplatinate	$K_2[Pt(C_2O_4)_2].2H_2O$	485.36	col, monocl pr	3.04^{12}	$-H_2O$, 100		sl s	s	
p403	oxalatouranate (IV)	$K_4[U(C_2O_4)_4].5H_2O$	836.59	yel, monocl	2.563					
p404	oxide, mon-	K_2O	94.20	col, cub, hygr	2.32^{0}	d 350		v s	v s	s al, eth
p405	*per*oxide	K_2O_2	110.20	wh, amorph, deliq		490	d			
p406	oxide, super-	KO_2	71.10	yel, cub leaf	2.14	380	d	v s, d		d al
p407	oxide, tri- (sesqui-)	K_2O_3	126.20	red		430		ev O_2		d dil H_2SO_4
p408	palladium chloride	$K_2(PdCl_4)$	326.42	yel-grnsh-br cr, tetr, 1.710, 1.523	2.67	524		s	s	sl s^{80} al
p409	palladium chloride	$K_2(PdCl_6)$	397.32	lt red, oct	2.738	d 170		v sl s	sl s	s HCl; i al
p410	palladium oxalate	$K_2[Pd(C_2O_4)_2].4H_2O$	432.71	yel need		dec in air, $-4H_2O$, 80		0.833^{27}	$9.98^{82.9}$	
p411	palmitate, acid (mixt)	$KC_{16}H_{31}O_2.C_{16}H_{32}O_2$	550.96	wh, fatty sol		138				0.198^{13} al
p412	1-phenol-2-sulfonate(*o*-)	$KC_6H_4(OH)SO_3.H_2O$	230.29	rhomb, 1.527, 1.568, 1.647	1.87	400		s		s al
p413	1-phenol-4-sulfonate(*p*-)	$KC_6H_4(OH)SO_3$	212.27	rhomb, 1.571, 1.608, 1.694	1.87	>260				
p414	phenylsulfate	$KC_6H_5SO_4$	212.27	rhomb leaf		d 150–160	d	14^{15}		v sl s al
p415	*meta*phosphate, hexa-	$(KPO_3)_6$	708.44	col mass, hygr	2.107	810	1320	s	s	i al
p416	*meta*phosphate, tetra-	$(KPO_3)_4.2H_2O$	508.33	col cr		$-2H_2O$, 100		100^{15}	s	
p417	*ortho*phosphate	K_3PO_4	212.28	col, rhomb, deliq	2.564^{17}	1340		90^{20}	s	i al
p418	*ortho*phosphate, di-H	KH_2PO_4	136.09	col, tetr, deliq, 1.510, 1.4864	2.338	252.6		33^{25}	83.5^{90}	i al
p419	*ortho*phosphate, mono-H	K_2HPO_4	174.18	wh, amorph, deliq		d		167^{20}	v s	v s al
p420	*pyro*phosphate	$K_4P_2O_7.3H_2O$	384.40	col, deliq	2.33	$-2H_2O$, 180	$-3H_2O$, 300	s	v s	i al
p421	*sub*phosphate	$K_2PO_3.4H_2O$	229.24	col, rhomb		40	$-4H_2O$, 150	v s	v s	

No.	Name	Synonyms and Formulae	Mol. wt.	Crystalline form, properties and index of refraction	Density or spec. gravity	Melting point, °C	Boiling point, °C	Solubility, in grams per 100 cc: Cold water	Hot water	Other solvents
	Potassium									
422	*hypo*phosphite	KH_2PO_2	104.09	wh, hex, deliq		d		200^{25}	330	v sl s abs al, NH_3; i eth; 11.1^{25} chl
423	*ortho*phosphite, di-H	KH_2PO_3	120.09	wh cr, deliq		d		220^{20}	v s	i al
424	phthalate, hydrogen	$KHC_8H_4O_4$	204.23	col, rhomb	1.636			10^{25}	33^{100}	
425	picrate	$KC_6H_2N_3O_7$	267.20	yel-redsh or grnsh, rhomb, 1.527, 1.903, 1.952	1.852	expl 310		0.5^{15}	25^{100}	0.184^{25} al
426	piperate	$KC_{12}H_9O_4$	256.31	lt yel cr powd				sl s	v s	
427	platinate, hydroxo-	$K_2Pt(OH)_6$	375.34	yel, rhomb		d 160		s		i al
428	platinorhodanide	$K_2[Pt(CNS)_6].2H_2O$	657.82	red, rhomb				s	8^{60}	s h al
429	platinum iodide, hexa-	K_2PtI_6	1034.72	blk, cub	4.963^{25}_{4}			s		sl s al
430	plumbate, hydroxo-	$K_2Pb(OH)_6$	387.44	col, rhomb				d	d	s KOH
431	praseodymium sulfate	$3K_2SO_4.Pr_2(SO_4)_3.H_2O$	1110.81	cr	3.275^{16}			sl s		s HCl, HNO_3
432	propionate	$KC_3H_5O_2.H_2O$	130.19	wh cr, hygr, or wh leaf, deliq		$-H_2O$, 120		207^{16}	359	22.2^{13} 95 % al
433	propyl sulfate	$KC_3H_7SO_4$	178.25	wh cr powd				v s		
434	*per*rhenate	$KReO_4$	289.30	wh, tetr, 1.643	4.887	550	1360–1370	1.21^{20}	14.0^{100}	v sl s al
435	rhenium (IV) chloride	$K_2[ReCl_6]$	477.12	yel-grn cr, oct	3.34	d		s	s	s a; i conc H_2SO_4; d h H_2SO_4
437	rhenium (V) oxychloride	$K_2[ReOCl_5]$	457.67	yel-grn cr, rhomb or monocl, 1.52				s		sl s HCl; s H_2SO_4; i al, eth
438	rhenium oxycyanide	$K_3[ReO_2(CN)_4]$	439.58	red cr, monocl	2.70^{25}_{4}	d 300–400, vac		s	s	v sl s al; i alk
439	rhodium cyanide	$K_3[Rh(CN)_6]$	376.32	pa yel, monocl, 1.5498, 1.5513, 1.5634				v s	v s	
440	rhodium oxalate	$K_3[Rh(C_2O_4)_3].4\frac{1}{2}H_2O$	565.34	col, tricl	2.171^{20}_{4}	$-4\frac{1}{2}H_2O$, 190		v s	v s	
441	rhodium sulfate	$KRh(SO_4)_2.12H_2O$	646.38	yel, cub	2.23			s		
442	ruthenate	$K_2RuO_4.H_2O$	261.29	blk, tetr		$-H_2O$, 200	d 400 vac	v s	d	d a, al
443	*per*ruthenate	$KRuO_4$	204.17	blk, tetr		d 44		sl s	s, d	
444	*d*-saccharate, acid	$KHC_6H_8O_8$	248.24	rhomb need				1.1^{6}	s	
445	salicylate	$KC_7H_5O_3$	176.22	wh powd				s		s al
446	santoninate	$KC_{15}H_{19}O_4$	302.42	wh cr powd, deliq				s		s al
447	selenate	K_2SeO_4	221.16	col, rhomb, hygr, 1.535, 1.539, 1.545	3.066			110.5^{0}	122.2^{100}	
448	selenide	K_2Se	157.16	wh cub, reddens in air, hygr	2.851^{15}			s d	s	
449	selenite	K_2SeO_3	205.16	wh, deliq		d 875		s		sl s al
450	selenocyanate	$KSeCN$	144.08	need, deliq	2.347	d 100		s	s	d a; s al
451	selenocyanoplatinate	$K_2Pt(SeCN)_6$	903.16	rhomb	$3.378^{12.5}$	d 80				
452	selenothionate	$K_2SeS_2O_6$	317.29	col, monocl pr		d 250		s, d		
453	*di*silicate	$K_2Si_2O_5$	214.37	col, rhomb, 1.502, 1.513	2.456^{25}_{4}	1015±10				
454	*meta*silicate	K_2SiO_3	154.29	col, rhomb (?), 1.520, 1.528		976		s	s	i al
455	*tetra*silicate	$K_2Si_4O_9.H_2O$	352.56	wh, rhomb, α 1.495, β 1.535	2.417	d 400		s	s	i al
456	*di*silicate, hydrogen	$KHSi_2O_5$	176.28	wh, rhomb, 1.480, 1.530	2.417^{15}_{4}	515				d HCl
457	silicotungstate	$K_4SiW_{12}O_{40}.18H_2O$	3354.95	col, hex		$-17H_2O$, 100		33.3^{20}	v s	v s acet; s MeOH; sl s al; eth, bz
458	silver carbonate	$KAgCO_3$	206.98	rect pl	3.769	d		d	d	
459	silver nitrate	$KNO_3.AgNO_3$	270.98	monocl	3.219	125		v s	v s	
460	sodium antimony tartrate	$KNaSbC_4H_2O_7$	346.97	wh, scales or powd				s		
461	sodium carbonate	$KNaCO_3.6H_2O$	230.19	monocl, hygr, effl	$1.61–1.63^{14}$	$-6H_2O$, 100		185.2^{15}		
462	sodium ferricyanide	$K_2Na[Fe(CN)_6]$	313.15	or-red, monocl		d		50^{25}	80^{80}	
463	sodium nitrocobaltate (III)	$K_2Na[Co(NO_2)_6].H_2O$	454.18	yel cr	1.633	135		0.07^{25}		i al
464	sodium sulfate	$3K_2SO_4.Na_2SO_4$	664.84	col, rhbdr	2.7				s	
465	sodium tartrate	Rochelle salt, seignette salt. $KNaC_4H_4O_6.4H_2O$	282.23	col, rhomb, 1.492, 1.493, 1.496	1.790	70–80	$-4H_2O$, 215	26^{0}	66^{26}	v sl s al
466	sorbate	$KC_6H_7O_2$	150.22	col cr	1.363^{25}_{20}	d 270		58.5^{25}		sl s MeOH

No.	Name	Synonyms and Formulae	Mol. wt.	Crystalline form, properties and index of refraction	Density or spec. gravity	Melting point, °C	Boiling point, °C	Solubility, in grams per 100 cc: Cold water	Hot water	Other solvents
	Potassium									
p467	stannate, hydroxo-	$K_2Sn(OH)_6$	298.94	col, trig	3.197			85^{10}	110.5^{20}	sl s KOH; i al, acet
p468	stearate	$KC_{18}H_{35}O_2$	322.58	wh cr powd					s	$0.145^{13.5}$ al; i eth, chl, CS_2
p469	stearate, acid (mixt)	$KC_{18}H_{35}O_2.C_{18}H_{36}O_2$	607.07	wh powd		153		s	s	$0.091^{13.5}$ al
p470	strontium chromium oxalate(ic)	$KSrCr(C_2O_4).6H_2O$	550.87	grnsh blk cr	2.155^{13}					
p471	styphnate	$KC_6H_2N_3O_8.H_2O$	301.22	yel, monocl pr		$-H_2O$, 120	expl	1.54^{20}		v sl s al
p472	succinate	$K_2C_4H_4O_4.3H_2O$	248.32	rhomb, hygr	1.564			s		
p473	succinate, hydrogen	$KHC_4H_4O_4$	156.18	monocl	1.767	d 242				
p474	succinate, hydrogen, dihydrate	$KHC_4H_4O_4.2H_2O$	192.21	rhomb, 1.417, 1.530, 1.533	1.616			s		s al
p475	succinate, hydrogen	$KHC_4H_4O_4.C_4H_6O_4$	274.27	monocl	1.56	162				
p476	sulfate	Nat. arcanite. K_2SO_4	174.27	col, rhomb or hex, 1.494, 1.495, 1.497	2.662	tr 588 m. p. 1069	1689	12^{25}	24.1^{100}	i al, acet, CS_2
p477	*peroxy*disulfate	$K_2S_2O_8$	270.33	col, tricl, 1.461, 1.467, 1.566	2.477	d<100		1.75^0	5.3^{20}	i al
p478	*pyro*sulfate	$K_2S_2O_7$	254.33	col, need	2.512^{25}_4	>300	d	s	d	
p479	sulfate, hydrogen	Nat. mercallite, misenite. $KHSO_4$	136.17	col, rhomb, deliq 1.480	2.322	214	d	36.3^0	121.6^{100}	i al, acet
p480	sulfide, di-	K_2S_2	142.33	red-yel cr		470		s	d	s al
p481	sulfide, hydro-	KHS	72.17	yel, rhomb deliq	1.68–1.70	455		d	d	s al
p482	sulfide, mono-	K_2S	110.27	yel-br, cub, deliq	1.805^{14}	840		s	v s	s al, glyc; i eth
p483	sulfide, mono-, pentahydrate	$K_2S.5H_2O$	200.34	col, rhomb		60	$-3H_2O$, 150	s		s al, glyc; i eth
p484	sulfide, penta-	K_2S_5	238.52	or cr, hygr		206	d 300	v s	v s	sl s al
p485	sulfide, tetra-	K_2S_4	206.46	red-brn cr		145	d 850	s		s al
p486	sulfide, tetra dihydrate	$K_2S_4.2H_2O$	242.49	yel				s	s	sl s al
p487	sulfide, tri-	K_2S_3	174.40	br-yel cr		252		s	d	s al
p488	sulfide, di-, trihydrate	$K_2S_2.3H_2O$	196.38	yel				v s	v s	s al
p489	sulfite	$K_2SO_3.2H_2O$	194.30	wh-yelsh, hex		d		100	<100	sl s al; i NH_3; d dil a
p490	*pyro*sulfite	Potassium *meta*bisulphite. $K_2S_2O_5$	222.33	col, monocl pl	2.34	d 190		sl s		sl s al; i eth
p491	sulfite, hydrogen	$KHSO_3$	120.17	col cr		d 190		s	s	i al
p492	*d*-tartrate	$K_2C_4H_4O_6.\frac{1}{2}H_2O$	235.28	col, monocl, β 1.526	1.98^{20}_4	$-H_2O$, 155	d 200–220	150^{14}	278^{100}	sl s al
p493	*dl*-tartrate	$K_2C_4H_4O_6.2H_2O$	262.31	col, monocl	1.984	$-2H_2O$, 100		100^{25}		
p494	*d*-tartrate, hydrogen	$KHC_4H_4O_6$	188.18	col, rhomb, 1.511, 1.550, 1.590	1.984^{18}			0.37	6.1^{100}	s a, alk; i al, ac a
p495	*dl*-tartrate, hydrogen	$KHC_4H_4O_6$	188.18	col, monocl	1.954			0.42^{15}	7.0^{100}	i al; s min a
p496	*meta*tellurate	K_2TeO_4	269.80	soft glutinous mass		d 200		d		
p497	*ortho*tellurate	$K_2H_4TeO_6.3H_2O$	359.88	col, rhomb, deliq		$-H_2O$	−O, 300	sl s	s	i al; sl s KOH
p498	telluride	K_2Te	205.80	col, cub, hygr	2.51			s, d	s	
p499	tellurite	K_2TeO_3	253.80	wh cr, deliq		d 460–470		v s	v s	s h K_2CO_3, KOH
p500	tellurium chloride	K_2TeCl_6	418.52	yel, oct, hygr	2.645			s d,		s, d al; s dil HCl
p501	thioantimonate	$2K_3SbS_4.4\frac{1}{2}H_2O$	815.68	yel cr, deliq				300^0	400^{80}	i al
p502	thioarsenate	K_3AsS_4	320.48	wh cr, deliq			d	v s		i al
p503	thioarsenite	K_3AsS_3	288.42			d		s		i al
p504	thiocarbonate, tri-	K_2CS_3	186.41	yel-red-brn cr, deliq		d		v s	s	s NH_3; sl s al; i eth
p505	thiocyanate	KNCS	97.18	col, rhomb pr, deliq	1.886^{14}	173.2	d 500	177.2^0	217^{20}	s al; 20.75^{22} acet; 0.18^{13} amyl al; v s liq NH_3
p506	*di*thionate	$K_2S_2O_6$	238.32	col, trig, 1.455, 1.515	2.278	d		6	66^{100}	i al
p507	*penta*thionate	$K_2S_5O_6.1\frac{1}{2}H_2O$	361.55	col, rhomb, —1.63	2.112	d		s	d	i al
p508	*tetra*thionate	$K_2S_4O_6$	302.46	col, monocl, 1.6057	2.296			v s		i al
p509	*tri*thionate	$K_2S_3O_6$	270.39	col, rhomb, 1.475, 1.480, 1.487	2.304	d 30–40		s	d	i al
p510	thioplatinate	$K_2Pt_4S_6$	1050.95	bl-gray cr	6.44^{15}	d, ign		i		d HCl
p511	thiosulfate, pentahydrate	$3K_2S_2O_3.5H_2O$	661.07	col, rhomb		d		$150.2^{17.2}$		
p512	*meta*thiostannate	$K_2SnS_3.3H_2O$	347.13	dk brn oil	1.847^{18}	$-3H_2O$, 100		s		i al
p513	thiosulfate	$K_2S_2O_3\frac{1}{3}.H_2O$	196.34	col, monocl, deliq	2.590 anhydr 2.23	$-H_2O$, 200	d	96.1^0	312^{90}	i al
p514	tungstate	$K_2WO_4.2H_2O$	362.08	col, monocl, deliq	3.113	tr 388, 921		51.5	151.5	d a; i al
p515	*meta*tungstate	$K_6[H_2W_{12}O_{40}].18H_2O$	3407.08	hex		*ca* 930		s	v s	d a
p516	*meta*uranate	K_2UO_4	380.23	or-yel, rhomb					i	v s a
p517	uranyl acetate	$KUO_2(C_2H_3O_2)_3.H_2O$	504.28	tetr	3.296^{15} ($\frac{1}{2}H_2O$)	$-H_2O$, 275		s		
p518	uranyl carbonate	$2K_2CO_3.UO_2CO_3$	606.46	yel, hex		$-CO_2$		7.4^{15}	d	s K_2CO_3 sol; i a

No.	Name	Synonyms and Formulae	Mol. wt.	Crystalline form, properties and index of refraction	Density or spec. gravity	Melting point, °C	Boiling point, °C	Solubility, in grams per 100 cc		
								Cold water	Hot water	Other solvents
	Potassium									
p519	uranyl sulfate	$K_2SO_4.UO_2SO_4.2H_2O$	576.39	yel, monocl	$3.363^{19.1}$	$-2H_2O$, 120		s		
p520	urate, acid	$KHC_5H_2N_4O_3$	206.21	wh powd				sl s		
p521	*meta*vanadate	KVO_3	138.04	col cr				sl s	s	sl s KOH; i al
p522	vanadium sulfate	Potassium vanadium alum. $KV(SO_4)_2.12H_2O$	498.35	violet cubic	1.783^{20}_4	20	$-12H_2O$, 230	1984^{10}		
p523	*ethyl*xanthate	KC_2H_5OCSS	160.30	wh to pa yel cr or cr powd	$1.558^{21.5}$	d>200		v s	d	s al; i eth
p524	**Praseodymium** (α form)	Pr	140.907	pa yel, met, hex up to 798	6.782	935, tr to β form 798	3127	d		s a
p525	(β form)	Pr	140.907	cub	6.64	935	3127	d		s a
p526	acetate	$Pr(C_2H_3O_2)_3.3H_2O$	372.09	grn need				v s		
p527	acetylacetonate	$Pr(C_5H_7O_2)_3$	438.24	cr ppt		146				s CS_2
p528	bromate	$Pr(BrO_3)_3.9H_2O$	686.77	grn, hex		56.5	$-7H_2O$, 170	196^{25}		
p529	bromide	$PrBr_3$	380.63	grn cr powd		691	1547	d, sl s		
p530	carbide	PrC_2	164.93	yel cr	5.10	d		d	d	s dil a
p531	carbonate	$Pr_2(CO_3)_3.8H_2O$	605.96	grn silky pl		$-6H_2O$, 100		i		s a
p532	chloride	$PrCl_3$	247.27	bl grn need	4.02^{25}	786	1700	103.9^{13}	∞^{100}	v s al; 2.4 pyr; i chl, eth
p533	chloride, hepta-hydrate	$PrCl_3.7H_2O$	373.37	grn, tricl	2.25^{17}	115		334^{13}	∞^{100}	s al, HCl
p534	hexantipyrine *per*chlorate	$[Pr(C_{11}H_{12}N_2O)_6].(ClO_4)_3$	1568.65	grn, hex leaf		d 286–291				
p535	iodide	PrI_3	521.62	gr cr, hygr		737		v s		
p536	molybdate	$Pr_2(MoO_4)_3$	761.63	grass-green tetr	4.84	1030				
p537	oxalate	$Pr_2(C_2O_4)_3.10H_2O$	726.03	lt grn cr				i		s a
p538	oxide, di-	PrO_2	172.91	br-bl powd	6.82	>350 tr to Pr_6O_{11}				
p539	oxide, sesqui-	Praseodymia. Pr_2O_3	329.81	yel-grn, amorph	7.07	d		0.000020^{29}		s a
p540	selenate	$Pr_2(SeO_4)_3$	710.69		4.30^{15}			36^{0}	3^{92}	
p541	sulfate	$Pr_2(SO_4)_3$	570.00	lt grn powd	3.72^{16}			23.7^{0} 17.7^{20}	1.02^{95}	
p542	sulfate, octahydrate	$Pr_2(SO_4)_3.8H_2O$	714.12	grn, monocl, 1.540, 1.549, 1.561	$2.827^{15.2}$	–		17.4^{20}	sl s	
p543	sulfate, pentahydrate	$Pr_2(SO_4)_3.5H_2O$	660.08	monocl pr	3.176^{16}				1.85^{35}	
p544	sulfide	Pr_2S_3	378.01	br powd	5.042^{11}	d		i	d	s dil a
p545	**Protactinium**	Pa	231.10	gray met, tetrag	15.37	~1230(?)				
p546	chloride	$PaCl_4$	372.91	yel-grn, tetrag		subl 400 in vac		s		s dil HCl
p547	fluoride	PaF_4	307.09	monocl				i		
p548	oxide, di-	PaO_2	263.10	blk, cub						
p549	oxide, pent-	Pa_2O_5	542.20	wh, cub						s dil HF
r1	**Radium**	Ra	226.00	silver wh met	5(?)	700	<1737	d, ev H_2O		d a
r2	bromide	$RaBr_2$	385.82	col-yelsh, monocl.	5.79	728	subl 900	s	s	s al
r3	bromide dihydrate	$RaBr_2.2H_2O$	421.85	wh, monocl		$-2H_2O$, 100		s	s	
r4	carbonate	$RaCO_3$	286.01	wh, or sl brnsh, monocl				i		d a
r5	chloride	$RaCl_2$	296.91	col-yelsh, monocl.	4.91	1000		s	s	s al
r6	chloride, dihydrate	$RaCl_2.2H_2O$	332.94	wh, monocl, discol		$-2H_2O$, 100		s	s	s HCl
r7	iodate	$Ra(IO_3)_2$	575.81					0.0175^{0}	0.170^{100}	
r8	nitrate	$Ra(NO_3)_2$	350.01	cr				13.9^{20}		
r9	sulfate	$RaSO_4$	322.06	col, rhomb				0.000002^{25}	0.000005^{45}	i a
r10	**Radon**	Niton, Radium emanation. Rn	222.00	col gas, opaque cr	gas 9.73 g/l liq 4.4^{-62} sol 4.0	−71	−61.8	51.0^{0} cm³ 22.4^{25} cm³	13.0^{50} cm³	sl s al, org liqu
r11	**Rhenium**	Re	186.2	met lust, hex	20.53	3180	5627	i	i	s dil HNO_3, H_2O_2; sl s H_2SO_4; i HCl
r12	bromide	$ReBr_3$	425.93	grn-blk cr		subl 500, vac				s dil H_2SO_4, HBr, liq NH_3
r13	carbonyl, penta-	$[Re(CO)_5]_2$	652.51	col, cub cr		d 250				v sl s org solv
r14	chloride, penta-	$ReCl_5$	363.47	dk grn to blk	4.9	d	d	d	d	s HCl, alk
r15	chloride, tetra-	$ReCl_4$	328.01	blk (exist. ?)			500	s, d	s, d	s HCl
r16	chloride, tri-	$ReCl_3$	292.56	dk red, hex			>550	s	s	s a, alk, liq NH_3, al; sl s eth
r17	fluoride	ReF_4	262.19	dk grn	5.383^{25}	124.5	d 500	d		s a
r18	fluoride, hexa-	ReF_6	300.19	pa yel, v hygr	liq 6.1573, sld $3.616^{18.8}$	18.8	47.6	s, d	s, d	d HNO_3, H_2SO_4
r19	iodide pentacarbonyl	$ReI.5CO$	453.16	yel, rhombd		200	d 400, subl vac 90	i		s bz
r20	oxide, di-	ReO_2	218.20	blk	11.4^{25}_4	d 1000		i	i	s conc HCl, H_2O_2
r21	oxide, hept-	Re_2O_7	484.40	yel pl or hex or powd, hygr	6.103	*ca* 297	subl 250	v s	v s	v s al; s alk, a
r22	oxide, per-	Re_2O_8(?)	500.40	wh	8.4	145		v s	v s	s alk; sl s eth

PHYSICAL CONSTANTS OF INORGANIC COMPOUNDS (Continued)

No.	Name	Synonyms and Formulae	Mol. wt.	Crystalline form, properties and index of refraction	Density or spec. gravity	Melting point, °C	Boiling point, °C	Solubility, in grams per 100 cc: Cold water	Hot water	Other solvents
	Rhenium									
r23	oxide, sesqui-	Dirhenium trioxide $Re_2O_3.xH_2O$		unstable				d, ev H_2		
r24	oxide, tri-	ReO_3	234.20	red, blue, cub	6.9–7.4		d 400	i	i	s H_2O_2, HNO_3
r25	oxybromide, tri-	ReO_3Br	314.11	wh		39.5	163			
r26	oxychloride, tri-	ReO_3Cl	269.65	col liq	3.867^{20}_{4}	4.5	131^{760}	d	d	s CCl_4
r27	oxy*tetra*chloride	$ReOCl_4$	344.01	or need		29.3	223.00	d	d	
r28	oxy*tetra*fluoride	$ReOF_4$	278.19	wh	lq 3.717 sol 4.032	39.7	62.7			
r29	oxy*di*fluoride, di-	ReO_2F_2	256.20	col		156				
r30	sulfide, di-	ReS_2	250.33	blk tr leaf	7.506^{20}_{4}		d	i	i	s HNO_3; i al, alk, HCl
r31	sulfide, hepta-	Re_2S_7	596.85	blk powd	4.866		d	i	i	s HNO_3, H_2O_2, alk; i HCl
r32	**Rhodium**	Rh	102.905	gray-wh, cub	12.4	1966±3	3727±100	i	i	s H_2SO_4+HCl, conc H_2SO_4; sl s a, aq reg
r33	amminechloride, hexa-	$[Rh(NH_3)_6]Cl_3$	311.45	rhomb pl	2.008^{25}_{4}	$-NH_3$ 210, d		12.5^{8}	s	
r34	carbonylchloride, basic	$RhCl_2.RhO.3CO$	376.75	ruby red need		subl 125.5		sl s	d	s CCl_4, ac a, bz
r35	chloride, tri-,	$RhCl_3$	209.26	br-red powd, deliq		d 450–500	subl 800	i	i	i a, aq reg
r36	chloride, tri-	$RhCl_3.xH_2O$		dk red, deliq		d 100		v s		s al, HCl; i eth
r37	fluoride, tri-	RhF_3	159.90	red, rhomb	5.38	>600 subl		i	i	i a, alk
r38	iodide, tri-	RhI_3	483.62	blk				i		
r39	nitrate	$Rh(NO_3)_3.2H_2O$	324.93	red, deliq				s	s	i al
r40	oxide, di-	RhO_2	134.90	br				i	i	i a, alk
r41	oxide, di-, dihydrate	$RhO_2.2H_2O$	170.93	olive grn		d		i		s HCl, acet a, alk
r42	oxide, sesqui-	Rh_2O_3	253.81	gray cr or amorph	8.20	d 1100–1150		i	i	i a, aq reg, KOH
r43	oxide, sesqui-, pentahydrate	$Rh_2O_3.5H_2O$	343.88	yel ppt		d		i	s	s a
r44	sulfate	$Rh_2(SO_4)_3.4H_2O$	566.05	red		d		s	s	
r45	sulfate	$Rh_2(SO_4)_3.12H_2O$	710.18	lt yel cr				v s	d	i al
r46	sulfate	$Rh_2(SO_4)_3.15H_2O$	764.22	pa yel cr		d		v s	d	i al, eth
r47	sulfide, hydro-	$Rh(HS)_3$	202.12	blk		d		i	d	s aq reg, aq Br; i Na_2S
r48	sulfide, mono-	RhS	134.97	gray-blk cr		d		i	i	i a, aq reg
r49	sulfide, sesqui-	Rh_2S_3	302.00	blk	6.40^{25}_{4}	d		i'	i	i a, aq reg, aq Br
r50	sulfite	$Rh_2(SO_3)_3.6H_2O$	554.09	yel cr		d		s		i al
r51	**Rubidium**	Rb	85.47	soft, silver wh met	1.532 liq: $1.475^{38.5}$	38.89	688	d	d	d al; s a
r52	acetate	$RbC_2H_3O_2$	144.52	col, nacreous leaf, hygr		246		$86^{44.7}$	$89.3^{99.4}$	
r53	aluminium sulfate	$RbAl(SO_4)_2.12H_2O$	520.76	col, cub, oct, 1.457, 1.45232, 1.46618	1.867^{0}	99		2.59^{20}	43.25^{80}	
r54	azide	RbN_3	127.49	col need or plates	2.7876	d *ca* 310		107.1^{16}		0.182^{16} al; i eth
r55	borofluoride	$RbBF_4$	172.27	very sm rhomb cr, 1.333	2.820^{20}	590	d 500	0.6^{17}	10^{100}	
r56	borohydride	$RbBH_4$	100.31	white, cubic	1.920	1.487		v s		sl s al, i ether, bz
r57	bromate	$RbBrO_3$	213.38	cub	3.68	430		2.93^{25}	5.08^{40}	
r58	bromide	RbBr	165.38	col, cub, 1.5530	3.35 liq: 2.79^{730}	682	1340	98^{5}	$205.2^{113.5}$	i al, sl s acet
r59	bromide, tri-	$RbBr_3$	325.20	red, rhomb		d 140				
r60	bromochloroiodide	RbIBrCl	327.74	rhomb		d 200				s al, d eth
r61	bromoiodide, di-	$RbIBr_2$	372.13	red, rhomb	3.84	225	d 265	s		s al, d eth
r62	carbonate	Rb_2CO_3	230.95	col cr, deliq		837	d 740	450^{20}	s	0.7 abs al
r63	carbonate, acid	$RbHCO_3$	146.49	wh, rhomb		d 175		53.73^{20}		2.0 al
r64	chlorate	$RbClO_3$	168.92	trim	3.19			5.0^{19}	62.8^{100}	
r65	perchlorate	$RbClO_4$	184.92	rhomb, 1.4701	2.80	fus	d	0.5^{0}	18^{100}	0.009^{25} al, 0.06^{25} MeOH
r66	chloride	RbCl	120.92	col, cub, 1.493^{25}	2.80: liq: 2.088^{750}	715	1390	77^{0}	138.9^{100}	0.08^{25} al, 1.41^{25} MeOH; v sl s NH_3
r67	chlorobromide, di-	$RbBrCl_2$	236.29	rhomb		d 110				
r68	chloro*di*bromide	$RbBr_2Cl$	280.74	rhomb		76				
r69	chloroiodide, di-	$RbICl_2$	283.28	dk orange, rhomb		180–200	d 265	v s	v s	
r70	chloroplatinate	Rb_2PtCl_6	578.75	yel, cub	$3.94^{17.5}$	d		0.184^{0}	0.634^{100}	i al
r71	chloroplatinate, hexa-	$Rb_2[PtCl_6]$	578.75	yel, cub	$3.94^{17.5}$	d		0.014^{0}	0.33^{100}	i al
r72	chromate	Rb_2CrO_4	286.93	yel, rhomb, ~1.71	3.518			62^{0}	95.7^{60}	
r73	*di*chromate	$Rb_2Cr_2O_7$	386.93	tricl or monocl, >1.95, 1.70	3.02 monocl 3.125 tricl			tricl: 4.96^{18} monocl: 5.42^{18}	27.3^{60} 28.1^{60}	
r74	chromiumsulfate	$RbCr(SO_4)_2.12H_2O$	545.77	vlt cub, 1.482	1.946	107		43.4^{25}		

No.	Name	Synonyms and Formulae	Mol. wt.	Crystalline form, properties and index of refraction	Density or spec. gravity	Melting point, °C	Boiling point, °C	Solubility, in grams per 100 cc Cold water	Hot water	Other solvents
	Rubidium									
r75	cobalt (II) sulfate	$RbSO_4.CoSO_4.6H_2O$	449.22	rubyred, monocl, 1.486, 1.491, 1.501	2.56^{15}			9.3^{25}	s	
r76	coppersulfate	$Rb_2SO_4.CuSO_4.6H_2O$	534.70	monocl, 1.489, 1.491, 1.504	2.57			10.28^{25}		
r77	cyanide	$RbCN$	111.49	col cr powder	2.32			s	s	i al, eth
r78	galliumsulfate	$RbGa(SO_4)_2.12H_2O$	563.50	col cr, 1.46579	1.962			s		
r79	fluoride	RbF	104.47	col, cubic, 1.398	3.557	775	1410	130.6^{18}		s dil HF; i al, eth, NH_3
r80	fluorogermanate	Rb_2GeF_6	357.52	wh cr				sl s	v s	
r81	fluosilicate	Rb_2SiF_6	313.02	cub, oct	3.332^{20}			0.16^{20}	1.35^{100}	s a, i al
r82	fluosulfonate	$RbFSO_3$	184.53	need		304				
r83	iodate	$RbIO_3$	260.37	monocl or cub	$4.33^{19.5}$	d		2.1^{23}		v s HCl
r84	*meta*periodate	$RbIO_4$	276.37	tetr	3.918^{16}			0.65^{13}		
r85	iodide	RbI	212.37	col, cub, 1.6474	3.55; liq: 2.87^{825}	642	1300	152^{17}	163^{25}	s liq NH_3; 0.674^{25} acet
r86	iodide, tri-	RbI_3	466.18	blk, rhomb	4.03^{22}	190		s		
r87	iodide, cmpd. with SO_2	$RbI.4SO_2$	468.63	lemon yel		13.5				
r88	iron (II) selenate	$Rb_2SeO_4.FeSeO_4.6H_2O$	620.79	blue-grn monocl, prism, 1.513, 1.520, 1.532	2.819					
r89	iron (III) selenate	$RbFe(SeO_4)_2.12H_2O$	643.42	cub, 1.507^{18}	2.31^{11}	45	$-12H_2O$, 100			
r90	iron (II) sulfate	$Rb_2SO_4.FeSO_4.6H_2O$	527.00	gr monocl, prism, 1.4815, 1.4874, 1.4977	2.516	d 60		24.2^{25} (anhydr)		
r91	iron (III) sulfate	$RbFe(SO_4)_2.12H_2O$	549.62	cub, 1.4823	1.91–1.95	48–53	$4.55^{6.6}$	52.6^{90}		
r92	hydride	RbH	86.48	col need	2.60	d 300		d	d	d a
r93	hydroxide	$RbOH$	102.48	gray-wh, deliq	3.203^{11}	301 ± 0.9		180^{15}	v s	s al
r94	neodymium nitrate	$2(?)RbNO_3.Nd(NO_3)_3.4H_2O$	697.26	redsh-vlt pl	2.56	47	$-4H_2O$, 60			
r95	nitrate	$RbNO_3$	147.47	col, hex cub, rhomb or tricl, hygr, 1.51, 1.52, 1.524	3.11; liq: 2.395^{100}	tr cub 161.4 m.p. 310	tricl rhom 219	44.28^{16}	452^{100}	sl s acet; v s HNO_3
r96	nitrate, hydrogen	$RbNO_3.HNO_3$	210.49	tetr		62				
r97	nitrate, hydrogen	$RbNO_3.2HNO_3$	273.50	col need		45				
r98	magnesium sulfate	$Rb_2SO_4.MgSO_4.6H_2O$	495.47	col, monocl, 1.467, 1.469, 1.478	2.386_4^{20}			20.2^{25} (anhydr)		
r99	permanganate	$RbMnO_4$	204.41	cr	$3.235^{10.4}$	d 295		0.5^0	4.7^{60}	
r100	oxide, mon-	Rb_2O	186.94	col-yel, cub	3.72	d 400		s d	s d	s liq NH_3
r101	oxide, per-	Rb_2O_2	202.94	yel, cub	3.65^0	570	d 1011^{760} mm	dec to RbOH $+H_2$		
r102	oxide, super	RbO_2, unstable	117.47	yel plates	3.80	432	d 1157^1 atm			
r103	oxide (tetr-)	Rb_2O_4	234.94	dk orange cr, deliq		dec 500 vac				
r104	oxide, tri- (sesqui)	Rb_2O_3 (or Rb_4O_6)	218.94	blk cub	3.53^0	489		s d		
r105	praseodymium nitrate	$2RbNO_3.Pr(NO_3)_3.4H_2O$	693.93	grnsh, monocl, need hygr	2.50	63.5	$-4H_2O$, 60			
r106	selenate	Rb_2SeO_4	313.90	col, rhomb, 1.5515, 1.5537, 1.5582	3.90			159^{12}		
r107	silicofluoride	Rb_2SiF_6	313.02	col, oct or hex	3.3383_4^{20}			sl s	s	s a, i al
r108	sulfate	Rb_2SO_4	267.00	col, rhomb hex, 1.513, 1.513, 1.514	3.613_4^{20} liq 2.53^{1100}	1060 trig 653	*ca* 1700	42.4^{10}	81.8^{100}	i acet; sl s NH_3
r109	sulfate, hydrogen	$RbHSO_4$	182.54	rhomb, 1.473	2.892^{16}	< red heat				
r110	sulfide, di-	Rb_2S_2	235.07	dk red		420	volat >850			
r111	sulfide, hexa-	Rb_2S_6	363.32	brn-red		201				
r112	sulfide, mono-	Rb_2S	203.00	wh-pale yel	2.912	530 d vac		v s	v s	
r113	sulfide, mono-tetrahydrate	$Rb_2S.4H_2O$	275.07	cr, deliq				v s	v s	
r114	sulfide, penta-	Rb_2S_5	331.26	red, rhomb, deliq	2.618^{15}	225		d		s 70 % al; i eth, chl
r115	sulfide, tri-	Rb_2S_3	267.13	redsh yel		213				
r116	tartrate, d & l	$Rb_2C_4H_4O_6$	319.01	trig	2.658_4^{20}			200^{25}		i toluol
r117	*dl*-tartrate, hydrogen	$RbHC_4H_4O_6$	234.55	trim pr	2.282	201 d		1.18^{25}	11.7^{100}	
r118	vanadium sulfate	Rubidium vanadium alum. $RbV(SO_4)_2.12H_2O$	544.72	yellow, cubic, 1.4689	1.915_4^{20}	64	230 $-12H_2O$, 300 dec	2.56^{10}		

No.	Name	Synonyms and Formulae	Mol. wt.	Crystalline form, properties and index of refraction	Density or spec. gravity	Melting point, °C	Boiling point, °C	Solubility, in grams per 100 cc		
								Cold water	Hot water	Other solvents
	Ruthenium									
r119	**Ruthenium**	Ru	101.07	gray-wh or silv brittle met, hex	12.30	2250	3900	i	i	i aq reg, a, al; s fus alk
r120	carbonyl, penta-	$Ru(CO)_5$	241.12	col liq		−22				s al, bz
r121	chloride, tetra-	$RuCl_4.5H_2O$	332.96	rdsh-br cr, hygr		d		s		s al
r122	chloride, tri-	$RuCl_3$	207.43	br cr, deliq	3.11	d >500		i	d	sl s al; s HCl; i CS_2
r123	fluoride, penta-	RuF_5	196.06	dk grn cr	$2.963^{16.5}$	101	250	d	d	
r124	hydroxide	$Ru(OH)_3$	152.09	blk powd				v sl s		s a; i alk
r125	oxide, di-	RuO_2	133.07	dk bl, tetr	6.97	d		i	i	i a; s fus alk
r126	oxide, tetr-	RuO_4	165.07	yel, rhomb need	3.29^{21}	25.5	d 108	2.033^{20}	2.249^{74}	s al, a, alk, CCl_4
r127	oxychloride ammoniated	Ruthenium red. $Ru_2(OH)_2Cl_4.7NH_3.3H_2O$	551.19	brn-red powd				s		
r128	silicide	RuSi	129.16	met pr	5.40^{4}			i	i	s HNO_3+HF
r129	sulfide	Nat. laurite. RuS_2	165.20	gray-blk, cub	6.99	d 1000		i	i	i a; s fus alk
s1	**Samarium**	Sm	150.35	wh-gray met, hex	7.536	1072	1900	i	i	s a
s2	acetate	$Sm(C_2H_3O_2)_3.3H_2O$	381.53		1.94			15^{25}		
s3	acetylacetonate	$Sm(C_5H_7O_2)_3$	447.68	cr mass		146–147		i		
s4	benzylacetonate	$Sm(C_{10}H_9O_2)_3.2H_2O$	669.92	straw color		103–105		i		s org solv
s5	bromate	$Sm(BrO_3)_3.9H_2O$	696.21	yel, hex		75	$-9H_2O$, 150	114^{25}		v sl s al
s6	(II) bromide	$SmBr_2$	310.17	dk brn	5.1	508	1880	d		
s7	(III) bromide	$SmBr_3.6H_2O$	498.17	yel cr, deliq	2.971^{22}	640				
s8	carbide	SmC_2	174.37	yel, hex	5.86			d	d	s, d a
s9	(II) chloride	$SmCl_2$	221.26	red-brn cr	4.56^{25}	740		s, d		i al, CS_2
s10	(III) chloride	$SmCl_3$	256.71	yelsh-wh cr, hygr	4.46^{18}	678±2	d	92.4^{10}	99.9^{80}	v s al; 6.4^{25} pyr
s11	(III) chloride, hexahydrate	$SmCl_3.6H_2O$	364.80	grn-yel, tricl, hygr	2.383	$-5H_2O$, 110				
s12	chromate	$Sm_2(CrO_4)_3.8H_2O$	792.80	yel				0.043^{25}		
s13	(II) fluoride	SmF_2	188.35			1306	>2400			
s14	(III) fluoride	SmF_3	207.35			1306	2323	i	i	
s15	hydroxide	$Sm(OH)_3$	201.37	pa yel powd				i		s a; i alk
s16	(II) iodide	SmI_2	404.16	dk brn		527	1580	d		
s17	(III) iodide	SmI_3	531.06	or-yel cr		850				
s18	kojate	$Sm(C_6H_5O_4)_3$	573.66			d 275		i		
s19	(III) methyl-phosphate, di	$Sm[(CH_3)_2PO_4]_3$	525.47	cream col, hex pr				35.2^{25}	10.8^{95}	
s20	(III) molybdate	$Sm_2(MoO_4)_3$	780.51	vlt, rhomb oct	5.36					
s21	(III) nitrate	$Sm(NO_3)_3.6H_2O$	444.46	pa yel, tricl	2.375	78–79		v s		
s22	(III) oxalate	$Sm_2(C_2O_4)_3.10H_2O$	744.91	wh cr				0.000054		s H_2SO_4
s23	oxide, sesqui-	Samaria. Sm_2O_3	348.70	wh-yelsh powd	8.347			i		v s a
s24	(III) sulfate	$Sm_2(SO_4)_3.8H_2O$	733.01	lt yel, monocl, 1.543, 1.552, 1.563	2.930	$-8H_2O$, 450		2.67^{20} 4.4^{25}	1.99^{40}	
s25	(III) sulfide	Sm_2S_3	396.89	yelsh-pink	5.729	1900			d	d dil a
s26	sulfate, basic	$Sm_2O_2SO_4$	428.76	yel powd		d 1100		i		i dil H_2SO_4
s27	**Scandium**	Sc	44.956	silv met, cubic or hex	2.992	1539	2727	d, ev H_2		
s28	acetylacetonate	$Sc(C_5H_7O_2)_3$	342.29	col pl		187.5	subl 210–215			s al, bz, chl
s29	bromide	$ScBr_3$	284.68		3.914	subl >1000				
s30	chloride	$ScCl_3$	151.32	col cr	2.39^{25}_{4}	939	subl 800–850	v s	v s	i abs al
s31	hydroxide	$Sc(OH)_3$	95.98	col amorph				i		s dil a
s32	nitrate	$Sc(NO_3)_3$	230.97	col, deliq		150		s		s al
s33	nitrate, tetrahydrate	$Sc(NO_3)_3.4H_2O$	303.03	col, pr, deliq		$-4H_2O$, 100		v s		
s34	oxalate	$Sc_2(C_2O_4)_3.5H_2O$	444.05	cr		$-4H_2O$, 140				
s35	oxide	Scandia. Sc_2O_3	137.91	wh powd	3.864			i	i	s h a
s36	sulfate	$Sc_2(SO_4)_3$	378.10	col cr	2.579	d		10.3^{25}	v s	
s37	sulfate, hexahydrate	$Sc_2(SO_4)_3.6H_2O$	486.19			$-4H_2O$, 100 $-6H_2O$, 250		v s		
s38	sulfate, pentahydrate	$Sc_2(SO_4)_3.5H_2O$	468.17		2.519	$-3H_2O$, 100		54.6^{25}		
s39	**Selenic acid**	H_2SeO_4	144.97	wh, hex prism, hygr	3.004^{15}_{4}	58 eas undercools	d 260	1300^{30}	∞^{60}	s H_2SO_4; d al; i NH_3
s40	monohydrate	$H_2SeO_4.H_2O$	162.99	wh, need	2.627^{15}_{4} liq 2.3564^{15}_{4}	26 eas undercools	205	v s	v s	
s41	tetrahydrate	$H_2SeO_4.4H_2O$	217.03	col liq		51.7 eas undercools	$-H_2O$, 172^{85}	∞		s H_2SO_4; d org sol
s42	**Selenium**	Se	78.96	blsh-gray, met hex	4.81^{20}_{4}	217	684.9±1.0	i	i	s H_2SO_4, $CHCl_3$; i al; v sl s CS_2
s43	"	Se	78.96	red, monocl prism	4.50	170–180 trsf to hex	684.8	i	i	$0.1^{46.6}$ CS_2; s H_2SO_4, HNO_3
s44	"	Se	78.96	red amorph, blk vitr	red 4.26 blk 4.28	tr to hex, 60–80	684.8	i	i	s H_2SO_4, CS_2, bz

No.	Name	Synonyms and Formulae	Mol. wt.	Crystalline form, properties and index of refraction	Density or spec. gravity	Melting point, °C	Boiling point, °C	Solubility, in grams per 100 cc: Cold water	Hot water	Other solvents
	Selenium									
s45	bromide, "mono-"	Diselenium dibromide. Se_2Br_2	317.74	dk red liq	3.604^{15}		227 d	d	d	d al; s CS_2, $CHCl_3$, C_2H_5Br
s46	bromide, tetra-	$SeBr_4$	398.60	or-red-brn cr		d 75		d	d	s CS_2, chl, C_2H_5Br, HCl
s47	bromide (mono-) chloride, tri-,	$SeBrCl_3$	265.23	yel br cr		190				i CS_2
s48	bromide (tri-) chloride	$SeBr_3Cl$	354.14	or cr, hygr		d				v sl s CS_2
s49	carbide	SeC_2	102.98	yel liq, 1.845	2.682^{20}_{4}	45.5	$125–126^{760}$	i		s CS_2, eth, CCl_4, bz, al
s50	chloride "mono"-	Diselenium dichloride. Se_2Cl_2	228.83	br red liq, 1.596	2.77^{25}_{4}	−85	d 130	d	d	d al, eth; s CS_2, chl, CCl_4, bz
s51	chloride, tetra-	$SeCl_4$	220.77	wh-yel, cub, deliq, 1.807	$3.78–3.85^{360}$	305, subl 170–196	d 288	d	d	i al, eth, CS_2; s $POCl_3$; d a, alk
s52	fluoride, hexa-	SeF_6	192.95	col gas, 1.895	3.25^{-28} g/l	−39, subl −46.6	−34.5	s d		
s53	fluoride, tetra-	SeF_4	154.95	col liq or wh cr		m.p. −13.8 frz −90	>100	d	d	
s54	hydride	H_2Se	80.98	col gas, pois	gas 3.664^{760} air; liq: $2.004^{-41.5}$	-60.4	−41.5	3.77^{4}	$270^{22.5}$	s CS_2, $COCl_2$
s55	iodide, "mono"-	Diselenium diiodide. Se_2I_2	411.73	steelgray cr (exist ?)		68–70	d 100	d	d	
s56	nitride	Se_4N_4	371.87	amorph, or yel-brickred, hygr		expl 160–200	d	i	sl d	i al, eth; v sl s acet ac, bz
s57	oxide, di-	SeO_2	110.96	wh, monocl, col, tetr, pois, >1.76	3.95^{15}_{15}	340–350 subl 315–317		38.4^{14}	82.5^{65}	6.67^{14} al; $4.35^{15.3}$ acet; $1.11^{13.9}$ ac a; s bz
s58	oxide, tri-	SeO_3	126.96	pa yel cub or fibre, deliq	3.6	118	d 180	d, v s	d, v s	s al, conc H_2SO_4; i eth, bz, chl, CCl_4
s59	oxybromide	Selenyl bromide. $SeOBr_2$	254.78	red yel cr	liq 3.38^{50}	41.6	217^{710} d	d		s CS_2, CCl_4, chl, H_2SO_4, bz
s60	oxychloride	$SeOCl_2$	165.87	col-yel, liq 1.651^{20}	2.42^{22}	8.5	176.4	d		s CS_2, CCl_4, chl, bz
s61	sulfur oxy*tetra*-chloride	$SeSO_3Cl_4$	300.83	hex pr		165	183	d		
s62	oxyfluoride	Selenyl fluoride. $SeOF_2$	132.96	col liq	2.67	4.6	124	d		s al, CCl_4
s63	sulfide, di-	SeS_2	143.09	br red-yel		<100	d	i		d aq reg, HNO_3; s $(NH_4)_2S$
s64	sulfide, "mono"-	SeS	111.02	or-yel tabl or powd	3.056^{0}	d 118–119		i	i	s CS_2; i eth; d al
s65	sulfur oxide	$SeSO_3$	159.02	grn pr or yel powd		$-SO_2$, 40		d, 118		s H_2SO_4; i SO_3
s67	**Selenious acid**	H_2SeO_3	128.97	col, hex, deliq	3.004^{15}_{4}	d 70	$-H_2O$	167^{20}	v s	v s al; i NH_3
s68	**Silane,** bromo-	SiH_3Br	111.02	col gas, expl in air	1.72^{-80} 1.533^{0}	−94	1.9			
s69	bromotrichloro-	$SiBrCl_3$	214.35	col liq	1.826	−62	80.3	d	d	
s70	chloro-	SiH_3Cl	66.56	col gas	gas: 3.033 g/l liq: 1.145^{-113}	−118.1	−30.4			
s71	dibromo-	SiH_2Br_2	189.92	col liq, inflam	2.17^{0}	−70.1	66	d		d alk
s72	dibromo-	Silicobromoform. $SiHBr_3$	268.82	col liq, inflam	2.7^{17}	−73	109	d	d	d NH_3
s73	dibromodichloro-	$SiBr_2Cl_2$	258.81	col liq	2.172^{25}_{4}	−45.5	104	d	d	
s74	dichloro-	SiH_2Cl_2	101.01	gas	gas 4.599 g/l liq 1.42^{-122}	−122	8.3	d	d	
s75	dichloro difluoro-	$SiCl_2F_2$	136.99	gas	6.2784 g/l	−144 ± 2	−31.7 ± 0.2	d	d	
s76	(hexa-)hexaoxocyclo-	Siloxane. $Si_6O_3H_6$	222.56	wh, pl	1.32^{20}	d 140, inflam		sl d	d	
s77	monochloro trifluoro-	$SiClF_3$	120.53	gas	5.455 g/l	−138.0 ± 2	−70.0 ± 0.2	d	d	
s78	monoiodo-	SiH_3I	158.01	col liq	$2.035^{14.8}$	−57.0	45.5	d		
s79	(tri-)nitrilo-	Silicylamine, tri-$(SiH_3)_3N$	107.34	col inflam liq	0.895^{-106}	−105.6				
s81	tribromochloro-	$SiBr_3Cl$	303.27	col liq	2.497^{25}_{4}	−20.8 ± 1	126–128	d	d	
s82	trichloro-	Silicochloroform. $SiHCl_3$	135.45	col liq	1.34	−126.5	33^{758} mm	d	d	s CS_2, CCl_4, chl, bz
s83	trichloroiodo-	$SiCl_3I$	261.35	col liq		>−60	113.5	d		
s84	trifluoro-	Silicofluoroform. $SiHF_3$	86.09	col gas	3.86^{0} g/l	−131.4	*ca* −95	d	d	d al, eth, alk; s tol
s85	triiodo-	Silicoiodoform. $SiHI_3$	409.81	red liq	3.314^{20}	8	220	d	d	s CS_2, bz

No.	Name	Synonyms and Formulae	Mol. wt.	Crystalline form, properties and index of refraction	Density or spec. gravity	Melting point, °C	Boiling point, °C	Solubility, in grams per 100 cc		
								Cold water	Hot water	Other solvents
	Silicane cyanate									
s86	**Silicane cyanate**	$Si(OCN)_4$	196.16	sol or liq	1.414^{20}_{4}	34.5±0.5	247.2±0.5 0.5^{760}	d	...	...
s87	diimide	$Si(NH)_2$	58.12	wh powd	...	d 900	...	...	...	...
s88	isocyanate	$Si(NCO)_4$	196.16	sol or liq	1.434^{25}_{4}	26.0±0.5	185.6±0.3^{760}	d	...	i acet; s bz, CCl_4, CS_2
s89	**Silicic acid, di-**	H_2SiO_5	138.18	col cr	...	d 150	...	i	i	s NH_3,HF
s90	meta-	H_2SiO_3	78.10	col, amorph	...	d room temp	...	i	i	s NH_3, HF, h alk
s91	**Silicon**	Si	28.086	steel gray, large to micr cr, cub	2.32–2.34	1410	2355	i	i	s HF+HNO_3; i HF
s92	acetate, tetra-	$Si(C_2H_3O_2)_4$	264.27	col cr, hygr	...	subl 110 d 160–170	148^{6} mm	d	...	d al; sl s acet, bz
s93	bromide, tetra-	Tetrabromosilane. $SiBr_4$	347.72	col fum liq, sol cub	liq: 2.7715^{25}_{4} sol: 3.292^{-79}	5.4	154	d	d	d H_2SO_4
s94	(di-)bromide, hexa-	Si_2Br_6	535.62	wh, rhomb	...	95	240	d	d	s CS_2; d KOH
s95	bromide(di-) sulfide	$SiSBr_2$	219.97	col pl	...	93	$150^{18.3}$ mm	d	d	s bz, CS_2
s96	carbide	SiC	40.10	col-blk, hex or cub, 2.654, 2.697	3.217	~2700, subl, d	...	i	i	i a; s fus KOH
s97	chloride, tetra-	Tetrachlorosilane. $SiCl_4$	169.90	col fum liq	liq: 1.483^{20}, sol: 1.90^{-97} gas: 7.59 g/l	−70	57.57	d	d	d al
s98	(di-)chloride, hexa-	Hexachlorodisilane. Si_2Cl_6	268.89	col liq, 1.4748^{18}	1.58^{0}	−1	145^{769}	d	d	d al
s99	chloride(di-) sulfide	$SiSCl_2$	131.06	col pr	...	75	$92^{22.5}$	d	d	s CCl_4, CS_2 bz
s100	chloride(tri-) sulfide, hydro-	$SiCl_3HS$	167.52	col liq	1.45	...	96–100	d	d	d al
s101	fluoride, tetra-	Tetrafluorosilane. SiF_4	104.08	col gas	gas: 4.69 g/l^{760} liq: 1.66^{-95}	−90.2	−86	d	d	s abs al, HF; i eth
s102	(di-)fluoride, hexa	Hexa-fluorodisilane. Si_2F_6	170.16	gas	7.759 g/l	−18.7	−18.5	d	d	...
s103	hydride	Silane, silicane. SiH_4	32.12	col gas	liq: 0.68^{-85} gas: 1.44 g/l	−185	-111.8^{760} mm	i	...	d KOH
s104	hydride	Disilane, disilicane. Si_2H_6	62.22	col gas	gas: 2.865 g/l liq: 0.0686^{-20}	−132.5	−14.5	sl d	...	s al, bz, CS_2
s105	hydride	Trisilane, trisilanepropane. Si_3H_8	92.32	col liq	liq: 0.743^{0}; gas: 4.15 g/l^{0}_{760}	−117.4	52.9	d	d	d CCl_4
s106	hydride	Tetrasilane, tetrasilane butane. Si_4H_{10}	122.42	col liq	liq: 0.79^{0} gas 5.48 g/l^{0}_{760}	−108	84.3	d	...	...
s107	iodide, tetra-	Tetraiodosilane. SiI_4	535.70	col, cub	4.198	120.5	287.5	d	...	2.2^{27} CS_2
s108	(di-)iodide, hexa-	Hexaiodosilane. Si_2I_6	817.60	col, hex	...	d 250	d	d	d	19^{19} CS_2
s109	nitride	Si_3N_4	140.28	gray-wh amorph powd	3.44	1900 press	...	i	i	s HF
s110	oxide, di-	Nat. cristobalite. SiO_2	60.08	col, cub or tetr, 1.487, 1.484	2.32	1713±5	2230 (2590)	i	i	s HF; v sl s alk
s111	oxide, di-	Nat. lechatelierite. SiO_2	60.08	col, amorph, vitr, 1.4588	2.19	...	2230 (2590)	i	i	s HF; v sl s alk
s112	oxide, di-	Nat. opal. $SiO_2.xH_2O$	...	col, amorph 1.41–1.46	2.17–2.20	>1600	...	i	i	s HF; v sl s alk
s113	oxide, di-	Nat. tridymite. SiO_2	60.08	col, rhomb, 1.469, 1.470, 1.471	2.26^{25}_{4}	1703	2230 (2590)	i	i	s HF; v sl s alk
s114	oxide, di-	Nat. quartz. SiO_2	60.08	col, hex, 1.544, 1.553	2.635–2.660	1610	2230 (2590)	i	i	s HF; v sl s alk
s115	oxide, mon-	SiO	44.09	wh, cub	2.13	>1702	1880	i	i	s dil HF+HNO_3
s116	oxychloride	Chlorosiloxane. Si_2OCl_6	284.89	col liq	...	28.1±0.2	137	d	...	∞ CS_2, CCl_4
s117	oxyfluoride	Si_2OF_6	186.16	col gas	1.358 liq	−47.8±0.5	−23.3	d	d	d alk
s118	sulfide, di-	SiS_2	92.21	wh need, rhomb	2.02	subl 1090	white heat	d	...	d al, liq NH_3; s dil alk; i bz
s119	sulfide, mono-	SiS	60.15	yel need	1.853^{15}	subl 940^{20}	...	d	d	d al, alk
s120	thiocyanate	$Si(CNS)_4$	260.41	wh, rhomb need	1.409^{20}_{4}	143.8	314.2	d	...	d al, a, alk; i eth CS_2, $CHCl_3$
s121	**Silicotungstic acid**	$H_4[Si(W_3O_{10})_4]\cdot 26H_2O$	3346.47	wh-sl yel cr, deliq	...	...	...	v s	v s	v s al
s122	**Silicyl oxide**	Disiloxane. $(SiH_3)_2O$	78.22	col gas	gas: 3.491 g/l liq: 0.881^{-80}	−144	−15.2	v sl s	sl d	...

No.	Name	Synonyms and Formulae	Mol. wt.	Crystalline form, properties and index of refraction	Density or spec. gravity	Melting point, °C	Boiling point, °C	Solubility, in grams per 100 cc		
								Cold water	Hot water	Other solvents
	Siloxane									
s123	**Siloxane, (di-), oxide**	$[H(O)Si]_2.O$	106.19	wh volum subst		expl *ca* 300		sl s		s, d HF; d al
s124	**Silver**	Ag	107.868	wh met, cub 0.54	10.5^{20}	960.8	2212	i	i	s HNO_3, h H_2SO_4, KCN; i alk
s125	acetate	$AgC_2H_3O_2$	166.92	wh pl	3.259^{15}	d		1.02^{20}	2.52^{80}	s dil HNO_3
s126	acetylide	Ag_2C_2	239.76	wh ppt		expl		i		s a; sl s al
s127	*ortho*arsenate	Ag_3AsO_4	462.53	dk red, cub	6.657^{25}	d		0.00085^{20}		s NH_4OH, ac a
s128	*ortho*arsenite	Ag_3AsO_3	446.53	yel, powd		d 150		0.00115^{20}	i	s ac a, NH_4OH, HNO_3; i al
s129	azide	AgN_3	149.89	wh rhomb pr, expl		252	297	i	0.01^{100}	s KCN, dil HNO_3; sl s NH_4OH
s130	benzoate	$AgC_7H_5O_2$	228.99	wh powd				0.262^{25}	s	0.017 al
s131	*tetra*borate	$Ag_2B_4O_7.2H_2O$	407.01	wh cr				sl s		s a
s132	bromate	$AgBrO_3$	235.78	col, tetr, 1.874, 1.920	5.206	d		0.196^{25}	1.33^{90}	s NH_4OH; sl s HNO_3
s133	bromide	Bromyrite: AgBr	187.78	pa yel, 2.253	6.473^{25}	432	d>1300	8.4×10^{-6}	0.00037^{100}	s KCN, $Na_2S_2O_3$, NaCl sol; sl s NH_4OH; i al
s134	carbonate	Ag_2CO_3	275.75	yel powd	6.077	d 218		0.0032^{20}	0.05^{100}	s NH_4OH, $Na_2S_2O_3$; i al
s135	chlorate	$AgClO_3$	191.32	wh, tetr	4.430^{20}_{4}	230	d 270	10^{15}	50^{80}	sl s al
s136	*per*chlorate	$AgClO_4$	207.32	wh, cr, deliq	2.806^{25}	d 486		557^{25}	s	s al; 101 tol; 5.28 bz
s137	chloride	Nat. cerargyrite. AgCl	143.32	wh, cub, 2.071	5.56	455	1550	0.000089^{10}	0.0021^{100}	s NH_4OH, $Na_2S_2O_3$, KCN
s138	chlorite	$AgClO_2$	175.32	yel cr		105 expl		0.45^{25}	2.13^{100}	
s139	chromate	Ag_2CrO_4	331.73	red, monocl	5.625			0.0014^{0}	0.008^{70}	s NH_4OH, KCN
s140	*di*chromate	$Ag_2Cr_2O_7$	413.73	red, tricl	4.770	d		0.0083^{15}	d	s a, NH_4OH, KCN
s141	citrate	$Ag_3C_6H_5O_7$	512.71	wh need		d		0.028^{18}	sl s	s a, NH_4OH, KCN, $Na_2S_2O_3$
s142	cyanate	AgOCN	149.89	col	4.00	d		sl s	s	s KCN, HNO_3, NH_4OH
s143	cyanide	AgCN	133.84	wh, hex	3.95	d 320		0.000023^{20}		s HNO_3, NH_4OH, KCN, $Na_2S_2O_3$
s144	ferricyanide	$Ag_3Fe(CN)_6$	535.56					0.000066^{20}		i a; s NH_4OH, h $(NH_4)_2CO_3$
s145	ferrocyanide	$Ag_4Fe(CN)_6.H_2O$	661.45	wh				i	i	s KCN; i a, NH_4 salts, NH_4OH
s146	fluogallate	$Ag_3[GaF_6].10H_2O$	687.47	col, orthorhomb cr, 1.493	2.90			v s		i al
s147	fluoride	AgF	126.87	yel, cub, deliq	$5.852^{15.5}$	435	*ca* 1159	$182^{15.5}$	205^{108}	sl s NH_4OH
s148	fluoride, di-	AgF_2	145.87	brn, rhomb	4.57–4.58	690	d 700	d		
s149	(di-)fluoride	Ag_2F	234.74	yel, hex	8.57	d 90		d		
s150	fluosilicate	$Ag_2SiF_6.4H_2O$	429.88	wh powd or col cr, deliq		>100	d	v s		
s151	fulminate	$Ag_2C_2N_2O_2$	299.77	need		expl		0.075^{13}	s	i HNO_3; s NH_4OH
s152	iodate	$AgIO_3$	282.77	col, rhomb	$5.525^{16.5}$	>200	d	0.003^{10}	0.019^{60}	s HNO_3, NH_4OH, KI
s153	*per*iodate	$AgIO_4$	298.77	or yel, tetrag	5.57	d 180		d		s HNO_3
s154	iodide(α)	Nat. iodyrite. AgI	234.77	yel, hex 2.21, 2.22	5.683^{30}_{4}	tr 146 to β		$2.8\times10^{-7.25}$	$2.5\times10^{-6.60}$	s KCN, $Na_2S_2O_3$, KI; sl s NH_4OH
s155	iodide (β)	AgI	234.77	or, cub	$6.010^{14.6}_{4}$	558	1506			
s156	iodomercurate (α)	Ag_2HgI_4	923.95	yel, tetrag	6.02	tr to β 50.7		i		s KI, KCN; i dil a
s157	iodomercurate (β)	Ag_2HgI_4	923.95	red, cub	5.90	d 158		i		s KI, KCN; i dil a
s158	hydrogen(tri-) *paraper*iodate	$Ag_2H_3IO_6$	441.69	yel, rhomb	5.68^{25}	60 d		1.68^{25}		s HNO_3
s159	hyponitrite	$Ag_2N_2O_2$	275.75	yel	5.75^{30}	d 110		v sl s		d HNO_3, H_2SO_4
s160	lactate	$AgC_3H_5O_3.H_2O$	214.96	wh or sl gray cr, powd				*ca* 7.7		
s161	laurate	$AgC_{12}H_{23}O_2$	307.19	wh, greasy powd		212.5				0.007^{25} al; 0.008^{15} eth
s162	levunilate	$AgC_5H_7O_3$	222.98	leaf				0.67^{17}	d	
s163	*per*manganate	$AgMnO_4$	226.81	dk vlt, monocl	4.27^{25}	d		0.55^{0}	$1.69^{28.5}$	d al
s164	mercury iodide (α)	Ag_2HgI_4	923.98	yel, tetrag	6.02	trst 50.7		i		
s165	mercury iodide (β)	Ag_2HgI_4	923.98	red, cub	5.90	158 d		i		
s166	myristate	$AgC_{14}H_{27}O_2$	335.24			211		0.007^{25}		0.006^{25} al; 0.007^{15} eth
s167	nitrate	$AgNO_3$	169.87	col, rhomb, 1.729, 1.744, 1.788	4.352^{19}	212	d 444	122^{0}	952^{190}	s eth, glyc; v sl s abs al

No.	Name	Synonyms and Formulae	Mol. wt.	Crystalline form, properties and index of refraction	Density or spec. gravity	Melting point, °C	Boiling point, °C	Solubility, in grams per 100 cc		
								Cold water	Hot water	Other solvents
	Silver									
s168	nitrite	$AgNO_2$	153.88	wh, rhomb	4.453^{26}	d 140	...	0.155^{0}	1.363^{60}	s ac a, NH_4OH; i al
s169	nitroplatinite	$Ag_2[Pt(NO_2)_4]$	594.85	yel-brn monocl pr	...	d 100	...	sl s	s	...
s170	nitroprusside	$Ag_2[FeNO(CN)_5]$	431.68	lt pink	...	...	...	i	...	s NH_4OH; i al, HNO_3
s171	oxalate	$Ag_2C_2O_4$	303.76	col cr	5.029^{4}	expl 140	...	0.00339^{18}	...	s KCN, NH_4OH, a
s172	oxide	Ag_2O	231.74	br-blk, cub	$7.143^{16.6}$	d 300	...	0.0013^{20}	0.0053^{80}	s a, KCN, NH_4OH, al
s173	oxide, per	Ag_2O_2 (or AgO)	247.74	gray-blk, cub	7.44	d>100	...	i	...	s H_2SO_4, HNO_3, NH_4OH
s174	palmitate	$AgC_{16}H_{31}O_2$	363.29	wh, greasy powd	...	209	...	0.0012^{20}	0.006^{25}	0.007^{15} eth; 0.006^{25} al
s175	*meta*phosphate	$AgPO_3$	186.84	wh, amorph	6.37	*ca* 482	...	i	...	s HNO_3, NH_4OH
s176	*ortho*phosphate	Ag_3PO_4	418.58	yel, cub	6.370^{25}	849	...	$0.00065^{19.5}$	...	s a, KCN, NH_4OH, NH_3
s177	*ortho*phosphate, mono-H	Ag_2HPO_4	311.75	wh, trig	1.8036	d 110	...	...	...	...
s178	*pyro*phosphate	$Ag_4P_2O_7$	605.42	wh	$5.306^{7.5}$	585	...	i	i	s a, NH_4OH, KCN, ac a
s179	propionate	$AgC_3H_5O_2$	180.94	wh leaf or need.	2.687^{25}_{4}	...	...	0.842^{20}	2.04^{80}	...
s180	*per*rhenate	$AgReO_4$	358.07	wh cr, tetrag or rhomb	7.05	430		0.32^{20}	...	...
s181	salicylate	$AgC_7H_5O_3$	244.99	wh to redsh-wh cr	...	...	...	sl s	...	s al
s182	selenate	Ag_2SeO_4	358.73	wh, orthorhomb cr	5.72	...	...	0.118^{20}	...	...
s183	selenide	Ag_2Se	294.70	thin gray pl, cub	8.0	880	d	i	...	s NH_4OH, h HNO_3
s184	stearate	$AgC_{18}H_{35}O_2$	391.35	wh powd amorph	...	205	...	0.006^{20}	...	0.006^{25} al; 0.006^{25} eth
s185	sulfate	Ag_2SO_4	311.80	wh, rhomb, 1.7583, 1.7748, 1.7852	$5.45^{29.2}$	652	d 1085	0.57^{0}	1.41^{100}	s a, NH_4OH; i al
s186	sulfide	Nat. acanthite. Ag_2S	247.80	gray-blk, rhomb	7.326	tr 175	d	0.00002	v sl s	s KCN, conc H_2SO_4, HNO_3
s187	sulfide	Nat. argentite. Ag_2S	247.80	blk, cub	7.317	825	d	0.000014^{20}		s KCN, a
s188	sulfite	Ag_2SO_3	295.80	wh cr	...	d 100	...	v sl s	...	s a, NH_4OH, KCN; i HNO_3
s189	*d*-tartrate	$Ag_2C_4H_4O_6$	363.81	wh, scales	3.423^{15}	d	...	0.2^{18}	0.203^{25}	s a, KCN, NH_4OH
s190	*ortho*tellurate, tetra-H	$Ag_2H_4TeO_6$	443.40	straw yel, rhomb bipyr	...	d>200	...	i	i	s KCN, NH_4OH
s191	telluride	Nat. hessite, Ag_2Te	343.34	gray, cub	8.5	955	...	i	i	s KCN, NH_4OH
s192	tellurite	Ag_2TeO_3	391.36	yel-wh ppt	...	250-bl 450-pa yel	...	i	i	s KCN, NH_3
s193[1]	thioantimonite	Nat. pyrargyrite. Ag_3SbS_3	541.55	red, trig, 3.084 2.881 (Li)	5.76	486	...	i	i	s HNO_3
s193[2]	thioarsenite	Nat. proustite. Ag_3AsS_3	494.72	scarlet red, trig, 3.088, 2.792	5.49	490	...	i	i	s HNO_3
s194	thiocyanate	AgSCN	165.95	col cr	...	d	...	0.000021^{25}	0.00064^{100}	s NH_4OH; i a
s195	*di*-thionate	$Ag_2S_2O_6.2H_2O$	411.90	rhomb cr, ~1.662	3.61	...	...	...	...	...
s196	thiosulfate	$Ag_2S_2O_3$	327.87	wh cr	...	d	...	sl s	...	s $Na_2S_2O_3$, NH_4OH
s197	tungstate	Ag_2WO_4	463.59	pa yel cr	...	...	...	0.05^{15}	...	s KCN, NH_4OH, HNO_3
	Silver complex									
s198	diamminesilver *per*rhenate	$[Ag(NH_3)_2]ReO_4$	392.13	col monocl cr	3.901	...	...	...	...	1.618 conc NH_4OH
s199	**Sodium**	Na	22.9898	silv, met cub, 4.22	0.97	97.81±0.03	892	d to $NaOH+H_2$	...	d al; i eth, bz
s200	acetate	$NaC_2H_3O_2$	82.03	wh gr powd, monocl, 1.464	1.528	324	...	119^{0}	170.15^{100}	sl s al
s201	acetate trihydrate	$NaC_2H_3O_2.3H_2O$	136.08	col, monocl pr, effl, β 1.464	1.45	58	123, $-3H_2O$, 120	76.2^{0}	138.8^{50}	2.1^{18} al; s eth
s202	alumina trisilicate	Nat. albite. $NaAlSi_3O_8$. (or $Na_2O.Al_2O_3.6SiO_2$)	262.22	col, tricl, 1.525, 1.529, 1.536	2.61	1100	...	...	sl d	s HCl; d dil al
s203	*meta*aluminate	$NaAlO_2$	81.97	wh amorph powd, hygr, 1.566, 1.595, 1.580	...	1800	...	s	v s	i al
s204	aluminum chloride	$NaCl.AlCl_3$	191.78	wh-yelsh cr powd, hygr	...	185	...	s	s	...
s205	aluminum *meta*-silicate	Nat. jadeite. $Na_2O.Al_2O_3.4SiO_2$	404.28	col, monocl	3.3	1000–1060	...	i	i	d HCl

No.	Name	Synonyms and Formulae	Mol. wt.	Crystalline form, properties and index of refraction	Density or spec. gravity	Melting point, °C	Boiling point, °C	Solubility, in grams per 100 cc: Cold water	Hot water	Other solvents
	Sodium									
s206	aluminum *ortho*-silicate	Nat. nephelite. $Na_2O.Al_2O_3.2SiO_2$	284.11	col, hex, 1.537 ± 0.002	2.619^{21}	1526	......	i	d	d a
s207	aluminum sulfate	$NaAl(SO_4)_2.12H_2O$	458.28	col, cub oct, 1.4388	1.6754^{20}	61	......	110^{15} (anhydr)	146^{30} (anhydr)	......
s208	amide	Sodamide. $NaNH_2$	39.01	wh, conchoid fract	......	210	400	d	d	d hot al; 0.1 liq NH_3
s209	ammonium phosphate	Microcosmic salt, stercorite. $NaNH_4HPO_4.4H_2O$	209.07	col, monoc, 1.439, 1.441, 1.469	1.554	d 79	......	16.7	100	i al, acet
s210	ammonmium sulfate	$NaNH_4SO_4.2H_2O$	173.12	wh, rhomb	1.63^{15}	d 80	......	s	s	......
s211	ammonium tartrate	$NaNH_4C_4H_4O_6.4H_2O$	261.16	wh, rhomb	1.590	......	......	21.09^{0}	......	......
s212	*meta*antimonate	Leuconine. $NaSbO_3$	192.74	wh powd	......	......	......	i	s	s Na_2S sol
s213	antimonate, hydroxy	"Pyroantimonate". $NaSb(OH)_6$	246.78	pseudo cub	......	......	......	$0.03^{12.3}$	0.3^{100}	sl s al
s214	*pyro*antimonate, dihydro-	$Na_2H_2Sb_2O_7.6H_2O$(?)	511.58	wh, tetrag	......	d 280	......	i	0.28^{100}, d	......
s215	antimonide	Na_3Sb	190.72	blk powd or bl cr, inflamm	......	856	......	d	......	sl s NH_3
s216	*meta*antimonite	$NaSbO_2.3H_2O$	230.78	col, rhomb	2.864	d	......	d	......	......
s217	*meta*arsenate	$NaAsO_3$	145.91	rhomb, effl, 1.479, 1.502, 1.527	2.301	615	......	v s	......	......
s218	*ortho*arsenate	$Na_3AsO_4.12H_2O$	423.93	col, trig or hex prism, 1.457, 1.466	1.752–1.804	86.3	......	$38.9^{15.5}$	......	1.67 al; 50^{15} glyc
s219	*ortho*arsenate, di-H	$NaH_2AsO_4.H_2O$	181.94	col, rhomb or monocl, 1.583, 1.553, 1.507	2.53	130, $-H_2O$, 100	d 200–280	s	......	......
s220	*ortho*arsenate, mono-H	$Na_2HAsO_4.7H_2O$	312.01	col, monocl, pois, 1.462, 1.466, 1.478	1.88	130, $-5H_2O$, 50	d 180	5.46^{0}	100^{100}	s glyc; sl s al
s221	*ortho*arsenate, mono-H	$Na_2HAsO_4.12H_2O$	402.09	col, monocl, effl, 1.445, 1.466, 1.451	1.736	28	$-12H_2O$, 100	56^{14}	140.7^{30}	sl s al; i liq Cl
s222	*pyro*arsenate	$Na_4As_2O_7$	353.79	wh cr	2.205	850	d 1000	v s	......	......
s223	arsenate fluoride	$2Na_3AsO_4.NaF.19H_2O$	800.06	wh, cub, 1.4657, 1.4693, 1.4726	2.849^{25}	......	......	10^{75}	......	......
s224	arsenite	Sodium metaarsenite (?) (com'l) $NaAsO_2$ (or mixt with Na_3AsO_3)	129.91	gray-wh powd, pois	1.87	......	......	v s	v s	sl s al
s225	arsenotartrate	$Na(AsO)C_4H_4O_6.2\frac{1}{2}H_2O$	307.02	shiny cr, pois	......	$-2\frac{1}{2}H_2O$, 275	d 275	6.5^{19}	......	i al
s226	azide	NaN_3	65.01	col, hex	1.846^{20}	d Na+N	d in vac	41.7^{17}	......	0.314^{16} al; s liq NH_3, i eth
s227	barbital	$NaC_8H_{11}N_2O_3$	206.18	wh powd	......	......	......	20^{25}	40^{100}	sl s al; i eth
s228	benzenesulfonate	$NaC_6H_5SO_3$	180.16	wh cr	......	......	......	35.8^{30}	v s	......
s229	benzoate	$NaC_7H_5O_2$	144.11	col cr, or wh amorph, or gran powd	......	......	......	66^{20}	74.2^{100}	1.64^{25} al
s230	*meta*bismuthate	$NaBiO_3$	279.97	yel-brn powd (com'l), yel (pure)	......	......	......	i	d	d a
s231	*meta*borate	$NaBO_2$	65.80	col, hex pr	2.464	966	1434	26^{20}	36^{35}	......
s232	*meta*borate, tetrahydrate	$NaBO_2.4H_2O$	137.86	tricl, coll	......	57	$-H_2O$, 120	v s	v s	......
s233	*meta*borate, peroxyhydrate	Sodium perborate (com'l). $NaBO_2.H_2O_2.3H_2O$	153.86	col, monocl	......	63.0	$-H_2O$, 130–150	2.55^{15}	3.75^{32}	s a, al, glyc
s234	*tetra*borate	$Na_2B_4O_7$	201.22	cr, 1.5010	2.367	741	d 1575	1.06^{0}	8.79^{40}	i al
s235	*tetra*borate, decahydrate	Borax. $Na_2B_4O_7.10H_2O$	381.37	col, monocl, effl, 1.447, 1.469, 1.472	1.73	75, $-8H_2O$, 60	$-10H_2O$, 320	2.01^{0}	170^{100}	v sl s al; s glyc; i a
s236	*tetra*borate, pentahydrate	$Na_2B_4O_7.5H_2O$	291.30	col, cub or hex, deliq, 1.461	1.815	$-H_2O$, 120	......	22.65^{65} (anhydr)	52.3^{100}	......
s237	borohydride	$NaBH_4$	37.83	white cub, 1.542	1.074	400 dec	......	55^{25}	v s	4 al; 16.4 MeOH; s pyr; i eth
s238	bromate	$NaBrO_3$	150.90	col, cub, 1.594	$3.339^{17.5}$	381	......	27.5^{0}	90.9^{100}	i al
s239	bromide	NaBr	102.90	col, cub, hygr, 1.6412	3.203^{25}_{4}	755	1390	116.0^{50}	121^{100}	sl s al
s240	bromide, dihydrate	$NaBr.2H_2O$	138.93	col, monocl pr	2.176	$-2H_2O$, 51	......	79.5^{0}	$118.6^{80.5}$	2.31^{25} al; s liq NH_3; 17.42^{15} MeOH
s241	bromoaurate	$NaAuBr_4.2H_2O$	575.62	br-blk cr	......	......	......	s	......	......
s242	bromoiridite	$Na_3IrBr_6.12H_2O$	956.81	dk grn, rhomb, effl	......	100	$-H_2O$, 150	......	......	s NH_4OH

No.	Name	Synonyms and Formulae	Mol. wt.	Crystalline form, properties and index of refraction	Density or spec. gravity	Melting point, °C	Boiling point, °C	Solubility, in grams per 100 cc: Cold water	Hot water	Other solvents
	Sodium									
s243	bromoplatinate	$Na_2PtBr_6.6H_2O$	828.62	dk red, tricl	3.323	d 150		v s	v s	v s al
s244	cacodylate	$Na[(CH_3)_2AsO_2].3H_2O$	214.03	wh	*ca* 60	$-H_2O$, 120		200^{15-20}		40^{25} al; 100^{15-20} 90% al
s245	calcium sulfate	$Na_2Ca(SO_4)_2.2H_2O$	314.21	col, monocl need.	2.64	$-2H_2O$, 80		d	d	
s246	*d*-camphorate	$Na_2C_{10}H_{14}O_4.3H_2O$	298.25	wh need, hygr		$-3H_2O$, 100		122^{14}		s al
s247	carbide	Na_2C_2	70.00	wh powd	1.575^{15}	*ca* 700		d	d	s a; d al
s248	carbonate	Na_2CO_3	105.99	wh powd, hygr, 1.535	2.532	851	d	7.1^0	45.5^{100}	sl s abs al; i acet
s249	carbonate, decahydrate	Washing soda. $Na_2CO_3.10H_2O$	286.14	wh, monocl, 1.405, 1.425, 1.440	1.44^{15}	32.5–34.5	$-H_2O$, 33.5	21.52^0	421^{104}	i al
s250	carbonate, heptahydrate	$Na_2CO_3.7H_2O$	232.10	rhomb bipyr, effl	1.51	$-H_2O$, 32		16.90	33.9^{35}	
s251	carbonate, monohydrate	Crystal carbonate, thermonatrite. $Na_2CO_3.H_2O$	124.00	col, rhomb, deliq, 1.506, 1.509	2.25	$-H_2O$, 100		33	52.08	14^{25} glyc; i al, eth
s252	carbonate, sesqui-	$Na_2CO_3.NaHCO_3.2H_2O$	226.03	col, monocl, 1.5073	2.112	d		13^0	42^{100}	
s253	carbonate hydrogen	$NaHCO_3$	84.00	wh, monocl pr, 1.500	2.159	$-CO_2$, 270		6.9^0	16.4^{60}	sl s al
s254	chlorate	$NaClO_3$	106.44	col, cub or trig, 1.513	2.490^{15}	248–261	d	79^0	230^{100}	s al, liq NH_3, glyc
s255	*per*chlorate	$NaClO_4$	122.44	wh, rhomb, deliq, 1.4606, 1.4617, 1.4731	d 482	d		s	v s	s al
s256	*per*chlorate, hydrate	$NaClO_4.H_2O$	140.46	col rhbdr, deliq	2.02	130	d 482	209^{15}	284^{50}	s al
s257	chloride	Common salt, nat. halite. NaCl	58.44	col, cub, 1.5442	2.165^{25}_{4}	801	1413	35.7^0	39.12^{100}	sl s al, liq, NH_3; s glyc; i HCl
s258	chlorite	$NaClO_2$	90.44	wh, cr, hygr		d 180–200		39^{17}	55^{60}	
s259	chlorite, pentahydrate	$NaOCl.5H_2O$	164.52	col		24.5		29.3^0	94.2^{23}	
s260	*hypo*chlorite	NaOCl	74.44	in solution only						
s261	*hypo*chlorite, dihydrate	$NaOCl.2\frac{1}{2}H_2O$	119.48	col, hygr		57.5		v s		
s262	chloroaurate	$NaAuCl_4.2H_2O$	397.80	yel, rhomb, ω 1.545 ϵ>1.75		d 100		150^{10}	990^{60}	v s al, eth
s263	chloroiridate	$Na_2IrCl_6.6H_2O$	558.99	dull red-blk, tricl		d 600		v s	v s	sl s al
s264	chloroiridite	$Na_3IrCl_6.12H_2O$	690.07	dk grn cr		$-H_2O$, 50		31.46^{15}	307.26^{85}	
s265	chloroosmate	$Na_2OsCl_6.2H_2O$	484.93	or-red, rhomb pr				v s		s al
s266	chloropalladite	$Na_2PdCl_4.3H_2O$	348.24	br-red cr, deliq				v s		s al
s267	chloroplatinate	Na_2PtCl_6	453.79	or-yel powd, hygr		tr 150–160		s	v s	s al
s268	*hexa*chloroplatinate	$Na_2PtCl_6.6H_2O$	561.88	or-red, tricl	2.500	$-6H_2O$, 100		66^{15}	v s	11.9 al, MeOH; i eth
s269	chloroplatinite	$Na_2PtCl_4.4H_2O$	454.98	red pr		100	$-H_2O$, 150	s		s al
s270	chlororhodite, hexa-	Na_3RhCl_6	384.59	red, tricl		d>550		v s		
s271	chlororhodite, hexa-, hydrate	$Na_3RhCl_6.18H_2O$	708.87	garnet red, oct, effl		d 904, effl		v s		i al
s272	chromate	Na_2CrO_4	161.97	yel, rhomb bipyram	2.710–2.736			87.3^{30}		sl s al; s MeOH
s273	chromate decahydrate	$Na_2CrO_4.10H_2O$	342.13	yel, monocl, deliq	1.483	19.92		50^{10}	126^{100}	sl s al; i ac a
s274	*di*chromate	$Na_2Cr_2O_7.2H_2O$	298.00	red, monocl pr, deliq, 1.661, 1.699, 1.751	2.52^{13}	$-2H_2O$, 100 356.7 (anhydr)	d 400 (anhydr)	238^0 (anhydr) 180^{20}	508^{80} (anhydr) 433^{98}	i al
s275	*peroxy*chromate	Na_3CrO_8	248.96	or pl		d 115		sl s		i al, eth
s276	cinnamate	$NaC_9H_7O_2$	170.14	wh cr powd				9.1	5^{100}	0.625 90% al; s glyc
s277	citrate, dihydrate	$Na_3C_6H_5O_7.2H_2O$	294.10	wh cr, gran or powd		$-2H_2O$, 150		72^{25}	167^{100}	0.625 90% al; s glyc
s278	citrate, pentahydrate	$Na_3C_6H_5O_7.5(\text{or } 5\frac{1}{2})H_2O$	348.15	wh, rhomb	$1.857^{23.5}$	$-5H_2O$, 150	d	92.6^{25}	250^{100}	sl s al
s279	cobaltinitrite	$Na_3Co(NO_2)_6$	403.94	yelsh-brnsh cr powd				v s, d		sl s al; d min a; i dil ac a
s280	cyanamide, mono-	$NaHCN_2$	64.02	wh cr powd, hygr				v s		
s281	cyanate	NaOCN	65.01	col need	1.937^{20}	d 700 vac		s	s	v sl s eth, bz
s282	cyanide	NaCN	49.01	col, cub, deliq, pois, 1.452		563.7	1496	48^{10}	82^{35}	sl s al; s NH_3
s283	cyanoaurite	Sodium aurocyanide. $NaAu(CN)_2$	271.99					s		
s284	cyanocuprate (I)	$NaCu(CN)_2$	138.57		1.013^{20}	d 100		s		
s285	cyanoplatinite	$Na_2[Pt(CN)_4].3H_2O$	399.19	col, tricl	2.646	$-3H_2O$, 120–125		s	s	s al
s286	enanthate	Sodium heptanoate. $NaC_7H_{13}O_2$	152.17	wh cr powd or leaf		240–350		s		s al
s287	ethyl acetoacetate	$NaC_6H_9O_3$	152.13	need		d		d		s eth
s288	ethyl sulfate	$NaC_2H_5SO_4.H_2O$	166.13	wh, hex pl, deliq		d		164^{17}		d alk, H_2SO_4; 142 al

No.	Name	Synonyms and Formulae	Mol. wt.	Crystalline form, properties and index of refraction	Density or spec. gravity	Melting point, °C	Boiling point, °C	Solubility, in grams per 100 cc		
								Cold water	Hot water	Other solvents
	Sodium									
s289	ferrate (III)	Ferrite. $Na_2Fe_2O_4$	221.67	br, hex pl or need	4.05			d		v s dil HCl
s290	ferricyanide	$Na_3Fe(CN)_6.H_2O$	298.92	red cr, deliq				18.9^{0}	67^{100}	i al
s291	ferrocyanate	Yellow prussiate of soda. $Na_4Fe(CN)_6.10H_2O$	484.04	pa yel, monocl, 1.519, 1.530, 1.544	1.458			31.85^{20}	156.5^{98}	i al
s292	fluoaluminate	Na_3AlF_6	209.94	col, monocl, β 1.364	2.90	1000		sl s		i HCl,; d alk
s293	fluoantimonate	$NaSbF_6$	258.73	rhomb	3.375^{18}	<1360		128.6^{20}		s al, acet
s294	fluoberyllate	Na_2BeF_4	130.99	wh, rhomb or monocl		d		1.47^{18}	2.94^{100}	
s295	fluoborate	$NaBF_4$	109.79	wh, rhomb	2.47^{20}	sl d 384	d	108^{26}	210^{100}	sl s al; d H_2SO_4
s296	fluoride	Nat. villiaumite. NaF	41.99	col, cub or tetr, 1.336	2.558^{41}	988	1695	4.22^{18}		s HF; v sl s al
s297	fluoride, hydrogen	NaF.HF	61.99	col, or wh cr powd, rhdr	2.08			s	s	
s298	fluoride *ortho*phosphate	$NaF.Na_3PO_4.12H_2O$	422.11		2.2165			12^{25}	57.5^{80}	
s299	fluoroacetate, mono-	$NaC_2H_2FO_2$	100.02	wh powd		200		111^{25}		1.4^{25} al; 5^{25} MeOH; 0.04^{25} acet; 0.0049^{25} CCl_4
s300	fluorophosphate, hexa-	$NaPF_6.H_2O$	185.97		2.369^{19}			103.2^{0}		
s301	fluorophosphate, mono-	Na_2PO_3F	143.95	col		*ca* 625		25		
s302	fluosilicate	Na_2SiF_6	188.06	col, hex, 1.312, 1.309	2.679	d		0.652^{17}	2.46^{100}	i al
s303	fluosulfonate	$NaSO_3F$	122.05	shiny leaf, hygr		d red heat		s		s al, acet; i eth
s304	formaldehyde-sulfoxylate	$NaHSO_2.CH_2O.2H_2O$	154.12	rhomb pr, hygr		64	d	v s		d a; s al, alk
s305	formate	$NaCHO_2$	68.01	col, monocl, deliq	1.92^{20}	253	d	97.2^{20}	160^{100}	sl s al; i eth
s306	2-furanacrylate	$NaC_7H_5O_3$	160.10	lt brn powd	1.919	d		s	s	sl s al; i eth
s307	*meta*germanate	Na_2GeO_3	166.57	wh, monocl, deliq, 1.59	3.31^{22}	1083			d	s a
s308	*meta*germanate, heptahydrate	$Na_2GeO_3.7H_2O$	292.68	col, rhomb		83		24.6^{0} 45.5^{25}		s a
s309	(mono-) *d*-glutamate	$NaC_5H_8NO_4$	169.11	wh cr		d		v s		sl s al
s310	glycerophosphate, monohydrate	$Na_2C_3H_7O_6P.H_2O$	234.05	yelsh visc liq; wh cr or powd				s		s al
s311	glycerophosphate, pentahydrate	$Na_2C_3H_7O_6P.5\frac{1}{2}H_2O$	315.12	wh pl, sc or powd		>130		v s		i al
s312	gold sulfide	$NaAuS.4H_2O$	324.08	col, monocl		d		s		s al
s313	hydride	NaH	24.00	silver need, 1.470	0.92	d 800		d	d	s molten Na; i CS_2, CCl_4, NH_3, bz
s314	hydroxide	NaOH	40.00	wh, deliq, 1.3576	2.130	318.4	1390	42^{0}	347^{100}	v s al, glyc; i acet, eth
s315	iodate	$NaIO_3$	197.89	wh, rhomb	$4.277^{17.5}$	d		9^{20}	34^{100}	i al; s ac a
s316	*meta*periodate	$NaIO_4$	213.89	col, tetr	3.865^{16}	d 300		14.44^{25}	$38.9^{51.5}$	s H_2SO_4, HNO_3, ac a
s317	*meta*periodate, trihydrate	$NaIO_4.3H_2O$	267.94	col, rhombdsh, effl	3.219^{18}_{4}	d 175		18.78^{25}	$36.4^{34.5}$	
s318	*para*periodate	Na_5IO_6	337.85	wh		800 d		d		
s319	(tri-)*para*periodate	$Na_3H_2IO_6$	293.88	col, hexag				sl s		s con NaOH sol
s320	iodide	NaI	149.89	col, cub, 1.7745	3.667^{25}_{4}	651	1304	184^{25}	302^{100}	42.57^{25} al; 39.9^{25} acet; s glyc
s321	iodide, dihydrate	$NaI.2H_2O$	185.92	col, monocl	$2.448^{20.8}$	752		317.9^{0}	1550^{100}	v s NH_3
s322	iodoplatinate	$Na_2PtI_6.6H_2O$	1110.59	brn, monocl	3.707			v s		s al
s323	iridium chloride	Sodium hexachloroiridate. $Na_3IrCl_6.12H_2O$	690.07	olive cr, rhomb or trig-rhomb		50		s	s	i al
s324	iron (III) nitrosopentacyanide	$Na_2[Fe(CN)_5NO].2H_2O$	297.95	ruby red, rhomb, 1.605, 1.575, 1.56	1.687^{25}	$-H_2O$, 100	d 160	40^{16}	s	
s325	iron (III) oxalate	$Na_3[Fe(C_2O_4)_3].5\frac{1}{2}H_2O$	487.96	grn, monocl	$1.973^{17.5}$	$-4H_2O$	d 300	32^{0}	182^{100}	
s326	iron (III) sulfate	$3Na_2SO_4.Fe_2(SO_4)_3.6H_2O$	934.09	wh, trig, 1.558, 1.613	2.5	$-6H_2O$, 100		d v sl		i al
s327	lactate	$NaC_3H_5O_3$	112.06	col or yelsh liq, very hygr		17	d 140	v s		s al; i eth
s328	lithium sulfate	$Na_3Li(SO_4)_2.6H_2O$	376.12	col, ditrig	2.009	$-6H_2O$, 50		s	s	
s329	magnesium carbonate	$Na_2CO_3.MgCO_3$	190.31	wh, rhomb	2.729^{15}	677 CO_2 1240/atm		d	d	
s330	magnesium sulfate	Nat. bloedite. $Na_2SO_4.MgSO_4.4H_2O$	334.48	col, monocl, 1.486, 1.488, 1.489	2.23			s		
s331	magnesium tartrate	$Na_2Mg(C_4H_4O_6)_2.10H_2O$	546.59	wh, monocl pr or powd				s	s	

No.	Name	Synonyms and Formulae	Mol. wt.	Crystalline form, properties and index of refraction	Density or spec. gravity	Melting point, °C	Boiling point, °C	Solubility, in grams per 100 cc		
								Cold water	Hot water	Other solvents
	Sodium									
s332	manganate	$Na_2MnO_4.10H_2O$	345.07	grn, monocl		17		s	d	
s333	*per*manganate	$NaMnO_4$	141.93	red cr, deliq		d		v s	v s	
s334	*per*manganate, trihydrate	$NaMnO_4.3H_2O$	195.97	purp cr, deliq	2.47	d 170		v s	v s	s NH_3; d alk
s335	methanearsenate	$Na_2CH_3AsO_3.6H_2O$	292.03	wh cr powd		130–140		*ca* 100		sl s al; i bz, eth, oils
s336	methoxide	$CH_3ONa.2CH_3OH$	118.11	wh powd		d, $-CH_3OH$		s, d		s CH_3OH
s337	methylsulfate	$NaCH_3SO_4.H_2O$	152.10	col cr, hygr				s		s al
s338	molybdate	Na_2MoO_4	205.92	opaque wh	3.28^{18}	687		s $44.3'$	84^{100}	
s339	molybdate, dihydrate	$Na_2MoO_4.2H_2O$	241.95	wh, rhbdr	3.28(?)	$-2H_2O$, 100		56.2^{0}	115.5^{100}	i meth acet
s340	*deca*molybdate	$Na_2Mo_{10}O_{31}.21H_2O$	1879.68	wh, monocl pr				sl s	0.842^{100}	
s341	*di*molybdate	$Na_2Mo_2O_7$	349.86	wh need		612		sl s	sl s	
s342	*octa*molybdate	$Na_2MoO_{25}.17H_2O$	1519.75	monocl cr		$-H_2O$, 20		v s	v s	
s343	*para*molybdate	$Na_6Mo_7O_{24}.22H_2O$	1589.84	col, monocl, effl		700 $-H_2O$, 100–120		117.9^{30} (anhydr)	s	
s344	*tetra*molybdate	$Na_2Mo_4O_{13}.6H_2O$	745.82	yel need				39.8^{21}	v s	
s345	*tri*molybdate	$Na_2Mo_3O_{10}.7H_2O$	619.96	need acicular		528 $-6H_2O$, 100–120		3.878^{20}	13.7^{100}	
s346	*meta*niobate	$Na_2Nb_2O_6.7H_2O$	453.90	pseudo-cub	4.512–4.559	$-H_2O$, 100		s		
s347	nitrate	Soda niter. $NaNO_3$	84.99	col, trig or rhbdr, 1.587, 1.336	2.261	306.8	d 380	92.1^{25}	180^{100}	s al, MeOH; v s NH_3; v sl s acet; sl s glyc
s348	nitride	Na_3N	82.98	dk gray		d 300		d		
s349	nitrite	$NaNO_2$	69.00	col-yel, rhomb pr, hygr	2.168^{0}	271	d 320	81.5^{15}	163^{100}	0.3^{20} eth; 4.4^{20} MeOH; 3 abs al; v s NH_3
s350	*hypo*nitrite	$Na_2N_2O_2$	105.99		1.728^{25}	d 300		d		i al
s351	*p*-nitrophenoxide	$NaOC_6H_4NO_2.4H_2O$	233.15	yel, monocl pr		$-2H_2O$, 36 $-4H_2O$, 120	d	5.97^{25}		sl s al
s352	nitroplatinite	$Na_2Pt(NO_2)_4$	425.09	pa yel rhomb or monocl pr, effl				s	s	
s353	nitroprusside	$Na_2[Fe(NO)(CN)_5].2H_2O$	297.95	red, rhomb	1.72			40^{16}		s al
s354	oleate	$Na_2C_{18}H_{32}O_2$	304.45	wh cr, or yel amorph gran		232–235		10^{12}		s al; sl s eth
s355	oxalate	$Na_2C_2O_4$	134.00	col cr, or wh powd	2.34	d 250–270		3.7^{20}	6.33^{100}	i al, eth
s356	oxalate, hydrogen	$NaHC_2O_4.H_2O$	130.03	wh, monocl		$-H_2O$, 100	d 200	1.7^{15}	21^{100}	
s357	oxalatoferrate (III)	$Na_3[Fe(C_2O_4)_3].xH_2O$	365.89 $+xH_2O$	grn, monocl cr	$1.973^{17.5}$	$-H_2O$, 100–120		32.5	182^{100}	
s358	oxide, mon-	Na_2O	61.98	wh-gray, deliq	2.27	subl 1275		d	d	d al
s359	oxide, per-	Na_2O_2	77.98	yel-wh powd	2.805	d 460	d 657	s	d	d al, NH_3; s dil a; i alk
s360	oxide, per-, octahydrate	$Na_2O_2.8H_2O$	222.10	wh, hex		d 30	d	s	d	i al
s361	palmitate	$NaC_{16}H_{31}O_2$	278.41	wh cr		270				
s362	pentobarbital	$NaC_{11}H_{17}N_2O_3$	248.26					s	s, d	s al
s363	phenobarbital	$NaC_{12}H_{11}N_2O_3$	254.22	wh				v s		s al; i eth, chl
s364	1-phenol-4-sulfonate(*p*-)	$NaC_6H_4(OH)SO_3.2H_2O$	232.19	col, monocl or gran, sl effl		d		23.8^{25}	125^{100}	0.75^{25} al; 20^{25} glyc
s365	phenoxide	$NaOC_6H_5$	116.10	wh cr need, deliq				v s		s al, acet; d a
s366	phenylcarbonate	$NaC_7H_5O_3$	160.11	col powd		d 120		d		d acet
s367	*hypo*phosphate	$Na_4P_2O_6.10H_2O$	430.06	col, monocl, 1.477, 1.482, 1.504	1.823	d		1.49^{25}	5.46^{50}	
s368	*hypo*phosphate, di-H.	$Na_2H_2P_2O_6.6H_2O$	314.03	col, monocl, 1.468, 1.490, 1.504	1.849	250 (anhydr)	$-6H_2O$, 100	2.35	25	s dil H_2SO_4, NH_4OH; i al
s369	*meta*phosphate, hexa-	Graham's salt. $(NaPO_3)_6$	611.17	col glass, 1.482 ± 0.002				v s		
s370	*meta*phosphate, tri-, hexahydrate	Knorre's salt. $(NaPO_3)_3.6H_2O$	413.98	col, tricl, effl, 1.433, 1.442, 1.446		53; $-6H_2O$, 50		s		
s371	*ortho*phosphate	$Na_3PO_4.10H_2O$	344.09	col, oct	$2.536^{17.5}$ (anhydr)	100		8.8 (anhydr)		
s372	*ortho*phosphate	$Na_3PO_4.12H_2O$	380.12	col, trig, 1.446, 1.452	1.62^{20}	d 73.3–76.7	$-12H_2O$, 100	1.5^{0}	157^{70}	i CS_2, al
s373	*ortho*phosphate, di-H	$NaH_2PO_4.H_2O$	137.99	col, rhomb, 1.456, 1.458, 1.487	2.040	$-H_2O$, 100	d 204	59.9^{0}	427^{100}	v sl s eth, chl, tol; i al
s374	*ortho*phosphate, di-H	$NaH_2PO_4.2H_2O$	156.01	col, rhomb, 1.4629	1.91	60		v s	v s	—
s375	*ortho*phosphate, mono-H	Sörensen's sodium phosphate. $Na_2HPO_4.2H_2O$	177.99	rhomb bispheroidal, 1.463	2.066^{16}	$-2H_2O$, 95		100^{50}	117^{80}	
s376	*ortho*phosphate, mono-H	$Na_2HPO_4.7H_2O$	268.07	col, monocl pr, 1.442	1.679	$-5H_2O$, 48.1		104^{40}		i al
s377	*ortho*phosphate, mono-H	$Na_2HPO_4.12H_2O$	358.14	col, rhomb or monocl, eff, wh powd, 1.432, 1.436, 1.437	1.52	$-5H_2O$, 35.1	$-12H_2O$, 100	4.15	87.4^{34}	i al

No.	Name	Synonyms and Formulae	Mol. wt.	Crystalline form, properties and index of refraction	Density or spec. gravity	Melting point, °C	Boiling point, °C	Solubility, in grams per 100 cc		
								Cold water	Hot water	Other solvents
	Sodium									
s378	*pyro*phosphate	$Na_4P_2O_7$	265.90	wh cr, 1.425	2.534	880		3.16^{0}	40.26^{100}	
s379	*pyro*phosphate	$Na_4P_2O_7.10H_2O$	446.06	col, monocl, 1.450, 1.453, 1.460	1.815–1.836	$-H_2O$, 93.8 m.p. 880		5.41^{0}	93.11^{100}	i al, NH_3
s380	*pyro*phosphate, di-H	$Na_2H_2P_2O_7.6H_2O$	330.03	monocl, 1.4599, 1.4645, 1.4649	1.85	$-H_2O$, 220		6.9^{0}	35^{40}	
s381	phosphide	Na_3P	99.94	red		d		d, PH_3		
s382	*hypo*phosphite	$NaH_2PO_2.H_2O$	105.99	col, monocl, deliq		d viol		100^{25}	667^{100}	v s al; s glyc; sl s NH_3, NH_4OH
s383	*ortho*phosphite, di-H	$NaH_2PO_3.2\frac{1}{2}H_2O$	149.01	col, monocl, 1.419, 1.431, 1.449		42	$-2\frac{1}{2}H_2O$, 100	56^{0}	193^{42}	
s384	*ortho*phosphite, mono-H	$Na_2HPO_3.5H_2O$	216.04	wh, rhomb deliq, β, 1.443	53	d 200–250		s	v s	i al, NH_4OH
s385	*tri*phosphate	Sodium tripoly-phosphate. $Na_5P_3O_{10}$	367.86	powd and gran				14.5^{25}	32.5^{100}	
s386	phthalate	$Na_2C_8H_4O_4$	210.10	wh powd or pearly pl		$-H_2O$, 150				
s387	platinate, hydroxo-	$Na_2Pt(OH)_6$	343.11	yel or red-brn, hex		$-3H_2O$, 150–170	d	s		i al; sl s HCl
s388	platinum cyanide	$Na_2[Pt(CN)_4].3H_2O$	399.19	col, tricl	2.646	$-H_2O$, 120–125		s		s al
s389	plumbate, hydroxo-	$Na_2Pb(OH)_6$	355.21	yel-wh lumps, hygr				d to PbO_2		d a; s alk
s390	potassium(*dl*)-tartrate	$NaKC_4H_4O_6$	210.14	col, tricl		90–100	d 200	47.4^{6} (anhydr)	v s	
s391	propionate	$NaC_3H_5O_2$	96.06	wh, gran powd				s		s al
s392	*per*rhenate	$NaReO_4$	273.19	col, hex pl, hygr	5.39	300 (in O_2) d 440 (vac)		100^{20}		s al
s393	*pyrohypo*rhenate	$Na_4Re_2O_7.H_2O$	594.37	sandy yel cr				0.004		
s394	rhodiumchloride	$Na_3RhCl_6.12H_2O$	600.78	dk-red cr, monocl pr		$-12H_2O$, 120		v s	v s	i al
s395	rhodiumnitrite	$Na_3[Rh(NO_2)_6]$	447.91	wh cr		d 360		40^{17}	s	i al; d a
s396	*per*ruthenate	$NaRuO_4.H_2O$	206.07	blk cr, lamellar		d 440 vac		v s	d	
s397	salicylate	$NaC_7H_5O_3$	160.11	wh cr powd				111^{15}	125^{25}	17^{15} al; 25 glyc
s398	selenate	Na_2SeO_4	188.94	col, rhomb	$3.213^{17.4}$			84^{35}	72.8^{100}	
s399	selenate, decahydrate	$Na_2SeO_4.10H_2O$	369.09	col, monocl	1.603–1.620	*ca* 32 trans		43.5^{20}	340^{100}	
s400	selenide	Na_2Se	124.94	wh to red, cr, deliq	2.625^{10}	>875		d		i NH_3
s401	selenite	$Na_2SeO_3.5H_2O$	263.01	wh cr, tetrag				s	s	i al
s402	silicate	Waterglass. $Na_2O.xSiO_2(x=3-5)$		col, amorph, deliq				s	s	i al, K and Na salts
s403	*di*silicate	$Na_2Si_2O_5$	182.15	rhomb pearly luster 1.500, 1.510		874		s	s	
s404	*meta*silicate	Na_2SiO_3	122.06	col, monocl, α 1.518, γ 1.527	2.4	1088		s	s, d	i al, K and Na salts
s405	*meta*silicate	$NaSiO_3.9H_2O$	284.20	col, rhomb bi-pyramid, effl		40–48	$-6H_2O$, 100	v s	v s	s dil NaOH; i al, a
s406	*ortho*silicate	Na_4SiO_4	184.04	col, hex, 1.530		1018		s	s	
s407	silicotungstate, dodeca-	$Na_4[Si(W_3O_{10})_4].20H_2O$ Sodium tungstosilicate	3326.53	col, tricl		$-7H_2O$, 100	d	v s	v s	s a, sl s al
s408	stannate, hydroxo-	$Na_2Sn(OH)_6$	266.71	col, hex or wh powd, or lumps		$-3H_2O$, 140		$61.3^{15.5}$	50^{100}	i al, acet
s409	stearate	$NaC_{18}H_{35}O_2$	306.47	wh fatty powd				s	s	s h al
s410	succinate	$Na_2C_4H_4O_4.6H_2O$	270.15	wh, gran or powd		$-6H_2O$, 120		21.45^{0}	86.63^{75}	v sl s al
s411	succinate, tetra-hydroxy-	Sodium dihydroxy tartrate. $Na_2C_4H_4O_8.3H_2O$	280.10			d		0.032^{0}	d	d min a; i al, eth
s412	sulfanilate	$NaC_6H_4(NH_2)SO_3$	195.17	wh lust cr leaf				s		
s413	sulfate, anhydr	Na_2SO_4	142.04	monocl (between *ca* 160–185), 1.480		884; tr to hex *ca* 241		s	$42-5^{100}$	s HI
s414	sulfate, anhydr	Nat. thenardite. Na_2SO_4	142.04	orthorhomb, 1.484, 1.477, 1.471	2.68			4.76^{0}	42.7^{100}	s glyc; i al
s415	sulfate, decahydrate	Glauber's salt, mirabilite. $Na_2SO_4.10H_2O$	322.19	col, monocl, effl, 1.394, 1.396, 1.398	1.464	32.38	$-10H_2O$, 100	11^{0}	92.7^{30}	i al
s416	sulfate, heptahydrate	$Na_2SO_4.7H_2O$	268.15	wh, rhomb or tetrag		tr to anhydr 24.4		19.5^{0}	44^{20}	i al
s417	*pyro*sulfate	Sodium metabisulfite. $Na_2S_2O_5$	190.10	wh powd or cr $(+7H_2O)$	1.4	>d 150		54^{20}	81.7^{100}	sl s al; s glyc
s418	*pyro*sulfate	$Na_2S_2O_7$	222.16	wh, transluc cr, deliq	2.658^{25}	400.9	d 460	s		s fum H_2SO_4
s419	sulfate hydrogen	$NaHSO_4$	120.06	col, tricl	2.435^{13}	>315	d	28.6^{25}	100^{100}	sl s al; i NH_3
s420	sulfate hydrogen, monohydrate	$NaHSO_4.H_2O$	138.07	col, monocl, deliq ~1.46	$2.103^{13.5}_{4}$	58.54±0.5		*ca* 67, d	d	d al
s421	sulfide, hydro-	NaHS	56.06	col, rhomb or wh gran cr, deliq		350		vs		s al

No.	Name	Synonyms and Formulae	Mol. wt.	Crystalline form, properties and index of refraction	Density or spec. gravity	Melting point, °C	Boiling point, °C	Solubility, in grams per 100 cc: Cold water	Hot water	Other solvents
	Sodium									
s422	sulfide, hydrodihydrate	$NaHS.2H_2O$	92.09	col need, deliq		d		s	s	d a; s al
s423	sulfide, mono-	Na_2S	78.04	wh cr, deliq	1.856^{14}	1180		15.4^{10}	57.2^{90}	d a; sl s al; i eth
s424	sulfide, mono-hydrate	$Na_2S.9H_2O$	240.18	col, tetr, deliq	1.427^{16}_{4}	d 920		47.5^{10}	96.7^{10}	d, sl s al
s425	sulfide, penta-	Na_2S_5	206.30	yel (exist ?)		251.8		s	s	s al
s426	sulfide, tetra-	Na_2S_4	174.24	yel, cub, hygr		275	d	s		s al
s427	sulfide, hydrotrihydrate	$NaHS.3H_2O$	110.11	col, lust rhomb cr		22	d	s	s	s al
s428	sulfite	Na_2SO_3	126.04	wh powd or hex, prism 1.565, 1.515	$2.633^{15.4}$	d red heat	d	12.54^{0}	28.3^{80}	sl s al; i liq Cl_2, NH_3
s429	sulfite hydrate	$Na_2SO_3.7H_2O$	252.15	col, monocl, effl	1.539^{15}	$-7H_2O$, 150	d	32.8^{0}	196^{40}	sl s al
s430	*hydro*sulfite	Dithionite, hyposulfate $Na_2S_2O_4.2H_2O$	210.14	col, monocl(?) cr, or yel-wh powd		d 52		25.4^{20}	d	d a; s alk; i al
s431	sulfite, hydrogen	$NaHSO_3$	104.06	wh, monocl, yel in sol, 1.526	1.48	d		v s	v s	sl s al
s432	*d*(& *l*)-tartrate	$Na_2C_4H_4O_6.2H_2O$	230.08	col, rhomb, 1.545, 1.49	1.818	$-2H_2O$, 150		29^{6}	66^{43}	i al
s433	*d*-tartrate, hydrogen	$NaHC_4H_4O_6.H_2O$	190.09	wh cr powd, rhomb, 1.53, 1.54, 1.60		$-H_2O$, 100	d 234	6.7^{18}	9.2^{30}	
s434	*dl*-tartrate hydrogen	$NaHC_4H_4O_6.H_2O$	190.09	col, monocl or tricl, 1.53, 1.54, 1.60		$-H_2O$, 100	d 219	8.9^{19}		
s435	*ortho*tellurate, tetra-H	$Na_2H_4TeO_6$	273.61	col, hex pl		d	d	0.77^{18}	2^{100}	s h dil HNO_3: i NaOH
s436	telluride	Na_2Te	173.58	wh cr powd very hygr, d in air	2.90	953		v s, d	v s, d	
s437	tellurite	Na_2TeO_3	221.58	wh, rhomb pr				sl s	s	
s438	thioantimonate	Schlippe's salt. $Na_3SbS_4.9H_2O$	481.11	pa yel, cub	1.806	87	d 234	20.15^{0}	100^{100}	i al, eth
s439	thioarsenate	$Na_3AsS_4.8H_2O$	416.27	yel, monocl, β 1.6802		d		v s	d	i al
s440	thiocarbonate, tri	$Na_2CS_3.H_2O$	172.20	yel need, deliq		d 75		s	d	s al; i eth, bz
s441	thiocyanate	NaSCN	81.07	col, rhomb deliq pois, ~1.625		287		$139.31^{21.3}$	225^{100}	v s al, acet
s442	*di*thionate	$Na_2S_2O_6.2H_2O$	242.13	col, rhomb, 1.482, 1.495, 1.519	2.189	$-2H_2O$, 110	$-SO_2$, 267	47.6^{16}	90.9^{100}	s HCl; i al
s443	thiosulfate	$Na_2S_2O_3$	158.11	col, monocl	1.667			50	231^{100}	i al
s444	thiosulfate, pentahydrate	"Hypo", sodium hyposulfite. $Na_2S_2O_3.5H_2O$	248.18	col, monocl, effle, 1.489, 1.508, 1.536	1.729^{17}	40–45 d 48	$-5H_2O$, 100	79.4^{0}	291.1^{45}	s NH_3; i al
s445	thiosulfoaurate (I)	$Na_3[Au(S_2O_3)_2].2H_2O$	526.22	wh cr, monocl	3.09	$-H_2O$, 150	d	50		i al
s446	*tri*titanate	$Na_2Ti_3O_7$	301.68	wh need, monocl	3.35–3.50	1128		i		s h HCl
s447	tungstate	Na_2WO_4	293.83	wh, rhomb	4.179	698		57.5^{0} 73.2^{21}	96.9^{100}	
s448	tungstate, dihydrate	$Na_2WO_4.2H_2O$	329.86	col pl, rhomb, 1.5533	3.23–3.25	$-2H_2O$, 100 anhydr 698		41^{0}	123.5^{100}	sl s NH_3; i al, a
s449	*meta*tungstate	$Na_2O.4WO_3.10H_2O$	1169.53	col, oct		706.6		s	v s	i a
s450	*para*tungstate	$Na_6W_7O_{24}.16H_2O$	2097.12	col, tricl	3.987	$-12H_2O$, 100; $-16H_2O$, 300		8	d	
s451	*meta*uranate	Na_2UO_4	348.01	gr yel or red pl, rhomb pr				i	i	s dil a, alk carb
s452	uranyl acetate	$NaUO_2(C_2H_3O_2)_3$	470.15	yel, tetr pr, 1.501	2.56					
s453	uranyl carbonate	$2Na_2CO_3.UO_2CO_3$	542.02	yel cr		d 400		sl s		i al
s454	urate	$Na_2C_5H_2N_4O_3.H_2O$	230.09	wh gran powd or hard cr nodules					1.3^{100}	v sl s 90 % al
s455	urate, acid	$NaHC_5H_2N_4O_3$	190.09	wh gran powd				0.083	0.8^{100}	
s456	valerate	$NaC_5H_9O_2$	124.12	wh cr or mass, hygr		140		s		s al
s457	*meta*vanadate	$NaVO_3$	121.93	col, monocl pr		630		21.1^{25}	38.8^{75}	
s458	*ortho*vanadate	Na_3VO_4	183.94	col, hex pr		850–866		s		i al
s459	*ortho*vanadate, decahydrate	$Na_3VO_4.10H_2O$	364.06	wh, cub or hex cr, 1.5305, 1.5398, 1.5475				s	s	
s460	*ortho*vanadate, hexadecylhydrate	$Na_3VO_4.16H_2O$	472.15	col need		866 (anhydr)		v s	d	i al
s461	*pyro*vanadate	$Na_4V_2O_7$	305.84	col, hex		632–654		s	s	i al
s462	ethylxanthate	NaC_2H_5OCSS	144.19	yelsh powd				s		s al
s463	zinc uranyl acetate	$NaZn(UO_2)_3(C_2H_3O_2)_9.9H_2O$	1591.98	monocl cr, α 1.475, γ 1.480				i		s al
	Stannous	See under tin								
	Stannic	See under tin								

No.	Name	Synonyms and Formulae	Mol. wt.	Crystalline form, properties and index of refraction	Density or spec. gravity	Melting point, °C	Boiling point, °C	Solubility, in grams per 100 cc: Cold water	Hot water	Other solvents
	Strontium									
s464	**Strontium**	Sr	87.62	silv wh to pa yel met	2.6^{20}	769	1384	d	d	s a, al, liq NH_3
s465	acetate	$Sr(C_2H_3O_2)_2$	205.71	wh cr	2.099	d		36.9	36.4^{97}	0.26^{15} MeOH
s466	acetate	$Sr(C_2H_3O_2)_2.\frac{1}{2}H_2O$	214.72	wh cr powd		$-\frac{1}{2}H_2O$, 150		s		sl s al
s467	*ortho*arsenate, acid	$SrHAsO_4.H_2O$	245.56	rhomb, need	3.606^{15}; 4.035 (anhydr)	$-H_2O$, 125		$0.284^{15.5}$	d	s a
s468	*ortho*arsenite	$Sr_3(AsO_3)_2.4H_2O$	580.76	cr, or wh powd				sl s		s a; sl s al
s469	borate, *tetra*-	$SrB_4O_7.4H_2O$	314.92						77^{100}	s HNO_3; NH_4 salts
s470	boride, hexa-	SrB_6	152.49	blk, cub	3.39^{15}	2235		i	i	s HNO_3; i HCl
s471	bromate	$Sr(BrO_3)_2.H_2O$	361.45	col yelsh, monocl, hygr	3.773	$-H_2O$, 120	d 240	33^{18}		
s472	bromide	$SrBr_2$	247.44	wh, hex need, hygr, 1.575	4.216^{24}	643	d	100^{20}	222.5^{100}	s al, amyl al
s473	bromide, hexahydrate	$SrBr_2.6H_2O$	355.53	col, hex, hygr	2.386^{25}_{4}	tr to $2H_2O$, 88.6	$-6H_2O$, >180	204.2^{0}	∞	63.9^{20} al; 113.4^{20} MeOH; 0.6^{20} acet; i eth
s474	carbide	SrC_2	111.64	blk, tetr	3.2	>1700		d	d	d a
s475	carbonate	Nat. strontianite. $SrCO_3$	147.63	col, rhomb, or wh powd trfrs to −hex at 926, 1.516, 1.664, 1.666	3.70	1497^{69atm}	$-CO_2$, 1340	0.0011^{18}	0.065^{100}	0.12 aq CO_2; s a, NH_4 salts
s476	chlorate	$Sr(ClO_3)_2$	254.52	col, rhomb, or wh powd, 1.516, 1.605, 1.626	3.152	d 120		174.9^{18}	v s	s dil al; i abs al
s477	*per*chlorate	$Sr(ClO_4)_2$	286.52	col cr, hygr				310^{25}	v s	212 MeOH; 181 al; i eth
s478	chloride	$SrCl_2$	158.53	col, cub 1.650^{25}	3.052	873	1250	53.8^{20}	100.8^{100}	v sl s abs al, acet; i NH_3
s479	chloride, dihydrate	$SrCl_2.2H_2O$	194.56	transp leaf, 1.594, 1.595, 1.617	2.6715^{25}					
s480	chloride, fluoride	$SrCl_2.SrF_2$	284.14	col, tetr, 1.651, 1.627	4.18	962		d	d	s conc HNO_3, conc HCl, i al
s481	chloride, hexahydrate	$SrCl_2.6H_2O$	266.62	col, trig, 1.536, 1.487	1.93	115, $-4H_2O$, 60	$-6H_2O$, 100	106.2^{0}	205.8^{40}	3.8^{6} al
s482	chromate	$SrCrO_4$	203.61	yel, monocl	3.895^{15}			0.12^{15}	3^{100}	s HCl, HNO_3, ac a, NH_4 salts
s483	cyanide	$Sr(CN)_2.4H_2O$	211.72	wh, rhomb, deliq		d		v s		
s484	cyanoplatinite	$Sr[Pt(CN)_4].5H_2O$	476.86	col, monocl pr, 1.696		$-5H_2O$, 150				s abs al
s485	glycerophosphate	$SrC_3H_7O_6P$	257.68	wh powd				sl s		i al
s486	ferrocyanide	$Sr_2Fe(CN)_6.15H_2O$	657.42	yel, monocl				50	100	
s487	fluoride	SrF_2	125.62	col, cub or wh powd, 1.442	4.24	1450+	2489^{760}	0.011^{0}	0.012^{27}	s hot HCl; i HF, al, acet
s488	fluosilicate	$SrSiF_6.2H_2O$	265.73	monocl	$2.99^{17.5}$	d		3.2^{15}	v s	s HCl; 0.065^{15} 50% al
s489	formate	$Sr(CHO_2)_2$	177.66	col, rhomb, 1.559, 1.547, 1.598	2.693	71.9		9.1^{0}	34.4^{100}	
s490	formate, dihydrate	$Sr(CHO_2)_2.2H_2O$	213.69	col, rhomb, 1.484, 1.521, 1.538	2.25	d, $-2H_2O$, 100		$11.62^{28.6}$	26.57^{100}	i al, eth
s491	hydride	SrH_2(?)	89.64	wh, rhomb, hygr	3.72	d 675	subl 1000 (in H_2)	d	d	d al
s492	hydroxide	$Sr(OH)_2$	121.63	wh, deliq	3.625	375 (in H_2)	$-H_2O$, 710	0.41^{0}	21.83^{100}	s a, NH_4Cl
s493	hydroxide, octahydrate	$Sr(OH)_2.8H_2O$	265.76	col, tetr, deliq, 1.499, 1.476	1.90	$-8H_2O$, 100		0.90^{0}	47.71^{100}	s a, NH_4Cl; i acet
s494	iodate	$Sr(IO_3)_2$	437.43	tricl	5.045^{15}			0.03^{15}	0.8^{100}	
s495	iodide	SrI_2	341.43	col pl	4.549^{25}_{4}	515	d	165.3^{0}	383^{100}	4.5^{20} al; 0.31^{0} NH_4OH; s MeOH
s496	iodide, hexahydrate	$SrI_2.6H_2O$	449.52	col-yelsh, hex, deliq	2.672^{25}	d 90		448.9^{0}	∞	s al; i eth
s497	lactate	$Sr(C_3H_5O_3)_2.3H_2O$	319.81	wh cr or gran powd		$-3H_2O$, 120		25	200^{100}	sl s al
s498	*per*manganate	$Sr(MnO_4)_2.3H_2O$	379.54	purpl, cub	2.75	d 175		270^{0}	291^{18}	
s499	molybdate	$SrMoO_4$	247.56	col, tetr ~1.91	4.54^{25}_{25}	d		0.0104^{17}		s a
s500	nitrate	$Sr(NO_3)_2$	211.63	col, cub	2.986	570		70.9^{18}	100^{90}	0.012 abs al; v s NH_3; sl s acet
s501	nitrate, tetrahydrate	$Sr(NO_3)_2.4H_2O$	283.69	col, monocl	2.2	$-4H_2O$, 100	1100 tr SrO	60.43^{0}	206.5^{100}	s liq NH_3; v sl s abs al, acet; i HNO_3

No.	Name	Synonyms and Formulae	Mol. wt.	Crystalline form, properties and index of refraction	Density or spec. gravity	Melting point, °C	Boiling point, °C	Solubility, in grams per 100 cc: Cold water	Hot water	Other solvents
	Strontium									
s502	nitride	Sr_3N_2	290.87					d	d	s HCl
s503	nitrite	$Sr(NO_2)_2.H_2O$	197.65	col, hex, 1.588	2.408_0^0	$-H_2O>100$	d 240	58.9^{0}	182^{100}	0.42^{20} 90% al
s504	*hypo*nitrite	$SrN_2O_2.5H_2O$	237.71	wh need	2.173_4^{25}			v sl s	sl s	v sl s NH_3
s505	oxalate	$SrC_2O_4.H_2O$	193.64	col cr		$-H_2O$, 150		0.0051^{18}	5^{100}	s HCl, HNO_3
s506	oxide	SrO	103.62	gray-wh, cub, 1.810	4.7	2430	~3000	0.69^{20}	22.85^{100}	30 fus KOH; sl s al; i eth, acet
s507	oxide, per-	SrO_2	119.62	wh powd	4.56	d 215^{760}		0.018^{20}	d	v s al, NH_4Cl; i acet
s508	oxide, per-, octahydrate	$SrO_2.8H_2O$	263.74	col cr	1.951	$-8H_2O$, 100	d	0.018^{20}	d	s NH_4Cl; i al, acet, NH_4OH
s509	*ortho*phosphate, di-	$SrHPO_4$	183.60	col, rhomb	3.544^{15}	1.62		i	i	s a, NH_4 salts
s510	salicylate	$Sr(C_7H_5O_3)_2.2H_2O$	397.88	col cr		d		5.6^{25}	28.6^{100}	1.5^{25}, 9.5^{78} al
s511	selenate	$SrSeO_4$	230.58	col, rhomb	4.23			i	i	s hot HCl; i HNO_3
s512	selenide	SrSe	166.58	wh, cub, 2.220	4.38			d	d	s HCl
s513	*meta*silicate	$SrSiO_3$	163.70	col, pr monocl, 1.599, 1.637	3.65	1580		i	i	
s514	*ortho*silicate	$SrSiO_4$	179.70	monocl, 1.728, 1.732, 1.758	3.84	>1750				
s515	sulfate	Nat. celestite. $SrSO_4$	183.68	col, rhomb, 1.622, 1.624, 1.631	3.96	1605		0.0113^{0}	0.014^{30}	sl s a; i al, dil H_2SO_4
s516	sulfate, hydrogen	$Sr(HSO_4)_2$	281.76	col		d		d		14^{70} H_2SO_4
s517	sulfide, hydro	$Sr(HS)_2$	153.76	col, cub need, 2.107		d		s	d	
s518	sulfide, mono	SrS	119.68	col, lt gray, cub, 2.107	3.70^{15}	>2000		i	d	d a
s519	sulfide, tetra-	$SrS_4.6H_2O$	323.97	redsh cr, hygr		25	$-4H_2O$, 100	s	s	s al
s520	sulfite	$SrSO_3$	167.68	col cr		d		0.0033^{17}		v s H_2SO_4; s a, HCl
s521	tartrate	$SrC_4H_4O_6.4H_2O$	307.75	wh, monocl	1.966			0.112^{0}	0.755^{85}	s dil HCl, dil HNO_3
s522	telluride	SrTe	215.22	wh, cub, 2.408	4.83					
s523	thiocyanate	$Sr(SCN)_2.3H_2O$	257.83	deliq		$-3H_2O$, 100	d 160–170	v s		v s al
s524	thiosulfate	$SrS_2O_3.5H_2O$	289.82	monocl need	2.17^{17}	$-4H_2O$, 100		2.5^{13}	57^{100}	i al
s525	*di*thionate	$SrS_2O_6.4H_2O$	319.81	trig, 1.530, 1.525	2.373	$-4H_2O$, 78		22^{16}	67^{100}	i al
s526	tungstate	$SrWO_4$	335.47	col, tetr	6.187	d		0.14^{15}		id a; i al
s527	**Sulfamic acid**	Amidosulfuric, aminosulfonic acid. NH_2SO_3H	97.09	col, rhomb	2.126^{25}	200 d	d	14.68	47.08^{80}	v sl s al, eth, acet; i CS_2, CCl
s528	**Sulfamide**	Sulfuryl amide. $SO_2(NH_2)_2$	96.11	rhomb pl	1.611	91.5	d 250	s		s al
s529	**Sulfur**(α)	S_8	256.512	yel, rhomb, 1.957	2.07^{20}	112.8 95.5 (revers.) 444.6	444.6	i	i	23^{0} CS_2; sl s tol, al, bz, eth, liq NH_3; s CCl_4
s530	(β)	S_8	256.512	pa yel, monocl	1.96	119.0	444.6	i	i	70 CS_2; s al, bz
s531	(γ)	S_8	256.512	pa yel, amorph	1.92	*ca* 120	444.6	i	i	i CS_2
s532	bromide, mono-	S_2Br_2	223.95	red liq, 1.730	2.63	−40	$54^{0.2}$	d	d	s CS_2
s533	chloride, di-	SCl_2	102.97	dk red liq, 1.557^{11}	1.621_{15}^{15}	−78	d 59			s CCl_4, bz; d al, et
s534	chloride, mono-	S_2Cl_2	135.03	yel-red liq, 1.666^{14}	1.678	−80	135.6	d	d	s bz, eth, CS_2
s535	chloride, tetra-	SCl_4	173.88	yel-br liq		−30	d −15	d	d	
s536	fluoride, hexa-	SF_6	146.05	col gas	gas 6.602 g/l liq $1.88^{-50.5}$	−50.5	63.8	sl s	sl s	s al, KOH
s537	fluoride, mono-	S_2F_2	102.12	col gas	liq 1.5^{-100}	−120.5	−38.4	d	d	d KOH
s538	fluoride, tetra-	SF_4	108.06	gas (exist ?)		−124	−40	d	d	
s539	(di) fluoride, deca-	S_2F_{10}	254.11	col liq	2.08_4^0	−92	29			d fus caust
s540	(tetra-) nitride, di-	S_4N_2	156.27	red liq or gray solid	1.901^{18}	23	d 100 expl	i		s eth; sl s al, CS_2
s541	(tetra-) nitride, tetra-	S_4N_4	184.28	or red, monocl	2.22^{15}	subl 179	expl 160	d		s CS_2, chl, bz, NH_3; sl s al, eth
s542	(tri-)*di*nitrogen dioxide	$S_3N_2O_2$	156.20	pa yel cr		100.7	d	i		s al, bz
s543	oxide, di-	SO_2	64.06	col gas or liq suffoc odor	gas 2.927 g/l liq 1.434	−72.7	−10	22.8^{0}	0.58^{90}	s al, ac a, H_2SO_4
s544	oxide, hept-	Sulfur oxide, per- S_2O_7	176.12	visc liq, or need		0	subl 10	d	d	s H_2SO_4
s545	oxide, mono-	SO (or S_2O_2)	48.06	col gas		d	d	d		
s546	oxide, sesqui-	S_2O_3	112.13	bl-grn cr		d 70–95		d	d	s al, eth, fum H_2SO_4
s547	oxide, tetra-	Sulfurperoxide. SO_4	96.06	wh		d 0–3		s, d		d dil H_2SO_4
s548	oxide, tri-(α)	SO_3	80.06	silky fibr need, stable modific	1.97^{20}	62.3	44.8	d	d	forms fum H_2SO_4
s549	oxide, tri-(β)	SO_3	80.06	asbestos like fibre, metastable		32.5	44.8	d	d	forms fum H_2SO_4

No.	Name	Synonyms and Formulae	Mol. wt.	Crystalline form, properties and index of refraction	Density or spec. gravity	Melting point, °C	Boiling point, °C	Solubility, in grams per 100 cc		
								Cold water	Hot water	Other solvents
	Sulfur(α)									
s550	oxide, tri-(γ)	SO_3	80.06	vitreous, orthorhomb, metastable	liq 1.920^{20}_{4} sld 2.29^{-10}	16.8	44.8	d	d	forms fum H_2SO_4
s551	oxy*tetra*chloride, mono-	S_2OCl_4	221.94	dk red liq	1.656^{0}		60	d	d	d al
s552	oxy*tetra*chloride, tri-	$S_2O_3Cl_4$	253.94	wh, rhomb need or pl		d 57		d	d	d al
s553	*tri*thiazyl chloride	S_4N_3Cl	205.73	pa yel cr		d 170 (vac)		d	d	
s554	**Sulfuric acid**	H_2SO_4	98.08	col liq	1.841 (96–98 %)	10.36 (100 %) 3.0 (98 %)	338 (98.3 %)	∞ ev heat	∞	d al
s555	dihydrate	$H_2SO_4.2H_2O$	134.11	col liq, 1.405	1.650^{0}	−38.9	167	∞	∞	d al, eth
s556	hexahydrate	$H_2SO_4.6H_2O$	206.17	liq		−54		v s	v s	
s557	monohydrate	$H_2SO_4.H_2O$	116.09	col liq or monocl cr, 1.438	1.788	8.62	290	∞	∞	d al
s558	octahydrate	$H_2SO_4.8H_2O$	242.20	liq		−62		v s	v s	
s559	peroxidi-	Per(di-)sulfuric acid. $H_2S_2O_8$	194.14	hygr cr		d 65	d	d	d	s al, eth, H_2SO_4
s560	peroximono-	Permonosulfuric acid. Caro's acid H_2SO_5	114.08			d 45		d	d	s H_3PO_4
s561	pyro-	$H_2S_2O_7$	178.14	col cr, hygr	1.9^{20}	35	d	d	d	d al
s562	tetrahydrate	$H_2SO_4.4H_2O$	170.14			−27		∞	∞	d al, eth
s563	**Sulfurous acid**	H_2SO_3	82.08	in sol only	*ca* 1.03			s		s al, eth, ac a
s564	**Sulfuryl** chloride	SO_2Cl_2	134.97	col liq, 1.444	1.6674^{20}_{4}	−54.1	69.1	d	d	s bz, ac a
s565	chloride fluoride	SO_2ClF	118.52	col gas	1.623^{0} g/l	−124.7	7.1	d		
s566	fluoride	SO_2F_2	102.06	col gas	gas 3.72 g/l liq 1.7	−136.7	−55.4	10^{9}		s al, CCl_4; sl s alk
s567	pyro-, chloride	$S_2O_5Cl_2$	215.03	col liq, 1.937^{20}	gas 9.6 g/l liq 1.818^{11}_{4}	−39 to −37	152.5	d	d	d a
t1	**Tantalum**	Ta	180.948	gray black hard metal, cub or powd	met 16.6 powd 14.401	2996	5425 ± 100	i	i	s HF, fus alk; i a
t2	boride, di-	TaB_2	202.57		11.15	3000(?)				
t3	bromide	$TaBr_5$	580.49	yel cr	4.67	265	348.8	d	d	s abs al, eth
t4	carbide	TaC	192.96	blk, cub	13.9	3880	5500	i	i	sl s H_2SO_4. HF
t5	chloride, penta-	$TaCl_5$	358.21	lt yel, vitr cr powd	3.68^{27}	216	242	d		s abs al, H_2SO_4
t6	fluoride	TaF_5	275.94	col, tetrag, deliq	4.74	96.8	229.5	s		s HF, eth
t7	nitride	TaN	194.95	br bronze or blk, hex	16.30	3360 ± 50		i	i	sl s aq reg, HF, HNO_3
t8	oxide, pent-	Ta_2O_5	441.89	col, rhomb	8.2	1800		i	i	s fus $KHSO_4$, HF; i a
t9	oxide, pent- hydrate	Tantalic acid $Ta_2O_5xH_2O$	......	col gel				s		s alk, exc conc HNO_3; i a
t10	oxide, tetr-	Ta_2O_4 (or TaO_2)	425.89	dk gray powd		oxidizes		i	i	i a
t11	sulfide	Ta_2S_4 (or TaS_2)	490.15	blk powd or cr		>1300		i	i	sl s HF+HNO_3; i HCl
t12	**Telluric acid, ortho-**	$Te(OH)_6$ or $H_2TeO_4.2H_2O$	229.64	wh, monocl pr	3.071	136		s	s	sl s dil a, HNO_3; i abs al, acet, eth
t13	**Telluric acid,**	$Te(OH)_6$ or H_6TeO_6	229.64	wh, cub	3.158	136		s	s	sl s dil a, HNO_3; i abs al, acet, eth
t14	**Tellurium**	Te	127.60	br blk, amorph, 1.0025	6.00	449.5 ± 0.3	989.8 ± 3.8	i	i	s H_2SO_4, HNO_3, aq reg, KCN, KOH; i HCl, CS_2
t15	**Tellurium**	Te	127.60	rhomb silv wh met, 1.0025	6.25	452	1390	i	i	s H_2SO_4, HNO_3, aq reg, KCN KOH; i HCl, CS_2
t16	bromide, di-	$TeBr_2$	287.42	brn to gray grn, need, unstable		210	339	d		s eth; sl s a; d NaOH
t17	bromide, tetra-	$TeBr_4$	447.27	or cr	4.31^{15}_{4}	380 ± 6	d 421	sl s	d	s eth, a, tart a, NaOH
t18	chloride, di-	$TeCl_2$	198.50	blk cr or amorph, unstable	7.05	209 ± 5	327	d	d	s min a, tart a; d NaOH
t19	chloride, tetra	$TeCl_4$	269.41	wh to yel cr, deliq	3.26^{18} 2.559^{232}	224	380^{760}	s d	s d	s HCl, bz, al, chl, CCl_4; i CS_2
t20	ethoxide	$Te(OC_2H_5)_4$	307.85			20	$107–107.5^{8.5}$			
t21	fluoride, hexa-	TeF_6	241.59	col gas unpleas odor	sol 4.006^{-191} liq 2.56^{-35}	−36	+35.5	d	d	d a, alk
t22	fluoride, tetra-	TeF_4	203.59	wh cr hygr		subl	>97	d	d	
t23	hydride	H_2Te	129.62	col gas pois	4.49	−48.9	-2.2^{760}	v s	s	d al

PHYSICAL CONSTANTS OF INORGANIC COMPOUNDS (Continued)

No.	Name	Synonyms and Formulae	Mol. wt.	Crystalline form, properties and index of refraction	Density or spec. gravity	Melting point, °C	Boiling point, °C	Solubility, in grams per 100 cc Cold water	Hot water	Other solvents
	Tellurium									
t24	iodide, di-	TeI_2	381.41	blk cr (exist ?)	...	subl	...	i	i	...
t25	iodide, tetra,-	TeI_4	635.22	blk cr	5.403^{15}_{4}	280	d	sl s	d	s alk, aq NH_3, HI
t26	methoxide	$Te(OCH_3)_4$	251.74	solid	...	...	123–124	...	...	...
t27	oxide, di-	Tellurite. TeO_2	159.60	wh, tetr or rhomb, 2.00, 2.18(Li), 2.35	tetr 5.67^{15} rhomb 5.91^{0}	733	1245	i	i	s HCl, hot HNO_3, alk; i NH_4OH
t28	oxide, mon-	TeO	143.60	blk, amorph (exist ?)	5.682	d 370 (in CO_2)	d	i	i	s dil a, H_2SO_4, KOH
t29	oxide, tri-	TeO_3	175.60	α yel amorph β gray cr	α 5.075^{105}_{4} β 6.21	d 395	...	i	i	d conc HCl; s hot KOH; i a, al
t30	sulfide	TeS_2	191.72	red-blk powd amorph (exist ?)	...	...	...	i	...	i a; s alk sulf
t31	sulfoxide	$TeSO_3$	207.66	deep red amorph	...	d 30	d	d	d	s H_2SO_4
t32	**Tellurous acid**	H_2TeO_3(?)	177.61	wh flocks, indef not isolated	3.05	d 40	...	0.00067	d	s a, NaOH; sl s NH_4OH; i al
t33	**Terbium**	Tb	158.924	silv-gray met, hex	8.272	1356	2800	i	i	s a
t34	bromide	$TbBr_3$	398.65	...	...	827	1490	s	s	...
t35	chloride hexa-hydrate	$TbCl_3.6H_2O$	373.38	col pr cr, deliq	4.35 (anhydr)	588 (anhydr)	$-H_2O$, 180–200 (in HCl gas)	v s	...	...
t36	fluoride	TbF_3	215.92	...	...	1172	2280(?)	i	i	i dil a
t37	iodide	TbI_3	539.64	...	...	946	>1300	s	s	...
t38	*di*methylphosphate	$Tb[(CH_3)_2PO_4]_3$	534.05	...	...	...	...	12.6^{25}	8.07^{40}	...
t39	nitrate	$Tb(NO_3)_3.6H_2O$	453.03	col, monocl cr	...	893	...	s	...	...
t40	oxalate	$Tb_2(C_2O_4)_3.10H_2O$	762.06	wh cr	2.60	$-H_2O$, 40	...	i	i	i dil a
t41	oxide	Terbia. Tb_2O_3	365.85	wh solid	...	...	...	...	...	s dil a
t42	oxide, per-	Tb_4O_7	747.69	dk-brn or blk solid	...	$-O_2$	...	i	i	s hot conc a
t43	sulfate	$Tb_2(SO_4)_3.8H_2O$	750.16	wh cr	...	$-8H_2O$, 360	...	3.561^{20}	2.51^{40}	...
t44	**Thallium**	Tl	204.37	bl-wh met, tetr	11.85	303.5	1457 ± 10	i	i	s HNO_3. H_2SO_4; sl s HCl
t45	acetate	$TlC_2H_3O_2$	263.42	silk wh cr, deliq	3.765^{137}	131	...	v s	...	v s al, $CHCl_3$; i acet
t46	aluminium sulfate	$TlAl(SO_4)_2.12H_2O$	639.66	oct, 1.488	2.306^{20}	91	...	11.78^{25}	...	...
t47	azide	TlN_3	246.39	yel, tetr	...	330 (vac)	...	0.1712^{0}	0.3^{16}	i al, eth
t48	bromate	$TlBrO_3$	332.28	col, need	...	...	...	0.35^{20}	s	s dil al
t49	bromide, di-	Bromothallate(ous). Tl_2Br_4 or $Tl^{I}_{3}[Tl^{III}Br_4]$	728.38	yel need	...	...	...	d	d	...
t50	bromide, mono-	TlBr	284.28	yel-wh, cub, 2.4–2.8	$7.557^{17.3}$	480	815	0.05^{25}	0.25^{68}	s al; i HB_2, acet
t51	bromide, tri-	$TlBr_3$	444.10	yel, deliq, unstable	...	d	...	s	v s	v s al
t52	carbonate	Tl_2CO_3	468.75	col, monocl	7.11	273	...	$4.03^{15.5}$	27.2^{100}	i abs al, eth, acet
t53	chlorate	$TlClO_3$	287.82	need (rhomb ?)	5.047^{9}	...	...	2^{0}	57.31^{100}	...
t54	*per*chlorate	$TlClO_4$	303.82	col, rhomb	4.89	501	d	20.5^{30}	167^{100}	sl s al
t55	chloride	TlCl	239.82	wh reg discol in air, 2.247	7.004^{30}_{4}	430	720	$0.29^{15.6}$	$2.41^{99.35}$	i al, acet; d a
t56	chloride, tri-	$TlCl_3$	310.73	hex pl, hygr	...	25	d	v s	...	s al, eth
t57	chloride, tri-	$TlCl_3.H_2O$	328.74	col, need	...	$-H_2O$, 60	d 100	v s	d	v s al, eth
t58	chloride, tri-	$TlCl_3.4H_2O$	382.79	col, need	...	37	$-4H_2O$, 100	86.2^{17}	d	s al, eth
t59	chloroplatinate	Tl_2PtCl_6	816.55	pale or cr	5.76^{17}	...	...	0.0064^{15}	0.05^{100}	...
t60	chromate	Tl_2CrO_4	524.73	yel	...	...	...	0.03^{60}	0.2^{100}	sl s a, alk; i ac a
t61	*di*chromate	$Tl_2Cr_2O_7$	624.73	red	...	...	...	i	...	d a
t62	chromium sulfate	Thallium chromium alum. $Tl[Cr(H_2O)_6](SO_4)_2.6H_2O$	664.67	vlt cr	2.394	92	...	163.8^{25}	...	...
t63	cyanate	TlCNO	246.39	col, need	5.487^{20}_{4}	...	...	s	v s	sl s al
t64	cyanide	TlCN	230.39	tabl	6.523	d	...	$16.8^{28.5}$	...	s a
t65	ethoxide	$(TlOC_2H_5)_4$	997.73	col liq	3.522	−3	d 80	s d	...	9.11^{25} al; s bz; i liq NH_3
t66	ethylate	$TlOC_2H_5$	249.43	liq, 1.6714^{20}	3.493^{20}_{4}	−3	d 130	...	...	sl s al; s eth
t67	ferrocyanide	$Tl_4Fe(CN)_6.2H_2O$	1065.46	yel, tricl	4.641	...	...	0.37^{18}	3.93^{101}	...
t68	fluogallate	$Tl_3(GaF_6H_2O)$	591.47	col, orthorhomb	6.44	...	...	s	...	...
t69	fluoride, mono-	TlF	223.37	col, cub, oct	8.23^{4}	327	655	78.6^{15}	d	sl s al
t70	fluoride, tri-	TlF_3	261.37	olive grn	8.36^{25}	d 550	...	d	...	i conc HCl
t71	fluosilicate	$Tl_2SiF_6.2H_2O$	586.85	hex pl	5.72	...	...	v s	...	...
t72	formate	$TlHCO_2$	249.39	col, need, hygr	4.967^{104}	101	...	500^{10}	...	v s MeOH, sl s al; i $ChCl_3$

No.	Name	Synonyms and Formulae	Mol. wt.	Crystalline form, properties and index of refraction	Density or spec. gravity	Melting point, °C	Boiling point, °C	Solubility, in grams per 100 cc		
								Cold water	Hot water	Other solvents
	Thallium									
t73	(I) hydroxide	TlOH	221.38	pa yel, need		d 139		25.9^{0}	52^{40}	s al
t74	iodate	$TlIO_3$	379.27	wh need				0.058^{20}	sl s	sl s HNO_3
t75	iodide (α)	TlI	331.27	yel, rhomb	7.29	tr to (β) 170		0.0006^{20}	0.12^{100}	s liq NH_3
t76	iodide (β)	TlI	331.27	red, cub	$7.098^{14.7}$	440	823	i	i	i al
t77	iodide, tri-	TlI_3	585.08	blk, lust rhomb		d		s		
t78	iron (III) sulfate	$TlFe(SO_4)_2.12H_2O$	668.52	pink, oct, n_D^{17} 1.524	2.351^{15}	$-H_2O$, *ca* 100		36.15^{25} (anhydr)		
t79	magnesium sulfate	$Tl_2SO_4.MgSO_4.6H_2O$	733.26	wh dull cr, 1.5660, 1.5836, 1.5900	3.573^{20}_{4}	$-6H_2O$, 40		d 0		
t80	methoxide	$TlOCH_3$	235.40	wh cr powd		d>120		s d		1.70^{25} CH_3OH; 3.16^{25} bz
t81	molybdate	Tl_2MoO_4	568.68	wh powd or cr		vol red heat		i	v sl s	i al; s alk carb, conc NH_4OH, HF
t82	myristate	$TlC_{14}H_{27}O_2$	431.74	wh powd		120–3				0.52^{25} 50 % al
t83	(I) nitrate (α)	$TlNO_3$	266.37	cubic		206	430	9.55^{20}	4.13^{100}	i al, s acet
t84	(I) nitrate (β)	$TlNO_3$	266.37	trig		tr 145 to (α)				
t85	(I) nitrate (γ)	$TlNO_3$	266.37	rhomb, α 1.817	$5.556^{21.4}$	tr 75 to (β)		3.91^{0}	414^{100}	i al; s acet
t86	(III) nitrate	$Tl(NO_3)_3$	390.38	cr				s		
t87	(III) nitrate	$Tl(NO_3)_3.3H_2O$	444.43	col, rhomb, deliq			s 100	d	d	
t88	nitrite	$TlNO_2$	250.38	yel micro cr		182		32.10^{25}	95.78^{98}	i a
t89	oleate	$TlC_{18}H_{33}O_2$	485.83	wh cr clusters		131–2		0.05^{15}	0.3^{80}	3.0^{25} al
t90	oxalate	$Tl_2C_2O_4$	496.76	monocl pr	6.31			1.48^{15}	9.02^{100}	
t91	oxalate, tetra-	$TlH_3(C_2O_4)_2.2H_2O$	419.46	tricl, leaf, 1.5097, 1.6319, 1.6538	2.992^{17}	d 100		76.9^{23}	v s	i cold al; s hot al
t92	(I) oxide	Tl_2O	424.74	blk, deliq	9.52^{16}	300	1080^{760} −O, 1865	v s d to TlOH		s a, al
t93	(III) oxide	Tl_2O_3	456.74	col, amorph pr, hex	hex 10.19^{22} am 9.65^{21}	717 ± 5	−2O, 875	i	i	s a; i alk
t94	palmitate	$TlC_{16}H_{31}O_2$	459.79	cr need		115–117		0.01^{15}	0.07^{60}	1.04^{45} al
t95	phenoxide	$TlOC_6H_5$	297.48	wh cr		233–5		d		s hot bz; sl s lgr
t96	*ortho*phosphate	Tl_3PO_4	708.08	col, need	6.89^{10}			0.5^{15}	0.67^{100}	i al, s NH_4 salts
t97	*ortho*phosphate, (di)-β	TlH_2PO_4	301.36	monocl	4.726	*ca* 190		sl s	sl s	i al
t98	*pyro*phosphate	$Tl_4P_2O_7$	991.42	monocl pr	6.786^{20}	>120		40		
t99	picrate	$TlC_6H_2N_3O_7$	432.47	red, monocl or yel tricl	red 3.164^{17} yel 2.993^{17}	expl 723–725		0.135^{0}	2.43^{70}	0.40 CH_3OH
t100	rhodanide	TlCNS	262.45	glossy leaflets, rhomb, tetr cr	4.954^{30}_{4}	d low temp		0.393^{25}		0.024^{0} liq SO_2; i acet; s MeOH
t101	selenate	Tl_2SeO_4	551.70	rhomb need, 1.949, 1.959, 1.964	>400			2.13^{10}	8.5^{50}	i al, eth
t102	selenide	Tl_2Se	487.70	gray leaf	9.05^{25}_{4}	340		i		s a; i acet a
t103	silver nitrate	$TlNO_3.AgNO_3$	436.25	wh cr powd		75		s		
t104	stearate	$TlC_{18}H_{35}O_2$	487.85	need		119		0.005^{15}	0.095^{75}	0.18^{15} al, 0.60^{60} al
t105	(I) sulfate	Tl_2SO_4	504.80	col, rhomb, 1.860, 1.867, 1.885	6.77	632	d	4.87^{20}	19.14^{100}	
t106	(I) sulfate hydrogen	$TlHSO_4$	301.44	pr need		120 d				v sl s dil H_2SO_4
t107	(III) sulfate	$Tl_2(SO_4)_3.7H_2O$	823.03	col leaf		$-6H_2O$, 220		d	d	s dil H_2SO_4
t108	(I) sulfide	Tl_2S	440.80	bl-blk tetr	8.46	448.5	d	0.02^{20}	sl s	s a; i alk, acet
t109	(III) sulfide	Tl_2S_3	504.93	blk, amorph		260 (in N_2)	d	i	i	s hot H_2SO_4
t110	sulfite	Tl_2SO_3	488.80	wh cr	6.427			3.34^{15}	v s	i al
t111	tartrate(dl)	$Tl_2C_4H_4O_6$	556.81	monocl	4.659	d 165		13.3^{15}		
t112	*meta*tellurate	Tl_2TeO_4	600.34	heavy wh ppt	$6.760^{17.6}$	red heat		sl s	sl s	
t113	thiocyanate	TlSCN	262.45	col, tetr	4.956^{20}			0.315^{20}	0.727^{40}	i al
t114	*di*thionate	$Tl_2S_2O_6$	568.86	monocl	5.57^{20}_{4}	d		41.8^{19}		
t115	thiosulfate	$Tl_2S_2O_3$	520.87	wh rhomb cr		d 130		sl s	v s	
t116	*meta*vanadate	$TlVO_3$	303.41	gray cr	6.09^{17}	424		0.87^{11}	0.21^{100}	
t117	*pyro*vanadate	$Tl_4V_2O_7$	1031.36	light yel	8.21^{19}	454		0.2^{14}	0.26^{100}	
t118	**Thiocarbonyl chloride**	Thiophosgene. $CSCl_2$	114.98	red yel liq, 1.5442	1.509^{15}		73.5	d		d al; s eth
t119	**Thiocarbonyl chloride, tetra-**	$CSCl_4$	185.89	yel	1.712^{13}		146–147		d	
t120	**Thiocyanic acid(iso)**	HSCN(HNCS)	59.09	col mass or gas		>−110	polym to solid −90	v s		v s al, eth, bz
t121	**Thiocyanogen**	$(SCN)_2$	116.16	liq, or yel solid		−2 to −3		d		s al, eth, CS_2, CCl_4

No.	Name	Synonyms and Formulae	Mol. wt.	Crystalline form, properties and index of refraction	Density or spec. gravity	Melting point, °C	Boiling point, °C	Solubility, in grams per 100 cc		
								Cold water	Hot water	Other solvents
	Thionyl bromide									
t122	**Thionyl** bromide	$SOBr_2$	207.88	or, yel liq	2.68^{18}	−52	138^{773}, 68^{40}	d	d	s bz, chl, CS_2, CCl_4
t123	chloride	$SOCl_2$	118.97	col, or yel liq; 1.527^{10}	1.655^{10}_{4}	−105	78.8^{746} d 140	d	d	d a, al, alk; s bz, chl
s124	chloride fluoride	$SOClF$	102.51	gas		−139.5	12.2			
t125	fluoride	SOF_2	86.06	col, gas	gas 2.93 g/l liq 1.780^{-100}	−110.5	−43.8	d	d	s eth, bz, chl, acet, $AsCl_3$
t126	**Thiophosphoramide**	Thiophosphorylamide. $PS(NH_2)_3$	111.11	yel wh amorph	1.7^{13}	d 200		sl s	d	
t127	**Thiophosphoryl** bromide	$PSBr_3$	302.76	yel, cub	2.85^{17}	37.8	125^{25}	d		s eth, CS_2, PCl_3
t128	thiophosphoryl bromide, hydrate	$PSBr_3.H_2O$	320.78	yel cr	2.794^{18}	35				
t129	thiophosphoryl bromide (mono-) chloride, di-	$PSBrCl_2$	213.85	yel liq	2.12^{0}	−30	d 150	d		
t130	thiophosphoryl bromide (di-) chloride	$PSBr_2Cl$	258.31	pa grn fum liq	2.48^{0}	−60	95^{60}	d		
t131	thiophosphoryl chloride	$PSCl_3$	169.40	col liq, 1.563 (c)	1.635	−35	125	d		s bz, CS_2, CCl_4
t132	thiophosphoryl fluoride	PSF_3	120.03	gas		$3.8^{7.6 atm}$	d	sl s d		s eth; i bz, CS_2
t133	**Thiosulfuric acid**	$H_2S_2O_3$	114.14	in sol only				s		
t134	**Thorium**	Th	232.038	gray, cub radioactive	11.7	~1700	~4000	i	i	s HCl, H_2SO_4, aq reg; sl s HNO_3
t135	boride, hexa-	ThB_6	296.92	dk viol-blk met, cub	6.4^{15}	2195		i	i	s HNO_3; i H_2SO_4, HCl, HF, aq alk
t136	boride, tetra	ThB_4	275.28	tetr pr	7.5^{15}			i	i	s HNO_3 HCl, hot H_2SO_4
t137	bromide	$ThBr_4$	551.67	col cr, hygr	5.67	subl 610	725	s	s	
t138	carbide	ThC_2	256.06	yel, tetr	8.96^{18}	2655±25	*ca* 5000 (?)	d		v sl s conc a
t139	carbonate	$Th(CO_3)_2$	352.06	exist?				i	d	s conc Na_2CO_3
t140	chloride	$ThCl_4$	373.85	wh, rhomb, deliq	4.59	770±2 subl 820	d 928	v s	v s	s al, a, KCl; sl s eth
t142	*tetra*cyanoplatinate	$Th[Pt(CN)_4]_2.16H_2O$	1118.61	yel-grn, rhomb	2.460			sl s	s	
t143	fluoride	ThF_4	308.03	wh cub powd	6.32^{24}	>900				sl d dil H_2SO_4, HCl; i conc H_2SO_4
t144	fluoride	$ThF_4.4H_2O$	380.09	cr		−H_2O, 100	−$2H_2O$, 140–200	0.017^{25}		i HF
t145	hydroxide	$Th(OH)_4$	300.02	wh gelat		d		i	i	s a; i alk, HF
t146	iodate	$Th(IO_3)_4$	931.65					i	i d	s dil H_2SO_4; i dil HNO_3
t147	iodide, tetra-	ThI_4	739.66	yel		566	839	s		
t148	nitrate	$Th(NO_3)_4$	480.06	plates, deliq		d 500		v s		s al
t149	nitrate	$Th(NO_3)_4.4H_2O$	552.12	col cr		swells		v s		v s al; sl s acet; 36.9 eth
t150	nitrate	$Th(NO_3)_4.12H_2O$	696.24	col leaf, deliq		d		v s		v s al, a
t151	nitride	Th_3N_4	752.14	dk brn powd, or blk cr				sl d	d	s HCl
t152	oxalate	$Th(C_2O_4)_2$	408.08	wh cr	4.637^{16}	d		0.0017^{17}	0.0017^{50}	s h aq $(NH_4)_2C_2O_4$; sl s a
t153	oxalate	$Th(C_2O_4)_2.6H_2O$	516.17	wh amorph powd				i		s Na_2CO_3, $(NH_4)_2C_2O_4$ sol; i HNO_3
t154	oxide, di-	Thorianite. ThO_2	264.04	wh cub, 2.20 (liq)	9.86	3050	4400	i	i	s hot H_2SO_4; i dil a, alk
t155	oxysulfide	$ThOS$	280.10	yel cr	6.44^{0}	d		i		s aq reg; sl s HNO_3
t156	2,4-pentanedione	Thorium acetylacetonate. $Th(C_5H_7O_2)_4$	628.48	col cr		171 subl 160^{10}	$260–270^{10}$	sl s		v s al, chl; s eth
t157	*hypo*phosphate	$ThP_2O_6.11H_2O$	588.15	wh amorph ppt.		−$11H_2O$, 160		i	i	i a, alk
t158	*meta*phosphate	$Th(PO_3)_4$	547.93	col rhomb pr	$4.08^{16.4}$					
t159	*ortho*phosphate	$Th_3(PO_4)_4.4H_2O$	1148.06	wh gelat				i	i	s 30^{0} HCl; i a
t160	picrate	$Th(C_6H_2N_3O_7)_4.10H_2O$	1324.59					0.305^{25}		
t161	selenate	$Th(SeO_4)_2.9H_2O$	680.09	col, monocl	3.026	−$8H_2O$, 200	d 1500	0.5^{0}	2.0^{100}	i a
t162	*ortho*silicate	Thorite $ThSiO_4$	324.12	col, tetr, 1.80, 1.81	6.82^{16}			v sl s		
t163	silicide	$ThSi_2$	288.21	blk, tetr	7.96^{16}					s hot HCl; sl s H_2SO_4
t164	sulfate	$Th(SO_4)_2$	424.16	wh cr, hygr	4.225^{17}			s	s	i a; v s $NH_4C_2H_3$
t165	sulfate	$Th(SO_4)_2.4H_2O$	496.22	wh need, or cr powd		−$4H_2O$, 400		9.41^{17} (anhydr)	2.54^{50} (anhydr)	i a

No.	Name	Synonyms and Formulae	Mol. wt.	Crystalline form, properties and index of refraction	Density or spec. gravity	Melting point, °C	Boiling point, °C	Solubility, in grams per 100 cc		
								Cold water	Hot water	Other solvents
	Thorium									
t166	sulfate	$Th(SO_4)_2.6H_2O$	532.26					1.63^{15}	6.64^{60}	i a
t167	sulfate	$Th(SO_4)_2.8H_2O$	568.29	monocl, prism, 1.5168		$-4H_2O$, 42		1.88^{25}	3.71^{44}	i a
t168	sulfate	$Th(SO_4)_2.9H_2O$	586.31	wh monocl	2.77	$-9H_2O$, 400		1.57^{20}	6.67^{55}	i a
t169	sulfide	ThS_2	296.17	dk brn-blk cr	7.30_{4}^{25}	1925 ± 50 (vac)		i	d 200	s hot aq reg; sl s a
t170	*pyro*vanadate	$ThV_2O_7.6H_2O$	554.01	yel				i	i	s conc a
t171	**Thulium**	Tm	168.934	silv wh met, hex.	9.332	1545	1727	i	i	
t172	bromide	$TmBr_3$	408.66			952	1440	s	s	
t173	chloride	$TmCl_3.7H_2O$	401.40	grn cr, deliq		824	1440	v s		v s al
t174	fluoride	TmF_3	225.93		1158	>2200		i	i	s dil a
t175	iodide	TmI_3	549.65	brt yel cr		1015	1260	s	s	
t176	oxalate	$Tm_2(C_2O_4)_3.6H_2O$	710.02	grn-wh ppt		$-H_2O$, 50		i		s alk oxal sol; i dil a
t177	oxide	Thulia. Tm_2O_3	385.87	grn-wh powd						sl s min a
t178	**Tin** gray	Sn	118.69	gray, cub	5.75	231.89	2270	i	i	s HCl, H_2SO_4, aq reg, alk; sl s dil HNO_3
t179	**Tin** white	Sn	118.69	wh met, tetr	7.28	231.88 stable 13.2 − 161	2260	i	i	s HCl, H_2SO_4, aq reg, alk; sl s dil HNO_3
t180	**Tin** brittle	Sn	118.69	wh, rhomb	6.52–56	231.89 stable >161	2260	i	i	s HCl, H_2SO_4, aq reg, alk; sl s dil HNO_3
t181	(II) acetate	$Sn(C_2H_3O_2)_2$	236.78	yelsh powd		182	240	d		s dil HCl
t182	*pyro*arsenate	$Sn_2As_2O_7$	499.22	flocculent ppt		d AS_2O_3 + SnO_2		i	i	i conc ac a
t183	(II) bromide	$SnBr_2$	278.51	pa yel, rhomb	5.117^{17}	215.5	620	85.2^{0}	222.5^{100}	s al, eth, acet
t184	(IV) bromide	$SnBr_4$	438.33	col, rhomb pyr, deliq	liq 3.34^{35}	31	202^{734}	s d	d	s acet, PCl_3, $AsBr_3$
t185	bromide chloride(tri-)	$SnBrCl_3$	304.96	col liq	2.51^{13}	−31	50^{30}			
t186	bromide(di-) chloride(di-)	$SnBr_2Cl_2$	349.41		2.82^{13}	−20	65^{30} d 191	d	d	
t187	bromide(tri-) chloride	$SnBr_3Cl$	393.87	liq	3.12^{13}	1	73^{30}			
t188	bromide(di-) iodode(di-)	$SnBr_2I_2$	532.32	or-red, hex pl	3.631^{15}	50	225	s	d<80	
t189	(II) chloride	$SnCl_2$	189.60	wh, rhomb	3.95_{4}^{25}	246	652	83.9^{0}	269.8^{15} d	s al, eth, acet, et acet, me acet, pyr
t190	(II) chloride dihydrate	$SnCl_2.2H_2O$	225.63	wh, monocl	$2.710^{15.5}$	37.7	d	d	d	s al, eth, acet, glac ac a
t191	(IV) chloride	$SnCl_4$	260.50	col liq, solid cub, 1.512	liq 2.226	−33	114.1	s	d	s eth
t192	(IV) chloride pentahydrate	$SnCl_4.5H_2O$	350.58	monocl cr		stable 19–56		s		
t193	(IV) chloride tetrahydrate	$SnCl_4.4H_2O$	332.56	opaque			stable 56–83	s		
t194	(IV) chloride trihydrate	$SnCl_4.3H_2O$	314.55	col, monocl cr		80	stable 64–83	s		
t195	(IV) chloride *di*ammine	$SnCl_4.2NH_3$	294.56	cr				s		d HCl
t196	chloride(tri-) bromide	$SnCl_3Br$	304.96	col liq	2.51^{13}	−31	50^{30}			
t197	chloride (di-) iodide(di-)	$SnCl_2I_2$	443.40	red mobile liq	3.287^{15}		297	s	d	s chl, bz CS_2
t198	(IV) chloride nitrosyl-chloride	$SnCl_4.2NOCl$	391.42	pa yel, oct cr	2.60	180		d	d	
t199	(IV) chromate	$Sn(CrO_4)_2$	350.68	br yel cr powd		d		s		
t200	(II) ferricyanide	$Sn_3[Fe(CN)_6]_2$	779.98	wh		d		i	i	s HCl
t201	(II) ferrocyanide	$Sn_2Fe(CN)_6$	449.33	wh gel				i	i	d HCl
t202	(IV) ferrocyanide	$SnFe(CN)_6$	330.64					i	i	d h HCl
t203	(II) fluoride	Fluoristan. SnF_2	156.69	wh, monocl cr				s		
t204	(IV) fluoride	SnF_4	194.68	wh, monocl cr, hygr	4.780^{19}	705 subl		v s	d	
t205	hydride	Stannane. SnH_4	122.72	gas		d − 150	−52			s $AgNO_3$, $HgCl_2$, conc alk, conc H_2SO_4
t206	iodide	SnI_2	372.50	yelsh-red to red monocl need	5.285	320	717	0.98^{20}	4.03^{100}	v s NH_4OH, HI soln
t207	(IV) iodide	SnI_4	626.31	or red cub, 2.106	4.473^{0}	144.5	364.5	s	d	141.1^{25} CS_2; 6.03^{15} CCl_4; 17.88^{25} bz

No.	Name	Synonyms and Formulae	Mol. wt.	Crystalline form, properties and index of refraction	Density or spec. gravity	Melting point, °C	Boiling point, °C	Solubility, in grams per 100 cc		
								Cold water	Hot water	Other solvents
	Tin									
t208	(II) nitrate	$Sn(NO_3)_2.20H_2O$	603.07	col leaf		−20		d	d	d HNO_3
t209	(II) nitrate, basic	$SnO.Sn(NO_3)_2$	377.39	wh cr mass		d >100 expl		d	d	
t210	(IV) nitrate	$Sn(NO_3)_4$	366.71	silky need		d 50		d		
t211	(II) oxide, mon-	SnO	134.69	blk, cub (tetr)	6.446^{0}	d 1080^{600}		i	i	s a, alk; sl s NH_4Cl
t212	oxide, mon-hydrate	$SnO.xH_2O$		wh powd or yellow-brn cr					d to SnO	d a; alk; s alk carb; i NH_4OH
t213	(IV) oxide, di-	Nat. cussiterite. SnO_2	150.69	wh, tetr, (also hex or rhomb), 1.997, 2.093	6.95	1127	subl 1800–1900	i	i	d KOH, NaOH; i aq reg
t214	oxide, di-hydrate	α-Stannic acid or "ordinary" stannic acid. $SnO_2.xH_2O$		amorph or gel				i	i	s a, alk, K_2CO_3
t215	oxide, di-hydrate	β-Stannic acid or "meta" stannic acid. $SnO_2.xH_2O$		wh, amorph or gel				i	i	i a, K_2CO_3; sol alk
t216	(II) *meta*phosphate	$Sn(PO_3)_2$	276.63	amorph mass	$3.380^{22.8}$					
t217	(II) *ortho*phosphate	$Sn_3(PO_4)_2$	546.01	wh, amorph	3.823^{17}			i	i	d a, alk
t218	(II) *ortho*phosphate, di-H	$Sn(H_2PO_4)_2$	312.66	wh, rhomb cr	$3.167^{22.8}$	d	d		d	
t219	(II) *ortho*phosphate, mono-H	$SnHPO_4$	214.67	cr	$3.476^{15.5}$	stabl >100	d	i	i	s dil min a
t220	(II) *pyro*phosphate	$Sn_2P_2O_7$	411.32	amorph powd	$4.009^{16.4}$					s conc a
t221	phosphide, mono-	SnP	149.66	silv wh	6.56	d	d	i	i	s HCl; i HNO_3
t222	phosphide, tri-	SnP_3	211.61	cr	4.10^{0}	<415 d to Sn_4P_3		i	i	d HNO_3; i HCl
t223	*tetra*phosphide, tri-	Sn_4P_3	567.68	wh cr	5.181	d <480		i	i	d fixed alk hydr, HCl
t224	phosphorus chloride	$SnCl_4.PCl_5$	468.74	col cr		subl 200		d	d	
t225	(II) selenide	$SnSe$	197.65	steelgray cr	6.179^{0}	861		i	i	d HCl, HNO_3, aq reg, alk sulf
t226	(II) sulfate	$SnSO_4$	214.75	wh-yelsh cr powd		>360 (SO_2)		33^{25}		s H_2SO_4
t227	(IV) sulfate	$Sn(SO_4)_2.2H_2O$	346.84	wh, hex pr, deliq				v s	d	s eth, dil H_2SO_4, HCl
t228	(II) sulfide	SnS	150.75	gray-blk cub, monocl	5.22^{25}	882	1230	0.000002^{18}		d HCl, alk, $(NH_4)_2S$
t229	(IV) sulfide	Mosaic gold. SnS_2	182.82	gold yel, hex	4.5	d 600		0.0002^{18}		d alk sulf, aq reg alk hydr, PCl_5, $SnCl_2$; i a
t230	(IV) sulfur chloride	$SnCl_4.2SCl_4$	608.25	yel cr		37	d <40	d	d	s eth, bz, CS_2, ethyl acet; d HNO_3
t231	tartrate	$SnC_4H_4O_6$	266.76	heavy wh powd				s		v s dil HCl
t232	(II) telluride	$SnTe$	246.29	gray cr	6.48	780	d	i	i	d alk sulf
t233	(IV) telluride	$SnTe_2$	373.89	blk, flocc ppt				i	i	d dil a, alk
t234	**Titanic acid, ortho-**	α Titanic acid. H_2TiO_4	113.91	wh		d		v sl s d		s dil HCl, dil H_2SO_4, conc alk
t235	**Titanium**	Ti	47.90	α hex, tr β cub 838, silv gray	4.5^{20}	1675	3260	i	i	s dil a
t236	boride, di-	TiB_2	69.52	hex	4.50	2900				
t237	bromide, di-	$TiBr_2$	207.72	blk powd	4.31	d >500		s ev H_2		
t238	bromide, tetra-	$TiBr_4$	367.54	or yel, deliq	2.6	39	230	d		s abs al, abs eth
t239	bromide, tri-	$TiBr_3.6H_2O$	395.72	redsh-viol or dk blue cr, deliq		115	d 400	v s		v s al, acet
t240	carbide	TiC	59.91	gr met, cub	4.93	3140±90	4820	i	i	s aq reg, HNO_3
t241	chloride, di-	$TiCl_2$	118.81	lt br-blk, hex, deliq	3.13	subl H_2	d 475 vac	d		s al, i eth, chl, CS_2
t242	chloride, tetra-	$TiCl_4$	189.71	lt yel liq, $1.61^{10.5}$	liq 1.726 sol 2.06^{-79}	−25	136.4	s	d	s dil HCl, al
t243	chloride, tri-	$TiCl_3$	154.26	dk viol, deliq	2.64	d 440	660^{108}	s	s	v s al; s HCl; i eth
t244	fluoride, tetra	TiF_4	123.89	wh powd, hygr	$2.798^{20.5}$	>400 (pressure)	284 (subl.)	s d		s H_2SO_4, al, C_5H_5N; i eth
t245	fluoride, tri-	TiF_3	104.90	purp-red or vlt	3.40	1200	1400	red s vlt i		
t246	hydride	TiH_2	49.92	gray powd	3.9^{12}	d 400				
t247	iodide, di-	TiI_2	301.71	blk, hygr	4.99	600	1000	d		d alk; s conc HF, conc HCl
t248	iodide, tetra-	TiI_4	555.52	red, cub	4.3	150	377.1	v s	d	
t249	nitride	TiN	61.91	yel-bronze, cub	5.22	2930		i	i	sl s hot aq reg +HF
t250	oxalate	$Ti_2(C_2O_4)_3.10H_2O$	540.01	yel pr				s	s	i al, eth
t251	oxide, di-	Nat. brookite. TiO_2	79.90	wh, rhomb, 2.583, 2.586, 2.741	4.17	1825		i	i	s H_2SO_4, alk; i a
t252	oxide, di-	Nat. octahedrite, anatase. TiO_2	79.90	br-blk, tetr, 2.554, 2.493	3.84			i	i	s H_2SO_4, alk; i a

No.	Name	Synonyms and Formulae	Mol. wt.	Crystalline form, properties and index of refraction	Density or spec. gravity	Melting point, °C	Boiling point, °C	Solubility, in grams per 100 cc		
								Cold water	Hot water	Other solvents
	Titanium									
t253	oxide, di-	Nat. rutile. TiO_2	79.90	col, tetr, 2.616, 2.903	4.26	1830–1850	2500–3000	i	i	s H_2SO_4, alk; i a
t254	oxide, mon-	TiO	63.90	yel blk, pr	4.93	1750	>3000			s dil H_2SO_4; i HNO_3
t255	oxide, sesqui-	Ti_2O_3	143.80	vlt blk, trig	4.6	2130 d		i	i	s H_2SO_4; i HCl, HNO_3
t256	phosphide	TiP	78.87	gray, met	3.95^{25}			i	i	i a
t257	sulfate	$Ti_2(SO_4)_3$	383.98	green powd				i	i	s dil a; i al, eth, conc H_2SO_4
t258	sulfate, basic	$TiOSO_4$	159.96	wh or sl yelsh powd, 1.80–1.89				d		
t259	sulfide, di-	TiS_2	112.03	yel sc	3.22^{20}			hyd sl	d in steam	d HCl; s dil HNO_3, H_2SO_4
t260	sulfide, mono-	TiS	79.96	redsh solid	4.12			i		s conc H_2SO_4; i HCl, HF, dil H_2SO_4
t261	sulfide, sesqui	Ti_2S_3	191.99	grayish-blk cr	3.584			i	i	s conc H_2SO_4, conc HNO_3; i dil H_2SO_4, dil HCl
t262	**Tungsten**	Wolfram. W	183.85	gray-blk, cub	19.35^{20}_{4}	3410±20	5927	i	i	v sl s HNO_3, H_2SO_4, aq reg; s HNO_3+HF, fus NaOH+$NaNO_3$; i HF, KOH
t263	arsenide	WAs_2	333.69	blk cr	6.9^{18}	d red heat		i		d hot HNO_3, hot H_2SO_4
t264	boride, di-	WB_2	205.47	silvery, oct	10.77	*ca* 2900		i	i	s aq reg
t265	bromide, di-	WBr_2	343.67	bl-blk need		d 400		d		
t266	bromide, penta-	WBr_5	583.40	vlt-brn need, hygr		276	333	d		s abs al, chl, eth, alk
t267	bromide, hexa-	WBr_6	663.30	bl-blk, need	6.9	232		i	d	s abs a, eth, CS_2, NH_4OH
t268	carbide	WC	195.86	blk, hex	15.63^{18}	2870±50	6000	i		s HNO_3+HF, aq reg
t269	(di-)carbide	W_2C	379.71	blk, hex	17.15	2860	6000	i		s HNO_3+HCl
t270	carbonyl	$W(CO)_6$	351.91	col, rhomb cr	2.65	d~150	175^{766}	i	i	s fum HNO_3; v sl s al, eth, bz
t271	chloride, di-	WCl_2	254.76	gray, amorph	5.436			d		
t272	chloride, hexa-	WCl_6	396.57	dk bl, cub	3.52^{25}_{4}	275	346.7		d^{60}	s al, eth, bz, CCl_4; v s CS_2, POCl
t273	chloride, penta-	WCl_5	361.12	blk, deliq	3.875^{25}_{4}	248	275.6		d to W_2O_5	v sl s CS_2
t274	chloride, tetra-	WCl_4	325.66	gray, deliq	4.624^{25}_{4}	d		d		
t275	fluoride, hexa-	WF_6	297.84	col gas, or lt yel liq	liq 3.44 gas 12.9 g/l	2.5^{420}	17.5	d	d	s alk
t276	iodide, di-	WI_2	437.66	br-gr, amorph	6.799^{25}_{4}	d		i	d	s alk; i al CS_2
t277	iodide, tetra-	WI_4	691.47	blk, cr	5.2^{18}	d		i	d	s abs al; i eth, chl, turp
t278	nitride, di-	WN_2	211.86	brn, cub		above 400 (vac)		d	d	
t279	oxide, di-	WO_2	215.85	br, cub	12.11	1500–1600 (in N_2)	*ca* 1430 subl 800	i	i	s a, KOH
t280	oxide, pent-	Mineral blue. W_2O_5 or W_4O_{11}	447.70 or 911.39	blue-vlt, tricl		subl 800–900	*ca* 1530 d 2000	i	i	i a
t281	oxide, tri-	Nat. wolframite. WO_3	231.85	yel, rhomb, or yel-or powd	7.16	1473		i	i	s hot alk; sl s HF; i a
t282	oxy*di*bromide, di-	WO_2Br_2	375.67	red, prism		d				
t283	oxy*tetra*bromide	$WOBr_4$	519.49	blk, deliq		277	327	d	d	
t284	oxy*tetra*chloride	$WOCl_4$	341.66	red, need		211	227.5	d	d	s CS_2, S_2Cl_2, bz
t285	oxy*di*chloride, di-	WO_2Cl_2	286.75	lt yel tabl		266		s	d	i al; s NH_4OH, alk
t286	oxy*tetra*fluoride-	WOF_4	275.84	col pl, hygr		110	187.5	d		sl s CS_2; i CCl_4
t287	phosphide	WP	214.82	gray, prism	8.5			i		s HNO_3+HF; i alk, HCl
t288	phosphide	WP_2	245.80	blk cr	5.8	d		i	i	s HNO_3+HF, aq reg; i al, eth
t289	phosphide	W_2P	398.67	dk gray prism	5.21	d				s fus Na_2CO_3+$NaNO_3$; i a, aq reg

PHYSICAL CONSTANTS OF INORGANIC COMPOUNDS (Continued)

No.	Name	Synonyms and Formulae	Mol. wt.	Crystalline form, properties and index of refraction	Density or spec. gravity	Melting point, °C	Boiling point, °C	Solubility, in grams per 100 cc		
								Cold water	Hot water	Other solvents
	Tungsten									
t290	silicide	WSi_2	240.02	blue, gray, tetrag.	9.4	above 900	......	i	i	s HNO_3+HF; i aq reg
t291	sulfide, di-	Nat. tungstenite. WS_2	247.98	dk gray, hexag	7.5^{10}	d 1250	......	i	......	s HNO_3+HF, fus alk; i al
t292	sulfide, tri-	WS_3	280.04	choc brn powd	......	......	......	sl s	s	s alk
t293	**Tungstic acid,** meta-	$H_2W_4O_{13}.9H_2O$	1107.55	col, tetrag	3.93	d 50	......	88.57^{22}	$111.87^{43.5}$	$110.76^{24.3}$ eth; s al
t294	**Tungstic acid,** ortho-	H_2WO_4	249.86	yel powd, 2.24	5.5	$-H_2O$, 100	1473	i	sl s	s alk, HF, NH_3; i most a
t295	**Tungstic acid,** ortho-	$H_2WO_4.H_2O$	267.88	wh	......	$H_2W_2O_7$ *at* 100	......	sl s	......	s alk
u1	**Uranic acid** meta-	Uranyl hydroxide. H_2UO_4(or $UO_2(OH)_2$)	304.04	yel, rhomb, or powd	5.926	$-H_2O$ 250–300	......	i	i	s a, alk carb
u2	**Uranium**	U	238.03	silvery, cubic, radioactive	19.05 ± 0.02^{25}	1132.3 ± 0.8	3818	i	i	s a; i alk, al
u3	boride, di-	UB_2	259.65	hex	12.70	2365	......	......	......	......
u4	bromide tetra-	UBr_4	557.67	br leaf, deliq	5.35	516	792^{760}	v s	v s	s d al, MeOH; i bz; s liq NH_3
u5	bromide tri-	UBr_3	477.76	dk brn need, hygr	6.53	730	volat	s	......	d al
u6	*di*carbide	UC_2	262.05	met cr	11.28^{16}	2350–2400	4370^{760}	d	d	i al; d dil inorg a
u7	chloride, penta-	UCl_5	415.30	dk green, gray need, red by trans light, hydr	3.81(?)	d 300	......	d	......	s abs als, a acet, NH_4Cl; d ac a; i bz, eth
u8	chloride, tetra-	UCl_4	379.84	dk grn met, cub oct, hygr	4.87	590 ± 1	792^{760}	v s	s	s al, acet, ac a; i eth, $CHCl_3$
u9	chloride, tri-	UCl_3	344.39	dk red need, hygr	5.44_{4}^{25}	842 ± 5	......	s	s	s MeOH, acet, glac acet a; i eth
u10	fluoride, hexa-	UF_6	352.02	col cr, deliq, monocl	4.68^{21}	64.5–64.8	56.2^{765}	d	......	d al, eth; s CCl_4, chl; i CS_2
u11	fluoride, tetra-	UF_4	314.02	green, tricl need	6.70 ± 0.10	960 ± 5	......	v sl s	......	i dil a, alk; s conc a, conc alk
u12	fluoride, tri-	UF_3	295.03	blk cr or fused	......	d above 1000	......	sl d	......	v sl s dil inorg a
u13	hydride	UH_3	241.05	blk-brn powd	10.95	......	......	i	i	i al, acet, liq NH_3; sl s dil HCl; d HNO_3
u14	hydride	UH_3	241.05	blk powd, cub	11.4	......	......	......	......	......
u15	iodide, tetra-	UI_4	745.65	blk, need	5.6^{15}	506	759	s	s d	......
u16	nitride, mono-	UN	252.04	br powd	14.31	*ca* 2630 ± 50	......	......	......	i HCl, H_2SO_4
u17	oxide, di-	UO_2	270.03	br-blk rhomb, or cub	10.96	2500	......	i	i	s HNO_3, conc H_2SO_4
u18	oxide, per-	$UO_4.2H_2O$	338.06	pa yel cr, hygr	......	d 115	......	0.0006^{20}	0.008^{90}	d HCl
u19	oxide, tri-	Uranyl oxide. UO_3	286.03	yel-red powd	7.29	d	......	i	......	s HNO_3, HCl
u20	*tri*-oxide, oct-	U_3O_8	842.09	olive green-blk	8.30	d 1300 to UO_2	......	i	i	s HNO_3, H_2SO_4
u21	(IV) sulfate	$U(SO_4)_2.4H_2O$	502.21	grn, rhomb	......	$-4H_2O$, 300	......	23^{11}	9^{63} (anhydr)	s dil a
u22	(IV) sulfate	$U(SO_4)_2.8H_2O$	574.28	......	......	d 90	......	11.3^{18}	58.2^{62}	i al; s dil a
u23	(IV) sulfate	$U(SO_4)_2.9H_2O$	592.29	grnsh, monocl	......	$-7H_2O$, 230	$-9H_2O$ red heat	......	......	s dil H_2SO_4
u24	sulfide, di-	US_2	302.16	gray-blk, tetr	7.96^{25}	>1100	oxidizes	sl d	......	s conc HCl; d HNO_3 v s al
u25	sulfide, mono-	US	270.09	blk amorph powd	10.87	above 2000	......	......	......	i HCl, HNO_3
u26	sulfide, sesqui-	U_2S_3	572.25	gray blk, rhomb need	......	ign	......	......	......	s+O aq reg conc HNO_3; i dil a
u27	**Uranyl** acetate	$UO_2(C_2H_3O_2)_2.2H_2O$	422.13	yel, rhomb	2.893^{15}	$-2H_2O$, 110	d 275	7.694^{15}	d	v s al
u28	benzoate	$UO_2(C_7H_5O_2)_2$	512.26	yel powd	......	......	......	sl s	......	sl s al
u29	bromide	UO_2Br_2	429.85	grn-yel need, hygr	......	......	......	s d	......	s al, eth
u30	*per*chlorate	$UO_2(ClO_4)_2.6H_2O$	577.02	yel cr, deliq, rhomb	......	90 d 110	......	......	......	......
u31	chloride	UO_2Cl_2	340.93	yel, deliq	......	578	d	320^{18}	v s	s al, amyl al, eth
u32	formate	$UO_2(CHO_2)_2.H_2O$	378.08	yel, oct	3.695^{19}	$-H_2O$, 110	......	7.2^{15}	......	sl s form a; 0.74 MeOH, 2.37^{15} acet
u33	iodate	$UO_2(IO_3)_2$	619.83	yel, rhomb	5.2	d 250	......	s	s	i HNO_3
u34	iodate	$UO_2(IO_3)_2.H_2O$	637.85	α prismatic, stable, β pyramidal	α 5.220^{18} β 5.052^{18}	......	......	α 0.1049^{18} β 0.1214^{18}	......	......
u35	iodide	UO_2I_2	523.84	red, deliq	......	d in air	......	......	......	s al, eth, bz
u36	nitrate	$UO_2(NO_3)_2.6H_2O$	502.13	yel, rhomb, deliq, 1.4967	2.807^{13}	60.2 d 100	118	......	∞ 60	v s al, eth, ac a, acet, MeOH
u37	oxalate	$UO_2C_2O_4.3H_2O$	412.09	yel cr	......	$-H_2O$, 110	......	0.8^{14}	3.3^{100}	s inorg a, alk, oxal a
u38	phosphate, mono-H	$UO_2HPO_4.4H_2O$	438.07	yel pl, tetr	......	......	......	i	i	s HNO_3, aq Na_2CO_3; i ac a
u39	potassium carbonate	$UO_2CO_3.2K_2CO_3$	606.46	yel cr	......	$-CO_2$, 300	......	7.4^{15}	d	i al

No.	Name	Synonyms and Formulae	Mol. wt.	Crystalline form, properties and index of refraction	Density or spec. gravity	Melting point, °C	Boiling point, °C	Solubility, in grams per 100 cc Cold water	Hot water	Other solvents
	Uranyl									
u40	sodium carbonate	$UO_2CO_3.2Na_2CO_3$	542.02	yel cr				sl s		i al
u41	sulfate	$UO_2SO_4.3H_2O$	420.14	yel-grn cr	$3.28^{16.5}$	d 100		$20.5^{15.5}$	22.2^{100}	24.3^{13} conc H_2SO_4; 30^{13} conc HCl
u42	sulfate	$2(UO_2SO_4).7H_2O$	858.29	yel		anh 300		sl s		s H_2SO_3
u43	sulfide	UO_2S	302.09	brn-blk, tetr		d 40–50		sl s	s d	s dil a, dil al, $(NH_4)_2CO_3$; i abs al
u44	sulfite	$UO_2SO_3.4H_2O$	422.15	pa-gr cr						s H_2SO_3
v1	**Vanadic acid,** meta	HVO_3	99.95	yel sc				i		s a, alk; i NH_4OH
v2	tetra-	$H_2V_4O_{11}$	381.78	br amorph				i		s a, alk, NH_4OH
v3	**Vanadium**	V	50.942	lt gray met, cub, 3.03	5.96	1890 ± 10	~3000	i	i	s aq reg, HNO_3, H_2SO_4, HF; i HCl, alk
v4	boride, di-	VB_2	72.56	hex	5.10					
v5[1]	bromide, tri-	VBr_3	290.67	grn-blk, deliq	4.00^{18}	d		s		s al, eth; i HBr
v5[2]	carbide	VC	62.95	blk, cub	5.77	2810	3900	i		s HNO_3, fus KNO_3; i HCl, H_2SO_4
v6	chloride, di-	VCl_2	121.85	grn, hex, deliq	3.23^{18}			s d	s d	s al, eth
v7	chloride, tetra-	VCl_4	192.75	red-br liq	1.816^{30}	−28 ± 2	148.5^{755}	s d		s abs al, eth, chl, acet a
v8	chloride, tri-	VCl_3	157.30	pink cr, deliq	3.00^{18}	d		s d	s d	s abs al, eth
v9	fluoride, penta	VF_5	145.93		2.177^{19}		111.2^{758}			s al
v10	fluoride, tetra-	VF_4	126.94	br yel	2.975^{23}	d 325		s		s acet; sl s al, chl
v11	fluoride, tri-	VF_3	107.94	grn, rhomb	3.363^{19}	>800	subl	i		s al, chl, CS_2
v12	fluoride, tri-	$VF_3.3H_2O$	161.98	dk gr, rhomb		$-3H_2O$, 100		s	v s d	i abs al
v13	iodide, di-	VI_2	304.71	vlt-rose, hex	5.44	750–800 subl vac		s		i al, CCl_4, CS_2, bz
v14	iodide, tri-	$VI_3.6H_2O$	539.75	gr cr, deliq		d		v s		s al
v15	nitride	VN	64.95	blk, cub	6.13	2320		i		sl s aq reg
v16	oxide, di-	Vanadium dioxide. VO (or V_2O_2)	66.94	lt gray cr	5.758^{14}	ign		i	i	s a
v17	oxide, di- (or tetr)-	VO_2 (or V_2O_4)	82.94	bl cr	4.339	1967		i	i	s a, alk
v18	oxide, pent-	V_2O_5	181.88	yel-red, rhomb, 1.46, 1.52, 1.76	3.357^{18}	690	d 1750	0.8^{20}		s a, alk; i abs al
v19	oxide, sesqui	Vanadium trioxide. V_2O_3	149.88	blk cr	4.87^{18}_{4}	1970		sl s	s	s HNO_3, HF, alk
v20	oxybromide	VOBr	146.85	vlt, oct	4.00^{18}	d 480		v sl s		s acet, anhydr eth, acet
v21	oxy di-bromide	$VOBr_2$	226.76	br powd, deliq		d 180		s		
v22	oxy*tri*bromide	$VOBr_3$	306.67	red liq	$2.933^{14.5}$	d 180	130^{100}	s		
v23	oxychloride	VOCl	102.39	yel brn powd	2.824, 3.64^{20}		127	i		v s HNO_3
v24	oxy*di*chloride	$VOCl_2$	137.85	grn, deliq	2.88^{13}			d		s dil HNO_3
v25	oxy*tri*chloride	$VOCl_3$	173.30	yel liq	1.829	−77 ± 2	126.7	s d		s al, eth, ac a
v26	oxy*di*fluoride	VOF_2	104.94	yel	3.396^{19}	d				
v27	oxy*tri*fluoride	VOF_3	123.94	yel-wh, hygr	2.459^{19}	300	480			sl s acet
v28	silicide, di-	VSi_2	107.11	met pr	4.42			i	i	s HF; i al, eth, a
v29	(*dl*-)silicide	V_2Si	129.97	silv wh pr	5.48^{17}			i	i	s HF; i al, eth, a
v30	sulfate (hypovanadous)	$VSO_4.7H_2O$	273.11	vlt, monocl		d in air				
v31	sulfide, mono- or (di-)	VS (or V_2S_2)	83.01	blk pl (exist ?)	4.20	d				s hot H_2SO_4, HNO_3; sl s KSH; i HCl, alk
v32	sulfide, penta-	V_2S_5	262.20	blk-grn powd	3.0	d		i		s HNO_3, alk sulf, alk
v33	sulfide, sesqui- or (tri-)	V_2S_3	198.08	grn-blk pl, or powd	4.72^{21}	d>600		i		s alk sulf; sl s, alk, HCl, HNO_3, H_2SO_4
v34	**Vanadyl sulfate**	$VOSO_4$	163.00	bl.				v s		
w1	**Water**	H_2O	18.01534	col liq, or col hex cr	liq 1.000^{4} sld 0.9168^{0}	0.00	100.00			s al
w2	**Water** heavy	Deuterium oxide. D_2O	20.03	col liq or hex cr, 1.33844^{20}	1.105^{20}_{4}	3.82	101.42	∞	∞	∞ al; sl s eth
w3	**Wolfram**	See tungsten								
x1	**Xenon**	Xe	131.30	col inert gas	gas 5.887 g/l ± 0.009 liq 3.52^{-109} solid 2.7^{-140}	−111.9	−107.1 ± 3	24.1^{0}, 11.9^{25}	8.4^{50}, 7.12^{80}	
y1	**Ytterbium**	Yb	173.04	cub	6.977 up to 789 6.54 above 789	824	1427	i		s a
y2	(III) acetate	$Yb(C_2H_3O_2)_3.4H_2O$	422.24	hex pl	2.09	$-4H_2O$, 100		v s	v s	
y3	(II) bromide	$YbBr_2$	332.86		5.91^{25}_{4}	677	1800	s	s	s dil a

No.	Name	Synonyms and Formulae	Mol. wt.	Crystalline form, properties and index of refraction	Density or spec. gravity	Melting point, °C	Boiling point, °C	Solubility, in grams per 100 cc: Cold water	Hot water	Other solvents
	Ytterbium									
y4	(III) bromide	$YbBr_3$	412.77	col cr		956	d	s	s	
y5	(II) chloride	$YbCl_2$	243.95	grn-yel cr	5.08	702	1900	s	s	s dil a
y6	(III) chloride	$YbCl_3.6H_2O$	387.49	grn, rhomb cr, deliq	2.575	865 $-6H_2O$, 180		v s	v s	s abs al
y7	(II) fluoride	YbF_2	211.04			1052	2380	i	i	
y8	fluoride	YbF_3	230.04			1157	2200	i	i	i dil a
y9	(II) iodide	YbI_2	426.85	lt yel, hex cr	5.40^{25}_{4}	780 ± 4	1300 d(700) vac	s	s	s dil a
y10	(III) iodide	YbI_3	553.75	gold yel cr		d 700	d	s	s	s dil a
y11	(III) oxalate	$Yb_2(C_2O_4)_3.10H_2O$	790.29	col cr	2.644			0.00033^{25}		sl s dil a
y12	(III) oxide	Ytterbia. Yb_2O_3	394.08	col	9.17			i	i	s h dil a
y13	(III) selenate	$Yb_2(SeO_4)_3.8H_2O$	919.08	hex pl	3.30			s d	s	
y14	(III) selenite	$Yb_2(SeO_3)_3$	726.95					i		
y15	(III) sulfate	$Yb_2(SO_4)_3$	634.26	col cr	3.793	d 900		44.2^{0}	4.7^{100}	
y16	(III) sulfate, octohydrate	$Yb_2(SO_4)_3.8H_2O$	778.39	prism	3.286			35.9^{25}	21.1^{40}	
y17	**Yttrium**	Y	88.905	gray-blk met, hex	4.34	1495 ± 5	2927	sl d	d	v s dil a; s h KOH
y18	acetate	$Y(C_2H_3O_2)_3.4H_2O$	338.10	col, tricl					9.03^{25}	
y19	bromate	$Y(BrO_3)_3.9H_2O$	634.76	hex pr		74	$-6H_2O$, 100	168^{25}		sl s al; i eth
y20	bromide	YBr_3	328.63	deliq		904		v s		s al; i eth
y21	bromide hydrate	$YBr_3.9H_2O$	490.77	col tabl, deliq				v s		sl s al; i eth
y22	carbide	YC_2	112.93	yel., microcr	4.13^{18}			d		
y23	carbonate	$Y_2(CO_3)_3.3H_2O$	411.88	wh-redsh powd						s dil min a, $(NH_4)CO_3$; sl s aq CO_2; i al, eth
y24	chloride	YCl_3	195.26	shiny wh leaf	2.67	721	1507	78^{10}	82^{80}	60.1^{15} al; 60.6^{15} pyr
y25	chloride, hexahydrate	$YCl_3.6H_2O$	303.36	redsh-wh, rhomb, deliq	2.18^{18}	$-5H_2O$, 100		217^{20}	235^{80}	s al; i eth
y26	chloride, monohydrate	$YCl_3.H_2O$	213.28	col cr		$-H_2O$, 160		v s		
y27	fluoride	YF_3	145.90	gelat	4.01	1387		i		v sl s dil a
y28	hydroxide	$Y(OH)_3$	139.93	wh-yel gelat or powd		d		i	i	s a, NH_4Cl; i alk
y29	iodide	YI_3	469.62	wh, cr, deliq		1004	650–700^{0.02}	v s		s al, acet; sl s eth
y30	molybdate	$Y_2(MoO_4)_3.4H_2O$	729.68	grayish or yelsh, tetr pl, 2.03	4.79^{16}_{16}	1347				
y31	nitrate, hexahydrate	$Y(NO_3)_3.6H_2O$	383.01	col, redsh cr, deliq	2.68	$-3H_2O$, 100		$134.7^{22.5}$		v s al, eth, HNO_3
y32	nitrate, tetrahydrate	$Y(NO_3)_3.4H_2O$	346.98	redsh-wh pr	2.682			s		s al, HNO_3
y33	oxalate	$Y_2(C_2O_4)_3.9H_2O$	604.01	wh cr powd		d		0.0001		sl s HCl
y34	oxide	Yttria. Y_2O_3	225.81	col-yelsh, cub or powd	5.01	2410		0.00018^{29}		s a; i alk
y35	sulfate	$Y_2(SO_4)_3$	465.99	wh powd	2.52	d 1000		5.38^{25}	s	s sat K_2SO_4 sol
y36	sulfate, octahydrate	$Y_2(SO_4)_3.8H_2O$	610.12	col-redsh, monocl, 1.543, 1.549, 1.576	2.558	$-8H_2O$, 120	d 700	7.47^{16} (anhydr)	1.99^{95} (anhydr)	i al, alk; s conc H_2SO_4
y37	sulfide	Y_2S_3	273.99	yel-gr powd						d a
y38	**Yttrium hexaantipyrine** *per*chlorate	$[Y(C_{11}H_{12}N_2O)_6](ClO_4)_3$	1516.60	col, hex cr		d 293–296		0.55^{20}		
y39	hexaantipyrine iodide	$[Y(C_{11}H_{12}N_2O)_6]I_3$	1598.96	col cr		280–282		4.65^{20}		
z1	**Zinc**	Zn	65.37	bluish-wh met, hex	7.14	419.4	907	i	i	s a, alk, ac a
z2	acetate	$Zn(C_2H_3O_2)_2$	183.46	col, monocl	1.84	d 200	subl vac	30^{20}	44.6^{100}	2.8^{25} al; 166.7^{79} al
z3	acetate, dihydrate	$Zn(C_2H_3O_2)_2.2H_2O$	219.49	col, monocl, β 1.494	1.735	237	$-2H_2O$, 100	31.1^{20}	66.6^{100}	2 al
z4	acetylacetonate	$Zn(C_5H_7O_2)_2$	263.59	need		138	subl	v s d		v s bz, acet; s al
z5	aluminate	Nat. gahnite. $ZnAl_2O_4$	183.33	cub, grn 1.78	4.58			i	i	i a; sl s alk
z6	amide	$Zn(NH_2)_2$	97.42	wh powd, amorph	2.13^{25}	d 200 vac		d	d	i al, eth
z7	antimonide	Zn_3Sb_2	439.61	silv wh, rhomb pr	6.33	570		d		
z8	*ortho*arsenate	Nat. koettigite. $Zn_3(AsO_4)_2.8H_2O$	618.08	monocl, 1.662, 1.683, 1.717	3.309^{15}	$-1H_2O$, 100		i	i	s HNO_3, H_3PO_4, alk
z9	*ortho*arsenate, basic	Nat. adamite. $Zn_3(AsO_4)_2.Zn(OH)_2$	573.34	col, rhomb	4.475^{15}	d 250				
z10	*ortho*arsenate, hydrogen	$ZnHAsO_4.4H_2O$	277.36	wh, rhomb		$-H_2O$, 327		d	d	
z11	arsenide	Zn_3As_2	345.95	met-gray, tetr	5.528	1015		i		d a
z12	benzoate	$Zn(C_7H_5O_2)_2$	307.60	wh powd				2.46^{20}	1.44^{20}	
z13	borate	$3ZnO.2B_2O_3$	383.35	wh tricl cr, or amorph powd	cr 4.22 powd 3.64	980		s		cr i HCl; amorph s HCl
z14	bromate	$Zn(BrO_3)_2.6H_2O$	429.28	wh, cub, 1.5452	2.566	100	$-6H_2O$, 200	v s	∞	
z15	bromide	$ZnBr_2$	225.19	col, rhomb, hygr n^{18}_{D} 1.5452	4.201^{25}_{4}	394	650	447^{20}	675^{100}	v s al, eth, acet; s NH_4OH

No.	Name	Synonyms and Formulae	Mol. wt.	Crystalline form, properties and index of refraction	Density or spec. gravity	Melting point, °C	Boiling point, °C	Solubility, in grams per 100 cc Cold water	Hot water	Other solvents
	Zinc									
z16	butyrate	$Zn(C_4H_7O_2)_2.2H_2O$	275.60	wh pr	...	...	...	10.7^{16}	d	...
z17	caproate	$Zn(C_6H_{11}O_2)_2$	295.68	...	...	...	...	$1.03^{24.5}$	...	...
z18	carbonate	Nat. smithsonite. $ZnCO_3$	125.39	col, trig, 1.818, 1.618	4.398	$-CO_2$, 300	...	0.001^{15}	...	s a, alk, NH_4 salts; i NH_3, acet, pyr
z19	chlorate	$Zn(ClO_3)_2.4H_2O$	304.33	col yelsh, cub, deliq	2.15	d 60	d	262^{20}	v s	167 al; s acet, eth, glyc
z20	chlorate, per-	$Zn(ClO_4)_2.6H_2O$	372.36	wh, rhomb, deliq, 1.508, 1.480	2.252 ± 0.01	105–107	d 200	s	...	s al
z21	chloride	$ZnCl_2$	136.28	wh, hex, deliq, 1.681, 1.713	2.91^{25}	283	732	432^{25}	615^{100}	$100^{12.5}$ al; v s eth; i NH_3
z22	chloroplatinate	$ZnPtCl_6.6H_2O$	581.27	yel, trig, hygr	2.717^{12}	d 160	...	v s	v s	v s al; d H_2SO_4
z23	chromate	$ZnCrO_4$	181.36	lem-yel pr	3.40	...	...	i	d	s a, liq NH_3; i acet
z24	chromate	$ZnCr_2O_4$	233.36	dk grn to black, cub	5.30^{15}	...	...	...	...	...
z25	*di*chromate	$ZnCr_2O_7.3H_2O$	335.40	redsh-brn cr, or or-yel powd, hygr	...	...	...	v s	d	i al, eth; s a
z26	citrate	$Zn_3(C_6H_5O_7)_2.2H_2O$	610.35	...	...	...	...	sl s	...	...
z27	cyanide	$Zn(CN)_2$	117.41	col, rhomb	1.852	d 800	...	0.0005^{20}	...	s alk, KCN, NH_3; i al
z28	ferrate (III)	Ferrite. $ZnFe_2O_4$	241.06	blk, oct	5.33^{20}	1590	...	...	...	s conc HCl; i dil a, alk
z29	ferrocyanide	$Zn_2Fe(CN)_6$	342.69	wh powd	1.85^{25}_{4}	...	...	i	...	s excess alk; i dil a
z30	ferrocyanide, trihydrate	$Zn_2Fe(CN)_6.3H_2O$	396.74	wh powd	...	d	...	i	i	i al, HCl; d NaOH; s NH_4OH; v sl s NH_3
z31	fluoride	ZnF_2	103.37	col, monocl or tricl	4.95^{25}_{4}	872	*ca* 1500	1.62^{20}	s	s hot a, NH_4OH; i al, NH_3
z32	fluoride, tetrahydrate	$ZnF_2.4H_2O$	175.43	col, rhomb	2.255	$-4H_2O$, 100	tr to ZNO, 3000	1.6^{18}	s	s a, alk, NH_4OH
z33	fluosilicate	$ZnSiF_6.6H_2O$	315.54	col, hex pr, 1.3824, 1.3956	2.104	d 100	...	v s	...	...
z34	formaldehyde-sulfoxylate	$Zn(HSO_2.CH_2O)_2$	255.56	rhomb pr	...	d	...	v s	v s	d a; i al
z35	formaldehyde-sulfoxylate, basic	$Zn(OH)HSO_2.CH_2O$	177.47	rhomb pr	...	d	...	i	i	d a; i al
z36	formate	$Zn(CHO_2)_2$	155.41	col, cr	2.368	d	...	3.80	62^{100}	...
z37	formate	$Zn(CHO)_2.2H_2O$	191.44	wh, monocl, 1.513, 1.526, 1.566	2.207^{20}	$-2H_2O$, 140	d	5.2^{20}	38^{100}	i al
z38	gallate	$ZnGa_2O_4$	268.81	wh fine cr, 1.74	6.15 calc	<800	...	i	i	i org solv; s dil a, NH_4OH
z39	glycerophosphate	$ZnC_3H_7O_6P$	235.43	wh amorph powd	...	...	...	s	...	i al, eth
z40	hydroxide(ε)	$Zn(OH)_2$	99.38	col, rhomb	3.053	d 125	...	v sl s	...	s a, alk
z41	iodate	$Zn(IO_3)_2$	415.18	wh, need	5.0632^{5}	d	...	0.87	1.31	s alk, HNO_3
z42	iodate, dihydrate	$Zn(IO_3)_2.2H_2O$	451.21	wh, cr powd	4.223^{25}_{4}	$-H_2O$, 200	...	0.877	1.32	s HNO_3, NH_4OH
z43	iodide	ZnI_2	319.18	col, hexag	4.7364^{25}_{4}	446	d 624	432^{18}	511^{100}	s a, al, eth, NH_3, $(NH_4)_2CO_3$
z44	*d*-lactate	$Zn(C_3H_5O_3)_2.2H_2O$	279.45	...	...	...	...	5.7^{15}	9^{33}	0.104 h 98 % al
z45	*dl*-lactate	$Zn(C_3H_5O_3)_2.3H_2O$	297.47	wh, rhomb cr	...	...	...	1.67^{105}	16.7^{100}	v sl s al
z46	laurate	$Zn(C_{12}H_{23}O_2)_2$	464.00	wh powd	...	128	...	0.01^{15}	0.019^{100}	0.010^{15} al
z47	*per*manganate	$Zn(MnO_4)_2.6H_2O$	411.33	vlt-br or bl, deliq	2.47	$-5H_2O$, 100	...	33.3	v s	d al, a
z48	nitrate, trihydrate	$Zn(NO_3)_2.3H_2O$	243.43	col, need	...	45.5	...	327.3^{40}	...	...
z49	nitrate, hexahydrate	$Zn(NO_3)_2.6H_2O$	297.47	col, tetrag	2.065^{14}	36.4	$-6H_2O$, 105–131	184.3^{20}	∞	v s al
z50	nitride	Zn_3N_2	224.12	gray	6.22^{25}_{4}	...	...	d	...	s HCl
z51	oleate	$Zn(C_{18}H_{33}O_2)_2$	628.30	wax-like solid	...	70	...	i	...	s al, eth, bz, CS_2; sl s acet
z52	oxalate	$ZnC_2O_4.2H_2O$	189.42	wh powd	3.28^{25}_{4}	d 100	...	0.00079^{18}	...	s a, alk
z53	oxide	Nat. zincite. ZnO	81.37	wh, hex, 2.008, 2.029	5.606	1975	...	0.00016^{29}	...	s a, alk, NH_4Cl; i al, NH_3
z54	oxide, per-	$ZnO_2.\frac{1}{2}H_2O$	106.38	yelsh, powd	3.00 ± 0.08	$-O_2$, vac	...	sl d	d	d al, eth, acet
z55	1-phenol-4-sulfonate(p)	$Zn(C_6H_5SO_4)_2.8H_2O$	555.83	col cr or fine wh powd, effl	...	$-8H_2O$, 125	...	62.5	250^{100}	55.6^{25} al
z56	*ortho*phosphate	$Zn_3(PO_4)_2$	386.05	col, rhomb	3.998^{15}	900	...	i	i	s a, NH_4OH; i al
z57	*ortho*phosphate, dihydrogen	$Zn(H_2PO_4)_2.2H_2O$	295.38	tricl	...	d 100	...	d	...	...
z58	*ortho*phosphate, octahydrate	$Zn_3(PO_4)_2.8H_2O$	530.18	rhomb pl	3.109^{15}	...	...	i	...	s alk

No.	Name	Synonyms and Formulae	Mol. wt.	Crystalline form, properties and index of refraction	Density or spec. gravity	Melting point, °C	Boiling point, °C	Solubility, in grams per 100 cc: Cold water	Hot water	Other solvents
	Zinc									
z59	*ortho*phosphate, tetrahydrate	α—Hopeite. $Zn_3(PO_4)_2.4H_2O$	458.11	col, rhomb, 1.572, 1.591, 1.59	3.04	tr >105		i	i	v s a, NH_4OH, NH_4 salts
z60	*ortho*phosphate tetrahydrate	β-Hopeite. $Zn_3(PO_4)_2.4H_2O$	458.11	col, rhomb, 1.574, 1.582, 1.582	3.03	tr >140		i	i	v s a, NH_4OH, NH_4 salts
z61	*ortho*phosphate tetrahydrate	Parahopeite. $Zn_3(PO_4)_2.4H_2O$	458.11	col, tricl, 1.614, 1.625, 1.665	3.75	tr >163		i	i	v s a, NH_4OH, NH_4 salts
z62	*pyro*phosphate	$Zn_2P_2O_7$	304.68	wh powd	3.75^{23}			i	i	s a, alk, NH_4OH
z63	phosphide	Zn_3P_2	258.06	dk gray, tetrag, pois	4.55^{13}	>420	1100; subl in H_2	d		d H_2SO_4 ev H_3P s HNO_3; s (viol) dil a; i a s alk
z64	hypophosphite	$Zn(H_2PO_2)_2.H_2O$	213.36	col, cr powd, hygr				s		
z65	picrate	$Zn(C_6H_2N_3O_7)_2.8H_2O$	665.69	yel cr powd, expl		expl		s		
z66	salicylate	$Zn(C_7H_5O_3)_2.3H_2O$	393.65	need				5^{20}		s al
z67	selenate	$ZnSeO_4.5H_2O$	298.40	wh, tricl	2.591_4^{20}	d >50		s		
z68	selenide	ZnSe	144.33	yelsh to redsh, cub, 2.89	5.42_4^{15}	>1100		i		s a; d HNO_3
z69	silicate	Nat. hemimorphite. $2ZnO.SiO_2.H_2O$	240.84	rhomb, or trigon 1.614, 1.617, 1.636	3.45			i	i	
z70	*meta*silicate	$ZnSiO_3$	141.45	col, rhomb	3.42	1437		i		i a
z71	*ortho*silicate	Nat. willemite. Zn_2SiO_4	222.82	trig, 1.694, 1.723	4.103	1509		i	i	s acet a
z72	stearate	$Zn(C_{18}H_{35}O_2)_2$	632.33	light powd		130		i		i al, eth
z73	sulfate	Nat. zinkosite. $ZnSO_4$	161.43	col, rhomb, 1.658, 1.669, 1.670	3.54_4^{25}	d 600		s	s	sl s al; s MeOH, glyc
z74	sulfate, heptahydrate	Nat. goslarite. $ZnSO_4.7H_2O$	287.54	col, rhomb, effl, 1.457, 1.480, 1.484	1.957_4^{25}	100	$-7H_2O$, 280	96.5^{20}	663.6^{100}	sl s al, glyc
z75	sulfate, hexahydrate	$ZnSO_4.6H_2O$	269.52	col, monocl or tetrag	2.072^{15}	$-5H_2O$, 70		s	117.5^{40}	
z76	sulfide,(α)	Nat. wurtzite. ZnS	97.43	col, hex, 2.356, 2.378	3.98	$1850^{150\ atm}$	subl 1185	0.00069^{18}		v s a; i ac a
z77	sulfide,(β)	Nat. sphalerite. ZnS	97.43	col, cub, 2.368	4.102^{25}	tr 1020		0.000065^{18}		v s a
z78	sulfide, monohydrate	$ZnS.H_2O$	115.45	yelsh-wh powd	3.98	1049		i		s a
z79	sulfite	$ZnSO_3.2H_2O$	181.46	wh, cr powd		$-2H_2O$, 100	d 200	0.16	d	i al; s H_2SO_3
z80	tartrate	$ZnC_4H_4O_6.H_2O$ (or $2H_2O$)	231.46	wh powd				0.055^{20}		s KOH, NaOH
z81	tellurate	Zn_3TeO_6	419.71	wh, gran ppt				i	i	s a
z82	telluride	ZnTe	192.97	red, cub, 3.56	6.34^{15}	1238.5		d		s d a
z83	thiocyanate	$Zn(SCN)_2$	181.53	wh powd, deliq				s		s al, NH_4OH
z84	valerate	$Zn(C_5H_9O_2)_2.2H_2O$	303.65	wh glist sc or powd				2.6^{24-25}	s	*ca* 2.5 al; v sl s et
	Zinc complexes									
z85	diamminezinc chloride	$[Zn(6H_3)_2]Cl_2$	170.34	col, rhomb, 1.625, 1.590	2.10	210.8	d 271	d		
z86	tetrammine perrhenate	$[Zn(NH_3)_4](ReO_4)_2$	633.89	wh, cub cr	3.608_4^{25}					0.1852 conc NH_4OH
z87	tetrapyridine fluosilicate	$[Zn(C_5H_5N)_4]SiF_6$	523.86	wh, rhomb	2.197					
z88	**Zirconium**	Zr	91.22	silver gray, met	6.49	1852 ± 2	3578	i	i	s HF, aq reg; sl s a
z89	bromide, di-	$ZrBr_2$	251.04	blk powd, ign in air		d >350		d ev H_2		
z90	boride, di-	ZrB_2	112.84	hex	6.085	*ca* 3000				
z93	bromide, tetra-	$ZrBr_4$	410.86	wh cr powd, deliq		$450 \pm 1^{15 atm}$	357 subl	i d		s liq NH_3, aceton i bz, CCl_4
z94	bromide, tri-	$ZrBr_3$	330.95	bl-blk powd		d 350		d ev H_2		
z95	carbide	ZrC	103.23	gray met, cub	6.73	3540	5100	i		sl s conc H_2SO_4
z96	carbonate, basic	$3ZrO_2.CO_2.H_2O$	431.68	wh, amorph powd				i		s a
z97	chloride, di-	$ZrCl_2$	162.13	blk	3.6^{16}	d 350		d ev H_2		
z98	chloride, tetra-	$ZrCl_4$	233.03	wh cr	2.803^{15}	$437^{25 atm}$	subl 331	s	d	s al, eth, conc H
z99	chloride, tri-	$ZrCl_3$	197.58	br cr	3.00^{16}	d 350		d ev H_2		s $-H_2$ conc al; i org cpd
z100	fluoride	ZrF_4	167.21	wh hex, 1.59	4.43	subl ~ 600		1.388^{25}	d	sl s HF
z101	hydride	ZrH_2	93.24	gray-blk powd						s dif HF, conc a
z102	hydroxide	$Zr(OH)_4$	159.25	wh amorp powd	3.25	$-2H_2O$, 500		0.02	i	s min a
z103	iodide	ZrI_4	598.84	wh need, hygr		499 ± 2 6.3atm	d ~600	s d	s	d al; s eth; v sl s CS_2, bz; i liq NH_3
z104	nitrate	$Zr(NO_3)_4.5H_2O$	429.32	col cr, deliq, 1.60, 1.61				v s		s al

No.	Name	Synonyms and Formulae	Mol. wt.	Crystalline form, properties and index of refraction	Density or spec. gravity	Melting point, °C	Boiling point, °C	Solubility, in grams per 100 cc: Cold water	Hot water	Other solvents
	Zirconium									
z105	nitride	ZrN	105.23	yel-brn cr	7.09	2980 ± 50		i	i	sl s inorg ac; s conc H_2SO_4, HF, aq reg
z106	oxide	Nat. baddeleyite. ZrO_2	123.22	col-yel-brn, monocl, 2.13, 2.19, 2.20	5.89	*ca* 2700	*ca* 5000	i	i	s H_2SO_4, HF
z107	oxide	Zirconia. ZrO_2 $HfO_2 < 2\%$	123.22	wh, monocl below 1000°, cub, above	5.6	2715		i	i	s H_2SO_4, HF
z108	oxide	Zirconium hydroxide, zirconic acid. $ZrO_2.xH_2O$		gel or wh amorph powd	3.25	$-2H_2O$, 550		0.02		s acids; i al, alk
z109	phosphide	ZrP_2	153.17	gray, brittle	4.77^{25}_{4}			i		v s conc hot H_2SO_4
z110	selenate	$Zr(SeO_4)_2.4H_2O$	449.20	hex trsp cr				s		sl s al, conc a
z111	selenite	$Zr(SeO_3)_2$	345.14	wh sm cr	4.3	d ~ 400		i	i	sl s H_2SO_4
z112	*ortho*silicate	Zircon, hyacinth. $ZrSiO_4$	183.30	tetr, var colors, 1.92–96; 1.97–2.02	4.56	2550		i	i	i a, aq reg, alk
z113	silicide	$ZrSi_2$	147.39	steel gray rhomb, lust met	4.88^{22}			i	i	s HF; i inorg a, aq reg
z114	sulfate	$Zr(SO_4)_2$	283.35	microcr powd, hygr	3.22^{16}	410 d		s	v s	
z115	sulfate	$Zr(SO_4)_2.4H_2O$	355.41	wh cr powd, rhomb	3.22^{16}	$-3H_2O$, 135–150		v s		i al
z116	sulfide	ZrS_2	155.35	steelgray cr, hexag	3.87	~ 1550		i	i	i a
z117	**Zirconyl bromide**	$ZrOBr_2.xH_2O$		brill need, deliq		$-H_2O$, 120		s		s hot conc HBr
z118	chloride	$ZrOCl_2.8H_2O$	322.25	wh, need, tetr, effl, 1.552, 1.563		$-6H_2O$, 150	$-8H_2O$, 210	s	d	s al, eth; sl s HCl
z119	iodide	$ZrOI_2.8H_2O$	505.15	col, need, hygr		d		v s	v s	s al; v s eth
z120	sulfide	Zirconium sulfoxide. ZrOS	139.28	yel powd	4.87	ign in air		i	i	

GRAVIMETRIC FACTORS AND THEIR LOGARITHMS

Rudolf Loebel

Compiled from International atomic weights of 1964. To facilitate use of this table the group of substances weighed under each element as well as the substance sought under each substance weighed are arranged in alphabetical order of their formulas.

Weighed	Sought	Factor	Log of Factor +10	Reciprocal of Factor	Log of Reciprocal of Factor +10
Aluminum					
Al	Al_2O_3	1.88946	10.27634	0.52925	9.72366
	$AlPO_4$	4.51987	10.65513	0.22125	9.34488
Al_4C_3	Al_2O_3	1.41653	10.15122	0.70595	9.84877
$Al(C_9H_6ON)_3$	Al	0.05873	8.76886	17.02811	11.23114
	Al_2O_3	0.11096	9.04516	9.01226	10.95483
$AlCl_3$	Al_2O_3	0.38233	9.58244	2.61554	10.41756
AlF_3	CaF_2	1.39464	10.14446	0.71703	9.85554
Al_2O_3	Al	0.52925	9.72366	1.88946	10.27634
	Al_4C_3	0.70595	9.84877	1.41653	10.15122
	$AlCl_3$	2.61552	10.41756	0.38233	9.58244
	$AlPO_4$	2.39214	10.37879	0.41804	9.62122
	$Al_2(SO_4)_3$	3.35567	10.52578	0.29800	9.47422
	$Al_2(SO_4)_3.18H_2O$	6.53605	10.81531	0.15300	9.18469
	$K_2SO_4.Al_2(SO_4)_3.24H_2O$	9.30532	10.96873	0.10747	9.03129
	$(NH_4)_2SO_4.Al_2(SO_4)_3.24H_2O$	8.89216	10.94901	0.11246	9.05100
$AlPO_4$	Al	0.22125	9.34488	4.51977	10.65512
	Al_2O_3	0.41804	9.62122	2.39211	10.37878
	P_2O_5	0.58196	9.76489	1.71833	10.23511
$Al_2(SO_4)_3$	Al_2O_3	0.29800	9.47422	3.35570	10.52578
$Al_2(SO_4)_3.18H_2O$	Al_2O_3	0.15300	9.18469	6.53595	10.81531
CaF_2	AlF_3	0.71703	9.85554	1.39464	10.14446
$K_2SO_4.Al_2(SO_4)_3.24H_2O$	Al_2O_3	0.10747	9.03129	9.30492	10.96872
$(NH_4)_2SO_4.Al_2(SO_4)_3.24H_2O$	Al_2O_3	0.11246	9.05100	8.89205	10.94900
P_2O_5	$AlPO_4$	1.71831	10.23510	0.58196	9.76489
Ammonium					
Ag	NH_4Br	0.90802	9.95810	1.10130	10.04191
	NH_4Cl	0.49589	9.69538	2.01657	10.30462
	NH_4I	1.34366	10.12829	0.74424	9.87171
AgBr	NH_4Br	0.52161	9.71735	1.91714	10.28265
	NH_4Cl	0.37323	9.57198	2.67931	10.42802
AgI	NH_4I	0.61737	9.79055	1.61977	10.20945
$BaSO_4$	$(NH_4)_2SO_4$	0.56615	9.75293	1.76632	10.24707
Br	NH_4Br	1.22574	10.08840	0.81583	9.91160
Cl	NH_4	0.50880	9.70655	1.96541	10.29345
	NH_4Cl	1.50881	10.17864	0.66277	9.82136
HCl	NH_4Cl	1.46710	10.16646	0.68162	9.83354
I	NH_4I	1.14214	10.05772	0.87555	9.94229
$MgNH_4PO_4.6H_2O$	NH_3	0.06941	8.84142	14.40714	11.15857
	NH_4	0.07352	8.86641	13.60174	11.13360
	$(NH_4)_2O$	0.10613	9.02584	9.42241	10.97416
N	NH_3	1.21589	10.08489	0.82244	9.91510
	NH_4	1.28785	10.10987	0.77649	9.89014
	NH_4Cl	3.81903	10.58195	0.26184	9.41804
	NH_4NO_3	5.71466	10.75699	0.17499	9.24302
	$(NH_4)_2O$	1.85899	10.26928	0.53793	9.73072
	$(NH_4)_2SO_4$	4.71699	10.67367	0.21200	9.32634
NH_3	$MgNH_4PO_4.6H_2O$	14.40648	11.15855	0.06941	8.84142
	N	0.82244	9.91510	1.21589	10.08489
	NH_4	1.05919	10.02498	0.94412	9.97503
	NH_4Cl	3.14093	10.49706	0.31837	9.50293
	$(NH_4)_2CO_3$	2.82099	10.45040	0.35448	9.54959
	NH_4HCO_3	4.64199	10.66671	0.21542	9.33329
	NH_4NO_3	4.69998	10.67210	0.21277	9.32791
	$(NH_4)_2O$	1.52891	10.18438	0.65406	9.81562
	NH_4OH	2.05783	10.31341	0.48595	9.68659
	$(NH_4)_2PtCl_6$	13.03213	11.11501	0.07673	8.88497
	$(NH_4)_2SO_4$	3.87945	10.58877	0.25777	9.41123
	N_2O_5	3.17106	10.50121	0.31535	9.49879
	Pt	5.72763	10.75797	0.17459	9.24202
	SO_3	2.35053	10.37117	0.42544	9.62884
NH_4	Cl	1.96540	10.29345	0.50880	9.70655
	$MgNH_4PO_4.6H_2O$	13.60144	11.13359	0.07352	8.86641
	N	0.77648	9.89013	1.28786	10.10986
	NH_3	0.94412	9.97503	1.05919	10.02498

Weighed	Sought	Factor	Log of Factor +10	Reciprocal of Factor	Log of Reciprocal of Factor +10
Ammonium (contd.)					
NH_3	NH_4Cl	2.96542	10.47208	0.33722	9.52792
	$(NH_4)_2PtCl_6$	12.30389	11.09005	0.08128	8.90998
	Pt	5.40757	10.73301	0.18493	9.26701
NH_4Br	Ag	1.10130	10.04191	0.90802	9.95810
	AgBr	1.91713	10.28265	0.52161	9.71735
	Br	0.81583	9.91160	1.22575	10.08840
NH_4Cl	Ag	2.01656	10.30461	0.49589	9.69538
	AgCl	2.67934	10.42802	0.37323	9.57198
	Cl	0.66277	9.82136	1.50882	10.17864
	HCl	0.68162	9.83354	1.46709	10.16646
	N	0.26185	9.41805	3.81898	10.58194
	NH_3	0.31838	9.50294	3.14090	10.49706
	NH_4	0.33722	9.52791	2.96542	10.47208
	$(NH_4)_2O$	0.48677	9.68732	2.05436	10.31268
	NH_4OH	0.65516	9.81635	1.52634	10.18365
	$(NH_4)_2PtCl_6$	4.14913	10.61795	0.24101	9.38204
	Pt	1.82354	10.26091	0.54838	9.73908
$(NH_4)_2CO_3$	NH_3	0.35448	9.54959	2.82103	10.45040
NH_4HCO_3	NH_3	0.21542	9.33329	4.64209	10.66672
NH_4I	Ag	0.74422	9.87170	1.34369	10.12830
	AgI	1.61977	10.20946	0.61737	9.79055
	I	0.87555	9.94228	1.14214	10.05772
NH_4NO_3	NH_3	0.21277	9.32791	4.69991	10.67209
	$(NH_4)_2PtCl_6$	2.77280	10.44292	0.36064	9.55709
	N_2O_5	0.67470	9.82911	1.48214	10.17089
	Pt	1.21865	10.08588	0.82058	9.91412
$(NH_4)_2O$	$MgNH_4PO_4.6H_2O$	9.42281	10.97418	0.10613	9.02584
	N	0.53733	9.73069	1.86105	10.26931
	NH_3	0.65407	9.81562	1.52889	10.18438
	NH_4Cl	2.05437	10.31268	0.48677	9.68732
	$(NH_4)_2PtCl_6$	8.52378	10.93063	.0.11732	9.06937
	N_2O_5	2.07406	10.31682	0.48215	9.68319
	Pt	3.74621	10.57359	0.26694	9.42641
NH_4OH	N	0.39967	9.60170	2.50206	10.39830
	NH_3	0.48595	9.68660	2.05782	10.31341
	NH_4	0.51471	9.71156	1.94284	10.28843
	NH_4Cl	1.52633	10.18365	0.65517	9.81635
	$(NH_4)_2PtCl_6$	6.33297	10.80158	0.15790	9.19841
	Pt	2.78334	10.44456	0.35928	9.55544
$(NH_4)_2PtCl_6$	NH_3	0.07674	8.88502	13.03101	11.11497
	NH_4	0.08128	8.90998	12.30315	11.09002
	NH_4Cl	0.24102	9.38206	4.14903	10.61794
	NH_4NO_3	0.36065	9.55709	2.77277	10.44291
	$(NH_4)_2O$	0.11732	9.06937	8.52370	10.93063
	NH_4OH	0.15791	9.19841	6.33272	10.80159
	$(NH_4)_2SO_4$	0.29768	9.47375	3.35931	10.52625
$(NH_4)_2SO_4$	$BaSO_4$	1.76632	10.24709	0.56615	9.75291
	H_2SO_4	0.74223	9.87054	1.34729	10.12947
	N	0.21200	9.32634	4.71698	10.67367
	NH_3	0.25777	9.41123	3.87943	10.58877
	$(NH_4)_2PtCl_6$	3.35927	10.52625	0.29768	9.47375
	$(NH_4)_2SO_4$	1.47640	10.16921	0.67732	9.83079
	SO_3	0.60589	9.78239	1.65046	10.21761
N_2O_5	NH_3	0.31535	9.49879	3.17108	10.50121
	NH_4NO_3	1.48214	10.17089	0.67470	9.82911
	$(NH_4)_2O$	0.48214	9.68318	2.07409	10.31682
Pt	NH_3	0.17459	9.24202	5.72770	10.75798
	NH_4	0.18493	9.26701	5.40745	10.73300
	NH_4Cl	0.54838	9.73908	1.82355	10.26092
	NH_4NO_3	0.82058	9.91412	1.21865	10.08588
	$(NH_4)_2O$	0.26694	9.42641	3.74616	10.57364
	NH_4OH	0.35928	9.55544	2.78334	10.44456
	$(NH_4)_2SO_4$	0.67732	9.83079	1.47641	10.16921
SO_3	NH_3	0.42543	9.62883	2.35056	10.37118
	$(NH_4)_2SO_4$	1.65046	10.21760	0.60589	9.78239
Antimony					
$K(SbO)C_4H_4O_6.\frac{1}{2}H_2O$	Sb	0.36460	9.56182	2.74273	10.43819
	Sb_2O_3	0.43647	9.63995	2.29111	10.36005

Weighed	Sought	Factor	Log of Factor +10	Reciprocal of Factor	Log of Reciprocal of Factor +10
Antimony (contd.)					
$K(SbO)C_4H_4O_6.\frac{1}{2}H_2O$	Sb_2O_4	0.46043	9.66317	2.17188	10.33684
	Sb_2S_3	0.50862	9.70640	1.96610	10.29360
Sb	$K(SbO)C_4H_4O_6.\frac{1}{2}H_2O$	2.74275	10.43819	0.36460	9.56182
	Sb_2O_3	1.19713	10.07814	0.83533	9.92186
	Sb_2O_4	1.26283	10.10137	0.79187	9.89866
	Sb_2O_5	1.32854	10.12340	0.75271	9.87659
	Sb_2S_3	1.39503	10.14458	0.71683	9.85540
	Sb_2S_5	1.65840	10.21969	0.60299	9.78031
Sb_2O_3	$K(SbO)C_4H_4O_6.\frac{1}{2}H_2O$	2.29111	10.36005	0.43647	9.63995
	Sb	0.83533	9.92186	1.19713	10.07814
	Sb_2O_4	1.05489	10.02320	0.94797	9.97680
	Sb_2O_5	1.10978	10.04523	0.90108	9.95476
	Sb_2S_3	1.16532	10.06645	0.85813	9.93356
	Sb_2S_5	1.38532	10.14155	0.72185	9.85845
Sb_2O_4	$K(SbO)C_4H_4O_6.\frac{1}{2}H_2O$	2.17190	10.33684	0.46043	9.66317
	Sb	0.79187	9.89866	1.26283	10.10137
	Sb_2O_3	0.94797	9.97680	1.05489	10.02300
	Sb_2O_5	1.05203	10.02203	0.95054	9.97797
	Sb_2S_3	1.10468	10.04324	0.90523	9.95671
	Sb_2S_5	1.31324	10.11834	0.76148	9.88166
Sb_2O_5	Sb	0.75270	9.87663	1.32853	10.12337
	Sb_2O_3	0.90108	9.95477	1.10977	10.04523
	Sb_2O_4	0.95054	9.97798	1.05202	10.02202
	Sb_2S_5	1.24828	10.09632	0.80109	9.90368
Sb_2S_3	$K(SbO)C_4H_4O_6.\frac{1}{2}H_2O)$	1.96603	10.29360	0.50863	9.70640
	Sb	0.71681	9.85542	1.39503	10.14458
	Sb_2O_3	0.85814	9.93356	1.16532	10.06645
	Sb_2O_4	0.90524	9.95676	1.10469	10.04332
	Sb_2O_5	0.95234	9.97879	1.05006	10.02121
Sb_2S_5	Sb	0.60299	9.78031	1.65840	10.21969
	Sb_2O_3	0.72186	9.85834	1.38570	10.14167
	Sb_2O_4	0.76148	9.88166	1.31323	10.11834
	Sb_2O_5	0.80110	9.90369	1.24828	10.09631
Arsenic					
As	As_2O_3	1.32032	10.12068	0.75739	9.87932
	As_2O_5	1.53387	10.18579	0.65195	9.81422
	As_2S_3	1.64194	10.21535	0.60904	9.78465
	As_2S_5	2.06991	10.31597	0.48311	9.68405
	$BaSO_4$	4.67291	10.66959	0.21400	9.33041
	$Mg_2As_2O_7$	2.07191	10.31637	0.48265	9.68364
	$MgNH_4AsO_4.\frac{1}{2}H_2O$	2.53968	10.40478	0.39375	9.59523
AsO_3	$BaSO_4$	2.84821	10.45457	0.35110	9.54543
	$Mg_2As_2O_7$	1.26286	10.10135	0.79185	9.89865
	$MgNH_4AsO_4.\frac{1}{2}H_2O$	1.54797	10.18976	0.64601	9.81024
AsO_4	$BaSO_4$	2.52018	10.40143	0.39680	9.59857
	$Mg_2As_2O_7$	1.11742	10.04821	0.89492	9.95178
	$MgNH_4AsO_4.\frac{1}{2}H_2O$	1.36969	10.13662	0.73009	9.86337
As_2O_3	As	0.75739	9.87932	1.32032	10.12068
	As_2O_5	1.16173	10.06511	0.86079	9.93490
	As_2S_3	1.24359	10.09468	0.80412	9.90532
	As_2S_5	1.56773	10.19527	0.63781	9.80472
	$BaSO_4$	3.53922	10.54891	0.28255	9.45110
	$Mg_2As_2O_7$	1.56925	10.19569	0.63725	9.80431
	$MgNH_4AsO_4.\frac{1}{2}H_2O$	1.92353	10.28410	0.51988	9.71590
As_2O_5	As	0.65195	9.81422	1.53386	10.18579
	As_2O_3	0.86077	9.93489	1.16175	10.06511
	As_2S_3	1.07046	10.02957	0.93418	9.97043
	As_2S_5	1.34947	10.13016	0.74103	9.86984
	$BaSO_4$	3.04648	10.48380	0.32825	9.51621
	$Mg_2As_2O_7$	1.35077	10.13058	0.74032	9.86942
	$MgNH_4AsO_4.\frac{1}{2}H_2O$	1.65573	10.21899	0.60396	9.78101
As_2S_3	As	0.60903	9.78464	1.64196	10.21536
	As_2O_3	0.80412	9.90532	1.24360	10.09468
	As_2O_5	0.93418	9.97043	1.07046	10.02957
	As_2S_5	1.26065	10.10060	0.79324	9.89939
	$Mg_2As_2O_7$	1.26187	10.10102	0.79247	9.89898
As_2S_5	As	0.48311	9.68405	2.06992	10.31595
	As_2O_3	0.63786	9.80472	1.56774	10.19527

Weighed	Sought	Factor	Log of Factor +10	Reciprocal of Factor	Log of Reciprocal of Factor +10
Arsenic (contd.)					
As_2S_5	As_2O_5	0.74103	9.86984	1.34947	10.13016
	As_2S_3	0.79324	9.89940	1.26065	10.10060
$BaSO_4$	As	0.21399	9.33039	4.67312	10.66961
	AsO_3	0.35110	9.54543	2.84819	10.45457
	AsO_4	0.39679	9.59856	2.52022	10.40143
	As_2O_3	0.28255	9.45110	3.53920	10.54890
	As_2O_5	0.32825	9.51621	3.04646	10.48379
$Ca_2As_2O_7$	As	0.43814	9.64161	2.28238	10.35839
	As_2O_3	0.57849	9.76230	1.72864	10.23770
$Mg_2As_2O_7$	As	0.48264	9.68363	2.07194	10.31638
	AsO_3	0.79184	9.89864	1.26288	10.10136
	AsO_4	0.89491	9.95178	1.11743	10.04822
	As_2O_3	0.63725	9.80431	1.56924	10.19569
	As_2O_5	0.74031	9.86942	1.35079	10.13059
	As_2S_3	0.79248	9.89899	1.26186	10.10102
$MgNH_4AsO_4.\frac{1}{2}H_2O$	As	0.39375	9.59523	2.53968	10.40478
	AsO_3	0.64600	9.81023	1.54799	10.18977
	AsO_4	0.73009	9.86337	1.36969	10.13662
	As_2O_3	0.51988	9.71590	1.92352	10.28410
	As_2O_5	0.60396	9.78101	1.65574	10.21899
Barium					
Ba	$BaCO_3$	1.43694	10.15744	0.69592	9.84256
	$BaCrO_4$	1.84455	10.26589	0.54214	9.73411
	$BaSiF_6$	2.03444	10.30844	0.49154	9.69156
	$BaSO_4$	1.69943	10.23030	0.58843	9.76969
$BaCl_2$	$BaCO_3$	0.94766	9.97665	1.05523	10.02334
	$BaCrO_4$	1.21647	10.08510	0.82205	9.91490
	$BaSO_4$	1.12077	10.04952	0.89224	9.95048
$BaCl_2.2H_2O$	$BaSO_4$	0.95546	9.98021	1.04662	10.01979
$BaCO_3$	Ba	0.69592	9.84256	1.43695	10.15747
	$BaCl_2$	1.05523	10.02334	0.94766	9.97665
	$BaCrO_4$	1.28366	10.10845	0.77902	9.89155
	$Ba(HCO_3)_2$	1.31426	10.11869	0.76088	9.88132
	BaO	0.77700	9.89042	1.28700	10.10958
	$BaSO_4$	1.18267	10.07286	0.84554	9.92713
	CO_2	0.22300	9.34830	4.48430	10.65170
$BaCrO_4^-$	Ba	0.54214	9.73411	1.84455	10.26589
	$BaCl_2$	0.82205	9.91490	1.21647	10.08510
	$BaCO_3$	0.77902	9.89155	1.28366	10.10846
	BaO	0.60530	9.78197	1.65207	10.21803
BaF_2	$BaSiF_6$	1.59359	10.20238	0.62751	9.79762
$Ba(HCO_3)_2$	$BaCO_3$	0.76088	9.88132	1.31427	10.11869
$Ba(NO_3)_2$	$BaSO_4$	0.89306	9.95088	1.11975	10.04912
BaO	$BaCO_3$	1.28701	10.10958	0.77699	9.89042
	$BaCrO_4$	1.65208	10.21803	0.60530	9.78197
	$BaSiF_6$	1.82233	10.26061	0.54878	9.73934
	$BaSO_4$	1.52211	10.18244	0.65698	9.81756
	CO_2	0.28701	9.45790	3.48420	10.54210
BaO_2	$BaSO_4$	1.37829	10.13934	0.72554	9.86066
BaS	$BaSO_4$	1.37780	10.13919	0.72579	9.86081
$BaSiF_6$	Ba	0.49152	9.69154	2.03451	10.30846
	BaF_2	0.62751	9.79762	1.59360	10.20238
	BaO	0.54878	9.73939	1.82222	10.26060
$BaSO_4$	Ba	0.58843	9.76969	1.69944	10.23030
	$BaCl_2$	0.89225	9.95049	1.12076	10.04952
	$BaCl_2.2H_2O$	1.04662	10.01979	0.95546	9.98021
	$BaCO_3$	0.84554	9.92713	1.18268	10.07286
	$Ba(NO_3)_2$	1.11975	10.04912	0.89306	9.95088
	BaO	0.65698	9.81756	1.52212	10.18244
	BaO_2	0.72554	9.86066	1.37828	10.13934
	BaS	0.72579	9.86081	1.37781	10.13911
CO_2	$BaCO_3$	4.48421	10.65169	0.22300	9.34830
	BaO	3.48421	10.54211	0.28701	9.45790
Beryllium					
Be	BeO	2.77530	10.44331	0.36032	9.55668
$BeCl_2$	BeO	0.31297	9.49551	3.19519	10.50450
BeO	Be	0.36032	9.55668	2.77531	10.44331
	$BeCl_2$	3.19525	10.50451	0.31296	9.49549

Weighed	Sought	Factor	Log of Factor +10	Reciprocal of Factor	Log of Reciprocal of Factor +10
Beryllium (contd.)					
BeO	$BeSO_4.4H_2O$	7.08211	10.85017	0.14120	9.14983
$Be_2P_2O_7$	Be	0.09389	8.97262	10.65076	11.02738
$BeSO_4.4H_2O$	BeO	0.14120	9.14983	7.08215	10.85017
Bismuth					
Bi	$BiAsO_4$	1.66475	10.22135	0.60069	9.77865
	Bi_2O_3	1.11484	10.04721	0.89699	9.95279
	BiOCl	1.24620	10.09559	0.80244	9.90441
	Bi_2S_3	1.23014	10.08996	0.81292	9.91005
$BiAsO_4$	Bi	0.60069	9.77865	1.66475	10.22135
	Bi_2O_3	0.66968	9.82587	1.49325	10.17413
$Bi(NO_3)_3.5H_2O$	Bi_2O_3	0.48030	9.68151	2.08203	10.31849
	BiOCl	0.53689	9.72988	1.86258	10.27011
Bi_2O_3	Bi	0.89699	9.95279	1.11484	10.04722
	$BiAsO_4$	1.49326	10.17414	0.66968	9.82587
	$Bi(NO_3)_3.5H_2O$	2.08204	10.31849	0.48030	9.68151
	BiOCl	1.11783	10.04837	0.89459	9.95163
	$BiONO_3$	1.23180	10.09054	0.81182	9.90946
	Bi_2S_3	1.10343	10.04275	0.90627	9.95726
BiOCl	Bi	0.80244	9.90441	1.24619	10.09559
	$Bi(NO_3)_3.5H_2O$	1.86256	10.27011	0.53690	9.72989
	Bi_2O_3	0.89458	9.95162	1.11784	10.04838
	$BiONO_3$	1.10195	10.04215	0.90748	9.95784
$BiONO_3$	Bi_2O_3	0.81182	9.90946	1.23180	10.09054
	BiOCl	0.90748	9.95784	1.10195	10.04216
$BiPO_4$	Bi	0.68754	9.83730	1.45446	10.16270
	Bi_2O_3	0.76651	9.88452	1.30461	10.11548
Bi_2S_3	Bi	0.81291	9.91005	1.23015	10.08996
	Bi_2O_3	0.90627	9.95726	1.10342	10.04274
Boron					
B	B_2O_3	3.21987	10.50784	0.31057	9.49216
	KBF_4	11.64619	11.06618	0.08586	8.93379
BO_2	B_2O_3	0.81314	9.91016	1.22981	10.08984
BO_3	B_2O_3	0.59192	9.77226	1.68942	10.22774
B_2O_3	B	0.31057	9.49216	3.21990	10.50785
	BO_2	1.22982	10.08985	0.81313	9.91016
	BO_3	1.68943	10.22774	0.59192	9.77226
	B_4O_7	1.11491	10.04724	0.89693	9.95276
	H_3BO_3	1.77630	10.24952	0.56297	9.75049
	KBF_4	3.61678	10.55832	0.27647	9.44165
	$Na_2B_4O_7.10H_2O$	2.73896	10.43758	0.36510	9.56241
B_4O_7	B_2O_3	0.89693	9.95276	1.11491	10.04724
$C_{20}H_{16}N_4.HBF_4$	B	0.02702	8.43169	37.00962	11.56832
	B_2O_3	0.08698	8.93942	11.49690	11.06058
H_3BO_3	B_2O_3	0.56297	9.75049	1.77629	10.24951
	KBF_4	2.03624	10.30883	0.49110	9.69117
KBF_4	B	0.08587	8.93384	11.64551	11.06616
	B_2O_3	0.27648	9.44167	3.61690	10.55834
	H_3BO_3	0.49110	9.69117	2.03625	10.30885
	$Na_2B_4O_7.10H_2O$	0.75725	9.87924	1.32057	10.12076
$Na_2B_4O_7.10H_2O$	B_2O_3	0.36510	9.56241	2.73898	10.43759
	KBF_4	1.32056	10.12075	0.75725	9.87924
Bromine					
Ag	Br	0.74079	9.86969	1.34991	10.13030
	BrO_3	1.18575	10.07399	0.84335	9.92601
	HBr	0.75013	9.87514	1.33310	10.12487
AgBr	Br	0.42555	9.62895	2.34990	10.37105
	BrO_3	0.68116	9.83325	1.46808	10.16675
	HBr	0.43092	9.63440	2.32062	10.36561
AgCl	Br	0.55754	9.74627	1.79359	10.25372
Br	Ag	1.34991	10.13030	0.74079	9.86969
	AgBr	2.34991	10.37105	0.42555	9.62895
	AgCl	1.79358	10.25372	0.55754	9.74627
	O	0.10011	9.00047	9.98901	10.99952
BrO_3	Ag	0.84335	9.92601	1.18575	10.07399
	AgBr	1.46809	10.16676	0.68116	9.83325
O	Br	9.98899	10.99952	0.10011	9.00047
HBr	Ag	1.33309	10.12486	0.75014	9.87514
	AgBr	2.32064	10.36561	0.43092	9.63440

Weighed	Sought	Factor	Log of Factor +10	Reciprocal of Factor	Log of Reciprocal of Factor +10
Cadmium					
Cd	$CdCl_2$	1.63087	10.21242	0.61317	9.78758
	$Cd(NO_3)_2$	2.10329	10.32290	0.47545	9.67711
	CdO	1.14235	10.05780	0.87539	9.94221
	CdS	1.28523	10.10898	0.77807	9.89102
	$CdSO_4$	1.85463	10.26825	0.53919	9.73174
$Cd(C_{13}H_8O_2N)_2$	Cd	0.21095	9.32418	4.74046	10.67582
	CdO	0.24098	9.38198	4.14972	10.61802
$CdCl_2$	Cd	0.61317	9.78758	1.63087	10.21242
	CdO	0.70045	9.84538	1.42765	10.15462
	CdS	0.78807	9.89657	1.26892	10.10343
	$CdSO_4$	1.13720	10.05584	0.87935	9.94417
$Cd(C_9H_6NO)_2$	Cd	0.28050	9.44793	3.56506	10.55207
	CdO	0.32043	9.50573	3.12081	10.49426
CdMoO4	Cd	0.41272	9.61566	2.42295	10.38434
	CdO	0.47147	9.67345	2.12103	10.32655
$Cd(NO_3)_2$	Cd	0.47545	9.67711	2.10327	10.32290
	CdO	0.54312	9.73490	1.84121	10.26510
	CdS	0.61106	9.78608	1.63650	10.21392
	$CdSO_4$	0.88177	9.94536	1.13408	10.05464
CdO	Cd	0.87539	9.94221	1.14235	10.05781
	$CdCl_2$	1.42765	10.15462	0.70045	9.85438
	$Cd(NO_3)_2$	1.84120	10.26510	0.54312	9.73490
	CdS	1.12508	10.05118	0.88883	9.94882
	$CdSO_4$	1.62352	10.21046	0.61595	9.78955
CdS	Cd	0.77807	9.89102	1.28523	10.10898
	$CdCl_2$	1.26893	10.10343	0.78807	9.89657
	$Cd(NO_3)_2$	1.63651	10.21392	0.61106	9.78608
	CdO	0.88883	9.94882	1.12507	10.05118
	$CdSO_4$	1.44303	10.15928	0.69299	9.84072
$CdSO_4$	Cd	0.53919	9.73174	1.85463	10.26825
	$CdCl_2$	0.87935	9.94417	1.13720	10.05584
	$Cd(NO_3)_2$	1.13408	10.05464	0.88177	9.94536
	CdO	0.61595	9.78955	1.62351	10.21046
	CdS	0.69299	9.84072	1.44302	10.15928
Calcium					
$BaSO_4$	CaS	0.30908	9.49007	3.23541	10.50993
	$CaSO_4$	0.58329	9.76588	1.71441	10.23411
	$CaSO_4.2H_2O$	0.73766	9.86786	1.35564	10.13214
Ca	$CaCl_2$	2.76921	10.44235	0.36111	9.55764
	$CaCO_3$	2.49726	10.39746	0.40044	9.60253
	CaF_2	1.94810	10.28961	0.51332	9.71039
	CaO	1.39920	10.14588	0.71469	9.85411
	$CaSO_4$	3.39671	10.53106	0.29440	9.46894
	Cl.	1.76911	10.24776	0.56526	9.75225
$Ca_3(AsO_4)_2$	$Mg_2As_2O_7$	0.77990	9.89204	1.28222	10.10798
$Ca_2As_2O_7$	Ca	0.23439	9.36994	4.26639	10.63006
	CaO	0.32795	9.51581	3.04925	10.48420
$CaC_2O_4.H_2O$	Ca	0.27430	9.43823	3.64564	10.56178
	CaO	0.38379	9.58410	2.60550	10.41589
$CaCl_2$	Ca	0.36111	9.55764	2.76924	10.44236
	$CaCO_3$	0.90179	9.95511	1.10891	10.04489
	CaO	0.50527	9.70352	1.97914	10.29648
	$CaSO_4$	1.22660	10.08870	0.81526	9.91130
	Cl	0.63885	9.80540	1.56531	10.19460
$CaCO_3$	Ca	0.40044	9.60253	2.49725	10.39751
	$CaCl_2$	1.10890	10.04489	0.90179	9.95511
	$Ca(HCO_3)_2$	1.61964	10.20942	0.61742	9.79058
	CaO	0.56030	9.74842	1.78476	10.25158
	$CaSO_4$	1.36018	10.13360	0.73520	9.86641
	$CaSO_4.2H_2O$	1.72015	10.23557	0.58134	9.76442
	CO_2	0.43970	9.64316	2.27428	10.35684
	HCl	0.72854	9.86245	1.37261	10.13755
CaF_2	Ca	0.51332	9.71039	1.94810	10.28961
	$CaSO_4$	1.74360	10.24145	0.57353	9.75855
$Ca(HCO_3)$	$CaCO_3$	0.61742	9.79058	1.61964	10.20942
	CaO	0.34594	9.53900	2.89067	10.46100
$Ca(IO_3)_2$	Ca	0.10280	9.01199	9.72763	10.98801
	CaO	0.14384	9.15788	6.95217	10.84212

Weighed	Sought	Factor	Log of Factor +10	Reciprocal of Factor	Log of Reciprocal of Factor +10
Calcium (contd.)					
$Ca(NO_3)_2$	N_2O_5	0.65824	9.81838	1.51920	10.18162
CaO	Ca	0.71470	9.85412	1.39919	10.14588
	$CaCl_2$	1.97914	10.29648	0.50527	9.70352
	$CaCO_3$	1.78477	10.25158	0.56030	9.74842
	CaF_2	1.39230	10.14373	0.71824	9.85627
	$Ca(HCO_3)_2$	2.89069	10.46100	0.34594	9.53900
	$Ca_3(PO_4)_2$	1.84368	10.26569	0.54239	9.73431
	$CaSO_4$	2.42760	10.38518	0.41193	9.61482
	$CaSO_4.2H_2O$	3.07008	10.48715	0.32572	9.51285
	Cl	1.26437	10.10187	0.79091	9.89813
	CO_2	0.78477	9.89474	1.27426	10.10527
	MgO	0.71879	9.85660	1.39123	10.14340
	SO_3	1.42760	10.15461	0.70048	9.84540
$Ca_3(PO_4)_2$	CaO	0.54239	9.73431	1.84369	10.26569
	$CaSO_4$	1.31672	10.11950	0.75946	9.88051
	$Mg_2P_2O_7$	0.71755	9.85585	1.39363	10.14415
	$(NH_4)_3PO_4.12MoO_3$	12.09843	11.08273	0.08266	8.91730
	P_2O_5	0.45761	9.66050	2.18527	10.33950
CaS	$BaSO_4$	3.23538	10.50992	0.30908	9.49007
$CaSO_4$	$BaSO_4$	1.71441	10.23411	0.58329	9.76581
	Ca	0.29440	9.46894	3.39674	10.53107
	$CaCl_2$	0.81526	9.91130	1.22660	10.08870
	$CaCO_3$	0.73520	9.86641	1.36017	10.13360
	CaF_2	0.57353	9.75855	1.74359	10.24145
	CaO	0.41193	9.61482	2.42760	10.38518
	$Ca_3(PO_4)_2$	0.75946	9.88051	1.31673	10.11950
	SO_3	0.58809	9.76944	1.70042	10.23056
$CaSO_4.2H_2O$	$BaSO_4$	1.35564	10.13213	0.73766	9.86786
	$CaCO_3$	0.58134	9.76443	1.72016	10.23557
	CaO	0.32572	9.51285	3.07012	10.48716
	SO_3	0.46502	9.66747	2.15045	10.33252
$CaWO_4$	WO_3	0.80523	9.90592	1.24188	10.09408
Cl	Ca	0.56526	9.75225	1.76910	10.24775
	$CaCl_2$	1.56531	10.19460	0.63885	9.80540
	CaO	0.79091	9.89813	1.26437	10.10188
CO_2	$CaCO_3$	2.27428	10.35684	0.43970	9.64316
	CaO	1.27427	10.10526	0.78476	9.89474
HCl	$CaCO_3$	1.37256	10.13753	0.72857	9.86247
$Mg_2As_2O_7$	$Ca_3(AsO_4)_2$	1.28221	10.10796	0.77990	9.89204
MgO	CaO	1.39118	10.14339	0.71881	9.85661
$Mg_2P_2O_7$	$Ca_3(PO_4)_2$	1.39365	10.14415	0.71754	9.85585
$(NH_4)_3PO_4.12MoO_3$	$Ca_3(PO_4)_2$	0.08265	8.91725	12.09892	11.08275
N_2O_5	$Ca(NO_3)_2$	1.51920	10.18162	0.65824	9.81838
P_2O_5	$Ca_3(PO_4)_2$	2.18521	10.33949	0.45762	9.66051
SO_3	CaO	0.70046	9.84539	1.42763	10.15462
	$CaSO_4$	1.70043	10.23056	0.58809	9.76944
	$CaSO_4.2H_2O$	2.15046	10.33253	0.46502	9.66747
WO_3	$CaWO_4$	1.24188	10.09408	0.80523	9.90592
Carbon					
Ag	CN	0.24120	9.38238	4.14594	10.61762
	HCN	0.25054	9.39888	3.99138	10.60112
	KCN	0.60369	9.78081	1.65648	10.21918
AgCN	CN	0.19433	9.28854	5.14589	10.71146
	HCN	0.20185	9.30503	4.95417	10.69497
	KCN	0.48638	9.68698	2.05601	10.31302
AgCNS	CNS	0.34999	9.54406	2.85722	10.45594
$Ag_4Fe(CN)_6$	C	0.11200	9.04922	8.92857	10.95078
AgOCN	OCN	0.28033	9.44767	3.56722	10.55233
$BaCO_3$	C	0.06086	8.78433	16.43115	11.21567
	CO_2	0.22300	9.34830	4.48431	10.65170
	CO_3	0.30408	9.48299	3.28861	10.51701
BaO	CO_2	0.28701	9.45790	3.48420	10.54210
	CO_2 (bicarbonate)	0.57402	9.75893	1.74210	10.24107
$BaSO_4$	CNS	0.24885	9.39594	4.01849	10.60406
C	$BaCO_3$	16.4305	11.21565	0.06086	8.78433
	CO_2	3.66409	10.56397	0.27292	9.43604
$CaCO_3$	CO_2	0.43970	9.64316	2.27428	10.35684
$Ca(HCO_3)_2$	CO_2	0.54295	9.73476	1.84179	10.26524

Weighed	Sought	Factor	Log of Factor +10	Reciprocal of Factor	Log of Reciprocal of Factor +10
Carbon (contd.)					
CaO	CO_2	0.78477	9.89474	1.27426	10.10526
	CO_2 (bicarbonate)	1.56954	10.19577	0.63713	9.80423
CN	Ag	4.14599	10.61762	0.24120	9.38238
	AgCN	5.14599	10.71147	0.19433	9.28854
CNS	AgCNS	2.85720	10.45594	0.34999	9.54406
	$BaSO_4$	4.01846	10.60406	0.24885	9.39594
	CuCNS	2.09394	10.32097	0.47757	9.67904
CO_2	$BaCO_3$	4.48420	10.65169	0.22301	9.34832
	$Ba(HCO_3)_2$	2.94672	10.46934	0.33963	9.53066
	BaO	3.48421	10.54211	0.28701	9.45790
	C	0.27292	9.43603	3.66408	10.56397
	$CaCO_3$	2.27426	10.35684	0.43970	9.64316
	$Ca(HCO_3)_2$	1.84174	10.26523	0.54296	9.73477
	CaO	1.27426	10.10526	0.78477	9.89474
	CO_3	1.36354	10.13467	0.73339	9.86534
	Cs_2CO_3	7.40329	10.86943	0.13508	9.13059
	$CsHCO_3$	4.40632	10.64407	0.22695	9.35593
	$FeCO_3$	2.63249	10.42037	0.37987	9.57964
	$Fe(HCO_3)_2$	2.02092	10.30555	0.49482	9.69445
	K_2CO_3	3.14049	10.49700	0.31842	9.50300
	$KHCO_3$	2.27491	10.35696	0.43958	9.64304
	K_2O	2.14049	10.33050	0.46718	9.66948
	Li_2CO_3	1.67887	10.22502	0.59564	9.77498
	$LiHCO_3$	1.54410	10.18868	0.64763	9.81133
	Li_2O	0.67887	9.83179	1.47304	10.16821
	$MgCO_3$	1.91595	10.28239	0.52193	9.71761
	$Mg(HCO_3)_2$	1.66265	10.22080	0.60145	9.77920
	MgO	0.91595	9.96188	1.09176	10.03812
	$MnCO_3$	2.61185	10.41695	0.38287	9.58305
	$Mn(HCO_3)_2$	2.01059	10.30332	0.49737	9.69668
	MnO	1.61185	10.20733	0.62041	9.79268
	Na_2CO_3	2.40829	10.38171	0.41523	9.61829
	$NaHCO_3$	1.90882	10.28077	0.52388	9.71923
	Na_2O	1.40829	10.14869	0.71008	9.85131
	$(NH_4)_2CO_3$	2.18329	10.33911	0.45802	9.66088
	NH_4HCO_3	1.79632	10.25439	0.55669	9.74561
	$PbCO_3$	6.07135	10.78328	0.16471	9.21672
	Rb_2CO_3	5.24767	10.71996	0.19056	9.28003
	$RbHCO_3$	3.32856	10.52225	0.30043	9.47774
	Rb_2O	4.24767	10.62815	0.23542	9.37184
	$SrCO_3$	3.35446	10.52562	0.29811	9.47438
	$Sr(HCO_3)_2$	2.3818	10.37690	0.41985	9.62309
	SrO	2.35446	10.37189	0.42473	9.62811
CO_3	$BaCO_3$	3.2886	10.51701	0.30408	9.48299
	CO_2	0.73339	9.86533	1.36353	10.13466
Cs_2CO_3	CO_2	0.13507	9.13056	7.40357	10.86944
$CsHCO_3$	CO_2	0.22695	9.35593	4.40626	10.64407
CuCNS	CNS	0.47757	9.67904	2.09393	10.32097
$FeCO_3$	CO_2	0.37987	9.57964	2.63248	10.42037
$Fe(HCO_3)_2$	CO_2	0.49482	9.69445	2.02094	10.30556
HCN	Ag	3.99137	10.60112	0.25054	9.39888
	AgCN	4.95408	10.69497	0.20185	9.30503
KCN	Ag	1.65648	10.21918	0.60369	9.78081
	AgCN	2.05602	10.31302	0.48638	9.68698
K_2CO_3	CO_2	0.31842	9.50300	3.14051	10.49700
$KHCO_3$	CO_2	0.43958	9.64304	2.27490	10.35696
K_2O	CO_2	0.46718	9.66948	2.14050	10.33052
Li_2CO_3	CO_2	0.59564	9.77498	1.67887	10.22502
Li_2O	CO_2	1.47304	10.16821	0.67887	9.83179
$MgCO_3$	CO_2	0.52193	9.71761	1.91597	10.28239
$Mg(HCO_3)_2$	CO_2	0.60145	9.77920	1.66265	10.22080
MgO	CO_2	1.09176	10.03812	0.91595	9.96188
$MnCO_3$	CO_2	0.38287	9.58305	2.61185	10.41695
MnO	CO_2	0.62041	9.79268	1.61184	10.20733
Na_2CO_3	CO_2	0.41523	9.61829	2.40830	10.38171
$NaHCO_3$	CO_2	0.52388	9.71923	1.90883	10.28077
Na_2O	CO_2	0.71008	9.85131	1.40829	10.14869
$(NH_4)_2CO_3$	CO_2	0.45802	9.66088	2.18331	10.33911

Weighed	Sought	Factor	Log of Factor +10	Reciprocal of Factor	Log of Reciprocal of Factor +10
Carbon (contd.)					
NH_4HCO_3	CO_2	0.55669	9.74561	1.79633	10.25439
$PbCO_3$	CO_2	0.16471	9.21672	6.07128	10.78328
Rb_2CO_3	CO_2	0.19056	9.28003	5.24769	10.71996
$RbHCO_3$	CO_2	0.30043	9.47774	3.32856	10.52225
Rb_2O	CO_2	0.23542	9.37184	4.24773	10.62815
$SrCO_3$	CO_2	0.29811	9.47438	3.35447	10.52562
$Sr(HCO_3)_2$	CO_2	0.41984	9.62308	2.38186	10.37691
SrO	CO_2	0.42472	9.62810	2.35449	10.37190
Cerium					
Ce	$Ce_2(C_2O_4)_3.3H_2O$	2.13513	10.32943	0.46836	9.67058
	$Ce(NO_3)_4$	2.77005	10.44249	0.36100	9.55751
	$Ce(NO_3)_4(NH_4NO_3)_2.H_2O$	4.04111	10.60650	0.24746	9.39351
	CeO_2	1.22838	10.08933	0.81408	9.91067
	Ce_2O_3	1.17128	10.06866	0.85377	9.93134
	$Ce_2(SO_4)_3$	2.02833	10.30714	0.49302	9.69286
$Ce_2(C_2O_4)_3.3H_2O$	Ce	0.46835	9.67057	2.13516	10.32943
	$Ce_2(SO_4)_3$	0.95000	9.97772	1.05263	10.02228
$Ce(NO_3)_4$	Ce	0.36100	9.55751	2.77008	10.44249
	CeO_2	0.44345	9.64684	2.25505	10.35316
	Ce_2O_3	0.42284	9.62618	2.36496	10.37382
$Ce(NO_3)_4(NH_4NO_3)_2.H_2O$	Ce	0.24745	9.39349	4.04122	10.60651
	CeO_2	0.30396	9.48282	3.28991	10.51719
	Ce_2O_3	0.28984	9.46216	3.45018	10.53784
CeO_2	Ce	0.81408	9.91067	1.22838	10.08933
	$Ce(NO_3)_4$	2.25505	10.35316	0.44345	.9.64685
	$Ce(NO_3)_4(NH_4NO_3)_2.H_2O$	3.28986	10.51718	0.30396	9.48281
	Ce_2O_3	0.95352	9.97933	1.04875	10.02068
Ce_2O_3	Ce	0.85377	9.93134	1.17128	10.06866
	$Ce(NO_3)_4$	2.36498	10.37383	0.42284	9.62618
	$Ce(NO_3)_4(NH_4NO_3)_2.H_2O$	3.45016	10.53784	0.28984	9.46216
	CeO_2	1.04874	10.02067	0.95353	9.97933
	$Ce_2(SO_4)_3$	1.73172	10.23848	0.57746	9.76152
$Ce_2(SO_4)_3$	Ce	0.49302	9.69286	2.02832	10.30714
	$Ce_2(C_2O_4)_3.3H_2O$	1.05265	10.02220	0.94998	9.97771
	Ce_2O_3	0.57746	9.76152	1.73172	10.23848
Cesium					
$AgCl$	$CsCl$	1.17468	10.06992	0.85130	9.93008
Cl	Cs	3.74877	10.57389	0.26675	9.42610
	$CsCl$	4.74877	10.67658	0.21058	9.32342
Cs	Cl	0.26675	9.42610	3.74883	10.57390
	$CsCl$	1.26675	10.10269	0.78942	9.89731
	Cs_2O_3	1.22576	10.08840	0.81582	9.91159
	Cs_2O	1.06019	10.02539	0.94323	9.97462
	Cs_2PtCl_6	2.53422	10.40385	0.39460	9.59616
	Cs_2SO_4	1.36139	10.13398	0.73454	9.86602
$CsB(C_6H_5)_4$	Cs	0.29394	9.46826	3.40205	10.53174
	Cs_2O	0.31164	9.49365	3.20883	10.50635
$CsCl$	$AgCl$	0.85130	9.93008	1.17467	10.06993
	Cl	0.21058	9.32342	4.74879	10.67658
	Cs	0.78942	9.89731	1.26675	10.10269
	Cs_2O	0.83693	9.92269	1.19484	10.07731
	Cs_2PtCl_6	2.00055	10.30114	0.49986	9.69885
	Cs_2SO_4	1.07470	10.03129	0.93049	9.96871
Cs_2CO_3	Cs	0.81582	9.91159	1.22576	10.08841
	Cs_2PtCl_6	2.06746	10.31544	0.48369	9.68457
	Cs_2SO_4	1.11065	10.04557	0.90037	9.95442
$CsClO_4$	Cs	0.57199	9.75739	1.74828	10.24261
	$CsCl$	0.72457	9.86008	1.38013	10.13992
	Cs_2O_3	0.70110	9.84578	1.42631	10.15421
	Cs_2O	0.60642	9.78277	1.64902	10.21723
	Cs_2SO_4	0.77870	9.89137	1.28419	10.10863
Cs_2O	Cs	0.94323	9.97462	1.06019	10.02539
	$CsCl$	1.19483	10.07731	0.83694	9.92269
	Cs_2PtCl_6	2.39034	10.37846	0.41835	9.62154
	Cs_2SO_4	1.28410	10.10860	0.77876	9.89140
	SO_3	0.28410	9.45347	3.51989	10.54653
Cs_2PtCl_6	Cs	0.39460	9.59616	2.53421	10.40385
	$CsCl$	0.49986	9.69885	2.00056	10.30115

Weighed	Sought	Factor	Log of Factor +10	Reciprocal of Factor	Log of Reciprocal of Factor +10
Cesium (contd.)					
Cs_2PtCl_6	Cs_2CO_3	0.48368	9.68456	2.06748	10.31544
	Cs_2O	0.41835	9.62154	2.39034	10.37846
Cs_2SO_4	Cs	0.73454	9.86602	1.36140	10.13399
	CsCl	0.93048	9.96871	1.07471	10.03129
	Cs_2CO_3	0.90037	9.95442	1.11065	10.04557
	Cs_2O	0.77876	9.89140	1.28409	10.10860
SO_3	Cs_2O	3.51987	10.54652	0.28410	9.45347
	Cs_2SO_4	4.51988	10.65513	0.22124	9.34486
Chlorine					
Ag	Cl	0.32866	9.51675	3.04266	10.48325
	HCl	0.33801	9.52893	2.95850	10.47107
AgCl	Cl	0.24736	9.39333	4.04269	10.60667
	ClO_3	0.58226	9.76512	1.71745	10.23488
	ClO_4	0.69389	9.84129	1.44115	10.15871
	HCl	0.25440	9.40552	3.93082	10.59448
$BaCrO_4$	Cl	0.27990	9.44700	3.57270	10.55299
Ca	Cl	1.76911	10.24776	0.56526	9.75225
Cl	Ag	3.04262	10.48325	0.32866	9.51675
	AgCl	4.04262	10.60666	0.24736	9.39333
	$BaCrO_4$	3.57276	10.55300	0.27990	9.44700
	Ca	0.56526	9.75225	1.76910	10.24775
	HCl	1.02843	10.01217	0.97236	9.98783
	K	1.10292	10.04255	0.90668	9.95745
	KCl	2.10292	10.32282	0.47553	9.67718
	Li	0.19572	9.29164	5.10934	10.70837
	Mg	0.34288	9.53514	2.91647	10.46486
	$MgCl_2$	1.34288	10.12804	0.74467	9.87196
	MnO_2	1.22609	10.08852	0.81560	9.91148
	Na	0.64846	9.81188	1.54212	10.18811
	NaCl	1.64846	10.21708	0.60663	9.21708
Cl	NH_4	0.50880	9.70655	1.96541	10.29346
	$PbCrO_4$	4.55787	10.65876	0.21940	9.34124
ClO_3	AgCl	1.71745	10.23488	0.58226	9.76512
	KCl	0.89340	9.95105	1.11932	10.04895
	NaCl	0.70032	9.84530	1.42792	10.15471
ClO_4	AgCl	1.44114	10.15870	0.69390	9.84130
	KCl	0.74967	9.87487	1.33392	10.12513
	NaCl	0.58765	9.76912	1.70169	10.23088
HCl	Ag	2.95850	10.47107	0.33801	9.52893
	AgCl	3.93086	10.59448	0.25440	9.40552
	NH_4Cl	1.46710	10.16646	0.68162	9.83354
	$(NH_4)_2SO_4$	1.81206	10.25817	0.55186	9.74183
K	Cl	0.90668	9.95745	1.10292	10.04255
KCl	Cl	0.47553	9.67718	2.10292	10.32282
	ClO_3	1.11932	10.04895	0.89340	9.95105
	ClO_4	1.33393	10.12513	0.74966	9.87486
Li	Cl	5.10924	10.70836	0.19572	9.29164
Mg	Cl	2.91650	10.46487	0.34288	9.53514
$MgCl_2$	Cl	0.74467	9.87196	1.34288	10.12804
MnO_2	Cl	0.81560	9.91148	1.22610	10.08853
Na	Cl	1.54212	10.18811	0.64896	9.81188
NaCl	Cl	0.60663	9.78293	1.64845	10.21708
	ClO_3	1.42791	10.15470	0.70032	9.84530
	ClO_4	1.70168	10.23088	0.58765	9.76912
NH_4	Cl	1.96539	10.29345	0.50880	9.70655
NH_4Cl	HCl	0.68162	9.83354	1.46709	10.16646
$(NH_4)_2SO_4$	HCl	0.55186	9.74184	1.81205	10.25817
$PbCrO_4$	Cl	0.21940	9.34124	4.55789	10.65876
Ag_2CrO_4	Cr	0.15674	9.19518	6.37999	10.80482
	Cr_2O_3	0.22895	9.35974	4.36777	10.64026
	CrO_3	0.30143	9.47919	3.31752	10.52082
	CrO_4	0.34966	9.54365	2.85992	10.45635
$BaCrO_4$	Cr	0.20525	9.31228	4.87211	10.68772
	CrO_3	0.39472	9.59629	2.53344	10.40371
	CrO_4	0.45788	9.66075	2.18398	10.33924
	Cr_2O_3	0.29998	9.47709	3.33356	10.52291
	$Cr_2(SO_4)_3.18H_2O$	1.41404	10.15046	0.70719	9.84954

Weighed	Sought	Factor	Log of Factor +10	Reciprocal of Factor	Log of Reciprocal of Factor +10
Chromium (contd.)					
$BaCrO_4$	$PbCrO_4$	1.27573	10.10576	0.78386	9.89424
Cr	$BaCrO_4$	4.87210	10.68772	0.20525	9.31228
	Cr_2O_3	1.46155	10.16482	0.68421	9.83519
	$PbCrO_4$	6.21547	10.79347	0.16089	9.20653
$Cr(C_9H_6NO)_3$	Cr	0.10733	9.03072	9.31706	10.96928
	Cr_2O_3	0.15677	9.19526	6.37877	10.80473
CrO_3	$BaCrO_4$	2.53345	10.40372	0.39472	9.59629
	Cr	0.51999	9.71600	1.92311	10.28401
	Cr_2O_3	0.75999	9.88081	1.31581	10.11920
	K_2CrO_4	1.94210	10.28827	0.51491	9.71173
	$K_2Cr_2O_7$	1.47105	10.16762	0.67979	9.83237
	$PbCrO_4$	3.23199	10.50947	0.30941	9.49053
CrO_4	$BaCrO_4$	2.18399	10.33925	0.45788	9.66075
	$PbCrO_4$	2.78618	10.44501	0.35891	9.55499
Cr_2O_3	$BaCrO_4$	3.33351	10.52291	0.29998	9.47709
	Cr	0.68420	9.83518	1.46156	10.16482
	CrO_3	1.31580	10.11919	0.75999	9.88081
	CrO_4	1.52634	10.18364	0.65516	9.81635
	$PbCrO_4$	4.25265	10.62866	0.23515	9.37135
$Cr_2(SO_4)_3.18H_2O$	$BaCrO_4$	0.70718	9.84953	1.41407	10.15047
	$PbCrO_4$	0.90217	9.95529	1.10844	10.04471
K_2CrO_4	CrO_3	0.51491	9.71173	1.94209	10.28827
	$PbCrO_4$	1.66418	10.22120	0.60090	9.77880
$K_2Cr_2O_7$	CrO_3	0.67979	9.83237	1.47104	10.16762
	$PbCrO_4$	2.19707	10.34184	0.45515	9.65815
$PbCrO_4$	Cr	0.16089	9.20652	6.21543	10.79347
	CrO_3	0.30941	9.49053	3.23196	10.50946
	CrO_4	0.35891	9.55499	2.78621	10.44501
	Cr_2O_3	0.23514	9.37133	4.25279	10.62868
	$Cr_2(SO_4)_3.18H_2O$	1.10843	10.04471	0.90218	9.95529
	K_2CrO_4	0.60090	9.77880	1.66417	10.22119
	$K_2Cr_2O_7$	0.45514	9.65815	2.19713	10.34186
Cobalt					
Co	$Co(NO_3)_2.6H_2O$	4.93731	10.69349	0.20254	9.30651
	$Co(NO_2)_3.(KNO_2)_3$	7.67433	10.88504	0.13030	9.11494
	CoO	1.27148	10.10431	0.78649	9.89569
	Co_3O_4	1.36197	10.13417	0.73423	9.86583
	$CoSO_4$	2.63001	10.41996	0.38023	9.58005
	$CoSO_4.7H_2O$	4.76984	10.67851	0.20965	9.32149
	$(CoSO_4)_2.(K_2SO_4)_3$	7.06550	10.84914	0.14153	9.15085
$Co(C_{10}H_6O_2N).2H_2O$	Co	0.09638	8.98399	10.37560	11.01602
	CoO	0.12255	9.08831	8.15993	10.91169
$(Co[C_5H_5N]_4).(SCN)_2$	Co	0.11990	9.07882	8.34028	10.92118
	CoO	0.15246	9.18316	6.55910	10.81684
$Co(C_9H_6ON)_2$	Co	0.16972	9.22973	5.89206	10.77027
	Co_2O_3	0.23883	9.37809	4.18708	10.62191
$Co(NO_3)_2.6H_2O$	Co	0.20254	9.30651	4.93730	10.69349
$Co(NO_2)_3.(KNO_2)_3$	Co	0.13030	9.11494	7.67460	10.88506
	CoO	0.16568	9.21927	6.03573	10.78073
CoO	Co	0.78648	9.89569	1.27149	10.10432
	$Co(NO_2)_3.(KNO_2)_3$	6.03571	10.78073	0.16568	9.21927
	Co_3O_4	1.07117	10.02986	0.93356	9.97014
	$CoSO_4$	2.06845	10.31564	0.48345	9.68435
	$(CoSO_4)_2.(K_2SO_4)_3$	5.55690	10.74483	0.17996	9.25518
Co_3O_4	Co	0.73423	9.86583	1.36197	10.13417
	CoO	0.93356	9.97014	1.07117	10.02986
$Co_2P_2O_7$	Co	0.40392	9.60630	2.47574	10.39371
	CoO	0.51357	9.71060	1.94715	10.28940
	Co_3O_4	0.55012	9.74046	1.81779	10.25953
	$CoSO_4$	1.06230	10.02624	0.94135	9.97375
	$CoSO_4.7H_2O$	1.92661	10.28479	0.51905	9.71521
$CoSO_4$	Co	0.38023	9.58005	2.62999	10.41996
	CoO	0.48345	9.68435	2.06847	10.31565
$CoSO_4.7H_2O$	Co	0.20965	9.32150	4.76985	10.67851
	CoO	0.26657	9.42581	3.75136	10.57419
$(CoSO_4)_2.(K_2SO_4)_3$	Co	0.14153	9.15085	7.06553	10.84914
	CoO	0.17996	9.25518	5.55679	10.74482
Columbium	See Niobium				

Weighed	Sought	Factor	Log of Factor +10	Reciprocal of Factor	Log of Reciprocal of Factor +10
Copper					
Cu	$Cu_2C_2H_3O_2.(AsO_2)_3$	3.98875	10.60084	0.25071	9.39917
	CuCNS	1.91407	10.28196	0.52245	9.71804
	CuO	1.25181	10.09754	0.79884	9.90246
	Cu_2O	1.12590	10.05150	0.88818	9.94850
	Cu_2S	1.25228	10.09770	0.79854	9.90230
	$CuSO_4.5H_2O$	3.92949	10.59433	0.25449	9.40567
$Cu_2C_2H_3O_2.(AsO_2)_3$	Cu	0.25071	9.39917	3.98867	10.60083
CuCNS	Cu	0.52245	9.71804	1.91406	10.28195
	CuO	0.65400	9.81558	1.52905	10.18443
$(Cu[C_5H_5N]_2.(SCN)_2$	Cu	0.18804	9.27425	5.31802	10.72575
	CuO	0.23539	9.37179	4.24827	10.62821
$Cu(C_7H_6NO_2)_2$	Cu	0.18922	9.27697	5.28485	10.72303
	CuO	0.23687	9.37451	4.22172	10.62549
$Cu(C_9H_6NO)_2$	Cu	0.18059	9.25669	5.53741	10.74330
	CuO	0.22606	9.35422	4.42360	10.64578
$Cu(C_{12}H_{10}ONS)_2$	Cu	0.12795	9.10704	7.81555	10.89296
	CuO	0.16017	9.20458	6.24337	10.79542
$Cu(NH_2C_6H_4CO_2)_2$	Cu	0.18922	9.27697	5.28485	10.72303
	CuO	0.23687	9.37451	4.22172	10.62549
CuO	Cu	0.79884	9.90246	1.25182	10.09754
	CuCNS	1.52904	10.18442	0.65401	9.81558
	Cu_2S	1.00038	10.00016	0.99962	9.99983
	$CuSO_4.5H_2O$	3.13905	10.49680	0.31857	9.50320
Cu_2O	Cu	0.88817	9.94850	1.12591	10.05150
	CuO	1.11183	10.04603	0.89942	9.95396
	Cu_2S	1.11224	10.04620	0.89909	9.95380
Cu_2S	Cu	0.79854	9.90230	1.25229	10.09770
	CuO	0.99962	9.99983	1.00004	10.00002
	Cu_2O	0.89908	9.95380	1.11225	10.04620
	$CuSO_4.5H_2O$	3.13787	10.49663	0.31869	9.50337
$CuSO_4.5H_2O$	Cu	0.25449	9.40567	3.92943	10.59433
	CuO	0.31857	9.50320	3.13903	10.49679
	Cu_2S	0.31869	9.50337	3.13785	10.49663
$Mg_2As_2O_7$	$Cu_2C_2H_3O_2.(AsO_2)_3$	1.08844	10.03681	0.91875	9.96320
Dysprosium					
Dy_2O_3	Dy	0.87131	9.94017	1.14770	10.05983
Erbium					
Er	Er_2O_3	1.14349	10.05824	0.87452	9.94177
Er_2O_3	Er	0.87452	9.94177	1.14349	10.05824
Europium					
Eu_2O_3	Eu	0.86361	9.93632	1.15793	10.06368
Fluorine					
BaF_2	$BaSiF_6$	1.59359	10.20238	0.62751	9.79762
$BaSiF_6$	BaF_2	0.62751	9.79762	1.59360	10.20238
	F	0.40795	9.61061	2.45128	10.38939
	HF	0.42960	9.63306	2.32775	10.36694
	H_2SiF_6	0.51568	9.71238	1.93919	10.28762
	SiF_4	0.37248	9.57110	2.68471	10.42889
	SiF_6	0.50847	9.70627	1.96668	10.29373
CaF_2	F	0.48664	9.68721	2.05491	10.31279
	6HF	0.51248	9.70968	1.95130	10.29033
	H_2SiF_6	0.61517	9.78900	1.62557	10.21100
	SiF_6	0.60657	9.78288	1.64861	10.21712
$CaSO_4$	F	0.27910	9.44576	3.58295	10.55424
	HF	0.29391	9.46821	3.40240	10.53178
F	$BaSiF_6$	2.45125	10.38939	0.40796	9.61062
	CaF_2	2.05491	10.31279	0.48664	9.68721
	$CaSO_4$	3.58293	10.55433	0.27910	9.44576
	H_2SiF_6	1.26407	10.10177	0.79110	9.89823
	K_2SiF_6	1.93244	10.28611	0.51748	9.71389
2HF	H_2SiF_6	3.60115	10.55644	0.27769	9.44356
6HF	H_2SiF_6	1.20038	10.07932	0.83307	9.92068
H_2SiF_6	$BaSiF_6$	1.93917	10.28762	0.51568	9.71237
	CaF_2	1.62555	10.21100	0.61518	9.78900
	F	0.79109	9.89823	1.26408	10.10178
	2HF	0.27769	9.44356	3.60114	10.55644
	6HF	0.83307	9.92068	1.20038	10.07932
	K_2SiF_6	1.52875	10.18434	0.65413	9.81566

Weighed	Sought	Factor	Log of Factor +10	Reciprocal of Factor	Log of Reciprocal of Factor +10
Fluorine (contd.)					
H_2SiF_6	SiF_4	0.72232	9.85873	1.38443	10.14129
	SiF_6	0.98601	9.99388	1.01419	10.00612
2KF	K_2SiF_6	1.89568	10.27777	0.52752	9.72224
K_2SiF_6	F	0.51748	9.71389	1.93244	10.28611
	6HF	0.54493	9.73634	1.83510	10.26366
	H_2SiF_6	0.65413	9.81566	1.52875	10.18434
	KF	0.52751	9.72223	1.89570	10.27777
	SiF_6	0.64498	9.80955	1.55044	10.19045
Na_2SiF_6	F	0.60615	9.78258	1.64976	10.21742
	6HF	0.63831	9.80503	1.56664	10.19498
	H_2SiF_6	0.76622	9.88435	1.30511	10.11565
	2NaF	0.44655	9.64987	2.23939	10.35013
	SiF_4	0.55344	9.74307	1.80688	10.25693
	SiF_6	0.75550	9.87823	1.32363	10.12177
PbClF	F	0.07261	8.86100	13.77221	11.13900
SiF_4	$BaSiF_6$	2.68467	10.42889	0.37249	9.57111
	H_2SiF_6	1.38444	10.14128	0.72231	9.85872
SiF_6	$BaSiF_6$	1.96666	10.29373	0.50848	9.70627
	CaF_2	1.64862	10.21712	0.60657	9.78288
	H_2SiF_6	1.01419	10.00612	0.98601	9.99388
	K_2SiF_6	1.55043	10.19045	0.64498	9.80955
Gallium					
Ga	Ga_2O_3	1.34423	10.12847	0.74392	9.87153
$Ga(C_9H_4NOBr_2)_3$	Ga	0.07146	8.85406	13.99384	11.14594
	Ga_2O_3	0.09606	8.98254	10.41016	11.01746
Ga_2O_3	Ga	0.74392	9.87153	1.34423	10.12847
Ga_2S_3	Ga	0.59178	9.77216	1.68982	10.22785
Germanium					
Ge	GeO_2	1.44083	10.15861	0.69404	9.84138
$(C_9H_6NOH)_4.(GeO_2.12MoO_3)$	Ge	0.03009	8.47842	33.23363	11.52158
	GeO_2	0.04335	8.63699	23.06805	11.36301
GeO_2	Ge	0.69404	9.84139	1.44084	10.15861
K_2GeF_6	Ge	0.27415	9.43799	3.64764	10.56201
Mg_2GeO_4	Ge	0.39193	9.59321	2.55148	10.40679
	GeO_2	0.56471	9.75183	1.77082	10.24817
Gold					
Au	$AuCl_3$	1.53998	10.18749	0.64936	9.81249
	$HAuCl_4.4H_2O$	2.09095	10.32034	0.47825	9.67966
	$KAu(CN)_4.H_2O$	1.81836	10.25968	0.54995	9.74032
$AuCl_3$	Au	0.64938	9.81250	1.53993	10.18750
C_6H_5SAu	Au	0.64339	9.80847	1.55427	10.19153
$HAuCl_4.4H_2O$	Au	0.47825	9.67966	2.09096	10.32034
$KAu(CN)_4.H_2O$	Au	0.54995	9.74032	1.81835	10.25967
Hafnium					
HfO_2	Hf	0.84797	9.92838	1.17929	10.07162
$Hf(C_6H_5CHOHCO_2)_4$	Hf	0.22794	9.35782	4.38712	10.64218
	HfO_2	0.26880	9.42943	3.72024	10.57057
Holmium					
Ho_2O_3	Ho	0.87297	9.94100	1.14551	10.05900
Hydrogen					
AgCNS	HCNS	0.35606	9.55152	2.80852	10.44848
$BaSO_4$	HCNS	0.25317	9.40341	3.94992	10.59659
CuCNS	HCNS	0.48586	9.68651	2.05821	10.31349
H	H_2O	8.93644	10.95116	0.11190	9.04883
	O	7.93644	10.89963	0.12600	9.10037
HCNS	AgCNS	2.80846	10.44847	0.35607	9.55154
	$BaSO_4$	3.94991	10.59659	0.25317	9.40341
	CuCNS	2.05822	10.31350	0.48586	9.68651
H_2O	H	0.11190	9.04883	8.93655	10.95117
O	H	0.12600	9.10037	7.93651	10.89963
Indium					
In	In_2O_3	1.20902	10.08244	0.92712	9.91757
	In_2S_3	1.41888	10.15194	0.70478	9.84805
$In(C_9H_6ON)_3$	In	0.20980	9.32181	4.76644	10.67819
	In_2O_3	0.25365	9.40423	3.94244	10.59577
In_2O_3	In	0.82711	9.91756	1.20903	10.08244
In_2S_3	In	0.70478	9.84805	1.41888	10.15194

Weighed	Sought	Factor	Log of Factor +10	Reciprocal of Factor	Log of Reciprocal of Factor +10
Iodine					
Ag	HI	1.18580	10.07401	0.84331	9.92599
	I	1.17646	10.07058	0.85001	9.92942
AgCl	I	0.88544	9.94716	1.12938	10.05284
AgI	HI	0.54483	9.73626	1.83543	10.26374
	I	0.54054	9.73283	1.85000	10.26717
	IO_3	0.74498	9.87214	1.34232	10.12786
	IO_4	0.81313	9.91016	1.22982	10.08985
	I_2O_5	0.71091	9.85181	1.40665	10.14819
	I_2O_7	0.77906	9.89157	1.28360	10.10843
HI	Ag	0.84331	9.92599	1.18580	10.07401
	AgI	1.83542	10.26374	0.54483	9.73626
	Pd	0.41590	9.61899	2.40442	10.38101
	PdI_2	1.40799	10.14860	0.71023	9.85140
	TlI	2.58981	10.41327	0.38613	9.58673
I	Ag	0.85000	9.92942	1.17647	10.07058
	AgCl	1.12937	10.05283	0.88545	9.94716
	AgI	1.85000	10.26717	0.54054	9.73283
	Pd	0.41921	9.62243	2.38544	10.37757
	PdI_2	1.41917	10.15203	0.70464	9.84797
	TlI	2.61039	10.41671	0.38308	9.58329
IO_3	AgI	1.34231	10.12785	0.74498	9.87214
	PdI_2	1.02971	10.01272	0.97115	9.98729
	TlI	1.89402	10.27738	0.52798	9.72262
IO_4	AgI	1.22981	10.08984	0.81313	9.91016
	PdI_2	0.94341	9.97470	1.05998	10.02530
	TlI	1.73528	10.23937	0.57628	9.76063
I_2O_5	AgI	1.40665	10.14819	0.71091	9.95181
	PdI_2	1.07907	10.03305	0.92672	9.96695
	TlI	1.98481	10.29772	0.50383	9.70228
I_2O_7	AgI	1.28360	10.10843	0.77906	9.89157
	PdI_2	0.98467	9.99329	1.01557	10.00671
	TlI	1.81120	10.25797	0.55212	9.74203
Pd	HI	2.40437	10.38100	0.41591	9.61900
	I	2.38542	10.37757	0.41921	9.62243
PdI_2	HI	0.71023	9.85140	1.40799	10.14860
	I	0.70463	9.70463	1.41918	10.15204
	IO_3	0.97114	9.98728	1.02972	10.01272
	IO_4	1.05998	10.02530	0.94341	9.97470
	I_2O_5	0.92672	9.96695	1.07907	10.03305
	I_2O_7	1.01556	10.00671	0.98468	9.99330
TlI	HI	0.38613	9.58673	2.58980	10.41327
	I	0.38308	9.58329	2.61042	10.41671
	IO_3	0.52798	9.72262	1.89401	10.27738
	IO_4	0.57627	9.76063	1.73530	10.23937
	I_2O_5	0.50382	9.70228	1.98484	10.29772
	I_2O_7	0.55212	9.74203	1.81120	10.25797
Iron					
Ag	$Fe_7(CN)_{18}$ (Prussian blue)	0.44253	9.64594	2.25973	10.35406
CN	$Fe_7(CN)_{18}$	1.83474	10.26358	0.54504	9.73643
CO_2	$FeCO_3$	2.63249	10.42037	0.37987	9.57964
	$Fe(HCO_3)_2$	2.02092	10.30555	0.49482	9.69445
	FeO	1.63249	10.21287	0.61256	9.78715
Fe	$Fe(HCO_3)_2$	3.18517	10.50313	0.31395	9.49686
	FeO	1.28648	10.10940	0.77731	9.89059
	Fe_2O_3	1.42973	10.15526	0.69943	9.84474
	$FePO_4$	2.70056	10.43145	0.37029	9.56854
	FeS	1.57414	10.19704	0.63527	9.80296
	$FeSO_4$	2.72009	10.43458	0.36763	9.56541
	$FeSO_4.7H_2O$	4.97817	10.69707	0.20088	9.30294
	$FeSO_4.(NH_4)_2SO_4.6H_2O$	7.02054	10.84637	0.14244	9.15363
$FeAsO_4$	$Mg_2As_2O_7$	0.79702	9.90147	1.25467	10.09853
$Fe(C_6H_5N[NO]O)_3$	Fe	0.11953	9.07748	8.36610	10.92252
	Fe_2O_3	0.17090	9.23274	5.85138	10.76726
$Fe(C_9H_6NO)_3$	Fe	0.11437	9.05821	8.74355	10.94169
	Fe_2O_3	0.16351	9.21354	6.11583	10.78646
$FeCl_3$	Fe_2O_3	0.49225	9.69219	2.03149	10.30781
$Fe_7(CN)_{18}$	Ag	2.25971	10.35405	0.44253	9.64594
	CN	0.54503	9.73642	1.83476	10.26358

Weighed	Sought	Factor	Log of Factor +10	Reciprocal of Factor	Log of Reciprocal of Factor +10
Iron (contd.)					
$FeCO_3$	CO_2	0.37986	9.57962	2.63255	10.42038
	FeO	0.62013	9.79248	1.61257	10.20752
	Fe_2O_3	0.68918	9.83833	1.45100	10.16167
$Fe(HCO_3)_2$	CO_2	0.49482	9.69445	2.02094	10.30555
	Fe	0.31396	9.49687	3.18512	10.50313
	FeO	0.40390	9.60627	2.47586	10.39373
	Fe_2O_3	0.44887	9.65212	2.22782	10.34788
FeO	CO_2	0.61256	9.78715	1.63249	10.21285
	Fe	0.77732	9.89060	1.28647	10.10940
	$FeCO_3$	1.61256	10.20751	0.62013	9.79248
	$Fe(HCO_3)_2$	2.47588	10.39373	0.40390	9.60627
	Fe_2O_3	1.11134	10.04584	0.89981	9.95415
	$FePO_4$	2.09918	10.32205	0.47638	9.67795
	FeS	1.22360	10.08764	0.81726	9.91236
	SO_3	1.11436	10.04703	0.89738	9.95298
Fe_2O_3	Fe	0.69943	9.84474	1.42974	10.15524
	$FeCl_3$	2.03149	10.30781	0.49225	9.69219
	$FeCO_3$	1.45099	10.16167	0.68918	9.83833
	$Fe(HCO_3)_2$	2.22781	10.34788	0.44887	9.65212
	$Fe(HCO_3)_3$	2.99200	10.47596	0.33422	9.52403
	FeO	0.89981	9.95415	1.11135	10.04584
	Fe_3O_4	0.96660	9.98525	1.03455	10.01475
	$FePO_4$	1.88886	10.27620	0.52942	9.72380
	FeS	1.10101	10.04113	0.90826	9.95821
	$FeSO_4$	1.90252	10.27933	0.52562	9.72067
	$FeSO_4.7H_2O$	3.48190	10.54182	0.28720	9.45818
	$Fe_2(SO_4)_3$	2.50406	10.39864	0.39935	9.60135
	$FeSO_4.(NH_4)_2SO_4.6H_2O$	4.91040	10.69112	0.20365	9.30888
Fe_3O_4	Fe_2O_3	1.03455	10.01475	0.96660	9.98525
$FePO_4$	Fe	0.37029	9.56854	2.70059	10.43145
	FeO	0.47638	9.67795	2.09916	10.32204
	Fe_2O_3	0.52942	9.72380	1.88886	10.27620
FeS	Fe	0.63527	9.80296	1.57413	10.19704
	FeO	0.81726	9.91236	1.22360	10.08764
	Fe_2O_3	0.90826	9.95821	1.10101	10.04179
$FeSO_4$	Fe	0.36763	9.56541	2.72013	10.43459
	Fe_2O_3	0.52562	9.72067	1.90252	10.27933
	SO_3	0.52704	9.72184	1.89739	10.27817
$FeSO_4.7H_2O$	Fe	0.20088	9.30294	4.97810	10.69706
	Fe_2O_3	0.28720	9.45818	3.48189	10.54182
$FeSO_4.(NH_4)_2SO_4.6H_2O$	Fe	0.14244	9.15363	7.02050	10.84637
	Fe_2O_3	0.20365	9.30888	4.91039	10.69112
$Fe_2(SO_4)_3$	Fe_2O_3	0.39935	9.60135	2.50407	10.39864
$Mg_2As_2O_7$	$FeAsO_4$	1.25468	10.09853	0.79702	9.90147
SO_3	FeO	0.89738	9.95298	1.11436	10.04703
	$FeSO_4$	1.89739	10.27816	0.52704	9.72184
Lanthanum					
La	La_2O_3	1.17277	10.06921	0.85268	9.93079
La_2O_3	La	0.85268	9.93079	1.17277	10.06921
Lead					
$BaSO_4$	$PbSO_4$	1.29927	10.11370	0.76966	9.88630
Pb	$PbCl_2$	1.34225	10.12783	0.74502	9.87217
	$PbCO_3$	1.28958	10.11044	0.77545	9.88955
	$(PbCO_3)_2.Pb(OH)_2$	1.24781	10.09615	0.80140	9.90385
	$PbCrO_4$	1.55982	10.19307	0.64110	9.80693
	PbO	1.07722	10.03231	0.92832	9.96770
	PbO_2	1.15445	10.06238	0.86621	9.93762
	$Pb(OH)_2$	1.16415	10.06601	0.85900	9.93399
	PbS	1.15474	10.06248	0.86600	9.93752
	$PbSO_4$	1.46363	10.16543	0.68323	9.83457
$Pb(C_2H_3O_2).3H_2O$	$PbCrO_4$	0.85198	9.93043	1.17374	10.06957
	$PbSO_4$	0.79944	9.90279	1.25088	10.09722
$Pb(C_7H_5NO_2)_2$	Pb	0.43396	9.63746	2.30436	10.36255
	PbO	0.46747	9.66975	2.13971	10.33024
$PbCl_2$	Pb	0.74502	9.87218	1.34225	10.12783
	PbO	0.80255	9.90448	1.24603	10.09553
$PbCO_3$	Pb	0.77541	9.88954	1.28964	10.11047
	PbO	0.83529	9.92184	1.19719	10.07816

Weighed	Sought	Factor	Log of Factor +10	Reciprocal of Factor	Log of Reciprocal of Factor +10
Lead (contd.)					
$PbCO_3$	$PbSO_4$	1.13492	10.05497	0.88112	9.94504
$(PbCO_3)_2.Pb(OH)_2$	Pb	0.80141	9.90386	1.24780	10.09614
	$PbCrO_4$	1.25007	10.09693	0.79996	9.90307
	$PbSO_4$	1.17298	10.06929	0.85253	9.93017
$PbCrO_4$	Pb	0.64110	9.80692	1.55982	10.19307
	$Pb(C_2H_3O_2)_2.3H_2O$	1.17371	10.06956	0.85200	9.93044
	$(PbCO_3)_2.Pb(OH)_2$	0.79996	9.90307	1.25006	10.09693
	PbO	0.69061	9.83922	1.44800	10.16077
	Pb_3O_4	0.70710	9.84947	1.41423	10.15052
	$PbSO_4$	0.93833	9.97235	1.06572	10.02765
$Pb(NO_3)_2$	PbO	0.67388	9.82859	1.48394	10.17141
	PbO_2	0.72219	9.85865	1.38468	10.14135
	$PbSO_4$	0.91561	9.96171	1.09217	10.03829
PbO	Pb	0.92831	9.96770	1.07723	10.03231
	$PbCl_2$	1.24602	10.09553	0.80256	9.90448
	$PbCO_3$	1.19718	10.07816	0.83530	9.92184
	$PbCrO_4$	1.44800	10.16077	0.69061	9.83923
	$Pb(NO_3)_2$	1.48393	10.17141	0.67389	9.82859
	PbO_2	1.07169	10.03007	0.93311	9.96993
	PbS	1.07196	10.03018	0.93287	9.96982
	$PbSO_4$	1.35871	10.13313	0.73599	9.86687
PbO_2	Pb	0.86622	9.93763	1.15444	10.06237
	$Pb(NO_3)_2$	1.38467	10.14135	0.72219	9.85865
	PbO	0.93311	9.96993	1.07169	10.03007
	$PbSO_4$	1.26782	10.10306	0.78876	9.89694
Pb_3O_4	$PbCrO_4$	1.41423	10.15052	0.70710	9.84948
	$PbSO_4$	1.32702	10.12288	0.75357	9.87712
$Pb(OH)_2$	Pb	0.85900	9.93399	1.16414	10.06601
$Pb_2P_2O_7$	Pb	0.70434	9.84778	1.41977	10.15222
	PbO	0.75874	9.88009	1.31797	10.11991
PbS	Pb	0.86600	9.93752	1.15473	10.06249
	PbO	0.93287	9.96982	1.07196	10.03017
	$PbSO_4$	1.26750	10.10296	0.78895	9.89705
$PbSO_4$	$BaSO_4$	0.76966	9.88630	1.29928	10.11370
	Pb	0.68323	9.83457	1.46364	10.16543
	$Pb(C_2H_3O_2)_2.3H_2O$	1.25088	10.09722	0.79944	9.90279
	$PbCO_3$	0.88112	9.94504	1.13492	10.05497
	$(PbCO_3)_2.Pb(OH)_2$	0.85253	9.93071	1.17298	10.06929
	$PbCrO_4$	1.06572	10.02765	0.93833	9.97236
	$Pb(NO_3)_2$	1.09216	10.03828	0.91562	9.96172
	PbO	0.73599	9.86687	1.35871	10.13313
	PbO_2	0.78876	9.89694	1.26781	10.10305
	Pb_3O_4	0.75357	9.87712	1.32702	10.12288
	PbS	0.78895	9.89705	1.26751	10.10295
Lithium					
CO_2	Li_2CO_3	1.67887	10.22504	0.59564	9.77498
	$LiHCO_3$	1.54410	10.18868	0.64763	9.81133
	Li_2O	0.67887	9.83179	1.47304	10.16821
Li	LiCl	6.10924	10.78599	0.16369	9.21402
	Li_2CO_3	5.32404	10.72624	0.18783	9.27377
	Li_2O	2.15283	10.33201	0.46450	9.66699
	Li_3PO_4	5.56218	10.74524	0.17979	9.25477
	Li_2SO_4	7.92189	10.89883	0.12623	9.10115
LiCl	Li	0.16369	9.21402	6.10911	10.78598
	Li_2CO_3	0.87147	9.94025	1.14749	10.05975
	Li_2O	0.35239	9.54702	2.83776	10.45297
	Li_3PO_4	0.91045	9.95926	1.09836	10.04074
	Li_2SO_4	1.29670	10.11284	0.77119	9.88716
Li_2CO_3	CO_2	0.59564	9.77498	1.67887	10.22502
	Li	0.18782	9.27374	5.32425	10.72626
	LiCl	1.14747	10.05974	0.87148	9.94026
	$LiHCO_3$	1.83963	10.26473	0.54359	9.73527
	Li_2O	0.40436	9.60677	2.47304	10.39323
	Li_3PO_4	1.04473	10.01901	0.95719	9.98100
$LiHCO_3$	CO_2	0.64762	9.81132	1.54412	10.18868
	Li_2CO_3	0.54363	9.73530	1.83949	10.26470
	Li_2O	0.21982	9.34207	4.54918	10.65794
	Li_3PO_4	0.56795	9.75431	1.76072	10.24569

Weighed	Sought	Factor	Log of Factor +10	Reciprocal of Factor	Log of Reciprocal of Factor +10
Lithium (contd.)					
Li_2O	CO_2	1.47304	10.16821	0.67887	9.83179
	Li	0.46449	9.66698	2.15290	10.33303
	LiCl	2.83767	10.45296	0.35240	9.54704
	Li_2CO_3	2.47304	10.39323	0.40436	9.60677
	$LiHCO_3$	4.54890	10.65791	0.21983	9.34209
	Li_3PO_4	2.58364	10.41223	0.38705	9.58777
	Li_2SO_4	3.67975	10.56582	0.27175	9.43417
	SO_3	2.67872	10.42809	0.37317	9.57191
Li_3PO_4	Li	0.17979	9.25477	5.56204	10.74523
	LiCl	1.09835	10.04074	0.91046	9.95926
	Li_2CO_3	0.95718	9.98099	1.04474	10.01901
	$LiHCO_3$	1.76070	10.24569	0.56796	9.75432
	Li_2O	0.38705	9.58777	2.58365	10.41223
	$Li_2.SO_4.H_2O$	1.65761	10.21949	0.60328	9.78052
Li_2SO_4	Li	0.12623	9.10116	7.92205	10.89883
	LiCl	0.77118	9.88716	1.29671	10.11284
	Li_2O	0.27175	9.43417	3.67975	10.56582
	Li_3PO_4	0.70213	9.84642	1.42424	10.15358
	SO_3	0.72823	9.86227	1.37319	10.13773
$Li_2SO_4.H_2O$	Li_3PO_4	0.60328	9.78052	1.65761	10.22948
SO_3	Li_2O	0.37317	9.57191	2.67974	10.42809
	Li_2SO_4	1.37318	10.13773	0.72824	9.86227
Lutetium					
Lu_2O_3	Lu	0.87938	9.94418	1.13716	10.05582
Magnesium					
$BaSO_4$	$MgSO_4$	0.51574	9.71243	1.93896	10.28757
	$MgSO_4.7H_2O$	1.05605	10.02368	0.94692	9.97631
Br	Mg	0.15212	9.18219	6.57376	10.81781
	$MgBr_2$	1.15212	10.06150	0.86797	9.93850
	$MgBr_2.6H_2O$	1.82807	10.26200	0.54703	9.73801
Cl	Mg	0.34288	9.53514	2.91647	10.46486
	$MgCl_2$	1.34288	10.12804	0.74467	9.87196
	$MgCl_2.6H_2O$	2.86643	10.45734	0.34887	9.54266
CO_2	$MgCO_3$	1.91595	10.28239	0.52193	9.71761
	MgO	0.91595	9.96187	1.09176	10.03812
I	Mg	0.09579	8.98132	10.43950	11.01868
	MgI_2	1.09578	10.03972	0.91259	9.96028
Mg	Br	6.57262	10.81780	0.15212	9.18219
	Cl	2.91650	10.46487	0.34288	9.53514
	I	10.43965	11.01869	0.09579	8.98132
	$MgCO_3$	3.46829	10.54011	0.28833	9.45989
	MgO	1.65807	10.21960	0.60311	9.78040
	$Mg_2P_2O_7$	4.57731	10.66061	0.21847	9.33939
	$MgSO_4$	4.95122	10.69471	0.20197	9.30529
$MgBr_2$	Br	0.86796	9.93850	1.15213	10.06150
$MgBr.6H_2O$	Br	0.54702	9.73800	1.82809	10.26200
MgC_2O_4	Mg	0.21643	9.33532	4.62043	10.66468
	MgO	0.35886	9.55493	2.78660	10.44508
$Mg(C_9H_6NO)_2$	Mg	0.07777	8.89081	12.85843	11.10919
	$MgBr_2$	0.58898	9.77010	1.69785	10.22990
	$MgBr_2.6H_2O$	0.93455	9.97060	1.07003	10.02939
	$MgCO_3$	0.26972	9.43091	3.70755	10.56909
	$MgCl_2$	0.30458	9.48370	3.28321	10.51630
	$MgCl_2.6H_2O$	0.65014	9.81301	1.53813	10.18699
	$Mg(HCO_3)_2$	0.46813	9.67037	2.13616	10.32963
	MgO	0.12895	9.11042	7.75494	10.88958
	$MgSO_4$	0.38505	9.58552	2.59707	10.41448
	$MgSO_4.7H_2O$	0.78842	9.89676	1.26834	10.10324
$MgCl_2$	Cl	0.74467	9.87196	1.34288	10.12804
	$Mg_2P_2O_7$	1.16872	10.06771	0.85564	9.93229
$MgCl_2.6H_2O$	Cl	0.34887	9.54266	2.86640	10.45723
	$Mg_2P_2O_7$	0.54753	9.73841	1.82638	10.26159
$MgCl_2.KCl.6H_2O$	$Mg_2P_2O_7$	0.40059	9.60270	2.49632	10.39730
$MgCO_3$	CO_2	0.52193	9.71761	1.91597	10.28239
	Mg	0.28833	9.45989	3.46825	10.54011
	$Mg(HCO_3)_2$	1.73559	10.23945	0.57617	9.76055
	MgO	0.47807	9.67949	2.09174	10.32051
	$Mg_2P_2O_7$	1.31977	10.12049	0.75771	9.87950

Weighed	Sought	Factor	Log of Factor +10	Reciprocal of Factor	Log of Reciprocal of Factor +10
Magnesium (contd.)					
$Mg(HCO_3)_2$	$MgCO_3$	0.57617	9.76055	1.73560	10.23945
	MgO	0.27545	9.44004	3.63042	10.55996
	$Mg_2P_2O_7$	0.76041	9.88105	1.31508	10.11895
MgI_2	I	0.91258	9.96027	1.09579	10.03973
MgO	CO_2	1.09176	10.03812	0.91595	9.96187
	Mg	0.60311	9.78040	1.65807	10.21960
	$MgCO_3$	2.09176	10.32051	0.47807	9.67949
	$Mg(HCO_3)_2$	3.63045	10.55996	0.27545	9.44004
	$Mg_2P_2O_7$	2.76064	10.44101	0.36223	9.55898
	$MgSO_4$	2.98613	10.47511	0.33488	9.52489
	SO_3	1.98611	10.29800	0.50350	9.70200
$Mg_2P_2O_7$	Mg	0.21847	9.33939	4.57729	10.66061
	$MgCl_2$	0.85563	9.93229	1.16873	10.06771
	$MgCl_2.6H_2O$	1.82638	10.26159	0.54753	9.73841
	$MgCl_2.KCl.6H_2O$	2.49633	10.39730	0.40059	9.60270
	$MgCO_3$	0.75771	9.87950	1.31977	10.12049
	$Mg(HCO_3)_2$	1.31508	10.11896	0.76041	9.88105
	MgO	0.36224	9.55900	2.76060	10.44101
	$MgSO_4$	1.08168	10.03410	0.92449	9.96590
	$MgSO_4.7H_2O$	2.21488	10.34535	0.45149	9.65465
$MgSO_4$	$BaSO_4$	1.93896	10.28757	0.51574	9.71243
	Mg	0.20197	9.30529	4.85123	10.69471
	MgO	0.33488	9.52489	2.98614	10.47511
	$Mg_2P_2O_7$	0.92449	9.96590	1.08168	10.03410
	SO_3	0.66511	9.82289	1.50351	10.17711
$MgSO_4.7H_2O$	$BaSO_4$	0.94693	9.97632	1.05604	10.02368
	$Mg_2P_2O_7$	0.45149	9.65465	2.21489	10.34535
	SO_3	0.32482	9.51164	3.07863	10.48836
SO_3	MgO	0.50350	9.70200	1.98610	10.29800
	$MgSO_4$	1.50351	10.17711	0.66511	9.82289
	$MgSO_4.7H_2O$	3.07863	10.48836	0.32482	9.51164
Manganese					
$BaSO_4$	$MnSO_4$	0.64696	9.81088	1.54569	10.18912
CO_2	$MnCO_3$	2.61184	10.41694	0.38287	9.58305
	MnO	1.61184	10.20733	0.62041	9.79268
Mn	$MnCO_3$	2.09230	10.32062	0.47794	9.67937
	MnO	1.29122	10.11100	0.77446	9.88900
	MnO_2	1.58246	10.19933	0.63193	9.80067
	Mn_2O_3	1.43684	10.15741	0.69597	9.84259
	Mn_3O_4	1.38830	10.14248	0.72031	9.85752
	$Mn_2P_2O_7$	2.58308	10.41213	0.38713	9.58786
MnC_2O_4	Mn	0.38429	9.58466	2.60220	10.41534
	Mn_3O_4	0.53352	9.72715	1.87434	10.27285
$Mn(C_9H_6NO)_2$	Mn	0.16005	9.20426	6.24805	10.79574
	Mn_3O_4	0.22220	9.34674	4.50045	10.65326
$MnCO_3$	CO_2	0.38287	9.58305	2.61185	10.41695
	Mn	0.47794	9.67937	2.09231	10.32063
	$Mn(HCO_3)_2$	1.53961	10.18741	0.64952	9.81259
	MnO	0.61713	9.79038	1.62040	10.20962
	Mn_3O_4	0.66353	9.82186	1.50709	10.17814
	$Mn_2P_2O_7$	1.23456	10.09152	0.81001	9.90849
	MnS	0.75689	9.87903	1.32120	10.12097
	$MnSO_4$	1.31365	10.11848	0.76124	9.88152
$Mn(HCO_3)_2$	$MnCO_3$	0.64952	9.81259	1.53960	10.18741
	MnO	0.40084	9.60297	2.49476	10.39703
	Mn_3O_4	0.43098	9.63446	2.32029	10.36555
MnO	CO_2	0.62041	9.79268	1.61184	10.20733
	Mn	0.77446	9.88900	1.29122	10.11099
	$MnCO_3$	1.62041	10.20963	0.61713	9.79038
	$Mn(HCO_3)_2$	2.49479	10.39703	0.40084	9.60297
	Mn_2O_3	1.11277	10.04641	0.89866	9.95360
	Mn_3O_4	1.07518	10.03148	0.93008	9.96852
	$Mn_2P_2O_7$	2.00049	10.30114	0.49988	9.69887
	MnS	1.22647	10.08865	0.81535	9.91134
	$MnSO_4$	2.12865	10.32811	0.46978	9.67189
	SO_3	1.12864	10.05255	0.88602	9.94744
MnO_2	Mn	0.63193	9.80067	1.58245	10.19933
	Mn_3O_4	0.87731	9.94315	1.13985	10.05684

GRAVIMETRIC FACTORS AND THEIR LOGARITHMS (Continued)

Weighed	Sought	Factor	Log of Factor +10	Reciprocal of Factor	Log of Reciprocal of Factor +10
Manganese (contd.)					
MnO_2	$Mn_2P_2O_7$	1.63233	10.21281	0.61262	9.78719
Mn_2O_3	Mn	0.69597	9.84259	1.43684	10.15741
	MnO	0.89865	9.95359	1.11278	10.04641
	Mn_3O_4	0.96622	9.98508	1.03496	10.01492
Mn_3O_4	Mn	0.72030	9.85751	1.38831	10.14249
	$MnCO_3$	1.50709	10.17814	0.66353	9.82186
	$Mn(HCO_3)_2$	2.32032	10.36555	0.43098	9.63446
	MnO	0.93007	9.96852	1.07519	10.03149
	MnO_2	1.13984	10.05684	0.87732	9.94316
	Mn_2O_3	1.03496	10.01492	0.96622	9.98508
	$MnSO_4$	1.97978	10.29662	0.50511	9.70339
$Mn_2P_2O_7$	Mn	0.38713	9.58786	2.58311	10.41214
	$MnCO_3$	0.81000	9.90849	1.23457	10.09152
	MnO	0.49987	9.69886	2.00052	10.30114
	MnO_2	0.61262	9.78719	1.63233	10.21281
	$MnSO_4$	1.06405	10.02696	0.93981	9.97304
MnS	Mn	0.63146	9.80035	1.58363	10.19966
	$MnCO_3$	1.32120	10.12097	0.75689	9.87903
	MnO	0.81535	9.91134	1.22647	10.08865
	$MnSO_4$	1.73559	10.23945	0.57617	9.76055
$MnSO_4$	$BaSO_4$	1.54570	10.18913	0.64696	9.81088
	Mn	0.36383	9.56090	2.74854	10.43910
	MnO	0.46978	9.67189	2.12866	10.32811
	Mn_3O_4	0.50510	9.70338	1.97981	10.29663
	$Mn_2P_2O_7$	0.93980	9.97304	1.06406	10.02696
	MnS	0.57617	9.76055	1.73560	10.23945
	SO_3	0.53021	9.72445	1.88605	10.27555
SO_3	MnO	0.88603	9.94745	1.12863	10.05255
	$MnSO_4$	1.88604	10.27555	0.53021	9.72445
Mercury					
Hg	HgCl	1.17673	10.07068	0.84981	9.92932
	$HgCl_2$	1.35351	10.13146	0.73882	9.86854
	HgO	1.07976	10.03332	0.92613	9.96667
	HgS	1.15983	10.06440	0.86220	9.93561
$Hg_3(AsO_4)_2$	Hg	0.68413	9.83514	1.46171	10.16486
$Hg(C_{12}H_{10}ONS)_2$	Hg	0.31681	9.50080	3.15647	10.49920
HgCl	Hg	0.84981	9.92932	1.17673	10.07068
	$HgCl_2$	1.15023	10.06079	0.86939	9.93921
	$HgNO_3$	1.11248	10.04629	0.89889	9.95371
	HgO	0.91760	9.96265	1.08980	10.03735
	Hg_2O	0.88369	9.94630	1.13162	10.05370
	HgS	0.98564	9.99372	1.01457	10.00629
$HgCl_2$	Hg	0.73882	9.86854	1.35351	10.13146
	HgCl	0.86939	9.93921	1.15023	10.06079
	HgS	0.85691	9.93294	1.16698	10.06706
$Hg(CN)_2$	HgS	0.92091	9.96422	1.08588	10.03578
Hg_2CrO_4	Hg	0.77572	9.88971	1.28912	10.11029
$HgNO_3$	HgCl	0.89889	9.95371	1.11248	10.04629
	HgS	0.88598	9.94742	1.12869	10.05257
$Hg(NO_3)_2$	HgS	0.71673	9.85536	1.39523	10.14464
$Hg(NO_3)_2.H_2O$	HgS	0.67903	9.83189	1.47269	10.16812
HgO	Hg	0.92613	9.96667	1.07977	10.03333
	HgCl	1.08980	10.03735	0.91760	9.96265
	HgS	1.07415	10.03106	0.93097	9.96894
Hg_2O	HgCl	1.13160	10.05369	0.88370	9.94630
	HgS	1.11535	10.04741	0.89658	9.95259
HgS	Hg	0.86220	9.93561	1.15982	10.06440
	HgCl	1.01457	10.00629	0.98564	9.99372
	$HgCl_2$	1.16698	10.06706	0.85691	9.93294
	$Hg(CN)_2$	1.08588	10.03578	0.92091	9.96422
	$HgNO_3$	1.12869	10.05257	0.88598	9.94743
	$Hg(NO_3)_2$	1.39522	10.14464	0.71673	9.85536
	$Hg(NO_3)_2.H_2O$	1.47268	10.16811	0.67903	9.83189
	HgO	0.93097	9.96894	1.07415	10.03106
	Hg_2O	0.89656	9.95258	1.11537	10.04741
	$HgSO_4$	1.27509	10.10554	0.78426	9.89446
$HgSO_4$	HgS	0.78426	9.89446	1.27509	10.10554
Molybdenum					
Mo	MoO_3	1.50031	10.17618	0.66653	9.82382

Weighed	Sought	Factor	Log of Factor +10	Reciprocal of Factor	Log of Reciprocal of Factor +10
Molybdenum					
Mo	MoC	1.12518	10.05122	0.88875	9.94878
	MoS_3	2.00261	10.30159	0.49935	9.69841
	$PbMoO_4$	3.82666	10.58282	0.26132	9.41717
MoC	C	0.11127	9.04638	8.89715	10.95362
	Mo	0.88874	9.94877	1.12519	10.05122
$MoO_2(C_9H_6NO)_2$	Mo	0.23049	9.36265	4.33858	10.63735
	MoO_3	0.34580	9.53883	2.89184	10.46118
MoO_3	Mo	0.66653	9.82382	1.50031	10.17617
	MoS_3	1.33479	10.12541	0.74918	9.87459
	$(NH_4)_2MoO_4$	1.36175	10.13410	0.73435	9.86590
	$(NH_4)_3PO_4.12MoO_3$	1.08630	10.03595	0.92056	9.96405
	$PbMoO_4$	2.55058	10.40664	0.39207	9.59336
MoS_3	Mo	0.49935	9.69841	2.00260	10.30159
	MoO_3	0.74918	9.87459	1.33479	10.12541
	$(NH_4)_2MoO_4$	1.02019	10.00868	0.98021	9.99132
MoSi	Mo	0.77352	9.88847	1.29279	10.11152
	Si	0.22645	9.35497	4.41599	10.64503
$(NH_4)_2MoO_4$	MoO_3	0.73435	9.86590	1.36175	10.13410
	MoS_3	0.98021	9.99132	1.02019	10.00868
	$(NH_4)_3PO_4.12MoO_3$	0.79771	9.90185	1.25359	10.09816
	$PbMoO_4$	1.87302	10.27254	0.53390	9.72746
$(NH_4)_3PO_4.12MoO_3$	MoO_3	0.92054	9.96404	1.08632	10.03596
	$(NH_4)_2MoO_4$	1.25359	10.09816	0.79771	9.90185
$PbMoO_4$	Mo	0.26132	9.41717	3.82673	10.58283
	MoO_3	0.39207	9.59336	2.55056	10.40664
	$(NH_4)_2MoO_4$	0.53390	9.72746	1.87301	10.27254
Neodymium					
Nd	Nd_2O_3	1.16639	10.06684	0.85735	9.93315
Nd_2O_3	Nd	0.85735	9.93316	1.16638	10.06684
Nickel					
Ni	$Ni(C_4H_7N_2O_2)_2$	4.92148	10.69209	0.20319	9.30790
	$Ni(NO_3)_2.6H_2O$	4.95231	10.69481	0.20193	9.30520
	NiO	1.27254	10.10467	0.78584	9.89533
	$NiSO_4$	2.63618	10.42099	0.37934	9.57903
Ni	$NiSO_4.7H_2O$	4.78419	10.67981	0.20902	9.32019
$Ni(C_4H_7N_2O_2)_2$	Ni	0.20319	9.30790	4.92150	10.69210
	NiO	0.25857	9.41258	3.86742	10.58742
$(Ni[C_5H_5N]_4)(SCN)_2$	Ni	0.11950	9.07737	8.36820	10.92263
	NiO	0.15207	9.18204	6.57592	10.81776
$Ni(C_7H_6O_2N)_2$	Ni	0.17736	9.24886	5.63825	10.75114
	NiO	0.22573	9.35359	4.43007	10.64641
$Ni(C_9H_6NO)_2$	Ni	0.16918	9.22835	5.91086	10.77165
	NiO	0.21529	9.33303	4.64490	10.66698
$Ni(NO_3)_2.6H_2O$	Ni	0.20193	9.30520	4.95221	10.69480
	NiO	0.25696	9.40987	3.89166	10.59013
	$NiSO_4$	0.53231	9.72616	1.87860	10.27383
NiO	Ni	0.78584	9.89533	1.27252	10.10467
	$Ni(C_4H_7N_2O_2)_2$	3.86749	10.58743	0.25857	9.41258
	$Ni(NO_3)_2.6H_2O$	3.89171	10.59014	0.25696	9.40987
	$NiSO_4$	2.07161	10.31631	0.48272	9.68370
	$NiSO_4.7H_2O$	3.75960	10.57514	0.26599	9.42487
$NiSO_4$	Ni	0.37934	9.57903	2.63616	10.42098
	$Ni(NO_3)_2.6H_2O$	1.87859	10.27384	0.53231	9.72616
	NiO	0.48272	9.68370	2.07159	10.31630
	$NiSO_4.7H_2O$	1.81482	10.25884	0.55102	9.74117
$NiSO_4.7H_2O$	Ni	0.20902	9.32019	4.78423	10.67981
	NiO	0.26599	9.42487	3.75954	10.57514
	$NiSO_4$	0.55102	9.74117	1.81482	10.25884
Niobium					
C	Nb	7.73497	10.88846	0.12928	9.11153
	NbC	8.73493	10.94126	0.11448	9.05873
	$\frac{1}{2}Nb_2O_5$	11.06508	11.04396	0.09038	8.95607
Nb	C	0.12928	9.11153	7.73515	10.88848
	Nb_2O_5	1.43053	10.15550	0.69904	9.84450
NbC	C	0.11448	9.05873	8.73515	10.94127
	Nb	0.88552	9.94720	1.12928	10.05280
Nb_2O_5	Nb	0.69904	9.84450	1.43053	10.15550
Nitrogen					
$AgNO_2$	HNO_2	0.30552	9.48504	3.27311	10.51496

Weighed	Sought	Factor	Log of Factor +10	Reciprocal of Factor	Log of Reciprocal of Factor +10
Nitrogen (contd.)					
$AgNO_2$	N_2O_3	0.24698	9.39266	4.04891	10.60733
$C_{20}H_{16}N_4.HNO_3$	N	0.37312	9.57185	2.68010	10.42815
	NO_3	0.16517	9.21793	6.05437	10.78207
HNO_2	$AgNO_2$	3.27310	10.51496	0.30552	9.48504
HNO_3	N	0.22228	9.34690	4.49883	10.65310
	NH_3	0.27027	9.43180	3.70000	10.56820
	NH_4Cl	0.84889	9.92885	1.17801	10.07115
	$(NH_4)_2PtCl_6$	3.52221	10.54682	0.28391	9.45318
	NO	0.47619	9.67778	2.10000	10.32222
	Pt	1.54801	10.18977	0.64599	9.81023
	SO_3	0.63528	9.80297	1.57411	10.19703
	N_2O_5	0.53413	9.72765	1.87220	10.27235
N	HNO_3	4.49877	10.65310	0.22229	9.34692
	$NaNO_3$	6.06816	10.78306	0.16479	9.21693
	NH_3	1.21589	10.08489	0.82244	9.91510
	NH_4Cl	3.81902	10.58195	0.26185	9.41805
	$(NH_4)_2PtCl_6$	15.84563	11.19991	0.06311	8.80010
	$(NH_4)_2SO_4$	4.71699	10.67367	0.21200	9.32634
	NO_2	3.28454	10.51648	0.30446	9.48353
	NO_3	4.42680	10.64609	0.22590	9.35392
	N_2O_3	2.71341	10.43352	0.36854	9.56648
	N_2O_5	3.85566	10.58610	0.25936	9.41390
	Pt	6.96416	10.84287	0.14359	9.15712
	SO_3	2.85798	10.45606	0.34990	9.54394
$NaNO_3$	N	0.16479	9.21693	6.06833	10.78307
	N_2O_5	0.63539	9.80304	1.57384	10.19696
NH_3	HNO_3	3.69998	10.56820	0.27027	9.43180
	N	0.82244	9.91510	1.21589	10.08489
	NO_3	3.64079	10.56119	0.27467	9.43881
	N_2O_5	3.17107	10.50121	0.31535	9.49879
$NH_4B(C_6H_5)_4$	N	0.04153	8.61836	24.07898	11.38164
	NH_3	0.05049	8.70320	19.80590	11.29679
	NH_4	0.05348	8.72819	18.69858	11.27181
NH_4Cl	HNO_3	1.17799	10.07115	0.84890	9.92886
	N	0.26185	9.41805	3.81898	10.58195
	NO_3	1.15914	10.06413	0.86271	9.93586
	N_2O_5	1.00959	10.00415	0.99050	9.99585
$(NH_4)_2PtCl_6$	HNO_3	0.28392	9.45320	3.52212	10.54680
	N	0.06311	8.80010	15.84535	11.19990
	NO_3	0.27938	9.44620	3.57935	10.55380
	N_2O_5	0.24333	9.38620	4.10965	10.61380
$(NH_4)_2SO_4$	N	0.21200	9.32634	4.71698	10.67367
	N_2O_5	0.81740	9.91243	1.22339	10.08757
NO	HNO_3	2.10000	10.32222	0.47619	9.67778
	NO_2	1.53320	10.18560	0.65223	9.81440
	NO_3	2.06641	10.31522	0.48393	9.68478
	N_2O_3	1.26660	10.10264	0.78952	9.89736
	N_2O_5	1.79980	10.25522	0.55562	9.74478
NO_2	N	0.30446	9.48353	3.28450	10.51647
	NO	0.65223	9.81440	1.53320	10.18560
NO_3	N	0.22590	9.35392	4.42674	10.64608
	NH_3	0.27467	9.43881	3.64073	10.56119
	NH_4Cl	0.86270	9.93586	1.15915	10.06413
	NO	0.48393	9.68478	2.06641	10.31522
	Pt	1.57318	10.19678	0.63566	9.80322
N_2O_3	$AgNO_2$	4.04875	10.60732	0.24699	9.39268
	N	0.36854	9.56648	2.71341	10.43352
	NO	0.78951	9.89736	1.26661	10.10264
N_2O_5	KNO_3	1.87217	10.27235	0.53414	9.72766
	N	0.25936	9.41390	3.85564	10.58610
	$NaNO_3$	1.57382	10.19695	0.63540	9.80305
	NH_3	0.31535	9.49879	3.17108	10.50121
	NH_4Cl	0.99049	9.99585	1.00960	10.00415
	$(NH_4)_2PtCl_6$	4.10965	10.61380	0.24333	9.38620
	$(NH_4)_2SO_4$	1.22339	10.08757	0.81740	9.91243
	NO	0.55562	9.74478	1.79980	10.25522
	Pt	1.80621	10.25677	0.55365	9.74324
	SO_3	0.74124	9.86996	1.34909	10.13004

Weighed	Sought	Factor	Log of Factor +10	Reciprocal of Factor	Log of Reciprocal of Factor +10
Nitrogen (contd.)					
Pt	HNO_3	0.64595	9.81020	1.54811	10.18980
	N	0.14358	9.15709	6.96476	10.84291
	NO_3	0.63562	9.80320	1.57327	10.19680
	N_2O_5	0.55364	9.74323	1.80623	10.25678
SO_3	HNO_3	1.57410	10.19703	0.63528	9.80297
	N	0.34990	9.54394	2.85796	10.45605
	N_2O_5	1.34908	10.13004	0.74125	9.86996
Osmium					
Os	OsO_4	1.33649	10.12597	0.74823	9.87404
OsO_4	Os	0.74823	9.87404	1.33649	10.12597
Palladium					
K_2PdCl_6	Pd	0.26781	9.42783	3.73399	10.57217
	$PdCl_2.2H_2O$	0.53687	9.72987	1.86265	10.27013
Pd	K_2PdCl_6	3.73402	10.57217	0.26781	9.42783
	$PdCl_2.2H_2O$	2.00460	10.30205	0.49883	9.69795
	PdI_2	3.38534	10.52960	0.29539	9.47040
	$Pd(NO_3)_2$	2.16541	10.33554	0.46181	9.66446
$Pd(C_4H_7N_2O_2)_2$	Pd	0.31610	9.49982	3.16356	10.50018
$PdCl_2.2H_2O$	K_2PdCl_6	1.86263	10.27012	0.53688	9.72988
	Pd	0.49883	9.69795	2.00469	10.30204
$PdCl_2.C_{12}H_8N_2$	Pd	0.29762	9.47366	3.35999	10.52634
PdI_2	Pd	0.29539	9.47040	3.38536	10.52960
$Pd(NO_3)_2$	Pd	0.46181	9.66446	2.16539	10.33554
Phosphorus					
Ag_3PO_4	P	0.07400	8.86923	13.51351	11.13077
	PO_4	0.22689	9.35582	4.40742	10.64418
	P_2O_5	0.16954	9.22927	5.89831	10.77072
$Ag_4P_2O_7$	P	0.10232	9.00996	9.77326	10.99004
	PO_4	0.31374	9.49657	3.18735	10.50343
	P_2O_5	0.23446	9.37007	4.26512	10.62993
Al_2O_3	P_2O_5	1.39214	10.14368	0.71832	9.85632
$AlPO_4$	PO_4	0.77875	9.89140	1.28411	10.10860
	P_2O_5	0.58196	9.76489	1.71833	10.23511
$Ca_3(PO_4)_2$	P_2O_5	0.45762	9.66051	2.18522	10.33949
$FePO_4$	PO_4	0.62971	9.79914	1.58803	10.20086
	P_2O_5	0.47058	9.67263	2.12504	10.32737
$Mg_2P_2O_7$	H_3PO_4	0.88059	9.94471	1.13560	10.05523
	Na_2HPO_4	1.27565	10.10573	0.78391	9.89427
	$Na_2HPO_4.12H_2O$	3.21828	10.50763	0.31073	9.49238
	$NaNH_4.HPO_4.4H_2O$	1.87870	10.27386	0.53228	9.72614
	P	0.27833	9.44456	3.59286	10.55544
	PO_4	0.85341	9.93116	1.17177	10.06884
	P_2O_5	0.63776	9.80466	1.56799	10.19535
Na_2HPO_4	$Mg_2P_2O_7$	0.78392	9.89427	1.27564	10.10573
	P_2O_5	0.49995	9.69893	2.00020	10.30105
$Na_2HPO_4.12H_2O$	$Mg_2P_2O_7$	0.31072	9.49237	3.21833	10.50763
	P_2O_5	0.19816	9.29702	5.04643	10.70299
$NaNH_4HPO_4.4H_2O$	$Mg_2P_2O_7$	0.53228	9.72614	1.87871	10.27386
	P_2O_5	0.33946	9.53079	2.94586	10.46921
$(NH_4)_3PO_4.12MoO_3$	P	0.01651	8.21775	60.56935	11.78223
	PO_4	0.05061	8.70424	19.75894	11.29577
	P_2O_5	0.03782	8.57772	26.44104	11.42227
P	Ag_3PO_4	13.51403	11.13078	0.07400	8.86923
	$Ag_4P_2O_7$	9.77314	10.99004	0.10232	9.00996
	$Mg_2P_2O_7$	3.59282	10.55544	0.27833	9.44456
	$(NH_4)_3PO_4.12MoO_3$	60.5786	11.78234	0.01651	8.21775
	P_2O_5	2.29136	10.36009	0.43642	9.63990
	$P_2O_5.24MoO_3$	58.0564	11.76385	0.01723	8.23616
	$U_2P_2O_{11}$	11.52587	11.06167	0.08676	8.93832
PO_4	Ag_3PO_4	4.40744	10.64418	0.22689	9.35582
	$Ag_4P_2O_7$	3.18739	10.50343	0.31374	9.49657
	$AlPO_4$	1.28410	10.10860	0.77876	9.89140
	$FePO_4$	1.58803	10.20086	0.62971	9.79914
	$Mg_2P_2O_7$	1.17175	10.06884	0.85342	9.93116
	$(NH_4)_3PO_4.12MoO_3$	19.7570	11.29572	0.05062	8.70432
	$P_2O_5.24MoO_3$	18.9344	11.27725	0.05281	8.72272
	$U_2P_2O_{11}$	3.75902	10.57507	0.26603	9.42493
P_2O_5	Ag_3PO_4	5.89780	10.77069	0.16955	9.22930

Weighed	Sought	Factor	Log of Factor +10	Reciprocal of Factor	Log of Reciprocal of Factor +10
Phosphorus (contd.)					
P_2O_5	$Ag_4P_2O_7$	4.26520	10.62994	0.23446	9.37007
	Al_2O_3	0.71831	9.85631	1.39216	10.14369
	$AlPO_4$	1.71831	10.23510	0.58197	9.76440
	$Ca_3(PO_4)_2$	2.18521	10.33949	0.45762	9.66051
	$FePO_4$	2.12502	10.32736	0.47058	9.67263
	$Mg_2P_2O_7$	1.56798	10.19534	0.63776	9.80466
	Na_2HPO_4	2.00020	10.30107	0.49995	9.69893
	$Na_2HPO_4.12H_2O$	5.04623	10.70297	0.19817	9.29704
	$NaNH_4HPO_4.4H_2O$	2.94578	10.46920	0.33947	9.53080
	$(NH_4)_3PO_4.12MoO_3$	26.4377	11.42222	0.03783	8.57784
	P	0.43642	9.63990	2.29137	10.36010
	$P_2O_5.24MoO_3$	25.3370	11.40376	0.03947	8.59622
	$U_2P_2O_{11}$	5.03013	10.70158	0.19880	9.29842
$P_2O_5.24MoO_3$	P	0.01722	8.23603	58.07201	11.76397
	PO_4	0.05281	8.72272	18.93581	11.27728
	P_2O_5	0.03947	8.59627	25.33570	11.40373
$U_2P_2O_{11}$	P	0.08676	8.93832	11.52605	11.06176
	PO_4	0.26603	9.42493	3.75897	10.57507
	P_2O_5	0.19880	9.29842	5.03018	10.70158
Platinum					
$H_2PtCl_6.6H_2O$	K_2PtCl_6	0.93852	9.97244	1.06551	10.02756
	Pt	0.37673	9.57603	2.65442	10.42397
K_2PtCl_6	$H_2PtCl_6.6H_2O$	1.06551	10.02756	0.93852	9.97244
	Pt	0.40141	9.60359	2.49122	10.39541
	$PtCl_4$	0.69320	9.84086	1.44259	10.15915
	$PtCl_4.5H_2O$	0.87854	9.94376	1.13825	10.05624
$(NH_4)_2PtCl_6$	Pt	0.43950	9.64296	2.27531	10.35704
	$PtCl_4$	0.75897	9.88022	1.31758	10.11978
	$PtCl_6$	0.91854	9.96310	1.08868	10.03690
Pt	$H_2PtCl_6.6H_2O$	2.65442	10.42397	0.37673	9.57603
	K_2O	0.48287	9.68383	2.07095	10.31617
	K_2PtCl_6	2.49121	10.39641	0.40141	9.60359
	$(NH_4)_2PtCl_6$	2.27531	10.35704	0.43950	9.64296
	$PtCl_4$	1.72690	10.23727	0.57907	9.76273
	$PtCl_4.5H_2O$	2.18863	10.34017	0.45691	9.65983
$PtCl_4$	K_2PtCl_6	1.44259	10.15915	0.69320	9.84086
	$(NH_4)_2PtCl_6$	1.31757	10.11978	0.75897	9.88022
	Pt	0.57907	9.76273	1.72691	10.23727
$PtCl_4.5H_2O$	K_2PtCl_6	1.13825	10.05624	0.87854	9.94376
	Pt	0.45691	9.65983	2.18861	10.34017
$PtCl_6$	K_2PtCl_6	1.19199	10.07628	0.83893	9.92373
	$(NH_4)_2PtCl_6$	1.08869	10.03691	0.91854	9.96310
Potassium					
Ag	KBr	1.10328	10.04269	0.90639	9.95732
	KCl	0.69116	9.83958	1.44684	10.16042
	$KClO_3$	1.13612	10.05543	0.88019	9.94458
	$KClO_4$	1.28444	10.10872	0.77855	9.89129
	KCN	0.60369	9.78081	1.65648	10.21919
	KI	1.53895	10.18723	0.64979	9.81277
AgBr	KBr	0.63378	9.80194	1.57783	10.19806
	$KBrO_3$	0.88939	9.94909	1.12437	10.05091
AgCl	KCl	0.52019	9.71616	1.92237	10.28384
	$KClO_3$	0.85508	9.93201	1.16948	10.06799
	$KClO_4$	0.96672	9.98530	1.03443	10.01470
AgCN	KCN	0.48638	9.68698	2.05601	10.31302
AgI	KI	0.70709	9.84947	1.41425	10.15053
	KIO_3	0.91154	9.95978	1.09704	10.04022
$BaCrO_4$	K_2CrO_4	0.76658	9.88456	1.30450	10.11544
	$K_2Cr_2O_7$	0.58064	9.76391	1.72224	10.23609
$BaSO_4$	$KHSO_4$	0.58343	9.76599	1.71400	10.23401
	K_2S	0.47244	9.67435	2.11667	10.32565
	K_2SO_4	0.74664	9.87311	1.33933	10.12689
Br	K	0.48933	9.68960	2.04361	10.31040
	KBr	1.48933	10.17299	0.67144	9.82701
CaF_2	$KF.2H_2O$	2.41114	10.38223	0.41474	9.61778
$CaSO_4$	$KF.2H_2O$	1.38286	10.14078	0.72314	9.85922
Cl	K	1.10292	10.04255	0.90668	9.95745
	KCl	2.10292	10.32282	0.47553	9.67718

Weighed	Sought	Factor	Log of Factor +10	Reciprocal of Factor	Log of Reciprocal of Factor +10
Potassium (contd.)					
Cl	$KClO_3$	3.45677	10.53867	0.28929	9.46133
	$KClO_4$	3.90808	10.59196	0.25588	9.40804
	K_2O	1.32856	10.12338	0.75269	9.87662
CO_2	$KHCO_3$	2.27491	10.35696	0.43958	9.64304
	K_2CO_3	3.14049	10.49700	0.31842	9.50300
	K_2O	2.14049	10.33051	0.46718	9.66948
I	KI	1.30811	10.11665	0.76446	9.88335
	KIO_3	1.68634	10.22695	0.59300	9.77306
K	Br	2.04360	10.31040	0.48933	9.68960
	Cl	0.90668	9.95745	1.10292	10.04256
	KBr	3.04360	10.48339	0.32856	9.51661
	KCl	1.90668	10.28028	0.52447	9.71972
	$KClO_3$	3.13419	10.49614	0.31906	9.50387
	$KClO_4$	3.54337	10.54941	0.28222	9.45059
	KI	4.24546	10.62793	0.23555	9.37208
	KNO_3	2.58572	10.41259	0.38674	9.58742
	K_2O	1.20458	10.08084	0.83016	9.91916
	K_2PtCl_6	6.21467	10.79342	0.16091	9.20658
	K_2SO_4	2.22835	10.34799	0.44876	9.65201
	Pt	2.49463	10.39701	0.40086	9.60299
K_3AsO_4	$Mg_2As_2O_7$	0.60584	9.78236	1.65060	10.21764
$K(B[C_6H_5]_4)$	K	0.10912	9.03790	9.16422	10.96210
	K_2O	0.13144	9.11873	7.60803	10.88127
KBr	Ag	0.90639	9.95732	1.10328	10.04269
	AgBr	1.57783	10.19806	0.63378	9.80194
	Br	0.67144	9.82701	1.48934	10.17299
	K	0.32856	9.51662	3.04358	10.48338
	K_2O	0.39580	9.59748	2.52653	10.40253
$KBrO_3$	AgBr	1.12436	10.05091	0.88939	9.94909
$KC_{12}H_4O_{12}N_7$	K	0.08192	8.91339	12.20703	11.08661
	K_2O	0.09868	8.99423	10.13377	11.00577
KCl	Ag	1.44685	10.16043	0.69116	9.83958
	AgCl	1.92238	10.28384	0.52019	9.71616
	Cl	0.47553	9.67718	2.10292	10.32283
	K	0.52447	9.71972	1.90669	10.28028
	$KClO_3$	1.64379	10.21585	0.60835	9.78415
	$KClO_4$	1.85840	10.26914	0.53810	9.73086
	K_2CO_3	0.92692	9.96704	1.07884	10.03296
	$K_2Cr_2O_7$	1.97299	10.29512	0.50684	9.70487
	$KHCO_3$	1.34289	10.12804	0.74466	9.87196
	KNO_3	1.35614	10.13230	0.73739	9.86770
	K_2O	0.63180	9.80058	1.58278	10.19942
	K_2PtCl_6	3.25941	10.51314	0.30680	9.48686
	K_2SO_4	1.16871	10.06770	0.85564	9.93229
	Pt	1.30836	10.11673	0.76432	9.88328
$KClO_3$	Ag	0.88019	9.94458	1.13612	10.05543
	AgCl	1.16948	10.06799	0.85508	9.93201
	Cl	0.28929	9.46133	3.45674	10.53867
	KCl	0.60835	9.78415	1.64379	10.21584
$KClO_4$	Ag	0.77855	9.89129	1.28444	10.10872
	AgCl	1.03443	10.01470	0.96672	9.98530
	Cl	0.25588	9.40804	3.90808	10.59196
	K	0.28222	9.45059	3.54333	10.54941
	KCl	0.53810	9.73086	1.85839	10.26914
	K_2O	0.33995	9.53142	2.94161	10.46858
KCN	Ag	1.65648	10.21918	0.60369	9.78081
	AgCN	2.05602	10.31302	0.48638	9.68698
K_2CO_3	CO_2	0.31842	9.50300	3.14051	10.49700
	KCl	1.07883	10.03295	0.92693	9.96705
	K_2O	0.68158	9.83352	1.46718	10.16648
	KOH	0.81191	9.90951	1.23166	10.09049
	K_2PtCl_6	3.51638	10.54610	0.28438	9.45390
	K_2SO_4	1.26085	10.10067	0.79312	9.89934
K_2CrO_4	$BaCrO_4$	1.30449	10.11544	0.76658	9.88456
$K_2Cr_2O_7$	$BaCrO_4$	1.72220	10.23608	0.58065	9.76391
	KCl	0.50685	9.70488	1.97297	10.29512
	K_2O	0.32021	9.50543	3.12295	10.49456
$KF.2H_2O$	CaF_2	0.41474	9.61778	2.41115	10.38222

Weighed	Sought	Factor	Log of Factor +10	Reciprocal of Factor	Log of Reciprocal of Factor +10
Potassium (contd.)					
$KF.2H_2O$	$CaSO_4$	0.72314	9.85922	1.38286	10.14078
K_2HAsO_4	$Mg_2As_2O_7$	0.71165	9.85227	1.40519	10.14774
$KHCO_3$	KCl	0.74466	9.87196	1.34289	10.12804
	K_2O	0.47046	9.67252	2.12558	10.32748
	K_2PtCl_6	2.42716	10.38510	0.41200	9.61490
	K_2SO_4	0.87029	9.93966	1.14904	10.06034
$KHSO_4$	$BaSO_4$	1.71401	10.23401	0.58343	9.76599
	K_2SO_4	0.63988	9.80610	1.56279	10.19389
KI	Ag	0.64980	9.81278	1.53894	10.18722
	AgI	1.41425	10.15053	0.70709	9.84947
	I	0.76446	9.88335	1.30811	10.11665
	K	0.23555	9.37208	4.24538	10.62792
	K_2O	0.28374	9.45292	3.52435	10.54708
KIO_3	AgI	1.09705	10.04023	0.91154	9.95978
	I	0.59300	9.77305	1.68634	10.22695
$KMnO_4$	Mn_2O_3	0.49948	9.69852	2.00208	10.30148
	MnS	0.55051	9.74007	1.81650	10.25924
K_2MnO_4	Mn_2O_3	0.40041	9.60250	2.49744	10.39749
	MnS	0.44132	9.64475	2.26593	10.35525
KNO_2	K_2SO_4	1.02380	10.01022	0.97675	9.98978
	N_2O_3	0.44656	9.64988	2.23934	10.35012
KNO_3	K	0.38674	9.58742	2.58572	10.41260
	KCl	0.73739	9.86770	1.35613	10.13230
	K_2O	0.46586	9.66826	2.14657	10.33174
	K_2PtCl_6	2.40344	10.38083	0.41607	9.61917
	K_2SO_4	0.86179	9.93540	1.16038	10.06460
	N	0.13853	9.14154	7.21865	10.85846
	NH_3	0.16844	9.22645	5.93683	10.77356
	NO	0.29678	9.47243	3.36950	10.52757
	N_2O_5	0.53414	9.72766	1.87217	10.27235
K_2O	Cl	0.75269	9.87662	1.32857	10.12338
	CO_2	0.46718	9.66948	2.14050	10.33052
	K	0.83016	9.91916	1.20459	10.08085
	KBr	2.52667	10.40255	0.39578	9.59745
	KCl	1.58284	10.19944	0.63178	9.80057
	$KClO_3$	2.60186	10.41529	0.38434	9.58472
	$KClO_4$	2.94155	10.46858	0.33996	9.53143
	K_2CO_3	1.46718	10.16648	0.68158	9.83352
	$K_2Cr_2O_7$	3.12294	10.49456	0.32021	9.50543
	$KHCO_3$	2.12558	10.32748	0.47046	9.67252
	KI	3.52439	10.54709	0.28374	9.45292
	KNO_3	2.14655	10.33174	0.46586	9.66826
	KOH	1.19122	10.07599	0.83948	9.92401
	K_2PtCl_6	5.15918	10.71258	0.19383	9.28742
	K_2SO_4	1.84990	10.26715	0.54057	9.73285
	N_2O_5	1.14657	10.05940	0.87217	9.94060
KOH	K_2CO_3	1.23164	10.09048	0.81193	9.90952
	K_2O	0.83946	9.92400	1.19124	10.07600
K_2PtCl_6	K	0.16091	9.20658	6.21465	10.79342
	KCl	0.30680	9.48686	3.25948	10.51315
	K_2CO_3	0.28438	9.45390	3.51642	10.54610
	$KHCO_3$	0.41200	9.61490	2.42718	10.38510
	KNO_3	0.41606	9.61916	2.40350	10.38084
	K_2O	0.19383	9.28742	5.15916	10.71258
	K_2SO_4	0.35856	9.55456	2.78893	10.44544
	$K_2SO_4.Al_2(SO_4)_3.24H_2O$	1.95218	10.29052	0.51225	9.70948
	$K_2SO_4.Cr_2(SO_4)_3.24H_2O$	2.05512	10.31283	0.48659	9.68716
K_2S	$BaSO_4$	2.11666	10.32565	0.47244	9.67435
	K_2SO_4	1.58039	10.19877	0.63276	9.80124
K_2SiO_3	SiO_2	0.38943	9.59043	2.56786	10.40958
K_2SO_4	$BaSO_4$	1.33933	10.12689	0.74664	9.87311
	K	0.44876	9.65201	2.22836	10.34799
	KCl	0.85565	9.93230	1.16870	10.06770
	K_2CO_3	0.79312	9.89934	1.26084	10.10066
	$KHCO_3$	1.14904	10.06034	0.87029	9.93966
	$KHSO_4$	1.56281	10.19390	0.63987	9.80609
	KNO_2	0.97676	9.98979	1.02379	10.01021
	KNO_3	1.16038	10.06460	0.86179	9.93540

Weighed	Sought	Factor	Log of Factor +10	Reciprocal of Factor	Log of Reciprocal of Factor +10
Potassium (contd.)					
K_2SO_4	K_2O	0.54057	9.73285	1.84990	10.26715
	K_2PtCl_6	2.78889	10.44543	0.35857	9.55457
	K_2S	0.63276	9.80124	1.58038	10.19876
	SO_3	0.45942	9.66221	2.17666	10.33779
$K_2SO_4.Al_2(SO_4)_3.24H_2O$	K_2PtCl_6	0.51224	9.70947	1.95221	10.29053
$K_2SO_4.Cr_2(SO_4)_3.24H_2O$	K_2PtCl_6	0.48658	9.68715	2.05516	10.31284
$Mg_2As_2O_7$	K_3AsO_4	1.65059	10.21764	0.60584	9.78236
	K_2HAsO_4	1.40519	10.14774	0.71165	9.85227
Mn_2O_3	$KMnO_4$	2.00207	10.30148	0.49948	9.69852
	K_2MnO_4	2.49731	10.39747	0.40043	9.60253
MnS	$KMnO_4$	1.81649	10.25923	0.55051	9.74077
	K_2MnO_4	2.26592	10.35524	0.44132	9.64475
N	KNO_3	7.21847	10.85845	0.13853	9.14154
NH_3	KNO_3	5.93678	10.77355	0.16844	9.22645
NO	KNO_3	3.36954	10.52757	0.29678	9.47243
N_2O_3	KNO_2	2.23934	10.35012	0.44656	9.64988
N_2O_5	KNO_3	1.87217	10.27235	0.53414	9.72765
	K_2O	0.87217	9.94060	1.14657	10.05940
Pt	K	0.40086	9.60299	2.49464	10.39701
	KCl	0.76431	9.88327	1.30837	10.11673
SiO_2	K_2SiO_3	2.56783	10.40957	0.38943	9.59043
SO_3	K_2SO_4	2.17664	10.33779	0.45942	9.66221
Praseodymium					
Pr	Pr_2O_3	1.17032	10.06832	0.85447	9.93169
Pr_2O_3	Pr	0.85447	9.93169	1.17032	10.06831
Pr_6O_{11}	Pr	0.82770	9.91787	1.20817	10.08213
Rhenium					
$C_{20}H_{16}N_4.HReO_4$	Re	0.33038	9.51901	3.02682	10.48098
$(C_6H_5)_4AsReO_4$	Re	0.29392	9.46823	3.40229	10.53177
Rhodium					
Na_3RhCl_6	Rh	0.26757	9.42744	3.73734	10.57256
Rh	Na_3RhCl_6	3.73735	10.57256	0.26757	9.42744
$RhCl_3$	Rh	0.49175	9.69174	2.03355	10.30826
Rubidium					
AgCl	Rb	0.59635	9.77550	1.67687	10.22450
	RbCl	0.84368	9.92618	1.18528	10.07382
Cl	Rb	2.41080	10.38216	0.41480	9.61784
	RbCl	3.41071	10.53284	0.29319	9.46715
Rb	AgCl	1.67688	10.22450	0.59635	9.77550
	Cl	0.41480	9.61784	2.41080	10.38216
	RbCl	1.41477	10.15069	0.70683	9.84932
	Rb_2CO_3	1.35106	10.13068	0.74016	9.86933
	Rb_2O	1.09360	10.03886	0.91441	9.96114
	Rb_2PtCl_6	3.38569	10.52965	0.29536	9.47035
	Rb_2SO_4	1.56195	10.19367	0.64023	9.80634
$RbB(C_6H_5)_4$	Rb	0.21119	9.32467	4.73507	10.67533
	Rb_2O	0.23096	9.36354	4.32975	10.63647
RbCl	AgCl	1.18527	10.07382	0.84369	9.92618
	Cl	0.29319	9.46715	3.41076	10.53285
	Rb	0.70683	9.84932	1.41477	10.15069
	Rb_2CO_3	0.95493	9.97997	1.04720	10.02003
	Rb_2O	0.77296	9.08816	1.29373	10.11184
	Rb_2PtCl_6	2.39301	10.37896	0.41788	9.62105
	Rb_2SO_4	1.10399	10.04297	0.90581	9.95704
Rb_2CO_3	Rb	0.74019	9.86934	1.35100	10.12066
	RbCl	1.04720	10.02003	0.45493	9.97997
	$RbHCO_3$	1.26864	10.10335	0.78825	9.89666
	Rb_2PtCl_6	2.50595	10.39897	0.39905	9.60103
	Rb_2SO_4	1.15609	10.06299	0.86498	9.93701
$RbClO_4$	Rb	0.46220	9.66483	2.16357	10.33517
	Rb_2CO_3	0.62446	9.79550	1.60138	10.20449
	RbCl	0.65390	9.81551	1.52929	10.18448
	$RbHCO_3$	0.79218	9.89883	1.26234	10.10118
	Rb_2O	0.50546	9.70369	1.97840	10.29632
	Rb_2SO_4	0.72193	9 85850	1.38518	10.14151
$RbHCO_3$	Rb_2CO_3	0.78831	9.89670	1.26854	10.10330
	Rb_2PtCl_6	1.97546	10.29567	0.50621	9.70433
	Rb_2SO_4	0.91136	9.95969	1.09726	10.04031

Weighed	Sought	Factor	Log of Factor +10	Reciprocal of Factor	Log of Reciprocal of Factor +10
Rubidium (contd.)					
Rb_2O	Rb	0.91441	9.96114	1.09360	10.03887
	RbCl	1.29368	10.11182	0.77299	9.88817
	Rb_2PtCl_6	3.09591	10.49079	0.32301	9.50922
	Rb_2SO_4	1.42827	10.15481	0.70015	9.84519
Rb_2PtCl_6	Rb	0.29537	9.47037	3.38558	10.52963
	RbCl	0.41787	9.62103	2.39309	10.37896
	Rb_2CO_3	0.39905	9.60103	2.50595	10.39898
	$RbHCO_3$	0.50624	9.70436	1.97535	10.29565
	Rb_2O	0.32301	9.50922	3.09588	10.49079
Rb_2SO_4	Rb	0.64022	9.80633	1.56196	10.19367
	RbCl	0.90577	9.95701	1.10403	10.04298
	Rb_2CO_3	0.86498	9.93701	1.15610	10.06300
	$RbHCO_3$	1.09730	10.04033	0.91133	9.95968
	Rb_2O	0.70015	9.84520	1.42827	10.15481
Samarium					
Sm_2O_3	Sm	0.86235	9.93568	1.15962	10.06431
Scandium					
Sc_2O_3	Sc	0.65196	9.81422	1.53384	10.18578
$Sc(C_9H_6NO)_3.(C_9H_6NOH)$	Sc	0.07221	8.85860	13.84850	11.14140
	Sc_2O_3	0.11076	9.04438	9.02853	10.95562
Selenium					
H_2SeO_3	Se	0.61224	9.78692	1.63335	10.21308
H_2SeO_4	Se	0.54466	9.73613	1.83601	10.26389
$PbSeO_4$	Se	0.22550	9.35315	4.43459	10.64685
	SeO_2	0.31689	9.50091	3.15567	10.49909
Se	H_2SeO_3	1.63336	10.21308	0.61223	9.78691
	H_2SeO_4	1.83599	10.26387	0.54467	9.73613
	SeO_2	1.40527	10.14776	0.71161	9.85224
	SeO_3	1.60790	10.20626	0.62193	9.79374
SeO_2	Se	0.71161	9.85224	1.40526	10.14776
SeO_3	Se	0.62193	9.79374	1.60790	10.20626
Silicon					
$BaSiF_6$	SiF_4	0.37249	9.57111	2.68464	10.42888
	SiO_2	0.21503	9.33250	4.65051	10.66750
H_2SiO_3	SiO_2	0.76933	9.88611	1.29983	10.11388
K_2SiF_6	SiF_4	0.47249	9.67439	2.11645	10.32561
	SiO_2	0.27277	9.43580	3.66609	10.56420
Si	SiO_2	2.13932	10.33027	0.46744	9.66973
SiC	C	0.29955	9.47647	3.33834	10.52353
	Si	0.70045	9.84538	1.42765	10.15462
SiF_4	$BaSiF_6$	0.26847	9.42890	3.72481	10.57110
	K_2SiF_6	2.11645	10.32561	0.47249	9.67439
SiF_4	SiO_2	0.57730	9.76140	1.73220	10.23860
SiO_2	$BaSiF_6$	4.65041	10.66749	0.21503	9.33250
	H_2SiO_3	1.29983	10.11388	0.76933	9.88611
	K_2SiF_6	3.66614	10.56421	0.27277	9.43580
	Si	0.46744	9.66973	2.13931	10.33027
	SiF_4	1.73221	10.23861	0.57730	9.76140
	SiO_3	1.26627	10.10252	0.78972	9.89747
	SiO_4	1.53256	10.18542	0.65250	9.81458
	Si_2O	0.60058	9.77857	1.66506	10.22182
	$Si(OH)_4$	1.59965	10.20403	0.62514	9.79597
$SiO_2.12MoO_3$	Si	0.01571	8.19618	63.65372	11.80382
	SiO_2	0.03362	8.52660	29.74420	11.47340
SiO_3	SiO_2	0.78972	9.89747	1.26627	10.10252
SiO_4	SiO_2	0.65250	9.81458	1.53257	10.18542
Si_2O	SiO_2	1.66505	10.22142	0.60058	9.77857
$Si(OH)_4$	SiO_2	0.62514	9.79598	1.59964	10.20402
Silver					
Ag	AgBr	1.74079	10.24075	0.57445	9.75925
	AgCl	1.32866	10.12341	0.75264	9.87659
	AgCN	1.24120	10.09384	0.80567	9.90616
	AgI	2.17645	10.33775	0.45946	9.66225
	$AgNO_3$	1.57481	10.19723	0.63500	9.80277
	Ag_2O	1.07416	10.03107	0.93096	9.96893
	Ag_3PO_4	1.29347	10.11176	0.77311	9.88824

Weighed	Sought	Factor	Log of Factor +10	Reciprocal of Factor	Log of Reciprocal of Factor +10
Silver (contd.)					
Ag	$Ag_4P_2O_7$	1.40313	10.14710	0.71269	9.85290
	Br	0.74079	9.86970	1.34991	10.13030
	Cl	0.32866	9.51675	3.04266	10.48325
	I	1.17646	10.07058	0.85001	9.92942
AgBr	Ag	0.57445	9.75925	1.74080	10.24075
	Br	0.42555	9.62895	2.34990	10.37105
AgCl	Ag	0.75264	9.87659	1.32866	10.12342
	$AgNO_3$	1.18526	10.07382	0.84370	9.92619
	Ag_2O	0.80845	9.90765	1.23693	10.09235
	Br	0.55754	9.74628	1.79359	10.25372
	Cl	0.24736	9.39333	4.04269	10.60667
AgCN	Ag	0.80567	9.90616	1.24120	10.09384
Ag_2CrO_4	Ag	0.65034	9.81314	1.53766	10.18686
AgI	Ag	0.45946	9.66225	2.17647	10.33775
	I	0.54054	9.73283	1.85000	10.26717
$AgNO_3$	Ag	0.63499	9.80277	1.57483	10.19723
	AgCl	0.84370	9.92619	1.18526	10.07381
Ag_2O	Ag	0.93096	9.96893	1.07416	10.03107
	AgCl	1.23693	10.09235	0.80845	9.90765
Ag_3PO_4	Ag	0.77311	9.88824	1.29348	10.11176
$Ag_4P_2O_7$	Ag	0.71269	9.85290	1.40313	10.14710
Br	Ag	1.34991	10.13030	0.74079	9.86970
	AgBr	2.34991	10.37105	0.42555	9.62895
	AgCl	1.79358	10.25372	0.55754	9.74628
Cl	Ag	3.04261	10.48325	0.32867	9.51676
	AgCl	4.04262	10.60666	0.24736	9.39333
I	Ag	0.85000	9.92942	1.17647	10.07058
	AgI	1.85000	10.26717	0.54054	9.73283
Sodium					
Ag	NaBr	0.95392	9.97951	1.04831	10.02049
	NaCl	0.54179	9.73383	1.84573	10.26617
	NaI	1.38958	10.14288	0.71964	9.85712
AgBr	NaBr	0.54798	9.73876	1.82488	10.26123
AgCl	NaCl	0.40777	9.61042	2.45236	10.38959
	$NaClO_3$	0.74267	9.87080	1.34649	10.21921
	$NaClO_4$	0.85429	9.93161	1.17056	10.06839
AgI	NaI	0.63846	9.80513	1.56627	10.19487
$BaSO_4$	$NaHSO_4$	0.51439	9.71129	1.94405	10.28871
	$NaHSO_4.H_2O$	0.59157	9.77201	1.69042	10.22799
	Na_2S	0.33438	9.52424	2.99061	10.47577
	Na_2SO_3	0.54002	9.73241	1.85178	10.26759
	$Na_2SO_3.7H_2O$	1.08032	10.03355	0.92565	9.96645
	Na_2SO_4	0.60857	9.78431	1.64320	10.21569
	$Na_2SO_4.10H_2O$	1.38043	10.14001	0.72441	9.85998
B_2O_3	$Na_2B_4O_7$	1.44511	10.15990	0.69199	9.84010
	$Na_2B_4O_7.10H_2O$	2.73895	10.43758	0.36510	9.56241
Br	Na	0.28770	9.45894	3.47584	10.54106
	NaBr	1.28770	10.10981	0.77658	9.89019
	Na_2O	0.38781	9.58862	2.57858	10.41138
$CaCl_2$	NaCl	1.05321	10.02252	0.94948	9.97749
$CaCO_3$	Na_2CO_3	1.05894	10.02488	0.94434	9.97513
CaF_2	NaF	1.07551	10.03161	0.92979	9.96838
CaO	Na_2CO_3	1.88996	10.27645	0.52911	9.72355
$CaSO_4$	Na_2CO_3	0.77853	9.89128	1.28447	10.10873
Cl	Na	0.64846	9.81188	1.54212	10.18812
	NaCl	1.64846	10.21708	0.60663	9.78292
	Na_2O	0.87410	9.94156	1.14403	10.05844
CO_2	Na_2CO_3	2.40829	10.38171	0.41523	9.61829
	Na_2O	1.40829	10.14869	0.71008	9.85131
H_3BO_3	$Na_2B_4O_7$	0.81356	9.91039	1.22917	10.08961
	$Na_2B_4O_7.10H_2O$	1.54195	10.18807	0.64853	9.81193
I	Na	0.18116	9.25806	5.51998	10.74194
	NaI	1.18115	10.07231	0.84663	9.92769
	Na_2O	0.24419	9.38773	4.09517	10.61227
KBF_4	$Na_2B_4O_7$	0.39954	9.60156	2.50288	10.39844
	$Na_2B_4O_7.10H_2O$	0.75725	9.87924	1.32057	10.12076
$Mg_2As_2O_7$	Na_2HAsO_3	1.09454	10.03923	0.91363	9.96077
	Na_2HAsO_4	1.19761	10.07832	0.83500	9.92169

Weighed	Sought	Factor	Log of Factor +10	Reciprocal of Factor	Log of Reciprocal of Factor +10
Sodium (contd.)					
$MgCl_2$	NaCl	1.22756	10.08904	0.81462	9.91096
$Mg_2P_2O_7$	Na_2HPO_4	1.27565	10.10573	0.78391	9.89427
	$Na_2HPO_4.12H_2O$	3.21828	10.50763	0.31072	9.49237
	$NaNH_4HPO_4.4H_2O$	1.87870	10.27386	0.53228	9.72614
	Na_3PO_4	1.47318	10.16825	0.67880	9.83174
	$Na_4P_2O_7.10H_2O$	2.00414	10.30193	0.49897	9.69807
N	$NaNO_3$	6.06815	10.78306	0.16479	9.21693
Na	Br	3.47585	10.54106	0.28770	9.45894
	Cl	1.54212	10.18811	0.64846	9.81188
	I	5.52003	10.74194	0.18116	9.25806
	NaBr	4.47585	10.65088	0.22342	9.34912
	NaCl	2.54213	10.40520	0.39337	9.59480
	Na_2CO_3	2.30513	10.36269	0.43382	9.63731
	$NaHCO_3$	3.65410	10.56278	0.27367	9.43723
	NaI	6.52003	10.81425	0.15337	9.18574
	Na_2O	1.34797	10.12968	0.74186	9.87032
	Na_2SO_4	3.08921	10.48985	0.32371	9.51016
$Na_2B_4O_7$	B_2O_3	0.69198	9.84009	1.44513	10.15991
	H_3BO_3	1.22916	10.08961	0.81356	9.91039
	KBF_4	2.50287	10.39844	0.39954	9.60156
$Na_2B_4O_7.10H_2O$	B_2O_3	0.36510	9.56241	2.73898	10.43759
	H_3BO_3	0.64853	9.81193	1.54195	10.18807
	KBF_4	1.32057	10.12076	0.75725	9.87924
NaBr	Ag	1.04831	10.02048	0.95392	9.97951
	AgBr	1.82489	10.26123	0.54798	9.73876
	Br	0.77658	9.89019	1.28770	10.10981
	Na	0.22342	9.34912	4.47588	10.65088
	Na_2O	0.30116	9.47880	3.32049	10.52120
$NaC_2H_3O_2.Mg(C_2H_3O_2)_2.$ $3UO_2(C_2H_3O_2)_2.6\frac{1}{2}H_2O$	Na	0.01527	8.18375	65.50075	11.81625
	NaBr	0.06833	8.83464	14.63400	11.16536
	Na_2CO_3	0.03519	8.54646	28.41474	11.45354
	NaCl	0.03881	8.58895	25.76589	11.41104
	NaF	0.02788	8.44536	35.86286	11.55465
	$NaHCO_3$	0.05579	8.74654	17.92500	11.25346
	NaI	0.09954	8.99801	10.04590	11.00199
	$NaNO_3$	0.05644	8.75162	17.71667	11.24837
	Na_2O	0.02058	8.31345	48.59086	11.68655
	NaOH	0.02656	8.42426	37.64746	11.57574
	Na_2SO_4	0.04716	8.67361	21.20261	11.32638
$NaC_2H_3O_2.Zn(C_2H_3O_2)_2.$ $3UO_2(C_2H_3O_2)_2.6\frac{1}{2}H_2O$	Na	0.01495	8.17461	66.89410	11.85239
	NaBr	0.06691	8.82549	14.94545	11.17450
	Na_2CO_3	0.03446	8.53730	29.01999	11.46270
	NaCl	0.03800	8.57981	26.31440	11.42019
	NaF	0.02733	8.43621	36.62601	11.56379
	$NaHCO_3$	0.05466	8.73739	18.30663	11.26261
	NaI	0.09747	8.98886	10.25977	11.01114
	$NaNO_3$	0.05527	8.74247	18.09398	11.25753
	NaOH	0.02601	8.41511	38.44970	11.58489
	Na_2O	0.02015	8.30430	49.62532	11.69570
	Na_2SO_4	0.04618	8.66446	21.65392	11.33554
Na_2CO_3	$CaCO_3$	0.94434	9.97513	1.05894	10.02488
	CaO	0.52911	9.72355	1.88997	10.27645
	$CaSO_4$	1.28447	10.10873	0.77853	9.89128
	CO_2	0.41523	9.61829	2.40830	10.38171
	Na	0.43381	9.63730	2.30516	10.36270
	NaCl	1.10281	10.04250	0.90677	9.95750
	$NaHCO_3$	1.58520	10.20008	0.63084	9.79992
	NaOH	0.75474	9.87780	1.32496	10.12220
	Na_2O	0.58477	9.76699	1.71007	10.23301
	Na_2SO_4	1.34015	10.12715	0.74619	9.87285
$Na_2CO_3.10H_2O$	Na_2SO_4	0.49639	9.69582	2.01455	10.30417
NaCl	Ag	1.84573	10.26617	0.54179	9.73383
	AgCl	2.45236	10.38958	0.40777	9.61042
	Cl	0.60663	9.78292	1.64845	10.21707
	Na	0.39337	9.59480	2.54214	10.40520
	$NaClO_3$	1.82129	10.26038	0.54906	9.73962

Weighed	Sought	Factor	Log of Factor +10	Reciprocal of Factor	Log of Reciprocal of Factor +10
Sodium (contd.)					
	$NaClO_4$	2.09503	10.32118	0.47732	9.67881
	Na_2CO_3	0.90677	9.95750	1.10282	10.04251
	$NaHCO_3$	1.43742	10.15759	0.69569	9.84242
	Na_2HPO_4	1.21451	10.08440	0.82338	9.91560
	Na_2O	0.53025	9.72448	1.88590	10.27552
NaCl	Na_2SO_4	1.21521	10.08465	0.82290	9.91535
$NaClO_3$	AgCl	1.34650	10.12921	0.74267	9.87080
	NaCl	0.54906	9.73961	1.82129	10.26038
$NaClO_4$	AgCl	1.17055	10.06839	0.85430	9.93161
	NaCl	0.47732	9.67881	2.09503	10.32119
NaF	CaF_2	0.92979	9.96838	1.07551	10.03161
Na_2HAsO_3	$Mg_2As_2O_7$	0.91362	9.96077	1.09455	10.03923
Na_2HAsO_4	$Mg_2As_2O_7$	0.83499	9.92168	1.19762	10.07832
$NaHCO_3$	Na	0.27366	9.43721	3.65417	10.56279
	NaCl	0.69569	9.84242	1.43742	10.15759
	Na_2CO_3	0.63083	9.79991	1.58521	10.20009
	Na_2O	0.36889	9.56690	2.71084	10.43310
Na_2HPO_4	$Mg_2P_2O_7$	0.78392	9.89427	1.27564	10.10573
	NaCl	0.82338	9.91560	1.21451	10.08440
	Na_2O	0.43660	9.64008	2.29043	10.35992
	$Na_4P_2O_7$	0.93654	9.97153	1.06776	10.02847
	P_2O_5	0.49994	9.69892	2.00024	10.30108
$Na_2HPO_4.12H_2O$	$Mg_2P_2O_7$	0.31072	9.49237	3.21833	10.50763
	$Na_4P_2O_7$	0.37122	9.56963	2.69382	10.43037
	P_2O_5	0.19816	9.29702	5.04643	10.70298
$NaHSO_3$	SO_2	0.61564	9.78933	1.62433	10.21076
$NaHSO_4$	$BaSO_4$	1.94404	10.28871	0.51439	9.71129
$NaHSO_4.H_2O$	$BaSO_4$	1.69039	10.22799	0.59158	9.77201
NaI	Ag	0.71964	9.85712	1.38958	10.14288
	AgI	1.56626	10.19486	0.63846	9.80513
	I	0.84662	9.92769	1.18117	10.07231
	Na	0.15337	9.18574	6.52018	10.81426
	Na_2O	0.20674	9.31542	4.83699	10.68458
$NaNH_4HPO_4.4H_2O$	$Mg_2P_2O_7$	0.53228	9.72614	1.87871	10.27386
	NH_3	0.08146	8.91094	12.27596	11.08905
	P_2O_5	0.33946	9.53079	2.94586	10.46921
$NaNO_3$	N	0.16479	9.21693	6.06833	10.78307
	Na_2O	0.36461	9.56183	2.74266	10.43818
	NH_3	0.20037	9.30183	4.99077	10.69817
	NO	0.35303	9.54781	2.83262	10.45219
	N_2O_5	0.63539	9.80304	1.57384	10.19696
Na_2O	Br	2.57854	10.41137	0.38782	9.58863
	Cl	1.14401	10.05843	0.87412	9.94157
	CO_2	0.71008	9.85131	1.40829	10.14869
	I	4.09501	10.61225	0.24420	9.38775
	Na	0.74185	9.87032	1.34798	10.12968
	NaBr	3.32039	10.52119	0.30117	9.47881
	NaCl	1.88587	10.27551	0.53026	9.72449
	Na_2CO_3	1.71008	10.23302	0.58477	9.76699
	$NaHCO_3$	2.71078	10.43310	0.36890	9.56691
	Na_2HPO_4	2.29044	10.35992	0.43660	9.64008
	NaI	4.83686	10.68457	0.20675	9.31545
	$NaNO_3$	2.74265	10.43817	0.36461	9.56183
	NaOH	1.29065	10.11081	0.77480	9.88919
	Na_2SO_4	2.29176	10.36017	0.43635	9.63984
	N_2O_5	1.74269	10.24122	0.57383	9.75878
	SO_3	1.29176	10.11118	0.77414	9.88882
NaOH	Na_2CO_3	1.32495	10.12220	0.75475	9.87780
	Na_2O	0.77479	9.88918	1.29067	10.11082
$Na_4P_2O_7$	Na_2HPO_4	1.06775	10.02847	0.93655	9.97153
	$Na_2HPO_4.12H_2O$	0.26938	9.43037	3.71223	10.56963
$Na_4P_2O_7.10H_2O$	$Mg_2P_2O_7$	0.49897	9.69807	2.00413	10.30193
Na_2S	$BaSO_4$	2.99062	10.47576	0.33438	9.52424
Na_2SO_3	$BaSO_4$	1.85176	10.26758	0.54003	9.73242
	SO_2	0.50827	9.70609	1.96746	10.29391
$Na_2SO_3.7H_2O$	$BaSO_4$	0.92564	9.96644	1.08033	10.03356
	SO_2	0.25407	9.40495	3.93592	10.59505
Na_2SO_4	$BaSO_4$	1.64319	10.21569	0.60857	9.78431
	Na	0.32370	9.51014	3.08928	10.48986

Weighed	Sought	Factor	Log of Factor +10	Reciprocal of Factor	Log of Reciprocal of Factor +10
Sodium (contd.)					
	$NaCl$	0.82289	9.91534	1.21523	10.08466
	Na_2CO_3	0.74619	9.87285	1.34014	10.21715
	$Na_2CO_3.10H_2O$	2.01451	10.30417	0.49640	9.69583
	Na_2O	0.43635	9.63984	2.29174	10.36016
	SO_3	0.56365	9.75101	1.77415	10.24899
$Na_2SO_4.10H_2O$	$BaSO_4$	0.72441	9.85998	1.38043	10.14001
$Na_2U_2O_7.2ZnU_2O_7$	Na	0.02369	8.37457	42.21190	11.62544
	Na_2O	0.03193	8.50420	31.31851	11.49579
NH_3	$NaNH_4HPO_4.4H_2O$	12.27607	11.08906	0.08146	8.91094
	$NaNO_3$	4.99070	10.69816	0.20037	9.30183
NO	$NaNO_3$	2.83258	10.45218	0.35304	9.54782
N_2O_5	$NaNO_3$	1.57382	10.19695	0.63540	9.80305
	Na_2O	0.57382	9.75878	1.74271	10.24123
P_2O_5	Na_2HPO_4	2.00020	10.30107	0.49995	9.69893
	$Na_2HPO_4.12H_2O$	5.04623	10.70297	0.19817	9.29704
	$NaNH_4HPO_4.4H_2O$	2.94578	10.46920	0.33947	9.53080
SO_2	$NaHSO_3$	1.62442	10.21070	0.61560	9.78930
	Na_2SO_3	1.96756	10.29393	0.50824	9.70607
SO_3	Na_2O	0.77414	9.88882	1.29176	10.11118
	Na_2SO_4	1.77414	10.24899	0.56365	9.75101
Strontium					
CO_2	$SrCO_3$	3.35446	10.52562	0.29811	9.47438
SO_3	SrO	1.29424	10.11202	0.77265	9.88798
	$SrSO_4$	2.29422	10.36063	0.43588	9.63937
Sr	$SrCO_3$	1.68489	10.22657	0.59351	9.77343
	$Sr(NO_3)_2$	2.41532	10.38298	0.41402	9.61702
	SrO	1.18261	10.07284	0.84559	9.92716
	$SrSO_4$	2.09633	10.32146	0.47702	9.67854
$SrCl_2$	$SrCO_3$	0.93124	9.96906	1.07384	10.03094
	SrO	0.65363	9.81533	1.52992	10.18467
	$SrSO_4$	1.15865	10.06395	0.86307	9.93605
$SrCO_3$	CO_2	0.29811	9.47438	3.35447	10.52562
	Sr	0.59351	9.77343	1.68489	10.22657
	$SrCl_2$	1.07383	10.03095	0.93125	9.96907
	$Sr(HCO_3)_2$	1.42010	10.15232	0.70418	9.84768
	$Sr(NO_3)_2$	1.43352	10.15640	0.69758	9.84359
	SrO	0.70189	9.84627	1.42472	10.15373
	$SrSO_4$	1.24419	10.09489	0.80374	9.90511
SrC_2O_4	Sr	0.49886	9.69798	2.00457	10.30202
	SrO	0.58996	9.77082	1.69503	10.22918
$Sr(HCO_3)_2$	$SrCO_3$	0.70417	9.84768	1.42011	10.15232
	SrO	0.49425	9.69395	2.02327	10.30602
$Sr(IO_3)_2$	Sr	0.20031	9.30170	4.99226	10.69829
	SrO	0.23688	9.37453	4.22155	10.69829
$Sr(NO_3)_2$	Sr	0.41402	9.61702	2.41434	10.38298
	$SrCO_3$	0.69759	9.84360	1.43351	10.15640
	SrO	0.48963	9.68987	2.04236	10.31014
	$SrSO_4$	0.86793	9.93849	1.15217	10.06151
SrO	SO_3	0.77265	9.88798	1.29425	10.11202
	Sr	0.84559	9.92716	1.18261	10.07284
	$SrCl_2$	1.52992	10.18467	0.65363	9.81533
	$SrCO_3$	1.42472	10.15373	0.70189	9.84627
	$Sr(HCO_3)_2$	2.02326	10.30605	0.49425	9.69395
	$Sr(NO_3)_2$	2.04237	10.31014	0.48963	9.68987
	$SrSO_4$	1.77263	10.24862	0.56413	9.75138
$SrSO_4$	SO_3	0.43588	9.63937	2.29421	10.36063
	Sr	0.47703	9.67855	2.09630	10.32145
	$SrCl_2$	0.86308	9.93605	1.15864	10.06395
	$SrCO_3$	0.80373	9.90511	1.24420	10.09489
	$Sr(NO_3)_2$	1.15217	10.06151	0.86793	9.93848
	SrO	0.56413	9.75138	1.77264	10.24862
Sulfur					
As_2S_3	H_2S	0.41555	9.61862	2.40645	10.38138
	S	0.39097	9.59214	2.55774	10.40786
$BaSO_4$	FeS_2	0.25702	9.40997	3.89077	10.59003
	H_2S	0.14602	9.16441	6.84838	10.83559
	H_2SO_3	0.35166	9.54612	2.84366	10.45388
	H_2SO_4	0.42021	9.62347	2.37976	10.37653

Weighed	Sought	Factor	Log of Factor +10	Reciprocal of Factor	Log of Reciprocal of Factor +10
Sulfur (contd.)					
	S	0.13738	9.13792	7.27908	10.86208
	SO_2	0.27448	9.43851	3.64325	10.56149
	SO_3	0.34302	9.53532	2.91528	10.46468
	SO_4	0.41158	9.61445	2.42966	10.38555
$C_{12}H_{12}N_2.H_2SO_4$	S	0.11357	9.05526	8.80514	10.94474
	SO_4	0.34026	9.53181	2.93893	10.46819
CdS	H_2S	0.23591	9.37275	4.23890	10.62725
	S	0.22196	9.34627	4.50532	10.65372
FeS_2	$BaSO_4$	3.89081	10.59004	0.25702	9.40997
H_2S	As_2S_3	2.40646	10.38138	0.41555	9.61862
	$BaSO_4$	6.84859	10.83560	0.14602	9.16441
	CdS	4.23885	10.62725	0.23591	9.37275
	SO_3	2.34924	10.37093	0.42567	9.62907
H_2SO_3	$BaSO_4$	2.84363	10.45287	0.35166	9.54612
H_2SO_4	$BaSO_4$	2.37974	10.37653	0.42021	9.62347
	$(NH_4)_2SO_4$	1.34728	10.12946	0.74224	9.87054
	SO_3	0.81631	9.91186	1.22502	10.08814
$(NH_4)_2SO_4$	H_2SO_4	0.74223	9.87054	1.34729	10.12946
	SO_3	0.60589	9.78239	1.65046	10.21760
S	As_2S_3	2.55775	10.40786	0.39097	9.59214
	$BaSO_4$	7.27919	10.86208	0.13738	9.13792
	CdS	4.50536	10.65373	0.22196	9.34627
SO_2	$BaSO_4$	3.64329	10.56149	0.27448	9.43851
SO_3	$BaSO_4$	2.91524	10.46468	0.34302	9.53532
	H_2S	0.42567	9.62907	2.34924	10.37093
	$(NH_4)_2SO_4$	1.65047	10.21761	0.60589	9.78239
SO_4	$BaSO_4$	2.42968	10.38555	0.41158	9.61445
Tantalum					
Ta	$TaCl_5$	1.97965	10.29659	0.50514	9.70341
	Ta_2O_5	1.22105	10.08674	0.81897	9.91327
TaC	C	0.06225	8.79413	16.06451	11.20587
	Ta	0.93775	9.97209	1.06638	10.02791
$TaCl_5$	Ta	0.50514	9.70341	1.97965	10.29659
	Ta_2O_5	0.61680	9.79041	1.62127	10.20985
Ta_2O_4	Ta_2O_5	1.03757	10.01602	0.96379	9.98398
Ta_2O_5	Ta	0.81897	9.91327	1.22105	10.08673
	$TaCl_5$	1.62126	10.20985	0.61680	9.79014
	Ta_2O_4	0.96379	9.98398	1.03757	10.01602
Tellurium					
H_2TeO_4	Te	0.65906	9.81893	1.51731	10.18108
$H_2TeO_4.2H_2O$	Te	0.55565	9.74480	1.79969	10.25520
Te	H_2TeO_4	1.51732	10.18108	0.65906	9.81893
	$H_2TeO_4.2H_2O$	1.79969	10.25520	0.55565	9.74480
	TeO_2	1.25078	10.09718	0.79950	9.90282
	TeO_3	1.37618	10.13868	0.72665	9.86133
	$(TeO_2)_2SO_3$	1.56450	10.19438	0.63918	9.80562
TeO_2	Te	0.79950	9.90282	1.25078	10.09718
TeO_3	Te	0.72665	9.86133	1.37618	10.13868
$(TeO_2)_2SO_3$	Te	0.63918	9.80562	1.56450	10.19438
Terbium					
Tb_4O_7	Tb	0.85021	9.92953	1.17618	10.07048
Thallium					
$(C_6H_5)_4AsTlCl_4$	Tl	0.28014	9.44738	3.56964	10.55263
Tl	TlCl	1.17346	10.06947	0.85218	9.93053
	Tl_2CO_3	1.14682	10.05949	0.87198	9.94051
	Tl_2CrO_4	1.28377	10.10849	0.77896	9.89151
	$TlHSO_4$	1.47497	10.16878	0.67798	9.83122
	TlI	1.62093	10.20976	0.61693	9.79024
	$TlNO_3$	1.30337	10.11507	0.76724	9.88493
	Tl_2O	1.03914	10.01668	0.96233	9.98332
	Tl_2PtCl_6	1.99772	10.30055	0.50057	9.69946
	Tl_2SO_4	1.23501	10.09167	0.80971	9.90833
TlCl	Tl	0.85218	9.93053	1.17346	10.06947
	Tl_2PtCl_6	1.70240	10.23106	0.58741	9.76894
Tl_2CO_3	Tl	0.87198	9.94051	1.14682	10.05951
	Tl_2PtCl_6	1.74197	10.24104	0.57406	9.75896
Tl_2CrO_4	Tl	0.77895	9.89151	1.28378	10.10850
$TlHSO_4$	Tl	0.67798	9.83122	1.47497	10.16878

Weighed	Sought	Factor	Log of Factor +10	Reciprocal of Factor	Log of Reciprocal of Factor +10
Thallium (contd.)					
TlI	Tl	0.61693	9.79024	1.62093	10.20976
	Tl_2PtCl_6	1.23244	10.09076	0.81140	9.90924
$TlNO_3$	Tl	0.76724	9.88493	1.30338	10.11507
	Tl_2PtCl_6	1.53272	10.18546	0.65243	9.81453
Tl_2O	Tl	0.96223	9.98332	1.03914	10.01668
	Tl_2PtCl_6	1.92247	10.28386	0.52016	9.71614
Tl_2PtCl_6	Tl	0.50057	9.69946	1.99772	10.30054
	TlCl	0.58741	9.76894	1.70239	10.23106
	Tl_2CO_3	0.57406	9.75896	1.74198	10.24105
	TlI	0.81140	9.90924	1.23244	10.09076
	$TlNO_3$	0.65244	9.81454	1.53271	10.18546
	Tl_2O	0.52016	9.71614	1.92249	10.28386
	Tl_2SO_4	0.61821	9.79114	1.61757	10.20886
Tl_2SO_4	Tl	0.80971	9.90833	1.23501	10.09167
	Tl_2PtCl_6	1.61757	10.20886	0.61821	9.79114
Thorium					
Th	ThO_2	1.13790	10.05610	0.87881	9.94390
$Th(C_9H_6NO)_4.(C_9H_6NOH)$	Th	0.24327	9.38609	4.11066	10.61391
	ThO_2	0.27682	9.44220	3.61246	10.55781
$ThCl_4$	ThO_2	0.70626	9.84896	1.41590	10.15103
$Th(NO_3)_4.6H_2O$	ThO_2	0.44898	9.65223	2.22727	10.34777
ThO_2	Th	0.87881	9.94390	1.13790	10.05610
	$ThCl_4$	1.41590	10.15103	0.70626	9.84896
	$Th(NO_3)_4.6H_2O$	2.22729	10.34778	0.44898	9.65223
Thulium					
Tm_2O_3	Tm	0.87561	9.94231	1.14206	10.05769
Tin					
Sn	$SnCl_2$	1.59744	10.20342	0.62600	9.79657
	$SnCl_2.2H_2O$	1.90100	10.27898	0.52604	9.72102
	$SnCl_4$	2.19479	10.34140	0.45562	9.65860
	$SnCl_4.(NH_4Cl)_2$	3.09622	10.49083	0.32297	9.50916
	SnO	1.13480	10.05492	0.88121	9.94508
	SnO_2	1.26961	10.10367	0.78764	9.89633
$SnCl_2$	Sn	0.62600	9.79657	1.59744	10.20342
	SnO_2	0.79478	9.90025	1.25821	10.09975
$SnCl_2.2H_2O$	Sn	0.52604	9.72102	1.90100	10.27898
	SnO_2	0.66786	9.82469	1.49732	10.17531
$SnCl_4$	Sn	0.45562	9.65860	2.19481	10.34139
	SnO_2	0.57846	9.76227	1.72873	10.23772
$SnCl_4.(NH_4Cl)_2$	Sn	0.32297	9.50916	3.09626	10.49084
	SnO_2	0.41005	9.61284	2.43873	10.38716
SnO	Sn	0.88121	9.94508	1.13480	10.05491
	SnO_2	1.11879	10.04875	0.89382	9.95125
SnO_2	Sn	0.78764	9.89633	1.26962	10.10367
	$SnCl_2$	1.25821	10.09976	0.79474	9.90023
	$SnCl_2.2H_2O$	1.49731	10.17531	0.66786	9.82469
	$SnCl_4$	1.72871	10.23772	0.57847	9.76228
	$SnCl_4.(NH_4Cl)_2$	2.43872	10.38716	0.41005	9.61284
	SnO	0.89382	9.95125	1.11879	10.04875
Titanium					
K_2TiF_6	F	0.47472	9.67644	2.10650	10.32356
	K	0.32573	9.51286	3.07003	10.48714
	Ti	0.19951	9.29996	5.01228	10.70004
	TiO_2	0.33279	9.52217	3.00490	10.47783
Ti	K_2TiF_6	5.01232	10.70004	0.19951	9.29996
	TiC	1.25073	10.09717	0.79953	9.90283
	TiO_2	1.66806	10.22222	0.59950	9.77779
TiC	C	0.20049	9.30210	4.98778	10.69791
	Ti	0.79953	9.90283	1.25073	10.09717
$TiO(C_9H_4NOCl_2)_2$	Ti	0.09776	8.99016	10.22913	11.00984
	TiO_2	0.16306	9.21235	6.13271	10.78765
$TiO(C_9H_6ON)_2$	Ti	0.13600	9.13354	7.35294	10.86646
	TiO_2	0.22685	9.35574	4.40820	10.64426
TiO_2	K_2TiF_6	3.00488	10.47782	0.33279	9.52217
	Ti	0.59950	9.77779	1.66806	10.22221
	TiC	0.74969	9.87488	1.33388	10.12512

Weighed	Sought	Factor	Log of Factor +10	Reciprocal of Factor	Log of Reciprocal of Factor +10
Tungsten					
$FeWO_4$	W	0.60539	9.78203	1.65183	10.21797
	WO_3	0.76344	9.88277	1.30986	10.11722
$MgWO_4$	W	0.67552	9.82964	1.48034	10.17036
	WO_3	0.85189	9.93038	1.17386	10.06962
$MnWO_4$	W	0.60719	0.78332	1.64693	10.21668
	WO_3	0.76571	9.88406	1.30598	10.11593
$PbWO_4$	W	0.40403	9.60641	2.47506	10.39359
	WO_3	0.50952	9.70716	1.96263	10.29284
W	W_2C	1.03266	10.01396	0.96837	9.98604
	WC	1.06532	10.02748	0.93869	9.97252
	WO_2	1.17405	10.06969	0.85175	9.93031
	WO_3	1.26108	10.10075	0.79297	9.89925
W_2C	C	0.03163	8.50010	31.61555	11.49990
	W	0.96837	9.98604	1.03266	10.01396
WC	C	0.06133	8.78767	16.30523	11.21233
	W	0.93868	9.97252	1.06533	10.02749
WO_2	W	0.85175	9.93031	1.17405	10.06969
$WO_2(C_9H_6ON)_2$	W	0.36467	9.56190	2.74221	10.43810
	WO_3	0.45987	9.66264	2.17453	10.33736
WO_3	W	0.79297	9.89926	1.26108	10.10074
Uranium					
U	UO_2	1.13444	10.05478	0.88149	9.94522
	U_3O_8	1.17925	10.07160	0.84800	9.92840
	$U_2P_2O_{11}$	1.49981	10.17604	0.66675	9.82396
UO_2	U	0.88149	9.94522	1.13444	10.05478
	U_3O_8	1.03950	10.01682	0.96200	9.98318
	$U_2P_2O_{11}$	1.32268	10.12145	0.75604	9.87854
$UO_2(C_9H_6ON)_2.(C_9H_7ON)$	U	0.33835	9.52937	2.95552	10.47063
	UO_2	0.38384	9.58415	2.60525	10.41585
U_3O_8	U	0.84799	9.92839	1.17926	10.07161
	UO_2	0.96199	9.98317	1.03951	10.01683
	$UO_2(NO_3)_2.6H_2O$	1.78864	10.25252	0.55908	9.74747
$UO_2(NO_3)_2.6H_2O$	U_3O_8	0.55909	9.74748	1.78862	10.25252
$U_2P_2O_{11}$	U	0.66675	9.82396	1.49981	10.17603
	UO_2	0.75639	9.87875	1.32207	10.12125
Vanadium					
V	VC	1.23578	10.09194	0.80921	9.90806
	V_2O_5	1.78518	10.25168	0.56017	9.74832
VC	C	0.19080	9.28058	5.24109	10.71942
	V	0.80921	9.90806	1.23577	10.09194
VO_4	V_2O_5	0.79120	9.89829	1.26390	10.10171
V_2O_5	V	0.56017	9.74832	1.78517	10.25168
	VC	0.69225	9.84026	1.44456	10.15974
	VO_4	1.26390	10.10171	0.79120	9.89829
Ytterbium					
Yb	YbO_3	1.13870	10.05641	0.87819	9.94359
Yb_2O_3	Yb	0.87820	9.94359	1.13869	10.05640
Yttrium					
Y	Y_2O_3	1.26994	10.10378	0.78744	9.89622
Y_2O_3	Y	0.78744	9.89622	1.26994	10.10378
Zinc					
$BaSO_4$	ZnS	0.41744	9.62059	2.39555	10.37941
	$ZnSO_4.7H_2O$	1.23196	10.09059	0.81171	9.90940
Zn	$ZnNH_4PO_4$	2.72877	10.43596	0.36647	9.56404
	ZnO	1.24476	10.09508	0.80337	9.90492
	$Zn_2P_2O_7$	2.33043	10.36744	0.42911	9.63257
	ZnS	1.49044	10.17332	0.67094	9.82668
$Zn(C_9H_6NO)_2$	Zn	0.18483	9.26677	5.41038	10.73323
	ZnO	0.23007	9.36186	4.34650	10.63814
$ZnCl_2$	ZnO	0.59708	9.77603	1.67482	10.22397
$ZnCO_3$	ZnO	0.64899	9.81224	1.54086	10.18777
$(Zn[C_5H_5N]_2).(SCN)_2$	Zn	0.19241	9.28423	5.19724	10.71577
	ZnO	0.23951	9.37932	4.17519	10.62068
$ZnNH_4PO_4$	Zn	0.36646	9.56403	2.72881	10.43597
	ZnO	0.45616	9.65912	2.19221	10.34089
ZnO	Zn	0.80337	9.90492	1.24476	10.09509
	$ZnCl_2$	1.67482	10.22397	0.59708	9.77603

Weighed	Sought	Factor	Log of Factor +10	Reciprocal of Factor	Log of Reciprocal of Factor +10
Zinc (contd.)					
	$ZnCO_3$	1.54086	10.18776	0.64899	9.81224
	$ZnNH_4PO_4$	2.19221	10.34088	0.45616	9.65912
	$Zn_2P_2O_7$	1.87219	10.27235	0.53413	9.72765
	ZnS	1.19737	10.07823	0.83516	9.92177
	$ZnSO_4.7H_2O$	3.53373	10.54823	0.28299	9.45177
$Zn_2P_2O_7$	Zn	0.42911	9.63257	2.33040	10.36743
	ZnO	0.53413	9.72765	1.87220	10.27235
ZnS	$BaSO_4$	2.39556	10.37941	0.41744	9.62059
	Zn	0.67094	9.82668	1.49045	10.17332
	ZnO	0.83516	9.92177	1.19738	10.07822
	$ZnSO_4.7H_2O$	2.95125	10.47001	0.33884	9.52999
$ZnSO_4.7H_2O$	$BaSO_4$	0.81171	9.90940	1.23197	10.09060
	ZnO	0.28299	9.45177	3.53369	10.54823
	ZnS	0.33884	9.52999	2.95125	10.47001
Zirconium					
K_2ZrF_6	Zr	0.32187	9.50768	3.10684	10.49232
	ZrF_4	0.58999	9.77084	1.69494	10.22915
	ZrF_6	0.72407	9.85978	1.38108	10.14022
	ZrO_2	0.43478	9.63827	2.30001	10.36173
Zr	ZrO_2	1.35080	10.13059	0.74030	9.86941
	ZrC	1.13166	10.05372	0.88366	9.94629
$Zr(BrC_8H_6O_3)_4$	Zr	0.09020	8.95521	11.08647	11.04479
	ZrO_2	0.12184	9.08579	8.20749	10.91421
$Zr(C_8H_7O_3)_4$	Zr	0.13110	9.11760	7.62777	10.88240
	ZrO_2	0.17709	9.24819	5.64685	10.75181
ZrC	C	0.11635	9.06577	8.59476	10.93424
	Zr	0.88366	9.94629	1.13166	10.05372
ZrO_2	Zr	0.74030	9.86941	1.35080	10.13059
	ZrC	0.83777	9.92312	1.19365	10.07687
$Zr_2P_2O_7$	Zr	0.34402	9.53658	2.90681	10.46342
	ZrO_2	0.46470	9.66717	2.15193	10.33283

PHYSICAL CONSTANTS OF MINERALS

Compiled by Ralph Kretz

The following table presents data for many of the more common minerals.

In order to avoid duplication and save space, very few cross references are given in the body of the table. If the name sought is not found in the table, consult the **synonym index** given below.

Specific gravities are given at normal atmospheric temperatures, a more precise statement being valueless considering the large variations in natural minerals.

Hardness is given in terms of Mohs' scale. (See under Hardness.)

Indices of refraction for the sodium line, λ = 5893 Å, unless otherwise indicated. Li, λ = 6708 Å. Indices will invariably be given in the order ω, ϵ or α, β, γ. Uniaxial crystals are considered positive if $\epsilon > \omega$, negative if $\omega > \epsilon$. Biaxial crystals are considered positive if β is nearer α in value than it is γ and negative if β is nearer γ than α.

ABBREVIATIONS

Abbreviation	Meaning of abbreviation	Abbreviation	Meaning of abbreviation	Abbreviation	Meaning of abbreviation
bl	blue	grn	green	rhbdr	rhombohedral
blk	black	grnsh	greenish	rhomb	rhombic
blksh	blackish	hex	hexagonal	somet	sometimes
blsh	bluish	iridesc	iridescent	tarn	tarnishes
br	brown	monocl	monoclinic	tetr	tetragonal
brnsh	brownish	oft	often	tricl	triclinic
col	colorless	pa	pale	vlt	violet
cub	cubic	purp	purple	wh	white
dk	dark	(R)	radioactive	yel	yellow
Fe	Fe, ferrous iron	redsh	redish	yelsh	yellowish
Fe^{+3}	Fe, ferric iron				

SYNONYM INDEX

Compound sought	Listed
Acmite	Aegirine
Agate	Quartz (impure)
Aluminum hydroxide	Boehmite, Diaspore, Gibbsite
Amphibole	Actinolite, Anthophyllite, Cummingtonite, Glaucophane, Hornblende, Riebeckite, Tremolite
Antimony oxide	Senarmontite, Valentinite
Antimony sulfide	Stibnite
Arsenic oxide	Arsenolite, Claudetite
Arsenic sulfide	Orpiment, Realgar
Barium carbonate	Witherite
Barium sulfate	Barite
Barytes	Barite
Bauxite	Gibbsite, Boehmite, Diaspore
Brimstone	Sulfur
Bronzite	Orthopyroxene
Cadmium sulfide	Greenockite
Calamine	Hemimorphite
Calcium carbonate	Aragonite, Calcite, Vaterite
Calcium sulfate	Anhydrite, Gypsum
Calcium sulfide	Oldhamite
Carborundum	Moissanite
Chalcedony	Quartz (impure, fibrous)
Chinaclay	Kaolinite
Chloanthite	Skutterodite
Chromespinel	Chromite
Chrysolite	Serpentine
Clinoptolite	Heulandite
Clayminerals	Illite, Kaolinite, Montmorillonite
Clinochlore	Chlorite
Cobaltbloom	Erythrite
Copper chloride	Nantokite
Copper oxide	Cuprite
Copper sulfide	Chalcocite, Covellite, Digenite
Emerald	Beryl
Emery	Mixture of Corundum, Magnetite and other minerals
Epsom salt	Epsomite
Feldspar	Orthoclase, Microcline, Anorthoclase, Albite, Oligoclase, Andesine, Anorthite
Fibrolite	Sillimanite
Flint	Quartz (impure)
Fluorapatite	Apatite
Fluorspar	Fluorite
Garnet	Almandine, Pyrope, Spessartite, Andradite, Grossularite, Uvarovite, Hydrogrossularite
Garnierite	Serpentine (Ni-bearing)
Glauber salt	Mirabilite
Hyacinth	Zircon
Iceland spar	Calcite
Idocrase	Vesuvianite
Iron carbonate	Siderite
Iron hydroxide	Goethite, Lepidocrocite
Iron oxide	Hematite, Magnetite
Iron spinel	Hercynite
Iron sulfide	Marcasite, Pyrite, Pyrrhotite
Lapis lazuli	Lazurite
Lead carbonate	Cerussite
Lead chloride	Cotunnite
Lead chromate	Crocoite
Lead oxide	Litharge, Minium
Lead sulfate	Anglesite
Lead sulfide	Galena
Limonite	Goethite (impure)
Lithiophyllite	Triphylite
Lithium mica	Lepidolite
Lodestone	Magnetite
Magnesium carbonate	Magnesite
Magnesium hydroxide	Brucite
Magnesium oxide	Periclase
Magnesium sulfate	Kieserite
Manganese carbonate	Rhodochrosite
Manganese hydroxide	Pyrochroite
Manganese oxide	Hausmannite, Manganosite, Pyrolusite
Manganese sulfide	Alabandite
Meerschaum	Serpentine
Mica	Muscovite, Paragonite, Phlogopite, Biotite, Lepidolite
Native copper	Copper
Native gold	Gold
Nickel oxide	Bunsenite
Nickel sulfide	Millerite
Orthite	Allanite
Penninite	Chlorite
Peridote	Olivine
Pistacite	Epidote
Pitchblende	Uraninite
Plagioclase	Albite, Oligoclase, Andesine, Anorthite
Potassium chloride	Sylvite
Potassium sulfate	Arcanite
Pyroxene	Diopsite, Angite, Aegirine, Jadeite, Pigeonite, Eustatite, Orthopyroxene
Rocksalt	Halite
Ruby	Corundum
Sapphire	Corundum
Silica	Christobalite, Quartz, Tridymite
Silver chloride	Cerargyrite
Silver iodide	Jodyrite, Miersite
Silver sulfide	Acanthite, Argentite
Smalltite	Skutterotite
Soapstone	Mixture of Talc and other minerals
Sodium chloride	Halite
Sodium sulfate	Thenardite
Strontium carbonate	Strontianite
Strontium sulfate	Celestite
Thorium oxide	Thorianite
Tin oxide	Cassiterite
Titanite	Sphene
Titanium oxide	Anatase, Brookite, Rutile
Uranium oxide	Uraninite
Zeolite	Natrolite, Mesolite, Scolecite, Thomasonite, Harmatome, Eddingtonite, Heulandite, Stilbite, Phillipsite, Chabazite, Gmelinite, Levyn, Laumontite, Mordenite
Zincblende	Sphalerite
Zinc carbonate	Smithsonite
Zinc oxide	Zincite
Zinc spinel	Gahnite
Zinc sulfide	Sphalerite, Wurtzite
Zirconium oxide	Baddeleyite

PHYSICAL CONSTANTS OF MINERALS

Name	Formula	Sp. gr.	Hardness	Crystalline form and color	Index of refraction (Na) η; ω ε α β γ
Acanthite	AgS	7.2–7.3	2–2.5	rhomb.(?), iron-blk.	
Actinolite	$Ca_2((Mg,Fe)_5Si_8O_{22}(OH,F)_2$	3.02–3.44	5–6	monocl., pa. to dk. grn.	1.599–1.688, 1.612–1.697, 1.622–1.70
Aegirine	$NaFe^{+3}Si_2O_6$	3.55–3.60	6	monocl., dk. grn. to grnsh. blk.	1.750–1.776, 1.780–1.820, 1.800–1.83
Åkermanite	$Ca_2MgSi_2O_7$	2.944	5–6	tetr., col., gray-grn., br.	1.632, 1.640
Alabandite	MnS	4.050	3.5–4	cub., iron-blk., tarn., br.	
Albite	$NaAlSi_3O_8$	2.63	6–6.5	tricl., col., wh., somet. yel., pink, grn.	1.527, 1.531, 1.538
Allanite	$(Ca,Mn,Ce,La,Y,Th)_2(Fe,Fe^{+3},Ti)(Al,Fe^{+3})_2$ $Si_3O_{12}(OH)$	3.4–4.2	5–6.5	monocl., pa. br. to blk.	1.690–1.791, 1.700–1.815, 1.706–1.82
Allemontite	$AsSb$	5.8–6.2	3–4	hex., tin-wh. to redsh., gray, tarn. gray–brnsh. blk.	
Almandine	$Fe_3Al_2Si_3O_{12}$	4.318	6–7.5	cub., red, dk. red, blk.	1.830
Altaite	$PbTe$	8.15	3	cub., tin-wh., yelsh., tarn. bronze-yel.	
Aluminite	$Al_2(SO_4)(OH)_4.7H_2O$	1.66–1.82	1–2	monocl.(?), wh.	1.459, 1.464, 1.470
Alunite	$(K,Na)Al_3(SO_4)_2(OH)_6$	2.6–2.9	3.5–4	rhbdr., wh., gray, yel., redsh., br.	1.572, 1.592
Alunogen	$Al_2(SO_4)_3.18H_2O$	1.77	1.5–2	tricl., col., wh., yelsh. wh., redsh. wh.	1.459–1.475, 1.461–1.478, 1.470–1.48
Amblygonite	$(Li,Na)Al(PO_4)(F,OH)$	3.0–3.1	5.5–6	tricl., wh., yelsh. wh., grnsh. wh., blsh. wh., gray	1.591, 1.604, 1.613
Analcite	$NaAlSi_2O_6.H_2O$	2.24–2.29	5.5	cub., wh., pink, gray	1.479–1.493
Anatase	TiO_2	3.90	5.5–6	tetr., br., yelsh. br., redsh. br., bl., blk., grn., gray	2.5612, 2.4880
Andalusite	Al_2OSiO_4	3.13–3.16	6.5–7.5	rhomb., pink, wh., red	1.629–1.640, 1.633–1.644, 1.638–1.65
Andesine	$([NaSi]_{0.7-0.5}[CaAl]_{0.3-0.5})AlSi_2O_8$	2.65–2.68	6–6.5	tricl., wh., gray, grn.	1.544–1.555, 1.548–1.558, 1.551–1.56
Andorite	$PbAgSb_3S_6$	5.33–5.37	3–3.5	rhomb., dk. steel gray, somet. tarn. yel. or iridesc.	
Andradite	$Ca_3Fe_2^{+3}Si_3O_{12}$	3.859	6–7.5	cub., brnsh. red, blk., somet. yel., grn.	1.887
Anglesite	$PbSO_4$	6.37–6.39	2.5–3	rhomb., col., wh., somet. gray, yelsh., grn. tinge	1.8771, 1.8826, 1.8937
Anhydrite	$CaSO_4$	2.98	3.5	rhomb., col., blsh. wh., vlt.	1.5698, 1.5754, 1.6136
Ankerite	$Ca(Fe,Mg,Mn)(CO_3)_2$	2.8–3.1	3.5–4	rhbdr., br., yelsh. br., grnsh. br., pink	1.690–1.750, 1.510–1.548
Anorthite	$CaAl_2Si_2O_8$	2.76	6–6.5	tricl., wh., yel., grn., blk.	1.577, 1.585, 1.590
Anorthoclase	$(Na,K)AlSi_3O_8$	2.56–2.60	6	tricl., col., wh.	1.523, 1.528, 1.529
Anthophyllite	$(Mg,Fe)_7Si_8O_{22}(OH,F)_2$	2.85–3.57	5.5–6	rhomb., wh., gray, grn., br., yelsh. br., dk. br.	1.596–1.694, 1.605–1.710, 1.615–1.72
Antimony	Sb	6.61–6.72	3–3.5	hex., tin-wh.	
Apatite	$Ca_5(PO_4)_3(OH,F,Cl)$	3.1–3.35	5	hex., grn., wh., yel., br. red, bl.	1.629–1.667, 1.624–1.666
Apophyllite	$KFCa_4Si_8O_{20}.8H_2O$	2.33–2.37	4.5–5	tetr., col., wh., pink, pa. yel., pa. grn.	1.534–1.535, 1.535–1.537
Aragonite	$CaCO_3$	2.94–2.95	3.5–4	rhomb., col., wh.	1.530–1.531, 1.680–1.681, 1.685–1.68
Arcanite	K_2SO_4	2.663		rhom., col., wh.	1.4935, 1.4947, 1.4973
Argentite	Ag_2S	7.2–7.4	2–2.5	cub., blksh. lead gray	
Arsenic	As	5.63–5.78	3.5	hex., tin-wh., tarn. dk. gray	
Arsenolite	As_2O_3	3.86–3.88	1.5	cub., wh., somet. blsh., yelsh., redsh. tinge	1.755
Arsenopyrite	$FeAsS$	5.9–6.2	5.5–6	monocl., silver-wh., to steel gray	
Atacamite	$Cu_2(OH)_3Cl$	3.74–3.78	3–3.5	rhomb., grn., dk. grn., blksh. grn.	1.831, 1.861, 1.880
Augelite	$Al_2(PO_4)(OH)_3$	2.696	4.5–5	monocl., col., wh., yelsh. wh., rose	1.5736, 1.5759, 1.5877
Augite	$(Ca,Mg,Fe,Fe^{+3},Ti,Al)_2(Si,Al)_2O_6$	3.23–3.52	5.5–6	monocl., pa. br., br., purp. br., grn., blk.	1.671–1.735, 1.672–1.741, 1.703–1.76
Autunite	$Ca(UO_2)_2(PO_4)_2.10–12H_2O$	3.1–3.2	2–2.5	tetr., yel., somet. grnsh. yel. to pa. grn.	1.577, 1.553
Axinite	$(Ca,Mn,Fe)_3Al_2BO_3Si_4O_{12}(OH)$	3.26–3.36	6.5–7	tricl., br., yelsh.	1.674–1.693, 1.681–1.701, 1.684–1.70
Azurite	$Cu_3(OH)_2(CO_3)_2$	3.77	3.5–4	monocl., azure bl., dk. bl., pa. bl.	1.730, 1.758, 1.838
Baddeleyite	ZrO_2	5.4–6.02	6.5	monocl., col., yel., gr., redsh. br., br., blk.	2.13, 2.19, 2.20
Barite	$BaSO_4$	4.50	3–3.5	rhomb., col., wh., somet. br., dk. br., gray	1.6362, 1.6373, 1.6482
Benitoite	$BaTi(SiO_3)_3$	3.65	6–6.5	rhbdr., bl., purp., col.	1.757, 1.804
Bertrandite	$Be_4Si_2O_7(OH)_2$	2.6	6	rhomb., col.	1.589, 1.602, 1.613
Beryl	$Be_3Al_2Si_6O_{18}$	2.66–2.83	7.5–8	hex., col., wh., blsh. grn., grnsh. yel., yel., bl.	1.565–1.590, 1.567–1.598
Beryllonite	$NaBe(PO)_4$	2.81	5.5–6	monocl., col., wh., pa. yel.	1.5520, 1.5579, 1.561
Biotite	$K(Mg,Fe)_3AlSi_3O_{10}(OH,F)_2$	2.7–3.3	2.5–3	monocl., blk., dk. br., redsh. br.	1.565–1.625–1.605–1.696, 1.605–1.69
Bismuth	Bi	9.70–9.83	2–2.5	rhbdr., silver-wh. to redsh wh.	
Bismuthinite	Bi_2S_3	6.75–6.81	2	rhomb., lead gray to tin-wh., tarn. yel. or iridesc.	
Bixbyite	$(Mn,Fe)_2O_3$	4.945	6–6.5	cub., blk.	
Bloedite	$Na_2Mg(SO_4)_2.4N_2O$	2.22–2.28	2.5–3	monocl., col., somet. blsh.-grn. or redsh.	1.483, 1.486, 1.487
Boehmite	$AlO(OH)$	3.01–3.06	3.5–4	rhomb., wh.	1.64–1.65, 1.65–1.66, 1.65–1.67
Boracite	$Mg_3B_7O_{13}Cl$	2.91–2.97	7–7.5	rhomb., col., wh., gray, yel., blsh.-grn., grn.	1.66, 1.66, 1.67
Borax	$Na_2B_4O_7.10H_2O$	1.715	2–2.5	monocl., col., wh., gray, blsh. or grnsh-wh.	1.4466, 1.4687, 1.4717
Bornite	Cu_5FeS_4	5.06–5.08	3	cub., copper red to pinchbeck br., tarn. purp., iridesc.	
Boulangerite	$Pb_5Sb_4S_{11}$	6.0–6.2	2.5–3	monocl., blsh. lead gray, oft, with yel. spots	
Bournonite	$PbCuSbS_3$	5.80–5.86	2.5–3	rhomb., steel gray to blk.	
Braggite	PtS	10.0		tetr., steel gray	
Braunite	$(Mn,Si)_2O_3$	4.72–4.83	6–6.5	tetr., brns. blk. to steel gray	
Bravoite	$(Ni,Fe)S_2$	4.62	5.5–6	cub., steel gray	
Breithauptite	$NiSb$	8.23	5.5	hex., pa. copper red to vlt., tarn.	
Brochantite	$Cu_4(SO_4)(OH)_6$	3.79	3.5–4	monocl., emerald-grn. to blksh. grn., pa. grn.	1.728, 1.771, 1.800
Bromyrite	$AgBr$	6.47	2.5	cub., col., gray, yelsh., grnsh.-br.	2.253
Brookite	TiO_2	4.08–4.20	5.5–6	rhomb., br., yelsh. br., redsh. br., blk.	2.5831, 2.5843, 2.7004
Brucite	$Mg(OH)_2$	2.38–3.40	2.5	hex., wh., pa. grn., gray, bl., yel., br.	1.560–1.590, 1.580–1.600
Bunsenite	NiO	6.898	5.5	cub., dk. pistachio-grn.	(Li) 2.37
Cacoxenite	$Fe_4(PO_4)_3(OH)_3.12H_2O$	2.2–2.4	3–4	hex., yel. to brnsh.-yel., redsh. yel., somet. grnsh.	1.575–1.585, 1.635–1.656
Calcite	$CaCO_3$	2.715–2.94	3	rhbdr., col., wh., somet. gray, yel., pink, bl.	1.658–1.740, 1.486–1.550
Caledonite	$Cu_2Pb_5(SO_4)_3(CO_3)(OH)_6$	5.75–5.77	2.5–3	rhomb., dk. grn., blsh. grn.	1.815–1.821, 1.863–1.869, 1.906–1.91
Calomel	$HgCl$	7.15	1.5	tetr., col., wh., gray, yelsh. wh., br.	1.973, 2.656
Cancrinite	$(Na,Ca)_{7-8}Al_6Si_6O_{24}(CO_3SO_4Cl)_{1.5-2}.1–5H_2O$	2.51–2.42	5–6	hex., col., wh., pa. bl., pa. grn., yel., redsh.	1.528–1.507, 1.503–1.495
Carnalite	$KMgCl_3.6H_2O$	1.602	2.5	rhomb., col., wh., oft. redsh., somet. yel., bl.	1.466, 1.475, 1.494
Carnotite	$K_2(UO_2)_2(VO_4)_2.3H_2O$		1–2	rhomb. or monocl., bright yel., yel., grnsh. yel.	1.75, 1.92, 1.95
Cassiterite	SnO_2	6.99	6–7	tetr., yelsh. or redsh. br., brnsh.-blk.	2.006, 2.0972
Celestite	$SrSO_4$	3.96	3–3.5	rhomb., col., wh. pa. bl., redsh., grnsh., brnsh.	1.621–1.622, 1.623–1.624, 1.630–1.63
Celsian	$BaAl_2Si_2O_8$	3.10–3.39	6–6.5	monocl., col., wh., yel.	1.579–1.587, 1.583–1.593, 1.588–1.60
Cervantite	$Sb_2O_4(?)$	6.64	4–5	rhomb.(?)., yel., wh., somet. redsh.-wh.	
Cerargyrite	$AgCl$	5.55	2.5	cub., col., gray, grnsh.-br., tarn. purp., yelsh.	2.071
Cerussite	$PbCO_3$	6.53–6.57	3–3.5	rhomb., col., wh., gray, somet. bl., blk., grn.	1.8036, 2.0765, 2.0786
Chabazite	$(Ca,Na_2)Al_2Si_4O_{12}.6H_2O$	2.05–2.10	4.5	rhbdr., redsh.-wh., wh., yelsh., grnsh.	1.470–1.494
Chalcocite	Cu_2S	5.5–5.8	2.5–3	rhomb., blksh., lead gray	
Chalcanthite	$CuSO_4.5H_2O$	2.28	2.5	tricl., dk. bl. to sky bl., somet. grnsh.	1.514, 1.537, 1.543
Chalcopyrite	$CuFeS_2$	4.1–4.3	3.5–4	tetr., brass-yel., tarn., iridisc.	
Chiolite	$Na_5Al_3F_{14}$	3.00	3.5–4	tetr., wh. to col.	1.349, 1.342
Chlorite	$(Mg,Al,Fe)_{12}(Si,Al)_8O_{20}(OH)_{16}$	2.6–3.3	2–3	monocl., grn., wh., yel., pink, br., red	1.57–1.66, 1.57–1.67, 1.57–1.67
Chloritoid	$(Fe,Mg,Mn)_2(AlFe^{+3})Al_3O_2SiO_4(OH)_4$	3.51–3.80	6.5	monocl., tricl., dk. grn.	1.713–1.730, 1.719–1.734, 1.723–1.74

Name	Formula	Sp. gr.	Hard-ness	Crystalline form and color	Index of refraction (Na) η; ω ϵ α β γ
Chondrodite	$Mg(OH,F)_2.2Mg_2SiO_4$	3.16–3.26	6.5	monocl., yel., br., red	1.592–1.615, 1.602–1.627, 1.621–1.646
Chromite	$FeCr_2O_4$	4.5–5.1	5.5	cub., blk.	2.16
Chrysoberyl	$BeAl_2O_4$	3.65–3.85	8.5	rhomb., grn., yel.	1.746, 1.748, 1.756
Chrysocolla	$CuSiO_3.2H_2O$	~2.4	2	rhomb., (?)., grn., bl., br., blk.	1.575, 1.597, 1.598
Cinnabar	HgS	8.090	2–2.5	hex., red, brnsh. red., gray	(Li) 2.814, 3.143
Claudetite	As_2O_3	4.15	2.5	monocl., col. to wh.	1.87, 1.92, 2.01
Clinozoisite	$Ca_2Al_3Si_3O_{12}(OH)$	3.21–3.38	6.5	monocl., col., pa. yel., gray, grn.	1.670–1.715, 1.674–1.725, 1.690–1.734
Cobaltite	$CoAsS$	6.33	5.5	cub., silver wh., redsh., steel gray, blk.	....
Colemanite	$Ca_2B_6O_{11}.5H_2O$	2.42–2.43	4.5	monocl., col., wh., yelsh. wh., gray	1.586, 1.592, 1.614
Columbite	$(Fe,Mn)(Cb,Ta)_2O_6$	5.15–5.25	6	rhomb., iron blk. to br. blk.	....
Connellite	$Cu_{19}(SO_4)Cl_4(OH)_{32}.3H_2O(?)$	3.36	3	hex., azure bl.	1.724–1.738, 1.746–1.758
Copiapite	$(Fe,Mg)Fe_4{}^{+3}(SO_4)_6(OH)_2.20H_2O$	2.08–2.17	2.5–3	tricl., yel., grnsh. yel.	1.51–1.53, 1.53–1.55, 1.58–1.60
Copper	Cu	8.95	2.5–3	cub., red	....
Coquimbite	$Fe_2(SO_4)_3.9H_2O$	2.10–2.12	2.5	hex., pa. vlt. to dk. amethystine, yelsh., grnsh.	1.53–1.55, 1.55–1.57
Cordierite	$Al_3(Mg,Fe)_2Si_5AlO_{18}$	2.53–2.78	7	rhomb., gray-bl., bl., dk. bl.	1.522–1.558, 1.524–1.574, 1.527–1.578
Corundum	Al_2O_3	4.022	9	hex., col., bl., yel., purp., grn., pink, red	1.767–1.772, 1.759–1.763
Cotunnite	$PbCl_2$	5.80	2.5	rhomb., col. to wh., somet. yelsh., grnsh.	2.199, 2.217, 2.260
Covellite	CuS	4.6–4.76	1.5–2	hex., indigo bl., dk. bl., iridesc. brass yel. to red	....
Cristobalite	SiO_2	2.33	6–7	tetr.(?)., col., wh., yel.	1.487, 1.484
Crocoite	$PbCrO_4$	5.96–6.02	2.5–3	monocl., red, orange red, orange yel.	2.29, 2.36, 2.66
Cryolite	Na_3AlF_6	2.96–2.98	2.5	monocl., col. to wh., brnsh., redsh., blk.	1.338, 1.338, 1.339
Cryolithionite	$Na_3Li_3Al_2F_{12}$	2.77	2.5–3	cub., col. to wh.	1.3395
Cubanite	$CuFe_2S_3$	4.03–4.18	3.5	rhomb., brass to bronze yel.	....
Cummingtonite	$(Mg,Fe)_7Si_8O_{22}(OH)_2$	3.2–3.5	5–6	monocl., dk. grn., br.	1.635–1.665, 1.644–1.675, 1.655–1.698
Cuprite	Cu_2O	6.14	3.5–4	cub., red, somet. blk.	....
Danburite	$CaSi_2B_2O_8$	3.0	7	rhomb., pa. yel., col., dk. yel., yelsh. br.	1.63, 1.63–1.64, 1.63–1.64
Datolite	$CaBSiO_4(OH)$	2.96–3.00	5–5.5	monocl. col., wh., yelsh., grnsh., pinksh.	1.622–1.626, 1.649–1.654, 1.666–1.670
Daubreelite	Cr_2FeS_4	3.80–3.82	?	cub., blk.	....
Derbylite	$Fe_6Ti_6Sb_2O_{23}(?)$	4.53	5	rhomb., pitch blk.	2.45, 2.45, 2.51
Diamond	C	3.50–3.53	10	cub., col., pa. yel. to dk. yel., pa. br. to dk. br., wh., blsh. wh.	2.4175
Diaspore	$AlO(OH)$	3.3–3.5	6.5–7	rhomb., wh., graysh. wh., col.	1.682–1.706, 1.705–1.725, 1.730–1.752
Digenite	$Cu_{2x}S$	5.546	2.5–3	cub., bl. to blk.	....
Diopside	$CaMgSi_2O_6$	3.22–3.38	5.5–6.5	monocl., wh., pa. grn., dk. grn.	1.664–1.695, 1.672–1.701, 1.695–1.721
Dioptase	$Cu_6Si_6O_{18}.6H_2O$	3.5	5	rhbdr., emerald grn.	1.64–1.66, 1.70–1.71
Dolomite	$CaMg(CO_3)_2$	2.86	3.5–4	rhbdr., wh., oft. yel. or br. tinge, col.	1.679, 1.500
Douglasite	$K_2FeCl_4.2H_2O(?)$	2.16	....	pa. grn., tarn. brnsh. red	1.485–1.491, 1.497–1.503
Dyscrasite	Ag_3Sb	9.67–9.81	3.5–4	rhomb., silver wh., tarn. gray, yelsh. or blksh.	....
Edingtonite	$BaAl_2Si_3O_{10}.4H_2O$	2.7–2.8	....	rhomb. or monocl., col., pink, br. wh.	1.541, 1.553, 1.557
Eglestonite	Hg_4OCl_2	8.4	2.5	cub., yel., orange-yel. to dk. brnsh., tarn. bl.	2.47–2.51
Emplectite	$CuBiS_2$	6.38	2	rhomb., gray to tin wh.	....
Empressite	$AgTe$	7.510	3–3.5	pa. bronze	....
Enargite	Cu_3AsS_4	4.4–4.5	3	rhomb., gray-blk. to iron-blk.	....
Enstatite	$MgSiO_3$	3.209	5–6	rhomb., col., gray, grn., yel., brn.	1.650–1.662, 1.653–1.671, 1.658–1.680
Epidote	$Ca_2Fe^{+3}Al_2Si_3O_{12}(OH)$	3.38–3.49	6	monocl., grn., yel., gray	1.715–1.751, 1.725–1.784, 1.734–1.797
Epsomite	$MgSO_4.7H_2O$	1.675–1.679	2–2.5	rhomb., col., wh. pink, grn.	1.4325, 1.4554, 1.4609
Erythrite	$(Co,Ni)_3(AsO_4)_2.8H_2O$	3.06	1.5–2.5	monocl., crimson-red, red, pa. pink	1.626, 1.661, 1.699
Eucairite	$CuAgSe$	7.6–7.8	2.5	silver wh. to lead gray	....
Euclasite	$BeAlSiO_4(OH)$	3.0–3.1	7.5	monocl., col., pa. grn., bl.	1.651, 1.655, 1.671
Eudialite	$(Na,Ca,Fe)_6ZrSi_6O_{18}(OH,Cl)(?)$	2.8–3.1	5–6	hex., pa. pink, red, br.	1.59–1.61, 1.59–1.61
Eulytite	$Bi_4Si_3O_{12}$	6.6	4.5	cub., br., yel., gray	2.05
Euxenite	$(Y,Ca,Ce,U,Th)(Cb,Ta,Ti)_2O_6$	5.0–5.9	5.5–6.5	rhomb., blk., grnsh. or brnsh. tint.	~2.2
Fayalite	Fe_2SiO_4	4.392	6.5	rhomb., grnsh., yelsh.	1.827, 1.869, 1.879
Ferberite	$FeWO_4$	7.51	4–4.5	monocl., br. to blk.	(Li)2.37–2.43
Fergussonite	$(Y,Er,Ce,Fe)(Cb,Ta,Ti)O_4$	5.6–5.8	5.5–6.5	tetr., gray, yel., br., dk. br.	2.1
Fluorite	CaF_2	3.18	4	cub., bl., purp., wh., col., yel., grn.	1.433–1.435
Forsterite	Mg_2SiO_4	3.222	7	rhomb., wh., grnsh., yelsh.	1.635, 1.651, 1.670
Franklinite	$ZnFe_2{}^{+3}O_4$	5.07–5.34	5.5–6.5	Cub., blk. to br.-blk.	(Li) ~2.36
Gahnite	$ZnAl_2O_4$	4.62	7.5–8	cub., dk. bl.-grn., somet. yelsh. or brnsh.	1.79–1.81
Galena	PbS	7.57–7.59	2.5–2.75	cub. lead gray	....
Galenabismuthite	$PbBi_2S_4$	7.04	2.5–3.5	rhomb., pa. gray to tin-wh., lead gray, somet. tarn., yel. or irid.	....
Ganomalite	$(Ca,Pb)_{10}(OH,Cl)_2(Si_2O_7)_3$	5.4–5.7	3–4	hex., col., gray	1.910, 1.945
Gaylussite	$Na_2Ca(CO_3)_2.5H_2O$	1.991	2.5–3	monocl., col. to yelsh. wh., graysh. wh., wh.	1.4435, 1.5156, 1.5233
Gehlenite	$Ca_2Al_2SiO_7$	3.038	5–6	tetr., col., gray-grn., br.	1.669, 1.658
Geikielite	$MgTiO_3$	4.05	5–6	rhbdr., brnsh blk., blsh.	2.31, 1.95
Gibbsite	$Al(OH)_3$	2.38–2.42	2.5–3.5	monocl., wh., graysh., grnsh. or redsh.-wh.	1.56–1.58, 1.56–1.58, 1.58–1.60
Glauberite	$Na_2Ca(SO_4)_2$	2.75–2.85	2.5–3	monocl., gray, yelsh., somet. col., redsh.	1.515, 1.535, 1.536
Glauconite	$(K,Na,Ca)_{1.2-2}(Fe^{+3},Al,Fe,Mg)_4Si_{7-7.6}Al_{1-0.4}O_{20}(OH)_4.nH_2O$	2.4–2.95	2	monocl., col., ye.lsh. grn., grn., blsh. gray	1.592–1.610, 1.614–1.641, 1.614–1.641
Glaucophane	$Na_2Mg_3Al_2Si_8O_{22}(OH)_2$	3.08–3.30	6	monocl., gray, lavender bl.	1.606–1.661, 1.622–1.667, 1.627–1.670
Gmelinite	$(Ca,Na_2)Al_2Si_4O_{12}.6H_2O$	~2.1	4.5	rhbdr., wh., redsh.-wh., yelsh., grnsh.	1.476–1.494, 1.474–1.480
Goethite	$FeO(OH)$	3.3–4.3	5–5.5	rhomb., blksh.-br., yelsh. or redsh.-br., yel.	2.260–2.275, 2.393–2.409, 2.398–2.515
Gold	Au	19.3	2.5–3	cub., yel.	....
Goslarite	$ZnSO_4.7H_2O$	1.978	2–2.5	rhomb., col., wh., somet. br., grn., bl.	1.4568, 1.4801, 1.4844
Graphite	C	2.09–2.23	1–2	hex., iron-blk. to steel gray	....
Greenockite	CdS	4.9	3–3.5	hex., yel. to orange	2.506, 2.529
Grossularite	$Ca_3Al_2Si_3O_{12}$	3.594	6–7.5	cub., wh., yel., grn., br., red	1.734
Gummite (R)	$UO_3.H_2O$	3.9–6.4	2.5–5	yel., orange, redsh.-yel., red, br. blk.	....
Gypsum	$CaSO_4.2H_2O$	2.30–2.37	2	monocl., wh., col., somet. gray, red, yel., br.	1.519–1.521, 1.523–1.526, 1.529–1.531
Halite	$NaCl$	2.16–2.17	2.5	cub., col., wh., orange, red	1.544
Hambergite	$Be_2(OH)(BO_3)$	2.36	7.5	rhomb., col. to gray, wh., yel.	1.56, 1.59, 1.63
Hanksite	$Na_{22}K(SO_4)_9(CO_3)_2Cl$	2.562	3–3.5	hex., col., somet. pa.-yelsh. or gray	1.481, 1.461
Harmotome	$BaAl_2Si_6O_{16}.6H_2O$	2.41–2.47	4.5	monocl., or rhomb., col., wh., pink, gray, yel.	1.503–1.508, 1.505–1.509, 1.508–1.514
Hausmannite	Mn_3O_4	4.83–4.85	5.5	tetr., brnsh.-blk.	(Li) 2.46, 2.15
Haüyne	$(Na,Ca)_{4-8}Al_6Si_6O_{24}(SO_4,S)_{1-2}$	2.44–2.50	5.5–6	cub., wh., gray, grn., bl.	1.496–1.505
Hedenbergite	$CaFeSi_2O_6$	3.50–3.56	6	monocl., brnsh.-grn., dk. grn., blk.	1.716–1.726, 1.723–1.730, 1.741–1.751
Helvite	$Mn_4Be_3Si_3O_{12}S$	3.20–3.44	6	cub., yel., br., redsh.-brn.	1.728–1.749
Hematite	Fe_2O_3	5.26	5–6	rhbdr., steel gray, dull red to bright red	3.22, 2.94
Hemimorphite	$Zn_4Si_2O_7(OH)_2.H_2O$	3.45	5	rhomb., col., wh., pa. bl., pa. grn., br.	1.614, 1.617, 1.636
Hercynite	$FeAl_2O_4$	4.40	7.5–8	cub., blk.	1.835
Herderite	$CaBe(PO_4)(Fe,OH)$	2.95–3.01	5–5.5	monocl., col. to pa. yel. or grnsh.-wh.	1.592, 1.612, 1.621
Hessite	Ag_2Te	8.24–8.45	2–3	monocl., (<149.5°), cub. (>149.5°)., gray	....
Heulandite	$(Ca,Na_2)Al_2Si_7O_{18}.6H_2O$	2.1–2.2	3.5–4	pseudo-monocl., col., wh., yel., pink, red, gray, br.	1.491–1.505, 1.493–1.503, 1.500–1.512

PHYSICAL CONSTANTS OF MINERALS (Continued)

Name	Formula	Sp. gr.	Hardness	Crystalline form and color	Index of refraction (Na) η; ω ϵ / ω β γ
Hopeite	$Zn_3(PO_4)_2.4H_2O$	3.0–3.1	3.25	rhomb., col. to grayish-wh., pa. yel.	1.57–1.59, 1.58–1.60, 1.58–1.60
Hornblende	$(Ca,Na,K)_{2-3}(Mg,Fe,Fe^{+3}Al)_5Si_6(Si,Al)_2O_{22}(OH,F)_2$	3.02–3.45	5–6	monocl., grn., dk. grn., blk.	1.615–1.705, 1.618–1.714, 1.632–1.730
Huebnerite	$MnWO_4$	7.12	4–4.5	monocl., yel.-br. to red br., somet. br., blk.	2.17, 2.22, 2.32
Humite	$Mg(OH,F)_2.3Mg_2SiO_4$	3.2–3.32	6	rhomb., yel., orange	1.607–1.643, 1.619–1.653, 1.639–1.675
Huntite	$Mg_3Ca(CO_3)_4$	2.696	....	rhomb.(?), wh.	....
Hydrogrossularite	$Ca_3Al_2Si_2O_8(SiO_4)_{1-m}(OH)_{4m}$	3.594–3.13	6–7.5	cub., wh., buff, pa. grn., gray, pink	1.734–1.675
Hydromagnesite	$Mg_4(OH)_2(CO_3.3H_2O$	2.236	3.5	monocl., col. to wh.	1.520–1.526, 1.524–1.530, 1.544–1.546
Illite	$K_{1-1.5}Al_4Si_{7-6.5}Al_{1-1.5}O_{20}(OH)_4$	2.6–2.9	1–2	monocl., wh.	1.54–1.57, 1.57–1.61, 1.57–1.61
Ilmenite	$FeTiO_3$	4.68–4.76	5–6	rhbdr., iron-blk.	....
Iodyrite	AgI	5.69	1.5	hex., col. on exposure to light, yel., br.	2.21, 2.22
Jadeite	$NaAlSi_2O_6$	3.24–3.43	6	monocl., col., wh., grn., grnsh. bl.	1.640–1.658, 1.645–1.663, 1.652–1.673
Jamesonite	$Pb_4FeSb_6S_{14}$	5.63	2.5	monocl., gray-blk., somet. tarn. iridesc.	....
Jarosite	$KFe_3(SO_4)_2(OH)_6$	2.91–3.26	2.5–3.5	rhbdr., ocherous, amber yel. to dk. br.	1.820, 1.715
Kainite	$KMg(SO_4)Cl{\cdot}3H_2O$	2.15	2.5–3	monocl., col., gray bl., vlt., yelsh., redsh.	1.494, 1.505, 1.516
Kaliophyllite	$KAlSiO_4$	2.61	6	hex., col.	1.532, 1.537
Kaolinite	$Al_4Si_4O_{10}(OH)_8$	2.61–2.68	2–2.5	tricl. or monocl., wh., redsh.-wh., grnsh.-wh.	1.533–1.565, 1.559–1.569, 1.560–1.570
Kernite	$Na_2B_4O_7.4H_2O$	1.908	2.5	monocl., col., wh.	1.454, 1.472, 1.488
Kieserite	$MgSO_4.H_2O$	2.571	3.5	monocl., col., gray, wh., yelsh.	1.520, 1.533, 1.584
Kyanite	Al_2OSiO_4	3.53–3.65	5.5–7	tricl., bl., wh., gray, grn., yel., pink	1.712–1.718, 1.721–1.723, 1.727–1.734
Lanarkite	$Pb_2(SO_4)O$	6.92	2–2.5	monocl., gray to grnsh. wh., pa. yel.	1.925–1.931, 2.004–2.010, 2.033–2.039
Lanthanite	$(La,Ce)_2(CO_3)_38H_2O$	2.69–2.74	2.5–3	rhomb., col. to wh., pink, yelsh.	1.51–1.53, 1.584–1.590, 1.610–1.616
Laumontite	$CaAl_2Si_4O_{12}.4–3.5H_2O$	2.2–2.3	3–3.5	monocl., col., wh., red, yel., brn.	1.502–1.514, 1.512–1.522, 1.514–1.525
Laurionite	$Pb(OH)Cl$	6.24	3–3.5	rhomb., col. to wh.	2.08, 2.12, 2.16
Lawsonite	$CaAl_2(OH)_2Si_2O_7.H_2O$	3.05–3.10	6	rhomb., col., wh.	1.655, 1.674–1.675, 1.684–1.686
Lazulite	$(Mg,Fe)Al_2(PO_4)_2(OH)_2$	3.08–3.38	5.5–6	monocl., bl., blsh. wh., dk. bl., blsh. grn.	1.604–1.626, 1.626–1.654, 1.637–1.663
Lazurite	$Na_4SSi_3Al_3O_{12}$	2.38–2.45	5–5.5	cub., berlin bl., azure bl., grnsh. bl., vlt.	1.500
Leadhillite	$Pb_4(SO_4)(CO_3)_2(OH)_2$	6.55	2.5–3	monocl., col. to wh., gray, pa. grn., pa. bl., yelsh.	1.87, 2.00, 2.01
Lepidocrocite	$FeO(OH)$	4.05–4.31	5	rhomb., ruby-red to red-br.	1.94, 2.20, 2.51
Lepidolite	$K_2(Li,Al)_{5-6}Si_{6-7}Al_{2-1}O_{20}(OH,F)_4$	2.80–2.90	2.5–4	monocl., col., pa. pink, pa. purp.	1.525–1.548, 1.551–1.585, 1.554–1.587
Leucite	$KAlSi_2O_6$	2.47–2.50	5.5–6	tetr., (pseudo-cub.) wh., gray	1.508–1.511
Levyne	$(Ca,Na_2)Al_2Si_4O_{12}.6H_2O$	~2.1	4.5	rhbdr., wh., redsh. wh., yelsh., grnsh.	1–496–1.505, 1.491–1.500
Litharge	PbO	9.14	2	tetr., red	(Li) 2.665, 2.535
Loellingite	$FeAs_2$	7.39–7.41	5–5.5	rhomb., silver wh. to steel-gray	....
Magnesite	$MgCO_3$	2.98–3.44	3.5–4.5	rhbdr., wh., col., somet. yel., br.	1.700–1.782, 1.509–1.563
Magnetite	$FeF_2{}^{+3}O_4$	5.175	5.5–6.5	cub., blk. to br.-blk.	2.42
Malachite	$Cu_2(OH)_2(CO_3)$	4.03–4.07	3.5–4	monocl., bright grn. to dk. grn., blksh. grn.	1.652–1.658, 1.872–1.878, 1.906–1.912
Manganite	$MnO(OH)$	4.32–4.43	4	monocl., dk. steel-gray to iron-blk.	(Li) 2.25, 2.25, 2.53
Manganosite	MnO	5.364	5.5	cub., emerald grn., tarn. bl.	....
Marcasite	FeS_2	4.887	6–6.5	rhomb., pa. bronze-yel., tin-wh.	....
Marialite	$Na_4Al_3Si_9O_{24}Cl$	2.50–2.62	5–6	tetr., col., wh., pa. grnsh. yel., gray, br.	1.546–1.550, 1.540–1.541
Marshite	CuI	5.68	2.5	cub., col. to pa. yel., on exposure to light, red	2.346
Mascagnite	$(NH_4)_2SO_4$	1.768	2–2.5	rhomb., col., gray, yelsh.	1.5202, 1.5230, 1.5330
Matlockite	$PbFCl$	7.12	2.5–3	tetr., col. or yel. to pa. amber, grnsh.	2.145, 2.006
Meionite	$Ca_4Al_6Si_6O_{24}CO_3$	2.78	5–6	tetr., col., wh., pa. grnsh. yel., gray, br.	1.590–1.600, 1.556–1.562
Melanterite	$FeSO_4.7H_2O$	1.898	2	monocl., grn., grnsh. bl., grnsh. wh.	1.47, 1.48, 1.49
Melilite	$(Ca,Na,K)_2(Mg,Fe,Fe^{+3},Al,Si)_3O_7$	2.95–3.05	5–6	tetr., yelsh., br., grn.-br.	1.624–1.666, 1.616–1.661
Mellite	$Al_2C_{12}O_{12}.18H_2O$	1.64	2–2.5	tetr., yel., redsh., brnsh., somet. wh.	1.5393, 1.5110
Mendipite	$Pb_3O_2Cl_2$	7.24	2.5	rhomb., col. to wh., gray, oft. yel., red, bl. tinge	2.22–2.26, 2.25–2.29, 2.29–2.33
Mesolite	$Na_2Ca_2(Al_2Si_3O_{10}).8H_2O$	~2.26	5	monocl., col., wh., gray, yel., pink, red	β = 1.504–1.508
Metacinnabar	HgS	7.65	3	cub., graysh.-blk.	....
Microcline	$KAlSi_3O_8$	2.56–2.63	6–6.5	tricl., col., wh., pink, red, yel., grn.	1.514–1.529, 1.518–1.533, 1.521–1.53
Microlite	$(Na,Ca)_2Ta_2O_6(O,OH,F)$	4.2–6.4	5–5.5	cub., pa. yel. to br., somet. red, grn.	~2.0
Miersite	AgI	5.64–5.68	2.5	cub., canary-yel.	2.18–2.22
Millerite	NiS	5.3–5.7	3–3.5	hex., pa. brass-yel. to bronze-yel., gray, tarn. iridesc.	....
Mimetite	$Pb_5(AsO_4,PO_4)_3Cl$	7.24	3.5–4	hex. pa. yel. to yelsh. br., orange-yel., wh.	2.147, 2.128
Minium	Pb_3O_4	8.9–9.2	2.5	scarlet red, bl. red, somet. yel. tint.	(Li) 2.40–2.44
Mirabilite	$Na_2SO_4.10H_2O$	1.490	1.5–2	monocl., col. to wh.	1.391–1.397, 1.393–1.399, 1.395–1.40
Moissanite	SiC	3.218	9.5	hex., grn. to blk., somet. blsh., red	2.647–2.649, 2.689–2.693
Molybdenite	MoS_2	4.62–4.73	1–1.5	hex., lead-gray	....
Monazite	$(Ce,La,Th)PO_4$	5.0–5.3	5	monocl., yel., br., redsh. br.	1.774–1.800, 1.777–1–801, 1.828–1.85
Monetite	$CaH(PO_4)$	2.929	3.5	tricl., wh., pa. yelsh.-wh.	1.587, ~1.615, 1.640
Monticellite	$CaMgSiO_4$	3.08–3.27	5.5	rhomb., col.	1.639–1.654, 1.646–1.664, 1.653–1.67
Montmorillonite	$(0.5Ca,Na)_{0.7}(Al,Mh,Fe)_4(Si,Al)_8O_{20}(OH)_4.nH_2O$	2–3	1–2	monocl., wh., yel., grn.	1.48–1.61, 1.50–1.64, 1.50–1.64
Montroydite	HgO	11.23	2.5	rhomb., dk. red to brnsh. red, br.	(Li) 2.37, 2.5, 2.65
Mordenite	$(Na_2,K_2Ca)Al_2Si_{10}O_{24}.7H_2O$	2.12–2.15	3–4	rhomb., col., wh., red, yel., br.	1.472–1.483, 1.475–1.485, 1.477–1.48
Muscovite	$KAl_2Si_3AlO_{10}(OH,F)_2$	2.77–2.88	2.5–3	monocl., col., pa. grn., pa. red, pa. br.	1.552–1.574, 1.582–1.610, 1.587–1.61
Nantokite	$CuCl$	4.136	2.5	cub., col. to wh., grayish, grn.	1.925–1.935
Natrolite	$Na_2Al_2Si_3O_{10}.2H_2O$	2.20–2.26	5	rhomb., col., wh., gray, yel., pink, red	1.473–1.483, 1.476–1.486, 1.485–1.49
Nepheline	$Na_3KAl_4Si_4O_{16}$	2.56–2.665	5.5–6	hex., col., wh., gray	1.529–1.546, 1.526–1.542
Newberyite	$MgH(PO_4).3H_2O$	2.10	3.0–3.5	rhomb., col.	1.511–1.517, 1.514–1.520, 1.530–1.53
Niccolite	$NiAs$	7.784	5–5.5	hex., pa. copper-red, tarn. gray to blk.	....
Nosean	$Na_8Al_6Si_6O_{24}SO_4$	2.30–2.40	5.5	cub., gray, bl., br.	1.495
Oldhamite	CaS	2.58	4	cub., pa. chestnut-br.	2.137
Oligoclase	$([NaSi]_{0.9-0.7}[CaAl]_{0.1-0.3})AlSi_2O_8$	2.63–2.65	6–6.5	tricl., col., wh., gray, grnsh., pink	1.533–1.544, 1.537–1.548, 1.543–1.55
Olivenite	$Cu_2(AsO_4)(OH)$	3.9–4.5	3	rhomb., olive grn., grnsh.-br., br., gray	1.75–1.78, 1.79–1.82, 1.83–1.87
Olivine	$(Mg,Fe)SiO_4$	3.22–4.39	6.5–7	rhomb., olive grn., grayish grn. to yelsh. br.	1.63–1.83, 1.65–1.87, 1.67–1.88
Opal	$SiO_2.nH_2O$	1.73–2.16	~6	col., wh., yel., br., red, grn., bl., blk.	1.41–1.46
Orpiment	As_2S_3	3.49	1.5–2	monocl., yel., brnsh. yel.	(Li) 2.4, 2.81, 3.02
Orthoclase	$KAlSi_3O_8$	2.55–2.63	6–6.5	monocl., col., wh., pink, red, yel., grn.	1.518–1.529, 1.522–1.533, 1.522–1.53
Orthopyroxene	$(Mg,Fe)SiO_3$	3.209–3.96	5–6	rhomb., col., gray, grn., yel., dk. brn.	1.650–1.768, 1.653–1.770, 1.658–1.78
Paragonite	$NaAl_2Si_3AlO_{10}(OH)_2$	2.85	2.5	monocl., col., pa. yel.	1.564–1.580, 1.594–1.609, 1.600–1.60
Parisite	$(Ce,La,Na)FCO_3.CaCO_3$	4.42	4.5	hex., brnsh., yel.	1.672, 1.771
Pectolite	$Ca_2NaH(SiO_3)_3$	2.86–2.90	4.5–5	tricl., col., wh.	1.595–1.610, 1.605–1.615, 1.632–1.64
Penfieldite	$Pb_4Cl_6(OH)_2$	6.6	....	hex., wh.	2.13, 2.21
Pentlandite	$(Fe,Ni)_9S_8$	4.6–5.0	3.5–4	cub., pa. bronze-yel.	....
Percylite	$PbCuCl_2(OH)_2(?)$	?	2.5	cub(?), sky bl.	2.04–2.06
Periclase	MgO	3.55–3.68	5.5	cub., col. to gray-wh., yel., brnsh. yel., grn., bl.	1.7350
Pekovskite	$CaTiO_3$	3.97–4.26	5.5	pseudo cub., blk., gray-blk., brnsh. bl., redsh. br., br., yel.	2.30–2.38
Petalite	$LiAlSi_4O_{10}$	2.412–2.422	6.5	monocl., wh., gray, somet. pink, grn.	1.504–1.507, 1.510–1.513, 1.516–1.52

Name	Formula	Sp. gr.	Hard-ness	Crystalline form and color	Index of refraction (Na) η; ω ϵ α β γ
harmacosiderite	$Fe_3(AsO_4)_2(OH)_3.5H_2O$	2.797	2.5	cub., olive-grn. to yel., br., redsh.	1.676–1.704
henakite	Be_2SiO_4	2.98	7.5	rhbder., col., rose, yel., br.	1.654, 1.670
hillipsite	$(0.5Ca,Na,K)_3Al_3Si_5O_{16}.6H_2O$	2.2	4–4.5	monocl. or rhomb., col., wh., pink, gray, yel.	1.483–1.504, 1.484–1.509, 1.496–1.514
hlogopite	$KMg_3AlSi_3O_{10}(OH,F)_2$	2.76–2.90	2–2.5	monocl., col., yelsh.-br., grn., redsh.-br., br.	1.530–1.590, 1.557–1.637, 1.558–1.637
hosgenite	$Pb_2(CO_3)Cl_2$	6.133	2–3	tetr., yelsh. wh. to yelsh. br., br., somet. wh., rose, gray	2.1181, 2.1446
iemontite	$Ca_2(Mn,Fe^{+3},Al)_2AlSi_3O_{12}(OH)$	3.45–3.52	6	monocl., redsh. brn., blk.	1.732–1.794, 1.750–1.807, 1.762–1.829
igeonite	$(Mg,Fe,Ca)(Mg,Fe)Si_2O_6$	3.30–3.46	6	monocl., br., grnsh. br., blk.	1.682–1.722, 1.684–1.722, 1.705–1.751
latinum	Pt	14–19	4–4.5	cub., whitish, steel gray to dk. gray	....
ollucite	$CsAlSi_2O_6$	2.9	6.5	tetr., (pseudo-cub.) col.	1.507–1.527
olybasite	$(Ag,Cu)_{16}Sb_2S_{11}$	6.0–6.2	2–3	monocl., iron-blk.	....
owellite	$Ca(Mo,W)O_4$	4.21–4.25	3.5–4	tetr., straw-yel., br., grnsh., somet. gray, bl., blk.	1.959–1.982, 1.967–1.993
rehnite	$Ca_2Al_2Si_3O_{10}(OH)_2$	2.90–2.95	6–6.5	rhomb., pa. grn., yel., gray, wh.	1.611–1.632, 1.615–1.642, 1.632–1.665
roustite	Ag_3AsS_3	5.57	2–2.5	rhbdr., scarlet-vermillion	3.0877, 2.7924
seudobrookite	Fe_2TiO_5	4.33–4.39	6	rhomb., dk. red-br. to brnsh. blk. and blk.	(Li) 2.38, 2.39, 2.42
silomelane	$BaMn^{+2}Mn_8^{+4}O_{16}(OH)_4$	4.71	5–6	rhomb., iron-blk. to steel-gray	
umpellyite	$Ca_4(Mg,Fe,Mn)(Al,Fe^{+3},Ti)_5(OH)_3Si_6O_{23}.2H_2O$	3.18–3.23	6	monocl., grn., blsh. grn., br.	1.674–1.702, 1.675–1.715, 1.688–1.722
yrargyrite	Ag_3SbS_3	5.85	2.5	rhbdr., deep red	(Li) 3.084, 2.881
yrite	FeS_2	5.018	6–6.5	cub., pa. brass-yel., tarn. iridesc.	....
yrochlore	$NaCaCb_2O_6F$	4.2–6.4	5–5.5	cub., br. to blk., yelsh., redsh. or blksh. br.	....
yrochroite	$Mn(OH)_2$	3.23–3.27	2.5	hex., col. to pa. grn. or bl., tarn. br. to blk.	1.72, 1.68
yrolusite	MnO_2	5.04–5.08	6–6.5	tetr., pa. steel-gray, iron-gray, blk., blsh.	....
yromorphite	$Pb_5(PO_4,AsO_4)_3Cl$	7.00–7.08	3.5–4	hex., grn., yel., br., orange, brnsh. red., gray	2.058, 2.048
yrope	$Mg_3Al_2Si_3O_{12}$	3.582	6–7.5	cub., red, pink	1.714
yrophyllite	$Al_2Si_4O_{10}(OH)_2$	2.65–2.90	1–2	monocl., wh., yel., pa. bl., gray-grn., brnsh.-grn.	1.534–1.556, 1.568–1.589, 1.596–1.601
yrrhotite	$Fe_{1-0.8}S$	4.58–4.65	3.5–4.5	hex., bronze-yel. to br., tarn., somet. iridesc.	....
uartz	SiO_2	2.65	7	rhbdr., col., wh., blk., purp., grn., bl., rose	1.544, 1.553
ammelsbergite	$NiAs_2$	7.0–7.2	5.5–6	tin. wh., redsh. tinge	....
aspite	$PbWO_4$	8.46	2.5–3	monocl., yelsh. br., pa. yel., gray	1.25–1.29, 1.25–1.29, 1.28–1.32
ealgar	AsS	3.56	1.5–2	monocl., aurora-red to orange-yel.	2.538, 2.684, 2.704
iebeckite	$Na_2Fe_3Fe_2^{+3}Si_8O_{22}(OH,F)_2$	3.02–3.42	5	monocl., dk. bl., bl.	1.654–1.701, 1.662–1.711, 1.668–1.717
hodochrosite	$MnCO_3$	3.70	3.5–4	rhbdr., pink, red, br., brnsh.-yel.	1.816, 1.597
hodonite	$(Mn,Fe,Ca)SiO_3$	3.57–3.76	5.5–6.5	tricl., pink to brnsh. red	1.711–1.738, 1.716–1.741, 1.724–1.751
utile	TiO_2	4.23–5.5	6–6.5	tetr., redsh. brn. to red, somet. yelsh., blsh.	2.605–2.613, 2.899–2.901
afflorite	$(Co,Fe)As_2$	7.0–7.5	4.5–5	rhomb., tin-wh., tarn. dk. gray	....
amarskite	$(Y,Er,Ce,U,Ca,Fe,Pb,Th)(Cb,Ta,Ti,Sn)_2O_6$	5.69	5–6	rhomb., velvet blk., somet. brnsh. tint	~2.20
apphirine	$(Mg,Fe)_2Al_4O_6SiO_4$	3.40–3.58	7.5	monocl., pa. bl., pa. grn.	1.701–1.717, 1.703–1.720, 1.705–1.724
capolite	$(Na,Ca)_4Al_3(Al,Si)_3Si_6O_{24}(Cl,F,OH,CO_3,SO_4)$	2.50–2.78	5–6	tetr., col., wh., pa. grnsh. yel., gray, bl.	1.546–1.600, 1.540–1.562
cheelite	$CaWO_4$	6.08–6.12	4.5–5	tetr., yelsh. wh., pa. yel., brnsh., col., wh., gray	1.920, 1.936
colecite	$CaAl_2Si_3O_{10}.3H_2O$	2.25–2.29	5	monocl., col., wh., gray, yel., pink, red	1.507–1.513, 1.516–1.520, 1.517–1.521
corodite	$Fe^{+3}(AsO_4).2H_2O$	3.28	3.5–4	rhomb., pa. grn., gray grn., br. somet. col., blsh., yel.	1.784, 1.795, 1.814
ellaite	MgF_2	3.15	5	tetr., col. to wh.	1.378, 1.390
narmontite	Sb_2O_3	5.50	2–2.5	pseudo-cub., col., gray-wh.	2.087
rpentine	$Mg_3Si_2O_5(OH)_4$	~2.55	2.5–3.5	monocl., wh., yel., gray, grn., blsh. grn.	1.53–1.57, 1.56, 1.54–1.57
iderite	$FeCO_3$	3.96	4–4.5	rhbdr., yelsh. br., br., dk. br.	1.875. 1.635
illimanite	Al_2OSiO_4	3.23–3.27	6.5–7.5	rhomb., col., wh., yelsh., br., grnsh.	1.654–1.661, 1.658–1.662, 1.637–1.683
ilver	Ag	10.1–11.1	2.5–3	cub., wh., tarn. gray or blk.	....
kutterudite	$(Co,Ni)As_3$	6.1–6.9	5.5–6	cub., between tin-wh. and silver-gray, tarn. gray or iridesc.	....
mithsonite	$ZnCO_3$	4.42–4.44	4–4.5	rhbdr., grayish wh. to dk. gray, grnsh., brnsh. wh.	1.848, 1.621
odalite	$Na_8Al_6Si_6O_{24}Cl_2$	2.27–2.33	5.5–6	cub., bl., grn., yel., gray, pink	1.483–1.487
perrylite	$PtAs_2$	10.58	6–7	cub , tin-wh.	....
pessartite	$Mn_3Al_2Si_3O_{12}$	4.190	6–7.5	cub., blk., dk. red, brnsh. red., bl., yelsh. orange	1.800
phalerite	ZnS	3.9–4.1	3.5–4	cub., br., blk., yel., red, wh.	2.369
phene	$CaTiSiO_4(O,OH,F)$	3.45–3.55	5	monocl., col., yel., grn., br., blk.	1.843–1.950, 1.870–2.034, 1.943–2.110
pinel	$MgAl_2O_4$	3.55	7.5–8	cub., grn., red, bl., br. to col.	1.719
podumene	$LiAlSi_2O_6$	3.03–3.22	6.5–7	monocl., col., gray-wh., pa. bl., pa. grn., yelsh.	1.648–1.663, 1.655–1.669, 1.662–1.679
annite	Cu_2FeSn_4	4.3–4.5	4	tetr., steel gray to iron blk.	
aurolite	$(Fe,Mg)_2(AlFe^{+3})_9O_6SiO_4(O,OH)_2$	3.74–3.83	7.5	monocl., brn., redsh., yelsh.	1.739–1.747, 1.745–1.753, 1.752–1.761
tercorite	$Na(NH_4)H(PO_4).4H_2O$	1.615	2	tricl., wh., yelsh., brnsh.	1.439, 1.442, 1.469
tibiotantalite	$Sb(Ta,Cb)O_4$	5.7–7.5	5.5	rhomb., dk. br. to pa. yel.-br., red-br., grnsh.-yel.	2.38, 2.41, 2.46
tibnite	Sb_2S_3	4.61–4.65	2	rhomb., lead-gray to steel-gray	....
tilbite	$(Ca,Na_2K_2)Al_2Si_7O_{18}.7H_2O$	2.1–2.2	3.5–4	monocl., col., wh., yel., pink, red, gray, br.	1.484–1.500, 1.492–1.507, 1.494–1.513
tilpnomelane	$(K,Na,Ca)_{0-1.4}(Fe^{+3}Fe,Mg,Al)_{6-8}Si_8O_{20}(OH)_4(O,OH,H_2O)_{4-8}$	2.59–2.96	3–4	monocl., br., dk. br., redsh. br., blk., dk. grn.	1.543–1.634, 1.576–1.745, 1.576–1.745
tolzite	$PbWO_4$	7.9–8.4	2.5–3	tetr., redsh. br., yelsh. gray, straw-yel., grnsh.	2.26–2.28, 2.18–2.20
trengite	$Fe^{+3}(PO_4).2H_2O$	2.90	3.5–4.5	rhomb., red, carmine, vlt., near col.	1.707, 1.719, 1.741
trontianite	$SrCO_3$	3.72	3.5	rhomb., col., wh., yel., grnsh., brnsh.	1.516–1.520, 1.664–1.667, 1.666–1.669
truvite	$Mg(NH_4)(PO_4).6H_2O$	1.71	2	rhomb., col., somet. yelsh., brnsh.	1.495, 1.496, 1.504
ulfur	S	2.07	1.5–2.5	rhomb., yel., brnsh., grnsh., redsh., gray	1.9579, 2.0377, 2.2452
ylvanite	$(Ag,Au)Te_2$	8.161	1.5–2	monocl., steel-gray to silver-wh.	....
ylvite	KCl	1.99	2	cub., col., wh., somet. grayish, blsh., yelsh., red	1.49031
alc	$Mg_3Si_4O_{10}(OH)_2$	2.58–2.83	1	monocl., col., wh., pa. grn., dk. grn., br.	1.539–1.550, 1.589–1.594, 1.589–1.600
antalite	$(Fe,Mn)(Ta,Cb)_2O_6$	7.90–8.00	6.5	rhomb., iron-bl. to br.-blk.	2.26, 2.32, 2.43
apiolite	$FeTa_2O_6$	7.9	6–6.5	tetr., blk.	(Li) 2.27, 2.42
ellurobismuthite	Bi_2Te_3	7.800–7.830	1.5–2	rhbdr., pa. lead-gray	....
erlinguaite	Hg_2OCl	8.725	2.5	monocl., yel. to grnsh.-yel., somet. br.	(Li) 2.33–2.37, 2.62–2.66, 2.64–2.68
etrahedrite	$(Cu,Fe)_{12}Sb_4S_{13}$	4.6–5.1	3–4.5	cub., flint-gray to iron-blk. to dull-blk.	
henardite	Na_2SO_4	2.664	2.5–3	rhomb., col., grayish-wh., yelsh., yelsh. br., redsh.	1.464–1.471, 1.473–1.477, 1.481–1.485
hermonatrite	$Na_2CO_3.H_2O$	2.255	1–1.5	rhomb., col. to wh., grayish, yelsh.	1.420, 1.506, 1.524
homsenolite	$NaCaAlF_6.H_2O$	2.981	2	monocl., col. to wh., somet. brnsh., redsh.	1.4072, 1.4136, 1.4150
homsonite	$NaCa_2([Al,Si]_5O_{10})_2.6H_2O$	2.10–2.39	5–5.5	rhomb., col., wh., pink, br.	1.497–1.530, 1.513–1.533, 1.518–1.544
horianite (R)	ThO_2	9.7	6.5	cub., dk. gray to brnsh.-blk., blk.	~2.20
horite (R)	$ThSiO_4$	5.2–5.4	4.5–5	tetr., orange-yel., brnsh. to blk.	~1.8
opaz	$Al_2SiO_3(OH,F)_2$	3.49–3.57	8	rhomb., col., wh., yel., gray, grn., red, bl.	1.606–1.629, 1.609–1.631, 1.616–1.638

Name	Formula	Sp. gr.	Hardness	Crystalline form and color	Index of refraction (Na) η; ω ϵ / ω β γ
Torbernite (R)	$Cu(UO_2)_2(PO_4)_2.8\text{-}12H_2O$	3.22	2–2.5	tetr., various shades of grn.	1.592, 1.582
Tourmaline	$Na(Mg,Fe,Mn,Li,Al)_3Al_6Si_6O_{18}(BO_3)_3(OH,F)_4$	3.03–3.25	7	rhbdr., blk., bl., grn., yel., red, col., br.	1.635–1.675, 1.610–1.650
Tremolite	$Ca_2Mg_5Si_8O_{22}(OH,F)_2$	3.0	5–6	monocl., col., gray, wh.	1.599, 1.612, 1.622
Tridymite	SiO_2	2.27	7	rhomb., col., wh.	1.471–1.479, 1.472–1.480, 1.474–1.48
Triphyllite-Lithiophyllite	$Li(Fe,Mn)PO_4$	3.34–3.58	4–5	rhomb., blsh. or grnsh. gray to yelsh. br., br.	1.66–1.70, 1.67–1.70, 1.68–1.71
Troegerite (R)	$(UO_2)_3(AsO_4)_2.12H_2O$	....	2–3	tetr., lemon-yel.	1.58–1.59, 1.625–1.635
Trona	$Na_3H(CO_3)_2.2H_2O$	2.14	2.5–3	monocl.. gray or yelsh. wh., col.	1.412, 1.492, 1.540
Turquois	$Cu(Al,Fe^{+3})_6(PO_4)_4(OH)_8.4H_2O$	2.6–3.2	4.5–6	tricl., bl., grn., grnsh.-gray	1.61–1.78, 1.62–1.84, 1.65–1.84
Ullmannite	$NiSbS$	6.61–6.69	5–5.5	cub., steel-gray to silver-wh.	....
Uraninite (R)	UO_2	8.0–11	5–6	cub., steel-blk., brnsh.-blk., grayish, grn.	....
Uvarovite	$Ca_3Cr_2Si_3O_{12}$	3.90	6–7.5	cub., emerald-grn.	1.86
Valentinite	Sb_2O_3	5.76	2.5–3	rhomb., col. to wh., somet. yelsh., redsh., gray, br.	2.18, 2.35, 2.35
Vanadinite	$Pb_5(VO_4)_3Cl$	6.5–7.1	2.75–3	hex., orange-red, red, brnsh.-red, br., brnsh.-yel., yel.	2.416, 2.350
Variscite-Strengite	$(AlFe^{+3})(PO_4).2H_2O$	2.57–2.87	3.5–4.5	rhomb., pa. grn., grn., blsh.-grn., red, vlt., col.	1.563–1.707, 1.588–1.719, 1.594–1.7
Vaterite	$CaCO_3$	2.645	....	hex., col.	1.550, 1.640–1.650
Vermiculite	$(Mg,Ca)_{0.7}(Mg,Fe^{+3}Al)_6(Al,Si)_8O_{20}(OH)_4.8H_2O$	~2.3	~1.5	monocl., col., yel., grn., br.	1.525–1.564, 1.545–1.583, 1.545–1.5
Vesuvianite	$Ca_{10}(Mg,Fe)_2Al_4(Si_2O_7)_2(SiO_4)_5(OH,F)_4$	3.33–3.43	6–7	tetr., yel., grn., br.	1.700–1.746, 1.703–1.752
Villiaumite	NaF	2.79	2–2.5	cub., carmine, (nat.), col. (artif.)	1.327
Vivianite	$Fe_3(PO_4)_2.8H_2O$	2.67–2.69	1.5–2	monocl., col., tarn. pa. bl., grnsh. bl., dk. bl., blsh. blk.	1.579–1.616, 1.602–1.656, 1.629–1.6
Wagnerite	$Mg_2(PO_4)F$	3.15	5–5.5	monocl., yel., gray, somet. red, grn.	1.568, 1.572, 1.582
Wavellite	$Al_3(OH)_3(PO_4)_2.5H_2O$	2 36	3.25–4	rhomb., grnsh. wh., grn. to yel., somet. br., bl., wh.	1.520–1.535, 1.526–1.543, 1.545–1.5
Whewellite	$Ca(C_2O_4).H_2O$	2.23	2.5–3	monocl., col., somet. yelsh., brnsh.	1.491, 1.554, 1.650
Willemite	Zn_2SiO_4	3.9–4.1	5.5	rhbdr., wh., yel., grn., red, gray, br.	1.691, 1.719
Witherite	$BaCO_3$	4.29–4.30	3.5	rhomb., col., wh., gray, yelsh. br.	1.529, 1.676, 1.677
Wolframite	$(Fe,Mn)WO_4$	7.12–7.51	4–4.5	monocl., dk. gray, brnsh. blk. to iron blk.	(Li) ~2.26, 2.32, 2.42
Wollastonite	$CaSiO_3$	2.87–3.09	4.5–5	tricl., wh., col., gray, pa. grn.	1.616–1.640, 1.628–1.650, 1.631–1.6
Wulfenite	$PbMoO_4$	6.5–7.0	2.75–3	tetr., orange-yel. to yel., gray, grn., br., red	2.403, 2.283
Wurtzite	ZnS	3.98	3.5–4	hex., brnsh. blk.	2.356, 2.378
Xenotime	$Y(PO_4)$	4.4–5.1	4–5	tetr., yelsh. br. to redsh. br., somet. gray, wh., pa. yel., grnsh.	1.721, 1.816
Zeunerite (R)	$Cu(UO_2)_2(AsO_4)_2.10\text{-}16H_2O$	....	....	tetr.	1.602–1.610
Zincite	ZnO	5.64–5.68	4	hex., orange-yel. to dk. red, somet. yel.	2.013, 2.029
Zircon	$ZrSiO_4$	4.6–4.7	7.5	tetr., redsh. br., yel., gray, grn., col.	1.923–1.960, 1.968–2.015
Zoisite	$Ca_2Al_3Si_3O_{12}(OH)$	3.15–3.365	6	rhomb., gray, grnsh., brnsh.	1.685–1.705, 1.688–1.710, 1.697–1.7

SOLUBILITY PRODUCT

The solubility product (or ion product constant) is the product of the concentrations of the ions in the saturated solution of a difficultly soluble salt. The concentrations are expressed as moles per liter of solution. The number of cations (or anions) resulting from the dissociation of one molecule of the salt, appears in the formula for calculations of the solubility product as the exponent of the concentration of the cation (or anion).

If two solutions, each containing one of the ions of a difficultly soluble salt, are mixed, no precipitation takes place unless the product of the ion concentrations in the mixture is greater than the solubility product.

In a solution containing two salts which yield a common ion the ratio of solubilities of the two salts is the ratio of the solubility products.

Substance	Solubility product at temperature noted
Aluminum hydroxide	4×10^{-13} (15°)
Aluminum hydroxide	1.1×10^{-15} (18°)
Aluminum hydroxide	3.7×10^{-15} (25°)
Barium carbonate	7×10^{-9} (16°)
Barium carbonate	8.1×10^{-9} (25°)
Barium chromate	1.6×10^{-10} (18°)
Barium chromate	2.4×10^{-10} (28°)
Barium fluoride	1.6×10^{-6} (9.5°)
Barium fluoride	1.7×10^{-6} (18°)
Barium fluoride	1.73×10^{-6} (25.8°)
Barium iodate, $Ba(IO_3)..2H_2O$	8.4×10^{-11} (10°)
Barium iodate, $Ba(IO_3)..2H_2O$	6.5×10^{-10} (25°)
Barium oxalate, $BaC_2O_4.3\frac{1}{2}H_2O$	1.62×10^{-7} (18°)
Barium oxalate, $BaC_2O_4.2H_2O$	1.2×10^{-7} (18°)
Barium oxalate, $BaC_2O_4.\frac{1}{2}H_2O$	2.18×10^{-7} (18°)
Barium sulfate	0.87×10^{-10} (18°)
Barium sulfate	1.08×10^{-10} (25°)
Barium sulfate	1.98×10^{-10} (50°)
Cadmium oxalate $CdC_2O_4.3H_2O$	1.53×10^{-8} (18°)
Cadmium sulfide	3.6×10^{-29} (18°)
Calcium carbonate (calcite)	0.99×10^{-8} (15°)
Calcium carbonate (calcite)	0.87×10^{-8} (25°)
Calcium fluoride	3.4×10^{-11} (18°)
Calcium fluoride	3.95×10^{-11} (26°)
Calcium iodate, $Ca(IO_3)_2. 6H_2O$	22.2×10^{-8} (10°)
Calcium iodate, $Ca(IO_3)_2.6H_2O$	64.4×10^{-8} (18°)
Calcium oxalate, $CaC_2O_4.H_2O$	1.78×10^{-9} (18°)
Calcium oxalate, $CaC_2O_4.H_2O$	2.57×10^{-9} (25°)
Calcium sulfate	1.95×10^{-4} (10°)
Calcium tartrate, $CaC_4H_4O_6.2H_2O$	0.77×10^{-6} (18°)
Cobalt sulfide	3×10^{-26} (18°)
Cupric iodate	1.4×10^{-7} (25°)
Cupric oxalate	2.87×10^{-8} (25°)
Cupric sulfide	8.5×10^{-45} (18°)
Cuprous bromide	4.15×10^{-8} (18–20°)
Cuprous chloride	1.02×10^{-6} (18–20°)
Cuprous iodide	5.06×10^{-12} (18–20°)
Cuprous sulfide	2×10^{-47} (16–18°)
Cuprous thiocyanate	1.6×10^{-11} (18°)
Ferric hydroxide	1.1×10^{-36} (18°)
Ferrous hydroxide	1.64×10^{-14} (18°)
Ferrous oxalate	2.1×10^{-7} (25°)
Ferrous sulfide	3.7×10^{-19} (18°)
Lead carbonate	3.3×10^{-14} (18°)
Lead chromate	1.77×10^{-14} (18°)
Lead fluoride	2.7×10^{-8} (9°)
Lead fluoride	3.2×10^{-8} (18°)
Lead fluoride	3.7×10^{-8} (26.6°)
Lead iodate	5.3×10^{-14} (9.2°)
Lead iodate	1.2×10^{-13} (18°)
Lead iodate	2.6×10^{-13} (25.8°)
Lead iodide	7.47×10^{-9} (15°)
Lead iodide	1.39×10^{-8} (25°)
Lead oxalate	2.74×10^{-11} (18°)
Lead sulfate	1.06×10^{-8} (18°)
Lead sulfide	3.4×10^{-28} (18°)
Lithium carbonate	1.7×10^{-3} (25°)
Magnesium ammonium phosphate	2.5×10^{-13} (25°)
Magnesium carbonate	2.6×10^{-5} (12°)
Magnesium fluoride	7.1×10^{-9} (18°)
Magnesium fluoride	6.4×10^{-9} (27°)
Magnesium hydroxide	1.2×10^{-11} (18°)
Magnesium oxalate	8.57×10^{-5} (18°)
Manganese hydroxide	4×10^{-14} (18°)
Manganese sulfide	1.4×10^{-15} (18°)
Mercuric sulfide	4×10^{-53} to 2×10^{-49} (18°)
Mercurous bromide	1.3×10^{-21} (25°)
Mercurous chloride	2×10^{-18} (25°)
Mercurous iodide	1.2×10^{-28} (25°)
Nickel sulfide	1.4×10^{-24} (18°)
Potassium acid tartrate $[K^+][HC_4H_4O_6^-]$	3.8×10^{-4} (18°)
Silver bromate	3.97×10^{-5} (20°)
Silver bromate	5.77×10^{-5} (25°)
Silver bromide	4.1×10^{-13} (18°)
Silver bromide	7.7×10^{-13} (25°)
Silver carbonate	6.15×10^{-12} (25°)
Silver chloride	0.21×10^{-10} (4.7°)
Silver chloride	0.37×10^{-10} (9.7°)
Silver chloride	1.56×10^{-10} (25°)
Silver chloride	13.2×10^{-10} (50°)
Silver chloride	215×10^{-10} (100°)
Silver chromate	1.2×10^{-12} (14.8°)
Silver chromate	9×10^{-12} (25°)
Silver cyanide $[Ag^+][Ag(CN)^-_2]$	2.2×10^{-12} (20°)
Silver dichromate	2×10^{-7} (25°)
Silver hydroxide	1.52×10^{-8} (20°)
Silver iodate	0.92×10^{-8} (9.4°)
Silver iodide	0.32×10^{-16} (13°)
Silver iodide	1.5×10^{-16} (25°)
Silver sulfide	1.6×10^{-49} (18°)
Silver thiocyanate	0.49×10^{-12} (18°)
Silver thiocyanate	1.16×10^{-12} (25°)
Strontium carbonate	1.6×10^{-9} (25°)
Strontium fluoride	2.8×10^{-9} (18°)
Strontium oxalate	5.61×10^{-8} (18°)
Strontium sulfate	2.77×10^{-7} (2.9°)
Strontium sulfate	3.81×10^{-7} (17.4°)
Zinc hydroxide	1.8×10^{-14} (18–20°)
Zinc oxalate, $ZnC_2O_4.2H_2O$	1.35×10^{-9} (18°)
Zinc sulfide	1.2×10^{-23} (18°)

X-Ray Crystallographic Data, Molar Volumes, and Densities of Minerals and Related Substances

From U.S. Geological Survey Bulletin 1248 by
Richard A. Robie, Philip M. Bethke and Keith M. Beardsley

An extensive list of references and the bases for the calculations and the selection of data are given in the above referenced Bulletin. Bulletin 1248 may be obtained from the Superintendent of Documents, U.S. Government Printing Office, Washington, D.C., 20402.

X-Ray Crystallographic Data of Minerals

	Name and formula	Crystal system	Space group	Structure type	Z	a_o	b_o
	Elements						
1	Silver Ag	cubic	Fm3m(225)	face-centered cubic	4	4.0862 ±.0002	
2	Arsenic As	hex-R	$R\bar{3}m$(166)	arsenic	6	3.760 ±.001	
3	Gold Au	cubic	Fm3m(225)	face-centered cubic	4	4.0786 ±.0002	
4	Bismuth Bi	hex-R	$R\bar{3}m$(166)	arsenic	6	4.5459 ±.0010	
5	Diamond C*	cubic	Fd3m(227)	diamond	8	3.5670 ±.0001	
6	Graphite C*	hex.	C6/mmc(194)	graphite	4	2.4612 ±.0001	
7	Copper Cu	cubic	Fm3m(225)	face-centered cubic	4	3.6150 ±.0005	
8	α-Iron Fe	cubic	Im3m(229)	body-centered cubic	2	2.8664 .0005	
9	Nickel Ni	cubic	Fm3m(225)	face-centered cubic	4	3.5238 ±.0005	
10	Lead Pb	cubic	Fm3m(225)	face-centered cubic	4	4.9505 ±.0005	
11	Platinum Pt	cubic	Fm3m(225)	face-centered cubic	4	3.9231 ±.0005	
12	orthorhombic Sulfur S	orth.	Fddd(70)	S_8 ring molecules	128	10.4646 ±.0020	12.8660 ±.0020
13	monoclinic Sulfur S	mon.	$P2_1/c$(14)	S_8 ring molecules	48	11.04 ±.03	10.98 ±.03
14	rhombohedral Sulfur S	hex-R	$R\bar{3}$(148)	S_6 ring molecules	18	10.818 ±.002	
15	Antimony Sb	hex-R	$R\bar{3}m$(166)	arsenic	6	4.310 ±.001	
16	Selenium Se	hex.	$P3_121$(152) $P3_221$(154)		3	4.3642 ±.0008	
17	Silicon Si	cubic	Fd3m(227)	diamond	8	5.4305 ±.0003	
18	β-Tin (white) Sn	tet.	$I4_1$/amd(141)		4	5.8315 ±.0008	
19	Tellurium Te	hex.	$P3_121$(152) $P3_221$(154)		3	4.4570 ±.0008	
20	Zinc Zn	hex.	$P6_3$/mmc(194)	hexagonal close packed	2	2.665 ±.001	
	Sulfides, arsenides, tellurides, selenides, and sulfosalts						
21	Shandite β-$Ni_3Pb_2S_2$*	hex-R	$R\bar{3}m$(166)		3	5.576 ±.010	
22	High-Argentite Ag_2S I	cubic			4	6.269 ±.020	

X-Ray Crystallographic Data of Minerals

Z; The number of gram formula weights per unit cell.

r; Indicates the data were obtained at an unspecified room temperature and may be taken as 25°±5°C.

*; Indicates the measurements were made on a natural specimen which may have deviated slightly from the listed formula. Densities for these minerals were calculated using the formula weight for the stoichiometric phase.

hex-R; Rhombohedral symmetry. To distinguish from true hexagonal symmetry.

X-Ray Crystallographic Data of Minerals

c_o	α_o	β_o	γ_o	Cell volume 10^{-24} cm^3	Molar volume cm^3	Molar volume cal bar^{-1}	X-Ray density grams cm^{-3}	Temp. °C	
				Elements					
				68.227 ± .010	10.272 ± .002	.24556 ±.00008	10.501 ±.002	25	1
10.555 ±.003				129.23 ± .08	12.972 ± .002	.31007 ±.00023	5.776 ±.004	26	2
				67.847 ± .010	10.215 ± .002	.24420 ±.00008	19.282 ±.003	25	3
11.8622 ±.0030				212.29 ± .11	21.309 ± .011	.50934 ±.00030	9.8071 ±.0050	26	4
				45.385 ± .004	3.4166 ± .0003	.08170 ±.00005	3.5155 ±.0003	25	5
6.7079 ±.0010				35.189 ± .006	5.2982 ± .0009	.12668 ±.00007	2.2670 ±.0004	15	6
				47.242 ± .020	7.1128 ± .0030	.17005 ±.00012	8.9331 ±.0037	25	7
				23.551 ± .012	7.0918 ± .0037	.16954 ±.00013	7.8748 ±.0041	25	8
				43.756 ± .019	6.5880 ± .0028	.15750 ±.00011	8.9117 ±.0038	25	9
				121.32 ± .04	18.267 ± .006	.43663 ±.00018	11.342 ±.003	25	10
				60.379 ± .023	9.0909 ± .0035	.21732 ±.00013	21.460 ±.008	25	11
24.4860 ±.0040				3296.73 ± .97	15.511 ± .005	.37078 ±.00015	2.0671 ±.0006	25	12
10.92 ±.03		96.73 ±.50		1314.6 ± 6.4	16.49 ± .08	.3943 ±.0020	1.944 ±.009	103	13
4.280 ±.001				433.78 ± .19	14.514 ± .006	.34693 ±.00020	2.2092 ±.0010	r	14
11.279 ±.003				181.45 ± .09	18.213 ± .010	.43535 ±.00028	6.685 ±.004	26	15
4.9588 ±.0008				81.793 ± .033	16.420 ± .007	.39249 ±.00020	4.8088 ±.0019	26	16
				160.15 ± .03	12.056 ± .002	.28819 ±.00009	2.3296 ±.0004	25	17
3.1813 ±.0006				108.18 ± .04	16.289 ± .005	.38935 ±.00017	7.2867 ±.0024	26	18
5.9290 ±.0010				102.00 ± .04	20.476 ± .008	.48944 ±.00024	6.2316 ±.0025	25	19
4.947 ±.001				30.428 ± .024	9.162 ± .007	.2190 ±.0002	7.134 ±.006	25	20
				Sulfides, arsenides, tellurides, selenides, and sulfosalts					
13.658 ±.010				367.76 ± 1.35	73.83 ± .27	1.765 ±.007	8.867 ±.033	r	21
				246.4 ± 2.4	37.09 ± .36	.8866 ±.0085	6.680 ±.064	600	22

X-Ray Crystallographic Data of Minerals

	Name and formula	Crystal system	Space group	Structure type	Z	a_o	b_o
	Sulfides, arsenides, tellurides, selenides, and sulfosalts—Continued						
23	Argentite Ag_2S II	cubic			2	4.870 ±.008	
24	Acanthite Ag_2S III	mon.	$P2_1/c(14)$		4	4.228 ±.002	6.928 ±.005
25	High-Naumanite Ag_2Se	cubic			2	4.993 ±.016	
26	Ag_2Te I	cubic			2	5.29 ±.01	
27	Ag_2Te II	cubic			4	6.585 ±.010	
28	Hessite Ag_2 Te III	mon.	$P2_1/c(14)$		4	8.09 ±.02	4.48 ±.01
29	$Ag_{1.55}Cu_{.45}S$ I	cubic			4	6.110 ±.010	
30	$Ag_{1.55}Cu_{.45}S$ II	cubic			2	4.825 ±.005	
31	Jalpaite $Ag_{1.55}Cu_{.45}S$ III	tet.			16	8.673 ±.004	
32	$Ag_{.93}Cu_{1.07}S$ I	cubic			4	5.961 ±.009	
33	$Ag_{.93}Cu_{1.07}S$ II	hex.			2	4.138 ±.004	
34	Stromeyerite $Ag_{.93}Cu_{1.07}S$ III	orth.	Cmcm(63)		4	4.066 ±.002	6.628 ±.003
35	Eucairite AgCuSe	orth.	pseudo P4/nmm(129)		10	4.105 ±.010	20.35 ±.02
36	Petzite Ag_3AuTe_2*	cubic	$I4_132(214)$		8	10.38 ±.02	
37	Maldonite Au_2Bi	cubic	Fd3m(227)	Cu_2Mg	8	7.958 ±.002	
38	High-Digenite Cu_2S I	cubic			4	5.725 ±.010	
39	High-Chalcocite Cu_2S II	hex.			2	3.961 ±.004	
40	Chalcocite Cu_2S III	orth.	Ab2m(39)		96	11.881 ±.004	27.323 ±.010
41	Digenite $Cu_{1.79}S$ (Cu rich side)	cubic		deformed fluorite	4	5.5695 ±.0010	
42	Digenite $Cu_{1.77}S$ (S rich side)	cubic		deformed fluorite	4	5.5542 ±.0010	
43	Berzelianite Cu_2Se	cubic			4	5.85 ±.01	
44	High-Bornite Cu_5FeS_4*	cubic			1	5.50 ±.01	
45	Metastable Bornite Cu_5FeS_4	cubic			8	10.94 ±.02	
46	Low-Bornite Cu_5FeS_4*	tet.	$P\overline{4}2_1c(144)$		16	10.94 ±.02	
47	Umangite Cu_3Se_2	tet.	P4/mmm(123)		2	6.402 ±.010	
48	Heazelwoodite Ni_3S_2	hex-R	R32(155)		3	5.746 ±.001	
49	Maucherite $Ni_{11}As_8$	tet.	$P4_12_12(92)$		4	6.870 ±.001	

X-Ray Crystallographic Data of Minerals

c_o	α_o	β_o	γ_o	Cell volume 10^{-24} cm³	Molar volume cm³	Molar volume cal bar⁻¹	X-Ray density grams cm⁻³	Temp. °C	
Sulfides, arsenides, tellurides, selenides, and sulfosalts—Continued									
				115.5	34.78	.8313	7.125	189	23
				± .6	± .17	±.0041	±.035		
7.862		99.58		227.08	34.19	.8172	7.248	25	24
±.003		±.30		± .29	± .04	±.0011	±.009		
				124.48	37.48	.8959	7.862	170	25
				± 1.20	± .36	±.0087	±.076		
				148.0	44.58	1.065	7.702	825	26
				± .8	± .26	±.006	±.044		
				285.54	42.99	1.028	7.986	250	27
				± 1.30	± .20	±.005	±.036		
8.96		123.33		271.33	40.85	.9764	8.405	r	28
±.02		±.30		± 1.43	± .22	±.0052	±.044		
				228.10	34.34	.8209	6.635	300	29
				± 1.12	± .17	±.0041	±.033		
				112.33	33.83	.8085	6.736	116	30
				± .35	± .11	±.0026	±.021		
11.756				884.30	33.286	.79559	6.8455	r	31
±.006				± .93	± .035	±.00088	±.0072		
				211.82	31.89	.7623	6.283	196	32
				± .96	± .14	±.0035	±.029		
7.105				105.36	31.73	.7583	6.316	100	33
±.007				± .23	± .07	±.0017	±.014		
7.972				214.84	32.35	.7732	6.194	r	34
±.004				± .18	± .03	±.0007	±.005		
6.31				527.12	31.75	.7588	7.887	r	35
±.01				± 1.62	± .10	±.0024	±.024		
				1118.4	84.19	2.012	9.214	r	36
				± 6.5	± .49	±.012	±.053		
				503.98	37.94	.9068	15.891	r	37
				± .38	± .03	±.0007	±.012		
				187.64	28.25	.6753	5.633	465	38
				± .98	± .15	±.0036	±.030		
6.722				91.34	27.50	.6574	5.786	152	39
±.007				± .21	± .06	±.0015	±.013		
13.491				4379.5	27.475	.65671	5.7924	r	40
±.004				± 2.5	± .016	±.00043	±.0034		
				172.76	26.012	.6217	5.605	25	41
				± .09	± .014	±.0004	±.003		
				171.34	25.798	.6166	5.602	25	42
				± .09	± .014	±.0004	±.003		
				200.2	30.14	.7205	6.835	170	43
				± 1.0	± .15	±.0037	±.035		
				166.4	100.2	2.395	5.008	240	44
				± .9	± .5	±.013	±.027		
				1309.34	98.57	2.356	5.091	r	45
				± 7.18	± .54	±.013	±.028		
21.88				2618.7	98.57	2.356	5.091	r	46
±.04				±10.7	± .40	±.010	±.021		
4.276				175.25	52.77	1.261	6.604	r	47
±.010				± .68	± .21	±.005	±.026		
7.134				203.98	40.95	.9788	5.867	r	48
±.002				± .09	± .02	±.0005	±.003		
21.81				1029.36	154.98	3.7043	8.0343	r	49
±.01				± .56	± .08	±.0021	±.0044		

	Name and formula	Crystal system	Space group	Structure type	Z	a_o	b_o
	Sulfides, arsenides, tellurides, selenides, and sulfosalts—Continued						
50	Pentlandite $Fe_{5.25}Ni_{3.75}S_8$	cubic	Fm3m(225)		4	10.196 ±.010	
51	Pentlandite $Fe_{4.75}Ni_{5.25}S_8$	cubic	Fm3m(225)		4	10.095 ±.010	
52	Sternbergite $AgFe_2S_3$*	orth.	Ccmm(63)		8	11.60 ±.02	12.675 ±.020
53	Argentopyrite $AgFe_2S_3$*	orth.	Pmmm(47)		4	6.64 ±.01	11.47 ±.02
54	Realgar AsS*	mon.	$P2_1/m(11)$		16	9.29 ±.05	13.53 ±.05
55	Oldhamite CaS	cubic	Fm3m(225)	rock salt	4	5.689 ±.006	
56	Greenockite CdS	hex.	$P6_3mc(186)$	zincite	2	4.1354 ±.0010	
57	Hawleyite CdS	cubic	$F\bar{4}3m(216)$	sphalerite	4	5.833 ±.002	
58	(hypothetical) CdS	cubic	Fm3m(225)	rock salt	4	5.516 ±.002	
59	Cadmoselite CdSe	hex.	$P6_3mc(186)$	zincite	2	4.2977 ±.0010	
60	CdTe	cubic	$F\bar{4}3m(216)$	sphalerite	4	6.4805 ±.0006	
61	(hypothetical) CoS	cubic	$F\bar{4}3m(216)$	sphalerite	4	5.339 ±.001	
62	Chalcopyrite ($CuFeS_2$) $CuFeS_{1.90}$	tet.	I42d(122)		4	5.2988 ±.0010	
63	Cubanite $CuFe_2S_3$*	orth.	Pcmn(62)		4	6.46 ±.01	11.12 ±.01
64	Covellite CuS	hex.	$P6_3/mmc(194)$		6	3.792 ±.001	
65	Klockmannite CuSe	hex.		deformed covellite	78	14.206 ±.010	
66	Troilite FeS	hex.	$P6_3/mmc(194)$	niccolite	2	3.446 ±.003	
67	Pyrrhotite $Fe_{.980}S$	hex.	$P6_3/mmc(194)$	defect niccolite	2	3.446 ±.001	
68	Pyrrhotite $Fe_{.885}S$	hex.	$P6_3/mmc(194)$	defect niccolite	2	3.440 ±.001	
69	(hypothetical) FeS	cubic	$F\bar{4}3m(216)$	sphalerite	4	5.455 ±.001	
70	(hypothetical) FeS	hex.	$P6_3mc(186)$	zincite	2	3.872 ±.001	
71	Cinnabar HgS	hex.	$P3_121(152)$ $P3_21(154)$	cinnabar	3	4.149 ±.001	
72	Metacinnabar HgS	cubic	$F\bar{4}3m(216)$	sphalerite	4	5.8517 ±.0010	
73	Tiemannite HgSe	cubic	$F\bar{4}3m(216)$	sphalerite	4	6.0853 ±.0050	
74	Coloradoite HgTe	cubic	$F\bar{4}3m(216)$	sphalerite	4	6.4600 ±.0006	
75	Alabandite MnS	cubic	Fm3m(225)	rock salt	4	5.2234 ±.0005	
76	(hypothetical) MnS	cubic	$F\bar{4}3m(216)$	sphalerite	4	5.611 ±.002	

c_o	α_o	β_o	γ_o	Cell volume 10^{-24} cm³	Molar volume cm³	Molar volume cal bar⁻¹	X-Ray density grams cm⁻³	Temp. °C	
Sulfides, arsenides, tellurides, selenides, and sulfosalts—Continued									
				1059.96 ± 3.12	159.59 ± .47	3.8144 ±.0113	4.823 ±.014	r	50
				1028.77 ± 3.06	154.89 ± .46	3.702 ±.011	4.998 ±.015	r	51
6.63 ±.01				974.81 ± 2.71	73.39 ± .20	1.754 ±.005	4.303 ±.012	r	52
6.45 ±.02				491.2 ± 1.9	73.96 ± .29	1.768 ±.007	4.269 ±.017	r	53
6.57 ±.03		106.55 ±.30		791.6 ± 6.4	29.80 ± .24	.7122 ±.0058	3.591 ±.029	r	54
				184.12 ± .58	27.722 ± .088	.6626 ±.0021	2.602 ±.008	r	55
6.7120 ±.0010				99.407 ± .050	29.934 ± .015	.71549 ±.00041	4.8261 ±.0024	r	56
				198.46 ± .20	29.88 ± .03	.7142 ±.0008	4.835 ±.005	r	57
				167.83 ± .18	25.27 ± .03	.6040 ±.0007	5.717 ±.006	r	58
7.0021 ±.0010				112.00 ± .05	33.727 ± .016	.80614 ±.00044	5.6738 ±.0028	r	59
				272.16 ± .08	40.977 ± .012	.97943 ±.00032	5.8569 ±.0016	25	60
				152.19 ± .09	22.91 ± .02	.5477 ±.0004	3.971 ±.002	r	61
10.434 ±.005				292.96 ± .18	44.109 ± .027	1.0543 ±.0007	4.0878 ±.0025	r	62
6.23 ±.01				447.53 ± 1.08	67.38 ± .16	1.611 ±.004	4.026 ±.010	r	63
16.34 ±.01				203.48 ± .16	20.42 ± .02	.4882 ±.0005	4.682 ±.001	r	64
17.25 ±.05				3014.8 ± 9.7	23.28 ± .08	.5564 ±.0018	6.122 ±.020	r	65
5.877 ±.001				60.439 ± .106	18.20 ± .03	.4350 ±.0008	4.830 ±.009	28	66
5.848 ±.002				60.14 ± .04	18.11 ± .01	.4329 ±.0003	4.793 ±.003	28	67
5.709 ±.003				58.507 ± .046	17.62 ± .02	.4211 ±.0004	4.625 ±.004	28	68
				162.32 ± .09	24.44 ± .01	.5842 ±.0004	3.597 ±.002		69
6.345 ±.002				82.38 ± .05	24.81 ± .02	.5930 ±.0004	3.544 ±.002	r	70
9.495 ±.002				141.55 ± .07	28.416 ± .015	.6792 ±.0004	8.187 ±.004	r	71
				200.38 ± .10	30.169 ± .016	.7211 ±.0004	7.712 ±.004	r	72
				225.34 ± .56	33.928 ± .084	.8110 ±.0020	8.239 ±.020	r	73
				269.59 ± .08	40.590 ± .011	.97016 ±.00032	8.0855 ±.0023	r	74
				142.51 ± .04	21.457 ± .006	.51289 ±.00019	4.0546 ±.0012	r	75
				176.65 ± .19	26.60 ± .03	.6357 ±.0007	3.271 ±.004	r	76

	Name and formula	Crystal system	Space group	Structure type	Z	a_o	b_o
	Sulfides, arsenides, tellurides, selenides, and sulfosalts—Continued						
77	(hypothetical) MnS	hex.	$P6_3mc(186)$	zincite	2	3.986 ±.001	
78	Niccolite NiAs	hex.	$P6_3/mmc(194)$	niccolite	2	3.618 ±.001	
79	Millerite NiS	hex-R	R3m(160)		9	9.616 ±.001	
80	Breithauptite NiSb	hex.	$P6_3/mmc(194)$	niccolite	2	3.942 ±.001	
81	Galena PbS	cubic	Fm3m(225)	rock salt	4	5.9360 ±.0005	
82	Clausthalite PbSe	cubic	Fm3m(225)	rock salt	4	6.1255 ±.0005	
83	Teallite $PbSnS_2$	orth.	Pbnm(62)	GeS	2	4.266 ±.003	11.419 ±.007
84	Altaite PbTe	cubic	Fm3m(225)	rock salt	4	6.4606 ±.0005	
85	Cooperite PtS	tet.	$P4_2/mmc(131)$		2	3.4699 ±.0006	
86	Herzenbergite SnS	orth.	Pbnm(62)	GeS	4	4.328 ±.002	11.190 ±.004
87	Sphalerite ZnS	cubic	$F\overline{4}3m(216)$	sphalerite	4	5.4093 ±.0005	
88	Wurtzite ZnS	hex.	$P6_3mc(186)$	zincite	2	3.8230 ±.0010	
89	Stilleite ZnSe	cubic	$F\overline{4}3m(216)$	sphalerite	4	5.6685 ±.0005	
90	ZnTe	cubic	$F\overline{4}3m(216)$	sphalerite	4	6.1020 ±.0006	
91	Orpiment As_2S_3*	mon.	$P2_1/n(14)$		4	11.49 ±.02	9.59 ±.02
92	Bismuthinite Bi_2S_3	orth.	Pbnm(62)	stibnite	4	11.150 ±.004	11.300 ±.004
93	Tellurobismuthite Bi_2Te_3	hex-R	$R\overline{3}m(166)$	Bi_2Te_2S	3	4.3835 ±.0020	
94	Stibnite Sb_2S_3	orth.	Pbnm(62)	stibnite	4	11.229 ±.004	11.310 ±.004
95	Linnaeite Co_3S_4	cubic	Fd3m(227)	spinel	8	9.401 ±.001	
96	Greigite Fe_3S_4	cubic	Fd3m(227)	spinel	8	9.876 ±.002	
97	Daubreeite $FeCr_2S_4$	cubic	Fd3m(227)	spinel	8	9.966 ±.005	
98	Violarite $FeNi_2S_4$	cubic	Fd3m(227)	spinel	8	9.464 ±.005	
99	Polymidite Ni_3S_4	cubic	Fd3m(227)	spinel	8	9.480 ±.001	
100	Co-Safflorite $CoAs_2$	mon.		deformed marcasite	2	5.049 ±.002	5.872 ±.002
101	Safflorite $(CO_{.5}Fe_{.5})As_2$	orth.	Pnnm(58)	marcasite	2	5.231 ±.002	5.953 ±.002
102	Cobaltite CoAsS*	cubic	$P2_13(198)$	NiSbS	4	5.60 ±.05	
103	Glaucodot (Co,Fe)AsS*	orth.	Cmmm(65)		24	6.64 ±.05	28.39 ±.10

X-Ray Crystallographic Data of Minerals

c_o	a_o	β_o	γ_o	Cell volume 10^{-24} cm³	Molar volume cm³	Molar volume cal bar⁻¹	X-Ray density grams cm⁻³	Temp. °C	
Sulfides, arsenides, tellurides, selenides, and sulfosalts—Continued									
6.465 ±.002				88.96 ± .05	26.79 ± .02	.6403 ±.0004	3.248 ±.002	r	77
5.034 ±.001				57.07 ± .03	17.18 ± .01	.4108 ±.0003	7.776 ±.005	r	78
3.152 ±.001				252.41 ± .10	16.891 ± .006	.40374 ±.00020	5.3743 ±.0020	r	79
5.155 ±.001				69.37 ± .04	20.89 ± .01	.4994 ±.0004	8.639 ±.005	r	80
				209.16 ± .05	31.492 ± .008	.75272 ±.00024	7.5973 ±.0019	26	81
				229.84 ± .06	34.605 ± .009	.82713 ±.00025	8.2690 ±.0020	r	82
4.090 ±.002				199.24 ± .21	59.996 ± .063	1.4340 ±.0016	6.501 ±.007	r	83
				269.66 ± .06	40.601 ± .009	.97043 ±.00027	8.2459 ±.0019	r	84
6.1098 ±.0010				73.563 ± .028	22.152 ± .008	.5295 ±.0003	10.254 ±.004	r	85
3.978 ±.001				192.66 ± .12	29.01 ± .02	.6933 ±.0005	5.197 ±.003	r	86
				158.28 ± .04	23.831 ± .007	.56962 ±.00020	4.0885 ±.0011	r	87
6.2565 ±.0010				79.190 ± .043	23.846 ± .013	.56998 ±.00036	4.0859 ±.0022	r	88
				182.14 ± .05	27.424 ± .007	.65548 ±.00022	5.2630 ±.0014	r	89
				227.20 ± .07	34.209 ± .010	.81765 ±.00029	5.6410 ±.0017	r	90
4.25 ±.01		90.45 ±.30		468.3 ± 1.7	70.51 ± .25	1.685 ±.006	3.490 ±.013	r	91
3.981 ±.001				501.59 ± .28	75.520 ± .043	1.8050 ±.0011	6.8081 ±.0038	26	92
30.487 ±.003				507.33 ± .47	101.85 ± .09	2.4342 ±.0023	7.862 ±.007	25	93
3.8389 ±.0010				487.54 ± .28	73.406 ± .042	1.7545 ±.0010	4.6276 ±.0026	25	94
				830.85 ± .27	62.548 ± .020	1.4950 ±.0005	4.8772 ±.0016	r	95
				963.26 ± .59	72.52 ± .04	1.733 ±.001	4.079 ±.003	r	96
				989.83 ± 1.49	74.52 ± .11	1.781 ±.003	3.866 ±.006	r	97
				847.66 ± 1.34	63.81 ± .10	1.525 ±.002	4.725 ±.008	r	98
				851.97 ± .27	64.138 ± .020	1.5330 ±.0005	4.7458 ±.0015	r	99
3.127 ±.001		90.45 ±.20		92.706 ± .057	27.92 ± .02	.6672 ±.0005	7.479 ±.005	26	100
2.962 ±.002				92.237 ± .078	27.775 ± .024	.6639 ±.0006	7.461 ±.006	26	101
				175.62 ± 4.70	26.44 ± .71	.6320 ±.0170	6.275 ±.168	r	102
5.64 ±.05				1063.2 ±12.9	26.68 ± .32	.6377 ±.0078	6.161 ±.075	r	103

	Name and formula	Crystal system	Space group	Structure type	Z	a_o	b_o
	Sulfides, arsenides, tellurides, selenides, and sulfosalts—Continued						
104	Cattierite CoS_2	cubic	Pa3(205)	pyrite	4	5.5345 ±.0005	
105	Trogtalite $CoSe_2$	cubic	Pa3(205)	pyrite	4	5.8588 ±.0010	
106	Loellingite $FeAs_2$	orth.	Pnnm(58)	marcasite	2	5.300 ±.002	5.981 ±.002
107	Arsenopyrite FeAsS*	tri.	$P\overline{1}$(2)		4	5.760 ±.010	5.690 ±.005
108	Gudmundite FeSbS*	mon.	$B2_1/d$(14)		8	10.00 ±.05	5.93 ±.03
109	Pyrite FeS_2	cubic	Pa3(205)	pyrite	4	5.4175 ±.0005	
110	Marcasite FeS_2*	orth.	Pnnm(58)	marcasite	2	4.443 ±.002	5.423 ±.002
111	Ferroselite $FeSe_2$	orth.	Pnnm(58)	marcasite	2	4.801 ±.005	5.778 ±.005
112	Frohbergite $FeTe_2$	orth.	Pnnm(58)	marcasite	2	5.265 ±.005	6.265 ±.005
113	Hauerite MnS_2	cubic	Pa3(205)	pyrite	4	6.1014 ±.0006	
114	Molybdenite MoS_2	hex.	$P6_3$/mmc(194)	molybdenite	2	3.1604 ±.0010	
115	Rammelsbergite $NiAs_2$	orth.	Pnnm(58)	marcasite	2	4.757 ±.002	5.797 ±.004
116	Pararammelsbergite $NiAs_2$	orth.	Pbca(61)		8	5.75 ±.01	5.82 ±.01
117	Gersdorfite NiAsS	cubic	$P2_13$(198)		4	5.693 ±.001	
118	Vaesite NiS_2	cubic	Pa3(205)	pyrite	4	5.6873 ±.0005	
119	$NiSe_2$	cubic	Pa3(205)	pyrite	4	5.9604 ±.0010	
120	Melonite $NiTe_2$	hex.	$P\overline{3}$ml(164)	cadmium iodide	1	3.869 ±.010	
121	Sperrylite $PtAs_2$	cubic	Pa3(205)	pyrite	4	5.968 ±.005	
122	Laurite RuS_2	cubic	Pa3(205)	pyrite	4	5.60 ±.02	
123	Tungstenite WS_2	hex.	$P6_3$/mmc(194)	molybdenite	2	3.154 ±.001	
124	Co-Skutterudite $CoAs_{3-x}$ $CoAs_{2.95}$	cubic	Im3(204)		8	8.2060 ±.0010	
125	Fe-Skutterudite $FeAs_{3-x}$ $FeAs_{2.95}$	cubic	Im3(204)		8	8.1814 ±.0010	
126	Ni-Skutterudite $NiAs_{3-x}$ $NiAs_{2.95}$	cubic	Im3(204)		8	8.3300 ±.0010	
127	Tennantite $Cu_{12}As_4S_{13}$	cubic	$I\overline{4}3$m(217)	tetrahedrite	2 2	10.190 ±.004	
128	Tetrahedrite $Cu_{12}Sb_4S_{13}$	cubic	$I\overline{4}3$m(217)	tetrahedrite	2 2	10.327 ±.004	
129	Enargite Cu_3AsS_4	orth.	Pnn2(34)		2	6.426 ±.005	7.422 ±.005
130	Luzonite Cu_3AsS_4*	tet.	$I\overline{4}2$m(121)		2	5.289 ±.005	

c_o	α_o	β_o	γ_o	Cell volume 10^{-24} cm^3	Molar volume cm^3	Molar volume cal bar^{-1}	X-Ray density grams cm^{-3}	Temp. °C	
Sulfides, arsenides, tellurides, selenides, and sulfosalts—Continued									
				169.53 ± .05	25.524 ± .007	.61009 ±.00021	4.8213 ±.0013	r	104
				201.11 ± .10	30.279 ± .016	.72374 ±.00042	7.1618 ±.0037	r	105
2.882 ±.001				91.357 ± .056	27.51 ± .02	.6576 ±.0005	7.477 ±.005	26	106
5.785 ±.005	90.00 ±.20	112.23 ±.20	90.00 ±.20	175.51 ± .44	26.42 ± .07	.6316 ±.0016	6.162 ±.015	r	107
6.73 ±.03		90.00 ±.50		399.09 ± 3.35	30.04 ± .25	.7181 ±.0061	6.978 ±.059	r	108
				159.00 ± .04	23.940 ± .007	.57221 ±.00020	5.0116 ±.0014	r	109
3.3876 ±.0015				81.622 ± .060	24.579 ± .018	.58749 ±.00047	4.8813 ±.0036	25	110
3.587 ±.004				99.50 ± .17	29.96 ± .05	.7162 ±.0013	7.134 ±.013	r	111
3.869 ±.002				127.62 ± .17	38.43 ± .05	.9185 ±.0013	8.094 ±.011	r	112
				227.14 ± .07	34.198 ± .010	.81741 ±.00029	3.4816 ±.0010	28	113
12.295 ±.002				106.35 ± .07	32.025 ± .021	.76547 ±.00055	4.9982 ±.0033	26	114
3.542 ±.002				97.645 ± .096	29.41 ± .03	.7030 ±.0007	7.091 ±.007	26	115
11.428 ±.02				382.42 ± 1.15	28.79 ± .09	.6882 ±.0021	7.244 ±.022	r	116
				184.51 ± .10	27.78 ± .01	.6640 ±.0004	5.964 ±.003	26	117
				183.96 ± .05	27.697 ± .007	.66203 ±.00022	4.4350 ±.0012	r	118
				211.75 ± .11	31.882 ± .016	.76204 ±.00043	6.7948 ±.0034	20	119
5.308 ±.010				68.81 ± .38	41.44 ± .23	.9905 ±.0055	7.575 ±.042	84	120
				212.56 ± .53	32.00 ± .08	.7650 ±.0020	10.778 ±.027	r	121
				175.6 ± 1.9	26.44 ± .28	.6320 ±.0068	6.248 ±.067	r	122
12.362 ±.004				106.50 ± .08	32.069 ± .023	.76652 ±.00059	7.7325 ±.0055	26	123
				552.58 ± .20	41.599 ± .015	.99428 ±.00041	6.7298 ±.0025	r	124
				547.62 ± .20	41.226 ± .015	.98537 ±.00041	6.7158 ±.0025	r	125
				578.01 ± .21	43.513 ± .016	1.0400 ±.0004	6.4286 ±.0023	r	126
				1058.09 ± 1.25	318.62 ± .38	7.1652 ±.0090	4.642 ±.006	r	127
				1101.3 ± 1.3	331.64 ± .39	7.9266 ±.0094	5.024 ±.006	r	128
6.144 ±.005				293.03 ± .38	88.24 ± .12	2.109 ±.003	4.463 ±.006	26	129
10.440 ±.008				292.04 ± .60	87.94 ± .18	2.1019 ±.0043	4.478 ±.009	26	130

	Name and formula	Crystal system	Space group	Structure type	Z	a_o	b_o
	Sulfides, arsenides, tellurides, selenides, and sulfosalts—Continued						
131	Famatimite Cu_3SbS_4*	tet.	$I\bar{4}m(121)$		2	5.384 ±.005	
132	Proustite Ag_3AsS_3	hex-R	R3c(161)		6	10.816 ±.001	
133	Pyrargyrite Ag_3SbS_3	hex-R	R3c(161)		6	11.052 ±.002	
134	Miargyrite $AgSbS_2$*	mon.	Cc(9)		8	12.862 ±.013	4.111 ±.004
	Oxides and hydroxides						
135	Corundum Al_2O_3	hex-R	$R\bar{3}c(167)$	corundum	6	4.7591 ±.0004	
136	Boehmite AlO(OH)*	orth.	Cmcm(63)	lepidocrocite	4	2.868 ±.003	12.227 ±.003
137	Diaspore AlO(OH)*	orth.	Pbnm(62)		4	4.401 ±.005	9.421 ±.005
138	Gibbsite $Al(OH)_3$	mon.	$P2_1/n(14)$		8	9.719 ±.002	5.0705 ±.0010
139	Arsenolite As_2O_3	cubic	Fd3m(227)	diamond	16	11.074 ±.005	
140	Claudetite As_2O_3	mon.	$P2_1/n(14)$		4	5.339 ±.002	12.984 ±.005
141	Bromellite BeO	hex.	$P6_3mc(186)$	zincite	2	2.6979 ±.0005	
142	Bismite α-Bi_2O_3	mon.	$P2_1/c(14)$	pseudo orthorhombic	8	8.166 ±.005	13.827 ±.010
143	Lime CaO	cubic	Fm3m(225)	rock salt	4	4.8108 ±.0005	
144	Portlandite $Ca(OH)_2$	hex.	$P\bar{3}ml(164)$	CdI_2	1	3.5933 ±.0005	
145	Monteponite CdO	cubic	Fm3m(225)	rock salt	4	4.6953 ±.0010	
146	Cerianite CeO_2	cubic	Fm3m(225)	fluorite	4	5.4110 ±.0020	
147	CoO	cubic	Fm3m(225)	rock salt	4	4.260 ±.002	
148	Eskolaite Cr_2O_3	hex-R	$R\bar{3}c(167)$	corundum	6	4.9607 ±.0020	
149	Tenorite CuO	mon.	C2/c(15)		4	4.684 ±.005	3.425 ±.005
150	Cuprite Cu_2O	cubic	Pn3m(224)		2	4.2696 ±.0010	
151	Wustite $Fe_{.953}O$	cubic	Fm3m(225)	defect rock salt	4	4.3088 ±.0003	
152	Hematite Fe_2O_3	hex-R	$R\bar{3}c(167)$	corundum	6	5.0329 ±.0010	
153	Magnetite Fe_3O_4	cubic	Fd3m(227)	spinel	8	8.3940 ±.0005	
154	Goethite α-FeO(OH)*	orth.	Pbnm(62)		4	4.596 ±.005	9.957 ±.010
155	Lepidocrocite γ-FeO(OH)*	orth.	Amam(63)		4	3.868 ±.010	12.525 ±.010
156	α-Ga_2O_3	hex-R	$R\bar{3}c(167)$	corundum	6	4.9793 ±.0010	

c_o	α_o	β_o	γ_o	Cell volume 10^{-24} cm^3	Molar volume cm^3	Molar volume cal bar^{-1}	X-Ray density grams cm^{-3}	Temp. °C	
Sulfides, arsenides, tellurides, selenides, and sulfosalts—Continued									
10.770 ±.008				312.19 ± .62	94.01 ± .19	2.2469 ±.0045	4.687 ±.009	26	131
8.6948 ±.0013				880.89 ± .21	88.420 ± .021	2.1133 ±.0006	5.595 ±.001	26	132
8.7177 ±.0020				922.18 ± .40	92.564 ± .040	2.2124 ±.0010	5.8506 ±.0025	26	133
13.220 ±.010		98.63 ±.15		691.10 ± 1.14	52.027 ± .086	1.244 ±.002	5.646 ±.009	r	134
Oxides and hydroxides									
12.9894 ±.0030				254.78 ± .07	25.575 ± .007	.61128 ±.00022	3.9869 ±.0011	25	135
3.700 ±.003				129.75 ± .17	19.535 ± .026	.46695 ±.00067	3.071 ±.004	26	136
2.845 ±.002				117.96 ± .17	17.760 ± .026	.4245 ±.0007	3.378 ±.005	r	137
8.6412 ±.0010		94.57 ±.25		424.49 ± .20	31.956 ± .015	.7638 ±.0004	2.441 ±.001	r	138
				1358.0 ± 1.8	51.118 ± .069	1.2218 ±.0017	3.870 ±.005	25	139
4.5405 ±.0010		94.27 ±.10		313.88 ± .19	47.259 ± .028	1.1296 ±.0007	4.1863 ±.0025	25	140
4.3772 ±.0005				27.592 ± .011	8.3086 ± .0032	.19862 ±.00012	3.0104 ±.0012	26	141
5.850 ±.004		90.00 ±.20		660.53 ± .77	49.73 ± .06	1.1885 ±.0014	9.371 ±.011	25	142
				111.34 ± .03	16.764 ± .005	.40071 ±.00017	3.3453 ±.0010	26	143
4.9086 ±.0020				54.888 ± .027	33.056 ± .016	.79011 ±.00043	2.2415 ±.0011	26	144
				103.51 ± .07	15.585 ± .010	.37254 ±.00028	8.2386 ±.0053	27	145
				158.43 ± .18	23.853 ± .026	.57016 ±.00068	7.216 ±.008	26	146
				77.31 ± .11	11.64 ± .02	.2782 ±.0004	6.438 ±.009	26	147
13.599 ±.010				289.82 ± .32	29.090 ± .032	.6953 ±.0008	5.225 ±.006	r	148
5.129 ±.005		99.47 ±.17		81.16 ± .17	12.22 ± .03	.2921 ±.0007	6.509 ±.014	26	149
				77.833 ± .055	23.437 ± .016	.56021 ±.00044	6.1047 ±.0043	26	150
				79.996 ± .017	12.044 ± .003	.28791 ±.00011	5.7471 ±.0012	17	151
		13.7492 ±.0010		301.61 ± .12	30.274 ± .012	.72361 ±.00034	5.2749 ±.0021	25	152
				591.43 ± .11	44.524 ± .008	1.0642 ±.0002	5.2003 ±.0009	22	153
3.021 ±.003				138.2 ± .2	20.82 ± .04	.4975 ±.0009	4.269 ±.008	r	154
3.066 ±.003				148.54 ± .43	22.364 ± .064	.5346 ±.0016	3.973 ±.011	r	155
13.429 ±.003				288.34 ± .13	28.943 ± .013	.69179 ±.00036	6.4762 ±.0030	24	156

	Name and formula	Crystal system	Space group	Structure type	Z	a_o	b_o
	Oxides and hydroxides—Continued						
157	Low-germania GeO_2	tet.	P4/mnm(136)	rutile	2	4.3963 ±.0010	
158	High-germania GeO_2	hex.	$P3_121(152)$ $P3_221(154)$	α-quartz	3	4.987 ±.002	
159	Ice H_2O	hex.	$P6_3/mmc(194)$		4	4.5212 ±.0010	
160	Hafnia HfO_2	mon.	$P2_1/c(14)$	baddeleyite	4	5.1156 ±.0010	5.1722 ±.0010
161	Montroydite HgO	orth.	Pnma(62)		4	6.608 ±.003	5.518 ±.003
162	Periclase MgO	cubic	Fm3m(225)	rock salt	4	4.2117 ±.0005	
163	Brucite $Mg(OH)_2$	hex.	$P\bar{3}ml(164)$	CdI_2	1	3.147 ±.004	
164	Manganosite MnO	cubic	Fm3m(225)	rock salt	4	4.4448 ±.0005	
165	Pyrolusite MnO_2	tet.	P4/mnm(136)	rutile	2	4.388 ±.003	
166	Bixbyite Mn_2O_3	cubic	Ia3(206)	Tl_2O_3	16	9.411 ±.005	
167	Hausmanite Mn_3O_4	tet.	$I4_1/amd(141)$		8	8.136 ±.005	
168	Molybdite MoO_3	orth.	Pbnm(62)		4	3.962 ±.002	13.858 ±.005
169	Bunsenite NiO	cubic	Fm3m(225)	rock salt	4	4.177 ±.002	
170	Litharge PbO red	tet.	P4/nmm(129)		2	3.9759 ±.0040	
171	Massicot PbO yellow	orth.	Pb2a(32)		4	5.489 ±.003	4.755 ±.004
172	Minium Pb_3O_4	tet.	$P4_2/mbc(135)$		4	8.815 ±.005	
173	Senarmontite Sb_2O_3	cubic	Fm3m(225)	arsenic trioxide	16	11.152 ±.003	
174	Valentinite Sb_2O_3	orth.	Pccn(56)	antimony trioxide	4	4.914 ±.002	12.468 ±.005
175	Cervantite Sb_2O_4	cubic	Fd3m(227)		8	10.305 ±.005	
176	Selenolite SeO_2	tet.	$P4_2/mbc(135)$ $P4_2bc(106)$		8	8.35 ±.01	
177	α-Quartz SiO_2*	hex.	$P3_121(152)$ $P3_221(154)$		3	4.9136 ±.0001	
178	β-Quartz SiO_2*	hex.	$P6_422(181)$ $P6_222(180)$		3	4.999 ±.001	
179	α-Cristobalite SiO_2	tet.	$P4_12_12(92)$ $P4_32_12(96)$		4	4.971 ±.003	
180	β-Cristobalite SiO_2	cubic	Fd3m(227)		8	7.1382 ±.0010	
181	Keatite SiO_2	tet.	$P4_12_12(92)$ $P4_32_12(96)$		12	7.456 ±.003	
182	β-Tridymite SiO_2	hex.	$P\bar{6}2c(172)$ $P6_3/mmc(194)$		4	5.0463 ±.0020	
183	Coesite SiO_2*	mon.	B2/b(15)		16	7.152 ±.001	12.379 ±.002

X-Ray Crystallographic Data of Minerals

c_o	a_o	β_o	γ_o	Cell volume 10^{-24} cm³	Molar volume cm³	Molar volume cal bar⁻¹	X-Ray density grams cm⁻³	Temp. °C	
Oxides and hydroxides—Continued									
2.8626 ±.0010				55.327 ± .032	16.660 ± .010	.39824 ±.00027	6.2777 ±.0036	25	157
5.652 ±.002				121.73 ± .11	24.438 ± .021	.58413 ±.00056	4.2797 ±.0038	26	158
7.3666 ±.0010				130.41 ± .06	19.635 ± .009	.46932 ±.00026	.9175 ±.0004	0	159
5.2948 ±.0010		99.18 ±.08		138.30 ± .06	20.823 ± .008	.49772 ±.00025	10.108 ±.004	r	160
3.519 ±.003				128.3 ± .1	19.32 ± .02	.4618 ±.0006	11.21 ±.01	25	161
				74.709 ± .027	11.248 ± .004	.26889 ±.00014	3.5837 ±.0013	25	162
4.769 ±.004				40.90 ± .11	24.63 ± .07	.5888 ±.0016	2.368 ±.006	26	163
				87.813 ± .030	13.221 ± .004	.31604 ±.00015	5.3653 ±.0018	26	164
2.865 ±.002				55.16 ± .08	16.61 ± .02	.3971 ±.0007	5.234 ±.008	r	165
				833.5 ± 1.3	31.37 ± .05	.7499 ±.0012	5.032 ±.008	25	166
9.422 ±.005				623.68 ± .84	46.95 ± .06	1.1222 ±.0016	4.873 ±.007	20	167
3.697 ±.004				202.98 ± .25	30.56 ± .04	.7305 ±.0010	4.710 ±.006	26	168
				72.88 ± .10	10.97 ± .02	.2623 ±.0004	6.809 ±.010	26	169
5.023 ±.004				79.40 ± .17	23.91 ± .05	.5715 ±.0013	9.334 ±.020	27	170
5.891 ±.004				153.8 ± .2	23.15 ± .03	.5533 ±.0007	9.641 ±.012	27	171
6.565 ±.003				510.13 ± .62	76.81 ± .09	1.836 ±.002	8.926 ±.009	25	172
				1386.9 ± 1.1	52.206 ± .042	1.2478 ±.0011	5.5837 ±.0045	26	173
5.421 ±.004				332.13 ± .31	50.007 ± .047	1.1952 ±.0012	5.8292 ±.0054	25	174
				1094.3 ± 1.6	82.38 ± .12	1.9690 ±.0029	3.733 ±.005	26	175
5.08 ±.01				354.2 ± 1.1	26.66 ± .08	.6373 ±.0020	4.161 ±.013	26	176
5.4051 ±.0001				113.01 ± .01	22.688 ± .001	.54229 ±.00007	2.6483 ±.0001	25	177
5.4592 ±.0020				118.15 ± .06	23.718 ± .013	.5669 ±.0004	2.533 ±.002	575	178
6.918 ±.003				170.95 ± .22	25.739 ± .033	.61521 ±.00083	2.3344 ±.0030	25	179
				363.72 ± .15	27.381 ± .012	.65447 ±.00032	2.1944 ±.0009	405	180
8.604 ±.005				478.3 ± .5	24.01 ± .02	.5738 ±.0006	2.503 ±.003	r	181
8.2563 ±.0030				182.08 ± .16	27.414 ± .024	.65527 ±.00062	2.1917 ±.0019	405	182
7.152 ±.001		120.00 ±.17		548.37 ± .95	20.641 ± .036	.49338 ±.00090	2.9110 ±.0050	25	183

	Name and formula	Crystal system	Space group	Structure type	Z	a_o	b_o
	Oxides and hydroxides—Continued						
184	Stishovite SiO_2*	tet.	P4/mnm(136)	rutile	2	4.1790 ±.0010	
185	Melanophlogite SiO_2*	cubic	Pm3n(223)	clathrate type	46	13.402 ±.004	
186	Cassiterite SnO_2	tet.	P4/mnm(136)	rutile	2	4.738 ±.003	
187	Tellurite TeO_2*	orth.	Pbca(61)	tellurite	8	5.607 ±.003	12.034 ±.005
188	Paratellurite TeO_2	tet.	$P4_12_12(92)$ $P4_32_12(96)$		4	4.810 ±.002	
189	Thorianite ThO_2	cubic	Fm3m(225)	fluorite	4	5.5952 ±.0005	
190	Rutile TiO_2	tet.	P4/mnm(136)		2	4.5937 ±.0005	
191	Anatase TiO_2	tet.	$I4_1/amd(141)$		4	3.785 ±.002	
192	Brookite TiO_2*	orth.	Pcab(61)		8	5.456 ±.002	9.182 ±.005
193	Titanium sesquioxide Ti_2O_3	hex-R	$R\overline{3}c(167)$	corundum	6	5.149 ±.002	
194	Uraninite UO_2	cubic	Fm3m(225)	fluorite	4	5.4682 ±.0010	
195	Karelianite V_2O_3	hex-R	$R\overline{3}c(167)$	corundum	6	4.952 ±.002	
196	Zincite ZnO	hex.	$P6_3mc(186)$	zincite	2	3.2495 ±.0005	
197	Baddeleyite ZrO_2	mon.	$P2_1/c(14)$	baddeleyite	4	5.1454 ±.0010	5.2075 ±.0010
	Multiple oxides						
198	Spinel $MgAl_2O_4$	cubic	Fd3m(227)	spinel	8	8.080 ±.002	
199	Hercynite $FeAl_2O_4$	cubic	Fd3m(227)	spinel	8	8.150 ±.004	
200	Galaxite $MnAl_2O_4$	cubic	Fd3m(227)	spinel	8	8.258 ±.002	
201	Gahnite $ZnAl_2O_4$	cubic	Fd3m(227)	spinel	8	8.0848 ±.0020	
202	Magnetite $FeFe_2O_4$	cubic	Fd3m(227)	spinel	8	8.3940 ±.0005	
203	Jacobsite $MnFe_2O_4$	cubic	Fd3m(227)	spinel	8	8.499 ±.002	
204	Trevorite $NiFe_2O_4$	cubic	Fd3m(227)	spinel	8	8.339 ±.003	
205	Picrochromite $MgCr_2O_4$	cubic	Fd3m(227)	spinel	8	8.333 ±.003	
206	Ilmenite $FeTiO_3$	hex-R	$R\overline{3}(148)$	ilmenite	6	5.093 ±.005	
207	Geikielite $MgTiO_3$	hex-R	$R\overline{3}(148)$	ilmenite	6	5.054 ±.005	
208	Pyrophanite $MnTiO_3$	hex-R	$R\overline{3}(148)$	ilmenite	6	5.155 ±.005	
209	Cobalt Titanate $CoTiO_3$	hex-R	$R\overline{3}(148)$	ilmenite	6	5.066 ±.001	

X-Ray Crystallographic Data of Minerals

c_o	α_o	β_o	γ_o	Cell volume 10^{-24} cm^3	Molar volume cm^3	Molar volume cal bar^{-1}	X-Ray density grams cm^{-3}	Temp. °C	
				Oxides and hydroxides—Continued					
2.6649 ±.0010				46.540 ± .028	14.014 ± .009	.33500 ±.00025	4.2874 ±.0026	r	184
				2407.2 ± 2.2	31.516 ± .028	.75325 ±.00072	1.9065 ±.0017	r	185
3.188 ±.003				71.57 ± .11	21.55 ± .03	.5151 ±.0009	6.992 ±.011	26	186
5.463 ±.003				368.61 ± .32	27.750 ± .024	.66328 ±.00062	5.7514 ±.0050	25	187
7.613 ±.002				176.14 ± .15	26.52 ± .02	.6339 ±.0006	6.018 ±.005	25	188
				175.16 ± .05	26.373 ± .007	.63038 ±.00021	10.012 ±.003	25	189
2.9618 ±.0010				62.500 ± .025	18.820 ± .008	.44986 ±.00023	4.2453 ±.0017	25	190
9.514 ±.006				136.30 ± .17	20.522 ± .025	.4905 ±.0007	3.893 ±.005	r	191
5.143 ±.003				257.6 ± .2	19.40 ± .02	.4636 ±.0005	4.119 ±.004	r	192
13.642 ±.010				313.2 ± .3	31.44 ± .03	.7515 ±.0009	4.574 ±.005	r	193
				163.51 ± .09	24.618 ± .014	.58843 ±.00037	10.969 ±.006	26	194
14.002 ±.010				297.36 ± .32	29.848 ± .032	.71342 ±.00081	5.0216 ±.0054	r	195
5.2069 ±.0005				47.615 ± .015	14.338 ± .005	.34273 ±.00016	5.6750 ±.0018	25	196
5.3107 ±.0010		99.23 ±.08		140.46 ± .06	21.148 ± .009	.50548 ±.00025	5.8267 ±.0023	r	197
				Multiple oxides					
				527.5 ± .4	39.71 ± .03	.9492 ±.0008	3.583 ±.003	26	198
				541.3 ± .6	40.75 ± .05	.9740 ±.0011	4.265 ±.005	25	199
				563.2 ± .4	42.39 ± .03	1.013 ±.001	4.078 ±.003	25	200
				528.45 ± .39	39.783 ± .030	.95088 ±.00075	4.6083 ±.0034	26	201
				591.43 ± .11	44.524 ± .008	1.0642 ±.0002	5.2003 ±.0009	22	202
				613.9 ± .4	46.22 ± .03	1.105 ±.001	4.990 ±.004	25	203
				579.9 ± .6	43.65 ± .05	1.043 ±.001	5.370 ±.006	25	204
				578.6 ± .6	43.56 ± .05	1.041 ±.001	4.415 ±.005	26	205
14.055 ±.020				315.73 ± .75	31.69 ± .08	.7574 ±.0019	4.788 ±.012	r	206
13.898 ±.010				307.44 ± .65	30.86 ± .07	.7376 ±.0016	3.896 ±.008	26	207
14.18 ±.01				326.3 ± .7	32.76 ± .07	.7829 ±.0017	4.605 ±.010	r	208
13.918 ±.005				309.34 ± .17	31.05 ± .02	.7422 ±.0004	4.986 ±.003	r	209

	Name and formula	Crystal system	Space group	Structure type	Z	a_o	b_o
	Multiple oxides—Continued						
210	Perovskite $CaTiO_3$	orth.	Pcmn(62)	perovskite	4	5.3670 ±.0010	7.6438 ±.0010
211	Chrysoberyl $BeAl_2O_4$	orth.	Pmnb(62)	olivine	4	5.4756 ±.0020	9.4041 ±.0030
	Halides						
212	Halite NaCl	cubic	Fm3m(225)	rock salt	4	5.6402 ±.0002	
213	Sylvite KCl	cubic	Fm3m(225)	rock salt	4	6.2931 ±.0002	
214	Villiaumite NaF	cubic	Fm3m(225)	rock salt	4	4.6342 ±.0005	
215	Chlorargyrite AgCl	cubic	Fm3m(225)	rock salt	4	5.5491 ±.0005	
216	Bromargyrite AgBr	cubic	Fm3m(225)	rock salt	4	5.7745 ±.0005	
217	Nantockite CuCl	cubic	$F\bar{4}3m$(216)	sphalerite	4	5.416 ±.003	
218	Marshite CuI	cubic	$F\bar{4}3m$(216)	sphalerite	4	6.0507 ±.0010	
219	Miersite AgI	cubic	$F\bar{4}3m$(216)	sphalerite	4	6.4963 ±.0010	
220	Iodargyrite AgI	hex.	$P6_3mc$(186)	zincite	2	4.5955 ±.0010	
221	Calomel HgCl	tet.	14/mm(139)		4	4.478 ±.005	
222	Fluorite CaF_2	cubic	Fm3m(225)	fluorite	4	5.4638 ±.0004	
223	Sellaite MgF_2	tet.	$P4_2/mnm$(136)	rutile	2	4.621 ±.001	
224	Chloromagnesite $MgCl_2$	hex-R	$R\bar{3}m$(166)		3	3.632 ±.004	
225	Lawrencite $FeCl_2$	hex-R	$R\bar{3}m$(166)		3	3.593 ±.003	
226	Scacchite $MnCl_2$	hex-R	$R\bar{3}m$(166)		3	3.711 ±.002	
227	Cotunnite $PbCl_2$	orth.	Pnmb(62)		4	4.535 ±.005	7.62 ±.01
228	Matlockite PbFCl	tet.	P4/nmm(129)		2	4.106 ±.005	
229	Cryolite Na_3AlF_6*	mon.	$P2_1/n$(14)		2	5.40 ±.01	5.60 ±.01
230	Neighborite $NaMgF_3$	orth.	Pcmn(62)	perovskite	4	5.363 ±.001	7.676 ±.001
	Carbonates and nitrates						
231	Calcite $CaCO_3$	hex-R	$R\bar{3}c$(167)	calcite	6	4.9899 ±.0010	
232	Otavite $CdCO_3$	hex-R	$R\bar{3}c$(167)	calcite	6	4.9204 ±.0010	
233	Cobalticalcite $CoCO_3$	hex-R	$R\bar{3}c$(167)	calcite	6	4.6581 ±.0010	
234	Siderite $FeCO_3$	hex-R	$R\bar{3}c$(167)	calcite	6	4.6887 ±.0010	

X-Ray Crystallographic Data of Minerals

c_o	a_o	β_o	γ_o	Cell volume 10^{-24} cm³	Molar volume cm³	Molar volume cal bar⁻¹	X-Ray density grams cm⁻³	Temp. °C	
				Multiple oxides—Continued					
5.4439 ±.0010				223.33 ± .07	33.626 ± .010	.80371 ±.00028	4.0439 ±.0012	r	210
4.4267 ±.0020				227.94 ± .15	34.320 ± .023	.82031 ±.00059	3.6997 ±.0025	25	211
				Halides					
				179.43 ± .02	27.015 ± .003	.64571 ±.00011	2.1634 ±.0002	26	212
				249.23 ± .02	37.524 ± .004	.89690 ±.00013	1.9868 ±.0002	25	213
				99.523 ± .032	14.984 ± .005	.35818 ±.00016	2.8021 ±.0009	25	214
				170.87 ± .05	25.727 ± .007	.61493 ±.00021	5.5710 ±.0015	26	215
				192.55 ± .05	28.991 ± .008	.69294 ±.00022	6.4772 ±.0017	26	216
				158.87 ± .26	23.92 ± .04	.5717 ±.0010	4.139 ±.007	25	217
				221.52 ± .11	33.353 ± .017	.7972 ±.0004	5.710 ±.003	26	218
				274.16 ± .10	41.278 ± .020	.9866 ±.0004	5.688 ±.003	r	219
7.5005 ±.0033				137.18 ± .10	41.308 ± .030	.9873 ±.0009	5.683 ±.004	25	220
10.910 ±.005				218.77 ± .50	32.939 ± .075	.7873 ±.0018	7.166 ±.016	26	221
				163.11 ± .04	24.558 ± .005	.58701 ±.00017	3.1792 ±.0007	25	222
3.050 ±.001				65.13 ± .04	19.61 ± .01	.4688 ±.0003	3.177 ±.002	18	223
17.795 ±.016				203.29 ± .48	40.81 ± .10	.9754 ±.0024	2.333 ±.006	r	224
17.58 ±.09				196.55 ± 1.06	39.46 ± .21	.9431 ±.0051	3.212 ±.017	r	225
17.59 ±.07				209.79 ± .86	42.11 ± .17	1.007 ±.004	2.988 ±.012	r	226
9.05 ±.01				312.74 ± .64	47.09 ± .10	1.1254 ±.0023	5.906 ±.012	26	227
7.23 ±.01				121.89 ± .34	36.70 ± .10	.8773 ±.0025	9.853 ±.028	26	228
7.776 ±.010		90.18 ±.25		235.1 ± .7	70.81 ± .20	1.692 ±.005	2.965 ±.009	r	229
5.503 ±.001				226.54 ± .07	34.11 ± .01	.8152 ±.0003	3.058 ±.001	18	230
				Carbonates and nitrates					
17.064 ±.002				367.96 ± .15	36.934 ± .015	.88278 ±.00041	2.7100 ±.0011	26	231
16.298 ±.003				341.72 ± .15	34.300 ± .015	.81983 ±.00041	5.0265 ±.0022	26	232
14.958 ±.003				281.07 ± .13	28.213 ± .013	.67435 ±.00036	4.2159 ±.0020	26	233
15.373 ±.003				292.68 ± .14	29.378 ± .014	.70219 ±.00037	3.9436 ±.0018	26	234

	Name and formula	Crystal system	Space group	Structure type	Z	a_o	b_o
	Carbonates and nitrates—Continued						
235	Magnesite $MgCO_3$	hex-R	$R\bar{3}c$(167)	calcite	6	4.6330 ±.0010	
236	Rhodochrosite $MnCO_3$	hex-R	$R\bar{3}c$(167)	calcite	6	4.7771 ±.0010	
237	Nickelous Carbonate $NiCO_3$	hex-R	$R\bar{3}c$(167)	calcite	6	4.5975 ±.0010	
238	Smithsonite $ZnCO_3$	hex-R	$R\bar{3}$(167)	calcite	6	4.6528 ±.0010	
239	Dolomite $CaMg(CO_3)_2$*	hex-R	$R\bar{3}$(148)	calcite	3	4.8079 ±.0010	
240	Huntite $Mg_3Ca(CO_3)_4$*	hex-R	R32(155)	calcite	3	9.498 ±.003	
241	Norsethite $BaMg(CO_3)_2$*	hex-R	R32(155)	calcite	3	5.020 ±.005	
242	Vaterite $CaCO_3$	hex.			6	7.135 ±.005	
243	Witherite $BaCO_3$	orth.	Pnam(62)	aragonite	4	6.430 ±.005	8.904 ±.005
244	Aragonite $CaCO_3$	orth.	Pnam(62)	aragonite	4	5.741 ±.005	7.968 ±.005
245	Cerussite $PbCO_3$	orth.	Pnam(62)	aragonite	4	6.152 ±.005	8.436 ±.005
246	Strontianite $SrCO_3$	orth.	Pnam(62)	aragonite	4	6.029 ±.005	8.414 ±.005
247	Shortite $Na_2Ca_2(CO_3)_3$	orth.	Amm2(38)		2	4.961 ±.005	11.03 ±.02
248	Malachite $Cu_2(OH)_2CO_3$	mon.	$P2_1/a$(14)		4	9.502 ±.007	11.974 ±.007
249	Azurite $Cu_3(OH)_2(CO_3)_2$	mon.	$P2_1/a$(14)		2	5.008 ±.005	5.844 ±.005
250	Niter KNO_3	orth.	Pnam(62)	aragonite	4	6.431 ±.005	9.164 ±.005
251	Soda Niter $NaNO_3$	hex-R	$R\bar{3}c$(167)	calcite	6	5.0696 ±.0010	
252	Gerhardite $Cu_2(NO_3)(OH)_3$	orth.	$P2_12_12_1$(19)		4	6.075 ±.004	13.812 ±.008
	Sulfates and borates						
253	Barite $BaSO_4$	orth.	Pnma(62)	barite	4	8.878 ±.005	5.450 ±.005
254	Anhydrite $CaSO_4$	orth.	Amma(63) Ccmm(63)	anhydrite	4	6.991 ±.005	6.996 ±.005
255	Anglesite $PbSO_4$	orth.	Pnma(62)	barite	4	8.480 ±.005	5.398 ±.005
256	Celestite $SrSO_4$	orth.	Pnma(62)	barite	4	8.359 ±.005	5.352 ±.005
257	Zinkosite $ZnSO_4$	orth.	Pnma(62)	barite	4	8.588 ±.008	6.740 ±.006
258	Arcanite KS_2SO_4	orth.	Pnma(62)	arcanite	4	5.772 ±.005	10.072 ±.005
259	Mascagnite $(NH_4)_2SO_4$	orth.	Pnma(62)	arcanite	4	7.782 ±.005	5.993 ±.005
260	Thenardite Na_2SO_4	orth.	Fddd(70)	thenardite	8	5.863 ±.005	12.304 ±.005

c_o	a_o	β_o	γ_o	Cell volume 10^{-24} cm^3	Molar volume cm^3	Molar volume cal bar^{-1}	X-Ray density grams cm^{-3}	Temp. °C	
Carbonates and nitrates—Continued									
15.016 ±.003				279.13 ± .13	28.018 ± .013	.66969 ±.00036	3.0095 ±.0014	26	235
15.664 ±.003				309.57 ± .14	31.073 ± .014	.74272 ±.00039	3.6992 ±.0017	26	236
14.723 ±.002				269.51 ± .12	27.052 ± .012	.64660 ±.00034	4.3886 ±.0020	26	237
15.025 ±.003				281.69 ± .13	28.275 ± .013	.67583 ±.00037	4.4343 ±.0021	26	238
16.010 ±.003				320.50 ± .15	64.341 ± .029	1.5378 ±.0008	2.8661 ±.0013	26	239
7.816 ±.004				610.63 ± .50	122.58 ± .10	2.9299 ±.0024	2.880 ±.002	26	240
16.75 ±.02				365.6 ± .8	73.39 ± .17	1.754 ±.004	3.838 ±.009	r	241
8.524 ±.007				375.80 ± .61	37.72 ± .06	.9016 ±.0015	2.653 ±.004	r	242
5.314 ±.005				304.24 ± .41	45.81 ± .06	1.095 ±.002	4.308 ±.006	26	243
4.959 ±.005				226.85 ± .33	34.15 ± .05	.8164 ±.0012	2.930 ±.004	26	244
5.195 ±.005				269.61 ± .38	40.59 ± .06	.9702 ±.0014	6.582 ±.009	26	245
5.107 ±.005				259.07 ± .37	39.01 ± .06	.9323 ±.0014	3.785 ±.005	26	246
7.12 ±.01				389.6 ± 1.0	117.3 ± .3	2.804 ±.007	2.610 ±.007	r	247
3.240 ±.003		98.75 ±.25		364.35 ± .54	54.86 ± .08	1.311 ±.002	4.030 ±.006	25	248
10.336 ±.005		92.45 ±.25		302.22 ± .43	91.01 ± .13	2.1752 ±.0031	3.787 ±.005	25	249
5.414 ±.005				319.07 ± .42	48.04 ± .06	1.148 ±.002	2.105 ±.003	26	250
16.829 ±.005				374.57 ± .19	37.598 ± .019	.89866 ±.00049	2.2606 ±.0011	25	251
5.592 ±.004				469.21 ± .53	70.65 ± .08	1.689 ±.002	3.399 ±.004	r	252
Sulfates and borates									
7.152 ±.003				346.05 ± .40	52.10 ± .06	1.245 ±.002	4.480 ±.005	26	253
6.238 ±.005				305.09 ± .39	45.94 ± .06	1.098 ±.002	2.964 ±.004	26	254
6.958 ±.003				318.50 ± .38	47.95 ± .06	1.146 ±.002	6.324 ±.008	25	255
6.866 ±.005				307.17 ± .41	46.25 ± .06	1.105 ±.002	3.972 ±.005	26	256
4.770 ±.005				276.10 ± .46	41.57 ± .07	.9936 ±.0017	3.883 ±.006	25	257
7.483 ±.004				435.03 ± .49	65.50 ± .07	1.566 ±.002	2.661 ±.003	25	258
10.636 ±.005				496.04 ± .57	74.68 ± .09	1.7851 ±.0021	1.7693 ±.0020	25	259
9.821 ±.005				708.47 ± .76	53.33 ± .06	1.275 ±.002	2.663 ±.003	25	260

	Name and formula	Crystal system	Space group	Structure type	Z	a_o	b_o
	Sulfates and borates—Continued						
261	Gypsum $CaSO_4.2H_2O$*	mon.	C2/c(15)		4	5.68 ±.01	15.18 ±.01
262	Epsomite $MgSO_4.7H_2O$	orth.	$P2_12_12_1$(19)		4	11.86 ±.01	11.99 ±.01
263	Goslarite $ZnSO_4.7H_2O$	orth.	$P2_12_12_1$(19)	epsomite	4	11.779 ±.005	12.050 ±.005
264	Mirabilite $Na_2SO_4.10H_2O$	mon.	$P2_1/c$(14)		4	11.51 ±.01	10.38 ±.01
265	Chalcanthite $CuSO_4.5H_2O$	tri.	$P\bar{1}$(2)		2	6.1045 ±.0050	10.72 ±.01
266	Brochantite $Cu_4SO_4(OH)_6$*	mon.	$P2_1/c$(14)		4	13.066 ±.010	9.85 ±.01
267	Syngenite $K_2Ca(SO_4)_2.H_2O$	mon.	$P2_1/m$(11)		2	9.775 ±.005	7.156 ±.005
268	Alunite $KAl_3(SO_4)_2(OH)_6$	hex-R	R3m(160)		3	6.982 ±.005	
269	Natroalunite $NaAl_3(SO_4)_2(OH)_6$	hex-R	R3m(160)		3	6.974 ±.005	
270	Hexahydrite $MgSO_4.6H_2O$	mon.	C2/c(15)		8	10.110 ±.005	7.212 ±.004
271	Leonhardtite $MgSO_4.4H_2O$	mon.	$P2_1/n$(14)		4	5.922 ±.006	13.604 ±.004
272	Melanterite $FeSO_4.7H_2O$	mon.	$P2_1/c$(14)		4	14.072 ±.010	6.503 ±.007
273	Vanthoffite $MgSO_4.3Na_2SO_4$	mon.	$P2_1/c$(14)		2	9.797 ±.003	9.217 ±.003
274	Dolerophanite $Cu_2O(SO_4)$	mon.	C2/m(15)		4	9.355 ±.010	6.312 ±.005
275	Retgersite $NiSO_4.4H_2O$	tet.	$P4_12_12$(92) $P4_32_1$(96)		4	6.782 ±.004	
276	Colemanite $CaB_3O_4(OH)_3.H_2O$*	mon.	$P2_1/a$(14)		4	8.743 ±.004	11.264 ±.002
277	Borax $Na_2B_4O_7.10H_2O$	mon.	C2/c(15)		4	11.858 ±.005	10.674 ±.005
278	Kernite $Na_2B_4O_7.4H_2O$	mon.	$P2_1/c$(14)		4	7.022 ±.003	9.151 ±.004
279	Hambergite $Be_2BO_3.(OH,F)$*	orth.	Pbca(61)		8	9.755 ±.001	12.201 ±.001
	Phosphates, molybdates, and tungstates						
280	Berlinite $AlPO_4$	hex.	$P3_121$(152) $P3_221$(154)	α-quartz	3	4.942 ±.005	
281	Xenotime YPO_4	tet.	14_1/amd(141)	zircon	4	6.885 ±.005	
282	Hydroxylapatite $Ca_5(PO_4)_3OH$	hex.	$P6_3/m$(176)	apatite	2	9.418 ±.003	
283	Fluorapatite $Ca_5(PO_4)_3F$	hex.	$P6_3/m$(176)	apatite	2	9.3684 ±.0030	
284	Chlorapatite $Ca_5(PO_4)_3Cl$	hex.	$P6/_3m$(176)	apatite	2	9.629 ±.005	
285	Carbonate-apatite $Ca_{10}(PO_4)_6CO_3H_2O$	hex.	$P6_3/m$(176)	apatite	1	9.436 ±.010	
286	Turquois $CuAl_6(PO_4)_4(OH)_8.4H_2O$*	tri.	$P\bar{1}$(2)		1	7.424 ±.008	7.629 ±.008

c_o	α_o	β_o	γ_o	Cell volume 10^{-24} cm^3	Molar volume cm^3	Molar volume cal bar^{-1}	X-Ray density grams cm^{-3}	Temp. °C	
Sulfates and borates—Continued									
6.29		113.83		496.1	74.69	1.785	2.305	r	261
±.01		±.22		± 1.5	± .22	±.005	±.007		
6.858				975.22	146.83	3.5094	1.679	25	262
±.007				± 1.53	± .23	±.0055	±.003		
6.822				968.29	145.79	3.4845	1.9723	25	263
±.003				± .72	± .11	±.0026	±.0015		
12.83		107.75		1459.9	219.8	5.253	1.466	24	264
±.01		±.17		± 2.6	± .4	±.009	±.003		
5.949	97.57	107.28	77.43	361.88	108.97	2.6045	2.2912	r	265
±.007	±.17	±.17	±.17	± .72	± .22	±.0052	±.0046		
6.022		103.27		754.3	113.6	2.715	3.982	r	266
±.010		±.25		± 1.8	± .2	±.006	±.009		
6.251		104.00		424.27	127.76	3.0535	2.5707	r	267
±.005		±.25		± .68	± .20	±.0049	±.0041		
17.32				731.2	146.8	3.508	2.822	r	268
±.01				± 1.1	± .2	±.005	±.004		
16.69				702.99	141.1	3.373	2.821	r	269
±.01				± 1.09	± .2	±.005	±.004		
24.41		98.30		1761.2	132.58	3.1689	1.7232	r	270
±.01		±.10		± 1.6	± .12	±.0029	±.0015		
7.905		90.85		636.78	95.88	2.2915	2.0071	r	271
±.005		±.20		± .78	± .12	±.0029	±.0025		
11.041		105.57		973.29	146.54	3.5025	1.8972	r	272
±.010		±.15		± 1.69	± .25	±.0061	±.0033		
8.199		113.50		678.96	204.45	4.8866	2.6730	r	273
±.003		±.10		± .65	± .20	±.0047	±.0025		
7.628		122.29		380.77	57.33	1.3703	4.171	r	274
±.005		±.10		± .70	± .11	±.0026	±.008		
18.28				840.80	126.59	3.0257	2.076	25	275
±.01				± 1.09	± .16	±.0040	±.003		
6.102		110.12		564.26	84.957	2.0306	2.4194	r	276
±.003		±.08		± .49	± .073	±.0018	±.0021		
12.197		106.68		1478.8	222.66	5.3217	1.7128	r	277
±.005		±.03		± 1.1	± .17	±.0041	±.0013		
15.676		108.83		953.40	143.55	3.4309	1.9038	r	278
±.008		±.25		± 1.61	± .24	±.0058	±.0032		
4.426				526.79	39.658	.9479	2.3663	r	279
±.001				± .14	± .011	±.0003	±.0006		
Phosphates, molybdates, and tungstates									
10.97				232.03	46.58	1.113	2.618	25	280
±.007				± .50	± .10	±.002	±.006		
5.982				283.57	42.69	1.020	4.307	26	281
±.005				± .48	± .07	±.002	±.008		
6.883				528.7	159.2	3.805	3.155	r	282
±.003				± .5	± .2	±.004	±.004		
6.8841				523.25	157.56	3.7659	3.2007	25	283
±.0030				± .41	± .12	±.0030	±.0025		
6.777				544.16	163.86	3.916	3.178	r	284
±.003				± .61	± .19	±.004	±.004		
6.883				530.74	319.6	7.640	3.281	r	285
±.010				± 1.36	± .8	±.020	±.008		
9.910	68.61	69.71	65.08	461.40	277.9	6.6416	2.927	r	286
±.010	±.20	±.20	±.20	± 1.12	± .7	±.0162	±.007		

	Name and formula	Crystal system	Space group	Structure type	Z	a_o	b_o
	Phosphates, molybdates, and tungstates—Continued						
287	Powellite $CaMoO_4$	tet.	$I4_1/a(100)$	scheelite	4	5.226 ±.005	
288	Wulfenite $PbMoO_4$	tet.	$I4_1/a(100)$	scheelite	4	5.435 ±.005	
289	Scheelite $CaWO_4$	tet.	$I4_1/a(100)$	scheelite	4	5.242 ±.005	
290	Stolzite $PbWO_4$	tet.	$I4_1/a(100)$	scheelite	4	5.4616 ±.0030	
291	Ferberite $FeWO_4$	mon.	P2/c(13)	wolframite	2	4.732 ±.004	5.708 ±.003
292	Huebnerite $MnWO_4$	mon.	P2/c(13)	wolframite	2	4.834 ±.004	5.758 ±.005
293	Wolframite $Fe_{.5}Mn_{.5}WO_4$	mon.	P2/c(13)	wolframite	2	4.782 ±.004	5.731 ±.004
294	Sanmartinite $ZnWO_4$	mon.	P2/c(13)	wolframite	2	4.691 ±.003	5.720 ±.003
	Ortho and ring structure silicates						
295	Forsterite Mg_2SiO_4	orth.	Pbnm(62)	olivine	4	4.758 ±.002	10.214 ±.003
296	Fayalite Fe_2SiO_4	orth.	Pbnm(62)	olivine	4	4.817 ±.005	10.477 ±.005
297	Tephroite Mn_2SiO_4*	orth.	Pbnm(62)	olivine	4	4.871 ±.005	10.636 ±.005
298	Lime Olivine γCa_2SiO_4	orth.	Pbnm(62)	olivine	4	5.091 ±.010	11.371 ±.020
299	Nickel Olivine Ni_2SiO_4	orth.	Pbnm(62)	olivine	4	4.727 ±.002	10.121 ±.005
300	Cobalt Olivine Co_2SiO_4	orth.	Pbnm(62)	olivine	4	4.782 ±.002	10.301 ±.005
301	Monticellite $CaMgSiO_4$	orth.	Pbnm(62)	olivine	4	4.827 ±.005	11.084 ±.005
302	Kerschsteinite $CaFeSiO_4$	orth.	Pbnm(62)	olivine	4	4.886 ±.005	11.146 ±.005
303	Knebelite $MnFeSiO_4$*	orth.	Pbnm(62)	olivine	4	4.854 ±.010	10.602 ±.010
304	Glauchroite $CaMnSiO_4$	orth.	Pbnm(62)	olivine	4	4.944 ±.004	11.19 ±.01
305	Fluor-Norbergite $Mg_2SiO_4.MgF_2$	orth.	Pnmb(62)		4	8.727 ±.005	10.271 ±.010
306	Chondrodite $2Mg_2SiO_4.MgF_2$*	mon.	$P2_1/c(14)$		2	7.89 ±.03	4.743 ±.020
307	Fluor-Humite $3Mg_2SiO_4.MgF_2$	orth.	Pnma(62)		4	10.243 ±.005	20.72 ±.02
308	Clinohumite $4Mg_2SiO_4.MgF_2$*	mon.	$P2_1/c(14)$		2	13.68 ±.04	4.75 ±.02
309	Grossularite $Ca_3Al_2Si_3O_{12}$	cubic	Ia3d(230)	garnet	8	11.851 ±.001	
310	Uvarovite $Ca_3Cr_2Si_3O_{12}$	cubic	Ia3d(230)	garnet	8	11.999 ±.002	
311	Andradite $Ca_3Fe_2Si_3O_{12}$	cubic	Ia3d(230)	garnet	8	12.048 ±.001	
312	Goldmanite $Ca_3V_2Si_3O_{12}$	cubic	Ia3d(230)	garnet	8	12.070 ±.005	

X-Ray Crystallographic Data of Minerals

c_o	α_o	β_o	γ_o	Cell volume 10^{-24} cm³	Molar volume cm³	Molar volume cal bar⁻¹	X-Ray density grams cm⁻³	Temp. °C	
Phosphates, molybdates, and tungstates—Continued									
11.43 ±.007				312.17 ± .63	47.00 ± .09	1.1234 ±.0023	4.256 ±.009	25	287
12.110 ±.007				357.72 ± .69	53.859 ± .104	1.2873 ±.0025	6.816 ±.013	25	288
11.372 ±.005				312.49 ± .61	47.049 ± .092	1.1245 ±.0023	6.120 ±.012	25	289
12.046 ±.005				359.32 ± .42	54.100 ± .064	1.2931 ±.0016	8.4110 ±.0099	25	290
4.965 ±.004		90.00 ±.05		134.11 ± .17	40.38 ± .05	.9652 ±.0013	7.520 ±.010	r	291
4.999 ±.004		91.18 ±.10		139.11 ± .20	41.89 ± .06	1.001 ±.002	7.228 ±.010	r	292
4.982 ±.004		90.57 ±.10		136.53 ± .18	41.11 ± .06	.9826 ±.0014	7.376 ±.010	r	293
4 925 ±.003		89.36 ±.20		132.14 ± .14	39.79 ± .04	.9511 ±.0010	7.872 ±.008	25	294
Ortho and ring structure silicates									
5.984 ±.002				290.81 ± .18	43.786 ± .027	1.0465 ±.0007	3.2136 ±.0020	25	295
6.105 ±.010				308.11 ± .62	46.389 ± .093	1.1088 ±.0023	4.3928 ±.0088	r	296
6.232 ±.005				322.87 ± .45	48.612 ± .067	1.1619 ±.0017	4.1545 ±.0058	r	297
6.782 ±.010				392.61 ± 1.19	59.11 ± .18	1.4129 ±.0043	2.914 ±.009	r	298
5.915 ±.002				282.98 ± .21	42.61 ± .03	1.0184 ±.0008	4.917 ±.004	r	299
6.003 ±.002				295.70 ± .21	44.52 ± .03	1.0642 ±.0008	4.716 ±.003	r	300
6.376 ±.005				341.13 ± .47	51.362 ± .071	1.2276 ±.0017	3.046 ±.004	r	301
6.434 ±.010				350.39 ± .67	52.756 ± .101	1.2609 ±.0025	3.564 ±.007	r	302
6.162 ±.010				317.11 ± .88	47.74 ± .13	1.1412 ±.0032	4.249 ±.012	r	303
6.529 ±.005				361.2 ± .9	54.38 ± .14	1.2997 ±.0032	3.441 ±.009	r	304
4.709 ±.002				422.09 ± .51	63.551 ± .077	1.5190 ±.0019	3.194 ±.004	25	305
10.29 ±.03		109.03 ±.30		364.0 ± 2.4	109.6 ± .7	2.620 ±.017	3.136 ±.021	r	306
4.735 ±.002				1004.9 ± 1.2	151.31 ± .18	3.6163 ±.0042	3.2017 ±.0037	25	307
10.27 ±.02		100.83 ±.50		655.5 ± 3.8	197.4 ± 1.1	4.717 ±.027	3.167 ±.018	r	308
				1664.43 ± .42	125.30 ± .03	2.9948 ±.0008	3.595 ±.001	25	309
				1727.57 ± .86	130.05 ± .07	3.1084 ±.0016	3.848 ±.002	26	310
				1748.82 ± .44	131.65 ± .03	3.1466 ±.0008	3.860 ±.001	25	311
				1758.42 2.19	132.38 ± .16	3.1639 ±.0040	3.765 ±.005	r	312

	Name and formula	Crystal system	Space group	Structure type	Z	a_o	b_o
	Ortho and ring structure silicates—Continued						
313	Almandite $Fe_3Al_2Si_3O_{12}$	cubic	Ia3d(230)	garnet	8	11.526 ±.001	
314	Pyrope $Mg_3Al_2Si_3O_{12}$	cubic	Ia3d(230)	garnet	8	11.459 ±.001	
315	Spessartite $Mn_3Al_2Si_3O_{12}$	cubic	Ia3d(230)	garnet	8	11.621 ±.001	
316	Zircon $ZrSiO_4$*	tet.	I4/amd(141)	zircon	4	6.604 ±.005	
317	Thorite $ThSiO_4$	tet.	I4/amd(141)	zircon	4	7.143 ±.004	
318	Coffinite $USiO_4$	tet.	I4/amd(141)	zircon	4	6.995 ±.004	
319	Kyanite Al_2SiO_5*	tri.	$P\bar{1}$(2)		4	7.123 ±.001	7.848 ±.002
320	Andalusite Al_2SiO_5*	orth.	Pnnm(58)		4	7.7959 ±.0050	7.8983 ±.0020
321	Sillimanite Al_2SiO_5*	orth.	Pbnm(62) Pnma(62)		4	7.4843 ±.0030	7.6730 ±.0030
322	3.2 Mullite $3AL_2O_3.2SiO_2$	orth.			3/4	7.557 ±.002	7.6876 ±.0020
323	2.1 Mullite $2Al_2O_3.SiO_2$	orth.	Pbam(55)		6/5	7.5788 ±.0020	7.6909 ±.0020
324	Staurolite $Fe_2Al_9Si_4O_{22}(OH)_2$*	mon.	C2/m(15)		2	7.90 ±.10	16.65 ±.15
325	Topaz $Al_2(SiO_4)(OH)$*	orth.	Pmnb(62)		4	8.394 ±.005	8.792 ±.007
326	Phenacite Be_2SiO_4*	hex-R	$R\bar{3}$(148)	phenacite	18	12.472 ±.005	
327	Willemite Zn_2SiO_4	hex-R	$R\bar{3}$(148)	phenacite	18	13.94 ±.01	
328	Dioptase CuH_2SiO_4*	hex-R	$R\bar{3}$(148)	phenacite	18	14.61 ±.02	
329	Larnite β-Ca_2SiO_4*	mon.	$P2_1/n$(14)		4	5.48 ±.02	6.76 ±.02
330	Akermanite $Ca_2MgSi_2O_7$	tet.	$P\bar{4}2_1m$(113)	melilite	2	7.8435 ±.0030	
331	Gehlenite $Ca_2Al_2SiO_7$	tet.	$P\bar{4}2_1m$(113)	melilite	2	7.690 ±.003	
332	Fe-Gehlenite $Ca_2Fe_2SiO_7$	tet.	$P\bar{4}2_1m$(113)	melilite	2	7.54 ±.01	
333	Hardystonite $Ca_2ZnSi_2O_7$*	tet.	$P\bar{4}2_1m$(113)	melilite	2	7.87 ±.03	
334	Sodium Melilite $NaCaAlSi_2O_7$	tet.	$P\bar{4}2_1m$(113)	melilite	2	8.511 ±.005	
335	Beryl $Be_3Al_2(Si_6O_{18})$*	hex.	P6/mmc(192)	beryl	2	9.215 ±.005	
336	Indialite high Cordierite $Mg_2Al_3(AlSi_5O_{18})$	hex.	P6/mmc(192)	beryl	2	9.7698 ±.0030	
337	Low Cordierite $Mg_2Al_3(AlSi_5O_{18})$	orth.	Cccm(66)		4	9.721 ±.003	17.062 ±.006
338	Fe-Indialite $Fe_2Al_3(AlSi_5O_{18})$	hex.	P6/mmc(192)	beryl	2	9.860 ±.010	
339	Fe-Cordierite $Fe_2Al_3(AlSi_5O_{18})$	orth.	Cccm(66)	cordierite	4	9.726 ±.010	17.065 ±.010

X-Ray Crystallographic Data of Minerals

c_o	α_o	β_o	γ_o	Cell volume 10^{-24} cm³	Molar volume cm³	Molar volume cal bar⁻¹	X-Ray density grams cm⁻³	Temp. °C	
Ortho and ring structure silicates—Continued									
				1531.21 ± .40	115.27 ± .04	2.7551 ±.0008	4.318 ±.001	25	313
				1504.67 ± .39	113.27 ± .03	2.7074 ±.0008	3.559 ±.001	25	314
				1569.39 ± .41	118.15 ± .03	2.8238 ±.0008	4.190 ±.001	25	315
5.979 ±.005				260.76 ± .45	39.261 ± .068	.9384 ±.0017	4.669 ±.008	25	316
6.327 ±.003				322.82 ± .39	48.60 ± .06	1.1617 ±.0015	6.668 ±.008	r	317
6.263 ±.005				306.45 ± .43	46.140 ± .064	1.103 ±.002	7.155 ±.010	r	318
5.564 ±.008	89.92 ±.15	101.25 ±.08	105.97 ±.08	292.83 ± .45	44.09 ± .07	1.054 ±.002	3.675 ±.006	25	319
5.5583 ±.0020				342.25 ± .27	51.530 ± .040	1.2316 ±.0010	3.145 ±.002	25	320
5.7711 ±.0040				331.42 ± .30	49.899 ± .044	1.1927 ±.0011	3.248 ±.003	25	321
2.8842 ±.0010				167.56 ± .09	134.55 ± .07	3.2159 ±.0016	3.166 ±.002	r	322
2.8883 ±.0010				168.35 ± .09	84.492 ± .043	2.0195 ±.0011	3.125 ±.002	r	323
5.63 ±.10		90.00 ±.25		740.5 ±17.5	223.0 ± 5.3	5.330 ±.126	3.825 ±.090	r	324
4.649 ±.003				343.10 ± .41	51.66 ± .06	1.2347 ±.0015	3.563 ±.005	26	325
8.252 ±.005				1111.6 ± 1.1	37.194 ± .037	.8890 ±.0009	2.960 ±.003	25	326
9.309 ±.003				1566.6 ± 2.3	52.42 ± .08	1.253 ±.002	4.251 ±.006	25	327
7.80 ±.01				1441.9 ± 4.4	48.24 ± .15	1.153 ±.004	3.247 ±.010	r	328
9.28 ±.02		94.55 ±.33		342.7 ± 1.8	51.60 ± .27	1.233 ±.006	3.338 ±.017	r	329
5.010 ±.003				308.22 ± .30	92.812 ± .090	2.2183 ±.0022	2.9375 ±.0029	r	330
5.0675 ±.0030				299.67 ± .29	90.239 ± .088	2.1568 ±.0022	3.0387 ±.0030	r	331
4.855 ±.005				276.01 ± .79	83.12 ± .24	1.9865 ±.0057	3.994 ±.011	r	332
5.01 ±.02				310.3 ± 2.7	93.44 ± .80	2.233 ±.019	3.357 ±.029	r	333
4.809 ±.003				348.35 ± .46	104.90 ± .14	2.507 ±.003	2.462 ±.003	r	334
9.192 ±.005				675.98 ± .82	203.55 ± .25	4.8651 ±.0060	2.641 ±.003	25	335
9.3517 ±.0030				773.02 ± .54	232.78 ± .16	5.5636 ±.0039	2.513 ±.002	25	336
9.339 ±.003				1548.96 ± .88	233.22 ± .13	5.5741 ±.0032	2.508 ±.001	25	337
9.285 ±.010				781.75 ± 1.80	235.40 ± .54	5.6264 ±.0130	2.753 ±.006	r	338
9.287 ±.010				1541.40 ± 2.47	232.08 ± .37	5.5468 ±.0089	2.792 ±.005	r	339

	Name and formula	Crystal system	Space group	Structure type	Z	a_o	b_o
	Ortho and ring structure silicates—Continued						
340	Mn-Indialite $Mn_2Al_3(AlSi_5O_{18})$	hex.	P6/mmc(192)	beryl	2	9.925 ±.010	
341	Sapphirine $Mg_2Al_4O_6SiO_4$*	mon.	$P2_1/c(14)$		8	11.26 ±.03	14.46 ±.03
342	Elbaite $NaLiAl_{7.67}B_3Si_6O_{27}(OH)_4$*	hex-R	R3m(160)	tourmaline	3	15.842 ±.010	
343	Schorl $NaFe_3Al_6B_3Si_6O_{27}(OH)_4$*	hex-R	R3m(160)	tourmaline	3	16.032 ±.010	
344	Dravite $NaMg_3Al_6B_3Si_6O_{27}(OH)_4$	hex-R	R3m(160)	tourmaline	3	15.942 ±.010	
345	Uvite $CaMg_4Al_5B_3Si_6O_{27}(OH)_4$	hex-R	R3m(160)	tourmaline	3	15.86 ±.01	
346	Sphene $CaTiSiO_5$*	mon.	A2/a(15)		4	7.07 ±.01	8.72 ±.01
347	Datolite $CaBSiO_4(OH)$*	mon.	$P2_1/c(14)$		4	9.62 ±.03	7.60 ±.03
348	Euclase $AlBeSiO_4(OH)$*	mon.	$P2_1/a(14)$		4	4.763 ±.005	14.29 ±.02
349	Chloritoid $H_2FeAl_2SiO_7$*	mon.	C2/c(15)		8	9.48 ±.01	5.48 ±.01
350	Hemimorphite $Zn_4(OH)_2Si_2O_7.H_2O$*	orth.	Imm2(35)		2	8.370 ±.005	10.719 ±.005
351	Zoisite $Ca_2Al_3(SiO_4)_3OH$	orth.	Pnma(62)		4	16.15 ±.01	5.581 ±.005
352	Clinozoisite $Ca_2Al_3(SiO_4)_3OH$	mon.	$P2_1/m(11)$		2	8.887 ±.007	5.581 ±.005
353	Epidote $Ca_2Al_{1.5}Fe_{1.5}(SiO_4)_3OH$*	mon.	$P2_1/m(11)$		2	8.89 ±.02	5.63 ±.01
354	Piemontite $Ca_2Al_{1.5}Mn_{1.5}(SiO_4)_3OH$*	mon.	$P2_1/m(11)$		2	8.95 ±.02	5.70 ±.01
355	Lawsonite $CaAl_2Si_2O_7(OH)_2.H_2O$	orth.	Cccm(63)		4	8.787 ±.005	5.836 ±.005
	Chain and band structure silicates						
356	Enstatite $MgSiO_3$*	orth.	Pcab(61)		16	8.829 ±.010	18.22 ±.01
357	Clinoenstatite $MgSiO_3$	mon.	$P2_1/c(15)$		8	9.620 ±.005	8.825 ±.005
358	Protoenstatite $MgSiO_3$	orth.	Pbcn(60)		8	9.25 ±.01	8.74 ±.01
359	High Clinoenstatite $MgSiO_3$	tri.			8	10.000 ±.005	8.934 ±.004
360	Clinoferrosilite $FeSiO_3$	mon.	$P2_1/c(15)$		8	9.7085 ±.0010	9.0872 ±.0011
361	Orthoferrosilite $FeSiO_3$	orth.	Pcab(61)	enstatite	16	9.080 ±.002	18.431 ±.004
362	Diopside $CaMg(SiO_3)_2$	mon.	C2/c(15)	diopside	4	9.743 ±.005	8.923 ±.005
363	Hedenbergite $CaFe(SiO_3)_2$*	mon.	C2/c(15)	diopside	4	9.854 ±.010	9.024 ±.010
364	Johannsenite $CaMn(SiO_3)_2$*	mon.	C2/c(15)	diopside	4	9.83 ±.03	9.04 ±.03
365	Ureyite $NaCr(SiO_3)_2$	mon.	C2/c(15)	diopside	4	9.550 ±.016	8.712 ±.007

X-Ray Crystallographic Data of Minerals

c_o	α_o	β_o	γ_o	Cell volume 10^{-24} cm^3	Molar volume cm^3	Molar volume cal bar^{-1}	X-Ray density grams cm^{-3}	Temp. °C	
Ortho and ring structure silicates—Continued									
9.297				793.11	238.8	5.708	2.706	r	340
±.010				± .81	± .5	±.013	±.006		
9.95		125.33		1321.7	99.50	2.378	3.464	r	341
±.02		±.50		± 9.7	± .73	±.017	±.025		
7.009				1523.4	305.82	7.3093	3.271	r	342
±.010				± 2.9	± .58	±.0140	±.006		
7.149				1591.3	319.45	7.635	3.297	r	343
±.010				± 3.0	± .60	±.014	±.006		
7.224				1589.99	319.19	7.629	3.004	r	344
±.010				± 2.97	± .60	±.014	±.006		
7.19				1566.3	314.4	7.515	3.095	r	345
±.01				± 2.9	± .6	±.014	±.006		
6.56		113.95		369.61	55.65	1.330	3.523	r	346
±.01		±.25		± 1.13	± .17	±.004	±.011		
4.84		90.15		353.9	53.28	1.273	3.003	r	347
±.02		±.25		± 2.3	± .35	±.008	±.020		
4.618		100.25		309.30	46.57	1.113	3.116	r	348
±.005		±.10		± .64	± .10	±.002	±.007		
18.18		101.77		924.6	69.61	1.664	3.619	r	349
±.01		±.25		± 2.2	± .16	±.004	±.008		
5.120				459.36	138.32	3.306	3.482	25	350
±.005				± .57	± .17	±.004	±.004		
10.06				906.74	136.52	3.263	3.328	r	351
±.01				± 1.34	± .20	±.005	±.005		
10.14		115.93		452.30	136.20	3.255	3.336	r	352
±.01		±.33		± 1.45	± .44	±.010	±.011		
10.19		115.40		460.72	138.7	3.316	3.587	r	353
±.02		±.30		± 1.97	± .6	±.014	±.015		
9.41		115.70		432.56	130.3	3.113	3.810	r	354
±.02		±.50		± 2.38	± .7	±.017	±.021		
13.123				672.96	101.32	2.4217	3.101	r	355
±.008				± .80	± .12	±.0029	±.004		
Chain and band structure silicates									
5.192				835.21	31.44	.7514	3.194	r	356
±.005				± 1.32	± .05	±.0012	±.005		
5.188		108.33		418.10	31.47	.7523	3.190	r	357
±.005		±.17		± .66	± .05	±.0012	±.005		
5.32				430.10	32.38	.7739	3.101	r	358
±.01				± 1.05	± .08	±.0019	±.008		
5.170	88.27	70.03	91.01	433.72	32.65	.7804	3.075		359
±.003	±.05	±.04	±.04	± .40	± .03	±.0008	±.003		
5.2284		108.43		437.60	32.943	.7874	4.005	r	360
±.004		±.05		± .15	± .011	±.0003	±.002		
5.238				876.6	33.00	.7887	3.998	r	361
±.001				± .54	± .02	±.0008	±.004		
5.251		105.93		438.97	66.09	1.580	3.277	r	362
±.003		±.25		± .69	± .10	±.003	±.005		
5.263		104.23		453.64	68.30	1.632	3.632	r	363
±.010		±.33		± 1.28	± .19	±.005	±.010		
5.27		105.00		452.35	68.11	1.628	3.629	r	364
±.02		±.50		± 2.87	± .43	±.010	±.023		
5.273		107.44		418.6	63.02	1.506	3.605	r	365
±.008		±.16		± 1.1	± .16	±.004	±.009		

	Name and formula	Crystal system	Space group	Structure type	Z	a_o	b_o
	Chain and band structure silicates—Continued						
366	Jadeite $NaAl(SiO_3)_2$*	mon.	C2/c(15)	diopside	4	9.409 ±.005	8.564 ±.005
367	Acmite (Aegirine) $NaFe(SiO_3)_2$	mon.	C2/c(15)	diopside	4	9.658 ±.005	8.795 ±.005
368	Ca Tschermak Molecule $CaAl_2SiO_6$	mon.	C2/c(15)	diopside	4	9.615 ±.005	8.661 ±.005
369	Spodumene $LiAl(SiO_3)_2$	mon.	C2/c(15)	diopside	4	9.451 ±.002	8.387 ±.002
370	β-Spodumene $LiAl(SiO_3)_2$	tet.	$P4_32_12(96)$ $P4_12_12(92)$		4	7.5332 ±.0008	
371	Pectolite $Ca_2NaH(SiO_3)_3$*	tri.	$P\bar{1}(2)$		2	7.99 ±.01	7.04 ±.01
372	Wollastonite $CaSiO_3$*	tri.	$P\bar{1}(2)$		6	7.94 ±.01	7.32 ±.01
373	Parawollastonite $CaSiO_3$*	mon.	$P2_1(4)$		12	15.417 ±.004	7.321 ±.002
374	Pseudowollastonite $CaSiO_3$*	tri.			24	6.90 ±.02	11.78 ±.02
375	Rhodonite $MnSiO_3$*	tri.	$P\bar{1}(2)$		10	7.682 ±.002	11.818 ±.003
376	Bustamite $CaMn(SiO_3)_2$*	tri.	$A\bar{1}(2)$		6	7.736 ±.003	7.157 ±.003
377	Pyroxmangite $MnFe(SiO_3)_2$*	tri.	$P\bar{1}(2)$		7	7.56 ±.02	17.45 ±.05
378	Tremolite $Ca_2Mg_5[Si_8O_{22}](OH)_2$*	mon.	C2/m(12)	tremolite	2	9.840 ±.010	18.052 ±.020
379	Fluor-tremolite $Ca_2Mg_5[Si_8O_{22}]F_2$	mon.	C2/m(12)	tremolite	2	9.781 ±.007	18.01 ±.01
380	Ferrotremolite $Ca_2Fe_5[Si_8O_{22}](OH)_2$	mon.	C2/m(12)	tremolite	2	9.97 ±.01	18.34 ±.02
381	Grunerite $Fe_7[Si_8O_{22}](OH)_2$	mon.	C2/m(12)	tremolite	2	9.572 ±.005	18.44 ±.01
382	Cummingtonite (hypo.) $Mg_7[Si_8O_{22}](OH)_2$	mon.	C2/m(12)	tremolite	2	9.476 ±.010	17.935 ±.010
383	Riebeckite $Na_2Fe_3Fe_2[Si_8O_{22}](OH)_2$	mon.	C2/m(12)	tremolite	2	9.729 ±.020	18.065 ±.020
384	Magnesioriebeckite $Na_2Mg_3Fe_2[Si_8O_{22}](OH)_2$	mon.	C2/m(12)	tremolite	2	9.733 ±.010	17.946 ±.020
385	Gaucophane I $Na_2Mg_3Al_2[Si_8O_{22}](OH)_2$	mon.	C2/m(12)	tremolite	2	9.748 ±.010	17.915 ±.020
386	Glaucophane II $Na_2Mg_3Al_2[Si_8O_{22}](OH)_2$	mon.	C2/m(12)	tremolite	2	9.663 ±.010	17.696 ±.020
387	Fluor-edenite $NaCa_2Mg_5[AlSi_7O_{22}]F_2$	mon.	C2/m(12)	tremolite	2	9.847 ±.005	18.00 ±.01
388	Fluor-richterite $Na_2CaMg_5[Si_8O_{22}]F_2$	mon.	C2/m(12)	tremolite	2	9.823 ±.005	17.96 ±.01
389	Anthophyllite $Mg_7[Si_8O_{22}](OH)_2$	orth.	Pnma(62)		4	18.61 ±.02	18.01 ±.06
	Framework structure silicates						
390	Microcline $KAlSi_3O_8$	tri.	$C\bar{1}(2)$		4	8.582 ±.002	12.964 ±.005
391	High Sanidine $KAlSi_3O_8$	mon.	C2/m(12)		4	8.615 ±.002	13.031 ±.003

X-Ray Crystallographic Data of Minerals

c_o	α_o	β_o	γ_o	Cell volume 10^{-24} cm³	Molar volume cm³	Molar volume cal bar⁻¹	X-Ray density grams cm⁻³	Temp. °C	
Chain and band structure silicates—Continued									
5.220		107.50		401.15	60.40	1.444	3.347	r	366
±.005		±.20		± .67	± .10	±.002	±.006		
5.294		107.42		429.06	64.60	1.544	4.411	r	367
±.005		±.20		± .70	± .11	±.003	±.007		
5.272		106.12		421.77	63.50	1.518	3.435	r	368
±.003		±.20		± .59	± .09	±.002	±.005		
5.208		110.07		387.7	58.37	1.395	3.188	r	369
±.001		±.03		± .1	± .02	±.001	±.001		
9.1540				519.48	78.215	1.8694	2.379	r	370
±.0008				± .12	± .018	±.0005	±.001		
7.02	90.05	95.27	102.47	383.84	115.58	2.763	2.876	r	371
±.01	±.25	±.25	±.25	± .99	± .30	±.007	±.007		
7.07	90.03	95.37	103.43	397.82	39.93	.9544	2.909	r	372
±.01	±.25	±.25	±.25	± 1.03	± .10	±.0025	±.008		
7.066		95.40		793.98	39.85	.9524	2.915	r	373
±.003		±.10		± .47	± .02	±.0006	±.002		
19.65	90.00	90.80	90.00	1597.0	40.08	.9579	2.899	r	374
±.02	±.30	±.30	±.30	± 5.6	± .14	±.0034	±.010		
6.707	92.36	93.95	105.66	583.77	35.158	.8403	3.727	r	375
±.002	±.05	±.05	±.05	± .31	± .019	±.0005	±.002		
13.824	90.52	94.58	103.87	740.38	74.32	1.776	3.326	r	376
±.010	±.25	±.25	±.25	± 1.08	± .11	±.003	±.005		
6.67	84.00	94.30	113.70	800.77	68.90	1.647	3.817	r	377
±.02	±.30	±.30	±.30	± 4.29	± .36	±.009	±.020		
5.275		104.70		906.34	272.92	6.523	2.977	r	378
±.010		±.25		± 2.43	± .73	±.018	±.008		
5.267		104.52		898.18	270.46	6.464	3.018	20	379
±.005		±.25		± 1.56	± .47	±.011	±.005		
5.30		104.50		938.24	282.53	6.753	3.434	r	380
±.01		±.10		± 2.92	± .69	±.017	±.008		
5.342		101.77		923.08	277.96	6.644	3.603	r	381
±.007		±.25		± 1.63	± .49	±.012	±.006		
5.292		102.23		878.97	264.68	6.326	2.950	r	382
±.005		±.25		± 1.58	± .47	±.011	±.005		
5.334		103.31		912.29	274.71	6.566	3.407	r	383
±.010		±.25		± 2.89	± .87	±.021	±.011		
5.299		103.30		900.74	271.24	6.483	3.102	r	384
±.010		±.25		± 2.37	± .71	±.017	±.008		
5.273		102.78		898.04	270.42	6.463	2.898	r	385
±.010		±.25		± 2.35	± .71	±.017	±.008		
5.277		103.67		876.79	264.02	6.310	2.968	r	386
±.010		±.10		± 2.17	± .65	±.016	±.007		
5.282		104.83		905.03	272.53	6.514	3.076	r	387
±.005		±.25		± 1.51	± .46	±.011	±.005		
5.268		104.33		900.47	271.15	6.481	3.033	r	388
±.005		±.25		± 1.48	± .45	±.011	±.005		
5.24				1756.3	264.4	6.320	2.953	r	389
±.01				± 7.0	± 1.1	±.025	±.012		
Framework structure silicates									
7.222	90.62	115.92	87.68	722.06	108.72	2.5984	2.560	r	390
±.002	±.10	±.10	±.10	± .67	± .10	±.0025	±.002		
7.177		115.98		724.28	109.05	2.6064	2.552	r	391
±.002		±.10		± .69	± .10	±.0025	±.002		

	Name and formula	Crystal system	Space group	Structure type	Z	a_o	b_o
	Framework structure silicates—Continued						
392	Orthoclase $KAlSi_3O_8$*	mon.	C2/m(12)		4	8.562 ±.003	12.996 ±.004
393	Fe-Sanidine $KFeSi_3O_8$	mon.	C2/m(12)		4	8.689 ±.008	13.12 ±.01
394	Fe-Microcline $KFeSi_3O_8$	tri.	$C\bar{1}$(2)		4	8.68 ±.01	13.10 ±.01
395	Low Albite $NaAlSi_3O_8$	tri.	$C\bar{1}$(2)		4	8.139 ±.002	12.788 ±.003
396	High Albite (Analbite) $NaAlSi_3O_8$	tri.	$C\bar{1}$(2)		4	8.160 ±.002	12.870 ±.003
397	Anorthite $CaAl_2Si_2O_8$	tri.	$P\bar{1}$(2)	primitive cell	8	8.177 ±.002	12.877 ±.003
398	Synthetic $CaAl_2Si_2O_8$	hex.	$P6_3$/mcm(193)		2	5.10 ±.02	
399	Synthetic $CaAl_2Si_2O_8$	orth.	$P2_12_12$(18)		2	8.22 ±.02	8.60 ±.02
400	Celsian $BaAl_2Si_2O_8$*	mon.	$I2_1$/c(15)		8	8.627 ±.010	13.045 ±.010
401	Paracelsian $BaAl_2Si_2O_8$*	mon.	$P2_1$/a(14)		4	8.58 ±.02	9.583 ±.020
402	Banalsite $BaNa_2Al_4Si_4O_{16}$*	orth.			4	8.50 ±.02	9.97 ±.02
403	Danburite $CaB_2Si_2O_8$*	orth.	Pnam(62)		4	8.04 ±.02	8.77 ±.02
404	Low Nepheline $NaAlSiO_4$	hex.	$C6_3$(178)		8	9.986 ±.005	
405	High Carnegeite $NaAlSiO_4$	cubic			4	7.325 ±.004	
406	Kaliophilite natural $KAlSiO_4$*	hex.	$P6_322$(182)		54	26.930 ±.010	
407	Kaliophilite synthetic $KAlSiO_4$	hex.	$P6_3$(173) $P6_322$(182)		2	5.180 ±.002	
408	Kalsilite $KAlSiO_4$	hex.	$P6_3$(173)		2	5.1597 ±.0020	
409	Leucite $KAlSi_2O_6$	tet.	$I4_1$/a(100)		16	13.074 ±.003	
410	High Leucite $KAlSi_2O_6$*	cubic	Ia3d(230)		16	13.43 ±.05	
411	Fe-Leucite $KFeSi_2O_6$	tet.	$I4_1$/a(100)		16	13.205 ±.002	
412	Petalite $LiAlSi_4O_{10}$*	mon.	$P2_1$/n(14)		2	11.32 ±.03	5.14 ±.01
413	Marialite $Na_4Al_3Si_9O_{24}Cl$	tet.	I4/m(87) P4/m(83)		2	12.064 ±.008	
414	Meionite $Ca_4Al_6Si_6O_{24}CO_3$	tet.	I4/m(87) P4/m(83)		2	12.174 ±.008	
	Sheet structure silicates						
415	Muscovite $KAl_2[AlSi_3O_{10}](OH)_2$*	mon.	C2/c(15)	$2M_2$ mica	4	5.203 ±.005	8.995 ±.005
416	Paragonite $NaAl_2[AlSi_3O_{10}](OH)_2$*	mon.	C2/c(15)	$2M_1$ mica	4	5.13 ±.03	8.89 ±.05
417	Lepidolite $K_2Al_3Li_2[AlSi_7O_{20}](OH)_4$*	mon.	C2/c(15)	$2M_2$ mica	2	9.2 ±.1	5.3 ±.1

X-Ray Crystallographic Data of Minerals

c_o	α_o	β_o	γ_o	Cell volume 10^{-24} cm^3	Molar volume cm^3	Molar volume cal bar^{-1}	X-Ray density grams cm^{-3}	Temp. °C	
Framework structure silicates—Continued									
7.193 ±.003		116.02 ±.15		719.25 ± 1.02	108.29 ± .15	2.5883 ±.0037	2.570 ±.004	r	392
7.319 ±.007		116.10 ±.30		749.28 ± 2.24	112.81 ± .34	2.6964 ±.0081	2.723 ±.008	r	393
7.340 ±.007	90.75 ±.25	116.05 ±.25	86.23 ±.25	748.09 ± 1.92	112.63 ± .29	2.692 ±.007	2.727 ±.007	r	394
7.160 ±.002	94.27 ±.10	116.57 ±.10	87.68 ±.10	664.65 ± .60	100.07 ± .09	2.3918 ±.0022	2.620 ±.002	26	395
7.106 ±.002	93.54 ±.10	116.36 ±.10	90.19 ±.10	667.00 ± .60	100.43 ± .09	2.4003 ±.0022	2.611 ±.002	r	396
14.169 ±.003	93.17 ±.02	115.85 ±.02	91.22 ±.02	1338.9 ± .6	100.79 ± .04	2.4090 ±.0011	2.760 ±.001	r	397
14.72 ±.02				331.57 ± 2.64	99.85 ± .79	2.386 ±.019	2.786 ±.022	r	398
4.83 ±.01				341.44 ± 1.35	102.82 ± .41	2.457 ±.010	2.706 ±.011	r	399
14.408 ±.020		115.20 ±.25		1467.1 ± 4.2	110.45 ± .31	2.640 ±.008	3.400 ±.010	r	400
9.08 ±.02		90.00 ±.50		746.6 ± 2.9	112.4 ± .4	2.687 ±.010	3.340 ±.013	r	401
16.72 ±.03				1416.9 ± 5.1	213.3 ± .8	5.099 ±.018	3.092 ±.011	r	402
7.74 ±.02				545.8 ± 2.3	82.17 ± .35	1.964 ±.008	2.992 ±.013	r	403
8.330 ±.004				719.38 ± .80	54.16 ± .06	1.294 ±.002	2.623 ±.003	r	404
				393.03 ± .64	59.18 ± .10	1.414 ±.002	2.401 ±.004	750	405
8.522 ±.004				5352.4 ± 4.7	59.69 ± .05	1.427 ±.001	2.650 ±.002	r	406
8.559 ±.004				198.89 ± .18	59.89 ± .05	1.431 ±.001	2.641 ±.002	r	407
8.7032 ±.0030				200.66 ± .17	60.424 ± .051	1.4442 ±.0031	2.618 ±.002	r	408
13.738 ±.003				2348.23 ± 1.19	88.389 ± .045	2.1126 ±.0011	2.469 ±.001	25	409
				2422.3 ±27.1	91.18 ± 1.02	2.179 ±.024	2.394 ±.027	625	410
13.970 ±.003				2435.98 ± .91	91.692 ± .034	2.1915 ±.0009	2.695 ±.001	25	411
7.62 ±.01		105.90 ±.20		426.41 ± 1.57	128.4 ± .5	3.069 ±.011	2.385 ±.009	r	412
7.514 ±.004				1093.6 ± 1.6	329.3 ± .5	7.871 ±.011	2.566 ±.004	r	413
7.652 ±.015				1134.07 ± 2.68	341.5 ± .8	8.162 ±.019	2.737 ±.007	r	414
Sheet structure silicates									
20.030 ±.010		94.47 ±.33		934.57 ± 1.21	140.71 ± .18	3.363 ±.004	2.831 ±.004	r	415
19.32 ±.10		95.17 ±.50		877.52 ± 8.47	132.1 ± 1.3	3.158 ±.031	2.893 ±.028	r	416
20.0 ±.2		98.00 ±.50		965.7 ±23.2	290.8 ± 7.0	6.950 ±.167	2.698 ±.065	r	417

	Name and formula	Crystal system	Space group	Structure type	Z	a_o	b_o
	Sheet structure silicates—Continued						
418	Phlogopite $KMg_3[AlSi_3O_{10}](OH)_2$	mon.	Cm(8)	1M mica	2	5.326 ±.010	9.210 ±.010
419	Fluor-phlogopite $KMg_3[AlSi_3O_{10}]F_2$	mon.	Cm(8)	1M mica	2	5.299 ±.005	9.188 ±.005
420	Annite $KFe_3[AlSi_3O_{10}](OH)_2$	mon.	Cm(8)	1M mica	2	5.391 ±.010	9.350 ±.005
421	Ferriannite $KFe_3[FeSi_3O_{10}](OH)_2$	mon.	C2/m(12)		2	5.430 ±.002	9.404 ±.003
422	Margarite $CaAl_2[Al_2Si_2O_{10}](OH)_2$*	mon.	C2/c(15)	2M mica	4	5.13 ±.02	8.92 ±.03
423	Talc $Mg_3Si_4O_{10}(OH)_2$*	mon.	C2/c(15)	$2M_1$	4	5.287 ±.007	9.158 ±.010
424	Pyrophyllite $Al_2Si_4O_{10}(OH)_2$*	mon.	C2/c(15)	$2M_1$	4	5.14 ±.02	8.90 ±.02
425	Minnesotaite $Fe_3Si_4O_{10}(OH)_2$*	mon.	C2/c(15)		4	5.4 ±.1	9.42 ±.04
426	Dickite $Al_2Si_2O_5(OH)_4$*	mon.	Cc(9)		4	5.150 ±.002	8.940 ±.003
427	Kaolinite $Al_2Si_2O_5(OH)_4$*	tri.	P1(1)		2	5.155 ±.007	8.959 ±.010
428	Nacrite $Al_2Si_2O_5(OH)_4$*	mon.	Cc(9)		4	8.909 ±.010	5.146 ±.010
	Zeolites						
429	Analcite $NaAlSi_2O_6.H_2O$	cubic	Ia3d(230)		16	13.733 ±.005	
430	Natrolite $Na_2Al_2Si_3O_{10}.2H_2O$*	orth.	Fdd2(43)		8	18.30 ±.02	18.63 ±.02

c_o	α_o	β_o	γ_o	Cell volume 10^{-24} cm^3	Molar volume cm^3	Molar volume cal bar^{-1}	X-Ray density grams cm^{-3}	Temp. °C	
				Sheet structure silicates—Continued					
10.311		100.17		497.83	149.91	3.5830	2.784	r	418
±.010		±.10		± 1.19	± .36	±.0086	±.007		
10.135		99.92		486.07	146.37	3.498	2.878	r	419
±.005		±.10		± .60	± .18	±.004	±.004		
10.313		99.70		512.40	154.30	3.688	3.318	r	420
±.020		±.25		± 1.45	± .44	±.010	±.009		
10.341		100.07		519.92	156.56	3.7419	3.454	r	421
±.006		±.20		± .51	± .15	±.0037	±.003		
19.50		95.00		888.9	133.8	3.199	2.975	r	422
±.05		±.50		± 5.2	± .8	±.019	±.017		
18.95		99.50		904.94	136.25	3.2565	2.784	r	423
±.01		±.20		± 1.71	± .26	±.0062	±.005		
18.55		99.92		835.9	125.9	3.008	2.863	r	424
±.03		±.20		± 4.0	± .6	±.015	±.014		
19.4		100.00		971.8	146.3	3.497	3.239	r	425
±.1		±.50		±19.2	± 2.9	±.069	±.064		
14.736		103.58		659.49	99.30	2.3733	2.600	r	426
±.005		±.10		± .49	± .07	±.0018	±.002		
7.407	91.68	104.87	89.93	330.48	99.52	2.3785	2.594	r	427
±.008	±.35	±.35	±.35	± .86	± .26	±.0062	±.007		
15.697		113.70		658.9	99.21	2.3713	2.602	r	428
±.020		±.25		± 2.1	± .32	±.0076	±.008		
				Zeolites					
				2589.98	97.49	2.3301	2.258	r	429
				± 2.83	± .11	±.0026	±.003		
6.60				2250.1	169.39	4.049	2.245	r	430
±.01				± 4.8	± .37	±.009	±.005		

DEFINITIVE RULES FOR NOMENCLATURE OF ORGANIC CHEMISTRY

IUPAC 1957 Rules.

Section A. Hydrocarbons

Section B. Fundamental Heterocyclic Systems

These rules are taken from 'Definitive Rules for the Nomenclature of Organic Chemistry' which were adopted unanimously by the Commission on Nomenclature and by The Council of the International Union of Pure and Applied Chemistry at Paris 1957, and subsequently published by Butterworths Scientific Publications on behalf of the Union. The extracts are printed here by permission of the Union and of Butterworths Scientific Publications. Future 'tentative' rules will be published in the Bulletin of the Union, and when made 'definitive' in its Journal 'Pure and Applied Chemistry.'

RULES

A. HYDROCARBONS

Acyclic Hydrocarbons

A-1

1.1.—The first four saturated unbranched acyclic hydrocarbons are called methane, ethane, propane and butane. Names of the higher members of this series consist of a numerical prefix and the termination "-ane." Examples of these numerical prefixes are shown in the table below. The generic name of saturated acyclic hydrocarbons (branched or unbranched) is "alkane."

Examples:

n		*n*		*n*		*n*	
1	Methane	12	Dodecane	22	Docosane	32	Dotriacontane
2	Ethane	13	Tridecane	23	Tricosane	33	Tritriacontane
3	Propane	14	Tetradecane	24	Tetracosane	40	Tetracontane
4	Butane	15	Pentadecane	25	Pentacosane	50	Pentacontane
5	Pentane	16	Hexadecane	26	Hexacosane	60	Hexacontane
6	Hexane	17	Heptadecane	27	Heptacosane	70	Heptacontane
7	Heptane	18	Octadecane	28	Octacosane	80	Octacontane
8	Octane	19	Nonadecane	29	Nonacosane	90	Nonacontane
9	Nonane	20	Eicosane	30	Triacontane	100	Hectane
10	Decane	21	Heneicosane	31	Hentriacontane	132	Dotriacontahectane
11	Undecane						

1.2.—Univalent radicals derived from saturated unbranched acyclic hydrocarbons by removal of hydrogen from a terminal carbon atom are named by replacing the ending "-ane" of the name of the hydrocarbon by "-yl." The carbon atom with the free valence is numbered as 1. As a class, these radicals are called normal, or unbranched-chain, alkyls.

Examples:

Pentyl $\overset{5}{C}H_3—\overset{4}{C}H_2—\overset{3}{C}H_2—\overset{2}{C}H_2—\overset{1}{C}H_2—$

Undecyl $\overset{11}{C}H_3—\overset{10-2}{[CH_2]_9}—\overset{1}{C}H_2—$

A-2

2.1.—A saturated branched acyclic hydrocarbon is named by prefixing the designations of the side chains to the name of the longest chain present in the formula.

Example:

$$\overset{5}{C}H_3—\overset{4}{C}H_2—\underset{\displaystyle CH_3}{\underset{|}{\overset{3}{C}H}}—\overset{2}{C}H_2—\overset{1}{C}H_3$$

3-Methylpentane

These names are retained for unsubstituted hydrocarbons only

Isobutane	$(CH_3)_2CH—CH_3$
Isopentane	$(CH_3)_2CH—CH_2—CH_3$
Neopentane	$(CH_3)_4C$
Isohexane	$(CH_3)_2CH—CH_2—CH_2—CH_3$

***Chemical Abstracts* index names for the above compounds are 2-methylpropane, 2-methylbutane, 2,2-dimethylpropane, and 2-methylpentane.**

2.2.—The longest chain is numbered from one end to the other by arabic numerals, the direction being so chosen as to give the lowest numbers possible to the side chains. When series of locants containing the same number of terms are compared term by term, that series is "lowest" which contains the lowest number on the occasion of the first difference. This principle is applied irrespective of the nature of the substituents.

Examples:

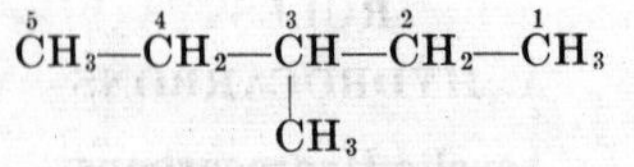

$$\overset{5}{C}H_3—\overset{4}{C}H_2—\overset{3}{C}H(CH_3)—\overset{2}{C}H_2—\overset{1}{C}H_3$$

3-Methylpentane

$$\overset{6}{C}H_3—\overset{5}{C}H(CH_3)—\overset{4}{C}H_2—\overset{3}{C}H(CH_3)—\overset{2}{C}H(CH_3)—\overset{1}{C}H_3$$

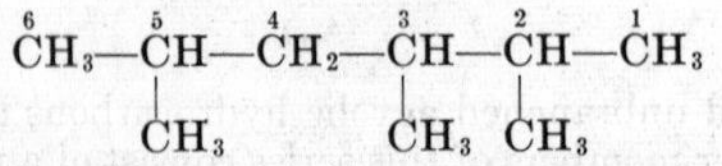

2,3,5-Trimethylhexane (not 2,4,5-Trimethylhexane)

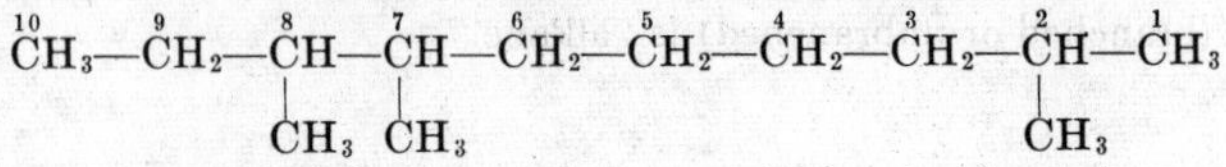

$$\overset{10}{C}H_3—\overset{9}{C}H_2—\overset{8}{C}H(CH_3)—\overset{7}{C}H(CH_3)—\overset{6}{C}H_2—\overset{5}{C}H_2—\overset{4}{C}H_2—\overset{3}{C}H_2—\overset{2}{C}H(CH_3)—\overset{1}{C}H_3$$

2,7,8-Trimethyldecane (not 3,4,9-Trimethyldecane)

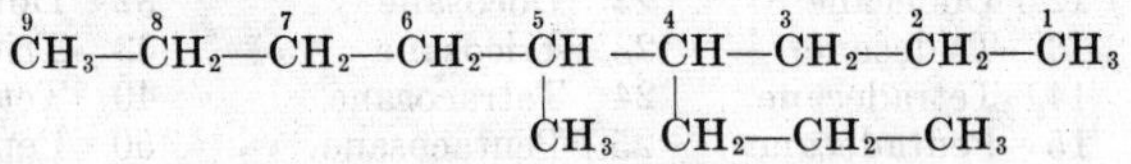

$$\overset{9}{C}H_3—\overset{8}{C}H_2—\overset{7}{C}H_2—\overset{6}{C}H_2—\overset{5}{C}H(CH_3)—\overset{4}{C}H(CH_2—CH_2—CH_3)—\overset{3}{C}H_2—\overset{2}{C}H_2—\overset{1}{C}H_3$$

5-Methyl-4-propylnonane (not 5-Methyl-6-propylnonane since 4,5 is lower than 5,6)

2.25.—Univalent branched radicals derived from alkanes are named by prefixing the designation of the side chains to the name of the unbranched alkyl radical possessing the longest possible chain starting from the carbon atom with the free valence, the starting atom being numbered as 1.

Examples:

1-Methylpentyl	$\overset{5}{C}H_3\overset{4}{C}H_2\overset{3}{C}H_2\overset{2}{C}H_2\overset{1}{C}H(CH_3)—$
2-Methylpentyl	$CH_3CH_2CH_2CH(CH_3)CH_2—$
5-Methylhexyl	$(CH_3)_2CHCH_2CH_2CH_2CH_2—$

These names are retained for the unsubstituted radicals only

Isopropyl	$(CH_3)_2CH—$
Isobutyl	$(CH_3)_2CHCH_2—$
sec-Butyl	$CH_3CH_2CH(CH_3)—$
tert-Butyl	$(CH_3)_3C—$
Isopentyl	$(CH_3)_2CHCH_2CH_2—$
Neopentyl	$(CH_3)_3CCH_2—$
tert-Pentyl	$CH_3CH_2C(CH_3)_2—$
Isohexyl	$(CH_3)_2CHCH_2CH_2CH_2—$

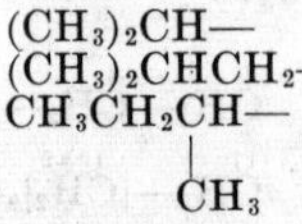

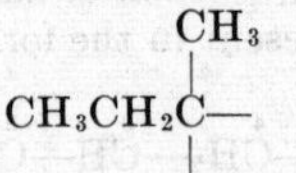

2.3.—If two or more side chains of different nature are present, they may be cited (a) in order of increasing complexity or (b) in alphabetical order.

(a) The side chains are arranged in order of increasing complexity by applying the following criteria in series until a decision is reached:

(i) The less complex is that containing the smaller total number of carbon atoms.

Examples:

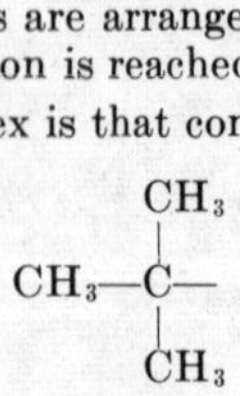

$$CH_3—C(CH_3)_2— \quad \text{less complex than} \quad \overset{5}{C}H_3—\overset{4}{C}H_2—\overset{3}{C}H_2—\overset{2}{C}H_2—\overset{1}{C}H_2—$$

(ii) The less complex is that containing the longer straight chain.

Example:

$$CH_3$$
$$CH_3—CH_2—CH—CH_2— \quad \text{less complex than} \quad CH_3—C(CH_3)_2—CH_2—$$

(iii) The less complex is that whose longest substituent has the lower locant.

Example:

$$\begin{array}{c} CH_3 \\ CH_3\ CH_2 \\ CH_3—CH_2—CH—CH—CH_2— \end{array} \quad \text{less complex than} \quad \begin{array}{c} CH_3 \\ CH_2\ CH_3 \\ CH_3—CH_2—CH—CH—CH_2— \end{array}$$

(iv) The less complex is that whose next longest substituent has the lower locant.

Example:

$$\begin{array}{c} CH_3 \\ CH_2\ CH_3 \\ CH_3—CH_2—CH—CH—CH_2— \end{array} \quad \text{less complex than} \quad \begin{array}{c} CH_3 \\ CH_3\ CH_2 \\ CH_3—CH—CH—CH_2—CH_2— \end{array}$$

(v) The less complex is that which is the more saturated.

Example:

$$CH_3—CH_2—CH_2— \quad \text{less complex than} \quad CH_3—CH{=}CH—$$

(vi) The less complex is that whose multiple linkage has the lower locant.

Example:

$$CH_3—CH{=}CH— \quad \text{less complex than} \quad CH_2{=}CH—CH_2—$$

(b) The alphabetical order is decided as follows:

(i) The names of simple radicals are first alphabetized and the multiplying prefixes are then inserted.

Example:

$$\begin{array}{c} CH_3—CH_2\ CH_3 \\ CH_3—CH_2—CH_2—CH—C—CH_2—CH_3 \\ CH_3 \end{array}$$

ethyl is cited before methyl, thus 4-Ethyl-3,3-dimethylheptane

(ii) The name of a complex radical is considered to begin with the first letter of its complete name.

Example:

$$\begin{array}{c} CH_3 \\ CH_3—CH—CH—CH_2—CH_2—CH_3 \\ CH_3—[CH_2]_5—CH—CH_2—CH—CH_2—CH_2—CH_2—CH_3 \\ CH_2—CH_3 \end{array}$$

dimethylpentyl (as a complete single substituent) is alphabetized under "d," thus 7-(1,2-Dimethylpentyl)-5-ethyltridecane

(iii) In cases where complex radicals are composed of identical words, priority for citation is given to that radical which contains the lowest locant at the first cited point of difference in the radical.

Example:

$$\begin{array}{c} CH_3 \qquad\qquad CH_3 \\ CH_3—CH_2—CH—CH_2 \qquad CH—CH_2—CH_2—CH_3 \\ CH_3—[CH_2]_4—CH—CH_2—CH—CH_2—CH_2—CH_2—CH_2—CH_3 \end{array}$$

6-(1-Methylbutyl)-8-(2-methylbutyl)tridecane

Chemical Abstracts has long been a proponent of the alphabetical order for prefixes. The alphabetical order as given in **A-2.3**(b) is the order used by *Chemical Abstracts*.

2.4.—If two or more side chains are in equivalent positions, the one to be assigned the lower number is that cited first in the name, whether the order of citation is based on complexity or on the alphabetical order.

Examples:

(a) Order based on complexity

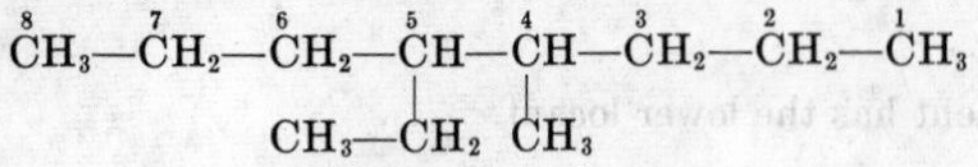

```
 8     7     6     5    4    3     2     1
CH3—CH2—CH2—CH—CH—CH2—CH2—CH3
                  |    |
           CH3—CH2   CH3
```

4-Methyl-5 ethyloctane

(b) Alphabetical order

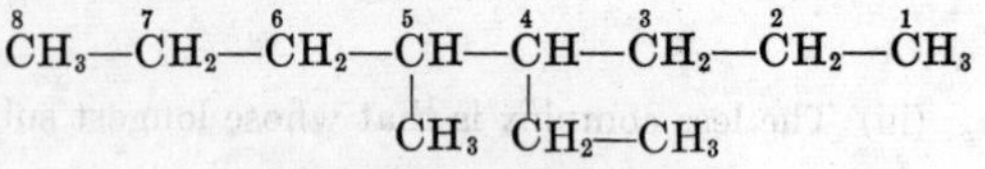

```
 8     7     6     5    4    3     2     1
CH3—CH2—CH2—CH—CH—CH2—CH2—CH3
                  |    |
                CH3  CH2—CH3
```

4 Ethyl-5-methyloctane

```
 8     7     6     5    4    3     2     1
CH3—CH2—CH2—CH—CH—CH2—CH2—CH3
                  |    |
            CH3—CH   CH2
                  |    |
                CH3  CH2—CH3
```

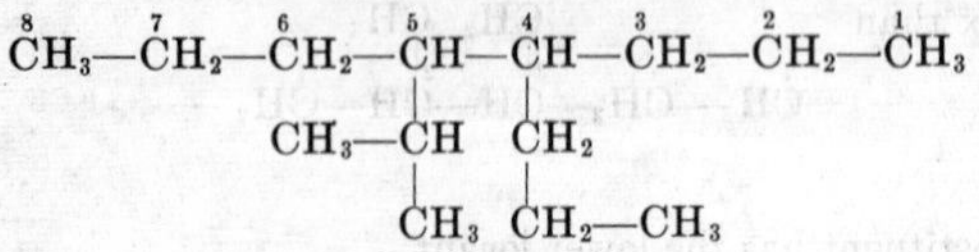

4-Propyl-5-isopropyloctane

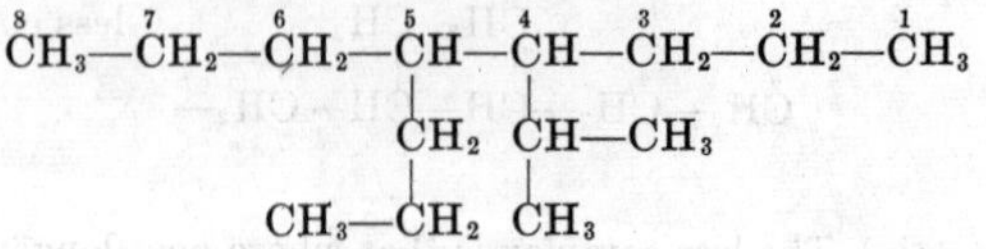

```
 8     7     6     5    4    3     2     1
CH3—CH2—CH2—CH—CH—CH2—CH2—CH3
                  |    |
                CH2  CH—CH3
                  |    |
           CH3—CH2   CH3
```

4-Isopropyl-5-propyloctane

2.5.—The presence of identical unsubstituted radicals is indicated by the appropriate multiplying prefix di-, tri-, tetra-, penta-, hexa-, hepta-, octa-, nona-, deca-, undeca-, *etc.*

Example:

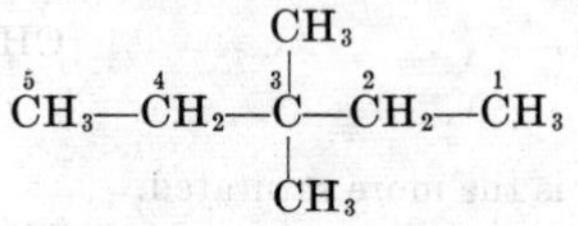

```
                CH3
 5     4     3|    2     1
CH3—CH2—C—CH2—CH3
              |
             CH3
```

3,3-Dimethylpentane

The presence of identical radicals each substituted in the same way may be indicated by the appropriate multiplying prefix bis-, tris-, tetrakis-, pentakis, *etc.* The complete expression denoting a side chain may be enclosed in parentheses or the carbon atoms in side chains may be indicated by primed numbers.

Example:

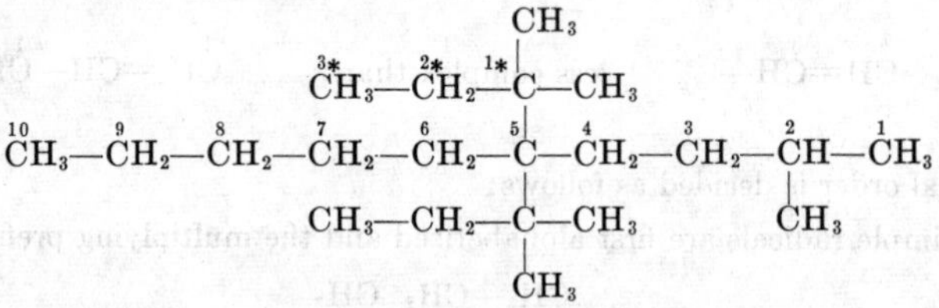

```
                                 CH3
                     3*    2*  1*|
                    CH3—CH2—C—CH3
                                 |
 10    9     8     7     6    5|    4     3     2    1
CH3—CH2—CH2—CH2—CH2—C—CH2—CH2—CH—CH3
                                 |                 |
                    CH3—CH2—C—CH3          CH3
                                 |
                                CH3
```

(a) Use of primes and order of complexity, * indicates primed numbers: 2-Methyl-5,5-bis-1′,1′-dimethylpropyldecane.
(b) Use of parentheses and alphabetical order, * indicates unprimed numbers: 5,5-Bis(1,1-dimethylpropyl)-2-methyldecane.
(c) Use of primes and alphabetical order, * indicates primed numbers: 5,5-Bis-1′,1-dimethylpropyl-2-methyldecane.

Chemical Abstracts uses parentheses and the alphabetical order without primes (**A-2.5**(b)). *Chemical Abstracts* would name the example in (b) 2-Methyl-5,5-di-*tert*-pentyldecane.

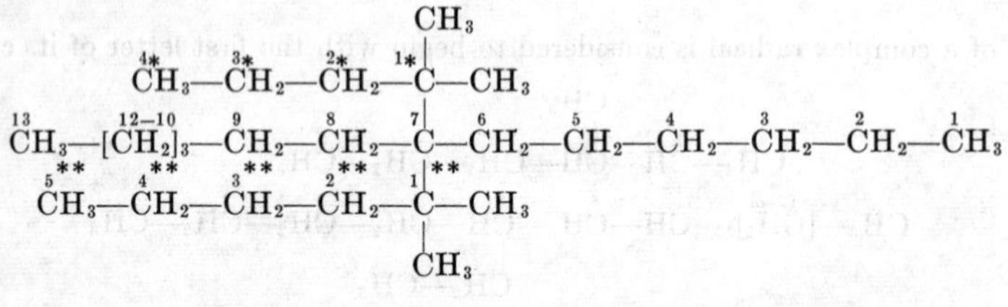

```
                                     CH3
             4*    3*     2*   1*|
            CH3—CH2—CH2—C—CH3
                                     |
 13   12—10      9     8     7|    6     5     4     3     2     1
CH3—[CH2]3—CH2—CH2—C—CH2—CH2—CH2—CH2—CH2—CH3
  5 **    4 **   3 **   2 **   1|**
 CH3—CH2—CH2—CH2—C—CH3
                                     |
                                    CH3
```

(a) Use of primes and order of complexity, * indicates primed numbers, ** indicates doubly primed numbers: 7-1′,1′-Dimethylbutyl-7-1″,1″-dimethylpentyltridecane.
(b) Use of parentheses and alphabetical order, * and ** indicate unprimed numbers: 7-(1,1-Dimethylbutyl)-7-(1,1-dimethylpentyl)-tridecane

2.6.—If chains of equal length are competing for selection as main chain in a saturated branched acyclic hydrocarbon, then the choice goes in series to:

(a) The chain which has the greatest number of side chains.

Example:

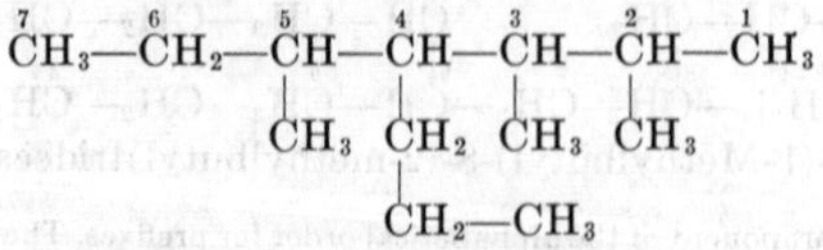

```
 7     6     5    4    3    2    1
CH3—CH2—CH—CH—CH—CH—CH3
              |    |    |    |
            CH3 CH2 CH3 CH3
                   |
                 CH2—CH3
```

2,3,5-Trimethyl-4-propylheptane

(b) The chain whose side chains have the lowest-numbered locants.

Example:

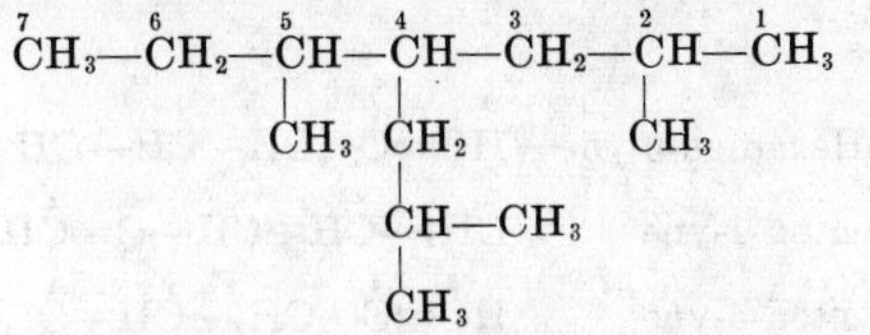

```
 7     6     5    4     3     2    1
CH3—CH2—CH—CH—CH2—CH—CH3
         |    |          |
        CH3  CH2        CH3
              |
              CH—CH3
              |
              CH3
```

4-Isobutyl-2,5-dimethylheptane

(c) The chain having the greatest number of carbon atoms in the smaller side chains.

```
          CH3          CH3              CH3        CH2—CH3
          |            |                |          |
CH3CH2CH—CH  —CH—CH2          CH2—CH—CH2—CHCH2CH3
13  12  11   10    9    8      7 \ / 6    5    4     3  2   1
                                  C
                                 / \
CH3CH2CH—CH2—CH—CH2          CH2—CH—CH2—CHCH2CH3
      |            |                |          |
      CH3          CH3              CH3        CH3
```

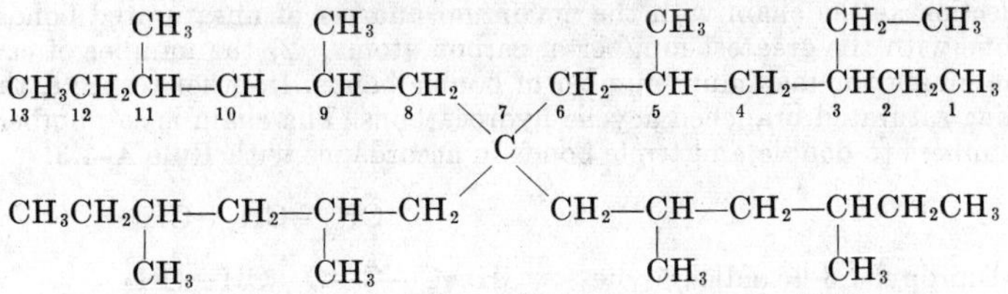

7,7-Bis(2,4-dimethylhexyl)-3-ethyl-5,9,11-trimethyltridecane

(d) The chain having the least branched side chains.

(1) Here the choice lies between two possible main chains of equal length, each containing six side chains in the same positions. Listing in increasing order, the number of carbon atoms in the several side chains of the first choice as shown and of the alternate second choice results as follows:

first choice	1, 1, 1, 2, 8, 8
second choice	1, 1, 1, 1, 8, 9

The expression, "the greatest number of carbon atoms in the smaller side chains," is taken to mean the largest side chain at the first point of difference when the size of the side chains is examined step by step. Thus, the selection in this case is made at the fourth step where 2 is greater than 1.

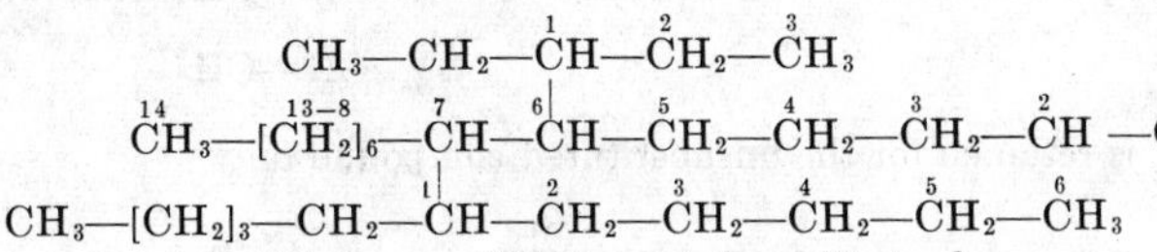

```
                              1      2      3
                  CH3—CH2—CH—CH2—CH3
 14    13—8      7      6|    5      4      3      2    1
CH3—[CH2]6—CH—CH—CH2—CH2—CH2—CH —CH3
                       1|    2      3      4      5      6
CH3—[CH2]3—CH2—CH—CH2—CH2—CH2—CH2—CH3
```

6-(1-Ethylpropyl)-7-(1-pentylhexyl)tetradecane

A-3

3.1.—Unsaturated unbranched acyclic hydrocarbons having one double bond are named by replacing the ending "-ane" of the name of the corresponding saturated hydrocarbon with the ending "-ene." If there are two or more double bonds, the ending will be "-adiene," "-atriene," *etc.* The generic names of these hydrocarbons (branched or unbranched) are "alkene," "alkadiene," "alkatriene," *etc.* The chain is so numbered as to give the lowest possible numbers to the double bonds.

Examples:

2-Hexene	$\overset{6}{C}H_3—\overset{5}{C}H_2—\overset{4}{C}H_2—\overset{3}{C}H=\overset{2}{C}H—\overset{1}{C}H_3$
1,4-Hexadiene	$\overset{6}{C}H_3—\overset{5}{C}H=\overset{4}{C}H—\overset{3}{C}H_2—\overset{2}{C}H=\overset{1}{C}H_2$

These non-systematic names are retained

Ethylene	$CH_2=CH_2$	Allene	$CH_2=C=CH_2$

Chemical Abstracts retains allene for the unsubstituted hydrocarbon only.

3.2.—Unsaturated unbranched acyclic hydrocarbons having one triple bond are named by replacing the ending "-ane" of the name of the corresponding saturated hydrocarbon with the ending "-yne." If there are two or more triple bonds, the ending will be "-adiyne," "-atriyne," *etc.* The generic names of these hydrocarbons (branched or unbranched) are "alkyne," "alkadiyne," "alkatriyne," *etc.* The chain is so numbered as to give the lowest possible numbers to the triple bonds.

The name "acetylene" for $HC\equiv CH$ is retained.

3.3.—Unsaturated unbranched acyclic hydrocarbons having both double and triple bonds are named by replacing the ending "-ane" of the name of the corresponding saturated hydrocarbon with the ending "-enyne," "-adienyne," "-atrienyne," "-enediyne," *etc.* Numbers as low as possible are given to double and triple bonds even though this may at times give "-yne" a lower number than "-ene." When there is a choice in numbering the double bonds are given the lowest numbers.

Examples:

```
                       6   5    4    3    2    1
1,3-Hexadien-5-yne    HC≡C—CH=CH—CH=CH2

                       5    4    3    2   1
3-Penten-1-yne        CH3—CH=CH—C≡CH

                       5   4    3    2    1
1-Penten-4-yne        HC≡C—CH2—CH=CH2
```

3.4.—Unsaturated branched acyclic hydrocarbons are named as derivatives of the unbranched hydrocarbons which contain the maximum number of double and triple bonds. If there are two or more chains competing for selection as the chain with the maximum number of unsaturated bonds, then the choice goes to (1) that one with the greatest number of carbon atoms; (2) the number of carbon atoms being equal, the one containing the maximum number of double bonds. In other respects, the same principles apply as for naming saturated branched acyclic hydrocarbons. The chain is so numbered as to give the lowest possible numbers to double and triple bonds in accordance with Rule **A-3.3.**

Examples:

```
                                           CH2—CH2—CH3
                                           |
                                           3   2   1
3,4-Dipropyl-1,3-hexadien-5-yne     CH≡C—C=C—CH=CH2
                                    6   5  4
                                       |
                                       CH2—CH2—CH3

                                    7    6    5    4    3    2
5-Ethynyl-1,3,6-heptatriene         CH2=CH—CH—CH=CH—CH=CH2
                                           |
                                           C≡CH

                                         CH3
                                    6    5|   4    3    2    1
5,5-Dimethyl-1-hexene               CH3—C—CH2—CH2—CH=CH2
                                         |
                                         CH3

                                    7    6  5   4    3    2    1
4-Vinyl-1-hepten-5-yne              CH3—C≡C—CH—CH2—CH=CH2
                                               |
                                               CH=CH2
```

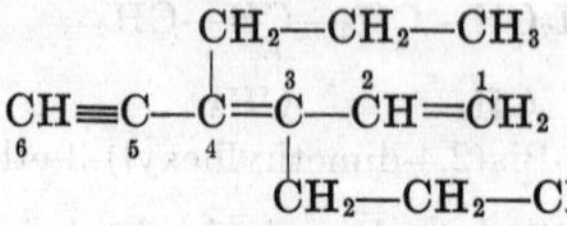

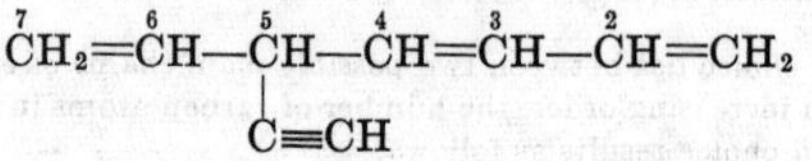

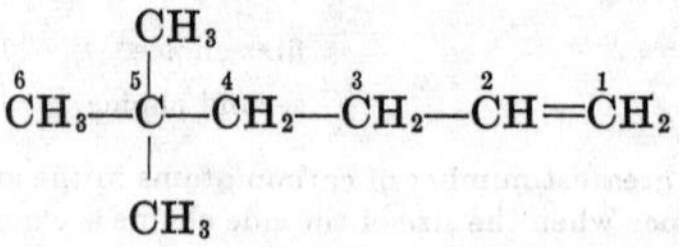

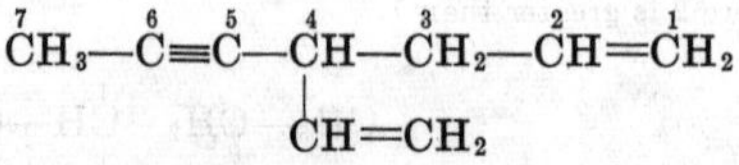

The name "isoprene" is retained for the unsubstituted compound only

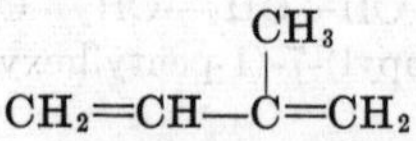

```
                CH3
                |
CH2=CH—C=CH2
```

3.5.—The names of univalent radicals derived from unsaturated acyclic hydrocarbons have the endings "-enyl," "-ynyl," "-dienyl," *etc.*, the positions of the double and triple bonds being indicated where necessary. The carbon atom with the free valence is numbered as 1.

Examples:

Ethynyl	CH≡C—
2-Propynyl	CH≡C—CH2—
1-Propenyl	CH3—CH=CH—
2-Butenyl	CH3—CH=CH—CH2—
1,3-Butadienyl	CH2=CH—CH=CH—
2-Pentenyl	CH3—CH2—CH=CH—CH2—
2-Penten-4-ynyl	CH≡C—CH=CH—CH2—

Exceptions: These names are retained

Vinyl (for ethenyl)	CH2=CH—	
Allyl (for 2-propenyl)	CH2=CH—CH2—	
Isopropenyl (for 1-methylvinyl)	CH2=C(CH3)—	(for unsubstituted radical only)

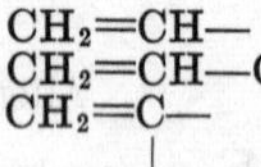

3.6.—When there is a choice for the fundamental chain of a radical, that chain is selected which contains (1) the maximum number of double and triple bonds; (2) the largest number of carbon atoms, and (3) the largest number of double bonds.

Examples:

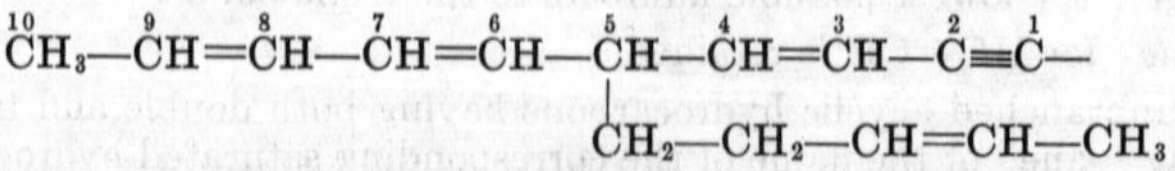

```
10   9    8    7    6    5    4    3   2  1
CH3—CH=CH—CH=CH—CH—CH=CH—C≡C—
                     |
                     CH2—CH2—CH=CH—CH3
```

5-(3-Pentenyl)-3,6,8-decatrien-1-ynyl

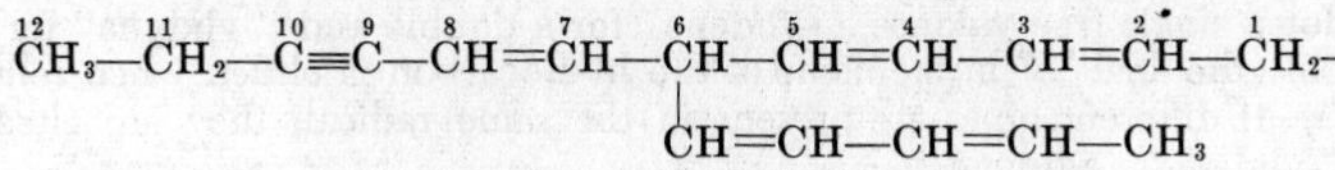

$$\overset{12}{CH_3}-\overset{11}{CH_2}-\overset{10}{C}\equiv\overset{9}{C}-\overset{8}{CH}=\overset{7}{CH}-\overset{6}{CH}(-CH=CH-CH=CH-CH_3)-\overset{5}{CH}=\overset{4}{CH}-\overset{3}{CH}=\overset{2}{CH}-\overset{1}{CH_2}-$$

6-(1,3-Pentadienyl)-2,4 7-dodecatrien-9-ynyl

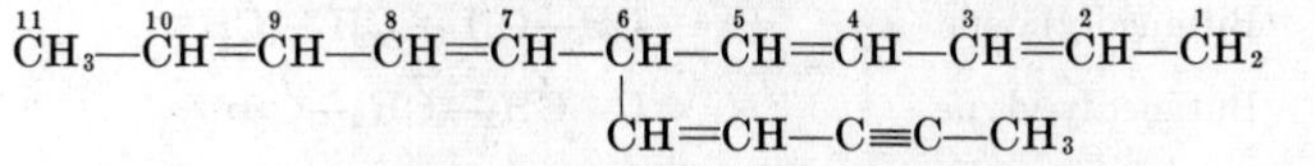

$$\overset{11}{CH_3}-\overset{10}{CH}=\overset{9}{CH}-\overset{8}{CH}=\overset{7}{CH}-\overset{6}{CH}(-CH=CH-C\equiv C-CH_3)-\overset{5}{CH}=\overset{4}{CH}-\overset{3}{CH}=\overset{2}{CH}-\overset{1}{CH_2}$$

6-(1-Penten-3-ynyl)-2,4,7,9-undecatetraenyl

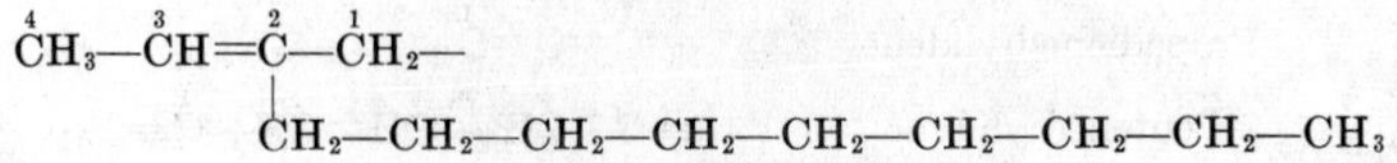

$$\overset{4}{CH_3}-\overset{3}{CH}=\overset{2}{C}(-CH_2-CH_2-CH_2-CH_2-CH_2-CH_2-CH_2-CH_2-CH_3)-\overset{1}{CH_2}-$$

2-Nonyl-2-butenyl

A-4

4.1.—Bivalent and trivalent radicals derived from univalent acyclic hydrocarbon radicals whose authorized names end in "-yl" by removal of one or two hydrogen atoms from the carbon atom with the free valences are named by adding "-idene" or "-idyne," respectively, to the name of the corresponding univalent radical. The carbon atom with the free valence is numbered as 1.

The name "methylene" is retained for the radical $CH_2=$.

Examples:

Methylidyne[1]	$CH\equiv$
Ethylidene	$CH_3-CH=$
Ethylidyne	$CH_3-C\equiv$
Vinylidene	$CH_2=C=$
Isopropylidene[2]	$(CH_3)_2C=$

4.2.—The names of bivalent radicals derived from normal alkanes by removal of a hydrogen atom from each of the two terminal carbon atoms of the chain are ethylene, trimethylene, tetramethylene, *etc.*

Examples:

Pentamethylene	$-CH_2-CH_2-CH_2-CH_2-CH_2-$
Hexamethylene	$-CH_2-CH_2-CH_2-CH_2-CH_2-CH_2-$

Names of the substituted bivalent radicals are derived in accordance with Rules **A-2.2** and **A-2.25.**

Example:

Ethylethylene $-\overset{2}{CH_2}-\overset{1}{CH}(-CH_2-CH_3)-$

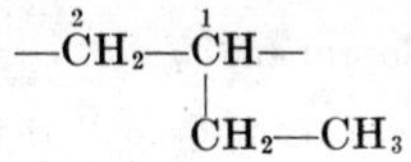

The name "propylene" is retained

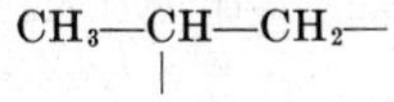

$$CH_3-CH(-)-CH_2-$$

4.3.—Bivalent radicals similarly derived from unbranched alkenes, alkadienes, alkynes, *etc.*, by removing a hydrogen atom from each of the terminals carbon atoms are named by replacing the endings "-ene," "-diene," "-yne," *etc.*, of the hydrocarbon name by "-enylene," "-dienylene," "-ynylene," *etc.*, the positions of the double and triple bonds being indicated where necessary.

Example:

Propylene $-\overset{3}{CH_2}-\overset{2}{CH}=\overset{1}{CH}-$

The name "vinylene" is retained (for ethenylene)

$$-CH=CH-$$

Names of the substituted bivalent radicals are derived in accordance with Rule **A-3.4.**

Examples:

4 Propyl-2-pentenylene $-\overset{5}{CH_2}-\overset{4}{CH}(-CH_2-CH_2-CH_3)-\overset{3}{CH}=\overset{2}{CH}-\overset{1}{CH_2}-$

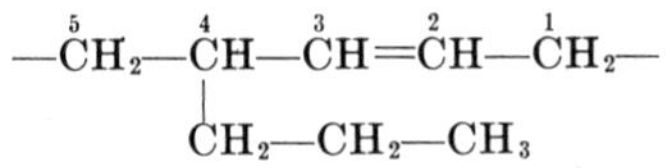

4.4.—Trivalent, quadrivalent and higher valent acyclic hydrocarbon radicals of two or more carbon atoms with the free valences at each end of a chain are named by adding to the hydrocarbon name the

(1) The group $=CH-$ may be referred to as the "methine" group.

(2) For unsubstituted radical only.

terminations "-yl" for a single free valence, "-ylidene" for a double and "-ylidyne" for a triple free valence on the same atom (the final "e" in the name of the hydrocarbon is elided when followed by a suffix beginning with "-yl"). If different types are present in the same radical, they are cited and numbered in the order of "-yl" "-ylidene," "-ylidyne."

Examples:

Butanediylidene	$=\overset{4}{C}H-\overset{3}{C}H_2-\overset{2}{C}H_2-\overset{1}{C}H=$
Butanediylidyne	$\equiv\overset{4}{C}-\overset{3}{C}H_2-\overset{2}{C}H_2-\overset{1}{C}\equiv$
1-Propanyl-3-ylidene	$=\overset{3}{C}H-\overset{2}{C}H_2-\overset{1}{C}H_2-$
Propadienediylidene	$=\overset{3}{C}=\overset{2}{C}=\overset{1}{C}=$
2-Pentenediylidyne	$\equiv\overset{5}{C}-\overset{4}{C}H_2-\overset{3}{C}H=\overset{2}{C}H-\overset{1}{C}\equiv$
1-Butanyliden-4-ylidyne	$\equiv\overset{4}{C}-\overset{3}{C}H_2-\overset{2}{C}H_2-\overset{1}{C}H=$

4.5.—Multivalent radicals containing three or more carbon atoms with free valences at each end of a chain and additional free valences at intermediate carbon atoms are named by adding the ending "-triyl," "-tetrayl," "-diylidene," "diylylidene," *etc.*, to the hydrocarbon name.

Examples:

$-\overset{3}{C}H_2-\overset{2}{C}H(-)-\overset{1}{C}H_2-$
1,2,3-Propanetriyl

$-\overset{3}{C}H_2-\overset{2}{C}(=)-\overset{1}{C}H_2-$
1,3-Propanediyl-2-ylidene

Monocyclic Hydrocarbons

A-11

11.1.—The names of saturated monocyclic hydrocarbons (with no side chains) are formed by attaching the prefix "cyclo" to the name of the acyclic saturated unbranched hydrocarbon with the same number of carbon atoms. The generic name of saturated monocyclic hydrocarbons (with or without side chains) is "cycloalkane."

Examples:

Cyclopropane (H_2C, CH_2, CH_2 ring)

Cyclohexane (H_2C, CH_2 ×6 ring)

11.2.—Univalent radicals derived from cycloalkanes (with no side chains) are named by replacing the ending "-ane" of the hydrocarbon name by "-yl," the carbon atom with the free valence being numbered as 1. The generic name of these radicals is "cycloalkyl."

Examples:

Cyclopropyl

Cyclohexyl

11.3.—The names of unsaturated monocyclic hydrocarbons (with no side chains) are formed by substituting "-ene," "-adiene," "-atriene," "-yne," "-adiyne," *etc.*, for "-ane" in the name of the corresponding cycloalkane. The double and triple bonds are given numbers as low as possible as in Rule **A-3.3.**

Examples:

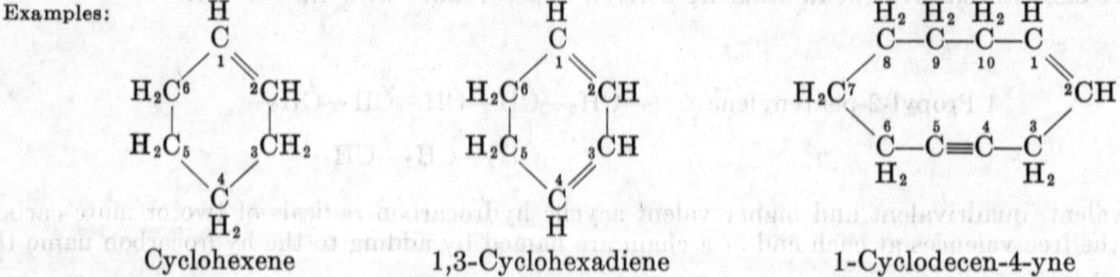

Cyclohexene 1,3-Cyclohexadiene 1-Cyclodecen-4-yne

The names "fulvene" (for methylenecyclopentadiene) and "benzene" are retained.

11.4.—The names of univalent radicals derived from unsaturated monocyclic hydrocarbons have the endings "-enyl," "-ynyl," "-dienyl," *etc.*, the positions of the double and triple bonds being indicated according to the principles of Rule **A-3.3.** The carbon atom with the free valence is numbered as 1, except as stated in the rules for terpenes(see Rules **A-72** to **A-75**).

Examples:

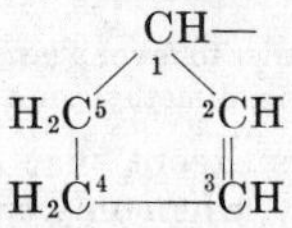

CH— (1); H_2C^5, ^2CH; H_2C^4—^3CH

2-Cyclopenten-1-yl

CH— (1); HC^5, ^2CH; HC^4—^3CH

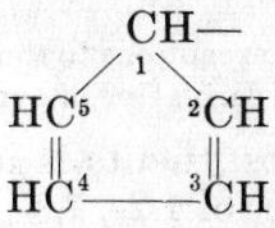

2,4-Cyclopentadien-1-yl

The radical name "phenyl" is retained.

The point of attachment is numbered 1 and the double and triple bonds are given numbers as low as possible. The number 1 for the point of attachment is given in the radical name to emphasize the fact that the point of attachment takes precedence over the double and triple bonds (except with terpenes which have fixed numberings).

11.5.—Names of bivalent radicals derived from saturated or unsaturated monocyclic hydrocarbons by removal of two atoms of hydrogen from the same carbon atom of the ring are obtained by replacing the endings "-ane," "-ene," "-yne," by "-ylidene," "-enylidene" and "-ynylidene," respectively. The carbon atom with the free valences is numbered as 1, except as stated in the rules for terpenes.

Examples:

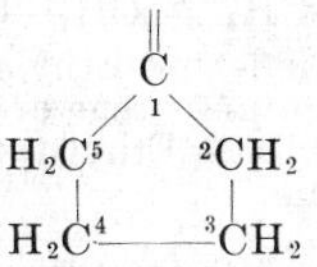

=C (1); H_2C^5, $^2CH_2$; H_2C^4—$^3CH_2$

Cyclopentylidene

=C (1); H_2C^6, ^2CH; HC_5, $_3CH$; C (4) H

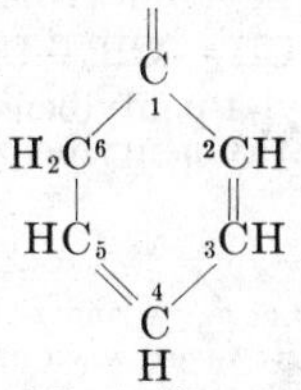

2,4-Cyclohexadien-1-ylidene

11.6.—Bivalent radicals derived from saturated or unsaturated monocyclic hydrocarbons by removing a hydrogen atom from each of two different carbon atoms of the ring are named by replacing the endings "-ane," "-ene," "-diene," "-yne," *etc.*, of the hydrocarbon name by "-ylene," "-enylene," "-dienylene," "-ynylene," *etc.*, the positions of the double and triple bonds and of the points of attachment being indicated. Preference in lowest numbers is given to the carbon atoms having the free valences.

Examples:

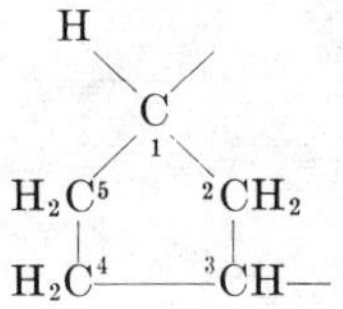

H—C (1); H_2C^5, $^2CH_2$; H_2C^4—^3CH—

1,3-Cyclopentylene

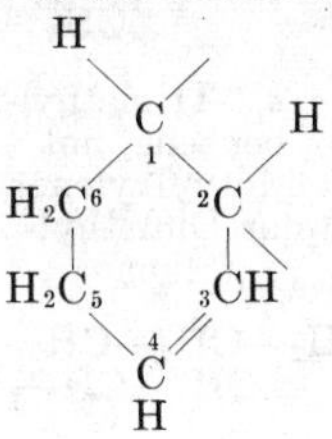

H—C (1); H; H_2C^6, ^2C; H_2C_5, $_3CH$; C (4) H

3-Cyclohexen-1,2-ylene

H—C (1); HC^6, ^2CH; HC_5, $_3CH$; C (4)—H

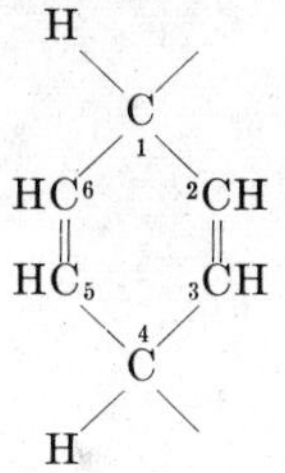

2,5-Cyclohexadien-1,4-ylene

The name "phenylene" is retained

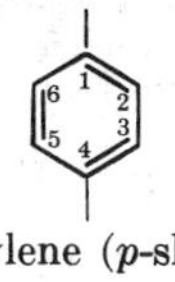

Phenylene (*p*-shown)

A-12

12.1.—These names for substituted monocyclic aromatic hydrocarbons are retained

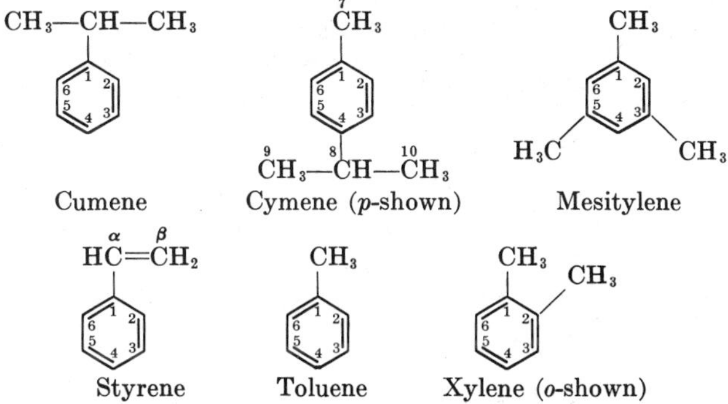

Cumene   Cymene (*p*-shown)   Mesitylene

Styrene   Toluene   Xylene (*o*-shown)

12.2.—Other substituted monocyclic aromatic hydrocarbons are named as derivatives of benzene or of one of the compounds listed in Part 1 of this rule. However, if the substituent introduced into such a compound, is identical with one already present in that compound, then the substituted compound is named as a derivative of benzene (see Rule **61.4**).

***Chemical Abstracts* makes certain exceptions to the last part of this rule in order to favor a larger parent, although still treating like groups alike, *e.g.*, an *ar*-methyl derivative of cymene is indexed at Cumene, dimethyl-, not at Benzene, isopropyldimethyl-.**

12.3.—The position of substituents is indicated by numbers except that *o*- (*ortho*-), *m*- (*meta*-) and *p*- (*para*-) may be used in place of 1,2-, 1,3-, and 1,4-, respectively, when only two substituents are present. The lowest numbers possible are given to substituents, choice between alternatives being governed by Rule **A-2** so far as applicable, except that when names are based on those of compounds listed in Part 1 of this rule the first priority for lowest numbers is given to the substituent(s) already present in those compounds.

Examples:

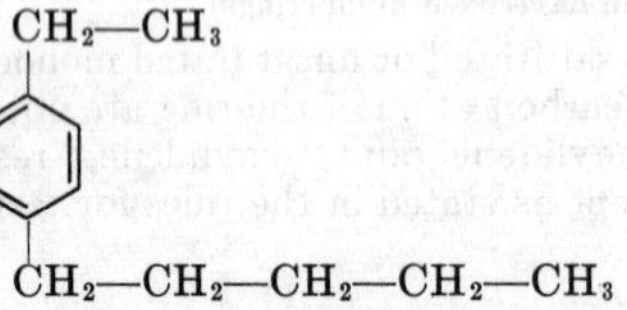

$CH_2—CH_3$ / $CH_2—CH_2—CH_2—CH_2—CH_3$
1-Ethyl-4-pentylbenzene or *p*-Ethylpentylbenzene

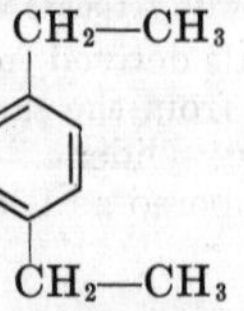

$CH_2—CH_3$ / $CH_2—CH_3$
1,4-Diethylbenzene or *p*-Diethylbenzene

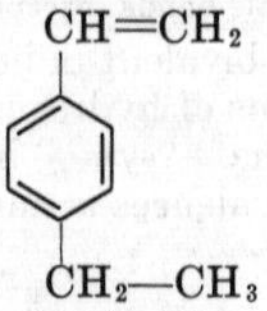

$CH=CH_2$ / $CH_2—CH_3$
4-Ethylstyrene or *p*-Ethylstyrene

***Chemical Abstracts* limits the use of *o*-, *m*- and *p*- to like substituents when both are expressed as prefixes, *e.g.*, *p*-diethylbenzene, 1-ethyl-4-pentylbenzene. However, when one substituent is expressed in the name, *m*-, *o*- or *p*- is used to indicate the position of the second and different substituent, *e.g.*, *o*-ethylcumene.**

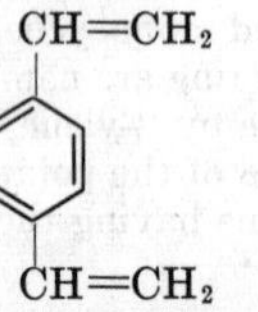

$CH=CH_2$ / $CH=CH_2$
1,4-Divinylbenzene or *p*-Divinylbenzene, not *p*-Vinylstyrene

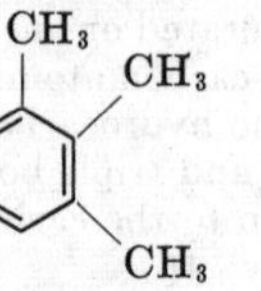

CH_3 / CH_3 / CH_3
1,2,3-Trimethylbenzene, not Methylxylene nor Dimethyltoluene

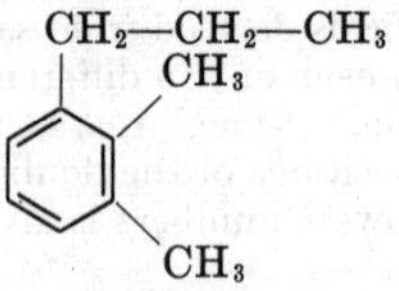

$CH_2—CH_2—CH_3$ / CH_3 / CH_3
1,2-Dimethyl-3-propylbenzene or 3-Propyl-*o*-xylene

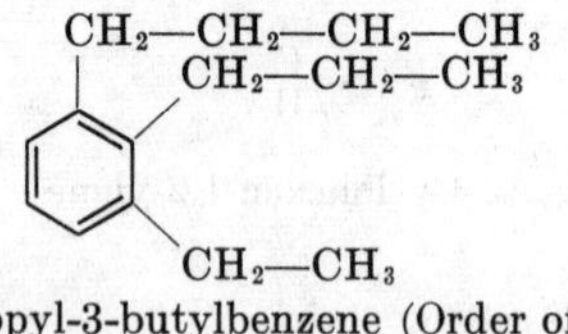

$CH_2—CH_2—CH_2—CH_3$ / $CH_2—CH_2—CH_3$ / $CH_2—CH_3$
1-Ethyl-2-propyl-3-butylbenzene (Order of complexity) or 1-Butyl-3-ethyl-2-propylbenzene (Alphabetical order)

12.4.—The generic name of monocyclic and polycyclic aromatic hydrocarbons is "arene."

A-13

13.1.—Univalent radicals derived from monocyclic aromatic hydrocarbons and having the free valence at a ring atom are given the names listed below. Such radicals not listed below are named as substituted phenyl radicals. The carbon atom having the free valence is numbered as 1.

Phenyl $C_6H_5—$

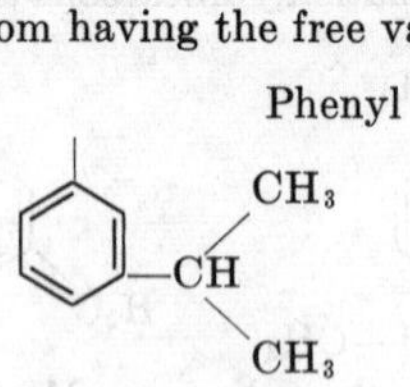

CH_3 / $—CH$ / CH_3
Cumenyl (*m*-shown)

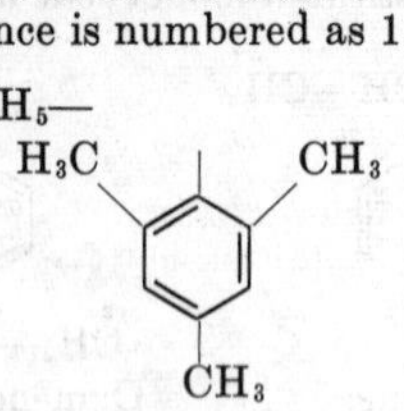

H_3C / CH_3 / CH_3
Mesityl

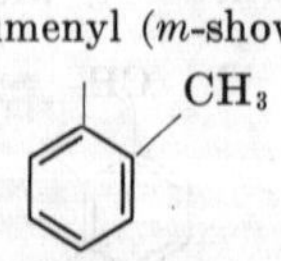

CH_3
Tolyl (*o*-shown)

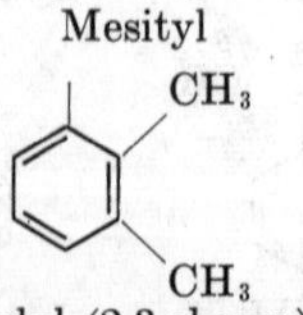

CH_3 / CH_3
Xylyl (2,3-shown)

13.2.—Since the name phenylene (*o*-, *m*-, or *p*-) is retained for the radical —C_6H_4— (exception to Rule **A-11.6**) bivalent radicals formed from substituted benzene derivatives and having the free valences at ring atoms are named as substituted phenylene radicals. The carbon atoms having the free valences are numbered 1,2-, 1,3-, or 1,4- as appropriate.

13.3.—These trivial names for radicals having a single free valence in the side chain are retained:

Benzyl	$C_6H_5—\overset{\alpha}{C}H_2—$
Cinnamyl	$C_6H_5—\overset{\gamma}{C}H=\overset{\beta}{C}H—\overset{\alpha}{C}H_2—$
Phenethyl	$C_6H_5—\overset{\beta}{C}H_2—\overset{\alpha}{C}H_2—$
Styryl	$C_6H_5—\overset{\beta}{C}H=\overset{\alpha}{C}H—$
Trityl	$(C_6H_5)_3C—$

Chemical Abstracts limits trityl to the unsubstituted radical.

13.4.—Multivalent radicals of aromatic hydrocarbons with free valences in the side chain are named in accordance with Rule **A-4.**

Examples:

Benzylidyne	$C_6H_5—C=$
Cinnamylidene	$C_6H_5—\overset{\gamma}{C}H=\overset{\beta}{C}H—\overset{\alpha}{C}H=$

A-14

14.1.—The generic names of univalent and bivalent aromatic hydrocarbon radicals are "aryl" and "arylene," respectively.

Fused Polycyclic Hydrocarbons

A-21

21.1.—The names of polycyclic hydrocarbons with maximum number of non-cumulative[1] double bonds end in "-ene." The names listed on pp. 11 and 12 are retained.

21.2.—The names of hydrocarbons containing five or more fused benzene rings in a straight linear arrangement are formed from a numerical prefix as specified in Rule **A-1.1** followed by "-acene."

Examples:

Pentacene　　Hexacene

The following list contains the names of polycyclic hydrocarbons which are retained (see Rule A-21.1).

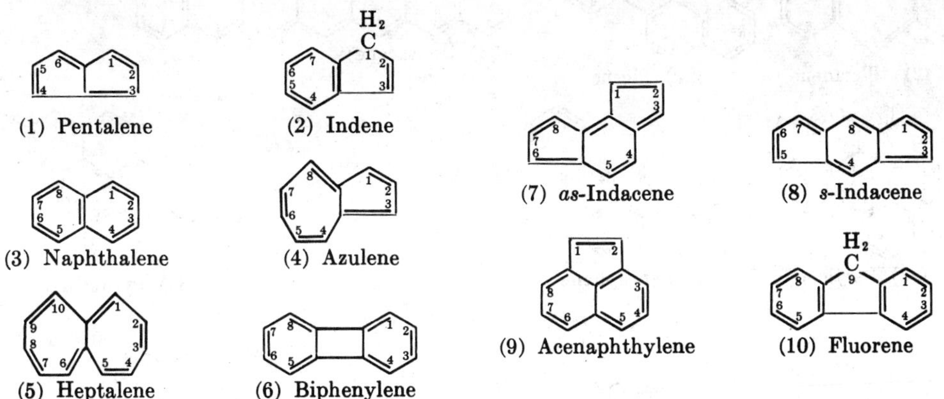

(1) Pentalene　(2) Indene　(3) Naphthalene　(4) Azulene　(5) Heptalene　(6) Biphenylene　(7) *as*-Indacene　(8) *s*-Indacene　(9) Acenaphthylene　(10) Fluorene

(1) Cumulative double bonds are those present in a chain in which at least three contiguous carbon atoms are joined by double bonds; non-cumulative double bonds comprise every other arrangement of two or more double bonds in a single structure. The generic name "cumulene" is given to compounds containing three or more cumulative double bonds.

$$CH_2=C=C=C=CH_2$$

Cumulative

$$CH_3—CH=CH—CH=CH—CH=CH_2$$

or

Non-cumulative

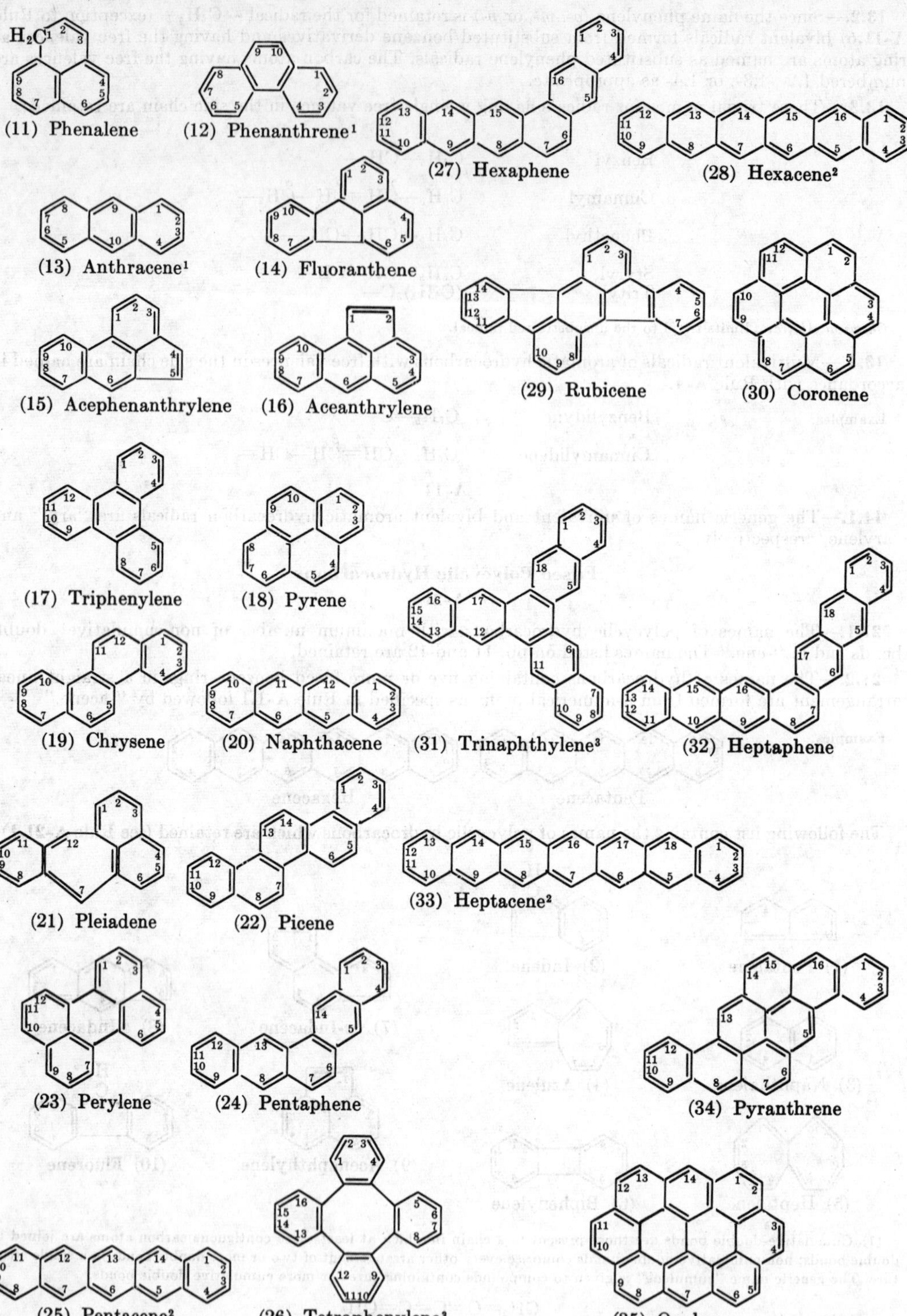

Beginning with Volume 51 Subject Index *Chemical Abstracts* uses biphenylene in place of cyclobutadibenzene and phenalene in place of benzonaphthene. The polycyclic hydrocarbons which have the trivial names listed above may be used as the base components in naming orthofused or ortho- and peri-fused polycyclic systems for which there are no acceptable trivial names.

(1) Denotes exception to systematic numbering. See Rule **A-22.5**.

(2) See Rule **A-21.2**.

(3) For isomer shown only.

21.3.—"Ortho-fused"[4] or "ortho- and peri-fused"[5] polycyclic hydrocarbons with maximum number of noncumulative double bonds which contain at least two rings of five or more members and which have no accepted trivial name such as those of **21.1** of this Rule, are named by prefixing to the name of a component ring or ring system (the base component) designations of the other components. The base component should contain as many rings as possible (provided it has a trivial name), and should occur as far as possible from the beginning of the list of Rule **A-21.1.** The attached components should be as simple as possible.

Example:

(not Naphthophenanthrene; benzo is "simpler" than naphtho, even though there are two benzo rings and only one naphtho)

Dibenzophenanthrene

(4) Polycyclic compounds in which two rings have two, and only two, atoms in common are said to be "ortho-fused." Such compounds have n common faces and $2n$ common atoms (Example I).

(5) Polycyclic compounds in which one ring contains two, and only two, atoms in common with each of two or more rings of a contiguous series of rings are said to be "ortho- and peri-fused." Such compounds have n common faces and fewer than $2n$ common atoms (Examples II and III).

Examples:

I — 3 common faces, 6 common atoms, "Ortho-fused" system

II (CH_2) — 7 common faces, 8 common atoms

III — 5 common faces, 6 common atoms

"Ortho- and peri-fused" systems

21.4.—The prefixes designating attached components are formed by changing the ending "-ene" of the name of the component hydrocarbon into "-eno," *e.g.*, "pyreno" (from pyrene). When more than one prefix is present, they are arranged in alphabetical order. These common abbreviated prefixes are recognized (see list in **21.1** of this Rule)

Acenaphtho	from	Acenaphthylene
Anthra	from	Anthracene
Benzo	from	Benzene
Naphtho	from	Naphthalene
Perylo	from	Perylene
Phenanthro	from	Phenanthrene

For monocyclic prefixes other than "benzo-," the following names are recognized, each to represent the form with the maximum number of non-cumulative double bonds: cyclopenta, cyclohepta, cycloocta, cyclonona, *etc.* When the base component is a monocyclic system, the ending "-ene" signifies the maximum number of noncumulative double bonds, and does not denote one double bond only.

Examples:

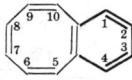

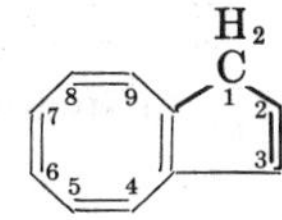

1*H*-Cyclopentacycloöctene

Benzocycloöctene

In forming names of polycyclic systems from the above carbocyclic prefixes the final "o" or "a" of the prefix is elided before another vowel, except in the case of anthra, phenanthro and prefixes ending in -eno from which the "a" and "o" are never elided. Examples: benz[*a*]anthracene, cyclopent[*a*]acenaphthylene, anthra[1,2 − *a*]anthracene, fluoreno[4,3,2-*de*]anthracene.

The ending "-ene" to signify the maximum number of noncumulative double bonds in all fused carbocyclic systems means that the endings -diene, triene, tetraene, *etc.*, are unnecessary when the base component is a monocyclic ring of seven or more carbon atoms. Higher stages of hydrogenation are indicated by the use of the prefixes dihydro-, tetrahydro-, *etc.* (*cf.* Rule **A-23.1**). *Chemical Abstracts* follows Rules **A-21.4** and **A-23.1** instead of Alternate Rule **A-23.5.**

21.5.—Isomers are distinguished by lettering the peripheral sides of the base component *a*, *b*, *c*, *etc.*, beginning with "*a*" for the side "1,2," "*b*" for "2,3" (or in certain cases "2,2a") and lettering every side around the periphery. To the letter as early in the alphabet as possible, denoting the side where fusion occurs, are prefixed, if necessary, the numbers of the positions of attachment of the other component. These numbers are chosen to be as low as is consistent with the numbering of the component, and their order conforms to the direction of lettering of the base component (see Examples II and IV). When two or more prefixes refer to equivalent positions so that there is a choice of letters, the prefixes are cited in alphabetical order according to Rule **A-21.4** and the location of the first cited prefix is indicated by a letter as early as possible in the alphabet (see Example V). The numbers and letters are enclosed in brackets and placed immediately after the designation of the attached component. This expression merely defines the manner of fusion of the components.

Examples:

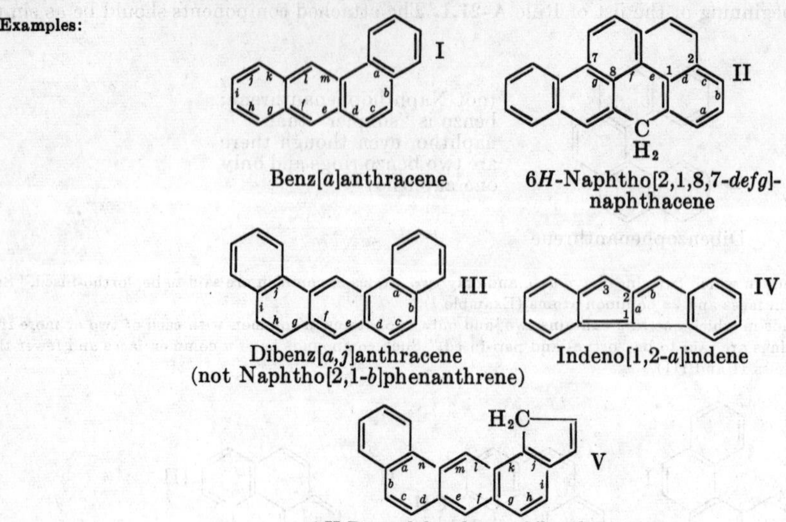

I Benz[*a*]anthracene

II 6*H*-Naphtho[2,1,8,7-*defg*]-naphthacene

III Dibenz[*a,j*]anthracene (not Naphtho[2,1-*b*]phenanthrene)

IV Indeno[1,2-*a*]indene

V 1*H*-Benzo[*a*]cyclopent[*j*]anthracene

The completed system consisting of the base component and the other components is then renumbered according to Rule **A-22,** the numbering of the component parts being ignored.

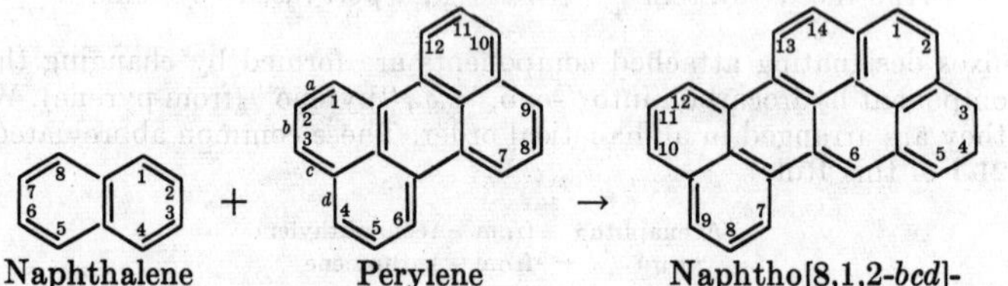

Naphthalene + Perylene → Naphtho[8,1,2-*bcd*]-perylene

21.6.—When a name applies equally to two or more isomeric condensed parent ring systems with the maximum number of non-cumulative double bonds and when the name can be made specific by indicating the position of one or more hydrogen atoms in the structure, this is accomplished by modifying the name with a locant, followed by italic capital *H* for each of these hydrogen atoms. Such symbols ordinarily precede the name. The said atom or atoms are called "indicated hydrogen." The same principle is applied to radicals and compounds derived from these systems.

Examples:

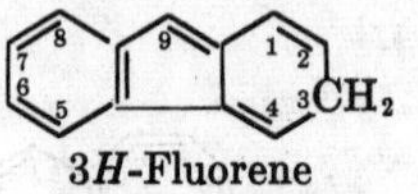

3*H*-Fluorene 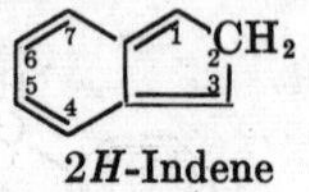 2*H*-Indene

A-22

22.1.—For the purposes of numbering, the individual rings of a polycyclic "ortho-fused" or "ortho- and peri-fused" hydrocarbon system usually are drawn as follows:

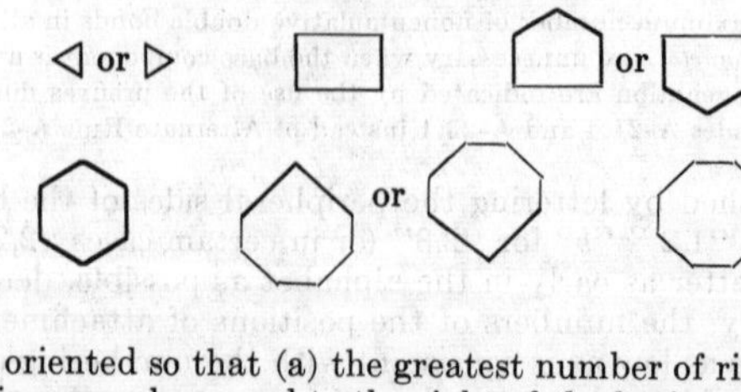

◁ or ▷ □ ⬠ or ⬠

⬡ ⬡ or ⬡ ⬡

and the polycyclic system is oriented so that (a) the greatest number of rings are in a horizontal row and (b) a maximum number of rings are above and to the right of the horizontal row (upper right quadrant).

If two or more orientations meet these requirements, the one is chosen which has as few rings as possible in the lower left quadrant.

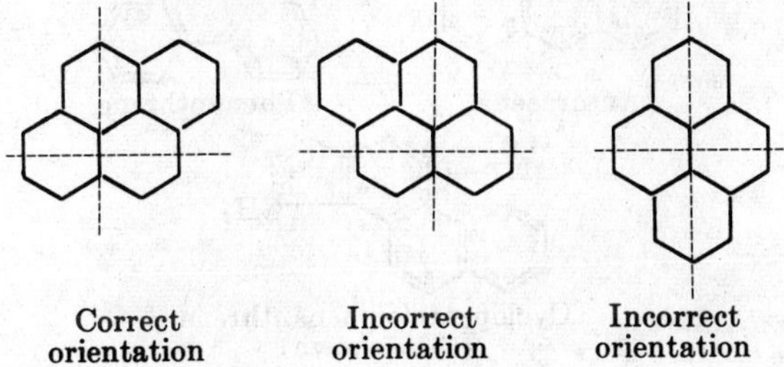

Correct orientation Incorrect orientation Incorrect orientation

The system thus oriented is numbered in a clockwise direction commencing with the carbon atom not engaged in ring-fusion in the most counterclockwise position of the uppermost ring, or if there is a choice, of the uppermost ring farthest to the right, and omitting atoms common to two or more rings.

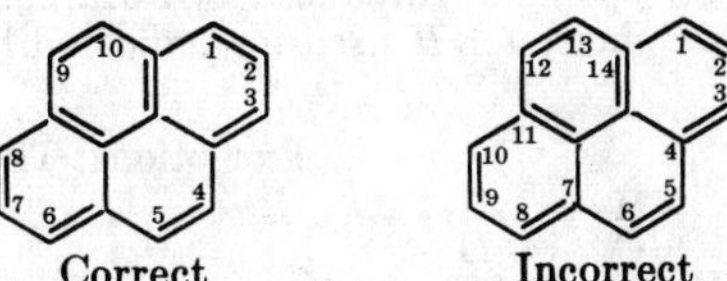

Correct Incorrect

22.2.—Atoms common to two or more rings are designated by adding roman letters "a," "b," "c," *etc.*, to the number of the position immediately preceding. Interior atoms follow the highest number, taking a clockwise sequence wherever there is a choice.

Examples:

[*cf.* Note below]

Correct Incorrect

22.3.—When there is a choice, carbon atoms common to two or more rings follow the lowest possible numbers.

Examples:

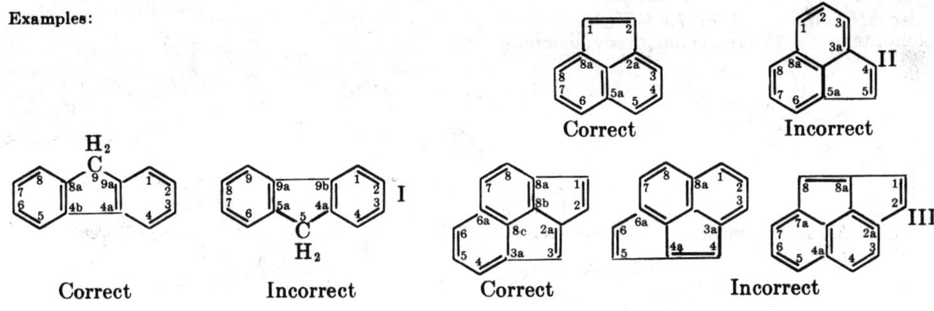

Correct Incorrect Correct Incorrect

Note: I. 4, 4, 8, 9 is lower than 4, 5, 9, 9.
II. 2, 5, 8 is lower than 3, 5, 8.
III. 2, 3, 6, 8 is lower than 3, 4, 6, 8 or 2, 4, 7, 8.

22.4.—When there is a choice, the carbon atoms which carry an indicated hydrogen atom are numbered as low as possible.

Examples:

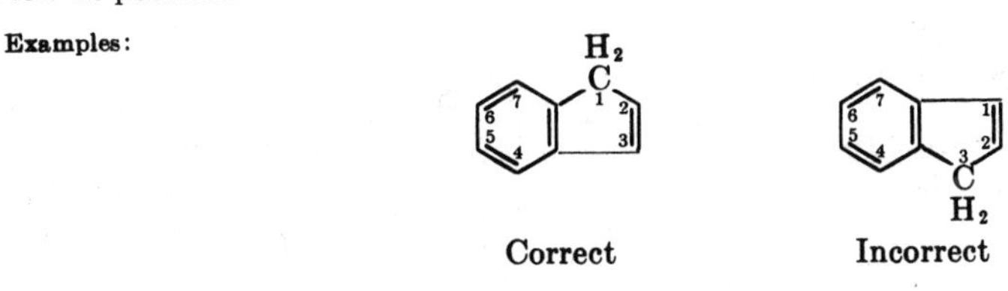

Correct Incorrect

22.5.—The following are recommended exceptions to the above rules on numbering

Anthracene

Phenanthrene

Cyclopenta[*a*]phenanthrene
(15*H*- shown)
See also rules on steroids

A-23

23.1.—The names of "ortho-fused" or "ortho- and peri-fused" polycyclic hydrocarbons with less than maximum number of non-cumulative double bonds are formed from a prefix "dihydro-," "tetrahydro-," *etc.*, followed by the name of the corresponding unreduced hydrocarbon. The prefix "perhydro-" signifies full hydrogenation. When there is a choice for *H* used for indicated hydrogen it is assigned the lowest available number.

Examples:

Exceptions: These names are retained

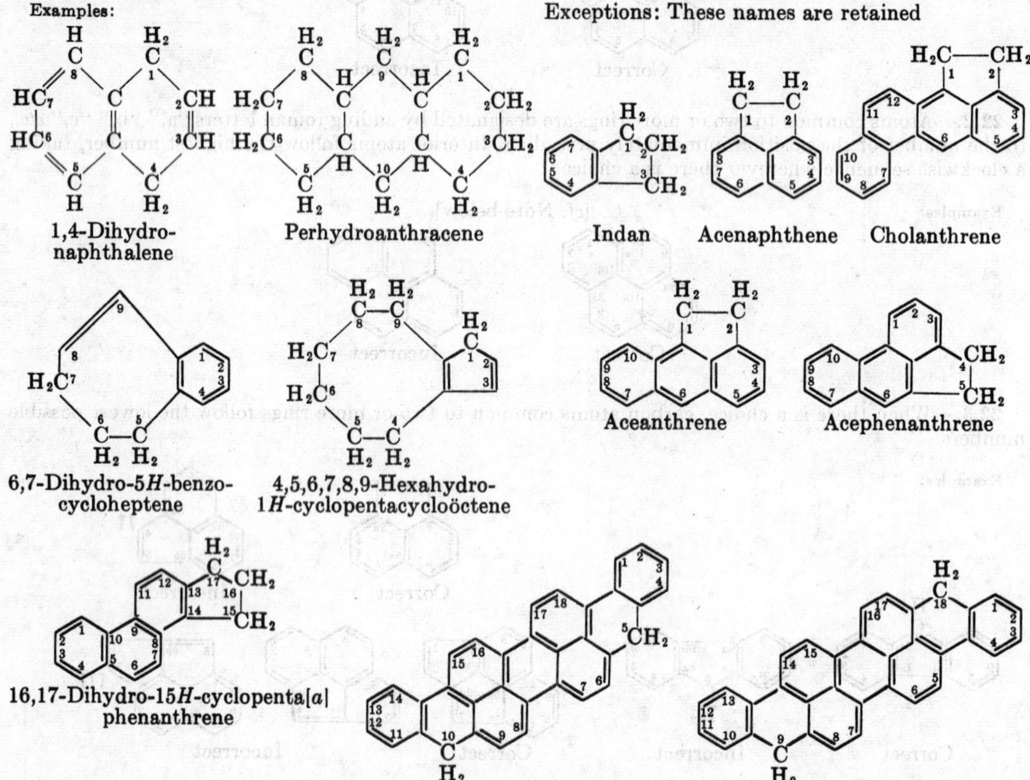

1,4-Dihydronaphthalene; Perhydroanthracene; Indan; Acenaphthene; Cholanthrene; 6,7-Dihydro-5*H*-benzocycloheptene; 4,5,6,7,8,9-Hexahydro-1*H*-cyclopentacycloöctene; Aceanthrene; Acephenanthrene; 16,17-Dihydro-15*H*-cyclopenta[*a*]phenanthrene; Violanthrene; Isoviolanthrene

Chemical Abstracts follows this Rule instead of the alternate Rule 23.5.

23.2.—When there is a choice, the carbon atoms to which hydrogen atoms are added are numbered as low as possible.

Example:

Correct

Incorrect

23.3.—Substituted polycyclic hydrocarbons are named according to the same principles as substituted monocyclic hydrocarbons (see Rules **A-12** and **A-61**).

23.5 (Alternate to part of Rule **A-23.1**).—The names of "ortho-fused" polycyclic hydrocarbons which have (a) less than the maximum number of non-cumulative double bonds, (b) at least one terminal unit which is most conveniently named as an unsaturated cycloalkane derivative, and (c) a double bond at the positions where rings are fused together, may be derived by joining the name of the terminal unit to that of the other component by means of a letter "o" with elision of a terminal "e". The abbreviations for fused aromatic systems laid down in Rule **A-21.4** are used, and the exceptions of Rule **A-23.1** apply.

Examples:

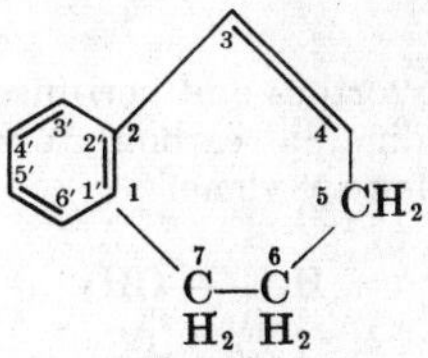

1,2-Benzo-1,3-cycloheptadiene

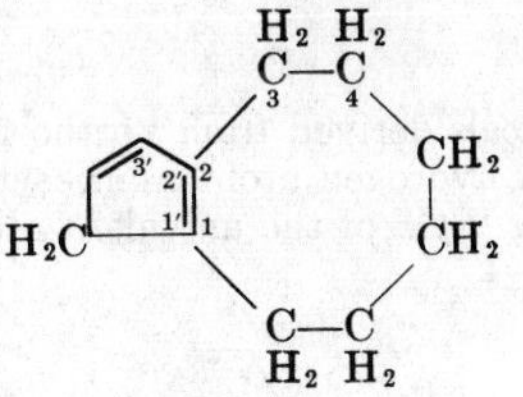

1,2-Cyclopenta-1′,3′-dienocyclooctene

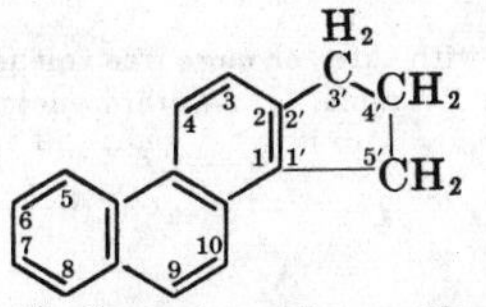

1,2-Cyclopentenophenanthrene

Chemical Abstracts follows Rules **A-21.4** and **A-23.1** instead of this alternate rule.

A-24

24.1.—For radicals derived from polycyclic hydrocarbons, the numbering of the hydrocarbon is retained. The point or points of attachment are given numbers as low as is consistent with the fixed numbering of the hydrocarbon.

24.2.—Univalent radicals derived from "ortho-fused" or "ortho- and peri-fused" polycyclic hydrocarbons with names ending in "-ene" by removal of a hydrogen atom from an aromatic or alicyclic ring are named in principle by changing the ending "-ene" of the names of the hydrocarbons to "-enyl."

Examples:

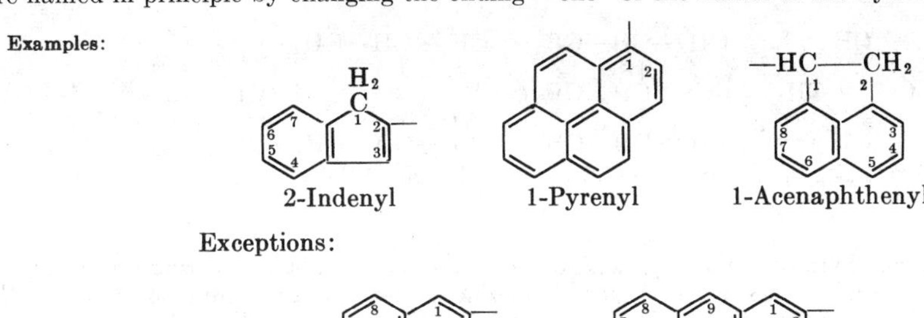

2-Indenyl   1-Pyrenyl   1-Acenaphthenyl

Exceptions:

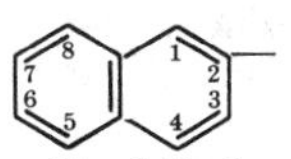

Naphthyl (2-shown)

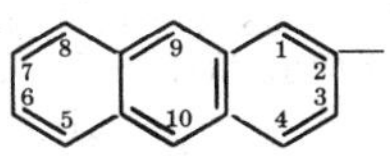

Anthryl (2-shown)

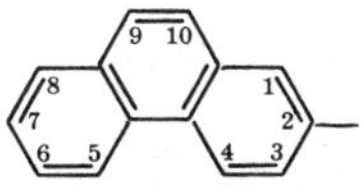

Phenanthryl (2-shown)

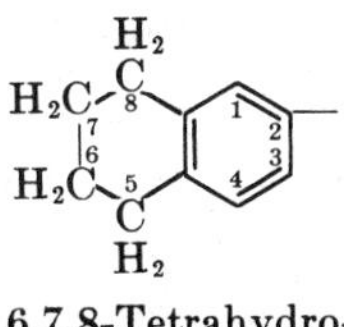

5,6,7,8-Tetrahydro-2-naphthyl

The above exceptions are limited to the simple rings as shown. Radicals from names of fused derivatives of these rings are formed according to the Rule, *e.g.*, benz[*a*]anthracenyl.

24.3.—Bivalent radicals derived from univalent polycyclic hydrocarbon radicals whose names end in "-yl", by removal of one hydrogen atom from the carbon atom with the free valence are named by adding "-idene" to the name of the corresponding univalent radical.

Example:

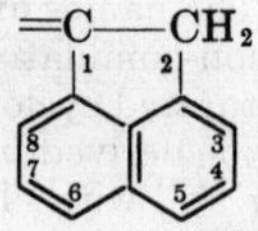

1-Acenaphthenylidene

24.4.—Bivalent radicals derived from "ortho-fused" or "ortho- and peri-fused" polycyclic hydrocarbons by removing of a hydrogen atom from each of two different carbon atoms of the ring are named by changing the ending "-yl" of the univalent radical name to "-ylene."

Examples:

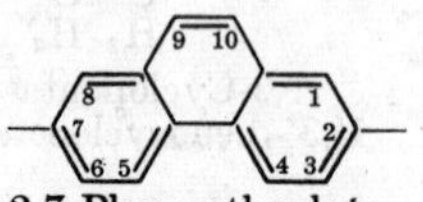

2,7-Phenanthrylene 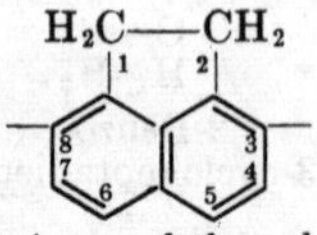   3,8-Acenaphthenylene

Chemical Abstracts names multivalent radicals with three or more free valences derived from polycyclic hydrocarbons by an extension of Rule **B-5.13** to carbocyclic systems. Example: 1,4,5-anthracenetriyl.

A-28

28.1.—Radicals formed from hydrocarbons consisting of polycyclic systems and side chains are named according to the principles of the preceding rules.

Bridged Hydrocarbons

A-31. von Baeyer System

31.1.—Saturated alicyclic hydrocarbon systems consisting of two rings only, having two or more atoms in common, take the name of an open chain hydrocarbon containing the same total number of carbon atoms preceded by the prefix "bicyclo-." The number of carbon atoms in each of the three bridges[1] connecting the two tertiary carbon atoms is indicated in brackets in descending order.

Examples:

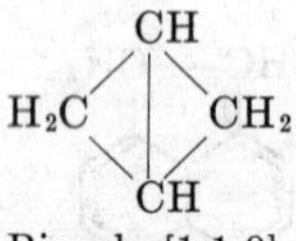

Bicyclo [1.1.0]-butane 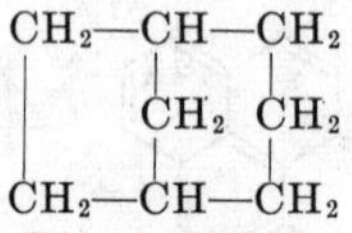   Bicyclo [3.2.1]-octane 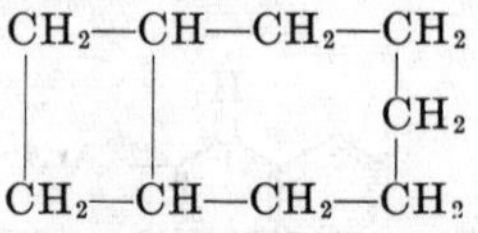   Bicyclo [5.2.0]nonane

31.2.—The system is numbered commencing with one of the bridge-heads, numbering proceeding by the longest possible path to the second bridge-head; numbering is then continued from this atom by the longer unnumbered path back to the first bridge-head and is completed by the shortest path.

Examples:

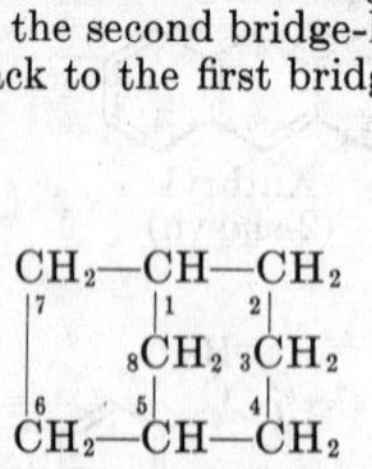

Bicyclo [3.2.1]octane 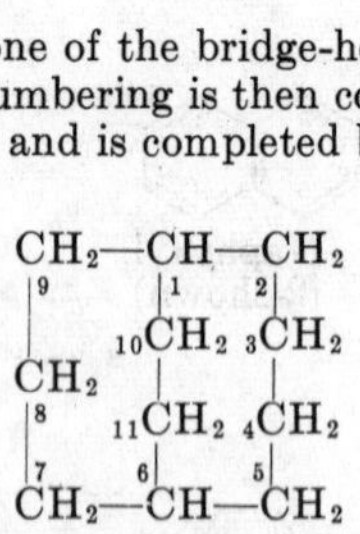   Bicyclo [4.3.2]undecane

Note: Longest path 1, 2, 3, 4, 5
Next longest path 5, 6, 7, 1
Shortest path 1, 8, 5

(1) A bridge is a valence bond or an atom or an unbranched chain of atoms connecting two different parts of a molecule. The two tertiary carbon atoms connected through the bridge are termed "bridge-heads."

31.3.—Unsaturated hydrocarbons are named in accordance with the principles set forth in Rule **A-11.3.** When there is a choice in numbering unsaturation is given the lowest numbers.

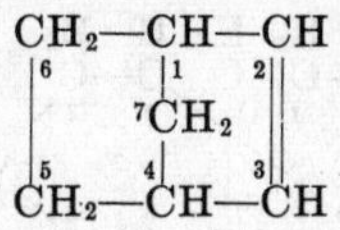

```
CH2—CH—CH
|6   |1  2||
    7CH2
|5   4|  3||
CH2—CH—CH
```

Bicyclo [2.2.1]hept-2-ene

Chemical Abstracts names the above ring system 2-norbornene by Rule **A-74.2,** but follows the above Rule for hydrocarbons which do not have terpene names.

31.4.—Radicals derived from these hydrocarbons are named in accordance with the principles set forth in Rule **A-11.4.** The numbering of the hydrocarbon is retained and the point or points of attachment are given numbers as low as is consistent with the fixed numbering of the hydrocarbon.

Example:

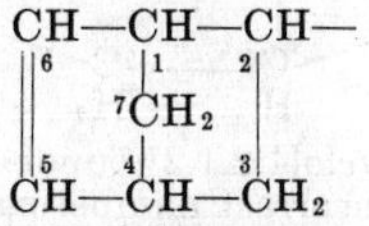

```
CH—CH—CH—
||6  |1  2|
    7CH2
||5  4|  3|
CH—CH—CH2
```

Bicyclo[2.2.1]hept-5-en-2-yl

Chemical Abstracts names the above radical 5-norbornen-2-yl by Rule **A-75.2,** but follows the above Rule for radicals derived from hydrocarbons which do not have terpene names. Radicals derived from the saturated hydrocarbons (except terpenes) are named by replacing the ending "-ane" of the hydrocarbon name by ,"-yl." The point of attachment is given a number as low as is consistent with the fixed numbering of the hydrocarbon.

A-32

32.11.—Cyclic hydrocarbon systems consisting of three or more rings may be named in accordance with the principles stated in Rule **A-31.** The appropriate prefix "tricyclo-," "tetracyclo-," *etc.*, is substituted for "bicyclo-" before the name of the open-chain hydrocarbon containing the same total number of carbon atoms.

32.12.—A polycyclic system is regarded as containing a number of rings equal to the number of scissions required to convert the system into an open-chain compound.

32.13.—The word "cyclo" is followed by brackets containing, in decreasing order, numbers indicating the number of carbon atoms in: the two branches of the main ring, the main bridge, the secondary bridges.

Examples:

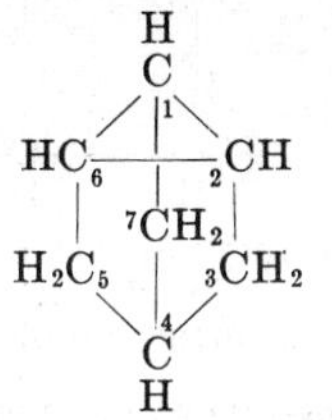

```
         H
         C
       / |1 \
   HC6———+———2CH
   |    7CH2   |
   H2C5  |   3CH2
       \ |4 /
         C
         H
```

Tricyclo[2.2.1.0[1]]heptane

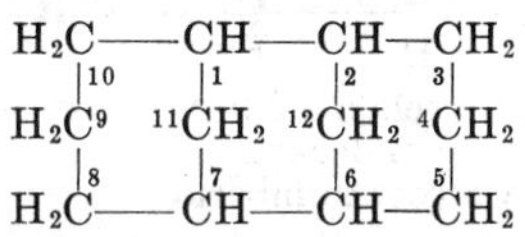

```
H2C——CH——CH—CH2
|10   |1   |2   3|
H2C9 11CH2 12CH2 4CH2
|8    |7   |6   5|
H2C——CH——CH—CH2
```

Tricyclo[5.3.1.1[1]]dodecane

(1) For location and numbering of the secondary bridge see Rules **A-32.22. A32.23, A-32.31.**

32.21.—The main ring and the main bridge form a bicyclic system whose numbering is made in compliance with Rule **A-31.**

32.22.—The location of the other or so-called secondary bridges is shown by superscripts following the number indicating the number of carbon atoms in the said bridges.

32.23.—For the purpose of numbering, the secondary bridges are considered in decreasing order. The numbering of any bridge follows from the part already numbered, proceeding from the highest-numbered bridge-head.

Numbering of the secondary bridges by proceeding from the highest-numbered bridge-head of each bridge is an improvement over the former practice of "following the shortest path from the highest previous number."

32.31.—When there is a choice, the following criteria are considered in turn until a decision is made:

(a) The main ring shall contain as many carbon atoms as possible, two of which must serve as bridgeheads for the main bridge.

Tricyclo[5.4.0.0^{2,9}]undecane
Correct numbering

Tricyclo[4.2.1.2^{7,9}]undecane
Incorrect numbering

Tricyclo[5.3.2.0^{4,9}]dodecane
Correct numbering

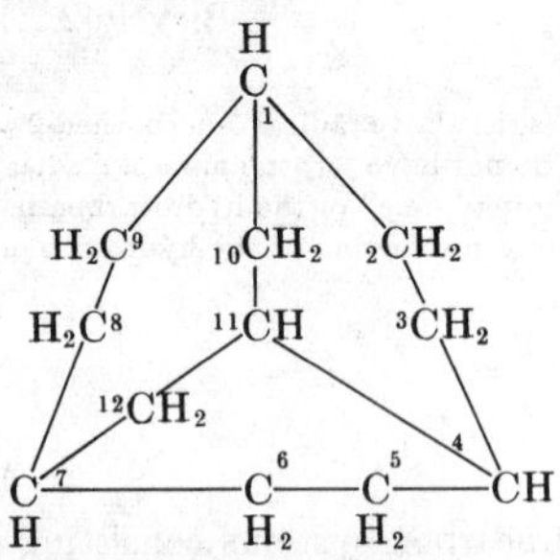

Tricyclo[5.2.3.0^{4,11}]dodecane
Incorrect numbering

(b) The main bridge shall be as large as possible.

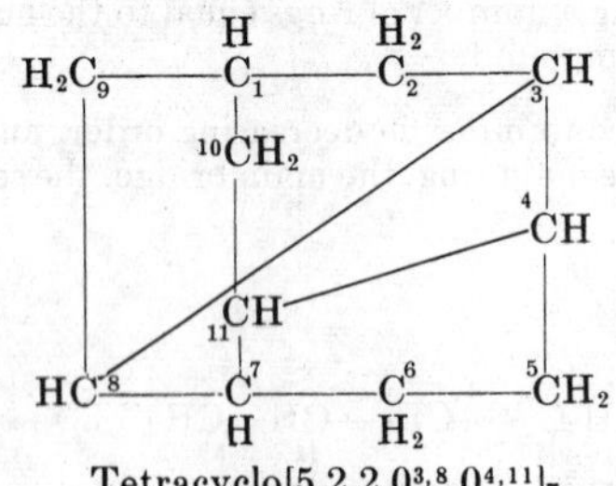

Tetracyclo[5.2.2.0^{3,8}.0^{4,11}]-undecane
Correct numbering

Tetracyclo[5.2.1.1^{4,10}.0^{2,6}]-undecane
Incorrect numbering

(c) The main ring shall be divided as symmetrically as possible by the main bridge.

Tricyclo[4.4.1.1^{1,5}]dodecane:
Correct numbering

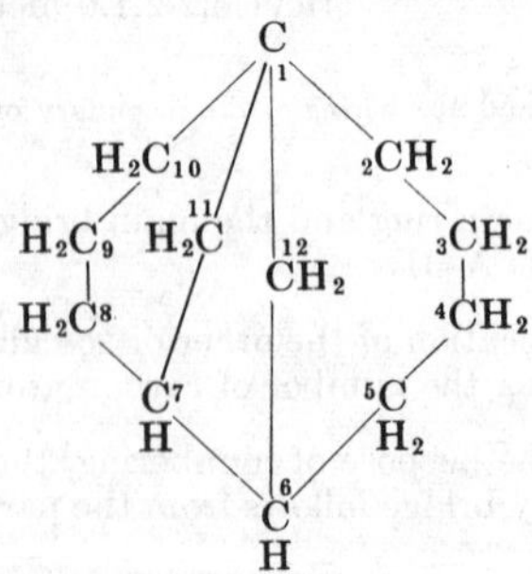

Tricyclo[5.3.1.1^{1,6}]dodecane:
Incorrect numbering

(d) The superscripts locating the other bridges shall be as small as possible (in the sense indicated in Rule **A-2.2**)

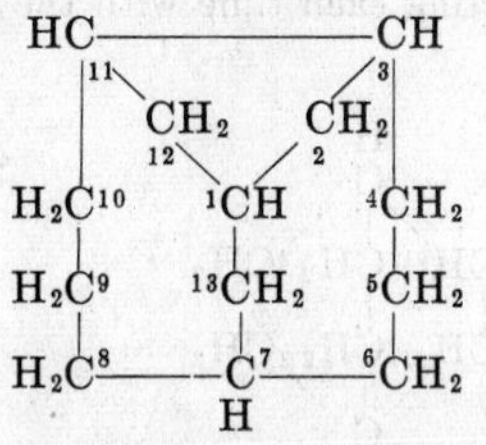

Tricyclo[$5.5.1.0^{3,11}$]tridecane
Correct numbering

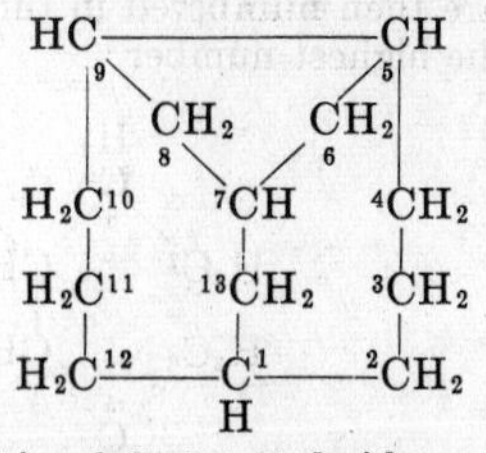

Tricyclo[$5.5.1.0^{5,9}$]tridecane
Incorrect numbering

A-34 (Alternate to Rule A-35)

34.1.—Polycyclic hydrocarbon systems which can be regarded as "ortho-fused" or "ortho- and peri-fused" systems according to Rule **A-21** and which, at the same time, have other bridges,[1] are first named as "ortho-fused" or "ortho- and peri-fused" systems. The other bridges are then indicated by prefixes derived from the corresponding hydrocarbon by replacing the final "-ane," "-ene," *etc.*, by "-ano," "-eno," *etc.*, and their positions are indicated by the points of attachment in the parent compound. If bridges of different types are present, they are cited in alphabetical order.

Examples of bridge names

Butano	$-CH_2-CH_2-CH_2-CH_2-$	Etheno	$-CH=CH-$
Benzeno (*o*-, *m*-, *p*-)	$-C_6H_4-$	Methano	$-CH_2-$
Ethano	$-CH_2-CH_2-$	Propano	$-CH_2-CH_2-CH_2-$

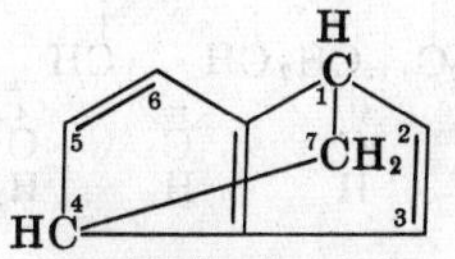

1,4-Dihydro-1,4-methanopentalene

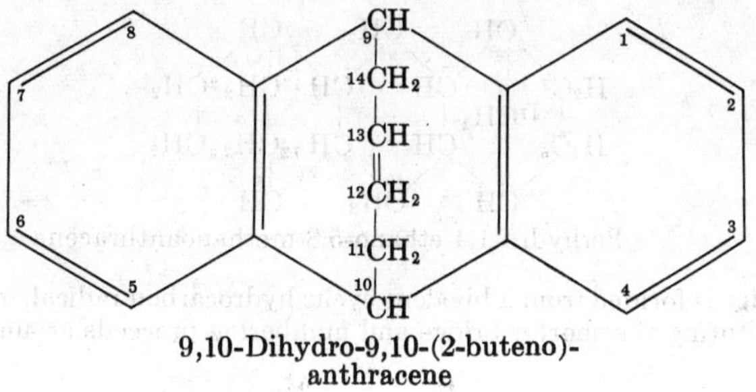

9,10-Dihydro-9,10-(2-buteno)-anthracene

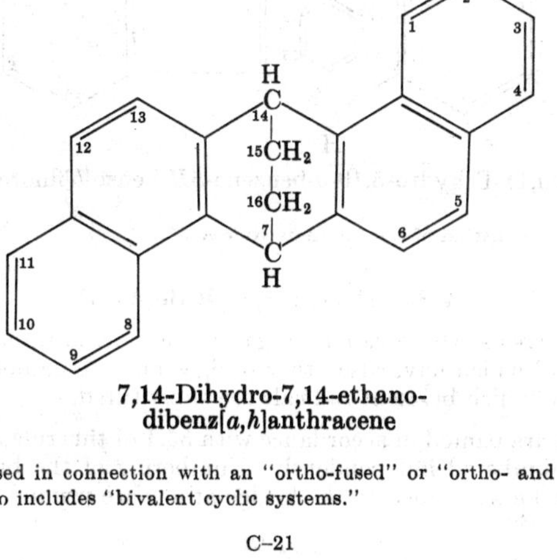

7,14-Dihydro-7,14-ethano-dibenz[*a,h*]anthracene

(1) The term "bridge," when used in connection with an "ortho-fused" or "ortho- and peri-fused" polycyclic system as defined in note to Rule **A-31.1**, also includes "bivalent cyclic systems."

34.2.—The parent "ortho-fused" "or ortho- and peri-fused" system is numbered as prescribed in Rule **A-22.** Where there is a choice, the position numbers of the bridge-heads should be as low as possible. The remaining bridges are then numbered in turn starting each time with the bridge atom next to the bridge-head possessing the highest number.

Example:

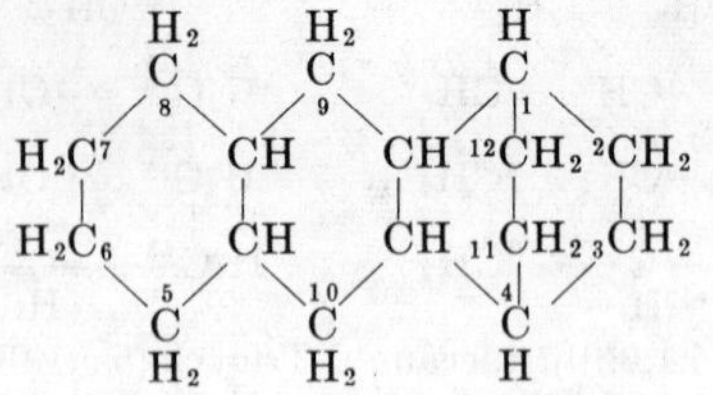

```
        H2     H2     H
        C      C      C
        8      9      1
H2C7    CH     CH  12CH2 2CH2
H2C6    CH     CH  11CH2 3CH2
        5      10     4
        C      C      C
        H2     H2     H
```

Perhydro-1,4-ethanoanthracene

```
        H2     H2     H
        C      C      C
        8      9      1
    H2C7    CH     CH  11CH2 2CH2
not
    H2C6    CH     CH  12CH2 3CH2
        5      10     4
        C      C      C
        H2     H2     H
```

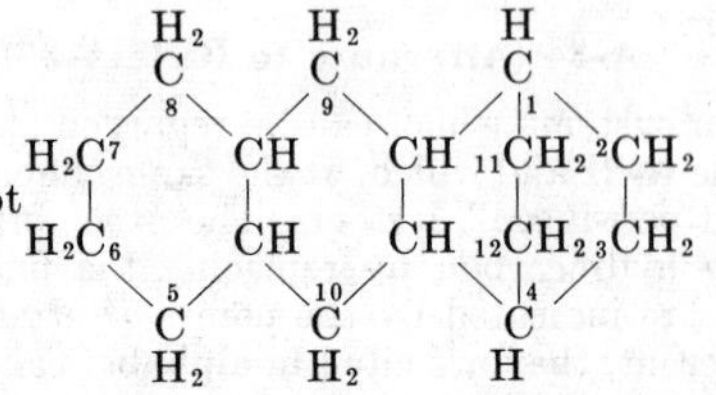

or

```
        H          H2        H2
        C          C         C
        8          9         1
H2C7 11CH2 CH      CH        2CH2
H2C6 12CH2 CH      CH        3CH2
        5          10        4
        C          C         C
        H          H2        H2
```

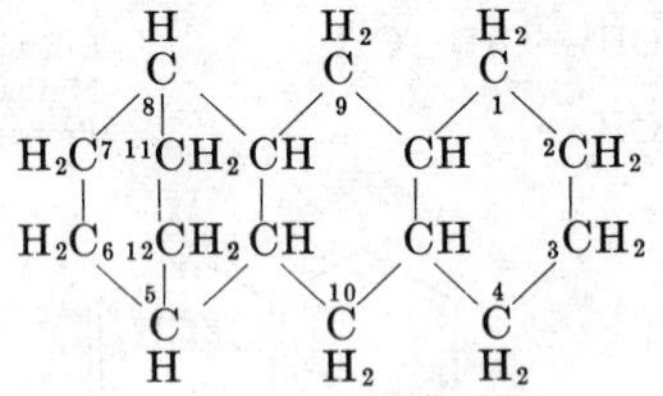

34.3.—When there is a choice of position numbers for the points of attachment for several individual bridges, the lowest numbers are assigned to the bridge-heads in the order of citation of the bridges and the bridge atoms are numbered according to the preceding rule.

Example:

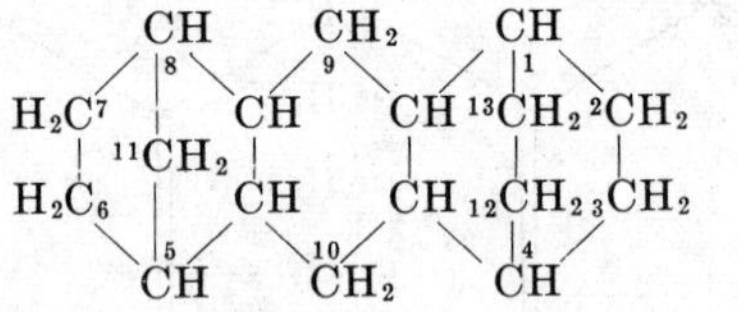

```
        CH      CH2     CH
        8       9       1
H2C7         CH      CH 13CH2 2CH2
     11CH2
H2C6         CH      CH 12CH2 3CH2
        5       10      4
        CH      CH2     CH
```

Perhydro-1,4-ethano-5,8-methanoanthracene

34.4.—When the bridge is formed from a bivalent cyclic hydrocarbon radical, low numbers are given to the carbon atoms constituting the shorter bridge and numbering proceeds around the ring.

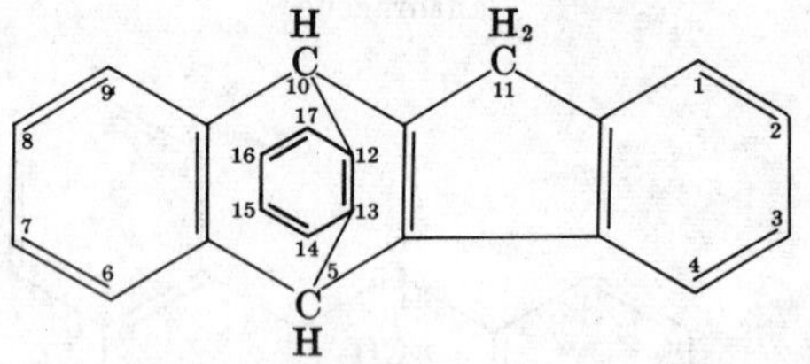

```
        H            H2
        C            C
        10           11
        17
9    16    12                    1
8                                   2
7    15    13                       3
6       14                       4
        5
        C
        H
```

10,11-Dihydro-5,10-*o*-benzeno-5*H*-benzo[*b*]fluorene

Chemical Abstracts follows this rule instead of the Alternate Rule **A-35.**

A-35 (Alternate to Rule A-34)

35.1.—Polycyclic hydrocarbons which can be regarded as "ortho-fused" or "ortho- and peri-fused" according to Rule **A-21** and which have also other bridges may be named as "ortho-fused" or "ortho- and peri-fused" systems into which bivalent radicals are substituted.

35.2.—Compounds which are named in accordance with **35.1** of this rule are numbered as prescribed in Rule **A-34.** The bridge is numbered independently. Numbering of the bridge is started at the lower-numbered bridge head. Numbers assigned to the bridge atoms are primed.

Examples:

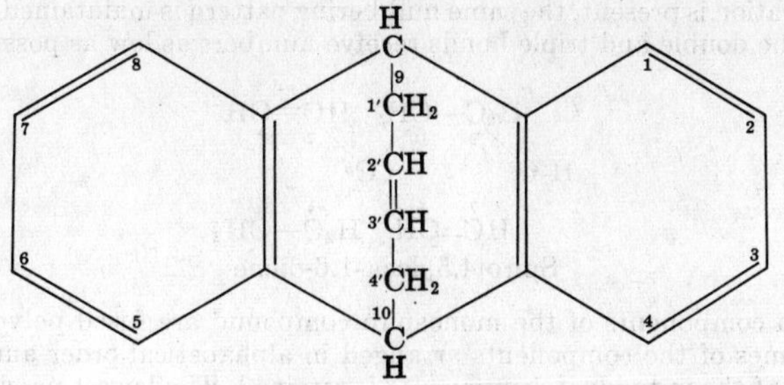

9,10-Dihydro-9,10-(2′-butylene)anthracene

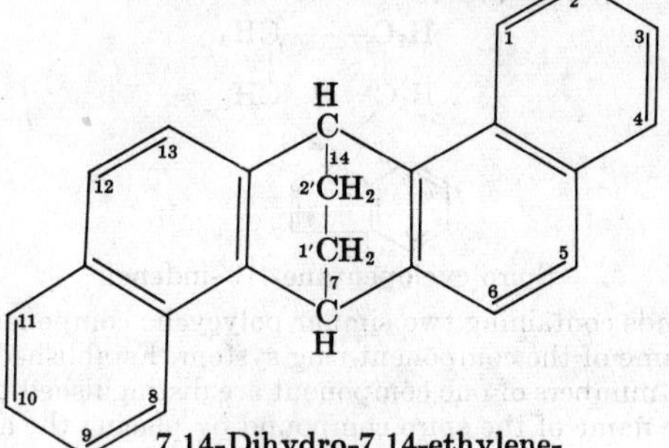

7,14-Dihydro-7,14-ethylene-dibenz[*a,h*]anthracene

Chemical Abstracts follows Rule A-34 instead of this Alternate Rule.

Spiro Hydrocarbons

A "spiro union" is one formed by a single atom which is the only common member of two rings. A "free spiro union" is one constituting the only union direct or indirect between two rings.[1] The common atom is designated as the "spiro atom." According to the number of spiro atoms present, the compounds are distinguished as monospiro, dispiro, trispiro compounds, *etc.* The following rules apply to the naming of compounds containing free spiro unions.

(1) An example of a compound where the spiro union is *not* free is:

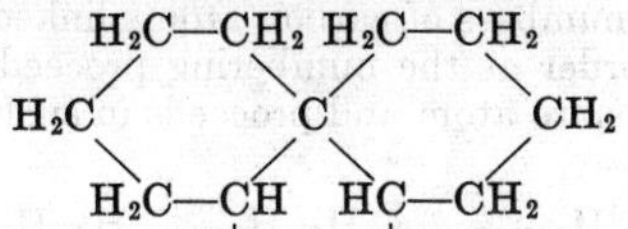

A-41 (Alternate to Rule A-42)

41.1.—Monospiro compounds consisting of only two alicyclic rings as components are named by placing "spiro" before the name of the normal acyclic hydrocarbon of the same total number of carbon atoms. The number of carbon atoms linked to the spiro atom in each ring is indicated in ascending order in brackets placed between the spiro prefix and the hydrocarbon name.

Examples:

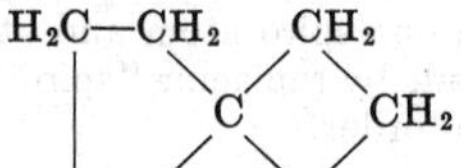

Spiro[3.4]octane

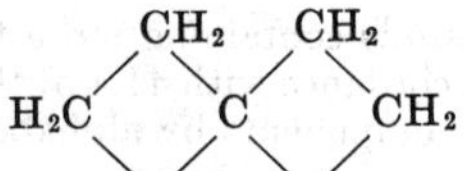

Spiro[3.3]heptane

41.2.—The carbon atoms in monospiro hydrocarbons are numbered consecutively starting with a ring atom next to the spiro atom, first through the smaller ring (if such be present) and then through the spiro atom and around the second ring.

Example:

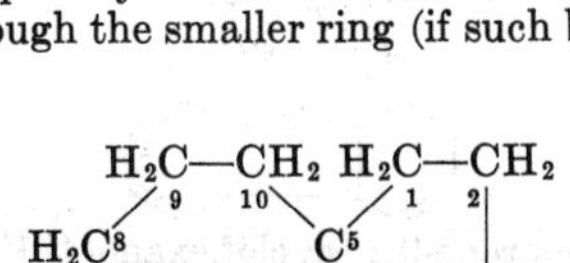

Spiro[4.5]decane

41.3.—When unsaturation is present, the same numbering pattern is maintained, but in such a direction around the rings that the double and triple bonds receive numbers as low as possible in accordance with Rule **A-11.**

Example:

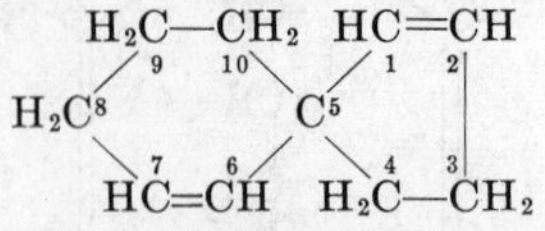

Spiro[4.5]deca-1,6-diene

41.4.—If one or both components of the monospiro compound are fused polycyclic systems, "spiro" is placed before the names of the components arranged in alphabetical order and enclosed in brackets. Established numbering of the individual components is retained. The lowest possible number is given to the spiro atom, and the numbers of the second component are marked with primes. The position of the spiro atom is indicated by placing the appropriate numbers between the names of the two components.

Example:

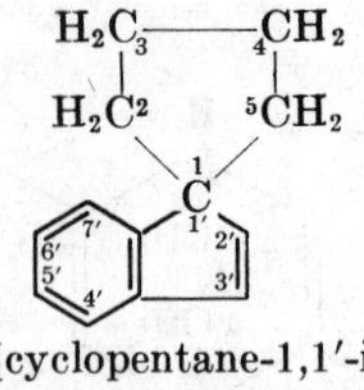

Spiro[cyclopentane-1,1′-indene]

41.5.—Monospiro compounds containing two similar polycyclic components are named by placing the prefix "spirobi" before the name of the component ring system. Established numbering of the polycyclic system is maintained and the numbers of one component are distinguished by primes. The position of the spiro atom is indicated in the name of the spiro compound by placing the appropriate locants before the name.

Example:

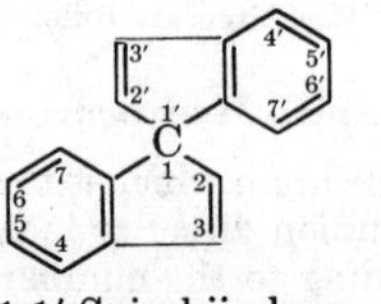

1,1′-Spirobiindene

41.6.—Polyspiro compounds consisting of three or more alicyclic systems are named by placing "dispiro-," "trispiro-," "tetraspiro-," *etc.*, before the name of the normal acyclic hydrocarbon of the same total number of carbon atoms. The numbers of carbon atoms linked to the spiro atoms in each ring are indicated in brackets in the same order as the numbering proceeds about the ring. Numbering starts with a ring atom next to a terminal spiro atom and proceeds in such a way as to give the spiro atoms as low numbers as possible.

Example:

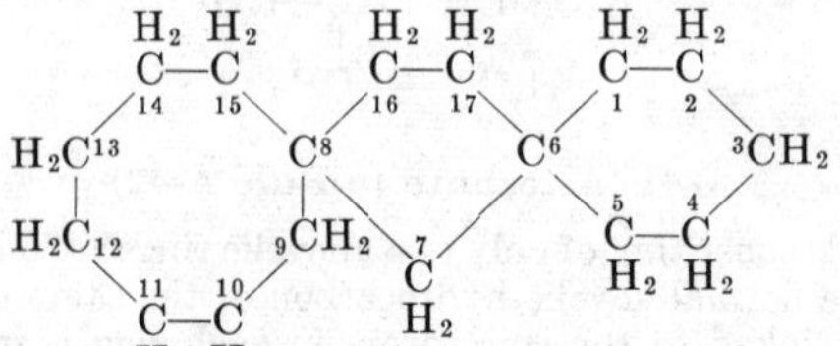

Dispiro[5.1.7.2]heptadecane

41.7.—Polycyclic compounds containing more than one spiro atom and at least one fused polycyclic component are named in accordance with **41.4** of this rule by replacing "spiro" with "dispiro," "trispiro," etc., and choosing the end components by alphabetical order.

Example:

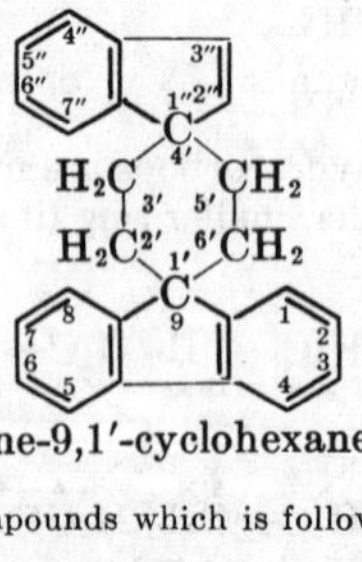

Dispiro[fluorene-9,1′-cyclohexane-4′,1″-indene]

This represents the method of naming spiro compounds which is followed by *Chemical Abstracts* instead of the Alternate Rule **A-42.**

A-42

42.1 (Alternate to **A-41.1** and **A-41.2**).—When two dissimilar cyclic components are united by a spiro union the name of the larger component is followed by the affix "spiro" which, in turn, is followed by the name of the smaller component. Between the affix "spiro" and the name of each component system is inserted the number denoting the spiro-position in the appropriate ring system, these numbers being as low as permitted by any fixed numbering of the component. The components retain their respective numberings but numbers for the component mentioned second are primed. Numbers 1 may be omitted when a free choice is available for a component.

Examples:

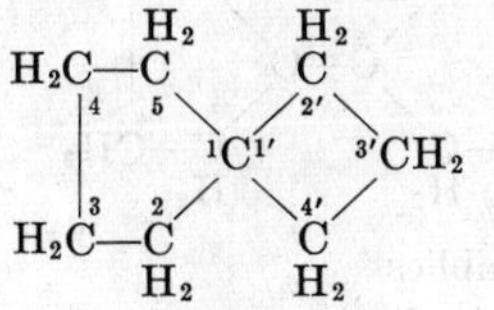

Cyclopentanespiro-cyclobutane

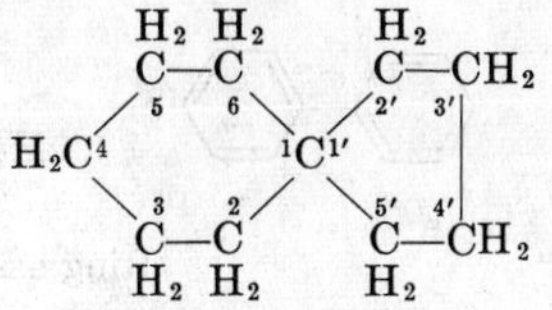

Cyclohexanespirocyclo-pentane

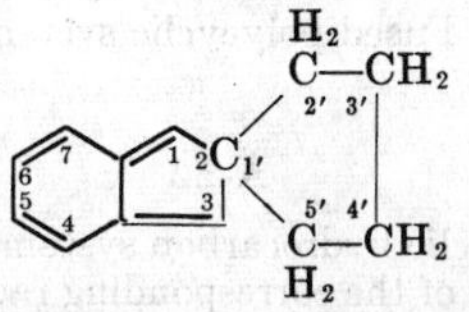

2*H*-Indene-2-spiro-1′-cyclopentane

42.2 (Alternate to **A-41.3**).—Rule **A-41.3** applies also with appropriate different numbering where nomenclature is according to Rule **A-42.1**, but the spiro-junction has priority for lowest numbers over unsaturation.

Example:

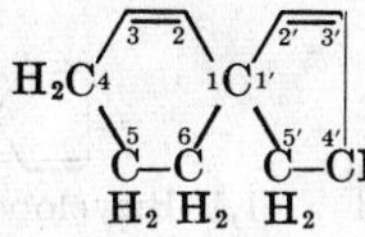

2-Cyclohexenespiro-(2′-cyclopentene)

42.3 (Alternate to **A-41.5**).—The nomenclature of Rule **A-41.5** is applied also to monocyclic components with identical saturation, the spiro-union being numbered 1.

Examples:

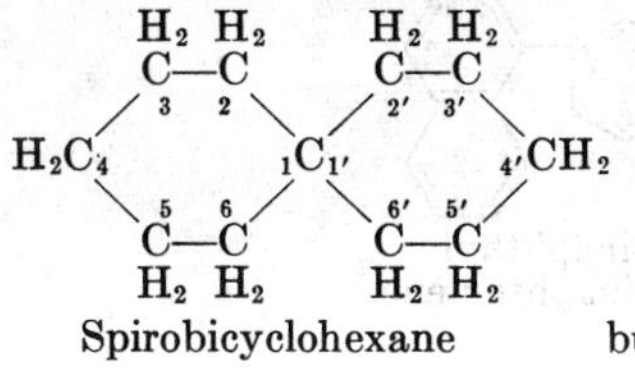

Spirobicyclohexane but 2-Cyclohexenespiro-(3′-cyclohexene)

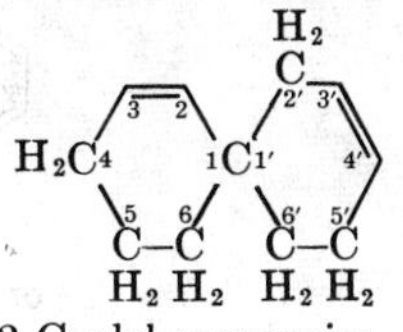

42.4 (Alternate to **A-41.6** and **A-41.7**).—Polycyclic compounds containing more than one spiro atom are named in accordance with Rule **A-42.1** starting from the senior[1] end-component irrespective of whether the components are simple or fused rings.

Examples:

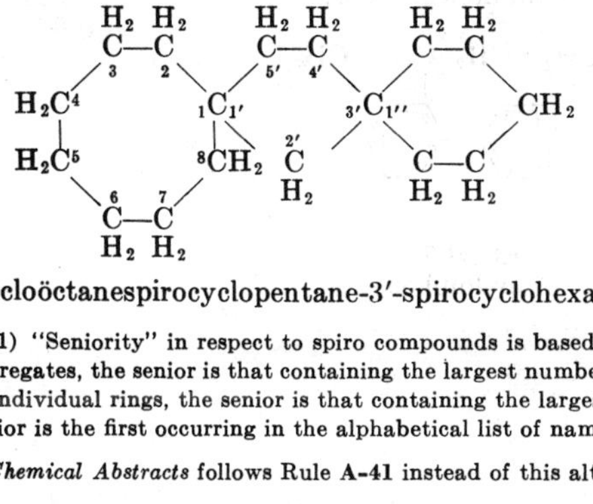

Cycloöctanespirocyclopentane-3′-spirocyclohexane

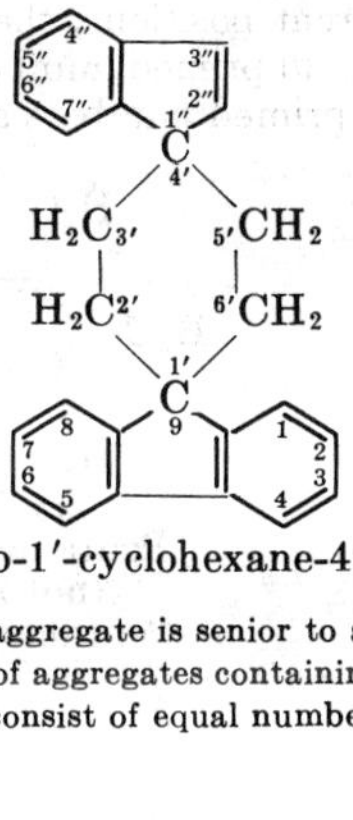

Fluorene-9-spiro-1′-cyclohexane-4′-spiro-1″-indene

(1) "Seniority" in respect to spiro compounds is based on the principles: (i) an aggregate is senior to a monocycle; (ii) of aggregates, the senior is that containing the largest number of individual rings; (iii) of aggregates containing the same number of individual rings, the senior is that containing the largest ring; (iv) if aggregates consist of equal number of equal rings the senior is the first occurring in the alphabetical list of names.

Chemical Abstracts follows Rule **A-41** instead of this alternate Rule.

Hydrocarbon Ring Assemblies

A-51

51.1.—Two or more cyclic systems (single rings or fused systems) which are directly joined to each other by double or single bonds are named "ring assemblies" when the number of such direct ring junctions is one less than the number of cyclic systems involved.

Examples:

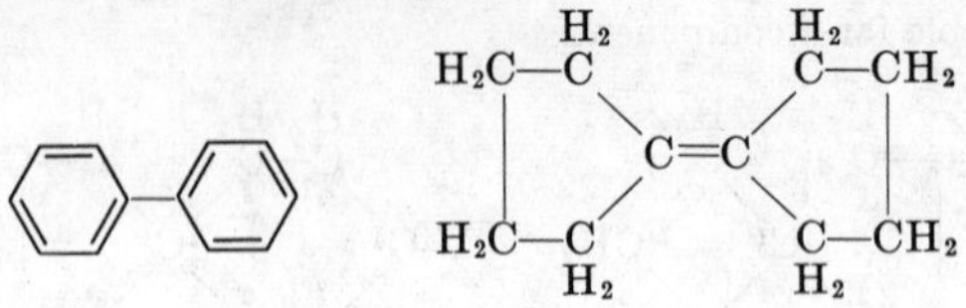

H_2C—$C(H_2)$ ... $C(H_2)$—CH_2; $C{=}C$; H_2C—$C(H_2)$ ... $C(H_2)$—CH_2

Ring assemblies

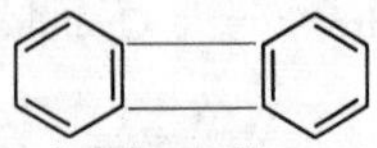

Fused polycyclic system

A-52

52.1.—Assemblies of two identical cyclic hydrocarbon systems are named in either of two ways: (a) by placing the prefix "bi-" before the name of the corresponding radical, or (b) for systems joined by a single bond by placing the prefix "bi-" before the name of the corresponding hydrocarbon. In each case, the numbering of the assembly is that of the corresponding radical or hydrocarbon, one system being assigned unprimed numbers and the other primed numbers. The points of attachment are indicated by placing the appropriate locants before the name.

Examples:

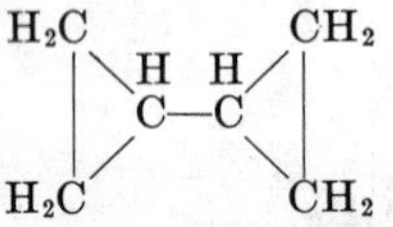

H_2C CH_2 H H C—C H_2C CH_2

1,1′-Bicyclopropyl
or 1,1′-Bicyclopropane

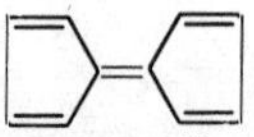

1,1′-Bicyclopentadienylidene

52.2.—If there is a choice in numbering, unprimed numbers are assigned to the system which has the lower-numbered point of attachment.

Example:

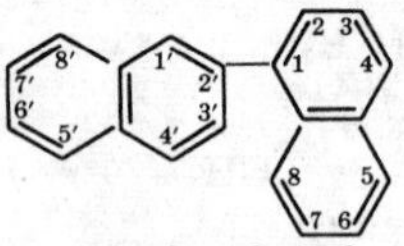

1,2′-Binaphthyl
or 1,2′-Binaphthalene

52.3.—If two identical hydrocarbon systems have the same point of attachment and contain substituents at different positions, the locants of these substituents are assigned according to Rule **A-2.2**; for this purpose an unprimed number is considered lower than the same number when primed. Assemblies of primed and unprimed numbers are arranged in ascending numerical order.

Examples:

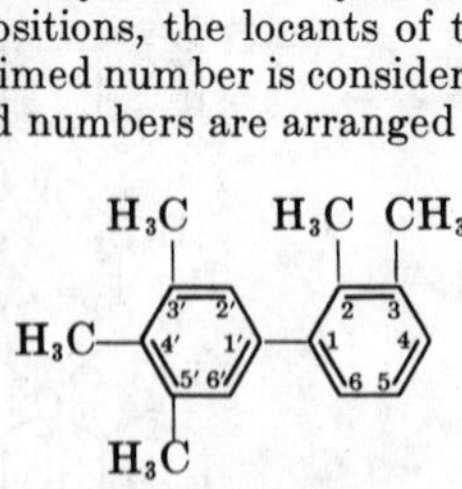

H_3C H_3C CH_3 H_3C H_3C

2,3,3′,4′,5′-
Pentamethylbiphenyl
(not 2′,3,3′,4,5-
Pentamethylbiphenyl)

CH_3—CH_2—CH_2 CH_2—CH_3

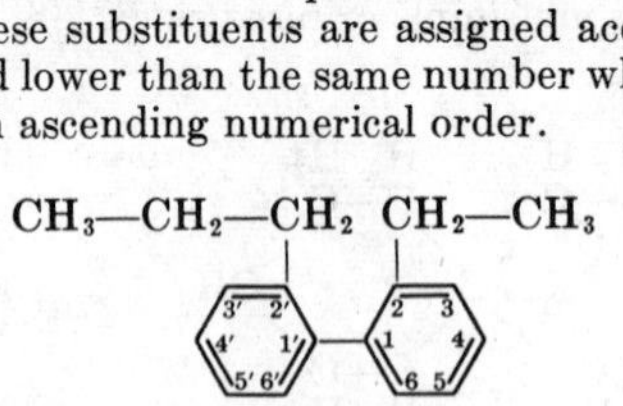

2-Ethyl-2′-
propylbiphenyl

52.4.—The name "biphenyl" is used for the assembly consisting of two benzene rings:

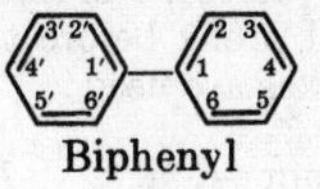

Biphenyl

A-53

53.1.—Other hydrocarbon ring assemblies are named by selecting one ring system as the base component and considering the other system as substituents of the base component. Such substituents are arranged in alphabetical order. The base component is assigned unprimed numbers and the substituents are assigned numbers with primes.

53.2.—The base component is chosen by considering these characteristics in turn until a decision is reached:

(a) The system containing the larger number of rings.

Examples:

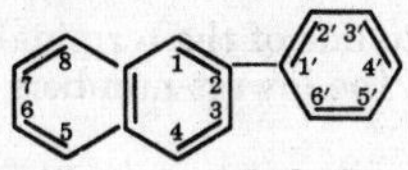

2-Phenylnaphthalene

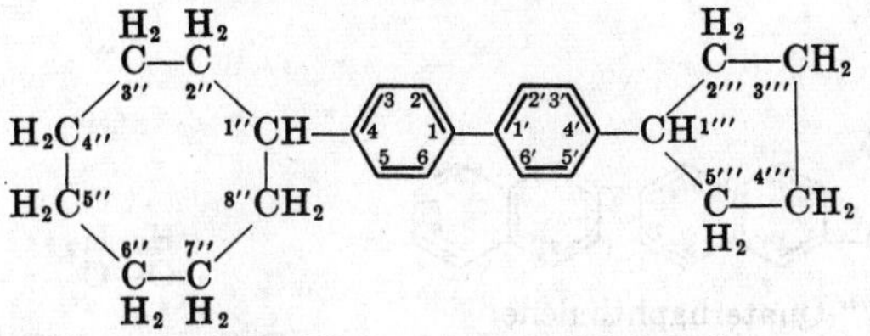

4-Cycloöctyl-4′-cyclopentylbiphenyl

(b) The system containing the larger ring.

Examples:

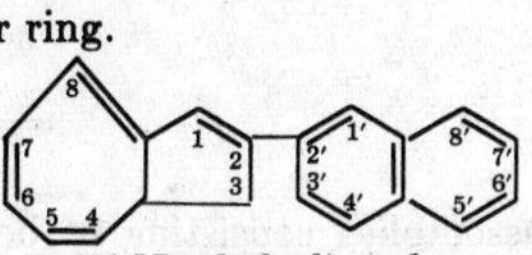

2-(2′-Naphthyl)azulene

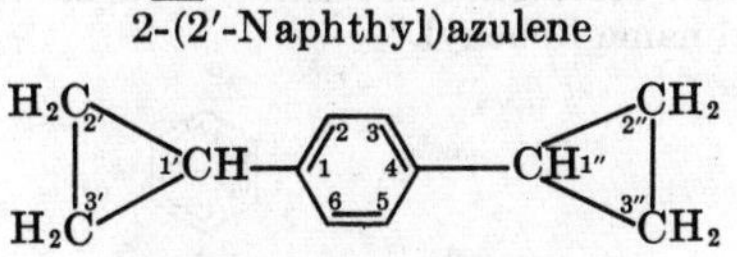

1,4-Dicyclopropylbenzene
or *p*-Dicyclopropylbenzene

(c) The system in the lowest state of hydrogenation (see also **53.3** of this rule).

Example:

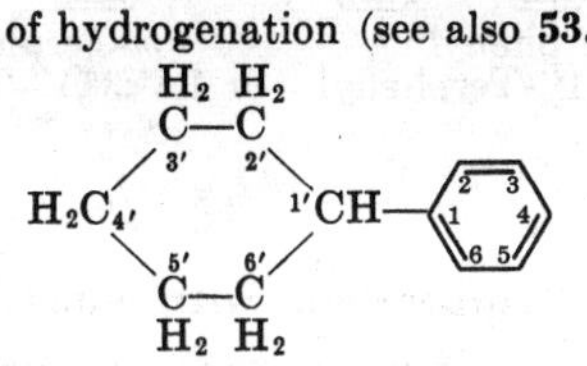

Cyclohexylbenzene

(d) The order of ring systems as set forth in the list of Rule **A-21.1.**

53.3.—Compounds covered by **53.2**(c) of this rule may also be named as hydrogenation products according to Rule **A-23.**

Example:

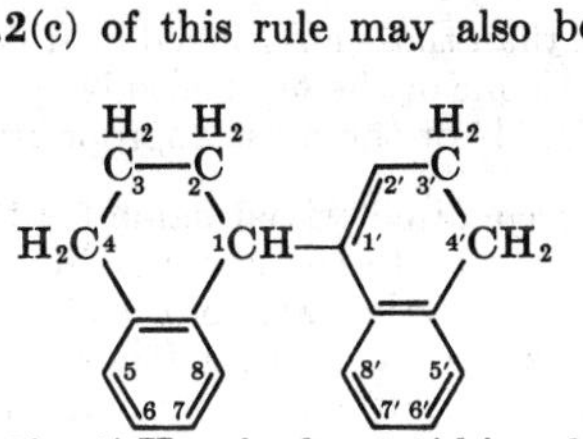

1,2,3,3′,4,4′-Hexahydro-1,1′-binaphthyl

The name of the second example in (a) which contains as a base component the ring assembly "biphenyl" indicates that the word system as used here includes ring assemblies as base components.

Chemical Abstracts does not use primed numbers with the groups which may be considered as substituent groups expressed as prefixes and attached to the name of the base component, Examples: 2-(2,4-dichlorophenyl)naphthalene; 2(2-naphthyl)-azulene.

A-54

54.1.—Unbranched assemblies consisting of three or more identical hydrocarbon ring systems are named by placing an appropriate numerical prefix before the name of the hydrocarbon corresponding to the repetitive unit. These numerical prefixes are used:

3. ter-	5. quinque-	7. septi-	9. novi-
4. quater-	6. sexi-	8. octi-	10. deci-

Example:

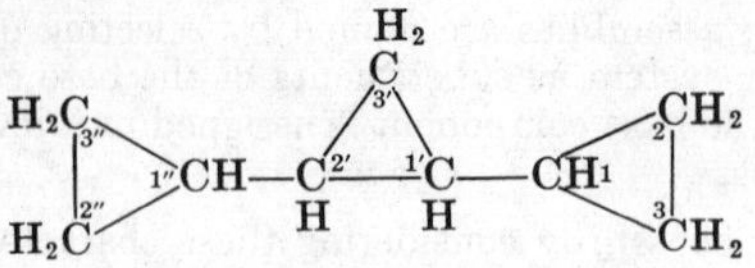

Tercyclopropane

54.2.—Unprimed numbers are assigned to one of the terminal systems, the other systems being primed serially. Points of attachment are assigned the lowest numbers possible.

Examples:

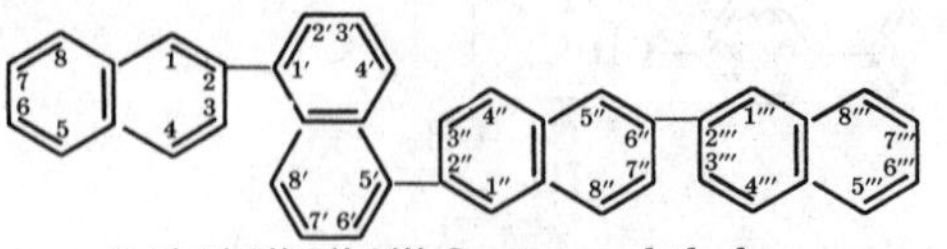

2,1′:5′,2″:6″,2‴-Quaternaphthalene

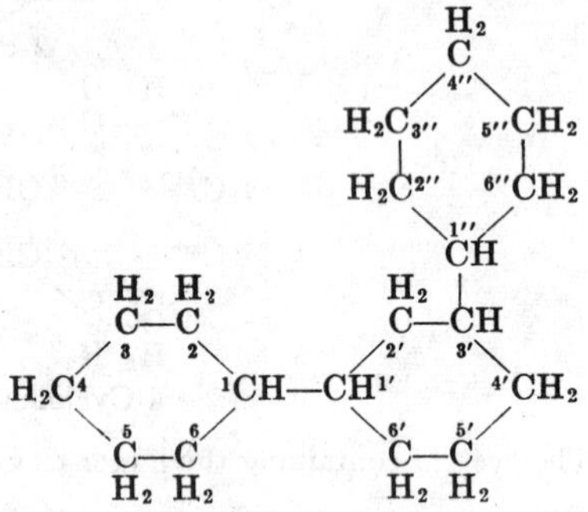

1,1′:3′,1″-Tercyclohexane

54.3.—As exceptions, unbranched assemblies consisting of benzene rings are named by using the appropriate prefix with the radical name "phenyl."

Examples:

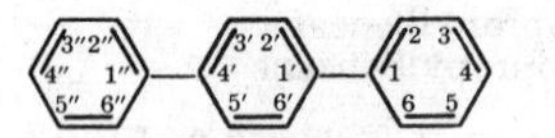

p-Terphenyl or 1,1′:4′,1″-Terphenyl

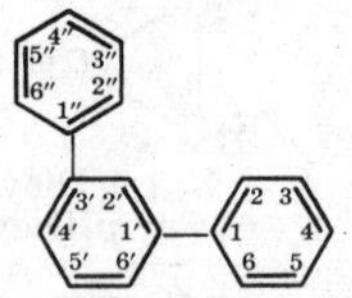

m-Terphenyl or 1,1′:3′,1″-Terphenyl

Cyclic Hydrocarbons with Side Chains

(Note: *cf.* Rules **A-12** and **A-13**)

A-61

61.1.—Hydrocarbons more complex than those envisioned in Rule **A-12,** composed of cyclic nuclei and aliphatic chains, are named according to one of the methods given below. Choice is made so as to provide the name which is the simplest permissible or the most appropriate for the chemical intent.

61.2.—When there is no generally recognized trivial name for the hydrocarbon, then (1) the radical name denoting the aliphatic chain is prefixed to the name of the cyclic hydrocarbon, or (2) the radical name for the cyclic hydrocarbon is prefixed to the name of the aliphatic compound. Choice between these methods is made according to the more appropriate of the following principles: (a) the maximum number of substitutions into a single unit of structure; (b) treatment of a smaller unit of structure as a substituent into a larger.

61.3.—In accordance with the principle (a) of **61.2** of this rule, hydrocarbons containing several chains attached to one cyclic nucleus generally are named as derivatives of the cyclic compound; and compounds containing several side chains and/or cyclic radicals attached to one chain are named as derivatives of the acyclic compound.

Examples:

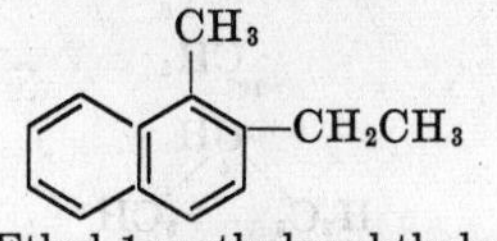

2-Ethyl-1-methylnaphthalene

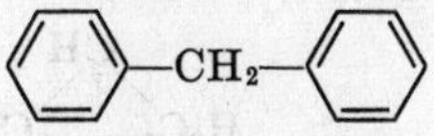

Diphenylmethane

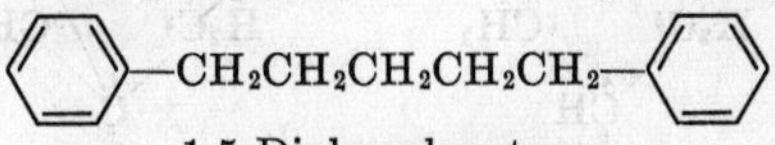

1,5-Diphenylpentane

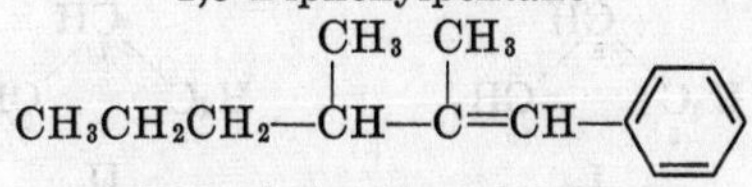

2,3-Dimethyl-1-phenyl-1-hexene

61.4.—In accordance with principle (b) of **61.2** of this rule, a hydrocarbon containing a small cyclic nucleus attached to a long chain is generally named as a derivative of the acyclic hydrocarbon; and a hydrocarbon containing a small group attached to a large cyclic nucleus is generally named as a derivative of the cyclic hydrocarbon.

Examples:

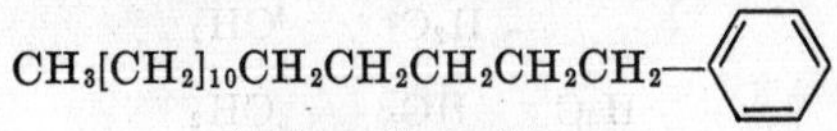

$CH_3[CH_2]_{10}CH_2CH_2CH_2CH_2CH_2$—$C_6H_5$

1-Phenylhexadecane

$CH_3[CH_2]_{10}CH_2CH_2CH_2CH{=}CH$—$C_6H_5$

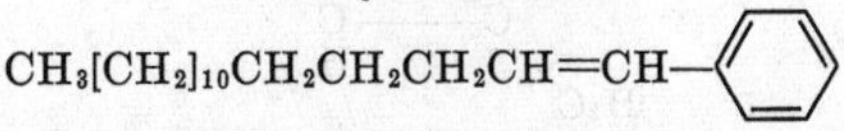

1-Phenyl-1-hexadecene

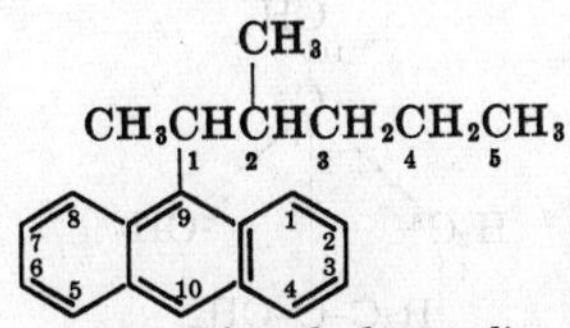

9-(1,2-Dimethylpentyl)-anthracene

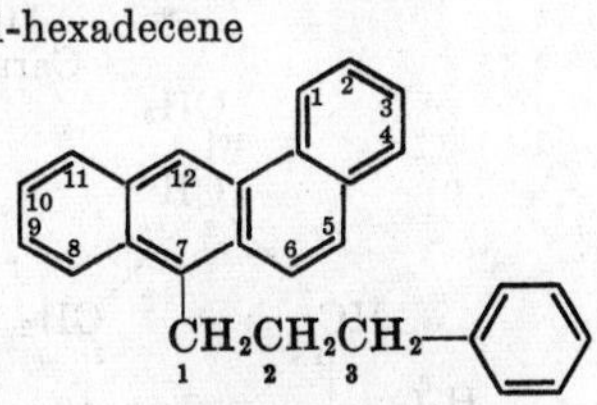

7-(3-Phenylpropyl)-benz[a]anthracene

61.5.—Recognized trivial names for composite radicals are used if they lead to simplifications in naming.

Examples:

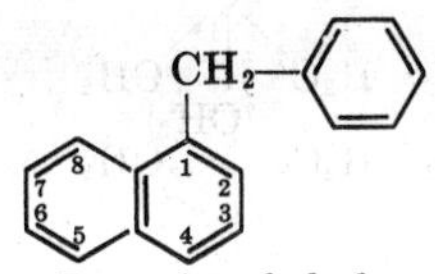

1-Benzylnaphthalene

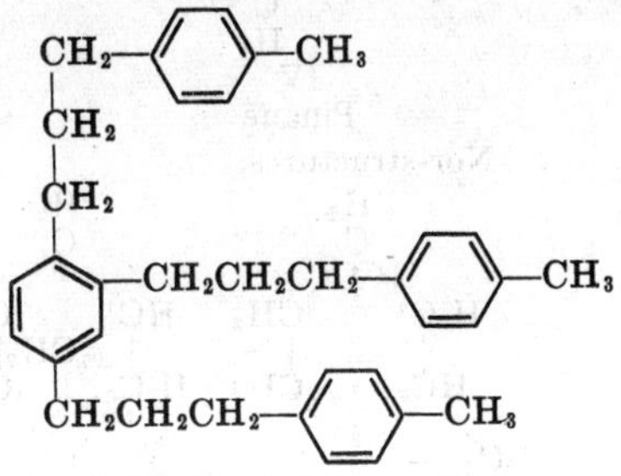

1,2,4-Tris(3-p-tolylyropyl)benzene

Terpene Hydrocarbons

Owing to long-established custom, terpenes are given exceptional treatment in these rules.

A-71. Acyclic Terpenes[1]

71.1.—The acyclic terpene hydrocarbons are named in a manner similar to that used for other unsaturated acyclic hydrocarbons when pure compounds are involved.

Example:

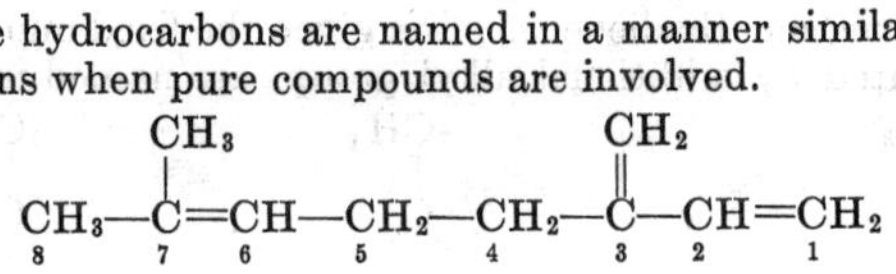

CH_3—$C(CH_3){=}CH$—CH_2—CH_2—$C({=}CH_2)$—$CH{=}CH_2$ (numbered 8 to 1)

7-Methyl-3-methylene-1,6-octadiene

A-72. Cyclic Terpenes

72.1.—The following structural types with their special names and special systems of numbering are used as the basis for the specialized nomenclature of monocyclic and bicyclic terpene hydrocarbons. The name "bornane" replaces camphane and bornylane; "norbornane" replaces norcamphane and norbornylane.

(1) For a more complete discussion of terpene nomenclature see "Nomenclature for Terpene Hydrocarbons," Advances in Chemistry Series No. 14 (American Chemical Society).

Fundamental terpene types:

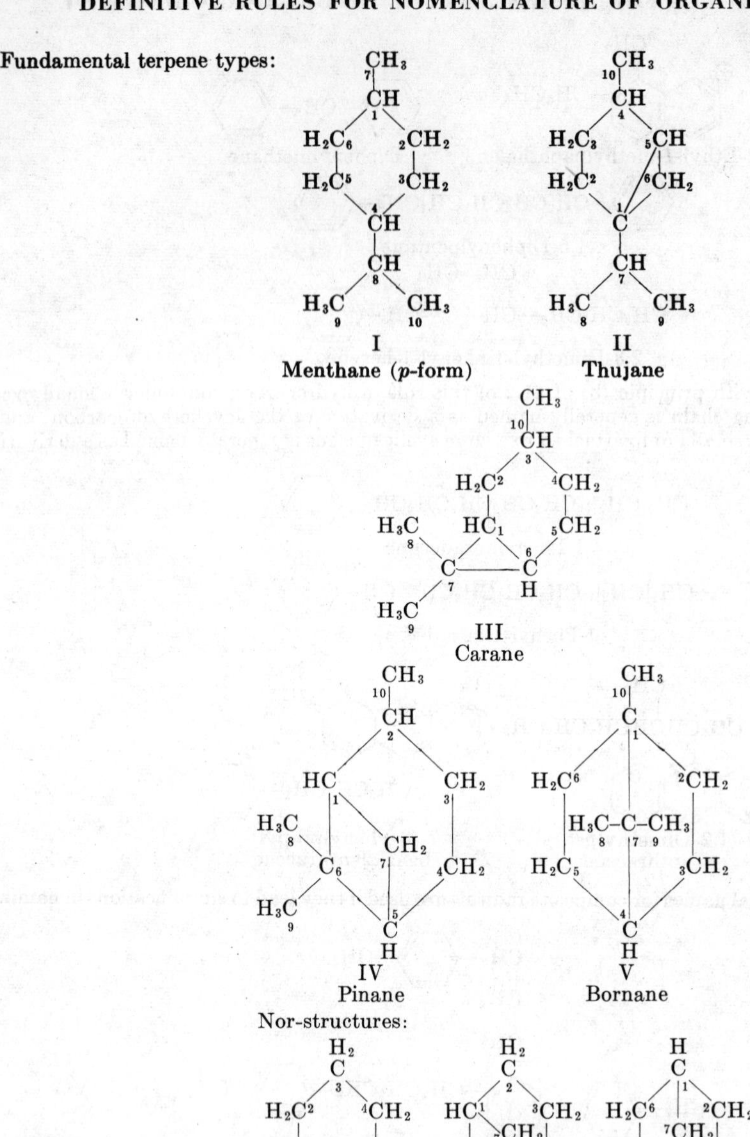

A-73. Monocyclic Terpenes

73.1.—Menthane Type: Monocyclic terpene hydrocarbons of this type (ortho, meta, and para isomers) are named menthane, menthene, menthadiene, *etc.*, are given the fixed numbering of menthane (Formula I). Such compounds substituted by additional alkyl groups are named in accordance with Rule **A-11.**

Examples:

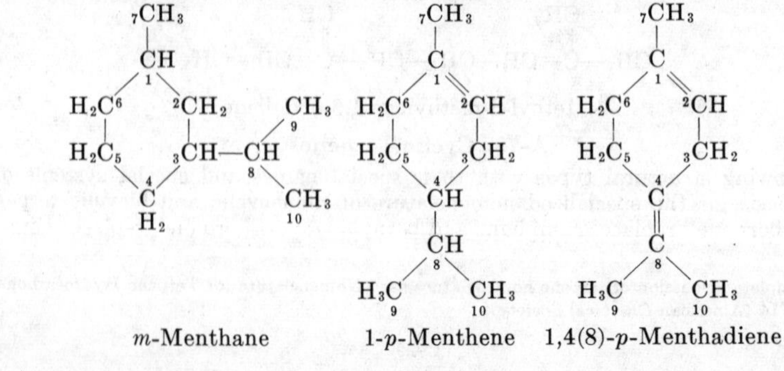

m-Menthane 1-*p*-Menthene 1,4(8)-*p*-Menthadiene

73.2.—Tetramethylcyclohexane Type: Monocyclic terpene hydrocarbons of this type are named systematically as derivatives of cyclohexane, cyclohexene, and cyclohexadiene (see Rule **A-11**).

Examples:

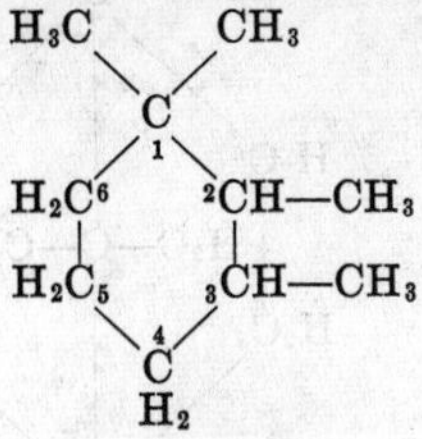

1,1,2,3-Tetramethylcyclohexane 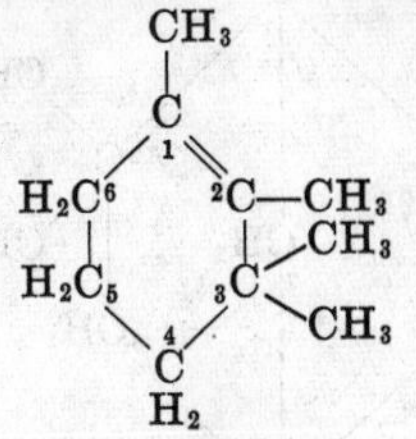 | 1,2,3,3-Tetramethylcyclohexene 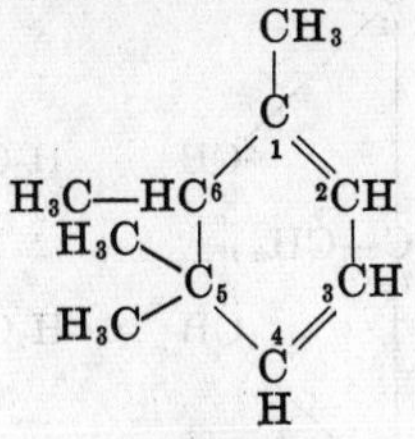 | 1,5,5,6-Tetramethyl-1,3-cyclohexadiene

A-74. Bicyclic Terpenes

74.1.—Bicyclic terpene hydrocarbons having the skeleton of Formula II or this skeleton and additional side chains except methyl or isopropyl (or methylene if one methylene group is already present) are named as thujane, thujene, thujadiene, *etc.*, and are given the fixed numbering shown for thujane (Formula II). Other hydrocarbons containing the thujane ring-skeleton are named from bicyclo[3.1.0]hexane and are given systematic bicyclo numbering (*cf.* Rule **A-31**).

Example:

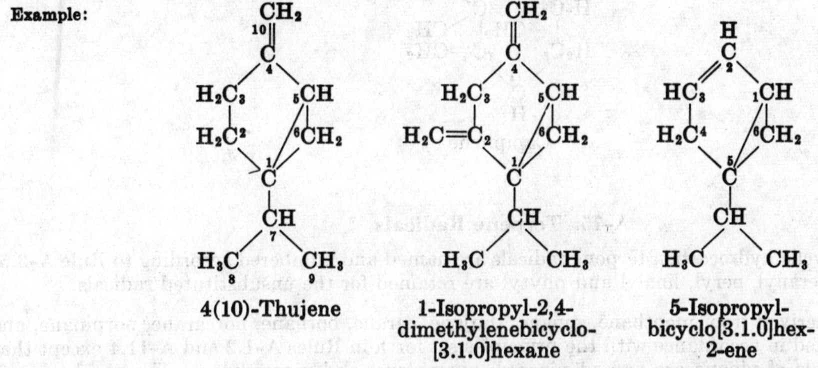

4(10)-Thujene | **1-Isopropyl-2,4-dimethylenebicyclo-[3.1.0]hexane** | **5-Isopropyl-bicyclo[3.1.0]hex-2-ene**

74.2.—Bicyclic terpene hydrocarbons having the skeleton of Formula III, IV, or V and additional side chains except methyl (or methylene if one methylene group is already present) are named, respectively, as carane, carene, caradiene, *etc.;* pinane, pinene, pinadiene, *etc.;* bornane, bornene, bornadiene, *etc.* They are given, respectively, the fixed numbering shown for carane (Formula III), pinane (Formula IV), and bornane (Formula V). Other hydrocarbons containing the ring-skeleton of carane, pinane, or bornane are named, respectively, from norcarane (Formula VI), norpinane (Formula VII), or norbornane (Formula VIII). These names are preferred to those from bicyclo [4.1.0]heptane, bicyclo[3.1.1]heptane, or bicyclo [2.2.1]heptane. The nor-names are given systematic bicyclo numbering (*cf.* Rule **A-31**).

Examples:

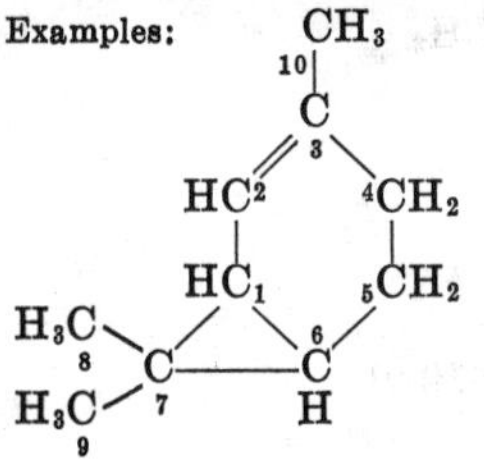

2-Carene 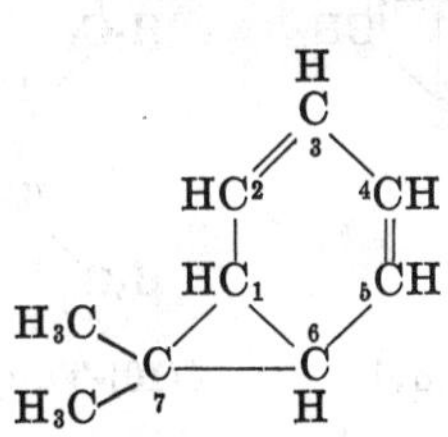 | 7,7-Dimethyl-2,4-norcaradiene 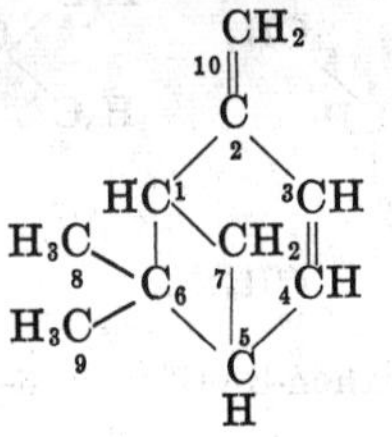 | 2(10),3-Pinadiene 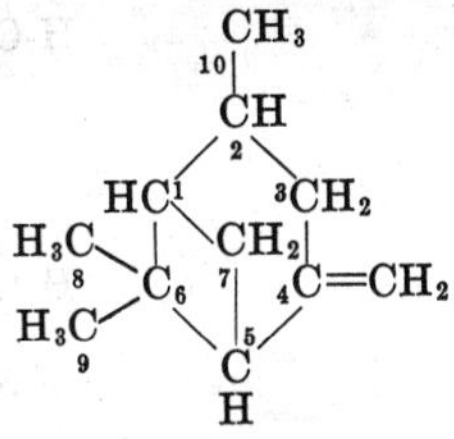 | 4-Methylenepinane

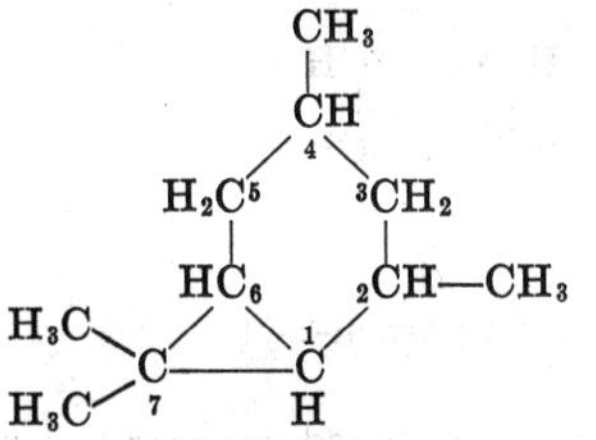

2,4,7,7-Tetramethylnorcarane 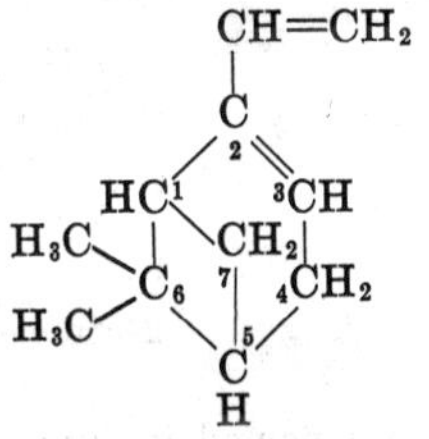 | 6,6-Dimethyl-2-vinyl-2-norpinene

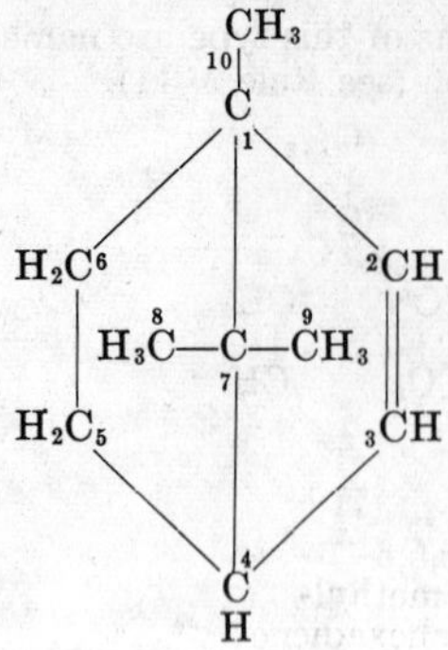

2-Bornene

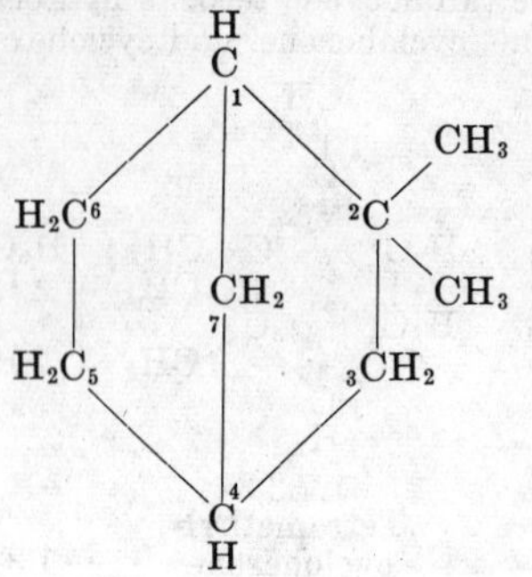

2,2-Dimethylnorbornane

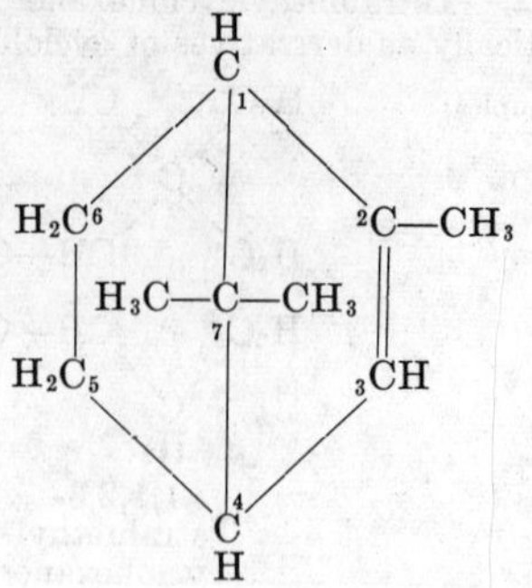

2,7,7-Trimethyl-2-norbornene

74.3.—The name "camphene" is retained for the unsubstituted compound 2,2-dimethyl-3-methylene-norbornane.

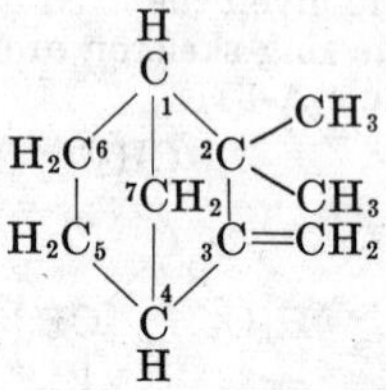

Camphene

A-75. Terpene Radicals

75.1.—Simple acyclic hydrocarbon terpene radicals are named and numbered according to Rule **A-3.5.** The trivial names geranyl, neryl, linalyl and phytyl are retained for the unsubstituted radicals.

75.2.—Radicals derived from menthane, pinane, thujane, carane, bornane, norcarane, norpinane, and norbornane are named in accordance with the principles set forth in Rules **A-1.2** and **A-11.4** except that the saturated radicals of pinane are named pinanyl, pinanylene, and pinanylidene. The numbering of the hydrocarbon is retained and the point or points of attachment, whether in the ring or side chain, are given numbers as low as is consistent with the fixed numbering of the hydrocarbon.

Examples:

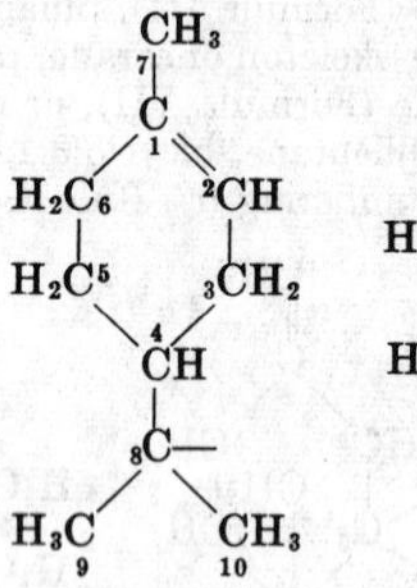

1-*p*-Menthen-8-yl

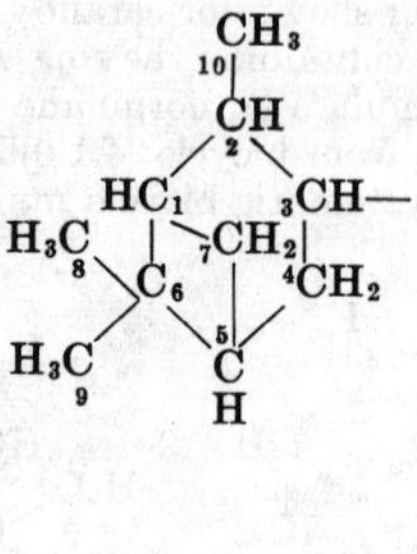

3-Pinanyl

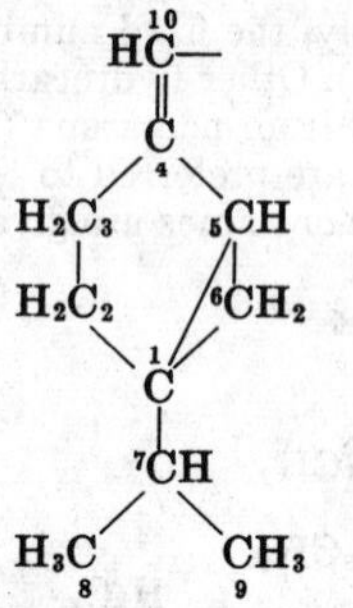

4(10)-Thujen-10-yl

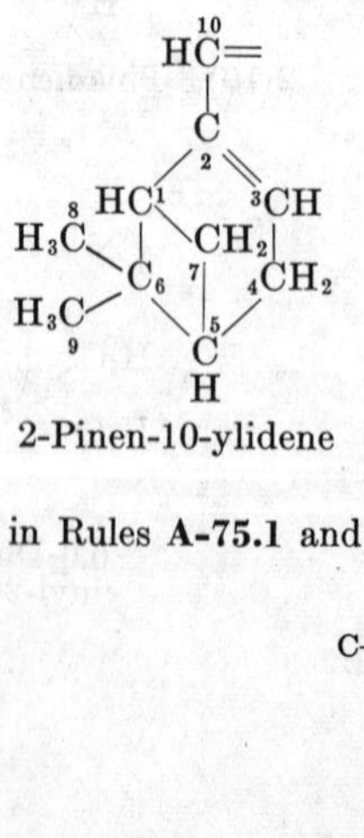

2-Pinen-10-ylidene

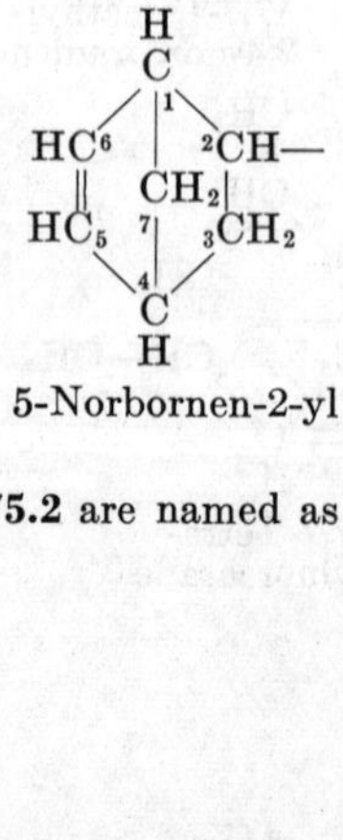

5-Norbornen-2-yl

75.3.—Radicals not named in Rules **A-75.1** and **A-75.2** are named as described in Rules **A-11** and **A-31.4.**

B. FUNDAMENTAL HETEROCYCLIC SYSTEMS

B-1. Extension of Hantzsch-Widman System

1.1.—Monocyclic compounds containing one or more hetero atoms in a three- to ten-membered ring are named by combining the appropriate prefix or prefixes from Table I (eliding "a" where necessary) with a stem from Table II. The state of hydrogenation is indicated either in the stem, as shown in Table II, or by the prefixes "dihydro-," "tetrahydro-," *etc.*, according to Rule **B-1.2.**

TABLE I

Element	Valence	Prefix
Oxygen	II	Oxa
Sulfur	II	Thia
Selenium	II	Selena
Tellurium	II	Tellura
Nitrogen	III	Aza
Phosphorus	III	Phospha[a]
Arsenic	III	Arsa[a]
Antimony	III	Stiba[a]
Bismuth	III	Bisma
Silicon	IV	Sila
Germanium	IV	Germa
Tin	IV	Stanna
Lead	IV	Plumba
Mercury	II	Mercura

[a] When immediately followed by "in-" or "-ine," "phospha-" should be replaced by "phosphor-," "arsa-" should be replaced by "arsen-" and "stiba" should be replaced by "antimon-".

"Bora" for boron is used frequently. *Chemical Abstracts* has placed boron with the valence of III between lead and mercury in Table I. This table shows the valence of tin and lead as IV instead of II (a correction of a typographical error).

TABLE II

No. of members in the ring	Rings containing nitrogen		Rings containing no nitrogen	
	Unsaturation[a]	Saturation	Unsaturation[a]	Saturation
3	-irine	-iridine	-irene	irane[e]
4	-ete	-etidine	-ete	-etane
5	-ole	-olidine	-ole	-olane
6	-ine[b]	[c]	-ine[b]	-ane[d]
7	-epine	[c]	-epine	-epane
8	-ocine	[c]	-ocin	-ocane
9	-onine	[c]	-onin	-onane
10	-ecine	[c]	-ecin	-ecane

[a] Corresponding to the maximum number of non-cumulative double bonds, the hetero elements having the normal valences shown in Table I. [b] For phosphorus, arsenic, antimony, see the special provisions of Table I. [c] Expressed by prefixing "perhydro" to the name of the corresponding unsaturated compound. [d] Not applicable to silicon, germanium, tin and lead. In this case, "perhydro-" is prefixed to the name of the corresponding unsaturated compound. [e] The syllables denoting the size of rings containing 3, 4 or 7–10 members are derived as follows: "ir" from *tri*-, "et" from *tet*ra, "ep" from *hep*ta, "oc" from *oc*ta, "on" from *non*a, and "ec" from *dec*a.

Examples:

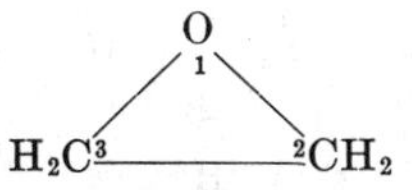

Oxirane

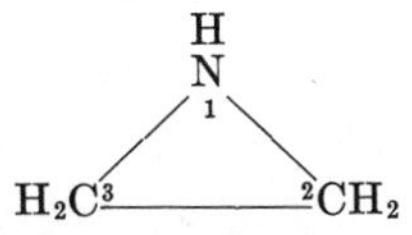

Aziridine

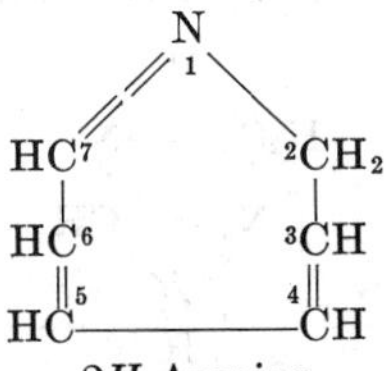

2*H*-Azepine

1.2.—Heterocyclic systems whose unsaturation is less than the one corresponding to the maximum number of noncumulative double bonds are named by using the prefixes "dihydro-," "tetrahydro-," etc.

In the case of 4- and 5-membered rings, a special termination is used for the structures containing one double bond, when there can be more than one noncumulative double bond.

No. of members of the partly saturated rings	Rings containing nitrogen	Rings containing no nitrogen
4	-etine	-etene
5	-oline	-olene

Examples:

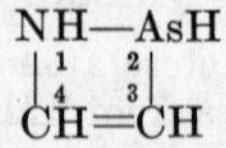

1,2-Azarset-3-ine

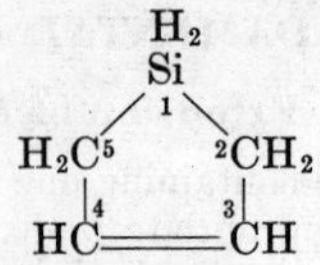

3-Silolene

1.3.—Multiplicity of the same hetero atom is indicated by a prefix "di," "tri-," etc., placed before the appropriate "a" term (Table I).

Example:

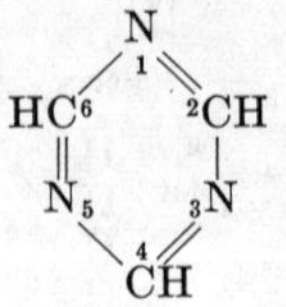

1,3,5-Triazine (or *s*-Triazine)

1.4.—If two or more kinds of "a" terms occur in the same name, their order of citation is by descending group number of the Periodic Table and increasing atomic number in the group as illustrated by the sequence in Table I.

Examples:

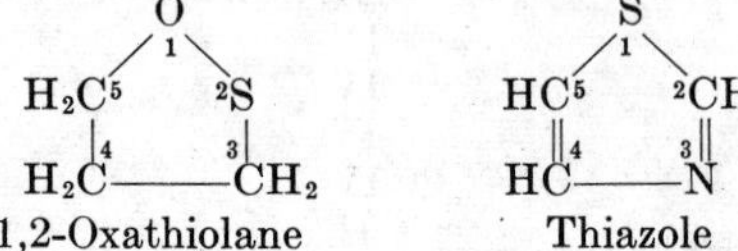

1,2-Oxathiolane

Thiazole

1.51.—The position of a single hetero atom determines the numbering in a monocyclic compound.

Example:

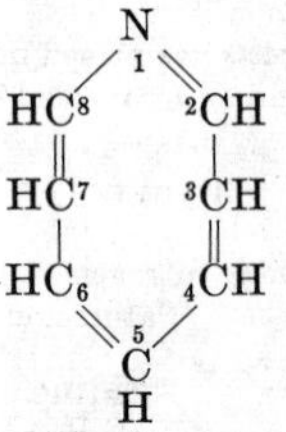

Azocine

1.52.—When the same hetero atom occurs more than once in a ring, the numbering is chosen to give the lowest locants to the hetero atoms.

Example:

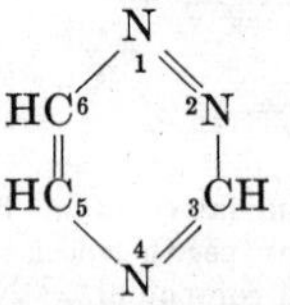

1,2,4-Triazine (or *as*-Triazine)

1.53.—When hetero atoms of different kinds are present, the locant 1 is given to a hetero atom which is as high as possible in Table I. The numbering is then chosen to give the lowest locants to the hetero atoms.

Examples:

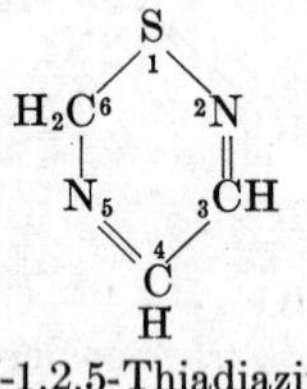

6*H*-1,2,5-Thiadiazine
(not 2,1,4-Thiadiazine)
(not 1,3,6-Thiadiazine)

The numbering must begin with the sulfur atom. This condition eliminates 2,1,4-thiadiazine. Then the nitrogen atoms receive the lowest possible locant, which eliminates 1,3,6-thiadiazine.

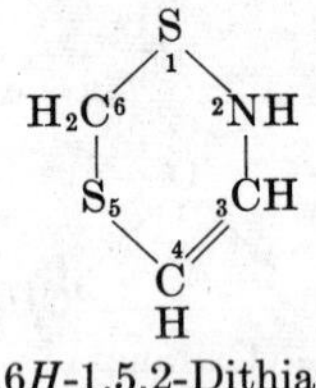

2*H*,6*H*-1,5,2-Dithiazine
(not 1,3,4-Dithiazine)
(not 1,3,6-Dithiazine)
(not 1,5,4-Dithiazine)

The numbering must begin with a sulfur atom. The choice of this atom is determined by the set of locants which can be attributed to the remaining hetero atoms of any kind. As the set 1,2,5 is lower than 1,3,4 or 1,3,6 or 1,5,4 in the usual sense, the name is 1,5,2-dithiazine.

B-2. Trivial and Semi-trivial Names

2.11.—The following trivial and semi-trivial names constitute a partial list of such names which are retained for the compound and as a basis of fusion names. They are arranged in the inverse order of the precedence prescribed in Rule **B-3.** The names of the radicals shown are formed according to Rule **B-5.**

	Parent Compound	Radical Name
(1)	Thiophene	Thienyl (2-shown)
(2)	Benzo[*b*]thiophene (replacing thianaphthene)	Benzo[*b*]thienyl (2-shown)
(3)	Naphtho[2,3-*b*]thiophene (replacing thiophanthrene)	Naphtho[2,3-*b*] thienyl (2-shown)
(4)	Thianthrene	Thianthrenyl (2-shown)
(5)	Furan	Furyl (3-shown)
(6)	Pyran (2*H*-shown)	Pyranyl (2*H*-Pyran-3-yl shown)
(7)	Isobenzofuran	Isobenzofuranyl (1-shown)
(8)	Chromene (2*H*- shown)	Chromenyl (2*H*-Chromen-3-yl shown)
(9)	Xanthene[1]	Xanthenyl[1] (2- shown)
(10)	Phenoxathiin	Phenoxathiinyl (2- shown)
(11)	2*H*-Pyrrole	2*H*-Pyrrolyl (2*H*-Pyrrol-3-yl shown)
(12)	Pyrrole	Pyrrolyl (3- shown)
(13)	Imidazole	Imidazolyl (2- shown)
(14)	Pyrazole	Pyrazolyl (1- shown)
(15)	Pyridine	Pyridyl (3- shown)
(16)	Pyrazine	Pyrazinyl
(17)	Pyrimidine	Pyrimidinyl (2-shown)
(18)	Pyridazine	Pyridazinyl (3- shown)
(19)	Indolizine	Indolizinyl (2-shown)
(20)	Isoindole	Isoindolyl (2- shown)
(21)	3*H*-Indole	3*H*-Indolyl (3*H*-Indol-2-yl shown)

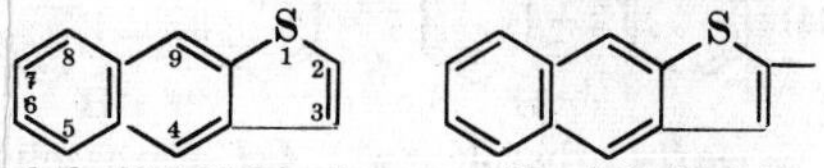

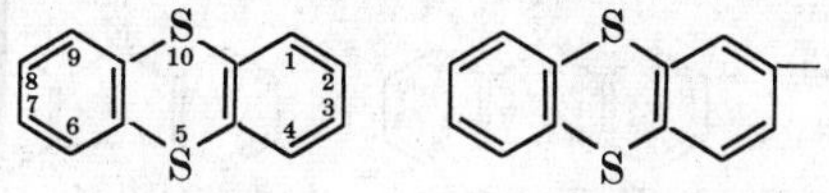

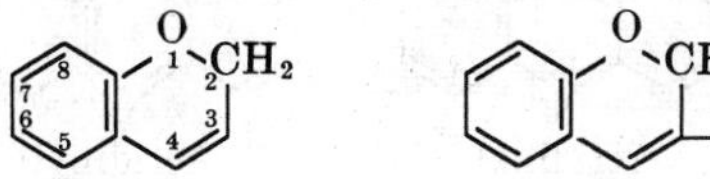

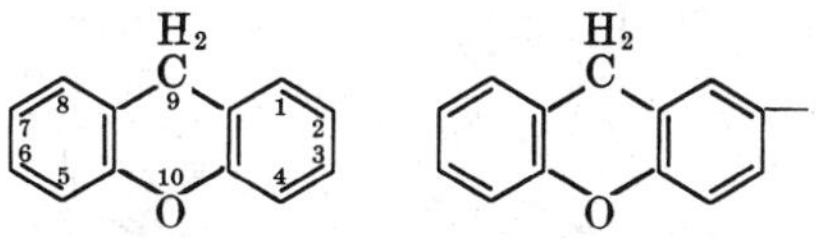

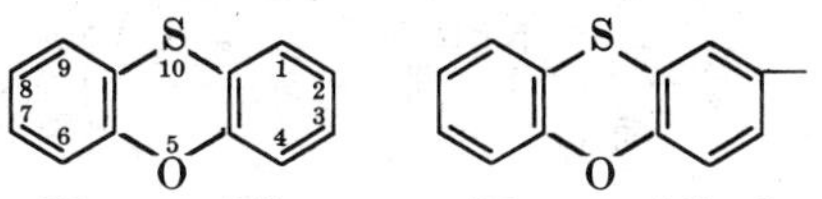

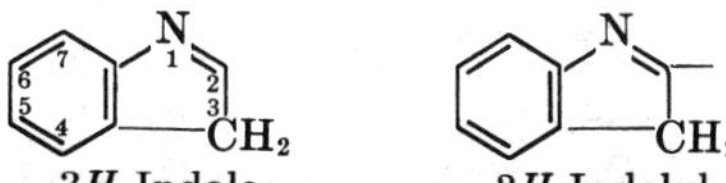

(1) Denotes exceptions to systematic numbering.

	Parent Compound	Radical Name
(22)	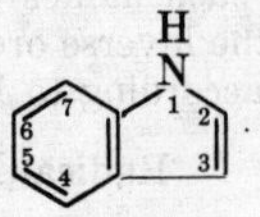 Indole	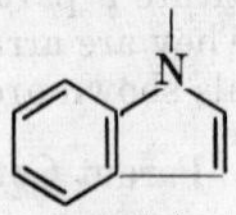 Indolyl (1- shown)
(23)	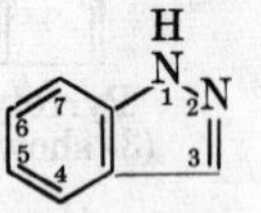 1*H*-Indazole	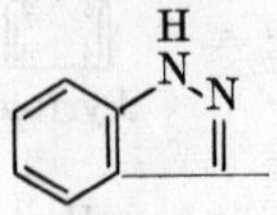 Indazolyl (1*H*-Indazol-3-yl shown)
(24)	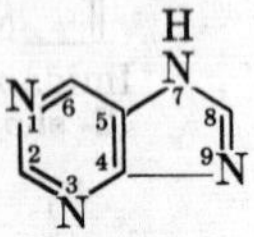 Purine[1]	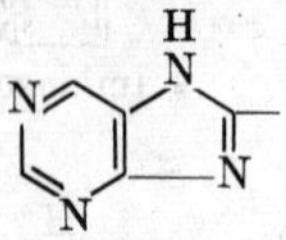 Purinyl[1] (8- shown)
(25)	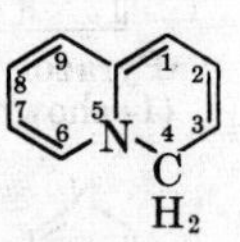 4*H*-Quinolizine	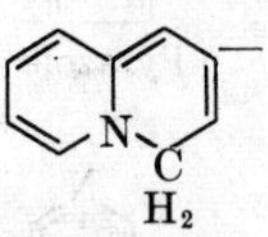 4*H*-Quinolizinyl (4*H*-Quinolizin-2-yl shown)
(26)	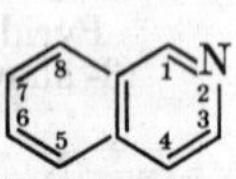 Isoquinoline	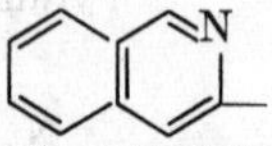 Isoquinolyl (3- shown)
(27)	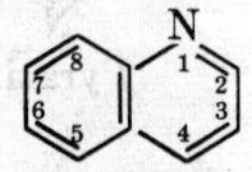 Quinoline	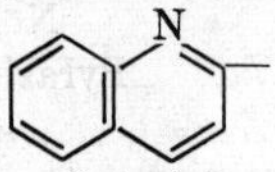 Quinolyl (2- shown)
(28)	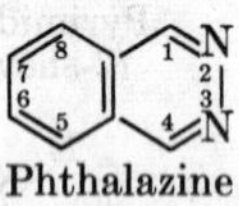 Phthalazine	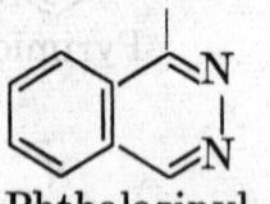 Phthalazinyl (1- shown)
(29)	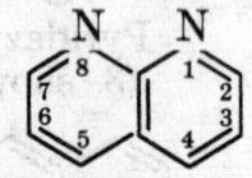 Naphthyridine (1,8- shown)	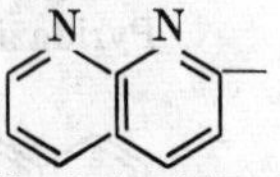 Naphthyridinyl (1,8-Naphthyridin-2-yl shown)
(30)	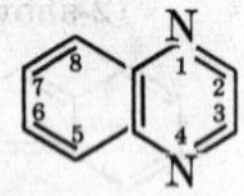 Quinoxaline	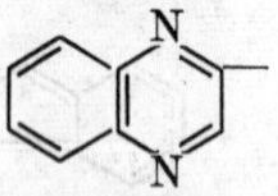 Quinoxalinyl (2- shown)
(31)	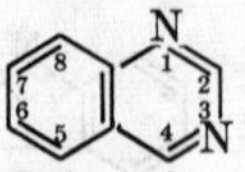 Quinazoline	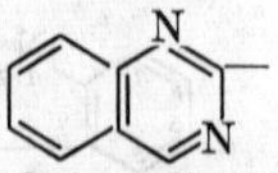 Quinazolinyl (2- shown)
(32)	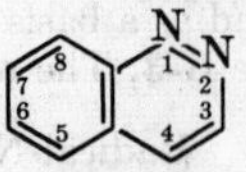 Cinnoline	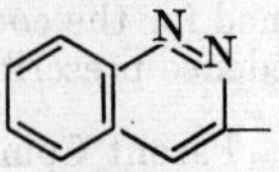 Cinnolinyl (3- shown)
(33)	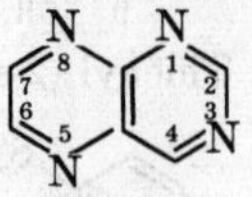 Pteridine	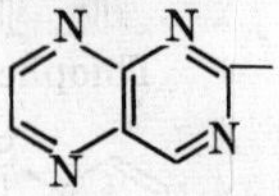 Pteridinyl (2- shown)
(34)	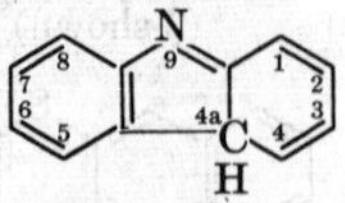 4a*H*-Carbazole[1]	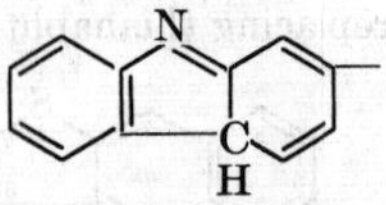 4a*H*-Carbazolyl[1] (4a*H*-Carbazol-2-yl shown)
(35)	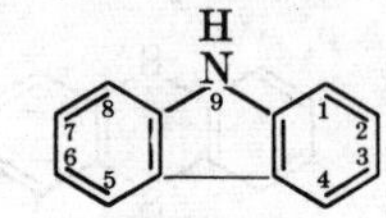 Carbazole[1]	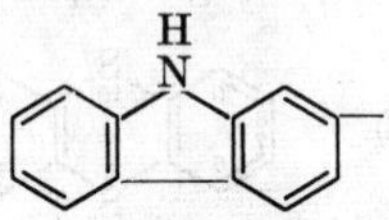 Carbazolyl[1] (2- shown)
(36)	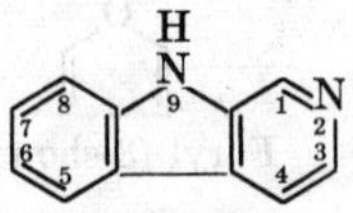 β-Carboline	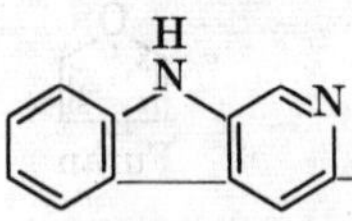 β-Carbolinyl (β-Carbolin-3-yl shown)
(37)	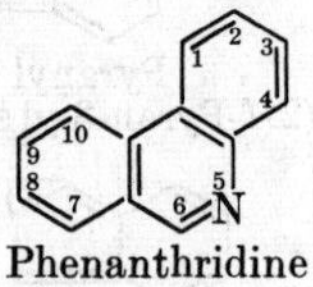 Phenanthridine	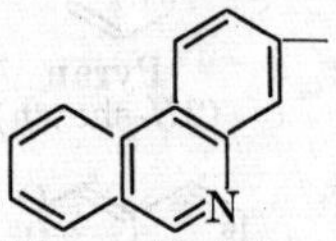 Phenanthridinyl (3- shown)
(38)	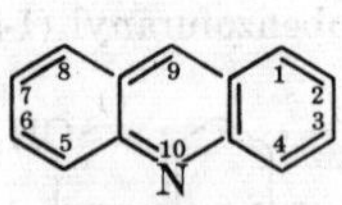 Acridine[1]	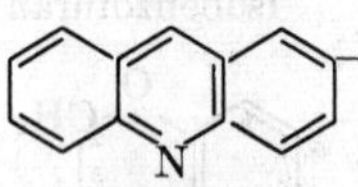 Acridinyl[1] (2- shown)
(39)	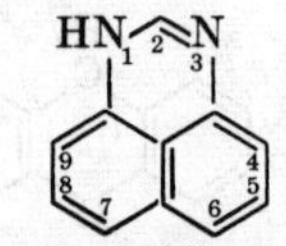 Perimidine	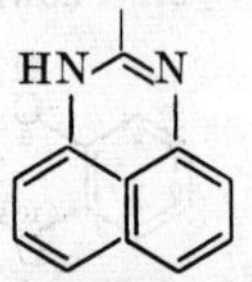 Perimidinyl (2- shown)
(40)	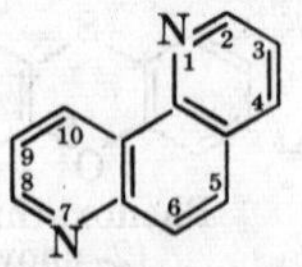 Phenanthroline (1,7- shown)	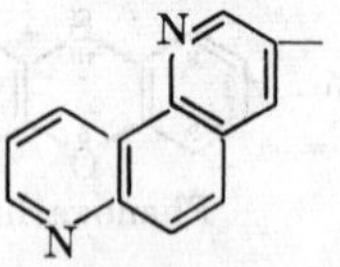 Phenanthrolinyl (1,7-Phenanthrolin-3-yl shown)

	Parent Compound	Radical Name
(41)	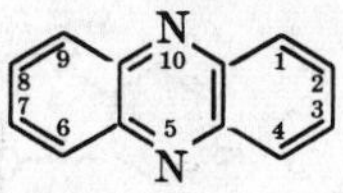 Phenazine	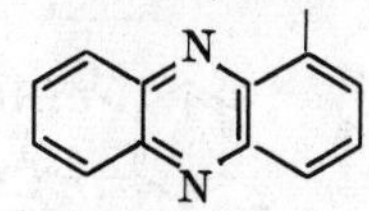 Phenazinyl (1- shown)
(42)	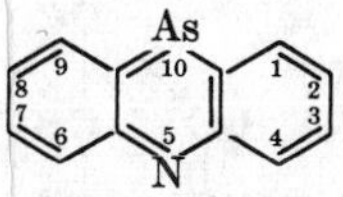 Phenarsazine	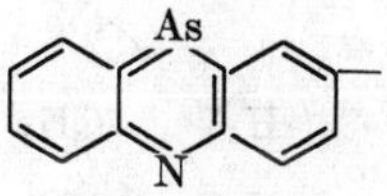 Phenarsazinyl (2- shown)
(43)	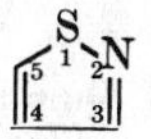 Isothiazole	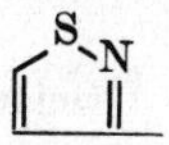 Isothiazolyl (3- shown)
(44)	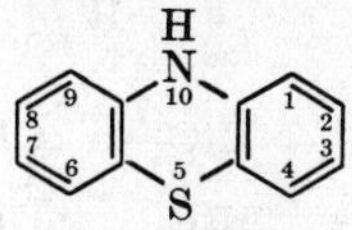 Phenothiazine	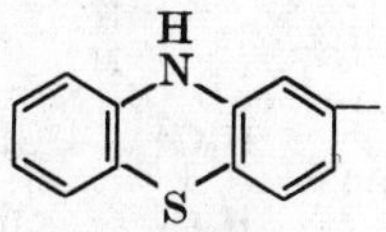 Phenothiazinyl (2- shown)
(45)	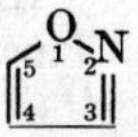 Isoxazole	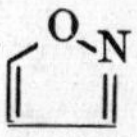 Isoxazolyl (3- shown)
(46)	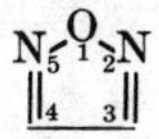 Furazan	 Furazanyl (3- shown)
(47)	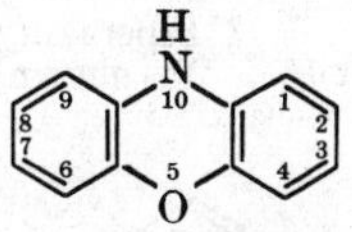 Phenoxazine	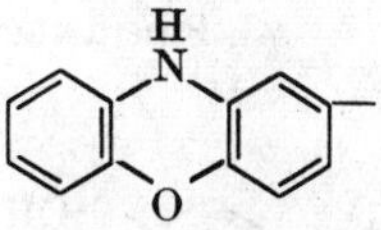 Phenoxazinyl (2- shown)

Chemical Abstracts uses 2*H*-1-benzopyran instead of chromene, and 9*H*-pyrido[3,4-*b*]indole instead of β-carboline.

B-2.12.—The following trivial and semi-trivial names are retained but are not recommended for use in fusion names. The names of the radicals shown are formed according to Rule **B-5.**

	Parent Compound	Radical Name
(1)	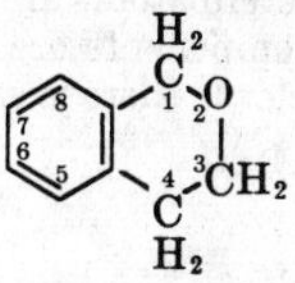 Isochroman	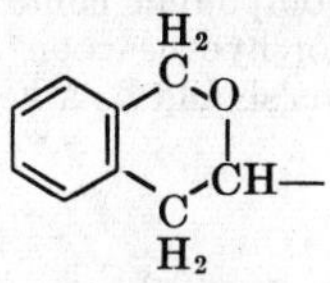 Isochromanyl (3- shown)
(2)	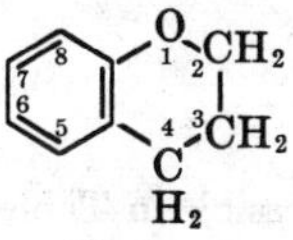 Chroman	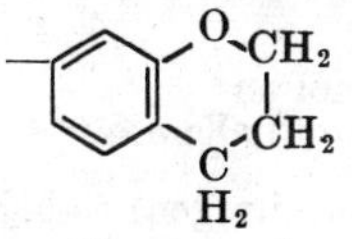 Chromanyl (7- shown)
(3)	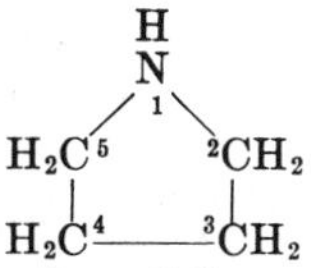 Pyrrolidine	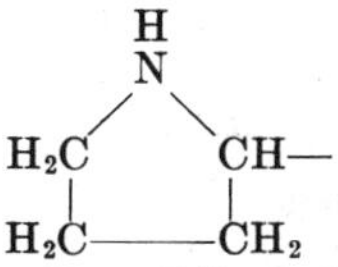 Pyrrolidinyl (2- shown)
(4)	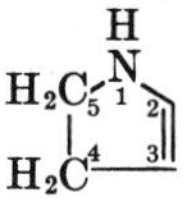 Pyrroline (2- shown[1])	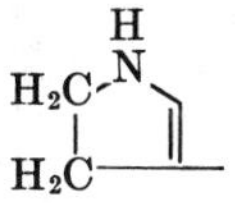 Pyrrolinyl (2-Pyrrolin-3-yl[1] shown)
(5)	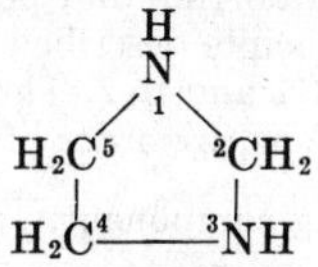 Imidazolidine	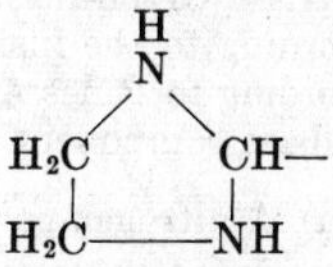 Imidazolidinyl (2- shown)
(6)	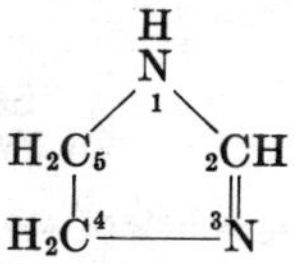 Imidazoline (2- shown[1])	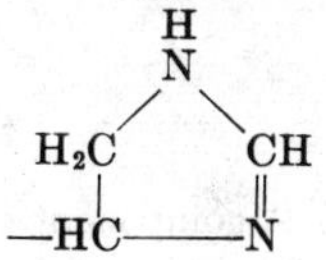 Imidazolinyl (2-Imidazolin-4-yl[1] shown)
(7)	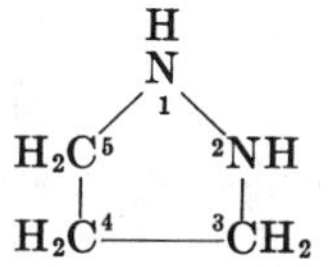 Pyrazolidine	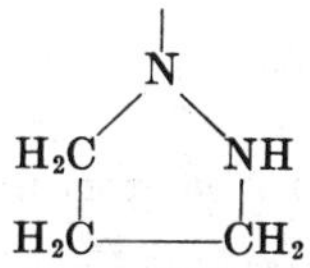 Pyrazolidinyl (1- shown)
(8)	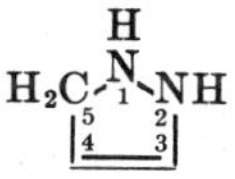 Pyrazoline (3- shown[2])	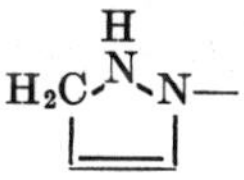 Pyrazolinyl (3-Pyrazolin-2-yl[2] shown)

(1) The "2-" denotes the position of the double bond.
(2) The "3-" denotes the position of the double bond.
(3) For 1-piperidyl use piperidino.

	Parent Compound	Radical Name
(9)	Piperidine	Piperidyl[3] (2- shown)
(10)	Piperazine	Piperazinyl (1- shown)
(11)	Indoline	Indolinyl (1- shown)
(12)	Isoindoline	Isoindolinyl (1- shown)
(13)	Quinuclidine	Quinuclidinyl (2- shown)
(14)	Morpholine	Morpholinyl[1] (3- shown)

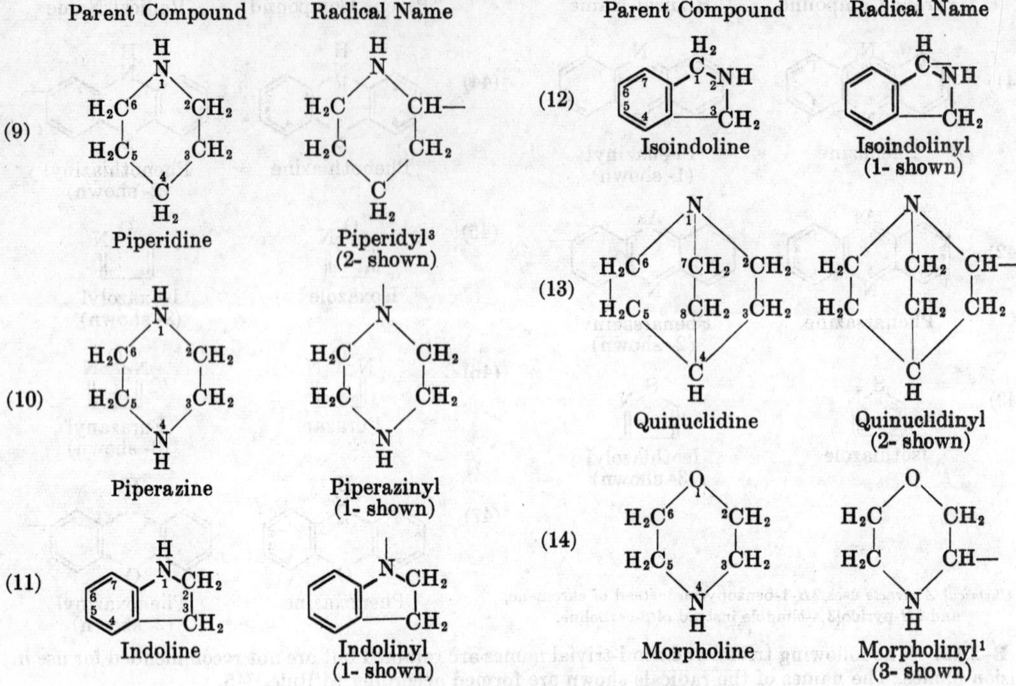

B-3. Fused Heterocyclic Systems

3.1.—"Ortho-fused" and "ortho- and peri-fused" ring compounds containing hetero atoms are named according to the fusion principle described in Rule **A-21** for hydrocarbons. The components are named according to Rules **A-21, B-1** and **B-2.** The base component should be a heterocycle. If there is a choice, the base component should be, by order of preference:

(a) A nitrogen-containing component.

Example:

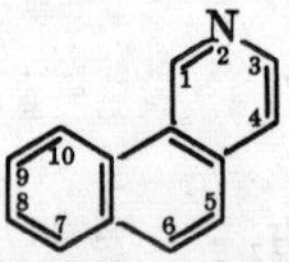

Benzo[*h*]isoquinoline
not Pyrido[3,4-*a*]naphthalene

(b) A component containing a hetero atom (other than nitrogen) as high as possible in Table I.

Example:

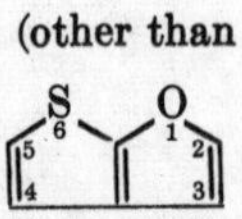

Thieno[2,3-*b*]furan
not Furo[2,3-*b*]thiophene

(c) A component containing the greatest number of rings.

Example:

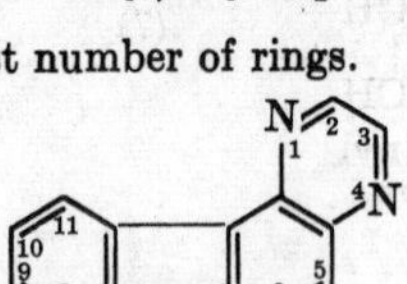

7*H*-Pyrazino[2,3-*c*]carbazole
not 7*H*-Indolo[3,2-*f*]quinoxaline

(d) A component containing the largest possible individual ring.

Example:

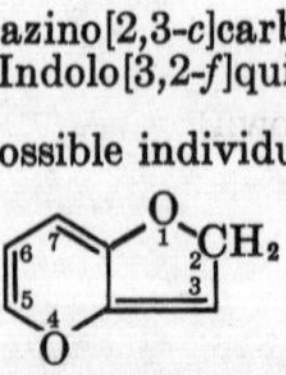

2*H*-Furo[3,2-*b*]pyran
not 2*H*-Pyrano[3,2-*b*]furan

(1) For 4-morpholinyl use morpholino.

(e) A component containing the greatest number of hetero atoms of any kind.

Example:

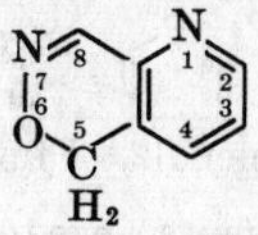

5*H*-Pyrido[2,3-*d*][1,2]oxazine
not [1,2]Oxazino[4,5-*b*]pyridine

(f) A component containing the greatest variety of hetero atoms.

Examples:

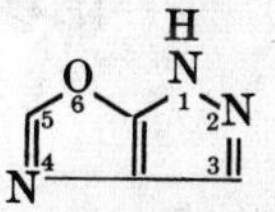

1*H*-Pyrazolo[4,3-*d*]oxazole not 1*H*-Oxazolo[5,4-*c*]-pyrazole

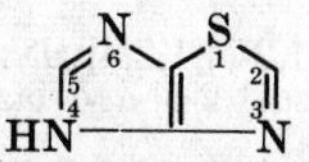

4*H*-Imidazo[4,5-*d*]thiazole not 4*H*-Thiazolo[4,5-*d*]-imidazole

(g) A component containing the greatest number of hetero atoms first listed in Table I.

Example:

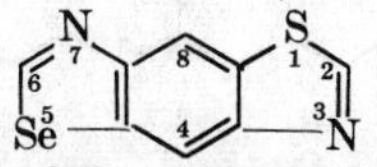

Selenazolo[5,4-*f*]benzothiazole[1]
not Thiazolo[5,4-*f*]benzoselenazole

(1) In this example the hetero atom first listed in Table I is sulfur and the greatest number of sulfur atoms in a ring is one.

(h) If there is a choice between components of the same size containing the same number and kind of hetero atoms choose as the base component that one with the lower numbers for the hetero atoms before fusion.

Example:

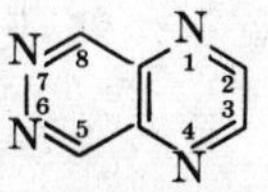

Pyrazino[2,3-*d*]pyridazine

3.2.—If a position of fusion is occupied by a hetero atom, the names of the component rings to be fused are so chosen as both to contain the hetero atom.

Example:

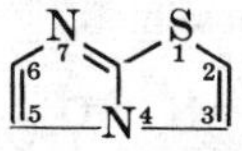

Imidazo[2,1-*b*]thiazole

3.3.—These contracted fusion prefixes may be used: furo, imidazo, isoquino, pyrido, pyrimido, quino and thieno.

Examples:

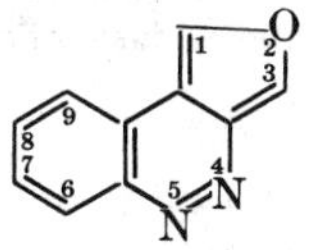

Furo[3,4-*c*]cinnoline

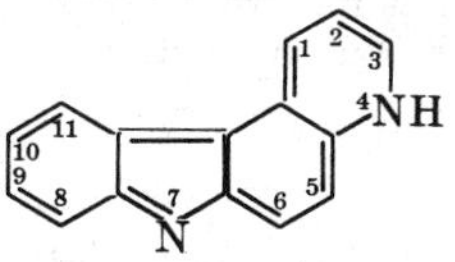

4*H*-Pyrido[2,3-*c*]carbazole

3.4.—In peripheral numbering of the complete fused systems, the ring system is oriented and numbered according to the principles of Rule **A-22.** When there is a choice of orientations, it is made in the following sequence in order to:

(a) Give low numbers to hetero atoms.

Examples:

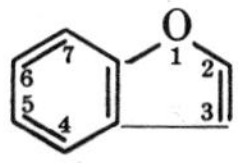

Benzo[*b*]furan

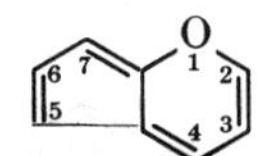

Cyclopenta[*b*]pyran

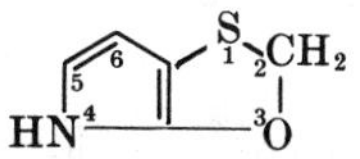

4*H*-[1,3]Oxathiolo-[5,4-*b*]pyrrole (N.B. 1,3,4 lower than 1,3,6)

(b) Give low numbers to hetero atoms in order of Table I.

Example:

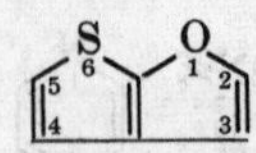

Thieno[2,3-*b*]furan

(c) Allow carbon atoms common to two or more rings to follow the lowest possible numbers (see Rules **A-22.2** and **A-22.3**). [A hetero atom common to the two rings is numbered according to Rule **B-3.4**(e).]

Examples:

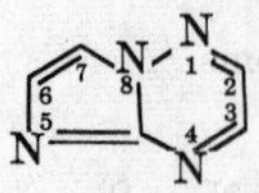

not 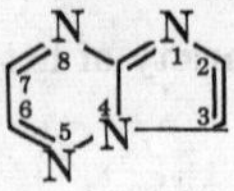 or

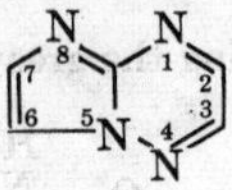

Imidazo[1,2-*b*][1,2,4]triazine
(or Imidazo[1,2-*b*]-*as*-triazine)

In a compound name for a fusion prefix (*i.e.*, when more than one pair of brackets is required), the points of fusion in the compound prefix are indicated by the use of unprimed and primed numbers, the unprimed numbers being assigned to the ring attached directly to the base component, thus

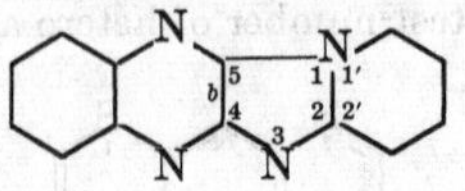

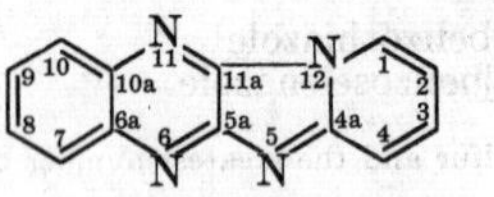

not

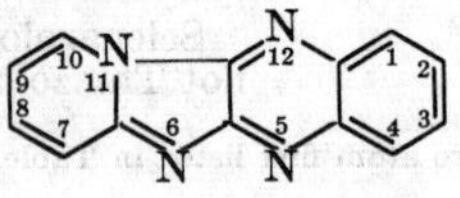

Pyrido[1′,2′:1,2]imidazo-
[4,5-*b*]quinoxaline

or

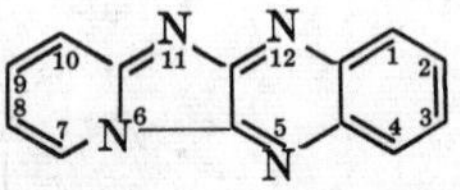

or

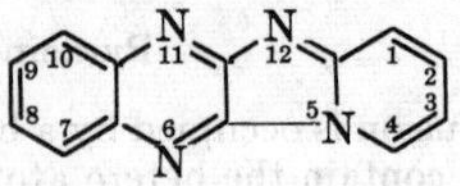

(d) Give hydrogen atoms lowest numbers possible.

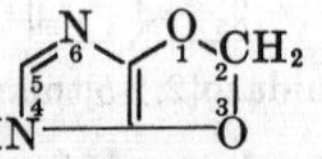

4*H*-1,3-Dioxolo[4,5-*d*]imidazole

(e) The ring is numbered as for hydrocarbons but numbers are given to all hetero atoms even when common to two or more rings. Interior hetero atoms are numbered last following the shortest path from the highest previous number.

B-4. "a" Nomenclature

4.1.—Names of heterocyclic compounds also may be formed by prefixing "a" terms (see Table I of Rule **B-1.1**) to the name of the corresponding homocyclic compound. The letter "a" should not be elided. There are two methods of applying this principle:

4.1(a). Stelzner Method.—In this method, the "a" term name relates to that of the hydrocarbon with the same distribution of bonds in the rings. Thus, I is not so related to benzene but to 1,4-cyclohexadiene, and II is not so related to naphthalene but to 1,4-dihydronaphthalene.

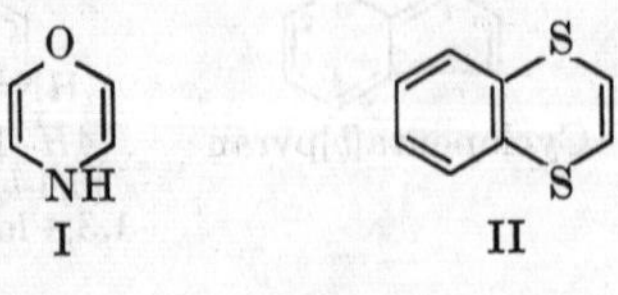

I II

4.1(b). Chemical Abstracts Method.—If the corresponding homocyclic compound is partially or completely hydrogenated and if this state of hydrogenation is denoted in its name without the use of hydro prefixes, as indan and cyclohexane, the procedure is the same as in (a). In other cases, positions in the skeleton of the corresponding homocyclic compound which are occupied by hetero atoms are denoted by the "a" terms, and the parent heterocyclic compound is considered to be that which contains the maximum number of conjugated or isolated[1] double bonds; hydrogen is added, as necessary, as hydro prefixes and/or as *H* to the "a" name thus obtained.

Examples:

Stelzner Method	*Chemical Abstracts* Method
Sila-2,4-cyclopentadiene	Sila-2,4-cyclopentadiene
Sila-1,3-cyclopentadiene	Sila-1,3-cyclopentadiene
Silabenzene	Silabenzene
7-Azabicyclo[2.2.1]heptane	7-Azabicyclo[2.2.1]heptane
1,3-Dithia-1,2,3,4-tetrahydronaphthalene	4*H*-1,3-Dithianaphthalene
1,4-Dithia-1,4-dihydronaphthalene	1,4-Dithianaphthalene

(1) Isolated double bonds are those which are neither conjugated nor cumulative as in

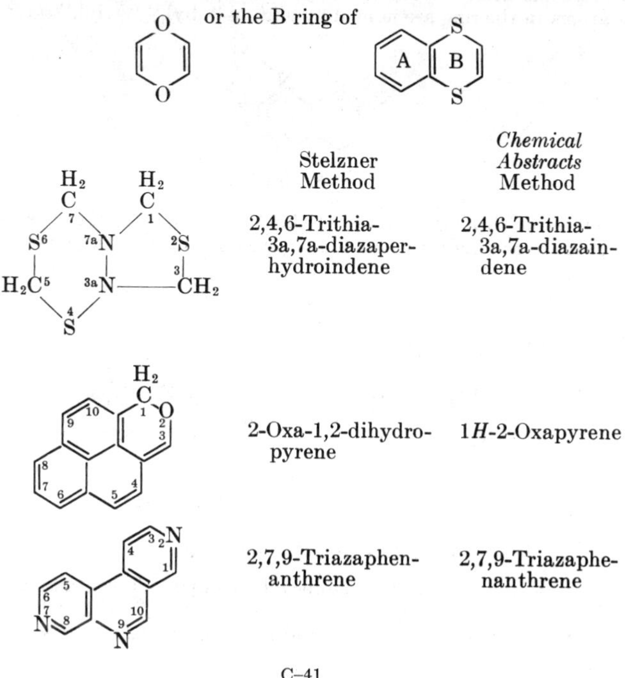

Stelzner Method	*Chemical Abstracts* Method
2,4,6-Trithia-3a,7a-diazaperhydroindene	2,4,6-Trithia-3a,7a-diazaindene
2-Oxa-1,2-dihydropyrene	1*H*-2-Oxapyrene
2,7,9-Triazaphenanthrene	2,7,9-Triazaphenanthrene

4.2.—In fusion names, the "a" terms precede the completed name of the parent hydrocarbon. If two or more kinds of "a" terms occur in the same name, the procedure described in Rule **B-1.4** applies. Prefixes denoting ordinary substitution procede the "a" terms.

Example:

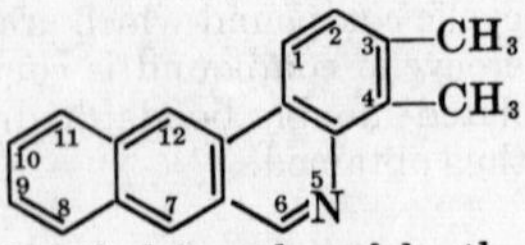

3,4-Dimethyl-5-azabenz[*a*]anthracene

Chemical Abstracts follows Rule **B-4.1** (b). Since "a" names for heterocyclic systems are based on the name of the carbocyclic system, the numbering of the "a" name must agree with the numbering of the carbocyclic parent. When there is a choice in direction of numbering the ring is numbered so as to give (1) lowest numbers to the hetero atoms, (2) lowest numbers to hetero atoms as high as possible in Table 1 (Rule **B-1.1**).

B-5. Radicals

5.11.—Univalent radicals derived from heterocyclic compounds by removal of hydrogen from a ring are in principle named by adding "yl" to the names of the parent compounds (with elision of final "e" if present).

Examples:

Indolyl from indole
Pyrrolinyl from pyrroline
Triazolyl from triazole
Triazinyl from triazine

(For further examples see Rule B-2.11).

These exceptions are retained: furyl, pyridyl, piperidyl, quinolyl, isoquinolyl and thienyl (from thiophene) (see also Rule **B-2.12**).

As exceptions, the names "piperidino" and "morpholino" are preferred to "1-piperidyl" and "4-morpholinyl."

5.12.—Bivalent radicals derived from univalent heterocyclic radicals whose names end in "-yl" by removal of one hydrogen atom from the atom with the free valence are named by adding "-idene" to to the name of the corresponding univalent radical.

Example:

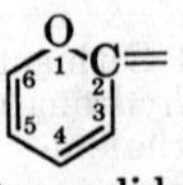

2-Pyranylidene

5.13.—Multivalent radicals derived from heterocyclic compounds by removal of two or more hydrogen atoms from different atoms in the ring are named by adding "-diyl," "-triyl," *etc.*, to the name of the ring system.

Example:

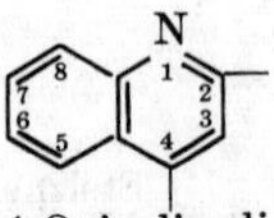

2,4-Quinolinediyl

5.21.—The use of "a" terms (Rule **B-4**) does not affect the formation of radical names. Such names are strictly analogous to those of the hydrocarbon analogs except that the "a" terms establish numbering in whole or in part.

Examples:

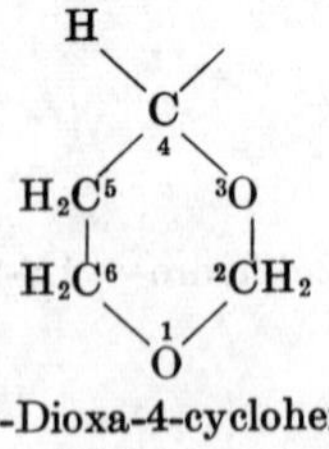

1,3-Dioxa-4-cyclohexyl 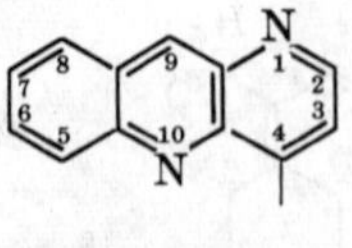   1,10-Diaza-4-anthryl

B-6. Cationic Hetero Atoms

6.1.—According to the "a" nomenclature, heterocyclic compounds containing cationic hetero atoms are named in conformity with the preceding rules by replacing "oxa-," "thia-," "aza-," *etc.*, by "oxonia-," "thionia-," "azonia-," *etc.*, the anion being designated in the usual way.

Examples:

Cl^- 1-Oxoniaanthracene chloride

Cl^- 4a-Azoniaanthracene chloride

Cl^- 1-Thioniabicyclo[2.2.1]heptane chloride

Cl^- 1-Methyl-1-oxoniacyclohexane chloride

HETEROCYCLIC SPIRO COMPOUNDS

B-10 (Alternate to B-11)

10.1.—Heterocyclic spiro compounds containing single-ring units only may be named by prefixing "a" terms (see Table I, Rule **B-1.1**) to the names of the spiro hydrocarbons formed according to Rules **A-41.1, A-41.2, A-41.3** and **A-41.6.** The numbering of the spiro hydrocarbon is retained and the hetero atoms in the order of Table I are given as low numbers as are consistent with the fixed numbering of the ring. When there is a choice, hetero atoms are given lower numbers than double bonds.

Examples:

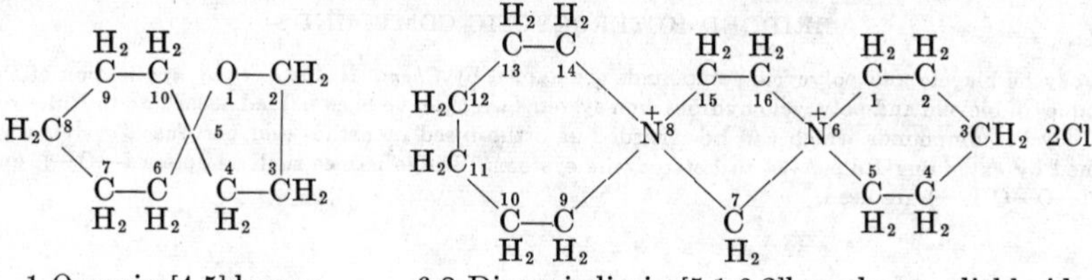

1-Oxaspiro[4.5]decane   6,8-Diazoniadispiro[5.1.6.2]hexadecane dichloride

10.2.—If at least one component of a mono- or polyspiro compound is a fused polycyclic system, the spiro compound is named according to Rule **A-41.4** or **A-41.7,** giving the spiro atom as low a number as possible consistent with the fixed numberings of the component systems.

Examples:

3,3′-Spirobi[3*H*-indole]   Spiro[piperidine-4,9′-xanthene]

Chemical Abstracts names heterocyclic spiro compounds according to Rules **B-10.1** and **B-10.2.**

B-11 (Alternate to B-10)

11.1.—Heterocyclic spiro compounds are named according to Rule **A-42,** the following criteria being applied where necessary: (a) spiro atoms have numbers as low as consistent with the numbering of the individual component systems; (b) heterocyclic components have priority over homocyclic components of the same size; (c) priority of heterocyclic components is decided according to Rule **B-3.** Parentheses are used where necessary for clarity in complex expressions.

Examples:

Cyclohexanespiro-2′-(tetrahydrofuran)

Tetrahydropyran-2-spirocyclohexane

3,3′-Spirobi(3*H*-indole)

1,2,3,4-Tetrahydroquinoline-4-spiro-4′-piperidine

Hexahydroazepinium-1-spiro-1′-imidazolidine-3′-spiro-1″-piperidinium dibromide

Chemical Abstract follows Rule **B-10.**

BRIDGED HETEROCYCLIC COMPOUNDS

Bridged heterocyclic bicyclo and polycyclo compounds are named by *Chemical Abstracts* by application of the "a" nomenclature to the names of bicyclo and polycyclo hydrocarbon systems which have been named according to Rules **A-31** and **A-32.**

Bridged heterocyclic compounds which can be regarded as ortho-fused or ortho- and peri-fused systems with additional bridges are named by extending Rule **A-34** to heterocyclic systems. Bridge names such as epoxy (—O—), epithio (—S—), epoxymethano (—O—CH_2—) are used.

INDEX

PREFIX NAMES OF ORGANIC RADICALS

This list is taken, by permission, from the Introduction to the Subject Index of *Chemical Abstracts*, Volumes 39 and 41.

acenaphthenyl $C_{12}H_9$— (*from acenaphthene*)
acetamido CH_3CONH—
acetenyl = ethynyl
acetimido $CH_3C(:NH)$—
acetonyl CH_3COCH_2—
acetonylidene $CH_3COCH=$
acetoxy CH_3COO—
acetyl CH_3CO—
acetylene $=CHCH=$
acridyl $C_{13}H_8N$— (*from acridine*)
acrylyl $CH_2:CHCO$—
adipyl —$OC(CH_2)_4CO$—
alanyl CH_3CHNH_2CO—
alkoxy RO— (*R = any alkyl radical*)
allyl $CH_2:CHCH_2$—
β-allyl = isopropenyl
amino H_2N—(*properly, in acid groups only; by some used synonymously with amino*)
amidoxalyl = oxamyl
amino H_2N—
amoxy $CH_3(CH_2)_4O$—
amyl $CH_3(CH_2)_4$—

tert-amyl: CH_3CH_2—C(=)—$(CH_3)_2$ [structure: CH_3CH_2 and $(CH_3)_2$ on C—]

amylidene $CH_3(CH_2)_3CH=$
anilino C_6H_5NH—
anisal = anisylidene
anisoyl *p*-$CH_3OC_6H_4CO$—
anisyl = methoxyphenyl or methoxybenzyl
anisylidene p-$CH_3OC_6H_4CH=$
anthranoyl *o*-$H_2NC_6H_4CO$—
anthraquinonyl $C_{14}H_7O_2$— (*from anthraquinone, 2 isomers*)
anthryl $C_{14}H_9$— (*from anthracene, 3 isomers*)
anthrylene —$C_{14}H_8$— (*from anthracene, 14 isomers*)
antimono —Sb:Sb—
antipyroyl (*from antipyric acid*)

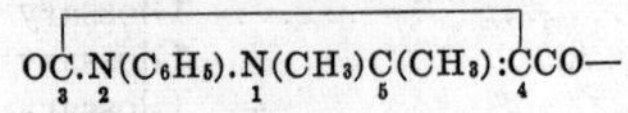

OC.N(C_6H_5).N(CH_3).C(CH_3):CCO— (ring positions 3 2 1 5 4)

antipyryl (*from antipyrine*)

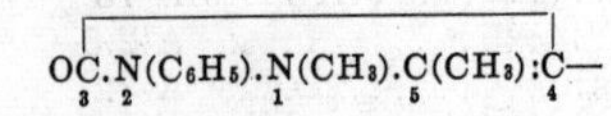

OC.N(C_6H_5).N(CH_3).C(CH_3):C— (ring positions 3 2 1 5 4)

arseno —As:As—
arsenoso O:As—
arsinico (*from arsinic acid*) $(HO)OAs=$
arsino H_2As—
arso O_2As—
arsono (*from arsonic acid*) $(HO)_2OAs$—
arsylene $HAs=$
asaryl 2,4,5-$(CH_3O)_3C_6H_2$—
asparagyl $H_2NCOCH_2CHNH_2CO$—
aspartyl —$COCH_2CHNH_2CO$—
auro Au—
azido N_3—
azino $=NN=$.
azo —N:N— (*as connective*)
azoxy —N(O)N—
benzal = benzylidene
benzamido C_6H_5CONH—
benzenyl = benzylidyne
benzidino $H_2NC_6H_4C_6H_4NH$— (*from benzidine*)
benziloyl $(C_6H_5)_2C(OH)CO$—
benzimidazolyl $C_7H_5N_2$— (*from benzimidazole*)
benzimido $C_6H_5C(:NH)$—
benzofuryl C_8H_5O— (*from benzofuran*)
benzhydryl $(C_6H_5)_2CH$—
benzhydrylidene $(C_6H_5)_2C=$
benzopyranyl C_9H_7O— (2-α-, etc.) (*from benzopyran*)
benzoxazolyl C_7H_4NO— (*from benzoxazole*)
benzoxy C_6H_5COO—
benzoyl C_6H_5CO—
benzoylene —C_6H_4CO—
benzyl $C_6H_5CH_2$—
benzylidene $C_6H_5CH=$
benzylidyne $C_6H_5C\equiv$
biphenylene —$C_6H_4C_6H_4$—
biphenylenebisazo —$N:NC_6H_4C_6H_4N:N$—
biphenylyl (2-, 3- or 4-) $C_6H_5C_6H_4$— (*from biphenyl*)
biphenylycarbonyl $C_6H_5C_6H_4CO$—
biphenylyloxy $C_6H_5C_6H_4O$—
bornyl (*from borneol*)

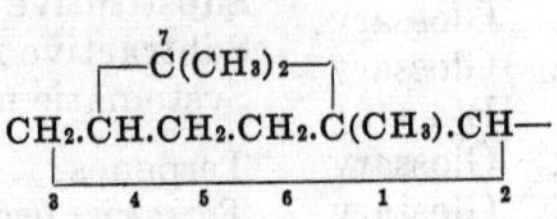

$CH_2.CH.CH_2.CH_2.C(CH_3).CH$— bridged by $C(CH_3)_2$ (positions 3 4 5 6 1 2; bridge 7)

boryl O:B—
bromo Br—
1-butenyl $CH_3CH_2CH:CH$—
2-butenyl $CH_3CH:CHCH_2$—
3-butenyl $CH_2:CH(CH_2)_2$—
butoxy $CH_3(CH_2)_3O$—
butyl $CH_3(CH_2)_3$

sec-butyl: CH_3CH_2 and CH_3 on CH—

tert-butyl $(CH_3)_3C$—
butylene (1,4) = tetramethylene
butylidene $CH_3(CH_2)_2CH=$
butyryl $CH_3(CH_2)_2CO$—
camphanyl $C_{10}H_{17}$— (*from camphane, 3 isomers*)
camphoroyl $C_{10}H_{14}O_2=$ (*from camphoric acid*)
camphoryl $C_{10}H_{15}O$— (*from camphor*)
camphorylidene $C_{10}H_{14}O=$ (*from camphor*)
caproyl $CH_3(CH_2)_4CO$—
capryl $CH_3(CH_2)_8CO$—
caprylyl $CH_3(CH_2)_6CO$—
carbamido = ureido
carbamyl H_2NCO—
carbanilino = phenylcarbamyl
carbazolyl $C_{12}H_8N$— (*from carbazole, 5 isomers*)
carbethoxy C_2H_5OOC—
carbomethoxy CH_3OOC—
carbonyl OC=
carbonyldioxy —OCOO—
carboxy HOOC—
carbyl —C—

carvacryl: $^4(CH_3)_2CH$ and $^1CH_3$ on C_6H_3—^{2}

cetyl = hexadecyl
chaulmoogroyl $C_5H_7(CH_2)_{12}CO$— (*from chaulmoogric acid*)
chaulmoogryl $C_5H_7(CH_2)_{12}CH_2$— (*from chaulmoogryl alcohol*)
chloro Cl—
chloromercuri ClHg—
chrysenyl $C_{18}H_{11}$— (*from chrysene*)
cinnamal = cinnamylidene
cinnamenyl = styryl
cinnamoyl $C_6H_5CH:CHCO$—
cinnamyl $C_6H_5CH:CHCH_2$—
cinnamylidene $C_6H_5CH:CHCH=$

cresotyl 2,3-$HO(CH_3)C_6H_3CO$— *(from cresotic acid)* (2,3-*shown*)
cresoxy = toloxy
cresyl = *ar*-hydroxytolyl = tolyl
cresylene = tolylene
crotonyl $CH_3CH{:}CHCO$—
crotyl = 2-butenyl
cumal = *p*-cuminylidene
cumidino $(CH_3)_2CHC_6H_4NH$—
cuminal = *p*-cuminylidene
cuminyl *p*-$(CH_3)_2CHC_6H_4CH_2$—
cuminylidene *p*-$(CH_3)_2CHC_6H_4CH$=
cumoyl *(from cumic acid)* *p*-$(CH_3)_2CHC_6H_4CO$
cumyl $(CH_3)_2CHC_6H_4$—
cyanato NCO—
cyano NC—
cyclobutyl *(from cyclobutane)* C_4H_7—
cyclohexadienyl (2,4- shown)

$CH_2.CH{:}CH.CH{:}CH.CH$—
6 5 4 3 2 1

cyclohexadienylidene (2,4- *or* 2,5-; 2,5- *shown*)

$CH{:}CH.CH_2CH{:}CH.C$=

cyclohexenyl C_6H_9— *(from cyclohexene. 3 isomers)*
cyclohexenylidene (2-shown)

$CH_2.CH_2.CH_2.CH{:}CH.C$=

cyclohexyl C_6H_{11}— *(from cyclohexane)*
cyclohexylidene

$CH_2.CH_2.CH_2.CH_2.CH_2.C$=

cyclopentenyl C_5H_7— *(from cyclopentene)*
cyclopentyl C_5H_9— *(from cyclopentane)*
cyclopropyl C_3H_5— *(from cyclopropane)*
cymyl $C_{10}H_{13}$— *(from cymene)*
2-*p*-cymyl = carvacryl
3-*p*-cymyl = thymyl
decyl $CH_3(CH_2)_9$—
desyl $C_6H_5COCH(C_6H_5)$—
diazo —N:N— or N:N=
diazoamino —N:NNH—
disilanoxy H_3SiSiH_2O—
disilanyl H_3SiSiH_2—
disilanylamino H_3SiSiH_2NH—
disilanylene —SiH_2SiH_2—
disilazanoxy $H_3SiNHSiH_2O$—
disilazanyl $H_3SiNHSiH_2$—
disilazanylamino $H_3SiNHSiH_2NH$—
disiloxanoxy $H_3SiOSiH_2O$—
disiloxanyl $H_3SiOSiH_2$—
disiloxanylamino $H_3SiOSiH_2NH$—
dodecyl $CH_3(CH_2)_{10}CH_2$—
duryl 2,3,5,6-$(CH_3)_4C_6H$—
durylene

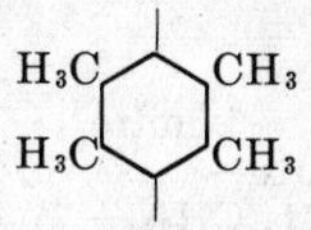

H_3C CH_3
H_3C CH_3

enanthyl $CH_3(CH_2)_5CO$—
epoxy —O— *(to different atoms already united in some other way)*
ethene = ethylene
ethenyl = ethylidyne = vinyl
ethinyl = ethynyl
ethoxalyl C_2H_5OOCCO—
ethoxy C_2H_5O—
ethyl CH_3CH_2—
ethylene —CH_2CH_2—
ethylenedioxy —$O(CH_2)_2O$—
ethylidene CH_3CH=
ethylidyne CH_3C≡
ethynyl CH:C—
ethynylene —C:C—
fenchyl $C_{10}H_{17}$— *(from fenchane)*
fluorenyl $C_{13}H_9$— *(from fluorene, 5 isomers)*
fluorenylidene $C_{13}H_8$= *(from fluorene)*
fluoro F—
formamido HCONH—
formazyl

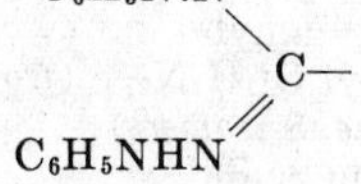

$C_6H_5N{:}N$
C—
C_6H_5NHN

formyl OCH—
furfural = furfurylidene
furfuryl $O.CH{:}CH.CH{:}CCH_2$—
furfurylidene

O.CH:CH.CH:CCH=

2-furoyl O.CH:CH.CH:CCO—
3-furoyl CH:CH.O.CH:CCO—
furyl (2 *isomers*, 2- *shown*)

O.CH:CH.CH:C—

furylidene

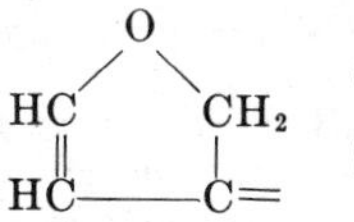

O
HC CH_2 3(2)- (also a 2(3)-form)
HC—C=

geranyl $C_{10}H_{17}$— *(from geraniol)*
glutamyl— —$OCCHNH_2(CH_2)_2CO$—
glutaryl $OC(CH_2)_2CO$—
glyceroyl $CH_2OHCHOHCO$—
glyceryl —CH_2CHCH_2—
glycolyl $HOCH_2CO$—
glycyl H_2NCH_2CO—
glyoxylyl OCHCO—
guaiacyl = *o*-methoxyphenyl
guanidino $H_2NC(:NH)NH$—
guanyl $H_2NC(:NH)$—
hendecyl $CH_3(CH_2)_{10}$—
heptyl $CH_3(CH_2)_6$—
hexadecyl $CH_3(CH_2)_{14}CH_2$—
hexyl $CH_3(CH_2)_5$—
hippuryl $C_6H_5CONHCH_2CO$—
homopiperonyl 3,4-$(CH_2O_2)C_6H_3CH_2CH_2$—
hydnocarpoyl $C_5H_7(CH_2)_{10}CO$— *(from hydnocarpic acid)*
hydnocarpyl $C_5H_7(CH_2)_{10}CH_2$— *(from hydnocarpyl alcohol)*
hydrazi

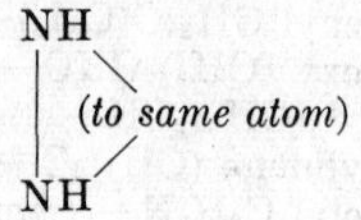

NH
(to same atom)
NH

hydrazino H_2NNH—
hydrazo —HNNH— *(to different atoms)*
hydrazono H_2NN=
hydrocinnamoyl $C_6H_5CH_2CH_2CO$—
hydroxamino HONH—
hydroximino = isonitroso
hydroxy (hydroxyl) HO—
-idene *added to any radical usually means a double bond at point of attachment*
imidazolidyl $C_3H_7N_2$— *(from imidazolidine)*
imidazolyl $C_3H_3N_2$— *(from imidazole, 4 isomers)*
imido NH= *(properly, in acid groups only); by some used synonymously with imino)*
imino NH=
indanyl C_9H_9— *(from indan, 4 isomers)*
indenyl C_9H_7— *(from indene, 7 isomers)*
indolyl C_8H_6N— *(from indole, 7 isomers)*
indolylidene

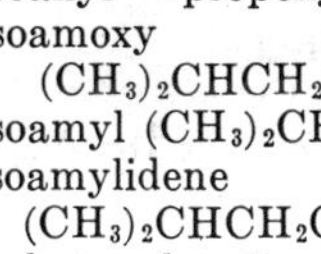

NH
C_6H_4 CH_2 3(2)-, etc.
C

indyl = indolyl
iodo I—
iodoso OI—
iodoxy O_2I—
isoallyl = propenyl
isoamoxy $(CH_3)_2CHCH_2CH_2O$—
isoamyl $(CH_3)_2CHCH_2CH_2$—
isoamylidene $(CH_3)_2CHCH_2CH$=
isobutenyl = 2-methylpropenyl
isobutoxy $(CH_3)_2CHCH_2O$—
isobutyl $(CH_3)_2CHCH_2$—
isobutyryl $(CH_3)_2CHCO$—
isocyanato OCN—
isocyano C:N—
isohexyl $(CH_3)_2CH(CH_2)_3$—

isoindolyl C_8H_6N— (*from isoindole, 4 isomers*)
isoleucyl $CH_3CH_2CH(CH_3)$-$CHNH_2CO$—
isonitro HOON=
isonitroso HON=
1-isopentenyl = 3-methyl-1-butenyl
isophthalal = isophthalylidene
isophthaloyl —OCC_6H_4CO— (*m*)
isophthalylidene =HCC_6H_4CH= (*m*)
isopropenyl CH_2:C(CH_3)—
isopropoxy $(CH_3)_2CHO$—
isopropyl $(CH_3)_2CH$—
isopropylidene $(CH_3)_2C$=
isoquinolyl C_9H_6N— (*from isoquinoline, 7 isomers*)
isothiocyano S:C:N—
isovaleryl $(CH_3)_2CHCH_2CO$—
isoxazolyl C_3H_2NO— (*from isoxazole, 5 isomers*)
keto = oxo
lauroyl (*from lauric acid*) $CH_3(CH_2)_{10}CO$—
leucyl (*from leucine*) $(CH_3)_2CHCH_2CHNH_2CO$—
malonyl —$OCCH_2CO$—
menthyl
$CH_3CH.(CH_2)_2.CH[CH(CH_3)_2].CH_2CH$— (1, 4, 3, 2)
(*from menthane; 2-p- shown*)
mercapto HS—
mercuri —Hg—
mesityl 2,4,6-$(CH_3)_3C_6H_2$—
α-mesityl 3,5-$(CH_3)_2C_6H_3CH_2$—
mesyl = methylsulfonyl
methene = methylene
methenyl = methylidyne
methionyl $CH_2(SO_2)_{22}$=
methoxy CH_3O—
methyl CH_3—
methylene CH_2=
methylenedioxy —OCH_2O—
methylidyne CH≡
methylol = (hydroxymethyl)
morpholinyl (*from morpholine*) = C_4H_8NO—
myristoyl (*from myristic acid*) $CH_3(CH_2)_{12}CO$—
naphthal = naphthylmethylene
naphthalimido (*from naphthalimide*) $C_{10}H_6(CO)_2N$—
naphthenyl = naphthylmethylidyne
naphthoxy $C_{10}H_7O$—
naphthoyl $C_{10}H_7CO$—
naphthyl (1- or 2-) $C_{10}H_7$—
naphthylene $C_{10}H_6$=
naphthylidene

1(4)-, etc.

nitramino O_2NNH—
nitrilo N≡
nitro O_2N—
aci-nitro = isonitro
nitroso ON—
nonyl $CH_3(CH_2)_8$—
norcamphanyl C_7H_{11}— (*from norcamphane*)
octyl $CH_3(CH_2)_7$—
oleoyl (*from oleic acid*) $C_{17}H_{33}CO$—
oxalyl —OCCO—
oxamido $H_2NCOCONH$—
oxamyl H_2NCOCO—
oxazolyl C_3H_2NO— (*from oxazole*)
oximido = isonitroso
oxo O= (*to same atom*)
oxy —O— (*used as a connective; cf. epoxy and oxo*)
palmitoyl $CH_3(CH_2)_{14}CO$— (*from palmitic acid*)
pelargonyl $CH_3(CH_2)_7CO$— (*from pelargonic acid*)
pentamethylene —$CH_2(CH_2)_3CH_2$—
pentazolyl N=N—N=N—N—
pentenyl (*like butenyl*) C_5H_9—
pentyl = amyl
perimidyl $C_{11}H_7N_2$— (*from perimidine, 8 isomers*)
perseleno Se:Se=
perthio (*replacing 0 only*) S=S=
phenacyl $C_6H_5COCH_2$—
phenacylidene C_6H_5COCH=
phenanthryl $C_{14}H_9$— (*from phenanthrene, 5 isomers*)
phenanthrylene —$C_{14}H_8$— (*several isomers*)
phenenyl C_6H_3≡ (*s-, as-, v-*)
phenethyl $C_6H_5CH_2CH_2$—
phenetidino $C_2H_5OC_6H_4NH$—
phenetyl (*o, m,* or *p*) $C_2H_5OC_6H_4$—
phenoxy C_6H_5O—
phenyl C_6H_5—
phenylacetyl $C_6H_5CH_2CO$—
phenylazo C_6H_5N:N—
phenylcarbamido = phenylureido
phenylene C_6H_4= (*o, m* or *p*)
phenylenebisazo —N:NC_6H_4N:N— (*o, m, p*)
phenylidene = cyclohexadienylidene
phenylsulfamyl $C_6H_5NHSO_2$—
phenylsulfonamido $C_6H_5SO_2NH$
phenylureido $C_6H_5NHCONH$—
phospharseno —P:As—
phosphazo —P:N—
phosphinico HOOP= (*from phosphinic acid*)
phosphino H_2P—
phospho O_2P—
phosphono $(HO)_2OP$—
phosphoro —P:P—
phosphoroso OP—

phthalal = phthalylidene
phthalidyl $C_6H_4CO.O.CH$— (*from phthalide*)
phthalidylidene $C_6H_4CO.O.C$= (*from phthalide*)
phthalimido $C_6H_4(CO)_2N$— (*o*)
phthaloyl —OCC_6H_4CO— (*o*)
phthalylidene =HCC_6H_4CH= (*o*)
phytyl
$(CH_3)_2CH(CH_2)_3CH(CH_3)(CH_2)_3$—
—$CH(CH_3)(CH_2)_3C(CH_3)$:$CHCH_2$—
picryl 2,4,6- $(NO_2)_3C_6H_2$—
piperidyl $C_5H_{10}N$— (*from piperidine, 4 isomers*)
piperonyl 3,4-$(CH_2O_2)C_6H_3CH_2$—
piperonylidene 3,4-$(CH_2O_2)C_6H_3CH$=
pivalyl $(CH_3)_3CCO$— (*from pivalic acid*)
prolyl (*from proline*)
$NH.CH_2.CH_2.CH_2.CH.CO$—
propargyl = 2-propynyl
propenyl CH_3CH:CH—
propenylidene CH_3CH:C=
propiolyl HC:CCO—
propionyl CH_3CH_2CO—
propoxy $CH_3CH_2CH_2O$—
propyl $CH_3CH_2CH_2$—
propylene —$CH(CH_3)CH_2$—
propylidene CH_3CH_2CH=
1-propynyl CH_3C:C—
2-propynyl CH:CCH_2—
pseudoallyl = isopropenyl
as-pseudocumyl 2,3,5-$(CH_3)_3C_6H_2$—
s-pseudocumyl 2,4,5-$(CH_3)C_6H_2$—
v-pseudocumyl 2,3,6-$(CH_3)_3C_6H_2$—
pseudoindolyl C_8H_6N— (*from pseudoindole, 7 isomers*)
pyranyl C_5H_5O— (2-α, 2-γ, 3-α-, etc.)
pyrazinyl $C_4H_3N_2$— (*from pyrazine*)
pyrazolidyl $C_3H_7N_2$— (*from pyrazolidine*)
pyrazolyl $C_3H_3N_2$— (*from pyrazole, 4 isomers*)
pyrenyl $C_{16}H_9$— (*from pyrene*)
pyridyl C_5H_4N— (*from pyridine, 3 isomers*)
pyridylidene

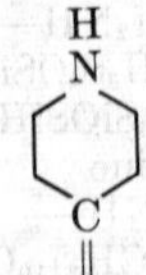

4(1), etc.

pyrimidyl $C_4H_3N_2$— *(from pyrimidine)*
pyromucyl = 2-furoyl
pyrrolidyl C_4H_8N— *(from pyrrolidine, 3 isomers)*
pyrroyl
CH:CH.CH:CH.N.CO— (ring)
pyrryl C_4H_4N— *(from pyrrole, 3 isomers)*
quinazolyl $C_8H_5N_2$— *(from quinazoline)*
quinolyl C_9H_6N— *(from quinoline, 7 isomers)*
quinonyl $C_6H_3O_2$— *(from quinone)*
quinoxalyl $C_8H_5N_2$— *(from quinoxaline)*
salicyl = *o*-hydroxyphenyl = *o*-hydroxybenzyl
salicylidene HOC_6H_4CH= *(o)*
salicyloyl HOC_6H_4CO— *(o)*
selenino (HO)OSe—
seleninyl OSe=
seleno Se=
selenocyano NCSe—
selenono HO_3Se—
selenonyl —SeO_2—
selenyl HSe—
semicarbazido $NH_2CONHNH$—
senecioyl $(CH_3)_2C:CHCO$— *(from senecioic acid)*
siloxy H_3SiO—
silyl H_3S—
silylamino H_3SiNH—
1-silyldisilanyl $(H_3Si)_2SiH$—
silylene H_2Si=
silylidene HSi≡
stannyl H_3Sn—
stearoyl $CH_3(CH_2)_{16}CO$—
stibarseno —Sb:As—
stibinico HOOSb=
stibino H_2Sb—
stibo O_2Sb—
stibono $(HO)_2OSb$—
stiboso O:Sb—
stibyl = stibino
stibylene HSb=
styrene —$CH(C_6H_5)CH_2$—
styrolene = styrene
styryl $C_6H_5CH:CH$—
succinamyl $H_2NCOCH_2CH_2CO$—
succinimido $(CH_2CO)_2N$—
succinyl —$OCCH_2CH_2CO$—
sulfamino HO_3SNH—
sulfamyl H_2NO_2S—
sulfanilamido *p*-$H_2NC_6H_4SO_2NH$— *(from sulfanilamide)*
sulfanilyl *p*-$H_2NC_6H_4SO_2$— *(from sulfanilic acid)*
sulfyhydryl = mercapto
sulfino HO_2S—
sulfinyl OS=
sulfo HO_3S—
sulfonamido —SO_2NH—
sulfonyl —SO_2—
sulfuryl = sulfonyl
tauryl $H_2NCH_2CH_2SO_2$—
telluro Te=
terephthalal = terephthalylidene
terephthaloyl —OCC_6H_4CO— *(p)*
terephthalylidene =HCC_6H_4CH= *(p)*
tetramethylene —$CH_2CH_2CH_2CH_2$—
tetrazolyl CHN_4— *(from tetrazole, 2 isomers)*
thenoyl C_4H_3SCO— *(from thiophenecarboxylic acid, 2 isomers)*
thenyl (2 *isomers*) $C_4H_3SCH_2$—
thenylidene (2 *isomers*) C_4H_3SCH=
thiazolidyl C_3H_6NS— *(from thiazolidine)*
thiazolyl C_3H_2NS— *(from thiazole, 3 isomers)*
thienyl C_4H_3S— *(from thiophene, 2 isomers)*
thio —S—
thiocarbamyl H_2NCS—
thiocarbonyl SC=
thiocyano NCS—
thiohydroxy = mercapto
thiol (S replacing O in OH) *Used in place of "thio" when required for distinction*
thiono (S replacing O in CO) *Used in place of "thio" when required for distinction*
thionyl = sulfinyl
thioxo (S replacing 2H in > CH_2)
thiuram = thiocarbamyl
thujyl $C_{10}H_{17}$— *(from sabinane, attached at 2-position)*
thymyl *(from thymol)*
$HC:C(CH_3).CH:CH.C[CH(CH_3)_2].C$— (ring)
toloxy (*o*, *m*, or *p*) $CH_3C_6H_4O$—
toluidino (*o*, *m* or *p*) $CH_3C_6H_4ONH$—
toluyl (*o*, *m* or *p*) $CH_3C_6H_4CO$—
α-toluyl = phenylacetyl
tolyl (*o*, *m* or *p*) $CH_3C_6H_4$—
α-tolyl = benzyl
tolylene (6 *isomers*) $CH_3C_6H_3$=
α-tolylene = benzylidene
tosyl = tolylsufonyl
triazeno $NH_2N:N$—
triazinyl $C_3H_2N_3$— *(from triazine)*
triazo = azido
triazolyl $C_2H_2N_3$— *(from triazole)*
trimethylene —$CH_2CH_2CH_2$—
trisilanyl $H_3SiSiH_2SiH_2$—
trisilanylene —$SiH_2SiH_2SiH_2$—
trityl = triphenylmethyl
tryptophyl $C_{11}H_{11}B_2O$— *(from tryptophan)*
tyrosyl *p*-$HOC_6H_4CH_2CHNH_2CO$— *(from tyrosine)*
undecyl = hendecyl (in sense $C_{11}H_{23}$—)
uramino = ureido
ureido H_2NCONH— *(by some used synonymously with ureylene)*
ureylene —HNCONH—
valeryl $CH_3(CH_2)_3CO$—
valyl $(CH_3)_2CHCHNH_2CO$— *(from valine)*
vanillal = vanillylidene
vanilloyl 3,4-$(CH_3O)(HO)C_6H_3CO$—
vanillyl 3,4-$(CH_3O)(HO)C_6H_3CH_2$—
vanillylidene 3,4-$(CH_3O)(HO)C_6H_3CH$=
veratral = veratrylidene
veratroyl 3,4-$(CH_3O)_2C_6H_3CO$—
veratryl 3,4-$(CH_3O)_2C_6H_3CH_2$—
veratrylidene 3,4-$(CH_3O)_2C_6H_3CH$=
vinyl $H_2C:CH$—
vinylene —CH:CH—
vinylidene $H_2C:C$=
xanthyl $C_{13}H_9O$— *(from xanthene, 6 isomers)*
xenyl = biphenylyl
xyloyl $(CH_3)_2C_6H_3CO$— *(from xylic acid, 7 isomers)*
xylyl $(CH_3)_2C_6H_3$—
xylylene —$H_2CC_6H_4CH_2$—

ORGANIC RING COMPOUNDS

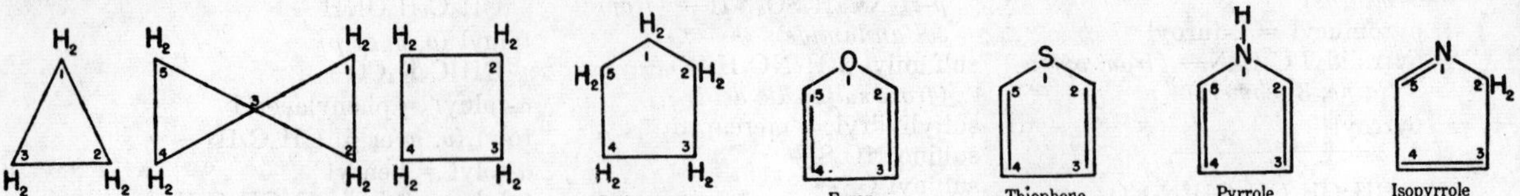

Cyclopropane

Spiropentane

Cyclobutane

Cyclopentane

Furan

Thiophene (Thiofuran)

Pyrrole (Azole)

Isopyrrole (Isoazole) (α-Pyrrolenine)

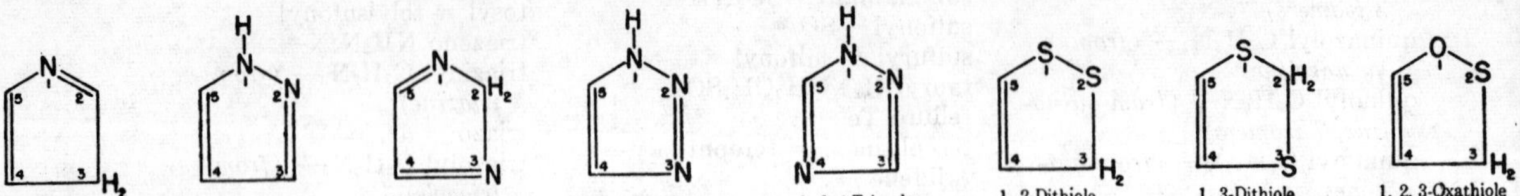

3-Isopyrrole (Isoazole) (β-Pyrrolenine)

Pyrazole (1, 2-Diazole)

2-Isoimidazole (1, 3-Isodiazole)

1, 2, 3-Triazole

1, 2, 4-Triazole

1, 2-Dithiole

1, 3-Dithiole

1, 2, 3-Oxathiole

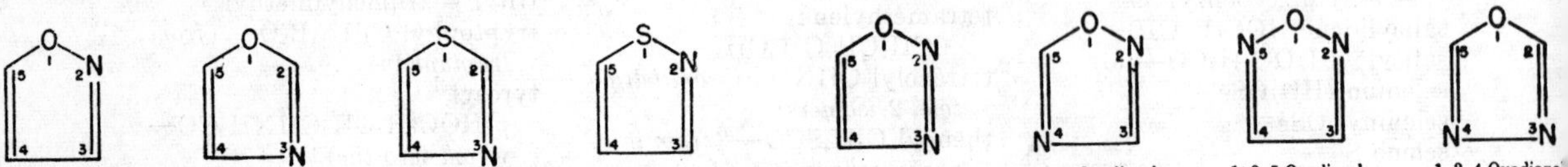

Isoxazole (Furo [α] monazole)

Oxazole (Furo [β] monazole)

Thiazole

Isothiazole

1, 2, 3-Oxadiazole

1, 2, 4-Oxadiazole (Azoxime)

1, 2, 5-Oxadiazole (Furazan)

1, 3, 4-Oxadiazole

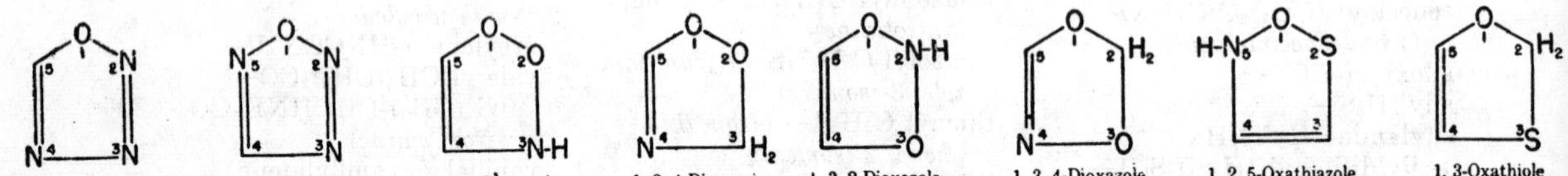

1, 2, 3, 4-Oxatriazole

1, 2, 3, 5-Oxatriazole

1, 2, 3-Dioxazole

1, 2, 4-Dioxazole

1, 3, 2-Dioxazole

1, 3, 4-Dioxazole

1, 2, 5-Oxathiazole

1, 3-Oxathiole

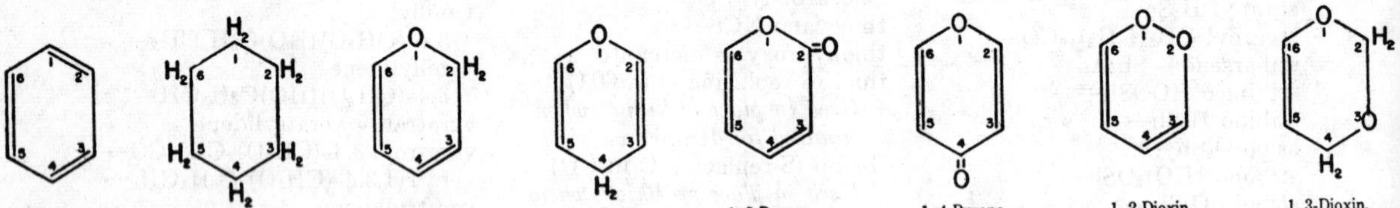

Benzene

Cyclohexane

1, 2-Pyran

1, 4-Pyran

1, 2-Pyrone (α-Pyrone)

1, 4-Pyrone (γ-Pyrone)

1, 2-Dioxin

1, 3-Dioxin, Dihydro form

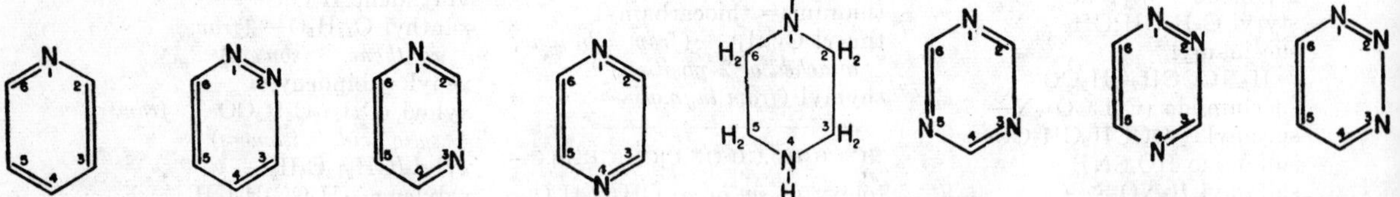

Pyridine (Azine)

Pyridazine (1, 2-Diazine)

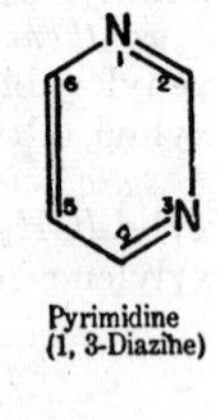

Pyrimidine (1, 3-Diazine)

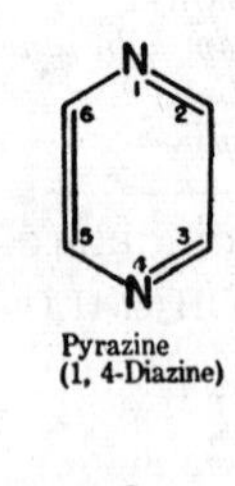

Pyrazine (1, 4-Diazine)

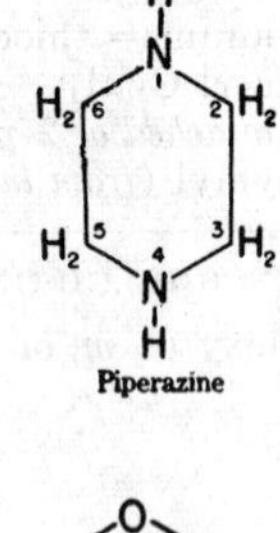

Piperazine

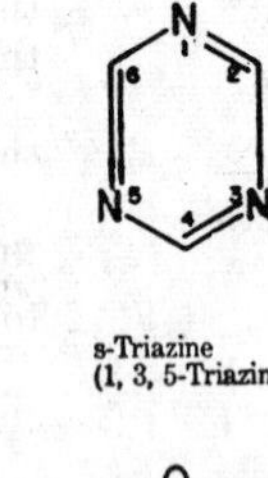

s-Triazine (1, 3, 5-Triazine)

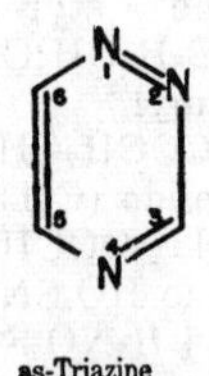

as-Triazine (1, 2, 4-Triazine)

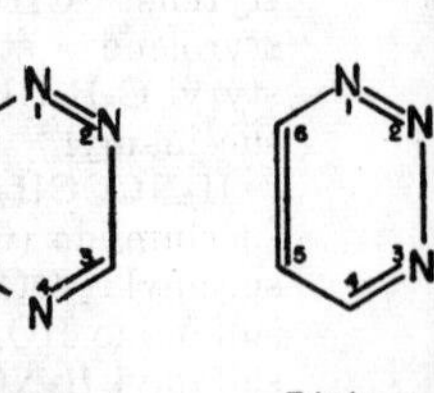

v-Triazine (1, 2, 3-Triazine)

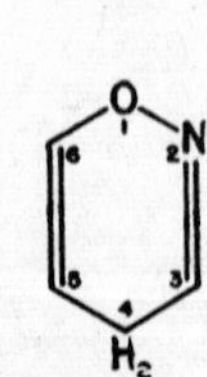

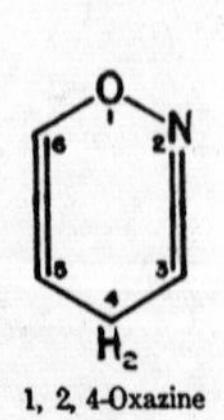

1, 2, 4-Oxazine

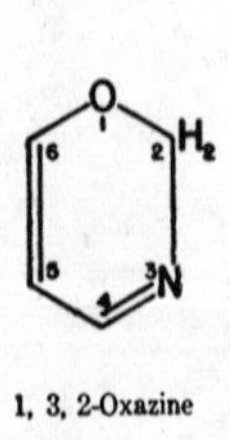

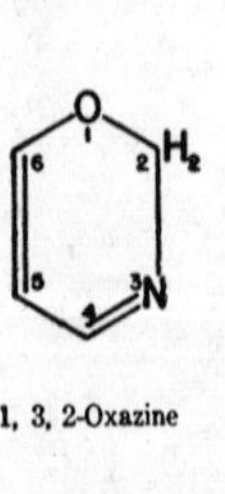

1, 3, 2-Oxazine

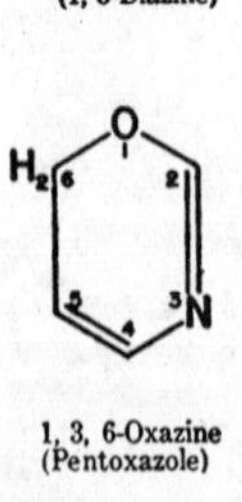

1, 3, 6-Oxazine (Pentoxazole)

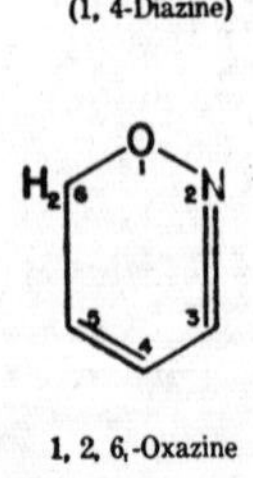

1, 2, 6,-Oxazine

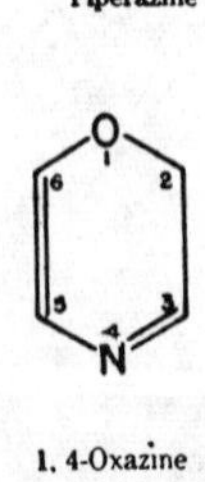

1, 4-Oxazine

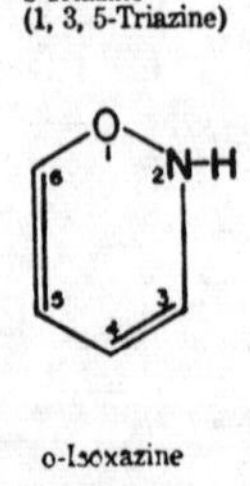

o-Isoxazine

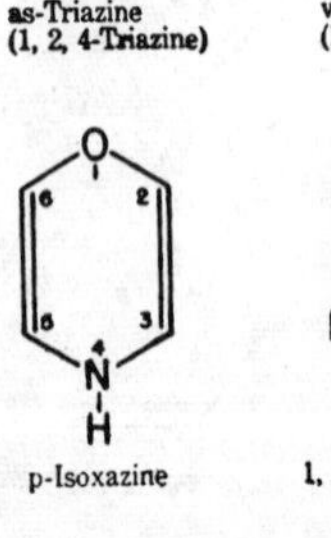

p-Isoxazine

1, 2, 5-Oxathiazine

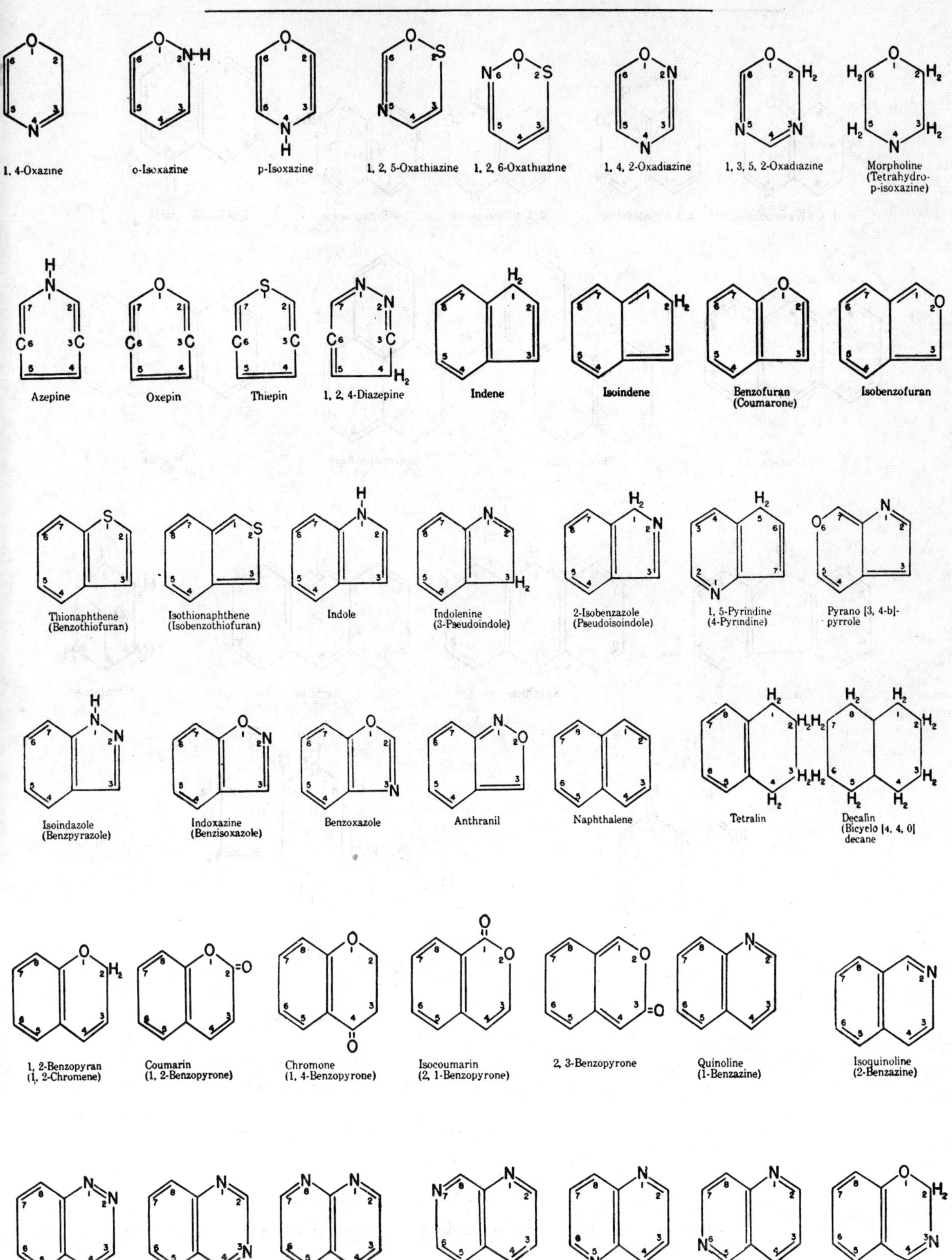
1, 4-Oxazine
o-Isoxazine
p-Isoxazine
1, 2, 5-Oxathiazine
1, 2, 6-Oxathiazine
1, 4, 2-Oxadiazine
1, 3, 5, 2-Oxadiazine
Morpholine (Tetrahydro-p-isoxazine)
Azepine
Oxepin
Thiepin
1, 2, 4-Diazepine
Indene
Isoindene
Benzofuran (Coumarone)
Isobenzofuran
Thionaphthene (Benzothiofuran)
Isothionaphthene (Isobenzothiofuran)
Indole
Indolenine (3-Pseudoindole)
2-Isobenzazole (Pseudoisoindole)
1, 5-Pyrindine (4-Pyrindine)
Pyrano [3, 4-b]-pyrrole
Isoindazole (Benzpyrazole)
Indoxazine (Benzisoxazole)
Benzoxazole
Anthranil
Naphthalene
Tetralin
Decalin (Bicyclo [4, 4, 0] decane
1, 2-Benzopyran (1, 2-Chromene)
Coumarin (1, 2-Benzopyrone)
Chromone (1, 4-Benzopyrone)
Isocoumarin (2, 1-Benzopyrone)
2, 3-Benzopyrone
Quinoline (1-Benzazine)
Isoquinoline (2-Benzazine)
Cinnoline (1, 2-Benzodiazine)
Quinazoline (1, 3-Benzodiazine)
Naphthyridine
Pyrido [3, 4-b]-pyridine
Pyrido [3, 2-b]-pyridine
Pyrido [4, 3-b]-pyridine
1, 3, 2-Benzoxazine

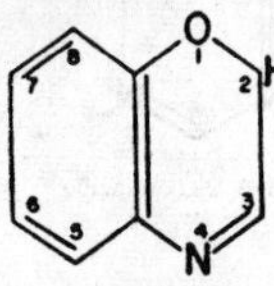

1, 4, 2-Benzoxazine

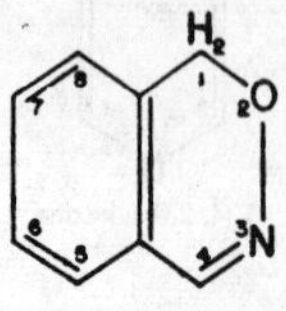

2, 3, 1-Benzoxazine

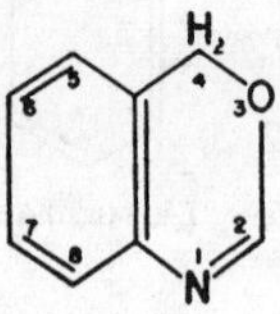

3, 1, 4-Benzoxazine

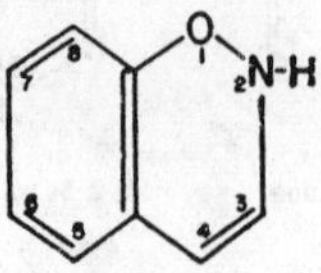

1, 2-Benzisoxazine

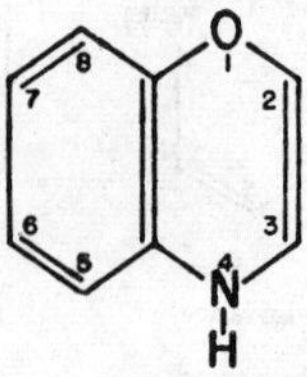

1, 4-Benzisoxazine

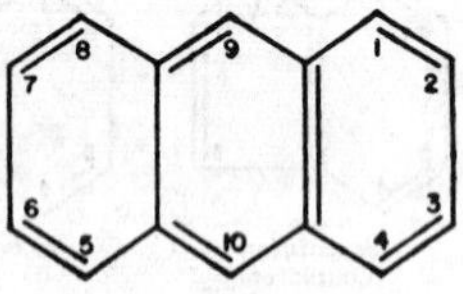

Anthracene

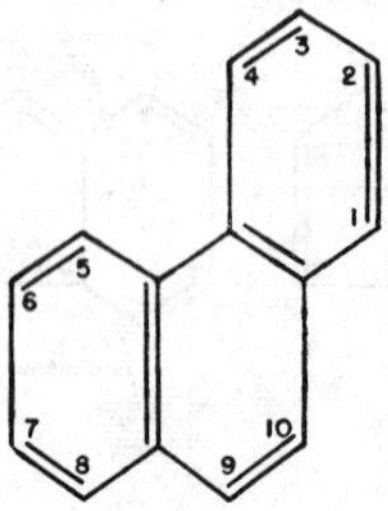

Phenanthrene

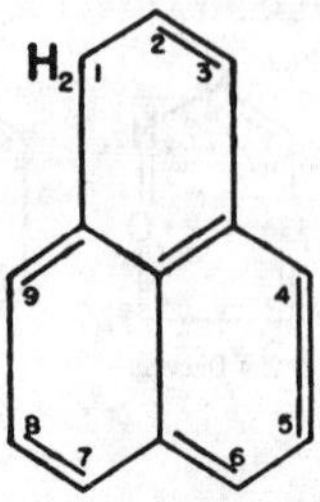

Benzonaphthene (Phenalene)

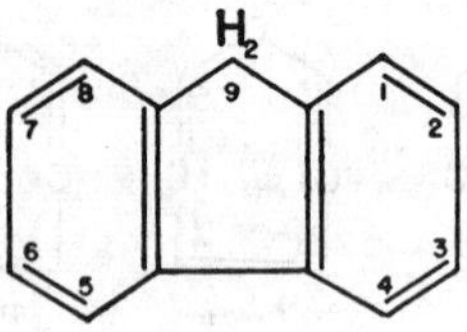

Fluorene

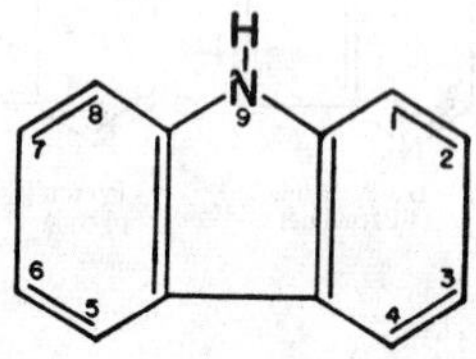

Carbazole

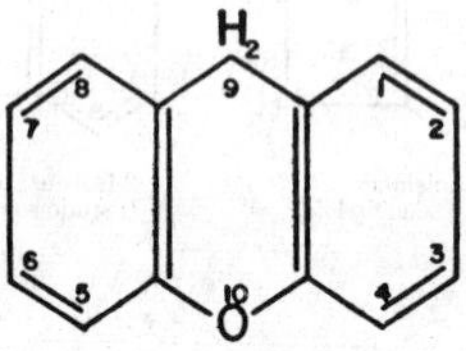

Xanthene

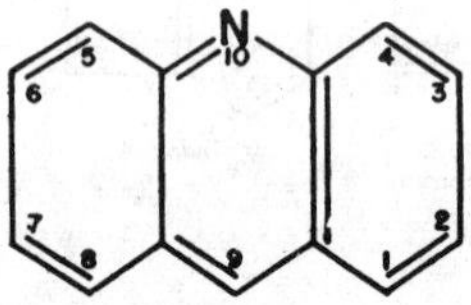

Acridine

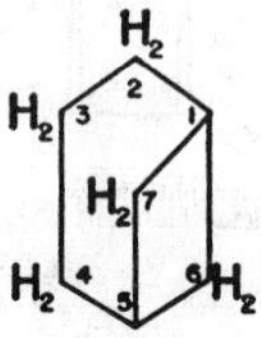

Norpinane

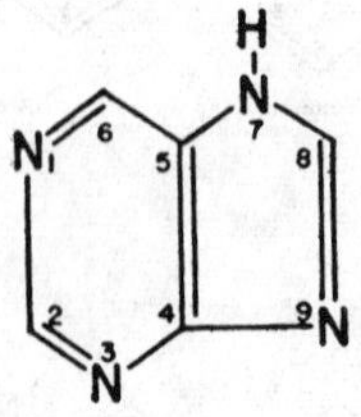

Purine

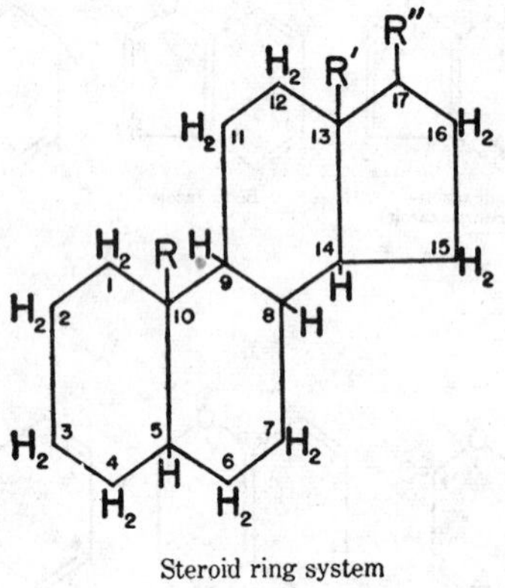

Steroid ring system

R; nearly always methyl
R′; usually methyl
R″; various groups

A more extensive listing of ring compounds and systems may be found in the following:

Chemical Abstracts, annual subject index.
The Ring Index, Patterson and Capell, Reinhold Publishing Company, 1940.
Lexikon der Kohlenstoffverbindungen, Richter, Leopold Voss, 1910.

The numbering system for the compounds listed above is that used in The Ring Index.

TABLE OF PHYSICAL PROPERTIES OF ORGANIC COMPOUNDS
EXPLANATION OF TABLE

The section of this book dealing with Physical Constants of Organic Compounds was prepared by Saul Patai, Ph.D. and J. Zabicky, Ph.D.

Key to the Explanation
(numbers refer to paragraphs)

EXPLANATION OF TABLE

1. The table of Physical Constants of Organic Compounds is a compilation of data on some 13600 organic compounds that include those of wide application in teaching, industry, medicine and research.

CONTENTS OF TABLE

2. **Number of entry.** Each entry has a number preceded by a letter corresponding to the first letter of the name of the parent compound (see paragraphs 17–26, 31, 32). Cross references are left unnumbered. The number of entry is used in the formula, melting point and boiling point indexes (see paragraphs 13–15).

3. **Name of compound.** See rules for naming of compounds (paragraphs 16 ff.).

4. **Synonyms and formula.**

a) *Synonyms.* For the sake of brevity, the number of synonyms in an entry is limited. Additional synonyms may readily be formed by derivation from the names given to simpler or less substituted compounds; *e.g.:* **Benzoic acid, 4-nitro-,** ethyl ester is given no synonyms but they may be derived from those appearing under **Benzoic acid,** for example, ethyl *p*-nitrobenzenecarboxylate.

b) *Formula.* Each entry has either a structural or an empirical formula. In the second case there are the following possibilities:

i) The structural formula was not known at the time of compilation.

iii) The compound is cyclic and appears among a series of compounds derived from the same cyclic parent compound; *e.g.:*

u133 **Uric acid, 7-methyl-** $C_6H_6N_4O_3$. *See* u131 means that in entry u131 appears the numbered skeleton of **Uric acid,** and the structural formula of u133 may be derived from that of u131.

Structural formulae in the Table give generally no configurational details, and when they do, they seldom give absolute configurations. These details are to be looked for either in the original literature or in specialized textbooks.

5. **Molecular weight.** Computed according to International Atomic Weight values of 1961.

6. **Crystalline form, color and specific rotation.**

a) *Crystalline form.* Solid compounds are to be considered as crystalline if not otherwise stated. Data on solvents used and solvents of crystallization are given in brackets. Some examples will explain the usage of the Table (see also list of abbreviations):

ABBREVIATION	MEANING
pl	plates
pl(al)	plates obtained from alcohol as the solvent
pl(al + 1½)	plates obtained from alcohol crystallizing with 1½ molecules of the **same solvent** per molecule of compound
pl(al + ?)	as above but the amount of solvent of crystallization per molecule of compound is undefined
pl(aq ace + 2w)	here the solvent of crystallization is one of the components of the solvent, *i.e.,* two molecules of water
(al)	compound crystallizes from alcohol but crystalline form is not reported

If no melting point is given in the Table the compound is a liquid at room temperature, unless otherwise stated.

b) *Color:* No special remark is made if the compound is colorless or no data on its color are available; otherwise an abbreviation appears before the crystalline form. (See table of abbreviations.)

c) *Specific rotation* appears according to the common usage, *e.g.:*

$$[a]_D^{25} - 25.8(w, c = 4)$$

Bracketted abbreviations indicate solvent and concentration, *i.e.,* in the above example, solvent water and concentration 4%.

7. **Melting point.** Remarks on this physical property appear as abbreviations after the m.p. Melting points of questionable accuracy are given in brackets.

8. **Boiling point.** The pressure at which this physical property was determined appears as a superscript. If no superscript is given, pressure is about one atmosphere. Remarks on this physical property appear as abbreviations. Boiling points of questionable accuracy are given in brackets.

9. **Density** is relative to water, otherwise it has the dimensions g/ml. A superscript indicates the temperature of the liquid and a subscript indicates the temperature of water to which the density is referred.

10. **Refractive index** is reported for the D line of the spectrum of sodium (n_D). The temperature of determination of this physical property appears as a superscript.

11. **Solubility.** Data are given for the most common solvents. As numerical data on this property scarcely appear in the literature, and when they appear their degree of approximation is not great, it was preferred to rely on the intuitive meaning that the following scale has for the chemist (corresponding abbreviations in brackets): insoluble (i), slightly soluble (δ), soluble (s), very soluble (v), miscible (∞), decomposes (d). If no special remark is made about the temperature, the reference is to room temperature; otherwise, a superscript appears.

12. **Literature references** are given to general sources of data such as *Beilstein's Handbook, Elsevier's Encyclopaedia,* etc. The superscript after the boldtype number indicates the number of the supplement (Ergenzungswerk) of *Beilstein's Handbook;* e.g.: $\mathbf{B7}^1$, 12 means page 12 of the first supplement to volume 7 of *Beilstein.*

INDEXES

Three indexes are found at the end of the Table:

13. **Empirical formula index** is arranged according to increasing C, H, and remaining elements in alphabetical order. Hydrates are entered under the formula of the *anhydrous* compound. Salts, complexes, etc., are found under their total formulae.

14. **Melting point index** is arranged according to increasing melting or freezing points. Only the lower temperature of the melting range is given. Remarks to the melting point in the main table, such as decomposition, sublimation, etc., are omitted. Melting points are rounded off to the nearest unit.

15. **Boiling point index** is arranged according to increasing boiling point of compounds of which the boiling point is given at a pressure near to 1 atm. (*i.e.,* 700–780 mm. Hg). Only the lower temperature of the boiling range is given. No special entries are made for remarks on the boiling point contained in the main table. Boiling points are rounded off to the nearest unit.

RULES FOR THE NAMING OF COMPOUNDS*

16. Naming of compounds is based mainly on *International Union of Chemistry* (*IUC*) and *Chemical Abstracts* usages, with variations adapted to the scope and range of the Table, in order to enable the user to find compounds even when their trivial names are not known.

17. A suitable systematic nomenclature is used in order to bring as near as possible compounds that are isomeric derivatives of the same parent compound, *e.g.,* physical properties appear under the entries **Benzene, 1,2-dihydroxy-, Benzene, 1,3-dihydroxy-,** and **Benzene, 1,4-dihydroxy-;** under **Hydroquinone, Pyrocatechol** and **Resorcinol** a cross reference directs the user towards the systematic names. No special entry or cross-reference is made for *e.g.,* **Bromoresorcinol;** this will be found under **Benzene, bromo(dihydroxy)-,** with the appropriate numbering before the names of the substituents. Another example: **4-Chloroquinaldine;** the trivial name of the parent compound is **Quinaldine,** it appears as: **Quinaldine** *see* **Quinoline, 2-methyl-;** accordingly the properties of **4-Chloroquinaldine** will appear under **Quinoline, 4-chloro-2-methyl-.**

18. Physical constants of a compound having widely used trivial and systematic names will appear under one of them only. The remaining names will appear with a cross-reference to the assigned name, *e.g.:*

Resorcinol *see* **Benzene, 1,3-dihydroxy-**

Propanoic acid, 2-amino- *see* **Alanine**

Very complicated compounds such as some natural products, dyes, etc. appear almost exclusively under their trivial names.

---

* The rules for nomenclature followed by the Table were adapted to the present, relatively limited scope of some thousands of compounds. For other use the IUPAC 1957 rules, which appear in this Handbook, may be consulted.

EXPLANATION OF TABLE (Continued)

19. Rules adopted by the Table for the naming of compounds according to their functional groups appear in paragraphs 27 ff.

20. The naming of a compound has three phases:

a) Determination of the parent compound.
b) Determination and numbering of the substituents in the skeleton of the parent compound.
c) Determination of the derivatives of the principal functions of the compound.

e.g.:

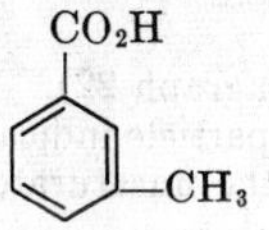

CO_2H / CH_3

Phase a) **Benzoic acid**
Phase b) **3-methyl-**

Accordingly it appears as **Benzoic acid, 3-methyl-**

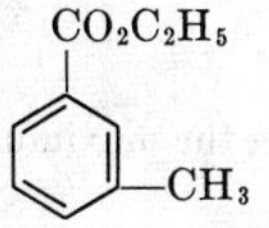

$CO_2C_2H_5$ / CH_3

Phase a) **Benzoic acid**
Phase b) **3-methyl-**
Phase c) ethyl ester

therefore it appears as **Benzoic acid, 3-methyl-**, ethyl ester.

21. The parent compound has at most one kind of principal function, (*e.g.:* acid) but this function may be multiple, (*e.g.:* diacid, triacid, etc.); examples:

$CH_3COCH_2CH_2CO_2H$ appears as **Pentanoic acid, 4-oxo-**
not as: **Pentanon-4-on-1-oic acid**
$HO_2CCH_2CH(CH_3)COCH_2CO_2H$ appears as **Hexanedioic acid, 3-methyl-4-oxo-**

Unsaturation is not considered as a function and, when possible, it is included in the name of the parent compound; *e.g.:*

$CH_2{:}CHCH_2CO_2H$ appears as **3-Butenoic acid.**

Exceptions to this rule are some cyclic ketones that may be designated as ring systems (see paragraph 83) and some natural products, such as **Pregnenolone** (the functions alcohol and ketone appear in the parent compound).

22. Entries in the Table are made first according to alphabetical order of parent compounds (paragraph 20, phase a), and then according to alphabetical order of the substituents in the skeleton of the parent compound (paragraph 20, phase b). It should be emphasized that numbers or single letters that are used in the designation of the compound are not considered for the alphabetisation. If the designation of a compound includes phase c of paragraph 20 this compound appears immediately after the compound that is designated without phase c and has the same substituents bearing the same numbers; *e.g.:*

Benzoic acid
Benzoic acid, 2-chloro-
Benzoic acid, 2-chloro-, ethyl ester
Benzoic acid, 3-chloro-
Benzoic acid, 3-chloro-, allyl ester

23. Prefixes indicating multiplicity of substituents as **di-**, **tri-**, **bis-**, **tris**, etc. and the prefix **iso-**, as in **dimethyl-**, **tripropyl-**, **bis**(**nitrophenyl**)**-**, **tris**(**aminomethyl**)**-**, **isopropyl-**, etc., are indexed under **d**, **t**, **b**, **t**, **i** respectively. The prefixes *sec-* and *tert-* are regarded as numbers, and therefore, they are not considered for the alphabetization.

24. If the designation of a compound is difficult, the use of the empirical formula index is suggested (paragraph 13).

25. When a compound is designated by the name of its principal functional group that alone is not a compound, as is the case with some amines, ketones, azo-compounds, etc., after the name of the functions the skeleton is designated by alphabetical order, and then the substituents on each of the groups attached to the principal function; *e.g.:*

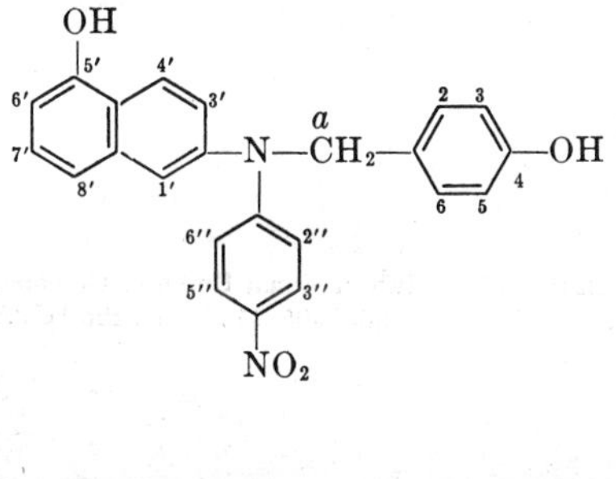

OH, N—CH₂—, —OH, NO_2

Amine, benzyl 2-naphthyl phenyl, 4,5′-dihydroxy-4″-nitro-

It is to be noted that substituents of benzyl (the first radical in the name) are unprimed; substituents of naphthyl (the second radical in the name) are primed; substituents of phenyl (the third radical in the name) are doubly primed.

26. The principal function of the parent compound has a number as low as possible.

HYDROCARBONS*

27. **Aliphatic hydrocarbons.** IUC rules 4–10 are followed, but some remarks are added:

a) For C_{11} hendecane (and not undecane) is used (see IUC rule 5).

b) From the possibilities in IUC rule 7 the alphabetical order is used always (see paragraph 22).

c) Regarding IUC rules 8 and 9: numbers before the name of the hydrocarbon or the particle indicating unsaturation indicate the carbon with the lower number of the two connected by the unsaturated link, *e.g.:*

$CH_3CH_2CH_2CH{:}CH_2$	**1-Pentene**
$CH_2{:}C{:}CHCH_3$	**1,2-Butadiene**
$CH_3C{\vdots}CCH{:}CHCH{:}CHCH_3$	**2,4-Octadien-6-yne**

d) Regarding IUC rule 10: in an unsaturated hydrocarbon the longest chain containing the maximum number of double and triple bonds will be considered; *e.g.:*

$CH_2{:}C{-}CH{:}CH_2$ (with CH_2CH_3 on C-2) appears as **1-3-Butadiene, 2-ethyl-**
not as **1-Pentene, 3-methylene-**
not as **1-Butene, 2-ethenyl-**

Preference is given to the double bond over the triple bond in the fundamental chain; *e.g.:*

$CH_2{:}CHCHCH{:}CH_2$ (with $C{\vdots}CH$ on C-3) Appears as **1,4-Pentadiene, 3-ethenyl-**
not as **1-Penten-4-yne, 3-ethenyl-**

28. **Cyclic hydrocarbons:**

a) *Saturated monocyclic* hydrocarbons follow IUC rule 11.

b) *Unsaturated cyclic* hydrocarbons follow IUC rule 12 with the following comments: the names **tetralin** for 1,2,3, 4-tetrahydronaphthalene, **decalin** for decahydronaphthalene, **indan** for 2,3-dihydroindene, and special names for steroidal skeletons are retained.

c) *Aromatic parent hydrocarbons.* The names **benzene, naphthalene, anthracene** are used. For linear arrangements of four or more fused benzene rings names are formed by a numeric prefix denoting the number of fused benzene rings and the suffix -cene: **tetracene, pentacene,** etc. The usual trivial names are given to other arrangements: **phenanthrene, chrysene, indene, fluorene, benzanthrene, dibenzophenanthrene,** etc.

d) *Bridged hydrocarbon ring systems* appear under the headings: **bicyclo-** or **tricyclo-** according to Baeyer's system; *e.g.:*

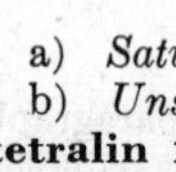

Bicyclo[3.1.0]hexane

However in most cases compounds having such structures appear only under their trivial names.

e) *Spiro, dispiro, etc.* compounds appear under these headings; *e.g.:*

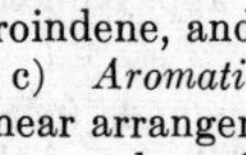

Spiro[4,5]decane

f) Cyclic hydrocarbons with side chains. IUC rules 49aI and 49aII are followed with some variations: the *Chemical Abstracts* usage is followed in conjunction with IUC rule 49aI: the parent compound is (when possible) the one containing the higher number of carbon atoms; *e.g.:* **Benzene, pentyl-,** *not* Pentane, 1-phenyl-; **Docosane, 1(1-naphthyl)-,** *not* Naphthalene, 1-docosyl-. The form **Hexane, phenyl-** is preferred. When the side chain is unsaturated and the cyclic compound is benzene or a benzene derivative, preference is given to the unsaturated aliphatic hydrocarbon; *e.g.:* $C_6H_5CH_2CH_2CH{:}CH_2$ is **1-Butene, 4-phenyl-;** $C_6H_5C{\vdots}CCH_2CH_3$ is **1-Butyne, 1-phenyl-.**

The following are exceptions:

$C_6H_5C{\vdots}CH$	**Benzene, ethynyl-**
$C_6H_5CH_2CH{:}CH_2$	**Benzene, allyl-**
$C_6H_5CH{:}CHCH_3$	**Benzene, propenyl-**
$C_6H_5CH{:}CH_2$	**Styrene**
$C_6H_5CH{:}CHC_6H_5$	**Stilbene**

* In order to avoid unnecessary repetition, the 1930 Liége rules for nomenclature of the International Union of Chemistry are referred to only by number, *e.g.* IUC rule 1. These rules may be found in *J. Am. Chem. Soc.* **55**, 3905–25 (1933); the IUPAC 1957 rules appeared after preparation of this table.

For alkyl derivatives of benzene the only trivial name retained is **toluene** for $C_6H_5CH_3$. Toluene derivatives appear under the heading **Toluene** if no other carbon atom is attached either to the methyl group or to the ring; otherwise the name will appear under **Benzene** or elsewhere as suitable; *e.g.:*

$H_2N-C_6H_4-CH_3$	**Toluene, 4-amino-**
$(CH_3)_2CH-C_6H_4-CH_3$	**Benzene, 1-isopropyl-4-methyl-**
$C_6H_5-C_6H_4-CH_3$	**Biphenyl, 4-methyl-**

Substitution in the methyl group of toluene will be labelled *a* and the word benzyl will not appear as a parent compound (only as a substituent); *e.g.:*

$Cl-C_6H_4-CH_2Br$	**Toluene, *a*-bromo-4-chloro-**
$C_6H_5-CH_2OH$	**Toluene, *a*-hydroxy-** (cross-reference under **Benzyl alcohol**).

Regarding IUC rule 49aII the trivial name **Stilbene** is used instead of **Ethene, 1,2-diphenyl-**.

29. **Heterocyclic systems.** Trivial names are used, or when these are lacking systematic names are formed by using the homocyclic system name and the particles **oxa-, aza-, thia-**, etc.

The following trivial names are used for simple heterocyclic compounds: **furan, pyran, pyrrole, pyrazole, imidazole, piperidine, pyridine, pyrazine, pyrimidine, pyrazidine, thiophene,** etc. Fused heterocyclic systems follow the same rule; trivial names such as acridine, indole, quinoline, isoquinoline, quinazoline, quinoxaline, xanthene, etc., or juxtaposed names such as **1,2-benzopyran,** etc. are used. All names of fused ring systems beginning with **benzo-, dibenzo-,** etc. appear under **b, d,** etc. respectively.

30. **Radicals.** Normal alkyl radicals are named according to the Geneva designation. The following radical names are also used: **phenyl, benzyl, tolyl, styryl, allyl, isopropyl, isobutyl, *sec*-butyl, *tert*-butyl, ethenyl, ethynyl.** Larger non-normal alkyl radicals are designated by forming the longest chain containing the radical linkage and numbering the carbon on which the radical links itself to other groups according to the nearest end of the longest chain; *e.g.:*

$$\overset{8}{C}H_3\overset{7}{C}H_2\overset{6}{C}H(CH_3)\overset{5}{C}H_2\overset{4}{C}H(-)\overset{3}{C}H_2\overset{2}{C}H(CH_3)\overset{1}{C}H_3$$ **(2,6-dimethyl-4-octyl)-**

The following IUC rules are also used: 54–62, 64, 65, 67.

31. **Substituted Hydrocarbons.** The following substituents never appear in a parent compound: **arsono-**(—AsO), **arso-**(—AsO_2), **boryl-** (—BO), **bromo-** (—Br), **chloro-** (—Cl), **fluoro-** (—F), **iodo-** (—I), **iodoso-** (—IO), **iodoxy-** (—IO_2), **nitro-** (—NO_2), **nitroso-** (—NO), **phospho-** (—PO_2), **phosphoroso-** (—PO), **stibo-** (—SbO_2), **triazo-** (—N_3). These are always treated as substituents (paragraph 20, phase b).

COMPOUNDS CONTAINING FUNCTIONAL GROUPS

32. **Order of precedence of functions.** When a compound bears one or more functions the parent compound is designated according to the following order of precedence: onium compound, acid (and derivatives of the acidic function), aldehyde, ketone, alcohol, amine, ether, sulfide, sulfone, sulfoxide, etc.

ONIUM COMPOUNDS

33. Onium compounds appear mainly under their trivial names.

34. **Ammonium compounds.** When four organic radicals are attached to the N atom, the compound appears under **Ammonium,** otherwise it appears as the corresponding salt of an amine; *e.g.:*

$C_6H_5\overset{+}{N}(CH_3)_3\overset{-}{I}$	**Ammonium, phenyl trimethyl,** iodide
$(C_2H_5)_3\overset{+}{N}H\ \overset{-}{Cl}$	**Amine, trimethyl,** hydrochloride
$\overset{-}{Cl}\ \overset{+}{N}H_3CH_2CO_2H$	**Glycine,** hydrochloride

35. **Betaines.** The word betaine appears after the name of the corresponding acid ("betaine" alone is the name of a compound); *e.g.:*

$C_6H_5(CH_3)_2\overset{+}{N}CH_2CO\overset{-}{O}$	**Glycine, *N,N*-dimethyl-*N*-phenyl-,** betaine

EXPLANATION OF TABLE (Continued)

ACIDS AND ACID DERIVATIVES

36. **Aliphatic monocarboxylic acids.** The longest chain containing the carboxylic group is taken as the parent compound. The carbon of the carboxyl group is numbered 1. The Geneva nomenclature is followed. (IUC rule 29); *e.g.:*

$\overset{6}{C}H_3\overset{5}{C}H_2\overset{4}{C}H_2\overset{3}{C}H(CH{:}CH_2)\overset{2}{C}H_2\overset{1}{C}O_2H$ — **Hexanoic acid, 3-ethenyl-**

$\overset{5}{C}H_3\overset{4}{C}H_2\overset{3}{C}H{:}\overset{2}{C}(CH_3)\overset{1}{C}O_2H$ — **2-Pentenoic acid, 2-methyl-**

In exception to this rule, the following trivial names are retained:

HCO_2H	**Formic acid**
CH_3CO_2H	**Acetic acid**
$C_6H_5CH{:}CHCO_2H$	**Cinnamic acid**

37. **Cyclic acids.** After the name of the ring the suffix **-carboxylic acid** is attached. The atom to which the carboxyl is attached is numbered 1 unless the ring system has a fixed numbering, in which case the carboxylic group takes a number as low as possible; *e.g.:*

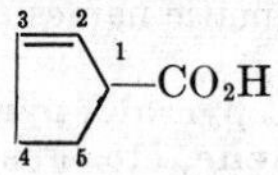

[cyclopentene ring with $-CO_2H$ at position 1] — **2-Cyclopentenecarboxylic acid**

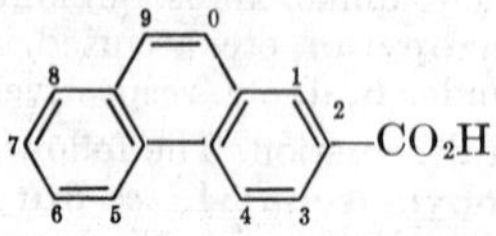

[phenanthrene with $-CO_2H$ at position 2] — **2-Phenanthrenecarboxylic acid**

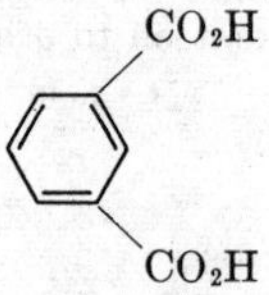

[benzene ring with two CO_2H groups, 1,3-] — **1,3-Benzenedicarboxylic acid** (cross reference under **Isophthalic acid**)

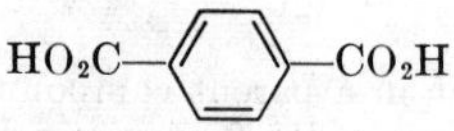

$HO_2C-C_6H_4-CO_2H$ — **1,4-Benzenedicarboxylic acid** (cross reference under **Terephthalic acid**)

Linear acids are given preference over cyclic acids; *e.g.:*

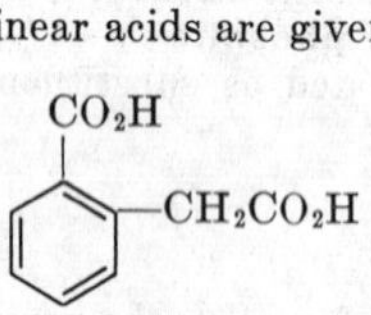

[benzene ring with CO_2H and $-CH_2CO_2H$ ortho] — **Acetic acid, (2-carboxyphenyl)-** (cross reference under **Homophthalic acid**)

Exceptions: **benzoic acid, 1-naphthoic acid, 2-naphthoic acid, phthalic acid.**

38. **Aliphatic, dicarboxylic acids** are found under the Geneva name of the chain containing both carboxylic groups; *e.g.:*

$HO_2\overset{5}{C}\overset{4}{C}H_2\overset{3}{C}(CH_2(CH_2)_6CH_3){:}\overset{2}{C}H\overset{1}{C}O_2H$ — **2-Pentenedioic acid, 3-octyl-**

Trivial names are used for the following acids: **oxalic, malonic, succinic, maleic, fumaric, tartaric, citric.**

39. **Aliphatic polycarboxylic acids** are designated using the **-carboxylic acid** suffix; *e.g.:*

$HO_2CCH_2CH_2CH(CO_2H)CH_2CO_2H$ — **1,2,4-Butanetricarboxylic acid**

40. **Thioacids.** After the name of the acid one of the following particles will appear: **thio-, dithio-, thiolo-, thiono-.** The last two appear only when it is certain that the ketonic or the hydroxylic oxygen is substituted by a sulfur atom.

41. **Imidic, hydroxamic, nitrolic and nitrosolic acids.** Only simple compounds of this type appear in the Table. The various cases arising are exemplified as follows:

$CH_3CH_2CH_2C(:NH)OH$	**Butanimidic acid**
$C_6H_5C(:NH)OH$	**Benzimidic acid**
(pyrrole ring, N—H)—C(:NH)OH	**2-Pyrrolecarboxymidic acid**
$CH_3CH_2C(:NOH)OH$	**Propanehydroxamic acid**
$C_6H_5C(:NOH)OH$	**Benzohydroxamic acid**
(furan ring, O)—C(:NOH)OH	**2-Furancarbohydroxamic acid**
$CH_3C(:NOH)NO_2$	**Acetonitrolic acid**
$CH_3C(:NOH)NO$	**Acetonitrosolic acid**

42. **Sulfonic and sulfinic acids.** According to IUC rule 47, with the following comment: onium and carboxylic acids are functions that precede the sulfonic and sulfinic acids, and so when the former functions are present in the compound the particles **sulfo-** and **sulfino-** will be used; *e.g.*:

$HO_3S—C_6H_4—CO_2H$ **Benzoic acid, 4-sulfo-**

43. **Sulfenic acids** are found under the corresponding hydrocarbon structure followed by the suffix **sulfenic acid;** *e.g.*:

$C_6H_5—S—OC_2H_5$ **Benzenesulfenic acid,** ethyl ester.

44. **Aldehydic and ketonic acids.** When the carbonyl group is present in the principal chain the particle **oxo-** is used; *e.g.*:

$CH_3C(=O)CH_2COOH$ **Butanoic acid, 3-oxo-**
(cross reference under **Acetoacetic acid**)

$CH_3CH(CHO)CH_2CH_2CO_2H$ **Pentanoic acid, 4-methyl-5-oxo-**

If the aldehyde group is not in the principal chain the particles **formyl-** or **oxo-** are used; *e.g.*:

$C_6H_4(CO_2H)(CHO)$ (benzene ring with CO_2H and CHO) **Benzoic acid, 3-formyl-**

$CH_3CH_2COCH_2CHCO_2H$ with $CH_3(CH_2)_3CH_2$ on CH **Dodecanoic acid, 2(2-oxobutyl)-**

See however paragraphs 83, 84.

45. **Hydroxyacids.** The hydroxy group is considered as a substituent; *e.g.*:

$CH_3CHOHCO_2H$ **Propanoic acid, 2-hydroxy-**
(cross reference under **Lactic acid**).

Exceptions: **tartaric, citric,** and acids derived from carbohydrates such as gluconic, mannonic, etc.

46. **Aminoacids.** Most of them appear under their trivial names with cross references under their systematic names.

47. **Orthoacids** appear under the heading **Ortho-;** *e.g.*:

$CH_3C(OCH_3)_3$ **Orthoacetic acid,** trimethyl ester.

$C_6H_5C(OC_2H_5)_3$ **Orthobenzoic acid,** triethyl ester.

48. **Acids attached by a non-carbon link.** The following examples explain the usage of the Table where the same acid appears attached twice to a non-carbon atom:

$NH(CH_2CO_2H)_2$	**Acetic acid, iminodi-**
HO_2C–C_6H_4–O–C_6H_4–CO_2H	**Benzoic acid, 2,3′-oxydi-**
HO_3S–C_6H_4–N:N–C_6H_4–SO_3H	**Benzenesulfonic acid, 2,3′-azodi-**

If the acids are different the name of the compound may appear under one of them or under the name of the linking functions.

49. After the name of the parent acid the particle **per-** is used; *e.g.:*

CH_3CO_3H	**Acetic acid, per-**

50. **Organic acids other than carboxylic, sulfonic, sulfinic and sulfenic.** See IUC rule 34.

51. **Inorganic acids.** Their derivatives may appear either under the name of the acid or under the name of the organic base; *e.g.:*

$SO_2(OCH_3)_2$	**Sulfuric acid,** dimethyl ester
$C_6H_5NH_2.HNO_3$	**Aniline** nitrate.

52. **Isocyanides.** The prefix **isocyano-** is used for the substituent —NC; *e.g.:*

C_6H_5—NC	**Benzene, isocyano-**

53. **Anhydrides.**

a) *Symmetrical anhydrides of monobasic acids.* After the name of the acid comes the word anhydride; *e.g.:*

$(CH_3CHBrCO)_2O$	**Propanoic acid, 2-bromo-,** anhydride.

b) *Mixed anhydrides:* Both possible names appear (one of them as a cross-reference); *e.g.:*

$CH_3CHClCOOCOCH_2CH_2Cl$	**Propanoic acid, 2-chloro-,** anhydride with 3-chloropropanoic acid; and also: **Propanoic acid, 3-chloro-,** anhydride with 2-chloropropanoic acid.

c) *Cyclic anhydrides of polyacids* are named by placing the word anhydride after the name of the acid; *e.g.:*

1,4,5,8-Naphthalenetetracarboxylic acid, 1:8,4:5-dianhydride

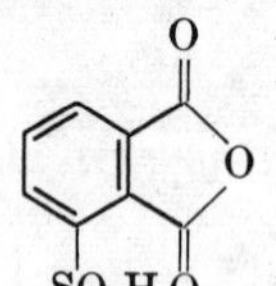

Phthalic acid, 3-sulfo-, anhydride.

54. **Acid halides and azides.** The name of the substituent of the hydroxyl group appears after the name of the corresponding acid; *e.g.:*

$CH_3CHClCOCl$	**Propanoic acid, 2-chloro-,** chloride
C_6H_5—CON_3	**Benzoic acid,** azide

55. **Amides.**

a) *Primary amides* appear under the name of the corresponding acid followed by the word amide; *e.g.:*

CH_3CONH_2	**Acetic acid,** amide (cross reference under **Acetamide**)

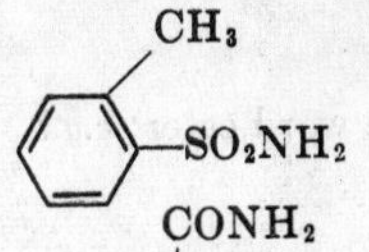

(structure: 2-methylbenzenesulfonamide, CH_3, $-SO_2NH_2$) **2-Toluenesulfonic acid,** amide

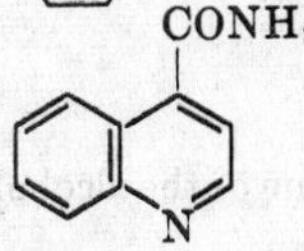

(structure: quinoline-4-carboxamide, $CONH_2$, N) **4-Quinolinecarboxylic acid,** amide

Nevertheless, when in forming the name of the corresponding acid the particle **carboxy-** is used, and this group turns to an amide group, the prefix **carbamyl-** is used; *e.g.:*

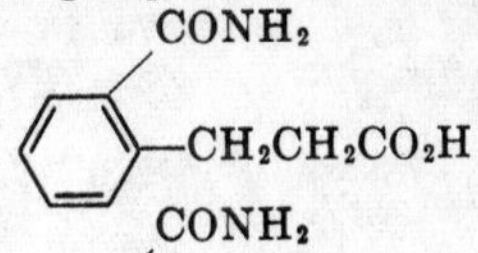

(structure: $CONH_2$, $-CH_2CH_2CO_2H$) **Propanoic acid, 3(2-carbamylphenyl)-** (see paragraph 37),

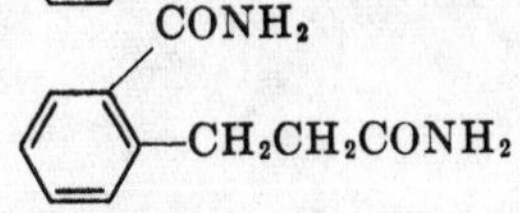

(structure: $CONH_2$, $-CH_2CH_2CONH_2$) **Propanoic acid, 3(2-carbamylphenyl)-,** amide

b) *N-Alkyl, N-Aryl and other N-derivatives of amides.* Substituents in the N atom appear in alphabetical order after the word amide; *e.g.:*

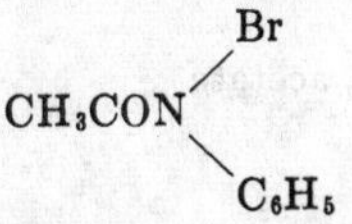

$CH_3CON(Br)(C_6H_5)$ **Acetic acid,** amide, *N*-bromo-*N*-phenyl-

Compounds with heterocyclic radicals attached to the *N*-atom of formamide, acetamide and benzamide are named with the heterocycle as parent compound using the particles **formamido-, acetamido-** and **benzamido;** *e.g.:*

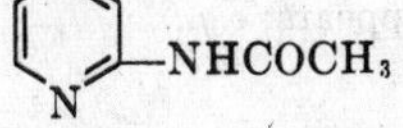

(structure: pyridine, $-NHCOCH_3$) **Pyridine, 2-acetamido-**

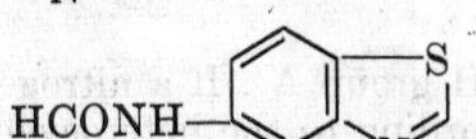

(structure: $HCONH-$ benzothiazole, S, N) **Benzothiazole, 5-formamido-**

Derivatives of other amides follow the general rule stated above, *e.g.:*

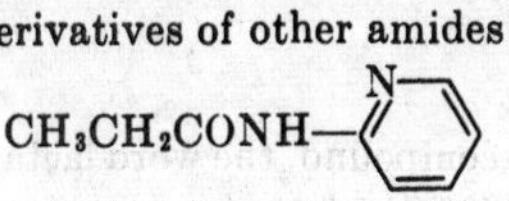

CH_3CH_2CONH-(2-pyridyl) **Propanoic acid,** amide, *N*(2-pyridyl)-

Amides in which the amide nitrogen is part of a heterocyclic system are named as acyl derivatives of that heterocycle:

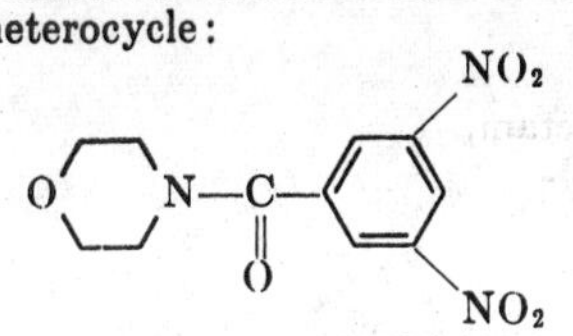

(structure: morpholine N–C(=O)– 3,5-dinitrophenyl, NO_2, NO_2) **Morpholine, 4(3,5-dinitrobenzoyl)-**

c) *Secondary and tertiary amides* are named according to the largest carboxylic acid from which it can be derived; *e.g.:*

$CH_3CONHCOCH_3$ **Acetic acid,** amide, *N*-acetyl-
(cross reference under **Diacetamide**)

$C_6H_5CONHCOCH_3$ **Benzoic acid,** amide, *N*-acetyl-

$(C_6H_5CO)_3N$ **Benzoic acid,** amide, *N,N*-dibenzoyl-
(cross reference under **Tribenzamide**)

A mixed amide of a carboxylic and a sulfonic acid is named as a derivative of the amide of the carboxylic acid *e.g.:*

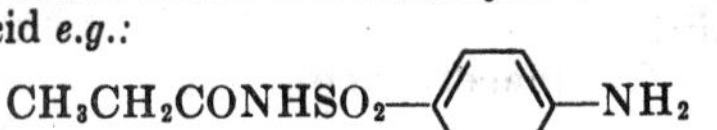

$CH_3CH_2CONHSO_2-C_6H_4-NH_2$ **Propanoic acid,** amide, *N*(4-aminobenzenesulfonyl)-

d) *Amides of hydroxamic acids.* As in paragraph 52a.

56. **Polypeptides.** A small number of polypeptides appear in the Table. They are found under their trivial name, or under the name of the amino acid that retains its carboxylic group and shares its amino group; *e.g.:*

$CH_3CH(NH_2)CONHCH_2CONHCH_2CO_2H$

Glycine, *N*(*N*-alanylglycyl)-

57. **Hydrazides of acids.** See paragraph 110.

58. **Esters.** After the name of the corresponding acid comes the radical and the word ester; *e.g.:*

$C_6H_5COOC_6H_5$ — **Benzoic acid,** phenyl ester

$HOSO_2OCH_3$ — **Sulfuric acid,** monomethyl ester

Nevertheless, esters of complicated and polyvalent alcohols may appear under the name of the alcohol; *e.g.:*

$CH_2O_2C(CH_2)_{15}CH_3$
$CHO_2C(CH_2)_{15}CH_3$
$CH_2O_2C(CH_2)_{15}CH_3$ — **Glycerol,** triheptadecenoate

CH_3C— — **Menthol,** acetate

CO_2H
CH_3CO_2— — **Aspirin,** *see* **Benzoic acid, 2-hydroxy-,** acetate

59. **Orthoesters.** See paragraph 47.

60. **Amidoximes** are treated as amides of hydroxamic acid (see paragraphs 41, 55).

61. **Amidines.** After the name of the corresponding acid the word **amidine** appears; *e.g.:*

$CH_3(CH_2)_6C(:NH)NH_2$ — **Octanoic acid,** amidine

The nitrogen atom of the $—NH_2$ group is designated N and that of the $=NH$ group N'. If a nitrogen atom of the amidine function is a part of a ring the compound is named according to the heterocyclic system.

62. **Betaines.** See paragraph 35.

63. **Lactams and lactones.** If there is no name for the ring system of the compound, the word lactam or lactone appears after the name of the corresponding aminoacid or hydroxyacid; *e.g.:*

CH_3— ... =O, N—H — **Hexanoic acid, 5-amino-,** lactam, *see* **2-Piperidone, 6-methyl-**

64. **Nitriles.** After the name of the corresponding acid the word nitrile appears; *e.g.:*

$CH_3CH(NH_2)CN$ — **Alanine,** nitrile

—CN — **2-Naphthoic acid,** nitrile

$CH_3(CH_2)_5CN$ — **Heptanoic acid,** nitrile

$NCCH_2CO_2H$ — **Malonic acid,** mononitrile (cross-reference under **Cyanoacetic acid**)

If the group —CN comes instead of a $—CO_2H$ group that is designated as **carboxy-,** the particle **cyano-** is used; *e.g.:*

CN
—CH_2CH_2COOH — **Propanoic acid, 3(2-cyanophenyl)-**

65. **Imides.** The word imide appears after the name of the corresponding dicarboxylic acid:

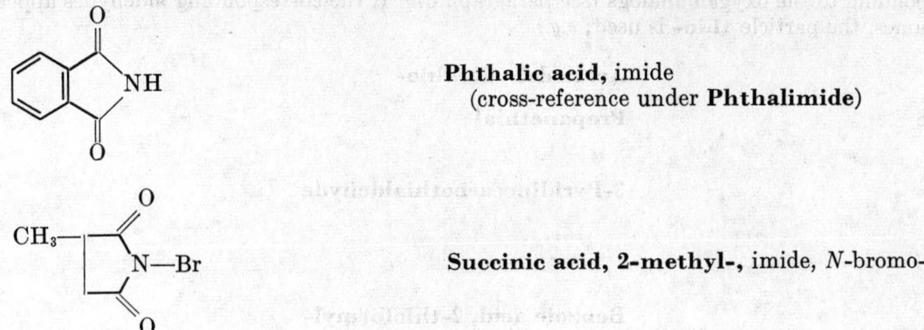

Phthalic acid, imide
(cross-reference under **Phthalimide**)

Succinic acid, 2-methyl-, imide, *N*-bromo-

For imides of monobasic acids see paragraph 55c.

66. **Cyanates, isocyanates, thiocyanates and isothiocyanates,** appear under the headings **Cyanic acid, Ioscyanic acid, Thiocyanic acid** and **Isothiocyanic acid** respectively, or, when the radical attached to these groups is complicated, the compound appears under the name of the corresponding alcohol (these rules are identical to those given for esters, amides, etc. in paragraphs 58, ff.).

ALDEHYDES

67. Geneva names for saturated and unsaturated aliphatic mono- and di-aldehydes are formed by dropping the final **e** of the name of the hydrocarbon and adding -al or (without dropping the **e**) **-dial.** The largest chain having the aldehydic group (or groups) as terminals is chosen.

Examples:

$CH_3(CH_2)_9CHO$	**Hendecanal**
$\overset{5}{C}H_3\overset{4}{C}H_2\overset{3}{C}H{:}\overset{2}{C}H\overset{1}{C}HO$	**2-Pentenal**
$OCHCH_2CH{:}CHCH{:}CHCH_2CHO$	**3,5-Octadienedial**

Some trivial names are used whenever the trivial name of the corresponding acid is used, *e.g.:* **formaldehyde, acetaldehyde, benzaldehyde, 1- and 2-naphthaldehyde, cinnamaldehyde,** etc.

When the **-carboxylic acid** nomenclature is used for the corresponding acid (see paragraphs 37, 39), the name of the aldehyde is formed by the use of the suffix **-carboxaldehyde;** *e.g.:*

(cyclohexenyl)—CHO	**1-Cyclohexanecarboxaldehyde**
OCH—(benzene ring)—CHO	**1,4-Benzenedicarboxaldehyde**
$OCH(CH_2)_4CH(CHO)_2$	**1,1,5-Pentanetricarboxaldehyde**

When a function of higher rank (see paragraph 32) appears in the molecule the —CHO group is designated as **oxo-** or as **formyl-** as the case demands; *e.g.:*

OCH—(benzene ring)—CO_2H	**Benzoic acid, 4-formyl-**
$OCHCH_2CH_2CO_2H$	**Butanoic acid, 4-oxo-**
$CH_3CH_2CH_2CH(CHO)CH_2CO_2H$	**Hexanoic acid, 3-formyl-**

68. **Acetals, hemiacetals and hydrates of aldehydes or ketones** appear in the Table immediately after the corresponding aldehyde or ketone. The following examples illustrate the nomenclature:

$CH_3CH_2CH(OC_2H_5)$	**Propanal,** diethyl acetal
$CH_3CH_2CH(OCH_3)OH$	**Propanal,** methyl hemiacetal
$Cl_3CCH(OH)_2$	**Acetaldehyde, trichloro-,** hydrate

69. **Thioaldehydes** are named by using the particles **-thial, -carbothialdehyde, thioformyl-** or **thioxo-** corresponding to the oxygen analogs (see paragraph 67). If the corresponding aldehydes appear under trivial names, the particle **thio-** is used; *e.g.:*

CH_3CHS	**Acetaldehyde, thio-**
CH_3CH_2CHS	**Propanethial**
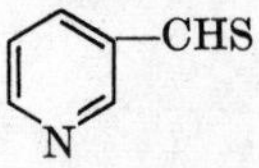	**3-Pyridinecarbothialdehyde**
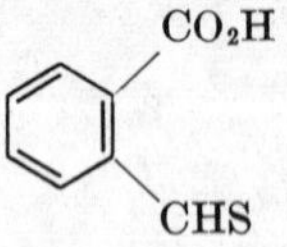	**Benzoic acid, 2-thioformyl-**
$SCHCH_2CO_2H$	**Propanoic acid, 3-thioxo-**

70. **Azines.** See paragraph 109.

71. **Imines.** See paragraph 99.

72. **Oximes.** The word oxime appears after the name of the corresponding aldehyde (or ketone).

KETONES

73. Ketones with no ring attached directly to the carbonyl group are named according to IUC rule 27.

74. Ketones with one aromatic ring attached to the carbonyl group are named according to IUC rule 27; *e.g.:*

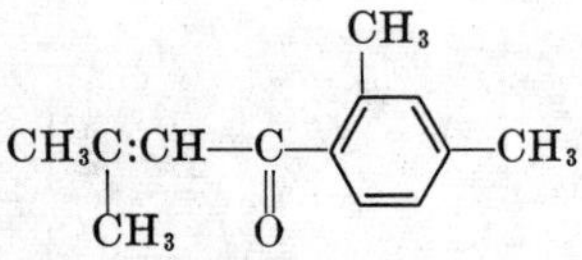	**2-Buten-1-one, 1(2,4-dimethylphenyl)-3-methyl-**
$CH_3CH_2COC_6H_5$	**1-Propanone, 1-phenyl-** (cross-reference under **Propiophenone**)

The following are exceptions:

$C_6H_5COCH_3$	**Acetophenone**
$C_6H_5CH{:}CHCOC_6H_5$	**Chalcone**
$C_6H_5COC_6H_5$	**Benzophenone.**

75. Monoketones with two rings attached to both sides of the carbonyl group appear under the heading **Ketone,** followed by the two rings in alphabetical order and then followed by the substituents in the rings (see paragraph 25); *e.g.:*

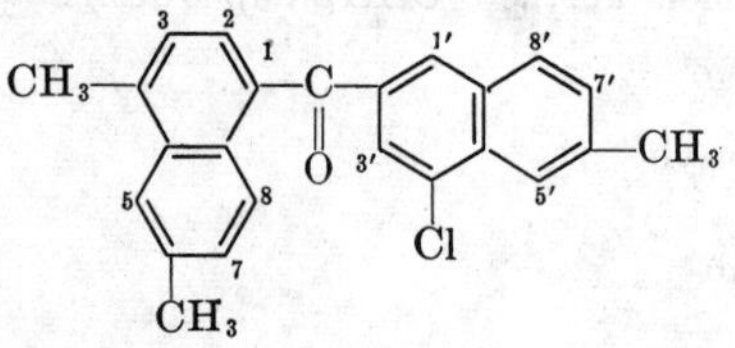	**Ketone, 1,2′-dinaphthyl, 4′-chloro-4,6,6′-trimethyl-**
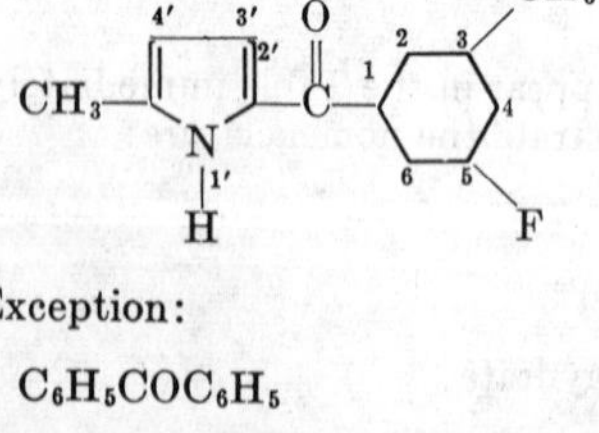	**Ketone, cyclohexyl 2-pyrryl, 3,5′-dimethyl-5-fluoro-**

Exception:

$C_6H_5COC_6H_5$	**Benzophenone**

76. Acetyl derivatives of cyclic compounds other than acetophenone are named according to the name of the cyclic compound:

O ‖ —CCH$_3$ (O in ring) **Furan, 2-acetyl-**

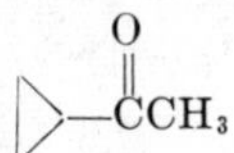

Cyclopropane, acetyl-

77. **Acetals, hemiacetals and hydrates of ketones.** See paragraph 68.

78. **Thiones (thioketones)** follow the Geneva system and IUC rule 27. When the corresponding ketone was given a trivial name the thione has the same name followed by the particle **thio-;** *e.g.:*

$C_6H_5CSCH_3$ **Acetophenone, thio-**

The particle **thioacetyl-** is used to designate the group CH_3CS— and its use is similar to that of the particle **acetyl-** (see paragraph 76). The heading corresponding to **Ketone** is **Thione;** and it is used as described in paragraph 75; *e.g.:*

Thione, cyclohexyl 3-pyridyl

The particle corresponding to **oxo-** is **thioxo-;** *e.g.:*

Cyclohexanecarboxylic acid, 5-methyl-2-thioxo-, amide, *N*-methyl-

79. **Ketenes.** According to IUC rule 28.

80. **Azines.** See paragraph 109.

81. **Imines.** See paragraph 99.

82. **Oximes.** See paragraph 72.

83. **Cyclic ketones and thiones.** When a cyclic system whose leading function is ketone or thione originally had two hydrogens in the site where O= or S= are attached, the compound gets the name of the parent compound with the ending **-one** or **-thione** *e.g.:*

Tetralin

2-Tetralone

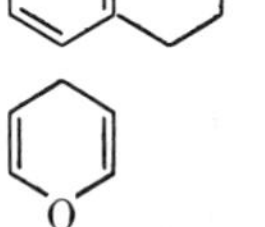

Pyran

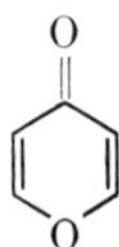

4-Pyrone

The ring system so obtained (including oxo- or thioxo- functions) may be used as a new ring system for the naming of compounds containing higher order functions (see paragraph 21); *e.g.:*

4-Pyrone-3-carboxylic acid

If the compound originally had only one hydrogen at the site of the ketonic function, the particles **oxo-** or **thioxo-** together with **dihydro-**, **tetrahydro-**, etc. are used; *e.g.:*

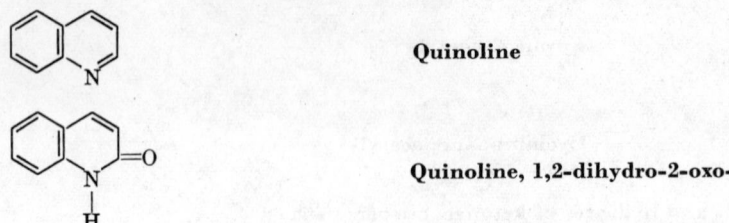

Quinoline

Quinoline, 1,2-dihydro-2-oxo-

84. **Quinones.** IUC rule 46 is followed. The quinone ring system may be used as a basis for the naming of compounds containing higher order functions.

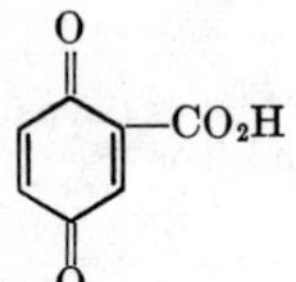

1,4-Benzoquinone-2-carboxylic acid

ALCOHOLS

85. **Aliphatic alcohols** appear under their Geneva names (IUC rule 20).

86. **Monocyclic aliphatic alcohols** appear under their Geneva names (IUC rule 20); *e.g.:*

Cyclohexanol, 2-methyl-

3-Cyclopenten-1-ol

87. **Heterocyclic and aromatic alcohols (phenols).** The name **Phenol** is used for C_6H_5OH and ring substituted derivatives with **no carbon** attached to the ring and no other hydroxyl group or higher order of precedence functions; *e.g.:*

Phenol, 3-amino-

Phenol, 4-methoxy-

Otherwise the compound appears under benzene, toluene or elsewhere; *e.g.:*

Toluene, 3-hydroxy-

Benzene, 1-hydroxy-3-isopropyl-

OH, –OH (structure) — **Benzene, 1,3-dihydroxy-** (cross-reference under **Resorcinol)**

CH_2OH, –OH (structure) — **Toluene, *a*,2-dihydroxy-**

NH_2, HO–, –OH (structure) — **Benzene, 1-amino-, 3,5-dihydroxy-**

Alcohols precede phenols in order; *e.g.:*

HO–C6H4–CH(OH)–C6H5 — **Methanol, (4-hydroxyphenyl)phenyl-**

$CH_3CHOHCH_2$–C6H4–OH — **2-Propanol, 1(2-hydroxyphenyl)-**

Exception: **Toluene, *a*-hydroxy-**, *not* **Methanol, phenyl-**
Other aromatic and heterocyclic phenols or alcohols will be named considering the hydroxyl group as an ordinary substituent; *e.g.:*

OH, –OH (structure) — **Naphthalene, 1,2-dihydroxy-**

–OH, N–H (structure) — **Piperidine, 3-hydroxy-**

88. **Polyalcohols** such as **glycerol, erythrol,** etc. appear under their trivial names. Exception: **1,2-ethanediol.**

89. **Sugars** appear under their trivial names.

90. **Esters of complicated alcohols.** See paragraph 58.

91. **Hemiacetals and hydrates of aldehydes and ketones.** See paragraph 68.

92. The **carbinol** nomenclature for alcohols **is not used.**

93. **Thiols.** Aliphatic thiols follows IUC rule 22. Aromatic and heterocyclic thiols and molecules containing higher order functions together with the SH group will be named using the particle **mercapto.** Thiol is lower than alcohol but higher than amine in the order of precedence; *e.g.:*

CH_3CH_2SH	**Ethanethiol**
HS–C6H4–CO_2H	**Benzoic acid, 4-mercapto-**
$HSCH_2CH_2OH$	**Ethanol, 2-mercapto-**

94. **Hydroperoxides.** See paragraph 112.

AMINES AND IMINES

95. The amine function is generally considered as an ordinary substituent (like **chloro-, nitro-,** etc.) *e.g.:*

$CH_3CH_2NH_2$	**Ethane, amino-**
–NH_2, N (structure)	**Pyridine, 3-amino-**

96. The name **aniline** is used for the compound $C_6H_5NH_2$ and as a parent compound for its derivatives unless a carbon atom is attached to the benzene ring or a function of rank higher than amino is present in the molecule; *e.g.:*

$O_2N—C_6H_4—NH_2$ **Aniline, 4-nitro-**

$H_2N—C_6H_4—OH$ **Phenol, 4-amino-**

$C_2H_5—C_6H_4—N(CH_2CH{:}CH_2)_2$ **Benzene, 1-diallylamino-4-ethyl-**

$CH_3CONH—C_6H_5$ **Acetic acid,** amide, *N*-phenyl-
(cross reference under **Acetanilide**)

$C_6H_5—C_6H_4—N(CH_3)(C_2H_5)$ **Biphenyl, 4(ethylmethylamino)-**

97. A list of names beginning with the word *Amine* appears in the Table; sometimes a cross-reference is given to the preferred name based on the largest radical present. If the choice is between a saturated and an unsaturated radical of the same number of carbon atoms, the unsaturated one is preferred. Regarding this nomenclature see paragraph 25.

$NH(CH_2C_6H_5)_2$ **Amine, dibenzyl-**

$C_2H_5—N(C_6H_5)—CH_3$ **Amine, ethyl methyl phenyl,**
see **Aniline, *N*-ethyl-*N*-methyl-**

$C_{10}H_7—NHC_6H_5$ **Naphthalene, 2(phenylamino)-,**
see **Amine, 2-naphthyl phenyl**

$C_{10}H_7—NH—C_6H_4—Cl$ **Amine, 2-naphthyl phenyl, 1,4′-dichloro-**
(see paragraph 25)

98. **Hydroxylamine derivatives.** See paragraph 111.

99. **Imines.** The particle **imino-** is used for the group NH= when it is attached to a single carbon atom; *e.g.:*

$CH_3CH{=}NH$ **Ethane, imino-**

In some instances when the *imine* group comes in place of the oxygen of a keto-group the word imine appears after the name of the ketone; *e.g.:*

O=C₆H₃(Cl)=NH **1,4-Benzoquinone, 2-chloro,** 4-imine

(This is similar to the usage in the case of oximes, see paragraph 72.)

ETHERS

100. Simple and complex monoethers are found under the name **Ether** followed by the two radicals (without substituents) attached to the oxygen in alphabetical order and finally followed by the substituents in the skeleton of the ether (see paragraph 25); *e.g.:*

CH_3OCH_3 **Ether, dimethyl**

$BrCH_2OCH_2Br$ **Ether, dimethyl, 1,1′-dibromo-**

$(C_6H_5)_2CHOCH_3$ **Ether, (diphenylmethyl) methyl**

$C_6H_5CH_2OC_6H_5$ **Ether, benzyl phenyl**

$HC{\vdots}COC_6H_5$ **Ether, ethynyl phenyl**

101. Alkoxy derivatives of benzene (linear, saturated and with less than 6 carbons) and alkoxy derivatives of polycyclic or heterocyclic hydrocarbons appear under the cyclic hydrocarbon as a parent compound; *e.g.:*

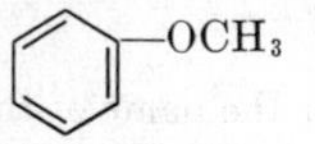

$-OCH_3$ **Benzene, methoxy** (cross-reference under **Anisole**)

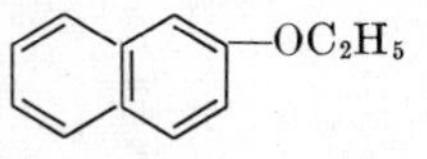

$-OC_2H_5$ **Isoquinoline, 3-ethoxy-**

102. **Epoxy compounds.** Follow IUC rule 24, but cyclic ethers such as **furan, dioxan, dioxolan,** etc. are designated by their heterocyclic system name.

103. **Polyethers.** Some appear under their trivial names. Ethers of polyalcohols sometimes appear after the name of the corresponding alcohol.

104. **Peroxides.** See paragraph 112.

105. **Thioethers.** See paragraph 114.

106. **Orthoesters.** See paragraph 47.

MISCELLANEOUS FUNCTIONS

107. **Azo and azoxy compounds. Azo** and **Azoxy** are the headings for these compounds. After the word **azo** or **azoxy** comes the names of the hydrocarbon (if symmetrical) or hydrocarbons (if unsymmetrical) attached to the azo or azoxy group in alphabetical order; places of substitution are given with unprimed numbers for the first hydrocarbon and primed numbers for the second hydrocarbon (see paragraph 25); *e.g.:*

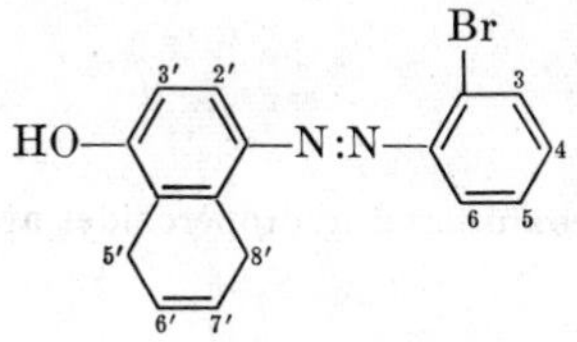

Br; 3′ 2′; 3; HO—; —N:N—; 4; 6 5; 5′; 8′; 6′ 7′ **Azo, benzene 1-naphthalene, 2-bromo$_2$, 4′-hydroxy-**

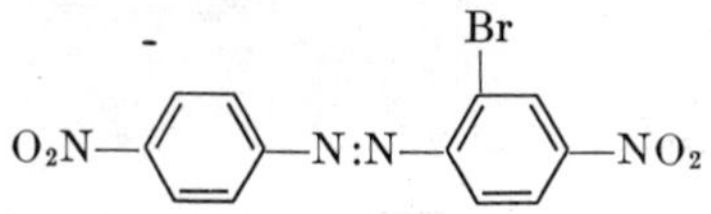

Br; O_2N- —N:N— $-NO_2$ **Azobenzene, 2-bromo-4,4′-dinitro-**

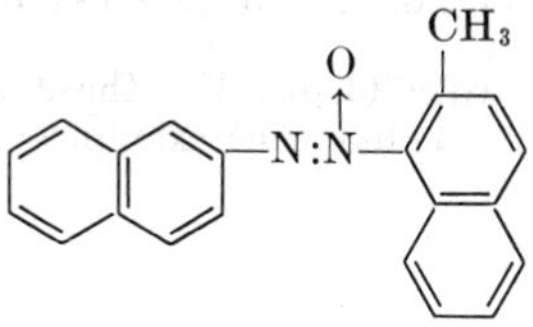

CH_3; O; ↑; —N:N— **1,2′-Azoxynaphthalene, 4′-methoxy-2-methyl-**

In the case of **azoxy** no attempt is made to distinguish between the nitrogen atoms.
Some of the azo compounds are cross-referred to other parent compounds (see paragraph 48).
Some multiple azo compounds are named by derivation from the most highly substituted nucleus; *e.g.:*

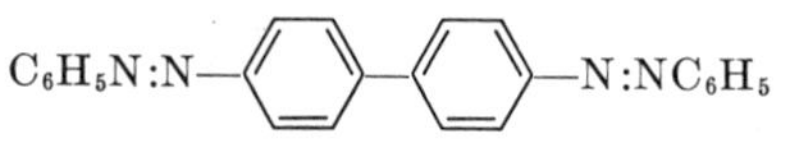

$C_6H_5N:N-$ $-N:NC_6H_5$ **Biphenyl, 4,4′-bis(phenylazo)-**

108. **Azides.** Azides of acids according to paragraph 54.
For other azides the particle **triazo-** is used (see paragraph 31); *e.g.:*

$CH_3CH_2N_3$ **Ethane, triazo-**

109. **Azines.** Appear under the name of the corresponding aldehyde or ketone if they are symmetrical; if not, under two headings with cross references; *e.g.:*

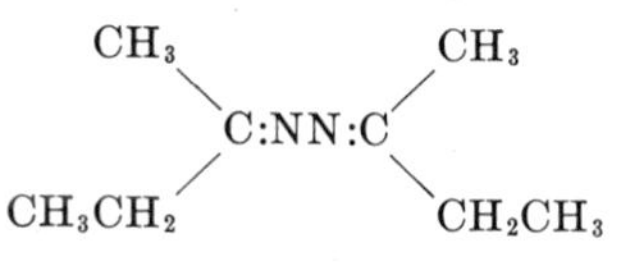

$(CH_3)(CH_3CH_2)C:NN:C(CH_3)(CH_2CH_3)$ **2-Butanone,** azine

CH_3 \ C:NN:CHCH$_3$ / CH_3	**2-Propanone,** azine with acetaldehyde *see* **Acetaldehyde,** azine with 2-propanone

110. **Hydrazine derivatives.** Monoacyl derivatives of hydrazine are entered under the name of the acid followed by the word hydrazide; *e.g.:*

$CH_3CONHNH_2$	**Acetic acid,** hydrazine

Diacyl hydrazines appear under **Hydrazine,** *e.g.:*

$CH_3CONHNHCOC_6H_5$ (1 2)	**Hydrazine, 1-acetyl-2-benzoyl-**

When a function of higher order of precedence (amino or higher) is also present, it may be convenient to use the particle **hydrazino-;** *e.g.:*

$NH_2NHCH_2CH_2CN$	**Propanoic acid, 3-hydrazino,** nitrile
$NH_2NHCONH_2$ (1 2 3 4)	is **Semicarbazide,** with derivatives under this name, but semicarbazones appear under the corresponding ketone or aldehyde.
$NH_2CONHNHCONH_2$ (1 2 3 4 5 6)	is **Biurea,** with all its derivatives appearing under this name.

111. **Hydroxylamine derivatives. Hydroxylamino-** is used for HONH—, and **aminooxy** for NH_2O—. Names of *N*-acyl derivatives of hydroxylamine are derived from that of the corresponding amide; *e.g.:*

$CH_3CONHOH$	**Acetic acid,** amide, *N*-hydroxy-

Oximes are explained in paragraph 72.

112. **Peroxides and hydroperoxides.** Peroxides are listed under **Peroxide** and hydroperoxides are listed under **Hydroperoxide;** *e.g.:*

C_2H_5OOH	**Hydroperoxide, ethyl**
$C_6H_5CH_2OOCH_2C_6H_5$	**Peroxide, dibenzyl**
$C_6H_5OOCH_3$	**Peroxide, methyl phenyl**

113. **Miscellaneous Phosphorus, Arsenic and Antimony Compounds.** IUC rule 34 is followed.

114. **Sulfides, disulfides, trisulfides sulfoxides and sulfones** follow rules identical to those of **ethers** see (paragraphs 100, 101), using an appropriate word or particle, as in the following examples:

CH_3OCH_3	**Ether,**	**dimethyl**	
CH_3SCH_3	**Sulfide,**	,,	
CH_3SSCH_3	**Disulfide,**	,,	
CH_3SSSCH_3	**Trisulfide,**	,,	
$CH_3S(O)CH_3$	**Sulfoxide,**	,,	
$CH_3S(O_2)CH_3$	**Sulfone,**	,,	
$CH_3OC_2H_5$	**Ether,**	**ethyl**	**methyl**
$CH_3SC_2H_5$	**Sulfide,**	,,	,,
$CH_3SSC_2H_5$	**Disulfide,**	,,	,,
$CH_3SSSC_2H_5$	**Trisulfide,**	,,	,,
$CH_3S(O)C_2H_5$	**Sulfoxide,**	,,	,,
$CH_3S(O_2)C_2H_5$	**Sulfone,**	,,	,,
$C_6H_5OCH_3$	**Benzene, methoxy-**		

$C_6H_5SCH_3$	**Benzene, methylthio-**
$C_6H_5SSCH_3$	,, **methyldithio**
$C_6H_5SSSCH_3$	,, **methyltrithio-**
$C_6H_5S(O)CH_3$	,, **methylsulfinyl-**
$C_6H_5S(O_2)CH_3$	,, **methylsulfonyl-**

115. **Urea derivatives.** According to IUC rule 36.

Cyclic urea derivatives are named according to their ring system or trivial names:

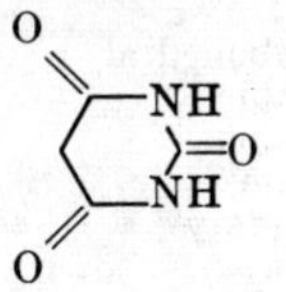

Barbituric acid

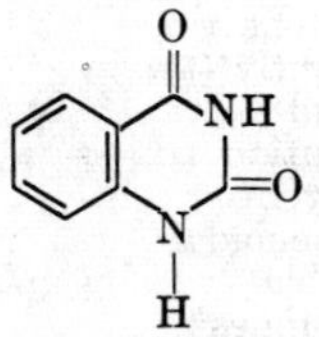

Quinazoline, 2,4-dioxo-1,2,3,4-tetrahydro-

The ureido function follows in order of precedence after the onium and the acid functions. The particle **ureido-** is used for the group NH_2CONH— and **thioureido-** for NH_2CSNH—; *e.g.:*

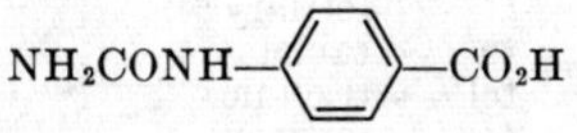

NH_2CONH— —CO_2H **Benzoic acid, 4-ureido-**

SYMBOLS AND ABBREVIATIONS

[*α*]	specific rotation
δ	slightly
>	above, more than
<	below, less than
∞	soluble in all proportions
*	name approved by the International Union of Chemists (I.U.C)[1]
?	unknown
Am	J. Am. Chem. Soc.
aa	acetic acid
abs	absolute
ac	acid
Ac	acetyl (CH_3CO—)
ace	acetone
al	alcohol[2]
alk	alkali
Am	amyl (pentyl)
amor	amorphous
anh	anhydrous
aqu	aqueous
as	asymmetric
atm	atmospheres
b	boiling
B	Beilstein
bipym	bipyramidal
bk	black[3]
bl	blue[3]
br	brown[3]
bt	bright
Bu	butyl
bz	Benzene
C	Chem. Abs.
c	percentage concentration
ca	about (circa)
chl	chloroform
co	columns
col	colorless
con	concentrated
cor	corrected
cr	crystals
d	decomposes
D	line in the spectrum of sodium (subscript)
D, d	dextro[4]
δd	slight decomposition
dil	diluted
diox	dioxane
distb	distillable
dk	dark
Dl, dl	racemic[4]
dlq	deliquescent
E	Elsevier's
eff	efflorescent
Et	ethyl
eth	ether[5]
exp	explodes
extrap	extrapolated
fl	flakes
flr	fluorescent
fr	freezes
fr. p.	freezing point
fum	fuming
gel	gelatinous
gl	glacial
gr	green[3]
gran	granular
gy	gray[3]
h	hot
H	Helvetica
hex	hexagonal
hyd	hydrate
hyg	hygroscopic
i	insoluble
i-	iso-
ign	ignites
in	inactive
inflam	inflammable
infus	infusible
irid	iridescent
J	J. Chem. Soc.
L, l	levo[4]
la	large
lf	leaf
li	little
lig	ligroin
liq	liquid
lo	long
lt	light
m	melting
m	meta
m-	meta-
M	molar (concentration)
mcl	monoclinic
Me	methyl
met	metallic
micr	microscopic
min	mineral
mod	modification
mut	mutarotatory
n	normal chain, refractive index
N	normal (concentration)
N	nitrogen[6]
Nat.	natural product[8]
nd	needles
o-	ortho-
oct	octahedral
og	orange[3]
or	or
oos	ordinary organic solvents
ord	ordinary
org	organic
orh	orthorhombic
os	organic solvents
p-	para-
pa	pale
par	partial
peth	petroleum ether
pk	pink[3]
Ph	phenyl
pl	plates
pr	prisms
Pr	propyl
purp	purple[3]
pw	powder
Py	pyridine
pym	pyramids
rac	racemic
rect	rectangular
red	red
rhd	rhombohedral
rh	rhombic
res	resinous
s	soluble
s	secondary[8]
sc	scales
sec	secondary[8]
sf	softens
sl	slightly (δ)
so	solid
sol	solution
solv	solvent
sph	sphenoidal
st	stable
sub	sublimes
suc	supercooled
sulf	sulfuric acid
sym	symmetrical
syr	syrup
t	tertiary[8]
ta	tablets
tcl	triclinic
tert	tertiary[8]
tetr	tetragonal
to	toluene
tr	transparent
trg	trigonal
undil	undiluted
uns	unsymmetrical
unst	unstable
v	very
vac	vacuum
var	variable
vic	vicinal
visc	viscous
volat	volatile or volatilises
vt	violet[3]
w	water
wh	white[3]
wr	warm
wx	waxy
ye	yellow[3]
xyl	xylene

1 For I.U.C. rules of nomenclature see General Index.

2 Generally means ethyl alcohol.

3 The abbreviation of a color ending in "sh" is to be read as ending with the suffix "-ish," *e.g.*, grsh means greenish.

4 *D, L* generally mean configuration and *d, l* generally means optical rotation, but there are many examples in the chemical literature for which the meaning of these symbols is ambigous and/or interchangeable.

5 Generally means diethyl ether.

6 *N* indicates a position in the molecule.

7 For additional information, see Table of Natural Products, according to section and number of entry indicated, *e.g.*, Nat. E2.

8 *s* and *sec.* or *t* and *tert*, are used as convenient.

No.	Name	Synonyms and Formula	Mol. wt.	Crystalline form, color and specific rotation	m.p. °C	b.p. °C	Density	n_D	Solubility w	al	eth	ace	bz	other solvents	Ref.
	Abietic acid														
a1	**Abietic acid**	Sylvic acid. $C_{20}H_{30}O_2$	302.46	pl$[\alpha]_D^{19}$ −104 (al, c = 10)	172–3	250^9			i	s	v	s	v	CS_2 s	**B9²**, 424
a2	—,methyl ester	$C_{21}H_{32}O_2$	316.49			$225–6^{16}$	1.049_4^{20}							aa s	**B9²**, 430
—	**Abacetin**	*see* **Flavone, 5,7-dihydroxy-4′-methoxy-**													
a3	**Acenaphtha-anthracene**	Naphtho-2′,3′:4,5-acenaphthene.	254.31	pa ye lf (lig)	192.5–3.5				i	.	.	.	s	con sulf s aa s lig s^h	**E14s**, 565
a4	**Acenaphthene**	Naphthyleneethylene.	154.21	nd (al)	96	278^{760}	1.0242_4^{99} 1.225_4^0	1.6066^{99}	i	δ s^h			v	aa s	**B5²**, 495
a5	—,**1-amino-**	$C_{12}H_{11}N$. *See* a4	169.23		135	sub			δ	v				os s	**B12²**, 764
a6	—,**3-amino-**	$C_{12}H_{11}N$. *See* a4	169.23	pl (al), nd (peth) pk in air	81.5					s				chl s (red sol)	**E13**, 148
a7	—,**4-amino-**	$C_{12}H_{11}N$. *See* a4	169.23	nd (al, w)	87				s^h					os v	**B12²**, 764
a8	—,**5-amino-**	$C_{12}H_{11}N$. *See* a4	169.23	nd (lig) red in air	108					s					**E13**, 149
a9	—,**5-bromo-**	$C_{12}H_9Br$. *See* a4	233.11	pl (al)	51.2	335^{760}	1.4392_4^{52}	1.6565^{54}	i	δ s^h					**B5¹**, 276
a10	—,**5-chloro-**	$C_{12}H_9Cl$. *See* a4	188.66	pl or nd (al)	70.5	319.2^{770}	1.954_4^{70}	1.6288^7	i	δ s^h					**B5¹**, 276
a11	—,**5-iodo-**	$C_{12}H_9I$. *See* a4	280.11	nd (al)	63–3.5		1.6687_4^{67}	1.6909		s			s	aa s	**E13**, 145
a12	—,**3-nitro-**	$C_{12}H_9NO_2$. *See* a4	199.21	gr-ye nd (aa)	151.5									sulf s (gy-bl→red) con alk s (red)	**E13**, 147
—	—,**1-oxo-**	*see* **1-Acenaphthenone**													
a14	**5-Acenaphthene-carboxylic acid**	5-Acenaphthoic acid.	198.22	(lig)	217								δ		**B9¹**, 280
a15	**Acenaphthene-quinone**		182.18	ye nd (aa)	261				i	δ			δ	lig s^h	**E13**, 169
a16	**3-Acenaphthene sulfonic acid**		234.28	hyg nd (bz)	87–9				s						**E13**, 181
a17	**1-Acenaph-thenone**	1-Oxoacenaphthene.	168.20	nd (al)	121				s^h	s^h			v	sulf s (ye-gr) NaOH s (vt)	**E13**, 164
a18	**Acenaphthylene**		152.20	pr (eth), pl (al)	92–3	265–75 par d	0.8988_2^{16}		i	v	v		v		**B5²**, 530
a19	**Acetaldehyde**	Acetic aldehyde. Ethanal*. CH_3CHO	44.05		−124.6	20.8^{760}	0.7834_4^{18}	1.3316^{20}	∞^h	∞	∞		∞		**B1³**, 321
a20	—,bis(2-chloroethyl)	Bis(2-chloroethoxy)ethane*. $CH_3CH(OCH_2CH_2Cl)_2$	187.07			194–6d	1.1712_{19}^{19}	1.4532^{16}							
a21	—,diacetate	Ethylidene diacetate. $CH_3CH(O_2CCH_3)_2$	146.14		18.9	169^{760}	1.3985^{25}	1.070^{25}		s					**B2²**, 167
a22	—,diethyl acetal	Acetal. 1,1-Diethoxyethane*. Ethylidene diethyl ether. $CH_3CH(OC_2H_5)_2$	118.18	volat		102.2^{760}	0.8314_4^{20}	1.3819	s	∞	∞				**B1³**, 326
a22¹	—,dimethyl acetal	1,1-Dimethoxyethane*. Ethylidene dimethyl ether. $CH_3CH(OCH_3)_2$	90.12		−113.2	64.5	0.8501_4^{20}	1.3668^{20}	s	s	s			chl s	**B1²**, 671

For explanations, symbols and abbreviations see beginning of table.

No.	Name	Synonyms and Formula	Mol. wt.	Crystalline form, color and specific rotation	m.p. °C	b.p. °C	Density	n_D	Solubility w	al	eth	ace	bz	other solvents	Ref.
	Acetaldehyde														
a23	—,2,4-dinitrophenylhydrazone (stable form)	$CH_3CHN{:}NH-C_6H_3(NO_2)-NO_2$	224.18	ye sc (al)	168.5				i	s^h	s	...	v	chl v	**B15**[1], 490
a24	—,—(unstable form)	$C_8H_8N_4O_4$. *See* a23	224.18	og-red	157										**B15**[1], 490
a25	—,oxime	Acetaldoxime. $CH_3CH{:}NOH$	59.07	nd	46.5	$114–5^{760}$	0.9656^{20}_{4}	1.4257^{20}	s^{wr}	∞	∞				**B1**[3], 675
a26	—,phenylhydrazone	*N*-Ethylidene-*N'*-phenylhydrazine. $CH_3CH{:}NNHC_6H_5$	134.18	nd or ta	98–101	$236–7^{21}$								lig δ^h	**B15**[2], 54
a27	—,semicarbazone	$CH_3CH{:}NNHCONH_2$	101.08	nd (w or al)	162		1.0300^{0}_{4}		δ	s					**B3**[3], 48
a28	—,**amino-**, diethyl acetal	Acetalylamine. Aminoacetal. β,β-Diethoxyethylamine. $H_2NCH_2CH(OC_2H_5)_2$	133.19			163	0.9159^{25}	1.4123^{25}	v	v	v			chl v	**B4**[2], 758
—	—,**ammonia**	*see* **Ethanol, 1-amino-**													
a29	—,**bromo-**, diethyl acetal	2-Bromoacetal. $BrCH_2CH(OC_2H_5)_2$	197.08			180^{760} d 66^{18}				s^h					**B1**[2], 682
a30	—,**chloro-**	$ClCH_2CHO$	78.50			$85–5.5^{748}$									**B1**[2], 675
a31	—,—,diethyl acetal	Chloroacetal. $ClCH_2CH(OC_2H_5)_2$	152.62			157.4^{760}	1.0180^{20}_{4}	1.4170^{20}	δ^h	∞	∞			dil sulf d	**B1**[2], 676
a32	—,—,dimethyl acetal	$ClCH_2CH(OCH_3)_2$	124.57			124.5–6.5	1.094^{20}_{20}	1.4150^{20}							
a33	—,**dichloro-**	Cl_2CHCHO	112.94			$87–8^{760}$				s					**B1**[2], 676
a35	—,—,diethyl acetal	Dichloroacetal. $Cl_2CHCH(OC_2H_5)_2$	187.07			185–6	1.1383^{14}								**B1**[2], 677
a36	—,—,ethyl hemiacetal	$Cl_2CHCH(OH)OC_2H_5$	159.01			109.5–111	1.314^{26}		δ	∞			∞	lig δ	**B1**[2], 677
a37	—,—,hydrate	$Cl_2CHCH(OH)_2$	130.96	cr (bz), ta	56–7	118–21			v	...	v			CS_2 s	**B1**[1], 614
a39	—,**dimethylamino-**, diethyl acetal	Dimethylaminoacetal. $(CH_3)_2NCH_2CH(OC_2H_5)_2$	161.25	ye		170–1	0.885^{7}		v	s	s			os s	**B4**[1], 308
a40	—,**diphenyl-**	2,2-Diphenylethanal*. $(C_6H_5)_2CHCHO$	196.25			315 δd 157.5^{7}	1.1061^{21}_{4}	1.5920^{21}	i	v	v	...	v		**B7**[2], 370
a41	—,**ethoxy-**	$C_2H_5OCH_2CHO$	88.11			105–6	0.942^{20}_{4}	1.3956^{20}	s	s					**B1**[1], 818
a42	—,**ethylmethylamino-**, diethyl acetal	$CH_3(C_2H_5)NCH_2CH(OC_2H_5)_2$	175.27			179–80			i					os v	**B4**[1], 309
a43	—,**hydroxy-**	Glycolaldehyde. Glycolic aldehyde. $HOCH_2CHO$	60.05	pl	97		1.366^{100}	1.4772^{19}	v	v	δ				**B1**[2], 863
a44	—,—,diethyl acetal	2,2-Diethoxyethanol*. $HOCH_2CH(OC_2H_5)_2$	134.18			167	0.888^{24}_{4}	1.4073^{20}		s	s				**B1**[2], 864
a45	—,**methoxy-**	CH_3OCH_2CHO	74.08			92.3	1.005^{25}_{4}	1.3950^{20}	s	s	s				
a45[1]	—,**2-oxo-2-phenyl-**, hydrate	Benzoyl formaldehyde, hydrate. Phenyl glyoxal, hydrate. $C_6H_5COCH(OH)_2$	152.15	nd (w, chl, alig, eth-liq)		93–4			δ	s	s			chl, CS_2 s	
a46	—,—,dioxime	Phenyl glyoxime. $C_6H_5C({:}NOH)CH{:}NOH$	164.17	nd (chl)	180				v^h	s	s			chl δ	**B7**[2], 602
a47	—,—,1-oxime	Isonitrosoacetophenone. $C_6H_5COCH{:}NOH$	149.15	mcl pr or lf (chl, w)	126–8				δ s^h					chl s^h	**B7**[2], 600
a48	—,**phenyl-**	α-Tolualdehyde. α-Toluic aldehyde. $C_6H_5CH_2CHO$	120.15	(w)	33–4	193–4	1.0272^{20}_{4}	1.5255^{20}	δ	∞	∞				**B7**[1], 292
a49	—,**tribromo-**	Bromal. Br_3CCHO	280.76			174^{760}	2.6650^{25}_{4}		d	s	s				**B1**[2], 683
a50	—,—,hydrate	$Br_3CCH(OH)_2$	298.77	mcl pr	53.5	d	2.5662^{40}_{4}		s	s	s			chl s	**B1**[2], 683
a51	—,—,monoethyl hemiacetal	Bromal monoethyl acetal. $Br_3CCHOHOC_2H_5$	326.83	nd	44	d			δ	v	v				**B1**[2], 684
a52	—,**trichloro-**	Chloral. Cl_3CCHO	147.39		−57.5	97.75^{760}	1.5121^{20}_{4}	1.45572^{20}	v^h	s^h	s^h				**B1**[2], 677
a53	—,—,butyl hemiacetal	Chloral *n*-butyl alcoholate. $Cl_3CCHOH(OC_4H_9^n)$	221.51	nd	49–50	$129–30^{743}$				v^h	s				**B1**[3], 267
a53[1]	—,—,diethyl acetal	Trichloroacetal. $Cl_3CCH(OC_2H_5)_2$	221.51			205^{760}	1.266^{25}_{4}		δ	∞	∞			glycerol ∞	**B1**, 621
a54	—,—,ethyl hemiacetal	Chloral alcoholate. $Cl_3CCHOH(OC_2H_5)$	193.46	nd	56–7	$115–6^{760}$	1.143^{40}		s^h	s^h	s^h				**B1**[2], 681

For explanations, symbols and abbreviations see beginning of table.

No.	Name	Synonyms and Formula	Mol. wt.	Crystalline form, color and specific rotation	m.p. °C	b.p. °C	Density	n_D	Solubility w	al	eth	ace	bz	other solvents	Ref.
	Acetaldehyde														
a55	—,—,hydrate	Trichloroethylidene glycol. $Cl_3CCH(OH)_2$	165.40	mcl pl	51.6–.7	96.3^{764}	1.9081^{20}_{4}		v	v	v			chl v	**B1**², 680
a57	—,—,β,β β-trichloro-*tert*-butyl hemiacetal	Chloralacetone chloroform. $Cl_3CCHOHOC(CCl_3)(CH_3)_2$	324.85	nd (bz)	65	sub			δ^h	s	s		s^h		
—	**Acetaldoxime**	*see* **Acetaldehyde, oxime**													
—	**Acetamide**	*see* **Acetic acid,** amide													
—	—,**α-cyano**	*see* **Malonic acid,** monoamide, mononitrile													
—	—,**diisobutyl-**	*see* **Pentanoic acid, 2-isobutyl-4-methyl-,** amide													
	Acetamidine	*see* **Acetic acid,** amidine													
—	**Acetanilide**	*see* **Acetic acid,** amide, *N*-phenyl-													
—	—,thio-	*see* **Acetic acid, thiono-,** amide, *N*-phenyl-													
a58	**Acetic acid**	Ethanoic acid*. CH_3CO_2H	60.05	rh (hyg)	16.604	118.5^{760} $0^{3.5}$	1.0491^{20}_{4}	1.3721^{20}	∞	∞	∞			os s	**B2**², 91 **B5**², 761
a59	—,allyl ester	Allyl acetate. $CH_3CO_2CH_2CH{:}CH_2$	100.12			103.5^{760}	0.9276^{20}_{4}	1.4049^{20}	δ	∞	∞				**B2**², 150
a60	—,amide	Acetamide. Ethanamide*. CH_3CONH_2	59.07	trg mcl	82.3	221.2^{760}	0.9986^{85}_{4}		s	v	i			chl s	**B2**², 177
a61	—,—(labile form)	CH_3CONH_2	59.07	rh	48.5	221.2^{760}			v	v					**B2,** 175
a62	—,—,*N*-acetyl-	*N*-Acetylacetamide. Diacetamide. $(CH_3CO)_2NH$	101.10	nd (eth)	79	223.5^{760}			s	s	s			lig s	**B2**², 180
a63	—,—,*N*-acetyl-*N*-(4-chlorophenyl)-	*p*-Chloroacetylacetanilide. $(CH_3CO)_2N-C_6H_4-Cl$	211.66	(al, eth)	66–7					v	v		v		**B12,** 612
a64	—,—,*N*-acetyl-*N*(4-ethoxy-phenyl	*p*-Ethoxydiacetanilide. $(CH_3CO)_2N-C_6H_4-OC_2H_5$	221.25	nd (lig)	53.5–4	182^{12}			δ	v	δ		δ	aa v lig δ^h	**B13,** 468
a65	—,—,*N*-acetyl-*N*-ethyl-	Diacethylethylamine. $(CH_3CO)_2NC_2H_5$	129.16			195–9	1.0092^{20}		i	s					**B4**¹, 352
a66	—,—,*N*-acetyl-*N*-methyl-	Diacetylmethylamine. $(CH_3CO)_2NCH_3$	115.13			192			∞		i				**B4**¹, 329
a67	—,—,*N*-acetyl-*N*-phenyl-	Diacetanilide. $(CH_3CO)_2NC_6H_5$	177.20	ta (lig)	37–8	199^{100}			δ	s			s		**B12**², 145
a68	—,—,*N*(4-acetyl-phenyl)-	*p*-Acetylacetanilide. 4-Acetylaminoacetophenone. $CH_3CONH-C_6H_4-COCH_3$	177.20	nd (w)	166–7				δ v^h	v	δ		v		**B14**², 33
a69	—,—,*N*(2-aminoethyl)-	*N*-Acetylethylenediamine. $CH_3CONHCH_2CH_2NH_2$	102.14	wh	51	128^3			s	s	i		s^h		
a70	—,—,*N*(2-aminophenyl)-	*o*-Aminoacetanilide. $CH_3CONH-C_6H_4-NH_2$ (ortho)	150.18	nd	132(145)				s	s	δ		s^h		**B13**², 15
a71	—,—,*N*(3-amino-phenyl)-	*m*-Aminoacetanilide. $CH_3CONH-C_6H_4-NH_2$ (meta)	150.18	nd or pl (bz)	86.5–7.5				v	v	δ	v	i		**B13**², 27
a72	—,—,—, hydrochloride	$C_8H_{10}N_2O\cdot HCl$	186.65	redsh pl (al)	248–51				v	s	i		i		**B13**², 27
a73	—,—,*N*(4-aminophenyl)-	*p*-Aminoacetanilide. $CH_3CONH-C_6H_4-NH_2$	150.18	nd (w)	162–2.5	267			δ	v	v				**B13**², 50
a74	—,—,*N*-benzyl-	*N*-Acetylbenzylamine. $CH_3CONHCH_2C_6H_5$	149.20	lf (eth)	61	$>300^{760}$ 157^2				v	v				**B12**², 588
a75	—,—,*N*-bromo-, hydrate	Acetobromamide. $CH_3CONHBr.\ H_2O$	155.99	bt ye pl (w + 1)	70–80	$39–40^{13}$			s^h	s	s			chl δ^c	**B2,** 181
a76	—,—,*N*-bromo-*N*-phenyl-	*N*-Bromoacetanilide. $CH_3CONBrC_6H_5$	214.07	ye pl (peth)	88									chl v lig δ	**B12**², 296
a77	—,—,*N*(2-bromophenyl)-	*o*-Bromoacetanilide. $CH_3CONH-C_6H_4-Br$ (ortho)	214.07	nd (al)	99				i	s	s				**B12**², 342

For explanations, symbols and abbreviations see beginning of table.

No.	Name	Synonyms and Formula	Mol. wt.	Crystalline form, color and specific rotation	m.p. °C	b.p. °C	Density	n_D	Solubility w	al	eth	ace	bz	other solvents	Ref.
	Acetic acid														
a78	—,—,*N*(3-bromophenyl)-	*m*-Bromoacetanilide. 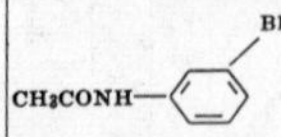 Br CH_3CONH—	214.07	nd (aq al)	87.5	...	...	...	...	v	v	...	...	...	**B12²**, 634
a79	—,—,*N*(4-bromophenyl)-	*p*-Bromoacetanilide. 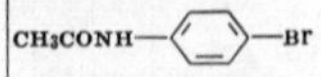 CH_3CONH— —Br	214.07	nd (labile, 60% al) mcl pr	168	...	1.717	...	i^h δ^h	δ	δ	...	δ	chl s	**B12²**, 348
a80	—,—,*N*-butyl-	$CH_3CONH(CH_2)_3CH_3$	115.18	...	...	229	...	...	...	...	...	...	...	...	**B4²**, 634
a81	—,—,*N*-butyl-*N*-phenyl-	$CH_3CON(C_6H_5)(CH_2)_3CH_3$	191.29	...	...	$273–5^{718}$	...	...	...	...	...	...	...	...	**B12**, 247
a82	—,—,*N*(4-butylphenyl)-	*p*-Butylacetanilide.  CH_3CONH— —$(CH_2)_3CH_3$	191.29	wh pl	105	...	...	...	...	s	...	...	...	...	**B12²**, 634
a83	—,—,*N*(2-chloro-4-nitrophenyl)-	2-Chloro-4-nitroacetanilide. 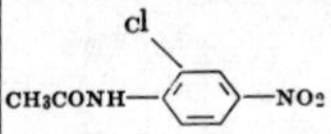 Cl CH_3CONH— —NO_2	214.62	nd	139–40	...	...	...	...	δ	...	...	...	...	**B12**, 733
a83¹	—,—,*N*(2-chloro-5-nitrophenyl)-	6-Chloro-3-nitroacetanilide. 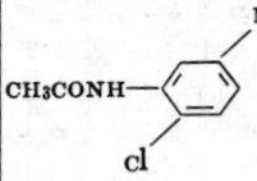 NO_2 CH_3CONH— Cl	214.62	nd (al)	153–4	...	...	...	...	δ	...	...	...	...	**B12**, 732
a84	—,—,*N*(3-chloro-4-nitrophenyl)-	3-Chloro-4-nitroacetanilide. 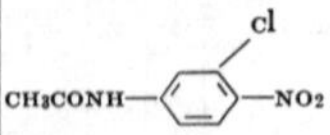 Cl CH_3CONH— —NO_2	214.62	pa ye nd	141–2	...	...	...	...	...	...	...	δ	...	**B12²**, 399
a85	—,—,*N*(4-chloro-2-nitrophenyl)-	4-Chloro-2-nitroacetanilide. 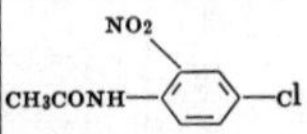 NO_2 CH_3CONH— —Cl	214.62	ye nd (al)	103	...	...	...	...	...	...	...	...	...	**B12²**, 397
a86	—,—,*N*(4-chloro-3-nitrophenyl)-	4-Chloro-3-nitroacetanilide. 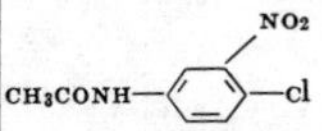 NO_2 CH_3CONH— —Cl	214.62	ye nd (al)	150	...	...	...	...	s^h	...	...	...	...	**B12**, 732
a88	—,—,*N*-chloro-*N*-phenyl-	*N*-Chloroacetanilide. $CH_3CON(Cl)C_6H_5$	169.91	nd (dil aa), pl (peth-chl)	87–8	...	...	...	δ	...	...	...	...	...	**B12²**, 295
a89	—,—,*N*(2-Chlorophenyl)-	*o*-Chloroacetanilide. 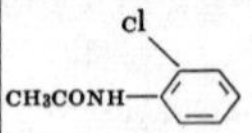 Cl CH_3CONH—	169.61	nd (dil aa)	86–8 (sub 50–60)	...	...	...	i	s	v	...	s	...	**B12²**, 317
a90	—,—,*N*(3-chlorophenyl)-	*m*-Chloroacetanilide. 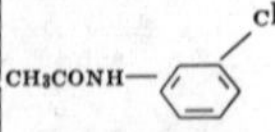 Cl CH_3CONH—	169.61	nd(50% aa)	76.6	...	...	...	δ	v	v	...	v	CS_2 v	**B12²**, 321
a91	—,—,*N*(4-chlorophenyl)-	*p*-Chloroacetanilide. 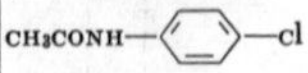 CH_3CONH— —Cl	169.61	nd (aq aa), ta (al, ace), cr (w)	179	...	1.385^{22}_{4}	...	i	s v^h	v	...	...	CCl_4 δ	**B12²**, 327
a92	—,—,*N*(2-chloro-3-tolyl)-	2-Chloro-3-acetaminotoluene. 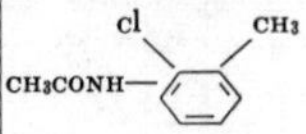 Cl CH_3 CH_3CONH—	183.64	nd	132	...	...	...	...	s	...	s	...	lig δ	**B12**, 871
a93	—,—,*N*(3-chloro-2-tolyl)-	6-Chloro-2-acetaminotoluene. 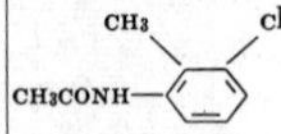 CH_3 Cl CH_3CONH—	183.64	nd	156–7	...	...	...	δ^h	s	...	...	s	...	**B12**, 836
a94	—,—,*N*(3-chloro-4-tolyl)-	3-Chloro-4-acetaminotoluene. 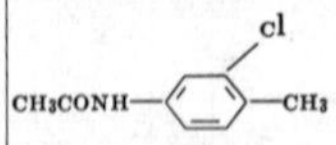 Cl CH_3CONH— —CH_3	183.64	tcl cr	113	...	...	...	...	...	...	...	...	...	**B12**, 989
a95	—,—,*N*(4-chloro-2-tolyl)-	5-Chloro-2-acetaminotoluene. 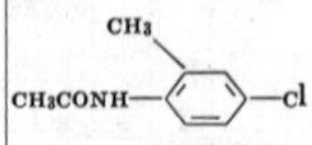 CH_3 CH_3CONH— —Cl	183.64	pl (al)	140	...	...	...	...	s^h	...	...	...	...	**B12**, 836
a96	—,—,*N*(4-chloro-3-tolyl)-	2-Chloro-5-acetaminotoluene. 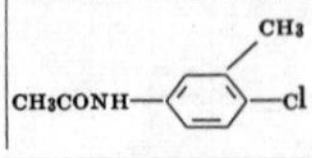 CH_3 CH_3CONH— —Cl	183.64	pl	91.2–.7	...	...	...	...	s	...	s	...	...	**B12**, 871

For explanations, symbols and abbreviations see beginning of table.

No.	Name	Synonyms and Formula	Mol. wt.	Crystalline form, color and specific rotation	m.p. °C	b.p. °C	Density	n_D	Solubility w	al	eth	ace	bz	other solvents	Ref.
	Acetic acid														
a97	—,—,*N*(5-chloro-2-tolyl)-	4-Chloro-2-acetaminotoluene. 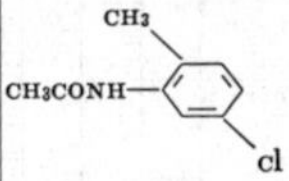	183.64	nd(w)	136–7	...	...	...	s^h	s	s	...	...	...	**B12**^{1}, 389
a98	—,—,*N*(5-chloro-3-tolyl)-	5-Chloro-3-acetaminotoluene. 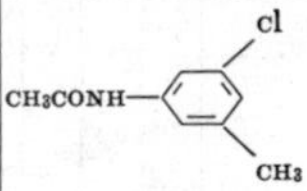	183.64	nd	146	...	...	...	...	...	...	...	...	...	**B12**, 871
a99	—,—,*N*(4-cyclo-hexylphenyl)-	4-Cyclohexylacetanilide. 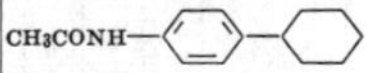	217.31	...	68–9	...	...	...	δ^h	v	v	...	...	chl v	...
a100	—,—,*N*,*N*-diacetyl-	Triacetamide. $(CH_3CO)_3N$	143.14	nd (eth)	79	...	...	...	...	...	s	...	...	...	**B2**^{1}, 82
a101	—,—,*N*,*N*-diethyl	$CH_3CON(C_2H_5)_2$	115.18	...	...	185–6	0.9248^9	...	s	s	...	...	...	...	**B4**^{2}, 602
a102	—,—,*N*,*N*-dimethyl-	$CH_3CON(CH_3)_2$	87.12	...	–20	165^{758}	0.9366^{25}_4	1.4351^{25}	∞	∞	∞	...	∞	chl ∞	**B4**^{2}, 564
a103	—,—,*N*(2,4-dimethyl-6-nitrophenyl)-	4-Acetamino-5-nitro-*m*-xylene. 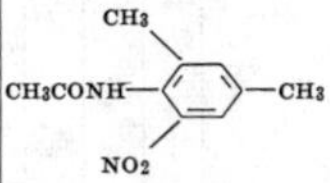	208.21	yesh nd (w)	172–3	...	...	...	s^h	s	...	...	s	...	**B12**^{2}, 612
a104	—,—,*N*(2,4-dimethylphenyl)-	2,4-Acetoxylide. 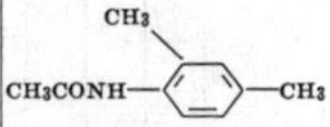	163.21	nd (al)	128–9	170^{10}	...	...	δ	v	...	...	...	...	**B12**^{2}, 608
a105	—,—,*N*(2,3-dinitrophenyl)-	2,3-Dinitroacetanilide. 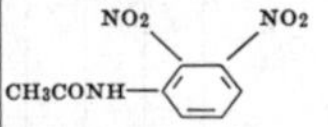	225.16	nd (al)	187	...	...	...	...	δ	δ	...	δ	chl δ	**B12**^{2}, 405
a106	—,—,*N*(2,4-dinitrophenyl)-	2,4-Dinitroacetanilide. 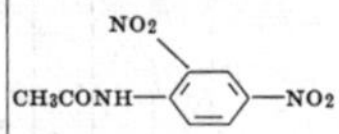	225.16	ye nd (al or bz)	125–6	...	...	...	∞	v	s	...	...	...	**B12**^{2}, 410
a107	—,—,*N*(2,5-dinitrophenyl)-	2,5-Dinitroacetanilide. 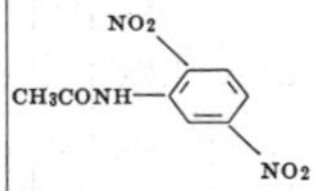	225.16	nd (al)	121	...	...	...	...	v	...	...	...	...	**B12**, 758
a108	—,—,*N*(2,6-dinitrophenyl)-	2,6-Dinitroacetanilide. 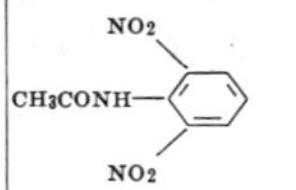	225.16	nd (aa)	197	...	...	...	...	...	...	...	...	...	**B12**, 758
a109	—,—,*N*(3,4-dinitrophenyl)-	3,4-Dinitroacetanilide. 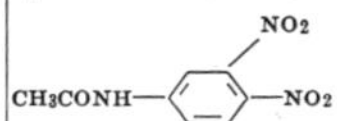	225.16	ye cr (al), nd (w)	144	...	...	...	...	v	...	...	...	...	**B12**^{2}, 414
a110	—,—,*N*(3,5-dinitrophenyl)-	3,5-Dinitroacetanilide. 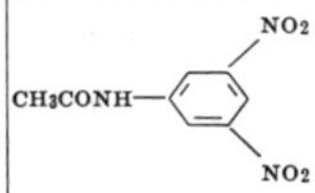	225.16	ye nd (dil aa, w)	191	...	...	...	δ	v^h	i	...	...	...	**B12**, 759
a111	—,—,*N*,*N*-diphenyl-	*N*-Acetyldiphenylamine. *N*-Phenylacetanilide. $CH_3CON(C_6H_5)_2$	211.25	rh or nd (w)	103	sub	...	...	δ	s	δ	...	...	...	**B12**, 247
a112	—,—,*N*,*N*-dipropyl-	$CH_3CON(CH_2CH_2CH_3)_2$	143.23	...	...	209–210	...	...	...	s	...	...	...	...	**B4**, 142
a113	—,—,*N*(2-ethoxy-4-nitrophenyl)-	2-Acetamino-5-nitrophenetole. 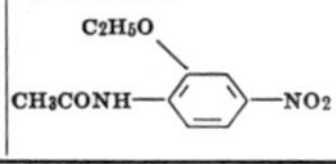	224.22	ye nd	165	...	...	...	...	s	...	...	...	...	**B13**^{2}, 194

For explanations, symbols and abbreviations see beginning of table.

No.	Name	Synonyms and Formula	Mol. wt.	Crystalline form, color and specific rotation	m.p. °C	b.p. °C	Density	n_D	Solubility w	al	eth	ace	bz	other solvents	Ref.
	Acetic acid														
a114	—,—,*N*(2-ethoxy-5-nitrophenyl)-	2-Acetamino-4-nitrophenetole. [C_2H_5O; CH_3CONH—; NO_2] 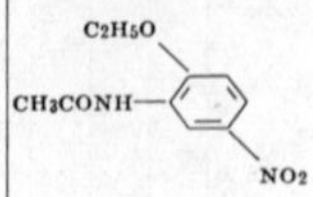	224.22	ye nd (al)	199					v^h					**B13²**, 193
a115	—,—,*N*(4-ethoxy-2-nitrophenyl)-	4-Acetamino-3-nitrophenetole. [NO_2; CH_3CONH—; —OC_2H_5] 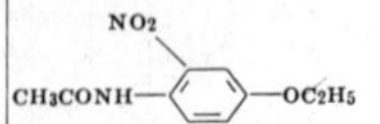	224.22	ye nd	104					v	v			chl v	**B13²**, 287
a116	—,—,*N*(4-ethoxy-3-nitrophenyl)-	4-Acetamino-2-nitrophenetole. [NO_2; CH_3CONH—; —OC_2H_5] 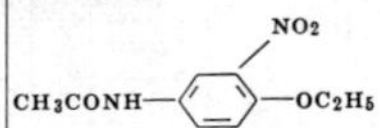	224.22	pr (dil al)	123					v	δ		v	chl v lig δ	**B13²**, 285
a117	—,—,*N*(5-ethoxy-2-nitrophenyl)-	3-Acetamino-4-nitrophenetole. [NO_2; CH_3CONH—; OC_2H_5] 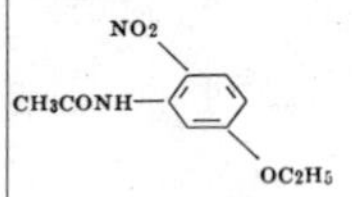	224.22	nd (al)	95					v		v		chl v lig s^h	**B13¹**, 136
a118	—,—,*N*(2-ethoxy phenyl)-	*o*-Acetophenetidide. 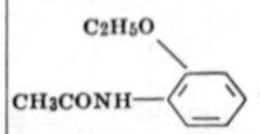 [C_2H_5O; CH_3CONH—]	179.21	lf (dil al)	78–9	>250			i	s	s			os s	**B13²**, 172
a119	—,—,*N*(3-ethoxyphenyl)-	*m*-Acetophenetidide. [OC_2H_5; CH_3CONH—] 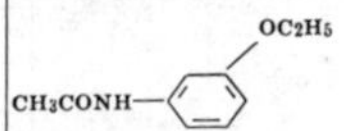	179.21	lf (w)	96.7										**B13¹**, 133
a120	—,—,*N*(4-ethoxyphenyl)-	*p*-Acetophenetidide. Phenacetin. 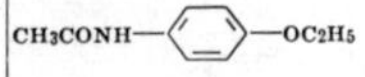 [CH_3CONH—; —OC_2H_5]	179.21	mcl pr	134.7	d		1.571	δ	s	δ				**B13²**,244
a121	—,—,*N*-ethyl-	Acetoethylamide. $CH_3CONHC_2H_5$	87.12			205	0.942_4^4		∞	∞				aa s	**B4**, 109
a122	—,—,*N*-ethyl-*N*(3-nitrophenyl)-	*N*-Ethyl-3-nitroacetanilide. 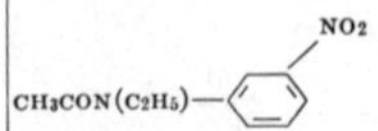 [NO_2; $CH_3CON(C_2H_5)$—]	208.22	pa ye nd	88–9					v			s		**B12**, 704
a123	—,—,*N*-ethyl-*N*-(4-nitrophenyl)-	*N*-Ethyl-4-nitroacetanilide. 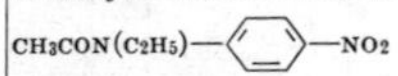 [$CH_3CON(C_2H_5)$—; —NO_2]	208.22	lf or pr	118–9				δ	v	δ		v		**B12**, 720
a124	—,—,*N*-ethyl-*N*-phenyl-	*N*-Ethylacetanilide. $CH_3CON(C_2H_5)C_6H_5$	163.22	(w), rh (eth)	51–3(54.5)	258^{731}			s		s				**B12²**, 143
—	—,—,*N*(formylphenyl)-	*see* **Benzaldehyde, acetamido-**													
a125	—,—,*N*(2-hydroxyethyl)-	*N*-Acetylethanolamine. $CH_3CONHCN_2CH_2OH$	103.12	(ace)	63–5	d^{10}	1.1223_{20}^{20}	1.4730^{20}	∞			s^h	δ	lig δ	**B4¹**, 430
a126	—,—,*N*(2-hydroxy-1-naphthyl)-	1-Acetamino-2-naphthol. 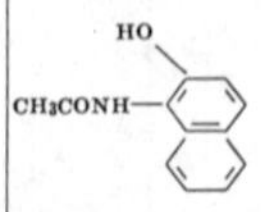 [HO; CH_3CONH—]	201.22	lf (w, dil al)	235d	sub				s^h	s^h	s^h	s^h	NaOH v	**B13²**, 414
a127	—,—,*N*(4-hydroxy-3-nitrophenyl)-*N*-methyl-	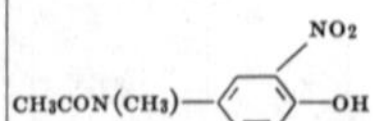 [NO_2; $CH_3CON(CH_3)$—; —OH]	210.20	(al)	161–2					s					**B13¹**, 186
a128	—,—,*N*(2hydroxyphenyl)-	2-Acetaminophenol. [HO; CH_3CONH—] 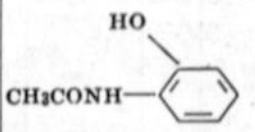	151.16	pl (dil al)	209				δ v^h	v	v		v	alk v	**B13²**, 171

For explanations, symbols and abbreviations see beginning of table.

No.	Name	Synonyms and Formula	Mol. wt.	Crystalline form, color and specific rotation	m.p. °C	b.p. °C	Density	n_D	Solubility w	al	eth	ace	bz	other solvents	Ref.
	Acetic acid														
a129	—,—,N(3-hydroxy-phenyl)-	3-Acetaminophenol. OH CH_3CONH—	151.16	nd (w)	148–9				v	v	δ		δ	chl δ	$B13^2$, 213
a130	—,—,N(4-hydroxy-phenyl)-	4-Acetaminophenol. CH_3CONH— —OH	151.16	mcl pr (w, al)	168–9		1.293^{21}		i^h v^h	v					$B13^2$, 243
a131	—,—,N(2-hydroxy-phenyl)-N-methyl-	2-(Acetylmethylamino)-phenol. HO $CH_3CON(CH_3)$—	165.20	nd (bz-peth)	151				δ	v	v		v		$B13^2$, 173
a132	—,—,N(4-hydroxy-phenyl)-N-methyl-	4-(Acetylmethylamino)-phenol. $CH_3CON(CH_3)$— —OH	165.20	sc (w)	245				δ	v	s				$B13^1$, 162
a133	—,—,N(4-iodo-phenyl)-	p-Iodoacetanilide. CH_3CONH— —I	261.07	ta (w), pr (w, al)	184.5				δ v^h	v	v		v		$B12^2$, 362
a134	—,—,Nisopropyl-Nphenyl-		177.28	lf (lig)	39	$262–3^{712}$									B12, 246
a135	—,—,N(2-methoxy-3-nitrophenyl)-	CH_3O NO_2 CH_3CONH—	210.20	pa ye pr (dil aa, me al)	103–4					s				MeOH s^h aa s^h	$B13^2$, 195
a136	—,—,N(2-methoxy-4-nitrophenyl)-	CH_3O CH_3CONH— —NO_2	210.20	pa ye cr AcOEt	153–4					s			s^h		$B13^2$, 194
a137	—,—,N(2-methoxy-5-nitrophenyl)-	CH_3O CH_3CONH— NO_2	210.20	nd (w)	175–6				δ s^h				δ		$B13^2$, 193
a138	—,—,N(2-methoxy-6-nitrophenyl)-	CH_3O CH_3CONH— NO_2	210.20	pa ye cr (dil al)	158–9					s^h			s		$B13^2$, 192
a139	—,—,N(3-methoxy-2-nitrophenyl)-	NO_2 OCH_3 CH_3CONH—	210.20	br amor	265 subl					s			s		$B13^1$, 136
a140	—,—,N(3-methoxy-4-nitrophenyl)-	OCH_3 CH_3CONH— —NO_2	210.20	ye nd (w)	165				s	v				aa v	$B13^1$, 137
a141	—,—,N(3-methoxy-5-nitrophenyl)-	OCH_3 CH_3CONH— NO_2	210.20	nd (aa)	201				δ	s	δ		s	chl δ aa s lig δ	$B13^2$, 216
a142	—,—,N(4-methoxy-2-nitrophenyl)-	NO_2 CH_3CONH— —OCH_3	210.20	ye nd (al)	116.5–7				v^h	v	v		v	aa v	$B13^2$, 287
a143	—,—,N(4-methoxy-3-nitrophenyl)-	NO_2 CH_3CONH— —OCH_3	210.20	og-ye nd (w or dil al)	153				s	s^h			δ	lig δ	$B13^1$, 186
a144	—,—,N(5-methoxy-2-nitrophenyl)-	NO_2 CH_3CONH— OCH_3	210.20	wh nd (al)	124					s^h					$B13^1$, 136
a145	—,—,N(2-meth-oxyphenyl)-	o-Acetaniside. CH_3O CH_3CONH—	165.20	nd (w)	87–8	303–5			v^h	v	s			os s	$B13^2$, 172
a146	—,—,N(3-meth-oxyphenyl)-	m-Acetaniside. OCH_3 CH_3CONH—	165.20	nd or pl (w)	80–1					v					$B13^1$, 133

For explanations, symbols and abbreviations see beginning of table.

No.	Name	Synonyms and Formula	Mol. wt.	Crystalline form, color and specific rotation	m.p. °C	b.p. °C	Density	n_D	Solubility w	al	eth	ace	bz	other solvents	Ref.
	Acetic acid														
a147	—,—,*N*(4-methoxyphenyl)-	*p*-Acetaniside, CH_3CONH— —OCH_3	165.20	ta (w)	130–2	...	...	...	δ	s	δ	s	...	chl s	**$B13^2$, 243**
a148	—,—,*N*-methyl-...	$CH_3CONHCH_3$	73.09	nd	27–8	$202–4^{726}$	0.9571^{25}_{4}	1.4301^{20}	v	v	v	...	v	chl v	**$B4^2$, 563**
a151	—,—,*N*-methyl-*N*-(1-naphthyl)-	$CH_3CON(CH_3)C_{10}H^{\alpha}_{7}$	199.24	nd (lig)	134	...	...	...	δ	s	s	...	...	...	**$B12^2$, 741**
a152	—,—,*N*-methyl-*N*-(4-nitrophenyl)-	$CH_3CON(CH_3)$— —NO_2	194.20	lf (w)	153	...	...	...	...	s	s	...	...	...	**$B12^1$, 352**
a153	—,—,*N*-methyl-*N*-phenyl-	Exalgin. $CH_3CON(CH_3)C_6H_5$	149.20	ta (eth), pr (al), lf (lig)	102–4	253^{712}	1.0036^{105}_{4}	1.576	s	s	...	...	...	...	**$B12^2$, 142**
153[1]	—,—,*N*-methyl-*N*-tolyl-	*N*-Methyl-*o*-acetotoluide. CH_3 $CH_3CON(CH_3)$—	163.22	...	55–6	260	...	...	...	s	...	...	...	...	**B12, 793**
a154	—,—,*N*-methyl-*N*(3-tolyl)-	CH_3 $CH_3CON(CH_3)$—	163.22	...	66	...	...	...	...	s	...	...	...	...	**B12, 861**
a155	—,—,*N*-methyl-*N*(4-tolyl)	$CH_3CON(CH_3)$— —CH_3	163.22	lf (eth-al)	83	283^{760}	...	...	...	...	...	...	...	...	**$B12^2$, 502**
a156	—,—,*N*(1-naphthyl)-	Aceto-1-naphthalide. $CH_3CONHC_{10}H^{\alpha}_{7}$	185.22	(al)	159–60	...	...	...	s^h	s^h	δ	...	...	...	**B12, 1231**
a157	—,—,*N*(2-naphthyl)-	Aceto-2-naphthalide. $CH_3CONHC_{10}H^{\beta}_{7}$	185.22	lf (al)	134	...	...	...	s^h	s^h	...	...	...	...	**B12, 1265**
a158	—,—,*N*(1-nitro-2-naphthyl)-	CH_3CONH NO_2—	230.23	ye nd (al)	126 (123.5)	...	...	...	...	s	s	...	s	aa s	**E12B, 759**
a159	—,—,*N*(2-nitrophenyl)-	*o*-Nitroacetanilide. NO_2 CH_3CONH—	180.17	ye pr (peth), lf (dil al)	93	$100^{0.1}$	1.419^{15}	...	s v^h	s	v	...	...	chl s	**$B12^2$, 371**
a160	—,—,*N*(3-nitrophenyl)-	*m*-Nitroacetanilide. NO_2 CH_3CONH—	180.17	wh lf (al)	154–60	$100^{0.0074}$	...	...	s^h	s	i	...	...	alk s	**$B12^2$, 380**
a161	—,—,*N*(4-nitrophenyl)-	*p*-Nitroacetanilide. CH_3CONH— —NO_2	180.17	ye pr	216	$100^{0.0084}$	...	...	δ	s	s	...	...	aa δ	**$B12^2$, 389**
a162	—,—,*N*(2-nitro-4-tolyl)-	2-Nitroacetoluidide. NO_2 CH_3CONH— —CH_3	194.19	dk ye (metastable)	95 (metastable form 93.5)	...	...	...	δ	δ	...	...	...	...	**$B12^2$, 536**
a163	—,—,*N*(4-octylphenyl)-	*p*-Octylacetanilide. CH_3CONH— —$(CH_2)_7CH_3$	247.37	lf or pl	93–4	...	...	...	i	v^h	v	...	...	...	**B12, 1185**
a164	—,—,*N*-phenyl-....	Acetanilide. $CH_3CONHC_6H_5$	135.16	rh or lf(w)	113–4	307^{760}	1.2105^{4}_{4}	...	δ^h s^h	v	s	...	...	chl v	**$B12^2$, 137**
a165	—,—,*N*-phenyl-*N*-propyl-	$CH_3CON(C_6H_5)CH_2CH_2CH_3$	177.25	mcl lf (eth, lig)	56(49)	266^{712}	...	...	i	v	v	...	...	...	**B12, 246**
a166	—,—,*N*(2-tolyl)-...	*o*-Acetotoluide. CH_3 CH_3CONH—	149.19	nd	110	296	1.168^{15}	...	δ	s	s	...	s	chl s aa s	**$B12^2$, 439**
a167	—,—,*N*(3-tolyl)-...	*m*-Acetotoluide. CH_3 CH_3CONH—	149.19	nd (w)	65.5	303	1.141^{15}	...	δ	v	v	...	...	...	**$B12^2$, 468**
a168	—,—,*N*(4-tolyl)-...	*p*-Acetotoluide. CH_3CONH— —CH_3	149.19	mcl cr or nd	153	307 sub	1.212^{15}	...	δ s^h	s	s	...	δ	aa s lig δ	**$B12^2$, 501**

For explanations, symbols and abbreviations see beginning of table.

No.	Name	Synonyms and Formula	Mol. wt.	Crystalline form, color and specific rotation	m.p. °C	b.p. °C	Density	n_D	Solubility w	al	eth	ace	bz	other solvents	Ref.
	Acetic acid														
a169	—,amidine	Acetamidine. Ethanamidine.* $CH_3(C:NH)NH_2$	58.09	...	−35	...	...	...	d^h δ	s	...	...	...	ac s	**B2²**, 183
a170	—,—,hydrochloride	$CH_3C(:NH)NH_2.HCl$	94.54	nd or, pr (al)	177	...	...	...	...	...	...	...	...	...	**B2²**, 185
—	—,—,*N,N'*-bis(4-ethoxyphenyl)-	*see* **Halocaine base**													
a171	—,—,*N,N'*-diphenyl-	$CH_3C(:NC_6H_5)NHC_6H_5$	210.28	nd (al)	131–2	...	...	...	δ v^h	...	v	...	...	ac s	**B12²**, 144
a172	—,anhydride	Acetic anhydride. Ethanoic anhydride*. $(CH_3CO)_2O$	102.09	...	−73.1	136.4	1.0820^{20}_{4}	1.3906^{20}	v	s	∞	...	s	chl s	**B2²**, 171
a173	—,benzyl ester	$CH_3CO_2CH_2C_6H_5$	150.18	...	−51.5	215.5^{760}	1.0563^{18}_{4}	1.5032^{20}	δ	∞	...	...	...	...	**B6²**, 415
a174	—,bromide	Acetyl bromide. Ethanoyl bromide*. CH_3COBr	122.95	ye in air	−96	76.7	1.663^{16}_{4}	1.4538^{16}	d	d	∞	...	∞	chl ∞	**B2²**, 176
a175	—,2-bromoethyl ester	$CH_3CO_2CH_2CH_2Br$	167.02	...	−13.8	162–3	1.514^{20}_{4}	...	v	∞	∞	...	...	...	**B2¹**, 57
a176	—,bromomethyl ester	$CH_3CO_2CH_2Br$	152.98	...	...	$130–3^{750}$	1.6560^{12}	...	i	s	v	...	...	chl v	**B2²**, 166
a177	—,2-buten-1-yl ester	Crotonyl acetate. $CH_3CO_2CH_2CH:CHCH_3$	114.15	...	...	128–9	0.9192^{20}_{4}	1.4181^{20}	δ	s	s	...	...	...	...
a178	—,butyl ester	$CH_3CO_2(CH_2)_3CH_3$	116.16	...	−77.9	126.5	0.8824^{18}_{4}	1.3970^{18}	δ	∞	∞	...	...	...	**B2²**, 140
a179	—,*sec*-butyl ester (*d*)	$CH_3CO_2CH(CH_3)CH_2CH_3$	116.16	$[\alpha]^{20}_{D}$ +25.43	...	112^{760}	0.8758^{16}_{4}	1.3877^{20}	...	...	...	...	...	...	**B2¹**, 59
a180	—,—(*dl*)	$CH_3CO_2CH(CH_3)CH_2CH_3$	116.16	...	...	112.2^{760}	0.8701^{20}_{4}	1.3882^{18}	i	s	s	...	...	...	**B2²**, 141
a181	——,—(*l*)	$CH_3CO_2CH(CH_3)CH_2CH_3$	116.16	$[\alpha]^{19}_{546}$ −20.2	...	$116–7^{760}$	0.873^{19}_{4}	1.3899^{18}	...	...	...	...	...	...	...
a182	—,*tert*-butylester	$CH_3CO_2C(CH_3)_3$	116.16	...	...	95^{750}	0.8620^{25}_{4}	1.3840^{25}	...	s	s	...	...	aa s	**B2²**, 142
a183	—,chloride	Acetyl chloride. Ethanoyl chloride*. CH_3COCl	78.50	fum	−112	51–2	1.1039^{21}_{4}	1.3898^{20}	d	d	∞	∞	∞	chl ∞	**B2²**, 175
a184	—,1-chloroethyl ester	$CH_3CO_2CHClCH_3$	122.55	...	...	121.5^{746} δd	1.114^{15}_{15}	...	d	...	s	...	...	...	**B2²**, 71
a185	—,2-chloroethyl ester	$CH_3CO_2CH_2CH_2Cl$	122.55	...	...	145	1.1783^{6}	...	i	∞	∞	...	...	...	**B2²**, 136
a186	—,2-chloroisopropyl ester	$CH_3CO_2CH(CH_3)CH_2Cl$	136.58	...	...	149–50	1.0788	1.4223	...	...	...	...	...	...	**B2**, 130
a187	—,chloromethyl ester	$CH_3CO_2CH_2Cl$	108.53	...	...	$115–6^{757}$	1.1953^{14}_{14}	...	i	s	s	...	...	con sulf s	**B2²**, 166
a188	—,3-chloropropyl ester	$CH_3CO_2CH_2CH_2CH_2Cl$	136.58	...	...	$163–5^{747}$	1.250^{19}	...	...	s	s	s	...	...	**B2²**, 139
a189	—,cyclohexyl ester	CH_3CO_2—(cyclohexane ring)	142.19	...	...	174.5	0.9688^{20}_{4}	1.4401^{20}	i	∞	∞	...	...	...	**B6²**, 10
a190	—,decyl ester	$CH_3CO_2(CH_2)_9CH_3$	220.32	...	−15.05	244	0.8671^{20}_{4}	1.4268^{20}	i	s	s	...	s	aa s	**B2**, 135
a191	—,1,3-dichloroisopropyl ester	$CH_3CO_2CH(CH_2Cl)_2$	171.03	...	...	202–8	1.281^{20}	1.4542^{20}	...	...	...	...	...	...	**B2²**, 140
a193	—,2,2-dimethylbutyl ester	$CH_3CO_2CH_2C(CH_3)_2CH_2CH_3$	144.22	...	34.5	$152–3^{755}$	0.8798^{18}_{4}	1.4068^{18}	...	...	...	...	...	...	**B2²**, 145
a194	—,1,1-dimethylpropyl ester	*tert*-Amyl acetate. $CH_3CO_2C(CH_3)_2CH_2CH_3$	130.19	...	...	124–4.5	0.8740^{18}_{4}	1.392^{20}	δ	s	s	...	...	aa s	**B2¹**, 160
a195	—,2,4-dinitrophenyl ester	NO_2; CH_3CO_2—(benzene ring)—NO_2	226.15	...	72	...	...	...	i	...	...	...	...	...	**B6¹**, 127
a196	—,3,5-dinitrophenyl ester	NO_2; CH_3CO_2—(benzene ring); NO_2	226.15	(bz-peth)	126–7	...	...	...	i	...	...	...	v	aa v^h	**B6**, 258
a197	—,ethenyl ester	Vinyl acetate. $CH_3CO_2CH:CH_2$	86.09	wh (unst), bl-gr(st)	−100.2	$71–2^{728}$	...	1.3941	i	∞	...	...	...	os s	**B2²**, 147

For explanations, symbols and abbreviations see beginning of table.

No.	Name	Synonyms and Formula	Mol. wt.	Crystalline form, color and specific rotation	m.p. °C	b.p. °C	Density	n_D	Solubility w	al	eth	ace	bz	other solvents	Ref.
	Acetic acid														
a198	—,2-ethoxyethyl ester	Cellosolve acetate. $CH_3CO_2CH_2CH_2OCH_2CH_3$	132.16		−61.7	156.4	0.9749^{20}_{4}	1.4058^{20}	v	∞	∞			os v	**B2**², 155
a199	—,ethyl ester	Ethyl acetate. $CH_3CO_2CH_2CH_3$	88.11		−83.578	77.06^{760}	0.9005^{20}_{4}	1.3701^{25}	s	∞	∞			chl ∞, os v	**B2**², 129
a200	—,2-ethylbutyl ester	$CH_3CO_2CH_2CH(C_2H_5)_2$	144.22		<−100	63^{20}	0.879^{20}_{4}	1.4109^{20}	i						
a201	—,2-ethylhexyl ester	$CH_3CO_2CH_2CH(C_2H_5)C_4H_9^n$	172.27		−93	199	0.8734^{20}_{20}	1.4204^{20}	i						
a202	— 3-ethyl-3-pentyl ester	$CH_3CO_2C(C_2H_5)_3$	158.24			160–3									**B2**, 134
a204	—,fluoride	Acetyl fluoride. Ethanoyl fluoride*. CH_3COF	62.04	fum	<−60	20.8^{760}	1.002^{15}_{4}		d	∞ ^{d}	∞		s	CS_2 δ	**B2**², 175
a205	—,furfuryl ester	$CH_3CO_2CH_2$— (furan ring, O)	140.14			177	1.1175^{20}_{4}	1.4627^{20}	i	s	s				**B17**², 115
a206	—,heptyl ester	$CH_3CO_2(CH_2)_6CH_3$	158.24			191.5	0.874^{16}_{16}	1.4153	i	s	s				**B2**, 134
206¹	—,4-heptyl ester	$CH_3CO_2CH(C_3H_7)_2$	158.24			170–2	0.8742^{0}_{0}								**B2**, 134
a207	—,hexadecyl ester	Cetyl acetate. $CH_3CO_2(CH_2)_{15}CH_3$	284.49	α)nd β) wax	α)18.5 β) 22.3	$200–1^{15}$	0.859^{25}_{4}	1.4358^{34}							**B2**², 146
a208	—,hexyl ester	$CH_3CO_2(CH_2)_5CH_3$	144.22			169.2	0.8092^{0}_{0}		i	v	v				**B2**², 145
a209	—,2-hexyl ester (*d*)	$CH_3CO_2CH(CH_3)C_4H_9^n$	144.22			57^{20}	0.8658^{18}_{4}								**B2**², 145
a210	—,—(*dl*)	$CH_3COCH(CH_3)C_4H_9^n$	144.22			157–8	0.875^{15}_{15}								
a211	—,3-hexyl ester (*d*)	$CH_3CO_2CH(C_2H_5)CH_2CH_2CH_3$	144.22	$[\alpha]^{20}_{D}+0.55$		149–51	0.8672^{20}_{4}	1.4037^{20}							**B2**¹, 61
a212	—,2-hydroxyethyl ester	Glycol monoacetate. $CH_3CO_2CH_2CH_2OH$	104.11			187–9	1.108^{15}		∞	∞	∞				**B2**², 154
a213	—,3-hydroxyphenyl ester	Eurisol. Resorcinol monoacetate. CH_3CO_2— (benzene ring, OH)	152.15	ye		283d			i	s					**B6**², 817
a214	—,iodide	Acetyl iodide. Ethanoyl iodide*. CH_3COI	169.95	turn br (separate I_2) fum		108	1.98^{17}		d	d	s				**B2**², 177
a215	—,isobutyl ester	$CH_3CO_2CH_2CH(CH_3)_2$	116.16		−98.58	117.2^{760}	0.8747^{20}_{4}	1.3901^{20}	δ	∞	∞				**B2**², 142
a216	—,isopropenyl ester	$CH_3CO_2C(CH_3){:}CH_2$	100.12		−92.9	$92–4^{732}$	0.9090^{20}	1.4033^{20}							
a217	—,isopropyl ester	$CH_3CO_2CH(CH_3)_2$	102.13		−73.4	93	0.8732^{18}_{4}	1.3740^{25}	s	s	∞				**B2**², 139
a218	—,menthyl ester (*l*)	$C_3H_7^i$, CH_3CO_2— (cyclohexane ring), CH_3	198.31	$[\alpha]^{20}_{D}-79.42$		109^{10}	0.9185^{20}_{4}	1.4469^{20}							**B6**², 42, 50
a219	—,2-methoxyethyl ester	Methyl cellosolve acetate. $CH_3CO_2CH_2CH_2OCH_3$	118.36			144.5–5	1.0090^{19}_{19}		s						**B2**², 154
a220	—,2-methoxyphenyl ester	CH_3O, CH_3CO_2— (benzene ring)	166.18		31–2	$134–5^{18}$			i	s	s				**B6**¹, 416
a221	—,methyl ester	Methyl acetate. $CH_3CO_2CH_3$	74.08		−98.1	57	0.9723^{20}_{4}	1.3617^{20}	v	∞	∞				**B2**², 125
a225	—,2-methylbutyl ester	*tert*-Amyl acetate. $CH_3CO_2C(CH_3)_2CH_2CH_3$	130.18			124–4.5	0.8740^{20}	1.4010^{20}						diox s	**B2**², 143
a226	—,3-methylbutyl ester	Isomyl acetate. $CH_3CO_2CH_2CH_2CH(CH_3)_2$	130.18		−78.5	142	0.8670^{20}_{4}	1.4003^{20}	δ	∞	∞			AmOH s	**B2**², 144
a227	—,methylene diester	Methanediol diacetate. $(CH_3CO_2)_2CH_2$	132.12		−23	164–5	1.136^{20}_{4}	1.4025^{24}	δ	∞	∞			os v	**B2**², 163

For explanations, symbols and abbreviations see beginning of table.

No.	Name	Synonyms and Formula	Mol. wt.	Crystalline form, color and specific rotation	m.p. °C	b.p. °C	Density	n_D	Solubility w	al	eth	ace	bz	other solvents	Ref.
	Acetic acid														
a228	—,2-methyl-3-heptyl ester	$CH_3CO_2CH(C_4H_9^n)CH(CH_3)_2$	172.27			172	0.875^{20}	1.4166	i	s					**B2**, 134
a229	—,3-methyl-2-heptyl ester	$CH_3CO_2CH(CH_3)CH(CH_3)C_4H_9^n$	172.27			185	0.8545_4^{21}	1.418^{21}	i	s					**B2**[2], 146
a230	—,6-methyl-2-heptyl ester	$CH_3CO_2CH(CH_3)(CH_2)_3CH(CH_3)$	172.27			187–8	0.8494^{20}	1.4137	i	s					**B2**, 135
a231	—,6-methyl-3-heptyl ester	$CH_3CO_2CH(C_2H_5)(CH_2)_2CH(CH_3)_2$	172.27				0.8554^{20}	1.4160	i	s					**B2**, 146
a232	—,2-methyl-1-naphthyl ester	1-Acetoxy-2-methyl naphthalene. 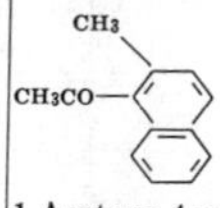 CH_3CO— ; CH_3	200.24			168–70					s				**B9**[2], 457
a234	—,4-methyl-1-naphthyl ester	1-Acetoxy-4-methyl-naphthalene. 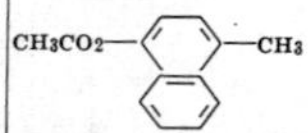 CH_3CO_2— ; —CH_3	200.24			$192–4^{12}$			i						**B9**[2], 458
a235	—,4-methyl-2-naphthyl ester	3-Acetoxy-1-methyl-naphthalene. 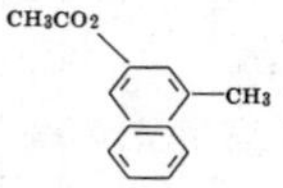 CH_3CO_2 ; —CH_3	200.24		39	188^{15}			i						**B9**[2], 458
a236	—,6-methyl-1-naphthyl ester	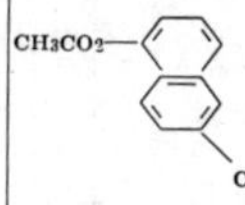 CH_3CO_2— ; CH_3	200.24			$176–9^{18}$	1.1400_4^{15}	1.6080^{15}	i						**B9**[2], 458
236[1]	—,2-methyl-3-pentyl ester	$CH_3CO_2CH(C_3H_7^i)C_2H_5$	144.22			148.5^{747}	0.8688^{20}								**B2**, 133
236[2]	—,3-methyl-3-pentyl ester	$CH_3CO_2C(C_2H_5)_2CH_3$	144.22			148	$0{,}8834_{20}^{20}$	$1{,}4109^{18}$							**B2**[2], 145
a237	—,4-methyl-2-pentyl ester	$CH_3CO_2CH(C_4H_9^i)CH_3$	144.22			147–8	0.8805^0								**B2**, 133
a238	—,2-methyl pentyl ester	$CH_3CO_2CH_2CH(CH_3)(CH_2)_3CH_3$	144.22			162.2^{746}	0.8717_{25}^{25}								**B2**, 133
a239	—,1-naphthyl ester	α-Naphthyl acetate. 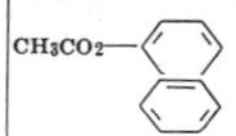 CH_3CO_2—	186.21	nd or pl (al)	49 (46)					s	s				**B6**[2], 580
a240	—,2-naphthyl ester	β-Naphthyl acetate. 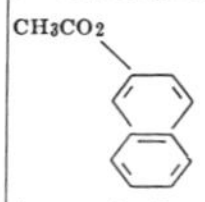 CH_3CO_2	186.21	nd (al)	70					s	s			chl s	**B6**[2], 600
a241	—,nitrile	Acetonitrile. Cyanomethane. Ethanenitrile*. Methyl cyanide. CH_3CN	41.05		−45.72	80.06	0.7856^{20}	1.3441^{20}	∞	∞	∞				**B2**[2], 181
a242	—,nitrilotri-	Triglycolamic acid. Trimethylamine-α,α′,α″-tricarboxylic acid. $N(CH_2CO_2H)_3$	191.14	pr	258–9				δ	s^h					**B4**[2], 801
a243	—,2-nitrophenyl ester	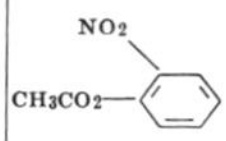 NO_2 ; CH_3CO_2—	181.15	nd or pr (lig)	40–1	253d			s	v	v		v	lig δ s^h	**B6**[2], 210
a244	—,3-nitrophenyl ester	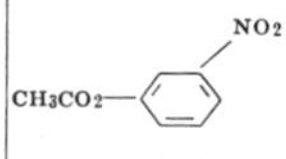 NO_2 ; CH_3CO_2—	181.15	nd (peth)	55–6				s	s				lig s	**B6**[2], 214
a245	—,4-nitrophenyl ester	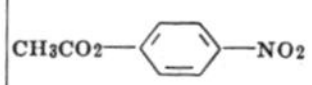 CH_3CO_2— ; —NO_2	181.15	lf (dil al)	79–80				v^h	s			v	os s lig s	**B6**[2], 223
a246	—,octadecyl ester	$CH_3CO_2(CH_2)_{17}CH_3$	312.54		34.5	$222–3^{15}$			i	s					**B2**[2], 147

For explanations, symbols and abbreviations see beginning of table.

No.	Name	Synonyms and Formula	Mol. wt.	Crystalline form, color and specific rotation	m.p. °C	b.p. °C	Density	n_D	Solubility w	al	eth	ace	bz	other solvents	Ref.
	Acetic acid														
a247	—,octyl ester	$CH_3CO_2(CH_2)_7CH_3$	172.27	...	−38.5	210	0.8847^{0}_{0}	...	i	s	...	...	...	os s	**B2**[2], 146
a248	—,2-octyl ester (*d*)	$CH_3CO_2CH(CH_3)(CH_2)_5CH_3$	172.27	$[\alpha]^{20}_{D}+6.84$ $[\alpha]^{17}_{D}+5.64$	...	84^{15}	0.8606^{19}_{4}	1.4141^{20}	...	s	...	...	s	CS_2, Py s	**B2**[2], 146
a249	—,—(*dl*)	$CH_3CO_2CH(CH_3)(CH_2)_5CH_3$	172.27	...	...	194.5^{744}	0.8626^{14}_{4}	...	...	s	...	...	...	...	**B2**[2], 146
a250	—,—(*l*)	$CH_3CO_2CH(CH_3)(CH_2)_5CH_3$	172.27	$[\alpha]^{14.5}_{D}-6.0$	...	$89–90^{17}$	0.8570^{25}_{4}	...	...	s	...	...	...	...	**B2**[2], 146
a251	—,3-octyl ester (*l*)	$CH_3CO_2CH(C_2H_5)(CH_2)_4CH_3$	172.27	$[\alpha]^{20}_{D}-4.30$	...	194–4.5	0.8693^{15}_{15}	1.4154^{20}	i	s	...	...	...	CS_2 s	**B2**[1], 62
a252	—,2-oxopropyl ester	Acetonyl acetate. Acetoxyacetone. $CH_3CO_2CH_2COCH_3$	116.11	...	...	$170–1^{755}$	1.0749^{20}_{7}	1.4150^{20}	v	v	v	...	...	...	**B2**[2], 168
a253	—,pentyl ester	Amyl acetate. $CH_3CO_2(CH_2)_4CH_3$	130.18	...	−70.8	149.25	0.8756^{20}_{4}	1.4031^{20}	δ	∞	∞	...	...	...	**B2**[2], 143
253[1]	—,2-pentyl ester (*d*)	*d*-2-Pentyl acetate. $CH_3CO_2CH(CH_3)CH_2CH_2CH_3$	130.18	$[\alpha]^{20}_{D}+17.16$ (undil) $[\alpha]^{20}_{D}+15.41$ (al)	...	130–1	0.8692^{18}_{4}	1.3960^{20}	...	s	s	...	...	...	**B2**[1], 160
253[2]	—,—(*dl*)	$CH_3CO_2CH(CH_3)CH_2CH_2CH_3$	130.18	...	...	133.5	0.8692^{18}_{4}	1.3960^{20}	i	s	s	...	...	...	**B2**[2], 143
253[3]	—,—(*l*)	$CH_3CO_2CH(CH_3)CH_2CH_2CH_3$	130.18	$[\alpha]^{20}_{D}+3.30$	...	142	0.8803^{13}	1.4012^{20}	...	...	...	...	...	...	**B2**, 131
253[4]	—,3-pentyl ester	$CH_3CO_2CH(C_2H_5)_2$	130.18	...	...	132^{741}	0.8712^{20}_{4}	1.3966^{20}	...	...	...	...	...	...	**B2**[2], 143
a254	—,phenacyl ester	ω-Acetoxyacetophenone. $CH_3CO_2CH_2COC_6H_5$	178.19	rh pl	49–9.5	270	1.1169^{65}_{4}	1.5036^{65}	i	v	v	...	δ	chl v lig δ	**B8**[2], 89
a255	—,phenyl ester	*o*-Acetylphenol. $CH_3CO_2C_6H_5$	136.15	...	...	195.7	1.0927^{0}_{4}	1.5088^{20}	δ	∞	∞	...	...	chl ∞	**B6**[2], 153
a256	—,2-phenylethyl ester	β-Phenethyl acetate. $CH_3CO_2CH_2CH_2C_6H_5$	164.20		...	232(224)	1.038^{15}	1.5108	i	s	s	...	...	...	**B6**[2], 451
a257	—,piperazinium salt	$2CH_3CO_2H.C_4H_{10}N_2$	206.24	...	208.5–9	...	...	...	s	s	i	...	...	...	...
a258	—,propyl ester	$CH_3CO_2CH_2CH_2CH_3$	102.13	...	−95	101.6	0.8884^{20}_{4}	1.3847^{20}	δ	∞	∞	...	...	...	**B2**[2], 137
a260	—,tetrahydrofurfuryl ester	$CH_3CO_2CH_2$— (O)	144.17	...	...	$204–7^{756}$	1.0624^{25}_{4}	1.4350^{25}	∞	∞	∞	...	...	chl ∞	**B17**[2], 107
a261	—,2-tolyl ester	*o*-Cresyl acetate. CH_3 / CH_3CO_2—	150.18	...	...	208	...	...	δ	v	v	...	...	...	**B6**[2], 330
a262	—,3-tolyl ester	*m*-Cresyl acetate. CH_3 / CH_3CO_2—	150.18	...	...	212	1.048^{26}_{4}	...	i	∞	∞	...	∞	chl ∞ lig ∞	**B6**[2], 352
a263	—,4-tolyl ester	*p*-Cresyl acetate. CH_3CO_2— —CH_3	150.18	...	...	212.5	1.0512^{17}_{4}	1.4991^{23}	δ	s	s	...	...	chl s	**B6**[2], 378
a264	—,1,2,2-trimethylpropyl ester (*d*)	$CH_3CO_2CH[C(CH_3)_3]CH_3$	144.22	$[\alpha]^{25}_{0}+9.63$	...	141	0.856^{25}_{4}	1.4001^{25}	...	...	...	...	...	...	...
a265	—,—(*dl*)	$CH_3CO_2CH[C(CH_3)_3]CH_3$	144.22	...	...	143^{757}	...	...	...	...	...	...	...	...	**B2**, 133
a266	**—,4-acetylaminophenoxy,** amide	H_2NCOCH_2O— —$NHCOCH_3$	208.22	nd (w)	208	...	...	...	δ v^h	v	i	...	i	...	**B13**, 465
—	**—,amino-**	*see* **Glycine**													
a267	—,**—,phenyl-(*dl*)-**	α-Amino-α-toluic acid. $C_6H_5CH(NH_2)CO_2H$	151.17	rh	242–4	sub	...	...	i	δ	...	...	...	os δ	**B14**[2], 282
a268	—,—,—,nitrile (*dl*)	α-Cyanobenzylamine. $C_6H_5CH(NH_2)CN$	132.17	hyg pl (lig)	55	...	...	...	...	...	...	...	...	lig δ	**B14**[2], 285
a269	**—,2-aminophenyl-,** nitrile	2-Aminobenzyl cyanide. NH_2 / —CH_2CN	132.17	lf (dil al)	72	...	...	...	δ	s	δ	...	v	...	**B14**[2], 275
a270	**—,4-aminophenyl-**	*p*-Amino-α-toluic acid. H_2N— —CH_2CO_2H	151.17	lf	199–200d	...	...	...	i s^h	δ	...	...	...	...	**B14**[2], 280

For explanations, symbols and abbreviations see beginning of table.

No.	Name	Synonyms and Formula	Mol. wt.	Crystalline form, color and specific rotation	m.p. °C	b.p. °C	Density	n_D	Solubility w	al	eth	ace	bz	other solvents	Ref.
	Acetic acid														
a271	—,—,nitrile	$H_2N-C_6H_4-CH_2CN$	132.17	pl (w)	46	312			δ^h	s	...	...	...	CS_2 δ^h os s	**B14²**, 281
—	—,**(2-amino-phenyl)-oxo-**	*see* **Isatic acid**													
a272	—,**arsono-**	$H_2O_3AsCH_2CO_2H$	183.97	pl (w or aa)	152d				v	v	i	...	i	AcOEt i	**B4²**, 999
—	—,**benzoyl-**	*see* **Propanoic acid, 3-oxo-3-phenyl-**													
a273	—,**bis(2,4-dinitro-phenyl)-**, ethyl ester	$(NO_2-C_6H_3(NO_2)-)_2CHCO_2CH_2CH_3$	420.30	lf (al), nd (bz or aa)	154				i	i	δ	...	δ	chl, CS_2 δ	**B9**, 675
a274	—,—,methyl ester	$(NO_2-C_6H_3(NO_2)-)_2CHCO_2CH_3$	406.27	lf (chl-MeOH)	159				i	...	...	...	...	MeOH i chl v	**B9**, 675
a275	—,**bromo-**	$BrCH_2CO_2H$	138.96	hex or rh	50	203^{760}	1.9335^{50}_{4}	1.4804^{50}	∞	∞	∞	...	...		**B2²**, 201
a276	—,—,amide	$BrCH_2CONH_2$	137.98	nd (al or bz)	91				v	δ	i	...	v^h		**B2¹**, 97
a277	—,—,bromide	$BrCH_2COBr$	201.87			147–50	2.317^{22}_{22}		d	d	...	...	...		
a278	—,—,*tert*-butyl ester	$BrCH_2CO_2C(CH_3)_3$	195.07			50^{10}			i	...	...	...	...	os v	**B2¹**, 96
a279	—,—,ethyl ester	$BrCH_2CO_2CH_2CH_3$	167.01			159	1.5059^{20}_{20}	1.4542^{13}	i	∞	∞	...	...		**B2²**, 202
a280	—,—,isobutyl seter	$BrCH_2CO_2CH_2CH(CH_3)_2$	195.07			188^{752}	1.3269^{20}_{4}		i	v	v	...	...		**B2¹**, 96
a281	—,—,methyl ester	$BrCH_2CO_2CH_3$	152.99			144d			...	s	...	...	...		**B2²**, 202
a282	—,—,nitrile	$BrCH_2CN$	119.96	pa ye		$150–1^{752}$			...	δ	s	...	...		**B2**, 216
a283	—,—,phenyl ester	$BrCH_2CO_2C_6H_5$	215.06	pl (al)	32	140^{20}			...	s	v	...	...		**B6¹**, 187
a284	—,—,propyl ester	$BrCH_2CO_2CH_2CH_2CH_3$	181.04			178	1.4099^{20}_{4}		i	s	s	...	...		**B2²**, 202
a285	—,**bromochloro-**	$BrClCHCO_2H$	173.41		38	210–2	1.9848^{31}_{4}	1.5014^{31}	v	v	v	v	...		**B2²**, 204
a286	—,—,ethyl ester	$BrClCHCO_2CH_2CH_3$	201.46			174	1.5890^{22}_{4}	1.4659^{24}	...	s	s	...	...		**B2²**, 204
a287	—,**bromodifluoro-**	F_2BrCCO_2H	174.95	lf (chl)	40	145–60			v	s	...	...	...	chl s^h	**B2**, 217
a288	—,**bromo-(diphenyl)-**	$(C_6H_5)_2CBrCO_2H$	291.16	(chl-peth)	133–4				...	...	...	...	...	to v^h, MeOH δ	**B9²**, 471
a289	—,—,bromide	$(C_6H_5)_2CBrCOBr$	354.07	nd (lig)	65–6				d	v d^h	v	...	v	chl v	**B9¹**, 283
a290	—,**bromofluoro-**	$BrFCHCO_2H$	156.95		48	183			s	v	...	...	...	chl v	**B2**, 216
a291	—,—,amide	$BrFCHCONH_2$	155.97	nd (CCl_4)	44				v	v	v	...	...	CCl_4 s^h	**B2**, 217
a292	—,**bromo-(phenyl)-**, nitrile	$C_6H_5CHBrCN$	196.05	yesh cr (dil al)	29(25.4)	242d	1.539^{29}_{4}		δ	v	v	v	v	chl v	**B9²**, 311
a293	—,**(2-bromo-phenoxy)-**	$2-Br-C_6H_4-OCH_2CO_2H$	231.06	nd (dil al)	141.15–3.0				...	v	v	...	...	chl v	**B6**, 198
a294	—,**(4-bromo-phenoxy)-**	$Br-C_6H_4-OCH_2CO_2H$	231.06	pr (al)	153–4				δ	v	v	...	...	CS_2 δ	**B6**, 200
a295	—,**(2-bromo-phenyl)-**	*o*-Bromotoluic acid. $2-Br-C_6H_4-CH_2CO_2H$	215.05	(aa)	105–6				...	v	v	...	...	CS_2 v aa v^h lig δ v^h	**B9**, 450
a296	—,—,nitrile	2-Bromobenzylcyanide. $2-Br-C_6H_4-CH_2CN$	196.06		0–1	$145–7^{14}$			...	v	...	...	...		**B9¹**, 181
a297	—,**(3-bromo-phenyl)-**	*m*-Bromotoluic acid. $3-Br-C_6H_4-CH_2CO_2H$	215.06	nd (w)	100				δ v^h	...	...	...	...		**B9¹**, 181
a298	—,**(4-bromo-phenyl)-**	$Br-C_6H_4-CH_2CO_2H$	215.06	nd	116				δ s^h	v	v	...	...	CS_2 v	**B9²**, 309
a299	—,—,nitrile	4-Bromobenzylcyanide. $Br-C_6H_4-CH_2CN$	196.06	pa ye cr (al)	47				i	δ	...	...	v	CS_2 v	**B9¹**, 181
a300	—,—,**hydroxy-**(*dl*)	*p*-Bromomandelic acid. $Br-C_6H_4-CH(OH)CO_2H$	231.06	nd (bz)	117–8			K	v	v	v	...	v	chl v	**B10²**, 125
—	—,**(4-carboxy-phenylthio)-**	*see* **Benzoic acid, 4-carboxy-methylthio-**													

For explanations, symbols and abbreviations see beginning of table.

No.	Name	Synonyms and Formula	Mol. wt.	Crystalline form, color and specific rotation	m.p. °C	b.p. °C	Density	n_D	Solubility w	al	eth	ace	bz	other solvents	Ref.
	Acetic acid														
a301	**—,(2-carboxy-methoxy)-**	*o*-(Carboxymethoxy) benzoic acid. Salicylacetic acid. CO_2H —OCH_2CO_2H	196.15	nd (w), lf (bz)	190–2	...	...	...	δ s^h	s	s	s	s^h	chl δ aa s lig δ	**B10**[1], 31
a302	**—,(3-carboxy-phenoxy)-**	*m*-(Carboxymethoxy) benzoic acid. HO_2C —OCH_2CO_2H	196.16	...	206–7	...	...	...	...	...	...	...	...	...	**B10**[1], 65
a303	**—,(4-carboxy-phenoxy)-**	*p*-(Carboxymethoxy) benzoic acid. HO_2C—⟨⟩—OCH_2CO_2H	196.16	nd (ace or w)	179	...	...	...	δ	v	v	v	s	chl v	**B10**[2], 94
a304	**—,(2-carboxy-phenyl)-**	*o*-Carboxy-α-toluic acid. Homophthalic acid. CO_2H —CH_2CO_2H	180.15	(w, eth)	180–1	...	...	...	s^h	s	δ	..	i	chl i	**B9**[2],617
a305	—,—,nitrile........	*o*-Carboxyphenylacetonitrile. CO_2H —CH_2CN	161.16	(aa), nd (w or bz)	126, 116d	...	...	...	δ s^h	v	v	...	v	CCl_4 v	**B9**[2], 618
a306	**—,(3-carboxy-phenyl)-**	*m*-Carboxy-α-toluic acid. Homoisophthalic acid. HO_2C —CH_2CO_2H	180.15	nd or pl (w)	184–5	...	...	...	δ s^h	s	s	...	δ	chl δ	**B9**, 860
a307	**—,(4-carboxy-phenyl)-**	*p*-Carboxy-α-toluic acid. Homoterephthalic acid. HO_2C—⟨⟩—CH_2CO_2H	180.15	mcl	239–41	...	...	...	s	s	δ		δ	....	**B9**, 861
a308	**—,(2-carboxy-phenyl)-2-oxo-**	*o*-Carboxyphenylglyoxylic acid. Phthalonic acid. CO_2H —$COCO_2H$	194.15	cr (+2w)	146(anh)	...	...	...	...	s	s	...	...	chl δ	**B10**[2], 60
—	—,**carboxyphenyl-thio)-**	*see* **Benzoic acid, carboxy-methylthio-**													
a309	**—,chloro-,(α)**.....	$ClCH_2CO_2H$..............	94.50	mcl pr	63	189	1.4043^{40}_4	1.4297^{6}	v	s	s	...	s	chl, CS_2 s	**B2**[2], 187
a310	—,—,(β).........	$ClCH_2CO_2H$..............	94.50	mcl pr	55–6	189	1.4043^{40}_4	1.4297^{65}	v	s	s	...	s	chl, CS_2 s	**B2**[2], 187
a311	—,—,(γ).........	$ClCH_2CO_2H$..............	94.50	mcl	50	189	0.3978^{65} 1.4043^{40}_4	1.4297^{65}	v	s	s	...	s	chl, CS_2 s	**B2**[2], 187
a312	—,—,amide.......	$ClCH_2CONH_2$..............	93.52	mcl pr	119.5	$224–5^{743}$	....	....	s	v	δ	...	...	....	**B2**, 199
a313	—,—,—,*N*-allyl-N-phenyl-	$ClCH_2CON(CH_2CH{:}CH_2)C_6H_5$	217.74	....	....	$119^{0.5}$	....	1.5079^{25}	...	...	...	...	...	....	**Am 78,** 255
a314	—,—,—,*N*,*N*-bis(2-chloroallyl)-	$ClCH_2CON(CH_2CCl{:}CH_2)_2$..	242.54	....	....	$161–3^{12}$	....	1.5220^{25}	...	...	...	...	...	....	**Am 78,** 255
a315	—,—,—,*N*,*N*-bis(3-chloroallyl)-	$ClCH_2CON(CH_2CH{:}CHCl)_2$.	242.54	....	....	$140–5^{1}$	....	1.5220^{25}	...	...	...	...	...	....	**Am 78,** 255
a316	—,—,—,*N*,*N*-bis(2-chloropropyl)-	$ClCH_2CON(CH_2CHClCH_3)_2$..	246.58	....	....	$134^{0.7}$	....	1.5018^{25}	...	...	...	...	...	....	**Am 78,** 255
a318	—,—,—,*N*,*N*-bis(2-ethylhexyl)-	$ClCH_2CON(CH_2CH(C_2H_5)C_4H^n_9)_2$	317.95	....	....	$154^{0.8}$	....	1.4622^{25}	...	...	...	...	...	....	**Am 78,** 255
a319	—,—,—,*N*,*N*-bis(2-methylallyl)-	$ClCH_2CON(CH_2C(CH_3){:}CH_2)_2$	201.70	....	....	$133–5^{20}$	....	1.4882^{2}	...	...	...	...	...	....	**Am 78,** 255
a320	—,—,—,*N*,*N*-bis(3-methylbutyl)-	$ClCH_2CON[CH_2CH_2CH(CH_3)_2]_2$	233.79	....	....	$109^{0.6}$	....	1.4625^{25}	...	...	...	...	...	....	**Am 78,** 255
a321	—,—,— *N*-butyl-..	$ClCH_2COHNC_4H^n_9$..........	149.63	....	....	110^{1}	....	1.4665^{25}	...	...	...	...	...	....	**Am 78,** 255
a322	—,—,—,*N-sec*-butyl-	$ClCH_2CONHC_4H^s_9$..........	149.63	....	45–5.5	$68^{0.7}$	....	....	...	...	...	...	...	....	**Am 78,** 255

For explanations, symbols and abbreviations see beginning of table.

No.	Name	Synonyms and Formula	Mol. wt.	Crystalline form, color and specific rotation	m.p. °C	b.p. °C	Density	n_D	Solubility w	al	eth	ace	bz	other solvents	Ref.
	Acetic acid														
a323	—,—,—,*N*-*tert*-butyl-	$ClCH_2CONHC_4H_9^t$	149.63		82.3										**Am 78,** 2556
a324	—,—,—,*N*-butyl-*N*-ethyl-	$ClCH_2CON(C_4H_9^n)C_2H_5$	177.63			$90^{1.5}$		1.4665^{25}							**Am 78,** 2556
a325	—,—,—,*N*-butyl-*N*-*iso*-propyl-	$ClCH_2CON(C_4H_9^n)C_3H_7^i$	191.70			$82^{0.5}$		1.4665^{25}							**Am 78,** 2556
a326	—,—,—,*N*(2-chloroallyl)-	$ClCH_2CONHCH_2CCl{:}CH_2$	168.02			$101^{1.4}$		1.5078^{25}							**Am 8,** 2556
a327	—,—,—,*N*(3-chloroallyl)-	$ClCH_2CONHCH_2CH{:}CHCl$	168.02		52–3.5	$112^{0.5}$									**Am 78,** 2556
a328	—,—,—,*N*(4-chlorobenzyl)-*N*-ethyl-	$ClCH_2CON(C_2H_5)$—C_6H_4—Cl (4-)	323.11		70–1										**Am 78,** 2556
a329	—,—,—,*N*-(2-chloro-4-nitrobenzyl)-	$ClCH_2CONHCH_2$—C_6H_3(2-Cl)—NO_2	263.08		118–9										**Am 78,** 2556
a330	—,—,—,*N*(2-chloropropenyl)-N-phenyl	$ClCH_2CON(CH{:}CClCH_2)C_6H_5$	243.11			$138^{0.7}$		1.5602^{25}							**Am 78,** 2556
a331	—,—,—,*N*(2-chloropropyl)-	$ClCH_2CONHCH_2CHClCH_3$	170.04			$88^{1.5}$		1.4942^{25}							**Am 78,** 2556
a332	—,—,—,*N*(3-chloropropyl)-	$ClCH_2CONHCH_2CH_2CH_2Cl$	170.04		36–7										**Am 78,** 2556
a333	—,—,—,*N*,*N*-diallyl-	$ClCH_2CON(CH_2CH{:}CH_2)_2$	173.64			92^2		1.4932							**Am 78,** 2556
a334	—,—,—,*N*,*N*-dibenzyl-	$ClCH_2CON(CH_2C_6H_5)_2$	273.76			$190^{1.8}$		1.5837^{25}							**Am 78,** 2556
a336	—,—,—,*N*,*N*-di-*sec*-butyl-	$ClCH_2CON(C_4H_9^s)_2$	205.73			$92^{0.7}$		1.4681^{25}							**Am 78,** 2556
a337	—,—,—,*N*(2,3-dichloroallyl)-	$ClCH_2CONHCH_2CCl{:}CHCl$	202.47			$131^{1.8}$		1.5311^{25}							**Am 78,** 2556
a338	—,—,—,*N*(2,4-dichloro-benzyl)-	$ClCH_2CONHCH_2$—C_6H_3(2-Cl)—Cl (4-)	252.53		96.5–7.5										**Am 78,** 2556
a339	—,—,—,*N*(2,5-dichlorobenzyl)-	$ClCH_2CONHCH_2$—C_6H_3(2-Cl, 5-Cl)	252.53		116–7										**Am 78,** 2556
a340	—,—,—,*N*(3,4-dichlorobenzyl)-	$ClCH_2CONHCH_2$—C_6H_3(3-Cl)—Cl (4-)	252.53		105–6										**Am 78,** 2556
a341	—,—,—,*N*(2,4-dichlorophenyl)-	$ClCH_2CONH$—C_6H_3(2-Cl)—Cl (4-)	238.50		101–2										**Am 78,** 2556
a342	—,—,—,*N*-(2,3-dichloropropyl)-	$ClCH_2CONHCH_2CHClCH_2Cl$	204.48		65–6										**Am 78,** 2502
a343	—,—,—,*N*,*N*-diethyl-	$ClCH_2CON(C_2H_5)_2$	149.62			$190–5^{25}$			v						**B4**[2], 656
a344	—,—,—,*N*,*N*-dihexyl-	$ClCH_2CON[(CH_2)_5CH_3]_2$	261.84		114.5–5.5										**Am 78,** 2556
a345	—,—,—,*N*-*N*-diisobutyl-	$ClCH_2CON(C_4H_9^i)_2$	205.73			99^2		1.4642^{25}							**Am 78,** 2556
a346	—,—,—,*N*,*N*-diisopropyl-	$ClCH_2CON(C_3H_7^i)_2$	177.68		48.5–9.5	$86^{2.7}$		1.4619^{25}							**Am 78,** 2556
a347	—,—,—,*N*(2,4-dinitrophenyl)-	$ClCH_2CONH$—C_6H_3(2-NO_2)—NO_2 (4-)	259.61	nd (al)	114–4.5										**B9**[2], 315
a348	—,—,—,*N* *N*-dipentyl-	$ClCH_2CON(C_5H_{11}^n)_2$	233.78			126^1		1.4651^{25}							**Am 78,** 2556
a349	—,—,—,*N*,*N*-dipropyl-	$ClCH_2CON(C_3H_7^n)_2$	177.68			125^8		1.4670^{25}							**Am 78,** 2556

For explanations, symbols and abbreviations see beginning of table.

No.	Name	Synonyms and Formula	Mol. wt.	Crystalline form, color and specific rotation	m.p. °C	b.p. °C	Density	n_D	Solubility w	al	eth	ace	bz	other solvents	Ref.
	Acetic acid														
a350	—,—,—,*N*-ethyl-*N*-hexyl-	$ClCH_2CON[(CH_2)_5CH_3]C_2H_5$	205.73			$120^{1.1}$		1.4978^{25}							**Am 78,** 2556
a351	—,—,—,*N*-furfuryl-	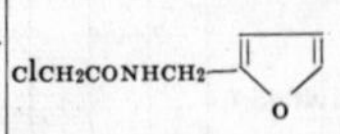 $ClCH_2CONHCH_2$— O	173.60		58–8.5										**Am 78,** 2556
a352	—,—,—,*N*-hexyl-	$ClCH_2CONH(CH_2)_5CH_3$	177.68		108.5–9.5										**Am 78,** 2556
352[1]	—,—,—,*N*-hexyl-*N*-methyl-	$ClCH_2CON[(CH_2)_5CH_3]CH_3$	191.70			$134^{3.8}$		1.5005^{25}							**Am 78,** 2556
a353	—,—,—,*N*(3-isopropoxypropyl)-	$ClCH_2CONHCH_2CH_2CH_2OCH(CH_3)_2$	193.68			$92^{0.5}$		1.4625^{25}							**Am 78,** 2556
a354	—,—,—,*N*-isopropyl-	$ClCH_2CONHCH_2(CH_3)_2$	136.60		62–2.5										**Am 78,** 2556
a355	—,—,—,*N*(3-methoxypropyl)-	$ClCH_2CONHCH_2CH_2CH_2OCH_3$	165.62		30	$88^{0.5}$		1.4712^{25}							**Am 78,** 2556
a357	—,—,—,*N*(2-methylallyl)-	$ClCH_2CONHCH_2C(CH_3){:}CH_2$	147.61			96^{1}		1.4860^{25}							**Am 78,** 2556
a358	—,—,—,*N*(3-methylbutyl)-	$ClCH_2CONHCH_2CH_2CH(CH_3)_2$	163.65		−15	$134–5^{13}$									**B4**[2], 647
a359	—,—,—,*N*-pentyl-	$ClCH_2CONH(CH_2)_4CH_3$	163.65			$82^{0.5}$		1.4665^{25}							**Am 78,** 2556
a360	—,—,—,*N*-propyl-	$ClCH_2CONHCH_2CH_2CH_3$	135.59			$105–6^{10.5}$			s	s	s	s		chl s	**B4**[1], 365
a361	—,—,—,*N*-tetradecyl-	$ClCH_2CONH(CH_2)_{13}CH_3$	289.90		64–5										**Am, 78** 2556
a362	—,—,—,*N*-tetrahydrofurfuryl-	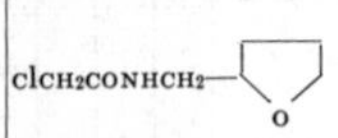 $ClCH_2CONHCH_2$— O	177.64		62.5–3.5										**Am 78,** 2556
a363	—,—,anhydride	$(ClCH_2CO)_2O$	170.99	pr (bz)	45.2	126^{24}			d	d^h	δ			chl δ	**B2**[2], 193
a364	—,—,benzyl ester	$ClCH_2CO_2CH_2C_6H_5$	184.63			147.5^{9}	1.2223^{4}_{4}	1.5426^{18}		s	s			chl s	**B6,** 435
a365	—,—,butyl ester	$ClCH_2CO_2(CH_2)_3CH_3$	150.61			183	1.081^{15}								**B2**[2], 192
a366	—,—,*sec*-butyl ester	$ClCH_2CO_2CH(CH_3)CH_2CH_3$	150.61			167.5	1.062^{20}_{20}	1.4251^{19}			s				
a367	—,—,chloride	Chloroacetyl chloride. $ClCH_2COCl$	112.95			$108–10^{764}$	1.4177^{20}_{4}	1.4535^{20}	d	d	∞				**B2**[2], 193
a368	—,—,2-chloroethyl ester	$ClCH_2CO_2CH_2CH_2Cl$	157.01			197–8	1.317		d^h		s				**B2,** 198
a369	—,—,ethyl ester	$ClCH_2CO_2CH_2CH_3$	122.55		−26	144^{740}	1.2570^{4}_{4}	1.4227^{20}	i	∞	∞				**B2**[2], 191
a370	—,—,2-hydroxyethyl ester	$ClCH_2CO_2CH_2CH_2OH$	138.56			240d $86^{1.6}$	1.330^{20}_{4}	1.4609^{20} 1.4585^{25}	∞	∞					
a371	—,—,isobutyl ester	$ClCH_2CO_2CH_2CH(CH_3)_2$	150.61			170	1.0675^{15}_{4}		s					os s	**B2**[1], 89
a372	—,—,isopropyl ester	$ClCH_2CO_2CH(CH_3)_2$	136.58			150.4–1.6	1.0888^{20}_{4}	1.4192^{20}	i	s	s				**B2**[2], 192
a373	—,—,2-methoxyethyl ester	Methyl cellosolve chloroacetate. $ClCH_2CO_2CH_2CH_2OCH_3$	152.58			$85–6^{9}$ $60^{1.3}$	1.2015^{20}_{4}	1.4382^{20}	d		v				
a374	—,—,methyl ester	$ClCH_2CO_2CH_3$	108.53		−32.7	131.5	1.2358^{20}_{4}	1.4329^{20}	δ	∞	∞				**B2**[2], 191
a375	—,—,nitrile	Chloromethyl cyanide. $ClCH_2CN$	75.50			123–4	1.193^{20}								**B2**[2], 194
a376	—,—,phenyl ester	$ClCH_2CO_2C_6H_5$	170.60	nd or pl (al)	44–5	230–5	1.2202^{44}_{4}	1.5146^{44}	i	v	v				**B6**[2], 154
a377	—,—,4-phenyl phenacyl ester	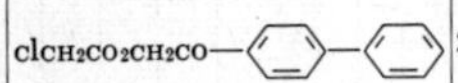 $ClCH_2CO_2CH_2CO$—	288.72		116										**C24,** 503
a378	—,—,piperazinium salt	$2ClCH_2CO_2H \cdot C_4H_{10}N_2$	275.14		145–6				s	s^h	i				
a379	—,—,propyl ester	$ClCH_2CO_2CH_2CH_2CH_3$	136.58			162–9	1.1033^{20}_{4}	1.4261^{20}	i		s				**B2**[1], 89
a380	—,—,4-tolyl ester	 $ClCH_2CO_2$— —CH_3	184.63	pl	32	162^{45}								os s	**B6**[2], 378
a381	—,**chloro-(diphenyl)-**	$(C_6H_5)_2CClCO_2H$	246.70	pl (bz-lig)	118–9d									aa v^h lig v^h	**B9**[2], 471
a382	—,—,amide	$(C_6H_5)_2CClCONH_2$	245.72	(to)	115					s	s		s	chl v	**B9**[1], 283
a383	—,—,chloride	$(C_6H_5)_2CClCOCl$	265.15	(lig)	50–1	180^{14}			d	d	δ				**B9**[2], 471
a384	—,—,ethyl ester	$(C_6H_5)_2CClCO_2C_2H_5$	274.75	pl (chl), cr (al)	43–4	185^{14}				s	s			chl s^h	**B9**[1], 282
a385	—,**(4-chloro-2-methylphenoxy)-**	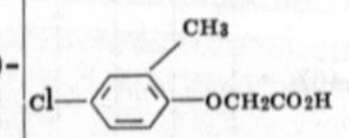 CH_3 Cl— —OCH_2CO_2H	200.63	pr	176				$δ^h$	v	v				**B6**[1], 188

For explanations, symbols and abbreviations see beginning of table.

No.	Name	Synonyms and Formula	Mol. wt.	Crystalline form, color and specific rotation	m.p. °C	b.p. °C	Density	n_D	Solubility w	al	eth	ace	bz	other solvents	Ref.
	Acetic acid														
a386	—,**(4-chloro-2-methyl-5-pyrimidyl)-**,ethyl ester	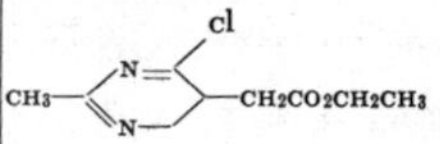 cl; CH_3—; N; N; —$CH_2CO_2CH_2CH_3$	214.65	...	40–1	110^{4}	...	...	i	...	s	...	s	...	...
a387	—,**(2-chloro-phenoxy)-**	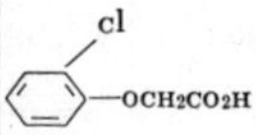 cl; —OCH_2CO_2H	186.60	nd (w or al)	145–6	...	...	...	s^h	s^h	...	...	...	...	**B6**², 172
a388	—,**(3-chloro-phenoxy)-**	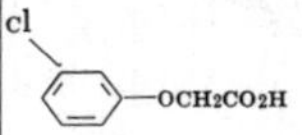 cl; —OCH_2CO_2H	186.60	...	110	...	...	...	...	...	...	...	...	...	...
a389	—,**(4-chloro-phenoxy)-**	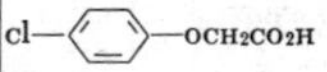 cl—; —OCH_2CO_2H	186.60	pr	155–6	...	...	...	δ	...	...	...	...	...	**B6**², 177
a390	—,**chloro-(phenyl)-**(*d*)	*d*-α-Chloro-α-toluic acid $C_6H_5CHClCO_2H$	170.60	(peth), $[\alpha]^{16}_D$ +191.2(bz)	60–1	...	...	...	...	s	...	...	s	lig s^h	**B9**², 306
a391	—,—(*dl*)	$C_6H_5CHClCO_2H$	170.60	pl	78	...	...	...	δ	v	v	...	...	lig δ	**B9**², 307
a392	—,—(*l*)	$C_6H_5CHClCO_2H$	170.60	nd (peth), $[\alpha]^{16}_D$ −191.8 (bz)	60–1	...	...	...	δ	v	v	...	v	chl v lig s^h	**B9**², 307
a393	—,**(2-chloro-phenyl)-**	*o*-Chloro-α-toluic acid. 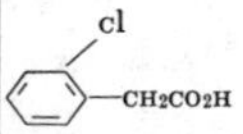 cl; —CH_2CO_2H	170.60	nd (w)	96	...	...	...	δ	v^h	...	...	...	...	**B9**¹, 178
a394	—,—,nitrile	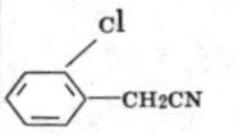 cl; —CH_2CN	151.66	grsh-ye	...	170^{120}	1.1737^{18}_{4}	1.15341^{18}	...	...	...	...	...	...	**B9**², 308
a395	—,**(3-chloro-phenyl)-**	*m*-Chloro-α-toluic acid. 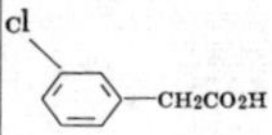 cl; —CH_2CO_2H	170.60	nd (*n*-C_7H_{16}), lf (aq al)	77.5–8.5	...	...	...	δ	δ	∞	...	δ	...	...
a396	—,**(4-chloro-phenyl)-**	*p*-Chloro-α-toluic acid. 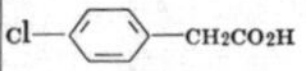 cl—; —CH_2CO_2H	170.60	nd (w)	103.5–4	...	...	...	v	v	v	...	s	...	**B9**², 306
a397	—,**(4-chloro-phenyl)hydroxy-**	*p*-Chloromandelic acid. 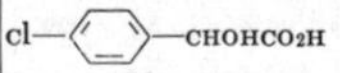 cl—; —$CHOHCO_2H$	186.60	nd	119–20	...	...	...	s	v	...	...	s^h	...	**B10**¹, 92
—	—,**cyano-**	*see* **Malonic acid,** mononitrile													
a398	—,**(3-cyano-phenyl)-**, nitrile	3-Cyanobenzyl cyanide. Homoisophthalonitrile. 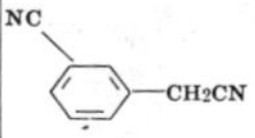 NC; —CH_2CN	142.15	nd (w)	84	...	...	...	s^h	s	s	...	s	chl s	**B9,** 860
a399	—,**1-cyclohexenyl-**, nitrile	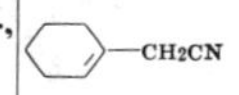 —CH_2CN	121.18	...	...	144^{90} 105^{22}	0.9473^{21}_{4}	1.4787^{21}	i	s	s	...	...	...	**B9**², 32
a400	—,**cyclohexyl-**	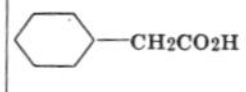 —CH_2CO_2H	142.19	...	30–1	242–4	0.9335^{20}_{4}	1.4775^{20}	δ	...	...	...	...	os s	**B9**², 9
a401	—,**cyclohexyl-idene-**, nitrile	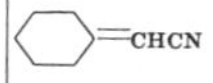 =CHCN	121.18	...	...	$107–8^{22}$	0.9483^{15}_{4}	1.4928^{15}	...	s	s	...	...	...	**B9**², 34
a402	—,**diazo-**, ethyl ester	Diazoacetic ester. Ethyl diazoacetate. $N_2CHCO_2CH_2CH_3$	114.10	ye rh	−22	$140–1^{720}$	1.0852^{18}_{4}	1.4588^{18}	δ	∞	∞	...	∞	lig ∞	**B3**², 390
a403	—,**dibromo-**	Br_2CHCO_2H	217.86	dlq cr	48	232 195^{250}	...	...	v	v	s	...	...	...	**B2**², 205
a404	—,—,amide,*N,N*-dimethyl-	$Br_2CHCON(CH_3)_2$	244.92	pr (w or eth)	79–80	128^{12}	...	...	...	...	δ	...	...	...	**B4,** 59
a405	—,—,ethyl ester	$Br_2CHCO_2CH_2CH_3$	245.91	...	...	194	1.903^{20}_{20}	1.5017^{13}	i	∞	∞	...	...	...	**B2**¹, 97
a406	—,—,methyl ester	$Br_2CHCO_2CH_3$	231.88	...	...	181.5–3.5	...	...	...	s	s	...	...	...	**B2**², 205
a407	—,**dichloro-**	Cl_2CHCO_2H	128.94	...	10.8	192–3 $91–2^{12}$	1.5634^{20}_{4}	1.4658^{20}	∞	∞	∞	...	...	...	**B2**², 194
a408	—,—,amide	$Cl_2CHCONH_2$	127.96	mcl pr	98.5	$233–4^{745}$	...	...	v^h	v	v	...	...	...	**B2**², 196
a409	—,—,anhydride	$(Cl_2CHCO)_2O$	239.87	...	...	214–6d $100–2^{16}$	1.574^{24}	...	d	d	...	...	...	...	**B2,** 204
a410	—,—,butyl ester	$Cl_2CHCO_2(CH_2)_3CH_3$	185.05	...	...	184	1.169^{15}	...	...	s	...	...	...	os s	**B2**², 194

For explanations, symbols and abbreviations see beginning of table.

No.	Name	Synonyms and Formula	Mol. wt.	Crystalline form, color and specific rotation	m.p. °C	b.p. °C	Density	n_D	Solubility w	al	eth	ace	bz	other solvents	Ref.
	Acetic acid														
a411	—,—,chloride	$Cl_2CHCOCl$	147.39			107–8	1.5315^{16}_{4}	1.4638^{16}	d	d	∞	...	...	...	**B2²**, 196
a412	—,—,ethyl ester	$Cl_2CHCO_2CH_2CH_3$	157.00			157	1.2821^{20}_{4}	1.4386^{20}	δ	∞	∞	...	...	...	**B2²**, 196
a413	—,—,(2-hydroxy-ethyl) ester	$Cl_2CHCO_2CH_2CH_2OH$	173.00			$106^{0.08}$	1.438^{20}_{4}	1.4730^{25}	δ	s	...	...	...	...	...
a414	—,—,isopropyl ester	$Cl_2CHCO_2CH(CH_3)_2$	173.00			163.8–4.8	1.2053^{20}_{4}	1.4328^{20}	...	...	s	...	...	...	...
a415	—,—,methyl ester	$Cl_2CHCO_2CH_3$	142.97			143^{764}	1.3808^{19}_{1}	...	i	s	...	...	...	...	**B2²**, 196
a416	—,—,nitrile	Cl_2CHCN	109.94			112–3	1.374^{11}	...	...	...	...	...	...	MeOH s	**B2¹**, 92
a417	—,—,propyl ester	$Cl_2CHCO_2CH_2CH_2CH_3$	171.02			177	1.2240^{20}	1.4392^{20}	...	s	s	...	...	...	**B2²**, 196
a418	—,**(2,4-dichloro-phenoxy)**	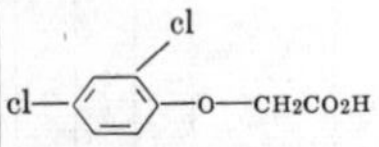 cl, cl—⟨ring⟩—O—CH_2CO_2H	221	(bz)	140–1	$160^{0.4}$	...	...	i	s	...	...	δ	...	...
a419	—,**diethoxy-**, ethyl ester	$(C_2H_5O)_2CHCO_2CH_2CH_3$	176.21			199 $75–7^{2-10}$	0.994^{18}	...	...	v	v	...	...	...	**B3²**, 389
a420	—,**difluoro-**	F_2CHCO_2H	96.02		–0.35	134.2^{766} $67–70^{20}$	1.5255^{20}	1.3420^{20}	∞	∞	∞	...	∞	...	**B2²**, 185
a421	—,**(2,5-dihydroxy-phenyl)-**	Homogentisic acid. 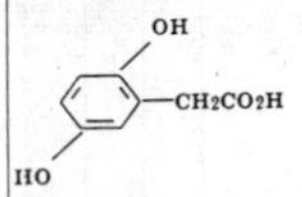 OH, —CH_2CO_2H, HO	168.15	pr (w+1), lf (al-chl)	152	...	...	...	v	v	v	...	i	chl i	**B10²**, 269
a422	—,**diiodo-**	I_2CHCO_2H	311.85	lt ye cr (naphtha), wh nd	110	...	...	...	s	s^h	s^h	...	s^h	...	**B2¹**, 99
a423	—,**(3,4-dimeth-oxyphenyl)-**	Homoveratric acid. 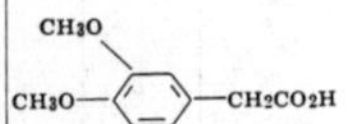 CH_3O, CH_3O—⟨ring⟩—CH_2CO_2H	196.20	nd (w+1), cr (bz-peth)	80–2 (hyd), 98–9 (anh)	...	...	...	s	v	v	...	...	...	**B10²**, 268
a424	—,—,amide	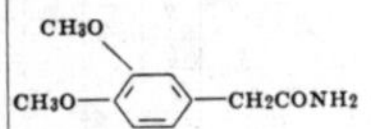 CH_3O, CH_3O—⟨ring⟩—CH_2CONH_2	195.22		145–7	...	...	...	...	s	s	...	...	...	**B10¹**, 198
a425	—,**(2,4-dimethyl-phenyl)hydroxy-**	2,4-Dimethylmandelic acid. 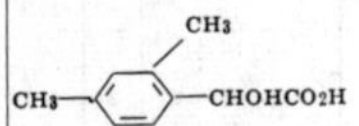 CH_3, CH_3—⟨ring⟩—$CHOHCO_2H$	180.21	rh (w), lf (to or peth-chl), nd (bz)	103 (to), 119 (bz or peth-chl)	...	...	...	v^h	v	v	...	...	chl, v	**B10**, 275
a426	—,**(2,5-dimethyl-phenyl)hydroxy-**	2,5-Dimethylmandelic acid. 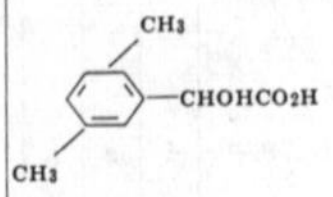 CH_3, —$CHOHCO_2H$, CH_3	180.21	nd or pr	114	...	...	...	δ	v	v	...	...	chl v̇	**B10²**, 268
a427	—,**(3,4-dimethyl-phenyl) hydroxy-**	3,4-Dimethylmandelic acid. 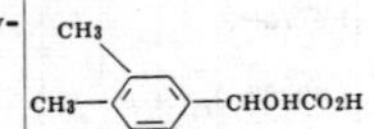 CH_3, CH_3—⟨ring⟩—$CHOHCO_2H$	180.21	lf (bz)	115–6	...	...	...	s	s^h	...	...	v^h	...	**B10²**, 167
a428	—,**(2,4-dinitro-phenoxy)-**	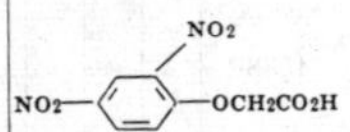 NO_2, NO_2—⟨ring⟩—OCH_2CO_2H	242.14	pa ye pr (w)	147–8	...	...	...	δ	s	...	...	i	...	**B6²**, 244
a429	—,**(3,5-dinitro-phenoxy)-**	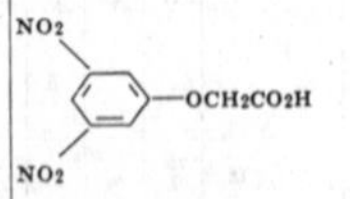 NO_2, —OCH_2CO_2H, NO_2	242.14	pa br cr	207	...	...	...	...	...	...	...	...	alk s	**B6**, 259
a430	—,**(2,4-dinitro-phenyl)-**	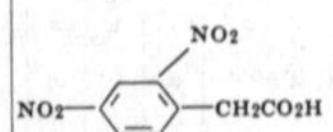 NO_2, NO_2—⟨ring⟩—CH_2CO_2H	226.14	nd (w)	179–80d	d	...	...	δ^h	s	s	...	...	...	**B9²**, 315
a431	—,—,chloride	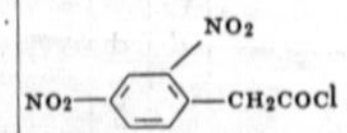 NO_2, NO_2—⟨ring⟩—CH_2COCl	244.60	lf (CS_2)	77	...	...	...	d	...	v	...	v	chl v	**B9¹**, 185

For explanations, symbols and abbreviations see beginning of table.

No.	Name	Synonyms and Formula	Mol. wt.	Crystalline form, color and specific rotation	m.p. °C	b.p. °C	Density	n_D	Solubility						Ref.
									w	al	eth	ace	bz	other solvents	
	Acetic acid														
a432	—,—,ethyl ester	NO_2; NO_2—⟨ring⟩—$CH_2CO_2CH_2CH_3$	254.20	nd (w)	35				δ	s	...	...	...	os v	**B9**, 459
a433	—,**(2,6-dinitrophenyl)-**	NO_2; ⟨ring⟩—CH_2CO_2H; NO_2	226.14	ye lf (aa)	201–2d				δ	s^h	...	...	...	aa δ	**B9**[1], 185
a434	—,**diphenyl-**	$(C_6H_5)_2CHCO_2H$	212.24	cr (aq), nd (w), lf (al)	146	195^{25} (sub)	1.258^{15}_{15}		δ v^h	v^h	v^h	...	...	chl v^h	**B9**[2], 466
a435	—,—,amide	$(C_6H_5)_2CHCONH_2$	211.25	pl (al)	167.5–8.5				...	s^h	...	...	...		**B9**[2], 468
a436	—,—,anhydride	$[(C_6H_5)_2CHCO]_2O$	406.48	nd (eth)	98	$220–5^{15}$			d^h	d^h	δ	...	v	chl v lig δ	**B9**[1], 281
a437	—,—,chloride	$(C_6H_5)_2CHCOCl$	230.70	pl (lig)	56–7	$170–1^{16}$			d	d	...	...	...	lig s^h	**B9**[1], 281
a438	—,—,ethyl ester	$(C_6H_5)_2CHCO_2CH_2CH_3$	240.29	nd (al), rh (AcOEt)	59	195^{25}			i	v	v	...	...	CS_2 v	**B9**[1], 277
a439	—,—,methyl ester	$(C_6H_5)_2CHCO_2CH_3$	226.26	mcl pl (AcOEt), lf (dil al)	60 (58.5)				i	s	s	...	...	AcOEt s^h	**B9**[2], 467
a440	—,—,nitrile	$(C_6H_5)_2CHCN$	193.25	pr (eth, peth), lf (dil al)	75–6	$181–4^{12}$			...	s^h	v	...	...	lig δ s^h	**B9**[2], 469
a441	—,**diphenyl-(hydroxy)-**	Benzilic acid. $(C_6H_5)_2COHCO_2H$	228.25	mcl nd (w)	151	d 180			δ v^h	v	v	...	...	con sulf s (red)	**B10**[2], 223
a442	—,—,ethyl ester	Ethyl benzylate. $(C_6H_5)_2COHCO_2CH_2CH_3$	256.30	pr or nd	34	201^{21}			i	s	s	...	...	con sulf s	**B10**[2], 225
a443	—,—,methyl ester	Methyl benzylate. $(C_6H_5)_2COHCO_2CH_3$	242.28	mcl or tcl cr,(al)	75 (73)	187^{13}			i	s^h	s	...	...	con sulf s aa s	**B10**[2],225
—	—,**diureido-**	*see* **Allantoic acid**													
a444	—,**ethoxy-**	Ethyloglycolic acid. $C_2H_5OCH_2CO_2H$	104.11			$206–7^{760}$ 102^{9}	1.1021^{20}_{4}	1.4194^{20}	v^h	v^h	v^h	...	...		**B3**[2], 170
a445	—,—,chloride	$C_2H_5OCH_2COCl$	122.55			127–8 $49–50^{37}$	1.1170^{20}_{4}	1.4204^{20}	d	d	s	...	...		**B3**[2], 173
a446	—,—,ethyl ester	$C_2H_5OCH_2CO_2CH_2CH_3$	132.16			$157.6–.8^{756}$ 52^{12}	0.9701^{20}_{4} 0.9945^{50}	1.4029^{20}	...	s	s	s	...		**B3**[2], 91
a447	—,—,*l*-menthyl ester	CH_3—⟨ring⟩—$C_3H_7^i$; $O_2CCH_2OC_2H_5$	242.36	$[\alpha]^{20}_{D}$ −66.35		155^{20}	0.9545^{20}_{4}		δ	s	s	...	...	chl s	**B6**[2], 47
a448	—,—,methyl ester	$C_2H_5OCH_2CO_2CH_3$	118.13			$147–8^{734}$	1.0112^{15}		...	s	s	s	...		**B3**[1], 191
a449	—,—,piperazinum salt	2 $C_2H_5OCH_2CO_2H$. $C_4H_{10}N_2$.	294.35		120–1				s	s	δ	...	...		
a450	—,**(2-ethoxyphenyl)-**	OC_2H_5; ⟨ring⟩—CH_2CO_2H	180.21	nd (lig), cr (w)	103–4				...	δ	...	...	...	lig s^h	**B10**[1], 82
a451	—,**(ethylthio)-**	$C_2H_5SCH_2CO_2H$	120.17		−8.7	164^{83} 117^{11}	1.1497^{20}_{4}		v	v	v	...	...		**B3**, 95
a452	—,**9-fluorenyl-**	⟨fluorene⟩; CH_2CO_2H	224.26	mcl	138–9	$218–20^{11}$			...	...	...	...	...		**E13**, 95
a453	—,**fluoro-**	FCH_2CO_2H	78.04		35.2	165^{760}			s^h	s^h	...	...	...		**B2**[2], 185
a456	—,**2-furyl-**	2-Furanacetic acid. ⟨furan⟩—CH_2CO_2H (O)	126.11	lf (bz)	66.8–7.5	$102–4^{0.4}$			s	...	...	...	...	MeOH s	
a457	—,—,nitrile	Furfuryl cyanide. ⟨furan⟩—CH_2CN (O)	107.11			$78–80^{20}$	1.0854^{25}_{4}	1.4715^{25}	δ	v	v	...	...		
—	—,**guanidino-**	*see* **Glycocyamine**													
a458	—,**hydroxy-**	Glycolic acid. $HOCH_2CO_2H$	76.05	rh, nd (w), lf(eth)	79	d			s	s	s	...	...		**B3**[2], 168
a459	—,—,acetate	Acetoxyacetic acid. $CH_3CO_2CH_2CO_2H$	118.09	nd(bz)	66–8	$144–5^{12}$			v	v	v	v	i	chl v	**B3**[2], 171
a460	—,—,—,ethyl ester	Ethylacetoxyacetate. Ethylacetylglycolate. $CH_3CO_2CH_2CO_2CH_2CH_3$	146.14			179	1.0993^{17}		δ	s	s	...	...	aa s	**B3**[2], 172
a461	—,—,amide	Glycolamide. Glycolic amide. $HOCH_2CONH_2$	75.07	lf (al), rh (AcOEt)	120				v	δ	δ	...	...		**B3**[2], 173
a462	—,—,anhydride	Glycolic anhydride. $(HOCH_2CO)_2O$	134.09	pw	130	d			i	i	i	...	...		**B3**[1], 92

For explanations, symbols and abbreviations see beginning of table.

No.	Name	Synonyms and Formula	Mol. wt.	Crystalline form, color and specific rotation	m.p. °C	b.p. °C	Density	n_D	Solubility w	al	eth	ace	bz	other solvents	Ref.
	Acetic acid														
a463	—,—,ethyl ester	Ethyl glycolate. $HOCH_2CO_2CH_2CH_3$	104.11			160	1.0826^{23}_{4}			v	v				**B3**^{2}, 171
a464	—,—,methyl ester	Methyl glycolate. $HOCH_2CO_2CH_3$	90.08		73.5–4	151.1	1.168^{18}_{4}		s	∞	∞				**B3**^{2}, 171
a465	—,—,nitrile	Ethanolnitrile. Formaldehyde. Cyanohydrin. Glycolonitrile. $HOCH_2CN$	57.05		<−72	183 δd 119^{24}		1.4117^{19}	v	v	v		i	chl i	**B3**^{2}, 174
a466	**—,hydroxy(4-iodophenyl)-**, (*dl*)	*p*-Iodomandelic acid (*dl*) I—C_6H_4—CHOHCO$_2$H	278.05		135				v	v	v			lig δ	**B10**, 210
a467	**—,hydroxy-(4-isopropyl-phenyl)-**, (*d*)	*p*-Isopropylmandelic acid (*dl*) $(CH_3)_2CH$—C_6H_4—CHOHCO$_2$H	194.23	ta (w), $[\alpha]^{17}_{D}+134.9$ (abs al, c=4.06)	153–4										**B10**, 279
a468	—,—,(*dl*)	$C_{11}H_{14}O_3$	194.23	nd (w)	159.2–60				δ	v	v				**B10**^{1}, 120
a469	—,—,(*l*)	$C_{11}H_{14}O_3$	194.23	ta (20% al), $[\alpha]^{17}_{D}-135$ (abs al, c=4.09)	153–4					v					**B10**, 279
a470	**—,hydroxy-(phenyl)-**,(*D*,−)	*d*(−)-Mandelic acid. *d*(−)-Phenylglycolic acid. *d*(−) $C_6H_5CHOHCO_2H$	152.15	ta, $[\alpha]^{20}_{D}-144.7$ (al, c=0.43)	132.8		1.341		s	s				aa s	**B10**^{2}, 114
a471	—,—,(*dl*)	*dl*-Mandelic acid. *dl*-Phenylglycolic acid. $C_6H_5CHOHCO_2H$	152.15	pl (w), rh (bz)	120.5	d	1.300^{20}_{4}		s	v	v				**B10**^{2}, 118
a472	—,—,(solvate),(*dl*)	$C_6H_5CHOHCO_2H$	152.15	mcl (bz+1)	32.6–.7										**B10**^{2}, 118
a473	—,—,(*L*,+)	*l*(+)-Mandelic acid. *l*(+)-Phenylglycolic acid. $C_6H_5CHOHCO_2H$	152.15	ta, $[\alpha]^{20}_{D}+156.2$ (w, p=3.2)	133.8				s v^{h}					chl δ	**B10**^{2}, 114
a474	—,—,acetate (*d*)	*d*(−)-*O*-Acetylmandelic acid. $C_6H_5CH(O_2CCH_3)CO_2H$	194.19	nd (w+1), $[\alpha]^{20}_{D}+156.4$ (ace)	96.5–8 (anh)				δ s^{h}	v	v	v	s	chl v CCl_4 δ lig δ	**B10**^{1}, 85
a475	—,—,—,(*dl*)	$C_6H_5CH(O_2CCH_3)CO_2H$	194.19	cr (bz), amor (chl-peth), cr (w+1), nd (w+½)	52–3(hyd), 76.8 (anh)				δ s^{h}	v	v		v	chl v	**B10**^{2}, 12(
a476	—,—,acetate chloride (*dl*)	*dl*-*O*-Acetylmandelyl chloride. $C_6H_5CH(O_2CCH_3)COCl$	212.64			129^{10} 142^{18}					v		v	chl v lig v	**B10**^{1}, 89
a477	—,—,amide (*dl*)	α-Hydroxy-α-toluamide. *dl*-Mandelamide. $C_6H_5CHOHCONH_2$	151.16	lf(bz)	133–4				δ	v^{h}	δ				**B10**^{2}, 122
a478	—,—,—,*N*-ethyl-(*d*)	*d*(−)-*N*-Ethylmandelamide. $C_6H_5CHOHCONHC_2H_5$	179.22	pl (chl-peth), $[\alpha]^{16}_{D}-34.4$ (al, c=3.7)	65.5–6.5				v	s	s	s	s	chl s	**B10**^{2}, 11?
a479	—,—,—,—,(*dl*)	$C_6H_5CHOHCONHC_2H_5$	179.22	pl (bz-peth)	53–4				v	v	v	v	v		**B10**^{1}, 89
a480	—,—,ethyl ester(*d*)	*d*(−)-Ethylmandelate. $C_6H_5CHOHCO_2CH_2CH_3$	180.20	(peth), $[\alpha]^{20}_{D}$ −128.4 (chl, c=6.67), $[\alpha]^{14}_{D}$ −200.2 (CS_2, c=2)	34–5	150^{21}	1.1270^{20}_{4}			s	s			CS_2 s	**B10**^{2}, 11(
a481	—,—,—(*dl*)	$C_6H_5CHOHCO_2CH_2CH_3$	180.20	nd (peth)	37	253–5				s	s			lig s	**B10**^{2}, 12
a482	—,—,—(*l*,+)	$C_6H_5CHOHCO_2CH_2CH_3$	180.20	(peth), $[\alpha]^{20}_{D}$ +205.1 (CS_2, c=0.7)	31.3					s	s			CS_2 s lig s^{h}	**B10**^{2}, 11
a483	—,—,methyl-ester (*D*,−)	*D*(−)-Methylmandelate. $C_6H_5CHOHCO_2CH_3$	166.18	$[\alpha]^{20}_{D}-131.5$(w)	55	160^{32}	1.1756^{20}		s	s		s	s	chl s	**B10**^{2}, 11
a484	—,—,—(*dl*)	$C_6H_5CHOHCO_2CH_3$	166.18	lf (bz-lig)		250 δd	1.1756^{20}		δ	s			s^{h}		**B10**^{2}, 12
a485	—,—,—(*L*,+)	$C_6H_5CHOHCO_2CH_3$	166.18	$[\alpha]^{20}_{D}+133.6$(w) $[\alpha]^{20}_{D}+252(CS_2)$	55.5				s	s					**B10**^{2}, 11

For explanations, symbols and abbreviations see beginning of table.

No.	Name	Synonyms and Formula	Mol. wt.	Crystalline form, color and specific rotation	m.p. °C	b.p. °C	Density	n_D	Solubility: w	al	eth	ace	bz	other solvents	Ref.
	Acetic acid														
a486	—,—,nitrile(*D*,−)	*D*(−)-Benzaldehyde cyanohydrin. *D*(−)-Mandelonitrile. $C_6H_5CHOHCN$	133.15												**B10²**, 117
a487	—,—,—(*dl*)	$C_6H_5CHOHCN$	133.15	ye pr	22	170d	1.1165^{20}_{4}		i	s	s				**B10²**, 123
a488	—,—,—(*L*,+)	$C_6H_5CHOHCN$	133.15	nd, $[\alpha]^{25}_{D}+46.9$											**B10¹**, 86
a489	—,**(3-hydroxy-4-methoxyphenyl)-**	Homoisoyanillic acid. 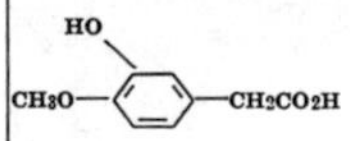 HO; CH_3O; CH_2CO_2H	182.17		130–1				s	s	s		s^h		**B10²**, 268
a490	—,**(2-hydroxy-5-nitrophenyl)-**	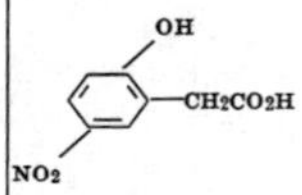 OH; CH_2CO_2H; NO_2	197.15	nd	160–2				v	s	s		δ	chl δ	**B10**, 189
a491	—,—,ethyl ester	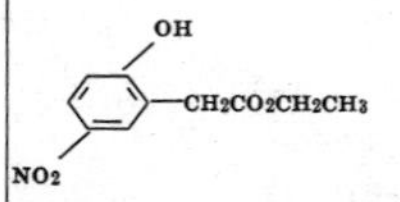 OH; $CH_2CO_2CH_2CH_3$; NO_2	225.11	pr or pl(al)	154–5						δ		s	chl s lig δ	**B10, 189**
a492	—,**(3-hydroxyphenoxy)**	Resorcinol-*O*-acetic acid. 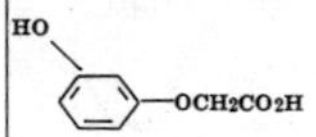 HO; OCH_2CO_2H	168.14	nd or pr (w, to)	158–9				s^h	s				to s^h	**B6**, 817
a493	—,—,ethyl ester	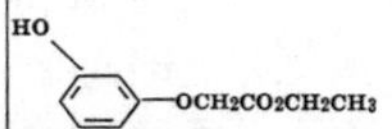 HO; $OCH_2CO_2CH_2CH_3$	192.20	pr (w, bz)	55	274 δd $170–3^{11}$			i s^h				s^h	lig δ	**B6¹**, 402
a494	—,**(4-hydroxyphenoxy)-**	Hydroquinone-O-acetic acid. 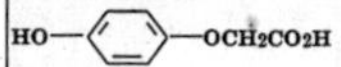 HO—; —OCH_2CO_2H	168.15	nd (to), pr (w+⅓)	152				s^h					to s^h	**B6**, 847
a495	—,**(2-hydroxyphenyl)-**	*o*-Hydroxy-α-toluic acid. 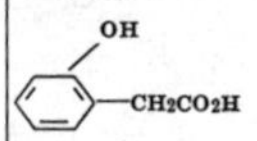 OH; CH_2CO_2H	152.15	nd (chl, eth)	147	240–3d			s		s			chl δ	**B10²**, 112
a496	—,—,amide	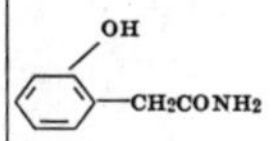 OH; CH_2CONH_2	151.17	lf (al, chl)	118					s^h				chl s^h	**B10**, 188
a497	—,—,hydrazide	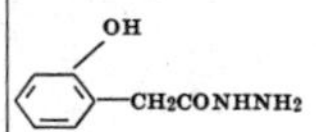 OH; $CH_2CONHNH_2$	166.18	lf (chl or bz), nd (al)	152					s^h			s	chl s^h	**B10²**, 122
a498	—,—,nitrile	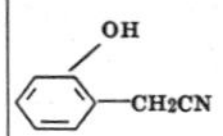 OH; CH_2CN	133.15	nd (lig-bz)	117–9				s^h					os v lig i	**B10**, 188
a499	—,**(3-hydroxyphenyl)-**	*m*-Hydroxy-α-toluic acid. 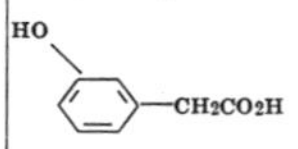 HO; CH_2CO_2H	152.15	nd (bz-lig)	129	190^{11}			v	v	v				**B10²**, 112
a500	—,—,nitrile	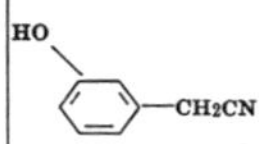 HO; CH_2CN	133.15	pl(w)	52–3				v^h	s	s				**B10**, 189
a501	—,**(4-hydroxyphenyl)-**	*p*-Hydroxy-α-toluic acid. 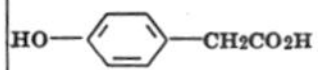 HO—; —CH_2CO_2H	152.14	nd	148	sub			δ v^h	v	v				**B10²**, 112
a502	—,—,nitrile	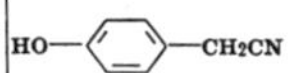 HO—; —CH_2CN	133.15	pl (w), mcl pr	69–70	330.5^{756} 210^{10}			δ	v	v				**B10²**, 113
a503	—,**iminodi-**	Diglycolamidic acid. $HN(CH_2CO_2H)_2$	133.10	rh pr	225				δ	i	i				**B4²**, 800
a504	—,—,dinitrile	α,α′-Dicyanodimethylamine. Iminodiacetonitrile. $HN(CH_2CN)_2$	95.10	lf(al)	78				s	s^h	δ		δ	chl δ	**B4²**, 481

For explanations, symbols and abbreviations see beginning of table.

No.	Name	Synonyms and Formula	Mol. wt.	Crystalline form, color and specific rotation	m.p. °C	b.p. °C	Density	n_D	Solubility w	al	eth	ace	bz	other solvents	Ref.
	Acetic acid														
a505	—,**(3-indolyl)-**	Heteroauxin. HN, CH_2CO_2H	175.18	lf(bz)	165				i	v	s	s		chl i	**B22^{2}**, 50
a506	—,**iodo-**	ICH_2CO_2H	185.95	pl (w, peth)	82	d			s^h	s^h	δ				**B2^{2}**, 206
a507	—,—,amide	ICH_2CONH_2	184.96	(w)	95				s^h						**B2^{1}**, 99
a508	—,—,ethyl ester	$ICH_2CO_2CH_2CH_3$	214.00			178–80 73^{16}	1.8173^{13}_{4}	1.5079^{13}		s	s				**B2^{2}**, 206
a509	—,**(2-isopropyl-5-methylphenoxy)-**	Thymoxyacetic acid. $CH(CH_3)_2$, OCH_2CO_2H, CH_3	208.26	nd (dil al)	148–50				δ s^h	v	v				**B6^{2}**, 499
a510	—,**(4-isopropylphenyl)-**, amide	$C_3H_7^i$—C_6H_4—CH_2CONH_2	177.25	pl (bz)		170			i	s	δ		s		**B9**, 561
a511	—,**isothiocyano-**	Mustard oil acetic acid. $SCNCH_2CO_2H$	117.13	pl(w), pr (eth)	128	$178.5–9^{19}$			v^h	s	s				**B27^{2}**, 28
a512	—,—,ethyl ester	Carbethoxymethyl isothiocyanate. $SCNCH_2CO_2CH_2CH_3$	145.18	red in air		215 δd 110^{10}	1.1649^{18}_{4}				s		s		**B4^{1}**, 480
a513	—,**menthoxy-(*l*)**	CH_3—C_6H_9—$C_3H_7^i$, OCH_2CO_2H	214.31	(eth), $[\alpha]^{20}_D$ −92.93	53–4					v				os v	**B6^{1}**, 25
a514	—,—,chloride(*l*)	CH_3—C_6H_9—C_3H_7, OCH_2COCl	232.75	$[\alpha]_D<0$		$128–31^{11}$			d^h	d^h	s				**B6^{1}**, 25
a515	—,**mercapto-**	Thioglycolic acid. $HSCH_2CO_2H$	92.12		−16.5	120^{20}	1.3253^{20}		∞	∞	∞				**B3^{2}**, 175
a516	—,—,acetate	Acetylthioglycolic acid. $CH_3COSCH_2CO_2H$	134.15	ye		$158–9^{17}$			s						**B3^{1}**, 96
516[1]	—,—,amide,*N*-(2-naphthyl)-	Thionalide. $C_{10}H_7^{\beta}NHCOCH_2SH$	217.29	nd	111–2				i	v				os v	
a517	—,—,—,*N*-phenyl-	Thioglycolic acid anilide. $HSCH_2CONHC_6H_5$	167.23	nd (w or al)	110–110.5				s^h	v	v		δ	lig δ	**B12^{1}**, 26
a518	—,—,ethyl ester	$HSCH_2CO_2CH_2CH_3$	120.17			156–8 55^{17}	1.0964^{15}			s	s				**B3^{2}**, 180
a519	—,**(2-mercaptophenyl)-**	SH, CH_2CO_2H	168.22	lf (w, bz-lig)	96–7				δ s^h	v	v		v	lig δ	**B10^{1}**, 82
a520	—,**methoxy-**	Methylglycolic acid. $CH_3OCH_2CO_2H$	90.08	hyg	203–4 96^{13}	1.4168^{20}		1.1768^{20}_{4}	s	s	s				**B3^{2}**, 170
a521	—,—,amide,*N*(4-ethoxyphenyl)-	*p*-Phenetitide methoxyacetic acid. C_2H_5O—C_6H_4—$NHCOCH_2OCH_3$	209.25	nd	99–9				δ^h	v	v				**B13**, 48
a522	—,—,chloride	CH_3OCH_2COCl	108.53			99	1.1871^{20}_{4}	1.4196^{20}	d	d				chl v	
a523	—,—,ethyl ester	$CH_3OCH_2CO_2CH_2CH_3$	118.13			143.9^{747}	1.0118^{15}		δ	v	v			os s	**B3^{2}**, 172
a524	—,—,methyl ester	$CH_3OCH_2CO_2CH_3$	104.11			131^{763} 57^{50}	1.0511^{20}_{4}	1.3964^{20}	δ	v	v			os s	**B3^{2}**, 171
a525	—,—,nitrile	CH_3OCH_2CN	71.08			120	0.9566^{15}_{4}	1.380^{28}	δ					ac, alk, os s	**B3^{2}**, 174
a526	—,—,piperazinium salt	$2CH_3OCH_2CO_2H.C_4H_{10}N_2$	266.30		155.7–6				s	s	i				
a527	—,**methoxy-(phenyl)-**(*D*,−)	$C_6H_5CH(OCH_3)CO_2H$	168.18	nd (peth), $[\alpha]^{13}_D$ −150.0 (al)	63–4				s	s		s		lig s^h	**B10^{1}**, 85
a528	—,—(*dl*-)	$C_6H_5CH(OCH_3)CO_2H$	166.18	pl (lig)	71–2				δ	v	v			lig δ s^h	**B10^{1}**, 87
a529	—,**(4-methoxy-2-nitrophenyl)-**	NO_2, CH_3O—C_6H_3—CH_2CO_2H	211.18	nd (50% al)	157–8									aa s^h	**B10^{2}**, 11

For explanations, symbols and abbreviations see beginning of table.

No.	Name	Synonyms and Formula	Mol. wt.	Crystalline form, color and specific rotation	m.p. °C	b.p. °C	Density	n_D	Solubility: w	al	eth	ace	bz	other solvents	Ref.
	Acetic acid														
a530	**—,(2-methoxy-phenoxy)-**	Pyrocatechol methylether *O*-acetic acid. 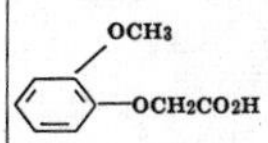 OCH_3 / OCH_2CO_2H	182.18	nd (w)	123–5 (129)	...	...	...	v^h	v	v	...	s	aa s lig i	**B6²**, 784
a531	**—,(3-methoxy-phenoxy)-**	Resorcinol methyl ether *O*-acetic acid. 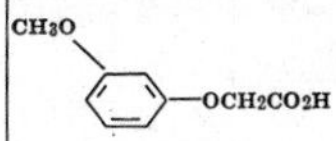 CH_3O / OCH_2CO_2H	182.18	nd (w)	118	...	...	...	δ v^h	...	...	...	...	...	**B6**, 817
a532	**—,(2-methoxy-phenyl)-**	*o*-Methoxytoluic acid. 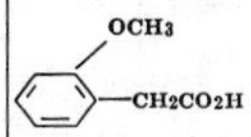 OCH_3 / CH_2CO_2H	166.18	nd (w)	123	...	...	...	...	v	v	v	v	to v^h chl v aa v lig δ	**B10**, 188
a533	—,—,nitrile	*o*-Methoxytolunitrile. 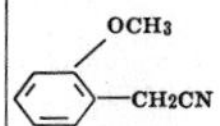 OCH_3 / CH_2CN	147.18	pr(bz-lig)	68	$141–3^{11}$	...	...	...	...	...	...	s^h	...	**B10²**, 112
a534	**—,(4-methoxy-phenyl)-**	Homoanisic acid. *p*-Methoxy-α-toluic acid 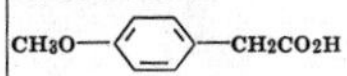 $CH_3O—C_6H_4—CH_2CO_2H$	166.18	lf	84–5.5	$138–40^{2-3}$	...	...	i s^h	v	s	...	s^h	...	**B10²**, 113
a535	—,—,nitrile	*p*-Methoxytolunitrile. 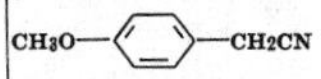 $CH_3O—C_6H_4—CH_2CN$	147.18	...	...	286–7 152^{16}	1.0845^{20}_{4}	1.5316^{17}	...	s	s	...	...	...	**B10²**, 113
—	**—,methylguanido-**	*see* **Creatine**													
a537	**—,(2-methyl-3-indolyl)-**	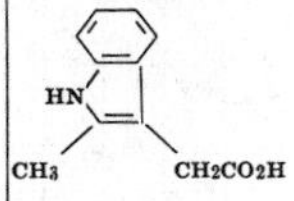 HN / CH_3 / CH_2CO_2H	189.22	(ace)	195–7d	...	...	...	δ^h	s^h	s	...	...	chl δ^h	**B22**, 68
a538	**—,(1-naphthoxy)-**	$C_{10}H_7^{\alpha}OCH_2CO_2H$	202.21	pr	190	...	...	...	δ	v	v	...	...	...	**B6²**, 580
a539	**—,(2-naphthoxy)-**	$C_{10}H_7^{\beta}OCH_2CO_2H$	202.21	pr (w)	156	...	...	...	s^h	s	s	...	...	aa s	**B6²**, 602
a540	**—,(1-naphthyl)-**	α-Naphthaleneacetic acid. $C_{10}H_7^{\alpha}CH_2CO_2H$	186.20	wh nd (w)	133	d	...	...	δ	δ	v	v	s	chl v aa s	**B9²**, 456
a541	—,—,amide	$C_{10}H_7^{\alpha}CH_2CONH_2$	185.23	nd (w)	180–1 sub	...	...	...	i s^h	...	...	...	s	CS_2 s aa s	**E12B**, 3267
a542	—,—,nitrile	$C_{10}H_7^{\alpha}CH_2CN$	167.21	wx so	32–3	$183–7^{13}$	...	...	...	s	...	...	...	...	**E12B**, 3268
a543	**—,(2-naphthyl)-**, amide	$C_{10}H_7^{\beta}CH_2CONH_2$	185.23	(al)	202–4.6d	...	...	...	δ^h	s	s	...	...	...	**E12B**, 3275
a544	—,—,nitrile	$C_{10}H_7^{\beta}CH_2CN$	167.21	(dil al or tetralin)	85.6–6.2	$145–50^{2}$ $202–5^{28}$	...	...	δ^h	v	v	...	v	chl v, CS_2 v	**E12B**, 3275
a545	**—,nitro-**	$O_2NCH_2CO_2H$	105.05	nd (chl)	87–9d	...	...	...	d	v	v	...	v^h	chl v^h	**B2²**, 207
a546	—,—,ethyl ester	$NO_2CH_2CO_2CH_2CH_3$	133.10	...	...	$105–7^{25}$	1.1922^{20}_{4}	...	δ	∞	v	...	...	dil alk s	**B2²**, 207
a547	**—,nitro(phenyl)-** nitrile	α-Nitrobenzyl cyanide. $C_6H_5CH(NO_2)CN$	162.14	...	39–40	...	...	...	d	s	...	...	...	...	**B9²**, 313
a548	**—,(2-nitro-phenoxy)-**	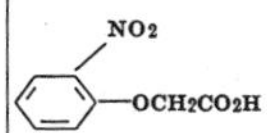 NO_2 / OCH_2CO_2H	197.15	...	156.5 (158)	...	...	...	δ	s	s	...	...	...	**B6²**, 211
a549	**—,(2-nitro-phenyl)-**	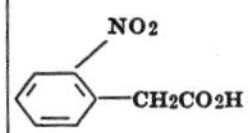 NO_2 / CH_2CO_2H	181.15	nd (w), pl (dil al)	141	...	...	...	s^h	s	...	...	...	...	**B9²**, 311
a550	—,—,nitrile	*o*-Nitrobenzyl cyanide. 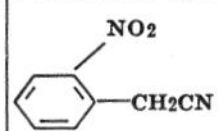 NO_2 / CH_2CN	162.14	nd (al-w) pr (aa or al)	82.5	178^{12}	...	...	s^h	s	s	...	...	...	**B9²**, 311
a551	**—,(3-nitro-phenyl)-**	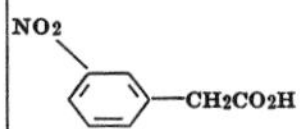 NO_2 / CH_2CO_2H	181.15	nd (w)	122	...	...	...	s^h	s	...	...	...	...	**B9²**, 311

For explanations, symbols and abbreviations see beginning of table.

No.	Name	Synonyms and Formula	Mol. wt.	Crystalline form, color and specific rotation	m.p. °C	b.p. °C	Density	n_D	Solubility w	al	eth	ace	bz	other solvents	Ref.
	Acetic acid														
a552	—,—,nitrile	*m*-Nitrobenzyl cyanide. NO_2 —CH_2CN	162.14	(eth-lig)	63	$175–80^{13}$			v		v		v	chl v lig δ	**B9**[2], 312
a553	—,**(4-nitro-phenyl)-**	NO_2—⟨⟩—CH_2CO_2H	181.14	pa ye nd (w)	151.5–2				δ	δ	δ				**B9**[2], 312
a554	—,—,nitrile	*p*-Nitrobenzyl cyanide. NO_2—⟨⟩—CH_2CN	162.14	lf	112	$195–7^{12}$			i	s	s				**B9**[2], 313
a555	—,—,piperazinium salt	$2(C_8H_7NO_4).C_4H_{10}N_2$. *See* a553	448.42		205.5–9.5				s^h	s^h	i				
a556	—,**(2-nitro-phenyl)oxo-**	*o*-Nitrobenzoylformic acid. *o*-Nitrophenylglyoxylic acid. NO_2 —$COCO_2H$	195.14	nd (w)	156–7				∞^h	s	δ	v		aa v lig δ	**B10**[1], 315
a559	—,**oxo-**,	Glyoxylic acid. Aldehydo-formic acid. $HCOCO_2H$	74.04	rh pr (w+½)	70–5 (+½w) 98				v						**B3**[2], 385
a560	—,**oxo(phenyl)-**	Benzoylformic acid. Phenyl-glyoxylic acid. $C_6H_5COCO_2H$	150.14	pr (CCl_4)	66	$147–51^{12}$			v	s	s			CS_2 i	**B10**[2], 454
a561	—,—,methyl ester	Methyl benzoyl formate. $C_6H_5COCO_2CH_2CH_3$	178.19	ye		246–8									**B10**[2], 455
a562	—,—,nitrile	Benzoyl cyanide. C_6H_5COCN	131.13	ta	32–3	206–8 99^{19}			i	v	v				**B10**[2], 457
a563	—,—,nitrile oxime	Phenylglyoxylonitrile oxime. $C_6H_5C(:NOH)CN$	146.14	lf(w), nd	126–7				s^h	v	s		v		**B10**[2], 457
a564	—,**oxo(2-thienyl)-**	2-Thienylglyoxylic acid. —$COCO_2H$ S	156.16	vlt	91.5				v		v				**B18**, 407
a565	—,**oxydi-**	Diglycolic acid. Oxydiethanoic acid. $O(CH_2CO_2H)_2.H_2O$	152.11	mcl pr (w+1)	148	d			s	s	s				**B3**[1], 90
565[1]	—,—,dichloride	Diglycolyl dichloride. $O(CH_2COCl)_2$	170.99			116^{12}			d	d				chl s	
a566	—,**per-**	Acetyl hydroperoxide. Peracetic acid. CH_3CO_2OH	76.05		0.1	105 (exp 110)	1.226^{15}_{4}		v	s	v			sulf v	
a567	—,**phenoxy-**	Glycolic acid phenyl ether. $C_6H_5OCH_2CO_2H$	152.14	nd or pl (w)	98–9	285			s	v	v		v	CS_2 v aa v	**B6**[2], 157
a568	—,—,amide	$C_6H_5OCH_2CONH_2$	151.17	nd (w, al)	101.5				i δ^h	v^h					**B6**, 162
a569	—,—,anhydride	$(C_6H_5OCH_2CO)_2O$	286.27	nd (eth)	67–9				i	δ	δ		v	lig δ	**B6**, 162
a570	—,—,chloride	$C_6H_5OCH_2COCl$	170.60			225–6 111^{13}			d	d	s				**B6**[2], 158
a571	—,—,ethyl ester	$C_6H_5OCH_2CO_2CH_2CH_3$	180.20			250–1 $145–50^{30}$	1.104^{18}		i	s	s				**B6**[2], 157
a572	—,—,methyl ester	$C_6H_5OCH_2CO_2CH_3$	166.17			245	1.150^{18}		i	∞	∞			CS_2 ∞	**B6**, 162
a573	—,—,nitrile	$C_6H_5OCH_2CN$	133.14			235^{745} 132^{30}	1.09^{18}			s	s				**B6**[2], 158
a574	—,**phenyl-**	Phenylethanoic acid*. α-Toluic acid $C_6H_5CH_2CO_2H$	136.14	lf, pl (peth)	76	266.5	1.091^{77}_{4} 1.228^{4}		s	v	v	s		CS_2 v lig i	**B9**[2], 264
a575	—,—,amide	α-Toluamide. $C_6H_5CH_2CONH_2$	135.16	pl		155	(240d)		δ s^h	δ s^h	δ		δ s^h		**B9**[1], 175
a576	—,—,—,*N*-methyl	$C_6H_5CH_2CONHCH_3$	149.20	(bz)	58				δ	v	v		δ	chl v	**B9**[2], 300
a577	—,—,—,*N*-phenyl-	α-Phenylacetanilide. $C_6H_5CH_2CONHC_6H_5$	211.23	pr (al)	117				i	s	v				**B12**, 275
a578	—,—,anhydride	α- Toluic anhydride. $(C_6H_5CH_2CO)_2O$	254.27	pr (eth)	71–2	$195–8^{12}$				s^h	s			chl s	**B9**[2], 299
a579	—,—,chloride	α-Toluyl chloride. $C_6H_5CH_2COCl$	154.59			$110–1^{23}$	1.1687^{20}_{4}		d	d	v				**B9**[2], 300
a580	—,—,ethyl ester	$C_6H_5CH_2CO_2CH_2CH_3$	164.20			227^{760} $120–1^{20}$	1.4023^{18}_{4}	1.4992^{18}	i	∞	∞				**B9**[2], 297
a581	—,—,isobutyl ester	Eglantine. $C_6H_5CH_2CO_2CH_2CH(CH_3)_2$	192.25			247	0.990		i	s	s				**B9**[2], 298

For explanations, symbols and abbreviations see beginning of table.

No.	Name	Synonyms and Formula	Mol. wt.	Crystalline form, color and specific rotation	m.p. °C	b.p. °C	Density	n_D	Solubility w	al	eth	ace	bz	other solvents	Ref.
	Acetic acid														
a582	—,—,methyl ester	$C_6H_5CH_2CO_2CH_3$	150.17			254 128–31^{20}	1.0808^0_0 1.044^{16}	1.5091^{16}	i	∞	∞				**B9**^{2}, 297
a583	—,—,nitrile	Benzylcyanide. α-Tolunitrile. $C_6H_5CH_2CN$	117.14		−23.8	234 85^8	1.0157^{20}_4	1.50144^{18} 1.5211^{25}	i	∞	∞				**B9**^{2}, 302
a584	—,—,2-phenylethyl ester	$C_6H_5CH_2CO_2CH_2CH_2C_6H_5$	240.29		26.5		1.080^{25}_{25}		i	s					
a585	—,—,piperazinium salt	2 $C_6H_5CH_2CO_2H$. $C_4H_{10}N_2$	358.43	nd	146.5–7.5				s^h	s^h	i				
a586	—,**3-pyrenyl-**	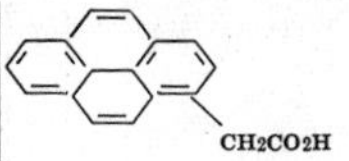 CH_2CO_2H	260.28	nd or pl (PhCl)	220d									dil alk s, con sulf s PhCl s	**E145**, 441
a587	—,**sulfo-**	Sulfoethanoic acid*. $HO_3SCH_2CO_2H$	140.11	hyg ta (w)	84–6	245^{760}d			s	s	i	s		chl s	**B4**^{2}, 531
a588	—,**2-thienyl-**	2-Thiophene acetic acid. 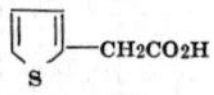 CH_2CO_2H	142.17		75–6				s^h	s	s				**B18**, 293
a589	—,**thiolo-**	Ethanethiolic acid. Thioacetic acid. CH_3COSH	76.11	ye	<−17	88–91.5^{760}	1.074^{10}_4		s v^h	v	∞				**B2**^{2}, 208
a591	—,—,ethyl ester	$CH_3COSCH_2CH_3$	104.16			116–6.2^{750}	0.9755^{25}_4	1.4503^{28}	i	v	v				**B2**^{2}, 209
a592	—,**thiono-**,amide	Thioacetamide. CH_3CSNH_2	75.14	cr (al) pl (eth)	115–6				v	v	δ		δ	lig δ	**B2**^{2}, 210
a593	—,—,—,*N*-phenyl-	Thioacetanilide. $CH_3CSHNC_6H_5$	151.22	nd (w)	74.5–6	d			i		i			alk s	**B12**^{2}, 142
a594	—,**thiono thiolo-**	Dithioacetic acid. CH_3CS_2H	92.19	ye-red		37^{15}	1.24^{20}		s	v	v	v	v	chl v aa v	**B2**^{2}, 212
a595	—,**2-tolyl-**	*o*-Methyl-α-toluic acid. 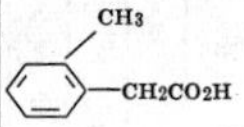	150.17	nd	88				v^h						**B9**^{1}, 207
a596	—,—,anhydride	*o*-Toluic anhydride. 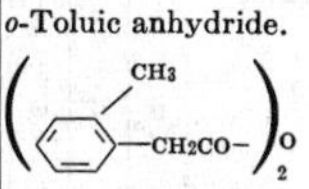	254.27	(eth)	38–9	>325^{760} 11^{220}			d	d	v				**B9**^{2}, 319
a597	—,—,nitrile	*o*-Methyl-α-tolunitrile. 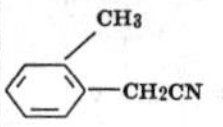	131.17			244	1.056^{22}		i	s	s		s		**B9**^{2}, 349
a598	—,**3-tolyl-**	*m*-Methyl-α-toluic acid. 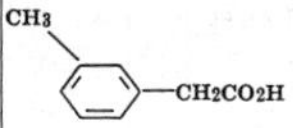	150.17	nd	60.5–1.5				v^h						**B9**^{1}, 208
a599	—,—,nitrile	*m*-Methyl-α-tolunitrile. 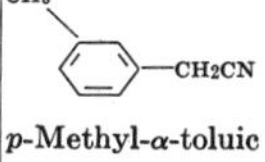	131.17			245–7^{745} δd 133^{15}	1.0022^{22}		i	s	s		s		**B9**^{2}, 349
a600	—,**4-tolyl-**	*p*-Methyl-α-toluic acid. 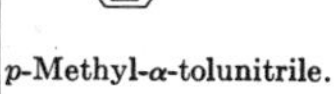	150.17	nd or lf (al)	91	265–7 (sub)			δ v^h	v	v		v	chl v	**B9**^{2}, 349
a601	—,—,nitrile	*p*-Methyl-α-tolunitrile. 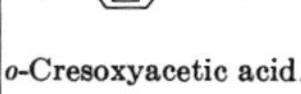	131.17		18	242–3 122^{13}	0.9922^{22}		i	s	s		s		**B9**^{2}, 349
a602	—,**2-tolyloxy-**	*o*-Cresoxyacetic acid. 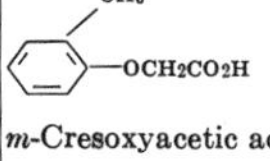	166.18	lf (w)	151–2				s^h				δ	os s con sulf s	**B6**^{2}, 331
a603	—,**3-tolyloxy-**	*m*-Cresoxyacetic acid. 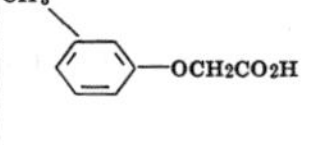	166.18	nd (w)	02–3				s^h	s			s		**B6**^{2}, 353

For explanations, symbols and abbreviations see beginning of table.

No.	Name	Synonyms and Formula	Mol. wt.	Crystalline form, color and specific rotation	m.p. °C	b.p. °C	Density	n_D	Solubility w	al	eth	ace	bz	other solvents	Ref.
	Acetic acid														
a604	—,**4-tolyloxy-**	*p*-Cresoxyacetic acid. $CH_3-C_6H_4-OCH_2CO_2H$	166.18	nd (w)	135–6				s^h	s			s	alk s	**B6²**, 380
a605	—,**tribromo-**	Br_3CCO_2H	296.78	mcl	135	245^{760}			v	v	v				**B2²**, 205
a606	—,—,amide	Br_3CCONH_2	295.80	mcl pr	121–2	sub				v^h	v		δ	chl δ	**B2²**, 206
a607	—,—,bromide	Br_3CCOBr	359.68			220–5 $88–90^{12}$			d	d	s		s	chl s	**B2²**, 206
a608	—,—,ethyl ester	$Br_3CCO_2CH_2CH_3$	324.83			225 148^{73}	2.2300^{20}_{20}	1.5438^{13}		s	s				**B2²**, 205
a609	—,**(2,4,6-tribromophenoxy)-**	$2,4,6-Br_3C_6H_2-OCH_2CO_2H$	388.87	nd (dil al)	200				i	v	v			CS_2 i lig i	**B6**, 205
a610	—,**trichloro-**	Cl_3CCO_2H	163.40	dlq cr	56.3	197.55^{760} $141–2^{25}$	1.62^{25}_{4} 1.6218^{61}_{4}	1.4603^{61}	v	s	s				**B2²**, 156
a611	—,—,amide	Cl_3CCONH_2	162.42	mcl pr (w)	141	$238–9^{746}$			δ	v	v				**B2²**, 94
a612	—,—,—,*N,N*-diethyl-	$Cl_3CCON(C_2H_5)_2$	218.52	pr	27				δ						**B4**, 110
a613	—,—,—,*N,N*-dimethyl-	$Cl_3CCON(CH_3)_2$	190.47		12	230–3	1.441^{15}		δ						**B4**, 59
a614	—,—,—,*N*-phenyl.	α-Trichloroacetanilide. $Cl_3CCONHC_6H_5$	238.51	lf (al)	95–7	$168–70^{760}$			i $δ^h$						**B12²**, 142
a615	—,—,anhydride	$(Cl_3CCO)_2O$	308.78			222–4d $98–100^{11}$	1.6908^{20}		d	d	s			aa s	**B2²**, 200
a616	—,—,bromide	Cl_3CCOBr	226.29			143	1.900^{15}_{14}		d	d	s		s		**B2**, 211
a617	—,—,butyl ester	$Cl_3CCO_2(CH_2)_3CH_3$	219.50			203–5 $100–1^{24}$	1.2778^{20}_{4}	1.4525^{25}	i	s	s		s		
a618	—,—,*sec*-butyl ester	$Cl_3CCO_2CH(CH_3)CH_2CH_3$	219.50			$88–9^{19}$	1.252^{25}_{4}	1.4440^{25}			s				
a619	—,—,*tert*-butyl ester	$Cl_3CCO_2C(CH_3)_3$	219.50	(pentane, MeOH)	25.5	37^{1}	1.2363^{25}_{4}	1.4398^{25}		v				MeOH v	
a620	—,—,chloride	Cl_3CCOCl	181.83			118	1.6564^{0}_{4} 1.6291^{16}		d	d	∞				**B2²**, 200
a621	—,—,2-chloroethyl ester	$Cl_3CCO_2CH_2CH_2Cl$	225.89			217^{766} 100^{14}	1.5357^{20}_{4}	1.4813^{20}	d		s				**B2**, 209
a622	—,—,2-hydroxyethyl ester	$Cl_3CCO_2CH_2CH_2OH$	207.44			$130–4^{12}$	1.532^{20}_{4}	1.4775^{20}	δ	v				CCl_4 v	
a623	—,—,ethyl ester	$Cl_3CCO_2CH_2CH_3$	191.44			167.5–8 62^{12}	1.3826^{20}_{4}	1.4507^{20}	i	s	s				**B2²**, 200
a624	—,—,isobutyl ester	$Cl_3CCO_2CH_2CH(CH_3)_2$	219.50			187–9 $93–4^{20}$	1.255^{25}_{4}	1.4456^{25}		s	s				**B2**, 209
a625	—,—,isopropyl ester	$Cl_3CCO_2CH(CH_3)_2$	205.47			174–5 $65.5–7^{15}$	1.2911^{25}_{4}	1.4409^{25}					s		
a626	—,—,2-methoxyethyl ester	$Cl_3CCO_2CH_2CH_2OCH_3$	221.47		14.6–4.8	$98.0–9.5^{17}$	1.3866^{20}_{4}	1.4563^{20}							
a627	—,—,methyl ester	$Cl_3CCO_2CH_3$	177.42		−17.5	153.8 44.5^{12}	1.4874^{20}_{4}	1.4572^{20}	i	v	v				**B2²**, 199
a628	—,—,2(2-methylbutyl) ester	$Cl_3CCO_2C(CH_3)_2CH_2CH_3$	233.52			191^{756} 105^{30}	1.2505^{20}_{20}		i		s				
a629	—,—,3-methylbutyl ester	$Cl_3CCO_2CH_2CH_2CH(CH_3)_2$	233.52			217 $92–5^{11}$	1.2314^{20}_{4}	1.4521^{20}	i	s	s				**B2²**, 200
a630	—,—,nitrile	Cl_3CCN	144.39		−44.14	84.6^{741}	1.4403^{25}_{4}	1.4409^{20}							**B2**, 212
a631	—,—,pentyl ester	$Cl_3CCO_2(CH_2)_4CH_3$	233.52			220.3–2.3 118^{30}	1.2475^{20}_{20}		i	s	s				
a632	—,—,piperazinium salt	$2Cl_3CCO_2H.\ C_4H_{10}N_2$	412.91		121				s	s^h	i				

For explanations, symbols and abbreviations see beginning of table.

No.	Name	Synonyms and Formula	Mol. wt.	Crystalline form, color and specific rotation	m.p. °C	b.p. °C	Density	n_D	Solubility w	al	eth	ace	bz	other solvents	Ref.
	Acetic acid														
a633	—,—,propyl ester	$Cl_3CCO_2CH_2CH_2CH_3$	205.47			187 69^{10}	1.3221^{20}_{4}	1.4501^{20}	i	s	s				
a634	—,—,trichloromethyl ester	$Cl_3CCO_2CCl_3$	280.75		34	191–2 $73\text{–}4^{10}$	1.6733^{35}_{4}				s		s	chl s lig s	**B3**[2], 17
a635	**—,(2,4,5-trichlorophenoxy)-**	cl, cl, cl ring; —OCH_2CO_2H	255.49	(bz)	157–8				δ				v^h		
a636	**—,(2,4,6-trichlorophenoxy-**	cl, cl, cl ring; —OCH_2CO_2H	255.49		177				δ^h	s^h			δ	lig δ	**B6**, 192
a637	**—,trifluoro-**	F_3CCO_2H	114.02		−15.25	72.4	1.5351^0		s						**B2**, 186
a638	—,—,amide,*N*-phenyl-	$F_3CCONHC_6H_5$	189.06	(60% al)	87.6					v					**B12**[2], 141
a639	—,—,anhydride	$(F_3CCO)_2O$	210.04		−65	39.5–40.1			d	d	s			aa s	**B2**[2], 186
a640	—,—,nitrile	F_3CCN	95.03			-63.9^{760}									
a641	**—,triiodo-**	I_3CCO_2H	437.76	ye lf	150d				s	s	s				**B2**, 225
a642	**—,trinitro-**,nitrile	$(NO_2)_3CCN$	176.05	wx	41.5	220 exp			d	d	s				**B2**, 229
a643	**—,triphenyl-**	$(C_6H_5)_3CCO_2H$	288.33	mcl pr (al), lf (aa)	271				i	s	δ		δ	MeOH s, CS_2, chl δ	**B9**[2], 500
—	**—,ureido-**	*see* **Hydantoic acid**													
—	**Acetic anhydride**	*see* **Acetic acid,** anhydride													
—	**Acetoacetic acid**	*see* **Butanoic acid, 3-oxo-***													
—	**Acetoin**	*see* **2-Butanone, 3-hydroxy-***													
—	**Acetol**	*see* **2-Propanone, 1-hydroxy-***													
—	**Acetonaphthone**	*see* **Naphthalene, acetyl-**													
—	**Acetone**	*see* **2-Propanone***													
—	**Acetonedicarboxylic acid**	*see* **Pentanedioic acid, 3-oxo-***													
—	**Acetonic acid**	*see* **Propanoic acid, 2-hydroxy-2-methyl-***													
—	**Acetonitrile**	*see* **Acetic acid,** nitrile													
a644	**Acetonitrolic acid**	Ethylnitrolic acid. $CH_3C(:NOH)NO_2$	104.07	ye rh (w, al)	84–5d				s	s	s				**B2**[2], 185
—	**Acetonylacetone,** dioxime	*see* **2,5-Hexanedione,** dioxime*													
a645	**Acetophenone**	Acetylbenzene. Methyl phenyl ketone ring positions 2, 3, 4, 5, 6; $COCH_3$	120.14	mcl pr or lf	19.655	202.0	1.0281^{20}_{4}	1.5363^{15}	i	s	s		s	chl s	**B7**[2], 208
a646	—,oxime	$C_6H_5C(:NOH)CH_3$	135.16	nd (w)	59	119^{20}			δ s^h	v	v	v	v	chl v lig v	**B7**[2], 216
—	**—,4-acetylamino-**	*see* **Acetic acid,** amide, *N*(4-Acetylphenyl)-													
a647	**—,2-amino-**	*o*-Acetylaniline. C_8H_9NO. *See* a645	135.16	ye	20	250–2 δd 135^{17}			i		s				**B14**[2], 28
a648	**—,3-amino-**	*m*-Acetylaniline. C_8H_9NO. *See* a645	135.16	ye pl (al), lf (eth)	98–9	289–90				s^h					**B14**[2], 30
a649	**—,4-amino-**	*p*-Acetylaniline C_8H_9NO. *See* a645	135.16	ye mcl pr	106	293–5 $195\text{–}200^{15}$			δ s^h	s	s		δ	chl δ lig δ	**B14**[2], 30
a650	**—,4-amino-α-chloro-**	*p*-Aminophenacyl chloride. C_8H_8ClNO. *See* a645	169.62	ye pl	148										**B14**[1], 367
a651	**—,4-amino-3-chloro-**	C_8H_8ClNO. *See* a645	169.62	pr (chl-peth)	92										**B14**, 49
a652	**—,3-amino-4-methoxy-**	$C_9H_{11}NO_2$. *See* a645	165.19	pr (al)	102					s	s		v		**B14**[2], 141
—	**—,benzal-**	*see* **Chalcone**													
—	**—,benzylidene**	*see* **Chalcone**													
a653	**—,α-bromo-**	Phenacyl bromide. $C_6H_5COCH_2Br$	199.05	nd (al), rh pr (al), pl (peth)	49.5–51	$133\text{–}5^{12}$	1.647^{20}_{4}		i	v^h	v		v	chl v	**B7**[2], 220
a654	**—,2-bromo-**	C_8H_7BrO. *See* a645	199.05	ye		112^{10} $131\text{–}5^{20}$		1.5678^{20}							**B7**[2], 220
a655	**—,4-bromo-**	C_8H_7BrO. *See* a645	199.05	lf (al)	49.5–51.5	255.5^{736}	1.647		i	s	s		s	CS_2 s	**B7**[2], 220
a656	**—,α-bromo-3-chloro-**	*m*-Chlorophenacyl bromide. C_8H_6BrClO. *See* a645	233.50	nd	96–6.5					v					**B7**[2], 221

For explanations, symbols and abbreviations see beginning of table.

No.	Name	Synonyms and Formula	Mol. wt.	Crystalline form, color and specific rotation	m.p. °C	b.p. °C	Density	n_D	Solubility						Ref.
									w	al	eth	ace	bz	other solvents	
	Acetophenone														
a657	**—,α-bromo-4-chloro-**	*p*-Chlorophenacyl bromide. C_8H_6BrClO. *See* a645	233.50	nd	96–6.5										**B7**, 285
a658	**—,4-bromo-α-chloro-**	*p*-Bromophenacyl chloride. C_8H_6BrClO. *See* a645	233.50	nd	116–7					v^h					**B7**, 285
a659	**—,α-bromo-4-methyl-**	*p*-Methylphenacyl bromide. C_9H_9BrO. *See* a645	213.08	nd, lf (al)	48–50	$155–9^{14}$			d	s	s				**B7**², 239
a660	**—,4-*tert*-butyl-2,6-dimethyl-3,5-dinitro-**	Musk ketone. $C_{13}H_{16}N_2O_5$. *See* a645	280.28	ye	134.5–6.5				i	δ					
a661	**—,α-chloro-**	Phenacyl chloride. $C_6H_5COCH_2Cl$	154.60	pl (dil al), rh, lf (peth), pr	56–5	247^{760} $133–4^{20}$	1.324_4^{15}		i	v	v	s	v	CS_2 v	**B7**², 219
a662	**—,2-chloro-**	C_8H_7ClO. *See* a645	154.60			$227–8^{738}$ 84^3	1.2016_4^{17}	1.685^{25}	δ		s				**B7**², 218
a663	**—,3-chloro-**	C_8H_7ClO. *See* a645	154.60			$241–5^{744}$ $127–31^{30}$	1.2130_4^{0}				s	s			
a664	**—,4-chloro-**	C_8H_7ClO. *See* a645	154.60		18.40	273^{760} $124–6^{30}$	1.1922_4^{20}	1.5550^{20}	i	∞	∞				**B7**², 219
a665	**—,α-chloro-2,4-dimethyl-**	2,4-Dimethylphenacyl chloride. $C_{10}H_{11}ClO$. *See* a645	182.65	nd	62					v	v		v		**B7**², 249
a666	**—,α-chloro-α-isonitroso-**	Benzoylformyl chloridoxime. $C_6H_5COCCl(:NOH)$	183.60	lf (bz), pr (chl)	132–3				i	v	v		v	chl v, CCl_4 δ	**B10**², 460
a667	**—,α-chloro-4-methyl-**	*p*-Methylphenacyl chloride. C_9H_9ClO. *See* a645	168.62	nd (al)	57–8	260–3 δd 113^4			d	v	v				**B7**¹, 165
a668	**—,α,4-dibromo-**	*p*-Bromophenacyl bromide. $C_8H_6Br_2O$. *See* a645	277.95	nd	110–1				i	s^h	s				**B7**², 222
a669	**—,α,α-dichloro-**	Phenacyl chloride. $C_6H_5COCHCl_2$	189.04	amor	20–1.5	247–8d 143^{25}	1.340^{16}			s			s		**B7**², 220
a670	**—,α,4-dichloro-**	*p*-Chlorophenacyl chloride. $C_8H_6Cl_2O$. *See* a645	189.04	nd (al)	101–2	270				s v^h			s	MeOH s	**B7**¹, 152
a671	**—,2,4-dichloro-**	$C_8H_6Cl_2O$. *See* a645	189.04		33–4	$140–50^{15}$			i						**B7**², 219
a672	**—,3,4-dichloro-**	$C_8H_6Cl_2O$. *See* a645	189.04	nd (peth)	76	135^{12}			i					lig s^h	**B7**², 219
a673	**—,2,3-dihydroxy-**	$C_8H_8O_3$. *See* a645	152.15	pr (bz-lig)	97–8					s				alk s	**B8**¹, 613
a674	**—,2,4-dihydroxy-**	Resacetophenone. $C_8H_8O_3$. *See* a645	152.15	nd or lf	142		1.1800^{141}		i	s^h	δ		δ	chl i, Py s aa s	**B8**², 294
a675	**—,2,5-dihydroxy-**	2-Acetylhydroquinone. Quinacetophenone. $C_8H_8O_3$	152.15	ye-gr nd (dil al)	203–4				δ	s	δ		δ		**B8**², 297
a676	**—,3,4-dihydroxy-**	$C_8H_8O_3$. *See* a645	152.15	nd (w or chl)	115–6				s^h					chl s^h	**B8**², 298
a677	**—,3,5-dihydroxy-**	$C_8H_8O_3$. *See* a645	152.15	(w)	147–8				v	v	v	v	δ		**B8**², 301
—	**—,3,4-dihydroxy-d-methylamino-**	*see* **Adrenalone**													
677¹	**—,3,4-dimethoxy-**	Acetoveratrone. $C_{10}H_{12}O_3$. *See* a645	180.20	pr (dil al)	51	$286–8^{760}$ $160–2^{10}$			s^h	s^h	s^h		s^h	chl s^h	**B8**², 298
677²	**—,3,5-dimethoxy-4-hydroxy-**	Acetosyringone. $C_{10}H_{12}O_4$. *See* a645	196.20	nd(w)	122				s^h	s	s				
a678	**—,2,4-dimethyl-**	$C_{10}H_{12}O$. *See* a645	148.21			228^{760} 110^{13}	1.0121^{15}		i	s					**B7**², 229
a679	**—,2,5-dimethyl-**	$C_{10}H_{12}O$. *See* a645	148.21			232–3 107^{13}	0.9963_4^{19}	1.530^{19}	i	v	v		v	CS_2 v	**B7**², 248
a680	**—,3,4-dimethyl-**	$C_{10}H_{12}O$. *See* a645	148.21			246–7 213^{310}	1.0090_4^{14}	1.5413^{15}	i	v	v		v	chl v aa v	**B7**², 248
a681	**—,3-dimethyl-amino-**	$C_{10}H_{13}NO$. *See* a645	163.22		42–3	148^{13}					v				**B14**, 45
a682	**—,4-dimethyl-amino-**	$C_{10}H_{13}NO$. *See* a645	163.22	nd (w, peth)	105.5	$172–5^{11}$			v^h		v			lig v^h	**B14**², 32
a683	**—,2-ethoxy-**	*o*-Acetylphenetole. $C_{10}H_{12}O_2$. *See* a645	164.21	pr (dil al), pl (lig)	43	243–4	1.0036^{78}		δ	v	v				**B8**², 85
a684	**—,4-ethoxy-**	*p*-Acetylphenetole. $C_{10}H_{12}O_2$. *See* a645	164.21	pl (eth)	36–7					v	v^h				**B8**², 85
a685	**—α-hydroxy-**	Benzoyl carbinol. Phenacyl alcohol. $C_6H_5COCH_2OH$	136.15	hex pl (al or eth), lf (w or dil al, +w)	86–7	$124–6^{12}$	1.0963_4^{99}		s^h	s	s			chl s lig δ	**B8**², 88
a686	**—,—,acetate**	Phenacyl acetate. $C_6H_5COCH_2O_2CCH_3$	178.19	rh pl (eth, lig or peth)	49–9.5	270 $150–2^{10}$	1.1169_4^{65}		i	v	v		s	chl v lig s	**B8**², 89
a687	**—,2-hydroxy-**	*o*-Acetylphenol. $C_8H_8O_2$. *See* a645	136.15			218^{760} $91–2^{13}$	1.1307_4^{21}	1.558^{21}	δ	∞	∞			aa ∞	**B8**², 81

For explanations, symbols and abbreviations see beginning of table.

No.	Name	Synonyms and Formula	Mol. wt.	Crystalline form, color and specific rotation	m.p. °C	b.p. °C	Density	n_D	Solubility w	al	eth	ace	bz	other solvents	Ref.
	Acetophenone														
a688	**—,3-hydroxy-**	*m*-Acetylphenol. $C_8H_8O_2$. *See* a645	136.15	nd or lf	94	296^{756} 153^{15}	1.0992^{109}	1.5348^{109}	v^h	v	v	...	v	chl v	**B8**^{2}, 84
a689	**—,4-hydroxy-**	*p*-Acetylphenol. $C_8H_8O_2$. *See* a645	136.15	nd (eth, dil al)	109	$147–8^{3}$	1.1090^{109}	...	δ s^h	v	v	...	...	...	**B8**^{2}, 84
a690	**—,α-hydroxy-4-methoxy-**	Anisoyl carbinol. $C_9H_{10}O_3$. *See* a645	166.18	lf	100	...	...	...	...	s	...	...	s	...	**B8**^{1}, 618
a691	**—,2-hydroxy-3-methoxy-**	*o*-Acetovanillon. $C_9H_{10}O_3$. *See* a645	166.18	pa ye nd (eth-pentane)	53–4	...	...	...	...	...	s	...	s	...	**B8**^{2}, 293
a692	**—,2-hydroxy-4-methoxy-**	Peonol. $C_9H_{10}O_3$. *See* a645	166.18	...	51.3	158^{20}	1.1310^{81}	1.5432^{81}	δ	s	s	...	s	chl s	**B8**^{2}, 294
a693	**—,2-hydroxy-5-methoxy-**	$C_9H_{10}O_3$. *See* a645	166.18	pa ye pr (dil al)	52	...	...	...	...	s^h	...	...	s	...	**B8**, 271
a694	**—,3-hydroxy-4-methoxy-**	Isoacetovanillon. $C_9H_{10}O_3$. *See* a645	166.18	anh cr (eth-lig), or (w+1)	67–8 (+1w) 91(anh)	...	...	...	s^h	...	s^h	...	...	...	**B8**^{2}, 298
a695	**—,4-hydroxy-α-methoxy-**	$C_9H_{10}O_3$. *See* a645	166.18	nd (bz), pr (w+¼)	130–1	...	...	...	s^h	v	s	v	s^h	lig δ	**B8**^{2}, 302
a696	**—,4-hydroxy-2-methoxy-**	Isopeonol. Resacetophenone 2-methyl ether. $C_9H_{10}O_3$. *See* a645	166.18	nd (w)	138	...	...	...	s^h	...	...	...	s	lig s	**B8**^{1}, 618
a697	**—,4-hydroxy-3-methoxy-**	Acetovanillon. Apocynin. $C_9H_{10}O_3$. *See* a645	166.18	pr (w)	115	295–300 $233–5^{15}$	...	...	δ v^h	s	v	...	s	chl v lig i	**B8**^{2}, 298
a698	**—,5-isopropyl-2-methyl-**	2-Acetyl-*p*-cymene. Carvacryl methyl ketone $C_{12}H_{16}O$. *See* a645	176.25	...	<−20	240	0.956^{20}_{4}	1.5181^{520}	...	...	...	...	...	...	**B7**^{2}, 262
a699	**—,2-methoxy-**	*o*-Acetoanisole. $C_9H_{10}O_2$. *See* a645	150.17	ye	...	239–40 $121–1^{13}$	1.0897^{20}_{4}	1.538^{24}	...	...	...	...	...	...	**B3**, 232
a700	**—,3-methoxy-**	*m*-Acetoanisole. $C_9H_{10}O_2$. *See* a645	150.17	...	95–6	240 $121–2^{12}$	1.0343^{19}	...	s	...	...	...	...	...	**B8**^{1}, 535
a701	**—,4-methoxy-**	*p*-Acetoanisole. $C_9H_{10}O_2$. *See* a645	150.17	pl (eth)	38–9	265	1.0818^{41}_{4}	1.53349^{20}	δ	s	s	...	...	...	**B8**^{2}, 84
a702	**—,4-methoxy-3-nitro-**	4-Acetyl-2-nitroanisole $C_9H_9NO_4$. *See* a645	195.17	nd (al)	99.5	...	...	...	...	s^h	s	s	s	...	**B8**^{1}, 538
a703	**—,2-methyl-**	*o*-Acetotoluene. $C_9H_{10}O$. *See* a645	134.17	...	...	211^{745} $89–92^{10}$	1.026^{20}_{4}	1.535^{13}	...	...	...	...	...	...	**B7**^{2}, 237
a704	**—,3-methyl-**	*m*-Acetotoluene. $C_9H_{10}O$. *See* a645	134.17	...	...	221^{745} $127.5–8^{30}$	1.007^{20}_{4}	1.5306^{20}	...	s	s	s	...	...	**B7**^{2}, 237
a705	**—,4-methyl-**	*p*-Acetotoluene. $C_9H_{10}O$. *See* a645	134.17	nd	28	226^{760}	1.0051	1.5335^{20}	i	s	s	...	s	chl s	**B7**^{1}, 238
a706	**—,2-nitro-**	$C_8H_7NO_3$. *See* a645	165.15	(al)	28–9	158^{16}	...	...	δ	v	v	...	...	chl v	**B7**^{2}, 222
a707	**—,3-nitro-**	$C_8H_7NO_3$. *See* a645	165.15	nd (al)	81	202 167^{18}	...	...	v^h	δ	v v	...	...	...	**B7**^{2}, 223
a708	**—,4-nitro-**	$C_8H_7NO_3$. *See* a645	165.15	yesh pr	80–1	...	...	...	...	s	...	...	...	...	**B7**^{2}, 223
a709	**—,α,α,α-trichloro-**	$C_6H_5COCCl_3$	223.50	...	...	256–7 145^{25}	1.425^{16}	...	i	s	v	...	...	...	**B7**^{1}, 152
a710	**—,2,3,4-tri-hydroxy-**	4-Acetylpyrogallol. Gallacetophenone. $C_8H_8O_4$. *See* a645	168.15	pa ye nd or lf	173	...	...	...	s^h	v	s	v	δ	chl δ aa v lig v	**B8**^{2}, 493
a711	**—,2,4,6-tri-hydroxy-**	2-Acetylphloroglucinol. Phloroacetophenone. $C_8H_8O_4$. *See* a645	168.15	ye (in alk), nd (w+1)	218–9	...	...	...	δ s^h	v	v	v	$δ^h$	chl δ aa v	**B8**^{2}, 442
a712	**—,2,4,5-tri-methyl-**	5-Acetylpseudocumene. $C_{11}H_{14}O$. *See* a645	162.22	...	10–1	247–8 $137–8^{20}$	1.001^{18}_{4}	1.541^{15}	i	v	v	...	v	CS_2 v aa v	**B7**^{2}, 256
a713	**—,2,4,6-tri-methyl-**	Acetylmesitylene. $C_{11}H_{14}O$. *See* a645	162.22	...	...	240.5^{735} 120^{12}	0.9754^{20}_{4}	1.5175^{20}	i	s	s	s	s	...	**B7**^{2}, 257
a714	**—,α,α,α-triphenyl-** (α form)	α-Benzopinacolone. $C_6H_5COC(C_6H_5)_3$	348.45	nd	204–5	...	...	...	...	i	...	...	s	chl, CS_2 s aa i	...
a715	**—,—,**(β form)	β-Benzopinacolone. $C_6H_5COC(C_6H_5)_3$	348.45	nd (al)	182	...	...	...	i	δ s^h	s	...	s	chl, CS_2 s aa s^h	**B7**^{2}, 513
—	**Acetopyruvic acid**	*see* **Pentanoic acid, 2,4-dioxo-***													
—	**Acetosyringone**	*see* **Acetophenone, 3,5-dimethoxy-4-hydroxy-**													
—	**Acetovanillon**	*see* **Acetophenone, 4-hydroxy 3-methoxy-**													
—	**Acetoveratrone**	*see* **Acetophenone, 3,4-dimethoxy-**													
—	**Acetoxime**	*see* **2-Propanone, oxime***													

For explanations, symbols and abbreviations see beginning of table.

No.	Name	Synonyms and Formula	Mol. wt.	Crystalline form, color and specific rotation	m.p. °C	b.p. °C	Density	n_D	Solubility w	al	eth	ace	bz	other solvents	Ref.
	Aceturic acid														
—	**Aceturic acid**	see **Glycine,** *N*-acetyl-													
—	**Acetylacetone**	see **2,4-Pentanedione***													
—	**Acetylbenzoyl**	see **1,2-Pentanedione, 1-phenyl-***													
—	**Acetyl bromide**	see **Acetic acid,** bromide													
—	**Acetyl chloride**	see **Acetic acid,** chloride													
—	**Acetyl fluoride**	see **Acetic acid,** fluoride													
—	**Acetyl iodide**	see **Acetic acid,** iodide													
—	**Acetylene**	see **Ethyne***													
—	**Acetylenedi-carboxylic acid**	see **Butynedioic acid***													
a718	**Achroodextrin**	$C_{36}H_{62}O_{31}$	990.88	amor	...	...	...	...	s	i	...	...	...	...	...
a719	**Aconic acid**	(structure)	128.08	lf (eth), rh	164	d	...	...	s	s	...	...	...	MeOH s	**B18,** 395
a720	**Aconine**	$C_{25}H_{41}NO_9$ $=C_{19}H_{19}(OCH_3)_4(OH)_5(NC_2H_5)$	499.61	amor,[α]+23	132	...	...	...	s	s	δ	...	...	chl s lig δ	...
—	**Aconitic acid**	see 1,2,3-Propenetri-carboxylic acid*													
a721	**Aconitine**	$C_{34}H_{47}NO_{11}=C_{19}H_{19}(OCH_3)_4(OH)_3$ $(NC_2H_5)(O_2CCH_3)(O_2CC_6H_5)$	645.75	rh lf, $[\alpha]_D^8+18.01$	204	...	...	...	i	s	δ	...	s	ch s	...
a722	—,diacetate	$C_{34}H_{47}NO_{11}(COCH_3)_2$. *See* a721	731.85	...	158	...	...	...	s	...	...	...	...	...	...
a723	—,hydrobromide	$C_{34}H_{47}NO_{11}$.HBr. 1½ H_2O	753.71	yesh pr (w), $[\alpha]_D-28.42$(w)	176–80	...	...	...	s^h	s	s	...	...	...	...
a724	—,hydrochloride(*l*)	$C_{34}H_{47}NO_{11}$HCl. 3½ H_2O	745.27	(w+3½) $[\alpha]_D-29.66$(w)	170–2	...	...	...	s	s	s	...	...	...	...
a725	—,nitrate(*l*)	$C_{34}H_{47}NO_{11}.HNO_3$	708.77	$[\alpha]_D^{20}-35$(2% aq sol)	ca. 200d	...	...	...	s	s	...	...	...	...	...
a726	—,sulfate(*l*)	$(C_{34}H_{47}O_{11}N)_2$. H_2SO_4	1389.60	yesh amor pw	...	...	...	...	s	s	...	...	...	...	...
a727	**Acridine**	2,3,5,6-Dibenzopyridine. (structure)	179.21	rh nd or pr (al), mcl, orh (2 forms)	111 sub	345–6	1.005_4^{20}	...	δ^h	v	v	...	v	CS_2 v	**B20²,** 300
a728	—,**2-amino-**	$C_{13}H_{10}N_2$. *See* a727	194.24	nd (w)	209	...	...	...	s^h	v	v	...	...	dil ac s	**B22,** 462
a729	—,**3-amino-**	$C_{13}H_{10}N_2$. *See* a727	194.24	og amor (bz-lig)	221–2	...	...	...	s	...	...	...	s	dil ac v	**B22²,** 376
a730	—,**4-amino-**	$C_{13}H_{10}N_2$. *See* a727	194.24	ye br pr (peth), og nd (MeOH)	105–6	183–4	...	...	...	v	v	...	v	MeOH s^h con sulf s dil HCl s	**B22²,** 376
a731	—,—hydrochloride.	$C_{13}H_{10}N_2$. HCl *See* a727	230.70	ye nd (dil al)	234	...	...	...	v	s	...	...	...	...	**B22²,** 376
a732	—,**9-amino-**	$C_{13}H_{10}N_2$. *See* a727	194.24	ye nd	233	...	...	...	...	s^h	...	...	...	dil HCl v	**B21²,** 280
a733	—,**2-amino-5(4-aminophenyl)-**	Chrysaniline. $C_{19}H_{15}N_3.2H_2O$ *See* a727	321.38	ye nd (95% al +2w)	260–7	...	...	...	i	δ	...	...	...	...	**B22²,** 403
a734	—,**9-chloro-**	$C_{13}H_8ClN$. *See* a727	213.67	nd (al)	122 sub	...	...	...	δ	v	...	...	...	...	**B20²,** 301
a735	—,**3,6-diamino**	$C_{13}H_{11}N_3$. *See* a727	209.25	nd (al or w)	283	...	...	...	s^h	v	δ	...	δ	...	**B22²,** 397
a736	—,**6,9-diamino-2-ethoxy-**	Rivanol. $C_{15}H_{15}N_3O$. *See* a727	253.31	ye nd	123–4d	...	...	...	...	...	...	...	...	...	**B22²,** 458
a737	—,**9,10-dihydro-**	Acridan. Carbazine. Dihydroacridine. $C_{13}H_{11}N$. *See* a727	181.24	pl or pr (al)	169	d300 sub	...	...	i	v^h	s	...	...	...	**B20²,** 291
a738	—,**9,10-dihydro-9-oxo-**	Acridone. $C_{13}H_9NO$. *See* a727	195.22	ye nd	354	>354 (no d)	...	...	i	δ v^h	i	...	i	alk s^h con sulf s chl i	**B21²,** 280
a739	—,**2-methyl-**	$C_{14}H_{11}N$. *See* a727	193.25	ye nd (al)	134	...	...	...	δ	v	v	...	v	con sulf s	**B20¹,** 173
a740	—,**9-phenyl-**	$C_{19}H_{13}N$. *See* a727	255.32	lf, ye nd (al)	182–3	403–4	...	...	i	δ	s	...	v	con sulf v	**B20²,** 332
—	*ms*-**Acridone**	see **Acridine, 9,10-dihydro-9-oxo-**													
—	**Acrolein**	see **Propenal***													
—	**β-Acrose**	see **Sorbose** (*L*)													
—	**Acrylamide**	see **Propenoic acid,** amide*													
—	**Acrylic acid**	see **Propenoic acid***													
—	**Acrylyl chloride**	see **Propenoic acid,** chloride													
a741	**Actidione**	(structure)	281.34	pl (AmOH, or 30% MeOH) $[\alpha]_D^{25}+6.8$(w), $[\alpha]_D^{25}-3.0$ (MeOH)	115.5–7.0	...	...	...	δ	v	...	...	...	os v lig i	...

For explanations, symbols and abbreviations see beginning of table.

No.	Name	Synonyms and Formula	Mol. wt.	Crystalline form, color and specific rotation	m.p. °C	b.p. °C	Density	n_D	Solubility w	al	eth	ace	bz	other solvents	Ref.
	Adabine														
—	**Adabine**	*see* **Urea, 1(2-bromo-2-ethylbutanoyl)-**													
a742	**Adamantane**	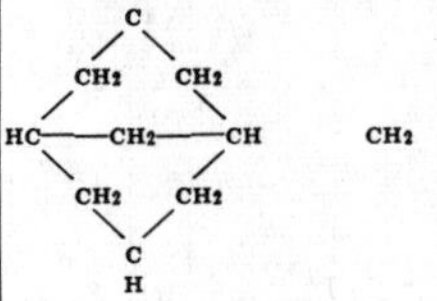	136.24	nd	268	....	1.07	1.568	...	...	...	...	...	...	E13, 1058
—	**Adenine**	*see* **Purine, 6-amino-**													
a743	**Adenosine**	Adenine-9-*d*-ribofuranoside. 6-Amino-ribofuransoide purine 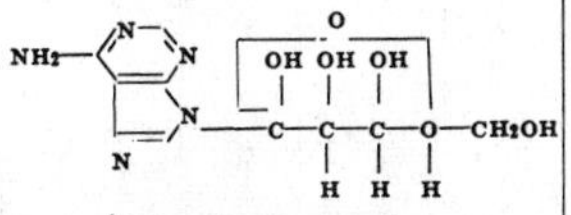	267.24	nd (w+1½), $[\alpha]^{20}_D -60.0$ (w, c=1), −43.5 (10% HCl, c=2)	229	....	....	....	s v^h	i	...	...	...	...	B31, 27
a744	**Adenylic acid**	Adenosine-5-phosphate. Muscle adenylic acid. Synadenylic acid. 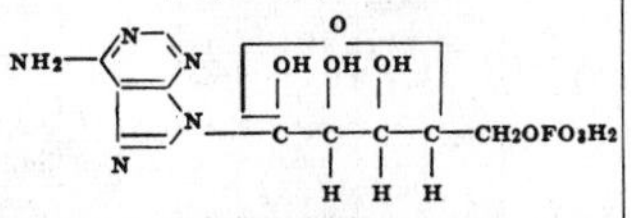	347.23	pw, nd (w, aq al) $[\alpha]^{50}_D -41.78$	195d	....	....	....	v^h	...	...	...	...	10% HCl s	....
—	**Adipamic acid**	*see* **Hexanedioic acid, monoamide***													
—	**Adipic acid**	*see* **Hexanedioic acid***													
—	**Adipoin**	*see* **Cyclohexanone-2-hydroxy-**													
a745	**Adonitol**	Adonite. Ribitol. 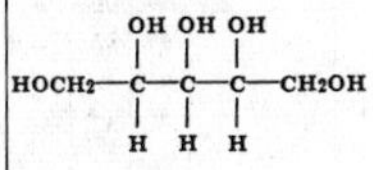	152.15	pr (w), nd (al)	102	....	....	....	s	s^h	i	...	...	lig i	B1^2, 604
a746	**Adrenaline(*d*)**	*d*-Epinephrine, *d*-2-methylamino-1-hydroxy-1-(3,4-dihydroxyphenyl)ethane. $C_9H_{13}NO_3$.	183.20	$[\alpha]^{20}_D +50.5$ (HCl)	215(d)	....	....	....	δ	i	...	...	...	...	B13^2, 523
a747	—(*l*)	$C_9H_{13}NO_3$	183.20	br (in air), pw $[\alpha]^{20}_D -50.72$ (aq HCl)	211–2	....	....	....	δ	i	...	...	...	...	B13^2, 523
a748	**Adrenalone**	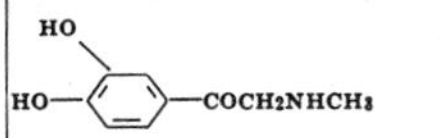	181.19	nd	235–6d	....	....	....	i	i	i	...	...	...	B14^2, 157
a749	**Adrenochrome**	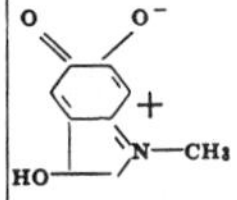	179.17	red (MeOH-HCO_2H, +½w)	115–20d	....	....	....	v	s	i	...	i	...	....
a750	**Adrenosterone**	D^4-Androstrene-3,11,17-trione. 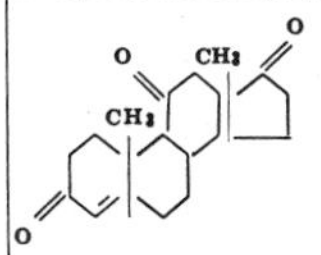	300.28	nd (al), lf (eth) $[\alpha]^{20}_D +262$ (abs al) $[\alpha]^{25}_D +281$ (ace)	220–3	sub (vac)	....	....	δ	s	s	s	...	chl s	E14s, 2979
—	**Adurol**	*see* **Benzene, 2-bromo-1,4-dihydroxy-***													
—	**Agaric acid**	*see* **1,2,3-Nonadecanetricarboxylic acid, 2-hydroxy-***													
a751	**Ajmalicine**	*Py*-tetrahydroserpentine. δ-Yohimbine. $C_{21}H_{24}N_2O_3$.	352.42	nd, $[\alpha]^{24}_D -58$ (chl)	....	....	....	....	i	...	...	...	...	...	....
a752	**Ajmaline**	$C_{20}H_{26}N_2O_2$	326.42	nd, $[\alpha]+131$ (chl)	158–60	....	....	....	i	s	...	...	...	chl s	....

For explanations, symbols and abbreviations see beginning of table.

No.	Name	Synonyms and Formula	Mol. wt.	Crystalline form, color and specific rotation	m.p. °C	b.p. °C	Density	n_D	Solubility w	al	eth	ace	bz	other solvents	Ref.
	α-Alanine														
a753	**α-Alanine** (*D*)	*l*-α-Aminopropionic acid. $CH_3CHNH_2CO_2H$	89.09	pr (w-al), $[\alpha]_D$ −9.68 (HCl)	297(d)	sub			δ v^h	δ	i				**B4²**, 812
a754	—(*DL*)	$CH_3CHNH_2CO_2H$	89.09	orh pr or nd (w)	280d (295–6)	sub 200	1.402		s v^h	δ	i	i			**B4²**, 815
a755	—(*L*)	$CH_3CHNH_2CO_2H$	89.09	rh (w), $[\alpha]^{13}_{5461}$ +3.1 (w, p=6)	297d				v	δ	i	i			**B4²**, 809
a756	—,ethyl ester hydrochloride (*dl*)	$CH_3CHNH_2CO_2CH_2CH_3.HCl$	153.61	hyg nd or not hyg pr (al)	70–5	d			v	v	v				**B4²**, 819
756¹	**β-Alanine**	β-Aminopropionic acid. $H_2NCH_2CH_2CO_2H$	89.09	nd, rh pr (al)	200d		1.437^{-5}		δ	δ	i	i			**B4²**, 827
a757	**α-Alanine, *N*-alanyl-** (*l*)	$H_2NCH(CH_3)CONHCH(CH_3)CO_2H$	160.17	lf, $[\alpha]^{20}_D$ +71.2 (w)	275–6d				v	δ	i				**B4²**, 813
—	**—,3-anthraniloyl-** (*l*)	*see* **Kynurenine** (*l*)													
a758	**—,*N*-benzoyl-**(*d*)	$CH_3CH(NHCOC_6H_5)CO_2H$	193.21	pl(w), $[\alpha]^{20}_D$, −2.4 (w,c=1)	151										**B9²**, 179
a759	—,—(*dl*)	$CH_3CH(NHCOC_6H_5)CO_2H$	193.21	pl or pr, lf (eth)	163–5	d			s	s	δ				**B9²**, 179
a760	—,—(*l*)	$CH_3CH(NHCOC_6H_5)CO_2H$	193.21	$[\alpha]^{20}_D$+2.4 (c=1, w)	151				δ						**B9²**, 179
a761	**—,*N*(carboxymethyl)-**(*dl*)	$HO_2CCH_2NHCH(CH_3)CO_2H$	147.14	(dil al)	222–3				s	i	i				**B4¹**, 497
a762	**—,*N*(4-chlorophenyl)**, nitrile.	Cl–⟨ ⟩–NHCH(CH₃)CN	180.69	lf (eth-peth)	114.5					s	s		v^h	chl v lig i	**B12**, 617
a763	**—,*N*,*N*-diethyl**, nitrile	$(C_2H_5)_2NCH(CH_3)CN$	126.21			81^{27} 47–9^{7}	0.857^{16}_{4}		s	s	s				**B4²**, 822
a764	**—,2(3,4-dihydroxyphenyl)-**(*L*)	*l*-Dopa; HO, HO–⟨ ⟩–CH₂CH(NH)₂CO₂H	197.19	pa br (w) $[\alpha]^{15}_D$ *ca.* −39.5 (w,p=1.3)	285.5				s	i	i	i	i	MeOH s alk s aa i	**B14²**, 399
a765	**—,*N*-fumaryl-**(*DL*)	Fumaroalanide. $HO_2CCH{:}CHCONHCH(CH_3)CO_2H$	187.15	nd	229d				δ s^h	s	δ			$NaHCO_3$ s lig i	
a766	**—,*N*-methyl-**(*dl*)	$CH_3CH(NHCH_3)CO_2H$	103.12	rh pr (abs al)	260d	292 sub			s^h	i					**B4²**, 821
a767	**β-Alanine, *N*-methyl-**	$CH_3NHCH_2CH_2CO_2H$	103.12	pl (al, +1w)	99–100 (+1w)				v	v^h					**B4²**, 828
a768	—,—,ethyl ester	$CH_3NHCH_2CH_2CO_2CH_2CH_3$	131.18			59.6^{14}	1.0082^{20}_{20}	1.4443^{20}		s	s			dil HCl s	**B4²**, 828
a769	—,—,nitrile	2-Methylaminopropanenitrile*. $CH_3NHCH_2CH_2CN$	83.11			82^{22}	0.8992^{20}_{4}	1.4312^{20}	s			s	s	MeOH, chl s	
—	**α-Alanine,-3-phenyl-**	*see* **Phenylalanine**													
a770	**—,*N*-phenyl-**, nitrile	α-Cyanoethyl aniline. $C_6H_5NH(CH_3)CN$	146.19	nd (bz or dil al), lf(dil al, bz, eth)	92				i $δ^h$	v	v		v	chl v	**B14**, 500
—	**Alantolactone**	*see* **Helenin**													
—	**Aldehydin**	*see* **Pyridine, 5-ethyl-2-methyl-**													
—	**Aldol**	*see* **Butanal, 3-hydroxy-***													
a772	**Aldrin**	Octalene. CCl₂; Cl–CH₂–Cl; Cl, Cl	364.93		104–4.5				i	s	s	s	s		
—	**Aleuritic acid**	*see* **Hexadecanoic acid, 9,10,16-trihydroxy-***													
—	**Aloemodine**	*see* **9,10-Anthraquinone, 1,8-dihydroxy-3-hydroxymethyl-***													
—	**Alizarin**	*see* **9,10-Anthraquinone, 1,2-dihydroxy-***													
—	**Alizarin cyanine R**	*see* **9,10-Anthraquinone, 1,2,4,5,8-pentahydroxy-***													
—	**Alkanine** (*d*)	*see* **Shikonine**													
a773	—(*l*)	CHOHCH₂CH:C(CH₃)₂; O=, =O; HO–, –OH	288.30	red br pr (bz or sub in vac), $[\alpha]^{20}_D$ 157 (bz)	149				i	δ	δ			alk s chl δ aa s	**B8²**, 544

For explanations, symbols and abbreviations see beginning of table.

No.	Name	Synonyms and Formula	Mol. wt.	Crystalline form, color and specific rotation	m.p. °C	b.p. °C	Density	n_D	Solubility w	al	eth	ace	bz	other solvents	Ref.
	Allatonic acid														
a774	**Allantoic acid**	Dicarbamidoacetic acid. Diureidoacetic acid. $(H_2NCONH)_2CHCO_2H$	176.13	nd, lf(MeOH)	173				δ	s				os δ	**B3**[2], 389
a775	**Allantoin**	Glyoxyldiureide. 5-Ureidohydantoin. O=C—NH CHNHCONH₂ HN —C=O	158.12	mcl p (w)	240				δ	i	i			MeOH i NaOH s	**B25**[2], 380
a776	**Allanturic acid**	Glyoxalyl urea. 5-Hydroxy-2,4-Imidazoledione. O=C—NH CHOH HN —C=O	116.08	amor hyg pw	turns br at 180				δ	i				hot alk d os i	**B25**[2], 388
—	**Allene**	*see* **Propadiene***													
a777	**Allicin**	*S*-Oxodiallyl disulfide. $C_6H_{10}OS_2$.	162.27			d	1.112[20]	1.561[20]	δ	∞	∞		∞		
a778	**Allitol**(*D*)	*D*-Allodulcitol. OH OH OH OH HOCH₂C—C—C—C—CH₂OH H H H H	182.17		150–1										
—	**Allocholesterol**	*see* **Coprostenol**													
—	**Allocinnamic acid**	*see* **Cinnamic acid** (*cis*)													
a779	**Alloisoleucine**(*D*)	$CH_3CH_2CH(CH_3)CH(NH_2)CO_2H$	131.18	lf, $[\alpha]_D^{20}$ 14.1 (w, p=2.92)	280–1d				s	δ					**B4**,457
a780	**Allomucic acid**	Tetrahydroxyadipic acid. $HO_2C(CHOH)_4CO_2H$	210.14	nd (w)	176				s	δ					**B3**[2], 376
a781	**Allonic acid,** γ-lactone(*D*)	O—CHOHCH₂OH O= HO— —OH	178.14	pr(al), $[\alpha]_D^{20}$ −6.8(w)	120				s	δ s[h]					**B18**[1], 406
—	**Alloocimene**	*see* **2,4,6-octatriene, 2,6-dimethyl-***													
—	**Allophanamide**	*see* **Biuret**													
—	**Allophanic acid**	*see* **1-Ureacarboxylic acid**													
a782	**Allose** (*D*)	OH OH OH OH HOH₂C C—C—C—C—CHO H H H H	180.16	$[\alpha]_D$+0.58	128–8.5					i					**B1**, 443
a783	—(*L*)	$C_6H_{12}O_6$. *See* a782	180.16	pr(al), $[\alpha]_D^{20}$ −0.58(al), −1.90 to −13.88 (w)	128				s	δ					
a784	**Alloxan**	Mesoxalylurea. Pyrimidinetetrone. H O N O= NH O O	142.07	wh tcl, rh pr (pk in air)dk ye cr(w +1,2,3,4)	256 (anh)	sub (vac)			v	s		s	s	aa s	**B24**[2], 301
a785	**Alloxanic acid**	O HN— OH —NH CO₂H O	160.09	tcl pr (eth)	162–3d				s	s	δ				**B3**[2], 266
a786	**Alloxantin**	O O OH OH —NH— —NH— O= =O —NH— —NH— O O	286.16	ta (w)	270–5				δ s[h]	δ	δ				**B26**[2], 335
—	**Allylacetic acid**	*see* **4-Pentenoic acid***													
—	**Allyl alcohol**	*see* **2-Propenol***													
—	**Allyl amine**	*see* **Propene, 3-amino-***													
—	**Allyl bromide**	*see* **Propene, 3-bromo***													
—	**Allyl cellosolve**	*see* **Ethanol, 2-allyloxy-**													

For explanations, symbols and abbreviations see beginning of table.

No.	Name	Synonyms and Formula	Mol. wt.	Crystalline form, color and specific rotation	m.p. °C	b.p. °C	Density	n_D	Solubility w	al	eth	ace	bz	other solvents	Ref.
	Allyl chloride														
—	**Allyl chloride**	*see* **Propene, 3-chloro-***													
—	**Allyl cyanide**	*see* **3-Butenoic acid,** nitrile													
—	**Allyl fluoride**	*see* **Propene, 3-fluoro-***													
—	**Allyl iodide**	*see* **Propene, 3-iodo-***													
—	**Allyl mustard oil**	*see* **Isothiocyanic acid,** allyl ester													
a787	**Aloetic acid**	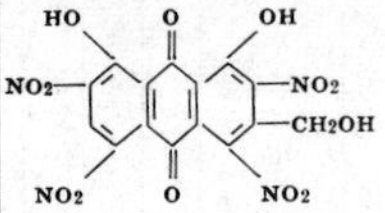	450.23	og-ye nd (aa)	285d >285 exp				δ s^h					NH_4OH v alk s (red)	**E13,** 575
a788	**Aloin**	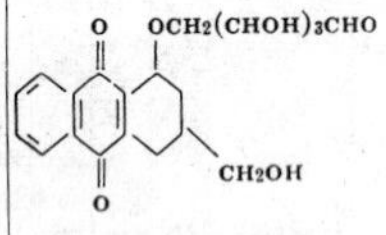	386.36	ye nd	147–9				δ	s	i	s	δ	KOH s chl i chl i	
—	**Alphol**	*see* **Benzoic acid, 2-hydroxy-,**1-naphthyl ester													
a789	**Alstonine**	$C_{21}H_{20}N_2O_3.3\frac{1}{2}H_2O$	348.41	br amor	205–10(d)										**B27**^{2}, 824
a790	**Altrose**(*d*)	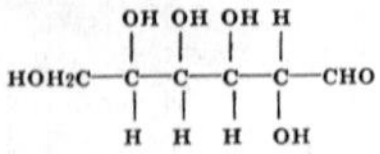	180.16	pr (MeOH-al) $[\alpha]^{20}_D$+11.7 to 33.1(30 min)	103–5				s	i					**B31**^{1}, 83
a791	—(*l,β*)	$C_6H_{12}O_6$. *See* a790	180.16	pr (al) $[\alpha]^{20}_D$−32.3											
a792	**Amalic acid**	Tetramethyl alloxantin. 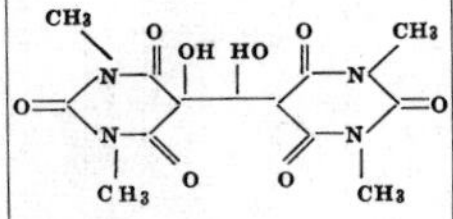	324.25	(w)	245d				δ^h	i				KOH s	
a793	**Amarine**	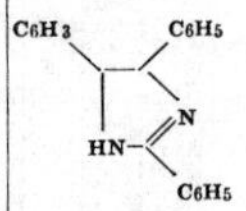	298.39	pr (eth bz-lig)	136	198d			i	s	s				**B23**^{2}, 274
a794	—,hydrate	$C_{21}H_{18}N_2.\frac{1}{2}H_2O$	307.40	pr (aq al+$\frac{1}{2}$w)	106										**B23**^{2}, 274
a795	**Amaron**	Benzoin imide. Ditolan azotide. Tetraphenylpyrazine. 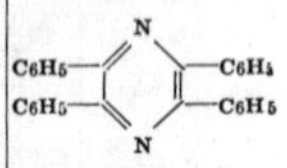	384.49	tcl nd or pr (ace, al)	252	sub			i	δ	v		s^h	chl, CS_2 s	**B23**^{2}, 304
a796	**Amine, allyl methyl**	CH_2:$CHCH_2NHCH_3$	71.12			64–6			∞	∞					**B4,** 206
a797	**—,benzyl, diphenyl**	$C_6H_5CH_2N(C_6H_5)_2$	259.35	nd	85–6				δ	δ v^h	v	v	v	aa δ lig v	**B12**^{2}, 551
a798	**—,benzyl ethyl phenyl**	*N*-Benzyl-*N*-ethyl aniline. $C_6H_5CH_2N(C_2H_5)C_6H_5$	211.31	yesh		285–6^{710}	1.034^{19}_4	1.5975^{15}	i	s					**B12**^{2}, 550
a799	**—,benzyl ethyl 2-tolyl**	*N*-Benzyl-*N*-ethyl-*o*-toluidine. 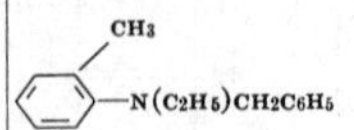	225.34	ye		230$^{20-5}$			i						**B12,** 1033
a800	**—,benzyl methyl 2-tolyl**	*N*-Benzyl-*N*-methyl-*o*-toluidine. 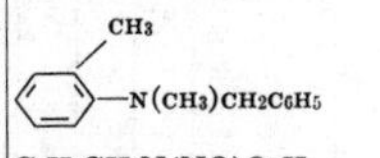	211.31	ye		167^{13}			i	s				os s	**B12,** 1033
a801	**—,benzyl *N*-nitroso phenyl**	$C_6H_5CH_2N(NO)C_6H_5$	212.25	ye nd (al)	58				i	s	s			chl s lig s	**B12**^{2}, 572
a802	**—,benzyl phenyl**	*N*-Benzylaniline. $C_6H_5CH_2NHC_6H_5$	183.26	pr (MeOH)	36.5–.8	321^{775} 171.5^{10}	1.0298^{65}_4 1.149^{14}	1.6118^{25}	i	s	s			MeOH s^h	**B12**^{2}, 548

For explanations, symbols and abbreviations see beginning of table.

No.	Name	Synonyms and Formula	Mol. wt.	Crystalline form, color and specific rotation	m.p. °C	b.p. °C	Density	n_D	Solubility w	al	eth	ace	bz	other solvents	Ref.
	Amine														
a803	—,**benzyl 2-tolyl**	*N*-Benzyl-*o*-toluidine. CH_3 / $NHCH_2C_6H_5$	197.28	(al, eth)	60	300–5 176^{10}	1.0142^{65}_{4}	1.5861^{65}	i	s	...	...	...	os s	**B12²**, 551
a804	—,**benzyl 3-tolyl-**	*N*-Benzyl-*m*-toluidine. CH_3 / $NHCH_2C_6H_5$	197.28	ye		315–7 180^{10}	1.0083^{65}_{4}	1.5845^{65}	...	δ	v	...	v	chl v	**B12²**, 552
a805	—,**benzyl 4-tolyl**	*N*-Benzyl-*p*-toluidine. CH_3—C_6H_4—$NHCH_2C_6H_5$	197.28	ye		319^{765} 181^{10}	1.0064^{65}_{4}	1.5832^{65}	i	v	v	...	...		**B12²**, 552
a806	—,**benzylidene ethyl**	$C_6H_5CH{:}NC_2H_5$	133.19			195^{749} 117^{12}	0.9411^{16}_{4}		i	s	s	...	...		**B7²**, 163
a807	—,**benzylidene methyl**	$C_6H_5CH{:}NCH_3$	119.17			185 $90\text{–}1^{30}$	0.9672^{14}_{4}	1.5519^{17}	...	s	s	...	...		**B7²**, 162
a808	—,**bis(dimethyl-phosphino)**	Amino-bis(dimethyl phosphine). $(CH_3)_2PNHP(CH_3)_2$	137.10		39.5	d $33.5^{5.4}$ subl			d	d	...	...	...	os ∞	
—	—,*tert*-**butyl**	*see* **Propane, 2-amino-2-methyl-***													
a809	—,**butyl diethyl,-2,2″-dihydroxy-**	Bis(2-hydroxyethyl)butyl-amine. $CH_3(CH_2)_3N(CH_2CH_2OH)_2$	161.25			$273\text{–}5d^{741}$ $176\text{–}80^{35}$	0.9692^{20}_{4}	1.4625^{20}	∞	∞	∞	s	∞	lig i	**B4**, 285
a810	—,**butyl dimethyl**	$CH_3(CH_2)_3N(CH_3)_2$	101.19			94.1– -5.2^{749}		1.3970^{20}	∞						**B4²**, 632
a811	—,**butyl ethyl**	$CH_3(CH_2)_3NHCH_2CH_3$	101.19			108–9	0.7398^{20}_{4}	1.4011^{20}	...	...	...	...	...		**B4**, 157
a812	—,*sec*-**butyl ethyl**(*d*)	$CH_3CH_2CH(CH_3)NHCH_2CH_3$	101.19	$[\alpha]^{15}_{D}+18$		98	0.7396^{15}_{4}	1.4043	...	...	...	...	...		
a813	—,—(*dl*)	$CH_3CH_2CH(CH_3)NHCH_2CH_3$	101.19		−104.5	$97\text{–}8^{741}$	0.7358^{20}_{4}		...	...	...	...	...		**B4²**, 636
a814	—,**butyl ethyl,-2,2′dihydroxy-**	$HOCH_2CH_2NHCH_2CHOHCH_2CH_3$	133.19	yesh		$118\text{–}22^{1}$	1.031^{20}_{4}	1.4690^{20}	s	s	...	...	...	lig i	
a815	—,—,**3,2′-dihydroxy-**	$HOCH_2CH_2NHCH(CH_3)CHOHCH_3$	133.19	yesh		$120\text{–}4^{2\text{-}5}$	1.0331^{20}_{4}	1.4718^{20}	s	...	...	s	...	lig i	
—	—,**cyano diphenyl**	*see* **Cyanamide, diphenyl-**													
—	—,**cyclohexyl phenyl**	*see* **Aniline,** *N*-cyclohexyl-													
a816	—,**diallyl**	$NH(CH_2\,CH{:}CH_2)_2$	97.16			111^{760}			...	...	...	...	...		**B4²**, 663
a817	—,**dibenzyl**	$NH(CH_2C_6H_5)_2$	197.28		−26	300d 270^{250}	1.0256^{22}_{4}	1.5143^{22}	i	v	v	...	...		**B12²**, 553
a818	—,**dibenzyl ethyl**	$CH_3CH_2N(CH_2C_6H_5)_2$	225.34			306			i	v	v	...	...	ac s	**B12²**, 553
a819	—,**dibenzyl phenyl**	$C_6H_5N(CH_2C_6H_5)_2$	273.38	nd or pr (al)	71–2	>300d	1.0444^{80}_{4}	1.6065^{80}	i	δ s^h	s	...	s	aa δ v^h	**B12²**, 554
a820	—,**dibutyl**	$NH(CH_2CH_2CH_2CH_3)_2$	129.25			159^{761}	0.767^{20}_{4}		s	v	v	...	...		**B4²**, 633
a821	—,**di-*sec*-butyl**	$[CH_3CH_2CH(CH_3)]_2NH$	129.25			132^{758}	0.7833^{0}_{0}		v	s	...	...	...	os s	**B4²**, 636
a822	—,—,**3,3′di-hydroxy-**	$[CH_3CHOHCH(CH_3)]_2NH$	161.25	yesh		$112\text{–}5^{3}$	0.9775^{20}_{4}	1.4162^{20}	v	v	δ	v	...	lig δ	
a823	—,**dibutyl,3,3′-dimethyl-**	Diisoamyl amine. $[(CH_3)_2CHCH_2CH_2]_2NH$	154.28		−44	188	0.7672^{21}_{4}	1.4229^{21}	i	s	∞	...	...		**B4²**, 646
a824	—,**dibutyl ethyl, 2,2′,2″-tri-hydroxy-**	$HOCH_2CH_2N(CH_2CHOHCH_2CH_3)_2$	205.30			$155\text{–}60^{1}$	1.0216^{20}_{4}	1.4705^{20}	v	v	...	v	...	lig δ	
a825	—,**dicyclohexyl**	C_6H_{11}—NH—C_6H_{11}	181.32		−0.10	255.8^{760}	0.925^{18}	1.488^{18}	δ^h	s	s	...	s		**B12²**, 7
a826	—,**didodecyl**	$[CH_3(CH_2)_{10}CH_2]_2NH$	353.68		α)46.95 β)51.8				...	v	v	δ	v	chl v	
a827	—,**diethyl**	$(CH_3CH_2)_2NH$	73.14		−48	56.3^{760}	0.7108^{18}_{4}	1.3871^{19}	v	∞	s	...	...		**B4²**, 590

For explanations, symbols and abbreviations see beginning of table.

No.	Name	Synonyms and Formula	Mol. wt.	Crystalline form, color and specific rotation	m.p. °C	b.p. °C	Density	n_D	Solubility						Ref.
									w	al	eth	ace	bz	other solvents	
	Amine														
a828	—,—,hydrochloride	$(CH_3CH_2)_2NH.HCl$	109.60	lf(al-eth)	223.5	320–30			v	δ	i	δ	...	chl δ	**B4**[2], 592
a829	—,**2-amino-2′-hydroxy-**	*N*(2-Hydroxyethyl)ethylenediamine. $H_2NCH_2CH_2NHCH_2CH_2OH$	104.15			238–40; 162^{50}	1.0304^{20}_{20}	1.4863^{20}	∞	∞	...	...	δ	lig δ	**B4**, 286
—	—,—,**2,2′-diamino-**	*see* **Diethylenetriamine**													
a830	—,—,**2,2′-dihydroxy**	Diethanolamine. Diethylolamine, $NH(CH_2CH_2OH)_2$	105.14	pr	28	270^{748}	1.09664^{20}_{4}	1.4776^{20}	v	v	δ	...	δ		**B4**[2], 729
a831	—,—,**1,1′-diphenyl-**	$NH[CH(CH_3)C_6H_5]_2$	225.34	ye		295–8; $61–2^{2}$	1.018^{13}	1.573	...	...	...	...	...		**B12**[2], 589
a832	—,—,**2,2′-diphenyl-**	$NH(CH_2CH_2C_6H_5)_2$	225.34		28–30	$335–7^{603}$; 195^{18}			i	v	v	...	...	ac s	**B12**[2], 593
a833	—,**diethyl methyl**	$(CH_3CH_2)_2NCH_3$	87.16			66^{760}			v	v	v	...	...		**B4**[2], 593
a834	—,—,**2,2′-dichloro-**, hydrochloride	$(ClCHC_2H_2)_2NCH_3.HCl$	192.53	hyg nd	109–10				v	s	...	...	...		
—	—,—,**2,2-diethoxy-**	*see* **Ethanal,2(ethylmethylamino)-** diethyl acetal													
a835	—,—,**2,2′-dihydroxy-**	$(HOCH_2CH_2)_2NCH_3$	119.17			$246–8^{747}$	1.0377^{20}	1.4678^{20}	∞	∞	δ	...	...		**B4**[2], 729
a836	—,—,**1′′′-methoxy-**	$(CH_3O)CH_2N(CH_2CH_3)_2$	117.19			$116.5–7^{763}$; 53^{68}			s	s	s	...	...	HCl s	**B4**[2], 598
a837	—,**diethyl *N*-nitro**	$(CH_3CH_2)_2NNO_2$	118.14			206.5^{757}; 93^{16}	1.057^{15}		δ	∞	∞	...	...		**B4**, 130
a838	—,**diethyl *N*-nitroso**	$(CH_3CH_2)_2NNO$	102.14	ye		176.9^{760}	0.9422^{20}_{4}	1.4386^{20}	s	s	s	...	...		**B4**[2], 617
a839	—,**diethyl phenyl, 2,2′-dihydroxy-**	*N*-Phenyldiethanolamine. $(HOCH_2CH_2)_2NC_6H_5$	181.24	pl (al)	58	228^{15}			δ	v	v	v	v		**B12**[2], 109
a840	—,**difurfuryl-**	(furyl)$—CH_2NHCH_2—$(furyl)	177.21			$135–42^{15}$; $102–3^{1}$			i	...	s	...	...		**B18**[2], 418
a841	—,**diheptyl-**	$[CH_3(CH_2)_6]_2NH$	213.41	nd	30	271^{750}			δ	s	v	...	...		**B4**[2], 652
a843	—,**dihexyl-**	$[CH_3(CH_2)_5]_2NH$	185.36			192–5			...	s	s	...	...		**B4**[2], 650
a844	—,**di-2-hydrindyl,1,1′-dihydroxy-**(*d*)	(indanyl, OH)—NH—(indanyl, HO)	281.36	nd (al), $[\alpha]^{18}_{D}$ +83.2 (dil HCl c=0.3)	223				...	s^h	...	...	...	os δ	**B13**[1], 265
a845	—,—,(*dl*)	$C_{18}H_{19}NO_2$	281.36	nd or pl (dil al)	205				δ	v	v	...	v		**B13**[1], 267
a846	—,—(*l*)	$C_{18}H_{19}NO_2$	281.36	ncl (al), $[\alpha]^{18}_{0}$ −83.3 (dil HCl, c=0.3)	223				...	s^h	...	...	...	os δ	**B13**[1], 266
a847	—,**diisobutyl**	$[(CH_3)_2CHCH_2]_2NH$	129.24		−73.5	139–40	0.7457^{20}	1.4093^{20}	δ	s	s	...	...		**B4**[2], 638
a848	—,**diisopropyl**	$[(CH_3)_2CH]_2NH$	101.19		−61	$83.5–4^{743}$	0.722^{22}		δ	...	...	...	...		**B4**[2], 630
a849	—,**diisopropyl-*N*-nitroso**	Nitrous diisopropylamine. $[(CH_3)_2CH]_2NNO$	130.19	(eth)	48	194.5	0.9422^{20}_{4}		δ	s	s	...	s		**B4**, 156
a850	—,**dimethyl**	$(CH_3)_2NH$	45.08		−96	7.4	0.6804^{0}_{4}	1.350^{17}	v	s	s	...	...		**B4**[2], 550
a851	—,—,hydrochloride	Dimethylammonium chloride. $(CH_3)_2NH.HCl$	81.55	rh, nd (al)	170				v	s	i	...	...	chl v	**B4**[2], 552
a852	—,—.**hexafluoro-**	Bis(trifluoromethyl)amine. $(CF_3)_2NH$	153.04		−130	-6.7^{760}			d	...	...	...	...		
a853	—,—,**perfluoro-**	$(CF_3)_2NF$	171.03			−37			i	...	...	...	...		
a854	—,**dimethyl isobutyl**	$(CH_3)_2CHCH_2N(CH_3)_2$	101.19			$80–1^{760}$			s	...	...	...	...	ac v	**B4**[2], 638
a855	—,**dimethyl *N*-nitro**	Dimethylnitramine. Nitric dimethylamide. $(CH_3)_2NNO_2$	90.08	nd (eth)	58	187^{759}	1.1090^{72}_{4}	1.4462^{12}	s	s	s	...	s		**B4**[2], 342
a856	—,**dimethyl *N*-nitroso**	Dimethylnitrosamine. Nitrous dimethylamide. $(CH_3)_2NNO$	74.08	ye		153^{774}	1.0059^{20}_{4}	1.4374^{18}	s	s	s	...	...		**B4**[2], 585
a857	—,**dimethyl pentyl**	$CH_3(CH_2)_4N(CH_3)_2$	115.22			123	0.743^{20}_{4}	1.4083^{20}	δ^h	...	∞	...	...		**B4**[2], 642
a858	—,**di-2-naphthyl**	$(C_{11}H^{\beta}_{7})_2NH$	269.33	pl (bz)	172.2	471			i	δ^h	s	...	δ	aa s^h	**B12**[2], 717
a859	—,**dioctadecyl**	$[CH_3(CH_2)_{16}CH_2]_2NH$	521.97		72.3				i	i	δ	i	δ	chl v	
a860	—,**dioctyl**	$[CH_3(CH_2)_6CH_2]_2NH$	241.46	nd	35.6	297–8; 175^{14}			i	v	v	...	...		**B4**[2], 655
a860[1]	—,**di-2-octyl**	$[CH_3(CH_2)_5CH(CH_3)]_2NH$	241.46			281.5^{739}	0.7948^{20}_{4}		i	v	v	...	...		**B4**, 197

For explanations, symbols and abbreviations see beginning of table.

No.	Name	Synonyms and Formula	Mol. wt.	Crystalline form, color and specific rotation	m.p. °C	b.p. °C	Density	n_D	Solubility w	al	eth	ace	bz	other solvents	Ref.
	Amine														
a861	**—,dipentyl**	$[CH_3(CH_2)_4]_2NH$	157.29	...	...	$202–3^{745}$ $91–3^{14}$	0.7771^{20}_{4}	1.4272^{20}	δ	v	∞	...	...	...	**B4**², 642
a862	**—,—,2,2′-dihydroxy-2,2′-dimethyl-**	$[CH_3CH_2CH_2C(CH_3)OHCH_2]_2NH$	217.36	...	...	$165–70^{15}$	0.9264^{20}_{4}	1.4584^{20}	v	v	s	...	...	lig i	...
a863	**—,diphenyl**	3 2 2′ 3′ 4 ⟨ ⟩—NH—⟨ ⟩ 4′ 5 6 6′ 5′	169.22	mcl lf	52.8	302	1.160^{20}_{20}	...	i	v	s	...	s	Py v aa s lig s	**B12**², 101
a864	**—,—, hydrobromide**	$(C_6H_5)_2NH.HBr$	250.15	pl	230	...	...	...	s	δ	i	...	...	...	**B12**, 180
a865	**—,—,2-amino-**	*N*-Phenyl-*o*-phenylenediamine. $C_{12}H_{12}N_2$. *See* a863	184.24	nd (w)	79–80	312.5^{744}	...	...	δ	...	...	s	s	chl s	**B13**², 13
a866	**—,—,4-amino-**	Diphenyl black base *P*. $C_{12}H_{12}N_2$. *See* a863	184.24	nd (al), (lig)	66 (al), 75 (lig)	364 $155^{0.026}$	...	...	δ	v^h	...	...	...	lig s^h	**B13**, 76
a867	**—,—,4,4′-bis(dimethylamino)-**	Leuco base of Bindschedler green. $C_{16}H_{21}N_3$. *See* a863	255.35	tetr pl (CS_2)	119	...	...	...	δ	δ	s	...	...	lig δ	**B13**², 56
a868	**—,—,4,4′-diamino-**	*p,p′*-Iminodianiline. $C_{12}H_{13}N_3$. *See* a683	199.24	lf (w)	158	d	...	...	δ s^h	s	s	...	...	...	**B13**², 55
a869	**—,—,2,2′-dinitro-**	$C_{12}H_9N_3O_4$. *See* a863	259.23	lf (al, ace MeOH), (aa)	169	...	...	...	...	v^h	...	s	...	MeOH s^h aa v^h	**B12**, 341
a870	**—,—,2,4-dinitro-**	$C_{12}H_9N_3O_4$. *See* a863	259.23	red nd (al)	156–6.5	...	...	...	i	$δ^h$	δ	s	δ	chl s	**B12**², 407
a871	**—,—,2,4′-dinitro-**	$C_{12}H_9N_3O_4$. *See* a863	259.23	ye red nd (aa)	222–3	...	...	...	i	δ	...	δ	...	to δ, chl s	**B12**², 387
a872	**—,—,2,6-dinitro-**	$C_{12}H_9N_3O_4$. *See* a863	259.23	og lf (al, aa)	107–8	...	...	...	i	s^h	...	...	...	aa s	**B12**², 413
a873	**—,—,3,4′-dinitro-**	$C_{12}H_9N_3O_4$. *See* a863	259.23	pa ye (chl)	217	...	...	...	...	δ	...	v	...	chl s con ac s	**B12**², 387
a874	**—,—,4,4′-dinitro-**	$C_{12}H_9N_3O_4$. *See* a863	259.23	ye nd (al)	216–6.5	...	...	...	i	δ	...	s	δ	to i aa s	**B12**², 387
a875	**—,—,2,4-dinitro-4′-hydroxy-**	$C_{12}H_9N_3O_5$. *See* a863	275.22	red lf	190	...	...	...	...	...	...	...	...	...	**B13**¹, 150
a876	**—,—,2,4-dinitro-5-hydroxy-**	$C_{12}H_9N_3O_5$. *See* a863	275.22	dk ye nd (al)	139	...	...	...	...	...	...	...	...	...	**B13**¹138
a877	**—,—,2,6-dinitro-2′-hydroxy-**	$C_{12}H_9N_3O_5$. *See* a863	275.22	red vt nd (al)	191	...	...	...	...	...	v^h	...	v	aa v	**B13**, 365
a878	**—,—,2,6-dinitro-3-hydroxy-**	$C_{12}H_9N_3O_5$. *See* a863	275.22	ye br nd (MeOH)	124–5	...	...	...	...	...	...	...	...	MeOH s^h	**B13**², 216
a880	**—,2,2′,4,4′,6,6′-hexanitro-**	Dipicrylamine. $C_{12}H_5N_7O_{12}$ *See* a863	439.21	pa ye pr (aa), (93% HNO_3)	239.5–40	...	...	...	i	i	δ	δ	...	aq alk s Py s	**B12**², 422
a881	**—,—,2-hydroxy-**	2-Anilinophenol. $C_{12}H_{11}NO$. *See* a863	185.23	pr (w)	69–70	$180–9^{20}$	...	...	$δ^h$	s	s	...	δ	...	**B13**, 365
a882	**—,—,3-hydroxy-**	3-Anilinophenol. $C_{12}H_{11}NO$. *See* a863	185.23	pl (w)	81.5–2	340	...	...	s^h	v	s	s	s	dil ac s lig δ	**B13**², 213
a883	**—,—,4-hydroxy-**	4-Anilinophenol. $C_{12}H_{11}NO$. *See* a863	185.23	pl	73	330 $215–6^{12}$	...	...	δ	v	v	...	v	chl v ac, alk s	**B13**², 231
a885	**—,—,2-nitro-**	$C_{12}H_{10}N_2O_2$. *See* a863	214.23	og lf (al-w), rh bipym	75	...	...	...	i	s	...	...	...	...	**B12**², 369
a886	**—,—,3-nitro-**	$C_{12}H_{10}N_2O_2$. *See* a863	214.23	lf or nd (dil al)	110–2	...	...	...	...	v	v	...	v	lig δ	**B12**², 377
a887	**—,—,4-nitro-**	$C_{12}H_{10}N_2O_2$. *See* a863	214.23	ye nd, tab (CCl_4)	133–4	211^{30}		...	i	v	...	...	...	aa v	**B12**², 386
a889	**—,diphenyl,4-nitroso-**	$C_{12}H_{10}N_2O$. *See* a863	198.23	blsh bk (al), ye pw, pl (bz)	143	...	...	...	δ	v^h	v	...	v	chl v lig δ	**B12**², 122
a890	**—,—,4-sulfo-**	*see* **Benzenesulfonic acid, 4-phenylamino-***													
a891	**—,—,2,2′,4,4′-tetrabromo-**	$C_{12}H_7Br_4N$. *See* a863	484.83	nd (chl or bz)	187.5	...	...	...	i	$δ^h$	...	...	s	...	**B12**², 356
a892	**—,—,2,2′,4,4′-tetranitro-**	$C_{12}H_7N_5O_8$. *See* a863	349.22	ye cr (aa)	201–1.5	...	...	...	i^h	$δ^h$	δ	...	$δ^h$	chl $δ^h$	**B12**², 409
892¹	**—,diphenyl ethyl**	$(C_6H_5)_2NCH_2CH_3$	197.28	...	...	295–7 150^{10}	1.0479^{14}_{4}	...	i	s	s	...	...	...	**B12**², 105
892²	**—,diphenyl methyl**	$(C_6H_5)_2NCH_3$	183.26	...	−7.55	293–4 145^{10}	1.0476^{20}_{4}	1.6193^{20}	i	δ	...	...	...	MeOH δ	**B12**², 105
892³	**—,diphenyl *N*-nitroso**	Nitrous diphenylamide. $(C_6H_5)_2NNO$	198.23	ye me pr (lig)	66.5	...	...	...	...	δ v^h	...	...	v^h	...	**B12**², 310
—	**—,dipycryl-**	*see* **Amine, diphenyl, 2,2′, 4,4′,6,6′-hexanitro-**													
a893	**—,dipropyl-**	$(CH_3CH_2CH_2)_2NH$	101.19		−39.6	109.4–10.4	0.7384^{20}_{4}	1.4046^{20}	s	s	∞	...	...	...	**B4**², 622
a894	**—,—,3,3′-diallyloxy-2,2′-dihydroxy-**	$[CH_2:CHCH_2OCH_2CHOHCH_2]_2NH$	245.32	...	73	$180–5^{2\cdot3}$	...	...	s	s	...	s	...	lig i	...
a895	**—,—,2,2′-dihydroxy-**	Diisopropanolamine. $[CH_3CHOHCH_2]_2NH$	133.19	...	42	$170–80^{30}$	0.989^{20}_{4}	...	s	s	...	...	...	...	**B4**², 737

For explanations, symbols and abbreviations see beginning of table.

No.	Name	Synonyms and Formula	Mol. wt.	Crystalline form, color and specific rotation	m.p. °C	b.p. °C	Density	n_D	Solubility w	al	eth	ace	bz	other solvents	Ref.
	Amine														
a896	**—,dipropyl ethyl, 2,2′,2″-trihydroxy-**	$(CH_3CHOHCH_2)_2NCH_2CH_2OH$	177.25	...	...	155–6[1]	1.0458^{20}_{4}	1.4708^{20}	v	v	...	v	...	lig δ	...
a897	**—,dipropyl isobutyl, perfluoro-**	$(CF_3)_2CFCF_2N(CF_2CF_2CF_3)_2$	571.08	...	...	146–8[760]	1.84^{25}_{4}	1.283^{25}	...	...	...	...	...	...	...
a898	**—,dipropyl methyl, 3,3′-dihydroxy-**	$(HOCH_2CH_2CH_2)_2NCH_3$	147.22	hyg	...	164–5[13]	...	...	∞	∞	δ	...	v	...	**B4²**, 735
a899	**—,dipropyl *N*-nitroso**	Dipropyl nitrosamine. Nitrous dipropylamide. $(CH_3CH_2CH_2)_2NNO$	130.19	golden	...	206[760] 104–5[22]	0.9163^{20}_{4}	1.4446^{20}	δ	∞	∞	...	...	...	**B4²**, 628
a900	**—,di-2-tolyl**	$[2\text{-}CH_3C_6H_4]_2NH$	197.27	bl flr, wh cr	52–3	312[727] 185[13]	...	...	δ	...	...	...	...	...	**B12²**, 437
a901	**—,di-3-tolyl**	$[3\text{-}CH_3C_6H_4]_2HN$	197.27	pa ye	<−12	319–20	...	...	δ	v	v	...	...	...	**B12²**, 467
a902	**—,di-4-tolyl**	$[CH_3\text{—}C_6H_4\text{—}]_2NH$	197.27	nd	79	330.5[760]	...	...	δ	...	...	...	...	...	**B12²**, 494
a903	**—,3,4-ditolyl** hydrochloride. $CH_3C_6H_4\text{—}NH\text{—}C_6H_4\text{—}CH_3.HCl$		233.75	...	202–3	...	...	...	s	v	i	v	v	chl v lig i	**B12¹**, 414
a904	**—,ethyl methyl**	$CH_3NHCH_2CH_3$	59.11	...	...	34–5	...	...	v	v	...	...	...	os s	**B4²**, 589
a905	—,—,hydrochloride	$CH_3NHCH_2CH_3.HCl$	95.58	(al-eth)	126–30	...	...	...	v	v	i	...	...	chl s	**B4²**, 589
a906	**—,ethyl *N*-nitrosophenyl**	*N*-Ethylphenylnitrosamine. $C_6H_5N(NO)CH_2CH_3$	150.18	yesh	...	119.5–20[15]	1.0874^{20}_{4}	...	i	...	...	...	...	aa s	**B12²**, 310
a907	**—,ethyl propyl, 2,2′-dihydroxy-**	$HOCH_2CH_2NHCH_2CHOHCH_3$	119.17	yesh	...	120–3[1]	$1\ 4695^{20}_{4}$	1.0455^{20}	s	s	...	s	...	...	...
a908	**—,ethyl 2-tolyl, 2-hydroxy-**	$2\text{-}CH_3C_6H_4\text{—}NHCH_2CH_2OH$	151.21	...	...	285–6 145–50[3]	1.0962^{25}_{20}	1.5675^{20}	...	s	s	...	...	chl s	**B12²**, 437
a909	**—,ethyl 4-tolyl, 2-hydroxy-**	$CH_3\text{—}C_6H_4\text{—}NHCH_2CH_2OH$	151.21	pl (eth-lig)	42–3	286–8 153–5[4]	...	...	δ d^h	v	v	...	v	chl v ac v	**B12²**, 495
a910	**—,methyl propyl**	$CH_3NHCH_2CH_2CH_3$	73.14	...	...	62–4[760]	0.7204^{17}	...	s	s	...	...	...	...	**B4²**, 621
a911	**—,1-naphthyl phenyl**	$C_{10}H_7^{\alpha}NHC_6H_5$	219.27	lf (lig), pr, nd (al)	62	335 200[10]	...	...	δ	s	s	...	s	chl s aa s	...
a912	**—,2-naphthyl phenyl**	$C_{10}H_7^{\beta}NHC_6H_5$	219.27	bl flr in sol	108	395–9.5 237[13]	...	...	i	i	s	...	...	...	**B12**, 1275
a913	**—,1-naphthyl propyl**	$C_{10}H_7^{\alpha}NHCH_2CH_2CH_3$	185.26	ye	...	316–8[771]	...	...	i	...	...	...	...	...	...
a914	**—,1-naphthyl 2-tolyl**	$C_{10}H_7^{\alpha}NH\text{—}C_6H_4CH_3\ (2)$	233.30	nd	94–5	198–202[9]	...	...	i	s	s	...	s	...	...
a915	**—,1-naphthyl 4-tolyl**	$C_{10}H_7^{\alpha}NH\text{—}C_6H_4\text{—}CH_3$	233.30	pr (al)	79	230[10]	...	...	δ	...	s	...	s	...	...
a916	**—,2-naphthyl 2-tolyl**	Yellow OB. $C_{10}H_7^{\beta}NH\text{—}C_6H_4CH_3\ (2)$	233.30	lf (lig)	95–6	398–402	...	...	...	s	s	v	v	chl v lig v	**B12**, 1277
a917	**—,2-naphthyl 4-tolyl**	$C_{10}H_7^{\beta}NH\text{—}C_6H_4\text{—}CH_3$	233.32	red lf	103–3.5	...	...	...	...	δ	...	...	...	PrOH s	**B12**, 1277
a918	**—,triallyl**	$(CH_2{:}CHCH_2)_3N$	137.23	...	...	155–6	0.894^{14}_{3}	...	...	...	...	...	...	ac s	**B4**, 208
a919	**—,tribenzyl**	$(C_6H_5CH_2)_3N$	287.41	pl (eth), mcl (al)	91	380–90	0.9912^{95}_{4}	...	δ	δ s^h	s	...	...	...	**B12²**, 555
a920	**—,tributyl**	$(CH_3CH_2CH_2CH_2)_3N$	185.36	hyg	...	216–7	0.7782^{20}_{20}	...	δ	v	v	...	...	...	**B4²**, 633
a921	**—,—,perfluoro-**	$[CF_3(CF_2)_3]_3N$	671.10	...	...	179[760]	1.873^{25}_{4}	1.291^{25}	...	...	...	...	...	...	...

For explanations, symbols and abbreviations see beginning of table.

No.	Name	Synonyms and Formula	Mol. wt.	Crystalline form, color and specific rotation	m.p. °C	b.p. °C	Density	n_D	Solubility w	al	eth	ace	bz	other solvents	Ref.
	Amine, tributyl														
a922	—,—,**2,2′,2″-tri-methyl-**	Tri-*active*-amylamine. $[CH_3CH_2CH(CH_3)CH_2]_3N$	227.44			230–7 94^{4}	0.7964^{13}	1.43305							**B4**^{2}, 642
a923	—,—,**3,3′,3″-tri-methyl-**	Triisoamylamine. $[(CH_3)_2CHCH_2CH_2]_3N$	227.44			235	0.7859$^{20}_{4}$		i	v	∞		∞	CCl_4 ∞	**B4**^{2}, 646
a924	—,**tridodecyl**	$[CH_3(CH_2)_{10}CH_2]_3N$	522.01		fr 15.7					δ	∞	δ	∞	chl ∞	
a925	—,**triethyl**	$(CH_3CH_2)_3N$	101.19		−115	89–90	0.7255$^{25}_{4}$	1.4003^{20}	s δ^{h}	s	s				**B4**^{2}, 593
a926	—,—,hydrochloride	$(CH_3CH_2)_3N.HCl$	137.65	hex (al)	253–4	sub 245	1.0689$^{21}_{4}$		v	s	i		δ	chl v	**B4**^{1}, 348
a927	—,—,hydroiodide	$(CH_3CH_2)_3N.HI$	229.11	pr (90–95% al)	263d		1.924		v	δ	i			chl δ	**B4**, 43
a928	—,—,**2-bromo-**, hydrobromide	$BrCH_2CH_2N(CH_2CH_3)_2.HBr$	261.01	nd (al-eth)	209				v	v^{h} δ					**B4**^{2}, 619
a929	—,—,**2,2-di-ethoxy-**	Diethylaminoacetal. $(CH_3CH_2)_2NCH_2CH(OCH_2CH_3)_2$	189.30			194–5	0.863^{16}		s	s	s			os s	**B4**, 309
a930	—,—,**2,2′-dihy-droxy-**	2,2′-Ethyliminodiethanol. $(HOCH_2CH_2)_2NC_2H_5$	133.19	ye		251–2^{750}	1.0135$^{20}_{4}$	1.4663^{20}	v	v	δ				**B4**, 284
a931	—,—,**perfluoro-**	$(CF_3CF_2)_3N$	371.05			70.3^{760}	1.736$^{25}_{4}$	1.262^{25}							
a932	—,—,**2,2′,2″-tri-chloro-**	Nitrogen mustard gas. $(ClCH_2CH_2)_3N$	204.53	pa ye	−4	143–4^{15}			δ	∞	∞		∞	HCl s	
a933	—,—,**2,2′,2″-tri-hydroxy-**	Triethanolamine. $(HOCH_2CH_2)_3N$	149.19	hyg	20–1.2	277^{150} 206–7^{15}	1.124$^{20}_{4}$	1.4852^{20}	∞	∞	δ		δ	chl s lig δ	**B4**^{2}, 729
a934	—,—,—,hydro-chloride	$(HOCH_2CH_2)_3N.HCl$	185.65	(dil al)	*ca.* 177				δ	i					**B4**, 285
a935	—,**trifurfuryl**	$[C_4H_3O{-}CH_2{-}]_3N$	257.29		fr in liq air	136–8^{1}			i		s				**B18**^{2}, 418
a936	—,**triheptyl**	$[CH_3(CH_2)_6]_3N$	311.60			330^{762}								ac s	**B4**, 193
a937	—,**trihexyl**	$[CH_3(CH_2)_5]_3N$	269.52			263–4			i	v	v			ac s	**B4**^{2}, 650
a938	—,**triisobutyl**	$[(CH_3)_2CHCH_2]_3N$	185.36		−21.8	191.5	0.764$^{25}_{4}$	1.4252^{17}	i	v	∞				**B4**^{2}, 639
a939	—,**trimethyl**	$(CH_3)_3N$	59.11		−117	3.5	0.6079$^{0}_{4}$		v	s	s		s	chl, to v	**B4**^{2}, 553
a940	—,—,hydrochloride	$(CH_3)_3N.HCl$	95.57	mcl dlq (al)	271–2	sub			s	s	i			chl δ	**B4**^{2}, 555
a941	—,—,oxide	$(CH_3)_3NO$	75.11	hyg nd	208				v	v	i		i	chl s^{h}	**B4**^{2}, 556
a942	—,—,oxide hydrochloride	$(CH_3)_3NO.HCl$	111.57	nd (al)	217–20d				v	s^{h}				MeOH v^{h}	**B4**^{2}, 556
a943	—,—,oxide hydroiodide	$(CH_3)_3NO.\ HI$	203.02	pr (al)	130d				v	v	i				**B4**, 50
a944	—,—,**perfluoro-**	$(CF_3)_3N$	221.03			−7 to −6									
—	—,—,**1,1′,1″-tricarboxy-**	*see* **Acetic acid, nitrilotri-**													
a945	—,**trioctadecyl**	$[CH_3(CH_2)_{16}CH_2]_3N$	774.49		fr 54					i	v		v	chl s	
a946	—,**tripentyl**	$[CH_3(CH_2)_4]_3N$	227.44			240–5 130^{14}				s				ac s	**B4**^{2}, 642
a947	—,**triphenyl**	$(C_6H_5)_3N$	245.31	mcl (MeOH, AcOEt, bz)	126.4	365	0.774$^{0}_{0}$	1.353^{16}	i	δ s^{h}	v		v	MeOH s^{h}	**B12**^{2}, 106
a948	—,**tripropyl**	$(CH_3CH_2CH_2)_3N$	143.27		−93.5	156^{760}	0.753$^{25}_{4}$	1.4176^{19}	δ	∞	s				**B4**^{2}, 623
a949	—,—,**perfluoro-**	$(CF_3CF_2CF_2)_3N$	521.11			130^{760}	1.822$^{25}_{4}$	1.279^{25}							
a950	—,—,**2,2′,2″-trihydroxy-**	Triisopropanolamine. $(CH_3CHOHCH_2)_3N$	227.31		45	170–80^{10}	1.0$^{20}_{4}$		s	s					
a951	—,**3,3′,4″-tritolyl**	$[CH_3{-}C_6H_4{-}]_2N{-}C_6H_4{-}CH_3$	287.41	nd (al)	89–90				i	s				os v con sulf s	**B12**^{1}, 415
a952	—,**tri(4-tolyl)**	$[CH_3{-}C_6H_4{-}]_3N$	287.41	(aa)	117					δ	s	s	v	chl v sulf s lig δ	**B12**^{2}, 494
—	**Aminomethanamidine**	*see* **Guanidine**													
—	**Aminonaphthosulfonic acid**	*see* **1-Naphthalenesulfonic acid, 4-amino-5-hydroxy-***													
a953	**Ammelide**	6-Amino-*s*-triazine-2,4-diol*. Cyanuramide. $C_3N_3(OH)_2NH_2$	128.09	mcl pr (w)	d				δ^{h}	i				dil sulf δ ac s aa δ	**B26**^{2}, 132

For explanations, symbols and abbreviations see beginning of table.

No.	Name	Synonyms and Formula	Mol. wt.	Crystalline form, color and specific rotation	m.p. °C	b.p. °C	Density	n_D	Solubility w	al	eth	ace	bz	other solvents	Ref.
	Ammeline														
a954	**Ammeline**	4,6-Diamino-*s*-triazin-2-ol. Cyanurodiamide. 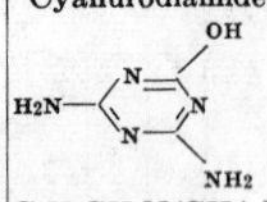	127.10	nd aq (Na_2CO_3)	d				i	i	i			ac s alk s aa i	**B26**[2], 132
a955	**Ammonium, benzyltrimethyl,** bromide	$C_6H_5CH_2N(CH_3)_3Br$	230.15	pl (al-lig), (w)	235				s^h	s^h					**B12**[2],546
a956	—,—,chloride	$C_6H_5CH_2N(CH_3)_3Cl$	185.70	(ace)	243				v			s^h			**B12**[1], 448
a957	—,—,iodide	$C_6H_5CH_2N(CH_3)_3I$	277.15	(al)	178–9					s					**B12**[1], 448
a958	—,—,nitrate	$C_6H_5CH_2N(CH_3)_3NO_3$	212.25		151–60				s	s					**B12**[2], 546
a959	—,**dodecyl trimethyl,** chloride	$CH_3(CH_2)_{11}N(CH_3)_3Cl$	263.90		246d				v	v	i	v	i	MeOH v chl v	
a960	—,**phenyl trimethyl,** bromide	$C_6H_5N(CH_3)_3Br$	216.13	hyg pr (al or al-eth)	213–4				v	s^h	i			aa s^h	**B12**[2], 88
a961	—,—,chloride	$C_6H_5N(CH_3)_3Cl$	171.67	hyg nd (al-eth)	d				s	v				chl δ	**B12**[2], 88
a962	—,—,iodide	$C_6H_5N(CH_3)_3I$	263.12	lf (al)	224 (228–9d)				v	s		δ		chl i, $δ^h$ aa s	**B12**[2], 98
a963	—,**propyl trimethyl-, 2-acetoxy,** chloride (*D*)	*O*-Acetyl-β-methylcholine chloride. $CH_3CH(O_2CCH_3)CH_2N(CH_3)_3Cl$	195.69	hyg, $[\alpha]_D+41.9$	172–3	d			s	s	i				
a964	—,—,—,—(*L*)	$CH_3CH(O_2CCH_3)CH_2N(CH_3)_3Cl$	195.69	$[\alpha]_D-41.3$					s						
a965	—,—,**2-hydroxy-,** chloride	β-Methylcholine chloride. $CH_3CHOHCH_2N(CH_3)_3Cl$	153.65	pr	165	d			s	s	i				
—	—,**purpurate**	*see* **Murexide**													
a966	—,**tetrabutyl,** iodide	$[CH_3(CH_2)_3]_4NI$	369.38	lf (w or bz)	144–5				δ s^h	v				os s	**B4**, 157
a967	—,**tetraethyl,** bromide	$(CH_3CH_2)_4NBr$	210.16	dlq (al)			1.397^{20}_4		v	v				chl v	**B4**[2], 597
a968	—,—,hydroxide, tetrahydrate	$(CH_3CH_2)_4NOH.4H_2O$	219.32	nd (w+4)	49–50	d (vac)			v	s					**B4**[2], 596
a969	—,**tetramethyl,** bromide	$(CH_3)_4NBr$	154.06	dlq ditetr bipym	230d	(360 sub in vac)	1.56		v	δ	i			chl i	**B4**[2], 558
a970	—,—,chloride	$(CH_3)_4NCl$	109.60	dlq ditetr, bipym	d>230		1.169^{20}_4		s	δ	i			chl i	**B4**[2], 557
a972	—,—,hydroxide, monohydrate	$(CH_3)_4NOH.H_2O$	109.17		130–5d without melting										**B4**, 50
a973	—,—,hydroxide, trihydrate	$(CH_3)_4NOH.3H_2O$	145.20		59–60										**B4**, 50
a974	—,—,hydroxide, pentahydrate	$(CH_3)_4NOH.5H_2O$	181.23	hyg nd	62–3	d			v ∞^h	v	i				**B4**[2], 557
a975	—,—,iodide	$(CH_3)_4NI$	201.05	ditetr bipym	230d		1.829		δ	δ	i	δ		chl i, alk δ liq SO_2 s	**B4**[2], 557
a976	—,**tetrapropyl,** chloride	$(CH_3CH_2CH_2)_4NCl$	221.82				1.0296^{13}		s					chl s	**B4**[2], 623
a977	—,—,iodide	$(CH_3CH_2CH_2)_4NI$	313.27	rh bipym	280d		1.3138^{25}_4		v	s	δ			chl v aa s	**B4**[2], 624
a978	**Amphetamine** (*d*)	Dexedrine. $C_6H_5CH_2CH(NH_2)CH_3$	135.21	$[\alpha]^{15}_D+37.6$ (bz, c=8%)		203–4 $83–4^{10}$	0.940^{15}_4	1.517^{26}	δ	s	s				
a979	—,sulfate (*d*)	Dexedrine sulfate. $2(C_6H_5CH_2CH(NH_2)CH_3).H_2SO_4$	368.50	$[\alpha]^{20}_D+22$ (w, c=8%)	d>300		1.15^{25}_4		v	δ	i				
a980	—(*dl*)	Benzedrine. $C_6H_5CH_2CH(NH_2)CH_3$	135.21			$200–3^{760}$ 116^{50}	0.9306^{25}_4	1.518^{26}	δ	s				chl s	
a981	—,**sulfate** (*dl*)	Benzedrine sulfate. $2(C_6H_5CH_2CH(NH_2)CH_3).H_2SO_4$	368.50		d>300		1.15^{25}_4		v	δ	i				
a982	**Amygdalin**	Amigdaloside. Mandelonitrile. β-Gentiobioside. $C_{20}H_{27}NO_{11}$.	457.44	$[\alpha]^{18}_D+41.6$ (w,p=1.6)	214–6				v^h	$δ^h$	i				**B31**, 401
a983	—,trihydrate	$C_{20}H_{27}NO_{11}.3H_2O$	511.48	rh (w+3), $[\alpha]^{20}_D$ 35 (w,c=3)	120–5				s ∞^h	s^h	i				**B31**, 400
—	**Amylamine**	*see* **Pentane, 1-amino-***													
—	**Amyl bromide**	*see* **Pentane, 1-bromo-***													

For explanations, symbols and abbreviations see beginning of table.

No.	Name	Synonyms and Formula	Mol. wt.	Crystalline form, color and specific rotation	m.p. °C	b.p. °C	Density	n_D	Solubility: w	a	eth	ace	bz	other solvents	Ref.
	Amyl chloride														
—	**Amyl chloride**	*see* **Pentane, 1-chloro-***													
—	**Amyl iodide**	*see* **Pentane, 1-iodo-***													
—	**Amytal**	*see* **Barbituric acid, 5-ethyl-5(3-methylbutyl)-**													
a984	**α-Amyrin**	α-Amyernol. $C_{30}H_{50}O$.	426.70	nd(al), $[\alpha]_D^{17}$ +91.6 (bz)	186	243^{07}									**E14s**, 1067
a985	**β-Amyrin**	β-Amyrenol. $C_{30}H_{50}O$.	426.70	nd (lig or al), $[\alpha]_D^{19}+99.8$(bz)	199.5–200	$260^{0.8}$			i	δ	s	...	s	chl s aa s lig δ	**B14s**, 946
a986	**Anabasine** (*l*)	Neonicotine. *l*-2(3-pyridyl)-piperidine. $C_{10}H_{14}N_2$.	162.23	darkens in air, $[\alpha]_D-83.1$	9	276^{760} 105^2	1.046_{20}^{20}	1.5430^{20}	∞	s	s	...	s		**B23²**, 113
a987	**Anacardic acid**	*o*-Pentadecadienylsalicylic acid. (structure: $C_{15}H_{27}$?, CO_2H, OH)	344.48		26			1.564^{28}	δ	s	s	...	...		**B10²**, 197
a988	**Anagyrine**	$C_{15}H_{20}NO_2$.	244.33	pa ye glass, $[\alpha]_D^{25}-168$ (al)		$260–70^{12}$			s $δ^h$	v	s	...	s	chl v dil HCl v lig i	**B24²**, 84
a989	**Analgene**	Quinolagen. Chinalgen. Labordin. Benzanalgen. (structure: CH_3CH_2O—, —$NHCOC_6H_5$, N)	292.33	ye nd (al)	206				i	δ s^h	δ	...	...	ac s	**B22**, 503
a990	**Anatabine**	2(3-Pyridyl)-1,2,3,6-tetrahydropyridine. $C_{10}H_{12}N_2$.	160.21	$[\alpha]_D^{17}-177.8$		$145–6^{10}$	1.091_4^{19}	1.5676^{20}	∞	s	s	...	s		
a991	**Androstane**	Etioallocholane. (structure: CH_3, CH_3)	260.45	lf (ace-MeOH or chl-ace) $[\alpha]_D^{16}+2$(chl)	50	$75–80^{0.01}$			i	s	s	s	...	MeOH s chl s lig s	**E14s**, 1396
a992	**Androsterone**	3(α)-Hydroxy-17-ketoandrostane. $C_{19}H_{30}O_2$.	290.43	lf or nd (al), $[\alpha]_D^{20}+94.6$ (abs al, c=0.7)	185–5.5				δ	s	s	...	...	os s	
—	**Δ⁴-Androstrene-3,11,17-trione**	*see* **Andrenosterone**													
a993	**Anemonin**	Anemone camphor. $C_{10}H_8O_4$.	192.16	rh pl (chl), nd (al or bz)	158				δ s^h	δ s^h	δ	...	...	chl s alk s lig δ	**B19²**, 182
—	**Anethole**	*see* **Benzene,1-methoxy-4-propenyl-***													
—	**Angelic acid**	*see* **2-Butenoic acid, 2-methyl- (*cis*)***													
a994	**Anhalamine**	$C_{11}H_{15}NO_3$.	209.25	nd (al)	187–8				s^h i	v^h i	δ	v	δ	chl $δ^h$ lig i	**B21²**, 157
a995	**Anhalonidine**	$C_{12}H_{17}NO_3$.	223.27	(bz, eth)	160				s	s	i	...	s^h	chl s lig i	**B21²**, 157
a996	**Anhalonine** (*d*)	$C_{12}H_{15}NO_3$.	221.26	$[\alpha]_D^{25}+56.7$	84.5–5.5				...	...	...	...	...		**B27²**, 542
a997	—(*dl*)	$C_{12}H_{15}NO_3$.	221.26	rh nd (peth)	85–5.5				s	v	v	...	v	chl v	**B27²**, 542
a998	—(*l*)	$C_{12}H_{15}NO_3$.	221.26	$[\alpha]_D^{25}-56.3$ (chl,c=4)	85–6				...	...	...	...	...		**B27²**, 542
a999	—,hydrochloride	$C_{12}H_{15}NO_3.HCl$.	257.72	rh pr, $[\alpha]_D^{25}$ −40.5(al)	254–5				δ v^h	s	δ	...	...	chl δ	**B27²**, 542
a1000	**Anhydroecgonine** (*dl*)	Ecgonidine (structure: CO_2H, CH_3—N)	185.23	(MeOH, w, MeOH-eth)	226–30d				v	δ	δ	...	δ	chl δ lig δ	**B22²**, 26
a1001	—,hydrochloride	Ecgonine hydrochloride. $C_9H_{16}ClNO_3$	221.69	rh nd(al +1w),ta (w), $[\alpha]_D^{15}-62.7$	240–1				...	s	s	...	...		**B22**, 31

For explanations, symbols and abbreviations see beginning of table.

No.	Name	Synonyms and Formula	Mol. wt.	Crystalline form, color and specific rotation	m.p. °C	b.p. °C	Density	n_D	Solubility w	al	eth	ace	bz	other solvents	Ref.
	Aniline														
a1002	**Aniline**	Aminobenzene. Phenylamine. 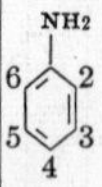	93.13		−6.2	184.32^{760} 69.2^{10}	1.0216^{20}_{4}	1.5863^{20}	s ∞^{h}	∞	∞		∞		B12², 44
a1003	—,benzenesulfonate	$C_6H_5NH_2.C_6H_5SO_3H$	251.31	nd	240 δd				v	v	i		i	CS_2 i	B12², 75
a1004	—,hydrochloride	$C_6H_5NH_2.HCl$	129.59	lf or nd	198	245^{760}	1.2215^{4}		v	v	i			chl i	B12², 65
a1005	—,nitrate	$C_6H_5NH_2.HNO_3$	156.14	rh	d>190		1.356^{4}		v	v	v				B12², 66
a1006	—,oxalate (acid)	$C_6H_5NH_2.H_2C_2O_4$	183.17	ta (aq al)	150–1				v	v		v			B12², 73
a1007	—,—(neutral)	$2(C_6H_5NH_2).H_2C_2O_4$	276.29	tcl pr (w)	174–5d				v	δ	i	δ			B12², 73
a1008	—,phosphate	$C_6H_5NH_2.H_3PO_4$	191.12	nd (al), lf(w-al)	179				v	v	i				B12, 117
a1009	—,picrate	$C_6H_5NH_2$ 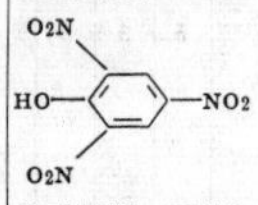	322.24	ye or red mcl pr (w), dk gr	181		1.558		δ	s			i		B12², 72
a1010	—,sulphate(acid)	$C_6H_5NH_2.H_2SO_4.\frac{1}{2}H_2O$	200.21	lf(w+½)	162										B12², 66
a1011	—,—(neutral)	$2(C_6H_5NH_2).H_2SO_4$	284.34	lf(al)	d		1.377^{4}		s	δ	i				B12², 66
a1012	—,**3-acetamido-**	*N*-Acetyl-*m*-phenylene-diamine. $C_8H_{10}N_2O$. *See* a1002	150.18	lf(ace-bz), nd or pr (bz)	87–9	100d			v	v	s	v	i		B13², 27
a1013	—,**4-acetamido-, sulfate**	*N*-Acetyl-*p*-phenylenediamine sulfate. $C_8H_{12}N_2O_5S$. *See* a1002	248.26	nd (eth-al)	285d				v	v	i				B13, 95
a1014	—,*N*-**allyl-**	$CH_2:CHCH_2NHC_6H_5$	133.19	ye		$217–9^{726}$ $105–8^{12}$	0.982^{25}_{4}		δ	s	∞				B12², 96
—	—,*N*-**benzyl-**	*see* **Amine, benzyl phenyl**													
a1015	—,*N*-**benzylidene-**	Benzalaniline. $C_6H_5CH:NC_6H_5$	181.24	ye nd (CS_2), pl (dil al)	54	310	1.038^{55}_{4}	1.600^{99}	i	s	s			liq SO_2, NH_3 s	B12², 113
a1016	—,**2-bromo-**	C_6H_6BrN. *See* a1002	172.03		32	229^{760} 110.5^{19}	1.578^{20}_{4}	1.6113^{20}_{4}	i	s	s				B12², 341
a1017	—,**3-bromo-**	C_6H_6BrN. *See* a1002	172.03		18.5 (fr 16.7)	251^{760}, $122–4^{10}$	1.5793^{20}_{4}	1.6260^{20}	δ	s	s				B12², 342
a1018	—,**4-bromo-**	C_6H_6BrN. *See* a1002	172.03	rh bipym, nd (60% al)	66.4	d	1.4970^{100}_{4}		i	s	s				B12², 344
a1019	—,**2-bromo-4,6-dichloro-**	$C_6H_4BrCl_2N$. *See* a1002	240.92	nd (al)	83.5	273^{760}				v			v	chl v	B12², 355
a1020	—,**4-bromo-2,3-dichloro-**	$C_6H_4BrCl_2N$. *See* a1002	240.92	nd	77.5					v	v		v		B12, 653
a1021	—,**4-bromo-2,5-dichloro-**	$C_6H_4BrCl_2N$. *See* a1002	240.92	nd	91					v	v				B12, 654
a1022	—,**4-bromo-3,5-dichloro-**	$C_6H_4BrCl_2N$ *See* a1002	240.92	nd(w+al)	129				s^{h}	v			v	chl v	B12, 654
a1023	—,**3-bromo-*N*,*N*-diethyl-**	$C_{10}H_{14}BrN$. *See* a1002	228.14			$139.5–42^{9}$				s		s			B12¹, 315
a1024	—,**4-bromo-*N*,*N*-diethyl-**	$C_{10}H_{11}BrN$. *See* a1002	228.15	nd or pr	33	270			i	v	v				B12², 347
a1025	—,**2-bromo-*N*,*N*-dimethyl-**	$C_8H_{10}BrN$. *See* a1002	200.08			$107–8,^{14}$				v					B12¹, 313
a1026	—,**3-bromo-*N*,*N*,-dimethyl-**	$C_8H_{10}BrN$. *See* a1002	200.08		11	259 (cor) 125^{10}				v				con ac s aa v	B12², 343
a1027	—,**4-bromo-*N*,*N*-dimethyl-**	$C_8H_{10}BrN$. *See* a1002	200.08	pl (al)	58	264			i	s	v			os s	B12², 345
a1028	—,**2-bromo-3,5-dinitro-**	$C_6H_4BrN_3O_4$. *See* a1002	262.03	ye lf (al)	181					δ v^{h}				HNO_3 s	B12, 762
a1029	—,**2-bromo-4,5-dinitro-**	$C_6H_4BrN_3O_4$. *See* a1002	262.03	pa ye (al)	186					v^{h}					B12, 762
a1030	—,**2-bromo-4,6-dinitro-**	$C_6H_4BrN_3O_4$. *See* a1002	262.03	ye nd (aa or al)	153–4	sub				v^{h}		v^{h}			B12², 418
a1030¹	—,**4-bromo-2,5-dinitro-**	$C_6H_4BrN_3O_4$. *See* a1002	262.03	ye (al)	186					v^{h}					B12, 761
a1031	—,**4-bromo-2,6-dinitro-**	$C_6H_4BrN_3O_4$. *See* a1002	262.03	og-red nd (abs al)	163 cor					v^{h}					B12², 418
a1032	—,**5-bromo-2,4-dinitro-**	$C_6H_4BrN_3O_4$. *See* a1002	262.03	pa ye nd	178.4					v	v		v	chl v	B12², 417
a1033	—,**6-bromo-2,3-dinitro-**	$C_6H_4BrN_3O_4$. *See* a1002	262.03	dk red (al)	158					v^{h}					B12, 760

For explanations, symbols and abbreviations see beginning of table.

No.	Name	Synonyms and Formula	Mol. wt.	Crystalline form, color and specific rotation	m.p. °C	b.p. °C	Density	n_D	Solubility w	al	eth	ace	bz	other solvents	Ref.
	Aniline														
a1034	—,**2-bromo-4-nitro-**	$C_6H_5BrN_2O_2$. *See* a1002	217.03	ye nd	104.5									aa v	**B12**[2], 403
a1035	—,**2-bromo-5-nitro-**	$C_6H_5BrN_2O_2$. *See* a1002	217.03	ye nd (al)	141					s	v		v	chl v lig δ	**B12**[2], 402
a1036	—,**3-bromo-4-nitro-**	$C_6H_5BrN_2O_2$. *See* a1002	217.03	ye nd(al)	175–6					v			i	chl i lig i	**B12**[2], 403
a1037	—,**4-bromo-2-nitro-**	$C_6H_5BrN_2O_2$. *See* a1002	217.03	og-ye nd (w)	111.5	sub			δ	v					**B12**[2], 401
a1038	—,**4-bromo-3-nitro-**	$C_6H_5BrN_2O_2$. *See* a1002	217.03	nd (al)	132					v	v			chl v aa v	**B12**[2], 402
a1039	—,**5-bromo-2-nitro-**	$C_6H_5BrN_2O_2$. *See* a1002	217.03	red ye nd	151–2					v				aa v	**B12**[2], 402
a1040	—,**6-bromo-2-nitro-**	$C_6H_5BrN_2O_2$. *See* a1002	217.03	ye nd (dil al)	74–5		1.988			v					**B12**[2], 402
a1041	—,**N-butyl-**	*N*-Phenyl butylamine. $C_6H_5NHC_4H_9^n$	149.24			241.6^{760} 68^{1}	0.93226^{20}_{4}	1.53412^{20}	δ	v	s				**B12**[2], 95
a1042	—,*N-sec*-**butyl-**	$C_6H_5NHC_4H_9^i$	149.24			225^{759} $112–4^{22}$		1.5318^{26}						os s	**B12**[2], 96
a1043	—,**2-chloro-(α)**	C_6H_6ClN. *See* a1002	127.58		—14	208.84	1.21256^{20}_{4}	1.5881^{20}	i	∞	s			os s	**B12**[2], 314
a1044	—,—(β)	C_6H_6ClN. *See* a1002	127.58		−3.5	208.84	1.21256^{20}_{4}	1.5881^{20}	i	∞	s			os	**B12**[2], 314
a1045	—,—,hydrochloride	$C_6H_6ClN.HCl$ *See* a1002	164.04	pl (w, aq al)	235		1.505^{18}		s	δ					**B12**[2], 314
a1046	—,**3-chloro-**	C_6H_6ClN. *See* a1002	127.58		−10.4	229.92 $99–100^{10}$	1.21606^{20}_{4}	1.59414^{20}		∞	∞		∞	CCl_4 ∞	**B12**[2], 319
a1047	—,—,hydrochloride	C_6H_6ClN,HCl. *See* a1002	164.04	pl	221.5–2.5				v	v					**B12**[2], 320
a1048	—,**4-chloro-**	C_6H_6ClN. *See* a1002	127.58	rh pr	72.5	232^{760}	1.429^{19}_{4}	1.5546^{87}	s^h	s	s			os s	**B12**[2], 322
a1049	—,**4-chloro-*N,N*-diethyl-**	$C_{10}H_{14}ClN$. *See* a1002	183.69	nd (al)	39	$95–6^{1.5}$				s^h					**B12**[2], 324
a1050	—,**4-chloro-*N,N*-dimethyl-**	$C_8H_{10}ClN$. *See* a1002	155.62	nd (al)	35–5	230–1				s^h					**B12**[1], 304
a1051	—,**4-chloro-*N,N*-dimethyl-3-nitro-**	$C_8H_9ClN_2O_2$. *See* a1002	200.63	ye nd	815–2.5					s				lig s	**B12**, 732
a1052	—,**4-chloro-2,6-dinitro-**	$C_6H_4ClN_3O_4$. *See* a1002	217.58	og-ye nd (al)	147 cor					s^h					**B12**[2], 416
a1053	—,**3-chloro-2,6-dinitro-**	$C_6H_4ClN_3O_4$. *See* a1002	217.58	og-ye nd (al)	112					s^h					**B12**[2], 416
a1054	—,**5-chloro-2-ethoxy-**	$C_8H_{10}ClNO$. *See* a1002	171.61	nd	42				δ	s			s	chl s	**B13**, 383
a1055	—,**2-chloro-*N*-ethyl-**	$C_8H_{10}ClN$. *See* a1002	155.62			219^{726}	1.1104^{25}_{4}								**B12**[2], 316
a1056	—,**4-chloro-2-methoxy-**	C_7H_8ClNO. *See* a1002	157.60	pr	52	260				s	s		s		**B13**[2], 184
a1057	—,**5-chloro-2-methoxy-**	2-Amino-4-chloroanisole. C_7H_8ClNO. *See* a1002	157.60	nd	82					s				lig δ	**B13**[2], 183
a1058	—,**2-chloro-4-nitro-**	$C_6H_5ClN_2O_2$. *See* a1002	172.58	ye nd (lig-CS_2, w, 20% aa)	104–5				δ	v	v			CS_2 v lig i	**B12**[2], 398
a1059	—,**3-chloro-4-nitro-**	$C_6H_5ClN_2O_2$. *See* a1002	172.58	ye pl (bz)	156–7					s				aa s lig i	**B12**[2], 398
a1060	—,**4-chloro-2-nitro-**	$C_6H_5ClN_2O_2$. *See* a1002	172.58	dk og-ye pr (50% al), nd (lig)	116–7					v	v			aa v lig δ	**B12**[2], 396
a1061	—,**4-chloro-3-nitro-**	$C_6H_5ClN_2O_2$. *See* a1002	172.58	ye nd or pr (w), nd (peth)	102–3				s^h	v	s			chl s lig δ	**B12**[2], 397
a1062	—,**5-chloro-2-nitro-**	$C_6H_5ClN_2O_2$. *See* a1002	172.58	gold-ye nd, pl	124–5					s	s			lig δ	**B12**[2], 397
a1063	—,**5-chloro-3-nitro**	$C_6H_5ClN_2O_2$. *See* a1002	172.58	og-ye nd (al)	133–4					s^h					**B12**, 732
a1064	—,**6-chloro-2-nitro-**	$C_6H_5ClN_2O_2$. *See* a1002	172.58	ye nd	76									50% al s	**B12**[2], 397
a1065	—,**6-chloro-3-nitro-**	$C_6H_5ClN_2O_2$. *See* a1002	172.58	ye nd (lig)	116–8					v	v	v		CS_2 δ aa v	**B12**[2], 398
a1066	—,***N*-cyclohexyl-**	Cyclohexylphenylamine. —NH—	175.26	mcl pr	16	279^{764}	1.0155^{20}_{4}	1.5610^{20}							**B12**[2], 98
a1067	—,**2,3-dibromo-**	$C_6H_5Br_2N$. *See* a1002	250.95	pl (dil al)	43				δ	v	v			aa v	**B12**, 655

For explanations, symbols and abbreviations see beginning of table.

No.	Name	Synonyms and Formula	Mol. wt.	Crystalline form, color and specific rotation	m.p. °C	b.p. °C	Density	n_D	Solubility						Ref.
									w	al	eth	ace	bz	other solvents	
	Aniline														
a1068	**—,2,4-dibromo-**	$C_6H_5Br_2N$. *See* a1002	250.95	rh bipym (chl), nd or lf (dil al)	79.5–80.5	156^{24}	2.260^{20}			s				chl s^h aa s	**B12**^{2}, 356
a1069	**—,2,5-dibromo-**	$C_6H_5Br_2N$. *See* a1002	250.95	pr (al)	51–2					v	s				**B12**^{2}, 357
a1070	**—,2,6-dibromo-**	$C_6H_5Br_2N$. *See* a1002	250.95	nd (al)	83–4	262–4				v	v		v	chl v	**B12**^{2}, 327
a1071	**—,3,4-dibromo-**	$C_6H_5Br_2N$. *See* a1002	250.95	lf (dil al)	81	100 sub				s^h					**B12**^{1}, 329
a1072	**—,3,5-dibromo-**	$C_6H_5Br_2N$. *See* a1002	250.95	nd	57					v	v		v		**B12**^{2}, 357
a1073	**—,2,4-dibromo-6-nitro-**	$C_6H_4Br_2N_2O_2$. *See* a1002	295.94	ye	128										**B12**^{2}, 403
a1074	**—,2,6-dibromo-4-nitro-**	$C_6H_4Br_2N_2O_2$. *See* a1002	295.94	ye nd (al or aa)	207				δ					aa s	**B12**^{2}, 404
a1075	**—,*N*,*N*-dibutyl-**	$C_6H_5N(CH_2CH_2CH_2CH_3)_2$	205.34		271^{760} 148.5^{4}	0.907			i	∞	∞			ac s	**B12**^{2}, 95
a1076	**—,2,3-dichloro-**	$C_6H_5Cl_2N$. *See* a1002	162.03	nd (lig)	24	252				s	v		δ	lig δ	**B12**, 621
a1077	**—,2,4-dichloro-**	$C_6H_5Cl_2N$. *See* a1002	162.03	pr (ace), nd (dil al), (lig)	63	245^{759}	1.567^{20}		δ	s	s				**B12**^{2}, 333
a1078	**—,2,5-dichloro-**	$C_6H_5Cl_2N$. *See* a1002	162.03	nd (lig)	50	251			δ	s	s		s	CS_2 s	**B12**^{2}, 336
a1079	**—,3,4-dichloro-**	$C_6H_5Cl_2N$. *See* a1002	162.03	nd (lig)	72	272 145^{15}				s	s		δ		**B12**^{2}, 337
a1080	**—,3,5-dichloro-**	$C_6H_5Cl_2N$. *See* a1002	162.03	nd	50.5	260^{741}			i	s	s				**B12**^{2}, 337
a1081	**—,2,6-dichloro-4-ethoxy-**	4-Amino-3,5-dichlorophenetole. $C_8H_9Cl_2NO$. *See* a1002	206.08	nd (dil al)	46	275			δ	v	v		v	chl v	**B13**^{2}, 276
a1082	**—,3,5-dichloro-4-ethoxy-**	4-Amino-2,6-dichlorophenetole. $C_8H_9Cl_2NO$. *See* a1002	206.08	nd (peth)	105–7					v	v		v	aa δ	**B13**^{2}, 275
a1083	**—,*N*-dichloromethylene-**	Phenyliminophosgene. Phenylisocyanide dichloride. $C_6H_5N{:}CCl_2$	174.04	oil	19.5	210^{760} 96^{19}	1.285^{15}								**B12**^{2}, 245
a1084	**—,2,6-dichloro-4-nitro-**	$C_6H_4Cl_2N_2O_2$. *See* a1002	207.03	ye nd (al, aa)	191					s				ac s	**B12**^{2}, 400
a1085	**—,*N*,*N*-diethyl-**	Diethylphenylamine. $C_6H_5N(C_2H_5)_2$	149.24	ye oil	−38.8	216.27 50^{16}	0.93507^{20}_{4}	1.5413	δ	s	v			chl v	**B12**^{2}, 92
a1086	**—,*N*,*N*-diethyl-2-ethoxy-**	*N*,*N*-Diethyl-*o*-phenetidine. $C_{12}H_{19}NO$. *See* a1002	193.29			231–3			i	∞	∞		∞	chl ∞ CS_2 ∞	**B13**, 365
a1087	**—,*N*,*N*-diethyl-4-ethoxy-**	*N*,*N*-Diethyl-*m*-phenetidine. $C_{12}H_{19}NO$. *See* a1002	193.29			286				s			s	aa s	**B13**^{1}, 131
a1088	**—,*N*,*N*-diethyl-3-nitro-**	$C_{10}H_{14}N_2O_2$. *See* a1002	194.24	ye		288–90			i						**B12**^{1}, 346
a1089	**—,*N*,*N*-diethyl-4-nitro-**	$C_{10}H_{14}N_2O_2$. *See* a1002	194.24	ye nd (lig), pl (al)	77–8		1.225			v^h				lig δ	**B12**^{1}, 351
a1090	**—,*N*,*N*-diethyl-4-nitroso-**	$C_{10}H_{14}N_2O$. *See* a1002	178.24	gr mcl pr (eth), lf (ace)	85		1.24^{15}_{4}		δ	s	s				**B12**^{2}, 365
a1091	**— 2,4-diiodo-**	$C_6H_5I_2N$ *See* a1002	344.93	br nd or rh cr (al)	95–6		2.748		δ s^h	s v^h	v	v		chl, CS_2 v lig s	**B12**^{2}, 63
a1092	**—,2,5-diiodo-**	$C_6H_5I_2N$. *See* a1002	344.93	nd	88–9									os s	**B12**, 675
a1093	**—,2,6-diiodo-**	$C_6H_5I_2N$. *See* a1002	344.93	nd (al)	122					s				os s	**B12**, 675
a1094	**—,3,4-diiodo-**	$C_6H_5I_2N$. *Sce* a1002	344.93	pa ye lf or pl (bz-peth)	74.5					v	v		v	lig δ	**B12**, 675
a1095	**—,3,5-diiodo-**	$C_6H_5I_2N$. *See* a1002	344.93	nd (al)	110 (107)					v	v			chl v	**B12**^{1}, 377
a1096	**—,2,4-diiodo-3-nitro-**	$C_6H_4I_2N_2O_2$. *See* a1002	389.93	pa ye mcl pl	125					v				os v lig δ	**B12**, 747
a1097	**—,2,6-diiodo-3-nitro-**	$C_6H_4I_2N_2O_2$. *See* a1002	389.93	ye nd	145.5(149)					v				os s^h, AcOEt δ	**B12**, 747
a1098	**—,2,6-diiodo-4-nitro-**	$C_6H_4I_2N_2O_2$. *See* a1002	389.93	pa ye lf or nd (bz)	245 (248)					$δ^h$			s^h		**B12**^{2}, 405
a1099	**—,4,6-diiodo-2-nitro-**	$C_6H_4I_2N_2O_2$. *See* a1002	389.93	ye nd (ace)	154				δ	δ	s	s	s	chl s aa s	**B12**^{1}, 361
a1100	**—,4,6-diiodo-3-nitro-**	$C_6H_4I_2N_2O_2$. *See* a1002	389.93	pa ye nd	149					$δ^h$	s				**B12**, 747
a1101	**—,2,3-dimethoxy-**	3-Aminoveratrole. $C_8H_{11}NO_2$. *See* a1002	153.18			137^{10}			s						**B13**^{2}, 464
a1102	**—,2,4-dimethoxy-**	4-Aminoresorcinol dimethyl ether. $C_8H_{11}NO_2$. *See* a1002	153.18	(lig)	39–40				δ	s	s		s		**B13**^{2}, 470
a1103	**—,2,6-dimethoxy-**	2-Aminoresorcinol dimethyl ether. $C_8H_{11}NO_2$. *See* a1002	153.18	pl (al)	75	146^{23}			δ	v	v		v	aa v lig s	**B13**^{2}, 468
a1104	**—,3,4-dimethoxy-**	4-Aminoveratrole. $C_8H_{11}NO_2$. *See* a1002	153.18	lf (eth)	87–8 (90)	$174–6^{22}$					s^h				**B13**^{2}, 465
a1105	**—,*N*.*N*-dimethyl-**	$C_6H_5N(CH_3)_2$	121.18	p ye	2.45	194.15^{760} $40^{2.4}$	0.9563^{20}_{4}	1.5587^{20}	δ	s	s		s		**B12**^{2}, 82
a1106	**—,—,hydrochloride**	$C_6H_5N(CH_3)_2.HCl$	157.64	hyg pl (w), (bz)	85		1.1156^{19}_{4}		s	s	i		δ	chl s	**B12**^{2}, 86

For explanations, symbols and abbreviations see beginning of table.

No.	Name	Synonyms and Formula	Mol. wt.	Crystalline form, color and specific rotation	m.p. °C	b.p. °C	Density	n_D	Solubility w	al	eth	ace	bz	other solvents	Ref.
	Aniline														
a1107	**—,—,N,*N*-dimethyl-**, *N*-oxide	$C_6H_5N(:O)(CH_3)_2$	137.18	pr	152–3	...	...	...	v	v	δ	...	...	chl v	**B12**^{2}, 88
a1108	**—,*N*,*N*-dimethyl-2-nitro-**	$C_8H_{10}N_2O_2$. *See* a1002	166.18	ye og	ca.–20	154^{24}	1.1794^{20}_{4}	1.6102^{20}	s	v^h	s	...	...	chl v	**B12**^{2}, 369
a1109	**—*N*,*N*-dimethyl-3-nitro-**	$C_8H_{10}N_2O_2$. *See* a1002	166.18	og-ye or red mcl pr (eth, or eth-al)	58–9	280–5^{760}	1.313^{17}	...	i	s	s	...	...	...	**B12**^{2}, 377
a1110	**—*N*,*N*-dimethyl-4-nitro-**	$C_8H_{10}N_2O_2$. *See* a1002	166.18	ye (bl) nd	164	...	...	...	i	s^h	s	...	...	aa s^h	**B12**^{2}, 386
a1111	**—,*N*,*N*-dimethyl 4-nitroso-**	$C_8H_{10}N_2O$. *See* a1002	150.18	gr pl (eth)	92.5–3.5	...	1.145^{20}	...	δ	s	s	...	...	$HCONH_2$ s	**B12**^{2}, 364
a1112	**—,4-dimethyl amino-*N*(2,4,6-trinitrobenzal)-**	O_2N / $(CH_3)_2N$—⟨⟩—N:CH—⟨⟩—NO_2 / O_2N	359.29	bk gr lf ($PhCO_2Et$), nd($PhNO_2$+1)	268 exp	...	...	...	i	...	...	δ	δ	chl, δ AcOEt δ aa δ	**B13**, 85
a1113	**—,2,4-dinitro-**....	$C_6H_5N_3O_4$. *See* a1002	183.12	ye nd (ace), gr-ye ta (al)	187.5–8	...	1.615^{14}	...	i δ^h	δ	...	...	...	HCl s^h	**B12**^{2}, 405
a1114	**—,2,6-dinitro-**....	$C_6H_5N_3O_4$. *See* a1002	183.12	gold lf (50% aa), nd (al)	137	...	...	...	i	δ	s	...	s^h	lig i	**B12**^{2}, 413
a1115	**—,*N*,*N*-dipropyl-**.	$C_6H_5N(CH_2CH_2CH_3)_2$	177.28	lt ye	...	240–3^{760} 127^{10}	0.9104^{20}	...	i	s	s	...	...	...	**B12**^{2}, 95
a1116	**—,2-ethoxy-**......	*o*-Aminophenetole. *o*-Phenetidine. $C_8H_{11}NO$. *See* a1002	137.18	...	<–21	232.5^{760}	...	...	δ	s	s	...	...	...	**B13**^{2}, 166
a1117	**—,3-ethoxy-**......	*m*-Aminophenetole. *m*-Phenetidine. $C_8H_{11}NO$. *See* a1002	137.18	...	...	248^{760} 127–8^{11}	...	...	δ	s	s	...	...	...	**B13**^{2}, 211
a1118	**—,4-ethoxy-**......	*p*-Aminophenetole. *p*-Phenetidine. $C_8H_{11}NO$. *See* a1002	137.18	...	2.4	249.9	1.0652^{15}_{4}	1.5632^{16}	δ	s	s	...	...	...	**B13**^{2}, 224
a1119	**—,4-ethoxy-*N*-(2-hydroxy-benzylidene)-**	*N*-Salicylidene-*p*-phenetidine. Malakin. OH / ⟨⟩—CH:N—⟨⟩—OC_2H_5	241.28	pa ye or grsh pl or nd (al)	94 (90)	...	...	...	i	v	v	...	v	...	**B13**^{2}, 241
a1120	**—,2-ethoxy-4-nitro-**	4-Nitro-*o*-phenetidine. $C_8H_{10}N_2O_3$. *See* a1002	182.18	ye nd (dil al)	91	...	...	...	...	v	v	v	...	lig δ	**B13**, 390
a1121	**—,2-ethoxy-5-nitro-**	5-Nitro-*o*-phenetidine. $C_8H_{10}N_2O_3$. *See* a1002	182.18	ye nd (dil al)	96–7	...	...	...	δ	s	...	...	...	...	**B13**^{2}, 192
a1122	**—,2-ethoxy-6-nitro-**	6-Nitro-*o*-phenetidine. $C_8H_{10}N_2O_3$. *See* a1002	182.18	ye, og (w)	60	...	...	...	...	...	...	...	...	...	**B13**, 388
a1123	**—,3-ethoxy-4-nitro-**	4-Nitro-*m*-phenetidine. $C_8H_{10}N_2O_3$. *See* a1002	182.18	nd (dil al)	122–3	...	...	...	δ	...	...	...	...	os s lig i	**B13**^{1}, 137
a1124	**—,4-ethoxy-2-nitro-**	2-Nitro-*p*-phenetidine. $C_8H_{10}N_2O_3$. *See* a1002	182.18	red pr (al)	112.7–3.2	...	...	...	...	δ s^h	v	...	...	chl v	**B13**^{2}, 286
a1125	**—,4-ethoxy-3-nitro-**	3-Nitro-*p*-phenetidine. $C_8H_{10}N_2O_3$. *See* a1002	182.18	og-ye nd (bz or to-lig)	41	...	...	...	δ^h	...	...	...	...	os v lig i	**B13**^{2}, 284
a1126	**—,5-ethoxy-2-nitro-**	6-Nitro-*m*-phenetidine. $C_8H_{10}N_2O_3$. *See* a1002	182.18	ye nd (dil al)	105–6	...	...	...	i	...	...	...	...	os s	**B13**^{1}, 136
a1127	**—,5-ethoxy-3-nitro-**	5-Nitro-*m*-phenetidine. $C_8H_{10}N_2O_3$. *See* a1002	182.18	ye, og-red nd (al)	115	...	...	...	δ v^h	v	...	...	...	...	**B13**, 422
a1128	**—,*N*-ethyl-**......	Ethyl phenylamine. $C_6H_5NHC_2H_5$	121.18	...	–63.5	204.72^{760} 97.5–8^{18}	0.9625^{20}_{4}	1.5559^{20}	i	∞	∞	...	...	os s	**B12**^{2}, 90
a1129	—,—,hydrochloride	$C_6H_5NHC_2H_5.HCl$	157.65	nd	178.5	...	1.0085^{182}_{4}	...	v	s	...	...	...	chl s	**B12**^{2}, 91
a1130	**—,*N*-ethyl-*N*-methyl-**	$C_6H_5N(CH_3)C_2H_5$	135.20	...	...	203–5^{760} 93–5^{12}	0.9193^{55}_{4}	...	i	∞	∞	...	...	...	**B12**^{2}, 91
a1131	**—,*N*-ethyl-*N*-nitroso-**	$C_6H_5N(NO)C_2H_5$	150.18	yesh	...	119.5–20^{15}	1.0874^{20}_{4}	1.55977^{20}	i	...	...	...	...	aa s	**B12**^{2}, 140
a1132	**—,2-fluoro-**.......	C_6H_6FN. *See* a1002	111.12	pa ye	–28.5	174.5–6^{757} 58^{11}	1.1513^{21}	1.5467^{18}	i	s	s	...	...	...	**B12**^{2}, 314
a1133	**—,3-fluoro-**.......	C_6H_6FN. *See* a1002	111.12	pa ye	...	187–9^{760} 82.3^{18}	1.1561^{19}	1.54528^{19}	δ	s	s	...	...	...	**B12**^{2}, 314
a1134	**—,4-fluoro-**.......	C_6H_6FN. *See* a1002	111.12	pa ye	–1.9	180.5––2.5^{757} 85^{19}	1.1725^{20}_{4}	1.5185^{20}	δ	...	...	...	...	...	**B12**^{2}, 314
a1135	**—,*N*-formyl-*N*-methyl-**	$C_6H_5N(CH:O)CH_3$	135.17	...	13.6–.7	253^{716} 215^{13}	1.0948^{20}_{4}	1.5589^{20}	δ	s	...	...	...	...	**B12**^{2}, 136

For explanations, symbols and abbreviations see beginning of table.

No.	Name	Synonyms and Formula	Mol. wt.	Crystalline form, color and specific rotation	m.p. °C	b.p. °C	Density	n_D	Solubility w	al	eth	ace	bz	other solvents	Ref.
	Aniline														
—	—,*N*(2-hydroxy-ethyl)-*N*-methyl	*see* **Ethanol, 2[methyl-(phenyl)amino]-***													
—	—,iminodi-	*see* **Amine, diphenyl, diamino-**													
a1136	**—,2-iodo-**	C_6H_6IN. *See* a1002	219.04	nd	60–1				δ	v	v			os s	**B12**[2], 360
a1137	**—,3-iodo-**	C_6H_6IN. *See* a1002	219.04	pl or nd	33	145–6[15]			i	s				chl s	**B12**[2], 360
a1138	**—,4-iodo-**	C_6H_6IN. *See* a1002	219.04	nd (w)	62–3				δ	s	s			os s lig i	**B12**[2], 361
a1139	**—,*N*-isobutyl-**	$C_6H_5NHCH_2CH(CH_3)_2$	149.24			231–2[760] 109–10[13]	0.940^{18}_4		i		v		v		**B12**[2], 96
a1140	**—,*N*-isopropyl-**	$C_6H_5NHCH(CH_3)_2$	135.21			203–4[760]									**B12**[2], 95
a1141	**—,2-mercapto-**	C_6H_7NS. *See* a1002	125.20	nd	26	234									**B13**[2], 198
a1142	**—,3-mercapto-**	C_6H_7NS. *See* a1002	125.20			180–90[16]				v	s			aa s	**B13**[1], 140
a1143	—,—,hydrochloride	C_6H_7NS. HCl. *See* a1002	161.66		232	546			v	v	i				**B13**, 425
a1144	**—,4-mercapto-**	C_6H_7NS. *See* a1002	125.20		46	140–5[15]			s						**B13**[2], 296
a1145	**—,2-methoxy-**	*o*-Anisidine. C_7H_9NO. *See* a1002	123.15		6.22	224[760] 90[4]	1.0923^{20}_4	1.5793[20]	δ	s	s				**B13**[2], 165
a1146	**—,3-methoxy-**	*m*-Anisidine. C_7H_9NO. *See* a1002	123.15		<−12	251 cor	1.096^{20}_4		δ	s	s				**B13**[2], 211
a1147	**—,4-methoxy-**	*p*-Anisidine. C_7H_9NO. *See* a1002	123.15	ta (w), rh pl	57.2	243 cor 115[13]	1.0605^{67}_4	1.5559[67]	s	v	v				**B13**[2], 223
a1148	—,—,hydrochloride	$C_7H_9NO.HCl$. *See* a1002	159.62	lf, nd	215				s						**B13**[2], 223
a1149	**—,2-methoxy-4-nitro-**	$C_7H_8N_2O_3$. *See* a1002	168.16	pa ye nd	139–40		1.2112[156]								**B13**[2], 194
a1150	**—,2-methoxy-5-nitro-**	$C_7H_8N_2O_3$. *See* a1002	168.16	og-red nd (al, eth, w)	118		1.2068[156]		s^h	v	s^h	v	v	AcOEt v lig δ	**B13**[2], 192
a1151	**—,2-methoxy-6-nitro-**	$C_7H_8N_2O_3$. *See* a1002	168.16	ye, pa red nd (al)	75–6					s^h					**B13**[2], 191
a1152	**—,3-methoxy-2-nitro-**	$C_7H_8N_2O_3$. *See* a1002	168.16	ye nd	143										**B13**[1], 136
a1153	**—,3-methoxy-4-nitro-**	$C_7H_8N_2O_3$. *See* a1002	168.16	ye nd (al)	169 sub					s		s		aa s	**B13**[1], 136
a1154	**—,3-methoxy-5-nitro-**	$C_7H_8N_2O_3$. *See* a1002	168.16	og cr	120		1.2034[156]		v^h	s			s	lig i	**B13**[2], 216
a1155	**—,4-methoxy-2-nitro-**	$C_7H_8N_2O_3$. *See* a1002	168.16	dk red pr (w or al)	123				s	s	s		δ		**B13**[2], 286
a1156	**—,4-methoxy-3-nitro-**	$C_7H_8N_2O_3$. *See* a1002	168.16	red (eth), og pr or pl (eth-lig)	57–7.5				v^h	v	v	v	s	to δ	**B13**[2], 284
a1157	**—,*N*-methyl-**	Methyl phenylamine. $C_6H_5NHCH_3$	107.15		−57	196.25[760] 86[15]	0.98912^{20}_4	1.5702[21]	i	s	s			chl s	**B12**[2], 79
a1158	—,—,hydrochloride	$C_6H_5NHCH_3.HCl$	143.62	nd(chl-eth)		122.5–3 cor	1.0660^{131}_4		v	s	i		i	chl v	**B12**[2], 81
a1159	**—,*N*-methyl-2-nitro-**	$C_7H_8N_2O_2$. *See* a1002	152.15	red (og) nd (peth)	38	d			$δ^h$	s	s	s		con ac s lig δ	**B12**[2], 369
a1160	**—,*N*-methyl-3-nitro-**	$C_7H_8N_2O_2$. *See* a1002	152.15	red ye nd (al), (lig)	68				s^h	s	s		s		**B12**[2], 377
a1161	**—,*N*-methyl-4-nitro-**	$C_7H_8N_2O_2$. *See* a1002	152.15	ye pr (al), (eth)	151–2	d	1.201^{155}_4		i	s	δ		s	lig δ	**B12**[2], 385
a1162	**—,*N*-methyl-*N*-nitroso-**	Methylnitroso phenylamine. $C_6H_5N(NO)CH_3$	136.15	ye	14.7	225 d 136[13]	1.1288^{20}_4	1.5769[20]	i	s	s				**B12**[2], 309
a1163	**—,*N*-methyl-4-nitroso-**	$C_7H_8N_2O$. *See* a1002	136.15	bl n	118				δ	s	s		δ	chl s lig δ	**B7**[2], 575
a1164	**—,*N*-methyl-*N*,2,4,6-tetranitro**	Tetryl. [structure: O_2N–ring(2,6-NO_2)–N(CH_3)(NO_2)]	287.15	ye pr	129	exp 187	1.57[19]		i	δ	δ			chl δ	**B12**, 770
a1165	**—,*N*-3-methyl-butyl-**	*N*-Isoamylaniline. $C_6H_5NHCH_2CH_2CH(CH_3)_2$	163.26			254.5[760] 126–7[14]	0.8912^{55}_4		i	∞	∞				**B12**[2], 96
a1166	**—,4-methylthio-**	4-Aminothioanisole. C_7H_9NS. *See* a1002	139.23			111[2.5]	1.1379^{20}_4	1.6395[20]							**B13**[2], 297
a1167	**—,*N*-nitro-**	Diazobenzolic acid. Nitranilide. Phenyl-nitramine. $C_6H_5NHNO_2$	138.12	lf (peth)	46–7	exp on heahg			v	s				lig δ	**B16**[2], 343

For explanations, symbols and abbreviations see beginning of table.

No.	Name	Synonyms and Formula	Mol. wt.	Crystalline form, color and specific rotation	m.p. °C	b.p. °C	Density	n_D	Solubility w	al	eth	ace	bz	other solvents	Ref.
	Aniline														
a1168	—,**2-nitro-**	$C_6H_6N_2O_2$. *See* a1002	138.12	ye lf or nd (w)	71.5 cor	284 165–6^{28}	1.442^{15}		δ	v	v	...	v	chl v	**B12**^{2}, 367
a1169	—,**3-nitro-**	$C_6H_6N_2O_2$. *See* a1002	138.12	ye nd, rh bipym (w)	112.5	305–7d 100$^{0.16}$	1.1747^{160}_{4}		δ	s	s	...	s	MeOH v chl s	**B12**^{2}, 374
a1170	—,**4-nitro-**	$C_6H_6N_2O_2$. *See* a1002	138.12	ye mcl nd (w)	147.8 cor	331.73 106$^{0.03}$	1.437^{14}		i δ^h	s	s	...	δ	MeOH v to s	**B12**^{2}, 383
a1171	—,**5-nitro-2-propoxy-**	$C_9H_{12}N_2O_3$. *See* a1002	196.20	og (PrOH-peth)	47.5–8.5				i	s	...	...	...	lig i	
a1172	—,**4-nitroso-**	$C_6H_6N_2O$. *See* a1002	122.12	bl nd (bz)	173–4	d on heating			s	s	...	...	δ		**B7**^{2}, 575
a1173	—,**2,3,4,5,6-penta-bromo-**	$C_6H_2Br_5N$. *See* a1002	487.66	nd (to-al)	265–6					s					**B12**, 669
a1174	—,**2,3,4,5,6-penta-chloro-**	$C_6H_2Cl_5N$. *See* a1002	265.37	nd (al)	232					s	s	...	...	lig δ	**B12**^{2}, 341
a1175	—,**2-phenoxy-**	2-Aminodiphenyl ether. $C_{12}H_{11}NO$. *See* a1002	185.23	(lig)	44–5	307–8^{728} 172–3^{14}								os v	**B13**^{2}, 167
a1176	—,**3-phenoxy-**	3-Aminodiphenyl ether. $C_{12}H_{11}NO$. *See* a1002	185.23	pr (bz-lig)	37	315								os s lig δ	**B13**, 404
a1177	—,**4-phenoxy-**	4-Aminodiphenyl ether. $C_{12}H_{11}NO$. *See* a1002	185.23	nd (w), (dil l)	82 cor	315–20^{30} 187–9^{14}			s^h	v	v	...	...	lig δ	**B13**^{2}, 227
—	—,***N*-phenyl-**	*see* **Amine, diphenyl**													
a1178	—,***N*-propyl-**	Phenylpropylamine. $C_6H_5NHCH_2CH_2CH_3$	135.20			222 cor 100^{4}	0.9443^{20}_{4}	1.5428^{20}	i	v	s	...	...	lig s	**B12**^{2}, 94
a1179	—,**2,3,4,5-tetra-chloro-**	$C_6H_3Cl_4N$. *See* a1002	230.92	nd (al)	118					s	s	...	s	aa s	**B12**^{2}, 340
a1180	—,**2,3,5,6-tetra-chloro-**	$C_6H_3Cl_4N$. *See* a1002	230.92	nd (lig, al)	110				i	s	v	...	...	CS_2 v lig s	**B12**^{2}, 340
a1181	—,**2,4,5-tribromo-**	$C_6H_4Br_3N$. *See* a1002	329.85	nd (al)	85–6					v	v	...	v	dil ac s	**B12**, 662
a1182	—,**2,4,6-tribromo-**	$C_6H_4Br_3N$. *See* a1002	329.85	nd (al, bz), rh bipym	122	300	2.35^{20}_{20}		i	δ s^h	s	...	...	chl s	**B12**^{2}, 358
a1183	—,—,hydrobromide	$C_6H_4Br_3N.HBr$. *See* a1002	410.77	nd	195–6d	sub			d	i	i	...	i	lig i	**B12**^{1}, 330
a1184	—,**3,4,5-tribromo-**	$C_6H_4Br_3N$. *See* a1002	329.85	nd (al)	123				i	s	s	...	...		**B12**, 668
a1185	—,**2,3,4-trichloro-**	$C_6H_4Cl_3N$. *See* a1002	196.47	nd (lig)	73	292^{774}				v	...	...	...	lig s	**B12**, 626
a1186	—,**2,4,5-trichloro-**	$C_6H_4Cl_3N$. *See* a1002	196.47	nd (lig or 50% al)	96.5	*ca.* 270				v	...	...	...	CS_2 v lig δ	**B12**^{2}, 338
a1187	—,**2,4,6-trichloro-**	*sym*-Trichloroaniline. $C_6H_4Cl_3N$. *See* a1002	196.47	(al), nd (lig or peth)	78.5	262^{746}			i	v	v	...	...	CS_2 v lig v	**B12**^{2}, 339
a1188	—,**2,3,5-triiodo-**	$C_6H_4I_3N$. *See* a1002	470.82	nd	116					δ	i	...	δ	lig i aa δ	**B12**, 676
a1189	—,**2,3,6-triiodo-**	$C_6H_4I_3N$. *See* a1002	470.82	nd (al or al-eth)	116.8					s^h	i	...	...		**B12**, 676
a1190	—,**2,4,6-triiodo-**	$C_6H_4I_3N$. *See* a1002	470.82	ye nd or pl (al), pr (aa)	185.5					i δ^h	...	...	...	CS_2 v AcOEt v	**B12**^{2}, 364
a1191	—,**3,4,5-triiodo-**	$C_6H_4I_3N$. See a1002	470.82	nd (al-ace)	174.5					δ	δ	v	s	chl v	**B12**, 676
a1192	—,**2,4,6-trinitro-**	Picramide, TNA. $C_6H_4N_4O_6$. See a1002	228.12	dk ye ta (aa)	190–1	exp	1.762^{14}		i	δ	δ	s	s	chl δ aa s^h	**B12**^{2}, 421
—	**Anisaldehyde**	*see* **Benzaldehyde, 4-methoxy-**													
—	**Anisic acid**	*see* **Benzoic acid, 4-methoxy-**													
—	**Anisidine**	*see* **Aniline, methoxy-**													
—	**Anisoin**	*see* **Benzoin, 4,4′-dimethoxy-**													
—	**Anisole**	*see* **Benzene, methoxy-***													
—	**Anol**	*see* **Benzene, 1-hydroxy-4(1-propenyl)-***													
a1193	**Anthracene**		178.23	ta (al)	216.2–.4	340 cor 226.5^{53}	1.25^{27}_{4}		i	i	δ	...	s	to δ chl δ	**B5**^{2}, 569 **B7**^{2}, 941
a1194	—,**1-acetyl-**	$C_{16}H_{12}O$. *See* a1193	220.27	pa ye (al)	103–5					s^h	...	...	...		**B7**^{2}, 450
a1195	—,**2-acetyl-**	$C_{16}H_{12}O$. *See* a1193	220.27	ye (al)	183–5					s^h	...	...	...		**B7**^{2}, 450
a1196	—,**9-acetyl-**	$C_{16}H_{12}O$. *See* a1193	220.27	pa ye (al)	ca. 80					s^h	...	...	...	aa s	**B7**^{2}, 450
a1197	—,**1-amino-**	1-Anthrylamine. $C_{14}H_{11}N$. *See* a1193	193.24	gold-ye nd (al)	127					v	...	...	...	con HCl i	**B12**^{2}, 785
a1198	—,**2-amino-**	2-Anthrylamine. $C_{14}H_{11}N$. *See* a1193	193.24	ye lf (al)	236–7	sub			i	s	...	...	...	con sulf i os δ	**B12**^{2}, 786
a1199	—,**9-amino-**	9-Anthrylamine. $C_{14}H_{11}N$. *See* a1193	193.24	ye lf (dil al), br (bz)	145–50					s	s	...	s	chl s	**B7**^{2}, 416
a1200	—,**9-benzoyl-**	9-Anthrophenone. $C_{21}H_{14}O$. *See* a1193	282.34	ye nd (AcOEt)	148							s	...	CCl_4, CS_2, Ac_2O s	**E13**, 321

For explanations, symbols and abbreviations see beginning of table.

No.	Name	Synonyms and Formula	Mol. wt.	Crystalline form, color and specific rotation	m.p. °C	b.p. °C	Density	n_D	Solubility: w	al	eth	ace	bz	other solvents	Ref.
	Anthracene														
a1201	**—,1-chloro-**	$C_{14}H_9Cl$. *See* a1193	212.68	lf (aa)	83.5		1.1707^{100}_{4}	1.6959^{100}	i				s	CCl_4 s	**E13**, 235
a1202	**—,2-chloro-**	$C_{14}H_9Cl$. *See* a1193	212.68		223				i				s	CCl_4, CS_2 s	**E13**, 236
a1203	**—,9-chloro-**	$C_{14}H_9Cl$. *See* a1193	212.68	gold-ye nd (al)	106				i	s^h				con sulf s (gr)	**E13**, 236
a1204	**—,10-chloro-9,10-dihydro-9-nitro-**	$C_{14}H_{11}ClNO_2$. *See* a1193	259.70	nd (bz)	*ca.* 163					δ			s^h	chl δ	**B5**^{2}, 549
a1205	**—,1,4-diamino-**	$C_{14}H_{12}N_2$. *See* a1193	208.27	unst as free base											**B13**, 269
a1206	**—,2,6-diamino-**	$C_{14}H_{12}N_2$. *See* a1193	208.27												**B13**^{2}, 130
a1207	**—,9,10-diamino-**	$C_{14}H_{12}N_2$. *See* a1193	208.27	red cr	196				i						
a1208	**—,9,10-dibromo-**	$C_{14}H_8Br_2$. *See* a1193	336.04	ye nd (to or xyl)	226	sub			i	δ	δ		δ	chl s	**B5**^{2}, 577
a1209	**—,9,10-dichloro-**	$C_{14}H_8Cl_2$. *See* a1193	247.13	ye nd (MeCOEt or CCl_4)	210					δ	δ		s		**B5**^{2}, 575
a1210	**—,9,10-dihydro-**	$C_{14}H_{12}$. *See* a1193	180.25	ta	108	sub 313	0.8976^{10}_{4}		i	s	s		s		**B5**^{2}, 545
—	**—,9,10-dihydro-9,10-dioxo-**	*see* **Anthraquinone**													
a1211	**—,9,10-dihydro-9-ethyl-**	$C_{16}H_{16}$. *See* a1193	208.30			320–3 δd (cor)	1.049^{18}_{18}		i	∞	∞		∞	aa s	**B5**, 649
a1212	**—,9,10-dihydro-9-hydroxy-**	Hydroanthrol. $C_{14}H_{12}O$. *See* a1193	196.24	nd (peth)	76				s^h	s	s		s	CS_2 s lig s	**B6**^{2}, 660
a1213	**—,9,10-dihydro-10-nitro-9-oxo-**	10-Nitroanthrone. $C_{14}H_9NO_3$	239.23	(bz-lig), nd (CS_2)	148d					s			v^h	alk s CS_2 s^h	**B7**^{2}, 240
a1214	**—,9,10-dihydro-9-oxo-**	Anthrone. $C_{14}H_{10}O$. *See* a1193	194.22	nd (bz-lig)	156								s^h	con sulf s dil alk s^h	**B7**^{2}, 414
a1215	**—,1,2-dihydroxy-**	1,2-Anthradiol. $C_{14}H_{10}O_2$. *See* a1193	210.22	v pa gr lf	160–2					v	v			alk s (dk) aa v	**B6**^{2}, 998
a1216	**—,1,5-dihydroxy-**	1,5-Anthradiol. Rufol. $C_{14}H_{10}O_2$. *See* a1193	210.22	ye nd	265d				v	s				alk s	**B6**, 1032
a1217	**—,1,8-dihydroxy-**	1,8-Anthradiol. Chrysazol. $C_{14}H_{10}O_2$. *See* a1193	210.22	ye nd (dil al), lf (al-aa)	225d				i	s	s		s	alk s AcOEt s	**B6**, 1033
a1218	**—,2,6-dihydroxy-**	2,6-Anthradiol. Flavol. $C_{14}H_{10}O_2$. *See* a1193	210.22	wh-lt gr lf	295–300 (dk at 270)				i	v	v			aa s	**B6**^{2}, 999
a1219	**—,9,10-dihydroxy-**	9,10-Anthradiol. Oxanthranol. $C_{14}H_{10}O_2$. *See* a1193	210.22	br nd	180						s		δ	chl δ	**B8**^{2}, 214
a1220	**—,1,3-dimethyl-**	$C_{16}H_{14}$. *See* a1193	206.27	pa bl flr lf (eth)	202–3				i	v	v				**B5**^{2}, 592
a1221	**—,2,3-dimethyl-**	$C_{16}H_{14}$. *See* a1193	206.27	bl gr flr lf (bz)	25.2				i	s			v		**B5**^{2}, 592
a1222	**—,9-ethyl-**	$C_{16}H_{14}$. *See* a1193	206.27	bl flr lf (al)	59		1.0413^{99}_{4}	1.6762^{99}	i	s	s				**B5**^{2}, 591
a1223	**—,1,2,3,4,5,6-hexahydro-**	$C_{14}H_{16}$. *See* a1193	184.27	lf (MeOH)	70					δ			v		**B5**^{2}, 472
a1224	**—,1-hydroxy-**	1-Anthrol. $C_{14}H_{10}O$. *See* a1193	194.22	(bz), nd or lf (al or aa)	158	224^{13}			i	v	v			NaOH s os s	**B6**^{2}, 669
a1255	**—,2-hydroxy-**	2-Anthrol. $C_{14}H_{10}O$. *See* a1193	194.22	ye (bz), nd, lf (al)	255				i	v	v	s		KOH s	**B6**^{2}, 669
a1226	**—,9-hydroxy-**	Anthranol. $C_{14}H_{10}O$. *See* a1193	194.22	ye red lf (dil al), nd (aa)	152				i	s			s^h	aq alk s os v	**B7**^{2}, 414
a1227	**—,1-hydroxy-9,10-dihydro-**	$C_{14}H_{12}O$. *See* a1193	196.24	sl grsh flr lf or nd (bz, peth)	94					s	s			aa s	**B6**^{2}, 660
a1228	**—,2-hydroxy-9,10-dihydro-**	$C_{14}H_{12}O$. *See* a1193	196.24	(bz-peth), lf (dil al)	129				δ					os v	**B6**^{2}, 660
a1229	**—,1-methyl-**	$C_{15}H_{12}$. *See* a1193	192.25	bl nd (MeOH), lf (al)	85–6	199–200	1.0471^{99}_{4}	1.6802^{99}	i	s	s		s	al (bl flr) chl, sulf s	**B5**^{2}, 585
a1230	**—,2-methyl-**	$C_{15}H_{12}$. *See* a1193	192.25	gr-bl flr lf (sub)	207	sub	1.81^{0}_{4}		i	δ	δ	δ	s	MeOH δ chl, CS_2 s aa δ	**B5**^{2}, 586
a1231	**—,9-methyl-**	$C_{15}H_{12}$. *See* a1193	192.26	yesh nd (aq al), pr (bz, al, MeOH)	81.5	$196–7^{12}$	1.065^{99}_{4}	1.6959^{99}						os v	**B5**^{2}, 586
a1232	**—,9-nitro-**	$C_{14}H_9NO_2$. *See* a1193	223.23	ye nd (al), pr (aa or xyl)	146	>360 ca. 275^{17}			i	δ			v	CS_2 v aa s	**B6**^{2}, 1245
a1233	**—,1,2,3,4,5,6,7,8-octahydro-**	Octhracene. $C_{14}H_{18}$. *See* a1193	186.30	pl (al)	73–4	$293–5^{760}$ 167^{12}	1.131^{0}_{4}	1.5363^{89}	i	v^h			v	aa s v^h	**B5**^{2}, 422
a1234	**—,9-phenyl-**	$C_{20}H_{14}$. *See* a1193	254.33	bl flr in sol, lf (al), (aa or al)	155–7	417			i	s^h	s^h		s^h	CS_2, chl s^h	**B5**^{2}, 639
a1235	**—,1,2,9-trihydroxy-**	$C_{14}H_{10}O_3$. *See* a1193	226.23	og ye lf	149–51									con sulf, alk s	**B8**^{2}, 371
a1236	**—,1,2,10-trihydroxy-**	$C_{14}H_{10}O_3$. *See* a1193	226.23	ye lf, nd (al-w)	208				δ	v	v	v	s	aa v	**B8**^{2}, 372

For explanations, symbols and abbreviations see beginning of table.

No.	Name	Synonyms and Formula	Mol. wt.	Crystalline form, color and specific rotation	m.p. °C	b.p. °C	Density	n_D	Solubility w	al	eth	ace	bz	other solvents	Ref.
	Anthracene														
a1237	**—,1,4,9-trihydroxy-**	$C_{14}H_{10}O_3$. *See* a1193	226.23	og-red nd (al)	156	...	...	...	...	s^h	...	...	...	...	**B8**, 330
a1238	**—,1,5,9-trihydroxy-**	$C_{14}H_{10}O_3$. *See* a1193	226.23	gold lf (al)	200d without m	...	...	...	...	s^h	...	...	...	...	**B8²**, 371
a1239	**—,1,8,9-trihydroxy-**	Anthralin. $C_{14}H_{10}O_3$. *See* a1193	226.23	yel lf or nd (lig)	176–7	...	...	...	i	δ	δ	s	s	chl, Py v dil NaOH s(ye)	**E13**, 372
a1240	**—,1,9,10-trihydroxy-(enol form)**	$C_{14}H_{10}O_3$. *See* a1193	226.23	gr nd (eth)	204–6	...	...	...	...	s	s	...	...	...	**B8²**, 372
a1241	**—,1,9,10-trihydroxy-(keto form)**	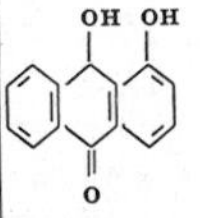 OH OH ... O	226.23	ye nd (lig)	135–7	...	...	...	...	s	...	...	...	os δ	**B8²**, 372
a1242	**—,2,3,9-trihydroxy-**	$C_{14}H_{10}O_3$. *See* a1193	226.23	ye br nd (al)	288–9	...	...	...	...	s	s	s	...	aa s	**B8²**, 372
a1243	**9-Anthracenecarboxaldehyde**	9-Anthraldehyde. 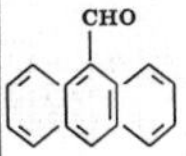 CHO	206.25	og nd (dil aa)	104–5	...	...	...	i	...	...	...	s	...	**E13**, 320
a1244	**1-Anthracenecarboxylic acid**	1-Anthroic acid. 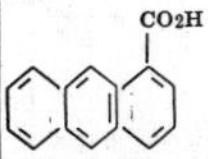 CO_2H	222.25	ye nd (gl aa), pr(al or AcOEt bt ye nd (sub)	246	sub	...	...	i	s	s	...	δ	chl, δ aa s	**B9²**, 493
a1245	**2-Anthracenecarboxylic acid**	2-Anthroic acid. 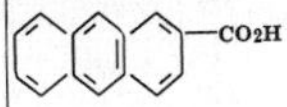 —CO_2H	222.25	ye lf (al)	280	sub	...	...	i	s^h	δ	...	δ	chl, δ	**B9²**, 494
a1246	**9-Anthracenecarboxylic acid**	9-Anthroic acid. 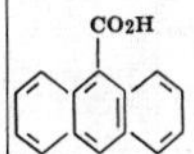 CO_2H	222.23	ye nd (bz, al)	217d	sub	...	...	i $δ^h$	s	...	...	...	...	**B9²**, 494
—	**Anthrachrysone**	*see* **9,10-Anthraquinone, 1,3,5,7-tetrahydroxy-***													
—	**Anthragallol**	*see* **9,10-Anthraquinone, 1,2,3-trihydroxy***													
—	**Anthraflavin**	*see* **9,10-Anthraquinone, 2,6-dihydroxy-***													
—	**Anthralin**	*see* **Anthracene, 1,8,9-trihydroxy-***													
a1247	**Anthranil**	3,4-Benzoisoxazol. 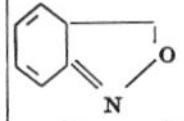 O N	119.12	...	<−18	215^{760} 99^{13}	1.8127^{20}_{4}	1.5845^{20}	$δ^h$	s	...	...	...	ac s os s	**B27²**, 17
—	**Anthranilaldehyde**	*see* **Benzaldehyde, 2-amino-**													
—	**Anthranilic acid**	*see* **Benzoic acid, 2-amino-**													
—	**Anthranol**	*see* **Anthracene, hydroxy-***													
—	**Anthrapurpurin**	*see* **9,10-Anthraquinone, 1,2,7-trihydroxy-***													
a1248	**9,10-Anthraquinone***	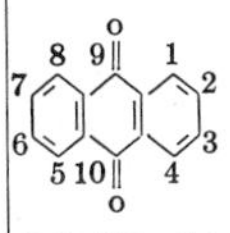 O 8 9 1 7 2 6 3 5 10 4 O	208.20	ye rh nd (al, bz)	286 (sub)	379.8^{760}	1.438^4	...	i	δ	i	...	$δ^h$	chl δ to $δ^h$	**B7²**, 709
a1249	**—,1-amino-***	$C_{14}H_9NO_2$. *See* a1248	223.22	red nd (al), (gl aa)	252–3	sub	...	...	i	δ	s	s	s	con HCl s chl s	**B22²**, 618
a1250	**—,2-amino-***	$C_{14}H_9NO_2$. *See* a1248	223.22	red nd (al, aa, sub)	302	sub	...	...	i	s	i	s	s	chl s	**B14²**, 107
a1251	**—,1-amino-2-benzoyl-**	$C_{21}H_{13}NO_3$. *See* a1248	327.34	red nd (aa)	190	...	...	...	...	...	...	...	...	aa s^h	**B14¹**, 482
a1252	**—,3-amino-2-benzoyl-**	$C_{21}H_{13}NO_3$. *See* a1248	327.34	ye pl (Py)	331	...	...	...	...	δ	...	...	δ	Py v aa δ	**B14¹**, 482
a1253	**—,1-amino-2-bromo-***	$C_{14}H_8BrNO_2$. *See* a1248	302.14	ye-red nd (aa), nd (xyl)	182	...	...	...	i	δ	δ	δ	δ	Py v lig δ	**B14¹**, 446

For explanations, symbols and abbreviations see beginning of table.

No.	Name	Synonyms and Formula	Mol. wt.	Crystalline form, color and specific rotation	m.p. °C	b.p. °C	Density	n_D	Solubility w	al	eth	ace	bz	other solvents	Ref.
	Anthraquinone														
a1254	**—,1-amino-3-bromo-***	$C_{14}H_8BrNO_2$. *See* a1248	302.14	red nd (to)	243					i	i	i	δ	to δ, Py, $PhNO_2$ v	**B14**[1], 446
a1255	**—,3-amino-1,2-dihydroxy-***	3-Aminoalizarin. $C_{14}H_9NO_4$. *See* a1248	255.23	deep red pr (aa)	>300	sub δd				δ				aq HCl δ aq NH_3 s	**E13**, 565
a1256	**—,4-amino-1,2-dihydroxy-***	4-Aminoalizarin. $C_{14}H_9NO_4$. *See* a1248	255.23	bk nd (al)	d					s^h				$PhNO_2$ s^h alk s (crimson)	**E13**, 565
a1257	**—,2-amino-1 hydroxy-***	β-Alizarinamide. $C_{14}H_9NO_3$. *See* a1248	239.22	red br nd (al)	226–7	sub			i		s		s	con sulf s Py v, aq NH_3 δ	**B14**[2], 167
a1258	**—,1-bromo-***	$C_{14}H_7BrO_2$. *See* a1248	287.11	ye nd (bz)	188	sub				s			s	$PhNO_2$ s con sulf s	**E11**, 413
a1259	**—,2-bromo-***	$C_{14}H_7BrO_2$. *See* a1248	287.11	ye nd (to), (gl aa)	204–5	sub				δ			s^h	to s AmOH s	**B7**[2], 717
a1260	**—,3-bromo-1,2-dihydroxy-***	3-Bromoalizarin. $C_{14}H_7BrO_4$. *See* s1248	319.13	br-red nd (to)	260–1				s					Py s dil alk s (bl-vt) con alk s (red) con sulf s (red)	**E13**, 555
a1261	**—,1-bromo-4-methylamino-**	$C_{15}H_{10}BrNO_2$. *See* a1248	316.17	br-red nd (Py)	194									Py, ac s	**B14**[1], 447
a1262	**—,2-bromo-1-methylamino*-**	$C_{15}H_{10}BrNO_2$. *See* a1248	316.17	br nd (aa)	170–2					δ				Py v^h aa v^h	**B14**[1], 446
a1263	**—,1-chloro-***	$C_{14}H_7ClO_2$. *See* a1248	242.65	ye nd (to or al)	162	sub			i	δ	∞		s^h	AmOH s^h $PhNO_2$ s	**B7**[2], 714
a1264	**—,2-chloro-***	$C_{14}H_7ClO_2$. *See* a1248	242.65	pa ye nd (aa or al)	212.4	sub			*i*	δ	i		δ v^h	to v^h $PhNO_2$ s	**B7**[2], 714
a1265	**—,1,2-diamino-***	$C_{14}H_{10}N_2O_2$. *See* a1248	238.24	vt nd ($PhNO_2$)	303–4					δ	δ			con sulf s con HCl i dil HCl δ Py s, chl δ	**B14**[2], 112
a1266	**—,1,3-diamino-***	$C_{14}H_{10}N_2O_2$. *See* a1248	238.24	red ($PhNO_2$)	290					δ	i			$PhNO_2$ δ, s^h con sulf s Py v	**B14**[2], 112
a1267	**—,1,4-diamino-***	$C_{14}H_{10}N_2O_2$. *See* a1248	238.24	dk vt nd (Py), (al)	268				δ v^h	δ			v	$PhNO_2$, Py v	**B14**[2], 113
a1268	**—,1.5-diamino-***	$C_{14}H_{10}N_2O_2$. *See* a1248	238.24	deep red nd (al)	319 cor	sub			i	δ	δ	δ	δ	con sulf s $PhNO_2$ s^h chl δ	**B14**[2], 116
a1269	**—,1,6-diamino-***	$C_{14}H_{10}N_2O_2$. *See* a1248	238.24	red nd (aa), lf (MeOPh)	292									$PhNO_2$ s^h aa s^h	**B14**[2], 110
a1270	**—,1,7-diamino-***	$C_{14}H_{10}N_2O_2$. *See* a1248	238.24	red nd ($PhNO_2$)	290									$PhNO_2$ s^h os δ	**B14**[1], 470
a1271	**—,1,8-diamino-***	$C_{14}H_{10}N_2O_2$. *See* a1248	238.24	red (al or aa)	262				i	δ	δ			Py, $PhNO_2$ s aa δ	**B14**[2], 119
a1272	**—,2,3-diamino-***	$C_{14}H_{10}N_2O_2$. *See* a1248	238.24	red ($PhNO_2$)	353									sulf, $PhNO_2$, Py s, chl, xyl δ	**B14**[2], 120
a1273	**—,2,6-diamino-***	$C_{14}H_{10}N_2O_2$. *See* a1248	238.24	red-br pr (Py)	310–20d				$δ^h$					con sulf s, Py s^h chl δ	**B14**[2], 120
a1274	**—,2,7-diamino-***	$C_{14}H_{10}N_2O_2$. *See* a1248	238.24	og-ye nd (al or $PhNO_2$), dk red nd (sub)	>330	sub			i	δ	δ			dil sulf, con ac s	**B14**[1], 473
a1275	**—,2,3-dibromo-***	$C_{14}H_6Br_2O_2$. *See* a1248	366.02	ye nd (to)	283	sub				i			s	chl, con sulf s	
a1276	**—,2,7-dibromo-***	$C_{14}H_6Br_2O_2$. *See* a1248	366.02	lt ye lf (anisole), nd, pl	236.5	sub				$δ^h$			s	con sulf s aa s^h	**B7**[2], 718
a1277	**—,1,3-dichloro-***	$C_{14}H_6Cl_2O_2$. *See* a1248	277.12	ye nd	208					$δ^h$	$δ^h$	$δ^h$		$PhNO_2$ s	**B7**[1], 411
a1278	**—,1,4-dichloro-***	$C_{14}H_6Cl_2O_2$. *See* a1248	277.12	ye nd	187.5–8					δ	δ		δ	$PhNO_2$, Py v	**B7**[2], 715
a1279	**—,1,5-dichloro-***	$C_{14}H_6Cl_2O_2$. *See* a1248	277.12	yesh (to), nd (aa)	252					δ		δ	δ	$PhNO_2$, con sulf s	**B7**[2], 715
a1280	**—,1,6-dichloro-***	$C_{14}H_6Cl_2O_2$. *See* a1248	277.12	ye nd (anisole-aa)	202–4					δ	δ	v		$PhNO_2$, to, anisole v	**B7**[2], 715
a1281	**—,1,7-dichloro-***	$C_{14}H_6Cl_2O_2$. *See* a1248	277.12	ye nd	213–4									$PhNO_2$ s^h	**B7**[2], 716
a1282	**—,1,8-dichloro-***	$C_{14}H_6Cl_2O_2$. *See* a1248	277.12	ye nd	202–3					δ				to v^h	**B7**[2], 716
a1283	**—,2,3-dichloro-***	$C_{14}H_6Cl_2O_2$. *See* a1248	277.12	ye nd (aa)	267–8					δ			v^h	aa $δ^h$	**B7**[2], 716

For explanations, symbols and abbreviations see beginning of table.

No.	Name	Synonyms and Formula	Mol. wt.	Crystalline form, color and specific rotation	m.p. °C	b.p. °C	Density	n_D	Solubility w	al	eth	ace	bz	other solvents	Ref.
	Anthraquinone														
a1284	**—,2,6-dichloro-***	$C_{14}H_6Cl_2O_2$. *See* a1248	277.12	ye nd (aa or al)	287	...	...	...	...	s^h	...	...	v^h	aa s^h	**B7**^{2}, 716
a1285	**—,2,7-dichloro-***	$C_{14}H_6Cl_2O_2$. *See* a1248	277.12	yesh nd (anisole)	212	...	...	...	...	...	...	...	...	anisole s^h	**B7**^{2}, 716
a1286	**—,1,2-dihydroxy-***	Alizarin. $C_{14}H_8O_4$. *See* a1248	240.20	og or red tcl or rh(al)	289–90 cor	430	...	...	δ	v	s	...	s	MeOH s^h CS_2 s	**B8**^{2}, 487
a1287	**—,1,3-dihydroxy-***	Purpuroxanthin. Xanthopurpurin. $C_{14}H_8O_4$. *See* a1248	240.20	ye-red nd (sub) pr(aa+2)	268–70	...	...	...	i	s	...	v	s	alk s(red) $PhNO_2$ s aa s^h	**E13**, 526
a1288	**—,1,4-dihydroxy-***	Quinizarin. $C_{14}H_8O_4$. *See* a1248	240.20	ye red lf(eth), deep red nd (al)	194–5	...	...	...	s^h	s^h	s^h	...	s^h	KOH, sulf s	**B9**^{2}, 889
a1289	**—,1,5-dihydroxy-***	Anthrarufin. $C_{14}H_8O_4$. *See* a1248	240.20	lf ye lf (gl aa, sub)	280	sub	...	...	i	δ	δ	δ	s	con sulf s CS_2 δ	**B8**^{2}, 496
a1290	**—,1,6-dihydroxy-***	$C_{14}H_8O_4$. *See* a1248	240.20	ye nd (gl aa)	269	...	...	...	...	δ	...	s	$δ^h$	dil alk s $PhNO_2$ s	**B8**^{1}, 721
a1291	**—,1,7-dihydroxy-***	$C_{14}H_8O_4$. *See* a1248	240.22	ye nd (sub)	291	sub	...	...	...	δ	s	...	s	chl s CS_2 aa s	**B8**^{1}, 721
a1292	**—,1,8-dihydroxy-***	Chrysazin. Istizin. $C_{14}H_8O_4$. *See* a1248	240.22	red or redsh-ye nd or lf(al)	192	sub	...	...	i	s	s	...	...	alk s chl s, $PhNO_2$ δ	**B8**^{2}, 500
a1293	**—,2,3-dihydroxy-***	Hystazarin. Hystazin $C_{14}H_8O_4$. *See* a1248	240.22	ye-br nd (aa) ye nd (sub)	>330	sub	...	...	...	δ	δ	δ	i	sulf, NH_3, NaOH s	**B3**^{2}, 504
a1294	**—,2,6-dihydroxy-***	Anthraflavin. Anthraflavic acid. $C_{14}H_8O_4$. *See* a1248	240.22	ye nd (al)	360d	...	...	...	δ	δ	i	...	i	con sulf, alk s chl i	**B3**^{2}, 504 **B9**^{2}, 890
a1295	**—,2,7-dihydroxy-***	Isoanthraflavin. Isoanthraflavic acid. $C_{14}H_8O_4$. *See* a1248	240.22	ye nd (dil al, sub)	>330	sub	...	...	...	s	i	...	δ	chl δ	**B8**^{2}, 505
a1296	**—,1,8-dihydroxy-3-hydroxymethyl-***	Aloeemodin. Rhabarberone. $C_{15}H_{10}O_5$. *See* a1248	270.23	og-ye nd (to)	223–4	sub	...	...	...	v^h	v	...	v	NH_4OH, sulf, Na_2CO_3 s (red)	**E13**, 571
—	**—,1,8-dihydroxy-3-hydroxymethyl-2,4,5,7-tetranitro-***	*see* Aloetic acid													
a1296^{1}	**—,1,2-dihydroxy-3-iodo-***	β-Iodoalizarin. $C_{14}H_7IO_4$. *See* a1248	366.12	og-red nd (xyl)	227–9	...	...	...	s	...	...	...	...	sulf s(og-red)	**E13**, 555
a1297	**—,1,2-dihydroxy-3-methyl-***	β-Methyl alizarin. $C_{15}H_{10}O_4$. *See* a1248	254.23	og nd	229	sub	...	...	...	s	s	s	...	...	...
a1298	**—,1,8-dihydroxy-3-methyl-***	Chrysophanic acid. Chrysophanol. $C_{15}H_{10}O_4$. *See* a1248	254.23	ye hex or mcl nd (sub)	196	sub	0.92	...	δ	δ	s	s	s^h	chl s aa s	**B8**^{2}, 510
a1299	**—,1,2-dihydroxy-3-nitro-***	Alizarin orange. β-Nitroalizarin. $C_{14}H_7NO_6$. *See* a1248	285.20	og ye nd (bz), ye lf (gl aa, al-chl)	244d	sub d	...	...	δ	...	...	...	s	sulf, chl s aa s	**B8**^{2}, 491
a1300	**—,1,2-dihydroxy-4-nitro***	α-Nitroalizarin. $C_{14}H_7NO_6$. *See* a1248	285.20	gold ye nd (aa or al)	289d	sub d	...	...	δ	s	...	...	s	sulf, chl s	**B8**^{2}, 491
a1301	**—,1,8-dihydroxy-2,4,5,7-tetrabromo-***	2,4,5,7-Tetrabromochrysazin. $C_{14}H_4Br_4O_4$. *See* a1248	555.85	og-ye nd (bz)	312	...	...	...	...	...	...	...	δ	dil KOH s aa δ	**B8**^{1}, 722
a1302	**—,1,4-dihydroxy-5,6,7,8-tetrachloro-***	5,6,7,8-Tetrachloroquinizarine. $C_{14}H_4Cl_4O_4$. *See* a1248	378.00	red pl (aa)	270	...	...	...	...	δ	δ	...	s	NaOH s aa s^h	**B8**^{1}, 716
a1303	**—,1,8-dihydroxy-2,4,5,7-tetranitro-***	Chrysammic acid. Chrysamminic acid. $C_{14}H_4N_4O_{12}$. *See* a1248	420.20	ye pl or lf	exp	d	...	...	δ	s	s	...	...	ac s	**B8**^{1}, 723
a1304	**—,1,2-dimethyl-***	$C_{16}H_{12}O_2$. *See* a1248	236.27	nd (ace or aa)	150	...	...	...	...	s	s	s^h	s	aa s^h	**B7**^{2}, 743
a1305	**—,1,3-dimethyl-***	$C_{16}H_{12}O_2$. *See* a1248	236.27	nd	162	...	...	...	...	δ	...	...	δ	...	**B7**^{2}, 743
a1306	**—,1,4-dimethyl-***	$C_{16}H_{12}O_2$. *See* a1248	236.27	ye nd (al, sub)	140–1	sub	...	...	...	δ s^h	...	...	s	xyl s aa v	**B7**^{2}, 743
a1307	**—,2,3-dimethyl-***	$C_{16}H_{12}O_2$. *See* a1248	236.27	ye nd (al or xyl) (aa)	208	sub	...	...	...	s^h	...	...	...	xyl s^h aa s^h	**B7**^{2}, 744
a1308	**—,2,6-dimethyl-***	$C_{16}H_{12}O_2$. *See* a1248	236.27	ye nd (aa or al)	242	sub	...	...	...	$δ^h$	...	...	...	to, $PhNO_2$ s aa s^h	**B7**^{2}, 744
a1309	**—,2,7-dimethyl-***	$C_{16}H_{12}O_2$. *See* a1248	236.27	yesh nd (al)	170	...	...	...	...	s^h	...	...	...	...	**B7**^{2}, 744
a1310	**—,1,3-dinitro-***	$C_{14}H_6N_2O_6$. *See* a1248	298.20	ye nd (HNO_3)	246–50	...	...	...	...	...	...	...	...	HNO_3 s	**E11**, 436
a1311	**—,1,5-dinitro-***	$C_{14}H_6N_2O_6$. *See* a1248	298.20	ye nd (xyl or $PhNO_2$), ye (sub)	422	sub	...	...	i	i	i	...	i	con sulf δ $PhNO_2$ v aa δ	**B7**^{2}, 721
a1312	**—,2-ethyl-1-nitro-***	$C_{16}H_{11}NO_4$. *See* a1248	281.27	yesh br (aa)	226	...	...	...	...	...	...	...	...	aa δ	**B7**^{2}, 743
a1313	**—,1,2,3,5,6,7-hexahydroxy-**	Rufigallic acid. $C_{14}H_8O_8$. *See* a1248	304.22	red rh, red-ye nd (sub)	...	sub	...	...	i	δ	δ	s	...	...	**E13**, 613

For explanations, symbols and abbreviations see beginning of table.

No.	Name	Synonyms and Formula	Mol. wt.	Crystalline form, color and specific rotation	m.p. °C	b.p. °C	Density	n_D	Solubility w	al	eth	ace	bz	other solvents	Ref.
	Anthraquinone														
a1314	**—,1-hydroxy-***	Erythrohydroxyanthraquinone. $C_{14}H_8O_3$. *See* a1248	224.20	red-og nd (al, sub)	193	sub			i	s	v	...	v	liq NH_3 δ	**B8²**, 388
a1315	**—,2-hydroxy-***	$C_{14}H_8O_3$. *See* a1248	224.20	ye lf or nd (al or aa), ye lf (sub)	306	sub			i	s	s	...	...	aq NH_3, KOH s	**B8²**, 393
a1316	**—,2-methyl-***	$C_{15}H_{10}O_2$. *See* a1248	222.23	yesh nd (al, aa or sub)	182–3	sub			...	v	δ	...	s	con sulf v aa s	**B7²**, 133 **B24²**, 713
a1317	**—,2-methyl-1-nitro-***	$C_{15}H_9NO_4$. *See* a1248	267.25	(aa)	270–1				...	i^h	$δ^h$	...	$δ^h$	aa $δ^h$	**E13**, 433
a1318	**—,6-methyl-1,2,5-trihydroxy-***	Morindone. $C_{15}H_{10}O_5$. *See* a1248	270.25	og-red nd (to)	281–2				i	v	v	...	v	sulf s(bl-vt) Py v	**E13**, 588
a1319	**—,6-methyl-1,3,8-trihydroxy-***	Emodin. $C_{15}H_{10}O_5$. *See* a1248	270.24	og-red mcl nd (aa), (dil aa+1w)	256–7	sub			i	s	δ	...	i	alk s, chl, CS_2, i aa δ	**B8²**, 563
a1320	**—,1-methylamino-***	$C_{15}H_{11}NO_2$. *See* a1248	237.26	ye-red nd	167				...	...	...	...	...	chl s aa s	**B14²**, 100
a1321	**—,2-methylamino-***	$C_{15}H_{11}NO_2$. *See* a1248	237.26	red nd (aa)	226–7				...	s	δ	...	...	chl s, to v aa v	**B14²**, 108
a1322	**—,1-nitro-***	$C_{14}H_7NO_4$. *See* a1248	253.21	yesh pr (ace), nd (aa)	232.5–3.5	$270-1^7$			i	δ	i	...	s	chl s aa s	**B7²**, 719
a1323	**—,2-nitro-***	$C_{14}H_7NO_4$. *See* a1248	253.21	ye nd (aa or al),	184.5–5	sub			i	δ	δ	s	...	sulf s	**B7²**, 720
a1324	**—,1,2,4,5,8-pentahydroxy-***	Alizarin cyanin R. Alizarin pentacyanin. $C_{14}H_8O_7$. *See* a1248	288.20	br lf ($PhNO_2$)	d	sub			...	...	...	...	...	os, alk, sulf s (bl with red flr)	**E13**, 613
a1325	**—,1,2,4,6-tetrahydroxy-***	Hydroxyflavopurpurin. $C_{14}H_8O_6$. *See* a1248	272.22	dk red nd (sub)					...	v	...	...	...	Py v aa δ	**B8²**, 582
a1326	**—,1,2,4,7-tetrahydroxy-***	4-Hydroxyanthrapurpurin. $C_{14}H_8O_6$. *See* a1248	272.22	red-ye (al, Py or aa)					...	...	...	...	...	con sulf s NH_3 s	**B8²**, 582
a1327	**—,1,2,5,6-tetrahydroxy-***	Rufiopin. $C_{14}H_8O_6$. *See* a1248	272.22	ye-red nd	sub	d			$δ^h$	s	δ	...	δ	chl δ	**B8²**, 583
a1328	**—,1,2,5,8-tetrahydroxy-***	Quinalizarin. $C_{14}H_8O_6$. *See* a1248	272.22	red nd ($PhNO_2$)	>275	sub			i	δ	δ	...	...	alk os δ aa δ	**B8²**, 584
a1329	**—,1,2,6,7-tetrahydroxy-***	$C_{14}H_8O_6$. *See* a1248	272.22	og nd ($PhNO_2$)	>330				...	$δ^h$	...	...	$δ^h$	xyl $δ^h$, $PhNO_2$ δ	**B8²**, 584
a1330	**—,1,2,7,8-tetrahydroxy-***	$C_{14}H_8O_6$. *See* a1248	272.22	red nd (aa)	318				...	δ	...	...	...	sulf, Py, NaOH s, $PhNO_2$ δ	**B8²**, 585
a1331	**—,1,3,5,7-tetrahydroxy-***	Anthrachrysone. $C_{14}H_8O_6$. *See* a1248	272.22	yesh nd (al+2w)	150–60d (+2w) >360 anh	sub			i	δ	δ	s	s	CS_2 i, chl s lig s	**B8²**, 585
a1332	**—,1,4,5,8-tetrahydroxy-***	$C_{14}H_8O_6$. *See* a1248	272.22	gr nd (aa), br nd (bz-lig)	>300	sub			i	...	...	...	...	alk s, Ac_2O δ	**B8²**, 586
a1333	**—,1,2,3-trihydroxy-***	Anthragallol. $C_{14}H_8O_5$. *See* a1248	256.22	ye nd (dil al), br (aa), og nd (sub)	313	sub			δ	s	s	...	...	sulf, alk δ, CS_2 s aa s	**B8²**, 549
a1334	**—,1,2,4-trihydroxy-***	Purpurin. $C_{14}H_8O_5$. *See* a1248	256.22	og-red, dk red or og-ye nd, (al)	259	sub			$δ^h$	v^h	s	...	v^h	aa v^h	**B8**, 509
a1335	**—,1,2,5-trihydroxy-***	2-Hydroxyanthrarufin. $C_{14}H_8O_5$. *See* a1248	256.22	red nd (gl aa, sub)	278	sub			i	...	s	...	...	...	**B8²**, 554
a1336	**—,1,2,6-trihydroxy-***	Flavopurpurin. $C_{14}H_8O_5$. *See* a1248	256.22	ye nd (al)	>330 (sub 160)	459 par d			$δ^h$	s	δ	...	s	...	**B8²**, 555
a1337	**—,1,2,7-trihydroxy-***	Anthrapurpurin. $C_{14}H_8O_5$. *See* a1248	256.22	ye nd (al)	374 (sub 170)	462^{760} par d			$δ^h$	s^h	δ	...	i	chl s aa s^h	**B8²**, 555
a1338	**—,1,2,8-trihydroxy-***	2-Hydroxychrysazin. $C_{14}H_8O_5$. *See* a1248	256.22	og nd (gl aa), red nd (sub)	235–6	sub			i	δ	...	...	...	...	**B8²**, 557
a1339	**—,1,3,8-trihydroxy-***	$C_{14}H_8O_5$. *See* a1248	256.22	bt red-br nd, (bz)	278				...	...	...	...	...	con alk s	**B8²**, 557
a1340	**—,1,4,5-trihydroxy-***	5-Hydroxyquinizarin. $C_{14}H_8O_5$. *See* a1248	256.22	red-br nd or lf ($PhNO_2$), deep red nd, (Py)	271				...	δ	...	...	...	aa s^h	**B8²**, 557
a1341	**—,1,4,6-trihydroxy-***	6-Hydroxyquinizarin. $C_{14}H_8O_5$. *See* a1248	256.22	bt red nd (al)	256				...	s	...	...	...	alk s Py s	**B8²**, 558
a1342	**9,10-Anthraquinone-2-carboxylic acid**	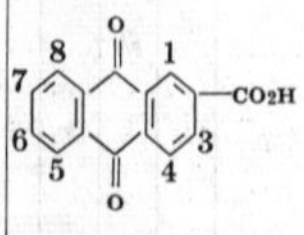 O 8 1 7 CO_2H 6 3 5 4 O	252.23												
a1342¹	**—,1,3-dihydroxy-**	Munjistin. $C_{15}H_8O_6$. *See* a1342	284.21	ye nd (al-w +1w), lf (aa)	230–1	sub			δ s^h	v^h	s	...	...	chl s sulf, alk s	**B10²**, 761
a1343	**—,1,4-dihydroxy-**	$C_{15}H_8O_6$. *See* a1342	284.21	ye br or red nd ($PhNO_2$)	249–50				...	δ	δ	s	δ	to, alk s aa s^h	**B10¹**, 505

For explanations, symbols and abbreviations see beginning of table.

No.	Name	Synonyms and Formula	Mol. wt.	Crystalline form, color and specific rotation	m.p. °C	b.p. °C	Density	n_D	Solubility w	al	eth	ace	bz	other solvents	Ref.
	Anthraquinone														
a1344	—,**4,5-dihydroxy-**	Rhein. $C_{15}H_8O_6$. *See* a1342	284.21	ye or og nd (MeOH, Py)	321–1.5	sub	.....	.....	δ^h	i	i	i	i	Py v, chl i, sulf, alk s	**B10**, 510
a1345	—,**4,5-dihydroxy-7-methoxy-**	Emodic acid monomethyl ether. $C_{16}H_{10}O_7$. *See* **a1342**	314.24	red-br nd (sub)	300	.....	.....	.....	...	...	...	...	...	os δ	**B10**^{2}, 767
a1346	—,**1-nitro-**	$C_{15}H_7NO_6$. *See* a1342	297.21	(aa)	283–4	.....	.....	.....	...	v	i	...	...	aa s^h	**B10**^{2}, 586
a1347	—,—,chloride	$C_{15}H_6ClNO_5$. *See* a1342	315.67	.....	243–4	.....	.....	.....	d^h	d^h	...	...	...	.....	**E13**, 665
a1348	—,**5-nitro-**	$C_{15}H_7NO_6$. *See* a1342	297.23	yesh nd (aa)	.....	.....	.....	.....	...	...	...	...	...	sulf s	**E13**, 665
a1349	**9,10-Anthraquinone-1,5-disulfonic acid**	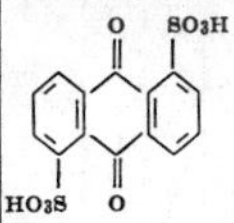 O SO3H / HO3S O	368.34	ye nd (HCl +4w), pl (dil aa+4w)	310–1d	.....	.....	.....	v	...	...	δ	δ	sulf s chl δ aa s	**B13**, 719
a1350	**9,10-Anthraquinone-1,6-disulfonic acid**	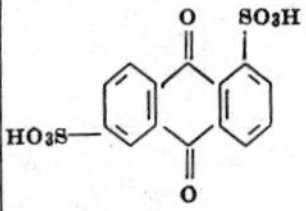 O SO3H / HO3S— / O	368.34	ye nd (HCl+5w) pr (dil aa +5w)	215–7d	.....	.....	.....	v	...	...	...	...	aa s	**B13**, 720
a1351	**9,10-Anthraquinone-1,7-disulfonic acid**	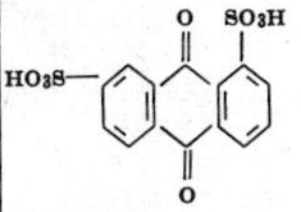 O SO3H / HO3S— / O	368.34	ye hyg pw (dil aa+4w)	d	.....	.....	.....	v	v	...	...	...	aa v	**B13**, 720
a1352	**9,10-Anthraquinone-1,8-disulfonic acid**	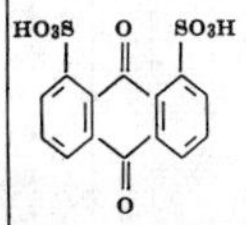 HO3S O SO3H / O	368.34	ye nd(+5w)	293–4d	.....	.....	.....	δ	...	...	...	...	.....	**E13**, 719
a1353	**9,10-Anthraquinone-1-sulfonic acid**	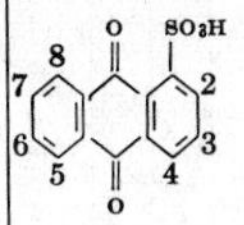 O SO3H / 8 7 2 6 3 5 4 / O	288.28	lf (aa), ye lf (con HCl +3w)	214 cor (anh) 218(hyd)	.....	.....	.....	v	...	...	...	...	aa v^h	**B11**^{2}, 192
a1354	**9,10-Anthraquinone-2-sulfonic acid**	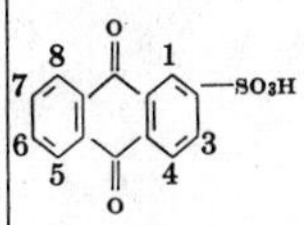 O / 8 1 7 —SO3H 6 3 5 4 / O	288.28	ye lf (+3w)	.....	.....	.....	.....	v	s	i	...	...	.....	**B11**^{2}, 193
a1355	—,amide	$C_{14}H_9NO_4S$. *See* a1354	287.30	ye nd (aa)	261	.....	.....	.....	...	δ	...	...	...	CS_2. to, chl δ	**B11**, 339
a1356	**9,10-Anthraquinone-1-sulfonic acid, 5-chloro-**	$C_{14}H_7ClO_5S$. *See* a1353	322.73	ye rh pr (HCl or aa +4w)	236–7	.....	.....	.....	v	...	...	...	...	sulf s (ye)	**E13**, 707
a1357	**9,10-Anthraquinone-2-sulfonic acid, 5-nitro-**	$C_{14}H_7NO_7S$. *See* a1354	333.28	yesh lf (dil HNO_3)	255d	.....	.....	.....	δ v^h	...	...	...	...	.....	**B11**^{2}, 195
—	**Anthrarufin**	*see* **9,10-Anthraquinone, 1,5-dihydroxy-***													
—	**Anthroic acid**	*see* **Anthracene carboxylic acid**													
—	**9-Anthrone**	*see* **Anthracene,9,10-dihydro-9-oxo-**													
a1358	**Antimalarine**	Plasmocid. 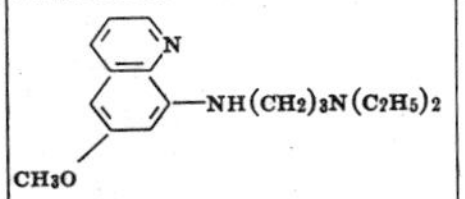 N / —NH(CH2)3N(C2H5)2 / CH3O	287.39	.....	.....	$182^{1.0}$	1.0569^{24}_{4}	1.5855^{24}	...	...	δ	...	...	dil HCl s	.....
a1359	**Antipyrine** (α-form)	Analgesine. 1,5-Dimethyl-2-phenyl-3-pyrazolone. Phenazone. 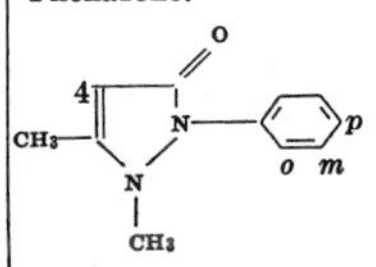 O / 4 / CH3— N— p / N o m / CH3	188.23	mcl lf (w, bz or eth)	113	319^{174} $211–2^{10}$	1.0747^{130}_{4}	1.5697	v	v	δ	...	s	liq SO_2 δ, chl s lig δ	**B24**^{2}, 11

For explanations, symbols and abbreviations see beginning of table.

No.	Name	Synonyms and Formula	Mol. wt.	Crystalline form, color and specific rotation	m.p. °C	b.p. °C	Density	n_D	Solubility w	al	eth	ace	bz	other solvents	Ref.
	Antipyrine														
a1360	—(β-form)	$C_{11}H_{12}N_2O$. *See* a1359	188.23	...	109 (18β→α)	...	...	...	...	...	...	...	...	...	**B24^2**, 11
a1361	—,2-hydroxy-benzoate	Salazolon. Salipyrazolone. Salipyrine. $C_{11}H_{12}N_2O.C_7H_6O_3$. *See* a1359	326.36	pw	89	...	...	...	δ s^h	v^h	δ	...	...	chl v^h	**B24^1**, 197
a1362	—,***m*-amino-**	$C_{11}H_{13}N_3O$. *See* a1359	203.25	redsh in air (bz)	148	...	...	...	s	s	i	...	...	chl s	**B24^1**, 210
a1363	—,***o*-amino-**	$C_{11}H_{13}N_3O$. *See* a1359	203.25	nd (AcOEt-eth)	165	...	...	...	s	s	...	...	...	...	**B24^1**, 210
a1364	—,***p*-bromo-**	$C_{11}H_{11}BrN_2O$. *See* a1359	267.13	nd (w)	117	...	...	...	δ s^h	v	δ	...	...	chl, to v	**B24^2**, 24
a1365	—,***p*-dimethyl-amino-**	$C_{13}H_{17}N_3O$. *See* a1359	231.30	pr or pl (lig or AcOEt)	134–5	...	...	...	s	s	δ	...	s	lig δ	**B24**, 46
—	**Antipyrine chloral**	*see* **Hypnal**													
a1366	**Aphanin**	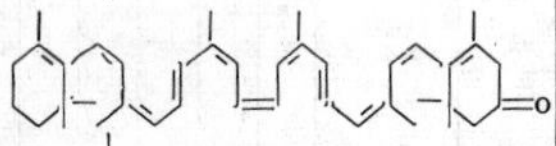	550.86	bl-bk lf (bz-MeOH)	178	...	...	...	i	δ	δ	...	v	CS_2, chl v	...
a1367	**Aphylline**	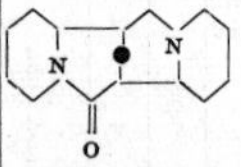	248.37	$[\alpha]^{20}_D$ +10.3 (MeOH)	52–7	200^{4}	...	...	...	...	...	...	...	os s dil HCl v	...
—	**Apigenin**	*see* **Flavone,4′5,7-tri-hydroxy-**													
a1368	**Apiol**	2,5-Dimethoxysafrole. Parsley camphor. 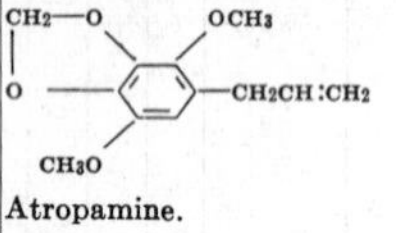	222.24	nd	29.5	294 179^{33}	1.176^{14}	1.536^{20}	i	s	s	s	s	chl s lig s	**B19^2**, 98
a1369	**Apoatropine**	Atropamine. 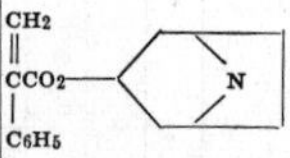	271.36	pr (chl)	62	...	...	...	δ	v	v	...	v	chl, CS_2 v lig δ	**B21^2**, 18
a1370	—,hydrochloride	$C_{17}H_{21}NO_2.HCl$. *See* a1369	307.82	lf	239	...	...	...	δ s^h	δ	i	δ	...	...	**B21^1**, 197
a1371	**Apocinchonidine**	$C_{19}H_{22}N_2O$. *See* a1372	294.40	lf (al), $[\alpha]^{20}_D$ −139.3 (chl-al, c=2)	252	...	...	...	...	v	δ	...	...	chl v	**B23^1**, 131
a1372	**Apocinchonine**	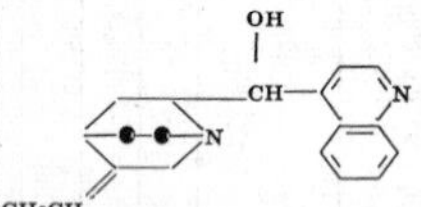	294.40	pr (al), $[\alpha]^{20}_D$ +167.4 (abs al, c=3)	219	...	...	...	i	s v^h	δ	...	δ	chl δ lig δ	**B23^1**, 131
a1373	—,hydrochloride	$C_{19}H_{22}N_2O.HCl.2H_2O$	366.89	nd (+2w) $[\alpha]^{16}_D$ +139 (w, c=0.006)	...	...	...	...	...	...	...	...	...	...	**B23**, 418
a1374	**Apocodeine**	Apomorphine 3-methyl ether. Pseudoapocodeine. 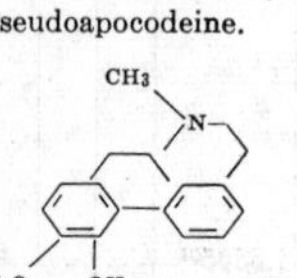	281.36	pr (MeOH) $[\alpha]^{24}_D$ −9 (abs al, c=0.449)	122.5–4.5	...	...	...	...	δ	s	s	s	lig s	**B27^2**, 88
a1375	—,(ethanol solvate)	$C_{18}H_{19}NO_2.C_2H_5OH$	327.43	lf(al+1)	105	...	...	...	...	...	...	...	...	...	**B21**, 188
a1376	—,(methanol solvate)	$C_{18}H_{19}NO_2.CH_3OH$	313.40	nd(MeOH+1)	85	...	...	...	...	...	...	...	...	...	**B21**, 188
—	**Apocupreine**	*see* **Apoquinine**													
a1377	**Apocyclene**	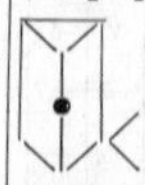	122.21	(al)	42.5–43	138–9^{764}	0.8710^{40}_4	1.4514^{40}	...	s	...	...	...	aa s	**E13**, 1042
—	**Apocynin**	*see* **Acetophenone,4-hydroxy-3-methoxy-**													
—	**Apofencho-camphoric acid**	*see* **1,3-Cyclopentane dicarboxylic acid, 4,4-dimethyl-***													

For explanations, symbols and abbreviations see beginning of table.

No.	Name	Synonyms and Formula	Mol. wt.	Crystalline form, color and specific rotation	m.p. °C	b.p. °C	Density	n_D	Solubility w	al	eth	ace	bz	other solvents	Ref.
	Aponal														
—	**Aponal**	*see* **Carbamic acid,** 2-methyl-2-butyl ester.													
a1378	**Apomorphine**	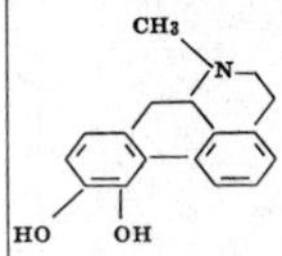 CH3, N, HO OH (structure)	267.33	pl (chl-peth), rods (eth+1), gr in air	195d	...	...	...	δ	...	δ	s	δ	HCl δ chl s lig δ	**B21**^{2}, 141
a1379	—,hydrochloride	$C_{17}H_{17}NO_2.NCl.\frac{3}{4}H_2O$	317.31	gr in air, mcl pr $[\alpha]^{25}_D -48$ (c=1.2)	200–10d	...	...	...	s^h	s	δ	...	...	chl δ	**B21**^{2}, 143
a1380	**Apoquinine** (α-form)	Apocupreine. 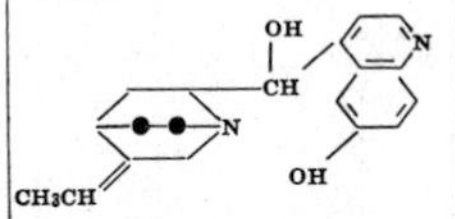 OH, CH, N, OH, CH_3CH (structure)	310.40	pr (eth) $[\alpha]^{20}_D$ −214.8 (al)	190d	...	...	...	i s^h	s	δ	δ	s	chl δ KOH s	**B23**^{2}, 412
a1381	—(β-form)	$C_{19}H_{22}N_2O_2.2H_2O$	346.43	$[\alpha]^{20}_D +194$	190d	...	...	...	...	...	...	...	...	...	**B23**^{2}, 412
a1382	**Aposafranone**	Benzeneindone. 10-Phenyl-2(10)-phenazinone. 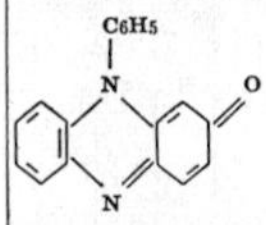 C_6H_5, N, O, N (structure)	272.31	br, gr (red in sol), nd (al)	248–9	...	...	...	δ^h	s	...	...	s	dil ac s	...
a1383	**Arabinose** (*D*)(α-form)	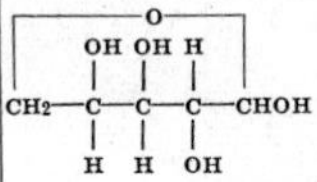 —O— OH OH H CH_2—C—C—C—CHOH H H OH	150.13	(MeOH), $[\alpha]^{20}_D$ +105	155.5–6.5 cor	...	1.585	...	v	δ	i	i	...	MeOH i	**B31**, 32
a1384	—(*D*)(β-form)	$C_5H_{10}O_5$. *See* a1383	150.13	(aq MeOH), $[\alpha]^{20}_D -175$ to −105	155.5–6.5	...	1.625	...	v	δ	i	i	...	MeOH i	**B31**, 32
a1385	—(*D*), diphenyl-hydrazone	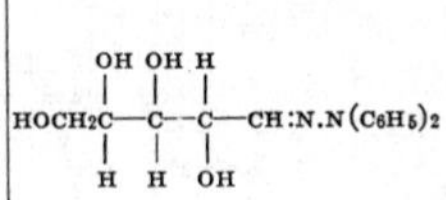 OH OH H $HOCH_2C$—C—C—$CH{:}N.N(C_6H_5)_2$ H H OH	316.36	orh pr	207	...	...	...	δ	δ	...	...	...	...	**B31**, 33
a1386	—(*DL*)	Pectinose. $C_5H_{10}O_5$. *See* a1383	150.13	pr, nd (al)	164.5 cor	...	1.585^{20}_4	...	v	δ	i	...	...	...	**B31**, 46
a1387	—(*DL*), diphenyl-hydrazone	$C_{17}H_{20}N_2O_4$. *See* a1385	316.36	nd (aq Py)	206	...	...	...	i δ^h	i δ^h	i	...	...	CS_2 i Py v	**B31**, 47
a1388	—(*L*)(α-form)	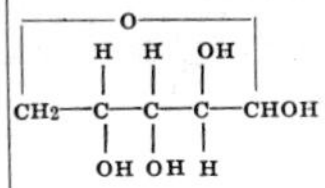 —O— H H OH CH_2—C—C—C—CHOH OH OH H	150.13	...	159–60	...	1.585^{20}_4	...	v	δ	i	...	...	...	**B31**, 34
a1389	—(*L*)(β-form)	$C_5H_{10}O_5$. *See* a1388	150.13	orh, $[\alpha]^{20}_D +190.6$ to +104.5	159–60	...	1.625^{20}_4	...	v	δ	i	...	...	...	**B31**, 34
a1390	—(*L*), diphenyl-hydrazone	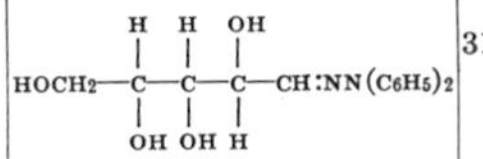 H H OH $HOCH_2$—C—C—C—$CH{:}NN(C_6H_5)_2$ OH OH H	316.36	orh pr, nd (dil) al) $[\alpha]^{20}_D$ +18.5 (Py)	204–5	...	...	...	δ	s^h	...	...	...	...	**B31**, 44
a1391	**Arabitol** (*D*)	Arabite *D*-Lyxitol. 1,2,3,4,5-Pentanepentol. 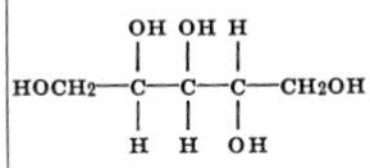 OH OH H $HOCH_2$—C—C—C—CH_2OH H H OH	152.15	pr, $[\alpha]^{20}_D +7.7$ (boraxsolution c=9.26)	103	...	...	...	v	δ	i	...	...	...	**B1**, 531
a1392	—(*DL*)	$C_5H_{12}O_5$. *See* a1391	152.15	pr (90% al)	105–6	...	...	...	...	δ	...	...	...	...	**B1**, 531
a1393	—(*L*)	$C_5H_{12}O_5$. *See* a1391	152.15	$[\alpha]_D -5.4$ (borax sol)	102	...	...	...	...	...	...	...	...	...	**B1**^{2}, 604
a1394	**Arabonic acid** (*D*)	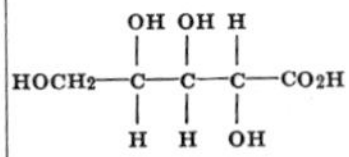 OH OH H $HOCH_2$—C—C—C—CO_2H H H OH	166.13	(dil aa), $[\alpha]^{25}_D$ +10.5 (c=6.0)	114–16	...	...	...	v	s	i	...	...	aa δ	**B3**^{2}, 303

For explanations, symbols and abbreviations see beginning of table.

No.	Name	Synonyms and Formula	Mol. wt.	Crystalline form, color and specific rotation	m.p. °C	b.p. °C	Density	n_D	Solubility w	al	eth	ace	bz	other solvents	Ref.
	Arabonic acid														
a1395	—(*DL*)	$C_5H_{10}O_6$. *See* a1394	166.13						s						
a1396	—(*L*)	$C_5H_{10}O_6$. *See* a1394	166.13	(al), $[\alpha]_D^{20} -9.8$	118–9				v						**B3**^{2}, 303
—	**Arachidic acid**	*see* **Eicosanoic acid***													
—	**Arachidonic acid**	*see* **6,10,14,18-Eicosatetraenoic acid***													
a1397	**Aramite**	$C_{15}H_{23}ClO_4S$	334.86			195^{2}	1.145_{20}^{20}	1.5100^{20}	i	s				os v	
a1398	**Arbutin**	Hydroquinone β-d-glucopyranoside $C_{12}H_{16}O_7$.	272.26	nd (w+1) $[\alpha]_D^{25} -64$ (c=3)	199.5–200 (anh)				v	s	i			chl, CS_2 i	**B31**, 210
a1399	—,hydrate	$C_{12}H_{16}O_7.H_2O$	290.27	$[\alpha]_D^{17} -60.3$ (p=5, w)	142										**B31**, 210
a1400	**Arecaidine**	Arecaine. 1-Methylguvacine. 1,2,5,6-tetrahydro-1-methylnicotinic acid. $C_7H_{11}NO_2$. H_2O.	159.18	pl (dil al), ta (dil al+1w)	223–4				v	i	i		i	dil al v, chl i	**B22**^{2}, 12
a1401	**Arecoline**	Arecaidine methyl ester. Methyl 1,2,5,6-tetrahydro-1-methylnicotinate. $C_8H_{13}NO_2$.	155.20			209 74^{7}	1.0504_{20}^{20}	1.4860^{20}	∞	∞	∞			chl s	**B22**^{2}, 12
a1402	—,hydrobromide	$C_8H_{13}NO_2.HBr$	236.12	mcl pr (al)	177.4–9				v	s v^{h}	δ			chl δ	**B22**^{2}, 12
a1403	—,hydrochloride	$C_8H_{13}NO_2.HCl$	191.66	nd	157–8				s	s					**B22**^{1}, 490
a1404	**Arginine** (*dl*)	*DL*-2-Amino-5-guanidopentanoic acid*. $H_2NC(:NH)NH(CH_2)_3CHNH_2CO_2H$	174.21		217–8										**B4**^{2}, 850
a1405	—(*l*)	$H_2NC(:NH)NH(CH_2)_3CHNH_2CO_2H$	174.21	pr (+2w), anh ta (60% al) $[\alpha]_D^{20} +12.1$ (w, c=2)	105d (hyd) 238d				s	i	i				**B4**^{2}, 845
a1406	—,diflavianate (*l*)	$C_6H_{14}N_4O_2(C_{10}H_6N_2O_8S)_2$	802.67	ye nd	202d				d	d				ac i	**B4**^{2}, 848
a1407	—,dipicrate (*dl*)	$C_6H_{14}N_4O_2.2\ C_6H_3N_3O_7$	632.45		196d										**B4**^{2}, 850
a1408	—,—(*l*)	$C_6H_{14}N_4O_2.2C_6H_3N_3O_7$	632.45		200d										**B4**^{2}, 848
a1409	—,flavianate (*l*)	$C_6H_{14}N_4O_2.C_{10}H_6N_2O_8S$	488.47	ye-og lf	258–60d				i	i	i				**B4**^{2}, 848
a1410	—,picrate (*d*)	$C_6H_{14}N_4O_2.C_6H_3N_3O_7.2H_2O$	439.34	nd (w)	217				δ						**B6, 287**
a1411	—,—(*dl*)	$C_6H_{14}N_4O_2.C_6H_3N_3O_7$	403.31	pr (w)	223d				i						**B4**^{2}, 851
a1412	—,—(*l*)	$C_6H_{14}N_4O_2.C_6H_3N_3O_7.2H_2O$	439.34	ye nd (w)	217–8d				δ	i	i				**B4**^{2}, 848
a1413	—,benzylidene (*l*)	$C_{13}H_{18}N_4O_2$.	262.31	lf(w)	204–5				s^{h}					MeOH s, alk i ac s	**B4**^{2}, 848
—	**Arsanilic acid**	*see* **Benzenearsonic acid, 4-amino-***													
a1414	**Arsenic acid, triethyl ester***	Ethyl arsenate*. $(C_2H_5O)_3AsO$	226.11			235–8	1.3264_4^0		d						**B2**, 332
a1415	**Arenous acid, triethyl ester*.**	Ethyl arsenite*. $(C_2H_5O)_3As$	210.09		165–6		1.2118_4^{13}	1.4369^{13}	d						**B1**^{2}, 332
a1416	**Arsine, bis(pentafluoroethyl)-iodo-***	$(CF_3CF_2)_2AsI$	439.86			120^{760}									
a1416[1]	**—,bis(trifluoromethyl)-***	$(CF_3)_2AsH$	213.94			19^{760} -25^{108}									
a1417	**—,bis(trifluoromethyl) bromo-***	$(CF_3)_2AsBr$	292.84			59.5^{745}		1.398^{20}	i						
a1418	**—,bis(trifluoromethyl) chloro-***	$(CF_3)_2AsCl$	248.38			46^{760}		1.351^{19}							
a1419	**—,bis(trifluoromethyl) fluoro-***	$(CF_3)_2AsF$	231.93			25^{760}									
a1420	**—,bis(trifluoromethyl)iodo-**	$(CF_3)_2AsI$	339.84	ye oil		92^{760}		1.425^{25}	i δ^{h}d		s				
a1421	**—,bis(trifluoromethyl)methyl-***	$(CF_3)_2AsCH_3$	227.97			52^{760}									
a1422	**—,chloro-(diphenyl)-***	$(C_6H_5)_2AsCl$	264.59	rh pl (peth)	39–40	333^{760} 172^{7}	1.387_4^{42}	1.6332^{56}	i	v	s	s			**B16**^{1}, 437
a1423	**—,dibromo(trifluoromethyl)-**	CF_3AsBr_2	303.75			118^{745}		1.528^{20}							
a1424	**—,dichloro-(methyl)-***	CH_3AsCl_2	160.86		−59	37^{25}, 133	1.838^{20}	1.5677^{15}	δ	v				os s	**B4**^{2}, 979

For explanations, symbols and abbreviations see beginning of table.

No.	Name	Synonyms and Formula	Mol. wt.	Crystalline form, color and specific rotation	m.p. °C	b.p. °C	Density	n_D	Solubility: w	al	eth	ace	bz	other solvents	Ref.
	Arsine														
a1425	**—,dichloro-(phenyl)-***	$C_6H_5AsCl_2$	222.93	...	...	254.4–7.6 131^{14}	1.6516^{19}_{4}	1.6386^{15}	i	v	s	...	s	...	**B16²**, 411
a1426	**—,dichloro(trifluoromethyl)-**	CF_3AsCl_2	214.83	...	...	71	1.431^{20}	...	i	...	...	...	...	...	...
a1427	**—,diethyl-***	$(C_2H_5)_2AsH$	134.05	...	...	105	1.1338^{24}_{4}	1.4709	δ	s	...	...	...	...	**B4²**, 980
a1428	**—,(difluoro)-ethenyl, 2-chloro-***	$ClCH{:}CHAsF_2$	174.41	...	...	>140d 43.5^{15}	...	...	d	...	...	...	...	...	...
a1429	**—,difluoro-(ethyl)-***	$C_2H_5AsF_2$	141.98	fum in air	−38.7	94.3 74^{100}	1.708^{17}	...	d	...	...	...	...	...	...
a1430	**—,difluoro-(methyl)-**	CH_3AsF_2	127.95	fum in air	−29.7	76.5	1.924^{18}	...	d	...	...	...	...	...	...
a1431	**—,difluoro-(phenyl)-***	$C_6H_5AsF_2$	190.02	wax	42	110^{48}	...	...	d	...	...	...	...	...	...
a1432	**—,diiodo(trifluoromethyl)***	CF_3AsI_2	397.74	...	...	183 δd 100^{48}	...	...	d	...	...	...	...	...	...
a1433	**—,dimethyl-***	Cacodyl hydride. $(CH_3)_2AsH$.	106.00	ign in air	...	35.6^{747}	1.213^{29}_{29}	...	δ	∞	∞	...	...	chl, CS_2, ∞ aa ∞	**B4²**, 978
a1434	**—,dimethyl(trifluoromethyl)-***	$(CH_3)_2AsCF_3$	174.00	...	...	58	...	...	...	...	...	...	...	...	...
a1435	**—,diphenyl-***	$(C_6H_5)_2AsH$	230.14	oil	...	$161–2^{20}$	...	...	i	s	s	...	...	...	**B16²**, 406
a1436	**—,ethyl-***	$C_2H_5AsH_2$	106.00	...	...	36	1.217^{22}_{22}	...	i	s	s	...	...	...	**B4²**, 980
a1437	**—,methyl-***	CH_3AsH_2	91.97	...	...	2	...	...	i	∞	∞	...	...	CS_2 ∞	**B4²**, 978
a1438	**—,triethyl-***	$(C_2H_5)_3As$	162.11	...	...	140^{736}	1.467^{20}	1.150^{20}_{4}	i	∞	∞	...	...	...	**B4²**, 980
a1440	**—,(trifluoro-methyl)-***	CF_3AsH_2	145.94	...	...	-11.6^{781}	...	...	...	...	...	...	...	...	...
a1441	**—,trimethyl-***	$(CH_3)_3As$	120.03	...	...	51.95^{736}	1.144^{15}	...	δ	s	s	...	s	...	**B4²**, 978
a1443¹	**—,triphenyl-***	$(C_6H_5)_3As$	306.24	tcl pr (bz, eth-peth)	59–60	360(CO_2)	1.2634^{48}_{4}	1.6888^{21}	i	δ	v	...	v	...	**B16²**, 407
a1443²	**—,tris(trifluoro-methyl)-***	$(CF_3)_3As$	281.94	...	...	33.3	...	...	i	...	s	...	...	...	...
a1443³	**—,tris(penta-fluoroethyl)-***	$(CF_3CF_2)_3As$	431.96	...	...	96–3	...	...	...	...	...	...	...	...	...
a1444	**Arsine oxide, 4-methoxy-phenyl-**	$CH_3O-C_6H_4-AsO$	198.05	(chl-eth, or bz-peth, solvate)	114–6 93–5 (solvate)	...	...	...	...	...	...	...	s	chl s	**B16²**, 448
a1445	**—,methyl-***	Methyl arsinic acid anhydride. CH_3AsO	105.96	pr (CS_2), (al)	95	275d	...	...	i	s	...	...	s	...	**B4²**, 992
a1446	**—,phenyl-**	C_6H_5AsO	168.03	(a, bz-eth)	144–6	...	...	...	i	δ v^h	i	...	v^h	chl v	**B16²**, 442
a1447	**—,2-tolyl-**	$C_6H_4CH_3AsO$	182.05	pw	145–6	...	...	...	v^h	...	i	...	...	alk δ	**B16**, 861
a1448	**—,4-tolyl-**	$C_6H_4CH_3AsO$	182.05	pw	156	...	...	...	v^h	...	i	...	...	alk δ	**B16**, 861
a1449	**Arsinic acid, bis(trifluoro-methyl)-***	Perfluorocacodylic acid. $(CF_3)_2AsO_2H$	245.94	nd (subl),rh (w+?)	150d 10^{-3} sub	...	...	...	s	...	...	...	...	chl δ^h	...
a1450	**—,diphenyl-***	$(C_6H_5)_2AsO_2H$	262.13	nd, pr	178	...	...	...	δ s^h	s^h	δ	...	...	chl s	**B16²**, 443
a1151	**Arsonium, tetra-phenyl-**, bromide	$(C_6H_5)_4AsBr$	463.25	...	281–4	...	...	...	δ	s	...	δ	...	MeOH s	...
a1452	—,—,chloride	$(C_6H_5)_4AsCl$	418.79	...	258–60	...	...	...	v	s	...	δ	...	MeOH s	...
a1453	**Artemisic acid**	$HO_2CCH(CH_3)$—[naphthalene ring with CH_3, —OH, CH_3]	244.29	nd (dil aa) $[\alpha]^{18.5}_{D}$ +70.4 (al)	135–6	...	...	...	...	v	v	...	...	aa v lig v	**E12B**, 3469 3469
—	**Arterenol**(*l*)	*see* **Noradrenalin**(*l*)													
—	**Artificial musk**	*see* **Benzene,1-*tert*-butyl-3-methyl-2,4,6-trinitro-***													
—	**Asaron**	*see* **Benzene,1-propenyl-2,4,5-trimethoxy-***													
—	**Asaronic acid**	*see* **Benzoic acid, 2,4,5-tri-methoxy-**													
a1454	**Ascaridole**	$C_{10}H_{16}O_2$	168.24	unst, $[\alpha]_D$ −4.14	...	exp	0.9985^{20}_{20}	1.4743^{20}	i d^h	d	...	...	...	os s	**B19²**, 18

For explanations, symbols and abbreviations see beginning of table.

No.	Name	Synonyms and Formula	Mol. wt.	Crystalline form, color and specific rotation	m.p. °C	b.p. °C	Density	n_D	Solubility w	al	eth	ace	bz	other solvents	Ref.
	Ascorbic acid														
a1455	**Ascorbic acid** (*d*)	Antiscorbutin. Vitamin C. $C_6H_8O_6$.	176.13	$[\alpha]^{18}$–48 (MeOH) $[\alpha]_D^{20}-23.8$ (w, c=3)	190d										
a1456	—(*dl*)	$C_6H_8O_6$	176.13		168–9										
a1457	—(*l*)	$C_6H_8O_6$	176.13	pw, $[\alpha]_D^{20}+24$(w)	190d				v	s	i	...	i	chl i	
a1458	**—,6-desoxy-(*L*)**	$H_3C-C(H)(OH)-C(H)-C(OH)=C(OH)-C=O$ (ring –O–)	160.13	pr (AcOEt) $[\alpha]_D^{22}+36.7$ (0.1N HCl)	160	$160^{0.01}$ sub			v	...	i	s	...	AcOEt δ	
a1459	**Asparagine** (*D*)	$H_2NCOCH_2CH(NH_2)CO_2H.1H_2O$	150.14	$(\alpha)_D^{18-20}+5.41$ (w)	234.5		1.543_4^{15}		δ						**B4**[1], 531
a1460	—(*DL*)	$C_4H_8N_2O_3.H_2O$. *See* a1459	150.14	tcl cr (w+1)	182–3	213–5d	1.4540_4^{15}		δ s^h	i	i	...	...		**B4**[2], 900
a1461	—(*L*)	$C_4H_8N_2O_3$. *See* a1459	132.12	rh (w+1)	235d		1.543_4^{15}		s v^h	i	i	...	...	MeOH i	**B4**[2], 896
a1462	**Aspartic acid** (*D*)	Aminosuccinic acid. Asparacemic acid. $HO_2CCH_2CH(NH_2)CO_2H$	133.10	$[\alpha]_D^{20}-25.5$ (HCl)	269–71		1.6613_{13}^{13}		s^h	i	i	...	...	dil HCl s	**B4**[1], 531
a1463	—(*DL*)	$C_4H_7NO_4$. *See* a1462	133.10	mcl pr	338–9		1.6632_{13}^{13}		δ s^h	i	i	...	...		**B4**[2], 900
a1464	—(*L*)	$C_4H_7NO_4$. *See* a1462	133.10	rh lf, $[\alpha]_D^{20}+4.36$ (w)	270–1 (sealed tube)		1.6613_{13}^{13}		δ s^h	i	...	...	...	dil HCl s	**B4**[2], 892 **B14**[2], 653
a1465	**—,*N*-benzoyl-**	$HO_2CCH_2CH(NHCOC_6H_5)CO_2H$	237.21	nd or lf, $[\alpha]_D^{19}+22.4$ (dil KOH)	182				δ^h	δ	i	...	...	chl i	**B9**[2], 185
a1466	**Aspidospermine**	$C_{22}H_{30}N_2O_2$	354.48	nd (al), $[\alpha]_D^{15}-100.2$ (al)	208				δ	s	δ	...	s	chl s	
—	**Aspirin**	*see* **Benzoic acid, 2-hydroxy-**, acetate													
—	**Atabrin**	*see* **Quinacrine**													
a1467	**Atisine**	Anthorine. $C_{22}H_{31}NO_2$	341.48	rh bipym	88.5–9				δ	v	v	...	...	MeOH-KOH s chl s	
a1468	—,hydrochloride	$C_{22}H_{31}NO_2.HCl$	377.94	nd,$[\alpha]_D+26.9$	136	340			v	v	i	...	...		
—	**Atophan**	*see* **4-Quinolinecarboxylic acid, 2-phenyl-**													
—	**Atoquinol**	*see* **4-Quinolinecarboxylic acid, 2-phenyl-**, allyl ester													
—	**Atranol**	*see* **Benzaldeyde, 2,6-dihydroxy-4-methyl-**													
—	**Atrolactic acid**	*see* **Propanoic acid, 2-hydroxy-2-phenyl-***													
—	**Atropamine**	*see* **Apoatropine**													
—	**Atropic acid**	*see* **Propenoic acid, 2-phenyl-***													
a1469	**Atropine**	*dl*-Daturine. *dl*-Hyoscyamine. Tropic acid, tropine ester. $C_{17}H_{23}NO_3$.	289.36	rh nd (dil al)	115–6	93–110 vac sub			δ	s	s	...	s	chl v lig i	**B21**[2], 19
a1470	—,hydrochloride	$C_{17}H_{23}NO_3.HCl$	325.83	nd	165										**B21**, 30
a1471	—,pentanoate	$C_{17}H_{23}NO_3.C_5H_{10}O_2.\frac{1}{2}H_2O$	400.52		42				∞	∞	δ				
a1472	—,sulfate	$2(C_{17}H_{23}NO_3).H_2SO_4$	676.80	nd (al-eth, or al-ace)	194	sub			v	v	i	...	...	chl δ	**B21**[2], 20
—	**Auligen**	*see* **Xantogen, diethyl-**													
a1473	**Auramine** (base)	Bis (*p*-dimethylaminophenyl)-methyleneimine. $[(CH_3)_2N-C_6H_4-]_2C{:}NH$	267.36	ye or colorless lf (al)	136				i	s	δ	...	...		**B14**[2], 58
a1474	—(dye)	Auramine hydrochloride. $C_{17}H_{22}ClN_3.H_2O$. *See* a1473	321.85	ye nd	267				δ	s	δ	...	...	chl v	**B14**[2], 58
a1475	—,*N*-methyl-	$C_{18}H_{23}N_3$. *See* a1473	281.39	ye (al)	133				i	s	...	s	δ	aa s	**B14**, 93

For explanations, symbols and abbreviations see beginning of table.

No.	Name	Synonyms and Formula	Mol. wt.	Crystalline form, color and specific rotation	m.p. °C	b.p. °C	Density	n_D	Solubility w	al	eth	ace	bz	other solvents	Ref.
	Aurin														
a1476	**Aurin**	Pararosolic acid, Rosolic acid. 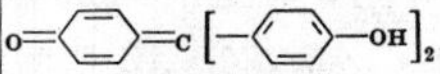	290.30	dk red lf or rh	>220				i	s	i	...	i	alk s chl δ aa s	**B8**[2], 417
a1477	**Aureomycin**	Biomycine. Chlorotetracycline. CTC. Duomycine. $C_{22}H_{23}ClN_2O_8$.	478.88	gold ye, ye flr $[\alpha]_D^{23} -275$ (MeOH)	168–9				i	δ	i	...	...	Cellosolve s	
a1478	—,hydrochloride	$C_{22}H_{23}ClN_2O_8$. HCl	515.35	ye orh $[\alpha]_D^{20} -106.5$ (dil HCl)	216d				δ	...	...	...	...		
a1479	**Auxin** (a)	Auxenetriolic acid. $C_{18}H_{32}O_5$.	328.44	hex (al-lig) $[\alpha]_D^{20} -3.19$ (al)	196				δ	s	δ	...	...	MeOH s	
a1480	—(b)	Auxenolonic acid. $C_{18}H_{30}O_4$.	310.42	(al-lig), $[\alpha]_D^{20} -2.8$	183				...	...	...	...	...		
a1481	**1-Azacyclooctane, 2-methyl-**	α-Methyl heptamethyleneimine. 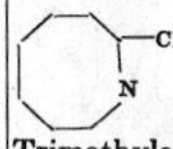	127.23	nd (ace)	156–7	162–3[746]	0.853^{20}	1.4620^{21}	...	...	...	...	...		**B20**[1], 30
a1482	**Azetidine**	Trimethyleneimine. 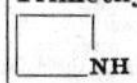	57.09			63	0.8436^{20}	1.4282^{25}	∞	∞	...	...	...		**B20**, 2
—	**Azelaic acid**	*see* **Nonanedioic acid***													
—	**Azibenzyl**	*see* **Ketone, benzyl phenyl, α-diazo-**													
a1483	**Azirane**	Dihydroazirine. Ethyleneimine. 	43.07			56	0.832_4^{20}		∞	s	...	...	...	os ∞	**B20**[2], 3
a1484	**Azo, benzene ethane**	Ethylphenyl diimide. $C_6H_5N{:}NC_2H_5$	134.18	bt ye		175–85 82.5–3[20]	0.9628_4^{22}	1.5313 (α) 1.5579 (β)	i	s	s	...	s	dil ac δ	**B16**[2], 3
a1485	—,**benzene methane**	Methylphenyl diimide. $C_6H_5N{:}NCH_3$	120.15	ye		60[15] 150d			...	s	s	...	...		**B16**, 7
a1485[1]	—,**benzene 1-naphthalene**	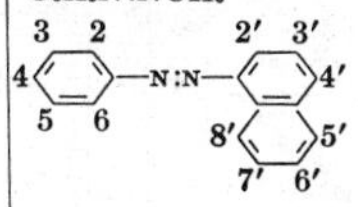	231.30						...	...	...	...	...		
a1485[2]	—,**benzene 2-naphthalene**	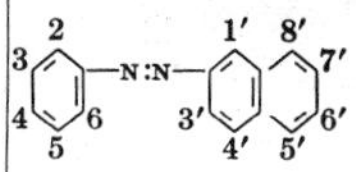	231.30						...	...	...	...	...		
a1486	—,**benzene 1-naphthalene, 2′-amino-**	Yellow AB. $C_{16}H_{13}N_3$. *See* a1485[1]	247.30	red pl (al)	102–4				i	v	...	...	...	aa v	**B16**[2], 193 **B25**[2], 545
a1487	—,—,**4′-amino-**	Naphthyl red. $C_{16}H_{13}N_3$. *Se* a1485[1]	247.30	red lf (dil al), nd (dil al)	125–5.5				...	s	s	...	s		**B16**[2], 186
a1488	—,—,—,hydrochloride	$C_{16}H_{13}N_3$. HCl. *See* a1485[1]	283.77	gr nd, pr (al or aa)	204–6				δ	s^h	...	...	...	aa s^h	**B16**[1], 324
a1489	—,—,**2,3-dimethyl-2′-hydroxy-**	$C_{18}H_{16}N_2O$. *See* a1485[1]	276.34	pa ye amor	125–30				i	...	...	...	...	to, CCl_4 s	**B16**[2], 71
a1490	—,—,**2,4-dimethyl-2′-hydroxy-**	$C_{18}H_{16}N_2O$. *See* a1485[1]	276.34	red nd	166				i	s	s	...	...		**B16**[2], 72
a1491	—,—,**2,5-dimethyl-2′-hydroxy-**	$C_{18}H_{16}N_2O$. *See* a1485[1]	276.34	nd (al)	150–1				...	s^h	...	...	...		**B16**, 168
a1492	—,—,**3,4-dimethyl-2′-hydroxy-**	$C_{18}H_{16}N_2O$. *See* a1485[1]	276.34	red nd (al)	146				...	δ s^h	δ ...	...	v	chl v	**B16**[1], 260
a1493	—,**benzene 2-naphthalene, 2,4-dimethyl-1′-hydroxy-**	$C_{18}H_{16}N_2O$. *See* a1485[2]	276.34	gold-red lf or nd (al-chl)	186				...	δ^h	...	...	s	alk i chl, CS_2 s lig i	**B16**[1], 143

For explanations, symbols and abbreviations see beginning of table.

No.	Name	Synonyms and Formula	Mol. wt.	Crystalline form, color and specific rotation	m.p. °C	b.p. °C	Density	n_D	Solubility w	al	eth	ace	bz	other solvents	Ref.
	Azo														
a1494	**—,—,4-dimethyl-amino-**	$C_{18}H_{17}N_3$. *See* 1458[2]	275.36	ye-br (bz+lig)	174								s^h		**B16**, 321
a1495	**—,benzene 1-naphthalene, 4,4′-dinitro-2-hydroxy-**	$C_{16}H_{10}N_4O_5$. *See* a1485[1]	338.29	purp bl sol in al-NaOH, (gl aa)	231									os δ aq alk i aa s^h	**B16**[1], 268
a1496	**—,benzene 1-naphthalene 4,8′-dinitro-4′-hydroxy-**	$C_{16}H_{10}N_4O_5$. *See* a1485[1]	338.29	red ($PhNO_2$)	252–60d						i			to $δ^h$ $PhNO_2$ s^h alk s lig i	**B16**, 161
a1497	**—,benzene 2-naphthalene, 4,4′-dinitro-1′-hydroxy-**	$C_{16}H_{10}N_4O_5$. *See* a1485[2]	338.29	red nd ($PhNO_2$-aa)	255d					δ		δ		$PhNO_2$ s aa δ	**B16**[2], 67
a1498	**—,—,4,5′-dinitro-1′-hydroxy-**	$C_{16}H_{10}N_4O_5$. *See* a1485[2]	338.29	(bz)	210	235d				s	s		s^h	chl s $PhNO_2$ v	**B16**, 154
a1499	**—,—,1′-hydroxy-**	$C_{16}H_{12}N_2O$. *See* 1485[2]	248.29	nd (al)	138				i	v				con alk i	**B16**[1], 248
a1500	**—,benzene 1-naphthalene, 2′-hydroxy-**	Sudan Yellow. $C_{16}H_{12}N_2O$. *See* a1458[1]	248,29	red-gold lf or nd (al)	133–4					s^h	s		s	lig s	**B16**[2], 70
a1501	**—,—,4′-hydroxy-**	$C_{16}H_{12}N_2O$. *See* a1485[1]	248.29	vt-br lf (bz)	205–6					v^h			v	dil NaOH, con sulf v	**B16**[2], 67
a1502	**—,—,2′-hydroxy-2-nitro-**	$C_{16}H_{11}N_3O_3$. *See* a1485[1]	293.29	og-red nd (gl aa)	210					δ				al-NaOH s^h aa s^h	**B16**[2], 70
a1503	**—,—,2′-hydroxy-4-nitro-**	Paranitraniline red. $C_{16}H_{11}N_3O_3$. *See* a1485[1]	293.27	br-og pl (to)	250				i	s		s			**B16**[2], 70
a1504	**—,—,2′-hydroxy-4′-nitro-**	$C_{16}H_{11}N_3O_3$. *See* a1485[1]	293.27	dk brsh-red nd (bz)	206–7					i			s^h	dil NaOH i	**B16**[2], 70
a1505	**—,—,3-nitro-2-hydroxy-**	$C_{16}H_{11}N_3O_3$. *See* a1485[1]	293.27	og pl (to)	194									xyl s al-KOH s to s^h	**B16**[2], 70
a1506	**—,—,4′-hydroxy-2′-nitro-**	$C_{16}H_{11}N_3O_3$. *See* a1485[1]	293.27	og-red nd (bz)	164								s^h	dil KOH s^h aa s	**E12B**, 1915
a1507	**—,—,3-nitro-4′-hydroxy-**	$C_{16}H_{11}N_3O_3$. *See* a1485[1]	293.27	dk red nd (al)	242–3d					s				NaOH s	**E12B**, 1762
a1508	**—,—,4′-hydroxy-4-nitro-**	$C_{16}H_{11}N_3O_3$. *See* a1485[1]	293.27	red nd	277–9d				δ					$PhNO_2$ s^h	
a1509	**—,benzene 2-naphthalene, 1′-hydroxy-4′-nitro-**	$C_{16}H_{11}N_3O_3$. *See* a1485[2]	293.27	dk red nd (gl aa)	180					$δ^h$			s^h	aa s^h	**E12B**, 1914
—	**Azo, benzene toluene**	*see* **Azobenzene, methyl-**													
a1510	**Azo,1-naphthalene 2-pyridine, 2-hydroxy-**	HO N:N N	249.28	og-red nd (al)	137				i	s	s			dil alk s^h dil ac s	**B22**[1], 694
a1511	**Azo, 1-naphthalene 2-pyridine, 4-hydroxy-**	N:N OH N	249.28	gr-red nd					i					dil ac s dil alk s con sulf s	**B22**[1], 694
a1512	**Azobenzene** (*cis*)	Azobenzidine. Diphenyldiimide. Phenylazobenzene*. 3′ 2′ 2 3 4′ N:N 4 5′ 6′ 6 5	182.22		71										
a1513	—(*trans*)	$C_{12}H_{10}N_2$. *See* a1512	182.22	og red mcl lf (al)	68.5	295–7^{749}	1.203_4	1.6266^{78}	δ	s	s			MeOH s lig δ	**B16**[2], 4
a1514	**—,4-acetamido-**	$C_{14}H_{13}N_3O$. *See* a1512	239.27	gold ye nd (al)	144–6				δ		δ		δ		**B16**[2], 155
a1515	**—,2-acetamido-4′,5-dimethyl-**	$C_{16}H_{17}N_3O$. *See* a1512	267.32	ye nd (al-aa)	157				i	s	s			chl s	**B16**[2], 182
a1516	**—,4-acetamido-2′,3-dimethyl-**	$C_{16}H_{17}N_3O$. *See* a1512	267.32	red nd (al)	186–7				i	δ	s			chl s	**B16**[2], 179
a1517	**—,4-acetoxy-2′-methyl-**	$C_{15}H_{14}N_2O_2$. *See* a1512	254.29	redsh ye lf (al)	68					s^h					**B16**, 105
a1518	**—,4-acetoxy-3-methyl-**	$C_{15}H_{14}N_2O_2$. *See* a1512	254.29	ye pl (dil al)	81–2					s	s		s	chl s	**B16**, 130
a1519	**—,4-acetoxy-4′-methyl-**	$C_{15}H_{14}N_2O_2$. *See* a1512	254.29	og nd (al)	97–8					s^h					**B16**[2], 42
a1520	**—,2-amino-**	2-Benzeneazoaniline. $C_{12}H_{11}N_3$. *See* a1512	197.23	red nd (al)	59				i	δ	δ			os v	**B16**[2], 147
a1521	**—,3-amino-**	3-Benzeneazoaniline. $C_{12}H_{11}N_3$. *See* a1512	197.23	og-ye nd (peth)	68				i	s	s	s	s	chl s	**B16**, 304

For explanations, symbols and abbreviations see beginning of table.

No.	Name	Synonyms and Formula	Mol. wt.	Crystalline form, color and specific rotation	m.p. °C	b.p. °C	Density	n_D	Solubility w	al	eth	ace	bz	other solvents	Ref.
	Azobenzene														
a1522	**—,4-amino-**	4-Benzeneazoaniline, $C_{12}H_{11}N_3$. *See* a1512	197.23	og mcl nd (al)	127	>360			δ^h	s^h	s	...	s^h	chl s lig δ	**B16**^{2}, 149
a1523	—,—,hydrobromide	$C_{12}H_{11}N_3$. HBr. *See* a1512	278.04	bk vt	206–7				...	δ	δ	...	...		**B16**, 307
a1524	—,—,hydrochloride	$C_{12}H_{11}N_3$. HCl. *See* a1512	233.71	bl vt or pa red nd or pw (w or HCl)	240				s^h	s^h	...	...	...	HCl s^h	**B16**^{2}, 149
a1525	**—,4-amino-2,2′-dimethyl-**	$C_{14}H_{15}N_3$. *See* a1512	225.29	ye nd (lig)	116–7				...	...	...	...	...	lig v^h	**B16**^{2}, 180
a1526	**—,4-amino-2,3′-dimethyl-**	$C_{14}H_{15}N_3$. *See* a1512	225.29	ye-gold nd (al), ye-br (peth)	80				δ	s	...	...	...		**B16**^{2}, 181
a1527	**—,4-amino-2′,3-dimethyl**	$C_{14}H_{15}N_3$. *See* a1512	225.29	ye ta (al)	101.5–3				i	s	s	...	...	chl s	**B16**^{2}, 178
a1528	**—,4-amino-2,4′-dimethyl-**	$C_{14}H_{15}N_3$. *See* a1512	225.29	ye pl (al), nd, (lig)	127				...	s^h	...	...	...	lig s^h	**B16**, 348
a1529	**—,4-amino-3,3′-dimethyl-**	$C_{14}H_{15}N_3$. *See* a1512	225.29	ye br pl or nd (lig)	124				...	...	...	...	...	lig s^h	**B16**, 345
a1530	**—,4-amino-3,4′-dimethyl-**	$C_{14}H_{15}N_3$. *See* a1512	225.29	og-ye nd (lig), ye pl (al)	128				i	δ	...	...	...	lig δ	**B16**^{1}, 322
a1531	**—,4-benzoxy-4′-methyl-**	4-*p*-Tolueneazophenol benzyl ether. $C_{20}H_{18}N_2O$. *See* a1512	302.38	pa ye lf (lig)	128				...	s	v	...	v	lig s^h	**B16**. 107
a1532	**—,3,3′-bis(dimethylamino)-**	3,3′-Azo-bis(*N,N*-dimethylaniline). $C_{16}H_{20}N_4$. *See* a1512	268.37	red nd (al)	118				i	s	...	...	s	lig δ	**B16**, 305
a1533	**—4,4′-bis(dimethylamino)-**	4,4′-Azo-bis(*N,N*-dimethylaniline). $C_{16}H_{20}N_4$. *See* a1512	268.37	og nd (bz)	273	sub			...	δ	δ	...	v	chl v	**B16**^{2}, 174
a1534	**—,4-bromo-**	$C_{12}H_9BrN_2$. *See* a1512	261.12	og nd	90–1				i	s	s	s	...		**B16**^{2}, 14
a1535	**—,4-diacetamido-2,3′-dimethyl-**	$C_{18}H_{19}N_3O_2$. *See* a1512	309.37	red nd (lig), cr	65 (nd), 75 (cr)				...	...	...	...	...	os v	**B16**^{2}, 179
a1536	**—,2,2′-diamino-**	*o,o′*-Azodianiline. $C_{12}H_{12}N_4$. *See* a1512	212.25	red lf (al or bz)	134				δ	δ s^h	s	s	δ		**B16**^{2}, 148
a1537	**—,2,4-diamino-**	Chrysoidine (base). $C_{12}H_{12}N_4$. *See* a1512.	212.25	pa ye nd (w)	117.5				δ^h	s	s	...	s	chl s	**B16**^{2}, 203
a1538	**—,4,4′-diamino-**	*p,p′*-Azodianiline. $C_{12}H_{12}N_4$. *See* a1512.	212.25	gold ye nd (al), or-ye pr (al or bz)	249 cor				δ	s	...	...	v	chl v lig δ	**B16**^{2}, 174
a1539	**—,2,2′-diethoxy-**	*o*-Azophenetole. $C_{16}H_{18}N_2O_2$. *See* a1512	270.32	red pr (al)	131	240d			i	s	s	...	...	HCl s	**B16**, 92
a1540	**—,4,4′-diethoxy-**	*p*-Azophenetole. $C_{16}H_{18}N_2O_2$. *See* a1512	270.32	ye-br pl	162	d			i	δ v^h	s	...	s	chl s	**B16**^{2}, 44
a1541	**—,2,2′-dihydroxy-**	*o*-Azophenol. $C_{12}H_{10}N_2O_2$. *See* a1512	214.22	gold ye pl (bz), nd (al)	172	sub			i	δ	v	...	δ	con alk s	**B16**^{2}, 33
a1541[1]	**—,2,4-dihydroxy-**	4-Phenylazoresorcinol. Sudan G. $C_{12}H_{10}N_2O_2$. *See* a1512	214.22	dk red nd (dil al)	161 ($\frac{1}{2}$w), 170 (anh)				i	v	v	...	v	chl v aa v	**B16**^{2}, 80
a1542	**—,3,3′-dihydroxy-**	*m*-Azophenol. $C_{12}H_{10}N_2O_2$. *See* a1512	214.22	ye lf (dil al)	205				δ	v^h	v	v	δ	alk s^h lig δ	**B16**^{2}, 37
a1543	**—,4,4′-dihydroxy-**	*p*-Azophenol. $C_{12}H_{10}N_2O_2.H_2O$. *See* a1512	232.33	og-ye pl (dil al +1w), gr (st), red (metast), ta (eth)	216–6.5				δ	s	s	s	δ	aa δ lig δ	**B16**^{2}, 43
a1544	**—,2,4-dihydroxy-4′-nitro-**	4-(*p*-Nitrophenylazo)resorcinol. $C_{12}H_9N_3O_4$. *See* a1512	259.23	red (al)	199–200				i	δ^h	...	...	...	aa δ	**B16**^{2}, 81
a1545	**—,2,2′-dimethyl-**	*o*-Azotoluene. $C_{14}H_{14}N_2$. *See* a1512	210.27	dk red mcl (eth)	55–6		1.0215^{65}_{4}	1.6180^{65}	i	s	v	...	v	CS_2, chl s	**B16**^{2}, 19
a1546	**—,3,3′-dimethyl-** (*cis*)	*m*-Azotoluene. $C_{14}H_{14}N_2$. *See* a1512	210.27	red					...	...	...	...	...		
a1547	—,—(*trans*)	$C_{14}H_{14}N_2$. *See* a1512	210.27	og-red orh	54–4.5		1.0123^{66}	1.6152^{66}	...	v	v	...	v		**B16**^{2}, 20
a1548	**—,4,4′-dimethyl** (*cis*)	*p*-Azotoluene. $C_{14}H_{14}N_2$. *See* a1512	210.27	deep red	104				...	...	...	...	...		
a1549	—,—,(*trans*)	$C_{14}H_{14}N_2$. *See* a1512	210.27	og ye mcl (peth-al)	144				i	s	s	...	s	dil ac i lig s	**B16**^{2}, 21
a1549[1]	**—,4-dimethylamino-**	Butter yellow. $C_{14}H_{15}N_3$. *See* a1512	225.29	ye lf (al)	117	d			i	v	s	...	...	Py v, ac s lig δ^h	**B16**^{2}, 151
a1550	**—,2,2′-dimethyl-4-ethoxy-**	$C_{16}H_{18}N_2O$. *See* a1512	254.34	deep red nd (al)	64				...	v	v	...	v	lig v	**B16**, 134
a1551	**—,2,3′-dimethyl-4-ethoxy-**	$C_{16}H_{18}N_2O$. *See* a1512	254.34	pr (al)	73				...	v	...	...	...	os v	**B16**, 135
a1552	**—,2′,3-dimethyl-4-ethoxy-**	$C_{16}H_{18}N_2O$. *See* a1512	254.34	(lig)	35–7				...	v	v	...	...	lig s^h	**B16**, 131
a1553	**—,2,4′-dimethyl-4-ethoxy-**	$C_{16}H_{18}N_2O$. *See* a1512	254.34	og-red pl (al)	64				...	v	v	...	v	lig v	**B16**, 135
a1554	**—,3,3′-dimethyl-4-ethoxy-**	$C_{16}H_{18}N_2O$. *See* a1512	254.34	red-ye pl (al)	46–7				...	v	v	...	v		**B16**, 131
a1555	**—,3,4′-dimethyl-4-ethoxy-**	$C_{16}H_{18}N_2O$. *See* a1512	254.34	og-ye nd (al)	73–4	251^{42}			...	...	...	...	...	os v	**B16**, 131

For explanations, symbols and abbreviations see beginning of table.

PHYSICAL CONSTANTS OF ORGANIC COMPOUNDS (Continued)

No.	Name	Synonyms and Formula	Mol. wt.	Crystalline form, color and specific rotation	m.p. °C	b.p. °C	Density	n_D	Solubility w	al	eth	ace	bz	other solvents	Ref.
	Azobenzene														
a1556	**—,4,5′-dimethyl-2-ethoxy-**	$C_{16}H_{18}N_2O$. *See* a1512	254.34	pa red nd (abs al)	43	253–4^{63}				s^h					**B16**, 141
a1557	**—,2,2′-dimethyl-4-hydroxy-**	$C_{14}H_{14}N_2O$. *See* a1512	226.28	og-red pl (bz)	113					v	v		v		**B16**^{2}, 61
a1558	**—,2,3′-dimethyl-4-hydroxy-**	$C_{14}H_{14}N_2O$. *See* a1512	226.28	og-ye pl (bz)	106–7					s	v		v	lig v	**B16**^{2}, 61
a1559	**—,2,3-dimethyl-4-hydroxy-**	$C_{14}H_{14}N_2O$. *See* a1512	226.28	red pr (al)	132				i	s	s		s		**B16**^{2}, 60
a1560	**—,2,4-dimethyl-5-hydroxy-**	$C_{14}H_{14}N_2O$. *See* a1512	226.28	og-ye nd (lig-peth)	113.5–4					v		v	v	MeOH v peth v	**B16**, 146
a1561	**—,2,4′-dimethyl-4-hydroxy-**	$C_{14}H_{14}N_2O$. *See* a1512	226.28	og-ye pr (bz)	135					v	v		v	lig δ	**B16**^{2}, 62
a1562	**—,3,3′-dimethyl-4-hydroxy-**	$C_{14}H_{14}N_2O$. *See* a1512	226.28	gold-ye nd (bz)	115					v	v		v		**B16**^{2}, 60
a1563	**—,3,4′-dimethyl-4-hydroxy-**	$C_{14}H_{14}N_2O$. *See* a1512	226.28	og nd	163									os v	**B16**^{2}, 60
a1564	**—,3,5-dimethyl-2-hydroxy-**	$C_{14}H_{14}N_2O$. *See* a1512	226.28	dk red nd (al, lig)	90					v	v		v	aa v lig v	**B16**, 145
a1565	**—,3,5-dimethyl-4-hydroxy-**	$C_{14}H_{14}N_2O$. *See* a1512	226.28	ye nd, pr (lig)	95–6									os v	**B16**, 145
a1566	**—,4,4′-dimethyl-2-hydroxy-**	$C_{14}H_{14}N_2O$. *See* a1512	226.28	og-ye (lig)	150–1									lig v^h	**B16**^{2}, 61
a1567	**—,4′,5-dimethyl-2-hydroxy-**	$C_{14}H_{14}N_2O$. *See* a1512	226.28	red cr, ye pl (to)	112–3					δ v^h	v		v	chl v to s	**B16**^{2}, 63
a1569	**—,2-ethoxy-**	$C_{14}H_{14}N_2O$. *See* a1512	226.28	red pl (peth), or mcl pr	43–4									os v	**B16**, 91
a1570	**—,3-ethoxy-**	$C_{14}H_{14}N_2O$. *See* a1512	226.28	pl	63.5–4	*ca.* 200^{22}								os v	**B16**, 95
a1571	**—,4-ethoxy-**	$C_{14}H_{14}N_2O$. *See* a1512	226.28	og nd (60–70% al)	77–8	325–6	1.0400^{100}_4								
a1572	**—,4-ethoxy-2-methyl-**	$C_{15}H_{16}N_2O$. *See* a1512	240.31	og-red nd	51.5					v	v			lig v	**B16**, 134
a1573	**—,4-ethoxy-2′-methyl-**	$C_{15}H_{16}N_2O$. *See* a1512	240.31	og lf (al)	53				δ	v	v		v	chl v	**B16**, 105
a1574	**—,4-ethoxy-3-methyl-**	$C_{15}H_{16}N_2O$. *See* a1512	240.31	nd, mcl pr (al)	59					s	s		s		**B16**, 130
a1575	**—,4-ethoxy-3′-methyl-**	$C_{15}H_{16}N_2O$. *See* a1512	240.31	og-red pr (al)	65					v	v		v		**B16**, 106
a1576	**—,4-ethoxy-4′-methyl-**	$C_{15}H_{16}N_2O$. *See* a1512	240.31	gold-ye lf (al)	121–2	251^{47}				s v^h				chl v	**B16**, 107
a1577	**—,2-hydroxy-**	*o*-Phenylazophenol. $C_{12}H_{10}N_2O$. *See* a1512	198.23	og-red nd (eth)	82.5–3				δ	s	s			os s dil alk s	**B16**^{2}, 32
a1578	—,—,benzoate	$C_{19}H_{14}N_2O_2$. *See* a1512	302.34	og nd (lig)	93					s^h			s^h	lig s^h	**B16**, 91
a1579	**—,3-hydroxy-**	*m*-Phenylazophenol. $C_{12}H_{10}N_2O$. *See* a1512	198.23	ye nd (w, lig), pr (bz)	116.5–7				i	s	s	s	s	dil alk s lig $δ^h$	**B16**, 94
a1580	—,—,acetate	$C_{14}H_{12}N_2O_2$. *See* a1512	240.26	og-pl (peth)	67.5									peth s^h	**B16**, 95
a1581	—,—,benzoate	$C_{19}H_{14}N_2O_2$. *See* a1512	302.34	og-red nd (peth)	91.5–2									peth s^h	**B16**, 95
a1582	**—,4-hydroxy-**	*p*-Phenylazophenol. $C_{12}H_{10}N_2O$. *See* a1512	198.23	ye lf (bz)	154.5–5.5	220–30$^{20}_{0}$ δd			i	v	v			con sulfs	**B16**^{2}, 38
a1583	—,—,acetate	$C_{14}H_{12}N_2O_2$. *See* a1512	240.26	og lf (al), nd (al, lig)	89	>360d				s^h				lig s^h	**B16**^{1}, 236
a1584	—,—,benzoate	$C_{19}H_{14}N_2O_2$. *See* a1512	302.34	og-ye nd (al, lig), ye-red pr (eth, al) lf (al)	138					δ s^h	δ s^h			to v lig s^h	**B16**^{2}, 41
a1585	**—,4-hydroxy-2′-methoxy-2-methyl-**	$C_{14}H_{14}N_2O_2$. *See* a1512	242.28	og	161				δ	v	s				**B16**, 135
a1586	**—,4-hydroxy-2′-methoxy-3-methyl-**	$C_{14}H_{14}N_2O_2$. *See* a1512	242.28	og red pl (dil al)	68					v	v		v		**B16**, 131
a1587	**—,2-hydroxy-4-methyl-**	$C_{13}H_{12}N_2O$. *See* a1512	212.25	red pl (lig)	122					v		v	v	chl v lig s	**B16**^{1}, 241
a1588	—,—,benzoate	$C_{20}H_{16}N_2O_2$. *See* a1512	316.36	og-ye nd (lig)	98					δ			v	MeOH δ	**B16**^{1}, 241
a1589	**—,2-hydroxy-5-methyl-**	$C_{13}H_{12}N_2O$. *See* a1512	212.25	og-ye lf (bz, bz-lig), gold lf (w-al)	107–9	sub				v			s		**B16**^{2}, 62
a1590	**—,4-hydroxy-2-methyl-**	$C_{13}H_{12}N_2O$. *See* a1512	212.25	ye nd (lig)	109					v	v		v	chl v	**B16**^{2}, 61
a1591	**—,4-hydroxy-2′-methyl-**	$C_{13}H_{12}N_2O$. *See* a1512	212.25	red pl or lf (bz-lig), og-ye nd (bz)	107–8					v	v		v	chl v lig δ	**B16**^{2}, 41
a1592	**—,4-hydroxy-3-methyl-**	$C_{13}H_{12}N_2O$. *See* a1512	212.25	gold-ye lf or nd	128–30				i $δ^h$	v	v		v	chl v lig s	**B16**^{2}, 59

For explanations, symbols and abbreviations see beginning of table.

No.	Name	Synonyms and Formula	Mol. wt.	Crystalline form, color and specific rotation	m.p. °C	b.p. °C	Density	n_D	Solubility w	al	eth	ace	bz	other solvents	Ref.
	Azobenzene														
a1593	—,—,benzoate	$C_{20}H_{16}N_2O_2$. *See* a1512	316.36	ye nd	110–1				...	δ v^h	v	v	...	chl v	**B16,** 130
a1594	**—,4-hydroxy-3′-methyl-**	$C_{13}H_{12}N_2O$. *See* a1513	212.25	ye pl (al), pr (bz-lig)	144–5				...	s^h	...	...	s	lig δ	**B16**[2], 41
a1595	**—,4-hydroxy-4′-methyl-**	$C_{13}H_{12}N_2O$. *See* a1513	212.25	og-red mcl, ye (Na_2CO_3 sol or bz-lig), og-ye (bz-lig)	152				s^h	v	v	...	v		**B16**[2], 42
a1596	—,—benzoate	$C_{20}H_{16}N_2O_2$. *See* a1513	316.36	og-red pr (bz), redsh ye nd (lig)	159				...	...	...	...	s^h	lig s^h	**B16**[1], 237
a1597	**—,2-methoxy-**	$C_{13}H_{12}N_2O$. *See* a1513	212.26	og-red nd (dil al)	40–1	195–7[14]	1.0728^{100}_{4}		...	s^h	...	...	...	os v	**B16**[2], 33
a1598	**—,3-methoxy-**	$C_{13}H_{12}N_2O$. *See* a1513	212.26	og-red pl (MeOH)	32.5–3.5	193–3.5[15]			...	...	...	...	...	os v MeOH s^h	**B16,** 95
a1599	**—,4-methoxy-**	$C_{13}H_{12}N_2O$. *See* a1513	212.26	og-red pl, lf (al), (peth)	64	340	1.12^{75-8}		i	s^h	...	...	...	os v	**B16**[2], 40
a1600	**—,4-methoxy-4′-methyl-**	$C_{14}H_{14}N_2O$. *See* a1513	226.28	og-ye pr (al)	110–1				...	s^h	...	...	...	os s	**B16**[1], 236
a1601	**—,4-nitro-**	$C_{12}H_9N_3O_2$. *See* a1513	227.22	lf, nd (al, lig)	134–5				...	s^h	...	s	s	chl s	**B16**[2], 17
a1602	**—,4-phenyl-amino-**	4-Anilineazobenzene. $C_{18}H_{15}N_3$. *See* a1513	273.34	ye pl or pr	82				i	v	v	...	...	lig v	**B16,** 314
a1603	**—,2,3,4-triamino-**	Bismark's brown. $C_{12}H_{15}N_3$. *See* a1513	227.27		143.5				i	v	v	...	...		**B16,** 386
a1604	**2-Azobenzene-carboxylic acid**	*o*-Phenylazobenzoic acid. 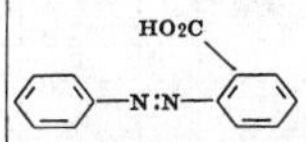 HO2C —N:N—	226.24	og-red nd or pl	97–8				...	v	v	...	v	aa v	**B16**[2], 97
a1605	**3-Azobenzene-carboxylic acid**	*m*-Phenylazobenzoic acid. 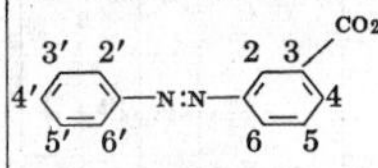 CO2H 3′ 2′ 2 3 4′ —N:N— 4 5′ 6′ 6 5	226.24	og-red pl (w)	170–1				δ^h	s	v	...	v	chl v lig δ	**B16,** 229
a1606	**4-Azobenzene-carboxylic acid**	*p*-Phenylazobenzoic acid. 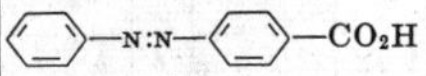 —N:N— —CO2H	226.24	red pl or lf (al)	248.5–9.5				...	s^h	s	...	s^h	chl s aa s^h lig δ	**B16**[2], 98
a1607	**3-Azobenzene-carboxylic acid, 4-hydroxy-**	$C_{13}H_{10}N_2O_3$. *See* a1605	242.24	nd (bz)	220d				...	...	...	...	...		
a1608	**—,4-hydroxy-2′-nitro-**	$C_{13}H_9N_3O_5$. *See* a1605	287.09	br-red cr	273				δ	s	...	...	...	chl δ	**B16**[2], 101
a1609	**—,4-hydroxy-3′-nitro-**	$C_{13}H_9N_3O_5$. *See* a1605	287.09	red br nd (al)	237				i	v	v	...	v	chl v aa v	**B16**[2], 101
a1610	**—,4-hydroxy-4′-nitro-**	$C_{13}H_9N_3O_5$. *See* a1605	287.09	og-br nd (dil aa)	253–4d				...	v	...	...	...	to s^h aa v	**B16**[2], 101
a1611	**—,3′-sulfo-**	$C_{13}H_{10}N_2O_5S$. *See* a1605	306.31	ye lf (al)	d				v	v	i	...	...		**B16,** 268
—	**2-Azobenzene-carboxylic acid, 4′-dimethyl-amino-**	*see* **Methyl red**													
a1612	**2,2′-Azobenzene-dicarboxylic acid**	*o*-Azobenzoic acid. 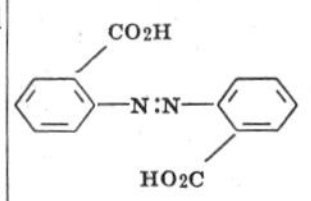 CO2H —N:N— HO2C	270.24	dk ye nd (al)	245–5.5 cor				δ^h	s v^h	v	...	i		**B16**[1], 287
a1613	**3,3′-Azobenzene-dicarboxylic acid**	*m*-Azobenzoic acid. 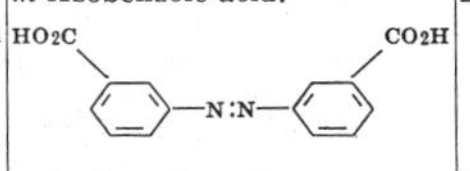 HO2C CO2H —N:N—	270.24	ye nd (aa)	*ca.* 340d				i	δ^h	δ	...	...		**B16,** 233
a1614	**4,4′-Azobenzene-dicarboxylic acid**	*p*-Azobenzoic acid. 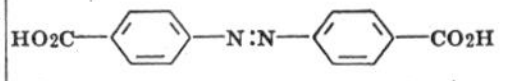 HO2C— —N:N— —CO2H	270.24	dk ye nd (al) or pk pw	>400d without melting				i	i	i	...	...	os i aa i^h	**B16,** 236
a1614[1]	**3,3′-Azobenzene-disulfonic acid**	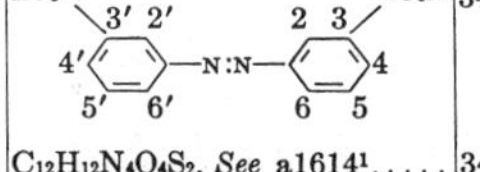 HO3S SO3H 3′ 2′ 2 3 4′ —N:N— 4 5′ 6′ 6 5	342.35												
a1615	—,diamide	$C_{12}H_{12}N_4O_4S_2$. *See* a1614[1]	340.39	ye cr (al)	295				...	s^h	...	...	...	os δ	**B16,** 268
a1616	—,dichloride	$C_{12}H_8Cl_2N_2O_4S_2$. *See* a1614[1]	379.26	red nd (eth)	166–7				d	d^h	s^h	...	...		**B16,** 268
a1617	—,diethyl ester	$C_{16}H_{18}N_2O_6S_2$. *See* a1614[1]	398.47	gold-ye nd (eth)	100				δ	v	v	...	...		**B16,** 268

For explanations, symbols and abbreviations see beginning of table.

No.	Name	Synonyms and Formula	Mol. wt.	Crystalline form, color and specific rotation	m.p. °C	b.p. °C	Density	n_D	Solubility w	al	eth	ace	bz	other solvents	Ref.
	3,4′-Azobenzenesulfonic acid														
a1617[1]	**3,4′-Azobenzene-disulfonic acid**	HO_3S–(3′ 2′ 5′ 6′)–N:N–(2 3 4 6 5)–SO_3H	342.35												
a1618	—,diamide	$C_{12}H_{12}N_4O_4S_2$. *See* a1617[1]	340.39	ye nd (al)	258				δ	s^h					**B16,** 279
a1619	—,dichloride	$C_{12}H_8Cl_2N_2O_4S_2$. *See* a1617[1]	379.26	red nd (eth, CS_2)	123–5				d^h	d^h	s^h			CS_2 s^h	**B16,** 279
a1619[1]	**4,4′-Azobenzene-disulfonic acid**	HO_3S–(3′ 2′ 5′ 6′)–N:N–(2 3 6 5)–SO_3H	342.35												
a1620	—,diamide	$C_{12}H_{12}N_4O_4S_2$. *See* a1619[1]	340.38	og nd	312d				δ	δ^h	i		i		**B16,** 280
a1621	—,dichloride	$C_{12}H_8Cl_2N_2O_4S_2$. *See* a1619[1]	379.26	br-red nd (eth, bz)	222				d^h	d^h	δ s^h		v	chl v	**B16,** 280
a1622	**3,3′-Azobenzene-disulfonic acid, 4,4′-dimethyl-,** dichloride	$C_{14}H_{12}Cl_2N_2O_4S_2$. *See* a1641′	407.31	deep red (bz)	194				d^h	d^h			s^h		**B16,** 283
a1622[1]	**—,6,6′-dimethyl.** dichloride	$C_{14}H_{12}Cl_2N_2O_4S_2$. *See* a1614[1]	407.31	red pr (bz, +2)	220				d^h	d^h	δ		s^h		**B16,** 285
a1623	**4,4′-Azobenzene-disulfonic acid, 2,2′-dimethyl-,** diamide	$C_{14}H_{16}N_4O_4S_2$. *See* a1619[1]	368.45	pl(NH_3)	>250				δ					NH_3 v	**B16,** 284
a1624	—,—,dichloride	$C_{14}H_{12}Cl_2N_2O_4S_2$ *See* a1619[1]	407.31	deep red nd (bz)	218				d^h	d^h	s		v^h		**B16,** 284
a1626	**3,3′-Azobenzene-disulfonic acid, 2,2′,4,4′,6,6′-hexabromo-,** dichloride	$C_{12}H_2Br_6Cl_2N_2O_4S_2$. *See* a1614[1]	852.71	vt pl (bz)	222–4				d		δ		δ s^h		**B16,** 269
a1627	**—,4,4′,6,6′-tetrabromo-,** dichloride	$C_{12}H_4Br_4Cl_2N_2O_4S_2$. *See* a1614[1]	694.85	red nd (bz)	232–3				d	d^h	δ		s^h		**B16,** 269
a1628	**4,4′-Azobenzene-disulfonic acid, 2,2′,6,6′-tetrabromo-,**	$C_{12}H_6O_6N_2Br_4S_2$. *See* a1619[1]	657.96	ye-br lf (bz)	258–62				d^h	d^h	δ		v		**B16,** 281
a1629	**4-Azobenzene-sulfonic acid**	4-Phenylazobenzenesulfonic acid* (3′ 2′ 4′ 5′ 6′)–N:N–(2 3 6 5)–SO_3H	262.29	og-red pl (w+3)	127(+3w)				δ v^h	δ	δ				**B16[2],** 115
a1630	**4-Azobenzene-sulfonic acid, chloride**	$C_{12}H_9ClN_2O_2S$. *See* a1629	280.73	og-red nd or pl (bz)	82				i	s			s^h		**B16,** 272
a1631	—,4′-chloro-	$C_{12}H_9ClN_2O_3S$. *See* a1629	296.73	br nd (w)	148				v	v					**B16,** 271
a1632	—,—,amide	$C_{12}H_{10}ClN_3O_2S$. *See* a1629	295.75	ye-br pr (al)	211					δ s^h	δ				**B16,** 272
a1633	—,—,chloride	$C_{12}H_8Cl_2N_2O_2S$. *See* a1629	315.18	red pr (eth)	130				d^h	s	s				**B16,** 272
—	**2-Azobenzene-sulfonic acid, 4′-dimethylamino-,** sodium salt	*see* **Methyl orange**													
a1634	**3-Azobenzene-sulfonic acid, 4′-formyl-3′-hydroxy-**	HO, OCH–(ring)–N:N–(ring)–SO_3H	306.30	red lf	>270				v	v	i			HCl δ	**B16,** 267
a1635	**4-Azobenzene-sulfonic acid, 4′-hydroxy-**	$C_{12}H_{10}N_2O_4S$. *See* a1629	278.29	ye red pr (dil al), amor (w)	200d				s	v^h	δ				**B16[2],** 115
a1636	**3-Azobenzene-sulfonic acid, 4′-hydroxy-3′-nitro-**	O_2N, HO–(ring)–N:N–(ring)–SO_3H	323.29	gold-ye pr (dil HCl+w)	116(+w), 235d (anh)				v	v					**B16,** 267
a1637	**4-Azobenzene-sulfonic acid, 4′-methyl-,** chloride	$C_{13}H_{11}ClN_2O_2S$. *See* a1629	294.76	red pr (bz)	130–2				d^h	d^h	v		s^h	chl v	**B16,** 272
—	**Azobenzoic acid**	*see* **Azobenzenedicarboxylic acid**													
a1638	***p,p′*-Azobiphenyl**	Di-*p*-xenyldiimide. (ring)–(ring)–N:N–(ring)–(ring)	334.24	og-red lf (bz)	256				i	i	v			con sulf s aa i	**B16[2],** 27

For explanations, symbols and abbreviations see beginning of table.

No.	Name	Synonyms and Formula	Mol. wt.	Crystalline form, color and specific rotation	m.p. °C	b.p. °C	Density	n_D	Solubility w	al	eth	ace	bz	other solvents	Ref.
	Azodicarboxylic acid														
a1639	**Azodicarboxylic acid,** diamide	Azobisformamide. Azodicarbonamide. $H_2NCON:NCONH_2$	116.08	og nd (w)	180–4d	....	....	....	i	i	δ	...	...	os i	**B3**^{2}, 99
a1640	—,diethyl ester	$C_2H_5O_2CN:NCO_2C_2H_5$	174.16	og-ye	....	106^{13}	....	....	...	...	...	...	...	....	**B3**^{2}, 98
a1641	**Azomethane** (*trans*)	Dimethyldiazene. Dimethyldiimide. $CH_3N:NCH_3$	58.08	col or pa ye gas	−78	1.5	....	....	...	...	...	...	...	C_7H_{16} s	....
a1642	—,**hexafluoro-**	$F_3CN:NCF_3$	166.03	pa grsh gas	−133	−31.6	....	....	δ	...	...	...	...	....	....
a1643	**1,1′-Azonaphthalene**	Di-α-naphthyldiimide. 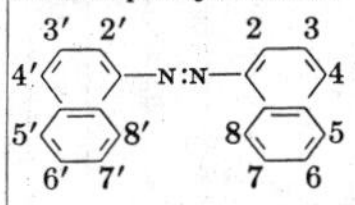	282.35	red nd (gl aa)	190	sub	....	....	i	δ	...	s	v	con sulf s AmOH s aa s^h	**B16**^{2}, 26
a1644	**1,2′-Azonaphthalene**	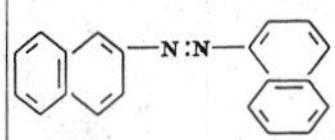	282.35	red nd, br lf (aa)	144–5 (136)	....	....	....	i	s	...	...	s	con sulf s aa v	**E12B,** 936
a1645	**2,2′,Azonaphthalene**	Di-β-naphthyldiimide. 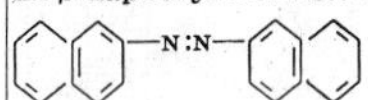	282.35	red lf (bz), og-ye nd (chl, al)	208	sub	....	....	i	δ	i	...	δ s^h	con sulf s chl s^h	**E12B,** 936
a1646	**1,1′-Azonaphthalene, 4-amino-**	$C_{20}H_{15}N_3$. *See* a1643	297.36	red-br nd of gr lustre	173–5	....	....	....	i	δ	δ	...	δ	....	....
—	**Azophenol**	*see* **Azobenzene, dihydroxy-**													
—	**Azotoluene**	*see* **Azobenzene, dimethyl-**													
a1647	**Azoxybenzene** (*cis*)	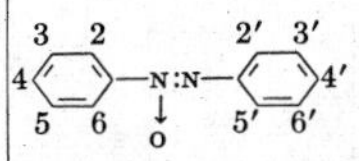	198.23	....	87	....	1.166^{20}_{4}	1.633^{20}	...	...	...	...	...	....	....
a1648	—,(*trans*)	$C_{12}H_{10}N_2O$. *See* a1647	198.23	lt ye rh	36	d	1.159^{20}_{4}	1.652^{20}	i	v	v	...	...	lig v	**B16**^{2}, 313
a1649	—,**4-bromo-**	$C_{12}H_7BrN_2O$. *See* a1647	277.13	deep ye pl	93.5–4.5	....	1.4138^{100}_{4}	....	i	s	...	...	s^h	MeOH s^h	**B16**^{2}, 315
a1650	—,**2,2′-dimethoxy-**	$C_{14}H_{14}N_2O_3$. *See* a1647	258.27	ye-og pr(MeOH)	81	....	....	....	...	v	v	v	v	MeOH s^h, chl v	**B16,** 635
a1651	—,**3,3′-dimethoxy-**	$C_{14}H_{14}N_2O_3$. *See* a1647	258.27	ye (al)	51	....	....	....	...	s^h	s^h	...	...	....	**B16**^{2}, 325
a1652	—,**4,4′-dimethoxy-**	$C_{14}H_{14}N_2O_3$. *See* a1647	258.27	ye mc	118.5–.6	1.171^{115}_{4}	....	....	...	s^h	...	...	s	....	**B16**^{2}, 326
a1653	**2,2′-Azoxybenzenedicarboxylic acid**	*o*-Azoxybenzoic acid. 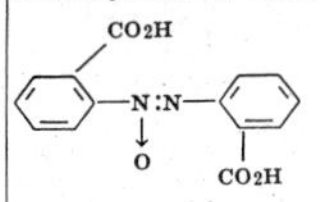	286.25	pa ye tcl pr or lf	254	....	....	....	δ	s^h	δ	s^h	δ	chl δ Py s aa s^h lig δ	**B16**^{2}, 335
a1654	**3,3′-Azoxybenzenedicarboxylic acid**	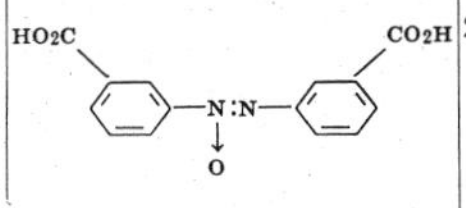	286.25	pa ye nd or lf (gl aa)	310–20d	....	....	....	i	...	δ	...	...	....	**B16**^{1}, 388
a1655	**4,4′-Azoxybenzenedicarboxylic acid**	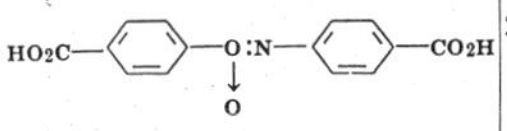	286.25	ye amor pw	240d	....	....	....	i	i	i	...	...	Py s oos i	**B16**^{2}, 336
—	**Azoxybenzoic acid**	*see* **Azoxybenzenedicarboxylic acid**													
a1656	**1,1′-Azoxynaphthalene**	$C_{10}H_7^{\alpha}N(O):NC_{10}H_7^{\alpha}$	298.35	ye or red orh pl (al)	127	....	....	....	i	s^h	δ	...	...	con sulf s	**E12B,** 1038
a1657	**2,2′-Azoxynaphthalene**	$C_{10}H_7^{\beta}N(O):NC_{10}H_7^{\beta}$	298.35	ye or red nd (al, gl aa)	164	....	....	....	i	δ	δ	...	s	chl s MeOH δ	**E12B,** 1039
a1658	**Azulene**	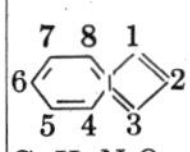	128.17	bl or grsh-bk pl	98.5–9	270d 115–35^{10}	....	....	i	s^h	...	...	...	ac s os s	**E12A,** 421
a1659	—,**1,3,5-trinitro-**	$C_{16}H_{11}N_3O_6$. *See* a1658	341.28	br nd	166.5–7.5	....	....	....	...	s^h	...	...	...	....	....
a1660	—,**1,4-dimethyl-1-isopropyl-**	Guaiazulene. $C_{15}H_{18}$. *See* a1658	198.31	bl-vt pl (al)	31.5	....	0.973^{20}_{4}	....	...	s^h	s	...	...	AcOEt s	**E12A,** 424
a1661	—,**1,5-dimethyl-8-isopropyl-**	Chamazulene. $C_{15}H_{18}$. *See* a1658	198.31	bl	....	161^{12}	0.9883^{20}_{4}	....	i	s	s	...	...	AcOEt s peth s	**B5**^{2}, 474
a1662	—,**4,8-dimethyl-2-isopropyl-**	Vetivazulene. $C_{15}H_{18}$. *See* a1658	198.31	red-vt nd (al)	31.5–2	140–60^2	....	....	...	s^h	...	...	...	....	**E12A,** 427
a1663	—,Δ^4-**Octahydro-** (*trans*)	*trans*-Bicyclo-(5.3.0)-Δ^2-decene. $C_{10}H_{16}$. *See* a1658	136.24	....	....	63.5^8	0.8996^{21}_{4}	....	i	s	...	...	...	aa s	**E12A,** 421

For explanations, symbols and abbreviations see beginning of table.

No.	Name	Synonyms and Formula	Mol. wt.	Crystalline form, color and specific rotation	m.p. °C	b.p. °C	Density	n_D	Solubility w	al	eth	ace	bz	other solvents	Ref.
	Baeyer's acid														
—	**Baeyer's acid**	*see* **2-Naphthalenesulfonic acid, 7-amino-**													
—	**Baicalin**	*see* **Flavone, 5, 6, 7-trihydroxy-**													
—	**B.A.L.**	*see* **1-Propanol, 2,3-dimercapto-**													
b1	**Bandrowski's base**	1,4-Benzoquinonebis(2,5-diaminoanil). 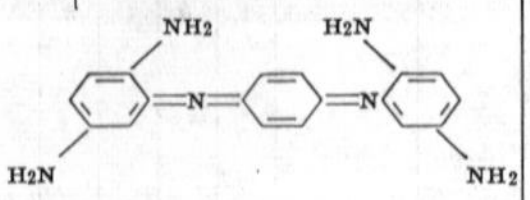	318.38	dk red or br lf	238				δ					os δ dil HCl s	**B13**^{2}, 146
—	**Banthine bromide**	*see* **Methantheline bromide**													
—	**Barbital**	*see* **Barbituric acid, 5,5-diethyl-**													
b2	**Barbituric acid**	Pyrimidinetroine. 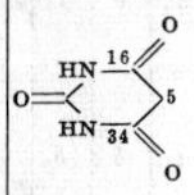	128.09	rh pr (w^h+2)	248	260d			δ	δ	s				**B24**^{2}, 267
b3	**—,5-allyl-5-butyl-**	$C_{11}H_{16}N_2O_3$. *See* b2	224.26		128				s^h	s^h					**B24**^{2}, 292
b4	**—,5-allyl-5(2-cyclopenten-1-yl)-**	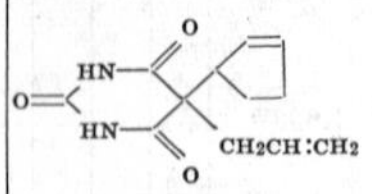	234.25		139–40				δ s^h	v					
b5	**—,5-allyl-5-isobutyl-**	Sandoptal. $C_{11}H_{16}N_2O_3$. *See* b3	224.26		138				δ v^{wr}	s	s	s		chl s lig i	**B24**^{2}, 292
b6	**—,5-allyl-5-isopropyl-1-methyl-**	$C_{11}H_{16}N_2O_3$. *See* b2	224.26		56–7	$176–8^{12}$				s	s		s		
b7	**—,5-allyl-5-phenyl-**	$C_{13}H_{12}N_2O_3$. *See* b2	244.25		156–7.5				δ	v	s		δ	lig i	
b8	**—,5-amino-**	Uramil. $C_4H_5N_3O_3$. *See* b2	143.10	lf (w^h)	>400				i δ^h		i		i		**B25**^{1}, 704
b9	**—,5-benzylidene-**	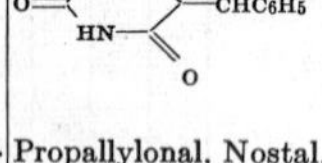	216.20	pr (åa)	256				i	i	i	s^h			**B24**^{2}, 299
b10	**—,5(2-bromoallyl)-5-isopropyl-**	Propallylonal. Nostal. $C_{10}H_{13}BrN_2O_3$. *See* b2	289.14		181				δ	v	δ	v	δ	chl δ aa v	**B24**^{2}, 291
b11	**—,5(1-cyclohexenyl)-5-ethyl-**	Cyclobarbital, 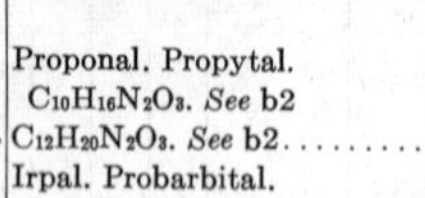	236.27	lf (w^h)	176.4				i s^h	v	s			dil alk s aa δ	**B24**^{2}, 294
b12	**—,5,5-diallyl-**	Allobarbitone. Curral. Dial. $C_{10}H_{12}N_2O_3$. *See* b2	208.22	lf (w)	174				δ	s	s		v		**B24**^{2}, 293
b13	**—,5,5-dibromo-**	Dibromin. $C_4H_2Br_2N_2O_3$. *See* b2	285.89	pl (MeOH-bz) lf (dil HNO_3)	234				δ	v	s			sulf s	**B24**^{2}, 272
b14	**—,5,5-diethyl**	Barbital. Barbitone. Veronal. $C_8H_{12}N_2O_3$. *See* b2	184.20	mcl, tcl, rh, bipym	191		1.220		δ s^h	s^h	s	s		chl s lig s	**B24**^{2}, 279
b15	**—,5,5-diethyl-1-methyl**	$C_9H_{14}N_2O_3$. *See* b2	198.22	nd	150–1				s^h						**B24**^{2}, 281
b16	**—,5,5-dipropyl-**	Proponal. Propytal. $C_{10}H_{16}N_2O_3$. *See* b2	212.25		146				i δ^h	s	s		s	dil alk s chl s	**B24**^{2}, 286
b17	**—,5-ethyl-5-hexyl-**	$C_{12}H_{20}N_2O_3$. *See* b2	240.30		112–3				δ	s	s		s		**B24**^{2}, 288
b18	**—,5-ethyl-5-isopropyl-**	Irpal. Probarbital. $C_9H_{14}N_2O_3$. *See* b2	198.13	nd	203				δ	i	s				**B24**^{2}, 284
b19	**—,5-ethyl-5(2-methyl-allyl)-2-thio-**	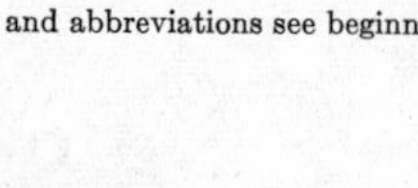	226.29		160–1				i					dil alk s	
b20	**—,5-ethyl-5(3-methylbutyl)-**	Amobarbital. Amytal. $C_{11}H_{18}N_2O_3$. *See* b2	226.28	lf (w)	156–8				δ	v	v				**B24**^{2}, 286

For explanations, symbols and abbreviations see beginning of table.

No.	Name	Synonyms and Formula	Mol. wt.	Crystalline form, color and specific rotation	m.p. °C	b.p. °C	Density	n_D	Solubility w	al	eth	ace	bz	other solvents	Ref.
	Barbituric acid														
b22	**—,5-ethyl-1-methyl-5-phenyl-**	Mephobarbital. Prominal. $C_{13}H_4N_2O_3$. *See* b2	246.27		176				δ s^h	v	δ			chl s	
b23	**—,5-ethyl-5-pentyl-**	$C_{11}H_{18}N_2O_3$. *See* b2	226.28		135–6				δ s^h	v	v			chl s	**B24**[2], 286
b23[1]	**—,5-ethyl-5(2-pentyl)**	Nembutal. Pentobarbital. $C_{11}H_{18}N_2O_3$. *See* b2	226.28	nd (w)	130				δ	s	s				**B24**[2], 287
b24	**—,5-ethyl-5(1-pyperidyl)-**	Eldoral. 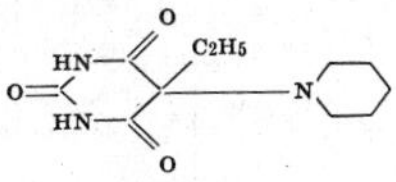	239.28		213–4				δ s^h	s	s		i		**C36**,3160
b25	**—,5-ethyl-5-phenyl-**	Luminal. Phenobarbital. $C_{12}H_{12}N_2O_3$. *See* b2	232.24		156–7 (unst) 175.3				i s^h	s	s				**B24**[2], 297
b26	**—,5(2-furfurylidene)-2-thio-**	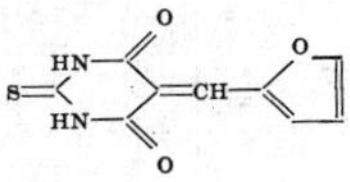	222.22	ye pl	>280d				i	i					**B27**, 607
b27	**—,5-hydroxy-**	Dialuric acid. $C_4H_4N_2O_4$. *See* b2	144.09	tetr	224, 214–5 (+w)				δ v^h						**B25**[2], 61
b28	**—,5-methyl-5-phenyl-**	Rutonal. $C_{11}H_{10}N_2O_3$. *See* b2	218.21		220				i	s	s			alk s	**B24**[2], 296
b29	**—,5-nitro-**	Dilituric acid. $C_4H_3N_3O_5$ *See* b2	173.09	pr, lf(w+3)	180–1				s^h	s	i				**B24**[2], 273
b30	**—,5(2-phenylethyl)-**	$C_{12}H_{12}N_2O_3$. *See* b2	232.24		212–3				δ	v					**B24**[2], 296
b31	**—,2-thio-**	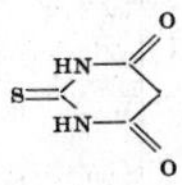	144.15	lf (w)	235(d)				δ s^h	s				dil alk s dil HCl s	**B24**[2], 275
—	**Batyl alcohol**	*see* **Glycerol, 1-octadecyl ether**													
b32	**Bebeerine(d)**	Chondrodendrin. 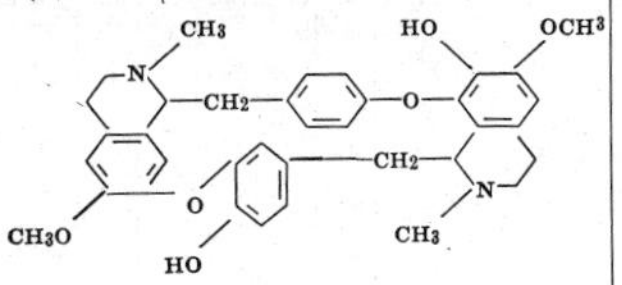	594.68	$[\alpha]^{25}_D$+297 (al)	221–1.5, 161 (+bz)							v		chl v	**B27**[2], 896
b33	—(*l*)	Curine. $C_{36}H_{38}N_2O_6$	594.68	pr, nd $[\alpha]^{20}_D$ −328 (Py) $[\alpha]^{28}_D$−298(al)	221,165–7 (+bz), 159–63 (+al)				δ	δ s^h	δ	s	s	chl s Py s ac s	**B27**[2], 896
—	**Behenic acid**	*see* **Docosanoic acid**													
—	**Behenolic acid**	*see* **13-Docosynoic acid**													
b35	**Benzaldehyde**	Benzenecarbonal*. Benzenecarboxaldehyde. 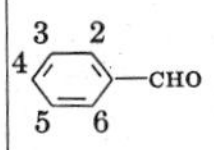	106.12		−55.6, −56.9	178.1^{760}	1.0415^{25}_4	1.5463^{20}	δ	∞	∞			liq NH_3 s lig v	**B7**[2], 145
b36	—,azine	Dibenzalhydrazine. $C_6H_5CH{:}NN{:}CHC_6H_5$	208.26	ye	93				i	s^h	s		s	chl s	**B7**[2], 171
b37	—,diacetate	Benzylidene diacetate. $C_6H_5CH(O_2CCH_3)_2$	208.21	pl (eth)	45–6	220^{760}	1.11^{20}			v	v				**B9**[2], 161
b38	—,hydrazone	$C_6H_5CH{:}NNH_2$	120.15	lf	16	140^{14}			d	s					**B7**[2], 171
b39	—,imide, *N*-ethyl-	*N*-Benzalethylamine. $C_6N_5CH{:}NC_2H_5$	133.19			195^{779}	0.937^{20}_4		i	s	s				**Am 66,** 1467
b40	—,—,*N*-methyl-	*N*-Benzalmethylamine. $C_6H_5CH{:}NCH_3$	119.16			180–1	0.9672^{15}_4	1.5540^{20}	δ				s		**B7**[2], 162
b41	—,—,*N*(4-tolyl)-	*N*-Benzal-*p*-toluidine. 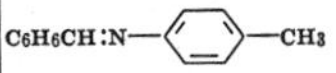 $C_6H_5CH{:}N-C_6H_4-CH_3$	195.27		35	326^{723}			i						**B12**[2], 496

For explanations, symbols and abbreviations see beginning of table.

No.	Name	Synonyms and Formula	Mol. wt.	Crystalline form, color and specific rotation	m.p. °C	b.p. °C	Density	n_D	Solubility w	al	eth	ace	bz	other solvents	Ref.
	Benzaldehyde														
b42	—,oxime(*anti*)	*anti*-Benzaldoxime. $C_6H_5CH{:}NOH$	121.13	nd (eth)	130		1.145_4^{20}		s^h	v	v				**B7**[2], 169
b43	—,—(*syn*)	*syn*-Benzaldoxime. $C_6H_5CH{:}NOH$	121.13	pr	34	200	1.1111_4^{20}	1.5908^{20}	δ	s	v		v		**B7**[2], 167
b44	—,phenylhydrazone	$C_6H_5CH{:}NNHC_6H_5$	196.24	nd (lig), pr	156					δ s^h	δ	s^h	s^h	liq NH_3 s	**B15**[2], 57
b45	**—,2-acetamido-**	$C_9H_9NO_2$ *See* b35	163.18	nd (w)	70–1				s^h					os s	**B14**, 26
b46	**—,3-acetamido**	$C_9H_9NO_2$ *See* b35	163.18	pl (bz)	84				δ	v			s^h		**B14**, 29
b47	**—,4-acetamido-**	$C_9H_9NO_2$ *See* b35	163.18	nd	155				s	δ			s		**B14**, 38
b48	**—,2-amino-**	Anthranilaldehyde. C_7H_7NO. *See* b35	121.13	lf	39–40				δ	v	v		s	chl s lig i	**B14**, 21
b49	—,—,oxime	NH_2 / CH:NOH	136.16	nd (bz)	135–6				δ	v	v		δ	CS_2 v lig δ	**B14**, 24
b50	**—,3-amino-**	C_7H_7NO *See* b35	121.13	ye (exists only in solution)							s			ac s	**B14**[2], 21
b51	—,—,oxime	H_2N / CH:NOH	136.16	nd (bz)	88					v	v		s	lig δ	**B14**, 28
b52	**—,4-amino-**	C_7H_7NO *See* b35	121.13	pl (w)	71				s	s	s			ac s	**B14**[2], 22
b53	—,—,oxime	$C_7H_7N_2O$. *See* b42	136.16	ye	124					v	v			ac s	**B14**, 31
b54	**—,2-bromo-**	C_7H_5BrO. *See* b35	185.03		21–2	230			i	v			v		**B7**[2], 181
b55	—,—,diacetate	Br / $CH(O_2CCH_3)_2$	287.13		84–6					s^h	s				**B7**[2], 181
b56	**—,3-bromo-**	C_7H_5BrO *See* b35	185.03			$228–30^{726}$			i	v	v				**B7**[2], 182
b57	**—,4-bromo-**	C_7H_5BrO *See* b35	185.03		57				i	v			v		**B7**[2], 182
b58	—,—,diacetate	Br— —$CH_2(O_2CCH_3)_2$	287.13	ye	94–5					s^h	s				**B7**[2], 182
b59	**—,2-bromo-4-hydroxy-**	$C_7H_5BrO_2$ *See* b35	201.03	nd (w)	159.5				δ s^h						**B8**[2], 74
b60	**—,2-bromo-5-hydroxy-**	$C_7H_5BrO_2$. *See* b35	201.03	nd (w)	135				i	v	v		v	chl s	**B8**[2], 57
b61	**—,3-bromo-2-hydroxy-**	$C_7H_5BrO_2$. *See* b35	201.03	nd (dil al)	49				δ					alk s, os s	**B8**, 54
b62	**—,3-bromo-4-hydroxy-**	$C_7H_5BrO_2$. *See* b35	201.03	lf (w)	124				δ	v	v	v	s	chl s lig δ	**B8**, 82
b63	**—,4-bromo-2-hydroxy-**	$C_7H_5BrO_2$. *See* b35	201.03	nd (dil al)	52				δ					os s	**B8**, 54
b64	**—,4-bromo-3-hydroxy-**	$C_7H_5BrO_2$. *See* b35	201.03		131.5									chl v	**B8**[2], 56
b65	**—,5-bromo-2-hydroxy-**	$C_7H_5BrO_2$. *See* b35	201.03	nd (al), lf (eth)	105				i	s	s				**B8**[2], 45
b66	**—,5-bromo-4-hydroxy-3-methoxy-**	$C_8H_7BrO_3$. *See* b35	231.06		164				i	s^h	δ		δ		**B8**[2], 286
b67	**—,2-chloro-**	C_7H_5ClO. *See* b35	140.57	nd	11	$213–4^{760}$	1.2483_4^{20}	1.5662^{20}	δ	s	s		s		**B7**[2], 177
b68	**—,3-chloro-**	C_7H_5ClO. *See* b35	140.57	pr	17–8	$213–4^{760}$	1.2410_4^{20}	1.5650^{20}	δ	s	s		s		**B7**[2], 178
b69	**—,4-chloro-**	C_7H_5ClO. *See* b35	140.57	pl	47	$213–4^{760}$	1.196_4^{61}	1.5552^{61}	s	v	v		v	CS_2 s aa s	**B7**[2], 178
b70	**—,2-chloro-3-hydroxy-**	$C_7H_5ClO_2$. *See* b35	156.57		139.5					v				chl δ aa v	
b71	**—,2-chloro-4-hydroxy-**	$C_7H_5ClO_2$. *See* b35	156.57	nd (w)	147–8				v					aa v^h	**B8**, 81
b72	**—,2-chloro-5-hydroxy-**	$C_7H_5ClO_2$. *See* b35	156.57	nd (w)	110.5–11.5					s				aa v	**B8**[2], 526
b73	**—,3-chloro-2-hydroxy-**	$C_7H_5ClO_2$. *See* b35	156.57	nd (MeOH)	55									os v	**B8**[1], 523
b74	**—,3-chloro-4-hydroxy-**	$C_7H_5ClO_2$. *See* b35	156.57	nd (w)	139	$149–50^{14}$			δ v^h	v	v				**B8**, 81
b75	**—,4-chloro-2-hydroxy-**	$C_7H_5ClO_2$. *See* b35	156.57	nd (al)	52.5				s					os v	
b76	**—,4-chloro-3-hydroxy-**	$C_7H_5ClO_2$. *See* b35	156.57	nd	121				s						
b77	**—,5-chloro-2-hydroxy-**	$C_7H_5ClO_2$ *See* b35	156.57	pl	98–9				i	v	s				**B8**[2], 45
b78	—,—,oxime	$C_7H_6ClNO_2$ *See* b42	171.58	nd	128					s^h					**B8**, 53

For explanations, symbols and abbreviations see beginning of table.

No.	Name	Synonyms and Formula	Mol. wt.	Crystalline form, color and specific rotation	m.p. °C	b.p. °C	Density	n_D	Solubility w	al	eth	ace	bz	other solvents	Ref.
	Benzaldehyde														
b79	**—,5-chloro-4-hydroxy-3-methoxy-**	$C_8H_7ClO_3$ *See* b35	186.60	tetr	165	...	...	...	i	s^h	...	...	...	aa s^h	**B8**[2], 22
b80	**—,2,3-dichloro-**	$C_7H_4Cl_2O$. *See* b35	175.02	...	65–7	...	...	...	...	v	v	...	...	...	...
b81	**—,2,4-dichloro-**	$C_7H_4Cl_2O$ *See* b35	175.02	pr	70–1	...	...	...	i	s	...	...	...	...	**B7**[2], 180
b82	**—,2,5-dichloro-**	$C_7H_4Cl_2O$ *See* b35	175.02	nd (al)	57–8	230–3	...	...	i	v	v	...	v	chl v	**B7**, 237
b83	**—,2,6-dichloro-**	$C_7H_4Cl_2O$ *See* b35	175.02	nd (lig)	70–1	...	...	...	i	s^h	s^h	...	...	lig s^h	**B7**, 237
b84	**—,3,4-dichloro-**	$C_7H_4Cl_2O$ *See* b35	175.02	...	43–4	247–8	...	...	i	s^h	s^h	...	...	...	**B7**, 238
b85	**—,3,5-dichloro-**	$C_7H_4Cl_2O$. *See* b35	175.02	nd or lf	65	$235–40^{748}$	...	...	δ	...	...	...	...	os v	...
b86	**—,2,4-dichloro-3-hydroxy-**	$C_7H_4Cl_2O_2$ *See* b35	191.01	...	141	...	...	...	...	...	...	...	...	lig δ v^h	...
b87	**—,2,6-dichloro-3-hydroxy-**	$C_7H_4Cl_2O_2$ *See* b35	191.01	...	140.5	...	...	...	δ^h	...	...	...	...	aa s	...
b88	**—,3,5-dichloro-2-hydroxy-**	$C_7H_4Cl_2O_2$ *See* b35	191.01	ye rh	95	...	...	...	i	...	...	...	...	...	**B8**, 54
b89	**—,—,oxime**	$C_7H_5O_2NCl_2$ *See* b35	206.02	nd	195–6	...	...	...	i	v	v	...	v	...	**B8**, 54
b90	**—,3,5-dichloro-4-hydroxy-**	$C_7H_4Cl_2O_2$ *See* b35	191.01	nd (chl)	158–9	...	...	...	...	v	v	...	δ	chl δ lig δ	**B8**, 81
b91	**—,4,6-dichloro-3-hydroxy-**	$C_7H_4Cl_2O_2$ *See* b35	191.01	...	130	...	...	...	v^h	...	...	...	v	lig i	**B8**[1], 526
b92	**—,2,3-diethoxy-**	$C_{11}H_{14}O_3$ *See* b35	194.23	...	...	169^{37}	...	...	...	s	s	...	...	...	**B8**[2], 268
b93	**—,3,4-diethoxy-**	Protocatechaldehyde. $C_{11}H_{14}O_3$ *See* b35	194.23	...	...	278–80	...	...	...	s	...	...	...	...	**B8**, 256
b94	**—,4-diethyl-amino-**	$C_{11}H_{15}NO$. *See* b35	177.25	nd (w)	41	174^7	...	...	v	s	...	...	s	...	**B14**[2], 25
b95	**—,2,4-dihydroxy-**	β-Resorcylaldehyde. $C_7H_6O_3$ *See* b35	138.12	nd (w)	135–6	$220–8^{22}$	...	...	s	v	v	...	δ	chl v	**B8**[2], 272
b96	**—,—,oxime**	$C_7H_7NO_3$ *See* b42	153.14	...	192	...	...	...	s	v	v	...	...	...	**B8**[2], 274
b97	**—,2,5-dihydroxy-**	Gentisinaldehyde. $C_7H_6O_3$ *See* b35	138.12	nd (bz)	99	...	...	...	v	v	...	...	s^h	chl v lig δ	**B8**[2], 276
b98	**—,3,4-dihydroxy-**	Protocatechualdehyde. $C_7H_6O_3$ *See* b35	138.12	lf (w)	153–4	...	...	...	s v^h	v	v	...	...	...	**B8**[2], 277
b99	**—,—,oxime**	$C_7H_7NO_3$ *See* b42	153.14	nd	149–51	...	...	...	v	v	δ	...	δ	chl δ	**B8**[2], 285
b100	**—,2,6-dihydroxy-4-methyl**	Atranol. $C_8H_8O_3$ *See* b35	152.15	...	124 118–9 (+½w)	...	...	...	v^h	v	v	...	s^h	chl v lig s^h	**B8**[2], 304
b101	**—,2,5-diiodo-4-hydroxy-3-methoxy-**	$C_8H_6I_2O_3$ *See* b35	403.94	nd	155–6	...	...	...	i	s	δ	...	...	...	...
b102	**—,2,4-dimethoxy-**	$C_9H_{10}O_3$ *See* b35	166.18	nd, (al or lig)	71	165^{10}	...	...	i	s	s	...	s	lig s	**B8**[2], 273
b103	**—,3,4-dimethoxy-**	Veratraldehyde. $C_9H_{10}O_3$ *See* b35	166.18	nd	42–3	281	...	...	δ	v	v	...	...	...	**B8**[2], 282
b104	**—,2,4-dimethoxy-6-hydroxy-**	$C_9H_{10}O_4$ *See* b35	182.18	nd, pl,	71	$190–5^{25}$	...	...	i	v	v	..	v	chl v aa v	**B8**[2], 436
b105	**—,2,6-dimethoxy-4-hydroxy-**	$C_9H_{10}O_4$ *See* b35	182.18	pr (bz)	69–71	...	...	...	...	...	...	...	...	...	**B8**[2], 436
b106	**—,3,4-dimethoxy-5-hydroxy-**	$C_9H_{10}O_4$ *See* b35	182.18	...	62–3	$177–80^{12}$	...	...	...	v	...	...	v	aa v lig v^h	**B8**[2], 437
b106[1]	**—,3,5-dimethoxy-4-hydroxy-**	Syringaldehyde. $C_9H_{10}O_4$ *See* b35	182.18	br nd (lig)	113	$192–3^{14}$	...	...	δ	v	v	...	v^h	chl v aa v lig δ	**B8**[2], 437
b107	**—,2,4-dimethyl-**	$C_9H_{10}O$. *See* b35	134.18	...	−9 to −8	218	...	...	...	s	s	s	s	...	**B7**[2], 240
b108	**—,2,5-dimethyl-**	$C_9H_{10}O$. *See* b35	134.18	...	...	$219–20^{738}$	...	...	...	v	...	...	...	os s	**B7**[1], 166
b109	**—,3,4-dimethyl-**	$C_9H_{10}O$. *See* b35	134.18	...	...	$223–5^{760}$	...	...	...	s	s	s	...	...	**B7**[2], 240
b110	**—,3,5-dimethyl-**	$C_9H_{10}O$. *See* b35	134.18	...	9	220–2	...	...	...	s	s	s	s	...	**B7**[2], 240
b111	**—,4-dimethyl-amino-**	$C_9H_{11}NO$. *See* b35	149.19	lf (w)	74	$176–7^{17}$	...	...	δ	s	s	s	...	...	**B14**[2], 23
b112	**—,2,4-dinitro-**	$C_7H_4N_2O_5$. *See* b35	196.12	pr (al), pl (bz)	72	$190–210^{10-20}$	...	...	δ	s	s	...	s	lig δ	**B7**[2], 205
b113	**—,2,6-dinitro-**	$C_7H_4N_2O_5$. *See* b35	196.12	lf	123	...	...	...	s^h	s	s	...	s	chl s	**B7**[2], 206
b114	**—,2-ethoxy-**	$C_9H_{10}O_2$. *See* b35	150.18	...	20–2	247–9	...	...	...	∞	∞	...	...	...	**B8**, 43
b115	**—,3-ethoxy-**	$C_9H_{10}O_2$. *See* b35	150.18	...	...	245.5^{760}	1.0768^{20}_{4}	1.5408^{20}	...	s	s	...	s	...	**B8**, 60
b116	**—,4-ethoxy-**	$C_9H_{10}O_2$. *See* b35	150.18	...	13–4	255	1.08^{21}_{21}	...	...	v	v	...	v	...	**B8**, 73
b117	**—,3-ethoxy-2-hydroxy-**	$C_9H_{10}O_3$ *See* b35	166.18	...	64–5	$263–4^{740}$	...	...	δ	...	...	...	...	os s	**B8**, 267
b118	**—,3-ethoxy-4-hydroxy-**	Bourbonal. Ethylvanillin. $C_9H_{10}O_3$. *See* b35	166.18	pl or sc (w)	77–8	...	...	...	δ	s^h	s	...	s	chl s	**B8**[2], 282
b119	**—,4-ethoxy-2-hydroxy-**	$C_9H_{10}O_3$. *See* b35	166.18	nd (dil al)	35	...	...	...	...	v	...	...	...	...	**B8**, 242
b120	**—,5-ethoxy-2-hydroxy-**	$C_9H_{10}O_3$. *See* b35	166.18	pr	51.5	230	...	...	δ	v	v	...	...	chl v	**B8**, 245
b121	**—,4-ethoxy-3-methoxy-**	$C_{10}H_{12}O_3$. *See* b35	180.21	mcl pr	73–4	...	...	...	δ^h	s	s	...	s	chl s aa s	**B8**[2], 283

For explanations, symbols and abbreviations see beginning of table.

No.	Name	Synonyms and Formula	wt.	Crystalline form, color and specific rotation	m.p. °C	b.p. °C	Density	n_D	Solubility w	al	eth	ace	bz	other solvents	Ref.
	Benzaldehyde														
b122	**—,2-hydroxy-**	Salicylaldehyde. $C_7H_6O_2$. *See* b35	122.12		−11 to −10	197^{760}	1.1461^{25}_{25}	1.5702^{20}	δ	∞	∞				**B8**, 31
b123	—,—,azine	$C_{14}H_{12}N_2O_2$. *See* b36	240.26	nd (al)	213–4				δ	$δ^h$			v	chl δ, alk v	**B8**², 43
b124	—,—,oxime	Salicylaldoxime. $C_7H_7NO_2$. *See* b42	137.14	pr (bz)	57				δ	v	v		v	lig i	**B8**², 42
b125	**—,3-hydroxy-**	$C_7H_6O_2$. *See* b35	122.12	nd (w)	106	240^{160}			δ	s			s	lig i	**B8**², 52
b126	—,—,azine	$C_{14}H_{12}N_2O_2$. *See* b36	240.26	ye pl (al)	205				i	δ			δ	chl δ	**B8**¹, 526
b127	—,—,oxime	$C_7H_7NO_2$ *See* b42	137.14		90				v	v			i	chl δ lig i	**B8**¹, 525
b128	**—,4-hydroxy-**	$C_7H_6O_2$ *See* b35	122.12	nd (w)	116		1.129^{130}_{4}		δ s^h	v	v		s		B8², 63
b129	—,—,azine	$C_{14}H_{12}N_2O_2$. *See* b36	240.26	ye	268d 273(+w)				$δ^h$	v			v	CCl_4 δ lig δ	**B8**¹, 531
b131	**—,4-hydroxy-2-iodo-3-methoxy-**	2-Iodovanillin. $C_8H_7IO_3$. *See* b35	278.05		155–6				i	s	δ				
b132	**—,4-hydroxy-5-iodo-3-methoxy-**	5-Iodovanillin. $C_8H_7IO_3$ *See* b35	278.05	pa ye	180				i	s	s				
b133	**—,2-hydroxy-3-methoxy-**	*o*-Vanillin. $C_8H_8O_3$. *See* b35	152.15	lt ye-lt gr nd (w, lig)	44–5	265–6^{760}			δ	v	v			CCl_4 v lig s	**B8**², 267
b134	**—,2-hydroxy-4-methoxy-**	$C_8H_8O_3$. *See* b35	152.15	nd (w), cr (al)	62–3				δ s^h	s v^h	v		v	lig v	**B8**², 272
b135	**—,2-hydroxy-5-methoxy-**	$C_8H_8O_3$ *See* b35	152.15	ye	4	247–8^{760}			δ	v	v				**B8**², 276
b136	**—,3-hydroxy-4-methoxy-**	Isovanillin. $C_8H_8O_3$ *See* b35	152.15	pl (w)	116–7	163–6^{10}	1.196		δ s^h	s	s		s^h	chl v aa s	**B8**², 282
b137	**—,3-hydroxy-5-methoxy-**	$C_8H_8O_3$ *See* b35	152.15		130–1					v			v	aa v lig δ	**B8**², 291
b138	**—,4-hydroxy-2-methoxy-**	$C_8H_8O_3$ *See* b35	152.15	lf (bz), nd (w)	153				δ	s^h	s		δ	chl s lig δ	**B8**², 272
b139	**—,4-hydroxy-3-methoxy-**	Vanillin. $C_8H_8O_3$. *See* b35	152.15	nd (w, lig)	80–1	285	1.056		δ v^h	v	v		v	os v	**B8**², 272
b140	—,—,acetate	Vanillin acetate. CH_3O; CH_3CO_2— —CHO	194.19	nd	77				δ	v	v				
b141	**—,2-hydroxy-3-nitro-**	$C_7H_5NO_4$ *See* b35	167.12		109–10								s		**B8**², 48
b142	**—,2-hydroxy-5-nitro-**	$C_7H_5NO_4$. *See* b35	167.12		126										**B8**², 48
b143	—,—,oxime	$C_7H_6N_2O_4$ *See* b42	182.14		225										**B8**², 48
b144	**—,3-hydroxy-2-nitro-**	$C_7H_5NO_4$. *See* b35	167.12	nd or pl (bz, lig)	157					s			s^h	lig δ	**B8**², 58
b145	—,—,oxime	$C_7H_6N_2O_4$. *See* b42	182.14	ye nd	172–5					s	s				**B8**², 58
b146	**—,3-hydroxy-4-nitro-**	$C_7H_5NO_4$. *See* b35	167.12	ye pl	134				δ	v	v			chl s lig δ	**B8**², 58
b147	—,—,oxime	$C_7H_6N_2O_4$ *See* b42	182.14	ye nd (chl)	161					v			δ	chl s	**B8**², 58
b148	**—,4-hydroxy-2-nitro-**	$C_7H_5NO_4$. *See* b35	167.12	ye nd	67				δ	v	v		v		**B8**, 83
b149	**—,4-hydroxy-3-nitro-**	$C_7H_5NO_4$. *See* b35	167.12	ye nd (al, w)	143–4				s^h	s v^h	δ			chl δ	**B8**², 77
b150	—,—,oxime	$C_7H_6N_2O_4$. *See* b42	182.14	pr or nd	169					v	v		δ	chl δ lig δ	**B8**², 77
b151	**—,4-isopropyl-**	Cumaldehyde. Cuminal. $C_{10}H_{12}O$. *See* b35	148.21			235–6	0.978^{20}_{4}	1.5301^{20}	i	s	s				**B7**², 247
b152	**—,2-methoxy-**	*o*-Anisaldehyde. $C_8H_8O_2$. *See* b35	136.15	pr	37–8	243–4^{760}	1.1326^{20}_{4}	1.5598^{20}	i	s	v			chl v	**B8**², 40
b153	**—,3-methoxy-**	*m*-Anisaldehyde. $C_8H_8O_2$ *See* b35	136.15			230	1.1187^{20}_{4}	1.5538^{20}		s	s				**B8**², 53
b154	**—,4-methoxy-**	Anisaldehyde. $C_8H_8O_2$ *See* b35	136.15		2.5	249.5^{760}	1.1192^{25}_{4}	1.5730^{20}	i	∞	∞				**B8**², 64
b155	—,—,oxime	$C_8H_9O_2N$. *See* b42	151.17	nd	134								δ		**B8**², 69
b156	**—,2-methoxy-3-nitro-**	$C_8H_7NO_4$. *See* b35	181.15	nd (w)	89–90				i	s	s				**B8**, 57
b157	**—,2-methyl-**	*o*-Tolualdehyde. C_8H_8O. *See* b35	120.15			197	1.0386^{19}_{4}	1.549^{19}	$δ^h$	s	s				**B7**, 295
b158	**—,3-methyl-**	*m*-Tolualdehyde. C_8H_8O. *See* b35	120.15			199	1.0189^{21}_{4}	1.541^{21}	δ	∞	∞				**B7**, 296

For explanations, symbols and abbreviations see beginning of table.

No.	Name	Synonyms and Formula	Mol. wt.	Crystalline form, color and specific rotation	m.p. °C	b.p. °C	Density	n_D	Solubility w	al	eth	ace	bz	other solvents	Ref.
	Benzaldehyde														
b159	**—,4-methyl-**	*p*-Tolualdehyde. C_8H_8O. *See* b35	120.15	...	...	204–5^{760}	1.0194^{17}_{4}	1.547^{17}	δ	∞	∞	...	...	...	**B7**, 297
b160	**—,3,4-methyl-enedioxy-**	Piperonal. 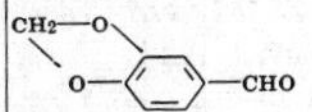 CH_2—O O— —CHO	150.14	wh-ye	37	263^{760}	...	...	δ s^h	v	∞	...	...	...	**B19**^{2}, 141
b161	—,—,oxime	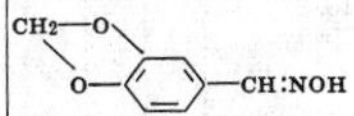 CH_2—O O— —CH:NOH	165.14	...	146	...	...	...	...	...	s	...	...	...	**B19**^{2}, 147
b162	**—,2-nitro-**	$C_7H_5NO_3$. *See* b35	151.12	ye	43.5–4	156^{23}	1.2844^{60}_{4}	...	δ	v	v	...	v	os v	**B7**^{2}, 185
b163	—,—,diacetate	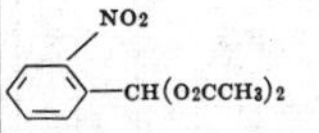 NO_2 —$CH(O_2CCH_3)_2$	253.22	...	88–9	...	...	...	...	s^h	...	...	...	...	**B7**^{2}, 187
b164	—,—,dimethyl acetal	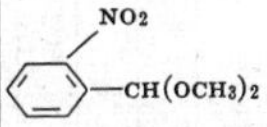 NO_2 —$CH(OCH_3)_2$	197.20	gr-ye	...	274–6^{762}	...	...	...	...	...	...	s	...	**B7**^{1}, 137
b165	**—,3-nitro-**	$C_7H_5NO_3$. *See* b35	151.12	lt ye hd (w)	58	164^{23}	1.2792^{20}_{4}	...	δ	s	s	...	...	os s	**B7**^{2}, 190
b166	—,—,diacetate	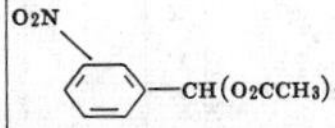 O_2N —$CH(O_2CCH_3)_2$	253.22	pr	66–7	...	1.393	...	...	s	...	...	...	os s	**B7**^{2}, 192
b167	—,—,dimethyl acetal	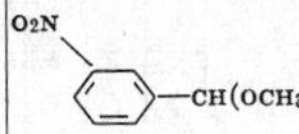 O_2N —$CH(OCH_3)_2$	197.20	...	...	162–4^{19}	1.209	...	...	...	...	...	s	...	**B7**, 253
b168	**—,4-nitro-**	$C_7H_5NO_3$. *See* b35	151.12	lf, pr (w)	106	...	1.496	...	δ	v	δ	...	s	lig δ	**B7**^{2}, 196
b169	—,—,diacetate	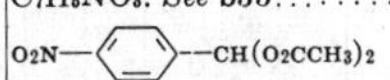 O_2N— —$CH(O_2CCH_3)_2$	253.22	pr (al)	127	...	...	...	...	s	...	...	...	os s	**B7**^{2}, 198
b170	—,—,dimethyl acetal	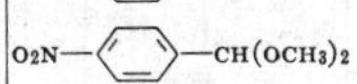 O_2N— —$CH(OCH_3)_2$	197.20	...	28–9	294–6^{774}	...	...	...	...	...	...	s	...	**B7**^{2}, 198
b171	**—,pentachloro-**	C_7HCl_5O. *See* b35	278.37	nd (bz, al)	202.5	...	...	...	...	δ	v	...	v	CS_2 v	**B7**^{1}, 134
b172	**—,2,3,4,5-tetra-chloro-**	$C_7H_2Cl_4O$. *See* b35	243.92	...	106	...	...	...	...	...	...	...	...	...	...
b173	**—,2,3,4-trichloro-**	$C_7H_3Cl_3O$, *See* b35	209.47	nd (al)	91	...	...	...	...	δ v^h	...	...	...	...	**B7**, 238
b174	**—,2,3,5-trichloro-**	$C_7H_3Cl_3O$. *See* b35	209.47	nd (al)	56	...	...	...	...	v	...	...	...	...	...
b175	**—,2,3,6-trichloro-**	$C_7H_3Cl_3O$. *See* b35	209.47	nd (lig)	86–7	...	...	...	...	...	v	...	...	lig δ v^h	**B7**, 238
b176	**—,2,4,5-trichloro-**	$C_7H_3Cl_3O$. *See* b35	209.47	nd (al)	112–3	...	...	...	δ	v	v	...	v	chl v CS_2 v	**B7**, 238
b177	**—,2,4,6-trichloro-**	$C_7H_3Cl_3O$. *See* b35	209.47	...	58–9	...	...	...	...	...	v	...	...	lig δ v^h	**B7**, 238
b178	**—,3,4,5-trichloro-**	$C_7H_3Cl_3O$. *See* b35	209.47	nd (al)	90–1	...	...	...	δ^h	v^h	v	v	v	chl v	...
b179	**—,2,4,6-tri-methyl-**	Mesitylaldehyde. $C_{10}H_{12}O$. *See* b35	148.20	...	14	237^{760}	...	...	i	s	s	s	s	...	**B7**^{2}, 250
b180	**—,2,4,6-trinitro-**	$C_7H_3N_3O_7$	241.12	dk gr pl (bz)	119	...	...	...	i	s	s	s	s	chl s aa s	**B7**^{2}, 207
—	**Benzaldoxime**	*see* **Benzaldehyde**, oxime													
—	**Benzamarone**	*see* **1,5-Pentanedione, 1,2,3,4,5-pentaphenyl-**													
—	**Benzamide**	*see* **Benzoic acid**, amide													
—	**Benzamidine**	*see* **Benzoic acid**, amidine													
—	**Benzanilide**	*see* **Benzoic acid**, amide, *N*-phenyl-													
b181	**1,2-Benzanthra-cene**	2,3-Benzophenanthrene. 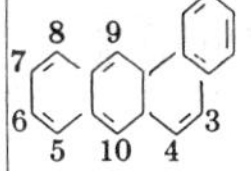 (ring numbering 3, 4, 5, 6, 7, 8, 9, 10)	228.29	ye br flr pl or lf (al, aa)	162	435^{760}	...	...	i	δ	...	...	v	aa δ	**E14**, 329
—	**2,3-Benzanthra-cene**	*see* **Naphthacene**													
b182	**1,2-Benzanthra-cene, 9,10-di-hydro-**	$C_{18}H_{14}$. *See* b181	230.31	lf (bz, al, aa)	112	...	...	...	i	δ	...	...	δ	aa δ	**E14**, 329
b183	**—,9,10-dimethyl-**	$C_{20}H_{16}$ *See* b181	256.35	pa ye pl (al, aa)	122–3	...	...	...	i	δ	...	s	v	CS_2, to s	**E14s**, 139
b184	**— 4-hydroxy-**	$C_{18}H_{12}O$. *See* b181	244.30	pa ye or og nd (bz)	230–1	...	...	...	...	...	...	...	s	Py, to s aa v	**E14s**, 171
b185	**—,3-methyl-**	$C_{19}H_{14}$ *See* b181	242.32	pl (bz-al), nd (bz-lig)	155	160$^{0.1}$ sub	...	...	i	...	...	...	s	xyl s	**E14s**, 116

For explanations, symbols and abbreviations see beginning of table.

No.	Name	Synonyms and Formula	Mol. wt.	Crystalline form, color and specific rotation	m.p. °C	b.p. °C	Density	n_D	Solubility w	al	eth	ace	bz	other solvents	Ref.
	Benzanthracene														
b186	—,**4-methyl-**	$C_{19}H_{14}$. *See* b181	242.32		124				i	s					**E14s**, 117
b187	—,**5-methyl-**	$C_{19}H_{14}$. *See* b181	242.32	ye mcl pl (bz-al)	158–9				i	s	s	...	s	xyl s	**E14s**, 118
b188	—,**6-methyl-**	$C_{19}H_{14}$ *See* b181	242.32		149–51				i	...	...	...	...	chl s xyl s^h CS_2 s^h	**E14s**, 118
b189	—,**7-methyl-**	$C_{19}H_{14}$ *See* b181	242.32		179–81				i	...	...	...	...	aa s^h	**E14s**, 119
b190	—,**8-methyl-**	$C_{19}H_{14}$. *See* b181	242.32	pl or nd (al, aa)	118				i	s	...	...	...	aa s	**E14s**, 120
b191	—,**9-methyl-**	$C_{19}H_{14}$. *See* b181	242.32	pa ye nd or pl (aa, MeOH)	138–9				i	s	...	...	...	CS_2 s aa s	**E14s**, 121
b192	—,**10-methyl-**	$C_{19}H_{14}$ *See* b181	242.32		141				i	s	...	s	...	CS_2, CCl_4 s aa s	**E14s**, 122
b193	**1,2-Benzanthracene-10-carboxaldehyde**	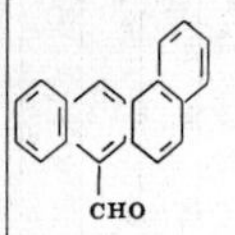	256.31	ye pr or nd	148				...	s	s	s	s	MeOH s	**E14s**, 186
b194	**1,2-Benz-3,4-anthraquinone**	2,3-Benzo-9,10-phenanthraquinone. 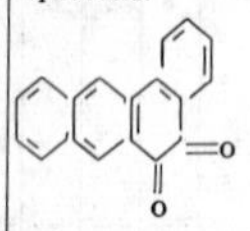	258.28	red nd (to)	262–3d				...	δ	...	...	δ	sulf s to δ aa δ	**B7**[2], 759
b195	**1,2-Benz-9,10-anthraquinone**	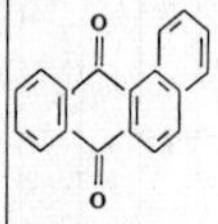	258.28	ye pr	168	sub			...	δ	δ	...	s	chl s aa δ lig δ	**B7**[2], 760
b196	**Benzanthrene**	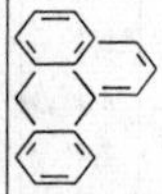	216.28	lf (al)	84				...	s	...	...	...	sulf s aa s xyl s	**E14s**, 343
b197	**1,9-Benzanthr-10-one**	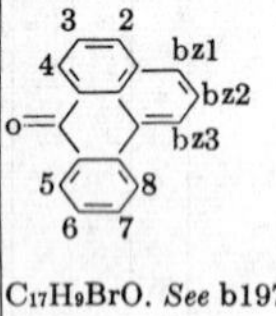	230.27	ye nd	170–1				i	...	...	...	...	sulf s^h	**B7**[2], 368
b198	—,*bz*-**1-bromo-**	$C_{17}H_9BrO$. *See* b197	309.19	ye nd (aa)	178				...	s	...	...	...	MeOH, sulf s	**E14s**, 355
b199	—,*bz*-**1-chloro-**	$C_{17}H_9ClO$ *See* b197	264.73	ye nd (aa)	182–3				...	...	...	...	δ	PhCl δ aa δ	**E14s**, 354
b200	—,*bz*-**2-hydroxy-**	$C_{17}H_{10}O_2$ *See* b197	246.27	ye nd	298				...	...	...	...	...	sulf s, os δ alk s	**E14s**, 373
b201	—,**4-hydroxy-**	$C_{17}H_{10}O_2$ *See* b197	246,27	ye nd (aa)	178–9				...	...	s	...	v^h	sulf s	**E14s**, 371
—	**Benzaurine**	*see* **Methanol, bis(4-hydroxyphenyl) phenyl-**													
—	**Benzedrine**	*see* **Amphetamine**													
b202	**Benzene***	Benzol. Phene* 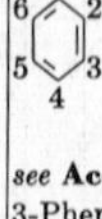	78.11		5.5	80.1	0.8787^{15}_{4}	1.5011^{20}	δ	∞	∞	∞	...	chl ∞ aa∞	**B5**[2], 119
—	—,**acetyl-**	*see* **Acetophenone**													
b203	—,**allyl-**	3-Phenylpropene* C_9H_{10}. *See* b202	118.17			156^{760}	0.8929^{20}_{0}	1.5143	i	s	s	...	s	chl s, CCl_4s	**B5**[2], 373
b204	—,**1-allyl-4-bromo-**	C_9H_9Br. *See* b202	197.08			$222–3^{730}$	1.324^{15}_{4}	1.559^{15}	i	...	s	...	s	chl s	**B5**[2], 373
b205	—,**2-allyl-4-chloro-1-hydroxy-**	C_9H_9ClO. *See* b202	168.62	pr (lig)	48	137^{18}	1.171^{15}		...	s	...	...	s	chl s alk s	**B6**[1], 282

For explanations, symbols and abbreviations see beginning of table.

PHYSICAL CONSTANTS OF ORGANIC COMPOUNDS (Continued)

No.	Name	Synonyms and Formula	Mol. wt.	Crystalline form, color and specific rotation	m.p. °C	b.p. °C	Density	n_D	Solubility w	al	eth	ace	bz	other solvents	Ref.
	Benzene														
b206	**—,1-allyl-3,4-dimethoxy-**	$C_{11}H_{14}O_2$. *See* b202	178.22			248–9	1.055^{15}	1.532^{20}	i	s	s	...	...	...	...
b207	**—,1-allyl-2-hydroxy-**	$C_9H_{10}O$. *See* b202	134.18		−6	220^{760}	1.0255^{15}_{15}	...	...	...	...	...	...	...	**B6,** 282
b208	**—,1-allyl-4-hydroxy-**	Chavicol. $C_9H_{10}O$. *See* b202	134.18		16	$230–1^{750}$	1.033^{18}_4	1.5441^{18}	s^h	∞	∞	...	...	chl ∞	**B6**[2], 529
b209	**—,4-allyl-1-hydroxy-2-methoxy-**	Betel phenol. Chavibetol. $C_{10}H_{12}O_2$. *See* b202	164.21		8	254–5	1.0690^{15}_{15}	1.5413^{20}_4	...	s	s	...	...	...	**B6**[2], 923
—	**—,amino-***	*see* **Aniline**													
b210	**—,1-amino-2-butyl-***	*o*-Butyl aniline. $C_{10}H_{15}N$. *See* b202	149.24	ye liq		$122–5^{12}$	0.953^{20}_4	...	...	...	...	...	...	oss ac s	**B7**[2], 633
b211	**—,1-amino-4-butyl-***	$C_{10}H_{15}N$. *See* b202	149.24	pa ye liq		248–50	0.945^{20}_4	...	i	...	...	...	...	os, ac s	**B7**[2], 633
b212	**—,1-amino-4-*sec*-butyl-(*dl*)**	$C_{10}H_{15}N$. *See* b202	149.24			238^{762}	0.949^{15}_4	...	...	...	...	...	...	...	**B12**[2], 635
b213	**—,1-amino-2-*tert*-butyl-**	$C_{10}H_{15}N$. *See* b202	149.24			233–5	0.977^{15}	...	...	v	v	...	v	ac s	**B12**[2], 636
b214	**—,1-amino-3-*tert*-butyl-**	$C_{10}H_{15}N$. *See* b202	149.24			229^{708}	...	...	...	v	v	...	v	...	**B12**[2], 637
b215	**—,1-amino-4-*tert*-butyl-**	$C_{10}H_{15}N$. *See* b202	149.24	ye to red	17	240^{740}	0.9525^{15}_4	1.5380^{20}	δ	∞	∞	...	...	...	**B12**[3], 637
b216	**—,1-amino-4-cyclohexyl-***	[structure]–NH_2	175.28	pl (lig)	55		...	...	...	s	...	...	s	...	**B12**[2], 667
b217	**—,2-amino-3,5-dibromo-1-ethoxy-***	$C_8H_9Br_2NO$. *See* b202	294.99	pr (dil al)	52		...	...	...	s	...	...	...	lig s^h	**B13,** 387
b218	**—,5-amino-1,3-dibromo-2-methoxy-***	$C_7H_7Br_2NO$. *See* b202	280.96	pl (lig)	66		...	...	...	s	s	s	s	chl s aa s	**B13,** 517
b219	**—,1-amino-2-diethylamino-***	$C_{10}H_{16}N_2$. *See* b202	164.25			312^{744}	...	...	δ	...	...	...	...	os s	**B13**[1], 6
b220	**—,1-amino-3-diethylamino-***	$C_{10}H_{16}N_2$. *See* b202	164.25			$276–8^{760}$	...	...	...	s	...	...	...	...	**B13**[2], 26
b221	**—,1-amino-4-diethylamino-***	$C_{10}H_{16}N_2$. *See* b202	164.25			260–2	...	...	...	...	...	...	...	...	**B13**[2], 40
b222	**—,1-amino-3,5-dihydroxy-***	Phloramin. $C_6H_7NO_2$. *See* b202	125.13	nd	146–52		...	...	δ^h	s	v	...	...	...	**B13,** 787
b223	**—,4-amino-1,2-dihydroxy-**	$C_6H_7NO_2$. *See* b202	125.13	br to vt	124–5d		...	...	s	s	...	v	δ	...	**B13**[2], 464
b224	**—,1-amino-2,3-dimethyl-***	2,3-Xylidine. $C_8H_{11}N$. *See* b202	121.18		<−15	224^{728}	0.9931^{20}	1.5684^{20}	δ	v	v	...	...	...	**B12**[2], 601
b225	**—,1-amino-2,4-dimethyl-***	2,4-Xylidine. $C_8H_{11}N$. *See* b202	121.18			214^{761}	0.9783^{20}_4	1.5607^{20}	δ	s	s	...	s	...	**B12**[2], 602
b226	**—,1-amino-3,5-dimethyl***	3,5-Xylidine. $C_8H_{11}N$. *See* b202	121.18			220–1	0.972^{20}_4	1.5581^{20}	δ	...	...	...	...	...	**B12**[2], 613
b227	**—,2-amino-1,3-dimethyl-***	2,6-Xylidine. $C_8H_{11}N$. *See* b202	121.18		11	214^{739}	0.9842^{20}	1.5610^{20}	i	s	s	...	...	...	**B12**[2], 604
b228	**—,2-amino-1,4-dimethyl-***	2,5-Xylidine. $C_8H_{11}N$. *See* b202	121.18	ye lf	16	214^{760}	0.9790^{21}_4	1.5591^{21}	δ	...	s	...	...	...	**B12**[2], 614
b229	**—,4-amino-1,2-dimethyl-***	3,4-Xylidine. $C_8H_{11}N$. *See* b202	121.18	pl or pr (lig)	49	224^{728}	1.076^{18}	...	δ	...	...	...	...	lig v	**B12**[2], 602
b230	**—,1-amino-2-dimethylamino-**	$C_8H_{12}N_2$. *See* b202	136.20			218^{751}	0.995^{25}	...	δ	v	v	...	...	...	**B8,** 15
b231	**—,1-amino-3-dimethylamino-**	$C_8H_{12}N_2$. *See* b202	136.20		<−20	$268–70^{740}$	0.995^{25}	...	δ	v	v	...	...	...	**B8,** 38
b232	**—,1-amino-4-dimethylamino-**	$C_8H_{12}N_2$. *See* b202	136.20	nd	37	257	1.036^{20}_4	...	δ	v	v	...	...	chl s	**B8,** 72
b233	**—,1-amino-2,3-dimethyl-4-nitro-***	$C_8H_{10}N_2O_2$. *See* b202	166.18	ye pr (al)	114		...	...	...	v^h	...	...	...	...	**B12,** 1103
b234	**—,1-amino-2,3-dimethyl-5-nitro-***	$C_8H_{10}N_2O_2$. *See* b202	166.18	pa ye nd (al)	111–2		...	...	...	v^h	...	...	...	...	**B12**[1], 429

For explanations, symbols and abbreviations see beginning of table.

No.	Name	Synonyms and Formula	Mol. wt.	Crystalline form, color and specific rotation	m.p. °C	b.p. °C	Density	n_D	Solubility w	al	eth	ace	bz	other solvents	Ref.
	Benzene														
b235	**—,1-amino-2,4-dimethyl-3-nitro-***	$C_8H_{10}N_2O_2$. *See* b202	166.18	ye nd	81–2				s^h	s				lig s	**B12^2**, 612
b236,	**—1-amino-2,4-dimethyl-5-nitro-***	$C_8H_{10}N_2O_2$. *See* b202	166.18	og ye nd	123					v^h					**B12^2**, 617
b237	**—,1-amino-3,4-dimethyl-2-nitro-***	$C_8H_{10}N_2O_2$. *See* b202	166.18	red pr (al)	65–6				i	v^h					**B12^1**, 481
b238	**—,1-amino-3,5-dimethyl-2-nitro-***	$C_8H_{10}N_2O_2$. *See* b202	166.18	ye nd	56										**B12^2**, 613
b239	**—,1-amino-4,5-dimethyl-2-nitro-***	$C_8H_{10}N_2O_2$. *See* b202	166.18	red-br nd (al)	140						δ			os v lig i	**B12**, 1106
b240	**—,2-amino-1,3-dimethyl-4-nitro-***	$C_8H_{10}N_2O_2$. *See* b202	166.18	ye nd (dil al)	81–2					s					**B12^2**, 605
b241	**—,2-amino-1,3-dimethyl-5-nitro-***	$C_8H_{10}N_2O_2$. *See* b202	166.18	og-red nd	158					s					**B12^2**, 606
b242	**—,2-amino-1,5-dimethyl-3-nitro-***	$C_8H_{10}N_2O_2$. *See* 202	166.18	og-red nd	76					s					**B12^2**, 612
b243	**—,2-amino-3,4-dimethyl-1-nitro-***	$C_8H_{10}N_2O_2$ *See* b202	166.18	red pl (al)	118–9					v^h					**B12**, 1102
b244	**—,5-amino-1,2-dimethyl-3-nitro-***	$C_8H_{10}N_2O_2$. *See* b202	166.18	og lf (al)	74–5					v^h					**B12**, 1106
b245	**—,5-amino-1,3-dimethyl-2-nitro-***	$C_8H_{10}N_2O_2$. *See* b202	166.18	og pr (bz) lf (lig)	133				s^h	v			s	lig δ	**B12^2**, 613
b245^1	**—,(1-amino-ethyl)-(*d*)***	d-α-Phenethylamine. $C_6H_5CH(CH_3)NH_2$	121.18	$[\alpha]_D^{15}+40.8$ (undil)		187^{760}	0.9651^{15}								**B12^2**, 586
b245^2	**—,—,(*dl*)**	$C_6H_5CH(CH_3)NH_2$	121.18			187^{760}	0.9651^{15}	1.5291^{15}	s	∞	∞				**B12^2**, 586
b245^3	**—,—,(*l*)**	$C_6H_5CH(CH_3)NH_2$	121.18	$[\alpha]_D^{21}-38.5$ (undil)		77^{16}	0.9520_4^{20}		s	v	v				**B12^2**, 586
b246	**—,1-amino-2-ethyl-***	*o*-Ethylaniline. $C_8H_{11}N$. *See* b202	121.18		−43	$209–10^{760}$	0.983_4^{22}	1.5584^{22}	δ	v	v				**B12^2**, 584
b247	**—,1-amino-3-ethyl-***	*m*-Ethylaniline. $C_8H_{11}N$. *See* b202	121.18		−64	$214–5^{764}$	0 9896^0		s^h	v	v				**B12^2**, 468
b248	**—,1-amino-4-ethyl-***	*p*-Ethylaniline. $C_8H_{11}N$. *See* b202	121.18		−5	216^{760}	0.975_4^{22}	1.5529^{22}	δ	v	v				**B12^2**, 584
b248^1	**—,(2-amino-ethyl)-***	β-Phenethylamine. $C_6H_5CH_2CH_2NH_2$	121.18			$194–5^{760}$	$0,9580_4^{24}$	1.5267	s	v	v				**B12^2**, 591
b249	**—,—,hydrochloride***	$C_6H_5CH_2CH_2NH_2.HCl$	157.65	pl or lf	218–9				v	v	i				**B12^2**, 592
b250	**—,4(2-amino-ethyl)-1,3-dihydroxy-, hydrochloride***	$C_8H_{11}NO_2HCl$. *See* b202	189.64	nd(w)	237d				v	v	δ				**B13^2**, 486
b251	**—,1(2-amino-ethyl)-2-hydroxy-***	$C_8H_{11}NO$. *See* b202	137.18		113–5										**B13^1**, 233
b252	**—,1(2-amino-ethyl)-3-hydroxy-***	$C_8H_{11}NO$. *See* b202	137.18		103–4d										**B13^1**, 233
—	**—,1(2-amino-ethyl)-4-hydroxy-***	*see* **Tyramine**													
b253	**—,5-amino-2-hydroxy-4-isopropyl-1-methyl-**	4-Aminocarvacrol. $C_{10}H_{15}NO$. *See* b202	165.24		134									MeOH s^h	**B13^2**, 392
b254	**—,6-amino-3-hydroxy-4-isopropyl-1-methyl-**	4-Aminothymol. $C_{10}H_{15}NO$. *See* b202	165.24	nd (bz)	178–9					s	δ			os s	**B13^2**, 393
b255	**—,1-amino-4-isobutyl-(*dl*)**	$C_{10}H_{15}N$. *See* b202	149.24	pa ye		238^{762}	0.949_4^{15}		i	∞				os ∞	**B12^2**, 635

For explanations, symbols and abbreviations see beginning of table.

No.	Name	Synonyms and Formula	Mol. wt.	Crystalline form, color and specific rotation	m.p. °C	b.p. °C	Density	n_D	Solubility w	al	eth	ace	bz	other solvents	Ref.
	Benzene														
b256	**—,1-amino-2-isopropyl-**	$C_9H_{13}N$. *See* b202	135.21			220–1[745]	0.9760^{12}_4								**B12²**, 624
b257	**—,1-amino-4-isopropyl-**	Cumidine. $C_9H_{13}N$. *See* b202	135.21		−63	225[760]	0.957^{20}_4	1.5415[20]	i v[h]	s[h]	s				**B12²**, 625
b258	**—,2-amino-1-isopropyl-4-methyl-**	Thymylamine. $C_{10}H_{15}N$. *See* b202	149.24			238–42[760]			δ	v	v				**B12²**, 642
b259	**—,2-amino-4-isopropyl-1-methyl-**	Carvacrylamine. $C_{10}H_{15}N$. *See* b202	149.24		−16	241[760]	0.9942^{20}_4	1.5387[20]	δ	s	s				**B12²**, 638
b260	**—,1-amino-2-mercapto-***	*o*-Aminothiophene. C_6H_7NS. *See* b202	125.19	nd	26	234									**B13**, 397
b261	**—,1-amino-3-mercapto-***	C_6H_7NS. *See* b202	125.19	pa ye		180–90[16]			s	s	i				**B13**, 425
b262	**—,1-amino-4-mercapto-***	C_6H_7NS. *See* b202	125.19	wh gran	46	140–5[16]									**B13**, 533
b263	**—,1-amino-2-methoxy-3-nitro-***	$C_7H_8N_2O_3$ *See* b202	168.15	pa ye nd (lig)	67					s				lig s[h]	**B13²**, 195
b264	**—,2-amino-4-methoxy-1-nitro-***	$C_7H_8N_2O_3$ *See* b202	168.15	br nd	131										**B13²**, 215
b265	**—,1-amino-4-methylamino-***	$C_7H_{10}N_2$ *See* b202	122.17	lf	36	257–9			v	v	v				**B13**, 71
b266	**—,1(amino-methyl)-4-isopropyl-***	Cuminylamine. $C_{10}H_{15}N$. *See* b202	149.24			225–7[724]			i	v	v				**B12**, 1172
b267	**—,1(amino-methyl)-2,4,5-trimethyl-**	$C_{10}H_{15}N$. *See* b202	149.24	nd (dil al)	65				i δ[h]	s[h]				chl s	**B12**, 1177
b268	**—,5(amino-methyl)-1,2,3-trimethyl-**	$C_{10}H_{15}N$. *See* b202	149.24	lf (w)	123				s[h]						**B12**, 1176
b269	**—,1-amino-4-octyl-***	1(4-Aminophenyl)octane*. $C_{14}H_{23}N$. *See* b202	205.35		20	170–2[17]			i		∞				**B7**, 1185
b270	**—,1-amino-2,3,4,5,6-penta-methyl-***	$C_{11}H_7N$. *See* b202	163.26	mcl pr (al)	152–3	278–80			i	s	s				**B12²**, 646
b271	**—,1-amino-2-propyl-***	*o*-Propylaniline. $C_9H_{13}N$. *See* b202	135.21			226[758]	0.9602^{30}_4	1.5427[30]	i	s	s				**B12²**, 620
b272	**—,1-amino-4-propyl-***	*p*-Propylaniline. $C_9H_{13}N$. *See* b202	135.21			224–6[760]			δ						**B12**, 1143
b273	**—,1(2-amino-propyl)-4-hydroxy-(L)***	*L*-Paredrine. $C_9H_{13}NO$. *See* b202	151.20	$[\alpha]^{17}_D -52$ (al)	111					s[h]	s			chl s	**H34**, 2202
b274	**—,1-amino-2,3,4,5-tetramethyl-***	$C_{10}H_{15}N$. *See* b202	149.24	lf (w)	70	259–60			s[h]	v	v			lig v	**B12**, 1175
b275	**—,1-amino-2,3,4,6-tetramethyl-***	Isoduridine. $C_{10}H_{15}N$. *See* b202	149.24		23–4	255	0.978[24]			s					**B12¹**, 506
b276	**—,1-amino-2,4,5,-trimethyl-***	Pseudocumidine. $C_9H_{13}N$. *See* b202	135.21	nd (w)	62	234–5	0.957			i					**B12²**, 629
b277	**—,2-amino-1,3,5,trimethyl-***	Mesidine. $C_9H_{13}N$. *See* b202	135.21		−5	229–30	0.9633								**B12²**, 631
—	**—,azido**	*see* **Benzene, triazo-***													
b278	**—,1-benzyl-4-ethyl-**	*p*-Ethyldiphenylmethane. $C_{15}H_{16}$. *See* b202	196.29		−24	297[760]	0.9777[20]	1.5616[20]		s	s			chl s	**B5**, 614
b279	**—,1-benzyl-2-hydroxy-**	$C_{13}H_{12}O$. *See* b202	184.24		21	312									**B6**, 675
b280	**—,1-benzyl-4-hydroxy-**	$C_{13}H_{12}O$. *See* b202	184.24	nd (al)	83–4	320–2			s[h]	s	s				**B6**, 678
b281	**—,1,2-bis(bromo-methyl)-**	*o*-Xylylene bromide. $C_8H_8Br_2$. *See* b202	263.97		95		1.988[0]		i	δ	δ			chl δ lig s	**B5²**, 285
b282	**—,1,3-bis(bromo-ethyl)-***	*m*-Xylylene bromide. $C_8H_8Br_2$. *See* b202	263.97	nd (chl)	76–7	135–40[20]	1.959[0]		i	s				chl s lig s	**B5²**, 294
b283	**—,1,4-bis(bromo-methyl)-***	*p*-Xylylene bromide. $C_8H_8Br_2$. *See* b202	263.97	mcl pr	145–7	245	2.012[0]		i	v	s		s	chl v	**B5**, 385
b284	**—,1,3-bis(bromo-methyl)-5-methyl-***	$C_9H_{10}Br_2$. *See* b202	278.00	pr	66				i	δ v[h]	s		s	lig s	**B5²**, 315

For explanations, symbols and abbreviations see beginning of table.

No.	Name	Synonyms and Formula	Mol. wt.	Crystalline form, color and specific rotation	m.p. °C	b.p. °C	Density	n_D	Solubility w	al	eth	ace	bz	other solvents	Ref.
	Benzene														
b285	—,**1,3-bis(carboxymethoxy)-***	$C_{10}H_{10}O_6$. *See* b202	226.19	nd (aa)	195				s^h					aa s^h	**B6²**, 817
b286	—,—,diethyl ester*	$C_{14}H_{18}O_6$. *See* b202	282.30	nd (eth)	42	228^{32}			i					lig δ	**B6**, 818
b287	—,**1,2-bis(chloromethyl)-***	*o*-Xylylene chloride. $C_8H_8Cl_2$. *See* b202	175.06	mcl (lig)	55	239–41	1.393^0		i	v	v			chl v lig v	**B5²**, 283
b288	—,**1,3-bis(chloromethyl)-***	*m*-Xylylene chloride. $C_8H_8Cl_2$. *See* b202	175.06		34	250–5	1.302^{20}		i	v	v				**B5²**, 29
b289	—,**1,4-bis(chloromethyl)-***	*p*-Xylylene chloride. $C_8H_8Cl_2$. *See* b202	175.06	lf (al)	100	240–5	1.417^0		i	v^h	v	v		chl v aa δ	**B5²**, 300
b290	—,**1,2-bis(cyanomethyl)-***	$C_{10}H_8N_2$. *See* b202	156.19	nd or pr (al)	59–60				i	s^h	s			lig i	**B9²**, 623
b291	—,**1,3-bis(cyanomethyl)-***	$C_{10}H_8N_2$. *See* b202	156.19	nd (w), pr (eth)	30	$230–1^{20}$			i	s^h	s			chl s lig i	**B9²**, 624
b292	—,**1,4-bis(cyanomethyl)-***	$C_{10}H_8N_2$. *See* b202	156.19	pr (al), nd (w)	96				i s^h	s^h	s			chl s	**B9²**, 624
b293	—,**1,2-bis(dibromomethyl)-***	$C_8H_6Br_4$. *See* b202	421.77		116–7				δ s^h					chl v lig i	**B5**, 367
b294	—,**1,3-bis(dibromomethyl)-***	$C_8H_6Br_4$. *See* b202	421.77	nd (al), pl (chl)	107				i	s^h			v	chl v lig s	**B5**, 375
b295	—,**1,4-bis(dibromomethyl)-***	$C_8H_6Br_4$. *See* b202	421.77	mcl pr (chl)	169				$δ^h$	δ	δ		v	chl δ, s^h	**B5**, 386
b296	—,**1,2-bis(diethylamino)-***	$C_{14}H_{24}N_2$. *See* b202	220.36			119^{10}	0.9267^{13}_4	1.5213^{13}		s	s				**B13²**, 12
b297	—,**1,3-bis(diethylamino)-***	$C_{14}H_{24}N_2$. *See* b202	220.36			148^9	0.9522^{12}_4	1.5537^{12}		s	s				**B13²**, 26
b298	—,**1,4-bis(diethylamino)-***	$C_{14}H_{24}N_2$. *See* b202	220.36	mcl pr, pl (al-w)	52	280				v	v		v	chl v lig v	**B13²**, 40
b299	—,**1,4-bis-(dimethylamino)-***	$C_{10}H_{16}N_2$. *See* b202	164.25	lf (dil al or lig)	51	260			δ s^h	v	v		v	chl v lig s	**B13²**, 40
b301	—,**1,3-bis(1,1-dimethylpropyl)-2-hydroxy-5-methyl-**	$C_{17}H_{28}O$. *See* b202	248.41			283^{760}	0.931^{25}_4	1.4950^{20}						alk i	
b302	—,**1,2-bis-(hydroxymethyl)-***	Phthalyl alcohol. *o*-Xylylene glycol. $C_8H_{10}O_2$. *See* b202	138.17	pl (eth)	64–5				s	s	v		δ		**B6**, 910
b303	—,**1,3-bis(hydroxymethyl)-***	*m*-Xylylene glycol. $C_8H_{10}O_2$. *See* b202	138.17		46–7	$154–9^{13}$	1.1359^{53}		v		s			AcOEt v	**B6¹**, 446
b304	—,**1,4-bis(hydroxymethyl)-***	*p*-Xylylene glycol. $C_8H_{10}O_2$. *See* b202	138.17	nd	112–3				v	v	v				**B6²**, 891
b304¹	—,**1,2-bis(methylamino)-4-chloro-***	$C_8H_{11}ClN_2$. *See* b202	170.64	pr (lig)	61				s						**B13**, 25
b305	—,**1,3-bis(3-methylbutoxy)-***	Resorcinol diisoamyl ether. $C_{16}H_{26}O_2$. *See* b202	250.38		47										**B6**, 815
b306	—,**1,3-bis(phenylazo)-2,4-dihydroxy-***	$C_{18}H_{14}N_4O_2$. *See* b202	318.34	red nd (chl-al)	222					δ				con ac s, con alk s, chl v^h	**B16**, 185
b307	—,**1,5-bis(phenylazo)-2,4-dihydroxy-***	$C_{18}H_{14}N_4O_2$. *See* b202	318.34	br-red nd	217					δ	δ			sulf s chl s	**B16**, 186
b308	—,**1,3-bis(4-tolylamino)-***	$C_{20}H_{20}N_2$. *See* b202	288.40	nd (al)	38–9				i	δ	δ				**B13**, 42
b309	—,**bromo-***	Phenyl bromide. C_6H_5Br. *See* b202	157.02		−31	$155–6^{760}$	1.5219^0_4	1.5598^{20}	i	v	v		v	CCl_4 s	**B5²**, 158
b310	—,**(4-bromobutoxy)-***	$C_{10}H_{13}BrO$. *See* b202	229.12		41	$153–6^{18}$				δ s^h					**B6²**, 146
b311	—,**1-bromo-4-*tert*-butyl-**	$C_{10}H_{13}Br$. *See* b202	213.12			232–3			i		s		s	chl s	**B5²**, 320
b312	—,**1-bromo-2-chloro-***	C_6H_4BrCl. *See* b202	191.46			204^{765}	1.6382^{25}_4	1.5786^{25}	i				v		**B5²**, 161
b313	—,**1-bromo-3-chloro-***	C_6H_4BrCl. *See* b202	191.46		−21	196	1.6302^{20}_4	1.578^{15}	i	v	v				**B5²**, 161
b314	—,**1-bromo-4-chloro-***	C_6H_4BrCl. *See* b202	191.46	mcl or rh pr	67	196^{756}	1.576^{71}_4	1.5531^{70}	i	δ s^h	s		s	chl s	**B5²**, 162

For explanations, symbols and abbreviations see beginning of table.

No.	Name	Synonyms and Formula	Mol. wt.	Crystalline form, color and specific rotation	m.p. °C	b.p. °C	Density	n_D	Solubility: w	al	eth	ace	bz	other solvents	Ref.
	Benzene														
b315	**—,1-bromo-4-cyclohexyl-***	$C_{12}H_{15}Br$. *See* b202	239.16			160^{23}	1.283^{25}_{4}	1.5584^{20}	i				s	chl s	**B5²**, 396
b316	**—,1-bromo-2,3-dichloro-***	$C_6H_3BrCl_2$. *See* b202	225.91	pl (al)	60	243^{705}			i	δ vʰ			v		**B5**, 209
b318	**—,1-bromo-3,5-dichloro-***	$C_6H_3BrCl_2$. *See* b202	225.91		74	232^{757}			i	sʰ			v		**B5**, 210
b319	**—,2-bromo-1,3-dichloro-***	$C_6H_3BrCl_2$. *See* b202	225.91	pl	65	242^{765}			i		v		v	chl v	**B5**, 210
b320	**—,2-bromo-1,4-dichloro-***	$C_6H_3BrCl_2$. *See* b202	225.91	pr or nd (al)	35	235^{731}			i	δ vʰ	v		v	chl v lig v	**B5**, 210
b321	**—,4-bromo-1,2-dichloro-***	$C_6H_3BrCl_2$. *See* b202	225.91	pr	25	237^{757}			i	δ vʰ	v		v	chl v	**B5**, 210
b322	**—,1-bromo-2,4-dihydroxy-***	$C_6H_5BrO_2$. *See* b202	189.01		103				v	δ	v		δ	CS_2 δ, chl δ	**B6²**, 819
b323	**—,1-bromo-3,5-dihydroxy-***	$C_6H_5BrO_2$. *See* b202	189.01	nd(bz), pr(w+1)	79(+w), 87				vʰ				vʰ s		**B6²**, 820
b324	**—,2-bromo-1,3-dihydroxy-***	$C_6H_5BrO_2$. *See* b202	189.01	nd (chl)	102						s			chl s	**B6²**, 819
b325	**—,2-bromo-1,4-dihydroxy-***	$C_6H_5BrO_2$. *See* b202	189.01	lf (lig)	110				v	v	v		v	chl δ lig δ	**B6²**, 846
b326	**—,4-bromo-1,2-dihydroxy-***	$C_6H_5BrO_2$. *See* b202	189.01	pr or nd (chl)	87				v	v	v		vʰ	chl δ lig i	**B6²**, 781
b327	**—,1-bromo-2,3-dimethyl-***	C_8H_9Br. *See* b202	185.07			214	1.365^{20}_{4}		i				s		**B5¹**, 180
b328	**—,1-bromo-2,4-dimethyl-***	C_8H_9Br. *See* b202	185.07		0	214	1.3708^{18}_{4}	1.5571^{18}	i	v	v				**B5²**, 285
b329	**—,2-bromo-1,4-dimethyl-***	C_8H_9Br. *See* b202	185.07	lf or pl	9	$205–10^{770}$	1.3582^{18}	1.5514^{18}	i	v					**B5²**, 301
b330	**—,4-bromo-1,2-dimethyl-***	C_8H_9Br. *See* b202	185.07			204			i	v	v				**B5²**, 293
b331	**—,1-bromo-2,3-dinitro-***	$C_6H_3BrN_2O_4$. *See* b202	247.02	pa ye pl (al)	101	320				vʰ	δ				**B5¹**, 138
b332	**—,1-bromo-2,4-dinitro-***	$C_6H_3BrN_2O_4$. *See* b202	247.02	ye	74				dʰ	vʰ					**B5¹**, 138
b333	**—,2-bromo-1,3-dinitro-***	$C_6H_3BrN_2O_4$. *See* b202	247.02	ye pr (al)	107				dʰ	δ vʰ					**B5¹**, 138
b334	**—,2-bromo-1,4-dinitro-***	$C_6H_3BrN_2O_4$. *See* b202	247.02	nd (al), pr (al-eth)	70					δ vʰ	v				**B5¹**, 139
b335	**—,4-bromo-1,2-dinitro-***	$C_6H_3BrN_2O_4$. *See* b202	247.02	nd (al), pl (al-eth)	35(unst), 59					δ vʰ	v			MeOH δ	**B5¹**, 138
b336	**—,1-bromo-2-ethoxy-***	C_8H_9BrO. *See* b202	201.08			222–6			i	v	v				**B6**, 197
b337	**—,1-bromo-3-ethoxy-***	C_8H_9BrO. *See* b202	210.08			228–31			i	v	v				**B6²**, 184
b338	**—,1-bromo-4-ethoxy***	C_8H_9BrO. *See* b202	201.08		4	233			i	v	v				**B6²**, 185
b338¹	**—,(1-bromo-ethyl)-(*d*)***	C_8H_9Br. *See* b202	185.07	$[\alpha]^{14}_{D}+1.5$		$86–8^{15}$	1.3108^{23}	1.5612^{20}	i	s	s				**B5²**, 278
b339	**—,—,(*dl*)***	C_8H_9Br. *See* b202	185.07			$202–3^{760}$	1.3605^{20}_{4}	1.5612^{20}	dʰ		s		s		**B5²**, 278
b340	**—,(2-bromo-ethyl)-***	C_8H_9Br. *See* b202	185.07			$217–8^{734}$			i		s		s		**B5²**, 278
b341	**—,1-bromo-4-fluoro-***	C_6H_4BrF. *See* b202	175.01		−17	152^{764}	1.4946^{20}_{4}	1.5604^{20}	i	s	s				**B5²**, 161
b342	**—,1-bromo-2-iodo-***	C_6H_4BrI. *See* b202	282.92		5	257^{754}	2.2571^{25}_{4}	1.6618^{25}	i	δ				aa δ	**B5²**, 167
b343	**—,1-bromo-3-iodo-***	C_6H_4BrI. *See* b202	282.92		−9	252^{754}			i	δ				aa δ	**B5²**, 168
b344	**—,1-bromo-4-iodo-***	C_6H_4BrI. *See* b202	282.92		92	252^{754}			i	δ	s				**B5²**, 168
b345	**—,1-bromo-4-isopropyl-**	$C_9H_{11}Br$. *See* b202	199.09		<−20	$216–7^{729}$	1.3646^{22}_{4}	1.5501^{23}	i		s		s	chl s	**B5²**, 307
b346	**—,(β-bromo-isopropyl)-**	$C_9H_{11}Br$. *See* b202	199.09	$[\alpha]^{20}_{D}(d)+15.6$, $(l)-15.6$		$106–8^{18}$	1.3155^{20}		i				s	chl s	**B5¹**, 191
b347	**—,2-bromo-4-isopropyl-1-methyl-**	$C_{10}H_{13}Br$. *See* b202	213.12			233–5	1.269^{18}		i	v	s				**B5¹**, 205

For explanations, symbols and abbreviations see beginning of table.

No.	Name	Synonyms and Formula	Mol. wt.	Crystalline form, color and specific rotation	m.p. °C	b.p. °C	Density	n_D	Solubility w	al	eth	ace	bz	other solvents	Ref.
	Benzene														
b348	**—,1-bromo-4-mercapto-***	C_6H_5BrS. *See* b202	189.08	lf (al)	75	230–1			δ^h	δ v^h	v			sulf v, CCl_4 v chl v	**B6**^{2}, 300
b349	**—,1-bromo-2-methoxy-***	C_7H_7BrO. *See* b202	187.04		2.5	223	1.5018^{20}_4	1.5727^{20}	i	v	v				**B6**^{2}, 183
b350	**—,1-bromo-3-methoxy-***	C_7H_7BrO. *See* b202	187.04			$210–1^{752}$			i	s	s		s	CS_2 s	**B6**^{2}, 184
b351	**—,1-bromo-4-methoxy-***	C_7H_7BrO. *See* b202	187.04		13	223	1.4564^{20}_4	1.5605^{20}	δ	v	v			chl v	**B6**^{2}, 185
b352	**—,1(bromo-methyl)-3,5-dimethyl-***	Mesityl bromide. $C_9H_{11}Br$. *See* b202	199.10	nd (eth)	40	229–31 δd 118^{22}				v	v		s	chl v	**B5**^{1}, 200
b353	**—,1(bromo-methyl)-2-methyl-***	*o*-Xylyl bromide. C_8H_9Br. *See* b202	185.07	pr	21	$216–7^{742}$	1.3811^{23}		i	s	s			os s	**B5**^{2}, 285
b354	**—,1(bromo-methyl)-3-methyl-***	*m*-Xylyl bromide. C_8H_9Br. *See* b202	185.07			212–3 δd 105^{13}	1.37^{23}		i	v	v				**B5**^{2}, 293
b355	**—,1(bromo-methyl)-4-methyl-***	*p*-Xylylbromide. C_8H_9Br. *See* b202	185.07	nd (al)	35	$218–20^{740}$	1.324		i	s	v^h			chl v^h	**B5**^{2}, 301
b356	**—,1-bromo-2-nitro-***	$C_6H_4BrNO_2$. *See* b202	202.02	pa ye (al)	43	258^{756}	1.6245^{80}_4		i	v	s		s		**B5**^{2}, 188
b357	**—,1-bromo-3-nitro-***	$C_6H_4BrNO_2$. *See* b202	202.02	rh	17 (unst) 56	265^{760}	1.7036^{20}_4	1.5979^{20}	i	s	s		s		**B5**^{2}, 188
b358	**—,1-bromo-4-nitro-***	$C_6H_4BrNO_2$. *See* b202	202.02	rh or mcl pr	126–7	254^{758}	1.948		i	s	s		s		**B5**^{2}, 188
b359	**—,1-bromo-2-nitroso-***	C_6H_4BrNO. *See* b202	186.02	nd	98								s	chl s	**B5**, 232
b360	**—,1-bromo-4-nitroso***	C_6H_4BrNO. *See* b202	186.02	nd (al)	95					s^h			s	chl v lig s	**B5**^{2}, 171
b361	**—,(3-bromo-1-propenyl)-***	Cinnamyl bromide. C_9H_9Br. *See* b202	197.08	nd	30	$98^{1.5}$	1.3428^{30}	1.610–1.613^{20}	i	s^h				hexane s	**B5**^{2}, 372
b362	**—,(3-bromo-propoxy)-***	γ-Phenoxy propyl bromide. $C_9H_{11}BrO$. *See* b202	215.10		7–8	127^{18}	1.365^{16}_{16}				v				**B6**^{2}, 145
b363	**—,(1-bromo-propyl)-***	$C_9H_{11}Br$. *See* b202	199.10			$100–1^{10}$	1.3098^{19}_4	1.5517^{19}	i d^h		v		s		**B5**^{2}, 305
b364	**—,(2-bromo-propyl)-***	$C_9H_{11}Br$. *See* b202	199.10			$107–9^{16}$	1.2908^{16}		d^h					chl s	**B5**^{1}, 190
b365	**—,(3-bromo-propyl)-***	$C_9H_{11}Br$. *See* b202	199.10			110^{12}			i		v				**B5**^{2}, 305
b366	**—,1-bromo-4-triazo-***	$C_6H_4BrN_3$. *See* b202	198.04	pl	20	105^{10}			i	δ	v		v		**B5**^{2}, 208
b367	**—,1-bromo-2,3,5-trimethyl-***	$C_9H_{11}Br$. *See* b202	199.10		<−15	238^{760}			i				s	sulf s	**B5**^{1}, 196
b368	**—,1-bromo-2,4,5-trimethyl-***	$C_9H_{11}Br$. *See* b202	199.10	nd (al)	73	233–5			i	v^h					**B5**^{1}, 196
b639	**—,2-bromo-1,3,5-trimethyl-***	$C_9H_{11}Br$. *See* b202	199.10		1	225	1.3191^{10}		i		v		s		**B5**^{2}, 315
b370	**—,butoxy-***	Butyl phenyl ether. $C_{10}H_{14}O$. *See* b202	150.22		−19	210^{760}	0.9351^{20}_4	1.4969^{20}							**B6**^{2}, 145
b371	**—,*sec*-butoxy-**	*sec*-Butyl phenyl ether. $C_{10}H_{14}O$. *See* b202	150.22			$195–6^{763}$	0.9415^{20}_4		i						**B6**^{2}, 146
b372	**—,butyl-***	$C_{10}H_{14}$. *See* b202	134.21		−88	183^{760}	$0\ 8601^{20}_4$	1.4898^{20}	i	∞	∞				**B5**^{2}, 317
b373	**—,*sec*-butyl-(*d*)**	$C_{10}H_{14}$. *See* b202	134.21	$[\alpha]^{20}_D+27.3$ (undil)	−75	173^{760}	0.8621^{20}_4	1.4902^{20}	i	∞	∞				**B5**^{2}, 319
b374	—,—,(*dl*)	$C_{10}H_{14}$. *See* b202	134.21		−83	173^{760}	0.8625^{20}_4	1.4892^{20}							**B5**^{2}, 319
b375	—,—,(*l*)	$C_{10}H_{14}$. *See* b202	134.21	$[\alpha]^{23}_{5461}-21.6$ (al c=5)		61^{18}	0.868^{24}_4	1.4891^{20}							
b376	**—,*tert*-butyl-**	$C_{10}H_{14}$. *See* b202	134.21		−58	168^{760}	0.8671^{20}_4	1.4923^{20}	i	v	v				**B5**^{2}, 320

For explanations, symbols and abbreviations see beginning of table.

No.	Name	Synonyms and Formula	Mol. wt.	Crystalline form, color and specific rotation	m.p. °C	b.p. °C	Density	n_D	Solubility						Ref.
									w	al	eth	ace	bz	other solvents	
	Benzene														
b377	**—,1-butyl-2,4-dihydroxy-***	$C_{10}H_{14}O_2$. *See* b202	166.21		47–8	$164–6^{6-7}$			i	δ			δ	chl δ	**B6^2**, 898
b378	**—,1-*tert*-butyl-2,3-dimethyl-4-hydroxy-**	$C_{12}H_{18}O$. *See* b202	178.26			145^{20}									
b379	**—,1-*tert*-butyl-2,5-dimethyl-4-hydroxy-**	$C_{12}H_{18}O$. *See* b292	178.26		71	264^{760}	0.939^{80}_{4}	1.5311^{20}						alk s	
b380	**—,1-*tert*-butyl-3,5-dimethyl-2-hydroxy-**	$C_{12}H_{18}O$. *See* b202	178.26		22	249^{760}	0.917^{80}_{4}	1.5183^{20}						alk i	
b381	**—,1-*tert*-butyl-4,5-dimethyl-2-hydroxy-**	$C_{12}H_{18}O$. *See* b202	178.26		46	145^{20}	0.920^{80}_{4}	1.5222^{20}						alk i	
b382	**—,5-*tert*-butyl-1,3-dimethyl-2-hydroxy-**	$C_{12}H_{18}O$. *See* b202	178.26		82	248^{760}	0.916^{80}_{4}							alk s	
b383	**—,1-*tert*-butyl-3,5-dimethyl-2,4,6-trinitro-**	Musk xylene. $C_{12}H_{15}N_3O_6$. *See* b202	297.27	pl, nd (al)	105–6 (unst) 113				i	δ v^h	s				**B5^2**, 340
b384	**—,2-*tert*-butyl-3,5-dinitro-1-iso-propyl-4-methyl-**	Moskene. $C_{14}H_{20}N_2O_4$. *See* b202	280.32	pa ye cr	132				i	δ					
b385	**—,5-*tert*-butyl-1,3-dinitro-4-methoxy-2-methyl-**	Musk ambrette. $C_{12}H_{16}N_2O_5$. *See* b202	268.41	pa ye cr	84–6				i	δ	s				
b386	**—,2-*tert*-butyl-1,3-dinitro-4,5,6-trimethyl-2-methyl-**	Musk tibetine. $C_{13}H_{18}N_2O_4$. *See* b202	266.29	pa ye cr	135				i	δ	s				
b387	**—,2-*tert*-butyl-4-ethyl-1-hydroxy-**	$C_{12}H_{18}O$. *See* b202	178.26		23	250^{760}								alk i	
b388	**—,4-*tert*-butyl-2 ethyl-1-hydroxy-**	$C_{12}H_{18}O$. *See* b202	178.26			257^{760}									
b389	**—,1-butyl-2-hydroxy-***	$C_{10}H_{14}O$. *See* b202	150.21			234–7	0.975^{20}_{4}	1.496^{15}	i	s	s				
b390	**—,1-butyl-3-hydroxy-***	$C_{10}H_{14}O$. *See* b202	150.21			$247–9^{758}$	0.974^{20}_{4}		i	s	s				
b391	**—,1-butyl-4-hydroxy-***	$C_{10}H_{14}O$. *See* b202	150.21		22	248^{765}	0.978^{20}_{4}	1.5165^{22}	i	s	s			alk s	**B6^2**, 485
b392	**—,1-*sec*-butyl-4-hydroxy-**	$C_{10}H_{14}O$. *See* b202	150.21	nd, $[\alpha]^{20}_{D}$ +13.3 (*m*-xyl)	61–2	$240–2^{750}$	0.9883^{20}_{20}	1.518^{25}	i	s	v				**B6^2**, 487
b393	**—,1-*tert*-butyl-2-hydroxy-**	$C_{10}H_{14}O$. *See* b202	150.21			217–20	0.9783^{20}_{4}	1.5160^{20}							**B6^2**, 489
b394	**—,1-*tert*-butyl-3-hydroxy-**	$C_{10}H_{14}O$. *See* b202	150.21		41	130^{20}									
b395	**—,1-*tert*-butyl-4-hydroxy-**	$C_{10}H_{14}O$. *See* b202	150.21	nd (lig)	99–100	$236–8^{740}$	0.908^{114}_{4}	1.4787^{114}	s^h	s	s				**B6^2**, 489
b396	**—,1-*tert*-butyl-2-hydroxy-4-methyl-**	$C_{11}H_{16}O$. *See* b202	164.24		24 37(+w)	224^{760}	0.922^{80}	1.5250^{20}	i					os s	**B6^2**, 507
b397	**—,2-*tert*-butyl-1-hydroxy-4-methyl-**	$C_{11}H_{16}O$. *See* b202	164.24		55	126^{20}	0.9247^{75}_{4}	1.4969^{75}	i					os s	
b398	**—,4-*tert*-butyl-1-hydroxy-2-methyl-**	$C_{11}H_{16}O$. *See* b202	164.24	yesh	27–8	132^{20}	0.965^{20}_{4}	1.5230^{20}	i					os s	**B6^2**, 507
b399	**—,1-butyl-2-methyl-***	$C_{11}H_{16}$. *See* b202	148.24			201	0.8702^{18}_{4}	1.4966^{18}	i	δ	s				**B5**, 437
b400	**—,1-butyl-3-methyl-***	$C_{11}H_{16}$. *See* b202	148.24			197–8	0.8624^{18}_{4}	1.4932^{18}	i	δ	s				**B5**, 437

For explanations, symbols and abbreviations see beginning of table.

No.	Name	Synonyms and Formula	Mol. wt.	Crystalline form, color and specific rotation	m.p. °C	b.p. °C	Density	n_D	Solubility w	al	eth	ace	bz	other solvents	Ref.
	Benzene														
b401	**—,1-butyl-4-methyl-***	$C_{11}H_{16}$. *See* b202	148.24			198–9	0.8586^{20}_{4}	1.4916^{20}	i	δ	s	...	...		**B5**, 437
b402	**—,1-*sec*-butyl-4-methyl-**	$C_{11}H_{16}$. *See* b202	148.24			200–5	0.8640^{19}	1.4917	i	...	v	...	...	chl v	**B5**^{2}, 333
b403	**—,1-*tert*-butyl-4-methyl-**	$C_{11}H_{16}$. *See* b202	148.24		−52	193^{760}	0.8534^{25}_{4}	1.4936^{17}	i	δ	v	s	s	chl v	**B5**^{1}, 210
b404	**—,1-*tert*-butyl-3-methyl-2,4,6-trinitro-**	Artificial musk. $C_{11}H_{13}N_3O_6$. *See* b202	283.24	ye nd (al)	96–7				i	v	v	...	s	chl s	**B5**, 438
b405	**—,2-*tert*-butyl-4-methyl-1,3,5-trinitro-**	Musk baur. $C_{11}H_{13}N_3O_6$. *See* b202	283.24	ye w	97				...	i	s	...	...		
—	**—,(1-butynyl)-***	*see* **Butyne, phenyl-**													
b407	**—,chloro-***	Phenyl chloride. C_6H_5Cl. *See* b202	112.56		−45	132^{760}	1.1064^{20}_{4}	1.5248^{20}	i	∞	∞	...	v	chl v CCl_2 v, CS_2 v	**B5**^{2}, 148
b408	**—,2-chloro-1,4-diacetamido-***	$C_{10}H_{11}ClN_2O_2$. *See* b202	226.67	nd	196–7				...	s	...	...	δ		**B13**, 118
b409	**—,1-chloro-2,4-diamino-***	$C_6H_7ClN_2$. *See* b202	142.60	pl. nd	91				δ	s	...	...	...	lig s	**B13**^{2}, 29
b410	**—,1-chloro-3,5-diamino-***	$C_6H_7ClN_2$. *See* b202	142.60	rh	105–6				s^h	s^h	...	...	s^h		**B13**^{2}, 29
b411	**—,2-chloro-1,3-diamino-***	$C_6H_7ClN_2$. *See* b202	142.60		85–6				...	...	...	...	...		**B13**^{1}, 15
b412	**—,2-chloro-1,4-diamino-***	$C_6H_7ClN_2$. *See* b202	142.60	nd	63–4				s	...	...	...	s^h	lig s^h	**B13**^{2}, 58
b413	**—,4-chloro-1,2-diamino-***	$C_6H_7ClN_2$. *See* b202	142.60	pl (bz, lig) nd (w^h)	76				δ	v	v	...	s	lig s	**B13**, 25
b414	**—,1-chloro-2,3-dihydroxy-***	$C_6H_5ClO_2$. *See* b202	144.56		46–8	$110–1^{11}$			...	...	...	...	...	lig v	**B6**^{1}, 388
b415	**—,1-chloro-2,4-dihydroxy-***	$C_6H_5ClO_2$. *See* b202	144.56		89	$255–6^{760}$			v	v	v	...	v	CS_2 v	**B6**^{2}, 818
b416	**—,1-chloro-3,5-dihydroxy-***	$C_6H_5ClO_2$. *See* b202	144.56	hyg nd	117	sub			s	...	...	...	...	os s peth i	**B6**^{2}, 819
b417	**—,2-chloro-1,3-dihydroxy-***	$C_6H_5ClO_2$. *See* b202	144.56		97–8				...	...	...	...	...		**C40**, 565
b418	**—,2-chloro-1,4-dihydroxy-***	$C_6H_5ClO_2$. *See* b202	144.56	red lf (chl), nd (bz)	105–6	263			v	i	s	...	v^h	chl s^h	**B6**^{2}, 844
b419	**—,4-chloro-1,2-dihydroxy-***	$C_6H_5ClO_2$. *See* b202	144.56	lf (bz-peth)	90–1	$139^{10.5}$			v	v	v	v	...	aa v lig δ	**B6**^{2}, 787
b420	**—,1-chloro-2,4-dimethyl-***	C_8H_9Cl. *See* b202	140.62			192–3	1.0598^{20}_{20}	1.5230^{25}	i	...	...	...	v		**B5**^{2}, 291
b421	**—,1-chloro-3,5-dimethyl-***	C_8H_9Cl. *See* b202	140.62			$190–1^{775}$			i	...	...	...	s		**B5**, 373
b422	**—,2-chloro-1,4-dimethyl-***	C_8H_9Cl. *See* b202	140.62		2	187	1.0589^{15}_{4}		i	...	...	...	v		**B5**^{1}, 186
b423	**—,4-chloro-1,2-dimethyl-***	C_8H_9Cl. *See* b202	140.62		−6	194^{755}	1.0692^{15}_{15}		i	...	...	...	v		**B5**, 363
b424	**—,1-chloro-2,3-dimethyl-4-hydroxy-***	C_8H_9ClO. *See* b202	156.62	nd (peth)	85				...	...	...	...	...	os s peth δ	**B6**, 454
b425	**—,1-chloro-2,3-dimethyl-5-hydroxy-***	C_8H_9ClO. *See* b202	156.62	nd (peth)	98				...	...	...	...	...	chl v peth δ	**B6**^{2}, 456
b426	**—,1-chloro-2,4-dimethyl-5-hydroxy-***	C_8H_9ClO. *See* b202	156.62	nd (lig or w)	90–1				s^h	...	...	...	...	os s	**B6**^{1}, 242
b427	**—,1-chloro-2,5-dimethyl-4-hydroxy-***	C_8H_9ClO. *See* b202	156.62	nd (lig)	74–5				...	...	...	...	...	CS_2 v lig δ	**B6**^{2}, 467
b428	**—,1-chloro-3,4-dimethyl-2-hydroxy-***	C_8H_9ClO. *See* b202	156.62			$221–3^{760}$			...	...	...	...	...	os s	**B6**^{1}, 241
b429	**—,1-chloro-4,5-dimethyl-2-hydroxy-***	C_8H_9ClO. *See* b202	156.62	nd (peth)	72				...	...	...	...	...	peth δ CS_2 v aa s	**B6**^{2}, 456
b430	**—,2-chloro-1,3-dimethyl-5-hydroxy-***	C_8H_9ClO. *See* b202	156.62		115–6	246			δ	s	s	...	δ		**B6**^{2}, 463

For explanations, symbols and abbreviations see beginning of table.

No.	Name	Synonyms and Formula	Mol. wt.	Crystalline form, color and specific rotation	m.p. °C	b.p. °C	Density	n_D	Solubility						Ref.
									w	al	eth	ace	bz	other solvents	
	Benzene														
b431	**—,2-chloro-1,5-dimethyl-3-hydroxy-***	C_8H_9ClO. *See* b202	156.62		49–50				...	...	s	...	...		**B6**^{2}, 464
b432	**—,2-chloro-3,4-dimethyl-1-hydroxy-***	C_8H_9ClO. *See* b202	156.62		27				...	...	...	...	...	peth s^h	
b433	**—,5-chloro-1,3-dimethyl-2-hydroxy-***	C_8H_9ClO. *See* b202	156.62		83				...	...	...	...	...		
b434	**—,1-chloro-2,3-dinitro-***	$C_6H_3ClN_2O_4$. *See* b202	202.56	pr (al)	78				i	δ	s	...	...		**B5**^{2}, 196
b435	**—,1-chloro-2,4-dinitro-***	$C_6H_3ClN_2O_4$. *See* b202	202.56	ye rh (eth) nd (al)	53 (α-form) 43 (β-form) 27 (γ-form)	315^{762}	1.4982_{4}^{75}	1.5857^{60}	i $δ^h$	δ s^h	s	...	s	CS_2 s	**B5**^{2}, 196
b436	**—,1-chloro-3,5-dinitro-***	$C_6H_3ClN_2O_4$. *See* b202	202.56	nd (al)	59				i	s	v	...	...		**B5**, 264
b437	**—,2-chloro-1,3-dinitro-***	$C_6H_3ClN_2O_4$. *See* b202	202.56	ye nd (al)	88	315	1.6867^{16}		i	s	s	...	...	to s	**B5**^{1}, 137
b438	**—,4-chloro-1,2-dinitro-***	$C_6H_3ClN_2O_4$. *See* b202	202.56	mcl or rh nd (eth or lig)	40	315d	1.6867_{4}^{20}		i	δ s^h	v	...	v	CS_2 s	**B5**^{2}, 196
b439	**—,1-chloro-2-ethoxy-***	C_8H_9ClO. *See* b202	156.61			210	1.1299^{24}		...	s	s	...	s		**B6**^{2}, 171
b440	**—,1-chloro-3-ethoxy-***	C_8H_9ClO. *See* b202	156.61			$204–5^{717}$	1.1712_{4}^{20}		...	v	v	...	v	aa v	**B6**, 185
b441	**—,1-chloro-4-ethoxy-***	C_8H_9ClO. *See* b202	156.61		21	212–4	1.1231^{20}	1.5227^{19}	...	s	s	...	...	aa s	**B6**^{2}, 176
b442	**—,(2-chloroethoxy)-***	C_8H_9ClO. *See* b202	156.61	pr	28	$217–20^{760}$			i	v	v	v	v	lig v	**B6**^{2}, 144
b443	**—,(1-chloroethyl)-(*d*)***	*d*-α-Phenethyl chloride. C_8H_9Cl. *See* b202	140.61	$[α]_D^{20}+50.6$ (undil)		85^{20}	1.0631_{4}^{20}		d^h	s	v	...	...		**B5**^{2}, 277
b444	**—,—,(*dl*)***	α-Phenethyl chloride. C_8H_9Cl. *See* b202	140.61			$81–2^{17}$	1.0620_{4}^{20}	1.5276^{20}	d^h	s	v	...	s		**B5**^{2}, 277
b445	**—,—,(*l*)***	*l*-α-Phenethyl chloride. C_8H_9Cl. *See* b202	140.61	$[α]_D^{19}-50.3$		85^{20}	1.0632_{4}^{20}		d^h	s	v	...	...		**B5**^{1}, 176
b446	**—,(2-chloroethyl)-***	β-Phenethyl chloride. C_8H_9Cl. *See* b202	140.61			$91–2^{20}$	1.069_{4}^{25}	1.5294^{20}	i	s	s	s	...	CS_2 s lig s	**B5**^{2}, 277
b447	**—,1-chloro-4-ethyl-***	C_8H_9Cl. *See* b202	140.61			180–2	1.0575_{4}^{14}	1.5223^{18}	i	...	...	...	v	aa s	**B5**^{1}, 177
b448	**—,(2-chloroethylthio)-***	C_8H_9ClS. *See* b202	172.67			$117–8^{11}$	1.1820_{15}^{24}	1.5840^{20}	...	...	...	...	...	chl s	**B6**^{2}, 287
b449	**—,1-chloro-2-fluoro-***	C_6H_4ClF. *See* b202	130.56		−43	$138–40^{758}$			i	...	...	...	...		**B5**^{2}, 153
b450	**—,1-chloro-4-fluoro-***	C_6H_4ClF. *See* b202	130.56		−28	130^{757}	1.226_{4}^{20}	1.4990^{15}	i	s	s	...	...		**B5**^{2}, 153
b451	**—,1-chloro-4-hydroxy-5-isopropyl-2-methyl-***	6-Chlorothymol. $C_{10}H_{13}ClO$. *See* b202	184.67	nd	59–61	259–63			δ	s	s	...	...	alk s	**B6**^{2}, 449
b452	**—,1-chloro-3-hydroxylamino-***	C_6H_6ClNO. *See* b202	143.58	pl (bz)	49				s	...	...	...	s	peth s	**B15**^{2}, 8
b453	**—,1-chloro-4-hydroxylamino-***	C_6H_6ClNO. *See* b202	143.58	lf	87–8				δ s^h	s	...	...	v	chl v peth δ	**B15**^{2}, 9
b454	**—,1-chloro-2-iodo-***	C_6H_4ClI. *See* b202	238.47			$234–5^{760}$	1.928^{25}	1.6331^{25}	i	...	...	...	...		**B5**^{1}, 119
b455	**—,1-chloro-3-iodo-***	C_6H_4ClI. *See* b202	238.47			230			i	...	...	...	...		**B5**^{2}, 167
b456	**—,1-chloro-4-iodo-***	C_6H_4ClI. *See* b202	238.47	lf (ace)	56–7	226–7	1.886_{4}^{57}		i	s	...	...	...	$PhNO_2$ s	**B5**^{2}, 167
b457	**—,1-chloro-4-isopropyl-***	$C_9H_{11}Cl$. *See* b202	154.64			$195–7^{750}$	1.0257_{20}^{20}	1.5120^{20}	i	...	...	...	v		**B5**^{2}, 307

For explanations, symbols and abbreviations see beginning of table.

No.	Name	Synonyms and Formula	Mol. wt.	Crystalline form, color and specific rotation	m.p. °C	b.p. °C	Density	n_D	Solubility w	al	eth	ace	bz	other solvents	Ref.
	Benzene														
b458	**—,1-chloro-2-mercapto-***	C_6H_5ClS. *See* b202	144.62			205–6	1.2752^{20}		δ	δ					**B6²**, 297
b459	**—,1-chloro-3-mercapto-***	C_6H_5ClS. *See* b202	144.62			205–7	1.2637^{13}		δ	s				peth s	**B6²**, 297
b460	**—,1-chloro-4-mercapto-***	C_6H_5ClS. *See* b202	144.62	pr or lf (al)	61	205–7			i	v^h	v		v		**B6²**, 297
b461	**—,1-chloro-2-methoxy-***	C_7H_7ClO. *See* b202	142.59			$90–1^{16}$			i						**B6²**, 171
b462	**—,1-chloro-3-methoxy-***	C_7H_7ClO. *See* b202	142.59			193–4	1.1759^{12}_{4}			s	s				**B6¹**, 100
b463	**—,1-chloro-4-methoxy-***	C_7H_7ClO. *See* b202	142.59		< −18	198–202			i	v	v			chl v	**B6²**, 179
b464	**—,1-chloro-3-methoxy-2-nitro-***	$C_7H_6ClNO_3$. *See* b202	187.58		55					s^h					**B6²**, 226
b465	**—,4-chloro-1-methoxy-2-nitro-***	$C_7H_6ClNO_3$. *See* b202	187.58	ye nd (al)	98					s^h				MeOH s	**B6²**, 226
b466	**—,1(chloromethyl)-2,4-dimethyl-***	$C_9H_{11}Cl$. *See* b202	154.64			215–6			i d^h	v	v		v		**B5²**, 313
b467	**—,1(chloromethyl)-4-ethyl-***	$C_9H_{11}Cl$. *See* b202	154.64			$95–6^{15}$			d^h	s			s	chl s	**B5²**, 310
b468	**—,1(chloromethyl)-2-methyl-***	*o*-Xylyl chloride. C_8H_9Cl. *See* b202	140.61			197–9			i	∞	∞				**B5²**, 283
b469	**—,1(chloromethyl)-3-methyl-***	*m*-Xylyl chloride. C_8H_9Cl. *See* b202	140.61			195–6	1.064^{20}	1.5327^{20}	i	s	s				**B5²**, 291
b470	**—,1(chloromethyl)-4-methyl-***	*p*-Xylyl chloride. C_8H_9Cl. *See* b202	140.61			200–2			i	s	∞				**B5²**, 299
b471	**—,(chloro-*tert*-butyl)-**	$C_{10}H_{13}Cl$. *See* b202	168.67			$104–5^{18}$	1.043^{20}_{4}	1.5525^{20}	i	∞	∞	∞	∞	CCl_4 ∞ lig ∞	**B5²**, 320
b472	**—,1-chloro-2-nitro-***	$C_6H_4ClNO_2$. *See* b202	157.56	mcl nd	32–3	246^{760}	1.368^{22}_{4}		i	s	s		s		**B5²**, 180
b473	**—,1-chloro-3-nitro-***	$C_6H_4ClNO_2$. *See* b202	157.56	pa ye rh pr (al)	24 (unst) 44	$235–6^{760}$	1.343^{50}_{4}		i	δ v^h	s		s	CS_2 s chl s aa s	**B5²**, 182
b474	**—,1-chloro-4-nitro-***	$C_6H_4ClNO_2$. *See* b202	157.56	mcl pr	83	242^{760}	1.9279^{90}		i	δ v^h	s			CS_2 s	**B5²**, 183
b475	**—,1-chloro-2-nitroso-***	C_6H_4ClNO. *See* b202	141.56	nd (al)	56–7					v	v		v	chl v, peth v^h	**B5²**, 171
b476	**—,1-chloro-3-nitroso-***	C_6H_4ClNO. *See* b202	141.56	nd (bz)	72					v	v	v	v	chl v	**B5²**, 171
b477	**—,1-chloro-4-nitroso-***	C_6H_4ClNO. *See* b202	141.56	(aa or al)	92–3					s				chl v aa s^h	**B5²**, 171
b478	**—,(3-chloropropoxy)-***	$C_9H_{11}ClO$. *See* b202	170.64		12	139^{25} 245–55	1.1167^{20}				v				**B6**, 142
b479	**—,(3-chloropropylthio)-***	$C_9H_{11}ClS$. *See* b202	186.71			$116–7^{4}$	1.1536^{20}_{4}	1.5752^{20}				s		Py s	**B6²**, 288
b480	**—,1-chloro-4-triazo-***	$C_6H_4ClN_3$. *See* b202	153.57			$90–1^{15}$			i		s				**B5²**, 208
b481	**—,2-chloro-1,3,5-trimethyl-***	$C_9H_{11}Cl$. *See* b202	154.64		< −20	204–6 104^{25}	1.0337^{30}	1.5212^{30}	i	v	v				**B5²**, 315
b482	**—,1-chloro-2,4,5-trinitro-**	$C_6H_2ClN_3O_6$. *See* b202	247.55	ye cr (al)	116				i	v^h			s	aa s	**B5²**, 205
b483	**—,2-chloro-1,3,5-trinitro-***	Picryl chloride. $C_6H_2ClN_3O_6$. *See* b202	247.55	wh nd	83		1.797		i	s^h	δ		s	peth δ, chl s^h	**B5²**, 205
b484	**—,cyclohexyl-***	Phenylcyclohexane*. $C_{12}H_{16}$. *See* b202	160.26		7–8	238–9	0.9502^{20}_{4}	1.5329^{20}	i	v	v				**B5**, 503
b485	**—,4-cyclohexyl-1,3-dihydroxy-***	$C_{12}H_{16}O_2$. *See* b202	192.26		124–5				δ	δ					**B6²**, 929
b486	**—,1-cyclohexyl-2-hydroxy-***	$C_{12}H_{16}O$. *See* b202	176.25	nd (lig)	56–7	147^{17} $278–82^{758}$			i	s				aa s lig s^h	**B6²**, 548
b487	**—,1-cylohexyl-4-hydroxy-***	$C_{12}H_{16}O$. *See* b202	176.25	nd (bz)	133				i	v	v		s	CCl_4 v^h lig δ	**B6²**, 548

For explanations, symbols and abbreviations see beginning of table.

No.	Name	Synonyms and Formula	Mol. wt.	Crystalline form, color and specific rotation	m.p. °C	b.p. °C	Density	n_D	Solubility w	Solubility al	Solubility eth	Solubility ace	Solubility bz	Solubility other solvents	Ref.
	Benzene														
b488	**—,cylopentyl-***	Phenylcyclopentane*. $C_{11}H_{14}$. *See* b202	146.22			215–6^{750}	0.9504_4^{20}	1.5288^{25}	i		s				**B5²**, 393
b489	**—,1,2-diacetamido-***	$C_{10}H_{12}N_2O_2$. *See* b202	192.22	nd (w)	185–6				s^h	s		s		chl s	**B8**, 20
b490	**—,1,4-diacetyl-**	$C_{10}H_{10}O_2$. *See* b202	162.18	pr (al)	114					v^h					**B7²**, 624
b491	**—,1,5-diacetyl-2,4-dihydroxy-**	Resodiacetophenone. $C_{10}H_{10}O_4$. *See* b202	194.18	nd	180				i	δ				os s^h	**B8²**, 456
b492	**—,1,2-diamino-***	*o*-Phenylenediamine. $C_6H_8N_2$. *See* b202	108.14	brsh ye lf (w), pl (chl)	102–3	256–8^{760}			δ	v	s			chl s	**B8**, 6
b493	**—,1,3-diamino-***	*m*-Phenylenediamine. $C_6H_8N_2$. *See* b202	108.14	rh (al)	63–4	282–4^{760}	1.0696_4^{58}	1.6339^{58}	v	s	s				**B8¹**, 10
b494	**—,1,4-diamino-***	*p*-Phenylenediamine. $C_6H_8N_2$. *See* b202	108.14	wh pl (bz)	140	267			s^h	s^h	s			chl s	**B8**, 61
b495	**—,1,2-diamino-3,5-dichloro-***	$C_6H_6Cl_2N_2$. *See* b202	177.04	nd (al)	60					s^h					**B13**, 27
b496	**—,1,3-diamino-2,5-dichloro-***	$C_6H_6Cl_2N_2$. *See* b202	177.04	nd (w)	106				v^h	s			v		**B13²**, 29
b497	**—,1,4-diamino-2,5-dichloro-*-**	$C_6H_6Cl_2N_2$. *See* b202	177.04	pr (w)	170				δ	δ					**B13**, 118
b498	**—,1,5-diamino-2,4-dichloro-***	$C_6H_6Cl_2N_2$. *See* b202	177.04	nd (dil al)	136–7					s					**B13**, 54
b499	**—,2,3-diamino-1,4-dichloro-***	$C_6H_6Cl_2N_2$. *See* b202	177.04	nd (50% al)	100				s	s				os v lig s	**B13²**, 20
b500	**—,2,5-diamino-1,3-dichloro-***	$C_6H_6Cl_2N_2$ *See* b202	177.04	nd pr (dil al)	124					s					**B13¹**, 37
b501	**—,1,2-diamino-4,5-dimethoxy-***	$C_8H_{12}N_2O_2$. *See* b202	168.15	bl	131				s	s	i			to δ	**B13**, 732
b502	**—,1,2-diamino-4-methoxy-***	$C_7H_{10}N_2O$. *See* b202	138.17		48	167–70^{11}					∞				**Am 69**, 586
b503	**—,2,4-diamino 1-methoxy-***	$C_7H_{10}N_2O$. *See* b202	138.17	nd (eth)	67–8					s	s^h				**B13²**, 308
b504	**—,1,4-diamino-2-methoxy-5-methyl-***	$C_8H_{12}N_2O$. *See* b202	152.20		164				δ	s	s				**B13²**, 349
b505	**—,2,3-diamino-1-methoxyl-4-methyl-***	$C_8H_{12}N_2O$. *See* b202	152.20	pr (eth-bz)	75–6				δ	v	v		δ		**B13²**, 349
b506	**—,1,2-diamino-4-methyl-***	$C_7H_{10}N_2$. *See* b202	122.17	pl (lig)	88	265			v						**B13**, 148
b507	**—,1,2-diamino-3-nitro-***	$C_6H_7N_3O_2$. *See* b202	153.15	dk red	158–9				δ					ac s	**B13¹**, 10
b508	**—,1,2-diamino-4-nitro-***	$C_6H_7N_3O_2$. *See* b202	153.15	dk red nd	199–200									ac s	**B13**, 29
b509	**—,1,4-diamino-2,3,5,6-tetra-methyl-***	Diaminodurene. $C_{10}H_{16}N_2$. *See* b202	164.25	nd (w)	149				s^h	v	s			chl v	**B13²**, 76
b510	**—,1,2-dibenzoxy-**	Pyrocatechol dibenzyl ether. $C_{20}H_{18}O_2$. *See* b202	290.34	yesh nd or pr (al)	63–4				i	s^h	s			peth s	**B6**, 772
b511	**—,1,4-dibenzoxy-**	Hydroquinone dibenzyl ether. $C_{20}H_{18}O_2$. *See* b202	290.34	pl (al)	130				i	s^h	s			aa s	**B5**, 845
b512	**—,1,2-dibromo-***	$C_6H_4Br_2$. *See* b202	235.92		6.7	221	1.9557_4^{20}	1.6081^{20}	i	s	∞				**B5²**, 162
b513	**—,1,3-dibromo-***	$C_6H_4Br_2$. *See* b202	235.92		−7	220	1.9523_4^{20}	1.6083^{17}	i	s	∞				**B5²**, 162
b514	**—,1,4-dibromo-***	$C_6H_4Br_2$. *See* b202	235.92	pl	87	218–9	1.8322_4^{100}	1.5742	i	s	v	v		CS_2 v	**B5²**, 163
b515	**—,1,5-dibromo-2,4-dimethyl-***	$C_8H_8Br_2$. *See* b202	263.98		72	255–6			i	s^h					**B5²**, 294
b516	**—,1,2-dibromo-3-nitro-***	$C_6H_3Br_2NO_2$. *See* b202	280.93	mcl pr	85				i		s	v		chl v	**B5²**, 190
b517	**—,1,2-dibromo-4-nitro-***	$C_6H_3Br_2NO_2$. *See* b202	280.93	nd (al), mcl pr	38–9	296		1.9835^{111}	i	s			s	aa s	**B5**, 250
b518	**—,1,3-dibromo-2-nitro-***	$C_6H_3Br_2NO_2$. *See* b202	280.93	nd (al), mcl pr	84		1.9211^{111}		i	s^h			s		**B5**, 250
b519	**—,1,3-dibromo-5-nitro-***	$C_6H_3Br_2NO_2$. *See* b202	280.93	pl or pr (eth), lf or nd (al)	106		1.9341^{111}		i	s^h	s^h		s		**B5**, 250
b520	**—,1,4-dibromo-2-nitro-***	$C_6H_3Br_2NO_2$. *See* b202	280.93	yesh pl (ace)	85–6		1.9416^{111}		i			s^h	s		**B5**, 250
b521	**—,2,4-dibromo-1-nitro-***	$C_6H_3Br_2NO_2$. *See* b202	280.93	ye pl or pr (al)	62		1.9581^{111}		i	s^h			s		**B5**, 250
b522	**—,2,3-dibromo-1,4,5-trimethyl-***	$C_9H_{10}Br_2$. *See* b202	277.98	nd (al)	63–4	293–4			i	v	v		v	chl v	**B5**, 403
b523	**—,2,4-dibromo-1,3,5-trimethyl-***	$C_9H_{10}Br_2$. *See* b202	277.98	nd (al)	64	285			i	δ v^h			s		**B5**, 408

For explanations, symbols and abbreviations see beginning of table.

No.	Name	Synonyms and Formula	Mol. wt.	Crystalline form, color and specific rotation	m.p. °C	b.p. °C	Density	n_D	Solubility						Ref.
									w	al	eth	ace	bz	other solvents	
	Benzene														
b524	**—,1,2-dibutoxy-***	Pyrocatechol dibutyl ether. $C_{14}H_{22}O_2$. *See* b202	222.32	ye		241^{765}									
b525	**—,1,4-di-*tert*-butyl-**	$C_{14}H_{22}$. *See* b202	190.32	nd (MeOH)	78	109^{15}, 236			i	s	s				**B5**^{2}, 344
b526	**—,1,3-di-*tert*-butyl-5-ethyl-2-hydroxy-**	$C_{16}H_{26}O$. *See* b202	234.37		44	134^{10}								alk i	**Am 67,** 304
b527	**—,1,4-di-*tert*-butyl-2,5-dihydroxy-**	$C_{14}H_{22}O_2$. *See* b202	222.32		210–12										**Am 64,** 937
b529	**—,1,3-di-*tert*-butyl-2-hydroxy-**	$C_{14}H_{22}O$, *See* b202	206.32	pr (al)	39	$94–8^{5-6}$				s^h				alk i	**Am 73,** 3179
b530	**—,2,4-di-*tert*-butyl-1-hydroxy-**	$C_{14}H_{22}O$. *See* b202	206.32		56.5	146^{20}		$1,5080^{20}$						alk i	**Am 60,** 2496
b531	**—,1,3-di-*tert*-butyl-2-hydroxy-5-methyl-**	Ionol. $C_{15}H_{24}O$. *See* b202	220.34		70	265 133^{11}	0.8937_4^{75}	1.4859^{75}	i	s		s	s	peth s alk i	**Am 69,** 1624
b533	**—,1,5-di-*tert*-butyl-2-hydroxy-3-methyl-**	$C_{15}H_{24}O$. *See* b202	220.34		52.5	113^8	0.891_4^{80}							alk i	**Am 67,** 305
b534	**—,1,5-di-*tert*-butyl-2-hydroxy-4-methyl-**	$C_{15}H_{24}O$. *See* b202	220.34		62	282 167^{20}	0.912_4^{80}		i	s	s	s	s	CCl_4 s alk i	
b534^1	**—,1,3-di-*tert*-butyl-2-hydroxy-5(2-methyl-2-butyl)-**	$C_{19}H_{32}O$. *See* b202	276.45		47	$135–8^6$									**Am 67,** 304
b535	**—,1,2-dichloro-***	$C_6H_4Cl_2$. *See* b202	147.01		−17	179	1.3048_4^{20}	1.5485^{20}	i	s	s				**B5**^{2}, 153
b536	**—,1,3-dichloro-***	$C_6H_4Cl_2$. *See* b202	147.01		−25	172	1.2881_4^{20}	1.5457^{21}	i	s	s		s		**B5**^{2}, 154
b537	**—,1,4-dichloro-***	$C_6H_4Cl_2$. *See* b202	147.01	mcl pr, lf (ace)	53	174	1.533_4^0	1.5210^{80}	i	δ s^h	s		s	CS_2 s chl s	**B5**^{2}, 154
b538	**—,1,2-dichloro-4,5-dihydroxy-***	$C_6H_4Cl_2O_2$. *See* b202	179.00	pr (chl-CS_2), nd (bz-peth)	116–7				v	v			v		**B6**^{1}, 389
b539	**—,1,3-dichloro-2,5-dihydroxy-***	$C_6H_4Cl_2O_2$. *See* b202	179.00	pl (al), nd (w or bz)	161–2				s^h	s			s^h		**B6**^{2}, 845
b540	**—,1,4-dichloro-2,5-dihydroxy-***	$C_6H_4Cl_2O_2$. *See* b202	179.00	nd (w), pl (bz)	166–70				s	v	v	v			**B6**^{2}, 845
b541	**—,1,5-dichloro-2,3-dihydroxy-***	$C_6H_4Cl_2O_2$. *See* b202	179.00		83–4				δ v^h	s					**B6,** 783
b542	**—,1,5-dichloro-2,4-dihydroxy-***	$C_6H_4Cl_2O_2$. *See* b202	179.00		113				v	v	v			lig δ	**B6**^{2}, 819
b543	**—,2,3-dichloro-1,4-dihydroxy-***	$C_6H_4Cl_2O_2$. *See* b202	179.00	nd (w)	144–5					v				lig i	**B6**^{2}, 845
b544	**—,1,2-dichloro-4,5-dimethyl-3-hydroxy-***	$C_8H_8Cl_2O$. *See* b202	191.06		90									chl v peth δ	**B6**^{2}, 454
b545	**—,1,3-dichloro-4,5-dimethyl-2-hydroxy-***	$C_8H_8Cl_2O$. *See* b202	191.06		52						s			peth δ	**B6**^{2}, 456
b546	**—,1,4-dichloro-2,3-dimethyl-5-hydroxy-***	$C_8H_8Cl_2O$. *See* b202	191.06		84						s			peth, δ s^h	**B6**^{2}, 456
b547	**—,2,4-dichloro-1,3-dimethyl-5-hydroxy-***	$C_8H_8Cl_2O$. *See* b202	191.06		83						s			lig δ s^h	**B6**^{2}, 464
b548	**—,2,4-dichloro-1,5-dimethyl-3-hydroxy-***	$C_8H_8Cl_2O$. *See* b202	191.06		95–6									chl v, CCl_4 v peth δ	
b549	**—,3,4-dichloro-1,2-dimethyl-5-hydroxy-***	$C_8H_8Cl_2O$. *See* b202	191.06		102.5									os s	
b550	**—,2,4-dichloro-1-ethoxy-***	$C_8H_8Cl_2O$. *See* b202	191.06			236–7			i						**B6,** 189
b551	**—,(1,2-dichloro-ethyl)-***	Styrene dichloride. $C_8H_8Cl_2$. *See* b202	175.06			$233–4.9^{35}$	1.240_4^{15}	1.5544^{15}	i				s		**B5**^{2}, 278

For explanations, symbols and abbreviations see beginning of table.

No.	Name	Synonyms and Formula	Mol. wt.	Crystalline form, color and specific rotation	m.p. °C	b.p. °C	Density	n_D	Solubility w	al	eth	ace	bz	other solvents	Ref.
	Benzene														
b552	**—,1,4-dichloro-2-iodo-***	$C_6H_3Cl_2I$. *See* b202	272.90	pl (al)	21	250–1			i	s					**B5**, 221
b553	**—,2,4-dichloro-1-methoxy-***	$C_7H_6Cl_2O$. *See* b202	177.03	pr	27–8	$232–3^{745}$			i	s	s				**B6**^{1}, 103
b554	**—,1,2-dichloro-3-nitro-***	$C_6H_3Cl_2NO_2$. *See* b202	192.00	mcl nd (peth)	61–2	257–8	1.721^{14}							peth s, os v aa s	**B5**^{2}, 185
b555	**—,1,2-dichloro-4-nitro-***	$C_6H_3Cl_2NO_2$. *See* b202	192.00	nd (al)	43	255–6	1.4266^{100}_{4}		i	s^h	s				**B5**^{2}, 186
b556	**—,1,3-dichloro-2-nitro-***	$C_6H_3Cl_2NO_2$. *See* b202	192.00	nd or pr (al)	72.5	130^8	1.4094^{80}		i	s^h					**B5**^{2}, 186
b557	**—,1,3-dichloro-5-nitro-***	$C_6H_3Cl_2NO_2$. *See* b202	192.00	pl (al)	65				i	s^h					**B5**^{2}, 186
b558	**—,1,4-dichloro-2-nitro-***	$C_6H_3Cl_2NO_2$. *See* b202	192.00	pl or pr (al)	55	266	1.669^{22}		i	s^h				chl s	**B5**, 245
b559	**—,2,4-dichloro-1-nitro-***	$C_6H_3Cl_2NO_2$. *See* b202	192.00		32	258.5	1.4390^{80}			s					**B5**^{2}, 185
b560	**—,1-dichloro-phosphino-4-isopropyl-***	$C_9H_{11}Cl_2P$. *See* b202	221.07			268–70	1.190^{12}		d^h						**B16**, 773
b561	**—,2,4-dichloro-1-triazo-***	$C_6H_3Cl_2N_3$. *See* b202	188.02	ye nd (al), pr (ace or bz)	54				i	s	s		s	peth s chl s	**B5**^{2}, 208
b562	**—,1,2-diethoxy-***	Catechol diethyl ether. $C_{10}H_{14}O_2$. *See* b202	166.22	pr (peth)	43–5					s	v				**B6**^{2}, 780
b563	**—,1,3-diethoxy-***	Resorcinol diethyl ether. $C_{10}H_{14}O_2$. *See* b202	166.22	pr	12.4	235			i	s	s				**B6**^{2}, 814
b564	**—,1,4-diethoxy-***	Hydroquinone diethyl ether. $C_{10}H_{14}O_2$. *See* b202	166.21	pl	71–2	246				v	s		s	chl s	**B6**^{2}, 840
b565	**—,1,2-diethyl-***	$C_{10}H_{14}$. *See* b202	134.21		<-20	183.5	0.8800^{20}	1.5035^{20}	i	s	s				**B5**^{2}, 327
b566	**—,1,3-diethyl-***	$C_{10}H_{14}$. *See* b202	134.21		<-20	181–2	0.8639^{20}_{4}	1.4955^{20}	i	s	s				**B5**^{2}, 327
b567	**—,1,4-diethyl-***	$C_{10}H_{14}$. *See* b202	134.21		−35	182–3	0.8620^{20}_{4}	1.4967^{20}	i	s	s				**B5**^{2}, 327
b568	**—,1,3-diethyl-5-methyl-***	$C_{11}H_{16}$. *See* b202	148.24			198–200	0.879^{20}_{4}		i	∞	∞				**B5**, 441
b569	**—,1,2-dihydroxy-***	Catechol. Pyrocatechol. $C_6H_6O_2$. *See* b202	110.11		105	240	1.371	1.604	s	s	s		s	alk s	
b570	—,—,carbonate	[structure: benzene ring fused to O–CO–O]	136.11	pr (ace or al)	120	225–30				s^h			s^h		**B19**^{1}, 660
b571	—,—,diacetate	$C_{10}H_{10}O_4$. *See* b202	194.19	nd (al)	64				i	v	v			peth s chl v	**B6**^{2}, 784
b572	—,—,dibenzoate	$C_{20}H_{14}O_4$. *See* b202	318.31	lf (eth-al)	86				i	s	s		s		**B9**^{2}, 112
b573	—,—,monoacetate	$C_8H_8O_3$. *See* b202	152.15	pl	57–8	$189–91^{102}$			s	s					**B6**^{2}, 783
—	—,—,monobenzoate	*see* **Benzoic acid,** 2-hydroxyphenyl ester													
b574	**—,1,3-dihydroxy-***	Resorcinol. $C_6H_6O_2$. *See* b202	110.11	nd	111	281	1.2717		s	s	s			chl δ	**B6**^{2}, 802
b575	—,—,diacetate	$C_{10}H_{10}O_4$. *See* b202	194.19			278			i	v	v				**B6**^{2}, 817
b576	—,—,dibenzoate	$C_{20}H_{14}O_4$. *See* b202	318.31	pl	117					s	s				**B9**^{2}, 113
b577	**—,1,4-dihydroxy-***	Hydroquinone. Quinol. $C_6H_6O_2$. *See* b202	110.11	mcl pr (sub), nd (w), pr (MeOH)	170	285^{730}	1.328^{15}		s v^h	v					**B6**^{2}, 832
b578	—,—,diacetate	$C_{10}H_{10}O_4$. *See* b202	194.19	pl (w)	122		0.8731^{25}_{4}		s^h	v^h	v			aa v^h lig v	**B6**^{2}, 843
b579	—,—,dibenzoate	$C_{20}H_{14}O_4$. *See* b202	318.31	mcl (al or to)	204				i	s^h	i			to s^h	**B9**^{2}, 114
b580	**—,1,2-dihydroxy-3,5-dimethyl-***	$C_8H_{10}O_2$. *See* b202	138.16	pr (w)	73–4				v	v	v				**B6**, 911
b581	**—,1,2-dihydroxy-4,5-dimethyl-***	$C_8H_{10}O_2$. *See* b202	138.16	mcl pr or nd (peth)	87–8	sub			v	v	v		δ	sulf s peth δ^h	**B6**^{1}, 444
b582	**—,1,3-dihydroxy-2,4-dimethyl-***	$C_8H_{10}O_2$. *See* b202	138.16	nd	149–50	sub			s	v	v				**B6**^{1}, 444
b583	**—,1,3-dihydroxy-2,5-dimethyl-***	$C_8H_{10}O$. *See* b202	138.16	nd (bz), pr (w)	163	277–80			s	s	s				**B6**^{2}, 891
b584	**—,1,4-dihydroxy-2,3-dimethyl-***	$C_8H_{10}O_2$. *See* b202	138.16		221				s	s	s				**B6**, 908
b585	**—,1,4-dihydroxy-2,5-dimethyl-***	$C_8H_{10}O_2$. *See* b202	138.16	lf (w)	217 sub				s^h	s	s		δ	CS_2 δ chl s	**B6**, 915
b586	**—,1,5-dihydroxy-2,3-dimethyl-***	$C_8H_{10}O_2$. *See* b202	138.16	nd (bz), pr (w+1)	136–7 115–7 (+1w)	sub			s	v			δ	peth δ CS_2 δ chl δ	**B6**, 908
b587	**—,1,5-dihydroxy-2,4-dimethyl-***	$C_8H_{10}O_2$. *See* b202	138.16	mcl pr, (w+1)	125	276–9 sub			s	s	s				**B6**^{2}, 889
b588	**—,2,5-dihydroxy-1,3-dimethyl-***	$C_8H_{10}O_2$. *See* b202	138.17	nd (xyl)	150					s	s				**B6**^{2}, 888

For explanations, symbols and abbreviations see beginning of table.

No.	Name	Synonyms and Formula	Mol. wt.	Crystalline form, color and specific rotation	m.p. °C	b.p. °C	Density	n_D	Solubility w	al	eth	ac	bz	other solvents	Ref.
	Benzene														
b589	**—,1,3-dihydroxy-2,4-dinitro-***	$C_6H_4N_2O_6$. *See* b202	200.11	ye nd	147–8				δ	δ					**B6**^{2}, 823
b590	**—,1,3-dihydroxy-2,4-dinitroso-***	$C_6H_4N_2O_4$. *See* b202	168.11	ye rh pl	162–3d				s	s					**B7**^{2}, 851
b591	**—,1,3-dihydroxy-2,4-di(2-tolylazo)**	$C_{20}H_{18}N_4O_2$. *See* b202	346.39	red nd (chl-al)	212					δ			s	chl s	**B16,** 186
b592	**—,1,3-dihydroxy-2,4-di(4-tolylazo)-**	$C_{20}H_{18}N_4O_2$. *See* b202	346.39	nd (chl-al)	230.5				δ					os s	**B16,** 186
b593	**—,2,4-dihydroxy-1,5-di(4-tolylazo)-**	$C_{20}H_{18}N_4O_2$. *See* b202	346.39	ye nd (chl-al)	255–6				s	δ				chl v^h	**B16,** 187
b594	**—,2,4-dihydroxy-1-ethyl-***	$C_8H_{10}O_2$. *See* b202	138.17	pr (chl or bz)	98–9	131^{15}			δ	δ	δ				**B6**^{2}, 885
b595	**—,2,4-dihydroxy-1-hexyl-***	$C_{12}H_{18}O_2$. *See* b202	194.28	pa ye nd	68–71	333, 178^8			δ	s	s		s	chl s	**B6**^{2}, 904
b596	**—,1,2-dihydroxy-4-iodo-***	$C_6H_5IO_2$. *See* b202	236.01	lf (CCl_4)	92	sub			δ	s	s	s	s	chl δ, peth δ	
b597	**—,1,3-dihydroxy-5-iodo-***	$C_6H_5IO_2$. *See* b202	236.01	nd (bz)	92.3	sub			i	v				peth i	**B6**^{2}, 821
b598	**—,1,4-dihydroxy-2-iodo-***	$C_6H_5IO_2$. *See* b202	236.01		115–6										**Am 56,** 667
b599	**—,2,4-dihydroxy-1-iodo-***	$C_6H_5IO_2$. *See* b202	236.01	pr	67										**B6**^{2}, 821
b600	**—,2,4-dihydroxy-1-isobutyl-**	$C_{10}H_{14}O_2$. *See* b202	166.22		62–3	$166–8^{6–7}$			δ	s	s				
b601	**—,1,2-dihydroxy-4-isopropyl-**	$C_9H_{12}O_2$. *See* b202	152.19	lf (peth)	78	270–2								peth s^h	**B6,** 929
b602	**—,1,4-dihydroxy-2-isopropyl-**	$C_9H_{12}O_2$. *See* b202	152.19	nd	130–1									lig s^h	**B6,** 929
b603	**—,2,4-dihydroxy-1-isopropyl-**	$C_9H_{12}O_2$. *See* b202	152.19		105	265–81			δ	s	s				
b604	**—,1,4-dihydroxy-2-isopropyl-5-methyl-**	Thymohydroquinone. $C_{10}H_{14}O_2$. *See* b202	166.22	pr	142	290 sub			δ	s	s		s	peth i	**B6**^{2}, 901
b605	**—,2,3-dihydroxy-1-isopropyl-4-methyl-**	$C_{10}H_{14}O_2$. *See* b202	166.22		48	270			δ					os s peth i	**B6**^{2}, 900
b606	**—,1,2-dihydroxy-3-methoxy-***	$C_7H_8O_3$. *See* b202	140.14		38–41										
b607	**—,1,3-dihydroxy-2-methoxy-***	$C_7H_8O_3$. *See* b202	140.14		85–7	$154–5^{24}$									
b608	**—,1,3-dihydroxy-5-methoxy-***	$C_7H_8O_3$. *See* b202	140.14	pl (bz)	78	213^{16}			δ	v	v				
b609	**—,1,2-dihydroxy-3-methyl-***	$C_7H_8O_2$. *See* b202	124.14	lf (bz)	68	241			s	s			s	chl s	**B6**^{2}, 858
b610	**—,1,3-dihydroxy-2-methyl-***	$C_7H_8O_2$. *See* b202	124.14	pr (bz)	119–20	264			s	s	s		s		**B6,** 878
b611	**—,2,4-dihydroxy-1(3-methylbutyl)-***	$C_{11}H_{16}O_2$. *See* b202	180.25		61–2	$177–8^6$			δ	s	s		s		**B6**^{2}, 903
b612	**—,2,4-dihydroxy-1(4-methylpentyl)-***	$C_{12}H_{18}O_2$. *See* b202	194.27		70–1	$182–3^{67}$			δ	s	s				**B6,** 904
b613	**—,1,3-dihydroxy-2-nitro-***	$C_6H_5NO_4$. *See* b202	155.11	og-red pr (al)	83.5					s^h					**B6**^{2}, 822
b614	**—,1,4-dihydroxy-2-nitro-***	$C_6H_5NO_4$. *See* b202	155.11	og-red rh	133–4				s^h	v	v				**B6**^{2}, 848
b615	**—,2,4-dihydroxy-1-nitroso-***	2-Hydroxy-1,4-benzoquinone oxime. $C_6H_5NO_3$. *See* b202	139.11	ye nd (+w)	148d				s	v	s	v	i	CS_2 i, chl s	**B8,** 253
b616	**—,1,3-dihydroxy-4-pentyl-***	$C_{11}H_{16}O_2$. *See* b202	180.24		72–3	$168–70^{6–7}$			i	s	s		s		**B6**^{2}, 902
b617	**—,1,3-dihydroxy-5-pentyl-***	Olivetol. $C_{11}H_{16}O_2$. *See* b202	180.24	nd (+w) pr(bz-lig)	49 40–1 (+w)	164^5			i	s	s		s		**J1945,** 311
b618	**—,1,2-dihydroxy-4-propyl-***	$C_9H_{12}O_2$. *See* b202	152.19	pr	58–60	152^{13}	1.100^{18}_4	1.4440^{18}	δ	s	s		s^h		**B6**^{2}, 892
b619	**—,1,3-dihydroxy-5-propyl-***	$C_9H_{12}O_2$. *See* b202	152.19	nd	83–4				s^h	s	s		s		**B6**^{2}, 893
b620	**—,2,4-dihydroxy-1-propyl-***	$C_9H_{12}O_2$. *See* b202	152.19	pr (bz)	107–8	$172–4^{14–15}$			s	s	s				
b621	**—,1,2-dihydroxy-3,4,5,6-tetrabromo-***	$C_6H_2Br_4O_2$. *See* b202	425.74	nd	192–3					s^h					**B6**^{2}, 788

For explanations, symbols and abbreviations see beginning of table.

PHYSICAL CONSTANTS OF ORGANIC COMPOUNDS (Continued)

No.	Name	Synonyms and Formula	Mol. wt.	Crystalline form, color and specific rotation	m.p. °C	b.p. °C	Density	n_D	Solubility						Ref.
									w	al	eth	ace	bz	other solvents	
	Benzene														
b622	**—,1,4-dihydroxy-2,3,5,6-tetra-bromo-***	$C_6H_2Br_4O_2$. *See* b202	425.74	mcl pr (al-eth)	244	3.023^{21}			i	v	v			aa s	**B6^2, 848**
b623	**—,1,2-dihydroxy-3,4,5,6-tetra-chloro-***	$C_6H_2Cl_4O_2$. *See* b202	247.91	nd (+3w)	194–5 94(+3w)								δ		**B6^2, 787**
b625	**—,1,3-dihydroxy-2,4,5,6-tetra-chloro-***	$C_6H_2Cl_4O_2$. *See* b202	247.91	nd (w)	141				v^h	v	v		v	aa v	**B6^2, 819**
b626	**—,1,4-dihydroxy-2,3,5,6-tetra-chloro-***	$C_6H_2Cl_4O_2$. *See* b202	247.91	nd (aa)	230–1				i	v	v		δ	CS_2 δ CCl_4 δ aa δ	**B6^2, 846**
b627	**—,1,4-dihydroxy-2,3,5,6-tetra-iodo-***	$C_6H_2I_4O_2$. *See* b202	613.75	(aa)	258					δ	s		δ	alk s^h, chl s aa δ	**B6^1, 417**
b628	**—,1,4-dihydroxy-2,3,5,6-tetra-methyl-***	Durohydroquinone. $C_{10}H_{14}O_2$. *See* b202	166.21	nd (al)	233					s^h	δ				
b629	**—,1,3-dihydroxy-2,4,6-tribromo-***	$C_6H_3Br_3O_2$. *See* b202	346.83	nd (w)	112				δ	s	s				**B6, 822**
b630	**—,1,4-dihydroxy-2,3,5-tribromo-***	$C_6H_3Br_3O_2$. *See* b202	346.83	nd (chl)	136–7				v^h	v	v		v	aa v lig s	**B6^2, 848**
b631	**—,1,2-dihydroxy-3,4,5-trichloro-***	$C_6H_3Cl_3O_2$. *See* b202	213.46	pr	115 (+w) 134–5 (+½w)				i	s	s			aa s	**B6^1, 389**
b632	**—,2,4-dihydroxy-1,3,5-trichloro-***	$C_6H_3Cl_3O_2$. *See* b202	213.46	nd	83				δ	v	v				**B6, 82**
b633	**—,1,3-dihydroxy-2,4,5-trimethyl-***	$C_9H_{12}O_2$. *See* b202	152.19		156										**B6, 931**
b634	**—,1,3-dihydroxy-4,5,6-trimethyl-***	$C_9H_{12}O_2$. *See* b202	152.19	nd or lf	163–4					v	v	v	δ		**B6, 930**
b635	**—,1,4-dihydroxy-2,3,5-trimethyl-***	$C_9H_{12}O_2$. *See* b202	152.19	nd (w)	167–8				v^h	v	v		v		**B6, 930**
b636	**—,2,4-dihydroxy-1,3,5-trimethyl-***	$C_9H_{12}O_2$. *See* b202	152.19	pl (al), lf	149–50	275			δ	v	v		δ		**B6, 939**
b637	**—,2,4-dihydroxy-1,3,5-trinitro-***	Styphnic acid. $C_6H_3N_3O_8$	245.11	hex (al)	179–80	sub			δ	s	s				**B6^2, 825**
b638	**—,1,2-diiodo-***	$C_6H_4I_2$. *See* b202	329.92	pl or pr (lig)	27	286^{750}	2.54^{20}	1.7179^{20}	i	δ	v				**B5^2, 168**
b639	**—,1,3-diiodo-***	$C_6H_4I_2$. *See* b202	329.92	rh pl or pr (eth-al)	40.4	285^{756}	2.47^{25}		i	s	s			chl s	**B5^2, 168**
b640	**—,1,4-diiodo-***	$C_6H_4I_2$. *See* b202	329.92	rh lf (al)	129.4	285 sub			i	s	v				**B5^2, 168**
b641	**—,1,2-diisopropyl-**	$C_{12}H_{18}$. *See* b202	162.26		−57	204	0.8771_4^{20}	1.4960^{20}	i	∞	∞	∞	∞	CCl_4 ∞	**B5, 447**
b642	**—,1,3-diisopropyl-**	$C_{12}H_{18}$. *See* b202	162.26		−63	203	0.8559_4^{20}	1.4883^{20}	i	∞	∞	∞	∞	CCl_4 ∞	**B5, 447**
b643	**—,1,4-diisopropyl-**	$C_{12}H_{18}$. *See* b202	162.26		−17	210	0.8568_4^{20}	1.4898^{20}	i	∞	∞	∞	∞	CCl_4 ∞	**B5^2, 339**
b644	**—,1,2-di-mercapto-***	$C_6H_6S_2$. *See* b202	142.25		28–9	238–9 120^{17}				v	v		v		**B6^2, 799**
b645	**—,1,3-di-mercapto-***	$C_6H_6S_2$. *See* b202	142.25	lf	25–7	245 141^{28}			δ					os v alk v	**B6^2, 829**
b646	**—,1,4-di-mercapto-***	$C_6H_6S_2$. *See* b202	142.25	lf (al)	98					v			v	aa v lig δ	**B6^2, 854**
b647	**—,1,2-di-methoxy-***	Veratrole. $C_8H_{10}O_2$. *See* b202	138.16	(lig)	22.5	206–7	1.084_{25}^{25}		δ	s	s				**B6^2, 779**
b648	**—,1,3-di-methoxy-***	Resorcinol dimethyl ether. $C_8H_{10}O_2$. *See* b202	138.16		−55	217–8	1.0552_{25}^{25}		δ	s	s		s	sulf s	**B6^2, 813**
b649	**—,1,4-di-methoxy-***	Hydroquinone dimethyl ether. $C_8H_{10}O_2$. *See* b202	138.16	lf (w)	56	212.6	1.036_4^{66}		δ	s	v		v		**B6^2, 839**
b650	**—,1,2-dimethoxy-4,5-dinitro-***	$C_8H_8N_2O_6$. *See* b202	228.16	nd (al)	130		1.3164_4^{140}		i	δ				MeOH δ	**B6^2, 794**
b651	**—,1,2-dimethoxy-4-iodo-***	$C_8H_9IO_2$. *See* b202	264.07	nd (dil MeOH)	35					s			s	MeOH s^h	**B6^1, 390**
b652	**—,1,4-dimethoxy-2-isopropyl-**	$C_{11}H_{16}O_2$. *See* b202	180.24			$114–6^{15}$	1.0129_4^{17}	1.5105^{17}			s		s		**B6, 929**
b653	**—,1,2-dimethoxy-3-nitro-***	$C_8H_9NO_4$. *See* b202	183.16	nd (al)	64–5		1.1404_4^{133}		i	v	v		v	aa s lig δ	**B6^2, 790**
b654	**—,1,2-dimethoxy-4-nitro-***	$C_8H_9NO_4$. *See* b202	183.16	ye nd (al-w)	98	230^{15-20}	1.1888_4^{133}		i	v	v			chl s lig δ	**B6^2, 790**

For explanations, symbols and abbreviations see beginning of table.

No.	Name	Synonyms and Formula	Mol. wt.	Crystalline form, color and specific rotation	m.p. °C	b.p. °C	Density	n_D	Solubility						Ref.
									w	al	eth	ace	bz	other solvents	
	Benzene														
b655	**—,1,3-dimethoxy-2-nitro-***	$C_8H_9NO_4$. *See* b202	183.16	ye nd (al)	131					s^h			s	chl s aa s^h	**B6**^{1}, 404
b656	**—,1,3-dimethoxy-5-nitro-***	$C_8H_9NO_4$. *See* b202	183.16	pa ye nd	89		1.1693^{132}_{4}		i	s				os s lig i	**B6**^{2}, 822
b657	**—,1,4-dimethoxy-2-nitro-***	$C_8H_9NO_4$. *See* b202	183.16	ye nd (dil al)	72–3	169^{13}	1.1666^{132}_{4}		i	v			s	chl s sulf s	**B6**^{2}, 849
b658	**—,2,4-dimethoxy-1-nitro-**	$C_8H_9NO_4$. *See* b202	183.16	nd (al)	75–6		1.1876^{132}_{4}		i	s				os s lig i	**B6**^{2}, 822
b659	**—,1,2-dimethoxy-4-propenyl- (*cis*)***	Isoeugenol methyl ether. $C_{11}H_{14}O_2$. *See* b202	178.22			$138–40^{12}$	1.0521^{20}_{4}	1.5616^{20}							**B6**, 956
b660	**—,1,2-dimethyl-***	*o*-Xylene. C_8H_{10}. *See* b202	106.16		−25	144	0.8968^{20}_{4}	1.5058^{20}	i	∞	∞				**B5**^{2}, 281
b661	**—,1,3-dimethyl-***	*m*-Xylene. C_8H_{10}. *See* b202	106.16		−47.4	139	0.8684^{15}_{4}	1.4973^{20}	i	∞	∞			os ∞	**B5**^{2}, 287
b662	**—,1,4-dimethyl-***	*p*-Xylene. C_8H_{10}. *See* b202	106.16		13–4	138			i	∞	∞			os ∞	**B5**^{2}, 296
b663	**—,1,2-dimethyl-3,4-dinitro-***	$C_8H_8N_2O_4$. *See* b202	196.16	nd (al)	82					δ				os δ	**B5**^{2}, 287
b664	**—,1,2-dimethyl-3,5-dinitro-***	$C_8H_8N_2O_4$. *See* b202	196.16	nd (al or peth)	75–6					s		δ	δ	chl δ	**B5**^{1}, 181
b665	**—,1,2-dimethyl-4,5-dinitro-***	$C_8H_8N_2O_4$. *See* b202	196.16	nd (al)	115–6				$δ^h$	δ				peth δ os δ	**B5**^{1}, 181
b666	**—,1,3-dimethyl-2,5-dinitro-***	$C_8H_8N_2O_4$. *See* b202	196.16	ye	101					s					**B5**^{2}, 295
b667	**—,1,4-dimethyl-2,3-dinitro-***	$C_8H_8N_2O_4$. *See* b202	196.16	mcl pr (al)	93				i	v^h					**B5**^{2}, 302
b668	**—,1,4-dimethyl-2,5-dinitro-***	$C_8H_8N_2O_4$. *See* b202	196.16	ye nd (al)	147				i	s^h	δ				**B5**^{1}, 188
b669	**—,2,3-dimethyl-1,4-dinitro-***	$C_8H_8N_2O_4$. *See* b202	196.16	(al)	89–90				i	s				os s	**B5**^{1}, 188
b670	**—,2,5-dimethyl-1,3-dinitro-***	$C_8N_8N_2O_4$. *See* b202	196.16	nd (al)	123				i	s					**B5**^{2}, 302
b671	**—,1,2-dimethyl-4-ethyl-***	$C_{10}H_{14}$. *See* b202	134.22		<−20	189	0.869^{20}_{4}		i	v	s				
b672	**—,1,3-dimethyl-5-ethyl-***	$C_{10}H_{14}$. *See* b202	134.22		<−20	185	0.861^{20}		i	s	s				**B5**^{2}, 328
b673	**—,1,4-dimethyl-2-ethyl-***	$C_{10}H_{14}$. *See* b202	134.22		<−20	185	0.8570^{22}	1.5051	i	s	s				**B5**^{2}, 328
b674	**—,2,4-dimethyl-1-ethyl-***	$C_{10}H_{14}$. *See* b202	134.22		<−20	185.5	0.8824^{17}_{4}		i	s	s				**B5**^{2}, 328
b675	**—,1,2-dimethyl-3-hydroxy-***	Xylenol. $C_8H_{10}O$. *See* b202	122.17	nd (w)	75	218			δ	s					**B6**^{1}, 239
b676	**—,1,2-dimethyl-4-hydroxy-***	$C_8H_{10}O$. *See* b202	122.17	nd (w)	62.5	225	1.023^{17}_{15}		δ	s	∞				**B6**^{2}, 455
b677	**—,—,acetate**	$C_{10}H_{12}O_2$. *See* b202	164.21		22	235			i	s	s		s		**B6**^{2}, 455
b678	**—,1,3-dimethyl-2-hydroxy-***	$C_8H_{10}O$. *See* b202	122.17		49	212			s	s					**B6**^{2}, 457
b679	**—,1,3-dimethyl-5-hydroxy-***	$C_8H_{10}O$. *See* b202	122.17	nd (w or peth)	68	219 sub			s	s					**B6**^{2}, 462
b680	**—,1,4-dimethyl-2-hydroxy-***	$C_8H_{10}O$. *See* b202	122.17	nd (w)	75	242			s	s	v				**B6**^{2}, 466
b681	**—,—,acetate**	$C_{10}H_{12}O_2$. *See* b202	164.21		<−20	237^{768}	1.0624^{15}		i	s	s				**B6**, 495
b682	**—,2,4-dimethyl-1-hydroxy-***	$C_8H_{10}O$. *See* b202	122.17	nd	26	$211–2^{766}$	1.0276^{14}_{4}	1.5420^{14}	δ	s	∞				**B6**^{2}, 458
b683	**—,—,acetate**	$C_{10}H_{12}O_2$. *See* b202	164.21			226	$1.0298^{15.5}_{4}$	1.4990^{15}	i	s	s				**B6**, 487
b684	**—,1,5-dimethyl-2-hydroxy-3-(hydroxymethyl)-***	$C_9H_{12}O_2$. *See* b202	152.19	nd	56–7					s	s			peth s^h	**B6**, 939
b685	**—,1,2-dimethyl-3-hydroxy-5-nitro-***	$C_8H_9NO_3$. *See* b202	167.17	og-ye nd (bz or w)	120–1				s^h	v		v	s^h	chl s^h peth i aa v	**B6**^{2}, 455
b686	**—,1,2-dimethyl-4-hydroxy-5-nitro-***	$C_8H_9NO_3$. *See* b202	167.17	ye rh (al)	87					v	v		v	chl v peth δ	**B6**, 484
b687	**—,1,3-dimethyl-2-hydroxy-4-nitro-***	$C_8H_9NO_3$. *See* b202	167.17	lf pr (bz), nd (lig)	99–100					v			s^h	chl v lig s^h	**B6**, 485

For explanations, symbols and abbreviations see beginning of table.

No.	Name	Synonyms and Formula	Mol. wt.	Crystalline form, color and specific rotation	m.p. °C	b.p. °C	Density	n_D	Solubility w	al	eth	ace	bz	other solvents	Ref.
	Benzene														
b688	**—,1,3-dimethyl-2-hydroxy-5-nitro-***	$C_8H_9NO_3$. *See* b202	167.17	pr (MeOH)	169–70					s			δ	MeOH s^h chl v lig δ	**B6**, 486
b689	**—,1,4-dimethyl-2-hydroxy-3-nitro-***	$C_8H_9NO_3$. *See* b202	167.17	nd (peth)	34–5	236d 150^{15}				v				os v	**B6**¹, 246
b690	**—,1,4-dimethyl-2-hydroxy-5-nitro-***	$C_8H_9NO_3$ *See* b202	167.17	pa ye nd	116–7				s^h	s	s				**B6**¹, 246
b691	**—,1,5-dimethyl-2-hydroxy-3-nitro-***	$C_8H_9NO_3$. *See* b202	167.17	ye nd	78				s^h						**B6**², 461
b692	**—,1,5-dimethyl-3-hydroxy-2-nitro-***	$C_8H_9NO_3$. *See* b202	167.17	ye nd (lig or dil MeOH)	66					v				os s peth i lig s^h	**B6**¹, 244
b693	**—,2,5-dimethyl-1-hydroxy-3-nitro-***	$C_8H_9NO_3$. *See* b202	167.17	ye lf (peth)	91				δ	v	v			peth s^h	**B6**, 497
b694	**—,1,2-dimethyl-4-hydroxy-3,5,6-tribromo-***	$C_8H_7Br_3O$. *See* b202	358.87	nd (al)	172–3				i	s					**B6**¹, 240
b695	**—,1,3-dimethyl-5-hydroxy-2,4,6-tribromo-***	$C_8H_7Br_3O$. *See* b202	358.87	nd (al)	162–4				s	s					**B6**², 465
b696	**—,1,2-dimethyl-3-hydroxy-4,5,6-trichloro-***	$C_8H_7Cl_3O$. *See* b202	225.50	nd (dil al or peth)	180–1					v				peth δ	**B6**², 454
b697	**—,1,2-dimethyl-4-hydroxy-3,5,6-trichloro-***	$C_8H_7Cl_3O$. *See* b202	225.50	nd (peth)	182.5									peth v^h	**B6**², 456
b698	**—,1,3-dimethyl-4-hydroxy-2,5,6-trichloro-***	$C_8H_7Cl_3O$. *See* b202	225.50	pa ye nd	174						s				**B6**², 460
b699	**—,1,3-dimethyl-5-hydroxy-2,4,6-trichloro-***	$C_8H_7Cl_3O$. *See* b202	225.50	ye nd (peth)	177–8									peth v^h	
b700	**—,1,4-dimethyl-2-hydroxy-3,5,6-trichloro-***	$C_8H_7Cl_3O$. *See* b202	225.50	pa gr nd (al)	175				i	s	v		v	chl v	**B6**², 467
b701	**—,1,2-dimethyl-3-iodo-***	C_8H_9I. *See* b202	232.07			$125–6^{15}$	1.6395^{20}_{4}	1.6074^{20}							**B5**², 286
b702	**—,1,2-dimethyl-4-iodo-***	C_8H_9I. *See* b202	232.07			111^{11} 228–32d	1.6334^{18}_{4}	1.6049^{18}							**B5**², 286
b703	**—,1,3-dimethyl-2-iodo-***	C_8H_9I. *See* b202	232.07			228–30			i				s		**B5**, 375
b704	**—,1,3-dimethyl-5-iodo-***	C_8H_9I. *See* b202	232.07			234–5 117^{27}	$1.6085^{18.5}_{4}$	$1.5967^{18.5}$	i					os s	**B5**², 294
b705	**—,1,4-dimethyl-2-iodo-***	C_8H_9I. *See* b202	232.07			$230–5^{770}$ $106–8^{13}$	1.6168^{17}_{4}	1.5992^{17}	i				s		**B5**², 302
b706	**—,2,4-dimethyl-1-iodo-***	C_8H_9I. *See* b202	232.07			232 111^{14}	1.6282^{16}_{4}	1.6008^{16}	i				s		**B5**², 294
b707	**—,1,2-dimethyl-3-methoxy-***	$C_9H_{12}O$. *See* b202	136.20		29	199	0.9596^{40}	1.5120^{40}	i	v	v		v		**B6**², 453
b708	**—,1,2-dimethyl-4-methoxy-***	$C_9H_{12}O$. *See* b202	136.20			204–5	0.9744^{14}_{4}	1.5198^{14}	i	s	s		s		**B6**², 455
b709	**—,1,3-dimethyl-2-methoxy-***	$C_9H_{12}O$. *See* b202	136.20			182–3	0.9619^{14}_{4}	1.5053^{14}	i	s	s		s		**B6**², 457
b710	**—,1,3-dimethyl-5-methoxy-***	$C_9H_{12}O$. *See* b202	136.20			194.5	0.9627^{15}_{4}	1.5106^{15}	i	s	s		s	CS_2 s aa s	**B6**², 462
b711	**—,1,4-dimethyl-2-methoxy-***	$C_9H_{12}O$. *See* b202	136.20			194^{772}	0.9693^{13}_{4}	1.5182^{13}	i	s	s		s	CS_2 s peth s	**B6**², 466
b712	**—,2,4-dimethyl-1-methoxy-***	$C_9H_{12}O$. *See* b202	136.20			192	0.9691^{13}_{4}	1.517^{13}	i	s	s		s	CS_2 s	**B6**², 459
b713	**—,1,2-dimethyl-3-nitro-***	$C_8H_9NO_2$. *See* b202	151.16	nd (al)	15	245–6 13^{20}			i	s					**B5**², 286

For explanations, symbols and abbreviations see beginning of table.

No.	Name	Synonyms and Formula	wt.	Crystalline form, color and specific rotation	m.p. °C	b.p. °C	Density	n_D	Solubility w	al	eth	ace	bz	other solvents	Ref.
	Benzene														
b714	**—,1,2-dimethyl-4-nitro-**	$C_8H_9NO_2$. *See* b202	151.16		29–30	254^{750} 143^{21}			i	∞^h					**B5²**, 286
b715	**—,1,3-dimethyl-2-nitro-***	$C_8H_9NO_2$. *See* b202	151.16		13	225^{774}			i	s					**B5²**, 294
b716	**—,1,3-dimethyl-5-nitro-***	$C_8H_9NO_2$. *See* b202	151.16		75	273^{740}			i	v	v				**B5²**, 295
b717	**—,1,4-dimethyl-2-nitro-***	$C_8H_9NO_2$. *See* b202	151.16			234–7			i	s					**B5²**, 302
b718	**—,2,4-dimethyl-1-nitro-***	$C_8H_9NO_2$. *See* b202	151.16		2	245^{744}			i	s	s				**B5²**, 294
b721	**—,1,2-dimethyl-3,4,5,6-tetrabromo-***	$C_8H_6Br_4$. *See* b202	421.79	nd (bz)	262	374–5			i	δ^h			s^h		**B5²**, 285
b722	**—,1,3-dimethyl-2,4,5,6-tetrabromo-***	$C_8H_6Br_4$. *See* b202	421.79	nd (xyl)	248				i	i			v	xyl s^h	**B5²**, 294
b723	**—,1,3-dimethyl-2,4,6-trinitro-***	$C_8H_7N_3O_6$. *See* b202	241.16	pa ye pr or lf (al-bz)	182				i	δ	δ				**B5²**, 295
b724	**—,1,4-dimethyl-2,3,5-trinitro-***	$C_8H_7N_3O_6$. *See* b202	241.16	mcl nd (al)	138–9	410 exp			i	s					**B5²**, 303
b725	**—,1,2-dinitro-***	$C_6H_4N_2O_4$. *See* b202	168.11	nd (bz)	118	319^{773} 194^{30}	1.3119^{120}_{4}		δ	s v^h				chl v	**B5²**, 193
b726	**—,1,3-dinitro-***	$C_6H_4N_2O_4$. *See* b202	168.11	rh pl	90	291^{756} 188^{33}	1.575^{18}_{4}		i	v^h	s		v	chl s	**B5²**, 193
b727	**—,1,4-dinitro-***	$C_6H_4N_2O_4$. *See* b202	168.11	nd (al)	172	299^{777} 183^{34}	1.625^{18}_{4}		i	δ				chl s aa s	**B5²**, 195
b728	**—,1,3-dinitro-2-ethoxy-***	$C_8H_8N_2O_5$. *See* b202	212.17	nd	57–8				i	δ	s				**B6¹**, 127
b729	**—,1,3-dinitro-5-ethoxy-***	$C_8H_8N_2O_5$. *See* b202	212.17		96–7				δ	δ	s				**B6**, 258
b730	**—,1,4-dinitro-2-ethoxy-***	$C_8H_8N_2O_5$. *See* b202	212.17		96–8				i		δ				**B6²**, 245
b731	**—,2,4-dinitro-1-ethoxy-***	$C_8H_8N_2O_5$. *See* b202	212.17		86				i	δ			δ		
b732	**—,2,4-dinitro-1-fluoro-***	$C_6H_3FN_2O_4$. *See* b202	186.10	pa ye (al)	25.8	296, 178^{25}	1.4718^{84}			s^h					**B5**, 262
b733	**—,1,3-dinitro-5-isopropyl-4-hydroxy-6-methyl-**	2,6-Dinitrothymol. $C_{10}H_{12}N_2O_5$. *See* b202	240.22	pr (peth)	55				i	s	s				**B6**, 543
b734	**—,1,2-dinitro-3-methoxy-***	$C_7H_6N_2O_5$. *See* b202	198.13	nd (al), pl (to)	118		1.2290^{137}		i	δ				lig δ^h	**B6²**, 239
b735	**—,1,2-dinitro-4-methoxy-***	$C_7H_6N_2O_5$. *See* b202	198.13	nd	70		1.3332^{110}			s				CS_2 s MeOH s	**B6¹**, 127
b736	**—,1,3-dinitro-2-methoxy-***	$C_7H_6N_2O_5$. *See* b202	198.13	nd (al)	118		1.3000^{128}			s				CS_2 s MeOH s	**B6²**, 248
b737	**—,1,3-dinitro-5-methoxy-***	$C_7H_6N_2O_5$. *See* b202	198.13	nd (al)	105.5		1.558^{12}		s^h	s				CS_2 s MeOH s	
b738	**—,1,4-dinitro-2-methoxy-***	$C_7H_6N_2O_5$. *See* b202	198.13	nd (bz-lig)	96–7	>360			δ	s				MeOH s CS_2 s	**B6¹**, 127
b739	**—,2,4-dinitro-1-methoxy-***	$C_7H_6N_2O_5$. *See* b202	198.13	nd (al or w)	94.5	sub	1.3364^{131}		δ	s^h	s				**B6²**, 241
b740	**—,1,4-dinitro-2,3,5,6-tetramethyl-***	$C_{10}H_{12}N_2O_4$. *See* b202	224.22	pr (al)	207–8	sub			i	δ^h	v		s		**B5²**, 330
b741	**—,2,4-dinitro-1,3,5-trimethyl-***	$C_9H_{10}N_2O_4$. *See* b202	210.19	rh (al)	86	418 exp			i	s^h					**B5²**, 316
—	**—,diphenyl-***	*see* **Terphenyl**													
b742	**—,1,2-dipropoxy-***	Catechol dipropyl ether. $C_{12}H_{18}O_2$. *See* b202	194.28		234–7 $117–20^{12}$										
b743	**—,1,3-dipropoxy-***	Resorcinol dibutyl ether. $C_{12}H_{18}O_2$. *See* b202	194.28		251 $127–8^{12}$	1.035^{20}_{21}	1.5138^{83}								**B6²**, 815
b744	**—,1,3-dipropyl-2-hydroxy-***	$C_{12}H_{18}O$. *See* b202	178.28		28	256					s				**B6²**, 511
b745	**—,(1,2-epoxyethyl)-***	Styrene oxide. C_8H_8O. *See* b202	120.15			191–2	1.0523^{16}_{4}		i	s	s				**B17**, 49
—	**—,ethenyl-***	*see* **Styrene**													
b746	**—,ethoxy-***	Phenetole. $C_8H_{10}O$. *See* b202	122.17		−30	172	0.9792^{4}_{4}	1.5076^{21}	i	s	s				**B6**, 140

For explanations, symbols and abbreviations see beginning of table.

No.	Name	Synonyms and Formula	Mol. wt.	Crystalline form, color and specific rotation	m.p. °C	b.p. °C	Density	n_D	Solubility w	al	eth	ace	bz	other solvents	Ref.
	Benzene														
b747	**—,1-ethoxy-4-ethyl-***	$C_{10}H_{14}O$. *See* b202	150.22			211, 92–3[12]	0.9585^{17}_{4}		i	s					**B6**[1], 234
b748	**—,1-ethoxy-2-fluoro-***	C_8H_9FO. *See* b202	140.16		−16.7	171.4	1.0872[13]	1.4933[13]							**B6**[2], 169
b749	**—,1-ethoxy-3-fluoro-***	C_8H_9FO. *See* b202	140.16		−27.5	171.4[755]	1.0716^{16}_{4}	1.4849[17]							**B6**[2], 170
b750	**—,1-ethoxy-4-fluoro-***	C_8H_9FO. *See* b202	140.16		−8.5	173[766]		1.4826[18]							**B6**[2], 170
b751	**—,1-ethoxy-2-iodo-***	C_8H_9IO. *See* b202	248.06			245[735]				s				os s	**B6**, 207
b752	**—,1-ethoxy-4-iodo-***	C_8H_9IO. *See* b202	248.06		29	249–50[729]					v			chl v	**B6**, 208
b753	**—,1-ethoxy-3-mercapto-***	$C_8H_{10}OS$. *See* b202	154.23			238–9				s	s			CS_2, alk s	**B6**, 833
b754	**—,1-ethoxy-4-mercapto-***	$C_8H_{10}OS$. *See* b202	154.23		1.6	238				s	s		s		**B6**[2], 852
b755	**—,1-ethoxy-2-nitro-***	$C_8H_9NO_3$. *See* b202	167.17	br ye	2	275, 134[8]	1.1903[15]	1.5425[20]	i	v	v				**B6**[2], 210
b756	**—,1-ethoxy-3-nitro-***	$C_8H_9NO_3$. *See* b202	167.17	br ye	34	284			i	s	s				**B6**[2], 214
b757	**—,1-ethoxy-4-nitro-***	$C_8H_9NO_3$. *See* b202	167.17	pr (al-w or peth)	60	283	1.1176^{125}_{4}		δ	v^h	v			peth s^h	**B6**[2], 221
b758	**—,ethyl-***	C_8H_{10}. *See* b202	106.17		−94	136	0.8672^{20}_{4}	1.4959[20]	i	∞	∞				**B5**[2], 274
b759	**—,1-ethyl-2-hydroxy-***	Phlorol. $C_8H_{10}O$. *See* b202	122.17		<−18	206	1.0371[0]		i	∞	∞		s		**B6**, 442
b760	**—,1-ethyl-3-hydroxy-***	$C_8H_{10}O$. *See* b202	122.17		−4	214[752]	1.0250[0]		δ	v	v				**B6**, 471
b761	**—,1-ethyl-4-hydroxy-***	$C_8H_{10}O$. *See* b202	122.17	nd	45–6	219			δ	v	v		v	CS_2 v	**B6**, 472
b762	**—,1-ethyl-2-iodo-***	C_8H_9I. *See* b202	232.07			226	1.6189^{16}_{4}	1.5941[22]	i					os s	**B5**[1], 177
b763	**—,1-ethyl-4-iodo-***	C_8H_9I. *See* b202	232.07		−17	209	1.6095^{16}_{4}	1.5909[22]	i					os s	**B5**[1], 178
b764	**—,1-ethyl-4-isobutyl-**	$C_{12}H_{18}$. *See* b202	162.28			200			i		s				
b765	**—,1-ethyl-3-isopropyl-**	$C_{11}H_{16}$. *See* b202	148.25		<−20	190–2			i	s	s				**B5**, 440
b766	**—,1-ethyl-4-isopropyl-**	$C_{11}H_{16}$. *See* b202	148.25		<−20	198	0.864[20]	1.4928[16]	i		s				**B5**[2], 334
b767	**—,1-ethyl-2-methoxy-***	$C_9H_{12}O$. *See* b202	136.20			186–8	0.9636^{19}_{4}	1.512	i		v		s		**B6**, 471
b768	**—,1-ethyl-3-methoxy-***	$C_9H_{12}O$. *See* b202	136.20			196–7	0.9575^{18}_{4}	1.5102	i		v		s		**B6**, 472
b769	**—,1-ethyl-4-methoxy-***	$C_9H_{12}O$. *See* b202	136.20			193–4	0.9624^{15}_{4}	1.5094	i		v		s		**B6**, 472
b770	**—,1-ethyl-2-methyl-***	C_9H_{12}. *See* b202	120.20		<−17	165	0.887^{20}_{4}	1.5042[20]	i	∞	∞				**B5**, 396
b771	**—,1-ethyl-3-methyl-***	C_9H_{12}. *See* b202	120.20		−95.5	161.3	0.8645[20]	1.4966[20]	i	v	v				**B5**[2], 309
b772	**—,1-ethyl-4-methyl-***	C_9H_{12}. *See* b202	120.20		−62.3	162	0.8612[20]	1.4950[20]	i	v	v				**B5**[2], 310
b773	**—,1-ethyl-2-nitro-***	$C_8H_9NO_2$. *See* b202	151.17		−23	228	1,126[25]	1.5354[20]	i	v	v				**B5**[2], 279
b774	**—,1-ethyl-3-nitro-***	$C_8H_9NO_2$. *See* b202	151.17			242–3	1.1345[0]		i	v	v				**B5**, 358
b775	**—,1-ethyl-4-nitro-***	$C_8H_9NO_2$. *See* b202	151.17		−12.3	245–6	1.1183^{20}_{4}	1.5455[20]	i	v	v				**B5**[2], 279
b776	**—,1-ethyl-4-propyl-***	$C_{11}H_{16}$. *See* b202	148.25			202–5	0.897[19]		i		s				**B5**, 439
b777	**—,(ethylthio-)***	Ethyl phenyl sulfide. $C_8H_{10}S$. *See* b202	138.23			203–4	1.024^{15}_{4}	1.5740[20]		s					**B6**[2], 287

For explanations, symbols and abbreviations see beginning of table.

No.	Name	Synonyms and Formula	Mol. wt.	Crystalline form, color and specific rotation	m.p. °C	b.p. °C	Density	n_D	Solubility w	al	eth	ace	bz	other solvents	Ref.
	Benzene														
b778	**—,ethynyl-***	Phenylacetylene. C_8H_6. *See* b202	102.14			142–3	0.9295^{20}_{4}	1.548^{20}	i	∞	∞				**B5^2, 406**
b779	**—,fluoro-***	C_6H_5F. *See* b202	96.11		−39.2	85	1.0244^{20}_{4}	1.4677^{20}	i	∞	∞				**B5^2, 147**
b780	**—,1-fluoro-4-iodo-***	C_6H_4FI. *See* b202	222.00		−20	182	1.9523^{15}		i	s	s				**B5^2, 167**
b781	**—,1-fluoro-2-methoxy-***	C_7H_7FO. *See* b202	126.13			154–5			i		s				**B6^2, 169**
b782	**—,1-fluoro-4-methoxy-***	C_7H_7FO. *See* b202	126.13		−43.5	157					s				**B6^1, 98**
b783	**—,1-fluoro-2-nitro-***	$C_6H_4FNO_2$. *See* b202	141.10	ye	−6	214d 86–7^{11}	1.3375^{17}	1.5323^{17}	i	s	s				**B5^2, 180**
b784	**—,1-fluoro-3-nitro-***	$C_6H_4FNO_2$. *See* b202	141.10	ye	4.1	197.5 86^{19}	1.3254^{19}_{4}	1.5280^{17}	i	s	s				**B5^2, 180**
b785	**—,1-fluoro-4-nitro-***	$C_6H_4FNO_2$. *See* b202	141.10	ye	26	206–7 87^{14}	1.3300^{20}_{4}	1.5316^{20}	i	s	s				**B5^2, 180**
b786	**—,hexabromo-***	C_6Br_6. *See* b202	551.52	mcl nd (bz)	310				i	s			s^h	chl δ aa δ	**B5^2, 164**
b787	**—,hexachloro-***	C_6Cl_6. *See* b202	284.78	mcl pr	226–7	322	1.569^{236}		i	δ			v^h	chl δ	**B5^2, 157**
b788	**—,hexaethyl-***	$C_{18}H_{30}$. *See* b202	246.44	mcl pr (al)	129	298	0.8305^{130}	1.4736^{130}	i	s^h	v		v	sulf s	**B5^2, 358**
b789	**—,hexahydroxy-***	$C_6H_6O_6$. *See* b202	174.11	nd	200d				δ	δ	δ		δ		**B5^2, 116**
b790	**—,hexaiodo-***	C_6I_6. *See* b202	833.49	red-br nd or mcl pr (bz)	350d				i	i	i			$PhNO_2$ s^h	**B5^2, 169**
b791	**—,hexamethyl-***	Mellitene. $C_{12}H_{18}$. *See* b202	162.28	rh pr (al)	165	265			i	δ	s	s	v	aa s	**B5^2, 34**
—	**—,hydroxy-***	*see* **Phenol***													
b792	**—,1-hydroxy-2-hydroxy-methyl-4-methyl-***	Homosaligenine. $C_8H_{10}O_2$. *See* b202	138.17	lf (w)	106.5				v^h	s	s				**B6^2, 889**
b793	**—,1-hydroxy-4-isobutyl-**	$C_{10}H_{14}O$. *See* b202	150.22			235–9	0.9796^{20}_{20}	1.5319^{15}		s	s		s		
b794	**—,1-hydroxy-2-isopropyl-**	*o*-Cumenol. $C_9H_{12}O$. *See* b202	136.20			212^{732}	1.012^{20}_{4}	1.5315^{20}	δ	s	s				**B6, 504**
b795	**—,1-hydroxy-3-isopropyl-**	$C_9H_{12}O$. *See* b202	136.20		26	228		1.5261^{20}	δ		s				**B6^2, 476**
b796	**—,1-hydroxy-4-isopropyl-**	$C_9H_{12}O$. *See* b202	136.20		61	223–5	0.990^{20}	1.5228^{20}	δ	s					**B6, 505**
b797	**—,2-hydroxy-1-isopropyl-4-methyl-**	Thymol. $C_{10}H_{14}O$. *See* b202	150.22		51	233	0.925^{80}_{4}	1.5227^{20}	i	v	v			EtOAc v chl v aa v	**B6^2, 494**
b798	**—,—,acetate**	$C_{12}H_{16}O_2$. *See* b202	192.26			245 131^{21}	1.009^{0}		i	∞	∞		∞	chl ∞	**B6^2, 499**
b799	**—,—,carbonate**	Thymotal. Tyratol. $C_{21}H_{26}O_3$. *See* b202	326.44	nd or pr	49				i	v^h	s			chl s	**B6^2, 499**
b800	**—,2-hydroxy-4-isopropyl-1-methyl-**	Carvacrol. $C_{10}H_{14}O$. *See* b202	150.22		0	237.7	0.9760^{20}_{4}	1.5230^{20}	i	s	s				**B6^2, 492**
b801	**—,—,acetate**	$C_{12}H_{16}O_2$. *See* b202	192.26			245	0.9896^{25}	1.4913^{28}		s	s				**B6^2, 494**
b802	**—,1-hydroxy-2-isopropyl-5-methyl-4-nitroso-**	$C_{10}H_{13}NO_2$. *See* b202	179.22	yesh red	175				i	v	v			chl v	**B7^2, 596**
b803	**—,1-hydroxy-5-isopropyl-2-methyl-3-nitroso-**	$C_{10}H_{13}NO_2$. *See* b202	179.22	yesh pr (bz), nd (dil al)	153				i	v	v		v	chl v	**B7^2, 596**
b804	**—,hydroxyl-amino-***	Phenylhydroxylamine. C_6H_7NO. *See* b202	109.13	nd (w, bz or peth)	83–4				s^h	v	v		v^h	chl v lig i	**B15^2, 4**
b805	**—,—,*N*-nitroso-**	$C_6H_6N_2O_2$. *See* b202	138.13	nd (lig)	59				i	v	v			os v	**B16^2, 34**
b806	**—,1-hydroxy-2-methoxy-4-propenyl-(*cis*)***	Isoeugenol (*cis*). $C_{10}H_{12}O_2$. *See* b202	164.21			134–5^{13}	1.0851^{20}_{4}	1.5700^{20}	δ	s	s				**B6, 955**
b807	**—,—,acetate**	$C_{12}H_{14}O_3$. *See* b202	206.24			160–2^{13}		1.5418^{20}	i		s				**B6^2, 919**
b808	**—,(*trans*)***	Isoeugenol (*trans*). $C_{10}H_{12}O_2$. *See* b202	164.21		33–4	141–2^{112}	1.0852^{20}_{4}	1.5782^{20}	δ	s	s				**B6, 955**
b809	**—,—,acetate**	$C_{12}H_{14}O_3$. *See* b202	206.24	nd (al)	79	282–3			i		s				

For explanations, symbols and abbreviations see beginning of table.

No.	Name	Synonyms and Formula	Mol. wt.	Crystalline form, color and specific rotation	m.p. °C	b.p. °C	Density	n_D	Solubility w	al	eth	ace	bz	other solvents	Ref.
	Benzene														
b809[1]	**—,1-hydroxy-4(2-methyl-2-butyl)-**	$C_{11}H_{16}O$. *See* b202	164.24	nd	92–3	248–50									**B6**, 548
b810	**—,1-hydroxy-2-methyl-4(3-methyl-3-pentyl)-***	$C_{13}H_{20}O$. *See* b202	192.30		35	160^{12}			i					os s	
b811	**—,1-hydroxy-4(3-methyl-butyl)-***	$C_{11}H_{16}O$. *See* b202	164.24	nd (w)	93	255			i	v	v				
b812	**—,1-hydroxy-methyl-4-isopropyl**	$C_{10}H_{14}O$. *See* b202	150.21			249	0.9818^{18}_{4}	1.5210^{20}	δ	∞	∞				**B6**, 543
b813	**—,1-hydroxy-2-methyl-3,4,5-trichloro-***	$C_7H_5Cl_3O$. *See* b202	211.48	nd (lig)	77				δ					os v alk s lig s^h	**B6**^{1}, 175
b814	**—,1-hydroxy-4-methyl-2,3,5-trichloro-***	$C_7H_5Cl_3O$. *See* b202	211.48		66–7				δ					os v alk s	**B6**, 404
b815	**—,2-hydroxy-4-methyl-1,2,5-trichloro-***	$C_7H_5Cl_3O$. *See* b202	211.48	nd (w), pl (peth)	45–7	265			s^h	v			δ	peth δ aa s^h	**B6**^{2}, 356
b816	**—,hydroxy-(pentamethyl)-***	$C_{11}H_{16}O$. *See* b202	164.24	nd (al or peth)	126	267			i	s					**B6**^{1}, 270
b817	**—,1-hydroxy-4-pentyl-***	$C_{11}H_{16}O$. *See* b202	164.24		<0	262			i	s	s				
b818	**—,1-hydroxy-4-propenyl-***	Anol. $C_9H_{10}O$. *See* b202	134.17	pl (w), lf	93–4	250d $140–5^{15}$			δ^h	s	s			os s	**B6**, 566
b819	**—,1-hydroxy-2-propyl-***	$C_9H_{12}O$. *See* b202	136.19			220			δ	s	s				**B6**^{2}, 469
b820	**—,1-hydroxy-3-propyl-***	$C_9H_{12}O$. *See* b202	136.19		26	228			δ	s	s				**B6**, 499
b821	**—,1-hydroxy-4-propyl-***	$C_9H_{12}O$. *See* b202	136.19		21–2	230–3			δ	s					**B6**^{2}, 469
b822	**—,1-hydroxy-2,3,4,6-tetra-methyl-***	Isodurenol. $C_{10}H_{14}O$. *See* b202	150.22		80–1	230–50				s					**B6**, 546
b823	**—,1-hydroxy-2,3,5,6-tetra-methyl-***	Durenol. $C_{10}H_{14}O$. *See* b202	150.22		118	249								peth s aa s	
b824	**—,2-hydroxy-1,3,5-tri-*tert*-butyl-**	$C_{18}H_{30}O$. *See* b202	262.44		131–2	130^{15}			i	s^h				os s alk i	
b825	**—,1-hydroxy-2,4,5-trimethyl-***	Pseudoemenol. $C_9H_{12}O$. *See* b202	136.20	nd	71–2	234–5			i	v	v				**B6**, 509
b826	**—,—,acetate**	$C_{11}H_{14}O_2$. *See* b202	178.23	nd (peth)	34	245–6				s	s			peth s^h	**B6**^{2}, 482
b827	**—,2-hydroxy-1,3,5-trimethyl-***	Mesitol. $C_9H_{12}O$. *See* b202	136.20		72	221			i	v	v				**B6**^{2}, 483
b828	**—,iodo-***	Phenyl iodide. C_6H_5I. *See* b202	204.01		−31.4	189	1.8230^{25}_{4}	1.6197^{20}	i	s	∞			chl s	**B5**^{2}, 165
b829	**—,(2-iodoethoxy)-***	C_8H_9IO. *See* b202	248.06		31–2									os s	**B6**^{1}, 81
b830	**—,1-iodo-2-methoxy-***	C_7H_7IO. *See* b202	234.04			$239–40^{739}$	1.8^{20}			v	v		v	chl v lig v	**B6**^{2}, 198
b831	**—,1-iodo-3-methoxy-***	C_7H_7IO. *See* b202	234.04			244–5				v	v				**B6**^{1}, 109
b832	**—,1-iodo-4-methoxy-***	C_7H_7IO. *See* b202	234.04	lf (al)	51–2	237^{726}				s^h	s				**B6**^{2}, 199
b833	**—,1-iodomethyl-2-methyl-***	C_8H_9I. *See* b202	232.07	nd (eth), (peth)	33–4						s^h			peth s^h	**B5**^{2}, 286
b834	**—,1-iodomethyl-4-methyl-***	C_8H_9I. *See* b202	232.07	nd (eth), (peth)	46–7				i		s^h		s	peth s^h	**B5**^{2}, 302
b835	**—,1-iodo-2-nitro-***	$C_6H_4INO_2$. *See* b202	249.01	ye rh nd	49.2	$288–9^{729}$ 162^{18}			i	s	s				**B5**^{2}, 190
b836	**—,1-iodo-3-nitro-***	$C_6H_4INO_2$. *See* b202	249.01	mcl pr	36–7	280 $145–55^{15}$	1.9477^{50}_{4}		i	s	s				**B5**^{2}, 191
b837	**—,1-iodo-4-nitro-***	$C_6H_4INO_2$. *See* b202	249.01	nd (al)	174	289^{772}			i	s				aa s	**B5**^{2}, 191
b838	**—,iodoso-***	C_6H_5IO. *See* b202	220.01		225 exp				s^h	s^h	i	i	i	peth i	**B5**^{2}, 166
b839	**—,1-iodo-2-triazo-***	$C_6H_4IN_3$. *See* b202	246.03	ye		$90–10^{0.9}$	1.8893^{25}_{4}	1.6631^{25}		s				os s	**B5**^{2}, 142

For explanations, symbols and abbreviations see beginning of table.

No.	Name	Synonyms and Formula	Mol. wt.	Crystalline form, color and specific rotation	m.p. °C	b.p. °C	Density	n_D	Solubility						Ref
									w	al	eth	ace	bz	other solvents	
	Benzene														
b840	**—,iodoxy-***	$C_6H_5IO_2$. *See* b202	236.01	nd (w)	236–7 exp				s^h	i			i	chl i aa s	**$B5^2$, 167**
b841	**—,isobutoxy-***	$C_{10}H_{14}O$. *See* b202	150.22			170	0.939^{16}_{4}	1.4932^{24}							**$B6^2$, 146**
b842	**—,1-isobutoxy-2-nitro-***	$C_{10}H_{13}NO_3$. *See* b202	195.22	ye		275–80	1.1361^{20}				s				**B6, 218**
b843	**—,isobutyl-***	$C_{10}H_{14}$. *See* b202	134.22		−51.5	173	0.8534^{20}_{4}	1.4866^{20}	i	∞	∞				**B5, 319**
b844	**—,isocyano-**	Phenyl isocyanate. C_7H_5NO. *See i*-149	119.12												
b845	**—,isopropenyl-**	C_9H_{10}. *See* b202	118.18			$163–4^{752}$	0.9139^{20}_{4}	1.5350^{22}	i	s	s				**$B5^2$, 374**
b846	**—,isopropoxy-**	$C_9H_{12}O$. *See* b202	136.20			177	0.943^{18}_{4}	1.4992^{20}	s	s					**$B6^2$, 145**
b847	**—,isopropyl-**	Cumene. C_9H_{12}. *See* b202	120.20		−96	152–3	0.864^{20}_{4}	1.4911^{20}	i	s	s		s	CCl_4 s	**$B5^2$, 306**
b848	**—,1-isopropyl-2-methyl-**	*o*-Cymene. $C_{10}H_{14}$. *See* b202	134.22			175	0.876^{20}_{4}	1.500^{20}	i	∞				os ∞	**$B5^1$, 322**
b849	**—,1-isopropyl-3-methyl-**	*m*-Cymene. $C_{10}H_{14}$. *See* b202	134.22		−25	176	0.8619^{20}_{4}	1.4922^{20}	i	∞	∞			chl s	**$B5^2$, 322**
b850	**—,1-isopropyl-4-methyl-**	*p*-Cymene. $C_{10}H_{14}$. *See* b202	134.22		−73.5	177	0.8569^{20}_{4}	1.4904^{20}	i	∞	∞			chl s	**$B5^2$, 322**
b851	**—,4-isopropyl-1-methyl-2-nitro-**	$C_{10}H_{13}NO_2$. *See* b202	179.22			$129–32^{15}$	1.0744^{20}_{4}	1.5309^{20}	i	v	v				**$B5^2$, 326**
b852	**—,1-isopropyl-2-nitro-**	$C_9H_{11}NO_2$. *See* b202	165.19	pa ye		$106–7^{9}$	1.101^{12}	1.5286^{12}	i					aa s	**$B5^2$, 307**
b853	**—,1-isopropyl-4-nitro-**	$C_9H_{11}NO_2$. *See* b202	165.19	pa ye		$124–5^{11}$	1.096^{12}	1.5400^{12}	i					lig s	**$B5^2$, 308**
b854	**—,mercapto-***	Thiophenol. Benzenethiol. C_6H_6S. *See* b202	110.18		−14.9	169.5	1.0728^{25}_{4}	1.5879^{24}	i	s	s		s		**$B6^2$, 284**
b855	**—,1-mercapto-4-methoxy-***	C_7H_8OS. *See* b202	140.21			227–9	1.140^{17}	1.5872^{15}		s	s		s		**$B6^2$, 852**
b856	**—,1-mercapto-2-nitro-***	$C_6H_5NO_2S$. *See* b202	155.18	ye nd (peth)	61					s	s		s	alk s	**$B6^2$, 303**
b857	**—,1-mercapto-4-nitro-***	$C_6H_5NO_2S$. *See* b202	155.18		77				s^h	s	v	v		alk s aa δ	**$B6^2$, 309**
b858	**—,2-mercapto-1,3,5-trinitro-***	Thiopicric acid. $C_6H_3N_3O_6S$. *See* b202	245.18	ye nd	114				v d^h	v d^h	v	v	v	chl v	**$B6^2$, 316**
b859	**—,methoxy-***	Anisole. C_7H_8O. *See* b202	108.13			153.8	0.9954^{20}_{4}	1.5179^{20}	i	s	s				**$B6^2$, 139**
b860	**—,1-methoxy-4-(3-methylbutyl)-***	$C_{12}H_{18}O$. *See* b202	178.28			$126–7^{18}$			i				s		**$B6^1$, 269**
b861	**—,1-methoxy-2-nitro-***	$C_7H_7NO_3$. *See* b202	153.13		10	273	1.2536^{20}_{4}	1.5161^{20}	i	∞	∞				**$B6^2$, 209**
b862	**—,1-methoxy-3-nitro-***	$C_7H_7NO_3$. *See* b202	153.13	nd (al)	38	258	1.373^{18}		i	s	v				**$B6^2$, 214**
b863	**—,1-methoxy-4-nitro-***	$C_7H_7NO_3$. *See* b202	153.13	pr (al)	52	258–60	1.2192^{60}_{4}	1.5070^{60}	i	v	v			peth δ	**$B6^2$, 220**
b864	**—,1-methoxy-4-propenyl-***	Anethole. $C_{10}H_{12}O$. *See* b202	148.20	lf (al)	22.5	253.3	0.988^{25}_{4}	1.5591^{22}	δ					os ∞	**$B6^2$, 523**
b865	**—,1-methoxy-2,3,4,6-tetrabromo-***	$C_7H_4Br_4O$. *See* b202	423 77	nd (al)	105–6	240d			i	s					**$B6^2$, 196**
b866	**—,1-methoxy-2,3,4,5-tetrachloro-***	$C_7H_4Cl_4O$. *See* b202	245.93		83				i	v			v	MeOH v	**$B6^2$, 182**
b867	**—,1-methoxy-2,3,5,6-tetrachloro-***	$C_7H_4Cl_4O$. *See* b202	244.93	nd (al)	89–90				i	s			v		**$B6^2$, 182**
b868	**—,2-methoxy-1,3,4,5-tetrachloro-***	$C_7H_4Cl_4O$ *See* b202	245.93	nd (MeOH), pr (al)	64–5				i	s	v		v	MeOH s^h	**$B6^2$, 182**
b869	**—,1-methoxy-2,3,4-tribromo-***	$C_7H_5Br_3O$. *See* b202	344.87	nd (al)	106				i	v				os v	**$B6^2$, 192**
b870	**—,1-methoxy-2,3,5-tribromo-***	$C_7H_5Br_3O$. *See* b202	344.87	pr (dil al)	82	305–12			i	v				os v	**$B6^2$, 192**

For explanations, symbols and abbreviations see beginning of table.

No.	Name	Synonyms and Formula	Mol. wt.	Crystalline form, color and specific rotation	m.p. °C	b.p. °C	Density	n_D	Solubility						Ref.
									w	al	eth	ace	bz	other solvents	
	Benzene														
b871	**—,1-methoxy-2,4,5-tribromo-***	$C_7H_5Br_3O$. *See* b202	344.87	nd (al)	105				i	s	...	...	...	os s	**B6²**, 192
b872	**—,2-methoxy-1,3,5-tribromo-***	$C_7H_5Br_3O$. *See* b202	244.87	nd (al)	87–8	297–9	2.491		i	δ	...	...	...	os s	**B6²**, 193
b873	**—,5-methoxy-1,2,3-tribromo-***	$C_7H_5Br_3O$. *See* b202	344.87	(al)	91–4	300–10			i	s	...	...	...	os s	**B6²**, 195
b874	**—,1-methoxy-2,3,5-trichloro-***	$C_7H_5Cl_3O$. *See* b202	211.48	nd (al)	84				i	s	...	v	...	os s	**B6²**, 180
b875	**—,1-methoxy-2,4,5-trichloro-***	$C_7H_5Cl_3O$. *See* b202	211.48	nd (dil al)	75				...	v	...	v	...	os s	**B6²**, 180
b876	**—,2-methoxy-1,3,4-trichloro-***	$C_7H_5Cl_3O$. *See* b202	211.48	pr (al)	42.8				...	s	...	...	...	os s	**B6²**, 180
b877	**—,2-methoxy-1,3,5-trichloro-***	$C_7H_5Cl_3O$. *See* b202	211.48	mcl nd (al)	60	240^{738}	1.640		...	s	...	...	...	os s	**B6²**, 181
b878	**—,2-methoxy-1,3,5-triiodo-***	$C_7H_5I_3O$. *See* b202	485.87	lf (bz) nd (eth or al)	98–9				i	v	...	...	...	os s	**B6²**, 204
b879	**—,1-methoxy-2,3,4-trinitro-***	$C_7H_5N_3O_7$. *See* b202	243.14	pa ye lf (al)	155	exp			...	s^h s^h	...	v	δ	aa v lig i	**B6¹**, 129
b880	**—,1-methoxy-2,3,5-trinitro-***	$C_7H_5N_3O_7$. *See* b202	243.14	ye nd (al)	119–20		1.618^{15}		s^h	v	...	s	v		**B6²**, 253
b881	**—,1-methoxy-2,4,5-trinitro-***	$C_7H_5N_3O_7$. *See* b202	243.14	ye (al)	106–7				v^h	...	s	...	s	chl s aa s lig i	**B6²**, 253
b882	**—,2-methoxy-1,3,5-trinitro-***	$C_7H_5N_3O_7$. *See* b202	243.14	ye pl	67–8		1.4947^{80}		i	v^h	s	...	v	chl v	**B6²**, 280
—	**—,methyl-***	*see* **Toluene**													
b883	**—,(3-methyl-butoxy)-***	$C_{11}H_{16}O$. *See* b202	164.25			$215–7^{718}$	0.9198^{22}_{4}	1.4872^{-0}	s	...	...	...	...		**B6²**, 146
b884	**—,(2-methyl-2-butyl)-**	$C_{11}H_{16}$. *See* b202	148.24			189–91	0.8740^{20}	1.4915^{23}	i	∞	∞	...	...		**B5²**, 333
b885	**—,(3-methyl-butyl)-***	$C_{11}H_{16}$. *See* b202	148.24			$197–9^{760}$	0.856^{20}_{4}	1.4853^{20}	i	s	s	...	...		**B5²**, 332
b886	**—,(4-methyl-pentyl)-***	$C_{12}H_{18}$. *See* b202	162.27			$214–5^{740}$	0.8568^{16}		i	s	s	...	...		**B5**, 444
b887	**—,1-methyl-2-propyl-***	$C_{10}H_{14}$. *See* b202	134.21		−60	185^{760}	0.8744^{20}	1.4998^{20}	...	...	...	...	...		**B5²**, 321
b888	**—,1-methyl-3-propyl-***	$C_{10}H_{14}$. *See* b202	134.21			182^{760}	0.8610^{20}	1.4936^{20}	...	...	...	...	...		**B5²**, 321
b889	**—,1-methyl-4-propyl-***	$C_{10}H_{14}$. *See* b202	134.21		−63.6	183^{760}	0.8584^{20}	1.4919^{20}	i	s	s	...	...		**B5²**, 321
b890	**—,methylthio-***	Thioanisole. Methyl phenyl sulfide. C_7H_8S. *See* b202	124.14			190^{760} $58–60^{6}$	1.0576^{20}_{4}	1.5869^{20}	i	s	...	...	...	os s	**B6²**, 287
b891	**—,nitro-***	$C_6H_5NO_2$. *See* b202	123.11		5.7	210.8^{760}	1.2037^{20}_{4}	1.5562^{20}	δ	v	v	...	v		**B5²**, 171
b892	**—,(α-nitro-isopropyl)-**	$C_9H_{11}NO_2$. *See* b202	165.19			224d $125–7^{15}$	1.1025^{20}_{0}	1.5209^{20}	i	...	...	...	...		**B5²**, 308
b893	**—,nitro(penta-chloro)*-**	Brassicol. Tritisan. $C_6Cl_5NO_2$. *See* b202	295 35		146		1.718^{25}_{4}		i	δ	...	...	s	chl s	
b894	**—,nitroso-***	C_6H_5NO. *See* b202	107.11	rh or mcl (al-eth)	67–8	$57–9^{18}$			i	s	s	...	...	lig v	**B5²**, 169
b895	**—,1-nitro-2,3,5,6-tetrachloro-***	$C_6HCl_4NO_2$. *See* b202	260.90		99		1.744^{25}_{4}		i	s	...	...	s	chl s	
b896	**—,1-nitro-2-triazo-***	$C_6H_4N_4O_2$. *See* b202	164.13	ye nd (bz-al)	51–2d				...	v	...	...	v	chl s aa v	**B5**, 278
b897	**—,1-nitro-3-triazo-***	$C_6H_4N_4O_2$. *See* b202	164.13	wh nd (dil al)	55				i	v	v	...	v	CS_2 v	**B5**, 278
b898	**—,1-nitro-4-triazo-***	$C_6H_4N_4O_2$. *See* b202	164.13	wh pl (dil al)	74				i	v^h	v^h	...	v^h	CS_2 v^h aa v^h	**B5**, 278
b899	**—,1-nitro-2,3,4-tribromo-***	$C_6H_2Br_3NO_2$. *See* b202	359.83	(al)	85.4				i	δ	v	...	v		**B5**, 251
b900	**—,1-nitro-2,3,5-tribromo-***	$C_6H_2Br_3NO_2$. *See* b202	359.83	nd	119.5				i	...	v	...	v		**B5**, 251
b901	**—,1-nitro-2,4,5-tribromo-***	$C_6H_2Br_3NO_2$. *See* b202	359.83	nd (al)	93.5	sub			i	v^h	v	...	...	CS_2 v	**B5**, 251
b902	**—,2-nitro-1,3,4-tribromo-***	$C_6H_2Br_3NO_2$ *See* b202	359.83	pl or pr (eth-al)	187 sub				i	s^h	v	...	v		**B5**, 251
b903	**—,2-nitro-1,3,5-tribromo-***	$C_6H_2Br_3NO_2$. *See* b202	359.83	mcl pr (chl)	125	177^{11}			i	$δ^h$	δ	...	...	chl v^h aa $δ^h$	**B5²**, 190
b904	**—,5-nitro-1,2,3-tribromo-***	$C_6H_2Br_3NO_2$. *See* b202	359.83	td (eth-al)	112	sub	2.645		i	...	v	...	v		**B5¹**, 133

For explanations, symbols and abbreviations see beginning of table.

No.	Name	Synonyms and Formula	Mol. wt.	Crystalline form, color and specific rotation	m.p. °C	b.p. °C	Density	n_D	Solubility						Ref.
									w	al	eth	ace	bz	other solvents	
	Benzene														
b905	**—,1-nitro-2,3,4-trichloro-***	$C_6H_2Cl_3NO_2$. *See* b202	226.46	nd	56				i	δ				CS_2 v	**B5^2**, 186
b906	**—,1-nitro-2,4,5-trichloro-***	$C_6H_2Cl_3NO_2$. *See* b202	226.46	pr (al or CS_2), nd (al)	57–8	288	1.790^{20}		i	s^h	v		v	CS_2 v	**B5^2**, 187
b907	**—,2-nitro-1,3,4-trichloro-***	$C_6H_2Cl_3NO_2$. *See* b202	226.46	nd (al)	89				i	s				lig δ	**B5^2**, 187
b908	**—,2-nitro-1,3,5-trichloro-***	$C_6H_2Cl_3NO_2$. *See* b202	226.46	nd (al)	68				i	v^h				CS_2 v lig v	**B5^2**, 187
b909	**—,5-nitro-1,2,3-trichloro-***	$C_6H_2Cl_3NO_2$. *See* b202	226.46		72.5		1.807		i	s^h					**B5^2**, 187
b910	**—,1-nitro-2,3,5-triiodo-***	$C_6H_2I_3NO_2$. *See* b202	500.84	ye pr	124				i	s^h				os s	**B5**, 256
b911	**—,1-nitro-2,4,5-triiodo-***	$C_6H_2I_3NO_2$. *See* b202	500.84	pa ye nd (CS_2)	178				i	δ	δ	δ		CS_2 s^h, chl δ	**B5**, 256
b912	**—,2-nitro-1,3,4-triiodo-***	$C_6H_2I_3NO_2$. *See* b202	500.84	nd (aa), pr (CS_2)	137				i					CS_2 s^h aa s^h	**B5**, 256
b913	**—,5-nitro-1,2,3-triiodo-***	$C_6H_2I_3NO_2$. *See* b202	500.84	ye pr (chl), nd (al)	167		3.256		i	δ	v		v	chl s aa v^h	**B5^2**, 192
b914	**—,1-nitro-2,3,5-trimethyl-***	$C_9H_{11}NO_2$. *See* b202	165.19	pr	20					s					**B5**, 404
b915	**—,1-nitro-2,4,5-trimethyl-***	$C_9H_{11}NO_2$. *See* b202	165.19	nd	70	265				s				peth s	**B5^2**, 313
b916	**—,2-nitro-1,3,5-trimethyl-***	$C_9H_{11}NO_2$. *See* b202	165.19	rh pr (al)	44	255^{760}	1.51			s					**B5^2**, 316
b917	**—,pentaamino-***	$C_6H_{11}N_5$. *See* b202	153.19	dk br nd	228				s	i	i			os i	**B13^2**, 155
b918	**—,pentabromo-***	C_6HBr_5. *See* b202	472.65	wh nd (aa or al)	160–1	sub				s^h			v	chl v lig δ	**B5^1**, 117
b919	**—,pentachloro-***	C_6HCl_5. *See* b202	250.35	nd	85–6	277	1.8342^{16}		i	s^h	v		v	CS_2 v, chl v	**B5^2**, 157
b920	**—,pentaethyl-***	$C_{16}H_{26}$. *See* b202	218.37		<-20	277	0.8985$^{19}_{19}$	1.5127^{20}	i						**B5^2**, 358
b921	**—,pentaiodo-***	C_6HI_5. *See* b202	707.62	nd (al)	172	sub			i	δ	δ		s	chl s aa s	**B5**, 229
b922	**—,pentamethyl-***	$C_{11}H_{16}$. *See* b202	148.24	pr (al)	54.3	232^{760}	0.917$^{20}_{4}$	1.527^{20}	i	v			v		**B5^2**, 336
b923	**—,pentoxy-***	$C_{11}H_{16}O$. *See* b202	164.25			111^{17}									**B6^1**, 82
b924	**—,pentyl-***	$C_{11}H_{16}$. *See* b202	148.24		−78	205.3^{760}	0.8662$^{20}_{4}$	1.4943^{20}	i	s	∞		∞		**B5^2**, 331
b924^1	**—,(2-pentyl)-***	$C_{11}H_{16}$. *See* b202	148.24			198–9^{757}	0.8594$^{21}_{4}$	1.4875^{21}	i	s	s				**B5**, 434
b925	**—,propenyl-***	C_9H_{10}. *See* b202	118.17		−52	175^{760}	0.911$^{20}_{4}$	1.549^{20}	i	∞	∞	∞	∞	peth ∞	**B5^2**, 371
b926	**—,1-propenyl-2,4,5-tri-methoxy-***	Asaron. $C_{12}H_{16}O_3$. *See* b202	208.25	mcl nd (w)	67	296^{760}	1.091^{11}	1.5719^{11}	s^h	s	s			peth s chl s CCl_4 s	**B6^2**, 1092
b927	**—,propoxy-***	Phenyl propyl ether. $C_9H_{12}O$. *See* b202	136.19			189.3^{760}	0.9350$^{15}_{15}$	1.5011^{20}		s	s				**B6^2**, 145
b928	**—,propyl-***	C_9H_{12}. *See* b202	120.19		−99.5	159.2^{760}	0.8620$^{20}_{4}$	1.4920^{20}	i	∞	∞	∞	∞	peth ∞ CCl_4 ∞	**B5^2**, 303
b929	**—,1,2,3,5-tetra-bromo-***	$C_6H_2Br_4$. *See* b202	393.74	nd (al)	98.5	329			i	s^h	v		v	CS_2 v	**B5^1**, 117
b930	**—,1,2,4,5-tetra-bromo-***	$C_6H_2Br_4$. *See* b202	393.74	mcl pr (CS_2)	182		3.072^{20}		i	δ	v				**B5^2**, 164
b931	**—,1,2,3,4-tetra-chloro-***	$C_6H_2Cl_4$. *See* b202	215.90	nd	47.5	254^{761}			i	δ	v			CS_2 v lig v	**B5^2**, 156
b932	**—,1,2,3,5-tetra-chloro-***	$C_6H_2Cl_4$. *See* b202	215.90	nd (al)	51	246			s^h	δ	s		s	CS_2 v lig v	**B5^2**, 157
b933	**—,1,2,4,5-tetra-chloro-***	$C_6H_2Cl_4$. *See* b202	215.90	nd, mcl pr (eth)	139–40	243–6	1.858^{22}		i	δ^h	s		s	chl s, CS_2 s	**B5^2**, 157
b934	**—,1,2,3,4-tetra-ethyl-***	$C_{14}H_{22}$. *See* b202	190.32		11.6	251^{734}	0.8875$^{20}_{4}$	1.5125^{20}	i	v	s				**B5^2**, 344
b935	**—,1,2,4,5-tetra-ethyl-***	$C_{14}H_{22}$. *See* b202	190.32		10	250	0.8788$^{20}_{4}$	1.5054^{20}	i	v	v				**B5**, 455
b936	**—,1,2,3,5-tetra-hydroxy-***	$C_6H_6O_4$. *See* b202	142.11	nd (w)	165				v	v		v		chl i aa i	**B6^1**, 570
b937	**—,1,2,4,5-tetra-hydroxy-***	$C_6H_6O_4$. *See* b202	142.11	lf (w)	215–20				v	v	v			aa δ	**B6^2**, 112
b938	**—,1,2,3,4-tetra-iodo-***	$C_6H_2I_4$. *See* b202	581.72	pr (CS_2)	136	sub				v	v			chl v	**B5**, 229
b939	**—,1,2,3,5-tetra-iodo-***	$C_6H_2I_4$. *See* b202	581.72	pr (eth or aa)	148	sub			i	δ	δ			chl δ aa v^h	**B5**, 229
b940	**—,1,2,4,5-tetra-iodo-***	$C_6H_2I_4$. *See* b202	581.62	pr (bz), nd (eth)	254	sub			i	δ	δ			CS_2 v aa s	**B5**, 229

For explanations, symbols and abbreviations see beginning of table.

No.	Name	Synonyms and Formula	Mol. wt.	Crystalline form, color and specific rotation	m.p. °C	b.p. °C	Density	n_D	Solubility w	al	eth	ace	bz	other solvents	Ref.
	Benzene														
b941	**—,1,2,3,4-tetramethyl-***	Prehnitene. $C_{10}H_{14}$. *See* b202	134.21		−64	$203–4^{760}$	0.901^{20}_{4}	1.5187^{20}	i	∞	∞				**B5**^{2}, 329
b942	**—,1,2,3,5-tetramethyl-***	Isodurene. $C_{10}H_{14}$. *See* b202	134.21		−24.1	$195–7^{760}$	0.8906^{20}_{4}	1.5134^{20}	i	s	v				
b943	**—,1,2,4,5-tetramethyl-***	Durene. $C_{10}H_{14}$. *See* b202	134.21		79	$190–1^{760}$	0.8380^{81}_{4}	1.4790^{81}	i	s	s		s		**B**^{2}**5**, 329
b944	**—,1,3,5-triacetyl-**	$C_{12}H_{12}O_3$. *See* b202	204.23	nd (w, aa or al)	163				s^h	s^h	δ			aa v	**B7**^{2}, 831
b945	**—,1,2,3-triamino-***	$C_6H_9N_3$. *See* b202	123.16		103	336			v	v	v				**B13**^{2}, 145
b946	**—,1,2,4-triamino-***	$C_6H_9N_3$. *See* b202	123.16	pl or lf (chl)	98–100	340			v	v	δ			chl s	**B13**^{2}, 146
b947	**—,triazo-***	Azidobenzene. $C_6H_5N_3$. *See* b202	119.12			70^{11}	1.0880^{20}_{20}	1.5589^{25}	i	δ	δ				**B5**^{2}, 207
b948	**—,1,2,3-tribromo-***	$C_6H_3Br_3$. *See* b202	314.83	pl (al)	87.8		2.658		i	$δ^h$	v				**B5**^{1}, 117
b949	**—,1,2,4-tribromo-***	$C_6H_3Br_3$. *See* b202	314.83	nd (al or eth)	44	275			i	s^h	v	v	v	CCl_4 v CS_2 v	**B5**^{2}, 164
b950	**—,1,3,5-tribromo-***	$C_6H_3Br_3$. *See* b202	314.83	nd (al), pr (eth)	121.5–2.5	271^{765}			i	$δ^h$	s		s	chl s	**B5**^{2}, 164
b951	**—,2,4,6-tribromo-1,3,5-trimethyl-***	$C_9H_9Br_3$. *See* b202	356.91	tcl nd (al or bz)	226				i	i			s		**B5**^{2}, 315
b952	**—,1,2,3-trichloro-***	$C_6H_3Cl_3$. *See* b202	181.46	pl (al)	53–4	218–9			i	δ	v		v	CS_2 v	**B5**^{2}, 156
b953	**—,1,2,4-trichloro-***	$C_6H_3Cl_3$. *See* b202	181.46	rh	16.9	213.5^{760}	1.4542^{20}_{4}	1.5717^{20}	i	δ	v				**B5**^{2}, 156
b954	**—,1,3,5-trichloro-***	$C_6H_3Cl_3$. *See* b202	181.46	nd	63–4	208^{763}			i	δ	v		v	CS_2 v lig v	**B5**^{2}, 156
b955	**—,1,2,3-trichloro-4,5,6-trihydroxy-***	$C_6H_3Cl_3O_3$. *See* b202	229.46	nd (al or bz)	185d				v^h	v	v		δ	chl δ CCl_4 δ CS_2 δ	**B6**, 1084
b956	**—,1,2,4-trichloro-3,5,6-trihydroxy-***	$C_6H_3Cl_3O_3$. *See* b202	229.46	nd (bz or aa)	160				δ	v	v			peth δ	**B6**^{2}, 1072
b957	**—,1,3,5-trichloro-2,4,6-trihydroxy-**	$C_6H_3Cl_3O_3$. *See* b202	229.46	(al)	134	sub			δ	s			i		**B6**, 1104
b958	**—,1,3,5-triethoxy-***	Phloroglucinol triethyl ether. $C_{12}H_{18}O_3$. *See* b202	210.27		43.5	175^{24}			i	v	v				**B6**^{2}, 1079
b959	**—,1,2,4-triethyl-***	$C_{12}H_{18}$. *See* b202	162.27			218^{760}	0.8738^{20}_{4}	1.5024^{20}	i	s	s				**B5**^{2}, 340
b960	**—,1,3,5-triethyl-***	$C_{12}H_{18}$. *See* b202	162.27		−66.5	216^{760}	0.8621^{20}_{4}	1.4958^{20}	i	v	v				**B5**^{2}, 340
b961	**—,1,2,3-trihydroxy-***	Pyrogallol. $C_6H_6O_3$. *See* b202	126.11	lf or nd	132.8	309^{760} 171^{12}	1.453^{4}_{4}		v	v	v		i		**B6**^{2}, 1059
b962	—,—,triacetate	$C_{12}H_{12}O_6$. *See* b202	252.22		165				i	s					**B6**^{2}, 1066
b963	**—,1,2,4-trihydroxy-***	Hydroxyquinol. $C_6H_6O_3$. *See* b202	126.11	pl (eth), lf or pl (w)	140.5				v	v	v		i	CS_2 i chl i	**B6**^{2}, 1071
b964	—,—,triacetate	$C_{12}H_{12}O_6$. *See* b202	252.22	nd (MeOH)	97	>300				s^h				MeOH s^h	**B6**^{2}, 1072
b965	**—,1,3,5-trihydroxy-***	Phloroglucinol. $C_6H_6O_3$. *See* b202	126.11		217–9	sub			δ	v	v				**B6**, 1075
b966	—,—,diacetate	$C_{10}H_{10}O_5$. *See* b202	210.18	pl (w)	104				s^h	s	s	s		lig δ	**B6**^{1}, 547
b967	—,—,triacetate	$C_{12}H_{12}O_6$. *See* b202	252.22	pr (w)	104–6					s^h					**B6**^{2}, 1079
b968	**—,1,2,3-triiodo-***	$C_6H_3I_3$. *See* b202	455.82	nd (al), pr (bz)	116	sub			i	s	s			chl s	**B5**^{1}, 122
b969	**—,1,2,4-triiodo-***	$C_6H_3I_3$. *See* b202	455.82	nd (al)	91.5	sub			i	s^h	s			chl s	**B5**^{1}, 122
b970	**—,1,3,5-triiodo-***	$C_6H_3I_3$. *See* b202	455.82	nd (aa)	184.2	sub			i	δ	δ		δ	chl δ aa s	**B5**^{1}, 122
b971	**—,1,2,3-trimethoxy-***	Pyrogallol trimethyl ether. $C_9H_{12}O_3$. *See* b202	168.19	rh nd (al)	47	241^{760}	1.1118^{46}_{45}		i	s	s		s		**B6**^{2}, 1066
b972	**—,1,3,5-trimethoxy-***	Phloroglucinol trimethyl ether. $C_9H_{12}O_3$. *See* b202	168.19	pr (al), lf (peth)	54–5	255.5^{760}			i	s	s		s		**B6**^{2}, 1078
b973	**—,1,2,3-trimethyl-***	Hemellitene. C_9H_{12}. *See* b202	120.20		−25.5	176	0.8944^{20}_{4}	1.5139^{20}	i	s	s				**B5**^{2}, 311
b974	**—,1,2,4-trimethyl-***	Pseudocumene. C_9H_{12}. *See* b202	120.20		−60.5	169.5^{760}	0.889^{4}_{4}	1.5044^{21}	i	s	s		s		**B5**^{2}, 312
b975	**—,1,3,5-trimethyl-***	Mesitylene. C_9H_{12}. *See* b202	120.20		−52.7	164.7^{760}	0.8642^{20}_{4}	1.4998^{20}	i	s	s				**B5**^{2}, 313
b976	**—,1,2,3-trimethyl-4,5,6-trinitro-***	$C_9H_9N_3O_6$. *See* b202	255.19	pr (al)	209				i	s					**B5**, 400

For explanations, symbols and abbreviations see beginning of table.

No.	Name	Synonyms and Formula	Mol. wt.	Crystalline form, color and specific rotation	m.p. °C	b.p. °C	Density	n_D	Solubility w	al	eth	ace	bz	other solvents	Ref.
	Benzene														
b977	**—,1,2,4-trimethyl-3,5,6-trinitro-***	$C_9H_9N_3O_6$. *See* b202	255.19	rh pr	185				i	δ			v^h		**B5²**, 316
b978	**—,1,3,5-trimethyl-2,4,6-trinitro-***	$C_9H_9N_3O_6$. *See* b202	255.19	tcl nd (al)	234	415 exp				$δ^h$	$δ^h$	$δ^h$			**B5²**, 316
b979	**—,1,2,3-trinitro-***	$C_6H_3N_3O_6$. *See* b202	213.11	pl (al)	127				i	s^h					**B5²**, 203
b980	**—,1,2,4-trinitro-***	$C_6H_3N_3O_6$. *See* b202	213.11	lf (eth), pr (al)	61				$δ^h$	s	s		v	chl v	**B5²**, 203
b981	**—,1,3,5-trinitro-***	$C_6H_3N_3O_6$. *See* b202	213.11	rh pl (bz), lf (w)	121–2		1.4775^{152}_{4}		$δ^h$	δ	δ	v	s		**B5²**, 203
b982	**—,1,3,5-triphenyl-***	$C_{24}H_{18}$. *See* b202	306.41	nd (al or aa)	172	459^{717}	1.199^{20}_{4}		i	i	s			CS_2 s aa s^h	**B5²**, 670
b983	**—,1,3,5-tris-(phenylamino)-***	$C_{24}H_{21}N_3$. *See* b202	351.46	nd (al)	193					δ	v		δ		**B13²**, 147
b984	**—,1,3,5-tris-(4-tolylamino)-**	$C_{27}H_{27}N_3$. *See* b202	393.54	nd (al)	186–7					δ				sulf s	**B13**, 299
b985	**Benzenearsonic acid***	3 2 4 ⟨ ⟩—AsO(OH)$_2$ 5 6	202.02		154–8d				s	v					**B16²**, 457
b986	**—,2-amino-***	$C_6H_8AsNO_3$. *See* b985	217.06	nd	153–4				v	v	δ			alk s aa s	**B16²**, 482
b987	**—,5-amino-**	$C_6H_8AsNO_3$. *See* b985	217.06	pr (w)	214				s^h					alk s os s	**B16²**, 489
b988	**—,4-amino-***	Arsanilic acid. $C_6H_8AsNO_3$. *See* b985	217.06	mcl nd (w or al)	231–2		1.9571^{20}		s^h	s^h		i	i	chl i aa δ	**B16²**, 491
b989	**—,3-amino-4-hydroxy-***	$C_6H_8AsNO_4$. *See* b985	233.06	pr (w+½)	290d				i	δ				os δ	**B16²**, 521
b990	**—,2-bromo-***	$C_6H_6AsBrO_3$. *See* b985	280.94	lf (w), pr (dil al)	201d				s^h	v					**B16²**, 458
b991	**—,2-chloro-***	$C_6H_6AsClO_3$. *See* b985	236.49	(w or dil al)	186–7				s^h	s	i		i	MeOH s, chl i	**B16²**, 457
b992	**—,3-chloro-***	$C_6H_6AsClO_3$. *See* b985	236.49	(w)	175				s	s				MeOH s	**B16²**, 457
b993	**—,4-chloro-***	$C_6H_6AsClO_3$. *See* b985	236.49	(al)	348					s				alk v	**B16²**, 457
b994	**—,2-hydroxy-***	$C_6H_7AsO_4$. *See* b985	218.04	nd (w)	190–1				v^h	v	i	v^h		chl δ aa v^h	**B16²**, 464
b995	**—,3-hydroxy-***	$C_6H_7AsO_4$. *See* b985	218.04	(w)	159–73				v	v		$δ^h$	i	chl i aa v^h	**B16¹**, 454
b996	**—,4-hydroxy-***	$C_6H_7AsO_4$. *See* b985	218.04	nd (ace)	170–4				v	v	δ			aa δ	**B16²**, 466
b997	**—,2-hydroxy-3-nitro-***	$C_6H_6AsNO_6$. *See* b985	263.04	pr (w)	252–4				s^h	s^h					**B16²**, 465
b998	**—,2-hydroxy-5-nitro-***	$C_6H_6AsNO_6$. *See* b985	263.04	pa ye (w)	250d				v^h	v^h	i	s			**B16²**, 465
b999	**—,3-hydroxy-2-nitro-***	$C_6H_6AsNO_6$. *See* b985	263.04	ye pl	208d					s^h					**B16²**, 466
b1000	**—,4-hydroxy-2-nitro-***	$C_6H_6AsNO_6$. *See* b985	263.04	ye nd (w)	228				v^h	s		s			**B16²**, 469
—	**—,methyl-***	*see* **Toluenearsonic acid**													
b1001	**—,2-nitro-***	$C_6H_6AsNO_5$. *See* b985	246.03	nd (w)	235–40d				δ s^h	δ s^h		δ		aa s	**B16²**, 449
b1002	**—,3-nitro-***	$C_6H_6AsNO_5$. *See* b985	246.03	lf	200				δ		i		δ	chl δ	**B16²**, 458
b1003	**—,4-nitro-***	$C_6H_6AsNO_5$. *See* b985	246.03	lf or nd (w)	>300				δ s^h	δ s^h					**B16¹**, 450
b1004	**—,4-nitroso-***	$C_6H_6AsNO_4$. *See* b985	230.03	ye nd	180d				δ s^h	δ	i			chl i aa v	**B16¹**, 449
b1005	**—,4-ureido**	$C_7H_9AsN_2O_4$. *See* b985	260.08	nd (w)	174				δ v^h	δ	i			alk s chl i	**B16²**, 497
b1006	**Benzeneboronic acid***	Phenylboric acid. 3 2 4 ⟨ ⟩—B(OH)$_2$ 5 6	121.93	nd (w)	216–9d				δ v^h	s	s		s		**B16²**, 638
b1007	**—,4-bromo-***	$C_6H_6BBrO_2$. *See* b1006	200.83	(w)	266				δ		s				**B16²**, 638
b1008	**—,2-chloro-***	$C_6H_6BClO_2$. *See* b1006	156.38		149				δ	s	s		s	lig s	
b1009	**—,4-chloro-***	$C_6H_6BClO_2$. *See* b1006	156.38	nd (w)	306–7				δ		s				**B16²**, 638
—	**Benzenecarboxylic acid***	*see* **Benzoic acid**													
b1010	**Benzene-diazonium chloride***	Diazobenzene chloride. $C_6H_5N(\vdots N)Cl$	140.57	nd (al)	exp				δ	s		s	i	chl i aa v	**B16¹**, 35
b1011	**Benzene-diazonium cyanide***	$C_6H_5N(\vdots N)CN$	131.14	pr	69				δ						**B16¹**, 43
b1012	**Benzene-diazonium nitrate***	$C_6H_5N(\vdots N)NO_3$	167.13	nd (al eth)	90 exp				δ	δ	i		i	chl i	**B16²**, 26
b1013	**Benzene-diazonium tribromide***	$C_6H_5N(\vdots N)Br_3$	344.85	pl (al)	63.5				i	δ	i				**B16**, 43

For explanations, symbols and abbreviations see beginning of table.

No.	Name	Synonyms and Formula	Mol. wt.	Crystalline form, color and specific rotation	m.p. °C	b.p. °C	Density	n_D	Solubility w	al	eth	ace	bz	other solvents	Ref.
	Benzenedicarboxaldehyde														
—	**1,2-Benzenedi-carboxaldehyde**	*see* **Phthalaldehyde**													
1014	**1,3-Benzenedi-carboxaldehyde**	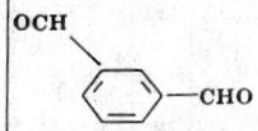 Isophthalaldehyde. OCH, —CHO	134.14	nd	89–90	$245–8^{771}$	...	...	δ	v	s	...	...	...	**B7**[2], 606
1015	**1,4-Benzenedi-carboxaldehyde**	 Terephthalaldehyde. OCH— —CHO	134.14	nd	116	247^{771}	...	...	δ	v	s	...	...	...	**B7**, 675
—	**1,2-Benzenedi-carboxylic acid***	*see* **Phthalic acid**													
1016	**1,3-Benzenedi-carboxylic acid***	Isophthalic acid. 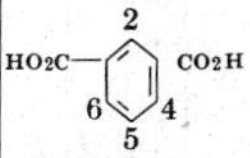 HO2C— 2 CO2H 6 4 5	166.13	nd (w or al)	345–7	sub	...	...	i	δ	i	...	i	aa s lig i	**B9**[2], 608
1016[1]	—,diamide	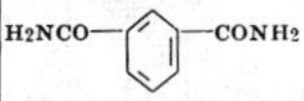 H2NCO— —CONH2	164.17	pl	270	...	...	...	δ^h	δ^h	...	...	...	os i	**B9**, 834
1017	—,—,*N,N,N′,N′*-**tetraethyl-**	$C_{16}H_{24}N_2O_2$, *See* b1016[1]	276.38	(eth-bz)	85	242^{12}	...	...	i	v	δ	v	v	lig i	**B9**[2], 609
1018	—,dichloride	$C_8H_4Cl_2O_2$. *See* b1016	203.03	pr (eth)	42–3	276	1.3380_4^{17}	1.570^{47}	δ	δ	s	...	...	...	**B9**[2], 609
1019	—,diethyl ester*	$C_{12}H_{14}O_4$. *See* b1016	222.24	...	11.5	302^{760}	1.1239_4^{17}	1.508^{18}	i	...	...	...	...	...	**B9**[1], 372
1020	—,dimethyl ester*	$C_{10}H_{10}O_4$. *See* b1016	194.19	nd (dil al)	79	282^{760}	1.194_4^{20}	1.5168^{20}	δ	...	...	...	...	...	**B9**[2], 1609
1021	—,dinitrile	1,3-Dicyanobenzene. 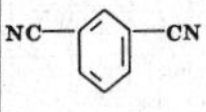 NC— —CN	128.13	nd	162	...	...	...	δ^h	v^h	s	...	s	peth i chl s	**B9**, 836
1022	—,mononitrile	3-Cyanobenzoic acid. 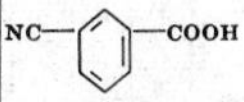 NC— —COOH	147.13	nd (w)	217	sub d	...	...	v^h	v	v	...	...	...	**B9**[2], 609
1023	—,piperazinium salt	$C_{12}H_{16}N_2O_4$. *See* b1016	252.27	...	252	...	...	...	s^h	s^h	s^h	...	...	...	...
1024	**1,4-Benzenedi-carboxylic acid***	Terephthalic acid. 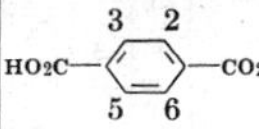 3 2 HO2C— —CO2H 5 6	166.13	nd	sub	300^{760}	...	...	i	s^h	δ	...	...	chl δ	**B9**, 841
1025	—,diamide	 H2NCO— —CONH2	164.17	nd (w), pl (aa)	>250	...	...	...	δ^h	i	i	...	i	chl i	**B9**[1], 376
1026	—,—,*N,N,N′,N′*-**tetraethyl**	$C_{16}H_{24}N_2O_2$. *See* b1025	276.38	cr (eth-al)	127	...	...	...	i	v	δ	...	s	lig i	**B9**[2], 613
1027	—,dichloride	 ClCO— —COCl	203.03	nd	82	...	...	...	d	d	s	...	...	...	**B9**, 844
1028	—,diethyl ester*	$C_{12}H_{14}O_4$. *See* b1024	222.24	mcl pr	44	302^{760}	...	...	...	v	s	...	...	...	**B9**, 844
1029	—,dimethyl ester*	$C_{10}H_{10}O_4$. *See* b1024	194.19	nd	141–2	sub	...	...	δ^h	v^h	s	...	...	MeOH δ	**B9**, 843
1030	—,dinitrile	1,4-Dicyanobenzene.  NC— —CN	128.13	nd (w)	222	sub	...	...	...	δ s^h	δ^h	...	s^h	aa v^h	**B9**[2], 613
1031	—,monoamide	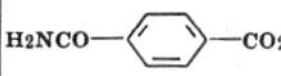 H2NCO— —CO2H	165.15	...	>300	sub	...	...	i	i	i	...	i	chl i	**B9**, 845
1032	—,mononitrile	4-Cyanobenzoic acid. 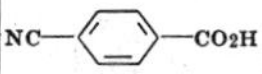 NC— —CO2H	147.13	pl or lf (w)	219	...	...	...	δ^h	s	s	...	...	aa s	**B9**[2], 613
1033	—,piperazinium salt	$C_{12}H_{16}N_2O_4$. *See* b1024	252.27	...	350d	...	...	...	δ^h	s^h	δ^h	...	...	...	...
1034	**1,3-Benzenedicar-boxylic acid, 2-amino-***	$C_8H_7NO_4$. *See* b1016	181.15	pl (al), nd (aa)	>340	...	...	...	i	δ	s	...	...	...	**B14**[2], 337
1035	—,—,dimethyl ester*	$C_{10}H_{11}NO_4$. *See* b1016	209.20	nd (al)	103–4	...	...	...	...	δ	...	...	...	...	**B14**[2], 337
1036	—,**4-amino-***	$C_8H_7NO_4$. *See* b1016	181.15	nd (w)	336–7	...	...	...	i	v^h	δ	...	...	aa v^h	**B14**[1], 633
1037	—,—,dimethyl ester*	$C_{10}H_{11}NO_4$. *See* b1016	209.20	nd (al)	127–9	...	...	...	...	δ	...	...	...	MeOH δ	**B14**[2], 337
1038	—,**5-amino-***	$C_8H_7NO_4$. *See* b1016	181.15	pr (al), pl (w)	>360	sub	...	...	i	δ	...	...	...	...	**B14**[1], 636
1039	—,—,dimethyl ester*	$C_{10}H_{11}NO_4$. *See* b1016	209.20	lf or pl (MeOH)	176	...	...	...	i	...	s	...	...	...	**B14**, 556
1040	**1,4-Benzenedicar-boxylic acid, 2-amino-***	$C_8H_7NO_4$. *See* b1024	181.15	ye	324–5d	...	...	...	δ	δ	δ	i	i	peth i chl i	**B14**[1], 637

For explanations, symbols and abbreviations see beginning of table.

No.	Name	Synonyms and Formula	Mol. wt.	Crystalline form, color and specific rotation	m.p. °C	b.p. °C	Density	n_D	Solubility w	al	eth	ace	bz	other solvents	Ref.
	Benzenedicarboxylic acid														
b1041	—,—,dimethyl ester*	$C_{10}H_{11}NO_4$. *See* b1024	209.20	nd (bz)	133–4					s^h	s^h	s		chl s	**B14**, 55
b1042	**—,2-benzoyl-**	$C_{15}H_{10}O_5$. *See* b1024	270.24	nd	285				i	s	s				**B10**, 88
b1043	**1,3-Benzenedicarboxylic acid, 4-bromo-***	$C_8H_5BrO_4$. *See* b1016	245.04	nd (al)	287					v^h					**B9**, 838
b1044	**1,4-Benzenedicarboxylic acid, 2-bromo-***	$C_8H_5BrO_4$. *See* b1024	245.04	nd (w or al)	299				δ v^h	v	i		i		**B9²**, 61
b1045	**1,3-Benzenedicarboxylic acid, 4-chloro-***	$C_8H_5ClO_4$. *See* b1016	200.58	nd (w)	295				δ	v	δ		i	chl i	**B9¹**, 37
b1046	**—,5-chloro-***	$C_8H_5ClO_4$. *See* b1016	200.58	nd (w+½)	278				δ v^h						**B9**, 838
b1047	**1,4-Benzenedicarboxylic acid, 2-chloro-***	$C_8H_5ClO_4$. *See* b1024	200.58	(w)	320				$δ^h$	v	v				**B9**, 847
b1048	**1,3-Benzenedicarboxylic acid, 4,6-dichloro-***	$C_8H_4Cl_2O_4$. *See* b1016	235.03	nd (w or dil al)	279–81				i	v	v			chl v	**B9**, 838
b1049	**1,4-Benzenedicarboxylic acid, 2,5-dichloro-***	$C_8H_4Cl_2O_4$. *See* b1024	235.03	nd (w)	306				$δ^h$		v				**B9**, 847
b1050	**—,2,5-dihydroxy-***	$C_8H_6O_6$. *See* b1024	198.14		d				s^h	s^h	s^h				**B10**, 55
b1051	**1,3-Benzenedicarboxylic acid, 4,5-dimethoxy-***	Isohemipinic acid. $C_{10}H_{10}O_6$. *See* b1016	226.18	nd (w)	245–6				$δ^h$	s	s				**B10**, 55
b1052	**—,4,6-dimethyl-***	α-Cumidic acid. $C_{10}H_{10}O_4$. *See* b1016	194.18	nd (w), pr (al-bz), lf (sub)	355				i $δ^h$						**B9²**, 62
b1053	**—,2-hydroxy-***	$C_8H_6O_5$. *See* b1016	182.14	nd (w+1)	244.5				$δ^h$	v	v			chl δ	**B9²**, 35
b1054	**—,4-hydroxy-***	$C_8H_6O_5$. *See* b1016	182.14	nd (w), lf, (dil al)	310				i	v	v			chl i aa s^h	**B9²**, 35
b1055	**—,5-hydroxy-***	$C_8H_6O_5$. *See* b1016	182.14	nd (w+2)	308.5				i v^h	v	v		s		**B9¹**, 25
b1056	**—,5-methyl-***	Livitic acid. $C_9H_8O_4$. *See* b1016	180.15	nd (w)	298				$δ^h$	s	s				**B9¹**, 38
b1057	**—,2-nitro-, dimethyl ester***	$C_{10}H_9NO_6$. *See* b1016	239.19	nd (w or al)	135				δ					peth i MeOH δ	**B9¹**, 37
b1058	**—,5-nitro-***	$C_8H_5NO_6$. *See* b1016	211.14	gr lf (+1½w)	250				i v^h	v	s				**B9¹**, 37
b1059	—,—,diethyl ester*	$C_{12}H_{13}NO_6$. *See* b1016	267.24	nd (al)	83.5				i	δ v^h					**B9**, 840
b1060	—,—,dimethyl ester*	$C_{10}H_9NO_6$. *See* b1016	239.19	nd	123				i	s	s				**B9²**, 61
b1061	**1,4-Benzenedicarboxylic acid, 2-nitro-***	$C_8H_5NO_6$. *See* b1024	211.14	nd	262–3				s^h	v^h					**B9**, 851
b1062	**1,3-Benzenedicarboxylic acid, tetrabromo-***	$C_8H_2Br_4O_4$. *See* b1016	481.76	nd (w)	288–92				s^h				i		**B9**, 839
b1063	**1,4-Benzenedicarboxylic acid, tetrabromo-***	$C_8H_2Br_4O_4$. *See* b1024	481.76	nd (w)	300d				i s^h	δ	δ		δ	aa δ	**B9**, 850
b1064	**1,3-Benzene-disulfonic acid, 4-amino-***	HO_3S, SO_3H, NH_2	253.26	nd (w+2)	120d				s	s	i				**B14²**, 47
b1065	**—,4-hydroxy-***	HO_3S, SO_3H, OH	254.18	nd	>100d				v	v	i				**B11²**, 13
b1066	**Benzenehexacarboxylic acid***	Mellitic acid. HO_2C, CO_2H, HO_2C, CO_2H, HO_2C, CO_2H	342.17	nd (al)	286–8				v	s				sulf s^h	**B9²**, 74
b1067	**Benzenepentacarboxylic acid***	HO_2C, CO_2H, HO_2C, HO_2C, CO_2H	298.16	nd (w+5)	228–30 (+5w), 238 (anh)				v^h	s	δ		i	peth i	**B9²**, 73
b1068	**Benzenephosphinic acid***	$C_6H_5P(OH)_2$	142.10	lf	83–6				s v^h	v	δ				**B16**, 79

For explanations, symbols and abbreviations see beginning of table.

No.	Name	Synonyms and Formula	Mol. wt.	Crystalline form, color and specific rotation	m.p. °C	b.p. °C	Density	n_D	Solubility w	al	eth	ace	bz	other solvents	Ref.
	Benzenephosphinic acid														
b1069	—,diethyl ester*	$C_6H_5P(OC_2H_5)_2$	198.20			235	1.032^{16}		i						**B16**, 791
b1070	**Benzenephosphonic acid***	$C_6H_5PO_3H_2$	158.09	lf (w)	160				v	s	s		i		**B16**[2], 390
b1071	—,dichloride	$C_6H_5POCl_2$	194.99			258	1.375^{20}		d						**B16**, 804
b1072	—,tetrachloride	$C_6H_5PCl_4$	249.89		73				d	d^h					**B16**, 804
b1073	**Benzenephosphothionic acid,** ethyl 4-nitrophenyl ester*	EPN. OC2H5 —PS —O— —NO2	323.31		36		1.27^{25}_{4}	1.5978^{30}	i	s	s		s		
b1074	**Benzeneseleninic acid***	$C_6H_5SeO_2H$	189.07	pl (w)	124–5		1.652^{123}_{4}		δ s^h					alk v	**B11**[2], 241
b1075	**Benzeneselenonic acid***	$C_6H_5SeO_3H$	205.07	nd	142	d			v	v	i		i		**B11**[1], 111
b1076	**Benzenesiliconic acid***	$C_6H_5SiO_2H$	138.20	glassy so (eth)	92				i	s	s				**B16**, 911
b1077	**Benzenestibonic acid***	$C_6H_5SbO_3H_2$	248.87	amor	>285				i	δ		δ		C_2Cl_4 s	**B16**[2], 584
b1078	**Benzenesulfenic acid,** chloride	Benzenesulfenyl chloride. 3 2 4 —SCl 5 6	144.62	red liq		$73–5^{9}$							s		**B6**[2], 295
b1079	—,**2,4-dinitro-,** chloride	$C_6H_3ClN_2O_4S$. *See* b1078	234.62	ye pr (bz-peth or CCl_4)	99								v	chl v aa v peth δ	**B6**[2], 316
b1080	—,**2-nitro-,** chloride	$C_6H_4ClNO_2S$. *See* b1078	189.62	ye nd (bz)	75						v		v	chl v	**B6**[2], 308
b1081	—,**4-nitro-,** chloride	$C_6H_4ClNO_2S$. *See* b1078	189.62	ye lf (peth)	52				d	d			v	aa d	**B6**[1], 160
b1082	**Benzenesulfinic acid**	3 2 4 —SO2H 5 6	142.18	pr (w)	85	100d			s^h	s	s		s	peth i	**B11**[2], 3
b1083	—,chloride	C_6H_5SOCl	160.62	pl (peth)	38	$85–90^{2}$			d	d	s			chl s^h	**B11**[2], 4
b1084	—,ethyl ester*	$C_6H_5SO_2C_2H_5$	170.23	liq		d			i	∞	∞		∞		**B11**, 6
b1085	—,**3-acetamido-**	$C_8H_9NO_3S$. *See* b1082	199.23	(w)	145				s^h						**B14**[2], 427
b1086	—,**4-acetamido-**	$C_8H_9NO_3S$. *See* b1082	199.23	(w)	160				s^h						**B14**[2], 427
b1087	—,**4-bromo-***	$C_6H_5BrO_2S$. *See* b1082	221.08	nd (w)	114				d^h	v	v				
b1088	—,**4-chloro-***	$C_6H_5ClO_2S$. *See* b1082	176.62	lf or nd (w)	99				s^h	v	v				**B11**[2], 4
—	—,**methyl-***	*see* **Toluenesulfinic acid**													
b1089	—,**3-nitro-***	$C_6H_5NO_4S$. *See* b1082	187.18	nd	95–6				s	s	s	s	i		**B11**[2], 5
b1090	—,**4-nitro-***	$C_6H_5NO_4S$. *See* b1082	187.18	pl or nd (w)	136d				δ	v	s			aa v	**B11**[2], 8
b1091	**Benzenesulfonic acid***	3 2 4 —SO3H 5 6	158.18	nd (bz)	52.5				s	s	i		δ	CS_2 i	**B11**[2], 18
b1092	—,hydrate*	$C_6H_5SO_3H.H_2O$	176.19	pl	45–6				s	s	i		δ	CS_2 i	**B11**[2], 18
b1093	—,amide	Benzenesulfonamide. 3 2 4 —SO2NH2 5 6	157.19	lf, nd (w)	150–1				i	s^h	s				**B11**[2], 24
b1094	—,—,*N,N*-dichloro-	Dichloramine B. $C_6H_5SO_2NCl_2$	226.08	ye mcl or pl	76					s					**B11**[2], 28
b1095	—,—,*N*-hydroxy-	$C_6H_5SO_2NHOH$	173.19	pl (w), rh	126d				v^h	v	v	v	δ	chl δ aa v	**B11**[1], 14
b1096	—,—,*N*-phenyl-	$C_6H_5SO_2NHC_6H_5$	233.29	tetr pr (al)	110				δ	v	v				**B12**[2], 298
b1097	—,chloride	$C_6H_5SO_2Cl$	176.62		14.5	251–2d $120–1^{17}$	1.3766^{25}_{25}		i	v	s				**B11**[2], 23
b1098	—,ethyl ester*	$C_6H_5SO_3C_2H_5$	186.23			156^{15}	1.2346^{0}_{4}	1.5081^{20}	δ d^h	s	v			chl v	**B11**[2], 20
b1099	—,fluoride	$C_6H_5SO_2F$	160.17			$90–1^{4}$	1.3286^{20}_{4}	1.4932^{18}		s	s				**B11**[2], 23
b1100	—,isopropyl ester	$C_6H_5SO_3C_3H_7^{i}$	200.26					1.5003^{20}	δ	v	v				**B11**[2], 20
b1101	—,methyl ester*	$C_6H_5SO_3CH_3$	172.20			154^{20}	1.2889^{0}_{4}	1.5151^{20}	δ	v	v			chl v	**B11**[2], 20
b1102	—,propyl ester*	$C_6H_5SO_3C_3H_7^{n}$	200.26			100d $162–3^{15}$	1.1804^{17}_{4}	1.5035^{25}	δ	s	v			chl v	**B11**[2], 20
b1103	—,sodium salt*	$C_6H_5SO_3Na$	180.16	nd (w)	450d				s v^h	$δ^h$					**B11**[2], 18
b1104	—,**3-acetamido-,** amide	$C_8H_{10}N_2O_3S$. *See* b1093	214.24	(aa)	216–9							s		aa s	**B14**, 682

For explanations, symbols and abbreviations see beginning of table.

No.	Name	Synonyms and Formula	Mol. wt.	Crystalline form, color and specific rotation	m.p. °C	b.p. °C	Density	n_D	Solubility w	al	eth	ace	bz	other solvents	Ref.
	Benzenesulfonic acid														
b1105	—,—,chloride	$C_8H_8ClNO_3S$. *See* b1091	233.67	nd (bz-peth)	88	...		...	d	s	s	...	δ	aa s	**B14², 43**
b1106	—,**4-acetamido-**, amide	$C_8H_{10}N_2O_3S$. *See* b1093	214.24	nd (aa)	219–20	...		...	s^h	s^h	...	s^h	...	...	**B14, 702**
b1107	—,—,chloride	$C_8H_8ClNO_3S$. *See* b1091	233.67	nd (bz), pr (bz-chl)	149	...		...	d	v	v	...	s^h	chl s^h	**B14, 439**
b1108	—,**2-amino-***	Orthanilic acid. $C_6H_7NO_3S$. *See* b1091	173.19	pr(+½w)	188	...		...	δ	i s^h	i	...	...	...	**B14², 42**
b1109	—,—,amide	$C_6H_8N_2O_2S$. *See* b1093	172.21	mcl pl or pr (w)	152.5–3.5	...		...	δ v^h	v	...	v	i	aa s	**B14², 42**
b1110	—**3-amino***	Metanilic acid. $C_6H_7NO_3S$. *See* b1091	173.19	nd, pr (w+1½)	d	...		...	δ	δ	i	...	...	...	**B14², 43**
b1111	—,—,amide	$C_6H_8N_2O_2S$. *See* b1093	172.21	lf or nd	142	...		...	δ s^h	v	...	...	...	...	**B14, 690**
b1112	—**4-amino-***	$C_6H_7NO_3S$. *See* b1091	173.19	rh pl or mcl (+w)	288	...	1.485^{25}_{4}	...	s^h	i	i	...	...	...	**B14², 43**
b1113	—,—,amide	Sulfanilamide. $C_6H_8N_2O_2S$. *See* b1093	172.21	lf (al)	163	...	1.08	...	s^h	s	s	s	...	...	**B14², 69**
b1114	—,—,—,*N*-acetyl-	$C_8H_{10}N_2O_3S$. *See* b1093	214.24	pr	182–4	...		...	δ	s	i	v	...	alk v	...
—	—,—,—,*N*(4,6-dimethyl-2-pyrimidyl)-	*see* **Sulfamethazine**													
—	—,—,—,*N*-guanido-	*see* **Sulfauganidine**													
—	—,—,—,*N*(4-methyl-2-pyrimidyl)-	*see* **Sulfamerazine**													
—	—,—,—,*N*(2-pyridyl)-	*see* **Sulfapyridine**													
—	—,—,—,*N*(2-quinoxalinyl)-	*see* **Sulfaquinoxaline**													
—	—,—,—,*N*(2-thiazolyl)-	*see* **Sulfathiazole**													
b1115	—,**2-amino-4-chloro-***	$C_6H_6ClNO_3S$. *See* b1091	207.65	nd or pl (w)	250d	...		...	δ s^h	δ	...	...	...	aa δ	**B14², 43**
b1116	—,**2-amino-5-chloro-**	$C_6H_6ClNO_3S$. *See* b1091	207.65	nd (w)	280d	...		...	δ s^h	...	...	...	...	...	**B14², 43**
b1117	—,**2-amino-3,4-dimethyl-***	$C_8H_{11}NO_3S$. *See* b1091	201.25	pr (w+1)	300d	...		...	δ	...	...	...	...	...	**B14¹, 731**
b1118	—,**2-amino-3,5-dimethyl-***	$C_8H_{11}NO_3S$. *See* b1091	201.25	pr or pl (w)	d	...		...	i	i	i	...	i	...	**B14², 45**
b1119	—,**2-amino-3,6-dimethyl-***	$C_8H_{11}NO_3S$. *See* b1091	201.25	amor	260	...		...	δ	...	i	...	i	...	**B14¹, 73**
b1120	—,**2-amino-4,5-dimethyl-***	$C_8H_{11}NO_3S$. *See* b1091	201.25	pl	>300	...		...	$δ^h$	...	...	...	...	...	**B14¹, 73**
b1121	—,**3-amino-2,5-dimethyl-***	$C_8H_{11}NO_3S$. *See* b1091	201.25	nd (w+1)	...	...		...	δ	...	...	...	...	...	**B14¹, 73**
b1122	—,**3-amino-4,5-dimethyl-***	$C_8H_{11}NO_3S$. *See* b1091	201.25	red nd (w+1)	315	...		...	δ	...	...	...	...	...	**B14¹, 73**
b1123	—,**4-amino-2,3-dimethyl-***	$C_8H_{11}NO_3S$. *See* b1091	201.25	nd	305	...		...	...	...	...	...	...	...	**B14¹, 73**
b1124	—,**4-amino-2,5-dimethyl-***	$C_8H_{11}NO_3S$. *See* b1091	201.25	pl or nd	>300	...		...	s	...	...	...	...	...	**B14¹, 73**
b1125	—,**5-amino-2,3-dimethyl-***	$C_8H_{11}NO_3S$. *See* b1091	201.25	pl (w+2)	294d	...		...	s	δ	...	...	δ	...	**B14¹, 73**
b1126	—,**5-amino-2,4-dimethyl-***	$C_8H_{11}NO_3S$. *See* b1091	201.25	pr or nd	290d	...		...	δ	i	...	...	i	chl i	**B14, 734**
b1127	—,**3-amino-4-hydroxy-***	$C_6H_7NO_4S$. *See* b1091	189.19	rh	155–6	...		...	δ	i	i	...	...	...	**B14, 814**
b1128	—,**3-amino-4-hydroxy-5-nitro-***	$C_6H_6N_2O_6S$. *See* b1091	234.20	pr	...	...		...	s^h	δ		...	...	...	**B14¹, 74**
b1129	—,**4(2-amino-1-napthylazo)-***	NH_2 — N:N — SO_3H	327.35	dk gr nd	...	...		...	s^h	v	i	...	...	...	**B16², 19**
b1130	—,**4(4-amino-1-naphthylazo)-***	NH_2 — N:N — SO_3H	327.35	dk gr cr	...	...		...	i	δ	...	...	...	...	**B16², 19**
b1131	—,**4(4-amino-phenyl-sulfon-amido)-,amide**	NH_2 — SO_2NH — SO_2NH_2	327.38	...	137	...		...	δ s^h	s	s	s	...	chl i peth i	...
b1132	—,**2-bromo-***	$C_6H_5BrO_3S$. *See* 61091	237.08	nd	...	...		...	v	v	...	...	...	...	**B11, 56**
b1133	—,—,amide	$C_6H_6BrNO_2S$. *See* b1093	236.09	nd (w), pr (al)	186	...		...	δ	δ	...	...	...	...	**B11, 56**

For explanations, symbols and abbreviations see beginning of table.

No.	Name	Synonyms and Formula	Mol. wt.	Crystalline form, color and specific rotation	m.p. °C	b.p. °C	Density	n_D	Solubility w	al	eth	ace	bz	other solvents	Ref.
	Benzenesulfonic acid														
b1134	—,—,chloride	$C_6H_4BrClO_2S$. *See* b1091	255.52	pr (eth)	51				d^h	d^h	δ				**B11**, 56
b1135	—,**3-bromo-**, amide	$C_6H_6BrNO_2S$. *See* b1093	236.09	nd or lf (w), pr (al)	153–4				δ s^h	δ s^h					**B11**, 56
b1136	—,**4-bromo-***	$C_6H_5BrO_3S$. *See* b1091	237.08	nd	102–3	155^{25}			s	s					**B11**, 57
b1137	—,—,amide	$C_6H_6BrNO_2S$. *See* b1093	236.09	nd (w or dil al)	166				δ s^h	s					**B11**, 56
b1138	—,—,bromide	$C_6H_4Br_2O_2S$. *See* b1091	299.98	(eth or bz)	76.5				d^h	d^h			s		**B11**^{1}, 56
b1139	—,—,chloride	$C_6H_4BrClO_2S$. *See* b1091	255.52	tcl or mcl pr (eth)	75–6	153^{15}			i	d	v				**B11**, 57
b1140	—,**2-chloro-**, amide	$C_6H_6ClNO_2S$. *See* b1093	191.64	lf (al)	188					s					**B11**, 54
b1141	—,**3-chloro-**,amide	$C_6H_6ClNO_2S$. *See* b1093	191.64	lf (dil al)	148				v^h	v	v				**B11**, 54
b1142	—,**4-chloro-***	$C_6H_5ClO_3S$. *See* b1091	192.62	nd, (w+1)	67	147–8^{25}			s	s	i		i		**B11**, 54
b1143	—,—,amide	$C_6H_6ClNO_2S$. *See* b1093	191.64	nd (w)	143.5				v^h	v^h	v^h				**B11**, 54
b1144	—,—,chloride	$C_6H_4Cl_2O_2S$. *See* b1091	211.07	pr or pl (eth)	55	141^{15}			d^h	d^h	v		v		**B11**, 55
b1145	—,—,4-chloro-phenyl ester*	Ovotran. $C_{12}H_8Cl_2O_3S$. *See* b1091	303.17		86.5				i	δ		s			
b1146	—,**3,4-diamino-***	$C_6H_8N_2O_3S$. *See* b1091	188.21	nd	d				v^h						**B14**, 717
b1147	—,**3,5-diamino-***	$C_6H_8N_2O_3S$. *See* b1091	188.21		d				δ v^h	δ	i				**B14**^{2}, 446
b1148	—,**2,5-dichloro**, dihydrate*	$C_6H_4Cl_2SO_3.2H_2O$. *See* b1091	263.10		<100				v		δ				**B11**^{2}, 30
b1149	—,—,chloride	$C_6H_3Cl_3SO_2$. *See* b1091	245.51	(bz), mcl pr	38				d^h	d^h					**B11**^{2}, 30
b1150	—,**3,4-dichloro-**, dihydrate*	$C_6H_4Cl_2SO_3.2H_2O$. *See* b1091	263.10	nd	71–2				v	v	v				**B11**, 55
b1151	—,—,chloride	$C_6H_3Cl_3SO_2$. *See* b1091	245.51	mcl pr	224				d^h	d^h					**B11**^{1}, 16
b1152	—,**3,5-diiodo-4-hydroxy-**, trihydrate*	Sozoiodolic acid. $C_6H_4I_2SO_4.3H_2O$. *See* b1091	480.02	nd	120, 190d				v	v	v				
b1153	—,**2,3-dimethyl-**, chloride	$C_8H_9ClO_2S$. *See* b1091	204.68	pr (peth)	47				d	d^h				peth v^h	**B11**, 120
b1154	—,**3,4-dimethyl-***	$C_8H_{10}O_3S$. *See* b1091	186.23	pl or pr (chl)	63–4				v						**B11**^{2}, 78
b1155	—,—,chloride	$C_8H_9ClO_2S$. *See* b1091	204.68	pr (eth)	51–2				d	d^h	v				**B11**, 121
b1156	—,**3,5-dimethyl-**, chloride	$C_8H_9ClO_2S$. *See* b1091	204.68	(eth)	89–90				d	d^h	v				**B11**, 126
b1157	—,**2,4-dinitro-***	$C_6H_4N_2O_7S$. *See* b1091	248.17	nd (w+3)	108				v	v	δ		i	peth i	**B11**^{2}, 36
b1158	—,**3 ,5-dinitro-***	$C_6H_4N_2O_7S$. *See* b1091	248.17	ye red (dil al)	235				i s^h	s^h	δ		δ	aa δ lig δ	**B11**, 79
b1159	—,**3-ethylamino-***	$C_8H_{11}NO_3S$. *See* b1091	201.25	nd (w)	294d				δ						**B14**^{2}, 435
b1160	—,**4-ethylamino-**	$C_8H_{11}NO_3S$. *See* b1091	201.25	pl (w)	258d				v						**B14**^{1}, 721
b1161	—,**4-fluoro-**, amide	$C_6H_6FNO_2S$. *See* b1093	175.18	pl or nd (w)	123				δ	v	v	v	δ		**B11**, 54
b1162	—,—,chloride	$C_6H_4ClFO_2S$. *See* b1091	194.61	pl or nd	36				d^h	d^h	v		v	chl v	**B11**, 53
b1163	—,**2-formyl-**	2-Benzaldehyde sulfonic acid. $C_7H_6O_4S$. *See* b1091	186.19	pr or lf	114				s	δ					**B11**^{2}, 185
b1164	—,**4-hydrazino-***	Phenylhydrazine-4-sulfonic acid. $C_6H_8N_2O_3S$. *See* b1091	188.21	nd or lf	286				δ						**B15**^{2}, 305
b1165	—,**2-hydroxy-***	*o*-Phenolsulfonic acid. $C_6H_6O_4S$. *See* b1091	174.17	(+¾w)	50, 145d (+w)				v	v					
b1166	—,**4-hydroxy-***	*p*-Phenolsulfonic acid. $C_6H_6O_4S$. *See* b1091	174.17	nd					s	s					**B11**^{2}, 134
b1167	—,—,amide	$C_6H_7NO_3S$. *See* b1093	173.20	(al or w)	176–7				s^h	s^h					**B11**, 243
b1168	—,**4-hydroxy-5-isopropyl-2-methyl-**	*α*-Thymolsulfonic acid. $C_{10}H_{14}O_4S$. *See* b1091	230.29	pl (+1w)	91–2				v						**B11**^{2}, 151
b1169	—,**4-hydroxy-3-nitro-***	$C_6H_5NO_6S$. *See* b1091	219.17	ye pl (AcOEt-bz), nd (w)	51.5 (+w), 141–2				s	v				AcOEt v chl v^h	**B11**, 246
b1170	—,**2-isopropyl-5-methyl-**	$C_{10}H_{14}O_3S$. *See* b1091	214.28		130–1				v	s	i				**B11**^{2}, 84
b1171	—,**4-isopropyl-2-**mcthyl	$C_{10}H_{14}O_3S$. *See* b1091	214.28	pl or pr	88–90				v	δ	δ		δ	chl δ	**B11**, 139
b1172	—,**5-isopropyl-2-methyl-**	$C_{10}H_{14}O_3S$. *See* b1091	214.28	pl (+2 w), mcl pr	220, 78–9 (+2 w)				v						**B11**^{2}, 83
b1173	—,**2-methoxy-**, chloride	$C_7H_7ClO_3S$. *See* b1091	202.65	nd (peth)	56				d	d	s			peth s^h	**B11**, 235
b1174	—,**4-methoxy-**, chloride	$C_7H_7ClO_3S$. *See* b1091	206.65	nd or pr (bz)	42–3				d	s d^h	s		s^h		**B11**^{2}, 136
—	—,**methyl-***	*see* **Toluenesulfonic acid**													
b1175	—,**2-nitro-***	$C_6H_5NO_5S$. *See* b1091	203.17		85				v	s	i				**B11**^{2}, 31
b1176	—,—,chloride	$C_6H_4ClNO_4S$. *See* b1091	221.63	pr (eth-peth)	68.5–9.5				d^h	d^h	s			peth δ	**B11**^{2}, 32
b1177	—,**3-nitro-***	$C_6H_5NO_5S$. *See* b1091	203.17	pl					v^h						**B11**, 68

For explanations, symbols and abbreviations see beginning of table.

No.	Name	Synonyms and Formula	Mol. wt.	Crystalline form, color and specific rotation	m.p. °C	b.p. °C	Density	n_D	Solubility w	al	eth	ace	bz	other solvents	Ref.
	Benzenesulfonic acid														
b1178	—,—,chloride	$C_6H_4ClNO_4S$. *See* b1091	221.63	mcl pr (eth), nd (lig)	64				i d^h	s d^h	...	...	...		**B11**², 33
b1179	—,**4-nitro-***	$C_6H_5NO_5S$. *See* b1091	203.17	(+2w)	95				s	...	...	...	...		**B11**², 33
b1180	—,—,chloride	$C_6H_4ClNO_4S$. *See* b1091	221.63	mcl nd (peth)	79.5–80.5				d^h	d^h	...	...	...	peth s^h	**B11**¹, 21
b1180¹	—,**4-phenyl-amino-**	4-Sulfodiphenylamine. $C_{12}H_{11}NO_3S$. *See* b1091	249.29	pl	d				v	v	i	...	...		**B14**¹, 72
b1181	—,**4-ureido,** amide	$C_7H_9N_3O_3S$. *See* b1091	215.24	nd	206–7				s^h	s^h	...	...	...		
b1182	**1,2,3,4-Benzene-tetracarboxylic acid***	Mellophanic acid. HO2C, CO2H, HO2C—, —CO2H	254.15	pr (+6 w)	238–42				s	...	...	s	...		**B9**², 730
b1183	—,tetramethyl ester*	$C_{14}H_{14}O_8$. *See* b1182	310.25	nd (MeOH or w)	133–5				i δ^h	v	δ	...	v	MeOH δ	**B9**², 730
b1184	**1,2,3,5-Benzene-tetra-carboxylic acid,** tetramethyl ester*	Tetramethyl prehintate. CH3O2C, CO2CH3, —CO2CH3, CH3O2C	310.26	nd (al)	108.4				i	s^h	...	...	...		**B9**¹, 435
b1185	**1,2,4,5-Benzene-tetra-carboxylic acid***	Pyromellitic acid. CO2H, HO2C—, —CO2H, HO2C	254.15	tcl pl (+2w)	275, 242 (+2w)				δ	s	...	...	...		**B9**, 997
b1186	—,tetraethyl ester*	$C_{18}H_{22}O_8$. *See* b1185	366.37	nd (al)	54	sub			i	s^h	...	...	...		**B9**, 998
b1187	—,tetramethyl ester*	$C_{14}H_{14}O_8$. *See* b1185	310.26	pl (al)	141.5				i	δ^h	...	...	...		**B9**, 998
—	**Benzenethiol***	*see* **Benzene, mercapto-***													
b1188	**1,2,3-Benzenetri-carboxylic acid***	Hemimellitic acid. HO2C, CO2H, —CO2H	210.14	tcl pl (+2w), or nd (w)	197, 223–4 (+2w)				v^h	...	v	...	...		**B9**², 712
b1190	**1,2,4-Benzenetri-carboxylic acid***	Trimellitic acid. CO2H, HO2C—, —CO2H	210.14	nd (w), (aa or al)	215–7				v	v	v	δ	δ	CCl_4 δ chl δ CS_2 δ aa s^h	**B9**², 712
b1191	**1,3,5-Benzenetri-carboxylic acid***	Trimesic acid. HO2C, —CO2H, HO2C	210.14	pr nd (w+1)	345–50				δ	v	s	...	...		**B9**², 712
b1192	—,triethyl ester*	$C_{15}H_{18}O_6$. *See* b1191	294.31	pr or nd (al)	133.5–4.5				i	s^h	v	...	s	CS_2 v aa s	**B9**², 712
b1193	—,trimethyl ester*	$C_{12}H_{12}O_6$. *See* b1191	252.23	nd	144				i	s	...	...	...		**B9**¹, 430
b1194	—,**2-chloro-***	$C_9H_5ClO_6$. *See* b1191	244.59	nd or pl (+1w)	285, 278(+w)				v	v	v	...	...	chl i	**B9**, 980
b1195	**1,2,4-Benzenetri-carboxylic acid, 6-hydroxy-***	$C_9H_6O_7$. *See* b1190	226.14	pr (w+2)	241d				s	s	...	...	...		**B10**, 580
b1196	**1,3,5-Benzenetri-carboxylic acid, 2-hydroxy-***	$C_9H_6O_7$. *See* b1191	226.14	pr (+1w), nd (+2w)	306 (anh)				s^h	v^h	δ	...	...	chl i	**B10**, 580
b1197	**1,3,5-Benzenetri-sulfonic acid***	HO3S, —SO3H, HO3S	318.30	(+3w)	100d				s	...	...	...	...		**B11**², 12
—	**Benzestrol**	*see* **Hexane,2,4-bis(4-hydroxyphenyl)-3-ethyl-***													
—	**Benzhydrol**	*see* **Methanol, diphenyl-***													
—	**Benzidine**	*see* **Biphenyl, 4,4′-diamino-**													
—	**Benzidinesulfone**	*see* **Dibenzothiophene, 2,7-diamino,** 9,9-dioxide													
b1198	**Benzil**	Diphenylglyoxal. Dibenzoyl. 1,2-Diphenyl-1,2-ethan-dione.* $C_6H_5COCOC_6H_5$	210.23	ye pr (al)	95	346–8d 188¹²	1.084^{102}_{4}		i	v	v	...	v		**B7**², 764
b1199	—,dioxime(*anti*)	α-Benzil dioxime. HON NOH ‖ ‖ C6H5C—CC6H5	240.16	lf (al or ace)	238d				i	δ	δ	...	...	alk s aa δ	**B7**², 680

For explanations, symbols and abbreviations see beginning of table.

No.	Name	Synonyms and Formula	Mol. wt.	Crystalline form, color and specific rotation	m.p. °C	b.p. °C	Density	n_D	Solubility w	al	eth	ace	bz	other solvents	Ref.
	Benzil														
b1200	—,—,(*syn*)	β-Benzil dioxime. NOH NOH ‖ ‖ C_6H_5C——CC_6H_5	240.26	nd (al)	207d	...	...	...	δ^h	s	s	...	...	alk s aa s	**B7²**, 681
b1201	—,—,(*amphi*)	γ-Benzil dioxime. NOH NOH ‖ ‖ C_6H_5C——CC_6H_5	240.26	nd (al), (aa)	164–6	...	...	...	i	s	...	s	s	peth δ alk s CCl_4 δ	**B7²**, 681
b1202	—,monooxime(α)	$C_6H_5C(:NOH)COC_6H_5$	225.25	lf (dil al bz, MeOH)	137–8	200d	...	...	δ	v	v	...	δ	chl δ aa v lig δ	**B7²**, 678
b1203	—,—,(β)	$C_6H_5C(:NOH)COC_6H_5$	225.25	pr or nd (bz+½)	70(+½bz) 113–4	...	...	...	δ	v	v	...	...	lig δ	**B7²**, 679
b1204	—,osazone(*anti*)	C_6H_5NHN $NNHC_6H_5$ ‖ ‖ C_6H_5C—CC_6H_5	390.49	ye nd	227	...	...	...	i	δ	...	δ	v^h	chl v^h	**B15²**, 72
b1205	—,—,(*syn*)	$NNHC_6H_5$ C_6H_5NHN ‖ ‖ C_6H_5C——CC_6H_5	390.49	ye nd (bz-al)	210	...	...	...	δ	s	...	δ	s	aa δ	**B15²**, 72
—	**Benzilic acid**	*see* **Acetic acid, diphenyl (hydroxy)-**													
b1206	**Benzimidazole**	Benzoglioxaline. 1,3-Benzodiazole. 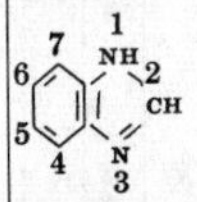 (ring positions 1–7; NH, CH, N)	118.14	rh bipym pl	170	>360	...	...	δ v^h	v	δ	...	i	dil alk s dil HCl v lig i	**B23²**, 151
b1207	—,2-amino-	$C_7H_7N_3$. *See* b1206	133.15	pl (w)	224	...	...	...	s	s	δ	s	δ	...	**B24¹**, 240
b1208	—,2-hydroxy-	*o*-Phenylene urea. $C_7H_6ON_2$. *See* b1206	134.14	lf (w or al)	310–2	...	...	...	δ	v	...	...	δ	alk δ con HCl s	**B24²**, 62
b1209	—,5,6-dimethyl-	$C_9H_{10}N_2$. *See* b1206	146.19	(eth)	205–6	140^3 sub	...	...	s	s	s	...	...	chl s	...
b1210	—,2-mercapto-	$C_7H_6N_2S$. *See* b1206	150.20	lf (dil al)	303–4	...	...	...	δ	v	...	...	...	...	**B24²**, 65
b1211	—,1-methyl-	$C_8H_8N_2$. *See* b1206	132.17	pr (peth), pl (al)	64	286^{756}	1.1283^{17}_4	1.6013^7	...	s^h	...	...	...	peth s^h	**B23²**, 152
b1212	—,2-methyl-	$C_8H_8N_2$. *See* b1206	132.17	pr (w)	176	...	...	...	s^h	s	s	...	...	...	**B23²**. 160
b1213	—,4-methyl-	$C_8H_8N_2$. *See* b1206	132.17	pl (w)	145	...	...	...	s^h	s	...	...	...	...	**B23¹**, 38
b1214	—,5-methyl-	$C_8H_8N_2$. *See* b1206	132.17	(w)	114	...	...	...	s^h	...	...	...	...	...	**B23¹**, 38
b1215	—,2-methyl-1(2-naphthyl)-5-nitro-	$C_{18}H_{13}N_3O_2$. *See* b1206	303.33	nd (al)	162	...	...	...	...	s	s	...	v	aa s	**B23**, 150
b1216	—,2-methyl-5-nitro-1-phenyl-	$C_{14}H_{11}N_3O_2$. *See* b1206	253.27	nd (al)	170–1	...	...	...	...	s^h	...	...	...	...	**B23²**, 161
b1217	—,6-nitro-	$C_7H_5O_2N_3$. *See* b1206	163.14	nd	203	...	...	...	δ	v	δ	...	δ	...	**B23²**, 154
b1218	—,2-phenyl-	$C_{13}H_{10}N_2$. *See* b1206	194.23	pl (aa, al w), nd (bz)	291	...	...	...	δ	s	...	...	δ	chl δ aa s	**B23²**, 238
b1219	**4,5-Benzindane**	1,2-Cyclopentanonaphthalene. 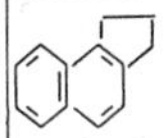	168.24	...	...	$244–5^{757}$ 170^5	1.066^{20}_4	1.6290^{20}	i	...	...	...	...	aa s	**E13**, 13
b1220	**1,2-Benzoisothiazole, 5-hydroxy-3-phenyl-**	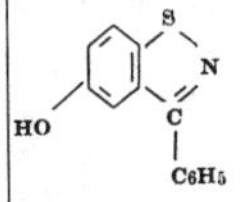 (S, N, HO, C, C_6H_5)	227.29	nd (dil al or aa)	159–60	...	...	...	...	v	...	v	δ	alk v aa v lig i	**B27²**, 87
b1221	**Benzisoxazole, 3-methyl-**	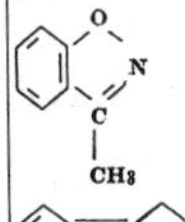 (O, N, C, CH_3)	133.14	...	...	$108–10^{16}$	...	...	i	...	s	...	...	...	...
b1222	**1,2-Benzocarbnzole,3,4-dihydro-**	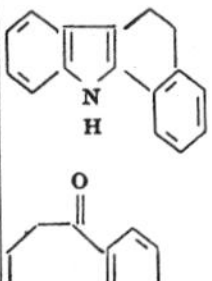 (N, H)	219.29	(lig or MeOH)	163–4	...	...	...	...	...	...	...	...	MeOH s^h lig s^h	**B20²**, 315
b1223	**1,2-Benzo-1-cyclooctene-3-one**	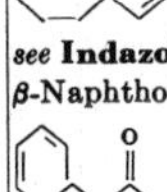 (O)	174.23	...	...	$87–8^{0.001}$	...	1.5577^{25}	...	...	...	...	...	...	**Am 76**, 5462
—	**1,2-Benzodiazole**	*see* **Indazole**													
b1224	**5,6-Benzoflavone**	β-Naphthoflavone. 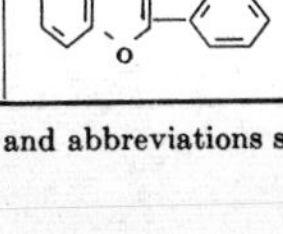 (O, O)	272.31	nd (al)	164–5	...	...	...	...	δ^h	s	..	v^h	...	**B17²**, 414

For explanations, symbols and abbreviations see beginning of table.

No.	Name	Synonyms and Formula	Mol. wt.	Crystalline form, color and specific rotation	m.p. °C	b.p. °C	Density	n_D	Solubility w	al	eth	ace	bz	other solvents	Ref.
	Benzoflavone														
b1225	**7,8-Benzoflavone**	α-Naphthoflavone. 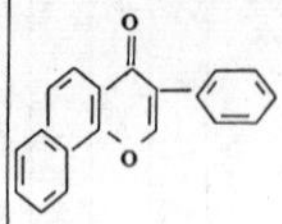	272.31	lf or nd (dil al)	155–6					δ				sulf s	**B17**[2], 414
b1226	**3,4-Benzofluoranthene**	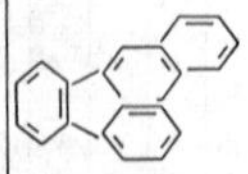	252.32	nd (bz)	167				i				δ		**E14s**, 569
b1227	**10,11-Benzofluoranthene**	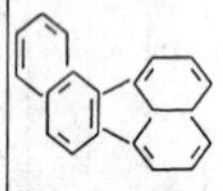	252.32	ye cr (al), nd (aa)	166				i	δ				aa δ	**E14s**, 570
b1228	**11,12-Benzofluoranthene**	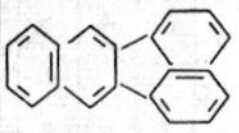	252.32	pa ye nd (bz)	217				i	s			s	aa s	**E14s**, 573
b1229	**1,2-Benzofluorene**	Chrysofluorene. 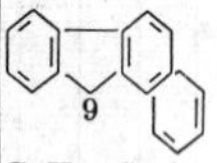	216.29	pl (ace or aa)	189–90	398^{758}				δ			s	chl s	**E14**, 299
b1230	**—,9-phenyl-**	$C_{23}H_{16}$. *See* b1229	292.38	nd (aa)	195.5					$δ^h$	v		v	sulf s	**E14**, 299
b1231	**Benzofuran**	Coumaron. 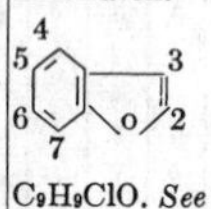	118.14		<−18	$173–5^{760}$ $62–3^{15}$	1.0913^{23}	1.565^{23}	i	s	s				**B17**[2], 57
b1232	**—,2(chloromethyl)-2,3-dihydro-**	C_9H_9ClO. *See* b1231	168.62			$118–9^{11}$	$1.2196^{7.5}_{16}$	$1.5620^{7.5}$							
—	**—,2,3-dihydro-**	*see* **Coumaran**													
b1233	**—,2-methyl-**	C_9H_8O. *See* b1231	132.16			197–8 $93–4^{20}$	1.050^{24}	1.5495^{22}							**B17**[2], 59
b1234	**—,3-methyl-**	C_9H_8O. *See* b1231	132.16			$196–7^{742}$	1.0540^{23}_{4}	1.5536^{16}	i						**B17**[1], 26
b1235	**—,5-methyl-**	C_9H_8O. *See* b1231	132.16			197–9	1.0467^{15}	1.5470^{16}						sulf s	**B17**[1], 27
b1236	**—,7-methyl-**	C_9H_8O. *See* b1231	132.16			190–1	1.0490^{19}	1.5525^{17}						sulf s	**B17**, 61
b1237	**Benzohydroxamic acid**	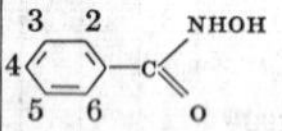	137.14	rh pl, lf	128–9	exp			s^h	s	δ		i		**B9**[2], 213
—	**—,amide**	*see* **Benzoic acid**, amidoxime													
b1238	**—,2-hydroxy-**	Salicylhydroxamic acid. $C_7H_7NO_3$. *See* b1237	153.14	nd (aa)	168				δ	v	v			aa s^h	**B10**[2], 60
b1239	**Benzoic acid**	Benzenecarboxylic acid. 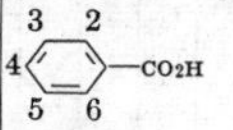	122.12	mcl lf or nd	122.4	249^{760}	$1.3211^{23.3}$	1.504^{132}	δ	v	v		v	CCl_4 s lig δ	**B9**[2], 72
b1240	—,4-acetamidophenyl ester	*p*-Benzoyloxyacetanilide. $C_{15}H_{13}NO_3$. *See* b1239	255.28	nd (al)	171				$δ^h$	v			v	aa v lig i	
b1241	—,2-acetylphenyl ester	*o*-Benzyloxyacetophenone. $C_{15}H_{12}O_3$. *See* b1239	240.26	nd (al)	88				i	v	v			aa v	**B9**[2], 133
b1242	—,4-acetylphenyl ester	*p*-Benzyloxyacetophenone. $C_{15}H_{12}O_3$. *See* b1239	240.26	nd (al)	134					s^h	s		s		**B9**[2], 133
b1243	—,allyl ester	$C_6H_5CO_2CH_2CH{:}CH_2$	162.19	ye		242	1.0578^{15}_{15}		i	s	s			MeOH s	**B9**[2], 93
b1244	—,amide	Benzamide. 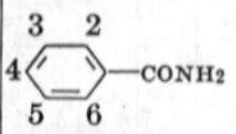	121.14	mcl pr or pl	132.5–3.5	290	1.0792^{130}_{4}		δ v^h	v	δ		s v^h	CCl_4 v CS_2 v	**B9**[2], 163
b1245	—,—,*N*(3-chloro-2-methylphenyl)-	$C_{14}H_{12}ClNO$ *See* b1244	245.71	nd	170–1				i	s	s	v	δ	aa δ	**B12**[2], 455
b1246	—,*N,N*-diphenyl-	Benzoyldiphenylamine. $C_6H_5CON(C_6H_5)_2$	273.34	rh pr (al), nd	180				δ	δ s^h	δ				**B12**[2], 155
b1247	—,—,*N*(1-naphthyl)-	$C_6H_5CONHC_{10}H_7^{α}$	247.30	nd (al or aa)	159–60					δ			s		**B12**[2], 525

For explanations, symbols and abbreviations see beginning of table.

No.	Name	Synonyms and Formula	Mol. wt.	Crystalline form, color and specific rotation	m.p. °C	b.p. °C	Density	n_D	Solubility						Ref.
									w	al	eth	ace	bz	other solvents	
	Benzoic acid														
b1248	—,—,N(2-nitrophenyl)-	$C_{13}H_{10}N_2O_3$. *See* b1244	242.24	ye nd (al)	98				δ^h	s	v				**B12^1, 342**
b1249	—,—,N(3-nitrophenyl)-	$C_{13}H_{10}N_2O_3$ *See* b1244	242.24	lf	157				i	δ	v			chl v	**B12^2, 380**
b1250	—,—,N(4-nitrophenyl)-	$C_{13}H_{10}N_2O_3$. *See* b1244	242.24	nd	199				i	δ^h				chl δ	**B12^2, 391**
b1251	—,—,N-phenyl-	Benzanilide $C_6H_5CONHC_6H_5$	197.24	lf (al)	163	$117–9^{10}$			i	δ v^h	δ			aa δ	**B12^2, 152**
b1252	—,—,N(2-tolyl)-	$C_{14}H_{13}NO$ *See* b1244	211.27	rh nd (aa-ace)	145–6		1.205^{15}		δ^h	s	s				**B12^2, 441**
b1253	—,—,N(3-tolyl)-	$C_{14}H_{13}NO$ *See* b1244	211.27	mcl pr (al)	125		1.170^{15}			s					**B12^1, 400**
b1254	—,—,N(4-tolyl)-	$C_{14}H_{13}NO$ *See* b1244	211.27	rh nd (al)	158	232	1.202^{15}		i	s	s				**B12^2, 505**
b1255	—,amidine	Benzamidine $C_6H_5C(:NH)NH_2$	120.16	lf (al)	80				s	v	δ				**B9^2, 199**
b1256	—,—,hydrochloride	$C_6H_5C(:NH)NH_2.HCl$	156.62	rh pr (w+2)	169, 72(+2w)				s^h	s^h					**B9^2, 199**
b1257	—,—,N,N-diphenyl-	$C_6H_5C(:NH)N(C_6H_5)_2$	272.35	rh pl (eth)	112					s	s		s		**B12^2, 155**
b1258	—,—,—,hydrochloride	$C_6H_5C(:NH)N(C_6H_5)_2$. HCl	308.31	mcl pr or nd	223d				v	v					**B12, 270**
b1259	—,—,N,N′-diphenyl-	$C_6H_5C(:NC_6H_5)NHC_6H_5$	272.35	nd (bz or al) pr (al)	146–7				δ	v	v				**B12^2, 157**
b1260	—,—,N(1-naphthyl)-	$C_6H_5C(:NH)NHC_{10}H_7^{\alpha}$	246.31	pl (al)	141				i	s	s				**B12, 1233**
b1261	—,amidoxime	Benzamidoxime $C_6H_5C(NH_2):NOH$	136.15	mcl pr (w)	79–80				δ	s	v		s	chl s lig i	**B9^2, 214**
b1262	—,anhydride	Benzoic anhydride. $(C_6H_5CO)_2O$	226.23	pr (eth)	42–3	360^{760}			i	s	s			lig i	**B9^2, 147**
b1263	—,azide	Benzoylazide. $C_6H_5CON_3$	147.14	pl (ace)	32	exp			i	s	s				**B9^2, 219**
b1264	—,benzoylmethyl ester	ω-Benzoyloxyacetophenone. $C_6H_5CO_2CH_2COC_6H_5$	240.26	pl (dil al)	118.5					s^h	v		v	chl v	**B9^2, 133**
b1265	—,benzyl ester	$C_6H_5CO_2CH_2C_6H_5$	212.24	nd or lf	19.4	327^{760} $170–1^{11}$	1.1121_4^{25}	1.5681^{21}	i	s	s	s	s	MeOH s chl s peth s	**B9^2, 100**
b1266	—,bromide	Benzoyl bromide. C_6H_5COBr	185.03		−24	218–9	1.570^{15}		d	d	∞				
b1267	—,butyl ester	$C_6H_5CO_2C_4H_9^n$	178.22		−22.4	247^{760}	1.000^{20}		i	∞	∞				**B9^2, 90**
b1268	—,chloride	Benzoyl chloride. C_6H_5COCl	140.57		−1	197.2^{760} 71^9	1.2105_4^{21}	1.5537^{20}	d	d	∞		s	CS_2 s	**B9^2, 159**
b1269	—,1-chloroethyl ester	$C_6H_5CO_2CHClCH_3$	184.63			134^{30}	1.172^{20}		i	s	s				**B9^2, 127**
b1270	—,2-chloroethyl ester	$C_6H_5CO_2CH_2CH_2Cl$	184.63			254^{729} $118–20^2$			i	v	v				**B9^2, 90**
b1271	—,cyclohexyl ester	$C_{13}H_{16}O_2$. *See* b1239	204.26		<−10	285			i	s	s				**B9^1, 65**
b1272	—,2-ethoxyethyl ester	$C_6H_5CO_2(CH_2)_2OC_2H_5$	194.22			$260–1^{740}$	1.0585_{25}^{25}	1.4969^{25}	i	v				os v	
b1273	—,ethyl ester	$C_6H_5CO_2C_2H_5$	150.17		−34.6	213^{760} 85^{10}	1.0458_4^{25}	1.5057^{19}	i	s	∞			chl s peth s	**B9^2, 88**
b1274	—,fluoride	Benzoyl fluoride. C_6H_5COF	124.11			154–5	>1		d^h	v	v				**B9^1, 94**
b1275	—,hexyl ester	$C_6H_5CO_2(CH_2)_5CH_3$	206.29			272^{770} $139–40^8$			i	s					**B9^2, 93**
b1276	—,hydrazide	Benzoyl hydrazine. $C_6H_5CONHNH_2$	136.15	pl (w)	112	267d			s	s	δ	δ		chl δ	**B9^2, 214**
b1277	—,2-hydroxyethyl ester	1,2-Ethanediol monobenzoate. $C_6H_5CO_2CH_2CH_2OH$	166.17		45	$150–1^{10}$			i	s					**B9^2, 108**
b1278	—,2-hydroxyphenyl ester	Pyrocatechol monobenzoate. $C_{13}H_{10}O_3$. *See* b1239	214.22	nd (w)	130–1				s^h	s					**B9, 130**
b1279	—,3-hydroxyphenyl ester	Resorcinol monobenzoate. $C_{13}H_{10}O_3$. *See* b1239	214.22	pr (dil al)	135–6				i					chl v aa v	**B9^2, 113**
b1280	—,4-hydroxyphenyl ester	Hydroquinone monobenzoate. $C_{13}H_{10}O_3$. *See* b1239	214.22	nd (al)	156–4d				i	s	v		v	chl v	**B9^2, 113**
b1281	—,iodide	Benzoyl iodide. C_6H_5COI	232.01	nd	3	135^{25}			d	s	∞				**B9^2, 162**
b1282	—,isobutyl ester	$C_6H_5CO_2C_4H_9^i$	178.22			242^{760}	0.9896_4^{30}		i	∞	∞				**B9^2, 191**
b1283	—,isopropyl ester	$C_6H_5CO_2C_3H_7^i$	164.21			218^{760}	1.0122_{15}^{15}		i	s	s				**B9^2, 90**
b1284	—,2-methoxyethyl ester	$C_6H_5CO_2(CH_2)_2OCH_3$	180.20			$254–6^{760}$	1.0891_{25}^{25}	1.5040^{25}	δ	s	s		s		**B9, 129**
b1285	—,2-methoxyphenyl ester	Guaiacyl benzoate. $C_{14}H_{12}O_3$. *See* b1239	228.25	(al)	58				δ^h	v^h	v			chl v aa δ	**B9^2, 112**

For explanations, symbols and abbreviations see beginning of table.

No.	Name	Synonyms and Formula	Mol. wt.	Crystalline form, color and specific rotation	m.p. °C	b.p. °C	Density	n_D	Solubility w	al	eth	ace	bz	other solvents	Ref.
	Benzoic acid														
b1286	—,methyl ester	$C_6H_5CO_2CH_3$	136.14		−12.3	195.5^{760}	1.0937^{15}_{4}	1.5205^{15}	i	s	∞	...	...	MeOH s	**B9^{2}**, 87
b1287	—,3-methylbutyl ester	Isoamyl benzoate. $C_6H_5CO_2(CH_2)_2CH(CH_3)_2$	192.55			263^{760} 133^{14}	$0.9910^{17.5}_{4}$	1.4950^{17}	i	s	∞	...	...		**B9^{2}**, 92
b1288	—,methylene diester	Methanediol dibenzoate. $(C_6H_5CO_2)_2CH_2$	256.25	nd or pr	99	225d	1.275^{22}		i	δ	s	...	s	MeOH v^h CCl$_4$ v peth δ	**B9^{2}**, 127
b1289	—,1-naphthyl ester	$C_6H_5CO_2C_{10}H_7^{\alpha}$	248.26	pl or pr (al-eth)	56				i	...	v	...	...		**B9**, 125
b1290	—,2-naphthyl ester	$C_6H_5CO_2C_{10}H_7^{\beta}$	248.26	nd or pr (al)	107				i	s^h	δ	...	...	aa δ v^h	**B9**, 125
b1291	—,nitrile	Phenyl cyanide. Benzonitrile. 3 2 4 —CN 5 6	103.12		−13	190.7	1.0102^{15}_{15}	1.5289^{20}	δ^h	∞	∞	...	...		**B9^{2}**, 196
b1292	—,2-nitrobenzyl ester	$C_{14}H_{11}NO_4$. *See* b1239	257.25	nd (dil al)	94				...	s^h	...	...	s	aa s lig δ	**B9**, 121
b1293	—,3-nitrobenzyl ester	$C_{14}H_{11}NO_4$. *See* b1239	257.25		71-2				i	s	s	...	...		
b1294	—,4-nitrobenzyl ester	$C_{14}H_{11}NO_4$. *See* b1239	257.25		89				...	...	...	...	...		**B9^{1}**, 68
b1295	—,phenyl ester	$C_6H_5CO_2C_6H_5$	198.21	mcl pr (eth-al)	71	314^{760}	1.235^{20}_{4}		i	s^h	s^h	...	...		
b1296	—,1-phenylethyl ester	$C_6H_5CO_2CH(CH_3)C_6H_5$	226.28			189^{21}	1.1108^{18}		i	s	...	...	...		**B9**, 121
b1297	—,2-phenylhydrazide	1-Benzoyl-2-phenylhydrazine. $C_6H_5CONHNHC_6H_5$	212.24	pr (al), nd (w), lf (dil al)	168	314			δ^h	s^h	δ	...	...	chl s	**B9^{2}**, 97
b1298	—,4-phenylphenacyl ester	—CO_2CH_2CO—	316.34		167				...	...	...	...	...		
b1299	—,propyl ester	$C_6H_5CO_2C_3H_7^{n}$	164.20		−51.6	231^{760}	1.0276^{15}_{15}	1.5014^{15}	i	∞	∞	...	...		**B9^{2}**, 90
b1300	—,tetrahydrofurfuryl ester	—CO_2CH_2— O	206.24			$300–2^{750}$ $138–40^{2}$			i	∞	∞	...	...	chl ∞	**B17^{2}**, 107
b1301	—,2-tolyl ester	$C_{14}H_{12}O_2$. *See* b1239	212.25			$307–8^{728}$			i	s	v	...	...		**B9^{2}**, 98
b1302	—,3-tolyl ester	$C_{14}H_{12}O_2$. *See* b1239	212.25		55–6	314			i	s	s	...	...	CS_2 s	**B9^{2}**, 99
b1303	—,4-tolyl ester	$C_{14}H_{12}O_2$. *See* b1239	212.25	pl (eth-al)	71.5	316			i	s	s	...	...		**B9^{2}**, 99
b1304	—,**2-acetamido-**	$C_9H_9NO_3$. *See* b1239	179.17	nd (aa)	185				δ v^h	v	v	v	v	aa v^h	**B14^{2}**, 219
b1305	—,**3-acetamido-**	$C_9H_9NO_3$. *See* b1239	179.17	nd (al)	248–50				i δ^h	v^h	δ	...	...		**B14^{2}**, 241
b1306	—,**4-acetamido-**	$C_9H_9NO_3$. *See* b1239	179.17	nd	256.5				i	s	δ	...	...		**B14^{2}**, 264
b1307	—,**2-acetamido-4-ethoxy-**	$C_{11}H_{13}NO_4$. *See* b1239	223.22	nd (al or MeOH)	199d				δ^h	v^h	δ^h	...	δ^h	chl δ^h MeOH v^h	**B14^{1}**, 657
b1308	—,**5-acetamido-2-ethoxy-**	$C_{11}H_{13}NO_4$. *See* b1239	223.22	nd (w)	190				δ^h	s	...	...	...		**B14**, 583
b1309	—,**2-acetyl-**	$C_9H_8O_3$. *See* b1239	164.15	nd (w), pr (bz)	114–5	$110–2^{2}$			s^h	∞	...	...	...		**B10^{2}**, 479
b1310	—,**4-acetyl-**	$C_9H_8O_3$. *See* b1239	164.15	nd (w)	210				s^h	δ	δ	...	...	chl δ lig i	**B10^{2}**, 480
b1311	—,—,methyl ester	$C_{10}H_{10}O_3$. *See* b1239	178.19	nd (w)	92				s^h	...	...	...	...		**B10**, 695
b1312	—,**5-allyl-2-hydroxy-2-methoxy-**	Eugenic acid. $C_{11}H_{12}O_4$. *See* b1239	208.21	pl (w+1)	85–8 (+1w) 127 (anh)				i s^h	s	s	...	...		**B10^{1}**, 215
b1314	—,**2-amino-**	Anthranilic acid. $C_7H_7NO_2$. *See* b1239	137.13	lf (al)	146		1.412^{20}		i s^h	s v^h	...	...	δ v^h	chl v^h	**B14^{2}**, 205
b1315	—,—,amide	$C_7H_8N_2O$. *See* b1244	136.15	lf (chl or w)	110	300			s^h	s	δ	...	...	AcOEt v	**B14^{2}**, 210
b1316	—,—,butyl ester	$C_{11}H_{15}NO_2$. *See* b1239	193.24		<0	182^{760}			...	...	...	...	...		**B14^{2}**, 209
b1317	—,—,ethyl ester	$C_9H_{11}NO_2$. *See* b1239	165.19		13	268^{760} $145–7^{15}$	1.088^{15}		i	s	s	...	...		**B14^{2}**, 209
b1318	—,—,isobutyl ester	$C_{11}H_{15}NO_2$. *See* b1239	193.24			$156–7^{13.5}$			...	...	...	...	...		**B14^{2}**, 209
b1319	—,—,methyl ester	$C_8H_9NO_2$. *See* b1239	151.16		24.5	256^{760}	1.1682^{19}_{4}		δ	v	v	...	...		**B14^{2}**, 208
b1320	—,—,nitrile	*o*-Cyanoaniline. $C_7H_6N_2$. *See* b1291	118.13	pr (CS_2), nd (peth)	49.5	263^{751}			s	v	v	v	v	Py v chl v peth i CS_2 v	**B14^{2}**, 210
b1321	—,—,phenyl ester	$C_{13}H_{11}NO_2$. *See* b1239	213.23	nd (al)	70				...	v	v	...	...		**B14^{2}**, 210
b1322	—,—,propyl ester	$C_{10}H_{13}NO_2$. *See* b1239	179.21			270^{760}			...	...	...	...	...		**B14^{2}**, 209
b1323	—,**3-amino-**	$C_7H_7NO_2$. *See* b1239	137.13	ye nd	180				δ	δ	δ	...	i		**B14^{2}**, 237
b1324	—,—,amide	$C_7H_8N_2O$. *See* b1244	136.15	ye mcl nd (+1w), nd (bz)	79–80 (+1w) 113–14				s	s	s	...	δ	chl δ	**B14^{1}**, 559
b1326	—,—,ethyl ester	$C_9H_{11}NO_2$. *See* b1239	165.20			294			δ	v	v	...	...		**B14^{2}**, 239

For explanations, symbols and abbreviations see beginning of table.

No.	Name	Synonyms and Formula	Mol. wt.	Crystalline form, color and specific rotation	m.p. °C	b.p. °C	Density	n_D	Solubility w	al	eth	ace	bz	other solvents	Ref.
	Benzoic acid														
b1327	—,—,methyl ester	$C_8H_9NO_2$. *See* b1239	151.17	lf	39		1.232[20]		...	v	v	...	v	chl v lig s peth s	**B14**[2], 238
b1328	—,—,nitrile	*m*-Cyanoaniline. $C_7H_6N_2$. *See* b1291	118.13	nd (dil al or CCl_4)	53	288–90			δ s^h	v	v	v	...	CS_2 s^h chl v	**B14**[2], 240
b1329	—,**4-amino-**	$C_7H_7NO_2$. *See* b1239	137.13	mcl pr	187	1.374^{25}_4			δ	s v^h	s	...	s		**B14**[2], 246
b1330	—,—,amide	$C_7H_8N_2O$. *See* b1244	136.15	(+½w)	183				δ	s	s	...	...		**B14**, 425
b1331	—,—,—,*N*(2-diethylamino-ethyl)-, hydrochloride	Procaine amide hydrochloride. $C_{13}H_{22}ClN_3O$. *See* b1244	271.79		165–9				v	s	i	...	i	chl δ	
b1332	—,—,—,*N*-phenyl-	$C_{13}H_{12}O$. *See* b1244	212.24		135–6				...	s	...	...	...		**B14**, 425
b1333	—,—,butyl ester	Butesin. $C_{11}H_{15}NO_2$. *See* b1239	193.24	(al or bz)	58	173–4[8]			i	s	s	...	s	chl s	**B14**[2], 249
b1334	—,—,—,picrate	$C_{23}H_{33}N_5O_{11}$. *See* b1239	615.58	ye	109–10				...	s	s	...	s	chl s	**B14**[2], 249
b1335	—,.2-diethyl-aminoethyl ester	Procaine. $C_{13}H_{20}N_2O_2$. *See* b1239	236.30	nd (w+2), pl (lig or eth)	51 (+2w) 61				δ	s	s	...	s	chl s	
b1336	—,—,—,hydro-chloride	Procaine hydrocloride. $C_{13}H_{21}ClN_2O_2$. *See* b1239	272.78	nd (al)	156		0.707[17]		v	s	i	...	...	chl δ	**B14**[2], 251
b1337	—,—,ethyl ester	Benzocaine. $C_9H_{11}NO_2$. *See* b1239	165.19	nd (w)	90	310			i	v	v	...	...	chl s	**B14**[2], 248
b1338	—,—,methyl ester	$C_8H_9NO_2$. *See* b1239	151.16	lf or nd	114				...	...	...	...	...		**B14**[2], 247
b1339	—,—,nitrile	$C_7H_6N_2$. *See* b1291	118.13	pr or pl	86				δ v^h	v	v	v	v	chl v lig δ CS_2δ CCl_4 δ	**B14**[1], 570
b1340	—,—,phenyl ester	$C_{13}H_{11}NO_2$. *See* b1239	213.23	nd	173				...	...	...	...	...		**B14**[1], 568
b1341	—,—,propyl ester	$C_{10}H_{13}NO_2$. *See* b1239	179.21		75				δ	v	v	...	v	chl v	**B14**[2], 248
b1342	—,—,2,2,2-trichlo-roethyl ester	$C_9H_8Cl_3NO_2$. *See* b1239	268.54	nd (peth)	87				...	...	...	...	...		
b1343	—,**2-amino-3,4-dichloro-**	$C_7H_5Cl_2NO_2$. *See* b1239	206.04	nd (aa)	240–2				δ	v	...	...	δ	chl s aa δ s^h	**B14**[2], 231
b1344	—,**2-amino-3,5-dichloro-**	$C_7H_5Cl_2NO_2$. *See* b1239	206.04	nd or lf	231–2d				i	v	...	...	...	os v	**B14**[1], 549
b1345	—,**2-amino-3,6-dichloro-**	$C_7H_5Cl_2NO_2$. *See* b1239	206.04	nd (w or aa)	155				v^h	v	...	...	v^h	aa s^h	**B14**, 367
b1346	—,**2-amino-4,5-dichloro-**	$C_7H_5Cl_2NO_2$. *See* b1239	206.04	nd (aa)	213–4				δ	s	...	...	...	aa s	**B14**[1], 549
b1347	—,**4-amino-3,5-dichloro-**	$C_7H_5Cl_2NO_2$. *See* b1239	206.04	(al)	291				i	δ s^h	δ	...	δ	chl δ aa δ peth δ	**B14**[2], 271
b1348	—,**6-amino-2,3-dichloro-**	$C_7H_5Cl_2NO_2$. *See* b1239	206.04	nd (MeOH)	176–7d				...	v	v	...	...	MeOH v aa v	**B14**, 368
b1349	—,**2-amino-3,5-diiodo-**	$C_7H_5I_2NO_2$. *See* b1239	388.94	pr (al)	232–3				i^h	δ	v	...	δ		**B14**[2], 233
b1350	—,—,ethyl ester	$C_9H_9I_2NO_2$. *See* b1239	417.00	pr (al)	101				...	δ s^h	...	...	...		**B14**[1], 554
b1351	—,**2-amino-4,5-diiodo-**	$C_7H_5I_2NO_2$. *See* b1239	388.94		210–20d				$δ^h$	s	s	...	s		**B14**[1], 555
b1352	—,—,ethyl ester	$C_9H_9I_2NO_2$. *See* b1239	417.00	pr (al)	137				$δ^h$	s^h	s	...	s		**B14**[1], 555
b1353	—,**4-amino-3,5-diiodo-**	$C_7H_5I_2NO_2$. *See* b1239	388.94	nd	>350				i	i	...	...	...	$PhNO_2$ δ	**B14**, 439
b1354	—,—,ethyl ester	$C_9H_9I_2NO_2$. *See* b1239	417.00	nd (al)	148				...	δ	...	...	...		**B14**, 439
b1355	—,**2-amino-3-hydroxy-**	$C_7H_7NO_3$. *See* b1239	153.14	lf (w)	164				s^h	s	s	...	...	chl s	**B14**[2], 355
b1356	—,**2-amino-4-hydroxy-**	$C_7H_7NO_3$. *See* b1239	153.14		148d				δ	s^h	...	...	s^h		**B14**[2], 359
b1357	—,**2-amino-5-hydroxy-**	$C_7H_7NO_3$. *See* b1239	153.14	pr	252				s^h	...	...	...	...	os δ	**B14**[2], 357
b1358	—,**3-amino-2-hydroxy-**	$C_7H_7NO_3$. *See* b1239	153.14		253d				i	i	i	...	...		**B14**[2], 350
b1359	—,**3-amino-4-hydroxy-**	$C_7H_7NO_3$. *See* b1239	153.14	pr (w+1)	210				s^h	$δ^h$	i	...	i	chl i aa s^h	**B14**, 593
b1360	—,—,methyl ester	$C_8H_9NO_3$. *See* b1239	167.16	st: nd (bz or aa) unst: (chl)	st 143 unst 111				i	v	δ	...	δ	ac s alk s	**B14**[2], 360
b1361	—,**4-amino-2-hydroxy-**	$C_7H_7NO_3$. *See* b1239	153.14	nd, pl (al-eth)	150–1				δ	s	δ	δ	i		**B14**, 579
b1362	—,**4-amino-3-hydroxy-**	$C_7H_7NO_3$. *See* b1239	153.14	lf (dil al)	226				δ	v	...	v	...		**B14**[2], 356
b1363	—,—,methyl ester	Orthoform $C_8H_9NO_3$. *See* b1239	167.16	lf	142				δ	v	v	...	...	aa v	**B14**[2], 356
b1364	—,**5-amino-2-hydroxy-**	$C_7H_7NO_3$. *See* b1239	153.14	(w)	283				$δ^h$	i	...	...	...		**B14**[2], 352
b1365	—,**2-amino-4-iodo-**	$C_7H_6INO_2$. *See* b1239	263.04	pr (dil al)	208d				δ	v	v	...	δ		**B14**[1], 554

For explanations, symbols and abbreviations see beginning of table.

No.	Name	Synonyms and Formula	Mol. wt.	Crystalline form, color and specific rotation	m.p. °C	b.p. °C	Density	n_D	Solubility						Ref.
									w	al	eth	ace	bz	other solvents	
	Benzoic acid														
b1366	—,**2-amino-5-iodo-**	$C_7H_6INO_2$. *See* b1239	263.04	nd or pr (dil al)	210.5d				δ	v				os v peth δ	**B14**^{2}, 23
b1367	—,**2-amino-3-methyl-**	$C_8H_9NO_2$. *See* b1239	151.17	nd (al), lf (w)	172				δ	v	v				**B14**^{2}, 29
b1368	—,**2-amino-4-methyl-**, nitrile	$C_8H_8N_2$. *See* b1291	132.17	lf	136				i	s		s	s	chl s	**B14**^{2}, 29
b1369	—,**2-amino-5-methyl-**	$C_8H_9NO_2$. *See* b1239	151.17	pl (al), nd (w)	175				δ	s	s				**B14**^{2}, 29
b1370	—,—,nitrile	$C_8H_8N_2$. *See* b1291	132.17	(dil al)	63				s^h	s				os s lig δ	**B14**, 48
b1371	—,**2-amino-6-methyl-**	$C_8H_9NO_2$. *See* b1239	151.17	nd (MeOH)	125–6d									MeOH s^h	**B14**^{2}, 29
b1372	—,—,nitrile	$C_8H_8N_2$. *See* b1291	132.17	ye pr (bz), (w)	131				δ s^h				δ		**B14**^{2}, 29
b1373	—,**3-amino-2-methyl-**, nitrile	$C_8H_8N_2$. *See* b1291	132.17	nd (w)	95.5				δ						**B14**, 47
b1374	—,**3-amino-4-methyl**	$C_8H_9NO_2$. *See* b1239	151.17	nd	164–5				s						**B14**, 48
b1375	—,—,nitrite	$C_8H_8N_2$. *See* b1291	132.17	pr (al)	81–2				δ	s				os v	**B14**, 48
b1376	—,**3-amino-5-methyl-**, nitrile	$C_8H_8N_2$. *See* b1291	132.17	nd (lig)	75					s				lig δ	**B14**^{1}, 60
b1377	—,**4-amino-2-methyl-**	$C_8H_9NO_2$. *See* b1239	151.17	nd (al)	153					s^h					**B14**, 47
b1378	—,—,nitrile	$C_8H_8N_2$. *See* b1291	132.17	(al)	90					s^h					**B14**^{1}, 59
b1379	—,**4-amino-3-methyl**	$C_8H_9NO_2$. *See* b1239	151.17	nd (w)	170				v^h						**B14**^{2}, 29
b1380	—,—,nitrile	$C_8H_8N_2$. *See* b1291	132.17	nd (w)	95				s^h						**B14**^{1}, 60
b1381	—,**5-amino-2-methyl**·	$C_8H_9NO_2$. *See* b1239	151.17	pr (w)	196				s^h	v^h					**B14**^{2}, 29
b1382	—,—,nitrile	$C_8H_8N_2$. *See* b1291	132.17	nd (peth)	88	100–10^{22}			$δ^h$	s				peth $δ^h$	**B14**^{2}, 29
b1383	—,**4-aminomethyl-**, nitrile, hydrochloride.	$C_8H_9ClN_2$. *See* b1291	168.63	pl (al)	274					δ					**B14**, 48
b1384	—,**2-amino-3-nitro-**	$C_7H_6N_2O_4$. *See* b1239	182.13	ye nd (w)	208–9				i	v	v		δ	chl δ	**B14**^{2}, 23
b1385	—,**2-amino-4-nitro-**	$C_7H_6N_2O_4$. *See* b1239	182.13	og pr	269				i $δ^h$	v	v				**B14**^{2}, 23
b1386	—,**2-amino-5-nitro-**	$C_7H_6N_2O_4$. *See* b1239	182.13	lf (al), nd (w)	278				i s^h	δ	δ		i	chl i	**B14**^{2}, 23
b1387	—,**2-amino-6-nitro-**	$C_7H_6N_2O_4$. *See* b1239	182.13	ye nd or lf (w)	184				v^h	v	v	v	δ	chl δ aa v	**B14**^{1}, 557
b1388	—,**3-amino-2-nitro-**	$C_7H_6N_2O_4$. *See* b1239	182.13	ye nd (w)	156–7				δ s^h	s	s			aa s^h lig i	**B14**, 414
b1389	—,**3-amino-4-nitro-**	$C_7H_6N_2O_4$. *See* b1239	182.13	red pl or nd (al)	298d				δ	s	s				**B14**, 415
b1390	—,**3-amino-5-nitro-**	$C_7H_6N_2O_4$. *See* b1239	182.13	ye pr	209–10				δ	s^h	δ		s	CS_2 δ aa v^h	**B14**, 415
b1391	—,**4-amino-2-nitro-**	$C_7H_6N_2O_4$. *See* b1239	182.13	red nd (w), pr (dil aa)	239.5d				s^h	s	δ		δ	aa s	**B14**^{1}, 583
b1392	—,**4-amino-3-nitro-**	$C_7H_6N_2O_4$. *See* b1239	182.13	ye nd	284d				i	δ				aa s	**B14**^{1}, 583
b1393	—,**5-amino-2-nitro-**	$C_7H_6N_2O_4$. *See* b1239	182.13	ye nd or pr (w)	235d				$δ^h$	s^h	δ				**B14**^{2}, 245
b1394	—,**3-amino-2,4,6-tribromo-**	$C_7H_4Br_3NO_2$. *See* b1239	373.87	nd (w)	170.5				s^h	v					**B14**, 413
b1395	—,**2-benzamido-**	$C_{14}H_{11}NO_3$. *See* b1239	241.24	nd (al or bz)	181				i	v	v				**B14**^{2}, 221
b1396	—,**3-benzamido-**	$C_{14}H_{11}NO_3$. *See* b1239	241.24	red pr (al)	248		1.5105$^{4}_{4}$		δ	s	δ				**B14**^{1}, 562
b1397	—,**4-benzamido-**	$C_{14}H_{11}NO_3$. *See* b1239	241.24	nd (al)	278				$δ^h$	s	s			aa s	**B14**^{1}, 557
b1398	—,**2-benzoyl-**	2-Benzophenonecarboxylic acid. $C_{14}H_{10}O_3$. *See* b1239	226.22	tcl nd (w)	127–8, 93.4 (+w)					v	v		s^h		**B10**^{2}, 517
b1399	—,—,amide	$C_{14}H_{11}NO_2$. *See* b1244	225.25	nd (to)	165				i	v			v	to s^h	**B10**, 749
b1400	—,—,ethyl ester	$C_{16}H_{14}O_3$. *See* b1239	254.27	rh pl	58		1.221$^{64}_{4}$	1.560^{64}	i	v	v			sulf s	**B10**^{2}, 517
b1401	—,—,methyl ester	$C_{15}H_{12}O_3$. *See* b1239	240.26	pl or mcl pr	52	350–2	1.1903$^{19}_{4}$	1.591^{20}	i					sulf s	**B10**^{2}, 517
b1402	—,**3-benzoyl-**	3-Benzophenonecarboxylic acid. $C_{14}H_{10}O_3$. *See* b1239	226.23	nd (w), fl (dil al)	161–2	sub			δ s^h	s	s		δ	to δ	**B10**^{2}, 521
b1403	—,**4-benzoyl-**	4-Benzophenonecarboxylic acid. $C_{14}H_{10}O_3$. *See* b1239	226.23	nd (dil aa), mcl lf (w)	194–5	sub			δ	s	s		δ	chl δ aa s	**B10**^{2}, 521
b1404	—,**2-benzoyl-4-chloro-**	$C_{14}H_9ClO_3$. *See* b1239	260.68	(xyl)	170					s	s			chl s	**B10**, 356
b1405	—,**2-benzoyl-3,4,5,6-tetrachloro-**, methyl ester	$C_{15}H_8Cl_4O_3$. *See* b1239	378.04	nd (MeOH)	92				i	s^h				MeOH s^h	**B10**, 750

For explanations, symbols and abbreviations see beginning of table.

No	Name	Synonyms and Formula	Mol. wt.	Crystalline form, color and specific rotation	m.p. °C	b.p. °C	Density	n_D	Solubility w	al	eth	ace	bz	other solvents	Ref.
	Benzoic acid														
b1406	**—,2-benzyl-**	$C_{14}H_{12}O_2$. *See* b1239	212.25	nd (dil al)	118				δ	s	s		s	chl s	**B9**[2], 471
b1407	**—,3-benzyl**	$C_{14}H_{12}O_2$. *See* b1239	212.25	nd (w), lf (al)	108	sub			δ	v	v			chl v	**B9**, 676
b1408	**—,4-benzyl-**	$C_{14}H_{12}O_2$. *See* b1239	212.25	nd (w), lf (dil al)	157–8	sub			δ	s	s		s	chl s	**B9**[1], 284
b1409	**—,4-(benzyl-sulfonamido)-**	[benzene ring]—CH_2SO_2NH—[benzene ring]—CO_2H	291.33		229–30				δ	s			δ	chl δ aa s	
b1410	**—,2-bromo-**	$C_7H_5BrO_2$. *See* b1239	201.03	mcl pr (w)	150	sub	1.929^{25}_{4}		δ s^h	s	s	s		chl s	**B9**[2], 230
b1411	—,—,amide	C_7H_6BrNO. *See* b1244	200.04	nd (w)	155				δ		δ				**B9**, 350
b1412	—,—,chloride	C_7H_4BrClO. *See* b1239	219.47	nd	11	245^{760}			d	d^h					**B9**[2], 231
b1413	—,—,ethyl ester	$C_9H_9BrO_2$. *See* b1239	229.08			254–5 135^{15}	1.4438^{15}_{4}	1.5455^{15}		s				os s	**B9**, 348
b1414	—,—,methyl ester	$C_8H_7BrO_2$. *See* b1239	215.05			246–7 122^{17}			i	s					**B9**, 348
b1415	—,—,nitrile	C_7H_4BrN. *See* b1291	182.03	nd (w)	53	$251–3^{754}$			v^h	v^h					**B9**[2], 232
b1416	—,—,piperazinium salt	$C_{18}H_{20}Br_2N_2O_4$. *See* b1239	488.19		227–30				$δ^h$	s^h	i				
b1417	**—,3-bromo-**	$C_7H_5BrO_2$. *See* b1239	201.03	mcl nd	155	>280	1.845^{20}		i	s	s				**B9**[2], 232
b1418	—,—,amide	C_7H_6BrNO. *See* b1244	200.04	pl (w or al)	155.3				δ	v					**B9**[2], 232
b1419	—,—,ethyl ester	$C_9H_9BrO_2$. *See* b1239	229.08			261^{760} 131^{17}	1.4308^{19}_{4}	1.5430^{19}						os s	**B9**[2], 232
b1420	—,—,methyl ester	$C_8H_7BrO_2$. *See* b1239	215.05	pl	31–2	122.5^{15}				s					**B9**[1], 143
b1421	—,—,nitrile	C_7H_4BrN. *See* b1291	182.03	(al)	38	225				v	v				**B9**[2], 233
b1422	—,—,piperazinium salt	$C_{18}H_{20}Br_2N_2O_4$. *See* b1239	488.19		169–71				$δ^h$	s^h	i				
b1423	**—,4-bromo-**	$C_7H_5BrO_2$. *See* b1239	201.03	nd (eth), lf (w), mcl pr	254.5	1.894^{20}			δ s^h	δ	s				**B9**[2], 233
b1424	—,—,amide-	C_7H_6BrNO. *See* b1244	200.04	nd or pl (w)	189.5				s	v	δ		s	CS_2 s lig s	**B9**, 353
b1425	—,—,chloride	C_7H_4BrClO. *See* b1239	219.47	nd (peth)	42	245–7 $117–20^{15}$			d	v	v		v	lig v	**B9**[2], 235
b1426	—,—,ethyl ester	$C_9H_9BrO_2$. *See* b1239	229.08			262^{737} 129^{15}	1.4332^{17}_{4}	1.5438^{17}		s				os s	**B9**[2], 234
b1427	—,—,methyl ester	$C_8H_7BrO_2$. *See* b1239	215.05	lf (dil al), nd (eth)	79–81		1.689			s	s	s	v	chl v peth s	**B9**[2], 234
b1428	—,—,nitrile	C_7H_4BrN. *See* b1291	182.03	nd (w or al)	113	235–7			s^h	s^h	s				**B9**[2], 236
b1429	—,—,piperazinium salt	$C_{18}H_{20}Br_2N_2O_4$. *See* b1239	488.19		224–6				$δ^h$	s^h	i				
b1430	**—,2-bromo-3-chloro-**	$C_7H_4BrClO_2$. *See* b1239	235.47	(bz)	143–4								v^h		**B9**[2], 236
b1431	**—,2-bromo-5-chloro-**	$C_7H_4BrClO_2$. *See* b1239	235.47	(bz)	153					v			v^h		**B9**, 355
b1432	**—,3-bromo-2-chloro-**	$C_7H_4BrClO_2$. *See* b1239	235.47	(al)	165					v					**B9**, 356
b1433	**—,3-bromo-4-chloro-**	$C_7H_4BrClO_2$. *See* b1239	235.47	pl (dil aa), (al)	215–6					s				aa v	**B9**[2], 236
b1434	**—,4-bromo-3-chloro-**	$C_7H_4BrClO_2$. *See* b1239	235.47	pl (dil aa), (al)	219					v				aa v^h	**B9**[2], 236
b1435	**—,5-bromo-2-chloro-**	$C_7H_4BrClO_2$. *See* b1239	235.47	nd (w), (aa)	155–6				δ	v				aa v^h	**B9**, 356
b1436	**—,3-bromo-2,4-dihydroxy-**	$C_7H_5BrO_4$. *See* b1239	233.03	br or ye nd (w)	202				s^h						**B10**[2], 254
b1437	**—,5-bromo-2,3-dihydroxy-**	$C_7H_5BrO_4$. *See* b1239	233.03	pr (w+1), nd	187 (pr), 215 (nd)				δ s^h	v	v		δ	aa v	**B10**[1], 175
b1438	**—,5-bromo-2,4-dihydroxy-**	$C_7H_5BrO_4$. *See* b1239	233.03	micr pr (w+1)	212				δ s^h	v	v				**B10**[2], 254
b1439	**—,5-bromo-3,4-dihydroxy-**	$C_7H_5BrO_4$. *See* b1239	233.03	nd (w or aa)	227–9				s^h					aa s^h	**B10**[1], 192
b1440	**—,2-bromo-3,5-dinitro-**	$C_7H_3BrN_2O_6$. *See* b1239	291.03	nd (w)	213					v			s	aa v lig s	**B9**[2], 284
b1441	**—,2-bromo-3-hydroxy-**	$C_7H_5BrO_3$. *See* b1239	217.03	nd (w)	160–1				δ v^h						**B10**[2], 83
b1442	**—,2-bromo-4-hydroxy-**	$C_7H_5BrO_3$. *See* b1239	217.03	nd (w)	151				δ v^h						**B10**[2], 103
b1443	**—,2-bromo-5-hydroxy-**	$C_7H_5BrO_3$. *See* b1239	217.03	(w)	185d				δ s^h						**B10**[2], 84
b1444	**—,3-bromo-2-hydroxy-**	$C_1H_5BrO_3$. *See* b1239	217.03	nd (dil al)	184.5				$δ^h$	v	v	v	δ	MeOH v, chl δ	**B10**[2], 63
b1445	**—,3-bromo-4-hydroxy-**	$C_7H_5BrO_3$. *See* b1239	217 03	nd or pr+w (w)	177				δ v^h	v	v			aa v	**B10**[2], 103
b1446	**—,4-bromo-2-hydroxy-**	$C_7H_5BrO_3$. *See* b1239	217.03	pl, nd (w)	214				δ v^h	v					**B10**[2], 63

For explanations, symbols and abbreviations see beginning of table.

No.	Name	Synonyms and Formula	Mol. wt.	Crystalline form, color and specific rotation	m.p. °C	b.p. °C	Density	n_D	Solubility						Ref.
									w	al	eth	ace	bz	other solvents	
	Benzoic acid														
b1447	**—,4-bromo-3-hydroxy-**	$C_7H_5BrO_3$. *See* b1239	217.03	nd (w)	227				δ s^h						**B10**^{2}, 83
b1448	**—,5-bromo-2-hydroxy-**	$C_7H_5BrO_3$. *See* b1239	217.03	nd (w or dil al)	165–6	sub >100			v^h	v	v				**B10**^{2}, 63
b1449	**—,2-bromo-3-nitro-**	$C_7H_4BrNO_4$. *See* b1239	246.03	(dil al)	191				δ s^h	v					**B9**^{2}, 277
b1450	**—,2-bromo-4-nitro-**	$C_7H_4BrNO_4$. *See* b1239	246.03	nd (w or dil al)	166–7	sub >155			δ s^h	v	v				**B9**^{2}, 277
b1451	**—,2-bromo-5-nitro-**	$C_7H_4BrNO_4$. *See* b1239	246.03	nd (w)	180–1	sub			δ s^h	v	v			chl v	**B9**^{2}, 277
b1452	**—,3-bromo-2-nitro-**	$C_7H_4BrNO_4$. *See* b1239	246.03	(eth)	250				δ		v^h		v^h		**B9**^{2}, 276
b1453	**—,3-bromo-4-nitro-**	$C_7H_4BrNO_4$. *See* b1239	246.03	nd	197				δ	v	v			chl v	**B9**, 408
b1454	**—,3-bromo-5-nitro-**	$C_7H_4BrNO_4$. *See* b1239	246.03	nd (w, bz or eth), pl (al)	159–60				δ s^h	v	v		s	CS_2 s aa v chl s	**B9**^{2}, 277
b1455	**—,4-bromo-2-nitro-**	$C_7H_4BrNO_4$. *See* b1239	246.03	nd (w)	163				s^h	v	v		v	chl v	**B9**^{2}, 276
b1456	**—,4-bromo-3-nitro-**	$C_7H_4BrNO_4$. *See* b1239	246.03	(dil aa)	203	sub			i	v					**B9**^{2}, 277
b1457	**—,5-bromo-2-nitro-**	$C_7H_4BrNO_4$. *See* b1239	246.03	(w or al)	140		1.920^{18}		s^h	s			v^h		**B9**, 406
b1458	**—,2-*tert*-butyl**	$C_{11}H_{14}O_2$. *See* b1239	178.23	pl (dil al)	68.5				i	v					**B9**^{2}, 365
b1459	**—,3-*tert*-butyl-**	$C_{11}H_{14}O_2$. *See* b1239	178.23	nd (peth)	127					v				peth v^h	**B9**, 560
b1460	**—,4-*tert*-butyl-**	$C_{11}H_{14}O_2$. *See* b1239	178.23	nd (dil al)	164				i	v			v		**B9**^{2}, 365
b1461	**—,4-butylamino-2-dimethylamino,** ethyl ester, hydrochloride	Tetracaine hydrochloride. $C_{15}H_{25}ClN_2O_2$. *See* b1239	300.83		147–50				v	s	i		i		
—	**—,(carboxymethoxy)-**	*see* **Acetic acid, (carboxyphenoxy)-**													
b1462	**—,4(4-carboxybenzoyl)-**	*p,p'*-Benzophenonedicarboxylic acid. $C_{15}H_{10}O_5$. *See* b1239	270.24	nd (al)	>360				i	δ	δ	δ	s	aa δ	**B10**^{2}, 618
b1465	**—,2-chloro-**	$C_7H_5ClO_2$. *See* b1239	156.57	mcl pr	142	sub			s^h	v	v		s	CS_2 δ peth s^h	**B9**^{2}, 221
b1466	—,—,amide	C_7H_6ClNO. *See* b1244	155.58	rh nd (w)	142.4				s^h	s	s				**B9**, 336
b1467	—,—,anhydride	$C_{14}H_8Cl_2O_3$. *See* b1239	295.12	nd (al)	79.6					v			v	chl v lig δ	**B9**^{2}, 223
b1468	—,—,chloride	$C_7H_4Cl_2O$. *See* b1239	175.02		−4	235–8			d	d					**B9**^{2}, 223
b1469	—,—,ethyl ester	$C_9H_9ClO_2$. *See* b1239	184.62			243–4 125^{20}	1.1942$^{15}_{4}$	1.5247^{15}	i	s	v				**B9**, 336
b1470	—,—,methyl ester	$C_8H_7ClO_2$. *See* b1239	170.60			234–5^{762}				s					**B9**^{2}, 222
b1471	—,—,nitrile	C_7H_4ClN. *See* b1291	137.57	nd	42–3	232			δ^h	s	s				**B9**^{2}, 223
b1472	—,—,piperazinium salt	$C_{18}H_{20}Cl_2N_2O_4$. *See* b1239	399.28		217–8d				s	s	s				
b1473	**—,3-chloro-**	$C_7H_5ClO_2$. *See* b1239	156.57	pr	155	sub	1.496$^{25}_{4}$		v^h	s	v		δ	CS_2 δ lig δ	**B9**^{2}, 223
b1474	—,—,amide	C_7H_6ClNO. *See* b1244	155.58	nd	134.5				δ s^h	s	s				**B9**, 338
b1475	—,—,anhydride	$C_{14}H_8Cl_2O_3$. *See* b1239	295.12	nd (al or peth), lf (bz)	95.5					v			v	peth s^h chl v lig δ	**B9**^{2}, 224
b1476	—,—,chloride	$C_7H_4Cl_2O$. *See* b1239	175.02			222–5			d	d					**B9**^{2}, 224
b1477	—,—,ethyl ester	$C_9H_9ClO_2$. *See* b1239	184.62			239–42 121^{20}	1.1859$^{16}_{4}$	1.5223^{15}	i	s	s				**B9**^{1}, 139
b1478	—,—,methyl ester	$C_8H_7ClO_2$. *See* b1239	170.60		21	231^{763} 114^{18}				s					**B9**, 338
b1479	—,—,nitrile	C_7H_4NCl. *See* b1239	137.57	nd	39–40				i	s	s				**B9**^{2}, 225
b1480	**—,4-chloro-**	$C_7H_5ClO_2$. *See* b1239	156.57	tcl pr	241.5				i	v	δ	δ	i	lig i	**B9**^{2}, 225
b1481	—,—,amide	C_7H_6ClNO. *See* b1244	155.58	nd (eth)	179–80				s^h	v	v				**B9**^{2}, 227
b1482	—,—,anhydride	$C_{14}H_8Cl_2O_3$. *See* b1239	295.12	nd or lf (bz)	193–4					δ	i		δ	CS_2 i chl δ peth i	**B9**^{2}, 227
b1483	—,—,chloride	$C_7H_4Cl_2O$. *See* b1239	175.02		14–6	220–2	1.377								**B9**^{2}, 227
b1484	—,—,ethyl ester	$C_9H_9ClO_2$. *See* b1239	184.62			237–8 122^{15}				s					**B9**^{1}, 140
b1485	—,—,methyl ester	$C_8H_7ClO_2$. *See* b1239	170.60	nd or mcl pr	42–4		1.382^{20}			s					**B9**^{2}, 226
b1486	—,—,nitrile	C_7H_4ClN. *See* b1291	137.57	nd (al)	94–6	223^{750} 95^{5}			δ^h	s	s		s	chl s lig δ	**B9**^{2}, 288
b1487	—,—,piperazinium salt	$C_{18}H_{20}Cl_2N_2O_4$. *See* b1239	399.28		219–20d				s	s	δ				
b1488	**—,2(4-chlorobenzoyl)-**	$C_{14}H_9ClO_3$. *See* b1239	260.68		150				δ^h	s	s		s		**B10**^{2}, 518

For explanations, symbols and abbreviations see beginning of table.

No.	Name	Synonyms and Formula	Mol. wt.	Crystalline form, color and specific rotation	m.p. °C	b.p. °C	Density	n_D	Solubility						Ref.
									w	al	eth	ace	bz	other solvents	
	Benzoic acid														
b1489	**—,2-chloro-3-hydroxy-**	$C_7H_5ClO_3$. *See* b1239	172.57	lf (w or bz)	157.5–8.5				δ v^h				v^h		**B10**, 142
b1490	**—,2-chloro-4-hydroxy-**	$C_7H_5ClO_3$ *See* b1239	172.57	nd (w)	159				δ v^h						
b1491	**—,2-chloro-5-hydroxy-**	$C_7H_5ClO_3$. *See* b1239	172.57	(w)	178–9				δ	v^h					**B10**, 143
b1492	**—,2-chloro-6-hydroxy-**	$C_7H_5ClO_3$. *See* b1239	172.57	nd (w)	166				s					os s	**B10**, 104
b1493	**—,3-chloro-2-hydroxy-**	$C_7H_5ClO_3$. *See* b1239	172.57	nd (w)	178–80				δ	s				chl s	**B10²**, 61
b1494	—,—,ethyl ester	$C_9H_9ClO_3$. *See* b1239	200.62	nd	21	269–70				s					**B10**, 101
b1495	—,—,methyl ester	$C_8H_7ClO_3$. *See* b1239	186.60	nd (MeOH or al)	38	259–60d									**B10**, 101
b1496	**—,3-chloro-4-hydroxy-**	$C_7H_5ClO_3$. *See* b1239	172.57	nd (w)	169–70				δ v^h	v	v	v	δ	chl δ lig δ	**B10**, 175
b1497	**—,4-chloro-2-hydroxy-**	$C_7H_5ClO_3$. *See* b1239	172.57	nd (w)	206–7				δ	s			s	chl s	**B10²**,161
b1498	**—,4-chloro-3-hydroxy-**	$C_7H_5ClO_3$. *See* b1239	172.57	nd (w)	219.5–20.5				δ v^h						
b1499	**—,5-chloro-2-hydroxy-**	$C_7H_5ClO_3$. *See* b1239	172.57	nd (w or al)	173–4				v	v			v	chl v	**B10²**, 62
b1500	—,—,ethyl ester	$C_9H_9ClO_3$. *See* b1239	200.62	nd (al)	25					s^h					**B10**, 103
b1501	—,—,methyl ester	$C_8H_7ClO_3$. *See* b1239	186.60	nd (al)	48	249d				v					**B10**, 103
b1502	**—,5-chloro-2-methoxy-**	$C_8H_7ClO_3$. *See* b1239	186.60	nd (w)	81–2				v						**B10**, 103
b1503	**—,2-chloro-4-methyl-**	$C_8H_7ClO_2$. *See* b1239	170.60	nd (al)	155				δ v^h	v	v		v^h	chl v	**B9**, 497
b1504	**—,2-chloro-5-methyl-**	$C_8H_7ClO_2$. *See* b1239	170.60	nd (w or al)	167				v^h	v^h					**B9**, 479
b1505	**—,2-chloro-6-methyl-**	$C_8H_7ClO_2$. *See* b1239	170.60	nd (w)	108				δ^h						
b1506	**—,3-chloro-2-methyl-**	$C_8H_7ClO_2$. *See* b1239	170.60	nd (al)	159					v	s				**B9**, 467
b1507	**—,3-chloro-4-methyl-**	$C_8H_7ClO_2$. *See* b1239	170.60	nd (dil al)	200–2				δ^h	v					**B9**, 498
b1508	**—,3-chloro-5-methyl-**	$C_8H_7ClO_2$. *See* b1239	170.60	nd (dil al)	178				δ	v					**B9**, 479
b1509	**—,4-chloro-2-methyl-**	$C_8H_7ClO_2$. *See* b1239	170.60	nd (w, al, dil aa, bz)	172				v^h	v^h			δ v^h	aa v	**B9**, 468
b1510	**—,4-chloro-3-methyl-**	$C_8H_7ClO_2$. *See* b1239	170.60	nd (w)	209.5				i δ^h						**B9**, 478
b1511	**—,5-chloro-2-methyl-**	$C_8H_7ClO_2$. *See* b1239	170.60	nd (al)	168.5–9.5					v^h					
b1512	**—,3-chloro-methyl-**, nitrile	C_8H_6ClN. *See* b1291	151.60	pr (al)	67	258–60[760]									**B9**, 479
b1513	**—,4-chloro-methyl-**, nitrile	C_8H_6ClN. *See* b1291	151.60	pr (al)	79.5	263[756]									**B9**, 498
b1514	**—,2-chloro-4-nitro-**	$C_7H_4ClNO_4$. *See* b1239	201.57	nd (w)	139–41				s^h				s^h		**B9²**, 276
b1515	**—,2-chloro-5-nitro-**	$C_7H_4ClNO_4$. *See* b1239	201.57	nd or pr (w)	164–5		1.608^{18}		δ	s	s		s		**B9²**, 276
b1516	**—,3-chloro-2-nitro-**	$C_7H_4ClNO_4$. *See* b1239	201.57	nd or pl	235		1.566^{18}		δ^h		s		i		**B9**, 400
b1517	**—,3-chloro-5-nitro-**	$C_7H_4ClNO_4$. *See* b1239	201.57	nd	147				δ	s	s			aa s	**B9**, 403
b1518	**—,4-chloro-2-nitro-**	$C_7H_4ClNO_4$. *See* b1239	201.57	pl, pr, nd	140–2				s^h				s^h		**B9²**, 274
b1519	—,—,nitrile	$C_7H_3ClN_2O_2$. *See* b1291	182.57	nd	100–1				v^h	s	s				**B9²**, 275
b1520	**—,4-chloro-3-nitro-**	$C_7H_4ClNO_4$. *See* 1239	201.57	nd or pl	180–2		1.645^{18}		i s^h	δ					**B9²**, 275
b1521	—,—,ethyl ester	$C_9H_8ClNO_4$. *See* b1239	229.62	ye nd	59					v			s	aa s	**B9**, 402
b1522	—,—,methyl ester	$C_8H_6ClNO_4$. *See* b1239	215.60	nd	83		1.522^{18}								**B9**, 402
b1523	**—,5-chloro-2-nitro-**,methyl ester	$C_8H_6ClNO_4$. *See* b1239	215.60	pl	48.5		1.453^{18}							MeOH v	**B9**, 401
b1524	**—,2,3-diamino-**	$C_7H_8N_2O_2$. *See* b1239	152.15	nd	190–1	d			δ	v				aa v	**B14¹**, 585
b1525	**—,2,4-diamino-**	$C_7H_8N_2O_2$. *See* b1239	152.15		140	>200d			s^h	v				aa v	**B14**, 448
b1526	**—,2,5-diamino-**	$C_7H_8N_2O_2$. *See* b1239	152.15	br pr (w)	200d				δ	δ	δ				**B14**, 448
b1527	**—,3,4-diamino-**	$C_7H_8N_2O_2$. *See* b1239	152.15	lf	210–11d				δ s^h						**B14¹**, 586
b1528	**—,3,5-diamino-**	$C_7H_8N_2O_2$. *See* b1239	152.15	nd (+1w)	228				δ s^h	s	v				**B14**, 453
b1529	**—,3-diazo-2-hydroxy-**	$C_7H_4N_2O_3$. *See* b1239	164.12	ye nd (ace)	155d							s			**B16**, 553
b1530	**—,2,3-dibromo-**	$C_7H_4Br_2O_2$. *See* b1239	279.93	nd (w)	149–50				δ^h					lig δ s^h	**B9¹**, 146
b1531	**—,2,4-dibromo-**	$C_7H_4Br_2O_2$. *See* b1239	279.93	lf (w)	174	sub			δ	s	s				**B9²**, 237

For explanations, symbols and abbreviations see beginning of table.

No.	Name	Synonyms and Formula	Mol. wt.	Crystalline form, color and specific rotation	m.p. °C	b.p. °C	Density	n_D	Solubility w	al	eth	ace	bz	other solvents	Ref.
	Benzoic acid														
b1532	**—,2,5-dibromo-**	$C_7H_4Br_2O_2$. *See* b1239	279.93	nd (al or w)	153	sub	...	...	s^h	s	s	...	...	chl s aa s	**B9**[2], 237
b1533	**—,2,6-dibromo-**	$C_7H_4Br_2O_2$. *See* b1239	279.93	nd (w), (lig)	150–1	335 209–10[16]	...	...	s^h	v	v	s	...	peth δ chl v	**B9**[2], 237
b1534	**—,3,4-dibromo-**	$C_7H_4Br_2O_2$. *See* b1239	279.93	nd (w), pl (al)	232–4	...	...	...	s^h	s	s	...	...	MeOH s	**B9**[2], 237
b1535	**—,3,5-dibromo-2,4-dihydroxy-6-methyl-**,ethyl ester	$C_{10}H_{10}Br_2O_4$. *See* b1239	354.01	pr (al)	144	...	...	...	...	δ v^h	s	...	...	...	**B10**[2], 275
b1536	**—,2,6-dibromo-3,4,5-trihydroxy-**	Gallobromol. $C_7H_4Br_2O_5$. *See* b1239	327.94	nd, pr or lf (w+1)	150	...	...	...	v	v	v	...	...	chl i	**B10**[2], 347
b1537	**—,2,3-dichloro-**	$C_7H_4Cl_2O_2$. *See* b1239	191.02	nd	168.3	...	...	...	s	s	s	...	...	...	**B9**[2], 228
b1538	**—,2,4-dichloro-**	$C_7H_4Cl_2O_2$. *See* b1239	191.02	nd (w or bz)	164.2	...	...	...	s^h	s	s	...	s	chl s	**B9**[2], 228
b1539	—,—,chloride	$C_7H_3Cl_3O$. *See* b1239	209.47	liq	...	150[34]	...	...	d^h	d^h	...	...	...	...	**B9**[2], 229
b1540	**—,2,5-dichloro-**	$C_7H_4Cl_2O_2$. *See* b1239	191.02	nd (w)	154.4	301	...	...	δ s^h	s	s	...	...	...	**B9**[2], 229
b1541	**—,2,6-dichloro-**	$C_7H_4Cl_2O_2$. *See* b1239	191.02	nd (al), pr (w)	144	sub	...	...	s^h	s	s	...	s	...	**B9**[2], 229
b1542	**—,3,4-dichloro-**	$C_7H_4Cl_2O_2$. *See* b1239	191.02	nd (w, al, bz)	207–9	...	...	...	s^h	v	s	...	...	...	**B9**[1], 141
b1543	—,—,chloride	$C_7H_3Cl_3O$. *See* b1239	209.47	liq	...	242	...	...	d^h	d^h	...	...	...	...	**B9**, 344
b1544	**—,3,5-dichloro-**	$C_7H_4Cl_2O_2$. *See* b1239	191.02	nd (al or w)	188	sub	...	...	δ	s	s	...	...	lig δ	**B9**[2], 229
b1545	**—,3,5-dichloro-2-hydroxy-**	$C_7H_4Cl_2O_3$. *See* b1239	207.02	nd (dil al)	220–1	sub	...	...	$δ^h$	v	...	...	...	...	**B10**[1], 48
b1546	**—,3,5-dichloro-4-hydroxy-**	$C_7H_4Cl_2O_3$. *See* b1239	207.02	nd (dil al or dil aa)	268–9	sub	...	...	δ s^h	v	v	...	...	...	**B10**[1], 78
b1547	**—,2(dichloro-sulfamyl)-**	SO_2NCl_2 / CO_2H (structure)	270.11	yesh gr pl (chl)	146–8 exp	...	...	...	$δ^h$	...	...	...	...	chl $δ^h$	**B11**, 377
b1548	**—,4(dichloro-sulfamyl)-**	Halazone. $HO_2C—C_6H_4—SO_2NCl_2$	270.11	pr (aa)	213	...	...	...	$δ^h$	...	...	...	...	chl $δ^h$ aa v	**B11**[2], 220
b1549	**—,2,3-dihydroxy-**	*o*-Pyrocatechuic acid. $C_7H_6O_4$. *See* b1239	154.12	pr or nd (w)	203–4	...	1.542^{20}_{4}	...	s	s	s	...	...	...	**B10**[2], 248
b1550	**—,2,4-dihydroxy-**	β-Resorcylic acid. $C_7H_6O_4$. *See* b1239	154.12	cr+w(w)	235–6	...	...	...	s^h	s	s	...	s	CS_2 i	**B10**[2], 251
b1551	**—,2,5-dihydroxy-**	Gentisic acid. $C_7H_6O_4$. *See* b1239	154.12	nd or pr (w)	205	...	...	...	v	v	v	...	δ	chl δ CS_2 i	**B10**[2], 257
b1552	**—,2,6-dihydroxy-**	γ-Resorcylic acid. $C_7H_6O_4$. *See* b1239	154.12	nd (w)	167d	...	...	...	s^h	s	s	...	...	...	**B10**[2], 259
b1553	**—,3,4-dihydroxy-**	Protocatechuic acid. $C_7H_6O_4$. *See* b1239.	154.12	mcl nd (w)	199d	...	1.524^4	...	v^h δ	v	s	...	i	...	**B10**, 389
b1554	**—,3,5-dihydroxy-**	α-Resorcylic acid. $C_7H_6O_4$. *See* b1239	154.12	pr or nd (+1½w)	237–40, 232–3 (+w)	...	...	...	δ v^h	v	v	...	...	...	**B10**[2], 266
b1555	**—,2,4-dihydroxy-6-methyl-**	*o*-Orsellinic acid. $C_8H_8O_4$. *See* b1239	168.15	nd (dil aa+lw)	176d	...	...	...	...	s	s	...	...	...	...
b1556	—,—,ethyl ester	$C_{10}H_{12}O_4$. *See* b1239	196.20	lf (aa), pr (al)	132	sub	...	...	$δ^h$	v	v	...	δ	chl δ lig δ	**B10**[2], 274
b1557	**—,2,4-dihydroxy-6-pentyl-**	Olivetolcarboxylic acid. $C_{12}H_{16}O_4$. *See* b1239	224.20	nd	147	...	...	...	...	...	...	...	...	...	...
b1558	**—,2-[(2,4-dihydroxyphenyl)-phenylmethyl]-**	HO, HO—, —CH—, CO_2H (structure)	320.35	(aa)	184	...	...	...	...	...	...	...	...	aa s^h	**B10**, 455
b1559	**—,3,5-diiodo-2-hydroxy-**	$C_7H_4I_2O_3$. *See* b1239	389.93	nd (al or aa)	235–6	...	...	...	i	v	v	...	i	chl i	**B10**[2], 65
b1560	—,—,ethyl ester	$C_9H_8I_2O_3$. *See* b1239	417.99	lf (al)	133	...	...	...	i	s	s	...	s	...	**B10**, 114
b1561	**—,3,5-diiodo-4-hydroxy-**	$C_7H_4I_2O_3$. *See* b1239	389.93	nd (dil al)	237	...	...	...	i	v	v	...	δ	chl δ lig δ,	**B10**[2], 105
b1562	—,—,ethyl ester	$C_9H_8I_2O_3$. *See* b1239	417.99	nd	123	...	...	...	...	s	...	...	...	os s	**B10**[2], 105
b1563	**—,2,4-dimethoxy-, nitrile**	$C_9H_9NO_2$. *See* b1291	163.18	lf ($PhNO_2$)	179	...	...	...	...	...	...	...	δ	to δ aa v $PhNO_2$ s^h	**B10**[2], 253
b1564	**—,2,5-dimethoxy-, amide**	$C_9H_{11}NO_3$. *See* b1244	181.19	lf (bz-peth), nd (w)	141–2	...	...	...	δ v^h	v	v^h	v	v	chl v	**B10**[1], 184
b1565	—,—,nitrile	$C_9H_9NO_2$. *See* b1291	163.18	nd (al)	82	...	...	...	...	s^h	...	...	v	chl v lig δ	**B10**[2], 258
b1566	**—,2,6-dimethoxy-, nitrile**	$C_9H_9NO_2$. *See* b1291	163.18	nd or pl	118	310	...	...	δ	v^h	δ	s	v	CS_2 δ lig δ chl v	**B10**[2], 260
b1567	**—,3,4-dimethoxy-**	Veratric acid. $C_9H_{10}O_4$. *See* b1239	182.18	nd (w)	180–1	...	...	...	i	v	v	...	...	...	**B10**[2], 261
b1568	—,—,amide	Veratramide. $C_9H_{11}NO_3$. *See* b1244	181.19	...	164	...	...	...	v^h	...	s	...	s	...	**B10**[2], 264
b1569	—,—,nitrile	Veratronitrile. $C_9H_9NO_2$. *See* b1291	163.18	nd (w)	67–8	...	...	...	s^h	s^h	...	...	s	...	**B10**[2], 264
b1570	**—,3,5-dimethoxy-, amide**	$C_9H_{11}NO_3$. *See* b1244	181.19	nd (bz)	148–9	...	...	...	...	v	v	...	v^h	lig v^h	**B10**[2], 267

For explanations, symbols and abbreviations see beginning of table.

No.	Name	Synonyms and Formula	Mol. wt.	Crystalline form, color and specific rotation	m.p. °C	b.p. °C	Density	n_D	Solubility w	al	eth	ace	bz	other solvents	Ref.
	Benzoic acid														
b1571	**—,2,3-dimeth-oxy-4-hydroxy-**	$C_9H_{10}O_5$. *See* b1239	198.18	pl or lf (w)	154–5				δ s^h	s		v		chl v lig i MeOH v	**B10**^{2}, 332
b1572	**—,2,3-dimeth-oxy-5-hydroxy-**	$C_9H_{10}O_5$. *See* b1239	198.18	pl or lf (w)	186–8				s^h	s^h					**B10**^{2}, 333
b1573	**—,2,4-dimeth-oxy-6-hydroxy-**	$C_9H_{10}O_5$. *See* b1239	198.18	nd (eth-bz)	153–5				s^h	s	s				**B10**^{2}, 334
b1574	**—,2,6-dimeth-oxy-4-hydroxy-**	$C_9H_{10}O_5$. *See* b1239	198.18	pl	175					δ	δ			Py v	**B10**^{1}, 235
b1575	**—,3,4-dimeth-oxy-2-hydroxy-**	$C_9H_{10}O_5$. *See* b1239	198.18	nd (w)	169–72				s^h						**B10**, 465
b1576	**—,3,4-dimeth-oxy-5-hydroxy-**	$C_9H_{10}O_5$. *See* b1239	198.18	nd (aa or w)	193–4				s	s				aa s	**B10**^{2}, 340
b1577	**—,3,5-dimeth-oxy-4-hydroxy-**	Syringic acid. $C_9H_{10}O_5$. *See* b1239	198.18	nd (w)	204–5				δ	s	v			chl v	**B10**, 480
b1578	**—,4,5-dimeth-oxy-2-hydroxy-**	$C_9H_{10}O_5$. *See* b1239	198.18	nd (w)	202d				δ s^h	s					**B10**^{1}, 234
b1579	—,—,methyl ester	$C_{10}H_{12}O_5$. *See* b1239	212.20	nd (w)	95				δ						**B10**^{1}, 234
b1580	**—,2,3-dimethyl-**	Hemellitic acid. 2,3-Xylic acid. $C_9H_{10}O_2$. *See* b1239	150.18	pr (al)	144				δ^h	s	s				**B9**^{1}, 209
b1581	**—,2,4-dimethyl-**	2,4-Xylic acid. $C_9H_{10}O_2$. *See* b1239	150.18	mcl or tcl nd (w)	127	267^{727}			δ^h	s^h					**B9**^{2}, 350
b1582	**—,2,5-dimethyl-**	Isoxylic acid. 2,5-Xylic acid. $C_9H_{10}O_2$. *See* b1239	150.18	nd (al)	132	268	1.069^{21}_4		i	s	s	s	s		**B9**^{1}, 210
b1583	**—,2,6-dimethyl-**	2,6-Xylic acid. $C_9H_{10}O_2$. *See* b1239	150.18	nd (lig)	116	274.5			δ	s	s			lig δ	**B9**^{2}, 350
b1584	**—,3,4-dimethyl-**	Paraxylic acid. 3,4-Xylic acid. $C_9H_{10}O_2$. *See* b1239	150.18	pr (al)	163	sub			i δ^h	s			s		**B9**^{2}, 353
b1585	**—,3,5-dimethyl-**	Mesitylenic acid. 3,5-Xylic acid. $C_9H_{10}O_2$. *See* b1239	150.18	nd (w)	170	sub			δ^h	v	v				**B9**^{2}, 354
b1586	**—,2(dimethyl-amino)-**	$C_9H_{11}NO_2$. *See* b1239	165.19	pr, nd (eth)	72	sub			v	v	s^h				**B14**^{2}, 213
b1587	**—,3(dimethyl-amino)-**	$C_9H_{11}NO_2$. *See* b1239	165.19	nd	151				v^h	s	s				**B14**, 392
b1588	**—,4(dimethyl-amino)-**	$C_9H_{11}NO_2$. *See* b1239	165.19	nd (al)	242.5–3.5					s	δ				**B14**^{2}, 259
b1589	**—,2,4-dinitro-**	$C_7H_4N_2O_6$. *See* b1239	212.12	nd (w)	182–3				δ s^h	δ s^h			δ v^h		**B9**^{2}, 279
b1590	**—,2,5-dinitro-**	$C_7H_4N_2O_6$. *See* b1239	212.12	pr (w)	177				δ s^h	s	s		δ		**B9**^{2}, 279
b1591	**—,2,6-dinitro-**	$C_7H_4N_2O_6$. *See* b1239	212.12	nd (w)	202–3				v^h	s	s				**B9**^{2}, 279
b1592	**—,3,4-dinitro-**	$C_7H_4N_2O_6$. *See* b1239	212.12	nd	165				δ s^h	v	v				**B9**^{1}, 167
b1593	**—,3,5-dinitro-**	$C_7H_4N_2O_6$. *See* b1239	212.12	mcl pr (al)	205				δ	s	δ		i	aa s	**B9**^{2}, 279
b1594	—,—,benzyl ester	$C_{14}H_{10}N_2O_6$. *See* b1239	302.25		112										
b1595	—,—,butyl ester	$C_{11}H_{12}N_2O_6$. *See* b1239	268.23	mcl nd	62.5			1.488		s^h					
b1596	—,—,chloride	$C_7H_3ClN_2O_5$. *See* b1239	230.57	ye nd (bz)	74	196^{12}			d	d	s				**B9**^{2}, 283
b1597	—,—,ethyl ester	$C_9H_8N_2O_6$. *See* b1239	240.17	nd (al)	92.7			1.560		s^h					**B9**, 414
b1598	—,—,furfuryl ester	NO_2 CO_2CH_2 O NO_2	292.21		78–81										**B17**^{2}, 115
b1599	—,—,isobutyl ester	$C_{11}H_{12}N_2O_6$. *See* b1239	268.23	mcl pl or nd	87–8										**B9**^{2}, 281
b1600	—,—,isopropyl ester	$C_{10}H_{10}N_2O_6$. *See* b1239	254.20	nd	122					s^h					
b1601	—,—,methyl ester	$C_8H_6N_2O_6$. *See* b1239	226.15	nd (w)	112				s^h	s^h					**B9**, 414
b1602	—,—,pentyl ester	$C_{12}H_{14}N_2O_6$. *See* b1239	282.26		46.4					s^h					
b1603	—,—,phenyl ester	$C_{13}H_8N_2O_6$. *See* b1239	288.22	rods	145–6				i	s^h	δ		v		**B9**, 119
b1604	—,—,propyl ester	$C_{10}H_{10}N_2O_6$. *See* b1239	254.20	mcl pl	73					s^h					
b1605	—,—,tetrahydro-furfuryl ester	NO_2 CO_2CH_2 O NO_2	296.24	nd (al)	83–4										
b1606	**—,2,4-dinitro-3-hydroxy-**	$C_7H_4N_2O_7$. *See* b1239	228.12	(w)	204				δ v^h						**B10**^{2}, 85
b1607	**—,3,5-dinitro-2-hydroxy-**	$C_7H_4N_2O_7$. *See* b1239	228.12	ye nd (+1w)	173–4				δ	s			s		**B10**, 122
b1608	**—,3,5-dinitro-4-hydroxy-**	$C_7H_4N_2O_7$. *See* b1239	228.12	ye lf (al)	244–6				δ s^h	v^h	v				**B10**^{2}, 108
b1609	**—,2-ethoxy-**	$C_9H_{10}O_3$. *See* b1239	166.18		19.5				δ s^h						**B10**^{2}, 40
b1610	—,—,amide	$C_9H_{11}NO_2$. *See* b1244	165.19	nd (w or al)	132.1				δ s^h	v^h	v^h				**B10**^{2}, 58
b1611	—,—,ethyl ester	$C_{11}H_{14}O_3$. *See* b1239	194.23			251^{760}	1.005		δ	v	s				**B10**, 74

For explanations, symbols and abbreviations see beginning of table.

No.	Name	Synonyms and Formula	Mol. wt.	Crystalline form, color and specific rotation	m.p. °C	b.p. °C	Density	n_D	Solubility w	al	eth	ace	bz	other solvents	Ref.
	Benzoic acid														
b1612	—,—,nitril	C_9H_9NO. *See* b1291	147.18		5	260.7 153^{15}				v	v			lig v	**B10**, 97
b1613	—,**3-ethoxy-**	$C_9H_{10}O_3$. *See* b1239	166.18	nd (w)	137	sub			δ	s	s		s		**B10**^{1}, 64
b1614	—,—,amide	$C_9H_{11}NO_2$. *See* b1244	165.19	nd (w)	139					v	s	v		peth δ chl v lig δ	**B10**, 411
b1615	—,—,ethyl ester	$C_{11}H_{14}O_3$. *See* b1239	194.23			264 $172–3^{50}$	1.0725^{20}_{20}		i $δ^h$	v	v				**B10**, 139
b1616	—,**4-ethoxy-**	$C_9H_{10}O_3$. *See* b1239	166.18	nd	198.5				$δ^h$						**B10**^{2}, 92
b1617	—,—,amide	$C_9H_{11}NO_2$. *See* b1244	165.19	pr (dil al)	206				δ	v					**B10**^{2}, 101
b1618	—,—,ethyl ester	$C_{11}H_{14}O_3$. *See* b1239	194.23			275 $148–9^{14}$	1.076^{12}		i	v	v				**B10**^{2}, 98
b1619	—,—,nitrile	C_9H_9NO. *See* b1291	147.18		16–2	258				v	v			lig v	**B10**^{2}, 101
b1620	—,—,piperazinium salt	$C_{22}H_{30}N_2O_6$. *See* b1239	418.49		176–7d				δ	s^h	i				
b1621	—,**4-ethoxy-2-hydroxy-**	$C_9H_{10}O_4$. *See* b1239	182.18	nd (w or bz)	154				δ s^h	v	v		v		**B10**, 379
b1622	—,**2-ethyl-**	$C_9H_{10}O_2$. *See* b1239	150.18	nd (w)	68	259^{760}	1.0413^{100}_{4}	1.5099^{100}	δ	v	s			lig δ	**B9**^{2}, 349
b1623	—,**3-ethyl-**	$C_9H_{10}O_2$. *See* b1239	150.18	nd (w or dil al)	47		1.042^{100}	1.5345^{100}	δ	s	v				**B9**^{1}, 208
b1624	—,**4-ethyl-**	$C_9H_{10}O_2$. *See* b1239	150.18	pr (al), pl or lf (w)	113.5				δ v^h	s			s	chl s	**B9**^{2}, 349
b1625	—,**2(ethylamino)-**	$C_9H_{11}NO_2$. *See* b1239	165.19	pr (dil al)	154					s	s				**B14**^{2}, 213
b1626	—,**3(ethylamino)-**	$C_9H_{11}NO_2$. *See* b1239	165.19	nd or pr	112	sub			$δ^h$	v	v				**B14**, 393
b1627	—,**4(ethylamino)-**	$C_9H_{11}NO_2$ *See* b1239	165.19		177–8					s	s			os s	**B14**^{1}, 572
b1628	—,**2-fluoro-**	$C_7H_5FO_2$. *See* b1239	140.11	nd (w)	126		1.460^{25}_{4}		δ v^h	v	v		i	chl s CS_2 i	**B9**^{2}, 220
b1629	—,**3-fluoro-**	$C_7H_5FO_2$. *See* b1239	140.11	lf (w)	124		1.474^{25}_{4}		δ						**B9**^{2}, 220
b1630	—,**4-fluoro-**	$C_7H_5FO_2$. *See* b1239	140.11	pr (w)	182.6		1.479^{25}_{4}		δ s^h	s	s				**B9**^{2}, 220
b1631	—,—,nitrile	C_7H_4FN. *See* b1291	121.12	nd (peth)	34.8	188.8^{760}	1.1070^{55}	1.4925^{55}						peth s^h	**B9**^{2}, 221
b1632	—,**2-formamido-**	$C_8H_7NO_3$ *See* b1291	165.15	nd (w+½)	162–3								s		**B14**^{2}, 219
b1633	—,**2-formyl-**	Phthalaldehydic acid. $C_8H_6O_3$. *See* b1239	150.13	lf (w)	96		1.404		s	v	v				**B10**, 666
b1634	—,**3-formyl-**	Isophthalaldehydic acid. $C_8H_6O_3$. *See* b1239	150.13	nd (w)	175				s^h	v	v				**B10**^{2}, 465
b1635	—,—,nitrile	3-Cyanobenzaldehyde. C_8H_5NO. *See* b1291	131.14	nd (eth)	79–81	210			v^h	v	v			chl v	**B10**, 671
b1636	—,**4-formyl-**	Terephthalaldehydic acid. $C_8H_6O_3$. *See* b1239	150.13	nd	248–50				$δ^h$	v	s			chl s	**B10**, 671
b1637	—,—,nitrile	4-Cyanobenzaldehyde. C_8H_5NO. *See* b1291	131.14	nd (w), pr (eth or dil al)	101–2	133^{12}			s^h	v	v			chl v	**B10**^{2}, 465
b1638	—,**3-formyl-2-hydroxy-**	$C_8H_6O_4$. *See* b1239	166.13	nd (w+1)	179				δ s^h						**B10**^{2}, 675
b1639	—,**3-formyl-4-hydroxy-**	$C_8H_6O_4$. *See* b1239	166.13	pr (w)	244	sub			δ s^h	s	s			chl δ	**B10**^{2}, 675
b1640	—,**4-formyl-3-hydroxy-**	$C_8H_6O_4$. *See* b1239	166.13	nd	234				$δ^h$	s	s				**B10**, 954
b1641	—,**5-formyl-2-hydroxy-**	$C_8H_6O_4$. *See* b1239	166.13	nd	248–9				$δ^h$	s^h	s			chl i	**B10**^{2}, 675
b1642	—,**2-hydrazino-**	$C_7H_8N_2O_2$. *See* b1239	152.16	nd (w)	250–1				s^h	δ s^h	δ				**B15**^{2}, 295
b1643	—,—,hydrochloride	$C_7H_9ClN_2O_2$ *See* b1239	188.62	nd	194–5d				v^h						**B15**^{2}, 295
b1644	—,**3-hydrazino-**	$C_7H_8N_2O_2$. *See* b1239	152.16	lf	186d				$δ^h$	δ	i				**B15**^{1}, 205
b1645	—,**4-hydrazino-**	$C_7H_8N_2O_2$. *See* b1239	152.16	nd or pl	220–5				δ s^h						**B15**^{2}, 297
b1646	—,**2,2′-hydrazodi-**	CO_2H HO_2C / C6H4—NHNH—C6H4	272.26	lf or pr (al)	205				i	s^h					**B15**^{2}, 296
b1647	—,**2,3′-hydrazodi-**	HO_2C CO_2H / C6H4—NHNH—C6H4	272.26	nd (al)	206d									os s	**B15**, 629
b1648	—,**4,4′-hydrazodi-**	HO_2C—C6H4—NHNH—C6H4—CO_2H	272.26	nd (al)	286d				i	$δ^h$				aa s^h	**B15**, 632
b1649	—,**2-hydroxy-**	Salicylic acid. $C_7H_6O_3$. *See* b1239	138.12	nd (w), mcl pr (al)	159	211^{760}	1.443^{20}_{4}	1.565	δ v^h	v	v	v	δ	chl δ	**B10**^{2}, 25
b1650	—,—,4-acetamidophenyl ester	$C_{15}H_{13}NO_4$. *See* b1239	271.28	pl (w), lf (al	187–8				i $δ^h$	s	s		s	peth i	**B14**^{2}, 247
b1651	—,—,acetate	Aspirin. $C_9H_8O_4$. *See* b1239	180.16	nd (w)	135				s^h		s		δ	chl s	**B10**^{2}, 41

For explanations, symbols and abbreviations see beginning of table.

No.	Name	Synonyms and Formula	Mol. wt.	Crystalline form, color and specific rotation	m.p. °C	b.p. °C	Density	n_D	Solubility w	al	eth	ace	bz	other solvents	Ref.
	Benzoic acid														
b1652	—,—,amide	$C_7H_7NO_2$. *See* b1244	137.14	lf	140	181.5^{14}	1.175^{140}_{4}		δ	s	δ				**B10.** 87
b1653	—,—,—,*N*-phenyl-	$C_{13}H_{11}NO_2$. *See* b1244	213.24	pr (w or al), lf (w)	134–5				s^h	δ	δ		δ	chl δ	**B12,** 500
b1654	—,—,—,*N*(2-tolyl)-	$C_{14}H_{13}NO_2$ *See* b1244	227.27	nd (al)	144					s^h					**B12**², 450
b1655	—,—,anhydride	$C_{14}H_{10}O_5$. *See* b1239	258.23	amor	200–20				i	v	v				
b1656	—,—,benzyl ester	$C_{14}H_{12}O_3$. *See* b1239	228.25			$186–8^{10}$	1.1755^{20}_{4}		δ	s	s				**B10**², 51
b1657	—,—,—,acetate	$C_{16}H_{14}O_4$. *See* b1239	270.29		26	$197–200^{7}$						v	v	os v lig δ peth δ	**B10**², 52
b1658	—,—,butyl ester	$C_{11}H_{14}O_3$. *See* b1239	194.23		−5.9	$259–60^{760}$ $136–8^{10}$									**B10**², 49
b1659	—,—,(2-carboxyphenyl) ester	Disalicylic acid. $C_{14}H_{10}O_5$. *See* b1239	258.23		148				v^h	v^h	s		$δ^h$		**B10**², 54
b1660	—,—,chloride	$C_7H_5ClO_2$. *See* b1239	156.57		18	92^{15}	1.3112^{20}	1.5812^{20}	d		s				**B10**¹, 43
b1661	—,—,2,4-dinitrophenyl ester	$C_{14}H_{10}N_2O_7$. *See* b1239	318.25	ye pl (aa)	167				i	δ	i		i	AcOEt v^h aa v^h	**B10**², 52
b1662	—,—,ethyl ester	$C_9H_{10}O_3$. *See* b1239	166.18		1.3	231–4 132.8^{37}	1.1362^{20}_{4}	1.5251^{20}	i	∞					**B10**², 47
b1663	—,—,—,acetate	Ethyl aspirin. $C_{11}H_{12}O_4$. *See* b1239	208.22			272 $148–50^{15}$	1.1566^{15}		i	s				os s	**B10**², 48
b1664	—,—,hydrazide	$C_7H_8N_2O_2$. *See* b1239	152.15	pl (al), pr (w)	147				i s^h	v	δ		s		**B10**², 61
b1665	—,—,2-hydroxyethyl ester	$C_9H_{10}O_4$. *See* b1239	182.18		37	173^{15}	1.2537^{15}_{15}		δ	v	v		v	chl v	**B10**², 53
b1666	—,—,isobutyl ester	$C_{11}H_{14}O_3$. *See* b1239	194.23		5.9	259–60 $136–8^{10}$			i	s	s				**B10**², 76
b1667	—,—,isopropyl ester	$C_{10}H_{12}O_3$. *See* b1239	180.21			240–2 118^{17}	1.0729^{20}_{4}	1.5065^{20}	i	∞	∞				
b1668	—,—,menthyl ester	$C_{17}H_{24}O_3$. *See* b1239	276.38			190^{15}	1.0467^{20}		i	∞			∞	os v	**B10**², 50
b1669	—,—,methoxymethyl ester	$C_9H_{10}O_4$. *See* b1239	182.18			153^{12}	1.2^{15}		i	∞	∞		∞	chl ∞	**B10**², 54
b1670	—,—,2-methoxyphenyl ester	$C_{14}H_{12}O_4$. *See* b1239	244.25		70				i	s	s			chl s	**B10,** 81
b1671	—,—,methyl ester	$C_8H_8O_3$. *See* b1239	152.15		−8.6	223.3^{760}	1.1738^{20}_{4}	1.5369^{20}	δ	v					**B10**², 44
b1672	—,—,—,benzoate	$C_{15}H_{12}O_4$. *See* b1239	256.26	pr (al or eth)	92	385d $270–80^{120}$			i $δ^h$	s	s		s	chl s	**B10**², 47
b1673	—,—,3-methylbutyl ester	Isoamyl salicylate. $C_{12}H_{16}O_3$. *See* b1239	208.26			$276–7^{743}$	1.042^{20}_{4}	1.506^{20}	i	v	s			chl s	**B10**², 49
b1674	—,—,1-naphthyl ester	Alphol. $C_{17}H_{12}O_3$. *See* b1239	264.28		83										**E12B,** 1183
b1675	—,—,2-naphthyl ester	Betol. $C_{17}H_{12}O_3$. *See* b1239	264.28		95		1.11^{116}		i	δ s^h	s		s		**B10**², 53
b1676	—,—,nitrile	2-Cyanophenol. C_7H_5NO. *See* b1291	119.12	pr (bz)	98	149^{14}	1.1052^{100}_{4}		δ	v	v		v	chl v	**B10**², 60
b1677	—,—,4-nitrobenzyl ester	$C_{14}H_{11}NO_2$. *See* b1239	273.25		97–8					v		v			**B10**², 52
b1678	—,—,pentyl ester	$C_{12}H_{16}O_3$. *See* b1239	208.25			$148–54^{760}$	1.065^{15}_{15}	1.506^{20}	δ	∞	∞				**B10**², 49
b1679	—,—,phenyl ester	Salol. $C_{13}H_{10}O_3$. *See* b1239	214.21		43	173^{12}	1.2614^{30}_{4}		i	v		v	v		**B10,** 76
b1680	—,—,propyl ester	$C_{10}H_{12}O_3$. *See* b1239	180.20		96–8	$236–40^{745}$	1.005^{25}	1.5100^{25}	δ	∞	∞				**B10,** 75
b1681	—,**3-hydroxy-**	$C_7H_6O_3$. *See* b1239	138.12	nd (w), pl or pr (al)	201.5				δ s^h	s^h	s		i		**B10**², 79
b1682	—,—,amide	$C_7H_7NO_2$. *See* b1244	137.13	lf (w)	170.5				δ s^h	s	s		i	chl i CS_2 i	**B10**², 82
b1683	—,—,—,*N*-phenyl	Salicylanilide. $C_{13}H_{11}NO_2$. *See* b1244	213.23	nd or pl (w)	156				δ	s	δ		δ	chl i	**B10**¹, 269
b1684	—,—,chloride	$C_7H_5ClO_2$. *See* b1239	156.57		<−15	$110–3^{0.5}$			d	d				chl s	**B10**¹, 66
b1685	—,—,ethyl ester	$C_9H_{10}O_3$. *See* b1239	166.18	nd (bz-peth)		238–40				s					**B10**², 81
b1686	—,—,methyl ester	$C_8H_8O_3$. *See* b1239	152.15	nd (w)	72.3	280^{709} 178^{17}				s			s^h	peth s^h	**B10**², 81
b1687	—,—,nitrile	3-Cyanophenol. C_7H_5NO. *See* b1291	119.12	pr (al or eth), lf (w)	82						v		v	chl v	**B10,** 141
b1688	—,**4-hydroxy-**	$C_7H_6O_3$. *See* b1239	138.12	pr	214.5–5.5				δ v^h	v^h	δ			CS_2 i	**B10**², 88
b1689	—,—,amide	$C_7H_7NO_2$. *See* b1244	137.13	nd (w+1)	162				δ s^h	s	δ			CS_2 i chl i	**B10**², 100
b1690	—,—,—,*N*-phenyl	$C_{13}H_{11}NO_2$. *See* b1244	213.24	pl or nd (w)	201–2				δ	v	δ		i	chl i	**B10**², 257

For explanations, symbols and abbreviations see beginning of table.

No.	Name	Synonyms and Formula	Mol. wt.	Crystalline form, color and specific rotation	m.p. °C	b.p. °C	Density	n_D	Solubility w	al	eth	ace	bz	other solvents	Ref.
	Benzoic acid														
b1691	—,—,butyl ester...	$C_{11}H_{14}O_3$. *See* b1239........	194.11	....	68–9	....	....	....	δ	s	...	...	...	....	....
b1692	—,—,ethyl ester...	$C_9H_{10}O_3$. *See* b1239........	166.18	....	116	297–8	....	....	δ	v	v	...	...	CS_2 i lig δ chl δ	**B10²**, 96
b1693	—,—,methyl ester .	$C_8H_8O_3$. *See* b1239........	152.14	nd (al)	131	270–80d	....	....	δ	v	v	v	...	....	**B10²**, 95
b1694	—,—,nitrile.......	4-Cyanophenol. C_7H_5NO. *See* b1291	119.12	lf (w)	113	....	....	....	δ s^h	v	v	...	...	chl v	**B10²**, 10
b1695	—,—,propyl ester..	$C_{10}H_{12}O_3$. *See* b1239........	180.21	pr (eth)	96.2	....	1.0630_4^{102}	1.5050^{102}	i s^h	s^h	s	...	...	....	**B10²**, 97
b1696	—,4(α-**hydroxy-benzyl)-**	$C_{14}H_{12}O_3$. *See* b1239........	228.34	nd (w)	164–5	....	....	....	s^h	v	v	...	...	chl δ to δ	**B10**, 34
b1697	—,**2-hydroxy-5-bromo-**, acetate	$C_9H_7BrO_4$. *See* b1239........	259.06	....	168–9	....	....	....	i	v	s	...	...	....	**B10²**, 64
b1699	—,**2-hydroxy-(dithio)-**	Dithiosalicylic acid. OH —CS_2H	170.25	og-ye nd (peth)	46–50	....	....	....	δ	s	s	...	s	MeOH s	**B10²**, 78
b1700	—,**2-hydroxy-3-iodo-**	$C_7H_5IO_3$. *See* b1239........	264.02	nd (w)	198–9	....	....	....	δ s^h	v	v	...	i	chl i	**B10²**, 64
b1701	—,**2-hydroxy-4-iodo-**	$C_7H_5IO_3$. *See* b1239........	264.02	pl or nd (al)	228–30d	....	....	....	...	s v^h	s	...	...	....	**B10²**, 65
b1702	—,**2-hydroxy-5-iodo-**	$C_7H_5IO_3$. *See* b1239........	264.02	nd (w)	196	....	....	....	δ s^h	v	...	...	i	chl i	**B10²**, 65
b1703	—,**3-hydroxy-2-iodo**	$C_7H_5IO_3$. *See* b1239........	264.02	nd (chl)	158–9	....	....	....	...	s	s	...	...	chl s^h	**B10²**, 84
b1704	—,**3-hydroxy-4-iodo-**	$C_7H_5IO_3$. *See* b1239........	264.02	nd (w)	226–8d	....	....	....	δ s^h	s	...	...	...	....	**B10²**, 84
b1705	—,**4-hydroxy-2-iodo**	$C_7H_5IO_3$. *See* b1239........	264.02	nd (w)	215d	....	....	....	s^h	s	...	...	δ	....	**B10²**, 10
b1706	—,**4-hydroxy-3-iodo**	$C_7H_5IO_3$. *See* b1239........	264.02	nd (w+$\frac{1}{2}$)	173.5–4.5	....	....	....	s^h	v	v	...	δ^h	chl δ aa s lig i	**B10²**, 10
b1707	—,**5-hydroxy-2-iodo**	$C_7H_5IO_3$. *See* b1239........	264.02	nd (w)	196–8d	....	....	....	s^h	s	s	...	...	....	**B10²**, 84
b1708	—,**2-hydroxy-3-isopropyl-6-methyl-**	*o*-Thymotic acid. $C_{11}H_{14}O_3$. *See* b1239	194.23	nd (w)	127	sub	....	....	i	s	s	...	s	....	**B10²**, 17
b1709	—,**2-hydroxy-5-isopropyl-3-methyl-**	$C_{11}H_{14}O_3$. *See* b1239........	194.23	nd (w)	147	....	....	....	i	v	...	...	...	....	**B10**, 28
b1710	—,**2-hydroxy-5-isopropyl-4-methyl-**	$C_{11}H_{14}O_3$. *See* b1239........	194.23	....	189–90	....	....	....	...	...	...	...	...	....	**B10²**, 17
b1711	—,**2-hydroxy-6-isopropyl-3-methyl-**	*o*-Carvacrotinic acid. .... $C_{11}H_{14}O_3$. *See* b1239	194.23	nd (w)	136	....	....	....	δ	v	v	...	...	....	**B10**, 28
b1712	—,**4-hydroxy-5-isopropyl-2-methyl-**	*p*-Thymotic acid. $C_{11}H_{14}O_3$. *See* b1239.......	194.23	lf (al)	156	....	....	....	δ^h	v	v	...	v	chl v	**B10²**, 17
b1713	—,**3-hydroxy-4-methoxy-**	Isovanillic acid. $C_8H_8O_4$. *See* b1239	168.15	nd, pr, pl	250	....	....	....	δ s^h	v	v	...	...	....	**B10²**, 26
b1714	—,**4-hydroxy-3-methoxy-**	Vanillic acid. $C_8H_8O_4$. *See* b1239	168.15	nd (w)	210	....	....	....	δ	v	s	...	...	....	**B10²**, 26
b1715	—,—,ethyl ester...	$C_{10}H_{12}O_4$. *See* b1239........	196.20	nd	44	291–3	....	....	i	v	v	...	...	....	....
b1716	—,—,methyl ester .	$C_9H_{10}O_4$. *See* b1239........	182.18	nd	64	118^2	....	....	...	s^h	...	...	...	peth s^h	**B10¹**, 18
b1717	—,**2-hydroxy-4-methoxy-6-methyl-**	Everninic acid. $C_9H_{10}O_4$. *See* b1239	182.18	nd (peth)	171–2	....	....	....	δ s^h	v ...	δ ...	v	δ	aa v lig δ	**B10²**, 27
b1718	—,—,methyl ester .	Sparassol. $C_{10}H_{12}O_4$. *See* b1239	196.20	pr (w), lf (MeOH)	67–8	....	....	....	δ^h	s	v	v	...	MeOH s peth s chl v	**B10²**, 27
b1719	—,**2-hydroxy-3-methyl**	2,3-Cresotic acid. $C_8H_8O_3$. *See* b1239	152.15	nd (w)	167	....	....	....	δ s^h	s	s	...	...	chl s	**B10²**, 13
b1720	—,—,chloride......	$C_8H_8ClO_2$. *See* b1239........	170.60	....	27–8	$87–9^{16}$	....	....	d	d^h	...	...	...	....	**B10²**, 132
b1721	—,**2-hydroxy-4-methyl**	$C_8H_8O_3$. *See* b1239........	152.15	nd (w), pr (al), pl (chl)	177.8	sub	....	....	δ	s	v	...	...	chl s	**B10²**, 137
b1722	—,**2-hydroxy-5-methyl-**	$C_8H_8O_3$. *See* b1239........	152.15	nd (w)	152.5	....	....	....	δ	s	s	...	...	chl s CS_4 i	**B10²**, 134
b1723	—,**2-hydroxy-6-methyl-**	$C_8H_8O_3$. *See* b1239........	152.15	nd (w)	173	....	....	....	δ s^h	v	v	...	...	chl s	**B10²**, 128
b1724	—,**3-hydroxy-2-methyl-**	$C_8H_8O_3$. *See* b1239........	152.15	nd (w)	145–6	....	....	....	δ s^h	v	v	...	...	....	**B10**, 214
b1725	—,**3-hydroxy-4-methyl-**	$C_8H_8O_3$. *See* b1239........	152.15	nd or pr (w)	208.5	sub	....	....	δ s^h	s	s	...	δ	peth δ chl i	**B10²**, 140

For explanations, symbols and abbreviations see beginning of table.

No.	Name	Synonyms and Formula	Mol. wt.	Crystalline form, color and specific rotation	m.p. °C	b.p. °C	Density	n_D	Solubility w	al	eth	ace	bz	other solvents	Ref.
	Benzoic acid														
b1726	—,**3-hydroxy-5-methyl-**	$C_8H_8O_3$. *See* b1239	152.15	nd (w)	210	sub			s^h	v	v	...	...		**B10**, 227
b1727	—,**4-hydroxy-2-methyl-**	$C_8H_8O_3$. *See* b1239	152.15	nd (w+½)	177–8	236–7 sub			δ s^h	s s	s	...	...	chl i	**B10^2**, 127
b1728	—,**4-hydroxy-3-methyl-**	$C_8H_8O_3$. *See* b1239	152.15	nd (w+½)	174–5				s^h	s	s	...	...	chl s^h	**B10^2**, 133
b1729	—,**5-hydroxy-2-methyl-**	$C_8H_8O_3$. *See* b1239	152.15	nd or pr	183–4				δ	v	v	...	...	chl δ	**B10**, 215
b1730	—,**2-(hydroxymethyl)-**	$C_8H_8O_3$. *See* b1239	152.15	nd	128				δ	v	v	...	...		**B10**, 218
—	—,—,lactone	*see* **Phthalide**													
b1731	—,**2-hydroxy-3-nitro-**	$C_7H_5NO_5$. *See* b1239	183.12	yesh nd (aa, w+1)	148–9 128–9 (+w)				δ	v	v	...	s	chl s	**B10^2**, 66
b1732	—,**2-hydroxy-4-nitro-**	$C_7H_5NO_5$. *See* b1239	183.12	nd (w or al)	232–3				i v^h	v	...	...	δ	chl δ aa s lig i	**B10^2**, 66
b1733	—,**2-hydroxy-5-nitro-**	$C_7H_5NO_5$. *See* b1239	183.12	nd (w)	229–30		1.650^{20}		δ s^h	v	...	...	...	os v	**B10^2**, 67
b1734	—,**2-hydroxy-6-nitro-**	$C_7H_5NO_5$. *See* b1239	183.12	pa ye nd					v	v	...	...	...		**J1950**, 2049
b1736	—,**3-hydroxy-2-nitro-**	$C_7H_5NO_5$. *See* b1239	183.12	pl or pr (w+1)	180.5–1.5				v	v	v	...	...		**B10^2**, 84
b1737	—,**3-hydroxy-4-nitro-**	$C_7H_5NO_5$. *See* b1239	183.12	lf (w)	235				δ	...	...	...	...		**B10^2**, 185
b1738	—,**3-hydroxy-5-nitro-**	$C_7H_5NO_5$. *See* b1239	183.12	ye lf or pl	194–5				s^h	s	s	...	...		**B10^2**, 85
b1739	—,**4-hydroxy-3-nitro-**	$C_7H_5NO_5$. *See* b1239	183.12	nd or lf (w)	183–4				v^h	v	v	...	...		**B10^2**, 106
b1740	—,**5-hydroxy-2-nitro-**	$C_7H_5NO_5$. *See* b1239	183.12	ye nd or pr (w+1)	172				v	v	v	...	...		**B10^2**, 85
b1741	—,**2-hydroxy-3-sulfo-**	$C_7H_6O_6S$. *See* b1239	218.18	pr (w+2)	120				v	v	s	...	...	aa s	**B11^2**, 231
b1742	—,**2-hydroxy-5-sulfo-**	$C_7H_6O_6S$. *See* b1239	218.18	hyg nd (w+2)	113, 180d (+2w)				∞	∞	v	...	...		**B11^2**, 232
b1743	—,**3-hydroxy-4-sulfo-**	$C_7H_6O_6S$. *See* b1239	218.18	nd (w+1½)	208				v	v	i	...	...		**B11**, 413
b1744	—,**3-hydroxy-5-sulfo-**	$C_7H_6O_6S$. *See* b1239	218.18	nd (w+1)	120d				s	v	v	...	...		**B11**, 413
b1745	—,**4-hydroxy-3-sulfo-**	$C_7H_6O_6S$. *See* b1239	218,18	nd or lf	d				v	s	i	...	...		**B11^2**, 235
b1746	—,**2-iodo-**	$C_7H_5IO_2$. *See* b1239	248.03	nd (w)	163	233 exp	2.249^{25}_{4}		i δ^h	v	v	...	...		**B9^2**, 239
b1747	—,—,methyl ester	$C_8H_7IO_2$. *See* b1239	262.06			277–8^{729} 167^{25}			...	s^h	...	...	...		**B9**, 364
b1748	—,**3-iodo-**	$C_7H_5IO_2$. *See* b1239	248.03	mcl pr	186.7				i	v	δ	...	...		**B9^2**, 240
b1749	—,—,methyl ester	$C_8H_7IO_2$. *See* b1239	262.06	nd	54–5	276–7^{739} 149–50^{18}			i	...	...	...	...	os v^h lig i	**B9**, 365
b1750	—,**4-iodo-**	$C_7H_5IO_2$. *See* b1239	248.03	mcl pr (dil al), lf (sub)	270	sub			i	δ	δ	...	...		**B9^2**, 240
b1751	—,—,methyl ester	$C_8H_7IO_2$. *See* b1239	262.06	nd (eth-al)	114 sub				...	s^h	s^h	...			**B9**, 367
b1752	—,**2-iodo-3-nitro-**	$C_7H_4INO_4$. *See* b1239	293.03	pr (w)	206				s^h	...	s	...	...		**B9^2**, 278
b1753	—,**2-iodo-4-nitro-**	$C_7H_4INO_4$. *See* b1239	293.03	pa ye nd (w)	146–7				δ v^h	...	...	...	δ	CCl_4 δ lig δ	**B9^1**, 166
b1754	—,**4-iodo-2-nitro-**	$C_7H_4INO_4$. *See* b1239	293.03	lf or pr (dil al)	192–3				δ^h	v	v	...	s^h		**B9^2**, 278
b1755	—,**4-iodo-3-nitro-**	$C_7H_4INO_4$. *See* b1239	293.03	pr (al)	213				i	v	...	...	...		**B9^2**, 278
b1756	—,**5-iodo-3-nitro-**	$C_7H_4INO_4$. *See* b1239	293.03	nd (al)	167				δ^h	v	...	...	...		**B9^2**, 278
b1757	—,**2-iodoso-**	$C_7H_5IO_3$. *See* b1239	264.03	lf (w)	223–5d				δ s^h	δ s^h	δ	...	...		**B9^2**, 239
b1758	—,**3-iodoso-**	$C_7H_5IO_3$. *See* b1239	264.03	ye amor	175–80				d^h	δ	δ	...	i	chl i	**B9**, 365
b1759	—,**4-iodoso-**	$C_7H_5IO_3$. *See* b1239	264.03	amor	212d				d^h	i	...	...	i	chl i	**B9**, 366
b1760	—,**2-isopropyl-**	*o*-Cuminic acid. $C_{10}H_{12}O_2$. *See* b1239	164.20	pr (w or peth)	59.5				δ s^h	v	v	...	...	peth s	**B9**, 546
b1761	—,**4-isopropyl-**	Cumic acid. *p*-Cuminic acid. $C_{10}H_{12}O_2$. *See* b1239	164.20	tcl (al)	116.5	sub			i	v	v	...	...		**B9**, 546
b1762	—,**2-mercapto-**	Thiosalicylic acid. $C_7H_6O_2S$. *See* b1239	154.18	nd (w)	168–9	sub			δ	s	s	...	...	aa s lig δ	**B10^2**, 70
b1763	—,**2-methoxy-**	*o*-Anisic acid. $C_8H_8O_3$. *See* b1239	152.14	pl (w), fl (al)	101	200			δ v^h	v	v	...	s	chl v CCl_4 s	**B10^2**, 39
b1764	—,—,amide	$C_8H_9NO_2$. *See* b1244	151.17	nd (bz), pr (eth), lf (w)	129				δ	...	v	...	v^h	sulf s	**B10^2**, 58
b1765	—,—,chloride	$C_8H_7ClO_2$. *See* b1239	170.60			254 145^{17}			d	d^h	...	...	...		**B10^2**, 55
b1766	—,—,ethyl ester	$C_{10}H_{12}O_3$. *See* b1239	180.20			246–8^{732} 135–6^{12}	1.1124^{20}_{4}	1.5224^{20}	i	s	s	...	...		**B10^2**, 48

For explanations, symbols and abbreviations see beginning of table.

No.	Name	Synonyms and Formula	Mol. wt.	Crystalline form, color and specific rotation	m.p. °C	b.p. °C	Density	n_D	Solubility w	al	eth	ace	bz	other solvents	Ref.
	Benzoic acid														
b1767	—,—,methyl ester	$C_9H_{10}O_3$. *See* b1239	166.18			228^{760} $129–30^{15}$	1.1571_4^{19}	$1.534^{19.5}$	i	s					**B10**², 4
b1768	—,—,nitrile	C_8H_7NO. *See* b1291	133.15		24.5	146^{20}	$1.1063_4^{20.5}$			s	v				**B10**², 6
b1769	—,—,piperazinium salt	$C_{20}H_{26}N_2O_6$. *See* b1239	390.43		190.5–1.5				s	s	i				
b1770	—,**3-methoxy-**	*m*-Anisic acid. $C_8H_8O_3$. *See* b1239	152.14	nd (w)	110	$170–2^{10}$			δ s^h	s	s		s	CCl_4 δ chl v	**B10**², 8
b1771	—,—,chloride	$C_8H_7ClO_2$. *See* b1239	170.60			$243–4^{760}$ $123–5^{15}$			d						**B10**, 14
b1772	—,—,ethyl ester	$C_{10}H_{12}O_3$. *See* b1239	180.20			$260–1^{760}$ 110^5			i	s	s				**B10**², 81
b1773	—,—,methyl ester	$C_9H_{10}O_3$. *See* b1239	166.18			236–8 $121–4^{10}$			i	s					**B10**², 8
b1774	—,—,piperazinium salt	$C_{20}H_{26}N_2O_6$. *See* b1239	390.43		137–8d				s^h	s	$δ^h$				
b1775	—,**4-methoxy-**	Anisic acid. $C_8H_8O_3$. *See* b1239	152.14	pr or nd (w)	185	275–80			i s^h	v	v			chl s	**B10**², 91
b1776	—,—,amide	$C_8H_9NO_2$. *See* b1244	151.16	nd or pl (w)	166.5–7.5	295			s	v	δ				**B10**², 10
b1777	—,—,butyl ester	$C_{12}H_{16}O_3$. *See* b1239	208.26			183^{40}	$1.054^{16.5}$	$1.5141^{18.5}$							**B10**², 97
b1778	—,—,chloride	$C_8H_7ClO_2$. *See* b1239	170.60	nd	24	262–3 91^1			d	d^h	s	s	v^h		**B10**¹, 17
b1779	—,—,ethyl ester	$C_{10}H_{12}O_3$. *See* b1239	180.21		7–8	$269–70^{760}$ $136–7^{13}$			i	s	s				**B10**², 9
b1780	—,—,methyl ester	$C_9H_{10}O_3$. *See* b1239	166.18		49	255			i	s	s				**B10**², 95
b1781	—,—,nitrile	C_8H_7NO. *See* b1291	133.15	nd (w), lf (al)	61–2	240.1^{76} $106–8^6$			i s^h	v^h	v		s		**B10**², 10
b1782	—,—,piperazinium salt	$C_{20}H_{26}N_2O_6$. *See* b1239	390.44		172–4				δ	s^h	i				
b1783	—,**5-methoxy 2(4-methoxy-phenoxy)-**	$C_{15}H_{14}O_5$. *See* b1239	274.28		95								s		**B10**¹, 18
b1784	—,**3-methoxy-2(methyl-amino)-**, methyl ester	Damascenine. $C_{10}H_{13}NO_3$. *See* b1239	195.22	pr (al)	27–9	270^{750}			i $δ^h$	v	v			CS_2 v lig v chl v	**B14**¹, 65
b1785	—,**2-methyl-**	*o*-Toluic acid $C_8H_8O_2$. *See* b1239	136.15	pr or nd (w)	107–8	$258–9^{751}$	1.062^{115}	1.512^{115}	i	v				chl s	**B9**, 462
b1786	—,—,amide	C_8H_9NO. *See* b1244	135.17	pl or nd (w)	147				δ v^h	v	δ		δ		**B9**, 465
b1787	—,—,2(4-biphenylyl)-2-oxoethyl ester	CH_3 —$COCH_2O_2C$—	330.39		94.5										
b1788	—,—,chloride	C_8H_7ClO. *See* b1239	154.60			$213–14^{760}$ $99–100^{12}$			d	d^h	v				**B9**², 319
b1789	—,—,ethyl ester	$C_{10}H_{12}O_2$. *See* b1239	164.21		<−10	221.3	1.038_4^{15}	1.507^{22}	i	∞	∞				**B9**, 463
b1790	—,—,methyl ester	$C_9H_{10}O_2$. *See* b1239	150.18		<−50	213^{760}	1.073^{15}		i	∞	∞				**B9**, 463
b1791	—,—,nitrile	C_8H_7N. *See* b1291	117.15		−14 to −13	205^{760}	0.9955_4^{20}	1.5270^{23}	i	∞	∞				**B9**², 319
b1792	—,**3-methyl-**	*m*-Toluic acid. $C_8H_8O_2$. *See* b1239	136.15	pr (w or al)	111–3	263	1.054^{112}	1.509	δ	v	v				**B9**, 475
b1793	—,—,amide	C_8H_9NO. *See* b1244	135.17	nd (eth)	97				δ	s	δ		δ		**B9**, 477
b1794	—,—,anhydride	$C_{16}H_{14}O_3$. *See* b1239	254.29		70	230^{17}					v		v	chl v	**B9**², 324
b1795	—,—,2(4-biphenylyl)-2-oxoethyl ester	CH_3 —$COCH_2O_2C$—	303.39		136.5										
b1796	—,—,chloride	C_8H_7ClO. *See* b1239	154.60		−25	$219–20^{766}$ 105^{20}			d	d^h	s				**B9**², 324
b1797	—,—,ethyl ester	$C_{10}H_{12}O_2$. *See* b1239	164.21			226.4	1.0301_{20}^{20}	1.505^{22}	i	∞	∞				**B9**, 476
b1798	—,—,methyl ester	$C_9H_{10}O_2$. *See* b1239	150.18			221^{758}	1.066^{15}		i	s					**B9**², 324
b1799	—,—,nitrile	C_8H_7N. *See* b1291	117.15		−23	213^{759}	1.0316_4^{20}		i $δ^h$	∞	∞				**B9**², 325
b1800	—,**4-methyl-**	*p*-Toluic acid. $C_8H_8O_2$. *See* b1239	136.15	nd (w)	179–80	274–5			i $δ^h$	v	v			MeOH v	**B9**, 483
b1801	—,—,amide	C_8H_9NO. *See* b1244	135.17	nd or pl(w)	155				δ v^h	v	v				**B9**, 486
b1802	—,—,anhydride	$C_{16}H_{14}O_3$. *See* b1239	254.27	lf (MeOH), nd (al)	98										**B9**²,329

For explanations, symbols and abbreviations see beginning of table.

No.	Name	Synonyms and Formula	wt.	Crystalline form, color and specific rotation	m.p. °C	b.p. °C	Density	n_D	Solubility						Ref.
									w	al	eth	ace	bz	other solvents	
	Benzoic acid														
b1803	—,—,2(4-biphenylyl)-2-oxoethyl ester	—$COCH_2O_2C$——CH_3	330.36	...	165	...	...	...	...	...	...	...	...	...	...
b1804	—,—,chloride	C_8H_7ClO. *See* b1239	154.60	...	−2	$225–7^{760}$	...	...	d	d^h	...	...	...	...	**B9**^{2}, 329
b1805	—,—,ethyl ester	$C_{10}H_{12}O_2$. *See* b1239	164.20	...	...	235.5^{760}	1.024^{25}_{25}	1.5089^{18}	i	∞	∞	...	...	...	**B9**, 484
b1806	—,—,methyl ester	$C_9H_{10}O_2$. *See* b1239	150.17	...	33.2	217	...	...	i	v	v	...	...	...	**B9**, 484
b1807	—,—,nitrile	C_8H_7N. *See* b1291	117.14	nd (al)	29.5	217.6^{760}	0.971^{43}_4	...	i	v	v	...	...	...	**B9**^{2}, 330
b1808	—,—,piperazinium salt	$C_{20}H_{26}N_2O_4$. *See* b1239	358.42	...	203	...	...	...	δ	s	δ^h	...	...	...	...
b1809	—,**2(methyl-amino)-**	$C_8H_9NO_2$. *See* b1239	151.17	lf (al or lig)	179	...	...	...	δ	v	v	...	v	chl v lig s	**B14**^{2}, 212
b1810	—,—,ethy ester	$C_{10}H_3NO_2$. *See* b1239	179.21	...	39	266^{760} $141–3^{15}$	...	...	i	...	v	...	...	...	**B14**^{2}, 213
b1811	—,—,methyl ester	$C_9H_{11}NO_2$. *See* b1239	165.09	...	18.5–9.5	255^{760} $130–1^{13}$	1.167^{15}_4	1.5839^{12}	i	s	s	...	...	...	**B14**^{2}, 212
b1812	—,**3(methyl-amino)-**	$C_8H_9NO_2$. *See* b1239	151.16	pl (peth)	147	...	...	...	i δ^h	v	i	...	v	chl v lig i	**B14**, 391
b1813	—,—,methyl ester	$C_9H_{11}NO_2$. *See* b1239	165.19	...	72	...	...	...	i	...	v	...	...	...	**B14**, 392
b1814	—,**4(methyl-amino)-**	$C_8H_9NO_2$. *See* b1239	151.16	nd (bz or w)	163	...	...	...	v^h	v	v	...	s^h	...	**B14**^{2}, 259
b1815	—,—,methyl ester	$C_9H_{11}NO_2$. *See* b1239	165.19	lf (dil al or lig)	95.5	...	...	...	i	s	s	...	...	...	**B14**^{1}, 571
b1816	—,**3(3-methyl butoxy)-**	*m*-Isoamyloxybenzoic acid. $C_{12}H_{16}O_3$. *See* b1239	208.26	...	74–5	...	...	...	...	δ s^h	...	...	...	...	**B10**^{1}, 64
b1817	—,**4(3-methyl-butoxy)-**	*p*-Isoamyloxybenzoic acid. $C_{12}H_{16}O_3$. *See* b1239	208.26	nd	141–2	...	...	...	...	...	...	...	...	...	**B10**^{1}, 70
b1818	—,—,chloride	$C_{12}H_{15}ClO_2$. *See* b1239	226.71	...	...	$180–2^{12}$	...	...	d	d^h	...	...	...	...	**B10**^{1}, 77
b1819	—,**3,4-methyl-enedioxy-**	Piperonylic acid. $C_8H_6O_4$. *See* b1239	166.13	...	229	sub	...	...	i	δ	δ	...	...	chl i	**B1**^{2}, 292
b1820	—,—,chloride	$C_8H_5ClO_3$. *See* b1239	184.58	...	80	155^{25}	...	...	...	...	...	...	...	...	**B19**, 270
b1821	—,—,ethyl ester	$C_{10}H_{10}O_4$. *See* b1239	194.18	...	18.5	285.5–6.5 $164–5^{11}$	...	...	i	v	v	...	...	peth v	**B19**^{2}, 293
b1822	—,—,methyl ester	$C_9H_8O_4$. *See* b1239	180.15	...	53	$270–1^{777}$	...	...	...	v	v	...	...	peth v CCl_4 v	**B19**^{2}, 293
b1823	—,**2-methyl-4-nitro-**, nitrile (*colorless*)	$C_8H_6N_2O_2$. *See* b1291	162.15	lf (sub)	100	sub	...	...	δ^h	δ	...	v	v	chl v	**B9**^{1}, 188
b1824	—,—,—,(*yellow*)	$C_8H_6N_2O_2$. *See* b1291	162.15	ye nd (al)	113–5	...	...	...	δ^h	δ	...	v	v	chl v	**B9**^{1}, 188
b1825	—,**2-methyl-5-nitro-**, nitrile	$C_8H_6N_2O_2$. *See* b1291	162.15	nd (al)	105	$174–5^{18}$	...	...	s^h	v^h	v	v	v	chl v	**B9**^{2}, 323
b1826	—,**2-methyl-6-nitro-**, nitrile	$C_8H_6N_2O_2$. *See* b1291	162.15	pl (al)	115	...	...	...	...	s	...	...	...	os s	**B9**^{2}, 323
b1827	—,**3-methyl-2-nitro-**, nitrile	$C_8H_6N_2O_2$. *See* b1291	162.15	nd (al)	84	...	...	...	...	s^h	...	...	...	...	**B9**^{1}, 191
b1828	—,**3-methyl-4-nitro-**, nitrile	$C_8H_6N_2O_2$. *See* b1291	162.15	pr (al), nd	80	...	...	...	δ	s^h	...	...	...	...	**B9**^{1}, 192
b1829	—,**3-methyl-5-nitro-**, nitrile	$C_8H_6N_2O_2$. *See* b1291	162.15	nd (lig)	104–5	...	...	...	...	s	...	...	...	lig s^h	**B9**^{1}, 192
b1830	—,**4-methyl-2-nitro-**, nitrile	$C_8H_6N_2O_2$. *See* b1291	162.15	nd (w)	99–101	...	...	...	s^h	v	...	...	v	chl v lig δ	**B9**^{2}, 334
b1831	—,**4-methyl-3-nitro-**, nitrile	$C_8H_6N_2O_2$. *See* b1291	162.15	ye nd (w)	107–8	171^{12}	...	...	s	s	v	v	v	chl v	**B9**^{2}, 334
b1832	—,**5-methyl-2-nitro-**, nitrile	$C_8H_6N_2O_2$ *See* b1291	162.15	nd (al)	92–3	...	...	...	v^h	v	...	...	v	aa v	**B9**^{1}, 192
b1833	—,**4(3-methyl-3-pentyl)-**	*p*-Isoamylbenzoic acid. $C_{13}H_{18}O_2$. *See* b1239	266.34	...	84	$153–4^1$	...	...	δ	s	...	...	δ	lig s	...
b1834	—,**2(1-naphth-oyl)-**	$C_{18}H_{12}O_3$. *See* b1239	276.29	...	174–6	...	...	...	δ^h	v	...	...	v	chl s	**E12B**, 3614
b1835	—,**2(2-naphth-oyl)-**	$C_{18}H_{12}O_3$. *See* b1239	276.29	nd (to)	168	...	...	...	...	...	...	...	...	os v aa v	**E12B**, 3621
b1836	—,**2-nitro-**	$C_7H_5NO_4$. *See* b1239	167.12	tcl nd (w)	147–8	...	...	...	δ^h	v	...	...	δ	chl δ lig δ	**B9**^{2}, 242
b1837	—,—,amide	$C_7H_6N_2O_3$. *See* b1244	166.14	nd (dil al)	176.6	317	...	...	s^h	s	s	...	...	...	**B9**^{2}, 246
b1838	—,—,—,*N*-phenyl-	$C_{13}H_{10}N_2O_3$. *See* b1244	242.24	nd (al or bz)	161–2	...	...	...	i	v	δ	...	s	chl s lig δ	**B12**^{2}, 153
b1839	—,—,azide	$C_7H_4N_4O_3$. *See* b1239	192.13	pr (eth)	37.5	...	...	...	...	...	v	...	v	chl v	**B9**, 376
b1840	—,—,ethyl ester	$C_9H_9NO_4$. *See* b1239	195.18	tcl	30	173^{18}	...	...	i	s	s	...	...	...	**B9**^{2}, 245
b1841	—,—,hydrazide	$C_7H_7N_3O_3$. *See* b1239	181.15	ye-br pr (w)	123	...	...	...	s	s	i	...	i	chl i	**B9**^{2}, 246
b1842	—,—,methyl ester	$C_8H_7NO_4$. *See* b1239	181.15	...	−13	275 176^{21}	...	...	i	s	s	...	s	MeOH s chl s	**B9**^{2}, 244
b1843	—,—,nitrile	$C_7H_4N_2O_2$. *See* b1291	148.12	nd (w or aa)	109.5	sub	...	...	δ s^h	s	s	v	...	CCl_4, aa s CS_2, peth δ chl v	**B9**^{2}, 246
b1844	—,**3-nitro-**	$C_7H_5NO_4$. *See* b1239	167.12	mcl pr (w)	140–1	...	1.494^{20}_4	...	δ	v	v	...	δ	chl s	**B9**^{2}, 247

For explanations, symbols and abbreviations see beginning of table.

No.	Name	Synonyms and Formula	Mol. wt.	Crystalline form, color and specific rotation	m.p. °C	b.p. °C	Density	n_D	Solubility w	al	eth	ace	bz	other solvents	Ref.
	Benzoic acid														
b1845	—,—,amide	$C_7H_6N_2O_3$. *See* b1244	166.14	ye mcl nd (w)	142.7	310–5			s^h	s	s	...	...		**B9**^{2}, 252
b1846	—,—,—,*N*-phenyl-	$C_{13}H_{10}N_2O_3$. *See* b1244	242.24	lf (w or al)	153–4	sub			i	s	s	...	s		**B12**^{2}, 153
b1847	—,—,azide	$C_7H_4N_4O_3$. *See* b1239	192.13	lf (dil al)	68				i	v	v	...	v	aa v	**B9**, 338
b1848	—,—,chloride	$C_7H_4ClNO_3$. *See* b1239	185.57	ye	35	275–8 154–5^{18}			d	d	v	...	...		**B9**^{2}, 252
b1849	—,—,ethyl ester	$C_9H_9NO_4$. *See* b1239	195.18	mcl pr	47	296–8 156^{10}			i	v	v	...	...		**B9**^{2}, 240
b1850	—,—,hydrazide	$C_7H_7N_3O_3$. *See* b1239	181.15	nd (w)	152				v^h	v^h	...	...	...		**B9**^{2}, 256
b1851	—,—,methyl ester	$C_8H_7NO_4$. *See* b1239	181.15	nd	78	279			i	δ	δ	...	...	MeOH δ	**B9**^{2}, 248
b1852	—,—,nitrile	$C_7H_4N_2O_2$. *See* b1239	148.12	nd (w)	117	sub			s^h	v^h	v	v	s^h	peth i aa v	**B9**^{2}, 254
b1853	**—,4-nitro-**	$C_7H_5NO_4$. *See* b1244	167.12	mcl lf (w)	242	sub	1.550^{32}_{4}		i s^h	δ	δ	s	δ	chl δ CS_2 δ	**B9**^{2}, 256
b1854	—,—,amide	$C_7H_6N_2O_3$. *See* b1244	166.14	nd (w)	201.4				i	s	s	...	...		**B9**^{2}, 271
b1855	—,—,*N*-phenyl-	$C_{13}H_{10}N_2O_3$. *See* b1239	242.24	lf (eth)	216				i	s	s	...	...		**B12**^{2}, 153
b1856	—,—,butyl ester	$C_{11}H_{13}NO_4$. *See* b1239	223.23	nd	35.3	160^8			...	s^h	v	...	v		**B9**^{2}, 259
b1857	—,—,chloride	$C_7H_4ClNO_3$. *See* b1239	185.57	nd (lig)	75	150–2^{15}			d	d	s	...	...		**B9**^{2}, 270
b1858	—,—,ethyl ester	$C_9H_9NO_4$. *See* b1239	195.18	tcl lf (al)	56.5				i	s	s	...	...		**B9**^{2}, 258
b1859	—,—,hydrazide	$C_7H_7N_3O_3$. *See* b1239	181.15	yesh nd (w)	217				$δ^h$	$δ^h$	i	...	i	chl i	**B9**^{2}, 274
b1860	—,—,methyl ester	$C_8H_7NO_4$. *See* b1239	181.15	ye mcl lf	96				i	s	s	...	...	chl s	**B9**^{2}, 258
b1861	—,—,nitrile	$C_7H_4N_2O_2$. *See* b1291	148.12	if (al), nd (bz)	147–8				δ s^h	δ s^h	...	...	...	chl s aa s	**B9**^{2}, 273
b1862	—,—,2,2,2-trichloroethyl ester	$C_9H_7Cl_3O_2$. *See* b1239	253.51	pr (al)	71				...	...	...	...	...		
b1863	**—,2-nitroso-**	$C_7H_5NO_3$. *See* b1239	151.12	(al or aa)	210d				...	$δ^h$	i	...	v	aa δ	**B9**^{2}, 241
b1864	**—,3-nitroso-**	$C_7H_5NO_3$. *See* b1239	151.12		230d				...	s	δ	...	δ		**B9**, 369
b1865	**—,4-nitroso-**	$C_7H_5NO_3$. *See* b1239	151.12	ye	250d				...	s^h	δ	...	δ	aa δ	**B9**^{2}, 242
b1866	**—,4-octyl-**	$C_{15}H_{22}O_2$. *See* b1239	234.34	lf (al)	99.5				...	s^h	...	...	...		**B9**, 571
b1867	**—,pentachloro-**	$C_7HCl_5O_2$. *See* b1239	294.35	nd or pl (bz or dil aa)	208				...	v	...	...	...	to v aa v	**B9**^{1}, 142
b1868	—,—,chloride	C_7Cl_6O. *See* b1239	312.80	pl (al)	87				...	v^h	...	...	...	MeOH v^h	
b1869	**—,pentamethyl-**	$C_{12}H_{16}O_2$. *See* b1239	192.26	nd (w), pl (dil al)	210.5	sub			i	v^h	...	...	...		**B9**, 569
b1870	**—,per-**	Perbenzoic acid, C_6H_5COOOH	138.12	mcl	42	97–110$^{13-5}$			i $δ^h$	v	v	...	...	chl s	**B9**^{2}, 157
b1871	**—,2-phenoxy-**	$C_{13}H_{10}O_3$. *See* b1239	214.22	fl (dil al)	113–4	355d			i $δ^h$	v	v	...	...	chl s	**B10**^{2}, 40
—	**—,phenyl-**	*see* **Biphenylcarboxylic acid**							...	...	...	...	...		
b1872	**—,2(phenylamino)-**	$C_{13}H_{11}NO_2$. *See* b1239	213.24	lf nd or pr (al)	183d				i $δ^h$	v^h	δ	...	$δ^h$		**B14**^{2}, 213
b1873	**—,2-phosphono-**	$C_7H_7O_5P$. *See* b1239	202.10	nd (w or al)	>300				v	δ	...	...	...		**B16**, 820
b1874	**—,2-propyl**	$C_{10}H_{12}O_2$. *See* b1239	164.21	lf (dil al)	58	272^{739}			s	v	v	...	...		**B9**^{1}, 213
b1875	**—,4-propyl-**	$C_{10}H_{12}O_2$. *See* b1239	164.21	pr or lf (w)	141				v^h	v	v	...	v	CS_2 v lig s chl v	**B9**, **545**
b1876	**—,3(1-semicarbazido)-**, amide	Cryogenine. $C_8H_{10}N_4O_2$. *See* b1239	194.19		172				i	s	s	s	...	chl s	**B15**^{2}, 297
—	**—,*x*-sulfamido-**	*see* **Benzoic acid, *x*-sulfo-, *x*-amide** ($x=2,3,4$)													
b1877	**—,2-sulfo-**	*o*-Carboxybenzenesulfonic acid. $C_7H_6O_5S$. *See* b1239	202.19	nd (w+3)	141, 70 (+3w)				v	v	...	...	...		**B11**^{2}, 215
b1878	—,—,1-amide	$C_7H_7NO_4S$. *See* b1244	201.20	pr(w+1)	193–4				v	v	...	...	...		**B11**^{2}, 215
b1879	—,—,2-amide	*o*-Sulfamylbenzoic acid. $C_7H_7NO_4S$. *See* b1239	201.20	pl or nd (w)	152				v	v	v	...	...		**B11**^{2}, 216
b1880	—,—,anhydride (endo)	CO O SO_2	184.17	pl or pr (bz)	129.5	184–6^{18}			i s^h	...	s	...	s	chl s	**B19**^{2}, 137
—	—,—,imide	*see* **Saccharine**													
b1881	—,—,nitrile	$C_7H_5NO_3S$. *See* b1291	183.19	nd (w)	279				s^h	s	δ	...	...	chl δ	**B11**, 372
b1882	**—,3-sulfo-**	$C_7H_6O_5S$. *See* b1239	202.19		141, 98(+w)				v	δ	v	...	i		**B11**^{2}, 217
b1883	—,—,3-amide	*m*-Sulfamylbenzoic acid. $C_7H_7NO_4S$. *See* b1239	201.20	pl (w)	246				s^h δ	s	δ	...	...		**B11**^{2}, 218
b1884	—,—,3-chloride	$C_7H_5ClO_4S$. *See* b1239	220.64	pr (bz)	131				d^h	d^h	s	...	s^h		**B11**^{2}, 218
b1885	—,—,dichloride	$C_7H_4Cl_2O_3S$ *See* b1239	239.08		20.4	153–4^7			...	...	...	...	...		**B11**, 386
b1886	**—,4-sulfo-**	$C_7H_6O_5S$. *See* b1239	202.11	nd (w+3)	259–60, 94 (+3w)				v	v	s	...	...		**B11**^{2}, 218
b1887	—,—,4-amide	*p*-Sulfamylbenzoic acid. $C_7H_7NO_4S$. *See* b1239	201.20	pr or fl (w)	280d				i $δ^h$	v	δ	...	i		**B11**^{2}, 219
b1888	—,—,4-chloride	$C_7H_5ClO_4S$. *See* b1239	220.64	nd (ace)		235–6			d^h	d^h	...	s	...	to s	**B11**^{2}, 219
b1889	**—,2,3,4,5-tetrachloro-**	$C_7H_2Cl_4O_2$. *See* b1239	259.92	nd (al)					δ	v	v	...	...		**B9**, 346

For explanations, symbols and abbreviations see beginning of table.

No.	Name	Synonyms and Formula	Mol. wt.	Crystalline form, color and specific rotation	m.p. °C	b.p. °C	Density	n_D	Solubility: w	al	eth	ace	bz	other solvents	Ref.
	Benzoic acid														
b1890	—,**2,3,4,5-tetra-hydroxy-**	$C_7H_6O_6$. *See* b1239	186.12	pr	84–5				v						**C50,** 14644
—	—,**2(3,4,5,6-tetra-hydroxy-xanthyl)-**	*see* **Gallin**													
b1891	—,**thiolo-**	$C_6H_5C(:O)SH$	138.19	ye pl (aa)	24	d			i	v	v			CS_2 v os s	**B9**^{2}, 286
b1892	—,**thiono-**, amide, *N*-phenyl-	$C_6H_5CSNHC_6H_5$	213.29	ye pl or pr	102	d			i	v	s		s^h	chl s lig δ	**B12**^{2}, 154
b1893	—,**2(2-toluyl)**	$C_{15}H_{12}O_3$. *See* b1239	240.25	nd (w+1)	130–2, 84(+w)				s	s^h	s			MeOH s	**B10**^{2}, 524
b1894	—,**2(3-toluyl)-**	$C_{15}H_{12}O_3$. *See* b1239	240.25	nd (w+1)	162				s^h						**B10**^{2}, 524
b1895	—,**2(4-toluyl)-**	$C_{15}H_{12}O_3$. *See* b1239	240.25	pr (w+1), nd (al)	146				δ^h	v	v	v	v	to v^h	**B10**^{2}, 525
b1896	—,—,methyl ester	$C_{16}H_{14}O_3$. *See* b1239	254.27	pl (MeOH)	66					v			v		**B10**, 759
b1897	—,**4(2-toluyl)-**	$C_{15}H_{12}O_3$. *See* b1239	240.25		177									xy s^h	**B10**^{2}, 527
b1898	—,**4(4-toluyl)-**	$C_{15}H_{12}O_3$. *See* b1239	240.25	nd	228				δ^h	v		v	δ	chl δ	**B10**^{2}, 528
b1899	—,**2,3,5-triamino-**	$C_7H_9N_3O_2$. *See* b1239	167.16						v^h	δ^h	i				**B14**, 455
b1900	—,**3,4,5-triamino-**	$C_7H_9N_3O_2$. *See* b1239	167.16	nd (w+½)					s^h	i	i				**B14**, 455
b1901	—,**2,3,4-tribromo-**	$C_7H_3Br_3O_2$. *See* b1239	358.85	nd (bz)	197–8				i	s	s		s^h		**B9**^{1}, 147
b1902	—,**2,3,5-tribromo-**	$C_7H_3Br_3O_2$. *See* b1239	358.85	nd (al)	193–4				i	v^h	v	v	δ	lig i	**B9**^{1}, 147
b1903	—,**2,4,5-tribromo-**	$C_7H_3Br_3O_2$. *See* b1239	358.85	nd (al or bz)	195–6				i	s^h				lig s^h	**B9**^{1}, 147
b1904	—,**2,4,6-tribromo-**	$C_7H_3Br_3O_2$. *See* b1239	358.85	pr (w)	194				δ	s			s		**B9**^{2}, 238
b1905	—,**3,4,5-tribromo-**	$C_7H_3Br_3O_2$ *See* b1239	358.85	nd (bz)	240				i	s				aa s^h	**B9**^{1}, 148
b1906	—,**2,3,4-trichloro-**	$C_7H_3Cl_3O_2$. *See* b1239	225.47	nd (w)	187–8										**B9**, 345
b1907	—,**2,3,5-trichloro-**	$C_7H_3Cl_3O_2$. *See* b1239	225.47	nd (w)	163				i					os s	**B9**, 345
b1908	—,**2,3,6-trichloro-**	$C_7H_3Cl_3O_2$. *See* b1239	225.47		163–4				δ		s				**B9**, 345
b1909	—,**2,4,5-trichloro-**	$C_7H_3Cl_3O_2$. *See* b1239	225.47	nd (w)	163				v^h s^h i	s	s				**B9**^{1}, 141
b1910	—,**2,4,6-trichloro-**	$C_7H_3Cl_3O_2$. *See* b1239	225.47	nd (w)	164				δ^h	v^h	v			chl v	**B9**, 345
b1911	—,**3,4,5-trichloro-**	$C_7H_3Cl_3O_2$. *See* b1239	225.47	nd (dil al)	210				δ^h i	s	s	s	s	CS_2 s	**B9**^{2}, 230
b1912	—,**2,3,5-triiodo-**	$C_7H_3I_3O_2$. *See* b1239	499.83	pr (al)	224–6				i	v^h	v		v^h		**B9**^{1}, 150
b1913	—,**2,4,5-triiodo-**	$C_7H_3I_3O_2$. *See* b1239	499.83	nd (al)	248				i	δ^h	δ		i		**B9**^{1}, 150
b1914	—,**3,4,5-triodo-**	$C_7H_3I_3O_2$. *See* b1239	499.83	pr (al)	288				i	v					**B9**, 367
b1915	—,**2,3,4-tri-hydroxy-**	$C_7H_6O_5$. *See* b1239	170.12	nd (w)	207–8				δ	s	s				**B10**^{2}, 331
b1916	—,—,ethyl ester	$C_9H_{10}O_5$. *See* b1239	198.17	(w+1)	102, 86(+w)				s^h i	v	v				**B10**, 467
b1917	—,—,methyl ester	$C_8H_8O_5$. *See* b1239	184.14	nd (w+2½)	151–2				s^h	s					**B10**, 466
b1918	—,**2,4,5-tri-hydroxy-**	$C_7H_6O_5$. *See* b1239	170.12	nd (w+½)	217–8d				v^h	v					**B10**^{2}, 334
b1919	—,**2,4,6-tri-hydroxy-**	$C_7H_6O_5$. *See* b1239	170.12	(w+1)	100				s^h δ	s	v		i		**B10**^{2}, 334
b1920	—,—,ethyl ester	$C_9H_{10}O_5$. *See* b1239	198.17	pr or nd (w+1)	129				s^h	v	v		v^h	lig δ	**B10**^{1}, 236
b1921	—,—,methyl ester	$C_8H_8O_5$. *See* b1239	184.14		174–6				δ	v	v		δ	CCl_4 δ	**B10**, 469
b1922	—,**3,4,5-tri-hydroxy-**	Gallic acid. $C_7H_6O_5$. *See* b1239	188.14	pr (w+1)	253d		1.694_4^6		v^h δ	v		s	i	chl i	**B10**^{2}, 335
b1923	—,—,amide	$C_7H_7NO_4$. *See* b1244	169.13	lf (w+1½)	244–5d				s^h δ	s				chl s aa s	**B10**^{2}, 346
b1924	—,—,—,*N*-phenyl-	$C_{13}H_{11}NO_4$. *See* b1244	245.24	lf (w+1), nd (w+2)	207				v^h i	s			i	chl i aa s	**B12**^{2}, 263
b1926	—,—,ethyl ester	$C_9H_{10}O_5$. *See* b1239	198.17	mcl pr (w+2½), nd (chl)	158				s^h δ	s	s			chl δ^h	**B10**^{2}, 343
b1927	—,—,isopropyl ester	$C_{10}H_{12}O_5$. *See* b1239	212.21		123–4				s	s	s		i	chl s^h	**B10**^{2}, 343
b1928	—,—,methyl ester	$C_8H_8O_5$. *See* b1239	184.15		202				δ	v				MeOH v^h	**B10**^{2}, 342
b1929	—,—,propyl ester	$C_{10}H_{12}O_5$. *See* b1239	212.21	nd (w)	147–8				δ s^h						**B10**^{2}, 343
b1930	—,**2,3,4-tri-methoxy-**	$C_{10}H_{12}O_5$. *See* b1239	212.21		100				s	s	s				**B10**^{2}, 332
b1931	—,**2,4,5-tri-methoxy-**	Asaronic acid. $C_{10}H_{12}O_5$. *See* b1239	212.21	nd (al)	144	300^{760}			s^h i	s			s	lig s	**B10**^{2}, 334
b1932	—,**3,4,5-tri-methoxy-**	$C_{10}H_{12}O_5$. *See* b1239	212.21	mcl nd (w)	166–8	$225–7^{10}$			δ	v	v			chl v	**B10**^{2}, 340
b1933	—,**2,3,4-tri-methyl-**	Prehnitilic acid. $C_{10}H_{12}O_2$. *See* b1239	164.21	pr (al)	167.5				s^h	s^h	s^h				**B9**, 552
b1934	—,**2,3,5-tri-methyl**	γ-Isodurilic acid. $C_{10}H_{12}O_2$. *See* b1239	164.21	pl (lig)	127					s					**B9**, 552
b1935	—,**2,3,6-tri-methyl-**	$C_{10}H_{12}O_2$. *See* b1239	164.21	nd (w or peth)	110–1				s	s	s				**B9**, 552
b1936	—,**2,4,5-tri-methyl-**	Durilic acid. $C_{10}H_{12}O_2$. *See* b1239	164.21	nd (bz)	152–3				δ^h	v	v		δ		**B9**^{2}, 361

For explanations, symbols and abbreviations see beginning of table.

No.	Name	Synonyms and Formula	Mol. wt.	Crystalline form, color and specific rotation	m.p. °C	b.p. °C	Density	n_D	Solubility w	al	eth	ace	bz	other solvents	Ref.
	Benzoic acid														
b1937	—,**2,4,6-trimethyl-**	β-Isodurilic acid. Mesitylene *eso*-carboxylic acid. $C_{10}H_{12}O_2$. *See* b1239	164.21	pr (lig)	155				δ	s	s	s		chl s	**B9,** 553
b1938	—,—,chloride	Mesiloyl chloride $C_{10}H_{11}ClO$. *See* b1239	182.65			$134–6^{60}$									**B9**¹, 360
b1939	—,**3,4,5-trimethyl-**	$C_{10}H_{12}O_2$. *See* b1239	164.21	nd (w)	215–6				$δ^h$ i	s	s				**B9,** 554
b1940	—,**2,3,6-trinitro-**	$C_7H_3N_3O_8$. *See* b1239	257.12	ye nd (w+2)	160, 55(+2w)				$δ^h$	s					**B9**¹, 168
b1941	—,**2,4,5-trinitro-**	$C_7H_3N_3O_8$. *See* b1239	257.12	ye lf (w)	194.5d				s^h	v	v		s	peth δ	**B9**¹, 168
b1942	—,**2,4,6-trinitro-**	$C_7H_3N_3O_8$. *See* b1239	257.12	ye nd (w)	228.7				δ	v	s	s		MeOH s	**B9**², 285
b1943	—,**3,4,5-trinitro-**	$C_7H_3N_3O_8$. *See* b1239	257.12	ye nd (eth+1)	168d				d^h		s				**B9**¹, 168
b1944	**Benzoin** (*d*)	Benzoylphenylcarbinol 4⟨⟩—CHOH—CO—⟨⟩4′	212.25	nd, $[\alpha]_D^{15}+92.8$ (Py, E=1)	133–4					v^h		v		MeOH v^h	**B8**², 193
b1945	—(*dl*)	$C_{14}H_{12}O_2$. *See* b1944	212.25	pr (al)	137	344^{768}	1.310_4^{20}		$δ^h$	s^h	δ			aa v^h	**B8**², 193
b1946	—(*l*)	$C_{14}H_{12}O_2$. *See* b1944	212.25	nd (MeOH), $[\alpha]_D^{12}-117.5$ (ace, c=1.25)	133–4					v^h		v		MeOH s^h	**B8,** 167
b1947	—,acetate	$C_6H_5CH(OCOCH_3)COC_6H_5$	254.29	pr or pl (eth)	83					v	v				**B8**², 196
b1948	—,ethyl ether	$C_6H_5CH(OC_2H_5)COC_6H_5$	240.30	nd (lig)	62	$194–5^{20}$	1.1016_4^{17}	1.5727^{17}		s	s		s	lig s^h	**B8**², 195
b1949	—,hydrazone	$C_6H_5CHOHC(:NNH_2)C_6H_5$	226.28	pr (al)	75				i	v^h					**B8,** 176
b1950	—,methyl ether	$C_6H_5CH(OCH_3)COC_6H_5$	226.28	nd (lig)	49–50	$188–9^{15}$	1.1278_4^{14}			v	v		v	lig v^h	**B8**², 195
b1951	—,oxime (*l*)	$C_6H_5CH(OH)C(:NOH)C_6H_5$	227.27	amor or pr (bz), $[\alpha]_D^{24}-3.2$ (chl, c=0.85)	163.5–4.5				i	s	s	s	$δ^h$		**B8,** 175
b1952	—,—,(*anti*)	$C_{14}H_{13}NO_2$. *See* b1951	227.27	pr (bz)	151–2										**B8**², 196
b1953	—,—,(*syn*)	$C_{14}H_{13}NO_2$. *See* b1951	227.27	pr (eth)	99										**B8**², 196
b1954	—,succinate	$[C_6H_5CH(COC_6H_5)O_2CCH-]_2$	506.56	lf (al)	129				i	s	s			CS_2 s	**B8,** 175
b1955	—,**4,4′-dimethoxy-**	Anisoin. $C_{16}H_{16}O_4$. *See* b1944	272.30	pr (dil al)	113				$δ^h$	v^h δ	δ				**B8**², 476
—	**Benzonitrile**	*see* **Benzoic acid,** nitrile													
—	**1,2-Benzophenanthrene**	*see* **Chrysene**													
b1956	**3,4-Benzophenanthrene**		228.30	nd or pl (lig or al)	167.5–8.3				i	δ				lig δ	**E14,** 356
b1957	**9,10-Benzophenanthrene**	Triphenylene.	228.30	nd (al, chl, bz)	199	425			i	s			v	chl v	**E14,** 357
b1958	**1,2-Benzophenazine**	Benzo[*a*]phenazine.	230.26	pr (al), nd (bz)	142.5	>360			i	s	s		δ	aa s	**B23**², 259
b1959	**Benzophenone**	Diphenyl ketone Benzoylbenzene	182.21	rh or mcl	48.1(α) 26(β)	305.9^{760}	α:1.146^{20} β:1.1076	α:1.6077^{19} β:1.6059^{23}	i	v				aa v	**B7**², 350
b1960	—,imine	$(C_6H_5)_2C:NH$	181.24			282 158^{12}	1.0847_4^{19}	1.6191^{19}	d		s				**B7**², 355
b1961	—,oxime	$(C_6H_5)_2C:NOH$	197.23	nd (al)	144				i	v^h	s	v		MeOH v^h	**B7**², 355
b1962	—,phenylhydrazone	$(C_6H_5)_2C:NNHC_6H_5$	272.34	pr or nd (al)	137					δ	s		s	aa s	**B15**², 63
b1963	—,**2-amino-**	$C_{13}H_{11}NO$. *See* b1959	197.23	lf or pr (al)	110–1					s^h	s				**B14**², 51
b1964	—,**3-amino-**	$C_{13}H_{11}NO$. *See* b1959	197.23	ye nd (w)	87				δ	s	s				**B14**¹, 388
b1965	—,**4-amino-**	$C_{13}H_{11}NO$. *See* b1959	197.23	lf (dil al)	124				s^h δ	s	s			aa s	**B14**², 54
b1966	—,**2-amino-4′-methyl-**	$C_{14}H_{13}NO$. *See* b1959	211.25	ye pr or pl (al)	96					v	v		v		**B14**², 63
b1967	—,**2-amino-5-methyl-**	$C_{14}H_{13}NO$. *See* b1959	211.25	ye nd or pl	66					v	v	v		chl v lig v aa v	**B14,** 106
b1968	—,**3-amino-4-methyl**	$C_{14}H_{13}NO$. *See* b1959	211.25	pa ye nd (MeOH)	108–10				$δ^h$	v	v	v	v	MeOH s^h CS_2 s^h chl v	**B14**², 63

For explanations, symbols and abbreviations see beginning of table.

No.	Name	Synonyms and Formula	Mol. wt.	Crystalline form, color and specific rotation	m.p. °C	b.p. °C	Density	n_D	Solubility						Ref.
									w	al	eth	ace	bz	other solvents	
	Benzophenone														
b1969	**—,3-amino-4′-methyl-**	$C_{14}H_{13}NO$. *See* b1959	211.25	pr (al)	111					v	v				**B14**, 107
b1970	**—,4-amino-3-methyl-**	$C_{14}H_{13}NO$. *See* b1959	211.25	pr (w)	112				δ s^h	v					**B14**, 105
b1971	**—,4-amino-4′-methyl-**	$C_{14}H_{13}NO$. *See* b1959	211.25	nd (bz)	179					v			s^h	chl v lig δ CS_2 v	**B14**, 107
b1973	**—,4,4′-bis-(diethylamino)-**	$C_{21}H_{28}N_2O$. *See* b1959	324.46	lf (al)	95–6					δ v^h					**B14**^{2}, 59
b1974	**—,4,4′-bis(dimethylamino)-**	Michler's ketone. $C_{17}H_{20}N_2O$. *See* b1959	268.35	lf (al), nd (bz)	174				i	δ	i		v^h	Py v	**B14**^{2}, 57
b1975	**—,4,4′-bis(dimethylamino), thio-**	$C_{17}H_{20}N_2S$. *See* b1959	284.23	pl	204				i	i	δ			chl s	**B14**^{2}, 60
b1976	**—,2-bromo-**	$C_{13}H_9BrO$. *See* b1959	261.13	pr (al), nd (lig)	42	345^{759}			i	s^h				lig s^h	**B7**^{2}, 360
b1977	**—,3-bromo-**	$C_{13}H_9BrO$. *See* b1959	261.13	nd	81.5				i	s^h					**B7**^{2}, 360
b1978	**—,4-bromo-**	$C_{13}H_9BrO$. *See* b1959	261.13	lf	82.5	350^{757}			i	δ^h	δ		δ		**B7**^{2}, 360
b1979	**—,2-chloro-**	$C_{13}H_9ClO$. *See* b1959	216.56	pl	45.5	330									**B7**^{1}, 227
b1980	**—,3-chloro-**	$C_{13}H_9ClO$. *See* b1959	216.56	nd	82–3					δ					**B7**^{1}, 227
b1981	**—,4-chloro-**	$C_{13}H_9ClO$. *See* b1959	216.56	nd (al)	76	332^{771}				s^h	s				**B7**^{2}, 359
b1982	**—,2-chloro-3,5-dinitro-**	$C_{13}H_7ClN_2O_5$. *See* b1959	306.66	ye nd (aa)	149					δ				chl s aa s^h lig i	**B7**, 428
b1983	**—,2,2′-diamino-**	$C_{13}H_{12}N_2O$. *See* b1259	212.24	lf (dil al), pr (bz)	134–5				i	s					**B14**, 87
b1984	**—,3,3′-diamino-**	$C_{13}H_{12}N_2O$. *See* b1259	212.24	nd (al)	171	285^{11}			i s^h	s	s				**B14**^{1}, 390
b1985	**—,4,4′-diamino-**	$C_{13}H_{12}N_2O$. *See* b1259	212.24	nd (al)	244–5				d^h	s	s				**B14**^{2}, 56
b1986	**—,4,4′-dibromo-**	$C_{13}H_8Br_2O$. *See* b1259	340.20	pl (al)	175				i	s^h			s	chl s CS_2 s	**B7**, 423
b1987	**—,2,4′-dichloro-**	$C_{13}H_8Cl_2O$. *See* b1259	251.12	pr (al)	67	$214–5^{22}$				s^h					**B7**, 420
b1988	**—,4,4′-dichloro-**	$C_{13}H_8Cl_2O$. *See* b1259	251.12	pl (al)	145	353^{757}			i	s^h	v			chl v aa v	**B7**, 420
b1989	**—,2,2′-dihydroxy-**	$C_{13}H_{10}O_3$. *See* b1259	214.21	lf or pr (lig)	59.5	330–40			i	s	s			chl s	**B8**^{2}, 354
b1990	**—,2,3′-dihydroxy-**	$C_{13}H_{10}O_3$. *See* b1259	214.21	nd (w)	126					s	s				**B8**, 315
b1991	**—,2,4-dihydroxy-**	$C_{13}H_{10}O_3$. *See* b1959	214.21	nd (w)	144				i	s	v		δ	aa v	**B8**^{2}, 352
b1992	**—,2,4′-dihydroxy-**	$C_{13}H_{10}O_3$. *See* b1959	214.21	pl (w)	150–1				δ^h	s	s		s^h		**B8**^{2}, 354
b1993	**—,2,5-dihydroxy-**	$C_{13}H_{10}O_3$. *See* b1959	214.21	ye nd (dil al)	125–6				s^h						**B8**^{2}, 353
b1994	**—,3,3′-dihydroxy-**	$C_{13}H_{10}O_3$. *See* b1959	214.21	nd (w)	163–4				s^h	s					**B8**, 316
b1995	**—,3,4′-dihydroxy-**	$C_{13}H_{10}O_3$. *See* b1959	214.21	nd (w)	206				s^h	s	s				**B8**, 316
b1996	**—4,4′-dihydroxy-**	$C_{13}H_{10}O_3$. *See* b1959	214.21	nd (lig)	212–3	1.133^{131}			δ s^h		s	s	i	CS_2, chl i, MeOH s	**B8**^{2},.355
b1997	**—,2,4-dihydroxy-6-methoxy-**	Isocotoin $C_{14}H_{12}O_4$. *See* b1959	244.25	ye nd (lig)	162									lig δ	**B8**^{2}, 467
b1998	**—,2,6-dihydroxy-4-methoxy-**	Cotoin. $C_{14}H_{12}O_4$. *See* b1959	244.25	yesh pr (chl), lf or nd (w)	130–1				δ^h	s	s	s	s^h	chl s CS_2 s	**B8**^{2}, 467
b1999	**—,4,4′-diodo-**	$C_{13}H_8I_2O$. *See* b1959	434.05	pl (to)	233–4				i	δ				to s^h	**B7**, 425
b2000	**—,2,4-dimethoxy-**	$C_{15}H_{14}O_3$. *See* b1959	242.28	pr (dil al)	87–8	218^{10}			i	v	δ			chl v lig δ	**B8**^{2}, 353
b2001	**—,2,4′-dimethoxy-**	$C_{15}H_{14}O_3$. *See* b1959	242.28	nd (al)	100				i	v^h	v		v	aa v	**B8**^{1}, 640
b2002	**—,2,5-dimethoxy**	$C_{15}H_{14}O_3$. *See* b1959	242.28		51	225^{28}			i	δ	δ			lig δ	**B8**^{2}, 354
b2003	**—,3,4-dimethoxy-**	$C_{15}H_{14}O_3$. *See* b1959	242 28	nd or pl (al)	103–4				i	v					**B8**^{2}, 354
b2004	**—,3,4′-dimethoxy-**	$C_{15}H_{14}O_3$. *See* b1959	242.28	pr (al)	55				i	v^h s					**B8**^{2}, 354
b2005	**—,4,4′-dimethoxy-**	$C_{15}H_{14}O_3$. *See* b1959	242.28	nd (al)	144					v^h			v	chl v	**B8**, 317
b2006	**—,4,4′-dimethyl-**	$C_{15}H_{14}O$. *See* b1959	210.26	rh (al)	95	333^{725}			i	v	v			chl v	**B7**^{2}, 387
b2007	**—,3(dimethylamino)-**	$C_{15}H_{15}NO$. *See* b1959	225.29	pa ye pl (al)	47	216^{15}				s^h					**B14**^{1}, 388
b2008	**—,4(dimethylamino)-**	$C_{15}H_{15}NO$. *See* b1959	225.29	ye lf (al), nd (peth)	92				i	δ v^h	v			peth s^h chl s	**B14**^{2}, 54
b2009	**—,2,2′-dinitro-**	$C_{13}H_8N_2O_5$ *See* b1959	272.21	nd (to or aa)	188–9				i					to s^h aa s^h	**B7**, 427
b2010	**—,2-hydroxy-**	$C_{13}H_{10}O_2$. *See* b1959	198.21	pl (dil al)	41				i	v	v		v	peth δ	**B8**, 155
b2011	**—,3-hydroxy-**	$C_{13}H_{10}O_2$. *See* b1959	198.21	lf (al)	116					v	v				**B8**, 157
b2012	**—,4-hydroxy-**	$C_{13}H_{10}O_2$. *See* b1959	198.21	nd (al)	135 (st), 122(unst)				s^h δ	v	v			aa v	**B8**^{2}, 184
b2013	**—,2-methoxy-**	$C_{14}H_{12}O_2$ *See* b1959	212.24		41	210^{27}			i	s			s	aa s	**B8**^{2}, 182
b2014	**—,3-methoxy-**	$C_{14}H_{12}O_2$ *See* b1959	212.24		44	$342–3^{730}$ 201^{17}			i	v			v	aa v	**B8**^{1}, 569
b2015	**—,4-methoxy-**	$C_{14}H_{12}O_2$. *See* b1959	212.24	pr (eth)	67–8	$354–5^{729}$			i	v	v		s	chl s aa s	**B8**^{1}, 569
b2016	**—,2-methyl-**	$C_{14}H_{12}O$. *See* b1959	196.25		<-18	309.5^{762}			i	v				os s	**B7**^{2}, 371
b2017	**—,3-methyl-**	$C_{14}H_{12}O$. *See* b1959	196.25			$314–5^{725}$ 170^8			i	s	s			chl s aa s	**B7**^{2}, 372
b2018	**—,4-methyl-**	$C_{14}H_{12}O$. *See* b1959	196.25	mcl pr	59–60				i	δ^h	s		s	lig δ	**B7**^{2}, 372
b2019	**—,—,imine**	$C_{14}H_{13}N$. *See* b1959	195.27		37	147^5	1.0617^{20}_4	1.6097^{20}	i						**B7**^{1}, 235
b2020	—,—,diphenyl-acetal	$C_{26}H_{22}O_2$. *See* b1959	366.43		134				i	i				os s peth i	**B7**^{2}, 372
b2021	**—,4-methyl-2-nitro-**	$C_{14}H_{11}NO_3$. *See* b1959	241.24	nd or pl (al)	126–7				i	s^h			s	chl s aa s^h	**B7**, 442
b2022	**—,4-methyl-2′-nitro-**	$C_{14}H_{11}NO_3$. *See* b1959	241.24	pr (aa)	155				i	δ^h			s^h	chl s aa s^h	**B7**, 442

For explanations, symbols and abbreviations see beginning of table.

No.	Name	Synonyms and Formula	Mol. wt.	Crystalline form, color and specific rotation	m.p. °C	b.p. °C	Density	n_D	Solubility w	al	eth	ace	bz	other solvents	Ref.
	Benzophenone														
b2023	**—,4-methyl-3-nitro-**	$C_{14}H_{11}NO_3$. *See* b1959	241.24	pr (al)	129–30				δ	s	s	v	v	CS_2 v aa v	**B7**², 374
b2024	**—,4-methyl-3′-nitro-**	$C_{14}H_{11}NO_3$. *See* b1959	241.24	lf	111				i	δ	s	...	s	chl s	**B7**², 375
b2025	**—,4-methyl-4-nitro-**	$C_{14}H_{11}NO_3$. *See* b1959	241.24	nd	122–4				i	s	s	...	...	chl s aa s	**B7**², 375
b2026	**—,—,oxime**	$C_{14}H_{12}N_2O_3$. *See* b1959	256.25	nd (eth-lig)	145				...	s	s	...	s	lig δ	**B7**, 443
b2027	**—,2-nitro-**	$C_{13}H_9NO_3$. *See* b1959	227.21	mcl (al)	105				...	δ	...	...	...		**B7**², 362
b2028	**—,3-nitro-**	$C_{13}H_9NO_3$. *See* b1959	227.21	ye nd (al)	94–5	234^{18}			...	sʰ	...	...	...		**B7**², 362
b2029	**—,4-nitro-**	$C_{13}H_9NO_3$. *See* b1959	227.21	nd or lf (al)	138		1.406^{0}_{4}		δ	δ sʰ	...	...	δ	CS_2 δ lig δ	**B7**², 362
b2030	**—,2,2′-3,4-tetra-hydroxy-**	$C_{13}H_{10}O_5$ *See* b1959	246.22	ye lf or pl (w+1)	149 102(+w)				sʰ	v	v	...	δ	aa v lig δ	**B8**², 539
b2031	**—,2,2′,4,4′-tetra-hydroxy-**	$C_{13}H_{10}O_5$. *See* b1959	246.22	ye nd (w+1½)	193–5				sʰ	v	v	v	sʰ	chl sʰ MeOH v	**B8**², 540
b2032	**—,2,2′,4,6′-tetra-hydroxy-**	Isoeuxanthonic acid. $C_{13}H_{10}O_5$. *See* b1959	246.22	(w+1)	200d				vʰ δ	s	s	...	...		**B8**, 496
b2033	**—,2,2′,5,6′-tetra-hydroxy-**	Euxanthoic acid. $C_{13}H_{10}O_5$ *See* b1959	246.22	nd (w)	200–2d				s vʰ	s	...	...	...		**B8**², 541
b2034	**—,2,3′,4,4′-tetra-hydroxy-**	$C_{13}H_{10}O_5$. *See* b1959	246.22	nd (w+2)	202				sʰ δ	v	v	v	δ	aa v lig i	**B8**², 541
b2035	**—,2,3′,4,6-tetra-hydroxy-**	$C_{13}H_{10}O_5$. *See* b1959	246.22	lf (w)	246d				vʰ	v	δ	...	...		**B8**², 540
b2036	**—,2,4,4′,6-tetra-hydroxy-**	$C_{13}H_{10}O_5$. *See* b1959	246.22	pr or nd (w+2)	210				sʰ	s	s	...	...		**B8**², 540
b2037	**—,3,3′,4,4′-tetra-hydroxy-**	$C_{13}H_{10}O_5$. *See* b1959	246.22		227–8				sʰ	s	...	...	s	peth δ	**B8**², 541
b2038	**—,2,2′,4,4′-tetra-methoxy-**	$C_{17}H_{18}O_5$. *See* b1959	302.31	lf (dil al)	130				...	v	...	...	...		**B7**², 540
b2039	**—,2,2′,5,5′-tetra-methoxy-**	$C_{17}H_{18}O_5$. *See* b1959	302.31		109				...	vʰ s	s	...	v	chl v aa s vʰ	**B8**, 497
b2040	**—,2,2′,6,6′-tetra-methoxy-**	$C_{17}H_{18}O_5$. *See* b1959	302.31	pl (bz)	204				...	δʰ	δ	...	δ	chl v aa δ	**B8**¹, 735
b2041	**—,2,3′,4,4′-tetra-methoxy-**	$C_{17}H_{18}O_5$. *See* b1959	302.31	nd or lf (al)	107				...	s	...	...	...		**B8**¹, 735
b2042	**—,2,3′,4,5′-tetra-methoxy-**	$C_{17}H_{18}O_5$. *See* b1959	302.31	nd (bz-peth)	73–4				...	v	v	...	vʰ	chl v aa δʰ peth i	**B8**¹, 735
b2043	**—,2,3′,4′,5-tetra-methoxy-**	$C_{17}H_{18}O_5$. *See* b1959	302.31	pr (dil al)	101–2				...	s	s	...	...		**B8**², 541
b2044	**—,2,3,4,6-tetra-methoxy-**	$C_{17}H_{18}O_5$. *See* b1959	302.31		125–6				i	v	...	v	v	lig sʰ	**B8**¹, 734
b2045	**—,2,4,4′,5-tetra-methoxy-**	$C_{17}H_{18}O_5$. *See* b1959	302.31		122–4				...	s	...	v	v	chl v	**B8**¹, 734
b2046	**—,2,4,4′,6-tetra-methoxy-**	$C_{17}H_{18}O_5$. *See* b1959	302.31	pr (al)	146				...	s	s	...	...		**B8**, 496
b2047	**—,3,3′,4,4′-tetra-methoxy-**	Veratrophenone $C_{17}H_{18}O_5$ *See* b1959	302.31	pr (al)	144.5				...	vʰ s	...	...	s	peth δ	**B8**², 541
b2048	**—,3,3′,4,5′-tetra-methoxy-**	$C_{17}H_{18}O_5$. *See* b1959	302.31	nd (bz)	114–5				...	v	v	...	vʰ	chl v lig i aa δ	**B8**¹, 735
b2049	**—,thio-**	$C_6H_5CSC_6H_5$	198.29	nd (peth)	51–2	174^{14}			...	δ	...	...	v	chl v peth δ os v	**B7**², 365
b2050	**—,2,2′,6-tri-hydroxy-**	$C_{13}H_{10}O_4$. *See* b1959	230.21	nd (al)	133–4				i sʰ	s	s	...	s		**B8**², 468
b2051	**—,2,3,4-tri-hydroxy-**	$C_{13}H_{10}O_4$. *See* b1959	230.21	ye nd (dil al)	140–1				sʰ δ	s	s	s	δ	aa s	**B8**², 466
b2052	**—,2,4,4′-tri-hydroxy-**	$C_{13}H_{10}O_4$. *See* b1959	230.21	nd (w+2)	200–1				sʰ	s	...	...	...		**B8**², 468
b2053	**—,2,4,6-tri-hydroxy-**	$C_{13}H_{10}O_4$. *See* b1959	230.21	nd (w+1)	165				vʰ δ	v	v	...	δ	chl δ lig i	**B8**², 466
b2054	**—,3,4,5-tri-hydroxy-**	$C_{13}H_{10}O_4$. *See* b1959	230.21	lf (w+1)	177–8				vʰ δ	v	v	v	δ	peth i	**B8**, 422
b2055	**—,2,3,4-tri-methoxy-**	$C_{16}H_{16}O_4$. *See* b1959	272.27	pr (dil al)	55				...	s	v	...	...		**B8**, 418
b2056	**—,2,4,4′-tri-methoxy-**	$C_{16}H_{16}O_4$. *See* b1959	272.27	nd (al)	73–4				...	sʰ	v	...	v	aa v lig i	**B8**¹, 702
b2057	**—,2,4,5-tri-methoxy-**	$C_{16}H_{16}O_4$. *See* b1959	272.27	nd (w)	97				δʰ i	v	...	v	v		**B8**¹, 701
b2058	**—,2,4,6-tri-methoxy-**	$C_{16}H_{16}O_4$. *See* b1959	272.27	mcl pr or rh pl (al)	115				δʰ i	vʰ	v	...	...	peth i aa δ vʰ chl v	**B8**, 420
b2059	**—,3,3′,4-tri-methoxy-**	$C_{16}H_{16}O_4$. *See* b1959	272.27	nd (MeOH)	83–4				...	s	...	...	...	MeOH sʰ	**B8**², 468
b2060	**—,3,4,4′-tri-methoxy-**	$C_{16}H_{16}O_4$. *See* b1959	272.27	nd (al)	98–9				...	sʰ	...	...	...		**B8**, 422

For explanations, symbols and abbreviations see beginning of table.

No.	Name	Synonyms and Formula	Mol. wt.	Crystalline form, color and specific rotation	m.p. °C	b.p. °C	Density	n_D	Solubility w	al	eth	ace	bz	other solvents	Ref.
	Benzophenone														
b2061	—,3,4′,5-trimethoxy-	$C_{16}H_{16}O_4$ *See* b1959	272.27	nd (bz)	97–8				...	v	v	...	v	chl v lig i aa δ	**B8**^{1}, 702
—	**Benzophenonecarboxylic acid**	*see* **Benzoic acid, benzoyl-**													
—	**Benzopinacol**	*see* **1,2-Ethanediol, 1,1,2,2-tetraphenyl-**													
—	**α-Benzopinacolone**	*see* **Acetophenone, α,α,α-triphenyl**													
—	**1,4-Benzopyran, 2,3-dihydro-**	*see* **Chroman**													
—	**1,2-Benzopyrazole**	*see* **Indazole**													
b2062	**3,4-Benzopyrene**	8 7 6 5 4 9 10 3 3′ 1 2 1′ 2′	252.30	mcl nd	179				i	...	...	...	...	...	**E14**, 457
b2063	—,5-amino-	$C_{20}H_{13}N$ *See* b2026	267.33	ye pl (bz-lig)	239–41				δ	s	δ	v	v	MeOH δ chl v	**E14s**, 697
b2064	—,5-hydroxy-	$C_{20}H_{12}O$ *See* b2026	268.32	nd (eth-lig)	207–9d				...	v	...	...	v	alk s^h aa s	**E14s**, 701
b2065	—,8-hydroxy-	$C_{20}H_{12}O$ *See* b2062	268.32	ye nd (bz-peth)	226–7d				...	s	s	s	s	...	**E14s**, 704
—	**1,2-Benzopyrone**	*see* **Isocoumarin**													
—	**2,1-Benzopyrone**	*see* **Coumarin**													
—	**1,4-Benzopyrone**	*see* **Chromone**													
—	**Benzo[α]pyrrole**	*see* **Indole**													
b2066	**5,6-Benzoquinoline**	N	179.22	lf (dil al or w)	93.5	350^{721} 202–5^8			$δ^h$	v	v	...	v	...	**B20**^{2}, 302
b2067	**7,8-Benzoquinoline**	N 2	179.22	lf (eth), pl (peth)	52	338^{719} 233^{47}			δ	s	s	...	s	os s	**B20**^{2}, 302
b2068	—,2-methyl-	$C_{14}H_{11}N$. *See* b2067	193.25			324–6	1.1464^{20}_4	1.6738^{20}	i	s	...	...	...	...	**B20**, 471
b2068[1]	**1,2-Benzoquinone**	3 O 4 =O 5 6	108.09	red pl or pr	60–70d				...	...	s	...	s^h	peth i	**B7**, 600
b2069	**1,4-Benzoquinone***	*p*-Quinone O= 4 5 6 1 =O 3 2	108.10	ye mcl pr (w)	115–7	sub			δ	s	...	...	...	peth δ	**B7**, 609
b2070	—,diimide	HN= =NH	106.13	ye nd	124				...	...	...	...	s^h	chl s	**B7**^{2}, 574
b2071	—,—,*N,N*-dichloro-	$C_6H_4Cl_2N_2$. *See* b2070	175.02	nd (w)	126d				s^h i	v^h	v	...	v^h	...	**B7**, 621
b2072	—,dioxime	HON= =NOH	138.13	ye nd	240d				s^h	...	...	...	...	...	**B7**, 627
b2073	—,monoimide, *N*-chloro-	O= —NCl	141.56	ye nd	85				v	$δ^h$	δ	...	...	chl $δ^h$	**B7**, 619
—	—,monooxime	*see* **Phenol, 4-nitroso-***													
b2074	—,2-bromo-6-methyl-*	$C_7H_5BrO_2$. *See* b2069	200.94	nd (al), pr (eth or lig)	95	sub			δ	v	v	...	...	chl v	**B7**^{2}, 591
b2075	—,5-bromo-2-methyl-*	$C_7H_5BrO_2$. *See* b2069	200.94	ye lf (lig)	106				...	v	...	...	...	lig δ v^h	**B7**^{2}, 591
b2076	1,2-Benzoquinone, 3-chloro-*	$C_6H_3ClO_2$. *See* b2068[1]	142.54	ye-red pr	68d				...	s	s	...	...	...	**B7**^{1}, 338
b2077	—,4-chloro-*	$C_6H_3ClO_2$. *See* b2068[1]	142.54	ye-red nd	78				...	...	s	...	...	lig s^h	**B7**^{1}, 338
b2078	1,4-Benzoquinone, 2-chloro-*	$C_6H_3ClO_2$. *See* b2069	142.54	ye-red	57				v	v	v	...	...	chl v	**B7**^{2}, 579
b2079	—,—,oxime	$C_6H_4NClO_2$ *See* b2069	157.56	gr-ye nd	184d				δ	s	s	...	...	...	**B7**^{2}, 579
b2080	—,2-chloro-3-methyl-*	$C_7H_5ClO_2$. *See* b2069	156.57		55				δ	v	...	...	...	chl v lig v	**B7**^{1}, 353
b2081	—,2-chloro-5-methyl-*	$C_7H_5ClO_2$ *See* b2069	156.57	ye nd (w or al)	104–5				s^h	v	v	...	...	chl v	**B7**^{2}, 590
b2082	—,2-chloro-6-methyl-*	$C_7H_5ClO_2$. *See* b2069	156.57	ye nd	90				δ	v	v	...	...	chl v	**B7**^{1}, 353
b2083	—,2,6-dibromo-, 1-imine, *N*-chloro-	$C_6H_2Br_2ClNO$. *See* b2073	299.36	ye pr	85–6				i	...	...	...	...	...	**B7**^{2}, 584
b2084	—,—,4-imine, *N*-chloro-	$C_6H_2Br_2ClNO$. *See* b2073	299.36	ye pr (al or aa)	83				...	s^h	...	...	...	...	**B7**^{2}, 584

For explanations, symbols and abbreviations see beginning of table.

No.	Name	Synonyms and Formula	Mol. wt.	Crystalline form, color and specific rotation	m.p. °C	b.p. °C	Density	n_D	Solubility w	al	eth	ace	bz	other solvents	Ref.
	Benzoquinone														
b2085	—,**3,5-dibromo-2,6-dimethyl-***	$C_8H_6Br_2O_2$. *See* b2069	293.95	ye lf	176	sub			i	s					B7^2, 593
b2086	—,**3,5-dibromo-2-methyl-***	$C_7H_4Br_2O_2$. *See* b2069	279.93	ye	117					v	v			chl v	B7^2, 591
b2087	—,**di-*tert*-butyl-**	$C_{14}H_{20}O_2$. *See* b2069	220.31	ye	152.5				i	s^h	s		s	aa s	B7^2, 670
b2088	**1,2-Benzoquinone, 4,5-dichloro-**	$C_6H_2Cl_2O_2$. *See* b2068[1]	176.99	yesh red pr or pl	94d						s		s		B7^1, 338
b2089	**1,4-Benzoquinone, 2,3-dichloro-**	$C_6H_2Cl_2O_2$. *See* b2069	176.99	ye lf	100–1						s		s		B7^2, 580
b2090	—,**2,5-dichloro-***	$C_6H_2Cl_2O_2$. *See* b2069	176.99	pa ye mcl pr	161				i	s^h δ	s			chl s	B7^2, 580
b2091	—,**2,6-dichloro-***	$C_6H_2Cl_2O_2$. *See* b2069	176.99	ye rh	120–1				δ^h	δ				chl s	B7, 633
b2092	—,—,1-imine *N*-chloro-	$C_6H_2Cl_3NO$. *See* b2073	210.44	ye nd (al)	67–8				δ	s^h	v			chl v	B7^2, 581
b2093	—,**2,5-dichloro-3,6-dihydroxy-***	Chloranilic acid. $C_6H_2Cl_2O_4$. *See* b2069	208.99	red lf (w+2)	283–4				s^h						B8^2, 433
b2094	—,**2,5-dihydroxy-***	$C_6H_4O_4$. *See* b2069	140.09	dk ye nd	211	215–20d			δ	s^h	i			AcOEt δ aa δ^h	B8^2, 432
b2095	—,**2,5-dihydroxy-3-dodecyl-***	Embelin. $C_{18}H_{28}O_4$. *See* b2069	308.42	og-red lf	143				i					os δ^h lig δ	
b2096	—,**2,5-dihydroxy-3-methoxy-6-methyl-***	Spinulosin. $C_8H_8O_5$. *See* b2069	184.15	red-bl	202				δ s^h					alk s	
—	—,**dihydroxy-**, monooxime	*see* **Benzene, trihydroxy-nitroso-**													
b2097	—,**2,6-dimethoxy-***	$C_8H_8O_4$. *See* b2069	168.15	ye mcl pr (aa)	256	sub			δ^h	δ	δ			alk v aa v^h	B8^2, 433
b2098	—,**2,3-dimethyl-***	o-Xyloquinone. $C_8H_8O_2$. *See* b2069	136.15	ye nd	55	sub			δ	s	s				B7^2, 593
b2099	—,**2,6-dimethyl-***	$C_8H_8O_2$. *See* b2069	136.15	ye nd	72–3		1.0479$^{78}_{4}$								B7^2, 593
b2100	—,**2,5-diphenyl-***	$C_{18}H_{12}O_2$. *See* b2069	260.30	ye lf (bz or aa)	214				i	δ			s^h	aa s^h	B7^2, 757
b2101	—,**3-hydroxy-2-methoxy-5-methyl-**	Fumigatin. $C_8H_8O_4$. *See* b2069	168.15	br nd or pl (peth)	116				δ	v	v	v	v	chl v peth δ	
b2102	**1,2-Benzoquinone, 4-isopropyl-5-methyl-**, 1-oxime	$C_{10}H_{13}NO_2$. *See* b2068	179.22	nd (bz-chl)	165–7d								s^h	chl s^h	B7^2, 595
b2103	**1,4-Benzoquinone, 2-isopropyl-5-methyl-, dioxime**	$C_{10}H_{14}N_2O_2$. *See* b2072	194.24		235d					v^h δ	δ		δ	chl δ aa δ v^h	B7, 665
b2104	**1,2-Benzoquinone, 3-methoxy-***	$C_7H_6O_3$. *See* b2068[1]	138.12	br-red pl, pr, nd	115–20				s	s	s		v	chl v	B8^2, 264
b2105	—,**4-methoxy-1-**oxime	$C_7H_7NO_3$. *See* b2068[1]	153.14	ye pr	158–9				δ	v			v	aa v	B8^2, 264
b2106	**1,4-Benzoquinone, 2-methoxy-***	$C_7H_6O_3$. *See* b2069	153.14	ye nd (w)	144–5				s^h	v				lig δ	B8^2, 265
b2107	—,**3-methoxy-5-propyl-**1-oxime	$C_{10}H_{13}NO_2$. *See* b2069	179.22	br nd (lig)	93–4					v			δ	aa v lig s^h	B7^2, 595
b2108	—,**2-methyl-***	Toluoquinone. $C_7H_6O_2$. *See* b2069	122.12	ye lf or nd	67	sub	1.08$^{75.5}_{4}$		δ s^h	s	s				B7^2, 588
b2109	—,**2-methyl-3,5,6-tribromo-***	$C_7H_3Br_3O_2$. *See* b2069	358.83	ye pl (al)	235–6				i	δ	s		s		B7^2, 592
b2110	—,**phenyl-***	$C_{12}H_8O_2$. *See* b2069	184.20	ye lf (peth)	114				δ^h				s	peth s^h chl v	B7, 740
b2111	—,**tetrachloro-***	Chloranil $C_6Cl_4O_2$. *See* b2069	245.88	ye mcl pr (bz), lf (aa)	290	sub			i	s^h δ	s			CS_2 v chl v	B7^2, 581
b2112	—,**tetrahydroxy-***	$C_6H_4O_6$. *See* b2069	172.10	bl-bk					v^h δ	v	δ				B8^2, 572
b2113	—,**tetramethyl-***	Duroquinone. $C_{10}H_{12}O_2$. *See* b2069	164.21	ye nd (al or lig)	111–2				i	s	s	s	s	sulf s aa s lig s^h δ	B7^2, 597
b2114	—,**trichloro-***	$C_6HCl_3O_2$. *See* b2069	211.43	ye fl (al)	168				i	s	v				B7^2, 581
b2115	**α-Benzosuberone**	1-Benzosuberanone. 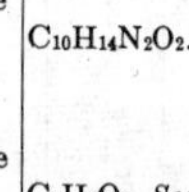	160.22			124–5^{7}	1.0780$^{20}_{4}$	1.5698^{20}		s					C50, 4850

For explanations, symbols and abbreviations see beginning of table.

No.	Name	Synonyms and Formula	Mol. wt.	Crystalline form, color and specific rotation	m.p. °C	b.p. °C	Density	n_D	Solubility w	al	eth	ace	bz	other solvents	Ref.
	Benzothiazole														
—	**Benzotetronic acid**	*see* **Coumarin,4-hydroxy-**													
b2116	**Benzothiazole**	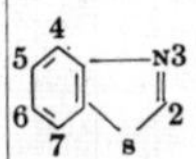	135.19			231^{760}	1.2384^{22}_{4}	1.6039^{22}_{α}	δ	v	v			CS_2 v	**B27²**, 17
b2117	**—,2-amino-**	$C_7H_6N_2S$. *See* b2116	150.20	pl (w)	132				δ	s	s			chl s	**B27²**, 225
b2118	**—,6-amino-**	$C_7H_6N_2S$. *See* b2116	150.20	pr (w)	87				δ	s	δ				**B27**, 366
b2119	**—,2-amino-6-ethoxy-**	$C_9H_{10}N_2OS$ *See* b2116	194.26	nd	163–4					s^h				MeOH s^h	**B27²**, 335
b2120	**—,5-amino-2-mercapto-**	$C_7H_6N_2S_2$. *See* b2116	182.26	nd	216									$PhNH_2$ s^h	**B27²**, 475
b2121	**—,6-amino-2-mercapto-**	$C_7H_6N_2S_2$. *See* b2116	182.26		263										**B27²**, 475
b2122	**—,2-amino-4-methyl-**	$C_8H_8N_2S$. *See* b2116	164.22	nd (w), pl (al)	136				δ	δ					**B27²**, 237
b2123	**—,2-amino-5-methyl-**	$C_8H_8N_2S$. *See* b2116	164.22	pl (dil al)	145					s					**B27²**, 240
b2124	**—,2-amino-6-methyl-**	$C_8H_8N_2S$. *See* b2116	164.22	nd (w), pr (dil al)	136				$δ^h$ s^h	s					**B27²**, 241
b2125	**—,2-chloro-**	C_7H_4ClNS. *See* b2116	169.63		24	248	1.3715^{19}_{4}	1.6338^{19}		s					**B27²**, 18
b2126	**—,5-chloro-2-methyl-**	C_8H_6ClNS. *See* b2116	183.66	pl	68					s				peth s	**B27²**, 22
b2127	**—,6-dimethyl-amino-2-mercapto-**	$C_9H_{10}N_2S_2$. *See* b2116	210.32	ye nd (bz)	230d						δ	s			**B27²**, 475
b2128	**—,2(2,4-dinitro-phenylthio)-**	$C_{13}H_7N_3O_4S_2$. *See* b2116	333.33	ye	162		1.24^{20}_{4}		i	δ s^h	δ				
b2129	**—,2-hydroxy-**	C_7H_5NOS. *See* b2116	151.19	pr (dil al), nd	138	360			i	v	v				**B27²**, 225
b2130	**—,6-hydroxy-2-phenyl-**	$C_{13}H_9NOS$. *See* b2116	227.29		228–9				i	s	i	s	s	aa s lig i	**B27²**, 88
b2131	**—,2(2-hydroxy-phenyl)-**	$C_{13}H_9NOS$. *See* b2116	227.29	nd or lf (al)	132					s^h					**B27²**, 91
b2132	**—,2-mercapto-**	$C_7H_5NS_2$. *See* b2116	167.25	nd (al)	179		1.42^{20}_{4}		i	s^h	δ			aa s	**B27²**, 233
b2133	**—,—,benzoate**	$C_{14}H_9NOS_2$. *See* b2116	271.35	ye	132				i	s^h δ	δ				
b2134	**—,2-mercapto-4-methyl-**	$C_8H_7NS_2$. *See* b2116	181.28	nd (dil aa)	186									aa s^h	**B27²**, 240
b2135	**—,2-mercapto-5-methyl-**	$C_8H_7NS_2$. *See* b2116	181.28	nd (to)	171–3									to s^h	**B27²**, 241
b2136	**—,2-mercapto-6-methyl-**	$C_8H_7NS_2$. *See* b2116	181.28		181				i	s		s	s^h		**B27²**, 242
b2137	**—,2-mercapto-7-methyl-**	$C_8H_7NS_2$. *See* b2116	181.28		184										**B27²**, 242
b2138	**—,2-mercapto-6-nitro-**	$C_7H_4N_2O_2S_2$. *See* b2116	212.25	nd (aa)	255–7									aa s^h	**B27²**, 234
b2139	**—,2-methyl-**	C_8H_7NS. *See* b2116	149.22		14	238 $150–1^{15}$	1.1763^{19}_{4}	1.6092^{19}	i	s					**B27²**, 21
b2140	**—,2(methylthio)-**	$C_8H_7NS_2$. *See* b2116	181.28	pr (dil al)	52					s^h					**B27²**, 71
b2141	**—,2-phenyl-**	$C_{13}H_9NS$. *See* b2116	211.29	nd (al)	114	>360			i	s^h	s			CS_2 s	**B27²**, 37
b2142	**—,2-phenyl-amino-**	$C_{13}H_{10}N_2S$. *See* b2116	226.30		161								δ	MeOH δ AcOEt s	**B27²**, 226
b2143	**Benzothiazoline, 3-methyl-2-imino-**	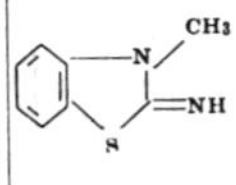	164.23	pl (w), nd (al)	128				δ v^h	v	v			chl v	**B27²**, 228
b2144	**—,3-methyl-2-thioxo-**	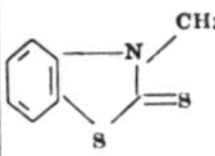	181.28	nd (al), pr (aa)	90	335^{757}			i	s^h δ	δ		v	CS_2 δ aa δ s^h chl v	**B27²**, 233
b2145	**Benzothiophene**	**Thionaphthene.** 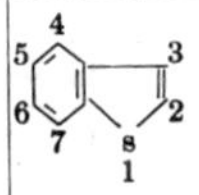	134.20	pl	32	221^{760}	1.1484^{32}_{4}	1.6374^{37}	i	v				os s	**B17²**, 59
b2146	**—,3-hydroxy-**	C_8H_6OS. *See* b2145	150.20	nd (w)	71				δ s^h	v				alk s peth δ	**B17²**, 128

For explanations, symbols and abbreviations see beginning of table.

No.	Name	Synonyms and Formula	Mol. wt.	Crystalline form, color and specific rotation	m.p. °C	b.p. °C	Density	n_D	Solubility w	al	eth	ace	bz	other solvents	Ref.
	Benzothiaphene														
b2147	—,**4-hydroxy-**	C_8H_6OS. *See* b2145	150.20	nd (sub)	72	sub	...	...	δ	s	...	...	...	alk s	**B17**, 121
b2148	**2,3-Benzothiophenequinone**	Thioisatin. 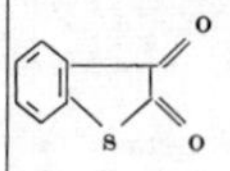	164.18	ye pr	121	247	...	...	i	s	...	...	s	aa s	**B17**, 467
b2149	**1,2,3-Benzotriazole**	Azimidobenzene. 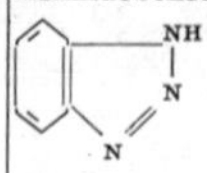	119.13	nd (chl or bz)	98.5	$201–4^{15}$	...	...	i	s	...	...	s	...	**B26**^{2}, 17
b2150	**1,2,3-Benzoxadiazole, 5,7-dinitro-**	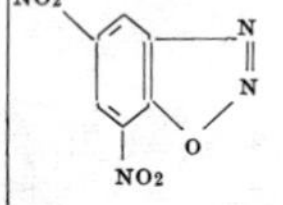	210.11	ye pl	158	...	...	...	...	s^h	...	...	...	...	**B16**^{2}, 287
b2151	**Benzoxazole**	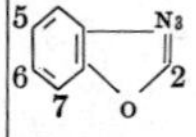	119.12	...	30.5	182.5^{760}	...	1.5594	δ	...	...	...		sulf s	**B27**^{2}, 17
b2152	—,**2-chloro-**	C_7H_4ClNO. *See* b2151	153.57	...	7	201–2	1.3453^{18}_{4}	1.5678^{20}	i	...	...	...	...	...	**B27**^{2} 17
b2153	—,**2-hydroxy-**	2(3)-Benzoxazolone. $C_7H_5NO_2$. *See* b2151	135.12	nd (bz or w+1)	141–2 97–8 (+w)	230^{30}	...	...	δ	s	s	...	...	...	**B27**^{2}, 223
b2154	—,**4-hydroxy-2-phenyl-**	$C_{13}H_9NO_2$. *See* b2151	211.22	nd (bz)	138–9	...	...	...	δ	v	v	...	s^h	aa v lig δ	**B27**^{2}, 88
b2155	—,**5-hydroxy-2-phenyl-**	$C_{13}H_9NO_2$. *See* b2151	211.12	nd (lig or dil al)	175	...	...	...	...	v	...	...	v	chl v aa v	**B27**^{2}, 88
b2156	—,**6-hydroxy-2-phenyl-**	$C_{13}H_9NO_2$. *See* b2151	211.12	...	216–7	...	...	...	...	δ	δ	v	s^h δ	aa v	**B27**, 117
b2157	—,**7-hydroxy-2-phenyl-**	$C_{13}H_9NO_2$. *See* b2151	211.12	nd (bz or dil al)	191–2	...	...	...	...	v	v	...	v^h	chl v lig v	**B27**^{2}, 91
b2158	—,**2(2-hydroxyphenyl)-**	$C_{13}H_9NO_2$. *See* b2151	211.12	nd (al)	123	338	...	...	...	...	...	...	...	os v	**B27**^{2}, 91
b2159	—,**2-mercapto-**	C_7H_5NOS. *See* b2151	151.19	nd (w)	192–3	...	...	...	δ s^h	s	v	...	...	aa v	**B27**^{2}, 224
b2160	—,**2-methyl-**	C_8H_7NO. *See* b2151	133.15	...	...	200–1	1.1211^{20}_{4}	1.5416^{20}	i	v	∞	...	...	...	**B27**^{2}, 20
b2161	**2,3-Benzoxazol-1-one**	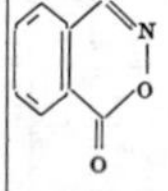	147.13	...	120d	...	...	...	...	...	...	...	...	...	**B27**^{2}, 249
—	**2(3)-Benzoxazolone**	*see* **Benzoxazole, 2-hydroxy-**													
—	**Benzoyl chloride**	*see* **Benzoic acid**, chloride													
—	**Benzyl alcohol**	*see* **Toluene, α-hydroxy-**													
—	**Benzylamine**	*see* **Toluene, α-amino-**													
—	**Benzylbromide**	*see* **Toluene, α-bromo-**													
—	**Benzyl chloride**	*see* **Toluene, α-chloro-**													
—	**Benzyl cyanide**	*see* **Acetic acid, phenyl-**, nitrile													
—	**Benzyl iodide**	*see* **Toluene, α-iodo-**													
b2162	**Berbamine**	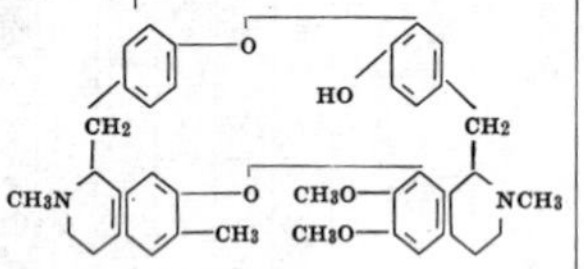	608.44	(w+4) $[\alpha]^{25}_{D}+106.3$ (chl)	170, 156 (+4w)	...	...	...	δ	s	s	...	...	peth s chl s	**B27**^{2}, 891
b2163	**Berberine**	$C_{20}H_{19}NO_5$	353.38	red ye (w+5½)	145, 110 (+5½w)	...	...	anh→ +5½w→	i s	s δ	s δ	...	s	...	**B27**^{2}, 567
b2164	—,compound with chloroform	$C_{21}H_{18}Cl_3NO_4$	454.74	tcl	179	...	...	...	δ	δ	...	...	...	chl v	**B27**, 492
b2165	—,hydrochloride	$C_{20}H_{22}ClNO_6$	407.8	(w+2), nd (w+4)	...	...	...	...	δ v^h	i	i	...	...	chl i	**B27**^{1}, 514
b2166	—,nitrate	$C_{20}H_{18}N_2O_7$	398.38	red ye	155d	...	...	...	v^h	δ	...	...	...	...	**B27**, 500
b2167	—,sulfate	$C_{40}H_{42}N_2O_{18}S$	822.85	red-ye	...	...	...	...	...	...	...	...	...	...	**B27**, 500
b2168	—,**tetrahydro-** (*dl*)	$C_{20}H_{21}NO_4$	339.38	mcl nd (al)	173–4	...	...	...	δ	s	...	...	...	CS_2 v chl v	**B27**^{2}, 557

For explanations, symbols and abbreviations see beginning of table.

No.	Name	Synonyms and Formula	Mol. wt.	Crystalline form, color and specific rotation	m.p. °C	b.p. °C	Density	n_D	Solubility w	al	eth	ace	bz	other solvents	Ref.
	Berbine														
—	**Berberonic acid**	*see* **2,4,5-Pyridinetricarboxylic acid**													
b2169	**Berbine** (*dl*)	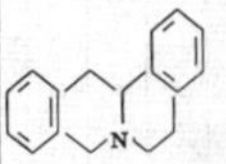	235.31	nd (eth or MeOH)	89				i	v				os v lig δ	**B20**[2], 311
b2170	**Betaine**	Lycine. Oxyneurine. $(CH_3)_3\overset{+}{N}CH_2COO^-$	117.15	(w+1), pr or pl (al)	239d				v	s	δ			MeOH v	**B4**[2], 785
—	**Betol**	**Benzoic acid, 2-hydroxy-2-naphthyl ester**													
b2172	**Betonicine**	$C_7H_{13}NO_3$	159.18	pr (dil al +1w) $[\alpha]_D^{21}$ −37	243d				δ	s					
b2173	**Betulin**	Lupenediol. $C_{30}H_{50}O_2$.	442.70	nd (al+1) $[\alpha]_D^{15}$ +20(Py)	261	170–80$^{0.08}$			i	δ s^h	δ		δ s^h	chl δ, s^h	**E14**, 568
b2174	**Betulinic acid**	$C_{30}H_{48}O_3$	456.71	pl or nd (al+1) $[\alpha]_{546}^{22}$ +7.9(Py)					δ	δ		δ	δ	chl δ Py s	**B6**[2], 939
b2175	**Biacene**	Biacenaphthylidene. 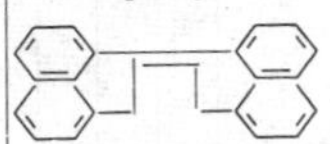	304.39	red-ye pl (bz)	277								s	CS_2 s^h	**E13**, 187
—	**Biacetyl**	*see* **2,3-Butanedione***													
—	**Biacetylene**	*see* **Butadiyne***													
—	**Bianiline**	*see* **Biphenyl, diamino-**													
—	**Bianisole**	*see* **Biphenyl, dimethoxy-**													
b2176	**10,10′-Bianthronyl**	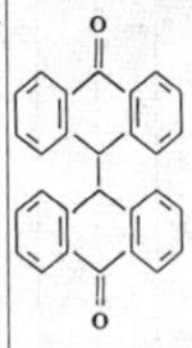	386.45	pl (ace)	250d					δ				chl v	**E13**, 766
b2177	**Biarsine, tetraethyl-**	Ethyl cacodyl. $(C_2H_5)_2AsAs(C_2H_5)_2$	266.06			185–90	1.1388_4^{24}	1.4709	i	s	s				**B4**, 616
b2178	**—,tetrakis(trifluoromethyl)-**	Perfluorocacodyl. $(CF_3)_2AsAs(CF_3)_2$	425.86			106–7^{760}		1.372^{19}	i						**J1952**, 2552
—	**—,tetramethyl-**	*see* **Cacodyl**													
—	**2,2-Bicamphane-2,2′-diol**	*see* **Camphor pinacole**													
b2179	**3,3′-Bicoumarin**	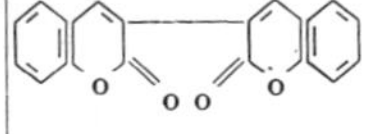	290.26	nd (aa)	>330					i	i		i	chl δ	**B19**, 183
—	**Bicresol**	*see* **Bipenyl, dihydroxy-dimethyl-**													
b2180	**Bicyclo[2,2,1]-hepta-2,5-diene.**	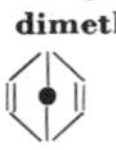	92.13		−19.9	89.5^{760}	0.909^{20}	1.4684^{25}	i					to ∞ lig ∞	
b2181	**Bicyclo[2,2,1]-heptane**	Norbornilane. Norcamphane. 	96.17		87.5–7.8				i	s				PrOH s	**B5**[2], 45
b2182	**Bicyclo[4,10]-heptane, 7-aza-**	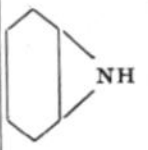	97.16		20–2	48–51^{22}				s	s				**B7**[1], 7
—	**Bicyclo[2,2,1]-heptane, 1,7,7-trimethyl-**	*see* **Camphane**													
b2183	**Bicyclo[2,2,1]-heptane-2,3-dicarboxylic acid**	2,3-Norcamphanedicarboxylic acid. 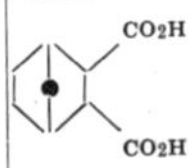	184.19		192–5				δ s^h	δ	δ				**E12A**, 982
b2184	**—,anhydride**	$C_9H_{10}O_3$. *See* b2183	166.12		165–7								s^h		**E12A**, 982

For explanations, symbols and abbreviations see beginning of table.

No.	Name	Synonyms and Formula	Mol. wt.	Crystalline form, color and specific rotation	m.p. °C	b.p. °C	Density	n_D	Solubility w	al	eth	ace	bz	other solvents	Ref.
	Bicyclo														
b2185	**Bicyclo[2,2,1]-heptane-2-carboxaldehyde**	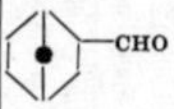	122.16	...	...	$70\text{–}2^{22}$	1.0227^{19}_{4}	1.4886^{19}	i	...	s	...	...	...	**E12A,** 704
b2186	**Bicyclohexyl**	Dodecahydrobiphenyl. 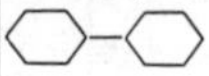	166.30	...	2.25	238	0.8917^{20}_{4}	1.4766^{20}	i	∞	∞	...	...	...	**B5**-108
b2188	**1,1′-Bicyclo-hexyldicarboxylic acid,** dinitrile	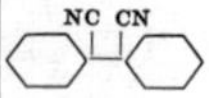	216.33	lf	220	...	...	...	...	...	s	...	...	...	**B9**[2], 571
b2189	**Bicyclo[3,3,1]-nonane**	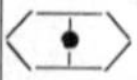	124.23	...	145–6	169–70	...	...	i	s	...	...	...	MeOH s^h aa s^h	**E12A,** 1057
b2190	**Bicyclo[3,3,0]-octane** (*cis*)		110.20	...	<−80	137^{765}	0.872^{18}_{4}	1.462^{18}	i	s	...	...	...	...	**E12A,** 80
b2191	—(*trans*)	C_8H_{14}. *See* b2190	110.20	...	−30	132^{755}	0.8626^{18}_{4}	1.4625^{18}	i	s	...	...	...	...	**E12A,** 80
b2192	**Bicyclo[3,2,1]-octane**	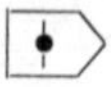	110.20	...	133–4	...	...	...	i	...	...	...	...	...	**E12A,** 1048
b2193	**Bicyclo[2,2,2]-octane**	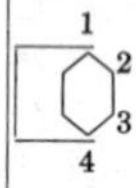	110.20	...	169–70	...	...	...	i	...	...	...	...	...	**E12A,** 1068
b2194	**Bicyclo[3,3,0]-octane-2,6-dione**	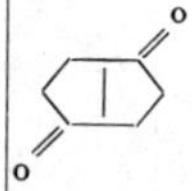	138.17	...	45	$86\text{–}8^{0.2}$	1.1290^{60}_{4}	1.4877^{54}	...	...	...	...	...	...	**E12A,** 84
b2195	**Bicyclo[2,2,2]-octane, 2-methyl-**	C_9H_{16}. *See* b2193	124.23	...	33–4	$157\text{–}8^{761}$	$0.8674^{40.5}_{4}$	$1.4613^{40.5}$	i	...	...	...	...	...	**E12A,** 1068
b2196	**Bicyclo[3,3,0]-octan-2-one**	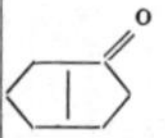	124.18	...	...	72^{12}	1.0097^{20}_{4}	1.4790^{20}	δ	s	...	...	...	...	**E12A,** 82
b2197	**9,9′-Bifluorenyl-**	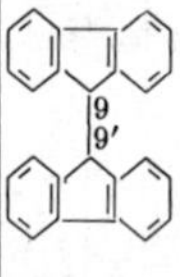	330.43	nd (bz-al)	247	...	...	...	...	δ	δ	...	...	Py s^h aa s^h	**E13,** 123
b2198	—,**9,9′-diphenyl-**	$C_{38}H_{26}$. *See* b2197	482.63	pl (bz)	205–30	1.266^{0}_{4}	...	...	i	...	...	...	δ	os δ	**B5**[2], 639
b2199	**Bifluorenylidene**	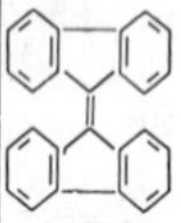	328.41	red nd	194–5	...	...	...	i	...	s	...	s	chl s	**E13,** 123
—	**Bigitaligenin**	*see* **Gitoxygenin**													
—	**Bigitaline**	*see* **Gitoxine**													
b2200	**Biguanide**	$H_2NC(:NH)NHC(:NH)NH_2$	101.12	pr or nd (al)	130	142d	...	...	v	s	...	...	i	chl i	**B3**[2], 76
b2201	—,**1(2-tolyl)**	$C_9H_{13}N_5$. *See* b2200	191.24	nd or pl (w+½)	144	...	...	...	δ v^h	v	i	v	i	chl i	**B12,** 803
b2202	**2,2′-Biindane, 1,1′,3,3′-tetraoxo-**	Bisdiketohydrindene. 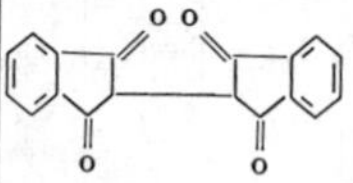	290.28	vt nd ($PhNO_2$)	297	...	...	...	...	...	...	...	...	$PhNO_2$ δ alk s	**B7**[2], 863
b2203	**Bikhaconitine**	$C_{36}H_{51}NO_{11}$	673.78	$[\alpha]_D+12$(al)	118–23, 113–16 (+w)	...	...	...	i	s	s	...	...	chl s peth i	...
b2204	**Bilifuscin**	$C_{16}H_{20}N_2O_4$	304.34	dk br	183	...	...	...	i	s	i	...	...	chl i aa s alk s	...
b2205	**Bilirubin**	$C_{33}H_{36}N_4O_6$	584.65	red mcl pr or pl (chl)	192	...	...	...	i	δ	δ	...	s	CS_2 s chl s	...
b2206	**Biliverdin**	Dehydrobilirubin. $C_{33}H_{34}N_4O_6$.	582.63	dk gr pl or pr (MeOH)	>300	...	...	...	i	s	δ	...	s	MeOH δ chl δ CS_2 δ alk s	...

For explanations, symbols and abbreviations see beginning of table.

No.	Name	Synonyms and Formula	Mol. wt.	Crystalline form, color and specific rotation	m.p. °C	b.p. °C	Density	n_D	Solubility w	al	eth	ace	bz	other solvents	Ref.
	Binaphthyl														
—	**Binaphthol**	*see* **Binaphthyl, dihydroxy-**													
b2207	**1,1-Binaphthyl**	α,α′-Dinaphthyl. (structure: 3′ 2′ 2 3; 4′ 4)	254.31	rh (al), lf (aa)	160.5	360d 240–4[12]			i	δ s^h	s		s	CS_2 s	**B5[2]**, 642
b2208	**2,2-Binaphthyl**	β,β′-Dinaphthyl.	254.31	pl	187–8	452[753]			i	δ	s		s^h	aa s	**B5[2]**, 643
b2209	**1,1-Binaphthyl, 4,4′-diamino-3,3′-dimethyl-**	$C_{22}H_{20}N_2$. *See* b2207	312.42		213					s			v		**B13[2]**, 140
b2210	**—,2,2′-dihydroxy-**	1,1′-Bi-2-naphthol. $C_{20}H_{14}O_2$. *See* b2207	286.31	nd (al)	218				i	s	s			chl δ alk s	**B6[2]**, 1026
b2211	**—,4,4′-dihydroxy-**	4,4′-Bi-1-naphthol. $C_{20}H_{14}O_2$. *See* b2207	286.31	pl	300				i	s	s		δ	chl δ	**B6**, 1053
b2212	**Biotin**	Vitamin H. Coenzyme R. $C_{10}H_{16}N_2O_3S$.	244.31	nd (w) $[\alpha]_D^{22}+92$	230–2				s	s	δ			chl δ	
b2213	—,methyl ester	$C_{11}H_{18}N_2O_3S$	258.34	$[\alpha]_D^{22}+57$ (chl)	166–7										
—	**Bioxirane**	*see* **Butane, 1,2,3,4-diepoxy-**													
—	**Biphenol**	*see* **Biphenyl, dihydroxy-**													
b2214	**Biphenyl**	Diphenyl. Phenylbenzene. (structure: 3 2 2′ 3′; 4 4′; 5 6 6′ 5′)	154.20	lf	70	255.9[760]	1.9896_4^{77}	1.588[77]	i	s	s				**B5[2]**, 479
b2215	**—,2-acetamido-**	$C_{14}H_{13}NO$. *See* b2214	211.27	pr or nd	121	355				v	v				**B12[2]**, 747
b2216	**—,3-acetamido-**	$C_{14}H_{13}NO$. *See* b2214	211.27	nd (al)	149					s^h			v^h		**B12[2]**, 751
b2217	**—,4-acetamido-**	$C_{14}H_{13}NO$. *See* b2214	211.27		172					v		v			**B12[2]**, 755
b2218	**—,4-acetamido-3-nitro-**	$C_{14}H_{12}N_2O_3$. *See* b2214	256.25	nd (al)	132					s^h	s			aa s^h	**B12[2]**, 760
b2219	**—,4-acetyl-**	4-Phenylacetophenone. $C_{14}H_{12}O$. *See* b2214	196.25	pr (ace)	121	325–7			i	v		v			**B7[2]**, 337
b2220	**—,2-amino-**	$C_{12}H_{11}N$. *See* b2214	169.22	lf (dil al)	49–50	299[760] 189–91[30]			i	s	s		s	peth δ	**B12[2]**, 747
b2221	**—,3-amino-**	$C_{12}H_{11}N$. *See* b2214	169.22	nd	30	195[15]			δ	s	s				**B12[2]**, 751
b2222	**—,4-amino-**	Xenylamine. $C_{12}H_{11}N$. *See* b2214	169.22	lf (dil al)	53	302 191[15]			δ^h	δ	δ			chl δ	**B12[2]**, 753
b2223	**—,3-amino-4-hydroxy-**	$C_{12}H_{11}NO$. *See* b2214	185.23	nd (chl)	208					v	v		v		**B13[2]**, 419
b2224	**—,4-amino-2′-hydroxy-**	$C_{12}H_{11}NO$. *See* b2214	185.23	nd (to)	181–2				δ	s				to δ aa s	**B13[2]**, 419
b2225	**—,4-amino-4′-hydroxy-**	$C_{12}H_{11}NO$. *See* b2214	185.23	pl (dil al)	273				i	δ	i	δ	δ	aa δ	**B13[2]**, 420
b2226	**—,5-amino-2-hydroxy-**	$C_{12}H_{11}NO$. *See* b2214	185.23	nd (al or bz)	198–9				i	s^h	δ		s	chl δ lig i	**B13[1]**, 280
b2227	**—,2-amino-4′-nitro-**	$C_{12}H_{10}N_2O_2$. *See* b2214	214.22	og-red nd (al)	158					s^h					**B12[2]**, 750
b2228	**—,2-amino-5-nitro-**	$C_{12}H_{10}N_2O_2$. *See* b2214	214.22	ye nd (al)	125					s^h					**B12[2]**, 750
b2229	**—,3-amino-4-nitro-**	$C_{12}H_{10}N_2O_2$. *See* b2214	214.22	og nd (dil al)	116					s					**B12[2]**, 752
b2230	**—,3-amino-4′-nitro-**	$C_{12}H_{10}N_2O_2$. *See* b2214	214.22	og nd (al)	137					s^h				aa s	**B12[2]**, 753
b2231	**—,4-amino-2′-nitro-**	$C_{12}H_{10}N_2O_2$. *See* b2214	214.22	red mcl pr (al)	97–8				δ^h	v					**B12[1]**, 547
b2232	**—,4-amino-3-nitro-**	$C_{12}H_{10}N_2O_2$. *See* b2214	214.22	red nd (al)	168–70					s^h	s			chl s aa s	**B12[1]**, 760
b2233	**—,4-amino-4′-nitro-**	$C_{12}H_{10}N_2O_2$. *See* b2214	214.22	red nd (al)	200				δ^h	v^h				aa s^h	**B12[2]**, 761
b2234	**—,2-benzyl-**	o-Biphenylylphenylmethane. $C_{19}H_{16}$. *See* b2214	244.32	mcl nd	54	283–7[110]			i	s	s		v		**B5**, 708
b2235	**—,4-benzyl-**	$C_{19}H_{16}$. *See* b2214	244.32	lf	85	285–6[110]	1.171_4^0		i	s	v		v		**B5[2]**, 618
b2236	**—,4,4′-bis-(diethylamino)-**	$C_{20}H_{28}N_2$. *See* b2214	296.44	nd (al)	85					s v^h	s				**B13[2]**, 98
b2237	**—,2,4′-bis(di-methylamino)-**	$C_{16}H_{20}N_2$. *See* b2214	240.34	pl (al)	51–2	206–7[11]				s v^h	s				**B13[2]**, 88
b2238	**—,4,4′bis(di-methylamino)-**	$C_{16}H_{20}N_2$. *See* b2214	240.34	nd (al or bz-lig)	198	>360				δ s^h	δ		v^h	chl v lig δ^h	**B13[2]**, 97
b2239	**—,4,4′-bis-(ethylamino)-**	$C_{16}H_{20}N_2$. *See* b2214	240.34	nd or pl (al)	120.5					v^h	v^h		v	lig δ	**B13**, 222

For explanations, symbols and abbreviations see beginning of table.

No.	Name	Synonyms and Formula	Mol. wt.	Crystalline form, color and specific rotation	m.p. °C	b.p. °C	Density	n_D	Solubility w	al	eth	ace	bz	other solvents	Ref.
	Biphenyl														
b2240	**—,4,4′-bis-(methylamino)-**	$C_{14}H_{16}N_2$. *See* b2214	212.28	lf (al or w)	146				δ^h	s^h					**B13²**, 97
b2241	**—,4,4′-bis-(phenylamino)-**	$C_{24}H_{20}N_2$. *See* b2214	336.42	lf (to)	244–5				i	δ	δ		δ	to v^h aa v^h	**B13²**, 98
b2242	**—,2-bromo-**	$C_{12}H_9Br$. *See* b2214	233.11		<−20	296–8 160^{11}			i	s	s				**B5²**, 485
b2243	**—,3-bromo-**	$C_{12}H_9Br$. *See* b2214	233.11			299–301			i						**B5²**, 485
b2244	**—,4-bromo-**	$C_{12}H_9Br$. *See* b2214	233.11	pl (al)	91.2	310	0.9327^{25}_{4}		i	s	s		s	CS_2 s	**B5²**, 485
b2245	**—,4(bromo-acetyl)-**	$C_{14}H_{11}BrO$. *See* b2214	275.15	nd	127					δ				CCl_4 s^h peth s^h	**Am52,** 3715
b2246	**—,3-bromo-4-hydroxy-**	$C_{12}H_9BrO$. *See* b2214	249.12	nd (chl-peth)	96					v				chl v^h aa v	**B6²**, 625
b2247	**—,4-bromo-4′-hydroxy-**	$C_{12}H_9BrO$. *See* b2214	249.12	pl (al)	164–6				δ	v	v	v	v	chl v	**B6²**, 625
b2248	**—,2-chloro-**	$C_{12}H_9Cl$. *See* b2214	188.65	mcl	32.4	274 154^{12}	$1.1499^{32.5}$		i	s	v			CCl_4 s lig v	**B5²**, 483
b2249	**—,3-chloro-**	$C_{12}H_9Cl$. *See* b2214	188.65		16	284–5	1.1579^{25}_{4}	1.6181^{25}	i	s	s				**B5²**, 483
b2250	**—,4-chloro-**	$C_{12}H_9Cl$. *See* b2214	188.65	lf (lig or al)	77.7	291			i	s	s			lig s	**B5²**, 483
b2251	**—,4-(chloro-acetyl)-**	$C_{14}H_{11}ClO$. *See* b2214	230.70	pl (al)	122–3					s^h					**B7**, 443
b2252	**—,3-chloro-2-hydroxy-**	$C_{12}H_9ClO$. *See* b2214	204.65		6	317–8d	1.24^{25}_{4}	1.6237^{30}	i	s				os s	
b2253	**—,4-chloro-4′-hydroxy-**	$C_{12}H_9ClO$. *See* b2214	204.65		146–7				δ	s	s	s	s	peth δ	**B6²**, 625
b2254	**—,2,2′-diaceta-mido-**	$C_{16}H_{16}N_2O_2$. *See* b2214	268.30	pr (al)	161				δ	v	δ		v	peth δ aa v	**B13**, 210
b2255	**—,2,4-diaceta-mido-**	$C_{16}H_{16}N_2O_2$. *See* b2214	268.30	nd (al)	202					s					**B13**, 212
b2256	**—,4,4′-diaceta-mido-**	$C_{16}H_{16}N_2O_2$. *See* b2214	268.30	nd (aa)	330–1				i	δ	δ				**B13²**, 102
b2257	**—,2,2′-diamino-**	*o*-Benzidine. *o,o′*-Bianiline. $C_{12}H_{12}N_2$. *See* b2214	184.23	mcl pr or nd (al)	81				s v^h				s		**B13²**,87
b2258	**—,2,4′-diamino-**	$C_{12}H_{12}N_2$. *See* b2214	184.13	nd (dil al)	54.5	363^{760}			δ	v	v				**B13²**, 88
b2259	**—,3,3′-diamino-**	*m*-Benzidine. $C_{12}H_{12}N_2$. *See* b2214	184.13	nd (w), pr (bz)	93.5				δ s^h		v		s^h		**B13²**, 90
b2260	**—,3,4-diamino-**	$C_{12}H_{12}N_2$. *See* b2214	184.23	lf (eth or al)	103					s v^h	s				**B13²**, 89
b2261	**—,4,4′-diamino-**	Benzidine. $C_{12}H_{12}N_2$. *See* b2214	184.23	nd (w)	128	400^{740}			δ^h	s	δ				**B13²**, 90
b2262	**—,4,4′-diamino-3,3′-dimethoxy-**	Bianisidine. $C_{14}H_{16}N_2O_2$. *See* b2214	244.29	lf or nd	137				i	v	v	v	v	chl v	**B13²**, 502
b2263	**—,4,4′-diamino-2,2′-dimethyl-**	Bitoluidine. *m*-Tolidine. $C_{14}H_{16}N_2$. *See* b2214	212.28	pr (w)	108–9				s^h	v	v				**B13**, 255
b2264	**—,4,4′-diamino-3,3′-dimethyl-**	*o*-Tolidine $C_{14}H_{16}N_2$. *See* b2214	212.28	lf	129				δ	v	v				**B13**, 256
b2265	**—,4,4′-diamino-3-ethoxy-**	$C_{14}H_{16}N_2O$. *See* b2214	228.28	nd (w)	134				δ	v^h	δ		δ		**B13²**, 419
b2266	**—,4,4′-dibromo-**	$C_{12}H_8Br_2$. *See* b2214	312.02	mcl pr (MeOH)	164	355–60			i	δ^h			s		**B5²**, 485
b2267	**—,4,4′-dichloro-**	$C_{12}H_8Cl_2$. *See* b2214	223.10	pr or nd (al)	148	315–9			i						**B5²**, 484
b2268	**—,4,4′-dichloro-2,2′-dinitro-**	$C_{12}H_6Cl_2N_2O_4$. *See* b2214	313.10	ye	139–40				i	δ s^h			s	aa s^h lig δ	**B5²** 584
b2269	**—,4,4′-diethoxy-3,3′-diethyl-**	$C_{20}H_{26}O_2$. *See* b2214	298.43	lf	120				i	δ					**B6**, 1015
b2270	**—,2,2′-diethoxy-3,3′-dimethyl**	$C_{18}H_{22}O_2$. *See* b2214	270.31	lf (al)	85				i	δ					**B6²**, 974
b2271	**—,2,4′-diethoxy-3,3′-dimethyl-**	$C_{18}H_{22}O_2$. *See* b2214	270.31	nd (al)	53				i	δ					**B6²**, 974
b2272	**—,4,4′-diethoxy-3,3′-dimethyl-**	$C_{18}H_{22}O_2$. *See* b2214	270.31	pl (al)	156				i	δ	δ				**B6**, 1010
b2273	**—,3,3′-diethyl-6,6′-dihydroxy-**	$C_{16}H_{18}O_2$. *See* b2214	242.31	nd (dil al)	131				i	δ					**B6²**, 981
b2274	**—,4,4′-difluoro-**	$C_{12}H_8F_2$. *See* b2214	190.20	mcl pr (al)	94–5	254–5 119^{14}			i s^h	v	v		v^h	chl v aa v^h MeOH v^h	**B5²**, 482
b2275	**—,2,2′-dihydroxy-**	*o,o′*-Biphenol. $C_{12}H_{10}O_2$. *See* b2214	186.20	lf (w+1), pr (to)	109–10, 71–3 (+w)				s^h	s	s		s	peth δ aa s	**B6²**, 960
b2276	**—,2,4′-dihydroxy-**	*o,p′*-Biphenol. $C_{12}H_{10}O_2$. *See* b2214	186.20	mcl pr or nd	162–3	342 206–10^{11}			δ^h	s	s				**B6²**, 961
b2277	**—,2,5-dihydroxy-**	Phenylhydroquinone. $C_{12}H_{10}O_2$. *See* b2214	186.20	nd	96–8				i	s					**B6**, 989
b2278	**—,3,3′-dihydroxy-**	*m,m′*-Biphenol. $C_{12}H_{10}O_2$. *See* b2214	186.20	nd (w)	123	247^{18}			s^h δ	s	s		s	chl s	**B6²**, 961

For explanations, symbols and abbreviations see beginning of table.

No.	Name	Synonyms and Formula	Mol. wt.	Crystalline form, color and specific rotation	m.p. °C	b.p. °C	Density	n_D	Solubility w	al	eth	ace	bz	other solvents	Ref.
	Biphenyl														
b2279	**—,3,4-dihydroxy-**	4-Phenylpyrocatechol. $C_{12}H_{10}O_2$. *See* b2214	186.20	...	136	>360	...	...	δ^h	v	v	s	s	CS_2 δ chl v	**B6**^{2}, 961
b2280	**—,4,4′-dihydroxy-**	*p,p′*-Biphenol. $C_{12}H_{10}O_2$. *See* b2214	186.20	nd or pl (al)	269–70	sub	...	...	δ	s	s	...	δ	...	**B6**^{2}, 962
b2281	**—,4,4′-dihydroxy-3,3′-diethyl-5,5′-dimethyl-**	$C_{18}H_{22}O_2$. *See* b2214	270.31	nd (aa)	148	...	...	...	...	v	...	v	v	peth δ aa v	**B6**^{2}, 984
b2282	**—,2,2′-dihydroxy-3,3′-dimethyl-**	$C_{14}H_{14}O_2$. *See* b2214	214.25	nd (peth)	113	sub	...	...	...	v	v	...	v	peth δ	**B6**^{2}, 974
b2283	**—,2,2′-dihydroxy-5,5′-dimethyl-**	$C_{14}H_{14}O_2$. *See* b2214	214.25	nd (bz)	153.5	sub	...	...	δ^h	s	v	...	...	os s	**B6**^{2}, 974
b2284	**—,2,2′-dihydroxy-6,6′-dimethyl-**	$C_{14}H_{14}O_2$. *See* b2214	214.25	pl (dil al)	164	...	...	...	i	v	...	...	...	...	...
b2285	**—,2,5′-dihydroxy-2′,5-dimethyl-**	$C_{14}H_{14}O_2$. *See* b2214	214.25	...	158	...	...	...	i	v	v	...	δ δ^h	...	**B6**^{2}, 973
b2286	**—,4,4′-dihydroxy-3,3′-dimethyl-**	$C_{14}H_{14}O_2$. *See* b2214	214.25	lf (w), nd	161	...	...	...	δ^h	v	v	...	...	...	**B6**^{2}, 974
b2287	**—,—,diacetate**	$C_{18}H_{18}O_4$. *See* b2214	298.33	nd (al)	135.5	...	...	...	i	δ	...	...	...	...	**B6**^{1}, 492
b2288	**—,5,5′-dihydroxy-2,2-dimethyl-**	$C_{14}H_{14}O_2$. *See* b2214	214.25	pr (al)	229	...	...	...	i δ^h	v	v	...	δ^h	peth i	**B6**^{2}, 973
b2289	**—,4,4′-dihydroxy-3,3′,5,5′-tetramethoxy-**	Hydrocerulignone. $C_{16}H_{18}O_6$. *See* b2214	306.31	mcl pr (al)	190	...	...	...	δ	s	δ	...	...	CS_2 i	**B6**^{1}, 593
b2290	**—,2,2′-dihydroxy-3,3′,5,5′-tetramethyl-**	$C_{16}H_{18}O_2$. *See* b2214	242.31	nd or pl (eth or lig)	137.5–8.5	...	...	...	i	v	v	...	...	peth δ^h lig v^h	**B6**^{2}, 981
b2291	**—,4,4′-dihydroxy-3,3′,5,5′-tetramethyl-**	$C_{16}H_{18}O_2$. *See* b2214	242.31	nd or pr (aa)	220–1	...	...	...	...	s^h	...	...	δ^h	aa s^h	**B6**, 1015
b2292	**—,4,4′-dihydroxy-3,3′,5,5′-tetranitro-**	$C_{12}H_6N_4O_{10}$. *See* b2214	366.20	ye nd	223	...	...	...	...	...	...	...	...	...	**B6**^{2}, 963
b2293	**—,2,2′-dimethoxy-**	*o,o′*-Bianisole. $C_{14}H_{14}O_2$. *See* b2214	214.25	rh bipyr (al)	155	$307–8^{766}$	1.268	...	i	v^h	δ^h	...	v^h	CCl_4, chl, CS_2 v^h	**B6**^{2}, 960
b2294	**—,3,3′-dimethoxy-**	*m,m′*-Bianisole. $C_{14}H_{14}O_2$. *See* b2214	214.25	nd (dil al)	36	328 $211–20^{15}$	...	...	i	v	v	...	v	CS_2 v aa v chl v	**B6**^{2}, 961
b2295	**—,4,4′-dimethoxy-**	*p,p′*-Bianisole. $C_{14}H_{14}O_2$. *See* b2214	214.25	lf (bz)	173	sub	...	...	i	v^h	δ	...	v	peth i chl v	**B6**^{2}, 962
b2296	**—,2,2′-dimethoxy-5,5′-dimethyl-**	$C_{16}H_{18}O_2$. *See* b2214	242.31	nd (dil al)	71	188^{12}	...	...	δ^h	s	...	...	...	os v	**B6**^{2}, 975
b2297	**—,2,5′-dimethoxy-2′,5-dimethyl-**	$C_{16}H_{18}O_2$. *See* b2214	242.31	pr (al)	86	168^4	...	...	i	δ v^h	...	...	s	peth s	**B6**^{2}, 973
b2298	**—,4,4′-dimethoxy-3,3′-dimethyl-**	$C_{16}H_{18}O_2$. *See* b2214	242.31	pr (al)	145.5	...	...	...	i	δ s^h	δ	...	...	...	**B6**, 1009
b2299	**—2,2′-dimethyl-**	*o,o′*-Bitolyl. $C_{14}H_{14}$. *See* b2214	182.25	...	18	256^{760}	0.955^{10}_{4}	...	i	v	v	...	v	...	**B5**^{2}, 512
b2300	**—,2,3′-dimethyl-**	$C_{14}H_{14}$. *See* b2214	182.25	...	...	273–4	0.9984^{22}	1.5848^{25}	i	v	v	...	v	...	**B5**^{2}, 512
b2301	**—,2,4′-dimethyl-**	$C_{14}H_{14}$. *See* b2214	182.25	...	...	$273–6^{760}$	...	...	...	...	...	...	...	...	**B5**^{2}, 512
b2302	**—,3,3′-dimethyl-**	*m,m′*-Bitolyl. $C_{14}H_{14}$. *See* b2214	182.25	...	9	280^{760}	0.9993^{16}_{4}	1.5962^{20}	i	v	v	...	v	...	**B5**^{2}, 513
b2303	**—,4,4′-dimethyl-**	*p,p′*Bitolyl. $C_{14}H_{14}$. *See* b2214	182.25	mcl pr (eth)	121–2	292^{752}	0.917^{121}_{4}	...	i	δ	s	...	s	CS_2 s	**B5**^{2}, 514
b2304	**—,3,3′-dimethyl-4,4′-dipropoxy-**	$C_{20}H_{26}O_2$. *See* b2214	298.43	lf	115	...	...	...	i	δ v^h	...	...	...	...	**B6**, 1010
b2305	**—,2,2′-dinitro-**	$C_{12}H_8N_2O_4$. *See* b2214	244.20	ye mcl pr	127–8	...	1.45(so)	...	i	v^h	s	...	s	aa s^h lig δ	**B5**^{2}, 490
b2306	**—,2,4′-dinitro-**	$C_{12}H_8N_2O_4$. *See* b2214	244.20	mcl pr	93.5	...	1.474	...	i	v^h	s	...	s^h	aa s^h	**B5**^{2}, 491
b2307	**—,3,3′-dinitro-**	$C_{12}H_8N_2O_4$. *See* b2214	244.20	ye-og nd (al or aa)	200	...	...	...	i	δ	δ	...	s^h	aa s^h	**B5**^{2}, 491
b2308	**—,4,4′-dinitro-**	$C_{12}H_8N_2O_4$. *See* b2214	244.20	nd (al)	238–9	...	...	...	i	δ s^h	...	...	s	aa s^h	**B5**^{2}, 491
—	**—,diphenyl-**	*see* **Quaterphenyl**													
—	**—,dodecahydro-**	*see* **Bicyclohexyl**													
b2310	**—,2-ethoxy-**	$C_{14}H_{14}O$. *See* b2214	198.25	pr (peth)	34	276	...	...	...	s	s	...	s	chl s	**B6**, 672
b2311	**—,3-ethoxy-**	$C_{14}H_{14}O$. *See* b2214	198.25	...	34	305	...	...	...	s	s	...	...	os s	**B6**, 674
b2312	**—,2-hydroxy-**	$C_{12}H_{10}O$. *See* b2214	170.20	nd (peth)	57	286^{760} 145^{14}	1.213^{25}_{4}	...	i	s	v	s	s	lig s	**B6**^{2}, 623
b2313	**—,3-hydroxy-**	$C_{12}H_{10}O$. *See* b2214	170.20	nd (w or peth)	78	>300	...	...	δ	v	...	...	v	peth s	**B6**^{2}, 624
b2314	**—,4-hydroxy-**	$C_{12}H_{10}O$. *See* b2214	170.20	nd (dil al)	164–5	305–8	...	...	δ	v	v	...	...	peth δ, chl v	**B6**^{2}, 624

For explanations, symbols and abbreviations see beginning of table.

No.	Name	Synonyms and Formula	Mol. wt.	Crystalline form, color and specific rotation	m.p. °C	b.p. °C	Density	n_D	Solubility w	al	eth	ace	bz	other solvents	Ref.
	Biphenyl														
b2315	**—,2-hydroxy-2′-methoxy-5,5′-dimethyl-**	$C_{15}H_{16}O_2$. *See* b2214	228.28			205^{12}				v	v		v	chl v	**B6**², 974
b2316	**—,2-iodo-**	$C_{12}H_9I$. *See* b2214	280.11			158^{6}	1.6038^{25}_{25}								**B5**², 486
b2317	**—,4-iodo-**	$C_{12}H_9I$. *See* b2214	280.11	nd	113–4	183^{11}			i	s^h	s		s	aa s	**B5**², 486
b2318	**—,2-methoxy-**	$C_{13}H_{12}O$. *See* b2214	184.24	pr (peth)	29	274 150^{13}	1.0233^{99}_{4}	1.5641^{99}	i	s^h				peth s	**B6**², 623
b2319	**—,4-methoxy-**	$C_{13}H_{12}O$. *See* b2214	184.24	pl (al)	90	157^{10}	1.0278^{100}_{4}	1.5744^{100}	i	s^h	s				**B6**², 625
b2320	**—,2-methyl-**	$C_{13}H_{12}$. *See* b2214	168.23		−0.2	255.3^{760}	1.010^{20}_{4}	1.5914^{20}	i	s	s				**B5**², 504
b2321	**—,3-methyl-**	$C_{13}H_{12}$. *See* b2214	168.23		4.5	272.7^{760}	1.0182^{17}_{4}	1.6039^{20}	i	s	s				**B5**², 504
b2322	**—,4-methyl-**	$C_{13}H_{12}$. *See* b2214	168.23	pl (lig)	49–50	267–8	1.015^{27}		i	s	s			os v	**B5**², 504
b2323	**—,2-nitro-**	$C_{12}H_9NO_2$. *See* b2214	199.20	lf (al)	37.2	201^{30}	1.44		i	s^h	s				**B5**², 487
b2324	**—,3-nitro-**	$C_{12}H_9NO_2$. *See* b2214	199.20	nd (al)	62	$225–30^{35}$			i	s				aa s	**B5**², 487
b2325	**—,4-nitro-**	$C_{12}H_9NO_2$. *See* b2214	199.20	nd (al)	113.7	340			i	δ v^h	s		s	chl s aa s	**B5**², 487
b2326	**—,3,3′,5,5′-tetra-hydroxy-**	$C_{12}H_{10}O_4$. *See* b2214	218.19	pl or nd (w+2)	310				s^h	s	s	i	i		**B6**², 1129
b2327	**—,2,2′,4,4′-tetra-nitro-**	$C_{12}H_6N_4O_8$. *See* b2214	334.20	ye pr (bz)	165				i	δ	δ		s	aa s	**B5**², 494
b2328	**—,2,4,4′-tri-amino-**	$C_{12}H_{13}N_3$. *See* b2214	199.25	nd	134										**B13**², 149
b2329	**2-Biphenylcar-boxylic acid**	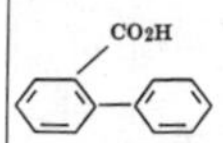	198.21	lf (dil al)	113.5–4.5	343–4 199^{10}			i	v			v	aa v	**B9**², 463
b2330	—,nitrile	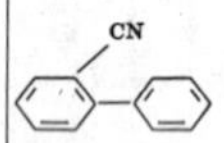	179.22	nd	41	175^{13}				s	v				**B9**², 463
b2331	**3-Biphenylcar-boxylic acid**	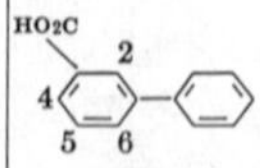	198.21	lf (al)	165–6				δ	v	v		v	aa v lig v	**B9**², 464
b2332	**4-Biphenylcar-boxylic acid**	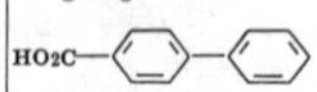	198.21	nd (bz or al)	228	sub			i	v	v		s		**B9**², 464
b2333	—,nitrile	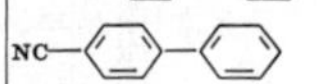	179.22		88				i	v	v				**B9**², 464
b2334	**3-Biphenylcar-boxylic acid, 2-hydroxy-**	$C_{13}H_{10}O_3$. *See* b2331	214.22		180					s^h					**B10**, 341
b2335	**2,2′-Biphenyldi-carboxylic acid**	Diphenic acid. 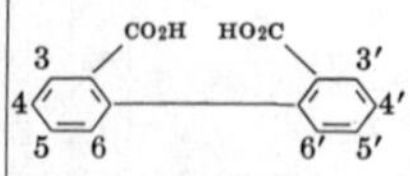	242.22	mcl pr (w)	228–9	sub			i $δ^h$	s	s			os s	**B9**², 655
b2336	—,anhydride	$C_{14}H_8O_3$. *See* b2335	224.20	nd (aa or bz)	217	sub			i		δ				**B17**², 497
b2337	—,dichloride	$C_{14}H_8Cl_2O_2$. *See* b2335	279.11		93–4					v			v	aa v	**B9**², 657
b2338	—,diethyl ester	$C_{18}H_{18}O_4$. *See* b2335	298.32		42										**B9**², 656
b2339	—,dimethyl ester	$C_{16}H_{14}O_4$. *See* b2335	270.27	mcl pr	73–4	$204–6^{14}$									**B9**², 656
b2340	—,imide	$C_{14}H_9NO_2$. *See* b2335	223.22	nd (al)	219–20				$δ^h$	δ s^h	δ			chl v	**B21**², 392
b2341	**2,3′-Biphenyldi-carboxylic acid**	Isodiphenic acid. 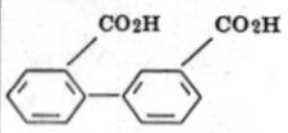	242.23	nd (w)	216				$δ^h$	v					**B9**², 663
b2342	**2,4′-Biphenyldi-carboxylic acid**	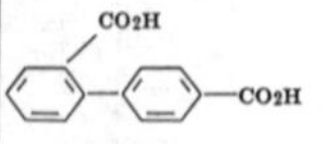	242.23	lf (al)	265d				δ	v^h			s	aa v	**B9**², 663
b2343	**3,3′-Biphenyldi-carboxylic acid**	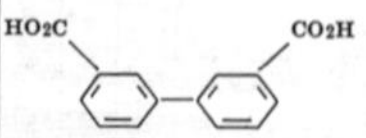	242.23		356–7				$δ^h$	δ s^h			i	aa δ lig i	**B9**², 664
b2344	—,dimethyl ester	$C_{16}H_{14}O_4$. *See* b2343	270.29	lf (MeOH)	104				i	v	v		v	lig δ	**B9**, 927
b2345	**3,4′-Biphenyldi-carboxylic acid**	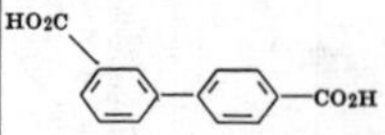	242.23	nd	333.5–4.5				i	δ		δ		$PhNO_2$ s	**B9**, 927

For explanations, symbols and abbreviations see beginning of table.

No.	Name	Synonyms and Formula	Mol. wt.	Crystalline form, color and specific rotation	m.p. °C	b.p. °C	Density	n_D	Solubility w	al	eth	ace	bz	other solvents	Ref.
	Biphenyldicarboxylic acid														
b2346	—,dimethyl ester	$C_{16}H_{14}O_4$. *See* b2345	270.29	nd (MeOH or lig)	98.5–9.5				i					MeOH s lig s	**B9**, 927
b2347	**3,5-Biphenyldicarboxylic acid**	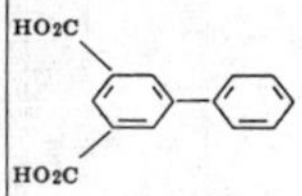	242.23		>310				i	v	v	v	v	aa s^h	**B9**, 926
b2348	—,dimethyl ester	$C_{16}H_{14}O_4$. *See* b2347	270.29	lf (MeOH)	214				i	δ			δ s^h		**B9**^{2}, 665
b2349	—,**2,2′-Biphenyldicarboxylic acid, 3,3′-dimethyl-**	$C_{16}H_{14}O_4$. *See* b2335	270.29		230					s	s		s		**B9**^{1}, 407
b2350	—,**4-nitro-**	$C_{14}H_9NO_6$. *See* b2335	287.23	pr, nd, lf	220–1				i s^h	v	v		δ	lig δ	**B9**^{2}, 659
b2351	—,**5-nitro-**	$C_{14}H_9NO_6$. *See* b2335	287.23	lf (w)	268				i v^h	v	v		v	MeOH v	**B9**^{2}, 659
b2352	—,**6-nitro-(*dl*)**	$C_{14}H_9NO_6$. *See* b2335	287.23	lf	248–50				i v^h	v	v	v	δ	chl δ lig δ aa v	**B9**^{2}, 659
b2353	—,**2,2′-Biphenyldisulfonic acid,** dichloride	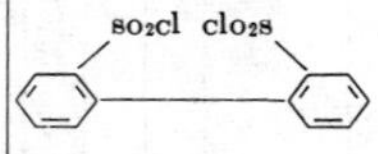	351.22	pr (chl)	138				d^h	d^h	v		v	chl v	**B11**^{2}, 123
b2354	**3,3-Biphenyldisulfonic acid,** diamide	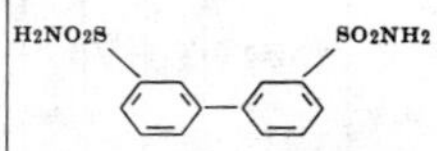	312.36	nd (ace)	285					s				MeOH s	**B11**, 219
b2355	—,dichloride	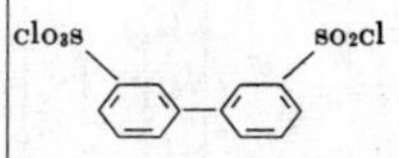	351.22	nd (chl)	128				d^h	d^h	s		s	AcOEt s	**B11**, 219
b2356	**4,4′-Biphenyldisulfonic acid**	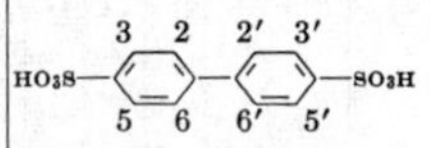	314.33	pr	72.5	>200			s						**B11**, 219
b2357	—,diamide	$C_{12}H_{12}N_2O_4S_2$. *See* b2356	312.36	nd (w)	300				v^h	δ	s		δ	CS_2 s	**B11**, 220
b2358	—,dichloride	$C_{12}H_8Cl_2O_4S_2$. *See* b2356	351.23	pr (aa)	203				d^h	d^h	s		s	CS_2 δ aa s	**B11**^{2}, 124
b2359	**2,2′-Biphenyldisulfonic acid, 4,4′-diamino-**	$C_{12}H_{12}N_2O_6S_2$. *See* b2353. 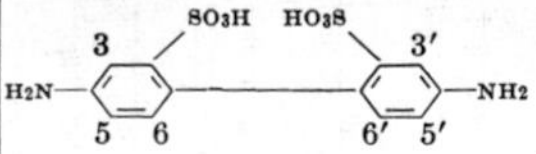	344.37	lf	175d				δ	i	i				**B14**^{1}, 743
b2360	—,**4,4′-diamino-5,5′-dimethyl-**	$C_{14}H_{16}N_2O_6S_2$. *See* b2359	372.43	nd (w+1½)					s^h	i	i			aa i	**B14**, 796
b2361	**2,2′-Bipyridyl**	2,2′-bipyridine 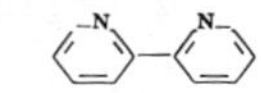	156.18	pr (peth)	70.1	273–5			δ	v	v		v	chl v lig v	**B23**^{2}, 211
b2362	**2,3′-Bipyridyl**	Isonicoteine. 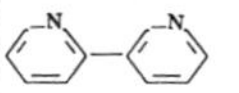	156.18			295.5–6.5	1.140^{20}_4	1.6223^{20}	δ	v	v		v	peth δ chl v	**B23**^{2}, 212
b2363	**2,4′-Bipyridyl**	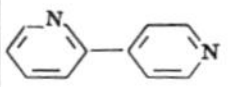	156.18			280–2			δ	v	v			chl v	**B23**, 200
b2364	**3,3′-Bipyridyl**	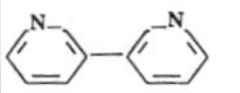	156.18		68	$291–2^{760}$	1.1635^{20}_{20}		v	v	δ				**B23**^{2}, 212
b2365	**3,4′-Bipyridyl**	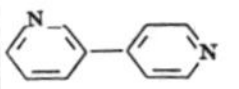	150.18	lf (peth)	61	297			v	s				peth s	**B23**^{2}, 212
b2366	**4,4′-Bipyridyl**	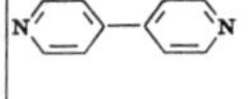	156.18	nd (w+2)	114, 73(+2w)	305^{760}			δ s^h	v	v		v	chl v lig v	**B23**^{2}, 212
b2367	**2,2′-Biquinolyl**	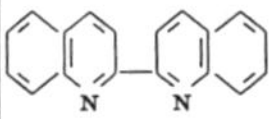	256.31	pl or lf (al)	196				i	v^h			s	os s	**B23**^{2}, 267
b2368	**2,3′-Biquinolyl**	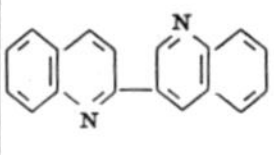	256.31	lf (al), pr (bz)	175–6				i	s^h	s		s	chl s	**B23**^{2}, 267

For explanations, symbols and abbreviations see beginning of table.

No.	Name	Synonyms and Formula	Mol. wt.	Crystalline form, color and specific rotation	m.p. °C	b.p. °C	Density	n_D	Solubility w	al	eth	ace	bz	other solvents	Ref.
	2,6-Biquinolyl														
b2369	**2,6-Biquinolyl**	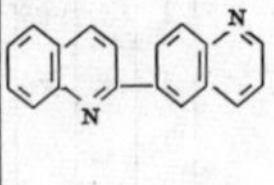	256.31	pl (al)	144					v	s		v		B23, 294
b2370	**2,7′-Biquinolyl** (higher melting)		256.31	mcl pr (al)	159–60				i	s^h	δ		δ		B23, 294
b2371	—,(lower melting)	$C_{18}H_{12}N_2$. *See* b2370	256.31	tcl	115					v	s		v	peth δ	B23, 294
b2372	**3,4′-Biquinolyl**	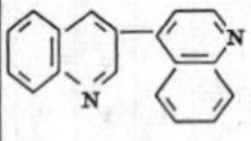	256.31		83–4					s			s	peth δ	B23[2], 26
b2373	**3,7′-Biquinolyl**	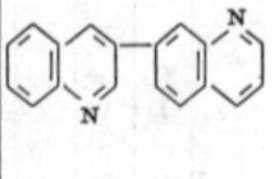	256.31	lf or nd (al or bz)	190				i	s^h	δ		s	chl s	B23[2], 26
b2374	**4,4′-Biquinolyl**	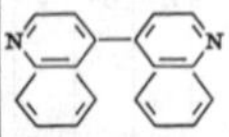	256.31	pr (peth)	166					v			v	peth δ	B23[2], 26
b2375	**4,6′-Biquinolyl**	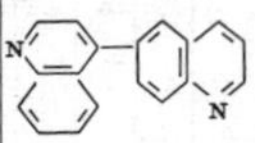	256.31		122					v	δ		v	AcOEt v chl v	B23, 294
b2376	**6,6′-Biquinolyl**	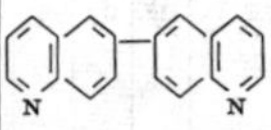	256.31	mcl pr (al)	181				i	s^h	s		s		B23, 295
b2377	**6,8′-Biquinolyl**	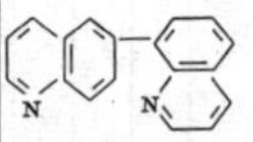	256.31	lf (al)	148				i	s^h	δ		v^h		B23, 296
b2378	**8,8′-Biquinolyl**	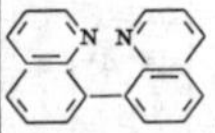	256.31	lf (al or aa)	205–7				i	s	δ	s	s	chl v lig δ	B23[2], 26
b2379	**Biquinone, 3,3′-dihydroxy-5,5′-dimethyl-**	Phenicin. 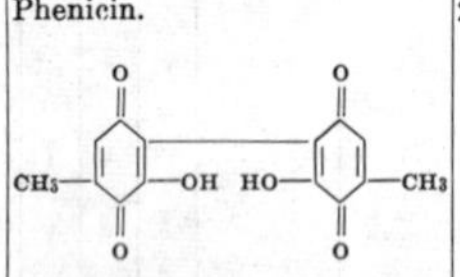	274.23	yesh-br	230–1				δ	v^h				chl v aa v	
—	**Bismark brown**	*see* **Azobenzene, 2,4,3′-triamino-**													
b2380	**Bistibine, tetramethyl-**	$(CH_3)_2SbSb(CH_3)_2$	303.64	br-red nd	17.5	190^{760}			i		s		s		J1935, 36
—	**Bistyryl**	*see* **1,3-Butadiene, 1,4-diphenyl-***													
b2381	**2,2′-Bithiophene**	2,2′-Bithienyl. 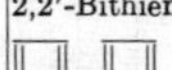	166.27	lf (al)	32–3	260^{760} 103^{3}			i	v	s			aa s	B19[2], 26
b2382	—,**hexabromo-**	$C_8Br_6S_2$ *See* b2381	639.66	nd (bz)	257–8				i				s^h		B19, 33
—	**Bitolyl**	*see* **Biphenyl, dimethyl-**													
b2383	**Biuret**	Allophanamide. Carbamylurea. $H_2NCONHCONH_2$	103.08	pl (al),nd(w+1)	193d				δ v^h	v	i				B3[2], 60
b2384	—,**acetyl-**	$CH_3CONHCONHCONH_2$	145.12	nd (w)	193–4				s	s	δ	i	δ		B3[1], 33
b2385	—,**1,5-diamino-**	$H_2NNHCONHCONHNH_2$	133.11	pr(dil al), nd (aa)	199–200				v	δ	δ			aa v^h	B3, 101
b2386	**Bixin**	$C_{25}H_{30}O_4$	394.52	pl (ace)	217d							s^h			
b2387	**Boric acid,** tributyl ester	Tributoxyborine. $B(OC_4H_9^n)_3$	230.16			230–1 $114\text{–}5^{15}$	0.8567^{20}_{4}	1.4080^{26}	d	s				MeOH v	B1[2], 398
b2388	—,triethyl ester	Triethoxyborine. $B(OC_2H_5)_3$	146.00			120^{760}	0.8546^{28}_{4}		d	∞	∞				B1[2], 333
b2389	—,triisopropyl ester	Triisopropoxyborine. $B(OC_3H_7^i)_3$	188.08			104^{760}			d	v			v	*i*-PrOH v	B1[2], 382
b2390	—,trimethyl ester	Trimethoxyborine. $B(OCH_3)_3$	103.91		−29.3	67–8	0.915^{20}		d		∞			MeOH ∞	B1[3], 121

For explanations, symbols and abbreviations see beginning of table.

No.	Name	Synonyms and Formula	Mol. wt.	Crystalline form, color and specific rotation	m.p. °C	b.p. °C	Density	n_D	Solubility						Ref.
									w	al	eth	ace	bz	other solvents	
	Boric acid														
b2391	—,tripropyl ester	$B(OC_3H_7^n)_3$	188.08			$179–80^{760}$ 64^9	0.8576_4^{20}	1.3948^{20}	d	v	∞			PrOH s	**B1**², 369
b2392	—,tris(3-methyl-butyl) ester	$B[OCH_2CH_2CH(CH_3)_2]_3$	272.24			254–5 $132–3^{12}$	0.8518_4^{20}	1.4156^{20}	d	∞	∞				**B1**³, 1645
b2393	—,**trithio-**, trimethyl ester	$B(SCH_3)_3$	152.11		5	218.2^{760}	1.126^{20}	1.5788^{20}	d	d				os ∞	
b2394	**Borine, bis(dimethylamino)-fluoro-**	$[(CH_3)_2N]_2BF$	117.96			106^{760}			d	d				os ∞	**Am76,** 3905
b2394¹	—,**bis(methylthio)methyl-**	$(CH_3S)_2BCH_3$	120.04		−59	100^{147}			d	d				os ∞	
b2395	—,**difluoro(dimethylamino)-**	$(CH_3)_2NBF_2$	92.88	rh	165d				d						**Am76,** 3904
b2396	—,**difluorophenyl-**	$C_6H_5BF_2$	125.91			70–5			d		s		s		**B16**², 638
b2397	—,**difluoro-(4-tolyl)-**	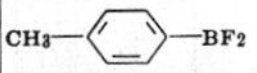 CH_3—⟨ ⟩—BF_2	139.94			95–7			d		s		s		
b2398	—,**dimethyl-arsino-**,(tetramer)	$[(CH_3)_2AsBH_2]_4$	471.27		150	185^{11}			i	s				os s	**Am 76,** 389
b2399	—,—,(trimer)	$[(CH_3)_2AsBH_2]_3$	353.46		50	146^{26}			i	s				os s	
b2400	—,**dimethyl(dimethylamino)-**	$(CH_3)_2NB(CH_3)_2$	84.99		−92	65^{760}								os ∞	
b2401	—,**dimethyl(dimethylphosphino)-**,(trimer)	$[(CH_3)_2PB(CH_3)_2]_3$	305.87		333	200^{10}			i	s					
b2402	—,**dimethyl-methoxy-**	$CH_3OB(CH_3)_2$	71.92			21^{760}			d	d				os ∞	
b2403	—,**dimethyl-(methylthio)-**	$CH_3SB(CH_3)_2$	87.98		−84	71^{760}			d	d				os ∞	
b2404	—,**(dimethylphosphino)-**,(tetramer)	$[(CH_3)_2PBH_2]_4$	295.48	nd	161	190^{29}			i	s				os s	
b2405	—,—,(trimer)	$[(CH_3)_2PBH_2]_3$	221.61	rh	86	90^4			i	s				os s	
b2407	—,**methylthio-**, (polymer)	$(CH_3SBH_2)_n$							d	d				os ∞	
b2408	—,**triethyl-**	$B(C_2H_5)_3$	98.00		−92.9	95	0.6961^{23}		δ	s	s				**B4**², 1022
b2409	—,**triisobutyl-**	$B(C_4H_9^i)_3$	182.16			188^{766} 86^{20}	0.7380_4^{25}	1.4188^{23}							**B4**², 1022
b2410	—,**trimethyl-**	$B(CH_3)_3$	55.92		−161.5	−20			δ	v	v				**B4**², 1022
b2411	—,**triphenyl-**	$B(C_6H_5)_3$	242.13		142	$245–50^{15}$			i	d	δ		s	lig s	**B16**², 636
b2412	—,**tripropyl-**	$B(C_3H_7^n)_3$	140.08		−56	156^{760}	0.7204_4^{25}	$1.4135^{22.5}$							**B4**², 1022
b2413	—,**tris(3-methyl-butyl)-**	$B[CH_2CH_2CH(CH_3)_2]_3$	224.24			119^{14}	0.7600_4^{23}	1.4321	v		s			os v	**B4**², 1023
—	—,**tris(methyl-thio)-**	*see* **Boric acid, trithio-,** trimethyl ester													
b2414	**Borneol**(*d*)	$C_{10}H_{18}O$	154.25	lf, $[\alpha]_D^{20}+37.7$ (al)	208	212^{760}	1.011_4^{20}		i	s	s		v	lig s	**B6**², 82
b2415	—(*dl*)	$C_{10}H_{18}O$	154.25	lf (lig)	210.5	sub	1.011_4^{20}		i	v	v		v		**B6**², 82
b2416	—(*l*)	$C_{10}H_{18}O$	154.25	pl, $[\alpha]_D^{20}-37.74$	208.6	210^{779}	1.1011_4^{20}		i	s	v		v		**B6**², 82
b2417	—,acetate(*d*)	$C_{12}H_{20}O_2$	196.28	rh, $[\alpha]_D^{20}+44.4$	29	$223–4^{760}$									**B6**², 83
b2418	—,—(*dl*)	$C_{12}H_{20}O$	196.28		<−17	223–4	0.9838_4^{20}	1.4630							**B6**², 86
b2419	—,—(*l*)	$C_{12}H_{20}O_2$	196.28	$[\alpha]_D^{20}-42.7$	29	225–6	0.9920_4^{20}	1.4634^{20}	δ	s	s				**B6**², 85
b2420	—,formate(*d*)	$C_{11}H_{18}O_2$	182.25	$[\alpha]_D+48.75$		90^{10}	1.017^{15}	1.4708^{15}							**B6**², 85
b2421	**Bornylamine**(*d*)	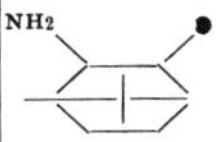 NH_2	153.26	$[N]_D^{30}+47.2$ (al)	163	200 sub			i	v	v			os v	**B12**², 39
b2422	**Bornyl chloride**(*d*)	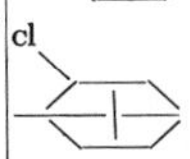 Cl	172.69	nd, $[\alpha]_D+34.5$ (eth)	149–51	207–8 sub			i	s	s			peth s	**B5**², 62
b2423	**Bornylene**(*d*)		136.23	$[\alpha]_D+19.3$	109–10	146^{760}									**B5**¹, 80
b2424	—(*l*)	$C_{10}H_{16}$. *See* b2423	136.23	$[\alpha]_D^{19}-23.9$ (bz)	113	146^{746}			i	s	s		s		**B5**², 105

For explanations, symbols and abbreviations see beginning of table.

No.	Name	Synonyms and Formula	Mol. wt.	Crystalline form, color and specific rotation	m.p. °C	b.p. °C	Density	n_D	Solubility						Ref.
									w	al	eth	ace	bz	other solvents	
	Bourbanal														
—	**Bourbonal**	*see* **Benzaldehyde, 3-ethoxy-4-hydroxy-**													
—	**Brassidic acid**	*see* **13-Decosenoic acid (*trans*)***													
b2425	**Brasilein**	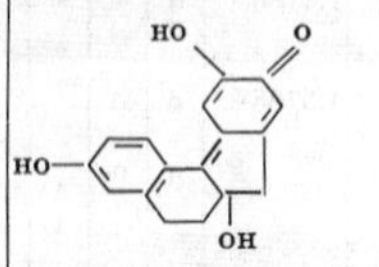	284.27	lf (w+1)	250				s^h	s	δ	...	s	chl s aa s	**B18²**, 194
—	**British antilewisite**	*see* **1-Propanol,2,3-dimercapto-***													
—	**Bromal**	*see* **Acetaldehyde,tribromo-**													
b2426	**Bromocresol green**	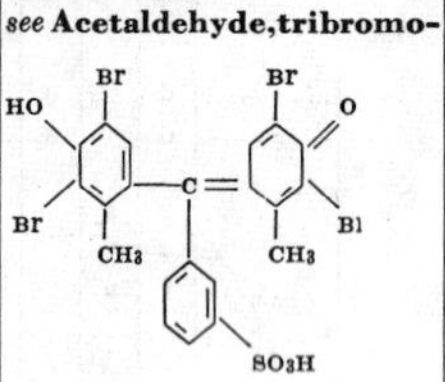	698.07	wh or red (+7w)	217–8				i	v	v	...	δ	AcOEt v aa s	**B19²**, 108
—	**Bromoform**	*see* **Methane, tribromo-***													
b2427	**Bromophenol blue**	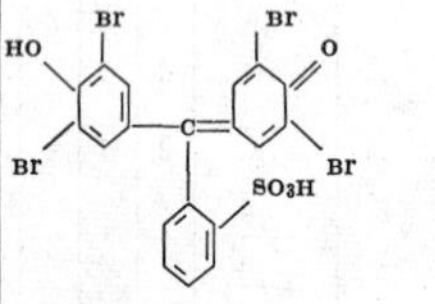	670.02	pr (aa)	270–1				i	s	...	...	s	aa s	**B19²**, 105
—	**Bromopicrin**	*see* **Methane nitrotribromo-***													
—	**Bromoprene**	*see* **1,3-butadiene, 2-bromo-***													
—	**Bromural**	*see* **Urea, 1(2-bromo-3-methylbutyl)-**													
b2428	**Brucine**	$C_{23}H_{26}N_2O_4$	394.44	pr (w+4), $[\alpha]^{20}_{5461}-149.5$ $[\alpha]^{20}_{5893}-120.5$	178				δ	v	δ	...	δ	chl v	**B27²**,797
b2429	—,hydrochloride	$C_{23}H_{27}ClN_2O_4$.	430.92	pr					s	δ	...	...	...	...	**B27²**, 797
b2430	—,nitrate	$C_{23}H_{31}N_3O_9$	493.51	pr	230				s	s	...	...	...	...	**B27²**, 797
b2431	—,sulfate	$C_{46}H_{68}N_4O_{19}S$	1013.1	nd, $[\alpha]_D-24.4$ (w)					δ v^h	δ	...	...	i	chl δ MeOH v	**B27²**, 797
b2432	**Bufotalin**	$C_{26}H_{36}O_6$	444.55	$[\alpha]^{20}_D+5.4$(chl)	223d				i	s	...	...	...	chl s	...
b2433	**Bulocapnine** (*d*)	$C_{19}H_{19}NO_4$	325.55	rh, $[\alpha]_D+237.1$ (chl, c=4)	199–200				i	s	...	...	...	os s chl v	**B27²**, 554
b2434	—(*dl*)	$C_{19}H_{19}NO_4$	325.55		209.10				...	...	...	...	...	...	**B27¹**, 468
b2435	**1,2-Butadiene***	Methylallene. $CH_2:C:CHCH_3$	54.09		−136.3	$18–19^{760}$	0.676^0_4	1.4205	i	∞	∞	...	...	...	**B1²**, 224
b2436	**1,3-Butadiene***	Bivinyl. $CH_2:CHCH:CH_2$	54.09		−108.9	−4.4		1.4292^{-25}	i	s	...	...	...	os s	**B1³**, 929
b2437	**1,2-Butadiene, 4-bromo-***	$CH_2:C:CHCH_2Br$	133.00			$109–11^{760}$	1.4255^{20}_4	1.5248^{20}	...	...	...	...	...	...	**B1³**, 929
b2438	**1,3-Butadiene, 2-bromo-***	Bromoprene. $CH_2:CBrCH:CH_2$	133.00	grsh-ye		$42–3^{165}$	1.397^{20}_4	1.4988^{20}	i	s	s	...	...	...	...
b2439	**1,2-Butadiene, 4-chloro-***	$CH_2:C:CHCH_2Cl$	88.54			88	0.9891^{20}_4	1.4775^{20}	δ	...	...	...	...	os v	**B1³**, 928
b2440	**1,3-Butadiene, 1-chloro-***	$CH_2:CHCH:CHCl$	88.54			68	0.9606^{20}_4	1.4712^{20}	...	...	...	...	...	chl v	**B1³**, 949
b2441	—,**2-chloro-***	Chloroprene. $CH_2:CClCH:CH_2$	88.54			59.4^{760}	0.9538^{20}_{20}	1.4538^{20}	δ	...	...	...	...	os s	...
b2442	—,**1,chloro-2-methyl-***	$CH_2:CHC(CH_3):CHCl$	102.57			99–101	0.9710^{20}_4	1.4792^{20}	...	v^h	...	s	...	...	**B1³**, 974

For explanations, symbols and abbreviations see beginning of table.

No.	Name	Synonyms and Formula	Mol. wt.	Crystalline form, color and specific rotation	m p. °C	b.p. °C	Density	n_D	Solubility w	al	eth	ace	bz	other solvents	Ref.
	1,3-Butadiene														
b2443	**—,1-chloro-3-methyl-***	$CH_2:C(CH_3)CH:CHCl$	102.57			99–100	0.9543^{20}_{4}				s			chl s	
b2444	**—,2-chloro-3-methyl-***	$CH_2:C(CH_3)CCl:CH_2$	102.57			93^{760}	0.9593^{20}_{4}	1.4686^{20}			v			chl v	**B1**[3], 975
b2445	**—,1,2-dichloro-***	$CH_2:CHCCl:CHCl$	122.99			$60–5^{105}$	1.199^{15}_{4}	1.5078^{15}						CCl_4 v	
b2446	**—,2,3-dichloro-***	$CH_2:CClCCl:CH_2$	122.99			98^{760}	1.1829^{20}_{4}	1.4890^{20}						chl v	
b2447	**—,2,3-dimethyl-***	Biisopropenyl. $CH_2:C(CH_3)C(CH_3):CH_2$	82.14		−76	75.9^{760}	0.7262^{20}_{4}	1.4393^{20}_{4}							**B1**[3], 991
b2448	**—,1,4-diphenyl-*** *(cis, cis)*	*cis, cis*-Bistyryl. $C_6H_5CH:CHCH:CHC_6H_5$	206.27	lf or nd (MeOH)	70.5		$0.9697^{100.7}$		i	s^h	v		v	peth s chl v aa s^h	**B5**[2], 589
b2449	—,*(trans,trans)**	*trans, trans*-Bistyryl. $C_6H_5CH:CHCH:CHC_6H_5$	206.27	lf (al or aa)	152.5	350^{720}			i	s	δ		s	peth s chl s	**B5**[2], 589
—	**—,1,4-epoxy-***	*see* **Furan**													
b2450	**—,1-ethoxy-***	$CH_2:CHCH:CHOC_2H_5$	98.15			109–12	0.8154^{20}_{4}	1.4529^{20}		s					**B1**[3], 1975
b2451	**—,2-ethoxy-***	$CH_2:CHC(OC_2H_5):CH_2$	98.15			$94.5–5.5^{760}$	0.8177^{20}_{4}	1.4400^{20}		v				os v	**B1**[3], 1977
b2452	**—,2-fluoro-***	Fluoroprene. $CH_2:CHCF:CH_2$	72.08			12^{760}	0.843^{4}_{4}	1.400^{4}							
b2453	**—,hexachloro-***	$CCl_2:CClCCl:CCl_2$	260.79		−21	215^{760} 101^{20}	1.6820^{20}_{4}	1.5542^{20}	i	s	s				**B1**[3], 955
b2454	**—,hexafluoro-***	$CF_2:CFCF:CF_2$	162.04		−132	6^{760}	1.553^{-20}_{4}	1.378^{-20}							**B1**[3], 948
b2455	**1,2-Butadiene, 4-iodo-***	$CH_2:C:CHCH_2I$	179.99			130^{760}	1.7129^{20}_{4}	1.5709^{20}							**B1**[3], 929
b2456	**1,3-Butadiene, 2-iodo-***	Iodoprene. $CH_2:CHCI:CH_2$	179.99			$111–3^{760}$	1.7278^{20}_{4}	1.5616							**B1**[3], 956
b2457	**1,2-Butadiene, 4-methoxy-***	$CH_2:C:CHCH_2OCH_3$	84.12			87–9	0.8286^{20}_{4}	1.435^{20}		s					**B1**[3], 1974
b2458	**1,3-Butadiene, 1-methoxy-***	$CH_2:CHCH:CHOCH_3$	84.12			$91–2^{760}$	0.8296^{20}_{4}	1.4590^{2}	s	s					**B1**[3], 1975
b2459	**—,2-methoxy-***	$CH_2:CHC(OCH_3):CH_2$	84.12			75^{760}	0.8272^{20}_{4}	1.4432^{20}		v				os v	**B1**[3], 1976
b2460	**1,2-Butadiene, 3-methyl-***	*asym*-Dimethylallene. $CH_2:C:C(CH_3)_2$	68.12		−120	40^{760}	0.6833^{20}_{4}	1.4166^{20}							**B1**[3], 1976
b2461	**1,3-Butadiene, 2-methyl-***	Isoprene. $CH_2:CHC(CH_3):CH_2$	68.12		−146	34^{760}	0.6810^{20}_{4}	1.4216^{20}	i	∞	∞				**B1**[3], 966
b2462	**—,Pentafluoro-2-trifluoro-methyl-***	Perfluoroisoprene. $F_2C:C(CF_3)CF:CF_2$	212.05			39^{760}	1.527^{0}_{4}	1.3000^{0}							
b2462[1]	**—,1-phenyl-,** *(trans)**	$CH_2:CHCH:CHC_6H_5$	130.19		4.5	86^{11}	0.9286^{20}_{4}	1.6128^{16}	i	s				os s	**B5**[2], 414
b2463	**1,2,3,4-tetra-chloro-,** (liquid)*	$ClCH:CClCCl:CHCl$	191.89		−4	188 67^{10}	1.516^{15}_{15}	1.5455	i					os s	**B1**[3], 954
b2464	—,—,(solid)*	$ClCH:CClCCl:CHCl$	191.89		50				i					chl v os s	**B1**[3], 954
b2465	**—,1,2,3-tri-chloro-***	$CH_2:CClCCl:CHCl$	157.44			$33–4^{7}$	1.4060^{20}_{4}	1.5262^{20}			s			chl s	**B1**[3], 954
b2466	**2,3-Butadien-1-ol***	$CH_2:C:CHCH_2OH$	70.09			$126–8^{756}$	0.9164^{20}_{4}	1.4759^{20}	s	v				os v	**B1**[3], 1974
b2468	**Butadiyne***	Biacetylene. CH⋮CC⋮CH	50.06		−36	10.3^{760}	0.7364^{0}_{4}	1.4189^{5}	v	s	v	v		chl s	**B1**[3], 1056
b2469	**—,1,4-bis(1-hydroxy-cyclo-hexyl-)***	OH HO ⬡—C⋮CC⋮C—⬡	246.35		174				i δ^h				δ^h	MeOH s chl δ^h	

For explanations, symbols and abbreviations see beginning of table.

No.	Name	Synonyms and Formula	Mol. wt.	Crystalline form, color and specific rotation	m.p. °C	b.p. °C	Density	n_D	Solubility w	al	eth	ace	bz	other solvents	Ref.
	Butadiyne														
b2470	—,**1,4-dichloro-***	$ClC{\vdots}CC{\vdots}CCl$	118.96			1–3								chl v	**B1**^{2}, 246
b2471	**Butanal***	Butyraldehyde. $CH_3CH_2CH_2CHO$	72.10		−99	75.7^{760}	0.8170^{20}_{4}	1.3843^{20}	s	∞	∞				**B1**^{2}, 721
b2472	—,oxime	$CH_3CH_2CH_2CH{:}NOH$	87.12		−29.5	152^{715}	0.923^{20}_{4}		v	∞	∞				**B1**^{2}, 724
b2473	—,phenylhydrazone	$CH_3CH_2CH_2CH{:}NNHC_6H_5$	162.23			152^{14}									**B15**^{2}, 55
b2474	—,**2-benzylidene-**	α-Ethylcinnamaldehyde. 2-Ethyl-3-phenylpropenal. $C_6H_5CH{:}C(C_2H_5)CHO$	160.21	ye		$157–8^{5}$	1.0201^{22}_{4}	1.578^{20}							
b2475	—,**3-chloro-**, diethylacetal	$CH_3CHClCH_2CH(OC_2H_5)_2$	180.68			$70–1^{12}$	0.9677^{20}_{4}	1.4210^{20}	i	∞				os ∞	**B1**^{2}, 724
b2476	—,**4-chloro-***	$ClCH_2CH_2CH_2CHO$	106.56			$50–1^{13}$	$1.107^{8.5}_{15}$	$1.4466^{8.5}$		v	v				
b2477	—,**2,3-dichloro-***	$CH_3CHClCHClCHO$	141.01			$58–60^{20}$	1.2666^{21}_{4}	1.4618^{21}	i	s	s			chl s	**B1**^{2}, 724
b2478	—,**2-ethyl-***	$(C_2H_5)_2CHCHO$	100.16			$115–7^{741}$	0.814^{20}_{4}	1.4040^{15}	δ	∞	∞				**B1**^{2}, 749
b2479	-,**3-hydroxy-***	Aldol. $CH_3CHOHCH_2CHO$	88.10			$79–80^{12}$ d85	1.103^{20}_{4}		∞	∞	s				**B1**^{2}, 868
b2480	—,**2-methyl-***	$CH_3CH_2CH(CH_3)CHO$	86.13			92–3	0.8029^{20}_{4}	1.3869^{20}	i	s	s				**B1**, 682
b2481	—,**3-methyl-***	Isovaleraldehyde. $(CH_3)_2CHCH_2CHO$	86.13		−51	92.5^{760}	0.803^{17}_{4}	1.3902^{20}	δ	s	s				**B1**^{2}, 742
b2482	—,—,oxime	$(CH_3)_2CHCH_2CH{:}NOH$	101.15		48	161.3^{759}	0.8934^{20}_{4}	1.4367^{20}	i						**B1**^{2}, 744
b2483	—,**2,2,3-trichloro-***	*n*-Butyrchloral. $CH_3CHClCCl_2CHO$	175.44			$164–5^{750}$	1.3956^{20}_{4}	1.4755^{20}	s	s	s				**B1**^{2}, 725
b2484	—,—,hydrate	$CH_3CHClCCl_2C(OH)_2$	192.45	rh pl or lf (w)	78	d	1.694^{20}_{4}		s^h	v	s				**B1**^{2}, 725
b2485	**Butane***	$CH_3CH_2CH_2CH_3$	58.12		−138.3	−0.5	0.6012^{0}_{4}	1.3543^{-13}	v	v	v			chl v	**B1**^{3}, 261
b2486	—,**1-amino-***	*n*-Butylamine. $CH_3CH_2CH_2CH_2NH_2$	73.14		−50.5	77.8^{760}	0.764^{25}_{4}	1.401^{20}	∞	s	s				**B4**^{2}, 631
b2487	—,**2-amino-(*d*)***	*sec*-Butylamine, $CH_3CH_2CH(CH_3)NH_2$	73.14	$[\alpha]^{20}_{D}+4.06$ (w)	−104.5	63^{760}	0.724^{20}_{4}	1.344^{20}	s	∞	∞				**B4**^{1}, 372
b2488	—,—(*dl*)*	$CH_3CH_2CH(CH_3)NH_2$	73.14		<−72	$66–8^{772}$	0.7271^{17}_{4}	1.3950^{17}							**B4**^{2}, 636
b2489	—,—(*l*)*	$CH_3CH_2CH(CH_3)NH_2$	73.14	$[\alpha]^{20}_{D}-5.00$ (w, c=4.7)		63	0.725^{20}_{4}		v	v					**B4**^{1}, 372
b2490	—,**2-amino-2,3-dimethyl-***	$(CH_3)_2CHC(CH_3)_2NH_2$	101.19			$104–5^{760}$	0.7683^{0}_{0}	1.4096^{17}							**B4**, 193
b2491	—,**3-amino-2,2-dimethyl-***	$(CH_3)_3CCH(CH_3)NH_2$	101.19		−20	$101.5–2.5^{751}$			v						**B4**, 193
b2492	—,—,hydrochloride*	$(CH_3)_3CCH(CH_3)NH_2.HCl$	137.65	nd	>245				v						**B4**, 193
b2494	—,**1-amino-3-methyl-***	$(CH_3)_2CHCH_2CH_2NH_2$	87.17			$95–7^{761}$	0.7505^{20}_{4}	1.4096^{18}	∞	∞	∞			chl s	**B4**^{2}, 644
b2495	—,**2-amino-2-methyl-***	$CH_3CH_2C(CH_3)_2NH_2$	87.17		−105	77^{760}	0.731^{25}_{4}		∞	∞	∞				**B4**^{2}, 644
b2496	—,**2-amino-3-methyl-***	$(CH_3)_2CHCH(CH_3)NH_2$	87.17			$84–7^{760}$	0.7574^{19}	1.4096^{18}	v	s					**B4**^{2}, 644
b2496^{1}	**1,4-bis(dicarbethoxyamino)-**	$C_2H_5O_2CNH(CH_2)_4NHCO_2C_2H_5$	232.28	nd (lig)	85–6				i	s	s		s	chl s lig s^h	**B4**^{1}, 421
b2496^{2}	—,**1,4-bis(dimethylamino)-**	$(CH_3)_3NCH_2CH_2CH_2CH_2N(CH_3)_2$	144.26			269	0.7942^{15}		s	s	s				**B4**^{2}, 702
b2497	—,**2,2-bis(ethylsulfonyl)-***	Trional. $CH_3CH_2C(SO_2C_2H_5)_2CH_3$	242.36	lf (w)	75.7	d	1.199^{85}_{4}		δ	v	s			peth s	**B1**^{2}, 731
b2498	—,**1,1-bis(4-hydroxyphenyl)-***	$CH_3CH_2CH_2CH(-C_6H_4-OH)_2$	242.32	nd (to)	136–7	270^{12}			i	s			s^h	to s^h	**B6**^{2}, 980
b2499	—,**2,2-bis(4-hydroxyphenyl-***	$CH_3CH_2C(CH_3)(-C_6H_4-OH)_2$	242.32	nd or pr	133–4	$250–3^{12}$			i	v	v	v	δ	MeOH v	**B6**^{2}, 980

For explanations, symbols and abbreviations see beginning of table.

No.	Name	Synonyms and Formula	Mol. wt.	Crystalline form, color and specific rotation	m.p. °C	b.p. °C	Density	n_D	Solubility w	al	eth	ace	bz	other solvents	Ref.
	Butane														
b2500	**—,1-bromo-***	*n*-Butyl bromide. $CH_3CH_2CH_2CH_2Br$	137.03		−112.3	101.3^{760}	1.2764^{20}_{4}	1.4398^{20}	i	∞	∞				**B1**^{3}, 290
b2501	**—,2-bromo-(*dl*)***	*sec*-Butyl bromide. $CH_3CH_2CH(CH_3)Br$	137.03		−112.1	91.2^{760}	1.2556^{20}_{4}	1.4366^{20}							**B1**^{3}, 293
b2502	—,—(*l*)*	$CH_3CH_2CH(CH_3)Br$	137.03	$[\alpha]^{22}_{D}-23.13$ (undil)		90–1	1.2536^{25}_{4}	$1.4359^{19.5}$							**B1**^{3}, 293
b2503	**—,1-bromo-4-chloro-***	$ClCH_2CH_2CH_2CH_2Br$	171.47			$174\text{–}5^{756}$	1.488^{20}_{4}	1.4885^{20}	i	s	s			chl s	**B1**^{3}, 294
b2504	**—,2-bromo-1-chloro-***	$CH_3CH_2CHBrCH_2Cl$	171.47			$146\text{–}7^{758}$	1.468^{20}_{4}	1.4880^{20}	i				s	chl s	**B1**^{2}, 83
b2505	**—,1-bromo-3,3-dimethyl-***	$(CH_3)_3CCH_2CH_2Br$	165.08			138	1.1556^{20}_{4}	1.4440	i		s			chl v	**B1**^{3}, 409
b2506	**—,2-bromo-2,3-dimethyl-***	$(CH_3)_2CHCBr(CH_3)_2$	165.08		24–5	$132\text{–}3^{742}$	1.1772^{20}	1.4517	d^{h}		s			chl v	**B1**^{3}, 414
b2507	**—,1-bromo-4-fluoro-***	$FCH_2CH_2CH_2CH_2Br$	155.01			$134\text{–}5^{740}$		1.4370^{25}	i	v	v				
b2508	**—,1-bromo-2-methyl-(*d*)***	$CH_3CH_2CH(CH_3)CH_2Br$	151.05	$[\alpha]_{5892}+4.04$		121.6^{760}	1.2234^{20}_{4}	1.4451^{20}	i	s	s				**B1**^{3}, 362
b2509	—,—(*dl*)*	$CH_3CH_2CH(CH_3)CH_2Br$	151.05			117–8	$1.222^{15.7}$	1.448^{21}							**B1**^{3}, 363
b2510	**—,1-bromo-3-methyl-***	Isoamyl bromide. $(CH_3)_2CHCH_2CH_2Br$	151.05		−112	121.5^{760}	1.2609^{20}_{4}	1.4420^{20}	i	s	s				**B1**^{3}, 363
b2511	**—,1-chloro-***	*n*-Butyl chloride. $CH_3CH_2CH_2CH_2Cl$	92.57		−123	78.4^{760}	0.8865^{20}_{4}	1.4021^{20}	i	∞	∞				**B1**^{3}, 275
b2512	**—,2-chloro-(*dl*)***	Isobutyl chloride. $CH_3CH_2CHClCH_3$	92.57		−131.3	68^{764}	0.8732^{20}_{4}	1.3971^{20}	i	∞	∞				**B1**^{3}, 278
b2513	—,—(*l*)*	$CH_3CH_2CHClCH_3$	92.57	$[\alpha]^{20}_{D}-8.48$	−140.5	68^{760}	0.8950^{0}_{4}								**B1**^{2}, 81
b2514	**—,1-chloro-2,2-dimethyl-***	*neo*-Hexyl chloride. $CH_3CH_2C(CH_3)_2CH_2Cl$	120.62			116^{735}		1.4200^{20}	i		v			chl v	
b2515	**—,1-chloro-3,3-dimethyl-***	$(CH_3)_3CCH_2CH_2Cl$	120.62			115	0.8670^{20}_{4}	1.4161^{20}			v			chl v	
b2516	**—,2-chloro-2,3-dimethyl-***	$(CH_3)_2CHCCl(CH_3)_2$	120.62		−7.2	118	0.8769^{22}	1.4162^{25}	d	s		s			**B1**^{3}, 414
b2517	**—,3-chloro-2,2-dimethyl-***	Pinacolyl chloride. $CH_3CHClC(CH_3)_3$	120.62			110^{734}	0.8767^{20}_{4}	1.4181			v				
b2519	**—,1-chloro-4-fluoro-***	$FCH_2CH_2CH_2CH_2Cl$	110.56			114.7		1.4020^{25}	i	v	v				
b2520	**—,1-chloro-2-methyl-(*d*)***	$CH_3CH_2CH(CH_3)CH_2Cl$	106.60	$[\alpha]^{20}_{5892}+1.64$		100.5^{760}	0.8857^{20}_{4}	1.4124	i	s	s				**B1**^{3}, 356
b2521	—,—(*dl*)*	$CH_3CH_2CH(CH_3)CH_2Cl$	106.60			99.9^{760}	0.8818^{15}_{15}								**B1**^{3}, 357
b2522	**—,1-chloro-3-methyl-***	Isoamyl chloride. $(CH_3)_2CHCH_2CH_2Cl$	106.60		−104.4	98.8^{760}	0.8704^{20}_{4}	1.4084^{20}	δ	∞	∞				**B1**^{3}, 359
b2523	**—,2-chloro-2-methyl-***	*tert*-Amyl chloride. $CH_3CH_2CCl(CH_3)$	106.60		−73.7	84.5^{759}	0.8650^{20}_{4}	1.4052^{20}	δ	s	s				**B1**^{3}, 357
b2524	**—,2-chloro-3-methyl-(*dl*)***	$(CH_3)_2CHCHClCH_3$	106.60			93^{760}	0.8685^{20}_{4}	1.4081^{20}			v				**B1**^{1}, 46
b2525	**—,2(chloro-methyl)-1,3-dichloro-(*dl*)***	$CH_3CHClCH(CH_2Cl)_2$	175.49			$79\text{–}81^{15}$	1.2793^{15}_{4}		i					chl s	
b2526	**—,2(chloro-methyl)-1,2,3-trichloro-***	$CH_3CHClCCl(CH_2Cl)_2$	209.93			$102-3^{18}$	1.3977^{18}_{4}		i		v			chl s	**B1**^{3}, 362
b2527	**—,1-chloro-2,2,3,3-tetra-methyl-***	$(CH_3)_3.C.C(CH_3)_2CH_2Cl$	148.68		52–3	$80\text{–}1^{40}$			i		v				**B1**^{3}, 502

For explanations, symbols and abbreviations see beginning of table.

No.	Name	Synonyms and Formula	Mol. wt.	Crystalline form, color and specific rotation	m.p. °C	b.p. °C	Density	n_D	Solubility						Ref.
									w	al	eth	ace	bz	other solvents	
	Butane														
b2528	**—,2-chloro-2,3,3-trimethyl-***	$(CH_3)_2CClC(CH_3)_3$	134.65		136				i		v			MeOH δ^h	**B1**[1], 59
b2529	**—,1,4-diamino-***	Putrescine. $H_2NCH_2CH_2CH_2CH_2NH_2$	88.15	lf	27–8	$158–9^{760}$	0.877^{25}_{4}		s						**B4**, 264
b2530	—,—,dihydro-chloride*	$H_2NCH_2CH_2CH_2CH_2NH_2.2HCl$	161.08	nd (w)	>290				v	s^h				MeOH i	**B4**[2], 701
b2533	**—,1,2-dibromo-***	$CH_3CH_2CHBrCH_2Br$	215.93			166.3^{760}	1.7951^{20}_{4}	1.5150^{20}	i		s			chl s	**B1**[3], 295
b2534	**—,1,3-dibromo-***	$CH_3CH_2BrCH_2CH_2Br$	215.93			175^{766}	1.80^{20}	1.507^{20}	i		s			chl s	**B1**[3], 295
b2535	**—,1,4-dibromo-***	Tetramethylene dibromide. $BrCH_2CH_2CH_2CH_2Br$	215.93		−26	$197–8^{760}$	1.8080^{20}_{4}	1.5175^{20}	i					chl s	**B1**[3], 295
b2536	**—,2,3-dibromo-***	$CH_3CHBrCHBrCH_3$	215.93		<−80	161^{760}	1.7893^{22}_{4}	1.5133^{22}	i		s			chl s	**B1**[3], 297
b2537	**—,1,1-dichloro-***	Butylidene dichloride. $CH_3CH_2CH_2CHCl_2$	127.01			$114–5^{763}$	1.0863^{20}_{4}	1.4355	i					chl s	**B1**[3], 280
b2538	**—,1,2-dichloro-***	$CH_3CH_2CHClCH_2Cl$	127.01			124^{760}	1.1116^{25}_{4}	1.4474^{15}	i		s			chl s	**B1**[3], 280
b2539	**—,1,3-dichloro-***	$CH_3CHClCH_2CH_2Cl$	127.01			$131–3^{758}$	1.1151^{20}_{4}	1.4445^{20}	i		s			chl s	**B1**[3], 281
b2540	**—,1,4-dichloro-***	Tetramethylene dichloride. $ClCH_2CH_2CH_2CH_2Cl$	127.01		−38.7	161–3	1.1598^{20}_{4}	1.4566^{20}	i					chl s	**B1**[2], 81
b2541	**—,2,2-dichloro-***	$CH_3CH_2CCl_2CH_3$	127.01		−74	104		1.4295	i					chl s	**B1**[3], 282
b2542	**—,2,3-dichloro-***	$CH_3CHClCHClCH_3$	127.01		−80.0	116^{760}	1.1134^{20}_{4}	1.4430^{20}	i					chl s	**B1**[3], 282
b2543	**—,2,2-dichloro-3,3-dimethyl-***	$(CH_3)_3CCCl_2CH_3$	155.08		151–2				i	v^h	s^h				**B1**[3], 409
b2544	**—,2,3-dichloro-2,3-dimethyl-***	$(CH_3)_2CClCCl(CH_3)_2$	155.08		164					v	v				**B1**, 152
b2545	**—,1,1-dichloro-3-methyl-(*dl*)***	$(CH_3)_2CHCH_2CHCl_2$	141.45			130	1.05^{24}_{24}		i	s	s				**B1**[1], 47
b2456	**—,1,2-dichloro-2 methyl-(*dl*)***	$CH_3CH_2CCl(CH_3)CH_2Cl$	141.05			133–5	1.0785^{20}_{4}	$1.4432^{23.5}$	d	s	s			chl s	**B1**[3], 360
b2547	**—,1,3-dichloro-3-methyl-***	$(CH_3)_2CClCH_2CH_2Cl$	141.05			145–6	1.0654^{20}_{4}	1.4455^{20}	i		s			chl s	**B1**[3], 361
b2548	**—,1,4-dichloro-2-methyl-***	$ClCH_2CH_2CH(CH_3)CH_2Cl$	141.05			$168–9^{760}$	1.1003^{25}_{4}	1.4562^{21}	i					chl s	**B1**[2], 101
b2549	**—,2,3-dichloro-2-methyl-***	$CH_3CHClCCl(CH_3)_2$	141.05			138^{760}	1.0696^{15}_{4}	1.4450^{18}	i	s	s				
b2550	**—,1:2,3:4-diepoxy-***	Butadiene dioxide. $CH_2CHCHCH_2$	86.09		−16	138^{767}	1.1157^{20}_{4}	1.4330^{20}_{4}	∞						**B19**[2], 14
b2551	—,—(*meso*)*	Bioxirane. $CH_2CHCHCH_2$	86.09		4	144^{760}	1.113^{20}	1.435^{20}	s	s					**B19**, 15
b2552	**—,1(diethyl-amino-methoxy)-***	Butyl(diethylaminomethyl) ether. $CH_3CH_2CH_2CH_2OCH_2N(C_2H_5)_2$	159.27			$172–4^{754}$				v				os s	**B4**[2], 598
b2553	**—,1,4-difluoro-octachloro-***	$FCCl_2CCl_2CCl_2CCl_2F$	369.70		4–5	152.5^{20}	1.9272^{20}_{4}	1.5256^{20}							**Am62**, 342
b2554	**—,2,2-di(2-furyl)-**	$(C_4H_3O)_2C(CH_3)CH_2CH_3$	190.23			$68–70^{2}$	1.033^{20}_{4}	1.4970^{20}	i	s	s			MeOH s	
b2555	**—,1,4-diiodo-***	Tetramethylene diiodide. $ICH_2CH_2CH_2CH_2I$	309.92		6	$125–6^{12}$	2.349^{26}_{4}	1.619^{25}	i					os s	**B1**[3], 302
b2556	**—,2,2-dimethyl-***	*neo*-Hexane. $CH_3CH_2C(CH_3)_3$	86.17			49.7^{760}	0.6492^{20}_{4}	1.3688^{20}	i	s	s				**B1**[3], 405
b2557	**—,2,3-dimethyl-***	$(CH_3)_2CHCH(CH_3)_2$	86.17		−128.8	58^{760}	0.6616^{20}_{4}	1.3750^{20}	i	s	s				**B1**[3], 410
b2558	**—,1,4-dinitro-***	$O_2NCH_2CH_2CH_2CH_2NO_2$	148.12	pr (al)	33–4	$176–8^{13}$			i	δ s^h	s		s	MeOH s^h	**B1**[3], 305
b2559	**—,2,3-dinitro-(*dl*)***	$CH_3CHNO_2CHNO_2CH_3$	148.12	pr (eth)	48–9					δ	s			MeOHs^h lig δ	**B1**[3], 305

For explanations, symbols and abbreviations see beginning of table.

No.	Name	Synonyms and Formula	Mol. wt.	Crystalline form, color and specific rotation	m.p. °C	b.p. °C	Density	n_D	Solubility w	al	eth	ace	bz	other solvents	Ref.
	Butane														
b2560	—,—(*meso*)*	$CH_3CHNO_2CHNO_2CH_3$	148.12	pr (eth)		$76–7^{1}$				δ	s			lig δ	**B1**³, 305
b2561	—,**1,2-epoxy-***	1,2-Butylene oxide. $CH_3CH_2CHCH_2$ (O bridged)	72.10			61–2	0.837^{17}_{4}	1.3855^{17}	d^h	v	∞			os v	**B17**², 17
—	—,**1,4-epoxy-***	*see* **Furan, tetrahydro-**													
b2562	—,**2,3-epoxy-**(*cis*)*	β-Butylene oxide. $CH_3CHCHCH_3$ (O bridged)	72.10		−80	59.7^{742}	0.8226^{25}_{4}	1.3828^{20}	d		v		v	os v	**B17**², 17
b2563	—,—(*trans*)*	$CH_3CHCHCH_3$ (O bridged)	72.10		−85	53.5^{742}	0.8010^{25}_{4}	1.3736^{20}	d		v		v	os v	**B17**², 17
b2564	—,**1-fluoro-***	Butyl fluoride. $CH_3CH_2CH_2CH_2F$	76.11			32^{760}	0.7761^{20}_{4}	1.3419^{15}		v					**B1**³, 273
b2565	—,**1,1,2,2,3,4,4-heptachloro-***	$Cl_2CHCHClCCl_2CHCl_2$	299.27			97.5^{2}	1.742^{20}_{20}	1.5407^{20}	i					CCl_4 v	
b2566	—,**1,1,2,3,4,4-hexachloro-*** (liquid)	$Cl_2CHCHClCHClCHCl_2$	264.82			111^{10}	1.6460^{20}_{4}	1.5258^{20}						CCl_4 s	**B1**³, 288
b2567	—,—(solid)	$Cl_2CHCHClCHClCHCl_2$	264.82	nd (al), pr (bz, aa)	109–10									CCl_4 s	**B1**³, 288
b2568	—,**1-iodo-***	Butyliodide. $CH_3CH_2CH_2CH_2I$	184.03		−103	130^{760}	1.6123^{20}_{4}	1.5001^{20}	i	∞	∞				**B1**³, 294
b2569	—,**2-iodo-**(*dl*)*	*sec*-Butyliodide. $CH_3CH_2CHICH_3$	184.03		−104.2	120^{760}	1.5984^{20}_{4}	1.4991^{20}	i	∞	∞				**B1**³, 305
b2570	—,—(*l*)*	$CH_3CH_2CHICH_3$	184.03	$[\alpha]_D$ −12.15 (al, c=20)		117.5–8.5	1.585^{20}_{4}	1.4945^{19}							**B1**³, 301
b2571	—,**1-iodo-2-methyl-**(*d*)*	$CH_3CH_2CH(CH_3)CH_2I$	198.05	$[\alpha]^{20}_{5982}$+5.69		148^{760}	1.5253^{20}_{4}	1.4977^{20}	i	s	s				**B1**³, 366
b2572	—,—(*dl*)*	$CH_3CH_2CH(CH_3)CH_2I$	198.05			144–7		1.497^{20}							**B1**³, 366
b2573	—,**1-iodo-3-methyl-***	Isoamyl iodide. $(CH_3)_2CHCH_2CH_2I$	198.05			147^{760}	1.5092^{20}_{4}	1.4939^{20}	δ	∞	∞				**B1**³, 367
b2574	—,**2-iodo-2-methyl-***	*tert*-Amyl iodide. $CH_3CH_2CI(CH_3)_2$	198.05			124.5^{760}	1.4937^{20}_{4}	1.4981^{20}	i	∞	∞				**B1**³, 366
b2575	—,**2-iodo-3-methyl-***	$(CH_3)_2CHCHICH_3$	198.05			$141–2^{57}$			i					os s	**B1**¹, 49
b2577	—,**1-isocyano-***	Butylcarbylamine. Butyl isocyanide. $CH_3CH_2CH_2CH_2NC$	83.13			118^{760}			i	∞	∞				
b2578	—,**1-isocyano-3-methyl-***	Isoamylcarbylamine. $(CH_3)_2CHCH_2CH_2NC$	97.16			$139–40^{760}$	0.806^{20}_{4}	1.406^{20}	i	∞	∞				**B4**, 184
b2579	—,**2-methyl-***	Isopentane. $(CH_3)_2CHCH_2CH_3$	72.15		−160	27.9^{760}	0.6197^{20}_{4}	1.3537^{20}	i	∞	∞				**B1**³, 352
b2580	—,**2-methyl-1,2,3-trichloro-***	$CH_3CHClCCl(CH_3)CH_2Cl$	175.50			$183–5^{762}$	1.2527^{20}_{4}		i					chl s	**B1**³, 361
b2581	—,**2-methyl-2,3,3-trichloro-***	$CH_3CCl_2CCl(CH_3)_2$	175.50			182–3			i					chl s, aa v^h	**B1**³, 361
b2582	—,**1-nitro-***	$CH_3CH_2CH_2CH_2NO_2$	103.12			153^{760}		1.4103^{20}	δ	∞	∞				**B1**³, 303
b2583	—,**2-nitro-***	$CH_3CH_2CHNO_2CH_3$	103.12		−132	139^{760}	0.9854^{17}_{4}	1.4057^{21}							**B1**³, 303
b2585	—,**1,1,2,2,3,3,4,4-octachloro-***	$Cl_2CHCCl_2CCl_2CHCl_2$	333.72		81					v^h					
b2586	—,**1,1,2,3,4-pentachloro-** (liquid)*	$ClCH_2CHClCHClCHCl_2$	230.37			95.5^{11}			i	s				CC_4 v	
b2587	—,—(solid)*	$ClCH_2CHClCHClCHCl_2$	230.37		49	230, 102^{11}	1.539^{53}	1.5065^{53}	i	v^h				CCl_4 v	
b2588	—,**1,2,2,3,4-pentachloro-***	$ClCH_2CHClCCl_2CH_2Cl$	230.37			85^{10}	1.5543^{20}_{4}	1.5157^{20}	i					chl s	**B1**³, 288
b2589	—,**1,1,4,4-tetrabromo-***	$Br_2CHCH_2CH_2CHBr_2$	373.74			$138–45^{10}$	2.529^{20}_{4}	1.6077^{20}	i	δ	∞		∞	chl ∞	**B1**, 121

For explanations, symbols and abbreviations see beginning of table.

No.	Name	Synonyms and Formula	Mol. wt.	Crystalline form, color and specific rotation	m.p. °C	b.p. °C	Density	n_D	Solubility						Ref.
									w	al	eth	ace	bz	other solvents	
	Butane														
b2590	**—,1,2,2,3-tetra-bromo-***	$CH_3CHBrCBr_2CH_2Br$	373.74			128–30^{14}	2.5085^{20}_{4}	1.6070^{20}	i						**B1**^{3}, 298
b2591	**—,1,2,2,4-tetra-bromo-***	$BrCH_2CH_2CBr_2CH_2Br$	373.74	nd (lig)	72–3				i					lig s^{h}	**B1**^{3}, 298
b2592	**—,1,2,3,4-tetra-bromo-(*dl*)***	$BrCH_2CHBrCHBrCH_2Br$	373.74	rh pl (peth)	39				i	v	v			lig s	**B1**^{3}, 298
b2593	—,—(*meso*)*	$BrCH_2CHBrCHBrCH_2Br$	373.74	nd (al or lig)	118	180–1^{60}			i	s^{h}				chl s lig i	**B1**^{3}, 298
b2594	**—,2,2,3,3-tetra-bromo-***	$CH_3CBr_2CBr_2CH_3$	373.74	lf (lig), pr (eth-lig)	242–3				i	δ	s		s	chl v	**B1**^{3}, 298
b2595	**—,1,1,1,2-tetra-chloro-***	$CH_3CH_2CHClCCl_3$	195.92			134–5^{742}	1.3932^{20}_{20}	1.4920^{20}	i					chl s	**B1**^{3}, 280
b2596	**—,1,2,2,3-tetra-chloro-***	$CH_3CHClCCl_2CH_2Cl$	195.92		−48, −46	182	1.4276^{18}_{4}	1.491^{20}						chl s	**B1**^{3}, 286
b2597	**—,1,2,3,3-tetra-chloro-***	$CH_3CCl_2CHClCH_2Cl$	195.92			90^{82}	1.4204^{20}_{4}	1.4958^{20}	i					chl s	**B1**^{3}, 286
b2601	**—,2,2,3,3-tetra-methyl-***	$CH_3C(CH_3)_2C(CH_3)_2CH_3$	114.23	lf (eth)	100.7	106.5^{760}	0.6485^{110}_{4}	1.4695^{20}	i		s				**B1**^{3}, 501
b2602	**—,1,1,2-tri-bromo-***	$CH_3CH_2CHBrCHBr_2$	294.85			216.2^{760}	2.1836^{20}_{4}	1.5626^{17}	i	s				chl s	**B1**^{2}, 84
b2603	**—,1,2,2-tri-bromo-***	$CH_3CH_2CBr_2CH_2Br$	294.83			213.8^{760}	2.1692^{20}_{4}	1.5624^{16}	i	s				chl s	**B1**^{2}, 85
b2604	**—,1,2,3-tri-bromo-***	$CH_3CHBrCHBrCH_2Br$	294.83			113–4^{21}	2.1908^{20}_{0}	1.5680^{20}	i	s	s			chl s	**B1**^{2}, 85
b2605	**—,1,2,4-tri-bromo-***	$BrCH_2CH_2CHBrCH_2Br$	294.83			112–3^{13}	2.234^{18}_{0}	1.574^{18}	i		s			chl s	**B1**^{2}, 85
b2606	**—,1,3,3-tri-bromo-***	$CH_3CBr_2CH_2CH_2Br$	294.83			200–5^{760}	2.1446^{20}_{4}	1.5564^{20}	i						**B1**^{3}, 298
b2607	**—,2,2,3-tri-bromo-***	$CH_3CHBrCBr_2CH_3$	294.83		1.8	206.5^{760}	2.1724^{20}_{4}	1.5628^{17}	i	s	s			chl s	**B1**^{2}, 85
b2608	**—,1,1,3-tri-chloro-***	$CH_3CHClCH_2CHCl_2$	161.46			150^{720}	1.317^{15}	1.4600^{15}	i					chl v	**B1**^{3}, 285
b2609	**—,1,2,3-tri-chloro-***	$CH_3CHClCHClCH_2Cl$	161.46			165–7^{760}	1.3164^{20}_{4}	1.4790^{20}	i		s			chl v	**B1**^{3}, 285
b2610	**—,2,2,3-tri-chloro-***	$CH_3CHClCCl_2CH_3$	161.46			143–5	1.2699^{20}_{4}	1.4645^{20}	i					chl v	**B1**^{3}, 285
b2611	**—,2,2,3-tri-methyl-***	Triptan. $(CH_3)_2CHC(CH_3)_3$	100.21		−25	80.9^{760}	0.6901^{20}_{4}	1.3894^{20}	i	s	s				**B1**^{3}, 454
b2612	**1-Butanearsonic acid***	*n*-Butylarsonic acid. $CH_3CH_2CH_2CH_2AsO_3H_2$	182.05		160				v	s	i				**B4**^{2}, 997
b2613	**1-Butaneboronic acid, 3-methyl-***	Isoamylboric acid. $(CH_3)_2CHCH_2CH_2B(OH)_2$	115.97	pl (w)	169				v^{h}	s				os s	**B4**^{2}, 1023
—	**Butanedial***	*see* **Succinaldehyde**													
—	**1,1-Butanedi-carboxylic acid***	*see* **Malonic acid, propyl-**													
—	**Butanedioic acid***	*see* **Succinic acid**													
—	**—,2,3-dihydroxy-***	*see* **Tartaric acid**													
b2614	**1,2-Butanediol(*d*)**	$CH_3CH_2CHOHCH_2OH$	90.12	$[\alpha]^{20}_{D}+14.5$ (al, c=6)		192–4	$1.0059^{17.5}_{0}$		v	∞					**B1**^{2}, 544
b2615	—(*dl*)*	α-Butyleneglycol. $CH_3CH_2CHOHCH_2OH$	90.12			189–91	1.001^{20}_{4}	1.435^{20}	s	s					**B1**^{2}, 544
b2616	—(*l*)*	$CH_3CH_2CHOHCH_2OH$	90.12	$[\alpha]^{22}_{D}-7.4$ (al, c=4)		94–6^{12}									**B1**^{2}, 544
b2617	**1,3-Butanediol(*d*)***	β-Butyleneglycol. $CH_3CHOHCH_2CH_2OH$	90.12	$[\alpha]^{22}_{D}+18.5$ (al, c=4)		204	1.0053^{20}_{4}	1.4418^{20}	s	s	i				**B1**^{2}, 545
b2618	—(*dl*)*	$CH_3CHOHCH_2CH_2OH$	90.12			205–8	1.0053^{20}_{4}	1.4418^{20}							**B1**^{2}, 545

For explanations, symbols and abbreviations see beginning of table.

No.	Name	Synonyms and Formula	Mol. wt.	Crystalline form, color and specific rotation	m.p. °C	b.p. °C	Density	n_D	Solubility w	al	eth	ace	bz	other solvents	Ref.
	1,3-Butanediol														
b2619	—(*l*)*	$CH_3CHOHCH_2CH_2OH$	90.12	$[\alpha]^{25}_D -18.8$ (al)		$107–10^{23}$									
b2620	—,sulfite	CH_3 (ring: O, SO, O)	135.16		5	185^{760}	1.2352^{20}_{4}	1.4461^{20}	d^h	v	v	v	v	aa v	
b2621	**1,4-Butanediol***	Tetramethyleneglycol. $HOCH_2CH_2CH_2CH_2OH$	90.12		19	235	1.0171^{20}_{4}	1.4467^{20}	∞	s	δ				**B1**², 545
b2622	**2,3-Butanediol**(*d*)*	$CH_3CHOHCHOHCH_3$	90.12	$[\alpha]^{25}_D +1.0$	34, 16.8 (+5w)	183–4	0.9872^{25}	1.4384^{25}	∞	∞	s				**B1**², 546
b2623	—(*dl*)*	$CH_3CHOHCHOHCH_3$	90.12			178–82	1.0033^{20}_{4}	1.4364^{25}	∞	∞					**B1**², 546
b2624	—(*l*)*	$CH_3CHOHCHOHCH_3$	90.12	$[\alpha]_D -5.5$	19	178–81	0.9869	1.4307							**B1**², 547
b2625	—(*meso*)*	$CH_3CHOHCHOHCH_3$	90.12		25		0.9939^{25}	1.4324^{35}							**B1**², 547
b2626	—,**2,3-dimethyl-***	Pinacol. $(CH_3)_2COHCOH(CH_3)_2$	118.17	nd (al or eth)	41.1	174.4^{760}			δ v^h	v					**B1**², 553
b2627	—,—,hexahydrate*	Pinacol hydrate. $(CH_3)_2COHCOH(CH_3)_2.6H_2O$	226.27	pl (w+6)	45.4		0.967^{15}		δ v^h	v					**B1**², 553
b2628	—,**2,3-diphenyl-***	Acetophenonepinacol. $CH_3COH(C_6H_5)COH(C_6H_5)CH_3$	242.31	pr	121–2				i	v	v			peth δ	**B6**², 979
b2629	**1,2-Butanediol, 3-methyl-***	α-Isoamylene glycol. $(CH_3)_2CHCHOHCH_2OH$	104.15			206	0.9987^{0}_{4}			s	s				**B1**², 550
b2630	**1,3-Butanediol, 3-methyl-***	γ-Isoamylene glycol. $(CH_3)_2COHCH_2CH_2OH$	104.15			202–3	0.9892^{20}		s	s					**B1**¹, 251
b2631	**2,3-Butanediol, 2-methyl-***	β-Isoamylene glycol. $CH_3CHOHCOH(CH_3)_2$	104.15			177–8	0.9893^{0}_{0}		s	s	s				**B1**², 550
b2632	**2,3-Butanedione***	Biacetyl. Dimethylglyoxal. $CH_3COCOCH_3$	86.09			89–90	$0.9808^{18.5}_{4}$	$1.3933^{18.5}$	v	∞	∞				**B1**², 824
b2633	—,dioxime	Dimethyl glyoxime. $CH_3C(:NOH)C(:NOH)CH_3$	116.12	nd (to)	245–6				δ	v	v			to δ	**B1**², 826
b2634	—,monooxime	Biacetyl monooxime. $CH_3COC(:NOH)CH_3$	101.10	pr (chl), lf (w)	74–5	$185–6^{760}$			δ	v	v			chl v	**B1**², 826
b2635	**1,3-Butanedione, 1-phenyl-***	Benzoylacetone. $CH_3COCH_2COC_6H_5$	162.18	pr	56	$261–2^{760}$	1.0599^{74}_{4}	1.5678^{78}	i		s				**B7**², 616
b2636	**1,4-Butane-dithiol-***	$HSCH_2CH_2CH_2CH_2SH$	122.26		−53.9	$195–6^{760}$	1.0621^{0}_{4}	1.5265^{25}	i	v					**B1**³, 2177
b2638	**1-Butanephos-phonic acid***	Butylphosphonic acid. $CH_3CH_2CH_2CH_2PO_3H_2$	138.11	pl	106	d			v	v	v		δ v^h		**Am 75**, 3379
b2639	**2-Butanephos-phonic acid***	*sec*-Butylphosphonic acid. $CH_3CH_2CH(CH_3)PO_3H_2$	138.11	pl	56	d			v	v	v		s	peth δ	**Am 75**, 3379
b2639¹	**1-Butanesulfonic acid,** amide	*n*-Butylsulfonamide. $CH_3CH_2CH_2CH_2SO_2NH_2$	137.21	lf (eth-lig)	45				v	v	v		v	lig i	**B4**, 8
b2640	—,chloride	$CH_3CH_2CH_2CH_2SO_2Cl$	156.64				$96–7^{18}$		d	d^h					**B4**, 8
b2641	**1,2,3,4-Butane-tetracarboxylic acid,** (high melting)*	$HO_2CCH_2[CH(CO_2H)]_2CH_2CO_2H$	234.17	lf (w)	236				v	v	δ		i	chl δ lig i	**B2**¹, 333
b2642	—,(low melting)*	$HO_2CCH_2[CH(CO_2H)]_2CH_2CO_2H$	234.17	nd or pr (w)	188–9				v						**B2**², 702
b2643	**1-Butanethiol***	*n*-Butylmercaptan. $CH_3CH_2CH_2CH_2SH$	90.19		−116	97–8			δ	v	v				**B1**², 398
b2644	**2-Butanethiol**(*d*)*	$CH_3CH_2CHSHCH_3$	90.19	$[\alpha]^{20}_D +15.7$		85–95	0.8299^{20}			s	s		s	peth s	**B1**³, 1549
b2645	—(*dl*)*	*sec*-Butylmercaptan. $CH_3CH_2CHSHCH_3$	90.19			85^{760}	0.8295^{20}_{4}	1.4366^{20}		s	s		s	peth s	**B1**³, 1548
b2646	—(*l*)*	$CH_3CH_2CHSHCH_3$	90.19	$[\alpha]^{17}_D -17.35$		83–4	0.8300^{17}_{4}			s	s		s	peth s	**B1**³, 1549
b2647	**1-Butanethiol, 2-methyl-**(*d*)*	$CH_3CH_2CH(CH_3)CH_2SH$	104.21	$[\alpha]^{23}_D +3.2$		119–21	0.8415^{23}								**B1**², 421
b2648	—,**3-methyl-***	Isoamylmercaptan. $(CH_3)_2CHCH_2CH_2SH$	104.21			120–2	0.8348^{20}_{4}	1.4412^{20}	i	∞	∞				**B1**², 434
—	**1,2,3-Butanetri-carboxylic acid, 2,3-dimethyl-***	*see* **Camphoronic acid**													

For explanations, symbols and abbreviations see beginning of table.

No.	Name	Synonyms and Formula	Mol. wt.	Crystalline form, color and specific rotation	m.p. °C	b.p. °C	Density	n_D	Solubility w	al	eth	ace	bz	other solvents	Ref.
	1,2,3-Butanetriol														
b2649	**1,2,3-Butanetriol***	1-Methylglycerol. $CH_3CH(OH)CH(OH)CH_2OH$	106.12			170^{20}		1.4622^{20}	v	v					$B1^{3}$, 2343
b2650	**1,2,4-Butanetriol***	$HOCH_2CHOHCH_2CH_2OH$	106.12			$172–4^{12}$	1.018^{20}	1.4688^{20}	v	v					$B1^{3}$, 2344
b2651	**Butanoic acid***	Butyric acid. $CH_3CH_2CH_2CO_2H$	88.10		−6.5	163.5^{760}	0.9640^{20}_{4}	1.3991^{20}	∞	∞	∞				$B2^{2}$, 237
b2652	—,allyl ester	$CH_3CH_2CH_2CO_2CH_2CH{:}CH_2$	128.17			$142–3^{772}$			i	∞	∞				B2, 272
b2653	—,amide	Butyramide. $CH_3CH_2CH_2CONH_2$	87.12	lf (bz)	115–6	216				s	δ				$B2^{2}$, 251
b2654	—,—,*N,N*-diethyl-	$CH_3CH_2CH_2CON(C_2H_5)_2$	143.23			206			∞	s					$B4^{2}$, 604
b2655	—,—,*N,N*-dimethyl-	$CH_3CH_2CH_2CON(CH_3)_2$	115.18			185–8				s					$B4^{2}$, 564
b2656	—,—,*N*-phenyl-	Butyranilide. $CH_3CH_2CH_2CONHC_6H_5$	163.21	mcl pr	97	189^{15}	1.134		i	v	v				$B12^{2}$, 146
b2657	—,anhydride*	Butyric anhydride. $(CH_3CH_2CH_2CO)_2O$	158.20		−75	198^{760}	0.9668^{20}_{4}	1.4124^{20}	d	d	s				$B2^{2}$, 251
b2658	—,benzyl ester	$CH_3CH_2CH_2CO_2CH_2C_6H_5$	178.23			235–42	1.014^{19}	1.4920^{20}	i	v	v				$B6^{2}$, 417
b2659	—,bromide	Butyryl bromide. $CH_3CH_2CH_2COBr$	151.01			128^{760}	1.4162^{17}_{4}	1.1596^{17}							$B2^{2}$, 251
b2660	—,(4-bromo-phenacyl) ester	$CH_3CH_2CH_2CO_2CH_2CO$— —Br	285.15		64										$B8^{2}$, 90
b2661	—,butyl ester*	$CH_3CH_2CH_2CO_2(CH_2)_3CH_3$	144.21		−91.5	166.6^{760}	0.8717^{20}_{20}	1.4045^{25}	δ	∞	∞				$B2^{2}$, 246
b2662	—,*sec*-butyl ester (*d*)	$CH_3CH_2CH_2CO_2CH_2CH(CH_3)_2$	144.21	$[\alpha]^{20}_{D}+22$	−91.5	151.5^{747} 54^{18}	0.8737^{13}_{4}	1.4011^{20}		s			s	CS_2 s Py s	$B2^{2}$, 247
b2663	—,—(*dl*)	$CH_3CH_2CH_2CO_2CH_2CH(CH_3)_2$	144.21			151.5^{747}	0.8572^{16}_{4}	1.4030^{16}_{4}							C33,5806
b2664	—,*tert*-butyl ester	$CH_3CH_2CH_2CO_2C(CH_3)_3$	144.21			145–7		$1.4007^{17.5}$		s	s	s			
b2665	—,chloride	Butyryl chloride. $CH_3CH_2CH_2COCl$	106.55		−89	102^{760}	1.0277^{20}_{4}	1.4121^{20}	d	d	∞				$B2^{2}$, 251
b2666	—,cyclohexyl ester*	$CH_3CH_2CH_2CO_2$—	170.24			212^{750}	0.9572^{0}_{4}		i	s				os s	$B6^{1}$, 6
b2668	—,2,2-dimethyl propyl ester*	*neo*-Pentyl butyrate. $CH_3CH_2CH_2CO_2CH_2C(CH_3)_3$	158.24			$165–6^{760}$	0.8719^{0}								B2, 272
b2669	—,ethyl ester*	$CH_3CH_2CH_2CO_2C_2H_5$	116.16		−97.8	120^{760}	0.8785^{20}_{4}	1.4000^{20}	δ	s	s				$B2^{2}$, 244
b2670	—,furfuryl ester	$CH_3CH_2CH_2CO_2CH_2$— O	168.19			$212–3^{764}$	1.0530^{20}_{4}		δ	s	∞				$B17^{2}$, 115
b2671	—,heptyl ester*	$CH_3CH_2CH_2CO_2(CH_2)_6CH_3$	186.29		−57.5	225–6	0.8637^{20}		i	s				os s	B2, 272
b2672	—,hexyl ester*	$CH_3CH_2CH_2CO_2(CH_2)_5CH_3$	172.26		−28	208	0.8652^{20}		i	s				os s	B2, 272
b2672[1]	—,isobutyl ester*	$CH_3CH_2CH_2CO_2CH_2CH(CH_3)_2$	144.21			157^{760}	0.8364^{18}_{4}	1.4030^{18}	δ	∞	∞				$B2^{2}$, 247
b2673	—,isopropyl ester*	$CH_3CH_2CH_2CO_2CH(CH_3)_2$	130.18			128	0.8652^{13}		i	s					B2, 271
b2674	—,methyl ester*	$CH_3CH_2CH_2CO_2CH_3$	102.13		<−95	102.6^{760}	0.8984^{20}_{4}	1.3870^{25}	δ	∞	∞				$B2^{2}$, 243
b2675	—,2-methyl butyl ester (*d*)*	$CH_3CH_2CH_2CO_2CH_2CH(CH_3)CH_2CH_3$	158.24	$[\alpha]^{20}_{D}+3.5$		179^{765}	0.8702^{18}_{4}	$1.4112^{20.5}$							$B2^{1}$, 120
b2676	—,—(*dl*)	$CH_3CH_2CH_2CO_2CH_2CH(CH_3)CH_2CH_3$	158.24			$178–9^{765}$	0.862^{20}_{4}								$B2^{2}$, 247
b2676[1]	—,2-methyl-2-butyl ester	*tert*-Amyl butyrate. $CH_3CH_2CH_2CO_2C(CH_3)_2CH_2CH_3$	158.24			164	0.8646^{15}_{0}								B2, 271
b2677	—,3-methylbutyl ester*	Isoamylbutyrate. $CH_3CH_2CH_2CO_2CH_2CH_2CH(CH_3)_2$	158.24			178.5^{760}	0.8823^{0}_{4}		i	v	v				$B2^{2}$, 247
b2678	—,2-methylpentyl ester(*d*)	$CH_3(CH_2)_2CO_2CH_2CH(CH_3)CH_2CH_2CH_3$	172.27	$[\alpha]_D+10.16$		98^{17}	0.8600^{17}_{4}	1.4160^{20}							
b2680	—,nitrile	Butyronitrile. Propyl cyanide. $CH_3CH_2CH_2CN$	69.11		−112	118^{760}	0.7936^{20}_{4}	1.3816^{24}	δ	∞	∞		s		$B2^{2}$, 252
b2681	—,octyl ester*	$CH_3CH_2CH_2CO_2(CH_2)_7CH_3$	200.32		−55.6	244	0.8629^{20}	1.4267^{15}_{He}	i	s				os s	$B2^{2}$, 248

For explanations, symbols and abbreviations see beginning of table.

No.	Name	Synonyms and Formula	wt.	Crystalline form, color and specific rotation	m.p. °C	b.p. °C	Density	n_D	Solubility w	al	eth	ace	bz	other solvents	Ref.
	Butanoic acid														
b2682	—,pentyl ester*	Amyl butyrate. $CH_3CH_2CH_2CO_2(CH_2)_4CH_3$	158.24		−73.2	186.4	0.8713^{15}_{4}	1.4139^{15}	i	v	v				**B2²**, 247
b2683	—,phenyl ester*	$CH_3CH_2CH_2CO_2C_6H_5$	164.21			$227\text{–}8^{760}$	1.0363^{4}_{4}	1.0267^{15}_{15}	i	s	s				**B6²**, 155
b2684	—,4-phenyl-phenacyl ester	$CH_3CH_2CH_2CO_2CH_2CO$—	282.34		97										**Am 73,** 5301
b2865	—,piperazinium salt	$(CH_3CH_2CH_2CO_2H)_2C_4H_{10}N_2$	262.35		121–2				s	s	i				**Am 56,** 1759
b2686	—,propyl ester*	$CH_3CH_2CH_2CO_2CH_2CH_2CH_3$	130.19		−97.2	143^{760}			δ	∞	∞				**B2²**, 246
b2687	—,**4(5-acenaphthyl)-4-oxo-**	—$COCH_2CH_2CO_2H$	254.29	nd (al or aa)	208d					s^h			s^h	xyl s^h	**B10²**, 532
b2688	—,**2-acetyl-3-oxo-**, ethyl ester	Ethyl diacetoacetate. Ethyl α-acetylacetoacetate. $(CH_3CO)_2CHCO_2C_2H_5$	172.18			109–11 104^{16}	1.104^{15}_{4}	1.4687^{20}	δ	v	v		v		**B3²**, 467
b2689	—,**2-amino-**(*d*)*	$CH_3CH_2CHNH_2CO_2H$	103.12	lf (dil al) $[\alpha]^{20}_{D}-7.9$ (w, p=4.5)	292d										**B4²**, 831
b2690	—,—(*d*)*l**	$CH_3CH_2CHNH_2CO_2H$	103.12	lf (w)	307d				v	δ	i				**B4²**, 831
b2691	—,—(*l*)*	$CH_3CH_2CHNH_2CO_2H$	103.12	lf (w-al) $[\alpha]_D+14.1$ (20% HCl)	270–80				s		δ			MeOH s	**B4²**, 831
b2692	—,**3-amino**(*d*)*	$CH_3CHNH_2CH_2CO_2H$	103.12	pr (MeOH) $[\alpha]^{20}_{D}+35.3$ (w, p=9.6)	220d										**B4¹**, 504
b2693	—,—(*dl*)*	$CH_3CHNH_2CH_2CO_2H$	103.12	nd (al)	193–4				v	i	i				**B4²**, 833
b2694	—,—(*l*)*	$CH_3CHNH_2CH_2CO_2H$	103.12	pr (MeOH) $[\alpha]^{20}_{D}-35.2$ (w, p=10)	220d										**B4¹**, 504
b2695	—,**4-amino-***	Piperidinic acid. $H_2NCH_2CH_2CH_2CO_2H$	103.12	pr (dil al)	203d				v	i	i		i	MeOH δ^h	**B4²**, 837
—	—,**2-amino-3-hydroxy-***	*see* **Threonine**													
b2696	—,**3-amino-2-hydroxy-***	$CH_3CHNH_2CHOHCO_2H$	119.12	pr (dil al)	200				s	δ					**B4**, 515
b2697	—,**4-amino-2-hydroxy-***	$H_2NCH_2CH_2CHOHCO_2H$	119.12	pr (w)	191–2				v	δ^h	i				**B4¹**, 548
b2698	—,**4-amino-3-hydroxy-***	$H_2NCH_2CHOHCH_2CO_2H$	119.12	pr (w)	214				s	δ	δ			MeOH δ	**B4²**, 937
—	—,**2-amino-2-methyl-***	*see* **Isovaline**													
—	—,**2-amino-3-methyl-***	*see* **Valine**													
b2699	—,**3-amino-3-methyl-***	$(CH_3)_2CNH_2CH_2CO_2H$	117.15	pr (w+1)	217				v	δ	i				**B4²**, 426
—	—,**2-amino-4(methylthio)-***	*see* **Methionine**													
b2700	—,**2-benzoyl-**, ethyl ester	Ethyl benzoyl ethyl acetate. $CH_3CH_2CH(COC_6H_5)CO_2C_2H_5$	220.27			152^7	1.0706^{15}_{4}	1.509^{15}	i		s				**B10²**, 488
—	—,**2-benzylidene-**	*see* **Cinnamic acid, α-ethyl-**													
b2701	—,**2-benzylidene-3-oxo-**, ethyl ester	Ethyl α-acetylcinnamate. Ethyl α-benzalacetoacetate. $C_6H_5CH{:}C(COCH_3)CO_2C_2H_5$	218.25	pl rh or pyr	60–1	$295\text{–}7^{760}$				δ	δ		δ	chl v CS_2 δ	**B10²**, 503
b2702	—,**2-bromo-**(*d*)*	$CH_3CH_2CHBrCO_2H$	167.01	$[\alpha]^{20}_{D}+35.2$ (eth, c=20), +12.9 (w, c=4)		$105\text{–}7^{15}$	1.216^{15}	1.4483^{15}		s					**B2²**, 255
b2703	—,—(*dl*)*	$CH_3CH_2CHBrCO_2H$	167.01		−4	$148\text{–}9^{74}$	1.5669^{20}_{20}		s	s	s				**B2²**, 255
b2704	—,—,amide	$CH_3CH_2CHBrCONH_2$	166.02	lf (bz), nd (ace)	112				v	v	v	v^h	v^h		**B2¹**, 125
b2705	—,—,bromide	$CH_3CH_2CHBrCOBr$	229.91			174–2			d	d	s				**B2²**, 255
b2706	—,—,chloride	$CH_3CH_2CHBrCOCl$	185.45			150–2	1.57^{11}		d^h	d^h	s				**B2¹**, 125
b2707	—,—,ethyl ester*	$CH_3CH_2CHBrCO_2C_2H_5$	195.06			177.5^{765}	1.321^{25}_{4}		i	∞	∞				**B2²**, 255
b2708	—,—,methyl ester*	$CH_3CH_2CHBrCO_2CH_3$	181.04			165–72				s					**B2¹**, 125
b2708¹	—,**4-bromo-***	$BrCH_2CH_2CH_2CO_2H$	167.01		32–3	$124\text{–}7^7$									**B2²**, 256

For explanations, symbols and abbreviations see beginning of table.

No.	Name	Synonyms and Formula	Mol. wt.	Crystalline form, color and specific rotation	m.p. °C	b.p. °C	Density	n_D	Solubility w	al	eth	ace	bz	other solvents	Ref.
	Butanoic acid														
b2709	—,—,methyl ester*	$BrCH_2CH_2CH_2CO_2CH_3$	181.04			186–7	1.450			s					**B2**[2], 283
b2710	—,—,nitrile	$BrCH_2CH_2CH_2CN$	148.01			205–7				s	s				**B2**[2], 256
b2711	—,**2-bromo-2-ethyl-**, amide	Neuronal. $(C_2H_5)_2CBrCONH_2$	194.08		67				δ v^h	v	v		v	peth δ	**B2**[2], 293
b2712	—,**2-bromo-2-ethyl-3-methyl-**, amide	2-Bromo-2-isopropylbutyramide. $(CH_3)_2CHCBr(C_2H_5)CONH_2$	208.11	nd (sub)	50–1	sub			δ	v	v		v	peth v	
b2713	—,**2-bromo-3-methyl-(*d*)***	$(CH_3)_2CHCHBrCO_2H$	181.04	pr (peth) $[\alpha]^{20}_D$+22.6 (bz, p=4), +4.5 (w,c=1.6), +7.5 (eth, c=7)	43.5	230 150^{40}			δ	v	s		s	peth v^h	**B2**[2], 279
b2714	—,—(*dl*)*	$(CH_3)_2CHCHBrCO_2H$	181.04	pr (eth or chl)	44	230			δ						**B2**[2], 279
b2715	—,—(*l*)*	$(CH_3)_2CHCHBrCO_2H$	181.04	$[\alpha]^{20}_D$−21.6 (bz, p=4)	40	11.9–20^{14}									**B2**[2], 279
b2716	—,—,amide	$(CH_3)_2CHCHBrCONH_2$	180.05	lf (bz)	133				s^h					lig s^h	**B2**, 318
b2717	—,—,ethyl ester*	$(CH_3)_2CHCHBrCO_2C_2H_5$	209.09			186	1.2776^{12}_{12}			v	v				**B2**[2], 279
b2718	—,—,methyl ester*	$(CH_3)_2CHCHBrCO_2CH_3$	195.06			174	1.353^{13}_{13}			s	s				**B2**, 317
b2719	—,**3-bromo-3-methyl-***	$(CH_3)_2CBrCH_2CO_2H$	181.04	nd (lig)	73.5					v	v		v	lig i	**B2**[2], 278
b2720	—,**2-bromo-3-oxo-**, amide, *N*-phenyl-	α-Bromoacetoacetanilide. $CH_3COCHBrCONHC_6H_5$	256.11	lf (al)	138d				δ	s^h	δ			chl δ	**B12**[1], 276
b2721	—,—,ethyl ester*	Ethyl α-bromoacetoacetate. $CH_3COCHBrCO_2C_2H_5$	209.04			210–5d 104–10^{15}	1.4294^{14}_4	1.463^{14}		s	s				**B3**[2], 427
b2722	—,**4-bromo-3-oxo-**, ethyl ester*	Ethyl γ-bromoacetoacetate. $BrCH_2COCH_2CO_2C_2H_5$	209.04			115–20^{15}	1.5278^{18}_4	1.483^{18}	δ	v	v				**B3**[2], 427
b2723	—,**2-chloro-***	$CH_3CH_2CHClCO_2H$	122.55			109.5^{24}	1.1796^{20}_4	1.4411^{20}	δ v^h	v	v				**B2**[2], 253
b2724	—,—,chloride	$CH_3CH_2CHClCOCl$	141.00			129–32			d	d	s				**B2**[1], 123
b2725	—,—,ethyl, ester*	$CH_3CH_2CHClCO_2C_2H_5$	150.61			163–4^{760}	1.063^{17}	1.4243		s	s				**B2**, 277
b2726	—,—,methyl ester*	$CH_3CH_2CHClCO_2CH_3$	136.58			145–6^{756}	1.0979^{14}	1.4253^{14}						MeOH s	**B2**, 277
b2727	—,**3-chloro-***	$CH_3CHClCH_2CO_2H$	122.55		16	116^{22}	1.1898^{20}_4	1.4421^{20}		s	v^h				**B2**[2], 253
b2728	—,—,chloride	$CH_3CHClCH_2COCl$	141.00			40–1^{12}	1.2163^{20}_4	1.4525^{20}	d	d				CS_2 s	**B2**[2], 253
b2729	—,—,ethyl ester*	$CH_3CHClCH_2CO_2C_2H_5$	150.61			169.5	1.0542^{20}_4	1.4246^{20}	i	s				CCl_4 s	**B2**[2], 253
b2730	—,—,methyl ester*	$CH_3CHClCH_2COCH_3$	136.58			155–6^{750}	1.0996^{20}_4	1.4258^{20}	i		v				**B2**, 277
b2731	—,**4-chloro-***	$ClCH_2CH_2CH_2CO_2H$	122.55		13–4	196^{22}	1.2236^{20}_4	1.4512^{20}	δ		v				**B2**[2], 253
b2732	—,—,chloride	$ClCH_2CH_2CH_2COCl$	141.00			173–4	1.2581^{20}_4	1.4616^{20}	d	d	s				**B2**[2], 254
b2733	—,—,ethyl ester*	$ClCH_2CH_2CH_2CO_2C_2H_5$	150.61			186^{760}	1.0754^{20}_4	1.4311^{20}		s	s	s			**B2**[2], 254
b2734	—,—,methyl ester*	$ClCH_2CH_2CH_2CO_2CH_3$	136.58			175–6^{764}	1.1268^{14}	1.4324^{20}	i	v^h	v	s			**B2**, 278
b2735	—,—,nitrile	$ClCH_2CH_2CH_2CN$	103.55			195–7			i	s	s				**B2**[2], 254
b2736	—,**2-chloro-2-methyl-***	$CH_3CH_2CCl(CH_3)CO_2H$	136.58			200–5^{754} 123–4^{36}	1.101^{10}	1.4508^{11}	i	s	s				**B2**, 306
b2737	—,—,chloride	$CH_3CH_2CCl(CH_3)COCl$	155.02			143–4^{739}	1.187^{14}		d		s				**B2**, 307
b2738	—,—,ethyl ester*	$CH_3CH_2CCl(CH_3)CO_2C_2H_5$	164.63			175^{747}	1.069^{14}	1.4368^{11}	i	s	s				**B2**, 305
b2739	—,**2-chloro-3-methyl-***	$(CH_3)_2CHCHClCO_2H$	136.58		16	210–2^{756} 125–6^{32}	1.135^{13}	1.4450^{11}	i	s	s				**B2**, 316
b2740	—,—,chloride	$(CH_3)_2CHCHClCOCl$	155.02			148–9^{760}	1.135^{13}		d	d	s				**B2**, 316
b2741	—,—,ethyl ester*	$(CH_3)_2CHCHClCO_2C_2H_5$	164.63			177–9^{756}	1.021^{13}		i	s	s				**B2**, 316
b2742	—,**2-chloro-2-methyl-3-oxo-**, ethyl ester*	Ethyl α-chloro-α-acetopropionate. $CH_3COCCl(CH_3)CO_2C_2H_5$	178.62			116–7^{75}	1.157^{18}_4	1.4382^{18}		s					**B3**[1], 237
b2743	—,**2-chloro-3-oxo-**, amide, *N*-phenyl-	α-Chloroacetoacetanilide. $CH_3COCHClCONHC_6H_5$	211.65	nd (al)	137.5				i	v			s	chl v MeOH v	**B12**[2], 267
b2744	—,—,ethyl ester*	Ethyl α-chloroacetoacetate. $CH_3COCHClCO_2C_2H_5$	164.59			86–9^{12}	1.19^{14}_{17}	1.4420^{17}	δ	v	v				**B3**[2], 426

For explanations, symbols and abbreviations see beginning of table.

No.	Name	Synonyms and Formula	Mol. wt.	Crystalline form, color and specific rotation	m.p. °C	b.p. °C	Density	n_D	Solubility w	al	eth	ace	bz	other solvents	Ref.
	Butanoic acid														
b2745	—,**4-chloro-3-oxo-**, chloride	$ClCH_2COCH_2CH_2COCl$	169.01			$117–9^{17}$	1.4397^{20}_{4}	1.4860^{20}	d	d	...	...	s	CCl_4 s	
b2746	—,—,ethyl ester*	Ethyl γ-chloroacetoacetate. $ClCH_2COCH_2CO_2C_2H_5$	164.59		−8	220d $95^{0.8}$	1.2176^{17}_{4}	1.4546^{17}	δ	s	s	...	s		**B3**², 426
b2747	—,**2(4-chloro-phenyl)-3-oxo-4-phenoxy-***	$Cl{-}C_6H_4{-}CH(CO_2H)COCH_2O{-}C_6H_5$	304.73	pl (al)	168				...	s^h	...	...	...		**B10**, 446
b2748	—,**2,3-dibromo-** (high m.p.)*	Dibromocrotonic acid. $CH_3CHBrCHBrCO_2H$	245.91	nd (eth), mcl (CS_2)	87	$100–10^{20}$			δ d^h	v	v	...	v		**B2**², 256
b2749	—,—,(low m.p.)*	Dibromocrotonic acid. $CH_3CHBrCHBrCO_2H$	245.91	nd (lig)	56–60				δ	v	.v	...	...		**B2**², 256
b2750	—,—,ethyl ester*	$CH_3CHBrCHBrCO_2C_2H_5$	273.96	nd	58–9				...	s	s	...	...		**B2**, 285
b2751	—,**2,4-dibromo-**, ethyl ester*	$BrCH_2CH_2CHBrCO_2C_2H_5$	273.96			$149–50^{52}$	1.6990^{20}_{0}	1.4960^{20}	...	s	s	...	...		**B2**, 285
b2752	—,**2,2-dibromo-3-oxo-**, ethyl ester*	Ethyl-α-α-dibromoaceto-acetate. $CH_3COCBr_2CO_2C_2H_5$	287.95			$120–4^{13}$			δ	s	s	...	...		**B3**², 427
b2753	—,**2,3-dichloro-** (high m.p.)*	Dichloroisocrotonic acid. $CH_3CHClCHClCO_2H$	157.00		78	131.5^{20}			δ	v	v	...	...	peth δ	**B2**, 279
b2754	—,—,(low m.p.)*	Dichlorocrotonic acid. $CH_3CHClCHClCO_2H$	157.00		63	124.5^{20}			...	v	v	...	v	CS_2 v lig δ chl v	**B1**¹, 124
b2755	—,**2,2-dichloro-3-oxo-**, ethyl ester	Ethyl α,α-dichloroacetoace-tate. $CH_3COCCl_2CO_2C_2H_5$	199.03			$205–7^{756}$ 91^{11}	1.293^{16}_{17}	1.4492^{17}	...	s	...	...	...		**B3**², 427
b2756	—,**2,2-diethyl-3-oxo-**, ethyl ester*	$CH_3COC(C_2H_5)_2CO_2C_2H_5$	186.25			210^{760}	0.9712^{17}_{4}	1.4326^{17}	i	∞	∞	...	...		**B3**², 445
b2757	—,**2,4-dihydroxy-3,3-dimethyl-**, amide, *N*(3-hy-droxy-propyl)-	$HOCH_2C(CH_3)_2CHOHCONH(CH_2)_2CH_2OH$	205.26	$[\alpha]^{20}_{D}+29.7$ (w, 3 %)		$118–20^{0.02}$			v	v	δ	...	...		**C41**, 6199
b2758	—,**2,2-dimethyl-***	$CH_3CH_2C(CH_3)_2CO_2H$	116.16		−13	186^{760}	0.9276^{20}_{4}	1.4145^{20}	δ	s	s	...	...		**B2**², 293
b2759	—,—,chloride	$CH_3CH_2C(CH_3)_2COCl$	134.61			132^{760}	0.9801^{20}_{4}	1.4245^{20}	d	d	s	...	...		**B2**, 336
b2760	—,**2,3-dimethyl-***	$(CH_3)_2CHCH(CH_3)CO_2H$	116.16		−1.5	$189–91^{760}$	0.9275^{20}_{4}	1.4146^{20}	s	s	s	...	...		**B2**², 294
b2761	—,—,amide	$(CH_3)_2CHCH(CH_3)CONH_2$	115.18		129				...	...	...	...	...		**B2**, 338
b2762	—,—,chloride	$(CH_3)_2CHCH(CH_3)COCl$	134.61			$135–6^{751}$	0.9795^{20}_{4}		d^h	d^h	s	...	...		
b2763	—,**3,3-dimethyl-***	$(CH_3)_3CCH_2CO_2H$	116.16		6–7	$183–4^{741}$	0.9124^{20}_{4}	1.4096^{20}	...	s	s	...	...		**B2**, 337
b2764	—,—,amide	$(CH_3)_3CCH_2CONH_2$	115.18	pl	132				...	s	...	...	...		**B2**¹, 144
b2765	—,—,chloride	$(CH_3)_3CCH_2COCl$	134.61			$128–30^{745}$	0.9696^{20}_{4}	1.4230^{20}	d	d	v	...	...		
b2766	—,**2,2-dimethyl-3-oxo-**, ethyl ester	$CH_3COC(CH_3)_2CO_2C_2H_5$	158.20			$180–2^{720}$	0.9773^{20}_{20}	1.418^{19}	s	s	s	...	...		**B3**², 439
b2767	—,**2,3-dioxo-**, 2-phenylhydrazone	$CH_3COC(:NNHC_6H_5)CO_2H$	206.20	ye nd or lf (al-eth)	161				i	s^h	s^h	...	...		**B15**², 133
b2768	—,**2,4-diphenyl-3-oxo-**, nitrile	Benzyl α-cyanobenzyl ketone $C_6H_5CH_2COCH(C_6H_5)CN$	235.29	nd (bz-peth)	86				...	s	s	...	s	aa s	**B10**, 762
b2769	—,**2,4-diphenyl-4-oxo-**, nitrile	$C_6H_5CH_2COCH(C_6H_5)CN$	235.29	pl (al)	127.5				...	v	v	...	...	chl v	**B10**², 528
b2770	—,**3,4-epoxy-**, nitrile	Epicyanohydrin. CH_2CHCH_2CN (with O bridging CH_2–CH)	83.09	pr	162				δ s^h	s	...	...	...		**B18**, 261
b2771	—,**2,3-epoxy-3-phenyl-** ethyl ester*	$CH_3C(C_6H_5)CHCO_2C_2H_5$ (with O bridging C–CH)	206.24			272–5 $147–9^{12}$	1.096^{15}	1.506^{17}	...	...	...	...	...		**B18**¹, 442
b2772	—,**2-ethyl-***	Diethylacetic acid. $(C_2H_5)_2CHCO_2H$	116.16		−31.8	194^{769}	0.9195^{18}_{4}	1.4179^{10}	δ	∞	∞	...	...		**B2**², 291
b2773	—,—,amide,*N*,*N*-diethyl-	$(C_2H_5)_2CHCON(C_2H_5)_2$	171.28			220–1 108^{12}			δ	s	v	...	v		**B4**, 111
b2774	—,—,chloride	$(C_2H_5)_2CHCOCl$	134.61			$138–9^{750}$	0.9825^{20}_{4}	1.4234^{20}	d	d^h	v	...	...		**B2**², 292
b2775	—,—,nitrile	3-Cyanopentane. $(C_2H_5)_2CHCN$	97.16			$144–6^{753}$			...	∞	∞	...	...		**B2**², 292

For explanations, symbols and abbreviations see beginning of table.

No.	Name	Synonyms and Formula	Mol. wt.	Crystalline form, color and specific rotation	m.p. °C	b.p. °C	Density	n_D	Solubility: w	al	eth	ace	bz	other solvents	Ref.
	Butanoic acid														
b2776	—,**2-ethyl-2-methyl-***	$(C_2H_5)_2C(CH_3)CO_2H$	130.19		<−20	203–4[760]			i	s					**B2**[2], 299
b2777	—,**2-ethyl-3-oxo-**, ethyl ester*	Ethyl ethylacetoacetate. $CH_3COCH(C_2H_5)CO_2C_2H_5$	158.20			190[760]	0.9924^{15}_{15}			v	s				**B3**[2], 438
b2778	—,—,methyl ester*	$CH_3COCH(C_2H_5)CO_2CH_3$	144.17			190	0.995^{14}			s	s	s			**B3**[2], 439
b2779	—,**4-fluoro-***	$FCH_2CH_2CH_2CO_2H$	105.09			78–9[6]		1.3993^{25}	s	v	v				
b2780	—,**2-hydroxy-***	$CH_3CH_2CHOHCO_2H$	104.11		43–4	260d 140[14]	1.125^{20}		s	s	s				**B3**[2], 216
b2781	—,—,ethyl ester (*d*)*	$CH_3CH_2CHOHCO_2C_2H_5$	132.16	$[\alpha]^{22}_D+8.4$		165–70	0.978^{15}	1.4101		s					**B3**, 215
b2782	—,—,—(*dl*)*	$CH_3CH_2CHOHCO_2C_2H_5$	132.16			167[760]	1.0044^{0}			s					**B3**[2], 217
b2783	—,**3-hydroxy-(*dl*)***	$CH_3CHOHCH_2CO_2H$	104.11		48–50	130[12–14]			v	v	v		i		**B3**[2], 220
b2784	—,—(*l*)*	$CH_3CHOHCH_2CO_2H$	104.11	$[\alpha]^{25}_D$−24.5 (w, c=5)	49–50				v	v	v		i		**B3**[2], 219
b2785	—,—,ethyl ester (*dl*)*	$CH_3CHOHCH_2CO_2C_2H_5$	132.16			184–5[755]	1.017^{20}_{4}	1.4182^{20}	s	s					**B3**[2], 221
b2786	—,—,methyl ester (*l*)*	$CH_3CHOHCH_2CO_2CH_3$	118.13	$[\alpha]^{20}_D-21.09$		70–2[17]	1.058^{20}		v	v	v		v	peth δ	**B3**[2], 219
b2787	—,**4-hydroxy-***	$HOCH_2CH_2CH_2H$	104.11		<−17	d									**B3**[2], 222
b2788	—,—,lactone*	Butyrolactone. (structure: five-membered ring with O and =O)	86.09		−42	206[760]	1.1286^{16}_{0}	1.4343^{26}	∞	v	v				**B17**[2], 286
b2789	—,**2-hydroxy-2-isopropyl-3-methyl-**, nitrile	Diisopropylketone cyanohydrin. $[(CH_3)_2CH]_2COHCN$	141.21	rh (eth or peth)	59	111[18]			i	s			s	os s	**B3**[2], 239
b2790	—,**2-hydroxy-3-methyl-(*d*)***	$(CH_3)_2CHCHOHCO_2H$	118.13	$[\alpha]^{25}_D-0.5$(w)					v	v		v			**B3**, 328
b2791	—,—(*dl*)*	$(CH_3)_2CHCHOHCO_2H$	118.13	rh bipyr	86				v	v	v				**B3**, 328
b2792	—,**3-hydroxy-3-methyl-***	$(CH_3)_2COHCH_2CO_2H$	118.13		<−32	162	0.9384^{20}	1.1581^{20}	v	v	v				**B3**, 327
b2793	—,**4-hydroxy-2-methylene-**, lactone*	(structure: five-membered ring with O, =O and =CH2)	98.10			57–60[2]			s	v				os s	
—	—,**4-hydroxy-3-oxo-**,lactone*	*see* **Furan, 2,4-dioxo-, tetrahydro-**													
b2794	—,**4(3-indolyl)-**	(structure: indole ring with 3-$CH_2CH_2CH_2CO_2H$)	203.24	pl (bz-peth)	123–4								v^h	peth i	**B22**[2], 54
b2795	—,**2-isopropyl-3-oxo-**, ethyl ester*	$CH_3COCH(C_3H_7^i)CO_2C_2H_5$	172.23			201[758]	0.957^{25}_{4}	1.4240^{0}	i	∞	∞				**B3**[2], 441
b2796	—,**2-methyl-(*d*)**	$CH_3CH_2CH(CH_3)CO_2H$	102.13	$[\alpha]^{21}_D+17.6$		176	0.9419^{20}_{4}								**B2**[2], 270
b2797	—,—(*dl*)*	$CH_3CH_2CH(CH_3)CO_2H$	102.13		<−80	177	0.941^{21}_{4}	1.4051	δ	∞	∞				**B2**[2], 270
b2798	—,—(*l*)*	$CH_3CH_2CH(CH_3)CO_2H$	102.13	$[\alpha]^{14}_D-10.1$ (al)		176–7	0.934^{20}_{4}								
b2799	—,—,chloride (*dl*)	$CH_3CH_2CH(CH_3)COCl$	120.58			115	0.9887^{20}_{4}	1.4156^{20}	d^h	d^h					**B2**, 315
b2800	—,—,ethyl ester (*d*)*	$CH_3CH_2CH(CH_3)CO_2C_2H_5$	130.19	$[\alpha]^{26}_{5893}+5.16$		130	0.860^{26}_{4}	1.3957^{24}		s			s		**B2**, 304
b2801	—,—,nitrile	$CH_3CH_2CH(CH_3)CN$	83.13			125[760]	0.8061^{0}_{4}			s	s				**B2**, 306
b2802	—,**3-methyl-***	Isovaleric acid. $(CH_3)_2CHCH_2CO_2H$	102.13		−37.6	176.7[760]	0.9373^{15}_{4}	1.4043^{20}	s	∞	∞			chl ∞	**B2**[2], 271
b2803	—,—,amide	$(CH_3)_2CHCH_2CONH_2$	101.15	mcl lf (al)	137	230			s	s	s				**B2**[2], 277
b2804	—,—,anhydride*	$[(CH_3)_2CHCH_2CO]_2O$	186.25			215[762] 102–3[15]	0.9327^{20}_{4}	1.4043^{20}	d^h	d^h	s				**B2**[2], 277
b2805	—,—,bromide	Isovaleryl bromide. $(CH_3)_2CHCH_2COBr$	165.04	mcl pl (al)		143			d	d^h	s				**B2**[2], 277
b2806	—,—,chloride	Isovaleryl chloride. $(CH_3)_2CHCH_2COCl$	120.58			114.5–5.5[771]	0.9887^{20}_{4}	1.4156^{10}	d	d	s				**B2**[2], 277
b2806[1]	—,—,ethyl ester	$(CH_3)_2CHCH_2CO_2C_2H_5$	130.19			134[760]	0.8656^{20}_{4}	1.3975^{25}		v	v				**B2**[2], 275

For explanations, symbols and abbreviations see beginning of table.

No.	Name	Synonyms and Formula	Mol. wt.	Crystalline form, color and specific rotation	m.p. °C	b.p. °C	Density	n_D	Solubility w	al	eth	ace	bz	other solvents	Ref.
	Butanoic acid														
b2807	—,—,isobutyl ester	$(CH_3)_2CHCH_2CO_2C_4H_9^i$	158.24			171.4^{760}	0.8736^{20}_{4}	1.4057^{20}	i	∞	∞				**B2**^{2}, 276
b2808	—,—,isopropyl ester*	$(CH_3)_2CHCH_2CO_2C_3H_7^i$	144.22			142^{756}	0.8538^{17}	1.3938^{25}	i	s	s				**B2**^{2}, 275
b2809	—,—,methyl ester*	$(CH_3)_2CHCH_2CO_2CH_3$	116.17			$116–7^{764}$	0.8787^{20}_{15}	1.3900^{25}	...	v	v				**B2**^{2}, 274
b2810	—,—,nitrile	Isovaleronitrile. $(CH_3)_2CHCH_2CN$	83.13		−101	130.5	0.7925^{19}_{4}	1.3927^{20}	δ	∞	∞				**B2**^{2}, 278
b2811	—,—,propyl ester*	$(CH_3)_2CHCH_2CO_2C_3H_7^n$	144.22			155.7^{760}	0.8643^{18}	1.4041^{18}	i	∞	∞				**B2**^{2}, 275
b2812	**—,2-methyl-3-oxo-**, ethyl ester*	Ethyl methylacetoacetate. $CH_3COCH(CH_3)CO_2C_2H_5$	144.17			187^{760}	1.0191^{20}_{4}	1.4207^{18}	δ	s	s				**B3**^{2}, 438
b2813	—,—,methyl ester*	$CH_3COCH(CH_3)CO_2CH_3$	130.14			177.4	1.0247^{25}_{4}	1.416^{24}	...	s	s				**B3**^{2}, 432
b2814	**—,3-methyl-2-phenyl-***	$(CH_3)_2CHCH(C_6H_5)CO_2H$	178.23	pr (lig)	63	$159–60^{14}$			...	s	v			lig δ	**B9**^{2}, 364
b2815	—,—,amide	$(CH_3)_2CHCH(C_6H_5)CONH_2$	177.25	nd (dil al)	111–2	$180–2^{14}$			i	s	s			lig δ	**B9**^{2}, 364
b2816	—,—,nitrile	$(CH_3)_2CHCH(C_6H_5)CN$	159.23			$245–9^{765}$ $123–4^{15}$	0.967^{15}		i	s			s		**B9**^{2}, 364
b2817	**—,2-oxo-***	α-Ketobutyric acid. $CH_3CH_2COCO_2H$	102.09	pl	31–2	$80–2^{16}$	1.200^{17}_{4}	1.3972^{20}	v	v	δ				**B3**^{2}, 411
b2818	—,—,oxime	$CH_3CH_2C(:NOH)CO_2H$	117.11	nd (w or bz)	161				δ	v	δ				**B3**^{2}, 412
b2819	**—,3-oxo-***	Acetoacetic acid. $CH_3COCH_2CO_2H$	102.09			<100d			∞	∞	s				**B3**^{2}, 413
b2820	—,—,allyl ester	Allyl acetoacetate. $CH_3COCH_2CO_2CH_2CH:CH_2$	142.16		−85	117^{50}	1.0385^{20}_{20}	1.4398^{20}	s	∞	...	...	∞	lig s	
b2821	—,—,amide,*N*(2-chlorophenyl)-	Cl; CH_3COCH_2CONH—⟨benzene ring⟩	211.65	nd	105				...	s	i	...		lig i	**B12**^{2}, 319
b2822	—,—,—,*N*-phenyl-	Acetoacetanilide. $CH_3COCH_2CONHC_6H_5$	177.20	pl or nd (bz or lig)	86				δ	s	s	...	s^h	chl s lig s	**B12**^{2}, 266
b2823	—,—,—,*N*(2-tolyl)-	CH_3; CH_3COCH_2CONH—⟨benzene ring⟩	191.23	pr (AcOEt)	107–8				...	s	...	...	s	lig δ	**B12**^{2}, 450
b2824	—,—,—,*N*(3-tolyl)-	CH_3; CH_3COCH_2CONH—⟨benzene ring⟩	191.23	pl (bz-peth)	57–8				...	s	...	...	s	lig δ	**B12**^{1}, 404
b2825	—,—,—,*N*(4-tolyl)-	CH_3COCH_2CONH—⟨benzene ring⟩—CH_3	191.23	pr (AcOEt)	95				...	s	...	...	s	lig δ	**B12**^{2}, 521
b2826	—,—,butyl ester*	$CH_3COCH_2CO_2C_4H_9^n$	158.20		−35.6	127^{50}	0.9694^{20}_{20}	1.4137^{20}	δ	∞	...	...	∞	lig ∞	
b2827	—,—,chloride	Acetoacetyl chloride. CH_3COCH_2COCl	120.54		−50, −51 (−20d)				d	d	v				
b2828	—,—,ethyl ester	Ethyl acetoacetate. $CH_3COCH_2CO_2C_2H_5$	130.15		<−80	180.4^{760}	1.025^{20}_{4}	1.4194^{20}	v	∞	∞	...	s	chl s	**B3**^{2}, 415
b2829	—,—,—(enol form)	$CH_3COH:CHCO_2C_2H_5$	130.15				1.0119^{10}	1.4432^{20}							**B3**^{2}, 415
b2830	—,—,—(keto form)	$CH_3COCH_2CO_2C_2H_5$	130.15		−39		1.0368^{10}_{4}	1.4171^{20}							**B3**^{2}, 415
b2831	—,—,isopropyl ester*	$CH_3COCH_2CO_2CH(CH_3)_2$	144.17		−27.3	185–7	0.9911^{20}_{20}	1.4173^{20}	s	∞	∞	...	...	lig ∞	**B3**, 659
b2832	—,—,methyl ester*	Methyl acetoacetate. $CH_3COCH_2CO_2CH_3$	116.11			169.5^{760}	1.0762^{20}_{4}	1.4184^{20}	v	∞	∞				**B3**^{2}, 414
b2833	—,—,nitrile	Acetoacetonitrile. Cyanoacetone. CH_3COCH_2CN	83.09			120–5				v	...	v			**B3**^{2}, 424
b2833^{1}	**—,4-oxo-4(4-methoxy-phenyl)**	CH_3O—⟨benzene ring⟩—$COCH_2CH_2CO_2H$	208.21	nd	150–1				s^h	s	...	...	s	chl s	**10**^{2}, 681
b2834	**—,3-oxo-2-phenyl-**, ethyl ester*	$CH_3COCH(C_6H_5)CO_2C_2H_5$	206.23			156^{22}			...	s	s				**B10**^{2}, 485
b2835	—,—,nitrile	$CH_3COCH(C_6H_5)CN$	159.19	pr (bz)	90				$δ^h$	v	v	...	δ s^h	chl v	**B10**^{2}, 485
b2836	**—,4-oxo-4-phenyl-***	β-Benzoylpropionic acid. $C_6H_5COCH_2CH_2CO_2H$	178.19		116				s^h	s	...				**B10**, 696

For explanations, symbols and abbreviations see beginning of table.

No.	Name	Synonyms and Formula	Mol. wt.	Crystalline form, color and specific rotation	m.p. °C	b.p. °C	Density	n_D	Solubility w	al	eth	ace	bz	other solvents	Ref.
	Butanoic acid														
b2838	—,**2-oxo-3-propyl-**, ethyl ester*	Ethyl-α-acetovalerate. $CH_3COCH(C_3H_7^n)CO_2C_2H_5$	172.22			224^{760}	0.9661_4^{20}	1.4255^{20}							**B3**^{2}, 441
b2839	—,**2-phenoxy-***...	$CH_3CH_2CH(OC_6H_5)CO_2H$	180.19	nd or pl (lig, w)	99	258			δ s^h					lig s^h	**B6**, 163
b2840	—,—,amide	$CH_3CH_2CH(OC_6H_5)CONH_2$	179.21	nd (w or al)	123				d^h	s^h	v	v	s^h	chl v	**B6**, 164
b2841	—,—,ethyl ester*	$CH_3CH_2CH(OC_6H_5)CO_2C_2H_5$	208.25			$250–1^{748}$	1.0388^{21}		i	∞	∞				**B6**, 164
b2842	—,—,nitrile	$CH_3CH_2CH(OC_6H_5)CN$	161.21			$228–30^{748}$				s	s				**B6**, 164
b2843	—,**4-phenoxy-***	$C_6H_5OCH_2CH_2CH_2CO_2H$	180.19	pl (lig)	64–5	$192–7^{15}$			i	v	v			CS_2 v lig s^h	**B6**^{2}, 158
b2844	—,—,amide	$C_6H_5OCH_2CH_2CH_2CONH_2$	179.21	pl (al)	80					v^h					**B6**, 164
b2845	—,—,ethyl ester*	$C_6H_5OCH_2CH_2CH_2CO_2C_2H_5$	208.25			$170–3^{25}$	1.045_{25}^{33}	1.491^{33}		s	s				**B6**^{2}, 159
b2846	—,—,nitrile	$C_6H_5OCH_2CH_2CH_2CN$	161.21	nd	45–6	$287–9^{765}$ $162–6^{22}$					s				**B6**^{2}, 159
b2847	—,**2-phenyl-***	$CH_3CH_2CH(C_6H_5)CO_2H$	164.21	pl	47.5	270–2					s		s		**B9**^{2}, 356
b2848	—,—,amide	$CH_3CH_2CH(C_6H_5)CONH_2$	163.22		86	185^{16}									**B9**^{2}, 356
b2849	—,—,methyl ester*	$CH_3CH_2CH(C_6H_5)CO_2CH_3$	178.22	nd	77–8	228			i	s	s				**B9**^{2}, 356
b2850	—,—,nitrile	$CH_3CH_2CH(C_6H_5)CN$	145.21			$238–40^{765}$ $141–3^{8}$			i	s	s		s		**B9**^{2}, 356
b2851	—,**3-phenyl-**, amide	$CH_3CH(C_6H_5)CH_2CONH_2$	163.22	nd (dil al)	106–7					s			δ		**B9**^{1}, 212
b2852	—,**4-phenyl-***	$C_6H_5CH_2CH_2CH_2CO_2H$	164.21	pl (w)	52	171^{15}			$δ^h$	s	s				**B9**^{2}, 354
b2853	—,—,amide	$C_6H_5CH_2CH_2CH_2CONH_2$	163.22	pl (w)	84.5				s^h	v	v				**B9**^{1}, 211
b2854	—,—,nitrile	$C_6H_5CH_2CH_2CH_2CN$	145.21			$142–5^{16}$									**B9**^{2}, 354
b2855	—,**4(3-pyrenyl)-**	—$CH_2CH_2CH_2CO_2H$	288.35	pl (aa)	190								s	aa s^h	**E14**, 392
b2856	—,**2,2,3,3-tetramethyl-**, amide	$(CH_3)_3CC(CH_3)_2CONH_2$	143.23	nd (peth-al)	200					s^h				lig i	**B2**^{2}, 305
b2857	—,**3-thioxo-**, ethyl ester	$CH_3CSCH_2CO_2C_2H_5$	146.21	dk red		75^{15}	1.0554_4^{31}	1.4712^{26}		s	s				**B3**^{2}, 427
b2858	—,**2,2,3-trichloro-***	$CH_3CHClCCl_2CO_2H$	191.45	lf or nd	60	236–8			v					peth v^h	**B2**^{2}, 255
b2859	—,—,ethyl ester*	$CH_3CHClCCl_2CO_2C_2H_5$	219.51			212	1.3138_{20}^{20}				v				**B2**, 281
b2860	—,**2,2,4-trichloro-***	$ClCH_2CH_2CCl_2CO_2H$	191.45		73–5				s	v					**B2**, 281
b2861	—,**2,3,3-trichloro-***	$CH_3CCl_2CHClCO_2H$	191.45	pl (lig)	52				δ	v	v		v	chl v	**B2**, 281
b2862	—,**4,4,4-trichloro-***	$Cl_3CCH_2CH_2CO_2H$	191.45	nd (w)	55				s^h						
—	—,**2,3,4-trihydroxy-**	*see* **Threonic acid**													
b2863	—,**2,2,3-trimethyl-**, chloride	$(CH_3)_2CHC(CH_3)_2COCl$	148.64			148–50			d		s				
b2864	**1-Butanol***	$CH_3CH_2CH_2CH_2OH$	74.12		−89.8	117.5^{760}	0.8098_4^{20}	1.3992^{20}	s	∞	∞				**B1**^{2}, 387
b2865	**2-Butanol** (*d*)	*sec*-Butyl alcohol. $CH_3CH_2CHOHCH_3$	74.12	$[α]_D^{20}$ +13.9 (undil)		99.5^{760}	0.8080_4^{20}	1.3954^{20}							**B1**^{2}, 404
b2866	—(*dl*)*	$CH_3CH_2CHOHCH_3$	74.12		<−100	99.5^{760}	0.8063_4^{20}	1.3978^{20}	v	∞	∞				**B1**^{2}, 400
b2867	—(*l*)*	$CH_3CH_2CHOHCH_3$	74.12	$[α]_D^{20}$ −13.9		99.5^{760}	0.8109_4^{15}								**B1**^{2}, 405
—	*tert*-**Butanol**	*see* **2-Propanol, 2-methyl-***													
b2868	**1-Butanol, 2-amino** (*d*)*	$CH_3CH_2CHNH_2CH_2OH$	89.14	$[α]_D^{20}$ +9.8		80^{11}	0.947^{20}		∞	∞	∞				**B4**^{1}, 438
b2869	—,—(*dl*)*	$CH_3CH_2CHNH_2CH_2OH$	89.14		−2	172–4	0.947^{20}	1.453^{20}	∞	∞	∞				**B4**^{1}, 438
b2870	—,**3-amino-***	$CH_3CHNH_2CH_2CH_2OH$	89.14			$82–5^{19}$			s						**B4**^{1}, 438
b2871	—,**4-amino-***	$H_2NCH_2CH_2CH_2CH_2OH$	89.14			206^{776} 100^{15}	0.967^{12}		s	s	i				**B4**^{1}, 439
b2872	**2-Butanol, 3-amino-***	$CH_3CHNH_2CHOHCH_3$	89.14		44	$83–5^{30}$		1.4502^{20}	s	s		s		lig i	
b2873	**1-Butanol, 2 amino-1-phenyl-***	$CH_3CH_2CHNH_2CH(C_6H_5)OH$	165.23	pl (bz-peth)	79–80					v			s	chl s peth δ	**B13**^{2}, 390
b2874	**2-Butanol, 3-bromo-***	$CH_3CHBrCHOHCH_3$	153.03			$46–50^{8}$	1.4500_{20}^{20}	1.4780^{20}		s	s				

For explanations, symbols and abbreviations see beginning of table.

No.	Name	Synonyms and Formula	Mol. wt.	Crystalline form, color and specific rotation	m.p. °C	b.p. °C	Density	n_D	Solubility w	al	eth	ace	bz	other solvents	Ref.
	1,Butanol														
b2875	**1-Butanol, 2-chloro-***	$CH_3CH_2CHClCH_2OH$	108.57			$74–6^{25}$	1.062^{25}_{4}	1.4410^{25}		v	v				
b2876	**—,4-chloro-***	Tetramethylene chlorohydrin. $ClCH_2CH_2CH_2CH_2OH$	108.57			$84–5^{16}$	1.0883^{20}_{4}	1.4518^{20}							**B1²**, 398
b2877	**2-Butanol, 1-chloro-***	$CH_3CH_2CHOHCH_2Cl$	108.57			141	1.068^{25}_{4}	1.4410^{25}		s	s				**B1²**, 402
b2878	**—,3-chloro-***	$CH_3CHClCHOHCH_3$	108.57			138–40	1.0692^{18}_{0}	1.4422^{20}							**B1²**, 403
b2879	**—,—(*erythro, dl*)***	$CH_3CHClCHOHCH_3$	108.57			135.4^{748}	1.0610^{25}_{4}	1.4397^{25}		v				chl v	
b2880	**—,4-chloro-***	$ClCH_2CH_2CHOHCH_3$	108.57			70^{13}				v	v				**B1¹**, 188
b2881	**—,1-chloro-2-methyl-***	$CH_3CH_2COH(CH_3)CH_2Cl$	122.60			152–3	1.068^{0}			v					**B1²**, 424
b2882	**—,3-chloro-2-methyl-***	$CH_3CHClCOH(CH_3)_2$	122.60			141–2	1.0355^{22}_{22}		s	v	v				**B1²**, 424
b2883	**—,1,3-dichloro-***	$CH_3CHClCHOHCH_2Cl$	143.02			$63–4^{10}$	1.2870^{15}_{4}	1.4766^{20}		v	v				**B1**, 373
b2884	**1-Butanol, 2,2-dimethyl-***	$CH_3CH_2C(CH_3)_2CH_2OH$	102.17		<-15	136.7^{760}	0.8283^{20}_{4}	1.4208^{20}	δ	s	s				**B1³**, 1675
b2885	**—,2,3-dimethyl-(*d*)***	$(CH_3)_2CHCH(CH_3)CH_2OH$	102.17	$[\alpha]^{25}_{D}+1.9$		142^{760}	0.823^{25}_{4}			s	s				**B1³**, 1677
b2886	**—,—(*dl*)***	$(CH_3)_2CHCH(CH_3)CH_2OH$	102.17			$144–5^{761}$	0.8297^{20}_{4}	$1.4195^{20.5}$		s					**B1³**, 1677
b2887	**—,3,3-dimethyl-***	$(CH_3)_3CCH_2CH_2OH$	102.17		−60, −65	143	0.844^{15}_{4}	1.4323^{15}		s	s				**B1³**, 1677
b2888	**2-Butanol, 2,3-dimethyl-***	$(CH_3)_2CHCOH(CH_3)_2$	102.17		−14	$120–1^{762}$	0.8232^{19}		s	∞	∞				**B1²**, 441
b2889	**—,3,3-dimethyl-***	$(CH_3)_3CCHOHCH_3$	102.17		5.5	120–1	0.8122^{25}		δ	s	∞				**B1²**, 441
b2890	**1-Butanol 2-ethyl-***	$(C_2H_5)_2CHCH_2OH$	102.17		<-15	146	0.8296^{25}	1.4205^{25}	δ	s	s				**B1**, 412
b2891	**—,4-fluoro-***	Tetramethylene fluorohydrin. $FCH_2CH_2CH_2CH_2OH$	92.11			58^{15}		1.3942^{25}	s	v	v				
b2892	**—,2-methyl-(*d*)***	$CH_3CH_2CH(CH_3)CH_2OH$	88.15	$[\alpha]^{20}_{D}-5.9$		129^{772}	0.8152^{25}_{4}	1.4087^{25}	s	∞	∞				**B1²**, 421
b2893	**—,—(*dl*)***	$CH_3CH_2CH(CH_3)CH_2OH$	88.15			129	0.8152^{25}_{4}	1.4087^{25}							**B1²**, 422
b2894	**—,—(*l*)***	$CH_3CH_2CH(CH_3)CH_2OH$	88.15	$[\alpha]^{18}_{D}+3.75$		130	0.816^{18}_{4}								**B1²**, 421
b2895	**—,3-methyl-***	$(CH_3)_2CHCH_2CH_2OH$	88.15			$131–2^{760}$	0.8092^{20}_{4}	1.4053^{20}_{4}	δ	∞	∞				**B1³**, 1633
b2896	**2-Butanol, 2 methyl-***	$CH_3CH_2COH(CH_3)_2$	88.15			102^{760}	0.8048^{25}_{4}	1.4052^{20}	s	∞	∞		s	chl s	**B1²**, 422
b2897	**—,3-methyl-(*d*)***	$(CH_3)_2CHCHOHCH_3$	88.15	$[\alpha]^{20}_{D}+5.34$ (al)		112^{734}	0.8225^{16}_{4}	1.3973^{20}		s	s			chl s	**B1¹**, 196
b2898	**—,—(*dl*)***	$(CH_3)_2CHCHOHCH_3$	88.15			$113–1^{760}$	0.8180^{20}_{4}	1.3973^{20}	δ	∞	∞			CCl_4 s	**B1²**, 425
b2899	**1-Butanol, 2-methyl-1-phenyl-***	$CH_3CH_2CH(CH_3)CH(C_6H_5)OH$	164.25			$125–6^{15}$			i	v				os v	**B6²**, 505
b2900	**—,2-methyl-4-phenyl-***	$C_6H_5CH_2CH_2CH(CH_3)CH_2OH$	164.25			$145–8^{18}$	0.9719^{20}_{4}		i	v				os v	**B6¹**, 269
b2901	**—,3-methyl-1-phenyl-***	$(CH_3)_2CHCH_2CHOHC_6H_5$	164.25			$235–6^{746}$	0.9537^{19}_{4}	1.5080^{18}	i	v	v			os v	**B6²**, 505
b2902	**—,3-methyl-2-phenyl-***	$(CH_3)_2CHCH(C_6H_5)CH_2OH$	164.25			127^{15}			i	v				os v	**B6**, 548
b2903	**2-Butanol, 2-methyl-1-phenyl-***	$CH_3CH_2COH(CH_3)CH_2C_6H_5$	164.25			$215–25^{747}$ $103–5^{11}$	0.9754^{20}_{0}	1.5182^{20}	i	s	s			lig s	**B6²**, 505
b2904	**—,2-methyl-3-phenyl-***	$CH_3CH(C_6H_5)COH(CH_3)_2$	164.25			105.7^{12}	0.9794^{20}_{4}	1.5193^{20}	i	s				os s	**B6¹**, 270
b2905	**—,2-methyl-4-phenyl-***	$C_6H_5CH_2CH_2COH(CH_3)_2$	164.25			121^{13}	0.9626^{21}_{4}	1.5077^{21}	i	v				os v	**B6²**, 506

For explanations, symbols and abbreviations see beginning of table.

No.	Name	Synonyms and Formula	Mol. wt.	Crystalline form, color and specfiic rotation	m.p. °C	b.p. °C	Density	n_D	Solubility						Ref.
									w	al	eth	ace	bz	other solvents	
	2-Butanol														
b2906	—,**3-methyl-3-phenyl-***	$(CH_3)_2CHCOH(C_6H_5)CH_3$	164.25			196–8^{760}	0.9653^{13}_{4}	1.5161^{13}	i	s				os s	**B6**^{1}, 269
b2907	**1-Butanol-, 2-nitro-***	$CH_3CH_2CHNO_2CH_2OH$	119.12		−47	105^{10}	1.1365^{11}		v	∞	∞				**B1**, 370
b2908	—,**1-phenyl-***	$CH_3CH_2CH_2CHOHC_6H_5$	150.22		14.5	113–5^{12}	1.0212^{18}_{4}		i	s	s				**B6**^{2}, 486
b2909	—,**2,2,3-trichloro-***	$CH_3CHClCCl_2CH_2OH$	177.47		62	199–200			δ^h	v	v				**B1**^{2}, 398
b2910	**2-Butanol, 1,1,1-trichloro-***	$CH_3CH_2CHOHCCl_3$	177.47			169–71^{738} 82–4^{22}	1.3670^{25}_{25}	1.4901^{25}		v	v	v	v	chl v CS_2 v	**B1**^{2}, 403
b2911	—,**2,3,3-trimethyl-***	$(CH_3)_3CCOH(CH_3)_2$	116.20		15–7	131–2^{760}	0.817^{22}	1.4233^{22}		∞	∞				**B1**^{2}, 447
b2912	**2-Butanone***	Ethyl methyl ketone. $CH_3CH_2COCH_3$	72.11		−87	79.6^{760}	0.8054^{20}_{7}	1.3814^{15}	v	∞	∞		∞		**B1**^{2}, 726
b2913	—,oxime	$CH_3CH_2C(:NOH)CH_3$	87.12		−29.5	152–3^{760}	0.9232^{20}_{4}	1.4428	s	∞	∞				**B1**^{2}, 730
b2914	—,**1-chloro-***	$CH_3CH_2COCH_2Cl$	106.55			139^{755}	1.080^{13}	1.4270^{10}	i					MeOH v^h	**B1**^{2}, 731
b2915	—,**3-chloro-***	$CH_3CHClCOCH_3$	106.55			117–9^{760}	1.032^{0}		i	v	v				**B1**^{2}, 731
b2916	—,**4-chloro-***	$ClCH_2CH_2COCH_3$	106.55			120–1^{760}				v	v			MeOH v	**B1**^{2}, 731
b2917	—,**3-chloro-3-methyl-***	$(CH_3)_2CClCOCH_3$	120.58			117.2^{758}	1.0083^{20}_{4}	1.4204^{20}			s				**B1**^{3}, 2818
b2918	—,**3,4-dibromo-4-phenyl-***	Benzalacetone dibromide. $C_6H_5CHBrCHBrCOCH_3$	306.01	nd (al)	124–5					δ s^h				chl v	**B7**^{2}, 244
b2919	—,**1,3-dichloro-***	$CH_3CHClCOCH_2Cl$	141.00			165^{753}	1.3116^{20}_{4}	1.4686^{20}	i	s	s			os s	**B1**^{1}, 348
b2920	—,**4(diethylamino)-***	$(C_2H_5)_2NCH_2CH_2COCH_3$	143.23			72–6^{16}			δ	s	s	s			**B4**^{1}, 452
b2921	—,**3,3-dimethyl-***	Pinacolone. $(CH_3)_3CCOCH_3$	100.16		−52.5	106^{760}	0.8012^{25}_{4}	1.3952^{20}	δ	s	s	s			**B1**^{3}, 354
b2922	—,**3,3-diphenyl-***	$CH_3C(C_6H_5)_2COCH_3$	224.31	pr (al)	41	310–1 176^{16}	1.069^{20}_{4}	1.5748^{20}	i	s v^h	v			chl v aa v	**B7**^{2}, 393
b2923	—,**1-hydroxy-***	$CH_3CH_2COCH_2OH$	88.11			153–4^{760}	1.0272^{20}_{4}	1.4189^{20}	v	v	v				**B1**^{2}, 870
b2924	—,**3-hydroxy-***	Acetoin. $CH_3CHOHCOCH_3$	88.11		15	143^{760}	1.0062^{20}_{20}	1.4190^{17}	∞	δ	δ			lig i	**B1**^{2}, 870
b2925	—,**1(1-hydroxy-2-naphthyl)-***	OH $CH_3CH_2COCH_2$— (2-naphthyl, 1-OH)	214.27	ye nd (eth)	85–6	145–52^{1}			i	s	s				**B8**, 152
b2926	—,**3-methyl-***	Isopropyl methyl ketone. $(CH_3)_2CHCOCH_3$	86.13		−92	95^{760}	0.8046^{16}_{4}	1.3879^{16}	δ	∞	∞				**B1**^{2}, 741
b2927	—,—,oxime	$(CH_3)_2.C.CH(:NOH)CH_3$	101.15			157–8			s	∞	∞				**B1**, 683
b2928	**1-Butanone, 3-methyl-1-phenyl-***	Isovalerophenone. $(CH_3)_2CHCH_2COC_6H_5$	162.23			236.5^{764}	$0.9701^{16.4}_{4}$	1.5139$^{15.3}$	i	∞	∞				**B7**^{2}, 252
b2929	—,**1-phenyl-***	Butyrophenone. $CH_3CH_2CH_2COC_6H_5$	148.21		11	228–9^{760}	0.988^{20}_{4}	1.532^{20}	i	∞	∞			peth δ	**B7**^{2}, 241
b2930	**2-Butanone, 1-phenyl-***	Ethyl benzyl ketone. $CH_3CH_2COCH_2C_6H_5$	148.21			230^{755}	1.002^{0}_{4}		i	s	∞				**B7**^{2}, 243
b2931	—,**4-phenyl-***	Benzylacetone. $C_6H_5CH_2CH_2COCH_3$	148.21			233–4	0.9894^{22}_{4}	1.511^{22}		s	s				**B7**^{2}, 243
b2932	**1-Butanone, 1(4-tolyl)-**	*p*-Methylbutyrophenone. $CH_3CH_2CH_2CO-C_6H_4-CH_3$	162.23		12	251.5^{758}	0.9745^{20}_{4}	1.5215^{20}	i	v	v				**B7**^{2}, 254
b2933	**2-Butenal** (*trans*)*	Crotonaldehyde. $CH_3CH:CHCHO$	70.09		−74	104	0.858^{16}_{4}	1.4373^{20}	v	∞	∞		∞		**B1**^{2}, 787
b2934	—,diethyl acetal	$CH_3CH:CHCH(OC_2H_5)_2$	144.22			146–8	0.8473^{18}_{4}	1.4162^{18}	i	∞	∞		∞	peth ∞	**B1**^{2}, 789
b2935	—,**2-bromo-**, diethyl acetal	$CH_3CH:CBrCH(OC_2H_5)_2$	223.12			86^{15}	1.2255^{21}	1.4565^{21}							**B1**^{1}, 380

For explanations, symbols and abbreviations see beginning of table.

No.	Name	Synonyms and Formula	Mol. wt.	Crystalline form, color and specific rotation	m.p. °C	b.p. °C	Density	n_D	Solubility w	al	eth	ace	bz	other solvents	Ref.
	2-Butenal														
b2936	**—,2-chloro-***	$CH_3CH{:}CClCHO$	104.54			147–8	1.1404_4^{23}	1.478^{23}	δ	s	s			CCl_4 s chl s	**B1**[2], 789
b2937	**—,3-ethoxy-,** diethyl acetal	1,1,3-Triethoxy-2-butene*. $CH_3C(OC_2H_5){:}CHCH(OC_2H_5)_2$	188.27			190–5 $79–82^{10}$	0.908_0^{21}	1.430^{21}		s					**B1**[3], 3267
b2938	**—,2-methyl-***	Tiglaldehyde. $CH_3CH{:}C(CH_3)CHO$	84.11			116.5	0.8710_4^{20}	1.4475^{20}	δ	∞	∞				**B1**[2], 792
b2939	**—,3-methyl-***	β,β-Dimethylacrolein. Senecialdehyde. $(CH_3)_2C{:}CHCHO$	84.11			133^{730}	0.8722_4^{20}	1.4526^{20}	s	s	s				
b2940	**1-Butene-***	α-Butylene. $CH_3CH_2CH{:}CH_2$	56.10		−185.4	-6.3^{760}	0.5946_4^{20}	1.3962^{20}	i	v	v				**B1**[3], 715
b2941	**2-Butene(*cis*)***	*cis*-β-Butylene. $CH_3CH{:}CHCH_3$	56.10		−138.9	3.7^{760}	0.6213_4^{20}	1.3931^{-25}	i	v	v				**B1**[3], 728
b2942	—(*trans*)*	*trans*-β-Butylene. $CH_3CH{:}CHCH_3$	56.10		−105.6	0.9	0.6041_4^{20}	1.3848^{-25}							**B1**[3], 730
b2943	**1-Butene, 4-bromo-***	$BrCH_2CH_2CH{:}CH_2$	135.01			98.5^{750}	1.3230_4^{20}	1.4622^{20}	i	s	s				**B1**[3], 727
b2944	**—,2-bromo-3-methyl-***	$(CH_3)_2CHCBr{:}CH_2$	149.04			105^{757}	1.2328_4^{20}	1.4504^{20}	i		s		s	chl s	**B1**[3], 800
b2945	**2-Butene, 1-bromo-3-methyl-***	$(CH_3)_2C{:}CHCH_2Br$	149.04			129–33d $50–1^{40}$	1.2819_0^{20}	1.4930^{15}	i d^h	s	s	s	s	CS_2 s chl s	**B1**[3], 796
b2946	**—,2-bromo-3-methyl-***	$(CH_3)_2C{:}CBrCH_3$	149.04			$119–20^{766}$	1.2773_0^{20}	1.4738^{20}	i		s	s		chl s	**B1**[3], 796
b2947	**1-Butene, 2-bromo-4-phenyl-***	$C_6H_5CH_2CH_2CBr{:}CH_2$	211.11			$117–8^{21}$	1.2901_4^{20}	1.5450^{20}							**B5**[2], 380
b2948	**2-Butene, 1 bromo-4-phenyl-***	$C_6H_5CH_2CH{:}CHCH_2Br$	211.11			$112–5^{14}$					s				**B5**[2], 379
b2949	**—,2-bromo-3-phenyl-***	$CH_3C(C_6H_5){:}CBrCH_3$	211.11			$114–6^{13}$					s		s		**B5, 488**
b2950	**1-Butene, 1-chloro-(*cis*)***	$CH_3CH_2CH{:}CHCl$	90.56			63.5^{760}	0.9153_4^{15}	1.4194^{15}	i	∞					
b2951	—,—(*trans*)*	$CH_3CH_2CH{:}CHCl$	90.56			68^{760}	0.9205_4^{15}	1.4225^{15}	i	∞					
b2952	**—,2-chloro-***	$CH_3CH_2CCl{:}CH_2$	90.56			58.5^{760}	0.9107_4^{15}	1.4115^{21}	i	∞					**B1, 204**
b2953	**—,3-chloro-***	$CH_3CHClCH{:}CH_2$	90.56			$64–5^{766}$	0.9001_4^{20}	1.4151^{20}				v			**B1**[3], 723
b2954	**—,4-chloro-***	$ClCH_2CH_2CH{:}CH_2$	90.56			75^{733}	0.9211_4^{20}	1.4233^{20}	i					chl v	
b2955	**2-Butene, 1-chloro-(*cis*)**	$CH_3CH{:}CHCH_2Cl$	90.56			85^{752}	0.9426_4^{20}	1.4390^{20}		s		s			**B1**[2], 176
b2956	—,—(*trans*)*	$CH_3CH{:}CHCH_2Cl$	90.56			$76–8^{760}$	0.9282_4^{20}	1.4359^{20}							**B1**[2], 176
b2957	**—,2-chloro-(*cis*)***	$CH_3CH{:}CClCH_3$	90.56		−117.3	70.6^{760}	0.9239_4^{20}	1.4240^{20}		∞					
b2958	—,—(*trans*)*	$CH_3CH{:}CClCH_3$	90.56		−105.8	62.8	0.9138_4^{20}	1.4190^{20}		∞					
b2959	**1-Butene, 3-chloro-2-chloromethyl-***	$CH_3CHClC(CH_2Cl){:}CH_2$	139.03			155^{760}	1.1328_4^{18}		i					chl s	**B1**[3], 787
b2960	**2-Butene, 1-chloro-2,3-dimethyl-***	$(CH_3)_2C{:}C(CH_3)CH_2Cl$	118.61			$111–2^{756}$	0.8895_4^{19}				v			chl v	**B1**[3], 819
b2961	**1-Butene, 1-chloro-2-methyl-***	$CH_3CH_2C(CH_3){:}CHCl$	104.58			96–7	0.9170_4^{20}		i		v				**B1**[2], 187

For explanations, symbols and abbreviations see beginning of table.

No.	Name	Synonyms and Formula	Mol. wt.	Crystalline form, color and specific rotation	m.p. °C	b.p. °C	Density	n_D	Solubility						Ref.
									w	al	eth	ace	bz	other solvents	
	1-Butene														
b2962	**—,1-chloro-3-methyl-***	$(CH_3)_2CHCH:CHCl$	104.58			$86\text{–}8^{756}$		1.4229^{20}	i		s			chl s	**B1³**, 799
b2963	**—,3-chloro-2-methyl-***	$CH_3CHClC(CH_3):CH_2$	104.58			94^{760}	0.9088^{20}_{4}	1.4304^{20}			v			chl s	**B1³**, 787
b2964	**2-Butene, 1-chloro-2-methyl-***	$CH_3CH:C(CH_3)CH_2Cl$	104.58			110^{760}	0.9327^{20}_{4}	1.4481^{20}	d^h	v		v			**B1²**, 189
b2965	**—,1-chloro-3-methyl-***	$(CH_3)_2C:CHCH_2C$	104.58			109	0.9335^{20}_{4}	1.4398^{20}						chl v	**B1²**, 191
b2966	**—,2-chloro-3-methyl-***	$(CH_3)_2C:CClCH_3$	104.58			97–8	0.925^{20}_{4}	1.4380^{20}			v				**B1²**, 189
b2967	**1-Butene, 2(chloromethyl)-1,3-dichloro-***	$CH_3CHClC(CH_2Cl):CH_2Cl$	174.48			$68\text{–}70^{8}$	1.2775^{17}_{4}		i					CCl_4 s	**B1³**, 788
b2968	**2-Butene, 1,4-dibromo-(*trans*)***	$BrCH_2CH:CHCH_2Br$	213.91	pl (peth)	53–4	205			δ	v				MeOH v peth v	**B1²**, 206
b2970	**1-Butene, 1,3-dichloro-***	$CH_3CHClCH:CHCl$	125.00			125^{730}	1.1341^{24}_{4}	1.464_{20}	i	s				chl s	**B1³**, 724
b2971	**—,2,3-dichloro-***	$CH_3CHClCCl:CH_2$	125.00			112	1.1411^{20}_{4}	1.4571^{20}	i					chl s	**B1³**, 724
b2972	**2-Butene, 1,1-dichloro-***	$CH_3CH:CHCHCl_2$	125.00			124–5	1.140^{18}_{18}	1.466^{18}			s			chl s	**B1³**, 741
b2973	**—,1,2-dichloro-** (high b.p.)*	$CH_3CH:CClCH_2Cl$	125.00			$132\text{–}4^{752}$	1.1597^{20}_{4}	1.4590^{20}	i			v		chl v	
b2974	—,—(low b.p.)*	$CH_3CH:CClCH_2Cl$	125.00			$116\text{–}8^{765}$	1.1544^{20}_{4}	1.4642^{20}				v		CCl_4 v	
b2975	**—,1,3-dichloro-** (*cis*)*	$CH_3CCl:CHCH_2Cl$	125.00			128^{745}	1.1528^{25}_{4}	1.4695^{25}	i					os s	**B1³**, 742
b2976	—,—(*trans*)*	$CH_3CCl:CHCH_2Cl$	125.00			130^{745}	1.1542^{25}_{4}	1.4711^{25}	i					os s	**B1³**, 742
b2976¹	**—,1,4-dichloro-** (*cis*)*	$ClCH_2CH:CHCH_2Cl$	125.00		−48	152.5^{758}	1.188^{25}_{4}	1.4887^{25}	i					os s	**B1³**, 743
b2977	—,—(*trans*)*	$ClCH_2CH:CHCH_2Cl$	125.00		1–3	155.5^{758}	1.183^{25}_{4}	1.4861^{25}	i					os s	**B1³**, 743
b2978	**—,2,3-dichloro-** (*cis*)*	$CH_3CCl:CClCH_3$	125.00			$125\text{–}6^{753}$	1.1618^{20}_{4}	1.4590^{20}	i					os s	**B1³**, 744
b2979	—,—(*trans*)*	$CH_3CCl:CClCH_3$	125.01			101^{753}	1.1416^{20}_{4}	1.4582^{20}	i						**B1³**, 744
b2980	**1-Butene, 3,3-dichloro-2-methyl-***	$CH_3CCl_2C(CH_3):CH_2$	139.03			$124\text{–}6^{762}$	1.085^{19}_{4}		i					chl s	**B1³**, 787
b2981	**2-Butene, 1,3-dichloro-2-methyl-***	$CH_3CCl:C(CH_3)CH_2Cl$	139.03			$151\text{–}3^{760}$	1.1293^{20}_{4}		i			v		chl s	**B1³**, 795
b2982	**—,1,4-dichloro-2-methyl-***	$ClCH_2CH:C(CH_3)CH_2Cl$	139.03			56^{10}	1.1526^{20}_{4}	1.4932^{20}	i			v		chl v	**B1³**, 795
—	**—,1,1-diethoxy-**	*see* **2-Butenal,** diethyl acetal*													
b2983	**1-Butene, 2,3-dimethyl-***	$(CH_3)_2CHC(CH_3):CH_2$	84.16		−157.3	56^{762}	0.6779^{20}_{4}	1.3904^{20}		s	s			CS_2 s	**B1³**, 816
b2984	**—,3,3-dimethyl-***	$(CH_3)_3CCH:CH_2$	84.16		−148.2	41.2^{760}	0.6529^{20}_{4}	1.3760^{20}							**B1³**, 814
b2985	**2-Butene, 2,3-dimethyl-***	$(CH_3)_2C:C(CH_3)_2$	84.16		−74.3	73.2^{760}	0.7081^{20}_{4}	1.4115^{20}		s	s				**B1³**, 817
b2986	**1-Butene, 3,4-epoxy-***	Vinyloxirane. $\overset{O}{CH_2CHCH:CH_2}$	70.09			70^{760}	0.9006^{0}	1.416^{20}						os s	**B17¹**, 13

For explanations, symbols and abbreviations see beginning of table.

No.	Name	Synonyms and Formula	Mol. wt.	Crystalline form, color and specific rotation	m.p. °C	b.p. °C	Density	n_D	Solubility w	al	eth	ace	bz	other solvents	Ref.
	1-Butene														
b2987	**—,2-ethyl-***	3-Methylenepentane. $(C_2H_5)_2C:CH_2$	84.16	......	−131.5	64.7^{760}	0.6894_4^{20}	1.3969^{20}	...	...	...	...	...	...	**B1**^{3}, 814
b2988	**—,2-ethyl-3-methyl-***	$(CH_3)_2CHC(C_2H_5):CH_2$	98.19	......	......	89	0.7186_4^{20}	1.410^{20}	...	...	...	...	...	...	**B1**^{3}, 833
b2989	**2-Butene, 1,1,2,3,4,4-hexachloro-** (liquid)*	$Cl_2CHCCl:CClCHCl_2$	262.78	......	......	$97–8^{10}$	1.651_{15}^{15}	1.5331	i	...	...	...	...	chl s	......
b2990	—,—(solid)*	$Cl_2CHCCl:CClCHCl_2$	262.78	lf (al)	80	......	......	......	...	v^h	v	...	v	chl v CCl_4 v	......
b2991	**1-Butene, 2-methyl-***	$CH_3CH_2C(CH_3):CH_2$	70.14	......	−133.8	38.6^{760}	0.6623_4^{20}	1.3874^{20}	...	...	...	...	...	...	**B1**^{3}, 788
b2992	**—,3-methyl-***	$(CH_3)_2CHCH:CH_2$	70.14	......	−68.5	20^{760}	0.6272_4^{20}	1.3640^{21}	i	∞	∞	...	...	...	**B1**^{3}, 797
b2993	**2-Butene, 2-methyl-***	$(CH_3)_2C:CHCH_3$	70.14	......	−123	38.4^{760}	0.6620_4^{20}	1.3878^{20}	i	s	s	...	...	lig v	**B1**^{2}, 187
b2994	**1-Butene, octafluoro-***	Perfluoro-α-butylene. $F_3CCF_2CF:CF_2$	200.03	......	......	4.8^{764}	1.615_4^{-20}	1.5443^0	...	...	...	...	...	...	**B1**^{3}, 722
b2995	**2-Butene, octafluoro-***	Perfluoro-β-butylene. $F_3CCF:CFCF_3$	200.03	......	−139, −129	$0–3^{760}$	......	1.5297^0	...	...	...	...	...	...	**B1**^{3}, 739
b2996	**—,1,1,1,4,4-pentachloro-***	$Cl_2CHCH:CHCCl_3$	228.33	......	......	$78–80^{11}$	1.612_{21}^{21}	1.5538^{21}	...	s^h	...	...	...	chl s	......
b2997	**1-Butene, 1,3,4,4-tetrachloro-***	$ClCH.CH.CHCl.CHCl_2$	193.89	......	......	$82–3^{17}$	......	......	i	...	...	...	...	chl s	**B1**^{3}, 726
b2998	**—,2,3,3,4-tetrachloro-***	$ClCH_2CCl_2CCl:CH_2$	193.89	......	......	$41–2^7$	1.4602_4^{20}	1.5133^{20}	i	...	...	...	...	chl s	**B1**^{3}, 726
b2999	**—,2,3,4-trichloro-***	$ClCH_2CHClCCl:CH_2$	159.44	......	......	60^{20}	1.3430_4^{20}	1.4944^{20}	i	...	...	...	...	chl s	**B1**^{2}, 275
b3000	**2-Butene, 1,2,4-trichloro-***	$ClCH_2CH:CClCH_2Cl$	159.44	......	......	$67–9^{10}$	1.3843_4^{20}	1.5175^{20}	i	s	...	...	...	chl s	......
—	—,1,1,3-triethoxy-*	*see* **2-Butenal, 3-ethoxy-diethyl acetal**													
b3001	**1-Butene, 2,3,3-trimethyl-***	Triptene. $(CH_3)_3CC(CH_3):CH_2$	98.19	......	−109.9	77.8^{760}	0.7050_4^{20}	1.4029^{20}	...	...	...	...	...	MeOH s	**B1**^{3}, 384
—	**3-Butene-1,1-dicarboxylic acid***	*see* **Malonic acid, allyl-**													
—	**Butenedioic acid** (*cis*)*	*see* **Maleic acid**													
—	—(*trans*)*	*see* **Fumaric acid**													
b3002	**2-Butene-1,4-diol** (*cis*)*	$HOCH_2CH:CHCH_2OH$	88.11	......	4	235^{760} 132^{16}	1.0698_4^{20}	1.4782^{20}	s	v	...	...	...	...	**B1**^{3}, 2255
b3003	—(*trans*)	$HOCH_2CH:CHCH_2OH$	88.11	......	25	131^{13}	1.070_4^{20}	1.4755^{20}	v	v	...	...	...	...	**B1**^{3}, 2256
b3004	**3-Butene-1,2-diol***	Vinylethylene glycol. $CH_2:CHCHOHCH_2OH$	88.11	......	......	169.5	1.0470_4^{20}	1.4068^{20}	s	s	...	...	...	...	**B1**^{3}, 2252
b3005	**2-Buten-1,4-dione, 1,4-diphenyl-**(*cis*)*	Dibenzoyl ethylene. $C_6H_5COCH:CHCOC_6H_5$	236.27	nd (al)	134	......	......	......	...	s^h	s	...	s	chl s	**B7**^{2}, 741
b3006	—,— (*trans*)*	$C_6H_5COCH:CHCOC_6H_5$	236.27	nd (al)	111	......	......	......	...	δ	...	...	s	chl v aa s lig i	**B7**^{2}, 741
b3007	**2-Butenoic acid** (*cis*)*	Isocrotonic acid. $CH_3CH:CHCO_2H$	86.09	nd or pr (peth)	14.5	169.3^{760}	1.0267_4^{20}	1.4483^{14}	v	s	...	...	...	...	**B2**^{2}, 394
b3009	—(*trans*)*	Crotonic acid. $CH_3CH:CHCO_2H$	86.09	mcl nd or pr (w or lig)	71.6	185^{760}	1.018_4^{15}	1.4228^{80}	v	v	...	...	...	...	**B2**^{2}, 290
b3010	—,amide (*trans*)	Crotonamide. $CH_3CH:CHCONH_2$	85.11	nd (ace)	160	......	......	......	δ	s	δ	...	s	...	**B2**^{2}, 392
b3011	—,anhydride (*trans*)*	Crotonic anhydride. $(CH_3CH:CHCO)_2O$	154.17	......	......	246–8 129^{19}	1.0397^{20}	1.4745^{20}	d	d	s	...	...	...	......
b3012	—,chloride (*trans*)	Crotonyl chloride. $CH_3CH:CHCOCl$	104.54	......	......	$124–5^{760}$	1.0905^{20}	1.460^{18}	d	d	...	...	...	...	**B2**^{2}, 392

For explanations, symbols and abbreviations see beginning of table.

No.	Name	Synonyms and Formula	Mol. wt.	Crystalline form, color and specific rotation	m.p. °C	b.p. °C	Density	n_D	Solubility w	al	eth	ace	bz	other solvents	Ref.
	2-Butenoic acid														
b3012[1]	—,ethyl ester (*cis*)*	Ethyl isocrotonate. $CH_3CH:CHCO_2C_2H_5$	114.15			126^{750}	0.9182^{20}_{4}	1.4242^{20}	...	s	...	...	...	os s	**B2**^{2}, 394
b3013	—,—(*trans*)*	Ethyl crotonate. $CH_3CH:CHCO_2C_2H_5$	114.15			143–7	0.9183^{20}_{4}	1.425^{14}	...	s	s	...	...		**B2**^{2}, 392
b3014	—,methyl ester (*trans*)*	Methyl crotonate. $CH_3CH:CHCO_2CH_3$	100.12			121	0.9444^{20}_{4}	1.4242^{20}	i	v	v	...	...		**B2**^{2}, 392
b3015	—,nitrile (*trans*)	Crotononitrile. $CH_3CH:CHCN$	67.09		−51.5	122	0.8239^{20}_{4}	1.4225^{20}	...	...	...	...	...		**B2**^{2}, 393
b3016	**3-Butenoic acid***.	Vinylacetic acid. $CH_2:CHCH_2CO_2H$	86.09		−3.9	169^{764}	1.0091^{20}_{4}	1.4252^{20}	s	∞	∞	...	...		**B2**^{2}, 389
b3017	—,ethyl ester	$CH_2:CHCH_2CO_2C_2H_5$	114.14			119		1.4218^{22}	...	s	...	...	...		**B2**, 407
b3018	—,nitrile	Allyl cyanide. $CH_2:CHCH_2CN$	67.09		−84	119	0.8318^{20}_{4}	1.4060^{20}	δ	∞	∞	...	...		**B2**^{2}, 389
b3019	**2-Butenoic acid, 3-amino-**, ethyl ester (*trans*)*	$CH_3CNH_2:CHCO_2C_2H_5$	129.16	mcl pr	34 20–1 unst	210–5d^{760} 105^{15}	1.0219^{19}_{4}	1.4988^{22}	i	s	s	...	s	chl s CS_2 s lig s	**B3**^{2}, 423
b3020	—,**3-bromo-** (*trans*)*	$CH_3CBr:CHCO_2H$	165.00	nd (lig), lf (w)	97				δ	s	s	...	s	CS_2 s aa s	**B2**, 419
b3021	—,**4-bromo-**, ethyl ester (*trans*)*	$BrCH_2CH:CHCO_2C_2H_5$	193.05			105–10^{15}	1.402^{16}_{4}	1.4925^{20}	...	v	...	...	...		**C50**, 10804
b3022	—,**2-chloro-** (*cis*)*.	$CH_3CH:CClCO_2H$	120.54	nd (w)	67				s	v	...	...	...	CS_2 v lig δ v^h	**B2**^{2}, 396
b3023	—,—(*trans*)*	$CH_3CH:CClCO_2H$	120.54	nd (w or peth)	99	212			δ	s	s	...	...		**B2**^{2}, 395
b3024	—,—,ethyl ester (*cis*)*	$CH_3CH:CClCO_2C_2H_5$	148.60			75^{30}	1.1021^{18}_{4}		...	s	s				**B2**^{2}, 396
b3025	—,—,—(*trans*)*	$CH_3CH:CClCO_2C_2H_5$	148.60			176–8	1.1090^{20}_{20}	1.4538^{20}	...	s	s	...	...		**B2**^{2}, 395
b3026	—,—,methyl ester (*trans*)*	$CH_3CH:CClCO_2CH_3$	134.57			161.5^{762}	1.160^{20}_{4}	1.4569^{23}	...	...	s	...	...		**B2**^{2}, 395
b3027	—,**3-chloro-** (*cis*)*.	$CH_3CCl:CHCO_2H$	120.54		61	195	1.1995^{66}_{4}	1.4704^{66}	δ	v	...	...	...	peth v^h CS_2 v	**B2**^{2}, 396
b3028	—,—(*trans*)*	$CH_3CCl:CHCO_2H$	120.54		94–5	206–11			δ	v	...	...	...	CS_2 v	**B2**^{2}, 396
b3029	—,—,ethyl ester (*cis*)*	$CH_3CCl:CHCO_2C_2H_5$	148.60			161.4	1.0924^{19}_{4}	1.4542^{19}_{4}	...	s	s	...	...		**B2**^{1}, 190
b3030	—,—,—(*trans*)*	$CH_3CCl:CHCO_2C_2H_5$	148.60			184	1.1062^{20}_{4}	1.459^{20}	...	s	s	...	...		**B2**^{2}, 396
b3031	—,—,methyl ester (*cis*)*	$CH_3CCl:CHCO_2CH_3$	134.57			142.4	1.138^{20}_{4}	1.4573^{19}	...	...	v	...	...	MeOH v	**B2**^{2}, 396
b3032	—,—,—(*trans*)*	$CH_3CCl:CHCO_2CH_3$	134.57			64–7^{14}	1.157^{20}_{4}	1.4628^{21}	...	v	v	...	...		**B2**^{2}, 396
b3033	—,**4-chloro-** (*trans*)*	$ClCH_2CH:CHCO_2H$	120.54		83	117–8^{13}			δ^h	...	v	...	...	peth δ aa v	**B2**, 418
b3034	—,**2,3-dibromo-4-oxo-***	Mucobromic acid. $O:CHCBr:CBrCO_2H$	257.89	pl (eth-lig)	125				s v^h	v	v	...	δ	chl δ CS_2 δ	**B3**^{2}, 460
b3035	—,**2,3-dichloro-4-oxo-***	Mucochloric acid. $O:CHCCl:CClCO_2H$	168.97	mcl pr (eth-lig)	127				δ s^h	s	s	...	s		**B3**, 727
b3036	—,**2-ethyl-** (*trans*)*	$CH_3CH:C(C_2H_5)CO_2H$	114.14	mcl pr	45	209	0.9484^{56}_{4}	1.4475^{50}	δ	s	s	...	...		**B2**^{2}, 408
b3037	**3-Butenoic acid, 2-hydroxy-**, ethyl ester*	$CH_2:CHCHOHCO_2C_2H_5$	130.14			173^{756}d 68^{15}	1.0470^{15}_{4}		v	∞	∞	...	...		**B3**, 271
b3038	—,**2-hydroxy-4-phenyl-***	Benzallactic acid. $C_6H_5CH:CHCHOHCO_2H$	178.18	nd (w)	46				s^h	...	δ	...	δ	CS_2 δ lig δ	
b3039	**2-Butenoic acid, 2-methyl-***	Angelic acid. $CH_3CH:C(CH_3)CO_2H$	100.11	mcl pr or nd	45–6	185	0.983^{49}_{4}	1.4434^{47}	δ s^h	s	v	...	...		**B2**^{2}, 401
b3040	—,—(*trans*)	Tiglic acid. $CH_3CH:C(CH_3)CO_2H$	100.11	tcl pr	64	198.5^{760}	0.9641^{76}_{4}		δ s^h	s	s	...	...		**B2**^{2}, 401
b3041	—,—,chloride	Tiglyl chloride. $CH_3CH:C(CH_3)COCl$	118.57			64^{35}			d	d	s	...	...		**B2**, 431

For explanations, symbols and abbreviations see beginning of table.

No.	Name	Synonyms and Formula	Mol. wt.	Crystalline form, color and specific rotation	m.p. °C	b.p. °C	Density	n_D	Solubility w	al	eth	ace	bz	other solvents	Ref.
	2-Butenoic acid														
b3042	—,—,ethyl ester	$CH_3CH:C(CH_3)CO_2C_2H_5$	128.17			152–4	0.9200^{20}_{4}	1.4340^{20}		s			s		**B2**^{2}, 401
b3042^{1}	—,3(4-nitro-phenyl)-*	$O_2N-C_6H_4-C(CH_3):CHCO_2H$	207.19	pa ye nd (aa)	168–9									aa s^h	**B9**, 615
b3043	—,4-oxo-4-phenyl-*	β-Benzoylacrylic acid. $C_6H_5COCH:CHCO_2H$	176.17	nd or pr (to)	99				δ s^h	s	s			to s^h lig δ	**B10**^{2}, 499
b3043^{1}	—,—,hydrate*	$C_6H_5COCH:CHCO_2H.H_2O$	194.19	f (w+1)	65				s^h	s	s				**B10**, 726
b3044	3-Butenoic acid, 4-phenyl-*	$C_6H_5CH:CHCH_2CO_2H$	162.19	nd (w), pr (CS_2)	88	302			i δ^h	v	v			CS_2 v	**B9**^{2}, 407
b3045	2-Buten-1-ol (*trans*)*	Crotyl alcohol. $CH_3CH:CHCH_2OH$	72.11		<−30	121^{754}	0.8454^{25}_{4}	1.4262^{25}	v	∞	∞				**B1**^{2}, 480
—	—,acetate	*see* **Acetic acid,** 2-buten-1-yl ester													
b3046	3-Buten-1-ol*	$CH_2:CHCH_2CH_2OH$	72.11			113^{755}	0.8379^{17}_{4}	1.4146^{17}	s	∞	∞				**B1**^{2}, 480
b3047	3-Buten-2-ol(*d*)*	$CH_3CHOHCH:CH_2$	72.11	$[\alpha]^{20}_{D}+33.9$ (undil)			0.8367^{15}_{4}	1.4120^{20}							**B1**^{2}, 480
b3048	—(*dl*)*	$CH_3CHOHCH:CH_2$	72.11		<−100	97.3^{760}	0.8318^{20}_{4}	1.4127^{20}	δ						**B1**^{2}, 479
b3049	2-Buten-1-ol, 2-chloro-*	$CH_3CH:CClCH_2OH$	106.55			159^{760}	1.0950^{23}_{4}	1.4682^{20}		v					**B1**^{2}, 481
b3050	—,4-chloro-*	$ClCH_2CH:CHCH_2OH$	106.55			$64–5^{2}$		1.4845^{20}	d^h	v	v				
b3051	3-Buten-1-ol, 2-chloro-*	$CH_2:CHCHClCH_2OH$	106.55			70^{30}		1.4665^{20}		v	v				
b3052	3-Buten-2-ol, 1-chloro-*	$CH_2:CHCHOHCH_2Cl$	106.55			144–7	1.111^{20}_{4}	1.4660^{20}						chl v	
b3053	—,3-chloro-*	$CH_3CHOHCCl:CH_2$	106.55			$53–7^{19}$	1.1138^{23}_{4}			v	v				
b3054	3-Butene-2-one*	Methyl vinyl ketone. $CH_3COCH:CH_2$	70.09			$79–80^{755}$	0.8636^{20}_{4}	1.4086^{20}	δ	s	v	v			**B1**^{2}, 786
b3055	—,4-bromo-4-phenyl-*	(α-Bromobenzal)acetone. $CH_3COCH:CBrC_6H_5$	225.09			$150–1^{10}$							s		**B7**, 367
b3056	2-Buten-1-one, 1,3-diphenyl-*	Dypnone. β-Methylchalcone. $CH_3C(C_6H_5):CHCOC_6H_5$	222.29			340–5 225^{22}	1.108^{20}_{0}				s				**B7**^{2}, 433
b3057	3-Buten-2-one 4(2-furyl)-	Furfuralacetone. (2-furyl)—CH:CHCOCH$_3$	136.15	nd	39–40	229d $112–5^{10}$	1.0496^{57}_{4}	1.5788^{45}	i	v	v			chl v peth s	**B17**^{2}, 326
b3058	—,(4-hydroxy-3-methoxy-phenyl)-	Vanillalacetone. $C_{11}H_{12}O_3$. *See* b2065	192.22	nd (al)	129				δ	v	v		v		
b3059	— 4(2-hydroxy-phenyl)-	Salicylideneacetone. $C_{10}H_{10}O_2$. *See* b3065	162.19	nd (al), pr (bz)	139				δ	s^h	v		s^h		**B8**^{2}, 153
b3060	—,4(3-hydroxy-phenyl)-	$C_{10}H_{10}O_2$. *See* b3065	162.19		97–8								s^h		**B8**^{2}, 155
b3061	—,4(4-hydroxy-phenyl)-	$C_{10}H_{10}O_2$. *See* b3065	162.19	nd (w)	114–5				δ s^h	v				aa v	**B8**^{2}, 155
b3062	—,4(4-methyloxy-phenyl)-	Anisilideneacetone. $C_{11}H_{12}O_2$. *See* b3065	176.22	lf (al)	73				i	v	v		v	aa s	**B8**^{2}, 155
b3063	—,3-methyl-	Isopropenyl methyl ketone. $CH_3COC(CH_3):CH_2$	84.11			98^{760}	0.8530^{20}_{4}	1.4236^{20}	δ	∞					**B1**, 733
b3064	—,4(3,4-methyl-enedioxy-phenyl)-	Piperonalacetone. $C_{11}H_{10}O_3$. *See* b3065	190.20		111				i δ^h	δ v^h	s		s	CCl_4 v peth i	**B19**^{2}, 157
b3065	—,4-phenyl-*	Benzalacetone. $CH_3COCH:CH-C_6H_5$ (ring positions 2, 3, 4, 5, 6)	146.19	pl	41–2	262^{760} 140^{16}	1.0097^{45}	1.5824^{47}	i	v	s		s	chl s peth δ	**B7**^{2}, 287
b3066	1-Buten-3-yne*	Vinylacetylene. $HC\vdots CCH:CH_2$	52.08			5.1^{760}	0.7095^{0}_{0}	1.4161^{1}							**B1**^{3}, 1032
b3067	—,4-chloro-*	$ClC\vdots CCH:CH_2$	86.52			$55–7^{760}$	1.0032^{20}_{4}	1.4663^{20}						chl s	**B1**^{3}, 1038
b3068	—,1-methoxy-*	$HC\vdots CCH:CHOCH_3$	82.10			$122–5d^{760}$ $30–2^{13}$	0.906^{20}_{4}	1.4818^{20}	i					os v	

For explanations, symbols and abbreviations see beginning of table.

No.	Name	Synonyms and Formula	Mol. wt.	Crystalline form, color and specific rotation	m.p. °C	b.p. °C	Density	n_D	Solubility w	al	eth	ace	bz	other solvents	Ref.
	1-Buten-3-yne														
b3069	—,**2-methyl-***	Valylene. $CH\vdots CC(CH_3):CH_2$	66.10			32^{760}	0.6801^{11}_{4}	1.4158^{20}							
—	**Butesin**	*see* **Benzoic acid, 4-amino-**, butyl ester													
—	***tert*-Butyl bromide**	*see* **Propane, 2-bromo-2-methyl-***													
—	***tert*-Butyl chloride**	*see* **Propane, 2-chloro-2-methyl-***													
—	**Butylene**	*see* **Butene***													
—	**Butylene oxide**	*see* **Butane, epoxy-***													
b3070	**1-Butyne***	$CH_3CH_2C\vdots CH$	54.09		−122.5	8.1^{760}	0.6784^{0}_{4}	1.3962^{20}	i	s	s				**B1**², 223
b3071	**2-Butyne***	Dimethyl acetylene. $CH_3C\vdots CCH_3$	54.09		−32.3	27^{760}	0.6913^{20}_{4}	1.3921^{20}	i	s	s				**B1**³, 925
b3072	—,**1-chloro-***	$CH_3C\vdots CCH_2Cl$	88.54			81–4					v	v			
b3073	**1-Butyne, 3-chloro-3-methyl-***	$(CH_3)_2CClC\vdots CH$	102.56			77–9	0.9061^{20}_{4}								**B1**³, 965
b3074	**2-Butyne, 1,4-dibromo-***	$BrCH_2C\vdots CCH_2Br$	211.89			92^{15}	2.014^{18}	1.588^{18}							**B1**³, 927
b3075	—,**1,4-dichloro-***	$CH_2ClC\vdots CCH_2Cl$	122.98			$165–6^{760}$ 73^{24}	1.258^{20}_{4}	1.5072^{20}							**B1**³, 927
b3076	—,**1,4-diiodo-***	$ICH_2C\vdots CCH_2I$	305.89	nd (al)	53	$70–2^{0.1}$				s^{h}				chl v	**B1**³, 927
b3077	**1-Butyne, 3,3-dimethyl-***	*tert*-Butylacetylene. $(CH_3)_3CC\vdots CH$	82.15		−81.2	$39–40^{760}$	0.6695^{20}_{4}	1.3749^{20}							
b3078	**2-Butyne, hexafluoro-***	Perfluoro-2-butyne*. $CF_3C\vdots CCF_3$	162.04		−117.4	-24.6^{760}				s	s	s		CCl_4 s aa s	**B1**³, 926
b3079	**1-Butyne, 3-methyl-***	Isopropyl acetylene. $(CH_3)_2CHC\vdots CH$	68.12		−134.3	29.3^{760}	0.7107^{20}_{4}	1.4039^{20}	i	∞	∞				**B1**³, 958
b3080	—,**4-phenyl-***	$C_6H_5CH_2CH_2C\vdots CH$	130.19			$189–91^{758}$	0.9218^{20}	1.5212^{20}	i						**B5**², 413
b3081	**2-Butynedioic acid***	Acetylenedicarboxylic acid. $HO_2CC\vdots CCO_2H$	114.06	pl (eth)	175–6				v	v	v				**B2**², 670
b3082	—,diethyl ester*	$C_2H_5O_2CC\vdots CCO_2C_2H_5$	170.17		1–2	$120–1^{20}$	1.0653^{25}_{4}	1.4405^{25}		s	s			CCl_4 s	**B2**², 671
b3083	—,dimethyl ester*	$CH_3O_2CC\vdots CCO_2CH_3$	142.11			$195–8^{760}$ 102^{20}	1.5638^{20}_{4}	1.4466^{20}		s	s				**B2**², 671
b3084	**2-Butyne-1,4-diol***	$HOCH_2C\vdots CCH_2OH$	86.09		58	140^{10}			v	v	δ	v	i	chl δ	**B1**¹, 261
b3085	**3-Butyne-1,2-diol***	$HC\vdots CCHOHCH_2OH$	86.09		40				s	s	δ				**B1**², 569
b3086	**2-Butynoic acid***	Tetrolic acid. $CH_3C\vdots CCO_2H$	84.07	pl (eth or CS_2)	77–8	203^{760} $99–100^{18}$	0.9641^{20}_{4}		v	v					**B2**², 451
b3087	—,ethyl ester*	$CH_3C\vdots CCO_2C_2H_5$	112.12			163^{760}	0.9641^{20}_{4}								**B2**¹, 208
b3088	**2-Butyn-1-ol***	$CH_3C\vdots CCH_2OH$	70.09		−2.2	143^{760}	0.9373^{20}_{4}	1.4530^{20}		s	∞				**B1**³, 1973
b3089	**3-Butyn-1-ol***	$HOCH_2CH_2C\vdots CH$	70.09		−63.6	129^{760}	0.9257^{20}_{4}	1.4409^{20}	s	v				MeOH v	**B1**³, 1972
b3090	**3-Butyn-2-ol**	$CH\vdots CCHOHCH_3$	70.09			107^{760}	0.8948^{20}	1.4526^{20}	s	v	s				**B1**³, 1971
b3091	—,**2-methyl-***	$(CH_3)_2COHC\vdots CH$	84.12		−3	102–4	0.8678^{16}_{4}	1.4187^{23}	s	s					**B1**², 505
—	**Butyraldehyde**	*see* **Butanal***													
—	**Butyramide**	*see* **Butanoic acid,** amide													
—	**Butyric acid**	*see* **Butanoic acid***													
—	**Butyroin**	*see* **4-Octanone, 5-hydroxy-***													
—	**Butyrolactone**	*see* **Butanoic acid, hydroxy-,** lactone*													
—	**Butyrophenone**	*see* **1-Butanone, 1-phenyl-***													

For explanations, symbols and abbreviations see beginning of table.

No.	Name	Synonyms and Formula	Mol. wt.	Crystalline form, color and specific rotation	m.p. °C	b.p. °C	Density	n_D	Solubility w	al	eth	ace	bz	other solvents	Ref.
	Cacodyl														
c1	**Cacodyl**	Diarsenic tetramethyl. Dimethylarsenic. Tetramethylbiarsine. $(CH_3)_2AsAs(CH_3)_2$	209.94		−5	163^{760}	1.447^{15}			s	s				**B4**^{2}, 1002
c2	**Cacodyl chloride**	Dimethylarsenic monochloride. Dimethylchlorarsine. $(CH_3)_2AsCl$	140.44		<−45	109	1.5046^{12}_4	1.5203^{12}	i	s	i				**B4**^{2}, 987
c3	**Cacodyl oxide**	Alkarsine. Bis-dimethylarsenic oxide. Dicacodyl oxide. $(CH_3)_2AsOAs(CH_3)_2$	225.94		−25	150^{760}	1.4943^9_4	1.5255^9	δ	s	s				**B4**^{2}, 989
c4	—,**perfluoro-**	Bis(trifluoromethyl) arsenous oxide. $(CF_3)_2AsOAs(CF_3)_2$	441.87			100^{760}		1.3540^{20}							**J 1953,** 1561
c5	**Cacodyl sulfide**	Bis-dimethylarsine sulfide. Dicacodyl sulfide. $(CH_3)_2AsSAs(CH_3)_2$	242.05		<−40	211^{760}			δ	s	s				**B4,** 608
c6	**Cacodyl trichloride**	Dimethylarsenic trichloride. Dimethylorthoarsenic acid trichloride. $(CH_3)_2AsCl_3$	211.35	(eth)	50d				d	d	s			CS_2 s	**B4,** 612
—	**Cacodylic acid**	*see* **Dimethylarsinic acid**													
—	**Cadalene**	*see* **Naphthalene, 1,6-dimethyl-4-isopropyl-**													
—	**Cadaverine**	*see* **Pentane, 1,5-diamino-***													
c7	**Cadinene**(*l*)	$C_{15}H_{24}$	204.36	$[\alpha]^{20}_D -130$		274^{760} 149^{20}	0.9176^{20}_4	1.5032^{20}	i	δ	s			chl s^h lig s	**B5**^{2}, 347
—	**Caffeic acid**	*see* **Cinnamic acid, 3,4-dihydroxy-**													
c8	**Caffeine**	Theine. 1,3,7-Trimethylxanthine. 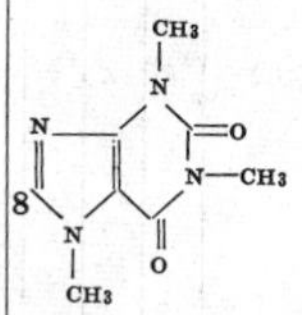	194.19	wh nd (w+1)	237	sub 178 sub 89^{15}	1.23^{19}		δ v^h	δ	i			chl s^h	**B26**^{2}, 266
c9	—,benzoate	$C_8H_{10}N_4O_2.C_6H_5COOH$. *See* c8	316.32	wh so					s	s					**B26**^{2}, 268
c10	—,citrate	$C_8H_{10}N_4O_2.C_6H_8O_7$. *See* c8	386.32	mcl					d	d					**B26**^{2}, 269
c11	—,hydrobromide	$C_8H_{10}N_4O_2.HBr.2H_2O$. *See* c8	311.14	ye	170				s	d	i			chl i	**B26**^{1}, 137
c12	—,hydrochloride	$C_8H_{10}N_4O_2.HCl.2H_2O$. *See* c8	266.69	mcl pr	80 (to) 100				d	d					**B26**^{2}, 268
c13	—,hydroiodide diiodide	Caffeine iodide. Diiodocaffeine hydroiodide. Caffeine triiodide. $C_8H_{10}N_4O_2.I_2.HI.1\frac{1}{2}H_2O$. *See* c8	602.94	gr pr	171d				d	v	i				**B26,** 466
c14	—,2-hydroxybenzoate	Caffeine salicylate. $C_8H_{10}N_4O_2.C_7H_6O_3$. *See* c8	332.32	wh nd (w)	137				δ	s					**B26**^{2}, 269
c16	—,3-methylbutanoate	Caffeine isovalerate. $C_8H_{10}N_4O_2.C_5H_{10}O_2$. *See* c8	296.32	unst nd					s						**B26,** 467
c17	—,sulfate	$C_8H_{10}N_4O_2.H_2SO_4$. *See* c8	292.27	wh nd					d	d					**B26,** 466
c18	—,**8-ethoxy-**	1,3,7-Trimethyl-2,6-dioxo-8-ethoxy-purine. $C_{10}H_{14}N_4O_3$. *See* c8	238.24	wh or yesh nd (w)	143				δ	v^h	δ				**B26**^{2}, 322
c19	—,**8-methoxy-**	$C_9H_{12}N_4O_3$. *See* c8	224.22	wh nd (al or w)	176				s^h	v			s		**B26**^{2}, 322
c20	**Calciferol**	Irradiated ergosterol. Vitamin D2. $C_{28}H_{44}O$	396.64	pr (ace) $[\alpha]^{20}_D +120.5$ (al), +81 (ace)	115–6				i	s	s	s			**E12A,** 170
c21	**Camphane**	Bornylane. Dihydrocamphene 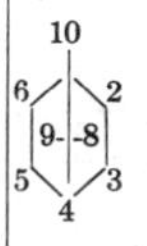	138.25	hex pl (al) pr (MeOH)	158–9	sub 161			i	s^h	s			MeOH s^h	**E12A,** 570
—	—,**2,3-dioxo-**	*see* **Camphorquinone**													

For explanations, symbols and abbreviations see beginning of table.

No.	Name	Synonyms and Formula	Mol. wt.	Crystalline form, color and specific rotation	m.p. °C	b.p. °C	Density	n_D	Solubility w	al	eth	ace	bz	other solvents	Ref.
	3-Camphane-carboxylic acid														
c22	**3-Camphane-carboxylic acid**	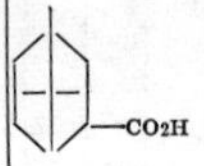	182.27	pl (dil aa) $[\alpha]^{20}_D$+56 (al)	90–1	153^{13}				s				aa v	E12A, 942
—	**2,3-Camphanedione**	*see* **Camphorquinone**													
c23	**Camphene**(*d*)	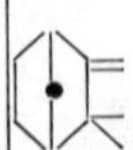	136.23	nd $[\alpha]^{17}_D$+103.9 (eth), +104.7 (al)	45–6	158^{760}	0.8446^{50}_4	1.4564^{b}	i	δ	s				E12A, 551
c24	—(*dl*)	$C_{10}H_{16}$. *See* c23	136.23	nd	50	$160–2^{760}$	0.879^{20}_4	1.4412	i	v	v				E12A, 551
c25	—(*l*)	$C_{10}H_{16}$. *See* c23	136.23	$[\alpha]^{19}_D$−80.7	45–6	158^{760}	0.8446^{50}_4	1.4451^{45}			s				E12A, 551
c26	**Camphenilone** (*d*)	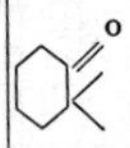	138.20	$[\alpha]^{20}_D$+70.4 (al)	39	193^{751} 78^{12}			i		s				E12A, 716
—	**Camphenol**	*See* **1-Epiborneol**													
—	**Camphol**	*See* **Isoborneol**													
c27	**Campholic acid** (*d*)	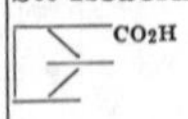	170.25	$[\alpha]^{20}_D$+59.3 (bz)	106	255^{758}			δ	s					$B9^2$, 17
c28	—(*dl*)	$C_{10}H_{18}O_2$. *See* c27	170.25	tcl pr	109				i	v	s				$B9^2$, 17
c29	—(*l*)	$C_{10}H_{18}O_2$. *See* c27	170.25	pr (dil al)	106.7	250^{760}				s					B9, 36
c30	Campholytic acid (α,*l*)	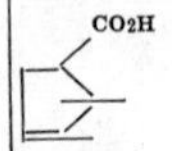	154.20	$[\alpha]^{18}_D$−59.6		140^{15}	1.0145^{18}	1.4712^{17}						lig v	$B9^1$, 33
c31	**Camphor** (*d*)	*d*-2-Camphanone. Formosa camphor. Laurel camphor. $C_{10}H_{16}O$ 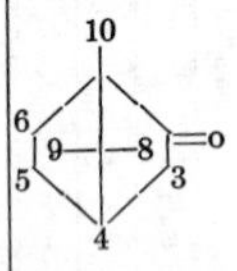	152.23	pl $[\alpha]^{20}_D$+44.26 (al)	176	209^{760} sub	0.990^{25}_4	1.5462	δ	v	v	s	s	chl v	$B7^2$, 93
c32	—(*dl*)	$C_{10}H_{16}O$. *See* c31	152.23	wh	174	sub 179									$B7^2$, 104
c33	—(*l*)	$C_{10}H_{16}O$. *See* c31	152.23	$[\alpha]^{16}_D$−43.6 (al, c=16.5)	178	204	0.9853^{18}								$B7^2$, 103
c34	**Camphor**, oxime (*d*)	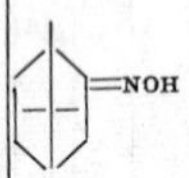	167.25	pr (lig-eth) $[\alpha]^{22}_D$+41.7 (al)	115										$B7^1$, 84
c35	—,—(*dl*)	$C_{10}H_{17}NO$. *See* c34	167.25	(peth)	118										$B7^1$, 85
c36	—,—(*l*)	$C_{10}H_{17}NO$. *See* c34	167.25	mcl nd (dil al) $[\alpha]^{20}_D$−42.4 (al)	118	249–54d	1.01^{116}_4		i	v	s				$B7^2$, 104
c37	—,**3-amino-**(*d*)	3-Aminocamphor. 3-Camphorylamine. $C_{10}H_{17}NO$. *See* c31	167.25	wx	110	244			i	s	s				$B14^2$, 6
c38	—,**3-bromo-**(*d*) (one stereoisomer)	$C_{10}H_{15}BrO$. *See* c31	231.14	nd (dil al) $[\alpha]^{20}_D$+29.4	78	265d				v				chl v CS_2 v aa v	$B7^2$, 101
c39	—,—(*d*) (one stereoisomer)	$C_{10}H_{15}BrO$. *See* c31	231.14	pr (al) $[\alpha]^{20}_D$+129.3 (MeOH, c=4.6)	76	274d	1.449^{20}_4		i	s	v		s	chl s	$B7^2$, 100
c40	—,—(*dl*)	$C_{10}H_{15}BrO$. *See* c31	231.14		51					v					$B7^2$, 105
c41	—,—(*l*)	$C_{10}H_{15}BrO$. *See* c31	231.14	mcl nd (al) $[\alpha]^{18}_D$−138.8 (ace, c=6)	76										$B7^2$, 104

For explanations, symbols and abbreviations see beginning of table.

No	Name	Synonyms and Formula	Mol. wt.	Crystalline form, color and specific rotation	m.p. °C	b.p. °C	Density	n_D	Solubility w	al	eth	ace	bz	other solvents	Ref.
	Camphor														
c42	—,**10-bromo-**(*d*)	$C_{10}H_{15}BrO$. *See* c31	231.14	pr (peth) $[\alpha]^{20}_D+19.1$ (abs al)	79	265d			i	v^h	v	...	v	chl v CS_2 v	**B7**², 101
c43	—,—(*dl*)	$C_{10}H_{15}BrO$. *See* c31	231.14		77										**B7**², 105
c44	—,**8-bromo-**(*d*)	$C_{10}H_{15}BrO$. *See* c31	231.14	tetr pr (lig) $[\alpha]^{19}_D+122.2$ (chl)	93	sub									**B7**, 123
c45	—,—(*dl*)	$C_{10}H_{15}BrO$. *See* c31	231.14	pr (eth-peth), pym (eth)	93	sub					v^h	...		peth s^h os s	**B7**, 136
c46	—,**3-chloro-**(*d*) (one stereoisomer)	$C_{10}H_{15}ClO$. *See* c31	186.68	lf $[\alpha]^{20}_D+71.1$ (bz, c=9.2)	94	244–7d			s^h	s^h	s	...	s	CS_2 s chl s	**B7**², 100
c47	—,—(*d*)(one stereoisomer)	$C_{10}H_{15}ClO$. *See* c31	186.68	$[\alpha]^{20}_D+35$ (al, c=5)	117	231d			δ	δ	s	...	...	CS_2 s chl s	**B7**², 100
c48	—,**10-chloro-**(*d*)	$C_{10}H_{15}ClO$. *See* c31	186.68	pr (al) $[\alpha]^{19}_D+40.7$ (al)	131–2								s	chl s peth s aa s	**B7**², 100
c49	—,**8-chloro-**(*d*)	$C_{10}H_{15}ClO$. *See* c31	186.68	pr $[\alpha]_D+99.9$ (chl)	139										**B7**, 136
c50	—,—(*dl*)	$C_{10}H_{15}ClO$. *See* c31	186.68		138	sub								os v	**B7**, 136
c51	—,**3,3-dibromo-** (*d*)	$C_{10}H_{14}Br_2O$. *See* c31	310.04	wh-ye rh pr (al, peth) $[\alpha]^{20}_D$ +40 (chl) +39.2 (al)	64	δ sub δ d	1.854		i	v	v	...	v	chl v peth v	**B7**², 101
c52	—,**3-nitro-**(*l*)	$C_{10}H_{15}NO_3$. *See* c31	197.23	mcl (bz) $[\alpha]^{15}_D$ −26 to −9 mut (al)	104				i	s	s	...	s	peth δ chl s	**B7**², 103
—	**Camphoramic acid**	*See* **Camphoric acid,** monoamide													
c53	**3-Camphorcarboxylic acid** (*d*)		196.25	pr (eth, 50 % al) $[\alpha]^{20}_D+34.9$ (bz)	128d				s	v	s	...	v	lig δ	**E12A**, 264
c54	—(*dl*)	$C_{11}H_{16}O_3$. *See* c53	196.25	(bz)	124				s	v	s^h	...	v^h	AcOEt s	**E12A**, 964
c55	—(*l*)	$C_{11}H_{16}O_3$. *See* c53	196.25	(bz) $[\alpha]^{20}_D$ −64 (al) −57.4 (AcOEt)	128				s	v	s^h	...	v^h	AcOEt s	**E12A**, 964
c56	**Camphoric acid** (*d*)		200.23	pr lf (w^h) $[\alpha]^{20}_D+47.7$ (al)	188		1.186^{20}_4		s^h	v	v	s	...		**B9**², 534
c57	—(*dl*)	$C_{10}H_{16}O_4$. *See* c56	200.23	pr (al, aa) mcl nd	202		1.228^{20}_4		v^h	s	v	...	...	chl δ	**B9**¹, 332
c58	—(*l*)	$C_{10}H_{16}O_4$. *See* c56	200.23	(w) $[\alpha]^{16}_D-48.1$ (abs al, c=8)	187		1.190		δ	s	s	s	...	MeOH s aa δ	**B9**², 539
c59	**Camphoric acid,** anhydride (*d*)		182.21	rh (al), pr (bz) $[\alpha]_D-7$ (bz)	222	270d	1.194^{20}		δ	δ	δ	...	v	chl δ	**B17**¹, 238
c60	—,—(*dl*)	$C_{10}H_{14}O_3$. *See* c59	182.21	rh (al)	221	270	1.194^{20}_4		δ	δ	δ	...	s	AcOEt s CS_2 s chl δ	**B17**, 459
c61	—,—(*l*)	$C_{10}H_{14}O_3$. *See* c59	182.21		221	>270				δ			s		**B17**, 459
c62	—,diethyl ester (*d*)	$C_{14}H_{24}O_4$. *See* c56	256.33	$[\alpha]^{15}_D+7.5$ (al), +9 (bz)		286^{752}	1.0298^{20}_4	1.4535^{26}	i	s	s	...	s	CCl_4 s	**B9**², 536
c62¹	—,dimethyl ester (*d*)	$C_{12}H_{18}O_4$. *See* c56	228.28		<−16	264^{738}	1.0747^{20}_4	1.4627^{19}	i	v	s	...	...		**B9**², 535
c63	—,1-monoamide (*dl*)	β-Camphoramic acid.	199.24	pl, nd $[\alpha]^{20}_D+60$ (6 % al)	183				s^h	s^h	δ	s	δ	lig δ	**B9**², 536

For explanations, symbols and abbreviations see beginning of table.

No.	Name	Synonyms and Formula	Mol. wt.	Crystalline form, color and specific rotation	m.p. °C	b.p. °C	Density	n_D	Solubility w	al	eth	ace	bz	other solvents	Ref.
	Camphoric acid														
c64	—,3-monoamide (*d*)	α-Camphoramic acid. (structure: CO_2H, $CONH_2$)	199.24	nd lf (w) $[\alpha]_D^{20}+45$ (6% al)	176–7				s^h	s^h	s^h	s^h	δ^h	MeOH s^h	**B9**[2], 536
c65	—,—(*dl*)	$C_{10}H_{17}NO_3$. *See* c64	199.29		198										
c67	**Camphoronic acid** (*d*)	$(CH_3)_2C(CO_2H)C(CH_3)(CO_2H)CH_2CO_2H$	218.20	nd (w) $[\alpha]^{19}+27.05$	159d				v	...	v	...	...		**B2**, 837
c68	—(*dl*)	$(CH_3)_2C(CO_2H)C(CH_3)(CO_2H)CH_2CO_2H$	218.20		172d				δ	...					**B2**, 839
c69	—(*l*)	$(CH_3)_2C(CO_2H)C(CH_3)(CO_2H)CH_2CO_2H$	218.20	nd (w) $[\alpha]_D^{19}-26.9$	164				v	v	s	v	δ	chl v	**B2**[2], 690
c70	**Camphor pinacol** (*l*)	(structure: OH HO)	306.48	rh $[\alpha]_D-27.2$ (bz)	158				i	s	s	...			
c71	**Camphor quinone**	3-Ketocamphor. (structure: =O, =O)	166.21	ye nd (dil al, w), pr (eth) $[\alpha]_D^{20}$ −113.2 (bz), −105.4 (chl)	199	sub			s^h	v	v	...	s	chl s	**B7**, 581
c72	**3-Camphorsulfonic acid**, methyl ester (*d*)	(structure: =O, SO_3CH_3)	246.33	(MeOH), nd (peth) $[\alpha]_D^{20}$ +98.6 (chl, c=4)	76	201[20]			i	v	...	...	...	os s MeOH v	**B2**, 179
c73	**10-Camphorsulfonic acid** (*d*)	(structure: CH_2SO_3H, =O)	232.30	pr (aa) $[\alpha]_D^{20}$ +32.8 (AcOEt c=3)	195d				v	...	i	...	...	aa δ	**B2**[2], 180
c74	—(*dl*)	$C_{10}H_{16}O_4S$. *See* c73	232.30	(aa)	202d				v	...	...	...	...	aa δ	**B2**[2], 182
c75	—(*l*)	$C_{10}H_{16}O_4S$. *See* c73	232.30	(aa), nd (AcOEt) $[\alpha]_D^{20}$ −20.75 (w)	194				v	...	i	...	...	AcOEt δ aa δ	**B2**[2], 182
c76	**α-Camphylamine**	(structure: $CH_2CH_2NH_2$)	153.26	$[\alpha]_D^{20}+3.83$		202–4	0.8688[20]	1.4728[18]	...						**B12**[2], 35
c77	**β-Camphylamine**	(structure: $CH_2CH_2NH_2$)	153.26	$[\alpha]_D+6$		206	0.8736_4^{18}	1.4728[18]	...						**B12**, 40
c78	**Canadine** (*d*)	*d*-Tetrahydroberberine. (structure: O—CH_2—O, CH_3O, CH_3O, N)	339.38	ye nd (dil al) $[\alpha]_D^{20}+297.4$ (chl, c=1)	140										**B27**[2], 557
c79	—(*dl*)	$C_{20}H_{21}NO_4$. *See* c78	339.38	mcl nd (al)	174				δ	δ	...	...	...	chl v CS_2 v	**B27**[2], 557
c80	—(*l*)	$C_{20}H_{21}NO_4$. *See* c78	339.38	ye nd (al) $]\alpha]_D^{20}$ −299.2 (chl)	134				i	v	s	...	s	chl s lig δ^h	**B27**[2], 557
c81	**Canaline**	$H_2NOCH_2CH_2CH(NH_2)CO_2H$	134.14	nd (al) $[\alpha]_D^{21}-8.31$	214d				...	s^h	...				
c82	**Canavanine**	$NH{:}C(NH_2)NHOCH_2CH_2(NH_2)CO_2H$	176.18	(al) $[\alpha]_D^{20}+7.9$ (w)	184										
—	**Canescine**	*see* **Deserpidine**													
c83	**Cannabidiol**	(structure: CH_3, OH, $(CH_2)_4CH_3$, $C_3H_5^{\prime}OH$)	314.45		67	187–190[2]			i	s	s	...	s	chl s	

For explanations, symbols and abbreviations see beginning of table.

No.	Name	Synonyms and Formula	Mol. wt.	Crystalline form, color and specific rotation	m.p. °C	b.p. °C	Density	n_D	Solubility w	al	eth	ace	bz	other solvents	Ref.
	Cannabinol														
c84	**Cannabinol**	CH_3; H_3C; OH; H_3C; $(CH_2)_4CH_3$	310.42	pl, lf (peth) $[\alpha]^{20}_D -148$ (al)	77	185^{105}			i	s				alk s	
c85	**Cantharidin**	$C_{10}H_{12}O_4$	196.22	rh pl	218	sub 84			i	i	δ	δ	δ	aa s	**E19²**, 179
—	**Capraldehyde**	*see* **Decanal***													
—	**Capric acid**	*see* **Decanoic acid***													
—	**Caproaldehyde**	*see* **Hexanal***													
—	**Caproic acid**	*see* **Hexanoic acid***													
—	**ε-Caprolactam**	*see* **Hexanoic acid, 6-amino-**, lactam*													
—	**Caprophenone**	*see* **1-Hexanone, 1-phenyl-***													
—	**Caprylaldehyde**	*see* **Octanal***													
—	**Caprylic acid**	*see* **Octanoic acid***													
c86	**Capsaicin**	$C_{18}H_{27}NO_3$	305.42	pl sc (peth)	64	$210^{0.01}$			i	v	s		s	peth s	**B13²**, 482
—	**Captan**	*see* **4-Cyclohexene-1,2-dicarboxylic acid**, imide, *N*(trichloromethylthio)-													
c87	**Carbamic acid,** benzyl ester	$NH_2CO_2CH_2C_6H_5$	151.17	pl (to), lf (w)	91	220d			$δ^h$	v	δ			to s	**B6**, 437
c88	—,4-benzylphenyl ester	NH_2CO_2–C_6H_4–CH_2–C_6H_5	227.27		145				$δ^h$	v^h			v	NaOH s	**B6²**, 630
c89	—,butyl ester	$NH_2CO_2C_4H_9^n$	117.15	pr	54	204d				v					**B3²**, 25
c90	—,chloride	NH_2COCl	79.49			62^{760}			d	d					**B3¹**, 15
c92	—,ethyl ester	Urethane. $NH_2CO_2C_2H_5$	89.09		48	185^{760}	0.9862^{21}_4	1.4144^{52}	v	v	v		v	chl v lig δ	**B3²**, 19
c93	—,isobutyl ester	$NH_2CO_2C_4H_9^i$	117.15	lf	67	207		1.4098^{76}	i	s	s				**B3²**, 26
c94	—,isopropyl ester	$NH_2CO_2C_3H_7^i$	103.12	nd	93	181^{711}	0.9951^{66}								**B3²**, 25
c95	—,methyl ester	Urethytan. $NH_2CO_2CH_3$	75.07	nd	54	177	1.1361^{56}_4	1.4125^{56}	v	v	s				**B3²**, 18
c95[1]	—,2-methyl-2-butyl ester	*tert*-Amyl carbamate. $NH_2CO_2C(CH_3)_2C_2H_5$	131.18	nd (dil al)	85–7				δ	δ				os v lig δ	**B3¹**, 14
c96	—,3-methylbutyl ester	$NH_2CO_2CH_2CH_2CH(CH_3)_2$	131.18	nd (w^h)	64	220^{760}	0.9438^{71}_4	1.4175^{71}	s^h	s	s				**B3²**, 26
—	—,nitrile	*see* **Cyanamide**													
c97	—,propyl ester	$NH_2CO_2CH_2CH_2CH_3$	103.12	pr	60	196			s	s	s				**B3²**, 25
c98	—,**N-benzyl-,** ethyl ester	$C_6H_5CH_2NHCO_2C_2H_5$	179.22	lf (lig)	49	230δd			δ	v	v		v	chl v lig s^h	**B12²**, 563
c99	—,**N-benzyl-N-nitro-,** ethyl ester	$C_6H_5CH_2N(NO_2)CO_2C_2H_5$	224.21	ye		d	1.213^{20}_{20}	1.5203^{20}		∞	∞				
c100	—,**N-butyl-,** butyl ester	$C_4H_9^nNHCO_2C_4H_9^n$	173.26			88^3	0.9238^{20}_{20}	1.4359^{20}		∞	∞				
c101	—,**N-butyl-N-nitro-,** butyl ester	$C_4H_9^nN(NO_2)CO_2C_4H_9^n$	218.25			98^3	1.048^{20}_{20}	1.448^{20}		∞	∞				**Am 73**, 5549
c102	—,**N-tert-butyl-N-nitro-,** ethyl ester	$C_4H_9^tN(NO_2)CO_2C_2H_5$	190.20			56^2d	1.051^{20}_{20}	1.4331^{20}		∞	∞				**Am 73**, 5449
c103	—,**N-cyclohexyl (N-ethyl) dithio-,** hexyl(ethyl)-, ammonium salt	$CS_2HN_2(CH_2)_5CH_3$; C_2H_5; C_2H_5	330.59	ye	93				s^h	v	s				**B12²**, 12
c104	—,**N-dibenzyldithio-,** dibenzylammonium salt	$(C_6H_5CH_2)_2NCS_2NH_2(CH_2C_6H_5)_2$	470.70	ye (eth)	82				s	v					**B12**, 1058
c105	—,**N,N-diethyl-**	$(C_2H_5)_2NCO_2H$	117.15	nd (eth)	−15d	171	0.9276^{20}_4	1.4206^{20}	v	v	δ				**B4²**, 611
c106	—,**N,N-diethyl-,** chloride	$(C_2H_5)_2NCOCl$	135.59			187–90			d^h	d^h					**B4²**, 611
c107	—,**N,N-diethyldithio-,** benzylidene ester	$[(C_2H_5)_2NCS_2]_2CHC_6H_5$	386.67	ye	110				i	s^h					
c108	—,—,diethylammonium salt	$(C_2H_5)_2NCS_2NH_2(C_2H_5)_2$	222.42	ye pl	82				v	v	δ				**B4**, 121

For explanations, symbols and abbreviations see beginning of table.

No.	Name	Synonyms and Formula	Mol. wt.	Crystalline form, color and specific rotation	m.p. °C	b.p. °C	Density	n_D	Solubility w	al	eth	ace	bz	other solvents	Ref.
	Carbamic acid														
c109	—,*N*,*N*-**diethyl-thio-**, chloride	$(C_2H_5)_2NCSCl$	151.66	pr	46	108^{10}									**B4**, 121
c110	—,*N*,*N*-**dimethyl-dithio-**, dimethyl-ammonium salt	$(CH_3)_2NCS_2NH_2(CH_3)_2\cdot\frac{1}{2}C_6H_6$	207.37	ye pl (bz+$\frac{1}{2}$)	125	142			v	v	δ				**B4²**, 577
c111	—,—,2,4-dinitro-phenyl ester	$(CH_3)_2NCS_2$—⬡—NO_2 (O_2N on ring)	287.32	ye	139		1.54^{20}_{4}		i	s^h		s	s		
c112	—,*N*,*N*-**dimethyl-thio-**, chloride	$(CH_3)_2NCSCl$	123.60	pr	42	98^{10}					v			peth s chl s	**B4²**, 576
c113	—,*N*,*N*-**diphenyl-**, chloride	$(C_6H_5)_2NCOCl$	231.68	lf (al)	85										**B12²**, 241
c114	—,—,ethyl ester	Diphenylurethan. $(C_6H_5)_2NCO_2C_2H_5$	241.29	pr (lig)	74	360^{760}			s	δ	s		s	peth s	**B12²**, 240
c115	—,**dithio-**	Aminodithioformic acid. Aminomethanethionothiolic acid* NH_2CS_2H	93.17	nd					d	v	v				**B3²**, 155
c116	—,*N*-**ethyl-**, butyl ester	$(C_2H_5)NHCO_2C_4H_9^n$	145.20			66^3	0.9413^{20}_{20}	1.4301^{20}		∞	∞				
c117	—,—,ethyl ester	Ethylurethan. $C_2H_5NHCO_2C_2H_5$	117.15			176^{760} 75^{14}	0.9813^{20}_{4}	1.4219^{20}	s						**B4²**, 607
c118	—,*N*-**ethyl-*N*-nitro-**, butyl ester	$C_2H_5N(NO_2)CO_2C_4H_9^n$	190.20			79^3	1.091^{20}_{20}	1.4455^{20}		∞	∞				
c119	—,—,ethyl ester	$C_2H_5N(NO_2)CO_2C_2H_5$	162.15			107^{31}	1.163^{20}_{20}	1.4432^{20}		∞	∞				
c119¹	—,—,methyl ester	$C_2H_5N(NO_2)CO_2CH_3$	148.12			72^{11}	1.233^{20}_{20}	1.4483^{20}		∞	∞				
c120	—,*N*-**ethyl-idenedi-**diethyl ester	Ethylidenediurethan. $CH_3CH(NHCO_2C_2H_5)_2$	204.33	nd (eth)	126	170–8^{20}			v	v	δ	v	δ	to δ lig δ	**B3¹**, 11
c121	—,*N*-**isobutyl-**, ethyl ester	Isobutylurethan. $C_4H_9^iNHCO_2C_2H_5$	145.20		>−65	96^{17}	0.9432^{20}_{4}	1.4288^{20}	δ						**B4²**, 640
c122	—,*N*-**isopropyl-**, ethyl ester	Isopropylurethan. $C_3H_7^iNHCO_2C_2H_5$	131.18			64^7	0.9548^{20}_{20}	1.4229^{20}		∞	∞				
c123	—,*N*-**isopropyl-*N*-nitro-**, ethyl ester	$C_3H_7^iN(NO_2)CO_2C_2H_5$	176.17			72^7	1.112^{20}_{20}	1.4381^{20}		∞	∞				**Am 73**, 5449
c124	—,*N*-**methyl-**, ethyl ester	Methylurethan. $CH_3NHCO_2C_2H_5$	103.12			170^{760}	1.035^{15}	1.4200^{18}	v	s					**B4²**, 567
c124¹	—,*N*-**methyl-*N*-phenyl-**, chloride	$CH_3N(C_6H_5)COCl$	169.61		88	280				v	v		s^h		**B12²**, 235
c126	—,*N*-**nitro-**, ethyl ester	$NO_2NHCO_2C_2H_5$	134.09	lf (lig)	64	140d	1.0074^{20}_{4}		v	v	v			lig δ	**B3²**, 99
c127	—,*N*-**nitro-*N*-propyl-**, ethyl ester	$C_3H_7^nN(NO_2)CO_2C_2H_5$	176.17			66^3	1.123^{20}_{20}	1.4431^{20}		∞	∞				**Am 73**, 5449
c128	—,—,methyl ester	$C_3H_7^nN(NO_2)CO_2CH_3$	162.15				1.1585^{15}_{15}			∞	∞				**B4**, 146
c129	—,*N*-**phenyl-**, ethyl ester	$C_6H_5NHCO_2C_2H_5$	165.19	wh nd (w)	52	238^{760}	1.1064^{30}_{4}	1.5376^{30}	s^h	s	s		s		**B12²**, 184
c130	—,—,isobutyl ester	$C_6H_5NHCO_2C_4H_9^i$	193.25	nd	86	216^{760}			δ	v	v				**B12¹**, 219
c131	—,—,isopropyl ester	$C_6H_5NHCO_2C_3H_7^i$	179.22	wh cr	90		1.09^{20}	1.4989^{91}	i	s			s		
c132	—,—,propyl ester	$C_6H_5NHCO_2C_3H_7^n$	179.22	wh nd	57–9										**B12**, 321
c133	—,*N*-propyl-, ethyl ester	$C_3H_7^nNHCO_2C_2H_5$	131.18			192^{758}	0.9921^{15}		s						**B4²**, 626
c134	—,**thiolo-**, ethyl ester	$NH_2COSC_2H_5$	105.16	lf	108	sub d			s^h	s	s				**B3¹**, 64

For explanations, symbols and abbreviations see beginning of table.

No.	Name	Synonyms and Formula	Mol. wt.	Crystalline form, color and specific rotation	m.p. °C	b.p. °C	Density	n_D	Solubility						Ref.
									w	al	eth	ace	bz	other solvent	
	Carbamic acid														
c135	—,**thiono-**, ethyl ester	Thiourethan. Xanthogenamide. $NH_2CSOC_2H_5$	105.16	mcl lf	41	d	1.069^{20}_{4}	1.520^{20}	δ	∞	∞	...	...	...	**B3²**, 107
—	**Carbamyl chloride**	*see* **Carbamic acid,** chloride													
—	**Carbamyl guanidine**	*see* **Guanidine, 1-ureido-**													
—	**Carbanilic acid**	*see* **Carbamiic acid,** ***N*-phenyl-**													
—	**Carbanilide**	*see* **Urea, 1,3-diphenyl-**													
c136	**Carbazide**	$NH_2NHCONHNH_2$	90.09	nd (dil al)	154	...	1.616^{-5}	...	v	v	...	...	...	...	**B3²**, 96
c137	—,**1,5-diphenyl-**	$CO(NHNHC_6H_5)_2$	242.28	(al+1)	170	d	...	...	$δ^h$	s^h	δ	s^h	s	aa s	**B14²**, 107
c138	—,**1-phenyl-**	$C_6H_5NHNHCONHNH_2$	166.18	nd (al)	151	...	...	...	s	s^h	...	...	...	...	**B15²**, 107
c139	—,**1,1,5,5-tetraphenyl-**	$[(C_6H_5)_2NNH]_2CO$	394.48	(al), nd (aa)	242	...	...	...	...	v^h	...	...	...	aa v^h	**B15²**, 115
c140	—,**3-thio-**	$NH_2NHCSNHNH_2$	106.15	nd, pl (w)	170d	...	...	...	v^h	$δ^h$	i	...	...	alk v	**B3²**, 137
—	**Carbazine**	*see* **Acridine, 9,10-dihydro-**													
c141	**Carbazole**	Dibenzopyrrole. Diphenylenimine.	167.21	pl lf	247	355^{760}	...	...	i	s	δ	...	s^h	chl δ to δ	**B20²**, 279
c142	—,**9-acetyl-**	$C_{14}H_{11}NO$. *See* c141	209.25	(eth), nd (w)	69	190^{6}	1.161^{100}_{4}	1.640^{100}	$δ^h$	...	...	...	...	...	**B20²**, 283
c143	—,**9-benzoyl-**	$C_{19}H_{13}NO$. *See* c141	271.32	nd or pr (al)	98	...	...	...	...	...	...	...	...	...	**B20¹**, 165
c144	—,**9-benzyl-**	$C_{19}H_{15}N$. *See* c141	257.34	nd (al)	118–120	...	...	...	i	δ	...	...	v	...	**B20²**, 282
c145	—,**9-butyl-**	$C_{16}H_{17}N$. *See* c141	223.32	nd (al)	58	...	...	...	...	...	...	...	...	...	**B20¹**, 164
c146	—,**9-ethenyl-**	$C_{14}H_{11}N$. *See* c141	193.25	(al)	68	190^{10}	1.059^{80}_{4}	1.6292^{80}	...	...	...	...	...	...	**B20²**, 282
c147	—,**9-ethyl-**	$C_{14}H_{13}N$. *See* c141	195.27	...	68	190^{10}	1.059^{80}_{4}	1.6394^{80}	i	v^h	v	...	...	...	**B20²**, 282
c148	—,**9-methyl-**	$C_{13}H_{11}N$. *See* c141	181.24	nd, lf (al)	88	195^{12}	...	...	i	δ	v	...	...	...	**B20²**, 281
c149	—,**1-nitro-**	$C_{12}H_8N_2O_2$. *See* c141	212.21	ye nd (aa)	187	...	...	...	i	...	...	...	...	CS_2 s aa v	**B20²**, 288
c150	—,**3-nitro-**	$C_{12}H_8N_2O_2$. *See* c141	212.21	ye	214	...	...	...	i	...	δ	...	δ	peth δ	**B20²**, 288
c151	—,**3-nitro-9-nitroso-**	$C_{12}H_7N_3O_3$. *See* c141	241.21	ye nd (al)	169	...	...	...	...	s^h	...	...	...	chl s lig δ	**B20²**, 289
c152	—,**1-oxo-1,2,3,4-tetrahydro-**	$C_{12}H_{11}NO$. *See* c141	185.23	nd	170	...	...	...	...	s	...	...	s	aa s	**B21²**, 275
c153	—,**9-phenyl-**	$C_{18}H_{13}N$. *See* c141	243.31	nd (al)	95	...	...	...	...	v^h	v	...	v	aa v	**B20²**, 282
c154	—,**9-propionyl-**	$C_{15}H_{13}NO$. *See* c141	223.28	...	90	...	...	...	...	v	v	...	...	peth δ	**B20¹**, 165
c155	—,**9-propyl-**	$C_{15}H_{15}N$. *See* c141	209.29	nd (al)	50	...	...	...	...	...	...	...	...	...	**B20¹**, 164
c156	—,**1,2,3,4-tetrahydro-**	$C_{12}H_{13}N$. *See* c141	171.24	lf (dil al)	120	190^{10}	...	...	...	s	v	...	v	MeOH v	**B20²**, 257
c157	**Carbazone, 1,5-diphenyl-**	$C_6H_5N{:}NCONHNHC_6H_5$	240.27	og nd (bz), pr (al)	157d	...	...	...	i	v	...	...	v	chl v	**B16²**, 9
—	**Carbinol**	see **Methanol***													
—	**Carbitol**	*see* **Diethyleneglycol,** monoethyl ether													
—	α-**Carbocinchomeronic acid**	*see* **2,3,4-Pyridinetricarboxylic acid**													
c158	**Carbodiimide, diphenyl-**	$C_6H_5N{:}C{:}NC_6H_5$	194.24	...	168–170	331	...	...	δ	δ	δ	...	s	...	**B12²**, 246
—	**Carbohydrazide**	*see* **Carbazide**													
c159	**Carbon dioxide**	CO_2	44.01	...	−56.6 (at 5.2 atm)	−78.5 sub	0.00198	...	...	...	...	...	...	...	...
c160	**Carbon diselenide**	CSe_2	169.93	ye	−45	126^{760}	2.6626^{25}_{4}	1.845^{20}	...	...	...	...	...	...	**J 1947,** 1080
c161	**Carbon disulfide**	CS_2	76.14	...	−112	45^{760}	1.2628^{20}_{4}	1.6255^{20}	sl	∞	∞	...	...	chl s	**B3²**, 139
c162	**Carbonic acid,** bis(2-chloroethyl) ester*	$OC(OCH_2CH_2Cl)_2$	187.02	...	8	241	1.3506^{20}_{4}	1.461^{20}	i	...	...	...	...	...	**B3²**, 5
c163	—,bis(3-chloropropyl) ester*	$OC(OCH_2CH_2CH_2Cl)_2$	215.08	...	...	$265–70^{740}$	...	...	...	...	...	...	...	...	**B3²**, 5
c164	—,bis(2-ethoxyethyl) ester*	$OC(OCH_2CH_2OC_2H_5)_2$	206.24	...	...	245	1.0635^{25}_{4}	1.4239^{25}	...	v	v	v	...	...	...
c165	—,bis(2-methoxyethyl) ester*	$OC(OCH_2CH_2OCH_3)_2$	178.19	...	...	232	1.0936^{25}_{4}	1.4193^{25}	...	s	s	s	...	...	...

For explanations, symbols and abbreviations see beginning of table.

No.	Name	Synonyms and Formula	Mol. wt.	Crystalline form, color and specific rotation	m.p. °C	b.p. °C	Density	n_D	Solubility w	al	eth	ace	bz	other solvents	Ref.
	Carbonic acid														
c166	—,bis(2-methoxy-phenyl) ester*	OC (—O—[ring], CH_3O)$_2$	274.28		88				i	δ	δ				**B6**[2], 784
c166[1]	—,bis(3-methyl-butyl) ester	$OC[OCH_2CH_2CH(CH_3)_2]_2$	202.30			233	0.9067^{20}_{4}	1.4174^{20}							**B3**[2], 7
c167	—,bis(trichloro-methyl) ester*	$OC(OCCl_3)_2$	296.75	(eth, peth)	79	203^{760} δd					δ^h			peth δ^h	**B3**[1], 8
c168	—,dibutyl ester*	$OC(OC_4H_9^n)_2$	174.24			207^{760}	0.9238^{20}_{4}	1.4117^{20}	i	s	s				**B3**[2], 6
c169	—,di-*tert*-butyl ester*	$OC(OC_4H_9^t)_2$	174.24	(al)	40 sub	158			i	s^h					
—	—,dichloride*	*see* **Phosgene**													
c170	—,diethyl ester*	$OC(OC_2H_5)_2$	118.13		−43	126	0.9752^{20}_{4}	1.3845^{20}	i	s	s				**B5**[2], 4
—	—,dihydrazide	*see* **Carbazide**													
c171	—,diisobutyl ester	$OC(OC_4H_9^i)_2$	174.24			190^{760}	0.9138^{20}_{4}	1.4072^{20}	i	∞	∞				**B3**[2], 6
c172	—,diisopropyl ester*	$OC(OC_3H_7^i)_2$	146.19			147	0.9162^{20}_{4}	1.3932^{20}		s					**B3**[2], 6
c173	—,dimethyl ester*	$OC(OCH_3)_2$	90.08		1	90	1.0694^{20}_{4}	1.3687^{20}	i	s	s				**B3**[2], 3
c175	—,diphenyl ester*	$OC(OC_6H_5)_2$	214.22	nd (al)	80	312^{760}	1.1215^{87}_{4}		i	s^h	s			CCl_4 s aa s^h	**B6**[2], 156
c176	—,dipropyl ester*	$CO(OC_3H_7^n)_2$	146.19			168^{760}	0.9435^{20}_{4}	1.4008^{20}	δ	∞	∞				**B3**[2], 5
c177	—,di-2-tolyl ester	OC (—O—[ring], H_3C)$_2$	242.28	nd (al)	60					s^h				aa v	**B6**[2], 330
c178	—,di-3-tolyl ester	OC (—O—[ring]—CH_3)$_2$	242.28	(al)	111				i	s^h					**B6**[2], 353
c179	—,di-4-tolyl ester	OC (—O—[ring]—CH_3)$_2$	242.28	nd (al)	115				δ	δ					**B6**[2], 380
c179[1]	—,**dithiolo-**, diethyl ester	$OC(SC_2H_5)_2$	150.26	ye		197	1.085^{19}	1.5237^{18}	i	s	s				**B3**[1], 84
—	—,ethylene ester	*see* **1,3-dioxolan-2-one**													
c180	—,monoethyl mono-2-butoxyethyl ester	$C_2H_5OCOOCH_2CH_2OC_4H_9^n$	190.23			224	0.9756^{25}_{4}	1.4143^{25}		s	s	s			
c181	—,monoethyl monomethyl ester	$CH_3OCOOC_2H_5$	104.10		−14	109	1.012^{20}_{4}	1.3378^{20}	i	∞	∞				**B3**[2], 4
c183	—,**trithio-**	$SC(SH)_2$	110.21	red	−30	57d	1.47^{17}_{4}		d	s				chl, to s	**B3**[2], 161
c184	**Carbon monoxide**	CO	28.01		−205	-191^{760}	0.8142	0.00125 at 0°/1 atm					s	aa s	**B1**, 720
—	**Carbon suboxide**	*see* **Propadiene, 1,3-dioxo-***													
—	**Carbon tetra-bromide**	*see* **Methane, tetrabromo-***													
—	**Carbon tetra-chloride**	*see* **Methane, tetrachloro-***													
—	**Carbon tetra-fluoride**	*see* **Methane, tetrafluoro-***													
—	**Carbon tetra-iodide**	*see* **Methane, tetraiodo-***													
c185	**Carbonyl fluoride**	F_2CO	66.01		−114	-83^{760}			d	d					**B3**[2], 9
c186	**Carbonyl sulfide**	Carbon oxysulfide. OCS	60.07		−138	-50^{760}	0.0021 at 1 atm 1.24^{-87}								**B3**[2], 104
—	**Carbostyril**	*see* **Quinoline, 2-hydroxy-**													
c187	**Carbothialdine**	[ring: NH—CS, H_3C—CH, S, NH—CH—CH_3]	162.27	(al)	120d				i	v^h	i			ac s	**B27**[2], 687
—	**Cardiazol**	*see* **Metrazol**													
c188	α-Δ^3-**Carene**(*d*)	$C_{10}H_{16}$	136.24	$[\alpha]_D$ +17.7		172^{760}	0.861^{20}_{4}	1.473^{20}				s	s	aa s	**E12A**, 32

For explanations, symbols and abbreviations see beginning of table.

No.	Name	Synonyms and Formula	Mol. wt.	Crystalline form, color and specific rotation	m.p. °C	b.p. °C	Density	n_D	Solubility w	al	eth	ace	bz	other solvents	Ref.
	Δ³-Carene														
c189	**Δ³-Carene**(*l*)	$C_{10}H_{16}$	136.24	$[\alpha]_D^{30}$ −5.72	...	167^{685}	0.8606^{30}_{30}	1.4684^{30}	...	...	...	s	s	aa s	**E12A**, 34
c190	**Δ⁴-Carene**(*l*)	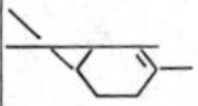 (structure)	136.24	$[\alpha]_D^{30}$ +62.2	...	167^{707}	0.8441^{30}_{30}	1.4717^{30}	...	...	...	s	...	aa s	**E12A**, 34
—	**Caricaxanthin**	*see* **Cryptoxanthin**													
c191	**Carminic acid**	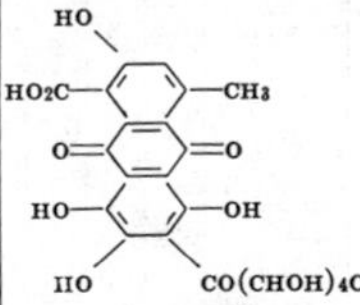 HO; HO_2C; CH_3; O; O; HO; OH; HO; $CO(CHOH)_4CH_3$ (structure)	492.38	red mcl pr $[\alpha]_{6450}^{15}$ +51.6(w)	136d	...	...	...	s	s	δ	...	i	chl i	**B10**^{2}, 776
c192	**Carnaubyl alcohol**	$CH_3(CH_2)_{29}CH_2OH$?	?	lf (al)	69	...	...	...	δ	s	...	...	...	...	**B1**^{2}, 472
—	**Carnegine**	*see* **Isoquinoline, 6,7-dimethoxy-1,2-dimethyl-1,2,3,4-tetrahydro-**													
c193	**Carnosine**(*D*,−)	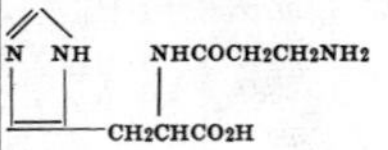 N NH; $NHCOCH_2CH_2NH_2$; CH_2CHCO_2H (structure)	226.23	$[\alpha]_D^{28}$ −20.4	260	...	...	...	s	...	...	...	...	...	**B25**^{2}, 408
c194	—,(*L*,+)	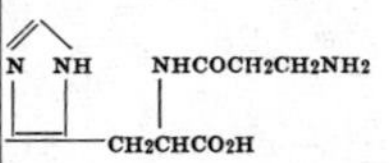 N NH; $NHCOCH_2CH_2NH_2$; CH_2CHCO_2H (structure)	226.24	nd (w-al) $[\alpha]_D^{20}$ +21(w)	246–50	...	...	...	v	i	...	...	...	...	**B25**^{2}, 408
c195	**α-Carotene**	α-Carotin. $C_{40}H_{56}$	536.89	red pl or pr (peth) $[\alpha]_{Cd}$ +380 (bz)	184	...	1.00^{20}_{20}	...	...	...	...	...	...	...	**B30**, 91
c196	**β-Carotene**	β-Carotin. $C_{40}H_{56}$	536.89	(bz-MeOH)	184	...	1.00^{20}_{20}	...	i	δ	δ	...	s	chl δ peth s	**B30**^{2}, 638
c197	**γ-Carotene**	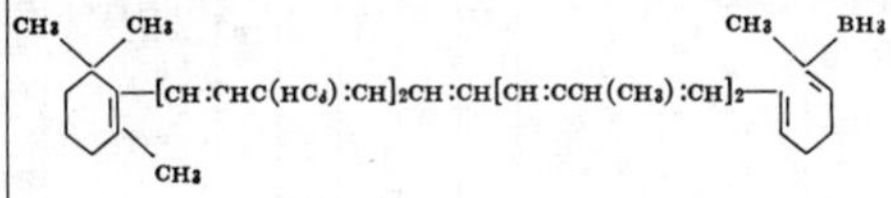	536.89	red pr (bz-MeOH)	177	...	...	...	i	i	δ	...	s	peth δ chl s	**B30**, 92
	CH_3 CH_3 —[CH:CHC(HC_4):CH]$_2$CH:CH[CH:CCH(CH_3):CH]$_2$— CH_3 BH_3; CH_3 (structure)														
—	**β-Carotene, dehydro-**	*see* **Isocarotene**													
c199	**Carpaine**(*d*)	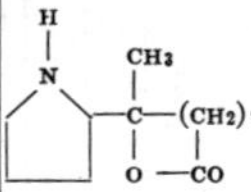 H; N; CH_3; C—$(CH_2)_7$; O —CO (structure)	239.36	mcl pr (al) $[\alpha]$+21.5 (al)	121	...	...	...	i	s	s	...	s	chl s	**B27**^{2}, 209
c200	hydrochloride(*d*)	$C_{14}H_{25}NO_2$.HCl. *See* c199	275.82	wh mcl nd	225d	...	...	...	v	s	s	...	...	...	**B27**^{2}, 210
c201	**Carpiline**	Carpidine. Pilosine. 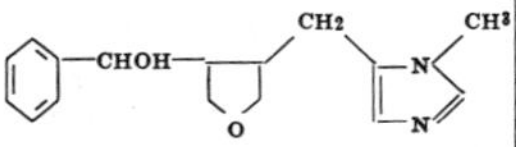 —CHOH—; CH_2; CH^3; N; N; O (structure)	286.33	pl (al), pr (dil al or w) $[\alpha]_D^{20}$ +35.9 (al)	187	...	...	...	δ	v^h	δ	δ	δ	chl δ dil ac v	**B27**^{1}, 612
—	**Carvacrol**	*see* **Benzene, 2-hydroxy-4-isopropyl-1-methyl-**													
—	*o*-**Carvacrotonic acid**	*see* **Benzoic acid, 2-hydroxy-6-isopropyl-3-methyl-**													
—	**Carvacrylamine**	*see* **Benzene, 2-amino-4-isopropyl-1-methyl-**													
c202	**Carvenone**(*dl*)	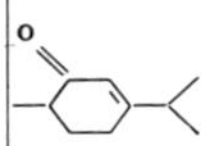 O (structure)	152.24	...	...	233	0.9263^{20}_{4}	1.4826^{20}	i	...	...	...	...	...	**B7**^{2}, 79
c203	—(*l*)	$C_{10}H_{16}O$. *See* c202	152.25	$[\alpha]_D$ −2.08	...	232–4	0.9290^{20}_{4}	1.4805	...	...	...	...	...	...	**B7**^{1}, 66
c204	**Carveol, dihydro-**(*d*)	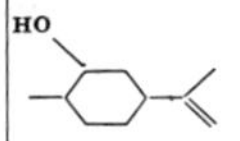 HO (structure)	154.25	$[\alpha]_D^{16}$ +34.2	...	225^{760}	0.9274^{20}_{4}	...	...	...	...	...	...	...	**B6**^{2}, 70
c205	—,—(*l*)	$C_{10}H_{18}O$. *See* c204	154.25	$[\alpha]_D^{20}$ −33.6	...	$89–90^{4}$	0.9368^{15}	1.4836^{20}	...	...	...	...	...	...	**B6**^{1}, 43

For explanations, symbols and abbreviations see beginning of table.

No.	Name	Synonyms and Formula	Mol. wt.	Crystalline form, color and specific rotation	m.p. °C	b.p. °C	Density	n_D	Solubility w	al	eth	ace	bz	other solvents	Ref.
	β-Carveol														
c206	**β-Carveol, dihydro-**	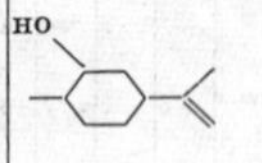 HO	154.25	$[\alpha]_D+7.64$		120^{20}	0.9266^{20}_{4}	1.4809^{20}							**B6**, 64
c207	**Carvomenthene**	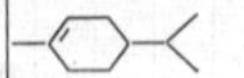	138.24			175^{760}	0.829^{20}_{4}	1.4601^{20}	i	s					**B5**², 52
c208	**Carvomenthol**	Hexahydrocarvacrol. 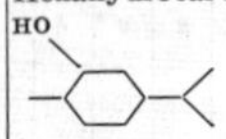 HO	156.26	$[\alpha]^{21}_D+31.40$		222	0.9056^{13}_{4}	1.4629^{17}		v	s				**B6**², 39
c209	—(*l*)	$C_{10}H_{20}O$. *See* c208	156.26			217–8	0.908^{20}_{4}								**B6**¹, 19
c210	**Carvomenthone**(*d*)	$C_{10}H_{18}O$	154.25	$[\alpha]^{21}_D+17.15$		$218–20^{745}$ $95–9^{15}$	0.9075^{15}	1.4544^{20}		v					**B7**², 37
c211	**Carvone**(*d*)	Carvol. $C_{10}H_{14}O$	150.21	$[\alpha]^{20}_D+69.1$		231^{760}	0.9608^{20}_{4}	1.4999^{18}	v^h	v	s			chl s	**B7**², 128
c212	—(*dl*)	$C_{10}H_{14}O$. *See* c211	150.21			230	0.9645^{15}_{15}	1.5003^{20}							**B7**², 130
c213	—(*l*)	$C_{10}H_{14}O$. *See* c211	150.21	$[\alpha]^{2}_D-62.46$		230–1	0.9593^{20}_{4}	1.4988^{20}							**B7**², 129
c214	—,oxime(*α,d*)	*D*-Carvoxime(*α*). $C_{10}H_{15}NO$. *See* c211	165.23	lf (al) $[\alpha]^{17}_D+39.71$ (al, p=8.45)	70–71										**B7**², 129
c215	—,—(*α,l*)	*L*-Carvoxime(*α*). $C_{10}H_{15}NO$. *See* c211	165.23	mcl (dil al) $[\alpha]^{18}_D-39.43$ (al, p=4.33)	73.5		1.0140^{73}		s	s					**B7**², 129
c216	—,—(*β,d*)	$C_{10}H_{15}NO$. *See* c211	165.23	nd (dil al) $[\alpha]_D+68.3$ (bz)	57–8									oos vs	**B7**¹, 102
c217	—,—(*β,l*)	$C_{10}H_{15}NO$. *See* c211	165.23	(al)	56–7										**B7**¹, 102
c218	—,—(*dl*)	*dl*-Carvoxime. $C_{10}H_{15}NO$. *See* c211	165.23		93–4										**B7**¹, 103
c219	—,**dihydro-**(*d*)	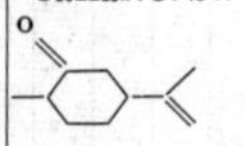 O	152.23	$[\alpha]_D+17.5$		221–4	0.928^{19}	1.4724							**B7**¹, 69
c220	—,—(*l*)	$C_{10}H_{16}O$. *See* c219	152.23	$[\alpha]^{20}_D-18.28$ $[\alpha]^{31}_D-46.63$		221–2	0.9253^{20}_{4}	1.4711^{20}	i						**B7**², 81
—	—,**tetrahydro-**	*see* **Carvomenthone**													
c221	**α-Caryophyllene**	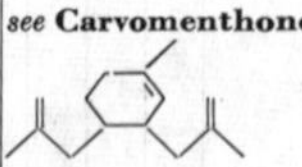	204.34	$[\alpha]^{20}_4+1.0\pm0.3$ (chl, c=9.26)		$155–7^{36}$	0.8966^{20}_{4}	1.5035^{20}							**Am72**, 5350
c222	**β-Caryophyllene**	$C_{15}H_{24}$	204.34	$[\alpha]^{20}_D-9.08$		$122^{13.5}$	0.9048^{22}_{22}	1.4965^{28}							**C47**, 8697
c223	**γ-Caryophyllene**	$C_{15}H_{24}$	204.34	$[\alpha]_{5461}-29.7$		$130-1^{24}$	0.8923^{25}_{25}	1.4942^{25}							**C32**, 8398
c224	**Caryophyllenic acid**(*dl,cis*)	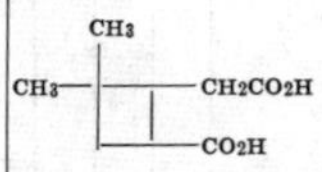 CH_3; CH_3—; CH_2CO_2H; CO_2H	186.20	pr	120						s				**B9**², 531
c225	—(*l,trans*)	$C_9H_{14}O_4$. *See* c224	186.21	$[\alpha]^{20}_D-28$ (bz)	81–2								s		**B9**², 531
c226	**Caryophyllin**	Oleanolic acid. 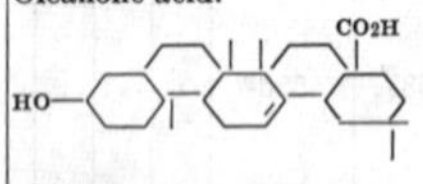 CO_2H; HO—	456.71	nd or pr (al) $[\alpha]^{20}_D+83.3$ (chl)	310d	280–308 sub (vac)			i	δ	δ		δ	Py v, CCl_4 δ, aa v^h	**B10**², 198; **E14**, 539
c227	**Catechin**,(*cis,d*)	*cis,d*-Catechol. 3,5,7,3′,4′-Flavanpentanol (*one form*) $C_{15}H_{14}O_6$	290.28	nd $[\alpha]^{18}_D+18.4$ (w, p=0.9)	174–5	240–5	1.344^{4}_{4}		v^h	v	δ	v	δ	chl δ, aa v, lig δ	**B17**², 254
c228	—(*cis,dl*)	$C_{15}H_{14}O_6$	290.28	nd (w+3)	212–4d					v	δ	v			**B17**², 256
c229	—(*cis,l*)	$C_{15}H_{14}O_6$	290.28	$[\alpha]_{5780}-16.8$ (w-ace, p=3)	174–5										**B17**², 255
—	**Catechol**	*see* **Benzene, 1,2-dihydroxy-***													

For explanations, symbols and abbreviations see beginning of table.

No.	Name	Synonyms and Formula	Mol. wt.	Crystalline form, color and specific rotation	m.p. °C	b.p. °C	Density	n_D	Solubility w	al	eth	ace	bz	other solvents	Ref.
	Catechol														
—	**Catechol**	*see* **Catechin**													
—	**Caulophylline**	*see* **Cytisine, *N*-methyl-**													
c230	**Cedrene**	$C_{15}H_{24}$	204.36	$[\alpha]^{22}_D -66.4$		$262-3^{760}$	0.9342^{20}_4	1.5034^{20}							**B5²**, 350
c231	**Cellobiose(β)**	**Glucose-β-glucoside.** 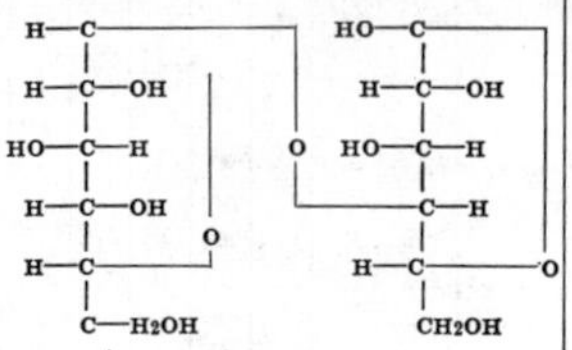	342.30	pw $[\alpha]^{20}_D +26.1$	255δd				s	δ	δ	i			**B31**, 380
c232	—,octaacetate(α)	Octaacetylcellobiose. $C_{28}H_{38}O_{19}$. *See* c231	678.60	nd (al) $[\alpha]^{20}_D +43.6$ (chl, p=z)	229.5				i	i	i			chl v aa v	**B31**, 382
c233	—,—(β)	$C_{28}H_{38}O_{19}$. *See* c231	678.60	nd (a') $[\alpha]^{20}_D -14.5$ (chl, c=11)	202					s^h			δ	chl v	**B31**, 382
—	**Cellosolve** acetate	*see* **Acetic acid,** 2-ethoxyethyl ester													
c234	**Cellulose**	Polycellobiose. $(C_6H_{10}O_5)_x$	$(162.14)_x$	wh amor	260–70d		1.27-1.61		i	i	i			oos i Cu(OH)₂.(NH₃)4 s	
c235	—,hexanitrate	Chief constituent of gun-cotton. $(C_{12}H_{14}N_6O_{22})_x$	$(594.27)_x$	wh amor	160–70 ign		1.66		i	i	i		i	$PhNO_2$ s	
c236	—,pentanitrate	$(C_{12}H_{15}N_5O_{20})_x$	$(549.27)_x$	wh amor			1.66		i	i	i		i	MeOH s eth-al s	
c237	—,tetranitrate	Constituent of collodion. $(C_{12}H_{16}N_4O_{18})_x$	$(504.27)_x$				1.66		i	i	i		i	MeOH s ethl-al s	
c238	—,triacetate	$(C_{12}H_{16}O_8)_x$	$(288.25)_x$	non-inflam yesh flakes $[\alpha]_D -22.5$ (chl)					i	i	i	i		$ClCH_2CH_2Cl$ s chl s $PhNO_2$ s aa s	
c239	—,triethyl ether	Ethyl cellulose. Triethyl cellulose. $(C_{12}H_{22}O_5)_x$	$(246.25)_x$	wh nd (bz) $[\alpha]^{20}_D +26.1$ (bz)	240–55				i	δ	s^h				
c240	—,trinitrate	Constituent of collodion. $(C_{12}H_{17}N_3O_{16})_x$	$(459.28)_x$				1.66		i	i		s		MeOH s aa s^h	
c241	**Cepharanthine**	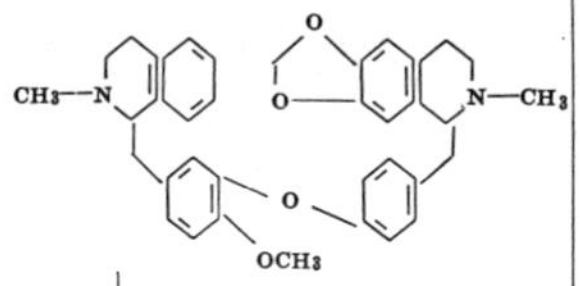	606.69	ye amor $[\alpha]^{20}_D +277$ (chl)	145–55					s	s	s		lig i	**C49**, 1745
c242	**Cerane**	Isohexacosane. $C_{26}H_{54}$	366.70	pl (eth)	61	$207^{0.7}$			i	s	s				**B1²**, 143
—	**Cerotic acid**	*see* **Hexacosanoic acid***													
—	**Ceryl alcohol**	*see* **1-Hexacosanol***													
c243	**Cerulignone**	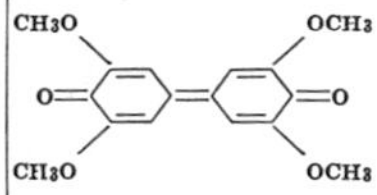	304.29	bl gr					i	i				oos i $PhNO_2$ s sulf s PhOH s	**B8²**, 573
—	**Cetene**	*see* **1-Hexadecene***													
—	**Cetyl amine**	*see* **Hexadecane, 1-amino-***													
—	**Cetyl bromide**	*see* **Hexadecane, 1-bromo-***													
—	**Cevadine**	*see* **Veratrine**													
c244	**Cevagenine**	$C_{27}H_{43}NO_8$	509.62	nd $[\alpha]^{20}_D -52$ (chl) $[\alpha]^{20}_D -47.5$ (al)	246–8										**H35**, 1270
c245	**Chalcone**	Benzylidene acetophenone. Benzalacetophenone. 1,3-Diphenyl-2-propen-1-one. 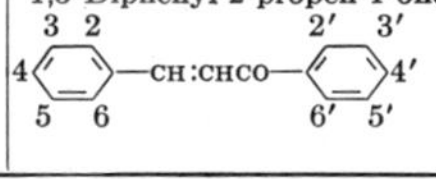	208.25	pa ye lf, pr, nd	62		1.0712^{62}_4			δ	s		s	chl s CS_2 s sulf s lig i	**B7²**, 423

For explanations, symbols and abbreviations see beginning of table.

No.	Name	Synonyms and Formula	Mol. wt.	Crystalline form, color and specific rotation	m.p. °C	b.p. °C	Density	n_D	Solubility						Ref.
									w	al	eth	ace	bz	other solvents	
	Chalcone														
c246	**—,4,4′-dimethyl-**	$C_{17}H_{16}O$. *See* c245	236.30	(MeOH)	127–9				i	s				MeOH s^h	**B7**[2], 441
c247	**—,3,3′-dinitro-**	$C_{15}H_{10}N_2O_5$. *See* c245	298.26	pa ye (Ac_2O)	210				i	i	i		s	Ac_2O s^h	**B7**[2], 429
c248	**—,2-methoxy-**	*o*-Anisylideneacetophenone. $C_{16}H_{14}O_2$. *See* c245	238.29	yesh nd (peth or eth-lig)	64–5					v	v		v	chl v: lig δ con sulf s	**B8**[2], 218
c249	**—,3-methoxy-**	$C_{16}H_{14}O_2$. *See* c245	238.29	yesh lf or pr (MeOH)	65	247[12]			i	s				oos s	**B8**[1], 579
c250	**—,4-methoxy-**	$C_{16}H_{14}O_2$. *See* c245	238.29	ye nd (al)	79				i	v^h	s			chl s aa s con sulf s MeOH v^h	**B8**[2], 218
c251	**—,3,4-methylenedioxy-**	Piperonylidenacetophenone. $C_{16}H_{12}O_3$. *See* c245	252.27	ye nd (al)	122				i	s^h				con sulf s aa s	**B9**, 141
c252	**—,2-nitro-**	$C_{15}H_{11}NO_3$. *See* c245	253.26	nd (al)	124					s^h	s			aa s	**B7**[2], 428
c253	**—,2′-nitro-**	$C_{15}H_{11}NO_3$. *See* c245	253.26	(al)	128–9					s^h	s				**B7**[2], 429
c254	**—,3-nitro-**	$C_{15}H_{11}NO_3$. *See* c245	253.26	ye (al)	145					s			s		**B7**[2], 429
c255	**—,4-nitro-**	$C_{15}H_{11}NO_3$. *See* c245	253.26	yesh (al)	162.5					s^h	s			chl s	**B7**[2], 429
c256	**—,α-nitro-**	$C_6H_5CH{:}C(NO_2)COC_6H_5$	253.26	ye pl (eth or bz-lig, (aa)	90									oos vs	**B7**, 483
—	**Chamazulene**	*see* **Azulene, 1,5-dimethyl-8-isopropyl-**													
c257	**Chaulmoogric acid (*d*)**	*d*-13(2-Cyclopentenyl)tridecanoic acid. $C_{18}H_{32}O_2$	280.45	pl or lf (al, aa) $[\alpha]_D+62$ (chl)	68–8.5	247–8[20]			i	δ	s			chl s	**B9**[2], 58
c258	**—(*dl*)**	$C_{18}H_{32}O_2$	280.45	(peth)	68.5										**B9**[2], 61
—	**Chavibetol**	*see* **Benzene, 4-allyl-1-hydroxy-2-methoxy-**													
—	**Chavicol**	*see* **Benzene, 1-allyl-4-hydroxy-**													
c259	**Chelerythrine,**	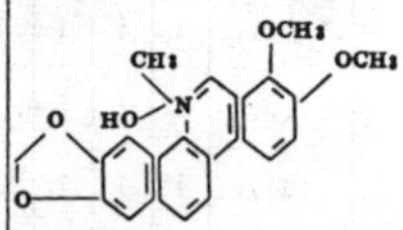	365.39	(chl-MeOH)	207				δ	δ	δ^h	s^h		chl s	**B27**[2], 563
c260	**Chelidonic acid**	Jervasic acid. 4-Pyrone-2,6-dicarboxylic acid. $C_7H_4O_6$	184.11	rose mcl (al-w+1w)	262				s^h	δ					
c261	**—,diethyl ester**	$C_{11}H_{12}O_6$	240.21	pr, nd	69				δ^h	s^h	s				**B18**[2], 367
c262	**Chelidonine(*d*)**	$C_{20}H_{19}NO_5$	353.38	mcl pl (al+1w) $[\alpha]_D+122.10$ (al, c=1)	136–40d				i^h	v^h	v^h			chl s^h	**B27**[2], 615
c263	**—,hydrochloride(*d*)**	$C_{20}H_{19}NO_5.HCl$	389.84	wh (w)					i	δ^h	δ^h				**B27**, 557
—	**Chenodeoxycholic acid**	*see* **Cholanic acid, 3(α),7(α)-dihydroxy-**													
—	**Chloral**	*see* **Acetaldehyde, trichloro-**													
—	**Chloralacetonechloroform**	*see* **Acetaldehyde, trichloro-**, β,β,β-trichloro-*tert*-butyl hemiacetal													
—	**Chloral-formamide**	*see* **Formic acid,** amide, *N*(1-hydroxy-2,2,2-trichloroethyl)-													
—	**Chloral hydrate**	*see* **Acetaldehyde,** trichloro-hydrate													
c264	**Chloralide**	2,5-Bis(trichloromethyl)-1.3-dioxolan-4-one. 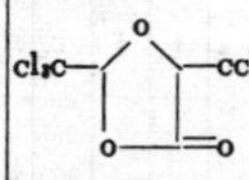	322.81	(al or eth)	116	275 147–8[12]			i	v^h	v			aa v	**B19**[1], 656
—	**Chloramphenicol**	*see* **Chlormycetin**													
—	**Chloranil**	*see* **1,4-Benzoquinone, tetrachloro-***													
—	**Chloranilic acid**	*see* **1,4-Benzoquinone, 2,5-dichloro-3,6-dihydroxy-***													
—	**Chloretone**	*see* **2-Propanol, 2-methyl-1,1,1-trichloro-***													
—	**Chloroform**	*see* **Methane, trichloro-**													
—	**Chloroform-*d***	*see* **Methane, trichlorodeutero-***													
c265	**Chloromycetin**	Chloramphenicol. Syntomycetin. $C_{11}H_{12}Cl_2N_2O_5$	323.14	pl or nd $[\alpha]_D^{25}+19$ (al, c=5)	151–2	sub vac			δ	v	s	v	i	chl s	**C42**, 4233
c266	**Chlorophyll a**	$C_{55}H_{72}MgN_4O_5$	893.48	bl bk hex pl	150–3				i	v^h	v^h			lig s	
c267	**Chlorophyll b**	$C_{55}H_{70}MgN_4O_6$	907.47	bl bk	120–30				i^h	v^h	v^h			MeOH s Py v^h lig s	

For explanations, symbols and abbreviations see beginning of table.

No.	Name	Synonyms and Formula	Mol. wt.	Crystalline form, color and specific rotation	m.p. °C	b.p. °C	Density	n_D	Solubility w	al	eth	ace	bz	other solvents	Ref.
	Chloropicrin														
—	**Chloropicrin**	*see* **Methane, nitrotrichloro-***													
—	**Chloroprene**	*see* **1,3-Butadiene, 2-chloro-***													
—	**Chlorosulfonic acid**	*see* **Sulfuric acid,** monochloride													
c268	**Chlorpromazine**	Largactil. Thorazine. 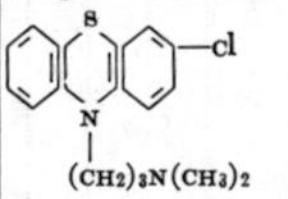	318.86		59–60				i	v	v		v	chl v dil HCl s	
c269	—,hydrochloride	$C_{17}H_{19}ClN_2S.HCl$	355.32		194–7d				s	v				MeOH v PrOH v chl v	
—	**Chlortetracyline**	*see* **Aureomycin**													
c270	**Cholanic acid**	Ursocholanic acid 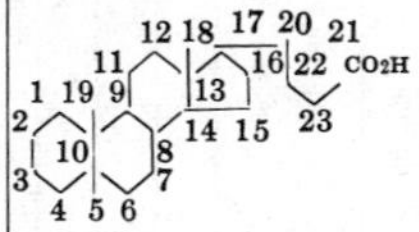	360.59	nd (al) $[\alpha]^{20}_D+215$ (chl)	163–4					s				chl s aa s	**E14,** 170
c271	—,ethyl ester	$C_{26}H_{44}O_2$. *See* c270	388.64	lf $[\alpha]_D+19$ (chl)	93–4									chl s	**C50,** 4175
c272	—,methyl ester	$C_{25}H_{42}O_2$. *See* c270	374.58	nd $[\alpha]_D+23\pm2$ (diox)	87–8									diox s	**C50,** 4175
c273	**—,3(α),6-dihydroxy-**	Hyodeoxycholic acid. $C_{24}H_{40}O_4$. *See* c270	392.56	$[\alpha]^{20}_D+37.2$ (MeOH)	196										**C50,** 5067
c274	**—,3(α),7(α)-dihydroxy-**	Chenodeoxycholic acid. $C_{24}H_{40}O_4$. *See* c270	392.56	nd $[\alpha]^{25}_D+11\pm2$ (diox)	140–2				i	v	s	v			**C48,** 3377
—	**—,3,12-dihydroxy-**	*see* **Desoxycholic acid**													
—	**—,3,7,12-trihydroxy-**	*see* **Cholic acid**													
—	**—,3,7,12-trioxo-**	*see* **Dehydrocholic acid**													
c275	**Cholanthrene**	Benz(*j*)aceanthrene. 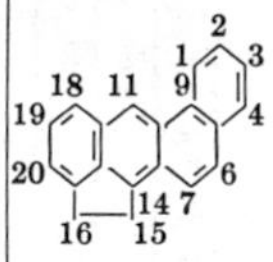	254.31	pa ye lf (bz-al)	174.5–5.0d	sub $210^{0.2}$			i	s				aa s lig s	**E14s,** 525
c276	**—,6,7-dihydro-20-methyl-**	$C_{21}H_{18}$. *See* c275	270.37	lf (MeOH)	155									MeOH s^h	**E14,** 433
c277	**—,11,14-dihydro-20-methyl-**	$C_{21}H_{18}$. *See* c276	270.37	nd (PrOH)	138–9									PrOH s^h	**E14,** 433
c278	**—,20-methyl-**	$C_{21}H_{16}$. *See* c275	268.34	yesh nd (bz)	177				i						
c279	**$\Delta^{2,4}$-Cholestadiene**	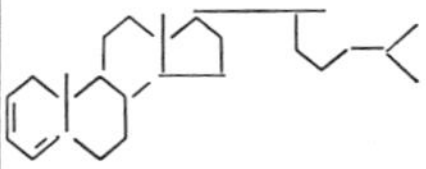	368.65	(eth-ace) $[\alpha]_D+114$ (chl)	63				i	s	s			chl s	**E14,** 18
c280	**$\Delta^{3,5}$-Cholestadiene**	Cholesterilene. 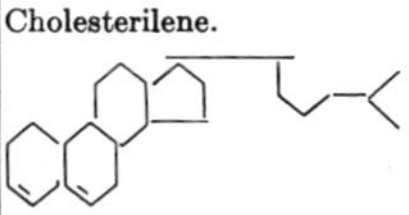	368.65	wh nd (al)	80	260^{13}	$0..925^{100}_4$		i^h	s^h	∞		∞	chl ∞ lig v	**E14s,** 1411
c281	**$\Delta^{5,7}$-Cholestadien-3(β)-ol**	7-Dehydrocholesterol. Provitan in D_3. 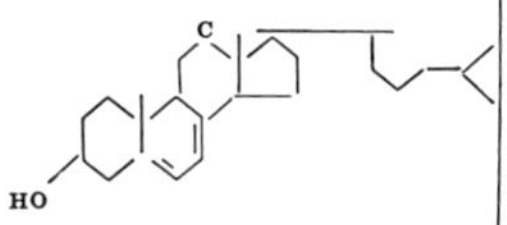	384.65	pl (eth-MeOH) $[\alpha]^{25}_D-115$ (chl)	150–1				i	δ				oos s	**E14s,** 1551
c282	**$\Delta^{4,6}$-Cholestadien-3-one**	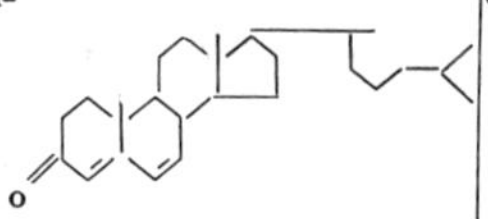	382.64	ye pr $[\alpha]^{20}_D+31$ (chl)	78										**J1956,** 627

For explanations, symbols and abbreviations see beginning of table.

No.	Name	Synonyms and Formula	Mol. wt.	Crystalline form, color and specific rotation	m.p. °C	b.p. °C	Density	n_D	Solubility w	al	eth	ace	bz	other solvents	Ref.
	Cholestane														
c283	**Cholestane**...	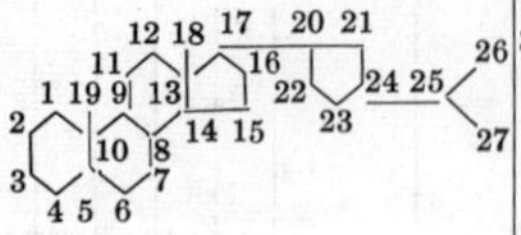	372.65	lf or pl (eth-al, ace, AcOEt) $[\alpha]_4^{20}+24.5$ (chl)	80	250^1	0.9090_4^{88}	1.4887^{88}	i	δ	v	...	v	chl v	**E14s,** 1429
—	—,3(β)-**hydroxy-**..	*see* **Cholestanol**													
c284	3(β)-**Cholestane-carboxylic acid**	$C_{28}H_{48}O_2$. *See* c283	416.66	nd $[\alpha]_D^{25}+28.8$ (chl, c=1.70)	210–1										**Am 75,** 6234
c285	3,6-**Cholestane-dione**	$C_{27}H_{44}O_2$. *See* c283	400.62	nd $[\alpha]_D^{20}+8.9$ (c=0.98)	171–2										**H37,** 258
c286	**Cholestanol**......	Dihydrocholesterol. 3(β)-Hydroxycholestane. $C_{27}H_{48}O$. *See* c283	388.65	lf (al+1 w) pr, pl (MeOH) $[\alpha]_D^{22}+34.2$ (chl)	146–7				i	v^h	v	...	...	MeOH δ chl v con sulf s	**E14s,** 1695
c287	3(α)-**Cholestanol**..	Epicholestanol. Epidihydro-cholesterol. $C_{27}H_{48}O$.	388.65	nd $[\alpha]_D^{20}+34$ (chl, c=1.1)	185–6										**Am77,** 4925
c288	3-**Cholestanone**...	Zymostanone. $C_{27}H_{46}O$. *See* c283	386.64	nd $[\alpha]_D+42$ (chl, c=2.12)	128–30										**Am77,** 190
c289	6-**Cholestanone,** 3-(β)-**hydroxy-**	6-Ketocholestanol. $C_{27}H_{46}O_2$. *See* c283	402.64	nd $[\alpha]_D-3.0$ (chl)	150–1										**Am76,** 532
c290	7-**Cholestanone-,** 3-(β)-**hydroxy-**	7-Oxocholestanol. 7-Keto-cholestanol. $C_{27}H_{46}O_2$. *See* c283	402.64	pl $[\alpha]_D^{22}-32$ (chl)	165–8										**Am76,** 532
c291	2-**Cholestene**.....	Neocholestene. 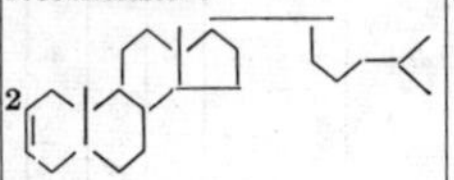	370.64	nd $[\alpha]_D+66$ (c=1.65)	74–5										**J1954,** 4284
c292	5-**Cholestene**.....	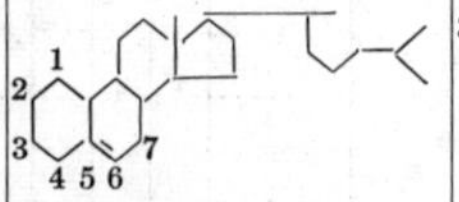	370.64	wh pr $[\alpha]_D-49$ (chl, c=1.43)	90–2										**Am 78,** 635
c293	—,3(β)-**bromo-**....	Cholesteryl bromide. $C_{27}H_{45}Br$. *See* c292	449.55	mcl lf (al) $[\alpha]_D^{20}-23.6$ (c=3.09)	98–9				...	s^h	...	...	s	chl s	**E14s,** 1454
c294	—,3(β)-**chloro-**....	Cholesteryl chloride. $C_{27}H_{45}Cl$. *See* c292	405.09	nd (al or ace) $[\alpha]_D^{20}-26.4$ (bz)	92–3				...	s^h	...	s^h	s	chl s aa v^h CS_2 v	**E14s,** 1450
c295	—,3(β)-**iodo**......	Cholesteryl iodide. $C_{27}H_{45}I$. *See* c292	496.55	nd (ace or AcOEt) $[\alpha]_D^{20}-11.9$ (chl)	106–7				i	δ	...	δ	s	chl s EtOAc s^h lig s	**E14s,** 1455
c296	5-**Cholestene-** 3(β), 7(β)-**diol**	7(β)-Hydroxycholesterol. $C_{27}H_{46}O_2$. *See* c292	402.64	wh nd $[\alpha]_D^{20}-84.1$	175–7										**C49,** 14042
c297	1-**Cholesten-3-one**	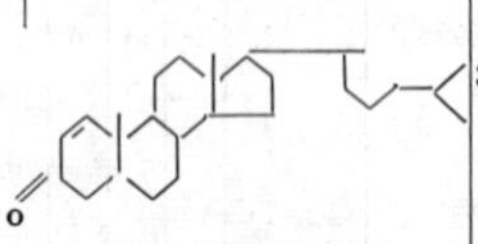	384.62	$[\alpha]_D^{25}+88.2$ (chl)	99–101								s		**Am75,** 3513
c298	4-**Cholesten-3-one**	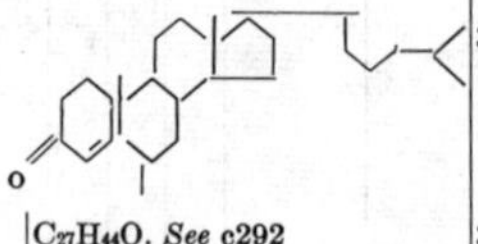	384.62	nd $[\alpha]_D^{25}+92$ (chl, c=2.01)	81–2				...	s	...	...	s		**C50,** 1568
c299	5-**Cholesten-3-one**	$C_{27}H_{44}O$. *See* c292	384.62	pr $[\alpha]_D^{25}-2.5$ (chl, c=2.03)	124–9										**C50,** 1568
—	**Cholesterilene**....	*see* **3,5-Cholestadiene**													
c300	**Cholesterol**......	Δ^5-3(β)-Cholestenol. Cho-lesterin. $C_{27}H_{46}O$	386.66	rh or tcl lf (al) $[\alpha]_D^{20}-31.61$ (chl)	148.5	360d	1.067_4^{20}		δ	δ	v	...	s	diox v chl v	**E14,** 38

For explanations, symbols and abbreviations see beginning of table.

No.	Name	Synonyms and Formula	Mol. wt.	Crystalline form, color and specific rotation	m.p. °C	b.p. °C	Density	n_D	Solubility w	al	eth	ace	bz	other solvents	Ref.
	Cholesterol														
c301	—,acetate	Cholesteryl acetate. $C_{29}H_{48}O_2$ Cholesteryl benzoate	428.68	wh nd (ace or al) $[\alpha]^{20}_D -47$ (chl)	115				δ	...	s	s	s	chl s	**E14s,** 1627
c302	—,benzoate	$C_{34}H_{50}O_2$	490.75	wh nd $[\alpha]-14.7$ (chl)	150–51				...	i	s	...	...	chl s	**E14,** 53
c303	—,hexadecanoate	Cholesteryl palmitate. $C_{43}H_{76}O_2$	625.08	wh nd (eth or al) $[\alpha]^{20}-23$ (chl)	90				i^h	s^h	s^h	s^h	s	chl s	**E14s,** 1627
—	**—,7(β)-hydroxy-**	*see* **5-Cholestene-3(β), 7(β)-diol**													
—	**—,7-dehydro-**	*see* **5,7-Cholestadien-3(β)-ol**													
—	**Cholesteryl bromide**	*see* **5-Cholestene, 3(β)-bromo-**													
—	**Cholesteryl chloride**	*see* **5-Cholestene, 3(β)-chloro-**													
c304	**Cholestrophane**	Dimethylparabanic acid. Oxalyldimethylurea. (structure: CH_3—N, N—CH_3 ring with three C=O)	142.11	lf or pl (w or al)	155	275–7 $148–50^{13}$			v^h	δ	s	...	...	...	**B24²,** 265
c305	**Cholic acid**	Cholanic acid. 3(α), 7(α), 12-Trihydroxycholalic acid. $C_{24}H_{40}O_5$	408.56	rh (eth), tetr rh (w or dil al+1w) $[\alpha]^{20}_D +37$ (al)	198 (anh)				i	s	v	s	...	chl v alk s aa s	**E14,** 191
c306	**Choline**	Trimethyl (2-hydroxyethyl) ammonium hydroxide. $(CH_3)_3\overset{+}{N}CH_2CH_2OH.OH^-$	121.18						v	v	i	...	...	...	**B4²,** 720
c307	**—,*O*-acetyl-,** bromide	Acetylcholine bromide. Progmoline. $(CH_3)_3\overset{+}{N}CH_2CH_2O_2CCH_3.Br^-$	226.12	hex pr (dil al)	143	d			v	s	i	...	...	MeOH s	**B4¹,** 428
c308	—,—,chloride	$(CH_3)_3\overset{+}{N}CH_2CH_2O_2CCH_3.Cl^-$	181.67	yesh nd	256–7				s^h	s^h	...	...	...	...	**B4²,** 723
c309	**—,*O*-benzoyl-,** chloride	$(CH_3)_3\overset{+}{N}CH_2CH_2O_2CC_6H_5Cl^-$	243.74	pr (ace-al)	200				...	s^h	...	s^h	...	...	**B9¹,** 90
c310	**—,carbamate,** chloride	$(CH_3)_3\overset{+}{N}CH_2CH_2O_2CNH_2.Cl^-$	182.65	hyg pr	210–2				v	δ	i	...	...	MeOH v chl i	...
—	**Chondrodendrin**	*see* **Bebeerine**(*d*)													
c311	**Chroman**	(structure)	134.17	...	...	214–5 $98–9^{18}$	1.0610^{20}	1.544^{20}	s^h	...	...	...	...	oos ∞	**B17,** 52
c312	**Chromone**	α-Benzopyrone. 4-Oxo-1,4-chromene. (structure)	146.14	nd (peth or w)	59	sub			$δ^h$	s	s	...	s	chl s	**B17¹,** 170
—	**—,2-phenyl-**	*see* **Flavone**													
—	**Chromotropic acid**	*see* **2,7-Naphthalene disulfonic acid, 4,5-dihydroxy-***													
—	**Chrysammic acid**	*see* **9,10-Anthraquinone, 1,8-dihydroxy-2,4,5,7-tetranitro-***													
—	**Chrysaniline**	*see* **Acridine, 2-amino-5(4-aminophenyl)-**													
—	**Chrysanthemum-monocarboxylic acid**	*see* **Cyclopropanecarboxylic acid, 2,2-dimethyl-3(2-methylpropenyl)-***													
—	**Chrysanthemunic acid**	*see* **Cyclopropanecarboxylic acid, 2,2-dimethyl-3(2-methylpropenyl)***													
—	**Chrysazin**	*see* **9,10-Anthraquinone, 1,8-dihydroxy-***													

For explanations, symbols and abbreviations see beginning of table.

No.	Name	Synonyms and Formula	Mol. wt.	Crystalline form, color and specific rotation	m.p. °C	b.p. °C	Density	n_D	Solubility w	al	eth	ace	bz	other solvents	Ref.
	Chrysazol														
—	**Chrysazol**	*see* **Anthracene, 1,8-dihydroxy-***													
c313	**Chrysene**	1,2-Benzophenanthrene. Benzo(*a*)-phenanthrene. 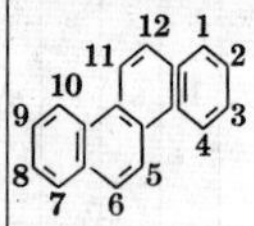	228.28	red bl flr rh pl (bz)	254	448			i	i	δ	...	δ	CS_2 δ aa δ	**E14**, 343
c314	—,1,2-dimethyl-	$C_{20}H_{16}$. *See* c313	256.35	pl or nd (bz-al)	128–9	$200^{0.5}$ (sub 140 vac)			i	s	...	...	...	CS_2 s aa s	**E14s**, 254
c315	—1,-methyl-	$C_{19}H_{14}$. *See* c313	242.32	nd (al-ace)	118.0–8.8				i	s	...	...	...		**E14s**, 246
c316	—,2-methyl-	$C_{19}H_{14}$. *See* c313	242.32	nd (bz-al or peth)	161.0–1.4	sub $140^{1.5}$			i	s	...	...	...	aa s	**E14s**, 247
c317	—,3-methyl-	$C_{19}H_{14}$. *See* c313	242.32	lf (hexane, al or to)	254–5	sub 130–40 vac			...	s	...	...	...		**E14s**, 246
—	**Chrysin**	*see* **Flavone, 5,7-dihydroxy-**													
—	**Chrysoidine**	*see* **Azobenzene, 2,4-diamino-**													
—	**Chrysophanic acid**	*see* **9,10-Anthraquinone, 1,8-dihdroxy-3-methyl-***													
c318	**1,2-Chrysoquinone**	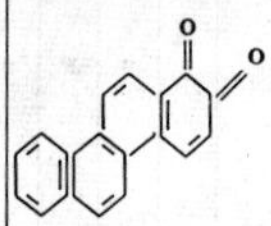	258.26	red nd (bz or to), lf or pl (aa)	239.5	sub			i	s^h	δ	...	s^h	sulf s aa δ	**B7**[2], 760
c319	**2,8-Chrysoquinone**	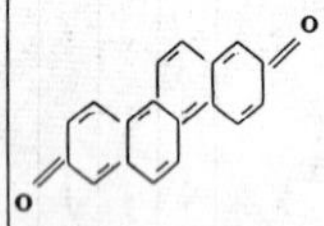	258.26	red ye nd (aa)	288–90d				...	...	...	...	...		**B7**[1], 441
—	**Chymyl alcohol**	*see* **Glycerol**, 1-hexadecyl ether													
—	**Cinchomeronic acid**	*see* **3,4-Pyridinedicarboxylic acid**													
c320	**Cinchonamine**	$C_{19}H_{24}N_2O$	296.40	rh nd (al) $[\alpha]_D^{20}+123$ (al, c=0.33)	186				i	v^h	v^h	...	δ	chl s	**B23**[2], 358
c321	**Cinchonicine**	Cinchotoxine. 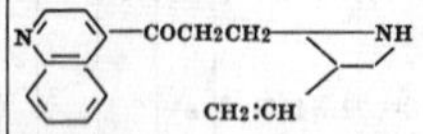	294.38	nd or pr (eth) $[\alpha]_D^{15}+48$ (al, c=1)	58–9				i	v	v	v	v	chl v	**B24**[2], 100
c322	**Cinchonidine**	Cinchovatine. α-Quinidine. $C_{19}H_{22}N_2O$	294.38	pr (al) $[\alpha]_D^{20}-109.20$ (al, p=1)	207.2	sub			i	s	δ	...	i	chl s MeOH s	**B23**[2], 373
c323	—,bisulfate	$C_{19}H_{22}N_2O.H_2SO_4.5H_2O$	482.54	mcl pr $[\alpha]_D-110$					v	v	...	...	...		**B23**, 441
c324	—,hydrochloride	$C_{19}H_{22}N_2O.HCl.H_2O$	348.88	wh pr (w) $[\alpha]_D^{20}-117.6$ (w, c=1.2)	242 (anh)				s	v	δ	...	...	chl v	**B23**[2], 373
c325	—,sulfate	$(C_{19}H_{22}N_2O)_2.H_2SO_4.6H_2O$	794.98	mcl eff nd (al+2w) $[\alpha]_D^{18.5}-97.9$ (w, c=1.2)	205 (anh)				δ	δ	i	...	...	chl δ	**B23**[2], 373
c326	**β-Cinchonidine**	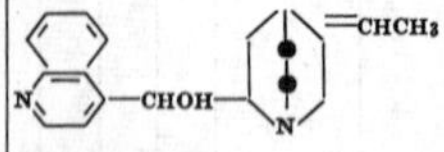	294.38	pr or lf (al or dil al) $[\alpha]_D^{25}-126.6$ (al, c=0.5)	241				i	v	δ	...	...	chl v	**B23**[1], 131
c327	**Cinchonine**	$C_{19}H_{22}N_2O$	294.38	nd or mcl cr (al) $[\alpha]_D^{17}+229$ (al)	265	sub >220			i	δ	δ	...	v	chl s	**B23**[2], 369
c328	—,bisulfate	$C_{19}H_{22}N_2O.H_2SO_4.4H_2O$	464.53	rh oct (w) $[\alpha]_D^{18}+195$ (w, c=2)					v	v	i	...	...		**B23**[2], 371

For explanations, symbols and abbreviations see beginning of table.

No.	Name	Synonyms and Formula	Mol. wt.	Crystalline form, color and specific rotation	m.p. °C	b.p. °C	Density	n_D	Solubility w	al	eth	ace	bz	other solvents	Ref.
	Cinchonine														
c329	—,dihydrochloride	$C_{19}H_{22}N_2O.2HCl$	367.33	pl $[\alpha]_D^{21}+205.5$ (w, c=3.6)					v	s	i	...			**B23**[1], 133
c330	—,hydrochloride	$C_{19}H_{22}N_2O.HCl.2H_2O$	366.88	mcl $[\alpha]_D^{25}+133.6$ (chl, c=1.4)	181		1.234		v^h	v	δ	...	...	chl s	**B23**[2], 370
c331	—,nitrate	$C_{19}H_{22}N_2O.HNO_3.\frac{1}{2}H_2O$	366.41	mcl (w)					s	s	...	...	...		**B23**, 420
c332	—,sulfate	$(C_{19}H_{22}N_2O)_2.H_2SO_4.2H_2O$	722.88	rh pr (w) $[\alpha]_D^{15}+169$ (w, c=1.4)	206–7				s^h	s	i	...	...	MeOH v chl δ	**B23**[2], 371
—	**Cinchoninic acid**	*see* **4-Quinolinecarboxylic acid**													
—	**Cinchopene**	*see* **4-Quinolinecarboxylic acid, 2-phenyl-**													
c333	**Cinchotine**	Hydrocinchonine. Pseudo-cinchonine. $C_{19}H_{24}N_2O$	296.42	pr $[\alpha]_D^{21}+203.4$	286				s^h	δ	i	...			**B23**[2], 356
c334	**1,4-Cineole**	*p*-Cineole. 1,4-Epoxy-*p*-menthane.	154.25		1	173–4	0.8997^{20}	1.4479^{25}	δ	∞	∞	...	...		**B17**[2], 32
c335	**1,8-Cineole**	Cajeputol. Eucalyptol. $C_{10}H_{18}O$	154.25	inactive	1	176.4 61^{14}	0.9267^{20}	1.4586^{20}	i	s	s	...	...	chl s	**B17**[2], 32
c336	**Cineolic acid**(*d*)	HO_2C, CH_3, CO_2H, CH_3, O, CH_3	216.24	(w+1) $[\alpha]_D^{20}+18.6$ (w, p=8.21)	79(+w) 138–9 (anh)				δ	v	δ	...	δ	chl v AcOEt v lig δ	**B18**, 332
c337	—(*dl*)	$C_{10}H_{16}O_5$	216.24		203–4d				δ	v^h	v	...	...	chl δ	**B18**[2], 285
c338	—(*l*)	$C_{10}H_{16}O_5.H_2O$	234.25	rh (+1 w)	79 (+w) 138–9 (anh)				δ	$δ^h$	δ	...	...	chl δ	**B18**, 322
c339	**Cinnamaldehyde**	Cinnamic aldehyde. β-Phenyl acrolein. 3-Phenylpropenal-(*trans*)* 6 β α 5 —CH:CHCHO 4 2 3	132.16	yesh	−7.5	253^{760}	1.0497_4^{20}	1.6195^{20}	δ	s	s	...	...	chl s lig i	**B7**[2], 273
c340	—,oxime(β)	$C_6H_5CH{:}CHCH{:}NOH$	147.18	nd	138.5				δ	δ	δ	...	s		**B7**[2], 278
c341	—,**4-hydroxy-3-methoxy-**	Coniferaldehyde. Ferulaldehyde. $C_{10}H_{10}O_3$. *See* c393	178.19		84				δ	s	s	...	s		**B8**, 288
c342	—,**4-methyl-**	$C_{10}H_{10}O$. *See* c339	146.19	ye lf (dil al)	41.5	$154–9^{25}$			...	s^h	...	...	...		**B7**, 369
c343	—,**α-methyl-**	$C_6H_5CH{:}C(CH_3)CHO$	146.19	ye		150^{100} $131–2^{16}$	1.0407_4^{17}	1.6057^{17}	...	...	...	...	...		**B7**[2], 291
c344	—,**2-nitro-**	$C_9H_7NO_3$. *See* c339	177.16	nd (eth or al)	127.0–7.5				v^h	s^h	s	...	...	chl v	**B7**[2], 281
c345	—,**3-nitro-**	$C_9H_7NO_3$. *See* c339	177.16	ye nd (w), pr (al), (aa)	116				$δ^h$	s^h	δ	...	v	aa v	**B7**[2], 282
c346	—,**4-nitro-**	$C_9H_7NO_3$. *See* c339	177.16	nd (w or al)	141-2				s^h	v	...	...	...	oos v	**B7**[2], 282
—	—,**α-pentyl-**	*see* **Heptanal, 2-benzylidene-***													
—	**Cinnamalonic acid**	*see* **Malonic acid, cinnamylidene** (*trans*)													
—	**Cinnamein**	*see* **Cinnamic acid**, benzyl ester													
c347	**Cinnamic acid** (*cis*) (m.p. 42)	*cis*-β-Phenylacrylic acid. 3 2 H H 4 —C=C—CO2H 5 6 β α	148.16	mcl pr (w)	42				δ	...	...	...	...	aa v lig v	**B9**[2], 393
c348	—(*cis*) (m.p. 58)	$C_9H_8O_2$. *See* c347	148.16	mcl pr	58	265			δ	s	v	v	...	chl s lig s	B9[2], 393
c349	—(*cis*) (m.p. 68)	Allocinnamic acid. Isocinnamic acid. $C_9H_8O_2$. *See* c347	148.16	mcl pr	68				δ	v^h	v^h	...	...	lig s	**B9**[2], 393
c350	—(*trans*)	*trans*-β-Phenylacrylic acid. *trans*-3-Phenylpropenoic acid* 3 2 H CO2H 4 —C=C—H 5 6 β α	148.16	mcl pr	135–6 cor	300^{760} cor	1.2475_4		i	v	s	...	s	chl s lig i	**B9**[2], 377

For explanations, symbols and abbreviations see beginning of table.

No.	Name	Synonyms and Formula	Mol. wt.	Crystalline form, color and specific rotation	m.p. °C	b.p. °C	Density	n_D	Solubility w	al	eth	ace	bz	other solvents	Ref.
	Cinnamic acid														
c351	—,allyl ester	Allyl cinnamate. $C_6H_5CH:CHCO_2CH_2CH:CH_2$	188.22			286d	1.048^{23}_{4}	1.530	i	v	∞				**B9²**, 387
c352	—,amide(*trans*)	Cinnamamide. $C_6H_5CH:CHCONH_2$	147.17	nd (bz)	148.0–8.5 cor				δ^h	v	v			CS_2 v	**B9²**, 391
c353	—,anhydride(*trans*)	$(C_6H_5CH:CHCO)_2O$	278.29	nd (bz or al), pr (al)	138				i	δ			v^h		**B9²**, 390
c354	—,benzyl ester (*trans*)	Benzyl cinnamate. Cinnamein. $C_6H_5CH:CHCO_2CH_2C_6H_5$	238.27	pr	39	244^5 350d	1.109^{15}			s	s				**B9²**, 388
c355	—,chloride(*trans*)	Cinnamoyl chloride $C_6H_5CH:CHCOCl$	166.60		36	257.5^{760} 101^2	1.1617^{45}_{4}	$1.614^{42.5}$	i	s^h				CCl_4 s lig s	**B9²**, 390
c356	—,ethyl ester(*trans*)	Ethyl cinnamate. $C_6H_5CH:CHCO_2C_2H_5$	176.21		7.5	271.5^{760} $128–33^6$	1.0491^{20}_{4}	1.5598^{20}	i	s	v			oos v	**B9²**, 385
c357	—,isopropyl ester (*trans*)	$C_6H_5CH:CHCO_2CH(CH_3)_2$	190.24			268–70	1.017^{18}		i	s					**B9²**, 387
c358	—,2-methoxyphenyl ester(*trans*)	*o*-Anisyl cinnamate. Guaiacol cinnamate. [structure: C_6H_5—CH:CHCO₂—C_6H_4(CH₃O)]	254.27	wh nd (al)	130				i			s	s	chl s	**B9²**, 389
c359	—,methyl ester (*trans*)	Methyl cinnamate. $C_6H_5CH:CHCO_2CH_3$	162.18		36.5	261.9	1.0911^{20}_{4}	1.5766^{21}	i	v	v			oos v	**B9²**, 385
c360	—,nitrile(*cis*)	Allocinnamonitrile. *cis*-Cinnamonitrile. *cis*-Styryl cyanide. $C_6H_5CH:CHCN$	129.15		−4.4	249 139^{30}							v		**Am 58,** 2428
c361	—,—(*trans*)	$C_6H_5CH:CHCN$	129.15		22	263.8 $134–6^{12}$	1.0283^{20}_{4}	1.6043^{23}							
c362	—,2-octyl ester (*trans, d*)	$C_6H_5CH:CHCO_2CH(CH_3)(CH_2)_5CH_3$	260.38	$[\alpha]^{17}_{D}+40.19$		218^{28}	0.9645^{20}_{4}	1.5145^{20}	i				s	chl s	**B9¹**, 229
c363	—,—(*trans, dl*)	$C_6H_5CH:CHCO_2CH(CH_3)(CH_2)_5CH_3$	260.38			240^{60} 213^{28}	0.9715^{17}_{4}		i				s	chl s	**B9¹**, 230
c364	—,—(*trans, l*)	$C_6H_5CH:CHCO_2CH(CH_3)(CH_2)_5CH_3$	260.38	$[\alpha]^{17}_{D}-39.78$		211^{23}	0.9692^{17}_{4}		i				s	chl s	**B9¹**, 229
c365	—,phenyl ester (*trans*)	$C_6H_5CH:CHCO_2C_6H_5$	224.26		72.5	$205–7^{12}$					s		s		**B9²**, 387
c366	—,*p*-phenylphenacyl ester(*trans*)	[structure: C_6H_5—CH:CHCO₂CH₂CO—C_6H_4—C_6H_5]	342.38		182.5										**Am 52,** 3719
c367	—,propyl ester (*trans*)	$C_6H_5CH:CHCO_2CH_2CH_2CH_3$	190.24			279^{750}	1.0435^{0}_{0}		i						**B9²**, 387
c368	—,4-ureidophenyl ester(*trans*)	[structure: C_6H_5—CH:CHCO₂—C_6H_4—NHCONH₂]	282.29	wh	204				i	s		s			
c369	—,**α-acetamido-** (*trans*)	$C_6H_5CH:C(NHCOCH_3)CO_2H$	205.22	(+2w)	185–6 (+2w) 193–4 (anh)				s						**B10²**, 471
c370	—,**2-amino-**(*trans*)	$C_9H_9NO_2$. *See* c350	163.17	ye nd	158–9d				s^h	s	s				**B14²**, 316
c371	—,**3-amino-**(*trans*)	$C_9H_9NO_2$. *See* c350	163.17	pa ye nd (w or al)	185				s^h	v	v			ac s alk s	**B14²**, 316
c372	—,**4-amino-**(*trans*)	$C_9H_9NO_2$. *See* c350	163.17	ye nd	175–6d				s^h	s	s				**B14²**, 317
c373	—,**2-benzamido-** (*trans*)	$C_{16}H_{13}NO_3$. *See* c350	267.29	nd (al)	262d					δ^h				oos δ aa s^h	**B14²**, 617
c374	—,**3-benzamido-**	$C_{16}H_{13}NO_3$. *See* c350	267.29	nd (AcOEt)	229										**B14¹**, 418
c375	—,**4-benzamido-**	$C_{16}H_{13}NO_3$. *See* c350	267.29	lf (aa), nd (ace)	274d							v^h		aa v^h	**B14¹**, 619
c376	—,α-bromo-(*cis*)	$C_6H_5CH:CBrCO_2H$	227.06	lf (w), pr (chl)	159–60				s^h	s			δ	CS_2 s	**B9²**, 399
c377	—,—(*trans*)	$C_6H_5CH:CBrCO_2H$	227.06	nd (w)	131–2				v^h	∞	s		s		**B9²**, 398
c378	—,**β-bromo**(*cis*)	$C_6H_5CBr:CHCO_2H$	227.06	nd (bz), pl (al)	159–60				δ^h	δ	s		δ	chl s	**B9²**, 398
c379	—,—(*trans*)	$C_6H_5CBr:CHCO_2H$	227.06	pa ye nd or pl (w), pr (chl)	178–9				δ^h	s			s^h	CS_2 δ	B9², 397
c380	—,**2-carboxy-** (*trans*)	β(2-Carboxyphenyl)acrylic acid. *o*,β-Styrenedicarboxylic acid. $C_{10}H_8O_4$. *See* c350	192.16	pr or nd (w)	205d				δ	v	δ		i	chl i	**B9²**, 641
c381	—,**3-carboxy-** (*trans*)	$C_{10}H_8O_4$. *See* c350	192.16	nd (ace-lig)	275				i	v	δ		i	chl i aa v	**B9²**, 642
c382	—,**4-carboxy-** (*trans*)	$C_{10}H_8O_4$. *See* c350	192.16	pw	358d	sub >350								oos i aa s^h	**B9²**, 642

For explanations, symbols and abbreviations see beginning of table.

No.	Name	Synonyms and Formula	Mol. wt.	Crystalline form, color and specific rotation	m.p. °C	b.p. °C	Density	n_D	Solubility w	al	eth	ace	bz	other solvents	Ref.
	Citrulline														
c444	**Citrulline**	α-Amino-δ-ureidovaleric acid. *N*-δ-Carbamylornithine. δ-Ureidonorvaline. $H_2NCONH(CH_2)_3CH(NH_2)CO_2H$	175.19	pr (aq MeOH) $[\alpha]_D^{20}+3.7$ (w)	222				s	i					
—	**Civetone**	*see* **9-Cycloheptadecen-1-one***													
—	**Cleve's acid**	*see* **1-Naphthalenesulfonic acid, 3-amino-***													
—	**Clupanodonic acid**	*see* **Docosapentaenoic acid***													
—	**Cocaethyline**	*see* **Ecgonine benzoate,** ethyl ester													
c445	**Cocaine**(*d*)	$C_{17}H_{21}NO_4$	303.36	mcl pr (eth) $[\alpha]_D^{20}+15.8$ (chl, p=10)	98										**B22²**, 150
c446	—(*dl*)	$C_{17}H_{21}NO_4$	303.36	rh bipym pr (peth)	79–80				i	v	v			chl v lig v	**B22²**, 156
c447	—(*l*)	$C_{17}H_{21}NO_4$	303.36	mcl pr (al) $[\alpha]_D^{20}-16.3$	98	sub 75–90 (vac)		1.5022^{98}	δ^h	v	v		s	chl v CS_2 s	**B22²**, 151
c448	—,chromate	$C_{17}H_{21}NO_4.H_2CrO_4.N_2O$.	439.39	og ye lf	127				δ						**B22**, 200
c449	—,hydrochloride(*d*)	$C_{17}H_{21}NO_4.HCl$	339.82	pl (al)	187 cor				v	v					**B22²**, 156
c450	—,—(*l*)	$C_{17}H_{21}NO_4.HCl$	339.82	mcl pr (al), (w+2) $[\alpha]_D^{20}-69.9$ (dil al, p=1.3)	197				v	s	i			chl s	**B22²**, 153
c451	**"Cocaine cinnamate"**(*d*)	Ecgonine cinnamate methyl ester. 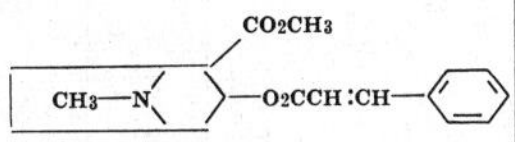	329.40	pr $[\alpha]_D+2$ (al)	68										
c452	—(*l*)	$C_{19}H_{23}O_4N$. *See* c451	329.40	mcl pr, nd (bz) $[\alpha]_D^{15}-4.7$ (chl, c=10)	121				i	s	s	s	s	chl s lig s	**B22²**, 154
c453	**Coclaurine**	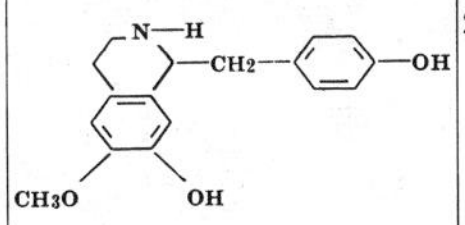	285.33	pl (al) $[\alpha]_D^{29}-17.01$	220–1				δ	v^h		s^h	i	chl δ	**C47**, 6430
c454	**Codamine**	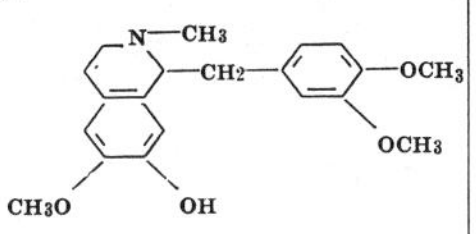	343.43	pr (bz)	126–7				s^h	v	s			chl v	**B21²**, 184
c455	**Codeine**	Morphine 3-methyl ether. $C_{18}H_{21}NO_3$	299.37	rh oct (eth or CS_2) $[\alpha]_D^{15}$ −137.8	157–8.5	250^{12}	1.32		s^h	v	s		s	chl s to s	**B27²**, 137
c456	—,hydrate	$C_{18}H_{21}NO_3.H_2O$	317.39	rh oct (w, aq al) $[\alpha]_D^{25}-136$ (al, c=2.8)			1.31		δ	v	s	v	s	chl v	**B27²**, 137
c457	—,hydrochloride	$C_{18}H_{21}NO_3.HCl.2H_2O$	371.86	nd $[\alpha]_D^{15}-108.2$ (w)	287d				v	s					**B27²**, 143
c458	—,phosphate	$C_{18}H_{21}NO_3.H_3PO_4.1\frac{1}{2}H_2O$	424.39	lf or pr (dil al)	220–35d				v	s^h	s			chl s	**B27²**, 144
c459	—,sulfate	$(C_{18}H_{21}NO_3)_2.H_2SO_4.5H_2O$	786.90	pr $[\alpha]_D^{15}-100.9$ (w, p=3)	278 (anh)				s	δ	i			chl i	**B27²**, 144
c460	**β-Codeine**	Neopine. 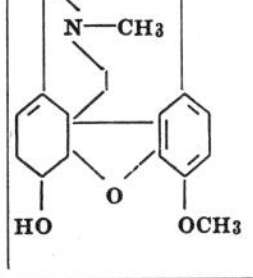	299.37	nd (peth) $[\alpha]_D^{23}-28$ (chl)	127.5				δ	s	s		s	MeOH v chl v lig δ	

For explanations, symbols and abbreviations see beginning of table.

No.	Name	Synonyms and Formula	Mol. wt.	Crystalline form, color and specific rotation	m.p. °C	b.p. °C	Density	n_D	Solubility w	al	eth	ace	bz	other solvents	Ref.
	Codeine														
c461	**Codeine, dihydro-**	Dihydroneopine. Paracodin. 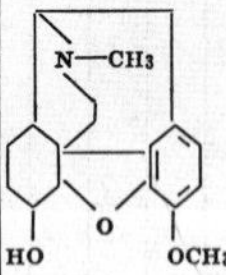	301.39	$[\alpha]_D +146.4$	110–2	248^{15}	......	......	...	...	...	...	...	......	B27^2, 103
c462	**Colchiceine**	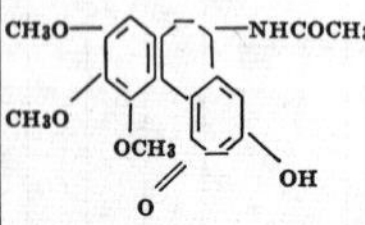	385.40	ye nd $[\alpha]^{20}_D -25.6$	175–7	......	1.24	......	δ	v	i	...	...	chl v	B14^2, 190
c463	**Colchicine**(*l*)	$C_{22}H_{25}NO_6$	399.43	ye pl (w+1½) $[\alpha]^{16.5}_D -121$ (chl)	155–7 (anh)	......	......	......	s	s	s	...	s	chl s MeOH s	B14^2, 191
c464	**Colchicinic acid, trimethyl-**	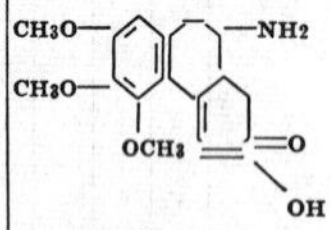	343.37	$[\alpha]^{20}_D -220$	155–7	......	......	......	s	s	...	...	...	......	Am 75, 5292
—	**α-Collidine**	*see* **Pyridine, 4-ethyl-2-methyl-**													
—	**β-Collidine**	*see* **Pyridine, 3-ethyl-4-methyl-**													
—	**γ-Collidine**	*see* **Pyridine, 2,4,6-trimethyl-**													
—	**Comenic acid**	*see* **4-Pyrone-2-carboxylic acid, 5-hydroxy-**													
c465	**Conessine**	Neriine. Wrightine. $C_{24}H_{40}N_2$	356.58	pl or nd (ace) $[\alpha]^{20}_D +25.3$	125–6	$165\text{–}7^{0.1}$	......	......	δ	...	...	...	...	aa s	C28, 5073
c466	**Congo red**		696.66	pw	......	......	......	......	δ	s	i	...	...	......	B26^2, 222
	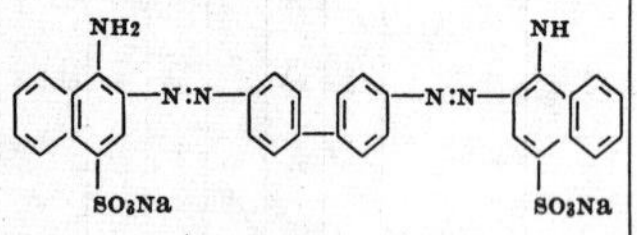														
c467	**Conhydrine**	α-Hydroxyconiine. 2(α-Hydroxypropyl)-piperidine. $C_8H_{17}NO$.	143.23	lf (eth) $[\alpha]^{21}_D +7.10$ (al, p=5)	120.6	226^{760}	......	......	s	v	v	...	...	......	B21^2, 11
c467[1]	—(*dl*)(lower m.p.)	$C_8H_{17}NO$	143.23	nd (peth)	69–70	sub	......	......	v	v	v	...	v	......	B21^2, 12
c467[2]	—(*dl*)(higher m.p.)	$C_8H_{17}NO$	143.23	nd (eth)	98–9	sub	......	......	v	v	v	...	v	......	B21^2, 11
c468	**α-Coniceine**(*d*)	2-Methylconidine 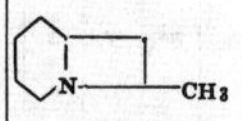	125.21	$[\alpha]^{15}_D +18.4$ (al, c=2)	−16	158^{760}	$0.891^{15.5}_{4}$	......	δ	s	...	...	...	......	B20, 152
c469	—(*inactire*)	$C_8H_{15}N$. *See* c468	125.21	......	47–9	156–9	$0.890^{15.5}_{4}$	......	δ	...	...	...	...	......	B20, 151
c470	**β-Coniceine**(*d*)	2-Propenylpiperidine 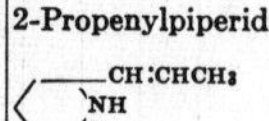	125.21	nd $[\alpha]^{45}_D +49.9$	38.5–9.0	168–9	......	......	δ	s	s	...	...	......	B20, 146
c471	—(*dl*)	$C_8H_{15}N$. *See* c470	125.22	nd	18	168.5–70^{752}	0.8716^{15}_{4}	......	...	...	...	...	...	......	B20, 146
c472	—(*l*)	$C_8H_{15}N$. *See* c470	125.22	nd $[\alpha]^{45}_D -50.5$	41	168.5–9^{760}	0.8520^{50}_{4}	......	δ	...	s	...	...	......	B20, 146
c473	**γ-Coniceine**	$C_8H_{15}N$	125.22	......	......	$173\text{–}4^{760}$ 76.8^{24}	0.8720^{20}_{4}	1.4607^{18}	δ	s	...	...	...	......	B20, 144
—	**δ-Coniceine**	*see* **Piperolidine**													
c474	**ε-Coniceine**(*d*)	*d*-2-Methylconidine. 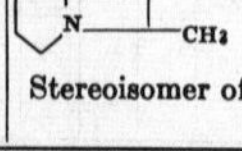 Stereoisomer of c468	125.22	$[\alpha]^{15}_D +67.4$	......	151.5–4.0	0.8856^{15}_{4}	......	...	...	...	...	...	......	......

For explanations, symbols and abbreviations see beginning of table.

No.	Name	Synonyms and Formula	Mol. wt.	Crystalline form, color and specific rotation	m.p. °C	b.p. °C	Density	n_D	Solubility w	al	eth	ace	bz	other solvents	Ref.
	ε-Coniceine														
c475	—(*dl*)	$C_8H_{15}N$. *See* c474	125.22			158–61	0.8836^{15}_{4}								
c476	—(*l*)	$C_8H_{15}N$. *See* c474	125.22	$[\alpha]^{15}_{D}-41.6$		151–3.5	0.8842^{15}_{4}								
c477	**Conidendrin**	Tsugalactone. Tsugaresinol. $C_{20}H_{20}O_6$	356.37	$[\alpha]^{22}_{D}-53.7$ (ace, 2.12%)	255–6				i	δ^h	δ^h				
—	**Conidine, 2-methyl-**	*see* **α-Coniceine**													
c478	**Conidine, 3-methyl-(*d*)**	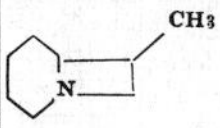 CH_3, N	125.22	$[\alpha]^{18}_{D}+16$	151–4		0.8856^{15}_{4}								**B20**, 153
c479	—,—(*dl*)	$C_8H_{15}N$. *See* c478	125.22		158^{760}		0.8836^{15}_{4}		δ	s	s				**B20**, 153
c480	—,—(*l*)	$C_8H_{15}N$. *See* c478	125.22	$[\alpha]^{17}_{D}-17.1$	143–5		0.8856^{15}_{4}								**B20**, 153
—	**Coniferaldehyde**	*see* **Cinnamaldehyde, 4-hydroxy-3-methoxy-**													
c481	**Coniferin**	$C_{16}H_{22}O_8$	342.35	nd (w+2) $[\alpha]^{20}_{D}-66.9$ (w, c=6)	185 (anh)				s^h	δ	i				**B31**, 221
c482	**Coniferyl alcohol**	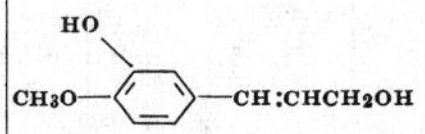 HO, CH_3O–, –CH:CHCH$_2$OH	180.21	pr	73–4	$163\text{–}5^{3}$			δ^h	s	s			alk s	**B6**^{2}, 1093
c483	**Coniine(*d*)**	Conicine. 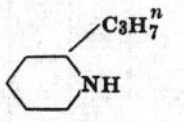 $C_3H_7^n$, NH	127.23	$[\alpha]^{20}_{D}+15.6$	−2.5	168^{760}	0.8430^{22}_{4}	$1.4512^{21.9}$	δ	∞	∞		s	chl δ	**B20**^{2}, 61
c484	—(*dl*)	$C_8H_{17}N$. *See* c483	127.23			$166\text{–}7^{745}$ $59\text{–}63^{17}$	0.8447^{20}	1.4581^{15}	δ					oos s	**B20**^{2}, 62
c485	—(*l*)	$C_8H_{17}N$. *See* c483	127.23	$[\alpha]^{15}_{D}-15.6$		166^{760}	0.845^{15}_{4}		δ	v				oos s	**B20**^{2}, 62
c486	—,hydrobromide(*d*)	$C_8H_{17}N.HBr$. *See* c483	208.15	pr	211				v	v				oos v	**B20**, 112
c487	—,hydrochloride(*d*)	$C_8H_{17}N.HCl$. *See* c483	163.69	orh (w) $[\alpha]^{20}_{D}+10.1$ (liq NH_3. c=8)	221				v	v	δ	δ		chl s	**B20**^{2}, 61
c488	—,—(*dl*)	$C_8H_{17}N.HCl$. *See* c483	163.69	nd (al-eth)	215–6				v	s	δ				**B20**^{1}, 31
c489	—,—(*l*)	$C_8H_{17}N.HCl$. *See* c483	163.69	nd	220–1				v	v	i				**B20**, 118
c490	—,picrate(*d*)	$C_8H_{17}N.C_6H_3N_3O_7$. *See* c483	356.33	ye pr (w)	75				s^h	s	s				**B20**, 112
c491	—,—(*l*)	$C_8H_{17}N.C_6H_3N_3O_7$. *See* c483	356.33		74										**B20**^{2}, 62
c492	—,*N*-**methyl-(*d*)**	$C_9H_{19}N$. *See* c483	141.25	$[\alpha]^{24}_{D}+81$		173–4	0.8318^{24}	1.4538^{15}	δ	s				oos s	
c493	**Conquinamine**	$C_{19}H_{24}N_2O_2$	312.40	ye tet r $[\alpha]^{15}_{D}+200$(al)	123				δ	s	s			chl s	
—	**Conyrine**	*see* **Pyridine, 2-propyl-**													
c496	**Copaene**	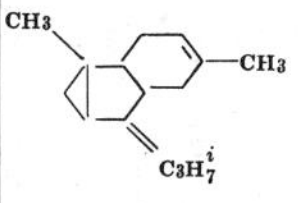 CH_3, –CH_3, $C_3H_7^i$	204.34	$[\alpha]_D$-13.3		246–51 $119\text{–}20^{10}$	0.9077^{15}_{15}	1.4894^{20}	i		s	s		aa s lig s	**B5**^{2}, 349
c497	**Coproergostane**	Pseudoergostane. 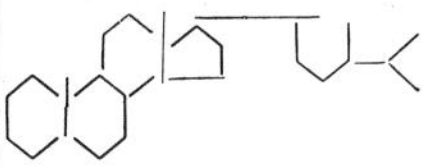	386.68	nd (ace) $[\alpha]^{19}_{D}+25.3$ (al)	64				i		s	s^h		chl s	**E14s**, 1439
c498	**Coprostane**	Pseudocholestane. 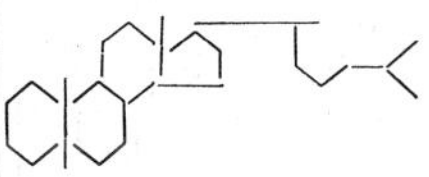	372.65	orh nd (al, ace or eth-al) $[\alpha]^{20}_{D}+25.5$ (chl)	72		$0.912^{87.7}_{4}$	1.4884^{88}_{4}	i	δ	v			chl v oos v	**E14s**, 1431
c499	3-(β)-**Coprostanol**	Coprosterol. Strecorin. $C_{27}H_{48}O$	388.65	nd (MeOH) $[\alpha]^{18}_{D}+28$ (chl)	102				i	v	v		v	chl v MeOH δ	**E14s**, 1704

For explanations, symbols and abbreviations see beginning of table.

No.	Name	Synonyms and Formula	Mol. wt.	Crystalline form, color and specific rotation	m.p. °C	b.p. °C	Density	n_D	Solubility w	al	eth	ace	bz	other solvents	Ref.
	Coprostenol														
c500	**Coprostenol**	Allocholesterol. 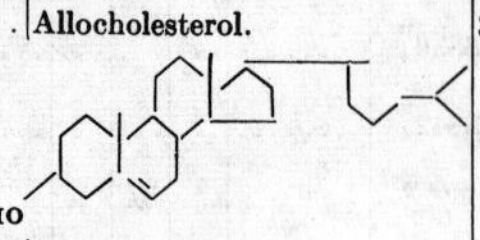	386.67	nd (eth) $[\alpha]_D +43.7$ (bz)	132	...	...	...	i	s	v	v	v	chl v	**E14**, 37
c501	**Coprostenone**	Δ^4-Cholesten-3-one. 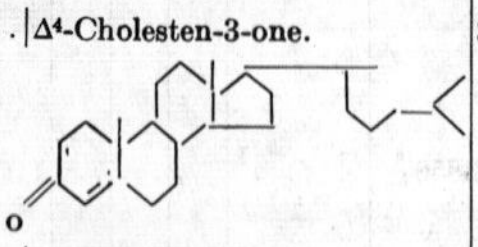	384.65	(MeOH) $[\alpha]_D +88.6$ (chl)	81–2	...	...	...	...	s	v	...	v	lig v	**E14**, 120
—	**Coprosterol**	*see* **Coprostanol**													
c502	**Coronene**	Hexabenzobenzene. 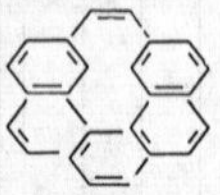	300.34	ye nd (bz)	438–40	525^{760}	...	...	...	...	...	...	s	...	
c503	**Corticosterone**	11,21-Dihydroprogesterone. $C_{21}H_{30}O_4$	346.45	nd (al+?), pl (ace) $[\alpha]_D^{15} +223$ (al)	180–22	sub $190^{00.1}$	...	...	...	...	...	...	...	...	
c504	—,**11-dehydro-**	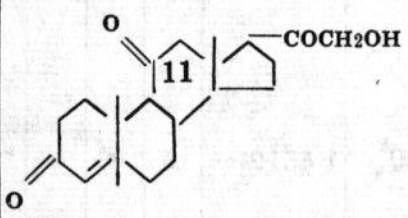	344.43	pr (ace-w, al, ace or ace-eth) $[\alpha]_D^{25} +258$ (al)	177–9	distb vac	...	...	i	s	...	s	s	...	**E14s**, 2972
c505	—,**17-hydroxy-**	Cortisol. Hydrocortisone. $C_{21}H_{30}O_5$	362.45	pl (al or *i*-PrOH) $[\alpha]_D^{22} +167$ (al)	217–20	...	...	...	δ^h	s^h	...	...	...	diox s aa s	**E14s**, 2868
—	**Cortisol**	*see* **Corticosterone, 17-hydroxy-**													
c506	**Cortisone**	17-hydroxy-11-dehydro-corticosterone. $C_{21}H_{28}O_5$.	360.43	rhd nd (al) $[\alpha]_D^{25} +209$ (al)	217	...	...	...	i	v	δ	v	δ	chl δ, MeOH v sulf s	**E14s**, 2977
c507	—,21-acetate	$C_{23}H_{30}O_6$	402.47	...	240–5	...	...	...	i	...	i	...	...	...	...
c508	**Corybulbine**(*d*)	Corydalis-G. 3-Hydroxy-13-methyl-2,9,10-trimethoxyberbine. 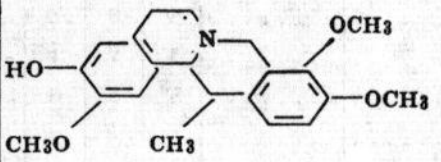	355.42	nd (al) $[\alpha]_D^{20} +303$ (chl, c=1.4)	242–3	...	...	...	i	δ	δ	...	s^h	chl s HCl s	**B21**^{2}, 200
c509	—(*dl*)	$C_{21}H_{25}NO_4$. *See* c508	355.42	(chl-al)	220–2	...	...	...	i	δ	δ	...	s^h	...	**B21**, 217
c510	**Corycavamine**	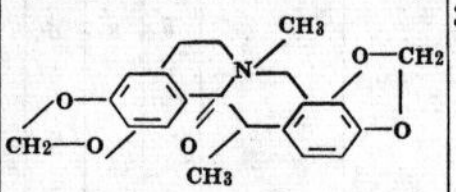	367.40	pr (eth) $[\alpha]_D^{20} +166.6$ (chl, c=2.2)	149	...	...	...	...	...	...	...	...	...	**B27**^{2}, 621
c511	**Corycavine**	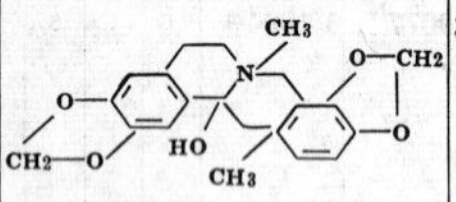	367.40	orh pl (al)	218	...	...	...	δ	v	...	...	...	chl s alk i	**B27**^{2}, 621
c512	**Corydaldine**	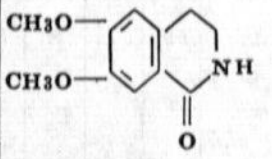	207.23	mcl pr (w or al)	175	...	...	...	s	s	s	...	s	chl s lig i	...
c513	**Corydaline**(*d*)	$C_{22}H_{27}NO_4$	369.46	pr (al) $[\alpha]_D^{20} +311$	135	...	...	...	i	s^h	s	...	s	chl s CS_2 s	**B21**^{2}, 200
c514	—(*dl*)	$C_{22}H_{27}NO_4$	369.46	(al)	135–6	...	...	...	...	...	...	...	...	...	**B21**^{2}, 200
c515	—(*meso, d*)	$C_{22}H_{27}NO_4$	369.46	pr (eth) $[\alpha]_D +180$ (chl, c=3)	155–6	...	...	...	...	...	...	...	...	...	**B21**^{1}, 257
c516	—(*mesco,dl*)	$C_{22}H_{27}NO_4$	369.46	(al)	163–4	...	...	...	...	...	...	...	...	...	**B21**^{2}, 200
c517	—(*mesco,l*)	$C_{22}H_{27}NO_4$	369.46	pr (eth) $[\alpha]_D -181$ (chl, c=3)	155–6	...	...	...	...	...	...	...	...	...	**B21**^{1}, 257

For explanations, symbols and abbreviations see beginning of table.

No.	Name	Synonyms and Formula	Mol. wt.	Crystalline form, color and specific rotation	m.p. °C	b.p. °C	Density	n_D	Solubility w	al	eth	ace	bz	other solvents	Ref.
	Corynantheine														
c518	**Corynantheine(β)**	$C_{22}H_{28}N_2O_4$	384.48	$[\alpha]_D^{18}$+28.5 (MeOH, c=1.0)	165–6	...	...	...	...	...	...	...	...	...	**C47**,6601
—	**Corynine**	*see* **Yohimbine**													
c519	**Cotarnine**	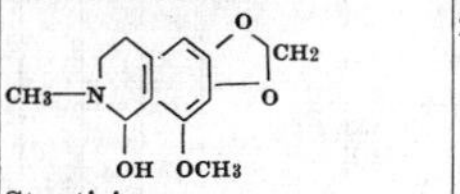	237.26	nd (bz), (eth)	132–3d	...	...	...	δ	s	s	...	s	chl s Na_2CO_3 s NH_4OH s NaOH i	**B27²**, 543
c520	—,chloride	Stypticin. $C_{12}H_{14}ClNO_3.2H_2O$	291.73	pw or nd (al-AcOEt)	197	...	...	...	s	s	...	...	...	...	**B27¹**, 456
c521	—,*O*(hydrogen phthalate)	$C_{12}H_{14}NO_3.O_2CC_6H_4CO_2H$. *See* c519	385.38	ye	115	...	...	...	δ	...	...	...	...	...	**B27**, 476
c522	—,*O*-phthalate	Styptol. $(C_{12}H_{14}NO_3)_2.C_6H_4(CO_2)_2$. *See* c519	604.62	ye	103	...	...	...	s	...	...	...	...	...	**B27**, 476
—	**Cotoin**	*see* **Benzophenone, 2,6-dihydroxy-4-methoxy-**													
c523	**Coumalic acid**	α-Pyrone-5-carboxylic acid. 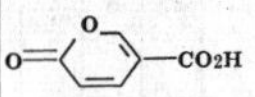	140.10	pr (MeOH)	205–10δd	218^{120}	...	...	δ	s	δ	δ	i	chl i aa s	**B18²**, 326
c524	**Coumaran**	2.3-Dihydrobenzofuran. Dihydrocoumarone. 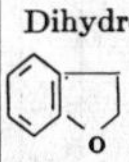	120.15	...	−21.5	188–9 76^{14}	1.058_4^{24}	1.5426^{20}	i	s	s	...	...	chl s CS_2 s	**B17¹**, 22
c525	**3-Coumaranone**	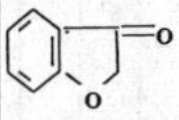	136.15	red nd (on standing)	102	$152–4^{16}$	...	...	δ	...	...	...	s	lig δ	**B17²**, 126
—	**Coumaric acid**	*see* **Cinnamic acid, hydroxy-**													
—	***o*-Coumaric acid, lactone**	*see* **Coumarin**													
—	**Coumarilic acid**	*see* **Coumarone-2-carboxylic acid**													
c526	**Coumarin**	*o*-Coumaric acid lactone. Coumarinic lactone. 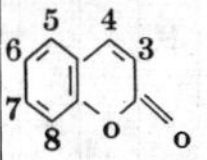	146.14	rhd pym (eth)	71	301.72^{760}	0.935_4^{20}	...	s^h	...	v	...	...	Py v chl v	**B17²**, 357
c527	—,**6-chloro-**	$C_9H_5O_2Cl$. *See* c526	180.59	nd (al)	165	...	...	...	...	v^h	v^h	...	s	CS_2 v	**B17²**, 359
c528	—,**7-diethyl-amino-4-methyl-**	$C_{14}H_{17}NO_2$. *See* c526	231.30	(al), (bz-lig)	89 (al) 135 (bz-lig)	...	...	...	δ^h	...	...	...	...	oos s	**B18**, 612
c529	—,**3,4-dihydro-**	Hydrocoumarin. Melilotic lactone. Melilotol. $C_9H_8O_2$. *See* c526	148.15	lf	25	272 145^{13}	1.169^{18}	1.5495^{18}	i	δ	δ	...	...	chl s	**B17²**, 334
c530	—,**7,8-dihydroxy-**	Daphnetin. $C_9H_6O_4$. *See* c526	178.14	yesh (dil al)	261–3	sub	...	...	s^h	s^h	δ	...	δ	chl δ CS_2 δ	**B18²**, 69
c531	—,**6,7-dihydroxy-**	Aesculetin. Esculetin. $C_9H_6O_4$. *See* c526	178.14	nd (aa+1w), lf (sub)	276	sub	...	...	δ	s^h	δ	s	...	alk, chl s AcOEt s	**B18²**, 68
c532	—,**5,7-dihydroxy-4-methyl-**	$C_{10}H_8O_4$. *See* c526	192.17	nd (al), lf (aa)	282–4	...	...	...	δ	v^h	δ	...	δ	alk v, chl δ	**B18¹**, 351
c533	—,**6,7-dihydroxy-4-methyl-**	4-Methylesculetin. $C_{10}H_8O_4$. *See* c526	192.17	ye nd (dil al)	269–70	...	...	...	s^h	s^h	...	...	...	aa s^h	**B18¹**, 351
c534	—,**7,8-dihydroxy-4-methyl-**	4-Methyldaphnetin. $C_{10}H_8O_4$. *See* c526	192.17	nd (w), pr (al)	235	...	...	...	i	s	...	...	...	...	**B18¹**, 352
c534¹	—,**7,8-dihydroxy-6-methoxy-**	**Fraxetin.** $C_{10}H_8O_5$. *See* c536	208.16	pl (al)	228	...	...	...	δ^h	δ s^h	δ	...	...	...	**B18²**, 152
c535	—,**5,7-dimethoxy-**	Limettin. $C_{11}H_{10}O_4$. *See* c526	206.19	pr or nd (al)	147.5	200δd	...	...	δ^h	v^h	i	v	...	chl v lig δ	**B18¹**, 348
c536	—,**7-ethoxy-4-methyl-**	Maraniol. $C_{12}H_{12}O_3$. *See* c526	204.22	wh	113–4	...	...	...	i	s	...	...	...	...	...
c537	—,**4-hydroxy-**	Benzotetronic acid. $C_9H_6O_3$. *See* c526	162.14	nd	212–3	...	...	...	s	s	s	...	...	...	**B17²**, 469
c538	—,**6-hydroxy-**	$C_9H_6O_3$. *See* c526	162.14	nd (dil HCl)	248–50	...	1.25	...	δ	s	...	...	...	AcOEt s	**B18¹**, 306
c539	—,**7-hydroxy-**	Umbelliferone. $C_9H_6O_3$. *See* c526	162.14	nd (w)	225–8	sub	...	...	δ^h	∞	∞	...	...	chl ∞ aa ∞	**B18²**, 16
c540	—,**8-hydroxy-**	$C_9H_6O_3$. *See* c526	162.14	nd (dil al)	280–5	...	...	...	s^h	s	i	...	i	chl i aa s	**B18¹**, 307
c541	—,**7-hydroxy-6-methoxy-**	Chrysatropic acid. Gelseminic acid. Scopoletin. $C_{10}H_8O_4$. *See* c526	192.16	nd or pr	204	...	...	...	δ	s^h	δ	...	i	chl s CS_2 i aa s^h	**B18²**, 68

For explanations, symbols and abbreviations see beginning of table.

No.	Name	Synonyms and Formula	Mol. wt.	Crystalline form, color and specific rotation	m.p. °C	b.p. °C	Density	n_D	Solubility w	al	eth	ace	bz	other solvents	Ref.
	Coumarin														
c542	—,7-hydroxy-4-methyl-	4-Methylumbelliferone. $C_{10}H_8O_3$. *See* c526	176.17	nd (al)	185–6				δ^h	s	δ			aa s	$B18^2$, 19
c543	—,4-hydroxy-3(1-phenyl-3-oxobutyl)-	Warfarin.	308.32		157–61				i	s		s		diox s	
c544	—,3-methyl-	$C_{10}H_8O_2$. *See* c526	160.16	rh bipym (al)	91	292.5			δ^h	δ				aq KOH i	$B17^2$, 362
c545	—,4-methyl-	$C_{10}H_8O_2$. *See* c526	160.16	nd (w), pr (bz)	83–4					s			s	alk s^h	$B17^2$, 362
c545[1]	—,7-methoxy-	Herniarin. $C_{10}H_8O_3$. *See* c526	176.17	lf	117				δ	s					
c546	—,6,7,8-trimethoxy-	Fraxetin dimethyl ether. $C_{12}H_{12}O_5$. *See* c526	236.22	rh bipym pl (dil al)	103–4	$90–100^{0.2}$				s	s				$B18^2$, 152
c547	**3-Coumarincarboxylic acid**		190.15	nd (w or bz)	190				s^h	s	i		i	alk s lig i	$B18^2$, 336
c548	**2-Coumaronecarboxylic acid**	Coumarilic acid.	162.14	nd (w)	192–3	310–5δd			δ^h	s				CS_2, chl δ	$B18^2$, 276
c549	—,ethyl ester	$C_{11}H_{10}O_3$	190.20		30–1	274^{760} 161^{15}	$1.1656^{28.5}_{4}$	$1.564^{27.6}$							$B18^1$, 442
c550	—,3-methyl-	$C_{10}H_8O_3$. *See* c548	176.17	nd (dil al)	188–9					s^h					B18, 309
—	**Coumaryl alcohol**	*see* **2-Propen-1-ol, 3(hydroxyphenyl)-**													
c551	**Coumestrol**	3,9-Dihydroxy-6H-benzofuro-[3,2-c] [1]-benzopyran-6-one.	268.21	gy micr rods	385d				i^h	δ	i	δ^h			
c552	—,diacetate	$C_{19}H_{12}O_7$. *See* c551	352.29	pl	234				i^h	i	i	δ^h			Am80, 4381
c553	**Creatine**	(α-Methylguanido)acetic acid. $NH_2C(:NH)N(CH_3)CH_2CO_2H$	131.14	mcl pr (w+1)	303				s	i	i				$B4^2$, 796
c554	**Creatinine**	1-Methylglycocyamidine.	113.12	rh pr (w+2)	260d				s	δ					$B24^2$, 128
—	**Cresol**	*see* **Toluene, hydroxy-**													
—	**Cresolol**	*see* **Toluene, 4-hydroxy-3-methoxy-**													
—	**Cresorcinol**	*see* **Toluene, 2,4-dihydroxy-**													
—	**Cresotic acid**	*see* **Benzoic acid, 2-hydroxy-3-methyl-**													
—	**Croceic acid**	*see* **1-Naphthalenesulfonic acid, 7-hydroxy-**													
c555	**Crocetin**(*trans*)	$C_{20}H_{24}O_4$	328.41	brick red rh (Ac_2O)	285d				δ	δ				Py s oos δ NaOH v	B30, 106
—	**Crocic acid**	*see* **Croconic acid**													
c556	**Croconic acid**	Corcic acid.	142.07	pa ye nd	eff 100 sub				v	s					$B8^2$, 532
—	**Crotonaldehyde**	*see* **2-Butenal***													
—	**Crotonic acid**	*see* **2-Butenoic acid***													
—	**Crotyl alcohol**	*see* **2-Buten-1-ol***													
—	**Cryogenine**	*see* **Benzoic acid, 3(1-semicarbazido)-, amide**													
c557	**Cryptopine**	$C_{21}H_{23}NO_5$	369.42	pr or pl (bz)	220–1		1.315^{20}_{4}		i	δ	δ		δ	chl s aa s	$B27^2$, 578

For explanations, symbols and abbreviations see beginning of table.

No.	Name	Synonyms and Formula	Mol. wt.	Crystalline form, color and specific rotation	m.p. °C	b.p. °C	Density	n_D	Solubility w	al	eth	ace	bz	other solvents	Ref.
	Cryptoxanthin														
c558	**Cryptoxanthin**	Carioxanthin. β-Caroten-3-ol. Cryptoxanthol. $C_{40}H_{56}O$	552.89	garnet red pr (bz-MeOH) $[\alpha]^{18}_{643.5} \pm 6$ (bz)	169				i	δ			v	chl v Py v lig δ	**B30**, 93
c559	**Cubebin**	$C_{20}H_{20}O_6$	356.38	nd (al or bz) $[\alpha]^{25}_D -45.6$ (chl, c = 5)	131–2				i	s	s			chl s	**B19**², 470
—	**Cumaldehyde**	*see* **Benzaldehyde, 4-isopropyl-**													
—	**Cumene**	*see* **Benzene, isopropyl-***													
—	**Cumenol**	*see* **Benzene, hydroxy-(isopropyl)-***													
—	**Cumic acid**	*see* **Benzoic acid, 4-isopropyl-**													
—	**Cumic alcohol**	*see* **Benzene, 1(hydroxymethyl)-4-isopropyl-***													
—	*m*-**Cumidic acid**	*see* **1,3-Benzenedicarboxylic acid, 4,6-dimethyl-***													
—	**Cumidine**	*see* **Benzene, 1-amino-4-isopropyl-***													
—	**Cuminic acid**	*see* **Benzoic acid, isopropyl-**													
—	**Cuminylamine**	*see* **Benzene, 1-amino-methyl-4-isopropyl-***													
c563	**Cupreine**	Hydroxycinchonine. 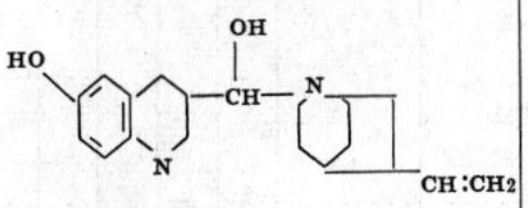	310.40	pr (eth) $[\alpha]^{17}_D -175.5$	198 (anh)				i	s	δ		δ	chl δ	**B23**, 510
c564	**Curarine**	$C_{19}H_{26}N_2O$	298.43	lf	161				i	s					
c565	*ar*-**Curcumene**	$C_{15}H_{22}$	202.34	$[\alpha]^{18}_D +35.8$		117–8⁷	0.8797^{15}_{15}	1.5029^{15}							**B5**², 402
c566	**Curcumin**	$C_{21}H_{20}O_6$	368.39	or ye pr, rh pr (MeOH)	183				i	δ				alk s lig i CS_2 δ	**B8**, 554
—	**Curine**	*see* **Bebeerine**(*l*)													
c567	**Cuscohygrine**	α,α′-Bis(*N*-methyl-α-pyrrolidyl)acetone. Cuskhygrine. $C_{13}H_{24}N_2O$	224.35			185^{32} 170^{23}	0.9782^{16}_4		∞						**B24**, 78
c568	—,hydrate	$C_{13}H_{24}N_2O.3\frac{1}{2}H_2O$	287.40	nd	40–1						s		s		**B24**, 78
c569	**Cusparine**	$C_{19}H_{17}NO_3$	307.35	nd (lig)	92					s	s	v	v	chl v lig δ	**B27**, 483
c570	**Cyamelide**	*sym*-Trioxanetriimine. 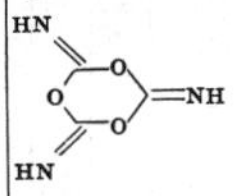	129.08	amor pw		d	1.127^{15}_4			i	i			oos i NH_4OH δ con sulf s	**B3**, 35
—	**Cyanacetamide**	*see* **Malonic acid**, mono-amide mononitrile													
c571	**Cyanamide**	Carbamic acid nitrile. Carbamonitrile. Carbodiimide. H_2NCN	42.04	nd	44	140^{19}	1.0729^{18}_4	1.4418^{48}	v	v	s		s	chl s CS_2 δ	**B3**, 76
c572	—,**benzyl-**	$C_6H_5CH_2NHCN$	132.17	pl (eth)	43				i	s	s				**B12**, 1051
c573	—,**diallyl-**	$(CH_2{:}CHCH_2)_2NCN$	122.17			140–5⁹⁰ 95⁹			i	s			s	oos s	**B4**², 666
c574	—,**dibutyl-**	$(C_4H_9^n)_2NCN$	154.26			187–91¹⁹⁰ 146–51³⁵			i					oos s	**B4**², 635
c575	—,**diethyl-**	*N*-Cyanodiethylamine. $(C_2H_5)_2NCN$	98.15			190^{760} 68^{10}	0.854^{20}_4	1.4126^{48}	i	s	s			HCl d	**B4**, 121
c576	—,**dimethyl-**	*N*-Cyanodimethylamine. $(CH_3)_2NCN$	70.09			163.5^{760} 68^{15}			v	s	s	s			**B4**², 474
c577	—,**diphenyl-**	*N*-cyanodiphenylamine. $(C_6H_5)_2NCN$	194.23	pr (al)	73–4	235–40⁶⁰			i	s^h				MeOH s^h lig s	**B7**, 430
c578	—,**methyl(1-naphthyl)-**	*N*-methyl-N-cyano-α-naphthylamine. $C_{10}H_7^{\alpha}N(CH_3)CN$	182.23	yesh		189–91⁵				s	s				**B12**², 697

For explanations, symbols and abbreviations see beginning of table.

No.	Name	Synonyms and Formula	Mol. wt.	Crystalline form, color and specific rotation	m.p. °C	b.p. °C	Density	n_D	Solubility w	al	eth	ace	bz	other solvents	Ref.
	Cyanamide														
c579	**—,phenyl-**	Carbanilonitrile. Cyananilide. *N*-Cyanoaniline. $C_6H_5NHCN.\frac{1}{2}H_2O$	127.15	lf (aa, dil KOH)	47(+3½w)				δ	v	v			KOH s	**B12**, 368
c580	**Cyanic acid**	Carbonimid. Isocyanic acid. HOCN	43.03	gas	−79 to −81	<0	1.140^{0}_{0} (hq)		s		s		s	chl s aa s	**B3**, 31
c581	—,ethyl ester	C_2H_5OCN	71.08			162d	0.89^{20}_{4}		i	∞	∞				
—	**Cyanoacetic acid**	*see* **Malonic acid,** mononitrile													
—	**Cyanoacetanilide**	*see* **Malonic acid,** monoamide mononitrile, *N*-phenyl-													
c581[1]	**Cyanogen**	Oxalic acid dinitrile. NCCN	52.04	gas	−34.4	-21.17^{760}		$0.953^{-21.17}_{4}$	s	s	s				**B2²**, 511
c582	**Cyanogen bromide**	Bromine cyanide. BrCN	105.93	nd	52	61.4^{760}	2.015^{20}_{4}		s	s	s				**B3**, 39
c583	**Cyanogen chloride**	Chlorine cyanide. ClCN	61.48		−6.9	12.66^{760}	1.186^{20}_{4}		δ	s					**B3**, 38
c584	**Cyanogen iodide**	Iodine cyadine. ICN	152.94	nd (al)	146–7	sub >45	2.59		s		s				**B3**, 41
c585	**Cyanogen sulfide**	Cyanogen thiocyanate. $S(CN)_2$	84.10	rh pl	60	sub 30–40			s		s				**B3**, 180
c586	**Cyanuric acid**	2,4,6-Triazinetriol*. Tricyanic acid. Trihydroxycyanidine. (structure: triazine ring with three OH . $2H_2O$)	129.08	mcl (w+2)	d	d	1.768^{0}		δ	δ	i	i	i	chl i aa i	**B26²**, 132
—	—,triamide	*see* **Melamine.**													
c587	—,tribenzyl ester	Benzyl cyanurate. $C_{24}H_{21}N_3O_3$. *See* c586	399.43	nd (al)	157	>320			i	s	δ				**B26¹**, 76
c588	—,trichloride	Cyanuric chloride. Trichloro-*s*-triazine*. Trichlorocyanidine. Tricyanogen chloride. (structure: triazine ring with three Cl)	184.43	(eth or bz)	145	190^{720}			i	s					**B26²**, 116
c589	**—,thio-**	Trithiocyanuric acid. (structure: triazine ring with three SH)	177.28	ye pr	200d				δ	i	δ				**B26**, 259
—	**Cyanurotriamide**	*see* **Melamine**													
—	**Cyclamen aldehyde**	*see* **Propanal, 3(4-isopropyl-phenyl)-2-methyl-***													
—	**Cyclobarbital**	*see* **Barbituric acid, 5(1-cyclohexenyl)-5-ethyl-**													
c590	**Cyclobutane***	Tetramethylene (structure: cyclobutane ring)	56.10		−50	13^{760}	0.703^{0}_{4}	1.4260^{20}	i	v	∞	v			**B5**, 17
c591	**—,ethyl*-**	C_6H_{12}. *See* c590	84.16		−143.2	72.5^{754}	0.7284^{20}_{4}	1.4004^{20}							**B5¹**, 11
c592	**—,methyl-***	C_5H_{10}. *See* c590	70.13			39–42	0.6931^{20}_{0}	1.3836^{20}	i	∞	∞				**B5¹**, 5
c593	**—,octafluoro-***	Perfluorocyclobutane. C_4F_8. *See* c590	200.04		−40.7	-4^{764}	1.654^{-20}_{4}				s				**Am70**, 4090
c594	**Cyclobutane-carboxylic acid***	Tetramethylenecarboxylic acid. (structure: cyclobutane ring with CO_2H)	100.11		−2	190^{754} $74–5^{2}$	1.0470^{20}_{4}	1.4400^{20}	δ	∞	∞				**B9²**, 5
c595	**1,1-Cyclobutane-dicarboxylic acid***	(structure: cyclobutane ring with two CO_2H)	144.13	pr (eth or w)	156.5 cor				v		s		s	lig δ	**B9²**, 514

For explanations, symbols and abbreviations see beginning of table.

No.	Name	Synonyms and Formula	Mol. wt.	Crystalline form, color and specific rotation	m.p. °C	b.p. °C	Density	n_D	Solubility w	al	eth	ace	bz	other solvents	Ref.
	1,1-Cyclobutanedicarboxylic acid														
c596	—,diethyl ester*	$C_{10}H_{16}O_4$. *See* c595	200.24			228.5–9.5^{755} 96–8^{8}	1.0456$^{20}_{4}$	1.4330^{26}		v					**B9**², 514
c597	**1,2-Cyclobutanedicarboxylic acid** (*cis, dl*)*	Ethylenesuccinic acid. 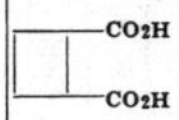	144.12	pl (w), pr (bz)	137–8				v	v	v		δ	lig δ	**B9**², 515
c598	—(*trans, d*)*	$C_6H_8O_4$. *See* c597	144.12	$[\alpha]^{30}_{D}$ +123.3 (w, c = 1.2)	105										**B9**², 515
c599	—(*trans, dl*)*	$C_6H_8O_4$. *See* c597	144.12	rh nd (bz)	130.5–31.0				s					con HCl δ	**B9**², 515
c600	—(*trans, l*)*	$C_6H_8O_4$. *See* c597	144.12	nd (HCl) $[\alpha]^{20}_{D}$ −124.3 (w, c = 0.8)	105										**B9**², 515
c601	**1,3-Cyclobutanedicarboxylic acid** (*cis*)*	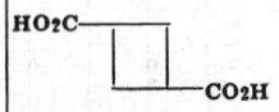 HO_2C–…–CO_2H	144.12	pr (w)	143.0–3.5	252			v	v	δ				**B9**, 726
c602	—(*trans*)*	$C_6H_8O_4$. *See* c601	144.12	pr (w), nd (sub)	170–1	sub			v^{h}		δ				**B9**, 726
c603	**Cyclobutene***	Cyclobutylene. 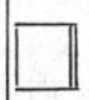	54.09			2	0.733$^{0}_{4}$					v			**B5**¹, 29
c604	—,**hexafluoro-***	Perfluorocyclobutene. C_4F_6. *See* c603	162.04		−61.1, −60.9	0^{764}	1.602$^{-20}_{4}$	1.298^{-20}							**C46**, 7988
c605	**Cyclocamphane**	Epicyclene. Isocyclene. β-Pericyclocamphane. 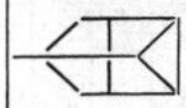	136.24		117–8	150–1^{743}	0.7948^{121}			s				aa s	**E13**, 1044
c606	**1,6-Cyclodecanediol***	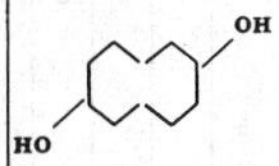	172.27	(AcOEt, chl)	148						s			AcOEt s^{h} chl s^{h}	**B6**², 753
c607	**1,6-Cyclodecanedione***	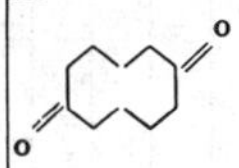	168.24	(eth)	100						s^{h}			aa s	**B7**², 540
c608	**Cyclodecanol***	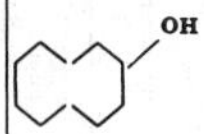	156.27		40–1	118–9$^{10.5}$	0.9626$^{20}_{4}$	1.4926^{20}							**Am74**, 3636
c609	**Cyclodecanone***	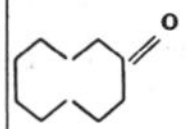	154.25	amor pw	28	106–7^{13}		1.4802^{25}			s		s	chl s	**B7**², 36
c610	**9-Cycloheptadecen-1-one***	Civetone. $C_{17}H_{30}O$	250.43		32.5	342^{745}	0.9135$^{37}_{4}$	1.4820^{37}	δ	s			s		**B7**², 121
c611	**Cycloheptane***	Heptamethylene. Suberane. 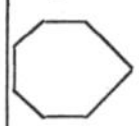	98.19		−12	118.48	0.8109$^{20}_{4}$	1.4449^{20}	i	v	v				**B5**², 15
c612	—,**1-aza-***	Hexamethylenimine. 	99.18			138.0–8.2^{749}	0.8841$^{15}_{4}$	1.4654^{23}							**B20**, 94
c613	—,**bromo-***	Suberyl bromide. $C_7H_{13}Br$. *See* c611	177.09			101.5^{40} 75^{12}	1.2887$^{22}_{4}$	1.4996^{20}	i		v			chl v	**B5**², 15
c614	—**methyl-***	C_8H_{16}. *See* c611	112.22			133–5^{479}	0.7984$^{20}_{4}$	1.4405^{20}							**B5**², 20
c615	**1,2-Cycloheptanedione***	$C_7H_{10}O_2$. *See* c611	126.16	ye	−40	107–9^{17}				s					**C50**, 6327
c616	**Cycloheptanol***	Suberol. Suberyl alcohol. $C_7H_{14}O$. *See* c611	114.19		2	185^{760} 95^{24}	0.9478$^{20}_{4}$	1.4447^{20}	δ	v	v				**B6**², 16
c617	**Cycloheptanone***	Suberone. $C_7H_{12}O$. *See* c611	112.17			178.5–9.5^{760} 71^{19}	0.9508$^{20}_{4}$	1.4608^{20}	δ	v	s				**B7**², 14

For explanations, symbols and abbreviations see beginning of table.

No.	Name	Synonyms and Formula	Mol. wt.	Crystalline form, color and specific rotation	m.p. °C	b.p. °C	Density	n_D	Solubility w	al	eth	ace	bz	other solvents	Ref.
	Cycloheptasiloxane														
c618	**Cycloheptasiloxane, tetradecamethyl-***	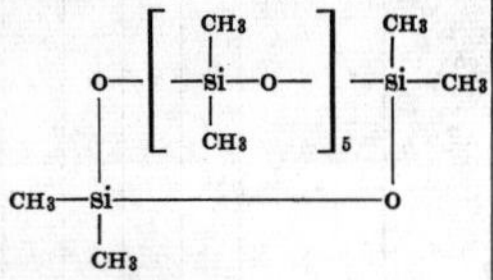	519.09		−26	154^{20}	0.9730	1.4040^{20}							
c619	**1,3,5-Cycloheptatriene***	Tropilidene. 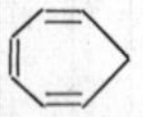	92.14	cubic (at −80)	−79.49	$116–8^{760}$ 60.5^{122}		1.5208^{25}	i	s	s		v	chl v	**B5**, 280
c620	**2,4,6-Cyclohepta-trien-1-one***	Tropone. 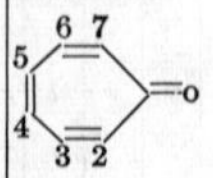	106.13		−7	113^{15} $84–5^{6}$	1.095^{22}_{4}	1.6070^{25}	∞		δ		δ	hexane s	**Am 73**, 876
c621	—,2-amino-*	C_7H_7NO. *See* c620	121.14	ye pl	106–7				s^h	s			s	chl s	**C46**, 7559
c622	—,3-bromo-2-hydroxy-*	$C_7H_5BrO_2$. *See* c620	201.03	ye pl or nd	107–8					s	s				**C49**, 2405
c623	—,2-hydroxy-*	Tropolone. $C_7H_6O_2$. *See* c620	122.12	nd	51	sub 40^{4}			s					oos s	**Am 74**, 4456
c624	—,2-hydroxy-4-isopropyl-*	$C_{10}H_{12}O_2$. *See* c620	164.21	pa ye (peth)	50–1				δ				δ	lig δ	**C45**, 2884
c625	—,2-hydroxy-4-methyl-*	$C_8H_8O_2$. *See* c620	136.15	nd	75–6						s			chl s	**C46**, 5035
c626	—,2-methoxy-*	$C_8H_8O_2$. *See* c620	136.15	pa ye nd (+½w)	41 (+½w)	128^{5}				s			s		**C46**, 4521
c627	**Cycloheptene***	Suberene. Suberylene. 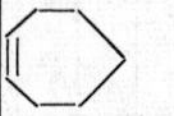	96.17			115^{760}	0.8239^{20}_{4}	1.4545^{20}	i	s	s				**B5²**, 42
c628	**1,3-Cyclohexadiene***	1,2-Dihydrobenzene. 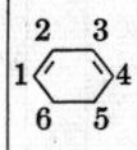	80.13		−98	80.5^{760}	0.8405^{20}_{4}	1.4153^{20}	i	s	v				**B5²**, 79
c629	**1,4-Cyclohexadiene***	1,4-Dihydrobenzene. 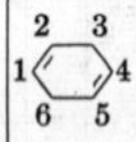	80.13			85.6^{760}	0.8471^{20}_{4}	1.4736^{20}	i	∞	∞				**B5²**, 80
c630	**1,3-Cyclohexadiene, 5-methyl-***	1,2-Dihydrotoluene. C_7H_{10}. *See* c628	94.16			110.0–10.5	0.8354^{20}_{4}	1.4763	i	v	s				**B5¹**, 62
c631	—,octafluoro-*	C_6F_8. *See* c628	224.05			63–4	1.601^{20}	1.3149^{20}							
c632	**1,4-Cyclohexadiene, octafluoro-***	C_6F_8. *See* c629	224.05			57–8		1.318^{18}							**J1954**, 3780
c633	**1,4-Cyclohexadiene-1,2-dicarboxylic acid***	3,6-Dihydrophthalic acid. 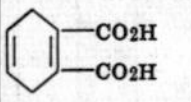	168.15	mcl pr (w)	153				δ						**B9**, 781
c634	**2,4-Cyclohexadiene-1,2-dicarboxylic acid***	2,3-Dihydrophthalic acid. 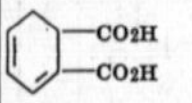	168.15	pr (w or al)	179–80				$δ^h$	s	δ				**B9**, 781
c635	**2,6-Cyclohexadiene-1,2-dicarboxylic acid***	4,5-Dihydrophthalic acid. 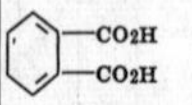	168.15	tcl	215				$δ^h$	s		s			**B9**, 782
c636	**2,5-Cyclohexadien-1-one, hexachloro-***	Hexachlorophenol. 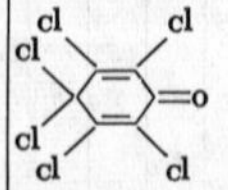	300.78		106								s	CCl_4 s	**B7¹**, 96

For explanations, symbols and abbreviations see beginning of table.

No.	Name	Synonyms and Formula	Mol. wt.	Crystalline form, color and specific rotation	m.p. °C	b.p. °C	Density	n_D	Solubility w	al	eth	ace	bz	other solvents	Ref.
	Cyclohexane														
c637	**Cyclohexane***	Hexahydrobenzene. Hexamethylene.	84.16		6.5	81^{755}	0.7791^{20}_{4}	$1.4266^{19.5}$	i	∞	∞				**B5**, 21
c638	—,**acetyl-**	Cyclohexyl methyl ketone. Hexahydroacetophenone. $C_8H_{14}O$. *See* c637	126.20			$179–80^{780}$	$0.9176^{20.7}_{4}$	$1.4496^{25.9}$	i		s				**B7**^{2}, 23
c639	—,**allyl-**	3-Cyclohexylpropene.* C_9H_{16}. *See* c637	124.23			152–3	0.8135^{20}	1.4500^{20}			s	s			**B5**^{2}, 49
c640	—,**amino-***	Cyclohexylamine. Hexahydroaniline. $C_6H_{13}N$. *See* c637	99.18		−17.7	134.5^{760} 30.5^{15}	0.8668^{20}_{4}	1.4592^{20}	s					oos ∞	**B12**^{2}, 4
c641	—,—,hydrochloride*	$C_6H_{13}N.HCl$. *See* c637	135.64	nd (w or al-eth)	206–7				v	v	s^h				**B12**^{2}, 4
c642	—,**bromo-***	Cyclohexyl bromide. $C_6H_{11}Br$. *See* c637	163.06			$163–5^{760}$ $61–2^{20}$	1.3264^{15}_{4}	1.4956^{15}	i	∞	∞				**B5**, 24
c643	—,**1-bromo-1-methyl-***	$C_7H_{13}Br$. *See* c637	177.09			156–60	1.2544^{18}_{18}	1.4817^{18}		s				chl s	**B5**^{1}, 12
c644	—,**1-bromo-3-methyl-***	$C_7H_{13}Br$. *See* c637	177.09			181.0–1.2 δd 60^{11}	1.268^{15}_{15}	1.4979^{20}	i		v				**B5**^{2}, 18
c645	—,**1-bromo-4-methyl-***	$C_7H_{13}Br$. *See* c637	177.09			130^{200} 55^{15}			i		v				**B5**^{2}, 18
c646	—,**(bromomethyl)-***	$C_7H_{13}Br$. *See* c637	177.09			$76–7^{26}$	1.2690^{25}_{4}	1.4889^{25}	i		s			chl s	**B5**^{2}, 18
c647	—,**butyl-***	1-Cyclohexylbutane.* $C_{10}H_{20}$. *See* c637	140.27		−78.6	178–80 68^{16}	0.8178^{20}_{4}	1.440^{20}	i						**B5**^{2}, 25
c648	—,*sec*-**butyl-**	2-Cyclohexylbutane.* $C_{10}H_{20}$. *See* c637	140.27			177.2^{760}	0.8156^{11}_{4}	1.4487^{11}	i			s			**B5**^{2}, 26
c649	—,*tert*-**butyl-**	$C_{10}H_{20}$. *See* c637	140.27			167–9	0.8305^{16}_{4}	1.4556^{16}	i						**B5**^{2}, 26
c650	—,**(butylamino)-***	*N*-Butylcyclohexylamine. $C_{10}H_{21}N$. *See* c637	155.29			207			δ	v	v				**B12**^{2}, 7
c651	—,**chloro-***	Cyclohexyl chloride. $C_6H_{11}Cl$. *See* c637	118.61		−43.9	142^{750}	1.000^{20}_{4}	1.4626^{20}	i	∞	∞		∞		**B5**, 21
c652	—,**cyclopentyl-***		152.28			209–10.5^{750}	0.8749^{20}_{4}	1.4725^{20}							
c653	—,**1,2-dibromo-** *(cis)**	$C_6H_{10}Br_2$. *See* c637	241.96		6.5	$100–4^{8}$	1.803^{25}_{25}	1.5514^{25}						chl, CCl_4 s lig s	**B5**^{2}, 12
c654	—,—*(trans)**	$C_6H_{10}Br_2$. *See* c637	241.96		−4, −2	$145–6^{100}$ $115–7^{28}$	1.784^{25}_{25}	1.5495^{25}	i					oos s	**B5**^{2}, 12
c655	—,**1,3-dibromo-** *(cis)**	$C_6H_{10}Br_2$. *See* c637	241.96	rods (al)	112				δ	s^h			v	lig δ	**B5**^{2}, 13
c656	—,—*(trans)**	$C_6H_{10}Br_2$. *See* c637	241.96		1	116^{16}				s^h			v		**B5**^{2}, 13
c657	—,**1,4-dibromo-*** (liquid)	$C_6H_{10}Br_2$. *See* c637	241.96			$137–8^{25}$	1.7834^{20}_{4}	1.5531^{20}	i		s				**B5**^{2}, 13
c658	—,—*(solid)	$C_6H_{10}Br_2$. *See* c637	241.96	(eth)	113				i		s^h				**B5**^{2}, 13
c659	—,**1,2-dichloro-** *(cis)**	$C_6H_{10}Cl_2$. *See* c637	153.05		−6, −5	93.5^{22}	1.2018^{20}_{4}	1.4963^{20}					s	CCl_4 s	**B5**^{1}, 8
c660	—,—*(trans)*	$C_6H_{10}Cl_2$. *See* c637	153.05		−7, −6	$74–5^{15}$	1.1842^{20}_{4}	1.4907^{20}							**B5**^{1}, 8
c661	—,**(diethylamino)-***	$C_{10}H_{21}N$. *See* c637	155.29			$192–3^{740}$	0.872^{0}_{0}			s					**B12**^{2}, 6
c662	—,**(difluoroamino)-decafluoro-***	Perfluorocyclohexylamine. $C_6F_{13}N$. *See* c637	333.06			$75–6^{760}$	1.787^{25}_{4}	1.286^{25}							

For explanations, symbols and abbreviations see beginning of table.

No.	Name	Synonyms and Formula	Mol. wt.	Crystalline form, color and specific rotation	m.p. °C	b.p. °C	Density	n_D	Solubility w	al	eth	ace	bz	other solvents	Ref.
	Cyclohexane														
c663	**—,1,1-dimethyl-***	C_8H_{16}. *See* c637	112.22		−33.54	119.2–9.4^{748}	0.7768^{25}_{4}	1.4292^{20}	i	s	s	s	s		$B5^{2}$, 20
c664	**—,1,2-dimethyl-** (*cis*)*	C_8H_{16}. *See* c637	112.22		−50.1	130.04^{760}	0.7962^{20}_{4}	1.4352^{20}	i	s			s	CCl_4 s lig v	$B5^{2}$, 21
c665	—,—(*trans*)*	C_8H_{16}. *See* c637	112.22		−89.4	123.7^{760}	0.7772^{20}_{0}	1.4273^{20}	i	s	s			CCl_4 s lig v	$B5^{2}$, 21
c666	**—,1,3-dimethyl-** (*cis*)*	C_8H_{16}. *See* c637	112.22		−85	124.9	0.7835^{20}_{4}	1.4269^{20}	i	∞	∞				B5, 36
c667	—,—(*trans*)*	C_8H_{16}. *See* c637	112.22	$[\alpha]_{549}+1.33$	−79.4	123.5	0.7663^{20}_{4}	1.4254^{20}							C40, 6427^{4}
c668	**—,1,4-dimethyl-** (*cis*)*	C_8H_{16}. *See* c637	112.22		−87	124.6	0.7827^{20}_{4}	1.4253^{10}							B5, 38
c669	—,—(*trans*)	C_8H_{16}. *See* c637	112.22		−37.2	119.63	0.7626^{20}_{0}								C30, 2180^{8}
c670	**—,1,2-dimethyl-ene-***	(ring)=CH_2, =CH_2	108.18			124^{740} 60–1^{90}	0.8229^{25}_{4}	1.4718^{25}							Am 75, 4780
c671	**—,1,2-epoxy-***	Cyclohexene oxide. 7 Oxabicyclo[4,1,0] heptane. $C_6H_{10}O$. *See* c637	98.15		<−10	131.5^{760}	0.9663^{20}	1.4519^{20}	i	v	v	v	v		$B17^{2}$, 29
c672	**—,1,2-epoxy-4(epoxy-ethyl)-***	4-Vinylcyclohexene dioxide. $C_8H_{12}O_2$. *See* c637	140.18		<−55	227^{760} 20<$^{0.1}$	1.0986^{20}_{20}	1.4787^{20}	v						
c673	**—,1,2-epoxy-4-ethenyl-**	$C_8H_{12}O$. *See* c637	124.18		<−100	169^{760} 20^{2}	0.9598^{20}_{20}	1.4700^{20}	δ						
c674	**—,ethyl-***	C_8H_{16}. *See* c637	112.22		−111.30	131.78^{760}	0.7880^{20}_{4}	1.4332^{20}		s	s	s	s	lig v	$B5^{2}$, 20
c675	**—,(ethylamino)-***	$C_8H_{17}N$. *See* c637	127.23			164^{760} 62–5^{15}	0.868^{0}_{0}		δ	∞	∞				$B12^{2}$, 6
c676	**—,fluoro-***	$C_6H_{11}F$. *See* c637	102.15		12–3	73–5^{300}		1.4142^{25}	i					Py s	Am 77, 3099
c677	**—,hendecafluoro-*(trifluoro-methyl)-**	Perfluoromethylcyclohexane. C_7F_{14}. *See* c637	350.06		−44.72	76.14^{760}	1.7878^{25}_{4}	1.285^{17}						CCl_4 s	J1955, 1749
c678	**—,1,1,2,3,4,5,6-heptachloro-,** (ε, *dl*)	$C_6H_5Cl_7$. *See* c636	325.28	rods	55.0–5.5										C46, 441
c679	**—,1,2,3,4,5,6-hexabromo-** (β or *cis*)*	Benzene-β-hexabromide. $C_6H_6Br_6$. *See* c637	557.57	pr	253d				i	δ	δ		i		B5, 25
c679′	—,—(α or *trans*)	$C_6H_6Br_6$. *See* c637	557.57	mcl pr (xyl)	212				i	δ	δ				B5, 25
c681	**—,1,2,3,4,5,6-hexachloro-(α)***	Benzene-*trans*-hexachloride. $C_6H_6Cl_6$. *See* c637	290.83	mcl pr (al or aa)	158.2	288	1.87^{20}		i	v^{h}			s	chl s PhNH$_2$ v	$B5^{2}$, 11
c682	—,—(β)*	Benzene-*cis*-hexachloride. $C_6H_6Cl_6$. *See* c637	290.83	(bz, al or xyl)	312 sub>312	60$^{0.58}$	1.89^{19}		i	δ			δ	chl δ aa δ	$B5^{2}$, 11
c683	—,—(γ)*	Benzeneγ-hexachloride. Gammexane. Lindane. $C_6H_6Cl_6$. *See* c637	290.83	nd (al)	112.9	60$^{0.48}$			i			s	v		B5, 8
c684	—,—(δ)*	$C_6H_6Cl_6$. *See* c637	290.83	pl	138–9	60$^{0.34}$									B5, 8
c685	**—,iodo-***	Cyclohexyl iodide. $C_6H_{11}I$. *See* c637	210.06			180 δd 68.5–9^{10}	1.626^{15}_{15}	1.547^{20}	i		s			oos v	$B5^{2}$, 13
c686	**—,isobutyl-**	$C_{10}H_{20}$. *See* c637	140.27			169^{754}	0.7950^{20}_{4}		i						$B5^{2}$, 26
c687	**—,isopropyl-***	Hexahydrocumene. Normenthane. C_9H_{18}. *See* c637	126.24		−89.8	154.5^{760}	0.8023^{20}_{4}	1.4409^{20}	i	v	v				B5, 41
—	**—,isopropyl-(methyl)-***	*see* **Menthane**													
c688	**—,methyl-***	Hexahydrotoluene. C_7H_{14}. *See* c637	98.18		−126.35	100.4	0.7695^{20}_{4}	1.4253^{15}	i	s	s				B5, 29
c689	**—,(methyl-amino)-**	$C_7H_{15}N$. *See* c637	113.20			145–7^{748} 76–7^{18}	0.868^{23}	1.4530^{23}	δ	v	∞				$B12^{2}$, 5

For explanations, symbols and abbreviations see beginning of table.

No.	Name	Synonyms and Formula	Mol. wt.	Crystalline form, color and specific rotation	m.p. °C	b.p. °C	Density	n_D	Solubility w	al	eth	ace	bz	other solvents	Ref.
	Cyclohexane														
c690	—,**(3-methyl-butyl)-***	1-Cyclohexyl-3-methylbutane.* Isoamylcyclohexane. $C_{11}H_{22}$. *See* c637	154.29	....	....	193	0.8023^{20}_{4}	1.4423^{20}	i	...	...	...	...	....	**B5**[2], 32
c691	—,**methylene-***...	=CH_2	96.17	....	−106.7	$101\text{–}2^{738}$	0.8004^{25}_{25}	1.4484^{25}	...	...	s	...	...	....	**B5**[2], 44
c692	—,**nitro-***........	$C_6H_{11}NO_2$. *See* c637........	126.16	....	fr. p. −34	205.5^{768} $106\text{–}8^{40}$	1.061^{20}_{4}	1.4612^{19}	...	s	...	...	...	lig s	**B5**[1], 10
c693	—,**pentyl-***.......	$C_{11}H_{22}$. *See* c637........	154.29	....	....	201.4–1.9 $84\text{–}5^{16}$	0.8018^{20}_{4}	1.4438^{20}	i	...	...	...	...	....	**B5**[2], 32
—	—,**phenyl-***.......	*see* **Benzene, cyclohexyl-***													
c694	—,**propyl-***.......	C_9H_{18}. *See* c637........	126.24	....	−94.5	$156\text{–}7^{772}$ $50\text{–}1^{26}$	0.7929^{20}_{4}	1.4370^{20}	i	...	s	...	s	peth v aa s	**B5**[2], 23
—	—,**(2-propyn-1-yl)***	*see* **Propyne, 3-cyclohexyl-***													
c695	—,**1,3,5-tri-methyl-(*cis*)***	Hexahydromesitylene. C_9H_{18}. *See* c637	126.24	....	....	140.5^{752}	0.7773^{20}_{4}	....	...	...	...	...	...	....	**B5**, 45
c696	—,—(*trans*)*......	C_9H_{18}. *See* c637........	126.24	....	....	138.5–9^{754}	0.7720^{20}_{4}	....	...	...	...	...	...	....	....
c697	**Cyclohexanecar-boxaldeyde**	Hexahydrobenzaldehyde. —CHO	112.17	....	....	$155\text{–}8^{760}$ 36^{10}	0.9405^{4}_{0}	1.4496^{20}	s	...	s	...	...	....	**B7**[2], 20
c698	**Cyclohexanecar-boxylic acid***	Hexahydrobenzoic acid. (ring positions 1–6) —CO_2H	128.17	mcl pr	30–1	232–3 100^{1}	1.0334^{22}_{4}	1.4599^{22}	δ	v	...	...	v	chl v	**B9**, 7
c699	—,chloride........	$C_7H_{11}ClO$. *See* c698........	146.62	....	....	184 $75\text{–}7^{15}$	1.0962^{15}_{4}	1.4711^{20}	d	d	...	...	...	....	**B9**, 9
c700	—,ethyl ester*.....	$C_9H_{16}O_2$. *See* c698........	156.22	....	....	196 $74^{4.5}$	0.9362^{20}_{4}	1.4396^{25}	...	v	v	...	...	AcOEt v chl v	**B9**, 8
c701	—,methyl ester*...	$C_8H_{14}O_2$. *See* c698........	142.19	....	....	$181\text{–}3^{750}$ $65\text{–}6^{12}$	0.9954^{15}_{4}	1.4537^{15}	i	...	...	...	...	....	**B9**[1], 5
c702	—,propyl ester*....	$C_{10}H_{18}O_2$. *See* c698........	170.25	....	....	215.5	0.9530^{15}_{4}	1.4486^{15}	i	...	...	...	...	....	**B9**, 8
c703	—,**1,1-azobis-,** dinitrile	1,1 Bicyanoazocyclohexane. (CN, NC; —N:N—)	244.37	(lig)	ca. 100	....	....	....	i	...	...	...	...	lig s[h]	**B16**[2], 97
c704	—,**epoxy-,**nitrile...	C_7H_9NO. *See* c698........	123.16	....	−33	244.5^{760} 110^{10}	1.0929^{20}_{20}	1.4763^{20}	s	...	...	...	...	....	....
c705	—,**1-hydroxy-,** nitrile	Cyclohexanone cyanohydrin. $C_7H_{11}NO$. *See* c698	125.17	....	34–6	$109\text{–}13^{9}$	....	1.4643^{20}	...	...	s	...	...	....	**Am** 77, 4571
c706	—,**2-hydroxy-***...	Hexahydrosalicylic acid. $C_7H_{12}O_3$. *See* c698	144.17	nd (AcOEt)	111	....	....	....	v	v	v	...	v	....	**B10**, 5
c706[1]	—,**1,3,4,5-tetra-hydroxy-(*d*)**	*d* Quinic acid. $C_7H_{12}O_6$. *See* c698	192.17	mcl pr $[\alpha]^{20}_{D}+44$ (w,c=10)	164	....	1.637	....	v^h	δ	δ	...	...	....	**B10**, 538
c706[2]	—(*dl*)...........	$C_7H_{12}O_6$. *See* c698........	192.17	....	174	....	....	....	...	...	...	...	...	....	**B10**, 538
c706[3]	—(*l*)............	$C_7H_{12}O_6$. *See* c698........	192.17	pr (w) $[\alpha]^{13}_{D}$ −44.1 (w, c=12)	183–4	d	....	....	v	δ	δ	...	...	ac s, alk s	**B10**[3], 317
c707	**1,2-Cyclohexane-dicarboxylic acid (*cis*)***	Hexahydrophthalic acid. —CO_2H, CO_2H	172.18	tcl nd (al)	192	d	....	....	δ	s	s	...	s	chl i lig δ	**B9**, 730
c708	—(*trans, d*)*......	$C_8H_{12}O_4$. *See* c707........	172.18	pw (w) $[\alpha]_D$ +18.2	179–83	....	....	....	v^h	...	δ	...	δ	....	....
c709	—(*trans, dl*)*.....	$C_8H_{12}O_4$. *See* c707........	172.18	lf or pr (w)	221	....	....	....	v^h	...	δ	...	δ	chl, peth i	**B7**, 730

For explanations, symbols and abbreviations see beginning of table.

No.	Name	Synonyms and Formula	Mol. wt.	Crystalline form, color, and specific rotation	m.p. °C	b.p. °C	Density	n_D	Solubility w	al	eth	ace	bz	other solvents	Ref.
	1,2-Cyclohexanedicarboxylic acid														
c710	—(*trans, l*)*	$C_8H_{12}O_4$. *See* c707	172.18	pw (w) $[\alpha]_D - 18.5$	178–83										
c711	—,diethyl ester (*cis*)*	$C_{12}H_{20}O_4$. *See* c707	228.29			133^{10}	1.0595^{14}_{4}	1.4551^{14}			s				**B9**², 521
c712	—,—(*trans*)	$C_{12}H_{20}O_4$. *See* c707	228.29			135^{11}	$1.0471^{13.1}_{4}$	$1.4522^{13.8}$							**B9**², 522
c713	**1,3-Cyclohexanedicarboxylic acid** (*cis*)*	(*cis*)-Hexahydroisophthalic acid. CO2H CO2H	172.18	nd (con HCl)	161–3				δ	v	s		v	lig δ	**B9**², 522
c714	—(*trans, d*)*	$C_8H_{12}O_4$. *See* c713	172.18	(w) $[\alpha]^{22}_D + 23.8$ (w, c=4)	134				s						**B9**², 523
c715	—(*trans, dl*)*	$C_8H_{12}O_4$. *See* c713	172.18	nd (w)	148				v^h						**B9**², 523
c716	—(*trans, l*)*	$C_8H_{12}O_4$. *See* c713	172.18	(w) $[\alpha]^{22}_D - 23.2$ (w, c=2)	134										**B9**², 523
c717	—,diethyl ester (*cis*)*	$C_{12}H_{20}O_4$. *See* c713	228.29			142^{11}	1.0505^{14}_{4}								**B9**², 523
c718	—,—(*trans*)*	$C_{12}H_{20}O_4$. *See* c713	228.29			142^{12}	1.0485^{21}_{4}								**B9**², 523
c719	**1,4-Cyclohexanedicarboxylic acid** (*cis*)*	*cis*-Hexahydroterephthalic acid. HO2C—⬡—CO2H	178.18	lf (w)	170–1				s^h	v	v			chl v	**B9**², 523
c720	—(*trans*)*	$C_8H_{12}O_4$. *See* c719	172.18	pr (w), pl (ace)	312–3	300 sub			δ^h	v	δ			chl i	**B9**², 524
c721	—,diethyl ester (*cis*)*	$C_{12}H_{20}O_4$. *See* c719	228.29			151^{13}	1.0516^{21}_{4}								**B9**², 524
c722	—,—(*trans*)*	$C_{12}H_{20}O_4$. *See* c719	228.29	nd	43–4		1.0105^{64}_{4}								**B9**², 529
c723	**1,2-Cyclohexanediol** (*cis*)*	OH OH	116.16		98	116^{13}	1.0297^{101}_{4}					s	s		**B6**², 743
c724	—(*trans*)*	$C_6H_{12}O_2$. *See* c723	116.16	(ace)	103–4	117^{13}			s			s	s		**B6**², 744
c725	**1,4-Cyclohexanediol**(*cis*)*	*cis*-Quinitol. HO—⬡—OH	116.16	tcl (ace)	113–4				s		i	s		MeOH s chl i	**B6**², 747
c726	—(*trans*)*	$C_6H_{12}O_2$. *See* c725	116.16	mcl (ace)	142		1.18^{20}_{4}				i	s		MeOH s	**B6**², 748
c727	**1,2-Cyclohexanedione***	Dihydropyrocatechol. 6 5 4 3 =O =O	112.12		38	193–5^{760} 96–7^{25}			s	s	s		s		**B7**², 526
c728	—,dioxime*	Nioxime. $C_6H_{10}N_2O_2$. *See* c727	142.16	(w or ace)	189–90				s^h			s^h			**B7**², 526
c729	**1,3-Cyclohexanedione***	Dihydroresorcinol. 6 5 4 2 O O	112.12	pr (bz or AcOEt)	105–6		1.0861^{91}	1.4576^{102}	s	s	δ	s	δ	chl s	**B7**², 526
c730	—,dioxime*	$C_6H_{10}N_2O_2$. *See* c729	142.16		156–7				v^h	v				aa v^h	**B7**, 555
c731	**1,4-Cyclohexanedione***	Tetrahydroquinone. O=⬡=O	112.12	mcl (w), nd (peth)	78	sub 100			s	s	s				**B7**², 526
c732	—,dioxime	$C_6H_{10}N_2O_2$. *See* c729	142.16	(w)	188				s						**B7**², 526
c733	**1,3-Cyclohexanedione,2-bromo-***	$C_6H_7BrO_2$. *See* c729	191.04	micr nd	166				i	s^h					**B7**, 556
c734	**1,2-Cyclohexanedione, 3,5-dimethyl-***	$C_8H_{12}O_2$. *See* c727	140.18	(dil MeOH)	71–2				s					MeOH s^h alk v	**B7**¹, 314
c735	**1,3-Cyclohexanedione, 5,5-dimethyl-***	Dimedone. Methone. $C_8H_{12}O_2$. *See* c729	140.18	yesh nd (w, aq ace), mcl pr (al-eth)	148–9				δ^h		δ	s		CCl_4 s chl v aa v	**B7**², 531

For explanations, symbols and abbreviations see beginning of table.

No.	Name	Synonyms and Formula	Mol. wt.	Crystalline form, color and specific rotation	m.p. °C	b.p. °C	Density	n_D	Solubility w	al	eth	ace	bz	other solvents	Ref.
	1,2,3,4,5,6-Cyclohexane-hexacarboxylic acid														
c736	**1,2,3,4,5,6-Cyclohexane-hexacarboxylic acid***	Hexahydromellitic acid. 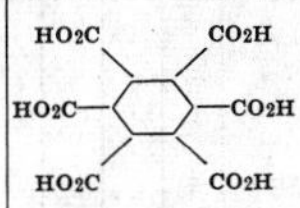	348.22		d				v	v	δ	...	...		**B9**, 1007
c736[1]	**Cyclohexanehexone**, octahydrate*	Cyclohexone hydrate. 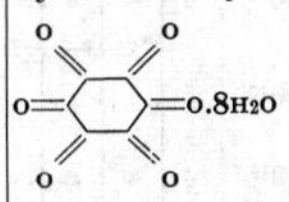	312.19	mic nd (dil HNO_3)	100–1				s^h	δ	δ	...	...	alk s	**B7^{2}**, 882
c737	**1,2,3,4,5-Cyclohexanepentol**(*d*)*	Protoquercitol. *d*-Quercite. *d*-Quercitol. 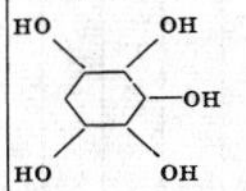	164.16	pr (w or al) $[\alpha]^{15}_{D}+25$ (c=10)	234		1.5845^{13}		v	δ^h	i	...	...	MeOH δ	**B6^{2}**, 1151
c738	—(*l*)*	$C_6H_{12}O_5$. *See* c737	164.16	pr (w), nd (al) $[\alpha]_D-49.5$(w)	174				v	δ	i	...	...		**B6**, 1188
c739	**Cyclohexanethiol***	Cyclohexyl mercaptant. 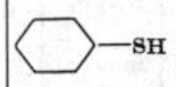	116.23			$158–60^{760}$	0.9486^{20}_{4}	1.4933^{20}	i	s	s	...	s	chl s	**B6^{2}**, 14
c740	**1,3,5-Cyclohexanetrione**, trioxime*	Phloroglucinol trioxime. 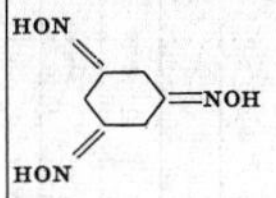	171.16	pw (aa)	155				δ	δ	...	s	...	chl s alk s ac s	**B15**, 34
c741	—,**2,2-dimethyl-***	Filicin. Filicinic acid. Filigic acid. 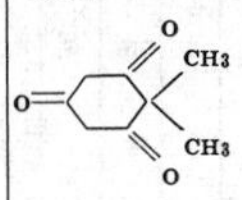	154.17	cubic or rhd (al)	213–5δd				δ^h	v^h	δ	...	δ	peth i Na_2CO_3 v	**B7^{1}**, 470
c742	**Cyclohexanol***	Hexahydrophenol. Hexalin. 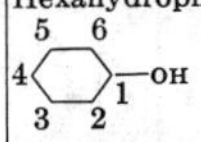	100.16	hyg nd	25.15	161.1	0.9624^{20}_{4}	1.4650^{22}	s	s	s	...	∞	CS_2 ∞	**B6^{2}**, 5
c743	—,**1-acetyl-**	$C_8H_{14}O_2$. *See* c742	142.20			$125–6^{50}$ 91^{11}	1.0257^{20}_{0}	1.4726^{11}	...	s	s	...	...		**B8^{2}**, 7
c744	—,**2-allyl-**(*trans*)	$C_9H_{16}O$. *See* c742	140.23			$94–6^{15}$	0.943^{20}_{4}	1.4778^{20}	i	...	...	...	...	MeOH s aa s	**Am74**, 399
c745	—,**2-amino-**(*trans, dl*)*	$C_6H_{13}NO$. *See* c742	115.18	hyg	66–8	105^{10}			...	...	...	...	s	chl v, ac s	**B13^{2}**, 157
c746	—,**2-butyl-***	$C_{10}H_{20}O$. *See* c742	156.27			$75–6^{3}$	0.902^{20}_{4}	1.4641^{20}	i	...	...	...	...	oos v	
c747	—,**2-chloro-**(*cis, dl*)*	*cis*-Cyclohexene chlorohydrin. $C_6H_{11}ClO$. *See* c742	134.61			$93–4^{26}$	1.1261^{25}	1.4860^{25}	...	s	...	...	...		...
c748	—,—(*cis, l*)*	$C_6H_{11}ClO$. *See* c742	134.61	hyg $[\alpha]_{546}-19.5$	36–7	87^{17}		1.4894^{25}	s	s	...	...	s		**B6**, 7
c749	—,—(*trans*)*	$C_6H_{11}ClO$. *See* c742	134.61	pr (bz-lig)	27.5–8	93^{26}	1.146^{16}_{4}	1.4899^{20}	...	v	s	...	...		**B6^{2}**, 12
c750	—,**4-chloro-***	$C_6H_{11}ClO$. *See* c742	134.61			106^{14}	1.1435^{17}_{4}	1.4930^{17}	...	s	...	...	...		**B6^{2}**, 12
c751	—,**3(dimethylamino)-***	$C_8H_{17}NO$. *See* c742	143.23			231^{740}	0.9766^{25}_{25}	1.4846^{22}	...	s	...	...	...		**B13^{2}**, 159
c752	—,**1-ethyl-***	$C_8H_{16}O$. *See* c742	128.22	pr	34.5	166, 67^{16}	0.927^{21}_{5}	1.4638^{18}	δ	...	...	...	...		**B6^{2}**, 26
c753	—,**2-ethyl-**(*cis*)*	$C_8H_{16}O$. *See* c742	128.22			74^{12}	0.9274^{21}_{4}	1.4655^{21}	i	...	...	...	...	oos v	**B6^{2}**, 26
c754	—,—(*trans*)*	$C_8H_{16}O$. *See* c742	128.22			79^{12}	0.9193^{21}_{4}	1.4640^{21}	i	...	...	...	...	oos v	**B6^{2}**, 26
c755	—,**1-ethynyl-***	$C_8H_{12}O$. *See* c742	124.18		22	73^{12}		1.4830^{20}	i	s	...	...	s		**B6^{2}**, 100
c756	—,**2(1-hydroxybutyl)-***	$C_{10}H_{20}O_2$. *See* c742	172.27			$115–8^{2}$	0.949^{20}_{4}	1.4750^{20}	i	...	...	...	...	oos v	...

For explanations, symbols and abbreviations see beginning of table.

No.	Name	Synonyms and Formula	Mol. wt.	Crystalline form, color and specific rotation	m.p. °C	b.p. °C	Density	n_D	Solubility w	al	eth	ace	bz	other solvents	Ref.
	Cyclohexanol														
c757	**—,2(1-hydroxy-ethyl)-***	$C_8H_{16}O_2$. *See* c742	144.22	...	...	$108\text{–}10^{2}$	0.976^{20}_{4}	1.4812^{22}	i	...	...	...	...	oos v	
—	**—,2-isopropyl-5-methyl-***	*see* **Neoisomenthol**													
c758	**—,1-methyl-***	$C_7H_{14}O$. *See* c742	114.19	...	25	155^{760}	0.9194^{26}_{4}	1.4558^{26}	i	...	...	...	s	chl s	**B6^2**, 16
c759	**—,2-methyl-(*cis, dl*)***	$C_7H_{14}O$. *See* c742	114.19	...	−9.5, −9.2	165	0.937^{20}	1.4640^{20}	δ	∞	s	...	...	...	**B6^2**, 20
c760	**—,—(*trans, d*)***	$C_7H_{14}O$. *See* c742	114.19	$[\alpha]^{20.1}_{D}+17.19$	...	78^{20}	...	1.4610^{20}	...	...	...	...	...	...	**B6^2**, 18
c761	**—,—(*trans, dl*)***	$C_7H_{14}O$. *See* c742	114.19	...	−21.2 −20.5	167.2–7.6	$0.9241^{19.7}_{4}$	$1.4613^{19.7}$	δ	∞	s	...	...	...	**B6^2**, 18
c762	**—,—(*trans, l*)***	$C_7H_{14}O$. *See* c742	114.19	$[\alpha]^{20}_{D}-17.81$	...	78^{20}	...	...	...	...	...	...	...	...	**B6^2**, 18
c763	**—,3-methyl-(*cis, l*)***	$C_7H_{14}O$. *See* c742	114.19	$[\alpha]^{22}_{D}-4.75$	−4.7	174–5	0.9135^{20}_{4}	1.4581^{20}	δ	∞	∞	...	...	...	**B6^2**, 20
c764	**—,—(*trans, l*)***	$C_7H_{14}O$. *See* c742	114.19	$[\alpha]^{29}_{D}-4.16$	...	$174\text{–}5^{762}$	0.9145^{22}_{4}	1.4550^{22}	...	...	...	...	...	...	**B6^2**, 20
c765	**—,4-methyl-(*cis*)***	$C_7H_{14}O$. *See* c742	114.19	...	−9.5, −9.2	$173\text{–}4^{760}$	0.9170^{20}_{4}	$1.4543^{21.5}$	δ	∞	s	...	...	...	**B6^2**, 22
c766	**—,—(*trans*)***	$C_7H_{14}O$. *See* c742	114.19	...	...	$173\text{–}4.5^{745}$	0.9118^{21}_{4}	1.4543^{21}	δ	∞	s	...	...	...	**B6^2**, 22
c767	**—,2-phenyl-(*cis*)***	$C_{12}H_{16}O$. *See* c742	176.26	...	41–2	$140\text{–}1^{16}$	1.035^{16}	1.5415^{16}	...	...	...	...	...	...	**B6^2**, 548
c768	**—,—(*trans*)***	$C_{12}H_{16}O$. *See* c742	176.26	...	56–7	$152\text{–}5^{16}$	...	...	...	s	...	...	...	chl s	**B6^2**, 548
c769	**—,2,2,6,6-tetrakis-(hydroxy-methyl)-***	$C_{10}H_{20}O_5$. *See* c742	220.27	pl (al)	131	...	...	...	v	v	i	i	i	MeOH v Py s	**B6^2**, 1151
c770	**—,1,2,2-tri-methyl-***	$C_9H_{18}O$. *See* c742	142.24	(+½w)	41(+½w)	$81.4\text{–}1.8^{20}$	0.9274^{18}_{4}	1.469^{18}	i	s	...	...	...	oos s	**B6^1**, 16
c771	**—,1,2,6-tri-methyl-***	$C_9H_{18}O$. *See* c742	142.24	...	...	78^{23}	0.9126^{15}_{4}	1.4598^{15}	i	s	s	...	...	...	**B6^1**, 17
c772	**—,1,3,3-tri methyl-***	$C_9H_{18}O$. *See* c742	142.24	pr	72.5–4.0	...	...	...	i	v	...	...	...	oos v	**B6^1**, 16
c773	**—,1,3,5-tri-methyl-***	$C_9H_{18}O$. *See* c742	142.24	...	...	181 $82\text{–}3^{19}$	0.8876^{17}_{4}	1.454^{17}	i	s	s	...	...	...	**B6^1**, 17
c774	**—,1,4,4-tri-methyl-***	$C_9H_{18}O$. *See* c742	142.24	hyg nd	58	$79\text{–}80^{15}$	...	...	i	s	s	...	...	...	**B6^1**, 16
c775	**—,2,2,3-tri-methyl-***	$C_9H_{18}O$. *See* c742	142.24	...	...	$85\text{–}7^{15}$	...	...	i	s	s	...	...	...	**B6^1**, 16
c776	**—,2,2,5-tri-methyl-***	$C_9H_{18}O$. *See* c742	142.24	...	...	$185\text{–}7^{748}$ $90\text{–}2^{23}$	0.9082^{16}_{4}	1.4632^{16}	...	s	...	...	...	oos s	**B6**, 22
c777	**—,2,2,6-tri-methyl-(liquid)***	$C_9H_{18}O$. *See* c742	142.24	...	...	$186\text{–}7^{753}$	0.9128^{20}_{4}	1.4600^{20}	i	s	s	...	...	...	**B6^1**, 16
c778	**—,—(solid)***	$C_9H_{18}O$. *See* c742	142.24	(peth or al)	51	87^{28}	...	...	i	s	s	...	...	...	**B6^1**, 16
c779	**—,2,3,3-tri-methyl-***	$C_9H_{18}O$. *See* c742	142.24	nd	28	97^{19}	...	...	i	v	...	...	...	oos v	**B6^1**, 16
c780	**—,2,3,6-tri-methyl-***	$C_9H_{18}O$. *See* c742	142.24	...	...	$193\text{–}5^{747}$	0.9117^{17}_{4}	...	i	s	...	...	...	...	**B6**, 22
c781	**—,2,4,5-tri-methyl-(*cis*)***	$C_9H_{18}O$. *See* c742	142.24	hyg	...	$191\text{–}3^{760}$ 84^{17}	0.912^{20}_{4}	1.463^{20}	i	s	s	...	...	...	**B6^2**, 36
c782	**—,—(*trans*)***	$C_9H_{18}O$. *See* c742	142.24	hyg	...	196^{760} 112^{35}	0.906^{20}_{4}	1.461^{20}	i	s	s	...	...	...	**B6^2**, 36
c783	**—,3,3,5-tri-methyl-(*cis*)***	*cis*-Dihydroisophorole. $C_9H_{18}O$. *See* c742	142.24	...	...	$201\text{–}3^{750}$ 92^{12}	0.9006^{16}_{4}	1.4550^{16}	i	s	s	...	...	...	**B6^1**, 16
c784	**—,—(*trans*)***	$C_9H_{18}O$. *See* c742	142.24	(eth)	52	196.5^{770} 95^{15}	0.8778^{40}_{4}	...	i	s	s	...	...	...	**B6^1**, 16
c785	**Cyclohexanone***	Ketohexamethylene. Pimelic ketone. (ring structure: 4, 5, 6, =O, 3, 2)	98.14	...	−16.4	155.65^{760}	0.9978^{20}_{4}	1.4522^{19}	δ	s	s	...	...	...	**B7^2**, 5

For explanations, symbols and abbreviations see beginning of table.

No.	Name	Synonyms and Formula	Mol. wt.	Crystalline form, color and specific rotation	m.p. °C	b.p. °C	Density	n_D	Solubility w	al	eth	ace	bz	other solvents	Ref.
	Cyclohexanone														
c786	—,oxime*	$C_6H_{11}NO$. *See* c785	113.16	hex pr	90	206–10			s	δ	s			MeOH s	**B7**[2], 10
c787	—,**2-acetyl-**	$C_8H_{12}O_2$. *See* c785	140.18			111–2^{18}	1.0782^{20}	1.5138							**B7**[2], 530
c788	—,**2-butyl-***	$C_{10}H_{18}O$. *See* c785	154.28			70^{2}	0.905$^{20}_{4}$	1.4545^{20}	i					oos v	
c789	—,**2-butylidene-**	$C_{10}H_{16}O$. *See* c785	152.26			98–100^{10}	0.935$^{20}_{4}$	1.4831^{20}	i					oos v	
c790	—,**2-chloro-***	C_6H_9ClO. *See* c785	132.59		23	82^{13}	1.161$^{20}_{15}$	1.4830^{20}			s		s	diox s	**B7**[2], 11
c790[1]	—,**3-chloro-***	C_6H_9ClO. *See* c785	132.59			91–2^{14}					s				**B7**, 10
c791	—,**4-chloro-***	C_6H_9ClO. *See* c785	132.59			95^{17}		1.4867^{20}			s				**B7**[2], 11
c793	—,**2,6-dibenzyl-idene-**	$C_{20}H_{18}O$. *See* c785	274.35	ye	117–8	185–95^{20}				δ			s	aa s	**B7**[2], 465
c794	—,**2,6-dibromo-***	$C_6H_8Br_2O$. *See* c785	255.96	(eth or aa)	106–7						s^{h}	s^{h}			**B7**[2], 12
c795	—,**2,4-dimethyl-,** oxime*	$C_8H_{15}NO$. *See* c785	141.21	nd (al)	96				i	s^{h}					**B7**[2], 26
c797	—,**2,5-di-methyl-(*d*)***	$C_8H_{14}O$. *See* c785	126.19	$[\alpha]^{20}+11.6$		172–4^{750} 51^{10}	0.8985$^{20}_{4}$	1.4445^{20}	i	s	s				**B7**[1], 19
c798	—,—(*trans, dl*)*	$C_8H_{14}O$. *See* c785	126.19			178	0.9025^{20}	1.4446							**B7**[2], 27
c799	—,**2(dimethyl-amino-methyl)-***	$C_9H_{17}NO$. *See* c785	155.23			92$^{10.5}$	0.9504$^{20}_{4}$	1.4672^{20}	δ	s	s				**B14**[2], 3
c800	—,**2-ethylidene-***	$C_8H_{12}O$. *See* c785	124.18			92^{20}	0.962$^{20}_{4}$	1.4882^{20}	i					oos v	**B7**[2], 58
c801	—,**2-hydroxy-***	Adipoin. $C_6H_{10}O_2$. *See* c785	114.14	nd (MeOH)	113			1.4785^{21}	v^{h}	v^{h}	i		i	peth i	**B8**[2], 3
c802	—,**2-isopropyl-***	$C_9H_{16}O$. *See* c785	140.23			72–3^{9}	0.923^{12}	1.4585^{12}	i					oos v	**B7**[2], 30
c803	—,**2-methyl-(*d*)***	$C_7H_{12}O$. *See* c785	112.17	$[\alpha]^{18}_{D}+14.21$			0.9262$^{18}_{4}$								**B7**[2], 15
c804	—,—(*dl*)*	$C_7H_{12}O$. *See* c785	112.17			165^{757}	0.9240$^{20}_{4}$	1.4493^{20}	i	s	s				**B7**[2], 16
c805	—,—(*l*)*	$C_7H_{12}O$. *See* c785	112.17	$[\alpha]^{25}_{D}-15.22$		59–60^{20}	0.9230$^{25}_{4}$								**B7**[2], 15
c806	—,**3-methyl-(*d*)***	$C_7H_{12}O$. *See* c785	112.17	$[\alpha]^{20}_{D}+12.7$ (undil)		169–70	0.9155$^{20}_{4}$	1.4493^{20}	i	s	s				**B7**[2], 17
c807	—,—(*dl*)*	$C_7H_{12}O$. *See* c785	112.17			168–9^{738} 60–0.2^{15}	0.9136$^{20}_{4}$	1.4430^{20}	i	s	s				**B7**[2], 18
c808	—,**4-methyl-***	$C_7H_{12}O$. *See* c785	112.17			169.5^{765}	0.9132$^{20}_{4}$	1.4458^{20}	i	s	s				**B7**[2], 19
c809	—,**2-propyl-***	$C_9H_{16}O$. *See* c785	140.23			199^{76} 70^{4}	0.927$^{20}_{4}$	1.4538^{20}	i					oos v	**B7**[2], 30
c810	**Cyclohexa-siloxane, dodecamethyl-**	$-O-[Si(CH_3)_2-O]_4-Si(CH_3)_2-O-Si(CH_3)_2-$ (ring)	444.71		−3	245 128^{20}	0.9672	1.4015^{20}							
c811	**Cyclohexene***	1,2,3,4-Tetrahydrobenzene. (ring numbered 6, 5, 1, 4, 2, 3)	82.14		−103.50	82.98^{760}	0.8110$^{20}_{4}$	1.4465^{20}	i	∞	∞	∞	∞	CCl_4 ∞ ig ∞	**B5**[2], 37
c812	—,**1-acetyl-**	$C_8H_{12}O$. *See* c811	124.18			200^{720} 63–4^{6}	0.9694$^{20}_{4}$	1.4904^{20}		s	s				**B7**[2], 58
c813	—,**3-bromo-***	C_6H_9Br. *See* c811	161.05			57.5–58^{12}	1.3890$^{20}_{4}$	1.5292^{20}							**B5**[2], 40
c814	—,**decafluoro-***	Perfluorocyclohexene. C_6F_{10}. *See* c811	262.06			52–3^{750}		1.296^{15}							**J 1952,** 1251
c815	—,**1-ethenyl-***	C_8H_{12}. *See* c811	108.18			142–4^{760} 50–2^{22}	0.8701$^{20}_{0}$	1.4915^{20}						MeOH v	**B5**[1], 62
c816	—,**4-ethenyl-***	C_8H_{12}. *See* c811	108.18			128.9^{760}	0.8299$^{20}_{4}$	1.4639^{20}							**B5**[2], 81
—	—,**1-isopropyl-4-methyl-**	*see* **Menthene**													
c817	—,**1-methyl-***	C_7H_{12}. *See* c811	96.17		−121	110^{760}	0.8127$^{20}_{4}$	1.4508^{20}						CCl_4 s	**B5**[2], 42
c818	—,**3-methyl-(*d*)***	C_7H_{12}. *See* c811	96.17	$[\alpha]^{20}_{D}+110$		104^{760}	0.8010$^{20}_{4}$	1.4444^{20}							**B5**[2], 43
c819	—,—(*dl*)*	C_7H_{12}. *See* c811	96.17			104	0.799^{25}	1.4432^{25}							**B5**[2], 43

For explanations, symbols and abbreviations see beginning of table.

No.	Name	Synonyms and Formula	Mol. wt.	Crystalline form, color and specific rotation	m.p. °C	b.p. °C	Density	n_D	Solubility w	al	eth	ace	bz	other solvents	Ref.
	Cyclohexene														
c820	**—,4-methyl-***	C_7H_{12}. *See* c811	96.17		−115.5	102.74^{760}	0.7991^{20}_{4}	1.4414^{20}	i	s	s				**B5**[2], 43
c821	**—,1,3,4,5,6-penta-chloro-**(γ)*	$C_6H_5Cl_5$. *See* c811	254.39			$115–6^{4}$		1.5630^{20}							**Am 76,** 1244
c822	—,—(δ)	$C_6H_5Cl_5$. *See* c811	254.39		68.2–8.6		1.80			s					**Am 76,** 1244
c823	**—,1-phenyl-***	1,2,3,4 Tetrahydrobiphenyl. $C_{12}H_{14}$. *See* c811	158.23		−18, −12	$251–3^{760}$ $69^{0.5}$	0.9931^{20}_{4}	1.5690^{20}						MeOH v	**B5**[2], 419
c824	**—,1-Cyclohexene-1-carboxaldehyde**	—CHO	114.15			90^{10}	0.9694^{20}_{4}	1.5005^{20}		s	s				**J 1955,** 320
c825	**3-Cyclohexene-1-carboxaldehyde**	—CHO	110.15		fr −96.1	164^{760}	0.9709^{20}_{4}	1.4725^{20}				s		MeOH s	**B7**[2], 57
c826	**1-Cyclohexene-1-carboxylic acid***	—CO_2H	126.15		33.5–35.8	$240–2^{760}$ $131–2^{13}$	1.109^{20}_{4}	1.4902^{20}	δ						**B9**[2], 30
c827	**3-Cyclohexene-1-carboxylic acid***	—CO_2H	126.15		17	132.5–3.0^{20}	1.0815^{20}_{4}	1.4812^{20}	v						**B9**[2], 30
c828	**2-Cyclohexene-1-carboxylic acid, 2,6-dimethyl-4-oxo,** ethyl ester*	3,5 dimethyl-4-carboxy-2-cyclohexene-1-one; CH_3, O=, —$CO_2C_2H_5$, CH_3	196.25			$157–8^{18}$	1.0493^{20}_{4}	1.4773^{20}							**B10**[2], 436
c829	**1-Cyclohexene-1,2-dicarboxylic acid***	Δ^1 Tetrahydrophthalic acid. —CO_2H, —CO_2H	170.16	mcl lf (w)	126				v						**B9**[2], 556
c830	—,anhydride*	O=C, O, C=O	152.14	pl (eth)	74				d^h	s	v	s		chl s	**B17**[2], 45
c831	**3-Cyclohexene-1,2-dicarboxylic acid, anhydride***	Δ^2 Tetrahydrophthalic anhydride. O=C, O, C=O	152.14	pr (eth)	78–9				d^h	s	s			chl s	**B17**[2], 457
c832	**4-Cyclohexene-1,2-dicarboxylic acid,** anhydride (*cis*)*	*cis*-Δ^4 Tetrahydrophthalic anhydride. O=C, O, C=O	152.14	pl (eth)	103–4				d^h	s				chl s lig δ	**B17**[2], 457
c833	—,—(*trans, d*)*	$C_8H_8O_3$. *See* c832	152.14	lf$[\alpha]^{25}_{D}$+6.6 (al)	128				d^h	v			v		**B17,** 462
c834	—,—(*trans, dl*)*	$C_8H_8O_3$. *See* c832	152.14	(bz-lig)	140				d^h	s			s	chl s	**B17,** 462
c835	—,imide, *N*(tri-chloromethylthio)-	Captan. O=C, $NSCCl_3$, C=O	300.60		172		1.74					δ	δ		
—	**Cyclohexene oxide**	*see* **Cyclohexene, 1,2-epoxy-***													
c836	**2-Cyclo-hexen-1-ol***	5 6 4 —OH 1 3 2	98.14			$164–6^{760}$	1.00^{26}_{26}	1.4790^{25}							**B6**[2], 60

For explanations, symbols and abbreviations see beginning of table.

No.	Name	Synonyms and Formula	Mol. wt.	Crystalline form, color and specific rotation	m.p. °C	b.p. °C	Density	n_D	Solubility w	al	eth	ace	bz	other solvents	Ref.
	2-Cyclohexen-1-ol														
c837	—,5-methyl- (*cis, d*)*	$C_7H_{12}O$. *See* c836	112.17	$[\alpha]_D^{30}+6.95$		83^{25}	0.9391^{25}_{4}	1.4727^{25}			s			lig s	**Am 77,** 4042
c838	—,—(*cis,l*)	$C_7H_{12}O$. *See* c836	112.17	$[\alpha]_D^{25}-7$		82^{25}	0.9391^{25}_{4}	1.4727^{25}							**Am 77,** 4042
c839	—,—(*trans, d*)	$C_7H_{12}O$. *See* c836	112.17	$[\alpha]_D^{27}+127$ (ace, c=19.4)		$68–9^{24}$	0.9430^{20}_{4}	1.4737^{25}			s			lig s	**Am 77,** 4042
c840	—,—(*trans, l*)	$C_7H_{12}O$. *See* c836	112.17	$[\alpha]_D^{27}-163.9$		$82–3^{24}$	0.9430^{20}_{4}	1.4737^{25}			s			lig s	**Am 77,** 4042
c841	**2-Cyclohexen-1-one***	(structure: ring positions 2–6, =O)	96.12			$169–71^{760}$ $61–2^{10}$		1.4883^{20}		v					**B7²**, 55
c842	—,2,5-dimethyl-*	$C_8H_{12}O$. *See* c841	124.18			189–90	0.938^{22}	1.4753^{22}			s				**B7¹**, 51
c842¹	—,3,5-dimethyl-*	$C_8H_{12}O$. *See* c841	124.18			211–2 $88–90^{2}$	0.9462^{12}_{4}	1.4819^{22}		s	s				**B7²**, 59
c843	—,3,6-dimethyl-*	$C_8H_{12}O$. *See* c841	124.18			75^{19}					s				**B7¹**, 51
c844	**3-Cyclo-hexen-1-one, 4,6-dimethyl-***	(structure: CH_3, CH_3, =O)	124.18			194	0.9539^{0}				s				**B7**, 61
—	**2-Cyclo-hexen-1-one, 3-hydroxy-***	*see* **1,3-Cyclohexanedione***													
c846	—,5-isopropyl-3-methyl-*	*m*-6-Menten-5-one. Hexeton. $C_{10}H_{16}O$. *See* c841	152.23	pa ye		244 124^{15}	0.9340^{21}	1.4865^{21}	δ	v				oos ∞	**B7²**, 74
c847	—,2-methyl-*	$C_7H_{10}O$. *See* c841	110.15			$178–9^{760}$ 56^{9}	0.9667^{20}_{4}	1.4865^{20}					s		**B7²**, 57
c848	—,3-methyl-*	$C_7H_{10}O$. *See* c841	110.15			$200–5^{760}$	0.9711^{20}_{4}	1.4946^{20}							**B7²**, 56
—	—,3,5,5-tri-methyl-*	*see* **Isophorone**													
—	**Cyclohexone,** hydrate	*see* **Cyclohexanehexone,** octahydrate.*													
—	**Cyclone**	*see* **Cyclopentadienone, tetraphenyl-**													
—	**Cyclonite**	*see* **1,3,5-Triazine, hexa-hydro-1,3,5-trinitro-**													
c850	**Cyclononanone***	(structure: O)	140.22		34	$90–6^{10-13}$	0.9591^{25}_{4}	1.4760^{20}							**B7²**, 29
c851	**Cyclonona-siloxane, octa-decamethyl-**	(structure: CH_3–Si(–CH_3)–O–[–Si(CH_3)$_2$–O–]$_7$–Si(CH_3)$_2$–O)	667.15			188^{20}		1.4070^{20}					s	ig s	
c852	**Cyclononene**(*cis*)*	(structure)	124.22			$167–9^{760}$ $71–2^{31}$	0.8671^{20}_{4}	1.4805^{20}							**Am 74,** 3643
c853	—(*trans*)	C_9H_{16}. *See* c852	124.22			$94–6^{80}$	0.8615^{20}_{4}	1.4799^{20}							**Am 74,** 3643
c854	**1,5-Cycloocta-diene**(*cis, cis*)*	(structure)	108.18		−70	150.8^{757}	0.8818^{25}_{4}	1.4905^{25}						CCl_4 s	**B5**, 116
c855	**Cyclooctane***	Octamethylene. (structure)	112.21		13.5	148.5–9.5^{749} 63^{45}	0.8337^{20}_{4}	1.1568^{20}							**B5²**, 20
c856	**Cyclooctanol***	(structure: OH)	128.22		14–5	$100–1^{15}$	0.9740^{20}_{4}	1.4871^{20}							**B6²**, 25

For explanations, symbols and abbreviations see beginning of table.

No.	Name	Synonyms and Formula	Mol. wt.	Crystalline form, color and specific rotation	m.p. °C	b.p. °C	Density	n_D	w	al	eth	ace	bz	other solvents	Ref.
	Cyclooctanone														
c857	**Cyclooctanone**	Azelaone. 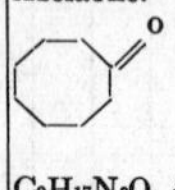	126.20		43.8	200–2^{713} 115–5.5^{60}	0.5984^{20}_{4}	1.6494^{20}							**B7^2**, 22
c858	—,semicarbazone	$C_9H_{17}N_3O$. *See* c857	183.26	lf	170–1									MeOH s^h	**B7^2**, 22
c859	**Cycloocta-siloxane, hexa-decamethyl-**	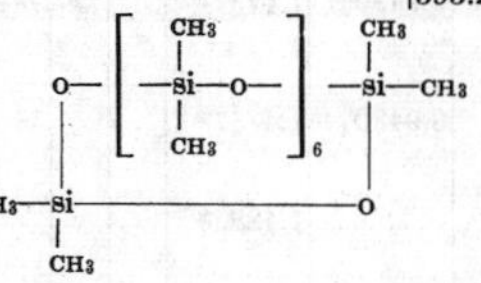	593.24		*ca.* 30	175^{20}		1.4060^{20}	i				s	lig s	
c860	**Cycloocta-tetraene***	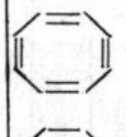	104.15	ye or wh	−27 (−5.4)	142–3^{760} 42–2.5^{17}	0.925^{20}_{4}	1.5394^{20}		s	s	s	s		**B5^1**, 228
c861	**Cyclooctene**(*cis*)	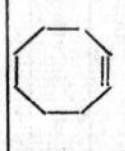	110.20		−12	138^{760} 42^{18}	0.8448^{25}_{4}	1.4682^{25}		s	s			CCl_4 s	**B5^2**, 45
c862	—(*trans*)	C_8H_{14}. *See* c861	110.20		−5.9	143^{760} 75^{78}	0.8456^{25}_{4}	1.4741^{25}		s				chl s	**B5^2**, 45
c863	**Cyclopenta-decanone***	Exaltone. 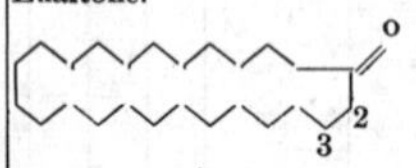	224.39		63	$120^{0.3}$	0.8895	1.4637^{66}	δ	s					
c864	—,3-methyl-*	Muscone. Muskone. $C_{16}H_{30}O$. *See* c863	238.42	ye $[\alpha]^{17}_{D}-13$ (undil)		327–30^{752} $130^{0.5}$	0.9221^{17}_{4}	1.4844^{15}	δ	∞	s	s			**B7^2**, 51
c865	**1,3-Cyclopenta-diene***	Cyclopentadiene* 	66.10		−97.2	40.83^{772}	0.8021^{20}_{4}	1.4429^{20}	i	∞	∞		∞		**B5^2**, 77
c866	—,hexachloro-	Perchlorocyclopentadiene. C_5Cl_6	272.79	ye gr	−10, −9	239^{753} 48–9$^{0.3}$	1.7019^{25}_{4}	1.5658^{20}							**Am 69,** 1918
—	**Cyclopentadiene-benzoquinone**	*see* **Naphthalene, 5,8-dioxo-1,4,5,8,9,10-hexahydro-1,4-methylene-***													
c867	**Cyclopenta-dienone, tetra-phenyl-***	Cyclone. Tetracyclone. 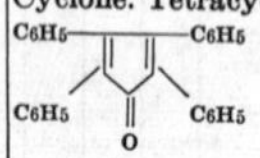	384.45	bk vt lf (aa or xyl)	220–1					s			s	xyl s^h iso-octane s aa s^h	**B7^2**, 521
c868	**Cyclopentane***	Pentamethylene. 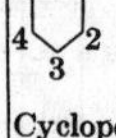	70.13		−93.879	49.262^{760}	0.7510^{20}_{4}	1.4064^{20}	i	∞	∞				**B5^2**, 4
c869	—,acetyl-	Cyclopentyl methyl ketone. $C_7H_{12}O$. *See* c868	112.17			162–3		1.4409^{25}			s				**B7^2**, 20
c870	—,bromo-*	C_5H_9Br. *See* c868	149.04			136.7–7.7^{760} 56^{48}	1.3900^{20}_{4}	1.4882^{20}							**B5^2**, 4
c871	—,butyl-*	C_9H_{18}. *See* c868	126.23		−108.2	156.7^{760}	0.7840^{20}_{4}	1.4318^{20}							**B5^2**, 25
c872	—,chloro-*	Cyclopentyl chloride. C_5H_9Cl. *See* c868	104.58			114.5–5.0	1.0051^{20}_{4}	1.4510^{20}	i					oos s	**B5^2**, 4
c873	—,1,2-diethyl-(*trans*)*	C_9H_{18}. *See* c868	126.23		−95.6	153.58^{760}	0.7959^{20}_{4}	1.4295^{20}							
c874	—,ethyl-*	C_7H_{14}. *See* c868	98.19		−137.9	103.6^{760}	0.7632^{20}_{4}	1.4196^{20}	i				s	to s, peth s	**B5^2**, 119
c875	—,isopropyl-*	C_8H_{16}. *See* c868	112.22		−112.7	126.8^{760}	0.7764^{20}_{4}	1.4261^{20}			s		s	peth v	**B5^2**, 22
c876	—,methyl-*	Methylpentamethylene. C_6H_{12}. *See* c868.	84.16		−142.4	72.0–2.2	0.7489^{20}_{4}	1.4096^{20}	i	∞	∞		∞		**B5**, 27

For explanations, symbols and abbreviations see beginning of table.

No.	Name	Synonyms and Formula	Mol. wt.	Crystalline form, color and specific rotation	m.p. °C	b.p. °C	Density	n_D	Solubility w	al	eth	ace	bz	other solvents	Ref.
	Cyclopentane														
c877	**—,1-methyl-2-propyl-(*trans*)***	C_9H_{18}	126.23		−104.9	152.58^{760}	0.7921_4^{20}	1.4321^{20}	...	...	...	...	...		
c878	**—,propyl-***	1-Cyclopentylpropane. C_8H_{16}. *See* c868	112.21		−121.7	131.3–1.5^{760}	0.7718_4^{20}	1.4266^{20}	i	...	...	...	s	peth v	**B5²**, 22
c879	**Cyclopentanecarboxaldehyde**	—CHO	98.14			133–4	0.9371_4^{20}	1.1432^{20}	s	...	s	...	...		**B7²**, 14
c880	**Cyclopentanecarboxylic acid***	—CO_2H	114.14		−8	212.5–13.5^{752} 106–8^{12}	1.0527_4^{20}	1.4532^{20}	δ	...	...	...	...	MeOH s	**B9²**, 6
c881	**—,3-formyl-2,2,3-trimethyl-,** methyl ester	$C_{11}H_{18}O_3$. *See* c880	198.25	$[\alpha]_D$+51.4 (al)		130–2^8	1.048_4^{20}	1.4160^{25}	...	s	...	...	...	MeOH s	**Am 64,** 1416
c882	**—,2-oxo-,** ethyl ester	$C_8H_{12}O_3$. *See* c880	156.18			218^{704} 110^{15}	1.0798_4^{20}	1.4519^{20}	...	...	s	...	s		**B10²**, 419
—	**3-Cyclopentanecarboxylic acid, 2,2,3-trimethyl-***	*see* **Campholytic acid**													
c883	**1,2-Cyclopentanedicarboxylic acid** (*cis*)*	—CO_2H —CO_2H	158.16	nd (w)	140										**B9**, 728
c884	—(*trans, d*)*	$C_7H_{10}O_4$. *See* c883	158.16	(w)$[\alpha]_D$+87.6 (w, c=0.9)	181										**B9¹**, 316
c885	—(*trans, dl*)*	$C_7H_{10}O_4$. *See* c883	158.16	(w)	162–3				v	v^h	δ	...	v	chl, peth v	**B9²**, 518
c886	—(*trans, l*)*	$C_7H_{10}O_4$. *See* c883	158.16	(w)$[\alpha]_D$−85.9 (w, c=1.2)	180–1										**B9¹**, 316
c887	**1,3-Cyclopentanedicarboxylic acid** (*cis*)*	Norcamphoric acid. —CO_2H CO_2H	158.16	pr (w)	119.9–120.6	>300d			v^h	s	s	s	s	peth i	**B9²**, 518
c888	—(*trans, d*)*	$C_7H_{10}O_4$. *See* c887	158.16	(CCl_4) $[\alpha]_D$+5.9 (w, c=5)	93.5										**B9²**, 519
c889	—(*trans, dl*)*	$C_7H_{10}O_4$. *See* c887	158.16	pr (CCl_4)	88										**B9²**, 519
c890	—,(*trans, l*)*	$C_7H_{10}O_4$. *See* c887	158.16	(CCl_4) $[\alpha]_D$−5.3 (w, c=5)	90–3										**B9²**, 519
c891	**—,4,4-dimethyl-***	Apofenchocamphoric acid. $C_9H_{14}O_4$. *See* c875	186.20	mcl	144–5				s^h	v	v	...	δ		**B9²**, 529
—	**Cyclopentanedicarboxylic acid, 1,2,2,3-tetramethyl-***	*see* **Campholic acid**													
—	**1,3-Cyclopentanedicarboxylic acid, 1,2,2-trimethyl-***	*see* **Camphoric acid**													
c892	**Cyclopentanol***	5 1 —OH 4 2 3	86.13		−19	140.85^{760}	0.9488_4^{20}	1.4530^{20}	δ	s	s	...	...		**B6²**, 3
c893	**—,3-acetyl-1,1,2,2,4-pentamethyl-(α)**	Desoxymesityl oxide. $C_{12}H_{22}O_2$. *See* c892	198.31	(peth-eth)	45				i	...	...	...	...		**B8²**, 11
c894	**Cyclopentanone***	Adipic ketone. Ketopentamethylene. 5 1 =O 4 2 3	84.11		−51.3	130.65^{760}	$0.9509_4^{18.2}$	1.4366^{20}	i	s	∞	...	...	MeOH s hexane s	**B7²**, 3
c895	**—,2-methyl-***	$C_6H_{10}O$. *See* c894	98.14		−76.1	139.5^{760}	0.9200_4^{20}	1.4347^{20}	s	v	v	v	...		**B7²**, 13
c896	**—,3-methyl-(*d*)***	$C_6H_{10}O$. *See* c894	98.14	$[\alpha]_D^{12}$+132.96		143.5^{742} 42.5–44^{12}	0.9140_4^{19}	1.4340^{19}	v	v	v	v	...		**B7²**, 13
c897	—,—(*dl*)*	$C_6H_{10}O$. *See* c894	98.14			143.7–3.9^{747} $37.5^{13.5}$	0.913^{22}	1.4329	s	v	v	v	...	aa v	**B7²**, 15

For explanations, symbols and abbreviations see beginning of table.

No.	Name	Synonyms and Formula	Mol. wt.	Crystalline form, color and specific rotation	m.p. °C	b.p. °C	Density	n_D	Solubility						Ref.
									w	al	eth	ace	bz	other solvents	
	Cyclopentasiloxane														
c898	**Cyclopentasiloxane,decamethyl-**	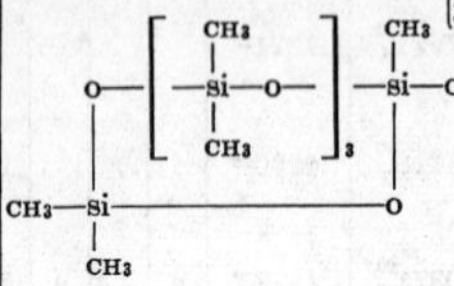	370.64		−38	210 101^{20}	0.9593	1.3982^{20}	i						
c899	**Cyclopentene***		68.11		−135.076	44.242^{760}	0.7743_4^{18}	1.4225^{20}	i	s	s				B5², 35
c900	**—,3-chloro-***	C_5H_7Cl. *See* c899	102.57			$25–31^{30}$	1.0577^{15}				s				B5², 36
c901	**—,octachloro-**	Perchlorocyclopentene. C_5Cl_8. *See* c899		(al)	41	$283–4^{733}$ 140^{10}		1.5660^{50}	i	v^h					B5, 62
c902	**3-Cyclopenten-1-one, 3,4-bis(4-methoxyphenyl)-***	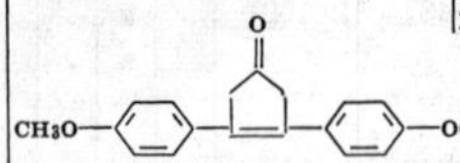	294.35	ye br	129									oos s	B8, 355
c903	**1,2-Cyclopentenophenanthrene**	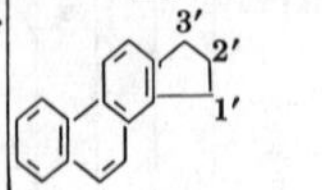	218.30	nd (al)	135–6				i	s					E14, 15
c904	**2,3-Cyclopentenophenanthrene**	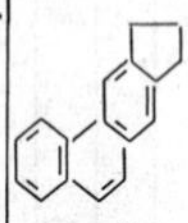	218.30	pl or pr (al), nd (MeOH)	84–5				i	s				MeOH s^h	E14s, 19
c905	**9,10-Cyclopenteno-phenanthrene**	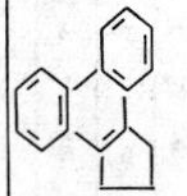	218.30	pl (xyl), nd (*i*-PrOH)	155–6				i	s			s		E14s, 25
c906	**1,2-Cyclopentenophenanthrene, 3-methyl-**	$C_{18}H_{16}$. *See* c903	232.33	(aa)	126–7				i					aa s^h	E14, 16
c907	**Cyclopropane***	Trimethylene. 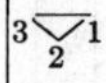	42.08		−126.6	−33	0.720_4^{-79}		δ	v	v				B5², 3
c908	**—,acetyl-**	Cyclopropyl methyl ketone. C_5H_8O. *See* c907	84.11			114^{772}	0.8993_4^{20}	1.4244^{20}	s	s	s				B7², 5
c909	**—,1,1-dimethyl-***	C_5H_{10}. *See* c907	70.13			21	0.6604_4^{20}	1.366^{20}	i	s	v				B5, 20
c910	**—,methoxy-***	Cyprome ether. C_4H_8O. *See* c907	72.10		−119	44.7^{760}	0.786_4^{25}	1.3802^{20}	s	s	v			oos s	
c911	**—,methyl-***	C_4H_8. *See* c907	56.10			4–5	0.6912_4^{20}		δ	v	v				B5², 3
c912	**Cyclopropanecarboxylic acid***	Ethyleneacetic acid. 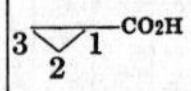	86.09		18–9	182–4	1.0885_4^{20}	1.4390^{20}	$δ^h$	s	s				B9², 3
c913	**—,nitrile**	Cyclopropanecarbonitrile*. Cyclopropyl cyanide. 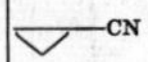	67.09			135^{760} $69–70^{88}$	0.911^{16}				s			hexane s	B9², 4
c914	**—,2,2-dimethyl-3(2-methyl-propenyl)-(*cis, d*)***	*d, cis*-Chrysanthemumic acid. $C_{10}H_{16}O_2$. *See* c912	168.24	pr $[\alpha]_D^{22}+83.3$ (chl, 1.6%)	40–2	$95^{0.1}$				s				chl s	C47, 3247
c915	**—,—(*cis, dl*)***	$C_{10}H_{16}O_2$. *See* c912	168.24	pr (AcOEt), (peth)	115–6					s				oos s	B9², 47
c916	**—,—(*cis, l*)***	$C_{10}H_{16}O_2$. *See* c912	168.24	pr $[\alpha]_D^{19}-83.3$ (chl, 1.56%)	41–3					s				chl s	C47, 3247[6]
c917	**—,—(*trans, d*)***	*d, trans*-Chrysanthemumic acid. $C_{10}H_{16}O_2$. *See* c912	168.24	pr $[\alpha]_D^{20}+25.8$ (chl, 2.48%)	17–21	245 δd $90^{0.1}$			δ	v				chl s oos v	B9², 45
c918	**—,—(*trans, dl*)***	$C_{10}H_{16}O_2$. *See* c912	168.24	pr	54	$145–6^{13}$				s				chl s AcOEt v	B9², 46

For explanations, symbols and abbreviations see beginning of table.

No.	Name	Synonyms and Formula	Mol. wt.	Crystalline form, color and specific rotation	m.p. °C	b.p. °C	Density	n_D	Solubility: w	al	eth	ace	bz	other solvents	Ref.
	Cyclopropanecarboxylic acid														
c919	—,—(*trans, l*)*	$C_{10}H_{16}O_2$. *See* c912	168.24	pr	17–21	99–$100^{0.2-0.3}$				s				chl s AcOEt v	
c920	**1,1-Cyclopropanedicarboxylic acid***	Ethylenemalonic acid. Vinaconic acid. (ring with two CO_2H)	130.10		140–1										
c921	—,diethyl ester*	$C_9H_{14}O_4$. *See* c920	186.21			$214–6^{748}$ $99–100^{11}$	1.0566^{25}_{25}	1.4345^{18}		v	v				B9^2, 512
c922	**1,2-Cyclopropanedicarboxylic acid** (*cis*)*	(ring with —CO_2H, —CO_2H)	130.10	pr (eth or w)	139				δ	v	v		δ	con HCl v chl v	B9^2, 513
c923	—(*trans, d*)*	$C_5H_6O_4$. *See* c922	130.10	$[\alpha]^{27}_D$+84.87	175				v	v	v		δ	chl δ	B9, 724
c924	—,(*trans, dl*)*	$C_5H_6O_4$. *See* c922	130.10	nd (eth), pl (ace-bz)		175	210^{30}		v	v	v		δ	chl δ	B9^2, 514
c925	—(*trans, l*)*	$C_5H_6O_4$. *See* c922	130.10	$[\alpha]^{27}_D$−84.40	175					v	v		δ	chl δ	B9, 724
c926	—,diethyl ester (*cis*)*	$C_9H_{14}O_4$. *See* c922	186.21			106.5–7.5	1.062^{13}_4			s	s				B9^2, 513
c927	—,dimethyl ester (*cis*)*	$C_7H_{10}O_4$. *See* c922	158.15			$219–20^{760}$ 110^3	1.1584^{16}_4	1.4472^{14}							B9^2, 513
c928	—,1-bromo-*	$C_5H_5BrO_4$. *See* c922	209.00	pr (eth-chl), cr (ace-bz)	175						v	v		chl δ	B9^2, 514
c929	**1,2,3-Cyclopropane-tricarboxylic acid***	HO_2C—(ring)—CO_2H, CO_2H	174.11	nd (w), (HCl)	220				s	s				chl δ	B9^2, 702
c930	**Cyclotetrasiloxane, octamethyl-**	$Si(CH_3)_2$—O—$Si(CH_3)_2$ / O O / $Si(CH_3)_2$—O—$Si(CH_3)_2$	296.62		17.5	175 74^{20}	0.9558	1.3968^{20}							
c931	—,octaphenyl-	$Si(C_6H_5)_2$—O—$Si(C_6H_5)_2$ / O O / $Si(C_6H_5)_2$—O—$Si(C_6H_5)_2$	791.02	nd (bz-al or aa)					i				s	aa s^h	
c932	—,2,4,6,8-tetramethyl, 2,4,6,8-tetraphenyl-	CH_3, CH_3 / $Si(C_6H_5)$—O—$Si(C_6H_5)$ / O O / $Si(C_6H_5)$—O—$Si(C_6H_5)$ / CH_3, CH_3	544.78				1.1183^{20}_4	1.5461^{20}				∞		heptane ∞	
c933	**Cyclotrisiloxane, hexaphenyl-**	$Si(C_6H_5)_2$—O / O $Si(C_6H_5)_2$ / $Si(C_6H_5)_2$—O	594.78	pl (bz-al or aa)	190	$290–300^1$			i				s	aa s	
c934	—,2,4,6-triethyl-2,4,6-triphenyl-	C_2H_5 / $Si(C_6H_5)$—O / O Si C_6H_5, C_2H_5 / $Si(C_6H_5)$—O / C_2H_5	450.66		177.5				i			s			
c935	—,2,4,6-trimethyl-2,4,6-triphenyl-(*cis*)	CH_3 / $Si(C_6H_5)$—O / O Si C_6H_5, CH_3 / $Si(C_6H_5)$—O / CH_3	408.58	pl (al)	99.5	$165–85^{1.5}$				s^h					
c936	—,—(*trans*)	$C_{21}H_{24}O_3Si_3$. *See* c935	408.58	nd (al)	39.5	$190^{1.5}$	1.1062^{20}_4	1.5397^{20}		s^h					
c937	**Cymarose**	4,5-Dihydroxy-3-methoxyhexanal*. 3-Methyldigitoxose. $CH_3(CHOH)_2CH(OCH_3)CH_2CHO$	162.18	pr (eth-peth), nd (ace) $[\alpha]^{21}_D$+53.4 (mut)	100–2				v		δ	v	δ	chl δ	

For explanations, symbols and abbreviations see beginning of table.

No.	Name	Synonyms and Formula	Mol. wt.	Crystalline form, color and specific rotation	m.p. °C	b.p. °C	Density	n_D	Solubility w	al	eth	ace	bz	other solvents	Ref.
	Cymediol														
—	**Cymediol**	see **Benzene, dihydroxy-(isopropyl)-methyl-***													
—	**Cymene**	see **Benzene, isopropyl-(methyl)-***													
c938	**Cysteic acid**(*d*)	*L*-2-Amino-3-sulfopropanoic acid*. *β*-Sulfoalanine. $HO_2CCH(NH_2)CH_2SO_3H$	169.16	oct cr or nd (dil al), pr or nd (w+1) $[\alpha]^{20}_D+8.66$	260d (anh)				s^h	i					
c939	—,(*dl*)	$HO_2CCH(NH_2)CH_2SO_3H$	169.16	pr (w)	272–4d				s^h	i					B4², 951
c940	**Cysteine**(*l*)	*L*-*β*-Mercaptoalanine. $HSCH_2CH(NH_2)CO_2H$	121.15	$[\alpha]^{30}_D+9.8$ (c=1.3)					v	v	i	i	i	aa v	B4², 920
c941	**Cystine**(*D*)	*D*-Dicysteine. $[HO_2CCH(NH_2)CH_2S\text{-}]_2$	240.29	$[\alpha]^{20}_D+223$ (1N HCl, c=1.00)	247–9				i	i				ac s alk s	B4², 919
c942	—(*D,L*)	$[HO_2CCH(NH_2)CH_2S\text{-}]_2$	240.29		260				i						B4², 936
c943	—(*L*)	$[HO_2CCH(NH_2)CH_2S\text{-}]_2$	240.29	hex pl (w) $[\alpha]^{20}_D-223.4$ (1N HCl, c=1.00)	260–1d				i	i	i		i	ac s alk s	B4², 925
c944	—(*meso*)	$C_6H_{12}N_2O_4S_2$. *See* c941	240.29		200–218d				i						B4², 936
c945	**Cytidilic acid**	Cytosilic acid. H OPO₃H OH H HO HOCH₂—C—C——C—C—N(pyrimidine ring, N)—NH₂ H H O	323.21	orh nd $[\alpha]^{23}_D+36.5$ (w, c=2.2)	231–3d				s^h	s^h					B31, 25
c946	**Cytidine**	3(*α*-Ribosido)cytosine. H OH OH H HO HOCH₂—C—C——C—C—N(pyrimidine ring, N)—NH₂ O	243.22	orh $[\alpha]^{20.5}_D$ +35.3 (w, c=1)	230–1d				v	δ					B31, 24
c947	**Cytisine**(*l*)	Baptitoxine. Sophorine. Ulexine, $C_{11}H_{14}N_2O$	190.24	orh pr (aa) $[\alpha]^{17}_D-119$ (w)	154.5	218^2			s	s			v	peth δ	B24², 70
c948	—,*N* **methyl-**	Canlophylline. $C_{12}H_{16}N_2O$	204.26	(w+2), nd (al, bz or lig), $[\alpha]^{18.5}_D-230$ (w)	137				s	v		v	v	chl v	B24², 70
—	**Cytosilic acid**	see **Cytidilic acid**													
c949	**Cytosine**	4-Amino-1,2-dihydro-1,3-diazin-2-one*. 4-Amino-2(1)pyrimidone. 1 N, 6, =O, NH 3, 5, NH₂	111.10	mcl or tcl pl (w+1)	320–25d				s^h	δ	i			ac s	
c950	—,**5-methyl-**	$C_5H_7N_3O$. *See* c949	125.13	pr (w)	270d				s					ac s	
—	—,**3(*α*-ribosido)-**	see **Cytidine**													

For explanations, symbols and abbreviations see beginning of table.

No.	Name	Synonyms and Formula	Mol. wt.	Crystalline form, color and specific rotation	m.p. °C	b.p. °C	Density	n_D	Solubility w	al	eth	ace	bz	other solvents	Ref.
	Daidzein														
—	**Daidzein**	*see* **Isoflavone, 4′,7-dihydroxy-**													
—	**Damascenine**	*see* **Benzoic acid, 3-methoxy-2(methylamino)-**, methyl ester													
—	**Danilone**	*see* **1,3-Indanedione, 2-phenyl-**													
—	**Daphnetin**	*see* **Coumarin, 2,8-dihydroxy-**													
—	**Datiscetin**	*see* **Flavone, 2′,3,5,7-tetrahydroxy-**													
—	**D.D.D.**	*see* **Ethane, 1,1-bis(4-chlorophenyl)-2,2-dichloro-**													
—	**D.D.T.**	*see* **Ethane, 1,1-bis(4-chlorophenyl)-2,2,2-trichloro-**													
d1	**1,3-Decadiene***	$CH_3(CH_2)_5CH{:}CHCH{:}CH_2$	138.25	...	...	168–70^{760}	0.750^{20}		δ	...	...	...	...	...	**B1**, 260
d2	**3,7-Decadien-5-yne,4,7-dipropyl***	$C_2H_5CH{:}C(C_3H_7^n)C{\vdots}C(C_3H_7^n)CH{:}CHC_2H_5$	218.37	...	...	125–7^{18}	0.8131^{19}_{4}	1.4890^{20}	...	...	...	...	...	...	...
d4	**4,6-Decadiyne***	$CH_3CH_2CH_2C{\vdots}CC{\vdots}CCH_2CH_2CH_3$	134.21	...	...	88^{12}	0.8695^{19}_{4}	...	...	...	...	...	...	...	...
d5	**4,6-Decadiyne-3,8-diol,3,8-dimethyl-***	$C_2H_5COH(CH_3)C{\vdots}CC{\vdots}CCOH(CH_3)C_2H_5$	194.28	...	89–91	...	...	...	δ s^h	...	...	...	δ s^h	CCl_4 δ MeOH s	...
d6	**Decalin**(*cis*)	Bicyclo[4,4,0]decane. Decahydronaphthalene. Naphthalene. 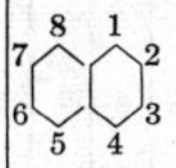	138.25	...	−45.4	195^{760}	0.8967^{20}_{4}	1.4811^{20}	i	∞	v	...	...	chl v	**E12B**, 78
d7	—(*trans*)	$C_{10}H_{18}$. *See* d6	138.25	...	−32.5	185.5^{760}	0.8700^{20}_{4}	1.4696^{20}	i	v	...	...	...	chl v MeOH δ	**E12B**, 78
d8	—,**1-amino**(*cis*)	*cis*-Decalylamine. $C_{10}H_{19}N$. *See* d6	153.27	...	8 (form I) −2 (form II)	100^{12}	...	...	...	...	...	...	...	os, ac s	**E12B**, 674
d10	—,—(*trans*)	$C_{10}H_{19}N$. *See* d6	153.27	...	−18	99^{11}	...	...	...	...	...	...	...	os, ac s	**E12B**, 674
d11	—,**1-chloro-**	$C_{10}H_{17}Cl$. *See* d6	172.70	...	...	d^{760} 114–6^{20}	...	...	...	...	...	...	...	...	...
d11[1]	—,**2-methylene-** (*trans*)	$C_{11}H_{18}$. *See* d6	150.25	...	...	200–1^{756}	0.8897^{20}_{4}	1.4841^{22}	...	...	...	...	...	...	**E12B**, 106
d12	**1,3-Decalindione** (*cis*)	1,3-Diketodecalin. 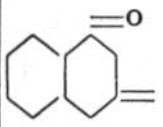	166.22	nd (bz, ace, dil al)	124–5	...	...	...	...	s	...	s	s^h	alk s	**E12B**, 2804
d13	—(*trans*)	$C_{10}H_{14}O_2$. *See* d12	166.22	nd (bz, dil al)	152–3	...	...	...	...	s	...	...	...	alk s	**E12B**, 2805
d14	**2,3-Decalindione** (*cis*)	2,3-Diketodecalin. 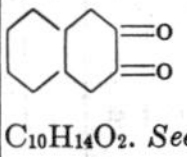	166.22	rh (al, ace, lig)	88–9	...	...	...	δ	s	...	...	...	alk s	**E12B**, 2812
d15	—(*trans*)	$C_{10}H_{14}O_2$. *See* d14	166.22	nd (w), lf (dil al)	100–1	...	...	...	δ	s	...	...	...	os s lig s^h	**E12B**, 2804
d17	**2-Decalincarboxylic acid**(*cis*)	Decahydro-2-naphthoic acid. 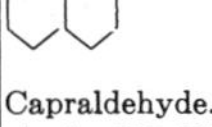	182.26	cr (hexane)	81	...	...	...	...	s	s	...	s	chl s	**E12B**, 4082
d18	**Decanal***	Capraldehyde. *n*-Decylaldehyde. $CH_3(CH_2)_8CHO$	156.27	...	*ca.* −5	208–9	0.830^{15}_{4}	1.4287^{20}	i	s	s	...	...	...	**B1**[2], 764
d19	—,oxime	Capraldoxime. $CH_3(CH_2)_8CH{:}NOH$	171.28	lf (dil MeOH)	69	...	...	...	i	s	s	...	...	...	**B1**, 711
d20	**Decane***	$CH_3(CH_2)_8CH_3$	142.29	...	−29.7	174.1^{760}	0.7300^{20}_{4}	1.4119^{20}	i	∞	s	...	...	...	**B1**[3], 519
d21	—,**1-amino-***	*n*-Decylamine. $CH_3(CH_2)_9NH_2$	157.30	...	15 fr. 16.1	220.5^{760}	0.951^{0}_{4}	1.4333^{30}	δ	∞	∞	∞	∞	chl ∞	**C49**, 15809[d]

For explanations, symbols and abbreviations see beginning of table.

PHYSICAL CONSTANTS OF ORGANIC COMPOUNDS (Continued)

No.	Name	Synonyms and Formula	Mol. wt.	Crystalline form, color and specific rotation	m.p. °C	b.p. °C	Density	n_D	Solubility: w	al	eth	ace	bz	other solvents	Ref.
	Decane														
d22	—,**1-bromo-***	*n* Decyl bromide. $CH_3(CH_2)_9Br$	221.18		−29.6	118^{16}	1.0658^{20}_{4}	1.4553^{20}	i		v			chl v	**B1**[3], 523
d23	—,**2-bromo-(*dl*)***	$CH_3(CH_2)_8CHBrCH_3$	221.18			$124\text{–}5^{30}$	1.0512^{20}	1.4526^{25}	i					chl v	**B1**[3], 523
d24	—,**1-bromo-10-fluoro-***	$F(CH_2)_{10}Br$	239.18			$131\text{–}2^{11}$	1.152^{20}_{4}	1.4512^{25}	i	v	v				
d25	—,**1-chloro-***	Decyl chloride. $CH_3(CH_2)_9Cl$	176.73			222–3	0.8683^{20}_{4}	1.4373^{20}	i						**B1**[3], 522
d26	—,**1-chloro-10-fluoro-***	$F(CH_2)_{10}Cl$	194.72			115^{9}	0.957^{20}_{4}	1.4333^{25}	i	v	v				
d27	—,**1,10-diamino***	Decamethylene diamine. $H_2N(CH_2)_{10}NH_2$	172.32		61.5	140^{12}									**B4**[2], 712
d28	—,**1,10-dibromo-***	Decamethylene dibromide. $Br(CH_2)_{10}Br$	300.09	pl (al)	27	160^{15}d	1.335^{30}	1.4905^{20}	i	δ, v^h	s				**B1**[2], 130
d29	—,**1-fluoro-***	*n*-Decyl fluoride. $CH_3(CH_2)_9F$	160.28			184	0.792^{10}				s				**B1**[2], 129
d30	—,**1-iodo-***	*n*-Decyl iodide. $CH_3(CH_2)_9I$	268.18			132^{15}	1.2567^{20}_{4}	1.4859^{20}	i	s	s				**B1**[3], 523
d31	**Decanedioic acid***	Sebacylic acid. Sebalic acid. $HO_2C(CH_2)_8CO_2H$	202.25	lf	134.5	295^{100}		1.422^{133}	δ	s					**B2**[2], 608
d32	—,diamide	Decanediamide. Sebacamide*. $H_2NCO(CH_2)_8CONH_2$	200.28	pr or pl (aa)	210				i, $δ^h$	i, v^h				aa v	**B2**, 720
d33	—,dibutyl ester	Di-*n*-butyl sebacate. $C_4H_9^nO_2C(CH_2)_8CO_2C_4H_9^n$	314.47			344–5	0.9329^{15}		i		s				**B2**, 719
d34	—,dichloride	Sebacyl chloride. $ClCO(CH_2)_8COCl$	239.14			220^{75}, 165^{11}	1.1212^{20}_{4}	1.4684^{18}	d	d					**B2**[2], 610
d35	—,diethyl ester*	Diethyl sebacate. $C_2H_5O_2C(CH_2)_8CO_2C_2H_5$	258.36		5(1)	306^{773}	0.9646^{20}_{4}	1.4359^{20}	δ	s			i		**B2**[1], 293
d36	—,di(2-ethylbutyl) ester*	$(C_2H_5)_2CHCH_2O_2C(CH_2)_8CO_2CH_2CH(C_2H_5)_2$	370.58		−22	$344\text{–}6^{760}$	0.920^{25}_{4}		i	s		s	s		**B2**, 719
d37	—,di(2-ethylhexyl) ester*	$C_4H_9^nCH(C_2H_5)CH_2O_2C(CH_2)_8CO_2CH_2CH(C_2H_5)(C_4H_9^n)$	426.66		−48	256^{5}	0.912^{25}_{4}	1.451^{25}	i	s		s	s		
d38	—,dimethyl ester	Dimethyl sebacate. $CH_3O_2C(CH_2)_8CO_2CH_3$	230.31	lo pr	26.6	144^{5}	0.9882^{28}	1.4355^{28}	i	s	s				**B2**, 719
d39	—,dinitrile	1,8-Dicyanooctane. Sebaconitrile. $NC(CH_2)_8CN$	164.25			$199\text{–}200^{15}$			i						**B2**[2], 610
d39[1]	—,monomethyl ester, mononitrile	9-Carbomethoxynonanonitrile. $NC(CH_2)_8CO_2CH_3$	197.28		3–4	178^{16}		1.4398^{25}	i		s				**Am 69**, 1684
d40	**1,10-Decanediol***	Decamethyleneglycol. $HO(CH_2)_{10}OH$	174.29	nd	71.5	170^{8}			δ	v	δ, v^h			lig i	**B1**[2], 560
d41	**1-Decanethiol***	*n*-Decyl mercaptan. $CH_3(CH_2)_9SH$	174.35			$125\text{–}7^{19}$	0.8410^{20}_{20}	1.4537^{20}	i	s	s				**Am 55**, 1090
d42	**Decanoic acid***	Capric acid. *n*-Decylic acid. $CH_3(CH_2)_8CO_2H$	172.27	nd	fr. 31.5	270^{760}	0.8858^{40}_{4}	1.4285^{40}	i	$∞^h$	s	v	v		**B2**[2], 309
d43	—,amide	Capramide. Decanamide*. $CH_3(CH_2)_8CONH_2$	171.28	lf (eth)	108, fr. 98		0.999^{20}_{4}	1.4261^{110}	i	s	s	s	δ	CCl_4 δ	**B2**, 356
d44	—,anhydride*	Capric anhydride $[CH_3(CH_2)_8CO]_2O$	326.52	lf	24.7		0.8865^{25}_{4}	1.400^{25}	i	s	s				**B2**[2], 311
d45	—,chloride	Capryl chloride. $CH_3(CH_2)_8COCl$	190.72		−34.5	232^{760}	0.973^{8}_{4}		d	d	s				**B2**, 356
d46	—,decyl ester*	*n*-Decyl caprate. $CH_3(CH_2)_8CO_2(CH_2)_9CH_3$	312.54		9.7	219^{15}	0.8586^{20}	1.4478^{20}							**B2**, 356
d47	—,ethyl ester*	Ethyl caprate. $CH_3(CH_2)_8CO_2C_2H_5$	200.32		−20	241.5^{760}	0.8648^{20}_{4}	1.4256^{20}		∞	∞			chl ∞	**J1948**, 624
d47[1]	—,isopropyl ester*	Isopropyl caprate. $CH_3(CH_2)_8CO_2CH(CH_3)_2$	214.35			121^{10}	0.8543^{20}	1.4221^{25}							
d48	—,methyl ester*	Methyl caprate. $CH_3(CH_2)_8CO_2CH_3$	186.30		−18	224^{760}	0.8733^{20}_{4}	1.4256^{20}	i	v	v				**J1948**, 624
d49	—,nitrile	Caprinitrile. Nonyl cyanide. $CH_3(CH_2)_8CN$	153.27		fr. −14.5	243.7^{760}	0.8295^{15}_{4}	1.4276^{25}	i	∞	∞	∞		chl ∞	**B2**, 356
d51	—,piperazine salt	$C_4H_{10}N_2.2CH_3(CH_2)_8CO_2H$	430.68		93				$δ^h$	s^h	i				**Am 70**, 2758

For explanations, symbols and abbreviations see beginning of table.

No.	Name	Synonyms and Formula	Mol. wt.	Crystalline form, color and specific rotation	m.p. °C	b.p. °C	Density	n_D	Solubility w	al	eth	ace	bz	other solvents	Ref.
	Decanoic acid														
d52	—,propyl ester*	Propyl caprate. $CH_3(CH_2)_8CO_2C_3H_7^n$	214.35			128.5^{10}	0.8620^{20}	1.4260^{25}							
d53	**—,2-bromo-***	$CH_3(CH_2)_7CHBrCO_2H$	251.17		4	$140\text{–}1^{2}$	1.1912^{24}	1.4595^{24}			v				**B2**, 356
d54	**—,10-fluoro-***	$F(CH_2)_9CO_2H$	190.26		49	$90\text{–}5^{0.1}$			δ	v	v			lig s	
d55	**—,2-octyl-***	9-Heptadecanecarboxylic acid. $[CH_3(CH_2)_7]_2CHCO_2H$	284.49	nd or lf (al)	38.5	$270\text{–}5^{100}$			i	s^h	s				**B2**², 368
d56	**—,4-oxo-***	γ-Ketocapric acid. $CH_3(CH_2)_5COCH_2CH_2CO_2H$	186.25	(dil al)	71					s^h					**B3**², 449
d57	**1-Decanol***	Decyl alcohol. $CH_3(CH_2)_9OH$	158.29		fr. 7	229^{760}	0.8287^{20}_{4}	1.4366^{20}	i	∞	∞	∞	∞	chl ∞	**B1**², 459
d58	**2-Decanol**(*dl*)*	$CH_3(CH_2)_7CHOHCH_3$	158.29		−2.4	210–1		1.4326^{25}		s			s		**B1**³, 1762
d59	**4-Decanol***	$CH_3CH_2CH_2CHOH(CH_2)_5CH_3$	158.29			210–1	0.826^{20}_{0}	1.4340^{25}	i	s					**B1**, 426
d60	**1-Decanol, 10-chloro-***	$Cl(CH_2)_{10}OH$	192.73			$185\text{–}9^{15}$		1.4584^{25}	i	v	v				
d61	**—,10-fluoro-***	$F(CH_2)_{10}OH$	176.28		ca. 22	$136\text{–}7^{15}$	0.919^{20}_{4}	1.4322^{25}	i	v	v				
d62	**2-Decanone***	Methyl *n*-octyl ketone. $CH_3CO(CH_2)_7CH_3$	156.27	nd	14	$210\text{–}1^{767}$	0.8230^{22}_{4}	1.4621^{20}	i	s	s				**B1**², 764
d63	**3-Decanone***	Ethyl *n*-heptyl ketone. $CH_3CH_2CO(CH_2)_6CH_3$	156.27			211	0.8251^{20}_{4}	1.4252^{20}		s	s				**B1**¹, 367
d64	**4-Decanone***	*n*-Propyl *n*-hexyl ketone. $CH_3CH_2CH_2CO(CH_2)_5CH_3$	156.27	nd	−9	206–7	0.824^{20}_{0}	1.4240^{21}	i	∞	∞				**B1**, 711
d65	**Decasiloxane, dicosamethyl-**	$CH_3[Si(CH_3)_2O]_9Si(CH_3)_3$	755.35			183^{4}	0.925	1.3988^{20}	i	δ			s	lig s	
d66	**1-Decene***	*n*-Decylene. $CH_3(CH_2)_7CH{:}CH_2$	140.26		−66.3	170.3^{756}	0.7408^{20}_{4}	1.4215^{20}	i	∞	∞				**B1**³, 859
d67	**5-Decene**(*cis*)*	*cis*-1,2-Dibutylethylene. $CH_3(CH_2)_3CH{:}CH(CH_2)_3CH_3$	140.26		−112	170^{739}	0.7445	1.4525	i	∞	∞				**Am 63,** 216
d68	—(*trans*)*	$CH_3(CH_2)_3CH{:}CH(CH_2)_3CH_3$	140.26		−73	170.2^{739}	0.7401	1.4235	i	∞	∞				**Am 63,** 216
d69	**1-Decene, 2-bromo-***	$CH_3(CH_2)_7CBr{:}CH_2$	219.18			$115\text{–}6^{22}$	1.0844^{20}_{4}	1.4629^{20}							**B1**³, 859
d70	**2-Decene, 1-bromo-***	$CH_3(CH_2)_6CH{:}CHCH_2Br$	219.18			121^{17}	1.074^{18}_{4}	1.4716^{18}						lig s	**B1**³, 859
d71	**1-Decen-3-yne***	$CH_3(CH_2)_5C{\equiv}CCH{:}CH_2$	136.23			45^{4}	0.7872^{25}_{4}	1.4565^{25}							**Am 61,** 572
d72	**1-Decen-4-yne***	$CH_3(CH_2)_4C{\equiv}CCH_2CH{:}CH_2$	136.23			$73\text{–}4^{22}$	0.7850^{25}_{4}	1.444^{25}							**Am 58,** 612
d73	**2-Decen-4-yne***	$CH_3(CH_2)_4C{\equiv}CCH{:}CHCH_3$	136.23			55^{5}	0.7850^{25}_{4}	1.4609^{25}							**Am 61,** 572
d74	**1-Decyne***	*n*-Octyl acetylene. $CH_3(CH_2)_7C{\equiv}CH$	138.25		−44	174^{760}	0.7655^{20}_{4}	1.4269^{20}	i	s	s			os s	**B1**³, 1016
d75	**3-Decyne***	$CH_3CH_2C{\equiv}C(CH_2)_5CH_3$	138.25			$175\text{–}6^{760}$	0.765^{21}_{4}	1.433^{21}_{4}							
d76	**4-Decyne***	$CH_3(CH_2)_4C{\equiv}C(CH_2)_2CH_3$	138.25			74.5^{19}	0.772^{17}_{4}	1.436^{17}							
d77	**5-Decyne***	Dibutylacetylene. $CH_3(CH_2)_3C{\equiv}C(CH_2)_3CH_3$	138.25		−73	177^{751}	0.7690^{20}_{4}	1.4331^{20}	i	s	s				**B1**³, 1017
d78	**4-Decyne, 3,3-dimethyl-***	$CH_3CH_2C(CH_3)_2C{\equiv}C(CH_2)_4CH_3$	166.30			86^{20}	0.7731^{20}_{4}	1.4399^{20}							**Am 62,** 1800
—	**Dehydroacetic acid**	*see* **4-Hexenoic acid, 2-acetyl-5-hydroxy-3-oxo-**, lactone													
d79	**Dehydrocholic acid**	3,7,2-Trioxocholanic acid. 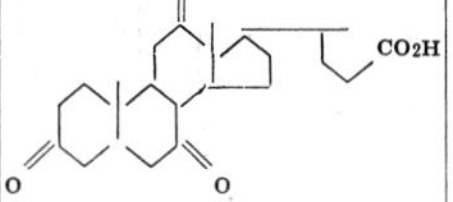	402.54	(ace), $[\alpha]_D$ +26 (al)	237				i	δ	i	s	δ	chl s AcOEt s	**E14**, 211

For explanations, symbols and abbreviations see beginning of table.

No.	Name	Synonyms and Formula	Mol. wt.	Crystalline form, color and specific rotation	m.p. °C	b.p. °C	Density	n_D	Solubility w	al	eth	ace	bz	other solvents	Ref.
	Dehydroergosterol														
d80	**Dehydro-ergosterol**	$\Delta^{5:6,\ 7:8,\ 9:11,\ 22:23}$-Ergo-statetraen-3-ol. HO— 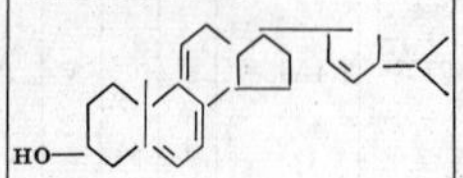	394.64	lf (al+1w), nd (eth) $[\alpha]_D^{15}+149.2$ (chl)	146	$230^{0.5}$	...	...	...	s^h	v	s	...	chl v AcOEt s	**E14**, 67
—	**Dehydromucic acid**	*see* **2,5-Furandicarboxylic acid**													
d81	**Delphinine**	$C_{33}H_{45}NO_9$	599.71	orh $[\alpha]_D^{25}+25$ (al)	198–200d	...	...	...	i	s	s	...	...	chl s	**J1952,** 1750
d82	**Delphinidine chloride**	3,3′,4′,5,5′,7′-Hexahydroxy-flavinium chloride. $C_{15}H_{11}ClO_7$.	338.70	br pr, nd or pl (HCl)	>350	...	...	...	v	v	...	...	...	MeOH v AcOEt s	**B18²**, 247
d83	**Derritol** 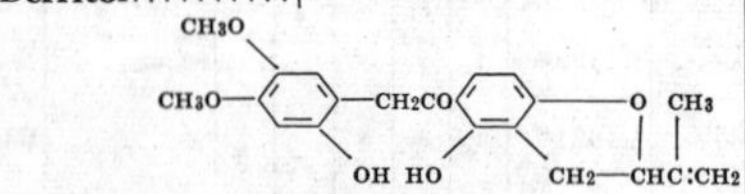	CH_3O, CH_3O—, —CH_2CO—, —O CH_3, OH HO, CH_2—$CHC:CH_2$	370.41	ye nd (MeOH) $[\alpha]_D^{20}-66.2$ (chl, c=3)	164	...	...	...	i	...	...	...	...	alk v	**B18²**, 223
d84	**Deserpidine**	Canescine. $C_{32}H_{38}N_2O_8$	578.67	nd or pr $[\alpha]_D-137$ (chl)	229–31	...	...	...	i	s^h	...	...	...	chl s	**Am** 77 4335
—	**Desmethyl-morphine**	*see* **Normorphine**													
—	**Desoxalic acid**	*see* **1,1,2-Ethane-tricarboxylic acid, 1,2-dihydroxy-***													
—	**Desoxybenzoin**	*see* **Ketone, benzyl phenyl**													
d85	**Desoxycholic acid**	3,12 Dihydroxycholanic acid. $C_{24}H_{40}O_4$	392.58	(al) $[\alpha]_D^{20}$ +57 (sl)	172–3	...	...	...	i	v	δ	δ	i	chl δ aa δ	**E14**, 183
d86	**Desoxycorti-costerone**	Δ^4-Pregnene-3,20-dion-21-ol. $C_{21}H_{30}O_3$	330.47	pl (eth) $[\alpha]_D^{20}+178$ (al)	141–2	...	...	...	...	v	s	v	...	...	**E14**, 153
d87	**Desthiobiotin**	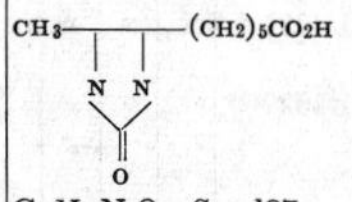 CH_3—, —$(CH_2)_5CO_2H$, N N, O	214.27	lo nd (w) $[\alpha]_D^{21}+10.7$	156–8	...	...	...	s	...	...	...	...	...	...
d88	—,methyl ester	$C_{11}H_{18}N_2O_3$. *See* d87	228.29	cr (MeOH) $[\alpha]_D^{28}+2.6$ (chl)	69–70	$194–7^{0.03}$	...	...	...	s^h	...	...	...	MeOH s chl v	...
—	**Desyl chloride**	*see* **Ketone, benzyl phenyl,** α-chloro-													
—	***O*-Deutero-methanol**	*see* **Methanol-*d***													
—	**Deuteroxy(tri-deutero)methane**	*see* **Methan-*d*₃-ol-*d***													
—	**Dexedrine**	*see* **Amphetamine**													
d89	**Dextrin** (starch)	$(C_6H_{10}O_5)_x$	$(162.14)_x$	amor $[\alpha]_D>+200$	chars	...	1.0384_4^{20}	...	s	i	i	...	...	...	...
—	**Dextronic acid**	*see* **Gluconic acid**(*D*)													
	Dextrose	*see* **Glucose**(*D*)													
d90	**Dextropimaric acid,** methyl ester	$C_{21}H_{32}O_2$	316.49	$[\alpha]_D+60.5$	69	$149^{0.03}$	1.030_4^{19}	1.5208^{19}	...	v	v	...	...	...	**B9²**, 453
—	**Diacetamide**	*see* **Acetic acid,** amide, *N*-acetyl-													
—	**Diacetanilide**	*see* **Acetic acid,** amide, *N*-acetyl-*N*-phenyl-													
—	**Diacetoacetic acid**	*see* **Butanoic acid, 2-acetyl-3-oxo-**													
—	**Diacetonamine**	*see* **2-Pentanone, 4-amino-4-methyl-***													
—	**Dialuric acid**	*see* **Barbituric acid, 5-hydroxy-**													
d91	**Diazoamino-benzene**	1,3 Diphenyltriazene. 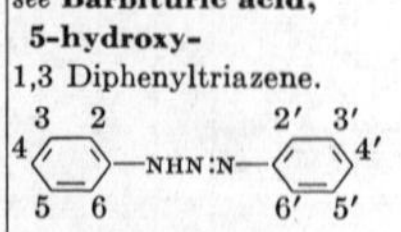 3 2 4 5 6 —NHN:N— 2′ 3′ 4′ 6′ 5′	197.24	ye lf (al)	100	...	...	...	i	v	v	...	v	Py v	**B16²**, 351
d92	—(isomer)	$C_{12}H_{11}N_3$. *See* d91	197.24	ye nd or pr (bz)	80–1	...	...	...	...	...	...	...	s	lig s	**B16²**, 351
d93	—,2,2′-dimethyl-	$C_{14}H_{15}N_3$. *See* d91	225.30	lt red nd (al)	121	...	...	...	...	s	s	...	...	lig s	**B16²**, 180

For explanations, symbols and abbreviations see beginning of table.

No.	Name	Synonyms and Formula	Mol. wt.	Crystalline form, color and specific rotation	m.p. °C	b.p. °C	Density	n_D	Solubility w	al	eth	ace	bz	other solvents	Ref.
	Diazoamino-benzene														
d94	**—,2,3′-dimethyl-**	$C_{14}H_{15}N_3$. *See* d91	225.30	ye cr (lig)	74									lig v	**B16**, 705
d95	**—,2,4′-dimethyl-**	$C_{14}H_{15}N_3$. *See* d91	225.30	ye nd (lig)	119–20									lig v	**B16**, 708
d96	**—,3,3′-dimethyl-**	$C_{14}H_{15}N_3$. *See* d91	225.30	ye nd (peth)	50–2									os v	**B16**[1], 407
d97	**—,3,4′-dimethyl-**	$C_{14}H_{15}N_3$. *See* d91	225.30	ye nd (lig)	96–7									os s lig s^h	**B16**, 708
d98	**—,4,4′-dimethyl-**	$C_{14}H_{15}N_3$. *See* d91	225.30	ye nd (lig) or pr (al)	116					s^h				sulf s lig s^h	**B16**[2], 355
d99	**—,4,4′-dinitro-**	$C_{12}H_9N_5O_4$. *See* d91	287.24	ye nd (al) or lf (bz)	220d				i	δ^h	s	δ	δ	chl δ	**B16**[2], 354
d100	**—,3-methyl-**	$C_{13}H_{13}N_3$. *See* d91	211.27	ye nd (lig)	86										**B16**, 714
d101	**—,4-methyl-**	$C_{13}H_{13}N_3$. *See* d91	211.27	ye pl	85 (91)				i						**B16**[2], 354
d102	**1,1′-Diazoamino-naphthalene**	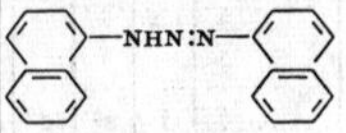 —NHN:N—	297.36	ye lf (al)	exp>100										**B16**, 716
d103	**2,2′-Diazoamino-naphthalene**	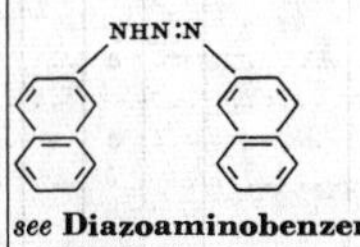 NHN:N	297.36	red nd (xyl)	156										
—	**Diazoamino-toluene**	*see* **Diazoaminobenzene, dimethyl-**													
—	**Diazomethane**	*see* **Methane, diazo-**													
d104	**1,2:3,4-Dibenz-anthracene**	Naphtho 2′,3′:9,10 phen-anthrene. 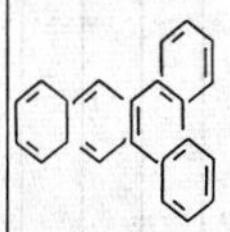	278.35	nd (aa)	205								δ		**B5**[2], 668
d105	**1,2:5,6-Dibenz-anthracene**	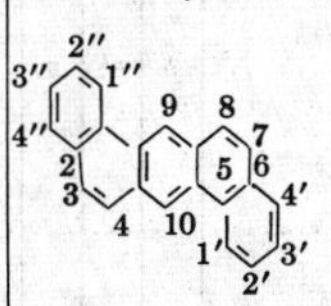	278.35	pl (dil ace)	269–70				i	δ		s	s	CS_2 s aa s	**E14s**, 604
d106	**1,2:6,7-Dibenz-anthracene**	1,2 Benzonaphthacene. Isopentaphene. 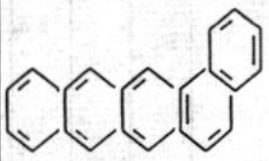	278.33	ye lf or nd (xyl)	263–4									con sulf s	**B5**[2], 667
d107	**1,2:7,8-Dibenz-anthracene**	Dinaphthanthracene. 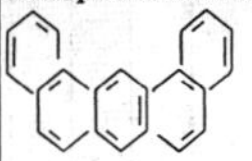	278.33	og lf (bz)	267.5				i	i	i		δ	aa i	**B5**[2], 668
d108	**1,2:5,6-Dibenz-anthracene, 4′,4″-dihydroxy-**	$C_{22}H_{14}O_2$. *See* d105	310.55	og	415–8	sub vac				δ^h	i^h				**E14s**, 620
d109	**Dibenzanthrone**	Violanthrone. 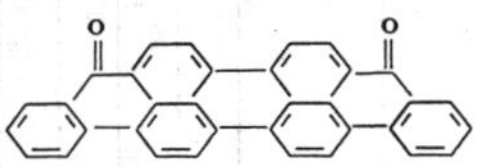	456.50	vt-bl or bk nd ($PhNO_2$ or quinoline)	490–5d					i			i	xyl, Py sulf s (vt) aa i	**E14s**, 907
d110	**1,2:5,6-Dibenzo-fluorene**	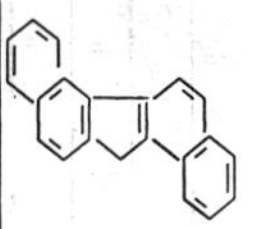	266.34	pl (bz-al or AcOEt)	174–5	$280^{0.3}$				δ			s	to, $PhNO_2$ s	**E14s**, 520
d111	**1,2:7,8-Dibenzo-fluorene**	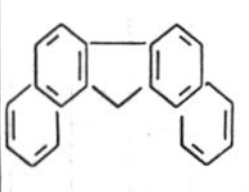	266.34	lf (bz)	230–1				i	δ	δ		s		**E14s**, 518
d112	**1,2:5,6-Dibenzo-9-fluorenone**	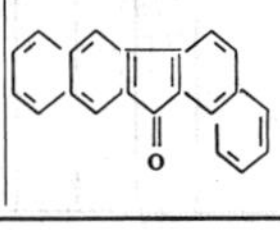	280.31	og-ye pl or pr (xyl or aa)	213	sub vac								to s^h sulf s (red)	**E14s**, 514

For explanations, symbols and abbreviations see beginning of table.

No.	Name	Synonyms and Formula	Mol. wt.	Crystalline form, color and specific rotation	m.p. °C	b.p. °C	Density	n_D	Solubility w	al	eth	ace	bz	other solvents	Ref.
	Dibenzofulvene														
—	**Dibenzofulvene**	*see* **Fluorene, 9-methylene-**													
d113	**Dibenzofuran**	Diphenylene oxide. 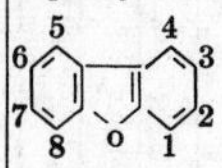	168.18	lf or sc (al)	86–7	287^{760}	1.0886^{99}_{4}	1.6079^{99}_{4}	i	s	v	...	v	aa v	**B17^2, 67**
d114	**—,1-amino-**	$C_{12}H_9NO$. *See* d113	183.21		85										
d115	**—,2-amino-**	$C_{12}H_9NO$. *See* d113	183.21	(dil al)	94										**B18^1, 557**
d116	**—,3-amino-**	$C_{12}H_9NO$. *See* d113	183.21	pl (dil al)	128				i	v^h	v				**B18^2, 423**
d117	**—,4-amino-**	$C_{12}H_9NO$. *See* d113	183.21	br nd (dil MeOH)	85										**Am 75, 4845**
d118	**—,1-bromo-**	$C_{12}H_7BrO$. *See* d113	247.09		67										
d119	**—,2-bromo-**	$C_{12}H_7BrO$. *See* d113	247.09	lf	110 (120)	220^{40}			i	s^h	s				**J1931, 529**
d120	**—,3-bromo-**	$C_{12}H_7BrO$. *See* d113	247.09	nd (al) lf (aa)	108–9	220^{40}				s				aa δ	**B17^2, 68**
d121	**—,2,8-dibromo-**	$C_{12}H_6Br_2O$. *See* d113	326.01		194										
d122	**—,3,6-dibromo-**	$C_{12}H_6Br_2O$. *See* d113	326.01	lf (al)	199–200					s	v	...	v	aa v	**B17^2, 69**
d123	**—,2,6-dinitro-**	$C_{12}H_6N_2O_5$. *See* d113	258.19		255–6					δ	...	s	i	xyl δ lig i	**B17^2, 69**
d124	**—,2,7-dinitro-**	$C_{12}H_6N_2O_5$. *See* d113	258.19		245										
d125	**—,1-nitro-**	$C_{12}H_7NO_3$. *See* d113	213.18	ye nd	138–9 (126)	$190–205^{15}$									
d126	**—,2-nitro-**	$C_{12}H_7NO_3$. *See* d113	213.18	ye nd (al), cr (aa)	159 (141)	$180–5^{3}$			i	δ	δ			aa s^h	**B17^2, 69**
d127	**—,4-nitro-**	$C_{12}H_7NO_3$. *See* d113	213.18	lt ye nd (al)	138–9 (120–1)										**Am 75, 4845**
d128	**1-Dibenzofuran-carboxylic acid**	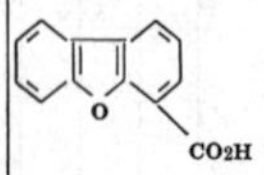	212.19	nd (al)	209–10										
d129	**3-Dibenzofuran-carboxylic acid**	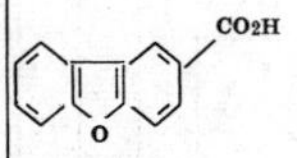	212.19	nd (dil al)	271–2				δ	s^h	s				**B18, 313**
d130	**4-Dibenzofuran-carboxylic acid**	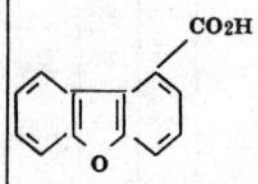	212.19	(50% al)	209–10										
d131	**1,2:6,7-Dibenzo-phenanthrene**	Naphtho-2′,1′:1,2-anthracene. Naphtho-2′,3′:1,2 phen-anthrene. 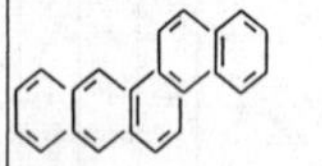	278.33	gr-ye lf	293–4										**B5^2, 668**
—	**1,2:7,8-Dibenzo-phenanthrene**	*see* **Picene**													
d132	**2,3:6,7-Dibenzo-phenanthrene**	Dibenzo[*b,h*]phenanthrene. Pentaphene. 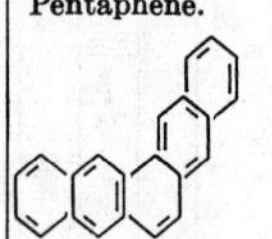	278.33	ye-gr nd or lf (xyl)	257				i	δ	δ	...	s	xyl δ	
—	**Dibenzo-1,4-pyran**	*see* **Xanthene**													
d133	**1,2:4,5-Dibenzo-pyrene**	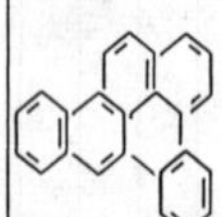	302.38	pa ye nd (to)	241–2					δ	...	δ	δ	to s^h con sulf s (ye-red) aa δ	**E14s, 805**
—	**Dibenzopyrrole**	*see* **Carbazole**													
d134	**Dibenzothiopene**	Diphenylene sulfide. 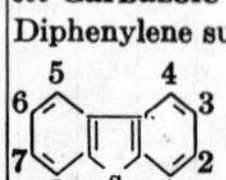	184.26	nd (dil al or lig)	97	332–3			s v^h	v	...	...	v	MeOH s^h	**B17^2, 70**
d135	**—,2-amino**	$C_{12}H_9NS$. *See* d134	199.28		129–31										**B18^2, 423**
d136	**—,3-amino**	$C_{12}H_9NS$. *See* d134	199.28	lf (dil al)	122–3										**B18^2, 423**

For explanations, symbols and abbreviations see beginning of table.

No.	Name	Synonyms and Formula	Mol. wt.	Crystalline form, color and specific rotation	m.p. °C	b.p. °C	Density	n_D	Solubility w	al	eth	ace	bz	other solvents	Ref.
	Dibenzothiopene														
d137	—,**1-bromo-**	$C_{12}H_7BrS$. *See* d134	263.16		84										
d138	—,—,dioxide	1 Bromodiphenylene sulfone. 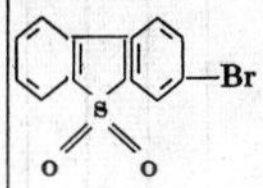	295.16		190										
d139	—,**2-bromo-**, dioxide	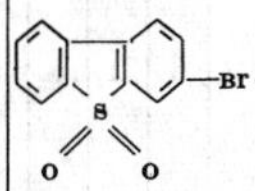	295.16		266–7										**Am 75, 3844**
d140	—,—,monoxide	2 Bromodiphenylene sulfonide. 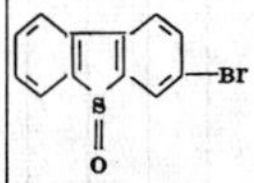	279.16		171–2										
d141	—,**3-bromo-**	$C_{12}H_7BrS$. *See* d134	263.16	nd	125–6 (98–9)										**$B17^2$, 70**
d142	—,—,monoxide	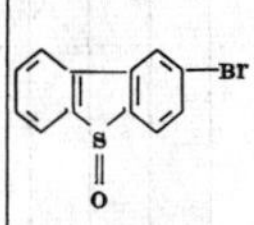	279.16		171–2										
d143	—,**4-bromo-**	$C_{12}H_7BrS$. *See* d134	263.16		83–4										**Am 76, 5786**
d144	—,**2,7-diamino-**, dioxide	Benzidine sulfone. 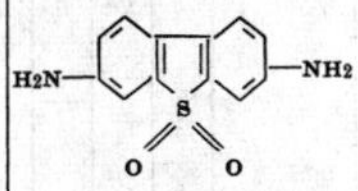	246.29	ye nd	327–8				i^h	i^h	i		i^h		**B18, 591**
d145	—,**3,6-diamino-**	$C_{12}H_{10}N_2S$. *See* d134	214.29	nd (al)	178					v^h					**$B18^2$, 427**
d146	—,**3,7-diamino-**	$C_{12}H_{10}N_2S$. *See* d134	214.29	pa ye cr	169–70										**Am 74, 1166**
d147	—,**2,8-dibromo-**	$C_{12}H_6Br_2S$. *See* d134	342.07		229										
d148	—,—,dioxide	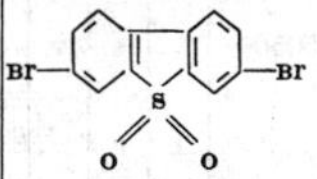	374.07		361–2										**Am 76, 5786**
d149	—,**3,6-dibromo-**	$C_{12}H_6Br_2S$. *See* d134	342.07		229										**$B17^2$, 71**
d150	—,**3,7-dinitro-**, dioxide	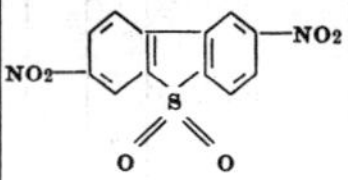	306.26		290										
d151	—,**2-nitro-**	$C_{12}H_7NO_2S$. *See* d134	229.26		99–100										
d152	—,**3-nitro-**	$C_{12}H_7NO_2S$. *See* d134	229.26	pa ye nd	186										**$B17^2$, 71**
d153	—,—,dioxide	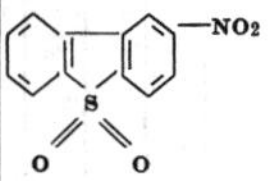	261.26		265										
d154	—,—,monoxide	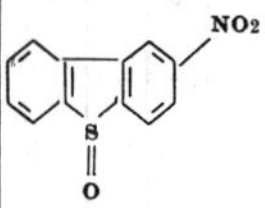	245.26		210										**Am 70, 1749**
d155	**3-Dibenzothiophenecarboxylic acid**	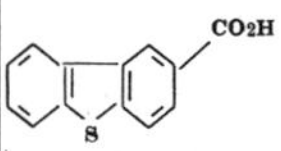	228.27		255										**$B18_2$, 279**
d156	**4-Dibenzothiophenecarboxylic acid**	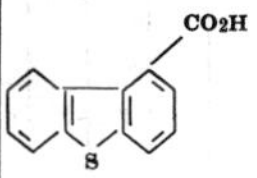	228.27		261–2										

For explanations, symbols and abbreviations see beginning of table.

No.	Name	Synonyms and Formula	Mol. wt.	Crystalline form, color and specific rotation	m.p. °C	b.p. °C	Density	n_D	Solubility w	al	eth	ace	bz	other solvents	Ref.
	4-Dibenzothiophencarboxylic acid														
d157	—,dioxide	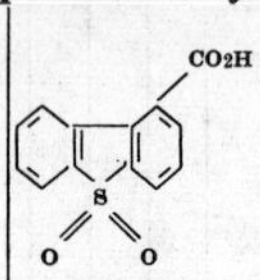	260.27		338–9										
—	**Dibenzoxazine**	*see* **Phenoxazine**													
—	**Dibenzyl**	*see* **Ethane, 1,2-diphenyl-**													
d158	**Diborane, methylthio-**	$CH_3SB_2H_5$	73.78		−101.5	53^{760} (extrap.) (-35^{9})			d	d				os ∞	**Am 76,** 3307
—	**Dichloramine B**	*see* **Benzenesulfonic acid,** amide, *N,N* dichloro-													
—	**Dichloramine T**	*see* **4-Toluenesulfonic acid,** amide, *N,N*-dichloro-													
—	**Dichlorophene**	*see* **Methane, bis(5-chloro-2-hydroxyphenyl)-**													
d159	**Dicoumarin**(*cis*)	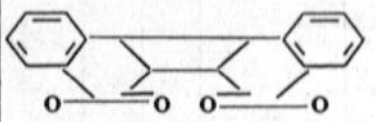	292.28	lf (aa)	262				i						**B19,** 181
d160	—(*trans*)	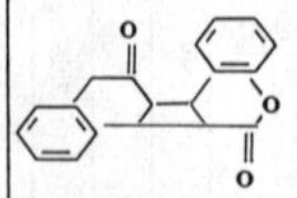	292.28	nd or pl (aa)	>275									aa δ	**B19,** 181
d161	**Dicoumarol**	4,4′-Dihydroxy-3,3′-methylene biscoumarin. Melitoxin. $C_{19}H_{12}O_6$	336.29	nd	288–92				i	i	i	i	δ	chl δ	**B19,** 197
d162	**Dictamine**	2,3-(Methoxy-2,3-Quinolino)-furan. 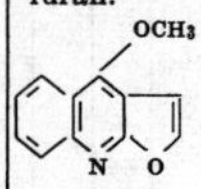	199.20	pr (al)	132–3				δ	v^h	s			chl, AcOEt con sulf s	**B27**[2]**,** 79
d163	**Dicyclohexadiene**	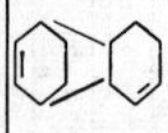	160.26			229–30	0.995_4^{14}	1.5288^{14}	i	δ	v	v	v	aa v	**E13,** 1038
d164	**α-Dicyclopentadiene**(*endo* form)		132.31		32	170 δd	0.9302_4^{35}	1.5050^{35}		s	v			CCl_4 s aa s	**E13,** 1018
d165	**—,3,4,5,6,7,8,8a-heptachloro-**	Heptachlor. $C_{10}H_5Cl_7$	373.34		95–6		$1.57–9^{9}$		i	s				lig s	
d166	**—,tetrahydro-**	Tricyclodecane. 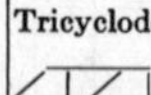	136.24	(al or aa)	77	193^{769}	0.9128^{79}	1.4726^{79}	i	s				aa s	**E13,** 1022
d167	**Dieldrin**	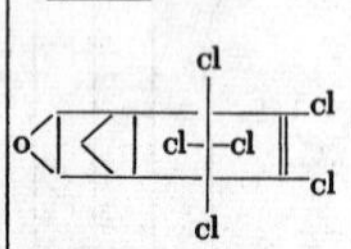	380.91		175–6		1.75		i	δ		s	s		
—	**Diethanolamine**	*see* **Amine, diethyl, 2,2′-dihydroxy-**													
d168	**Diethylene glycol**	2,2′-Dihydroxydiethyl ether. 2,2′-Oxydiethanol. $HOCH_2CH_2OCH_2CH_2OH$	106.12		−10.5	245^{760}	1.118_{20}^{20}	1.4488^{15}	s	s	s				**B1**[2]**,** 520
d169	—,diacetate	$CH_3CO_2CH_2CH_2OCH_2CH_2O_2CCH_3$	190.20			245–51	1.1078_{15}^{15}			v					**B2**[2]**,** 155
d170	—,diethyl ether	Diethyl carbitol. $C_2H_5OCH_2CH_2OCH_2CH_2OC_2H_5$	162.23			189^{760}	0.9063_4^{20}	1.4115^{20}	v	v	s			os v	**B1**[3]**,** 2098
d171	—,dioctadecanoate	Diethylene glycol distearate. Glycosterin. $[CH_3(CH_2)_{16}CO_2CH_2]_2O$	611.01	wx	54–5		0.9333_4^{20}			i	i				
d172	—,dioleate	$[CH_3(CH_2)_7CH:CH(CH_2)_7CO_2CH_2CH_2]_2O$	635.03	pa ye liq			0.9310_4^{20}			∞	∞				

For explanations, symbols and abbreviations see beginning of table.

No.	Name	Synonyms and Formula	Mol. wt.	Crystalline form, color and specific rotation	m.p. °C	b.p. °C	Density	n_D	Solubility w	al	eth	ace	bz	other solvents	Ref.
	Diethylene glycol														
d173	—,monobutyl ether	Butyl carbitol. $CH_3(CH_2)_3OCH_2CH_2OCH_2CH_2OH$	162.23		−78	231^{760}	0.9553^{20}_{4}	1.4321^{20}	∞	v	v				B1[2], 521
d174	—,—,acetate	$CH_3(CH_2)_3OCH_2CH_2OCH_2CH_2O_2CCH_3$	204.26		−32	247^{760}	0.985^{20}_{4}	1.4262^{20}	s	∞	∞			oos ∞	
d175	—,monododecanoate	Ethylene glycol monolaurate. Glaurin. $CH_3(CH_2)_{10}CO_2CH_2CH_2OCH_2CH_2OH$	288.42	lt ye	17–18	$>270^{760}$	0.96^{25}_{25}			∞		∞		to s	C37 3202[1]
d176	—,monoethyl ether	Carbitol. $CH_3CH_2OCH_2CH_2OCH_2CH_2OH$	134.18	hyg liq		201^{760}	0.9881^{20}_{4}	1.4273^{20}	∞				∞	oos ∞	B1[2], 520
d177	—,—,acetate	Carbitol acetate. $CH_3CH_2OCH_2CH_2OCH_2CH_2O_2CCH_3$	176.21	hyg liq	−25	218^{760}	1.0096^{20}_{4}	1.4213^{20}	∞	∞	∞			oos s	B2[2], 155
d178	—,mono(2-hydroxypropyl) ether	$CH_3CHOHCH_2OCH_2CH_2OCH_2CH_2OH$	164.20			$277–9^{760}$	1.0789^{20}_{4}	1.4498^{20}	∞	v			v	MeOH, CCl_4 v	
d179	—,monomethyl ether	Methyl carbitol. $CH_3OCH_2CH_2OCH_2CH_2OH$	120.15			193^{760}	1.0354^{20}_{4}	1.4264^{20}	∞	v					C32, 2217[5]
d180	**Diethylenetriamine**	2,2′-Diaminodiethylamine. $(NH_2CH_2CH_2)_2NH$	103.17	ye hyg liq	−39	208^{760}	0.9542^{20}_{20}	1.4810^{25}	∞	∞	i			lig s	B4[2], 695
d181	***m*-Digallic acid**	Gallic acid 3-monogallate. 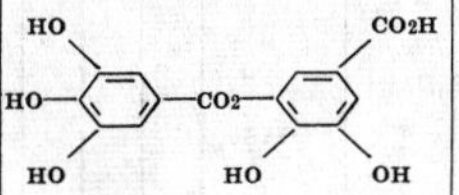	332.23	nd (dil al+1w)	280				i, s^h	s	δ	s		MeOH s aa δ	B10[2], 344
—	**Digine**	*see* **Gitogenin**													
d182	**Digitalose**	3-Methyl *D*-fucose. 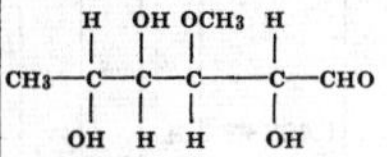	178.19	nd (AcOEt). mut $[\alpha]^{22}_D$ +106 (17 hrs.)	106 (fresh) 119 (after 4 months)				s					AcOEt s^h	
d183	**Digitogenin**	$C_{27}H_{44}O_5$	448.65	nd (al) $[\alpha]^{19}_D$ −18 (chl)	280–3				i	s^h				chl s	E14, 286
d184	**Digitoxigenin**	$C_{23}H_{34}O_4$	374.53	(dil MeOH) $[\alpha]^{20}_D$+18.1 (MeOH)	256					s				MeOH v	E14, 225
d185	**Digitoxin**	$C_{41}H_{64}O_{13}$	764.96	(chl+eth) $[\alpha]^{20}_D$+4.8 (diox)	252–3				δ	v				chl, MeOH Py s	E14, 230
—	**Digitoxit**	*see* **1,3,4,5-Hexanetetrol**													
d186	**Digitoxose**	2-Desoxy-*D*-altromethylose. $CH_3(CHOH)_3CH_2CHO$	148.16	cr (MeOH+eth or AcOEt) mut. $[\alpha]^{18}_D$ +27.9 to +43.3 (pyr)	112				v			v		Py s AcOEt s^h	
—	—,3-methyl-	*see* **Cymarose**													
—	**Diglycine**	*see* **Glycine, *N*-glycyl-**													
—	**Diglycolic acid**	*see* **Acetic acid, oxydi-**													
d187	**Digoxigenin**	$C_{23}H_{34}O_5$	390.53	pr (AcOEt). $[\alpha]^{20}_D$+23.1 (MeOH)	222					v				chl δ MeOH v	E14, 239
d190	**Dihydrosamidin**	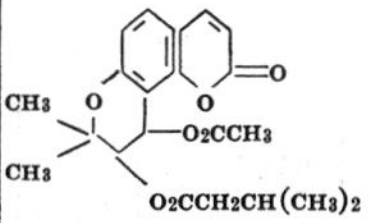	388.40	$[\alpha]_D$+19 (al)	117–9				i	s	s				Am 79, 3534
d191	**Diisoeugenol**	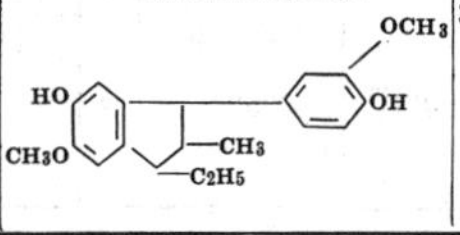	328.42	nd (bz lig or al)	180–1					s^h	v		δ	chl v	B6[2], 917

For explanations, symbols and abbreviations see beginning of table.

No.	Name	Synonyms and Formula	Mol. wt.	Crystalline form, color and specific rotation	m.p. °C	b.p. °C	Density	n_D	Solubility w	al	eth	ace	bz	other solvents	Ref.
	Dilactic acid														
d192	**Dilactic acid**	1,1′-Dicarboxydiethyl ether. Dilactylic acid. $HO_2CCH(CH_3)OCH(CH_3)CO_2H$	162.14	rh	112–3 (105–7)				s	δ	s		δ		**B3**², 205
—	**α,γ-Dilaurin**	*see* **Glycerol, 1,3-didodecanoate**													
—	**Dilirutic acid**	*see* **Barbituric acid, 5-nitro-**													
—	**Dimedone**	*see* **1,3-Cyclohexanedione, 5,5-dimethyl-**													
d193	**Dimethisoquin,** hydrochloride	(structure: isoquinoline with $C_4H_9^n$ and $OCH_2CH_2N(CH_3)_2$) .HCl	308.86		146				s	s					
d194	**Dimethylarsinic acid**	Alkargen. Cacodylic acid. $(CH_3)_2AsO_2H$	138.00	tcl	200				v	v	i				**B4**², 993
d195	**Dimethyl phosphinic acid,** ethyl ester	$(CH_3)_2PO_2C_2H_5$	122.10			89^{15}	1.0278^{25}_{4}	1.4261^{25}	δ	s	s				**Am 73,** 5466
—	**Dimite**	*see* **Ethanol, 1,1-bis(4-chlorophenyl)-**													
d196	**Dimissine**	$C_{50}H_{83}NO_{22}$	1050.21	[α]−20 (Py)	305–8					s^h				MeOH s^h dil ac s	
—	**Dinicotinic acid**	*see* **3,5-Pyridinedicarboxylic acid**													
d197	**1,3-Dioxane**	*m*-Dioxane. Trimethylene glycol methylene ether. (ring structure, positions 1–6)	88.11		−42	105^{755}	1.0342^{20}_{4}	1.4165^{20}_{α}	∞	∞	∞				**B19**², 4
d198	**1,4-Dioxane**	Diethylene dioxide. *p*-Dioxane. Glycol ethylene ether. (ring structure, positions 1–6)	88.11		11.8	101^{760}	1.0336^{20}_{4}	1.4224^{20}	∞	∞	∞	∞		os ∞ aa ∞	**B19**², 4
d199	**—,2,3-dichloro-**	$C_4H_6Cl_2O_2$. *See* d198	157.00		30	$80–2^{10}$	1.468^{20}_{4}	1.4928^{20}	i d^h		v	v	v	chl, CCl_4, diox v lig v	
d200	**1,3-Dioxane, 2,4-dimethyl-**	$C_6H_{12}O_2$. *See* d197	116.16			120 (180)	0.9392^{20}_{4}	1.4136^{20}	δ					os v	**B19**², 10
d201	**—,5-ethyl-4-propyl-**	$C_9H_{18}O_2$. *See* d197	158.24			196^{760}	0.9305^{20}_{4}	1.4370^{20}	i					os s	
d202	**1,4-Dioxane, heptachloro-**	$C_4HCl_7O_2$. *See* d198	329.22		54–6	$123–8^{8}$			i		v	v	v	chl, CCl_4 v lig v	
d203	**1,3-Dioxane, 5-hydroxy-2-methyl-**	$C_5H_{10}O_3$. *See* d197	118.13			176^{760}	1.0705^{17}_{4}	1.4375^{17}	∞						**B19**², 70
d204	**—,4-methyl-**	$C_5H_{10}O_2$. *See* d197	102.13			114^{760}	0.9953^{20}_{4}	1.4168^{20}	δ					os v	
d205	**—,4-methyl-4-phenyl-**	$C_{11}H_{14}O_2$. *See* d197	178.22		35–40	256^{760}	1.079^{30}_{4}		i					os s	
d206	**—,2-phenyl-**	$C_{10}H_{12}O_2$. *See* d197	164.19	nd (peth)	41	252–4			d^h	v	v				**B19**², 22
d207	**—,4-phenyl-**	$C_{10}H_{12}O_2$. *See* d197	164.19			245^{760}	1.1038^{20}_{4}	1.5306^{18}	i					os s	**B19**¹, 616
—	**2,5-*p*-Dioxanedione**	*see* **Glycolide**													
d208	**1,3-Dioxolane**	Glycol methylene ether (ring structure, positions 1–5)	74.08			78^{765}	1.0600^{20}_{4}	1.3974^{20}_{α}	∞	s	s	s			**B19**², 3
d209	**—,4-(hydroxymethyl)-2-methyl-**	Glycerol ethylidene ether. $C_5H_{10}O_3$. *See* d208	118.13			187^{760}	1.1243^{17}_{4}	1.4415^{17}	δ	s					**B19**², 71
d210	**—,2-methyl-**	$C_4H_8O_2$. *See* d208	88.10			172–3 (82–5)	0.9811^{20}_{4}	1.4035^{17}	v	∞	∞				**B19,** 8

For explanations, symbols and abbreviations see beginning of table.

No.	Name	Synonyms and Formula	Mol. wt.	Crystalline form, color and specific rotation	m.p. °C	b.p. °C	Density	n_D	Solubility w	al	eth	ace	bz	other solvents	Ref.
	1,3-dioxolane-4-carboxaldehyde														
d211	**1,3-Dioxolane-4-carboxaldehyde, 2,2-dimethyl-**	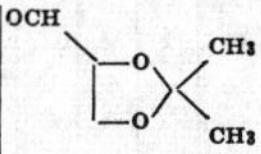 OCH, CH_3, CH_3 (structure)	130.14	......	......	74^{50}	......	1.4189^{25}	s^h	...	...	...	...	......	......
d212	**1,3-Dioxolane-2-one**	1,2-Ethanediol carbonate*. Ethylene. carbonate. Glycol carbonate. 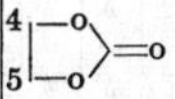 (structure)	88.06	mcl pl (al)	39–40	248^{760}	1.3218^{39}_{4}	1.4158^{50}	∞	∞	∞	...	∞	chl ∞ AcOEt ∞ aa ∞	**B19**, 100
d213	—,**4-methyl-**.....	Propylene carbonate. Isopropylene carbonate. $C_4H_6O_3$. *See* d212	102.09	......	−48.8	242^{760}	1.2069^{20}_{20}	1.4209^{20}	v	δ	s	...	s	......	......
—	α,γ-**Dipalmitin**...	*see* **Glycerol, 1,3-**dihexadecanoate													
—	**Dipentene**........	*see* **Limonene,**													
d214	**Diphenadione**....	2-Diphenylacetyl-1,2-indanedione. 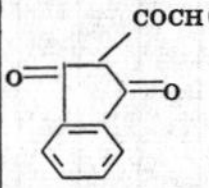 COCH$(C_6H_5)_2$ (structure)	340.36	ye mcl	144–5.5	......	......	1.670β	...	δ	...	...	...	Acetonitrile, cyclohexane, s	......
—	**Diphenic acid**....	*see* **2,2′-Biphenyldicarboxylic acid**													
—	**Diphenylene-imine**	*see* **Carbazole**													
—	**Diphenylene oxide**	*see* **Dibenzofuran**													
—	**Diphenylene sulfide**	*see* **Dibenzothiophen**													
—	**Diphenylmethane carboxylic acid**	*see* **Benzoic acid, benzyl-**													
—	**Diphosgene**......	*see* **Formic acid, chloro-,** trichloromethyl ester													
d215	**Diphosphine, tetrakis(trifluoromethyl)-**	$(F_3C)_2PP(CF_3)_2$........	338.04	......	......	84^{760}	>1.0	......	i δd^h	...	...	...	...	......	**J1953,** 1565
—	**Dipicolinic acid**..	*see* **2,6-Pyridinedicarboxylic acid**													
d216	**Diploicin**........	$C_{16}H_{10}Cl_4O_5$........	424.07	(bz)	232 (225d)	......	......	......	i	δ	δ	...	δ	lig δ	**B19**^{2}, 238
—	**Dipropionamide**..	*see* **Propanoic acid,** amide, *N*-propanoyl-													
d217	**Dipropylene glycol**	2,2′-Dihydroxydipropyl ether. $(CH_3CHOHCH_2)_2O$	......	......	......	229–32	1.0224^{20}_{20}	......	∞	s	...	...	...	......	......
—	**Disalicylic acid**...	*see* **Benzoic acid, 2-hydroxy-,** 2-carboxyphenyl ester													
d218	**Diselenide, diphenyl-**	$C_6H_5SeSeC_6H_5$........	312.13	ye nd	59–61	1557^{80}_{4}	......	......	...	s^h	...	...	...	xyl s MeOH s^h	**B6**^{2}, 319
d219	**Disiloxane, 1,3-diphenyl-1,1,3,3-tetramethyl-**	$C_6H_5Si(CH_3)_2OSi(CH_3)_2C_6H_5$..	286.53	......	......	110–1^1	0.9763^{20}_{4}	1.5122^{20}	...	...	...	...	...	......	......
d220	—,**hexaethyl-**.....	$[(C_2H_5)_3Si]_2O$........	246.54	......	......	233^{756}	0.8590^{0}_{0}	1.4345^{20}	...	...	...	...	...	......	**B4,** 627
d221	—,**hexakis(2-ethylbutoxy)-**	$([(C_2H_5)_2CHCH_2O]_3Si)_2O$....	679.19	......	<−54	220^1	0.9219^{20}_{20}	1.4330^{20}	i	δ	s	...	s	......	......
d222	—,**hexakis(2-ethylhexoxy)-**	$([CH_3(CH_2)_3CH(C_2H_5)CH_2O]_3Si)_2O$	847.52	......	......	$253^{0.9}$	0.9044^{20}_{20}	1.4402^{20}	i	δ	s	...	s	......	......
—	α,γ,-**Distearin**....	*see* **Glycerol,** 1,3-dioctadecanoate													
d223	**Distibine, tetrakis-(trifluoromethyl)-**	Perfluoro antimony cacodyl. $(F_3C)_2SbSb(CF_3)_2$	519.53	pa ye or col, slow d	......	*ca.* 136^{760}	......	......	i	...	...	...	...	alk d	**J1957,** 3708

For explanations, symbols and abbreviations see beginning of table.

No.	Name	Synonyms and Formula	Mol. wt.	Crystalline form, color and specific rotation	m.p. °C	b.p. °C	Density	n_D	Solubility w	al	eth	ace	bz	other solvents	Ref.
	Disulfide														
d224	**Disulfide, bis-(dibenzylthiocarbamyl)**	Tetrabenzylthiuram disulfide. $[(C_6H_5CH_2)_2NCS]_2S_2$	544.83	ye cr	132–3				i	δ	δ			chl s	
d225	**—,bis(dibutylthiocarbamyl)**	$[(C_4H_9^n)_2NCS]_2S_2$	408.76	red liq			1.03^{20}		i	δ	s				
d226	**—,bis(diethylthiocarbamyl)**	Antabuse. $[(C_2H_5)_2NCS]_2S_2$	296.54		69–70	1.17^{17}			i	s^h	δ			chl v	**B4²**, 613
d227	**—,bis(dimethylthiocarbamyl)**	$[(CH_3)_2NCS]_2S_2$	240.43	wh or ye	150–2	129^{20}			i	δ s^h	δ			chl s	**B4²**, 557
d228	**—,bis(ethylmethylthiocarbamyl)**	$[CH_3N(C_2H_5)CS]_2S_2$	268.49	ye	72				i	δ	δ			chl s	
d229	**—,bis(1-piperidylthiocarbamyl)**	Dicyclopentamethylenethiuram disulfide. 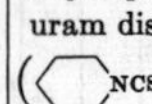	320.56		137–8 (129)				i	s^h	δ			chl s	**Am 65,** 1267
d230	**—,diacetyl**	$CH_3COSSCOCH_3$	150.22		20	d			i d^h	v	v			CS_2 v	**B2²**, 210
d231	**—,2,2′-dibenzothiazyl**	2,2′-Dithiobisbenzothiazole. 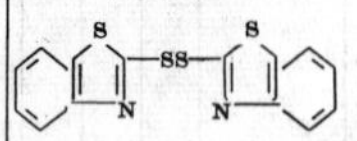	332.49	lt ye sc (bz)	180	d	1.50_4^{20}		i	i			v^h	chl δ	**B27¹**, 249
d232	**—,dibenzoyl**	$C_6H_5COSSCOC_6H_5$	274.36	pr (al) sc (chl-peth)	136	d				$δ^h$	δ			CS_2 s^h	**B9²**, 289
d233	**—,dibenzyl**	$C_6H_5CH_2SSCH_2C_6H_5$	246.40	lf (MeOH or al) nd (aa)	69 (72)				δ	s^h	s		s		**B6²**, 437
d234	**—,4,4′-dimethoxy-**	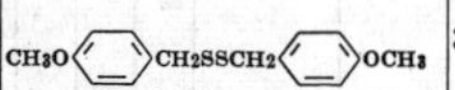	306.45	lf or nd (al)	101				i	i			v	chl v	**B6**, 901
d235	**—,dibutyl**	$CH_3(CH_2)_3SS(CH_2)_3CH_3$	178.36			226^{760}	0.9383_4^{20}	1.4926^{20}	i	∞	∞				**B1³**, 1524
d236	**—,—,3,3′-dimethyl**	Isoamyl sulfide. $(CH_3)_2CHCH_2CH_2SSCH_2CH_2CH(CH_3)_2$	206.41			250	0.9192_4^{20}	1.4864^{20}	d						**B1³**, 1647
d237	**—,diethoxy-**	Ethoxyl disulfide. $C_2H_5OSSOC_2H_5$	154.25			$67–8^{16}$	1.0913_4^{20}	1.4766^{20}							**B1³**, 1314
d238	**—,diethyl-**	$C_2H_5SSC_2H_5$	122.25			153	0.9982_4^{20}	1.5063^{20}	δ	∞	∞				**B1²**, 345
d239	**—,—,2,2′-diamino-**	Dithiobisethylamine. Cystamine. $NH_2CH_2CH_2SSCH_2CH_2NH_2$	152.28			d			s		δ				**B4²**, 731
d240	**—,di(2-furfuryl)**	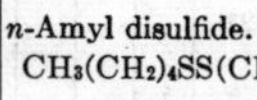	226.32		10	$112–3^{0.5}$				v				os v lig δ	**B17²**, 116
d241	**—,diisopropyl**	$(CH_3)_2CHSSCH(CH_3)_2$	150.31			176^{757}	0.9435_4^{20}	1.4916^{20}							**B1³**, 1480
d242	**—,dimethyl**	CH_3SSCH_3	94.20			116–8 (107)	1.0636_0^0	1.5208^{30}	i	∞	∞				**B1²**, 278
d243	**—,—,hexafluoro-**	F_3CSSCF_3	202.14			34.6^{760}	>1.26		i	s				CS_2 i peth s	**J1952,** 2198
d244	**—,—,1,1,1′,1′-tetraphenyl-**	$(C_6H_5)_2CHSSCH(C_6H_5)_2$	398.59	nd (al)	152–3					s^h				CS_2 v	**B6²**, 636
d245	**—,1,1′-dinaphthyl**	$(C_{10}H_7)^{\alpha}SS(C_{10}H_7)^{\alpha}$	318.46	pl (al) nd (lig)	91				i	δ	v				**B6²**, 588
d246	**—,2,2′-dinaphthyl**	$(C_{10}H_7)^{\beta}SS(C_{10}H_7)^{\beta}$	318.46	nd	139				i	v	v			lig i	**B6¹**, 317
d247	**—,dipentyl**	*n*-Amyl disulfide. $CH_3(CH_2)_4SS(CH_2)_4CH_3$	206.41			119^7	0.9221_4^{20}	1.4889^{20}	i		s				**B1³**, 1608
d248	**—,diphenyl**	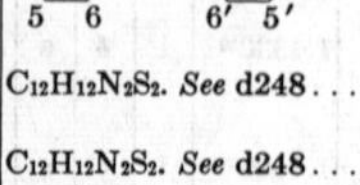	218.34	nd (al)	60–1	310^{760}	1.353_4^{20}		i	s	s		s		**B6**, 323
d249	**—,—,2,2′-diamino-**	$C_{12}H_{12}N_2S_2$. *See* d248	248.36	pa ye pl or nd (50 % al)	93				i	v^h	s				**B13²**, 201
d250	**—,—,3,3′-diamino-**	$C_{12}H_{12}N_2S_2$. *See* d248	248.36	nd (dil al)	59–60				i	v	v		v		**B13²**, 219
d251	**—,—,4,4′-diamino-**	$C_{12}H_{12}N_2S_2$. *See* d248	248.36	ye or col pr or nd (w or eth or dil al)	85 (77)				s^h	v	v		$δ^h$	chl v lig $δ^h$	**B13²**, 300
d252	**—,—,4,4′-dichloro-2,2′-dinitro-**	$C_{12}H_6Cl_2N_2O_4S_2$. *See* d248	377.23	ye (ace-al)	212–3					δ		δ	s	CCl_4 δ peth i lig δ	**B6²**, 313

For explanations, symbols and abbreviations see beginning of table.

No.	Name	Synonyms and Formula	Mol. wt.	Crystalline form, color and specific rotation	m.p. °C	b.p. °C	Density	n_D	Solubility w	al	eth	ace	bz	other solvents	Ref.
	Disulfide														
d253	—,—,5,5′-di-chloro-2,2′-dinitro-	$C_{12}H_6Cl_2N_2O_4S_2$. *See* d248	377.24	ye nd (90% aa)	171					δ			v	chl, CS_2 v lig i	**B6**², 314
d254	—,—,2,2′-dimethoxy-	$C_{16}H_{18}O_2S_2$. *See* d248	306.44	nd (al)	88–90					s^h					**B6**, 795
d255	—,—,4,4′-diethoxy-	$C_{16}H_{18}O_2S_2$. *See* d248	306.44	nd (al)	48–9				i	s^h					**B6**¹, 421
d257	—,—,2,2′-dimeth-oxy-5,5′-dimethyl-	$C_{16}H_{18}O_2S_2$. *See* d248	306.44	pa ye pr (al) or pl (al or eth)	67				i	s^h					**B6**², 874
d258	—,—,2,2′-dinitro-	$C_{12}H_8N_2O_4S_2$. *See* d248	308.34	ye nd (bz or aa) pl (aa)	195					δ	i	δ	δ	peth i aa δ	**B6**², 307
d259	—,—,3,3′-dinitro-	$C_{12}H_8N_2O_4S_2$. *See* d248	308.34	ye nd (al) or rh	84					δ	δ				**B6**, 339
d260	—,—,4,4′-dinitro-	$C_{12}H_8N_2O_4S_2$. *See* d248	308.34	nd (aa) pl (al)	181					δ				aa δ	**B6**², 312
d261	—,—,2,2′4,4′-tetranitro-	$C_{12}H_6N_4O_8S_2$. *See* d248	398.34	ye nd	303d				i					oos i $PhNO_2$, Py s	**B6**², 316
d262	—,dipropyl	$CH_3CH_2CH_2SSCH_2CH_2CH_3$	150.31			193.5^{750}	0.9599^{20}_{4}	1.4981^{20}							**B1**³, 1435
d263	—,di-2-tolyl	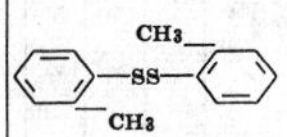	246.40	lf (al)	38–9				i	s				os s	**B6**², 342
d264	—,di-4-tolyl	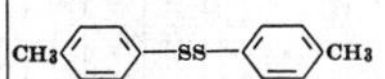	246.40	nd or lf (al)	45–6				i	s	v				**B6**², 400
—	**Ditaine**	*see* **Echitamine**													
—	**Ditan**	*see* **Methane, diphenyl-**													
d265	**1,4-Dithiane**	Diethylene disulfide. 	120.22	mcl pr	111–2	199–200			δ^h	s^h	s			CS_2 s aa s^h	**B19**², 6
d266	**1,3,5-Dithiazine, 4,5-dihydro-5-methyl-**	Methylthioformaldine. 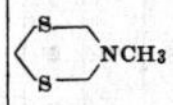	135.26	nd (eth)	65	ca. 185d				s	s			aa s	**B27**, 460
d267	**1,4-Dithiin, 2,5-diphenyl-**	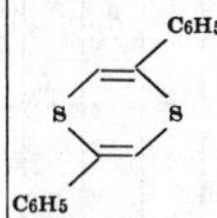	268.40	ye pr (al)	118–9				i	s			v	con sulf s alk s	**B19**², 46
—	**Dithioacetic acid**	*see* **Acetic acid, thiolo-thionothiolo-**													
d268	**1,3-Dithiolane**	Trimethylene 1,3-disulfide. 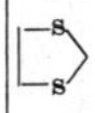	106.21		−51	175^{760}	1.259^{17}	1.5975^{15}		s	s			xyl s	**B19**², 3
d269	**Dithizone**	Diphenylthiocarbazone. Formazyl mercaptan. $C_6H_5N{:}NCSNHNHC_6H_5$	256.32	bl-bk (chl-al)	165–9d				i	δ	δ			chl, alk s	**B16**, 26
d270	**Diurea**	*p*-Urazine. 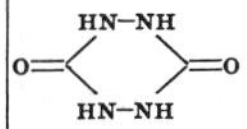	116.08	pr (w)	264–6				δ^h	δ				aa δ	**B26**², 258
—	**Divarinol**	*see* **Benzene, 1,3-dihydroxy-5-propyl-**													
d271	**Djenkolic acid**	β,β′-Methylenedithiodialanine. $HO_2CCH(NH_2)CH_2SCH_2SCH_2CH(NH_2)CO_2H$	254.13	nd (w, HCl) $[\alpha]^{26}_{D}-44.5$ (1% HCl)	300–50d				s^h					dil ac s dil alk s	
d272	**Docosane***	$CH_3(CH_2)_{20}CH_3$	310.59	pl (to) cr (eth)	44–1	327^{768}	0.7944^{20}_{4}	1.1455^{20}	i		v			chl s	**B1**³, 574
d273	**Docosanoic acid***	Behenic acid. Dodosoic acid. $CH_3(CH_2)_{20}CO_2H$	340.57	nd	80	306^{60}			δ	δ	δ				**B2**³, 373
d274	—,ethyl ester	Ethyl behenate. $CH_3(CH_2)_{20}CO_2C_2H_5$	368.63	nd (al) cr (ace)	49	$240–2^{10}$			i	s	s				**B2**², 374
d275	—,methyl ester	Methyl behenate. $CH_3(CH_2)_{20}CO_2CH_3$	354.60	nd (ace)	53–4	$224–5^{15}$		1.4339^{60}	i	s	s				**B2**², 373
d276	**1-Docosanol***	Docosyl alcohol. $CH_3(CH_2)_{21}OH$	326.61	(ace, chl)	71	$180^{0.22}$			δ	v	δ v^h			chl s MeOH v peth v^h	**B1**³, 1846
d277	**Docosapentaenoic acid**	Clupanodonic acid. $C_{21}H_{33}CO_2H$	330.50	pa ye	<−78	236^{5}	0.9356^{20}_{4}	1.5020^{20}	i		s				**B2**², 470

For explanations, symbols and abbreviations see beginning of table.

No.	Name	Synonyms and Formula	Mol. wt.	Crystalline form, color and specific rotation	m.p. °C	b.p. °C	Density	n_D	Solubility w	al	eth	ace	bz	other solvents	Ref.
	13-Docosenoic acid														
d278	**13-Docosenoic acid***(*cis*)	Erucic acid. $CH_3(CH_2)_7CH:CH(CH_2)_{11}CO_2H$	338.56	nd (al)	33–4	265^{15}	0.860_4^{55}	1.4758^{20}	i	s	v	...	...	MeOH v	**B2²**, 445
d279	—,(*trans*)*	Brassidic acid. Brassic acid. Isoerucic acid. $CH_3(CH_2)_7CH:CH(CH_2)_{11}CO_2H$	338.56	lf (al)	61.5	285^{30}	0.8585^{57}	1.4472^{64}	i	δ	δ	...	...	...	**B2²**, 447
d280	—,anhydride (*trans*)*	Brassidic anhydride. $[CH_3(CH_2)_7CH:CH(CH_2)_{11}CO]_2O$	659.14	nd (al) pl (eth, ace, peth)	64	0.835_4^{70}	1.4366^{100}	...	i	δ	s	s	...	peth s	**B2²**, 448
d281	—,13,14-diiodo-, ethyl ester*(*trans*)	Iodobrassid. $CH_3(CH_2)_7CI:CI(CH_2)_{11}CO_2C_2H_5$	618.43	nd or sc	37 (40)	d	...	...	i	s v^h	v	...	v	chl v	**B2²**, 448
d282	**13-Docosynoic acid***	Behenolic acid. $CH_3(CH_2)_7C⋮C(CH_2)_{11}CO_2H$	336.56	mcl pr or nd (al)	58	...	...	...	i	v	v	...	...	chl s	**B2²**, 462
d282¹	**Dodecanal***	Lauraldehyde. $CH_3(CH_2)_{10}CHO$	184.32	lf	44.5	185^{100}	0.8352_4^{15}	...	i	s	s	...	...	...	...
d283	—,dimethyl acetal	1,1-Dimethoxydodecane*. $CH_3(CH_2)_{10}CH(OCH_3)_2$	230.39	...	...	$132–4^{5}$	...	1.4310^{25}	...	v	v	...	...	MeOH v	...
d284	**Dodecane***	$CH_3(CH_2)_{10}CH_3$	170.34	...	−9.6	216.3^{760}	0.7487_4^{20}	1.4216^{20}	i	v	v	v	...	chl, CCl_4, v	**B1³**, 539
d285	—,**1-amino-***	Dodecylamine. Laurylamine. $CH_3(CH_2)_{11}NH_2$	185.36	...	28.3	$247–9^{760}$	...	...	δ	∞	∞	...	∞	chl, CCl_4 ∞	**B4**, 200
d286	—,—,acetate	$CH_3(CH_2)_{11}NH_2.CH_3CO_2H$	245.41	...	fr. 69.5	...	...	...	v	v	...	...	δ	...	...
d287	—,—,hydrochloride	$CH_3(CH_2)_{11}NH_2.HCl$	221.82	...	184–6	...	...	...	v	v	i	...	δ s^h	...	**Am 74**, 4287
d288	—,**1-bromo-***	Dodecyl bromide. Laurylbromide. $CH_3(CH_2)_{11}Br$	249.24	...	−9.6	$200–1^{100}$	1.0382_4^{20}	1.4581^{20}	i	s	s	...	...	...	**B1³**, 542
d289	—,**1-bromo-12-fluoro-***	$F(CH_2)_{12}Br$	267.23	...	...	$85–6^{0.15}$	...	1.4524^{25}	i	v	v	...	...	...	...
d290	—,**1-chloro-***	Lauryl chloride. $CH_3(CH_2)_{11}Cl$	204.79	...	fr. −9.4	116.5^{5}	0.8673_4^{20}	1.4426^{20}	i	v	...	...	s	CCl_4 ∞ lig ∞	**J1943**, 636
d291	—,**1,2-dibromo-***	$CH_3(CH_2)_9CHBrCH_2Br$	328.14	nd (aa, al)	41	215^{15}	...	...	i	v	s	...	...	chl v aa s	**B1²**, 543
d292	—,**1-iodo-***	$CH_3(CH_2)_{11}I$	296.24	...	...	$159–60^{15}$	1.2015_4^{20}	1.4844^{20}	i	s $∞^h$	∞	∞	...	chl, CCl_4 ∞ MeOH s	**B1¹**, 67
d293	**Dodecanedioic acid**, dimethyl ester*	$CH_3O_2C(CH_2)_{10}CO_2CH_3$	258.36	pr	31.3	150^{2}	...	...	i	...	...	...	...	AcOEt s^h	...
d294	**1-Dodecanethiol**	Lauryl mercaptan. $CH_3(CH_2)_{11}SH$	202.41	...	...	$165–6^{39}$	0.8450_{20}^{20}	1.4589^{20}	i	s	s	...	...	...	...
d295	**Dodecanoic acid***	Lauric acid. $CH_3(CH_2)_{10}CO_2H$	200.32	nd (al)	44	131^{1}	0.8679_4^{50}	1.4304^{50}	i	v	v	s	v	...	**Am 64**, 2739
d296	—,amine*	Lauramide. $CH_3(CH_2)_{10}CONH_2$	199.34	...	102	...	...	...	i	s	δ	s	δ	CCl_4 s^h	...
d297	—,—,*N*-phenyl-	Lauranilide. $CH_3(CH_2)_{10}CONHC_6H_5$	275.44	...	77	...	...	...	i	s	s	s	s	CCl_4, chl	...
d298	—,anhydride*	Lauric anhydride. $[CH_3(CH_2)_{10}CO]_2O$	382.63	lf (al)	41.8	166	0.8552_4^{70}	1.4292^{70}	d	s^h	...	...	...	...	**B2²**, 321
d299	—,benzyl ester*	Benzyl laurate. $CH_3(CH_2)_{10}CO_2CH_2C_6H_5$	290.45	...	8.5	$209–11^{12}$	0.9457_4^{20}	1.4812^{20}	i	s	v	...	v	chl v	**B6²**, 417
d300	—,chloride	Lauryl chloride. $CH_3(CH_2)_{10}COCl$	218.77	...	−17	145^{18}	...	1.4458^{20}	d	d	s	...	...	...	**B2²**, 321
d301	—,ethyl ester*	Ethyl laurate. $CH_3(CH_2)_{10}CO_2C_2H_5$	228.36	...	fr. −1.8	273	0.8618_4^{20}	1.4311^{20}	i	v	∞	...	...	...	**J1948**, 624
d302	—,isopropyl ester*	Isopropyl laurate. $CH_3(CH_2)_{10}CO_2C_3H_7^i$	242.39	...	...	117.4^{2}	0.8536^{20}	1.4280^{25}	...	...	...	...	...	...	...
d303	—,methyl, ester*	Methyl laurate. $CH_3(CH_2)_{10}CO_2CH_3$	214.34	...	fr. 5.2	262	0.8702_4^{20}	1.4319^{20}	i	∞	∞	∞	∞	MeOH, chl, CCl_4, AcOEt s	**J1948**, 624
d304	—,nitrile	Lauronitrile. Undecyl cyanide. $CH_3(CH_2)_{10}CN$	181.31	...	fr. 4	198^{100}	0.8373^{15}	1.4360^{20}	i	∞	∞	∞	∞	chl ∞	**Am 66**, 361
d305	—,phenyl ester*	Phenyl laurate. $CH_3(CH_2)_{10}CO_2C_6H_5$	276.42	lf (al)	24.5	210^{15}	...	...	i	s	s	...	...	...	**B6**, 154
d306	—,4-phenyl-phenacyl ester	$CH_3(CH_2)_{10}CO_2CH_2CO-C_6H_4-C_6H_5$	394.56	...	86	...	...	...	...	...	...	...	...	...	...
d307	—,piperazine salt	$2CH_3(CH_2)_{10}CO_2H.C_4H_{10}N_2$	486.76	...	92–2.5	...	...	...	s	s	i	...	...	...	**Am 70**, 2758

For explanations, symbols and abbreviations see beginning of table.

No.	Name	Synonyms and Formula	Mol. wt.	Crystalline form, color and specific rotation	m.p. °C	b.p. °C	Density	n_D	Solubility						Ref.
									w	al	eth	ace	bz	other solvents	
	Dodecanoic acid														
d308	—,propyl ester*	Propyl laurate. $CH_3(CH_2)_{10}CO_2CH_2CH_2CH_3$	242.39		123.7^{2}	1.8600^{20}	1.4311^{25}								
d309	**—,2-bromo-***	$CH_3(CH_2)_9CHBrCO_2H$	279.22	pl	30–1	$157–9^{2}$	1.1474^{24}	1.4585^{24}	i	v	...	...	v		**B2**, 363
d310	**—,12-fluoro***	$F(CH_2)_{11}CO_2H$	218.30		60–1				...	v	v	...	...	lig s^h	
d311	**1-Dodecanol***	Lauryl alcohol. $CH_3(CH_2)_{11}OH$	186.33	lf (dil al)	26 (22)	$255–9^{760}$	0.8309^{24}_{4}		i	s	s	...	...		**B1²**, 463
d312	**6-Dodecanol***	$CH_3(CH_2)_4CHOH(CH_2)_5CH_3$	186.33	(peth)	30	119^{9}			...	s	s	...	...		**B1**, 428
d313	**2-Dodecanone***	*n*-Decyl methyl ketone. $CH_3CO(CH_2)_9CH_3$	184.31		20.5	$246–7^{760}$	0.8198^{30}_{4}	1.4286^{30}	i	s	...	...	...	os s	**B1²**, 769
d314	**1-Dodecanone, 1-phenyl-***	Laurophenone. Lauroylbenzene. *n*-Undecyl phenyl ketone. $CH_3(CH_2)_{10}COC_6H_5$	260.42	og cr	46–7	$222–3^{21}$	0.8969^{52}_{4}	1.4850^{52}	...	...	...	...	...		**B7¹**, 186
—	**1,6,10-Dodecatrien-3-ol, 3,7,11-trimethyl-***	*see* **Nerolidol**													
d315	**1-Dodecene***	α-Dodecylene. $CH_3(CH_2)_9CH{:}CH_2$	168.31		−35.2	213.4^{760}	0.7584^{20}_{4}	1.4300^{20}	i	s	s	s	...	peth s	**B1³**, 869
d316	**2-Dodecenedioic acid**(*cis*)*	Traumatic acid. $HO_2CCH{:}CH(CH_2)_8CO_2H$	228.28	(al, ace)	67–8				δ	s	s	...	s	chl s	
d317	—,(*trans*)	$HO_2CCH{:}CH(CH_2)_8CO_2H$	228.28	(al, ace)	165–6				δ	s	s	...	...	chl s	
d318	**1-Dodecen-3-yne***	$CH_2{:}CHC{\vdots}C(CH_2)_7CH_3$	164.28			78^{4}	0.7858^{25}_{4}	1.4510^{25}	...	...	...	...	...		**Am 61**, 572
d319	**1-Dodecyne***	$CH{\vdots}C(CH_2)_9CH_3$	166.30			$95–8^{15}$	0.7758^{24}_{4}	1.4351^{26}	...	...	...	...	...		**Am 55**, 3453
d320	**2-Dodecyne***	$CH_3C{\vdots}C(CH_2)_8CH_3$	166.30		−9	105^{15}	0.7917^{15}_{4}		...	...	...	...	...		
d321	**3-Dodecyne***	$CH_3CH_2C{\vdots}C(CH_2)_7CH_3$	166.30			95^{12}	0.7871^{20}_{4}	1.4442^{20}	...	...	...	...	...		**Am 60**, 1884
d322	**6-Dodecyne***	Di *n*-amylacetylene. $CH_3(CH_2)_4C{\vdots}C(CH_2)_4CH_3$	166.30			209^{745}	0.7871^{20}_{4}	1.4442^{20}	i	s	s	...	...		**B1³**, 1025
d323	**Dotriacontane***	Dicetyl. $CH_3(CH_2)_{30}CH_3$	450.85	(bz, chl, aa, eth)	69.7	310^{15}	0.8124^{20}_{4} suc	1.4364^{70}	i	δ	s^h	...	δ	CCl_4, aa s^h chl δ	**B1³**, 587
d324	**1-Dotriacontanol***	$CH_3(CH_2)_{30}CH_2OH$	466.85	(bz)	89.4	sub $200–50^{1}$			...	...	...	...	...		**B1³**, 1851
—	**Dulcitol**	*see* **Galactitol**													
—	**Durene**	*see* **Benzene, 1,2,4,5-tetramethyl-**													
—	**Durenol**	*see* **Benzene, 1-hydroxy-2,3,5,6-tetramethyl-**													
—	**Durohydroquinone**	*see* **Benzene, 1,4-dihydroxy-2,3,5,6-tetramethyl-**													
—	**Duroquinone**	*see* **1,4-Benzoquinone, tetramethyl-**													
—	**Durylic acid**	*see* **Benzoic acid, 2,4,5-trimethyl-**													
—	**Dypnone**	*see* **2-Buten-1-one, 1,3-diphenyl-**													

For explanations, symbols and abbreviations see beginning of table.

No.	Name	Synonyms and Formula	Mol. wt.	Crystalline form, color and specific rotation	m.p. °C	b.p. °C	Density	n_D	Solubility w	al	eth	ace	bz	other solvents	Ref.
	Ecgonidine														
e1	**Ecgonidine(*l*)**	Anhydroecgonine(*l*) 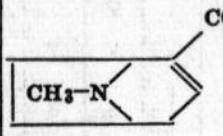	167.21	$[\alpha]_D^{14}-84.6$ (w, p=1.7)	235										B22², 26
e2	**Ecgonine(*dl*)**	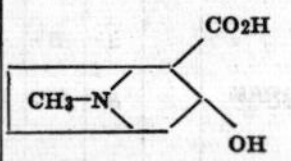	185.22	pl (w+3)	203 anh 93–118 (+3w)				v^h		i			abs al	B22², 156
e3	—(*l*)	Tropinecarboxylic acid. $C_9H_{15}NO_3$. *See* e2	185.22	mcl pr (al) $[\alpha]_D^{12}-45.5$ (w, p=5)	205				v					MeOH s	B22², 150
e4	—,hydrate(*l*)	$C_9H_{15}NO_3.H_2O$. *See* e2	203.33	mcl pr (al), eff 120–30	198				v					MeOH s	B22, 196
e5	—,benzoate(*l*)	*O*-Benzoyl-*l*-ecgonine. 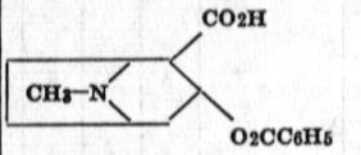	289.32	nd (w) $[\alpha]_D^{14}-63.5$ (w, p=1.7)	195				δ s^h	s	i		v^h	dil ac s dil alk s	B22, 150
e6	—,—,ethyl ester(*l*)	Cocaethyline. Homocaine. $C_{18}H_{23}NO_4$. *See* e5	317.37	pr (al)	109				δ	s	s				B22, 262
e7	—,—,tetra-hydrate(*l*)	$C_{16}H_{19}NO_4.4H_2O$. *See* e5	361.39	pr (w) $[\alpha]_D^{15}-44.6$ (abs al, c=3)	92				δ s^h	s	i		v^h		B22, 197
e8	—,hydrochloride(*l*)	$C_9H_{15}NO_3.HCl$. *See* e2	221.69	rh (al), nd (*i*-AmOH) $[\alpha]_D^{28}-48.2$ (w, p=44)	246				s	s				MeOH v *i*-AmOH δ	B22², 150
e8¹	**Echinochrome A**	$C_{12}H_{10}O_7$	266.20	dk red nd	220d										
e9	—,trimethyl ether	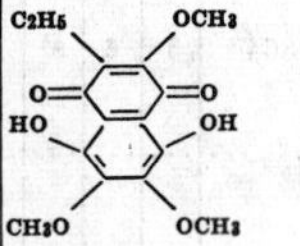	308.22	dk red lf (aq diox), nd	133 vac									MeOH s NaOH s (bl)	
e10	**Echinopsine**	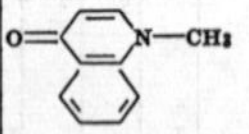	159.18	nd (bz), (w+1)	152 (143)				s v^h	v	δ		v^h	chl v MeOH v	B21², 259
e11	**Echitamidine**	$C_{20}H_{26}N_2O_3$	342.43	pl (eth) $[\alpha]_D^{16}-515$ (al)	244d				s	s					J 1932, 2628
e12	**Echitamine**	Ditaine. $C_{22}H_{28}N_2O_4.4H_2O$	456.52	pr (al+4w) eff 105 (−3w) $[\alpha]_D^{20}-29$ (al)	206 (+1w)				s	s	s			chl s	J 1925, 1640
e13	—,hydrochloride	$C_{22}H_{28}N_2O_4.HCl$	420.92	nd (w), $[\alpha]_D^{15}-58$ (w, % c=1)	292–5d				s	v	δ		δ		J 1925, 1640
e14	**Echitin**	$C_{32}H_{52}O_2$	468.77	lf $[\alpha]_D+73$ (eth)	170					s^h	s	s	s	peth s	
e15	**Egonol**	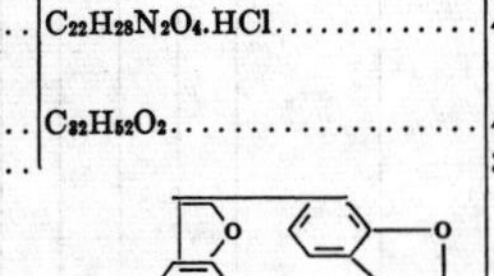	326.33	pl (BuOH)	118	$228–30^{0.15}$									
e16	**Eicosane***	$CH_3(CH_2)_{18}CH_3$	282.54	(al or bz)	36.8	343^{760}	0.7886_4^{20}	1.4425^{20}	i		s		s	peth s	B1³, 570
e17	**Eicosanedioic acid***	$HO_2C(CH_2)_{18}CO_2H$	342.52		124–5	$233–4^2$									
e18	**Eicosanoic acid***	Arachic acid. Eicosoic acid. $CH_3(CH_2)_{18}CO_2H$	312.52	pl (al)	75.4	328 δd $203–5^1$	0.824_4^{100}	1.425^{100}	i	δ	v		s	chl s, peth δ	B2², 369
e19	—,ethyl ester*	$CH_3(CH_2)_{18}(CO_2C_2H_5)$	340.58		49–50	$186–7^2$			i	s	s				B2², 370
e20	—,methyl ester*	$CH_3(CH_2)_{18}(CO_2CH_3)$	326.55	lf (MeOH)	54.5	$215–6^{10}$		1.4317^{60}	i	s	s		s	chl s	B2², 370

For explanations, symbols and abbreviations see beginning of table.

No.	Name	Synonyms and Formula	Mol. wt.	Crystalline form, color and specific rotation	m.p. °C	b.p. °C	Density	n_D	Solubility w	al	eth	ace	bz	other solvents	Ref.
	1-Eicosanol														
e21	**1-Eicosanol***	Arachidic alcohol. *pri-n*-Eicosyl alcohol. $CH_3(CH_2)_{18}CH_2OH$	298.54	wx (al)	65.5	369^{760}	0.8405^{2}_{4} suc	1.4550^{20} suc	i	δ			s^h	peth s^h	**B1**^{3}, 1843
e22	**2-Eicosanol***	2-Eicosyl alcohol. $CH_3(CH_2)_{17}CHOHCH_3$	298.54		50.7	357^{760}	0.8378^{20}_{4} suc	1.4524^{20} suc							
e23	**2-Eicosanone***	Methyl-*n*-octadecyl ketone. $CH_3(CH_2)_{17}COCH_3$	296.52		58					δ	δ	δ			
e24	**3-Eicosanone***	Ethyl-*n*-heptadecyl ketone. $CH_3(CH_2)_{16}COC_2H_5$	296.52	lf (al)	59.5					δ	s		s	chl s	**B1**^{2}, 774
e25	—,oxime*	$CH_3(CH_2)_{16}C(:NOH)C_2H_5$	311.54	nd (al)	55–6					δ	s	s		peth δ	**B1**, 719
e26	**7-Eicosanone***	Hexyl-*n*-tridecyl ketone. $CH_3(CH_2)_{12}CO(CH_2)_5CH_3$	296.52	so		$210–1^{11}$									**B1**^{2}, 774
e27	**6,10,14,18-Eicosatetraenoic acid***	Arachidonic acid. $CH_3[CH:CHCH_2CH_2]_4CH_2CH_2CO_2H$	304.46		−49.5	d		1.4824^{20} 1.5563^{23}	i	s				os s	**B2**^{2}, 469
e28	**1-Eicosene***	$CH_3(CH_2)_{17}CH:CH_2$	280.52		28.5	341^{760}	0.7882^{30}_{4}	1.4440^{30}							
e29	**1-Eicosyne***	$CH_3(CH_2)_{17}C{\vdots}CH$	278.50			$314–5^{760}$	0.8181^{24}_{4}								
—	**Elaidic acid**	*see* **9-Octadecenoic acid-(*trans*)***													
—	**Elaidyl alcohol**	*see* **9-Octadecen-1-ol-(*trans*)***													
e30	**α-Elaterin**	$C_{28}H_{38}O_7$	486.59	(al) $[\alpha]^{25}_D$ −58.9 (chl)	234				i	δ	s		s	chl s	**J95**, 1989
e31	**β-Elaterin**	$C_{20}H_{28}O_5$	348.42	nd (al) $[\alpha]^{25}_D$+13.9	195.5				i	i	δ		δ	chl s	
—	**Eldoral**	*see* **Barbituric acid, 5-ethyl-5-piperidyl-**													
e32	**Elemane**	Dihydroelemene. $C_{15}H_{28}$	206.36			115–9	0.8830^{17}_{4}	1.4950^{17}							
e33	**Elemenonic acid**	Dihydro-β-elemonic acid. $C_{30}H_{48}O_3$	456.69	nd (al or AcOEt) $[\alpha]_D$ +37.5 (chl)	245–6										
e34	**α-Elemol**	 OH, C_2H_5 (structure)	222.36		47	$142–3^{12}$	0.9345^{18}_{4}	1.4980^{18}							
—	**Eleostrearic acid**	*see* **9,11,13-Octadecatrienoic acid***													
e35	**Ellagene**	Indeno-2′,3′:2,3-fluorene. 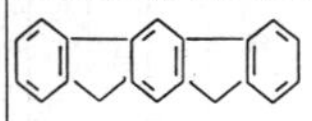	254.33	pl (bz)	197										**E14s**, 498
e36	**Ellagic acid,** dihydrate	$C_{14}H_6O_8.2H_2O$	338.23	pa ye pr	450–80 d without m				δ	δ		δ		os δ	**B19**^{2}, 284
e37	**Elliptic acid**	$C_{20}H_{18}O_8$	386.36	nd (aq al)	190										**J1942** 587
—	**Embelin**	*see* **1,4-Benzoquinone, 2,5-dihydroxy-3-dodecyl-**													
e38	**Emeraldine** 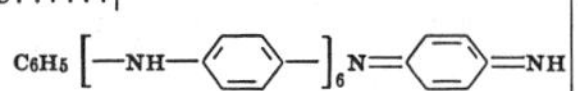	C_6H_5 [—NH—⟨⟩—]$_6$N=⟨⟩=NH	726.89	indigo-bl pw						i			i	chl i, gl aa i 80 % aa s	**B12**^{1}, 147
e39	**Emetine**(*l*)	$C_{29}H_{40}N_2O_4$	480.65	amor pw $[\alpha]^{15}_D$−22.10 (al, c=2)	74 cor				i	s		s			**B23**^{2}, 449
e40	—,hydrochloride(*l*)	$C_{29}H_{40}N_2O_4.2HCl.7H_2O$	679.68	nd (w^h) $[\alpha]$+53 (MeOH)	269–70				v	s	i				
e41	**Emicymarin**	$C_{30}H_{46}O_9$	550.70	nd or pr (MeOH) $[\alpha]^{20}_D$+12.8	ca. 207										
—	**Emodin**	*see* **9,10-Anthracenequinone, 6,methyl-1,3,8-trihydroxy-**													
—	**Enanthic acid**	*see* **Heptanoic acid***													
e42	**Enneaphyllin**	$C_{90}H_{154}$	1236.23	rods (bx)	98					s	δ	i	s	Py i lig δ	**C32**, 2686

For explanations, symbols and abbreviations see beginning of table.

No.	Name	Synonyms and Formula	Mol. wt.	Crystalline form, color and specific rotation	m.p. °C	b.p. °C	Density	n_D	Solubility w	al	eth	ace	bz	other solvents	Ref.
	Eosin														
e43	**Eosin**	2,4,5,7-Tetrabromofluorescein. 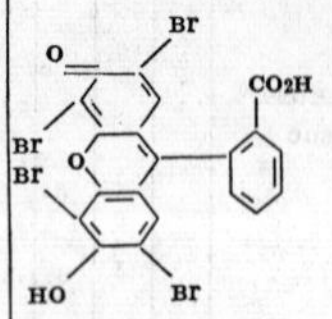 (structure: O, Br, CO_2H, Br, Br, O, HO, Br)	647.92	ye-red	295–6				i	s	i		δ	alk s, chl δ aa δ	**B19**[2], 254
e44	—,sodium salt decahydrate	Eosin dye. Sodium eosin. $C_{20}H_6Br_4Na_2O_5.10H_2O$. *See* e43	872.04	red nd (dil al) eff 150°					s	s	i				**B19**, 230
e45	**Ephedrine**(*d*)	$C_{10}H_{15}NO$	165.24	pl (w)	40										**Am 69**, 128
e46	—(*dl*)	Racephedrine. $C_{10}H_{15}NO$.	165.24	nd (eth or peth)	76–7	$135\text{–}7^{12}$			s	s	s		s	chl s	**B13**[2], 376
e47	—(*l*)	Natural ephedrine. $C_{10}H_{15}NO.H_2O$	183.25	pl (w+1) $[\alpha]_D+13.4$ (4% w) $[\alpha]_D^{20}-5.5$ (al)	40	225^{760}									
e48	—,hydrochloride(*d*)	$C_{10}H_{15}NO.HCl$	201.70	pl (abs al) $[\alpha]_D^{20}+35.8$ (w, c=11.5)	218				v	s	i				**B13**[2], 375
e49	—,—(*dl*)	Ephetonin. $C_{10}H_{15}NO.HCl$	201.70	pl	189–90				v	s	i				**B13**[2], 376
e50	—,—(*l*)	$C_{10}H_{15}NO.HCl$	201.70	orh $[\alpha]_D^{20}-36.6$	218				v	s	i				
e51	—,sulfate(*l*)	$(C_{10}H_{15}NO)_2.H_2SO_4$	428.55	hex pl $[\alpha]-31.5$	240–8				v	δ					
e52	—,*N*-**methyl-**,(*l*)	$C_6H_5CHOHCH(CH_3)N(CH_3)_2$	179.26	nd or pl (al or eth) $[\alpha]_D-29.5$ (MeOH, c=4.5)	87–8 cor	v volat			i	s	s^h			MeOH s	**B13**[2], 378
e53	—,*N*-(4-nitrobenzoyl)-,(*dl*)	$NO_2—C_6H_4—N(CH_3)CH(CH_3)CHOHC_6H_5$	281.33	pa ye pl (al)	162										**Am 69**, 128
e54	**Epi-β-amyrin**	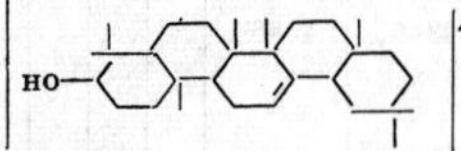 (structure: HO—)	426.10	(MeOH) $[\alpha]_D+73.3$ (chl)	225 cor										**E14s**, 948
e55	**Epi-α-amyrin**, acetate	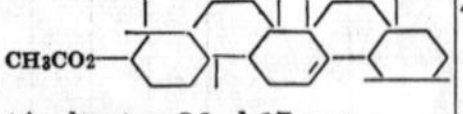 (structure: CH_3CO_2—)	468.43	nd (chl-MeOH) $[\alpha]_D+36$ to +39 (chl)	135 cor										**E14s**, 1070
e56	**Epiandrosterone**	Androstan-3β-ol-17-one. Isoandrosterone $C_{19}H_{30}O_2$	290.43	$[\alpha]_D+89$ (MeOH)	175–6										
e57	**Epiborneol**(*l*)	3-Camphenol. 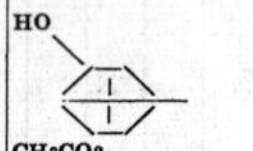 (structure: HO)	154.24	nd (C_5H_{12}) $[\alpha]_D^{17}+11.1$ (al)	181–2	213^{742}								peth s	**B6**[2], 91
e58	—,acetate	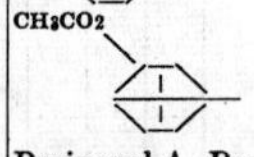 (structure: CH_3CO_2)	196.28	$[\alpha]_D^{16}+15.63$	<−15	101^{11}									**B6**[2], 92
e59	**Epibreinonol**	Breinonol A. Breienonol A. 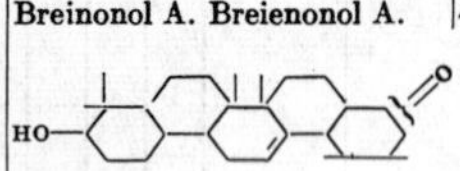 (structure: HO—, =O)	440.69	pl (MeOH), nd (chl) $[\alpha]_D^{17}+37$ (chl)	204										**E14s**, 1083
—	**Epibromohydrin**	*see* **Propane, 3-bromo-1,2-epoxy-***													
e60	**Epicamphor**(*d*)	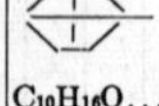 (structure: O=)	152.13	$[\alpha]_D^{17}+45.4$ (bz)	182				i	s					**B7**[1], 86
e61	—(*dl*)	$C_{10}H_{16}O$	152.13	(peth)	174.3 cor									peth v	**B7**[2], 104
e62	—(*l*)	β-Camphor. $C_{10}H_{16}O$	152.13	$[\alpha]_D^{19}+58.21$ (chl, c=13)	184 cor	213			i	v	v				**B7**[2], 105
e63	—,oxime(*d*)	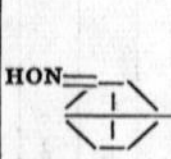 (structure: HON=)	167.25	nd (MeOH) $[\alpha]_D-98.9$	103										**B7**[1], 86
e64	—,—(*dl*)	$C_{10}H_{17}NO$	167.25	nd (dil al)	98–100										**B7**[1], 87

For explanations, symbols and abbreviations see beginning of table.

No.	Name	Synonyms and Formula	Mol. wt.	Crystalline form, color and specific rotation	m.p. °C	b.p. °C	Density	n_D	Solubility w	al	eth	ace	bz	other solvents	Ref.
	Epicamphor														
e65	—,—(*l*)	$C_{10}H_{17}NO$	167.25	nd (dil MeOH) $[\alpha]_D+100.5$ (bz, c=6.3)	103–4									os s	**B7**^{1}, 86
e66	—,*semicarbazone*(*d*)	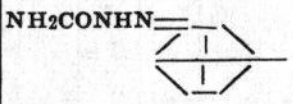 $NH_2CONHN=$	209.29	nd	237–8										**B7**^{1}, 86
e67	—,—(*l*)	$C_{11}H_{19}N_3O$	209.29	nd (al)	237–8d				i	s^h					**B7**^{1}, 86
e68	—,**bromo-**(*l*)	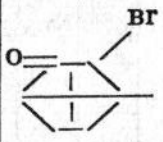 Br	231.14	nd or pr (peth) $[\alpha]_D-86.6$ (AcOEt, c=3.6)	133–4					δ			v	chl v lig s	**B7**^{1}, 86
e70	**α-Epicamphyl-amine**	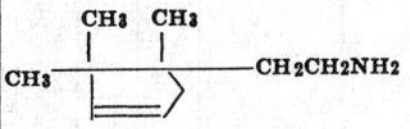 CH_3 CH_3 CH_3 $CH_2CH_2NH_2$	153.26	$[\alpha]_D+17.6$ (bz, c=6.5)		$127–8^{100}$			v						**B12**^{2}, 35
e71	**Epicatechin**(*dl*)	$C_{15}H_{14}O_6$	290.26	nd (w+1) or pr (w+4)	224–6 d anh cor					s	δ	s			**B17**^{2}, 258
e72	—(*l*)	$C_{15}H_{14}O_6$	290.26	(w+4) $[\alpha]^{25}_D-69$ (al)	245d cor				δ		δ	s			**B17**^{2}, 257
—	**β-Epichloro-hydrin**	*see* **Propane, 2-chloro-1,3-epoxy-***													
—	**Epicholestanol**	*see* **(3α)-Cholestanol**													
e74	**Epicholestan-4-ol**	4(α)-Hydroxycholestane. 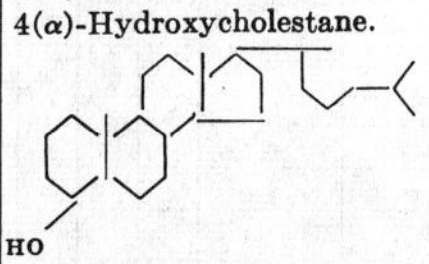 HO	388.65	lf (al) $[\alpha]^{21}_D+29.0$ (chl)	132 (185–6)										**E14**, 65
e75	**Epicholestanol**	ε-Cholestanol. 3(α)-Hydroxy-cholestane. 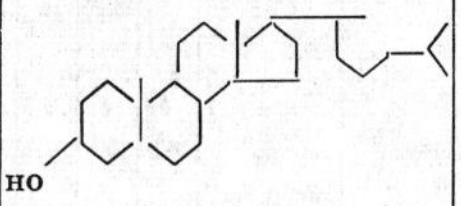 HO	388.65	nd (al) $[\alpha]^{20}_D+33.95$ (chl)	185–6				i	δ s^h	s			chl s MeOH δ	**E14**, 59
e76	**Epicholesterol**	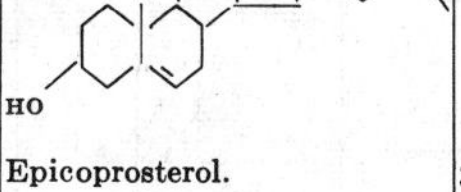 HO	386.64	(al) $[\alpha]_D-37.5$ (al)	141.5										**E14**, 56
e77	**Epicoprostanol**	Epicoprosterol. 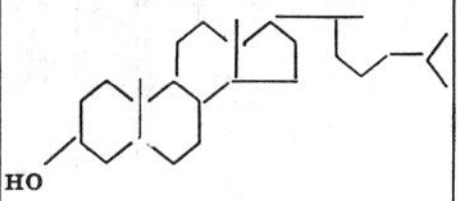 HO	388.65	(al, ace) $[\alpha]^{20}_D+31.6$ (chl)	117–8										**E14**, 60
e78	**Epicoprostenol**	Epiallocholesterol. 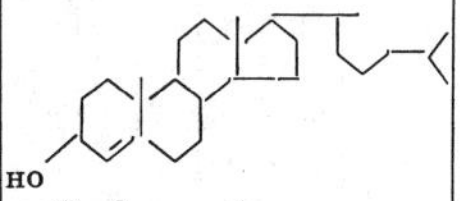 HO	386.64	nd (ace) $[\alpha]^{24}_D+120.8$ (bz)	84					s	v		v	chl v	**E14**, 37
—	**Epicylene**	*see* **Cyclocamphane**													
e79	**Epidicentrin**(*dl*)	*dl*-Domesticene methyl ether. 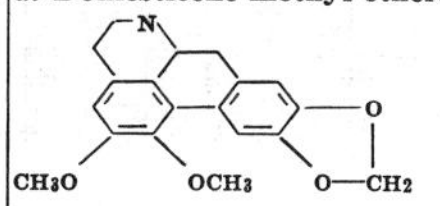 N O CH_3O OCH_3 O—CH_2	339.38	pr (MeOH)	142					v^h	s			chl v^h	**B27**^{2}, 553
e80	—(*l*)	$C_{20}H_{21}NO_4$	339.38	pr (MeOH) $[\alpha]^{18}_D-101.3$ (chl, c=0.5)	138–9										**B27**^{2}, 553
e81	**Epiergosterol**(*D*)	$\Delta^{7,9(11),22}$-Ergostatrien-3(α)-ol. 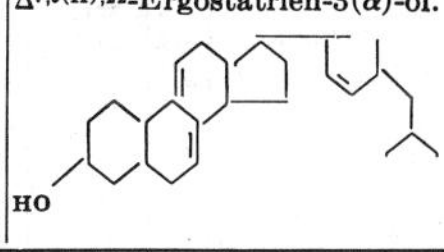 HO	396.63	nd (eth-MeOH) $[\alpha]^{19}_D+36.2$ (chl)	203–4										**E14s**, 1752

For explanations, symbols and abbreviations see beginning of table.

No.	Name	Synonyms and Formula	Mol. wt.	Crystalline form, color and specific rotation	m.p. °C	b.p. °C	Density	n_D	Solubility w	al	eth	ace	bz	other solvents	Ref.
	Epifucitol														
e82	**Epifucitol**(*d*)	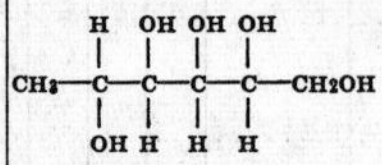 CH_3—C(H)(OH)—C(OH)(H)—C(OH)(H)—C(OH)(H)—CH_2OH	166.17	(w) $[\alpha]_D^{21}$+2 (w)	104										
e83	—(*l*)	$C_6H_{14}O_5$	166.17	(w) $[\alpha]_D$ −2.3 (w)	104				s						
e84	**Epifucose**(*l*)	*α-D*-Glucomethylose. *l*-Talomethylose. 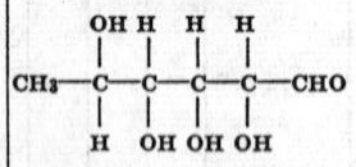 CH_3—C(OH)(H)—C(H)(OH)—C(H)(OH)—C(H)(OH)—CHO	164.16	(AcOEt) $[\alpha]_D$ −36.9 (w, c=6)	135–45				s	...	i	i	...	abs al s	
—	**Epiiodohydrin**	*see* **Propane, 1,2-epoxy-3-iodo-***													
e85	**Epiisofenchol**	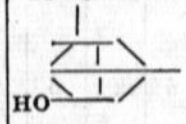	154.25	nd (sub)	71–2										E12A, 645
e86	**Epiisofenchone**(*d*)	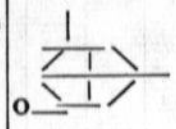	152.23	$[\alpha]_D$ +13.6		195	0.934_4^{20}	1.459^{20}							E12A, 732
e87	—(*dl*)	$C_{10}H_{16}O$	152.23			195–8									E12A, 732
e88	**Epilupinine**	*d*-Isolupinine. 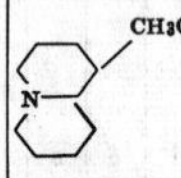 CH_2OH	169.26	nd (peth) $[\alpha]_D^{15}$+38.2	76–8				s					peth δ os s	
—	**Epimethylin**	*see* **Ether, methyl propyl, 2′,3′-epoxy-.**													
—	**Epinephrine**	*see* **Adrenaline**													
e89	**Epinine**	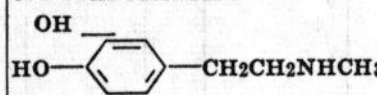 OH, HO—C_6H_3—$CH_2CH_2NHCH_3$	167.20	nd (al)	188–9 cor				δ s^h	δ δ^h				os δ	B13², 487
e90	**Epiquinidine**	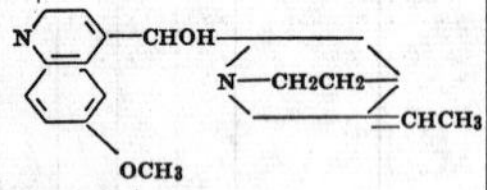 —CHOH—, N—CH_2CH_2—, =CHCH$_3$, OCH_3	324.41	(AcOEt) $[\alpha]_D^{20}$+103.7 (al)	113					s	δ				C31, 181
e91	**Epirhodanhydrin**	2,3-Epoxypropylrhodanine. CH_2CHCH_2SCN (O)	115.16	dk red liq. garlic smell		d			i	s				chl s	
e92	**Episarsapogenin**	$C_{27}H_{44}O_3$	416.62	nd (ace) $[\alpha]_D$ −71	204–6										E14, 284
e93	**Epitruxillic acid**	2,4-*cis*-Diphenylcyclobutan-1,3-*trans*-dicarboxylic acid*. 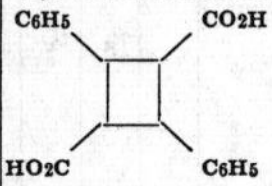 C_6H_5, CO_2H, HO_2C, C_6H_5	296.31	(dil al)							i		i		B9², 691
e94	**Equilenin**(*d*)	$C_{18}H_{18}O_2$	266.33	nd (dil al)	258–9	170–80$^{0.01}$ sub				δ s^h		δ		chl δ	E14, 134
e95	—(*dl*)	$C_{18}H_{18}O_2$	266.34	(bz)	276–8										Am 67, 2274
e96	—(*l*)	$C_{18}H_{18}O_2$	266.34		258–9 cor	170–80^1 sub				s^h					Am 67, 2274
e97	**Equilin**	$C_{18}H_{20}O_2$	268.36	orh sph pl (AcOEt) $[\alpha]_D^{25}$+308 (diox)	238–40	170–200 sub vac			δ	s		s		diox s	E14, 135
e98	—,*α*-**dihydro-**	$\Delta^{1,3,5(10),7}$-Estratetraene-3,17β-diol. HO, OH 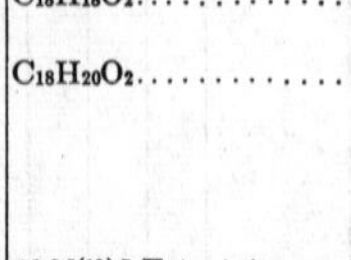	270.37	$[\alpha]_D$ +220 (diox)	174–6									sulf s	E14s, 1962

For explanations, symbols and abbreviations see beginning of table.

No.	Name	Synonyms and Formula	Mol. wt.	Crystalline form, color and specific rotation	m.p. °C	b.p. °C	Density	n_D	Solubility w	al	eth	ace	bz	other solvents	Ref.
	Equisterin														
e99	**Equisterin**	Kaempferol-7-diglucoside. $C_{27}H_{30}O_{16}$	610.53	ye nd (+2w)	195–6					s					
e100	**Equol**	$C_{15}H_{14}O_3$	242.28	(aq al) $[\alpha]_{5461}-21.5$	189–90										
e101	**Eremophilol**	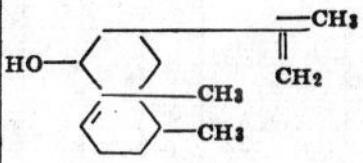 HO– CH₃ CH₂ CH₃ CH₃	220.36	visc oil $[\alpha]_{5461}-55.6$ (MeOH)		164.5^{13}		1.5202							**E12B,** 1432
e102	**Eremophilone**	$C_{15}H_{22}O$	218.34	nd (MeOH) $[\alpha]_D+171.6$ (al)	42–3	171^{15}	0.9994^{25}_{25}	1.5182^{25}							**E12C,** 2597
e103	—,8,9-epoxy-	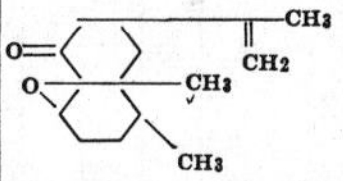 O= CH₃ CH₂ O CH₃ CH₃	234.34	nd (peth) $[\alpha]_{5461}-208$ (MeOH)	63–4									oos v	**E12B,** 2600
e104	**Ergine**	Lysergamide. 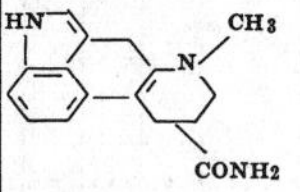 HN CH₃ N CONH₂	283.33	(MeOH+1), pr (aq ace+2w) $[\alpha]^{20}_{5461}+598$ (chl)	135d (115d)									os δ	**H32,** 506
—	**Ergobasine**	*see* **Ergometrine**													
—	**Ergobasinine**	*see* **Ergomatrinine**													
e107	**Ergocornine**	$C_{31}H_{39}N_5O_5$	561.69	$[\alpha]^{20}_D-188$ (chl), −105 (Py)	182–4d				i	δ		s		chl s	**H34,** 1544
e108	**Ergocorninine**	$C_{31}H_{39}N_5O_5$	561.69	lo pr (al) $[\alpha]^{20}_D+409$ (chl, c=1)	228d				i	s		v	s^h	chl v	**B27²,** 860
e109	**Ergocristine**	$C_{35}H_{39}N_5O_5$	609.70	rh (bz+2) $[\alpha]^{20}_D-183$ (chl)	165–7d				i	s		s		chl s	**H34,** 1544
e110	**Ergocristinine**	$C_{35}H_{39}N_5O_5$	609.70	pr (al, AcOEt) $[\alpha]^{20}_D+366$ (chl, c=1)	226d (214d)				i	δ^h		δ		chl δ	**B27²,** 860
e111	**Ergocryptine**	$C_{32}H_{41}N_5O_5$	575.69	pr (al) $[\alpha]^{20}_D-187$ (chl), −112 (Py)	212–4d				i	s				chl s	**H34,** 1544
e112	**Ergocryptinine**	$C_{32}H_{41}N_5O_5$	575.69	lo pr (al) $[\alpha]^{20}_D+408$ (chl, c=1)	240–2d		δ		δ	s^h		v		chl v	**B27²,** 860
e113	**Ergometrine**(*l*)	Ergobasine. *l*-Lysergic-*l*(β-hydroxyisopropylamide). $C_{19}H_{23}N_3O_2$.	325.39	nd (bz) $[\alpha]^{20}_D-89$ (w)	159–69d cor				δ	s		s		chl δ	**Am 69,** 1701
e114	**Ergometrinine**(*d*)	Ergobasinine. *d*-Isolysergic-*d*-(β-hydroxy-isopropylamide). HN CH₃ N CH₃ CONHCHCH₂OH 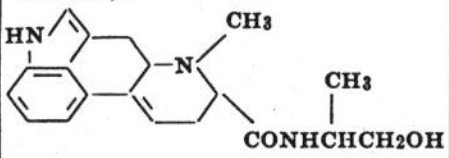	325.39	pr (ace) $[\alpha]^{20}_{5461}$ +520 (chl), +413 (MeOH)	195–7d				i	δ				Py s chl s	**Am 60,** 1701
e115	—(*l*)	$C_{19}H_{23}N_3O_2$	325.39	pr (ace) $[\alpha]^{20}_D-415$ (chl)	196d cor				i	δ				Py s	**Am 60,** 1701
e116	**Ergopinacol**	Bisergostadienol. $C_{56}H_{86}O_2$	791.27	nd (bz-al)	205										
e117	**Ergosine**	$C_{30}H_{37}N_5O_5$	547.64	pr (MeOH) $[\alpha]^{20}_D+16$ (ace) −161 (chl)	228d							s		chl s MeOH δ	**H34,** 1541
e118	**Ergosinine**	$C_{30}H_{37}N_5O_5$	547.64	pr (al, aq ace, bz) nd (MeOH +)½ $[\alpha]^{20}_D+420$	220d (228d)				i	δ		s		chl s	**J1937,** 396

For explanations, symbols and abbreviations see beginning of table.

No.	Name	Synonyms and Formula	Mol. wt.	Crystalline form, color and specific rotation	m.p. °C	b.p. °C	Density	n_D	Solubility						Ref.
									w	al	eth	ace	bz	other solvents	
	$\Delta^{5:6,7:8}$-Ergostadien-3β-ol														
e119	**$\Delta^{5:6,7:8}$-Ergostadien-3β-ol**	22,23-Dihydroergosterol. Provitamin Dy. 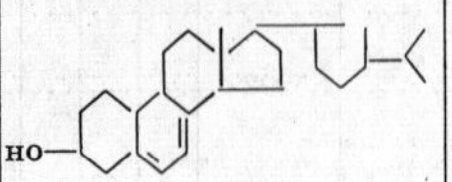	398.65	nd (MeOH-AcOEt) $[\alpha]^{19}_{D}-109$ (chl)	152–3										**E14s, 1457**
e120	**$\Delta^{14,22}$-Ergostadien-3β-ol**	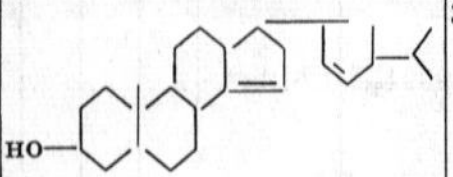	398.65	(MeOH) $[\alpha]^{20}_{D}\pm^{5}-9$ (chl)	116										**E14s, 1762**
e121	**α-Ergostadienone**	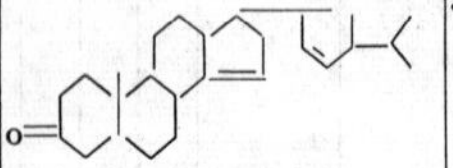	396.63	lf (al) $[\alpha]^{19}_{D}+2$ (chl)	182–3										**E14, 125**
e122	**Ergostane**	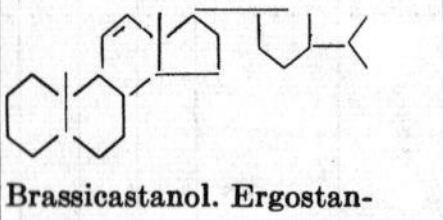	386.68	lf or pl (eth+MeOH or ace) $[\alpha]^{23}_{D}+21$ (chl)	85				i	δ	v	v	...	chl v	
e123	**Ergostanol**	Brassicastanol. Ergostan-3β-ol. 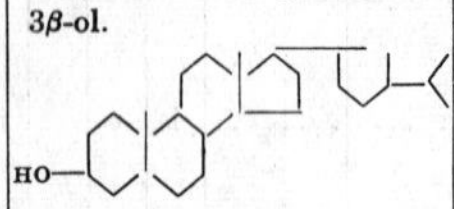	402.68	nd (MeOH-eth) $[\alpha]_{D}+15.94$ (chl)	144–5				i	...	s	...	...	chl s	**E14s, 1767**
e124	**$\Delta^{3,5,7,22}$-Ergostatetrraene**	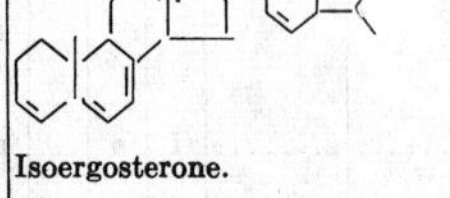	378.61	pl (Ac_2O or al) $[\alpha]^{20}_{D}-40.5$(chl)	104										**E14s, 1434**
e125	**$\Delta^{4,6,22}$-Ergostatrienone**	Isoergosterone. 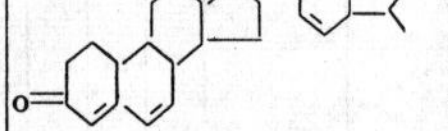	394.67	nd (ace-eth)											**E14, 124**
e126	**α-Ergostenol**	$\Delta^{8:14}$-Ergosten-3β-ol. α-Tetrahydroergostero 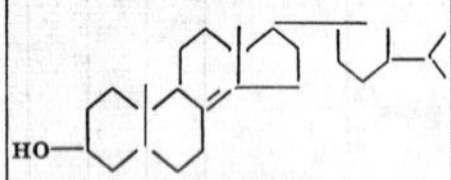	400.67	lf or nd (MeOH) nd (gl aa) $[\alpha]_{D}+17.86$ (chl)	134–5					δ	s	...	s	chl s	**E14s, 1765**
e127	**β-Ergostenol**	$\Delta^{14:15}$-Ergosten-3β-ol. β-Tetrahydroergosterol. 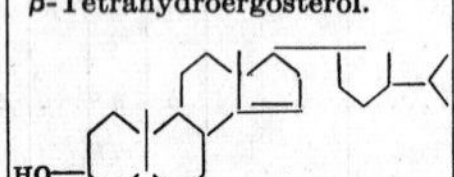	400.67	pl (al) $[\alpha]^{20}_{D}+19.4$ (chl)	141–2										**E14s, 1767**
e128	**γ-Ergostenol**	Δ^{7}-Ergosten-3β-ol. β-Tetrahydroergosterol. 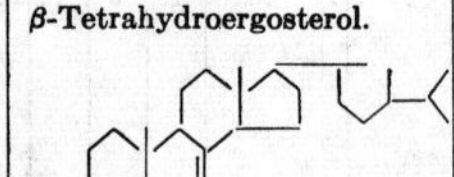	400.67	nd (MeOH) $[\alpha]^{19}_{D}\ 0$	148										**E14s, 1765**
e129	**δ-$\Delta^{8:9}$-Ergostenol**	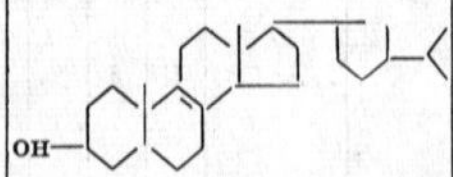	400.67	$[\alpha]_{D}+39$	155										**J1949, 214**
e130	**Ergosterol**	$\Delta^{5,7,22}$-Ergostatrien-3β-ol. Ergosterin. $C_{28}H_{44}O$.	396.63	pl (al), nd (eth) $[\alpha]^{20}_{D}-123.5$	162–4				i	δ s^h	δ s^h	...	...	chl s	**E14s, 1735**
e131	**Ergosterol D**	$\Delta^{7,9,11,22}$-Ergostatrien-3β-ol. 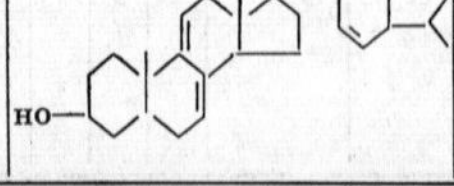	396.63	nd $[\alpha]^{17}_{D}+24.6$	167										**E14s, 1752**

For explanations, symbols and abbreviations see beginning of table.

No.	Name	Synonyms and Formula	Mol. wt.	Crystalline form, color and specific rotation	m.p. °C	b.p. °C	Density	n_D	Solubility w	al	eth	ace	bz	other solvents	Ref.
	Ergosterol														
e133	**Ergosterol,5,6-dihydro-**	α-Dihydroergosterol. HO—	398.65	pr lf (ace) $[\alpha]^{20}_D - 19$ (chl)	176										**E14s,** 1758
e134	**Ergosterone**	$\Delta^{4,1,22}$-Ergostatrienone. O=	394.62	nd (ace-MeOH) $[\alpha]^{20}_D - 0.52$ (chl)	132										**E14,** 124
e135	**Ergotamine**	$C_{33}H_{35}N_5O_5$	581.68	nd (al) pr (bz) $[\alpha]^{20}_D - 155$	212–4				i	δ		δ	δ	$C_6H_5NO_2$ chl v	**B27²,** 860
e136	**Ergotaminine**	$C_{33}H_{35}N_5O_5$	581.68	lf (MeOH) pl (al) $[\alpha]^{20}_D + 369$ (chl, c=0.5)	241–3 cor d					i		i	s	chl s	**B27²,** 862
e137	**Ergothioneine**	$C_9H_{15}O_2N_3S$	229.30	eff nd (w+2), mcl $[\alpha]_D + 115$ (w, c=1)	259–82 d (290d)				v^h	v^h	i			chl i	**B25²,** 413
—	**Eriodictol**	*see* **Flavanone, 3′,4′,5,7-tetrahydroxy-**													
—	**Erucic acid**	*see* **13-Docosenoic acid** (*cis*)													
e138	**Erysocine**	$C_{18}H_{21}NO_3$	299.37	nd (eth) $[\alpha]_D + 238.1$	162					s	s			chl s	**Am 62,** 1677
e139	**Erysodine**	$C_{18}H_{21}NO_3$	299.37	nd (al) $[\alpha]^{27}_D + 248$ (al)	204–5					s	s			chl s	**Am 73,** 589
e140	**Erysonine**	$C_{17}H_{19}NO_3$	285.35	(al) $[\alpha]^{25}_D$ +285−8 (aq HCl), +272 (morpholine)	236–7d					δ				os δ	**Am 36,** 1544
e141	**Erysopine**	$C_{17}H_{19}NO_3$	285.35	(al) $[\alpha]^{25}_D + 265.2$ (alcoholic glycerol)	241–2				δ					chl δ	**Am 73,** 589
e142	**Erysothiopine**	$C_{19}H_{21}NO_7S$	407.44	(al-w+w) $[\alpha]^{25}_D + 194$ (al)	168–9										**Am 66** 1083
e143	**Erysovine**	$C_{18}H_{21}NO_3$	299.37	pr (eth) $[\alpha]_D + 252$ (al)	178–9					s	s			chl s	**Am 73,** 589
e144	**Erytraline**	−4H; N—; OCH₃; O; CH_2—O	297.34	(al) $[\alpha]^{27}_D + 211.8$ (al)	106–7					s				chl s	**Am 73,** 589
e145	**Erythramine**	−2H; N—; OCH_3; O; CH_2—O	299.36	(eth-peth) $[\alpha]^{30}_D + 228$ mm. (al, c=0.19)	103–4	125 at 3.9×10^{-4} mm.				s^d	s		s	lig i	**Am 73,** 589
e146	**Erythratine**	−2H; N—; OCH_3, OH; O; CH_2—O	315.36	(eth-peth+½w) $[\alpha]^{28}_D + 145.5$ (al)	170										**Am 73,** 589
e147	**Erythritol** (*d*)	*d*-Erythryte *L*-Threite. $HOCH_2$—C(H)(OH)—C(OH)(H)—CH_2OH	122.12	pr (w) nd (al) $[\alpha]_D + 11.1$ (al, p=5), −4.4(w, p=5–10)	88–9										**B1³,** 2360
e148	—(*l*)	*l*-Erythryte. *D*-Threite. $HOCH_2$—C(OH)(H)—C(H)(OH)—CH_2OH	122.12	pr (w) nd (al) $[\alpha]_D - 11.1$ (al, p=5), −4.33 (w, p=6)	90–1										**B1³,** 2360

For explanations, symbols and abbreviations see beginning of table.

No.	Name	Synonyms and Formula	Mol. wt.	Crystalline form, color and specific rotation	m.p. °C	b.p. °C	Density	n_D	Solubility w	al	eth	ace	bz	other solvents	Ref.
	Erythritol														
e148[1]	—(*meso*)	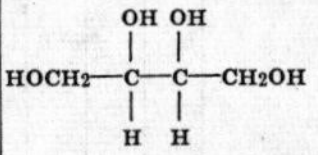 $HOCH_2$—C(OH)(H)—C(OH)(H)—CH_2OH	122.12	bipym tetr	126	329–31	1.451_4^{20}		v	δ s[h]	i			Py v	**B1**[3], 2356
e148[2]	—,tetranitrate (*meso*)	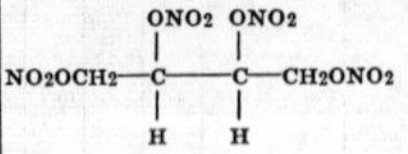 NO_2OCH_2—C(ONO_2)(H)—C(ONO_2)(H)—CH_2ONO_2	312.12	la lf (al)	61				i	s	s			glycerol s	**B1**[3], 2358
e149	**β-Erythroidine**	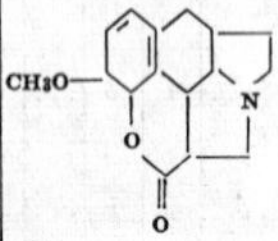 (structure: CH_3O—, N, O, =O)	273.32	(abs al) $[\alpha]_D^{25}+88.8$ (w)	98–9				s	s	δ	...	s	chl s	**Am 4,** 1866
e150	**Erythronic acid,** γ-lactone(*d*)	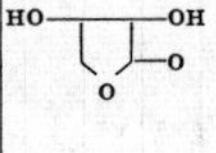 (structure: HO—, —OH, O, =O)	118.09	pr $[\alpha]_D^{20}-73.1$ (w, p=4)	102–3					s[h]					**B18**[3], 54
e151	—,—(*l*)	$C_4H_6O_4$	118.09	pr $[\alpha]_D^{20}-72.5$ (w, p=4)	104 (102–3)					s[h]					**B18**[3], 54
e152	**Erythrose**(*D*)	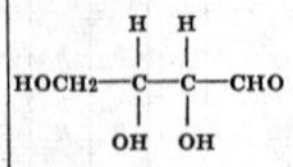 $HOCH_2$—C(H)(OH)—C(H)(OH)—CHO	120.11	syr $[\alpha]_D^{20}-14.8$ (w, 3 days, c=11)					s	v					**B31,** 12
e153	—(*L*)	$C_4H_8O_4$	120.11	syr $[\alpha]_D^{20}+21.5$ (w, 3 days, p=5)					s	v					**B31,** 12
e154	**Erythrosin**	2,4,5,7-Tetraiodofluorescein. 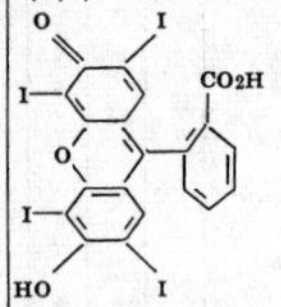 (structure: O, I, I—, CO_2H, O, I—, HO, I)	835.94	og-ye (eth)					δ	s	s	...	i	aq ace s lig i	**B19**[2], 258
e155	—(dye)	Disodium salt of erytrosin. Iodeosin B. $C_{20}H_6I_4Na_2O_5$	879.91	red-br pw					v	s					
e156	**Erythrulose**(*L*)	$HOCH_2CHOHCOCH_2OH$	120.10	syr $[\alpha]_D$ ca. +12 (at equilibrium)		d			s	s					**B31,** 14
e157	**Escholerine**	$C_{41}H_{61}NO_{13}$	775.91	pl $[\alpha]_D^{25}-30\pm2$ (Py)	235 vac d				i	...	...	s			**Am 76,** 1152
—	**Esculetin**	*see* **Coumarin,6,7-dihydroxy-**													
e158	**Esculin**	Aesculin. 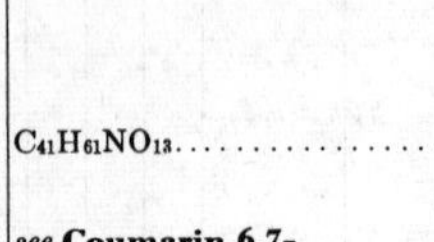 (structure: HO, O—CH, HCOH, O, HCOH, HCOH—CHCH2OH, O, =O)	340.28	nd (w+½) $[\alpha]_D^{18}-14.6$ (MeOH)	160d	230d			δ s[h]	δ s[h]	i			alk s chl s[h] aa s	**B31,** 246
e159	**Estradiol**	$C_{18}H_{24}O_2$. α-isomer β-isomer	272.37	wh or yesh pr (80% al) $[\alpha]_D+54$ (α isomer) $[\alpha]_D+81$ (β isomer)	178 (α) 222 (β)				i	s				(α) peth s	
—	**Estragole**	*see* **Benzene,1-allyl-4-methoxy-**													
e160	**Estriol**	$C_{18}H_{24}O_3$	288.37	lf (al) $[\alpha]_D+61$ (al), +30 (Py)	288d cor					s				Py v os δ	**E14s,** 2152

For explanations, symbols and abbreviations see beginning of table.

No.	Name	Synonyms and Formula	Mol. wt.	Crystalline form, color and specific rotation	m.p. °C	b.p. °C	Density	n_D	Solubility w	al	eth	ace	bz	other solvents	Ref.
	Estrone														
e161	**Estrone**	$C_{18}H_{22}O_2$	270.36	orh, mcl $[\alpha]^{25}_D+158$ to $+168$ (diox)	258–62				i	δ^h	δ	s^h	δ	diox Py s	
—	**Ethanal***	*see* **Acetaldehyde**													
e162	**Ethane***	CH_3CH_3	30.07	gas, hex cr	−182.8	−88.63	0.572^{-108}_{4}	1.0^3769 at 0°/546 mm.	i	δ					**B1³**, 120
e163	**—,amino-***	Ethylamine. $C_2H_5NH_2$	45.08		−84	16.6^{760}	0.6892^{15}_{15}		∞	∞	∞				**B4²**, 586
e164	**—,—,hydro-bromide***	$C_2H_5NH_2.HBr$	126.00	mcl nd or pl (al)	159.5					v				chl δ	**B4²**, 588
e165	**—,—,hydro-chloride***	$C_2H_5NH_2.HCl$	81.55	mcl pl (al)	109–10	d315			v	v	i	i			**B4²**, 588
e166	**—,—,hydroiodide***	$C_2H_5NH_2.HI$	173.00	mcl nd (w)	188.5		2.100		v	v	i				**B4²**, 588
e167	**—,1-amino-2-bromo,** hydro-chloride*	$BrCH_2CH_2NH_2.HCl$	160.45	lf (al-AcOEt)	174–5d cor				v				s		**B4²**, 618
e168	**—,1-amino-2-chloro-,** hydro-chloride*	$ClCH_2CH_2NH_2.HCl$	115.99	hyg (AmOH)	119–23				v	v				os v	**B4**, 133
—	**—,2-amino-1,1-diethoxy-***	*see* **Acetaldehyde, 2-amino-,** diethyl acetal													
e169	**—,1-amino-2-diethylamino-***	*N,N*-diethylethylenediamine $H_2NCH_2CH_2N(C_2H_5)_2$	116.21			145	0.8211^{20}_{20}		∞	s	s			to s	**B4²**, 690
e170	**—,1-amino-2-ethoxy-***	2-Aminodiethyl ether. $C_2H_5OCH_2CH_2NH_2$	89.14			108^{758} cor	0.8512^{20}_{4}	1.4101^{20}	∞	∞	∞				**B4²**, 718
e171	**—,1-amino-2-methoxy-***	$CH_3OCH_2CH_2NH_2$	75.11			95^{756}			v	v					**B4²**, 718
e176	**—,aminoxy-**	Ethoxyamine. *α*-Ethyl-hydroxylamine. $C_2H_5ONH_2$	61.08		68	0.8827^{8}_{8}			∞	∞				os ∞	**B1²**, 333
e177	**—,1,2-bis(4-aminophenyl)-***	*α,α′*-Bi-*p*-toluidine. 4,4′-Diaminobibenzyl. $H_2N-C_6H_4-CH_2CH_2-C_6H_4-NH_2$	212.30	pl (w)	135–6 sub	sub			i s^h	v					**B13¹**, 75
e178	**—,1,2-bis-(benzoyl-amino)-***	*N,N′*-Dibenzoyl-ethylenediamine. $C_6H_5CONHCH_2CH_2NHCOC_6H_5$	268.32	pr or nd (al)	247				i	i				aa v^h	**B9²**, 187
e179	**—,1,2-bis(2-bromophenyl)-***	2,2′-Dibromobibenzyl (Br)$C_6H_4-CH_2CH_2-C_6H_4$(Br)	340.07	pl (al)	84.5					s^h					**B5²**, 507
e180	**—,1,2-bis(4-bromophenyl)-***	4,4′-Dibromobibenzyl. $Br-C_6H_4-CH_2CH_2-C_6H_4-Br$	340.07	pr (al)	114–5	*ca* 198^{10}				i δ^h			δ		**B5²**, 507
e181	**—,1,2-bis(2-chloroethoxy)-***	Triglycol dichloride ether. $ClCH_2CH_2OCH_2CH_2OCH_2CH_2Cl$	187.07			230^{760} 118^{10}	1.197^{20}_{20}								
e182	**—,1,1-bis(4-chlorophenyl)-2,2-dichloro-***	DDD. Rothane $(Cl-C_6H_4-)_2CHCHCl_2$	320.05		109–10										
e183	**—,2,2-bis(4-chlorophenyl)-1,1,1-trichloro-***	DDT. Dichlorodiphenyl-trichloroethane. $(Cl-C_6H_4-)_2CHCCl_3$	354.49	nd (al)	108.5				i	δ	v	v			**Am 67**, 189
e184	**—,1,2-bis(ethyl-thio)-***	$C_2H_5SCH_2CH_2SC_2H_5$	150.31			210–3	0.9815^{20}_{4}	1.5118^{20}		s	s				**B1³**, 2131
e185	**—,2,2-bis(4-methoxyphenyl)-1,1,1-trichloro-***	Methoxychlor. $(CH_3C-C_6H_4-)_2CHCCl_3$	345.65		94				i	s	v		v	os v	**H27**, 892
e186	**—,1,2-bis(2-methoxy-phenoxy)-***	Guaiacol ethylene ether. (OCH_3)$C_6H_4-OCH_2CH_2O-C_6H_4$(CH_3O)	274.30	nd (al)	139–40				i	v	v				**B6**, 772
e187	**—,1,2-bis(methyl-amino)-***	*N,N′*-Dimethylenediamine. $CH_3NHCH_2CH_2NHCH_3$	88.15			119–20	0.828^{15}_{4}			s	s			dil HCl s	**B4³**, 689

For explanations, symbols and abbreviations see beginning of table.

No.	Name	Synonyms and Formula	Mol. wt.	Crystalline form, color and specific rotation	m.p. °C	b.p. °C	Density	n_D	Solubility: w	al	eth	ace	bz	other solvents	Ref.
	Ethane														
e189	**—,1,2-bis(methylthio)-***	$CH_3SCH_2CH_2SCH_3$	122.26			182.5^{760}	1.0371^{20}_{4}	1.5292^{20}	s	s				chl s	**B1**^{3}, 2130
e190	**—,1,2-bis(2-nitrophenyl)-***	2,2′-Dinitrobibenzyl NO_2 NO_2 —CH_2CH_2—	272.27	pr (aa)	127					δ	v		v	aa s^h	**B5**, 603
e191	**—,1,2-bis(4-nitrophenyl)-***	4,4′-Dinitrobibenzyl. NO_2— —CH_2CH_2— —NO_2	272.27	yesh nd (al or bz)	180–2					i $δ^h$	δ		δ s^h	chl δ	**B5**^{3}, 508
e192	**—,1,2-bis(phenylamino)-***	*N,N′*-Diphenylethylenediamine. $C_6H_5NHCH_2CH_2NHC_6H_5$	212.29	lf (dil al)	66.2	$178–82^{62}$			i	s	s				**B12**^{3}, 287
e193	**—,1,2-bis(phenylsulfonyl)-***	$C_6H_5SO_2CH_2CH_2SO_2C_6H_5$	310.40	nd of lf	180				$δ^h$	s^h			s	aa s^h	**B6**^{3}, 291
e194	**—,1,2-bis(phenylthio)-***	$C_6H_5SCH_2CH_2SC_6H_5$	246.39	ta (al)	69				i						**B6**^{3}, 291
e195	**—,bromo-***	Ethyl bromide. C_2H_5Br	108.98		−118.9	38.402^{760}	1.4604^{20}_{4}	1.4239^{20}	δ	∞	∞			chl ∞	**B1**^{3}, 171
e196	**—,1-bromo-2-chloro-***	Ethylene chlorobromide. $BrCH_2CH_2Cl$	143.43		−16.7	107^{760}	1.7392^{20}_{4}	1.4917^{20}	δ	s	s				**B1**^{3}, 179
e197	**—,1-bromo-2-fluoro-***	$BrCH_2CH_2F$	126.97			$70–1^{745}$	1.7044^{25}_{4}	1.4236^{20}	i	v	v				
e199	**—,chloro-***	Ethyl chloride. C_2H_5Cl	64.52		−138.7	13.1^{760}	0.9028^{15}_{4} (liq)	1.3742^{10}	δ	v	∞				**B1**^{3}, 133
e200	**—,1-chloro-2-diethylamino, hydrochloride***	β-Chlorotriethylammonium chloride. $ClCH_2CH_2N(C_2H_5)_2.HCl$	172.11	nd (al-eth)	210–1				s	s	i				**B4**^{2}, 618
e201	**—,1-chloro-1,2-dinitro-1,2,2-trifluoro-***	$F_2C(NO_2)CClFNO_2$	208.50			98.5–100^{760}									**J1953**, 2075
e202	**—,1-chloro-2-fluoro-***	$ClCH_2CH_2F$	82.51			53.2			i	v	v				
e203	**—,1-chloro-1-nitro-***	$CH_3CHClNO_2$	109.52			$124–5^{760}$	1.247^{8}	1.423	i	s				alk s	**B1**, 101
e204	**—,chloropentafluoro-***	$ClCF_2CF_3$	154.48	gas	−106	$-38s^{760}$			i	s	s				**B1**^{3}, 139
e205	**—,1-(2-chloroethoxy)-2-phenoxy-***	β-chloro-β′-phenoxy diethyl ether. $C_6H_5OCH_2CH_2OCH_2CH_2Cl$	200.67			149^{10}	1.149^{15}_{15}								**B6**^{2}, 150
e206	**—,2-(2-chlorophenyl)-2-4-chlorophonyl)-1,1,1-trichloro-***	*o,p′*-DDT. Cl— —CH(CCl₃)— (Cl)	354.51	(MeOH)	>4cor										**Am 67**, 1498
e207	**—,1,2-diamino-***	Ethylenediamine. 1,2-Ethanediamine. $H_2NCH_2CH_2NH_2$	60.10		8.5	116.5^{760}	0.8995^{20}_{20}	1.4499^{20}	v	∞	i		i		**B4**^{2}, 676
e208	**—,—,hydrate***	$H_2NCH_2CH_2NH_2.H_2O$	78.11		10	118^{760}	0.963^{21}_{4}	1.4500	∞						**B4**^{2}, 676
e209	**—,—,hydrochloride***	$H_2NCH_2CH_2NH_2.2HCl$	133.02	mcl pr (w)	>270 sub			1.633		i					**B4**^{2}, 677
e210	**—,2-diazo-1,1,1-trifluoro-***	Trifluorodiazoethane. N_2CHCF_3	110.04	ye		13^{752}			s		s				**Am 65**, 1458
e211	**—,1,1-dibromo-***	Ethylene bromide. CH_3CHBr_2	187.86		−63	110^{760}	2.0554^{20}_{4}	1.5122^{20}	i	s	v				**B1**^{3}, 61
e212	**—,1,2-dibromo-***	Ethylene bromide. $BrCH_2CH_2Br$	187.86		10.06	131^{760}	2.180^{20}_{4}	1.5389^{20}	δ	v	∞				**B1**^{3}, 182
e213	**—,1,2-dibromo-1,1-dichloro-***	$BrCH_2CBrCl_2$	256.76		−67	176–8d	2.2695^{15}_{4}	1.5593^{15}	i					os s	**B1**^{2}, 64
e214	**—,1,2-dibromo-1,2-dichloro-***	$BrCHClCHBrCl$	256.76		−27	195^{760}	2.135^{20}_{4}	1.5669^{20}	i					os s	**B1**^{3}, 190
e215	**—,1,1-dichloro-***	Ethylidene chloride. CH_3CHCl_2	98.96		−96	57 cor	1.1776^{20}_{4}	1.4160^{20}	δ	v	v				**B1**^{3}, 190

For explanations, symbols and abbreviations see beginning of table.

No.	Name	Synonyms and Formula	Mol. wt.	Crystalline form, color and specific rotation	m.p. °C	b.p. °C	Density	n_D	Solubility w	al	eth	ace	bz	other solvents	Ref.
	Ethane														
e216	**—,1,2-dichloro-***	Ethylene chloride. $ClCH_2CH_2Cl$	98.96		−35	84^{760}	1.256^{20}_{4}	1.4448^{20}	δ	v	∞			oos s	**B1**^{3}, 141
e217	**—,1,1-dichloro-2,2-difluoro-1,2-dinitro-**	$O_2NCF_2CCl_2NO_2$	224.94			$81–2^{103}$ 142^{760}d									**J1953,** 2075
e218	**—,1,1-dichloro-1,2,2,2-tetrafluoro-***	Freon 114. Cl_2CFCF_3	170.92		−94	3.8^{760} cor	1.455^{25}_{4}	1.3092^{0}	i	s	s				**B1**^{3}, 152
e219	**—,1,2-diethoxy-***	Glycol diethyl ether. $C_2H_5OCH_2CH_2OC_2H_5$	118.18			123^{760}	0.8484^{20}			s				os v	**B1**^{2}, 519
—	**—,2,2-diethoxy-1,1,1-trichloro-***	*see* **Acetaldehyde, trichloro-**, diethyl acetal													
e221	**—,1,1-difluoro-***	Ethylidene fluoride. CH_3CHF_2	66.05	gas		−24.7		1.3011^{-72}							**Am 64,** 2289
e222	**—,1,2-difluoro-***	Ethylene fluoride. FCH_2CH_2F	66.05			30.7^{760}									**B1**, 82
e223	**—,1,1-difluoro-1,2,2,2-tetrachloro-***	$F_2CClCCl_3$	203.82		40.6	91.5^{760} cor			i	s	s				**B1**^{3}, 164
e224	**—,1,2-difluoro-1,1,2,2-tetrachloro-***	$Cl_2CFCFCl_2$	203.82		25	93^{760} -37.5^{1}	1.6447^{25}_{4}	1.4130^{25}	i	s	s				**B1**^{3}, 165
e225	**—,(difluoroamino)pentafluoro-***	Perfluoroethylamine. $F_3CCF_2NF_2$	171.02			-35^{760}			i						**J1950,** 1966
e227	**—,1,1-diiodo-***	Ethylidene iodide. CH_3CHI_2	281.86			$179–80^{760}$	2.84^{0}	1.673^{20}	i	v	v			chl s	**B1**^{3}, 198
e228	**—,1,2-diiodo-***	Ethylene iodide. ICH_2CH_2I	281.86	ye pr or pl (eth), d in lt	81–2		2.132^{00}		δ	s	s				**B1**^{2}, 69
e229	**—,1,2-dimethoxy-***	Ethylene glycol dimethyl ether. $CH_3OCH_2CH_2OCH_3$	90.12		−58	85^{760}	0.8665^{20}_{4}	1.3796^{20}	s	s	s				**B1**^{3}, 2073
e230	**—,1,2-di-*N*-morpholyl-**	O(ring)N—CH_2CH_2—N(ring)O	200.28	wh-yesh (eth or lig)	75	$160–3^{25}$			v	v		s	s	MeOH s	**B27, 7**
e231	**—,1,1-dinitro-***	$O_2NCH_2CH_2NO_2$	120.07	ye mcl so	185–6 cor	1.3503^{24}_{24}			δ	s	s				**B1**^{2}, 70
e232	**—,1,2-dinitro-1,2-diphenyl-***, (low melting form)	α,α′-Dinitrobibenzyl. $C_6H_5CH(NO_2)CH(NO_2)C_6H_5$	272.26	(al), pr (aa)	150–2					v	v	v	v	os v aa v lig s	**B5**^{2}, 508
e233	—,—,(high melting form)*	$C_6H_5CH(NO_2)CH(NO_2)C_6H_5$	272.26	nd (aa)	235–6 (226d)					δ	δ	v		aa δ s^h	**B5**^{2}, 508
e234	**—,1,2-dinitro-1,1,2,2-tetrafluoro-***	$O_2NCF_2CF_2NO_2$	192.03			$58–9^{760}$									
e235	**—,1,2-diphenoxy-***	Ethylene diphenyl ether. $C_6H_5OCH_2CH_2OC_6H_5$	214.27	lf (al)	98	$180–5^{12}$			i	δ s^h	s			chl s	**B6**^{2}, 150
e236	**—,1,1-diphenyl-***	α-Methylditan. $CH_3CH(C_6H_5)$-	182.27		−21.5	286	0.9875^{20}_{4}	1.5761^{20}	i	∞	∞				**B5**^{2}, 653
e237	**—,1,2-diphenyl-***	Bibenzyl. Dibenzyl. $C_6H_5CH_2CH_2C_6H_5$	182.27	nd (al)	52.2	285^{760} $140–50^{13}$	0.995^{20}_{4}	1.5338α	i	s v^h	s			liq SO_2 s liq NH_3 i	**B5**^{2}, 506
e238	**—,2,2-diphenyl-1,1,1-trichloro-***	$(C_6H_5)_2CHCCl_3$	285.60	lf (al)	64					s v^h					**B35**^{2}, 510
e239	**—,1,1-di-4-tolyl-**	$(CH_3—C_6H_4—)_2CHCH_3$	210.32		<−20	295–300 $153–6^{11}$	0.974^{20}_{4}		i				s	aa s	**B5**^{2}, 502
e240	**—,1,2-di-4-tolyl-**	$CH_3—C_6H_4—CH_2CH_2—C_6H_4—CH_3$	210.32	lf (MeOH or dil al) pl (lig)	82–3 (85)	$296–8^{730}$ 178^{18}			i	δ			s	peth s	**B5**^{2}, 521
e240[1]	**—,2,2-di-4-tolyl-1,1,1-trichloro-**	$(CH_3—C_6H_4—)_2CHCCl_3$	313.66	mcl pr (al or eth-al)	89				i	s v^h	v				**B5**^{2}, 522
e241	**—,1,2-epoxy-***	Ethylene oxide. Oxirane. CH_2—O—CH_2 (ring)	44.05		−111	$13–4^{11}$	0.882^{10}_{10}	1.3597^{20}	s	s	s		s		**B17**^{2}, 9
e242	**—,1,2-epoxy-1-phenyl-***	Styrene oxide. C_6H_5CH—O—CH_2 (ring)	120.15		−35.6	$191–2^{760}$ $20^{0.3}$	1.0469^{25}_{4}	1.5350^{20}	δ	s	∞		∞		**B17**^{2}, 49

For explanations, symbols and abbreviations see beginning of table.

No.	Name	Synonyms and Formula	Mol. wt.	Crystalline form, color and specific rotation	m.p. °C	b.p. °C	Density	n_D	Solubility w	al	eth	ace	bz	other solvents	Ref.
	Ethane														
e243	**—,1-ethoxy-2-methoxy-***	$C_2H_5O.CH_2CH_2.OCH_3$	104.15			102	0.8529^{20}_{4}	1.3868^{20}	∞	∞	∞				
e244	**—,1-ethoxy-2-methylamino-***	Ethyl β-methylaminoethyl ether. $CH_3NHCH_2CH_2OC_2H_5$	103.17			$114–5^{744}$ cor	0.8363^{20}_{4}	1.414^{20}	∞	∞	∞				**B4**, 276
e245	**—,fluoro-***	Ethyl fluoride. C_2H_5F	48.06	gas	−143.2	−37.7	0.00220 at ca. 0°	1.303^{37}	s	v	v				**B1**^{3}, 130
e246	**—,fluoropentachloro-***	$FCCl_2CCl_3$	220.29		101.3	$134–6^{760}$			i	s	s				**B1**^{3}, 168
e247	**—,hexabromo-***	Perbromoethane. Br_3CCBr_3	503.48	rh pr	198–205	d	2.823^{20}_{4}	1.863	i	δ^h	δ			CS_2 s	**B1**^{3}, 193
e248	**—,hexachloro-***	Perchloroethane. Cl_3CCCl_3	236.74	rh (al, eth)	187	186^{777}	2.091^{20}_{4}		i	v	v			liq HF δ	**B1**^{3}, 168
e249	**—,hexafluoro-***	Perfluoroethane. F_3CCF_3	138.01	gas	−94	−79	1.590^{-78}		i	δ	δ				**B1**^{3}, 132
e250	**—,hexaphenyl-***	$(C_6H_5)_3CC(C_6H_5)_3$	486.66		145–7d	d				s^h				chl, CS_2 v CCl_4, lig δ MeOH v	**B1**^{2}, 626
e251	**—,hydroxylamino-***	β-Ethylhydroxylamine. C_2H_5NHOH	61.08	nd (liq)	59–60d		0.9079^{64}_{4}	1.4152^{64}	v	v	δ		δ	lig δ	**B4**^{2}, 953
e252	**—,2-imino-1,1,1-trichloro-***	Chloralimide. 2,2,2-Trichloroethylideneimide. $Cl_3CCH{:}NH$	146.40		150–5				δ	v	v				
e253	**—,iodo-***	Ethyliodide. CH_3CH_2I	155.97		−108	72^{760}	1.950^{20}_{20}	1.5168^{15}	δd	∞				os ∞	**B1**^{3}, 193
e254	**—,isocyano-**	Ethylcarbylamine. Ethylisocyanide. C_2H_5NC	55.08		<−66	79	0.7402^{20}_{4}	1.3659^{24}	v	∞	∞				**B4**^{2}, 600
254[1]	**—,1(4-methoxyphenyl)1-phenyl-***	1-p-Anisyl-1-phenylethane. $CH_3O—C_6H_4—CH(C_6H_5)CH_3$	212.29			$180–2^{19}$ $130^{0.5}$	1.0473^{20}_{4}	1.5725^{20}	i					os s	**B6**^{2}, 639
e255	**—,nitro-***	$CH_3CH_2NO_2$	75.07		−90	115^{760}	1.0448^{25}_{4}	1.3917^{20}	s	∞	∞			dil alk s	
e256	**—,nitropentafluoro-***	$F_3CCF_2NO_2$	165.02			0^{760}									**J1953**, 2075
e257	**—,2-nitro-1,1,1-trifluoro-***	$F_3CCH_2NO_2$	129.04			96^{760}									**Am 72**, 3579
e258	**—,1(2-nitrophenyl)2-(4-nitrophenyl)-***	2,4′-Dinitrobibenzyl. $(2\text{-}NO_2C_6H_4)—CH_2CH_2—C_6H_4—NO_2$	272.26	nd	74–5					v					**B5**, 603
e259	**—,nitrosopentafluoro-***	F_3CCF_2NO	149.02	bl		-42^{760}									**J1953**, 2075
e260	**—,pentabromo-***	Br_3CCHBr_2	424.58	mcl pr	56–7	210^{300}d	3.312^{20}_{4}		i	s	v				**B1**^{3}, 193
e261	**—,pentachloro-***	Pentalin. Cl_3CCHCl_2	202.30		−29	162^{760}	1.6796^{20}_{4}	1.5054^{15}	i	∞	∞				**B1**^{3}, 165
e262	**—,pentaiodo-***	I_3CCHI_2	659.55	mcl pr (aa)	182–4					s			s	aa s	**B1**^{1}, 31
e264	**—,1,1,1,2-tetrabromo-***	Br_3CCH_2Br	345.67		0.0	112^{18}d	2.8748^{20}_{4}	1.6277^{20}		s					**B1**^{3}, 191
e265	**—,1,1,2,2-tetrabromo-***	Acetylene tetrabromide. $Br_2CHCHBr_2$	345.67	yesh	−1	239–42d $137–8^{36}$	2.9672^{20}_{4}	1.6380^{20}	i	∞	∞			chl ∞ $PhNH_2$ ∞ aa ∞	**B1**^{3}, 192
e266	**—,1,1,1,2-tetrachloro-***	Cl_3CCH_2Cl	167.85	yesh red	−68.1	$129–30^{760}$	1.5532^{20}_{4}	1.4821^{20}	δ	∞	∞				**B1**^{3}, 158
e267	**—,1,1,2,2-tetrachloro-***	Acetylene tetrachloride. $Cl_2CHCHCl_2$	167.85		−43.8	146^{760}	1.5984^{20}_{4}	1.4944^{20}	δ	∞	∞				**B1**^{3}, 159
e268	**—,1,1,1,2-tetrafluoro-***	F_3CCH_2F	102.03			-26^{736}									**Am 77**, 4899
e269	**—,1,1,1,2-tetraphenyl-***	$(C_6H_5)_3CCH_2C_6H_5$	334.47	mcl (eth-peth)	143–4	$277–80^{21}$			i	δ		δ	s		**B5**^{2}, 674
e270	**—,1,1,2,2-tetraphenyl-***	$(C_6H_5)_2CHCH(C_6H_5)_2$	334.47	(bz+1), rh nd (chl)	212.5	358–62				δ^h			s^h	aa s	**B5**^{2}, 673
e271	**—,triazo-***	Azidoethane. Ethyl azide. $C_2H_5N_3$	71.08			48^{760}								peth s	**B1**^{2}, 71
e272	**—,1,1,2-tribromo-***	Br_2CHCH_2Br	266.77		−35.5	187–90	2.5789^{20}_{4}	1.5933^{20}		s					**B1**^{3}, 191

For explanations, symbols and abbreviations see beginning of table.

No.	Name	Synonyms and Formula	Mol. wt.	Crystalline form, color and specific rotation	m.p. °C	b.p. °C	Density	n_D	Solubility w	al	eth	ace	bz	other solvents	Ref.
	Ethane														
e273	—,**1,1,1-trichloro-***	Methylchloroform. CH_3CCl_3	133.41	...	−33	74^{760}	1.3492^{29}_4	1.4199^{21}_{α}	i	∞	∞	...	...	...	**B1²**, 55
e274	—,**1,1,2-trichloro-***	Cl_2CHCH_2Cl	133.41	...	−37.4	113^{760}	1.4405^{20}_4	1.4706^{20}	δ	...	∞	...	...	...	**B1³**, 154
e275	—,**1,1,1-trichloro-2,2,2-trifluoro-***	Cl_3CCF_3	187.38	...	14.2	45.8^{760} cor	1.5790^{20}_4	1.3610^{25}	i	s	s	...	...	...	**B1³**, 157
e276	—,**1,1,2-trichloro-1,2,2-trifluoro-***	$Cl_2CFCClF_2$	187.38	...	−36.4	47.7^{760} cor	1.5635^{25}_4	1.3557^{25}	i	s	∞	...	∞	...	**B1³**, 157
e277	—,**1,1,1-trifluoro-***	CH_3CF_3	84.04	gas	−111.3	-47.3^{760}	0.00378	...	...	...	...	...	...	...	**B1³**, 131
e278	—,**1,1,1-triiodo-***	Methyliodoform. CH_3CI_3	407.76	ye oct (al)	93	...	...	...	...	δ	v	...	v	CS_2 v lig δ	**B1³**, 99
e279	—,**1,1,1-triphenyl-***	α-Methyltritan. $(C_6H_5)_3CCH_3$	258.37	nd (al, eth)	95	...	...	...	i	δ s^h	v	...	...	...	**B5¹**, 350
e280	—,**1,1,2-triphenyl-***	$(C_6H_5)_2CHCH_2C_6H_5$	258.37	mcl lf (dil al)	56	$348–9^{751}$ cor	...	...	i	v	v	...	...	...	**B5²**, 620
e281	**Ethanearsonic acid***	Ethylarsonic acid. $C_2H_5AsO_3H_2$	154.00	nd (al)	99.5	...	...	...	v	v	...	...	...	alk v	...
e282	**Ethaneboronic acid***	Ethylboric acid. $C_2H_5B(OH)_2$	73.89	volat pl (eth)	40 subl	d	...	...	v	v	v	...	...	...	**B4**, 642
—	**Ethanedial***......	*see* **Glyoxal**													
—	**Ethanedioic acid***	*see* **Oxalic acid**													
e283	**1,2-Ethanediol***	Dihydroxyethane*. Ethylene glycol. Glycol. $HOCH_2CH_2OH$	62.07	...	−13.2	198^{760} cor	1.1088^{20}_4	1.4314^{20}	∞	∞	s	∞	...	aa ∞	**B1³**, 2053
e284	—,bis(chloroacetate)	$ClCH_2CO_2CH_2CH_2O_2CCH_2Cl$	215.03	(eth-peth)	45–6	$142–4^2$	...	...	i	d^h	v	...	...	...	...
—	—,carbonate......	*see* **1,3-Dioxolan-2-one**													
—	**1,1-Ethanediol,** diacetate	*see* **Acetaldehyde,** diacetate.													
e285	**1,2-Ethanediol,** diacetate	Ethylene acetate. $CH_3CO_2CH_2CH_2O_2CCH_3$	146.14	...	−31	190^{760}	1.104^{20}_4	1.4150^{20}	v	∞	∞	∞	∞	CS_2, CCl_4 ∞ aa ∞	
e286	—,dibenzoate	Ethylene benzoate. $C_6H_5CO_2CH_2CH_2O_2CC_6H_5$	270.29	rh pr (eth)	73–4	>360d	...		i	...	s	...	...	...	**B9²**, 109
e287	—,dibutanoate*	Ethylene butyrate. $C_3H^n_7CO_2CH_2CH_2O_2CC_3H^n_7$	202.25	...	...	240	1.024^0_4	...	i	v	v	...	...	...	**B2²**, 272
e288	—,didodecanoate*	Ethylene laurate. $CH_3(CH_2)_{10}CO_2CH_2CH_2O_2C(CH_2)_{10}CH_3$	426.69	pl (al)	56.6	188^{20}	...	...	i	v	v	...	...	...	**B2²**, 320
e289	—,diethyl ether	1,2-Diethoxyethane*. $C_2H_5OCH_2CH_2OC_2H_5$	118.18	...	...	123^{758}	0.8484^{20}	1.3860^{20}	...	...	...	...	...	...	**B1²**, 519
e290	—,diformate	$HCO_2CH_2CH_2O_2CH$	118.09	...	...	174	1.193^0_4	1.3580	δ	s	s	...	...	...	**B2**, 23
e291	—dihexadecilate*	Ethylene palmitate. $CH_3(CH_2)_{14}CO_2CH_2CH_2O_2C(CH_2)_{14}CH_3$	538.90	lf or nd (al-chl)	69–70	226	0.8594^{78}	...	i	δ	s^h	...	...	oos v	**B2²**, 338
e292	—,dimethyl ether	1,2-Dimethoxyethane* $CH_3OCH_2CH_2OCH_3$	90.12	...	−58	$83–4^{760}$	0.8664^{20}_4	1.3813^{20}	s	...	...	...	...	...	**B1²**, 518
e293	—,dinitrate*	Ethylene nitrate. $O_2NOCH_2CH_2ONO_2$	152.06	ye	−22.3	197 ± 3^{760}	1.4918^{20}_4	...	i	s	...	...	...	alk d	**B1³**, 2112
e294	—,dinitrite*	Ethylene nitrite. $ONOCH_2CH_2ONO$	120.07	...	<−15	98	1.2156^0_4	...	i	s	s	...	...	glycerol s	**B1**, 469
e295	—,dioctadecylate*	Ethylene stearate. $CH_3(CH_2)_{16}CO_2CH_2CH_2O_2C(CH_2)_{16}CH_3$	595.01	lf	76–7 (79)	241^{20}	0.8581^{78}	...	i	δ	v	...	...	...	**B2²**, 354
e296	—,dipropanoate*	Ethylene propionate. $C_2H_5CO_2CH_2CH_2O_2CC_2H_5$	174.20		...	211	1.020^{15}	...	δ	∞	∞	...	...	...	**B2**, 242
e297	—,ditetradecylate*	Ethylene myristate. $CH_3(CH_2)_{12}CO_2CH_2CH_2O_2C(CH_2)_{12}CH$	482.79		65	...	...	...	...	...	...	...	...	...	**B2²**, 327
e298	—,dithiocyanate	$NCSCH_2CH_2SCN$	144.22	rh pl or nd (w), ta (al or eth)	90	d	...	...	δ s^h	s	s	v	δ	...	**B3²**, 123
e301	—,**1,2-bis(4-methoxyphenyl)-*** (lower m.p. form)	*p,p′*-Dimethoxyhydrobenzoin, Isohydroanisoin. $CH_3O—C_6H_4—CHOHCHOH—C_6H_4—OCH_3$	274.32	(eth)	125–6	...	...	...	...	v	v	...	...	...	**B6²**, 1130
e302	—,—,(higher m.p. form)	Hydroanisoin. $C_{16}H_{18}O_4$. *See* e301	274.32	rh ta	170–1	...	...	...	i $δ^h$	δ s^h	i $δ^h$	... ...	... ...	...	**B6²**, 1129

For explanations, symbols and abbreviations see beginning of table.

No.	Name	Synonyms and Formula	Mol. wt.	Crystalline form, color and specific rotation	m.p. °C	b.p. °C	Density	n_D	Solubility w	al	eth	ace	bz	other solvents	Ref.
	1,2-Ethanediol														
e304	**—,1,2-dicyclo-hexyl-***	Cyclohexanonepinacol. Dodecahydrobenzoin. (ring)—CHOHCHOH—(ring)	226.32	nd	129–30	...	...	...	...	...	...	...	v	peth s	...
e305	**—,1,2-diphenyl-*** (*d*)	*d*-Hydrobenzoin. *d*-Isohydrobenzoin. (ring)—CHOHCHOH—(ring)	214.27	nd (w), lf or pr (abs al) $[\alpha]_D +92$ (abs al, c = 1.2), +128 (bz, c = 0.3)	149–50d	...	...	...	δ^h	s	...	...	δ^h	MeOH s AcOEt s	**B6**[2], 970
e306	—,—(*dl*)	$C_{14}H_{14}O_2$. *See* e305	214.27	nd (w or al) ta (eth)	117–8 (anh)	$133^{0.023}$	...	...	i δ^h	v	v	...	...	chl v lig i	**B6**[2], 969
e307	—,—(*l*)	$C_{14}H_{14}O_2$ *See* e305	214.27	lf (eth or abs al), pr (bz or abs al) $[\alpha]_D -92$ (abs al, c = 1.2)	149–50d	...	...	...	...	v	s^h	v	s	MeOH v chl s	**B6**[2], 970
e308	—,—(*meso*)	Tolylene glycol. $C_{14}H_{14}O_2$. *See* e305	214.27	nd or lf (w or bz-peth)	137–8	$139^{0.023}$ $>300^{760}$	...	...	i δ^h	v	...	...	...	lig i	**B6**[2], 967
e309	**—,phenyl-***	Phenyl ethylene glycol. $C_6H_5CHOHCH_2OH$	138.17	nd	67–8	$272–4^{755}$	...	...	v	v	v	...	v	lig δ v^h	**B6**[2], 887
309[1]	**—,tetraphenyl-***	Benzopinacol. $(C_6H_5)_2COHCOH(C_6H_5)_2$	366.46	pr (bz+1)	178–80	...	...	...	...	δ^h	s	...	s^h	chl s lig i CS_2 s	**B6**[2], 1034
—	**1,1-Ethanediol,** 2,2,2-**tribromo-***	*see* **Acetaldehyde, tri-bromo-,** hydrate													
e310	**1,2-Ethanedione, 1,2-di(2-furyl)-**	Bipyromucyl. Furyl. (furan ring)—COCO—(furan ring)	190.16	ye nd	165–6	...	...	...	i	δ	s	...	...	chl s	**B19**[2], 183
—	**—,1,2-diphenyl-***	*see* **Benzil**													
e311	**1,2-Ethanedisul-fonic acid***	Ethylenedisulfonic acid. $HO_3SCH_2CH_2SO_3H$	190.19	nd (gl aa)	104	...	...	...	v	v	...	...	...	...	...
e312	**1,2-Ethanedi-thiol***	Dithioglycol. Ethylene mercaptan. $HSCH_2CH_2SH$	94.20	...	−41.2	146^{760}	1.1192^{25}_{4}	1.5558^{25}	...	s	...	...	...	alk v	**B1**[2], 529
e313	**Ethanedithiolic acid, diethyl** ester*	Diethyl dithioloxalate. $C_2H_5SCOCOSC_2H_5$	178.27	ye nd (eth)	27	235 $80–2^{32}$	1.0565^{21}_{4}	...	...	...	s^h	...	...	...	**B2**[2], 515
e314	**Ethanephos-phonic acid***	Ethylphosphonic acid. $C_2H_5PO_3H_2$	110.05	pl	61–2	...	...	...	v	v	v	...	i	lig i	...
e315	—,diethyl ester*	$C_2H_5PO(OC_2H_5)_2$	166.16	...	...	198^{760}	1.0259^{20}_{4}	1.4163^{20}	δ	s	s	...	...	...	**B4**[2], 975
e316	—,dimethyl ester*	$C_2H_5PO(OCH_3)_2$	138.10	...	...	82^{18}	1.1029^{30}_{4}	1.4128^{30}	s	s	...	...	s	...	**J1954,** 3222
e317	**Ethanesulfinic acid***	Ethylsulfinic acid. $C_2H_5SO_2H$	94.13	syr	...	...	...	...	...	...	...	...	...	alk s	**B4**[2], 524
e318	**Ethanesulfonic acid***	Ethylsulfonic acid. $C_2H_5SO_3H$	110.13	hyg	−17	...	1.3341^{25}	1.4340^{16}	s	s	...	...	...	alk s	**B4**[2], 525
e319	—,chloride*	$C_2H_5SO_2Cl$	128.58	pa ye	...	65^{13} 171^{760}	1.357^{22}	1.4339^{20}	d	d	v	...	...	...	**B4**[2], 526
—	**—,2-amino-***	*see* **Taurine**													
e320	**—,2-bromo-,** chloride	$BrCH_2CH_2SO_2Cl$	207.48	pa ye	...	$119–21^{25}$	1.921^{20}	1.5242^{20}	d	d	...	...	...	...	**B4**[2], 526
e321	**—,1-chloro-,** chloride*	$CH_3CHClSO_2Cl$	163.02	...	...	$72–3^{13}$	...	...	...	...	...	...	...	...	...
e322	**—,2-chloro-*,** chloride	$ClCH_2CH_2SO_2Cl$	163.02	...	...	200–3 $93–7^{17}$	...	...	d	d	...	...	...	...	**B4**[2], 526
e323	**—,2-hydroxy-***	$HOCH_2CH_2SO_3H$	126.13	syr	100d	100d	...	...	∞	∞	...	...	...	...	**B4**[2], 529
e324	**1,1,2,2-Ethane-tetracarboxylic acid,** 1,2-diethyl-ester*	$C_2H_5O_2CCH(CO_2H)CH(CO_2H)CO_2C_2H_5$	262.22	hyg lf (+½w)	132–3d	...	...	...	s	v	v	...	...	chl d CS_2 δ	**B2**, 858
e325	—,tetraamide	Dimalonamide. Ethane-1,1,2,2-tetracarbonamide*. $(H_2NCO)_2CHCH(CONH_2)_2$	202.17	...	230d	...	...	...	δ^h	δ^h	...	...	...	os i	**B2**, 859
e326	—,tetraethyl ester*	Ethyl dimalonate. $(C_2H_5O_2C)_2CHCH(CO_2C_2H_5)_2$	318.33	tetr pr	76	305d	1.064^{80}	1.4105^{80}	...	s	...	...	...	...	**B2**[2], 699
e327	—,tetramethyl ester*	Methyl dimalonate. $(CH_3O_2C)_2CHCH(CO_2CH_3)_2$	262.22	(eth, al, bz)	138	...	...	...	...	v^h	δ	...	s^h	lig i	**B2**[2], 699
e328	**Ethanethiol***	Ethyl hydrosulfide. Ethyl mercaptan. Ethyl thio-alcohol. C_2H_5SH	62.13	...	−144.4	37^{760}	0.8391^{20}_{4} 0.8315^{25}_{4}	1.4351^{25} 1.4306^{20}	δ	s	s	...	...	dil alk s	**B1**[2], 341

For explanations, symbols and abbreviations see beginning of table.

No.	Name	Synonyms and Formula	Mol. wt.	Crystalline form, color and specific rotation	m.p. °C	b.p. °C	Density	n_D	Solubility						Ref.
									w	al	eth	ace	bz	other solvents	
	Ethanethiol														
e329	—,sodium salt	Sodium thioethylate. C_2H_5SNa	84.12	wh cr	d				v	s					**B1**, 341
e330	—,**2-amino-***	$NH_2CH_2CH_2SH$	77.15	(subl)	99–100	subl$_{vac}$ d^{760}			v	s				dil HCl v	**B4^1**, 431
e331	—,**2-chloro-***	$ClCH_2CH_2SH$	96.58			113^{760}	1.1826^{20}_{4}	1.4929^{20}	s	v	v			diox v	**B1^3**, 1381
e332	—,**1-phenyl-(*l*)***	$C_6H_5CH_2CH_2SH$	138.23	$[\alpha]^{20}_{D}$ −8.1 (eth)		199–200				s	s		s		**B6^2**, 445
e333	**1,1,1-Ethanetri-carboxylic acid***	$CH_3C(CO_2H)_3$	162.10	pr	159d				s	s	s				
e334	**1,1,2-Ethanetri-carboxylic acid***	Carboxysuccinic acid. $HO_2CCH_2CH(CO_2H)_2$	162.10	pr (w)	159d				v	v	v		δ^h		**B2^2**, 681
e335	—,**1,2-dihydroxy-***	Desoxalic acid. $HO_2CCHOHCOH(CO_2H)_2$	194.10	hyg (+1w)	d50				v d^h		δ				**B3**, 586
—	**Ethanoic acid***	*see* **Acetic acid**													
e336	**Ethanol***	Alcohol. Ethyl alcohol. Methyl carbinol. Spirit of wine. C_2H_5OH	46.07		−117.3	78.5	0.7893^{20}_{4}	1.3611^{20}	∞	∞	∞	∞		chl ∞ aa ∞	**B1^3**, 1223
e337	—,(*d*)	*o*-Deuteroethanol. Deuter-oxyethane. C_2H_5OD	47.08			78.8^{760}	0.801^{25}_{4}		∞	∞	∞	∞		aa ∞	**B1^3**, 1287
e338	—,**2-allyloxy-**	Glycol monoallyl ether. $CH_2{:}CHCH_2OCH_2CH_2OH$	102.13			$155–61^{760}$	0.9506^{29}_{4}	1.4358^{20}	∞	v			s	MeOH s CCl_4 s	**B1**, 468
e339	—,**1-amino-***	Acetaldehyde-ammonia. $CH_3CH(OH)NH_2$	61.08	rh	97	110 par d			s		δ				
e340	—,**2-amino-***	Colamine. Ethanolamine. $NH_2CH_2CH_2OH$	61.08		10.3	170^{760}	1.0180^{20}_{4}	1.4541^{20}	∞	∞	δ		δ	chl s lig δ glycerol ∞	**B4^2**, 717
—	—,**2-amino-1(3,4-dihydroxy-phenyl)-(*l*)***	*see* **Noradrenaline(*l*)**													
e341	—,**2-amino-1-phenyl-***	$NH_2CH_2CHOHC_6H_5$	137.18	nd	56–7	160			v	s					**B13^2**, 361
e342	—,**1-amino-2,2,2-trichloro-***	Chloral-ammonia. $Cl_3CCH(OH)NH_2$	164.42	nd	63–4	100d			δ	s	s		s		
e343	—,**2(2-amino-ethylamino)-***	$NH_2CH_2CH_2NHCH_2CH_2OH$	104.15	hyg liq		$238–40^{752}$ cor			v	v	δ				**B4**, 286
e344	—,**2(2-amino-phenyl)***	o-$NH_2C_6H_4CH_2CH_2OH$ (structure: benzene ring with NH_2 and CH_2CH_2OH)	137.18	ye in air		$152–3^{6}$			s						**B13^2**, 362
e345	—,**2(4-amino-phenyl)-***	NH_2—C_6H_4—CH_2CH_2OH	137.18	nd (al)	108					δ s^h					**B13^2**, 362
e346	—,**2-benzyloxy-**	Glycol monobenzyl ether. $C_6H_5CH_2OCH_2CH_2OH$	152.19		<−75	256^{760} 138^{15}	1.0640^{20}_{4}	1.5233^{20}	s	s	s				**B6^2**, 413
346^1	—,**1,1-bis(4-chlorophenyl)-***	4,4′-Dichloro-α-methyl-benzylhydrol. Dimite. $(Cl$—C_6H_4—$)_2COHCH_3$	267.16		69–70				i	i	s		s		
e347	—,**bromo-***	Ethylene bromohydrin. $BrCH_2CH_2OH$	124.97			$149–50^{750}$	1.7629^{20}_{4}	1.4915^{20}	∞	∞	∞			oos v lig δ	**B1^2**, 337
e348	—,**2-butoxy-***	Glycol monobutyl ether. $C_4H_9^nOCH_2CH_2OH$	118.18			171^{743}	0.9027^{20}_{4}	1.4177^{26}	∞	∞	∞				**B1^2**, 519
e349	—,**2-butyl-amino-***	$C_4H_9^nNHCH_2CH_2OH$	117.19			199–200^{756}	0.8907^{20}_{4}	1.4437^{20}	v	v	v				**B4**, 283
e350	—,**2-chloro-***	Ethylene chlorohydrin. $ClCH_2CH_2OH$	80.51		−67.5	128^{760} 44^{20}			∞	∞	δ				**B1^2**, 333
e351	—,**2-chloro-1-phenyl-***	Styrene chlorohydrin. $C_6H_5CHOHCH_2Cl$	156.61			128^{17}	1.165^{20}_{4}	1.5538^{25}		s	v				**B6^2**, 446
e352	—,**2(2-chloro-ethoxy)-***	β-Chloroethyl cellosolve. $ClCH_2CH_2OCH_2CH_2OH$	124.57			180–85 $92–100^{12}$		very high	v	∞	∞				**B1^2**, 519
e353	—,**1-cyclohexyl-***	Methylcyclohexylcarbinol. C_6H_{11}—$CHOHCH_3$	128.22			189	0.9278^{19}	1.4661^{19}		v	v				**B6^2**, 27
e354	—,**2-cyclohexyl-***	C_6H_{11}—CH_2CH_2OH	128.22			$207–9^{757}$ $97–9^{15}$	0.9165^{20}_{4}	1.4635^{20}		s	s		s		**B6^2**, 27

For explanations, symbols and abbreviations see beginning of table.

No.	Name	Synonyms and Formula	Mol. wt.	Crystalline form, color and specific rotation	m.p. °C	b.p. °C	Density	n_D	Solubility w	al	eth	ace	bz	other solvents	Ref.
	Ethanol														
e355	**—,2-cyclopentyl-***	—CH_2CH_2OH	114.19			96–7^{24}	0.9180^{20}_{4}	1.4577^{20}	i		s				**B6^2**, 25
e356	**—,2,2-dichloro-***	Cl_2CHCH_2OH	114.96			146^{760}	1.145^{15}_{15}		δ	s	s				**B1**, 338
—	**—,2,2-diethoxy-***	*see* **Acetaldehyde, 2-hydroxy-**, diethyl acetal													
e358	**—,2-diethyl-amino**	β-Hydroxytriethylamine. $(C_2H_5)_2NCH_2CH_2OH$	117.19			163^{760}	0.884^{20}_{4}	1.4400^{25}	∞	s	s	s	s	peth s	**B4^2**, 727
e359	**—,2-dimethyl-amino-***	$(CH_3)_2NCH_2CH_2OH$	89.14			135^{758} cor	0.8866^{20}_{4}	1.43^{20}	∞	∞	∞				**B4^2**, 719
e360	**—,1,2-diphenyl-** (*d*)*	$C_6H_5CH_2CHOHC_6H_5$	198.27	(eth-peth), nd (dil al) $[\alpha]^{60}_{D}+26.1$	67–8	167–70^{10}	1.0358^{70}_{4}		i	v	v				**B6^2**, 637
e361	—,—(*dl*)*	$C_6H_5CH_2CHOHC_6H_5$	198.27	nd (bz-peth)	67	177^{15}									**B6^2**, 638
e362	—,—(*l*)*	$C_6H_5CH_2CHOHC_6H_5$	198.27	nd (eth-peth) $[\alpha]^{20}_{D}-9.4$ (c=10)	67		1.0358^{70}_{4}								**B6^2**, 637
e363	**—,2-ethoxy-***	Ethyl cellosolve. Glycol monoethyl ester. $C_2H_5OCH_2CH_2OH$	90.12			135^{760}	0.9297^{20}_{4}	1.4080^{20}	∞	∞	∞				**B1^2**, 518
e364	**—,2-ethylamino-***	$C_2H_5NHCH_2CH_2OH$	89.14			169–70^{760}	0.914^{20}_{4}	1.444^{20}	v	v	v				**B4^2**, 727
e365	**—,2(ethylthio)-***	$C_2H_5SCH_2CH_2OH$	106.19		ca. −100	184	1.0166^{20}_{4}	1.4867^{20}	δ	s					**B1^3**, 2120
e366	**—,2-fluoro-***	Ethylene fluorohydrin. FCH_2CH_2OH	64.06		−43	103.5	1.040^{20}_{4}	1.3647^{18}	∞	∞	∞				**B1^2**, 333
e367	**—,2-hexyloxy-***	Hexyl cellosolve. $CH_3(CH_2)_5OCH_2CH_2OH$	146.23		−45.1	208^{760} 96^{13}	0.8894^{20}_{20}		δ	v	v				**B1^3**, 2086
e368	**—,2-iodo-***	Ethylene iodohydrine. ICH_2CH_2OH	171.97			176–7d 85–8^{25}	2.1968^{20}_{4} 2.2289^{0}_{4}	1.5713^{20}	s	v	v				**B1^3**, 1363
368^1	**—,isobutoxy-**	Isobutyl cellosolve. $C_4H_9^iOCH_2CH_2OH$	118.18			159^{746}	0.8900^{20}_{4}	1.4143^{20}		v	v				**B1^3**, 2084
e369	**—,2-isobutyl-amino-***	$C_4H_9^iNHCH_2CH_2OH$	117.19			199–200 cor	0.8818	1.4437^{20}	v	v	v				**B4**, 283
e370	**—,2-isopropoxy-***	Isopropyl cellosolve. $C_3H_7^iOCH_2CH_2OH$	104.15			141^{736}	0.9030^{20}_{4}	1.4095^{20}	∞	∞	∞				**B1^3**, 2080
e371	**—,2-isopropyl-amino-***	$C_3H_7^iNHCH_2CH_2OH$	103.17			171^{741}	0.8970^{20}_{4}	1.4395^{20}	∞	∞	∞				**B4**, 282
e372	**—,2-mercapto-***	Monothioethylene glycol. Thioglycol. $HSCH_2CH_2OH$	78.14			157–8^{742}δd 55^{13}	1.1143^{20}_{4}	1.4996^{20}	s	s	s				**B1^2**, 523
e373	**—,2-methoxy-***	Methyl cellosolve. $CH_3OCH_2CH_2OH$	76.10		−85.1	125^{768}	0.9647^{20}_{4}	1.4024^{20}	∞	v	∞		∞		**B1^3**, 2069
e374	**—,1(2-methoxy-phenyl)-***	*o*-Anisylmethylcarbinol. OCH_3 —CHOHCH$_3$	152.19			128^{17}	1.0862^{15}_{4}	1.5379	i	s	s				**B6^2**, 886
e375	**—,1(3-methoxy-phenyl)-***	CH_3O —CHOHCH$_3$	152.19			133^{15}	1.0781^{19}_{4}	1.5325		v	v				**B6^2**, 886
e376	**—,1(4-methoxy-phenyl)-***	CH_3O— —CHOHCH$_3$	152.19			ca 310^{760} δd	1.086^{16}_{4}	1.537		s	s			con sulf s	**B6^2**, 886

For explanations, symbols and abbreviations see beginning of table.

No.	Name	Synonyms and Formula	Mol. wt.	Crystalline form, color and specific rotation	m.p. °C	b.p. °C	Density	n_D	Solubility w	al	eth	ace	bz	other solvents	Ref.
	Ethanol														
e377	**—,2-methylamino-***	$CH_3NHCH_2CH_2OH$	75.11			159^{747} cor	0.937^{20}	1.4385^{20}	∞	∞	∞				**B4²**, 718
—	**—,2-methylamino-1-phenyl-(*l*)***	*see* **Halostachine**													
e378	**—,2(methylphenyl)amino-***	$C_6H_5N(CH_3)CH_2CH_2OH$	151.21	yesh		150^{14}	0.9995^{15}_{0}		s^h	v	v	v	v		**B12²**, 107
e380	**—,2(methylthio)-***	$CH_3SCH_2CH_2OH$	92.16			$68–70^{20}$	1.6640^{20}_{20}	1.4867^{30}	s	v	v				**B1²**, 524
—	**—,morpholinyl-**	*see* **Morpholine, (hydroxyethyl)-**													
e381	**—,1(1-naphthyl)-(*dl*)***	$C_{10}H_7^{\alpha}CHOHCH_3$	172.22	nd (peth)	66	178^{15}			i	s	...	s	s	chl s	**B6²**, 619
e382	**—,—(*l*)***	$C_{10}H_7^{\beta}CHOHCH_3$	172.22	nd. $[\alpha]^{20}_{D}-69.8$ (al)	47	166^{11}	1.1190^{14}_{4}		i	s	...	s	s	chl s	**B6²**, 619
e383	**—,2(2-naphthylamino)-***	$C_{10}H_7^{\beta}NHCH_2CH_2OH$	187.23	lf (eth)	51						v				**B12²**, 717
e384	**—,2-nitro-***	$O_2NCH_2CH_2OH$	91.07		−80	194^{765}	1.270^{15}_{4}	1.4438^{19}	∞	∞	∞	...	i		**B1³**, 1364
e385	**—,2-phenoxy-***	Phenoxytol. $C_6H_5OCH_2CH_2OH$	138.16			137^{25}	1.102^{22}_{4}	1.534^{20}	s	s					
e386	**—,1-phenyl-(*d*)***	$C_6H_5CHOHCH_3$	122.17	$[\alpha]_D+41.77$, +43.4		100^{18}	1.0191^{13}_{4}	1.5211^{20}	...	v	...	...	s	chl s	**B6²**, 444
e387	**—,—(*dl*)***	$C_6H_5CHOHCH_3$	122.17		20	204^{760}	1.0180^{0}_{4}	1.5244^{25}	i	∞	∞				**B6²**, 444
e388	**—,—(*l*)***	$C_6H_5CHOHCH_3$	122.17	$[\alpha]^{20}_{D}-54.9$		$202–4^{760}$	1.018^{0}_{4}		...	v	v				**B6²**, 444
e389	**—,2-phenyl-***	Benzylcarbinol. Phenethyl-alcohol. $C_6H_5CH_2CH_2OH$	122.17		−27fr	219^{760} $99–100^{10}$	1.023^{15}_{4}	1.5267^{18}	δ	∞	∞				**B6²**, 448
e390	**—,2-phenylamino-***	2-Anilinoethanol. $C_6H_5NHCH_2CH_2OH$	137.18			286^{760}	1.1129^{25}_{20}	1.5749^{20}	δ	v	v	...		chl v	**B12²**, 106
—	**—,piperidyl-**	*see* **Piperidine, hydroxyethyl-**													
e391	**—,2-propoxy-***	Propyl cellosolve. $C_3H_7^nOCH_2CH_2OH$	104.15			150^{736}	0.9112^{20}_{4}	1.4133^{20} 1.4112^{26}	s	v	v				
e393	**—,1(3-tolyl)-**	(structure: CH_3 on benzene ring with —CHOHCH₃)	136.19			112^{12}	0.9974^{15}_{4}	1.526^{15}	...	v	v				**B6²**, 478
e394	**—,1(4-tolyl)-**	$CH_3—C_6H_4—CHOHCH_3$	136.19			219^{756}	0.9668^{16}_{4}		i	v	v				**B6²**, 479
—	**—,tolylamino-**	*see* **Amine, ethyl tolyl, hydroxy-**													
e395	**—,2(4-tolylthio)-**	$CH_3—C_6H_4—SCH_2CH_2OH$	168.26			174^{30} 282–3d			...	s	s	...	s	chl δ aa s lig i	**B6²**, 396
e396	**—,2-triazo-***	$N_3CH_2CH_2OH$	87.08			60^{8}, 73^{20}	1.149^{24}_{24}	1.4578^{25}	∞						**B1¹**, 471
e397	**—,2,2,2-tribromo-***	Avertin. Bromethol. Ethobrom. Renarcol. Br_3CCH_2OH	282.79		79–82 δd 70	$92–3^{10}$			δ	s	s	...	s	lig s^h	
e398	**—,2,2,2-trichloro-***	Cl_3CCH_2OH	149.42	hyg rh ta	17.8	151^{737}			δ	∞	∞	...		alk s	**B1³**, 1358
e399	**—,1,1,2-triphenyl***	Benzyldiphenylcarbinol. $(C_6H_5)_2COHCH_2C_6H_5$	274.36	nd (bz-lig) pr (peth)	89–90	222^{11}			i	v	δ	...		peth δ lig δ	**B6²**, 696
e400	**—,2,2,2-triphenyl-***	Tritylcarbinol. $(C_6H_5)_3CCH_2OH$	274.36	(al, eth, lig)	107d				i	s	s	...	s	lig s	**B6²**, 696
e401	**Ethene***	Ethylene. $CH_2{:}CH_2$	28.05	gas, mcl pr	−169.15 fr −181	−104	0.00126 at 0° 760 mm.	1.363^{100}	i	δ	s				**B1¹**, 75
e402	**—,amino-***	Ethenylamine. Vinylamine. $CH_2{:}CHNH_2$	43.07			$55–6^{756}$	0.8321^{24}		∞	s	∞				**B4**, 203
e403	**—,bromo-***	Vinyl bromide. $CH_2{:}CHBr$	106.96		fr −138	16^{750}	1.5167^{14}_{4}	1.4462	i	i	i				**B1²**, 162

For explanations, symbols and abbreviations see beginning of table.

No.	Name	Synonyms and Formula	Mol. wt.	Crystalline form, color and specific rotation	m.p. °C	b.p. °C	Density	n_D	Solubility w	al	eth	ace	bz	other solvents	Ref.
	Ethene														
e404	**—,1-bromo-1,2,2-triphenyl-***	$(C_6H_5)_2C{:}CBrC_6H_5$	335.25	nd (aa)	116–7				i					aa s^h	**B5**[1], 355
e405	**—,chloro-***	Vinyl chloride. $CH_2{:}CHCl$	62.50	gas	−160	-13.9^{760}			δ	s	v				**B1**[2], 157
e406	**—,1-chloro-2-dichloroarsino-***	Lewisite. $ClCH{:}CHAsCl_2$	207.32		0.1	190^{760}d 93^{26}	1.888_4^{20}		i	s				os s alk d	
e407	**—,1-chloro-1,2,2-triphenyl-***	$(C_6H_5)_2C{:}CClC_6H_5$	290.80		117				i		δ			chl v	**B5**[1], 355
e408	**—,1,2-dibromo-** (*cis*)*	$BrCH{:}CHBr$	185.80		−53	112.5^{760}	2.2464^{20}	1.5428^{20}	i	v	v				**B1**[3], 672
e409	—,—(*trans*)*	$BrCH{:}CHBr$	185.80		−6.5	108^{760}	2.2308^{20}	1.5505^{18}							**B1**[3], 672
e410	**—,1,1-dibromo-2-ethoxy-***	$C_2H_5OCH{:}CBr_2$	229.91			$170–2^{747}$ cor	1.7697_4^{18}		i		v				**B1**[2], 273
e411	**—,1,1-dichloro-***	Vinylidene chloride. $CH_2{:}CCl_2$	96.95		−122.1	32^{760}	1.218^{20}	1.4249^{20}	i						**B1**[3], 647
e412	**—,1,2-dichloro-** (*cis*)*	$ClCH{:}CHCl$	96.95		−80.5	60.3^{760}	1.2837_4^{20}	1.4490^{20}	δ	∞	∞				**B1**[3], 651
e413	—,—(*trans*)*	$ClCH{:}CHCl$	96.95		−50	47.5	1.2565_4^{20}	1.4454^{20}	δ		∞				**B1**[3], 651
e414	**—,1,1-difluoro-***	Vinylidine fluoride. $CH_2{:}CF_2$	64.04	gas		<−84			i	s	v				**B1**[3], 638
e415	**—,1,1-dinitro-2,2-diphenyl-***	$(C_6H_5)_2C{:}C(NO_2)_2$	270.24	ye (al)	146−7				s^h	s	s				**B5**[2], 545
e416	**—,1,1-diphenyl-***	$(C_6H_5)_2C{:}CH_2$	180.24		8.2	277^{760}									**B5**[2], 543
—	**—,1,2-diphenyl-***	*see* **Stilbene**													
e417	**—,fluoro-***	Vinyl fluoride. $CH_2{:}CHF$	46.04	gas		−51			i	s		s			**B1**[2], 77
e418	**—,iodo-***	Vinyl iodide. $CH_2{:}CHI$	153.95			56	2.037^{20}	1.5385	i	∞	∞				**B1**, 199
e419	**—,tetrabromo-***	$Br_2C{:}CBr_2$	343.66	pl (dil al), nd (al)		56.5	226–7		i	s				chl v	
e420	**—,tetrachloro-***	Perchloroethylene. $Cl_2C{:}CCl_2$	165.83		−22	121^{760}	1.623_4^{20}	1.5044^{20}	i	∞	∞		∞		**B1**[3], 664
e421	**—,tetrafluoro-***	Perfluoroethylene. $F_2C{:}CF_2$	100.02	gas	−142.5	-76.3^{760}	$1.519^{76.3}$		i						**B1**[3], 638
e422	**—,tetraiodo-***	Periodoethylene. I_2CCI_2	531.64	lemon ye lf, pr (eth)	190	sub	2.983^{20}		i	δ	δ		s	CS_2 v chl s	**B1**[3], 676
e423	**—,tetraphenyl-***	$(C_6H_5)_2C{:}C(C_6H_5)_2$	332.45	mcl or rh (bz-eth) or (chl-al)	223–4	$415–25^{760}$	1.155_4^{0}		i	δ	δ		v^h		**B5**[2], 679
e424	**—,tribromo-***	$BrCH{:}CBr_2$	264.76			$163–4^{760}$	$2.708_4^{20.5}$	1.6045^{16}							**B1**[2], 164
e425	**—,trichloro-***	$ClCH{:}CCl_2$	131.39		−88	87^{760}	1.462_4^{20}	1.4784^{20}	δ	∞	∞				**B1**[3], 656
e426	**—,triphenyl-***	$(C_6H_5)CH{:}C(C_6H_5)_2$	256.35	lf (al or MeOH)	72–3	$220–1^{14}$	1.0373_4^{78}	1.6292^{78}	i	s	v			MeOH s	**B5**[2], 630
e427	**Ethenetetra carboxylic acid,** tetraethyl ester*	$(C_2H_5O_2C)_2C{:}C(CO_2C_2H_5)_2$	316.31	tcl pinacoidal pr (eth)	58	325–8 par d 210^{22}			i	v	v				**B2**[2], 709
e428	**Ether, allyl butyl, 3′-methyl-**	Allyl isoamyl ether. $(CH_3)_2CHCH_2CH_2OCH_2CH{:}CH_2$	128.22			120–2			i	∞	∞				**B1**[2], 477
e429	**—,allyl ethyl**	3-Ethoxypropene*. $C_2H_5OCH_2CH{:}CH_2$	86.13			66^{761}	0.7651_4^{20}	1.3881^{20}	i	∞	∞				**B1**[3], 1881
e430	**—,allyl isopropyl**	$C_3H_7^iOCH_2CH{:}CH_2$	100.16			83–4	0.7764^{20}	1.3946^{20}			∞				**B1**[3], 1882
e431	**—,allyl methyl**	$CH_3OCH_2CH{:}CH_2$	72.11			$42.5–3^{757}$	0.77_{11}^{11}	1.3803	i	∞	∞				**B1**[3], 1881
e432	**—,allyl 2-naphthyl**	$C_{10}H_7^{\beta}OCH_2CH{:}CH_2$	184.24			d 210			i						**B6**[1], 313
e433	**—,allyl phenyl**	$C_6H_5OCH_2CH{:}CH_2$	134.18			191.7^{760}	0.9832_4^{15}	1.5218^{20}	i	s	∞				**B6**[2], 146
e434	**—,—,4′-chloro-**	$Cl–C_6H_4–OCH_2CH{:}CH_2$	168.62			d^{760} $106–7^{12}$	1.131^{15}		i	s	s		s		**B6**[1], 101
434[1]	**—,—,2′,4′,6′-tribromo-**	$Br–C_6H_2Br_2–OCH_2CH{:}CH_2$ (Br at 2,6)	370.88	nd (al)	33–4										**B6**[2], 194
e435	**—,allyl propyl**	$CH_3CH_2CH_2OCH_2CH{:}CH_2$	100.16			90–2	0.7670_4^{20}	1.3919^{20}		s	∞				**B1**[3], 1882
e436	**—,allyl 2-tolyl**	$CH_3C_6H_4–OCH_2CH{:}CH_2$ (2-CH_3)	148.21			205–8 85^{12}	0.9698_4^{15}	1.5188^{15}							**B6**[2], 329

For explanations, symbols and abbreviations see beginning of table.

No.	Name	Synonyms and Formula	Mol. wt.	Crystalline form, color and specific rotation	m.p. °C	b.p. °C	Density	n_D	Solubility w	al	eth	ace	bz	other solvents	Ref.
	Ether														
e437	**—,allyl 3-tolyl**	CH_3-(3-position) C_6H_4—$OCH_2CH{:}CH_2$	148.21			211–4	0.965^{15}_{4}								**B6**[1], 178
e438	**—,allyl 4-tolyl**	CH_3—C_6H_4—$OCH_2CH{:}CH_2$	148.21			214.5^{760}	0.9728^{15}_{15}								**B6**[2], 377
e439	**—,benzyl butyl**	$C_6H_5CH_2O(CH_2)_3CH_3$	164.25			220.5^{744}	0.9310^{10}_{4}		i	∞	∞				**B6**[2], 410
e440	**—,—,2′-methyl-(*d*)**	$C_6H_5CH_2OCH_2CH(CH_3)CH_2CH_3$	178.28	$[\alpha]^{22}_{D}+1.82$		231^{722}	0.911^{22}_{4}	1.4854^{22}	i						**B6**, 431
e441	**—,—,3′-methyl-**	Benzyl isoamyl ether. $C_6H_5CH_2OCH_2CH_2CH(CH_3)_2$	178.28			$236–7^{748}$	0.9200^{10}_{4}		i				δ		**B6**[2], 410
e442	**—,benzyl ethyl**	α-Ethoxytoluene. $C_6H_5CH_2OC_2H_5$	136.20			186	0.9490^{20}_{4}	1.4955^{20}	i	∞	∞				**B6**[2], 409
e443	**—,benzyl isobutyl**	$C_6H_5CH_2OCH_2CH(CH_3)_2$	164.25			$211–2^{743}$	0.9250^{10}_{4}		i		v			chl v	**B6**[2], 410
e444	**—,benzyl methyl**	α-Methoxytoluene. $C_6H_5CH_2OCH_3$	122.17			174	0.9643^{25}_{25}		i	v	v		s	lig i	**B6**[2], 409
e445	**—,—,1′chloro-**	$C_6H_5CH_2OCH_2Cl$	156.60			103^{13}		1.5270^{17}							**B6**[2], 414
e446	**—,benzyl 1-naphthyl**	$C_6H_5CH_2OC_{10}H_7^{\alpha}$	234.30	(al)	77–8 d				i	δ v^h					**B6**[2], 579
e447	**—,benzyl 2-naphthyl-**	$C_6H_5CH_2OC_{10}H_7^{\beta}$	234.30	lf (al)	99				i	s	s		s	chl s	**B6**[2], 599
e448	**—,butyl ethenyl**	Butoxyethene.* $CH_3(CH_2)_3OCH{:}CH_2$	100.16		−92	93.8^{760}	0.7742^{25}_{4}	1.4017^{20}	i					oos ∞. glycol δ glycerol δ	**B1**[3], 1860
e449	**—,—,3-methyl-**	Isoamyl vinyl ether. $(CH_3)_2CHCH_2CH_2OCH{:}CH_2$	114.19			$112–3^{760}$	0.7826^{20}_{4}	1.4072^{20}		s	v				**B1**[3], 1863
e450	**—,butyl ethyl**	1-Ethoxybutane*. $CH_3(CH_2)_3OC_2H_5$	102.18		−124	92^{760}	0.7490^{20}_{4}	1.3818^{20}	i	∞	∞				**B1**[3], 1502
e451	**—,*sec*-butyl ethyl**	$CH_3CH(C_2H_5)OC_2H_5$	102.18			81^{760}	0.7503^{20}_{4}	1.3802^{20}			v				**B1**[3], 1533
e452	**—,*tert*-butyl ethyl**	$(CH_3)_3COC_2H_5$	102.18		−94	70^{758}	0.7519^{25}	1.3794^{20}	i	v	v				**B1**[3], 1577
e453	**—,butyl ethyl, 2′-chloro-**	$CH_3(CH_2)_3OCH_2CH_2Cl$	136.63			$49–50^{11}$ d^{760}	0.9335^{20}_{4}	1.4155^{20}			s				
e454	**—,—,3-methyl-**	Ethyl isoamyl ether. $(CH_3)_2CHCH_2CH_2OC_2H_5$	116.21			112	0.7695^{21}_{15}		i	∞	∞				**B1**[3], 432
e454[1]	**—,*sec*-butyl ethyl, 2-methyl**	Ethyl *tert*-amyl ether. $CH_3CH_2C(CH_3)_2OC_2H_5$	116.21			101	0.7657^{20}_{4}	1.3912^{20}	δ	v	∞				**B1**, 389
e455	**—,butyl ethynyl**	Butoxyacetylene. $CH_3(CH_2)_3OC{:}CH$	98.15			50.5^{110} $102–4^{760}$ exp ca. 100	0.8120^{20}_{4}	1.4053^{20}	d^h	v					**B1**[3], 1969
e456	**—,butyl furfuryl**	(furan ring)—$CH_2O(CH_2)_3CH_3$	154.21			189.90^{765}	0.9516^{20}_{4}	1.4522^{20}	i	s	v				**B17**[2], 114
e457	**—,butyl isobutyl**	$CH_3(CH_2)_3OCH_2CH(CH_3)_2$	130.22			131–2	0.763^{16}		i	s	∞				**B1**, 376
e458	**—,butyl isopropyl**	$CH_3(CH_2)_3OCH(CH_3)_2$	116.20		108^{738}	0.7594			i	s				os s con sulf s	**B1**[3], 1503
458[1]	**—,butyl methyl**	$CH_3(CH_2)_3OCH_3$	88.15		−115.5	71^{760}	0.7443^{20}_{4}	1.3736^{20}	i	∞	∞				**B1**[2], 395
e459	**—,*sec*-butyl methyl**	$CH_3CH_2CH(CH_3)OCH_3$	88.15			59cor	0.7415^{20}_{4}		δ	v	v				**B1**[3], 1532
459[1]	**—,—,2-methyl-**	$CH_3CH_2C(CH_3)_2OCH_3$	102.17			86.3	0.7703^{20}_{4}	1.3885^{20}	δ	v	∞			MeOH v	**B1**, 389
e460	**—,*tert*-butyl methyl**	$(CH_3)_3COCH_3$	88.15		−109	55.2^{760} cor			s	v	v				**B1**[3], 1576
e461	**—,butyl methyl, 3-methyl-**	Isoamyl methyl ether. $(CH_3)_2CHCH_2CH_2OCH_3$	102.17			91^{765}	0.6871^{11}_{4}			s	s				**B1**[2], 432

For explanations, symbols and abbreviations see beginning of table.

No.	Name	Synonyms and Formula	Mol. wt.	Crystalline form, color and specific rotation	m.p. °C	b.p. °C	Density	n_D	Solubility w	al	eth	ace	bz	other solvents	Ref.
	Ether														
e462	—,**butyl propyl**	$CH_3(CH_2)_3OCH_2CH_2CH_3$	116.20			117.1	0.7773^{0}_{0}		i	v	v				**B1**[3], 1503
e463	—,—,**3-methyl-**	Isoamyl propyl ether. $(CH_3)_2CHCH_2CH_2OCH_2CH_2CH_3$	130.23			130			i	s	∞				**B1**[2], 432
e464	—,**butyl 2-tolyl**	CH_3, $O(CH_2)_3CH_3$ (structure)	164.25			223	0.9437^{0}_{0}								**B6**, 353
e465	—,**2-butynyl methyl**	$CH_3C{:}CCH_2OCH_3$	84.12			$99–100^{760}$	0.8496^{20}_{4}	1.4262^{20}	i	s	s		s		**B1**[3], 1973
e466	—,**cyclohexyl 2-furyl**	(structure)	166.22			$118–9^{28}$	1.0200^{28}_{4}	1.4861^{28}							
e467	—,**cyclohexyl methyl**	Hexahydroanisole. (structure, $-OCH_3$)	114.19		−74.37	133^{760}	0.8790^{20}_{4}	1.4355^{20}			s			MeOH s	**B6**[2], 9
e468	—,—,**2-bromo-**	(structure, Br, $-OCH_3$)	193.09			$78–9^{12}$	1.3257^{20}_{4}	1.4871^{20}	i		s			MeOH s	**B6**[2], 13
e469	—,**diallyl**	Allyl ether. $CH_2{:}CHCH_2OCH_2CH{:}CH_2$	98.15			94	0.8260^{20}_{4}		i	∞	∞				**B1**[2], 477
e470	—,**dibenzyl**	Benzyl ether. $C_6H_5CH_2OCH_2C_6H_5$	198.27		3.60	298^{760} 184^{28}	1.0428^{20}_{4}		i	∞	∞				**B6**[2], 412
e471	—,**dibutyl**	Butyl ether. $CH_3(CH_2)_3O(CH_2)_3CH_3$	130.23		−97.9	142^{760}	0.7725^{15}_{4}	1.4010^{15}	i	∞	∞				**B1**[2], 396
e472	—,**di-*sec*-butyl**	*sec*-Butyl ether. $CH_3CH(C_2H_5)OCH(C_2H_5)CH_3$	130.23			$120–1^{760}$	0.756^{21}	1.393^{25}	i	∞	∞				**B1**[2], 402
e473	—,**dibutyl, 3,3′-dimethyl-**	Isoamyl ether $(CH_3)_2CHCH_2CH_2OCH_2CH_2CH(CH_3)_2$	158.29			$172–3^{760}$	0.7777^{20}_{4}	1.4085^{20}	i	∞	∞			chl ∞	**B1**[3], 1638
e474	—,—,**3-methyl-**	Butyl isoamyl ether. $CH_3(CH_2)_3OCH_2CH_2CH(CH_3)_2$	144.26			157^{756}			i	s	∞				**B1**, 401
e475	—,**diethenyl**	Ethenyloxyethene*. Vinyl ether. $CH_2{:}CHOCH_2{:}CH$	70.09			39	0.773^{20}_{4}	1.3989^{20}		∞	∞	∞		chl ∞	**B1**[2], 473
e476	—,—,**hexachloro-**	$Cl_2C{:}CClOCCl{:}CCl_2$	276.76			210	1.654^{21}								**B1**, 725
e477	—,**diethyl**	Ether. Ethyl ether. Ethoxyethane*. $CH_3CH_2OCH_2CH_3$	74.12		fr −116.2	34.6	0.714^{20}_{20}	1.3526^{20}	s	∞			∞	chl ∞ oils ∞ lig ∞	**B1**[2], 311
e478	—,—,borofluoride	$(C_2H_5)_2O.BF_3$	141.93		−60.4	$125–6^{760}$ d 60^{20}	1.125^{25}_{4}	1.348^{20}	d	d	s				**B1**[3], 1308
e479	—,—,**2-bromo-**	$BrCH_2CH_2OC_2H_5$	153.02	ye nd (al)		$127–8^{755}$	1.3572^{20}_{4}	1.447^{20}	δ	∞	∞				**B1**[3], 1361
e480	—,—,**1-chloro-**	$CH_3CHClOC_2H_5$	108.57			98 par d	0.950^{20}_{4}	1.4053^{20}	d	d^h					
e481	—,—,**2-chloro-**	$ClCH_2CH_2OC_2H_5$	108.57			107–8	0.9894^{20}_{4}	1.4113^{20}							**B1**[3], 1349
e482	—,—,**decachloro-**	Perchloroether. $Cl_3CCCl_2OCCl_2CCl_3$	418.57	oct. double pym, tetr sc	69	d									**B2**, 210
e483	—,—,**2,2′-di-benzoxy-**	Oxybis(β-ethylbenzoate). $C_6H_5CO_2CH_2CH_2OCH_2CH_2O_2CC_6H_5$	314.34			$279–81^{24}$	1.1701^{15}_{15}		s	s					**B9**[2], 108
e484	—,—,**1,2-dibromo-**	$BrCH_2CHBrOC_2H_5$	231.93			80^{20}	1.7320^{20}_{4}	1.5044^{20}		s				chl s	**B1**, 625
—	—,—,**1,1′-di-carboxy-**	*see* **Dilactic acid**													
e485	—,—,**1,1′-di-chloro-**	$CH_3CHClOCHClCH_3$	143.01			116–7	1.1376^{12}			v	∞				**B1**, 607
e486	—,—,**1,2-dichloro-**	$ClCH_2CHClOC_2H_5$	143.01			140–5	1.1870^{20}_{4}	1.4435^{20}	d	v	v				**B1**[2], 676
e487	—,—,**1,2′-di-chloro-**	$ClCH_2CH_2OCHClCH_3$	143.01			d^{760} $55–7^{17}$	1.1823^{19}_{19}	1.4497^{16}	d	v	v				**B1**[2], 674
e488	—,—,**2,2′-di-chloro-**	$ClCH_2CH_2OCH_2CH_2Cl$	143.01		−24.5	178^{760}	1.2199^{20}_{4}	1.4575^{20}	i $δ^h$		s		∞	MeOH ∞ oos s	**B1**[3], 1349

For explanations, symbols and abbreviations see beginning of table.

No.	Name	Synonyms and Formula	Mol. wt.	Crystalline form, color and specific rotation	m.p. °C	b.p. °C	Density	n_D	Solubility w	al	eth	ace	bz	other solvents	Ref.
	Ether														
—	—,—,**2,2′-dihydroxy-**	*see* **Diethylene glycol**													
e489	—,—,**2,2′-diphenoxy-**	$C_6H_5OCH_2CH_2OCH_2CH_2OC_6H_5$	258.32	nd (dil al)	66–7					s					**B6²**, 150
e490	—,—,**1,1′-diphenyl-**	$C_6H_5CH(CH_3)OCH(CH_3)C_6H_5$	226.32			280.2 167–8^{23}	1.0058^{15}_{4}		i		s			chl s	**B6²**, 445
e491	—,—,**2,2′-diphenyl-**	$C_6H_5CH_2CH_2OCH_2CH_2C_6H_5$	226.32	vt-bl flr		317–20^{760} 194.5^{20}	1.0178^{15}		i		s			chl s	**B6²**, 450
—	—,—,**2-methoxy-**	*see* **Ethane, 1-ethoxy-2-methoxy-***													
e492	—,—,**1,1′,2,2,2,2′,2′,2′-octachloro-**	$Cl_3CCHClOCHClCCl_3$	349.68	(al or MeOH)	40–2	130–1^{11}			i	δ			v	peth v MeOH v	**B1²**, 681
e493	—,**difurfuryl**	Furfuryl ether. (furan)—CH_2OCH_2—(furan)	178.19			101^{2} 88–9^{1}	1.1405^{20}_{4}	1.5088^{20}	i						**B17²**, 116
e494	—,**diheptyl**	Heptyl ether. $CH_3(CH_2)_6O(CH_2)_6CH_3$	214.39			258.5^{769}	0.8008^{20}_{4}	1.4275^{20}	i	s	s				**B1³**, 1683
e495	—,**dihexadecyl**	Dicetyl ether. Hexadecyl ether. $CH_3(CH_2)_{15}O(CH_2)_{15}CH_3$	466.88	lf (al)	55	270 d			i	s	s				**B1²**, 467
e496	—,**dihexyl**	Hexyl ether. $CH_3(CH_2)_5O(CH_2)_5CH_3$	186.35			223^{763}	0.7936^{20}_{4}	1.4204^{20}	i		s				**B1³**, 1656
e496^{1}	—,**diisobutyl, α,β-dichloro-**	$(CH_3)_2CClCHClOC_4H_9^i$	199.12			192.5 83^{15}	1.031^{5}_{4}							os v	**B1**, 675
e497	—,**diisopropyl**	Isopropyl ether. $(CH_3)_2CHOCH(CH_3)_2$	102.18		−85.89	69^{761}	0.7241^{20}_{4}	1.3679^{20}	δ	∞	∞			dil sulf i	**B1³**, 1459
e498	—,—,**β,β′-dichloro-**	$ClCH_2CH(CH_3)OCH(CH_3)CH_2Cl$	171.07			187^{760}	1.103^{20}_{4}	1.4505^{20}	i					oos ∞	**B1³**, 1470
e499	—,**dimethyl**	Methoxymethane*. Methyl ether. CH_3OCH_3	46.07	gas	−138	−24.9^{760}			s	s	s				**B1³**, 1188
e500	—,—,**borofluoride**	$(CH_3)_2O.BF_3$	113.88		−14, −12	127^{760} d	1.2348^{25}_{4}	1.302^{20}	d	d					**B1³**, 1192
e501	—,—,**chloro-**	$ClCH_2OCH_3$	80.51		−103.5	59.15^{760}	1.0605^{20}_{4}	1.3974^{20}	d	s	s				**B1²**, 645
e502	—,—,**1,1′-dichloro-**	$ClCH_2OCH_2Cl$	114.96			104^{760}	1.328^{15}_{4}	1.435^{21}	d	∞	∞				**B1²**, 646
e503	—,—,**1,1,1′,1′-tetraphenyl-**	Benzhydryl ether. $(C_6H_5)_2CHOCH(C_6H_5)_2$	350.46	mcl (bz)	110	315^{745} d 267^{15}				δ^{h}	δ		v		**B6²**, 634
e504	—,—,**1,1,1-tri-2-tolyl-**	(CH_3-phenyl)$_3COCH_3$	316.45		108				i	s				MeOH s	**Am 73**, 3644
e505	—,**1,1′-dinaphthyl**	α-Naphthyl ether. $C_{10}H_7^{\alpha}OC_{10}H_7^{\alpha}$	270.33	lf (al or al-eth)	110	280–5^{23}			i	δ s^{h}	s		s	aa δ	**E12B**, 1209
e506	—,**1,2′-dinaphthyl**	α,β′-Naphthyl ether. $C_{10}H_7^{\alpha}OC_{10}H_7^{\beta}$	270.33	nd (al or al-eth)	81	264^{15}			i	δ	s	s	s	chl s	**E12B**, 1305
e507	—,**2,2′-dinaphthyl**	β-Naphthyl ether. $C_{10}H_7^{\beta}OC_{10}H_7^{\beta}$	270.33	nd or lf (al)	105	250^{19}			i	δ s^{h}	v		v	aa δ	**E12B**, 1305
e508	—,**dioctyl**	Octyl ether. $CH_3(CH_2)_7O(CH_2)_7CH_3$	242.45			286–7	0.8204^{0}_{0}		δ	s	s				**B1**, 419
e509	—,**dipentyl**	Amyl ether. Amyl oxide. $CH_3(CH_2)_4O(CH_2)_4CH_3$	158.29		−69	190^{760} 70^{12}	0.774^{20}_{4}		i	∞	∞				**B1²**, 417
e510	—,**diphenyl**	Phenyl ether. 3 2 — 2′ 3′ 4 (phenyl)—O—(phenyl) 4′ 5 6 — 6′ 5′	170.21	mcl, rh	27–8	258–9^{760}	1.0863^{20}	1.5780^{27}	i	s	s		s		**B6**, 146
—	—,—,**amino-**	*see* **Aniline, phenoxy-**													
e511	—,—,**4-bromo-**	$C_{12}H_9BrO$. *See* e510	249.11		18	305 165^{16}	1.4225^{19}_{1}	1.6088^{19}	i		s				**B62**, 185

For explanations, symbols and abbreviations see beginning of table.

No.	Name	Synonyms and Formula	Mol. wt.	Crystalline form, color and specific rotation	m.p. °C	b.p. °C	Density	n_D	Solubility w	al	eth	ace	bz	other solvents	Ref.
	Ether														
e512	—,—,**4,4′-dibromo-**	$C_{12}H_8Br_2O$. *See* e510	328.01	lf (al)	60.5	338–40 210^{11}	...	...	i	s	v	...	s	aa δ	**B6²**, 185
e513	—,—,**4,4′-dichloro-**	$C_{12}H_8Cl_2O$. *See* e510	239.10	nd (al)	20.9	312–4	1.1231^{20}	1.611	i	...	...	...	...	...	**B6²**, 176
e514	—,—,**2,2′-dihydroxy-**	*o*-Diphenol ether. 2,2′-Oxydiphenol. $C_{12}H_{10}O_3$. *See* e510	202.21	nd (w)	121	...	...	...	δ^h	s	...	...	s	lig δ	**B6**, 773
e515	—,—,**4,4′-dihydroxy-**	*p*-Diphenol ether. 4,4′-Oxydiphenol. $C_{12}H_{10}O_3$. *See* e510	202.21	lf (w)	160–1	...	...	...	δ s^h	v	v	...	v^h	...	**B6**, 845
e516	—,—,**2,2′-dimethoxy-**	*o*-Anisyl ether. $C_{14}H_{14}O_3$. *See* e510	230.27	pl (lig)	78	330–1	...	...	i	v	v	...	...	lig i s^h	**B6**, 773
e517	—,—,**2,3′-dimethoxy-**	$C_{14}H_{14}O_3$. *See* e510	230.27	...	33–4	326–9	...	...	...	s	s	...	s	lig δ	**B6²**, 816
e518	—,—,**3,3′-dimethoxy-**	$C_{14}H_{14}O_3$. *See* e510	230.27	pa br liq	...	332–4	...	...	...	δ	v	...	v	lig δ	**B6²**, 816
e519	—,—,**2,2′-dinitro-**	$C_{12}H_8N_2O_5$. *See* e510	260.21	pl or nd (al)	114–5	...	...	...	...	δ v^h	...	...	...	...	**B6**, 219
e520	—,—,**2,4-dinitro-**	$C_{12}H_8N_2O_5$. *See* e510	260.21	pl (al), nd (al-ace)	70 cor	$230–50^{27}$	...	...	δ	v	s	...	...	...	**B6²**, 242
e521	—,—,**2,4′-dinitro-**	$C_{12}H_8N_2O_5$. *See* e510	260.21	nd (al)	103.5	...	...	...	...	δ s^h	...	...	...	...	**B6²**, 222
e522	—,—,**2,6-dinitro-**	$C_{12}H_8N_2O_5$. *See* e510	260.21	lf (al)	99–100	...	...	...	...	δ s^h	...	...	...	...	**B6²**, 245
e523	—,—,**3,4-dinitro-**	$C_{12}H_8N_2O_5$. *See* e510	260.21	pa ye	89	...	...	...	...	...	s	...	...	...	**B6¹**, 127
e524	—,—,**4,4′-dinitro-**	$C_{12}H_8N_2O_5$. *See* e510	260.21	ye nd (al)	142–3	...	...	...	i	δ s^h	δ	...	...	aa-bz s	**B6²**, 222
e525	—,—,**2-methoxy-**	$C_{13}H_{12}O_2$. *See* e510	200.24	(MeOH), nd (lig)	79	288 $91–2^7$	...	...	i	s	s	...	s	MeOH s^h lig s^h	**B6²**, 781
e526	—,—,**2-nitro-**	$C_{12}H_9NO_3$. *See* e510	215.21	ye	<−20	$183–5^8$	1.2539^{22}	1.575^{20}	i	s	s	...	...	...	**B6²**, 222
e527	—,—,**4-nitro-**	$C_{12}H_9NO_3$. *See* e510	215.21	pl	61 (56)	$188–90^8$	...	...	i	δ	s	...	s	...	**B6²**, 210
e528	—,**dipropyl**	Propyl ether. $CH_3CH_2CH_2OCH_2CH_2CH_3$	102.18	...	fr −122	91^{760}	0.7360^{20}_{4}	1.3832^{14}	δ	∞	∞	...	...	...	**B1²**, 367
e529	—,—,**1,2-dichloro-**	$CH_3CHClCHClOC_3H_7^n$	171.07	...	...	176^{760}	1.129^{15}_{4}	1.447^{16}	i	...	s	...	...	...	**B1¹**, 334
e530	—,—,**1,3-dichloro-**	$ClCH_2CH_2CHClOC_3H_7^n$	171.07	...	...	56^{12}	1.112^{20}	1.4423^{20}	i	s	s	...	...	...	**B1²**, 690
e531	—,—,**2,2′-dichloro-**	$CH_3CHClCH_2OCH_2CHClCH_3$	171.07	...	...	188^{762}	1.109^{20}_{4}	1.4467^{20}	...	s	s	...	...	...	**B1²**, 370
e532	—,—,**3,3′-dichloro-**	$Cl(CH_2)_3O(CH_2)_3Cl$	171.07	...	...	215^{745}	1.140^{20}_{20}	...	...	s	s	...	...	...	**B1²**, 370
e534	—,**ethenyl ethyl**	Ethyl vinyl ether. $C_2H_5OCH{:}CH_2$	72.11	...	−115.8	$35–6^{760}$	0.7589^{20}_{4}	1.3767^{20}	δ	s	∞	...	...	...	**B1³**, 1857
e535	—,—,**2′-chloro-**	$ClCH_2CH_2OCH{:}CH_2$	106.55	...	...	108^{760}	1.0475^{20}_{4}	1.4378^{20}	...	v	v	...	...	...	**B1³**, 1859
e536	—,—,**1,2-dichloro-**	$C_2H_5OCCl{:}CHCl$	141.00	...	...	128	1.1972^{25}_{4}	1.4558^{17}	d^h s	...	...	...	...	...	**B1²**, 780
e537	—,**ethenyl isobutyl**	Isobutyl vinyl ether. $(CH_3)_2CHCH_2OCH{:}CH_2$	100.16	...	−112	83^{760}	0.7645^{20}_{4}	1.3966^{20}	δ	...	...	...	...	oos ∞ glycol δ glycerol δ	**B1³**, 1862
e538	—,**ethenyl isopropyl**	Isopropyl vinyl ether. $(CH_3)_2CHOCH{:}CH_2$	86.13	...	...	$55–6^{760}$	0.7534^{20}_{4}	1.3840^{20}	...	v	v	...	...	...	**B1³**, 1859
e539	—,**ethenyl methyl**	Methoxyethene*. Methyl vinyl ether. $CH_3OCH{:}CH_2$	58.08	...	−122	8^{760}	0.7725^{0}_{4}	1.3730^{0}	δ	v	...	...	...	os v glycol i glycerol i	**B1³**, 1857
e540	—,**ethenyl phenyl**	Phenyl vinyl ether. $C_6H_5OCH{:}CH_2$	120.15	...	...	155–6	0.9770^{20}_{4}	1.5224^{20}	...	...	...	...	...	...	**B6²**, 146
e541	—,**ethyl ethynyl**	Ethoxyacetylene. $C_2H_5OC\vdots CH$	70.09	...	...	50^{760} exp 100	0.7929^{20}_{4}	1.3812^{20}	...	v	v	...	v	...	**B1³**, 1968
e542	—,**ethyl furfuryl**	O —$CH_2OC_2H_5$	126.16	...	...	$149–50^{770}$	0.9844^{20}_{4}	1.4523^{20}	i	s	s	...	...	...	**B17²**, 114
e543	—,**ethyl heptyl**	$CH_3(CH_2)_6OC_2H_5$	144.26	...	...	166.6	0.790^{16}_{4}	1.4111^{20}	i	s	s	...	...	...	**B1³**, 1682
e544	—,**ethyl hexyl**	$CH_3(CH_2)_5OC_2H_5$	130.23	...	...	$142–3^{773}$	0.7722^{20}_{4}	1.4008^{20}	i	v	v	...	...	...	**B1³**, 1656

For explanations, symbols and abbreviations see beginning of table.

No.	Name	Synonyms and Formula	Mol. wt.	Crystalline form, color and specific rotation	m.p. °C	b.p. °C	Density	n_D	Solubility w	al	eth	ace	bz	other solvents	Ref.
	Ether														
e545	**—,ethyl isobutyl**	$(CH_3)_2CHCH_2OC_2H_5$	102.18			81^{760}	0.751^{20}_{4}	1.3739^{25}	i	∞	∞				**B1**³, 1559
e546	**—,ethyl isopropyl.**	$(CH_3)_2CHOC_2H_5$	88.15			53–4	0.720^{25}_{4}		s	∞	∞				**B1**³, 1458
e547	**—,ethyl methyl**	$CH_3OC_2H_5$	60.10			7^{760}	0.7252^{0}_{0}		s	∞	∞				**B1**³, 1288
e548	**—,—,1′bromo-**	$BrCH_2OC_2H_5$	139.00			109^{746}	1.4402^{20}_{4}	1.4515^{20}			v				**B1**², 647
e549	**—,—,1-chloro-**	$CH_3OCHClCH_3$	94.54			$72–3^{751}$	0.9909^{20}_{4}	1.4004^{20}	d		v				**B1**³, 2654
e550	**—,—,1′-chloro-**	$ClCH_2OC_2H_5$	94.54			83^{763}	1.0372^{0}_{4}	1.4040^{20}		s	v				**B1**², 645
e551	**—,—,2-chloro-**	$CH_3OCH_2CH_2Cl$	94.54			92–3	1.0345^{20}_{4}	1.4111^{20}	s		v				**B1**², 335
e552	**—,—,1′-di-methylamino-**	$(CH_3)_2NCH_2OC_2H_5$	103.17			136^{730}			i	v				os v	**B4**², 598
—	**—,—,2-methyl-amino**	*see* **Ethane, 1-ethoxy-2-methylamino-**													
e553	**—,ethyl octyl**	$CH_3(CH_2)_7OC_2H_5$	158.29		12.5	182–4 $72–3^{8}$	0.794^{17}_{4}		i	s				AcOEt s	**B1**³, 1708
e554	**—,ethyl pentyl**	Amyl ethyl ether. $CH_3(CH_2)_4OC_2H_5$	116.21			117–8	0.7622^{20}_{4}		δ	∞	∞				**B1**³, 1602
e555	**—,—,1-chloro-**	$CH_3(CH_2)_4OCHClCH_3$	150.65			63^{8}	0.9200^{20}_{4}	1.4218^{20}	d		s				
e556	**—,ethyl phenyl, 2-bromo-**	$C_6H_5OCH_2CH_2Br$	201.08		39	240–50d $125–30^{20}$			i	v	v				**B6**², 145
e557	**—,ethyl propyl**	$CH_3CH_2CH_2OC_2H_5$	88.15		<−79	63.6^{760}	0.7330^{20}_{4}	1.3695^{20}	s	∞	∞			aa ∞	**B1**³, 1414
e558	**—,—,1-chloro-**	$C_3H_7^nOCHClCH_3$	122.60			$112–5^{731}$ δd	0.9322^{20}_{4}	1.4013^{20}	d		s				**B1**, 607
e560	**—,—,2′,3′-epoxy-**	$CH_2—CHCH_2OC_2H_5$ [ring O]	102.14			128^{760}	0.9646^{12}		s		s				**B17**, 105
e561	**—,ethyl 1-propynyl**	Methylethoxyacetylene. $CH_3C{\equiv}COC_2H_5$	84.12			84^{760}	0.8276^{20}_{4}	1.4130^{20}	i d^h					os v	**B13**, 1969
e562	**—,ethyl 3-propynyl**	Ethyl propargyl ether. $C_2H_5OCH_2C{\equiv}CH$	84.12			82	0.8326^{20}_{4}	1.4039	i	s	s				**B1**², 504
e563	**—,ethynyl methyl**	Methoxyacetylene. $CH_3OC{\equiv}CH$	56.06			$22–3^{760}$		1.3693^{16}		s	v				**B1**³, 1968
e564	**—,ethynyl phenyl**	Phenoxyacetylene. $C_6H_5OC{\equiv}CH$	118.13		−37, −36	$43–4^{10}$	1.0614^{20}_{4}	1.5125^{20}	i		s				**B6**, 145
e565	**—,ethynyl propyl**	Propoxyacetylene	84.12			75	0.8080^{20}_{4}	1.3935^{20}	d^h	v	v				**B1**³, 1969
e566	**—,furfuryl methyl**	[furan ring, O]—CH_2OCH_3	112.13			$131–3^{760}$	1.0163^{20}_{4}	1.4570^{20}	i	s	v				**B17**², 114
e567	**—,2-furyl octyl**	[furan ring, O]—O—$(CH_2)_7CH_3$	196.28			$129–30^{18}$	0.9214^{28}_{4}	1.4520^{28}							
e568	**—,2-furyl** phenyl	[furan ring, O]—O—[benzene ring]	160.16			$105–6^{18}$	1.1010^{23}_{4}	1.5418^{23}							
e569	**—,heptyl methyl**	$CH_3(CH_2)_6OCH_3$	130.23			151^{760}	0.7869^{15}_{15}	1.4073^{20}	i	∞	∞				**B1**³, 1682
e570	**—,heptyl phenyl**	$CH_3(CH_2)_6OC_6H_5$	192.30			267	0.9319^{0}_{0}								**B6**, 144
—	**—,—,4′-hydroxy-**	*see* **Phenol, 4-heptoxy-**	318.46	lf (al)	41.8	200^{1}	0.8434^{82}	1.4556^{82}							**B6**, 144
e572	**—,hexadecyl phenyl**	Cetyl phenyl ether. $CH_3(CH_2)_{15}OC_6H_5$													
e573	**—,hexyl phenyl**	$CH_3(CH_2)_5OC_6H_5$	178.28		−19	240^{760}	0.9174^{20}_{4}	1.4921^{20}							**B6**², 146
e575	**—,isobutyl methyl**	$(CH_3)_2CHCH_2OCH_3$	88.15			58^{760} cor	0.7311^{20}_{4}		i	s	s				**B1**³, 1559
e576	**—,isobutyl propyl**	$(CH_3)_2CHCH_2OCH_2CH_2CH_3$	116.20			105–6			δ	s	∞				**B1**², 410
e577	**—,isopropyl methyl**	$(CH_3)_2CHOCH_3$	74.12			32.5^{777}	0.7237^{15}_{4}	1.3576^{20}	s	∞	∞				**B1**³, 1458

For explanations, symbols and abbreviations see beginning of table.

No.	Name	Synonyms and Formula	Mol. wt.	Crystalline form, color and specific rotation	m.p. °C	b.p. °C	Density	n_D	Solubility w	al	eth	ace	bz	other solvents	Ref.
	Ether														
e578	**—,isopropyl propyl**	$(CH_3)_2CHOCH_2CH_2CH_3$	102.17			83	0.7474^{12}_{4}		δ						**B1**², 381
e579	**—,methyl 2-octyn-1-yl**	$(CH_3)OCH_2C≡C(CH_2)_4CH_3$	140.22			61^{12}	0.808^{21}_{4}	1.4280^{21}							
e580	**—,methyl pentyl**	$CH_3(CH_2)_4OCH_3$	102.18			$99–100^{760}$	0.767^{19}	1.3855^{19}							**B1**², 417
e581	**—,methyl propyl**	$CH_3CH_2CH_2OCH_3$	74.12			38.9^{760}	0.738^{20}_{4}	1.3579^{20}	s	∞	∞				**B1**³, 1413
e582	**—,—,1-chloro-**	$CH_3CH_2CH_2OCH_2Cl$	108.57			109^{760}	0.9884^{20}_{4}	1.4125^{20}	d	v	v				**B1**¹, 305
e583	**—,—,2′3′-dibromo-**	$CH_3CHBrCHBrOCH_3$	231.94			185 84^{15}	1.8320^{12}_{4}	1.5123^{20}			s				**B1**³, 1428
e585	**—,—,2′,3′-epoxy-**	Epimethylin. $CH_2CHCH_2OCH_3$ (epoxide ring via O)	88.10			115–8		1.4116	s						**B17**, 104
e586	**—,1-naphthyl pentyl**	$C_{10}H_7^{\alpha}O(CH_2)_4CH_3$	214.26		30	322			i	s	s		s		
e587	**—,2-naphthyl pentyl**	$C_{10}H_7^{\beta}O(CH_2)_4CH_3$	214.26		24.5	328		1.5587^{30}	i	s	s		s		
e588	**—,octyl phenyl**	$C_6H_5O(CH_2)_7CH_3$	206.33		8	285^{760}	0.9139^{15}_{15}								**B6**, 144
e590	**Ethionic acid,** anhydride	Carbyl sulfate. ring: O_2S, SO_2, O, O	188.19	dlq	ca. 80				d						**B19**, 433
—	**Ethyl acetate**	*see* **Acetic acid**, ethyl ester													
—	**Ethyl acetoacetate**	*see* **Butanoic acid, 3-oxo-,** ethyl ester*													
—	**Ethyl alcohol**	*see* **Ethanol***													
—	**Ethylamine***	*see* **Ethane, amino***													
—	**Ethyl cellosolve**	*see* **Ethanol, 2-ethoxy-***													
—	**Ethyl ether**	*see* **Ether, diethyl**													
—	**Ethyl vanillin**	*see* **Benzaldehyde, 3-ethoxy-4-hydroxy-**													
—	**Ethylene**	*see* **Ethene***													
—	**Ethylene bromide**	*see* **Ethane, 1,2-dibromo-***													
—	**Ethylene bromohydrin**	*see* **Ethanol, 2-bromo-***													
—	**Ethylene chloride**	*see* **Ethane, 1,2-dichloro-***													
—	**Ethylene chlorohydin**	*see* **Ethanol, 2-chloro-***													
—	**Ethylenediamine**	*see* **Ethane, 1,2-diamino-***													
—	**Ethylene dicyanide**	*see* **Succinic acid**, dinitrile													
—	**Ethylene fluoride**	*see* **Ethane, 1,2-difluoro-***													
—	**Ethylene fluorohydrin**	*see* **Ethanol, 2-fluoro-***													
—	**Ethylene glycol**	*see* **1,2-Ethanediol***													
—	—,dimethyl ether	*see* **Ethane, 1,2-dimethoxy-***													
—	—,monoallyl ether	*see* **Ethanol, 2-allyloxy-**													
—	—,monoethyl ether	*see* **Ethanol, 2-ethoxy-***													
—	**Ethylene iodide**	*see* **Ethane, 1,2-diiodo-***													
—	**Ethylene iodohydrin**	*see* **Ethanol, 2-iodo-***													
—	**Ethylene oxide**	*see* **Ethane, 1,2-epoxy-***													
—	**Ethylene urea**	*see* **2-Imidazolidone**													
—	**Ethylenimine**	*see* **Azirane**													
—	**Ethylidene acetate**	*see* **Acetaldehyde**, diacetate													
e591	**Ethyne***	Acetylene, Ethine, HC≡CH	26.04		−81.8	-83.6^{760} sub	0.6181^{-82}_{4}	1.00051^{0}	δ	δ		s	s	chl s	**B1**³, 887
e592	**—,bis(1-hydroxycyclohexyl)-***	OH, HO; cyclohexyl–C≡C–cyclohexyl	222.33	nd (CCl_4)	109–10					v				CCl_4 s^h lig δ	**B6**², 909

For explanations, symbols and abbreviations see beginning of table.

No.	Name	Synonyms and Formula	Mol. wt.	Crystalline form, color and specific rotation	m.p. °C	b.p. °C	Density	n_D	Solubility w	al	eth	ace	bz	other solvents	Ref.
	Ethyne														
e593	—,**bromo-***	Ethynyl bromide. HC⋮CBr	104.94			4.7	0.0047 (760 mm.)		δ		s			dil HNO_3 s dil HCl δ	**B1**[3], 919
e594	—,**chloro-***	Ethynyl chloride. HC⋮CCl	60.48		−126	−32	0.002 (760 mm.)		d	s					**B1**[3], 917
e599	—,**dibromo-***	BrC⋮CBr	183.84	nd	−25	exp			i	s	s			oos s	**B1**[3], 919
e600	—,**dichloro-***	ClC⋮CCl	94.93		−66	exp				s	s				**B1**[3], 918
e601	—,**diiodo-***	IC⋮CI	277.83	rh nd (lig)	78.5–8.9	exp				s		s		os s lig δ	**B1**[3], 919
e602	—,**diphenyl-***	$C_6H_5C⋮CC_6H_5$	178.24	mcl pr (al)	63.5	170^{19} 300^{760}	0.9657_4^{100}		i	δ v^h	v				**B5**[2], 568
e603	—,**1(2-naphthyl)-2-phenyl-***	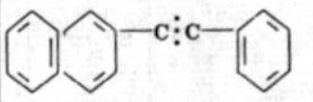 [structure]	228.30	(al)	117					s					**B5**[2], 628
e604	α-**Eucaine**	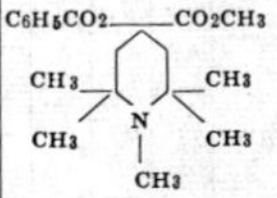 $C_6H_5CO_2$, CO_2CH_3, CH_3, CH_3, N, CH_3, CH_3, CH_3	333.43	pr (eth or al)	104–5				i	s	s		s	peth s lig s	**B22**, 194
e605	—,hydrochloride	$C_{19}H_{27}NO_4.HCl$. *See* e604	369.89	pl (w+1), pr (MeOH+2)	ca. 200 d				v	v	δ				**B22**, 194
e606	β-**Eucaine**(*d*)	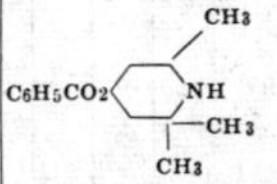 CH_3, $C_6H_5CO_2$, NH, CH_3, CH_3	247.34	pr (peth)	57–8										**B21**[2], 14
e607	—(*dl*)	Betacaine. $C_{15}H_{21}NO_2$. *See* e605	247.34	pl (peth)	70–1				i	s	s		s	chl s	**B21**[2], 13
e608	—(*l*)	$C_{15}H_{21}NO_2$. *See* e605	247.34	pr (peth)	57–8										**B21**[2], 14
e609	—,hydrochloride(*dl*)	$C_{15}H_{21}NO_2.HCl$ *See* e605	283.80	pl (w)	277–9				s	s	s			chl s	**B21**[2], 13
e610	—,lactate(*dl*)	$C_{15}H_{21}NO_2.CH_3CHOHCO_2H$. *See* e605	337.42		ca. 152				v	s	δ			chl s	**B21**[2], 13
—	**Eucaliptol**	*see* **1,8-Cineole**													
e611	**Eugenol**	4-Allylguaiacol Eugenic acid 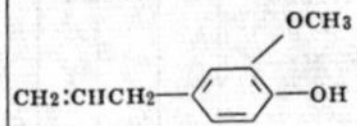 OCH_3, $CH_2{:}CHCH_2$—, —OH	164.20		9.2	255^{760}	1.0664_4^{20}	1.5410^{20}	i	∞	∞			chl s oils s	**B6**[2], 921
—	—,acetate	$C_{12}H_{14}O_3$. *See* e611	206.23		27		1.080_{25}^{25}	1.520^{20} (suc)	δ	s					
e612	**Eugetic acid**	*see* **Benzoic acid, 5-allyl-2-hydroxy-3-methoxy-**													
e613	**Eupittone**	Eupittonic acid. 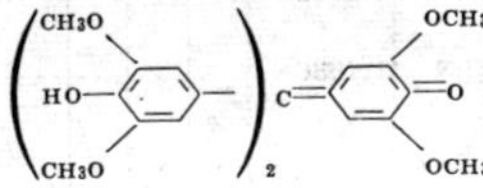 (CH_3O, HO—, CH_3O)$_2$C=, OCH_3, =O, OCH_3	470.46	nd (al-eth)	200					$δ^h$				alk s(bl) aa s	**B8**, 574
e614	**Euxanthic acid**	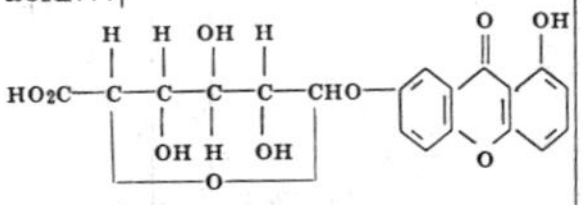 HO_2C—C(H)—C(H)(OH)—C(OH)(H)—C(H)(OH)—CHO—, O, OH	404.43	ye nd (+1w) [α]−108 (+1w)	130 d (+w) 162 (anh)				δ s^h	v^h				alk s	**B31**, 277
—	**Euxanthone**	*see* **Xanthone, 1,7-dihydroxy-**													
—	**Euxanthonic acid**	*see* **Benzophenone, 2,2′,5,6′-tetrahydroxy-**													
e615	**Evernic acid**	Orselinic acid 4-everitate. Lecanoric acid monomethyl ether. 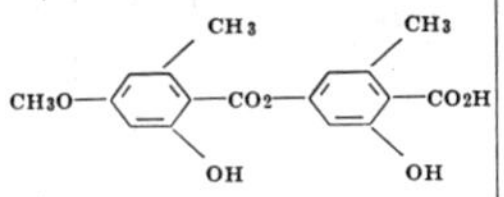 CH_3O—, CH_3, —CO_2—, CH_3, —CO_2H, OH, OH	332.30	nd (w or ace), pr (al)	170				i $δ^h$	δ s^h	i $δ^h$				**B10**[2], 274
—	**Everninic acid**	*see* **Benzoic acid, 2-hydroxy-4-methoxy-6-methyl-**													
e616	**Evodiamine**(*d*)	$C_{19}H_{17}N_3O$	303.35	yesh pl (al) $[α]_D^{15}+352$ (ace, c=0.5)	278				i	δ	δ	s	i	chl δ peth i aa δ	**B26**[2], 103
e617	—,hydrate(*d*)	$C_{19}H_{17}N_3O.H_2O$	321.37	pl (al)	146–7										**B24**[2], 72
—	**Exaltone**	*see* **Cyclopentadecanone***													

For explanations, symbols and abbreviations see beginning of table.

No.	Name	Synonyms and Formula	Mol. wt.	Crystalline form, color and specific rotation	m.p. °C	b.p. °C	Density	n_D	Solubility w	al	eth	ace	bz	other solvents	Ref.
	Fagaramide														
f1	**Fagaramide**	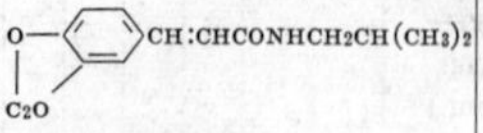	247.28	nd (bz, dil al or peth) pl (AcOEt)	119.5				i^h	s^h			s	peth δ AcOEt s^h	**B19²**, 299
f2	**β-Fagarine**	Skimmianine. 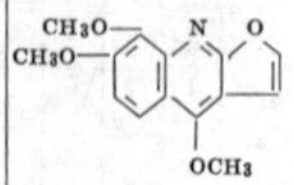	259.25	pym, oct (al)	178				i	s	δ			chl s CS_2 δ peth i	......
f3	**γ-Fagarine**	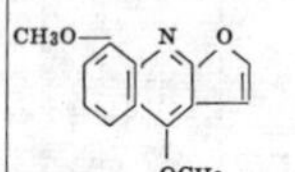	229.25	pr (al)	142				δ	s^h	s		s	chl s peth δ	......
f4	**α-Farnesene**	$(CH_3)_2C:CH(CH_2)_2C(CH_3):CHCH_2CH:C(CH_3)CH:CH_2$	204.34			129–32[12]	0.8410^{20}_{4}	1.4836^{20}	i					peth ∞ lig ∞	**B1³**, 1067
f5	**β-Farnesene**	$CH_2:CHC(:CH_2)(CH_2)_2CH:C(CH_3)(CH_2)_2CH:C(CH_3)_2$	204.34			121–2[9]	0.8363^{20}_{4}	1.4899^{20}	i					chl v aa s os s	**B1³**, 1067
f6	**Farnesol**	$(CH_3)_2C:CH(CH_2)_2C(CH_3):CH(CH_2)_2C(CH_3):CHCH_2OH$	222.36			160[10]	0.8883^{20}_{4}	1.4899^{20}	i	v				os s	**B1³**, 2041
f7	**α-Fenchene**(*d*)	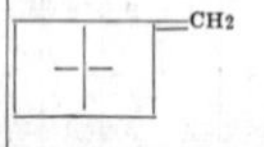	136.24	$[\alpha]^{14}_{D}+29$ (l = 10 cm)		155–6									**B5¹**, 86
f8	—(*dl*)	Isopinene. $C_{10}H_{16}$. *See* f7	136.24			154–6	0.8660^{20}_{4}	1.4705^{20}							**B5¹**, 86
f9	—(*l*)	$C_{10}H_{16}$. *See* f7	136.24	$[\alpha]^{20}_{D}-32.3$		157–9	0.8665^{20}_{4}	1.4713^{20}	i	∞	s				**B5²**, 109
f10	Fenchone(*d*)	*d*-2-Fenechanone, *d*-1,3,3-Trimethylnorcamphor. 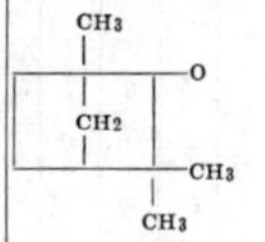	152.24	$[\alpha]^{20}_{D}+66.9$	6	193.5^{760}	0.9465^{20}_{4}	1.4623^{20}	i	v	v				**E12A** 724
f11	—(*dl*)	$C_{10}H_{16}O$. *See* f10	152.24		−18	193–4	0.9501^{15}_{15}	1.4702^{20}							**B7²**, 93
f12	—(*l*)	$C_{10}H_{16}O$. *See* f10	152.24	$[\alpha]^{19}_{578}-59.3$ (cyclohexane)	5		0.948^{20}	1.4636							**B7²**, 92
f13	**Fenchyl alcohol** (*dl*)	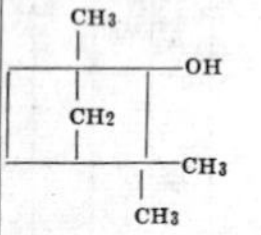	154.25		38–9	201.4			i	v	v			peth v	**B6**, 71
f14	**Ferron**	Loretin. 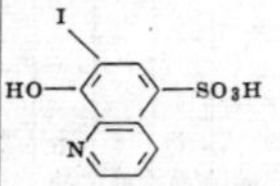	351.12		260–70				$δ^h$	δ					**B22**, 408
—	**Ferulaldehyde**	*see* **Cinnamaldehyde, 4-hydroxy-3-methoxy-**													
—	**Ferulic acid**	*see* **Cinnamic acid, 4-hydroxy-3-methoxy-**													
—	**Filicinic acid**	*see* **1,3,5-Cyclohexanetrione, 2,2-dimethyl-**													
f15	**Filicic acid**	Filicic acid. Filicin. 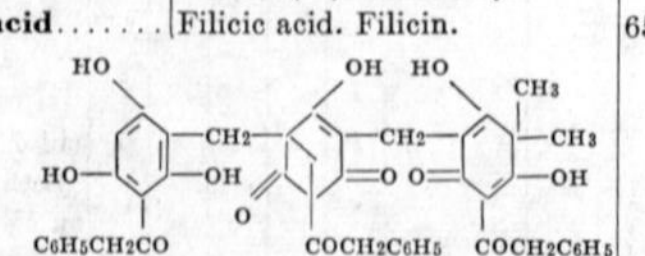	652.70	lt ye pl (AcOEt)	184–5				i	i d^h	δ		s	CS_2, chl, xyl s	**E12A**, 525
—	**Fisetin**	*see* **Flavone, 3,3′,4′,7-tetrahydroxy-**													

For explanations, symbols and abbreviations see beginning of table.

No.	Name	Synonyms and Formula	Mol. wt.	Crystalline form, color and specific rotation	m.p. °C	b.p. °C	Density	n_D	Solubility w	al	eth	ace	bz	other solvents	Ref.
	Flavaniline														
f16	**Flavaniline**	2(*p*-Aminophenyl) lepidine. 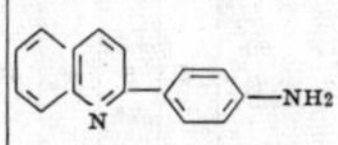	234.30	pr (bz)	97	...	...	...	δ	s	...	...	s	...	**B22**, 469
f17	**Flavanone**	2,3-Dihydro-2-phenyl-1,4-benzopyrone. 4-Oxo-2-phenylchroman. 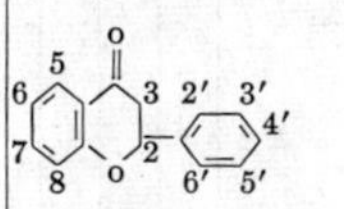	224.26	...	75–6	...	...	...	...	...	...	...	...	...	**B17^{2}**, 387
f18	**—,4′-methyl-3′,5,7-trihydroxy-**	Hesperetin. 4-methoxy-2′,3,4,6′,-tetrahydroxychalcone. (2 *tautomeric forms*). $C_{18}H_{34}O_5$ *See* f17	302.29	pl (dil al+½w), pl (AcOEt)	227d	subl $205^{0.004}$	...	...	δ	v	s	...	δ	chl δ dil alk s	**B8^{2}**, 580 **B18^{2}**, 204
f19	**—,3′,4′,5,7-tetrahydroxy-**	Eriodictol. 2′,3,4,4′,6′-Pentahydroxychalcone. (2 *tautomeric forms*). $C_{15}H_{12}O_6$. *See* f17	288.26	pa br nd (dil al +1.5 w or dil aa+2.5w)	267d 257d (1.5w)		...	...	$δ^h$	$δ^h$	δ	...	...	dil alk s os δ aa δ	**B8**, 543 **B18^{2}**,204
—	**Flavianic acid**	*see* **2-Naphthalenesulfonic acid, 2,4-dinitro-1-hydroxy-**													
—	**Flavinium, 3,4′,5,7-tetrahydroxy-, chloride**	*see* **Pelargonidine**, chloride													
—	**Flavol**	*see* **Anthracene, 2,6-dihydroxy-**													
f20	**Flavone**	2-Phenyl-γ-benzopyrone. 2-Phenylchromone. 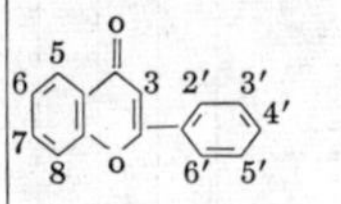	222.25	nd (lig)	99–100	...	...	...	i	s	s	...	...	...	**B17^{2}**, 395
f21	**—,6-bromo-**	$C_{15}H_9BrO_2$. *See* f20	301.15	nd (al)	189–90	...	...	...	...	s^h	...	...	...	...	**B17**, 373
f22	**—,5,7-dihydroxy-**	Chrysin. $C_{15}H_{10}O_4$. *See* f20	254.24	pa ye pl or pr (MeOH), nd (subl)	285	subl	...	...	i	s^h	s	...	δ	CS_2, chl s aa s^h lig δ	**B18^{2}**, 97
f23	**—,5,7-dihydroxy-4′-methoxy-**	Acacetin. $C_{16}H_{12}O_5$. *See* f20	284.27	ye nd (al)	262	...	...	...	i	δ s^h	i	v	δ	AcOEt δ alk s lig δ	**B18^{3}**, 173
f24	**—,5,7-dihydroxy-6-methoxy-**	Oroxylin. $C_{16}H_{12}O_5$. *See* f20	284.27	ye cr	230–2	...	...	...	...	s	s	s	s^h	alk s aa s	...
f25	**—,3-hydroxy-**	Flavonol. $C_{15}H_{10}O_3$. *See* f20	238.25	pa ye nd (al)	168–70	...	...	...	...	s^h	...	...	...	...	**B17^{2}**, 498
—	**—,7-methoxy-3,3′,4′,5-tetrahydroxy-**	*see* Rhamnetin													
f26	**—,2′,3,3′,5,7-pentahydroxy-**	$C_{15}H_{10}O_7$. *See* f20	302.24	ye nd (aa+1.5w)	295	...	...	...	...	...	...	...	...	...	**B18^{2}**, 235
f27	**—,2′,3,4′,5,7-pentahydroxy-**	Morin. $C_{15}H_{10}O_7$. *See* f20	302.24	pa ye nd	285	...	...	...	$δ^h$	v	s	...	s	CS_2 i alk s	**B18**, 239
f28	**—,3′,3,5,5′,7-pentahydroxy-**	$C_{15}H_{10}O_7$. *See* f20	302.24	bed-ye cr (dil al+1w)	317	...	...	...	...	...	...	...	...	...	**B18^{2}**, 236
f29	**—,3,3′,4′,5,7-pentahydroxy-**	Quercitin. Meletin. Sophoretin. $C_{15}H_{10}O_7$. *See* f20	302.24	ye nd (dil+2w)	316–7 (anh)	subl	...	...	$δ^h$	s	δ	s	...	MeOH δ Py s	**B18^{2}**, 236
f30	**—,3,3′,4′,7,8-pentahydroxy-**	$C_{15}H_{10}O_7$. *See* f20	302.24	ye nd (chl al+1w)	308	...	...	...	...	...	...	...	...	...	**B18**, 250
f31	**—,3,3′5,5′,7-pentahydroxy-**	$C_{15}H_{10}O_7$. *See* f20	302.24	ye nd	>300	...	...	...	...	...	...	...	...	...	**B18^{2}**, 239
f32	**—,3′,4′,5,5′,7-pentahydroxy-**	Tricetin. $C_{15}H_{10}O_7$. *See* f20	302.24	ye (dil al+?w)	>330 d	...	...	...	...	δ	i	...	i	...	**B18^{1}**, 423
f33	**—,2′,3,5,7-tetrahydroxy-**	Datiscetin. $C_{15}H_{10}O_6$. *See* f20	286.24	pa ye nd (al)	268–9 (276)	...	...	...	$δ^h$	v	v	...	...	os, alk, con sulf s	**B18^{2}**, 214
f34	**—,3,3′4′,7-tetrahydroxy-**	Fisetin. $C_{15}H_{10}O_6$. *See* f20	286.24	lt ye nd (dil al +1w)	330	sub^{vac}	...	...	i	s	δ	s	δ	peth δ	**B18^{2}**,216
f35	**—,3′,4′,5,7-tetrahydroxy-**	Luteolin. $C_{15}H_{10}O_6$. *See* f20	286.24	ye nd (dil al+1w)	329–30 (anh)	sub^{vac}	...	...	$δ^h$	s	δ	...	...	alk s con sulf s	**B18^{2}**, 212
f36	**—,3,4′,5,7-tetrahydroxy-**	Kaempferol. $C_{15}H_{10}O_6$. *See* f20	286.24	ye nd (al+1w) or aa)	276–7	...	...	...	i	v^h	v	v	i	chl δ aa s^h alk s	**B18^{2}**, 214

For explanations, symbols and abbreviations see beginning of table.

No.	Name	Synonyms and Formula	Mol. wt.	Crystalline form, color and specific rotation	m.p. °C	b.p. °C	Density	n_D	Solubility w	al	eth	ace	bz	other solvents	Ref.
	Flavone														
f37	**—,4′,5,7-tri-hydroxy-**	Apigenin. $C_{15}H_{10}O_5$. *See* f20	270.24	ye nd (Py-w or w, +½w), lf (al)	247–8	sub^{vac}	....	....	i	s^h	...	...	...	Py s dil alk v con sulf s	**B18²**, 172
f38	**—,5,6,7-tri-hydroxy-**	Baicalein. $C_{15}H_{10}O_5$. *See* f20...	270.24	ye pr (al, MeOH or aa)	264–5d	....	....	....	δ	s v^h	s^h	s^h	δ	AcOEt s^h alk s(og-red) aa s	**B18²**, 172
—	**Flavopurpurin**...	*see* **9,10-Anthraquinone, 1,2,6-trihydroxy-**													
f39	**Floridosine**.......	Glycerol-*D*-monogalactoside. $HOCH_2—C(OH)(H)—C(H)(OH)—C(H)(OH)—C(OH)(H)—CH_2OCH_2CHOHCH_2CH_3$	252.27	nd (al), pr (w)	128.5	....	....	....	v	δ	...	...	...	....	....
f40	**Fluoran**.........	9-Hydroxy-9-xanthene-*o*-benzoic acid lactone. 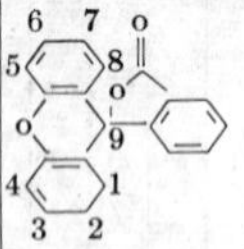	300.32	nd (al+2 al) w (al+½ al)	182	....	....	....	i	s	...	...	...	con sulf s	**B19²**, 173
f41	**—,1,6-dihydroxy-**	$C_{20}H_{12}O_5$. *See* f40...........	332.32	ye nd	>280	....	....	....	...	...	...	...	...	....	**B19²**, 247
f42	**—,2,6-dihydroxy-**	$C_{20}H_{12}O_5$. *See* f40...........	332.32	ye nd (al)	177	....	....	....	...	...	...	...	...	....	**B19²**, 247
f43	**—,3,5-dihydroxy-**	$C_{20}H_{12}O_5$. *See* f40...........	332.32	ye-gr nd (aa)	179	....	....	....	...	...	...	...	...	....	**B19²**, 248
f44	**Fluoranthene**....	1,2-Benzacenaphthene. Idryl. 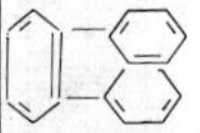	202.26	pa ye nd or pl (al)	111	*ca.* 375	1.252^0_4	....	i	δ s^h	s	...	s	chl, CS_2 s aa s	**E14s**, 54
f45	**Fluorene**.........	2,3-Benzindene. Diphenyl-enemethane. 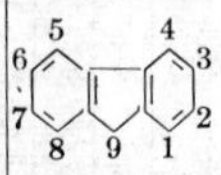	166.22	lf (al)	116–7	$293–5^{760}$	1.203^0_4	....	i	δ s^h	s	...	s	CS_2 s to s, v^h aa s v^h	**E13**, 25
f46	**—,2-acetamido-**..	*N*(2-Fluorenyl) acetamide. $C_{15}H_{13}NO$. *See* f45	223.28	nd (50% al or 50% aa)	187–8	....	....	....	i	s	...	...	...	aa s	**B12**, 1331
f47	**—,9-acetamido-**..	$C_{15}H_{13}NO$. *See* f45..........	223.28	nd (aa)	262	....	....	....	...	...	...	...	...	con sulf s	**B12²**, 781
f48	**—,2-amino-**......	$C_{13}H_{11}N$. *See* f45...........	181.24	lo pl or nd (dil al)	129	....	....	....	i	s	s	...	...	....	**B12²**, 779
f49	**—,9-amino-**......	$C_{13}H_{11}N$. *See* f45...........	181.24	nd (lig)	64	....	....	....	i	v	s	...	s	MeOH, chl v con sulf s (gr)	**B12²**, 780
f50	**—,1-amino-9-hydroxy-**	$C_{13}H_{11}NO$. *See* f45..........	197.24	dk red nd (w)	142	....	....	....	δ	v	v	...	s	con sulf s aa v	**B13²**, 435
f51	**—,2-amino-9-hydroxy-**	$C_{13}H_{11}NO$. *See* f45..........	197.24	irid nd (al)	200	....	....	....	...	s^h	δ	...	...	chl s^h con sulf s	**B13²**, 435
f52	**—,4-amino-9-hydroxy-**	$C_{13}H_{11}NO$. *See* f45..........	197.24	ye (60% al)	183–4	....	....	....	...	s	...	...	...	....	**B13²**, 436
f53	**—,9-benzhydryl-idene-**	β,β-Diphenyl-α,α-biphenylene-ethylene. ω,ω-Diphenyl-dibenzofulvene. $C_{26}H_{18}$. *See* f45	330.43	ye (bz)	229.5	....	....	....	...	δ	δ	...	...	chl s	**E13**, 37
f54	**—,9-benzylidene-**.	ω-Phenyldibenzofulvene. $C_{20}H_{14}$. *See* f45	254.33	lf (al)	76	....	....	....	...	s^h	...	...	s	....	**E13**, 35
f55	**—,2-bromo-**......	$C_{13}H_9Br$. *See* f45...........	245.13	nd (al)	111.5	*ca.* 185^{135}	....	....	i	s^h	...	...	...	chl v aa s	**B5²**, 534
f56	**—,9-bromo-**......	$C_{13}H_9Br$. *See* f45...........	245.13	(lig or al)	104	....	....	....	i	s^h	...	s	...	con sulf s^h	**B5²**, 534
f57	**—,9(3-bromo-benzylidene)-**	$C_{20}H_{13}Br$. *See* f45..........	333.24	ye nd (aa)	92–3	....	....	....	i	s	...	...	...	MeOH s con sulf s^h	**B5¹**, 358
f58	**—,9(4-bromo-benzylidene)-**	$C_{20}H_{13}Br$. *See* f45..........	333.24	ye nd (aa)	144	....	....	....	i	...	...	...	...	con sulf s^h aa s^h	**B5²**, 640
f59	**—,9(2-chloro-benzylidene)-**	$C_{20}H_{13}Cl$. *See* f45..........	288.78	ye nd (aa)	176	....	....	....	i	...	...	...	...	con sulf s^h aa s^h	**B5¹**, 358
f60	**—,9(3-chloro-benzylidene)-**	$C_{20}H_{13}Cl$. *See* f45..........	288.78	pa ye pr or pym (MeOH)	90.5	....	....	....	i	v	...	...	...	con sulf s^h lig v	**B5¹**, 358
f61	**—,9(4-chloro-benzylidene)-**	$C_{20}H_{13}Cl$. *See* f45..........	288.78	ye nd (aa)	149.5	....	....	....	i	...	...	...	...	con sulf s^h aa s^h	**B5¹**, 358
f62	**—,9-cinnamyl-idene-**	$C_{22}H_{16}$. *See* f45..........	280.37	pa ye nd (aa)	155	....	....	....	...	s^h	...	...	...	chl s^h sulf s (og)	**E13**, 36
f63	**—,2,7-diamino-**..	$C_{13}H_{12}N_2$. *See* f45..........	196.25	nd (w), pr (bz)	162	....	....	....	i s^h	s	...	...	...	chl s	**B13²**, 123
f64	**—,2,7-dichloro-**...	$C_{13}H_8Cl_2$. *See* f45..........	235.11	pl or nd (bz)	128	subl	....	....	i	...	...	...	s^h	chl, CCl_4 s	**B5²**, 533

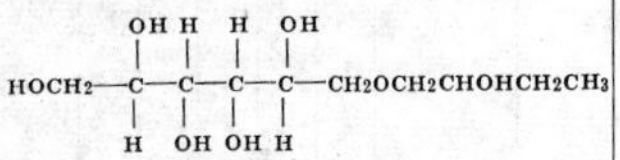

For explanations, symbols and abbreviations see beginning of table.

No.	Name	Synonyms and Formula	Mol. wt.	Crystalline form, color and specific rotation	m.p. °C	b.p. °C	Density	n_D	Solubility: w	al	eth	ace	bz	other solvents	Ref.
	Fluorene														
f65	**—,9,9-dichloro-**	$C_{13}H_8Cl_2$. *See* f45	235.11	rh pr (bz)	102.5				d^h	v				os v con sulf s	**B5^2, 534**
f65[1]	**—,1,8-dimethyl-9-hydroxy-9-(2-tolyl)-**	$C_{22}H_{20}O$. *See* f45	300.40		141–2					s	s		s		
f66	**—,1,8-dimethyl-9-(2-tolyl)-**	$C_{22}H_{19}O$. *See* f45	299.40		168–9					s	s	s			
f67	**—,2-hydroxy-**	$C_{13}H_{10}O$. *See* f45	182.22	lf (w) nd (chl or 60%)	171				i	v	v			alk v aa v lig i	**B6^2, 655**
f67[1]	**—,9-hydroxy-**	Diphenylenecarbinol. 9-Fluorenol. $C_{13}H_{10}O$. *See* f45	182.22	hex nd (w or peth)	158–9					δ	s		v	peth δ	**E13, 63**
f67[2]	**—,9-hydroxy-9-phenyl-**	$C_{19}H_{14}O$. *See* f45	258.12	ye or col pr (lig)	108–9								s	con sulf s (ye) aa s	**E13, 64**
f68	**—,9-methyl-**	$C_{14}H_{12}$. *See* f45	180.25	pr	46–7	$154–6^{15}$	1.0263^{66}_{4}	1.610^{66}	i					chl s	**E13, 29**
f69	**—,9(2-methyl-benzylidene)-**	$C_{21}H_{16}$. *See* f45	268.34	nd or pr (aa)	109.5				i	s^h			s^h	con sulf s^h	**B5^1, 359**
f70	**—,9(4-methyl-benzylidene)-**	$C_{21}H_{16}$. *See* f45	268.34	nd (aa), pr (al)	97.5				i	s^h			s^h	con sulf s^h	**B5^1, 239**
f71	**—,9-methylene-**	Biphenyleneethylene. Dibenzofulvene. $C_{14}H_{10}$. *See* f45	178.23		53				i					os v	**E13, 34**
f72	**—,2-nitro-**	$C_{13}H_9NO_2$. *See* f45	211.22	nd (50% aa or ace)	160 (15.4)				i			s^h	s^h		**B5^2, 535**
f73	**—,3-nitro-**	$C_{13}H_9NO_2$. *See* f45	211.22	pa ye nd (al)	106					s^h					**E13, 48**
f74	**—,9-nitro-**	$C_{13}H_9NO_2$. *See* f45	211.22	gr-ye nd (al)	ca. 150					s^h					**E13, 48**
—	**—,9-oxo-**	*see* **9-Fluorenone**													
f75	**—,9-phenyl-**	$C_{19}H_{14}$. *See* f45	242.32	nd or lf (al or bz)	147–8				i	s^h	s		s	aa s	**E13, 29**
—	**Fluorenecarboxylic acid, 9-oxo**	*see* **9-Fluorenonecarboxylic acid**													
f76	**2-Fluorenesulfonic acid**	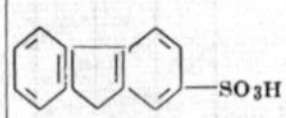 $-SO_3H$	246.29	nd (aa)	155 (+1w)				v					chl s alk s	**E13, 114**
—	**Fluorenol**	*see* **Fluorene, hydroxy-**													
f80	**9-Fluorenone**	Diphenylene ketone. 9-Oxofluorene. (5, 4, 6, 3, 7, 2, 8, 1, O) 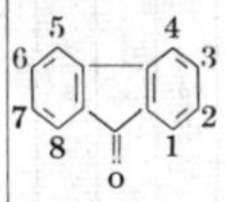	180.19	rh bipym (al)	83 (86)	341.5^{760}	1.1300^{99}_{4}	1.6369^{99}	i	s	v		s v^h	to v	**E13, 77**
f81	**—,oxime**	9-Isonitrosofluorene. 9-Oximinofluorene. 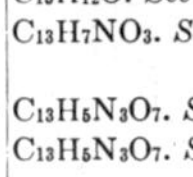 (NOH)	195.21	nd (chl-peth or bz)	195–6					s				chl s dil alk v	**B7^2, 407**
f82	**—,1-amino-**	$C_{13}H_9NO$. *See* f80	195.21	og nd (dil al)	110				s^h	s				oos vs	**B14, 113**
f83	**—,2-amino-**	$C_{13}H_9NO$. *See* f80	195.21	red-vt pr (al)	156 (163)				i	s^h	s^h		s^h	peth δ	**B14^2, 68**
f84	**—,3-amino-**	$C_{13}H_9NO$. *See* f80	195.21	ye nd (w or dil al)	158–9				s^h	s					**B14^2, 69**
f85	**—,4-amino-**	$C_{13}H_9NO$. *See* f80	195.21	og nd	138					v	v			chl v aa v	**B14^2, 69**
f86	**—,2-bromo-**	$C_{13}H_7BrO$. *See* f80	259.12	ye nd (al or aa)	142–3					s^h		s	s^h		**E13, 83**
f87	**—,2-chloro-**	$C_{13}H_7ClO$. *See* f80	214.66	og-ye nd (dil al)	125	subl				s				sulf s (vt-red)	**E13, 89**
f88	**—,1,8-dimethyl-**	$C_{15}H_{12}O$. *See* f80	208.55	ye	197–8					i				chl s aa s	
f89	**—,2-nitro-**	$C_{13}H_7NO_3$. *See* f80	225.21	ye nd or lf (aa)	22–3							s		sulj s aa s^h (red-ye)	**E13, 85**
f90	**—,2,3,7-trinitro-**	$C_{13}H_5N_3O_7$. *See* f80	315.21	pa ye nd(aa)	180–1					δ	δ		v	chl v	**B7^1, 254**
f91	**—,2,4,7-trinitro-**	$C_{13}H_5N_3O_7$. *See* f80	315.21	ye nd (aa or bz)	176				δ			v	v	chl v	**B7^2, 410**
f92	**9-Fluorenone-1-carboxylic acid**	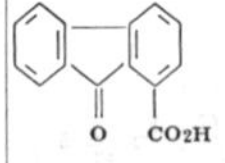 (O, CO_2H)	224.22	og-red nd (dil al)	191–2				δ	v	v			sulf s (red)	**B10^2, 534**
f93	**—,amide**	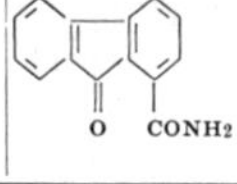 (O, $CONH_2$)	223.22	ye nd (al)	229–30					δ s^h					**B10, 774**

For explanations, symbols and abbreviations see beginning of table.

No.	Name	Synonyms and Formula	Mol. wt.	Crystalline form, color and specific rotation	m.p. °C	b.p. °C	Density	n_D	Solubility						Ref.
									w	al	eth	ace	bz	other solvents	
	9-Fluorenone-1-carboxylic acid														
f94	—,chloride	$C_{14}H_7ClO_2$. *See* f92	242.65	pa ye nd (bz)	140				d^h	s	s		v^h		**B10, 77**
f95	—,ethyl ester	$C_{16}H_{12}O_3$. *See* f92	252.26	ye nd (dil al)	84–6				i	s	s			sulf s	**B10, 77**
f96	**9-Fluorenone-2-carboxylic acid**	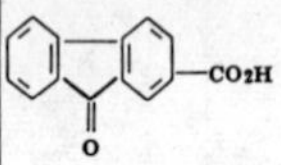	224.22	ye nd (al or aa)	338	subl				s^h				aa s^h	**E13, 10**
f97	—,methyl ester	$C_{15}H_{10}O_3$. *See* f96	238.23	ye nd (MeOH)	181				i	s	s	s			**B10, 77**
f98	**9-Fluorenone-3-carboxylic acid**	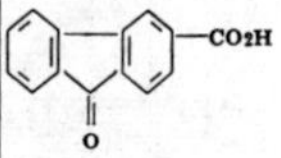	224.22	ye (aa)	285–6					s^h				aa s^h	**E13, 10**
f99	**9-Fluorenone-4-carboxylic acid**	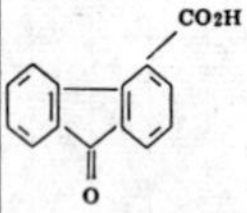	224.22	ye nd (al)	227				i	s^h	s			sulf s (red) aa s^h	**B10^2, 53**
f100	**Fluorescein**	3′,4′-Dehydroxyfluoran. Resorcinolphthalein. 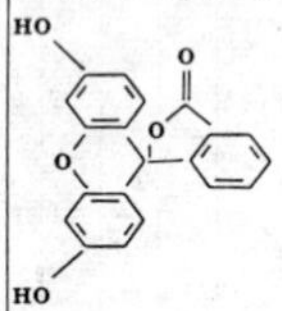	332.30	red rh pr (stable form) ye gran (labile form)	314–6 d (sealed tube)				δ^h	s	δ	s^h		peth i	**B19^2, 24**
—	—,sodium salt	*see* **Uranine**													
f102	**Fluorescin**	2(3,6-Dihydroxyxanthyl)-benzoic acid. 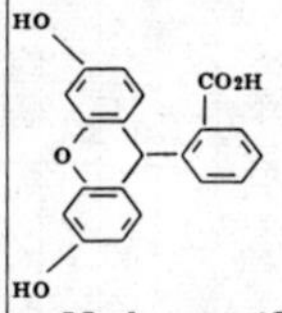	334.33	col or ye nd (aa or eth) pl (bz)	125–7						s		s^h	aa s^h	**B18^2, 30**
—	**Fluoroform**	*see* **Methane, trifluoro-**													
f103	**Fluorophosphoric acid,** diisopropyl ester	$(C_3H_7^iO)_2POF$	184.15			62^9	1.055	1.3830^{25}	δ d^h		s			oils v lig δ	
—	**Fluoroprene**	*see* **1,3-Butadiene, 2-fluoro-**													
f104	**Folic acid**	P.G.A. Pteroylglutamic acid. Vitamin Bc. $C_{19}H_{19}N_7O_6$	441.41	ye-og pl (w)	ca. 250 d				i δ^h	δ	i	i	i	Py δ chl i alk δ	
f105	**Folinic acid**	5-Formyl-5,6,7,8-tetrahydro-pteroyl-glutamic acid. Lencoforin 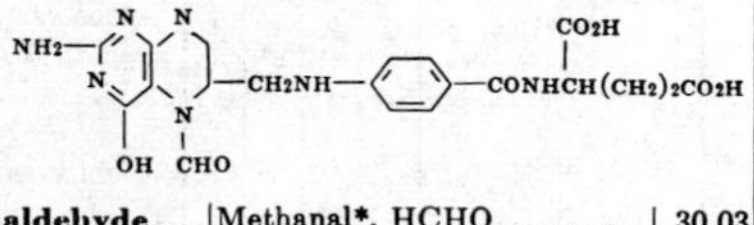	473.45	pl [α] − 15.1	248–50 d										**Am 73, 197**
f106	**Formaldehyde**	Methanal*. HCHO	30.03	gas	−92	−21^{760}	0.815^{-20}		s	s	s			chl s	**B1^2, 619 B6^2, 124**
f108	—,β,β′-Dichloro-isopropyl ethyl acetal	$(ClCH_2)_2CHOCH_2OC_2H_5$	187.07			96–8^{16}	1.182^{17}_{17}	1.4491^{17}			s				**B1^2, 640**
f109	—,diethyl acetal	Diethoxymethane*. Diethyl formal. Ethylal. $CH_2(OC_2H_5)_2$	104.15		−66.5	88	0.8319^{20}_4	1.3748^{18}	s						**B1^2, 639**
f111	—,dimethyl acetal	Methyal. $CH_2(OCH_3)_2$	76.10		−104.8	45.5^{760}	0.8876^{18}_4	1.3589^{18}	v	∞	∞			oos ∞	**B1^2, 638**
f111[1]	—,2,4-dinitro-phenylhydrazone		210.15	ye cr	166				i	s^h	δ				**B15, 490**
f111[2]	—,dipropyl acetal	Dipropoxymethane*. $CH_2(OCH_2CH_2CH_3)_2$	132.20		−97.3	140.5^{760}	0.8345^{20}	1.3939^{19}	v						**B1^2, 639**

For explanations, symbols and abbreviations see beginning of table.

No.	Name	Synonyms and Formula	Mol. wt.	Crystalline form, color and specific rotation	m.p. °C	b.p. °C	Density	n_D	Solubility w	al	eth	ace	bz	other solvents	Ref.
	Formaldehyde														
f112	—,oxime	Formaldoxime. CH_2:NOH	45.04			84			s δ^h	v	v				**B1**, 590
f113	—,**fluoro-**	Formyl fluoride. FCHO	48.02	gas		-24^{760}				d					**Am 77** 5278
—	—,**thio-, trimer**	*see* **1,3,5-Trithane**													
f114	**Foraldomedone**	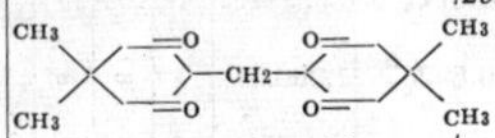	292.38		189–90				i						**B7**^{2}, 852
—	**Formamide**	*see* **Formic acid**, amide													
—	**Formamidine**	*see* **Formic acid**, amidine													
—	**Formanilide**	*see* **Formic acid**, amide, *N*-phenyl-													
f115	**Formic acid**	Methanoic acid*. HCO_2H	46.03		8.4	100.7	1.220^{20}_{4}	1.3714	∞	∞	∞				
f116	—,allyl ester	HCO_2CH_2CH:CH_2	86.09			83	0.948^{18}_{4}		δ	s	∞				
f117	—,amide	Formamide. $HCONH_2$	45.04		2.55	$105–6^{11}$	1.134^{20}_{4}	1.4453	∞	∞	δ		δ		**B2**, 26
f119	—,—,*N*,*N*-diethyl-	$HCON(C_2H_5)_2$	101.15			177–8	0.908		∞	v	v				**B4**, 109
f120	—,—,*N*,*N*-dimethyl-	$HCON(CH_3)_2$	73.10		−61	153^{758}	0.9445^{25}_{4}	1.4269^{25}	∞	∞	∞		∞	chl ∞ lig δ	**B4**, 58
f121	—,—,*N*,*N*-diphenyl-	*N*-Formyldiphenyl amine. $HCON(C_6H_5)_2$	197.24	rh (dil al)	72.4	337.5^{762}			i s^h	s	s		s		**B12**^{1}, 190
f122	—,—,*N*-ethyl-	$HCONHC_2H_5$	73.10			197–9	0.952^{21}_{4}		∞	∞	∞				**B4**, 109
f123	—,—,*N*(1-hydroxy-2,2,2-trichloroethyl)-	Chloral formamide. $HCONHCHOHCCl_3$	192.42			118			s	v	v	v			**B2**^{2}, 37
f124	—,—,*N*-methyl-	$HCONHCH_3$	59.07			180–5	1.011^{19}		s	s	i				**B4**^{2}, 563
f125	—,—,*N*-phenyl-	Formanilide. $HCONHC_6H_5$	121.14	mcl pr	47.5	271	1.1437^{17}_{4}		s	v	s				
f126	—,—,*N*-2-tolyl-	Form-*o*-toluidide. *N*-Formyl-*o*-toluidine 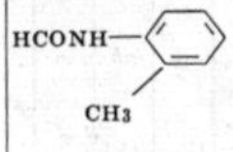	135.17	pl (al)	62	288^{760}	1.086^{55}_{4}			v					**B12**^{2}, 349
f127	—,—,*N*-3-tolyl-	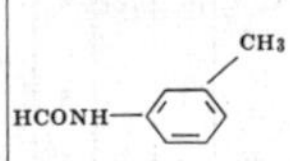	135.17		<−18	278^{724}			s					NaOH s	**B12**^{2}, 468
f128	—,—,*N*-4-Tolyl-	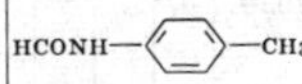	135.17	nd	53				s	v					**B12**^{1}, 419
f129	—,amidine	Formamidine. HN:$CHNH_2$	44.06	pr	81	d			v	s					**B2**, 90
f130	—,—,*N*,*N'*-diphenyl-	C_6H_5N:$CHNHC_6H_5$	196.24		143	>250			δ	s	v	s	s	chl s, lig δ CS_2 δ	**B12**, 236
f130[1]	—,amidoxime	Isoretin. HON:$CHNH_2$	60.06	rh nd (al)	114–5	d			v	δ	δ		i		**B2**^{2}, 89
f131	—,benzyl ester	Benzyl formate. $HCO_2CH_2C_6H_5$	136.14			203.4	1.081^{20}_{4}		i	s	∞				
f132	—,butyl ester	Butyl formate. $HCO_2CH_2CH_2CH_2CH_3$	102.13		−91.9	106.8	0.9102^{0}_{4}	1.3874^{25}	δ	∞	∞				
f133	—,*sec*-butyl ester	$HCO_2CH(CH_3)C_2H_5$	102.13			97	0.8846	1.3812^{25}	δ	∞	∞				
f134	—,cyclohexyl ester	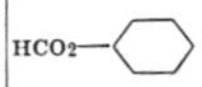	128.17			162.5	1.0057^{0}_{4}		i	s				HCO_2H s aa s	**B6**^{2}, 10
f135	—,ethyl ester	$HCO_2C_2H_5$	74.08		−80.5	54.3	0.9117	1.3598	s	∞	∞				
f136	—,heptyl ester	$HCO_2(CH_2)_6CH_3$	144.22			176.7	0.894^{0}_{4}		i	∞	∞				
f137	—,hexyl ester	$HCO_2(CH_2)_5CH_3$	130.19			153.6	0.898^{0}_{4}		i	∞	∞				
f138	—,hydrazide	Formyl hydrazide. $HCONHNH_2$	60.06	ye lf or nd (al)	54				i	v	v^h		v	chl v	**B2**, 93
f139	—,isobutyl ester	$HCO_2CH_2CH(CH_3)_2$	102.13		−94.5	98.4^{760}	0.8854^{20}_{4}	1.3855^{20}	δ	∞	∞				**B2**^{2}, 30
f140	—,isopropyl ester	Isopropyl formate. $HCO_2CH(CH_3)_2$	88.11			$68–71^{751}$	0.8728^{20}_{4}		δ	∞	∞				**B2**^{2}, 29
f141	—,methyl ester	Methyl formate. HCO_2CH_3	60.05		−99	31.5^{760}	0.9742^{20}_{4}	1.3433^{20}	v	∞	s			MeOH s	**B2**^{2}, 25

For explanations, symbols and abbreviations see beginning of table.

No.	Name	Synonyms and Formula	Mol. wt.	Crystalline form, color and specific rotation	m.p. °C	b.p. °C	Density	n_D	Solubility w	al	eth	ace	bz	other solvents	Ref.
	Formic acid														
f142	—,3-methylbutyl ester	Isoamyl formate. $HCO_2(CH_2)_2CH(CH_3)_2$	116.16			124.2	0.8820	1.3476^{20}	δ	s	∞				
f143	—,4-nitrobenzyl ester	$HCO_2CH_2—C_6H_4—NO_2$	181.15	(dil al)	31					s^h					**B6**^{1}, 223
f144	—,octyl ester	$HCO_2(CH_2)_7CH_3$	158.24			198.1	0.872^{12}_{4}		i						**B2**, 22
f145	—,pentyl ester	Amyl formate. $HCO_2(CH_2)_4CH_3$	116.16		−73.5	132.1	0.8926^{15}_{4}	1.3992^{20}	δ	∞	∞				
f146	—,propyl ester	Propyl formate. $HCO_2CH_2CH_2CH_3$	88.11		−92.9	81.3^{760}	0.9006^{20}_{4}	1.3769^{20}	δ	∞	∞				**B2**^{2}, 28
—	—,**amino-**	*see* **Carbamic acid**													
—	—,**azodi-**	*see* **Azodicarboxylic acid**													
—	—,**chloro-**, amide	*see* **Carbamic acid**, chloride													
f150	—,—,benzyl ester	Carbobenzoxy chloride. $ClCO_2CH_2C_6H_5$	170.60			103^{20}			d	d^h	s				**B6**, 437
f151	—,—,butyl ester	$ClCO_2(CH_2)_3CH_3$	136.58			142	1.074^{25}_{4}	1.417^{8}	d	d	∞				**B3**^{2}, 111
f152	—,—,2-chloroethyl ester	$ClCO_2CH_2CH_2Cl$	142.97			155–60	1.420^{18}		i	s				oos s	**B2**^{2}, 193
f153	—,—,chloromethyl ester	$ClCO_2CH_2Cl$	128.94			107^{760}	1.465^{15}	1.4286^{22}	d^h			v			**B3**^{2}, 11
f154	—,—,3-chloropropyl ester	$ClCO_2(CH_2)_3Cl$	157.00			177	1.2949^{25}_{20}	1.4456^{20}	i						**B3**^{2}, 10
f155	—,—,cyclohexyl ester	$ClCO_2—C_6H_{11}$	162.62			87.5^{27}					s				
f156	—,—,dichloromethyl ester	$ClCO_2CHCl_2$	163.39			$110–1^{760}$			d^h			v			**B3**^{2}, 11
f157	—,—,2-ethoxyethyl ester	Cellosolve chloroformate. $ClCO_2CH_2CH_2OC_2H_5$	152.58			67.2^{14}	1.1341^{25}	1.4169^{25}	i	s					
f158	—,—,ethyl ester	Ethyl chloroformate. $ClCO_2C_2H_5$	108.53		−80.6	95^{760}	1.3577^{20}_{4}	1.3955^{20}	d	d	s		s	chl s	**B3**^{2}, 10
f159	—,—,isobutyl ester	$ClCO_2CH_2CH(CH_3)_2$	136.58			128.8	1.0426^{18}_{4}		d	s^h	∞		s	chl s	**B3**^{2}, 11
f160	—,—,isopropenyl ester	$ClCO_2C(CH_3){:}CH$	119.53			100^{760}	1.103^{20}_{20}				s				
f161	—,—,isopropyl ester	$ClCO_2CH(CH_3)_2$	122.55			$103–4^{723}$			i		s				**B3**^{2}, 10
f162	—,—,2-methoxyethyl ester	$ClCO_2CH_2CH_2OCH_3$	138.55			58.7^{13}	1.1905^{25}	1.4163^{25}	i		v				
f163	—,—,methyl ester	Methyl chloroformate. $ClCO_2CH_3$	94.50			$72–3^{767}$	1.2231^{20}_{4}	1.3868^{20}	d	∞	∞		s	chl s	**B3**^{2}, 9
f164	—,—,3-methylbutyl ester	Isoamyl chloroformate. $ClCO_2CH_2CH_2CH(CH_3)_2$	150.61			154.3	1.0288^{17}_{4}		d	∞	∞				**B3**^{2}, 11
f165	—,—,pentyl ester	Amyl chloroformate. $ClCO_2(CH_2)_4CH_3$	150.61			$60–2^{15}$		1.4181^{15}			s				
f166	—,—,propyl ester	$ClCO_2CH_2CH_2CH_3$	122.55			115.2	1.0961^{20}_{4}	1.4035^{20}	d	∞	∞				**B3**^{2}, 10
f167	—,—,trichloromethyl ester	Diphosgene. Perchloromethyl formate. Superpalite. $ClCO_2CCl_3$	197.83		−57	128^{760}	1.6525^{14}	1.4566^{22}	i	s	v				**B3**^{2}, 8
—	—,**cyano-**	*see* **Oxalic acid**, mononitrile													
f168	—,**(ethoxy)dithio-**	Ethylxanthic acid. $C_2H_5OCS_2H$	122.21		ca. −53				δ					chl, CS_2 s $PhNO_2$ s	**B3**^{2}, 151
—	—,**fluoro-**, fluoride	*see* **Carbonyl fluoride**													
f169	**Formimidic acid, *N*-phenyl-**, ethyl ester	*N*-Phenylformininoethyl ether. $C_6H_5N{:}CHOC_2H_5$	149.19			$213–5^{761}$	1.0051^{20}_{4}	1.5279^{20}			s		s		**B12**, 239
f170	**Formonitrolic acid**	Methyl nitrolic acid. $HON{:}CHNO_2$	90.04	nd (eth)	64				s	s	s				**B2**, 92
f171	**Frangulin**	Franguloside. Rhamnoxanthin. $C_{21}H_{20}O_9$	416.39	ye or red (al or AcOEt) $[\alpha]_D$ −134 (80% aa, c=0.05)	249				i s^h	s^h			s^h	chl s^h aa s	**B31**, 74
—	**Fraxetin**	*see* **Coumarin 7,8-dihydroxy-6-methoxy-**													

For explanations, symbols and abbreviations see beginning of table.

No.	Name	Synonyms and Formula	Mol. wt.	Crystalline form, color and specific rotation	m.p. °C	b.p. °C	Density	n_D	Solubility: w	al	eth	ace	bz	other solvents	Ref.
	Fraxin														
f173	**Fraxin**	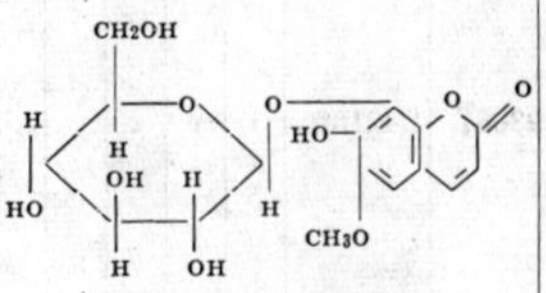	370.32	ye nd (al)	190 (205)				δ v^h	δ s^h	i				**B31**, 249
—	**Freon**	*see* **Methane, dichloro-* difluoro-***													
—	**Freon 114**	*see* **Ethane, 1,1-dichloro-1,2,2,2-tetra-fluoro-***													
f174	**1-Fructosamine** (*D*)	Isoglucosamine. 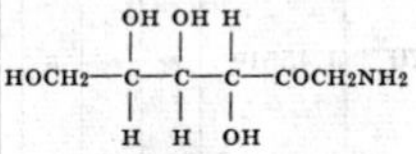	179.17	syr						i	i			dil ac s	**B31**, 342, 345
f175	**β-Fructose**(*D*)	Fruit sugar. Levulose. 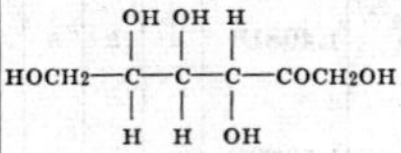	180.16	pr or nd (w) $[\alpha]^{20}_D -122$ (MeOH)	102–4				v	δ		s		MeOH s	**B31**, 321
f176	**—α-Fucose**(*L*)	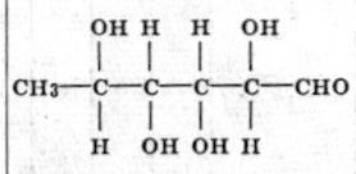	164.16	nd (al) $[\alpha]_D$ −124.1	145				v	δ	i				**B31**, 76, 78, 80
—	**3-methyl-**(*D*)	*see* Digitalose													
f177	**Fucoxanthin**	$C_{40}H_{56}O_6$	632.89	red-br pl(+2w) $[\alpha]^{18}_D +72.5$	166–8					s	δ			MeOH δ	**B30**, 105
—	**Fulminuric acid**	*see* **Malonicacid, nitro-, amide nitrile**													
f178	**Fumaric acid**	*trans*-Butenedioic acid*. $HO_2CCH:CHCO_2H$	116.07	nd, mcl pr or pl	286–7 (293–5)	$165^{1.7}$ sub	1.635^{20}_4		δ s^h	s	δ	δ		chl, CCl_4 δ con sulf s	**B2**[2], 631
f179	—,dichloride	Fumaryl chloride. $ClCOCH:CHCOCl$	152.97	pa ye liq		$158–60^{760}$	1.408^{20}								
f180	—,diiosbutyl ester	$(CH_3)_2CHCH_2O_2CCH:CHCO_2CH_2CH(CH_3)_2$	228.29			170^{160}			i	s					**B2**, 742
f181	—,diisopropyl ester	$(CH_3)_2CHO_2CCH:CHCO_2CH(CH_3)_2$	200.24			225–6				s	s				**B2**, 742
f182	—,dinitrile	*trans*-1,2-Dicyanoethylene. Fumaronitrile. $NCCH:CHCN$	78.07	nd	96	186^{760}	0.9416^{111}	1.4349^{111}	s	s	s				**B2**[1], 302
f183	—,diphenyl ester	$C_6H_5O_2CCH:CHCO_2C_6H_5$	268.27	nd (al)	161–2	219^{14}			i	δ s^h					**B6**, 156
f184	—,dipropyl ester	$CH_3CH_2CH_2O_2CCH:CHCO_2CH_2CH_2CH_3$	200.24				1.0120^{17}_4	1.4435^{20}		s	s				**B2**[2], 639
f185	—,piperazinium salt	$C_4H_{10}N_2.HO_2CCH:CHCO_2H$	202.21		240d				s^h	s	i				**Am70**, 2758
185[1]	—,**bromo-**	$HO_2CCBr:CHCO_2H$	194.98	pl	185–6	200d			s	s					
185[2]	—,**chloro-**	$HO_2CCCl:CHCO_2H$	150.52	pl (aa)	191–2	sub			v	v	v		δ		
f186	—,—,dichloride	$ClCOCCl:CHCOCl$	187.42	pa gr		184–7δd $73–5^{20}$	1.564^{20}_4	1.5206^{20}	d	d	v			aa s	**B2**[2], 640
f187	—,—,diethyl ester	$C_2H_5O_2CCH:CClCO_2C_2H_5$	206.63			250 δd 136.5^{19}	1.188^{20}_4	1.4571^{20}	i	v	v				**B2**[2], 640
f188	—,—,dimethyl ester	$CH_3O_2CCH:CClCO_2CH_3$	178.57			224	1.2899^{25}_4			s	s				**B2**[2], 640
f189	—,**dimethyl-**	$HO_2C(CH_3)C:C(CH_3)CO_2H$	144.12	nd (w)	241				δ s^h	δ			i	chl i	**B2**[2], 660
f190	—,**methyl-**	Mesaconic acid. $HO_2CC(CH_3):CHCO_2H$	130.10	rh nd or mcl pr (eth, AcOEt)	202–4	250d	1.466^{20}_4		δ v^h	v	s		δ	chl, CS_2 δ lig δ	**B2**[2], 651
f191	—,—,diethyl ester	$C_2H_5O_2CC(CH_3):CHCO_2C_2H_5$	186.20			229^{760}	1.0453^{20}_{20}	1.4488^{20}		s	s		s		**B2**[2], 652
f192	—,—,dimethyl ester	$CH_3O_2CC(CH_3):CHCO_2CH_3$	158.15			203.5	1.0914^{20}_4	1.4512^{20}	δ	s	s				**B2**[2], 652

For explanations, symbols and abbreviations see beginning of table.

No.	Name	Synonyms and Formula	Mol. wt.	Crystalline form, color and specific rotation	m.p. °C	b.p. °C	Density	n_D	Solubility						Ref.
									w	al	eth	ace	bz	other solvents	
	Fumigatin														
—	**Fumigatin**	see **1,4-Benzoquinone, 3-hydroxy-2-methoxy-5-methyl-***													
f193	**Furan**	1,4-Epoxy-1,3-butadiene*. Furfuran. (ring numbered 4, 3, 2, 5, O)	68.07	...	...	32^{758}	0.9366^{20}_{4}	1.4216^{20}	i	v	v	...	...	...	**B17^{2}**, 34
f194	**—,2-acetyl-**	2-Furyl methyl ketone. $C_6H_6O_2$. *See* f193	110.11	(peth)	28.5	173	1.098^{20}	1.5017^{20}	i	s	s	...	...	...	**B17^{2}**, 31
194^{1}	**—,2(amino-methyl)-**	Furfurylamine. C_5H_7NO. *See* f193	97.11	...	...	$145–6^{761}$	1.0995^{20}_{4}	1.4908^{20}	∞	s	s	...	...	...	**B18^{2}**, 41
f195	**—,2(amino-methyl)tetra-hydro-**	Terahydrofurfuryl amine. $C_5H_{11}NO$. *See* f220	101.15	...	...	$151–2^{735}$	0.9770^{20}_{20}	1.4551^{20}	∞	∞	s	...	...	...	**Am67**, 69
f196	**—,2-bromo-**	α-Furyl bromide. C_4H_3BrO. *See* f193	146.98	...	...	102^{744}	1.630	1.4980^{20}	δ	s	s	...	s	...	...
f197	**—,3-bromo-**	β-Furyl bromide. C_4H_3BrO. *See* f193	146.98	...	...	102.5^{745}	1.650^{20}_{4}	1.4981^{20}	i	s	s	...	s	...	**B17**, 27
f198	**—,2(bromo-methyl)-**	Furfuryl bromide. C_5H_5BrO. *See* f193	161.01	pa ye	...	$33–4^{2}$	1.560^{20}_{20}	1.5380^{20}	...	...	s	...	...	...	**B17^{2}**, 39
f199	**—,2(bromo-methyl)tetra-hydro-**	1-Bromo-2,5-epoxypentane*. C_5H_9BrO. *See* f220	165.04	...	...	$168–70^{744}$	...	...	...	s	s	...	...	...	**B17^{2}**, 21
f200	**—,2-*tert*-butyl-**	$C_8H_{12}O$. *See* f193	124.18	...	...	$116–7^{737}$	0.869^{20}_{4}	1.4373^{20}	i	s	s	...	...	...	...
f200^{1}	**—,2-chloro-**	α-Furyl chloride. C_4H_3ClO. *See* f193	102.52	...	77.5	1.1923^{20}_{4}	1.4571	...	i	s	...	...	...	...	...
f201	**—,3-chloro-**	β-Furyl chloride. C_4H_3ClO. *See* f193	102.52	...	79^{740}	1.2094^{20}_{4}	1.4601^{20}	...	δ	...	s	...	...	...	...
f201^{1}	**—,2-chloro-methyl-**	Furfuryl chloride. C_5H_5ClO. *See* f193	116.55	...	...	49^{26}	1.1783^{20}_{4}	1.4941	i	s	s	...	...	...	...
f202	**—,2,5-dibromo-**	$C_4H_2Br_2O$. *See* f193	225.88	pl	9–10	$164–5^{764}$	2.27^{20}_{20}	1.5455	...	...	...	...	...	...	**B17**, 28
f203	**—,2,5-di-*tert*-butyl-**	$C_{12}H_{20}O$. *See* f193	180.28	...	...	210^{760}	0.837^{20}_{4}	1.4369^{20}	i	s	s	...	...	...	...
f204	**—,2,5-dichloro-**	$C_4H_2Cl_2O$. *See* f193	136.96	...	...	115	1.371^{25}	...	...	...	...	...	...	...	...
f205	**—,2,2-diethyl-(tetrahydro)-**	$C_8H_{16}O$. *See* f220	128.21	...	...	145–50	0.8703^{20}_{4}	1.4317^{20}	i	...	...	...	...	os v	**B17^{2}**, 25
f206	**—,2,5-dimethyl-**	C_6H_8O. *See* f193	96.13	...	...	93^{764}	0.9000^{20}_{4}	1.4470^{20}	i	s	s	...	s	chl s aa s	**B17^{1}**, 20
f207	**—,2,4-dioxo-(tetrahydro-)**	Tetronic acid. β-Ketobutyro-lactone. $C_4H_4O_3$. *See* f220	100.07	pl (al-lig)	141	...	...	...	v^{h}	v^{h}	δ	...	δ	chl δ lig δ	**B17**, 403
f207^{1}	**—,2,5-dinitro-**	$C_4H_2N_2O_5$. *See* f193	158.07	nd (w), pr (al)	101	...	...	...	...	i	s	...	...	...	...
f207^{2}	**—,2,5-diphenyl-**	$C_{16}H_{12}O$. *See* f193	220.26	nd or lf (dil al)	91	243–5	...	...	i	v	v	...	...	oos s	...
f208	**—,2-ethoxy-**	Ethyl 2-furyl ether. $C_6H_8O_2$. *See* f193	112.12	...	...	125–6	0.9849^{23}_{4}	1.4500^{23}	...	...	...	...	...	...	...
f209	**—,2-ethyl-**	C_6H_8O. *See* f193	96.12	...	...	$92–3^{768}$	0.912^{13}_{15}	1.4466^{13}	...	...	...	...	...	...	...
f210	**—,2-ethyl-(tetrahydro)-**	$C_6H_{12}O$. *See* f220	100.16	...	...	107^{760}	0.8609^{16}_{16}	1.4169						os s	**B17^{2}**, 23
—	**—,(hydrox-methyl)-**	see **Methanol, furyl-**													
f210^{1}	**—,2-Iodo-**	C_4H_3IO. *See* f193	193.98	...	...	$43–5^{15}$	2.024^{20}_{4}	1.5661^{20}	...	...	s	...	...	...	...
f210^{2}	**—,3-Iodo-**	C_4H_3IO. *See* f193	193.98	...	...	132.2^{732}	2.045^{20}_{4}	...	i	...	s	...	...	...	...
f211	**—,2-isopropoxy-**	Isopropyl 2-furyl ether. $C_7H_{10}O_2$. *See* f193	126.15	...	...	135–6	0.9689^{20}_{4}	1.4419^{20}	...	...	...	...	...	...	...
f212	**—,2-methoxy-**	Methyl 2-furyl ether. $C_5H_6O_2$. *See* f193	98.10	...	...	110–1	1.0646^{25}_{4}	1.4468^{25}	...	...	...	...	...	...	...

For explanations, symbols and abbreviations see beginning of table.

No.	Name	Synonyms and Formula	Mol. wt.	Crystalline form, color and specific rotation	m.p. °C	b.p. °C	Density	n_D	Solubility: w	al	eth	ace	bz	other solvents	Ref.
	Furan														
f213	**—,2-methyl-**	Sylvan. C_5H_6O. *See* f193	82.10			63^{737}	0.9132^{20}_{4}	1.4342^{20}	i	s	s				**B17**[2], 39
f214	**—,3-methyl-**	C_5H_6O. *See* f193	82.10			65–6	0.923^{18}_{4}	1.4255^{18}	i	s	s				**B17**[2], 40
f215	**—,2-methyl-(tetrahydro)**	$C_5H_{10}O$. *See* f220	86.14			78^{750}	0.8552^{20}_{4}		s	v	v			chl, os v	**B17**[2], 21
f216	**—,3-methyl-(tetrahydro-)**	$C_5H_{10}O$. *See* f220	86.14			83^{738}	0.8642^{20}_{4}	1.4112^{20}	s					os v	**B17**[2], 21
f217	**—,2-nitro-**	$C_4H_3NO_3$. *See* f193	113.07	mcl (peth)	29	$133–5^{123}$			i		s				**B17**, 28
f218	**—,2-phenyl-**	$C_{10}H_8O$. *See* f193	144.17			$107–8^{18}$	1.083^{20}_{4}	1.5920^{20}							**B17**[2], 64
f219	**—,2-propyl-**	$C_7H_{10}O$. *See* f193	110.15			$114–5^{750}$	0.8905^{20}_{4}	1.4459^{20}							
f219[1]	**—,2-propyl(tetrahydro)-**	$C_7H_{14}O$. *See* f220	114.18			114^{750}	0.8547^{20}_{4}	1.4242^{20}	i						**B17**[1], 11
f220	**—,tetrahydro-**	1,4-Epoxybutane*. Tetramethylene oxide. (ring numbered 4, 3, 2, 5, O)	72.10		fr −65	64–5	0.888^{21}_{4}	1.4076^{21}	v	v	v			os v	**B17**[2], 15
f220[1]	**—,tetraiodo-**	C_4I_4O. *See* f193	571.69	nd	165				i	s				lig δ	**B17**[2], 34
f221	**—,2,3,5-trichloro-**	C_4HCl_3O. *See* f193	171.42			147	1.50^{25}								
—	**2-Furanacrylic acid**	*see* **Propenoic acid, 3(2-furyl)-**													
—	**2-Furancarboxaldehyde**	*see* Furfural													
f222	**2-Furancarboxylic acid**	α-Furoic acid. Pyromucic acid. (ring numbered 4, 3, 2, 5, O; 2-CO_2H)	112.09	mcl nd or lf (w)	133	230–2			s v^h	s	v				**B18**[2], 265
f223	—,allyl ester	$C_8H_8O_3$. *See* f222	152.15			$206–9^{760}$	1.118^{25}_{25}	1.4945^{20}							
f224	—,benzyl ester	$C_{12}H_{10}O_3$. *See* f222	202.21	ye		$179–81^{18}$	1.1632^{20}_{4}	1.5550^{20}	i						**B18**[2], 267
f225	—,butyl ester	$C_9H_{12}O_3$. *See* f222	168.19			233	1.0555^{20}_{4}	1.4740	i	s	s		s	peth s	**B18**[2], 266
f226	—,*sec*-butyl ester	$C_9H_{12}O_3$. *See* f222	168.19			$67–9^{1}$	1.0465^{20}_{4}		i	∞	∞				**B18**[2], 266
f227	—,chloride	$C_5H_3ClO_2$. *See* f222	130.53		−2 (+1)	173–4			i d^h	d	s			chl s	**B18**[2], 267
f228	—,ethyl ester	$C_7H_8O_3$. *See* f222	140.14	lf or pr	34–5	196.8^{760}	1.1174^{21}_{4}	1.4791^{21}	i	∞	∞		s	peth ∞	**B18**[2], 266
f229	—,furfuryl ester	(furyl)—CO_2CH_2—(furyl)	192.17	dimorphic	27.5	122^{2}	1.2384^{25}_{25}	1.5280^{20}	i	s	s		∞	chl ∞	**B18**[2], 267
f230	—,heptyl ester	$C_{12}H_{18}O_3$. *See* f222	210.28			$116–7^{1}$	1.0005^{20}_{4}		i	s					**B18**[2], 267
f231	—,hexyl ester	$C_{11}H_{16}O_3$. *See* f222	196.25			$105–7^{1}$	1.0170^{20}_{4}		i	s			∞	MeOH s	**B18**[2], 267
f232	—,isobutyl ester	$C_9H_{12}O_3$. *See* f222	168.19			$221–3^{760}$	1.0383^{28}_{4}	1.4676^{28}							**B18**[2], 266
f233	—,isopropyl ester	$C_8H_{10}O_3$. *See* f222	154.17			$198–9^{760}$	1.0655^{24}_{4}	1.4682^{24}							**B18**, 275
f234	—,methyl ester	$C_6H_6O_3$. *See* f222	126.11			181.3^{760}	1.1786^{21}_{4}	1.4860^{20}	i	s	s		s		**B18**[2], 266
f235	—,3-methylbutyl ester	Isoamyl α-furoate. $C_{10}H_{14}O_3$. *See* f222	182.22			282^{760}	1.030^{20}_{4}	1.4274^{20}	i	∞				peth ∞ aa s	**B18**[2], 266
f236	—,nitrile	α-Furyl cyanide. (furyl)—CN	93.09			146^{738}	1.0822^{20}_{4}	1.4715^{25}							**B18**[2], 268
f237	—,octyl ester	$C_{13}H_{20}O_3$. *See* f222	224.30			$126–7^{1}$	0.9885^{20}_{4}		i	s					**B18**[2], 267
f238	—,pentyl ester	Amyl furoate. $C_{10}H_{14}O_3$. *See* f222	182.22			$95–7^{1}$	1.0335^{20}_{4}		i	∞					**B18**[2], 266
f239	—,piperazinium salt	$2(C_5H_4O_3).C_4H_{10}N_2$. *See* f222	310.31		234–6				s	s	δ				**Am70**, 2758

For explanations, symbols and abbreviations see beginning of table.

No.	Name	Synonyms and Formula	Mol. wt.	Crystalline form, color and specific rotation	m.p. °C	b.p. °C	Density	n_D	Solubility w	al	eth	ace	bz	other solvents	Ref.
	2-Furancarboxylic acid														
f240	—,propyl ester	$C_8H_{10}O_3$. *See* f222	154.17			210.9	1.0745^{26}_{4}	1.4737^{26}	i	s	s	...	s	peth s	**B18**[2], 266
f241	**3-Furancarboxylic acid**	β-Furoic acid. (ring: 4 3 —CO_2H, 5 2, O 1)	112.09	nd (w)	122–3	$105–10^{12}$			δ s^h	s	v	...	...	AcOEt s	**B18**[1], 439
f242	—,methyl ester	$C_6H_6O_3$. *See* f241	126.11			160^{760}	1.1744^{15}_{15}	1.4676^{20}	...	s					**B18**[1], 439
f243	**2-Furancarboxylic acid, 3-bromo-**	$C_5H_3BrO_3$. *See* f222	190.99	nd (w)	128–9				δ	s	s	...	δ	chl δ	**B18**, 284
f244	—,**4-bromo-**	$C_5H_3BrO_3$. *See* f222	190.99	nd (w)	128–9				δ	s	s	...	s	chl s lig δ CS_2 δ	**B18**[2], 268
f245	—,**5-bromo-**	$C_5H_3BrO_3$. *See* f222	190.99	lf (w)	190–1 (185–6)				δ	s	v	...	s	chl s lig i peth i	**B18**, 284
f246	—,—,ethyl ester	$C_7H_7O_3Br$. *See* f222	219.04	pr	17	235^{767}	1.528^{20}		i	s	s				**B18**, 284
f247	—,**3-chloro-**	$C_5H_3ClO_3$. *See* f222	146.53	pl or pr (w)	153–4				δ	s	s	...	s^h	chl s^h	**B18**, 282
f248	—,**4-chloro-**	$C_5H_3ClO_3$. *See* f222	146.53		149										**B18**, 282
f249	—,**5-chloro-**	$C_5H_3ClO_3$. *See* f222	146.53	lf	179–80				δ s^h	s	s	...	s^h		**B18**, 282
f250	—,—,methyl ester	$C_6H_5ClO_3$. *See* f222	160.56		40–2										
f251	—,**5(cyclohexyloxy)**	$C_{11}H_{14}O_4$. *See* f222	210.22		138–9				δ d^h	s					
f252	**3-Furancarboxylic acid, 2,5-dimethyl-**	Pyrotritaric acid. $C_7H_8O_3$. *See* f241	140.14	nd (w)	134–5	sub			s^h	s	v				**B18**, 297
f253	**2-Furancarboxylic acid 5-ethoxy-**	$C_7H_8O_4$. *See* f222	156.14		140–1				δ d^h	s					
f254	—,**5-iodo-**	$C_5H_3IO_3$. *See* f222	237.98		192										**C25**, 4262[6]
f255	—,**5-methoxy-**	$C_6H_6O_4$. *See* f222	142.11		136–41				δ d^h	s					
f256	—,**3-methyl-**	$C_6H_6O_3$. *See* f222	126.11	nd (w)	136–7				s^h						**B18**[1], 439
f257	—,—,ethyl ester	$C_8H_{10}O_3$. *See* f222	154.17	pl	47–8	205^{760}									**B18**[2], 271
f258	—,—,methyl ester	$C_7H_8O_3$ *See* f222	140.41	pl	36–8										**B18**, 293
f259	—,**4-methyl-**	$C_6H_6O_3$. *See* f222	126.11	hd (bz-peth)	131–2										
f260	—,**5-methyl-**	$C_6H_6O_3$. *See* f222	126.11	pl or nd (w)	109–10	105^{1}			δ v^h	s	s	...	δ	chl s	**B18**[2], 272
f261	—,—,methyl ester	$C_7H_8O_3$. *See* f222	140.14			98^{15}									**B18**[2], 272
f262	**3-Furancarboxylic acid, 2-methyl-**	$C_6H_6O_3$. *See* f241	126.11	(w)	102–3				...	s	s	...	...	peth s aa s	**B18**[2], 271
f263	—,—,ethyl ester	$C_8H_{10}O_3$. *See* f241	154.17			$85–7^{20}$	1.0102^{25}_{4}		i	...	s				**B18**[1], 439
f264	—,**4-methyl-**	$C_6H_6O_3$. *See* f241	126.11	nd (bz-peth)	138–9				s						
f265	—,**5-methyl-**	$C_6H_6O_3$. *See* f241	126.11	(w)	119										
f266	**2-Furancarboxylic acid, 5-nitro-**	$C_5H_3NO_5$. *See* f222	157.08	pa ye pl (w)	185	sub			s^h	s	s	...	δ	chl i	**B18**, 287
f267	—,**5-octyloxy-**	$C_{13}H_{20}O_4$. *See* f222	240.30		125–6				δ d^h	s					
—	**3-Furancarboxylic acid, 5-oxo(tetrahydro)-**	*see* **Paraconic acid**													
f268	**2-Furancarboxylic acid, 5-phenoxy-**	$C_{11}H_8O_4$. *See* f222	204.18		122–3				δ d^h	s					
f269	—,**tetrahydro-**	(ring O)—CO_2H	116.12		21	$131–2^{14}$	1.1933^{20}_{4}	1.4612^{20}	s						**B18**[2], 262
f270	**2,3-Furandicarboxylic acid**	(ring O)—CO_2H, —CO_2H	156.10	pr (aa or sub)	224–5				v	v	δ	...	...	AcOEt v aa δ v^h	**B18**[2], 287
f271	—,dimethyl ester	$C_8H_8O_5$. *See* f270	184.15		37				i	s	v				**B18**[2], 288
f272	**2,4-Furandicarboxylic acid**	HO_2C—(ring O)—CO_2H	156.10	lf (w+1)	266 (anh)	sub			δ v^h	v	...	v	...	chl δ CS_2 δ aa δ	**B18**, 327
f273	—,dimethyl ester	$C_4H_2O(CH_3CO_2)_2$	184.14	pr (MeOH)	109–10										**B18**, 327
f273[1]	**2,5-Furandicarboxylic acid**	Dehydromucic acid. HO_2C—(ring O)—CO_2H	156.09	nd (w), lf (al)	>320	sub			δ	δ					**B18**[2], 288

For explanations, symbols and abbreviations see beginning of table.

No.	Name	Synonyms and Formula	Mol. wt.	Crystalline form, color and specific rotation	m.p. °C	b.p. °C	Density	n_D	Solubility w	al	eth	ace	bz	other solvents	Ref.
	2,5-Furandicarboxylic acid														
f274	—,dichloride	Dehydromucyl chloride. $C_6H_2Cl_2O_3$. *See* f273	192.99	nd (sub)	80	ca. 245			v	v	v			chl v	**B18**, 330
f275	—,dimethyl ester	$C_8H_8O_5$. *See* f273[1]	184.14	nd (w)	112	$154–6^{15}$			i	s	s				**B18**, 329
f275[1]	**3,4-Furandicarboxylic acid.**	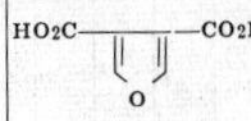 HO_2C——CO_2H O	156.09		217–8										**Am 66**, 51
f276	**Furazan, 3,4-dimethyl-**	3,4-Dimethyl-1,2,5-oxadiazole 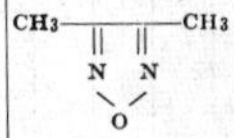 CH_3——CH_3 N N O	98.11		−7	156^{744}	1.0528^{14}_{4}	1.4271^{19}	δ	v	v				**B27²**, 622
f277	**Furfural**	Fural. 2-Furaldehyde. 2-Furancarboxaldehyde. 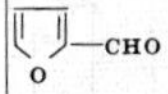 —CHO O	96.08		−38.7	161.7^{760}	1.1598^{20}_{4}	1.5261^{20}	s v^h	v	∞				**B17²**, 305
f278	—,diacetate	Furfurylidene diacetate. 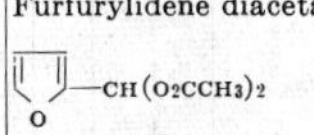 —$CH(O_2CCH_3)_2$ O	198.17	nd or pl (eth-peth)	52–3	220^{760}			δ	s	v		v	peth δ	**B17²**, 309
f279	—,diethyl acetal	$C_9H_{14}O_3$. *See* f277	170.21			$184–5^{740}$	0.9994^{20}_{20}	1.4451^{20}	i	s					**B17²**, 309
f280	—,oxime (*anti*)	α-Furfuraldoxime. $C_5H_5NO_2$. *See* f277	111.10	nd (lig)	75–6				s	v	v		v	lig δ s^h	**B17²**, 311
f281	—,—(*syn*)	β-Furfuraldoxime. $C_5H_5NO_2$. *See* f277	111.10	nd (lig)	91–2	201–8 d 98^9			δ	v	v		v	CS_2 v aa v lig s^h	**B17²**, 311
f282	—,phenylhydrazone	$C_{11}H_{10}N_2O$. *See* f277	186.21	ye lf (al)	96				i	v	v			lig δ	**B17²**, 312
f283	**—,5-bromo-**	$C_5H_3BrO_2$. *See* f277	174.99		82	112^{16}			δ	s	s				
f284	**—,5-chloro-**	$C_5H_3ClO_2$. *See* f277	130.53		31–3	70^1									
f285	**—,5-hydroxymethyl-**	$C_6H_6O_3$. *See* f277	126.11	nd (eth-peth)	35	$110^{0.02}$	1.2062^{25}_{4}	1.5627^{18}	s	s	s		s	chl s, CCl_4 δ	**B18²**, 6
f286	**—,5-methyl-**	$C_6H_6O_2$. *See* f277	110.11			187^{760}	1.1072^{18}_{4}		s	v	∞				**B17²**, 315
f287	**—,5-nitro-**	$C_5H_3NO_4$. *See* f277	141.08	pa ye	35–6	$128–32^{10}$			δ					peth s	**Am 52**, 2550
f288	—,—,semicarbazone	Furacin. Nitrofurazone. $C_6H_6N_4O_4$. *See* f277	198.14	pa ye nd, darkens in light	236–40 d				i	δ	i			alk s	
f289	—,tetrahydro-	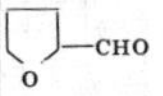 —CHO O	100.11			$142–3^{779}$	1.0727^{20}_{4}	1.4366^{20}	s		s				
f290	**Furfurine**	Furin. 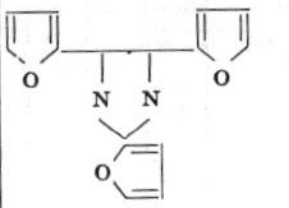 O N N O O	268.28	lt br nd or rh pr (w or eth)	116				$δ^h$	v	v				**B27²**, 918
—	**Furfuryl alcohol**	*see* **Methanol, 2-furyl-**													
—	**Furfurylamine**	*see* **Furane, 2(aminomethyl)-**													
—	**2-Furfurylfuran**	*see* **Methane, di-2-furyl-**													
—	**Furfuryl mercaptan**	*see* **Methanethiol, 2-furyl-**													
—	**Furil**	*see* **1,2-Ethanedione, 1,2-difuryl-**													
—	**Furoic acid**	*see* **Furancarboxylic acid**													
f291	**Furoin**	Furoylfurylcarbinol. 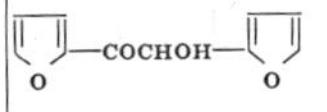 —COCHOH— O O	192.17	lt br pr	135				$δ^h$	δ	s				**B19²**, 224

For explanations, symbols and abbreviations see beginning of table.

No.	Name	Synonyms and Formula	Mol. wt.	Crystalline form, color and specific rotation	m.p. °C	b.p. °C	Density	n_D	Solubility w	al	eth	ace	bz	other solvents	Ref.
	Gaidic acid														
—	**Gaidic acid**	*see* **1-Hexadecanoic acid**													
g1	**Galactitol**(*D*)	*D*-Dulcitol. $HOCH_2—CH(OH)—CH(OH)—CH(OH)—CH(OH)—CH_2OH$	182.17		188.5	sub			s v^h	i	i				**B1**[2], 612
g2	**Galactonic acid, α-lactone**(*D*)	$HOCH_2—CH(OH)—CH(OH)—CH(OH)—CH(OH)—CO$ (ring —O—)	178.14	nd (al or AcOEt) $[\alpha]^{20}_D -70.1$ (w)	112				s					AcOEt v	**B18**[1], 408
g3	**Galactose**(*D*)	$HOCH_2—CH(OH)—CH(OH)—CH(OH)—CH(OH)—CHO$	180.16	pl or pr (al) $[\alpha]+81.5$	165.5				v^h	δ				Py s	**B1**, 909
g4	—,3,6-an-hydro-(*D*)	$CH_2—CH(OH)—CH(OH)—CH—CH(OH)—CHO$ (ring —O—)	162.14	lf (PrOH) $[\alpha]_D+24$(w)	123–5				s						
g5	—,2,3,4,6-tetra-O-methyl-(*D*)	$CH_3OCH_2—CH—CH(OCH_3)—CH(OCH_3)—CH(OCH_3)—CHOH$ (ring —O—)	236.27	$[\alpha]^{20}_D+102.2$ (w, p=3) (mut)	72	172^{12}			s	s	s				**B31**, 304
g6	**Galacturonic acid**(*D*)	$OHC—CH(OH)—CH(OH)—CH(OH)—CH(OH)—COOH$	194.14	$[\alpha]_D+53.6$ (mut)	159–60d				s	s^h	i				**B3**[1], 306
g7	**Galegine**	4-guanidino-2-methyl-2-butene. γ,γ-Dimethylallyl-guanidine. $C(CH_3)_2:CH_1CH_2N:C(NH_2)_2$	127.19	hyg	60–5	d			v	v	δ			chl δ lig δ	**B4**[2], 672
g8	**Galipine**	$C_{20}H_{21}NO_3$	323.40	pr (al, eth), nd (peth)	115.5				δ	v	v	v	v	chl v peth δ	**B21**[2], 171
—	**Gallamide**	*see* **Benzoic acid, 3,4,5-tri-hydroxy-, amide**													
—	**Gallanilide**	*see* **Benzoic acid, 3,4,5-tri-hydroxy-**, amide, *N*-phenyl-													
g9	**Gallein**	4,5-Dihydroxyfluorescein. (structure: HO, HO, +O, CO_2^-, HO, HO)	364.31	gr cr or br-red pw	>300				i	v	δ	δ	δ	chl δ alk s aa δ	**B19**[2], 282
—	**Gallic acid**	*see* **Benzoic acid, 3,4,5-tri-hydroxy-***													
g10	**Gallin**	(structure: HO, HO, O, CO_2H, HO, HO)	365.32	nd (eth), turns red in air					δ	v	δ	v		aa v	**B18**, 368
—	**Gammexane**	*see* **Cyclohexane, 1,2,3,4,5,6-hexachloro-***													
—	**Gelsaminic acid**	*see* **Coumarin, 7-hydroxy-6-methoxy-**													
g12	**Gelsemine**(*d*)	Gelseminin. (structure: $CH_3—N$, C, O, N—H)	322.41	(ace) $[\alpha]^{20}_D$ +17.8 (chl)					i	s	s		s	chl s	**B27**[2], 720

For explanations, symbols and abbreviations see beginning of table.

No.	Name	Synonyms and Formula	Mol. wt.	Crystalline form, color and specific rotation	m.p. °C	b.p. °C	Density	n_D	Solubility w	al	eth	ace	bz	other solvents	Ref.
	Gelsemine														
g13	—,hydrochloride(*d*)	$C_{20}H_{22}N_2O_2$.HCl. *See* g12	358.87	pr (w) $[\alpha]^{20}_D$+5 (w)	326				s	δ					**B27**, 720
—	**Genistein**	*see* **Isoflavone, 4',5,7-trihydroxy-**													
—	**Gentianin**	*see* **Xanthone, 1,7-dihydroxy-3-methoxy-**													
—	**Gentisic acid**	*see* **Benzoic acid, 2,5-dihydroxy-**													
—	**Gentisin**	*see* **Xanthone, 1,7-dihydroxy-3-methoxy-**													
—	**Gentisinaldehyde**	*see* **Benzaldehyde, 2,5-dihydroxy-**													
—	**Gentisyl alcohol**	*see* **Toluene, α,2,5-trihydroxy-**													
—	**Geranial**	*see* **Citral a**													
g14	**Geraniol**	3,7-Dimethyl-2,6-octadien-1-ol*. $C_{10}H_{18}O$	154.24			107^{8}	0.8830^{15}_{4}	1.4766^{20}_{D}	i	s					
g15	—,formate	$C_{11}H_{18}O_2$	182.26			229d, $113\text{–}4^{15}$	0.9086^{25}_{4}	1.4659^{20}	i	v	s				**B2**[2], 33
g16	—,**tetrahydro-(*dl*)**	3,5-Dimethyl-1-octanol*. $(CH_3)_2CH(CH_2)_3CH(CH_3)CH_2CH_2OH$	158.29			212–3	0.8302^{20}_{4}	1.4351^{20}		s	s		s		**B1**[2], 460
g17	**Geranyl bromide**	8-Bromo-2,6-dimethyl-2,6-octadiene*. $BrCH_2CH{:}C(CH_3)CH_2CH_2CH{:}C(CH_3)_2$	217.16			$101\text{–}2^{12}$	1.0940^{22}_{4}	1.4858^{22}	d^h	s	s				**B1**[2], 240
g18	**Germanidine**	$C_{37}H_{57}NO_{10}$	675.86	nd $[\alpha]^{24}_D$−30 ±2 (al)	221–2 (vac)				i						**Am 75,** 4925
g19	**Germanitrine**	$C_{39}H_{59}NO_{11}$	717.90	nd $[\alpha]^{24}_D$−61 ±2 (Py)	228–9 (vac)				i						**Am 75,** 4925
g20	**Germerine**	$C_{37}H_{59}NO_{11}$	693.88	$[\alpha]_D$+16 (chl), −7 (Py)	200–3								s	MeOH s^h, alk s, ac s	**Am 72,** 4625
g21	**Germidine**	$C_{34}H_{53}NO_{10}$	635.80	pl, $[\alpha]_D$+13 (chl), −11 (Py)	198–20 (230–1)					s^h			s	MeOH s^h, ac s	**Am 72,** 4625
g22	**Germine**	$C_{27}H_{43}NO_8$	509.65	pr $[\alpha]_D$+5 (95% al)	218–21								s	MeOH s^h, alk s, ac s	
g23	**Germinitrine**	$C_{39}H_{57}NO_{11}$	715.89	pr $[\alpha]^{24}_D$ −36 (Py)	175 (vac)				i						**Am 75,** 4925
g24	**Germitrine**	$C_{39}H_{61}NO_{12}$	735.92	pr $[\alpha]_D$−4 (chl), −69 (Py)	216–9				i	d^h		s^h	s	chl s, ac s	**Am 72,** 4625
g25	**Gitogenin**	Digine. $C_{27}H_{44}O_4$	432.65	lf (bz) $[\alpha]^{20}_D$ −61 (w)	271–2				i	s^h	δ			chl s	**Am 77,** 5661
g26	**Gitoxigenin**	Bigitaligenin. Hydroxy-digitoxin. $C_{23}H_{34}O_5$	390.52	$[\alpha]^{20}_{546}$+38.5 (MeOH, c=0.7	lf (MeOH) 224–5d										
g27	**Gitoxin**	Bigitalin. Anhydrogitalin. $C_{41}H_{74}O_{14}$	780.96	nd $[\alpha]^{20}_{546}$+3.5 (Py, c=1)					δ	δ				chl δ	**E14**, 243
—	**Glucitol**	*see* **Sorbitol**													
g28	**Glucoascorbic acid**(*D*)	$HOCH_2$—CH(OH)—CH(OH)—C(OH)=C(OH)—C(=O), ring —O— (structure: H H / HOCH2—C—C—C=C—C=O / OH OH OH OH OH)	206.15	rods (ace-MeOH-peth) $[\alpha]^{20}_D$ −22 (MeOH), −80 (NaOH)	190 (anh) 138 (+1w)				s	s				MeOH s	
g29	**Gluco-α-heptose**(*D*)	$HOCH_2$—CHOH—CHOH—CHOH—CHOH—CHOH—CHO (structure: OH OH H OH OH / HOCH2—C—C—C—C—C—CHO / H H OH H H)	210.19	rh pl (w) $[\alpha]^{20}_D$ max −19.7 (w) (mut)	210–15				v	δ					**B31**, 359

For explanations, symbols and abbreviations see beginning of table.

No.	Name	Synonyms and Formula	Mol. wt.	Crystalline form, color and specific rotation	m.p. °C	b.p. °C	Density	n_D	Solubility w	al	eth	ace	bz	other solvents	Ref.
	Glucomethylose														
g30	**Gluco-methylose**(*D*)	Epifucose. *D*-Isorhamnose. Quinovose. CH_3—C—C—C—C—CHO (H H OH H / OH OH H OH)	164.16	(AcOEt) (mut) $[\alpha]^{20}_D +73$ to $+29.7$	134–5	....	....	....	v	v	δ	δ	...	AcOEt δ^h	....
g31	**Gluconal**(*d*)	HO— (ring, O) —CH_2OH	146.14	hyg nd $[\alpha]^{22}_D -7$ (w)	60	....	....	....	δ	δ	i	...	...	MeOH δ	**B31**, 116 **B17**², 214
g32	**Gluconic acid**(*D*).	Dextronic acid. $HOCH_2$—C—C—C—C—CO_2H (OH OH H OH / H H OH H)	196.16	nd (al-eth) (mut) $[\alpha]^{25}_D -3.49$ to $+12.95$	125–6	....	....	....	s d^h	i	i	...	...	....	**B3**², 352
g33	—,γ-lactone	$HOCH_2$—C—C—C—C—CO (ring —O—) (OH H OH / H H OH H)	178.14	nd $[\alpha]_D +68$ (w)	130–5	....	....	....	...	v^h	...	...	...	....	**B18**. 203
g34	—,δ-lactone	$HOCH_2$—C—C—C—C—CO (ring —O—) (OH H OH / H H OH H)	178.14	nd (al) $[\alpha]^{25}_D$ (mut) $+66.7$ to $+6.2$	153	1.610^{-5}	....	....	d	...	...	...	...	....	**B18**², 190
g35	—,nitrile(*d*)	*d*-Glucononitrile. $HOCH_2$—C—C—C—C—CN (OH OH H OH / H H OH H)	177.16	pl (al) $[\alpha]^{21}_D +8.8$	120.5	....	....	....	v	δ	...	δ	...	MeOH s Py s	**B3**², 351
g36	—,phenyl-hydrazide(*D*)	$HOCH_2$—C—C—C—C—$CONHNHC_6H_5$ (OH OH H OH / H H OH H)	286.27	pr (w) $[\alpha] +18$ $(+13)$ (w)	195	....	....	....	s	δ^h	i	...	...	....	**B15**², 122
g37	—,**5-oxo-**(*D*)	$HOCH_2CO$—C—C—C—CO_2H (OH H OH / H OH H)	194.14	cr or syr $[\alpha]_D -14.5$ (w)	125–6	....	....	....	s	s	δ	...	...	....	**B3**, 883
—	**Glucosamine**.....	*see* **Glucose, amino-**													
g38	**Glucose**(*D*) (equilibrium mixture)	Dextrose. $HOCH_2$—C—C—C—C—CHO (H H OH H / OH OH H OH)	180.16	$[\alpha]^{20}_D +52.7$ (w)	146 (150)	....	....	....	s	s^h	...	...	...	Py s^h	**B1**, 1879
g39	—(*D*)(α)	$HOCH_2$—C—C—C—C—C—OH (ring —O—) (H OH H / H OH H OH H)	180.16	rods, cubes, orh, nd (al) $[\alpha]^{20}_D$ (mut) $+112.2$ to $+52.7$ (w, c=4)	146d	....	1.5620^{18}_4	....	v	δ	...	i	...	MeOH δ AcOEt i	**B31**, 83
g40	—,mono-hydrate(*D*)(α)	$C_6H_{12}O_6.H_2O$	198.17	lf or pl or orh $[\alpha]^{20}_D$ (mut) $+102$ to $+47.9$	83	....	....	....	v	δ	i	...	...	....	**B31**, 86
g41	—(*D*)(β)	$HOCH_2$—C—C—C—C—CH (ring —O—) (H OH H / OH OH H OH OH)	180.16	nd (al), (w+1) $[\alpha]^{20}_D$ (mut) $+18.7$ to $+527$ (w, c=4)	150	....	1.5620^{18}_4	....	v	δ	i	...	...	....	**B31**, 89
g43	—,penta-acetate(*D*)(α)	$C_{16}H_{22}O_{11}$. *See* g39..........	390.35	pl or nd (al) $[\alpha]^{20}_D +100.9$ (al, c=0.5)	112–3	sub	....	....	δ	δ	s	...	...	chl s aa s CS_2 δ lig δ	**B31**, 119
g44	—,—(*D*)(β)	$C_{16}H_{22}O_{11}$. *See* **g41**..........	390.35	nd (al) $[\alpha]_D +3.9$ (chl, c=6)	134	sub vac	....	....	i	δ	δ	...	s	peth δ chl ∞	**B31**, 120

For explanations, symbols and abbreviations see beginning of table.

No.	Name	Synonyms and Formula	Mol. wt.	Crystalline form, color and specific rotation	m.p. °C	b.p. °C	Density	n_D	Solubility w	al	eth	ace	bz	other solvents	Ref.
	Glucose														
g45	—,pentamethyl-ether(*D*)	$C_{11}H_{22}O_6$. *See* g38	250.29	$[\alpha]^{20}_D$+147.4 (w, p=10)		$108^{0.1}$	1.082^{20}_4	1.4464^{20}	s	s	s	s			**B31**, 180
g46	—,*α*-phenyl-hydrazone(*D*)	H H OH H HOCH$_2$—C—C—C—C—CH:NNHC$_6$H$_5$ OH OH H OH	270.29	pl (al) $[\alpha]^{25}_D$ (mut) −8.7 to −52.5 (w, c=2)	160				s	δ^h	δ^h				**B31**, 173
g47	—,*β*-phenyl-hydrazone(*D*)	$C_{12}H_{18}N_2O_5$. *See* g46	270.29	pr, nd $[\alpha]^{19}_D$ (mut) −4.5 to −53.7 (w-Py)	140–1				s	δ	i				**B31**, 174
g48	—,phenyl-osazone(*D*)	*D*-Fructosazone. *D*-Glucosazone. *D*-Mannosazone. H H OH NNHC$_6$H$_5$ HOCH$_2$—C—C—C—C—C:NNHC$_6$H$_5$ OH OH H	358.40	ye nd (dil al) $[\alpha]_D$−41 (MeOH), −0.35 (al-Py)	208d				i	s^h				lig SO_2 i	**B31**, 350
g49	—,—(*DL*)	$C_{18}H_{22}N_4O_4$. *See* g48	358.40	ye nd (al)					i	δ	i		i		**B31**, 355
g50	—,**2-amino-**(*D*)	Chitosamine. *D*-Glucosamine. H H OH H HOCH$_2$—C—C—C—C—CHO OH OH H NH$_2$	179.17	nd(al, MeOH) $[\alpha]_D$+47.1 (w, c=0.4)	110d				v	δ	i			MeOH δ chl i	**B31**, 167
g51	—,—,hydro-chloride(*D*)	$C_6H_{13}NO_5HCl_2$. *See* g50	215.64	mcl (w or dil al) $[\alpha]^{20}_D$+100					v	s^h					**B31**, 169
—	—,**6-benzoyl-**(*D*)	*see* **Vacciniin**													
g52	—,**2(methyl-amino)-**	*N*-Methyl-2-glucosamine. H H OH H HOCH$_2$—C—C—C—C—CHO OH OH H NHCH$_3$	193.20	gummy $[\alpha]^{25}_4$ (mut) −65 (MeOH, c=1)	130–2d				s						**B31**, 186
g53	—,**thio-**	Glucothiose. H H OH H HOCH$_2$—C—C—C—C—CHS OH OH H OH	196.22	hyg pw $[\alpha]^{30}_D$ (mut) +48.7 (w, c=1.4)	70				s						
g54	**Glucoside, *α*-methyl-**(*D*)	Methyl-*α*-*D*-glucopyranoside. H OH H HOCH$_2$—C—C—C—C—C—OCH$_3$ H OH H OH H	194.18	rh nd (al) $[\alpha]^{20}_D$+158	168	$200^{0.2}$	1.46^{30}_4		v	δ	i			MeOH s	**B31**, 179
g55	—,***β*-methyl-**(*D*)	Methyl-*β*-*D*-glucopyranoside. H OH H HOCH$_2$—C—C—C—C—CH H OH H OH OCH$_3$	194.18	tetr (al) $[\alpha]^{20}_D$ −34.2 (w, p=10)	108–10				v	δ	i				**B31**, 182
—	**Glucothiose**	*see* **Glucose, thio-**													
—	**Glucurone**	*see* **Glucuronic acid**, lactone.													
g56	**Glucuronic acid**(*D*)	H H OH HC—C—C—C—C—CO$_2$H OH OH OH H H	194.14	nd (al) $[\alpha]^{20}_D$ (mut) +12.4 to +35.9 (w)	165				s	s					**B31**, 261
g57	—,*α*-lactone	Glucurone. H H H O:C—C—C—C—C—CHO OH OH H OH	176.13	mcl $[\alpha]^{22}_4$ +19.4 (w)	177				s						**B31**, 264

For explanations, symbols and abbreviations see beginning of table.

No.	Name	Synonyms and Formula	Mol. wt.	Crystalline form, color and specific rotation	m.p. °C	b.p. °C	Density	n_D	Solubility w	al	eth	ace	bz	other solvents	Ref.
	Glutamic acid														
g58	**Glutamic acid**(*D*)	α-Aminoglutaric acid. 2-Aminopentanedioic acid.* $HO_2CCH_2CH_2CH(NH_2)CO_2H$	147.13	pl (w) $[\alpha]_D^{25}$ −31.7 (1.7N HCl)	206	....	1.538_4^{20}	....	δ	δ	i	...	...	....	**B4**², 910
g59	—(*DL*)........	$HO_2CCH_2CH_2CH(NH_2)CO_2H$	147.13	rh (al)	193–4	....	1.4601_4^{20}	....	δ s^h	i	i	...	...	CS_2 i lig i	**B4**², 911
g60	—(*L*,+)........	$HO_2CCH_2CH_2CH(NH_2)CO_2H$	147.13	orh $[\alpha]_D^{20}$+30.4 (HCl)	224–5	sub vac	1.538_4^{20} (vac)	....	δ s^h	i	...	i	...	MeOH i	**B4**², 902
g61	—,hydrochloride(*L*)	$HO_2CCH_2CH_2CH(NH_2)CO_2H.HCl$	183.59	rh, $[\alpha]_D^{19}$+31.1 (dil HCl)	213	....	....	....	s	s	...	...	...	....	....
g62	—,*N*-**acetyl**-(*L*)...	$HO_2CCH_2CH_2CH(NHCOCH_3)CO_2H$	189.17	$[\alpha]_D$−15.3 (w, c=2)	191	....	....	....	s^h	s^h	...	...	...	....	**B4**², 911
g63	—,**3-hydroxy**-(*D*)	$HO_2CCH_2CHOHCH(NH_2)CO_2H$	163.13	hyg pr (w)	100	d	....	....	v	i	i	...	...	aa v	**B4**², 946
g64	—,—(*dl*)........	$HO_2CCH_2CHOHCH(NH_2)CO_2H$	163.13	rh pr and nd	198d	....	....	....	v	i	...	...	...	os i	**B4**¹, 550
—	—,**pteroyl**-......	*see* **Folic acid**													
g65	**Glutamine**(*L*,+)..	α-Aminoglutaramic acid. $HO_2CCH(NH_2)CH_2CH_2CONH_2$	146.15	nd (w) $[\alpha]_D^{20}$ +32.5 (5%HCl)	184–5	....	....	....	s	i	i	...	i	MeOH i	....
—	**α-Glutamylcysteinglycine**	*see* **Glutathione**													
—	**Glutaraldehyde**...	*see* **Pentanedial**													
—	**Glutaric acid**.....	*see* **Pentanedioic acid***													
—	**Glutaronitrile**....	*see* **Pentanedioic acid,** dinitrile													
—	**Glutaryl chloride**	*see* **Pentanedioic acid,** dichloride													
g66	**Glutathione**......	α-Glutamylcystenylglycine. $HO_2CCH(NH_2)CH_2CH_2CONHCH(CH_2SH)CONHCH_2CO_2H$	307.32	$[\alpha]_{546}^{15}$−18.5 (w)	190–2d	....	....	....	s	i	i	...	...	....	**B4**², 931
g67	**Glyceraldehyde** (*D*)	*D*-2,3-Dihydroxypropanal*. $HOCH_2CHOHCHO$	90.08	syr $[\alpha]_D$+13.5	....	....	....	....	...	...	...	...	...	....	**B1**¹, 427
g68	—(*dl*)........	$HOCH_2CHOHCHO$........	90.08	nd or pr (40% MeOH, al-eth)	145	....	1.453_4^{18}	....	s	δ	δ	...	i	peth i lig i	**B1**², 888
g68¹	—(*L*)........	$HOCH_2CHOHCHO$........	90.08	[α]−13.8 (w, c=11)	....	....	....	....	...	...	...	...	...	....	**B1**¹, 427
g69	—,dimethyl acetal-(*dl*)	3,3-diethoxy-1,2-propanediol*. $HOCH_2CHOHCH(OC_2H_5)_2$	164.21	....	....	136^{27}	....	....	∞	∞	∞	...	...	....	**B1**², 889
—	**Glyceric acid**.....	*see* **Propanoic acid, 2,3-dihydroxy-***													
—	**Glycerin**........	*see* **Glycerol**													
g70	**Glycerol**........	Glycerin. 1,2,3-Propanetriol*. $HOCH_2CHOHCH_2OH$	92.10	syr	18.6	290d	1.2613_4^{20}	1.4746^{20}	∞	∞	δ	...	i	chl, CCl_4 CS_2, peth	**B1**³, 2297
g71	—,borate........	$(C_3H_5BO_3)t$........	(99.89)	glass	ca. 150	....	....	....	d^h	...	...	...	...	....	**B1**, 519
g72	—,1(2-chlorophenyl) ether	$HOCH_2CHOHCH_2O-C_6H_4Cl$ (2-Cl)	202.64	nd (bz)	65	250^{19}	....	....	s^h	...	v	...	s^h	....	**B6**¹, 99
g73	—,1(4-chlorophenyl) ether	$HOCH_2CHOHCH_2O-C_6H_4-Cl$ (4-Cl)	202.64	nd (eth+peth), lf (bz)	76	$214–5^{19}$	....	....	i	v	v	...	s^h	con H_2SO_4s	**B6**¹, 101
g74	—,1,3-diacetate...	Diacetin. $HOCH(CH_2O_2CCH_3)_2$	176.17	....	40	259–61	1.184_4^{16}	1.44	v	v	δ	...	...	CS_2 i	**B2**², 160
g75	—,1,2-dibutanoate	α,β-Dibutyrin. $C_3H_7^nCO_2CH_2CH(CH_2OH)O_2CC_3H_7^n$	232.28	....	....	273.5	....	....	i	s	...	...	...	....	**B2**, 273
g76	—,1,3-didodecanoate	α,γ-Dilaurin. Glycerol-1,3-dilaurate. $HOCH[CH_2O_2C(CH_2)_{10}CH_3]_2$	456.69	pl (al), nd (eth-al)	57	....	....	....	...	s	s	...	s	chl s lig s	**B2**², 320
g77	—,1,3-dihexadecanoate	α,γ-Dipalmitin. Glycerol-1,3-dipalmitate. $HOCH[CH_2O_2C(CH_2)_{14}CH_3]_2$	568.89	(al or chl)	70	....	....	....	δ s^h	s^h	...	...	...	chl s^h	....

For explanations, symbols and abbreviations see beginning of table.

No.	Name	Synonyms and Formula	Mol. wt.	Crystalline form, color and specific rotation	m.p. °C	b.p. °C	Density	n_D	Solubility w	al	eth	ace	bz	other solvents	Ref.
	Glycerol														
g78	—,1,2-dimethyl ether(*dl*)	2,3-Dimethoxy-1-propanol*. $CH_3OCH_2CH(OCH_3)CH_2OH$	120.15	...	...	180^{760}	1.016_4^{25}	1.4200^{20}	∞	s	...	...	...	...	**B1**^{3}, 2317
g79	—,1,3-dimethyl ether	1,3-Dimethoxy-2-propanol*. $CH_3OCH_2CHOHCH_2OCH_3$	120.15	...	...	169^{760}	1.0085_4^{20}	1.4192^{20}	∞	v	...	...	...	...	**B1**^{3}, 2318
g80	—,1,3-dinitrate, hemihydrate	$O_2NOCH_2CHOHCH_2ONO_2 \cdot \frac{1}{2}H_2O$	191.11	pr (w)	26	148^{15}	1.47_4^{15}	...	s	v	s	...	...	...	...
g81	—,1,3-dioctadecanoate	α,γ-Distearin. Glycerol 1,3-distearate. $[CH_2O_2C(CH_2)_{16}CH_3]_2$	625.03	nd or pl (eth, chl, lig)	29.1	...	...	...	...	δ s^h	δ s^h	...	...	os s^h	**B2**^{2}, 356
g82	—,1,3-diphenyl ether	$C_6H_5OCH_2CHOHCH_2OC_6H_5$	244.29	lf (al)	81–2	...	1.179_4^{24}	...	i	v^h	v	...	v	chl v	**B6**^{2}, 152
g83	—,—,2-acetate	$CH_3CO_2CH(CH_2OC_6H_5)_2$	286.33	(dil al)	70–1	...	...	...	i	v	v	...	v	chl v	**B6**, 149
g84	—,1,3-dipropanoate	α,γ-Dipropioin. $HOCH(CH_2O_2CCH_2CH_3)_2$	204.22	...	...	$170–3^{10}$	...	...	...	s	...	...	...	...	**B2**^{1}, 107
—	**—,1,2-dithio-**	*see* **1-Propanol, 2,3-dimercapto-***													
g85	—,¹-hexadecyl ether	Chimyl alcohol. $HOCH_2CHOHCH_2O(CH_2)_{15}CH_3$	316.53	lf (hexane) $[\alpha]_D^{20}+3$ (chl)	64	$120^{0.005}$	...	...	...	...	...	s	...	chl, s peth s	...
g86	—,1(2-methoxyphenyl)ether	Guaiacol α-glyceryl ether. $HOCH_2CHOHCH_2O$— (benzene ring with CH_3O)	198.22	pw	78–9	...	...	...	s	s	...	...	...	chl s	**B6**^{1}, 382
g87	—,1-methyl ether	3-Methoxy-1,2-propanediol*. $HOCH_2CHOHCH_2OCH_3$	106.12	hyg lig	...	220^{760} $110–2^{13}$	1.114_4^{20}	1.442^{25}	v	v	s	...	...	os s	**B1**^{3}, 2317
g88	—,2-methyl ether	2-Methoxy-1,3-propanediol*. $HOCH_2CH(OCH_3)CH_2OH$	106.12	hyg lig	...	232^{760} $119–20^9$	1.124_4^{25}	1.4505^{17}	v	v	...	...	...	os v	**B1**^{3}, 2317
g89	—,1-monoacetate(*dl*)	Monoacetin. $HOCH_2CHOHCH_2O_2CCH_3$	134.13	...	...	158^{165}	1.2060_4^{20}	1.4151^{20}	s	s	δ	...	i	...	**B2**^{2}, 159
g90	—,1-monobutanoate	α-Monobutyrin. $HOCH_2CHOHCH_2O_2CCH_2CH_2CH_3$	162.19	...	...	269.71	1.129^{18}	...	s	s	...	...	...	...	**B2**^{2}, 249
g91	—,1-monododecanoate(*dl*)	α-Monolaurin. $HOCH_2CHOHCH_2O_2C(CH_2)_{10}CH_3$	274.40	lf (CCl_4 or peth)	62–3	142 (vac)	...	...	...	δ s^h	v	v	s	chl v	**B2**^{2}, 320
g92	—,—(*l*)	$HOCH_2CHOHCH_2O_2C(CH_2)_{10}CH_3$	274.40	(eth or peth) $[\alpha]_D-4.9$ (Py)	54–5	...	...	...	...	...	...	...	...	...	...
g93	—,monohexadecanoate(*dl*)	α-Monopalmitin. $HOCH_2CHOHCH_2O_2C(CH_2)_{14}CH_3$	330.51	pl or lf (eth)	77	...	...	...	i	v	δ	...	...	...	**B2**^{2}, 338
g94	—,—(*l*)	$HOCH_2CHOHCH_2O_2C(CH_2)_{14}CH_3$	330.51	(eth or peth) $[\alpha]_D-4.37$ (Py)	71–2	...	...	...	...	...	...	...	...	...	...
g95	—,1-mono(2-hydroxy benzoate)	Glycosal. α-Glyceryl salicylate. $HOCH_2CHOHCH_2O_2C$— (benzene ring with HO)	212.20	nd (eth)	76	...	...	...	δ v^h	v	δ	...	v^h	chl δ peth δ	**B10**^{2}, 53
g96	—,1-mono(12-hydroxy-9-octadecenoate)	Glycerol-1-monoricinoleate. α-Monoricinolein. $HOCH_2CHOHCH_2O_2C(CH_2)_7CH{:}CHCH_2CHOH(CH_2)_5CH_3$	372.55	ye	...	...	1.028_4^{20}	...	...	δ	s	...	s	chl s lig δ MeOH, CS_2 δ	**B3**^{1}, 138
g97	—,1-mononitrate	$HOCH_2CHOHCH_2ONO_2$	137.09	pr (w, al or eth)	58–9	155–60	1.40_4^{15}	...	v	v	δ	...	...	...	**B1**^{2}, 591
g98	—,2-mononitrate	$HOCH_2CH(ONO_2)CH_2OH$	137.09	pr	58–9	155–60	1.40_4^{20}	...	s	s	s	...	...	...	**B1**^{2}, 591
g99	—,1-mono(9,12-octadecadienoate)	Glycerol-α-monolinoleate. Linolein. $HOCH_2CHOHCH_2O_2C(CH_2)_7CH{:}CHCH_2CH{:}CH(CH_2)_4CH_3$	354.53	...	8.9	...	...	...	...	δ	v	...	v	chl v lig δ MEOH, CS_2 δ	**B2**^{1}, 213
g100	—,1-monooctadecanoate(*dl*)	α-Monostearin. $HOCH_2CHOHCH_2O_2C(CH_2)_{16}CH_3$	358.57	pl (MeOH)	82	...	0.9841_4^{20}	...	i	δ s^h	δ v^h	...	...	peth δ lig s	**B2**^{2}, 354
g101	—,—(*l*)	$HOCH_2CHOHCH_2O_2C(CH_2)_{16}CH_3$	358.57	(eth or peth) $[\alpha]-3.58$ (Py)	76–7	...	...	...	...	...	...	...	...	...	...
g102	—,1-mono(9-octadecenoate)	Glycerol 1-monooleate. α-Monoolein. $HOCH_2CHOHCH_2O_2C(CH_2)_7CH{:}CH(CH_2)_7CH_3$	356.55	pl (al)	35.5	$238–40^3$	0.942_4^{21}	1.4626^{40}	i	s	s	...	...	chl s	**B2**^{2}, 439

For explanations, symbols and abbreviations see beginning of table.

No.	Name	Synonyms and Formula	Mol. wt.	Crystalline form, color and specific rotation	m.p. °C	b.p. °C	Density	n_D	Solubility w	al	eth	ace	bz	other solvents	Ref.
	Glycerol														
—	—,1-monooleate....	*see* **Glycerol,** 1-mono(9-octadecenoate)													
—	—,1-monoricinoleate	*see* **Glycerol,** 1-mono(12-hydroxy-9-octadecenoate)													
g103	—,1-octadecyl ether	Batyl alcohol. 3-Octadecyl oxy-1,2-propanediol*. $HOCH_2CHOHCH_2O(CH_2)_{17}CH_3$	344.58	....	70–1	....	....	....	i	s^h	s	...	...	....	**B1^2**, 590
g104	—,1-phenyl ether...	Antodyne. $HOCH_2CHOHCH_2OC_6H_5$	168.19	nd (eth, peth or bz-lig)	69–70	200^{22}	1.225^{20}_{4}	....	v	v	s^h	...	v	peth δ con sulf s	**B6^2**, 152
g105	—,1(2-tolyl)ether...	CH_3 $HOCH_2CHOHCH_2O$—	182.20	nd (bz-peth)	66 (71)	....	....	....	s	v	s	...	...	....	**B6**, 354
g106	—,triacetate.......	Triacetin. $CH_3CO_2CH(CH_2O_2CCH_3)_2$	218.21	....	3.2	258–60	1.1562^{20}_{4}	1.5064^{20}	δ	∞	∞	...	∞	chl ∞ lig δ CS_2 δ	**B2^2**, 160
g107	—,tribenzoate.....	Tribenzoin. $C_6H_5CO_2CH(CH_2O_2CC_6H_5)_2$	404.42	nd (MeOH)	76	....	1.228^{12}_{4}	....	i	s^h	v	v	v	chl v	**B9^2**, 122
g108	—,tributanoate....	Tributyrin. $CH_3CH_2CH_2CO_2CH(CH_2O_2CCH_2CH_2CH_3)_2$	302.37	....	<−75	315	1.0350^{20}_{4}	1.4359^{20}	i	s	v	...	s	....	**B2^2**, 249
g109	—,tridodecanoate..	Glycerol trilaurate. Trilaurin. $CH_3(CH_2)_{10}CO_2CH[CH_2O_2C(CH_2)_{10}CH_3]_2$	639.02	nd (al)	49	....	0.8944^{60}_{4}	1.4404^{60}	i	δ	s	...	v	chl s peth s	**B2^2**, 320
—	—,trielaidate......	*see* **Glycerol,** tri(*trans*-9-octadecenoate)													
g111	—,trihexadecanoate	Glycerol tripalmitate. Tripalmitin. $CH_3(CH_2)_{14}CO_2CH[CH_2O_2C(CH_2)_{14}CH_3]_2$	807.35	nd (eth)	65.5	310–20	0.8752^{20}_{4}	1.4381^{80}	i	δ	s	...		chl s	**B2^2**, 340
g112	—,trihexanoate....	Glycerol tricaproate. $CH_3(CH_2)_4CO_2CH[CH_2O_2C(CH_2)_4CH_3]_2$	386.45	....	−60 (−72)	>200	0.9867^{20}_{4}	1.4426^{20}	...	∞	∞	...	∞	peth, chl ∞	**B2^2**, 285
g113	—,tri(3-methyl-butanoate)	Glycerol isovalerate. Tri-isovalerin. $(CH_3)_2CHCH_2CO_2CH[CH_2O_2CCH_2CH(CH_3)_2]_2$	344.45	....	....	330 194^{15}	0.9984^{20}_{4}	1.4354^{20}	...	s	s	...	...	....	**B2^2**, 277
g114	—,trimethyl ether..	1,2,3-Trimethoxypropane*. $CH_3OCH_2CH(OCH_3)CH_2OCH_3$	134.18	....	148	0.946^{15}_{4}	1.4055^{15}	....	∞	...	...	...	...	sulf v	**B1^3**, 2318
g115	—,trinitrate.......	Nitroglycerin. Trinitrin. $O_2NOCH_2CH(ONO_2)CH_2ONO_2$	227.09	pa ye tcl or rh.	13(stable) 2(labile)	256 exp.	1.5918^{25}_{4}	1.4786^{12}	δ	s	∞	...	...	MeOH, CS_2, lig δ chl v peth δ	**B1^2**, 591
g116	—,trioctadecanoate.	Glycerol tristearate. Tristearin. $CH_3(CH_2)_{16}CO_2CH[CH_2O_2C(CH_2)_{16}CH_3]_2$	891.45	....	α 53.5 β′ 64.5 β 73	....	0.862^{80}_{40}	1.4399^{80}	i	i	...	i	i	CCl_4 δ chl s AcOEt i lig i	....
g117	—,tri(*cis*-9-octa-decenoate)	Glycerol trioleate. Triolein. $CH_3(CH_2)_7CH:CH(CH_2)_7CO_2CH[CH_2O_2C(CH_2)_7CH:CH(CH_2)_7CH_3]_2$	885.46	polymorphic	−5	$235–40^{18}$	0.915^{20}_{4}	1.4621^{40}	i	δ	s	...	...	chl s peth s	**B2^2**, 440
g117[1]	—,tri(*trans*-9-octadecenoate	Trielaidin. $CH_3(CH_2)_7CH:CH(CH_7CO_2CH[CH_2O_2C(CH_2)_7CH:CH(CH_2)_7CH_3]_2$	885.46	....	38(32)	....	....	....	...	δ	v	...	...	....	**B2**, 470
g118	—,trioctanoate.....	Glycerol tricaprylate. Tricaprylin. $CH_3(CH_2)_6CO_2CH[CH_2O_2C(CH_2)_6CH_3]_2$	470.70	....	8.3 (st) −21 (metast)	>200	0.9540^{20}_{4}	1.4482^{20}	...	∞	v	...	v	peth v chl v	**B2^2**, 303
g119	—,tripropanoate...	Tripropionin. $CH_3CH_2CO_2CH(CH_2O_2CCH_2CH_3)_2$	260.29	....	....	$177–82^{20}$	1.100^{20}_{18}	1.4318^{19}	i	s	...	...	...	....	**B2^2**, 222
g120	—,tritetradecanoate	Glycerol trimyristate. trimyristin. $CH_3(CH_2)_{12}CO_2CH[CH_2O_2C(CH_2)_{12}CH_3]_2$	723.19	polymorphic (al-eth)	56.5	311	0.8848^{60}_{4}	1.4428^{60}	i	δ	s	...	s	chl s lig δ CS_2 δ	**B2^2**, 328
g121	**Glycerophosphoric acid**(*L*)	$HOCH_2CHOHCH_2OPO(OH)_2$	172.08	syr $[\alpha]_D$ −1.45 (Ba salt in 2N HCl, c=10.3)	....	....	....	....	...	...	...	...	...	....	**B1^3**, 2335
—	**Glycidic acid**.....	*see* **Propanoic acid, 2,3-epoxy-***													

For explanations, symbols and abbreviations see beginning of table.

No.	Name	Synonyms and Formula	Mol. wt.	Crystalline form, color and specific rotation	m.p. °C	b.p. °C	Density	n_D	Solubility w	al	eth	ace	bz	other solvents	Ref.
	Glycocyamine														
g180	**Glycocyamine**	Guanidoacetic acid. *N*-Guanylglycine. $NH{:}C(NH_2)NHCH_2CO_2H$	117.11	lf or nd (w)	250–66				δ s^h	δ	δ				**B4²**, 793
g181	**Glycogen**	Animal starch. $(C_6H_{10}O_5)_x$		amor $[\alpha]^{25}_D$ +191 (w)	240				v	i s^h	i				
—	**Glycol**	*see* **1,2-Ethanediol**													
—	**Glycolaldehyde**	*see* **Acetaldehyde, hydroxy-**													
—	**Glycolic acid**	*see* **Acetic acid, hydroxy-**													
g182	**Glycolide**	2,5-*p*-Dioxanedione. 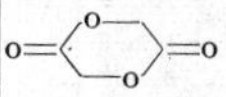	116.07	lf (al-chl)	86–7				s^h	δ s^h	δ	v		chl s^h	**B19²**, 175
g183	**Glycoluril**	Acetylenediurein. Glyoxaldiurene. 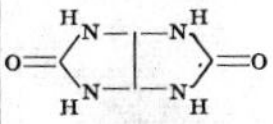	142.12	nd or pr	300d				$δ^h$	i	s			HCl s NH₄OH s	**B26²**, 260
—	**Glycolyurea**	*see* **Hydrantoin**													
—	**Glycylglycine**	*see* **Diglycine**													
g184	**18α-Glycyrrhetinic acid**	Glycyrrhetic acid. $C_{30}H_{46}O_4$	470.70	pl (chl-MeOH) $[\alpha]_D$ +98±4 (chl, c=0.13)	331–5									chl s	**E14**, 561
g185	—,glycoside	Glycyrrhizic acid. Glycyrrhizin. $C_{42}H_{62}O_{16}$	822.95	pl or pr (aa)	220d				v^h	i	i				
g186	**Glyoxal**	Ethanedial*. Biformyl. OCHCHO	58.04	ye pr	15	50.4	1.14²⁰	1.3626²⁰	v	s	s				**B1²**, 816
g187	—,dioxime	Diisonitroseothane. Glyoxime. HON:CHCH:NOH	88.07	rh pl (w)	178d	sub			v^h	v^h	v^h				**B1²**, 818
g188	**—,phenyl-**	Benzoylformaldehyde. $C_8H_6O_2$.	134.14	ye syr		142¹²⁵ 95–7²⁵			s d^h	s	s			chl s	**B7¹**, 360
—	**Glyoxaline**	*see* **Imidazole**													
—	**Glyoxime**	*see* **Glyoxal,** dioxime													
—	**Glyoxylic acid**	*see* **Acetic acid, oxo-**													
—	**Gnoscopine**	*see* **Narcotine**(*dl*)													
g189	**Gramine**	3(Dimethylaminomethyl)-indole. Donaxine. 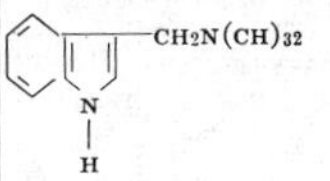	174.25	lf (eth), nd (ace)	134 (139)				δ	s	s			chl s peth i	
g190	**Griseofulvin**	$C_{17}H_{17}ClO_6$	352.77	oct (bz) $[\alpha]^{17}_D$ +376 (chl)	218–9				i	δ		δ	δ		
—	**Guaethol**	*see* **Phenol, 2-ethoxy-***													
—	**Guaiacol**	*see* **Phenol, 2-methoxy-***													
—	**Guaiacol, 4-propenyl-**	*see* **Isoeugenol**													
—	**Guaiazulene**	*see* **Azulene, 1,4-dimethyl-7-isopropyl-**													
—	**Guaiene**	*see* **Naphthalene, 2,3-dimethyl-***													
g191	**Guaiol**	3,8-Dimethyl-5(α-hydroxypropyl)-Δ⁹-octahydroazulene. $C_{15}H_{26}O$	222.37	trg pr (al) $[\alpha]^{20}_D$ −30 (al)	93	288 165¹⁷	0.9074^{100}_{20}	1.4716¹⁰⁰	i	s	s				**E12A**, 431
g192	**Guanidine**	Aminomethanamidine. Carbanamidine. $HN{:}C(NH_2)_2$	59.07	cr					v	v					**B3²**, 69
g193	—,acetate	$HN{:}C(NH_2)_2.CH_3CO_2H$	119.12	nd	229–30				v	v	i				**B3²**, 69
g194	—,carbonate	$2[HN{:}C(NH_2)_2].H_2CO_3$	180.17	oct tetr pr	197		1.24₄		v	i					**B3²**, 72
g195	—,hydrochloride	$HN{:}C(NH_2)_2.HCl$	95.53	cr					v	v					**B3²**, 71
g196	—,nitrate	$CH_5H_3.HNO_3$	122.08	lf (w)	215–6	d			v	s v^h		δ			**B3²**, 72
g197	—,picrate	$CH_5N_3.C_6H_3N_3O_7$	288.18	og-ye pl or nd (w)	333d				$δ^h$	δ	δ				**B6²**, 265
g198	—,thiocyanate	$CH_5N_3.HCNS$	118.16	lf	118				v						**B3²**, 121
g199	**—,1-amino-***	Guanylhydrazine. $HN{:}C(NH_2)NHNH_2$	74.09	cr	d				s	s	i				**B3²**, 95
g200	**—,1-cyano-***	Dicyanodiamide. Param. $NH{:}C(NH_2)NHCN$	84.08	rh lf or pl	208	d	1.404¹⁴		s	δ	i	s		AcOEt δ chl, CS_2, CCl_4 i	**B3²**, 75

For explanations, symbols and abbreviations see beginning of table.

No.	Name	Synonyms and Formula	Mol. wt.	Crystalline form, color and specific rotation	m.p. °C	b.p. °C	Density	n_D	Solubility w	al	eth	ace	bz	other solvents	Ref.
	Guanidine														
g201	—,**1,3-diphenyl-***	Melaniline. HN:C(NHC$_6$H$_5$)$_2$	211.27	mcl nd (al)	151.5	d	1.13_4^{20}	....	δ	s	v	...	...	CCl$_4$, chl s to s^h	**B12**², 216
g202	—,**1,3-di(2-tolyl)**	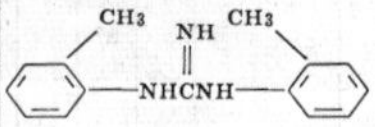	239.32	(dil al)	179	....	1.10_4^{20}	....	δ^h	δ s^h	v	...	...	....	**B12**, 803
g203	—,**1-methyl-**, sulfate*	[HN:C(NH$_2$)NHCH$_3$]$_2$.H$_2$SO$_4$	244.27	(w)	238–40	....	....	....	v	δ	...	...	...	os i	**B4**², 570
g204	—,**1-nitro-***	HN:C(NH$_2$)NHNO$_2$	104.07	nd (w)	246–7 (230)	....	....	....	δ s^h	δ	i	...	...	alk v	**B3**², 100
g205	—,**1-phenyl-3(2-tolyl)-**	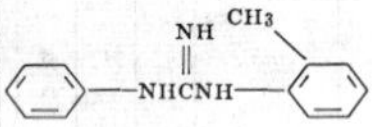	225.30	nd (aa)	123–5 (130)	....	....	....	s	v	...	...	v	chl v	**B12**², 445
g206	—,**1,1,3,3-tetra-phenyl-***	HN:C[N(C$_6$H$_5$)$_2$]$_2$	363.47	rh (lig)	130–1	....	....	....	i	v	v	...	v	....	**B12**, 430
g207	—,**1,1,3-tri-phenyl-***	HN:C(NHC$_6$H$_5$)N(C$_6$H$_5$)$_2$	287.37	ta (dil al)	134	....	....	....	δ	v	v	...	δ	....	**B12**², 242
g208	—,**1,2,3-tri-phenyl-***	C$_6$H$_5$N:C(NHC$_6$H$_5$)$_2$	287.37	nd or pr (al)	145	d	1.13_4^{20}	....	δ^h	s	...	...	...	....	**B12**², 246
g209	—,—,hydrochloride monohydrate	C$_6$H$_5$N:C(NHC$_6$H$_5$)$_2$.HCl.H$_2$O	341.84	pr	ca. 238d	....	....	....	δ	...	...	...	...	HCl δ	**B12**², 246
g210	—,**1-ureido-***(?)	1-Carbamylguanidine. Guanylurea. Dicyandiamidine. HN:C(NH$_2$)NHCONH$_2$	102.10	pr (al)	105	160d	....	....	s^h	δ	i	...	i	Py s chl, CS$_2$ i	**B3**, 89
g211	**Guanine**	2-Aminohydroxanthine. 6-hydroxy-2-aminopurine. 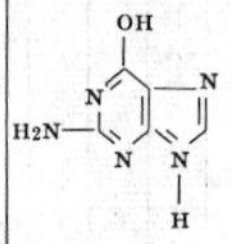	151.13	nd or pl (NH$_4$OH)	360d	....	....	....	i	δ	δ	...	...	NH$_4$OH δ alk s ac s aa i	**B26**², 262
g212	**Guanosine**	9-D-Ribosidoguanine. Vernine. 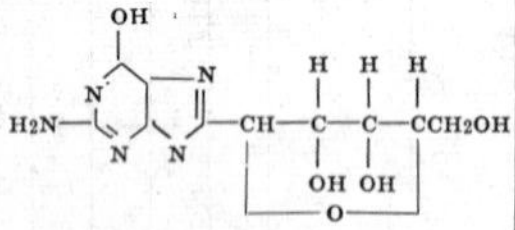	283.25	nd (w) $[\alpha]_D^{20}$ −60.5 (anh)	230–5	....	....	....	i s^h	i	i	...	...	aa v^h	**B31**, 28
—	**Guanosine phosphoric acid**	*see* **Guanylic acid**													
g213	**Guanylic acid**	Guanosine 3′-phosphate. Guanosine phosphoric acid. GMP. 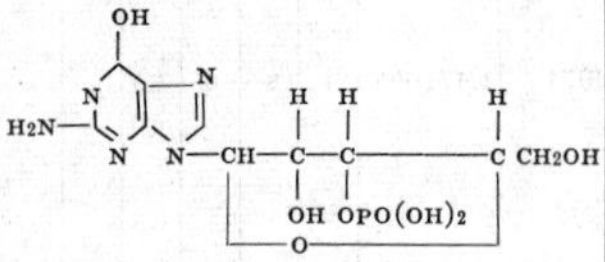	363.23	nd (w+2) $[\alpha]_D^{20}$ −135 (w, p=1)	208d	....	....	....	s v^h	...	...	...	...	....	**B31**, 29
—	**Gulitol**	*see* **Sorbitol**													
g214	**Gulonic acid,** α-lactone(*d*)	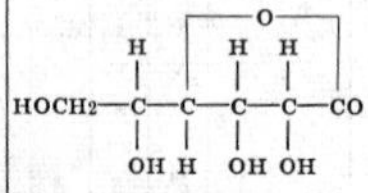	178.14	pr ta $[\alpha]_D^{20}$+55.1	180–1	....	....	....	s	δ	...	...	...	....	**B18**², 193
g215	—,—(*l*)	C$_6$H$_{10}$O$_6$. *See* g214	178.14	rh pr $[\alpha]_4^{20}$ −57.1	185	....	....	....	v^h	δ^h	...	...	...	....	**B18**², 193
g216	—,phenyl-hydrazide(*D*)	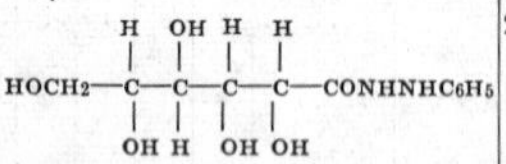 HOCH$_2$—C—C—C—C—CONHNHC$_6$H$_5$	286.29	....	147–9	....	....	....	v^h	v^h	...	...	...	....	**B15**, 332

For explanations, symbols and abbreviations see beginning of table.

No.	Name	Synonyms and Formula	Mol. wt.	Crystalline form, color and specific rotation	m.p. °C	b.p. °C	Density	n_D	Solubility w	al	eth	ace	bz	other solvents	Ref.
	Gulonic acid														
g217	—,**1-oxo-**(*L*)	$HOCH_2—CH(OH)—CH(OH)—CH(OH)—COCO_2H$ (H, OH H / OH H OH)	194.14	(w) $[\alpha]^{18}_D -48$	171δd				δ					alk s aa δ	
g218	**Gulose**(*D*,—)	$HOCH_2—C—C—C—C—CHO$ (H OH H H / OH H OH OH)	180.16	syr. $[\alpha]^{20}_D -20.4$		d			s	δ					**B23**, 282
g219	—(*L*,+)	$C_6H_{12}O_6$. *See* g218	180.16	syr $[\alpha]^{20}_D +61.6$		d			s	δ					**B31**, 282
g220	—,osazone(*D*)	D-Idosazone. $HOCH_2.C—C—C—C{:}NNHC_6H_5$ (H OH H $NHNC_6H_5$ / OH H OH)	358.39	$[\alpha]$+0.5 (al-Py)	168 (180d)										**B31**, 355
g221	**Guvacine**	1,2,5,6-Tetrahydronicotinic acid. (ring N)—CO_2H	127.14	pr (w)	293–5d				s	i	i		i	chl i	**B22**[1], 489

For explanations, symbols and abbreviations see beginning of table.

No.	Name	Synonyms and Formula	Mol. wt.	Crystalline form, color and specific rotation	m.p. °C	b.p. °C	Density	n_D	Solubility w	al	eth	ace	bz	other solvents	Ref.
	Halazone														
—	**Halazone**	*see* **Benzoic acid, 4(dichlorosulfamyl)-***													
h1	**Halostachine(*l*)**	$C_6H_5CHOHCH_2NHCH_3$	151.21	$[\alpha]-47$	43–5				s	s	s	i			
h2	**Harmaline**	3,4-Dihydroharmine. $C_{13}H_{14}N_2O$	214.27	ta (MeOH), rh (al)	239				i	s	δ			chl s Py s	**B23**^{2}, 345
h3	**Harmine**	Banisterine. Telepathine. Yageine. $C_{13}H_{12}N_2O$	212.55	rh (al), nd (MeOH)	257–9	sub			δ	δ	δ			Py s	**B23**^{2}, 348
h4	**Hecogenin**	$C_{27}H_{42}O_4$	430.63	pl (eth) $[\alpha]_D^{22}+7$	245 (253) (268)						s	s		lig s	**J1955**, 1966
h5	—,acetate	$C_{29}H_{44}O_5$	472.67	(MeOH)	243 (252)										**Am 77**, 5196
h6	**Hederagenin**	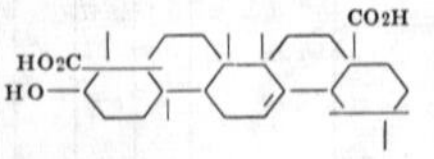	472.71	pr (al) $[\alpha]_D^{21}$ +70.1 (chl-MeOH), +80 (Py)	332				i	s				Py s aa s	**B10**^{2}, 305
h7	**Helenin**	Alantolactone. 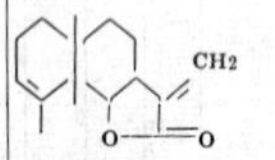	232.33	nd	76	275^{760}			δ^h	v	v		s	chl s aa s	**B17**^{2}, 356
h8	**Helicin(l)**	Salicyladehyde-*D*-glucoside. $C_{13}H_{16}O_7$	284.27	nd (w) $[\alpha]_D^{20}$ −60.4 (w)	175				v^h	s	i				**B31**, 223
h9	**Helvolic acid**	Fumigaein. $C_{32}H_{42}O_8$	554.69	$[\alpha]_D^{18}-138$ (chl)	212				i	s					
h10	**Hematein**	Haematein. 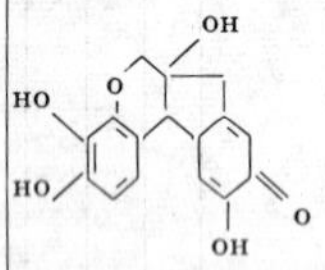	300.27	cr	250d				i	δ	i		i	aa δ	**B18**^{2}, 221
h11	**Hematin**	Hydroxyhemin. Ferriporphyrin hydroxide. $C_{34}H_{32}N_4O_4.FeOH$	633.51	br pw	>200				i	s^h	i			Py δ^h alk s	
h12	**Hematommic acid**	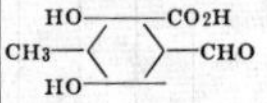	244.17	nd	172										**H16**, 282
h13	**Hematoporphyrin**	$C_{34}H_{38}N_4O_6$	598.71	red					i	s	δ				
h14	**Hematoxylin**	Hydroxybrasilin. Haematoxylin. $C_{16}H_{14}O_6.3H_2O$	356.33	yesh $[\alpha]+11$ (w, c=3.7)	140				v	s	s			alk s	**B17**^{2}, 273
—	**Hemellitone**	*see* **Benzene 1,2,3-trimethyl-***													
—	**Hemellitic acid**	*see* **Benzoic acid, 2,3-dimethyl-**													
—	**Hemimellitic acid**	*see* **1,2,3-benzenetricarboxylic acid***													
h15	**1,10-Hendecadiyne***	$HC{:}C(CH_2)_7C{:}CH$	148.24		−17.0	83^{12}	0.8182_4^{21}	1.453^{21}							
h16	**Hendecanal***	*n*-Undecylaldehyde. $CH_3(CH_2)_9CHO$	170.29		−4	117^{18}	0.8251_4^{23}	1.4520^{20}	i	s	s				**B1**, 712
h17	—,oxime*	$CH_3(CH_2)_9CH{:}NOH$	185.31	wh nd (al)	72				s	s	s				**B1**, 713
h18	—,**2-methyl-**	$CH_3(CH_2)_8CH(CH_3)CHO$	184.31			114^{10}	0.830_4^{15}	1.4382^{20}	δ	s	s				
h19	**Hendecane***	Undecane*. $CH_3(CH_2)_9CH_3$	156.31		−25	195^{760}	0.74017_4^{20}	1.4172^{20}	i	∞	∞				**B1**^{3}, 534
h20	—,**1-amino-***	$CH_3(CH_2)_{10}NH_2$	171.32		17	242^{760}			s^h	s	i				**B4**^{2}, 658
h21	—,**1-bromo-11-fluoro-***	$F(CH_2)_{11}Br$	253.21			$95^{0.6}$		1.4518^{25}	i	v	v				
h22	**Hendecanoic acid.**	Undecanoic acid*. Undecylic acid. $CH_3(CH_2)_9CO_2H$	186.29		28.6	280^{760} 162^{10}	0.8907	1.4294^{45}	i	v		v	∞		**B2**^{2}, 314
h23	—,ethyl ester*	$CH_3(CH_2)_9CO_2C_2H_5$	214.35			123^9								os s	**B2**^{1}, 154
h24	—,nitrile	Decyl cyanide. $CH_3(CH_2)_9CN$	167.29			253^{760} 125^{11}			i	s	s				**B2**, 358
h25	—,**2-bromo-***	$CH_3(CH_2)_8CHBrCO_2H$	265.19		10	$178\text{–}183^{14}$			i					alk s	**B2**^{1}, 155
h26	—,**10-bromo-***	$CH_3CHBr(CH_2)_8CO_2H$	265.19	(peth-eth)	35				i	v				os v	**B2**^{2}, 315

For explanations, symbols and abbreviations see beginning of table.

No.	Name	Synonyms and Formula	Mol. wt.	Crystalline form, color and specific rotation	m.p. °C	b.p. °C	Density	n_D	Solubility w	al	eth	ace	bz	other solvents	Ref.
	Heptane														
h93	**Heptane***	$CH_3(CH_2)_5CH_3$	100.21		−91	98.42	0.68376$^{20}_{4}$	1.38764^{20}	i	v	∞			chl ∞	**B1**^{3}, 415
h94	**—,1-amino-***	$CH_3(CH_2)_6NH_2$	115.22		−23	155^{760}	0.777^{20}	1.41954$^{26}_{\alpha}$	δ	∞	∞				**B4**^{2}, 652
h95	**—,2-amino-***	$CH_3(CH_2)_4CHNH_2CH_3$	115.22			142^{760}	0.7665^{18}	1.41997^{18}	δ	s	s			peth s	
h96	**—,1-bromo-***	$CH_3(CH_2)_6Br$	179.11		−58	180^{760}	1.1384$^{20}_{4}$	1.4505$^{20}_{D}$	i	v	v				**B1**^{3}, 431
h97	**—,1-bromo-7-fluoro-***	$F(CH_2)_7Br$	197.10			85^{11}		1.4463	i	v	v				
h98	**—,1-chloro-***	$CH_3(CH_2)_6Cl$	134.65		−69	159^{750}	0.881$^{16}_{0}$		i	∞	∞				**B1**^{2}, 117
h99	**—,2-chloro-***	$CH_3(CH_2)_4CHClCH_3$	134.65			46^{19}	0.8725$^{15}_{4}$	1.4221^{20}	i					chl s aa s	**B1**^{3}, 429
h100	**—,3-chloro-***	$CH_3(CH_2)_3CHClCH_2CH_3$	134.65			144^{751}	0.8960$^{20}_{4}$	1.4237^{20}	i					chl v	**B1**^{2}, 430
h101	**—,4-chloro-***	$(CH_3CH_2CH_2)_2CHCl$	134.65			144^{758} 48^{21}	0.8710$^{20}_{4}$	1.4237^{20}						chl s	**B1**^{2}, 117
h102	**—,3-chloro-2,3-diethyl-***	$CH_3(CH_2)_3CCl(CH_3)CH(CH_3)_2$	162.70			54^{8}	0.885$^{20}_{4}$	1.4391^{20}			s			chl s	
h103	**—,5-chloro-2,5-dimethyl-***	$CH_3CH_2CCl(CH_3)CH_2CH_2CH(CH_3)_2$	162.70			63^{15}	0.8692$^{18}_{4}$	1.43457^{15}	i				s	chl s	**B1**^{1}, 64
h104	**—,3-chloro-3-ethyl***	$CH_3(CH_2)_3CCl(C_2H_5)CH_2CH_3$	162.70			46^{3}	0.8856$^{20}_{4}$	1.4400^{20}			s			chl s	**B1**^{3}, 510
h105	**—,4-chloro-4-ethyl-***	$(CH_3CH_2CH_2)_2CClCH_2CH_3$	162.70			67^{12}	0.8821$^{20}_{4}$	1.4438^{20}	i				s	chl s	**B1**^{3}, 511
h106	**—,1-chloro-7-fluoro-***	$F(CH_2)_7Cl$	152.64			70^{10}	0.993^{20}	1.4222^{25}	i	v	v				
h107	**—,2-chloro-2-methyl-***	$CH_3(CH_2)_4CCl(CH_3)_2$	148.68			50^{15}	0.8568$^{25}_{4}$	1.4240^{25}							**B1**^{3}, 472
h108	**—,2-chloro-6-methyl-***	$(CH_3)_2CH(CH_2)_3CHClCH_3$	148.68			74^{35}		1.4260^{15}							**B1**^{3}, 472
h109	**—,3-chloro-3-methyl-***	$CH_3(CH_2)_3CCl(CH_3)CH_2CH_3$	148.68			64^{27}	0.8764$^{20}_{4}$	1.4317^{20}			v				**B1**^{3}, 474
h110	**—,4-chloro-4-methyl-***	$(CH_3CH_2CH_2)_2CClCH_3$	148.68			50^{12}	0.8690$^{20}_{4}$	1.43098^{15}	i		s		s		**B1**^{3}, 477
h110^1	**—,3(chloro-methyl)**	$CH_3(CH_2)_3CH(C_2H_5)CH_2Cl$	148.68			165–70^{760}	0.8769$^{20}_{4}$	1.4319^{20}	i					os s	
h111	**—,1,1-dichloro-***	$CH_3(CH_2)_5CHCl_2$	169.10			191	1.011$^{20}_{4}$	1.4440^{20}	i		s				**B1**^{3}, 430
h112	**—,1,2-dichloro-***	$CH_3(CH_2)_4CHClCH_2Cl$	169.10			68–72^{7}	1.064$^{20}_{4}$	1.4490^{20}						chl v	**B1**^{3}, 430
h113	**—,2,2-dichloro-***	$CH_3(CH_2)_4CCl_2CH_3$	169.10			77^{25}	1.012$^{20}_{4}$	1.4440^{20}	i				s	chl s	**B1**^{3}, 430
h114	**—,4,4-dichloro-***	$(CH_3CH_2CH_2)_2CCl_2$	169.10			86^{27}	1.008^{17}	1.448^{17}	i					chl s	**B1**, 154
h115	**—,2,6-dichloro-2,6-dimethyl-***	$(CH_3)_2CCl(CH_2)_3CCl(CH_3)_2$	197.16		43	93^{16}			i				s	chl s	**B1**^{3}, 513
h116	**—,2,3-dimethyl-***	$CH_3(CH_2)_3CH(CH_3)CH(CH_3)_2$	128.25		−116	141^{760}	0.7270$^{20}_{4}$	1.409^{20}						os v	**B1**^{3}, 511
h117	**—,2,4-dimethyl-***	$CH_3CH_2CH_2CH(CH_3)CH_2CH(CH_3)_2$	128.25			133^{760}	0.7158$^{20}_{4}$	1.4033^{20}	i					os ∞	**B1**^{3}, 512
h118	**—,2,5-dimethyl-** (*d*)	$CH_3CH_2CH(CH_3)CH_2CH_2CH(CH_3)_2$	128.25			136^{760}	0.7198$^{20}_{4}$	1.4033^{20}	i					os ∞	**B1**^{2}, 128
h119	**—,2,6-dimethyl-***	Diisobutylmethane. Isobutyl-isoamyl. $(CH_3)_2CH(CH_2)_3CH(CH_3)_2$	128.25		−103	135^{760}	0.712$^{20}_{0}$	1.4007^{20}							
h120	**—,3,3-dimethyl-***	$CH_3(CH_2)_3C(CH_3)_2CH_2CH_3$	128.25			137^{760}	0.7304$^{20}_{4}$	1.4095^{20}	i					os v	**B1**^{3}, 513
h122	**—,4-ethyl-***	Ethyldipropylmethane. $CH_3(CH_2)_2CH(C_2H_5)(CH_2)_2CH_3$	128.25			141	0.730$^{20}_{4}$	1.496^{20}	i	i	s				
h123	**—,1-fluoro-***	$CH_3(CH_2)_5CH_2F$	118.19		−76	119^{750}		1.3860^{20}							**B1**, 56
h124	**—,hexadeca-fluoro-***	Perfluoroheptane. $CF_3(CF_2)_5CF_3$	388.7		−78	82	1.7333	1.2618^{20}							**B1**^{3}, 429

For explanations, symbols and abbreviations see beginning of table.

PHYSICAL CONSTANTS OF ORGANIC COMPOUNDS (Continued)

No.	Name	Synonyms and Formula	Mol. wt.	Crystalline form, color and specific rotation	m.p. °C	b.p. °C	Density	n_D	Solubility						Ref.
									w	al	eth	ace	bz	other solvents	
	Heptane														
h125	—,**1-iodo-***	$CH_3(CH_2)_5CH_2I$	226.11		−48.2	204^{760}	1.3792^{20}_{4}	1.4897^{20}	i	s	s				**B1**³, 433
h126	—,**2-methyl-***	$CH_3(CH_2)_4CH(CH_3)_2$	114.23		−109	118^{760}	0.6979^{20}_{4}	1.39494^{20}	i	δ	s				**B1**³, 471
h127	—,**3-methyl-(*d*)***	$CH_3(CH_2)_3CH(CH_3)CH_2CH_3$	114.23	$[\alpha]^{26}_{D}+9.34$		115–118	0.7075^{18}_{4}	1.4002^{18}							**B1**³, 475
h128	—,—(*dl*)*	$CH_3(CH_2)_3CH(CH_3)CH_2CH_3$	114.23		−121	119	0.70583^{20}_{4}	1.3985^{20}_{D}	i	δ	s				**B1**³, 476
h129	—,—(*l*)*	$CH_3(CH_2)_3CH(CH_3)CH_2CH_3$	114.23	$[\alpha]^{25}_{D}-3.87$		115									
h130	—,**4-methyl-***	$(CH_3CH_2CH_2)_2CHCH_3$	114.23		−121	118^{760}	0.70463^{20}_{4}	1.39792^{20}_{D}	i	δ	s				**B1**³, 476
h131	—,**2(methyl amino)-**	$CH_3(CH_2)_4CH(NHCH_3)CH_3$	129.24			155^{760}			δ						
h132	**Heptanedioic acid***	Pimelic acid. $HO_2C(CH_2)_5CO_2H$	160.17		106	272^{100}sub	1.329^{15}		δ	s	s		i		**B2**², 586
h133	—,dichloride	$ClCO(CH_2)_5COCl$	197.07			137^{15}			d	d					**B2**, 671
h134	—,diethyl ester*	$C_2H_5O_2C(CH_2)_5CO_2C_2H_5$	216.27		−24	$252–5^{748}$	0.99448^{20}	1.43052^{20}	i	s	s			AcOEt s	**B2**¹, 282
h135	—,dimethyl ester*	$CH_3O_2C(CH_2)_5CO_2CH_3$	188.22		−21	80^{1}, 120^{10}	1.0625^{20}_{4}	1.4302^{20}		s					**B2**², 587
h136	—,dinitrile	1,5-Dicyanopentane. $NC(CH_2)_5CN$	122.17			175^{14}	0.949^{18}		i	∞	∞			chl ∞	**B2**², 587
h137	—,monoethyl ester	$HO_2C(CH_2)_5CO_2C_2H_5$	182.22		10	182^{18}		1.4415^{20}	δ	δ	δ		δ	chl δ	**B2**¹, 282
h138	—,**4-oxo-**	$OC(CH_2CH_2CO_2H)_2$	174.15	rh pl (w)	143				s^h	s	δ	δ	i		**B3**², 487
h139	**1,7-Heptanediol***	Heptamethyleneglycol. $HOCH_2(CH_2)_5CH_2OH$	132.20		22	262	0.9510^{25}_{4}	1.4558^{25}	s	s	δ				**B1**³, 221
h140	**2,4-Heptanediol, 3-methyl-***	$CH_3CH_2CH_2CHOHCH(CH_3)CHOHCH_3$	146.23			115^{2}	0.928^{20}_{4}	1.4459^{20}	i						**B1**, 491
h141	**2,4-Heptanedione***	$CH_3CH_2CH_2COCH_2COCH_3$	128.17			174^{760}	0.9411^{15}_{4}								**B1**, 792
h142	**1-Heptanethiol***	$CH_3(CH_2)_6SH$	132.27		−43	177^{760}	0.8427^{20}_{4}	1.4521^{20}	i	∞	∞				**B1**³, 168
h143	**2,4,6-Heptanetrione***	Diacetylacetone. $CH_3COCH_2COCH_2COCH_3$	142.16		49	121^{10}	1.0681^{40}_{4}	1.4930^{20}	s	s					
h144	**Heptanoic acid***	Enanthic acid. *n*-Heptylic acid. $CH_3(CH_2)_5CO_2H$	130.19		−10	223^{760}	0.9185^{20}_{4}	1.4216^{20}	δ	s	s				**B2**², 294
h145	—,amide	$CH_3(CH_2)_5CONH_2$	129.20	nd (al), lf (w)	94	250–8	0.8489^{11}_{4}		s	s	s				**B2**², 29
h146	—,anhydride*	$[CH_3(CH_2)_5CO]_2O$	242.36		−11	268–271	0.932^{20}_{4}	1.4335^{15}	i	s	s				**B2**, 340
h147	—,butyl ester*	$CH_3(CH_2)_5CO_2C_4H_9^n$	186.30			225^{760}	0.86382^{20}	1.4228^{15}	i	s				os s	**B2**, 340
h148	—,chloride*	$CH_3(CH_2)_5COCl$	148.63				0.9590^{20}	1.4345^{15}	d	d	s			lig v	**J87**, 93
h149	—,ethyl ester*	$CH_3(CH_2)_5CO_2C_2H_5$	158.24		−66	187^{760}	0.8685^{20}_{4}	1.4144^{15}	δ	s	s				**B2**², 29
h150	—,heptyl ester*	$CH_3(CH_2)_5CO_2(CH_2)_6CH_3$	228.38		−33	$276–8^{760}$	0.86492^{19}_{4}	1.4320^{20}	i	s	s				**B2**², 14
h151	—,hexyl ester*	$CH_3(CH_2)_5CO_2(CH_2)_5CH_3$	214.35		−48	261	0.86114^{20}	1.429^{15}	i	s				os s	**B2**², 29
h152	—,methyl ester*	$CH_3(CH_2)_5CO_2CH_3$	144.22		−56	172^{760}	0.8806^{15}_{4}	1.4137^{15}	δ	s	s				**B2**², 29
h153	—,2-methylpropyl ester	$CH_3(CH_2)_5CO_2CH_2CH(CH_3)_2$	186.30			209^{759}	0.8593^{20}		i	s				os s	**B2**¹, 14
h154	—,nitrile*	$CH_3(CH_2)_5CN$	111.19			183^{765}	0.895^{22}	1.4195	i		s		s	aa s	**B2**², 29
h155	—,octyl ester*	$CH_3(CH_2)_5CO_2(CH_2)_7CH_3$	242.41		−22	290^{760}	0.85961^{20}	1.43488^{15}	i	s				os s	**B2**, 340
h156	—,pentyl ester*	$CH_3(CH_2)_5CO_2(CH_2)_4CH_3$	200.32		−50	245	0.86232^{20}	1.42627^{15}	i	s				os s	
h157	—,*p*-phenylphenacyl ester	$CH_3(CH_2)_5CO_2CH_2CO—C_6H_4—C_6H_5$	324.41		62										
h158	—,propyl ester*	$CH_3(CH_2)_5CO_2CH_2CH_2CH_3$	172.27		−65	206^{760}	0.86556^{15}_{4}	1.41835^{15}	i					os s	**B2**, 340
h159	—,**7-amino-***	$NH_2(CH_2)_6CO_2H$	145.20	(al)	195				s^h	s^h	i	s^h			**B4**, 459
h160	—,**2-bromo-***	$CH_3(CH_2)_4CHBrCO_2H$	209.09			147^{12}, 250d	1.319^{18}	1.471^{18}							**B2**², 29
h161	—,**7-amino-***	$Br(CH_2)_6CO_2H$	209.09		31	280								os s lig i	**B2**, 341
h162	—,**7-fluoro-***	$F(CH_2)_6CO_2H$	148.18			133^{10}		1.4207^{25}	δ	δ	δ				
h163	—,**2-hydroxy-6-methyl-***	$(CH_3)_2CH(CH_2)_3CHOHCO_2H$	160.21	nd (eth)	152d	192d			δ	s	s				
h164	—,**7-iodo-***	$I(CH_2)_6CO_2H$	256.08	lf	49–51				δ					os s	**B2**, 341
h165	—,**6-oxo-***	$CH_3CO(CH_2)_4CO_2H$	144.17		31–2	181^{25}			v	v	v				**B3**¹, 24

For explanations, symbols and abbreviations see beginning of table.

No.	Name	Synonyms and Formula	Mol. wt.	Crystalline form, color and specific rotation	m.p. °C	b.p. °C	Density	n_D	Solubility w	al	eth	ace	bz	other solvents	Ref.
	1-Heptanol														
h166	**1-Heptanol***	$CH_3(CH_2)_6OH$	116.20		−35	177^{760}	0.8219^{20}_{4}	1.4241^{20}	δ	∞	∞				$B1^{3}$, 1679
h167	**2-Heptanol**(*d*)*	$CH_3(CH_2)_4CHOHCH_3$	116.20	$[\alpha]^{25}_{D}+11.4$ (al)		73^{20}	0.8190^{20}_{4}	1.4209^{20}							$B1^{3}$, 1687
h168	—(*dl*)	$CH_3(CH_2)_4CHOHCH_3$	116.20			160^{760}	0.8167^{20}_{4}	1.4210^{20}	δ	s	s				$B1^{3}$, 1686
h169	—(*l*)	$CH_3(CH_2)_4CHOHCH_3$	116.20	$[\alpha]^{12}_{D}-10.5$		74^{23}	0.8184^{29}_{4}								$B1^{3}$, 1687
h170	**3-Heptanol**(*d*)*	$CH_3(CH_2)_3CHOHCH_2CH_3$	116.20	$[\alpha]^{20}_{D}+6.7$ (w)	−70	157^{750}	0.8227^{20}_{4}	1.4201^{20}	δ						$B1^{2}$, 444
h171	**4-Heptanol***	$(CH_3CH_2CH_2)_2CHOH$	116.20		−82	161^{760}	0.8183^{20}_{4}	1.4258^{20}	i	s	s				$B1^{3}$, 1735
h172	**2-Heptanol, 6-amino-2-methyl-**, hydrochloride*	$CH_3CHNH_2(CH_2)_3COH(CH_3)_2.HCl$	181.71		150				v	s	i		i		
h173	**1-Heptanol, 7-chloro-***	$Cl(CH_2)_7OH$	150.65	(peth)	10	150^{20} 120^{13}	0.9998^{15}_{4}	1.4537^{25}	δ	v				peth s	
h174	**4-Heptanol, 2,6-dimethyl-***	$[(CH_3)_2CHCH_2]_2CHOH$	144.26			178^{760}	0.809^{21}_{4}	1.423^{20}	i	s	s				$B1^{3}$, 1753
h175	**4-Heptanol, 4-ethyl-***	$(CH_3CH_2CH_2)_2COHC_2H_5$	144.26			182^{760}	0.8350^{20}	1.4332^{20}	i	s	s				$B1^{2}$, 457
h176	**1-Heptanol, 7-fluoro-***	Heptamethylene fluorohydrin. $F(CH_2)_7OH$	134.19			89^{12}	0.956^{20}_{4}	1.4197^{25}	δ*	v	v				
h177	—,**4-methyl-***	$CH_3CH_2CH_2CH(CH_3)(CH_2)_3OH$	130.22			183^{760} 85^{20}	0.8065^{25}_{4}	1.4253^{25}	i	s				os s	$B1^{3}$, 1734
h178	—,**6-methyl-***	$(CH_3)_2CH(CH_2)_5OH$	130.22		−106	188^{764}	0.8176^{25}_{4}	1.4251^{25}	i						**Am 63,** 3100
h179	**2-Heptanol, 2-methyl-***	$CH_3(CH_2)_4COH(CH_3)_2$	130.23			156^{760}	0.8142^{20}_{4}	1.4238^{20}	i	s	s				$B1^{3}$, 1725
h180	**3-Heptanol, 3-methyl-***	$CH_3(CH_2)_3COH(CH_3)CH_2CH_3$	130.23		−83	163^{760}	0.8282^{20}_{4}	1.4279^{20}	i	s	s				$B1^{3}$, 1729
h181	**4-Heptanol, 4-methyl-***	$(CH_3CH_2CH_2)_2COHCH_3$	130.23		−82	161^{760}	0.8248^{20}_{4}	1.4258^{20}_{4}	i	s	s				$B1^{3}$, 1735
h182	**1-Heptanol, 1-phenyl-***	$CH_3(CH_2)_5CHOHC_6H_5$	192.30			275 $153–5^{18}$	0.946	1.501	i						$B6^{2}$, 513
h183	**4-Heptanol, 4-propyl-***	$(CH_3CH_2CH_2)_3COH$	158.28			190–2	0.8338^{21}_{0}	1.43551^{21}	i	s					$B1^{3}$, 1769
h184	**2-Heptanone***	$CH_3(CH_2)_4COCH_3$	114.18		35	151	0.8111^{20}_{4}	1.41156^{15}	v	s	s				$B1^{2}$, 753
h185	**3-Heptanone***	$CH_3(CH_2)_3COCH_2CH_3$	114.18		−39	150	0.8183^{20}_{4}		i	∞	∞				$B1^{2}$, 754
h186	**4-Heptanone***	$CH_3CH_2CH_2COCH_2CH_2CH_3$	114.18		−33	144	0.8174^{20}_{4}	1.4073^{20}	i	∞	∞				$B1^{2}$, 754
h187	**2-Heptanone, 1-chloro-***	$CH_3(CH_2)_4COCH_2Cl$	148.63			83^{16}	0.802^{20}	1.450^{20}	i		s				**Am 67,** 1944
h187[1]	**4-Heptanone, 2,6-dimethyl-***	$[(CH_3)_2CHCH_2]_2CO$	142.24			168^{760}	0.8053^{20}_{4}	1.412^{21}	i	∞	∞				$B1^{2}$, 763
h188	**3-Heptanone, 6-dimethyl-aminol-4,4-diphenyl-**(*l*)*	*l*-Amidon. *l*-Methadone. $(CH_3)_2NCH(CH_3)CH_2C(C_6H_5)_2COC_2H_5$	309.34	$[\alpha]^{20}_{D}-32$ (al)	99					s				dil HCl s	
h189	—,—hydrochloride (*dl*)*	$(CH_2)_2NCH(CH_3)CH_2C(C_6H_5)_2COC_2H_5.HCl$	345.90	pl (al-eth)	235				v	s	i			chl s	
h190	—,—,—(*l*)*	$(CH_3)_2NCH(CH_3)CH_2C(C_6H_5)_2COC_2H_5.HCl$	345.90	$[\alpha]^{20}_{D}-169$	241				s	s	i			chl s	
h191	**2-Heptanone, 3-isopropyl-***	$CH_3(CH_2)_3CH[CH(CH_3)_2]COCH_3$	156.27			78^{15}	0.8195^{20}_{4}	1.4250^{20}	δ					os v	
h192	—,**3-methyl-***	$CH_3(CH_2)_3CH(CH_3)COCH_3$	128.21			167^{760}	0.8218^{20}_{4}	1.4172^{20}	δ					os s	$B1^{2}$, 759
h193	**3-Heptanone, 6-methyl-***	$(CH_3)_2CHCH_2COCH_2CH_3$	128.21		163		0.8304^{20}	1.4209	i	s	s				$B1^{2}$, 759

For explanations, symbols and abbreviations see beginning of table.

No.	Name	Synonyms and Formula	Mol. wt.	Crystalline form, color and specific rotation	m.p. °C	b.p. °C	Density	n_D	Solubility w	al	eth	ace	bz	other solvents	Ref.
	4-Heptanone														
h194	**4-Heptanone, 2-methyl-***	$CH_3CH_2CH_2COCH_2CH(CH_3)_2$	128.21			155^{750}	0.813_{0}^{22}		i	s	s				**B1**[1], 363
194[1]	**1-Heptanone, 1-phenyl-***	Enanthiophenone. $CH_3(CH_2)_5COC_6H_5$	190.27	lf	16	283 155^{15}	0.9516_{4}^{20}		i	s	s	s			**B7**[2], 264
h195	**Heptasiloxane, hexadecamethyl-**	$CH_3[Si(CH_3)_2O-]_6Si(CH_3)_3$	532.96			165^{20}	0.911	1.3965^{20}	i	δ			s	peth s lig s	
h196	**1,4,6-Heptatrien-3-one, 1(4-methoxy-phenyl)-7-phenyl-***	CH_3O—	290.42	ye lf (al)	139				i	s^h	s		s	chl s lig s	**B8**, 208
h197	**1-Heptene***	α-Heptylene. $CH_3(CH_2)_4CH{:}CH_2$	98.18		−119	94^{760}	0.6970_{4}^{20}	1.3998^{20}	i	s	s				**B1**[3], 820
h198	**2-Heptene(*cis*)***	β-Heptylene. $CH_3(CH_2)_3CH{:}CHCH_3$	98.18			98^{760}	0.708_{4}^{20}	1.406^{20}							**B1**[3], 824
h199	**—,(*trans*)***	$CH_3(CH_2)_3CH{:}CHCH_3$	98.18		−109	98^{760}	0.7012_{4}^{20}	1.4045^{20}							**B1**[3], 825
h200	**3-Heptene(*cis*)***	$CH_3CH_2CH_2CH{:}CHCH_2CH_3$	98.18			96^{760}	0.7030_{4}^{20}	1.4059^{20}							**B1**[3], 826
h201	**—(*trans*)***	$CH_3CH_2CH_2CH{:}CHCH_2CH_3$	98.18		−137	96	0.6981_{4}^{20}	1.4043^{20}							**B1**[3], 827
h202	**1-Heptene, 1-chloro-***	$CH_3(CH_2)_4CH{:}CHCl$	132.63			155 $78\text{–}82^{75}$								chl v to v	**B1**[2], 196
h203	**—,2-chloro-***	$CH_3(CH_2)_4CCl{:}CH_2$	132.63			138^{748}	0.8895_{4}^{20}	1.4349^{20}	i					chl s	**B1**[3], 823
h204	**2-Heptene, 4-chloro-***	$CH_3CH_2CH_2CHClCH{:}CHCH_3$	132.63			140–145 49^{21}	0.879_{4}^{18}	1.4430^{21}	i					MeOH v chl v	**B1**[3], 825
h205	**3-Heptene, 4-chloro-***	$CH_3CH_2CH_2CCl{:}CHCH_2CH_3$	132.63			139	0.883^{14}	1.437^{14}	i		s			chl s	**B1**[3], 827
h206	**2-Heptene, 6-chloro-2-methyl-***	$CH_3CHCl(CH_2)_2CH{:}CH(CH_3)_2$	147.67			$60\text{–}61^{15}$	0.8931_{4}^{18}	1.4458^{18}	i					os s	**B1**[2], 20
h207	**2-Heptene, 2-methyl-6-methylamino-***	$CH_3CH(NHCH_3)CH_2CH_2CH{:}C(CH_3)_2$	141.26			176–178 $58\text{–}59^{7}$			i	v	v				
h208	**1-Heptene-2-carboxaldehyde, 1-phenyl-***	α-*m*-Amylcinnamaldehyde. Jasminaldehyde. $CH_3(CH_2)_4C(CHO){:}CHC_6H_5$	202.30			$174\text{–}5^{20}$	0.9718^{20}	1.5381^{20}	i	s	s				**B7**[2], 310
h209	**2-Hepten-4-ol***	$CH_3CH_2CH_2CH(OH)CH{:}CHCH_3$	114.19			$152\text{–}4^{760}$	0.8445_{4}^{20}	1.4373^{20}							**B1**, 447
h210	**1-Hepten-4-ol, 4-methyl-***	$CH_3CH_2CH_2COH(CH_3)CH_2CH{:}CH_2$	128.22			162	0.8525_{4}^{18}	1.4479^{18}							**B1**[3], 194
h211	**3-Hepten-2-one** (*trans*)*	$CH_3CH_2CH_2CH{:}CHCOCH_3$	112.17			62^{15}	0.8445_{4}^{2}	1.4430^{20}							
h212	**5-Hepten-2-one, 6-methyl-***	$(CH_3)_2C{:}CHCH_2CH_2COCH_3$	126.20		−67	173^{760}	0.8546_{4}^{16}	1.4400^{20}	i	∞	∞				**B1**[2], 797
h213	**1-Hepten-3-yne***	$CH_3CH_2CH_2C{\vdots}CCH{:}CH_2$	94.16			44^{75}	0.7671_{4}^{25}	1.4520^{25}							**Am 61**, 57
h214	**1-Hepten-4-yne, 6,6-dimethyl-***	$CH_2{:}CHCH_2C{\vdots}CC(CH_3)_3$	122.21			125^{760}	0.758^{20}	1.4305^{20}							
h215	**6-Hepten-4-yne-3-ol-3-ethyl-***	$CH_2{:}CHC{\vdots}CCOH(C_2H_5)CH_2CH_3$	138.21			62^{4}	0.8875_{4}^{20}	1.4800^{20}	i	v	v				**B1**[3], 203
—	**Heptyl alcohol**	*see* **Heptanol***													
—	**Heptyl thiocyanate**	*see* **Thiocyanic acid,** heptyl ester													
—	**Heptylamine***	*see* **Heptane, amino-***													
—	**Heptylene**	*see* **Heptene***													
h216	**1-Heptyne***	$CH_3(CH_2)_4C{\vdots}CH$	96.17		−81	100^{760}	0.7338_{4}^{20}	1.4084^{20}	δ	∞	∞				**B1**[3], 995
h217	**2-Heptyne***	$CH_3(CH_2)_3C{\vdots}CCH_3$	96.17			112^{760}	0.748_{4}^{20}	1.4230^{20}	i	∞					**B1**[3], 998
h218	**3-Heptyne***	$CH_3CH_2CH_2C{\vdots}CCH_2CH_3$	96.17			$105\text{–}6^{760}$	0.7527_{4}^{20}	1.4220^{20}	i	∞	∞				**B1**[3], 999

For explanations, symbols and abbreviations see beginning of table.

No.	Name	Synonyms and Formula	Mol. wt.	Crystalline form, color and specific rotation	m.p. °C	b.p. °C	Density	n_D	Solubility						Ref.
									w	al	eth	ace	bz	other solvents	
	1-Heptyne														
h219	**1-Heptyne, 1-bromo-***	$CH_3(CH_2)_4C\vdots CBr$	175.07			164^{758} 69^{25}	1.2120^{22}_{4}	1.4678^{22}							**B1**³, 998
h220	**1-Heptyne, 1-bromo-***	$CH_3(CH_2)_3C\vdots CCH_2Br$	175.07			104^{56}		1.4878^{25}		s	s	s			**B1**³, 998
h221	**2-Heptyne, 1-chloro-***	$CH_3(CH_2)_4C\vdots CCl$	130.62			141^{760}	0.9250^{24}_{4}	1.4411^{24}		v	v			MeOH v	
h222	**2-Heptyne, 1-chloro-***	$CH_3(CH_2)_3C\vdots CCH_2Cl$	130.62			167^{760}		1.4570^{25}		s	s				**B1**³, 998
h223	**—,7-chloro-***	$ClCH_2(CH_2)_3C\vdots CCH_3$	130.62			175^{760}		1.4599^{25}	i	s	s				**B1**³, 998
h224	**3-Heptyne, 1-chloro-***	$CH_3CH_2CH_2C\vdots CCH_2CH_2Cl$	130.62			162^{760} $90–93^{70}$		1.4520^{25}	i		s				**B1**³, 999
h225	**—,7-chloro-***	$Cl(CH_2)_3C\vdots CCH_2CH_3$	130.62			164^{760}		1.4540^{25}	i		s				**B1**³, 999
h226	**—,2,6-dimethyl-*.**	$(CH_3)_2CHCH_2C\vdots CCH(CH_3)_2$	124.23			$130–6^{760}$	0.785^{20}_{4}								
h227	**—,5,5-dimethyl-*.**	$CH_3CH_2C(CH_3)_2C\vdots CCH_2CH_3$	124.23			69^{100}	0.7610^{20}_{4}	1.4360^{20}							**Am 62,** 1800
h228	**—,5-ethyl-5-methyl-***	$(CH_3CH_2)_2C(CH_3)C\vdots CH_2CH_3$	138.25			88^{100}	0.7714^{20}_{4}	1.4386^{20}							**Am 62,** 1800
h229	**2-Heptyn-1-ol***	Butyl propargyl alcohol $CH_3(CH_2)_3C\vdots CCH_2OH$	112.17			94^{22}		1.4523^{25}							**Am 71,** 1292
—	**Herniarin**	*see* **Coumarin 7-methoxy-**													
—	**Heroin**	*see* **Morphine, diacetyl-**													
h231	**Hesperetin**	$C_{16}H_{14}O_6$	302.29	pl (al)	227				i	v	s		δ	chl δ	
—	**Hesperetol**	*see* **Styrene, 3-dihydroxy-4-methoxy-**													
h232	**Hesperidin**	$C_{28}H_{34}O_{15}$	610.39	wh nd	261–3				δ	s				aa s	**Am 68,** 2109
—	**Hespertin**	*see* **Flavanone, 4′-methoxy-,3′,5,7-trihydroxy-**													
—	**Heteroxanthine**	*see* **Xanthine, 7-methyl-**													
h233	**Hexacene**	Anthraceno-2′:3′,2:3-anthracene.	328.41	(sub)		sub									**E14S,** 789
—	**Hexachlorophene**	*see* **Methane, bis(2-hydroxy-3,5,6-trichlorophenyl)-***													
h234	**Hexacosane***	Cerane. $CH_3(CH_2)_{24}CH_3$	366.72	mcl, tcl or rh (bz)	56	262^{15}	0.7783^{60}	1.4357^{60}	i	s^h			s	chl s lig v	**B1**², 143
h235	**Hexacosanoic acid***	Cerotic acid. $CH_3(CH_2)_{24}CO_2H$	396.70	nd (al)	88		0.8359^{79}_{4}	1.4301^{100}	i	v^h	v				**B2**², 381
h236	**1-Hexacosanol***	Cerotin. Ceryl alcohol. $CH_3(CH_2)_{25}OH$	382.72	rh pl	80	305^{20}d			i	s	s				**B1**², 470
h237	**1,15-Hexadecadiyne***	$HC\vdots C(CH_2)_{12}C\vdots CH$	218.39		44–5	$152–5^{12}$									
h238	**6,9-Hexadecadiyne***	$CH_3(CH_2)_5C\vdots CCH_2C\vdots C(CH_2)_4CH_3$	218.39			169^{15}	0.845^{18}_{4}	1.4694^{18}							
h239	**6,10-Hexadecadiyne***	$CH_3(CH_2)_4C\vdots CCH_2CH_2C\vdots C(CH_2)_4CH_3$	218.39			157^{10}	0.846^{17}	1.4686^{22}							
h240	**Hexadecanal***	Palmitaldehyde. $CH_3(CH_2)_{14}CHO$	240.43	pl (eth) nd (peth)	34	$200–2^{29}$			i	s				os s	**B1**², 771
h241	—,dimethyl acetal	$CH_3(CH_2)_{14}CH(OCH_3)_2$	286.50			144^{2}		1.4382	v		v			MeOH v	
h242	—,oxime	$CH_3(CH_2)_{14}CH{:}NOH$	255.45	nd (al)	88										**B1**², 771
h243	**Hexadecane***	Cetane. $CH_3(CH_2)_{14}CH_3$	226.45	lf (ace)	18	287^{760}	0.77331^{20}_{4}	1.4345	i	δ^h	∞				**B1**³, 555
h244	**—,1-amino-***	$CH_3(CH_2)_{15}NH_2$	241.46	lf	46	322^{760} 144^{2}			i	v	v	s	v	chl s	**B4**², 660
h245	**—,1-bromo-***	Cetyl bromide. $CH_3(CH_2)_{15}Br$	305.35		15	149^{1}	1.9949	1.4608^{25}	i		s				**B1**², 138
h246	**—,1-chloro-***	Cetyl chloride. $CH_3(CH_2)_{15}Cl$	260.89		12	187^{16}	0.8520^{20}_{4}	1.4548^{20}	i						**B1**³, 559
h247	**—,1,16-dibromo-***	$Br(CH_2)_{16}Br$	384.25	lf (al)	56	204^{4}			i	δ v^h				chl v	**B1**², 138
h248	**—,1-fluoro-**	Cetyl fluoride. $CH_3(CH_2)_{15}F$	244.44			287 181^{24}	0.809^{18}				s			lig s	**B1**², 138
h249	**—,1-iodo-***	Cetyliodide. $CH_3(CH_2)_{15}I$	352.35		22	211^{15} 160–3	1.123^{20}_{4}	1.4806^{20}	i	δ	s	s	∞	chl v	**B1**³, 560

For explanations, symbols and abbreviations see beginning of table.

No.	Name	Synonyms and Formula	Mol. wt.	Crystalline form, color and specific rotation	m.p. °C	b.p. °C	Density	n_D	Solubility						Ref.
									w	al	eth	ace	bz	other solvents	
	Hexadecane														
h250	—,**1-phenyl-***	Cetylbenzene. $CH_3(CH_2)_{15}C_6H_5$	302.55	...	27	237^{16}	0.8560^{20}_{4}	1.4814^{20}	i	δ	v	...	v	CS_2 v lig v	**B5**, 472
h251	**Hexadecanedioic acid***	Thapsic acid. $HO_2C(CH_2)_{14}CO_2H$	286.42	pl (al)	123	...	...	...	i	s^h	...	s^h	s^h	AcOEt s^h	**B2**, 733
h252	**1-Hexadecanethiol***	Cetyl mercaptan. $CH_3(CH_2)_{15}SH$	258.51	(lig)	18	$123–8^1$	...	...	i	δ	s	...	...	...	...
h253	**Hexadecanoic acid***	Palmitic acid. $CH_3(CH_2)_{14}CO_2H$	256.43	nd	63	167^1	0.8474^{80}_{4}	1.42549^{60}	i	s	∞	s	s	chl v	**Am 64**, 2739
h254	—,amide	$CH_3(CH_2)_{14}CONH_2$	255.45	lf	107	236^{12}	...	...	i	δ	δ	δ	δ	...	...
h255	—,—,*N*-phenyl-*	Palmitanilide. $CH_3(CH_2)_{14}CONHC_6H_5$	331.55	nd (al)	89	282 132^{10}	...	...	i	v	v	v	v	chl v aa v^h	**B12²**, 148
h256	—,anhydride*	$[CH_3(CH_2)_{14}CO]_2O$	492.85	lf (peth)	63	...	0.8383^{83}_{4}	1.4357^{70}	i	δ	v	...	...	peth $δ^h$	**B2²**, 341
h257	—,benzyl ester*	$CH_3(CH_2)_{14}CO_2CH_2C_6H_5$	346.56	...	36	...	0.9136^{35}_{25}	1.4689^{50}	i	s	s	...	s	chl v	**B6²**, 417
h258	—,butyl ester*	$CH_3(CH_2)_{14}CO_2C_4H_9^n$	312.54	(dil al)	16	...	1.432^{50}	...	i	s	s	...	...	...	**B2²**, 336
h259	—,chloride	$CH_3(CH_2)_{14}COCl$	274.88	...	12	199^{20}	...	...	d	d	v	...	...	...	**B2²**, 341
h260	—,ethyl ester*	$CH_3(CH_2)_{14}CO_2C_2H_5$	284.49	nd	25	191^{10}	0.8577^{25}_{4}	1.4347^{34}	i	s	s	...	...	...	**B2²**, 336
h261	—,hentriacontyl ester*	Myricyl palmitate. $CH_3(CH_2)_{14}CO_2(CH_2)_{30}CH_3$	691.27	...	72	...	...	...	i	i	s	...	...	...	**B2**, 373
h262	—,hexadecyl ester*	Cetyl palmitate. $CH_3(CH_2)_{14}CO_2(CH_2)_{15}CH_3$	480.87	pl (eth)	53	360	0.8324^{50}_{4}	1.4425^{50}	i	s^h	s	...	...	...	**B2²**, 337
h262¹	—,2-hydroxyethyl ester*	$CH_3(CH_2)_{14}CO_2CH_2CH_2OH$	300.49	...	51	...	0.8786^{60}_{4}	...	...	v	s^h	...	...	...	...
h263	—,isopropyl ester*	$CH_3(CH_2)_{14}CO_2C_3H_7^i$	298.51	...	...	160^2	0.8404^{38}	1.4364^{25}	...	...	...	...	...	...	...
h264	—,methyl ester,*	$CH_3(CH_2)_{14}CO_2CH_3$	270.46	...	30	$415–8^{747}$ 148^2	...	...	...	v	s	v	v	chl v	**B2**, 372
h265	—,nitrile	$CH_3(CH_2)_{14}CN$	237.43	hex	31	251^{100} 142^1	0.822^{31}	1.4396^{35}	i	v	v	v	v	chl v	...
h266	—,propyl ester*	$CH_3(CH_2)_{14}CO_2C_3H_7^n$	298.51	...	...	166^2	0.8455^{38}	1.4392^{25}	...	...	...	...	...	...	...
h267	—,**16-hydroxy-***	$HO(CH_2)_{15}CO_2H$	272.43	...	95	...	...	...	i	s	$δ^h$	s	$δ^h$	...	**B3**, 362
h268	—,**9,10,16-trihydroxy***	Aleuritic acid. $HO(CH_2)_6CHOHCHOH(CH_2)_7CO_2H$	304.43	lf nd	100–1	...	...	...	δ	...	...	...	...	...	**B3²**, 272
h269	**1-Hexadecanol***	Cetyl alcohol. $CH_3(CH_2)_{15}OH$	242.45	...	50	344 190^{15}	0.8176^{50}_{4}	1.4283^{79}	i	δ	v	s	v	chl v	**B1²**, 466
h270	**1-Hexadecene***	Cetene, Cetylene. $CH_3(CH_2)_{13}CH{:}CH_2$	224.43	...	4	155^{15}	0.7825^{20}_{4}	1.4441^{20}	i	s	s	...	...	peth s	**B1²**, 206
h271	**Hexadecenoic acid***	Gaidic acid. $C_{15}H_{29}CO_2H$	254.42	lf (al)	39	...	...	...	...	v	...	...	...	peth s, chl v	**B2**, 461
h272	**2-Hexadecenoic acid***	Δ,α,β-Hypogeic acid. $CH_3(CH_2)_{12}CH{:}CHCO_2H$	254.42	...	49	...	...	...	...	...	v	...	...	chl v, peth v	**B2**, 424
h273	**7-Hexadecenoic acid***	Hypogeic acid (artificial), $CH_3(CH_2)_7CH{:}CH(CH_2)_5CO_2H$	254.42	...	33	230^{10}	...	...	i	v	s	...	...	...	**B2**, 460
h274	**1-Hexadecyne***	$CH_3(CH_2)_{13}C{:}CH$	222.42	...	20–5	284^{760}	0.7965^{20}_{4}	1.4440^{20}	i	...	...	...	...	...	**B1³**, 1028
h275	**2-Hexadecyne***	$CH_3(CH_2)_{12}C{:}CCH_3$	222.42	...	20	160^{15}	0.8039^{20}_{4}	...	...	...	...	...	...	...	...
h276	**7-Hexadecynoic acid***	$CH_3(CH_2)_7C{:}C(CH_2)_5CO_2H$	252.40	nd (w), (al)	47	214^{15}	...	...	i	v	v	...	...	...	**B2**, 494
h277	**2,4-Hexadien-1-al***	Sorbaldehyde. $CH_3CH{:}CHCH{:}CHCHO$	96.13	...	...	76^{30} 43^5	0.898^{20}	1.5384^{20}	...	...	...	...	...	...	**B1²**, 809
h278	**1,2-Hexadiene***	Propylallene. $CH_3CH_2CH_2CH{:}C{:}CH_2$	82.15	...	...	78	0.7149^{20}	1.4298^{17}	i	...	s	...	...	chl s	**B1³**, 189
h279	**1,3-Hexadiene***	$CH_3CH_2CH{:}CHCH{:}CH_2$	82.15	...	...	72	0.7152^{13}_{4}	1.4416^{12}	i	...	s	...	...	...	**B1³**, 981
h280	**1,4-Hexadiene***	Allylpropenyl. $CH_3CH{:}CHCH_2CH{:}CH_2$	82.15	...	...	66^{761}	0.710^{16}_{4}	1.4167^{16}	i	...	v	...	...	...	**B1³**, 982
h281	**1,5-Hexadiene***	Biallyl. $HC_2{:}CHCH_2CH_2CH{:}CH_2$	82.15	...	−141	60^{763}	0.6923^{20}_{4}	1.4044^{20}	i	s	s	...	...	...	**B1³**, 982
h282	**2,4-Hexadiene***	Bipropenyl. $CH_3CH{:}CHCH{:}CHCH_3$	82.15	...	−79	82^{760}	0.7196^{20}_{4}	1.4516^{20}	i	...	...	...	...	...	**B1³**, 985

For explanations, symbols and abbreviations see beginning of table.

No.	Name	Synonyms and Formula	Mol. wt.	Crystalline form, color and specific rotation	m.p. °C	b.p. °C	Density	n_D	Solubility w	al	eth	ace	bz	other solvents	Ref.
	1,3-Hexadiene														
h283	**1,3-Hexadiene,3-chloro-***	$CH_3CH_2CH:CClCH:CH_2$	116.59			68^{117}	0.9390^{20}_{4}	1.4770^{20}			s			chl s	
h284	**1,4-Hexadiene,5-chloro-2-isopropyl-***	$CH_3CCl:CHCH_2C[CH(CH_3)_2]:CH_2$	158.67			95^{18}	0.9310^{25}_{4}	1.4730^{25}	i					chl s	
h285	**2,4-Hexadiene, 6-chloro-2-methyl-***	$ClCH_2CH:CHCH:C(CH_3)_2$	130.62			57^{11}	0.9416^{20}_{4}	1.5120^{20}	i						**B1**^{3}, 1000
h286	**1,5-Hexadiene, decachloro-***	Perchlorodiallyl. $Cl_2C:CClCCl_2CCl_2CCl:CCl_2$	426.60		49		1.905^{51}_{4}	1.6012^{51}				s		chl s	**B1**^{3}, 984
h287	**2,4-Hexadiene 1,3-dichloro-***	$CH_3CH:CHCCl:CHCH_2Cl$	151.04			$80–2^{17}$	1.1456^{20}_{4}	1.5271^{20}						chl s	
h288	**1,5-Hexadiene, 2,5-dimethyl-***	Diisobutenyl. $CH_2:C(CH_3)CH_2CH_2C(CH_3):CH_2$	110.20		−76	137^{755}	0.7512^{20}	1.43995^{21}	i						**B1**^{2}, 239
h289	**2,4-Hexadiene, 2,5-dimethyl-***	Diisocrotyl. $(CH_3)_2C:CHCH:C(CH_3)_2$	110.20		15	134^{760}	0.7636^{20}_{4}	1.4785^{20}_{D}	i	s	s				**B1**^{2}, 1011
h290	—,**1,3,4,6-tetrachloro***	$ClCH_2CH:CClCCl:CHCH_2Cl$	219.93			$84–9^{2}$	1.4013^{20}_{4}	1.5465^{20}						chl, MeOH s	
h291	**1,4-Hexadiene, 3,3,6-trichloro-**	$ClCH_2CH:CHCCl_2CH:CH_2$	185.48			$100–3^{4}$	1.3036^{20}_{4}	1.5585^{20}			s				**B1**^{3}, 982
h292	**2,4-Hexadiene-1,6-dioic acid** *(cis)**	*cis*-Muconic acid. $HO_2CCH:CHCH:CHCO_2H$	142.11		188									aa s	**B2**, 803
h293	—*(trans)**	*trans*-Muconic acid. $HO_2CCH:CHCH:CHCO_2H$	142.11	nd (al)	305d	320			δ	δ				AcOEt s aa s	**B2**, 803
h294	—,dimethyl ester *(trans)**	$CH_3O_2CCH:CHCH:CHCO_2CH_3$	170.17	nd	159					s	s				**B2**^{2}, 672
h295	**1,5-Hexadiene-3,4-diol**(*d*)*	Divinyl glycol. $H_2C:CHCHOHCHOHCH:CH_2$	114.15	$[\alpha]^{17}_{4}$ +94.8 (al)	−60	198^{760} 97^{13}	1.006^{20}_{4}	1.4700^{20}	s	s	s			chl s	**B1**^{3}, 2272
h296	—(*dl*)	$H_2C:CHCHOHCHOHCH:CH_2$	114.15		21.7		1.017^{19}_{4}								**B1**^{3}, 2272
h297	—(*meso*)	$CH_2:CHCHOHCHOHCH:CH_2$	114.15		18	100^{14}	1.023^{19}_{4}	1.4810^{19}	v	v	v			chl v	**B1**^{3}, 2272
h298	**2,4-Hexadienoic acid**	Sorbic acid. $CH_3CH:CHCH:CHCO_2H$	112.13	nd	133	228d (153^{50})			δ^h	s	v				**B2**, 489
h299	—,amide	$CH_3CH:CHCH:CHCONH_2$	111.14	nd (w)	170				s	s					**B2**^{2}, 453
h300	—,chloride	$CH_3CH:CHCH:CHCOCl$	130.58			78^{15}	1.0666^{19}_{4}	1.5471^{19}							**B2**^{2}, 453
h301	—,ethyl ester*	$CH_3CH:CHCH:CHCO_2C_2H_5$	140.18			85^{20}	0.9560^{20}_{4}	1.502^{20}							**B2**^{2}, 452
h302	—,methyl ester*	$CH_3CH:CHCH:CHCO_2CH_3$	126.16	lf	5	180^{759}				s	s				**B2**^{2}, 452
h303	**2,4-Hexadien-1-ol***	Sorbyl alcohol. $CH_3CH:CHCH:CHCH_2OH$	98.15	nd	31	76^{12}	0.8967^{23}_{4}	1.4971^{23}	i	s	s				
h304	**3,5-Hexadien-2-ol***	$H_2C:CHCH:CHCHOHCH_3$	98.15			$77–8^{26}$ $44–5^{8}$				v					**B1**^{3}, 1987
h306	**3,5-Hexadien-2-one, 6-phenyl-**	Cinnamilideneacetone. $C_6H_5CH:CHCH:CHCOCH_3$	172.23	wh	68	$170–2^{15}$			δ	s				chl s MeOH s	**B7**^{2}, 321
h307	**1,3-Hexadien-5-yne***	$HC\vdots CCH:CHCH:CH_2$	78.11			32^{100}	0.7821^{20}_{20}	1.4900^{20}	i					sulf s	**B1**^{3}, 1058
h308	**1,5-Hexadien-3-yne***	Divinyl acetylene. $HC_2:CHC\vdots CCH:CH_2$	78.11		−88	84^{760}	0.7857^{20}_{4}	1.5035^{20}							**B1**^{3}, 1058
h309	**3,5-Hexadien-1-yne***	$CH\vdots CCH:CHCH:CH_2$	78.11			83^{760} 32^{100}	0.7734^{20}_{4}	1.5095^{20}	i						**B1**^{3}, 1058
h310	**1,5-Hexadien-3-yne, 2,5-dimethyl-***	Diisopropenyl acetylene. $CH_2:C(CH_3)C\vdots CC(CH_3):CH_2$	106.16	ye		123	0.7863^{25}_{4}	1.4845^{20}							**B1**^{3}, 1062
h311	**1,4-Hexadiyne***	$HC:CCH_2C:CCH_3$	78.11		<-80	$78–83^{760}$	0.825^{0}_{4}								

For explanations, symbols and abbreviations see beginning of table.

No.	Name	Synonyms and Formula	Mol. wt.	Crystalline form, color and specific rotation	m.p. °C	b.p. °C	Density	n_D	Solubility w	al	eth	ace	bz	other solvents	Ref.
	1,5-Hexadiyne														
h312	**1,5-Hexadiyne***	$HC{:}CCH_2CH_2C{:}CH$	78.11		−6	86^{760} 20^{46}	0.8049^{20}_{4}	1.4380^{23}	i	s	s			os s	$B1^{3}$, 1057
h313	**2,4-Hexadiyne***	$CH_3C{:}CC{:}CCH_3$	78.11	pr (sub)	64	129				v	v				$B1^{2}$, 242
h314	**—,1,6-di(4-morpholinyl)-**	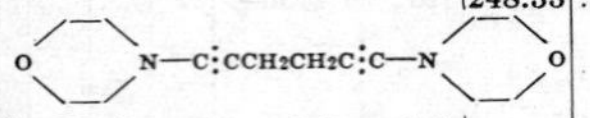 O N—C:CCH₂CH₂C:C—N O	248.33		108				s^{h}				s		
—	**Hexahydrobenzyl alcohol**	*see* **Methanol, cyclohexyl-***													
—	**Hexahydro-*p*-cymene**	*see* ***p*-Methane(*trans*)**													
—	**Hexahydrothymol**	*see* **Menthol**													
—	**Hexamethylene glycol**	*see* **1,6-Hexanediol***													
—	**Hexamethylenediamine**	*see* **Hexane, 1,6-diamino-***													
—	**Hexamethyleneimine**	*see* **Cycloheptane. 1-aza-***													
h315	**Hexamethylenetetramine**	Urotropine. 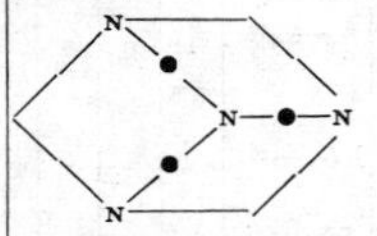	140.19	rh	280 sub		1.331^{-5}		v	s	δ	s	δ	chl s	$B1^{2}$, 200
h316	**Hexanal***	Caproaldehyde. *n*-Caproic aldehyde. $CH_3(CH_2)_4CHO$	100.16		−56	128^{760}			δ	v	v				$B1^{2}$, 745
h317	—,oxime*	$CH_3(CH_2)_4CH{:}NOH$	115.17	(MeOH)	51					s	s				B1, 689
h318	**—,2-ethyl-***	$CH_3(CH_2)_3CH(C_2H_5)CHO$	128.21		<−100	163^{760}			i	s	s				B1, 707
h320	**Hexane***	$CH_3(CH_2)_4CH_3$	86.18		−95	68^{760}	0.65937^{20}_{4}	1.37486^{20}	i	v	s			chl s	$B1^{3}$, 374
—	***neo*-Hexane**	*see* **Butane, 2,2-dimethyl-***													
h321	**Hexane, 1-amino-***	*n*-Hexylamine. $CH_3(CH_2)_5NH_2$	101.19		−19	129^{742}	0.763^{25}_{4}	1.4255^{17}	δ	∞	∞				$B4^{2}$, 649
h321[1]	**—,2-amino-*(*d*)**	*d*-α-Methylamylamine. $CH_3(CH_2)_3CH(NH_2)CH_3$	101.19	$[\alpha]^{27}_{D}+4.3$		64^{90}									
h322	**—,—(*dl*)***	$CH_3(CH_2)_3CH(NH_2)CH_3$	101.19		−19	116–8	0.767^{20}_{4}								B4, 190
h323	**—,2-amino-4-methyl-***	$CH_3CH_2CH(CH_3)CH_2CH(NH_2)CH_3$	115.21			$130–5^{760}$	0.7655	1.4150^{25}	δ	v	v			chl, dil ac v	
h324	**—,2,4-bis(4-hydroxyphenyl)-**	Benzestrol. 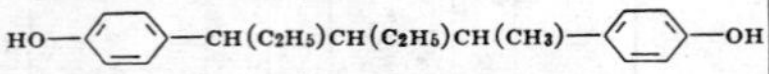 HO— —$CH(C_2H_5)CH(C_2H_5)CH(CH_3)$— —OH	298.41		162–6				i	v	v	v	δ	MeOH v chl δ aa s	
h325	**—,3,4-bis(4-hydroxy-3-methylphenyl)-**	Dimethylhexestrol. Promethestrol. 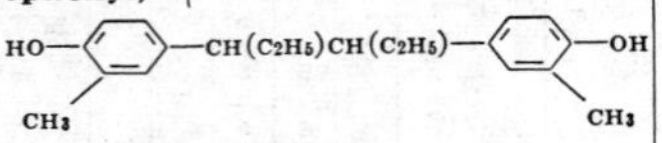 HO— —$CH(C_2H_5)CH(C_2H_5)$— —OH; CH_3, CH_3	298.41	(dil aa)	145									aa δ	
h326	**—,1-bromo-***	$CH_3(CH_2)_5Br$	165.08		−85	156^{759}	1.1763^{20}_{4}	1.44781^{20}	i	∞	∞				$B1^{3}$, 391
h327	**—,2-bromo-***	$CH_3(CH_2)_3CHBrCH_3$	165.09			144^{760}	1.1658^{20}_{4}	1.4432^{25}	i	v				chl v	$B1^{3}$, 392
h328	**—,3-bromo-***	$CH_3CH_2CH_2CHBrCH_2CH_3$	165.09			144^{744}	1.1799^{20}_{4}	1.4486^{20}	i					chl v	B1, 144
h329	**—,1-bromo-6-fluoro-***	$F(CH_2)_6Br$	183.08			$67–8^{11}$	1.293^{20}_{4}	1.4435^{25}	i	v	v				
h330	**—,1-chloro-***	$CH_3(CH_2)_5Cl$	120.62		−83	$132–3^{760}$	0.8784^{20}_{4}	1.41991^{20}	i						$B1^{8}$, 388
h331	**—,2-chloro-***	$CH_3(CH_2)_3CHClCH_3$	120.62			$122–3^{760}$	0.8694^{21}_{4}	1.4142^{22}							$B1^{3}$, 389
h332	**—,3-chloro-***	$CH_3CH_2CH_2CHClCH_2CH_3$	120.62			123	0.870^{20}	1.4163^{20}	i		v			chl v	
h333	**—,2-chloro-2,5-dimethyl-***	$(CH_3)_2CHCH_2CH_2CCl(CH_3)_2$	148.68			44^{14}	0.8476^{18}_{4}	1.4232^{20}	i		s		s		$B1^{3}$, 485
h334	**—,3-chloro-2,3-dimethyl-***	$CH_3CH_2CH_2CCl(CH_3)CH(CH_3)_2$	148.68			$41–3^{12}$	0.8869^{20}_{4}	1.4333^{25}	i	s					$B1^{3}$, 481
h336	**—,1-chloro-3-ethyl-***	$CH_3CH_2CH_2CH(C_2H_5)CH_2CH_2Cl$	148.68	$[\alpha]^{27}_{D}+1.15$		85^{40}	0.879^{21}_{4}	1.4335^{25}	i		s				$B1^{3}$, 478

For explanations, symbols and abbreviations see beginning of table.

No.	Name	Synonyms and Formula	Mol. wt.	Crystalline form, color and specific rotation	m.p. °C	b.p. °C	Density	n_D	Solubility w	al	eth	ace	bz	other solvents	Ref.
	Hexane														
h337	—,**3-chloro-3-ethyl-***	$CH_3CH_2CH_2CCl(C_2H_5)_2$	148.68			155d (62–3^{24})	0.9018^{0}	1.4358^{20}	i		s			chl s	**B1**^{3}, 479
h338	—,**1-chloro-6-fluoro-***	$F(CH_2)_6Cl$	138.62			62^{15}	1.015$^{20}_{4}$	1.4168^{25}	i	v	v				
h339	—,**1-chloro-3-methyl-***	$CH_3CH_2CH_2CH(CH_3)CH_2CH_2Cl$	134.65			150–2^{758}	0.8766$^{20}_{4}$	1.4274^{20}						chl s	**B1**^{2}, 119
h340	—,**2-chloro-2-methyl-***	$CH_3(CH_2)_3CCl(CH_3)_2$	134.65			135d	0.8661$^{25}_{4}$	1.4205^{20}	i	v				chl s	**B1**, 156
h341	—,**2-chloro-5-methyl-***	$(CH_3)_2CHCH_2CH_2CHClCH_3$	134.65			138^{735}d	0.863$^{20}_{4}$				v				**B1**, 156
h342	—,**3-chloro-3-methyl-***	$CH_3CH_2CH_2CCl(CH_3)CH_2CH_3$	134.65			135	0.8787$^{20}_{4}$	1.4250^{20}			v				**B1**^{2}, 119
h343	—,**3-chloro-2,2,3-trimethyl***	$CH_3(CH_2)_2CCl(CH_3)C(CH_3)_3$	162.71			64–5^{13}	0.8973$^{20}_{4}$	1.4468^{20}			s			chl s	**B1**^{3}, 515
h344	—,**1,6-diamino-***	Hexamethylenediamine. $H_2N(CH_2)_6NH_2$	116.21		41–2	204			v	s			s		**B4**^{2}, 710
h345	—,—,dihydro-chloride*	$H_2N(CH_2)_6NH_2.2HCl$	189.14	nd (al-eth)	248–50				v	δ	i			chl i	**B4**^{2}, 710
h346	—,**2,5-diamino-***	$CH_3CH(NH_2)CH_2CH_2CH(NH_2)CH_3$	116.21			175			v	v	v				**B4**, 269
h347	—,**1,2-dibromo-***	$CH_3(CH_2)_3CHBrCH_2Br$	243.99			87^{16}	1.5872$^{15}_{4}$	1.5012^{19}			∞				**B1**^{2}, 109
h348	—,**1,2-dichloro-***	$CH_3(CH_2)_3CHClCH_2Cl$	155.08			172–4	1.085^{15}							chl s	**B1**, 144
h349	—,**1,6-dichloro-***	$Cl(CH_2)_6Cl$	155.08			203–5 94^{22}	1.0677$^{20}_{4}$	1.4572^{20}	i					chl s	**B1**^{3}, 389
h350	—,**2,2-dichloro-***	$CH_3(CH_2)_3CCl_2CH_3$	155.08			68^{49}	1.0150$^{25}_{4}$	1.4353^{25}	i		s			chl s	**B1**^{3}, 390
h351	—,**2,3-dichloro-***	$CH_3CH_2CH_2CHClCHClCH_3$	155.08			162–5	1.0527^{11}				s			chl s	**B1**^{2}, 109
h352	—,**2,5-dichloro-***	$CH_3CHClCH_2CH_2CHClCH_3$	155.08		19	177^{751}	1.0431$^{25}_{4}$	1.4495^{12}						chl s	**B1**, 144
h353	—,**2,5-dichloro-*** (*meso*)	$CH_3CHClCH_2CH_2CHClCH_3$	155.08			178^{752}	1.0459$^{25}_{4}$							chl s	**B1**, 144
h354	—,**3,4-dichloro-***	$CH_3CH_2CHClCHClCH_2CH_3$	155.08			69–70^{30}	1.055^{20}	1.4490^{20}	i			s		chl s	
h355	—,**2,5-dichloro-2,5-dimethyl-***	$(CH_3)_2CClCH_2CH_2CCl(CH_3)_2$	183.13	lf, nd	66–7					s	s		s	chl s	**B1**^{2}, 127
h356	—,**3,4-dichloro-3,4-dimethyl-***	$CH_3CH_2CCl(CH_3)CCl(CH_3)CH_2CH_3$	183.13			165^{760}								chl s	**B1**^{2}, 62
h357	—,**1,6-diiodo-***	$I(CH_2)_6I$	338.00	nd	10	113^{3}	2.03$^{22}_{4}$	1.585^{20}	i	v	v				**B1**^{2}, 110
h358	—,**2,2-dimethyl-***	$CH_3(CH_2)_3C(CH_3)_3$	114.23		−121	107^{760}	0.69528$^{20}_{4}$	1.39349^{20}	i	v				os v lig v	**B1**^{3}, 479
h359	—,**2,3-dimethyl-** (*dl*)*	$CH_3CH_2CH_2CH(CH_3)CH(CH_3)_2$	114.32			116^{760}	0.71214$^{20}_{4}$	1.40113^{20}	i	δ	s				**B1**^{3}, 480
h360	—,—(*l*)*	$CH_3CH_2CH_2CH(CH_3)CH(CH_3)_2$	114.23	$[\alpha]^{25}_{D}-0.92$		113^{760}									**B1**^{3}, 482
h361	—,**2,4-dimethyl-** (*d*)*	$CH_3CH_2CH(CH_3)CH_2CH(CH_3)_2$	114.23	$[\alpha]^{30}_{D}+2.99$		111^{760}	0.696$^{30}_{4}$								**B1**^{3}, 483
h362	—,—(*dl*)*	$CH_3CH_2CH(CH_3)CH_2CH(CH_3)_2$	114.23			109^{760}	0.70036$^{20}_{4}$	1.39534^{20}	i	δ	s				**B1**^{2}, 482
h363	—,—(*l*)*	$CH_3CH_2CH(CH_3)CH_2CH(CH_3)_2$	114.23	$[\alpha]^{21}_{D}+0.8$		110^{760}									**B1**^{3}, 483
h364	—,**2,5-dimethyl-***	$(CH_3)_2CHCH_2CH_2CH(CH_3)_2$	114.23		−91	109^{760}	0.69354$^{20}_{4}$	1.39246^{20}	i	δ	s				**B1**^{3}, 283
h365	—,**3,3-dimethyl-***	$CH_3CH_2CH_2C(CH_3)_2CH_2CH_3$	114.23		−126	112^{760}	0.71000$^{20}_{4}$	1.40009^{20}	i	s				os s	**B1**^{3}, 486
h366	—,**3,4-dimethyl-***	$CH_3CH_2CH(CH_3)CH(CH_3)CH_2CH_3$	114.23			118^{760}	0.71923$^{20}_{4}$	1.40406^{20}	i	δ	s				**B1**^{3}, 488
h367	—,**1,6-dimitro-***	$O_2N(CH_2)_6NO_2$	176.17		37				i					aa s	**B1**^{3}, 396
h368	—,**3-ethyl-***	$CH_3CH_2CH_2CH(CH_2CH_3)_2$	114.23			119^{760}	0.713$^{20}_{4}$	1.4016^{20}	i	δ	s				**B1**^{3}, 478

For explanations, symbols and abbreviations see beginning of table.

No.	Name	Synonyms and Formula	Mol. wt.	Crystalline form, color and specific rotation	m.p. °C	b.p. °C	Density	n_D	Solubility w	al	eth	ace	bz	other solvents	Ref.
	Hexane														
h369	**—,1-fluoro-***	$CH_3(CH_2)_5F$	104.17		−104	93^{755}					s				**B1**^{3}, 387
h370	**—,1-iodo-***	$CH_3(CH_2)_5I$	212.07			181^{747}	1.4367^{20}_{4}	1.49290^{20}	i						**B1**^{3}, 394
h371	**—,2-methyl-***	Isoheptane. $CH_3(CH_2)_3CH(CH_3)_2$	100.21		−118	90^{760}	0.67869^{20}_{4}	1.38485^{20}	i	s	∞				**B1**^{3}, 434
h372	**—,3-methyl-(*d*)***	$CH_3CH_2CH_2CH(CH_3)CH_2CH_3$	100.21	$[\alpha]^{20}_{D}+9.5$	−119	92^{760}	0.6860^{20}_{4}	1.3887^{20}	i	s	∞				**B1**^{3}, 439
h373	—,—(*dl*)*	$CH_3CH_2CH_2CH(CH_3)CH_2CH_3$	100.21		*ca.* −173	92^{760}	0.6868^{20}	1.3886							**B1**^{3}, 437
h374	—,—(*l*)*	$CH_3CH_2CH_2CH(CH_3)CH_2CH_3$	100.21	$[\alpha]^{21}_{D}-7.75$		92^{760}	0.687^{21}_{4}	1.3854^{25}							**B1**^{3}, 440
h375	**—,1,1,1,2,2-pentachloro-***	$CH_3(CH_2)_3CCl_2CCl_3$	258.40			$129–31^{10}$	1.370^{25}	1.4980^{25}	i		s				
h376	**—,1-phenyl-***	Hexylbenzene. $CH_3(CH_2)_5C_6H_5$	162.28		−62	226^{760}			i		∞				**B5**^{2}, 337
h377	**—,1,1,2,2-tetra-chloro-***	$CH_3(CH_2)_3CCl_2CHCl_2$	223.96			99–101	1.3096^{25}_{4}	1.488^{25}							
h378	**—,2,2,5-tri-methyl-***	$(CH_3)_2CHCH_2CH_2C(CH_3)_3$	128.26		−106	124^{760}	0.7032^{25}_{4}	1.3997^{20}	i					os v lig ∞	**B1**^{3}, 516
h379	**1,6-Hexanedial***	Adipaldehyde. $OCH(CH_2)_4CHO$	114.15			$92–4^{9}$	1.003^{19}_{4}	1.4307^{19}	δ	v	v			aa s	**B1**^{2}, 839
h380	**Hexanedioic acid***	Adipic acid. $HO_2C(CH_2)_4CO_2H$	146.14	mcl pr (w, ace-lig)	153	267^{750}	1.360^{25}_{4}		δ	v	s			aa i lig i	**B2**^{2}, 572
h381	—,diamide	$H_2NCO(CH_2)_4CONH_2$	144.17	pl	220				δ	v	δ				**B2**^{2}, 576
h382	—,diazide	$N_3OC(CH_2)_4CON_3$	196.17		−1					v	v				**B2**^{1}, 278
h383	—,dibutyl ester*	$C_4H_9^nO_2C(CH_2)_4CO_2C_4H_9^n$	258.36		−37	145^{4}	0.9652^{20}_{4}	1.4369^{20}	i	∞	∞				**B2**^{2}, 575
h384	—,dichloride	$ClCO(CH_2)_4COCl$	183.04			132^{18}									**B2**^{2}, 575
h385	—,diethyl ester*	$C_2H_5O_2C(CH_2)_4CO_2C_2H_5$	202.25		−21	$239–41^{760}$ 127^{13}	1.0076^{20}_{4}	1.4272^{20}	i	s	s				**B2**^{2}, 574
h386	—,di(2-ethylbutyl) ester*	$(C_2H_5)_2CHCH_2O_2C(CH_2)_4CO_2CH_2CH(C_2H_5)_2$	314.47		−26	200^{10}	0.934^{25}_{4}	1.4434^{20}	i	s		s		aa s	
h387	—,di(2-ethylhexyl) ester	$C_3H_7^nCH(C_2H_5)CH_2O_2C(CH_2)_4CO_2CH_2CH(C_2H_5)C_3H_7^n$	370.58		−60	214^{5}	0.922^{25}_{4}	1.4474^{20}	i	s	s	s		aa s	
h388	—,dihydrazide	$H_2NNHOC(CH_2)_4CONHNH_2$	174.20	lf (w)	171				s	s^h	i		s	aa i	
h389	—,diisobutyl ester	$C_4H_9^iO_2C(CH_2)_4CO_2C_4H_9^i$	258.36		−17	282^{760}	0.953^{20}	1.4315^{25}	i	s		s		aa s	
h390	—,dimethyl ester*	$CH_3O_2C(CH_2)_4CO_2CH_3$	174.20		10	105^{7} 115^{13}	1.0600^{20}_{4}	1.4277	i	s	s				**B2**, 365
h391	—,dinitrile	$NC(CH_2)_4CN$	108.14		1	295^{760} (180^{20})	0.951^{19}	1.4597	δ	s	δ			chl s	**B2**^{2}, 576
h392	—,dipropyl ester*	$C_3H_7^nO_2C(CH_2)_4CO_2C_3H_7^n$	230.31		−20	155^{16}	0.9790^{20}_{4}	1.4314^{20}	i	s	s				**B2**^{2}, 574
h393	—,monoamide*	Adipamic acid. $HO_2C(CH_2)_4CONH_2$	145.16	nd (w)	161										**J 1934,** 1101
h394	—,monoethyl ester*	$HO_2C(CH_2)_4CO_2C_2H_5$	174.20	hyg (eth-peth)	28	285^{760} (160^{7})	1.081^{20}	1.4388^{20}		s	s^h			peth s^h	**B2**^{2}, 574
h395	—,monomethyl ester*	$HO_2C(CH_2)_4CO_2CH_3$	160.17		3	162^{10}				s					**B2**, 652
h396	—,piperazinium salt	$HO_2C(CH_2)_4CO_2H.C_4H_{10}N_2$	232.28		244d				s		i				**Am 56,** 1759
h397	**—,2-amino-***	$HO_2C(CH_2)_3CH(NH_2)CO_2H$	161.16	pl (w)	206				s^h	δ	δ				
h398	**—,2-methyl-***	$HO_2C(CH_2)_3CH(CH_3)CO_2H$	160.17	(peth-bz)	64				s	s	s		δ	chl s	**B2**^{2}, 588
h399	**—,3-methyl-(*d*)***	$HO_2CCH_2CH_2CH(CH_3)CH_2CO_2H$	160.17	(chl-b^{8}) $[\alpha]^{25}_{D}$ +9.81 (w)	93	230^{30}			v	v	v	v		chl v	**B2**^{2}, 588
h400	—,—(*dl*)*	$OH_2CCH_2CH_2CH(CH_3)CH_2CO_2H$	160.17	nd (bz), (ace-bz)	94	$190–200^{12}$			v	v	v	v	δ	chl v, lig δ AcOEt v	**B2**^{2}, 589
h401	**—,2-oxo-***	$OH_2C(CH_2)_3COCO_2H$	160.12	(al)	124				v	v	i	v	i	peth, chl i	**B3**^{2}, 485
h402	**—,3-oxo-***	$HO_2CCH_2CH_2COCH_2CO_2H$	160.12	(chl)	86–8				δ	δ					**B3**^{2}, 486
—	**—,2,3,4,5-tetra-hydroxy-***	*see* **Mucic acid.**													
h406	**1,6-Hexanediol***	Hexamethylene glycol. $HO(CH_2)_6OH$	118.17	nd (w)	43	250^{760}			s	s	δ		i		**B1**^{3}, 2201

For explanations, symbols and abbreviations see beginning of table.

No.	Name	Synonyms and Formula	Mol. wt.	Crystalline form, color and specific rotation	m.p. °C	b.p. °C	Density	n_D	Solubility w	al	eth	ace	bz	other solvents	Ref.
	2,3-Hexanediol														
h407	**2,3-Hexanediol***	$CH_3CH_2CH_2CHOHCHOHCH_3$	118.17		60	206	0.9900^{15}	1.4510^{15}	s	s	s				**B1**, 484
h408	**2,5-Hexanediol***	$CH_3CHOHCH_2CH_2CHOHCH_3$	118.17		43	$216–8^{750}$	0.9610^{20}_{4}	1.4475^{20}	s	s	s				**B1**², 551
h409	**3,4-Hexanediol, 3,4-diethyl-***	$(C_2H_5)_2COHCOH(C_2H_5)_2$	174.28	(eth)	27	95^{4} 230			i	v	v				**B1**³, 3225
h410	**2,5-Hexanediol, 2,5-dimethyl-***	$(CH_3)_2COHCH_2CH_2COH(CH_3)_2$	146.22		92	214	0.898^{20}		s	v			v^h	chl v	**B1**², 557
h411	**1,3-Hexanediol, 2-ethyl-***	$CH_3CH_2CH_2CHOHCH(C_2H_5)CH_2OH$	146.22		−40	244^{760}	0.9325^{22}_{4}	1.4530^{22}	δ	s	s				
h412	**2,3-Hexanedione,** 3-oxime*	$CH_3(CH_2)_2C(:NOH)COCH_3$	129.16	(al)	60										**B1**¹, 405
h413	**2,5-Hexanedione*.**	Acetonylacetone. $CH_3COCH_2CH_2COCH_3$	114.14		−8	194^{754}	0.7370^{20}_{4}	1.4232	v	v	v			peth $δ^h$	**B1**², 841
h414	—,dioxime*	$CH_3C(:NOH)CH_2CH_2C(:NOH)CH_3$	144.18	lf(bz)	134–5				v^h	v^h	v^h				
h415	**1,6-Hexanedione, 1,6-diphenyl-***	1,4-Dibenzoylbutane. $C_6H_5CO(CH_2)_4COC_6H_5$	266.34	nd (al), pr (chl-al), lf (dil al)	107					s^h				MeOH s^h chl s^h	**B7**², 705
h416	**2,5-Hexanedione, 3-hydroxy-***	$CH_3COCH_2CHOHCOCH_3$	130.14			$62–7^{0.5}$		1.4497^{25}	δ		s		s		**Am 71**, 3165
h417	**3,4-Hexanedione, 2,2,5,5-tetramethyl-***	Di-*tert*-butylglyoxal. Pivalil. $(CH_3)_3CCOCOC(CH_3)_3$	170.24		−2	168^{745}	0.8776^{20}	1.4157^{20}	i		∞				**B1**, 800
h418	**1,2,3,4,5,6-Hexanehexol***	Dulcite. Dulcitol. Melampyrin. $CH_2OH(CHOH)_4CH_2OH$	182.17	mcl pr	189	$290–5^{3-4}$	1.466^{15}		s^h	δ					**B1**², 612
h419	**1,3,4,5-Hexanetetrol***	Digitoxit. $CH_3(CHOH)_3CH_2CH_2OH$	150.11	pr $[α]^{15}_{D}$ −86.2	88				i	v	s^h				**B1**², 603
h420	**1-Hexanethiol***	*n* Hexyl mercaptan. $CH_3(CH_2)_5SH$	118.24		−80	151^{760}	0.8424^{20}_{4}	1.4496^{20}	i	v	v				**B1**³, 1659
h421	**2-Hexanethiol***	2-Mercaptohexane*. $CH_3(CH_2)_3CHSHCH_3$	118.24		−147	140^{760} $60–6^{50}$	0.8345^{20}_{4}	1.4451^{20}		s			s		**B1**³, 1662
h422	**3-Hexanethiol***	3-Mercaptohexane*. $CH_3CH_2CH_2CHSHCH_2CH_3$	118.24			57^{25}	0.832^{20}_{20}	1.4428^{20}		s					**B1**³, 1665
h423	**1,2,3-Hexanetriol***	$CH_3CH_2CH_2CHOHCHOHCH_2OH$	134.18	hyg	60–2	168^{14}			s	s					**B1**³, 2349
h424	**1,2,4-Hexanetriol***	$CH_3CH_2CHOHCH_2CHOHCH_2OH$	134.18			$190–2^{30}$			s	s					**B1**, 521
h425	**1,2,5-Hexanetriol***	$CH_3CHOHCH_2CH_2CHOHCH_2OH$	134.18			181^{10}	1.1012^{0}		s	s	i				**B1**³, 2349
h426	**2,3,4-Hexanetriol***	$CH_3CH_2(CHOH)_3CH_3$	134.18			256–7 $130–1^{4}$			s	s					**B1**³, 2349
h427	**Hexanoic acid***	*n*-Caproic acid. $CH_3(CH_2)_4CO_2H$	116.16			205^{760}	0.9274^{20}_{4}	1.4163^{20}	δ	s	s				**B2**², 281
h428	—,amide	Caproamide. $CH_3(CH_2)_4CONH_2$	115.18		101	255^{760}	0.999^{20}_{4}	1.4200^{110}	δ	s	δ		s		**B2**², 286
h429	—,—,*N*-phenyl-	Hexanilide. $CH_3(CH_2)_4CONHC_6H_5$	191.28	nd (peth), pr (al)	95		1.112		i	v	v			peth s^h	**B12**², 147
h430	—,anhydride*	$[CH_3(CH_2)_4CO]_2O$	214.30		−41	143^{15} $241d^{760}$	0.9240^{15}_{4}	1.4280^{25}		s					**B2**², 285
h431	—,butyl ester*	$CH_3(CH_2)_4CO_2(CH_2)_3CH_3$	172.26		−64	208^{760}	0.8623^{25}_{4}	1.4153^{25}	i	s	∞				**B2**, 323
h432	—,chloride	$CH_3(CH_2)_4COCl$	134.61		−87	153^{760}	0.9754^{20}_{4}	1.4264^{20}			s				**B2**², 286
h433	—,ethyl ester*	$CH_3(CH_2)_4CO_2C_2H_5$	144.21		−67	168^{760}	0.8710^{20}_{4}	1.4073^{20}	i	s	s				**B2**², 285
h434	—,heptyl ester*	$CH_3(CH_2)_4CO_2(CH_2)_6CH_3$	214.34		−34	259^{760}	0.8611^{20}	1.4293^{15}	i	s				os s	**B2**, 323
h435	—,hexyl ester*	$CH_3(CH_2)_4CO_2(CH_2)_5CH_3$	200.31		−55	246^{761}	0.865^{18}	1.4264^{15}	i	s				os s	**B2**, 323
h436	—,methyl ester*	$CH_3(CH_2)_4CO_2CH_3$	130.18		−71	151^{760}	0.9038^{0}_{4}	1.4038^{23}	i	v	v				**B2**², 284
h437	—,3-methylbutyl ester*	Isoamyl caproate. $CH_3(CH_2)_4CO_2CH_2CH_2CH(CH_3)_2$	186.29			$224–7^{760}$	0.861		i	s					
h438	—,nitrile	Amyl cyanide. $CH_3(CH_2)_4CN$	97.16		−45	160^{760}	0.8093^{20}_{20}	1.4115^{20}	i	s	s				**B2**², 286
h439	—,octyl ester*	$CH_3(CH_2)_4CO_2(CH_2)_7CH_3$	228.36		−28	275^{760}	0.8603^{20}	1.4326^{15}	i	s				os s	**B2**, 323

For explanations, symbols and abbreviations see beginning of table.

No.	Name	Synonyms and Formula	Mol. wt.	Crystalline form, color and specific rotation	m.p. °C	b.p. °C	Density	n_D	Solubility w	al	eth	ace	bz	other solvents	Ref.
	Hexanoic acid														
h440	—,pentyl ester*	$CH_3(CH_2)_4CO_2(CH_2)_4CH_3$	186.30		−47	226^{760}	0.8612^{25}_{4}	1.4202^{25}							**B2**², 285
h441	—,4-phenyl-phenacyl ester*	$CH_3(CH_2)_4CO_2CH_2CO$—(biphenyl) 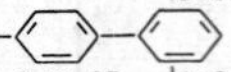	310.38		64–5										**Am 73,** 3087
h442	—,piperazinium salt	$2[CH_3(CH_2)_4CO_2H].C_4H_{10}N_2$	318.45		111				s	s	i	s^h			**Am 56,** 1759
h443	—,propyl ester	$CH_3(CH_2)_4CO_2CH_2CH_2CH_3$	158.23		−74	185^{760}	0.8844^{2}_{2}	1.4140^{15}	i	s	s				**B2**, 323
h444	**—,2-acetyl-,** ethyl ester	Ethyl α-butylacetoacetate $CH_3(CH_2)_3CH(COCH_3)CO_2C_2H_5$	186.24			219–24 104^{12}	0.9523^{20}_{4}	1.4301^{20}							**B1**³, 706
h445	**—,6-amino-***	$H_2N(CH_2)_5CO_2H$	131.17	lf (eth)	202				v	i				MeOH δ	**B4**², 856
h446	—,—,lactam*	ε-Caprolactam. 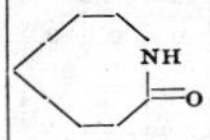 (ring: NH, =O)	113.16	(lig)	60–2	$140–2^{15}$			s	v			s	chl v	**B21**², 216
h447	**—,6-amino-3-methyl-, lactam***	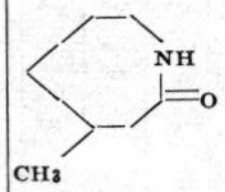 (ring: NH, =O, CH_3)	127.19	(bz-peth) $[\alpha]^{20}_{D}$ −36.1	105–6				v				s		**B21**, 243
h448	**—,6-amino-5-methyl-, lactam***	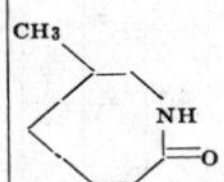 (ring: CH_3, NH, =O)	127.19	(peth, bz-lig) $[\alpha]^{20}_{D}$ −22.2	68–9				v		s				**B21**, 242
h449	**—,6-benzoyl-amino-**	$C_6H_5CONH(CH_2)_5CO_2H$	235.29	nd (al-eth)	79–80				δ^h				δ	AcOEt s lig δ	**B9**², 181
h450	**—,6-benzoyl-amino-2-bromo-** (*dl*)	$C_6H_5CONH(CH_2)_4CHBrCO_2H$	314.20	(al)	166					v^h					**B9**², 181
h451	—,—(*l*)	$C_6H_5CONH(CH_2)_4CHBrCO_2H$	314.20	nd (dil al) $[\alpha]^{18}_{D}$ −29.2 (al)	129					s^h					**B9**², 181
h452	**—,2-bromo-**(*dl*)*	$CH_3(CH_2)_3CHBrCO_2H$	195.07		4	240^{760} $140–2^{23}$				s	s				**B2**², 287
h453	—,—(*l*)*	$CH_3(CH_2)_3CHBrCO_2H$	195.07	$[\alpha]^{30}_{D}$ −27 (eth, c = 5)		129^{14}				s	s				**B2**², 287
h454	—,—,ethyl ester*	$CH_3(CH_2)_3CHBrCO_2C_2H_5$	223.12			205–10				s					**B2**¹, 141
h455	**—,3-bromo-,***	$CH_3CH_2CH_2CHBrCH_2CO_2H$	195.07	nd	35								v	chl v lig v	**B2**, 235
h456	**—,6-bromo-,***	$Br(CH_2)_5CO_2H$	195.07	(peth)	35	$165–70^{20}$								peth s^h	**B2**², 287
h457	**—,6-cyclohexyl-***	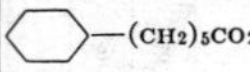 (cyclohexyl)—$(CH_2)_5CO_2H$	198.30		32	180^{11}	0.9626^{20}_{4}	1.4750^{20}			s				**B9**², 21
—	**—,2,6-diamino***	*see* **Lysine**													
h458	**—,2,4-dioxo-***	Propionylpyruvic acid. $CH_3CH_2COCH_2COCO_2H$	144.12	(al, w+1)	83				s	s	s		δ	peth i	**B3**², 437
h459	—,—,ethyl ester*	$CH_3CH_2COCH_2COCO_2C_2H_5$	172.18			163–5				v	v	v			**B3**², 437
h460	**—,2-ethyl-***	$CH_3(CH_2)_3CH(C_2H_5)CO_2H$	144.21			228^{755}	0.9031^{25}_{4}	1.4255^{28}		δ					**B2**², 304
h461	—,—,amide	$CH_3(CH_2)_3CH(C_2H_5)CONH_2$	143.23	nd (w)	101				s^h						**B2**, 350
h462	**—,6-fluoro-***	$F(CH_2)_5CO_2H$	134.15			130^{13}		1.4166^{25}	δ	v	v				
h463	**—,2-hydroxy-(*d*)***	$CH_3(CH_2)_3CHOHCO_2H$	132.16	$[\alpha]^{20}_{D}$ +0.7 (w, c = 14)											**B3**², 231
h464	—,—,—(*dl*)*	$CH_3(CH_2)_3CHOHCO_2H$	132.16	pr (eth-al, peth)	61	270				v	s			chl v	**B3**², 232
h465	—,—(*l*)*	$CH_3(CH_2)_3CHOHCO_2H$	132.16	pr (eth) $[\alpha]^{20}_{D}$ −3.8 (w, c = 4.5)	60–1				v	v	v			chl v	**B3**², 232
h466	**—,4-hydroxy-,** lactone*	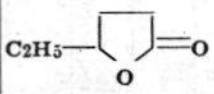 C_2H_5—(ring: O, =O)	114.14		−18	215–18 $107–9^{17}$			s	s					**B17**², 290
h467	**—,2-methyl-(*d*)*.**	$CH_3(CH_2)_3CH(CH_3)CO_2H$	130.18	$[\alpha]^{22}_{D}$ +19.7		105^{5}	0.909^{25}								**B2**², 296
h468	—,—(*dl*)*	$CH_3(CH_2)_3CH(CH_3)CO_2H$	130.18			210^{760}	0.9612^{20}_{4}	1.4193^{20}	∞	∞	∞		∞		**B2**², 297
h469	—,—(*l*)*	$CH_3(CH_2)_3CH(CH_3)CO_2H$	130.18	$[\alpha]^{25}_{D}$ −4.3 (w, c = 26)	121^{20}		0.909^{25}_{4}	1.4189^{25}							**B2**², 296

For explanations, symbols and abbreviations see beginning of table.

No.	Name	Synonyms and Formula	Mol. wt.	Crystalline form, color and specific rotation	m.p. °C	b.p. °C	Density	n_D	Solubility w	al	eth	ace	bz	other solvents	Ref.
	Hexanoic acid														
h470	**—,3-methyl-, chloride**	$CH_3CH_2CH_2CH(CH_3)CH_2COCl$	148.64			163^{751}	0.967^{20}_{4}							to δ^h	**B2**^{2}, 295
h471	**—,4-methyl-(*d*)***	$CH_3CH_2CH(CH_3)CH_2CH_2CO_2H$	130.18	$[\alpha]^{20}_{D}+8.44$	221^{760}		0.9228^{20}_{4}	1.4198	δ	s		s	s	MeOH s	**B2**^{1}, 146
h472	**—,—(*dl*)***	$CH_3CH_2CH(CH_3)CH_2CH_2CO_2H$	130.18		−80	217^{754}			δ	s	s			os s	**B2**^{2}, 298
h473	**—,—,chloride(*dl*)**	$CH_3CH_2CH(CH_3)CH_2CH_2COCl$	148.64			167^{767}	0.9677^{20}_{4}				s				
h474	**—,5-methyl-***	$(CH_3)_2CH(CH_2)_3CO_2H$	130.18		<−25	216	0.9138^{21}_{4}	1.4209^{19}	δ	s	s				**B2**^{2}, 297
h475	**—,—,chloride**	$(CH_3)_2CH(CH_2)_3COCl$	148.64			168^{739}			i		s				**B2**, 342
h476	**—,4-oxo-**	Homolevulinic acid. $CH_3CH_2COCH_2CH_2CO_2H$	130.14	hyg ta	32	160^{24}			s	s	s				**B3**^{2}, 435
—	**—,2,3,4,5,6-pentahydroxy-***	*see* **Idonic acid**													
h478	**1-Hexanol***	*n*-Hexyl alcohol. $CH_3(CH_2)_5OH$	102.18		−47	158^{760}	0.8136^{20}_{4}	1.4178^{20}	δ	s	∞		∞		**B1**^{3}, 1650
h479	**2-Hexanol(*d*)***	$CH_3(CH_2)_3CHOHCH_3$	102.18	$[\alpha]^{25}_{D}+14.1$ (eth, c=11)		138^{760}	0.81036^{25}_{4}	1.4126^{25}	δ	s	s				**B1**^{2}, 437
h480	**—(*dl*)***	$CH_3(CH_2)_3CHOHCH_3$	102.18			140^{760}	0.8159^{20}_{4}	1.4144^{20}	δ						**B1**^{3}, 1660
h481	**—(*l*)***	$CH_3(CH_2)_3CHOHCH_3$	102.18	$[\alpha]^{18}_{5780}-12.04$		$136–8^{754}$	0.8178^{18}_{4}								**B1**^{3}, 1663
h482	**3-Hexanol(*d*)***	$CH_3CH_2CH_2CHOHCH_2CH_3$	102.18	$[\alpha]^{20}_{D}+6.81$		131–3									**B1**^{3}, 1663
h483	**—(*dl*)***	$CH_3CH_2CH_2CHOHCH_2CH_3$	102.18			135^{760}	0.8193^{20}_{4}	1.4168^{20}	δ						**B1**^{3}, 1663
h484	**—(*l*)***	$CH_3CH_2CH_2CHOHCH_2CH_3$	102.18	$[\alpha]^{18}_{0}-7.17$		135^{760}	0.8213^{20}_{4}	1.4140^{22}	δ	s	∞				**B1**^{3}, 1665
h485	**1-Hexanol, 6-chloro-***	Hexamethylene chlorohydrin. $Cl(CH_2)_6OH$	136.63			107^{12}		1.4531^{25}	δ	v	v				
h486	**2-Hexanol, 1-chloro-***	$CH_3(CH_2)_3CHOHCH_2Cl$	136.63			$72–5^{12}$	1.0139^{20}_{4}	1.4478^{20}							**B1**^{2}, 1661
h487	**3-Hexanol, 1-chloro-***	$CH_3CH_2CH_2CHOHCH_2CH_2Cl$	136.63			120^{35}	1.003^{25}_{4}	1.446^{25}	δ					os v	**B1**^{3}, 1664
h488	**—,2-chloro-***	$CH_3CH_2CH_2CHOHCHClCH_3$	136.63			170^{753}	1.0143^{11}		i					os s	**B1**^{2}, 438
h489	**—,5-chloro-***	$CH_3CHClCH_2CHOHCH_2CH_3$	136.63			$78–9^{13}$	1.0012^{19}_{4}	1.4433^{19}	δ					os ∞	**B1**^{3}, 1664
h490	**1-Hexanol, 2,3-epoxy-2-ethyl-***	$CH_3CH_2CH_2\underset{\backslash O/}{CHC}(C_2H_5)CH_2OH$	144.21		<−65	131^{50} 96^{10}	0.9517^{20}_{20}	1.4422^{20}	δ						
h491	**—,2-ethyl-***	$CH_3(CH_2)_3CH(C_2H_5)CH_2OH$	130.23		<−76	185	0.8328^{20}_{4}	1.4318^{20}	i	s	s				**Am 58**, 586
h492	**3-Hexanol, 3-ethyl*-**	$CH_3CH_2CH_2COH(C_2H_5)_2$	130.23			160^{760}	0.8373^{20}_{4}	1.4300^{20}	i	s	s				**B1**^{3}, 1736
h493	**—,3-ethyl-5-methyl-***	$(CH_3)_2CHCH_2COH(C_2H_5)_2$	144.25			272	0.8396^{22}_{4}	1.43457^{13}	i	s	s				**B1**^{3}, 1775
h494	**1-Hexanol, 6-fluoro-***	Hexamethylene fluorohydrin. $F(CH_2)_6OH$	120.16			$85–6^{14}$	0.975^{20}_{4}	1.4141^{25}	δ	v	v				
h495	**—,2-isopropyl-5-methyl-***	$(CH_3)_2CHCH_2CH_2CH(C_3H_7^i)CH_2OH$	158.28			211^{760}	0.8366^{17}_{4}	1.4389^{17}	i	s	v				**B1**^{1}, 215
h496	**—,2-methyl-(*d*)***	$CH_3(CH_2)_3CH(CH_3)CH_2OH$	116.20	$[\alpha]^{25}_{D}+2.47$ (eth, c=22)		164–5	0.8313^{13}_{4}	1.4245^{17}	i	∞	s				**B1**^{2}, 444
h497	**—,—(*dl*)***	$CH_3(CH_2)_3CH(CH_3)CH_2OH$	116.20			164	0.8270^{20}_{4}	1.4250^{20}							**B1**^{3}, 415
h498	**—,4-methyl-(*d*)***	$CH_3CH_2CH(CH_3)(CH_2)_3OH$	116.20	$[\alpha]^{28}_{D}+2.2$		77^{20}	0.818^{23}_{4}	1.4232^{25}		s				os s	**B1**^{3}, 1695
h499	**—,— (*dl*)***	$CH_3CH_2CH(CH_3)(CH_2)_3OH$	116.20			168^{740} 83^{24}	0.8239^{20}_{4}	1.4219^{20}		s				os s	**B1**^{3}, 1695
h500	**—,5-methyl-***	$(CH_3)_2CH(CH_2)_4OH$	116.20			170^{755}	0.819^{24}_{4}	1.4251^{23}	δ	s	s				**B1**^{3}, 1692

For explanations, symbols and abbreviations see beginning of table.

No.	Name	Synonyms and Formula	Mol. wt.	Crystalline form, color and specific rotation	m.p. °C	b.p. C	Density	n_D	Solubility: w	al	eth	ace	bz	other solvents	Ref.
	2-Hexanol														
h501	**2-Hexanol, 2-methyl-***	$CH_3(CH_2)_3COH(CH_3)_2$	116.20			143^{760}	0.8119^{20}_{4}	1.4185^{20}	δ	∞	∞				$\mathbf{B1}^{3}$, 1690
h502	**—,5-methyl-***	$(CH_3)_2CHCH_2CH_2CHOHCH_3$	116.20			150^{744}	0.814^{20}_{4}	1.4194^{20}	i	s	s				$\mathbf{B1}^{2}$, 1692
h503	**3-Hexanol, 3-methyl-***	$CH_3CH_2CH_2COH(CH_3)CH_2CH_3$	116.20			143^{760}	0.8283^{18}_{4}	1.4231^{20}	δ	s	s				$\mathbf{B1}^{3}$, 1693
h504	**—,5-methyl-(*d*)***	$(CH_3)_2CHCH_2CHOHCH_2CH_3$	116.20	$[\alpha]^{35}_{D}+21.23$		81^{60}		1.4171^{25}							$\mathbf{B1}^{3}$, 1692
h505	—,—(*dl*)*	$(CH_3)_2CHCH_2CHOHCH_2CH_3$	116.20			147^{756}	0.827^{0}	1.4128^{20}	i	s	s				$\mathbf{B1}^{3}$, 1692
h506	—,—(*l*)*	$(CH_3)_2CHCH_2CHOHCH_2CH_3$	116.20	$[\alpha]^{23}_{D}-3.88$		63^{19}									$\mathbf{B1}^{3}$, 1692
h507	**1-Hexanol, 1-phenyl-***	$CH_3(CH_2)_4CHOH(C_6H_5)$	178.26			170^{50}	0.9477^{25}_{4}	1.5042		s	s				
h508	**3-Hexanol, 2,2,5,5-tetra-methyl-***	$(CH_3)_3CCH_2CHOHC(CH_3)_3$	158.28		52	166–70			i	s	s				$\mathbf{B1}^{3}$, 1772
h509	**2-Hexanol, 2,3,4-trimethyl-***	$CH_3CH_2CH(CH_3)CH(CH_3)COH(CH_3)_2$	144.26			57^{5}	0.853^{15}_{4}	1.4415^{15}		s	s				$\mathbf{B1}^{3}$, 1756
h510	**—,2,3,5-tri-methyl-***	$(CH_3)_2CHCH_2CH(CH_3)COH(CH_3)_2$	144.26			171^{755}	0.8316^{0}_{20}	1.4313		s	s				$\mathbf{B1}$, 212
h511	**3-Hexanol, 2,2,3-trimethyl-***	$CH_3CH_2CH_2COH(CH_3)C(CH_3)_3$	144.26			170^{760}	0.8531^{4}_{4}	1.4536							$\mathbf{B1}^{3}$, 1755
h512	**—,2,3,5-tri-methyl-***	$(CH_3)_2CHCH_2COH(CH_3)CH(CH_3)_2$	144.26			72^{21}	0.8271^{20}_{20}	1.4321^{20}		s	s				$\mathbf{B1}^{3}$, 1756
h513	**—,2,4,4-tri-methyl-***	$CH_3CH_2C(CH_3)_2CHOHCH(CH_3)_2$	144.26			170^{760}	0.8489^{20}	1.4395^{20}		s	s				$\mathbf{B1}^{2}$, 458
h514	**—,2,5,5-tri-methyl-***	$(CH_3)_3CCH_2CHOHCH(CH_3)_2$	144.26			77^{32}	0.825^{20}	1.4286^{20}		s	s				$\mathbf{B1}^{3}$, 1756
h515	**—,3,4,4-tri-methyl-***	$CH_3CH_2C(CH_3)_2COH(CH_3)CH_2CH_3$	144.26			$165–6^{760}$	0.8323^{21}_{0}	1.43407^{21}		s	s				$\mathbf{B1}$, 425
h516	**—,3,5,5-tri-methyl-***	$(CH_3)_3CCH_2COH(CH_3)CH_2CH_3$	144.26			62^{14}	0.8350^{20}	1.4352^{20}		s	s				$\mathbf{B1}^{3}$, 1755
h517	**2-Hexanone***	$CH_3(CH_2)_3COCH_3$	100.16		−57	126^{760}	0.81162^{20}_{4}	1.40152^{20}		∞	∞				$\mathbf{B1}^{2}$, 745
h518	**3-Hexanone***	$CH_3CH_2CH_2COCH_2CH_3$	100.16			123^{765}	0.81491^{22}_{4}	1.3990^{22}	δ	∞	∞				$\mathbf{B1}^{2}$, 746
h519	**1-Hexanone, 1(2,4-dihydroxy-phenyl)-***	HO; $CH_3(CH_2)_4CO$—⟨ring⟩—OH	208.25	pl (to-peth)	56	343d			i	s	s	s		chl s	$\mathbf{B8}^{2}$, 314
h520	**2-Hexanone, 3,3-dimethyl-***	$CH_3CH_2CH_2C(CH_3)_2COCH_3$	128.21			149^{745}	0.838^{0}_{4}			s	s				$\mathbf{B1}^{3}$, 760
h521	**—,3,4-dimethyl-***	$CH_3CH_2CH(CH_3)CH(CH_3)COCH_3$	128.21			158	0.8295^{22}_{4}	1.412^{24}		s	s				$\mathbf{B1}^{1}$, 760
h522	**3-Hexanone, 2,2-dimethyl-***	$CH_3CH_2CH_2COC(CH_3)_3$	128.21			$145–8^{745}$	0.8105^{25}_{4}	1.4148^{17}		s	s				$\mathbf{B1}^{2}$, 760
h523	—,—,oxime*	$CH_3CH_2CH_2C(:NOH)C(CH_3)_3$	143.23	nd (al)	76					δ					$\mathbf{B1}^{1}$, 364
h524	**—,2,5-dimethyl-***	$(CH_3)_2CHCH_2COCH(CH_3)_2$	128.21			147^{744}	0.8270^{0}_{0}			s	s	s			$\mathbf{B1}^{2}$, 760
h525	**—,4,4-dimethyl-***	$CH_3CH_2C(CH_3)_2COCH_2CH_3$	128.21			151	0.8298^{20}		s	s		s		chl s	$\mathbf{B1}^{2}$, 760
h526	**—,6-dimethyl-amino-4,4-diphenyl-5-methyl-(*l*)***	*l*-Isomethadone. *l*-Isoamidon. $(CH_3)_2NCH_2CH(CH_3)C(C_6H_5)_2COCH_2CH_3$	309.43	$[\alpha]^{25}_{D}-20$ (al)		$162–5^{0.6}$				s					
h527	**—,4,4-hydroxy-***	Propioin. $CH_3CH_2CHOHCOCH_2CH_3$	116.16			60^{12}	0.956^{21}_{4}	1.4340^{21}							$\mathbf{B1}$, 835
h528	**—,4-hydroxy-2,2,5,5-tetra-methyl-***	Pivaloin. $(CH_3)_3CCHOHCOC(CH_3)_3$	172.20		81 sub	80^{10}			s^{h}		s			peth i	$\mathbf{B1}^{3}$, 884
h529	**2-Hexanone, 3-methyl-***	$CH_3CH_2CH_2CH(CH_3)COCH_3$	114.18			136^{760}								os v	$\mathbf{B1}^{1}$, 360

For explanations, symbols and abbreviations see beginning of table.

No.	Name	Synonyms and Formula	Mol. wt.	Crystalline form, color and specific rotation	m.p. °C	b.p. °C	Density	n_D	Solubility w	al	eth	ace	bz	other solvents	Ref.
	2-Hexanone														
h530	**—,4-methyl-***	$CH_3CH_2CH(CH_3)CH_2COCH_3$	114.18			139^{762} $35–7^{11}$								os ∞	**B1²**, 756
h531	**—,5-methyl-***	$(CH_3)_2CHCH_2CH_2COCH_3$	114.18		144^{760}		0.888^{20}_{4}		δ	∞	∞				**B1²**, 756
h532	—,—,oxime*	$(CH_3)_2CHCH_2CH_2C(:NOH)CH_3$	129.20			$195–6^{761}$	0.8881^{20}_{4}	1.4448							**B1** 701
h533	**3-Hexanone, 4-methyl-***	$CH_3CH_2CH(CH_3)COCH_2CH_3$	114.19			134	0.8248^{19}							os v	**B1**, 701
h534	**—,5-methyl-***	$(CH_3)_2CHCH_2COCH_2CH_3$	114.19			134^{735}	0.815^{17}_{0}		i	∞	∞				**B1¹**, 760
h535	**1-Hexanone, 1-phenyl-***	Caprophenone. $CH_3(CH_2)_4COC_6H_5$	176.26		27	265^{760}	0.9576^{25}_{4}	1.5027^{25}	δ		s				**B7²**, 257
h536	**Hexasiloxane tetradeca-methyl-**	$(CH_3)_2SiO[-Si(CH_3)_2O-]_4Si(CH_3)_3$	443.97		<−100	142^{20}	0.8910	1.3948^{20}		δ			s		
h538	**1,2,3,5-Hexa-tetraene, 4-chloro-***	$CH_2:CHCCl:C:C:CH_2$	112.56			$127d^{760}$ 55^{54}	0.9997^{20}_{4}	1.5280^{20}						chl s	**B1³**, 1061
h539	**1,2,4,5-Hexa-tetraene, 3,4-dichloro-***	$CH_2:C:CCl.CCl:C:CH_2$	147.00			$38–40^{8}$	1.1819^{20}_{4}	1.5456^{20}						chl s	**B1³**, 1061
h540	**1,3,5-Hexatriene** (*cis*)	Divinylethylene. $CH_2:CHCH:CHCH:CH_2$	80.13		−12	78^{760}	0.7175^{20}_{4}	1.4577^{20}		s					**B1²**, 243
h541	—(*trans*)	$CH_2:CHCH:CHCH:CH_2$	80.13		−12	79	0.7369^{15}_{4}	1.5135^{15}				s			**B1³**, 1041
h542	**1,3,4-Hexatriene, 3,6-dichloro-***	$ClCH_2CH:C:CClCH:CH_2$	149.02			$45–6^{3}$	1.1807^{20}_{4}	1.5195^{20}		s				AcOEt s	
h543	**1,3,5-Hexatriene, 2,5-dimethyl-***	$CH_2:C(CH_3)CH:CHC(CH_3):CH_2$	108.18		−9	145^{747}	0.7822^{20}_{4}	1.5122^{20}						MeOH s lig s	**B1³**, 1047
h544	**—,1,6-diphenyl-***	$C_6H_5CH:CHCH:CHCH:CHC_6H_5$	232.33	lf (ace)	200					i	i		δ	chl δ aa i	**B5²**, 605
h545	**1,2,4-Hexatriene, 3,4,6-trichloro-***	$ClCH_2CH:CClCCl:C:CH_2$	183.47			50^{1}	1.3132^{20}_{4}	1.5517^{20}						chl s	
h546	**2-Hexanal***	$CH_3CH_2CH_2CH:CHCHO$	98.15			47^{17}	0.861^{13}_{4}	1.4407^{13}							
h547	**2-Hexene**(*cis*)*	$CH_3CH_2CH_2CH:CHCH_3$	84.16		−146	68^{749}	0.6845^{20}_{4}	1.3454^{20}	i	s	s				**Am 63,** 2683
h548	—(*trans*)*	$CH_3CH_2CH_2CH:CHCH_3$	84.16		−133	68^{750}	0.6780^{20}_{4}	1.3935^{20}	i	s	s				**Am 63,** 2683
h549	**3-Hexene**(*cis*)*	$CH_3CH_2CH:CHCH_2CH_3$	84.16		−135	67^{741}	0.6796^{20}_{4}	1.3934^{20}	i	s	s				**Am 63,** 2683
h550	—(*trans*)*	$CH_3CH_2CH:CHCH_2CH_3$	84.16		−113	67^{741}	0.677^{20}_{4}	1.3938^{20}	i	s	s				**Am 63,** 2683
h551	**1-Hexene, 5-amino-4-methyl-***	$CH_3CHNH_2CH(CH_3)CH_2CH:CH_2$	113.20			$133–6^{760}$	0.793^{15}_{0}		s						**B4**, 226
h552	**—,1-chloro-***	$CH_3(CH_2)_3CH:CHCl$	118.61			121	0.8872^{22}	1.4300^{22}	i		v			chl v	
h553	**—,2-chloro-***	$CH_3(CH_2)_3CCl:CH_2$	118.61			113^{740}	0.8886^{25}_{4}	1.4278^{25}	i					chl v	
h554	**—,5-chloro-***	$CH_3CHClCH_2CH_2CH:CH_2$	118.61			$121–5^{760}$ $28–30^{13}$	0.8891^{25}_{4}	1.4279^{25}	i		v			chl v	**B1³**, 803
h555	**2-Hexene, 4-chloro-***	$CH_3CH_2CHClCH:CHCH_3$	118.61			123^{760} 30^{10}	0.9148^{20}_{20}	1.4400^{20}			v	v		chl v	**B1²**, 192
h556	**3-Hexene, 1-chloro-***	$CH_3CH_2CH:CHCH_2CH_2Cl$	118.61			$59–61^{60}$	0.900^{24}_{4}	1.435^{24}	i		v	v		chl v	
h557	**—,3-chloro-***	$CH_3CH_2CH:CClCH_2CH_3$	118.61			113^{748}	0.8898^{25}_{4}	1.4320^{25}	i		s			chl s	
h558	**—,2-chloro-2,5-dimethyl-***	$(CH_3)_2CHCH:CHCCl(CH_3)_2$	146.66			$45–60^{15}$		1.45^{20}			s				

For explanations, symbols and abbreviations see beginning of table.

No.	Name	Synonyms and Formula	Mol. wt.	Crystalline form, color and specific rotation	m.p. °C	b.p. °C	Density	n_D	Solubility w	al	eth	ace	bz	other solvents	Ref.
	3-Hexene														
h559	—,**1-chloro-4-ethyl-***	$CH_3CH_2C(C_2H_5):CHCH_2CH_2Cl$	146.66	...	...	173	0.9102^{20}_{4}	1.4524^{20}	...	...	...	...	s	chl s	**B1**^{2}, 201
h560	**1-Hexene, 1,2-dichloro-** (*cis*)*	$CH_3(CH_2)_3CCl:CHCl$	153.05	...	...	88^{30}	...	...	...	...	...	...	...	CCl_4 s	...
h561	—,—(*trans*)*	$CH_3(CH_2)_3CCl:CHCl$	153.05	...	...	$63–5^{32}$	1.1167^{25}_{4}	1.4576^{25}	...	...	...	...	...	chl s	...
h562	—,**dodecafluoro-***	Perfluoro-1-hexene. $F_3C(CF_2)_3CF:CF_2$	300.05	...	57	...	...	...	...	...	...	...	...	...	...
h563	**3-Hexene, dodecafluoro-***	Perfluoro-2-hexene. $F_3CCF_2CF:CFCF_2CF_3$	300.05	...	...	49	...	...	...	...	...	...	...	...	...
h564	**1-Hexene, 2-ethyl-***	$CH_3(CH_2)_3C(C_2H_5):CH_2$	112.22	...	...	120	0.7274^{20}_{3}	1.4207^{20}	i	...	v	...	v	peth v	...
h565	**3-Hexene, 1,2,3,4,5,6-hexachloro-***	$ClCH_2CHClCCl:CClCHClCH_2Cl$	290.83	(peth)	58	$110–12^{2}$	...	...	...	...	...	...	...	MeOH, chl v^h	...
h566	—,**3(4-hydroxyphenyl)-4(4-methoxyphenyl)-***	Mestilbol. H_5C_2 C_2H_5 / CH_3O— —C : C— —OH	282.39	nd (bz-lig)	116–18	$185–95^{03}$	...	...	i	s	v	v	...	...	...
h567	**1-Hexene, 1,1,2-trichloro-***	$CH_3(CH_2)_3CCl:CCl_2$	187.50	...	...	$90–93^{10}$	1.225^{25}	1.4760^{25}	i	...	s	...	...	...	...
h568	**3-Hexenoic acid***	Hydrosorbic acid. $CH_3CH_2CH:CHCH_2CO_2H$	114.15	...	12	208	0.964	...	...	...	...	...	...	...	**B2**, 435
h569	**4-Hexenoic acid, 2-acetyl-5-hydroxy-3-oxo-**, lactone	O O / CH_3— —$COCH_3$ / O	168.15	nd (w)	109	270^{760}	...	...	v^h	δ	v	...	...	...	**B17**^{2}, 524
h570	**2-Hexenoic acid, 3-phenyl-**	$CH_3(CH_2)_2C(C_6H_5):CHCOOH$	190.24	(peth)	94	$183–4^{14}$	...	...	...	...	...	...	s	peth δ	**B9**^{2}, 477
h571	**1-Hexen-3-ol***	$CH_3CH_2CH_2CHOHCH:CH_2$	100.16	...	...	134	0.834^{23}_{4}	1.4215^{26}	...	...	...	...	...	...	...
h572	**3-Hexen-1-ol**(*cis*)*	$CH_3CH_2CH:CHCH_2CH_2OH$	100.16	...	...	156	0.846^{22}_{15}	1.4803^{20}	...	v	s	...	...	...	...
h573	**1-Hexen-3-yne***	$CH_3CH_2C{\vdots}CCH:CH_2$	80.13	...	...	85^{758}	0.7492^{20}_{4}	1.4522^{20}	...	...	...	...	...	...	**B1**^{3}, 1041
h574	**1-Hexen-5-yne***	Diallylene. $HC{\vdots}CCH_2CH_2CH:CH_2$	80.13	...	...	70^{760}	0.8579^{18}_{4}	...	...	...	...	...	...	...	...
h575	**1-Hexen-3-yne, 5-chloro-5-methyl-***	$(CH_3)_2CClC{\vdots}CCH:CH_2$	128.60	...	...	48^{28}	0.9375^{15}	1.4778^{20}	...	...	...	v^h	...	...	**B1**^{3}, 1045
h576	**3-Hexen-1-yne, 3-propyl-***	$CH_3CH_2CH:C(C_3H_7^n)C{\vdots}CH$	122.21	...	...	136^{760}	0.7799^{25}_{4}	1.4432^{25}	...	...	...	...	...	...	**Am 64**, 363
h577	**4-Hexen-1-yn-3-ol**(*d*)*	$CH_3CH:CHCHOHC{\vdots}CH$	96.13	$[\alpha]^{19}_{5593}$ +16.06	...	157–9	0.9030^{30}_{4}	1.4645^{16}	...	...	...	...	...	...	...
h578	—(*dl*)*	$CH_3CH:CHCHOHC{\vdots}CH$	96.13	...	...	154–6	0.9148^{23}_{4}	1.4651^{23}	...	...	...	...	...	...	...
—	**Hexestrol, dimethyl-**	*see* **Hexane, 3,4-bis(4-hydroxyphenyl)-3-methyl-***													
—	**Hexeton**	*see* **2-Cyclohexen-1-one, 5-isopropyl-3-methyl-***													
h579	**1-Hexyne***	*n*-Butylacetylene. $CH_3(CH_2)_3C{\vdots}CH$	82.15	...	−132	71^{760}	0.71518^{20}_{4}	1.39840^{20}	i	s	s	...	...	...	**B1**^{3}, 977
h580	**2-Hexyne***	$CH_3CH_2CH_2C{\vdots}CCH_3$	82.15	...	−88	84^{760}	0.7317^{20}_{4}	1.4135^{20}	i	∞	∞	...	...	...	**B1**^{3}, 980
h581	**3-Hexyne***	Diethylacetylene. $CH_3CH_2C{\vdots}CCH_2CH_3$	82.15	...	−105	81	0.7231^{20}_{4}	1.4115^{20}	i	s	s	...	...	...	**B1**^{3}, 980
h582	—,**2,5-dichloro-2,5-dimethyl-***	$(CH_3)_2CClCH:CHCCl(CH_3)_2$	181.11	...	29	$175–8^{745}$	...	...	...	...	...	...	...	chl s	**B1**^{3}, 847
h583	—,**1:2,5:6-diepoxy-**	$CH_2CHC{\vdots}CCHCH_2$ (O bridges)	110.11	...	−16	98^{20}	1.1189^{23}	1.4871^{23}	...	...	...	...	...	chl s	**B19**^{2}, 19
h584	**1-Hexyne, 5-methyl-***	Isoamylacetylene. $(CH_3)_2CHCH_2CH_2C{\vdots}CH$	96.19	...	−125	92^{760}	0.7274^{20}_{4}	1.4059^{20}	i	...	s	...	...	...	**B1**^{3}, 1000

For explanations, symbols and abbreviations see beginning of table.

No.	Name	Synonyms and Formula	Mol. wt.	Crystalline form, color and specific rotation	m.p. °C	b.p. °C	Density	n_D	Solubility w	al	eth	ace	bz	other solvents	Ref.
	2-Hexyne														
h585	**2-Hexyne, 5-methyl-***	$(CH_3)_2CHCH_2C⋮CCH_3$	96.19			$100–105^{760}$	0.745^{20}_{4}	1.418^{17}							
h586	**3-Hexyne, 2-methyl-***	$CH_3CH_2C⋮CCH(CH_3)_2$	96.19			$94–5^{760}$	0.7263^{20}_{4}	1.4114^{20}							
h587	**3-Hexyne-2,5-diol, 2,5-dimethyl-***	Acetylen-pinacon. $(CH_3)_2COHC⋮CCOH(CH_3)_2$	142.19	nd	95	205^{759}	0.949^{20}_{20}		s	v	v	v	v	chl s	**B1**^{2}, 572
h588	**1-Hexyn-3-ol, 3-methyl-***	$CH_3CH_2CH_2COH(CH_3)C⋮CH$	112.17			137^{760}	0.8620^{20}_{4}	1.4338^{20}	s	s	s				**B1**^{3}, 1994
h589	**3-Hexyn-2-ol, 2-methyl-***	$CH_3CH_2C⋮CCOH(CH_3)_2$	112.17			145–7	0.962^0	1.4411^0	s	s	s				**B1**^{3}, 1993
h589[1]	**Hippuric acid**	Benzamidoacetic acid. *N*-benzoylglycine. *p* (ring) —$CONHCH_2CO_2H$ *m o*	179.18	pr (w)	190		1.371^{20}_{4}		δ	s				AcOEt s	**B9**^{2}, 174
h589[2]	—,4-phenyl-phenacyl ester (ring)—$CONHCH_2CO_2CH_2CO$—(ring)—(ring)		373.41		163										**Am 52,** 3715
h589[3]	—,piperazinium salt	$2(C_6H_5CONHCH_2CO_2H).C_4H_{10}N_2$	444.48	wh	182–4d				s^h	s^h	i				**Am 70,** 2758
h589[4]	—,*p*-**amino-**	$C_9H_{10}N_2O_3$. *See* h589[1]	194.19	pr (w)	198–9				i	s	i				**B14**^{2}, 258
h589[5]	—,*m*-**bromo-**	$C_9H_8BrNO_3$. *See* h589[1]	258.09	nd (w)	146–7				s^h	s			δ	MeOH s	**B9**^{2}, 233
h589[6]	—,*o*-**bromo-**	$C_9H_8BrNO_3$. *See* h589[1]	258.09	nd (w)	192–3				s^h		i			AcOEt v	**B9**^{2}, 231
h589[7]	—,*p*-**bromo-**	$C_9H_8BrNO_3$. *See* h589[1]	258.09	nd (w)	162				s^h	v					**B9**, 354
h590	**Histamine**	4-Imidoazoleethylamine. β-Aminoethylglyoxaline. C_5H_9N	111.15	pl (chl) wh	86	209^{18}			s	s	i			chl s^h	**B25**^{2}, 302
h591	—,dihydrochloride	$C_5H_9N_3.2HCl$	184.08	(al-eth)	242				v	δ	i			MeOH v	**B25**^{2}, 303
h592	**Histidine**(*d*)	—$CH_2CH(NH_2)CO_2H$ N NH (ring)	155.16	ta (w) $[\alpha]^{20}_{D}+40.2$	287d				s	i	i	i		chl i	**B25**^{2}, 404
h593	—(*dl*)	$C_6H_9N_3O_2$. *See* h592	155.16	ta, pr	285d				s					os i	**B25**^{2}, 409
h594	—(*l*)	$C_6H_9N_3O_2$. *See* h592	155.16	lf (dil al) $[\alpha]^{20}_{D}-39.7$	285d				s	δ	i				**B25**^{2}, 409
h595	—,bis-3,4-dichloro-benzene-sulfonate(*l*)	$C_6H_9N_3O_2.2C_6H_4Cl_2O_3S$. *See* h592	609.29	rh nd	280d				s^h	i					
h596	—,diflavianate(*l*)	$C_6H_9N_3O_2.2C_{10}H_6N_2O_8S$. *See* h592	783.64	nd (w)	251–4d				δ	i	i				**B25**^{2}, 427
h597	—,dihydro-chloride(*dl*)	$C_6H_9N_3O_2.2HCl$. *See* h592	228.09		237d										**B25**^{1}, 718
h598	—,—(*l*)	$C_6H_9N_3O_2.2HCl$. *See* h592	228.09	rh pl	245d				s	s	i				**B25**, 513
h599	—,monohydro-chloride mono-hydrate	$C_6H_9N_3O_2.HCl.H_2O$. *See* h592	209.64	pl (w) $[\alpha]^{20}_{D}+1.7$	259				s	i	i				**B25**^{2}, 407
—	—,β-**alanyl-**	*see* **Carnosine**													
h600	**Holocaine** (base)	*N,N*-bis(*p*-Ethoxyphenyl) acetamidine. Phenacaine. C_2H_5O—(ring)—$NHC:N$—(ring)—OC_2H_5 CH_3	298.37	nd (al)	117–8				δ	v	v		v	lig s	**B13**^{2}, 247
h601	—,hydrochloride	$C_{18}H_{22}N_2O_2.HCl$. *See* h600	334.84		190–2				v^h	v	i				**B13**^{2}, 272
h602	**Homatropine**	Homotropine. Mandelyltropeine. CH_3—N(ring)—O_2CCHOH—(ring)	275.35	pr (al)	98				δ	s	s	s	δ	chl s	**B21**^{2}, 18
h603	—,hydrobromide	$C_{16}H_{21}NO_3.HBr$. *See* h602	356.28	rh pym (w)	217–8d				s	s	i			chl δ	**B21**, 23
h604	—,hydrochloride	$C_{16}H_{21}NO_3.HCl$. *See* h602	311.82	wh pr	216–7				s	s					
—	**Homoanisic acid**	*see* **Acetic acid, (4-methoxyphenyl)-**													
—	**Homocaine**	*see* **Ecgonine**, benzoate ethyl ester													
—	**Homogentisic acid**	*see* **Acetic acid, (2,5-dihydroxyphenyl)-**													

For explanations, symbols and abbreviations see beginning of table.

No.	Name	Synonyms and Formula	Mol. wt.	Crystalline form, color and specific rotation	m.p. °C	b.p. °C	Density	n_D	Solubility w	al	eth	ace	bz	other solvents	Ref.
	Homoisophthalic acid														
—	**Homoisophthalic acid**	*see* **Acetic acid, (3-carboxyphenyl)-**													
—	**Homoisovanillic acid**	*see* **Acetic acid, (3-hydroxy-4-methoxy-phenyl)-**													
—	**Homophthalic acid**	*see* **Acetic acid, (2-carboxyphenyl)-**													
—	**Homopyrocatechol**	*see* **Tóluene, 3,4-dihydroxy-**													
—	**Homosaligenine**	*see* **Benzene, 1-hydroxy-2-hydroxymethyl-4-methyl-**													
—	**Homoterephthalic acid**	*see* **Acetic acid, (4-carboxyphenyl)-**													
—	**Homoveratric acid**	*see* **Acetic acid, (3,4-dimethoxyphenyl)-**													
h605	**Hordenine**	Anhaline. 1(Dimethylamino)-2(4-hydroxyphenyl)ethane*. $C_{10}H_{15}NO$	165.23	rh pr	117	173^{11}			s	v	v	...	δ	chl s, lig s	$B13^2$, 356
h606	—,sulfate	$2(C_{10}H_{15}NO).H_2SO_4$	428.16		205				...	...	...	...	...		$B13^1$, 236
h607	—,sulfate, dihydrate	$2(C_{10}H_{15}NO).H_2SO_4.2H_2O$.	464.59	ta	197				s	δ	i	...	...		$B13^1$, 236
h608	**Humulon**	α-Lupulic acid. (structure: HO, $COCH_2CH(CH_3)_2$, $(CH_3)_2CHCH{:}CH$, HO, O, $CH_2CH{:}C(CH_3)_2$)	362.45	yesh (eth) $[\alpha]_D^{20}$ −232 (bz), −212 (al)	65				δ^h	s	...	...	...	os s, alk s	$B8^2$, 537
h609	**Hydantoic acid**	*N*-Carbamylglycine. Ureidoacetic acid. $H_2NCONHCH_2CO_2H$	118.10	mcl pr	180d				v^h	v^h	...	...	...		$B4^2$, 792
h610	—,ethyl ester	$H_2NCONHCH_2CO_2C_2H_5$	146.15	nd (w)	134				v^h	s^h	i	...	...		$B4^2$, 794
h611	**—,phenylthio-**	*N*-Phenylpseudothiohydantoic acid. $C_6H_5N{:}C(NH_2)SCH_2CO_2H$	210.25	wh	151–8	175^{760}			s^h	δ	δ	...	δ		$B12^1$, 248
h612	**Hydantoin**	Glycolylurea. (structure: 5 4, O, HN NH, 1 3, 2, O)	100.18	nd (al), lf (w)	220				v^h	s^h	i	...	...	peth i	$B24^2$, 127
h613	**—,1-acetyl-2-thio-**	$C_5H_6N_2O_2S$. *See* h612	158.19	(al)	175				...	s^h	...	...	...		$B24^1$, 293
h614	**—,1-benzoyl-2-thio-**	$C_{10}H_8N_2O_2S$. *See* h612	220.26	(al)	165d				...	s^h	...	...	...		$B24^1$, 294
h615	**—,5-benzylidene-2-thio-**	$C_{10}H_8N_2OS$. *See* h612	204.26	ye nd (al)	258d				...	s^h	...	...	...	aa δ	$B24^1$, 355
h616	**—,5,5-dimethyl-**	$C_5H_8N_2O_2$. *See* h612	128.14	pr (w-al)	178				v	v	v	v	v	chl v	$B24^2$, 157
h617	**—,5,5-diphenyl-**	$C_{15}H_{12}N_2O_2$. *See* h612	252.26	nd (al)	286				i	δ	...	s	δ	chl δ	$B24^2$, 227
h618	**—,3,5-diphenyl-2-thio-**	$C_{15}H_{12}N_2OS$. *See* h612	268.11		233				δ	v	v	...	v		B24, 385
h619	**—,5-ethyl-5-phenyl-(*d*)**	*d*-Nirvanol. $C_{11}H_{12}N_2O_2$. *See* h612	204.22	$[\alpha]_D$ +123 (al)	237				...	s	...	...	...		Am 54, 4697
h620	**—,5-ethyl-5-phenyl-(*dl*)**	*dl*-Nirvanol. $C_{11}H_{12}N_2O_2$. *See* h612	204.22	wh nd	199				δ^h	s	δ	...	...	aa s	$B24^2$, 206
h621	**—,5(2-hydroxybenzylidene)-2-thio-**	$C_{10}H_8N_2O_2S$. *See* h612	220.26	nd (aa)	248				...	...	...	...	...	aa s^h	$B25^1$, 502
h622	**—,1-methyl-**	$C_4H_6N_2O_2$. *See* h612	114.11	(w), pr (al)	157–8				s	s	s	...	...	chl s	$B24^2$, 128
h623	**—,5-methyl-(*dl*)**	α-Lactylurea. $C_4H_6N_2O_2$. *See* h612	114.11		145				...	...	...	...	...		
h624	—,—(*l*)	$C_4H_6N_2O_2$. *See* h612	114.11	(w) $[\alpha]_D^{20}$ −50.6	175				v	...	...	...	...		$B24^1$, 304
h625	—,—,hydrate(*dl*)	$C_4H_6N_2O_2.H_2O$. *See* h612	132.13	rh (w)	155–6				v	v	...	v	...		$B24^2$, 155
h626	**—,2-thio-**	Glycolylthiourea. $C_3H_4N_2OS$. *See* h612	116.15	ye nd (w)	227				v^h	...	s	...	...	alk s	$B24^2$, 138
—	**—,5-ureido-**	*see* **Allantoin**													
—	**Hydracrylic acid**	*see* **Propanoic acid, 3-hydroxy-***													
h627	**Hydrastine**	$C_{21}H_{21}NO_6$	383.41	yesh (al) $[\alpha]_D^{17}$ −67.8 (chl, c=2.5)	135				...	δ	δ	...	s	chl s	$B27^2$, 603

For explanations, symbols and abbreviations see beginning of table.

No.	Name	Synonyms and Formula	Mol. wt.	Crystalline form, color and specific rotation	m.p. °C	b.p. °C	Density	n_D	Solubility w	al	eth	ace	bz	other solvents	Ref.
	Hydrastine														
h628	—,hydrochloride	$C_{21}H_{21}NO_6.HCl$	419.87	micr pw $[\alpha]_D+158.0$ (w, c=2)	116	...	...	...	v^h	s	...	...	...	...	**B27**[2], 604
h629	**Hydrastininine**	O, CH_3, N, O, OH	207.23	nd (lig), (eth)	116	...	...	...	s^h	v	v	...	...	chl v	**B27**[2], 530
h630	—,bisulfate	$C_{11}H_{11}NO_2.H_2SO_4$. *See* h629	287.30	gr-flr ye cr (al)	216d	...	...	...	s	s	...	...	...	...	**B27**, 465
h631	—,hydrochloride	$C_{11}H_{11}NO_2.HCl$. *See* h629	225.68	ye nd	212d	...	...	...	v	v	...	...	...	chl s	**B27**[2], 530
—	**Hydratropaldehyde**	*see* **Propanal, 2-phenyl-***													
h633	**Hydrazine, 1-acetyl-2-phenyl-**	Acetic acid β-phenyl hydrazide. Hydracetin. $C_6H_5NHNHCOCH_3$	150.18	hex pr	129	...	...	...	v^h	v	δ	...	s	chl s	**B15**[2], 92
h634	—,allyl-	$CH_2{:}CHCH_2NHNH_2$	72.11	...	...	$122–4^{757}$	...	...	...	...	...	...	...	...	**B4**[1], 562
—	—,benzoyl-	*see* **Benzoic acid,** hydrazide													
h635	—,benzyl-	$C_6H_5CH_2NHNH_2$	122.17	...	26	103^{41}	...	...	∞	∞	∞	...	...	...	**B15**[2], 244
h636	—,1-benzyl-2(4-tolyl)-	H_3C— —$NHNHCH_2C_6H_5$	212.30			212^{17}	...	...	...	...	...	...	...	...	**B15**, 533
h637	—,1,2-bis(3-aminophenyl)-*	*m,m'*-Hydrazoaniline. $C_{12}H_{14}N_4$. *See* h669	214.27	pym	151	...	...	...	i	δ	i	...	...	...	**B15**[2], 309
h638	—,1,2-bis(4-aminophenyl)-*	*p,p'*-Hydrazodianiline. Diphenine. $C_{12}H_{14}N_4$. *See* h669	214.27	ye	145	...	...	...	s^h	v	v	...	...	...	**B15**, 653
—	—,1,2-bis(carboxyphenyl)-*	*see* **Benzoic acid, hydrazodi-**													
h640	—,1,2-bis(1-cyanocyclohexyl)-	CN, NC, NHNH	246.36	(al)	145d	...	...	...	...	s^h	...	...	...	...	**B15**[2], 294
h641	—,2-bromophenyl-*	$C_6H_7BrN_2$. *See* h694	187.05	nd	48	...	...	...	...	...	...	...	...	...	**B15**[1], 117
h643	—,4-bromophenyl-*	$C_6H_7BrN_2$ *See* h694	187.05	nd (w), lf (lig), (al)	108	...	...	...	s^h	s	s	...	...	peth δ lig v^h	**B15**[2], 160
h644	—,1-butyl-1-phenyl-*	$NH_2N(C_6H_5)C_4H_9^n$	164.25		...	250^{763}	...	...	...	...	...	...	...	...	**B15**[1], 28
h645	—,1,2-diallyl-	$CH_2{:}CHCH_2NHNHCH_2CH.CH_2$	112.18	...	...	145^{752}	...	...	...	...	...	...	...	...	**B4**[2], 963
h646	—,1,2-dibenzoyl-	$C_6H_5CONHNHCOC_6H_5$	240.26	nd (MeOH)	237–9	...	...	...	$δ^h$	δ	...	...	...	MeOH s^h	**B9**[2], 216
h647	—,1,2-dibenzoyl-1,2-dimethyl-	$C_6H_5CON(CH_3)N(CH_3)COC_6H_5$	268.32	pr (al)	85	...	...	...	i	s^h	i	...	...	...	**B9**[2], 217
h648	—,1,1-dibenzyl-	$NH_2N(CH_2C_6H_5)_2$	212.30	lf (dil al)	47	...	...	...	i	v	v	...	...	...	**B15**[2], 246
h649	—,1,2-dibenzyl-	$C_6H_5CH_2NHNHCH_2C_6H_5$	212.30	(peth)	65	...	...	...	...	...	δ	...	...	...	**B15**[2], 245
h650	—,(2,4-dichlorophenyl)-*	$C_6H_6Cl_2N_2$. *See* h694	177.03	nd (w^h)	...	105	...	...	s^h	v	v	...	...	aa v	B15, 431
h651	—,1,1-diethyl-*	$NH_2N(C_2H_5)_2$	88.15	...	...	99	...	...	v	v	v	...	v	chl v	**B4**, 550
h652	—,1,2-diethyl-*	$C_2H_5NHNHC_2H_5$	88.15	...	...	84–6	...	...	...	s	s	...	s	...	**B4**, 550
h653	—,1,2-diformyl-	Hydrazodiformic acid. OCHNHNHCHO	88.07	pr	159	...	...	...	v	δ	i	...	...	...	**B2**[1], 38
h654	—,1,2-diisobutyl-	$(C_4H_9^i)NHNH(C_4H_9^i)$	144.26	...	...	170^{735} 63^{10}	0.8002_4^{20}	1.4276	δ	...	...	...	...	os s	B4[2], 962
h655	—,1,2-diisopropyl-*	$(C_3H_7^i)NHNH(C_3H_7^i)$	116.21		...	125^{760} 63^{84}	0.7894_4^{20}	1.4173^{20}	...	...	...	...	...	os ∞	**B4**[2], 960
h656	—,1,1-dimethyl-*	$NH_2N(CH_3)_2$	60.10	...	...	63^{752}	0.7914^{22}	1.4075^{22}	v	v	v	...	...	MeOH v	**B4**[2], 958
h657	—,1,2-dimethyl-*	$CH_3NHNHCH_3$	60.10	...	...	81^{753}	0.8274_4^{20}	1.4209^{20}	∞	∞	∞	...	...	...	**B4**[2], 958
h658	—,—,dihydrochloride*	$CH_3NHNHCH_3.2HCl$	133.02	pr (w)	168d	...	...	...	v	v	...	...	...	...	**B4**[1], 560
h659	—,(2,3-dimethylphenyl)-*	$C_8H_{12}N_2$. *See* h694	136.20	nd (al)	108	...	...	...	...	v^h	s	...	...	...	**B15**[1], 171
h660	—,(2,4-dimethylphenyl)-*	$C_8H_{12}N_2$. *See* h694	136.20	nd (eth)	85	...	...	...	δ	v	s	...	...	...	**B15**[1], 173
h661	—,(2,5-dimethylphenyl)-*	$C_8H_{12}N_2$. *See* h694	136.20	nd	78	...	...	...	i	v	s	...	...	os s	**B15**[1], 175
h662	—,(2,6-dimethylphenyl)-*	$C_8H_{12}N_2$. *See* h694	136.20	nd (lig)	46	...	...	...	...	...	...	...	...	lig s	**B15**[1], 172
h663	—,(3,4-dimethylphenyl)-*	$C_8H_{12}N_2$. *See* h694	136.20	yesh nd (eth)	57	...	...	...	...	...	...	...	...	...	**B15**[1], 172
h664	—,1,2-di(1-naphthyl)-*	1,1'-Hydrazonaphthalene. $C_{10}H_7^{\alpha}NHNHC_{10}H_7^{\alpha}$	284.36	lf (bz), pl (peth)	153	...	...	...	i	...	s	...	s	peth δ	**B15**[2], 256

For explanations, symbols and abbreviations see beginning of table.

No.	Name	Synonyms and Formula	Mol. wt.	Crystalline form, color and specific rotation	m.p. °C	b.p. °C	Density	n_D	Solubility w	al	eth	ace	bz	other solvents	Ref.
	Hydrazine														
h665	**—,1,2-di(2-naphthyl)-***	2,2′-Hydrazonaphthalene. $C_{10}H_7^{\beta}NHNHC_{10}H_7^{\beta}$	284.36	pl (bz)	140–1					i	δ			os s	**B15**, 569
h666	**—,(2,4-dinitrio-phenyl)-***	$C_6H_6N_4O_4$. *See* h694	198.14	blsh-red	194				i	s^h	δ		δ	chl δ AcOEt s^h	
h667	**—,(2,6-dinitro-phenyl)-***	$C_6H_6N_4O_4$. *See* h694	198.14	nd (dil al)	145										
h668	**—,1,1-diphenyl**	$NH_2N(C_6H_5)_2$	184.24	ta (lig)	44	220^{40-50}	1.190^{16}_{4}		δ	v	v		v	chl v	**B15**², 52
h669	**—,1,2-diphenyl-***	Hydrazobenzene. 3′ 2′ 4′ 5′ 6′ C_6H_5—NHNH—C_6H_5 2 3 4 6 5	184.24	ta (al-eth)	131		1.158^{16}_{4}			v			δ	aa i	**B15**², 52
h670	**—,1,1-di(4-toyl)-**	$NH_2N(C_6H_4CH_3)_2$	212.30	lf (al)	93					s^h				sulf s	**B15**¹, 154
h671	**—,1,2-di(2-tolyl)-**	$C_6H_4(CH_3)$—NHNH—$C_6H_4(CH_3)$	212.30	lf (al)	165						s		s		**B15**², 223
h672	**—,1,2-di(3-tolyl)-**	*m*-Hydrazotoluene. $CH_3C_6H_4$—NHNH—$C_6H_4CH_3$	212.30	(peth)	38	224			i	s					**B15**, 506
h673	**—,1,2-di(4-tolyl)-**	*p*-Hydrazotoluene. CH_3—C_6H_4—NHNH—C_6H_4—CH_3	212.30	lf (lig), al, (bz-al)	135		0.957^{20}_{4}		i	s	s		s		**B15**², 234
h674	**—,ethyl-***	$NH_2NHC_2H_5$	60.10			101^{760}			v	v	v		v	chl v	**B4**², 959
h675	**—,1-ethyl-1-phenyl-**	$NH_2N(C_2H_5)C_6H_5$	136.20			237^{760} $115–9^{19}$	1.0181^{21}_{4}	1.5711^{21}							**B15**², 50
h676	**—,1-ethyl-2-phenyl-**	$C_6H_5NHNHC_2H_5$	136.20			240^{750} $100–104^{10}$	1.0103^{21}_{4}	1.5691^{16}	δ	v	v		v		**B15**², 50
h680	**—,1-isobutyl-1-phenyl-**	$NH_2N(C_6H_5)C_4H_9^{i}$	164.25			240–5	0.9633^{15}_{4}								**B15**, 121
h681	**—,isopropyl-***	$NH_2NHCH(CH_3)_2$	74.13			107^{750}			∞	∞	∞		∞	AcOEt ∞	**B4**², 960
h682	**—, methyl-***	NH_2NHCH_3	46.07		<−80	87^{745}			s	∞	s			lig i	**B4**², 957
h683	**—,1-methyl-2-isopropyl-***	$(CH_3)_2CHNHNHCH_3$	88.15	(eth)	76	100^{760}									**B4**², 960
h684	**—,1-methyl-1-phenyl-***	$NH_2N(C_6H_5)CH_3$	122.17			227^{745}	1.0404^{20}_{4}	1.5823^{22}	s	∞	∞		∞	chl ∞	**B15**², 49
h685	**—,1-methyl-2-phenyl-***	$C_6H_5NHNHCH_3$	122.17			230^{728}	1.0296^{23}_{4}	1.5715^{23}							**B15**², 50
h686	**—,1-methyl-2(3-tolyl)-**	H_3C—C_6H_4—$NHNHCH_3$	136.20	ye	59–61			1.0265^{100}_{4}	i	v	δ		v		**B15**², 229
h687	**—,1-methyl-2(4-tolyl)-**	CH_3—C_6H_4—$NHNHCH_3$	136.20	ye nd (eth), pl (lig)	91					s	s		s		**B15**¹, 154
h687¹	**—,1(3-methyl-butyl)1-phenyl-***	$NH_2N(C_6H_5)CH_2CH_2CH(CH_3)_2$	178.28			236	0.9588^{15}								**B15**, 121
h688	**—,(1-naphthyl)-***	$NH_2NHC_{10}H_7^{\alpha}$	158.20	lf (al), eth, sc (eth)	117	203^{20}			δ^h	v^h	s		v^h	chl v^h	**B15**², 256
h689	**—,(2-naphthyl)-***	$NH_2NHC_{10}H_7^{\beta}$	158.20	lf (w)	124–5				s^h	v^h	δ		v^h	chl δ	**E12B**, 885
h690	**—,(2-nitro-phenyl)-***	$C_6H_7N_3O_2$. *See* h694	153.14	red nd (bz)	90				v^h	δ	δ		δ	lig δ	**B15**², 177
h691	**—,(3-nitro-phenyl)-***	$C_6H_7N_3O_2$. *See* h694	153.14	red nd or pr (ace), ye nd (al)	93				δ^h	δ			δ	chl s aa s	**B15**², 182
h692	**—,(4-nitro-phenyl)-***	$C_6H_7N_3O_2$. *See* h694	153.14	og red lf or nd (al^h)	158d				s	v	s		s^h	chl s AcOEt s	**B15**², 183
h693	**—,1-phentyl-2-phenyl-(*d*)***	$C_6H_5NHNH(CH_2)_4CH_3$	178.28	$[\alpha]_D$ +4.45		$173–5^{50}$									**B15**, 121

For explanations, symbols and abbreviations see beginning of table.

No.	Name	Synonyms and Formula	Mol. wt.	Crystalline form, color and specific rotation	m.p. °C	b.p. °C	Density	n_D	Solubility w	al	eth	ace	bz	other solvents	Ref.
	Hydrazine														
h694	**—,phenyl-***	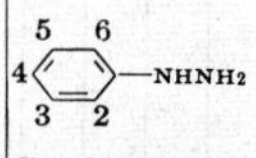 5 6 4 —$NHNH_2$ 3 2	108.14		20	243^{760} 144^{15}	1.099^{20}_{4}	1.6083^{18}	s^h	∞	∞	...	∞	chl ∞ lig δ	**B15²**, 44
h695	—,—,hemihydrate*	$C_6H_5NHNH_2.\frac{1}{2}H_2O$	117.15		24	120^{12}	1.0970^{25}_{25}	1.6081^{20}	δ	...	...	...	...		**B15**, 67
h696	—,—,hemihydro-chloride*	$2(C_6H_5NHNH_2).HCl$	252.75	nd	225				s	s	δ	...	...		**B15**, 108
h697	—,—,hydro-chloride*	$C_6H_5NHNH_2.HCl$	144.61	lf (al)	240				v	s	i	...	...		**B15²**, 48
h698	**—,1-phenyl-2(2-toyl)-**	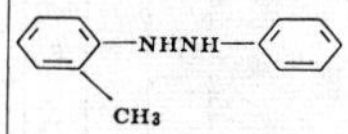 —NHNH— CH_3	198.27	lf (al)	101				i	δ	v	...	v	peth δ	**B15**, 497
h699	**—,1-phenyl-2(3-tolyl)-**	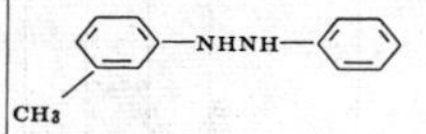 —NHNH— CH_3	198.27	ye (peth)	61		1.0265^{100}_{4}		...	v	δ	...	...	lig s	**B15²**, 229
h700	**—,1-phenyl-2(4-tolyl)-**	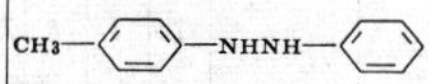 CH_3— —NHNH—	198.27	ta (lig), (al)	91				...	v	...	...	v		**B15¹**, 154
h701	**—,propyl-***	$NH_2NHCH_2CH_2CH_3$	74.13		119										**B4²**, 960
h702	**—,tetraphenyl-**	$(C_6H_5)_2NN(C_6H_5)_2$	336.44	pr	149d				i	$δ^h$	...	v	v	chl v	**B15¹**, 29
h703	**—,(2-tolyl)-**	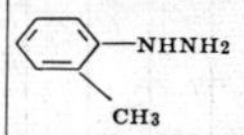 —$NHNH_2$ CH_3	122.17	nd	59				...	v	v	...	...	chl v	**B15²**, 222
h704	**—,3-tolyl-**	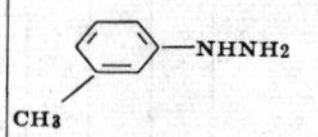 —$NHNH_2$ CH_3	122.17			244	1.057^{20}_{4}								**B15²**, 229
h705	**—,4-tolyl-**	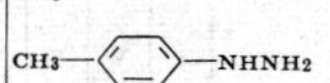 CH_3— —$NHNH_2$	122.17	lf (w)	66	244			δ	v	v	...	v		**B15²**, 233
h706	**—,1(3-tolyl)-2(4-tolyl)-**	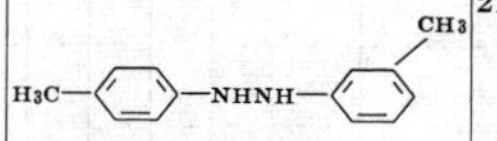 H_3C— —NHNH— CH_3	212.30	ta (lig)	74										**B15**, 511
h707	**—,(2,4,6-tri-bromophenyl)-***	$C_6H_5Br_3N_2$. *See* h694	344.85	nd (peth), (lig)	146				...	...	...	...	v	chl v lig s	**B15¹**, 126
h708	**—,(2,4,6-tri-chlorophenyl)-***	$C_6H_5Cl_3N_2$. *See* h694	211.48	(bz)	143				s^h	...	...	...	s^h		**B15²**, 156
h709	**—,(2,4,6-trinitro-phenyl)-***	$C_6H_5N_5O_6$. *See* h694	243.14	pr (al)	178				...	...	δ	...	δ	chl δ aa s	**B15²**, 221
h710	**—,triphenyl-***	$C_6H_5NHN(C_6H_5)_2$	260.34	nd (bz-peth)	142d		0.869^{70}_{4}		i	s	δ	...	v		**B15²**, 54
h711	**Hydrazinecar-boxylic acid,** ethyl ester*	*N*-Aminourethane. $NH_2NHCO_2C_2H_5$	104.11		45	198d 92^{13}			...	s	s	...	...		**B3²**, 79
h712	**1,1-Hydrazine-dicarboxylic acid,** diethyl ester*	$NH_2N(CO_2C_2H_5)_2$	176.17	pr	29	138^{12}			...	s	...	...	...		**B3²**, 79
h713	**1,2-Hydrazine-dicarboxylic acid,** diamide	Dicarbonamide. $H_2NCONHNHCONH_2$	118.10	pl (w)	257–9				s^h	i	i	...	...	30% KOH s	**B3²**, 95
h714	—,diethyl ester*	$C_2H_5O_2CNHNHCO_2C_2H_5$	176.17	nd (chl), pr (w)	135	250d			s^h	v	v	...	...	chl s^h	**B3²**, 79
—	**Hydrazobenzene**	*see* **Hydrazine, 1,2-diphenyl-***													
—	**Hydrazo compounds**	*see* **Hydrazine**													
h715	**Hydrindane** (*trans, l*)	Hexahydroindane. Octahydroindene. 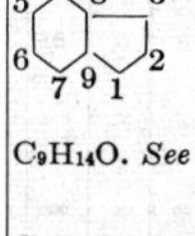 4 8 3 5 6 2 7 9 1	124.23	$[\alpha]^{21}_{D}-5.98$		159^{760}	0.863^{20}_{4}	1.4655^{18}	i	...	s	...	...		**B5²**, 50
h716	**2-Hydrindanone** (*cis*)	$C_9H_{14}O$. *See* h715	138.21		10	225^{754} 109^{23}			...	...	...	...	...		**E12A**, 221
h717	—(*trans*)	$C_9H_{14}O$. *See* h715	138.21		−12	218^{754}	0.9807^{17}_{4}	1.4769^{17}	...	s	...	...	...		**E12A**, 222
—	**Hydrindene**	*see* **Indan**													
—	**Hydrindene, 1,2,3,-triketo,** monohydrate	*see* **Ninhydrin**													

For explanations, symbols and abbreviations see beginning of table.

No.	Name	Synonyms and Formula	Mol. wt.	Crystalline form, color and specific rotation	m.p. °C	b.p. °C	Density	n_D	Solubility w	al	eth	ace	bz	other solvents	Ref.
	Hydrobenzamide														
h718	**Hydrobenzamide**	Tribenzaldiamine. $C_6H_5CH(N{:}CHC_6H_5)_2$	298.39	nd (bz), (al, w)	101	130^{760}			i	v	v				$B7^2$, 166
—	**Hydrobenzoin**	*see* **1,2-Ethanediol, 1,2-diphenyl-***													
h719	**Hydroberberine(*d*)**	*d*-Canadine(*D*). Tetrahydroberberine.	339.39	nd (dil al) $[\alpha]^{20}_D+297.4$ (chl, c=1)	132										$B27^2$, 556
h720	—(*l*)	$C_{20}H_{21}NO_4$. *See* h719	339.39	nd (al) $[\alpha]^{20}_D$ −298.2 (chl, c=1)	132				i	v	s		s	chl s lig δ^h	$B27^2$, 557
—	**Hydrocerulignone**	*see* **Biphenyl, 4,4′-dihydroxy-3,3′,5,5′-tetramethoxy-**													
h721	**Hydrocinchonidine**	Cinchamidine. Dihydrocinchonidine.	296.42	$[\alpha]^{20}_D-97.5$	231				i	δ	δ				$B23^2$, 357
—	**Hydrocinnamaldehyde**	*see* **Propanal, 3-phenyl-***													
—	**Hydrocinnamic acid**	*see* **Propanoic acid, 3-phenyl-***													
—	**Hydrocortisone**	*see* **Corticosterone, 17-hydroxy-**													
h722	**Hydrocotarnine,** hemihydrate	$C_{12}H_{15}NO_3 \cdot \frac{1}{2}H_2O$	230.27	pr (eth)	56				i	v	v	v		chl v aa v	$B27^2$, 541
h723	**Hydrocupreine**		312.42	pl (dil al) $[\alpha]^{20}_D$ +154.8 (al)	230				δ		s			chl s	$B23^1$, 151
h724	**Hydrocyanic acid***	**Hydrogen cyanide.** Formonitrile. HCN	27.03		−14	26^{760}	0.6884^{20}_4 0.925^{-40} (solid)	1.2619^{20}	∞	∞	∞				$B2^2$, 37
h725	**Hydrofuramide**	Furfuralhydramide. Furfuramide. Trifurfuraldiamine.	268.26	nd (al)	117				i	v	v				$B17^2$, 311
—	**Hydrogen cyanide**	*see* **Hydrocyanic acid***													
h726	**Hydrohydrastinine**		191.23	nd (lig)	66	$301–3^{752}$				v	v	v		AcOEt v CS_2 v aa v	$B27^2$, 528
h727	**Hydrolapachol**		244.29	ye nd (al), (peth)	94					s				peth s^h	E12B, 3082
—	**Hydroperoxide, acetyl**	*see* **Acetic acid, per-**													
h728	—,*tert*-**butyl-**	$(CH_3)_3COOH$	90.12		6	13^3	0.896^{20}	1.4010^{20}	s	s	s			chl s	
h729	—,**cyclohexyl-**		116.16		−20	$42^{0.1}$	1.019^{20}_4	1.4645^{25}						aa s	Am 72, 3333
h730	—,**ethyl-**	C_2H_5OOH	62.07		−100	$93–7^{760}$ exp>100	0.9332^{20}_4	1.3800^{20}	∞	∞	∞				$B1^3$, 1312
h731	—,**methyl-**	CH_3OOH	48.04			$38–40^{65}$	1.9967^{15}_4	1.3641^{15}	∞	∞	∞		s	chl δ aa δ	$B1^2$, 270
—	**Hydrophlorone**	*see* **Benzene, 1,4-dihydroxy-2,5-dimethyl-***													

For explanations, symbols and abbreviations see beginning of table.

No.	Name	Synonyms and Formula	Mol. wt.	Crystalline form, color and specific rotation	m.p. °C	b.p. °C	Density	n_D	Solubility w	al	eth	ace	bz	other solvents	Ref.
	Hydroquinidine														
h732	**Hydroquinidine**	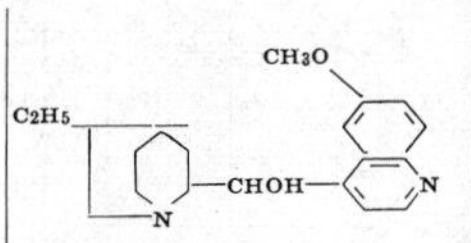 CH_3O, C_2H_5, —CHOH—, N	326.43	nd	168–9										
h733	**Hydroquinine**(*d*)	$C_{20}H_{26}N_2O_2$. *See* h732	326.43	$[\alpha]_D^{16}+143.5$	171										
h734	—(*dl*)	$C_{20}H_{26}N_2O_2$. *See* h732	326.43		175–7										
h735	—(*l*)	$C_{20}H_{26}N_2O_2$. *See* h732	326.43	$[\alpha]_D^{20}-142.2$											**B23**^{2}, 400
—	**Hydroquinol**	*see* **Benzene, 1,2,4-trihydroxy-***													
—	**Hydroquinone**	*see* **Benzene, 1,4-dihydroxy-***													
h736	**Hydro-quinoneph-thalein**	2,7-Dihydroxyfluoran. 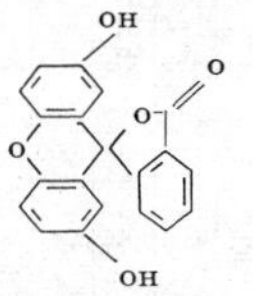 OH, O, O, OH	332.30	nd (eth)	228–9				δ^h	v	v	v		aa v	**B19**^{2}, 247
—	**Hydrosorbic acid**	*see* **3-Hexenoic acid***													
h737	**Hydroxy-amphetamine**	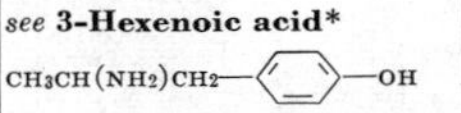 $CH_3CH(NH_2)CH_2$— —OH	151.20		125–6				δ	s				chl s	
h738	—,hydrobromide	$C_9H_{13}NO$. HBr. *See* h737	232.10		189				v	s					
h739	**Hydroxy-citronellal**	$(CH_3)_2COH(CH_2)_3CH(CH_3)CH_2CHO$	172.26			103^3	0.9220^{20}	1.4494^{20}	δ	s					
—	**Hydroxyditan**	*see* **Methane, (hydroxy-phenyl)phenyl-***													
—	**Hydroxyhemin**	*see* **Hematin**													
h742	**Hyenic Acid**	$C_{25}H_{50}O_2$	382.66	(eth), nd (bz)	77–8				i	δ	v				**B2**^{2}, 380
h743	**Hygrine**	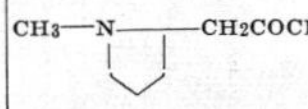 CH_3—N— —CH_2COCH_3	141.21			$193–5^{760}$ $(92–4^{20})$	0.935_4^{17}		δ	s				chl s	**B21**^{2}, 218
—	**Hyodeoxycholic Acid**	*see* **Cholanic Acid, 3(α),6(α)dihydroxy-**													
h744	**Hyoscine**(*dl*)	Atroscine. Scopolamine. 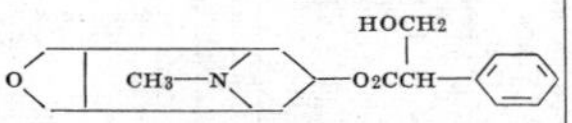 $HOCH_2$, O, CH_3—N, —O_2CCH—	303.35	pr	82				δ	v	v	v	v		**B27**^{1}, 248
h745	—,dihydrate(*dl*)	$C_{17}H_{21}NO_4.2H_2O$. *See* h744	339.38	nd (w+2)	36–7						δ				**B27**, 101
h746	—,hydrobromide(*d*)	$C_{17}H_{21}NO_4$.HBr. *See* h744	384.28		193										
h747	—,—(*dl*)	$C_{17}H_{21}NO_4$.HBr. *See* h744	384.28	eff (ace)	185				s	δ	i	δ			**B27**^{1}, 248
h748	—,—(*l*)	$C_{17}H_{21}NO_4$.HBr. *See* h744	384.28	lf (al) $[\alpha]_D-26$	209				v	δ	i				**B27**^{2}, 64
h749	—,hydrobromide trihydrate(*d*)	$C_{17}H_{21}NO_4HBr.3H_2O$. *See* h744	438.32	ta (w) $[\alpha]_D+23$ (w,c=3)	55										**B27**^{1}, 247
h750	—,—(*dl*)	$C_{17}H_{21}NO_4.HBr.3H_2O$. *See* h744	438.32	(w)	55–8										**B27**^{1}, 248
h751	—,—(*l*)	$C_{17}H_{21}NO_4.HBr.3H_2O$. *See* h744	438.32	ta (w+3) $[\alpha]_D$ −22.8 (w, c=2)					v	δ					**B27**^{2}, 64
h752	—,hydrochloride(*l*)	$C_{17}H_{21}NO_4$.HCl. *See* h744	339.82	(al)	200				v	v					**B27**^{2}, 64
h753	—,hydrocholride dihydrate(*l*)	$C_{17}H_{21}NO_4.HCl.2H_2O$. *See* h744	375.85	pr (w)	80				v	v					**B27**, 101
h754	—,monohydrate(*dl*)	$C_{17}H_{21}NO_4.H_2O$. *See* h744	321.38	(w+1)	56				δ	v	v			chl v	**B27**, 101
h755	—,—(*l*)	$C_{17}H_{21}NO_4.H_2O$. *See* h744	321.38	(w+1) $[\alpha]_D^{20}$ −28 (w), −18 (abs al, c=5.2)	59				v^h	v	v	v	v	chl v	**B27**^{2}, 63
h756	—,nitrate(*l*)	$C_{17}H_{21}NO_4.HNO_3$. *See* h744	366.37	nd (al)	213										**B27**^{2}, 64
h757	—,sulfate	$(C_{17}H_{21}NO_4)_2.H_2SO_4.2H_2O$. *See* h744	740.83	nd (w, ace)											**B27**, 101
h758	**Hyoscyamine**(*d*)	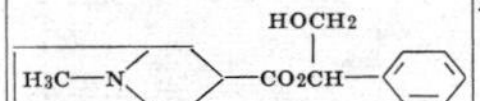 $HOCH_2$, H_3C—N, —CO_2CH—	289.38	nd (dil al) $[\alpha]_D+31.3$ (c=4)	106				δ	v	v		v	chl v	**B21**^{2}, 18
h759	—(*l*)	Daturine. *l*-Tropyltropeine. $C_{17}H_{23}NO_3$. *See* h758	289.38	nd (dil al) $[\alpha]_D-20.7$ (abs al c=4)	108						δ			chl δ	**B21**^{2}, 18
h760	—,hydrobromide(*l*)	$C_{17}H_{23}NO_3$.HBr. *See* h758	370.29	pr	149–50				v	v					**B21**^{1}, 198

For explanations, symbols and abbreviations see beginning of table.

No.	Name	Synonyms and Formula	Mol. wt.	Crystalline form, color and specific rotation	m.p. °C	b.p. °C	Density	n_D	Solubility w	al	eth	ace	bz	other solvents	Ref.
	Hyoscyamine														
h761	—,hydrochloride(*l*)	$C_{17}H_{23}NO_3.HCl$. *See* h758	325.84	$[\alpha]_D-23.2$ (w, c = 0.5)	149–51				s	s					**B21**, 26
h762	—,sulfate(*l*)	$2(C_{17}H_{23}NO_3).H_2SO_4.2H_2O$. *See* h758	712.86	dlq nd (al) $[\alpha]_D-28.3$	206 (anh)				v	v^h	i	v^h			**B21**², 19
h763	**Hypaphorine**(*d*)	*N,N*-Dimethyl-*L*-tryptophane methylbetaine.	246.31	(dil al) $[\alpha]^{25}_D+113.4$ (w, c = 1.6)	253–4				v	v				os i	**B22**², 469
h764	**Hypnal**	Antipyrine-chloralhydrate.	353.64	wh	68				v	v					
h765	**Hypochlorous acid,** *tert*-butyl ester	*tert*-Butyl hypochlorite. $(CH_3)_3COCl$	108.57			80^{750}	0.9583^{18}_4	1.403^{20}	i	d		s		aa d	**B1**³, 1581
h766	—,ethyl ester*	CH_3CH_2OCl	80.51			36^{0758}	1.013^{-6}_4		i	s	s		s	chl s	**B1**², 325
h767	—,methyl ester*	CH_3OCl	66.49			12^{726}			i	s					
h768	**Hypofluorous acid,** trifluoromethyl ester*	F_3COF	104.00		<−215	-95^{760}									**Am70,** 3986
—	$\Delta^{\alpha,\beta}$**-Hypogeic acid**	*see* **2-Hexadecenoic acid***													
h769	**Hypophosphoric acid,** tetraethyl ester*	Tetraethyl hypophosphate. $(C_2H_5O)_2POOP(OC_2H_5)_2$	274.20			$116–7^2$	1.1283^{18}	1.4284^{20}	d	d			s		**B1**³, 1330
h770	**Hypoxanthine**	6(1)-Purinone. Sarcine.	136.11	oct nd	150d				δ^h					alk s dil ac s	**B26**², 252
—	—,**2-amino-**	*see* **Guanine**													
—	—,riboside	*see* **Inosine**													
—	**Hystarazin**	*see* **9,10-Anthraquinone, 2,3-dihydroxy-***													

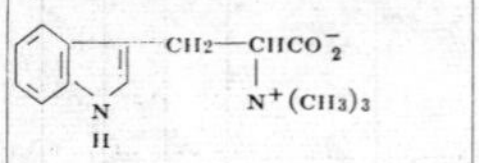

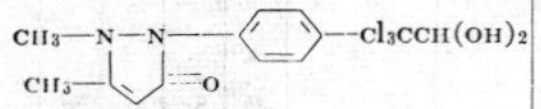

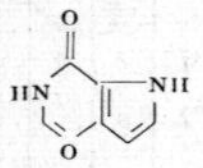

For explanations, symbols and abbreviations see beginning of table.

No.	Name	Synonyms and Formula	wt.	Crystalline form, color and specific rotation	m.p. °C	b.p. °C	Density	n_D	Solubility w	al	eth	ace	bz	other solvents	Ref.
	Iditol														
i1	**Iditol**	*D*-Idite. 1,2,3,4,5,6-Hexanehexol. 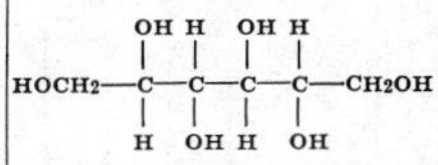	182.16	$[\alpha]_D$ +3.5 (w)	73.5										
i1[1]	**Idonic acid**(*l*)	2,3,4,5,6-pentahydroxyhexanoic acid. 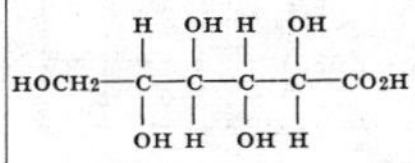	196.16	nd (w, al) $[\alpha]_D^{20}$ −3.25 (w,p=10.56)	205d				s						B3[2], 354
i2	—,α-lactone(*d*)	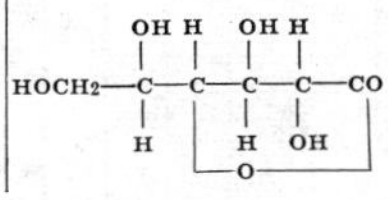	178.14	pl, $[\alpha]_D^{20}$ −52.6 (w)	174										
i3	—,phenylhydrazide (*L*)	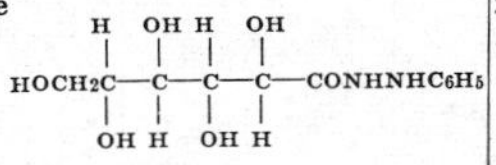	286.29	$[\alpha]_D^{20}$ −12.4	115				v	v^h				AcOEt δ^h	**B15**, 82
i4	Idose(*D*)	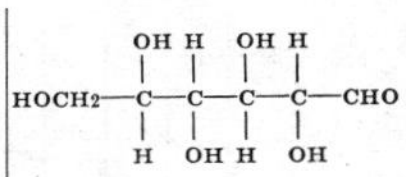	180.16	syr $[\alpha]_D$ +16					s						**B31**, 294 **B1**, 909
i5	—(*L*)	$C_6H_{12}O_6$. *See* i4	180.16	syr, $[\alpha]_D^{20}$ +52.7 (c=6.2)					s						**B1**, 909
—	**Ignotine**	*see* **Carnosine**													
i6	**Imesatin**	Isatin-3-imide. 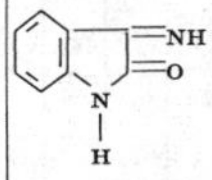	146.15	dk ye pr	175–6				i s^h	v	δ		i	lig i	**B21**[2], 332
i7	**Imidazole**	1,3-Diazole. Glyoxaline. Iminazole. 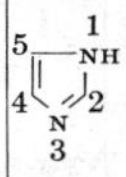	68.08	mcl pr (bz)	90	256	1.0303^{101}_4	1.4801^{101}	v	v	δ			Py s chl s	**B23**[2], 34
—	**—,4-ethylamino-**	*see* **Histamine**													
i8	**—,1-methyl-**	Oxalmethyline. $C_4H_6N_2$. *See* i7	82.10		−6	198	1.6325^{21}_4	1.4924^{21}							**B23**[2], 39
i9	**—,1-methyl-2-mercapto-**	Methimazol. Topazole. $C_4H_6N_2S$. *See* i7	114.17	lf (al)	142	280 δd			v	s	δ		δ	chl s lig δ	
—	**—,2,4,5-triphenyl-**	*see* **Lophine**													
i10	**4,5-Imidazoledicarboxylic acid**	1,3-Diazole-4,5-dicarboxylic acid. 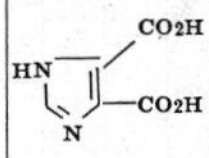	156.10	amor		288 d	1.749		δ^h	i	i			Py δ con ac s os i	**B25**[2], 159
i11	**2-Imidazolidinethione**	*N,N'*-Ethylenethiourea. 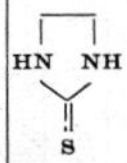	102.16	nd (al), pr (AmOH)	197–8				v	s	i		i	chl i lig i	**B24**[2], 4
i12	**2-Imidazolidone**	Ethylene urea. 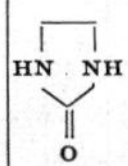	86.09	nd (chl)	131				v	v^h	δ				**B24**[2], 3

For explanations, symbols and abbreviations see beginning of table.

No.	Name	Synonyms and Formula	Mol. wt.	Crystalline form, color and specific rotation	m.p. °C	b.p. °C	Density	n_D	Solubility w	al	eth	ace	bz	other solvents	Ref.
	Δ²-Imidazoline														
i13	**Δ²-Imidazoline, 2-methyl-**	Lysidine. 2-Methyl-2-glyoxalidine. 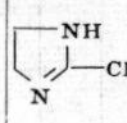	84.12	hyg	107	195–8 ca. 120^{65}			v	v	i			chl s	**B23²**, 26
i14	**Imperatorin**	$C_{16}H_{14}O_4$	270.27		102				i	s	s		s	os s alk s	**B19²**, 227
i15	**Indaconitine**	Acetylbenzoylpseudoconine. $C_{34}H_{47}NO_{10}$	629.67		202–3d				i	s	s				
i16	**Indan**	2,3-Dihydroindene. Hydrindene. 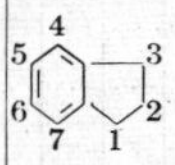	118.18		−51.4	117	0.9645^{20}_{4}	1.5351^{21}	i	∞	∞				**B5²**, 376
i17	**—,1-amino-**	*dl*-1-Hydrindamine. $C_9H_{11}N$. *See* i16	133.20			220.5^{747} $96–7^{8}$	1.038^{15}_{4}	1.5619^{15}	δ						**E12A**, 149
i18	**—,5-amino-**	5-Hydrindamine. $C_9H_{11}N$. *See* i16	133.20	nd (peth)	37–8	$247–9^{745}$ $146–7^{25}$			δ					ac s os v	**E12A**, 154
i19	**—,2,3-dibromo-**	Indene dibromide. $C_9H_8Br_2$. *See* i16	276.00	pr (peth)	31.5–2.5	$100–5^{1}$	1.747^{25}_{4}	1.6290^{25}	d^h	d^h	s			sulf s(red)	**E12A**, 142
i20	**—,2,3-dichloro-**	Indene dichloride. $C_9H_8Cl_2$. *See* i16	187.08			$83–5^{3}$	1.254^{25}_{4}	1.5690^{25}	d^h	d^h					**E12A**, 142
—	**—,hexahydro-**	*see* **Hydrindane**													
i21	**—,1-hydroxy-**	$C_9H_{10}O$. *See* i16	134.18	nd (peth)	55	255 128^{12}			$δ^h$	v	v				**B6²**, 530–1
i22	**—,4-hydroxy-**	$C_9H_{10}O$. *See* i16	134.18		47–51	120^{12}									**B6²**, 530–1
i23	**—,5-hydroxy-**	$C_9H_{10}O$. *See* i16	134.18	nd (peth)	55	255			δ	v	v			sulf s (ye)	**E12A**, 182
i24	**—,1-methyl-**	$C_{10}H_{12}$. *See* i16	132.21			186–7 60^{10}	0.9402^{20}_{4}	1.5222^{20}	i						**E12A**, 106
i25	**—,2-methyl-**	$C_{10}H_{12}$. *See* i16	132.21			70^{10}	0.9317^{25}_{4}	1.5189^{25}	i						**E12A**, 107
i26	**—,4-nitro-**	$C_9H_9NO_2$. *See* i16	163.18	(al)	44–4.5	139^{10}			i					os v	**E12A**, 146
i27	**—,1-phenyl-1,3,3-trimethyl-**	$C_{18}H_{20}$. *See* i16	236.36	tcl pr (al)	52–3	299–300 $118–20^{0.1}$	1.1183^{20}_{4}	1.5633^{20}	i	s^h			s	MeOH s	**E12A**, 117
i28	**1,2-Indandione**	α,β-Diketohydrindane. 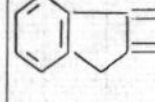	146.15	gold ye pl or lf (bz, eth)	95					v				chl v MeOH v	**B7²**, 631
i29	**—**,β-oxime	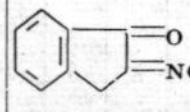	161.17	nd (al), (bz)	210d					s^h			s^h	dil alk s (ye)	**B7²**, 631
i30	**1,3-Indanedione**	1,3-Dioxohydrindene. Oxinolone. 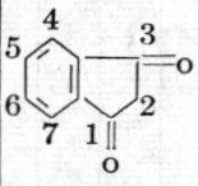	146.15	nd (HCl), (eth, lig)	129–31d		1.37^{21}		δ	v^h	s^h		v	alk s (ye) lig δ	**B7²**, 632
i31	**—**,dioxime	$C_9H_8N_2O_2$ *See* i30	176.18	nd (w)	ca. 225d				$δ^h$	δ	i				**B7**, 695
i32	**—,2-*tert*-butyl**	Pivalyl indandione. Pivalyl valone. $C_{14}H_{14}O_3$. *See* i30	230.25		108.5–10.5									aq alk s (bt ye)	
i33	**—,2(3-methylbutyl)-**	Valone. $C_{14}H_{14}O_3$. *See* i30	230.25	ye	67–8				i	s	s	s			
i34	**—,2-phenyl-**	Danilone. $C_{15}H_{10}O_2$. *See* i30	222.23	lf (al or bz)	149				i	s	s	s	s		**E1217**, 297
i35	**1-Indanone**	α-Hydrindone. 1-Keto-hydrindene. 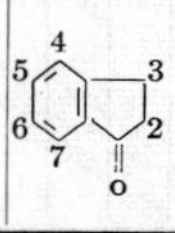	132.16	ta, nd (w+3)	40	243–5	1.0940^{45}_{4}	1.561^{45}	δ	s	s			chl s lig s	**B7²**, 283

For explanations, symbols and abbreviations see beginning of table.

No.	Name	Synonyms and Formula	Mol. wt.	Crystalline form, color and specific rotation	m.p. °C	b.p. °C	Density	n_D	Solubility w	al	eth	ace	bz	other solvents	Ref.
	2-Indanone														
i36	**2-Indanone**	β-Hydrindone. 2-Keto-hydrindene. 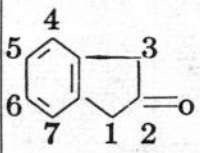	132.16	nd (al or eth)	59	220–5d	1.0712^{69}_{4}	1.538^{67}	i	v	v	v	...	chl v	**B7**², 286
i37	**1-Indanone, 2-nitro-**	$C_9H_7NO_3$. *See* i35	177.16	ye nd (bz-lig)	117d	...	...	...	s	...	...	...	...	os s, lig δ^h	**E12A,** 260
i38	**—,6-nitro-**	$C_9H_7NO_3$. *See* i35	177.16	ye lf and nd (peth)	74	...	...	...	δ	v	v	...	v	AcOEt v, chl v	**E12A,** 260
i39	**2-Indanone, 5-nitro-**	$C_9H_7NO_3$. *See* i36	177.16	br nd (al)	141–1.5	...	...	...	δ	v	v	...	...	aa v	**E12A,** 260
i40	**Indanthrene**	Dihydroanthraquinonazine. Indanthrone. 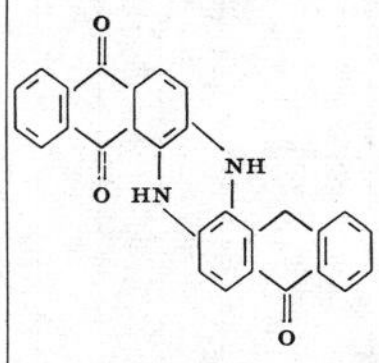	442.43	bl nd	470–500d	...	...	...	i	i	i	...	...	dil alk s, $PhNO_2$ s	**B24**², 317
i41	**Indazole**	1,2-Benzodiazole. 1,2-Benzopyrazole 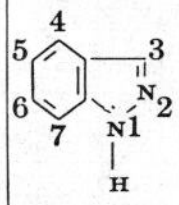	118.14	nd (al^h)	146–5	270^{743}	...	...	s^h	s	s	...	...	...	...
i42	**—,3-chloro-**	$C_7H_5ClN_2$. *See* i41	152.58	nd (w or lig)	148–8.5	sub	...	...	s^h	v	v	...	v	lig s^h	**B23**², 139
i43	**—,4-chloro-**	$C_7H_5ClN_2$. *See* i41	152.58	nd (to)	156	...	...	...	s	s	...	...	...	os s, to s^h	**B23**², 139
i44	**—,4-nitro-**	$C_7H_5N_3O_2$. *See* i41	163.16	...	203	...	...	...	...	s	...	v	...	aa v	**B23**², 144
i45	**—,5-nitro-**	$C_7H_5N_3O_2$. *See* i41	163.16	yesh nd (al)	208	...	...	...	...	s	s	v	s	aa v	**B23**², 145
i46	**—,6-nitro-**	$C_7H_5N_3O_2$. *See* i41	163.16	nd (w, al, aa or xyl)	181d	...	...	...	s^h	s	...	...	...	NaOH s^h, xyl s^h	**B23**², 146
i47	**—,7-nitro-**	$C_7H_5N_3O_2$. *See* i41	163.16	...	186.5–7.5 sub	...	...	...	...	...	...	...	...	2N NaOH v	**B23**², 150
i48	**3-Indazolinone**	Benzopyrazolone. Indazolone. 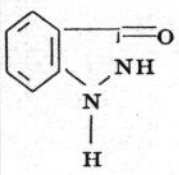	134.14	nd or lf (MeOH or w), pl or nd (al)	247	...	...	...	δ^h	s^h δ	δ	...	...	MeOH s^h	**B24**², 59
i49	**Indene**	Indonaphthene. 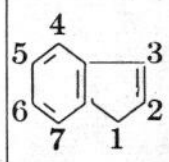	116.16	(aa)	−2	182.2–2.4	0.9915^{20}_{4}	1.5642^{20}	i	∞	∞	s	s	Py s, CS_2 s	**B5**², 410; **B16**², 871
i50	**—,1-benzhydrylidene-**	ω,ω-Diphenyl benzofulvene. 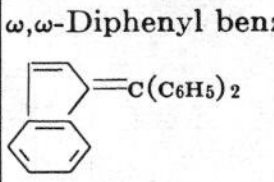	280.37	og-ye (al)	114.5	...	...	...	...	δ	...	...	...	os s, aa δ	**E12A,** 128
—	**—,2,3-dihydro-**	*see* **Indan**													
i51	**—,1,2-diphenyl-**	$C_{21}H_{16}$. *See* i49	268.36	nd (aa)	177–8	...	...	...	...	s^h	s	...	...	sulf s(gr), aa s^h	**E12A,** 118
i52	**—,1,3-diphenyl-**	$C_{21}H_{16}$. *See* i49	268.36	pym or nd (aa)	68 (nd), 85 (pym)	230^{15}	...	...	...	s^h	s	...	...	aa s^h	**E12A,** 119
i53	**—,2,3-diphenyl-**	$C_{21}H_{16}$. *See* i49	268.36	pr (aa)	108–9	$235–40^{12}$	...	...	...	s^h	...	...	...	sulf s(gr), aa s^h	**E12A,** 119
i54	**—,2-methyl-**	$C_{10}H_{10}$. *See* i49	130.19	...	...	$184–5^{741}$, 79^{10}	0.9734^{19}_{4}	1.5650^{19}	i	...	...	...	...	...	**E12A,** 107
i55	**—,3-methyl-**	$C_{10}H_{10}$. *See* i49	130.19	...	...	205–6d, 70^{10}	0.9682^{27}_{4}	1.5591^{27}	i	...	...	...	...	...	**E12A,** 107
i56	**1-Indenecarboxylic acid**	$C_{10}H_8O_2$. *See* i49	160.17	pa ye nd (bz)	161	$193–5^{12}$	...	...	...	v	δ	...	...	to, chl δ	**E12A,** 356

For explanations, symbols and abbreviations see beginning of table.

No.	Name	Synonyms and Formula	Mol. wt.	Crystalline form, color and specific rotation	m.p. °C	b.p. °C	Density	n_D	Solubility w	al	eth	ace	bz	other solvents	Ref.
	2-Indenecarboxylic acid														
i57	**2-Indenecarboxylic acid**	$C_{10}H_8O_2$. *See* i49	160.17	(bz)	234 sub	...	...	...	δ	v	v	...	s	...	**E12A**, 357
i58	**1-Indenone, 2,3-diphenyl-**	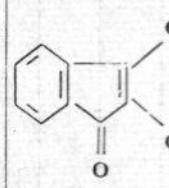	282.34	og-red (lig or al)	150–1	...	...	...	i	s	...	s	s	lig s^h	**B7^2**, 501
i59	**Indican**	Indoxyl-β-glucoside. $C_{14}H_{23}NO_9$	349.35	nd (w), $[\alpha]^{19}_{546}$ −65.6 (w,c=1)	57–8	...	...	...	v	v	δ	s	δ	MeOh s AcOEt, CS_2 δ	**B31**, 258
—	**Indigo carmine**	*see* **5,5′-Indigotindisulfonic acid.**													
—	**Indigo red**	*see* **Indirubin**													
i60	**Indigo white**	2,2′-Diindoxyl. Leucoindigo. 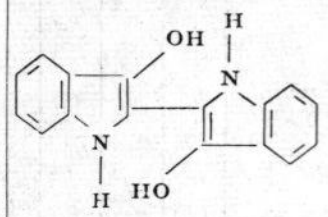	264.29	ye	...	...	...	...	δ	s	s	...	...	alk s	**B23^2**, 429
i61	**4,4′-Indigotindicarboxylic acid**	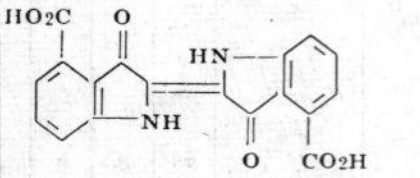	350.29	bl nd	...	...	...	...	i	i	i	...	...	sulf s chl i	**B25**, 273
i62	**5,5′-Indigotindisulfonic acid**	Indigo carmine. 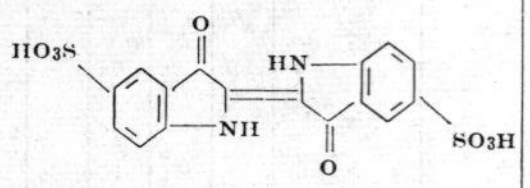	422.40	dk bl amor, or br-red cr	...	...	...	...	s	s	...	...	...	...	**B25^2**, 298
i63	**Indigotinsulfonic acid**	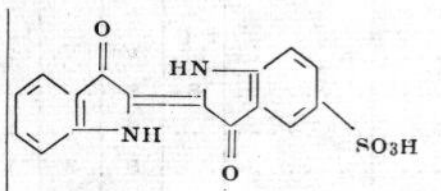	342.34	amor	200(d)	...	...	...	s	s	...	...	...	dil ac i	**B25^2**, 296
i64	**Indirurin**	Indigo red. 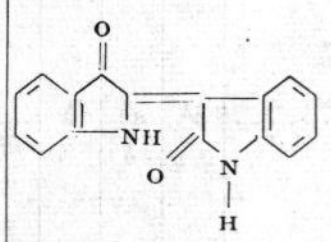	262.27	red or br rh nd (sub)	sub	...	...	...	i	δ	s	...	...	con sulf s aa v	**B24^2**, 246
—	**Indogenic acid**	*see* **2-Indolecarboxylic acid, 3-hydroxy-**													
i65	**Indole**	1-Benzo[b]pyrrole. 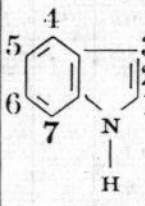	117.15	lf (peth), (eth)	52.5	254	...	...	s^h	v	v	...	s	lig s	**B21^2**, 56
i66	**—,1-acetyl-**	$C_{10}H_9NO$. *See* i65	159.19	...	...	$152–3^{14}$	...	...	...	...	...	...	...	...	**B20^2**, 200
i67	**—,3(2-aminoethyl)-**	Tryptamine. $C_{10}H_{12}N_2$. *See* i65	160.22	nd (al-bz)	120	...	...	...	i	s	i	s	i	chl i	**B22^2**, 346
—	**—,2,3-dihydro-**	*see* **Indoline**													
i69	**—,1,3-dimethyl-**	*N*-Methylskatole. $C_{10}H_{11}N$. *See* i65	145.20	nd	141–3	225–32	...	...	...	...	s	...	...	...	**B20^2**, 204
—	**—,3(dimethylaminomethyl)-**	*see* **Gramine**													
i69^1	**—,3-hydroxy-**	3-Indolol. Indoxyl. C_8H_7NO. *See* i65	133.15	bt ye pr	85	110	...	...	s	s	s	...	s	chl s lig δ	**B21^2**, 42
i69^2	**—,—,β-glucoside**	*see* **Indican**													
i70	**—,3-hydroxy-1-nitroso-**	Isatoxim. $C_8H_6N_2O_2$. *See* i65	162.15	ye nd	202	...	...	...	δ	s	...	...	...	KOH s	**B21**, 73
—	**—,2-methyl**	C_9H_9N. *See* i65	131.18	nd or lf (w)	61	272	1.07^{20}_{4}	...	δ^h	v	v	...	...	sulf s	**B20^2**, 201
i71	**—,3-methyl-**	Skatole. C_9H_9N. *See* i65	131.18	lf (lig)	95	265–6	...	...	s	s	s	...	...	chl s	**B20^2**, 203
i72	**2-Indolecarboxylic acid**	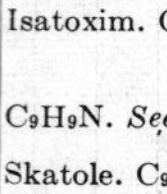	161.16	ye pl (bz)	203	...	...	...	δ	s	s	...	...	aa δ	**B22^2**, 45

For explanations, symbols and abbreviations see beginning of table.

No.	Name	Synonyms and Formula	Mol. wt.	Crystalline form, color and specific rotation	m.p. °C	b.p. °C	Density	n_D	Solubility w	al	eth	ace	bz	other solvents	Ref.
	2-Indolecarboxylic acid														
i73	**—,3-hydroxy**	Indogenic acid. Indoxylic acid. $C_9H_7NO_3$. *See* i72	177.16	pw		122–3 sub d			δ d^h						**B22**[2], 168
i74	**Indoline**	2,3-Dihydroindole. 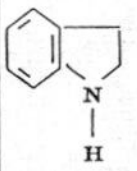	119.17			228–30	1.063_{4}^{20}	1.5923^{20}	δ						**B20**[2], 170
—	**2,3-Indolinedione**	*see* **Isatin**													
—	**Indolol**	*see* **Indole, hydroxy-**													
—	**Indonaphthene**	*see* **Indene**													
i75	**Indone, 2,3-dibromo-**	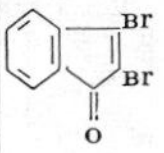	287.95	og-ye nd (al, aa)	123				i	s	v			chl v	**E12A**, 254
i76	**Indophenin**	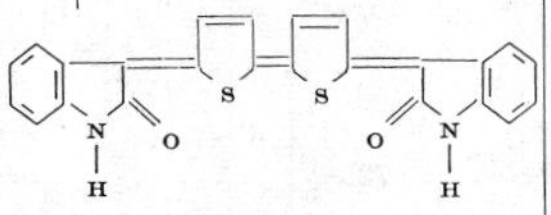	426.52	bl nd or pw	d				i	δ	δ		δ	chl δ	**B21**[2], 330
i77	**Indophenol**	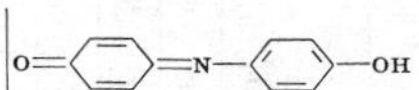	199.21	red br pl (peth)	160				s^h	s	s			chl s aa s	
i78	**Indoxazene**	4,5-Benzoisoazol. 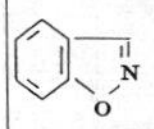	119.12			$82–3^{14}$	1.1727_{4}^{21}	1.5570^{20}							**B27**[2], 15
—	**Indoxylic acid**	*see* **2-Indolecarboxylic acid, 3-hydroxy-**													
i82	**Inosine**	Hypoxanthosine. Hypoxanthine riboside. 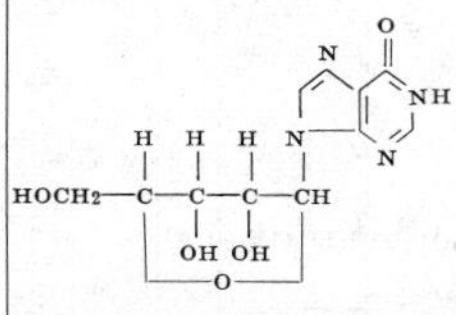	268.23	pl (w+2), nd (80% al) $[\alpha]_D^{20}-73$ (dil NaOH)	90 (+2w) 218d (anh)				δ s^h	s				dil alk s	**B31**, 25
i83	**Inositol**(*d*)	*d*-Inosite. 1,2,3,4,5,6-Cyclohexanehexol*. 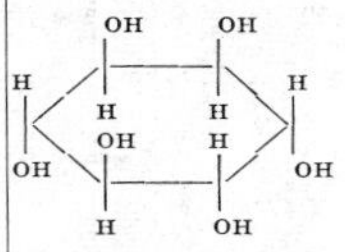	180.16	pr (w+2), (al) $[\alpha]_D+65.0$ (w, 12%)	247–8 cor				v	δ	i				**B6**[2], 1157
i84	—(*dl*)	*i*-Inosite. Mesoinosite. Phaseomannitol. $C_6H_{12}O_6$. *See* i83	180.16	mcl pr (w), cr (gl aa)	228	319 vac	1.752^{15}		v	δ	i			aa v	**B6**[2], 1158
i85	—(*l*)	*l*-Inosite. $C_6H_{12}O_6$. *See* i83	180.16	nd (w+2) $[\alpha]_D^{28}-64.1$ (w, 4%)	236	250 vac	1.598^{20}		v	δ	i				**B6**[2], 1157
i86	**Inulin**	Plant starch $[(C_6H_{10}O_5)_6.H_2O]$	5200	amor or pw, $[\alpha]_D^{21}-38.3$	178d		1.35_{4}^{20}		δ s^h	i				HCO_2H s	**E1**, 1318
—	**Iodisan**	*see* **2-Propanol, 1,3-bis(dimethylamino)-,** dimethyl iodide													
—	**Iodival**	*see* **Urea, 1(2-iodo-3-methyl butanoyl)-***													
—	**Iodoalphionic acid**	*see* **Propanoic acid, 3(4-hydroxy-3,5-diiodophenyl)2-phenyl-**													
—	**Iodoform**	*see* **Methane, triiodo-***													

For explanations, symbols and abbreviations see beginning of table.

No.	Name	Synonyms and Formula	Mol. wt.	Crystalline form, color and specific rotation	m.p. °C	b.p. °C	Density	n_D	Solubility w	al	eth	ace	bz	other solvents	Ref.
	Iodogorgoic acid														
i87	**Iodogorgoic acid**(*d*)	3,5-Diiodotyrosine(*d*). 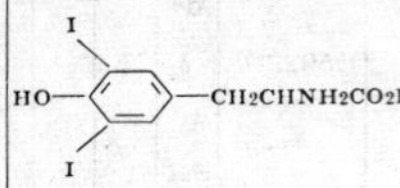 I, HO—, —$CH_2CHNH_2CO_2H$, I	432.99	yesh nd (w or 70% al), $[\alpha]^{23}_{546}+2.6$ (HCl, c=5)	204				δ						**B14**[2], 378
i88	—(*dl*)	$C_9H_9I_2NO_3$. *See* i87	432.99	nd (50% aa), (70% al)	198.4				s						**B14**[2], 384
i89	—(*l*)	$C_9H_9I_2NO_3$. *See* i87	432.99	nd (w or 70% al), $[\alpha]^{19}_D-2.9$ (4% HCl, c=5)	202										**B14**[2], 366
i90	**Iodonium fluoride, diphenyl-**	$(C_6H_5)_2IF$	300.12	dlq rh (ace)	85d				v	v	i	s[h] δ			
i91	**Iodonium iodide, diphenyl**	$(C_6H_5)_2II$	408.02	ye nd (al)	182					s[h]					
—	**Iodophen**	*see* **Phenolphthalein, 3,3″,5,5″-tetraiodo-**													
—	**Iodoprene**	*see* **1,3-Butadiene, 2-iodo-***													
—	**Ionol**	*see* **Benzene, 1,3-di-*tert*-butyl-2-hydroxy-5-methyl-**													
i92	**α-Ionol**	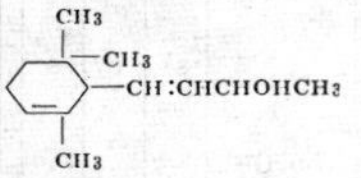 CH_3, CH_3, —CH:CHCHOHCH$_3$, CH_3	194.31			127^{15}	0.9474^{20}_4	1.4735^{20}							
i93	**β-Ionol**	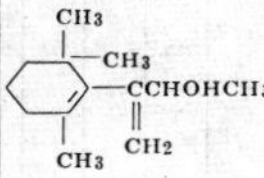 CH_3, CH_3, —CCHOHCH$_3$, ‖ CH_2, CH_3	194.31			131^{15}	0.9243^{20}_4	1.4969^{20}							
i94	**α-Ionone**(*d*)	4(2,6,6-Trimethyl-2-cyclohexenyl)-3-butene-2-one*. 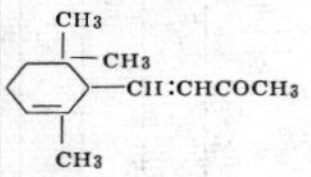 CH_3, CH_3, —CH:CHCOCH$_3$, CH_3	192.30	$[\alpha]^{23}_D+347$				1.5021^{22}							
i95	—(*dl*)	$C_{13}H_{20}O$. *See* i94	192.30			146^{28}	0.9301^{21}_4	1.5041^{20}	δ	∞	∞			chl i	**B7**[2], 140
i96	—(*l*)	$C_{13}H_{20}O$. *see* i94	192.30	$[\alpha]^{27}_D-406$				1.5000^{25}							
i97	—,semicarbazone	$C_{14}H_{23}N_3O$. *See* i94	249.36	(60% al)	143										**B7**, 169
i98	**β-Ionone**	$C_{13}H_{20}O$	192.30			140^{18}	0.9445^{20}_4	1.5210^{20}	δ	∞	∞				**B7**[2], 140
i99	—,semicarbazone	$C_{14}H_{13}N_3O$	239.28	nd (al)	148–9				i	s	s		s	chl s lig i	**B7**, 168
—	**Ipral**	*see* **Barbituric acid, 5-ethyl-5-isopropyl-**													
i100	**β-Irone**	4(2,2,3,6-Tetramethyl-1-cyclohexenyl)-3-butene-2-one. $C_{13}H_{20}O$	192.30	$[\alpha]_D+33.31$ (1=100 mm.)		$85–90^{0.1}$	0.9434^{21}_4	1.5017^{20}	δ	v	v		v	chl v lig v	**B7**[1], 110
i101	**Isatic acid**	2-Aminobenzoyl formic acid. 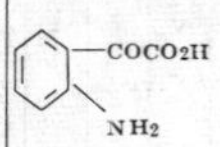 —$COCO_2H$, NH_2	165.15	pw	d				s d[h]						**B14**[2], 410
i102	**Isatin**	2,3-Indolinedione. Isatic acid lactam. 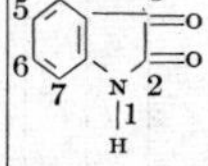 4, 3, 5, =O, =O, 6, 7, N 2, |1, H	147.13	red pr	203–5	sub			s[h] δ	v[h]	δ	s	s	alk s, d[h]	**B21**[2], 327
i103	—,chloride	2-Chloro-3-pseudoindozone. 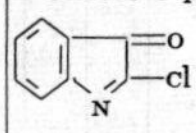 =O, —Cl, N	165.58	br nd	180d				i	v	v		δ v[h]	aa v lig δ	**B21**[2], 259

For explanations, symbols and abbreviations see beginning of table.

No.	Name	Synonyms and Formula	Mol. wt.	Crystalline form, color and specific rotation	m.p. °C	b.p. °C	Density	n_D	Solubility w	al	eth	ace	bz	other solvents	Ref.
	Isatin														
—	—,3-imide	*see* **Imesatin**													
i104	—,2-oxime	$C_8H_6N_2O_2$. *See* i102	162.15	ye-og nd (al or w)	198–200 (d ca. 200)				v^h	s^h	v	v	δ	alk s lig i chl δ	**B21**[1], 353
i105	—,3-oxime	$C_8H_6N_2O_2$. *See* i102	162.15	gold ye nd	220d				v^h δ	v	δ			alk, ac s lig i	**B21**[2], 334
i106	—,**1-acetyl**	$C_{10}H_7NO_3$. *See* i102	189.16	ye pr, nd	141				δ	s					**B21**[2], 338
i107	—,**1-methyl-**	$C_9H_7NO_2$. *See* i102	161.16	red-ye nd (w)	134				s^h	s		s	s		**B21**[2], 336 567
i108	—,**5-methyl-**	$C_9H_7NO_2$. *See* i102	161.16	red pl (w), nd (w or al)	187				δ	s	δ			HCl, alk s	**B21**[2], 375
i109	—,**5-nitro-**	$C_8H_4N_2O_4$. *See* i102	192.14	ye nd (al)	254–5d				δ	v				KOH s	**B21**[2], 346
i110	**Isatoic acid,** anhydride	*N*-Carboxyanthranilic anhydride.	163.14	pr (al or gl aa), (al or diox)	243d				δ	$δ^h$	i	δ	i	chl i	**B27**[2], 299
—	**α-Isatropic acid**	*see* **1,4-Tetralindicarboxylic acid, 1-phenyl-**													
—	**Isoanthraflavin**	*see* **9,10-Anthraquinone, 2,7-dihydroxy-***													
i111	**Isoapiol**	CH_2—O, OCH_3, CH_3O, CH:CHCH$_3$	222.23	mcl pr, lf or pl (al)	56	303–4, 189^{33}			i	v^h	v	v	v	con sulf s	**B19**[2], 97
—	**Isobarbituric acid**	*see* **Uracil, 5-hydroxy-**													
i112	**Isobergaptene**	$C_{12}H_8O_4$	216.18		222–3					s				diox, MeOH s	
i113	**Isoborneol**(*d*)	OH	154.24	(peth), $[\alpha]_D^{20} -34.6$ (al, c=5)	214				δ	s	s		δ	lig δ	**B6**[2], 89
i114	—(*dl*)	α,β-Camphol. $C_{10}H_{18}O$. *See* i113	154.24	ta (peth)	212 (sealed tube)	sub			i	v	v		δ	chl v	**B6,** 87
i115	—(*l*)	β-Camphol. $C_{10}H_{18}O$. *See* i113	154.24	(peth), $[\alpha]_D +33$ (al)	214				i	v	v		δ	chl v	**B6,** 87
i116	—,acetate(*d*)	$C_{12}H_{20}O_2$. *See* i113	196.28	$[\alpha]_4^{20} -50.2$ (al)			0.9905_4^{20}	1.4633^{20}							**B6**[2], 90
i117	—,—(*dl*)	$C_{12}H_{20}O_2$. *See* i113	196.28			107^{13}	0.9841_4^{20}	1.4619^{23}							**B6**[2], 91
i118	—,—(*l*)	$C_{12}H_{20}O_2$. *See* i113	196.28		<−50	225, $123–7^{35}$	1.002^0								**B6,** 89
i119	—,formate(*d*)	$C_{11}H_{18}O_2$. *See* i113	182.25	$[\alpha]_D^{20} +29.5$ (al, c=5)		94^{15}	1.0136_4^{20}	1.4678_{6708}^{22}		s					**E12A,** 681
—	**Isobornylamine**	*see* Neobornylamine													
—	**Isobutylphosphonic acid**	*see* **1-Propanephosphonic acid, 2-methyl-***													
—	**Isobutyranilide**	*see* **Propanoic acid, 2-methyl-,** amide, *N*-phenyl-													
—	**Isobutyric acid**	*see* **Propanoic acid, 2-methyl-***													
—	**Isobutyrophenone**	*see* **1-Propanone, 2-methyl-1-phenyl-***													
i120	**Isocalycanthine**	$C_{11}H_{14}N_2 \cdot \frac{1}{2}H_2O$	183.26	rh	235					s					
i121	**Isocamphane**(*d*)	Dihydrocamphene. 2,2,3-Trimethylnorcamphane.	138.25	(MeOH), $[\alpha]_D^{22} +8.68$ (bz, p=20)	62–3	$166–6.5^{750}$ cor									**B5**[2], 67
i122	—(*dl*)	$C_{10}H_{18}$. *See* i121	138.25	(al)	65–7 cor	165.5–5.7^{730}				v		v	v	MeOH δ AcOEt v	**B5**[1], 52
i123	—(*l*)	$C_{10}H_{18}$. *See* i121	138.25	$[\alpha]_D^{20} -10.8$	60–7	$164–5^{757}$	0.8276_4^{67}	1.4419^{67}		δ			δ	lig δ	**B5**[2], 67

For explanations, symbols and abbreviations see beginning of table.

No.	Name	Synonyms and Formula	Mol. wt.	Crystalline form, color and specific rotation	m.p. °C	b.p. °C	Density	n_D	Solubility w	al	eth	ace	bz	other solvents	Ref.
	Isocamphoric acid														
i124	**Isocamphoric acid**	*α-trans*-1,2,2-Trimethyl-1,3-cyclopentanedicarboxylic acid*. 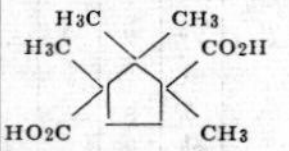	200.24	lf (w), $[\alpha]_D^{20}+48.6$	171–2	....	....	....	δ	v	...	...	...	aa v	**B9**, 762
i125	—(*dl*)	$C_{10}H_{16}O_4$. *See* i124	200.24	pr (al or gl aa) (w)	197	....	1.249	....	δ^h	δ	v	...	i^h	lig i	**B9**^{1}, 334
i126	—(*l*)	$C_{10}H_{16}O_4$. *See* i124	200.24	tetr $[\alpha]_D^{17}-48.4$ (MeOH, p=9.9)	173	....	....	....	δ	s	...	...	...	...	**B9**^{1}, 333
—	**Isocaproamide**	*see* **Pentanoic acid, 4-methyl-**, amide													
—	**Isocaproic acid**	*see* **Pentanoic acid, 4-methyl-***													
—	**—,α-amino-**	*see* **Leucine**													
—	**Isocarbostyril**	*see* **Isoquinoline, 1-hydroxy-**													
i127	**Isocarotene**	Dehydro-β-carotene. $C_{40}H_{54}$	534.87	vt pr (bz-MeOH), vt nd+lf (bz)	192–3 cor	....	....	....	...	...	...	...	...	...	**B30**, 88
i128	**Isocarvomenthol**(*d*)	$C_{10}H_{20}O$. 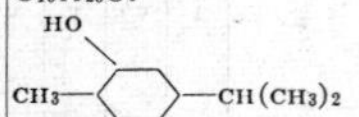	156.26	....	....	110^{20}	0.904_4^{20}	1.4669^{18}	...	...	...	...	...	...	...
i129	—(*l*)	$C_{10}H_{20}O$. *See* i128	156.26	$[\alpha]_D^{16}-17.7$	....	106^{17}	0.9109_4^{20}	1.4662^{20}	...	...	...	...	...	...	**B6**^{2}, 39
—	**Isocholesterol**	*see* **Lanosterol**													
—	**Isocinchomeronic acid**	*see* **2,5-Pyridinedicarboxylic acid**													
—	**Isocinnamic acid**	*see* **Cinnamic acid**(*cis*)													
—	**Isocitric acid**	*see* **1,2,3-Propanetricarboxylic acid, 1-hydroxy-***													
i130	**Isocodeine**	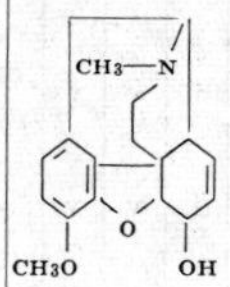	299.36	pl (bz), pr (AcOEt) $[\alpha]_D^{15}-152$ (chl, c=2)	171–2	d	1.87^{4}	1.675	...	...	...	...	...	...	**B27**^{2}, 175
i131	**Isocorybulbine**	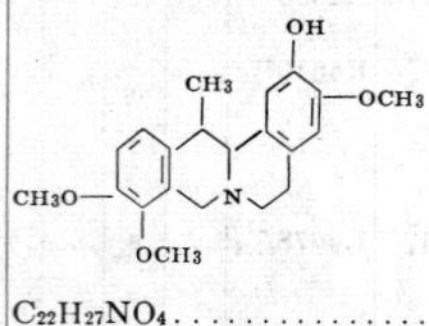	355.44	lf (al) $[\alpha]_D^{15}+299.8$ (chl, c=1)	187.5–8.5	....	1.045_4^{20}	....	i	s	...	...	...	chl s	**B21**^{2}, 200
i132	**Isocorydaline**	$C_{22}H_{27}NO_4$	369.47	....	136	....	....	....	...	...	...	...	...	...	...
i133	**Isocorydine**	Corytuberine methyl ether. Luteanine. 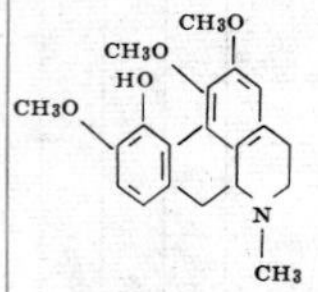	341.39	pl, $[\alpha]_D^{20}+195.3$ (chl)	185	....	....	....	...	...	δ	...	...	chl s	**B21**^{2}, 192
—	**Isocotin**	*see* **Benzophenone, 2,4-dihydroxy-6-methoxy**													
i134	**Isocoumarin**	2,1-Benzopyrone. *o*(β-Hydroxyvinyl)-benzoic acid lactone.	146.14	pl (bz)	46–7	285–6^{719}	....	....	i	v	v	...	v	CS_2 v	**B17**, 333

For explanations, symbols and abbreviations see beginning of table.

No.	Name	Synonyms and Formula	Mol. wt.	Crystalline form, color and specific rotation	m.p. °C	b.p. °C	Density	n_D	Solubility w	al	eth	ace	bz	other solvents	Ref.
	Isocrotonic acid														
—	**Isocrotonic acid**	*see* **2-Butenoic acid**(*cis*)*													
i135	**Isocyanic acid,** 4-bromo-phenyl ester	*p*-Bromophenyl isocyanate. C_7H_4BrNO. *See* i149	198.03		42	226 158^{14}			d^h	d^h	v				**B12**[1], 321
i136	—,*tert*-butyl ester	*tert*-Butyl isocyanate. $(CH_3)_3CNCO$	99.14		85.5 cor		0.8670^9							HCl d	**B4**, 175
i137	—,2-chlorophenyl ester	*o*-Chlorophenyl isocyanate. C_7H_4ClNO. *See* i149	153.57			$114–5^{43}$									**B12**, 601
i138	—,3-chlorophenyl ester	*m*-Chlorophenyl isocyanate. C_7H_4ClNO. *See* i149	153.57			$113–4^{43}$									**B12**, 606
i139	—,4-chlorophenyl ester	*p*-Chlorophenyl isocyanate. C_7H_4ClNO. *See* i149	153.57		30–1	$115–7^{45}$									**B12**, 616
i140	—,ethyl ester	Ethyl isocyanate. C_2H_5NCO	71.08			60	0.8981		i	∞	∞				**B4**[2], 613
i141	—,hendecyl ester	Undecyl isocyanate. $CH_3(CH_2)_{10}NCO$	197.32			103^3			d^h					lig v	**B4**[2], 658
i142	—,isobutyl ester	Isobutyl isocyanate. $(CH_3)_2CHCH_2NCO$	99.13			101.5									**B4**[2], 641
i143	—,methyl ester	Methylcarbylamine. Methyl isocyanate. CH_3NCO	57.05			43–45	0.967_4^{16}	1.3419^{18}	s						**B4**, 77
i144	—,1-naphthyl ester	α-Naphthylcarbylamine. α-Naphthylisocyanate. $C_{10}H_7^{\alpha}NCO$	169.18			269–70	1.774_4^{20}								**E12B**, 478
i145	—,2-naphthyl ester	β-Naphthylcarbylamine. β-Naphthylisocyanate. $C_{10}H_7^{\beta}NCO$	169.18	pl	55–6						v		v		**B12**, 1297
i146	—,2-nitrophenyl	*o*-Nitrophenylisocyanate. $C_7H_4N_2O_3$. *See* i149	164.12	wh nd	41				d	d	s		s	chl s	**B12**[2], 373
i147	—,3-nitrophenyl ester	*m*-Nitrophenylisocyanate. $C_7H_4N_2O_3$. *See* i149	164.12	wh lf (lig)	51				d	d	s		s	chl s lig s	**B12**[2], 382
i148	—,4-nitrophenyl ester	*p*-Nitrophenylisocyanate. $C_7H_4N_2O_3$. *See* i149	164.12	pa ye nd	56–7	$160–2^{18}$			d	d	v		v		**B12**[2], 394
i149	—,phenyl ester	Phenyl isocyanate. Phenylcarbylamine. (ring positions 5 6 4 3 2)—NCO	119.12			$162–3^{751}$ cor	1.0943_4^{20}	1.5368^{20}	d	d	v				**B12**[2], 244
i150	—,2-tolyl ester	*o*-Tolylisocyanate. CH_3 / —NCO	133.15			184–7			i d^h	d^h	s				**B12**, 812
i151	—,3-tolyl ester	*m*-Tolylisocyanate. CH_3 / —NCO	133.15			195–8			d	d	s		s		**B12**, 864
i152	—,4-tolyl ester	*p*-Tolylisocyanate. CH_3— —NCO	133.15			187^{751}			d	d	s		s		**B12**[1], 427
i153	**Isocyanuric acid,** trimethyl ester	Tricarbonimide trimethyl ester (structure: CH_3, O, N, O=, N—CH_3, N, CH_3, O)	171.16	mcl pr (w or al)	176–7	274			i δ^h	s					**B26**[2], 134
—	**Isocyclene**	*see* **Cyclocamphane**													
i154	**Isoderritol**	$C_{21}H_{22}O_6$	270.41	ye lf	148										
—	**Isodurene**	*see* **Benzene, 1,2,3,5-tetramethyl-***													
—	**Isodurenol**	*see* **Benzene, 2-hydroxy-1,3,4,5-tetramethyl-***													
—	**Isoduridine**	*see* **Benzene, 2-amino-1,3,4,5-tetramethyl-***													
—	**α-Isodurylic acid**	*see* **Benzoic acid, 3,4,5-trimethyl-**													
—	**β-Isodurylic acid**	*see* **Benzoic acid, 2,4,6-trimethyl-**													

For explanations, symbols and abbreviations see beginning of table.

γ-Isodurylic acid

No.	Name	Synonyms and Formula	Mol. wt.	Crystalline form, color and specific rotation	m.p. °C	b.p. °C	Density	n_D	Solubility w	al	eth	ace	bz	other solvents	Ref.
—	**γ-Isodurylic acid**	*see* **Benzoic acid, 2,3,5-trimethyl-**													
—	**Isoergosterone**	*see* **$\Delta^{4,6,22}$-Ergostatrienone**													
i155	**8-Isoestradiol**	$\Delta^{1,3,5}$-8-Epiestratrien-3,17(β)-diol. 8-Epiestradiol. 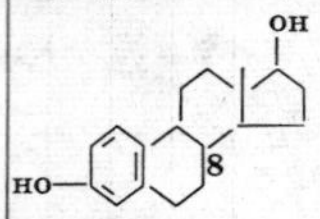	272.37	(dil MeOH-chl), $[\alpha]^{20}_D$+18 (diox)	181					s				diox s	**E14s**, 1982
i156	**8-Isoestrone**	8-Epiestrone. 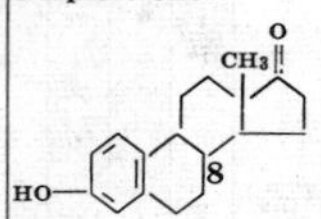	270.36	(MeOH or MeOH-eth), $[\alpha]^{20}_D$+94 (diox)	247				i		s			diox s	**E14s**, 2559
i157	**Iso-β-eucaine(*dl*)**	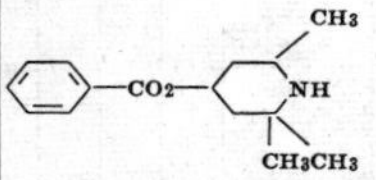	247.31		<−5	188^{19}	1.0467^{22}_{22}		i	s					**B21²**, 15
i158	—,hydrochloride(*d*)	$C_{15}H_{21}NO_2.HCl$. *See* i157	283.81	nd (w), $[\alpha]_{5461}$+14.9 (w, c=1)	271–3										**B21²**, 16
i159	—,hydrochloride(*dl*)	$C_{15}H_{21}NO_2.HCl$. *See* i157	283.81	unst nd or st ta (w), pl (aq al or al-eth)	269–71										**B21²**, 15
i160	—,hydrochloride(*l*)	$C_{15}H_{21}NO_2.HCl$. *See* i157	283.81	nd $[\alpha]_{5461}$−19.3 (w, c=1)	271–73										**B21²**, 15
—	**Isoeugenol**	*see* **Benzene, 1-hydroxy-2-methoxy-4-propenyl-***													
—	**Isoeuxanthone**	*see* **Xanthone 1,6-dihydroxy-**													
—	**Isoeuxanthonic acid**	*see* **Benzophenone, 2,2′,4,6′-tetrahydroxy-**													
i161	**Isoflavone, 4′,7-dihydroxy-**	Daidzein. 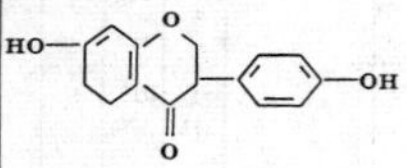	254.23	pa ye nd (50% al)	323d	sub				s	s				**B18²**, 100
i162	—,4′,5,7-tri-hydroxy-	Genistein. Prunetol. 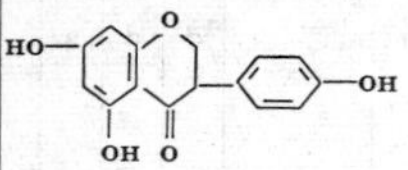	270.23	nd (eth) pl (60% al)	297–8δd				δ^h	δ				os s dil alk s aa δ	**B18²**, 176
i163	**Isofurfurine**	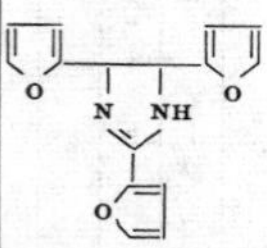	268.28	nd (w)	143										**B27**, 764
i164	**Isogeraniolene**	CH_2:C(CH_3)CH:CHCH$_2$CH(CH_3)$_2$	124.22			$143–4^{755}$	0.7561^{20}_4	1.4520^{20}	i						**B1²**, 1015
—	**Isohemipinic acid**	*see* **1,3-Benzenedicarboxylic acid, 4,5-dimethoxy-**													
—	**Isohydroanisoin**	*see* **1,2-Ethanediol, 1,2-bis(4-methoxyphenyl)-***													
i165	**Isolysergic acid**	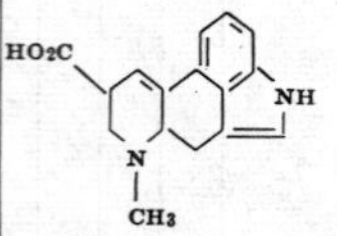	268.32	(w+2)	208d				δ	δ				Py s	
—	**α-Isomalic acid**	*see* **Malonic acid, hydroxy (methyl)-**													

For explanations, symbols and abbreviations see beginning of table.

No.	Name	Synonyms and Formula	Mol. wt.	Crystalline form, color and specific rotation	m.p. °C	b.p. °C	Density	n_D	Solubility w	al	eth	ace	bz	other solvents	Ref.
	Isomanide														
i166	**Isomanide**	1:4, 3:6-Dianhydromannitol. 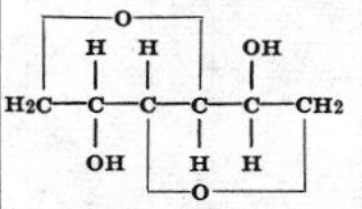	146.15	mcl $[\alpha]^{26.2}_D$ +62.2 (chl)	87–8	274d	...	...	v	δ	i	...	i	chl δ	**B1**³, 2402
—	**Isoindolinone**	*see* **Phthalic acid,** imidine													
i167	**Isomenthol**(*d*)	*p*-Menthanol-3. 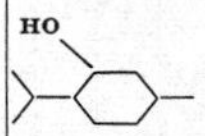	156.26	$[\alpha]_D$+25.9	82.5	218	...	...	...	...	...	...	...	...	**B6,** 41
i168	—(*dl*)	$C_{10}H_{20}O$. *See* i167	156.26	nd	53–4	218.5	0.8855^{55}_4	...	...	...	...	...	...	...	**B6**², 51
i169	—(*l*)	$C_{10}H_{20}O$. *See* i167	156.26	$[\alpha]^{15}_D$−24.1	82.5	...	...	...	...	...	...	...	...	...	**B6**², 51
—	**Isomethadone**	*see* **3-Hexanone, 6-dimethylamino-4,4-diphenyl-5-methyl-***													
i170	α-**Isomorphine**	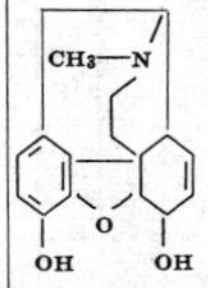	285.38	nd (MeOH-AcOEt) $[\alpha]^{15}_D$−167 (MeOH c=3)	248	...	...	...	s^h	...	...	...	...	MeOH v	**B27**², 174
i171	**Isonicoteine**	*see* **2,3′-Bipyridyl**													
	Isonicotinaldehyde	*see* **4-Pyridinecarboxaldehyde**													
	Isonicotine	4(4-Pyrydyl)piperidine. 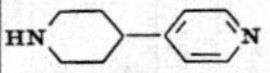	162.24	hyg nd	78	260d	...	...	δ	v	s	...	s	lig s	**B23,** 119
—	**Isonicotinic acid**	*see* **4-Pyridinecarboxylic acid**													
—	**Isonipecotic acid**	*see* **4-Piperidinecarboxylic acid**													
i172	**Isopapaverine, *N*-benzyl-**	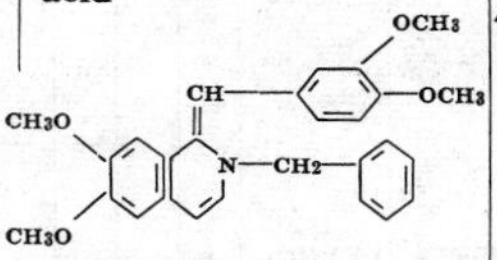	429.52	ye lf (al)	139–40	...	...	...	...	s^h	s	...	...	...	**B21,** 229
i173	—,*N*-**ethyl-**	$C_{22}H_{25}NO_4$. *See* i172	367.45	pr (al)	*ca.* 101	...	...	...	...	...	...	...	...	...	**B21,** 229
i174	—,*N*-**methyl-**	$C_{21}H_{23}NO_4$. *See* i172	353.42	ye hyg cr	129–31	...	...	...	s	δ	...	...	...	os s dil al v	**B21,** 229
i175	**Isopelletierine**(*dl*)	Isopimicine. 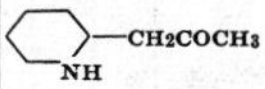	141.21	...	...	86^{10}	0.9624^{20}_4	1.4683^{20}	...	s	...	...	...	chl s dil ac s	**B21**², 219
i176	—,*N*-**methyl-**	$C_9H_{17}NO$. *See* i175	155.24	...	...	$96–8^{13}$	0.948^{20}_4	1.4674^{20}	s	...	...	...	...	dil ac s lig s	...
—	**Isopentane**	*see* **Butane, 2-methyl***													
—	**Isopeonol**	*see* **Acetophenone, 4-hydroxy-2-methoxy-**													
i177	**Isophenolphthalein**	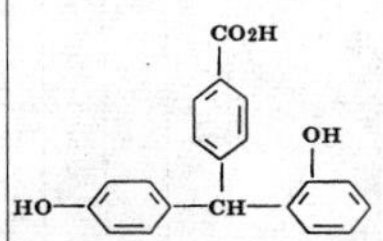	320.35	(dil aa)	189–90	...	...	...	...	v^h	...	v^h	...	...	**B10**², 322
i178	**3-Isophenothiazin-3-one**	Azthione. Thiazone. 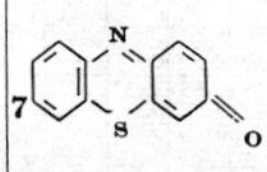	213.26	red	164	...	...	...	s^h	v	...	...	v^h	chl s	**B27**¹, 251
i179	—,**7-hydroxy-**	Thionol. $C_{12}H_7NO_2S$. *See* i178.	229.26	red br pw or nd	>360	...	...	...	i	s	...	...	...	chl i aa s^h	**B27**², 109

For explanations, symbols and abbreviations see beginning of table.

No.	Name	Synonyms and Formula	Mol. wt.	Crystalline form, color and specific rotation	m.p. °C	b.p. °C	Density	n_D	Solubility w	al	eth	ace	bz	other solvents	Ref.
	Isophorone														
i180	**Isophorone**	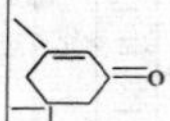	138.20			215.2	0.9229^{20}		δ						
—	**Isophthaldehyde**	*see* **1,3-Benzenedicarboxaldehyde**													
—	**Isophthaldehydic acid**	*see* **Benzoic acid, 3-formyl-**													
—	**Isophthalic acid**	*see* **1,3-Benzenedicarboxylic acid***													
i181	**Isopilocarpine**(*o*)	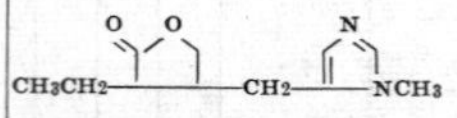	208.26	pr $[\alpha]+50$ (w)		261^{10}			v	v	i		δ	chl v lig i	
i182	**Isopimpinellin**	$C_{13}H_{10}O_5$	246.22	ye nd (MeOH)	149–51									MeOH s	
—	**Isopinene**	*see* **α-Fenchene**													
i183	**Isopomiferin**	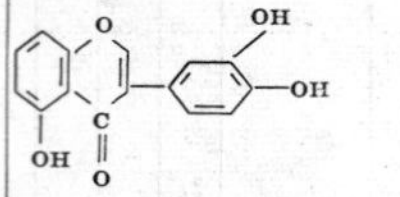	270.25	nd	265										
—	**Isoprene**	*see* **1,3-Butadiene, 2-methyl-***													
—	**Isopropyl alcohol**	*see* **2-Propanol***													
—	**Isopropyl bromide**	*see* **Propane, 2-bromo-***													
—	**Isopropyl chloride**	*see* **Propane, 2-chloro-***													
i184	**Isopulegol**(*d*)	$\Delta^{8(9)}$-Menthenol-3. 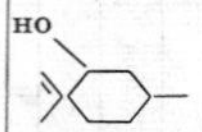	154.25			212 $93–4^{14}$	0.911^{20}_{4}	1.4723^{20}							**B6**[2], 70
i185	—(*l*)	$C_{10}H_{18}O$. *See* i184	154.25	$[\alpha]^{20}_{546}-25.90$		212 94^{14}	0.9110^{20}_{4}	1.4723^{20}							
i186	**α-Isoquinine**	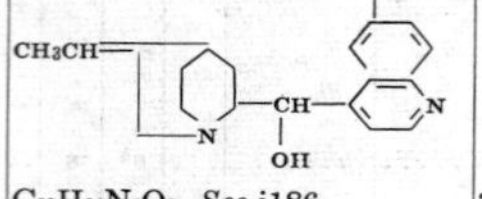	324.21	(bz-peth) $[\alpha]^{18}_{D}-245$ (al, c=1), −248 (al, c=0.5)	196.5				i	v	v				**B23**[2], 414,423
i187	**β-Isoquinine**	$C_{20}H_{24}N_2O_2$. *See* i186	324.43	pr (dil al), or amor, $[\alpha]^{17}_{D}-187$ (97% al, c=1)	190–1				i	v	δ		v	chl v lig s	**B23**[2], 413
i188	**Isoquinoline**	2-Benzazine*. Benzo(c)pyridine. Leucoline. 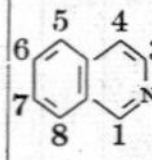	129.16	hyg	24.8	243.25	1.0986^{20}	1.6148^{20}	i						**B20**[2], 236
i189	—,**hydrochloride**	Isoquinolinium chloride. $C_9H_7N.HCl$. *See* i188	165.63	pr or pl (al)	209–9.5				s	s^h					**B20**[2], 237
i190	—,**hydrogen sulfate**	$C_9H_7N.H_2SO_4$. *See* i188	227.15	pr or pl (al)	209–9.5				v	s^h					**B20**[2], 236
i191	—,**1-amino-**	$C_9H_8N_2$. *See* i188	144.18	pl (w)	123				δ	v	δ				**B22**[1], 640
i192	—,**5-amino-**	$C_9H_8N_2$. *See* i188	144.18	nd (peth)	128	sub								lig s^h	**B22**[2], 359
i193	—,**4-bromo-**	C_9H_6BrN. *See* i188	208.07	(peth)	40	280–5					v^h				**B20**[2], 238
i194	—,**6,7-dimethoxy-1,2-dimethyl-1,2,3,4-tetrahydro-**	Carnegine. $C_{13}H_{19}NO_2$. *See* i188	221.29	pa br syr	170^1				s	s	s			chl s con sulf s, HCl s	**B21**[2], 110
i195	—,**1-hydroxy-**	Isocarbostyril. 1-Isoquinolinol. C_9H_7NO. *See* i188	145.19	mcl (bz), nd (bz, al, w)	209–10	240 sub			δ	v^h	δ s^h		δ s^h	lig i	**B21**[2]. 58
i196	—,**1-methyl-**	$C_{10}H_9N$. *See* i188	143.19		10.1–10.4	255.3 $124–6^{10}$	1.0777^{20}_{4}	1.6153^{20}							**B20**[2], 247
i197	—,**3-methyl-**	$C_{10}H_9N$. *See* i188	143.19	(eth)	68	252–3									**B20**[2], 247
i198	—,**4-methyl-**	$C_{10}H_9N$. *See* i188	143.19		<−75	256									**B20**[1], 152
i199	—,**6-methyl-**	$C_{10}H_9N$. *See* i188	143.19		85–6	265.5								os s	**B20**[2], 247
i200	—,**7-methyl-**	$C_{10}H_9N$. *See* i188	143.19		76	245									**B20**[2], 248
i201	—,**8-methyl-**	$C_{10}H_9N$. *See* i188	143.19			258									**B20**, 404
i202	—,**5-nitro-**	$C_9H_6N_2O_2$. *See* i188	174.16	nd (w+1)	110	sub			δ v^h	v	v			CS_2, chl v aa v	**B20**[2], 238

For explanations, symbols and abbreviations see beginning of table.

No.	Name	Synonyms and Formula	Mol. wt.	Crystalline form, color and specific rotation	m.p. °C	b.p. °C	Density	n_D	Solubility w	al	eth	ace	bz	other solvents	Ref.
	Isoquinoline														
i203	**—,1,2,3,4-tetra-hydro-**	2-Azatetralin. $C_9H_{11}N$. *See* i188	133.20		<−15	234	1.0642^{24}_{4}		i	s			s^h	dil ac s xyl s^h	**B20²**, 276
i204	**1-Isoquinoline-carboxylic acid,** nitrile	1-Cyanoisoquinoline. 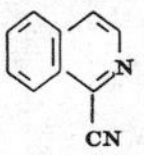	154.17	nd (peth), nd (MeOH)	93				δ	v	v		v	lig δ	**B22¹**, 511
i205	**5-Isoquinoline-carboxylic acid,** nitrile	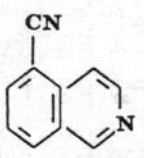	154.17	nd (w or dil al)	135	100–20 sub			δ	v				dil ac v	**B22²**, 58
i207	**Isoraunescine**	$C_{31}H_{36}N_2O_8$	564.62	wh nd $[\alpha]_D$ −70 (chl)	241–2.5				i	δ				chl s aa s	
i208	**Isoreserpiline**	$C_{23}H_{28}N_2O_5$	412.49	wh pr, $[\alpha]^{20}_{D}$ −82 (Py)	212–2				i						
—	**Isorhamnose**	*see* **Glucomethylose**													
i209	**Isorubijervine**	Δ^5-3β-18-Dihydroxysolanidene. $C_{27}H_{43}NO_2$ 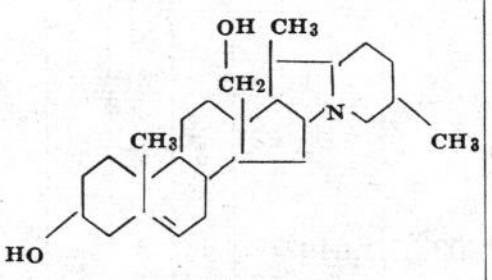	413.65	pr $[\alpha]_D$ +9.2 (al)	236–8					i s^h			s	ac s alk i chl s	
i210	**Isorubijervosine**	3-β-*D*-Glucosyl-Δ-5-solanidene-18-ol. $C_{33}H_{53}NO_7$	575.76	wh nd, $[\alpha]^{24}_{D}$ −20±2(Py)					i						
i211	**Isosaccharic acid**	Tetrahydro-3,4-dihydroxy-2,5-furandicarboxylic acid. 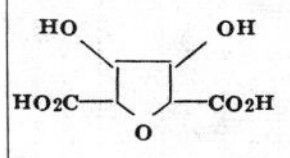	192.13	rh $[\alpha]^{20}_{D}$+46.1 (w, p=4.2)	185	d			v	v	δ				**B18²**, 309
i212	Isosafrole (trans)	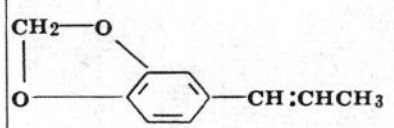	162.19			252	1.1224^{20}_{4}	1.5782	i	∞	∞		∞		**B19**, 27
i213	**Isoserine**(*l*)	$H_2NCH_2CH(OH)CO_2H$	105.10	$[\alpha]^{20}_{D}$ −32.58	199–201				s						**B4**, 503
—	**Isosuccinic acid**	*see* **Malonic acid, 2-methyl-**													
i214	**Isothebaine**(*d*)	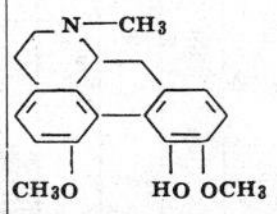	311.39	rh (al), $[\alpha]^{18}_{D}$ 285.1 (al, c=2)	203–4					∞	δ			MeOH s chl ∞	**B21²**, 1169
i215	—,sulfate(*d*)	$(C_{19}H_{21}NO_3)_2.H_2SO_4$	720.85	nd	120–1d				v						**B21¹**, 250
i216	**Isothiocyanic acid,** allyl ester	Allyl mustard oil. $CH_2{:}CHCH_2NCS$	99.16		−80	152	1.0126^{20}_{4}	1.5300^{20}	δ	∞	∞		v		**B4²**, 667
i217	—,benzyl ester	Benzyl mustard oil. $C_6H_5CH_2NCS$	149.22	ye		243	1.1246^{15}_{4}	1.6049^{15}	i	∞	s				**B12²**, 567
i218	—,butyl ester	Butyl mustard oil. *n*-Butyl thiocarbimide. $CH_3(CH_2)_2CH_2NCS$	115.20			167 $64\text{–}6^{12}$	0.946^{20}_{4}		i	v	v				**B4²**, 635
i219	—,*sec*-butyl ester(*d*)	*sec*-Butyl mustard oil. $CH_3CH_2CH(CH_3)NCS$	115.20	$[\alpha]^{20}_{D}$+61.88		159–63	0.943^{20}_{4}								**B4**, 161
i220	—,—(*dl*)	$CH_3CH_2CH(CH_3)NCS$	115.20			159.5	0.944^{12}		i	s	s				**B4**, 162
i221	—,—(*l*)	$CH_3CH_2CH(CH_3)NCS$	115.20	$[\alpha]^{20}_{D}$ −61.80		159	0.942^{20}_{4}								**B4**, 161
i222	—,*tert*-butyl ester	*tert*-Butyl mustard oil. $(CH_3)_3CNCS$	115.20		10.5	140 64^{12}	0.9187^{10}_{4}		i	i	s				**B4**, 175

For explanations, symbols and abbreviations see beginning of table.

No.	Name	Synonyms and Formula	Mol. wt.	Crystalline form, color and specific rotation	m.p. °C	b.p. °C	Density	n_D	Solubility w	al	eth	ace	bz	other solvents	Ref.
	Isothiocyanic acid														
i223	—,4-biphenylyl ester	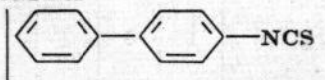 (biphenyl)—NCS	211.29	nd (eth)	58						v				**B12,** 1319
i224	—,4-bromophenyl ester	C_7H_4BrNS. *See* i236	214.09	(al)	56				i	s^h					**B6**², 301
i225	—,chlorophenyl ester	C_7H_4ClNS. *See* i236	169.63	nd (al)	45	249–50				s^h					**B12**², 330
i226	—,cyclohexyl ester	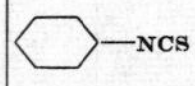 (cyclohexyl)—NCS	141.24			219^{746}									**B12**², 12
i227	—,4(dimethyl-amino)phenyl ester	$C_9H_{10}N_2S$. *See* i236	178.26	pr	67										**B13**³, 53
i228	—,ethyl ester	Ethyl mustard oil. C_2H_5NCS	87.14		−5.9	131–2 cor	0.995_4^{20}	1.5134^{20}	i	∞	∞				**B4**², 614
i229	—,isobutyl ester	Isobutyl mustard oil. $(CH_3)_2CHCH_2NCS$	115.20			160 cor	0.9638_4^{14}	1.5005^{14}	i	s	s				**B4**², 641
i230	—,isopropyl ester	Isopropyl mustard oil. $(CH_3)_2CHNCS$	101.17			137–7.5									**B4**, 155
i231	—,methyl ester	Methyl mustard oil. CH_3NCS	73.12		35	119	1.0691_4^{37}	1.5258	δ	∞	v				**B4**², 579
i232	—,3-methyl butyl ester	Isoamyl mustard oil. $(CH_3)_2CHCH_2CH_2NCS$	129.23			182–4	0.9419^{17}		δ	v	v				**B4**², 649
i233	—,1-naphthyl ester	α-Naphthyl mustard oil. $C_{10}H_7NCS$	185.25	nd (al)	58				i	v^h	v	v	v	chl v lig δ	**B12**², 698
i234	—,2-naphthyl ester	β-Naphthyl mustard oil. $C_{10}H_7NCS$	185.25	yesh nd (al)	62–3					v^h	v				**B12**², 724
i235	—,pentyl ester	*n*-Amyl mustard oil. $CH_3(CH_2)_4NCS$	129.23	bt ye		193.4 cor			δ	v	v				**B4**², 642
i236	—,phenyl ester	Phenyl mustard oil. 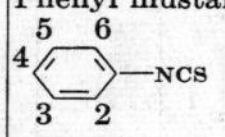 (phenyl, positions 2–6)—NCS	135.19		−21 cor	221 cor	1.1303_4^{20}	1.6492^{23}	i	s	s				**B12**², 247
i237	—,propyl ester	*n*-Propyl mustard oil. $CH_3CH_2CH_2NCS$	101.17			153 cor	0.9781_4^{16}	1.5085^{16}	δ	∞	∞				**B4**², 627
i238	—,4-propylphenyl ester	$C_{10}H_{11}NS$. *See* i236	177.27			263				v	v				**B12,** 1144
—	**Isothiourea,**	*see* Urea, **2-thio-**(*S*-substituted)													
—	**Isovaleraldehyde**	*see* **Butanol, 3-methyl-***													
—	**Isovaleric acid**	*see* **Butanoic acid, 3-methyl-***													
—	**Isovaleronitrile**	*see* **Butanoic acid, 3-methyl,** nitrile													
—	**Isovalerophenone.**	*see* **1-Butanone 3-methyl-1-phenyl-***													
—	**Isovaleryl chloride**	*see* **Butanoic acid, 3-methyl-**chloride													
i241	**Isovaline**(*d*)	2-Amino-2-methyl butanoic acid*. $CH_3CH_2C(CH_3)(NH_2)CO_2H$	117.15	nd (aq al+1w) $[\alpha]_D^{20}+10.7$ (w, c=4)		sub									**B4**², 851
i242	—(*dl*)	$CH_3CH_2C(CH_3)(NH_2)CO_2H$	117.15	rh nd (al-eth, +1w)	315	sub 300			s	δ	s				**B4**², 851
i243	—(*l*)	$CH_3CH_2C(CH_3)(NH_2)CO_2H$	117.15	$[\alpha]_D^{20}-7.8$ (w, c=2)	*ca.* 300	sub			v	v					**B4**², 851
—	**Isovanillic acid**	*see* **Benzoic acid, 3-hydroxy-4-methoxy-**													
—	**Isovanillin**	*see* **Benzaldehyde, 3-hydroxy-4-methoxy-***													
i244	**Isoxanthen-3-one, 2,6,7-trihydroxy-9-phenyl-**	2,6,7-Trihydroxy-9-phenylfluorone. 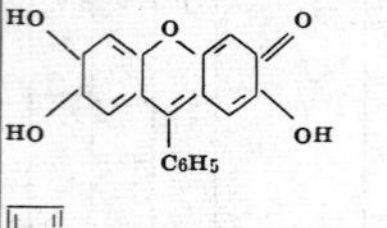 (structure: HO, HO, OH, O, C_6H_5)	320.21	og red	>300					δ				os δ alk s	**B18**¹, 404
i245	**Isoxazole**	(ring: N, O)	69.07			95–5.5	1.078_4^{20}		s						**B27**, 414
—	**Isoxylic acid**	*see* **Benzoic acid, 2,5-dimethyl-**													
—	**Itaconic acid**	*see* **Succinic acid, methylene-**													

For explanations, symbols and abbreviations see beginning of table.

PHYSICAL CONSTANTS OF ORGANIC COMPOUNDS (Continued)

No.	Name	Synonyms and Formula	Mol. wt.	Crystalline form, color and specific rotation	m.p. °C	b.p. °C	Density	n_D	Solubility: w	al	eth	ace	bz	other solvents	Ref.
	Jacareubin														
j1	**Jacareubin**	$C_{18}H_{14}O_6$	326.29	ye pr	256–7d	…	…	…	…	s	δ	s^h	δ	AcOEt s, MeOH s, Chl δ	**J.** 1953, 3932
j2	**Jaconecic acid**	$C_{10}H_{16}O_6$	232.24	nd (eth), $[\alpha]_D^{25}+28.1$	183–4	…	…	…	…	s	s^h	…	…	…	**JACS, 78,** 3518
j3	**Japaconitine-a**	Acetyl benzoyl aconitine. $C_{34}H_{47}NO_{11}$	645.76	rh $[\alpha]_D^{13}+20.7$ (chl)	202–3	…	…	…	i	s^h	s^h	v	…	chl s	…
j4	**—-al**	$C_{34}H_{47}NO_{11}$	645.76	rh (MeOH) $[\alpha]_D^{12}+26.4$ (chl)	208–9d	…	…	…	…	…	…	…	…	MeOH s	…
j5	**—-b**	$C_{34}H_{47}NO_{11}$	645.76	rh (MeOH) $[\alpha]_D^{11}+26.9$	208–9	…	…	…	…	s	…	…	…	…	…
—	**Jasminaldehyde**	*see* **1-Heptene-2-carboxaldehyde, 1-phenyl***													
j6	**Jasmone**	$C_{11}H_{16}O$	164.24	ye oil	…	$257–8^{755}$ $109–10^{5}$	0.9437_4^{22}	1.4979^{22}	δ	s	s	…	…	CCl_4 s lig s	…
j7	**Javanicin**	$C_{15}H_{14}O_6$	290.26	(red)	208d	…	…	…	…	…	…	…	…	alk s	**J. 1947,** 1021
—	**Jervasic acid**	*see* **Chelidonic acid**													
j8	**Jervine**	$C_{27}H_{43}NO_5$	461.65	nd (w+2) $[\alpha]_D-147$ (c=0.99, al)	240–1	…	…	…	i	s	δ	s	…	chl s	…
—	**Juglone**	*see* **1,4-Naphthoquinone, 5-hydroxy-***													
j9	**Julolidine**	N	173.26	…	40	280d $155–6^{17}$	1.003^{20}	1.568^{25}	…	…	…	…	…	…	**B20**[2], 214
j10	—,1,6-dioxo-	1,6-Diketojulolidine. $C_{12}H_{11}NO_2$	201.22	(EtOH)	145–6	$190–210^{0.3}$	…	…	…	s^h	…	…	…	MeOH s	**J.** 1951 1898
j11	**Junipal**	5(α-propynyl)-2-formyl thiophen. CH_3—C⋮C—(S)—CHO	150.19	nd (pet eth)	80	…	…	…	δ s^h	…	…	…	…	org. sol s	
j12	**Juniperol**	$C_{15}H_{24}O$	220.36	tcl (EtOH) $[\alpha]_D^{20}$ +17.8 (EtOH) +18.4 (Chl)	107	286–8d	1.0460_{20}^{20}	1.519	i	…	…	…	…	org sol s	**BG**[2], 516
j13	**Junipic acid**	5(α-propynyl)-2-thiophencarboxylic acid CH_3—C⋮C—(S)—CO_2H	166.19	ye nd (pet eth), (aq EtOH)	169–71	…	…	…	i	s	…	…	…	$NaHCO_3$ s lig s^h	…
j14	—,methyl ester	$C_9H_8O_2S$. *See* j13	180.22	(aq MeOH)	62	sub 50–5	…	…	…	…	…	s	…	…	…

For explanations, symbols and abbreviations see beginning of table.

No.	Name	Synonyms and Formula	Mol. wt.	Crystalline form, color and specific rotation	m.p. °C	b.p. °C	Density	n_D	Solubility w	al	eth	ace	bz	other solvents	Ref.
	Kaempferol														
—	**Kaempferol**	*see* **Flavone, 3,4′,5,7-tetrahydroxy-**													
—	**Kairoline**	*see* **Quinoline, 1-methyl-1,2,3,4-tetrahydro**													
—	**Kavahin**	*see* **Methysticin**													
k1	**Ketene**	Carbomethene. Ethonone*. $CH_2{:}CO$	42.04		−151	−56			d	d	δ	δ	...	NH_3 d	**B1**², 779
k2	**—,diethyl acetal**	1,1-Diethoxyethene*. $CH_2{:}C(OC_2H_5)_2$	116.16			124–6	0.8776_4^{25}	1.4101^{25}	d	d	...	...	...		
k3	**—,dimethyl-***	$(CH_3)_2C{:}CO$	70.09		−97.5	34			d	d	...	...	...		**B1**², 789
k3¹	**—,diphenyl-***	$(C_6H_5)_2C{:}CO$	193.23	ye-red		265–70d $119–21^{3.5}$	1.1107_4^{14}	1.615^{14}	...	...	s	...	s		**B7**², 412
k4	**—,methyl-**	$CH_3CH{:}CO$	56.06		−80										**B1**², 782
—	**Ketine**	*see* **Pyrazine, 2,5-dimethyl-**													
k5	**Ketone, benzyl 1-naphthyl**	α-Phenyl-1-acetonaphthone. $C_6H_5CH_2COC_{10}H_7^{\alpha}$	246.31	ta, lf (al)	66–7				i	δ v^h	s	...	...	chl s	**B7**², 461
k6	**—,benzyl 2-naphthyl**	β-Phenyl-2-acetonaphthone. $C_6H_5CH_2COC_{10}H_7^{\beta}$	246.31	nd (al)	99.5				...	s^h	s	...	s	chl s	**B7**², 461
k7	**—,benzyl phenyl**	Desoxybenzoin. α-Phenyl acetophenone. $C_6H_5CH_2COC_6H_5$	196.24	pl (al)	60	320	1.201_4^0		δ^h	s	s	...	...		**B7**², 368
k8	**—,—,α-chloro-**	Desyl chloride. $C_6H_5COCHClC_6H_5$	230.70	nd (al)	65	d			...	s^h	...	...	...	alk i	**B7**², 369
k9	**—,—,α-diazo**	Azibenzil. Diazodesoxy-benzoin. $N_2{:}C(C_6H_5)COC_6H_5$	222.25	red pl (eth)	79				...	δ	δ s^h	...	δ		**B7**², 683
k10	**—,cyclobutyl phenyl**	Benzoylcyclobutane. 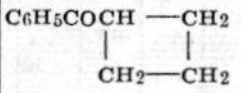 $C_6H_5COCH—CH_2$ / $CH_2—CH_2$	160.22			258	1.0457_{25}^{25}		...	...	...	...	...		**B7**, 374
k11	**—,1,1′-dinaphthyl**	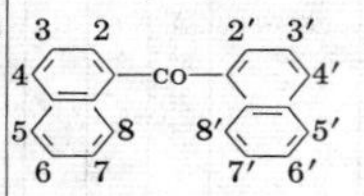 (structure: 1-naphthyl—CO—1′-naphthyl)	282.32	wh nd, yesh pr (aa)	100				...	v^h δ	δ	...	v	aa δ^h lig δ^h	**B7**², 503
k12	**—,1,2′-dinaphthyl**	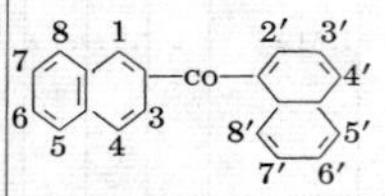 (structure: 2-naphthyl—CO—1′-naphthyl)	282.32	nd (al, bz-lig)	135	$235^{0.6}$			...	δ	...	...	v		**B7**², 504
k13	**—,2,2′-dinaphthyl**	$C_{10}H_7^{\beta}COC_{10}H_7^{\beta}$	282.32	nd (eth), lf (chl-eth)	125.5				...	δ	δ	...	...	chl s	**B7**², 504
k14	**—,1,1′-dinaph-thyl-, 2-methyl-**	$C_{22}H_{16}O$. *See* k11	296.35	gr-yesh nd (al)	171	$230–5^5$			i	s^h	...	...	...	con sulf s	**B7**², 506
k15	**—,1,2′-dinaph-thyl-, 2-methyl-**	$C_{22}H_{16}O$. *See* k12	296.35	gr-yesh (aa)	141–2				i	...	...	...	...	con sulf s aa s^h	**B7**², 506
—	**—,diphenyl**	*see* **Benzophenone**													
k16	**—di(2-thienyl)**	Thienone. 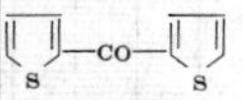 (structure: 2-thienyl—CO—2-thienyl)	194.26	nd	87–8	326			i	s^h	...	...	...		**B19**, 135
—	**—,ethyl methyl**	*see* **2-Butanone***													
—	**—,ethyl styryl**	*see* **1-Penten-3-one, 1-phenyl-***													
k18	**—,2-furyl phenyl**	2-Benzoylfuran. 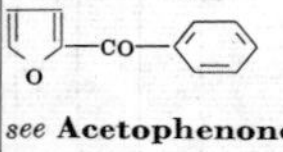 (structure: 2-furyl—CO—phenyl)	172.19	br	<−15	285	1.1839_{19}^{19}	1.6055^{20}	i	s	s	...	s		**B17**², 372
—	**—,methyl phenyl**	*see* **Acetophenone**													
k20	**—,1-naphthyl 2-tolyl**	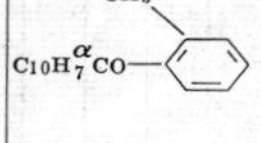 $C_{10}H_7^{\alpha}CO$—(2-CH_3-phenyl)	246.29	(al)	64	365			i	s^h	v	...	...	lig δ	**B7**², 462
k21	**—,1-naphthyl 3-tolyl**	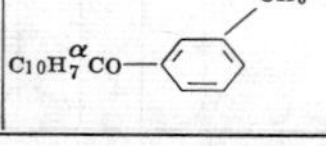 $C_{10}H_7^{\alpha}CO$—(3-CH_3-phenyl)	246.29	nd (al)	74–5				i	s	...	...	...	os s	**B7**¹, 284

For explanations, symbols and abbreviations see beginning of table.

No.	Name	Synonyms and Formula	Mol. wt.	Crystalline form, color and specific rotation	m.p. °C	b.p. °C	Density	n_D	Solubility w	al	eth	ace	bz	other solvents	Ref.
	Ketone														
k22	**—,phenyl 4-piperidyl, 1'-methyl-**	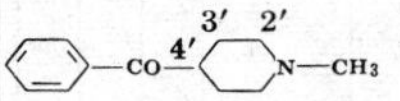 (phenyl)—CO—4'(piperidyl, 3' 2')N—CH_3	203.27	....	....	....	130–7[2]	1.5430^{23}	s^h	s	v	...	s	....	....
—	**—,phenyl pyridyl**	*see* **Pyridine, benzoyl**													
k23	**—,phenyl 2-thienyl**	2-Benzoylthiophene. 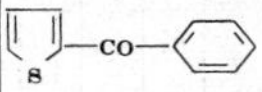 (thienyl)—CO—(phenyl)	188.25	nd (dil al)	56–7	300	1.1890^{54}_{4}	1.6181^{54}_{α}	i	v^h	v	...	...	....	**B17**[2], 372
k24	**Khellin**	$C_{14}H_{12}O_5$	260.24	(MeOH or w)	154–5	180–$200^{0.05}$	....	....	s^h i	...	δ	s	...	MeOH s, v^h	**B19**[2], 236
—	**Kojic acid**	*see* **4-Pyrone, 5-hydroxy-2-hydroxymethyl-**													
—	**Kynurenic acid**	*see* **2-Quinoline carboxylic acid, 4-hydroxy-**													
k25	**Kynurenine(*l*)**	3-Anthranylylalanine. 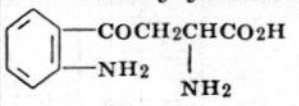 (benzene)—$COCH_2CHCO_2H$ (with NH_2 on CH), —NH_2	208.21	lf, $[\alpha]^{17}_{D}-28.5$ (w)	190d	....	....	....	...	...	...	...	...	....	**B14**[2], 415
—	**Kynurine**	*see* **Quinoline, 4-hydroxy-**													

For explanations, symbols and abbreviations see beginning of table.

No.	Name	Synonyms and Formula	Mol. wt.	Crystalline form, color and specific rotation	m.p. °C	b.p. °C	Density	n_D	Solubility w	al	eth	ace	bz	other solvents	Ref.
	Lactamide														
—	**Lactamide**	*see* **Propanoic acid 2-hydroxy-, amide**													
—	**Lactaric acid**	*see* **Octadecanoic acid, 6-oxo-***													
—	**Lactic acid**	*see* **Propanoic acid, 2-hydroxy-***													
—	**β-Lactic acid**	*see* **Propanoic acid, 3-hydroxy-***													
l1	**Lactide**(*d*)	O, O, O, O	144.13	hyg nd (eth)	95	150^{25}				s	s		s	chl s	**B19**, 154
l2	—(*dl*)	$C_6H_8O_4$. *See* l1	144.13	tcl (al)	124	255^{757}	0.862_4^{10}		δ	s	s	s	s	peth δ	**B19**[2], 176
l3	—(*l*)	$C_6H_8O_4$. *See* l1	144.13	(eth)$[\alpha]_D^{16}$ 281.6 (bz, c=0.82)	95	150^{25}									**B19**, 154
l4	**Lactobionic acid**	COOH, HCOH, HOCH, O, HC, HCOH, CH_2OH; CH, HCOH, HOCH, O, HOCH, HC, CH_2OH	358.29						v	δ	i			aa δ	**B31**, 415
—	**Lactobiose**	*see* **Lactose**													
—	**Lactoflavin**	*see* **Riboflavin**													
l5	**Lactose**(α)	Lactobiose. Milk sugar. HCOH, HCOH, O, HOCH, HC, HC, H_2COH; O, CH, HCOH, HOCH, HOCH, O, HC, H_2COH	342.31	pw $[\alpha]_D^{20}$ (mut) +92.6 (w, c=4.5) to +52.3 (final)	222–8				v						**B31**, 408
l6	—(β)	HOCH, HCOH, HOCH, HC, O, HC, H_2COH; O, CH, HCOH, HOCH, HOCH, O, HC, H_2COH	342.31	$[\alpha]_D^{20}$ (mut) +34.2 (5 min) to 56 (final) (w)	253		1.59^{20}		v						**B31**, 408
l7	—,monohydrate(α)	$C_{12}H_{22}O_{11}.H_2O$. *See* l5	360.32	rh $[\alpha]_D^{20}$+83.5 (w, 10 min)	201–2	d	1.525^{20}		v	i	i			MeOH i chl i	**B31**, 407
l8	**Lanosterol**	Isocholesterol. Zanosterol. $C_{27}H_{46}O$.	386.67	nd (eth)	137–8					s^h	v			aa s^h	
l9	—,benzoate	Isocholesterol benzoate. $C_{34}H_{50}O_2$.	490.78	pw or nd	191–5					s	s				
l10	**Lanthionine**(*D*)	*d*-β,β-Thiodialanine. $[HO_2CCH(NH_2)CH_2]_2S$	208.24	hex pl $[\alpha]_D^{21}$ −8.0 (2.4N NaOH)	293–5d darkens 245				δ					dil alk s dil ac s	
l11	—(*DL*)	$[HO_2CCH(NH_2)CH_2]_2S$	208.24	hex. pl.	282–95d chars 240				δ					dil alk s dil ac s	
l12	—(*L*)	$[HO_2CCH(NH_2)CH_2]_2S$	208.24	hex pl $[\alpha]_D^{22}$ +8.6 (2.4N NaOH)	293–5d darkens 245				δ					dil alk s dil ac s	
l13	—(*meso*)	$[HO_2CCH(NH_2)CH_2]_2S$	208.24	hex pl (dil NH_4OH)	304d softens 207				δ		i	i		dil alk s dil ac s	
l14	**Lanthopine**	$C_{23}H_{25}NO_4$	379.46	pw	200					δ	δ			chl s	

For explanations, symbols and abbreviations see beginning of table.

No.	Name	Synonyms and Formula	Mol. wt.	Crystalline form, color and specific rotation	m.p. °C	b.p. °C	Density	n_D	Solubility w	al	eth	ace	bz	other solvents	Ref.
	Lindane														
—	**Lindane**	*see* **Cyclohexane 1,2,3,4,5,6-hexachloro-**													
—	**Linoleic acid**	*see* **9,12-Octadecadienoic acid***													
—	**Linolenic acid**	*see* **9,12,15-Octadecatrienoic acid***													
—	**Linolenin**	*see* **Glycerol, 1-mono-(9,12-octadecadienoate)**													
149	**α-Lipoic acid**	6,8-Epidithioctanoic acid*. Thioctic acid. 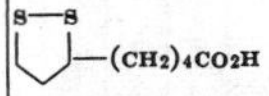	206.31	pa ye pl $[\alpha]^{25}_D$+96.7	47.5	...	...	...	δ	...	...	...	v	MeOH v AcOEt v	**Am 76,** 1270
150	**Lithocholic acid**	3(α)-Hydroxycholanic acid. $C_{24}H_{40}O_3$	376.58	hex lf (al), pr (aa) $[\alpha]^{20}_D$ +32.14 (al)	184–6	...	...	...	i	v^h	...	...	...	chl s aa s lig i	...
151	**Lithofellic acid**	$C_{20}H_{36}O_4$	340.51	...	206	d	...	...	i	s	...	...	...	...	...
152	**Lobelanidine**	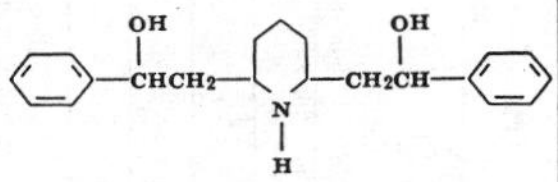	339.48	sc (al or eth)	150	distb vac	...	...	δ^h	s	δ	v	v	Py v chl v	**B21²**, 136
153	**Lobelanine**	$C_{22}H_{25}NO_2$	335.45	nd (peth or eth)	99	...	...	...	δ	v	s	v	v	MeOH, chl v	**B21²**, 394
154	**Lobeline**(*dl*)	$C_{22}H_{27}NO_2$	337.47	ye pl (eth), pr (al)	110	...	...	...	δ	v	v	...	s	chl s	**B21²**, 434
155	—(*l*)	$C_{22}H_{27}NO_2$	337.47	nd (al, eth or bz) $[\alpha]^{15}_D$−42.85 (al, c=1)	130–31	...	...	...	δ	s^h	δ	...	...	peth δ chl s	**B21²**, 434
156	**Longifolene**(*d*)	$C_{15}H_{24}$	204.34	$[\alpha]^{18}_D$+36	...	$254–6^{706}$ $150–1^{36}$	0.9284^{30}_{30}	1.495^{30}	...	...	...	...	...	...	**B5,** 226
157	**Lophine**	2,4,5-Triphenylimidazole. 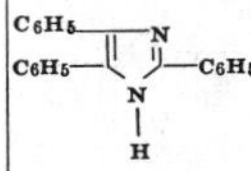	296.37	nd (al)	275	sub	...	...	i	δ	δ	...	...	...	**B23²**, 280
—	**Loretin**	*see* **Ferron**													
—	**Loretine**	*see* **5-Quinolonesulfonic acid, 8-hydroxy-7-iodo-**													
158	**Lucernic acid**	$C_{30}H_{46}O_7$	518.67	amor [α]+12	...	...	...	...	i	δ^h	i^h	...	...	...	...
159	—,triacetate	$C_{36}H_{52}O_{10}$	644.78	prismatic rods [α]+7.7	297–9	...	...	...	i^h	v^h	δ^h	...	...	...	...
—	**Luleolin**	*see* **Flavone, 3′,4′,5,7-tetrahydroxy-**													
—	**Luminal**	*see* **Barbituric acid, 5-ethyl-5-phenyl-**													
160	**Luminol**	5-Amino-2,3-dihydro-1,4-phthalazinedione. 3-Aminophthalhydrazide. 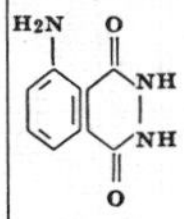	177.17	ye nd	332–3 (280)	...	...	...	i	δ	δ	...	...	alk v dil ac δ aa s^h	**B25²**, 389
161	**Lumisterol**	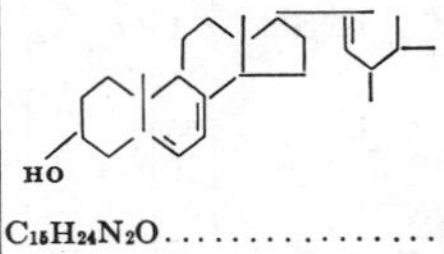	396.66	nd (ace-MeOH) $[\alpha]^{19}_D$+191.5	118	...	...	...	...	...	...	s	...	aa s	**E14,** 76
162	**Lumpanine**(*d*)	$C_{15}H_{24}N_2O$	248.37	nd $[\alpha]^{20}_D$+82.4 (w, c=3)	44	$185–6^{0.08}$	...	1.5444^{24}	s	v	v	...	...	chl v lig v	**B24²**, 53
163	—(*dl*)	$C_{15}H_{24}N_2O$	248.37	nd (peth)	98	$233–4^{18}$	...	...	v	v	v	...	...	peth v chl v	**B24²**, 55
164	**Lupeol**	Lupenol. 2-Hydroxy-$\Delta^{20,29}$-lupene. $C_{30}H_{50}O$.	426.70	nd (al or ace) $[\alpha]^{20}_D$+27.2 (chl)	215	...	0.9457^{218}_4	1.4910^{218}	i	v^h	v	...	v	chl v lig v	**E14s,** 1115

For explanations, symbols and abbreviations see beginning of table.

No.	Name	Synonyms and Formula	Mol. wt.	Crystalline form, color and specific rotation	m.p. °C	b.p. °C	Density	n_D	Solubility w	al	eth	ace	bz	other solvents	Ref.
	Lupetidine														
—	**Lupetidine**	see **Piperidine, dimethyl-**													
165	**Lupinine**(*l*)	$C_{10}H_{19}NO$	169.26	rh $[\alpha]_D^{17}-19$	68.5–9.2	269–70[754]			s	s	s			chl s	**B21**[2], 28
166	—,hydrochloride	$C_{10}H_{19}NO.HCl$	205.74	pr (aq al) $[\alpha]_D-14$ (w)	212–3				s	s					
167	—,des-N-methyl-	(ring structure: $-CH_2OH$, $-CH_2CH_2CH{:}CH_2$, N, CH_3)	183.29			145–6[15]			i	s	s				**B21**[2], 27
—	**Lupulic acid**	see **Humulon**													
168	**Lupulone**	β-Bitter acid. β-Lupulinic acid. $C_{26}H_{38}O_4$.	414.57	pr (al)	92–4				i	s				peth δ hexane s	**B7**[2], 852
—	**Lutein**	see **Xanthophyll**													
—	**Lutidine**	see **Pyridine, dimethyl-**													
—	**Luxitol**	see **Arabitol**													
169	**Lycaconitine**	$C_{36}H_{48}N_2O_{10}$	668.76	amor $[\alpha]_D+31.5$	111–14				δ	s	δ			peth, chl s	
170	**Lycomarasmine**	$HO_2CCH_2CH(CO_2H)NHCOCH_2NHCH(CO_2H)CH_2NH_2$	277.23	$[\alpha]_D^{20}-48$ (w)	227–9d				δ					dil ac, dil alk v	
171	**Lycopene**	Lycopin. Neolycopene. $C_{40}H_{56}$.	536.85	pr or nd	173					$δ^h$	s		v^h	chl v^h	**B30**, 81
172	**Lycorine**(*l*)	Amarylline. Narcissine. $C_{16}H_{17}NO_4$.	287.32	pr (al or Py) $[\alpha]_D^{26}-120$ (al)	280	sub			i	δ	δ			chl δ	**B27**[2], 547
173	**Lycoxanthin**	(structure: HO-ring–$(CH{:}CHC{:}CH)_2CH{:}CH(CH{:}CCH{:}CH)_2$–ring, with CH_3 and CH substituents)	552.85	purple or red br nd (CS_2), (bz-peth)	168				i	δ			s	CS_2 s peth δ	
—	**Lysergamide**	see **Ergine**													
174	**Lysergic acid**	$C_{16}H_{16}N_2O_2$	268.32	lf (w) $[\alpha]_D^{20}+40$ (Py)	238d					s				Py s	
—	**Lysidine**	see **Δ^2-Imidazoline, 2-methyl-**													
175	**Lysine**(*L*)	*L*-α,ε-Diaminocaproic acid. $H_2N(CH_2)_4CH(NH_2)CO_2H$	146.19	nd (w or dil al) $[\alpha]_D^{20}+14.6$ (w, c=6)	224–5 d				v	i					**B4**[2], 857
176	—,dihydro-chloride(*L*)	$H_2N(CH_2)_4CH(NH_2)CO_2H.2HCl$	219.12	(al-eth or aq HCl) $[\alpha]_D^{20}$ +15.29 (w)	200–201				v	s^h					**B4**[2], 857
177	—,N^α-**benzoyl-**(*dl*)	$H_2N(CH_2)_4CH(NHCOC_6H_5)CO_2H$	250.30	nd	235–49				v^h	δ	i		i	chl i	**B9**[2], 193
178	—,N^ϵ-**benzoyl-**(*dl*)	$C_6H_5CONH(CH_2)_4CH(NH_2)CO_2H$	250.30	cr (w)	268				s^h	i	i				**B9**[2], 193
179	—,—(*L*)	$C_6H_5CONH(CH_2)_4CH(NH_2)CO_2H$	250.30	lf (w) $[\alpha]_D^{19}$ +20.1	240d				s^h						**B9**[2], 193
	Malachite green														
—	**Malachite green, leuco-**	see **Methane, bis(4-dimethylaminophenyl)-(phenyl)-**													
m1	**Malathion**	*S*(1,2-Dicarboxyethyl)-*O*, *O*-dimethyldithiophosphate.	330.36	ye-br	2.85	156–7[0.7]	1.23_4^{25}	1.4985[25]	δ	s	s		s		**Am 73**, 557
m2	**Maleic acid**	*cis*-Butenedioic acid*. $HO_2CCH{:}CHCO_2H$	116.07	mcl pr	137–138		1.590[20]		v	v	s	v	δ	con sulf s aa s	**B2**[2], 641
m3	—,anhydride	(ring structure: O=, O, =O)	98.06	nd (chl or eth)	56 (52)	197–9	1.314[60]		s d^h		s	s		chl s lig δ	
m3[1]	—,diethyl ester	Ethyl maleate. $C_2H_5O_2CCH{:}CHCO_2C_2H_5$	172.18			225 105–6[14]	1.064[25]		i	s	s				
m3[2]	—,dimethyl ester	Methyl maleate. $CH_3O_2CCHCHCO_2CH_3$	144.12		−19	205 102[17]	1.1606_4^{20}		i		s				
m4	—,diphenyl ester	$C_6H_5O_2CCH{:}CHCO_2C_6H_5$	268.26	lf (lig)	73	226[15]			i	v	v	v	v	chl v	**B6**, 156

For explanations, symbols and abbreviations see beginning of table.

No.	Name	Synonyms and Formula	Mol. wt.	Crystalline form, color and specific rotation	m.p. °C	b.p. °C	Density	n_D	Solubility w	al	eth	ace	bz	other solvents	Ref.
	Maleic acid														
—	—,hydrazide	see **Pyridazine, 3,6-dioxo-1,2,3,6-tetrahydro-**													
m5	—, imide	Maleimide	97.07	pl (bz)	93	sub	...	...	s	s	s	...	...	...	**B21**[2],311
m6	—,—,*N*-ethyl-	$C_6H_7NO_2$. *See* m6	125.12	(bz)	45.5	...	...	...	δ	v	v	...	...	...	...
m6[1]	—,—,*N*-phenyl-	Maleanil. $C_{10}H_7NO_2$. *See* m6	173.16	yend (bz-lig)	90–1	162[12]	...	...	δ	s	s	...	s	chl s, CS_2 δ	**B21**, 400
m6[2]	—,monoamide	Maleamic acid $HO_2CCH{:}CHCONH_2$	115.09	pl	152–3	...	...	...	v	s[h]	i	...	...	...	...
m6[3]	—,**bromo-**	$HO_2CCH{:}CBrCO_2H$	194.98	nd or pr	128 (138–41)	d	...	...	v	v	v	...	...	...	...
m6[4]	—,**chloro**	$HO_2CCH{:}CClCO_2H$	150.52	pr (eth-chl)	108 (114)	...	...	...	s[h]	v	v	...	δ	chlδ, peth i aa s	...
m7	—,—,diethyl ester	$C_2H_5O_2CCH{:}CClCO_2C_2H_5$	206.63	...	...	235δd 125[18]	1.174^{20}_{4}	...	...	v	...	...	...	...	**B2**[2],646
m8	—,—,dimethyl ester	$CH_3O_2CCH{:}CClCO_2CH_3$	178.58	...	...	106.5[18]	1.2775^{25}_{4}	...	...	v	v	...	...	...	**B2**[2], 646
m9	—,**dichloro-**	$HO_2CCCl{:}CClCO_2H$	184.97	nd (lig or eth)	119–20	...	...	...	v	s	...	...	i	chl, CS_2 i	**B2**, 753
m10	—,**dihydroxy-**	$HO_2CCOH{:}COHCO_2H$	148.08	amor	155	...	...	...	δ	s	δ	...	...	MeOH δ	**B3**[2], 346
m11	—,**dimethyl-,** anhydride	Pyrocinchonic anhydride. $C_6H_6O_3$. *See* m3	126.11	rh	96	223 105[−12]	1.107^{100}_{4}	...	δ	v	v	...	v	...	**B17**[2], 450
m12	—,**methyl-**	Citraconic acid. $HO_2CCH{:}C(CH_3)CO_2H$	130.10	tcl pr or pl (eth-bz), nd (eth-lig)	92–3	...	1.617	...	v	...	δ	...	δ	chl δ CS_2i	**B2**[2], 652
m13	—,—,anhydride	Citraconic anhydride. $C_5H_4O_3$. *See* m3	112.08	...	7–8	213–4 99–100[15]	1.2469^{16}_{4}	1.4710[21]	...	v	s	...	...	...	**B17**[2], 448
m14	—,—,diethyl ester	Diethyl citraconate, $C_2H_5O_2CCH{:}C(CH_3)CO_2C_2H_5$	186.21	...	...	230.3 120[20]	1.0401^{20}_{4}	1.4442[20]	...	s	s	...	...	aa s	**B2**[2], 653
m15	—,—,dimethyl ester	Dimethyl citraconate. $CH_3O_2CCH{\cdot}CH(CH_3)CO_2CH_3$	158.16	...	...	210.5 94–5[11]	1.1153^{20}_{4}	1.4486[20]	δ	s	s	...	...	...	**B2**[2], 653
m16	—,**phenyl-,**	$C_{10}H_6O_3$. *See* m3	174.16	ye nd (CS_2)	119–20	...	...	...	...	s	s	...	...	chl s, CS_2 δ	**B17**[2], 481
m16[1]	**d-Malic acid** anhydride	Hydroxysuccinic acid $HO_2CCHOHCH_2CO_2H$	134.09	col. cryst., $[a]_D$ +2.92 (MeOH) $[a]^{20}_{D}$ +3.3 (H_2O, p = 4)	98–99	dec.	...	...	v	v	δ	...	...	MeOH	**B3**[2], 275
m16[2]	**dl-Malic acid**	Hydroxysuccinic acid $HO_2CCHOHCH_2CO_2H$	134.09	col. cryst.	130	dec.	D^{20}_{4} 1.601 (vac.)	...... 14.5[26] 411.5[79]	...	...	...	...		...	**B3**[2], 289 **B3**[3], 918
m16[3]	**l-Malic acid**	Hydroxysuccinic acid $HO_2CCHOHCH_2CO_2H$	134.09	col. need., (acet. + $CHCl_3$) $[a]^{18}_{D}$ −2.3 (H_2O, p = 7)	100–103	dec.	1.595[4]	...	27[10] 36[20]	v	3[20]	...	...	...	**B3**[2], 276 **B3**[3], 909
—	**Malonaldehydic acid**	see **Propanoic acid, 3-oxo-***													
—	**Malonanilic acid**	see **Malonic acid,** monoamide, *N*-phenyl-													
m17	**Malonic acid**	Propanedioic acid*. $CH_2(CO_2H)_2$	104.06	trc	135.6 δ sub	d	...	...	v	s	s	...	...	Py v	**B2**[2], 516
m19	—,diamide	Malonamide. $C_3H_6N_2O_2$	102.09	mcl st pr (w), unst tetr	170 (st)	...	...	...	s	i	i	...	...	...	**B2**[2], 530
m21	—,—,*N*,*N*-diphenyl-	Malonanilide. $CH_2(CONHC_6H_5)_2$	254.29	nd (al)	223–5	...	...	...	i	v[h]	i	...	v	...	**B12**[2], 167
m22	—,dibutyl ester	Dibutyl malonate. $CH_2(CO_2CH_2CH_2CH_2CH_3)_2$	216.28	...	−83	251–2 137[14]	0.9856^{15}_{4}	1.4262[20]	...	...	...	...	...	aa s	**B2**[2], 528
m22[1]	—,dichloride	Malonyl chloride, $CH_2(COCl)_2$	140.95	...	...	58[27]	1.4486^{19}_{4}	1.462^{18}_{4}	d[h]	d[h]	s	...	...	AcOEt s	**B2**[2], 529

For explanations, symbols and abbreviations see beginning of table.

No.	Name	Synonyms and Formula	Mol. wt.	Crystalline form, color and specific rotation	m.p. °C	b.p. °C	Density	n_D	Solubility w	al	eth	ace	bz	other solvents	Ref.
	Malonic acid														
m23	—,diethyl ester	Ethyl malonate. Malonic ester. $CH_2(CO_2C_2H_5)_2$	160.17		−50	199	1.0550^{20}_{4}	1.4143^{20}	δ	∞	∞			chl s aa s	B2^2, 524
m24	—,methyl diester	Methyl malonate. $CH_2(CO_2CH_3)_2$	132.12		−80	183	1.1544^{20}_{4}	1.4140^{20}	δ	∞	∞				B2^2, 522
m25	—,dinitrile	Malononitrile. Methylene cyanide. Dicyanomethane. $CH_2(CN)_2$	66.06		30–1	218–9	1.0494^{35}_{4}	1.4146^{34}	s	v	v		s	chl s aa s	B2^2, 535
m26	—,dipropyl ester	Propyl malonate. $CH_2(CO_2CH_2CH_2CH_3)_2$	188.23	glass		229	1.0088^{20}_{4}	1.4206^{20}							B2^2, 528
m27	—,monoamide, *N*-phenyl-	Malonanilic acid. $C_6H_5NHCOCH_2CO_2H$	179.18	(al)	131–2d				s^h	s^h	s^h		s^h		B12^1, 208
m28	—,monoamide mononitrile	Cyanoacetamide. H_2NCOCH_2CN	84.08	pl	118 (124)				s	δ					B2^2, 534
m29	—,—,*N*-phenyl-	α-Cyanoacetanilide. $C_6H_5NHCOCH_2CN$	160.18	nd (al)	199–200				δ						B12^2, 167
m30	—,monochloride monoethyl ester	Carbethoxyacetyl chloride. $ClCOCH_2CO_2C_2H_5$	150.56			170–80d $75–7^{17}$			d	d	v				B2^2, 529
m31	—,monochloride monomethyl ester	$ClCOCH_2CO_2CH_3$	136.54			$57–9^{12}$			d	d	v				B2^1, 252
m32	—,monoethyl ester mononitrile	Ethyl cyanoacetate. $NCCH_2CO_2C_2H_5$	113.12		−22.5	206 99^{15}	1.063^{20}_{4}	1.4179^{20}	i	∞	∞				B2^2, 531
m33	—,monoethyl ester piperazinium salt	$C_4H_{10}N_2.2HO_2CCH_2CO_2C_2H_5$	350.38		144				s	s^h	i				C28, 1704
m34	—,monomethyl ester mononitrile	Methyl cyanoacetate. $NCCH_2CO_2CH_3$	99.09		−22.5	200 115^{36}	1.123^{15}		i	∞	∞				B2^2, 530
m35	—,mononitrile	Cyanoacetic acid. $NCCH_2CO_2H$	85.06	dlq dimorphic cr	70 (66)	$108^{0.15}$δd			s	s	s			chl δ aa δ	B2^2, 530
m36	—,piperazinium salt	$C_4H_{10}N_2.CH_2(CO_2H)_2$	190.20		180d				s	s^h	i				C28, 1704
m37	—,**acetamido-,** diethyl ester	$CH_3CONHCH(CO_2C_2H_5)_2$	217.22	(al)	95	185^{20}			δ^h	s^h	δ				B4^2, 891
m38	—,**acetyl-, diethyl ester**	$CH_3COCH(CO_2C_2H_5)_2$	202.21			240 120^{17}	1.0834^{26}_{4}	1.4374^{26}						Na_2CO_3 s	B3^2, 485
m39	—,**allyl-**	3-Butene-1,1-dicarboxylic acid*. $CH_2:CHCH_2CH(CO_2H)_2$	144.13	tcl (eth)	105	>180d			s	s	s		s^h		B2^2, 658
m40	—,—,diethyl ester	$CH_2:CHCH_2CH(CO_2C_2H_5)_2$	200.24			222–3 93^6	1.0148^{14}		i	v	v				B2^2, 658
m41	—,**amino-,** monohydrate	$H_2NCH(CO_2H)_2.H_2O$	137.09	pr (w+1)	109				δ	δ					B4^2, 890
m42	—,—,diethyl ester	$H_2NCH(CO_2C_2H_5)_2$	175.19			$122–3^{16}$	1.100^{16}_{4}	1.4353^{16}	v	v	v			lig i	B4^2, 890
m43	—,**benzoylamino-,** diethyl ester	Ethyl benzamidomalonate. $C_6H_5CONHCH(CO_2C_2H_5)_2$	279.30	nd (peth)	61				i	v	s			peth δ	B9^2, 185
m44	—,**benzyl-**	$C_6H_5CH_2CH(CO_2H)_2$	194.19	pr (bz, eth, chl-peth)	121				v	v	v		v^h		B9^2, 619
m45	—,—,diamide	$C_6H_5CH_2CH(CONH_2)_2$	192.22	nd (al)	225					δ s^h	i				B9, 869
m46	—,—,diethyl ester	$C_6H_5CH_2CH(CO_2C_2H_5)_2$	250.30			300 169^{12}	1.0750^{20}_{4}	1.4872^{20}	i						B9^2, 620
m47	—,—,dinitrile	Benzylmalononitrile. $C_6H_5CH_2CH(CN)_2$	156.19	pl (al), nd (w or lig)	91	174^{23}				s	s		s	AcOEt s lig δ	B9, 870
m48	—,**benzyl-(hydroxy)-**	Benzyltartronic acid. $C_6H_5CH_2COH(CO_2H)_2$	210.19	pr	147				s	s	s				B10, 515
m49	—,**benzylidene-**	$C_6H_5CH:C(CO_2H)_2$	192.17	pr (w)	195–6d				δ s^h	s	δ	s	δ	AcOEt s CS_2, chl δ	B9^2, 638
m50	—,—,diethyl ester	$C_6H_5CH:C(CO_2C_2H_5)_2$	248.28		32(27)	308–12δd 180^{14}	1.1045^{20}_{4}	1.5389^{20}	i	s			s	os s	B9^2, 639
m51	—,—,monoethyl ester mononitrile	Ethyl α-cyanocinnamate. Ethyl benzalcyanoacetate. $C_6H_5CH:C(CN)CO_2C_2H_5$	201.23	nd	48(51)	188^{15}			i	δ s^h	s		v	chl s aa v	B9^2, 640
m52	—,—,mononitrile	Benzalcyanoacetic acid. α-Cyanocinnamic acid. $C_6H_5CH:C(CN)CO_2H$	173.17	(al)	180					s^h				dil NH_3 s	B9^2, 639
m53	—,**bromo-**	$BrCH(CO_2H)_2$	182.96	nd (eth), pl (ace-bz)	113d					v	v				B2^2, 538

For explanations, symbols and abbreviations see beginning of table.

No.	Name	Synonyms and Formula	Mol. wt.	Crystalline form, color and specific rotation	m.p. °C	b.p. °C	Density	n_D	Solubility w	al	eth	ace	bz	other solvents	Ref.
	Malonic acid														
m54	—,—,diethyl ester	$BrCH(CO_2C_2H_5)_2$	239.07		−54	233–5d $121–5^{16}$	1.4022^{25}_{4}		i	∞	∞				$B2^2$, 538
m55	—,**butyl-**	1,1-Pentanedicarboxylic acid.* $CH_3(CH_2)_3CH(CO_2H)_2$	160.17	pr (w)	96–102				v	s	s				$B2^2$, 588
m56	—,—,diethyl ester	$CH_3(CH_2)_3CH(CO_2C_2H_5)_2$	216.28			235–40 122^{12}				v	v				$B2^2$, 588
m57	—,—,monoethyl ester mononitrile	Ethyl α-cyanocaproate. $CH_3(CH_2)_3CH(CN)(CO_2C_2H_5)$	169.22	syr		$230–3^{734}$ $108–9^{9}$	0.9537^{25}_{4}	1.4242^{25}		s			s		$B2^2$, 588
m58	—,*sec*-**butyl-**, diethyl ester	$CH_3CH_2CH(CH_3)CH(CO_2C_2H_5)_2$	216.28			245–50	0.988^{15}	1.4248	i	v	v				$B2^2$, 590
m59	—,**chloro-**	$ClCH(CO_2H)_2$	138.51	pr	133	d			v	v	v				$B2^2$, 537
m60	—,—,diethyl ester	Ethyl chloromalonate. $ClCH(CO_2C_2H_5)_2$	194.62			222–3 110^{6}	1.2040^{20}_{4}	1.4327^{20}	i	∞	∞			chl ∞	$B2^2$, 537
m61	—,**(3-chloropropyl)-**, diethyl ester	$ClCH_2CH_2CH_2CH(CO_2C_2H_5)_2$	236.70			$154–7^{17}$			i	s	s			chl s	$B2^1$, 278
m62	—,**cinnamylidene-**,(*trans*)	Cinnamalmalonic acid. $C_6H_5CH{:}CHCH{:}C(CO_2H)_2$	218.21	dk ye nd	212d				i	s				chl s AcOEt s	$B9^2$, 649
m63	—,**1-cyclohexenyl-**, mononitrile	1-Cyclohexenylcyanoacetic acid (cyclohexenyl ring)—CH(CN)CO2H	165.19	nd (bz)	109–10				δ v^h				s^h		$B9^2$, 560
m64	—,**cyclohexyl-**, diethyl ester	(cyclohexyl ring)—CH(CO2C2H5)2	242.32			$163–5^{20}$	1.0281^{19}	1.4478^{25}	i	s				os s	$B9^2$, 526
m65	—,**2-cyclopentenyl-**, diethyl ester	(cyclopentenyl ring)—CH(CO2C2H5)2	226.27			141^{16}	1.0507^{20}_{4}	1.4536^{20}							$B9^2$, 559
m66	—,cyclopentylidene-, diethyl ester	(cyclopentylidene ring)=C(CO2C2H5)2	226.27			140^{10}	1.0616^{20}_{4}	1.4724^{20}							$B9^2$, 559
m67	—,**dibenzyl-**, diethyl ester	$(C_6H_5CH_2)_2C(CO_2C_2H_5)_2$	340.42		14	$234–5^{23}$	1.093^{20}_{4}			s	s				$B9^1$, 408
m68	—,**dibromo-** diethyl ester	$Br_2C(CO_2C_2H_5)_2$	317.97			250–6δd 154^{28}									$B2^2$, 539
m69	—,**dibutyl-**, diethyl ester	$(CH_3CH_2CH_2CH_2)_2C(CO_2C_2H_5)_2$	272.39			$153–4^{14}$			i	s	s				$B2^2$, 614
m70	—,**diethyl-**	3,3-Pentanedicarboxylic acid*. $(C_2H_5)_2C(CO_2H)_2$	160.17	pr (w or bz)	125	d			v	v	v		$δ^h$	chl δ	$B2^2$, 593
m71	—,—,diethyl ester	$(C_2H_5)_2C(CO_2C_2H_5)_2$	216.28			230 100^{12}	0.9917^{15}_{15}	1.4252^{17}	i	∞	∞				$B2^2$, 593
m72	—,—,monoethyl ester mononitrile	$(C_2H_5)_2C(CN)CO_2C_2H_5$	169.22			215–6 $100–1^{15}$			i	∞	∞				$B2^2$, 594
m73	—,—,piperazinium salt	$(C_2H_5)_2C(CO_2H)_2.C_4H_{10}N_2$	246.31		80–1				s	s	i	s^h			C28, 1704^8
—	—,**dihydroxy-**	*see* **Malonic acid, oxo-**, hydrate													
m74	—,**dimethyl-**, diethyl ester	$(CH_3)_2C(CO_2C_2H_5)_2$	188.23			196–7 $97–8^{22}$	0.9910^{25}_{4}	1.4105^{24}	i	∞	∞				$B2^2$, 572
m75	—,**(2,4-dinitrophenyl)-**, diethyl ester	O2N—(benzene ring, NO2)—CH(CO2C2H5)2	326.25	pr	51				i	v^h	v^h			alk s	$B9^1$, 378
m76	—,**(ethoxymethylene)-**, diethyl ester	$C_2H_5OCH{:}C(CO_2C_2H_5)_2$	216.24			279–81d 165.5^{19}			i	s	s				$B3^2$, 300
m77	—,**ethyl-**	1,1-Propanedicarboxylic acid*. $CH_3CH_2CH(CO_2H)_2$	132.12	pr (w+1)	111.5 anh	160d			v	δ	s	i	s	MeOH s AcOEt s chl s	$B2^2$, 569
m78	—,—,diethyl ester	$C_2H_5CH(CO_2C_2H_5)_2$	188.23			207–9	1.004^{20}_{20}	1.4170^{20}	δ	v	v				$B2^2$, 570
m79	—,**ethyl(isopropyl)-**, diethyl ester	$(CH_3)_2CHC(C_2H_5)(CO_2C_2H_5)_2$	230.31			232–3				s	s	s			B2, 706
m80	—,**ethyl(methyl)-**	2,2-Butanedicarboxylic acid*. $C_2H_5C(CH_3)(CO_2H)_2$	146.14	pr or nd	121–2				s	s	s				$B2^2$, 585
m81	—,**ethyl(phenyl)-**, diethyl ester	$C_6H_5C(C_2H_5)(CO_2C_2H_5)_2$	264.33			170^{19}	1.071^{20}_{4}		i	s	s				$B9^2$, 630

For explanations, symbols and abbreviations see beginning of table.

No.	Name	Synonyms and Formula	Mol. wt.	Crystalline form, color and specific rotation	m.p. °C	b.p. °C	Density	n_D	Solubility w	al	eth	ace	bz	other solvents	Ref.
	Malonic acid														
m82	—,**ethylidene-**, diethyl ester	$CH_3CH{:}C(CO_2C_2H_5)_2$	186.21			115–8^{17}	1.0194^{17}_{4}	1.4308^{17}		s	s				**B2**^{2}, 654
m83	—,(**formyl-amino**)-, diethyl ester	Diethyl formamidomalonate. $HCONHCH(CO_2C_2H_5)_2$	203.19		48–9	173–4^{11}				s					**B4**^{2}, 891
m84	—,**heptyl-**	1,1-Octanedicarboxylic acid*. $CH_3(CH_2)_6CH(CO_2H)_2$	202.25	pr (bz-peth)	96				i	v	v	v			**B2**^{2}, 610
m85	—,**hexadecyl-**	Cetylmalonic acid. 1,1-Heptadecanedicarboxylic acid*. $CH_3(CH_2)_{15}CH(CO_2H)_2$	328.50	nd (lig), lf (aa)	119–22				i	δ			v^h	peth δ, aa v^h, lig s^h	**B2**^{2}, 627
m86	—,—,diethyl ester	$CH_3(CH_2)_{15}CH(CO_2C_2H_5)_2$	384.61	amor	22	238–40^{14}				s					**B2**^{1}, 299
m87	—,**hexyl-** diethyl ester	1,1-Heptanedicarboxylic acid. $CH_3(CH_2)_5CH(CO_2C_2H_5)_2$	244.32			268–70	0.9556^{25}		i	s				os s	**B2**^{2}, 604
m88	—,**hydroxy-**	Tartronic acid. $HOCH(CO_2H)_2$	120.06	pr (w+1)	156–8 anh	sub			s	s	δ				**B3**, 415
m89	—,—,diethyl ester	$HOCH(CO_2C_2H_5)_2$	176.17		−25	222–5, 121^{15}				s				os s	**B3**^{1}, 148
m90	—,—,dimethyl ester	$HOCH(CO_2CH_3)_2$	148.11	nd (eth-peth or MeOH)	53.4 (44.5)	122^{19}			s	s	v			chl v, os s, peth δ	**B3**^{2}, 274
m91	—,(**2-hydroxy-cyclohexyl**)- lactone, monoethyl ester	$CHCO_2C_2H_5$	212.24	pa ye		199^{30}	1.0735^{17}								**B18**^{2}, 322
m92	—,**hydroxy-(methyl)-**	α-Isomalic acid. $CH_3COH(CO_2H)_2$	134.09		142	170d			v	v	v				**B3**, 440
m93	—,**isobutyl-**	$(CH_3)_2CHCH_2CH(CO_2H)_2$	160.17	(bz)	108	d			v	v	v				**B2**^{1}, 284
m94	—,—,diethyl ester	$(CH_3)_2CHCH_2CH(CO_2C_2H_5)_2$	216.28			225	0.983^{17}		i	v	v				**B2**, 683
m95	—,**isopropyl-**, diethyl ester	$(CH_3)_2CHCH(CO_2C_2H_5)_2$	202.25			214	0.9850^{25}_{25}	1.418	δ	v	v				**B2**^{2}, 586
m96	—,**isopropylidene-**	$(CH_3)_2C{:}C(CO_2H)_2$	144.12	(ace-chl)	170–1	d				s				alk v, chl i	**B2**^{1}, 312
m97	—,—,diethyl ester	$(CH_3)_2C{:}C(CO_2C_2H_5)_2$	200.23			175–8, 140–1^{20}	1.0284^{16}_{4}	1.4486^{17}		s	s			CCl_4 s	**B2**^{2}, 661
m98	—,**methyl-**	1,1-Ethanedicarboxylic acid*. Isosuccinic acid. $CH_3CH(CO_2H)_2$	118.09	nd (AcOEt-bz)	120–35d		1.455^{20}_{4}		v	v	v			aa v	**B2**^{2}, 562
m99	—,—,diethyl ester	$CH_3CH(CO_2C_2H_5)_2$	174.20			201	1.018^{20}_{4}	1.4137^{19}	δ	v	v				**B2**^{2}, 563
m100	—,—,dimethyl ester	Methyl isosuccinate. $CH_3CH(CO_2CH_3)_2$	146.15			178	1.0952^{20}_{4}	1.4144^{20}	δ	∞	∞				**B2**^{2}, 562
m101	—,(**3-methyl-butyl**)-, diethyl ester	Ethyl isoamylmalonate. $(CH_3)_2CHCH_2CH_2CH(CO_2C_2H_5)_2$	230.31			240–2, 130–2^{15}			i	v	v				**B2**^{2}, 599
m102	—,**nitro-**, monoamide mononitrile	Cyanonitroacetamide. Fulminuric acid. Isocyanuric acid. $NCCH(NO_2)CONH_2$	129.08	pr (al)	145 exp				s	s	δ		i	chl i, lig i	**B2**^{1}, 258
m103	—,**octyl-**	1,1-Nonanedicarboxylic acid*. $CH_3(CH_2)_7CH(CO_2H)_2$	216.27	pr	113–4								s^h	chl s^h	**B2**^{2}, 613
m104	—,**oxo-**, hydrate	Mesoxalic acid. Dihydroxy-malonic acid. $(HO)_2C(CO_2H)_2$	136.06	dlq nd					v	s	s				**B3**^{2}, 472
m105	—,—,diethyl ester	Ethyl mesoxalate. $OC(CO_2C_2H_5)_2$	174.16	pa ye gr	ca. −30	208–10^{220}	1.1419^{16}_{4}	1.419^{16}	v	s	s			chl s, CS_2 i	**B3**^{1}, 267
m106	—,—,diethyl ester hydrate	Ethyl dihydroxymalonate. $(HO)_2C(CO_2C_2H_5)_2$	192.17	pl (bz)	57	ca. 200			s	s	v		s	chl s, lig δ, CS_2 i	**B3**^{2}, 474
m107	—,—,diethyl ester, oxime	Ethyl isonitrosomalonate. $HON{:}C(CO_2C_2H_5)_2$	189.17			172^{12}	1.182^{18}_{4}	1.4544^{18}	i					os ∞, alk s	**B3**^{2}, 475
m108	—,**pentyl-**, diethyl ester	$CH_3(CH_2)_4CH(CO_2C_2H_5)_2$	230.31			134–6^{14}		1.4253	i	v	v				**B2**^{2}, 597
m109	—,(**3-pentyl**)-, diethyl ester	Ethyl *sec*-amylmalonate. $(C_2H_5)_2CHCH(CO_2C_2H_5)_2$	230.31			230^{736}, 85–6^{12}		1.4294^{20}	i	v	v				**B2**^{2}, 599
m111	—,**phenyl-**, diethyl ester	$C_6H_5CH(CO_2C_2H_5)_2$	236.27			205δd, 170–2^{14}	1.0950^{20}_{4}	1.4977^{20}	i	s					**B9**^{2}, 615
m112	—,—,dimethyl ester	$C_6H_5CH(CO_2CH_3)_2$	208.21		50(?)	145–7^{13}				s	s			lig δ^h	**B9**^{2}, 615
m113	—,—,dinitrile	$C_6H_5CH(CN)_2$	142.16	(dil al)	70–1	152–3^{21}			δ	v	δ, v^h			lig δ	**B9**^{2}, 616
m114	—,—,monoethyl ester mononitrile	Ethyl phenylcyanoacetate. $C_6H_5CH(CN)CO_2C_2H_5$	189.22			275δd, 165^{20}	1.091^{20}_{4}		i	∞				os ∞	**B9**^{2}, 615

For explanations, symbols and abbreviations see beginning of table.

No.	Name	Synonyms and Formula	Mol. wt.	Crystalline form, color and specific rotation	m.p. °C	b.p. °C	Density	n_D	Solubility w	al	eth	ace	bz	other solvents	Ref.
	Malonic acid														
m115	—,**(phenyl-amino)-**, diethyl ester	Ethyl anilinomalonate. $C_6H_5NHCH(CO_2C_2H_5)_2$	251.29	(al or lig)	45	....	....	....	i	v	v	...	v	CS_2, chl v	**B12**[1], 271
m116	—,**phthalimido-**, diethyl ester	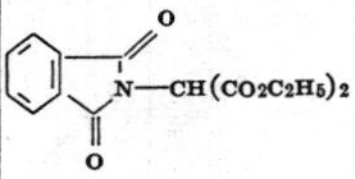 $N-CH(CO_2C_2H_5)_2$	305.29	pr (al)	74	....	....	....	δ	v^h	s	v	v	chl v CS_2 s peth δ	**B21**[1], 379
m117	—,**(4-phthali-midobutyl)** diethyl ester	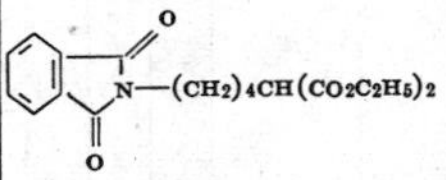 $N-(CH_2)_4CH(CO_2C_2H_5)_2$	361.50	nd (peth)	46	....	....	....	...	...	...	...	...	peth s^h	**B21**, 489
m118	—,**propyl-**........	1,1-Butanedicarboxylic acid*. $CH_3CH_2CH_2CH(CO_2H)_2$	146.15	pl (bz)	96	d	....	....	v	s	s	...	δ	chl s	**B2**[2], 581
m119	—,—,diethyl ester..	$CH_3CH_2CH_2CH(CO_2C_2H_5)_2$...	202.25	....	....	222	0.9860^{25}_{25}	....	δ	v	v	...	...	....	**B2**[2], 582
—	**Maltobiose**.......	*see* **Maltose**													
—	**Maltol**...........	*see* **4-Pyrone, 3-hydroxy-2-methyl-**													
m120	**α-Maltose,** anhydrous	Glucose-4-α-glucosido. Maltobiose. 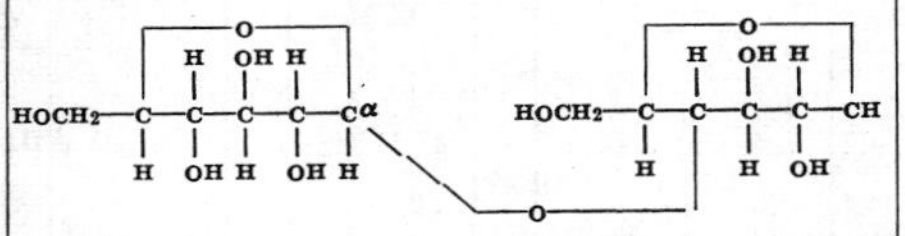	342.30	(abs al) $[\alpha]_D$ +140.7 (w, c=10)	160–5	....	....	....	s	s^h	...	...	...	....	....
m121	—,monohydrate....	$C_{12}H_{22}O_{11}.H_2O$. *See* m120.....	360.32	nd (w) $[\alpha]^{20}_D$ +111.7 (c=4)	102–3	....	1.540	....	v	δ	i	...	...	....	**B31**[2], 386
m122	**Malvidine chloride**	Syringidine chloride. 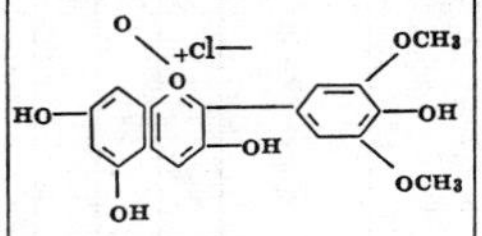	366.76	rh pr (al-aq HCl +1w)	300	....	....	....	δ	s	...	...	...	MeOH s	....
—	**Mandelic acid**....	*see* **Acetic acid, hydroxy-(phenyl)-**													
—	***l*-Mandelo-nitrile-β-gentio-bioside**	*see* **Neoamygdalin**													
m123	**Mannitane**.......	3,6(1,4)-Anhydro-*D*-mannitol. 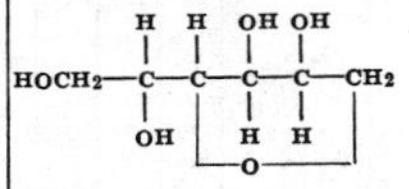	164.16	pl. $[\alpha]^{20}_D$ −26.2 (w)	146–7	....	....	....	v	i	...	...	...	....	**B1**[1], 284 **B17**[2], 235
—	**Mannite**.........	*see* **Mannitol**													
m124	**Mannitol,(*D*)**.....	*D*-Mannite. 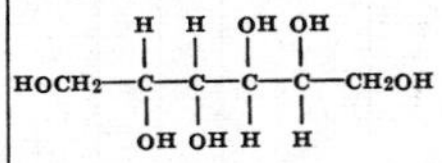	182.17	rh nd or pr $[\alpha]^{25}_D$ −0.208	168	$295^{3.5}$	1.489^{20}_4	1.3330	v	δ^h	i	...	...	Py δ	**B1**[3], 2392
m125	—(*dl*)............	α-Acritol. $C_6H_{14}O_6$. *See* m124	182.17	....	168	....	....	....	s^h	δ	...	...	...	....	**B1**[3], 2405
m126	—(*L*).............	*L*-Mannite. $C_6H_{14}O_6$. *See* m124	182.17	nd	162–3	....	....	....	v	δ	...	...	...	MeOH δ^h	**B1**[3], 2405
m127	—,hexanitrate.....	Manexin. Nitranitol. Nitro-mannitol. $C_6H_8N_6O_{18}$. *See* m124	452.16	nd $[\alpha]^{22}_{546}$+46.8 ($ClCH_2CH_2Cl$, c=0.33)	112	120 exp	1.8^{20}_4	....	i	s	s	...	s^h	....	**B1**[3], 2404
—	—,**1,4:3,6-dianhydro-**	*see* **Isomannide**													
m128	**Mannoheptose-** (*d*, α)	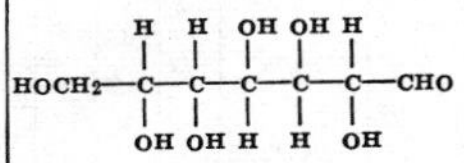	210.19	nd (al). $[\alpha]^{20}_D$ (mut) +85 to +68.6 (w,c=10)	134–5	....	....	....	v	δ	...	...	...	....	**B31**, 361
m129	—(*d*, β), mono-hydrate	$C_7H_{14}O_8.H_2O$. *See* m128......	228.21	(w+1) $[\alpha]^{20}_D$ +45.7	83	....	....	....	s	...	...	...	...	....	....

For explanations, symbols and abbreviations see beginning of table.

No.	Name	Synonyms and Formula	Mol. wt.	Crystalline form, color and specific rotation	m.p. °C	b.p. °C	Density	n_D	Solubility w	al	eth	ace	bz	other solvents	Ref.
	Mannonic acid														
m132	**Mannonic acid,** α-lactone(*D*)	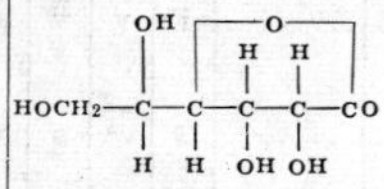	178.14	nd (96% al), pr (abs al) $[\alpha]_D^{20}+51.8$ (w)	151	...	...	...	v	δ^h	...	...	...	...	**B18¹, 407**
m133	—,—(*L*)	$C_6H_{10}O_6$. *See* m132	178.14	rh nd $[\alpha]_D-51.8$	151	...	...	...	s	δ	...	...	...	...	**B18², 193**
m134	**Mannose**(*D*)	Carubinose. Seminose. 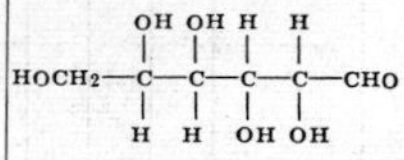	180.16	nd or pr (al) $[\alpha]_D^{20}$ (mut) −17 to +14.6 (w, p=3)	132	...	1.539_4^{20}	...	v	δ	i	...	...	MeOH δ	**B31, 284**
m135	—(*DL*)	$C_6H_{12}O_6$. *See* m134	180.16		132–3	...	...	...	...	δ	...	...	...	...	...
m136	—(*L*)	$C_6H_{12}O_6$. *See* m134	180.16	(al) $[\alpha]_D$ (mut) +14 to −14	132	...	...	...	v	δ	...	...	...	MeOH δ	...
m137	—,phenyl-hydrazone(*D*)	$C_{12}H_{18}N_2O_5$. *See* m134	270.29	ye pr (w), nd (al). $[\alpha]_D$ (mut) +26.3 to +6.3	199–200	...	...	...	i s^h	s^h	...	...	...	...	**B31, 290**
m138	—,—(*L*)	$C_{12}H_{18}N_2O_5$. *See* m134	270.29	...	195	...	...	...	...	...	...	...	...	...	...
m139	**Mannosaccharic acid**,γ,γ′-di-lactone(*d*)	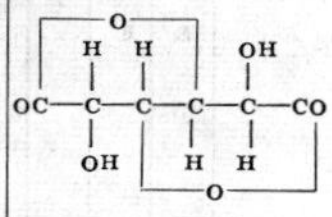	174.11	nd (w) $[\alpha]_D^{23}$ +202	190d	...	...	...	δ v^h	s^h	...	...	...	...	**B19², 266**
m140	—,—(*l*)	$C_6H_6O_6$. *See* m139	174.11	nd (w or al) $[\alpha]_D^{20}-202.5$	183–5	...	...	...	s v^h	δ	i	...	...	...	**B19², 266**
m141	**Mannuronic acid**(α)	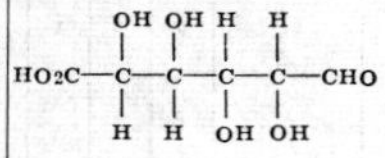	194.14	nd (al-eth) $[\alpha]_D^{25}$ (mut) +16 to −6 (w)	120–30d	...	...	...	s	...	...	...	...	...	...
m142	—(β)	$C_6H_{10}O_7$	194.14	(w+½) $[\alpha]_D^{25}$ (mut) −47.9 to −23.9 (w)	165–7	...	...	...	s	...	...	...	...	...	...
—	**Maraniol**	*See* **Coumarin, 7-ethoxy-4-methyl-**													
—	**Maretine**	*see* **Semicarbazide, 1(3-tolyl)-**													
—	**Margaraldehyde**	*see* **Heptadecanal***													
—	**Margaric acid**	*see* **Heptadecanoic acid***													
m144	**Matrine**	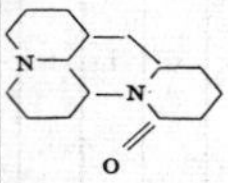	248.37	(4 forms) α: nd or pr, β: orh pr, γ: liq, δ: pr or lf (peth) $[\alpha]_D^{10}+31.9$ (w)	α 76–7 β 87 δ 84	223⁶	...	...	s δ^h	v	s	...	v	CS_2 v chl v peth δ	**B24², 58**
m145	**Meconic acid**	3-Hydroxy-α-pyrone-2,6-dicarboxylic acid. $C_7H_4O_7$.	200.11	rh pl (w or dil HCl+3w), pr (con HCl+1w)	100	d	...	...	δ v^h	s	δ	δ	s	MeOH δ	**B18², 372**
m146	**Meconidine**	$C_{21}H_{23}NO_4$	353.42	ye amor	58		...	...	i	...	...	...	...	...	...
m147	**Meconin**	6,7-Dimethoxyphthalide. 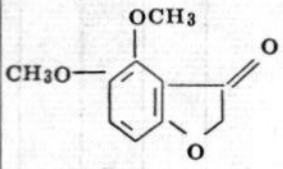	194.19	nd (w)	102–3	155 sub	...	...	s^h	s	s	...	s	chl s	**B18², 62**
m148	**Medicagenic acid**	2β,3β-Dihydroxy-Δ¹²-oleanene-23,28-dioic acid. $C_{30}H_{46}O_6$	502.67	pr $[\alpha]+111$	349–50	...	...	...	i^h	δ^h	δ	...	...	...	**Am, 76** 2271
m149	—,diacetate	$C_{34}H_{50}O_8$. *See* m148	586.77	mcl $[\alpha]+94$	210–2	...	1.190_{25}^{25}	...	i^h	δ^h	i	...	...	...	**Am, 79** 5292
m150	**Melam**	$C_6H_9N_{11}$	235.21	pw	...	...	...	...	i	δ	...	...	...	ac s	**B3², 121**
m151	**Melamine**	Cyanurotriamide. Cyanuramide. 2,4,6-Triamino-1,3,5-triazine*. 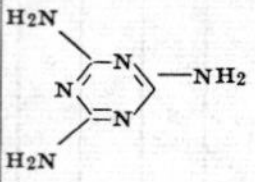	126.12	mcl pr (w)	354	sub	1.573^{14}	1.872^{20}	δ v^h	δ^h	i	...	...	...	**B26², 132**

For explanations, symbols and abbreviations see beginning of table.

No.	Name	Synonyms and Formula	Mol. wt.	Crystalline form, color and specific rotation	m.p. °C	b.p. °C	Density	n_D	Solubility w	al	eth	ace	bz	other solvents	Ref.
	Melaniline														
—	**Melaniline**	*see* **Guanidine, 1,3-diphenyl-**													
m152	**Melene**	$C_{30}H_{60}$	420.81	nd (ace)	62–3	380	0.9128^{25}_{25}	1.4228^{90}	i	δ^h	δ^h	δ^h	δ	to δ	**B1³, 885**
m153	**Melezitose**	Melizitose. 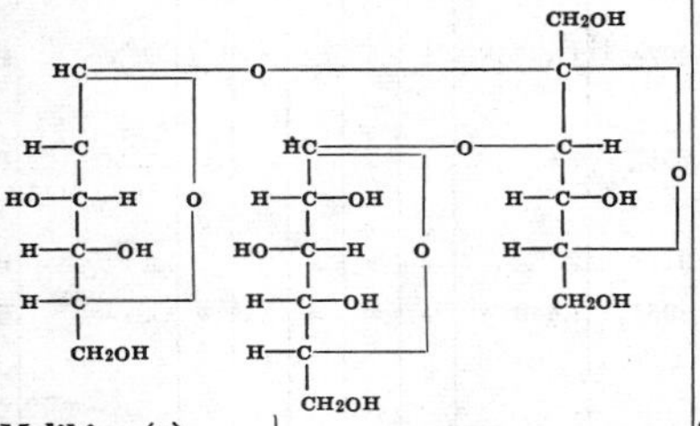	504.45	(w+2) $[\alpha]^{20}_D$ +88 (w, c=4)	153–4 (anh)		1.5565^0		s	δ					**B31, 466**
m154	**Melibiose**(α)	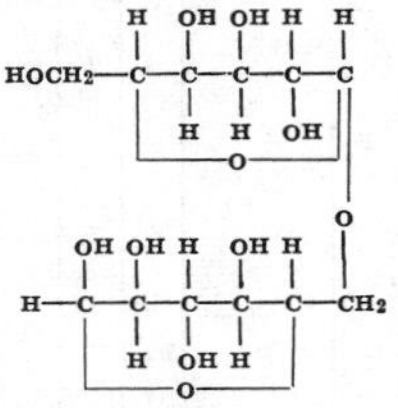	342.30	amor (anh) $[\alpha]^{15}$ (mut) +145.8 to +141.6 (w, c=2)											**B31, 422**
m155	—(β), dihydrate	Glucose-6-α-galactoside. $C_{12}H_{22}O_{11}.2H_2O$	378.33	mcl (w+2) $[\alpha]^0_D$ (mut) +115.4 to +129.5	84–5				v	δ					**B31, 421**
—	**Melilotic acid**	*see* **Propanoic acid, 3(2-hydroxyphenyl)-***													
—	**Melilotol**	*see* **Coumarin, 3,4-dihydro-**													
—	**Melissic acid**	*see* **Hentriacontanoic acid***													
—	**Melitoxin**	*see* **Dicoumarol**													
—	**Mellitene**	*see* **Benzene, hexamethyl-***													
—	**Mellitic acid**	*see* **Benzenehexacarboxylic acid***													
—	**Mellophanic acid**	*see* **1,2,3,4-Benzenetetracarboxylic acid***													
—	**Menadione**	*see* **1,4-Naphthoquinone, 2-methyl-***													
m157	*p*-**Menthane**(*cis*)	Hexahydro-*p*-cymene. 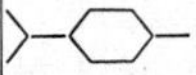	140.27			168.5	0.8039^{20}_4	1.4419^{20}	i	v	v				**B5², 26**
m158	—(*trans*)	$C_{10}H_{20}$. *See* m157	140.27			169–70	0.792^{20}_4	1.4393^{20}	i	v	v				**B5², 27**
—	**Menthanone**	*see* **Menthone**													
m159	Δ^3-*p*-**Menthene**(*d*) (one form)	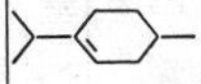	138.25	$[\alpha]^{20}_D$+115.6 (undil)		168	0.8118^{20}_4	1.4524^{20}		s	s			aa s	**B5², 52**
m160	—(*d*) (one form)	$C_{10}H_{18}$. *See* m159	138.25	$[\alpha]_D$ (mut) +29.6 to +54.4		167–8	0.8078^{20}_4								
m161	—(*dl*)	$C_{10}H_{18}$. *See* m159	138.25			168^{754}	0.8129^{20}_4	1.4540^{20}_4							**B5², 53**
m162	—(*l*)	$C_{10}H_{18}$. *See* m159	138.25	$[\alpha]_D$ −13.46 (al)		167–8	0.8112^{19}_{19}	1.4511		s					**B5, 89**
—	$\Delta^{8(9)}$-*p*-**Menthenol**	*see* **Isopulegol**													
—	**Menthenone**	*see* **Piperitone**													
—	**Menthofuran**	*see* **Coumarone, 3,6-dimethyl-4,5,6,7-tetrahydro-**													
m163	**Menthol**(*d*)	3-*p*-Menthanol. Hexahydrothymol. 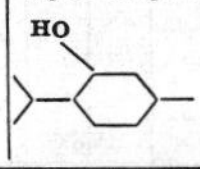	156.27	$[\alpha]_D$+48.3 (al)	38–40					s				os s	**B6², 49**

For explanations, symbols and abbreviations see beginning of table.

No.	Name	Synonyms and Formula	Mol. wt.	Crystalline form, color and specific rotation	m.p. °C	b.p. °C	Density	n_D	Solubility w	al	eth	ace	bz	other solvents	Ref.
	Menthol														
m164	—(*dl*)	$C_{10}H_{20}O$. *See* m163	156.27	nd (peth)	35–6	216^{760}	0.904_{15}^{15}	1.4615^{20}	i	v	v	v		os v	**B6²**, 49
m165	—(*l*)	$C_{10}H_{20}O$. *See* m163	156.27	nd (MeOH) $[\alpha]_D^{20}-48.92$ (p=9.06)	42.5	216.4^{760}	0.890_{15}^{15}	1.460^{22} suc	δ	v	v			chl v peth s aa s	**B6²**, 39
m166	—,3-methyl-butanoate	Menthyl isovalerate. $C_{15}H_{28}O_2$. *See* m163	240.39	$[\alpha]_D-64.02$ (bz, c=10)		129^{9}	0.907^{15}	1.4486^{20}	i	s					**B6¹**, 22
m167	Menthone(*d*)	*d*-3-*p*-Menthanone. $C_{10}H_{18}O$.	154.25	$[\alpha]_D^{18}+24.85$		204^{750}	0.895_{18}^{18}								**B7¹**, 35
m168	—(*dl*)	$C_{10}H_{18}O$	154.25			210.5	0.911^{0}		δ	s	s			aa s	**B7²**, 41
m169	—(*l*)	$C_{10}H_{18}O$	154.25	$[\alpha]_D^{20}-28.46$	−6.6	209.6^{760}	0.8954_{4}^{20}	1.4498^{20}	δ	∞	∞		∞	CS_2 ∞	**B7¹**, 35
—	**Mepheridine**	*see* **4-Piperidinecarboxylic acid, 1-methyl-4-phenyl-,** ethyl ester													
—	**Mesaconic acid**	*see* **Fumaric acid, methyl-**													
m170	**Mescaline**	Mezcaline. 3,4,5-Trimethoxy-β-phenethylamine. $C_{11}H_{17}NO_3$.	211.27		35–6	180^{12}			s	s	δ		s	chl s lig δ	**B13²**, 521
—	**Mesidine**	*see* **Benzene, 2-amino-1,3,5-trimethyl-***													
—	**Mesitaldehyde**	*see* **Benzaldehyde, 2,4,6-trimethyl-**													
—	**Mesitol**	*see* **Benzene, 2-hydroxy-1,3,5-trimethyl-***													
—	**Mesitoic acid**	*see* **Benzoic acid, 2,4,6-trimethyl-**													
—	**Mesitylene**	*see* **Benzene, 1,3,5-trimethyl-***													
—	**Mesitylenic acid**	*see* **Benzoic acid, 3,5-dimethyl-**													
—	**Mesityl oxide**	*see* **3-Penten-2-one, 4-methyl-***													
—	**Mesorcinol**	*see* **Benzene, 2,4-dihydroxy-1,3,5-trimethyl-***													
—	**Mesoxalic acid**	*see* **Malonic acid, oxo-**													
—	**Mestilbol**	*see* **3-Hexene, 3(4-hydroxyphenyl)4(4-methoxyphenyl)-***													
—	**Metacetaldehyde,** (*tetramer*)	*see* **Metaldehyde II**													
m171	**Metacrolein**	2,4,6-Triethenyl-1,3,5-trioxane. 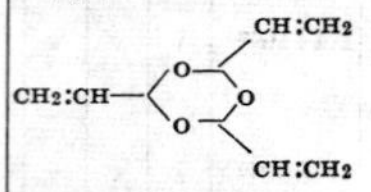	168.20	lf (al)	40	170^{760}			i δ^h	s	s				**B1¹**, 378
—	**Metahemipinic acid**	*see* **Phthalic acid, 4,5-dimethoxy-**													
m172	**Metaldehyde**	Metaacetaldehyde. $(C_2H_4O)_{4-6}$	$(44.05)_{4-6}$	tetr nd or pr (al)	246.2 (sealed tube)	112–5 sub			i	δ^h	δ^h	i		CS_2 i aa i	**B1²**, 671
m173	**Metaldehyde II**	**Metacetaldehyde** (tetramer). 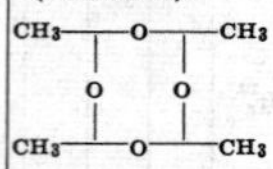	176.20		47	71^{760}								os s	
m174	**Metamekonin**	5,6-Dimethoxyphthalide. 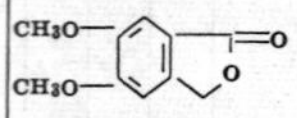	194.12	(dil al)	155–7				i s^h		δ			KOH v	**B18²**, 62
—	**Metanilic acid**	*see* **Benzenesulfonic acid, 3-amino-***													
—	**Methacrylic acid**	*see* **Propenoic acid, 2-methyl-***													
—	**Methadone**	*see* **3-Heptanone, 6-dimethylamino-4,4-diphenyl-***													

For explanations, symbols and abbreviations see beginning of table.

No.	Name	Synonyms and Formula	Mol. wt.	Crystalline form, color and specific rotation	m.p. °C	b.p. °C	Density	n_D	Solubility w	al	eth	ace	bz	other solvents	Ref.
	Methanal														
—	**Methanal***	see **Formaldehyde**													
m175	**Methane***	CH_4	16.04	gas	−182.48	-161.49^{760}	0.415^{-164}		δ	δ	δ		s	MeOH δ to δ	**B1**³, 1
m176	**—,amino-***	Methylamine. CH_3NH_2	31.06	gas	−93.5	-6.3^{760}	0.699^{-11}		v	s	∞				**B4**², 546
m177	—,—,hydrochloride	$CH_3NH_2.HCl$	67.52	dlq tetr ta (al)	227–8	sub^{760} $225\text{–}30^{15}$			s	s	i	i		AcOEt i chl i	
m178	**—,amino-(diphenyl)-***	Benzhydrylamine. $(C_6H_5)_2CHNH_2$	183.24	hex pl	34	304^{763} 176^{23}	1.0635^{20}_{0}	1.5963^{22}	δ						**B12**², 768
—	**—,amino(2-furyl)-**	see **Furan, 2(aminomethyl)-**													
m180	**—,aminooxy-,** hydrochloride.	*O*-Methylhydroxylamine hydrochloride. $CH_3ONH_2.HCl$	83.52	pr	149				s	s					**B1**², 275
m181	**—,(2-aminophenyl)bis-(4-aminophenyl)-***	$C_{19}H_{19}N_3$. *See* m312	289.39	(al)	165				s^h	v	i				**B13**², 150
m182	**—,(3-aminophenyl)bis-(4-aminophenyl)-***	$C_{19}H_{19}N_3$. *See* m312	289.39	nd (eth or eth-lig)	150				i	s	δ			lig i	**B13**, 312
m183	**—,(3-aminophenyl)-diphenyl***	3-Aminotritan. $C_{19}H_{17}N$. *See* m312	259.34	nd (eth)	120						s				**B12**², 790
m184	**—,(4-aminophenyl)-diphenyl-***	4-Aminotritan. *p*-Benzhydrylaniline. $C_{19}H_{17}N$. *See* m312	259.34	pr or lf (eth, lig) ta (peth)	84	ca. 248^{12}			i		s		s	lig s	**B12**², 790
m185	**—,(3-aminophenyl)phenyl-***	3-Aminoditan. *m*-Benzylaniline. $C_{13}H_{13}N$. *See* m271	183.24	(lig)	46									lig s	**B12**, 1323
m186	**—,(4-aminophenyl)phenyl-***	4-Aminoditan. *p*-Benzylaniline. $C_{13}H_{13}N$. *See* m271	183.24	mcl (lig)	34–5	300	1.038^{55}		i	v	v			lig s	**B12**, 1323
—	**—,biphenylyl-(phenyl)**	see **Biphenyl, benzyl-**													
m187	**—,bis(4-aminocyclohexyl)-** (*cis, cis*)	4,4′-Methylenebiscyclohexylamine. H_2N—⬡—CH_2—⬡—NH_2	210.37		61	139^{18}		1.5014^{27} (mixture of isomers)							**Am 73**, 741
m188	—,—(*cis, trans*)	$C_{13}H_{26}N_2$. *See* m187	210.37		36–7	$120.3^{0.8}$	0.9608^{25}_{4}	1.5046^{25}							**Am 73**, 741
m189	—,—(*trans, trans*)	$C_{13}H_{26}N_2$. *See* m187	210.37		64–5	$130\text{–}1^{0.8}$		1.5032^{25}							**Am 73**, 741
m190	**—,bis(2-amino-4-nitrophenyl)-**	$C_{13}H_{12}N_4O_4$. *See* m271	288.26	og pl (al)	205					v^h	δ		δ	aa v lig δ	**B13**², 113
m191	**—,bis(3-amino-4-nitrophenyl)-**	$C_{13}H_{12}N_4O_4$. *See* m271	288.26	red nd (PhOH-al)	228–30					i			i	$PhNO_2$ s^h chl i	**B13**², 113
m192	**—,bis(4-aminophenyl)-***	4,4′-Diaminoditan. $C_{13}H_{14}N_2$. *See* m271	198.27	pl or nd (w), pl (bz)	92–3	$398\text{–}9^{768}$ 257^{18}			δ	v	v		v		**B13**², 111
m193	**—,bis(4-aminophenyl)phenyl-***	4,4′-Diaminotritan. $C_{19}H_{18}N_2$. *See* m312	274.37	pr (bz or eth)	139–40				$δ^h$	v	v			chl s lig s	**B13**², 134
m194	**—,bis(3-carboxy-4-hydroxyphenyl)-***	5,5′-Methylenedisalicylic acid. HO_2C, CO_2H; HO—⬡—CH_2—⬡—OH	288.26	nd	243–4				δ	s	s	s	δ		**B10**², 397
m195	**—,bis(5-chloro-2-hydroxydienyl-***	Dichlorophene. Dihydroxane. $C_{13}H_{10}Cl_2O_2$. *See* m271	269.13		177–8				i	s		s			
m196	**—,bis(4-chlorophenoxy)-***	Neotran. Oxythane. Cl—⬡—OCH_2O—⬡—Cl	269.13		65				δ	δ	v	v			
m197	**—,bis(4-dimethylaminophenyl)-***	$C_{17}H_{22}N_2$. *See* m271	254.38	pl or ta (al, lig)	91–2	390d			i	δ s^h	v		v	CS_2 v ac s	**B15**², 434
m198	**—,bis(4-dimethylaminophenyl)(aminophenyl)-***	$C_{23}H_{27}N_3$. *See* m312	345.49	(al)	151–2					δ					**B13**, 314
m198ⁱ	**—,[2,4-bis(dimethylamino)phenyl]diphenyl-***	$C_{23}H_{26}N_2$. *See* m312	330.48	pl (peth)	122–3				i	δ				os s	**B13**, 273
m199	**—,bis(4-dimethylaminophenyl)(2-hydroxyphenyl)-***	$C_{23}H_{26}N_2O$. *See* m312	346.48	nd (al)	127–8				δ	δ v^h			v	lig δ	**B13**², 440
m200	**—,bis(4-dimethylaminophenyl)-(3-hydroxyphenyl)-***	$C_{23}H_{26}N_2O$. *See* m312	346.48		149				δ	δ					**B13**², 440

For explanations, symbols and abbreviations see beginning of table.

No.	Name	Synonyms and Formula	Mol. wt.	Crystalline form, color and specific rotation	m.p. °C	b.p. °C	Density	n_D	Solubility w	al	eth	ace	bz	other solvents	Ref.
	Methane														
m201	**—,bis(4-dimethyl-aminophenyl)-(4-hydroxy-phenyl)-***	$C_{23}H_{26}N_2O$. *See* m312	346.48	(al)	165	...	...	...	δ	v^h	...	...	v	alk s lig δ	**B13²**, 440
m202	**—,bis(4-dimethyl-aminophenyl) phenyl-**	Leucomalachite green. $C_{23}H_{26}N_2$. *See* m312	330.48	nd or lf (al, bz)	102(94)	...	...	...	i	s	v	...	v	to v lig δ	**B13²**, 135
m203	**—,bis(2,4-dinitro-phenyl)-***	2,2′4,4′-Tetranitroditan. $C_{13}H_8N_4O_8$. *See* m271	348.23	ye pr (aa)	173	...	...	...	...	i	i	...	δ	alk s aa δ	**B5²**, 503
m204	**—,bis(3-formyl-4-hydroxy-phenyl)-**	5,5′-Methylene-disalicylaldehyde. $C_{15}H_{12}O_4$. *See* m271	256.26	pa ye (aa)	140	...	...	...	...	...	...	...	...	aa s^h	**B8**, 436
m205	**—,bis(2-hydroxy-3,5,6-trichloro-phenyl)-***	Gamophen. Hexachlorophen. Hexosan. $C_{13}H_6Cl_6O_2$. *See* m271	406.91	(bz)	164–5	...	...	...	...	s	s	s	...	dil alk s chl s	...
m207	**—,bis(3-hydroxy-phenyl)-***	*m,m*′-Methylenediphenol. $C_{13}H_{12}O_2$. *See* m271	200.23	nd (dil aa)	102–3	...	...	...	s^h	v	v	...	δ	aa v lig δ	**B6**, 995
m208	**—,bis(4-hydroxy-phenyl)-***	*p,p*′-Methylenediphenol. $C_{13}H_{12}O_2$. *See* m271	200.23	lf or nd (w)	162	sub	...	...	...	s	s	...	...	chl, alk s CS_2 i	**B6²**, 964
m209	**—,bis(2-nitro-phenyl)-***	2,2′-Dinitroditan. $C_{13}H_{10}N_2O_4$. *See* m271	258.23	pl or nd (w)	159	...	...	...	δ	v	v	...	...	lig i	**B5**, 595
m210	**—,bis(3-nitro-phenyl)-***	3,3′-Dinitroditan. $C_{13}H_{10}N_2O_4$. *See* m271	258.23	lf (aa)	175.5	...	...	...	...	s	δ	...	v^h	aa v^h	**B5²**, 503
m211	**—,bis(4-nitro-phenyl)-***	4,4′-Dinitroditan. $C_{13}H_{10}N_2O_4$. *See* m271	258.23	nd (bz)	183–5	...	...	...	...	i	δ	...	v^h	aa v^h	**B5**, 595
—	**—,bis(phenyl-imino)-***	*see* **Carbodiimide, *N,N*′-diphenyl-**													
m212	**—,bis(trichloro-silyl)-**	$Cl_3SiCH_2SiCl_3$	272.89	...	...	64^{10}	...	...	...	...	v	...	...	...	**C50**, 9997
m213	**—,bromo-***	Methyl bromide. CH_3Br	94.95	...	−95	3.59	1.732_0^0	1.4234^{10}	δ	∞	∞	...	...	chl, CS_2 ∞	**B1³**, 1143
m214	**—,bromochloro-***	$ClCH_2Br$	129.40	...	−86.5	69	1.991^{19}	1.48^{25}	i	...	...	...	...	os s	...
m215	**—,bromochloro-dinitro-***	$BrCCl(NO_2)_2$	219.40	...	9.2–9.3	$75–6^{15}$	2.0394_4^{20}	1.4793	...	v	...	...	...	...	**B1³**, 115
m216	**—,bromochloro-fluoro-***	$BrCHClF$	147.47	...	−115	36.1^{756}	1.9771_4^0	...	i	...	...	...	...	...	**B1³**, 84
m217	**—,bromodi-chloro-***	$BrCHCl_2$	163.85	...	...	$89–90^{742}$	1.980_4^{20}	1.4964^{20}	i	...	...	...	...	chl ∞, os v	**B1³**, 84
m218	**—,bromodi-fluoro-***	$BrCHF_2$	130.93	...	−14.5	...	1.55^{16}	...	s	v	...	...	...	...	**B1³**, 83
m219	**—,bromodi-fluoronitroso-***	$BrCF_2NO$	159.94	bl gas	...	ca. -12^{760}	...	...	...	...	...	...	...	...	**J1953**, 2072
m220	**—,bromodiiodo-***	$BrCHI_2$	346.75	ye (peth)	60	110^{25}	...	...	...	...	...	...	...	peth δ	**B1**, 72
m221	**—,bromo-(diphenyl)-***	Benzhydryl bromide. α-Bromoditan. $BrCH(C_6H_5)_2$	247.13	tcl (peth)	45	193^{26}δd	...	...	d^h	s	...	...	v	...	**B5²**, 502
m222	**—,bromofluoro-***	$BrCH_2F$	112.94	...	...	18–20	...	...	...	s	...	...	...	chl v	**B1³**, 83
m223	**—,bromoiodo-***	$BrCH_2I$	220.85	...	...	$138–41^{760}$	2.926^{17}	1.6410^{20}	...	...	...	...	...	chl v	**B1³**, 99
m224	**—,bromonitro-***	$BrCH_2NO_2$	139.95	...	...	148–50	...	...	...	v	...	...	...	alk s	**B1³**, 115
m225	**—,bromotri-chloro-***	$BrCCl_3$	198.30	...	−21	104^{760}	2.012_4^{20}	1.5061^{20}	i	∞	∞	...	...	...	**B1³**, 85
m226	**—,bromotri-fluoro-***	$BrCF_3$	148.93	gas	...	−60, -59^{760}	...	...	...	...	...	...	...	chl v	**B1³**, 83
m227	**—,bromotri-nitro-***	$BrC(NO_2)_3$	229.95	...	17–8	56^{10}	...	...	δ	s	...	...	...	chl s	**B1³**, 110
m228	**—,chloro-***	Methyl chloride. CH_3Cl	50.49	gas	−97	-23.76^{760}	0.92^{20}	1.3661^{-10}	s	s	∞	...	...	chl ∞ aa ∞	**B1³**, 36
m229	**—,chlorodi-bromo-***	$ClCHBr_2$	208.31	...	...	$118–20^{748}$	1.451_4^{20}	1.5482^{20}	i	...	...	...	...	os s	**B1³**, 87
m230	**—,chlorodi-fluoro-***	$ClCHF_2$	86.48	gas	−146–7	−39.8	1.4909^{-69}	...	v	...	...	...	...	...	**B1³**, 41
m231	**—,chlorodifluoro-nitro-***	$ClCF_2(NO_2)$	131.48	...	...	25^{760}	...	...	...	...	...	...	...	...	**J1953**, 2075
m232	**—,chlorodifluoro-nitroso-***	$ClCF_2(NO)$	115.48	bl gas	...	ca. -35^{760}	...	...	...	...	...	...	...	...	**J1953**, 2075
m233	**—,chlorodiiodo-***	$ClCHI_2$	302.30	...	−4	$200d^{760}$ 88^{30}	...	...	...	...	...	...	...	chl v	**B1**, 72
m234	**—,chlorodinitro-***	$ClCH(NO_2)_2$	140.49	...	...	$34–6^{13}$	1.6123^{20}	1.4575	...	...	...	...	...	alk s	**B1³**, 115
m235	**—,chloro-(diphenyl)-***	Benzhydryl chloride. α-Chloroditan. $ClCH(C_6H_5)_2$	202.68	nd	20.5 (14)	173^{19}	1.1398_4^{20}	1.5959^{20}	...	...	...	...	...	...	**B5²**, 500
m237	**—,chlorofluoro-***	$ClCH_2F$	68.48	gas	...	−9.1	...	...	...	...	...	...	...	chl v	**B1³**, 41

For explanations, symbols and abbreviations see beginning of table.

No.	Name	Synonyms and Formula	Mol. wt.	Crystalline form, color and specific rotation	m.p. °C	b.p. °C	Density	n_D	Solubility w	al	eth	ace	bz	other solvents	Ref.
	Methane														
m238	**—,chloroiodo-***	$ClCH_2I$	176.38			109^{760}	2.422^{20}_{4}	1.5822^{20}	...	...	...	...	...	chl v liq HF δ	**B1**[2], 99
m239	**—,chloronitro-***	$ClCH_2NO_2$	95.49			122–3 (107)	1.466^{15}		s	...	...	...	...	alk v	**J1953,** 2075
m239[1]	**—,(4-chloro-phenyl)phenyl-**	4-Chloroditan. $C_{13}H_{11}Cl$. *See* m271	202.69		7.5	298^{742}	1.1247^{20}_{0}		...	...	...	...	...		**B5**[2], 500
m240	**—,chlorotri-bromo-***	$ClCBr_3$	287.19	lf (eth)	55	158–9	2.71^{15}		i	...	s	...	...		**B1**[3], 91
m241	**—,chlorotri-fluoro-***	$ClCF_3$	104.46	gas	−181	-81.1^{760}			...	...	...	...	...		**B1**[3], 42
m242	**—,chlorotri-nitro-***	$ClC(NO_2)_3$	185.48		4.5	133–5δd 52^{3}	1.6810^{15}_{4}	1.4560	δ	s	s	...	...	chl s	**B1**[3], 116
m243	**—,chloro(tri-phenyl)-***	Trityl chloride. $ClC(C_6H_5)_3$	278.78	nd or pr (bz-peth)	112–3	310^{760} $230–5^{20}$			i d^h	δ	v	...	v	chl, CCl_4 v CS_2 v	**B5**[2], 615
—	**—,cyano-**	*see* **Acetic acid,** nitrile													
m243[1]	**—,deuterotri-chloro-***	Chloroform-*d*. $CDCl_3$	120.39		−64.12				i	...	...	...	...		**B1**[3], 63
m244	**—,diazo-***	Azimethylene. CH_2N_2	42.04	ye gas	−145	−23			d	s^h	s	...	...		**B1**[2], 650
m245	**—,diazo-(diphenyl)-***	$(C_6H_5)_2CN_2$	194.24	bl-red nd (peth)	29–30	exp			...	s	...	...	...	os v	**B7**[2], 358
—	**—,dibenzoyl-**	*see* **1,3-Propanedione, 1,3-diphenyl-***													
m246	**—,dibromo-***	Methylene bromide. CH_2Br_2	173.85		−52 fr	97	2.4921	1.5419^{20}	δ	∞	∞	∞	...		**B1**[3], 85
m247	**—,dibromodi-chloro-***	Br_2CCl_2	242.76	nd	22	*ca.* 135	2.42^{25}_{0}		i	...	...	...	...	os s	**B1**[3], 88
m248	**—,dibromodi-fluoro-***	F_2CBr_2	209.83			24.5^{760}			s	...	...	...	...		**B1**[1], 16
m249	**—,dibromo-dinitro-***	$Br_2C(NO_2)_2$	263.84	nd	5.5	158d 77^{21}	2.4440^{20}_{4}	1.5215^{2}	i	∞	...	...	...		**B1**[3], 115
m250	**—,dibromo-fluoro-***	$FCHBr_2$	191.84			64.9^{757}	2.421^{20}_{4}	1.4685^{20}	...	...	...	...	...		**B1**[3], 87
m251	**—,dibromoiodo-***	$ICHBr_2$	299.74	pl (peth)	22.5	91^{42} $(101–4^{15})$			...	...	...	...	...	peth s	**B1**[3], 99
m252	**—,dichloro-***	Methylene chloride. CH_2Cl_2	84.93		fr −97	40–1	1.335^{15}_{4}	1.3348^{15}	δ	∞	∞	...	...		**B1**[2], 13
m253	**—,dichlorodi-fluoro-***	Freon. Cl_2CF_2	120.91		−155	-29^{760}			s	s	s	...	...		**B1**, 61
m254	**—,dichloro-dinitro-***	$Cl_2C(NO_2)_2$	174.93			46^{20}	1.6124^{20}_{4}	1.4575	...	...	...	...	...		**B1**[3], 115
m255	**—,dichloro-(diphenyl)-***	Benzophenone dichloride. $(C_6H_5)_2CCl_2$	237.13			305d 172^{16}	1.235^{18}		...	...	s	...	s		**B5**[2], 501
m256	**—,dichloro-fluoro-***	$FCHCl_2$	102.92		−135	9^{760}	1.426^{0}	1.3724^{9}	i	s	s	...	...		**B1**[3], 47
m257	**—,dichloroiodo-***	$ICHCl_2$	210.83			132^{760}	2.392^{20}_{4}	1.5840^{20}	...	...	...	...	...	chl v	**B1**[3], 99
m258	**—,dichloronitro-***	Cl_2CHNO_2	129.93			107^{760}			...	...	...	...	...		**J1953,** 2075
m259	**—,difluoro-***	Methylene fluoride. CH_2F_2	52.02			−51.6			i	s	...	...	...		**B1**, 59
m260	**—,(difluoro-amino)trifluoro-**	Perfluoromethylamine. F_3CNF_2	121.01			-75^{760}			...	...	...	...	...		**J1950,** 1966
m261	**—,difluoroiodo-***	$ICHF_2$	177.92		−122	21.6^{760}	3.238^{-19}		...	...	...	...	...		**B1**[3], 98
m262	**—,di(2-furyl)-**	2-Furfurylfuran. O —CH_2— O	148.15			$94^{22.5}$	1.102^{20}_{4}	1.5049^{20}	i	s	s	...	...	MeOH s	**Am 55,** 3302
m263	**—,diiodo-***	Methylene iodide. CH_2I_2	267.87	ye nd or lf	6	181^{760}δd $106–7^{70}$	3.3254^{20}_{4}	1.7559^{10}	δ	s	s	...	s	chl s	**B1**[2], 37
m264	**—,diiodofluoro-***	$FCHI_2$	285.84	pa ye	−34	$100–1^{760}$	3.1969^{22}		δ	s	s	...	...		**B1**[3], 102
m265	**—,[3(dimethyl-amino)-phenyl] [4(dimethyl-amino)-phenyl]-***	$C_{23}H_{26}N_2$. *See* m271	330.48	(abs al)	83–4				...	$δ^h$	$δ^h$	...	...		**B13,** 274
m267	**—,di(1-naphthyl)-**	$C_{10}H_7^{\alpha}CH_2C_{10}H_7^{\alpha}$	268.36	pr or nd (al)	109	>360 $270–2^{14}$			δ s^h	δ v^h	s	...	s	chl s peth δ	**B5**[2], 650

For explanations, symbols and abbreviations see beginning of table.

No.	Name	Synonyms and Formula	Mol. wt.	Crystalline form, color and specific rotation	m.p. °C	b.p. °C	Density	n_D	Solubility w	al	eth	ace	bz	other solvents	Ref
	Methane														
m268	**—,di(2-naphthyl)-**	$C_{10}H_7^{\beta}CH_2C_{10}H_7^{\beta}$	268.36	nd (al or eth)	93	...	...	...	i	δ	...	...	s	...	**B5**[1], 360
m269	**—,dinitro-***	$CH_2(NO_2)_2$	106.04	ye nd	<−15	100exp	...	...	i δ^h	s	s	...	...	...	**B1**[3], 115
m270	**—,dinitro-(diphenyl)-***	$(C_6H_5)_2C(NO_2)_2$	258.23	pl (dil al)	...	78	...	...	...	v^h	v	...	v	chl v	**B5**, 596
m271	**—,diphenyl-***	Ditan. 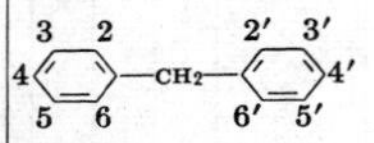	168.24	pr nd	26–7	265.6^{760}	1.0060_4^{20}	1.5768^{20}	i	s	s	...	...	chl s	**B5**[2], 498
m272	**—,diphenyl(3-tolyl)-**	3-Methyltritan. 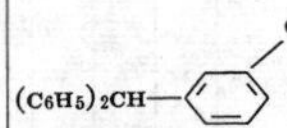	258.36	pr (al or MeOH)	62	354^{706}	1.07^{16}	...	...	δ	v	...	v	chl v aa v lig s	**B5**, 710
m273	**—,di(α-pyrryl)-**	2,2′-Methylenedipyrrole. 2,2′-Pyrromethane. 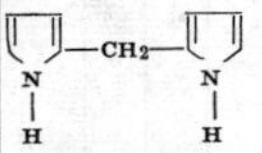	146.19	lf or nd	73	$163–7^{12}$	...	...	i	...	s	...	s	lig s^h	**B23**, 167
m274	**—,disilano-**	2-Carbatrisilane. Disilylmethane. $CH_2(SiH_3)_2$	76.26	...	...	14.7^{754}	0.6979_0^4	1.4117^4	...	...	...	...	...	...	**C48**, 5080
m275	**—,(di-4-tolyl)-phenyl-***	4,4′-Dimethyltritan. 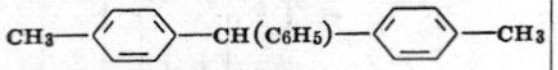	272.39	nd (MeOH)	56	$218–20^{12}$	...	...	...	s	v	...	v	chl v CS_2 v aa s lig s	**B5**[1], 352
m276	**—,fluoro-***	Methyl fluoride. CH_3F	34.03	...	−141.8	-78.4^{760}	0.8428^{-60}	...	v	v	v	...	...	...	**B1**[3], 33
m277	**—,fluoroiodo-***	FCH_2I	159.94	...	...	53.4	2.366_4^{20}	1.4911^{20}	...	...	...	...	...	chl v	**B1**[3], 98
m278	**—,fluorotri-bromo-***	$FCBr_3$	270.76	...	...	106^{760}	2.7648_4^{20}	1.5256^{20}	i	...	s	...	...	...	**B1**[3], 91
m279	**—,hydroxyl-amino-***	*N*-Methylhydroxylamine. CH_3NHOH	47.06	hyg nd	42	62.5^{15}	1.0003_4^{20}	1.4164^{20}	v	v	δ	...	δ	MeOH v lig δ	**B4**[2], 952
m280	**—,(2-hydroxy-phenyl)-(4-hydroxy-phenyl)-***	2,4′-Dihydroxyditan. $C_{13}H_{12}O_2$. *See* m271	200.24	nd (dil al)	119–20	...	...	...	δ	v	v	...	...	...	**B6**, 994
m281	**—,(4-hydroxy-phenyl)-phenyl-***	4-Benzylphenol. p-Hydroxyditan. $C_{13}H_{12}O$. *See* m271	184.24	nd or pl (al)	84	320–30 $198–200^{10}$	...	...	δ^h	s v^h	s	...	s	chl s aa s	**B6**[2], 629
m282	**—,iodo-***	Methyl iodide. CH_3I	141.94	...	−66.5	42.5	2.28_4^{20}	1.5293^{21}	δ	∞	∞	...	...	...	**B1**[2], 35
m283	**—,iodotrichloro-***	$ICCl_3$	245.27	...	...	142	2.355_4^{20}	1.5854^{20}	...	...	...	...	...	chl v	**B1**[3], 99
m284	**—,(1-naphthyl)-phenyl-***	α-Benzylnaphthalene. $C_{10}H_7^{\alpha}CH_2C_6H_5$	218.30	mcl lf or ta (al)	59	350^{760}	1.165^0	...	i	δ	v	...	s	chl s CS_2 v	**B5**[2], 604
m285	**—,(2-naphthyl) phenyl-***	β-Benzylnaphthalene. $C_{10}H_7^{\beta}CH_2C_6H_5$	218.30	mcl pr (al)	35.5	350^{760}	1.176^0	...	i	δ v^h	...	...	v	...	**B5**, 690
m286	**—,nitro-***	CH_3NO_2	61.04	mcl	−28.5	100.8^{760}	1.1354_4^{22}	1.3935^{20}	s	s	s	...	...	alk s	**B1**[2], 40
m287	**—,(2-nitro-phenyl)(4-nitro-phenyl)-***	2,4′-Dinitrodiian. $C_{13}H_{10}N_2O_4$. *See* m271	258.24	ye mcl pr (bz)	118	...	...	...	...	...	...	...	δ s^h	...	**B5**, 595
m288	**—,(3-nitro-phenyl)(4-nitro-phenyl)-***	3,4′-Dinitroditan. $C_{13}H_{10}N_2O_4$. *See* m271	258.24	nd (al)	103–4	...	...	...	...	δ s^h	...	...	...	...	**B5**, 595
m289	**—,nitrosotri-fluoro-***	F_3CNO	99.01	...	−197 (−205)	-84^{760} (-93^{760})	...	...	...	...	...	...	...	...	**J1953** 3755
m290	**—,nitrotri-bromo-***	Bromopicrin. Nitrobromoform. Br_3CNO_2	297.74	pr	10.25	exp^{760} $89–90^{20}$	2.7930_4^{20}	1.5790^{20}	i	s	s	...	...	...	**B1**[3], 115
m291	**—,nitrotri-chloro-***	Chloropicrin. Nitrochloroform. Cl_3CNO_2	164.38	...	−64	112^{757}	1.6558_4^{20}	1.4611^{20}	δ^h	∞	...	∞	∞	MeOH ∞ aa ∞	**B1**[3], 113
m291[1]	**—,nitrotrifluoro-***	Fuoropicrin. F_3CNO_2	115.01	...	...	−33.6 (−20)	...	...	...	...	...	...	...	...	**J1953**, 2075

For explanations, symbols and abbreviations see beginning of table.

No.	Name	Synonyms and Formula	Mol. wt.	Crystalline form, color and specific rotation	m.p. °C	b.p. °C	Density	n_D	Solubility w	al	eth	ace	bz	other solvents	Ref.
	Methane														
m292	**—,(pentafluoro-thio)trifluoro-**	$F_3C(SF_5)$	196.06	...	...	-20^{760}	...	...	i	i	...	...	...	CS_2 s	**J1953,** 2372
m293	**—,phenyl(3-tolyl)-**	*m*-Benzyltoluene. 3-Methylditan. (structure: C_6H_5–CH_2–C_6H_4–CH_3)	182.27	...	...	275^{747}	0.997^{18}	...	...	s	s	...	...	...	**B5,** 607
m294	**—,phenyl(4-tolyl)-**	*p*-Benzyltoluene. 4-Methylditan. (structure: C_6H_5–CH_2–C_6H_4–CH_3)	182.27	...	−30	286^{760}	0.9978^{19}	1.5710^{19}	...	s	s	...	...	chl s aa s	**B5^2,** 511
m295	**—,tetrabromo-***	Carbon tetrabromide. CBr_4	331.65	mcl ta	90–1	$189–90^{760}$	2.9609^{100}_{4}	...	i	s	s	...	...	chl s liq HF δ	**B1^3,** 92
m296	**—,tetrachloro-***	Carbon tetrachloride. CCl_4	153.82	...	−22.96	76.75^{760}	1.5942^{20}_{4}	1.4664^{20}	i	s	∞	...	∞	chl ∞	**B1^3,** 65
m297	**—,tetrafluoro-***	Carbon tetrafluoride. CF_4	88.00	...	−183.7	-129^{754}	3.034^{0}	...	δ	...	...	...	...	...	**B1^3,** 35
m298	**—,tetraiodo-***	Carbon tetraiodide. CI_4	519.63	red lf (bz or chl)	171d	$130–40^{1-2}$ sub	4.50^{0}	...	i	i	...	...	...	chl CCl_4 s Py s	**B1^3,** 104
—	**—,tetrakis (hydroxy-methyl)-***	*see* **Pentaerythritol**													
m299	**—,tetranitro-***	$C(NO_2)_4$	196.03	...	13	126	1.6372^{21}_{4}	1.4398^{17}	i	s	s	...	...	...	**B1^2,** 47
m300	**—,tetraphenyl-***	$C(C_6H_5)_4$	320.44	rh nd (bz or Ac_2O)	285	431^{760} (sub)	...	...	i	i	i	...	s^h	to s^h aa s^h lig i	**B5^2,** 672
m301	**—,tribenzoyl-** (enol form)	$(C_6H_5CO)_3CH$	328.37	...	155	...	...	...	...	...	...	...	...	chl v	**B7^1,** 485
m302	**—,—,(keto form)**	$(C_6H_5CO)_3CH$	328.37	nd (al or ace)	231	sub	...	...	i	i δ^h	...	δ	δ	chl δ	**B7^2,** 842
m303	**—,tribromo-***	Bromoform. $CHBr_3$	252.75	hex sc	8.3	149.5^{760}	2.8899^{20}_{4}	1.5976^{20}	δ	∞	∞	...	s	chl s lig s	**B1^3,** 88
m306	**—,trichloro-***	Chloroform. $CHCl_3$	119.38	...	−63.5	61.2^{760}	1.4916^{18}_{4}	1.4433^{25}	δ	∞	∞	s	∞	lig ∞	**B1^2,** 14
—	**—,—,*d***	*see* **Methane, deuterotri-chloro-***													
m308	**—,trifluoro-***	Fluoroform. CHF_3	70.01	...	−163	-82.2^{760}	1.52^{-100}	...	δ	v	...	...	...	chl δ	**B1^3,** 34
m310	**—,triiodo-***	Iodoform. CHI_3	393.73	ye hex pr or nd (ace)	119–21	...	4.008^{20}_{4}	...	i	s^h	s	...	...	chl, CS_2 s	**B1^3,** 102
m311	**—,trinitro-***	Nitroform. $CH(NO_2)_3$	151.04	...	15(19)	exp^{760} $45–7^{22}$	1.479^{20}_{4}	...	s	...	...	...	...	alk s	**B1^3,** 116
m312	**—,triphenyl-***	Tritan. (structure: $(C_6H_5)_3CH$, ring positions 2–6, 2′–6′, 2″–6″)	244.34	rh	94	$358–9^{754}$	1.014^{99}_{4}	1.5839^{99}	...	δ v^h	v	...	...	chl v	**B5^2,** 613
m314	**—,tri(2-tolyl)-**	(structure: $(CH_3C_6H_4)_3CH$)	286.42	nd	130–1	...	...	...	...	s	s	...	...	MeOH s	...
m315	**—,tris(4-amino-phenyl)-***	*p*-Leucaniline. $C_{19}H_{19}N_3$. *See* m312	289.38	lf (w, al or bz)	202.5	...	...	...	i	s	s	...	...	...	**B13^2,** 150
m316	**—,tris(4(di-methylamino)-phenyl)-***	Leucocrystal violet. $C_{25}H_{31}N_3$. *See* m312	373.55	lf (al), nd (bz or lig)	175	...	...	...	i	δ v^h	v	...	v	chl v aa v	**B13^2,** 150
—	**—,tris(ethyl-thio)-***	*see* **Orthoformic acid, trithio-,** triethyl ester													
m317	**—,tris(4-hydroxy-phenyl)-***	Leucoaurin. $C_{19}H_{16}O_3$. *See* m312	292.34	pr (aa)	240	...	...	...	δ	v	s	...	...	aa v	**B6^2,** 1106
m318	**—,tris(4-nitro-phenyl)-***	$C_{19}H_{13}N_3O_6$. *See* m312	379.33	sc (bz)	212.5	...	...	...	...	...	δ	...	δ	aa δ	**B5^2,** 618
m319	**Methanearsonic acid***	Methylarsonic acid. $CH_3AsO(OH)_2$	139.97	lf (al)	160–1	...	...	...	s	s	...	...	...	...	**B4^2,** 996
m320	**Methanedisul-fonic acid*,** dihydrate	Methionic acid. $CH_2(SO_3H)_2 \cdot 2H_2O$	212.20	hygr nd (w+2w)	...	220–70^{15-20} d	...	...	s	s^h	...	...	...	...	**B1^2,** 644
—	**Methaneortho-siliconic acid**	*see* **Silane, methyltri-hydroxy-**													

For explanations, symbols and abbreviations see beginning of table.

No.	Name	Synonyms and Formula	Mol. wt.	Crystalline form, color and specific rotation	m.p. °C	b.p. °C	Density	n_D	Solubility w	al	eth	ace	bz	other solvents	Ref.
	Methanephosphonic acid														
m323	**Methanephosphonic acid***	Methylphosphonic acid. $CH_3PO(OH)_2$	96.02	pl	104	d			v	v	v		i	peth i	
m324	—,diethyl ester*	Diethyl methylphosphonate. $CH_3PO(OC_2H_5)_2$	152.13			194^{763} 85^{15}	1.0406_4^{30}	1.4101^{30}	δ	s					**J1954,** 3222
m325	—,dimethyl ester*	Dimethyl methylphosphonate. $CH_3PO(OCH_3)_2$	124.08			79.5^{20}	1.1507_4^{20}	1.4099^{30}	s	s	s				**J1954,** 3222
m326	**—,trifluoro-,** diammonium salt*	$F_3CPO(ONH_4)_2$	184.06		212–6d				s						**J1954,** 3598
m328	**Methanesiliconic acid***	Silicoacetic acid. CH_3SiO_2H	76.13	amor					i		s			KOH s	**B4,** 629
m329	**Methanesulfenic acid, trichloro-,** chloride	Perchloromethylmercaptan. Cl_3CSCl	185.89	ye		149	1.6996_4^{20}	1.5484							**B3**^{2}, 106
m330	**—,trifluoro-,** chloride	Trifluoromethylsulfenyl chloride F_3CSCl	136.52	ye		-0.7^{760}			i d						**J1953,** 3219
m332	**Methanesulfonic acid***	Methylsulfonic acid. CH_3SO_3H	96.11		20	167^{10}	1.4812_4^{18}	1.4317^{16}	v	s	v				**B4**^{2}, 524
m333	—,chloride	Methanesulfonyl chloride*. CH_3SO_2Cl	114.55			161^{730}	1.480_4^{18}		i	s	s				**B4**^{2}, 525
m334	**—,phenyl-,** amide, *N*-methyl-	*N*-Methylbenzylsulfonamide. $C_6H_5CH_2SO_2NHCH_3$	185.25	nd or lf (aa-lig)	108–9				s^h	s	s			alk s	**B11**^{2}, 73
m335	**—,trichloro-,** chloride	Cl_3CSO_2Cl	217.89		135	170(δsub)			i d^h	s d^h	s			CS_2 s	**B3**^{2}, 16
m336	**—,trifluoro-***	F_3CSO_3H	150.08	hyg liq	34(+1w)	162^{760}				d	∞ d^h				**J1955,** 2901
m337	—,—,amide	$F_3CSO_2NH_2$	149.09		119				s					chl s	**J1956,** 173
m338	—,—,—,*N*,*N*-diethyl-	$F_3CSO_2N(C_2H_5)_2$	205.20			55^7			i		s				**J1956,** 173
m339	—,—,anhydride*	$(F_3CSO_2)_2O$	282.14			84^{760}			d	d					**J1957,** 4069
m340	—,—,chloride	F_3CSO_2Cl	168.52			31.6^{760}			i						**J1955,** 2901
m341	—,—,ethyl ester	$F_3CSO_3C_2H_5$	178.13			115^{760}			d		s				**J1956,** 175
m342	—,—,fluoride	F_3CSO_2F	152.07			-21.7^{760}			d	i^h					**J1956,** 175
m343	—,—,potassium salt	CF_3SO_3K	188.17		230										**J1957,** 4069
m344	—,—,sodium salt	CF_3SO_3Na	172.06		248							s			**J1957,** 4069
m345	**Methanethiol***	Methylmercaptan. CH_3SH	48.11		−123	6	0.8665_4^{20}		δ^h	v	v				**B1**^{3}, 1212
m346	—,**(2-furyl)-**	Furfurylmercaptan. (structure: furan ring with $-CH_2SH$; O)	114.17			155^{760}	1.1319_4^{20}	1.5324^{20}	i						**B17**^{2}, 116
m347	**Methanesulfinic acid, amino-(imino)-***	$HN{:}C(NH_2)SO_2H$	108.12	nd (al)	144d				v					os i	**B3**^{1}, 36
m348	**Methanetricarboxylic acid,** trimethyl ester*	Tricarboxymethoxymethane. $CH(CO_2CH_3)_3$	190.15	pr (MeOH)	45–6	243 128^{15}				v	v		v	chl v MeOH s^h	**B2**^{2}, 680
—	Methanal*	*see* **Formaldehyde**													
—	**Methanoic acid***	*see* **Formic acid**													
m349	**Methanol***	Carbinol. Methyl alcohol. Wood alcohol. CH_3OH	32.04		−97.8	64.96^{760}	0.7914_4^{20}	1.3288^{20}	∞	∞	∞	∞			**B1**^{3}, 1147
m349[1]	**Methanol-*d***	*O*-Deuteromethanol. CH_3OD	33.05		−100	65.5^{760}	0.8127_4^{20}		∞	∞	∞				**B1**^{3}, 1186
m349[2]	**Methan-*d*³-ol-*d***	Deuteroxy(trideutero)-methane*. Tetradeuteromethanol. D_3COD	36.07			65.4^{760}			∞	∞	∞				**B1**^{3}, 1187
m350	**Methanol, (2-amino-3-methylphenyl)-***	2-Amino-3-methylbenzyl alcohol (structure: CH_3, NH_2, $-CH_2OH$)	137.18	nd (bz)	71	$135–45^{12}$							δ s^h		**B13**^{2}, 367

For explanations, symbols and abbreviations see beginning of table.

No.	Name	Synonyms and Formula	Mol. wt.	Crystalline form, color and specific rotation	m.p. °C	b.p. °C	Density	n_D	Solubility w	al	eth	ace	bz	other solvents	Ref.
	Methanol														
m351	—,**(2-amino-4-methyl-phenyl)-***	2-Amino-4-methylbenzyl alcohol. NH2 CH3— —CH2OH	137.18	nd (bz)	141	140–50[13]							δ s[h]		**B13**[2], 369
m352	—,**(2-amino-5-methyl phenyl)-***	2-Amino-5-methylbenzyl alcohol. NH2 —CH2OH CH3	137.18	nd (bz)	123	145–50[13]							δ s[h]		**B13**[2], 367
m353	—,**(4-amino-3-methylphenyl)-bis(4-amino-phenyl)-***	Rosaniline. $C_{20}H_{21}N_3O$. *See* m381	319.39	nd(w)	186d				δ	s	i				**B13**, 763
m355	—,**(4-amino-phenyl)phenyl-*-**	*p*-Aminobenzhydrol. $C_{13}H_{13}NO$. *See* m366	199.24	nd (w or bz)	121					v	δ	s		MeOH v AcOEt v, peth δ	**B13**, 696
m356	—,**bis[4(dimethyl-amino)phenyl]-***	Michler's hydrol. $C_{17}H_{22}N_2O$. *See* m366	270.36	lt gr lf or pr (bz)	95–6				i	v[h]	s		s	aa s	**B13**[2], 423
m357	—,**bis[2(dimethyl-amino)-phenyl] phenyl-***	$C_{23}H_{26}N_2O$. *See* m381	346.48	pr (lig)	105						δ			ac s lig δ	**B13**, 741
m358	—,**bis[3(dimethyl-amino)-phenyl] phenyl-***	$C_{23}H_{26}N_2O$. *See* m381	346.48	(eth)	128–9						v[h]			aa s	**B13**, 742
m359	—,**bis[4(dimethyl-amino)phenyl] phenyl-***	$C_{23}H_{26}N_2O$. *See* m381	346.48	(eth, bz, lig, MeOH or peth)	121–3 (107)						v		v[h]	ac s lig v[h]	**B13**[2], 442
m360	—,**bis(4-hydroxy-phenyl)-phenyl-***	Benzaurin. $C_{19}H_{16}O_3$ *See* m381	292.34	ye-red powd	100				i	δ	δ		δ		**B8**[2], 245
m362	—,**cyclohexyl-**	Hexahydrobenzyl alcohol. Hydroxymethylcyclohexane. —CH2OH	114.18			184–6[784] 83[14]	0.9215^{25}_{4}	1.4640^{25}		s	s				
m363	—,**[2(dimethyl-amino)phenyl]-[3(dimethyl-amino)phenyl]-phenyl-***	$C_{23}H_{26}N_2O$. *See* m381	346.48	pl (bz)	183–4					δ	δ		s	HCl s	**B13**, 742
m364	—,**[2(dimethyl-amino)phenyl]-[4(dimethyl-aminophenyl]-phenyl-***	$C_{23}H_{26}N_2O$. *See* m381	346.48	(al)	169–70					δ	δ		v	ac s	**B13**, 742
m365	—,**[3(dimethyl-amino)phenyl]-[4(dimethyl-amino)phenyl]-phenyl-***	$C_{23}H_{26}N_2O$. *See* m381	346.48	pr (bz-al)	140					δ	δ		v	ac s	**B13**, 742
m366	—,**diphenyl-***	Benzhydrol. 3′ 2′ 4′ 5′ 6′ —CHOH— 2 3 4 6 5	184.23	nd (lig)	69	297–8[748] 176[13]			δ[h]	v	v			CCl4 v chl v aa s lig δ	**B6**[2], 631
m367	—,**diphenyl-(1-naphthyl)-***	$C_{10}H_7^{\alpha}COH(C_6H_5)_2$	310.37	(lig)	136.5	d			i	v[h]	v		s	lig i s[h]	**B6**[2], 721
m368	—,**diphenyl-(2-naphthyl)-***	$C_{10}H_7^{\beta}COH(C_6H_5)_2$	310.37	pr (eth-lig)	118					v			v	oos s lig i	**B6**[2], 722
m368[1]	—,**di(4-tolyl)-**	$(CH_3-C_6H_4-)_2CHOH$	212.28	nd (al)	68				i	s	s	s		chl s aa s	**B6**, 688
m369	—,**(2-furyl)-**	Furfuryl alcohol. —CH2OH O	98.10	colorless to ye		171[750] 82–3[25]	1.1296^{20}_{4}	1.4845^{20}	∞ d	v	v				**B17**[2], 113
m370	—,**(4-methoxy-phenyl)-phenyl-***	*p*-Anisylphenylcarbinol. *p*-Methoxybenzhydrol. $C_{14}H_{14}O_2$. *See* m366	214.25	nd (w, lig or dil al)	60(68)				s[h]	v			s	chl s lig s[h]	**B6**[2], 965

For explanations, symbols and abbreviations see beginning of table.

No.	Name	Synonyms and Formula	Mol. wt.	Crystalline form, color and specific rotation	m.p. °C	b.p. °C	Density	n_D	Solubility						Ref.
									w	al	eth	ace	bz	other solvents	
	Methanol														
m371	**—,(5-methyl-2-furyl)-**	5-Methylfurfuryl alcohol. 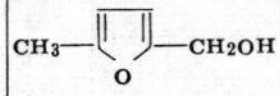	112.13			194–6^{744} 81^{13}	1.0769$^{20}_{4}$	1.4853^{20}		v	v				**B17^1, 56**
m375	**—,(1-naphthyl)-***.	1(Hydroxymethyl)-naphthalene*. $C_{10}H_7^{\alpha}CH_2OH$	158.20	nd (w or al)	60	301^{715} 163^{12}			δ^h	v	v				**B6^2, 617**
m376	**—,(1-naphthyl)-phenyl-***	$C_{10}H_7^{\alpha}CHOHC_6H_5$	234.30	(al or lig)	86.5	ca. 360			i	v	v		v	lig δ	**B6^2, 681**
m377	**—,(2-naphthyl)-phenyl-***	$C_{10}H_7^{\beta}CHOHC_6H_5$	234.30	nd (al or lig)	87–8				i	s	s		s	to s lig δ	**B6^2, 681**
m378	**—,(2-tetrahydro-furyl)-**	Tetrahydrofurfuryl alcohol. 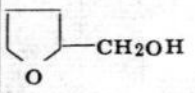	102.14	hyg		177–8^{750} 80–2^{20}	1.0544$^{20}_{4}$	1.4517^{20}							**B17^2, 106**
m379	**—,(2-thienyl)-**	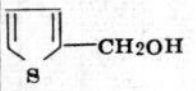	114.16			207^{760}			i	v	v				**B17, 113**
m379^1	**—,(2-tolyl)-**	*o*-Methylbenzyl alcohol. 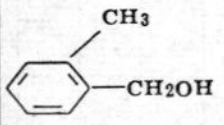	122.16	nd	35	223^{750}	1.023^{40}		δ v^h	v	v			chl v	**B6^2, 457**
m379^2	**—,(3-tolyl)-**	*m*-Methylbenzyl alcohol. 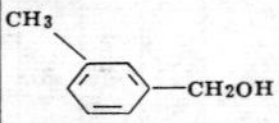	122.16		<−20	215–6^{760} 108–11^{10}	0.9157^{17}		δ s^h	v	v				**B6^2, 465**
m380	**—,(4-tolyl)-**	*p*-Methylbenzyl alcohol. 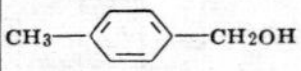	122.16	nd (heptane)	59	217^{760} 116–8^{20}			δ s^h	v	v				**B6^2, 469**
m381	**—,triphenyl-***	Triphenylcarbinol. Tritanol 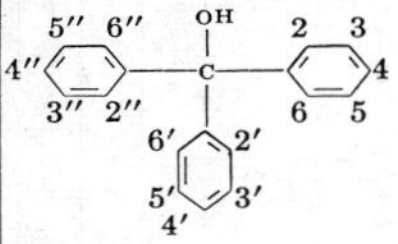	260.30	pl (al), trig (bz), rh (CCl_4)	164.2	380	1.199$^{0}_{4}$		i	v	v		v	peth i aa s	**B6^2, 686**
m382	**—,tris(4-amino-phenyl)-***	Pararosaniline. $C_{19}H_{19}N_3O$. *See* m381	305.37	colorless to red lf	189 (ca. 205)				δ	s	δ				**B13^2, 447**
m383	**—,tris-(4-biphenylyl)-**	Tri-p-xenylcarbinol. 	488.63	nd (aa or bz)	212					δ			v	aa s^h	**B6^2, 741**
m383^1	**—,tris(3-nitro-phenyl)-***	$C_{19}H_{13}N_3O_7$. *See* m381	395.32	rhd (MeOH, chl or AcOEt-lig)	167					δ^h	δ		s	CS_2 δ aa s	**B6^1, 352**
m384	**—,tris(4-nitro-phenyl)-***	$C_{19}H_{13}N_3O_7$. *See* m381	395.32	mcl pr (bz or aa)	189					δ^h	δ		s	CS_2δ aa s	**B6^1, 352**
m386	**Methantheline bromide**	Banthine bromide. 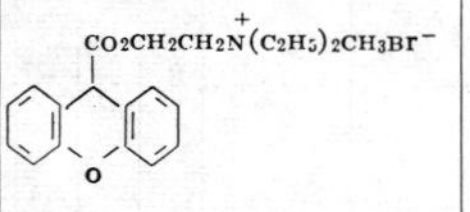	420.34	(*i*-PrOH)	171–2				s	s	i			chl s	
m387	**Methapryilene,** (base)	Histadyl base. Tenalin base. Thenylene base. 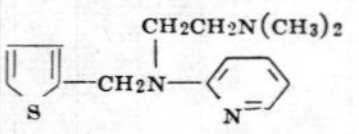	261.38			173–5^3		1.5842^{25}							
—	**Methionic acid**	*see* **Methanedisulfonic acid***													
m388	**Methionine**(*DL*)	*dl*-2-Amino-4(methylthio)-butanoic acid. $CH_3SCH_2CH_2CH(NH_2)CO_2H$	149.22	pl (al)	218d		1.340		v	δ	i				**B4^2, 938**

For explanations, symbols and abbreviations see beginning of table.

No.	Name	Synonyms and Formula	Mol. wt.	Crystalline form, color and specific rotation	m.p. °C	b.p. °C	Density	n_D	Solubility w	al	eth	ace	bz	other solvents	Ref.
	Methionine														
m389	—(*L*)	$CH_3SCH_2CH_2CH(NH_2)CO_2H$.	149.22	hex pl (dil al) $[\alpha]^{25}_D -8.2$ (c=1)	280–1d	...	...	...	s	i	i	i	i	peth i	**B4**², 938
—	**Methone**	*see* **1,3-Cyclohexanedione, 5,5-dimethyl-***													
—	**Methoxychlor**	*see* **Ethane, 1,1-bis(4-methoxyphenyl)-1,2,2-trichloro-***													
—	**Methyl alcohol**	*see* **Methanol***													
—	**Methylamine**	*see* **Methane, amino-***													
—	**Methyl cellosolve**	*see* **Ethanol, 2-methoxy-***													
m390	**Methyl green**	Paris green. Heptamethyl pararosaniline chloride. 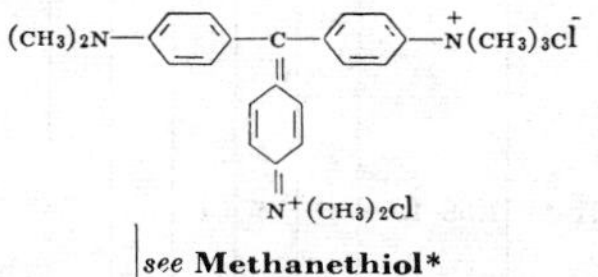	458.46	gr pd	...	...	...	...	s	δ	i	...	...	...	**B13**², 451
—	**Methyl mercaptan**	*see* **Methanethiol***													
m391	**Methyl orange**	Sodium 4′-dimethylaminoazobenzene-4-sulfonate. Helianthin. Orange III. 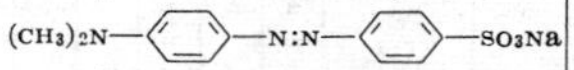	327.35	og ye pl	d	...	...	...	δ	s	i	...	...	Py δ	**B16**, 331
m392	**Methyl red**	4′-Dimethylaminoazobenzene-2-carboxylic acid. 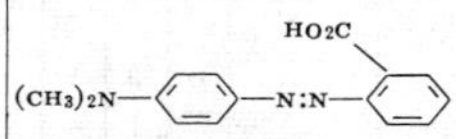	269.31	vt or red pr (to or bz), nd (aq aa), lf (dil al)	183	...	...	...	δ	s	...	v^h	v^h	chl v peth δ aa v lig δ	**B16**², 164
m393	**Methylene blue**	3,9-Bisdimethylamino-phenazothionium chloride. 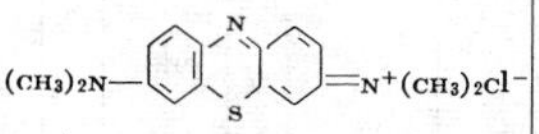	319.87	dk gr cr or powd	...	...	...	...	s	s	...	...	...	chl s	...
—	**Methylene bromide**	*see* **Methane, dibromo-***													
—	**Methylene chloride**	*see* **Methane, dichloro-***													
—	**Methylene cyanide**	*see* **Malonic acid,** dinitrile													
—	**5,5′-Methylene disalicylaldehyde**	*see* **Methane, bis(3-formyl-4-hydroxy-phenyl)-**													
—	**Methylene fluoride**	*see* **Methane, difluoro-***													
—	**Methylhydroxylamine**	*see* **Methane, hydroxylamino-***													
m394	**Methysticin**	Kavahin. Kavatin 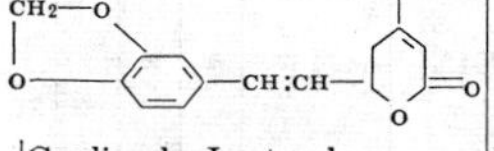	274.28	nd or pr (MeOH, ace) $[\alpha]^{20}_D +94.3$ (ace)	138–9	...	...	...	i δ^h	v^h	δ	s	s	chl s peth δ	**B19**², 431
m395	**Metrazol**	Cardiazole. Leptazole 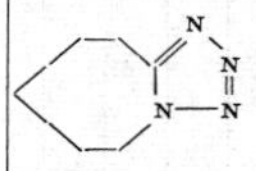	138.17	...	59	...	...	...	s	v	s	...	...	os s	...
m396	**Metycaine**	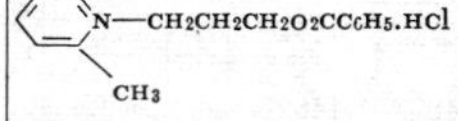	292.79	...	172–5	...	...	...	s	s	i	...	...	chl s	...
—	**Michler's ketone**	*see* **Benzophenone, 4,4′-bis(dimethylamino)-**													
m397	**Mimosine**(*dl*)	Leucenol. 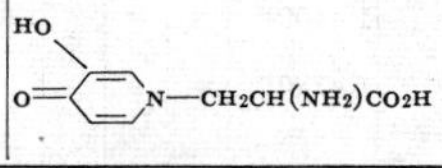	198.18	(w)	291d	...	...	...	δ	...	...	...	...	oos i dil ac s dil alk s	...

For explanations, symbols and abbreviations see beginning of table.

No.	Name	Synonyms and Formula	Mol. wt.	Crystalline form, color and specific rotation	m.p. °C	b.p. °C	Density	n_D	Solubility w	al	eth	ace	bz	other solvents	Ref.
	Mimosine														
m398	—(*l*)	*l*-Leucenol. $C_8H_{10}N_2O_4$. *See* m397	198.18	ta $[\alpha]_D^{22} -21$	227–8d				δ					oos i	
—	**Monolaurin**	*see* **Glycerol,** monododecanoate													
—	**Monoolein**	*see* **Glycerol,** mono(9-octadecoate)													
—	**Monopalmitin**	*see* **Glycerol,** monohexadecanoate													
—	**Monoricinolein**	*see* **Glycerol,** mono(12-hydroxy-9-octadecenoate)													
—	**Monostearin**	*see* **Glycerol,** monooctadecanoate													
—	**Morin**	*see* **Flavone, 2′,3,4′,5,7,8-pentahydroxy-**													
—	**Morindone**	*see* **9,10-Anthraquinone, 6-methyl-1,2,5-trihydroxy-***													
m399	**Morphine**	$C_{17}H_{19}NO_3$	285.35		254				δ^h	δ		i			
m400	—,monohydrate	$C_{17}H_{19}NO_3.H_2O$	303.36	orh pr $[\alpha]_D^{25}$ −132 (MeOH)	230d	d	1.32_4^{20}	1.58–1.64	s^h	δ	i				B27[2], 122
m401	—,acetate trihydrate(*l*)	$C_{17}H_{19}NO_3.CH_3CO_2.3H_2O$.	398.44	colorless to ye $[\alpha]_D^{15} -77$ (w)	200d				v	s v^h	i			chl δ	B27[2], 134
m402	—,hydrochloride trihydrate	$C_{17}H_{19}NO_3.HCl.3H_2O$.	375.85	nd or flakes $[\alpha]_D^{25} -111.5$ (w)	200d (250d)				s v^h	s	i			chl i	B27[2], 132
m403	—,*N*-oxide	Genomorphine. $C_{17}H_{19}NO_4$.	301.35	pr (50% al)	274–5				δ	δ		i	i	chl i NH_4OH v	B27[2], 159
m404	—,sulfate pentahydrate	$2(C_{17}H_{19}NO_3).H_2SO_4.5H_2O$.	758.82	powd or cubes $[\alpha]_D^{25} -94.5$ (w)	250d				s v^h	δ	i		i	chl i	B27[2], 133
m405	—,*O,O*-**diacetyl-**	Diamorphine. Heroin. $C_{21}H_{23}NO_5$.	369.40	rh $[\alpha]_D^{15} -166$ (MeOH)	171		1.56–1.61		δ	s	δ		s	chl v MeOH s	B27[2], 151
m406	—,—,hydrochloride monohydrate	$C_{21}H_{23}NO_5.HCl.H_2O$	423.90	$[\alpha]_D^{20} -153$ (w)	231–2 (243–4)				v	s	i			chl i	B27[2], 153
m407	—,**ethyl-**, hydrochloride dihydrate	Dionin. $C_{19}H_{23}NO_3.HCl.2H_2O$. *See* m399	385.88		123d (170 anh)				s	s	δ			chl δ	B27[2], 148
—	**Morphol**	*see* **Phenanthrene, 3,4-dihydroxy-***													
m408	**Morpholine**	Diethylenimide oxide. Tetrahydro-1,4-isoxazine. 1 2 3 4 / O NH / 6 5	87.12	hyg	−4.9	128^{760}	0.9994_4^{20}	1.4545^{20}	∞	s	s			os s	B17, 5
m409	—,**4-acetyl**	$C_6H_{11}NO_2$. *See* m408	129.16		14	152^{50}	1.1165_{20}^{20}	1.4838^{20}	∞						
m410	—,**4(2-aminoethyl)-**	$C_6H_{14}N_2O$. *See* m408	130.19		25.6	116^{50}	0.9915_{20}^{20}	1.4715^{20}	∞	∞			∞	lig ∞	
m411	—,**4(3-aminopropyl)-**	$C_7H_{16}N_2O$. *See* m408	144.21		−15	134^{50}	0.9872_{20}^{20}	1.4749^{30}	∞	∞			∞	lig ∞	
m412	—,**4-benzyl-**	$C_{11}H_{15}NO$. *See* m408	177.25			260–1 $128–9^{13}$	1.0387_4^{20}		δ					ac s	B27[1], 203
m413	—,**4-butyl-**	$C_8H_{17}NO$. *See* m408	143.23		−57.1	96^{50}	0.8957_{20}^{20}	1.4451^{20}	v						
m414	—,**2,6-dimethyl-**	$C_6H_{13}NO$. *See* m408	115.18		fr–85	146.6^{760}	0.9346_{20}^{20}	1.4462^{20}	∞	∞			∞	lig ∞	
m415	—,**4(2-ethyoxyethyl)-**	$C_8H_{17}NO_2$. *See* m408	159.23			206^{760}	0.963^{20}		∞						
m416	—,**4-ethyl-**	$C_6H_{13}NO$. *See* m408	115.18			$138–9^{763}$	0.8996_4^{20}	1.4400^{20}	∞	∞	∞				B27[1], 203
m417	—,**4(2-hydroxyethyl)-**	4-Morpholineethanol $C_6H_{13}NO_2$. *See* m408	131.17			227^{757}	1.071	1.4780^{20}	s	s					B27, 7

For explanations, symbols and abbreviations see beginning of table.

No.	Name	Synonyms and Formula	Mol. wt.	Crystalline form, color and specific rotation	m.p. °C	b.p. °C	Density	n_D	Solubility w	al	eth	ace	bz	other solvents	Ref.
	Morpholine														
m418	**—,4(2-hydroxy-propyl)-**	$C_7H_{15}NO_2$. *See* m408	145.21	....	....	$92–4^{13}$	1.0174^{20}_{4}	1.464^{20}	s	s	...	s	s	MeOH s	....
m419	**—,4-methyl-**	$C_5H_{11}NO$. *See* m408	101.15	....	....	$115–6^{750}$	0.9051^{20}_{4}	1.4332^{20}	s	s	s	...	...	....	**B27**, 6
m420	**—,4-phenyl-**	$C_{10}H_{13}NO$. *See* m408	163.22	(al-eth)	57–8	$259–60^{745}$ $165–70^{45}$	....	....	i	i	v	...	...	....	**B27**[2], 3
m421	**—,4(4-tolyl)-**	$C_{11}H_{15}NO$. *See* m408	177.25	(dil al)	51	167^{30}	....	....	...	v	s	...	...	....	**B27**[2], 4
m422	**Mucic acid**	2,3,4,5-Tetrahydroxy-hexanedioic acid*. $HO_2C–CH(OH)–CH(OH)–CH(OH)–CH(OH)–CO_2H$	210.14	cr or powd	206d (213–4)	255	....	....	δ	i	i	...	...	alk s	....
—	**Mucobromic acid**	*see* **2-Butenoic acid, 2,3-dibromo-4-oxo-***													
—	**Mucochloric acid**	*see* **2-Butenoic acid, 2,3-dichloro-4-oxo-***													
—	**Muconic acid**	*see* **2,4-Hexadienedioic acid***													
m423	**Murexide**	Ammonium purpurate. [O=C<(NH)(NH)>C(=O)... ring–N=ring ...]$.NH_4OH.H_2O$	302.22	red-gr pr	....	....	....	....	δ s^h	i	i	...	...	alk s	**B25**, 499
—	**Muscone**	*see* **Cyclopentadecanone, 3-methyl-***													
—	**Musk baur**	*see* **Benzene, 2-*tert*-butyl-4-methyl-1,3,5-trinitro-**													
—	**Musk ketone**	*see* **Acetophenone, 4-*tert*-butyl-2,6-dimethyl-3,5-dinitro-**													
—	**Musk xylene**	*see* **Benzene, 1-*tert*-butyl-3,5-dimethyl-2,4,6-trinitro-**													
—	**Mustard gas**	*see* **Sulfide, diethyl, 2,2′-dichloro-**													
m424	**Mycophenolic acid**	$C_{17}H_{20}O_6$	320.33	nd (w)	141	....	....	....	$δ^h$	v	v	...	δ	chl v to δ, v^h	**B18**[2], 393
m425	**Myrcene**	$C_{10}H_{16}$	136.23	....	....	167^{760} 65^{20}	0.8013^{15}_{0}	1.4706^{20}	i	s	s	...	...	chl s aa s	**B1**, 264
—	**Myricyl alcohol**	*see* **1-Triacontanol***													
—	**Myristaldehyde**	*see* **Tetradecanal***													
—	**Myristic acid**	*see* **Tetradecanoic acid***													
m426	**Myristicin**	CH_2—O (methylenedioxy) ring, —CH_2CH:CH_2, CH_3O	192.22	....	<−20	$276–7^{760}$ 157^{21}	1.1437^{20}_{20}	1.5403^{20}	...	δ	s	...	...	....	**B19**[2], 84
—	**Myristyl bromide**	*see* **Tetradecane, 1-bromo-***													

For explanations, symbols and abbreviations see beginning of table.

No.	Name	Synonyms and Formula	Mol. wt.	Crystalline form, color and specific rotation	m.p. °C	b.p. °C	Density	n_D	Solubility w	al	eth	ace	bz	other solvents	Ref.
	Naphthacene														
n1	**Naphthacene**	2,3-Benzanthracene. Tetracene. 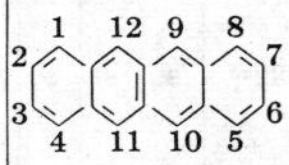	228.30	og-ye lf	ca. 335	sub							i		**B5**^{2}, 628
n2	**—,9,10-dihydro-**	$C_{18}H_{14}$. *See* n1	230.31	nd or lf (bz)	212	ca. 400					δ^h			sulf s (ye) PhNO$_2$ v	**E14s**, 765
n3	**—,9,10-diphenyl-**	$C_{30}H_{20}$. *See* n1	380.49	og (eth+½)	171–2									os v	**E14s**, 77
n4	**—,9,11-diphenyl-**	$C_{30}H_{20}$. *See* n1	380.49	ye	301–2				i		s		s	CS$_2$ s	**E14**, 313
n5	**—,9,10,11-triphenyl-**	$C_{36}H_{24}$. *See* n1	456.59	og (eth), (bz+1)	236–7 177–8 (+1 bz)				i		s		s	CS$_2$ s	**E14**, 313
n6	**9,10-Naphthacenequinone**	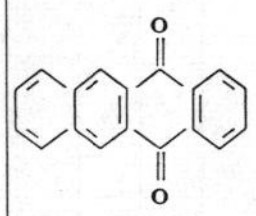	258.28	pa ye nd (PhNO$_2$ or aa)	294	sub				s		δ	δ	sulf s aa i	**E14**, 319
n7	**9,11-Naphthacenequinone**	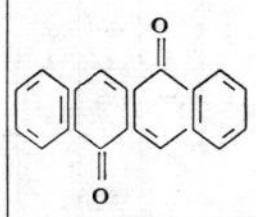	258.28	dk red (aa or xyl)	322d									os δ xyl δ	**E14s**, 86
n8	**1-Naphthaldehyde**	1-Formylnaphthalene. 1-Naphthalenecarbonal*. 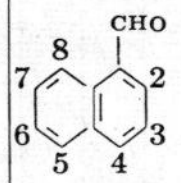	156.19	pa ye	33–4	291 150^{13}	1.1503^{20}_{4}	1.6546^{19}	i	s	s				**E12B**, 2197
n9	**2-Naphthaldehyde**	2-Formylnaphthalene. 2-Naphthalenecarbonal. 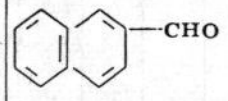	156.19	lf (w)	61	$155\text{–}60^{13}$	1.0775^{99}_{4}	1.6211^{99}							**E12B**, 2204
n10	**1-Naphthaldehyde, 2-ethoxy-**	$C_{13}H_{12}O_2$. *See* n8	200.24	yesh nd (al or aa)	112–3					s^h				aa s^h	**B8**^{1}, 564
n11	**—,4-ethoxy-**	$C_{13}H_{12}O_2$. *See* n8	200.24	yesh cr (aa)	75									aa s^h	**B8**^{2}, 174
n12	**—,2-hydroxy-**	β-Naphthol-1-aldehyde. $C_{11}H_8O_2$. *See* n8	172.19	pr (al), nd (aa)	82	192^{27}			i	s	s			aq alk s lig s	**B7**^{2}, 171
n13	**2-Naphthaldehyde, 1-hydroxy-**	α-Naphthol-2-aldehyde. $C_{11}H_8O_2$. *See* n9	172.19	grsh-ye nd (dil aa or lig)	60				δ s^h		s			os s	**E12B**, 2369
n14	**Naphthalene***	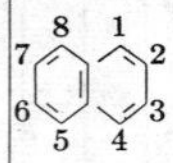	128.18	mcl	80.22	210.8^{720}	1.145^{24}_{24}	1.4003^{24}	i	s	v^h		v	CS$_2$ v chl v	**B5**, 531
n15	—,picrate	$C_{10}H_8.C_6H_3N_3O_7$. *See* n14	357.28	ye pr or pl (eth-aa), (eth)	151.5		1.42		δ		s		s		**B6**^{2}, 259
n16	**—,1-acetamido-2-nitro-**	$C_{12}H_{10}N_2O_3$. *See* n14	230.23	lt ye nd (aa)	202–3					s^h				aa s	**E12B**, 746
n17	**—,1-acetamido-4-nitro-***	$C_{12}H_{10}N_2O_3$. *See* n14	230.23	pa ye nd (ace)	192.5–3.5							s^h			**E12B**, 751
n18	**—,1-acetamido-5-nitro-***	$C_{12}H_{10}N_2O_3$. *See* n14	230.23	ye (al)	220				δ	s^h	δ		δ		**E12B**, 755
n19	**—,1-acetamido-8-nitro-***	$C_{12}H_{10}N_2O_3$. *See* n14	230.23	(w)	187–8				s^h						**E12B**, 756
n20	**—,2-acetamido-1-nitro-***	$C_{12}H_{10}N_2O_3$. *See* n14	230.23	ye rh bipym nd or pl (al)	123.5				δ	v	s		v	aa v lig s	**B12**^{2}, 731
n21	**—,2-acetamido-6-nitro-***	$C_{12}H_{10}N_2O_3$. *See* n14	230.23	lt ye nd	224				i	v	i		δ	aa v	**E12B**, 765
n22	**—,6-acetamido-1-nitro-***	$C_{12}H_{10}N_2O_3$. *See* n14	230.23	ye br rh (al), ye nd (bz)	185.5					s			δ	aa s	**E12B**, 764
n23	**—,7-acetamido-1-nitro-***	$C_{12}H_{10}N_2O_3$. *See* n14	230.23	ye nd (al)	195.5					δ			δ	aa s	**E12B**, 766
n24	**—,1-acetyl-**	1-Acetonaphthone. Methyl 1-naphthyl ketone. $C_{12}H_{10}O$. *See* n14	170.21		34	296–8 $170\text{–}0.5^{20}$	1.1336^{0}_{4}	1.6280^{22}	i	s	s			os s	**B7**^{2}, 337

For explanations, symbols and abbreviations see beginning of table.

No.	Name	Synonyms and Formula	Mol. wt.	Crystalline form, color and specific rotation	m.p. °C	b.p. °C	Density	n_D	Solubility						Ref.
									w	al	eth	ace	bz	other solvents	
	Naphthalene														
n25	—,**2-acetyl-**	2-Acetonaphthone. Methyl 2-naphthyl ketone. $C_{12}H_{10}O$. *See* n14	170.20	nd (lig)	56	301–3 171–3[17]	...	...	...	δ	...	...	...	CS_2 s lig δ	**B7**[2], 338
n26	—,**2-acetyl-1-amino-**	$C_{12}H_{10}O_2$. *See* n14	186.21	ye-gr nd (al)	103	325δd	...	...	i	δ	...	...	s	chl, CS_2 s aa s	**B8**[2], 178
n27	—,**2-acetyl-4-bromo-1-hydroxy-**	$C_{12}H_9BrO_2$. *See* n14	265.11	ye-gr nd (al)	126–7	...	...	...	i	s	s	...	s	CS_2, chl s lig s	**B8**[2], 178
n28	—,**1-acetyl-2-hydroxy-***	$C_{12}H_{10}O_2$. *See* n14	186.21	pa ye lf (peth), rh (lig), nd or pl (gasoline)	64–5	...	...	...	...	v	v	...	v	os, con sulf v	**B8**[2], 175
n29	—,**1-acetyl-4-hydroxy-**	$C_{12}H_{10}O_2$. *See* n14	186.21	pr (aa or to)	198	...	...	...	...	s	...	...	s	alk s aa s[h]	**B8**[2], 176
n30	—,**2-acetyl-1-hydroxy-**	$C_{12}H_{10}O_2$. *See* n14	186.21	ye nd (bz, lig)	98	325δd	...	...	i	δ	...	...	s	chl, CS_2 s aa s	**B8**[1], 567
n31	—,**2-acetyl-3-hydroxy-**	$C_{12}H_{10}O_2$. *See* n14	186.21	ye lf (al, peth)	112	...	...	...	...	δ	...	v	v	dil NaOH s lig δ	**B8**[2], 179
n32	—,**2-acetyl-6-hydroxy-**	$C_{12}H_{10}O_2$. *See* n14	186.21	pr (bz)	172	...	...	...	...	s	...	...	s[h]	NaOH s	**B8**[2], 179
n33	—,**3-acetyl-1-hydroxy-**	$C_{12}H_{10}O_2$. *See* n14	186.21	nd	173–4	...	...	...	...	v	...	...	δ	alk s aa v	**B8**, 150
n34	—,**2-acetyl-1-hydroxy-4-nitro-**	$C_{12}H_9NO_4$. *See* n14	231.21	ye nd (al)	159	...	...	...	i	δ s[h]	s	...	s	...	**B8**[2], 179
n35	—,**1-allyl-**	$C_{13}H_{12}$. *See* n14	168.24	...	...	256–7 129–30[10]	1.0228^{20}_{4}	1.6140[20]	...	s	...	...	s	chl s	**E12B**, 118
n36	—,**1-amino-***	1-Naphthylamine*. α-Naphthylamine. $C_{10}H_9N$. *See* n14	143.19	rh (al-w)	50	300.8 sub	1.123^{25}_{25}	1.6703	δ	v	v	...	...	...	**B12**[2], 675
n37	—,—,hydrochloride	$C_{10}H_9N.HCl$. *See* n14	179.65	nd	...	sub	...	...	s	s	s	...	...	...	**B12**, 1220
n38	—,**2-amino-***	2-Naphthylamine*. β-Naphthylamine. $C_{10}H_9N$. *See* n14	143.19	lf (w)	113	306.1	1.0614^{98}_{4}	1.6498[98]	s	s	s	...	...	...	**E12B**, 544
n39	—,—,hydrochloride	$C_{10}H_9N.HCl$. *See* n14	179.65	...	254	...	...	...	...	...	...	...	...	...	**B12**[2], 710
n42	—,**1-amino-4-bromo-***	$C_{10}H_8BrN$. *See* n14	222.09	nd (al, bz or peth)	102	...	...	...	...	s	...	s	s	lig s	**E12B**, 713
n43	—,**1-amino-5-bromo-***	$C_{10}H_8BrN$. *See* n14	222.09	lf or pl (w or lig)	69 (cor) (subl)	...	...	...	δ[h]	v	v	...	v	chl v lig s	**E12B**, 717
n44	—,**2-amino-1-bromo-***	$C_{10}H_8BrN$. *See* n14	222.09	rh nd (dil al or lig)	63–4	...	...	...	δ	v	s	...	v	chl v	**E12B**, 723
n45	—,**2-amino-3-bromo-***	$C_{10}H_8BrN$. *See* n14	222.09	pl (al)	169	...	...	...	...	s	...	...	...	aa s	**E12B**, 729
n46	—,**2-amino-6-bromo-***	$C_{10}H_8BrN$. *See* n14	222.09	lf (al, w or peth)	128	...	...	...	δ	v[h]	...	s	s	...	**E12B**, 731
n47	—,**3-amino-1-bromo-***	$C_{10}H_8BrN$. *See* n14	222.09	nd (bz-peth or 90% aa)	72	...	...	...	i	v	v	...	v	...	...
n48	—,**6-amino-1-bromo-***	$C_{10}H_8BrN$. *See* n14	222.09	...	38	...	...	...	i	v	...	...	...	os v lig δ	**E12B**, 730
n49	—,**1-amino-4-bromo-2-nitro-***	$C_{10}H_7BrN_2O_2$. *See* n14	207.09	og cr (al or aa)	200	...	...	...	...	v[h]	...	...	v	to, chl v	**E12B**, 775
n50	—,**1-amino-2-chloro-***	$C_{10}H_8ClN$. *See* n14	177.63	nd (peth or dil al)	60	...	...	...	δ[h]	v	...	s	...	...	**E12B**, 709
n51	—,**1-amino-4-chloro-***	$C_{10}H_8ClN$. *See* n14	177.63	nd (al, bz or lig)	98	...	...	...	...	v	v	...	...	...	**E12B**, 711
n52	—,**2-amino-1-chloro-***	$C_{10}H_8ClN$. *See* n14	177.63	nd (al or peth)	60	...	...	...	...	s	...	...	...	aa s lig s[h]	**E12B**, 720
n53	—,**1-amino-2,4-dibromo-***	$C_{10}H_7Br_2N$. *See* n14	300.99	nd (al)	118–9	...	...	...	i	v	v	...	v	chl v lig v	**E12B**, 735
n54	—,**2-amino-1,4-dibromo-***	$C_{10}H_7Br_2N$. *See* n14	300.99	nd (al or bz)	106–7	...	...	...	...	...	...	...	...	os v	**E12B**, 737
n55	—,**2-amino-1,6-dibromo-***	$C_{10}H_7Br_2N$. *See* n14	300.99	nd (al or peth)	122–3	...	...	...	δ	v	...	...	v	aa s	**E12B**, 737
n56	—,**1-amino-2,4-dichloro-***	$C_{10}H_7Cl_2N$. *See* n14	212.08	nd (al)	83–4	...	...	...	i	v	...	...	...	...	**E12B**, 734
n57	—,**5-amino-1,4-dihydro-***	$C_{10}H_{11}N$. *See* n14	145.21	pl or nd (to), turns pink in air	37.5 cor	247[408]	...	...	...	s	...	...	...	chl s ac s	**E12B**, 667
n58	—,**1-amino-2,4-dinitro-***	$C_{10}H_7N_3O_4$. *See* n14	233.18	ye nd (al), pr (aa)	242	...	...	...	δ	s	s	v	s	chl s lig δ	**E12B**, 784
n59	—,**2-amino-1,6-dinitro-***	$C_{10}H_7N_3O_4$. *See* n14	233.18	gold-ye nd (al or aa), pw (Py)	245–6	...	...	...	δ	δ	i	v	δ	chl δ lig i	**E12B**, 793

For explanations, symbols and abbreviations see beginning of table.

No.	Name	Synonyms and Formula	Mol. wt.	Crystalline form, color and specific rotation	m.p. °C	b.p. °C	Density	n_D	Solubility w	al	eth	ace	bz	other solvents	Ref.
	Naphthalene														
n60	**—,1-amino-4-fluoro-***	$C_{10}H_8FN$. *See* n14	161.18	lt ye	48	162^{16}								ac s	E12B, 710
n61	**—,1-amino-2-hydroxy-***	1-Amino-2-naphthol. $C_{10}H_9NO$. *See* n14	159.18	silvery lf (bz)	150d				δ^h	s	s^h			dil alk, dil ac v	E12B, 1657
n62	**—,1-amino-6-hydroxy-***	5-Amino-2-naphthol. $C_{10}H_9NO$. *See* n14	159.18	nd (w) or og pr (w)	190.6(wr)				s	s	s			NH_3 s	E12B, 1688
n63	**—,1-amino-7-hydroxy-***	8-Amino-2-naphthol. $C_{10}H_9NO$. *See* n14	159.18	nd (w, al)	205.7				v^h	v	s		δ		E12B, 1696
n64	**—,2-amino-3-hydroxy-***	3-Amino-2-naphthol. $C_{10}H_9NO$. *See* n14	159.18	silvery lf (bz), nd (al)	235				δ s^h	v^h	s^h		s		E12B, 1677
n65	**—,2-amino-6-hydroxy-***	6-Amino-2-naphthol. $C_{10}H_9NO$. *See* n14	159.18	br (dil al), pr (w)	212–3d (sealed tube)				s	s					E12B, 1689
n66	**—,2-amino-7-hydroxy-***	7-Amino-2-napthol. $C_{10}H_9NO$. *See* n14	159.18	nd or lf (al)	208				δ	v	v				E12B, 169
n67	**—,2-amino-3-iodo-***	$C_{10}H_8IN$. *See* n14	269.09	(50 % al)	137					v			s	chl v aa v	E12B, 729
n68	**—,1-amino-2-methyl-***	$C_{11}H_{11}N$. *See* n14	157.22	nd (peth), turns red in air	32				i					os v lig s	B12², 742
n69	**—,1-amino-3-methyl-***	$C_{11}H_{11}N$. *See* n14	157.22	(peth)	51–2					s				os s lig s	B12², 743
n70	**—,1-amino-4-methyl-***	$C_{11}N_{11}N$. *See* n14	157.22	nd (peth)	51–2				i					os v lig δ	B12², 740
n71	**—,2-amino-1-methyl-***	$C_{11}H_{11}N$. *See* n14	157.22	nd (lig), pr (peth)	51				δ^h	v	v		v	chl v lig s	B12², 740
n72	**—,2-amino-6-methyl-***	$C_{11}H_{11}N$. *See* n14	157.22	lf (peth), turns red in air	129–30									os v min ac v	B12², 743
n73	**—,1-amino-2-nitro-***	$C_{10}H_8N_2O_2$. *See* n14	188.19	ye-red mcl pr (al)	144					s^h					B12², 703
n74	**—,1-amino-3-nitro-***	$C_{10}H_8N_2O_2$. *See* n14	188.19	og-ye nd (50 % al)	137					v			s	chl s	E12B, 748
n75	**—,1-amino-4-nitro-***	$C_{10}H_8N_2O_2$. *See* n14	188.19	og-ye nd (al)	195				δ^h	s					B12², 704
n76	**—,1-amino-5-nitro-***	$C_{10}H_8N_2O_2$. *See* n14	188.19	red nd	118–9										B12², 705
n77	**—,1-amino-6-nitro-***	$C_{10}H_8N_2O_2$. *See* n14	188.19	og-red nd	143					s					
n78	**—,1-amino-8-nitro-***	$C_{10}H_8N_2O_2$. *See* n14	188.19	red lf (peth)	96–7									dil sulf s lig s^h	B12², 705
n79	**—,2-amino-1-nitro-***	$C_{10}H_8N_2O_2$. *See* n14	188.19	og-ye nd	126–7				δ^h	s				aa s	B12, 1313
n80	**—,2-amino-6-nitro-***	$C_{10}H_8N_2O_2$. *See* n14	188.19	lt og pl (al or aa)	206–7				δ				s	diox s	E12B, 765
n81	**—,6-amino-1-nitro-***	$C_{10}H_8N_2O_2$. *See* n14	188.19	red nd	143–5									aa s lig i	B12, 1314
n82	**—,7-amino-1-nitro-***	$C_{10}H_8N_2O_2$. *See* n14	188.19	red nd	105					s	s		s		B12, 1315
n83	**—,2-amino-1-nitroso-***	$C_{10}H_8N_2O$. *See* n14	172.19	nd (al)	150–2				s^h						B7, 717
n84	**—,2-amino-1,3,6-tribromo-***	$C_{10}H_6Br_3N$. *See* n14	379.89	(chl, al or al-eth)	143–4				δ^h		v			chl v lig s	E12B, 741
n85	**—,1(2-amino-ethyl)-***	$C_{12}H_{13}N$. *See* n14	171.24			$182–3^{18}$				v				xyl s	E12B, 425
n86	**—,2(2-amino-ethyl)-***	$C_{12}H_{13}N$. *See* n14	171.24			174^{25}				s				aa s	
n87	**—,1(2-amino-ethylamino)-***	*N*-α-Naphthylethylenediamine. $C_{10}H_7^{\alpha}NHCH_2CH_2NH_2$	186.26	ye		320d 204^9	1.114_4^{25}	1.6648^{25}	δ	s				lig i	E12B, 518
n88	**—,1(amino-methyl)-***	$C_{11}H_{11}N$. *See* n14	157.22			294–5 $162–3^2$				s				sulf s (bl)	E12B, 418
n89	**—,2(amino-methyl)-***	$C_{11}H_{11}N$. *See* n14	157.22	pr (eth)	59–60	180^{24}			δ	v	v				E12B, 421
n90	**—,1-benzyl-2-hydroxy-**	1-Benzyl-2-naphthol. $C_{17}H_{14}O$. *See* n14	234.30	nd (bz)	115				i	s	v	v	s	chl s	B6², 680
n91	**—,1-benzyl-4-hydroxy-**	4-Benzyl-1-naphthol. $C_{17}H_{14}O$. *See* n14	234.30	pl or nd (dil aa-lig)	125–6				δ					os s lig δ	B6², 680
n92	**—,2-benzyl-1-hydroxy-**	2-Benzyl-1-naphthol. $C_{17}H_{14}O$. *See* n14	234.30	nd or pr (lig)	73.5–4.0	$237–40^{12}$			i					lig δ	B6², 681
n92¹	**—,1(benzyldine-amino)-**	Benzaldehyde α-naphthylimide. $C_{17}H_{13}N$. *See* n14	231.30	ye lf (al)	73.5					s^h			s	MeOH s	E12B, 507
n92²	**—,2(benzylidene-amino)-**	Benzaldehyde β-naphthylimide. $C_{17}H_{13}N$. *See* n14	231.30	yesh nd (al)	102–3					s^h			s	chl s aa s	E12B, 602

For explanations, symbols and abbreviations see beginning of table.

No.	Name	Synonyms and Formula	Mol. wt.	Crystalline form, color and specific rotation	m.p. °C	b.p. °C	Density	n_D	Solubility						Ref.
									w	al	eth	ace	bz	other solvents	
	Naphthalene														
n93	**—,1-bromo-***	$C_{10}H_7Br$. *See* n14	207.08	pr (β form)	−6.2(α) 2–2.7(β)	281	1.4887^{17}_{4}	1.6588^{19}	s^h	∞	∞	...	∞	...	**B5**, 547
n94	**—,2-bromo-***	$C_{10}H_7Br$. *See* n14	207.08	pl or rh lf (al)	59	281–2	1.605^0	...	i	s	s	...	s	chl, CS_2, s	**B5²**, 548
n95	**—,1-bromo-2(bromo-methyl)-***	$C_{11}H_8Br_2$. *See* n14	300.01	nd (al), (peth)	107–8	...	...	...	d^h	s	...	...	s	lig s^h	**B5²**, 465
n96	**—,1-bromo-2-hydroxy-***	1-Bromo-2-naphthol. $C_{10}H_7BrO$. *See* n14	223.08	rh pr (bz-lig)	84	130d	...	...	i	s	s	...	s	aa v lig s	**E12B**, 1484
n97	**—,1-bromo-4-hydroxy-***	4-Bromo-1-naphthol. $C_{10}H_7BrO$. *See* n14	223.08	nd (dil al)	127–8	...	...	...	...	s^h	...	...	...	chl s aa s	**B6²**, 582
n98	**—,1-bromo-5-hydroxy-***	5-Bromo-1-naphthol. $C_{10}H_7BrO$. *See* n14	223.08	nd (w)	137	...	...	...	s^h	...	...	...	...	...	**B6²**, 582
n99	**—,1-bromo-6-hydroxy-***	5-Bromo-2-naphthol. $C_{10}H_7BrO$. *See* n14	223.08	nd (w)	105	...	...	...	s^h	s	...	...	...	...	**B6²**, 605
n100	**—,1-bromo-8-hydroxy-***	8-Bromo-1-naphthol. $C_{10}H_7BrO$. *See* n14	223.08	pl (peth)	60–1	...	...	...	...	...	...	...	...	lig s^h	**B6**, 614
n101	**—,2-bromo-3-hydroxy-***	3-Bromo-2-naphthol. $C_{10}H_7BrO$. *See* n14	223.08	nd (lig)	84–5	...	...	...	δ^h	v	...	...	v	lig s^h	**B6²**, 605
n102	**—,2-bromo-6-hydroxy-***	6-Bromo-2-naphthol. $C_{10}H_7BrO$. *See* n14	223.08	nd (bz)	129–30	...	...	...	...	s	...	...	s^h	...	**B6²**, 605
n103	**—,2-bromo-7-hydroxy-***	7-Bromo-2-naphthol. $C_{10}H_7BrO$. *See* n14	223.08	(peth)	132–3	...	...	...	...	...	...	...	...	lig s^h	**B6²**, 605
n104	**—,6-bromo-1-hydroxy-***	6-Bromo-1-naphthol. $C_{10}H_7BrO$. *See* n14	223.08	nd (w)	129–30	...	...	...	v^h	...	...	...	...	...	**B6²**, 583
n105	**—,7-bromo-1-hydroxy-***	7-Bromo-1-naphthol. $C_{10}H_7BrO$. *See* n14	223.08	(w)	105.5–6.5	...	...	...	s^h	...	...	...	...	...	**B6²**, 583
n106	**—,6-bromo-2-hydroxy-1-methyl-***	$C_{11}H_9BrO$. *See* n14	237.10	nd (bz)	129	...	...	...	i	v^h	v^h	...	δ^h	chl v aa v^h	**E12B**, 1501
n107	**—,1(bromo-methyl)-***	$C_{11}H_9Br$. *See* n14	221.10	(peth or al)	53.5–4.5	183^{18}	...	...	i^h	s	s	...	s	aa s^h	**B5²**, 462
n108	**—,2(bromo-methyl)-***	$C_{11}H_9Br$. *See* n14	221.10	lf (al)	56	213^{100} $165–9^{14}$	...	...	δd	s	s	...	...	chl s aa s^h	**B5²**, 464
n109	**—,1-chloro-***	$C_{10}H_7Cl$. *See* n14	162.62	...	...	263	1.1938^{20}_{4}	1.6332^{20}	i	s	s	...	s	CS_2 s	**B6**, 1248
n110	**—,2-chloro-***	$C_{10}H_7Cl$. *See* n14	162.62	pl (dil al), lf	59	256 $121–2^{12}$	1.1377^{71}_{4}	1.6079^{71}	i	s	s	...	s	chl s CS_2 s	**B5²**, 445
n111	**—,1-chloro-2-hydroxy-***	1-Chloro-2-naphthol. $C_{10}H_7ClO$. *See* n14	178.62	nd (lig), pr (chl) pl (w)	70	...	...	...	δ v^h	v^h	...	...	v^h	chl v lig s^h	**B6²**, 603
n112	**—,1-chloro-2-nitro-***	$C_{10}H_6ClNO_2$. *See* n14	207.62	pa ye nd	81	...	...	...	...	...	...	...	...	...	**E12B**, 374
n113	**—,1-chloro-3-nitro-***	$C_{10}H_6ClNO_2$. *See* n14	207.62	ye nd (HCO_2H)	127	...	...	...	...	...	...	...	...	...	**E12B**, 376
n114	**—,1-chloro-4-nitro-***	$C_{10}H_6ClNO_2$. *See* n14	207.62	lt ye nd (peth or al)	87–7.5	...	...	...	i	s	s	...	...	...	**E12B**, 370
n115	**—,1-chloro-5-nitro-***	$C_{10}H_6ClNO_2$. *See* n14	207.62	nd (dil al or aa)	111	>360 181^2	...	...	...	...	...	...	...	...	**E12B**, 372
n116	**—,1-chloro-6-nitro-***	$C_{10}H_6ClNO_2$. *See* n14	207.62	ye nd	120	...	...	...	...	...	...	...	...	...	**E12B**, 377
n117	**—,1-chloro-8-nitro-***	$C_{10}H_6ClNO_2$. *See* n14	207.62	lt ye nd (gl aa or bz)	94	>360	...	...	...	s	...	...	...	aa s lig δ	**E12B**, 373
n118	**—,2-chloro-1-nitro-***	$C_{10}H_6ClNO_2$. *See* n14	207.62	nd (peth)	95.5	175^2 >360	...	...	...	v	v	v	v	aa v	**E12B**, 368
n119	**—,2-chloro-3-nitro-***	$C_{10}H_6ClNO_2$. *See* n14	207.62	br	94.5	...	...	...	...	...	...	...	...	...	**E12B**, 375
n120	**—,2-chloro-6-nitro-***	$C_{10}H_6ClNO_2$. *See* n14	207.62	ye nd	170	$180–90^{15}$	...	...	...	...	...	...	...	...	...
n121	**—,2-chloro-7-nitro-***	$C_{10}H_6ClNO_2$. *See* n14	207.62	ye nd	136	...	...	...	...	...	...	...	...	...	...
n122	**—,3-chloro-1-nitro-***	$C_{10}H_6ClNO_2$. *See* n14	207.62	grsh-br nd ($PhNO_2$ or 90% HCO_2H)	105	...	...	...	...	...	...	...	...	...	**E12B**, 369
n123	**—,6-chloro-1-nitro-***	$C_{10}H_6ClNO_2$. *See* n14	207.62	nd (aq ace)	100.5	...	...	...	...	...	...	...	...	...	**E12B**, 373
n124	**—,7-chloro-1-nitro-***	$C_{10}H_6ClNO_2$. *See* n14	207.62	ye nd (al)	116	...	...	...	i	s	s	...	...	...	**E12B**, 373
n126	**—,1(chloro-methyl)-***	$C_{11}H_9Cl$. *See* n14	176.65	pr	32	291–2 $135–9^6$	...	...	...	...	...	...	...	...	**B5²**, 461
n127	**—,2(chloro-methyl)-***	$C_{11}H_9Cl$ *See* n14	176.65	lf (al)	48	168^{20}	...	...	δ^h	s^h	...	...	s	...	**B5²**, 464
—	**—,decahydro-***	*see* **Decalin**													
n128	**—,1,2-diamino-***	1,2-Naphthylenediamine $C_{10}H_{10}N_2$. *See* n14	158.20	lf (w), rose to br in air	94–8	$150–1^{0.5}$	...	...	δ^h	v	v	...	...	chl v	**E12B**, 805

For explanations, symbols and abbreviations see beginning of table.

No.	Name	Synonyms and Formula	Mol. wt.	Crystalline form, color and specific rotation	m.p. °C	b.p. °C	Density	n_D	Solubility						Ref.
									w	al	eth	ace	bz	other solvents	
	Naphthalene														
n129	**—,1,4-diamino-***	1,4-Naphthylenediamine. $C_{10}H_{10}N_2$. *See* n14	158.20	nd	120	...	...	1.6441^{18}	δ^h	v	v	...	v	chl v	**B13**[2], 82
n130	**—,1,5-diamino-***	1,5-Naphthylenediamine. $C_{10}H_{10}N_2$. *See* n14	158.20	pr (eth)	190	sub	1.4	...	δ^h	s	s	...	...	chl s	**E12B**, 824
n131	**—,1,6-diamino-***	1,6-Naphthylenediamine. $C_{10}H_{10}N_2$. *See* n14	158.20	nd (w)	77.5	...	1.1477^{99}_{4}	1.7083^{99}	s^h δ	s^h	s^h	...	s^h	...	**E12B**, 828
n132	**—,1,7-diamino-***	1,7-Naphthylenediamine. $C_{10}H_{10}N_2$. *See* n14	158.20	pl (bz), nd (w)	117.5	...	...	...	δ^h	s^h	δ	...	s^h	lig δ	**B13**, 205
n133	**—,1,8-diamino-***	1,8-Naphthylenediamine. $C_{10}H_{10}N_2$. *See* n14	158.20	(w-al)	66.5	205^{12} sub	1.1265^{99}_{4}	1.6828^{99}	δ s^h	v	v	...	...	...	**B13**[2], 85
n134	**—,2,3-diamino-***	2,3-Naphthylenediamine. $C_{10}H_{10}N_2$. *See* n14	158.20	lf (eth or w)	193–4	...	1.0968^{26}_{4}	1.6342^{26}	...	v	s	...	...	...	**B13**[2], 86
n135	**—,2,6-diamino-***	2,6-Naphthylenediamine. $C_{10}H_{10}N_2$. *See* n14	158.20	nd or lf (w)	222	...	...	...	δ^h	δ	δ	...	...	...	**B13**[2], 86
n136	**—,1,4-diamino-2-methyl-***	$C_{11}H_{12}N_2$. *See* n14	172.23	(peth)	113–4	...	...	...	...	...	...	...	...	ac s aa s	**E12B**, 846
n137	—,—,dihydro-chloride	Vitamin K_6. $C_{11}H_{12}N_2.2HCl$. *See* n14	245.15	(dil HCl)	300d	...	...	...	v	...	...	...	...	...	**E12B**, 846
n138	**—,1,5-diamino-2-methyl-***	$C_{11}H_{12}N_2$. *See* n14	172.23	red-ye lf (dil al)	136	...	...	...	δ	v	v	...	v	...	**E12B**, 847
n139	**—,1,3-diamino-2-phenyl-***	$C_{16}H_{14}N_2$. *See* n14	234.30	pl (MeOH or bz)	116	...	...	...	...	v	δ	...	v	lig i	**E12B**, 843
n140	**—,2,6-dibromo-1,5-dihydroxy-,** diacetate	$C_{14}H_{10}Br_2O_4$. *See* n14	402.05	nd	147.5	...	...	...	...	...	...	...	...	lig s^h	**B6**[2], 951
n141	**—,1,6-dibromo-2-hydroxy-***	1,6-Dibromo-2-naphthol. $C_{10}H_6Br_2O$. *See* n14	301.98	nd (peth or aa)	106	...	...	...	i	s	s	...	...	...	**E12B**, 1512
n142	**—,1,2-dichloro-***	$C_{10}H_6Cl_2$. *See* n14	197.07	pl (al)	35	295–8	1.3147^{49}_{4}	1.6338^{49}	...	s	s	...	...	...	**E12B**, 302
n143	**—,1,3-dichloro-***	α-Dichloronaphthalene. $C_{10}H_6Cl_2$. *See* n14	197.07	nd or pr (al)	61–2	291^{755}	...	...	...	s	...	...	...	...	**E12B**, 304
n144	**—,1,4-dichloro-***	β-Dichloronaphthalene. $C_{10}H_6Cl_2$. *See* n14	197.07	nd or pr (al)	68	286–7	1.2997^{76}_{4}	1.6228^{76}	i	δ	s	v	s	...	**E12B**, 306
n145	**—,1,5-dichloro-***	γ-Dichloronaphthalene. $C_{10}H_6Cl_2$. *See* n14	197.07	nd or lf (MeOH)	107	...	sub	...	i	δ	s	...	...	...	**E12B**, 309
n146	**—,1,6-dichloro-***	η-Dichloronaphthalene. $C_{10}H_6Cl_2$. *See* n14	197.07	nd or pr (al)	49	sub	...	...	...	...	...	...	...	...	**E12B**, 310
n147	**—,1,7-dichloro-***	δ-Dichloronaphthalene. $C_{10}H_6Cl_2$. *See* n14	197.07	nd or pr (al, aa)	63.5–4.5	285–6	1.2611^{100}_{4}	1.6092^{100}	...	s	s	...	s	aa s	**E12B**, 312
n148	**—,1,8-dichloro-***	ζ-Dichloronaphthalene. *peri*-Dichloronaphthalene. $C_{10}H_6Cl_2$. *See* n14	197.07	rh pl (hexane)	87	d	1.2924^{99}_{4}	1.6236^{100}	...	...	...	...	...	...	**E12B**, 313
n149	**—,2,3-dichloro-***	ι-Dichloronaphthalene. $C_{10}H_6Cl_2$. *See* n14	197.07	rh lf (al)	120	...	...	...	i	δ s^h	s	...	...	...	**E12B**, 315
n150	**—,2,6-dichloro-***	ϵ-Dichloronaphthalene. $C_{10}H_6Cl_2$. *See* n14	197.07	pr (aa), nd (al), lf (al), pl (PhCO$_2$Et)	140–1	285	...	...	...	δ	s	...	...	chl s aa s	**E12B**, 317
n151	**—,2,7-dichloro-***	δ-Dichloronaphthalene. $C_{10}H_6Cl_2$. *See* n14	197.07	pl or lf	116	...	...	...	...	v^h	...	...	...	hexane s^h aa s^h	**E12B**, 318
n152	**—,1(diethyl-amino)-***	*N,N*-Diethyl-1-naphthyl-amine*. $C_{14}H_{17}N$. *See* n14	199.30	...	...	$155–65^{30}$ 285	1.015^{20}_{20}	1.5961^{20}	i	s	s	...	...	aa s	...
n153	**—,2,3-di-hydrazino-***	$C_{10}H_{12}N_4$. *See* n14	188.23	red-br (al or w), colorless nd (bz)	167–8d (w) 155–6 (bz, -al)	...	...	...	...	v	...	...	...	dil ac s sulf s(red) aa δ	**E12B**, 908
n154	**—,1,2-dihydro-***	Δ^1-Dialin. Δ^1-Dihydro-naphthalene. $C_{10}H_{10}$. *See* n14	130.19	lf, pl on cooling	−9	206–7 78^{9}	0.9974^{20}_{4}	1.5817^{20}	...	...	...	...	...	...	**E12B**, 53
n155	**—,1,4-dihydro-***	Δ^2-Dialin. $C_{10}H_{10}$. *See* n14	130.19	pl	24	209 94^{17}	0.9928^{33}_{4}	1.5549^{33}	...	...	...	...	...	aa s^h	**E12B**, 54
n156	**—,1,2-dihydroxy-***	β-Naphthohydroquinone. 1,2-Naphthalenediol*. $C_{10}H_8O_2$. *See* n14	160.16	lf or nd (CS_2) lf (w+1), nd (lig)	103–4 (anh) 58–60 (+1w)	...	...	...	δ	...	s	...	...	alk s	**E12B**, 1962

For explanations, symbols and abbreviations see beginning of table.

No.	Name	Synonyms and Formula	Mol. wt.	Crystalline form, color and specific rotation	m.p. °C	b.p. °C	Density	n_D	Solubility						Ref.
									w	al	eth	ace	bz	other solvents	
	Naphthalene														
n157	**—,1,4-dihydroxy-***	1,4-Naphthalenediol*. α-Naphthohydroquinone. $C_{10}H_8O_2$. *See* n14	160.17	mcl nd (bz)	176	...	...	...	s^h	s	s	...	s	CS_2 i aa s^h lig δ	**E12B,** 1973
n158	**—,1,5-dihydroxy-***	1,5-Naphthalenediol*. $C_{10}H_8O_2$. *See* n14	160.17	nd, pr (w)	258	d	...	...	δ	s	s	s	s	aa i lig i	**R12B,** 1980
n159	—,—,diacetate	$C_{14}H_{12}O_4$. *See* n14	244.25	nd (bz)	161	...	...	...	...	...	...	...	...	aa s^h	**E12B,** 1982
n160	**—,1,6-dihydroxy-***	1,6-Naphthalenediol*. $C_{10}H_8O_2$. *See* n14	160.17	pr (bz)	138	...	...	...	δ	s	s	s	...	MeOH s lig i	**E12B,** 1987
n161	**—,1,7-dihydroxy-***	1,7-Naphthalenediol*. $C_{10}H_8O_2$. *See* n14	160.17	nd (bz or sub)	178	sub	...	...	δ	v	v	...	...	aa s	**E12B,** 1989
n162	**—,1,8-dihydroxy-***	1,8-Naphthalenediol*. $C_{10}H_8O_2$. *See* n14	160.17	unst nd	144	...	...	...	δ	s^h	v	...	...	to v lig δ	**E12B,** 1991
n163	**—,2,3-dihydroxy-***	2,3-Naphthalenediol*. $C_{10}H_8O_2$. *See* n14	160.17	lf (w)	159	...	...	...	s^h	v	v	...	v	aa s lig s	**E12B,** 1994
n164	**—,2,6-dihydroxy-***	2,6-Naphthalenediol*. $C_{10}H_8O_2$. *See* n14	160.17	rh pl (w)	218	sub	...	...	v^h δ	s	s	s	δ	aa s lig i	**E12B,** 2003
n165	**—,2,7-dihydroxy-***	2,7-Naphthelenediol*. $C_{10}H_8O_2$. *See* n14	160.17	nd (w or dil al)	190	sub	...	...	s	s	s	...	s	chl s lig i	**E12B,** 2007
n166	**—,1,4-dihydroxy-2-methyl-, diacetate**	$C_{15}H_{14}O_4$. *See* n14	258.28	pr	113	...	...	...	i	s^h	...	...	...	...	**B6**^{2}, 958
n167	**—,1,5-di-mercapto-***	1,5-Naphthalened ithiol*. $C_{10}H_8S_2$. *See* n14	192.30	lf (al, eth, bz)	103	...	...	...	...	v	v	...	v	dil HCl v	**B6**^{2}, 952
n168	**—,1,4-dimethyl-***	α-Dimethylnaphthalene. $C_{12}H_{12}$. *See* n14	156.22	...	>−18	$262\text{–}4^{751}$	1.0157^{20}	1.6158^{16}	i	v	∞	...	...	...	**B5**^{2}, 468
n169	**—,2,3-dimethyl-***	Guaiene. $C_{12}H_{12}$. *See* n14	156.22	lf (al)	104–4.5	265–6	1.008^{20}_{4}	...	i	s^h δ	s	...	s	...	**E12B,** 142
n170	**—,1(dimethyl-amino)-***	*N,N*-Dimethyl-1-naphthyl-amine*. $C_{12}H_{13}N$. *See* n14	171.24	vt flr	59.5–6.7	274.5 $139\text{–}40^{12}$	1.0446^{15}_{15}	1.624^{15}	i	s	s	...	...	...	...
n171	**—,2(dimethyl-amino)-***	*N,N*-Dimethyl-2-naphthyl-amine*. $C_{12}H_{13}N$. *See* n14	171.24	dk red nd	52–3	$212\text{–}3^{69}$ $160\text{–}1^{12}$	1.0387^{70}_{70}	1.6443	i	s	s	...	...	...	**B12,** 1273
n172	**—,1,6-dimethyl-4-isopropyl-***	Cadalene. $C_{15}H_{18}$. *See* n14	198.31	...	...	$291\text{–}2^{720}$ 146^{7}	0.9725^{24}_{4}	1.5852^{25}	...	...	...	...	...	...	**B5**^{2}, 473
n174	**—,1,3-dinitro-***	γ-Nitronaphthalene. $C_{10}H_6N_2O_4$. *See* n14	218.17	ye nd (bz)	144–5	sub	...	...	i	s	...	...	...	...	**B5**^{2}, 454
n175	**—,1,5-dinitro-***	$C_{10}H_6N_2O_4$. *See* n14	218.17	hex nd (aa)	217.5	sub	...	...	i	δ	v	...	s	CS_2 i, Py s	**B5**^{2}, 454
n176	**—,1,8-dinitro-***	$C_{10}H_6N_2O_4$. *See* n14	218.17	ye rh pl (chl)	173–3.5	445d	...	...	i	i	...	...	δ	Py si, chl δ	**B5**^{2}, 455 **B14**^{2}, 653
$n176^1$	**—,1,6-dinitro-2-hydroxy-***	$C_{10}H_6N_2O_5$. *See* n14	234.17	gr ye nd	208d	...	...	...	δ	s	...	...	...	Py, chl s lig i	**E12B,** 1581
n177	**—,2,4-dinitro-1-triazo-***	$C_{10}H_5N_5O_4$. *See* n14	259.18	ye rh nd (al)	105d	...	...	...	...	s^h	s	...	s	chl, to s lig s^h	**B5**^{2}, 460
n178	**—,5,8-dioxo-1,4,5,8,9,10-hexa-hydro-1,4-methylene-***	Cyclopentadienebenzo-quinone. $C_{11}H_{10}O_2$. *See* n14	174.20	gr-ye lf	77–8	...	...	...	...	s	s	...	s	...	**E13,** 1033
n179	**—,1-ethoxy-***	Ethyl α-naphthyl ether. $C_{12}H_{12}O$. *See* n14	172.23	nd	5.5	280.5	1.060^{20}_{4}	1.602^{20}	i	v	v	...	...	...	**B6**^{2}, 578
n180	**—,2-ethoxy-***	Ethyl β-naphthyl ether. Nerolin. $C_{12}H_{12}O$. *See* n14	172.23	nd	35.5–36	280	1.0606^{25}_{4}	1.5975^{36}	i	s	s	...	...	to s lig s CS_2 s	**B6**^{2}, 598
n181	**—,1-ethyl-***	$C_{12}H_{12}$. *See* n14	156.23	...	−13.88	258.67	1.00816^{20}_{4}	1.6062^{20}	i	∞	∞	...	...	...	**B5**^{2}, 467
n182	**—,2-ethyl-***	$C_{12}H_{12}$. *See* n14	156.23	...	−7.4	257.9	0.9922^{20}_{4}	1.5999^{20}	i	∞	∞	...	...	...	**B5**^{2}, 467
n183	**—,1(ethyl-amino)-***	$C_{12}H_{13}N$. *See* n14	171.24	...	...	191^{16}	1.060^{20}	1.6477	i	s	s	...	...	...	...
n184	**—,2(ethyl-amino)-***	$C_{12}H_{13}N$. *See* n14	171.24	...	<15	306–17 167^{10-2}	1.057	...	...	...	...	...	...	...	**B12,** 1274
n185	**—,1-fluoro-***	$C_{10}H_7F$. *See* n14	146.17	...	−13	215^{758}	1.134^{0}	1.5939^{20}	i	s	s	...	...	...	**B5**^{2}, 444
n186	**—,2-fluoro-***	$C_{10}H_7F$. *See* n14	146.17	nd	59	212.5	...	...	i	s	s	...	s	chl s	**B5,** 541
n187	**—,1(formyl-amino)-***	$C_{11}H_9NO$. *See* n14	171.20	nd	137.5	...	...	...	s^h	s	...	...	...	os s	**E12B,** 459
n188	**—,2(formyl-amino)-***	$C_{11}H_9NO$. *See* n14	171.20	lf (bz-peth)	129	...	...	...	...	s	...	...	...	...	**E12B,** 562
n189	**—,1,2,3,4,9,10-hexahydro-***	Naphthalene hexahydride. $C_{10}H_{14}$. *See* n14	134.22	...	...	195	0.934	1.5331	...	...	...	...	...	...	**B5,** 433
n190	**—,1-hydroxy-***	1-Naphthol. α-Naphthol. $C_{10}H_8O$. *See* n14	144.17	ye mcl pr	93.35	288^{760}	1.103^{101}	1.6206^{99}	$δ^h$ i	s	s	...	s	chl s CCl_4 δ	**E12B,** 1148

For explanations, symbols and abbreviations see beginning of table.

No.	Name	Synonyms and Formula	Mol. wt.	Crystalline form, color and specific rotation	m.p. °C	b.p. °C	Density	n_D	Solubility w	al	eth	ace	bz	other solvents	Ref.
	Naphthalene														
n191	—,2-hydroxy-*	2-Naphthol. β-Naphthol. $C_{10}H_8O$. *See* n14	144.17	mcl lf	122	295^{760}	1.28		i	v^h	v		v	SO_2 δ lig $δ^h$	E12B, 1210
n192	—,—,acetate	2-Acetoxynaphthalene. $C_{12}H_{10}O_2$. *See* n14	186.21	nd (al)	70	132–44			i	v	v			chl v	E12B, 1256
n193	—,—,benzoate	2-Benzoyloxynaphthalene. $C_{17}H_{12}O_2$. *See* n14	248.28	nd or pr (al)	108				i	v^h	δ				E12B, 1260
n197	—,2-hydroxy-1-methyl-*	1-Methyl-2-naphthol. $C_{11}H_{10}O$. *See* n14	158.20	nd (w, bz-lig)	110	180^{12}			δ	v	v		v	aa v	E12B, 1388
n198	—,1-hydroxy-8-nitro-*	8-Nitro-2-naphthol. $C_{10}H_7NO_3$. *See* n14	189.17	ye nd (al)	144–5				s	v	v	v	v		E12B, 1559
n199	—,2-hydroxy-1-nitro-*	1-Nitro-2-naphthol. $C_{10}H_7NO_3$. *See* n14	189.17	ye nd or lf (al)	103	$115^{0.05}$			s^h	v	v				E12B, 1547
n200	—,6-hydroxy-1-nitro-*	5-Nitro-2-naphthol. $C_{10}H_7NO_3$. *See* n14	189.17	lt ye nd (w)	147				v^h	δ	v				E12B, 1556
n201	—,1-hydroxy-4-nitroso-*	1,4-Naphthoquinone 1-oxime*. 4-Nitroso-1-naphthol. $C_{10}H_7NO_2$. *See* n14	173.17	bt ye nd (bz), (dil al)	198				i	v	v	v	$δ^h$	MeOH v	E12B, 1597
n202	—,2-hydroxy-1-nitroso-*	1,2-Naphthoquinone 1-oxime*. 1-Nitroso-2-naphthol. $C_{10}H_7NO_2$. *See* n14	173.17	ye nd (bz)	112				i	s v^h	v		v	aa v lig δ	E12B, 1597
—	—,1-hydroxy-1,2,3,4-tetrahydro*-	*see* **Tetralin, hydroxy-**													
n208	—,2-hydroxy-1,3,6-tribromo-*	1,3,6-Tribromo-2-naphthol. $C_{10}H_5Br_3O$. *See* n14	380.88	nd (aa or al)	133					s			s	MeOH s CCl_4 s	$B6^2$, 606
n209	—,2-hydroxy-1,4,6-tribromo-*	Providoform. 1,4,6-Tribromo-2-naphthol. $C_{10}H_5Br_3O$. *See* n14	380.88	nd (bz)	157–8					s			s^h	chl s aa s	E12B, 1522
n210	—,3-hydroxy-1,2,7-tribromo-*	3,4,6-Tribromo-β-naphthol. $C_{10}H_5Br_3O$. *See* n14	380.88	nd (bz)	127–8					s			v^h		$B6^2$, 607
n211	—,1-hydroxy-2,3,4-trichloro-*	2,3,4-Trichloro-α-naphthol. $C_{10}H_5Cl_3O$. *See* n14	247.51	nd (aa or lig)	159–60				i	s^h	v			aa $δ^h$ lig $δ^h$	$B6^2$, 582
n212	—,2-hydroxy-1,3,4-trichloro-*	1,3,4-Trichloro-β-naphthol. $C_{10}H_5Cl_3O$. *See* n14	247.51	nd	162					s				aa s	$B6^2$, 604
n213	—,1-iodo-*	$C_{10}H_7I$. *See* n14	254.07			305	1.7474_4^{14}	1.7054^{14}	i	∞	∞		∞	CS_2 ∞	$B5^2$, 449
n214	—,2-iodo-*	$C_{10}H_7I$. *See* n14	254.07	lf	54.5	308 172^{21}	1.6319_4^{99}	1.6662^{99}	i	v	v			aa v	$B5^2$, 450
n214[1]	—,1-mercapto-*	1-Naphthalenethiol*. $C_{10}H_8S$. *See* n14	160.24			208.5^{200} 285d	1.155_4^{20}		δ	v	v				$B6^2$, 588
n214[2]	—,2-mercapto-*	2-Naphthalenethiol*. $C_{10}H_8S$. *See* n14	160.24	pl (al)	81; 162.7^{20}	288	1.550		δ	v	v			lig v	$B6^2$, 610
n215	—,1-methoxy-*	Methyl α-naphthyl ether. $C_{11}H_{10}O$. *See* n14	158.20		<−10	269 cor	1.0964_2^{14}	1.6232^{14}	i	s	s			chl s, CS_2 v	$B6^2$, 578
n216	—,2-methyl-*	Methyl β-naphthyl ether. Nerolin. $C_{11}H_{10}O$. *See* n14	158.20	lf (eth)	72	274	sub		δ	δ	s		v	CS_2 s, chl v	$B6^2$, 598
n217	—,1-methoxy-*	$C_{11}H_{10}$. *See* n14	142.20		−22	$240–3^{759}$	1.0287^{12}	1.618	i	v	v				$B5^2$, 460
n218	—,2-methyl-*	$C_{11}H_{10}$. *See* n14	142.20	mcl (al)	37–8	240–2	1.029_4^{20}	1.6026^{40}	i	v	v				$B5^2$, 463
n219	—,1(methylamino)-*	$C_{11}H_{11}N$. *See* n14	157.22			293 $165–75^{15}$		1.6722^{20}	i	s	s			CS_2 s	
n220	—,2(methylamino)-*	$C_{11}H_{11}N$. *See* n14	157.22	dk in air		317 $165–70^{12}$									B12, 1273
m221	—,1(β-methylbutoxy)-*	Isoamyl α-naphthyl ether. $C_{15}H_{18}O$. *See* n14	214.31			$317–9^{742}$ cor	1.0069_4^{14}	1.5705^{14}							B6, 607
n222	—,2(3-methylbutoxy)-*	Isoamyl β-naphthyl ether. $C_{15}H_{18}O$. *See* n14	214.31	lf	26.5	323–6d	1.0155_4^{12}	1.5768^{12}	i	s	s				B6, 642
n223	—,1-methyl-2-nitro-*	$C_{11}H_9NO_2$. *See* n14	187.20	lt ye nd (al)	58–9				i	s					E12B, 363
n224	—,1-methyl-3-nitro-*	$C_{11}H_9NO_2$. *See* n14	187.20	ye nd (al)	81–2				i	s					E12B, 364
n225	—,1-methyl-4-nitro-*	$C_{11}H_9NO_2$. *See* n14	187.20	bt ye nd	71–2	$182–3^{18}$								os s	E12B, 362
n226	—,1-methyl-5-nitro-*	$C_{11}H_9NO_2$. *See* n14	187.20	br nd (al)	82–3				i	s					E12B, 363
n227	—,1-methyl-6-nitro-*	$C_{11}H_9NO_2$. *See* n14	187.20	nd (dil al)	76–7				i	s					E12B, 364

For explanations, symbols and abbreviations see beginning of table.

No.	Name	Synonyms and Formula	Mol. wt.	Crystalline form, color and specific rotation	m.p. °C	b.p. °C	Density	n_D	Solubility w	al	eth	ace	bz	other solvents	Ref.
	Naphthalene														
n228	**—,1-methyl-7-nitro-***	$C_{11}H_9NO_2$. *See* n14	187.20	ye nd (al)	98–9				i	s^h					**E12B,** 364
n229	**—,1-methyl-8-nitro-***	$C_{11}H_9NO_2$. *See* n14	187.20	lt ye (90% al)	65				i	s^h					**E12B,** 363
n230	**—,2-methyl-1-nitro-***	$C_{11}H_9NO_2$. *See* n14	187.20	yesh pr or nd (al)	81	188^{20}			i	s					**E12B,** 362
n231	**—,2-methyl-3-nitro-***	$C_{11}H_9NO_2$. *See* n14	187.20	yesh lf (al)	117–8				i	s^h					**E12B,** 364
n232	**—,2-methyl-6-nitro-***	$C_{11}H_9NO_2$. *See* n14	187.20	ye nd	119				i	s^h					**E12B,** 364
n233	**—,2-methyl-7-nitro-***	$C_{11}H_9NO_2$. *See* n14	187.20	yesh pl (al)	105				i	s^h					**E12B,** 364
n234	**—,3-methyl-1-nitro-***	$C_{11}H_9NO_2$. *See* n14	187.20	pa ye nd (al)	49–50				i	s^h					**E12B,** 362
n235	**—,6-methyl-1-nitro-***	$C_{11}H_9NO_2$. *See* n14	187.20	ye nd (al)	61–2				i	s^h					**E12B,** 363
n236	**—,7-methyl-1-nitro-***	$C_{11}H_9NO_2$. *See* n14	187.20	ye nd (al)	36–8				i	s^h					**E12B,** 363
n237	**—,1-nitramino-***	1-Diazonaphthalenic acid. $C_{10}H_7^{\alpha}NHNO_2$	188.19	lt ye nd (w)	123–4									alk, org solv v	**E12B,** 862
n238	**—,2-nitramino-***	2-Diazonaphthalenic acid. $C_{10}H_7^{\beta}NHNO_2$	188.19	lf or nd	131.5–6										**E12B,** 862
n239	**—,1-nitro-***	$C_{10}H_7NO_2$. *See* n14	173.17	ye nd (al)	58.5	304	1.332_4^{20}		i	s	v			chl, CS_2 v	**B5**², 450
n240	**—,2-nitro-***	$C_{10}H_7NO_2$. *See* n14	173.17	ye rh nd (al)	79.0	165^{15}			i	v	v				**B5**², 451
n241	**—,1-nitro-5-triazo-***	$C_{10}H_6N_4O_2$. *See* n14	214.18	gold-ye nd (al)	121				δ s^h						**B5**², 459
n242	**—,2-nitro-1-triazo-***	$C_{10}H_6N_4O_2$. *See* n14	214.18	ye nd (dil ace)	103–4d					v		v		aa v lig δ^h	**B5**, 565
n243	**—,1(nitroso-hydroxyl-amino)-***	$C_{10}H_7^{\alpha}N(NO)OH$. *See* n14	188.19	nd (peth)	54–5									chl, NH_3 s	**E12B,** 863
n244	**—,2(nitroso-hydroxyl-amino)-***	$C_{10}H_7^{\beta}N(NO)OH$	188.19	nd (aa-peth)	88–92						s			NH_3 s aa s	**E12B,** 863
n245	**—,octachloro-***	Perchloronaphthalene. $C_{10}Cl_8$. *See* n14	403.74	nd (CCl_4-bz)	197.5–8 cor	440–2d $246^{0.5}$				δ			v	chl v lig v	**B5**², 446
n246	**—,1-phenyl-***	$C_{16}H_{12}$. *See* n14	204.27		ca. 45	334^{770}			i	v	v		v		**B5**², 602
n247	**—,2-phenyl-***	$C_{16}H_{12}$. *See* n14	204.27	bl lf (al)	101–2	345–6				s	v		s	aa s	**B5**², 603
n248	**—,1-propoxy-***	α-Naphthyl propyl ether. $C_{13}H_{14}O$. *See* n14	186.26			$298–9^{762}$	1.0447_4^{18}	1.5928^{18}							**B6**, 607
n249	**—,2-propoxy-***	β-Naphthyl propyl ether. $C_{13}H_{14}O$. *See* n14	186.26	nd (al)	39.5–40					s^h					**B6**, 641
—	**—,1,2,3,4-tetra-hydro-***	*see* **Tetralin**													
n251	**—,1,3,5,8-tetra-nitro-***	γ-Tetranitronaphthalene. $C_{10}H_4N_4O_8$. *See* n14	308.17	lt ye tetr (ace)	194–5					δ		v		chl δ HNO_3 s	**E12B,** 411
n252	**—,1,3,6,8-tetra-nitro-***	β-Tetranitronaphthalene. $C_{10}H_4N_4O_8$. *See* n14	308.17	ye nd (al)	203	exp			i	δ					**B5**², 459
n253	**—,1-triazo-***	α-Naphthyl azide. $C_{10}H_7N_3$. *See* n14	169.19	pa ye pr	12		1.1713^{25}	1.6550^{25}		v	∞	∞			**E12B,** 1041
n254	**—,2-triazo-***	β-Naphthyl azide. $C_{10}H_7N_3$. *See* n14	169.19	pr (al), nd (peth)	33				δ	s				MeOH s, os v	**E12B,** 1041
n255	**—,1,2,5-tri-methyl-***	$C_{13}H_{14}$. *See* n14	170.26	nd (al)	33.5	140^{12}	1.011_4^{18}	1.6082^{18}	i	δ	v		v	MeOH δ	**B5**², 470
n256	**—,1,2,6-tri-methyl-***	$C_{13}H_{14}$. *See* n14	170.26			$145–6^{15}$					s		s		**B5**, 571
n257	**—,1,2,7-tri-methyl-***	$C_{13}H_{14}$. *See* n14	170.26			$147–8^{16}$	1.008_4^{15}	1.6093^{15}	i		s		s		**B5**², 470
n258	**—,2,3,6-tri-methyl-***	$C_{13}H_{14}$. *See* n14	170.26		92–3	$263–4^{760}$			i				s		**B5**, 572
n259	**—,1,2,5-trinitro-***	$C_{10}H_5N_3O_6$. *See* n14	263.17	lt ye nd (al)	112–3				i	s				CCl_4 s	**B5**², 457
n260	**—,1,3,5-trinitro-***	$C_{10}H_5N_3O_6$. *See* n14	263.17	ye rh (chl)	122.0	364 exp			i	v	δ	v		chl s	**B5**², 457
n261	**—,1,3,8-trinitro-***	$C_{10}H_5N_3O_6$. *See* n14	263.17	yesh mcl pr (ace or aa), wh nd (80% HNO_3)	218				i^h	i	s			to δ Py s	**E12B,** 405

For explanations, symbols and abbreviations see beginning of table.

No.	Name	Synonyms and Formula	Mol. wt.	Crystalline form, color and specific rotation	m.p. °C	b.p. °C	Density	n_D	Solubility w	al	eth	ace	bz	other solvents	Ref.
	Naphthalene														
n262	**—,1,4,5-trinitro-***	$C_{10}H_5N_3O_6$. *See* n14	263.17	ye lf, nd or rh pl	148–9				i	δ	δ	v^h		AcOEt s Py v^h	**E12B,** 406
—	**Naphthalene carboxylic acid***	*see* **Naphthoic acid**													
n263	**1,2-Naphthalenedicarboxylic acid***		216.20	nd (al)	175d				s^h	s	s			aa s	**E12B,** 4681
n264	**1,4-Naphthalenedicarboxylic acid***		216.20	(aa or $PhNO_2$)	320				$δ^h$	v				os δ	**E12B,** 4690
n265	**1,5-Naphthalenedicarboxylic acid***		216.20	nd ($PhNO_2$)	320–2 cor				i	$δ^h$	δ		i^h	chl $δ^h$ lig i^h	**E12B,** 4693
n266	**1,6-Naphthalenedicarboxylic acid***		216.20	nd (aa)	305 (sinters ca. 290)					s^h				os δ aa s^h	**E12B,** 4695
n267	**1,6-Naphthalenedicarboxylic acid***		216.20	ye (dil al or aa)	308d					s	s			aa s	**E12B,** 4696
n268	**1,8-Naphthalenedicarboxylic acid***		216.20	nd	270d				δ	δ s^h	δ				**E12B,** 4697
n269	—,anhydride	Naphthalic anhydride. $C_{12}H_6O_3$. *See* n268	198.18	nd	271.5–2				i	$δ^h$	δ		i	aa s	**E12B,** 4705
n270	—,dichloride	Naphthaloyl chloride. $C_{12}H_6Cl_2O_2$. *See* n268	253.09	pr (CS_2)	ca. 84–6	$195–200^{0.2}$							s	chl s	**E12B,** 4713
n271	—,diethyl ester	Diethyl naphthalate. $C_{16}H_{16}O_4$. *See* n268	272.30	yesh mcl pr or nd (dil al)	59–60	$238–9^{19}$	1.1399^{70}_4	1.5586^{70}	i	s	s			AcOEt s con sulf s	**B9², 652**
n272	—,dimethyl ester	Dimethyl naphthalate. $C_{14}H_{12}O_4$. *See* n268	244.25	nd or pr (al, MeOH)	104				i	s				MeOH v aa s	**E12B,** 4703
n273	—,imide	Naphthalimide $C_{12}H_7NO_2$. *See* n268	197.20	nd	300				i	δ	i		i		**E12B,** 4724
n274	**2,3-Naphthalenedicarboxylic acid***		216.20	pr (aa or w)	239–40d				$δ^h$	δ	δ		i	CS_2 i chl i lig i	**E12B,** 4723
n275	**2,6-Naphthalenedicarboxylic acid***		216.20	nd (al)	>300				$δ^h$	s^h			δ	to $δ^h$ aa δ	**E12B,** 4728
n276	**2,7-Naphthalenedicarboxylic acid***		216.20	nd (w, al or dil HCl)	>300d				i^h	s^h			i	aa i	**E12B,** 4730
n277	**1,2-Naphthalenedicarboxylic acid, 3,4-dihydro-,** anhydride	$C_{12}H_8O_3$. *See* n263	200.20	pa ye nd (lig)	166–7								v	lig δ s^h	**B17², 487**
n278	**1,8-Naphthalenedicarboxylic acid, 3,6-dinitro-***	$C_{12}H_6N_2O_8$. *See* n268	306.19	silvery lf (w)	212				$δ^h$	v	i		i	AcOEt v $PhNO_2$ v aa v	**E12B,** 4775
n279	—,3-nitro-, anhydride	$C_{12}H_5NO_5$. *See* n268	243.18	ye nd (aa or $PhNO_2$)	252–3				i^h	i^h			i^h		**E12B,** 4771
n280	**—,4-nitro-***	$C_{12}H_7NO_6$. *See* n268	261.19	ye nd	140–50d					i	i			aa s^h lig i	**E12B,** 4772
n281	**1,5-Naphthalenedisulfonic acid***		288.30	lf	d		1.493		v	s	i				**B11², 119**

For explanations, symbols and abbreviations see beginning of table.

No.	Name	Synonyms and Formula	Mol. wt.	Crystalline form, color and specific rotation	m.p. °C	b.p. °C	Density	n_D	Solubility w	al	eth	ace	bz	other solvents	Ref.
	Naphthalenedisulfonic acid														
n282	**1,6-Naphthalene-disulfonic acid***	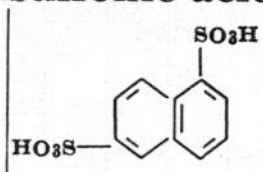	288.30	og	125d				v	s	i				E11², 121
n283	**2,7-Naphthalene-disulfonic acid***	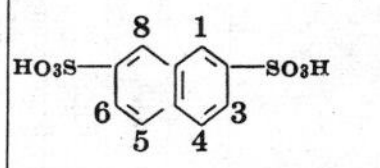	288.30	hyg nd	199				s					con HCl δ	E11², 122
n284	**1,3-Naphthalene-disulfonic acid, 7-amino-***	β,γ-Acid. Amino-G-acid. 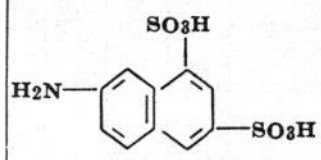	303.31	mcl pr (w)	273–5				s						B14, 784
n285	**2,7-Naphthalene-disulfonic acid, 4-amino-5-hydroxy-***	H acid. $C_{10}H_9NO_7S_2$. *See* n283	319.31						δ	δ	δ				B14², 502
n286	**—,4,5-dihydroxy-***	Chromotropic acid. $C_{10}H_8O_8S_2.2H_2O$. *See* n283	356.34	nd or lf (w+2)					s	i	i				B11², 72
n287	**1,3-Naphthalene-disulfonic acid, 7-hydroxy-***	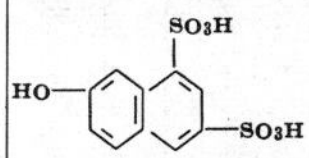	304.30						s						B11², 165
n288	**2,7-Naphthalene-disulfonic acid, 3-hydroxy-***	R-Acid. $C_{10}H_8O_7S_2$. *See* n283	304.30	dlq nd	d				s	i	i				B11², 164
n289	**1-Naphthalene-phosphonic acid***	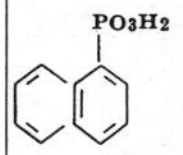	208.16	(w)	189				δ	v					B16², 392
n290	—,dichloride	$C_{10}H_7Cl_2OP$. *See* n289	245.05		ca. 60				d	d^h					B16², 392
n291	**1-Naphthalene-sulfinic acid***	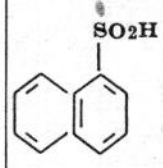	192.24	nd (w)	104				s	s	δ				B11¹, 5
n292	**2-Naphthalene-sulfinic acid***	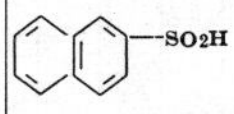	192.24	nd	105				s	s	s				B11², 10
n293	**1-Naphthalene-sulfonic acid***	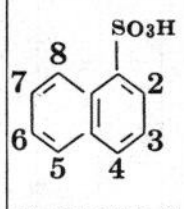	208.24		90				v	s	δ				B11², 91
n294	—,chloride	$C_{10}H_7ClO_2S$. *See* n293	226.68	lf (eth)	67.5	195^{13}			i	s	v				B11², 93
n295	**2-Naphthalene-sulfonic acid***	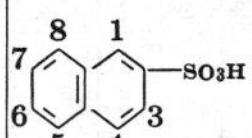	208.24	dlq pl	102	d	1.441^{25}_{4}		v	s	s		$δ^h$		B11², 96
n296	—,chloride	$C_{10}H_7ClO_2S$. *See* n295	226.68	pw or lf	76	201^{13}			i	s	v		s	chl, CS_2 s	B11², 99
n297	**1-Naphthalene-sulfonic acid, 3-amino-***	Cleve's acid. $C_{10}H_9NO_3S.H_2O$. *See* n293	241.27	nd (w+1)											E12B, 5088
n298	**—,4-amino-***	Naphtaionic acid. $C_{10}H_9NO_3S.\frac{1}{2}H_2O$. *See* n293	232.26	wh nd (w+½), red-br cr	d		1.6703^{25}_{4}		i	δ				MeOH s Py s	E12B, 5033
n299	**—,7-amino-***	Amino-F-acid. Bayer's acid. Cassela's acid. $C_{10}H_9NO_3S.H_2O$. *See* n293	241.27	nd (w+1), pl (aq ace)					i	δ	δ			aa s	E12B, 5103
n300	**—,4-amino-5-hydroxy-***	S-acid. $C_{10}H_9NO_4S$. *See* n293	239.25	nd					δ	i	i				B14, 835
n301	**2-Naphthalene-sulfonic acid, 5,7-dinitro-8-hydroxy-***	Flavianic acid. $C_{10}H_6N_2O_8S$. *See* n295	314.23	pa ye nd (con HCl, +3w), (w)	100(+3w) 150–1 (anh)				v	v				con HCl δ	B11², 156
n302	**1-Naphthalene-sulfonic acid, 4-hydroxy-***	Neville-Winther acid. $C_{10}H_8O_4S$. *See* n293	224.23	ta or pl (w)	137				v						B11, 271

For explanations, symbols and abbreviations see beginning of table.

No.	Name	Synonyms and Formula	Mol. wt.	Crystalline form, color and specific rotation	m.p. °C	b.p. °C	Density	n_D	Solubility w	al	eth	ace	bz	other solvents	Ref.
	1-Naphthalene-sulfonic acid														
n303	**—,5-hydroxy-***	$C_{10}H_8O_4S$. *See* n293	224.24	dlq	120				s					aa s	**B11**, 273
n304	**—,7-hydroxy-***	Croceic acid. $C_{10}H_8O_4S$. *See* n293	224.24												**B11**[2], 163
n305	**—,8-hydroxy-***	$C_{10}H_8O_4S$. *See* n293	224.24	(w+1)	106–7 (+1w)				v						**B11**, 275
n306	—,—,lactone	Naphthosultone. $C_{10}H_6O_3S$. *See* n293	206.22	pr (bz)	154	>360			δ	δ			s	CS_2 i, chl v^h	**B19**, 43
n307	**2-Naphthalene-sulfonic acid, 1-hydroxy-***	Schaffer acid. $C_{10}H_8O_4S$. *See* n295	224.24	pl (w)	>250				v^h s	s	i			dil HCl δ	**E12B**, 5262
n308	**—,6-hydroxy-***	$C_{10}H_8O_4S$. *See* n295	224.24	lf	167 (anh)				v	v	i			aa s	**B11**, 282
n309	**—,7-hydroxy-***	$C_{10}H_8O_4S$. *See* n295	224.24	nd (HCl)	115–6 (anh)	150d			v	v	i		i		**B11**, 285
n309[1]	**—,6-hydroxy-5-nitroso-***	$C_{10}H_7NO_5S$. *See* n295	253.23	og	d				v						**B11**[2], 190
—	**—,5,6,7,8-tetrahydro-**	*see* **Tetralinsulfonic acid**													
n310	**1,4,5,8-Naphthalenetetracarboxylic acid***	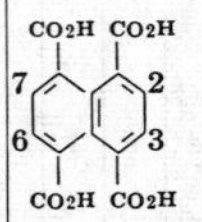	304.22	lf or nd	ca. 150d				s^h	δ		v	δ	CS_2 δ, chl δ, aa s^h	**E12B**, 4853
n311	**—,1,8:4,5-dianhydride***	$C_{14}H_4O_6$. *See* n310	268.18	nd (aa), pr ($PhNO_2$)	>300	sub 320[3]			i					$NaCO_3$ s (slowly d)	**E12B**, 4833
n312	**1,2,5-Naphthalenetricarboxylic acid***	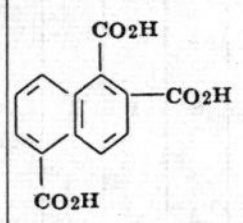	260.21	nd (MeOH)	270				s		δ			MeOH v	**E12B**, 4829
n313	**1,4,5-Naphthalenetricarboxylic acid***	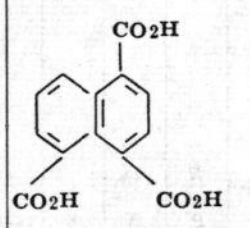	260.21	(eth)	266–8				δ	v					**E12B**, 4830
—	**Naphthalic acid**	*see* **1,8-Naphthalenedicarboxylic acid***													
—	**Naphthazarin**	*see* **1,4-Naphthoquinone, 5,8-dihydroxy-***													
—	**Naphthionic acid**	*see* **1-Naphthalenesulfonic acid, 4-amino-***													
n314	**Naphthocaine,** hydrochloride	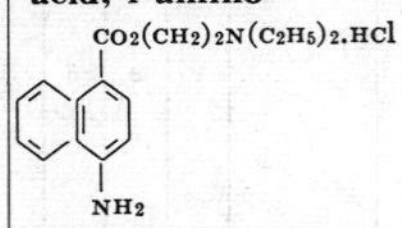	322.84	pa ye	212–6				δ						
—	**Naphthohydroquinone**	*see* **Naphthalene, dihydroxy-***													
n315	**1-Naphthoic acid**	α-Naphthalene carboxylic acid.* α-Naphthoic acid. 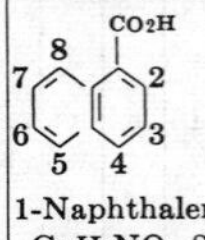	172.19	orh pl	160–1	>300	1.398		i	δ	s			chl s	**E12B**, 3966
n316	—,amide	1-Naphthalenecarbonamide*. $C_{11}H_9NO$. *See* n315	171.29	nd or pl (al or gl aa)	204–5	sub			δ	δ				con HCl s^h	**E12B**, 3992
n317	—,amide	1-Naphthamidine*. $C_{11}H_{10}N_2$. *See* n315	170.22	lf (dil al)	154				$δ^h$	s	δ	s	δ	chl s, lig δ	**E12B**, 4002
n318	—,anhydride	1-Naphthoic acid anhydride. $C_{22}H_{14}O_3$. *See* n315	326.36	pr (bz)	145–6				i	δ	s		s		**E12B**, 3988
n319	—,chloride	1-Naphthoyl chloride. $C_{11}H_7ClO$. *See* n315	190.63		20	297.5, 172^{15}			d^h	d^h					**B9**[2], 450
n320	—,ethyl ester	$C_{13}H_{12}O_2$. *See* n315	200.24			309, 221^{74}	1.1217^{25}_{25}	1.5966^{15}	i	s					**B9**[2], 449
n321	—,methyl ester	$C_{12}H_{10}O_2$. *See* n315	186.21		59.5	$167–9^{20}$	1.129^{20}_{4}	1.6086^{20}		s			s		**E12B**, 3981
n322	—,nitrile	1-Naphthonitrile. $C_{11}H_7N$. *See* n315	153.19	nd (lig)	38	296.5, $184–5^{37}$	1.1113^{25}_{25}	1.6298^{18}	i	v	v			lig s	**B9**, 649

For explanations, symbols and abbreviations see beginning of table.

No.	Name	Synonyms and Formula	Mol. wt.	Crystalline form, color and specific rotation	m.p. °C	b.p. °C	Density	n_D	Solubility w	al	eth	ace	bz	other solvents	Ref.
	1-Naphthoic acid														
n323	—,piperazinium salt	$2C_{11}H_8O_2.C_4H_{10}N_2$. *See* n315	430.51	wh	131.5–39d	...	...	...	δ	s	δ	...	...	...	...
n324	**2-Naphthoic acid**	2-Naphthalenecarboxylic acid*. β-Naphthoic acid.	172.19	mcl pl or nd (chl, eth, bz)	185	>300	1.077_4^{100}	...	i	s	s	...	...	lig δ^h	**E12B**, 4021
n325	—,amide	2-Naphthalenecarbonamide.* $C_{11}H_9NO$. *See* n324	171.20	lf (al)	195 cor	...	...	...	δ	s	s	s	...	chl, CS_2 s lig s	**E12B**, 4039
n326	—,amidine	2-Naphthamidine. $C_{11}H_{10}N_2$. *See* n324	170.22	(bz)	133–6	...	...	...	δ	v	δ	...	...	aa v^h	**E12B**, 4047
n327	—,anhydride	2-Naphthoic acid anhydride. $C_{22}H_{14}O_3$. *See* n324	326.36	nd (eth)	135	...	...	...	i	...	δ s^h	...	...	aa v	**E12B**, 4037
n328	—,chloride	2-Naphthoyl chloride. $C_{11}H_7ClO$. *See* n324	190.63	...	43	304–6	...	...	d^h i	...	s	...	...	chl s aa s	**B9**, 657
n329	—,ethyl ester	$C_{13}H_{12}O_2$. *See* n324	200.24	...	32	308–9 224^{74}	1.1143_4^{23}	1.5951^{23}	i	s	s	...	...	chl s aa s^h	**B9**[2], 453
n330	—,methyl ester	$C_{12}H_{10}O_2$. *See* n324	186.21	lf (MeOH)	73.2 cor	ca. 290	...	...	...	v	v	...	...	MeOH v chl v	**E12B**, 4031
n331	—,nitrile	2-Naphthonitrile. $C_{11}H_7N$. *See* n324	153.19	lf (lig)	...	306.5 $156–8^{12}$	1.0939_{60}^{60}	...	i	v	v	...	...	lig s^h	**B9**, 659
n332	—,piperazinium salt	$2C_{11}H_8O_2.C_4H_{10}N_2$. *See* n324	430.51	wh	194–5d	...	...	...	δ^h	s	δ	...	...	...	...
n333	**2-Naphthoic acid, 4-acetyl-3-hydroxy-**	$C_{13}H_{10}O_4$. *See* n324	230.22	ye pl (aa)	194	...	...	...	δ^h	v	...	...	...	chl δ aa δ	**E12B**, 4589
n334	**1-Naphthoic acid, 2-amino-**	$C_{11}H_9NO_2$. *See* n315	187.20	nd (dil al)	126d	...	...	...	δ	v	v	...	...	aa v^h	**E12B**, 4184
n335	—,**3-amino-**	$C_{11}H_9NO_2$. *See* n315	187.20	ye or pksh nd (eth)	178–9	...	...	...	...	s	...	...	...	MeOH s	**E12B**, 4185
n336	—,**4-amino-**	$C_{11}H_9NO_2$. *See* n315	187.20	brsh nd (w or al)	176	...	...	...	δ s^h	s	s	s	s	aa δ lig δ	**E12B**, 4188
n337	—,**5-amino-**	$C_{11}H_9NO_2$. *See* n315	187.20	ye nd (w or dil al)	211–2	...	...	...	δ^h	δ	i	...	s	...	**E12B**, 4198
n338	—,**6-amino-**	$C_{11}H_9NO_2$. *See* n315	187.20	ye (al or w)	205–6	...	...	...	δ	...	s	s	v	AcOEt s aa δ	**E12B**, 4200
n339	—,**7-amino-**	$C_{11}H_9NO_2$. *See* n315	187.20	pa br pr (al)	223–4	...	...	...	...	v	s	s	v	AcOEt s	**B14**[2], 322
n340	**2-Naphthoic acid, 1-amino-**	$C_{11}H_9NO_2$. *See* n324	187.20	nd (dil al or aa)	202–5d	...	...	...	δ	s	s	...	s	...	**B14**[2], 323
n341	—,**3-amino-**	$C_{11}H_9NO_2$. *See* n324	187.20	ye lf (dil al)	214	...	...	...	...	v	v	...	...	...	**B14**[2], 323
n342	—,**4-amino-**	$C_{11}H_9NO_2$. *See* n324	187.20	nd (dil al)	204–6	...	...	...	...	s^h	s	s	...	AcOEt s	**B14**[2], 323
n343	—,**5-amino-**	$C_{11}H_9NO_2$. *See* n324	187.20	lf (al)	228	...	...	...	δ	s^h	i	...	...	aa i	**B14**[2], 324
n344	—,**6-amino-**	$C_{11}H_9NO_2$. *See* n324	187.20	nd (w or dil al)	225	...	...	...	δ s^h	v	v	s	v	dil ac v aa δ	**B14**[2], 324
n345	—,**7-amino-**	$C_{11}H_9NO_2$. *See* n324	187.20	pa ye nd or lf (al)	240	...	...	...	...	v	s	s	δ	NH_3 v AcOEt s	**B14**[2], 324
n346	—,**8-amino-**	$C_{11}H_9NO_2$. See n324	187.20	nd (aa)	219	...	...	...	...	v	s	s	δ	AcOEt s	**B14**[2], 324
n347	**1-Naphthoic acid, 4-bromo-**	$C_{11}H_7BrO_2$. *See* n315	251.09	nd (aa, dil al or xyl)	220	...	...	...	δ^h	v	...	...	...	xyl s^h aa s^h	**E12B**, 4121
n348	—,**5-bromo-**	$C_{11}H_7BrO_2$. *See* n315	251.09	nd (aa or al)	246	...	...	...	δ^h	s	δ	...	v	...	**E12B**, 4123
n349	—,**8-bromo-**	$C_{11}H_7BrO_2$. *See* n315	251.09	pr (w or bz)	178	...	...	...	δ^h	v	δ	...	v	chl δ lig δ	**E12B**, 4130
n350	**2-Naphthoic acid, 1-bromo-**	$C_{11}H_7BrO_2$. *See* n324	251.09	nd (aa or bz)	191	...	...	...	δ	s	...	...	s^h	aa s	**E12B**, 4133
n351	—,**5-bromo-**	$C_{11}H_7BrO_2$. *See* n324	251.09	nd (al)	270 cor	...	...	...	δ^h	...	s	...	s	aa s	**E12B**, 4138
n352	**1-Naphthoic acid, 2-chloro-**	$C_{11}H_7ClO_2$. *See* n315	206.63	(w)	151–2	...	...	...	δ^h	v	v	...	...	...	**E12B**, 4120
n353	—,**4-chloro-**	$C_{11}H_7ClO_2$. *See* n315	206.63	nd (al)	210	...	...	...	δ	v	δ	...	...	chl δ aa v	**E12B**, 4121
n354	—,**5-chloro-**	$C_{11}H_7ClO_2$. *See* n315	206.63	nd	245 sub	...	...	...	δ	v	...	...	δ	aa δ	**E12B** 4122
n355	—,**8-chloro-**	$C_{11}H_7ClO_2$. *See* n315	206.63	pl (al, w or bz)	169–70	...	...	...	s^h	s^h	...	v	...	xyl s^h aa s	**E12B**, 4128
n356	**2-Naphthoic acid, 1-chloro-**	$C_{11}H_7ClO_2$. *See* n324	206.63	nd (bz)	196	...	...	...	δ	s	...	s	...	...	**E12B**, 4132
n357	—,**3-chloro-**	$C_{11}H_7ClO_2$. *See* n324	206.63	(dil MeOH)	216.5	...	...	...	...	v	...	v	v	MeOH v chl, AcOEt v	**E12B**, 4134
n358	—,**5-chloro-**	$C_{11}H_7ClO_2$. *See* n324	206.63	nd (al-aa)	270 cor	...	...	...	...	s^h	...	...	s	aa s	**E12B**, 4137

For explanations, symbols and abbreviations see beginning of table.

No.	Name	Synonyms and Formula	Mol. wt.	Crystalline form, color and specific rotation	m.p. °C	b.p. °C	Density	n_D	Solubility w	al	eth	ace	bz	other solvents	Ref.
	2-Naphthoic acid														
n359	**—,4-chloro-1-hydroxy-**	$C_{11}H_7ClO_3$. *See* n324	222.63	nd (al)	234				i^h	s	...	s	δ	...	**B10**[1], 146
n360	**—,4-chloro-3-hydroxy-**	$C_{11}H_7ClO_3$. *See* n324	222.63	ye nd	230 (233d)				...	...	...	...	...	...	**B10**, 336
n361	**1-Naphthoic acid, 1,2-dihydro-**	$C_{11}H_{10}O_2$. *See* n315	174.20	(50% al)	138				s	...	...	...	...	sulf s (red)	**E12B**, 4067
n362	**—,1,4-dihydro-**	$C_{11}H_{10}O_2$. *See* n315	174.20	nd or lf (lig)	91				δ	v	...	v	δ	AcOEt v CS_2 v	**E12B**, 4069
n363	**—,3,4-dihydro-**	$C_{11}H_{10}O_2$. *See* n315	174.20	nd (w), (peth)	125				δ	s	...	...	...	MeOH s chl, CS_2 s	**E12B**, 4070
n364	**2-Naphthoic acid, 1,2-dihydro-**	$C_{11}H_{10}O_2$. *See* n324	174.20	nd (dil al, w or peth)	101				i v^h	s	...	...	...	chl v aa s.	**E12B**, 4072
n365	**—,1,4-dihydro-**	$C_{11}H_{10}O_2$. *See* n324	174.20	lf (bz, w or dil al)	162–3				i s^h	v	v	...	δ	CS_2 i lig δ	**E12B**, 4073
n366	**—,3,4-dihydro-**	$C_{11}H_{10}O_2$. *See* n324	174.20	nd (dil aa or dil al)	119–20				δ s^h	...	...	...	s	chl v aa s	**E12B**, 4075
n367	**—,1,3-dihydroxy-,** ethyl ester	$C_{13}H_{12}O_4$. *See* n324	232.24	nd (dil al or dil aa)	83–4				i	v	v	...	...	alk s lig v^h	**B10**[2], 310
n368	**1-Naphthoic acid, 4,5-dinitro-**	$C_{11}H_6N_2O_6$. *See* n315	262.18	yesh nd or pl (al)	266.7 sub				δ^h	v^h	δ	...	δ	aa v	**E12B**, 4175
n369	**—,5-hydroxy-**	$C_{11}H_8O_3$. *See* n315	188.18	pr (w), (bz)	236				δ^h	v	s	...	...	chl s aa s	**E12B**, 4267
n370	**—,6-hydroxy-**	$C_{11}H_8O_3$. *See* n315	188.18	nd (w), pr (w)	208				s^h	v	s	...	...	chl s aa s	**E12B**, 4268
n371	**—,7-hydroxy-**	$C_{11}H_8O_3$. *See* n315	188.18	nd (w or gl aa)	255–6				s	s	δ	...	i	aa s	**E12B**, 4269
n372	**2-Naphthoic acid, 1-hydroxy-**	$C_{11}H_8O_3$. *See* n324	188.18	(dil al, w, aa), nd (al, eth, bz), pw (chl)	186–8				i	s	s	...	s	...	**E12B**, 4278
n273	—,—,phenyl ester	$C_{17}H_{12}O_3$. *See* n324	264.28		96				...	s^h	...	...	s	...	**B10**, 332
n374	**—,3-hydroxy-**	$C_{11}H_8O_3$. *See* n324	188.18	lf ye (dil al), ye lf (w), nd (dil al)	216				i	s	s	...	s	to chl s	**E12B**, 4295
n375	—,—,ethyl ester	$C_{13}H_{12}O_3$. *See* n324	216.24	nd or mlc pr (aa)	85	291			i^h	s^h	...	s	...	chl s aa s^h	**B10**[2], 213
n376	—,—,hexadecyl ester	$C_{27}H_{40}O_3$. *See* n324	412.62	grsh-wh pr	72–3				i	δ	...	...	s	lig s	...
n377	—,—,methyl ester	$C_{12}H_{10}O_3$. *See* n324	202.21	pa ye rh nd (dil MeOH)	75–6	205–7			i	s	...	...	...	os s	**B10**[2], 213
n378	**—,5-hydroxy-**	$C_{11}H_8O_3$. *See* n324	188.18	wh nd (w)	213				i s^h	s	s	s	i	chl i aa s	**E12B**, 4352
n379	**—,7-hydroxy-**	$C_{11}H_8O_3$. *See* n324	188.18	wh lf (aq al or dil al), pa ye nd (al), pl (aq al)	262				i s^h	s	s	s	i	chl i aa s	**E12B**, 4356
n380	**1-Naphthoic acid, 8-iodo-**	$C_{11}H_7IO_2$. *See* n315	298.08	pr (w)	164.5				...	v	v	...	v	aa v	**E12B**, 4131
n381	**—,3-nitro-**	$C_{11}H_7NO_4$. *See* n315	217.18	(al)	268–9				δ^h	s^h	...	...	...	...	**E12B**, 4154
n382	**—,4-nitro-**	$C_{11}H_7NO_4$. *See* n315	217.18	yesh nd (al or aa)	225–6				δ^h	v	...	...	δ	chl v lig δ	**E12B**, 156
n383	**—,5-nitro-**	$C_{11}H_7NO_4$. *See* n315	217.18	yesh nd (al)	239				i s^h	i v^h	s	...	s	chl s aa v	**E12B**, 4161
n384	**—,8-nitro-**	$C_{11}H_7NO_4$. *See* n315	217.18	nd (w), pr (al)	217				i	s	δ	...	δ	aa s	**E12B**, 4165
n385	**2-Naphthoic acid, 5-nitro-**	$C_{11}H_7NO_4$. *See* n324	217.18	yesh nd (al)	295				δ	δ^h	δ	...	δ	chl δ lig δ	**E12B**, 4167
n386	**—,8-nitro-**	$C_{11}H_7NO_4$. *See* n324	217.18	yesh nd (al)	291 sub				δ	...	...	...	...	...	**E12B**, 4170
—	**α-Naphthol**	*see* **Naphthalene, 1-hydroxy-***													
—	**β-Naphthol**	*see* **Naphthalene, 2-hydroxy-***													
—	**β-Naphthol-1-aldehyde**	*see* **1-Naphthaldehyde, 2-hydroxy-**													
—	**α-Naphtholsulfonic acid L**	*see* **1-Naphthalenesulfonic acid, 5-hydroxy-***													
—	**α-Naphtholsulfonic acid S**	*see* **1-Naphthalenesulfonic acid, 8-hydroxy-***													
—	**α-Naphtholsulfonic acid**	*see* **2-Naphthalenesulfonic acid, 1-hydroxy-***													
—	**β-Naphtholsulfonic acid**	*see* **2-Naphthalenesulfonic acid, 7-hydroxy-***													
—	**β-Naphtholsulfonic acid S**	*see* **2-Naphthalenesulfonic acid, 6-hydroxy-***													

For explanations, symbols and abbreviations see beginning of table.

No.	Name	Synonyms and Formula	Mol. wt.	Crystalline form, color and specific rotation	m.p. °C	b.p. °C	Density	n_D	Solubility w	al	eth	ace	bz	other solvents	Ref.
	1-Naphthol-8-sulfonic acid														
—	**1-Naphthol-8-sulfonic acid,** sultone	*see* **1-Naphthalenesulfonic acid, 8-hydroxy-, lactone***													
—	**Naphthonitrile**...	*see* **Naphthoic acid,** nitrile													
n387	**Naphtho-(2′,3′:5,6)quinoline**	β-Anthraquinoline. 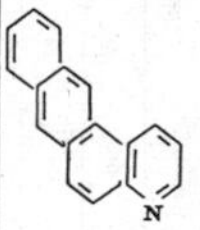	229.28	lf or ta	168.5–9.5	446	...	...	i	v^h	v	...	v	...	**B20,** 506
n388	**1,2-Naphthoquinone-***	1,2-Dihydro-1,2-diketonaphthalene*. 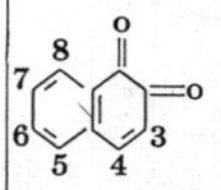	158.16	ye-red nd (eth), og lf (bz)	126	...	1.450	...	s	s	s	...	...	sulf s lig δ	**B7,** 709
n389	—,dioxime*.......	$C_{10}H_8N_2O_2$. *See* n388	188.19	nd (bz-lig or dil al)	152–4d	...	...	...	...	...	...	...	s	diox s sulf s(red) alk s (ye)	**E12B,** 2754
—	—,1-oxime*.......	*see* **Naphthalene, 2-hydroxy-1-nitroso-**													
n390	**1,4-Naphthoquinone***	1,4-Dihydro-1,2-diketonaphthalene*. 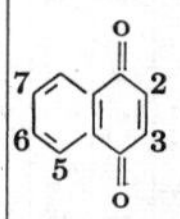	158.15	bt ye nd (al or peth), ye (sub)	128.5	sub	...	...	δ	v^h	δ	...	δ	CS_2, chl δ lig δ	**E12B,** 2781
n391	—,dioxime*.......	$C_{10}H_8N_2O_2$. *See* n390	188.19	nd (dil al)	207d	...	...	...	s	s	...	...	...	alk s	**E12B,** 2792
—	—,1-oxime*.......	*see* **Naphthalene, 1-hydroxy-4-nitroso-***													
n392	**2,6-Naphthoquinone***	2,6-Dihydro-2,6-diketonaphthalene.* 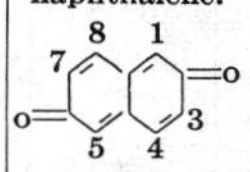	158.16	ye-red pr (bz or bz-peth)	135d	...	...	...	...	v	δ	d	δ	Py d MeOH v aa d	**E12B,** 2803
n393	**1,4-Naphthoquinone, 4-anil-, 2-anilino-**	Naphthoquinone dianilide. 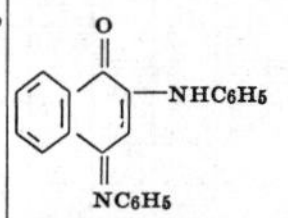	324.39	ye-red nd (bz- or al)	182.5–3	...	...	...	i	δ	...	...	...	dil ac i os s	**E12B,** 2957
n394	**1,2-Naphthoquinone, 3-bromo-***	$C_{10}H_5BrO_2$. *See* n388	237.06	red nd or pl (aa or al)	179	...	...	...	...	s^h	...	...	s^h	aa s^h	**E12B,** 2888
n395	**—,4-bromo-***.....	$C_{10}H_5BrO_2$. *See* n388	237.06	red nd (bz-lig)	154	...	...	...	...	s	...	...	v	sulf s (red) lig δ	**E12B,** 2890
n396	**—,6-bromo-***.....	$C_{10}H_5BrO_2$. *See* n388	237.06	og-red or ye pr or pl (bz or AcOEt), nd (w)	168d	...	...	...	...	s	s^h	...	...	xyl s^h aa s^h lig s^h	**E12B,** 2891
n397	**1,4-Naphthoquinone, 2-bromo-***	$C_{10}H_5BrO_2$. *See* n390	237.06	ye pl or nd (al or dil aa)	130.5	...	...	...	...	v^h	v	δ	v	chl v CS_2 δ	**E12B,** 2894
n398	**—,5-bromo-2,3-dichloro-***	$C_{10}H_3BrCl_2O_2$. *See* n390	305.95	ye pr (al)	180	...	...	...	...	s^h	...	v	v	...	**B7**[2], 655
n399	**—,2-bromo-3-hydroxy-***	$C_{10}H_5BrO_3$. *See* n390	253.06	ye mcl pr (al), nd (al, w or aa)	202	sub	...	...	i $δ^h$	s v^h	v	s	δ	...	**E12B,** 3123
n400	**—,2-bromo-3-methyl-***	$C_{11}H_7BrO_2$. *See* n390	251.09	ye-br nd (al)	151	sub 100	...	...	i	s	v	s	s	alk i aa s chl v	**E12B,** 2897
n401	**1,2-Naphthoquinone, 3-chloro-***	$C_{10}H_5ClO_2$. *See* n388	192.60	red nd (al, aa, bz or chl)	172d	...	...	...	...	s^h	...	...	...	Na_2CO_3 i aa s^h	**E12B,** 2882
n402	**1,4-Naphthoquinone, 2-chloro-***	$C_{10}H_5ClO_2$. *See* n390	192.60	ye nd (al or aa)	117–8	...	...	...	...	v	δ	...	v	Na_2CO_3 i	**E12B,** 2892

For explanations, symbols and abbreviations see beginning of table.

No.	Name	Synonyms and Formula	Mol. wt.	Crystalline form, color and specific rotation	m.p. °C	b.p. °C	Density	n_D	Solubility w	al	eth	ace	bz	other solvents	Ref.
	1 4-Naphthoquinone														
n403	**—,5-chloro-***	$C_{10}H_5ClO_2$. *See* n390	192.60	ye nd (lig)	163	sub	...	...	δ	s	...	...	...	aa s lig s	**E12B,** 2895
n404	**—,6-chloro-***	$C_{10}H_5ClO_2$. *See* n390	192.60	red-br (al or eth)	106–7	...	...	...	v	...	δ	...	v	sulf s(red) chl δ	**E12B,** 2896
n405	**—,2-chloro-3-hydroxy-***	$C_{10}H_5ClO_3$. *See* n390	208.60	ye nd (al or aa)	214–5	sub	...	...	s^h δ	s	s	...	s	...	**E12B,** 3119
n406	**1,2-Naphthoquinone, 3,4-dibromo-***	$C_{10}H_4Br_2O_2$. *See* n388	315.96	red lf or pl (aa or bz)	172–4	...	...	...	...	δ	δ	...	...	alk s	**E12B,** 2901
n407	**—,3,6-dibromo-***	$C_{10}H_4Br_2O_2$. *See* n388	315.96	red pr (AcOEt)	176	...	...	...	...	s	...	...	...	aa s^h	**E12B,** 2903
n408	**—.4,6-dibromo-***	$C_{10}H_4Br_2O_2$, *See* n388	315.96	og-red pr (bz), nd (AcOEt)	153	...	...	...	...	δ	...	...	v	aa v lig δ	**E12B,** 2904
n409	**1,4-Naphthoquinone, 2,3-dibromo-***	$C_{10}H_4Br_2O_2$. *See* n390	315.96	ye nd (aa)	218	...	...	...	...	...	δ	...	...	aa s^h lig δ	**B7,** 731
n410	**—,5,8-dibromo-***	$C_{10}H_4Br_2O_2$. *See* n390	315.96	ye nd (al)	171–3	...	...	...	...	s^h	...	...	...	...	**B7,** 732
n411	**1,2-Naphthoquinone, 3,4-dichloro-***	$C_{10}H_4Cl_2O_2$. *See* n388	227.05	red lf or pl (aa or bz), nd (bz or chl)	184	sub	...	...	...	δ	...	...	...	chl s	**E12B,** 2899
n412	**1,4-Naphthoquinone, 2,3-dichloro-***	$C_{10}H_4Cl_2O_2$. *See* n390	227.05	ye nd (al)	192.5–3	...	...	...	i	δ	δ	...	δ	...	**B7²,** 653
n413	**—,2,6-dichloro-***	$C_{10}H_4Cl_2O_2$. *See* n390	227.05	ye nd	148–9	...	...	...	i	s^h	...	...	...	...	**B7,** 730
n414	**—,5,6-dichloro-***	$C_{10}H_4Cl_2O_2$. *See* n390	227.05	ye nd	181	sub	...	...	...	δ	...	...	...	...	**B7,** 730
n415	**—,5,8-dichloro-**	$C_{10}H_4Cl_2O_2$. *See* n390	227.05	ye nd (al)	173–4	...	...	...	i	s^h	s	...	...	...	**B7,** 730
n416	**2,6-Naphthoquinone, 1,5-dichloro-**	$C_{10}H_4Cl_2O_2$. *See* n392	227.05	red ye pr (chl), gold-ye nd (al)	206d	...	...	...	i	v	δ	v	v	chl v aa v lig i	**E12B,** 2915
n417	**1,4-Naphthoquinone, 2,3-dihydroxy-**	Isonaphthazarin. $C_{10}H_6O_4$. *See* n390	190.16	red og nd (sub) br lf	282	sub	...	...	$δ^h$	δ	δ	s	δ	alk s chl δ	**E12B,** 3177
n418	**—,5,8-dihydroxy-***	Naphthazarin. $C_{10}H_6O_4$. *See* n390	190.16	dk red mcl pr (bz)	276–80	sub	...	...	$δ^h$	$δ^h$	$δ^h$	...	...	aa s^h	**E12B,** 3184
n419	**—,3,5-dihydroxy-2-methyl-***	Droserone. $C_{11}H_8O_4$. *See* n390	204.18	og-ye nd	180–1	...	...	...	...	...	...	...	...	aa s^h	**E12B,** 3194
n420	**—,5,8-dihydroxy-2-methyl-***	$C_{11}H_8O_4$. *See* n390	204.18	gr pl	173	...	...	...	i	s^h	...	...	...	...	**E12B,** 3195
n421	**—,2,3-dimethyl-***	$C_{12}H_{10}O_2$. *See* n390	186.21	ye pr (al)	127	...	...	...	i	s^h	...	...	s	aa s	**B7²,** 658
n422	**—,2,5-dimethyl-***	$C_{12}H_{10}O_2$. *See* n390	186.21	ye nd (peth or eth or MeOH)	...	...	...	...	...	...	...	...	...	diox s lig δ	**E12B,** 2848
n423	**—,2,6-dimethyl-***	$C_{12}H_{10}O_2$. *See* n390	186.21	ye pr or nd (AcOEt)	137–8	...	...	...	...	s	s	...	s	...	**E12B,** 2848
n424	**—,2,8-dimethyl-***	$C_{12}H_{10}O_2$. *See* n390	186.21	pr (peth), nd (MeOH)	135–6	...	...	...	$δ^h$	s	...	...	...	...	**E12B,** 2850
n425	**—,2-ethyl-***	$C_{12}H_{10}O_2$. *See* n390	186.21	pr (al), nd (aa, MeOH, peth)	87–8	...	...	...	...	s^h	...	...	...	MeOH s^h aa s^h	**E12B,** 2838
n426	**—,2-ethyl-3-hydroxy-***	$C_{12}H_{10}O_3$. *See* n390	202.21	ye nd	137–8	...	...	...	...	...	s	s^h	...	...	**E12B,** 3075
n427	**—,2-ethyl-3,5,6,7,8-pentahydroxy-***	Echinochrome A. $C_{12}H_{10}O_7$. *See* n350	266.21	red nd (diox-w)	220 δd	sub 120^{10}	...	...	δ	...	v	s	v	conc. sulf s	**E12B,** 3018
n428	**1,2-Naphthoquinone, 6-hydroxy-***	$C_{10}H_6O_3$. *See* n388	174.16	red lf (ace), og-ye nd	165d	...	...	...	δ	s	s	...	δ	aa s	**E12B,** 3020
n429	**—,7-hydroxy-***	$C_{10}H_6O_3$. *See* n388	174.16	br nd	194	...	...	...	...	v	i	...	i	chl i	**E12B,** 3020
n430	**1,4-Naphthoquinone, 2-hydroxy-***	Lawsone. $C_{10}H_6O_3$. *See* n390	174.16	redsh-br (aa)	190d	...	...	...	...	v	i	...	i	chl i	**B8,** 300
n431	**—,—,acetate**	$C_{12}H_8O_4$. *See* n390	216.20	ye lf (al)	131	...	...	...	δ	s	s	...	...	CS_2 s^h	**E12B,** 3054
n432	**—,5-hydroxy-***	Juglon. Nucin. $C_{10}H_6O_3$. *See* n390	174.16	redsh-ye nd or pr (chl or bz)	154	d	...	...	i	δ	δ	...	...	chl v	**B8²,** 347
n433	**—,6-hydroxy-***	$C_{10}H_6O_3$. *See* n390	174.16	gold-ye or red ye nd (w, bz-al, bz-ace)	170d	...	...	...	...	v	v	v	δ	sulf s (red) MeOH v lig δ	**E12B,** 3067
n434	**—,2-hydroxy-3-methyl-***	Phthiocol. $C_{11}H_8O_3$. *See* n390	188.16	ye pr (eth-peth)	173–4	sub	...	...	δ	...	...	...	...	os s lig i	...
n435	**—,5-hydroxy-2-methyl-***	Plumbagin. $C_{11}H_8O_3$. *See* n390	188.16	gold pr og-ye nd (dil al)	78–9	sub	...	...	δ	s v^h	v	v	v	chl, CS_2 v lig v	**E12B,** 3102
n436	**—,2-hydroxy-3-phenyl-***	$C_{16}H_{10}O_3$. *See* n390	250.26	gold-ye or og pr or nd (al, bz, MeOH)	147	...	...	...	...	v^h	v	...	v	alk s chl v lig s	**E12B,** 3091

For explanations, symbols and abbreviations see beginning of table.

No.	Name	Synonyms and Formula	Mol. wt.	Crystalline form, color and specific rotation	m.p. °C	b.p. °C	Density	n_D	Solubility w	al	eth	ace	bz	other solvents	Ref.
	1,2-Naphthoquinone														
n437	**1,2-Naphthoquinone, 3-methyl-***	$C_{11}H_8O_2$. *See* n388	172.19	red or og nd (abs al), lf (bz)	116 (122–2.5)					s				alk s sulf s (gr)	**E12B**, 2814
n438	**—,4-methyl-***	$C_{11}H_8O_2$. *See* n388	172.19	og nd (MeOH)	109d									MeOH s^h	**E12B**, 2814
n439	**1,4-Naphthoquinone, 2-methyl-***	Menadione. Vitamin K_3. $C_{11}H_8O_2$. *See* n390	172.19	ye nd (al, peth)	104–6				i	δ			s	aa δ lig δ	**B8**[1], 565
n440	**1,2-Naphthoquinone, 3-nitro-***	$C_{10}H_5NO_4$. *See* n388	203.16	crimson pl (aa)	156				s^h		δ		s	CS_2 δ lig i	**E12B**, 2932
n441	**1,4-Naphthoquinone, 2-phenyl-***	$C_{16}H_{10}O_2$. *See* n390	234.26	gold-ye nd (al)	111					v	v		v	sulf s (red) chl v lig δ	**E12B**, 2875
n442	**—,2(phenylamino)-***	Lawsone anilide. $C_{16}H_{11}NO_2$. *See* n390	249.27	red nd (dil al)	190–1	sub				s^h	s		s	sulf s (red) lig δ	**E12B**, 2955
n443	**—,5,6,7,8-tetrahydro-***	$C_{10}H_{10}O_2$. *See* n390	162.19	gold-ye nd (peth)	55–6					s	s			aa s	**E12B**, 2807
n444	**—,2,5,8-trihydroxy-***	Naphthopurpurin. $C_{10}H_6O_5$. *See* n390	206.16	red nd (bz or MeOH)	195				δ v^h	v				NH_4OH s sulf s (red)	**E12B**, 3222
—	**Naphthosultone**	*see* **1-Naphthalenesulfonic acid, 8-hydroxy-, lactone***													
—	**α-Naphthylamine**	*see* **Naphthalene, 1-amino-***													
—	**β-Naphthylamine**	*see* **Naphthalene, 2-amino-***													
—	**α-Naphthylazide**	*see* **Naphthalene, 1-triazo-***													
—	**Naphthyl red**	*see* **Azo, benzene 1-naphthalene, 4′-amino-**													
—	**Naphthylene-diamine**	*see* **Naphthalene, diamino-***													
n445	**Narceine**	Narcein. Narcéne. Pseudonarcéne. $C_{23}H_{27}NO_8 . 3H_2O$	499.52	nd (w+3)	138 (anh) 176 (+3w)				s^h δ	s^h	s^h		s^h		**B19**[2], 386
n447	—,bisulfate decahydrate	$C_{23}H_{27}NO_8 . H_2SO_4 . 10H_2O$.	723.70	nd (sulf), pw	d				s	s^h	s				**B19**[2], 386
n448	—,hydrochloride trihydrate	$C_{23}H_{27}NO_8 . HCl . 3H_2O$.	535.98	pr (HCl)	188 (anh)				$δ^h$	s^h	δ				**B19**[1], 372
n449	**α-Narcotine**(*dl*)	$C_{22}H_{23}NO_7$	413.43	nd (al, chl, MeOH)	232–3					δ			δ	chl v^h	**B22**[2], 607
n450	—(*l*)	$C_{22}H_{23}NO_7$	413.43	pr (al), $[\alpha]-198$ (chl)	176				i	s	δ	v	s		**B27**[2], 606
n451	—,hydrochloride	$C_{22}H_{23}NO_7 . HCl$	449.89	(w+3), $(\alpha)+100$	193 (anh)				v	δ					**B27**[2], 606
n452	**β-Narcotine**(*dl*)	β-Gnoscopine. $C_{22}H_{23}NO_7$.	413.43	nd, pr (MeOH, al)	229										**B27**[1], 559
n452[1]	—,hydrochloride	$C_{22}H_{23}NO_7 . HCl$	449.89	pr	224–6										**B27**[1], 559
n453	**Naringin**	Naringenin-7-rhamnoglucoside. Narinogoside. $C_{27}H_{32}O_{14} . 2H_2O$.	616.58	wh mcl hyg $[\alpha]_D^{19}-82.11$ (al)	171d				δ	δ v^h	i		i	chl i	
n454	**Neoamygdalin**	*l*-Mandelonitrile-β-gentobioside. $C_{20}H_{27}NO_{11}$	457.44	(al),$[\alpha]_D^{25}-61.4$ (w, c=8.5)	212				v	v					**B31**, 404
n455	**Neoabietic acid,** methyl ester	Methyl neoabiate. $C_{21}H_{32}O_2$.	316.49	(MeOH)	61.5–2									MeOH v	**B9**[2], 433
n456	**Neobornyl amine**	Isobornylamine.	153.27	pw, $[\alpha]_D-43.7$ (4% al)	184				i					os s	**B12**[2], 40
n457	**Neocarvomenthol**(*dl*)		156.27					1.4637^{20}							
n458	—(*l*)	$C_{10}H_{20}O$. *See* n457	156.27	$[\alpha]_D^{21}-41.7$		102^{18}	0.9012_4^{20}	1.4632^{20}							**B6**[2], 39
—	**Neocholestane**	*see* **2-Cholestane**													
n458[1]	**Neoergosterol**		380.62	pr (al), $[\alpha]_D^{17}$ −11.0 (chl)	151–2				i					os s	**E14**, 35

For explanations, symbols and abbreviations see beginning of table.

No.	Name	Synonyms and Formula	Mol. wt.	Crystalline form, color and specific rotation	m.p. °C	b.p. °C	Density	n_D	Solubility						Ref.
									w	al	eth	ace	bz	other solvents	
	Neogermitrine														
n459	**Neogermitrine**	$C_{36}H_{55}NO_{11}$	677.84	wh nd $[\alpha]_D^{25}$ −79.2 (Py)	234–4.8				i	d^h				chl s	
—	**Neohexane**	see **Butane, 2,2-dimethyl-***													
n460	**Neoisocarvomenthol(*l*)**		156.27	$[\alpha]_D^{17}$ −34.7	<−25	$87–8^{4}$	0.9102^{20}_{4}	1.4676^{20}							**B6**², 39
n461	**Neoisomenthol(*d*)**	*p*-Menthol-3.	156.27	$[\alpha]_D^{16}$ +2.2 (al)	−8	214.6	0.9131^{18}_{4}	1.4674							**B6**, 42
n462	—(*dl*)	$C_{10}H_{20}O$. *See* n461	156.27		14	214.5	0.8854^{55}_{4}	1.4676^{17}							**B6**², 51
n463	**Neomenthol(*d*)**	*d*-*β*-Pulegomenthol.	156.27	$[\alpha]_D^{16}$ +17.8	−22	212	0.897^{22}_{4}	1.4617							**B6**², 50
n464	—(*dl*)	$C_{10}H_{20}O$. *See* n463	156.27	pl or pr (peth)	51	212.3^{756}	0.903^{15}_{15}	1.4604^{20}	i						**B6**², 50
n465	—(*l*)	$C_{10}H_{20}O$. *See* n463	156.27	$[\alpha]_D^{18}$ −19.62 (al)				1.4603^{20}							**B6**², 29
	Neonicotine	see **Anabasine**													
—	**Neopentane**	see **Propane, 2,2-dimethyl-***													
—	**Neopentyl alcohol**	see **1-Propanol, 2,2-dimethyl-***													
—	**Neopentyl chloride**	see **Propane, 1-chloro-2,2-dimethyl-***													
—	**Neopine**	see ***β*-Codeine**													
—	**Neral**	see **Citral b**													
—	**Neriine**	see **Conessine**													
n466	**Nerol**	3,7-Dimethyl 2,6-octadien-1-ol*. $C_{10}H_{18}O$	154.25		<−15	$224–5^{745}$ 124^{25}	0.8813	1.475^{20}	i	s					**B1**², 510
n467	**Nerolidol(*d*)**	*α*-3,7,11-Trimethyl-1,6,10-dodecatrien-3-ol*. $C_{15}H_{26}O$.	222.37	$[\alpha]_D^{20}$ +142		276 $97^{0.2}$	1.8783^{20}_{4}	1.4898^{20}		v				os s	**B1**³, 2041
n468	—(*dl*)	$C_{15}H_{26}O$	222.37			$145–6^{12}$	0.8756^{19}_{4}	1.4801^{19}		s	s			os s aa s	**B1**³, 2042
n469	—(*l*)	$C_{15}H_{26}O$	222.37	$[\alpha]_D$ −6.5		$124–6^{3}$	0.8881^{15}_{15}	1.4799^{20}		s				os s	**B1**³, 2042
n470	**Neurine**	Trimethylethenylammonium hydroxide*. $CH_2{:}CHN^+(CH_3)_3OH^-$	103.17	syr					s	s	s				**B4**², 661
—	**Neuronal**	see **Butanoic acid, 2-bromo-2-ethyl-**, amide													
—	**Neville-winther acid**	see **1-Naphthalenesulfonic acid, 4-hydroxy-**													
—	**Niacinamide**	see **3-Pyridinecarboxylic acid**, amide													
—	**Nicotinamide**	see **3-Pyridinecarboxylic acid**, amine													
n471	**Nicotine(*d*)**	*d*-*β*-Pyridyl-*α*-*N*-methylpyrrolidine $C_{10}H_{14}N_2$	162.24	$[\alpha]_D^{20}$ +163.2		245.5–6.5^{729}	1.0094^{20}_{4}								**B23**, 117
n472	—(*dl*)	$C_{10}H_{14}N_2$	162.24				1.0084^{20}_{4}	1.5289^{20}							**B23**, 117
n473	—(*l*)	$C_{10}H_{14}N_2$	162.24	br in air, $[\alpha]_D$ −161.55	−79	246.7^{745}	1.0107^{20}	1.527^{20}	∞	v	v			chl v lig s	
n474	—,hydrochloride	$C_{10}H_{14}N_2.HCl$	198.70	dlq, $[\alpha]_D$ +14.44 (HCl)			1.0337								**B23**, 114
—	**Nicotinic acid**	see **3-Pyridinecarboxylic acid**													
—	**Nicotinonitrile**	see **3-Pyridinecarboxylic acid**, nitrile													
n475	**2,2′-Nicotyrine** (solid)	*N*-Methyl-2(2-pyridyl)-pyrrole. *α*-Nicotyrine. $C_{10}H_{10}N_2$	158.20	br in air	43.5–4.5				i	s	δ		s	dil HCl s	
n476	—(liquid)	$C_{10}H_{10}N_2$	158.20		−28	273^{764} $149–50^{22}$			δ					os s	**B23**², 191

For explanations, symbols and abbreviations see beginning of table.

No.	Name	Synonyms and Formula	Mol. wt.	Crystalline form, color and specific rotation	m.p. °C	b.p. °C	Density	n_D	Solubility: w	al	eth	ace	bz	other solvents	Ref.
	2′,3-Nicotyrine														
n477	**2′,3-Nicotyrine**	*N*-Methyl-2(3 pyridyl)-pyrrole. β-Nicotyrine. 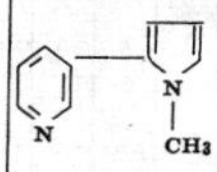	158.20	br in air	108	280–1[744] 150[15]	1.2411^{20}_{4}	1.6057^{20}	δ s^h	s				os s	**B23**[2], 192
n478	**Ninhydrin**	1,2,3-Triketohydrindene monohydrate. 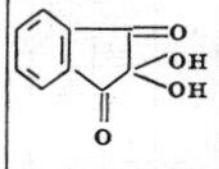	178.15	pr (w)					v^h	s	δ				**B7**[2], 831
—	**Nioxime**	*see* **1,2-Cylohexanedione, dioxime***.													
—	**Nirvanol**	*see* **Hydantoin, 5-ethyl-6-phenyl-**													
n479	**Nitranilic acid**	3,6-Dinitro-2,5-dihydroxy-*p*-benzoquinone. 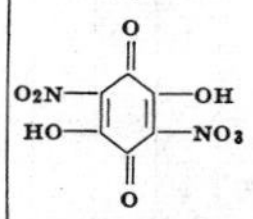	230.09	gold-ye pl (dil HNO_3+w)	86–7 (+w) (170d)				v	v	i				**B8**[3], 433
n480	**Nitric acid**, butyl ester*	*n*-Butyl nitrate. $CH_3(CH_2)_3ONO_2$	119.12			135.5[763]	1.0228^{30}	1.4013^{23}	i	s	s				**B1**[2], 397
n481	—,*sec*-butyl ester	*sec*-Butyl nitrate. $CH_3CH_2CH(CH_3)ONO_2$	119.12			124	1.0264^{20}_{4}	1.4016^{20}		∞	∞				
n482	—,decyl ester*	*n*-Decyl nitrate. $CH_3(CH_2)_9ONO_2$	203.28			127–8[11]	0.951^{0}_{4}								**B1**, 425
n483	—,ethyl ester*	Ethyl nitrate. $CH_3CH_2ONO_2$	91.07	inflam	−112	86.3–7[728]	1.105^{20}_{4}	1.3849^{22}	s	∞	∞				**B1**[2], 328
n484	—,isopropyl ester*	Isopropyl nitrate. $(CH_3)_2CHONO_2$	105.09			100–1.4	1.036^{19}_{19}	1.3912^{16}							**B1**[3], 1465
n485	—,methyl ester*	Methyl nitrate. CH_3ONO_2	77.04	explosive vapor	−108 exp	65 exp	1.2032^{25}		δ	s	s				**B1**[1], 284
n486	—,3-methylbutyl ester*	Isoamyl nitrate. $(CH_3)_2CHCH_2CH_2ONO_2$	133.15			147–8	0.9961^{22}_{4}	1.4122^{22}	δ	∞	∞				**B1**[3], 1642
n487	—,octyl ester*	*n*-Octyl nitrate. $CH_3(CH_2)_7ONO_2$	175.23			110–2[20]	0.8419^{17}_{17}								**B1**, 419
n488	—,propyl ester*	*n*-Propyl nitrate. $CH_3CH_2CH_2ONO_2$	105.09			110.5	1.0580^{20}_{4}	1.3972^{20}	δ	s	s				**B1**[2], 369
—	**Nitroform**	*see* **Methane, trinitro-***													
—	**Nitrofurazone**	*see* **Furfural, 5-nitro-**, semicarbazone													
—	**Nitrogen mustard**	*see* **Amine, triethyl, 2,2′2″-trichloro-**													
—	**Nitroglycerin**	*see* **Glycerol**, trinitrate													
—	***N*-Nitroguanidine**	*see* **Guanidine,1-nitro-***													
n489	**Nitron**	4,5-Dihydro-1,4-diphenyl-3,5-phenylimino-1,2,4-triazole. 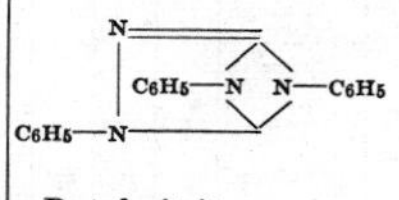	312.38	ye nd	180d				i	s	δ	s	s	chl s dil ac s	**B26**[2],76, 199
n490	**Nitrous acid,** butyl ester	*n*-Butyl nitrite. $CH_3(CH_2)_3ONO$	103.12			77.8 27[88]	0.8756^{26}_{4}	1.3768		∞	∞				**B1**[2], 397
n491	—,*sec*-butyl ester	*sec*-Butyl nitrite. $C_2H_5CH(CH_3)ONO$	103.12			68	0.8720^{20}_{4}	1.3710^{20}	δ	v	v			CCl_4 s	**B1**[2], 402
n492	—,*tert*-butyl ester	*tert*-Butyl nitrite. $(CH_3)_3CONO$	103.12	pa ye		62.8–3.2	0.8941^{0}		δ	s	s			chl, Cs_2 s	**B1**[2], 415
n493	—,decyl ester*	*n*-Decylnitrite. $CH_3(CH_2)_9ONO$	187.28	yesh		105–8[12]									**B1**, 425
n494	—,ethyl ester*	Ethyl nitrite. CH_3CH_2ONO	75.07	yesh or colorless		17	0.90^{15}_{15}		d	∞	∞				**B1**[2], 328

For explanations, symbols and abbreviations see beginning of table.

No.	Name	Synonyms and Formula	Mol. wt.	Crystalline form, color and specific rotation	m.p. °C	b.p. °C	Density	n_D	Solubility w	al	eth	ace	bz	other solvents	Ref.
	Nitrous acid														
n495	—,heptyl ester*	*n*-Heptyl nitrite. $CH_3(CH_2)_6ONO$	145.20			155–8	0.8939_4^0		i		s				B1^2, 443
n496	—,hexyl ester*	*n*-Hexyl nitrite. $CH_3(CH_2)_5ONO$	131.18	ye		$129–30^{774}$ cor	0.8778	1.3987	i	s	s				B1^1, 20
n497	—,isopropyl ester*	Isopropyl nitrite. $(CH_3)_2CHONO$	89.09			45^{762}	0.844_{25}^{25}								B1^3, 1464
n498	—,methyl ester*	Methyl nitrite. CH_3ONO	61.04	gas	−17	−12	0.991^{15} (liq)			s	s				B1, 284
n499	—,3-methyl butyl ester*	Isoamyl nitrite. $(CH_3)_2CHCH_2CH_2ONO$	117.15	yesh, inflam			0.8717_4^{20}	1.3871^{20}	δ	∞	∞				B1^3, 1641
n500	—,octyl ester	*n*-Octyl nitrite. $CH_3(CH_2)_7ONO$	159.22	ye		174–5	0.862^{17}								
n501	—,pentyl ester*	*n*-Amyl nitrite. $CH_3(CH_2)_4ONO$	117.15	ye		104.5^{763} cor	0.8817_4^{20}	1.3893^{20}	δ	∞	∞				B1^2, 1604
n502	—,propyl ester*	*n*-Propyl nitrite. $CH_3CH_2CH_2ONO$	89.09			57	0.935^{21}	1.3613^{20}		s	s				B1^2, 369
n502[1]	**Nonaconsane***	$CH_3(CH_2)_{27}CH_3$	408.80	rh	64–5	295^{15}	0.7630_4^{100}		i	v^h	v	v	s	chl δ	B1^2, 144
n503	**—,2-methyl-***	$(CH_3)_2CH(CH_2)_{26}CH_3$	422.83	pl (lig)	73–4	$222^{0.3}$									B1^1, 72
n504	**1-Nonacosanol***	$CH_3(CH_2)_{27}CH_2OH$	424.80	(al)	82.5										B1^2, 471
n505	**Nonadecane***	$CH_3(CH_2)_{17}CH_3$	268.53	wax	32	330 193^{15}	0.7774_4^{32}	1.4409^{20}	i	δ	s				B1^3, 568
n506	**1,2,3-Nonadecanetricarboxylic acid, 2-hydroxy-***	Agaric acid. Agaricin. $CH_3(CH_2)_{15}CH(CO_2H)C(OH)(CO_2H)CH_2(CO_2H)$	416.56	lf $[\alpha]_D^{19}$ −9.8	141–2				s^h	δ			i	chl i	B3^2, 372
n507	—,triethyl ester	$C_{25}H_{52}O_7$ *See* n506	500.72	nd	36–7								s		B3^2, 373
n508	—,trimethyl ester	$C_{25}H_{46}O_7$ *See* n506	458.64	nd	63–4					s^h			s		B3^2, 373
n509	**Nonadecanoic acid***	*n*-Nonadecylic acid. $CH_3(CH_2)_{17}CO_2H$	298.51	lf (al)	68.7	$297–8^{100}$			i	v^h	v		v	chl v lig v	B2^2, 368
n510	**1-Nonadecanol***	*n*-Nonadecyl alcohol. $CH_3(CH_2)_{17}CH_2OH$	284.53	(ace)	62–3	$166–7^{0.32}$									B1^3, 1841
n511	**2-Nonadecanone***	Methyl *n*-heptadecyl ketone. $CH_3(CH_2)_{16}COCH_3$	282.51	pl (al)	55–6	266.5^{110} 165^3	0.8108^{36}		i	δ	v	v		chl v	B1^2, 773
n512	—,oxime*	$CH_3(CH_2)_{16}C(:NOH)CH_3$	297.53	(al)	76.5–7.5					s^h					B1, 718
n513	**4-Nonadecanone***	*n*-Propyl *n*-pentadecyl ketone. $CH_3(CH_2)_{14}CO(CH_2)_2CH_3$	282.51		50.5	201^{11}			i	δ	v	v			B1^2, 773
n514	**10-Nonadecanone***	Caprinone. Dinonyl ketone. $[CH_3(CH_2)_7CH_2]_2CO$	282.51	lf (al)	58	>350			i	δ	s	s	v	chl s lig s	B1, 718
n515	**1-Nonadecyne***	$CH_3(CH_2)_{16}C\vdots CH$	264.50		37–8	$144.0^{1.5}$									
n516	**1,8-Nonadiyne***	$HC\vdots C(CH_2)_5C\vdots CH$	120.20		−21.0	55^{13}	0.8159_4^{21}	1.4521^{21}							
n517	**Nonanal***	*n*-Nonylaldehyde. Pelargonaldehyde. $CH_3(CH_2)_7CHO$	142.24			185 $80–2^{13}$	0.8268_{19}^{19}	1.4273^{20}							B1^2, 761
n518	—,oxime*	$CH_3(CH_2)_7CH{\cdot}NOH$	157.26	lf (dil al)	64				i	s	s				B1^2, 761
n519	**Nonane***	$CH_3(CH_2)_7CH_3$	128.26		−51	150.798	0.7176_4^{20}	1.4054^{220}	i	v	v				B1^3, 502
n520	**—,1-amino-***	*n*-Nonylamine $CH_3(CH_2)_7CH_2NH_2$	143.27			202.2			δ	s	s				
n521	**—,1-chloro-***	*n*-Nonyl chloride. $CH_3(CH_2)_7CH_2Cl$	162.73			202 $76–9^4$	0.8704_4^{20}	1.4369^{20}	i		s			chl s	B1^2, 128
n522	**—,2-chloro-***	2-Nonyl chloride. $CH_3(CH_2)_6CHClCH_3$	162.70			190^{764}	0.8563^{20}		i					chl s	B1, 166
n523	**—,5-chloro-***	5-Nonyl chloride. $(CH_3CH_2CH_2CH_2)_2CHCl$	162.70			$85–7^{14}$	0.8639_4^{15}	1.4314^{15}	i		v				B1^2, 128
n524	**—,1-chloro-9-fluoro-***	$FCH_2(CH_2)_7CH_2Cl$	180.70			$102–2.5^{11}$	0.966_4^{20}	1.4301^{25}	i	v	v				
n525	**Nonanedioic acid***	Azelaic acid. 1,7-Heptanedicarboxylic acid*. $HO_2C(CH_2)_7CO_2H$	188.23	lf or nd	106.5	>360d 225.5^{10}	1.225_4^{25}	1.4303^{111}	s^h δ	s	δ		v^h		B2^2, 602
n526	—,dichloride	Azelayl dichloride. $ClCO(CH_2)_7COCl$	225.12			$180–3^{35}$			d	d^h	s		v^h		B2, 709
n527	—,diethyl ester*	Ethyl azelate. $C_2H_5O_2C(CH_2)_7CO_2C_2H_5$	244.33		−18.5	291–2 $130–2^{15}$	0.9729_4^{20}	1.4351^{20}	i	s	s				B2^2, 603

For explanations, symbols and abbreviations see beginning of table.

No.	Name	Synonyms and Formula	Mol. wt.	Crystalline form, color and specific rotation	m.p. °C	b.p. °C	Density	n_D	Solubility w	al	eth	ace	bz	other solvents	Ref.
	Nonanedioic acid														
n528	—,di(2-ethylbutyl) ester*	Di-2-ethylbutyl azelate. $CH_2[(CH_2)_3CO_2CH_2CH(C_2H_5)_2]_2$	356.55		−45	230^{5}	0.928^{25}_{4}	1.443	i	s	...	s	s		
n529	—,di(2-ethylhexyl) ester*	Di-2-ethylhexyl azelate. $CH_2[(CH_2)_3CO_2CH_2CH(C_2H_5)C_4H^n_9]_2$	412.66		−78	237^{5}	0.915^{25}_{4}	1.446^{25}	i	s	...	s	s		
n530	—,dimethyl ester*	Dimethyl azelate. $CH_3O_2C(CH_2)_7CO_2CH_3$	216.28			156^{20}	1.0069^{20}_{4}	1.4360^{20}	i	s	...	...	...	os s	**B2**^{1}, 290
n531	—,dinitrile	Azelanitrile. $NC(CH_2)_7CN$	150.23			$195–6^{20}$	0.9405^{0}		i	v	v	...	v		**B2**^{1}, 290
n532	—,diphenyl ester*	Diphenyl azelate. $C_6H_5O_2C(CH_2)_7CO_2C_6H_5$	340.40	nd (al)	48–9				i	δ s^h	v	...	v		**B6**, 156
n533	**1,9 Nonanediol***	Enneamethylene glycol. Nonamethylene glycol. $HOCH_2(CH_2)_7CH_2OH$	160.26	(bz)	45.8	169^{14}			δ	v	v	...	v^h	lig i	**B1**^{3}, 2226
n534	**Nonanoic acid***	*n*-Nonylic acid. Pelargonic acid. $CH_3(CH_2)_7CO_2H$	158.24		15	253–5 150^{20}	0.9057^{20}_{4}	1.4343^{19}	i	s	s	...	...	chl s	**B2**^{2}, 360
n535	—,amide	Nonanamide*. Pelargonamide. $CH_3(CH_2)_7CONH_2$	157.26		98.9				i	δ	δ	...	...		
n536	—,chloride	Nonanyol chloride*. Pelargonyl chloride. $CH_3(CH_2)_7COCl$	176.69		−60.5	215.35	0.9463^{15}_{4}		d	d	s	...	...		**B2**^{2}, 308
n537	—,ethyl ester*	$CH_3(CH_2)_7CO_2C_2H_5$	186.30		−36.7	222 $96–8^{10}$	0.8657^{20}_{4}	1.4220^{20}	i	s	s	...	...		**B2**^{2}, 307
n538	—,methyl ester	$CH_3(CH_2)_7CO_2CH_3$	172.27			$213–4^{757}$ 80^{5}	0.8799^{15}	1.4214^{20}	i	s	s	...	...		**B2**^{2}, 307
n539	—,nitrile	Nonanenitrile*. Octyl cyanide. Pelargonitrile. $CH_3(CH_2)_7CN$	139.24		−34.2	224.0 55^{1}	0.786^{16}	1.4235^{25}	i	δ	s	...	...		
n540	—,piperazinium salt	$2[CH_3(CH_2)_7CO_2H].C_4H_{10}N_2$	402.62	wh	95.1–6.2				s^h	s	i	...	...		
n541	**—,9-amino-***	$H_2NCH_2(CH_2)_7CO_2H$	173.26		185.6–6.6				s	s	...	...	...		**B4**, 463
n542	**—,9-fluoro-***	ω-Fluoropelargonic acid. $FCH_2(CH_2)_7CO_2H$	176.23		ca. 18	$89–90^{0.2}$		1.4289^{25}	δ	v	v	...	...		
n543	**1-Nonanol***	Nonyl alcohol. $CH_3(CH_2)_7CH_2OH$	144.26			212 118^{15}	0.8273^{20}_{4}	1.4333	i	s	s	...	...		
n549	**5-Nonanol, 5-butyl-***	Tri-*n*-butyl carbinol. $[CH_3(CH_2)_3]_3COH$	200.37		20	230–5d 100^{4}	0.8408^{20}_{4}	1.4445^{20}	...	s	...	...	...		**B1**^{3}, 1802
n550	**1-Nonanol, 9-chloro-***	Nonamethylene. Chlorohydrin. $ClCH_2(CH_2)_7CH_2OH$	178.70		28	$146–8^{14}$		1.4575	i	v	v	...	...		
n551	**—,9-fluoro-***	Nonamethylene fluorohydrin. $FCH_2(CH_2)_7CH_2OH$	162.25			$125–6^{15}$	0.928^{20}_{4}	1.4279^{25}	i	v	v	...	...		
n552	**2-Nonanone***	Heptyl methyl ketone. $CH_3(CH_2)_6COCH_3$	142.24		−7.46	193^{749}	0.8208^{20}_{4}	1.42096^{20}	i	s	s	s	s		
n553	**4-Nonanone-***	*n*-Propyl *n*-amyl ketone. $CH_3(CH_2)_4CO(CH_2)_2CH_3$	142.24			187–8 $76–7^{11}$	0.837^{20}_{4}		...	...	...	...	...		**B1**^{1}, 365
n554	**5-Nonanone***	Dibutyl ketone. $CH_3(CH_2)_3CO(CH_2)_3CH_3$	142.24		−4.8	188.4	0.8270^{13}_{4}	1.421^{15}	i	s	v	...	...	chl v	**B1**^{2}, 762
n555	**Nonasiloxane, eicosamethyl-**	$CH_3[Si(CH_3)_2O-]_8Si(CH_3)_3$	681.47			173^{5}	0.918	1.3980^{20}	i	δ	...	...	s		
n556	**1,3,6,8-Nonatetraen-5-one, 1,9-diphenyl-***	Dicinnamylidene acetone. $(C_6H_5CH{:}CHCH{:}CH)_2CO$	286.38	ye nd (abs al)	144				i	δ s^h	δ	...	...	con sulf s	**B7**, 524
n557	**1-Nonen-3-yne***	$CH_3(CH_2)_4C{\vdots}CCH{:}CH_2$	122.21			$27.2–8.2^{4}$	0.7602^{25}_{4}	1.4487^{25}	...	...	...	...	...		
n558	**1-Nonen-4-yne***	$CH_3(CH_2)_3C{\vdots}CCH_2CH{:}CH_2$	122.21			58.0^{22}	0.777^{25}_{4}	1.4413^{25}	...	...	...	...	...		
n559	**2-Nonen-4-yne***	$CH_3(CH_2)_3C{\vdots}CCH{:}CHCH_3$	122.21			$70–0.5^{29}$	0.7832^{25}_{4}	1.4590^{25}	...	...	...	...	...		
n560	**1-Nonyne***	$CH_3(CH_2)_6C{\vdots}CH$	124.23		−65.0	160.67	0.760^{20}_{4}	1.123^{20}	...	...	...	...	...		
n561	**2-Nonyne***	$CH_3(CH_2)_5C{\vdots}CCH_3$	124.23			$155–6^{747}$	0.769^{20}_{4}	1.4331^{20}	...	...	s	...	...	lig s	**B1**^{3}, 1013
n562	**3-Nonyne***	$CH_3(CH_2)_4C{\vdots}CCH_2CH_3$	124.23			$153–5^{745}$ 92^{97}	0.7616^{20}_{4}	1.4299^{20}	i	...	s	...	...	lig s	**B1**^{3}, 1013

For explanations, symbols and abbreviations see beginning of table.

No.	Name	Synonyms and Formula	Mol. wt.	Crystalline form, color and specific rotation	m.p. °C	b.p. °C	Density	n_D	Solubility w	al	eth	ace	bz	other solvents	Ref.
	4-Nonyne														
n563	**4-Nonyne***	$CH_3(CH_2)_3C\vdots CCH_2CH_2CH_3$	124.22			150–4[752]	0.757_4^{25}	1.4296^{25}							
n564	**1-Nonyne, 1-chloro-***	$CH_3(CH_2)_6C\vdots CCl$	158.07			75–7[15]	0.906^{20}	1.450^{20}			v				**B1[3]**, 1012
n565	**4-Nonyne,3,3-dimethyl-***	$CH_3(CH_2)_3C\vdots CC(CH_3)_2CH_2CH_3$	152.28			82[40]	0.7658_4^{20}	1.4313^{20}							
n566	**—,8-methyl-***	$(CH_3)_2CHCH_2CH_2C\vdots CCH_2CH_2CH_3$	138.25			104.5	0.7681_4^{20}	1.4311^{20}							
—	**Nopinene**	*see* **β-Pinene**													
n567	**Nopinone**	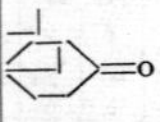	138.21	$[\alpha]_D^{24}+16.64$	−1 to 0	211.8 77–8.5[8]	0.977^{20}	1.4769^{25}	s	s	s				**E12A,** 510
n568	**Noradrenaline**(*l*)	*l*-Arterenol. *l*-Norepinephrine. 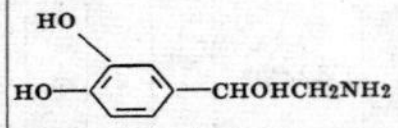	169.18	$[\alpha]_D^{25}-37.3$ (dil HCl)	216.5–8d				δ	δ	δ			alk v dil HCl v	**B13[2]**, 522
—	**Norbornylane**	*see* **Bicyclo(2,2,1)heptane**													
—	**Norcamphane**	*see* **Bicyclo(2,2,1)heptane**													
—	**Norcamphane dicarboxylic acid**	*see* **Bicyclo(2,2,1)heptane-2,3-dicarboxylic acid**													
—	**Norcamphoric acid**	*see* **1,3-Cyclopentane-dicarboxylic acid***													
n569	**Nordihydro-guaiaritic acid**	NDGA. 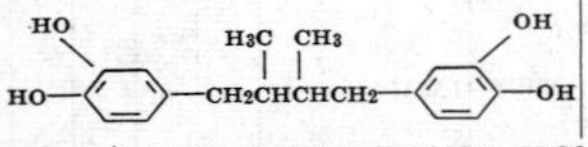	302.37	(dil aa or al)	184–5				δ^h	s	s		i	to i alk s chl δ	**B6[1]**, 577
n570	**Norephedrine,** hydrochloride	$C_6H_5CHOHCH(CH_3)NH_2.HCl$	187.67	pl (abs al), (dil HCl, al or AcOEt)	194				v	s	i		i	chl i	**B13[2]**, 371
n571	**—,*N,N*-diethyl-(*dl*)**	$C_6H_5CHOHCH(CH_3)N(C_2H_5)_2.HCl$	243.78	pr (lig)	63–4					s					**B13[2]**, 378
n572	**—,*N*-ethyl-**	$C_6H_5CHOHCH(CH_3)NCH_2H_5.HCl$	217.73	(lig)	47–8	138[18]							s		
n573	**Normorphine**	Desmethylmorphine. 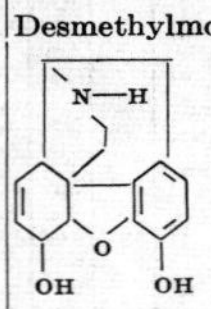	271.32	(w+1½), (MeOH+1)	273 (w+1½) 277 (anh)				i^h	i^h			i	chl i	**B27[2]**, 117
n574	**Nornicotine**	*l*-3(2-Pyrrolidyl)pyridine. $C_9H_{12}N_2$	148.21	hyg, $[\alpha]_D^{22}-88.8$		270 120[1]	1.0737^{19}	1.5378^{19}	s	v	v				**B23[2]**, 107
—	**Norvaline**	*see* **Pentanoic acid, 2-amino-***													
—	**Novocain**	*see* **Benzoic acid, 4-amino-2(diethylamino)-, ethyl ester hydrochloride**													
—	**Novonal**	*see* **4-Pentenoic acid, 2,3-diethyl, amide**													

For explanations, symbols and abbreviations see beginning of table.

No.	Name	Synonyms and Formula	Mol. wt.	Crystalline form, color and specific rotation	m.p. °C	b.p. °C	Density	n_D	Solubility w	al	eth	ace	bz	other solvents	Ref.
	Octacosane														
o1	**Octacosane**	$CH_3(CH_2)_{26}CH_3$	394.77	mcl or rh	61.5	242^3	0.7750^{70}	1.4330^{70}	i				s	chl s	**B1**^{3}, 582
o2	**Octacosanoic acid***	$CH_3(CH_2)_{26}CO_2H$	424.76		90.4			1.4313^{100}							
o3	**1-Octacosanol***	$CH_3(CH_2)_{27}OH$	410.77		83.3										
o4	**9,12-Octadecadienoic acid***	Linoleic acid. $CH_3(CH_2)_4CH{:}CHCH_2CH{:}CH(CH_2)_7CO_2H$	280.45	pa ye or colorless	−5	$229–30^{16}$	0.9025_4^{20}	1.4699^{20}	i	∞	∞	∞	∞	chl ∞	**B2**^{2}, 459
o5	—,ethyl ester*	Ethyl linoleate. $CH_3(CH_2)_4CH{:}CHCH_2CH{:}CH(CH_2)_7CO_2C_2H_5$	308.51	ye or colorless		$270–5^{180}$	0.8865_4^{20}		i	s	s				**B2**^{2}, 461
o6	—,methyl ester	Methyl linoleate. $CH_3(CH_2)_4CH{:}CHCH_2CH{:}CH(CH_2)_7CO_2CH_3$	294.48	ye or colorless		$211–2^{16}$	0.8886_4^{18}		i	s	s				**B2**^{2}, 461
o7	**10,12-Octadecadienoic acid***	10,12-Linoleic acid. $CH_3(CH_2)_4CH{:}CHCH{:}CH(CH_2)_8CO_2H$	280.45		56–7		0.8686_4^{70}	1.4689^{60}							**B2**^{1}, 212
o8	**7,11-Octadecadiyne***	$CH_3(CH_2)_5C{\vdots}CCH_2CH_2C{\vdots}C(CH_2)_5CH_3$	246.44			$167–8^7$	0.841_4^{19}	1.4698^{19}							
o9	**Octadecanal***	Stearaldehyde. $CH_3(CH_2)_{16}CHO$	268.49		38	261			i						**B2**, 772
o10	—,dimethyl acetal	1,1-Dimethoxyoctadecane*. $CH_3(CH_2)_{16}CH(OCH_3)_2$	314.56			$168–70^3$		1.4410^{25}		v	v			MeOH v	
o11	**Octadecane***	$CH_3(CH_2)_{16}CH_3$	254.50	nd (al or eth-MeOH)	28	305–7 (317)	0.7768_4^{28}	1.4349^{35}	i	δ	s	s		lig s	**B1**^{3}, 565
o12	—,1-amino-*	$CH_3(CH_2)_{17}NH_2$	269.52		fr 53.1	232^{32}			i	s	s	δ	s	chl v	
o13	—,—,acetate	$CH_3(CH_2)_{17}NH_2.CH_3CO_2H$	329.57		fr 84.5				i s^h	v			s		
o14	—,—,hydrochloride	$CH_3(CH_2)_{17}NH_2.HCl$	305.98	orh	>180d				i v^h	δ v^h	i		i		**C26**, 5469^4
o15	—,1-bromo-*	$CH_3(CH_2)_{17}Br$	333.40		30				i	s	s				**Am 73**, 4886
o16	—,1-chloro-*	$CH_3(CH_2)_{17}Cl$	288.95		21–3	$157–8^{1.5}$		1.4525^{20}	i						**B1**^{3}, 567
o17	—,1,18-dibromo-*	$Br(CH_2)_{18}Br$	412.31	nd or lf (al)	64	$205–7^{1.5}$			i	δ v^h				chl v	**B1**^{2}, 139
o18	—,1-iodo-*	$CH_3(CH_2)_{17}I$	380.40	lf (lig), nd(ace or al-ace)	34	$169^{0.5}$			i	δ	δ				**B1**^{3}, 567
o19	**Octadecanedioic acid; diethyl ester***	$C_2H_5O_2C(CH_2)_{16}CO_2C_2H_5$	370.58		48.2	240^{12}			i	s	s				**B2**^{2}, 626
o20	**1,18-Octadecanediol***	$HO(CH_2)_{18}OH$	286.50	lf (al, AcOEt or bz), nd (bz)	97–9	$210–1^2$			i						**B1**^{3}, 2247
o21	**1-Octadecanethiol***	*n*-Octadecyl mercaptan. $CH_3(CH_2)_{17}SH$	286.57		20–30	$169–71^1$		1.4648	i	δ	s				
o22	**Octadecanoic acid***	Stearic acid. $CH_3(CH_2)_{16}CO_2H$	284.49	mcl lf	70.1	358–83 183.5^1	0.9408_4^{20}	1.4299^{80}	i	δ	v	δ	δ	chl, CCl_4 s CS_2 s	**B2**^{2}, 346
o23	—,amide	Stearamide. $CH_3(CH_2)_{16}CONH_2$	283.50	lf	109	$250–1^{12}$			i	δ	δ	δ	δ	chl δ	**B2**, 384
o24	—,—,*N*-phenyl-	Stearanilide. $CH_3(CH_2)_{16}CONHC_6H_5$	359.60	nd (al)	94	153.5^{10}			i	s	v		v		**B12**^{2}, 148
o24[1]	—,anhydride*	Stearic anhydride. $[CH_3(CH_2)_{16}CO]_2O$	550.96		71.5		0.8368_4^{82}	1.4362^{80}	i	i	δ		δ		**B2**^{2}, 360
o25	—,benzyl ester	Benzyl stearate. $CH_3(CH_2)_{16}CO_2CH_2C_6H_5$	374.61	pa ye	45–6		0.9075_{25}^{50}	1.4663^{50}	i	δ	δ			chl δ	**B6**^{2}, 418
o26	—,butyl ester*	*n*-Butyl stearate. $CH_3(CH_2)_{16}CO_2(CH_2)_3CH_3$	340.59		27.5	223	0.855_4^{20}	1.4328^{50}	i	s		v			**B2**^{2}, 352
o27	—,chloride	Stearyl chloride. $CH_3(CH_2)_{16}COCl$	302.93		23	215^{15}				s^h					**B2**, 384
o28	—,cyclohexyl ester-*	$CH_3(CH_2)_{16}CO_2$—(cyclohexyl ring)	366.63		44 (28–9)		0.890_{25}^{25}		i	i	s				**B6**^{2}, 11
o29	—,ethyl ester*	Ethyl stearate. $CH_3(CH_2)_{16}CO_2C_2H_5$	312.54		33.9	$213–5^{15}$	1.057_4^{20}	1.4292^{40}	i	s	s	v			**B2**^{2}, 379
o30	—,hexadecyl ester*	Cetyl stearate. $CH_6(CH_2)_{16}CO_2(CH_2)_{15}CH_3$	508.92	la lf or pl (eth, aa)	56.6			1.4410^{70}	i	i	δ s^h	s	i s^h	chl s CS_2 s	**B2**^{2}, 353
o30[1]	—,2-hydroxyethyl ester*	Glycol monostearate. $CH_3(CH_2)_{16}CO_2CH_2CH_2OH$	328.54		58.5		0.8780_4^{60}			s	s^h				
o31	—,isobutyl ester	Isobutyl stearate. $CH_3(CH_2)_{16}CO_2CH_2CH(CH_3)_2$	340.59	wax	25	223^{15}			i		s				**B2**^{2}, 353

For explanations, symbols and abbreviations see beginning of table.

No.	Name	Synonyms and Formula	Mol. wt.	Crystalline form, color and specific rotation	m.p. °C	b.p. °C	Density	n_D	Solubility						Ref.
									w	al	eth	ace	bz	other solvents	
	Octadecanoic acid														
o32	—,isopropyl ester*	Isopropyl stearate. $CH_3(CH_2)_{16}CO_2CH(CH_3)_2$	326.57		28										B2^2, 352
o33	—,methyl ester*	Methyl stearate. $CH_3(CH_2)_{16}CO_2CH_3$	298.51		39.1	215^{15}		1.4290^{60}_{2}	i	s	s	v		chl v	
o34	—,3-methylbutyl ester*	Isomyl stearate. $CH_3(CH_2)_{16}CO_2CH_2CH_2CH(CH_3)_2$	354.62		23	192^2	0.855^{20}_4	1.433^{50}	i	δ	s				B2^2, 352
o35	—,nitrile	Stearonitrile. $CH_3(CH_2)_{16}CN$	265.49		fr 40.8	214^{13}	0.817^{20}_4	1.4389^{45}	i	s	v	v		chl v	B2, 384
o36	—,pentyl ester*	*n*-Amyl stearate. $CH_3(CH_2)_{16}CO_2(CH_2)_4CH_3$	354.62	pl	30			1.4342^{50}	i	s	v				B2^2, 353
o37	—,phenyl ester*	Phenyl stearate. $CH_3(CH_2)_{16}CO_2C_6H_5$	360.59		52	267^{15}			i	s	s				B6^2, 155
o38	—,*p*-phenylphenacyl ester	$CH_3(CH_2)_{16}CO_2CH_2CO$—$C_6H_4$—$C_6H_5$	478.72		97(91)										C32, 4944
o39	—,propyl ester*	Propyl stearate. $CH_3(CH_2)_{16}CO_2CH_2CH_2CH_3$	326.57	(eth or PrOH)	fr 28.9	186.8^2	0.8452^{38}	1.4400^{30}	i	s	v	v		chl, CCl_4 v	B2^2, 352
o40	—,tetrahydrofurfuryl ester	$CH_3(CH_2)_{16}CO_2CH_2$— (tetrahydrofuryl ring, O)	368.61		22		0.917^{25}_{25}		i	s	s				
o41	—,9,10-dibromo-* (*trans, dl*)	Elaidic acid dibromide. $CH_3(CH_2)_7CHBrCHBr(CH_2)_7CO_2H$	442.29	colorless or ye nd (eth)	29–30		1.2458^{30}_4	1.4893^{42}	i	i	s				B2^2, 362
o42	—,2,3-dihydroxy-*	$CH_3(CH_2)_{14}CHOHCHOHCO_2H$	316.49	nd (aa)	α:107 β:126				$δ^h$	δ	δ	δ		chl $δ^h$ peth δ	B3, 406
o43	—,9,10-dihydroxy-*	$CH_3(CH_2)_7CHOHCHOH(CH_2)_7CO_2H$	316.49	lf	93–4				i	δ	δ				Am 73, 2730
o44	—,9,10-dioxo-*	Stearoxylic acid. $CH_3(CH_2)_7COCO(CH_2)_7CO_2H$	312.45	lf	85				i	v^h	v^h				
o45	—,9,10,12,13,15,16-hexabromo-*	α-Linolenic acid hexabromide. $CH_3[CH_2CHBrCHBr]_3(CH_2)_7CO_2H$	757.89		181				i		i			chl i xyl s^h	B2^2, 364
o46	—,2-hydroxy-*	$CH_3(CH_2)_{15}CHOHCO_2H$	300.49	flat nd	91–2					s		s		chl, AcOEt s	B3, 364
o47	—,3-hydroxy-*	$CH_3(CH_2)_{14}CHOHCH_2CO_2H$	300.49	lf (chl)	89–90					s^h	s			chl s^h	B3^2, 249
o48	—,10-hydroxy-*	$CH_3(CH_2)_7CHOH(CH_2)_8CO_2H$	300.49	pl	83–4				i	s	δ				B3^2, 249
o49	—,11-hydroxy-*	$CH_3(CH_2)_6CHOH(CH_2)_9CO_2H$	300.49	ta (al)	84–5				i	δ	δ			lig δ	
o50	—,12-hydroxy-*	$CH_3(CH_2)_5CHOH(CH_2)_{10}CO_2H$	300.49	(al)	83				i	s	s			chl s	B3^2, 250
o51	—,10-methyl-* (*D*⁻)	Tuberculostearic acid. $CH_3(CH_2)_7CH(CH_3)(CH_2)_8CO_2H$	298.51	$[\alpha]^{22}_D -0.11$	12–3	$175–8^{0.7}$	0.8771^{24}_4	1.4512^{25}		s^h	s^h	s^h			B2^2, 369
o52	—,3-oxo-*	$CH_3(CH_2)_{14}COCO_2CO_2H$	298.47		64–5					s^h					B3^2, 457
o53	—,—,ethyl ester*	Ethyl palmitoylacetate. $CH_3(CH_2)_{14}COCH_2CO_2C_2H_5$	356.52	(al or peth)	37–8				i	s^h				peth s^h	B3^2, 457
o54	—,6-oxo-*	Lactaric acid. $CH_3(CH_2)_{11}CO(CH_2)_4CO_2H$	298.47	lf (al or peth-al)	87				i		δ		δ	chl s	B3^2, 457
o55	—,—,ethyl ester*	$CH_3(CH_2)_{11}CO(CH_2)_4CO_2C_2H_5$	326.52		41				i	v	v				B3^1, 253
o56	—,10-oxo-*	$CH_3(CH_2)_7CO(CH_2)_8CO_2H$	298.47	lf (al)	82					δ					B3^2, 457
o57	—,—,ethyl ester*	$CH_3(CH_2)_7CO(CH_2)_8CO_2C_2H_5$	326.52	lf (al)	41				i	s^h					B3, 725
o58	—,12-oxo-*	$CH_3(CH_2)_5CO(CH_2)_{10}CO_2H$	298.47	lf (aa)	82					s^h				aa s^h	B3^2, 457
o59	—,—,ethyl ester*	$CH_3(CH_2)_5CO(CH_2)_{10}CO_2C_2H_5$	326.52	lf (al)	38				i	s^h					B3^2, 458
o60	—,9,10,12,13-tetrabromo-*	Linoleic acid tetrabromide. $CH_3(CH_2)_4[CHBrCHBrCH_2]_2(CH_2)_6CO_2H$	600.12	pl or lf (aa)	114–5				i	v	v		v	chl v aa v	B2^2, 362
o61	—,—,ethyl ester*	$CH_3(CH_2)_4[CHBrCHBrCH_2]_2(CH_2)_6CO_2C_2H_5$	628.15	nd	58–9				i	δ				peth δ	B2^2, 363
o62	—,—,methyl ester*	$CH_3(CH_2)_4[CHBrCHBrCH_2]_2(CH_2)_6CO_2CH_3$	614.13	lf	39.1	215^{15}		1.4346^{45}	i	s	s			chl s	B2^2, 351
o63	**1-Octadecanol***	Stearyl alcohol. $CH_3(CH_2)_{17}OH$	270.48	lf (al)	58.5	210.5^{15}	0.8124^{59}_4		i	s	s	δ	δ	chl s	B1^3, 1833
o64	**9,11,13-Octadecatrienoic acid** (*cis*)*	α-Eleostearic acid. $CH_3(CH_2)_3[CH{:}CH]_3(CH_2)_7CO_2H$	278.44	nd (al)	49	235^{12}δd	0.8980^{56}_4	1.5112^{50}	i	s	s				B2^2, 465
o65	—(*trans*)*	β-Eleostearic acid. $CH_3(CH_2)_3[CH{:}CH]_3(CH_2)_7CO_2H$	278.44	lf (al, MeOH or CCl_4)	71–2		0.8839^{80}_4	1.5000^{74}	i	s^h	i			MeOH s	B2^2, 467
o66	**9,12,15-Octadecatrienoic acid** (*cis, cis, cis*)*	α-Linolenic acid. $CH_3[CH_2CH{:}CH]_3(CH_2)_7CO_2H$	278.44		−11	$230–2^{17}$	0.9157^{20}_4	1.480^{20}	i	s	s				B2^2, 463

For explanations, symbols and abbreviations see beginning of table.

No.	Name	Synonyms and Formula	Mol. wt.	Crystalline form, color and specific rotation	m.p. °C	b.p. °C	Density	n_D	Solubility: w	al	eth	ace	bz	other solvents	Ref.
	9,12,15-Octadecatrienoic acid														
o67	—,ethyl ester (*cis, cis, cis*)*	Ethyl linolenate. $CH_3[CH_2CH{:}CH]_3(CH_2)_7CO_2C_2H_5$	306.49	...	...	$132–3^{0.01}$	0.8919^{20}_{4}	1.4675^{20}	i	s	s	...	...	...	**B2**^{2}, 465
o68	**9-Octadecenol***	Olealdehyde. $CH_3(CH_2)_7CH{:}CH(CH_2)_7CHO$	266.47	ye nd	...	$168–9^{3-4}$	0.8513^{15}	1.4557^{20}	...	...	...	...	...	...	**B1**, 749
o69	**1-Octadecene***	$CH_3(CH_2)_{15}CH{:}CH_2$	252.49	...	17.5	179^{15}	0.7891^{20}_{4}	1.4448^{20}	i	...	...	s^h	...	...	**B1**^{3}, 878
o70	**9-Octadecene***	$CH_3(CH_2)_7CH{:}CH(CH_2)_7CH_3$	252.49	...	−30.5	162^{9}	0.7916^{20}_{4}	1.4470^{20}	i	...	...	...	...	...	**B1**^{3}, 879
o71	**9-Octadecenoic acid***(*cis*)	Oleic acid. $CH_3(CH_2)_7CH{:}CH(CH_2)_7CO_2H$	282.47	nd	16.3 (13.4)	286^{100}	0.895^{18}_{4}	1.4582^{20}	i	∞	∞	s	∞	chl ∞	...
o72	—(*trans*)	Elaidic acid. $CH_3(CH_2)_7CH.CH(CH_2)_7CO_2H$	282.47	pl (al)	45	288^{100}	0.851^{79}_{4}	1.4499^{45}	i	δ	s	...	s	chl s	**B2**^{2}, 441
o73	—,amide(*cis*)	Oleamide. $CH_3(CH_2)_7CH{:}CH(CH_2)_7CONH_2$	281.49	...	76	...	...	...	i	s^h	s	...	...	...	**Am 71,** 2215
o74	—,—,(*trans*)	Elaidamide. $CH_3(CH_2)_7CH{:}CH(CH_2)_7CONH_2$	281.49	...	90	...	...	...	i	s^h	...	...	...	...	**Am 71,** 3017
o75	—,—,*N*-phenyl-(*cis*)	Oleanilide. $CH_3(CH_2)_7CH{:}CH(CH_2)_7CONHC_6H_5$	357.58	nd	41	143.5^{10}	...	...	i	s^h	v	s^h	v^h	MeOH s aa s	**B12**^{1}, 198
o76	—,benzyl ester(*cis*)	$CH_3(CH_2)_7CH{:}CH(CH_2)_7CO_2CH_2C_6H_5$	372.60	...	...	237^{7}	0.9330^{25}_{25}	1.4875^{25}	i	s	v	..	...	...	**B6**^{2}, 418
o77	—,butyl ester(*cis*)*	Butyl oleate. $CH_3(CH_2)_7CH{:}CH(CH_2)_7CO_2(CH_2)_3CH_3$	338.58	ye	<−10	$227–8^{10}$	0.8657^{25}_{25}	1.4480^{25}	i	s	...	...	...	...	**B2**^{2}, 439
o78	—,ethyl ester(*cis*)*	Ethyl oleate. $CH_3(CH_2)_7CH{:}CH(CH_2)_7CO_2C_2H_5$	310.52	...	...	$216–7^{15}$	0.8671^{25}	1.4449^{25}	i	∞	∞	...	...	...	...
o79	—,—(*trans*)*	Ethyl elaidate. $CH_3(CH_2)_7CH{:}CH(CH_2)_7CO_2C_2H_5$	310.52	...	...	$217–9^{15}$	0.868^{18}	1.445^{25}	i	s	s	...	...	...	**B2**^{2}, 443
o80	—,methyl ester (*cis*)*	Methyl oleate. $CH_3(CH_2)_7CH{:}CH(CH_2)_7CO_2CH_3$	296.50	...	19.9	$216–7^{20}$	0.879^{18}	1.4466^{35}	i	∞	∞	...	...	...	...
o81	—,—(*trans*)*	Methyl elaidate. $CH_3(CH_2)_7CH{:}CH(CH_2)_7CO_2CH_3$	296.50	...	...	$213–5^{15}$	0.872^{18}	1.446^{25}	i	s	s	...	...	...	**B2**^{2}, 443
o82	—,3-methylbutyl ester(*cis*)*	Isoamyl oleate. $CH_3(CH_2)_7CH{:}CH(CH_2)_7CO_2CH_2CH_2CH(CH_3)_2$	352.61	...	...	$223–4^{10}$	0.897^{15}	...	i	s	v	...	...	...	**B2**^{2}, 439
o83	—,nitrile(*cis*)	Oleonitrile. $CH_3(CH_2)_7CH{:}CH(CH_2)_7CN$	263.47	...	−1	330–40d $213–4^{16}$	...	...	...	...	...	...	...	...	**B2**^{2}, 443
o84	**11-Octadecenoic acid**(*trans*)*	Vaccenic acid. $CH_3(CH_2)_5CH{:}CH(CH_2)_9CO_2H$	282.47	pl (ace)	43–4	...	...	1.4439^{60}	...	...	...	s^h	...	...	...
o85	**6-Octadecenoic acid, 6,7-dioido-***	6,7-Diiodotariric acid. Iodostarin. $CH_3(CH_2)_{10}CI{:}CI(CH_2)_4CO_2H$	534.26	nd (al)	48.5	d	...	...	i	δ v^h	v	...	v	chl, CS_2 v alk δ	**B2**^{2}, 428
o86	**9-Octadecenoic acid, 12-hydroxy-**(*cis*)*	Ricinoleic acid. $CH_3(CH_2)_5CHOHCH_2CH{:}CH(CH_2)_7CO_2H$	298.47	$[\alpha]^{22}_{D}+5.05$	α7.7 β16 γ5.5	$226–8^{10}$	0.9496^{15}	1.4716^{20}	i	s	s	...	...	...	**B3**, 385
o87	—,—,butyl ester (*cis*)*	Butyl ricinoleate. $CH_3(CH_2)_5CHOHCH_2CH{:}CH(CH_2)_7CO_2(CH_2)_3CH_3$	354.58	$[\alpha]+3.73$	...	275^{13}	0.9058^{22}	1.4566^{22}	i	...	s	...	...	...	**B3**, 388
o88	—,—,ethyl ester (*cis*)*	Ethyl ricinoleate. $CH_3(CH_2)_5CHOHCH_2CH{:}CH(CH_2)_7CO_2C_2H_5$	326.52	$[\alpha]^{22}_{D}+5.28$	...	258^{13}	0.9145^{22}	1.4618^{22}	i	s^h	...	...	...	...	**B3**^{2}, 259
o89	—,—(*trans*)	Ethyl ricinelaidate. $CH_3(CH_2)_5CHOHCH_2CH{:}CH(CH_2)_7CO_2C_2H_5$	326.52	...	16	...	...	...	i	s	...	...	...	...	**B3**, 389
o90	—,—,isobutyl ester(*cis*)*	Isobutyl ricinoleate. $CH_3(CH_2)_5CHOHCH_2CH{:}CH(CH_2)_7CO_2C_4H_9^i$	354.58	$[\alpha]+4.01$	...	262^{9}	0.9078^{22}	1.4538^{22}	i	...	s	...	...	...	**B3**, 388
o91	**9-Octadecen-1-ol**(*cis*)*	Olelyl alcohol. $CH_3(CH_2)_7CH{:}CH(CH_2)_8OH$	268.49	...	6–7	$205–10^{15}$	0.8489^{20}	1.4607^{20}	i	s	s	...	...	...	**B1**^{3}, 1962
o92	—(*trans*)*	Elaidyl alcohol. $CH_3(CH_2)_7CH{:}CH(CH_2)_8OH$	268.49	(al or ace)	36–7	ca. 333 198^{10}	0.8338^{40}_{4}	1.4522^{40}	i	s^h	...	s^h	...	dil NaOH s	**B1**^{3}, 1965
o93	**1-Octadecyne***	$HC{:}C(CH_2)_{15}CH_3$	250.47	(al)	22.5 fr 26	313^{760} 180^{15}	0.7955^{30}	...	...	...	...	...	...	...	**B1**^{3}, 1029
o94	**2-Octadecyne***	$CH_3(CH_2)_{14}C{\vdots}CCH_3$	250.47	...	30	184^{15}	0.8016^{30}_{4}	...	...	...	...	...	...	...	...
o95	**9-Octadecyne***	$CH_3(CH_2)_7C{\vdots}(CH_2)_7CH_3$	250.47	...	3	$163–4^{7}$	0.8012^{20}_{4}	1.4488^{25}	...	...	...	...	...	...	**Am 61,** 630
o96	**9-Octadecynoic acid***	Stearolic acid. $CH_3(CH_2)_7C{\vdots}C(CH_2)_7CO_2H$	280.45	pr (al)	48	260	...	1.4510^{54}	i	δ	v	...	...	...	...
o97	—,ethyl ester*	Ethyl stearolate. $CH_3(CH_2)_7C{\vdots}C(CH_2)_7CO_2C_2H_5$	308.51	...	...	$180^{2.5}$	...	1.4555^{20}	...	...	...	...	...	...	...

For explanations, symbols and abbreviations see beginning of table.

No.	Name	Synonyms and Formula	Mol. wt.	Crystalline form, color and specific rotation	m.p. °C	b.p. °C	Density	n_D	Solubility						Ref.
									w	al	eth	ace	bz	other solvents	
	9-Octadecynoic acid														
o98	—,methyl ester*	Methyl stearolate. $CH_3(CH_2)_7C\vdots C(CH_2)_7CO_2CH_3$	294.48	...	...	175^{3}	...	1.4562^{20}	...	...	...	...	...	...	...
o99	**1,3-Octadiene, 3-chloro-***	$CH_3(CH_2)_3CH{:}CClCH{:}CH_2$	144.65	...	...	$64–5^{18}$	0.9366_4^{20}	1.4794^{20}	i	...	s	...	...	...	**B1**[3], 1007
o100	**2,6-Octadiene, 2,6-dimethyl-***	Dihydroocimene. $CH_3CH{:}C(CH_3)CH_2CH_2CH{:}C(CH_3)_2$	138.25	...	...	172–3	0.775_4^{21}	1.4497^{19}	...	...	...	...	...	...	**B1**[3], 1018
o101	**2,4-Octadiene, 7-methyl-***	$(CH_3)_2CHCH_2CH{:}CHCH{:}CHCH_3$	124.23	...	...	149	0.7521_4^{18}	1.4543^{18}	i	...	...	...	...	...	**B1**[2], 239
o102	**1,7-Octadien-3,5-diyne, 1,8-dimethoxy-***	$CH_3OCH{:}CHC\vdots CC\vdots CCH{:}CHOCH_3$	162.19	...	33–4	exp	...	...	...	s	...	...	...	...	...
—	**1,6-Octadien-2-ol, 3,7-dimethyl-**	*see* **Linalool**													
—	**2,6-Octadien-1-ol, 3,7-dimethyl-**	*see* **Geraniol, Nerol**													
o103	**1,5-Octadien-3-yne, 5-propyl-***	$CH_3CH_2CH{:}C(C_3H_7^n)C\vdots CCH{:}CH_2$	148.25	...	...	$57–8^{6}$	0.8047_4^{20}	1.4949^{20}	...	...	...	...	...	...	**Am 55,** 1097
o104	**2,6-Octadien-4-yne, 3,6-dimethyl-***	$CH_3CH{:}C(C_2H_5)C\vdots C(C_2H_5)C{:}CHCH_3$	162.28	...	...	$169–71^{760}$	0.8157_4^{20}	1.4971^{20}	...	...	...	...	...	...	**C31,** 4268
o105	—,3,6-dimethyl-*	$CH_3CH{:}C(CH_3)C\vdots C(CH_3)C{:}CHCH_3$	134.22	...	−45	170^{760}	0.8071_4^{22}	1.4977^{20}	...	...	...	...	...	...	...
o106	**1,7-Octadiyne***	$HC\vdots C(CH_2)_4C\vdots CH$	106.17	...	...	$135–6^{760}$	0.8169_4^{21}	1.453^{21}	...	...	...	...	...	...	...
o107	**2,6-Octadiyne***	$CH_3C\vdots CCH_2CH_2C\vdots CCH_3$	106.17	...	27	62^{19}	0.828_4^{30}	1.4658^{30}	...	...	...	...	...	...	...
o108	**3,5-Octadiyne***	$CH_3CH_2C\vdots CC\vdots CCH_2CH_3$	106.17	...	...	$163–4^{760}$	0.826_4^{0}	1.4968^{0}	...	...	...	...	...	...	...
o109	—,2,7-dimethyl-*	$(CH_3)_2CHC\vdots CC\vdots CCH(CH_3)_2$	134.22	...	...	74^{12}	0.8090_4^{20}	...	...	...	...	...	...	...	...
—	**Octalene**	*see* **Aldrin**													
o110	**Octanal***	Capryl aldehyde. $CH_3(CH_2)_6CHO$	128.22	...	...	$167–70^{760}$	0.8211_4^{20}	1.4217^{20}	...	s	∞	...	...	oos s	**B1**[2], 757
o111	—,oxime	Caprylaldoxime. $CH_3(CH_2)_6CH{:}NOH$	143.23	nd (peth or dil al)	60	$120–5^{10}$	...	...	...	...	...	...	...	MeOH s	**B1**[2], 758
o112	**Octane***	$CH_3(CH_2)_6CH_3$	114.23	...	−56.5	125–6	0.7025_4^{20}	1.3975	i	δ	s	...	...	...	**B1**[3], 457
o113	—,1-amino-*	Octylamine, $CH_3(CH_2)_7NH_2$	129.25	...	...	179.6^{760}	0.7819^{20}	1.4292^{20}	δ	v	v	...	...	...	**J1948,** 1825
o114	—,2-amino-*(*d*)	$CH_3(CH_2)_5CHNH_2CH_3$	129.25	...	...	70^{25}	...	...	...	...	...	...	...	...	...
o115	—,—(*dl*)	$CH_3(CH_2)_5CHNH_2CH_3$	129.25	...	...	$163–4^{760}$	0.7745_0^{20}	1.4232^{25}	i	v	v	...	...	...	**B4,** 196
—	—,1(amino-phenyl)-	*see* **Benzene, amino(octyl)-**													
o116	—,1-bromo-*	$CH_3(CH_2)_7Br$	193.13	...	−55	202.3	1.118_4^{20}	1.4527^{20}	i	∞	∞	...	...	...	**B1**[2], 466
o117	—,2-bromo-*(*d*)	$CH_3(CH_2)_5CHBrCH_3$	193.13	$[\alpha]_D^{25}+34.2$	...	$76–7^{18}$	1.0982_4^{25}	1.4475^{25}	i	∞	∞	...	...	...	**B1**[2], 125
o118	—,—(*dl*)	$CH_3(CH_2)_5CHBrCH_3$	193.13	...	...	$188–9^{760}$	1.0878_4^{25}	1.4442^{25}	i	∞	∞	...	...	...	**B1**[2], 125
o119	—,—(*l*)	$CH_3(CH_2)_5CHBrCH_3$	193.13	$[\alpha]_D^{25}-27.47$	...	$72–3^{18}$	1.0914_4^{17}	1.4475^{25}	i	∞	∞	...	...	...	**B1**[2], 125
o120	—,1-bromo-8-fluoro*	$F(CH_2)_8Br$	211.12	...	...	$118–20^{22.5}$	...	1.4484^{25}	i	v	v	...	...	...	...
o121	—,1-chloro-*	$CH_3(CH_2)_7Cl$	148.68	...	...	181.5^{765}	0.8748_4^{20}	1.4306^{20}	i	v	v	...	...	...	**B1,** 159
o122	—,2-chloro-*(*d*)	$CH_3(CH_2)_5CHClCH_3$	148.68	$[\alpha]_D^{20}+33.7$	...	171–3	0.871_4^{15}	1.4273^{21}	i	v	s	...	...	...	**B1,** 160
o123	—,4-chloro-*(*d*)	$CH_3(CH_2)_3CHClCH_2CH_2CH_3$	148.68	$[\alpha]_D^{25}+0.28$ (undil. 1=10 cm)	...	92^{50}	...	...	i	...	s	...	...	chl s	**B1**[3], 466
o124	—,1-chloro-8-fluoro-*	$F(CH_2)_8Cl$	166.67	...	...	87^{10}	0.978_4^{20}	1.4266^{25}	i	v	v	...	...	...	...
o125	—,3-chloro-3-methyl-*	$CH_3(CH_2)_4CCl(CH_3)CH_2CH_3$	162.70	...	...	$73–4^{15}$	0.8680_4^{25}	1.4351^{20}	...	...	...	...	...	chl s	**B1**[3], 508

For explanations, symbols and abbreviations see beginning of table.

No.	Name	Synonyms and Formula	Mol. wt.	Crystalline form, color and specific rotation	m.p. °C	b.p. °C	Density	n_D	Solubility w	al	eth	ace	bz	other solvents	Ref.
	Octane														
o126	**—,4-chloro-4-methyl-***	$CH_3(CH_2)_3CCl(CH_3)CH_2CH_2CH_3$	162.70			$71^{14.5}$	0.8723^{20}_{4}	1.4360^{20}			s			chl s	B1³, 509
o127	**—,1,2-dibromo-***	$CH_3(CH_2)_5CHBrCH_2Br$	272.03			118.5^{15}	1.4580^{20}_{4}	1.4942^{20}							B1², 125
o128	**—,2,7-dimethyl-***	$(CH_3)_2CH(CH_2)_4CH(CH_3)_2$	142.29		−49.2	159.6^{760}	0.7278^{18}_{4}	1.4105^{15}							B1², 131
o129	**—,1-fluoro-***	$CH_3(CH_2)_7F$	132.22			$142–3^{760}$	0.8103^{20}_{4}	1.3935^{20}							B1², 124
o130	**—,1-iodo-***	$CH_3(CH_2)_7I$	240.13		−45.7	255.5^{760}	1.3297^{20}_{4}	1.4890^{20}		s	s				B1², 185
o131	**—,2-iodo-***(*D*,−)	$CH_3(CH_2)_5CHICH_3$	240.13	$[\alpha]^{20}_{D}-47.87$		$98–100^{18}$	1.3219^{20}_{4}	1.4863^{25}	i					lig s	B1³, 470
o132	—,—(*DL*)	$CH_3(CH_2)_5CHICH_3$	240.13			210	1.3251^{20}_{4}	1.4896^{20}	i					lig s	B1³, 470
o133	—,—(*L*,+)	$CH_3(CH_2)_5CHICH_3$	240.13	$[\alpha]^{24}_{D}+62.6$		101^{22}	1.3314^{17}_{4}		i					lig s	B1³, 470
o134	**—,2-methyl-***	$CH_3(CH_2)_5CH(CH_3)_2$	128.26		−80.5	142.8^{760}	0.7134^{20}_{4}	1.4032^{20}	i	s	s			peth v lig v	B1³, 507
o135	**—,3-methyl-***(*d*)	$CH_3(CH_2)_4CH(CH_3)CH_2CH_3$	128.26	$[\alpha]^{17}_{D}+9.38$	−107.6	$144–5^{756}$	0.7206^{17}	1.4068^{20}							B1, 166
o136	—,—(*l*)	$CH_3(CH_2)_4CH(CH_3)CH_2CH_3$	128.26	$[\alpha]^{27}_{D}-8.5$		143^{760}	0.714^{27}_{4}	1.4052^{25}							
o137	**—,4-methyl-***(*dl*)	$CH_3(CH_2)_3CH(CH_3)CH_2CH_2CH_3$	128.26		−119.1	142.4^{760}	0.7199^{20}_{4}	1.4061^{20}						os v	B1³, 509
o138	—,—(*l*)	$CH_3(CH_2)_3CH(CH_3)CH_2CH_2CH_3$	128.26	$[\alpha]^{19}_{D}-1.06$		141^{760}	0.714^{22}_{4}		i					os v	B1³, 510
o139	**—,1(4-nitrophenyl)-***	1-Nitro-4-octylbenzene*. NO_2—C_6H_4—$(CH_2)_7CH_3$	235.33		−5	200^{10}			i		∞				B5, 454
o140	**—,1-phenyl-***	Octylbenzene*. $C_6H_5(CH_2)_7CH_3$	190.33		−37	$264–5^{760}$	0.8572^{20}_{4}	1.4849^{20}	i		∞				B5², 343
o141	**Octanedial***	Suberaldehyde. $OCH(CH_2)_6CHO$	142.20			230–40d $140–5^{30}$			v	v					B1², 845
o142	—,dioxime	Suberaldoxime. $HON{:}CH(CH_2)_6CH{:}NOH$	172.23	pr (al)	150–5				δ^{h}	s^{h}					B1², 485
o143	**Octanedioic acid***	Suberic acid. $HO_2C(CH_2)_6CO_2H$	174.20	lo nd or pl	140	300 sub 219.5^{10}			δ s^{h}		δ				B2², 595
o144	—,diethyl ester	Diethyl suberate. $C_2H_5O_2C(CH_2)_6CO_2C_2H_5$	230.31		5.9	$282–6^{760}$	0.9822^{20}_{4}	1.4328^{20}	i	s	s				B2², 596
o145	—,dimethyl ester	Dimethyl suberate. $CH_3O_2C(CH_2)_6CO_2CH_3$	202.25		−4.8	268 124.3^{6}	1.0210^{20}_{4}	1.4337^{20}	i	s				os s	B2, 693
o146	**1,8-Octanediol***	$HO(CH_2)_8OH$	146.23	nd (bz-lig), pr	63	172^{20}			δ	v	δ		s	lig δ	B1², 556
o147	**4,5-Octanediol** (*dl*)*	1,2-Dipropylethylene glycol. $CH_3CH_2CH_2CHOHCHOHCH_2CH_2CH_3$	146.23		28	$117–8^{12}$		1.4419^{25}							B1³, 2219
o148	—(*meso*)*	$CH_3CH_2CH_2CHOHCHOHCH_2CH_2CH_3$	146.23	lf (bz or al)	123–4					δ	δ				B1³, 2219
o149	**2,3-Octanedione***	Acetylcaproyl. $CH_3COCO(CH_2)_4CH_3$	142.20			$172–3^{733}$				s					B1, 795
o150	—,dioxime*	Methylpentylglyoxime. $CH_3C({:}NOH)C({:}NOH)(CH_2)_4CH_3$	172.23	nd (al)	173					s					B1 795
o151	—,3-oxime*	$CH_3COC({:}NOH)(CH_2)_4CH_3$	157.21	(lig)	59	133^{11}					v			lig s^{h}	B1, 795
o152	**2,7-Octanedione***	1,4-Diacetylbutane. $CH_3CO(CH_2)_4COCH_3$	142.20	pl	43–4	114^{10}			δ	s					B1¹, 408
o153	—,dioxime*	$CH_3C({:}NOH)(CH_2)_4C({:}NOH)CH_3$	172.23	(al)	158				i	s^{h}					B1, 795
o154	**3,6-Octanedione***	$CH_3CH_2COCH_2CH_2COCH_2CH_3$	142.20	pl (al)	34–5	98^{14}			δ^{h}	s					B1², 845
o155	**4,5-Octanedione***	Dibutyryl. Dipropylglyoxal. $CH_3CH_2CH_2COCOCH_2CH_2CH_3$	142.20	ye		168^{160}	0.934^{0}_{4}			v	v	v			B1², 845
o156	—,dioxime*	Dipropylglyoxime. $CH_3CH_2CH_2C({:}NOH)C({:}NOH)CH_2CH_2CH_3$	172.23		186–7	sub			i	s	s			lig i	B1², 846
o157	**3,6-Octanedione, 2,2,7,7-tetramethyl-**	Dipivalylethane. $(CH_3)_3CCOCH_2CH_2COC(CH_3)_3$	198.31		22–5	$115–7^{17}$	0.900^{27}_{4}	1.4353^{25}	i	s	s				
o158	**1-Octanethiol***	Octyl mercaptan. $CH_3(CH_2)_7SH$	146.30		−49.2	199.1^{760} 86^{15}	0.8428^{20}_{4}	1.4540^{20}		s					B1³, 1710

For explanations, symbols and abbreviations see beginning of table.

No.	Name	Synonyms and Formula	Mol. wt.	Crystalline form, color and specific rotation	m.p. °C	b.p. °C	Density	n_D	Solubility w	al	eth	ace	bz	other solvents	Ref.
	2-Octanethiol														
o159	**2-Octanethiol** (*dl*)*	$CH_3(CH_2)_5CH(SH)CH_3$	146.30		−79	186.4^{760} 88.9^{30}	0.8366^{20}_{4}	1.4504^{20}		s	s		s		**B1**³, 1717
o160	—(*l*)*	$CH_3(CH_2)_5CH(SH)CH_3$	146.30	$[\alpha]^{25}_{546}$ −36.4		$78–80^{22}$	0.830^{25}_{4}			s	s		s		**B1**³, 1722
o161	**Octanoic acid***	Caprylic acid*. $CH_3(CH_2)_6CO_2H$	144.22		16.3	$110–1^{4}$	0.8615^{80}	1.4278^{20}	δ^h	∞				chl ∞ CH_3CN ∞	**B2**², 301
o162	—,amide	Caprylamide. $CH_3(CH_2)_6CONH_2$	143.23	lf	fr 105.9	239^{760}			δ^h	v		s	δ	chl δ	**B2**², 303
o163	—,anhydride*	Caprylic anhydride. $[CH_3(CH_2)_6CO]_2O$	270.42	lf, pl	0.9	280–5	0.9065^{18}_{4}	1.4358^{18}	d	s	∞			oos s	**B2**², 303
o164	—,butyl ester*	Butyl caprylate. $CH_3(CH_2)_6CO_2(CH_2)_3CH_3$	200.32		−42.9	245	0.8628^{20}		i	s				os s	**B2**, 348
o165	—,chloride	Caprylyl chloride. $CH_3(CH_2)_6COCl$	162.66		−63	195.6^{760}	0.9535^{15}_{4}		d	d	s				**B2**², 303
o166	—,ethyl ester*	Ethyl caprylate. $CH_3(CH_2)_6CO_2C_2H_5$	172.27		−43.1	208.5^{760}	0.8693^{20}_{4}	1.4180^{20}	i	v	v				**B2**², 302
o167	—,heptyl ester*	$CH_3(CH_2)_6CO_2(CH_2)_6CH_3$	242.41		−10.6	290.6	0.8596^{20}		i	s				os s	**B2**, 348
o168	—,hexyl ester*	$CH_3(CH_2)_6CO_2(CH_2)_5CH_3$	228.38		−30.6	277.4	0.8603^{20}		i	s				os s	
o169	—,isopropyl ester*	Isopropyl caprylate. $CH_3(CH_2)_6CO_2CH(CH_3)_2$	186.30			93.8^{10}	0.8555^{20}	1.4147^{25}							
o170	—,methyl ester*	Methyl caprylate. $CH_3(CH_2)_6CO_2CH_3$	158.24		−40, −41	192.9^{760}	0.8775^{20}_{4}	1.4160^{25}	i	v	v				**B2**², 302
o171	—,3-methylbutyl ester*	Isoamyl caprylate. $CH_3(CH_2)_6CO_2CH_2CH_2CH(CH_3)_2$	214.35			136^{10}			δ	s					
o172	—,nitrile	Caprylonitrile. $CH_3(CH_2)_6CN$	125.22		−45.6	205.2^{760}	0.8141^{20}_{4}	1.4204^{20}	i	δ	s				**B2**², 303
o173	—,octyl ester*	$CH_3(CH_2)_6CO_2(CH_2)_7CH_3$	256.43		−18.1	306.8^{760} 192.5^{30}	0.8745^{0}_{4}	1.4352^{20}	i	s				os s	**B2**, 348
o174	—,pentyl ester*	Amyl caprylate. $CH_3(CH_2)_6CO_2(CH_2)_4CH_3$	214.35		−34.8	260.2	1.8613^{20}		i	s				os s	
o175	—,piperazinium salt	$C_4H_{10}N_2.2[CH_3(CH_2)_6CO_2H]$	374.57		98				s	s	i				**Am 70,** 2758
o176	—,propyl ester*	Propyl caprylate. $CH_3(CH_2)_6CO_2CH_2CH_2CH_3$	186.30		−46.2	226.4 100^{10}	0.8638^{20}	1.4191^{25}	i	s				os s	**B2**, 348
o177	—,**2-amino-**(*d*)*	$CH_3(CH_2)_5CHNH_2CO_2H$	159.23	$[\alpha]^{26}_{D}$+23.5 (6N HCl c=1)						δ					**B4**², 886
o178	—,—,(*dl*)	$CH_3(CH_2)_5CHNH_2CO_2H$	159.23	lf (w)	230	sub d			δ	δ	δ		δ	aa s	**B4**², 886
o179	—,—(*l*)	$CH_3(CH_2)_5CHNH_2CO_2H$	159.23	$[\alpha]_D$ −13 (1N NaOH, c=2)	276										**B4**², 886
o180	—,**8-amino-***	$H_2N(CH_2)_7CO_2H$	159.23		172				δ v^h	v					**B4**¹, 527
o181	—,**2-bromo-***	$CH_3(CH_2)_5CHBrCO_2H$	223.12			$128–9^{2}$	1.2785^{24}	1.4613^{24}							**B2**², 303
o182	—,**8-fluoro-***	$F(CH_2)_7CO_2H$	162.21		35	$132–3^{4}$			δ	v	v			lig s^h	
o183	—,**2-hydroxy-***	$CH_3(CH_2)_5CHOHCO_2H$	160.21	pl	69.5				δ	v	v				**B3**², 237
o184	—,**4-hydroxy-,** lactone*	γ-Caprylolactone. $CH_3(CH_2)_3$— O =O	142.20			$132–3^{20}$				s					**B17**¹, 133
o185	—,**2-methyl-3-oxo-,** ethyl ester*	$CH_3(CH_2)_4COCH(CH_3)CO_2C_2H_5$	200.28			$128–9^{12}$	0.963^{0}_{4}			s					**B3**, 713
o186	**1-Octanol***	Octyl alcohol. $CH_3(CH_2)_7OH$	130.23		16.7	194–5	0.8270^{20}_{4}	1.4293^{20}	i	∞	∞				**J1952,** 514
o187	**2-Octanol***(*d*)	$CH_3(CH_2)_5CHOHCH_3$	130.23	$[\alpha]^{20}_{D}$+9.79		86^{20}	0.8216^{20}_{4}	1.4256^{20}							**B1**¹, 208
o188	—(*dl*)	$CH_3(CH_2)_5CHOHCH_3$	130.23		−38.6	180^{760}	0.8193^{20}_{4}	1.4203^{20}	δ						**B1**², 449
o189	—(*l*)	$CH_3(CH_2)_5CHOHCH_3$	130.23	$[\alpha]^{20}_{D}$ −9.84		80^{11}	0.8201^{20}_{4}	1.4256^{20}							**B1**², 451
o190	**1-Octanol, 8-chloro-***	Octamethylene chlorohydrin. $Cl(CH_2)_8OH$	164.68			$129–30^{11}$		1.4563^{25}	δ	v	v				

For explanations, symbols and abbreviations see beginning of table.

No.	Name	Synonyms and Formula	Mol. wt.	Crystalline form, color and specific rotation	m.p. °C	b.p. °C	Density	n_D	Solubility w	al	eth	ace	bz	other solvents	Ref.
	1-Octanol														
o191	—,**3,7-dimethyl-** (*d*)*	$(CH_3)_2CH(CH_2)_3CH(CH_3)CH_2CH_2OH$	158.29	$[\alpha]^{20}_{D}+4.09$		$105–6^{10}$	0.8288^{20}_{4}	1.4342							**B1**^{2}, 460
o192	—,—(*l*)	$(CH_3)_2CH(CH_2)_3CH(CH_3)CH_2CH_2OH$	158.29	$[\alpha]^{11}_{546}-3.67$		109^{15}	0.830^{18}	1.4370^{15}							**B1**^{2}, 461
o193	**3-Octanol, 3,7-dimethyl-***(*dl*)	Tetrahydrolinalool. $(CH_3)_2CH(CH_2)_3COH(CH_3)CH_2CH_3$	158.29		31–2	127–8	0.832^{25}_{25}		i	s					
o194	—,**3-ethyl-***	$CH_3(CH_2)_4COH(C_2H_5)_2$	158.29			199	0.8361^{25}_{4}	1.4390^{20}							**B1**, 426
o195	**1-Octanol, 8-fluoro-***	Octamethylene fluorohydrin. $F(CH_2)_8OH$	148.22			$106–7^{10}$	0.945^{20}_{4}	1.4248^{25}	δ	v	v				
o196	**2-Octanol, 2-methyl-***	$CH_3(CH_2)_5COH(CH_3)_2$	144.26			178	0.823^{19}_{19}	1.4299^{19}	i	s	s				**B1**^{2}, 457
o197	**2-Octanone***	Hexyl methyl ketone. $CH_3(CH_2)_5COCH_3$	128.22		−20.9	173^{760}	0.8185^{20}_{4}	1.4161^{20}	δ	∞	∞				**B1**^{2}, 758
o198	**3-Octanone***	Amyl ethyl ketone. $CH_3(CH_2)_4COC_2H_5$	128.22			$169–70^{738}$	0.8220^{20}_{4}	1.4156^{20}	i	∞	∞				**B1**^{1}, 362
o199	**4-Octanone***	Butyl propyl ketone. $CH_3(CH_2)_3COCH_2CH_2CH_3$	128.22			163^{760}		1.4173^{14}							**B1**^{2}, 759
o200	—,**5-hydroxy-***	Butyroin. $CH_3CH_2CH_2CHOHCOCH_2CH_2CH_3$	144.22			$180–90^{760}$ 95^{20}	0.9367^{0}_{4}	1.4346^{17}		s	s	s			**B1**^{2}, 880
o201	—,**7-methyl-***	Isoamyl propyl ketone. $(CH_3)_2CHCH_2CH_2COCH_2CH_2CH_3$	142.24			$177–9^{760}$	0.8239^{20}_{4}	1.4210^{20}							**B1**^{1}, 366
o202	**Octasiloxane, Octadeca-methyl-**	$CH_3[-Si(CH_3)_2O-]_7Si(CH_3)_3$	607.31			$153^{5.1}$	0.913	1.3970^{20}	i	δ			s	peth s lig s	
o203	**1,3,5,7-Octa-tetraene***	$CH_2:CHCH:CHCH:CHCH:CH_2$	106.17	(bz)	ca. 50									peth s^{h} aa s	**B1**^{3}, 1062
o204	**2,4,6-Octatriene***	$CH_3CH:CHCH:CHCH:CHCH_3$	103.18	lf		$147–8^{764}$	0.7961^{23}_{4}	1.5131^{27}		s				chl s lig s	**B1**^{3}, 1047
o205	**1,3,7-Octatriene, 3,7-dimethyl-***	Ocimene. $CH_2:C(CH_3)CH_2CH_2CH:C(CH_3)CH:CH_2$	136.24			176–8d 81^{30}	0.799^{21}_{4}	1.4778^{25}							**Am** 75, 5946
o206	**2,4,6-Octatriene, 2,6-dimethyl-** (*4-trans, 6-trans*)*	Alloocimene A. $CH_3CH:C(CH_3)CH:CHCH:C(CH_3)_2$	136.24		−35.4	91^{20}	0.8118^{20}_{4}	1.5446^{20}							**B1**^{3}, 1050
o207	—,—(*4-trans, 6-cis*)	Alloocimene B. $CH_3CH:C(CH_3)CH:CHCH:C(CH_3)_2$	136.24		−20.6	89^{20}	0.8060^{20}	1.5446^{20}							**B1**^{3}, 1050
o208	**1-Octene***	$CH_3(CH_2)_5CH:CH_2$	112.22		−101.7	121.3^{760}	0.7149^{20}_{4}	1.4091^{20}	i	∞				chl, os v	**B1**^{3}, 835
o209	**2-Octene**(*cis*)*	$CH_3(CH_2)_4CH:CHCH_3$	112.22		−100.5	125.6^{760}	0.7243^{20}_{4}	1.4150^{20}	i					chl s	**B1**^{3}, 839
o210	—(*trans*)*	$CH_3(CH_2)_4CH:CHCH_3$	112.22		−88	124.9^{760}	0.7184^{20}_{4}	1.4132^{20}	i					chl v	**B1**^{3}, 839
o211	**3-Octene***(*cis*)	$CH_3(CH_2)_3CH:CHCH_2CH_3$	112.22		−126	122.3^{741}	0.7189^{20}_{4}	1.4125^{20}	i	s	s	s		lig s	
o212	—(*trans*)	$CH_3(CH_2)_3CH:CHCH_2CH_3$	112.22		−108, −107	122.4^{741}	0.7163^{20}_{4}	1.4124^{20}	i	s	s	s		lig s	
o213	**4-Octene**(*cis*)*	$CH_3CH_2CH_2CH:CHCH_2CH_2CH_3$	112.22		−118	121.7^{739}	0.7186^{20}_{4}	1.4136^{20}	i	s	s	s		lig s	
o214	—(*trans*)*	$CH_3CH_2CH_2CH:CHCH_2CH_2CH_3$	112.22		−105	121.4^{739}	0.7147^{20}_{4}	1.4116^{20}	i	s	s	s		lig s	
o215	**1-Octene, 2-chloro-***	$CH_3(CH_2)_5CCl:CH_2$	146.66			168–70	0.9274^{0}_{0}				s				**B1**, 221
o216	**2-Octene, 2-chloro-***	$CH_3(CH_2)_4CH:CClCH_3$	146.66			167–8	0.8923^{16}_{16}	1.4424^{16}		s	s				**B1**^{2}, 200
o217	—,**4-chloro-***	$CH_3(CH_2)_3CHClCH:CHCH_3$	146.66			153	0.8924^{20}_{4}	1.4452^{20}	i			s		chl s	

For explanations, symbols and abbreviations see beginning of table.

No.	Name	Synonyms and Formula	Mol. wt.	Crystalline form, color and specific rotation	m.p. °C	b.p. °C	Density	n_D	Solubility w	al	eth	ace	bz	other solvents	Ref.
	4-Octene														
o218	**4-Octene, 4-chloro-***	$CH_3CH_2CH_2CH:CClCH_2CH_2CH_3$	146.66	...	...	$157–9^{750}$	0.8788^{25}_{4}	1.4394^{25}	...	...	...	...	...	chl s	...
o219	**1-Octen-3-yne***	$CH_3(CH_2)_3C\vdots CCH:CH_2$	108.18	...	...	62^{60}	0.7579^{25}_{4}	1.4568^{20}	...	...	...	...	...	...	**B1**^{3}, 1046
o220	**1-Octen-3-yn-5-ol, 5-methyl-***	$CH_2:CHC\vdots CC(CH_3)(OH)CH_2CH_2CH_3$	138.21	...	...	80^{13}	0.8851^{15}_{4}	1.4740^{18}	...	s	...	...	...	...	**B1**^{3}, 2034
o221	**1-Octyne***	$CH_3(CH_2)_5C\vdots CH$	110.20	...	−80	125.2^{760}	0.7457^{20}	1.4159^{20}	i	s	s	...	...	...	**B1**, 258
o222	**2-Octyne***	$CH_3(CH_2)_4C\vdots CCH_3$	110.20	...	−60.2	137.2^{760}	0.7591^{20}_{4}	1.4278^{20}	i	s	s	...	...	...	**B1**, 258
o223	**3-Octyne***	$CH_3(CH_2)_3C\vdots CC_2H_5$	110.20	...	−105	133^{760}	0.7529^{20}_{4}	1.4250^{20}	i	s	s	...	...	...	**Am63**, 2684
o224	**4-Octyne***	$CH_3CH_2CH_2C\vdots CCH_2CH_2CH_3$	110.20	...	−102	131^{760}	0.7506^{20}_{4}	1.4248^{20}	i	s	s	...	...	...	...
o225	**1-Octyne, 1-chloro-***	$CH_3(CH_2)_5C\vdots CCl$	144.65	...	...	$61–2^{17}$	0.912^{20}	1.445^{20}	...	v^h	v	...	...	...	**B1**^{3}, 1005
o226	**3-Octyne, 2-chloro-2-methyl-***	$CH_3(CH_2)_3C\vdots CCCl(CH_3)_2$	158.67	...	...	68^{15}	0.8929^{20}_{4}	1.4480^{20}	i	...	v	...	...	...	**B1**^{3}, 1014
o227	**—,2,2-dimethyl-***	$CH_3(CH_2)_3C\vdots CC(CH_3)_3$	138.25	...	...	79^{70}	0.7491^{20}_{4}	1.4270^{20}	...	...	...	...	...	...	**Am62**, 1800
o228	**—,7-methyl-***	$(CH_3)_2CH(CH_2)_2C\vdots CCH_2CH_3$	124.23	...	...	87^{99}	0.7599^{20}_{4}	1.4280^{20}	...	...	...	...	...	...	**Am61**, 2897
o229	**2-Octyn-1-ol***	$CH_3(CH_2)_4C\vdots CCH_2OH$	126.20	...	$−20^{-17}$	$76–8^2$	0.884^{17}_{4}	1.4550^{20}	...	...	∞	...	...	...	**B1**^{2}, 506
—	**Octhracene**	*see* **Anthracene, 1,2,3,4,5,6,7,8-octahydro-**													
—	**Olealdehyde**	*see* **9-Octadecenal***													
—	**Oleanilide**	*see* **9-Octadecenoic acid,** amide, *N*-phenyl-(*cis*)													
—	**Oleanolic acid**	*see* **Caryphyllin**													
—	**Oleic acid**	*see* **9-Octadecenoic acid**(*cis*)*													
—	**Oleyl alcohol**	*see* **9-Octadecen-1-ol**(*cis*)*													
—	**Olivetol**	*see* **Benzene, 1,3-dihydroxy-5-pentyl-***													
—	**Olivetolcarboxylic acid**	*see* **Benzoic acid, 2,4-dihydroxy-6-pentyl-**													
o230	**Orcein**	Orcin. $C_{28}H_{24}N_2O_7$	500.51	amor	...	...	...	...	i	s	i	s	i	chl, CS_2 i	**B6**^{2}, 876
—	**Orcinol**	*see* **Toluene, 2,5-dihydroxy-**													
o231	**Ornithine**(*L*)	*L*-2,5-Diaminopentanoic acid*. $H_2N(CH_2)_3CH(NH_2)CO_2H$	132.16	$[\alpha]+16.65$	...	...	...	...	s	s	δ	...	...	...	**B4**^{2}, 844
o232	—,monohydrochloride(*DL*)	$H_2N(CH_2)_3CH(NH_2)CO_2H.HCl$	168.62	nd	215	...	...	...	s	...	...	...	...	...	**B4**, 424
o233	—,sulfate(*DL*)	$H_2N(CH_2)_3CH(NH_2)CO_2H.H_2SO_4$	230.24	...	213	...	...	...	v	δ	i	...	...	...	**B4**, 424
—	**Orotic acid**	*see* **6-Uracilcarboxylic acid**													
—	**Oroxylin**	*see* **Flavone, 5,7-dihydroxy-6-methoxy-**													
—	**Orsellinic acid**	*see* **Benzoic acid, 2,4-dihydroxy-6-methyl-**													
—	**Orthanilic acid**	*see* **Benzenesulfonic acid, 2-amino-***													
o234	**Orthoacetic acid,** triethyl ester	1,1,1-Triethoxyethane*. $CH_3C(OC_2H_5)_3$	162.23	...	...	142	0.8847^{25}_{4}	1.3949^{25}	i	∞	∞	...	...	chl, CCl_4 ∞	**B2**^{2}, 137
o235	—,trimethyl ester	1,1,1,-Trimethoxyethane*. $CH_3C(OCH_3)_3$	120.15	...	...	107–9	0.9428^{25}_{4}	1.3859^{25}	...	v	v	...	...	...	**B2**^{2}, 128
o236	**Orthocarbonic acid,** tetraethyl ester	Tetraethoxy methane*. $C(OC_2H_5)_4$	192.26	...	...	158–9	0.9197^{19}_{4}	1.3935^{19}	...	∞	∞	...	...	...	**B3**^{1}, 4
o237	—,tetrapropyl ester	Tetrapropoxymethane*. $C(OCH_2CH_2CH_3)_4$	248.35	...	...	224.2	0.911^8	...	...	...	...	...	...	...	**B3**, 6
—	**Orthoform**	*see* **Benzoic acid, 4-amino-3-hydroxy-,** methyl ester													
o238	**Orthoformic acid,** triethyl ester	Triethoxymethane*. $HC(OC_2H_5)_3$	148.20	...	...	145–7	0.8909^{20}_{4}	1.3922^{20}	d	s	s	...	...	...	...

For explanations, symbols and abbreviations see beginning of table.

No.	Name	Synonyms and Formula	Mol. wt.	Crystalline form, color and specific rotation	m.p. °C	b.p. °C	Density	n_D	Solubility w	al	eth	ace	bz	other solvents	Ref.
	Orthoformic acid														
o239	—,triisobutyl ester	Triisobutoxymethane. $HC[OCH_2CH(CH_3)_2]_3$	232.37			220–2	0.861^{23}			s	s				**B2**[2], 31
o240	—,triisopropyl ester	Triisopropoxymethane*. $HC[OCH(CH_3)_2]_3$	190.29			166–8	0.8621^{20}_{4}	1.4000^{20}							
o241	—,trimethyl ester	Trimethoxymethane*. $HC(OCH_3)_3$	106.12			101–2	0.9676^{20}_{4}	1.3793^{20}							**B2**, 19
o242	—,tri(3-methylbutyl) ester	Triisoamyl orthoformate. $HC[OCH_2CH_2CH(CH_3)_2]_3$	274.45			265–7d	0.864^{23}			s	s				**B2**[2], 31
o243	—,triphenyl ester	Triphenoxymethane.* $HC(OC_6H_5)_3$	292.34		76–7	$269–70^{50}$				s^h	s		s^h	chl s	
o244	—,tripropyl ester	Tripropoxymethane.* $HC(OCH_2CH_2CH_3)_3$	190.29			$140–1^{745}$	0.8805^{20}_{4}	1.4072^{20}							
o245	**—,trithio-**, triethyl ester	Ethyl orthothioformate. Tris(ethylthio)methane*. $HC(SC_2H_5)_3$	196.40			235^{760}d 136.5^{23}	1.053^{20}_{4}			s	s				**B2**[1], 39
o246	**Orthopropionic acid,** triethyl ester	1,1,1-Triethoxypropane*. $CH_3CH_2C(OC_2H_5)_3$	176.26			161^{766}				v	v				**B2**[2], 220
o247	**Orthosilicic acid,** tetraethyl ester	$Si(OC_2H_5)_4$	208.33			165.5	0.933^{20}_{4}		d	v	∞				**B1**, 334
o248	—,tetrakis(2-ethylbutyl) ester	$[(C_2H_5)_2CHCH_2O]_4Si$	432.77	lig			0.8920^{20}_{4}	1.4307^{20}	i	δ	s		s		
o249	—,tetrakis(2-ethylhexyl)ester	$[CH_3(CH_2)_3CH(C_2H_5)CH_2O]_4Si$	544.98		−90	227^{5}	0.8803^{20}_{4}	1.4388^{20}	i	δ	s		s		
o250	—,tetramethyl ester	Methyl silicate $(CH_3O)_4Si$	152.22			$25–7^{12}$	1.0280^{22}_{4}	1.3677		s					**B1**[2], 274
o253	**Orthovanadic acid,** triisopropyl ester	$[(CH_3)_2CHO]_3VO$	244.22	pa ye to colorless	−180	143^{24}	1.088^{15}		d	s	s		s	to s	**B51**, 751
o254	**1,3,4-Oxadiazole, 2,5-dimethyl-**	(ring structure: CH_3–C=N–N=C–CH_3 with O bridging)	98.11			178–9			∞	∞	∞				**B24**, 565
o255	**Oxalic acid,**	Ethanedioic acid*. HO_2CCO_2H	90.04	mcl ta or pr (+2w) orth (anh)	189.5d 101–2 (−2w)	157 sub	1.90^{17}_{4}anh 1.653^{19}_{4} (+2w)		s	v	δ		i	chl i peth i	
o256	—,diallyl ester	$CH_2{:}CHCH_2O_2CCO_2CH_2CH{:}CH_2$	170.17			$206–7^{754}$	1.055		i	s					
o257	—,diamide	Oxamide. $H_2NCOCONH_2$	88.07	pw	419d				i	i	i				
o258	—,—,*N,N′*-diethyl-	$C_2H_5NHCOCONHC_2H_5$	144.17	nd (al)	175(180)		1.169^{4}		δ	s	i				**B4**[2], 605
o259	—,—,*N,N′*-diisopropyl-	$(CH_3)_2CHNHCOCONHCH(CH_3)_2$	172.23	nd	110					v^h					**B4**, 154
o260	—,—,*N,N′*-dimethyl-	$CH_3NHCOCONHCH_3$	116.12	pl	217		1.3^{4}_{4}		v^h	δ	i		s^h	chl s^h	**B4**[2], 564
o261	—,—,*N,N′*-diphenyl	Oxanilide. $C_6H_5NHCOCONHC_6H_5$	240.26	lf (bz or $PhNO_2$)	254	>360			i	$δ^h$	i		s		**B12**[2], 165
o262	—,dibutyl ester	$CH_3(CH_2)_3O_2CCO_2(CH_2)_3CH_3$	202.25		−29.5	243.4	1.0099^{0}_{0}	1.4232^{20}	i	s	s				**B2**[2], 507
o263	—,dichloride	Oxalyl chloride. $ClCOCOCl$	126.93		−16	63–4	1.488^{13}_{4}	1.4340^{13}	d	d	s				**B2**[2], 508
o264	—,di(2-chloroethyl) ester	$ClCH_2CH_2O_2CCO_2CH_2CH_2Cl$	215.03		45	132^{3}							s		
o265	—,dicyclohexyl ester	(cyclohexyl)–O_2CCO_2–(cyclohexyl)	254.33	(MeOH)	45	$190–1^{73}$				v	v			MeOH s	**B6**[1], 6
o266	—,diethyl ester	Oxalic ester. $C_2H_5O_2CCO_2C_2H_5$	146.14		−40.6	185.7^{760}	1.0785^{20}_{4}	1.4101^{20}	$δ^h$	∞	∞				**B2**[2], 504
o267	—,dihydrazide	$H_2NNHCOCONHNH_2$	118.10	nd (w)	243–4d				s^h	δ	δ		δ	chl δ	**B2**[2], 514
o268	—,diisobutyl ester	$(CH_3)_2CHCH_2O_2CCO_2CH_2CH(CH_3)_2$	202.25			229^{758}	1.002^{14}	1.4180^{20}	i	s	s				**B2**[2], 507
o269	—,diisopropyl ester	$(CH_3)_2CHO_2CCO_2CH(CH_3)_2$	174.18			189				s	s				**B2**[1], 234
o270	—,dimethyl ester	$CH_3O_2CCO_2CH_3$	118.09	mcl ta	54	163–4	1.148^{15}	1.379^{82}	δ	s	s				**B2**[2], 503
o271	—,di(3-methylbutyl) ester	$(CH_3)_2CHCH_2CH_2O_2CCO_2CH_2CH_2CH(CH_3)_2$ Isoamyl oxalate.	230.31			267–8	0.968^{11}_{11}		i	v	v				

For explanations, symbols and abbreviations see beginning of table.

No.	Name	Synonyms and Formula	Mol. wt.	Crystalline form, color and specific rotation	m.p. °C	b.p. °C	Density	n_D	Solubility w	al	eth	ace	bz	other solvents	Ref.
	Oxalic acid														
—	—,dinitrile	*see* **Cyanogen**													
o273	—,dipropyl ester	$CH_3CH_2CH_2O_2CCO_2CH_2CH_2CH_3$	174.20		−46.3	214–5	1.0169^{20}_{4}	1.4168^{20}	δ	∞	s				
o274	—,di(2-tolyl)ester	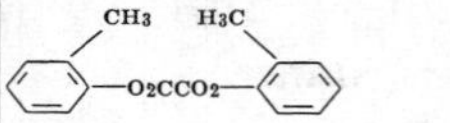 [structure: CH_3, H_3C, $-O_2CCO_2-$]	270.29	nd (al)	91	distb			i	v	v		v	chl, CS_2 v os v	**B6**[2], 330
o275	—,di(3-tolyl)ester	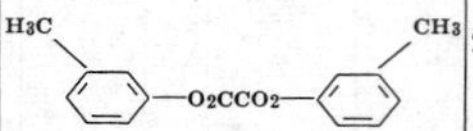 [structure: H_3C, CH_3, $-O_2CCO_2-$]	270.29	nd (al)	106	distb			i	v				os v	**B6**[2], 353
o276	—,di(4-tolyl)ester 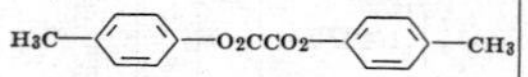	[structure: H_3C-, $-O_2CCO_2-$, $-CH_3$]	270.29	lf or pl (al-eth)						s				os s	**B6**[2], 397
o277	—,imide	Oximide. 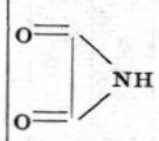 [structure: O=, NH, O=]	71.04	pr					d^h	δ					**B21**, 368
o278	—,monoamide	Oxamic acid. HO_2CCONH_2	89.05		210d				δ	i	i				**B2**[2], 509
o279	—,—,*N-sec*-butyl-	$CH_3CH_2CH(CH_3)NHCOCO_2H$	145.16	(eth)	88–9						s^h				**B4**, 162
o280	—,—,*N*-phenyl-	Oxanilic acid. $C_6H_5NHCOCO_2H$	165.15	nd	150				s^h	v	s			chl s	**B12**[2], 164
o281	—,monoamide monoethyl ester	Ethyl oxamate. $C_2H_5O_2CCONH_2$	117.11		114–5				s		s		i		**B2**[2], 509
o282	—,—,*N*-acetyl-	$C_2H_5O_2CCONHCOCH_3$	159.14	nd	54					s	s				**B2**[2], 509
o283	—,—,*N*-phenyl-	$C_2H_5O_2CCONHC_6H_5$	193.20	pl or pr (al), nd (w)	65–6	260–300			δ					oos v	**B12**[2], 164
o284	—,monoamide monohydrazide	Semioxamazide. $H_2NNHCOCONH_2$	103.08	lf	ca. 220d				δ s^h	i	i			alk v ac v	
o285	—,monoamide monoureide	Oxalam. Oxaluramide. $H_2NCOCONHCONH_2$	131.09		>310				i					sulf s	**B3**[2], 54
o286	—,monochloride monoethyl ester	$ClCOCO_2C_2H_5$	136.54	hyg		135–8	1.2223^{20}_{4}		d	d	s		s		**B2**[2], 508
o287	—,monoethyl-monomethyl ester	$CH_3O_2CCO_2C_2H_5$	132.12			173.7	1.5505^{0}_{0}		i	v	v				
o288	—,monoethyl ester mononitrile	Ethyl cyanoformate. $C_2H_5O_2CCN$	99.09			113–5	1.0034^{20}_{4}	1.3821^{20}	i	s	s				**B2**[2], 510
o289	—,monoureide	Oxaluric acid. $H_2NCONHCOCO_2H$	132.08		187	208–10d			s					oos δ	**B3**[2], 54
o290	—,piperazinum salt	$C_2H_2O_4.C_4H_{10}N_2$	176.18		>300				s^h	v	i				
o291	—,**dithiono-**, diamide	Dithiooxamide. $H_2NCSCSNH_2$	120.20	og-red	sub				δ	δ				con sulf s	**B2**[2], 515
—	**Oxaloacetic acid**	*see* **Succinic acid, 2-oxo**													
—	**Oxaluramide**	*see* **Oxalic acid,** monoamide monoureide													
—	**Oxaluric acid**	*see* **Oxalic acid,** monureide													
—	**Oxamic acid**	*see* **Oxalic acid,** monoamide													
—	**Oxanilic acid**	*see* **Oxalic acid,** monoamide, *N*-phenyl-													
—	**Oxanthranol**	*see* **Anthracene,9,10-dihydroxy-**													
o292	**Oxazole**	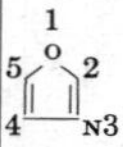 [structure: 1, O, 2, 5, 4, N3]	69.06												
o291[1]	—,**2,4-dimethyl-**	C_5H_7NO. *See* o292	97.12			108	0.9352^{15}_{4}		v	v	v				**B24**[2], 10
o293	—,**2,5-dimethyl-**	C_5H_7NO. *See* o292	97.12			$117–8^{755}$	0.9958^{21}_{4}		∞						**B24**[2], 10
o294	—,**2,4-diphenyl-**	$C_{15}H_{11}NO$. *See* o292	221.26	lf (al)	103	338–40				v^h	v		v		**B27**, 78
o295	—,**2,5-diphenyl-**	$C_{15}H_{11}NO$. *See* o292	221.26	nd (lig)	74	360	1.0906^{100}_{4}		δ	v	v				**B27**[2], 43
o296	—,**4,5-diphenyl-**	$C_{15}H_{11}NO$. *See* o292	221.26	pl or pr (lig)	44	$192–5^{15}$			δ	v	v			con ac s	**B27**, 79
o297	—,**2,4,5-triphenyl-**	Azobenzil. Benzilam. $C_{21}H_{15}NO$. *See* o292	297.36	pr	116					δ			s		
o298	**2,4-Oxazolidinedione 5,5-dipropyl-**	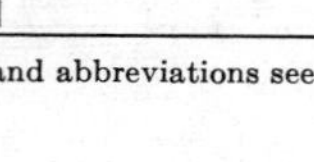 [structure: O=, NH, =O, $CH_3CH_2CH_2$, $CH_3CH_2CH_2$]	185.22		42–3	$148–50^{3}$			i						**Am67**, 522

For explanations, symbols and abbreviations see beginning of table.

No.	Name	Synonyms and Formula	Mol. wt.	Crystalline form, color and specific rotation	m.p. °C	b.p. °C	Density	n_D	Solubility w	al	eth	ace	bz	other solvents	Ref.
	Oxindole														
o299	**Oxindole**	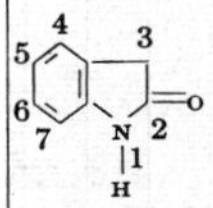	133.15	nd (w)	126	227^{73}			δ^h	s	s				**B21**², 249
o300	—,**1-ethyl-**	$C_{10}H_{11}NO$. *See* o299	161.21	nd (ace or w)	97–8				δ			s^h			**B21**¹, 291
o301	—,**3-ethyl-1-methyl-**	$C_{11}H_{13}NO$. *See* o299	175.23			$280–5^{745}$ $103–7^{0.5}$		1.557^{25}		s	s				**B21**², 258
o302	—,**3-hydroxy-**	Dioxindole. $C_8H_7NO_2$. *See* o299	149.15	(al)	180										**B21**², 415
o303	**Oxonium, dimethyl-, bromide**	$[(CH_3)_2OH]^+Br^-$	126.99		−13									liq HBr v	**B1**³, 1192
o304	—,—,chloride	$[(CH_3)_2OH]^+Cl^-$	82.54	gas	−97	−2					s			liq HCl v	**B1**³, 1192
o305	—,**triethyl-, borofluoride**	$[(C_2H_5)_3O]^+BF_4^-$	147.96	hyg nd	148d				d			v^h		chl, CH_3NO_2, $PhNO_2$ v	**B1**³, 1193
o306	**Oxyacanthine**	Vinetine. 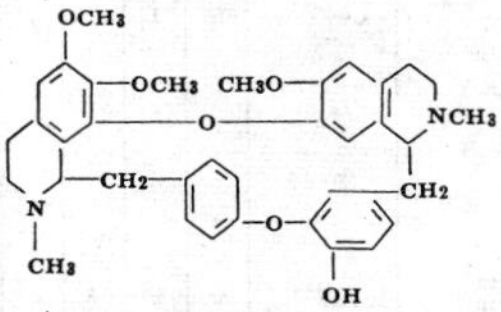	608.74	nd (al or eth)	202–14				s	s	s		s	chl s	**B27**², 892
o307	—,hydrochloride	$C_{16}H_{21}NO_3$. HCl. *See* o306	645.20	nd, $[\alpha]^{15}_D$ −163.6	270–1				s						
o308	—,nitrate dihydrate	$C_{19}H_{21}NO_3.HNO_3.\ 2H_2O$ *See* o306	671.75	nd	195–200				δ						
o309	**Oxynarcotine**	Narcotine-*N*-oxide. 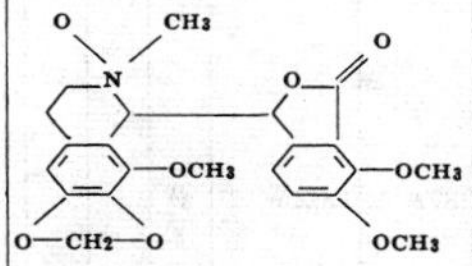	429.43	nd					δ	δ	i	δ			**B27**², 607
o311	**Oxysparteine**	Isolupamine. 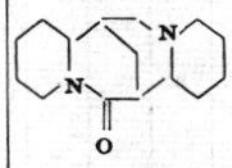	248.37	ye to colorless hyg nd	83–4	$209^{12.5}$			v	v	v			chl s	**B24**², 56
o312	—,monohydrochloride tetrahydrate	$C_{15}H_{24}N_2O.HCl.4H_2O$. *See* o311	356.90	(w)	48–50				s	s					**B24**², 57
—	**Oxytetracycline**	*see* **Terramycin**													

For explanations, symbols and abbreviations see beginning of table.

No.	Name	Synonyms and Formula	Mol. wt.	Crystalline form, color and specific rotation	m.p. °C	b.p. °C	Density	n_D	Solubility: w	al	eth	ace	bz	other solvents	Ref.
	Palmitaldehyde														
—	**Palmitaldehyde**	*see* **Hexadecanal***													
—	**Palmitanilide**	*see* **Hexadecanoic acid, amide,** *N*-phenyl-*													
—	**Palmitic acid**	*see* **Hexadecanoic acid***													
p1	**Paludrine**	Chloroguanide. Proguanil BPC Cl—C_6H_4—NHC(:NH)NHC(:NH)NHCH$(CH_3)_2$	253.74	pl	129									to s^h	
p2	—,hydrochloride	$C_{11}H_{16}N_5Cl$. HCl. *See* p1	290.20		248–52				δ	s	i			chl i	
p3	**Panthesin**	NH_2-C_6H_4-$CO_2CH_2CH(CH_2CH(CH_3)_2)N(C_2H_5)_2.CH_3SO_3H$	388.53	pa ye pw	157–9				v	s					
p4	**Pantothenic acid.**	Chick antidermatitis factor *N*(α,γ-dihydroxy-β,β-dimethylbutyryl)β-alanine. $C_9H_{17}NO_5$	219.24	ye visc oil					v		s		v	diox s, AmOH v aa v	
p5	—,calcium salt(*d*)	Calcium pantothenate. $[C_9H_{16}NO_5]_2Ca$	476.54	wh (MeOH), $[\alpha]_D^{25}+24.27$	195–6				s	i		i		MeOH s	
p6	—,—(*l*)	$[C_9H_{16}NO_5]_2Ca$	476.54	(MeOH), $(\alpha)_D^{26}$ −27.8 (w)	187.5–9										
p7	**Pantothenyl alcohol**	Penthenol. $HOCH_2C(CH_3)_2CHOHCONHC(CH_3)_2OH$	205.25	slightly hyg oil $[\alpha]_D^{20}+29.5$		d $118–20^{0.02}$	1.2_{20}^{20}	1.497^{20}	v	v	δ			MeOH v alk d	
p8	**Papaveraldine**	Xantholine. (CH_3O)(OCH_3)-isoquinolyl-C(=O)-C_6H_3(OCH_3)(OCH_3)	353.38	nd (al), (bz-peth)	210				i	δ	δ		s	chl s min ac v aa v^h lig δ	**B21**[2], 479
p9	**Papaverine**	Papaveroline tetramethyl ether. $C_{20}H_{21}NO_4$	339.39	wh pr (al-eth), nd (chl-peth)	147	d	1.337_4	1.625	s^h δ	v		s	s^h	chl δ lig δ	**B21**[2], 202
p10	—,hydrochloride	$C_{20}H_{21}NO_4.HCl$	375.86	wh mcl pr (w)	224–5				s	s	δ			chl s	**B21**[2], 203
—	—,**tetrahydro-*N*-methyl**	*see* **Landanosine**													
p11	**Parabanic acid**	Imidazoletrione. Oxalylurea. HN—C(=O)—C(=O)—NH—C(=O) (ring)	114.06	mcl nd (w)	243–5				s	v					**B24**[2], 263
p12	**Parabutyraldehyde**	$CH_3CH_2H_2C$-(1,3,5-trioxane)-($CH_2CH_2CH_3$)$_2$	216.32			$98–100^{35}$	0.918								**B19**[2], 402
p13	**Paraconic acid**	Hydroxymethylsuccinic acid γ-lactone. Itamalic acid γ-lactone. Tetrahydro-5-oxo-3-furancarboxylic acid. O=(lactone ring)-CO_2H	130.10	dlq	57–8				s						**B18**[2], 311
p14	**Paracyanogen**	$(CN)_x$		br pw		sub			i	i				KOH s	**B2**[1], 239
p14[1]	**Paraisobutyraldehyde**	2,4,6-Triisopropyl-1,3,5-trioxane	216.32	nd (al)	59–60	195 sub δd			i	s	v				**B19**, 390
p15	**Paraldehyde**	Paraacetaldehyde. 2,4,6-Trimethyl-1,3,5-trioxane. CH_3-(1,3,5-trioxane)-(CH_3)$_2$	132.16		12.45	128.0	0.9923_4^{20}	1.4049^{20}	v s^h	∞	∞			chl ∞	**B19**[2], 394

For explanations, symbols and abbreviations see beginning of table.

No.	Name	Synonyms and Formula	Mol. wt.	Crystalline form, color and specific rotation	m.p. °C	b.p. °C	Density	n_D	Solubility w	al	eth	ace	bz	other solvents	Ref.
	Paraldehyde														
p16	**—,trichloro-**	Chloroacetaldehyde trimer. 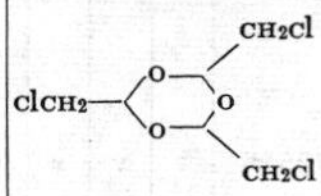	235.51	nd (eth)	87–7.5 cor	142–4[10]			i	v^h δ	v				**B19**[1], 807
p17	**Paraldol**	$C_8H_{16}O_4$	176.21	wh tcl pr	89–91	90[15]	1.116[20]	1.4610[20]	v	v	s				**B1**[2], 869 **B19**[2], 93
—	**Pararosaniline**	*see* **Methane, tris(4-aminophenyl)-***													
p18	**Parasorbic acid**	2-Hexen-1,5-olide. 5-Hydroxy-2-hexenoic acid lactone*. 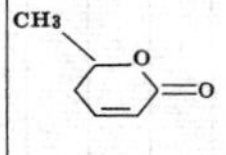	112.13			104–5[14]	1.079^{18}_4		s	v	v				
p19	**Parathion**	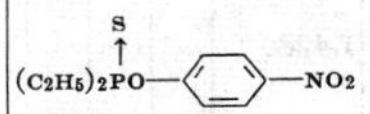 $(C_2H_5)_2P(S)O-C_6H_4-NO_2$	291.26	ye	6.1	375 157–62[0.6]	1.26^{25}_4	1.5370[25]	i	∞	s	s		chl v AcOEt v lig i	
—	**Paredrine**	*see* **Benzene, 1(2-aminopropyl)-4-hydroxy-***													
—	**Paris green**	*see* **Methyl green**													
—	**α-Parvoline**	*see* **Pyridine,2-ethyl-3,5-dimethyl-**													
—	**β-Parvoline**	*see* **Pyridine, tetramethyl-**													
p20	**Patulin**	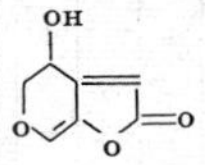	154.12	pr or pl (eth, chl)	111				s	s				os s lig i	
p21	**Paucine,** hydrate	$C_{27}H_{39}N_5O_5.6\frac{1}{2}H_2O$	630.74	ye lf	126d				s^h	s^h	i			alk s^h, chl i	
—	**Peganine**	*see* **Vasicine** *(DL)*													
—	**Pelargonaldehyde**	*see* **Nonanal***													
—	**Pelargonic acid**	*see* **Nonanoic acid**													
p22	**Pelargonidin,** chloride	$C_{15}H_{11}ClO_5$	306.70	red br hyg (anh), pr or pl (dil HCl), nd (al-HCl, +w)	>350 (anh)				s v^h	v				MeOH, chl v conc sulf s	**B18**[2], 200
p23	**Pelletierine**	β(2-Piperidyl)propionaldehyde. Punicine. $C_8H_{15}NO$	141.21	oil $[\alpha]_D -31.1$		195d 106[21]	0.988^{20}_4		s	s	s		s		**B21**[2], 220
p24	—sulfate	$2(C_8H_{15}NO).H_2SO_4$	380.50			133 (anh)			s	s					**B21**, 246
p25	**Pellotine**	*N*-Methylanhalonidine. $C_{13}H_{19}NO_3$	237.25	(peth)	110–2				δ	s	s				**B21**[1], 249
p26	**Penicillic acid**	$CH_2:C(CH_3)COC(OCH_3):CHCO_2H$	170.16	rh or hex pl (+1w), nd (peth h)	64–5 (+1w) 87 (anh)				v^h s	v	v		v	chl v lig δ^h	**B3**[2], 319
p27	**Pentacene***	2,3,6,7-Dibenzanthracene. 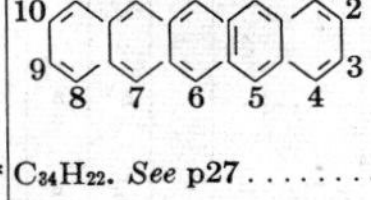	278.35	deep vt-bl nd or lf ($PhNO_2$)	high	290–300 sub			i					os δ	**E14s**, 582
p28	**—,6,13-diphenyl-***	$C_{34}H_{22}$. *See* p27	430.55	vt-bl nd ($PhNO_2$+1)		318–20				s^h			s	CCl_4 s^h aa s	**E14s**, 583
p29	**6,9-Pentadecadiyne***	$CH_3(CH_2)_4C:CCH_2C:C(CH_2)_4CH_3$	204.34			135–6[4]	0.840^{21}_4	1.4693[21]							
p30	**Pentadecanal***	*n*-Pentadecylaldehyde. $CH_3(CH_2)_{13}CHO$.	226.39	nd	24–5	185[25]			i	s	v	v		os v	**B1**[2], 770
p31	**—,oxime***	$CH_3(CH_2)_{13}CH:NOH$	241.41	nd (al)	86					δ	s				**B1**, 716
p32	**Pentadecane***	$CH_3(CH_2)_{13}CH_3$	212.41		10	270.5	0.7689^{20}_4	1.4315[20]	i	v	v				**B1**[2], 68
p33	**—,1-amino-***	*n*-Pentadecylamine. $CH_3(CH_2)_{14}NH_2$	227.43		36.5	307.6			i	s	δ				**B4**, 201
p34	**—,1-bromo-***	Pentadecyl bromide. $CH_3(CH_2)_{14}Br$	291.33		18.63	170–5[80] 127–8[0.5]		1.4592[20]	i					chl v	**B1**[3], 553
p35	**—,1,15-dibromo-***	Pentadecamethylene bromide. $Br(CH_2)_{15}Br$	370.24	lf (al)	25.5	197[2]			i	δ v^h				chl v	**B1**[3], 554

For explanations, symbols and abbreviations see beginning of table.

No.	Name	Synonyms and Formula	Mol. wt.	Crystalline form, color and specific rotation	m.p. °C	b.p. °C	Density	n_D	Solubility w	al	eth	ace	bz	other solvents	Ref.
	Pentadecane														
p36	**—,1(2,3-dihydroxyphenyl)-***	Tetrahydroxirushiol. 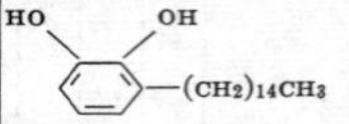 HO OH —$(CH_2)_{14}CH_3$	320.50	nd (to, xyl, eth or peth)	59	...	...	...	...	v	v	...	v	lig δ^h v^h	**B6**², 911
p37	**Pentadecanoic acid***	Pentadecylic acid. $CH_3(CH_2)_{13}CO_2H$	242.39	pl (aq al), (peth)	53–4	257^{100} 158^{1}	0.8423^{80}	1.4254^{80}	i	v	s	v	v	CS_2 s chl s	**B2**², 329
p38	—,methyl ester	$CH_3(CH_2)_{13}CO_2CH_3$	256.42	nd (dil al)	18.5	...	0.8618^{25}_{4}	1.4390^{25}	...	s	s	...	...	os s	**B2**², 330
p39	**8-Pentadecanone**	Caprilone. Diheptyl ketone. $CH_3(CH_2)_6CO(CH_2)_6CH_3$	226.39	...	40	178	...	...	...	s	...	...	...	...	**B1**, 717
p40	**1-Pentadecene***	$CH_3(CH_2)_{12}CH:CH_2$	210.41	...	2–8	144–5^{10}	0.7751^{25}_{4}	...	...	...	...	...	...	...	**B1**², 206
p41	**1-Pentadecyne***	$CH_3(CH_2)_{12}C\vdots CH$	208.37	...	...	112–3^{5}	0.8261^{20}_{4}	1.4410^{20}	...	...	...	...	...	...	...
p43	**2,4-Pentadienal, 5-phenyl-***	Cinnamylideneacetaldehyde. $C_6H_5CH:CHCH:CHCHO$	158.19	ye oil	...	155–65^{3}	...	...	i	∞	v	...	∞	...	**B7**², 320
p44	**1,2-Pentadiene***	Ethylallene. $CH_3CH_2CH:C:CH_2$	68.11	...	−137.26	44–5	0.6890^{20}	1.4149	...	...	...	...	...	...	**B1**, 251
p45	**1,3-Pentadiene***	Piperylene. $CH_3CH:CHCH:CH_2$	68.11	...	...	41.8 42.2^{748}	0.6830^{20}_{4}	1.4280^{20}	...	...	...	...	...	...	...
p46	**1,4-Pentadiene***	Allylethylene. Divinylmethane. $CH_2:CHCH_2CH:CH_2$	68.11	...	−148.28	25.8–26.2^{756}	0.6594^{20}_{4}	1.3883^{20}	...	...	...	...	...	...	**B1**, 250
p47	**2,3-Pentadiene***	*sym*-Dimethylallene. $CH_3CH:C:CHCH_3$	68.11	...	−125.65	49–51	0.7024^{20}_{0}	1.4284^{20}	...	...	...	...	...	...	**B1**, 250
p48	**1,3-Pentadiene, 3-chloro-***	3-Methylchloroprene. $CH_3CH:C(Cl)CH:CH_2$	102.57	...	...	99.5–101.5	0.9576^{20}_{4}	1.4785^{20}	...	...	...	...	...	...	...
p49	**1,2-Pentadiene, 1-chloro-3-ethyl-***	$(C_2H_5)_2C:C:CHCl$	130.62	...	...	85–8^{100}	0.9297^{19}_{4}	...	i	...	s	...	...	...	**B1**³, 1002
p50	**—,1-chloro-3-methyl-***	$CH_3CH_2C(CH_3):C:CHCl$	116.59	...	...	68–70^{100}	0.9562^{20}_{4}	...	...	...	s	...	...	...	...
p51	**1,3-Pentadiene, 1-chloro-3-methyl-***	$CH_3CH:C(CH_3)CH:CHCl$	116.59	...	...	62–3^{100}	0.9574^{20}_{4}	...	i	...	s	...	...	chl s	...
p52	**—,2-chloro-3-methyl-***	$CH_3CH:C(CH_3)CCl:CH_2$	116.59	...	...	57–60^{95}	0.9437^{20}_{4}	1.4671^{20}	...	...	s	...	...	chl s	...
p53	**2,4-Pentadienoic acid***	*β*-Vinylacrylic acid. $CH_2:CHCH:CHCO_2H$	98.10	pr (eth)	80	110–5d	...	...	s^h	v	v	...	...	lig δ	**B2**, 451
p54	**—,4-hydroxy-, lactone***	Protoanemonin. 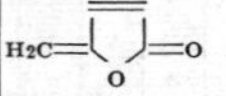 H_2C= =O O	96.08	pa ye oil	...	$45^{1.5}$	...	...	δ	...	...	...	...	chl s	...
p55	**—,5(3,4-methylenedioxiphenyl)-**	Piperic acid. 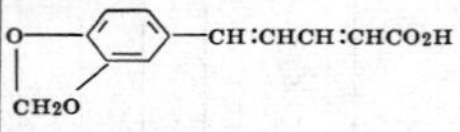 O— —CH:CHCH:CHCO₂H CH_2O	218.20	ye in light, nd (al), ye nd (sub)	215	sub	...	...	i	s^h	δ	...	δ	...	**B19**², 300
p56	**—,5-phenyl-***	Cinnamylideneacetic acid. $C_6H_5CH:CHCH:CHCO_2H$	174.19	pl or pr (al, bz)	166–7	...	...	...	...	s^h	s	...	s v^h	lig δ	**B9**², 441
p57	—,—,ethyl ester	$C_6H_5CH:CHCH:CHCO_2C_2H_5$	202.14	ye	25–6	...	1.0299^{42}_{4}	1.5768^{80}	...	s	s	...	...	...	**B9**², 441
p58	—,—,methyl ester	$C_6H_5CH:CHCH:CHCO_2CH_3$	188.22	lf or pl	71	185^{20}	...	...	i	v	...	...	...	MeOH v	**B9**², 441
p59	**1,4-Pentadien-3-one, 1,5-bis-(2-ethoxyphenyl)-***	$C_{21}H_{22}O_3$. *See* p72	322.41	ye (dil al)	89	...	...	...	i	v^h	s^h	...	...	...	**B8**, 352
p61	**—,1,5-bis(2-hydroxyphenyl)-**	Disalicylidene acetone. $C_{17}H_{14}O_3$. *See* p72	266.30	ye nd (dil al)	159	...	...	...	δ	v	s	v	s	ac s chl s Py v	**B8**², 404
p62	**—,1,5-bis(4-hydroxyphenyl)-***	Bis(4-hydroxystyryl) ketone. $C_{17}H_{14}O_3$. *See* p72	266.30	ye og nd or lf (dil al)	237–8	...	...	...	...	v	s	v	s	chl, MeOH s alk, ac s	**B8**², 405
p63	**—,1,5-bis(2-methoxyphenyl)-***	$C_{19}H_{18}O_3$. *See* p72	294.35	ye nd (al), pl (al)	127	...	...	...	i	δ s^h	...	...	...	...	**B8**², 405

For explanations, symbols and abbreviations see beginning of table.

No.	Name	Synonyms and Formula	Mol. wt.	Crystalline form, color and specific rotation	m.p. °C	b.p. °C	Density	n_D	Solubility w	al	eth	ace	bz	other solvents	Ref.
	1,4-Pentadien-3-one														
p64	**—,1,5-bis(3-methoxyphenyl)-***	$C_{19}H_{18}O_3$. *See* p72	294.35	nd (chl-MeOH)	55–6									os v lig δ	**B8^1**, 666
p65	**—,1,5-bis(4-methoxyphenyl)-***	$C_{19}H_{18}O_3$. *See* p72	294.35	ye pl (aa)	129–30								v	chl v aa s	**B8^2**, 406
p66	**—,1,5-bis(3,4-methylenedioxyphenyl)-***	$C_{19}H_{14}O_5$. *See* p72	322.30	ye nd (bz or AcOEt)	185				i	δ		v	s^h	chl v con sulf s lig i	**B19**, 446
p67	**—,1,5-bis(2-nitrophenyl)-***	$C_{17}H_{12}N_2O_5$. *See* p72	324.30	ye nd (aa)	170.5–1					δ				chl v con sulf s	**B7^2**, 455
p68	**—,1,5-bis(3-nitrophenyl)-***	$C_{17}H_{12}N_2O_5$. *See* p72	324.30	ye, br (Ac_2O)	238									con sulf δ, Ac_2O δ, os s	**B7^2**, 455
p69	**—,1,5-bis(4-nitrophenyl)-***	$C_{17}H_{12}N_2O_5$. *See* p72	324.30	ye (Ac_2O)	254									Ac_2O δ os v	**B7^2**, 455
p70	**—,1(2-chlorophenyl)-5(3-chlorophenyl)-**	$C_{17}H_{12}Cl_2O$. *See* p72	303.20	ye nd (dil al)	67–8				s		δ			CS_2 δ lig δ	**B7^2**, 454
p71	**—,1(2-chlorophenyl)-5(4-chlorophenyl)-**	$C_{17}H_{12}Cl_2O$. *See* p72	303.20	ye nd (al)	109				δ						**B7^2**, 454
p71^1	**—,1,5-di(2-furyl)-**	Difurfurylidene acetone. (furyl)—CH:CHCOCH:CH—(furyl)	214.21	dlq pr (peth)	60–1				i	v	v			chl v con HCl s sulf s lig δ^h	**B19^2**, 162
p72	**—,1,5-diphenyl-**	Dibenzalacetone. Cinnamone. Styryl ketone. C_6H_5(3,2,4,5,6)—CH:CHCOCH:CH—C_6H_5(2′,3′,4′,5′,6′)	234.30	pl or lf (ace)	113	d			i	δ	s	s		chl s	
p72^1	**2,4-Pentadiene-1-one, 1,5-diphenyl-**	Cinnamalacetophenone. $C_6H_5CH{:}CHCH{:}CHCOC_6H_5$	234.30	ye nd (al)	102–3				i	s^h					**B7^2**, 451
p73	**1,3-Pentadiyne***	$CH_3C{\vdots}CC{\vdots}CH$	64.08			55–6	0.7375^{21}_4	1.4431^{21}							
p74	**Pentaerythritol**	Tetrakis(hydroxymethyl)methane*. $C(CH_2OH)_4$	136.15		260	sub		1.548	s						**B1^2**, 601
p75	**—,tetraacetate**	$C(CH_2O_2CCH_3)_4$	304.30	tetr nd (w)	84–6		1.273^{18}_4		s	v	v				**B2^2**, 150
p76	**—,tetranitrate**	Pentrit. PETN. $C(CH_2ONO_2)_4$	316.15	tetr (ace). pr (ace-al)	140–1		1.773^{20}_4		δ	δ	δ	v	s		**B1^2**, 602 **B9^2**, 887
—	**Pentalin**	*see* **Ethane, pentachloro-***													
p77	**Pentanal***	Valeral. Valeraldehyde. $CH_3(CH_2)_3CHO$	86.13		−91.5	102–3	0.8095^{20}_4	1.3944^{20}	δ	s	s				**B1^2**, 735
p78	**—,oxime***	Valeraldoxime. $CH_3(CH_2)_3CH{:}NOH$	101.15		52										
p78^1	**—,4-hydroxy-(*dl*)***	$CH_3CHOHCH_2CH_2CHO$	102.14			$63–5^{10}$	1.019^{20}_4	1.4359^{17}	s					os s	**B1^2**, 871
p78^2	**—,4-hydroxy-(*l*)***	$CH_3CHOHCH_2CH_2CHO$	102.14	$[\alpha]^{23}_D -7.8$		$43–6^1$			s					os s	**B1^2**, 872
p79	**—,3-hydroxy-2-methyl-***	Propionaldol. $C_2H_5CHOHCH(CH_3)CHO$	116.16			$94–6^{23}$	0.986^{25}_4	1.4502	s	v	v	v			**B1^1**, 423
p80	**—,2-methyl-***	$CH_3CH_2CH_2CH(CH_3)CHO$	100.16			116^{737}			s		s	s			**B1^1**, 355
p81	**—,4-methyl-2-oxo-***	Formul isobutyl ketone. $(CH_3)_2CHCH_2COCHO$	114.15	ye-gr		$45–6^{12}$								os s	**B1^1**, 406
p82	**—,2-oxo-***	Butyrylformaldehyde. $CH_3(CH_2)_2COCHO$	100.12			36^{16}			∞	∞	∞				**B1^2**, 830
p83	**—,4-oxo-***	γ-Ketovaleraldehyde. Levulinaldehyde. Levulinic aldehyde. $CH_3COCH_2CH_2CHO$	100.12		<−21	186–8	1.0184^{21}_4	1.4257^{22}	∞	∞	∞				**B1^2**, 830
p84	**Pentane***	$CH_3(CH_2)_3CH_3$	72.15		−129.72	36	0.6262^{20}_4	1.3579^{20}	v						**B1^2**, 92
p85	**—,1-amino-***	*n*-Amylamine. Pentylamine*. $CH_3(CH_2)_4NH_2$	87.16		−55	103	0.7614^{20}_4	1.413^{20}	∞	∞	∞				**B4^2**, 641
p86	**—,2-amino-***	2-*sec*-Buylamine. $CH_3(CH_2)_2CHNH_2CH_3$	87.16			91.5^{755}	0.7384^{20}_4		∞	∞	∞				**B4^2**, 643

For explanations, symbols and abbreviations see beginning of table.

No.	Name	Synonyms and Formula	Mol. wt.	Crystalline form, color and specific rotation	m.p. °C	b.p. °C	Density	n_D	Solubility						Ref.
									w	al	eth	ace	bz	other solvents	
	Pentane														
p87	**—,3-amino-***	$(C_2H_5)_2CHNH_2$	87.16			89–91	0.7487^{20}_{4}	1.4063^{20}		s					**B4**[2], 643
p88	**—,3(amino-methyl)-***	$(C_2H_5)_2CHCH_2NH_2$	101.19			125.3									**B4**, 192
—	**—,2,2-bis-(hydroxy-methyl)-***	*see* **1,3-Propanediol, 2-methyl-2-propyl-***													
p89	**—,3,3-bis(ethyl-sulfonyl)-***	Tetronal. $(C_2H_5)_2C(SO_2C_2H_5)_2$	256.38	lf	85				δ	s	s				**B4**, 681
p90	**—,1-bromo-***	Amyl bromide. $CH_3(CH_2)_4Br$	151.06		−95	129.6	1.2177^{20}_{4}	1.1444^{20}	i	s	∞				**B1**[3], 344
p91	**—,2-bromo-***	$CH_3(CH_2)_2CHBrCH_3$	151.06			117–8 58.4^{100}	1.2039^{20}_{4}	1.4413^{20}							**B1**[3], 345
p92	**—,3-bromo-***	$(C_2H_5)_2CHBr$	151.06		118.2–.5 59^{100}	1.2170^{20}_{4}	1.4443^{20}		i						**B1**[3], 346
p93	**—,1-bromo-5-fluoro-***	$F(CH_2)_5Br$	169.05		162		1.4406^{25}		i	v	v				
p94	**—,1-bromo-2-methyl-***	$CH_3(CH_2)_2CH(CH_3)CH_2Br$	165.09			$142–5^{748}$ $51–3^{25}$	1.1624^{20}_{4}	1.4495^{20}	i		v			chl v	**B1**[3], 399
p95	**—,1-bromo-3-methyl-***	$CH_3CH_2CH(CH_3)CH_2CH_2Br$	165.09			148.6– 9.4^{766}	1.1829^{20}_{4}		i		s			chl v	**B1**[3], 403
p96	**—,1-bromo-4-methyl-***	Isohexyl bromide. $(CH_3)_2CH(CH_2)_3Br$	165.09			$147–8^{759}$	1.1683^{20}	1.4490	i		v			chl v	**B1**[3], 399
p97	**—,2-bromo-2-methyl-***	$CH_3CH_2CH_2CBr(CH_3)_2$	165.09			77.8^{145}		1.442^{23}	d^h		s			chl v	**B1**[3], 399
p98	**—,3-bromo-3-methyl-***	$(C_2H_5)_2CBrCH_3$	165.09			$82–3^{145}$	1.1835^{20}		d^h		s			chl v	**B1**[1], 54
p99	**—,1-chloro-***	*n*-Amyl chloride. Pentyl chloride. $CH_3(CH_2)_4Cl$	106.60		−99	108.2 10^{15}	0.8828^{20}_{4}	1.4128^{20}	i	∞	∞				**B1**[3], 339
p100	**—,2-chloro-(*d*)***	$CH_3CH_2CH_2CHClCH_3$	106.60	$[\alpha]_D$ +34.07	−137, −139	97^{770}	0.8732^{20}_{4}	1.4079^{20}	i	s	s				**B1**[2], 95
p101	**—,3-chloro-***	$(C_2H_5)_2CHCl$	106.60		−105, −106	98	0.8723^{20}_{4}	1.4082^{20}							**B1**[1], 95
p102	**—,2-chloro-2,3-dimethyl-***	$CH_3CH_2CH(CH_3)CCl(CH_3)_2$	134.65			$38–9^{20}$		1.4264^{20}	i		s		s		**B1**[3], 447
p103	**—,2-chloro-2,4-dimethyl-***	$(CH_3)_2CHCH_2CCl(CH_3)_2$	134.65			$127–8^{733}$d $33–4^{20}$	0.861^{20}_{4}	1.4180^{20}			v				**B1**[1], 59
p104	**—,3-chloro-2,3-dimethyl-***	$CH_3CH_2C(CH_3)ClCH(CH_3)_2$	134.65			$135–8^{757}$ δd $41–2^{20}$	0.884^{22}_{22}	1.4318^{20}	i		s			chl s	**B1**[3], 448
p105	**—,4-chloro-2,2-dimethyl-***	$CH_3CHClCH_2C(CH_3)_3$	134.65			93^{25}	0.855^{20}_{4}	1.4180^{20}			s				
p106	**—,3-chloro-2,2-dimethyl-3-ethyl-***	$(C_2H_5)_2CClC(CH_3)_3$	162.71			$d53–4^{6}$		1.4528^{25}	d		s			chl s	**B1**[3], 517
p107	**—,2-chloro-3-ethyl-***	$(C_2H_5)_2CHCHClCH_3$	134.65			$62.0–2.5^{50}$ cor	0.8911^{25}_{25}	1.4318^{20}	i					chl s	**B1**[1], 120
p108	**—,3-chloro-3-ethyl-***	$(C_2H_5)_3CCl$	134.65			143–4 $43–4^{20}$	0.8644^{25}_{4}	1.4329^{20}			v				**B1**[2], 120
p109	**—,3-chloro-3-ethyl-2-methyl-***	$(C_2H_5)_2CClCH(CH_3)_2$	148.68			150–5d		1.4405^{25}	d	s				chl s	**B1**[3], 490
p110	**—,1-chloro-5-fluoro-***	$F(CH_2)_5Cl$	124.59			143.2		1.4109^{25}	i	v	v				
p111	**—,2-chloro-2-methyl-***	$CH_3(CH_2)_2CCl(CH_3)_2$	120.63			$110–1^{734}$d $36–7^{15}$	0.863^{20}_{4}	1.4126^{20}	d^h		v				**B1**[2], 111
p111[1]	**—,2-chloro-4-methyl-***	$(CH_3)_2CHCH_2CHClCH_3$	120.63			$111–2^{773}$	0.861^{20}_{4}	1.4113^{20}			v				**B1**[2], 111
p112	**—,3-chloro-2-methyl-***	$CH_3CH_2CHClCH(CH_3)_2$	120.63			115– 6.5^{752}d			i		s			chl s	**B1**[2], 111

For explanations, symbols and abbreviations see beginning of table.

No.	Name	Synonyms and Formula	Mol. wt.	Crystalline form, color and specific rotation	m.p. °C	b.p. °C	Density	n_D	Solubility w	al	eth	ace	bz	other solvents	Ref.
	Pentane														
p113	**—,3-chloro-3-methyl-***	$(C_2H_5)_2CClCH_3$	120.63			116 35^{25}	0.8900^{20}_{4}	1.4210^{20}	d^h		v			chl v	**B1**^{2}, 112
p114	**—,3(chloro-methyl)-***	$(C_2H_5)_2CHCH_2Cl$	120.63			125–7	0.8914^{20}_{4}	1.4230^{20}	i		v			chl v	
p115	**—,2-chloro-2,4,4-trimethyl-***	$(CH_3)_2CClCH_2C(CH_3)_3$	148.68		−26	145–50d 44^{16}	0.8891^{0}_{0}	1.4307^{20}	d	s				alk d^h	**B1**^{3}, 498
p116	**—,1,5-diamino-***	Cadaverine. Pentamethylene diamine. $H_2N(CH_2)_5NH_2$	102.18		9	178–80	0.873^{25}_{4}		s	s	δ				**B4**^{2}, 708
p117	**—,1,5-dibromo-***	Pentamethylene bromide. $Br(CH_2)_5Br$	229.96		−39.5	222.2 99^{12}	1.6879^{15}_{4}	1.5091^{15}							**B1**^{2}, 97
p118	**—,1,2-dichloro-***	$CH_3CH_2CH_2CHClCH_2Cl$	141.05			148.4–.8 $58–9^{28}$	1.0872^{20}_{4}	1.4485^{20}	i	s				alk d chl v	**B1**^{2}, 95
p119	**—,1,3-dichloro-***	$CH_3CH_2CHClCH_2CH_2Cl$	141.05			80.4^{60}	1.0834^{20}_{4}	1.4485^{20}	i			v		chl v	
p120	**—,1,4-dichloro-***	$CH_3CHCl(CH_2)_2CH_2Cl$	141.05			88.1^{60}	1.0840^{20}_{4}	1.4503^{20}	d^h			s		chl s	**B1**, 131
p121	**—,1,5-dichloro-***	$Cl(CH_2)_5Cl$	141.05		−72.8	182.3	1.1058^{15}_{4}	1.4563^{20}	i	s	s				**B1**^{2}, 95
p122	**—,2,2-dichloro-***	$CH_3CH_2CH_2CCl_2CH_3$	141.05			128–9 cor $36–7^{20}$	1.040^{20}	1.434^{20}	i		s			chl s	**B1**^{3}, 342
p123	**—,2,4-dichloro-***	$CH_3CHClCH_2CHClCH_3$	141.05			147–50 62^{12}	1.0529^{12}_{4}	1.447^{18}			v			chl v alk d	**B1**^{1}, 43
p124	**—,3,3-dichloro-***	$CH_3CH_2CCl_2CH_2CH_3$	141.05			$131–2^{750}$ 32^{14}	1.053^{20}	1.442^{20}	i					chl s	**B1**^{3}, 343
p125	**—,1,2-dichloro-4,4-dimethyl-***	$(CH_3)_3CCH_2CHClCH_2Cl$	169.10			$173–5^{745}$ $58–9^{12}$	1.0259^{20}_{4}	1.4489^{20}						chl s	
p126	**—,1,5-dichloro-3,3-dimethyl-***	$(ClCH_2CH_2)_2C(CH_3)_2$	169.10			135^{80} $58–9^{8}$	1.0917^{15}_{4}	1.4899^{15}	i	s^h				chl s	**B1**^{3}, 454
p127	**—,2,4-dichloro-2,4-dimethyl-***	$(CH_3)_2CClCH_2CCl(CH_3)_2$	169.10		23–4	$51.5–.7^{8}$			i	δ				chl s	**B1**^{3}, 451
p128	**—,3,3-dichloro-2,4-dimethyl-***	$(CH_3)_2CHCCl_2CH(CH_3)_2$	169.10			118–20δd	0.9513^{9}		i		s			chl s	**B1**, 158
p129	**—,3,3-diethyl-***	Tetraethylmethane. $C(C_2H_5)_4$	128.25			138.2	0.7522^{20}_{4}	1.4206^{18}	i					os s	
p130	**—,1,5-diiodo-***	$I(CH_2)_5I$	323.97		9	149^{20}	2.1903^{15}	1.6046^{15}	i		s			chl s	**B1**^{3}, 350
p131	**—,2,2-dimethyl-***	$CH_3CH_2CH_2C(CH_3)_3$	100.21		−123.82	79.197	0.6739^{20}_{4}	1.3826^{20}	i	s	s				**B1**^{3}, 442
p132	**—,2,3-dimethyl-***	$CH_3CH_2CH(CH_3)CH(CH_3)_2$	100.21			89.8	0.6951^{20}_{4}	1.3919^{20}	i	s	s				**B1**^{3}, 445
p133	**—,2,4-dimethyl-***	$(CH_3)_2CHCH_2CH(CH_3)_2$	100.21		−119.24	80.5^{760}	0.6727^{20}_{4}	1.3815^{20}	i	s	s				**B1**^{3}, 449
p134	**—,3,3-dimethyl-***	$CH_3CH_2C(CH_3)_2CH_2CH_3$	100.21		−135.0	86.064	0.6933^{20}_{4}	1.3909^{20}	i	s	s				**B1**^{3}, 452
p135	**—,1,5-dinitro-***	$O_2NCH_2(CH_2)_3CH_2NO_2$	162.15			$134^{1.2}$		1.461^{20}	i				s		**B1**, 45
p136	**—,1,1-diphenyl-***	$CH_3(CH_2)_3CH(C_6H_5)_2$	224.35			277–8			i						**B5**^{2}, 523
p137	**—,1,5-diphenyl-***	$C_6H_5(CH_2)_5C_6H_5$	224.35			330.6 cor $187–9^{10}$	0.9814^{19}_{0}	1.559^{19}	i						**B5**^{2}, 523
p138	**—,1,5-dithio-cyanato-**	$NCS(CH_2)_5SCN$	186.31	yesh		221–2									**B3**^{1}, 72
p139	**—,1,2-epoxy-2,4,4-trimethyl-***	$(CH_3)_3CCH_2C(CH_3)CH_2$ (1,2-epoxide, O bridging C1–C2)	128.22		−64	140.9 $20^{5.4}$	0.8287^{20}_{20}	1.4097^{20}	δ						**B1**^{3}, 850
p140	**—,3-ethyl-***	Triethylmethane. $(C_2H_5)_3CH$	100.21		−118.604	93.5	0.6982^{20}_{4}	1.3934^{20}	i	s	s				**B1**^{3}, 441
p141	**—,3-ethyl-2-methyl-***	$(CH_3)_2CHCH(C_2H_5)_2$	114.23		−114.960	115.65	0.7193^{20}_{4}	1.4040^{20}	i	δ	s				**B1**^{3}, 489
p142	**—,3-ethyl-3-methyl-***	$(C_2H_5)_3CCH_3$	114.23		−90.9	118.4	0.7274^{20}_{4}	1.4081^{20}	i		s				
p143	**—,1-fluoro-***	*n*-Amyl fluoride. $CH_3(CH_2)_4F$	90.14		<−80	62–3 cor	0.7885^{20}_{4}	1.3573^{20}	i	v	v				**B1**^{3}, 338
p144	**—,1-iodo-***	*n*-Amyl iodide. $CH_3(CH_2)_4I$	198.06		−85.6	155^{740} 35^{7}	1.517^{20}_{4}	1.4955^{20}	δ	s	s				**B1**^{3}, 348

For explanations, symbols and abbreviations see beginning of table.

No.	Name	Synonyms and Formula	Mol. wt.	Crystalline form, color and specific rotation	m.p. °C	b.p. °C	Density	n_D	Solubility w	al	eth	ace	bz	other solvents	Ref.
	Pentane														
p145	**—,2-iodo-***	$CH_3(CH_2)_2CHICH_3$	198.05	...	...	144–5	1.539^{0}	...	i	...	...	...	...	os s	**$B1^{1}$**, 44
p146	**—,3-iodo-***	$(C_2H_5)_2CHI$	198.05	...	...	145–6	1.528^{0}	1.4967^{20}	i	...	...	...	...	os s	**$B1^{1}$**, 44
p147	**—,1-isocyano-**	*n*-Amylcarbylamine. Amyl isocyanide. $CH_3(CH_2)_4NC$	97.16	...	−51.1	155.5	0.806^{20}_{4}	...	i	s	...	...	...	...	...
p148	**—,2-methyl-***	$CH_3CH_2CH_2CH(CH_3)_2$	86.18	...	−153.67	60.3	0.6532^{20}_{4}	1.3714^{20}	i	s	s	...	...	...	**$B1^{3}$**, 396
p149	**—,3-methyl-***	$(C_2H_5)_2CHCH_3$	86.18	...	...	63.265	0.6643^{20}_{4}	1.3765^{20}	i	s	δ	...	...	...	**$B1^{3}$**, 401
p150	**—,3-nitro-***	$(C_2H_5)_2CHNO_2$	117.15	...	...	153–5	0.957^{0}_{4}	...	i	s	s	...	...	...	...
p151	**—,2,2,4,4-tetra-methyl-***	$(CH_3)_3CCH_2C(CH_3)_3$	128.25	...	−67.0	122.3	0.7185^{20}_{4}	1.4069^{20}	i	v	...	...	...	os v	**$B1^{3}$**, 519
p152	**—,2,2,3-tri-methyl-***	$CH_3CH_2CH(CH_3)C(CH_3)_3$	114.23	...	−112.27	109.8	0.7161^{20}_{4}	1.4028^{20}	i	∞	...	...	s	...	**$B1^{3}$**, 491
p153	**—,2,2,4-tri-methyl-***	Isooctane. $(CH_3)_2CHCH_2C(CH_3)_3$	114.23	...	−107.41	99.2	0.6918^{20}_{4}	1.3915^{20}	i	δ	s	...	...	...	**B1**, 164
p154	**—,2,3,3-tri-methyl-***	$CH_3CH_2C(CH_3)_2CH(CH_3)_2$	114.23	...	−100.7	114.8	0.7262^{20}_{4}	1.4075^{20}	i	v	...	...	...	os v	**$B1^{3}$**, 499
p155	**—,2,3,4-tri-methyl-***	$(CH_3)_2CHCH(CH_3)CH(CH_3)_2$	114.23	...	−109.21	113.4	0.7191^{20}_{4}	1.4042^{20}	i	v	...	...	...	to, os v	**$B1^{3}$**, 500
p156	**Pentanedial***	Glutaraldehyde. Glutaric aldehyde. $OCH(CH_2)_3CHO$	100.12	...	...	187–9d $71-2^{10}$	...	...	∞	∞	...	...	...	...	**$B1^{2}$**, 831
p157	—,dioxime	Glutaraldoxime. $HON{:}CH(CH_2)_3CH.NOH$	130.15	nd (w)	178	sub	...	...	s^{h}	...	...	...	...	Py s^{h}	**$B1^{2}$**, 831
p158	**Pentanedioic acid***	Glutaric acid. $HO_2C(CH_2)_3CO_2H$	132.11	mcl pr	97.5	302–4δd 200^{20}	1.424^{25}_{4}	1.4188^{106}	v	v	v	...	s	chl s con sulf s lig δ	**$B2^{2}$**, 564
p159	—,dichloride	Glutaryl chloride. $ClCO(CH_2)_3COCl$	169.02	...	...	216–8 cor 100^{15}	1.324^{20}_{4}	1.4728^{20}	d	d	s	...	...	...	**$B2^{2}$**, 566
p160	—,diethyl ester*	Diethyl glutarate. $C_2H_5O_2C(CH_2)_3CO_2C_2H_5$	188.23	syr	−24.1 cor	236.5–7	1.0220^{20}_{4}	1.4241^{20}	δ	δ	s	...	...	...	**$B2^{2}$**, 565
p161	—,dimethyl ester*	Dimethyl glutarate. $CH_3O_2C(CH_2)_3CO_2CH_3$	160.17	...	−42.5	214^{751} $84-5^{6}$	1.0876^{20}_{4}	1.4242^{20}	...	v	v	...	...	...	**$B2^{2}$**, 565
p162	—,dinitrile	Glutaronitrile. Pentanedinitrile*. $NC(CH_2)_3CN$	94.12	...	−29	286	0.995^{15}_{4}	1.4365^{23}	s	s	i	...	...	GS_2 i chl s	**$B2^{2}$**, 566
p163	—,diphenyl ester*	Diphenyl glutarate. $C_6H_5O_2C(CH_2)_3CO_2C_6H_5$	284.31	nd (lig)	54	236.5^{15}	...	...	i	s	...	...	...	os v lig v	**B6**, 156
p164	—,piperazinium salt	$2[HO_2C(CH_2)_3CO_2H].C_4H_{10}N_2$	350.37	...	152	...	...	...	s	s^{h}	i	...	...	...	...
p165	**—,2-acetyl-,** diethyl ester	$CH_3COCH(CO_2C_2H_5)CH_2CH_2CO_2C_2H_5$	230.26	...	...	271–2δd $110-2^{0.1}$	1.0712^{20}_{4}	1.4420^{19}	δ	s	s	...	...	...	**$B3^{2}$**, 488
—	**—,2-amino-***	*see* **Glutamic acid**													
p166	**—,2,2-dimethyl-***	$HO_2CCH_2CH_2C(CH_3)_2CO_2H$	160.17	nd (con HCl, bz-lig)	85	...	...	...	v	v	...	...	...	chl v lig δ	**$B2^{2}$**, 590
p167	**—,3,3-dimethyl-***	$HO_2CCH_2C(CH_3)_2CH_2CO_2H$	160.17	mcl pl, nd (bz)	103–4	...	...	...	v	v	v	...	δ s^{h}	lig i	**$B2^{2}$**, 592
p168	**—,3[2(3,5-di-methyl-2-oxo-cyclohexyl)2-hydroxy-ethyl]-, imide**	Actidone. Cycloheximide. CH_3 O O CH_3 —CHCH$_2$— NH OH O	281.35	wh	115–7	...	...	...	δ	s	s	s	...	...	...
p169	**—,2-ethyl-3-methyl-***	$HO_2CCH_3CH(CH_3)CH(C_2H_5)CO_2H$	175.21	pr (w)	100–1	...	...	...	v	...	v	...	...	...	**B2**, 701
p170	**—,2-ethyl-4-methyl-***(low m.p.)	Mesomethylethylglutaric acid. $HO_2CCH(CH_3)CH_2CH(C_2H_5)CO_2H$	175.21	nd (w)	61	...	...	...	s^{h}	...	...	...	...	os v	**B2**, 699
p171	—(high m.p.)	Paramethylethylglutaric acid. $HO_2CCH(CH_3)CH_2CH(C_2H_5)CO_2H$	175.21	nd (w)	105	...	...	...	s^{h}	...	...	...	i	os v lig s CS_2 i	**B2**, 699
p172	**—,3-ethyl-3-methyl-***	$HO_2CCH_2C(CH_3)(C_2H_5)CH_2CO_2H$	175.21	nd (w or bz), pl (bz-peth)	86	260^{740}	...	...	s^{h}	s	s	...	s^{h}	...	**$B2^{2}$**, 600
p173	**—,2-hydroxy-(*d*)***	$HO_2CCH_2CH_2CHOHCO_2H$	148.12	(eth), $[\alpha]^{19}_{D}$ 1.76 (w, 3%)	72	...	...	...	...	...	...	...	...	...	**$B3^{1}$**, 447

For explanations, symbols and abbreviations see beginning of table.

No.	Name	Synonyms and Formula	Mol. wt.	Crystalline form, color and specific rotation	m.p. °C	b.p. °C	Density	n_D	Solubility w	al	eth	ace	bz	other solvents	Ref.
	Pentanedioic acid														
p174	—,—(*dl*)*	$HO_2CCH_2CH_2CHOHCO_2H$	148.12	pr (AcOEt)	98–100d				s	s					**B3**², 293
p175	—,—(*l*)*	$HO_2CCH_2CH_2CHOHCO_2H$	148.12	($[\alpha]_D -1.98$ (18.8% weight /weight)	72–3										**B3**², 93
p177	**—,2-oxo-***	α-Ketoglutaric acid. $HO_2CCH_2CH_2COCO_2H$	146.10	(ace-bz)	115–6				v	v	v				**B3**², 481
p178	**—,3-oxo-***	Acetonedicarboxylic acid. β-Ketoglutaric acid. $HO_2CCH_2COCH_2CO_2H$	146.10	nd (al), rh (w)	138	d			v	v	δ		i	chl i lig i	**B3**², 482
p179	—,diethyl ester*	$C_2H_5O_2CCH_2COCH_2CO_2C_2H_5$	202.21			250	1.113^{20}_{4}		δ	∞					**B3**², 484
p181	**—,2-phenyl-***	$C_6H_5CH(CO_2H)CH_2CH_2CO_2H$	208.22	(bz or eth-peth)	82–3						s^h		s^h		**B9**, 877
p182	—,—,anhydride*	O O O C_6H_5	190.20	nd (eth)	95	$218–30^{13}$				s^h					**B17**, 494
p183	**—,3-phenyl-,** anhydride*	O C_6H_5— O O	190.20	(bz)	105	$217–9^{15}$					s		s	chl s lig i	**B17**², 476
p184	**—,2,3,4-tri-hydroxy-(*d*)***	*D*-Arabotrihydroxyglutaric acid. $HO_2C(CHOH)_3CO_2H$	180.11	(ace) $[\alpha]_D +22$	127				v	s		s^h			**B3**¹, 192
p185	—,—(*dl*)*	$HO_2C(CHOH)_3CO_2H$	180.11		154.5d				v	v	i	δ		chl i	**B3**, 553
p186	**1,2-Pentanediol** (*d*)*	1,2-Amylene glycol. $CH_3(CH_2)_2CHOHCH_2OH$	104.15	$[\alpha]^{20}_D +0.95$		210.5–211.5^{751}	0.9802^{20}_{20}	1.4412^{19}							
p187	**1,4-Pentanediol***	γ-Pentylene glycol. $CH_3CHOHCH_2CH_2CH_2OH$	104.15			211.8 134^{32}	0.9960^{17}_{4}	1.4439^{17}	∞	∞				chl ∞	**B1**¹, 250
p188	**1,5-Pentanediol***	Pentamethylene glycol. $HO(CH_2)_5OH$	104.15		−18	260	0.9939^{20}_{20}	1.4498	s	s	δ				**B1**, 481
p189	**2,3-Pentanediol***	β-*n*-Amylene glycol. $CH_3CH_2CHOHCHOHCH_3$	104.15			187.5	0.9800^{19}_{0}		s	s	δ				**B1**, 481
p190	**1,3-Pentanediol-, 2,2-dimethyl-***	$CH_3CH_2CHOHC(CH_3)_2CH_2OH$	132.21	(eth)	60–3	212–4 119^{21}				s					**B1**, 190
p191	**1,5-Pentanediol, 2,2-dimethyl-***	$HO(CH_2)_3C(CH_3)_2CH_2OH$	132.21			130^{12}				s	s				**B1**¹, 254
p192	**2,4-Pentanediol, 2-methyl-***	$CH_3CHOHCH_2COH(CH_3)_2$	118.19			$190–4^{740}$	0.9254^{17}_{4}	1.4311^{17}	s	s	s				**B1**¹, 252
p193	**—,3-methyl-***	$CH_3CHOHCH(CH_3)CHOHCH_3$	118.19			221–2 91^{3}	0.964^{20}	1.4433^{20}	s	s					**B1**³, 2209
p194	**1,2-Pentanediol, 2,4,4-trimethyl-***	$(CH_3)_3CCH_2COH(CH_3)CH_2OH$	146.23	pr or pl (bz)	62–3				s	v	v		s^h		**B1**³, 2226
p195	**1,3-Pentanediol, 2,2,4-trimethyl-***	$(CH_3)_2CHCHOHC(CH_3)_2CH_2OH$	146.23		51.8–2.2	234^{737} $81–2^{1}$	0.937^{15}_{15}	1.4513^{15}	δ	v	v				**B1**³, 2225
p196	**1,4-Pentanediol, 2,2,4-trimethyl-***	$(CH_3)_2COHCH_2C(CH_3)_2CH_2OH$	146.23	(eth)	86	209–11 $114–5^{13}$				v	s				**B1**, 493
p197	**2,3-Pentanediol, 2,4,4-trimethyl-***	$CH_3C(CH_3)_2CHOHC(CH_3)OHCH_3$	146.23	mcl pr (lig)	65–6					v	s			lig s^h	**B1**³, 2225
p198	**2,3-Pentane-dione***	Acetylpropionyl. Ethyl methylglyoxal. $CH_3CH_2COCOCH_3$	100.12	dk ye		108	0.9565^{19}_{4}	1.4014^{19}	s	∞	∞	∞			**B1**², 831
p199	—,dioxime*	2,3-Diisonitrosopentane. $CH_3CH_2C(:NOH)C(:NOH)CH_3$	130.15	ye nd or pl	172–3	sub			i	s	δ			to s^h	**B1**², 831
p200	—,2-oxime*	$CH_3CH_2COC(:NOH)CH_3$	115.14	pl	69–72					v	v				**B1**, 776
p201	—,3-oxime*	$CH_3CH_2C(:NOH)COCH_3$	115.14	pl (lig)	58–9	183–7			δ	v	v			chl v	**B1**², 831
p202	**2,4-Pentane-dione***	Acetylacetone. Diacetyl-methane. $CH_3COCH_2COCH_3$	100.11		−23	139^{746}	0.9721^{25}_{4}	1.4541^{17}	v	∞	∞			chl ∞	**B1**, 777
p203	—,dioxime*	2,4-Diisonitrosopentane. $CH_3C(:NOH)CH_2C(:NOH)CH_3$	130.15	pr (eth)	149–50				δ	s	δ				**B1**², 838
p204	—,monoimide	$CH_3COCH_2C(:NH)CH_3$	99.13		43	209			v		v				**B1**², 838
p205	**—,3,3-dimethyl-***	$CH_3COC(CH_3)_2COCH_3$	128.17		19	173 58^{10}	0.9564^{15}_{15}	1.4332^{20}			s				**B1**², 844
p206	**—,3-ethyl-***	$CH_3COCH(C_2H_5)COCH_3$	128.17			$69–70^{13}$	0.9531^{19}_{4}	1.4408^{19}_{4}		s	s			chl s	**B1**², 844

For explanations, symbols and abbreviations see beginning of table.

No.	Name	Synonyms and Formula	Mol. wt.	Crystalline form, color and specific rotation	m.p. °C	b.p. °C	Density	n_D	Solubility w	al	eth	ace	bz	other solvents	Ref.
	1,5-Pentanedione														
p207	**1,5-Pentanedione, 1,2,3,4,5-pentaphenyl-***	Benzamaron. Benzylidene-bis-desoxybenzoin. $C_6H_5CH[CH(C_6H_5)COC_6H_5]_2$	480.61	lf (w)	218	...	...	...	δ	δ	...	...	s	...	**B7**[2], 803
p208	**1,4-Pentanedione, 1-phenyl-***	γ-Oxovalerophenone. Phenacylacetone. $C_6H_5COCH_2CH_2COCH_3$	176.22	ye oil	...	162^{12}	...	...	s^h	...	...	...	...	alk i	**B7**[1], 368
p209	**1-Pentanethiol***	*n*-Amyl mercaptan. *n*-Thioamyl alcohol. $CH_3(CH_2)_3CH_2SH$	104.22	...	−75.7	126.5	0.8375^{25}	1.4459^{20}	i	∞	∞	...	...	...	**B1**[3], 1607
p210	**2-Pentanethiol***	(α-Methylbutyl) mercaptan. $CH_3CH_2CH_2CHSHCH_3$	104.22	...	−169	112.9 63.9^{150}	0.8327^{20}_{4}	1.4386^{25}	...	s	...	...	...	lig s	**B1**[3], 1615
p211	**3-Pentanethiol***	(α-Ethylpropyl) mercaptan. $(C_2H_5)_2CHSH$	104.22	...	−110.8	105	0.8410^{20}_{4}	1.4447^{20}	...	s	...	...	...	...	**B1**[3], 1619
p212	**1,3,5-Pentanetricarboxylic acid, 3-acetyl-3-ethyl ester, 1,5-dinitrile***	$CH_3COC(CH_2CH_2CN)_2CO_2C_2H_5$	236.27	...	81–2	$190–200^{2}$	...	...	i	s^h	...	...	...	...	**B3**[2], 512
p213	**1,2,3-Pentanetriol***	$CH_3CH_2CHOHCHOHCH_2OH$	120.15	syr	...	192^{63}	1.0851^{34}_{0}	...	∞	∞	s	...	...	...	**B1**[3], 2346
p214	**Pentanoic acid***	Propylacetic acid, Valerianic acid. Valeric acid. $CH_3(CH_2)_3CO_2H$	102.13	...	−34.5	186–7 $86–8^{15}$	0.939^{20}_{4}	1.4086^{20}	s	s	s	...	...	...	**B2**[2], 263
p215	—,amide	Pentanamide*. $CH_3(CH_2)_3CONH_2$	101.15	mica-like mcl pl (peth)	101	...	1.023	...	s	s	s	...	...	...	...
p216	—,—,*N,N*-dimethyl	$CH_3(CH_2)_3CON(CH_3)_2$	129.20	...	−51	141^{100}	1.4419^{25}	...	∞	∞	∞	...	...	...	...
p217	—,—,*N*-phenyl-	Valeranilide. $CH_3(CH_2)_3CONHC_6H_5$	177.24	mcl pr	63	...	...	...	...	...	s	...	...	...	...
p218	—,anhydride*	Valeric anhydride. $[CH_3(CH_2)_3CO]_2O$	186.25	...	−56.1	215	0.924^{20}_{4}	...	d^h	s	v	...	...	...	...
p219	—,butyl ester*	*n*-Butyl valerate. $CH_3(CH_2)_3CO_2(CH_2)_3CH_3$	158.24	...	−92.8	185.8	0.8700^{15}_{4}	1.4126^{15}	δ	s	s	...	...	...	**B2**[2], 266
p220	—,*sec*-butyl ester-(*d*)*	*d*-*sec*-Butyl valerate. $CH_3(CH_2)_3CO_2CH(CH_3)C_2H_5$	158.24	$[\alpha]^{20}_{D}+20.72$	...	174.5 67^{18}	0.8605^{20}_{4}	1.4081^{20}	i	s	s	...	s	Py s	**B2**[1], 131
p221	—,chloride	Pentanoyl chloride*. Valeryl chloride. $CH_3(CH_2)_3COCl$	120.58	...	−110.0	127	1.016^{15}	...	...	...	...	...	...	...	...
p222	—,ethyl ester*	Ethyl valerate. $CH_3(CH_2)_3CO_2C_2H_5$	130.19	...	−91.2	145–6	0.877^{20}_{4}	1.3732^{20}	i	∞	∞	...	...	...	...
p223	—,furfuryl ester*	Furfuryl valerate. $CH_3(CH_2)_3CO_2CH_2$—(furan ring, O)	182.22	...	...	$228–9^{764}$ $82–3^{1}$	1.0284^{20}_{4}	...	i	s	s	...	...	...	**B17**[2], 115
p224	—,heptyl ester*	*n*-Heptyl valerate. $CH_3(CH_2)_3CO_2(CH_2)_6CH_3$	200.32	...	−46.4	245.2	0.8623^{20}	...	i	s	...	...	...	os s	**B2**, 301
p225	—,hexyl ester*	*n*-Hexyl valerate. $CH_3(CH_2)_3CO_2(CH_2)_5CH_3$	186.30	...	−63.1	226.3	0.8635^{20}	...	i	s	...	...	...	os s	**B2**, 301
p226	—,isobutyl ester	Isobutyl valerate. $CH_3(CH_2)_3CO_2CH_2CH(CH_3)_2$	158.24	...	170–2	0.853^{20}	...	1.4046^{20}	i	∞	...	...	...	...	...
p227	—,isopropyl ester*	Isopropyl valerate. $CH_3(CH_2)_3CO_2CH(CH_3)_2$	144.22	...	...	153.5	0.8579^{20}_{4}	1.4009^{20}	i	s	s	...	...	...	**B2**[2], 266
p228	—,methyl ester*	Methyl valerate. $CH_3(CH_2)_3CO_2CH_3$	116.16	...	...	127.3	0.9097	...	δ	∞	∞	...	...	...	...
p229	—,nitrile	Pentanonitrile*. Valeronitrile. $CH_3(CH_2)_3CN$	83.13	...	−96	141.25	0.7949^{25}	1.3917^{18}	...	...	...	...	...	...	**B2**[2], 267
p230	—,octyl ester*	*n*-Octyl valerate. $CH_3(CH_2)_3CO_2(CH_2)_7CH_3$	214.35	...	−42.3	261.6	0.8615^{20}	1.4273^{15}	i	s	...	...	...	os s	**B2**, 301
p231	—,pentyl ester*	*n*-Amyl valerate. $CH_3(CH_2)_3CO_2(CH_2)_4CH_3$	172.27	...	−78.8	203.7	0.858^{19}_{4}	1.413^{19}	δ	∞	∞	...	...	...	...
p232	—,phenylphenacyl ester	$CH_3(CH_2)_3CO_2CH_2CO$—C_6H_4—C_6H_5	296.37	...	63–3.5	...	...	...	...	...	...	...	...	...	...
p233	—,piperazinium salt	$2C_4H_9CO_2H.C_4H_{10}N_2$	290.41	wh	112–5.3	...	...	...	s	s	i	...	...	diox s^h	...

For explanations, symbols and abbreviations see beginning of table.

No.	Name	Synonyms and Formula	Mol. wt.	Crystalline form, color and specific rotation	m.p. °C	b.p. °C	Density	n_D	Solubility: w	al	eth	ace	bz	other solvents	Ref.
	Pentanoic acid														
p234	—,propyl ester*	*n*-Propyl valerate. $CH_3(CH_2)_3CO_2CH_2CH_2CH_3$	144.22	...	...	167.5	0.8888^0	...	i	s	s	...	...	chl s	...
p235	**—,2-acetyl-**, ethyl ester	Ethyl propylacetoacetate. $CH_3COCH(C_3H_7^n)CO_2C_2H_5$	172.23	...	...	$90.2–.5^{10}$	0.9682^{15}_4	1.4271	i	s	...	...	...	...	**B3**², 441
p236	**—,2-amino-** (*D*,−)*	$CH_3CH_2CH_2CHNH_2CO_2H$	117.15	lf $[\alpha]^{20}_D -24.2$ (20% HCl, c=2)	ca. 307	...	...	...	v^h s	i	i	...	...	chl i lig i	...
p237	—,—(*DL*)*	Norvaline. $CH_3CH_2CH_2CHNH_2CO_2H$	117.15	lf (al)	303	sub	...	...	s v^h	i	i	...	...	chl i lig i	**B4**², 843
p238	—,—(*L*,+)*	$CH_3CH_2CH_2CHNH_2CO_2H$	117.15	(dil al) $[\alpha]^{20}_D$ +23 (20% HCl, c=10)	ca. 305	...	...	...	v^h	i	i	...	...	chl i lig i	...
p239	**—,4-amino-**(*D*)*	$CH_3CHNH_2CH_2CH_2CO_2H$	117.15	(aq al), $[\alpha]^{20}_D$ +12.0 (w, p=10)	214 cor	...	...	...	...	...	...	...	...	...	**B4**², 843
p240	—,—(*DL*)*	$CH_3CHNH_2CH_2CH_2CO_2H$	117.15	...	193d	d	...	...	v	i	i	...	i	lig i	**B4**², 843
p241	**—,5-amino-***	$H_2N(CH_2)_4CO_2H$	117.15	lf	157–8d	d	...	...	v	δ	i	...	...	...	**B4**², 844
—	—,—,lactam*	*see* **2-Piperidone**													
—	**—,2-amino-5-guanidino-***	*see* **Arginine**													
—	**—,2-amino-4-methyl-***	*see* **Leucine**													
—	**—,2-amino-3-sulfo-***	*see* **Cysteic acid**													
p242	**—,2-bromo-***	$CH_3CH_2CH_2CHBrCO_2H$	181.04	...	...	67^{10}	...	...	δ	v	s	...	...	...	**B2**², 268
p243	—,—,ethyl ester*	$CH_3CH_2CH_2CHBrCO_2C_2H_5$	209.09	...	...	190–2 $92–4^{18}$	1.226^{18}_4	...	i	s	s	...	...	...	**B2**, 302
p244	**—,2-bromo-4-methyl-**(*d*)*	$(CH_3)_2CHCH_2CHBrCO_2H$	195.07	oil $[\alpha]^{20}_D$ +29.8 (eth, c=6) 26.8 (20% dil al, c=8)	...	...	...	...	...	v	v	...	...	...	**B2**², 290
p245	—,—(*l*)*	$(CH_3)_2CHCH_2CHBrCO_2H$	195.07	oil $[\alpha]^{20}_D -121$	...	$94^{0.2-0.4}$ cor	...	...	...	v	v	...	...	...	**B2**², 291
p246	—,—,amide(*d*)*	$(CH_3)_2CHCH_2CHBrCONH_2$	194.09	(w) $[\alpha]^{20}_D -48.3$ (al, c=5.9)	118 cor	...	...	...	v^h	s	s	...	...	chl s AcOEt s	**B2**², 291
p247	—,—,ethyl ester*	$(CH_3)_2CHCH_2CHBrCO_2C_2H_5$	223.12	pa ye	...	202–4 $100–3^{17}$	...	...	...	...	...	...	...	...	**B2**¹, 142
p248	**—,2-chloro-**(*dl*)*	$CH_3CH_2CH_2CHClCO_2H$	136.58	...	−15	222^{763} $132–5^{32}$	1.141^{13}	1.4481^{11}	i	s	s	...	...	...	**B2**, 302
p249	—,—,chloride(*dl*)	$CH_3CH_2CH_2CHClCOCl$	155.03	...	...	$155–7^{763}$	1.246	...	d	d	s	...	...	...	**B2**, 302
p250	—,—,ethyl ester*	$CH_3CH_2CH_2CHClCO_2C_2H_5$	164.64	...	...	185^{752}	1.040^{12}	1.4307^{11}	i	s	s	...	...	...	**B2**, 302
p251	**—,3-chloro-***	$CH_3CH_2CHClCH_2CO_2H$	136.58	...	33	112^{10}	1.1484^{20}_4	1.4462^{20}	...	...	s	...	...	...	...
p252	—,—,ethyl ester*	$CH_3CH_2CHClCH_2CO_2C_2H_5$	164.64	...	...	189	1.0330^{20}_4	1.4278^{20}	i	s	s	...	...	...	...
p253	**—,4-chloro-***	$CH_3CHClCH_2CH_2CO_2H$	136.58	...	...	117^{10}	1.1514^{20}	1.4458^{20}	...	...	s	...	...	...	**B2**¹, 131
p254	—,—,chloride	$CH_3CHClCH_2CH_2COCl$	155.03	...	...	61^8	...	...	d	d	v	...	...	...	**B2**¹, 132
p255	—,—,ethyl ester*	$CH_3CHClCH_2CH_2CO_2C_2H_5$	164.64	...	...	196^{760} 70.5^9	1.0393^{20}_4	1.4310^{20}	i	s	s	...	...	...	**B2**¹, 131
p256	**—,5-chloro-***	$Cl(CH_2)_4CO_2H$	136.58		18	230^{760} $141–9^{12}$	1.3416^{25}_4	1.4859^{25}	s v^h	s	v	...	...	...	**B2**, 203
p257	—,—,chloride	$Cl(CH_2)_4COCl$	155.03	...	...	$75–80^{5-8}$	...	...	d	d	s	...	...	...	...
p258	—,—,ethyl ester	$Cl(CH_2)_4CO_2C_2H_5$	164.64	...	...	205–6 93^{16}	...	1.4355^{20}	i	s	s	...	...	...	**B2**², 268
—	**—,2,5-diamino-***	*see* **Ornithine**													
p259	**—,4,4-dimethyl-**, chloride	Neopentylacetyl chloride. $(CH_3)_3CCH_2CH_2COCl$	148.64	...	...	94^{100}	...	1.4294^{20}	d	d	s	...	...	...	...
p260	**—,2,4-dioxo-***	Acetoneoxalic acid. Acetopyruvic acid. $CH_3COCH_2COCO_2H$	130.10	nd (bz-chl), pr (bz)	55–63 (+1w) 98–101δd (anh)	...	...	...	s	s	s	s	s	AcOEt s chl s lig i	**B3**², 465
p261	—,—,ethyl ester*	Ethyl acetoneoxalate. $CH_3COCH_2COCO_2C_2H_5$	158.16	...	18	213–5 $113–6^{19}$	1.1251^{20}	1.4757^{17}	...	s	s	...	...	...	**B3**², 465

For explanations, symbols and abbreviations see beginning of table.

No.	Name	Synonyms and Formula	Mol. wt.	Crystalline form, color and specific rotation	m.p. °C	b.p. °C	Density	n_D	Solubility w	al	eth	ace	bz	other solvents	Ref.
	Pentanoic acid														
p262	—,**2-ethyl-***	3-Hexanecarboxylic acid*. $CH_3CH_2CH_2CH(C_2H_5)CO_2H$	130.18			209.2			i	s	s				**B2**², 299
p263	—,—,chloride	$CH_3CH_2CH_2CH(C_2H_5)COCl$	148.64			158–60 50^{11}			d		s				**B2**, 344
p264	—,**3-ethyl-***	$(C_2H_5)_2CHCH_2CO_2H$	130.18			212									**B2**, 344
p265	—,**5-fluoro-***	$F(CH_2)_4CO_2H$	120.12			83^{2}		1.4080^{25}	δ	v	v				
p266	—,**2-hydroxy-***	Valerolactic acid. $CH_3CH_2CH_2CHOHCO_2H$	118.13	hyg pl	31	sub			s	s	s				**B3**², 225
p267	—,**4-hydroxy-**, lactone(*d*)*	γ-Valerolactone. 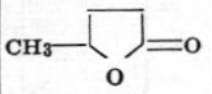	100.11	$[\alpha]^{20}_{D}+13.5$		$86–90^{14}$									**B17**², 288
p268	—,—,(*dl*)*	$C_5H_8O_2$. *See* p267	100.11		−31	206	1.0465^{25}	1.4305^{25}	∞						**B17**², 288
p269	—,—,(*l*)*	$C_5H_8O_2$. *See* p267	100.11	$[\alpha]^{20}_{D}+4.6$ (eth, c=10)		$78–80^{8}$									**B17**², 288
p270	—,**5-hydroxy-**, lactone	δ-Valerolactone. 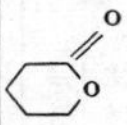	100.11		−12.5	219–22	1.0794^{20}	1.4503^{20}_{D}	δ	v	v				**B17**, 235
—	—,**2-hydroxy-4-methyl-***	*see* **Leucic acid**													
p271	—,**2-hydroxy-2-propyl-**, nitrile	4-Heptanone cyanohydrin. $(CH_3CH_2CH_2)_2C(OH)CN$	141.21			$119–20^{21}$	0.9077^{18}	1.4337^{18}		s					**B3**², 238
p282	—,**2-isobutyl-4-methyl-**, amide	Diisobutylacetamide. $[(CH_3)_2CHCH_2]_2CHCONH_2$	171.28	lf or nd (eth)	74–5				δ	v				os v lig δ	**B2**¹, 153
p283	—,**2-methyl-(*d*)***	Methylpropylacetic acid. $CH_3CH_2CH_2CH(CH_3)CO_2H$	116.16	$[\alpha]^{25}_{D}+5.58$ (eth)		96^{15}	0.908^{25}_{4}	1.4112^{25}							**B2**², 288
p284	—,—(*dl*)*	$CH_3CH_2CH_2CH(CH_3)CO_2H$	116.16			193^{748} cor	0.9230^{20}_{4}	1.4136	s	s	s				**B2**², 288
p285	—,—(*l*)*	$CH_3CH_2CH_2CH(CH_3)CO_2H$	116.16	$[\alpha]^{25}_{D}-7.08$ (eth)		96^{15}	0.920^{25}_{4}	1.4117^{25}							**B2**², 288
p286	—,—,chloride(*dl*)	$CH_3CH_2CH_2CH(CH_3)COCl$	134.61			140.0–0.8^{745}	0.9781^{20}_{4}		d^{h}	d^{h}	s				
p287	—,**3-methyl-(*d*)***	*sec*-Butylacetic acid. $CH_3CH_2CH(CH_3)CH_2CO_2H$	116.16	$[\alpha]^{26}_{D}+4.01$		197.4–8	0.930^{15}								**B2**², 241
p288	—,—(*dl*)*	$CH_3CH_2CH(CH_3)CH_2CO_2H$	116.16		−41.6	197.2–.8	0.9262^{20}	1.4159^{20}		s	s				**B2**², 291
p289	—,—(*l*)*	$CH_3CH_2CH(CH_3)CH_2CO_2H$	116.16	$[\alpha]^{26}_{D}-2.54$		110^{150}	0.923^{26}_{4}								
p290	—,—,chloride(*dl*)	*sec*-Butylacetyl chloride. $CH_3CH_2CH(CH_3)CH_2COCl$	134.61			142.5–3^{749}	0.9781^{20}_{4}		d	d	v			CS_2 v	
p291	—,**4-methyl-***	Isobutylacetic acid. Isocaproic acid. $(CH_3)_2CHCH_2CH_2CO_2H$	116.16		−33.0	200–1	0.9225^{20}_{4}	1.4144^{20}	δ	s	s				**B2**², 289
p292	—,—,amide	Isocaproamide. $(CH_3)_2CHCH_2CH_2CONH_2$	115.18	nd (al)	119–21				s	s^{h}					**B2**², 290
p293	—,—,—,*N*-phenyl-	Isocaproanilide. $(CH_3)_2CHCH_2CH_2CONHC_6H_5$	191.28	nd (bz, dil al)	110–2				i	s^{h}	s		s^{h}		**B12**², 147
p294	—,—,chloride	Isocaproyl chloride. $(CH_3)_2CHCH_2CH_2COCl$	134.61			141–2	0.9697^{20}		d^{h}	d^{h}					**B2**², 289
p295	—,—,nitrile	Isoamyl cyanide. Isocapronitrile. $(CH_3)_2CHCH_2CH_2CN$	97.16		−51	155–6	0.8038^{20}_{4}	1.4048^{22}	i	s	∞				**B2**², 290
p296	—,**4-methyl-2-phenyl-***	α-Phenylisocaproic acid. $(CH_3)_2CHCH_2CH(C_6H_5)CO_2H$	192.25	pr (peth)	78–9	$178–80^{30}$				s				alk s lig δ	**B9**², 369
p297	—,—,nitrile	α-Phenylisocapronitrile. $(CH_3)_2CHCH_2CH(C_6H_5)CN$	173.25			$263–6^{765}$ $136–8^{15}$	0.942^{16}		i	s	s		s	aa s	**B9**¹, 220
p298	—,**3-oxo-**, ethyl ester*	Ethyl propionylacetate. $CH_3CH_2COCH_2CO_2C_2H_5$	144.17			$91–3^{17}$		1.4183^{23}		s	s		s		**B3**², 430
p299	—,**4-oxo-***	β-Acetopropionic acid. Levulinic acid. $CH_3COCH_2CH_2CO_2H$	116.11	lf or pl	37.2	245–6 δd $137–9^{10}$	1.1335^{20}_{4}	1.4429^{17}	v	v	v				**B3**², 430
p299¹	—,—,benzyl ester	Benzyl levulinate. $CH_3COCH_2CH_2CO_2CH_2C_6H_5$	206.23			$132–4^{2}$	1.0935^{20}_{4}	1.5090^{20}						to s	

For explanations, symbols and abbreviations see beginning of table.

No.	Name	Synonyms and Formula	Mol. wt.	Crystalline form, color and specific rotation	m.p. °C	b.p. °C	Density	n_D	Solubility w	al	eth	ace	bz	other solvents	Ref.
	Pentanoic acid														
p300	—,—,butyl ester*	Butyl levulinate. $CH_3COCH_2CH_2CO_2(CH_2)_3CH_3$	172.22	...	...	237.8	0.9735^{20}_{4}	1.4201^{20}	...	s	s	s	...	...	...
p301	—,—,*sec*-butyl ester	*sec*-Butyl levulinate. $CH_3COCH_2CH_2CO_2CH(CH_3)CH_2CH_3$	172.22	...	...	225.8	0.9669^{20}_{4}	1.4249^{20}	...	s	s	s	...	...	...
p302	—,—,isobutyl ester	Isobutyl levulinate. $CH_3COCH_2CH_2CO_2CH_2CH(CH_3)_2$	172.22	...	...	230.9	0.9677^{20}_{4}	1.4268^{20}	...	s	s	s	...	...	...
p303	—,—,isopropyl ester*	Isopropyl levulinate. $CH_3COCH_2CH_2CO_2CH(CH_3)_2$	158.19	...	...	209.3	0.9872^{20}_{4}	1.4209^{20}	...	s	s	s	...	...	...
p304	—,—,methyl ester*	Methyl levulinate. $CH_3COCH_2CH_2CO_2CH_3$	130.14	...	...	191^{743} $85–6^{14}$	1.05113^{20}_{4}	1.4233^{20}	δ	s	∞	s	...	...	**B3**^{2}, 431
p305	—,—,propyl ester*	Propyl levulinate. $CH_3COCH_2CH_2CO_2CH_2CH_2CH_3$	158.19	...	...	221.2	0.9937^{20}_{20}	1.4258^{20}	...	s	s	s	...	...	**B3**, 675
p306	**—,5-oxo-5-phenyl-***	γ-Benzoylbutyric acid. $C_6H_5CO(CH_2)_3CO_2H$	192.22	(w)	128–9	...	...	...	s^h	...	...	...	...	...	**B10**^{2}, 488
p307	**—,2-phenyl-(*d*)***	α-Phenylvaleric acid. $CH_3CH_2CH_2CH(C_6H_5)CO_2H$	178.23	$[\alpha]_D$ +58.81 (chl)	...	165^{14}	1.047^{20}_{4}	...	...	s	s	...	...	chl s	**B9**, 557
p308	—,—(*dl*)*	$CH_3CH_2CH_2CH(C_6H_5)CO_2H$	178.23	nd (lig)	52	...	...	...	...	s	...	...	...	chl s lig s^h	**B9**, 557
p309	—,—,nitrile	$CH_3CH_2CH_2CH(C_6H_5)CN$	159.22	...	...	$254–5^{750}$ $125–8^{13}$	0.960^{15}	...	i	s	...	...	s	...	**B9**^{1}, 216
p310	**—,4-phenyl-***	γ-Phenylvaleric acid. $C_6H_5CH(CH_3)CH_2CH_2CO_2H$	178.23	...	ca. 13	210^{85} 147^{2}	1.0554^{15}_{4}	...	...	s	...	...	...	...	**B9**^{2}, 362
p311	**—,5-phenyl-***	δ-Phenylvaleric acid. $C_6H_5(CH_2)_4CO_2H$	178.23	pl (w), pr (peth)	57–8 (61)	$190–3^{30}$	...	...	$δ^h$	v	...	...	...	os s	**B9**^{1}, 216
p312	**—,2,2,4-tri-methyl-, amide**	$(CH_3)_2CHCH_2C(CH_3)_2CONH_2$	143.23	lf (lig)	71	...	...	...	...	s^h	...	...	...	lig δ	**B2**^{2}, 305
p313	**1-Pentanol***	Butyl carbinol. *n*-Amyl alcohol. $CH_3(CH_2)_4OH$	88.15	...	−79	137.3^{748}	0.8110^{25}_{4}	1.4101^{20}	i	∞	∞	...	...	...	**B1**, 385
p314	**2-Pentanol***	Methyl *n*-propyl carbinol. $CH_3CH_2CH_2CHOHCH_3$	88.15	...	...	118.5–9.5	0.8103^{20}_{4}	1.4053^{20}	v	s	s	...	...	...	**B1**, 384
p315	**3-Pentanol***	Diethyl carbinol. $(CH_3CH_2)_2CHOH$	88.15	...	...	115.5^{754}	0.8154^{25}	1.4077^{25}	δ	s	s	...	...	...	**B1**, 385
p316	**2-Pentanol, 1-allyloxy-3-amino-***	$CH_2:CHCH_2OCH_2CHOHCHNH_2CH_2CH_3$	159.23	...	...	$93–6^{2-3}$	1.0212^{20}_{4}	1.4714^{20}	s	s	...	s	...	lig δ	...
p317	**—,1-amino-2-methyl-***	$CH_3(CH_2)_2C(CH_3)(OH)CH_2NH_2$	117.19	...	...	$80–5^{15}$	0.9081^{20}_{4}	1.4510^{20}	s	s	...	s	...	lig i	...
p318	**1-Pentanol, 5-chloro-***	Pentamethylene chlorohydrin. $Cl(CH_2)_5OH$	122.60	...	...	$91–2^{11}$	...	1.4512^{25}	δ	v	v	...	...	...	...
p319	**2-Pentanol, 1-chloro-***	$CH_3CH_2CH_2CHOHCH_2Cl$	122.60	...	...	$157–60^{735}$ $59–62^{14}$	1.037^{20}_{20}	1.4404^{25}	...	s	v	...	...	...	**B1**^{2}, 419
p320	**3-Pentanol, 1-chloro-***	$CH_3CH_2CHOHCH_2CH_2Cl$	122.60	...	...	173	1.0327^{25}_{4}	1.466^{25}	δ s^h	v	v	...	...	...	**B1**^{2}, 421
p321	**1-Pentanol,2,4-dimethyl-(*dl*)***	$(CH_3)_2CHCH_2CH(CH_3)CH_2OH$	116.20	...	...	160–2 $65–7^{18}$	0.793^{20}_{4}	1.427^{20}	...	s	...	...	...	os s	**B1**^{3}, 1700
p322	—,—(*l*)*	$(CH_3)_2CHCH_2CH(CH_3)CH_2OH$	116.20	$[\alpha]^{22}_D$ −1.1	...	157	0.816^{25}_{4}	...	...	s	...	...	...	os s	**B1**^{3}, 1700
p323	**2-Pentanol, 2,4-dimethyl-***	Dimethyl isobutyl carbinol. $(CH_3)_2CHCH_2C(OH)(CH_3)_2$	116.20	...	<-20	133^{749}	0.8103^{20}_{4}	1.4172^{20}	i	s	s	...	...	...	**B1**, 417
p324	**3-Pentanol, 2,3-dimethyl-***	Ethyl isopropyl methyl carbinol. $(CH_3)_2CHC(OH)(CH_3)CH_2CH_3$	116.20	...	<-30	$138–40^{750}$	0.833^{20}_{4}	1.4287^{20}	i	s	s	...	...	...	**B1**, 417
p325	**3-Pentanol, 2,4-dimethyl-***	Diisopropyl carbinol. $(CH_3)_2CHCH(OH)CH(CH_3)_2$	116.20	...	<70	140	0.8288^{20}_{4}	1.4226^{20}	δ	s	s	...	...	...	**B1**,417
p326	**—,2,4-dimethyl-3-phenyl-***	Diisopropylphenyl carbinol. $[(CH_3)_2CH]_2COHC_6H_5$	192.30	ye	60.5	229 157^{60}	0.959	...	δ	i	s	...	...	...	**B6**, 273

For explanations, symbols and abbreviations see beginning of table.

No.	Name	Synonyms and Formula	Mol. wt.	Crystalline form, color and specific rotation	m.p. °C	b.p. °C	Density	n_D	Solubility w	al	eth	ace	bz	other solvents	Ref.
	3-Pentanol														
p327	**—,3-ethyl-***	Triethyl carbinol. $(C_2H_5)_3COH$	116.20			140–2	0.8389_0^{20}	1.4266^{22}	i	s	s				**B1**, 417
p328	**—,3-ethyl-2-methyl-***	Diethylisopropyl carbinol. $(CH_3)_2CHCOH(C_2H_5)_2$	130.23			$159–61^{750}$	0.8295^{20}		i	s	s				**B1**, 423
p329	**1-Pentanol, 5-fluoro-***	Pentamethylene fluorohydrin. $F(CH_2)_5OH$	106.14			$70–1^{11}$		1.4057^{25}	s	v	v				
p330	**—,2-methyl-***	Methyl *n*-amyl carbinol. $CH_3CH_2CH_2CH(CH_3)CH_2OH$	102.18			148^{762}	0.8192_4^{25}	1.4230^{19}							**B1**, 409
p331	**—,3-methyl-***	$CH_3CH_2CH(CH_3)CH_2CH_2OH$	102.18			153^{760}	0.822_4^{27}	1.4182^{25}	i	s	s				**B1**, 411
p332	**—,4-methyl-***	Isoamyl carbinol. Isohexyl alcohol. $(CH_3)_2CH(CH_2)_3OH$	100.18			152–3	0.8205_4^{25}	1.4177^{25}	i	s	s				**B1²**, 202
p334	**2-Pentanol, 2-methyl-***	Dimethyl *n*-propyl carbinol. $CH_3CH_2CH_2COH(CH_3)_2$	102.18	transparent mass	107–9	121.5	0.8350_4^{16}	1.4125^{16}	δ	s	s				**B1**, 409
p335	**—,3-methyl-***	*sec*-Butyl methyl carbinol. $CH_3CH_2CH(CH_3)CHOHCH_3$	102.18			134.32 75.6^{50}	0.8235_4^{25}	1.4179^{25}	δ	s	s				**B1³**, 1672
p336	**—,4-methyl-***	Methyl isobutyl carbinol. $(CH_3)_2CHCH_2CHOHCH_3$	102.18			130	0.8025_4^{25}	1.4090^{25}	δ	s	s				**B1**, 410
p337	—,—,acetate	4-Methyl-2-pentyl acetate. $(CH_3)_2CHCH_2CH(CH_3)O_2CCH_3$	144.22			147	0.8805_0^0								
p338	**3-Pentanol, 2-methyl-***	Ethyl isopropyl carbinol. $(CH_3)_2CHCHOHCH_2CH_3$	102.17			$129–30^{721}$	0.8230_4^{20}	1.4175^{20}	δ	∞	∞				**B16**, 1037
p339	**—,3-methyl-***	Diethyl methyl carbinol. $(CH_3CH_2)_2COHCH_3$	102.17		<−38	122.1–.9	0.8237_0^{20}	1.418^{21}	δ	∞	∞				**B1**, 411
p340	**1-Pentanol, 1-phenyl-(*d*)***	Butyl phenyl carbinol. $C_6H_5CH(OH)(CH_2)_3CH_3$	164.25	$[\alpha]_D^{20}+40.8$		$140–2^{25}$	0.9672_{20}^{20}	1.4086^{25}	i	s	s	s			**B6²**, 504
p341	**—,5-phenyl-***	$C_6H_5(CH_2)_5OH$	164.25			155^{20}			i	v	v				**B6²**, 505
p342	**2-Pentanol, 2-phenyl-***	Methyl phenyl propyl carbinol. $C_6H_5COH(CH_3)(CH_2)_2CH_3$	164.25			216 112^{14}	0.9723_4^{22}		i	v					**B6²**, 505
p343	**3-Pentanol, 1-phenyl-(*d*)***	$C_6H_5CH_2CH_2CHOHCH_2CH_3$	164.25	$[\alpha]_D^{20}+26.8$ (al), +30.2 (CS_2)	38	143^{19}	0.9687_4^{20}		i	v				CS_2 v	**B6²**, 404
p344	**—,3-phenyl-***	Diethyl phenyl carbinol. $(C_2H_5)_2COHC_6H_5$	164.25		<−17	$223–4^{762}$ 110^{12}	0.9831_4^{20}	1.5182^{20}	i	v					**B6²**, 506
p345	**2-Pentanol, 2,4,4-trimethyl-***	Isodibutol. $(CH_3)_3CCH_2COH(CH_3)_2$	130.22		−20	147.5	0.8417^0	1.4209	i	δ	s				**B1**, 423
p346	**2-Pentanone***	Methyl propyl ketone. $CH_3CH_2CH_2COCH_3$	86.13		−77.8	102	0.8124_{15}^{15}	1.3895^{20}	δ	∞	∞				**B1**, 676
p347	—,oxime*	Methyl propyl ketoxime. $CH_3CH_2CH_2C(:NOH)CH_3$	101.15			167^{748}	0.9095_4^{20}	1.4450^{20}	s	∞	∞				**B1**, 677
p348	**3-Pentanone***	Diethyl ketone. Propione $CH_3CH_2COCH_2CH_3$	86.13		−42	102.7	0.8159_4^{19}	1.3905^{25}	v	∞	∞				**B1**, 679
p349	**2-Pentanone, 4-amino-4-methyl-***	Diacetonamine. $(CH_3)_2CNH_2CH_2COCH_3$	115.17			$25^{0.14}$	<1		s	∞	∞				**B4²**, 767
p350	**1-Pentanone, 1(4-amino-phenyl)-***	3-Aminovalerophenone. H_2N—⬡—$CO(CH_2)_3CH_3$	177.24			$160–3^3$			i	s	s				**B14²**, 43
p351	**2-Pentanone, 1-chloro-***	$CH_3CH_2CH_2COCH_2Cl$	120.58			154.5–6δd $58–9^{17}$			δ					MeOH s	**B1²**, 738
p352	**—,5-chloro-***	$Cl(CH_2)_3COCH_3$	120.58			76^{34}	1.0571_4^{18}	1.4371^{18}			s	s			**B1²**, 738
p353	**3-Pentanone, 1-chloro-***	$CH_3CH_2COCH_2CH_2Cl$	120.58			68^{20}				s	v			MeOH s alk d^h	**B1**, 680
p354	**—,2-chloro-***	$CH_3CH_2COCHClCH_3$	120.58			135			i	s	s				**B1**, 681
p355	**—,2,4-dimethyl-***	Diisopropyl ketone. Tetramethyl acetone. $(CH_3)_2CHCOCH(CH_3)_2$	114.18			124–5	0.8062_4^{20}	1.4001^{20}	i	∞	∞		s		**B1**, 703

For explanations, symbols and abbreviations see beginning of table.

No.	Name	Synonyms and Formula	Mol. wt.	Crystalline form, color and specific rotation	m.p. °C	b.p. °C	Density	n_D	Solubility w	al	eth	ace	bz	other solvents	Ref.
	2-Pentanone														
p356	**2-Pentanone, 3-ethyl-4-methyl-***	$(CH_3)_2CHCH(C_2H_5)COCH_3$	128.22			155–7	0.812_4^{20}	1.4105^{20}	s	∞	∞	...	∞	os v aa ∞	**B1**, 707
p357	**—,1-hydroxy-***	$CH_3CH_2CH_2COCH_2OH$	102.14			$62–4^{18}$	0.9860_4^{20}	1.4234^{20}	s	s	s	...	...		**B1**^{2}, 872
p358	**—,3-hydroxy-***	$CH_3CH_2CH(OH)COCH_3$	102.14			152–3 59^{147}	0.9722_4^{17}		∞	s	s	...	...		**B1**^{2}, 872
p359	**—,4-hydroxy-***	$CH_3CH(OH)CH_2COCH_3$	102.14			$62–4^{12}$	1.0091_4^{4}	1.4265^{20}	s	s	s	...	...		**B1**^{2}, 872
p360	**—,5-hydroxy-***	$HOCH_2CH_2CH_2COCH_3$	102.14			$116–8^{33}$	1.0071_4^{20}	1.4415^{16}	∞	s	s	...	...		**B1**^{2}, 873
p361	**3-Pentanone, 2-hydroxy-***	$CH_3CH_2COCHOHCH_3$	102.14			152.5^{761} $45–8^{11}$			s	s	s	...	...		**B1**^{3}, 873
p362	**2-Pentanone, 4-hydroxy-4-methyl-***	Diacetone alcohol. $(CH_3)_2C(OH)CH_2COCH_3$	116.16		54–7	$164–6^{11}$	0.9306_4^{25}	1.4300^{9}	∞	∞	∞	...	...		**B1**, 836
p363	**1-Pentanone, 5-hydroxy-1-phenyl-***	δ-Hydroxy-*n*-valerophenone. $C_6H_5CO(CH_2)_4OH$	178.23	pl (w)	40–1				s^h	v	v	...	v	MeOH v lig δ	**B8**, 123
p364	**2-Pentanone, 3-methyl-***	*sec*-Butyl methyl ketone. $CH_3CH_2CH(CH_3)COCH_3$	100.16			118	0.8181_4^{14}	1.4002^{18}	δ	∞	∞	...	...		**B1**, 693
p365	**—,4-methyl-***	Isobutyl methyl ketone. Isopropyl acetone. $(CH_3)_2CHCH_2COCH_3$	100.16		−84.7	116.85	0.801_4^{20}	1.396^{20}	δ	∞	∞	...	∞		**B1**, 691
p366	**3-Pentanone, 2-methyl-***	Ethyl isopropyl ketone. $CH_3CH_2COCH(CH_3)_2$	100.16			$114.5–5^{745}$	0.830_0^{0}		δ	v	∞	...	...		**B1**, 691
p367	**1-Pentanone, 4-methyl-1-phenyl-***	Isocaprophenone. $(CH_3)_2CHCH_2CH_2COC_6H_5$	176.26		−10	255–6	0.9623_4^{15}	1.533^{20}	i	v	v	...	...		**B7**^{2}, 258
p368	**—,1-phenyl-***	Valerophenone. $CH_3(CH_2)_3COC_6H_5$	162.22			248.5 cor $131–3^{13}$	0.988_{20}^{20}	1.532^{20}	i	v	v	...	...		**B7**^{2}, 251
p369	**3-Pentanone, 2,2,4,4-tetramethyl-***	Hexamethyl acetone. Piavlone. $(CH_3)_3CCOC(CH_3)_3$	142.23			150–1	0.8199_4^{25}	1.4195^{18}	i	s	s	s	...	chl s aa s	**B7**^{2}, 251
—	**Pentaphene**	*see* **2:3, 6:7-Dibenzophenanthrene**													
p370	**Pentaquine**	CH_3O— —$NH(CH_2)_5NHCH(CH_3)_2$ N	301.42			$165–70^{0.02}$		1.5785^{25}	...	...	...	...	...	dil HCl s	
p371	**Pentasiloxane, dodecamethyl-**	$CH_3[—Si(CH_3)_2O—]_4Si(CH_3)_3$	384.71		ca. −80	229^{710}	0.8755	1.3925^{20}	δ	...	...	...	s	lig s	
p372	**Pentatriacontane***	$CH_3(CH_2)_{33}CH_3$	492.23		74.7	331^{15}	0.782_4^{75}		...	...	δ^h	...	...		**B1**, 177
p373	**18-Pentatriacontanone***	Diheptadecyl ketone. Stearone. $CH_3(CH_2)_{16}CO(CH_2)_{16}CH_3$	506.92	lf (lig)	87.8		0.793_4^{95}		i	δ^h	δ^h	δ^h	δ	chl δ lig δ	**B1**, 720
p374	**2-Pentenal, 2-methyl-***	Methylethyl acrolein. $CH_3CH_2CH:C(CH_3)CHO$	98.14			$137.3^{8.6}$ cor	0.8581_4^{20}	1.4488^{20}	i	s	s	...	...	MeOH s	**B1**^{2}, 793
p375	**1-Pentene***	1-Pentylene. Propylethylene. $CH_3CH_2CH_2CH:CH_2$	70.13		−138	29.2^{740}	0.6411_4^{20}	1.3711^{20}	i	∞	∞	...	...	dil sulf v	**B1**, 210
p376	**2-Pentene**(*cis*)*	*cis*-β-*n*-Amylene $CH_3CH_2CH:CHCH_3$	70.13		−180, −178	37.8	0.6586_4^{20}	1.3822^{20}	i	∞	∞	...	...	dil sulf s	**B1**, 210
p377	—(*trans*)	$CH_3CH_2CH:CHCH_3$	70.13		−136, −135	36.25	0.6486_4^{20}	1.3790^{20}	i	∞	∞	...	...	dil sulf v	**B1**, 210
p378	**—,1-Pentene, 1-bromo-***	$CH_3CH_2CH_2CH:CHBr$	149.04			121.5–22 43.5^{30}	1.2606_4^{25}	1.4777^{25}	i	...	v	...	...		**B1**^{3}, 783
p379	**—,2-bromo-***	$CH_3CH_2CH_2CBr:CH_2$	149.04			107–8	1.228^{20}	1.4535^{20}	i	...	...	...	s	chl s lig s	**B1**^{2}, 183

For explanations, symbols and abbreviations see beginning of table.

No.	Name	Synonyms and Formula	Mol. wt.	Crystalline form, color and specific rotation	m.p. °C	b.p. °C	Density	n_D	Solubility						Ref.
									w	al	eth	ace	bz	other solvents	
	2-Pentene														
p380	**—,3-bromo-***	$CH_3CH_2CHBrCH{:}CH_2$	149.04			30.5^{30}	1.2417^{25}_{4}	1.4626^{25}	i						**B1**^{2}, 183
p380^1	**2-Pentene, 1-bromo-***	$CH_3CH_2CH{:}CHCH_2Br$	149.04			123–4 35^{25}	1.2545^{20}	1.4731^{20}	i						**B1**^{2}, 185
p381	**—,2-bromo-***	$CH_3CH_2CH{:}CBrCH_3$	149.04			110.5^{750}	1.277^{20}_{20}	1.4580^{20}	i						**B1**^{3}, 784
p382	**—,3-bromo-***	$CH_3CH_2CBr{:}CHCH_3$	149.04			115.2^{750}	1.273^{20}_{20}	1.4628^{20}	i						**B1**^{3}, 784
p383	**—,4-bromo-***	$CH_3CHBrCH{:}CHCH_3$	149.04			116.7–9.2d 22^{8}	1.2312^{21}	1.4752^{21}	d					chl v	**B1**^{3}, 784
p384	**—,5-bromo-***	$BrCH_2CH_2CH{:}CHCH_3$	149.04			121.7^{621}	1.2715^{20}_{4}	1.4695^{20}	i						**B1**^{3}, 784
p385	**1-Pentene, 2-chloro-***	$CH_3CH_2CH_2CCl{:}CH_2$	104.58			95–7	0.872^{5}			s	v				**B1**^{3}, 774
p386	**—,3-chloro-***	$CH_3CH_2CHClCH{:}CH_2$	104.58			93–4	0.8978^{20}_{4}	1.4254^{20}		s	v	s			**B1**^{2}, 182
p387	**—,4-chloro-***	$CH_3CHClCH_2CH{:}CH_2$	104.58			97–100	0.934^{15}	1.417^{15}	i		v			chl v	**B1**^{3}, 774
p388	**—,5-chloro-***	$ClCH_2CH_2CH_2CH{:}CH_2$	104.58			$103.5–4.5^{773}$ 36.5^{61}	0.9125^{20}_{4}	1.4297^{20}	δ			v			
p389	**2-Pentene, 1-chloro-***	$CH_3CH_2CH{:}CHCH_2Cl$	104.58			109.5 62^{148}	0.908^{22}_{4}	1.4352^{22}		v^h	v	s			**B1**^{2}, 184
p390	**—,2-chloro-***	$CH_3CH_2CH{:}CClCH_3$	104.58			95–7 45^{130}	0.903^{24}	1.421^{24}						xyl v^h	**B1**^{2}, 185
p391	**—,3-chloro-***	$CH_3CH_2CCl{:}CHCH_3$	104.58			$90–2^{781}$	1.423^{20}	0.9125^{20}_{20}						xyl v^h	**B1**^{2}, 185
p392	**—,4-chloro-***	Piperylene hydrochloride. $CH_3CHClCH_2{:}CHCH_3$	104.58			103 $18–20^{12}$	0.9004^{20}_{20}	1.4322^{20}							**B1**^{3}, 782
p393	**—,5-chloro-***	$ClCH_2CH_2CH{:}CHCH_3$	104.58			$107.0–.6^{755}$	0.9043^{20}_{4}	1.4310^{20}			v	v			**B1**^{3}, 782
p394	**—,3-chloro-2,4-dimethyl-***	$(CH_3)_2CHCCl{:}C(CH_3)_2$	132.63			118–20	0.9513^{9}_{9}				v			chl v	**B1**, 221
p395	**1-Pentene, 2-chloro-3-ethyl-3-methyl-***	$(C_2H_5)_2C(CH_3)CCl{:}CH_2$	146.66			147^{743} 53^{20}	0.9147^{20}_{4}	1.4450^{25}	i					chl s	**B1**^{3}, 847
p396	**—,3-chloro-2-methyl-***	$CH_3CH_2CHClC(CH_3){:}CH_2$	118.61			121–4		1.4422^{20}			v			chl v	**B1**^{3}, 809
p397	**2-Pentene, 5-chloro-2-methyl-***	$ClCH_2CH_2CH{:}C(CH_3)_2$	118.61			$132–3^{758}$	0.9265^{20}_{4}	1.4419^{20}	i			s			**B1**^{3}, 810
p398	**1-Pentene, decafluoro-***	Perfluoropentene. $F_3CCF_2CF_2CF{:}CF_2$	250.04			$29–30^{740}$		1.2571^{25}							
p399	**2-Pentene,2 2,5-dichloro-***	$ClCH_2CH_2CH{:}CClCH_3$	139.03			$40–1^{8}$	1.1182^{15}_{4}		i					chl s	**B1**^{3}, 783
p400	**1-Pentene, 2,3-dimethyl-***	$CH_3CH_2CH(CH_3)C(CH_3){:}CH_2$	98.19			84.1–.3	0.7054^{20}_{4}	1.4022^{20}	i	∞	∞			dil sulf v	
p401	**—,2,4-dimethyl-***	$(CH_3)_2CHCH_2C(CH_3){:}CH_2$	98.19			80.9–1.3	0.6937^{20}_{4}	1.3970^{20}							**B1**, 220
p402	**—,3,3-dimethyl-***	$CH_3CH_2C(CH_3)_2CH{:}CH_2$	98.19			76.9	0.6961^{20}_{4}	1.3991^{20}							
p403	**—,4,4-dimethyl-***	Vinylneopentane. $(CH_3)_3CCH_2CH{:}CH_2$	98.19		−136.5	72.49	0.6827^{20}_{4}	1.3919^{20}	i					os s	**B1**^{2}, 832
p404	**2-Pentene, 2,3-dimethyl-***	$CH_3CH_2C(CH_3){:}C(CH_3)_2$	98.19			91–3	0.7190^{20}_{4}	1.4142^{29}	i	s	s				**B1**^{2}, 199
p405	**—,2,4-dimethyl-***	$(CH_3)_2CHCH{:}C(CH_3)_2$	98.19			82.4–.8	0.6954^{20}_{4}	1.4014^{20}	i	s	s				**B1**^{2}, 199
p406	**—,3,4-dimethyl-***	$(CH_3)_2CHC(CH_3){:}CHCH_3$	98.19			86.2–.4	0.7126^{20}_{4}	1.4052^{20}							
p407	**—,4,4-dimethyl-***	$(CH_3)_3CCH{:}CHCH_3$	98.19			76.0–.1	0.6881^{20}_{4}	1.3986^{20}							
p408	**1-Pentene, 2-ethyl-***	3-Vinylhexane. $CH_3CH_2CH_2C(C_2H_5)C{:}CH_2$	98.19			93.9–4.3	0.7079								
p409	**2-Pentene, 3-ethyl-***	$(C_2H_5)_2C{:}CHCH_3$	98.19			94.8–.9	0.7172^{20}_{4}	1.4120^{20}	i	s					**B1**^{2}, 198

For explanations, symbols and abbreviations see beginning of table.

No.	Name	Synonyms and Formula	Mol. wt.	Crystalline form, color and specific rotation	m.p. °C	b.p. °C	Density	n_D	Solubility						Ref.
									w	al	eth	ace	bz	other solvents	
	1-Pentene														
p410	**1-Pentene, 2-methyl-***	$CH_3CH_2CH_2C(CH_3):CH_2$	84.16		−135.76	61.5–2	0.6817^{20}_{4}	1.3921^{20}							**B1**, 90
p411	—,**3-methyl-***	$CH_3CH_2C(CH_3)CH:CH_2$	84.16			53.6–4	0.6700^{20}_{4}	1.3835^{20}							
p412	—,**4-methyl-***	1-Isohexene. $(CH_3)_2CHCH_2CH:CH_2$	84.16		−153.63	53.6–.9	0.6646^{20}_{4}	1.3825^{20}							
p413	**2-Pentene, 2-methyl-***	$CH_3CH_2CH:C(CH_3)_2$	84.16		−134.75	67.2–.5	0.6904^{20}_{4}	1.4005^{20}	i	s					**B1**², 193
p414	—,**3-methyl-(*cis*)***	$CH_3CH_2C(CH_3):CHCH_3$	84.16			65.7–6.2	0.6940^{20}_{4}	1.3994^{20}	i	s					**B1**², 194
p415	—,—(*trans*)*	$CH_3CH_2C(CH_3):CHCH_3$	84.16		−134.84	67.6–8.2	0.6956^{20}_{4}	1.4002^{20}	i	s					**B1**², 194
p416	—,**4-methyl-(*cis*)***	2-Isohexene. $(CH_3)_2CHCH:CHCH_3$	84.16		−134.43	57.7–8.5	0.6709^{20}_{4}	1.3885^{20}	i	s					**B1**², 194
p417	—,—(*trans*)*	$(CH_3)_2CHCH:CHCH_3$	84.16			54.2–5.2	0.6702^{20}_{4}	1.3881^{20}							**B1**², 194
p418	**1-Pentene, nonafluoro-2-trifluoromethyl-***	Perfluoro-2-methylpentene. $F_3C(CF_2)_2(CF_3)C:CF_2$	300.06			60									
p419	—,**2,4,4-trimethyl-***	Diisobutylene. $(CH_3)_3CCH_2C(CH_3):CH_2$	112.22		−93	101.44	0.7150^{20}_{4}	1.4112^{20}	i					lig s	**B1**³, 849
p420	**2-Pentene, 2,4,4-trimethyl-***	Diisobutylene. $(CH_3)_3CCH:C(CH_3)_2$	112.22		−106.33	112.3	0.7195^{20}_{4}	1.4135^{25}	i					lig v	**B1**³, 848
p421	**2-Pentenedioic acid,** diethyl ester(*trans*)*	Ethyl glutaconate. $C_2H_5O_2CCH_2CH:CHCO_2C_2H_5$	186.21			125^{12}	1.0540^{19}_{4}	1.4466^{19}		s	s				**B2**², 649
p422	—,**3-methyl,** monoethyl monomethyl ester*	Ethyl methyl 3-methyl-glutaconate. $C_2H_5O_2CCH_2C(CH_3):CHCO_2CH_3$	186.21			118^{12}				s					**B2**, 778
p423	**4-Pentenoic acid***	Allylacetic acid. $CH_2:CHCH_2CH_2CO_2H$	100.12		<−18	187–9	0.9843^{18}_{4}	$1.4341^{7.5}$	δ	v	v				**B2**², 399
p424	—,nitrile	Allylacetonitrile. Allylmethyl cyanide. $CH_2:CHCH_2CH_2CN$	81.12			140	1.1803^{13}		i	∞	∞				**B2**, 426
p425	—,**2-acetyl-,** ethyl ester	Ethyl allylacetoacetate. $CH_2:CHCH_2CH(COCH_3)CO_2C_2H_5$	170.21			211–2δd	0.9898^{20}_{4}	1.4388^{18}	i	∞	∞		∞		**B3**, 738
p426	—,**2,2-diethyl-,** amide	Novonal. $CH_2:CHCH_2C(C_2H_5)_2CONH_2$	155.24	wh pw, cr (eth-peth)	75–6	155^{10}			δ	v	v				**B2**², 418
p428	**2-Pentenoic acid, 4-hydroxy-,** lactone*	β-Angelica lactone. Δ¹-Angelica lactone. CH_3— =O O	98.10		<−17	$208–9^{751}$ 98^{15}	1.076^{20}_{4}		∞	s	s				**B17**², 297
p429	**3-Penteneoic acid, 2-hydroxy-,** nitrile(*cis*)	Angelactinonitrile. *cis*-Crotonaldehyde cyanohydrin. $CH_3CH:CHCHOHCN$	97.12			139^{70}	0.9675^{15}_{4}	1.4460^{21}	s	s	s		s	chl s CS_2 i lig i	**B3**², 156
p430	—,**4-hydroxy-,** lactone*	γ-Angelica lactone. Δ²-Angelica lactone. CH_3— =O O	98.10	nd	18	δd	1.084^{20}_{4}		s^d	s	s			CS_2 s	**B17**², 297
p431	**2-Pentenoic acid, 4-methyl-***	Isobutylideneacetic acid. α-Isohexenic acid. $(CH_3)_2CHCH:CHCO_2H$	114.15		35	204–8 $115–6^{20}$	0.9529^{21}_{4}	1.4489^{21}		v					**B2**², 406
p432	**1-Penten-3-ol***	Ethyl vinyl carbinol. $CH_2:CHCHOHCH_2CH_3$	86.13			114–6	0.8395^{22}_{4}	1.4183^{28}	δ	∞	∞				**B1**², 482
p433	**4-Penten-1-ol***	$CH_2:CHCH_2CH_2CH_2OH$	86.13			139–42	0.863^{0}_{4}								**B1**², 483

For explanations, symbols and abbreviations see beginning of table.

No.	Name	Synonyms and Formula	Mol. wt.	Crystalline form, color and specific rotation	m.p. °C	b.p. °C	Density	n_D	Solubility: w	al	eth	ace	bz	other solvents	Ref.
	4-Penten-2-ol														
p434	**4-Penten-2-ol***	Allyl methyl carbinol. $CH_2:CHCH_2CHOHCH_3$	86.13			115	0.8367^{20}_{4}	1.4225^{20}	v	∞	∞				**B1**, 443
p435	**3-Penten-2-ol, 2-methyl-***	Dimethyl propenyl carbinol. $CH_3CH:CHCOH(CH_3)_2$	100.16			$121.6–2^{757}$	0.8343^{20}_{4}	1.4295^{20}	s	∞	∞				**B1**[2], 486
p436	**4-Penten-2-ol, 2-methyl-***	Dimethyl allyl carbinol. $CH_2:CHCH_2COH(CH_3)_2$	100.16			117–9	0.8335^{20}_{4}	1.4302^{20}	δ						**B1**, 445
p437	**3-Penten-2-one***	Ethylideneacetone. Methyl propenyl ketone. $CH_3CH:CHCOCH_3$	84.12			122	0.8624^{20}_{4}	1.4350^{20}	s						**B1**[2], 791
p438	**1-Penten-3-one, 5,5-dimethyl-1-phenyl-***	Benzalpinacolone. $C_6H_5CH:CHCOCH_2CH(CH_3)_2$	188.27		43	154^{25}	0.9509^{46}	1.5523^{25}	δ	s			s	chl s	**B7**[2], 308
p439	**—,1(4-methoxy-phenyl)-***	α-Methylanisalacetone. $CH_3O—C_6H_4—CH:CHCOCH_2CH_3$	190.24	colorless-lt ye	60				i	s					
p440	**—,2-methyl-***	Ethyl isopropenyl ketone. $C_2H_5COC(CH_3):CH_2$	98.14		69.5	118.5	0.8560^{20}_{20}		δ	s					
p441	**—,4-methyl-***	Mesityl oxide. $(CH_3)_2C:CHCOCH_3$	98.14		−59	130–1	0.8578^{20}_{4}	1.4440^{20}	s	∞	∞				**B1**[2], 793
p441[1]	**—,1-phenyl-***	$C_6H_5CH:CHCOC_2H_5$	160.22	lf (lig)	38–9	142^{12}	0.8697^{20}_{4}	1.5684^{50}	δ	v	v		v		**B7**[2], 298
p442	**1-Penten-3-yne***	$CH_2:CHC\vdots CCH_3$	66.10			59.2	0.7401^{20}_{4}	1.4496^{20}							
p443	**1-Penten-4-yne***	$CH_2:CHCH_2C\vdots CH$	66.10			41–2	0.777^{22}_{22}	1.3653^{22}							
p444	**3-Pentene-1-yne, 3-ethyl-***	$CH_3CH:C(C_2H_5)C\vdots CH$	94.16			96.5	0.7886^{25}	1.4338^{25}							
p445	**1-Pentene-3-yne, 2-methyl-***	$CH_2:C(CH_3)C\vdots CCH_3$	80.13			75–7		1.4002^{20}							
p446	**1-Pentene-1-yne, 3-methyl-**	$CH_3CH:C(CH_3)C\vdots CH$	80.13			67–9		1.4332^{20}							
—	**Pentobarbital**	*see* **Barbituric acid, 5-ethyl-5(2-pentyl)-**													
p447	**1-Pentyne***	*n*-Propylacetylene. $CH_3CH_2CH_2C\vdots CH$	68.12		−90.0	39.3	0.6909^{25}_{4}	1.3827^{25}	i	v	∞				**B1**, 250
p448	**2-Pentyne***	Ethylmethyl acetylene. $CH_3CH_2C\vdots CCH_3$	68.12		−101	55.5	0.7127^{17}_{4}	1.4045^{17}	i	v	∞				**B1**[1], 110
p449	**1-Pentyne, 3-chloro-3-ethyl-***	$(CH_3CH_2)_2CClC\vdots CH$	130.62			$73–6^{100}$	0.9230^{19}_{4}	1.4437^{19}_{α}			s				**B1**[3], 1002
p450	**—,3-chloro-3-methyl-***	$CH_3CH_2CCl(CH_3)C\vdots CH$	116.59			55^{130}	0.9163^{20}_{4}	1.4330^{20}			s				
p451	**2-Pentyne, 4-chloro-4-methyl-***	Trimethylpropargyl chloride. $CH_3CCl(CH_3)C\vdots CCH_3$	116.59			$57–61^{47}$		1.4143^{20}				v^h			
p452	**1-Pentyne, 4,4-dimethyl-***	$(CH_3)_3CCH_2C\vdots CH$	96.17			73–5	0.7154^{20}_{4}	1.4028^{20}							
p453	**2-Pentyne, 4,4-dimethyl-***	$(CH_3)_3CC\vdots CCH_3$	96.17			82.9–3 .0	0.7176^{20}_{4}	1.4071^{20}							
p454	**1-Pentyne, 3-ethyl-***	$(C_2H_5)_2CHC\vdots CH$	96.17			87–8.5	0.7246^{25}_{4}	1.4043^{25}	i		s				**B1**[3], 1002
p455	**—,1-ethyl-3-methyl-***	$(C_2H_5)_2C(CH_3)C\vdots CH$	110.20			$98–100^{745}$	0.7360^{20}_{4}	1.4102^{20}							
p456	**1-Pentyne, 4-methyl-***	$(CH_3)_2CHCH_2C\vdots CH$	82.15		−105.1	61.1–2	0.7092^{15}_{4}								
p457	**2-Pentyne, 4-methyl-***	$(CH_3)_2CHC\vdots CCH_3$	82.15			72.0–.5	0.716^{19}_{4}	1.4078^{19}							
p458	**2-Pentynoic acid***	Ethylpropiolic acid. $CH_3CH_2C\vdots CCO_2H$	98.10	(peth)	50	122^{10}	0.978^{20}	1.4619^{20}	v						**B2**, 481

For explanations, symbols and abbreviations see beginning of table.

No.	Name	Synonyms and Formula	Mol. wt.	Crystalline form, color and specific rotation	m.p. °C	b.p. °C	Density	n_D	Solubility						Ref.
									w	al	eth	ace	bz	other solvents	
	4-Pentynoic acid														
p459	**4-Pentynoic acid*.**	Ethynylpropionic acid. HC⋮$CCH_2CH_2CO_2H$	98.10		57.7	102^{17}			v	v	v				B2, 481
p460	**1-Pentyn-3-ol, 3,4-dimethyl-***	Ethynyl methyl isopropyl carbinoi. $(CH_3)_2CHCOH(CH_3)C$⋮CH	112.17			133	0.876^{15}_{4}	1.459^{15}	s	s	s				B1, 506
p461	**—,3-methyl-***	$CH_3CH_2COH(CH_3)C$⋮CH	98.14		30–1	120–1 61^{70}	0.8665^{20}_{4}	1.4318^{20}							$B1^2$, 506
—	**Peonol**	*see* **Acetophenone, 2-hydroxy-4-methoxy-**													
—	**Peracetic acid**	*see* **Acetic acid, per-**													
—	**Perbenzoic acid**	*see* **Benzoic acid, per-**													
p462	**Perchloric acid, ethyl ester***	Ethyl perchlorate*. $CH_3CH_2OClO_3$	128.52			89			d	∞	∞				
p463	**—,methyl ester***	Methyl perchlorate*. CH_3OClO_3	114.49			ca. 52			d	∞	∞				
p464	**—,trichloromethyl ester***	Trichloromethyl perchlorate*. Cl_3COClO_3	217.83		−55	exp (40d)			d					bz, al exp CCl_4 ∞	
p465	**Pereirine**	$C_{19}H_{24}N_2O$	296.40	pa ye amorph pw, [α]+137.5 (al)	135d				i	v	v			chl v	
p466	**Perimidine**	Perinaphthimineazole. (structure: N, NH)	168.20	gr	222				i	s					$B23^1$, 53
p467	**Peroxide, acetyl benzoyl**	$C_6H_5COOOCOCH_3$	180.15	wh nd (lig)	40–1	85–100 exp 130^{19}			d	d	s			chl s oils s	$B9^2$, 157
p468	**—,diacetyl**	Acetyl peroxide. $CH_3COOOCOCH_3$	118.09	nd (eth), lf	30	63^{21}			δ	s	s^h			NaOH d CCl_4 v	$B2^2$, 174
p469	**—,dibenzoyl**	Benzoyl peroxide. (structure: 3 2, 4, —COOOCO—, 2′ 3′, 4′, 5 6, 6′ 5′)	242.22	rh (eth), pr	106–8	exp		1.545	δ	δ	s	s	s	MeOH δ CS_2 s	$B9^2$, 157
p470	**—,—,3,3′-dinitro-**	$C_{14}H_8N_2O_8$. *See* p469	332.23	nd (al)	139–40d						v	v	v	AcOEt s^h	$B9^2$, 252
p471	**—,—,4,4′-dinitro-**	$C_{14}H_8N_2O_8$. *See* p469	332.23	ye cr (ace), nd (to)	156d									AcOEt s^h to s^h	$B9^2$, 270
p472	**—,di(*tert*-butyl)**	$(CH_3)_3COOC(CH_3)_3$	146.22		−40	111^{760}	0.794^{20}	1.3890^{20}	i			∞		octane ∞	
p473	**—,dododecanoyl-**	Alperox c. Lauroyl peroxide. $CH_3(CH_2)_{10}COOOCO(CH_2)_{10}CH_3$	398.61	wh pl	49				i						
p474	**—,diethyl**	Ethyl peroxide. $C_2H_5OOC_2H_5$	90.12		−70	$58–9^{625}$	0.8240^{19}	1.3715^{17}	δ	∞	∞				$B1^3$, 1313
p475	**—,dimethyl, hexafluoro-**	Perfluorodimethyl peroxide. F_3COOCF_3	170.02			−40, −32			d					aqKOH i	
p476	**—,di(1-naphthoyl)**	$C_{10}H_7^{\alpha}COOOCOC_{10}H_7^{\alpha}$	342.35	yesh pr (chl, ace, eth)	91d exp					s		s	s	chl s	E12B, 3989
p477	**—,disuccinyl**	$(HO_2CCH_2CH_2CO)_2O_2$	234.16	cr pw	127d				s	s	δ	s	i	chl i	
p478	**Perseitol**	α Mannoheptitol. Perseite. (structure: $HOCH_2C$—C—C—C—$CH(OH)CH_2OH$; above: H H OH OH; below: OH OH H H)	212.20	nd, [α]+4.53	188 cor				s	δ s^h					B1, 548
p479	**Perylene**	(structure: 8 7 6 5, 9, 4, 10, 3, 11 12 1 2)	252.31	gold ye pl (bz, aa)	273–4	350–400 sub			i	δ	δ	v	s	chl v	$B5^2$, 655
p480	**3-Perylene-carboxylic acid**	$C_{21}H_{12}O_2$. *See* n479	296.13	og br nd ($PhNO_2$)	330									sulf s (vt, red flr) os s (gr flr)	E14s, 743
—	**Petn**	*see* **Pentaerythritol,** tetranitrate													
p481	**Peucedanin**	(structure: O, O, O, $CH(CH_3)_2$, OCH_3)	258.26	yesh (eth)	109	$276–81^{17}$			δ	s^h	s		δ	chl v	$B19^2$, 228

For explanations, symbols and abbreviations see beginning of table.

No.	Name	Synonyms and Formula	Mol. wt.	Crystalline form, color and specific rotation	m.p. °C	b.p. °C	Density	n_D	Solubility						Ref.
									w	al	eth	ace	bz	other solvents	
	α-Phellandrene														
p482	**α-Phellandrene**	5-Isopropyl-2-methyl-1,3-cyclohexadiene* $C_{10}H_{16}$	136.23	$[\alpha]_D$ +44.41	...	175–6	0.8463^{25}_{4}	1.477^{20}	...	i	s	...	...	...	**B5**, 129
p483	**β-Phellandrene**	3-Isopropyl-6-methylene-cyclohexene*. 1(7),2-*p*-Menthadiene. $C_{10}H_{16}$	136.23	$[\alpha]$ +17.64	...	171–2	0.8520^{20}_{4}	1.4788^{20}	i	i	s	...	...	...	**B5**, 131
—	**Phenacetin**	*see* **Acetic acid, amide,** *N*(4-ethoxyphenyl)-													
p483[2]	**Phenaceturic acid**	Glycine, *N*-(phenylacetyl)	193.2		143					v			δ	$CHCl_3$ δ	**B9**,493
—	**Phenacyl alcohol**	*see* **Acetophenone, α-hydroxy-**													
—	**Phenacyl bromide**	*see* **Acetophenone, α-bromo-**													
—	**Phenacyl chloride**	*see* **Acetophenone, α-chloro-**													
p484	**Phenanthrene***	10 9 1 11 14 8 2 7 3 4 12 13 5 6	178.22	bl flr (in sol)	101	340	1.182	1.5973	i	s	δ	...	...	CS_2 δ	**B5**, 667
p485	**—,2-acetyl-***	$C_{16}H_{12}O$. *See* p484	220.27	(MeOH)	143 cor.	...	...	...	i	s	...	...	...	...	**E13**, 878
p486	**—,3-acetyl-***	$C_{16}H_{12}O$. *See* p484	220.27	nd (MeOH)	72	...	...	...	i	...	...	...	...	MeOH v, chl s	**E13**, 879
p487	**—,9-acetyl-***	$C_{16}H_{12}O$. *See* p484	220.27	bl flr nd (MeOH), lf (al)	74–4.5	...	...	...	...	v	v	...	...	MeOH s^h	**E13**, 879
p488	**—,2-amino-***	2-Phenanthrylamine. $C_{14}H_{11}N$. *See* p484	193.25	lt ye (lig)	85	...	...	...	i	s^h i	s^h i	...	...	...	**E13**, 819
p489	**—,3-amino-***	3-Phenanthrylamine. $C_{14}H_{11}N$. *See* p484	193.25	(lig)	143(α) 87–5(β)	...	...	...	δ	v	...	...	...	HCl s xyl δ	**E13**, 820
p490	**—,9-amino-** (a form)*	9-Phenanthrylamine. $C_{14}H_{11}N$. *See* p484	193.25	lt ye cr (al)	137–8	sub	...	...	...	v	v	...	v	chl v	**E13**, 819
p491	**—,—**(b form)*	$C_{14}H_{11}N$. *See* p484	193.25	lt ye nd	104	sub	...	...	...	v	v	...	v	chl v	**E13**, 819
p492	**—,9-bromo-***	$C_{14}H_9Br$. *See* p484	257.14	pr (al)	63	>360	1.4093^{101}_{4}	...	i	s	s	...	...	CS_2 v	**B5**[2], 583
p493	**—,9,10-diamino-***	$C_{14}H_{12}N_2$. *See* p484	208.27	lf	166		...	...	...	...	...	...	...	...	**E13**, 822
p494	**—,9,10-dihydro-***	$C_{14}H_{12}$. *See* p484	180.25	nd (MeOH)	34.5–5	$168–9^{15}$	1.0757^{40}_{4}	...	...	s	v	...	...	MeOH δ	**E13**, 794
p495	**—,3,4-dihydroxy-***	3,4-Phenanthrenediol*. Morphol. $C_{14}H_{10}O_2$. *See* p484	210.23	colorless nd, dk in air	143		...	...	i	s	s	...	...	alk s	**E13**, 846
p496	**—,3,4-dimethoxy-***	Dimethylmorphol. $C_{16}H_{14}O_2$. *See* p484	238.29	bt ye lf	43–4	298–303	...	...	i	v	v	...	...	...	...
p497	**—,9,10-dimethyl-***	$C_{16}H_{14}$. *See* p484	206.29	lt red pr and nd (aa)	139	sub	...	...	...	δ	...	...	s	chl s aa s	**E13**, 803
p498	**—,9,10-diphenyl-***	$C_{26}H_{18}$. *See* p484	330.43	nd (eth, bz)	235	sub 270	...	...	i	δ	s	...	s	...	**E13**, 810
p499	**—,1,2,3,4,5,6,7,8,9,10,11,12-dodecahydro-***	$C_{14}H_{22}$. *See* p484	190.33	...	...	$81–2^{1.5}$	0.9674^{20}_{4}	1.5102^{20}	i	...	...	...	...	os s	**E13**, 795
p500	**—,9-ethyl-***	$C_{16}H_{14}$. *See* p484	206.29	...	61–2	199–200	1.0603^{78}_{4}	1.6582^{78}	i	s	...	...	...	...	**E13**, 799
p501	**—,1,2,3,4,11,12-hexahydro-***	$C_{14}H_{16}$. *See* p484	184.28	...	...	$125–6^{2.5}$	...	1.5810^{15}	i	...	...	...	...	os s	**E13**, 799
p502	**—,1-hydroxy-***	1-Phenanthrol*. $C_{14}H_{10}O$. *See* p484	194.24	nd (peth, bz-lig)	157		...	...	...	...	...	...	...	sulf s	**E13**, 839
p503	**—,2-hydroxy-***	2-Phenanthrol*. $C_{14}H_{10}O$. *See* p484	194.24	pl, lf (eth, lig)	168		...	...	δ	s	s	...	s	...	**B6**, 704
p504	**—,3-hydroxy-***	3-Phenanthrol*. $C_{14}H_{10}O$. *See* p484	194.24	nd (al, lig)	122–3		...	...	i s^h	v	s	...	s^h	lig s^h	**B6**, 705
p505	**—,9-hydroxy-***	9-Phenanthrol*. $C_{14}H_{10}O$. *See* p484	194.24	nd (lig)	152–3		...	...	δ	v	v	...	v	chl v lig i	**E13**, 831
p505[1]	**—,7-isopropyl-1-methyl-***	Retene. $C_{18}H_{18}$. *See* p484	234.34	pl (al)	100.5–1	390	1.035	...	...	s^h	...	...	s	CS_2 s lig s	**B5**, 683
p506	**—,1-methyl-***	$C_{15}H_{12}$. *See* p484	192.26	lf, pl (al)	123		...	...	i	s	...	...	...	...	**E13**, 798
p507	**—,3-methyl-***	$C_{15}H_{12}$. *See* p484	192.26	pr	65		...	...	i^h	s	...	...	...	os s	**B5**, 1667
p508	**—,4,5-dimethyl-***	$C_{16}H_{10}$. *See* p484	190.25	(al)	114.3–5.3		...	...	δ	...	...	...	...	$PhNO_2$ s	**E14**, 309
p509	**—,9-nitro-***	$C_{14}H_9NO_2$. *See* p484	223.23	ye nd (al)	116–7		...	...	...	s	...	...	...	sulf s (red) MeOH s	**E13**, 817
p510	**—,1,2,3,4,5,6,7,8-octahydro-***	$C_{14}H_{18}$. *See* p484	186.30	...	16.7	295 169^{15}	1.026^{20}_{4}	1.5701^{13}	i	...	...	...	s	CS_2 s aa s	**E13**, 795

For explanations, symbols and abbreviations see beginning of table.

No.	Name	Synonyms and Formula	Mol. wt.	Crystalline form, color and specific rotation	m.p. °C	b.p. °C	Density	n_D	Solubility w	al	eth	ace	bz	other solvents	Ref.
	Phenanthrene														
p511	**—,1,2,3,4,9,10,11, 12-octahydro-** (*cis*)*	$C_{14}H_{18}$. *See* p484	186.30			$88–90^{0.1}$	1.0072^{25}_{4}	1.5549^{21}	i				s		**E13**, 795
p512	—,—(*trans*)*	$C_{14}H_{18}$. *See* p484	186.30	nd	23–4			1.5528^{21}	i				s		**E13**, 795
p513	**—,tetra-decahydro-***	Perhydrophenanthrene. $C_{14}H_{24}$. *See* p484	192.35			$86–9^{2}$	0.9447^{20}_{4}	1.5011^{20}	i					os s	**E13**, 796
p514	**—,1,2,3,4-tetra-hydro-***	Tetranthrene. $C_{14}H_{14}$. *See* p484	182.27	lf (MeOH)	33–4	173^{11}		1.0601^{40}_{4}	i	s		s			**E13**, 794
p515	**—,1,3,4,5-tri-hydroxy-***	3,4,5-Phenanthrenetriol*. $C_{14}H_{10}O_3$	226.23	lf or pl (w)	148				i	v	s			chl s	**E13**, 861
p516	**1-Phenanthrene-carboxylic acid***	1-Phenanthroic acid. $C_{15}H_{10}O_2$. *See* p484	222.25	nd (al)	232–3					s					**E13**, 953
p517	**2-Phenanthrene-carboxylic acid***	2-Phenanthroic acid. $C_{15}H_{10}O_2$. *See* p484	222.25	(aa)	258.5–60					s				aa s	**E13**, 953
p518	—,nitrile	$C_{15}H_9N$. *See* p484	203.27	(bz-lig)	105					v				os v	**B9**, 706
p519	**3-Phenanthrene-carboxylic acid***	3-Phenanthroic acid. $C_{15}H_{10}O_2$. *See* p484	222.25	nd	270					s					**E13**, 954
p520	—,nitrile	$C_{15}H_9N$. *See* p484	203.27	nd (abs al)	102					s^h				os v	**B9**, 706
p521	**9-Phenanthrene-carboxylic acid***	9-Phenanthroic acid. $C_{15}H_{10}O_2$. *See* p484	222.25	nd (aa), lf (sub)	250–2	sub			δ	s	s		s		**E13**, 955
p522	—,nitrile	$C_{15}H_9N$. *See* p484	203.27	nd (al)	103					s^h				os v	**B9**, 707
p523	**2-Phenanthrene-sulfonic acid***	$C_{14}H_{10}O_3S$. *See* p484	258.30	(bz)	ca. 150				v	v			v^h	to s	**E13**, 995
p524	**3-Phenanthrene-sulfonic acid***	$C_{14}H_{10}O_3S$. *See* p484	258.30	lf (bz), (w+1)	88–9 (+2w) 120–1 (+1w) 175–7 (anh)				s				s^h		**B13**, 995
p525	**9-Phenanthrene-sulfonic acid***	$C_{14}H_{10}O_3S$. *See* p484	258.30	lf or nd (bz), (+2w)	134 (+2w) 174 (anh)				s	s			δ s^h	aa s	**B13**, 996
p526	**Phenanthridine**	9-Azaphenanthrene. 3,4-Benzoquinoline.	179.22	nd (al)	106	360			i	v	v		v	chl v CS_2 v	**B20**, 466
p527	—,9-hydroxy-	9-Phenanthridone. $C_{13}H_9NO$. *See* p526	195.21	(al), nd (sub)	293–4	sub			i	δ s^h	i			$PhNO_2$ s^h con sulf s	**B21**[2], 79
—	**Phenanthrol***	*see* **Phenanthrene, hydroxy-***													
p528	**1,5-Phenanthroline**	1,5-Diazaphenanthrene*. *m*-Phenanthroline.	180.20	pl (anh), nd (w+2)	78–8.5 (anh), 65.5 (+2w)	>360			s^h	v	δ		$δ^h$		**B23**[1], 61
p529	**1,8-Phenanthroline**	1,8-Diazaphenanthrene*. *p*-Phenanthroline.	180.20	nd (w)	177	sub 100			s v^h	v	δ			chl v lig s^h	**B23**[1], 61
p530	**4,5-Phenanthroline**	4,5-Diazaphenanthrene*. *o*-Phenanthroline.	180.20	wh nd	117				v^h	v					**B23**, 227
p531	—,hydrate	$C_{12}H_8N_2.H_2O$. *See* p530	198.22	wh nd (w)	93–4				δ s^h	δ	δ				**B23**[2], 235
p532	**1,5-Phenanthroline, 9-nitro**	$C_{12}H_7N_3O_2$. *See* p528	225.08	nd (dil al)	168				s^h δ	s^h				os s	**B23**[2], 236

For explanations, symbols and abbreviations see beginning of table.

No.	Name	Synonyms and Formula	Mol. wt.	Crystalline form, color and specific rotation	m.p. °C	b.p. °C	Density	n_D	Solubility w	al	eth	ace	bz	other solvents	Ref.
	9,10-Phenanthroquinone														
p533	**9,10-Phenanthroquinone**	9,10,Phenanthrenequinone*. 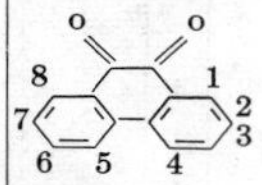	208.20	og nd	207	360	1.405^{20}_{4}		δ	δ	δ				**E13**, 910
p534	—,**2-bromo-***	$C_{14}H_7BrO_2$. *See* p533	287.13	(aa)		233–4				s				sulf s (gr)	**E13**, 924
p535	—,**3-bromo-***	$C_{14}H_7BrO_2$. *See* p533	287.13	nd (aa)	268–9				i					sulf s (dk br) PhNO₂ s	**E13**, 925
p536	—,**2-chloro-***	$C_{14}H_7ClO_2$. *See* p533	242.67	ye-red nd (aa)	252–3				i	s				sulf s (gr-br)	**E13**, 924
p537	—,**3-chloro-***	$C_{14}H_7ClO_2$. *See* p533	242.67	og-ye nd (aa)	264–5				i	s				aa s	**E13**, 925
p538	—,**1,2-dihydroxy-***	$C_{14}H_8O_4$. *See* p533	240.20	dk nd (ace)	d				δ	v		v		alk aa v	**E13**, 940
p539	—,**2,5-dihydroxy-***	$C_{14}H_8O_4$. *See* p533	240.20	dk red nd (w)	400d									alk, sulf s	**E13**, 941
p540	—,**2,7-dihydroxy-***	$C_{14}H_8O_4$ *See* p533	240.20	dk red nd	>400d					s	δ	s	δ	NaOH s	**E13**, 941
p541	—,**4,5-dihydroxy-***	$C_{14}H_8O_4$. *See* p533	240.20	dk red (al), nd (w)	>400										**B8**^{2}, 508
p542	—,**2,5-dinitro-***	$C_{14}H_6N_2O_6$. *See* p533	298.22	red ye pr (aa)	228									aa s	**E13**, 929
p543	—,**2,7-dinitro-***	$C_{14}H_6N_2O_6$. *See* p533	298.22	gold ye nd	301–3					δ				aa δ	**E13**, 929
p544	—,**2-hydroxy-***	$C_{14}H_8O_3$. *See* p533	224.22	br-red or bk-vt nd	283 cor				s					Ac_2O s al-KOH s (bl)	**E13**, 937
p545	—,**3-hydroxy-***	$C_{14}H_8O_3$. *See* p533	224.22	ye-red or red nd (aa, sub)	ca. 330d				s	s				alk s (red) AC_2O s	**E13**, 937
p546	—,**7-isopropyl-1-methyl-***	Retenoquinone. $C_{18}H_{16}O_2$. *See* p533	264.31	og nd (al)	197	sub				$δ^h$	$δ^h$		s	CS_2 v^h	**B7**, 819
p547	—,**2-nitro-***	$C_{14}H_7NO_4$. *See* p533	253.20	ye lf or nd (aa)	257					i					**E13**, 927
p548	—,**1,2,4-trihydroxy-***	$C_{14}H_8O_5.C_2H_5OH$. *See* p533	302.27	red (al+1)	d					v				alk, con sulf, Py s, os δ	**E13**, 943
p549	—,**2,3,4-trihydroxy-***	$C_{14}H_8O_5$. *See* p533	256.20	red br pw	185d				s	δ					**E13**, 943
—	**Phenanthrylamine***	*see* **Phenanthrene, amino-***													
p550	**Phenazine**	Azophenylene. 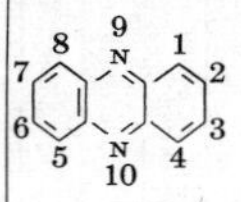	180.20	ye-red nd (aa)	171	>360 sub			δ	s^h	δ				**B23**, 223
p551	—,**9,10-dihydro-**	Hydrazophenylene. $C_{12}H_{10}N_2$ *See* p550	182.22	rh lf	212d	317			i	δ			i		**B23**, 209
p552	—,**1,4-dihydroxy-**, di(*N*-oxide)	$C_{12}H_8N_2O_4$. *See* p550	244.20	purp (chl)	236d				i					chl, dil alk, con sulf s	
p553	—,**1-Hydroxy-**	Hemipiocyanine. 1-Phenazinol. $C_{12}H_8N_2O$. *See* p550	196.20	ye nd (bz, dil MeOH), lf (dil MeOH)	158	sub			δ s^h	v				dil alk, con ac s os v	**B23**^{2}, 360
p554	—,**2-methyl-**	$C_{13}H_{10}N_2$. *See* p550	194.23	lt ye nd or pr	117	350d			$δ^h$	v	v			chl v lig δ	**B23**^{2}, 239
—	**Phenazinol**	*see* **Phenazine, hydroxy-**													
—	**Phenethyl alcohol**	*see* **Ethanol, 2-phenyl-***													
—	**Phenethyl amine**	*see* **Ethane, 1-amino-2-phenyl-***													
—	**Phenethyl chloride**	*see* **Benzene, (chloroethyl)-***													
—	**Phenetidine**	*see* **Aniline, ethoxy-***													
—	**Phenetole**	*see* **Benzene, ethoxy-**													
—	**Phenicin**	*see* **Biquinone, 3,3′-dihydroxy-5-5′-dimethyl-**													
—	**Phenobarbital**	*see* **Barbituric acid, 5-ethyl-5-phenyl-**													
—	**Phenocoll**	*see* **Glycine**, amide, N′-ethoxyphenyl													
p555	**Phenol***	Carbolic acid. Benzenol*. Hydroxybenzene*. 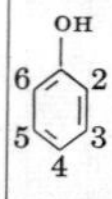	94.11		43 (fr 41)	182 90.2^{25}	1.0722^{20}_{20} 1.0710^{25}_{4}	1.5509^{21}	s ∞^h	s	v			chl, CS_2 s	**B6**^{2}, 116 **B11**^{2}, 285
p556	—,**2-amino-***	*o* Hydroxyaniline. C_6H_7NO. *See* p555	109.12	wh rh bipym nd	174	sub 153^{11}	1.328		v	v	v		$δ^h$		**B13**^{2}, 164

For explanations, symbols and abbreviations see beginning of table.

No.	Name	Synonyms and Formula	Mol. wt.	Crystalline form, color and specific rotation	m.p. °C	b.p. °C	Density	n_D	Solubility w	al	eth	ace	bz	other solvents	Ref.
	Phenol														
p557	—,**3-amino-***	*m*-Hydroxyaniline. C $_6H_7NO$. *See* p555	109.12	pr (to)	123	164^{11}	...	...	v^h	v	v	...	δ	lig δ	**B13**^{2}, 209
p558	—,—,hydro-bromide*	$C_6H_7NO.HBr$. *See* p555	190.05	pr (w)	...	...	...	...	...	...	...	...	...	...	...
p559	—,—,hydro-chloride*	$C_6H_7NO.HCl$. *See* p555	145.60	pr (w)	224 229	... ...	... ...	... ...	s^h s^h	... ...	... ...	... ...	... ...	... ...	**B13**, 403 **B13**, 403
p560	—,—,hydroiodide	$C_6H_7NO.HI$. *See* p555	237.00	pr	209	...	...	...	s^h	...	...	...	...	...	**B13**, 403
p561	—,**4-amino-***	*p*-Hydroxyaniline. C_6H_7NO. *See* p55	109.12	wh pl (w)	186	sub par d	...	...	v^h δ	v	...	...	i	chl i	**B13**^{2}, 220
p562	—,**2-amino-3-chloro-***	C_6H_6ClNO. *See* p555	143.58	nd (aq $NaHSO_3$)	122	...	...	...	s	...	...	...	...	...	**B13**^{2}, 182
p563	—,**2-amino-5-chloro-***	C_6H_6ClNO. *See* p555	143.58	nd (al), pr (aq $NaHSO_3$)	154	...	...	...	s^h	s	s	...	δ	...	**B13**^{2}, 184
p564	—,**3-amino-2-chloro-***	C_6H_6ClNO. *See* p555	143.58	...	85–7	...	...	...	δ	s	...	...	...	...	**B13**, 420
p565	—,**4-amino-2-chloro-***	C_6H_6ClNO. *See* p555	143.58	nd (al, eth, w)	153	...	...	...	δ	v	v	...	...	alk s	**B13**^{2}, 272
p566	—,**4-amino-3-chloro-***	C_6H_6ClNO. *See* p555	143.58	nd (dil $NaHSO_3$)	160	...	...	...	δ	s	...	...	...	...	**B13**^{2}, 273
p567	—,**2-amino-4-chloro-5-nitro-***	$C_6H_5ClN_2O_3$. *See* p555	188.58	ye nd	225d (dk at 200)	...	...	...	$δ^h$	v	...	...	...	alk s	**B13**^{2}, 196
p568	—,**2-amino-4-chloro-6-nitro-***	$C_6H_5ClN_2O_3$. *See* p555	188.58	...	152	...	...	...	$δ^h$	v	...	...	...	...	**B13**^{2}, 196
p569	—,**2-amino-6-chloro-4-nitro-***	$C_6H_5ClN_2O_3$. *See* p555	188.58	ye nd (w+1)	160	...	...	...	$δ^h$	v	v	...	...	...	**B13**^{2}, 195
p570	—,**4-amino-2-chloro-6-nitro-***	$C_6H_5ClN_2O_3$. *See* p555	188.58	...	130	...	...	...	$δ^h$	v	...	...	...	...	**B13**, 524
p571	—,**2-amino-3,5-dibromo-***	$C_6H_5Br_2NO$. *See* p555	266.95	nd (lig)	145	...	...	...	s^h	...	...	...	...	lig s^h	**B13**^{2}, 187
p572	—,**2-amino-4,6-dibromo-***	$C_6H_5Br_2NO$. *See* p555	266.95	nd (dil al)	99	...	...	...	δ	s	s	...	s	chl s	**B13**^{2}, 188
p573	—,**4-amino-2,6-dibromo-***	$C_6H_5Br_2NO$. *See* p555	266.95	nd (al)	192.5	...	...	...	i	s	δ	...	...	...	**B13**^{2}, 280
p574	—,**2-amino-3,5-dichloro-***	$C_6H_5Cl_2NO$. *See* p555	178.03	nd (bz, w)	132–3	...	...	...	...	s	...	...	s	chl s	**B13**^{2}, 185
p575	—,**3-amino-4,6-dichloro-***	$C_6H_5Cl_2NO$. *See* p555	178.03	ye br pr (w)	135–6	...	...	...	i	s	v	v	s	chl s	**B13**^{1}, 135
p576	—,**4-amino-2,5-dichloro-***	$C_6H_5Cl_2NO$. *See* p555	178.03	(bz)	178–9	...	...	...	d^h	s	v	...	...	chl i aa v alk s	**B13**^{2}, 274
p577	—,**4-amino-2,6-dichloro-***	$C_6H_5Cl_2NO$. *See* p555	178.03	nd (w, bz)	167	...	...	...	$δ^h$ i	v	v	...	...	aa δ	**B13**^{2}, 274
p578	—,**4-amino-3,5-dichloro-***	$C_6H_5Cl_2NO$. *See* p555	178.03	nd (w, bz)	154	...	...	...	s	s	s	...	...	CS_2 CCl_4 s aa s	**B13**^{2}, 276
p579	—,**2-amino-4,6-dinitro-***	Picramic acid. $C_6H_5N_3O_5$. *See* p555	199.13	red	169	...	...	...	s	s	δ	...	s	chl δ aa s	**B13**^{2}, 196
p580	—,**2-amino-3-nitro-***	$C_6H_6N_2O_3$. *See* p555	154.12	red nd (w)	212	sub	...	...	v^h	...	...	...	...	...	**B13**^{1}, 191
p581	—,**2-amino-4-nitro-***	$C_6H_6N_2O_3$. *See* p555	154.12	og pr (w+?)	80–90 (+w) 142–3 (anh)	...	...	...	δ	v	s	...	s^h	MeOH s aa s	**B13**^{2}, 192
p582	—,**2-amino-6-nitro-***	$C_6H_6N_2O_3$. *See* p555	154.12	red nd (al)	111–2	...	...	...	δ	s	v	...	v	chl v	**B13**^{1}, 122
p583	—,**3-amino-4-nitro-***	$C_6H_6N_2O_3$. *See* p555	154.12	og nd	185–6	...	...	...	$δ^h$	s	v	...	v	chl v	**B13**^{1}, 136
p584	—,**4-amino-2-nitro-***	$C_6H_6N_2O_3$. *See* p555	154.12	dk red pl or nd (w, al)	131	...	...	...	s^h	s	s	...	...	...	**B13**^{2}, 839
p585	—,**4-amino-3-nitro-***	$C_6H_6N_2O_3$. *See* p555	154.12	red pr (eth)	154	...	...	...	s	s	s	...	...	chl s	**B13**^{1}, 186
p586	—,**4(2-amino-propyl)-***	$C_9H_{13}NO$. *See* p555	151.20	(bz)	125–6	...	...	...	s	s	...	...	s^h	chl, AcOEt s	...
—	—,**anilino-**	*see* **Amine, diphenyl, hydroxy-**													
p587	—,**2-benzylidene-amino-**	$C_{13}H_{11}NO$. *See* p555	197.24	lf (al)	89	...	...	...	...	δ	v	...	v	con sulf s	**B13**^{1}, 112
p588	—,**2-benzyloxy-**	Pyrocatechol monobenzyl ether. $C_{13}H_{12}O_2$. *See* p555	200.23	...	...	$173–4^{13}$	1.154^{22}	1.1588^{22}	...	s	s	...	...	alk s	...
p589	—,**3-benzyloxy-**	Resorcinol monobenzyl ether. $C_{13}H_{12}O_2$. *See* p555	200.23	(CCl_4)	69.2	200^5	...	...	s	...	...	...	...	5 % KOH s CCl_4 s^h	...
p590	—,**4-benzyloxy-**	Hydroquinone monobenzyl ether. $C_{13}H_{12}O_2$. *See* p555	200.23	pl (w)	122–5.5	...	...	...	δ v^h	v	v	...	v	...	**B6**, 845
p591	—,**4[bis(2-hydroxyethyl) amino]-***	*N-p*-Phenoldiethanolamine. $C_{10}H_{15}NO_3$. *See* p555	197.24	(w or dil Na_2CO_3)	140	...	...	...	s	...	...	...	...	ac s alk, s	**B13**^{2}, 233

For explanations, symbols and abbreviations see beginning of table.

No	Name	Synonyms and Formula	Mol. wt.	Crystalline form, color and specific rotation	m.p. °C	b.p. °C	Density	n_D	Solubility						Ref.
									w	al	eth	ace	bz	other solvents	
	Phenol														
p592	**—,2-bromo-***	C_6H_5BrO. *See* p555	173.02		5.6	194–5	1.4924^{20}_{4}		δ	s	s			alk s	**B6**^{1}, 104
p593	**—,3-bromo-***	C_6H_5BrO. *See* p555	173.02	lf	33	236.5 $135–40^{12}$			δ	v	v			chl s alk s	**B6**^{1}, 105
p594	**—,4-bromo-***	C_6H_5BrO. *See* p555	173.02	tetr	66.4	238	1.840^{15}		s	v	v			chl s	**B6**^{1}, 105
p595	**—,3-bromo-5-chloro-***	C_6H_4BrClO. *See* p555	207.47	nd (peth)	70	$256–60^{756}$				s				10 % alk s aa s	**B6**^{2}, 187
p596	**—,4-bromo-2-chloro-***	C_6H_4BrClO. *See* p555	207.47	nd (lig)	29–30	$127–30^{12}$	1.6170^{20}_{4}	1.5859^{20}		v	v	v	v	to v lig $δ^h$	**B6**^{2}, 187
p597	**—,3-bromo-2,4-dinitro-***	$C_6H_3BrN_2O_5$. *See* p555	263.02	pa ye nd (w)	175				v						**B6**^{2}, 249
p598	**—,3-bromo-2,6-dinitro-***	$C_6H_3BrN_2O_5$. *See* p555	263.02	nd (peth)	131				δ					lig δ v^h	**B6**^{2}, 250
p599	**—,4-bromo-2,6-dinitro-***	$C_6H_3BrN_2O_5$. *See* p555	263.02	pa ye nd (w), nd (al),pr (aa)	78				i $δ^h$	s	v		v	chl v lig δ	**B6**^{2}, 250
p600	**—,5-bromo-2,4-dinitro-***	$C_6H_3BrN_2O_5$. *See* p555	263.02	pr (al, eth)	92				s	v	v			con HNO_3 s	**B6**^{2}, 249
p601	**—,6-bromo-2,4-dinitro-***	$C_6H_3BrN_2O_5$. *See* p555	263.02	pa ye (al), pr (eth), colorless nd (w, al)					$δ^h$	s^h	v		v	lig s	**B6**^{2}, 250
p602	**—,2-bromo-3-nitro-***	$C_6H_4BrNO_3$. *See* p555	218.02	pa ye nd (HCl)	147–8	sub			i $δ^h$	v	v			CS_2 δ lig δ	**B6**, 244
p603	**—,2-bromo-4-nitro-***	$C_6H_4BrNO_3$. *See* p555	218.02	(to w), nd (chl, eth, dil al)	113–4				i	v	v			CCl_4 s chl s	**B6**^{2}, 233
p604	**—,2-bromo-5-nitro-***	$C_6H_4BrNO_3$. *See* p555	218.02	pa ye nd (w)	124					v	v			lig δ	**B6**^{2}, 233
p605	**—,2-bromo-6-nitro-***	$C_6H_4BrNO_3$. *See* p555	218.02	pa ye nd (chl, al)	66.5				s^h	v				aa s	**B6**^{2}, 233
p606	**—,3-bromo-2-nitro-***	$C_6H_4BrNO_3$. *See* p555	218.02	ye (peth)	65–7 (anh) 35 (hyd)					v				:lig δ v^h	**B6**^{2}, 232
p607	**—,3-bromo-4-nitro-***	$C_6H_4BrNO_3$. *See* p55	218.02	pa ye nd (bz, peth)	129–30					v	v		s v^h	lig s^h	**B6**^{2}, 234
p608	**—,3-bromo-5-nitro-***	$C_6H_4BrNO_3$. *See* p555	218.02	(w)	145				δ s^h	v			δ		**B6**^{2}, 233
p609	**—,4-bromo-2-nitro-***	$C_6H_4BrNO_3$. *See* p555	218.02	nd, lf (al), pr (eth)	92				i	s	v		v	chl v lig δ	**B6**^{2}, 232
p610	**—,4-bromo-3-nitro-***	$C_6H_4BrNO_3$. *See* p555	218.02	nd (w^h)	147				$δ^h$ i						**B6**, 244
p611	**—,5-bromo-2-nitro-***	$C_6H_4BrNO_3$. *See* p555	218.02	ye pr or nd (lig)	42				s^h	v	v				**B6**^{2}, 233
p612	**—,2-butoxy-***	Pyrocatechol monobutyl ether. $C_{10}H_{14}O_2$. *See* p555	166.21			231–4 159^{69}	1.026^{25}	1.5113^{25}							**B6**, 772
—	**—,butyl-***	*see* **Benzene, butyl (hydroxy)-***													
p613	**—,2-chloro-***	C_6H_5ClO. *See* p555	128.56		8.7	175–6	1.2410^{18}_{15}	1.5473^{40}	v		s		v		**B6**^{2}, 170
p614	**—,3-chloro-***	C_6H_5ClO. *See* p555	128.56	nd	32.8	213–6	1.268^{25}	1.5565^{40}	s^h δ	s	s		v		**B6**^{2}, 172
p615	**—,4-chloro-***	C_6H_5ClO. *See* p555	128.56	nd	43	219.75	1.2651^{40}_{4}	1.5579^{40}	s	v	v		v	alk s	**B6**^{2}, 174
p616	**—,4-chloro-2,3-dinitro-***	$C_6H_3ClN_2O_5$. *See* p555	218.56	pr	127				$δ^h$	s	v		v		**B6**^{2}, 248
p617	**—,2-chloro-2,6-dinitro-***	$C_6H_3ClN_2O_5$. *See* p555	218.56	pr or pl	90		1.74^{22}		δ	v	v			chl s	**B6**^{2}, 248
p618	**—,2-chloro-4-nitro-***	$C_6H_4ClNO_3$. *See* p555	173.56	wh nd (w, 50 % al)	110–1				s^h	v	v		δ	chl v	**B6**^{2}, 228
p619	**—,2-chloro-5-nitro-***	$C_6H_4ClNO_3$. *See* p555	173.56	ye nd (w), rosettes (bz)	121–2				s^h				s^h		**B6**, 240
p620	**—,4-chloro-2-nitro-***	$C_6H_4ClNO_3$. *See* p555	173.56	ye mcl (al)	88–9				δ	s v^h	v			chl v	**B6**^{2}, 266
p621	**—,5-chloro-2-nitro-***	$C_6H_4ClNO_3$. *See* p555	173.56	ye pr (w)	39.5–40	sub			δ	s	s			aa s	**B6**^{2}, 227
p622	**—,2,4-diamino-***	$C_6H_8N_2O$. *See* p555	124.14	lf	78–80d					s	δ	s		chl δ lig δ ac, alk s	**B13**^{2}, 308
p623	**—,—,dihydro-chloride***	$C_6H_8N_2O.2HCl$. *See* p555	197.07	nd	230–40d				v	δ	δ				**B13**^{2}, 308
p624	**—,2,5-diamino-***	$C_6H_8N_2O$. *See* p555	124.14	nd	68				v						**B13**^{2}, 312
p625	**—,3,4-diamino-***	$C_6H_8N_2O$. *See* p555	124.14		170–2										**B13**^{1}, 210
p626	**—,3,5-diamino-***	$C_6H_8N_2O$. *See* p555	124.14	nd (chl)	180				v		δ				**B13**, 567
p627	**—,2,4-dibromo-***	$C_6H_4Br_2O$. *See* p555	251.92	nd (peth)	40	177^{17}			δ	v	v		v	CS_2 v	**B6**^{2}, 188
p628	**—,2,6-dibromo-***	$C_6H_4Br_2O$. *See* p555	251.92	nd (w)	56–7	162^{21}			s^h	v	v				**B6**^{2}, 189
p629	**—,2,6-dibromo-4-nitro-***	$C_6H_3Br_2NO_3$. *See* p555	296.92	pa ye pr or lf (al)	145–6	d>144			i	v^h	v^h		δ	CS_2 v^h lig i	**B6**^{2}, 234

For explanations, symbols and abbreviations see beginning of table.

No.	Name	Synonyms and Formula	Mol. wt.	Crystalline form, color and specific rotation	m.p. °C	b.p. °C	Density	n_D	Solubility: w	al	eth	ace	bz	other solvents	Ref.
	Phenol														
p630	**—,2,3-dichloro-***	$C_6H_4Cl_2O$. *See* p555	163.01	(lig, bz)	57	...	...	...	...	s	s	...	...	...	**B6**[1]**1**, 102
p631	**—,2,4-dichloro-***	$C_6H_4Cl_2O$. *See* p555	163.01	hex nd (bz)	45	$206–8^{753}$	...	...	δ	v	v	...	v	chl v	**B6**[2], 178
p632	**—,2,5-dichloro-***	$C_6H_4Cl_2O$. *See* p555	163.01	pr (bz)	59	211^{744}	...	...	δ	v	v	...	v	...	**B6**[2], 178
p633	**—,2,6-dichloro-***	$C_6H_4Cl_2O$. *See* p555	163.01	nd (peth)	68–9	$219–20^{740}$	...	...	...	v	v	...	...	...	**B6**[2], 179
p634	**—,3,4-dichloro-***	$C_6H_4Cl_2O$. *See* p555	163.01	nd (peth)	68	253.5^{767}	...	...	...	...	...	...	...	...	**B6**[2], 179
p635	**—,3,5-dichloro-***	$C_6H_4Cl_2O$. *See* p555	163.01	pr (peth)	68	233^{757} 122–48	...	...	δ	v	...	...	...	...	**B6**[2], 179
p637	**—,2,6-dichloro-4-nitro-***	$C_6H_3Cl_2NO_3$. *See* p555	208.01	nd (w)	exp<100 (125d)	...	1.822	...	s^h δ	s^h	v	...	δ	chl v	**B6**[2], 231
—	**—,diethyl-***	*see* **Benzene, diethyl-(hydroxy)-***													
p638	**—,3(diethyl-amino)-***	$C_{10}H_{15}NO$. *See* p555	165.23	rh bipym (CS_2-lig)	78	276–80 $174–5^{20}$	...	...	s	s	s	...	...	CS_2 s	**B13**[2], 212
p639	**—,2,4-diiodo-***	$C_6H_4I_2O$. *See* p555	345.93	nd (w)	72–3	sub 100	...	...	s^h δ	v	v	...	δ	chl δ	**B6**[2], 202
p640	**—,2,3-dimeth-oxy-***	Pyrogallol 1,2 dimethyl ether. $C_8H_{10}O_3$. *See* p555	154.16	...	65–70	233–4 $124–5^{17}$	...	...	...	...	...	...	...	...	**B6**, 1081
p641	**—,2,6-dimeth-oxy-***	Pyrogallol 1,3 dimethyl ether. $C_8H_{10}O_3$. *See* p555	154.16	mcl pr (w)	55–6	262.7	...	...	δ	s	v	...	...	aq alk s	**B6**[2], 1065
p642	**—,3,5-dimeth-oxy-***	Phloroglucinol dimethyl ether. $C_8H_{10}O_3$. *See* p555	154.16	(bz, lig)	36–8	$198–200^{35}$	...	...	...	...	s	...	...	...	**B6**[2], 1078
—	**—,dimethyl-***	*see* **Benzene, dimethyl-(hydroxy)-***													
p643	**—,3(dimethyl-amino)-***	$C_8H_{11}NO$. *See* p555	137.18	nd (lig)	87	265–8 $152–3^{15}$	...	1.5895^{26}	$δ^h$	v	v	v	v	CS_2 v	**B13**[2], 211
p644	**—,2,3-dinitro-***	$C_6H_4N_2O_5$. *See* p555	184.11	ye nd (w)	144	...	1.681^{20}	...	δ	v^h	v	...	...	...	**B6**, 251
p645	**—,2,4-dinitro-***	$C_6H_4N_2O_5$. *See* p555	184.11	ye pl or lf (w)	115–6 (111.6)	...	...	...	δ s^h	s	s	...	s	chl s	**B6**, 251
p646	**—,2,5-dinitro-***	$C_6H_4N_2O_5$. *See* p555	184.11	ye mcl pr or nd (dil al, w, lig)	108	...	...	...	δ s^h	δ v^h	v	...	s v^h	MeOH δ	**B6**[2], 244
p647	**—,2,6-dinitro-***	$C_6H_4N_2O_5$. *See* p555	184.11	pa ye rh nd or lf	63–4	...	...	...	s^h δ	v^h	v	...	s	chl s	**B6**, 257
p648	**—,3,4-dinitro-***	$C_6H_4N_2O_5$. *See* p555	184.11	tcl nd (w)	134	...	1.672	...	δ s^h	s	s	...	...	...	**B6**, 257
p649	**—,3,5-dinitro-***	$C_6H_4N_2O_5$. *See* p555	184.11	lf	126.1	...	1.702	...	...	s	...	...	s	chl s lig δ	**B6**, 258
p650	**—,2-ethoxy-***	Pyrocatechol monoethyl ether. $C_8H_{10}O_2$. *See* p555	138.16	...	29	217	...	...	δ	∞	∞	...	...	...	**B6**, 771
p651	**—,3-ethoxy-***	Resorcinol monoethyl ether. $C_8H_{10}O_2$. *See* p555	138.16	...	66	246–7 $117^{5.5}$	...	...	i	s	s	...	s	...	**B6**, 814
p652	**—,4-ethoxy-***	Hydroquinone monoethyl ether. $C_8H_{10}O_2$. *See* p555	138.16	pr or lf (w)	66–7	246–7	...	...	δ s^h	v	v	...	...	...	**B6**, 843
—	**—,ethyl-***	*see* **Benzene, ethyl-(hydroxy)-**													
p653	**—,2(ethyl-amino)-***	$C_8H_{11}NO$. *See* p555	137.18	pl (bz)	107.5	...	...	...	i	v	δ	...	s^h	...	**B13**, 364
p654	**—,3(ethyl-amino)-***	$C_8H_{11}NO$. *See* p555	137.18	(bz)	62	176^{12}	...	...	s^h	s	s	...	...	chl v	**B13**, 408
p655	**—,4(ethyl-amino)-***	$C_8H_{11}NO$. *See* p555	137.18	nd (w)	110–2	...	...	...	s^h	s	s	...	...	...	**B13**, 443
p655[1]	**—,4(heptyloxy)-***	Hydroquinone monoheptyl ether. $C_{13}H_{20}O_2$. *See* p555	208.29	...	60	...	...	...	...	...	...	...	...	...	...
p655[2]	**—,4(hexyloxy)-***	Hydroquinone monohexyl ether. $C_{12}H_{18}O_2$. *See* p555	194.26	...	48	...	...	...	...	...	...	...	...	...	...
p656	**—,2-iodo-***	C_6H_5IO. *See* p555	220.01	nd	43	$186–7^{160}$	1.8757^{80}	...	s	v	v	...	...	CS_2 v	**B6**, 201
p657	**—,3-iodo-***	C_6H_5IO. *See* p555	220.01	nd (lig)	40	d	...	...	δ	s	s	...	...	...	**B6**, 208
p658	**—,4-iodo-***	C_6H_5IO. *See* p555	220.01	nd (w)	93–4	$138–40^{5}$d	1.8573^{112}	...	δ	v	v	...	...	...	**B6**, 208
p659	**—,2-mercapto-***	Monothiopyrocatechol. C_6H_6OS. *See* p555	126.17	...	5–6	$216–7^{751}$	1.2373^{0}_{0}	...	δ	...	s	...	...	alk s	**B6**, 793
p660	**—,3-mercapto-***	Monothioresorcinol. C_6H_6OS. *See* p555	126.17	...	16–7	168^{35}	...	...	δ	s	...	...	...	os v alk, con sulf	**B6**[2], 285
p661	**—,4-mercapto-***	Monothiohydroquinone. C_6H_6OS. *See* p555	126.17	...	29–30	$166–8^{45}$	...	...	s	s	...	...	...	alk s con sulf s	**B6**[1], 419
p662	**—,2-methoxy-***	Pyrocatechol monomethyl ether. Guaiacol. $C_7H_8O_2$. *See* p555	124.14	hex pr	28.65	205.05	1.1287^{21}_{4}	1.5383^{21}	δ	s	s	...	...	os v chl s	**B6**[2], 776
p663	—,acetate	$C_9H_{10}O_3$. *See* p555	166.18	...	...	$123–4^{13}$	1.1285^{25}_{4}	1.5101^{25}	i	∞	∞	...	...	...	**B6**[2], 783
p664	**—,3-methoxy-***	Resorcinol monomethyl ether. $C_7H_8O_2$. *See* p555	124.14	...	<-17.5	244.3	...	...	δ	∞	∞	...	...	...	**B6**, 813
p665	**—,4-methoxy-***	Hydroquinone monomethyl ether. $C_7H_8O_2$. *See* p555	124.14	...	53	243	...	...	s	v	v	...	s	...	**B6**, 843

For explanations, symbols and abbreviations see beginning of table.

No.	Name	Synonyms and Formula	Mol. wt.	Crystalline form, color and specific rotation	m.p. °C	b.p. °C	Density	n_D	Solubility						Ref.
									w	al	eth	ace	bz	other solvents	
	Phenol														
p666	**—,2-methoxy-3-nitro-***	$C_7H_7NO_4$. *See* p555	169.14	yesh rh pr (CS_2, peth)	68.5–9.5				δ	v				CS_2 δ, s^h	**B6²**, 789
p667	**—,2-methoxy-4-nitro-***	$C_7H_7NO_4$. *See* p555	169.14	ye nd (w)	103–4				v^h	v	v				**B6²**, 790
p668	**—,2-methoxy-5-nitro-***	$C_7H_7NO_4$. *See* p555	169.14	pa ye nd (w)	104				s^h	s	s				**B6²**, 790
p669	**—,2-methoxy-6-nitro-***	$C_7H_7NO_4$. *See* p555	169.14	og-ye nd (sub)	62				v	s					**B6²**, 789
p670	**—,3-methoxy-4-nitro-***	$C_7H_7NO_4$. *See* p555	169.14	ye nd (al)	144					s^h			s^h		**B6²**, 822
p671	**—,3-methoxy-5-nitro-***	$C_7H_7NO_4$. *See* p555	169.14	ye	141–2					s					**B6**, 825
p672	**—,4-methoxy-2-nitro-***	$C_7H_7NO_4$. *See* p555	169.14	og ye nd or mcl cr (al, lig)	80					s^h					**B6²**, 849
p673	**—,4-methoxy-3-nitro-***	$C_7H_7NO_4$. *See* p555	169.14	pa ye nd (w), (bz)	98–100				δ v^h				δ v^h		**B6²**, 848
p674	**—,5-methoxy-2-nitro-***	$C_7H_7NO_4$. *See* p555	169.14	yesh nd (al)	95					s^h					**B6²**, 822
—	**—,methyl-***	*see* **Toluene, hydroxy-**													
p675	**—,2(methyl-amino)-***	C_7H_9NO. *See* p555	123.16	pl (bz)	86–7				i	s			s		**B13**, 362
p676	**—,4(methyl-amino)-, sulfate***	Metol. Photol. Pictol. $2(C_7H_9NO).H_2SO_4$. *See* p555	344.39	wh pw	250–60d				v	s					**B13**, 441
p677	**—,4(methyl-amino)2-nitro-***	$C_7H_8O_3N_2$. *See* p555	168.15	dk red-br nd (al)	113–4					s^h				alk s (red)	**B13¹**, 186
p678	**—,4(2-methyl-aminopropyl)-***	$C_{10}H_{15}NO$. *See* p555	165.24	(MeOH)	162–3				δ	s	s			MeOH s^h dil ac v	
p679	**—,2-nitro-***	$C_6H_5NO_3$. *See* p555	139.11	ye nd or pr (eth, al)	44.9	216	1.485^{14}		$δ^h$	v^h	v			alk s	**B6**, 213
p680	**—,3-nitro-***	$C_6H_5NO_3$. *See* p555	139.11	ye mcl (aq HCl, eth)	97	194^{70}	1.2797^{100}_{4}		δ v^h	v				dil ac s alk s	**B6**, 222
p681	**—,4-nitro-***	$C_6H_5NO_3$. *See* p555	139.11	ye mcl pr (to)	114	279d (sub)	1.479^{20}		δ v^h	v	v			chl s	**B6**, 226
p682	**—,4-nitroso-***	1,4 Benzoquinone monoxime. $C_6H_5NO_2$. *See* p555	123.11	pa ye rh nd (ace bz)	126d				δ	s	s	s		dil alk s	**B6²**, 205 **B7²**, 574
p682¹	**—,4(octyloxy)-***	Hydroquinone monooctyl ether. $C_{14}H_{22}O_2$. *See* p555	222.33		60–1										
p683	**—,pentabromo-***	C_6HBr_5O. *See* p555	488.62	mcl pr (aa), nd (al, CS_2)	229.5	sub			i	s^h	δ		s^h		**B6²**, 197
p684	**—,pentachloro-***	C_6HCl_5O. *See* p555	266.35	mcl pr (al+1w), nd (bz)	174 (+1w), 191 (anh)	$309–10^{754}$ d	1.978^{22}_{4}		δ	v	v				**B6²**, 182
p685	**—,2-propoxy-***	Pyrocatechol monopropyl ether. $C_9H_{12}O_2$. *See* p555	152.19			223–6 140^{68}	1.0523^{25}	1.5176^{25}							
p686	**—,3-propoxy-***	Resorcinol monopropyl ether. $C_9H_{12}O_2$. *See* p555	152.19			120^5									
p687	**—,4-propoxy-***	Hydroquinone monopropyl ether. $C_9H_{12}O_2$. *See* p555	152.19		56–7										
—	**—,propyl-***	*see* **Benzene, hydroxy-(propyl)-***													
p688	**—,seleno-***	Phenylselen mercaptan. C_6H_5SeH	157.07			183.6	1.4865^{15}		i	s	v			CCl_4 v	**B6²**, 317
p689	**—,2,3,4,6-tetra-bromo-***	$C_6H_2Br_4O$. *See* p555	409.74	nd (al)	120	sub				v			s	alk s	**B6²**, 196
p690	**—,2,3,4,5-tetra-chloro-***	$C_6H_2Cl_4O$. *See* p555	231.91	nd (peth, sub)	11.6	sub				v				alk, MeOH v	**B6²**, 182
p691	**—,2,3,4,6-tetra-chloro-***	$C_6H_2Cl_4O$. *See* p555	231.91	nd (lig, aa)	69–70	d 150^{15}			i $δ^h$	v				chl, NaOH v	**B6²**, 182
p692	**—,2,3,5,6-tetra-chloro-***	$C_6H_2Cl_4O$. *See* p555	231.91	(peth)	115				δ				v		**B6²**, 182
p693	**—,2,3,4,6-tetra-nitro-***	$C_6H_2N_4O_9$. *See* p555	274.11	lt ye nd (chl)	140d	exp			v						**B4**, 293
p694	**—,2,4,6-tri-amino-***	$C_6H_9N_3O$. *See* p555	139.16			257			v	s	s				**B13²**, 317
p695	**—,2,3,5-tri-bromo-***	$C_6H_3Br_3O$. *See* p555	330.83	nd or pl (al, lig)	94–5				δ	v	v	v		alk v lig v	**B6²**, 192
p696	**—,2,4,5-tri-bromo-***	$C_6H_3Br_3O$. *See* p555	330.83	nd (lig)	87					s				alk s lig s	**B6²**, 192
p697	**—,2,4,6-tri-bromo-**	$C_6H_3Br_3O$. *See* p555	330.83	nd (al), pr (bz)	95–6	$282–90^{764}$ sub	2.55^{20}_{20}		δ	v	s			chl s	**B6**, 203

For explanations, symbols and abbreviations see beginning of table.

No.	Name	Synonyms and Formula	Mol. wt.	Crystalline form, color and specific rotation	m.p. °C	b.p. °C	Density	n_D	Solubility w	al	eth	ace	bz	other solvents	Ref.
	Phenol														
p698	**—,3,4,5-tribromo-***	$C_6H_3Br_3O$. *See* p555	330.83	tcl (bz lig)	129	...	...	...	...	s	...	...	s	alk v aa s	**B6**², 195
p699	**—,2,3,4-trichloro-***	$C_6H_3Cl_3O$. *See* p555	197.46	nd (sub)	83.5	sub	...	...	...	s	s	...	...	alk s aa s	**B6**², 179
p700	**—,2,3,5-trichloro-***	$C_6H_3Cl_3O$. *See* p555	197.46	nd (al)	62	248.5–9.5^{250}	...	...	δ	s	s	...	...	...	...
p701	**—,2,3,6-trichloro-***	$C_6H_3Cl_3O$. *See* p555	197.46	nd (dil al, lig)	58	...	...	...	$δ^h$	v	v	...	v	aa s lig s	**B6**², 180
p702	**—,2,4,5-trichloro-***	$C_6H_3Cl_3O$. *See* p555	197.46	nd (al), (lig)	66–7	sub	...	...	δ	s	...	...	...	os s lig s	**B6**², 180
p703	**—,2,4,6-trichloro-***	$C_6H_3Cl_3O$. *See* p555	197.46	rh nd	68	246	1.4901^{75}_{4}	...	δ	v	v	...	...	...	**B6**², 179
p704	**—,3,4,5-trichloro-***	$C_6H_3Cl_3O$. *See* p555	197.46	nd (lig)	101	271–7^{746}	...	...	...	...	...	...	...	lig δ v^h	**B6**², 181
p705	**—,2,3,5-triiodo-***	$C_6H_3I_3O$. *See* p555	471.84	nd (peth)	114	...	...	...	...	s	...	...	...	os v	**B6**, 211
p706	**—,2,4,6-triiodo-***	$C_6H_3I_3O$. *See* p555	471.84	nd (al)	156–8	sub d	...	...	i	δ	s	s	...	...	**B6**, 211
p707	**—,2,3,6-trinitro-***	$C_6H_3N_3O_7$. *See* p555	229.11	nd (w)	119	...	...	...	v^h δ	v	v	...	...	aa s	**B6**, 265
p708	**—,2,4,5-trinitro-***	$C_6H_3N_3O_7$. *See* p555	229.11	wh nd	96	...	...	...	δ s^h	v	...	...	...	aa v	**B6**², 253
p709	**—,2,4,6-trinitro-***	Picric acid. $C_6H_3N_3O_7$. *See* p555	229.11	ye	122–3	...	exp>300	1.763	s^h δ	s	δ	...	...	chl s aa s	**B6**², 253
p710	**Phenolphthalein**	2,2 bis(*n* Hydroxyphenyl) phthalide. 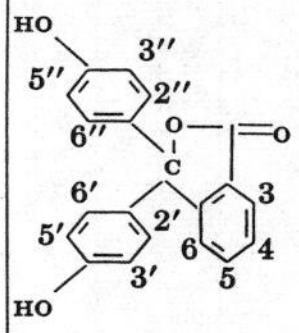	318.31	wh rh nd	261–2	...	1.277^{32}_{4}	...	δ	v	s	...	...	...	**B18**, 143
p711	**—,3′,3″,5′,5″-tetrabromo-**	$C_{20}H_{10}Br_4O_4$. *See* p710	633.96	nd (al, eth, aa)	293	...	...	...	i	i	v	...	...	alk s	**B18**², 124
p712	**—,3′3″,5′5″-tetrachloro-**	$C_{20}H_{10}Cl_4O_4$. *See* p710	456.12	(bz, aa)	225	...	...	...	i	s	...	s	v	$PhNO_2$ s	**B18**², 123
p713	**—,3′,3″,5′,5″-tetraiodo-**	Iodophen. Nosophen. $C_{20}H_{10}I_4O_4$. *See* p710	821.96	amor	268–9d	...	2.0246^{22}_{22}	...	i	i	i	...	...	chl δ	...
p714	**Phenolsulfonphthalein**	Phenol red. 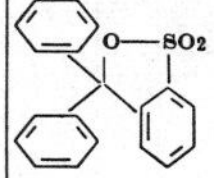	346.39	nd or pl	...	...	...	...	δ	s	δ	...	δ	KOH s	**B19**², 102
p715	**Phenolphthalin**	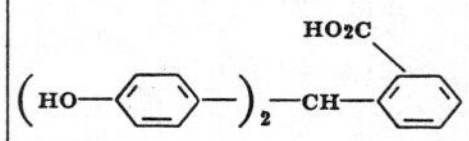	320.35	nd (w)	229–32	...	...	...	i	s	...	...	...	NaOH s	**B10**², 322
p716	**Phenothiazine**	Thiodiphenylamine. 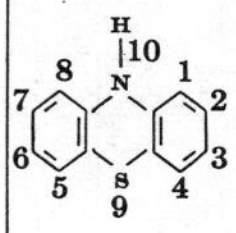	199.27	ye pr (al)	185.5–.9	371 290^{40}	...	...	δ	...	s	v	s	CCl_4 i	**B27**, 63
p717	**—,2,7-dinitro-,** 9-oxide	$C_{12}H_7N_3O_5S$. *See* p716	305.28	ye-red lf (aa), pr ($PhNH_2$)	d	...	...	...	...	...	...	...	...	os δ $PhNO_2$ s^h	**B27**¹, 229
p718	**—,10(2-diethylaminoethyl)-**	$C_{18}H_{22}N_2S$. *See* p716	298.44	...	...	195–208$^{4-5}$	...	...	i	...	...	...	...	dil HCl s	...
p719	**—,10(2-dimethylaminopropyl)-**	Phenergan base. $C_{17}H_{20}N_2S$. *See* p716	284.41	...	60	190–2^{3}	...	...	i	...	...	...	...	dil HCl v	...
p720	**—,—,hydrochloride**	$C_{17}H_{20}N_2S.HCl$. *See* p716	320.88	...	203–4	...	...	...	v	s	...	...	...	chl s	...
p721	**—,10-phenyl-**	$C_{18}H_{13}NS$. *See* p716	275.36	ye pr (al)	89–90	...	...	...	i	s^h	...	...	...	con sulf s	**B27**, 227
p722	**Phenoxanthin**	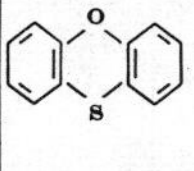	200.26	nd, cr (MeOH)	60–1	183–4^{12}	...	...	...	s	...	...	...	os v	**B19**², 33

For explanations, symbols and abbreviations see beginning of table.

No.	Name	Synonyms and Formula	Mol. wt.	Crystalline form, color and specific rotation	m.p. °C	b.p. °C	Density	n_D	Solubility w	al	eth	ace	bz	other solvents	Ref.
	Phenoxazine														
p723	**Phenoxazine**	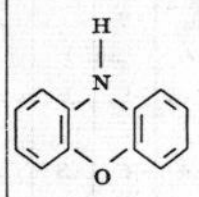	183.20	lf (dil al, bz)	156	δd				v	v			chl v con sulf s aa v lig s	**B27²**, 31
—	**Phenoxytol**	*see* **Ethanol, 2-phenoxy-***													
p724	**Phenylalanine**	*l*-α-Aminohydrocinnamic acid. *l*-β-Phenyl-α-aminopropionic acid. $C_6H_5CH_2CH(NH_2)CO_2H$	165.19	nd or lf (w), $[\alpha]_D^{18}$−48.4 (NaOH, p=6.4), $[\alpha]_D^{22}$ −35.03(w)	283	sub		1.600	s	δ^h	i			dil ac δ alk s os i	**B14²**, 297
p725	—(*DL*)	$C_6H_5CH_2CH(NH_2)CO_2H$	165.19	red-br lf (aq al), pr (w)	284–8d	sub par d			δ	δ	i				**B14²**, 299
p726	—(*L*)	$C_6H_5CH_2CH(NH_2)CO_2H$	165.19	lf (w), $[\alpha]_D^{25}$ +35.3	283–4				s	i	i			MeOH δ os i	**B14²**, 296
p727	—,*N*-**acetyl-(*d*,+)**	$C_6H_5CH_2CH(NHCOCH_3)CO_2H$	207.23	(w), $[\alpha]_D^{26}$−50.9	172				s^h	s					**B14²**, 297
p728	—,—(*dl*)	$C_6H_5CH_2CH(NHCOCH_3)CO_2H$	207.23	nd	152				s	s					**B14²**, 302
p729	—,—(*l*)	$C_6H_5CH_2CH(NHCOCH_3)CO_2H$	207.23	$[\alpha]_D^{26}$+51.4	172					s					**B14²**, 298
—	**Phenylenediamine**	*see* **Benzene, diamino-***													
—	**Phenylglycidol**	*see* **1-Propanol, 2,3-epoxy-3-phenyl-***													
—	**Phenylglycine**	*see* **Acetic acid, amino (phenyl)-***													
—	**Phloramin**	*see* **Benzene, 1-amino-3,5-dihydroxy-***													
p730	**Phloretin**	Dihydronaringenin. 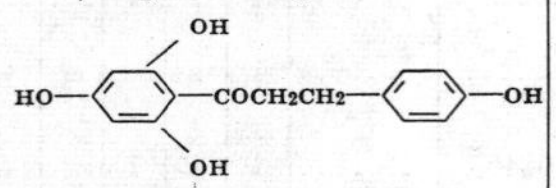	274.27	lf	262–4d				δ^h	∞	δ		∞		**B8**, 498
p731	**Phlorhizin**, dihydrate	Asebotin. Asebotoside. 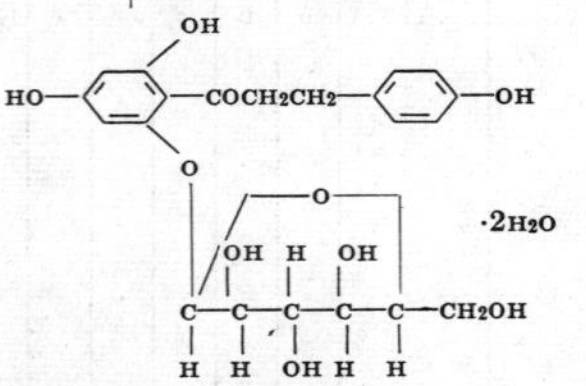	472.44	nd	110(+2w) 170d (anh)		1.4298		i						
—	**Phloroglucinol**	*see* **Benzene, 1,3,5-trihydroxy-***													
—	**Phlorol**	*see* **Benzene, 1-ethyl-2-hydroxy-***													
—	**Phorone**	*see* **2,5-Heptadiene-4-one, 2,6-dimethyl-***													
p732	**Phosgene**	Carbonic acid dichloride. Carbonyl chloride. Chloroformyl chloride. Cl_2CO.	98.92		−118	8.02	1.392_4^{19}		d	d				to s aa s	**B3**, 13
p733	—,**thio-**	Thiocarbonyl chloride. $SCCl_2$.	114.98	red		73	1.508^{15}	1.5442	i	s	s				**B3**, 105
p734	**Phosphine, bis-(trifluoro-methyl)-**	$(F_3C)_2PH$	169.99	spontaneously inflam		1									
p735	—,**bis(trifluoro-methyl)chloro-**	$(F_3C)_2PCl$	204.44	spontaneously inflam		21			d						
p736	—,**bis(trifluoro-methyl)cyano-**	$(F_3C)_2PCN$	195.00	spontaneously inflam		48		1.3248^{20}							
p737	—,**bis(trifluoro-methyl)iodo-**	$(F_3C)_2PI$	295.89			73		1.403^{15}	d					Bu_2O s	

For explanations, symbols and abbreviations see beginning of table.

No.	Name	Synonyms and Formula	Mol. wt.	Crystalline form, color and specific rotation	m.p. °C	b.p. °C	Density	n_D	Solubility w	al	eth	ace	bz	other solvents	Ref.
	Phosphine														
p738	**—,dichloro(2,4-dimethylphenyl)-**	$C_8H_9Cl_2P$. *See* p741	207.04			256–8			d^h						**B16**, 773
p739	**—,dichloro(2,5-dimethylphenyl)-**	$C_8H_9Cl_2P$. *See* p741	207.04		−30	253–4	1.25^{18}_{18}		d^h						**B16**, 773
p740	**—,dichloro(4-ethylphenyl)-**	$C_8H_9Cl_2O$. *See* p741	207.04			250–2	1.227^{17}		d^h						**B16**, 772
p741	**—,dichlorophenyl-**	3 2 / 4⟨ ⟩—PCl_2 / 5 6	178.99			224.6	1.319^{20}	1.6053^{7}	d				∞	CS_2 ∞	**B16**², 373
p742	**—,dichlorotrifluoromethyl-**	CF_3PCl_2	170.89			37			d						
p743	**—,diethyl-**	$(C_2H_5)_2PH$	90.11			85									**B4**, 582
p744	**—,diiodo(trifluoromethyl)-**	F_3CPI_2	353.79	ye fum		d^{760} 73^{37}		1.630^{20}	i d		s			Bu_2O s	
p745	**—,dimethyl-**	$(CH_3)_2PH$	62.05			25	<1		i	s	s				**B4**, 580
p746	**—,diphenyl-(ethyl)-**	$(C_6H_5)_2PC_2H_5$	214.25			293			δ				δ		**B16**², 371
p747	**—,ethyl-**	$CH_3CH_2PH_2$	62.05			25	<1		d						**B4**², 969
p748	**—,methyl-**	CH_3PH_2	48.02	gas		-14^{759}			i	δ	v				**B4**², 969
p749	**—,phenyl-**	$C_6H_5PH_2$	110.10			160–1	1.001^{15}								**B16**², 369
p750	**—,tris(chloromethyl)-**	$P(CH_2Cl)_3$	179.41			100^{7}									
p751	**—,triethyl-**	$(C_2H_5)_3P$	118.16			127.5^{741}	0.8006^{19}_{4}	1.458^{15}	i	∞	∞				**B4**², 969
p752	—,—,oxide	$(C_2H_5)_3PO$	134.16	wh hyg nd	50	243			s	s	s				**B4**, 592
p753	—,—,sulfide	$(C_2H_5)_3PS$	150.22		94				s						**B4**, 592
p754	**—,(trifluoromethyl)-**	F_3CPH_2	102.00	spontaneously inflam		−26.5									
p755	**—,trimethyl-**	$(CH_3)_3P$	76.08			40–2	<1		i		s				**B4**², 969
p756	**—,triphenyl-**	$(C_6H_5)_3P$	262.29		80	>360	1.194	1.5248^{65}	i	s	v		s	chl s	**B16**, 760
p757	—,—,oxide	$(C_6H_5)_3PO$	278.29	pr (hyd)	156	>360	1.2124^{23}_{4}		$δ^h$	v	δ		v		**B16**, 783
p758	**—,tris(trichloromethyl)-**	$(CCl_3)_3P$	386.08		53				d						
p759	**—,tris(trifluoromethyl)-**	$(CF_3)_3P$	237.99	spontaneously inflam		17.3			i					w d^{200}	
p760	—,—,oxide	$(CF_3)_3PO$	253.99			23.6			d						
p761	**Phosphinic acid, bis(trifluoromethyl)***	$(F_3C)_2PO_2H$	201.99	visc liq		182	<1.9		s					con sulf i	
p762	**—,diethyl-***	$(C_2H_5)_2PO_2H$	122.10		18.5	$92^{0.08}$			v	v	v				
p763	—,—,anhydride*	$[(C_2H_5)_2PO]_2O$	226.20			188^{14}	1.1053^{23}_{4}	1.4720^{23}	d	d	s				
p764	—,—,chloride	$(C_2H_5)_2POCl$	140.55				104^{15}		d	d					
p765	—,—,ethyl ester*	$(C_2H_5)_2PO(OC_2H_5)$	150.16			95^{14}	1.0022^{25}_{4}	1.4375^{25}	δ	s	s				
p766	**—,dimethyl-***	$(CH_3)_2PO_2H$	94.05		92	377			v	v	v		s		**B4**, 593
p767	—,—anhydride*	$[(CH_3)_2PO]_2O$	170.09	nd	119–21	192^{15}			d	d					
p768	—,—,chloride	$(CH_3)_2POCl$	112.51	nd	66.8–8.4	$220–4^{745}$			d	d					
—	**—,ethyl-***	*see* **Ethanephosphinic acid**													
—	**—,methyl-***	*see* **Methanephosphinic acid**													
—	**—,phenyl-***	*see* **Benzenephosphinic acid**													
p770	**Phosphoric acid,** diethyl ester	$(C_2H_5O)_2POOH$	154.10			203.3	1.175	1.4148^{25}	i		s				**B1**², 331
p771	—,dimethyl ester*	$(CH_3O)_2POOH$	126.05			172–6d	1.335^{25}	1.408^{25}	s	s	i	s		to, CCl_4 i	**B1**³, 1205
p772	—,monoethyl ester*	$(C_2H_5O)PO(OH)_2$	126.05	hyg cr		d	1.430^{25}_{4}	1.427	s d^h	s	s	s		CCl_4 i to i	**B1**³, 1327
p773	—,monophenyl ester*	$(C_6H_5O)PO(OH)_2$	174.09	pl (chl), nd (w)	99.5				v	v	v		v	chl s	**B6**², 125
p774	—,triamide, hexamethyl-*	$[(CH_3)_2N]_3PO$	179.20			$98–100^{6}$	1.024^{25}_{25}	1.4570^{25}							
p775	—,tributyl ester*	$(C_4H_9^nO)_3PO$	266.32			289 $160–2^{15}$	0.9727^{25}_{4}	1.4224^{25}	s	∞	s		s	CS_2 s	**B1**³, 1511
p776	—,triethyl ester*	$(C_2H_5O)_3PO$	182.16		−56.4	215–6	1.0725^{19}	1.4067^{17}	$s^{δd}$	s ∞	s		s		**B1**², 331
p777	—,triisobutyl ester	$(C_4H_9^iO)_3PO$	266.32			264	0.9617^{25}_{4}	1.4193^{20}	v						**B1**³, 1563

For explanations, symbols and abbreviations see beginning of table.

PHYSICAL CONSTANTS OF ORGANIC COMPOUNDS (Continued)

No.	Name	Synonyms and Formula	Mol. wt.	Crystalline form, color and specific rotation	m.p. °C	b.p. °C	Density	n_D	Solubility w	al	eth	ace	bz	other solvents	Ref
	Phosphoric acid														
p778	—,triisopropyl ester*	$(C_3H_7^iO)_3PO$	224.24			$218–20^{763}$ $95–6^{8}$	0.9867_4^{20}	1.4057^{20}	...	s	...	...	...		$B1^3$, 1466
p779	—,trimethyl ester*	$(CH_3O)_3PO$	140.08			197.2	1.2506_0^0		v	δ	s	...	...		B1, 286
p780	—,triphenyl ester*	$[C_6H_5\text{–}O\text{–}]_3PO$ (ring positions 2, 3, 4, 5, 6)	326.29	(abs al-lig), pr (al), nd (eth-lig)	50	245^{11}	1.2055_4^{50}		i	s	v	...	v	chl v	$B6^2$, 166
p781	—,tripentyl ester*	$[CH_3(CH_2)_4O]_3PO$	308.40			167^{5}		1.4319^{20}	i	s	s	...	...	to, CS_2 s	
p782	—,tripropyl ester*	$(C_3H_7^nO)PO$	224.24			252 107.5^{5}	1.0121_4^{20}	1.4165^{20}	δ	s	s	...	...	to, CS_2 s	$B1^1$, 179
p783	—,tris(2,4-dimethyl-phenyl) ester*	$C_{24}H_{27}O_4P$. *See* p780	410.45	glassy on fr		232–5	1.142_4^{38}	1.5550^{20}	i	...	...	...	s	hexane s	B6, 488
p784	—,tris(2,5-dimethyl-phenyl) ester*	$C_{24}H_{27}O_4P$. *See* p780	410.45		78.6–81	$260–5^{8}$			i	δ	...	...	s	hexane δ	
p785	—,tris(2,6-dimethyl-phenyl) ester*	$C_{24}H_{27}O_4P$. *See* p780	410.45		136–8	$262–4^{6}$			i	δ	...	...	s	hexane δ	
p786	—,tris(3,4-dimethyl-phenyl) ester*	$C_{24}H_{27}O_4P$. *See* p780	410.45	wax	71.5–2.5	$260–3^{7}$			i	δ	...	...	s	hexane δ	B6, 482
p787	—,tris(3,5-dimethyl-phenyl) ester*	$C_{24}H_{27}O_4P$. *See* p780	410.45		46–6.5	290^{10}			i	δ	...	...	...	hexane δ aa s	
p788	—,tri(2-tolyl) ester	$C_{21}H_{21}O_4P$. *See* p780	368.37	colorless or pa ye	11	410 $263–5^{20}$	1.183^{25}	1.5575^{20}	...	...	...	...	...	aa s	
p789	—,tri(3-tolyl) ester	$C_{21}H_{21}O_4P$. *See* p780	368.37	wax	25–6	$258–63^{4}$	1.150^{25}	1.5575^{20}	i	δ	s	...	...	aa s	
p790	—,tri(4-tolyl) ester	T.P.C. $C_{21}H_{21}O_4P$. *See* p780	368.37	nd (al), ta (eth)	77–8	$244^{3.5}$	1.247^{25}		...	s	v	...	s	chl s	B6, 401
p793	—,tris(2,2,2-tri-chloroethyl) ester*	$(Cl_3CCH_2O)_3PO$	492.16		73–4				...	...	v	...	...	lig δ	B1, 338
p794	**—,thiono-,** tri(2-tolyl) ester	$(2\text{-}CH_3C_6H_4\text{–}O)_3PS$	384.44	nd (al)	45–6				i	s	...	...	...	aa v	$B6^1$, 173
p795	—,—,tri(3-tolyl) ester	$[3\text{-}CH_3C_6H_4\text{–}O]_3PS$	384.44	nd (al)	40–1	$270–2^{12}$			i	s^h δ	...	...	...	MeOH δ aa v lig δ	$B6^2$, 354
p796	—,—,tri(4-tolyl) ester	$[CH_3\text{–}C_6H_4\text{–}O]_3PS$	384.44	nd (al)	93–4				i	s^h δ	...	...	δ	chl v aa v lig δ	$B6^2$, 382
p797	**Phosphorous acid,** diethyl ester*	$(C_2H_5O)_2POH$	138.10			87^{20}	1.0720_0^{20}	1.4101^{20}	d^h	s	s	...	...		$B1^3$, 1329
p798	—,dimethyl ester*	$(CH_3O)_2POH$	110.05			170–1 70^{25}	1.2004_0^{20}	1.4036^{20}	...	s	...	...	...	Py s	$B1^3$, 1203
p799	—,tributyl ester*	$[(CH_3(CH_2)_3O]_3P$	250.32			122^{12}	0.9259_4^{20}	1.4321^{19}	d^h	s	v	...	...		$B1^1$, 187
p800	—,triethyl ester*	$(C_2H_5O)_3P$	166.16			157.9 48.2	0.9687_4^{20}	1.4131^{20}	i	v	v	...	...		$B1^2$, 330
p801	—,triisopropyl ester*	$[(CH_3)_2CHO]_3P$	208.24			$60–1^{10}$	$0.9687_0^{18.5}$		...	s	s	...	...		$B1^2$, 382
p802	—,trimethyl ester*	$(CH_3O)_3P$	124.08			111–2 22^{23}	1.0520_0^{20}	1.4095^{20}	d^h	v	v	...	...		$B1^3$, 1205
p803	—,triphenyl ester*	$(C_6H_5O)_3P$	310.29		*ca.* 25	360 $200–1^{5}$	1.184_{18}^{18}		i d^h	v	...	...	...	os v	$B6^2$, 164
p804	—,tripropyl ester	$(CH_3CH_2CH_2O)_3P$	208.24			206–7 $83–4^{10}$	0.9705_0^0		...	s	s	...	...		$B1^2$, 369
p805	—,tri(2-tolyl) ester	$[2\text{-}CH_3C_6H_4\text{–}O]_3P$	352.37	ye		238^{11}			δd	δd	s	...	...		$B6^2$, 331
p806	—,tri(4-tolyl) ester	$[CH_3\text{–}C_6H_4\text{–}O]_3P$	352.37	pa ye		$250–5^{10}$			...	...	s	...	...		$B6^2$, 381
p807	—,tris(2,2,2-tri-chloroethyl) ester*	$(Cl_3CCH_2O)_3P$	476.16			263			...	...	...	...	...		B1, 338
p808	**Phthalaldehyde**	1,2 Benzene dicarbonal*. *o*-Phthalic aldehyde. $C_6H_4(CHO)_2$	134.14	ye, nd (peth)	56–7				δ	s	s	...	...	lig δ	B7, 674

For explanations, symbols and abbreviations see beginning of table.

No.	Name	Synonyms and Formula	Mol. wt.	Crystalline form, color and specific rotation	m.p. °C	b.p. °C	Density	n_D	Solubility w	al	eth	ace	bz	other solvents	Ref.
	Phthalaldehydic acid														
—	**Phthalaldehydic acid**	*see* **Benzoic acid, 2-formyl-**													
—	**Phthalamic acid**	*see* **Phthalic acid,** monoamide													
—	**Phthalamide**	*see* **Phthalic acid,** diamide													
—	**Phthalanil**	*see* **Phthalic acid,** imide, *N*-phenyl-													
—	**Phthalazine, 6-amino-1,4-dioxo-1,2,3,4-tetrahydro-**	*see* **Phthalic acid, 4-amino-,** hydrazide													
p810	**—,1,2-dihydro-1-oxo**	1(2H)-Phthalazinone. 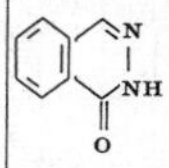	146.15	nd (w)	184–5	337^{755}			v^h	v^h	...	...	v^h	alk s ac s	**B24**[2], 70
p811	**—,1,4-dihydroxy-**	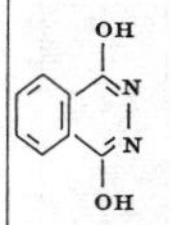	162.15	nd (w, dil al, aa)	341–4 cor				δ^h i	δ^h	i	...	i	chl i dil alk s aa s	**B24**[2], 194
p812	**Phthalic acid**	1,2-Benzene dicarboxylic acid*. 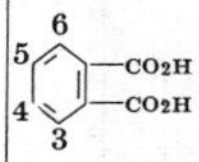	166.13	pl (w^h)	206–8	d>191 (sealed tube)	1.593		δ v^h	v	...	...	...	chl i	**B9**, 791 **B14**, 936
p813	—,anhydride	Phthalandione. Phthalic anhydride. $C_8H_4O_3$. *See* p812	148.12	wh nd	130.8	284.5 sub	1.527^4		δ	s	δ	...	...		**B17**[2], 463 **B21**[2], 466
p814	—,diamide	Phthalamide. Phthalic diamide. $C_8H_8N_2O_2$. *See* p812	164.17		219–20				i	i	i	...	...		**B9**, 814 **B20**, 565
p815	—,—,*N,N,N′,N′*,-tetraethyl-	$C_{16}H_{24}N_2O_2$. *See* p812	276.38		36	204^{16}			s	...	...	...	...	os v	**B9**[2], 602
p816	—,dibenzyl ester	$C_{22}H_{18}O_4$. *See* p812	346.39	pr (al)	42–3	277^{15}			i	v	v	...	...	lig i	**B9**, 802
p817	—,dibutyl ester	$C_{16}H_{22}O_4$. *See* p812	278.35			340	1.043^{21}		i	∞	∞	...	∞		
p818	—,dichloride	Phthalyl chloride. $C_8H_4Cl_2O_2$. *See* p812	203.03		11	$131–3^9$	1.4081^{15}_4	$1.571^{15.5}$	...	...	...	...	...		**B9**, 363
p819	—,dicyccohexyl ester	$C_{20}H_{26}O_4$. *See* p812	330.43	pr (al)	66				i	s	s	...	...		**B9**, 799
p820	—,di(2-ethoxy-ethyl) ester	$C_{16}H_{22}O_6$. *See* p812	310.35		34	345 $233–5^{23}$	1.1229^{21}		...	...	...	...	...		**B9**[2], 597
p821	—,diethyl ester	$C_{12}H_{14}O_4$. *See* p812	222.24			296	1.2321^{14}_4		i	∞	∞	...	s		**B9**, 798
p822	—,diisobutyl ester	$C_{16}H_{22}O_4$. *See* p812	278.35			295–8	1.0490^{15}		...	...	...	...	...		**B9**[2], 587
p823	—,di(2-methoxy-ethyl) ester	$C_{14}H_{18}O_6$. *See* p812	282.30			230^{10}	1.1708^{15}		...	...	...	...	...		**B9**[2], 597
p824	—,dimethyl ester	$C_{10}H_{10}O_4$. *See* p812	194.18	pa ye		282–5	$1.1905^{20.7}_4$	$1.515^{20.7}$	i	∞	∞	...	s		**B9**, 797
p825	—,di(3-methyl-butyl) ester	$C_{18}H_{26}O_4$. *See* p812	306.39			330–8d 225^{40}	1.0220^{16}_{16}	1.4871^{20}	i	s	...	...	...	os s	**B9**[2], 587
p826	—,dinitrile	Phthalonitrile. $C_8H_4N_2$. *See* p812	128.14	nd (w, lig)	141				s^h i	v	s	...	v	lig s^h	**B9**[2], 602
p827	—,diphenyl ester	$C_{20}H_{14}O_4$. *See* p812	318.31	pr (al, lig)	73	$250–7^{14}$			i	δ	δ	...	...		**B9**, 801
p828	—,dipropyl ester	$C_{14}H_{18}O_4$. *See* p812	250.30			304–5			i	s	s	...	...		**B9**[2], 586
p829	—,hydrazide	Phthalhydrazide. 1,4-Dioxo-1,2,3,4-tetra-hydrophthalazine. 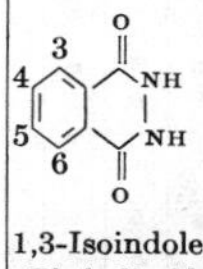	162.15	mcl nd (w, dil al)	341–4				i δ^h	i δ^h	...	...	...	aa s	**B24**[2], 194
p830	—,imide	1,3-Isoindoledione. Phthalimide. 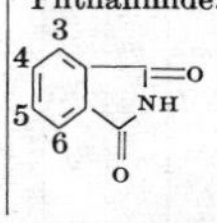	147.13	mcl pr (w, sub)	238				δ	...	...	...	i	alk s aa s lig i	**B21**[2], 348

For explanations, symbols and abbreviations see beginning of table.

No.	Name	Synonyms and Formula	Mol. wt.	Crystalline form, color and specific rotation	m.p. °C	b.p. °C	Density	n_D	Solubility w	Solubility al	Solubility eth	Solubility ace	Solubility bz	Solubility other solvents	Ref.
	Phthalic acid														
p831	—,—,*N*-acetyl-	$C_{10}H_7NO_3$. *See* p830	189.17	nd (bz), (al, aa)	135–6				i		s		v^h	chl v	**B21**^{2}, 357
p832	—,—,*N*-benzyl-	$C_{15}H_{11}NO_2$. *See* p830	237.26	ye nd (al), (aa)	116		1.343^{16}			s^h				aa s^h	**B21**^{2}, 351
p833	—,—,*N*(2-bromoethyl)-	$C_{10}H_8BrNO_2$. *See* p830	254.10	nd (w)	82.0–3.5				s^h d^h		v				**B21**^{2}, 349
p834	—,—,*N*(2-bromoisobutyl)-	$C_{12}H_{12}BrNO_2$. *See* p830	282.15	nd (al), lf (chl)	97					s v^h			v	AcOEt v chl v	**B21**^{2}, 349
p835	—,—,*N*(bromomethyl)-	$C_9H_6BrNO_2$. *See* p830	240.07	pr (chl, bz, aa, AcOEt)	149–50				d^h			s	i	chl δ AcOEt v^h	**B21**^{2}, 354
p836	—,—,*N*(2-bromopropyl)-	$C_{11}H_{10}BrNO_2$. *See* p830	268.12	nd (MeOH, al)	110–11				δ^h	v	v			lig δ	**B21**^{2}, 349
p837	—,—,*N*(3-bromopropyl)-	$C_{11}H_{10}BrNO_2$. *See* p830	268.12	nd (lig)	72–3				i	v^h	v^h			peth i lig δ	**B21**^{2}, 349
p838	—,—,*N*-carbethoxy-	Ethyl *N,N*-phthalyl-carbamate. $C_{11}H_9NO_4$. *See* p830	219.20	(bz-peth)	87–9						v		v	chl v con ac s	**B21**^{2}, 35
p839	—,—,*N*-ethyl-	$C_{10}H_9NO_2$. *See* p830	175.19	nd (al)	79	285.6^{758} cor					s				**B21**^{2}, 349
p840	—,—,*N*(2-hydroxyethyl)-	$C_{10}H_9NO_3$. *See* p830	191.19	nd, lf (w)	129.5 cor				δ						**B21**^{2}, 352
p841	—,—,*N*(hydroxymethyl)-	$C_9H_7NO_3$. *See* p830	177.16	lf, pr (to)	141–2				s^h i	δ s^h	i		δ s^h	to s^h CCl_4 i	**B21**^{2}, 354
p842	—,—,*N*-isobutyl-	$C_{12}H_{13}NO_2$. *See* p830	203.23		93	293.5									**B21**, 463
p843	—,—,*N*-methyl-	$C_9H_7NO_2$. *See* p830	161.16	nd (al), lf (sub)	134	285–7 cor			i	δ v^h					**B21**^{2}, 348
p844	—,—,*N*(1-naphthyl)-	$C_{18}H_{11}NO_2$. *See* p830	273.29	pr, pl (al-aa)	180–1					s^h	s^h				**B21**, 469
p845	—,—,*N*(2-naphthyl)-	$C_{18}H_{11}NO_2$. *See* p830	273.29	nd (aa)	216									aa v^h	**B21**^{2}, 352
p846	—,—,*N*(4-nitrophenyl)-	$C_{14}H_8N_2O_4$. *See* p830	268.23	(aa)	271–2					i^h				xyl i aa i s^h	**B21**^{2}, 350
p847	—,—,*N*-phenyl-	$C_{14}H_9NO_2$. *See* p830	223.22	wh nd (al)	207	sub			i	i				chl ∞	**B21**, 464
p849	—,monoamide	Phthalamic acid. $C_8H_7NO_3$. *See* p812	165.14	pr	148–9				δ	δ	i		i	lig i	**B9**^{2}, 600
p850	—,—,*N*(1-naphthyl)-	Alanap-1. $C_{18}H_{13}NO_3$. *See* p812	291.29		185				i	δ		δ	δ		
p851	—,monoamide mononitrile	Phthalamic nitrile. $C_8H_6N_2O$. *See* p812	146.14	nd (MeOH), (aa)	173				i	v				MeOH s^h AcOEt δ	**B9**, 815
p852	—,monobutyl ester	$C_{12}H_{14}O_4$. *See* p812	222.24	pl (al, ace)	73–4									os s	**B9**^{2}, 586
p853	—,mono-*sec*-butyl ester(*d*)	$C_{12}H_{14}O_4$. *See* p812	222.24	$[\alpha]^{20}_D$+38.4 (al)	48					s				chl s	**B9**^{2}, 586
p854	—,—(*dl*)	$C_{12}H_{14}O_4$. *See* p812	222.24	(peth)	63–3.5									peth s^h	**B9**^{2}, 587
p855	—,monoethyl ester	$C_{10}H_{10}O_4$. *See* p812	194.18		2	d	1.1877^{22}_4	1.509^{22}	δ	s	s				**B9**, 797
p856	—,mononitrile	$C_8H_5NO_2$. *See* p812	147.13	nd (al)	187d				v^h	v^h	δ		i	peth i	**B9**^{2}, 602
p857	—,mono(2-octyl) ester(*d*)	$C_{16}H_{22}O_4$. *See* p812	278.35	pr (peth) $[\alpha]^{20}_D$ +48.7 (al)	75		1.027^{72}_4			v			v	AcOEt, chl v CS_2 s	**B9**^{2}, 587
p858	—,—(*dl*)	$C_{16}H_{22}O_4$. *See* p812	278.35	(peth)	55					v		v	v	chl v peth s^h	**B9**^{1}, 353
p859	—,—(*l*)	$C_{16}H_{22}O_4$. *See* p812	278.35	pr (peth) $[\alpha]_D$ −48.26 (al)	75					v			v	chl v peth s^h	**B9**^{2}, 587
p860	**—,3-amino-**	$C_8H_7NO_4$. *See* p812	181.15	nd	191–2d				δ	δ	δ		i	chl i	**B14**^{2}, 336
p861	**—,4-amino-,** dimethyl ester	$C_{10}H_{11}NO_4$. *See* p812	209.20	pl (al, bz), pr (w)	84 cor				δ^h	s	δ			chl, Py s	**B14**, 554
p862	—,—,hydrazide	$C_8H_7N_3O_2$. *See* p829	177.17	ye nd (w+1)	>305					δ	i		i	chl i lig i	**B25**, 487
p863	**—,3-benzoyl-**	2,3-Benzophenone dicarboxylic acid $C_{15}H_{10}O_5$. *See* p812	270.23	pl, nd (+1w)	145–50 (anh)				s^h	s	δ		s		**B10**, 880
p864	**—,4-benzoyl-**	3,4-Benzophenone dicarboxylic acid. $C_{15}H_{10}O_5$. *See* p812	270.23	lf (xyl)	189				s	s					
p865	**—,3-bromo-**	$C_8H_5BrO_4$. *See* p812	245.04	nd	188				s	s	s			chl i	**B9**, 821
p866	**—,4-bromo-**	$C_8H_5BrO_4$. *See* p812	245.04		173–5				δ^h	v					**B9**^{2}, 605
p867	**—,3-chloro-**	$C_8H_5ClO_4$. *See* p812	200.58	nd (w)	186–7				δ	v	v				**B9**^{1}, 366
p868	—,—,anhydride	$C_8H_3ClO_3$. *See* p812	182.57	nd	122–3										**B17**^{2}, 466
p869	**—,4-chloro-**	$C_8H_5ClO_4$. *See* p812	200.58	nd	157				s	s	s				**B9**, 816
p870	—,—,anhydride	$C_8H_3ClO_3$. *See* p812	182.57	pr	95–7	294.5^{720}				s	s			chl s	**B17**^{2}, 466
p871	**—,3,4-dichloro-**	$C_8H_4Cl_2O_4$. *See* p812	235.03	ta (w)	195				v		v				**B9**, 817
p871^1	—,—,anhydride	$C_8H_2Cl_2O_3$. *See* p812	217.01	pl	120–1	329				s				chl, to s	**B17**^{2}, 467
p872	**—,3,5-dichloro-**	$C_8H_4Cl_2O_4$. *See* p812	235.03	nd, ta (aq HCl)	164d					v	v	v	δ	chl δ	**B9**, 817
p873	—,—,anhydride	$C_8H_2Cl_2O_3$. *See* p812	217.01	nd	89								s	chl s	**B17**, 483
p874	**—,3,6-dichloro-**	$C_8H_4Cl_2O_4$. *See* p812	235.03	ta (w)					v^h	v	v				**B9**, 817

For explanations, symbols and abbreviations see beginning of table.

No.	Name	Synonyms and Formula	Mol. wt.	Crystalline form, color and specific rotation	m.p. °C	b.p. °C	Density	n_D	Solubility w	al	eth	ace	bz	other solvents	Ref.
	Phthalic acid														
p875	—,—,anhydride	$C_8H_2Cl_2O_3$. *See* p812	217.01	nd	190–1	339									**B17**[2], 467
p876	—,**4,5-dichloro-**	$C_8H_4Cl_2O_4$. *See* p812	235.03	nd (w)	199–200				s v^h		v				**B9**[1], 366
p877	—,—,anhydride	$C_8H_2Cl_2O_3$. *See* p812	217.01	ta or pr (to^h, CCl_4)	187.5–8.8 cor	313							δ v^h	to v CCl_4 δ	**B18**[1], 254
—	—,**dihydro-**	*See* **Cyclohexadiene-dicarboxylic acid***													
p878	—,**3,6-dihydroxy-**, dinitrile	$C_8H_4N_2O_2$. *See* p812	160.14	yesh lf	230				δ	v	v		δ	chl δ	**B10**[2], 383
p879	—,—,imide	$C_8H_5NO_4$. *See* p830	179.14	ye nd (w+3)	273–4				v^h						**B21**[1], 478
p880	—,**4,5-dimethoxy-**	*m*-Hemipic acid. *m*-Hemipinic acid. $C_{10}H_{10}O_6$. *See* p812	226.18	nd	206				δ						**B10**[2], 382
p881	—,**3-hydroxy-**	$C_8H_6O_5$. *See* p812	182.13	nd, pr (eth-peth)	166–7	sub			$δ^h$	v	v				**B10**, 498
p882	—,**4-hydroxy-**	$C_8H_6O_5$. *See* p812	182.13	rosettes (w)	204–5				s	s	s		δ		**B10**, 499
p882[1]	—,**monoper-**	$C_8H_6O_5$. *See* p812	182.13	nd	110d				v	v			s	chl s	**B9**[2], 599
p883	—,**3-nitro-**	$C_8H_5NO_6$. *See* p812	211.13	pa ye pr (w)	218				δ	s^h	δ		i	chl i	**B9**, 823
p884	—,—,anhydride	$C_8H_3NO_5$. *See* p812	193.12	nd (al)	164				i	s^h			δ	aa s^h	**B17**[2], 468
p885	—,—,diethyl ester	$C_{12}H_{13}NO_6$. *See* p812	267.24	pr (al), nd (peth)	46				i	v	v				**B9**[2], 606
p887	—,**4-nitro-**	$C_8H_5NO_6$. *See* p812	211.13	pa ye nd (w, eth)	165				s	s			i	chl i	**B9**[2], 829
p888	—,—,imide	$C_8H_4N_2O_4$. *See* p830	192.13	colorless nd (w), ye lf (al-ace)	201–2				$δ^h$	s s^h		s		aa s	**B21**[2], 373
p889	—,—,dimethyl ester	$C_{10}H_9NO_6$. *See* p812	239.19	(dil al)	69–71				i	s				MeOH s	**B9**[2], 607
p890	—,—,piperazinium salt	$C_8H_5NO_6.C_4H_{10}N_2$. *See* p812	297.27	wh	201.5–4.5 d				s^h	s^h	δ				
p891	—,**tetrabromo-**	$C_8H_2Br_4O_4$. *See* p812	481.76	nd (w^h)	266d				δ s^h	δ				os δ	**B9**[1], 367
p892	—,—,anhydride	$C_8Br_4O_3$. *See* p812	463.75	nd (aa-xyl)	279.5–80				i	i			δ	os $δ^h$, $PhNO_2$ s	**B17**[2], 468
p893	—,**tetrachloro-**	$C_8H_2Cl_4O_4$. *See* p812	303.93	pl (w)	250d				δ	s		v	δ	chl δ	**B9**[2], 604
p894	—,—,anhydride	$C_8Cl_4O_3$. *See* p812	285.92	pr, nd (sub)	255–6.5	sub			d^h		δ				**B17**[2], 467
p895	—,—,monoethyl ester	$C_{10}H_6Cl_4O_4$. *See* p812	331.99		94–5	d 150			δ	v	v				**B9**[2], 604
p896	—,**tetraiodo-**, anhydride	$C_8I_4O_3$. *See* p812	651.73	ye pr, nd (aa, $PhNO_2$), nd (sub)	327–8	sub			i					oos i aa $δ^h$	**B17**[2], 468
—	**Phthalic anhydride**	*see* **Phthalic acid,** anhydride													
p897	**Phthalide**	α-Hydroxy-*o*-toluic acid lactone. 1(3)-Isobenzofuranone.	134.13	nd	73	290			s^h	v					**B17**[2], 332
p898	—,**3-benzylidene-**	$C_{15}H_{10}O_2$. *See* p897	222.23	mcl pr	108				i	s^h					
p899	—,**3,3-diphenyl-**	Phthalophenone. Triphenylcarbinol-*o*-carboxylic anhydride. $C_{20}H_{14}O_2$. *See* p897	286.31	lf (al)	118–20	235^{15}									**B17**, 391
p900	—,**6-nitro-**	$C_8H_5NO_4$. *See* p897	179.13	nd (al)	141				i	s	s			chl s^h	**B17**[2], 334
—	**Phthalimide**	*see* **Phthalic acid,** imide													
p901	**Phthalocyanine**	Tetrabenzoporphyrazine.	514.53	grsh-bl mcl with purp luster	does not melt									os i	
—	**Phthalonic acid**	*see* **Acetic acid, 2(2-carboxyphenyl)-2-oxo-***													
—	**Phthalyl chloride**	*see* **Phthalic acid,** dichloride													
—	**Phthiocol**	*see* **1,4-Naphthoquinone, 2-hydroxy-3-methyl-***													

For explanations, symbols and abbreviations see beginning of table.

No.	Name	Synonyms and Formula	Mol. wt.	Crystalline form, color and specific rotation	m.p. °C	b.p. °C	Density	n_D	Solubility w	al	eth	ace	bz	other solvents	Ref.
	Physostigmine														
p902	**Physostigmine**	Eserine. $C_{15}H_{21}N_3O_2$	275.34	orth pr (eth, bz), $[\alpha]^{17}_D -76$ (chl, c=1.3), $[\alpha]^{25}_D -120$ (bz, c=1)	105–6 86–7 (unst)				δ	s			s	chl s	B23², 330
p903	—,2-hydroxy-benzoate	Eserin salicylate. $C_{15}H_{21}N_3O_2.C_7H_6O_3$	413.46	lt, air, heat→ red	185–7				δ s^h	s v^h	δ			chl s	
p904	—,sulfate	$2(C_{15}H_{21}N_3O_2).H_2SO_4$	648.77	dlq sc	140				v	v	δ				
p905	**Phytadiene**	$C_{20}H_{38}$	278.51			$185–8^{22}$	0.826^0_4							peth, MeOH ∞ aa ∞	B1², 243
p906	**Phytol**	3,7,11,15-Tetramethyl-2-hexadecen-1-ol*. $C_{20}H_{40}O$	296.52			$202–4^{10}$	0.8497^{25}_4	1.4595^{25}	i					org solv s	B1², 503
p907	**Picein**	*p*-Hydroxyacetophenone-*D*-glucoside. $C_{14}H_{18}O_7$	298.28	nd (w+1), nd (MeOH)	195–6				v^h δ	δ	i			AcOEt δ chl i aa s	B31, 226
p908	**Picene**	Dibenzo[*a,i*]phenanthrene. 	278.34	lf, pl (xyl, cumene, sub)	364	518–20			i	δ			δ^h	chl δ^h cumene s aa δ^h	
—	**Picoline**	*see* **Pyridine, methyl-**													
—	**Picolinic acid**	*see* **2-Pyridinecarboxylic acid***													
—	**Picramic acid**	*see* **Phenol, 2-amino-4,6-dinitro-***													
—	**Picric acid**	*see* **Phenol, 2,4,6-trinitro-***													
p909	**Picrolonic acid**	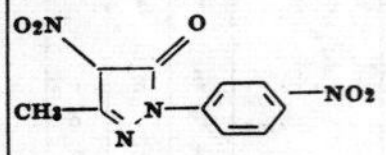	264.20	ye nd (al)	124d				δ^h	s^h	s^h			MeOH s^h	B24², 25
p910	**Picropodophyllin**	$C_{22}H_{22}O_8$	414.42	nd (al)	228				i	v^h	v	s		alk s chl, MeOH v	B19, 424
p911	**Picrotoxin**	Cocculin. $C_{30}H_{34}O_{13}$	602.60	rh lf, $[\alpha]^{16}_D -29.3$	203				δ v^h	s v^h	δ			chl δ	
—	**Picryl chloride**	*see* **Benzene, 2-chloro-1,3,5-trinitro-***													
p912	**Pilocarpidine** (*d*)	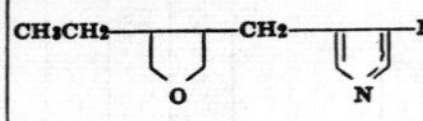	194.24	syr, $[\alpha]^{20}_D +81.3$ (w, al, chl)					s	v	δ				B27², 694
p913	—,nitrate(*d*)	$C_{10}H_{14}N_2O_2.HNO_3$	257.25	pr (w), $[\alpha]^{20}_D$ +73.2 (w)	137					s					
p914	**Pilocarpine**	$C_{11}H_{16}N_2O_2$	208.26	nd, $[\alpha]^{20}_D +100.5$	34	260^5			s	s	δ		δ	chl v lig i	B27², 694
p915	—,**hydrochloride**	$C_{11}H_{16}N_2O_2.HCl$	244.72	hyg cr	204–5				v	v	i			chl δ	
p916	—,2-hydroxy-benzoate	Pilocarpine salicylate. $C_{11}H_{16}N_2O_2.C_7H_6O_3$	346.39	nd or lf, $[\alpha]_D +63$	120					s	s				B27, 635
p917	—,nitrate	$C_{11}H_{16}N_2O_2.HNO_3$	271.28	wh pw, $[\alpha]_D +77–83$ (10% aq sol)	174–8				v	s^h δ	i			chl i	
p918	—,sulfate	$(C_{11}H_{16}N_2O_2)_2.H_2SO_4$	514.60	hyg, $[\alpha]^{20}_D +85$	132				s	s					B27, 635
—	**Pilosine**	*see* **Carpiline**													
p919	**Pimaric acid**	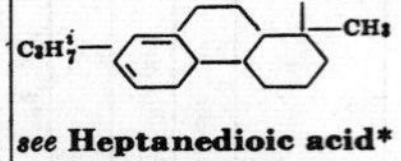	302.46	orh (ace), $[\alpha]^{18}_D +74.7$ (chl, c=4)	211–2				i	s	s				
—	**Pimelic acid**	*see* **Heptanedioic acid***													
—	**Pinacol**	*see* **2,3-Butanediol, 2,3-dimethyl-***													
—	**Pinacolone**	*see* **2-Butanone, 3,3-dimethyl-***													
p920	**Pinane** (*d, cis*)	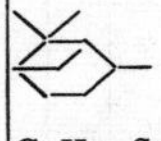	138.25	$[\alpha]_D +22.7$, $[\alpha]_{546} +28.5$	−50	$166–7^{755}$	0.861^{15}_{15}	1.4647^{15}						os s	E12A, 454
p921	—(*dl*)	$C_{10}H_{18}$ *See* p920	138.25			164.5–5	0.8551^{20}_4	1.4609^{20}							B5¹, 47

For explanations, symbols and abbreviations see beginning of table.

No.	Name	Synonyms and Formula	Mol. wt.	Crystalline form, color and specific rotation	m.p. °C	b.p. °C	Density	n_D	Solubility w	al	eth	ace	bz	other solvents	Ref.
	Pinane														
p922	—(*l, cis*)	$C_{10}H_{18}$. *See* p920	138.25	$[\alpha]_D -20.9$, $[\alpha]_{578} -23.8$		$167–8^{757}$	0.8556^{21}	1.4627^{21}						os s	**E12A,** 454
p923	**α-Pinene**	$C_{10}H_{16}$	136.24		−55	156.2	0.8582^{20}_{4}	1.4658^{20}	δ	∞	∞			chl ∞	
p924	**β-Pinene**(*d*)	Nopinene. 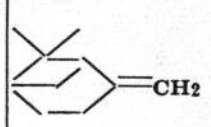 =CH_2	136.24	$[\alpha]_D +28.6$		164–6	0.8654^{20}_{4}	1.4739^{20}						os s	**E12A,** 478
p925	—(*l*)	$C_{10}H_{16}$. *See* p924	136.24	$[\alpha]_{578} -22.44$ $[\alpha]_{492} -23.10$		164	0.8740^{15}	1.4773^{20}						os s	**E12A,** 478
p926	**Pinic acid**(*dl*) (*a* form)	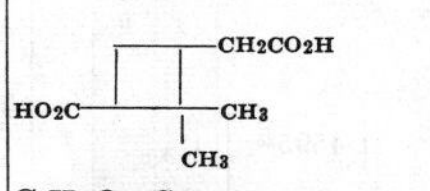 CH_2CO_2H HO_2C CH_3 CH_3	186.20	wh pr (w)	101–2.5	$214–6^{9}$	1.23		δ						**B9**^{2}, 529
p927	— (*dl*) (*b* form)	$C_9H_{14}O_4$. *See* p926	186.20		68	$214–6^{9}$									**B9**^{2}, 529
p928	—(*dl*) (*c* form)	$C_9H_{14}O_4$. *See* p926	186.20	pl	58	$214–6^{9}$									**B9**^{2}, 529
p929	—,—(*l*)	$C_9H_{14}O_4$. *See* p926	186.20	nd(w)	135										**B9**^{1}, 320
p930	**Pinol**(*dl*)	*dl*-Sobrerone. 6,8-Epoxy-1-*p*-menthene. 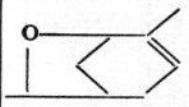 O	152.23			183–4 $76–7^{14}$	0.9420^{20}_{4}	1.4709^{20}		s	s				**B17**^{2}, 44
p931	—,hydrate(*dl*)	$C_{10}H_{16}O.H_2O$. *See* p930	170.25	pl or nd	131	270–1			s	v	v				
—	**Pinosylvin**	*see* **Stilbene, 3,5-dihydroxy-** (*trans*)													
—	**Pipecoline**	*see* **Piperidine, methyl-**													
—	**Pipecolyl carbinol**	*see* **Piperidine, 2(2-hydroxyethyl)-**													
p932	**Piperazine**	Diethylenediamine. Hexahydropyrazine. 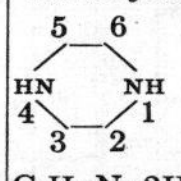 5 6 HN NH 4 1 3 2	86.14	lf (al^{h})	104	145–6		1.446^{113}	v	s	i			glycols v	**B24**^{2}, 3
p933	—,dihydrobromide	$C_4H_{10}N_2.2HBr$. *See* p932	247.99	wh nd	d				v	i	i				
p934	—,dihydrochloride	$C_4H_{10}N_2.2HCl$. *See* p932	159.07	nd (dil al)	335–40				v	i	i				**B23**^{2}, 4
p935	—,hexahydrate	$C_4H_{10}N_2.6H_2O$. *See* p932	194.24	wh (w)	44	125–30			v	s	v				
p936	—,**1-benzyl-**	$C_{11}H_{16}N_2$. *See* p932	176.25			$145–7^{12}$			s	s	s				
p937	—,**1,4-bis(4-methoxy-benzoyl)-**	1,4-Dianisoylpiperazine. $C_{20}H_{22}N_2O_4$. *See* p932	354.39	wh	192.5–3.5				i	s^{h}	i				
p938	—,**1,4-bis-(phenylacetyl)-**	*N,N′*-Di-α-toluylpiperazine. $C_{20}H_{22}N_2O_2$. *See* p932	322.39	wh	150–1				δ	s^{h}	δ				
p939	—,**1,4-bis-(3-phenyl-propanoyl)-**	1,4-Bis(hydrocinnamyl)-piperazine. $C_{22}H_{26}N_2O_2$. *See* p932	350.44	wh	122.5–3				i	s^{h}	δ				
p940	—,**1,4-dimethyl-**	$C_6H_{14}N_2$. *See* p932	114.19		247–50	$131–2^{750}$			v	v	v				**B23**, 7
p941	—,**2,5-dimethyl-** (*cis*)	$C_6H_{14}N_2$. *See* p932	114.19	rh bipym nd or pr (chl)	114	162 sub			v	v	δ		δ	chl v	**B23**^{2}, 21
p942	—,—(*trans*)	$C_6H_{14}N_2$. *See* p932	114.19	mcl pl or pr (bz, chl)	118	162 sub			v	v	δ				**B23**^{2}, 19
p943	—,**2,6-dimethyl-** (*cis*)	$C_6H_{14}N_2$. *See* p932	114.19	lf or pl	110–1	162			s	s	i		δ	chl s peth v	**B23**^{2}, 22
p944	—,**1,4-dinitro-**	$C_4H_8N_4O_2$. *See* p932	144.14	lt ye pl	158				δ	v^{h}	δ				**B23**^{1}, 7
p945	—,**1-ethyl-**	$C_6H_{14}N_2$. *See* p932	114.19			155–8			s	s	s				
p946	—,**1-methyl-**	$C_5H_{12}N_2$. *See* p932	100.17			134–6			v	v	v				
p947	—,**2-methyl-**	$C_5H_{12}N_2$. *See* p932	100.17	hyg lf	62	155^{763}			v	s	s		s	chl s	**B23**^{2}, 16
p948	—,**1-phenyl-**	$C_{10}H_{14}N_2$. *See* p932	162.23	pa ye oil		$156–7^{10}$		1.0621^{20}_{4}	i	∞	∞				
p949	**1-Piperazine-carboxylic acid,** ethyl ester	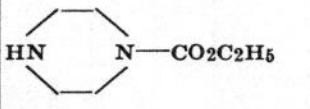 HN N—$CO_2C_2H_5$	158.20			237 $116–7^{12}$			s	s	s				
p950	**1,4-Piperazine-dicarboxylic acid,** diethyl ester	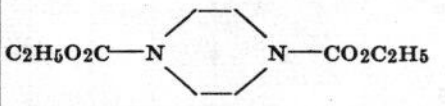 $C_2H_5O_2C$—N N—$CO_2C_2H_5$	230.26	nd	45	315				s	s				**B23**, 12
p951	**2,5-Piperazine-dione**	Glycine anhydride. 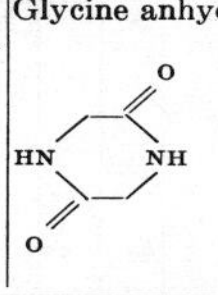 O HN NH O	114.11	ta or pl (w)	311–2 sub 260				δ	δ^{h}				HCl s	**B24**^{2}, 141

For explanations, symbols and abbreviations see beginning of table.

No.	Name	Synonyms and Formula	Mol. wt.	Crystalline form, color and specific rotation	m.p. °C	b.p. °C	Density	n_D	Solubility w	al	eth	ace	bz	other solvents	Ref.
	Piperic acid														
—	**Piperic acid**	*see* **2,4-Pentadienoic acid, 5(3,4-methylene-dioxyphenyl)-***													
p952	**Piperidine**	Hexahydropyridine. Pentamethyleneimine. 5 6 / 4⟨⟩NH / 3 2 1	85.15		−9	106.0 17.7^{20}	0.8606^{20}_{4}	1.4534^{20}	∞	∞			s	chl s	**B20**², 6
—	**—,2-acetonyl-**	*see* **Isopelletierine**													
p953	**—,1-acetyl-**	$C_7H_{13}NO$. *See* p952	127.18		107–9	226–7	1.011^{9}		∞	s					**B20**², 33
p954	**—,2-allyl-**	$C_8H_{15}NO$. *See* p952	125.22			170–1	0.8823^{15}_{4}								**B20**, 147
p955	**—,1-benzoyl-**	$C_{12}H_{15}NO$. *See* p952	189.25	tcl	49	320–1 180^{15}			i	s	s				**B20**², 34
—	**—,4-benzoxy-2,2,6-trimethyl-**	*see* **Iso-β-eucaine**													
p956	**—,1-butyl-**	$C_9H_{19}N$. *See* p952	141.25			175–7	0.8245^{20}_{4}	1.4467^{20}							**B20**², 13
p957	**—,2-butyl-(*d*)**	$C_9H_{19}N$. *See* p952	141.25	$[\alpha]_D +15.7$			0.8512								**B20**, 127
p958	**—,—(*dl*)**	$C_9H_{19}N$. *See* p952	141.25			186–9 75^{14}	0.8529^{15}_{4}		i					os s	**B20**, 128
p959	**—,—(*l*)**	$C_9H_{19}N$. *See* p952	141.25	$[\alpha]^{16}_D -18.7$			0.8533								**B20**, 127
p960	**—,3-butyl-**	$C_9H_{19}N$. *See* p952	141.25			196–7			i	s	s			chl s	**B20**¹, 30
p961	**—,1-*sec*-butyl-(*dl*)**	$C_9H_{19}N$. *See* p952	141.25			$175–6^{768}$	0.8378^{20}_{4}	1.4506^{20}							**B20**², 13
p962	**—,—(*l*)**	$C_9H_{19}N$. *See* p952	141.25	$[\alpha]^{25}_D -54.59$		$75–6^{25}$	0.835^{25}_{4}	1.4486^{21}							
p963	**—,1-*tert*-butyl-**	$C_9H_{19}N$. *See* p952	141.25			166^{763}	0.8465^{20}_{4}	1.4532^{20}	δ	∞	∞				**B20**², 13
p964	**—,2,4-diethyl-**	$C_9H_{19}N$. *See* p952	141.25			174–9	0.8722^{0}		δ						**B20**, 128
p965	**—,2,5-diethyl-**	$C_9H_{19}N$. *See* p952	141.25			190 $100–5^{22}$	0.8722^{0}				v			chl v	**B20**, 128
p966	**—,3,4-diethyl-(*cis*)**	$C_9H_{19}N$. *See* p952	141.25	$[\alpha]^{22}_D +26.0$ (95% al, c=4.35)		70^{12}									
p967	**—,—(*trans*)**	$C_9H_{19}N$. *See* p952	141.25			193^{720}									**B20**, 128
p969	**—,1,2-dimethyl-(*d*)**	$C_7H_{15}N$. *See* p952	113.20	$[\alpha]^{15}_D +68.8$		127	0.825^{16}								**B20**, 98
p970	**—,—(*dl*)**	$C_7H_{15}N$. *See* p952	113.20			127.9	0.824^{15}_{4}	1.4395^{20}	s	s	s			CS_2 s	**B20**, 95
p971	**—,1,3-dimethyl-(*dl*)**	$C_7H_{15}N$. *See* p952	113.20			124–6	0.818^{15}								**B20**, 100
p972	**—,2,3-dimethyl-**	α,β-Lupetidine. $C_7H_{15}N$. *See* p952	113.20			$138–40^{720}$			s	v					**B20**¹, 29
p973	**—,2,4-dimethyl-(*d*)**	α,γ-Lupetidine. $C_7H_{15}N$. *See* p952	113.20	$[\alpha]_D +23.2$			0.845								**B20**, 108
p974	**—,—(*dl*)**	$C_7H_{15}N$. *See* p952	113.20			140–2	0.8615^{20}			v	v			ac s	**B20**¹, 29
p975	**—,2,5-dimethyl-**	α,β′-Lupetidine. $C_7H_{15}N$. *See* p952	113.20			138–40 cor				v				ac s	**B20**, 108
p976	**—,2,6-dimethyl-**	α,α′-Lupetidine. $C_7H_{15}N$. *See* p952	113.20			127.5–8.3^{768}	0.8492^{0}		∞	∞	∞			ac s	**B20**², 60
p977	**—,3,3-dimethyl-**	β,β′Lupetidine. $C_7H_{15}N$. *See* p952	113.20			137				v					**B20**¹, 29
p978	**—,4,4-dimethyl-**	γ,γ′Lupetidine. $C_7H_{15}N$. *See* p952	113.20			145–6			∞					os v	**B20**¹, 29
p979	**—,4(dimethylamino)1-phenyl-**	Irenal. $C_{13}H_{20}N_2$. *See* p952	204.31	lf (bz)	47.5–8.5	$123–6^{0.5}$							s^h	dil HCl v	
p980	**—,1(2,2-dimethylpropyl)-**	*N*-Neopentylpiperidine. $C_{10}H_{21}N$. *See* p952	155.28			188	0.8608^{20}_{4}	1.4593^{20}							**B20**², 13
p981	**—,1-dodecyl-**	$C_{17}H_{35}N$. *See* p952	253.46	pa ye		161^{5}	0.8378^{20}_{4}	1.4588^{20}							
p982	**—,1-ethyl-**	$C_7H_{15}N$. *See* p952	113.20			130.8	0.8237^{20}_{4}	1.4445^{19}							**B20**², 13
p983	**—,2-ethyl-(*d*)**	$C_7H_{15}N$. *See* p952	113.20	$[\alpha]_D +17.1$		142–3.5	0.8680^{4}								**B20**, 104
p984	**—,—(*dl*)**	$C_7H_{15}N$. *See* p952	113.20			141.5–2.5	0.8666^{0}_{0}		s						**B20**¹, 28
p985	**—,—(*l*)**	$C_7H_{15}N$. *See* p952	113.20	$[\alpha]_D -14.9$		138–42									**B20**¹, 28

For explanations, symbols and abbreviations see beginning of table.

No.	Name	Synonyms and Formula	Mol. wt.	Crystalline form, color and specific rotation	m.p. °C	b.p. °C	Density	n_D	Solubility w	al	eth	ace	bz	other solvents	Ref.
	Piperidine														
p986	**—,3-ethyl-(*dl*)**	$C_7H_{15}N$. *See* p952	113.20	fum in air		152.6	0.8565^{23}_{4}	1.4531^{20}	δ						**B20**¹, 29
p987	—,—(*l*)	$C_7H_{15}N$. *See* p952	113.20	$[\alpha]^{15}_{D}-4.5$		155									**B20**, 106
p988	**—,4-ethyl-**	$C_7H_{15}N$. *See* p952	113.20			156–8	0.8759^{0}		δ						**B20**, 108
—	**—,1,4-ethylene-**	*see* **Quinuclidine**													
p989	**—,1-heptyl-**	$C_{12}H_{25}N$. *See* p952	183.33			239.5	0.8316^{20}	1.4531^{20}							
p990	**—,1-hexyl-**	$C_{11}H_{23}N$. *See* p952	169.30			219.2	0.830^{20}	1.4517^{20}							
p991	**—,1(2-hydroxy-ethyl)-**	2-*N*-Piperidylethanol. $C_7H_{15}NO$. *See* p952	129.20			$200–2^{742}$ $89–91^{20}$	0.9732^{25}_{25}	1.4749^{20}	∞	v					**B20**², 18
p992	**—,2(1-hydroxy-ethyl)-**	Methyl-2-piperidylcarbinol. $C_7H_{15}NO$. *See* p952	129.20			$106–10^{18}$									**B21**², 9
p993	**—,2(2-hydroxy-ethyl)-**	$C_7H_{15}NO$. *See* p952	129.20		39–40	234.5 $145–6^{36}$	1.01^{17}		v	v	v				**B21**², 9
p994	**—,3(2-hydroxy-ethyl)-**	$C_7H_{15}NO$. *See* p952	129.20			$121–3^{6}$	1.0106^{25}_{4}	1.4888^{25}							**B21**², 10
p995	**—,4(2-hydroxy-ethyl)-**	$C_7H_{15}NO$. *See* p952	129.20	syr	132–3.5	227–8 $120–5^{15}$	1.0059^{15}_{4}								**B21**², 10
—	**—,2(2-hydroxy-propyl)-**	*see* **Conyhydrine**													
p996	**—,1-isobutyl-**	$C_9H_{19}N$. *See* p952	141.25			$160–1^{740}$ 87^{43}	0.8161^{25}	1.4382^{25}	v						**B20**, 20
p997	**—,2-isobutyl-**	$C_9H_{19}N$. *See* p952	141.25			181–2	0.8510^{22}_{4}	1.4553^{22}		v	v				**B20**, 128
p998	**—,1-isopropyl-**	$C_8H_{17}N$. *See* p952	127.23			$149–50^{757}$	0.8389^{20}_{4}	1.4491^{20}	s						**B20**², 13
p999	**—,2-isopropyl-(*dl*)**	$C_8H_{17}N$. *See* p952	127.23			162	0.8668^{0}		δ i^h						**B20**, 120
p1000	—,—(*l*)	$C_8H_{17}N$. *See* p952	127.23	$[\alpha]_D-13.1$		161.5	0.8503^{19}								**B20**, 121
p1001	**—,4-isopropyl-**	$C_8H_{17}N$. *See* p952	127.23			168–71 $66–70^{15}$			δ i^h						**B20**, 121
p1002	**—,1-methyl-**	$C_6H_{13}N$. *See* p952	99.17			107	0.8159^{20}_{4}	1.4355^{20}	v s^h	∞	∞				**B20**², 12
p1003	**—,2-methyl-(*d*)**	$C_6H_{13}N$. *See* p952	99.17	$[\alpha]^{15}_{D}+18.7$ (chl)		117–7.5		1.4498^{15}							**B20**², 57
p1004	—,—(*dl*)	α-Pipecoline. $C_6H_{13}N$. *See* p952	99.17		−4.9	$117–8^{747}$	0.8436^{24}_{4}	1.4464^{24}	v	s	s			dil KOH i	**B20**², 57
p1005	**—,3-methyl-(*dl*)**	β-Pipecoline. $C_6H_{13}N$. *See* p952	99.17			$125–6^{763}$	0.8446^{24}_{4}	1.4463^{24}	v						**B20**², 58
p1006	—,—(*l*)	$C_6H_{13}N$. *See* p952	99.17	$[\alpha]^{25}_{D}-4$		124			v						**B20**², 58
p1007	**—,4-methyl-**	γ-Pipecoline. $C_6H_{13}N$. *See* p952	99.17			126.5–9	0.8674^{0}		v						**B20**, 101
p1008	**—,1(3-methyl-butyl)-**	*N*-Isoamylpiperidine. $C_{10}H_{21}N$. *See* p952	155.28			188–9			δ						**B20**², 14
p1009	**—,1(2-methyl-2-pentyl)-**	$C_{11}H_{23}N$. *See* p952	169.30			205–7	0.8517^{20}_{4}	1.4592^{20}							**B20**², 14
p1010	**—,1(3-methyl-3-pentyl)-**	$C_{11}H_{23}N$. *See* p952	169.30			214^{752}	0.8614^{20}_{4}	1.4637^{20}							**B20**², 14
—	**—,1-methyl-2-propyl-**	*see* **Coniine, *N*-methyl-**													
p1011	**—,1-nitroso-**	$C_5H_{10}N_2O$. *See* p952	114.15	pa ye		217^{721} 102^{13}	$1.0631^{18.5}_{4}$	$1.4933^{18.5}$	s					HCl s	**B20**¹, 24
p1012	**—,1-nonyl-**	$C_{14}H_{29}N$. *See* p952	211.38			$135–7^{11}$	0.8313^{25}	1.4538^{25}							
p1013	**—,1-octyl-**	$C_{13}H_{27}N$. *See* p952	197.36			122^{10} 89^{1}	0.8324^{20}_{4}	1.4544^{20}							
p1014	**—,1-pentyl-**	$C_{10}H_{21}N$. *See* p952	155.28		198.2 80^{8}	1.4498^{20}	0.8282^{20}								**B20**², 13
p1015	**—,1-phenyl-**	$C_{11}H_{15}N$. *See* p952	161.25			$257–8^{752}$ 126.5^{15}				v	v		v	chl v	**B20**², 14
p1016	**—,2-propenyl-(*d*)**	$C_8H_{15}N$. *See* p952	125.21	$[\alpha]^{15}_{D}+24.8$											**B20**, 147
p1017	—,—(*dl*)	$C_8H_{15}N$. *See* p952	125.21	nd	−15	166.5–8.5	0.8695^{15}_{4}								**B20**, 147

For explanations, symbols and abbreviations see beginning of table.

No.	Name	Synonyms and Formula	Mol. wt.	Crystalline form, color and specific rotation	m.p. °C	b.p. °C	Density	n_D	Solubility w	al	eth	ace	bz	other solvents	Ref.
	Piperidine														
p1018	—,—(*l*)	$C_8H_{15}N$. *See* p952	125.21	$[\alpha]_D^{15} -29.0$			0.8672_4^{15}								**B20**, 147
p1019	—,**1-propyl-**	$C_8H_{17}N$. *See* p952	127.23			151.2	0.8231^{20}	1.4446^{20}	s	v	v				**B20**^{2}, 13
—	—,**2-propyl-**	*see* **Coniine**													
p1020	—,**3-propyl-(*d*)**	$C_8H_{17}N$. *See* p952	127.23	$[\alpha]_D^{16} +5.9$											**B20**, 120
p1021	—,—(*dl*)	$C_8H_{17}N$. *See* p952	127.23			174^{758}	0.8475_4^{26}								**B20**, 119
p1022	—,—(*l*)	$C_8H_{17}N$. *See* p952	127.23	oil, $(\alpha)_D^{16} -6.6$		174^{752}	0.8517_4^{19}								**B20**, 120
p1023	—,**4-propyl-**	$C_8H_{17}N$. *See* p952	127.23			178–80									**B20**, 120
p1024	—,**2,2,6,6-tetramethyl-**	$C_9H_{19}N$. *See* p952	141.25			155.5–6.5	0.8367_4^{16}				s				**B20**, 129
p1025	—,**2,2,4-trimethyl-**	$C_8H_{17}N$. *See* p952	127.23			148	0.832^{15}		δ	s	s				**B20**, 124
p1026	—,**2,3,6-trimethyl-**	$C_8H_{17}N$. *See* p952	127.23			36^8	0.8302_4^{20}	1.4434^{20}							
p1027	—,**2,4,6-trimethyl-**	$C_8H_{17}N$. *See* p952	127.23			151–3	0.8315_4^{19}	1.4435^{19}	δ	∞	∞				**B20**^{2}, 64
p1028	**1-Piperidinecarboxaldehyde**	N-Formyl piperidine. N—CHO	113.16	pa ye		222 $106–10^{17}$	1.0205_4^{25}		∞	∞	∞		∞	chl ∞ aa ∞ lig ∞	**B20**^{2}, 33
p1029	**1-Piperidinecarbodithioic acid,** piperidinium salt	N—CS_2H.HN	246.44	pa ye lf (al), (CS_2-eth)	169–70				v^h	v			v		**B20**^{2}, 37
p1030	**4-Piperidinecarboxylic acid**	Hexahydronicotinic acid. Isonipecotic acid. 1 2 3 4 HN —CO_2H 6 5	129.16		ca. 320				s	i					**B22**^{1}, 410
p1031	—,ethyl ester	Ethyl isonipecotate. $C_8H_{15}NO_2$. *See* p1030	157.21			$95–8^7$		1.4569^{23}	s	s	s		s		
p1032	—,methyl ester	Methyl isonipecotate. $C_7H_{13}NO_2$. *See* p1030	143.19			$88–9^9$		1.4635^{25}	s	s	s		s		
p1033	—,**1-methyl-,** methyl ester	$C_8H_{15}NO_2$. *See* p1030	157.21			$96–100^{20}$		1.4510^{25}	s	s	s		s		**B22**, 486
p1034	—,**1-methyl-4-phenyl,** ethyl ester hydrochloride	Mepheridine hydrochloride. $C_{15}H_{22}ClNO_2$. *See* p1030	283.79		186–9				s	δ	i	s	i	AcOEt s	
p1034[1]	**2,4,-Piperidinedione, 3,3-diethyl-**	O= NH C_2H_5 C_2H_5 O	169.20		102–7				v	v				MeOH v	
—	**Piperidinic acid**	*see* **Butanoic acid 4-amino-***													
—	**Piperidinium,** cyclopentamethylene dithiocarbamate	*see* **Piperidinecarbodithoic acid,** piperidinium salt													
p1035	**2-Piperidone**	5-Aminopentanoic acid lactam*. δ-Valerolactam. O 3 1 4 NH 5 6	99.14	hyg	39–40	258–62 137^{14}			v		v			dil ac s, con alk i	**B21**, 238
p1036	**4-Piperidone,** hydrochloride	3 2 1 O= NH.HCl 5 6	135.60		139–41d				v						**B21**^{2}, 215
p1037	**2-Piperidone, 3-hydroxy-(*dl*)**	$C_5H_9NO_2$. *See* p1035	115.13	nd or pr (AcOEt)	141–2				v	v	δ			AcOEt, chl δ	**B21**^{2}, 411
p1038	—,**5-hydroxy-**	$C_5H_9NO_2$. *See* p1035	115.13	(al, al-AcOEt)	145–6 cor				v	v	i		i	AcOEt, chl δ	**B21**^{2}, 412
p1039	**4-Piperidone, 2,2,6,6-tetramethyl-**	Triacetoneamine. $C_9H_{17}NO.H_2O$. *See* p1036	173.25	rh ta (moist eth, +1w), nd (eth)	56 (hyd), 34.6 (anh)	205			v	v	v				**B21**^{2}, 222
p1040	**Piperine**	1-Piperylpiperidine. $C_{17}H_{19}NO_3$	285.33	pr (AcOEt), pl (al)	129.5				i	s v^h	v		v	Py s chl v	**B20**^{2}, 53

For explanations, symbols and abbreviations see beginning of table.

No.	Name	Synonyms and Formula	Mol. wt.	Crystalline form, color and specific rotation	m.p. °C	b.p. °C	Density	n_D	Solubility w	al	eth	ace	bz	other solvents	Ref.
	Piperine														
p1041	—,hydroiodide diiodide	$2(C_{17}H_{19}NO_3).HI.I_2$	952.43	bl nd	145	....	....	....	s	v^h	v	...	...	chl, CS_2 v	**B20**, 80
—	**Piperitenone**	*see* **3-Terpinolenone**													
p1042	**Piperitone**(*d*)	Δ^1-*p*-Methanone-3. 1-Methyl-4-isopropylcyclohexanone-3. $C_{10}H_{16}O$	152.24	yesh in air $[\alpha]^{20}_D +49.13$	....	222–30 116–8.5^{20}	0.9344^{20}_4	1.4848^{20}	...	...	s	...	...	....	**B7**^{2}, 74
p1043	—(*dl*)	$C_{10}H_{16}O$	152.24	....	....	232–3 113^{18}	0.9331^{20}_4	1.4845^{20}	...	...	...	...	...	....	**B7**^{1}, 64
p1044	—(*l*)	$C_{10}H_{16}O$	152.24	$[\alpha]^{20}_D -51.53$		235 109.5–105^{15}	0.9324^{20}_4	1.4848^{20}	...	...	...	...	...	....	**B7**^{2}, 75
p1045	**Piperolidine**(*dl*)	Octahydropyrrocoline. δ-Coniceine. 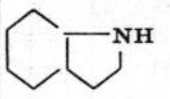	125.21	....	....	161.5^{750}	0.9012^{15}_4	....	...	...	...	...	...	....	**B20**, 150
p1046	—(*l*)	$C_8H_{15}N$. *See* p1045	125.21	$[\alpha]_D -7.8$ (c=1)	....	158^{729}	0.8976^{23}_4	....	...	...	...	...	...	....	**B20**, 150
—	**Piperonal**	*see* **Benzaldehyde, 3,4-methylenedioxy-**													
—	**4-Piperone**	*see* **Verbenone**													
—	**Piperonyl alcohol**	*see* **Toluene, α-hydroxy-3,4-methylenedioxy-**													
—	**Piperonylic acid**	*see* **Benzoic acid, 3,4-methylenedioxy-**													
—	**Pivaldehyde**	*see* **Propanal, 2,2-dimethyl-***													
—	**Pivalic acid**	*see* **Propanoic acid, 2,2-dimethyl-***													
—	**Pivalil**	*see* **3,4-Hexanedione, 2,2,5,5-tetramethyl-***													
—	**Pivalilindandione**	*see* **1,3-Indandione, 2-*tert*-butyl-**													
—	**Pivaloin**	*see* **3-Hexanone, 4-hydroxy-2,2,5,5-tetramethyl-***													
—	**Pivalophenone**	*see* **1-Propanone, 2,2-dimethyl-1-phenyl-***													
—	**Plasmocid**	*see* **Antimalarine**													
—	**Plumhagin**	*see* **1,4-Naphthoquinone, 5-hydroxy-2-methyl-***													
p1047	**Podophyllotoxin**	$C_{22}H_{22}O_8$	414.42	wh (w+1 or 2) $[\alpha]^{14}_D -101.3$	114–8	....	....	....	i	v	i	v	v	chl v	**B19**^{2}, 443
p1048	**Polyglycolid**	$[-O-CH_2CO-]x$	......	wh ($PhNO_2$)	223	....	....	....	...	...	...	...	...	....	**B19**^{1}, 679
p1049	**Populin**	Salicin benzoate. 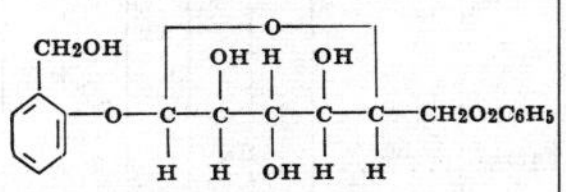	390.39	nd (w+2, al +2w) $[\alpha]_D -2.0$	180	....	....	....	δ	s^h	v	...	...	aa s	**B31**, 216
p1050	**Porphin**	Tetramethenetetrapyrrole. 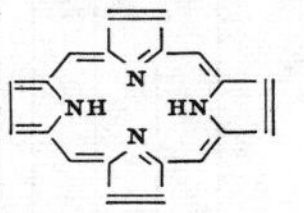	310.36	met red or og lf (chl-MeOH)	360	300^{12} sub	1.336	....	i	δ	...	...	...	chl δ Py δ, diox s	**B26**^{2}, 228
p1051	**Pregnane**	17(β)Ethyletiocholane. 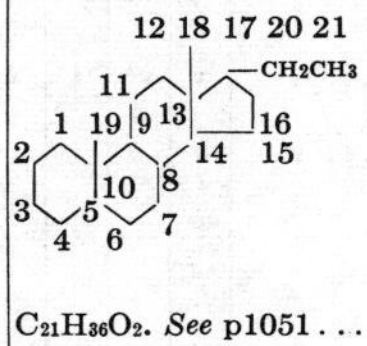	288.62	mcl sc or pl (MeOH) $[\alpha]^{19}_D +20$	....	83.5	1.032^{15}_4	....	...	...	...	...	...	chl s MeOH s^h	**E14**, 17
p1052	**Pregnane-3(α), 20(α)diol**	$C_{21}H_{36}O_2$. *See* p1051	320.52	pl (ace) $[\alpha]^{20}_D +27.4$	238	....	1.15	....	...	δ	...	s^h	...	os δ	**E14**, 95

For explanations, symbols and abbreviations see beginning of table.

No.	Name	Synonyms and Formula	Mol. wt.	Crystalline form, color and specific rotation	m.p. °C	b.p. °C	Density	n_D	Solubility						Ref.
									w	al	eth	ace	bz	other solvents	
	Pregnane-3(α)														
p1053	**Pregnane-3(α), 20(β)diol**	$C_{21}H_{36}O_2$. *See* p1051	320.52	(al)	232–4					s^h					**E14**, 95
p1054	**Pregnane-3(β), 20(α)diol**	$C_{21}H_{36}O_2$. *See* p1051	320.52	(al)	182					s^h					**E14**, 95
p1055	**Pregnane-3(β), 20(β)diol**	$C_{21}H_{36}O_2$. *See* p1051	320.52	cr (AcOEt-peth)	174–6									AcOEt s	**E14**, 95
p1056	**3,20-Pregnane-dione**	$C_{21}H_{32}O_2$. *See* p1051	316.49	nd (dil al)	123					v^h				os v	**E14**, 152
p1057	**Pregan-3(α)ol-20-one**	$C_{21}H_{34}O_2$. *See* p1051	318.50	(dil al)	149					s^h					**E14**, 115
p1058	**Pregnan-3(β)ol-20-one**	$C_{21}H_{34}O_2$. *See* p1051	318.50	(dil al)	149					s^h					**E14**, 115
p1059	**Pregnan-20(α)ol-3-one**	$C_{21}H_{34}O_2$. *See* p1051	318.50	pr (ace)	152							s^h			**E14**, 132
p1060	**Pregnan-20(β)ol-3-one**	$C_{21}H_{34}O_2$. *See* p1051	318.50	(dil MeOH)	172									MeOH s^h	**E14**, 132
p1061	**4-Pregnene-11(β), 17(α),20,21-tetrol-3-one**	$C_{21}H_{32}O_5$. *See* p1051	364.49	(aq ace+w) $[\alpha]_D^{20}+87$ (al)	ca. 120d					s		s^h			**E14**, 150
p1062	**4-Pregnene-17(α), 20(β),21-triol-3-one**	$C_{21}H_{32}O_4$. *See* p1051	348.49	(MeOH) $[\alpha]_D+63$	190									diox, chl s MeOH s	
p1063	**5-Pregnene-3(β)-ol-20-one**	$C_{21}H_{32}O_2$. *See* p1051	316.49	nd (dil al) $[\alpha]_D^{20}+28$	193				δ	δ		δ	δ	CCl_4 δ lig i	**E14**, 114
—	**Prehnitene**	*see* **Benzene, 1,2,3,4-tetramethyl-***													
—	**Prehnitic acid**	*see* **1,2,3,5-Benzenetetra-carboxylic acid***													
—	**Prehnitylic acid**	*see* **Benzoic acid, 2,3,4-trimethyl-**													
p1064	**Primeverose**	6(*β-D*-Xylosido)*D*-glucose. 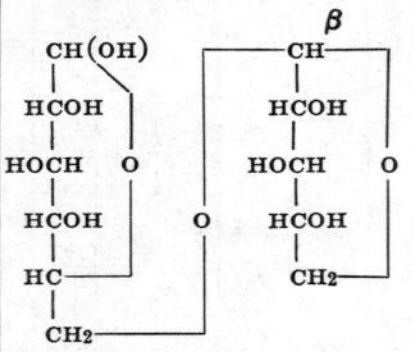	312.28	(MeOH, 80 % al), $[\alpha]_D^{20}$ (mut) +23 to −3.2					s	s				MeOH s	**B31**, 372
—	**Procaine**	*see* **Benzoic acid, 4-amino-2(diethylamino)**, ethyl ester													
p1065	**α-Progesterone**	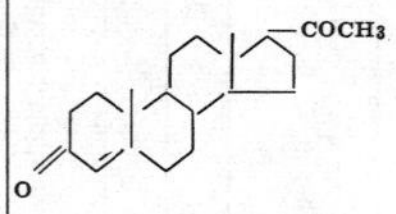	314.47	orb pr (dil al) $[\alpha]_D+192$	128.5		1.166^{23}		δ					os δ	
p1066	**β-Progesterone**	$C_{21}H_{30}O_2$. *See* p1065	314.47	$[\alpha]_D^{20}+172–82$ (diox, c=2)	121		1.171^{20}		i	s		s		con sulf s diox s	
p1067	**Progesterone**, dioxime	$C_{21}H_{32}N_2O_2$. *See* 1065	344.50	pl (dil al)	243					s^h					**E14**, 151
p1068	**Proline**(*D*)	*d*-2-Pyrrolidinecarboxylic acid. 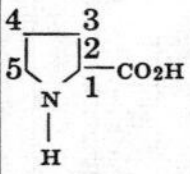	115.13	pr (al-eth) $[\alpha]_D+81.9$ (w)	215–20d										**B22**, 1
p1069	—(*DL*)	$C_5H_9NO_2$. *See* p1068	115.13	nd (al-eth)	205d				v	v^h s	i	δ	δ	chl δ	**B22**[2], 5
p1070	—(*L*)	$C_5H_9NO_2$. *See* p1068	115.13	nd (al), pr (w)	220–2				v	δ	i			BuOH, PrOH i	**B22**[2], 3
p1071	—,**4-hydroxy-** (*D, cis*)	$C_5H_9NO_3$. *See* p1068	131.13	nd (w+1) $[\alpha]_D^{18}+58.6$ (w, p=5)	237–41										**B22**, 546
p1072	—,—(*DL, cis*)	$C_5H_9NO_3$. *See* p1068	131.13	(w, dil al)	243–4				v					MeOH δ	**B22**[2], 144

For explanations, symbols and abbreviations see beginning of table.

No.	Name	Synonyms and Formula	Mol. wt.	Crystalline form, color and specific rotation	m.p. °C	b.p. °C	Density	n_D	Solubility w	al	eth	ace	bz	other solvents	Ref.
	Proline														
p1073	—,—(*L, cis*)	$C_5H_9NO_2$. *See* p1068	131.13	nd (w+1), $[\alpha]^{18}_D -58.1$ (w, p=5.2)	238–41	...	...	...	v	...	...	...	...	MeOH δ	**B22**, 546
p1074	—,—(*D, trans*)	$C_5H_9NO_2$. *See* p1068	131.13	lf (dil al) $[\alpha]^{21}_D +75.2$	274	...	...	...	...	...	...	...	...	...	**B22**^{1}, 545
p1074	—,—(*DL, trans*)	$C_5H_9NO_2$. *See* p1068	131.13	pl (MeOH)	270	...	...	...	v	δ	...	...	...	MeOH s^h	**B22**^{2}, 143
p1075	—,—(*L, trans*)	$C_5H_9NO_2$. *See* p1068	131.13	lf (dil al), pr (w) $[\alpha]_D -76.5$ (w, c=2.5)	274	...	...	...	...	...	...	...	...	...	**B22**^{1}, 143
p1077	—,—,betaine(*d*)	$C_7H_{13}NO_2$. *See* p1068	159.18	pr (w+1), $[\alpha]^{21}_D +36$	249d	...	...	...	δ s^h	δ	...	...	...	...	...
—	**Prominal**	*see* **Barbituric acid, 5-ethyl-3-methyl-5-phenyl-**													
p1078	**Prontisol**	H_2NSO_2—⟨C₆H₄⟩—N:N—⟨C₆H₃(H_2N)⟩—NH_2.HCl	327.8	og-red	248–50	...	...	...	δ	s	...	s	...	oils s fats s	...
p1079	**Propadiene***	Allene. Dimethylenemethane. $CH_2:C:CH_2$	40.07	gas	−136	−34.5	1.787	1.4168	...	...	...	...	...	...	**B1**^{3}, 922
p1080	—,**1,3-dioxo-***	Carbon suboxide. Dioxoallene. O:C:C:C:O	68.03	gas	−111.3	6.8	...	1.4538^0	d	d	s	...	s	CS_2 s HCO_2H, MeOH d	**B1**^{2}, 856
p1081	—,**tetrafluoro-***	Perfluoroallene. $F_2C:C:CF_2$	112.03	gas	...	−38	...	...	...	...	...	...	...	...	...
p1082	—,**tetraphenyl-***	$(C_6H_5)_2C:C:C(C_6H_5)_2$	344.43	nd, pr (dil ace, al),	166	...	...	...	i	δ	s	s	v	CS_2, chl v	**B5**^{2}, 692
—	**Propaesin**	*see* **Benzoic acid, 4-amino-,** propyl ester													
p1083	**Propanal***	Propionaldehyde. Propional. CH_3CH_2CHO	58.08	...	−81	48.8	0.807^{20}_{4}	1.3636^{20}	s	∞	∞	...	...	...	**B1**, 629
p1084	—,diethyl acetal	1,1-Diethoxypropane*. Propylal. $CH_3CH_2CH(OC_2H_5)_2$	132.20	...	...	122.8^{744}	0.8232^{20}_{4}	...	...	v	v	...	...	...	**B1**^{2}, 685
p1085	—,oxime*	Propionaldoxime. $CH_3CH_2CH:NOH$	73.09	...	40	131.5	0.9258^{20}_{4}	1.4287^{20}	...	...	...	...	...	...	**B1**, 631
p1086	—,**2-bromo-***	$CH_3CHBrCHO$	136.99	...	...	109–10 $52–4^{80}$	1.592^{20}	1.4813^{20}	...	...	v	...	...	...	**B1**^{2}, 691
p1087	—,**2-chloro-***	$CH_3CHClCHO$	92.53	...	...	86	1.182^{13}_{4}	1.431^{17}	δ	...	∞	...	∞	...	**B1**^{1}, 334
p1088	—,**3-chloro-***	$ClCH_2CH_2CHO$	92.53	...	...	130–1 40^{19}	1.268^{15}	1.475^{15}	i	s	s	...	...	...	**B1**^{2}, 690
p1089	—,—,diethyl acetal*	$ClCH_2CH_2CH(OC_2H_5)_2$	166.65	...	...	74^{20}	0.9951^{19}_{4}	1.4206^{19}	...	...	...	...	...	...	**B1**^{2}, 690
p1090	—,**2-chloro-2-methyl-***	$(CH_3)_2CClCHO$	106.56	...	...	90	1.053^{15}_{4}	...	...	v	v	...	...	...	**B1**, 675
p1091	—,**2,3-dibromo-***	Acrolein dibromide. $BrCH_2CHBrCHO$	215.90	...	...	$73–5^{10}$	2.198^{15}	...	...	...	v	...	...	...	**B1**^{2}, 691
p1092	—,**2,2-dichloro-***	CH_3CCl_2CHO	126.98	(peth)	38–9	77.8–8.0	...	...	...	...	...	...	...	peth δ	**B1**^{3}, 2692
p1093	—,**2,3-dichloro-***	$ClCH_2CHClCHO$	129.98	hyd (w)	...	73^{60}	1.400^{20}	1.4762^{20}	...	d	...	...	...	CCl_4 s	**B1**^{2}, 691
—	—,**2,3-dihydroxy-***	*see* **Glyceraldehyde**													
p1094	—,**2,2-dimethyl-***	Pivalaldehyde. $(CH_3)_3CCHO$	86.13	...	6	75	0.7927^{17}	1.3791^{20}	...	s	s	...	...	...	**B1**^{2}, 740
p1095	—,—,oxime*	Pivaldoxime. $(CH_3)_3CCH:NOH$	101.15	...	41	65^{20}	...	...	...	...	...	...	...	...	**B1**^{1}, 354
p1096	—,**3-ethoxy-,** diethyl acetal*	1, 1, 3-Triethoxypropane*. $C_2H_5OCH_2CH_2CH(OC_2H_5)_2$	176.26	...	...	184–6 sl d 78^{14}	0.898^{15}_{4}	1.4067^{20}	δ	s	...	...	...	...	**B1**^{3}, 3190
—	—,**2-ethyl-3-phenyl-***	*see* **Butanal, 2-benzylidene**													
p1097	—,**3-hydroxy-2-oxo-***	Propanolonal. Reductone. $HOCH_2COCHO$	88.07	ye	...	200–20d	...	...	s	s	...	...	...	...	**B1**^{2}, 895
p1098	—,3(4-isopropyl-phenyl)2-methyl	Cyclamen aldehyde. Cyclomal. $C_3H_7^i$—⟨C₆H₄⟩—$CH_2CH(CH_3)CHO$	190.27	...	...	127^9	0.950^{25}_{25}	1.509^{20}	i	s	s	...	s	...	...
p1099	—,**2-oxo-***	Methylglyoxal. CH_3COCHO	72.06	ye hyg	...	72	1.0455^{24}	1.4002^{18}	...	s	s	...	s	...	...
p1100	—,—,**dioxiime***	Methyl glyoxime. $CH_3C(:NOH)CH:NOH$	102.09	nd	157	...	...	...	δ s^h	s	s	...	...	...	...
p1101	—,—,**1-oxime***	Isonitrosoacetone. $CH_3COCH:NOH$	87.08	nd (CCl_4), lf (eth-peth)	68	sub	1.0744^{67}	...	s	...	s	...	s	chl s CCl_4 s	**B1**^{3}, 822

For explanations, symbols and abbreviations see beginning of table.

No.	Name	Synonyms and Formula	Mol. wt.	Crystalline form, color and specific rotation	m.p. °C	b.p. °C	Density	n_D	Solubility						Ref.
									w	al	eth	ace	bz	other solvents	
	Propanal														
p1102	**—,2-phenoxy-***	$C_6H_5OCH(CH_3)CHO$	150.18			$99–101^{16}$			i	v	v		v		**B6**, 151
p1103	**—,—,oxime***	$C_6H_5OCH(CH_3)CH:NOH$	165.19	nd	110				δ	v	v				**B6**, 151
p1104	**—,2-phenyl-***	Hydratropic aldehyde. $C_6H_5CH(CH_3)CHO$	134.17			92.5^{10}	1.0089^{20}_{4}	1.5221^{20}	i	s					
p1105	**—,3-phenyl-***	Hydrocinnamaldehyde. $C_6H_5CH_2CH_2CHO$	134.17	mcl	47	223			i	v	∞				**B7**², 236
p1106	**—,2,2,3-trichloro-***	$ClCH_2CCl_2CHO$	161.43			$63–5^{45}$	1.470^{25}	1.473^{25}			v			CCl_4 s	**B1**², 691
p1107	**Propane***	$CH_3CH_2CH_3$	44.09		−189.9	−44.5	0.5853^{-45}_{4}	1.2898^{20}	s	v	v		v	chl v	**B1**, 104
p1108	**—,1-amino-***	Propylamine. $CH_3CH_2CH_2NH_2$	59.11		−83	49	0.719^{20}_{20}	1.389^{30}	s	v	v				**B4**², 619
p1109	**—,2-amino-***	Isopropylamine. $(CH_3)_2CHNH_2$	59.11		−101	33.0	0.691^{18}_{4}	1.3770^{15}	∞	∞	∞				**B4**², 629
p1110	**—,1-amino-3(2-butyloctyloxy)-***	$CH_3(CH_2)_5CH(C_4H_9^n)CH_2O(CH_2)_3NH_2$	243.44	yesh		$105–15^{1}$	0.8455^{20}_{4}	1.4478^{20}	i			s	s	MeOH, chl s CCl_4 s	
p1111	**1-amino-2,2-dimethyl-***	Neopentylamine. $(CH_3)_3CCH_2NH_2$	87.16			80.5^{729}	0.7455^{20}_{4}	1.4023^{20}	δ						**B4**, 188
p1112	**—,1-amino-3-dodecyloxy-***	$CH_3(CH_2)_{11}O(CH_2)_3NH_2$	243.44			$115–25^{1}$	0.8439^{20}_{4}	1.4487^{20}	i			s	s	MeOH, chl s	
p1113	**—,1-amino-3-methoxy-***	$CH_3OCH_2CH_2CH_2NH_2$	89.14			116–9	0.8727^{20}_{4}	1.4191^{20}	s			s	s	MeOH, chl s CCl_4 s	
p1114	**—,2-amino-2-methyl-***	*tert*-Butylamine. $(CH_3)_3CNH_2$	73.14		−67.5±1	45.2	0.696^{20}_{4}	1.3786^{20}	∞	∞	∞				**B4**², 641
—	**—,2-amino-1-phenyl-***	*see* **Amphetamine**													
p1114¹	**—,2,2-bis(bromomethyl)1,3-dibromo-***	Pentaerythrityl bromide. $C(CH_2Br)_4$	387.78	(ace), nd (lig), tcl (bz+1)	163	305–6	2.596^{15}			s^h	δ		s^h	to s^h, os δ	**B1**², 105
p1115	**—,2,2-bis(chloromethyl)1,3-dichloro-***	Pentaerythrityl tetrachloride. $C(CH_2Cl)_4$	209.95		97	110^{12}			i		s			chl s	
p1116	**—,1,1-bis(ethylsulfonyl)2-methyl-***	$(C_2H_5SO_2)_2CHCH(CH_3)_2$	242.36	nd (al)	94				s^h						**B1**, 676
p1117	**—,1,1-bis(4-hydroxyphenyl)-***	4,4′-Propylidenediphenol. $[HO-C_6H_4-]_2CHCH_2CH_3$	228.29	nd (w)	130				i	v	v			alk v	**B6**², 977
p1118	**—,2,2-bis(4-hydroxyphenyl)-***	4,4′-Isopropylidenediphenol. $[HO-C_6H_4-]_2C(CH_3)_2$	228.29	pr (dil aa), nd (w)	152–3	$250–2^{13}$			i	v	v		v		**B6**², 978
p1118¹	**—,2,2-bis(iodomethyl)1,3-diiodo-***	Pentaerythryl iodide. $C(CH_2I)_4$	575.74	nd (to)	225					δ	δ			to s^h	**B1**³, 105
p1119	**—,1-bromo-***	*n*-Propylbromide. $CH_3CH_2CH_2Br$	123.00		−109.85	70.82	1.3539^{20}_{4}	1.4341^{20}	δ	s	s				**B1**², 74
p1120	**—,2-bromo-***	Isopropyl bromide. $CH_3CHBrCH_3$	123.00		−90.8	59.41	1.3097^{20}_{4}	1.4251^{20}	δ	∞	∞				**B1**², 242
p1121	**—,1-bromo-2-chloro-***	$CH_3CHClCH_2Br$	157.44			118	1.531^{20}_{4}	1.4745^{20}							
p1122	**—,1-bromo-3-chloro-***	$ClCH_2CH_2CH_2Br$	157.44			141.3–2.3 $68–70^{62}$	1.4718^{25}_{22}	1.4950^{20}	i	v	v			chl v	**B1**³, 245
p1123	**—,2-bromo-1-chloro-***	$CH_3CHBrCH_2Cl$	157.44			117.0^{756}	1.537^{20}_{4}	1.4776							
p1124	**—,2-bromo-2-chloro-***	$(CH_3)_2CBrCl$	157.44			$93.0–5.5^{745}$	1.474^{21}	1.4575^{20}	i		v			chl v	**B1**³, 246
p1125	**—,1-bromo-2,2-dimethyl-***	Neopenty bromide. $(CH_3)_3CCH_2Br$	151.05			34.6^{100} $89–91^{749}$	1.2604^{20}_{0}		i	s	s				

For explanations, symbols and abbreviations see beginning of table.

No.	Name	Synonyms and Formula	Mol. wt.	Crystalline form, color and specific rotation	m.p. °C	b.p. °C	Density	n_D	Solubility						Ref.
									w	al	eth	ace	bz	other solvents	
	Propane														
p1126	**—,1-bromo-2,3-epoxy-(*dl*)***	Epibromohydrin. O CH_2CHCH_2Br	136.98			138–40 $61–2^{50}$	1.615^{14}		i	s^h	s				**B17**^{2}, 14
p1127	**—,—(*l*)***	C_3H_5BrO. *See* p1126	136.98	$[\alpha]^{16}_D+23.1$ (undil); +6.5 (al, p=5.5)		$134–6^{50}$									**B17**^{2}, 14
p1128	**—,1-bromo-3-fluoro-***	$FCH_2CH_2CH_2Br$	140.99			101.4	1.542^{25}_4	1.4290^{25}	i	v	v			chl v	**B1**^{3}, 244
p1129	**—,2-bromo-2-methly-***	*tert*-Butylbromide. $(CH_3)_3CBr$	137.03		−20	73.3	1.2220^{20}_4	1.4283^{20}	i						**B1**^{3}, 322
p1130	**—,2(bromo-methyl)1,2,3-tribromo-***	$(BrCH_2)_3CBr$	373.73		25	$150–1^{14}$	2.5595^{20}_4	1.6246^{20}	i			s			**B1**^{2}, 91
p1131	**—,1-bromo-1-nitro-***	$CH_3CH_2CHBrNO_2$	168.00			160–5 82.5^{50}				s	s			KOH s	**B1**^{3}, 260
p1132	**—,2-bromo-2-nitro-***	$(CH_3)_2CBrNO_2$	168.00			151.7 $-.8^{745}$ $73–5^{50}$	1.6562^0			s	s			alk i	**B1**, 116
p1133	**—,1-chloro-***	*n*-Propylchloride. $CH_3CH_2CH_2Cl$	78.54		−122.8	46.60	0.8923^{20}_4	1.3886^{20}	δ	∞	∞				**B1**^{2}, 72
p1134	**—,2-chloro-***	Isopropyl chloride. $CH_3CHClCH_3$	78.54		−117	34.8	0.8590^{20}_4	1.3781^{20}	δ	∞	∞				**B1**^{3}, 221
p1135	**—,1-chloro-2,2-dimethyl-***	Neopentyl chloride. $(CH_3)_3CCH_2Cl$	106.60		−20±1	84.4	0.866^{20}_4	1.4042^{20}							
p1136	**—,1-chloro-2,3-epoxy-(*dl*)***	α-Epichlorohydrin. O CH_2CHCH_2Cl	92.53		−48	116.5	1.1801^{20}_4	1.438^{20}	δ d^h	∞	∞		s		**B17**^{2}, 73
p1137	**—,—(*l*)**	C_3H_5ClO. *See* p1136	92.53	$[\alpha]^{18}_D-25.6$		$92–3^{360}$	1.2007								**B17**^{1}, 4
p1138	**—,2-chloro-1,3-epoxy-***	3-Chloro-1-oxacyclobutane. β-Epichlorohydrin (Cl–, O ring)	92.53		132–4					v	v				**B17**, 6
p1139	**—,3-chloro-1,2-epoxy-2-methyl-***	O $CH_2CH(CH_3)CH_2Cl$	106.56			122.0	1.1025^{20}_4	1.4340^{20}	s						
p1140	**—,1-chloro-3-fluoro-***	$FCH_2CH_2CH_2Cl$	96.54			82–5		1.3871^{25}	i	v	v				
p1141	**—,2-chloro-2-methyl-***	*tert*-Butyl chloride. $(CH_3)_3CCl$	92.57		−27.1	50.7	0.8511^{20}_4	1.3903^{20}	δ	∞	∞				**B1**^{3}, 316
p1142	**—,(2-chloro-methyl)1,1,2,3-tetrachloro-***	$(ClCH_2)_2CClCHCl_2$	230.37			$226.3–.4^{737}$ 95^9	1.5686^{25}_4	1.5165^{25}						chl s	**B1**^{3}, 321
p1143	**—,(2-chloro-methyl)1,2,3-trichloro-***	$(ClCH_2)_3CCl$	195.92			209.8– 10.5^{737} 87^9	1.481^{25}_4	1.508^{20}						chl s	**B1**^{3}, 321
p1144	**—,1-chloro-1-nitro-***	$CH_3CH_2CHClNO_2$	123.54			141–3	1.209^{20}_{20}	1.430	δ	s	s			glycols s oils s	**B1**, 116
p1145	**—,1-chloro-2-nitro-***	$CH_3CH(NO_2)CH_2Cl$	123.54			172–3 94^{46}	1.2^{18}								**B1**, 116
p1146	**—,1-chloro-3-nitro-***	$O_2NCH_2CH_2CH_2Cl$	123.54			197 $115–6^{40}$	1.267^{20}								**B1**, 116
p1147	**—,2-chloro-1-nitro-***	$CH_3CHClCH_2NO_2$	123.54			172^{749} 75^{15}	1.2361^{15}		δ	s	s				**B1**, 116
p1148	**—,2-chloro-2-nitro-***	$CH_3CCl(NO_2)CH_3$	123.54						δ	s	s			KOH i glycerol, oils s	**B1**^{2}, 79
p1148[1]	**—,1-chloro-3-phenyl-***	$C_6H_5CH_2CH_2CH_2Cl$	154.64			219–20	1.056^{25}_4	1.5160^{25}							

For explanations, symbols and abbreviations see beginning of table.

No.	Name	Synonyms and Formula	Mol. wt.	Crystalline form, color and specific rotation	m.p. °C	b.p. °C	Density	n_D	Solubility w	al	eth	ace	bz	other solvents	Ref.
	Propane														
p1149	—,1,2-di-acetamido-	$CH_3CONHCH(CH_3)CH_2NHCOCH_3$	158.20	nd (bz)	138–9	190^{18}	...	...	v	v	i	...	δ s^{h}	chl v	**B4**, 261
p1150	—,1,2-di-amino-(*d*)*	1,2-Propanediamine*. $CH_3CH(NH_2)CH_2NH_2$	74.13	$[\alpha]^{25}_{D}+20.7$	...	117.35	0.8584^{25}_{4}	...	v	...	i	...	...	chl v	**B4²**, 697
p1151	—,1,3-diamino-*	Trimethylenediamine. $H_2NCH_2CH_2CH_2NH_2$	74.13	...	...	135.5	0.884^{25}_{4}	...	s	∞	∞	...	...	...	**B4²**, 697
p1152	—,1,1-dibromo-*	Propylidene bromide. $CH_3CH_2CHBr_2$	201.91	...	...	$135.3\text{–}.7^{74}$	...	1.5100^{20}	...	...	...	...	...	...	...
p1153	—,1,2-dibromo-*	Propylene bromide. $CH_3CHBrCH_2Br$	201.91	...	−55.5	141.4	1.9366^{20}_{4}	1.5206^{20}_{4}	...	s	s	...	...	...	...
p1154	—,1,3-dibromo-*	Trimethylene bromide. $BrCH_2CH_2CH_2Br$	201.91	...	−34.2	167.34	1.9893^{20}_{4}	1.523	...	...	...	...	...	...	...
p1155	—,2,2-dibromo-*	Isopropylidene bromide. $CH_3CBr_2CH_3$	201.91	...	...	114.5	1.7825^{20}_{4}		...	...	...	...	...	...	...
p1157	—,1,1-dibromo-2,2-dimethyl-*	Neopentylidene bromide. $(CH_3)_3CCHBr_2$	229.95	...	...	$64\text{–}5^{43}$	1.7614^{20}_{0}	1.5085	i	...	s	...	...	...	**B1³**, 371
p1158	—,1,3-dibromo-2,2-dimethyl-*	$(CH_3)_2C(CH_2Br)_2$	229.95	...	...	185–90 72^{14}	1.6934^{20}_{4}	1.5050^{20}	i	...	s	...	...	...	**B1³**, 371
p1159	—,1,2-dibromo-2-methyl-*	Isobutylene bromide. $(CH_3)_2CBrCH_2Br$	215.93	...	−70.3	149.0	1.759^{20}_{4}	1.509	...	...	...	...	...	...	...
p1160	—,1,1-dichloro-*	Propylidene dichloride. $CH_3CH_2CHCl_2$	112.99	...	...	88.3	1.1321^{20}_{4}	1.4289^{20}	...	...	...	...	...	...	**B1**, 105
p1161	—,1,2-dichloro-*	Propylene dichloride. $CH_3CHClCH_2Cl$	112.99	...	−100.42	96.20	1.1558^{20}_{4}	1.4390^{20}	δ	...	...	...	...	...	**B1**, 105
p1163	—,1,3-dichloro-*	Trimethylene chloride. $ClCH_2CH_2CH_2Cl$	112.99	...	...	120.4	1.1896^{20}	1.4469	δ	v	v	...	...	...	...
p1164	—,2,2-dichloro-*	Acetone dichloride. Isopropylidene chloride. $CH_3CCl_2CH_3$	112.99	...	−34.6	69.7	1.093^{20}	1.4093^{20}	i	s	∞	...	...	...	...
p1165	—,1,1-dichloro-2-methyl-*	Isobutylidene chloride. $(CH_3)_2CHCHCl_2$	127.01	...	...	105–6	1.0111^{12}_{12}	...	i	...	...	...	...	chl s	**B1³**, 319
p1166	—,1,2-dichloro-2-methyl-*	Isobutylene chloride. $(CH_3)_2CClCH_2Cl$	127.01	...	>−130	108 $38\text{–}9^{70}$	1.093^{20}_{4}	1.4370^{20}	i	∞	∞	∞	∞	CCl_4 ∞	**B1³**, 319
p1167	—,1,3-dichloro-2-methyl-*	$CH_3CH(CH_2Cl)_2$	127.01	...	...	136 60^{49}	1.1325^{25}_{4}	1.4488^{25}	i	s	...	...	...	CCl_4 s	**B1³**, 319
—	—,1,1-diethoxy-*	*see* **Propanal**, diethyl acetal*													
p1167[1]	—,2,2-bis(ethyl-sulfonyl)-*	Acetone diethyl sulfone. Sulfonal. $(CH_3)_2C(SO_2C_2H_5)_2$	228.33	mcl (w), pr (al)	125.8	300d	...	...	δ	v	...	...	s	...	**B1**, 345
p1168	—,2,2-difluoro-*	Isopropylidene fluoride. $CH_3CF_2CH_3$	80.08	gas	...	−0.6 (−0.2)	0.92^{0}	...	...	...	...	...	...	...	...
p1169	—,2,2-difuryl-*	$(CH_3)_2C(\text{—furyl})_2$	176.22	...	−12	$73\text{–}6^{5}$	1.043^{20}_{4}	1.4966^{20}	i	s	s	...	...	MeOH s	
p1170	—,1,2-diiodo-*	Propylene iodide. CH_3CHICH_2I	295.89	...	...	...	$2.490^{18.5}$	...	...	...	...	...	...	...	**B1**, 115
p1171	—,1,3-diiodo-*	Trimethylene iodide. $ICH_2CH_2CH_2I$	295.89	...		224^{768} 110^{19}	2.5755^{20}_{4}	1.6423^{20}	i	...	s	...	...	CCl_4 s chl s	**B1³**, 255
p1172	—,2,2-dioodo-*	Isopropylidene iodide. $CH_3CI_2CH_3$	295.89	...	...	147–8d 71^{12}	2.4458^{0}	...	i	...	s	...	...	CCl_4 s chl s	**B1³**, 225
p1174	—,2,2-dimethyl-*	Neopentane. $C(CH_3)_4$	72.15	gas	−20	9.45	0.61350^{20}	1.3476^{6}	i	s	s	...	...	...	**B1²**, 104
p1175	—,1(dimethyl-amino)-*	Dimethyl propylamine*. $CH_3CH_2CH_2N(CH_3)_2$	87.17	...	...	65.5^{752}	0.7152^{20}_{4}	1.3907^{20}	...	...	...	...	...	...	**B4²**, 621
p1176	—,1(dimethyl-amino)2-methyl	Dimethyl isobutylamine*. $(CH_3)_2CHCH_2N(CH_3)_2$	101.19	...	...	81.0^{753}	0.7097^{20}_{4}	1.3907^{20}	...	...	...	...	...	...	**B4²**, 638
p1177	—,1,1-dinitro-*	$CH_3CH_2CH(NO_2)_2$	134.09	...	−42	184	1.2610^{25}	1.4339^{20}	...	...	...	...	...	alk s	**B1³**, 260
p1178	—,1,3-dinitro-*	$O_2NCH_2CH_2CH_2NO_2$	134.09	...	−21.4	103^{1}	1.353^{26}_{4}	1.4654^{20}	i	...	s	...	...	...	**B1³**, 261
p1179	—,2,2-dinitro-*	$CH_3C(NO_2)_2CH_3$	134.09	...	53	185.5	1.30^{25}	...	δ	...	...	...	...	...	**B1³**, 261

For explanations, symbols and abbreviations see beginning of table.

No.	Name	Synonyms and Formula	Mol. wt.	Crystalline form, color and specific rotation	m.p. °C	b.p. °C	Density	n_D	Solubility w	al	eth	ace	bz	other solvents	Ref.
	Propane														
p1179[1]	**—,1,2-di-phenoxy-***	$CH_3CH(OC_6H_5)CH_2OC_6H_5$	228.29	rh (MeOH)	32	175–8[12]	$1.0748_4^{33.3}$	1.5542[33.3]	i	s	s	s	s	MeOH, chl s	**B6**[2], 151
p1179[2]	**—,1,3-di-phenoxy-***	$C_6H_5OCH_2CH_2CH_2OC_6H_5$	228.29	lf (al)	61	338–40 160[25]			i	s	s				**B6**[2], 151
p1180	**—,1,2-epoxy-***	Methyloxyrane. Propylene oxide. CH_2CHCH_3 (epoxide O)	58.08			35	0.859_4^0		v	∞	∞				**B17**, 6
p1181	**—,1,3-epoxy-***	Trimethylene oxide.	58.08			47–8 cor	0.8930_4^{25}	1.3897[25]	∞	∞	s			os v	**B17**[2], 12
p1182	**—,1,2-epoxy-3-iodo-***	Epiiodohydrin. CH_2CHCH_2I (epoxide O)	189.98			160–2	1.982[24]		i	s	s				**B17**[2], 15
p1183	**—,1,2-epoxy-2-methyl-***	Isobutylene oxide. $CH_2C(CH_3)_2$ (epoxide O)	72.11				0.8311[0]			s	s				**B17**[2], 17
p1184	**—,1,2-epoxy-3,3,3-trichloro-***	CH_2CHCCl_3 (epoxide O)	161.42			149[764]	1.495_4^{20}	1.4737[25]			v				
p1184[1]	**—,1-ethoxy-3-phenoxy-**	$C_6H_5OCH_2CH_2CH_2OC_2H_5$	180.25	oil		328–30									**B6**, 147
p1185	**—,1-fluoro-***	*n*-Propyl fluoride. $CH_3CH_2CH_2F$	62.09		−159	−2.5	0.7788^{-3}	1.3154[10]	δ	v	v				**B1**[3], 217
p1186	**—,1,1,1,2,2,3,3-heptachloro-***	$Cl_2CHCCl_2CCl_3$	285.22	amor	29.4	247–8 132[30]	1.8048_4^{31}								**B1**[3], 238
p1187	**—,1,1,1,2,3,3,3-heptachloro-***	$Cl_3CCHClCCl_3$	285.22		11–1.5	249 93[3]	1.7921_4^{34}	1.5427[21]							**B1**[3], 239
p1188	**—,heptafluoro-1-nitro-***	$F_3CCF_2CF_2NO_2$	215.03			25									
p1189	**—,heptafluoro-1-nitroso-***	$F_3CCF_2CF_2NO$	199.03	deep bl	−151	−12									
p1190	**—,1,1,1,2,2,3-hexafluoro-***	$FCH_2CF_2CF_3$	152.04			1.2									
p1191	**—,1(hydroxyl-amino)-***	$CH_3CH_2CH_2NHOH$	75.11	nd (eth)	46				v		s			lig δ	**B4**, 537
p1192	**—,1-iodo-***	*n*-Propyl iodide. $CH_3CH_2CH_2I$	169.99		−101.3	102.45	1.7454_4^{20}	1.5055[20]	δ	∞	∞				**B1**[3], 252
p1193	**—,2-iodo-***	Isopropyl iodide. CH_3CHICH_3	169.99		−90.1	89.45	1.7042_4^{20}	1.4997[20]	δ	∞	∞		∞	chl ∞	**B1**[3], 253
p1194	**—,1-iodo-2,2-dimethyl-***	Neopentyl iodide. $(CH_3)_3CCH_2I$	198.05			127–9d	1.5317[13]		i	s	s				
p1195	**—,2-iodo-2-methyl-***	*tert*-Butyl iodide. $(CH_3)_3CI$	184.02		−38.20	100 d 20.8[30]	1.571_0^0	1.4918[20]	d	∞	∞				**B1**[3], 326
p1196	**—,1-isocyano-***	Propylcarbilamine. $CH_3(CH_2)_2NC$	69.11			99.5			i	∞	∞				**B4**[2], 625
p1197	**—,2-isocyano-***	Isopropylcarbylamine. $(CH_3)_2CHNC$	84.14			87	0.7596[0]		i	∞	∞				**B4**, 154
p1198	**—,2-isocyano-2-methyl-***	*tert*-Butyl isocyanide. $(CH_3)_3CNC$	83.13			118				∞	∞			HCl d	**B4**[2], 641
p1198[1]	**—,methoxy-(phenoxy)-**	$CH_3OCH_2CH_2CH_2OC_6H_5$	166.22	oil		230–1									**B6**, 147
p1199	**—,1(2-methoxy-phenyl)-2(methyl-amino)-,** hydrochloride.	OCH_3 (ortho) C_6H_4—$CH_2CH(NHCH_3)CH_3.HCl$	215.72		124–8				v	v	δ		δ	chl v	
p1200	**—,2-methyl-1-nitro-***	Nitroisobutane. $(CH_3)_2CHCH_2NO_2$	103.12		158–9		0.9625_{25}^{25}		δ	∞	∞				
p1201	**—,2-methyl-1,1,1,2,3-penta-chloro-***	$ClCH_2CCl(CH_3)CCl_3$	230.35	(al)	73.5	213[737] 90–3[10]			i					chl s	**B1**[3], 321
p1202	**—,2-methyl-1,1,1,2-tetra-bromo-***	$(CH_3)_2CBrCBr_3$	373.74	lf	217				i	v	v				**B1**, 128

For explanations, symbols and abbreviations see beginning of table.

No.	Name	Synonyms and Formula	Mol. wt.	Crystalline form, color and specific rotation	m.p. °C	b.p. °C	Density	n_D	Solubility						Ref.
									w	al	eth	ace	bz	other solvents	
	Propane														
p1203	**—,2-methyl-1,1,2,3-tetrabromo-***	$BrCH_2CBr(CH_3)CHBr_2$	373.74			$134–5^{15}$	2.4545^{20}	1.5990^{20}	i						**B1**^{3}, 325
p1204	**—,2-methyl-1,1,1,2-tetrachloro-*-**	$(CH_3)_2CClCCl_3$	195.92	(al)	178.6–9.6	$192^{117.5}$ cor			i	v	v				**B1**, 126
p1205	**—,2-methyl-1,1,2,3-tetrachloro-***	$ClCH_2CCl(CH_3)CHCl_2$	195.92		−46	190.6–1.3 69^{12}	1.4393^{25}_{4}	1.4963^{20}	i					chl s	**B1**^{3}, 321
p1206	**—,2-methyl-1,1,2-tribromo-***	$(CH_3)_2CBrCHBr_2$	294.85			208–15d 96^{14}	2.0169^{20}_{4}		i						**B1**^{3}, 325
p1207	**—,2-methyl-1,2,3-tribromo-***	$(BrCH_2)_2CBrCH_3$	294.85			88.5^{9}	2.1750^{20}_{4}	1.5652^{20}							**B1**^{3}, 325
p1208	**—,2-methyl-1,1,2-trichloro-***	$(CH_3)_2CClCHCl_2$	161.47		−6.5, −6	144.5–5.5 $46–7^{18}$	1.2588^{20}_{4}	1.4666^{20}			s			chl v	**B1**^{3}, 320
p1209	**—,2-methyl-1,2,3-trichloro-***	$CH_3CCl(CH_2Cl)_2$	161.47			162–3 81^{50}	1.3012^{25}_{4}	1.4765^{20}	i					chl s	**B1**^{3}, 320
p1210	**—,1-nitro-***	$CH_3CH_2CH_2NO_2$	89.09		−108	130.5–1.5	1.0221^{24}_{4}	1.4003^{24}	δ	∞	∞				**B1**^{2}, 79
p1211	**—,2-nitro-***	$CH_3CH(NO_2)CH_3$	89.09		−93	120	1.024^{0}	1.3941^{20}	δ						**B1**^{2}, 79
p1212	**—,1(nitroamino)-***	*n*-Propylnitramine. $CH_3CH_2CH_2NHNO_2$	104.11		−21, −23	$128–9^{40}$	1.1046^{15}		δ	∞	∞				**B4**^{1}, 569
p1213	**—,3-nitro-1,1,1-trifluoro-***	$O_2NCH_2CH_2CF_3$	143.07			135									
p1214	**—,octachloro-***	Perchloropropane. $Cl_3CCCl_2CCl_3$	319.69	wh	160	$268–9^{734}$			i	v	v				**B1**^{1}, 35
p1215	**—,1,1,1,2,3-pentachloro-***	$ClCH_2CHClCCl_3$	216.34	nd (al^{h})	179–80	sub			i					os s	**B1**, 107
p1216	**—,1,1,2,3,3-pentachloro-***	$Cl_2CHCHClCHCl_2$	216.34			198–200 $98–100^{20}$	1.6086^{34}_{4}	1.5131^{16}	i	v					**B1**^{1}, 34
p1218	**—,1,1,1,2-tetrachloro-***	$CH_3CHClCCl_3$	181.89		−64	152–3 cor $87–8^{104}$	1.4695^{22}_{22}	1.4855^{20}	i	v					**B1**, 107
p1219	**—,1,1,2,2-tetrachloro-***	$CH_3CCl_2CHCl_2$	181.89			153	1.47^{13}_{13}		i	∞	∞				**B1**, 107
p1220	**—,1,1,2,3-tetrachloro-***	$ClCH_2CHClCHCl_2$	181.89			179–80 cor 61^{12}	1.522^{15}_{15}	1.5037^{17}						alk d^{h} chl s	
p1221	**—,1,2,2,3-tetrachloro-***	$ClCH_2CCl_2CH_2Cl$	181.89			164	1.496^{17}_{17}		i	v	v				**B1**, 107
p1222	**—,1,1,2-tribromo-***	$CH_3CHBrCHBr_2$	280.82			200–1 83^{6}	2.3548^{20}_{4}	1.5790^{20}	i	s				chl s aa s	**B1**^{3}, 251
p1223	**—,1,2,2-tribromo-***	$CH_3CBr_2CH_2Br$	280.82			190–1 81^{20}	2.2985^{20}_{4}	1.5670^{20}	i					chl s aa s	**B1**^{2}, 77
p1224	**—,1,2,3-tribromo-***	$BrCH_2CHBrCH_2Br$	280.82		16–17	219–21	2.4114^{15}	1.584^{18}	i	v	v				**B1**, 77
p1225	**—,1,1,1-trichloro-***	$CH_3CH_2CCl_3$	147.44			106.5–8.5	1.287^{23}_{4}		i					chl v	**B1**^{3}, 230
p1226	**—,1,1,2-trichloro-***	$CH_3CHClCHCl_2$	147.44			140	1.372^{25}		i					chl s	**B1**^{3}, 230
p1227	**—,1,1,3-trichloro-***	$CH_2ClCH_2CHCl_2$	147.44			146–8	1.351^{18}	1.474^{18}	i	s				chl v aa s	**B1**^{3}, 330
p1228	**—,1,2,2-trichloro-***	$CH_3CCl_2CH_2Cl$	147.44			123–5	1.318^{25}	1.4609^{25}	i	s				chl v	**B1**^{3}, 231
p1229	**—,1,2,3-trichloro-***	Trichlorohydrin. $ClCH_2CHClCH_2Cl$	147.44		−14.7	156	1.394^{20}	1.4858^{20}	δ	s	s				**B1**, 73
—	**—,1,3,3-triethoxy-***	*see* **Propanal, 3-ethoxy-, diethyl acetal***													
p1230	**—,1,2,3-triphenyl-***	$(C_6H_5)_3CCH_2CH_3$	272.39	pr	51										**B5**, 622
p1231	**1-Propanearsonic acid***	$CH_3CH_2CH_2AsO_3H_2$	168.02	nd (al), pl (w)	126–7				v	v	i				**B4**^{2}, 997
p1232	**1-Propaneboronic acid***	*n*-Propylboric acid. $CH_3CH_2CH_2B(OH)_2$	87.91	wh nd	107	d			v	s	s				**B4**^{2}, 1023

For explanations, symbols and abbreviations see beginning of table.

No.	Name	Synonyms and Formula	Mol. wt.	Crystalline form, color and specific rotation	m.p. °C	b.p. °C	Density	n_D	Solubility w	al	eth	ace	bz	other solvents	Ref.
	Propaneboronic acid														
p1233	**—,2-methyl-***	Isobutylboric acid. $(CH_3)_2CHCH_2B(OH)_2$	101.94	lo pl (w)	112				δ	s	s				**B4**[2], 1023
—	**Propanedioic acid***	*see* **Malonic acid**													
p1234	**1,2-Propanediol***	Propylene glycol. $CH_3CHOHCH_2OH$	76.10			189, 101.8^{24}	1.0361^{20}_{4}	1.4324^{20}	∞	∞	s				**B1**, 535
p1235	—,carbonate	CH_3-cyclic carbonate (ring with O, C=O)	102.09		<−70	240, 110^{10}	1.2041^{20}_{4}	1.4212^{20}	v	v	v	v	v		
p1236	—,diacetate	$CH_3CO_2CH(CH_3)CH_2O_2CCH_3$	160.17			186^{758}	1.109^{0}		v	s					**B2**[2], 156
p1243	—,sulfite	CH_3-cyclic sulfite (ring with O, SO)	122.14		<−60	175, 85^{25}	1.2960^{20}_{4}	1.4370^{20}	d^h v	v	v	v	v	AcOEt v	
p1244	**1,3-Propanediol-***	Trimethylene glycol. $HOCH_2CH_2CH_2OH$	76.10			213.5	1.0529^{20}_{4}	1.4389^{20}	∞	∞	v				**B1**[2], 540
p1245	**—,diacetate**	$CH_3CO_2CH_2CH_2CH_2O_2CCH_3$	160.17			209–10	1.070^{19}		v	s					**B2**[2], 156
p1250	**—,2-amino-2-ethyl-***	$CH_3CH_2CNH_2(CH_2OH)_2$	119.16	ye	108–10	$143\text{–}5^{10}$	1.099^{20}_{4}	1.490^{20}	∞						
p1251	**—,2-amino-2(hydroxy-methyl)-***	$H_2NC(CH_2OH)_3$	121.14	or nd	170.5–1.5	$219\text{–}20^{10}$			v						
p1252	**—,2-amino-2-methyl-***	$CH_3CNH_2(CH_2OH)_2$	105.14		109–11	151.2^{10}			v						
p1253	**—,2-butyl-2-ethyl-***	$CH_3CH_2CH_2CH_2C(C_2H_5)(CH_2OH)_2$	160.26	wh	43.8	262, 123^{15}	0.929^{50}_{20}	1.4587^{25}	δ						
p1254	**1,2-Propanediol, 3-chloro-***	α-Chlorohydrin. $ClCH_2CHOHCH_2OH$	110.54	yesh liq		213d, 116^{11}	1.326^{18}_{15}		s	s	s				**B1**, 537
p1255	—,—,diacetate	$ClH_2CH(O_2CCH_3)CH_2O_2CCH_3$	194.62		245, 116^{12}	245	1.199^{25}_{4}	1.4407^{20}							**B2**[1], 67
p1256	**1,3-Propanediol, 2-chloro-***	$HOCH_2CHClCH_2OH$	110.54			$124.5\text{–}5.0^{15}$	1.3219^{20}_{4}	1.4831^{20}	v	v		v			**B1**[2], 542
p1257	**1,2-Propanediol. 3-chloro-2-methyl-***	$ClCH_2COH(CH_3)CH_2OH$	124.57			$80^{1.6}$	1.2362^{20}_{4}	1.4748^{20}	∞	∞	∞				
—	**—,3,3-diethoxy-***	*see* **Glyceraldehyde,** diethyl acetal													
p1258	**1,3-Propanediol, 2,2-diethyl-***	$HOCH_2C(C_2H_5)_2CH_2OH$	132.20	wh	61.3	240–1	1.052^{20}_{20}		v	v	v			os s, lig i	
p1259	**1,2-Propanediol, 3(diethyl-amino)-***	$(C_2H_5)_2NCH_2CHOHCH_2OH$	147.22	syr	233–5				s	s	s			chl s	**B4**, 302
p1260	**1,3-Propanediol, 2,2-dimethyl-***	$HOCH_2C(CH_3)_2CH_2OH$	104.15	nd (bz)	127	206^{747}, $120\text{–}30^{15}$			s	v	v		s^h		**B1**[2], 550
p1261	**1,2-Propanediol, 3(dimethyl-amino)-***	$(CH_3)_2NCH_2CHOHCH_2OH$	119.16			220^{749}			s	s	s			chl s	**B4**, 302
p1262	**1,3-Propanediol, 2,2-dinitro-***	$HOCH_2C(NO_2)_2CH_2OH$	166.09		142				s^h	s^h			s^h	diox, $PhNO_2$ s^h	
p1263	**—,2-ethyl-2-hydroxymethyl-***	TMP. Trimethylolpropane. $CH_3CH_2C(CH_2OH)_3$	134.18	wh pw or pl	58	160^{5}			∞	∞			i	CCl_4 i	
p1264	**—,2-ethyl-2-nitro-***	$CH_3CH_2C(NO_2)(CH_2OH)_2$	149.15	nd (w)	57–8	d			v	v	v				**B1**, 483
—	**1,2-Propanediol, 3-hexadecyloxy-***	*see* **Glycerol,** 1-hexadecyl ether													
p1265	**1,3-Propanediol, 2(hydroxy-methyl)2-methyl-***	Pentaglycerol. Trimethyl-olethane. $CH_3C(CH_2OH)_3$	120.15	wh pw or nd (al)	201	$135\text{–}7^{15}$			∞	∞	i		i	aa v^h	**B1**[3], 2348
p1266	**—,2(hydroxy-methyl)2-nitro-**	$O_2NC(CH_2OH)_3$	151.12	nd or pr	165	d			v	v	s				**B1**[2], 596
p1267	**1,2-Propanediol, 3-mercapto-***	1-Thioglycerol. $HSCH_2CHOHCH_2OH$	108.16	visc		d	1.295^{14}_{14}	1.5268^{20}	δ	∞	δ	v	δ		**B1**[3], 2339
p1268	**—,2-methyl-***	Isobutylene glycol. $(CH_3)_2C(OH)CH_2OH$	90.12			177	1.003^{20}_{4}		s						**B1**[2], 547

For explanations, symbols and abbreviations see beginning of table.

No.	Name	Synonyms and Formula	Mol. wt.	Crystalline form, color and specific rotation	m.p. °C	b.p. °C	Density	n_D	Solubility w	al	eth	ace	bz	other solvents	Ref.
	1,3-Propanediol														
p1269	**1,3-Propanediol, 2-methyl-2-nitro-***	$CH_3C(NO_2)(CH_2OH)_2$	135.12	mcl	147–9	d	...	...	v	v	...	...	...	...	**B1²**, 547
p1270	—,**2-methyl-2-propyl-***	2,2-bis(hydroxymethyl)-pentane. $(HOCH_2)_2C(CH_3)CH_2CH_2CH_3$	132.20	...	56	234	...	...	s	...	...	...	...	os s	...
—	**1,2-Propanediol, 3-octadecyloxy-***	*see* **Glycerol**, 1-octadecyl ether													
p1271	**1,3-Propanedione, 2,2-dibromo-1,3 diphenyl-***	$C_6H_5COCBr_2COC_6H_5$	382.06	pr (eth)	95	...	...	...	...	δ	s^h	...	...	os δ	**B7**, 772
p1272	**1,2-Propanedione, (3,5-dimethoxy-4-hydroxy-phenyl)-**	Syrignoyl methyl ketone. (structure: ring with CH_3O, HO—, —$COCOCH_3$, CH_3O)	224.22	ye nd	80–1	...	...	...	δ^h	s	...	...	s	peth s^h	...
p1273	**1,3-Propanedione, 1,3-diphenyl**(one enol form)*	Dibenzoyl methane. $C_6H_5COCH_2COC_6H_5$	224.26	unst nd	70–1	...	...	...	...	...	s	...	...	peth, chl s dil NaOH s	**B7²**, 689
p1274	—,—(one enol form)*	$C_6H_5COCH_2COC_6H_5$	224.26	metast mcl	72–3	...	...	...	...	...	...	...	...	...	**B7²**, 689
p1275	—,—(one enol form)*	$C_6H_5COCH_2COC_6H_5$	224.26	st rh bipym	77.5–8	219–21[18]	...	...	...	s	v	...	...	Na_2CO_3 i NaOH s chl v	**B7²**, 689
p1276	—,—(keto form)	$C_6H_5COCH_2COC_6H_5$	224.26	metast nd (eth)	80–1	...	...	...	...	s	s	...	...	chl s dil NaOH s	**B7²**, 690
p1277	**1,2-Propanedione, 1-phenyl-***	Acetyl benzoyl. $C_6H_5COCOCH_3$	148.16	ye	...	222 101[12]	1.0065^{20}_{4}	1.537[10]	s	s	s	...	...	...	...
p1278	—,—,dioxime*	Methyl phenyl glyoxime. $C_6H_5C(:NOH)C(:NOH)CH_3$	178.20	nd (al)	238–40	...	...	...	i	s	...	...	...	...	...
p1279	—,—.1-oxime*	$C_6H_5C(:NOH)COCH_3$	163.18	pa ye nd or lf (aa, al)	164–5	...	...	...	δ	s^h	...	...	δ	aa s^h	**B7²**, 608
p1280	—,—,2-oxime*	α-Isonitrosopropiophenone. $C_6H_5COC(:NOH)CH_3$	163.18	wh nd (w)	115	...	...	...	s^h	...	...	...	...	to s^h	**B7²**, 608
p1281	**1,3-Propanedi-thiol***	$HSCH_2CH_2CH_2SH$	108.23	...	−79	172.9 63[15]	1.0783^{20}_{4}	1.5403[20]	δ	∞	∞	...	∞	chl ∞	**B1³**, 2164
p1282	**Propanephos-phonic acid***	*n*-Propylphosphonic acid. $CH_3CH_2CH_2PO_3H_2$	124.08	pl	73	d	...	...	v	v	v	...	δ	peth δ	...
p1283	**2-Propanephos-phonic acid***	Isopropylphosphonic acid. $(CH_3)_2CHPO_3H_2$	124.08	pl	74–5	d	...	...	v	v	v	...	δ s^h	...	...
p1284	**1-Propanephos-phonic acid, 2-methyl-***	Isobutylphosphonic acid. $(CH_3)_2CHCH_2PO_3H_2$	138.10	pl	119	d	...	...	v	v	s	...	δ	...	...
p1285	**2-Propanephos-phonic acid, 2-methyl-***	*tert*-Butylphosphonic acid. $(CH_3)_3CPO_3H_2$	138.10	nd	192	d	...	...	v	v	...	...	...	peth δ aa s	...
p1286	**1-Propanesul-fonic acid,** amide	1-Propansulfonamide. $CH_3CH_2CH_2SO_2NH_2$	123.17	pr (eth)	52	...	...	...	v	v	δ	...	...	...	**B4**, 8
p1287	—.chloride	1-Propanesulfonyl chloride*. $CH_3CH_2CH_2SO_2Cl$	142.61	...	...	180d 77[13]	1.2864^{15}_{4}	...	d^h	d^h	...	...	...	...	**B4**, 8
p1288	**2-Propanesul-fonic acid***	Isopropylsulfonic acid. $(CH_3)_2CHSO_3H$	124.16	...	d>100	...	...	...	v	...	...	...	...	...	**B4²**, 526
p1289	—,amide	2-Propanesulfonamide*. $(CH_3)_2CHSO_2NH_2$	123.17	(eth-peth)	60	...	...	...	v	s	s	...	s	peth i	**B4**, 8
p1290	**1-Propanesul-fonic acid, 2-methyl-**, chloride	Isobutanesulfonyl chloride*. $(CH_3)_2CHCH_2SO_2Cl$	156.63	...	...	189–91 80[13]	...	...	d	d^h	...	...	...	...	**B4**, 8
p1291	**1,1,2,3-Propane-tetracarboxylic acid**, tetraethyl ester*	$C_2H_5O_2CCH_2CH(CO_2C_2H_5)CH(CO_2C_2H_5)_2$	332.35	...	...	203–4	1.1184^{20}_{4}	...	...	s	...	...	...	...	**B2¹**, 332
p1292	**1,1,3,3-Propane-tetracarboxylic acid**. tetraethyl ester*	Ethyl methanedimalonate. $(C_2H_5O_2C)_2CHCH_2CH(CO_2C_2H_5)_2$	332.35	...	...	300–10d 150–8[3]	1.116[20]	...	...	s	...	...	...	...	**B2²**,701
p1293	**1-Propanethiol***	*n*-Propylmercaptan. $CH_3CH_2CH_2SH$	76.16	...	−111.5	67–8	0.8357^{25}_{4}	1.4351[25]	δ	s	s	...	...	...	**B1²**, 372

For explanations, symbols and abbreviations see beginning of table.

No.	Name	Synonyms and Formula	Mol. wt.	Crystalline form, color and specific rotation	m.p. °C	b.p. °C	Density	n_D	Solubility: w	al	eth	ace	bz	other solvents	Ref.
	2-Propanethiol														
p1294	**2-Propanethiol***	Isopropylmercaptan. $(CH_3)_2CHSH$	76.16		−130.7	57–60	0.8085^{25}_{4}	1.4223^{25}	δ	∞	∞				B1², 387
p1295	**1-Propanethiol, 2-methyl-***	Isobutylmercaptan. $(CH_3)_2CHCH_2SH$	90.19		<−79	88	0.8357^{20}_{4}	1.4386^{20}	δ	v	v				B1², 412
p1296	**2-Propanethiol, 2-methyl-***	*tert*-Butylmercaptan. $(CH_3)_3CSH$	90.19		1.26	64.22	0.7947^{25}	1.4232^{20}	i					heptane s	B1³, 1589
p1297	**1,2,3-Propanetricarboxylic acid***	Tricarballylic acid. $HO_2CCH_2CH(CO_2H)CH_2CO_2H$	176.13	orh (w, eth)					v	v	δ				B2,² 682
p1298	**—,1,2-dihydroxy-*(*l*)**	$HO_2CCH_2COH(CO_2H)CHOHCO_2H$	208.13	nd $[\alpha]_D$ −17.7 (ac)	159–60		1.39^{35}		v	δ	v				B3¹, 303
p1299	**—,1-hydroxy-***	Isocitric acid. $HO_2CCH_2CH(CO_2H)CH(OH)CO_2H$	192.13	yesh syr	105				δ	δ	δ				B3², 359
—	**—,2-hydroxy-***	*see* **Citric acid**													
—	**1,2,3-Propanetriol***	*see* **Glycerol**													
p1301	**Propanetrione, diphenyl-***	Diphenyltriketone. $C_6H_5COCOCOC_6H_5$	238.25	ye nd (lig)	69–70	289^{175}			i	δ	s				B7, 871
p1302	**Propanoic acid***	Propionic acid. $CH_3CH_2CO_2H$	74.08		−20.8	141.1	0.992^{20}_{4}	1.3874^{20}	∞	∞	s				B2, 234
p1303	—,allyl ester	$CH_3CH_2CO_2CH_2CH{:}CH_2$	114.15			$124–4.5^{774}$									B2, 241
p1304	—,amide	Propanamide*. Propionamide. $CH_3CH_2CONH_2$	73.09	rh, pl (bz)	81.3	213	1.042		v	v	v			chl v	B2, 243
p1305	—,—,*N*,*N*,-diethyl-	$CH_3CH_2CON(C_2H_5)_2$	129.21			191				v				ac v	B4², 603
p1306	—,—,*N*-phenyl-	Propionanilide. $CH_3CH_2CONHC_6H_5$	149.19	lf (eth, al, bz)	105	222.2	1.175		δ	v	v				B12, 250
p1307	—,—,*N*-propionyl-	Dipropionamide. $CH_3CH_2CONHCOCH_2CH_3$	129.16	nd (w, eth)	154 (sub 100)	210–20			δ		δ				B2², 224
p1308	—,—,*N*(2-tolyl)-	CH_3CH_2CONH—[benzene ring with CH_3]	163.22	nd (bz)	87	298–9				v	v		s^h	chl v aa v	B12², 440
p1309	—,—,*N*(3-tolyl)-	CH_3CH_2CONH—[benzene ring with CH_3]	163.22	nd (eth)	81					v	v				B12, 861
p1310	—,anhydride	Propionic anhydride. $(CH_3CH_2CO)_2O$	130.14		−45	$168.1–.4^{712}$	1.4038^{20}_{4}	1.0110^{20}	d	d	∞				B2², 223
p1311	—,bromide	Propionyl bromide. CH_3CH_2COBr	136.99		$103.0–.6^{770}$		1.5210^{16}_{4}	1.4578^{16}			s				B2², 108
p1312	—,butyl ester*	$CH_3CH_2CO_2(CH_2)_3CH_3$	130.18		−89.55	145.5	0.8818^{15}	1.3982^{25}	δ	∞	∞				B2, 241
p1313	—,*sec*-butyl ester(*d*)	$CH_3CH_2CO_2CH(CH_3)CH_2CH_3$	130.18	$[\alpha]^{20}_{D}$ +23.85		132.0–.5	0.8663^{20}_{4}	1.3952^{20}		s	s				B2², 221
p1314	—,chloride	Propionyl chloride. CH_3CH_2COCl	92.53		−94	80	1.0646^{20}	1.4057^{25}	d	d	s				B2², 223
p1315	—,cyclohexyl ester*	$CH_3CH_2CO_2$—[cyclohexane ring]	156.22			193^{750}	0.9718^{0}_{4}		i	s	s	s			B6¹, 6
p1316	—,ethyl ester*	$CH_3CH_2CO_2C_2H_5$	102.13		−73.9	99.10	0.8889^{20}_{4}	1.3839^{20}	δ	∞	∞				B2, 240
p1317	—,fluoride	Propionyl fluoride. CH_3CH_2COF	76.07			44	0.972^{15}_{4}								B2², 108
p1318	—,furfuryl ester	$CH_3CH_2CO_2CH_2$—[furan ring, O]	154.17			$195–6^{762}$	1.1085^{20}_{4}		δ	s	∞				B17², 115
p1319	—,heptyl ester*	$CH_3CH_2CO_2(CH_2)_6CH_3$	172.26		−50.9	210(208)	0.8679^{20}		i					os s	B2, 241
p1320	—,hexyl ester*	$CH_3CH_2CO_2(CH_2)_5CH_3$	158.23		−57.5	190	0.8698^{20}		i	s	s			AcOEt s	
p1321	—,iodide	Propionyl iodide. CH_3CH_2COI	183.99			127–8									B2, 243
p1322	—,isobutyl ester	$CH_3CH_2CO_2CH_2CH(CH_3)_2$	130.18		−71.4	136.8	0.8876^{20}_{4}	1.3975^{20}	δ						B2, 241
p1323	—,isopropyl ester*	$CH_3CH_2CO_2CH(CH_3)_2$	116.16			111.3	0.8931^{0}		δ	∞	∞				B2, 241
p1324	—,methyl ester*	$CH_3CH_2CO_2CH_3$	88.11		−87.5	78.7	0.9151^{20}_{4}	1.3779^{20}	δ					os s	B2, 21
p1325	—,3-methylbutyl ester*	Isoamyl propionate. $CH_3CH_2CO_2CH_2CH_2CH(CH_3)_2$	144.21			160–1	0.8580^{20}_{15}	1.4065^{20}	δ	s	s				B2, 241

For explanations, symbols and abbreviations see beginning of table.

No.	Name	Synonyms and Formula	Mol. wt.	Crystalline form, color and specific rotation	m.p. °C	b.p. °C	Density	n_D	Solubility w	al	eth	ace	bz	other solvents	Ref.
	Propanoic acid														
p1326	—,nitrile	Ethyl cyanide. Propionitrile. CH_3CH_2CN	55.08		−91.9	97.2	0.7720$^{20}_{4}$	1.3683^{15}	v	s	s				**B2**^{2}, 225
p1327	—,octyl ester*	$CH_3CH_2CO_2(CH_2)_7CH_3$	186.29		−41.6	227.9	0.8663^{20}		i	s				os s	**B2**, 241
p1328	—,pentyl ester*	$CH_3CH_2CO_2(CH_2)_4CH_3$	144.22		−73.1	168.65	0.8761$^{15}_{4}$	1.4096^{15}	i	∞	∞				**B2**, 221
p1329	—,phenyl ester*	$CH_3CH_2CO_2C_6H_5$	150.17		75–7	211,98–9^{18}	1.0467$^{25}_{25}$	1.5003^{25}	i	v	v		s		**B6**, 154
p1330	—,4-phenylphenacyl ester*	$CH_3CH_2CO_2CH_2CO$—C_6H_4—C_6H_5	268.30		102										
p1331	—,piperazinium salt	$2(CH_3CH_2CO_2H).C_4H_{10}N_2$	234.29	(diox)	124–5										
p1332	—,propyl ester*	$CH_3CH_2CO_2CH_2CH_2CH_3$	116.16		−75.9	123.0	0.8809$^{20}_{4}$	1.3935^{20}	δ	∞	∞				**B2**, 240
p1333	—,tetrahydrofurfuryl ester*	$CH_3CH_2CO_2CH_2$— (tetrahydrofuryl, O)	158.20			204–7^{756} 85–7^{3}	1.044$^{20}_{0}$		i	∞	∞			chl ∞	**B17**^{2}, 107
—	—,**2-amino-***	*see* **Alanine**													
—	—,**3-amino-***	*see* **β-Alanine**													
p1335	—,**2(2-aminoacetamido)2-methyl-**	$(CH_3)_2C(CO_2H)NHCOCH_2NH_2$	160.17	nd	260d					s^h					**B4**^{2}, 841
—	—,**2-amino-3-hydroxy-***	*see* **Serine**													
—	—,**2-amino-3(4-hydroxyphenyl)-***	*see* **Tyrosine**													
p1336	—,**2-amino-2-methyl-***	$(CH_3)_2C(NH_2)CO_2H$	103.12	ta or pr (w)	335 cor	sub 280			v	δ	i				**B4**^{2}, 839
—	—,**3(2-aminophenyl)-, lactam***	*see* **Quinoline, 2-oxo-1,2,3,4-tetrahydro-**													
—	—,**2-amino-3-phenyl-***	*see* **Phenylalanine**													
p1337	—,**3-amino-3-phenyl-(*d*)***	$C_6H_5CHNH_2CH_2CO_2H$	165.19	pl (w) $[\alpha]_D$ −9.2 (1N NaOH, p=9) $[\alpha]^{20}_D$ +7 (w, p=1)	234–5d				δ	δ^h					**B14**^{1}, 602
p1338	—,—(*dl*)*	$C_6H_5CHNH_2CH_2CO_2H$	165.19	w	231d				δ s^h	δ s^h	δ			dil ac s	**B14**^{2}, 295
p1339	—,—(*l*)*	$C_6H_5CHNH_2CH_2CO_2H$	165.19	(w) $[\alpha]^{25}_D$ −7.5 (w, c=1) $[\alpha]_D$ +8.9 (IN NaOH, p=10)	234–5d				δ	δ^h					**B14**^{1}, 602
—	—,**2-benzylidene-**	*see* **Cinnamic acid, α-methyl-**													
p1340	—,**2-benzyl-3-phenyl-**	Dibenzylacetic acid. $(C_6H_5CH_2)_2CHCO_2H$	240.30	pl (peth, dil aa) nd (w)	89	235^{18}			δ^h	v	v		v	chl v aa v	**B9**^{2}, 475
p1341	—,—,amide	$(C_6H_5CH_2)_2CHCONH_2$	239.32	nd (al, w)	128–9					v	v				**B9**, 683
p1342	—,—,methyl ester	$(C_6H_5CH_2)_2CHCO_2CH_3$	254.33	nd (al)	42–3				i	δ				peth δ, os v	**B9**^{2}, 475
p1343	—,—,nitrile	$(C_6H_5CH_2)_2CHCN$	221.30	lf or pl (al)	89–91				i	v	v				**B9**, 683
p1344	—,**2-bromo-***	$CH_3CHBrCO_2H$	152.99	pr	25.7	203.5 96^{10}	1.700^{20}	1.4753^{20}	v	v	v				**B2**, 254
p1345	—,—,amide, *N*(2-tolyl)-	$CH_3CHBrCONH$—C_6H_4—CH_3 (2-)	274.62	nd	131					s	v		s	chl v	**B12**, 794
p1346	—,—,bromide	$CH_3CHBrCOBr$	215.90			152–4 cor			d	d					**B2**^{2}, 230
p1347	—,—,chloride	$CH_3CHBrCOCl$	171.44			131–3	1.697^{11}		d	d^h	s			chl s	**B2**^{2}, 229
p1348	—,—,ethyl ester*	$CH_3CHBrCO_2C_2H_5$	181.04			159–61d	1.394$^{20}_{4}$		i	∞	∞				**B2**, 255
p1349	—,—,methyl ester(*d*)*	$CH_3CHBrCO_2CH_3$	167.01	$[\alpha]^{17}_D$ +42.65		93–6$^{120-35}$ 61–2^{36}	1.482^{17}			s					**B2**^{1}, 112
p1350	—,—,—(*dl*)*	$CH_3CHBrCO_2CH_3$	167.01			51.5^{19}	1.4966$^{21}_{4}$			s	s			MeOH s	**B2**^{1}, 112
p1351	—,—,—(*l*)*	$CH_3CHBrCO_2CH_3$	167.01	$[\alpha]^{20}_{578}$ −55.5		61–3^{32}	1.484^{20}			s					**B2**^{2}, 229
p1352	—,—,nitrile	$CH_3CHBrCN$	133.98			59^{24}	1.5505$^{20}_{4}$	1.4585^{20}							**B2**^{2}, 231

For explanations, symbols and abbreviations see beginning of table.

No.	Name	Synonyms and Formula	Mol. wt.	Crystalline form, color and specific rotation	m.p. °C	b.p. °C	Density	n_D	Solubility w	al	eth	ace	bz	other solvents	Ref.
	Propanoic acid														
p1353	—,—,piperazinium salt	$2(CH_3CHBrCOOH).C_4H_{10}N_2$	392.10	wh	195d				s	s	i	...	...		
p1354	—,**3-bromo-***	$BrCH_2CH_2CO_2H$	152.98	pl (CCl_4)	62.5		1.48		s	s	s	...	s	chl s	**B2**, 256
p1355	—,—,amide	$BrCH_2CH_2CONH_2$	152.00	(w)	110–1				s^h	...	...	...	...		**B2**[2], 231
p1356	—,—,butyl ester*	$BrCH_2CH_2CO_2(CH_2)_3CH_3$	209.09			130^{26}	1.2609^{15}_{4}	1.4577^{9}	...	...	...	...	...	os s	**B2**[2], 231
p1357	—,—,ethyl ester*	$BrCH_2CH_2CO_2C_2H_5$	181.04			112^{44}	1.4123^{18}_{4}	1.4569^{18}	...	...	...	...	...	os v	**B2**[2], 231
p1358	—,—,methyl ester*	$BrCH_2CH_2CO_2CH_3$	167.01			105.5^{60}	1.5122^{0}_{4}	1.4603^{17}	...	s	...	...	...	os v	**B2**[2], 231
p1359	—,—,3-methylbutyl ester*	Isoamyl β-bromopropionate. $BrCH_2CH_2CO_2CH_2CH_2CH(CH_3)_2$	223.12			$110\text{–}1^{11}$	1.2217^{15}_{4}	1.4556^{9}	...	s	s	...	...		**B2**[2], 231
p1360	—,—,nitrile	$BrCH_2CH_2CN$	133.98			92^{25}	1.6152^{20}_{4}	1.1470^{20}	...	v	v	...	...		**B2**[2], 231
p1361	—,**2-bromo-2-methyl-***	$(CH_3)_2CBrCO_2H$	167.01	(peth)	48	198–200	1.5225^{60}_{60}		v	s	s	...	...	os v	**B2**[2], 263
p1362	—,—,amide	$(CH_3)_2CBrCONH_2$	166.02	pr (chl)	148	145^{17}			s	s	...	...	...	chl v	**B2**[2], 263
p1363	—,—,—,*N*-methyl-*N*-phenyl-	$(CH_3)_2CBrCON(CH_3)C_6H_5$	256.15	(lig)	44				...	...	...	...	...	to s lig s^h	**B12**, 254
p1364	—,—,bromide	$(CH_3)_2CBrCOBr$	229.91			162–4	1.4067^{14}_{4}	1.4552^{14}	d	d^h	...	...	...	AcOEt, CS_2 s	**B2**[2], 263
p1365	—,—,ethyl ester*	$(CH_3)_2CBrCO_2C_2H_5$	195.06			164^{762}	1.3287^{20}_{20}		i	s	∞	...	...		**B2**[2], 263
p1366	—,—,nitrile	$(CH_3)_2CBrCN$	148.01			$61.2\text{–}.6^{5}$	1.4796^{15}_{4}	1.4739^{15}	...	s	s	...	s		**B2**[2], 263
p1367	—,**2-bromo-2-phenyl-***	$C_6H_5CBr(CH_3)CO_2H$	229.08	pl (CS_2)	93–4				i	...	...	...	v	CS_2 v lig δ	**B9**, 525
p1368	—,**3-bromo-2-phenyl-***	$C_6H_5CH(CH_2Br)CO_2H$	229.08	pr (CS_2)	93–4				i	v	v	...	v		**B9**, 526
p1369	—,**3-bromo-3-phenyl***	$C_6H_5CHBrCH_2CO_2$	229.08	pw [α]57.8–85.0 (w), −75.1 (al)	135				s	s	...	...	...		**B9**[2], 343
p1370	—,**3-bromo-3-phenyl-**, methyl ester*	$C_6H_5CHBrCH_2CO_2CH_3$	243.11	pr	37.5–8.5				s	s	δ	...	...	os s lig δ	**B9**[1], 201
p1371	—,**3(2-carboxy-phenyl)-***	CO_2H / $—CH_2CH_2CO_2H$ (benzene ring structure)	194.19	nd (w)	166				v	...	s	...	δ		**B9**[2], 622
p1372	—,**3(4-carboxy-phenyl)-***	HO_2C—(benzene ring)—$CH_2CH_2CO_2H$	194.19	nd (al)	294				v^h	v^h	...	...	...		**B9**[2], 622
p1373	—,**2-chloro-***	$CH_3CHClCO_2H$	108.53			186 84^{12}	1.28^{0}		∞	∞	∞	...	...		**B2**, 248
p1374	—,—,amide, *N*(2-tolyl)-	CH_3 / $CH_3CHClCONH$—(benzene ring structure)	197.67	nd (abs al)	111				...	v^h	...	...	...		**B12**, 794
p1375	—,—,chloride(*d*)	$CH_3CHClCOCl$	126.98	$[\alpha]^{18}_{D}+0.2$ (l=10)		110	$1.2394^{7.5}$		d	d	...	...	...		**B2**[1], 110
p1376	—,—,butyl ester*	$CH_3CHClCO_2C_4H_9^{n}$	164.64			183.5–5.0 71.6–2.6^{10}	1.0253^{20}_{4}	1.4263^{20}	...	...	s	...	...		
p1377	—,—,ethyl ester*	$CH_3CHClCO_2C_2H_5$	136.58			147–8	1.087^{20}_{4}	1.4185^{20}	i	∞	∞	...	...		**B2**, 249
p1378	—,—,isobutyl ester(*d*)*	$CH_3CHClCO_2CH_2CH(CH_3)_2$	164.64	$[\alpha]_D+5.21$		175–7	1.0312^{20}_{4}	1.4247^{20}	...	s	s	...	...	os s	**B2**, 248
p1379	—,—,isopropyl ester*	$CH_3CHClCO_2CH(CH_3)_2$	150.61			151.5–2.5 $46.1\text{–}.9^{12}$	1.0315^{20}_{4}	1.4149^{20}	i	s	s	...	...		
p1380	—,—,methyl ester(*d*)*	$CH_3CHClCO_2CH_3$	122.56	$[\alpha]_D+19.01$		132–4 $49\text{–}50^{35}$	1.1520^{20}_{4}		...	s	...	...	...		**B2**, 248
p1381	—,—,—(*dl*)*	$CH_3CHClCO_2CH_3$	122.56			132.5	1.0750_{4}		...	s	...	...	...	0.1N HCl s	**B2**[2], 226
p1382	—,—,—(*l*)*	$CH_3CHClCO_2CH_3$	122.56	$[\alpha]_D-26.83$			1.158^{5}_{4}		...	s	...	...	...		**B2**, 248
p1383	—,—,nitrile	$CH_3CHClCN$	89.53			123–4	1.0792^{10}		...	...	...		...		**B2**[1], 111
p1384	—,**3-chloro-***	$ClCH_2CH_2CO_2H$	108.53	lf (w)	61	204			s	s	∞	...	...		**B2**, 249

For explanations, symbols and abbreviations see beginning of table.

No.	Name	Synonyms and Formula	Mol. wt.	Crystalline form, color and specific rotation	m.p. °C	b.p. °C	Density	n_D	Solubility w	al	eth	ace	bz	other solvents	Ref.
	Propanoic acid														
p1385	—,—,amide, *N*(2-tolyl)-	CH_3 $ClCH_2CH_2CONH$—	197.67	(w, dil al)	78				s^h	v^h					B12², 441
p1386	—,—,chloride	$ClCH_2CH_2COCl$	126.98	yesh		$143–5^{763}$	1.3307^{13}		δ	v d^h	v			chl v	B2², 227
p1387	—,—,butyl ester*	$ClCH_2CH_2CO_2C_4H_9^n$	164.64			104^{22}	1.0708^{15}_{4}	1.4321^{20}	s		s			os s	B2², 227
p1388	—,—,ethyl ester*	$ClCH_2CH_2CO_2C_2H_5$	136.58			162	1.1086^{20}_{4}	1.4254^{20}	δ	∞	∞				B2, 250
p1389	—,—,isobutyl ester	$ClCH_2CH_2CO_2CH_2CH(CH_3)_2$	164.64			191–3	1.0323^{20}_{4}	1.4295^{20}	i	s	s				B2, 250
p1390	—,—,methyl ester*	$ClCH_2CH_2CO_2CH_3$	122.56			155–7 $40–2^{10}$	1.1861^{15}_{4}	1.4319^{12}		s					B2², 227
p1391	—,—,3-methyl-butyl ester*	$ClCH_2CH_2CO_2CH_2CH_2CH(CH_3)_2$	178.66			$207–8^{740}$ 87^{12}	1.0171^{20}_{4}	1.4343^{20}	i	s	s				B2², 227
p1392	—,—,nitrile	$ClCH_2CH_2CN$	89.53			173–4.5 $85–7^{20}$	1.1443^{18}								B2², 227
p1393	—,—,propyl ester*	$ClCH_2CH_2CO_2CH_2CH_2CH_3$	150.61			178–81	1.0656^{20}_{4}	1.4290^{20}	i	s	s				B2, 250
p1394	**—,3-chloro-2(chloromethyl)-2-hydroxy-***	$(ClCH_2)_2COHCO_2H$	173.01	ta (chl)	93					v	v			peth δ	B3², 224
p1395	**—,3-chloro-2,2-dimethyl-***	Chloropivalic acid. $ClCH_2C(CH_3)_2CO_2H$	136.58		40–2	$126–9^{30}$								CCl_4 v	
p1396	—,—,chloride	$ClCH_2C(CH_3)_2COCl$	155.03			$85–6^{60}$			d	d				CCl_4 s	
p1397	**—,3-chloro-2-hydroxy-2-methyl-, nitrile**	Chloroacetone cyanohydrin. $ClCH_2C(CH_3)OHCN$	119.15			110^{27}	1.2027^{15}		v	v				os v peth i	B3, 317
p1398	**—,2-chloro-2-methyl-***	$(CH_3)_2CClCO_2H$	122.56		31	118^{50}			v	v					B2, 294
p1399	—,—,chloride	$(CH_3)_2CClCOCl$	141.01			117–8		1.4369^{20}	d	d	s				B2², 263
p1400	—,—,ethyl ester*	$(CH_3)_2CClCO_2C_2H_5$	150.61			148.5–9 cor	1.062^{0}	1.4109^{16}			s				B2, 295
p1401	—,—,methyl ester*	$(CH_3)_2CClCO_2CH_3$	136.58			135 $42–4^{17}$	1.0893^{15}_{15}	1.4122^{21}			s				
p1402	**—,3-chloro-2-methyl-***	$ClCH_2CH(CH_3)CO_2H$	122.56			$128–33^{50}$				s	v			CCl_4 s	
p1403	—,—,chloride	$ClCH_2CH(CH_3)COCl$	141.01			$171–2^{765}$		1.4542^{20}	d	d				CCl_4 s	B2, 295
p1404	**—,3-cyclohexyl-***	—$CH_2CH_2CO_2H$	156.22		16	193^{750} $111–2^{4}$	0.9966^{20}_{4}	1.4658^{20}	s		s				B9², 13
p1405	**—,2,3-diamino-***	$H_2NCH_2CHNH_2CO_2H$	104.11	hyg rosettes	110–20				s	i	i				B4¹, 500
p1406	**—,2,3-dibromo-***	$BrCH_2CHBrCO_2H$	231.89	st pl, labile pr	64	220–40d			s	s			s	CS_2 s	B2, 258
p1407	—,—,ethyl ester*	$BrCH_2CHBrCO_2C_2H_5$	259.95			211–4 119.2^{30} cor	1.7882^{16}_{4}	1.5015^{16}		s	s				B2², 232
p1408	—,—,methyl ester*	$BrCH_2CHBrCO_2CH_3$	245.92			203 115^{25}	1.9605^{0}_{4}	1.5147^{17}		s					B2², 232
p1410	**—,2,3-dibromo-3-phenyl-(*d*)***	*d*-Cinnamic acid dibromide. $C_6H_5CHBrCHBrCO_2H$	307.99	pr (chl) $[\alpha]^{15}_{D}$ +45.8 (abs al)	182					v	v			chl s^h	B9², 344
p1411	—,—(*dl*)*	$C_6H_5CHBrCHBrCO_2H$	307.99	mcl pr (chl)	204	sub			d^h	v	v			CS_2 s	B9², 344
p1412	—,—(*meso*)*	$C_6H_5CHBrCHBrCO_2H$	307.99	nd	91–3				d^h				s	CS_2 s lig i	B9², 344
p1413	—,ethyl ester(*d*)*	$C_6H_5CHBrCHBrCO_2C_2H_5$	336.05	(CS_2) $[\alpha]_D$ +59.1	71					s				CS_2 s^h	B9, 518
p1414	—,—(*dl*)*	$C_6H_5CHBrCHBrCO_2C_2H_5$	336.05	mcl pr or pl	75–6				i	s	v			chl v	B9², 345
p1415	**—,2,2-dichloro-***	$CH_3CCl_2CO_2H$	142.98			185–90 $90–2^{14}$	$1.389^{22.8}_{4}$		v	v	s			alk v	B2², 228
p1416	—,—,chloride	CH_3CCl_2COCl	161.43			$117.4–.8^{753}$ $68–73^{89}$	1.4062^{20}_{4}	1.4524^{20}	d	d					B2², 228
p1417	**—,2,3-dichloro-***	$ClCH_2CHClCO_2H$	142.98	hyg nd (peth)	50	210d 113^{12}			s	s	s				B2², 228
p1418	—,—,chloride	$ClCH_2CHClCOCl$	161.43			$52–4^{16}$	1.4757^{20}_{4}	1.4764^{20}	d	d	s				
p1419	—,—,ethyl ester*	$ClCH_2CHClCO_2C_2H_5$	171.03			183–4 $76–7^{15}$	1.2461^{20}_{4}	1.4482^{20}		s	s				B2, 252

For explanations, symbols and abbreviations see beginning of table.

No.	Name	Synonyms and Formula	Mol. wt.	Crystalline form, color and specific rotation	m.p. °C	b.p. °C	Density	n_D	Solubility w	al	eth	ace	bz	other solvents	Ref.
	Propanoic acid														
p1420	—,—,methyl ester(*d*)*	$ClCH_2CHClCO_2CH_3$	157.01	$[\alpha]^{20}_D+1.70$	...	92^{50}	1.3282^{20}_4	...	...	s	...	...	...	...	**B2**^{1}, 111
p1421	**—,3,3-dichloro-***	$Cl_2CHCH_2CO_2H$	142.98	pr	56	...	...	...	v	v	v	...	v	chl v	**B2**, 252
p1422	—,—,chloride*	Cl_2CHCH_2COCl	161.43	...	...	$43–4^{10}$	1.4557^{20}_4	1.4738^{20}	d	d	s	...	...	diox s	...
p1423	**—,3,3-dichloro-2-hydroxy-2-methyl-***	$Cl_2CHC(CH_3)OHCO_2H$	173.01	pr (al, eth)	82–3	d	...	...	s	v	...	...	...	...	**B3**^{2}, 224
p1423^1	**—,2,3-dihydroxy-***	Glyceric acid. $HOCH_2CHOHCO_2H$	106.08	syr	...	...	...	...	∞	∞	i	v	...	...	**B3**^{2}, 261
p1424	—,—,ethyl ester(*D*)*	*D*-Ethyl glycerate. $HOCH_2CHOHCO_2C_2H_5$	134.14	$[\alpha]^{11}_D-22.73$ (1 = 19.84 cm)	...	...	...	...	...	...	...	...	...	...	**B3**^{1}, 141
p1425	—,—,—(*DL*)*	$HOCH_2CHOHCO_2C_2H_5$	134.14	...	...	230–40	1.1909^{15}_{15}	...	s	v	v	...	...	...	**B3**^{2}, 264
p1426	—,—,methyl ester(*D*)*	*D*-Methyl glycerate. $HOCH_2CHOHCO_2CH_3$	120.11	$[\alpha]^{15}_D-4.8$	...	$119–20^{14}$	1.2798^{15}_{15}	...	...	...	...	...	...	...	**B3**^{1}, 141
p1427	—,—,—(*DL*)	$HOCH_2CHOHCO_2CH_3$	120.11	...	...	239–44	1.2814^{15}_{15}	...	∞	∞	δ	...	...	...	**B3**^{1}, 142
p1428	**—,3(2,5-dimethoxyphenyl)-2-oxo-***	OCH_3 / CH_2COCO_2H / CH_3O (2,5-dimethoxyphenyl ring)	224.22	yesh (aa)	166–70d	...	...	...	...	...	...	s	δ	MeOH s aa s^h	**B10**^{2}, 723
p1429	**—,3(3,4-dimethoxyphenyl)-2-oxo-***	CH_3O / CH_3O—(ring)—CH_2COCO_2H	224.22	lf (aa)	*ca.* 187d	...	...	...	s^h	v	δ	v	δ	chl δ	**B10**^{2}, 723
p1430	**—,2,2-dimethyl-***	Pivalic acid. $(CH_3)_3CCO_2H$	102.13	nd	...	163.7–.8 70^{14}	0.905^{50}	$1.3931^{36.5}$	δ	v	v	...	...	...	**B2**^{2}, 280
p1431	—,—,amide,*N*,*N*-diethyl-	$(CH_3)_3CCON(C_2H_5)_2$	157.26	...	...	203	0.891^{15}	...	s δ^h	s	s	...	...	...	**B4**, 111
p1432	—,—,chloride	Pivalyl chloride. $(CH_3)_3CCOCl$	120.58	...	...	107 48^{100}	...	1.4126^{20}	d	d	v	...	...	...	**B2**^{2}, 280
p1433	—,—,ethyl ester*	$(CH_3)_3CCO_2C_2H_5$	130.19	...	...	118–8.2	0.856^{20}_4	1.3912^{20}	...	s	s	...	...	...	**B2**^{2}, 280
p1434	—,—,methyl ester*	$(CH_3)_3CCO_2CH_3$	116.16	...	...	101–3	0.891^0_4	1.3880^{20}	δ	∞	∞	...	...	...	**B2**^{1}, 139
p1435	—,—,nitrile	*tert*-Butyl cyanide. Pivalonitrile. $(CH_3)_3CCN$	83.13	...	15–6	105–6	...	...	...	...	...	...	...	...	**B2**^{2}, 181
p1436	**—,2(dimethylamino)-**,nitrile	Dimethyl-*dl*-alanine nitrile. $(CH_3)_2NCH(CH_3)CN$	98.15	...	...	144	...	...	...	s	s	...	...	...	**B4**, 392
p1437	**—,2,2-diphenyl-***	$(C_6H_5)_2C(CH_3)CO_2H$	226.28	pl (bz-peth), nd (w), lf (dil al)	173–4	sub >300	...	...	δ^h	s v^h	v	...	v	chl, to v peth s	**B9**^{2}, 474
p1438	**—,2,3-diphenyl-** (*d*)*	$C_6H_5CH_2CH(C_6H_5)CO_2H$	226.28	(dil al) $[\alpha]^{20}_D+94$ (bz)	83–9	...	...	...	δ^h	s	s	...	s	...	**B9**^{1}, 284
p1439	—,—(*dl*)*	$C_6H_5CH_2CH(C_6H_5)CO_2H$	226.28	a) pr (chl), b) pl (chl), c) (MeOH)	a) 88–9 b) 95–6 c) 82	330–40	a) 1.1481 b) 1.1495 c) 1.1430	...	i δ^h	v	v	...	s	CS_2 s chl s^h	**B9**^{1}, 285
p1440	—,—(*l*)*	$C_6H_5CH_2CH(C_6H_5)CO_2H$	226.28	nd (dil al) $[\alpha]^{20}_D-85.1$ (bz)	83–9	...	...	...	δ^h	s	s	...	s	...	**B9**^{1}, 285
p1441	**—,3,3-diphenyl-***	$(C_6H_5)_2CHCH_2CO_2H$	226.28	nd (dil al)	155	...	...	...	δ	v	s	...	...	...	**B9**^{2}, 478
p1442	**—,2,3-diphenyl-2-hydroxy-***	α-Benzylmandelic acid. $C_6H_5CH_2COH(C_6H_5)CO_2H$	242.28	nd (bz), (dil al)	165–6	...	...	...	δ^h	v	v	...	s^h	aa v	**B10**^{2}, 227
p1443	**—,3,3-diphenyl-2-hydroxy-***	β,β′ Diphenyllactic acid. $(C_6H_5)_2CHCHOHCO_2H$	242.28	nd (w)	159	...	...	...	δ s^h	s	δ	...	...	...	**B10**^{2}, 228
p1444	**—,3,3-diphenyl-3-hydroxy-***	$(C_6H_5)_2COHCH_2CO_2H$	242.28	nd (dil al)	212	...	...	...	δ^h	v	δ	v	δ	aa v	**B10**^{1}, 156
p1445	—,—,ethyl ester*	$(C_6H_5)_2COHCH_2CO_2C_2H_5$	270.33	pr (dil al)	87	...	...	...	δ s^h	v	...	...	...	con sulf s os v	**B10**^{2}, 228
p1446	**—,2,3-epoxy-***	Acrylic acid oxide. Glycidic acid. CH_2CHCO_2H (with O bridging CH_2 and CH)	88.06	...	...	...	...	...	∞	∞	∞	...	...	...	**B18**^{1}, 435
p1447	**—,2-ethoxy-**, nitrile	$CH_3CH(OC_2H_5)CN$	99.13	...	...	131^{765}	0.878^{16}_4	1.390^{16}	δ	v	v	...	...	aa v	**B3**, 385
p1448	**—,3-ethoxy-**, nitrile	$C_2H_5OCH_2CH_2CN$	99.13	...	...	171.3–.5	0.9285^{15}_4	...	...	v	v	...	...	ac d	**B3**^{1}, 113
p1449	**—,3-fluoro-***	$FCH_2CH_2CO_2H$	92.07	...	...	$83–4^{14}$	...	1.3889^{25}	s	v	v	...	...	...	...

For explanations, symbols and abbreviations see beginning of table.

No.	Name	Synonyms and Formula	Mol. wt.	Crystalline form, color and specific rotation	m.p. °C	b.p. °C	Density	n_D	Solubility w	al	eth	ace	bz	other solvents	Ref.
	Propanoic acid														
p1451	—,**3(2-furyl)-**	Furfurylacetic acid. (furan ring)—$CH_2CH_2CO_2H$	140.14	(w, peth), ye in HCl	56.5–8	229 $108–10^{10}$			s		s				$B18^2$, 272
p1452	—,—,ethyl ester	$C_9H_{12}O_3$. *See* p1451	168.19			209									$B18^2$, 272
p1452[1]	—,**3(2-furyl)3-oxo-**, ethyl ester	$C_9H_{10}O_4$. *See* p1451	182.18	lt ye		$142–3^{10}$	1.165^{17}		i	s	s				$B18^2$, 327
p1453	—,—,methyl ester	$C_8H_8O_4$. *See* p1451	168.15	ye in air		$144–5^{20}$				s	s				
p1454	—,**2-hydroxy-(*d*)***	Lactic acid. Sarcolactic acid. $CH_3CHOHCO_2H$	90.08	hyg pr $[\alpha]_D^{15}+3.82$ (w, c=10.5)	25–6	119^{12} d			∞	∞	∞				$B3^2$, 186
p1455	—,—(*dl*)*	$CH_3CHOHCO_2H$	90.08	ye	18	119^{12}	1.2060_4^{25}	1.4392^{20}	v	v	δ				$B3^2$, 192
p1456	—,—(*l*)*	$CH_3CHOHCO_2H$	90.08	pr lf $[\alpha]_D-2.26$ (w, c=1.24)	26–7	103^2			v	v	v				$B3^2$, 186
p1457	—,—,acetate(*dl*)	$CH_3CO_2CH(CH_3)CO_2H$	132.12	dlq	57–60	$167–70^{78}$ 127^{11}			d^h	s			s	peth δ	$B3^2$, 205
p1458	—,—,acetate chloride(*d*,+)	$CH_3CO_2CH(CH_3)COCl$	150.56	$[\alpha]_{578}^{18}+32.4$		$51–3^{11}$	1.177^{20}		d^h	d^h					$B3^2$, 189
p1459	—,—,—(*dl*)	$CH_3CO_2CH(CH_3)COCl$	150.56			56^{11}	1.1920_4^{17}	1.4241^{17}	d^h	d^h					B3, 283
p1460	—,—,allyl ester	Allyl lactate. $CH_3CHOHCO_2CH_2CH{:}CH_2$	130.14			$56–60^8$	1.0452_4^{20}	1.4369^{20}	i					Py s	
p1461	—,—,amide(*d*)	Lactamide. Lactic amide. $CH_3CHOHCONH_2$	89.09	$[\alpha]_{578}^{18}+22.2$	49–51										
p1462	—,—,—(*dl*)	$CH_3CHOHCONH_2$	89.09	pl (AcOEt)	73.5		1.1381_4^{80}		v	v					$B3^2$, 208
p1463	—,—,—,*N*(4-ethoxyphenyl)-	*N*-Lactyl-β-phenetidide. $CH_3CHOHCONH$—(benzene ring)—OC_2H_5	209.25	(w)	117.8				s^h	v	δ		δ	peth δ	$B13^2$, 262
p1464	—,—,anhydride*	Lactic anhydride. $(CH_3CHOHCO)_2O$	162.14	pa ye amor or syr	260d	d			δ	v	v				B3, 282
p1465	—,—,butyl ester(*d*)*	*d*-Butyl lactate. $CH_3CHOHCO_2C_4H_9^n$	146.19	$[\alpha]_D^{27.3}+13.63$		77^{10}	$0.9744_4^{27.6}$								$B3^2$, 188
p1466	—,—,—(*dl*)*	$CH_3CHOHCO_2C_4H_9^n$	146.19		−43	83^{13}	0.9803_4^{22}	1.4217^{20}	δ	∞	∞				B3, 207
p1467	—,—,ethyl ester(*d*)*	Ethyl lactate. $CH_3CHOHCO_2C_2H_5$	118.13	$[\alpha]_D^{19}-11.26$		$69–70^{36}$	1.0415_4^{14}	1.4157^{13}							$B3^2$, 185
p1468	—,—,—(*dl*)*	$CH_3CHOHCO_2C_2H_5$	118.13			154.5 cor	1.0299_4^{25}		∞	v	v				$B3^2$, 205
p1469	—,—,—(*l*)*	$CH_3CHOHCO_2C_2H_5$	118.13	$[\alpha]_D^{19}+14.52$		53^{15}	$1.0324_4^{20.4}$								$B3^2$, 187
p1470	—,—,isopropyl ester*	Isopropyl lactate. $CH_3CHOHCO_2CH(CH_3)_2$	132.16			166–8			s	s	s		s		
p1471	—,—,methyl ester(*d*)*	L(+)Methyl lactate. $CH_3CHOHCO_2CH_3$	104.11	$[\alpha]_D^{20}+7.46$		48^{16}	1.0857_4^{26}								$B3^2$, 187
p1472	—,—,—(*dl*)*	$CH_3CHOHCO_2CH_3$	104.11			144.8 cor	1.0898^{19}	1.4156^{16}	∞	s	s				$B3^2$, 205
p1473	—,—,—(*l*)*	*D*(−)Methyl lactate. $CH_3CHOHCO_2CH_3$	104.11	$[\alpha]_D^{20}-8.25$		58^{19}	1.096_4^{18}	1.4139^{20}							$B3^2$, 184
p1474	—,—,3-methyl-butyl ester*	Isoamyl lactate. $CH_3CHOHCO_2CH_2CH_2CH(CH_3)_2$	160.21			202.4	0.9768_0^0		i	s					$B3^2$, 207
p1475	—,—,nitrile	Acetaldehyde cyanohydrin. Lactonitrile. $CH_3CH(OH)CN$	71.08	ye liq	−40	182–4d	0.9877_4^{20}	1.4058^{18}	∞	∞	s			CS_2, peth i	$B3^2$, 209
p1476	—,—,—,acetate	$CH_3CO_2CH(CH_3)CN$	113.12			172–3	1.032_4^{14}		s	δ	δ			aa δ	B3, 285
p1477	—,—,4-phenyl-phenacyl ester*	$CH_3CHOHCO_2CH_2CO$—(benzene ring)—(benzene ring)	284.31		145										
p1478	—,—,piperazinium salt	$2(CH_3CHOHCO_2H).C_4H_{10}N_2$	266.30		96				s	s^h	i			cellosolve s^h	
p1479	—,**3-hydroxy-***	Hydracrylic acid. β-Lactic acid. $HOCH_2CH_2CO_2H$	90.08	syr		d			v	s	∞				$B3^2$, 212

For explanations, symbols and abbreviations see beginning of table.

No.	Name	Synonyms and Formula	Mol. wt.	Crystalline form, color and specific rotation	m.p. °C	b.p. °C	Density	n_D	Solubility w	al	eth	ace	bz	other solvents	Ref.
	Propanoic acid														
p1480	—,—,lactone*	β-Propiolactone. O—O (ring)	72.06		−33.4	155d 51^{10}	1.1460^{20}_{5}	1.4131^{20}	d	d	∞	...	...	chl s	**B17**[1], 130
p1481	—,—,nitrile	Ethylene cyanohydrin. Hydracrylonitrile. $HOCH_2CH_2CN$	71.08		220–2^{723} 116–8^{20}	1.0588^{0}			∞	∞	δ	...	...	CS_2 i	**B3**[2], 213
p1482	**—,3(4-hydroxy-3,5-diiodo-phenyl)2-phenyl-***	Iodoalphionic acid. I, HO—, —$CH_2CH(C_6H_5)CO_2H$, I	494.07	pa ye (aa)	163.5–.8				i	s	s	...	δ	chl δ alk s	
p1483	**—,3(4-hydroxy-3-methoxy-phenyl)-***	Hydroferulic acid. HO—, —$CH_2CH_2CO_2H$, CH_3O	196.20	pl (w)	89–90				s	s	s	...	...		**B10**, 424
p1484	**—,2-hydroxy-2-methyl-***	Acetonic acid. $(CH_3)_2COHCO_2H$	104.11	hyg pr or nd (bz)	79–80	214–5			v	v	v	...	v^h δ		**B3**[2], 223
p1485	—,—,ethyl ester*	$(CH_3)_2COHCO_2C_2H_5$	132.16			150 cor			∞	∞	...	...	...		**B3**[2], 223
p1486	—,—,methyl ester*	$(CH_3)_2COHCO_2CH_3$	118.13			137			v	v	...	...	...		**B3**[2], 223
p1487	—,—,nitrile	Acetone cyanohydrin. $(CH_3)_2C(OH)CN$	85.11		−19	82^{23}	0.932^{19}	1.3996^{20}	v	v	v	...	...	peth i, os v	**B3**[2], 224
p1488	**—,2-hydroxy-3-oxo-3-phenyl-**	Benzoylglycolic acid. $C_6H_5COCH(OH)CO_2H$	180.16	lo pr (lig)	112				δ s^h	s	s	...	...	chl s	**B9**[2], 147
p1490	**—,2-hydroxy-2-phenyl-(*D*)***	Atrolactic acid. $C_6H_5COH(CH_3)CO_2H$	166.18	pr (w) $[\alpha]^{10.5}_{D}+37.7$ (al, c=3.5)	116.5–7				s	...	...	...	...		**B10**[2], 155
p1491	—,—(*DL*)*	$C_6H_5COH(CH_3)CO_2H$	166.18	nd, pl (lig)	93–5				δ	...	...	...	...		**B10**[2], 156
p1492	—,—(*L*)*	$C_6H_5COH(CH_3)CO_2H$	166.18	nd (bz, w) $[\alpha]^{13.8}_{D}-37.7$ (al, c=3.4)	116–7				s	s^h v	...	v	v^h		**B10**[2], 156
p1493	—,—,hemi-hydrate(*Dl*)*	$C_6H_5COH(CH_3)CO_2H.\frac{1}{2}H_2O$	175.19	nd	67–8				s^h	s^h	...	...	...		**B10**[2], 157
p1494	**—,2-hydroxy-3-phenyl-(*d*)***	$C_6H_5CH_2CHOHCO_2H$	166.18	nd (w) $[\alpha]^{20}_{D}+22.2$ (w, p=2.2)	124–6				v^h	s	s^h	s	s^h	MeOH s CS_2, chl δ	**B10**[2], 152
p1495	—,—(*dl*)*	$C_6H_5CH_2CHOHCO_2H$	166.18	(chl bz), pr (w)	98	148–50^{15}			δ	...	s	...	...	CCl_4 s^h	**B10**[2], 154
p1496	—,—(*l*)*	$C_6H_5CH_2CHOHCO_2H$	166.18	nd (w) $[\alpha]_D-19.9$ (w, c=3.2)					v^h	v	v	...	...		**B10**[2], 153
p1497	**—,3-hydroxy-2-phenyl-(*d*)***	$HOCH_2CH(C_6H_5)CO_2H$	166.18	nd (w), pr (eth) $[\alpha]^{16}_{D}+72.2$ (al, c=2.7)	129–30				...	...	...	...	δ		**B10**[2], 158
p1498	—,—(*dl*)*	Tropaic acid. Tropic acid. $HOCH_2CH(C_6H_5)CO_2H$	166.18	nd, pl (al, bz)	118	d			∞^h s	s	s	...	δ	peth i	**B10**[2], 158 **B23**[2], 534
p1499	—,—(*l*)*	$HOCH_2CH(C_6H_5)CO_2H$	166.18	ta (AcOEt), nd (w) $[\alpha]^{15}_{D}-79.0$ (w, c=1.5)	130				...	...	...	...	δ	AcOEt s	**B10**[2], 158
p1500	**—,3-hydroxy-3-phenyl-(*d*)***	$C_6H_5CHOHCH_2CO_2H$	166.18	$[\alpha]^{18}_{D}+20.6$ (MeOH, c=5.04)	116				δ	s	s	...	...	MeOH s chl δ	**B10**[2], 148
p1501	—,—(*dl*)*	$C_6H_5CHOHCH_2CO_2H$	166.18	pr (w)	96				δ ∞^h	v	...	v	δ	chl s, MeOH v	**B10**[2], 148
p1502	—,—(*l*)*	$C_6H_5CHOHCH_2CO_2H$	166.18	nd (bz) $[\alpha]_D-39.5$ (al, c=4.8)	115–6				δ	...	...	...	δ	chl δ	**B10**[2], 147
p1502[1]	**—,3(2-hydroxy-phenyl)-***	*o*-Hydrocoumaric acid. Melilotic acid. OH, —$CH_2CH_2CO_2H$	166.18	pr (w)	82–3				s v^h	...	s	...	...		**B10**[2], 143
p1503	**—,2-hydroxy-3,3,3-trichloro-***	$Cl_3CCHOHCO_2H$	193.41	pr (eth)	124	140–70^{45}			v	v	v	...	...	chl v	**B3**[2], 210
p1504	—,—,monohydrate	$Cl_3CCHOHCO_2H.H_2O$	211.43		105–10				...	...	...	...	...		**B3**, 287

For explanations, symbols and abbreviations see beginning of table.

No.	Name	Synonyms and Formula	Mol. wt.	Crystalline form, color and specific rotation	m.p. °C	b.p. °C	Density	n_D	Solubility w	al	eth	ace	bz	other solvents	Ref.
	Propanoic acid														
p1505	—,—,nitrile	Chloral cyanohydrin. $Cl_3CCHOHCN$	174.41	pl (w, CS_2)	61	215–20d			v	v	v			CS_2 s	**B3**[2], 210
p1506	**—,2,2′-iminodi-***	$HN[CH(CH_3)CO_2H]_2$	161.16	pr (eth), nd (w)	234–5d				v	i	i s^h	i			**B4**[2], 824
p1507	—,—,dinitrile	$HN[CH(CH_3)CN]_2$	123.16	nd (eth)	67–8d				δ	s	s				**B4**[2], 825
p1508	**—,3,3′-iminodi-,** *N*-methyl-, diethyl ester*	$(C_2H_5O_2CCH_2CH_2)_2NCH_3$	231.29			$136–8^{4}$	1.0190^{20}_{20}	1.4421^{20}		s	s				**B4**[2], 829
p1509	**—,β(3-indolyl)-***	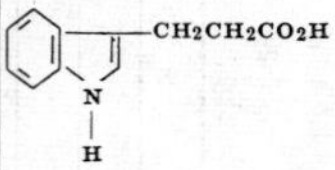 $CH_2CH_2CO_2H$ N H	189.22	pl (w)	133–4				s^h δ	v	v	v	v	AcOEt v chl v	**B22**[2], 53
p1510	—,—,methyl ester	$C_{12}H_{13}NO_2$. *See* p1509	203.24	pr (MeOH)	79–80					s				MeOH s^h	**B22**[2], 53
p1511	**—,2-iodo-***	CH_3CHICO_2H	199.98	nd, $[\alpha]_D$: +50.7 *L* −29.9 (eth)	45–5		2.073^{18}_{4}		δ	v	v				**B2**[2], 233
p1512	**—,3-iodo-***	$ICH_2CH_2CO_2H$	199.98	lf	81.2				s^h δ	v	s	s			**B2**, 261
p1513	—,—,methyl ester*	$ICH_2CH_2CO_2CH_3$	214.00			188^{756}	1.8408^{7}			s					**B2**, 262
p1513[1]	**—,3-mercapto-***	$HSCH_2CH_2CO_2H$	106.15	amor	16.8	110.5–1.5[15]	1.218^{21}		s	s	s				**B3**[2], 214
p1514	**—,2-methoxy-,** nitrile	$CH_3CH(OCH_3)CN$	85.11			118^{729}	0.893^{20}_{4}	1.382^{20}		s					**B3**, 285
p1515	**—,3-methoxy-,** nitrile	$CH_3OCH_2CH_2CN$	85.11			165.5^{763}	0.9463^{15}_{4}			s	s				**B3**[1], 113
p1516	**—,2-methoxy-3-methyl-,** 3-*p*-menthyl ester*	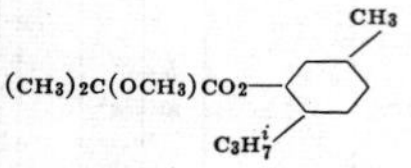 CH_3 $(CH_3)_2C(OCH_3)CO_2$— $C_3H_7^i$	256.39			$124–6^{10}$	0.9466			s	s				
p1518	**—,2-methyl-***	Isobutyric acid. $(CH_3)_2CHCO_2H$	88.11		−47	154.3	0.9504^{20}_{4}	1.3930^{20}	v	∞	∞				**B2**[2], 257
p1519	—,—,allyl ester	Allyl isobutyrate. $(CH_3)_2CHCO_2CH_2CH{:}CH_2$	128.17			$133-5^{766}$			δ	∞	∞				**B2**, 292
p1521	—,—,amide, *N*-phenyl-	Isobutyranilide. $(CH_3)_2CHCONHC_6H_5$	163.21	mcl pr (al, eth), nd (lig),	106–7				δ^h	v	v				**B12**[2], 147
p1522	—,—,anhydride*	Isobutyric anhydride. $[(CH_3)_2CHCO]_2O$	158.20		−53.5	181.5^{734}	0.9540^{20}_{15}		d	d	∞				**B2**[2], 262
p1523	—,—,bromide	Isobutyryl bromide. $(CH_3)_2CHCOBr$	151.01			116–8	1.4067^{15}_{4}	1.4552^{15}	d	d					**B2**[2], 262
p1524	—,—,*tert*-butyl ester	*tert*-Butyl isobutyrate. $(CH_3)_2CHCO_2C(CH_3)_3$	144.22			126.7		1.3921^{20}	i	s	s				**B2**[1], 128
p1525	—,—,chloride*	Isobutyryl chloride. $(CH_3)_2CHCOCl$	106.55		−90.0	90–3	1.0174^{20}_{4}	1.4079^{20}	d	d	s				**B2**[2], 262
p1526	—,—,cyclohexyl ester*	Cyclohexyl isobutyrate. 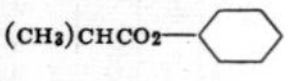 $(CH_3)CHCO_2$—	170.24			204^{750} $87–8^{15}$	0.9489^{0}_{4}		i	s				os s	**B6**[2], 11
p1527	—,—,ethyl ester*	Ethyl isobutyrate. $(CH_3)_2CHCO_2C_2H_5$	116.16		−88.2	111.0	0.8693^{20}_{4}	1.3903	δ	∞	∞				**B2**[2], 260
p1528	—,—,furfuryl ester	Furfuryl isobutyrate. 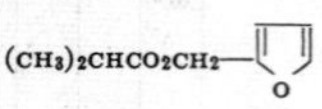 $(CH_3)_2CHCO_2CH_2$— O	168.19			$85–6^{15}$	1.0313^{20}_{4}								**B17**[2], 115
p1529	**—,—,α,α′,1,2-hydrazodi-** dinitrile	*α,α′*-Hydrazodiisobutyronitrile. $(CH_3)_2C(CN)NHNHC(CN)(CH_3)_2$	166.23	pl (eth)	92–3				i	v	v				**B4**, 561
p1530	—,—,isobutyl ester	Isobutyl isobutyrate. $(CH_3)_2CHCO_2CH_2CH(CH_3)_2$	144.22		−80.66	148.6	0.8750^{0}_{4}	1.3999	δ	s	∞				**B2**[2], 260
p1531	—,—,isopropyl ester*	Isopropyl isobutyrate. $(CH_3)_2CHCO_2CH(CH_3)_2$	130.19			120.8	0.8687^{0}_{4}		i	s	s				**B2**[1], 128

For explanations, symbols and abbreviations see beginning of table.

No.	Name	Synonyms and Formula	Mol. wt.	Crystalline form, color and specific rotation	m.p. °C	b.p. °C	Density	n_D	Solubility						Ref.
									w	al	eth	ace	bz	other solvents	
	Propanoic acid														
p1532	—,—,methyl ester*	Methyl isobutyrate. $(CH_3)_2CHCO_2CH_3$	102.13		−84.7	92.3	0.8906^{20}_{4}	1.3840	δ	∞	∞				**B2²**, 260
p1533	—,—,3-methylbutyl ester*	Isoamyl isobutyrate. $(CH_3)_2CHCO_2CH_2CH_2CH(CH_3)_2$	158.24			168.9	0.8627^{20}		δ	s	s				**B2²**, 261
p1534	—,—,nitrile	Isobutyronitrile. Isopropyl cyanide. $(CH_3)_2CHCN$	69.11		−71.5	103.8	0.7608^{30}_{4}	1.3723^{25}	δ	v	v				**B2²**, 262
p1535	—,—,pentyl ester*	Amyl isobutyrate. $(CH_3)_2CHCO_2(CH_2)_4CH_3$	158.24	wh		155	0.8592^{13}	1.4076	δ	∞	∞				
p1536	—,— piperazinium salt	$2[(CH_3)_2CHCO_2H]. C_4H_{10}N_2$	262.35	wh	89.5–90				s	s	i^h			diox s^h	
p1537	—,—,propyl ester*	*n*-Propyl isobutyrate. $(CH_3)_2CHCO_2CH_2CH_2CH_3$	130.19			134.0	0.8843^{0}_{4}		δ	s	v				**B2²**, 260
p1538	**—,2-methy-3-phenyl-***	$C_6H_5CH_2CH(CH_3)CO_2H$	164.21		36.5	272			δ v^h	v	v				
p1539	**—,3(1-naphthyl)-***	$C_{10}H_7^{\alpha}CH_2CH_2CO_2H$	200.24	(bz), nd (al)	156–6.7	179^{11}			s^h	s					**E12B**, 3283
p1540	**—,3(2-naphthyl)-***	$C_7H_{10}^{\beta}CH_2CH_2CO_2H$	200.24	lf or nd (w, al)	134–5				v^h	s					**E12B**, 3285
p1541	**—,3(2-nitro-phenyl)2-oxo-***	NO_2 / $—CH_2COCO_2H$ (structure)	209.16	nd or lf (w)	121				$δ^h$	v	v		δ	chl δ	**B10²**, 476
p1542	**—,2-oxo-***	Acetyl formic acid. Pyruvic acid. CH_3COCO_2H	88.06		13.6	115d 48^2	1.2272^{20}_{4}	1.4138^{20}	∞	∞	∞				**B3²**, 393
p1543	—,—,ethyl ester*	Ethyl pyruvate. $CH_3COCO_2C_2H_5$	116.12			155 $69–71^{42}$	$1.0596^{15.6}_{4}$	$1.408^{15.6}$	δ	∞	∞				**B3²**, 403
p1544	—,—,methyl ester*	Methyl pyruvate. $CH_3COCO_2CH_3$	102.09			134–7 53^{15}	1.154^{0}_{4}		δ	∞	∞				**B3²**, 402
p1545	—,—,nitrile	Acetyl cyanide. Pyruvonitrile. CH_3COCN	69.06	rh		92.3	0.9745^{20}_{4}	1.3764^{20}	d	d	s			CH_3CN s	**B3²**, 404
p1545¹	**—,3(2-oxocyclo-hexyl)-, nitrile**	O / $—CH_2CH_2CN$ (structure)	151.21			$138–42^{10}$	1.0181^{20}_{4}	1.4755^{20}							**Am73**, 724
p1546	**—,2-oxo-3-phenyl-***	Phenylpyruvic acid. $C_6H_5CH_2COCO_2H$	164.16	lf (chl, bz)	157–8				$δ^h$	v	v		s^h	chl s^h	**B10²**, 471
p1546¹	**—,3-oxo-2-phenyl-, ethyl ester***	Ethyl phenylmalonaldate. $C_6H_5CH(CHO)CO_2C_2H_5$	192.22		70–1	$135–40^{10}$	1.2045^{20}_{20}	1.532^{21}	i	∞	v				**B10²**, 478
p1547	**—,3-oxo-3-phenyl-***	Benzoylacetic acid. $C_6H_5COCH_2CO_2H$	164.16	nd (bz-peth)	103–4d				δ v^h	s	s			lig δ	**B10²**, 466
p1548	—,—,amide,*N*-phenyl-*	α-Benzoylacetanilide. $C_6H_5COCH_2CONHC_6H_5$	239.28	lf (bz)	108				δ	v			δ v^h	alk v, chl v	**B12²**, 270
p1549	—,—,ethyl ester*	$C_6H_5COCH_2CO_2C_2H_5$	192.22		<0	265–70d 165^{14}	1.122^{20}_{4}	1.5312^{16}	δ	s	s				**B10²**, 467
p1550	—,—,methyl ester (enol form)*	$C_6H_5COCH_2CO_2CH_3$	178.19		ca. 40			1.5620^{16}							**B10²**, 466
p1551	—,—,—(keto+enol forms)*	$C_6H_5COCH_2CO_2CH_3$	178.19	colorless to ye		265d 151.5^{13}	1.158^{29}_{4}	1.537^{25}	i	∞	∞	s		dil alk s	**B10²**, 466
p1552	—,—,nitrile	Benzoylacetonitrile. α-Cyanoacetophenone. $C_6H_5COCH_2CN$	145.16	pr or lf (w^h)	80–1	160^{10}			δ	s	s		s	chl, aq KCN, alk s	**B10²**, 468
p1553	**—,pentachloro-***	$Cl_3CCCl_2CO_2H$	246.31	cr(CCl_4)		200–15d			v					CCl_4 δ, v^h	**B2²**, 228
p1554	—,—,chloride*	Cl_3CCCl_2COCl	264.75	nd	42				d	d			s		**B2¹**, 112
p1555	**—,2-phenoxy-(*D*)***	$CH_3CH(OC_6H_5)CO_2H$	166.18	nd (w) $[\alpha]^{21}_{D}+39.3$ (al, c=1.2)	87	$265–6^{758}$			δ v^h	v	v				**B6²**, 158
p1556	—,—(*DL*)*	$CH_3CH(OC_6H_5)CO_2H$	166.18	nd	115–6	265–6			δ	s	s				**B6²**, 158
p1557	—,—,amide*	$CH_3CH(OC_6H_5)CONH_2$	165.19	nd or pl (to, w)	132–3				s^h	v	v			to s^h lig v^h	**B6**, 163
p1558	—,—,chloride	$CH_3CH(OC_6H_5)COCl$	184.62			$146–7^{55}$ $115–7^{10}$			d	d	s				**B6**, 163
p1559	—,—,ethyl ester*	$CH_3CH(OC_6H_5)CO_2C_2H_5$	194.23			243–4 $120–5^6$	1.360^{17}_{4}		i	s	s				

For explanations, symbols and abbreviations see beginning of table.

No.	Name	Synonyms and Formula	Mol. wt.	Crystalline form, color and specific rotation	m.p. °C	b.p. °C	Density	n_D	Solubility w	al	eth	ace	bz	other solvents	Ref.
	Propanoic acid														
p1560	—,**3-phenoxy-***	$C_6H_5OCH_2CH_2CO_2H$	166.18	nd (w), lf (lig)	97.5–8	$234–45^{771}$			s^h					lig s^h	**B6²**, 158
p1561	—,—,amide	$C_6H_5OCH_2CH_2CONH_2$	165.19	nd (w)	119				s^h	v^h	v^h				**B6²**, 158
p1562	—,—,ethyl ester*	$C_6H_5OCH_2CH_2CO_2C_2H_5$	194.23			170^{40}			i	s	s				**B6²**, 158
p1563	—,**2-phenyl-(*d*)***	Hydratropic acid. $C_6H_5CH(CH_3)CO_2H$	150.18	$[\alpha]^{20}_D+81.1$ (al, c=3)											**B9²**, 347
p1564	—,—(*dl*)*	$C_6H_5CH(CH_3)CO_2H$	150.18		<−20	264–5 160^{25}	1.1^0_4		δ						**B9²**, 348
p1565	—,—(*l*)*	$C_6H_5CH(CH_3)CO_2H$	150.18	$[\alpha]^{20}_D-7.0$		152^{16}									
p1566	—,—,amide	Hydratropamide. $C_6H_5CH(CH_3)CONH_2$	149.19	lo nd (w) $[\alpha]^{28}_D+57.9$ (chl, c=1.6)	100.5				i δ^h	s				chl s	**B9**, 525
p1567	—,—,nitrile	Hydratroponitrile. $C_6H_5CH(CH_3)CN$	131.18			$116–7^{20}$	0.9854^{20}_4	1.5095^{25}	i		s				**B9²**, 348
p1568	—,—,piperazinium salt	$2[C_6H_5CH(CH_3)CO_2H].C_4H_{10}N_2$	386.50	wh	122.5–3				δ	s^h	i				
p1569	—,**3-phenyl-***	Hydrocinnamic acid. $C_6H_5CH_2CH_2CO_2H$	150.18	(lig)	48.6	279.8 $169–70^{28}$	1.047^{100}_4		s	s	s		v	CCl_4 s chl s	**B9²**, 337
p1570	—,—,amide	Hydrocinnamamide. $C_6H_5CH_2CH_2CONH_2$	149.19	nd	106–8				s^h	v	v				**B9²**, 340
p1571	—,—,benzyl ester	Benzyl hydrocinnamate. $C_6H_5CH_2CH_2CO_2CH_2C_6H_5$	240.30			$198–9^{20}$	1.090^{15}								**B9²**, 339
p1572	—,—,chloride	Hydrocinnamyl chloride. $C_6H_5CH_2CH_2COCl$	168.22			225 d 107^{11-2}	1.135^{21}		d	d^h	s			CS_2 s	**B9²**, 340
p1573	—,—,ethyl ester*	Ethyl hydrocinnamate. $C_6H_5CH_2CH_2CO_2C_2H_5$	178.23			247.2 123^{16}	1.0147^{20}	1.4954^{20}	i	s	s				**B9²**, 339
p1574	—,—,isopropyl ester*	Isopropyl hydrocinnamate. $C_6H_5CH_2CH_2CO_2CH(CH_3)_2$	192.26			126^{11}	0.9860^{25}_4								**B9²**, 339
p1575	—,—,methyl ester*	Methyl hydrocinnamate. $C_6H_5CH_2CH_2CO_2CH_3$	164.21			236.6	1.0455^0		i	s	s		s	AcOEt s	**B9²**, 338
p1576	—,—,nitrile	Hydrocinnamonitrile. $C_6H_5CH_2CH_2CN$	131.17			$125–6^{15}$	1.0016^{20}	1.5266^{20}							**B9²**, 341
p1577	—,—,propyl ester*	Propyl hydrocinnamate. $C_6H_5CH_2CH_2CO_2CH_2CH_2CH_3$	192.26			262.1 135^{16}	1.008^{12}								**B9²**. 339
p1578	—,**2-propoxy-***	$CH_3CH_2CH_2OCH(CH_3)CN$	113.16			150^{727}	0.866^{20}_4	1.398^{20}							**B3**, 285
p1579	—,**3(3-pyrenyl)-**	$CH_2CH_2CO_2H$ (pyrene structure)	274.30	pl(aa)	180				i					con sulf s dil alk s aa s	**E14s**, 441
p1580	—,**2,2,3,3-tetrachloro-***	$Cl_2CHCCl_2CO_2H$	211.08	(CS_2-chl)	76				s					CS_2 v	**B2**, 253
p1581	—,**3(2-tetrahydrofuryl)-**	(tetrahydrofuran ring, O)—$CH_2CH_2CO_2H$	144.17			263 $118–20^2$	1.1155^{20}_{20}	1.4578^{25}		s				alk v	**B18²**, 263
p1582	—,—,ethyl ester*	$C_9H_{16}O_3$. *See* p1581	172.22			221–2 73^2	1.024^7_{15}	1.440^{20}		s					**B18²**, 263
p1584	—,**2,2,3-trichloro-***	$ClCH_2CCl_2CO_2H$	177.43	hyg (CS_2)	60				v	v			v		**B2²**, 228
p1585	—,**2,2,3-triphenyl-***	$(C_6H_5)_2CHCH(C_6H_5)CO_2H$	302.38	nd (dil al, peth)	222–3				i	v	v			peth s	**B9**, 715
p1586	—,**3,3,3-triphenyl-***	$(C_6H_5)_3CCH_2CO_2H$	302.38	pr (al)	179–80					v	v				**B9²**, 504
p1587	**1-Propanol-***	*n*-Propyl alcohol. $CH_3CH_2CH_2OH$	60.09		−127	97.1	0.7796^{20}_4	1.3850^{20}	v	v	v		v		**B1²**, 360
p1588	**2-Propanol***	Isopropanol. Isopropyl alcohol. $CH_3CHOHCH_3$	60.09		−89.5	82.4	0.7851^{20}_4	1.3776^{20}	∞	∞	∞				**B1³**, 1439
p1589	**1-Propanol, 2-amino-(*dl*)***	$CH_3CH(NH_2)CH_2OH$	75.11			173–6			v	v	v				**B4²**, 733
p1590	—,**3-amino-***	$H_2NCH_2CH_2CH_2OH$	75.11			$187–8^{756}$ cor	1.4570^{26}	0.9824^{26}_4	s	s					**B4²**, 734
p1591	**2-Propanol, 1-amino-(*dl*)***	$CH_3CHOHCH_2NH_2$	75.11			$160–1^{750}$ cor	0.973^{18}	1.45^{18}	s	s	i				**B4²**, 736
p1592	—,—(*l*)*	$CH_3CHOHCH_2NH_2$	75.11			$156–8^{758}$	0.973^{18}		s	s	i				**B4²**, 736

For explanations, symbols and abbreviations see beginning of table.

No.	Name	Synonyms and Formula	Mol. wt.	Crystalline form, color and specific rotation	m.p. °C	b.p. °C	Density	n_D	Solubility w	al	eth	ace	bz	other solvents	Ref.
	1-Propanol														
p1593	**1-Propanol, 2-amino-1(4-aminophenyl)-,** dihydrochloride*	p-Aminoneoephedrine dihydrochloride. $H_2N—C_6H_4—CHOHCH(NH_2)CH_3.2HCl$	239.15	lf(al-eth)	192–3d				v	s	i				
p1594	**2-Propanol, 1-amino-3-(diethylamino)-***	$H_2NCH_2CH(OH)CH_2N(C_2H_5)_2$	146.24			223	0.937^{20}_{4}	1.465^{20}		s	s				**B4**², 740
p1595	**1-Propanol, 2-amino-2-methyl-***	$(CH_3)_2CNH_2CH_2OH$	89.14		25–6	165.5	0.934	1.449^{20}	∞						
—	—,2-amino-1-phenyl-*	see **Norephedrin**													
p1595	**2-Propanol, 2-benzyl-**	$C_6H_5CH_2COH(CH_3)_2$	150.23			$103–5^{10}$	0.9790^{20}_{25}	1.5174^{20}	δ	s					
p1596	—,1,3-bis(dimethylamino)-*	$(CH_3)_2NCH_2CHOHCH_2N(CH_3)_2$	146.24			178–85 $79–81^{18}$			v						**B4**, 290
p1597	—,—,dimethiodide	Iodisan. $HOCH[CH_2N(CH_3)_3I]_2$	430.14	wh cr pw	ca. 275d				v	δ	i	i			
p1598	**1-Propanol, 3-bromo-***	Trimethylene bromohydrin. $BrCH_2CH_2CH_2OH$	139.00			$98–112^{185}$	1.5374^{20}_{4}		s	∞	∞				**B1**, 356
p1599	**2-Propanol, 1-bromo-***	Propylene bromohydrin. $CH_3CHOHCH_2Br$	139.00			145–8 49.6^{12}	1.5988^{0}	1.4801^{20}	s	v	v				**B1**³, 1474
p1600	—,1-butoxy-*	1,2-Propyleneglycol, 1-monobutyl ether. $CH_3CHOHCH_2O(CH_2)_3CH_3$	132.21			168–75	1.0035^{20}_{4}	1.4168^{20}		s			s	MeOH s CCl_4 s	**B1**², 537
p1601	**1-Propanol, 2-chloro-***	Propylene chlorohydrin. $CH_3CHClCH_2OH$	94.54			133–4	1.103^{20}	1.4362^{20}	s	s				os s	
p1603	—,3-chloro-*	Trimethylene chlorohydrin. $ClCH_2CH_2CH_2OH$	94.54			161–2	1.1309^{20}_{4}	1.4469^{20}	v	s	s				**B1**, 356
p1604	**2-Propanol, 1-chloro-***	Propylene chlorohydrin. $CH_3CHOHCH_2Cl$	94.54			126–7 cor	1.11^{20}_{20}	1.4392^{20}	∞	∞	∞				**B1**, 363
p1605	—,1-chloro-3-isopropoxy-*	$ClCH_2CHOHCH_2OCH(CH_3)_2$	152.63			$87–7.5^{20}$	1.0530^{25}	1.4370^{25}							**B1**³, 2153
p1606	**1-Propanol, 2-chloro-2-methyl-**	β-Isobutylene chlorohydrin. $(CH_3)_2CClCH_2OH$	108.57	visc		132–3δd			d					con HCl s	**B1**, 378
p1607	—,3-chloro-2-methyl-*	$ClCH_2CH(CH_3)CH_2OH$	108.57			$76–8^{21}$	1.083^{25}_{4}	1.4460^{25}		v	v				
p1608	**2-Propanol, 1-chloro-2-methyl-***	$(CH_3)_2COHCH_2Cl$	108.57			128–9 71^{100}			v δd^h	v					**B1**², 415
p1609	—,1-chloro-3-propoxy-*	$CH_3CH_2CH_2OCH_2CHOHCH_2Cl$	152.63			$92–5^{15}$	1.0526^{25}_{4}	1.4378^{25}							**B1**³, 2153
p1610	—,1,3-diamino-*	$H_2NCH_2CHOHCH_2NH_2$	90.13		42	235									**B4**², 739
p1611	—,—,dihydrochloride	$H_2NCH_2CHOHCH_2NH_2.2HCl$	163.06	pr	175–7				v	v	i				**B4**², 739
p1612	**1-Propanol, 2,3-dibromo-(d)***	$BrCH_2CHBrCH_2OH$	217.91	$[\alpha]_D$ +7.27			2.11		δ	∞	∞	∞	∞		**B1**², 371
p1613	—,—(dl)*	$BrCH_2CHBrCH_2OH$	217.91			219δd 118^{17}	2.1197^{20}_{3}		δ	∞	∞	∞	∞		**B1**, 357
p1614	**2-Propanol, 1,3-dibromo-***	$BrCH_2CHOHCH_2Br$	217.91	yesh liq		219	2.1202^{25}_{4}	1.5495^{25}	s	s	s				**B1**³, 1475
p1615	**1-Propanol, 2,3-dichloro-***	$ClCH_2CHClCH_2OH$	128.99	visc		182 $70–80.5^{17}$	1.3534^{20}_{4}	1.4875^{18}	δ	∞	∞	∞	∞	lig δ	**B1**², 370
p1617	**2-Propanol, 1,1-dichloro-***	$CH_3CHOHCHCl_2$	128.99			$146–8^{765}$	1.3334^{22}_{22}		δ	v	v				**B1**², 383
p1618	—,1,3-dichloro-*	$ClCH_2CHOHCH_2Cl$	128.99			175.8– 6.3^{765} 69^{12}	1.359^{25}_{4}	1.4837^{20}	v	v	∞				**B1**², 383
p1620	—,1,1-dichloro-2-methyl-*	$(CH_3)_2COHCHCl_2$	143.02			150–5 52^{10}	1.2363^{19}_{4}	1.4598^{19}		v			v^h		**B1**, 382

For explanations, symbols and abbreviations see beginning of table.

No.	Name	Synonyms and Formula	Mol. wt.	Crystalline form, color and specific rotation	m.p. °C	b.p. °C	Density	n_D	Solubility w	al	eth	ace	bz	other solvents	Ref.
	2-Propanol														
p1621	**—,1,3-dichloro-2-methyl-***	$(ClCH_2)_2COHCH_3$	143.02			174–5 55–6^{10}	1.2758^{20}_{4}	1.4744^{21}	s	v					**B1**, 382
p1622	**—,1(diethyl-amino)-***	$CH_3CHOHCH_2N(C_2H_5)_2$	131.22			158–9				s					**B4**^{2}, 737
p1623	**1-Propanol, 2,3-dimercapto-***	BAL. British antilewisite. 1,2-Dithioglycerol. $HSCH_2CHSHCH_2OH$	124.21	visc liq		120^{15}	1.2385^{25}_{4}	1.5720^{25}	d	s	s			oils s	
p1624	**—,3(3,5-di-methoxy-4-hydroxy-phenyl)-***	Hydrosinapyl alcohol. 2-Syringylethanol. CH_3O, HO—⟨⟩—$CH_2CH_2CH_2OH$, CH_3O	212.24	wh	75.5–6.5				s				δ	lig i	
p1625	**—,2,2-dimethyl-***	*tert*-Butylcarbinol. Neopentyl alcohol. $(CH_3)_3CCH_2OH$	88.15		52–3	113–4	0.812		δ	v	v				**B1**, 406
p1626	**—,2,2-dimethyl-1-phenyl-***	$C_6H_5CHOHC(CH_3)_3$	164.25	nd	45	114–6^{16}			i	s				os s	**B6**^{1}, 270
p1627	**—,2,2-dimethyl-3-phenyl-***	$C_6H_5CH_2C(CH_3)_2CH_2OH$	164.25	nd	34–5	125–6$^{14.5}$			i	s				os s	**B6**^{2}, 507
p1628	**—,2,3-epoxy-***	Glycide. Glycidol. 3-Hydroxy-propylene oxide. CH_2CHCH_2OH (with O bridge)	74.08			166–7	1.111^{22}	1.4350^{16}	∞	∞	∞	s	i	chl s	**B17**^{2}, 104
p1629	**—,2,3-epoxy-3-phenyl-***	Phenyl glycidol. $C_6H_5CH—CHCH_2OH$ (with O bridge)	150.18		26.5	138^3	1.512^{27}	1.5432^{27}						CCl_4 s	
p1630	**—,2-ethoxy-***	$CH_3CH(OC_2H_5)CH_2OH$	104.15			140–1	0.9044^{20}_{4}	1.4122^{20}	s						
p1630^1	**2-Propanol, 1-ethoxy-***	$CH_3CHOHCH_2OC_2H_5$	104.15			131	0.9028^{20}_{4}	1.4075^{20}							
p1631	**1-Propanol, 3-fluoro-***	Trimethylene fluorohydrin. $FCH_2CH_2CH_2OH$	78.09			127.8	1.0390^{25}_{4}	1.3771^{25}	s	v	v				
p1632	**—,3(4-hydroxy-3-methoxy-phenyl)-***	Hydroconieryl alcohol. 2-Vanillyl alcohol. CH_3O, HO—⟨⟩—$CH_2CH_2CH_2OH$	182.22			197^{15}		1.5545^{25}		s	s				
p1632^1	**2-Propanol, 1-methoxy-***	$CH_3CHOHCH_2OCH_3$	90.12			118–8.5	0.9620^{20}_{4}	1.4070^{20}							
p1633	**—,2-methyl-***	*tert*-Butyl alcohol. $(CH_3)_3COH$	74.12	rh pr or pl	25.5	82.2–.3 20^{31}	0.7856^{20}_{4}	1.3838^{20}	∞	∞	∞				**B1**^{2}, 413
p1634	**1-Propanol, 2-methyl-2-nitro-***	$(CH_3)_2C(NO_2)CH_2OH$	119.12	nd or pl (MeOH)	82				δ	v	v				**B1**, 378
p1635	**—,2-methyl-1-phenyl-***	α-Isopropylbenzyl alcohol. $C_6H_5CH(OH)CH(CH_3)_2$	150.22	d:$[\alpha]^{20}_{D}+47.7$ l:$[\alpha]^{20}_{D}-25.2$		222–4 112–3^{15}	0.9869^{14}	1.5193^{14}	i	s					**B6**^{2}, 488
p1636	**2-Propanol, 2-methyl-1-phenyl-***	$(CH_3)_2COHCH_2C_6H_5$	150.22	nd	24	214–6 103–5^{10}	0.9774^{19}_{4}	1.5201^{15}	i	s	s		s		**B6**, 523
p1637	**—,2-methyl-1,1,1-tribromo-***	Brometone. $(CH_3)_2COHCBr_3$	310.83	nd (lig), (dil al)	168–70	sub			δ	s	s				**B1**^{3}, 1588
p1638	**—,2-methyl-1,1,1-trichloro-***	Chloretone. $(CH_3)COHCCl_3$	177.46	hyg nd (w+1)	98.5–9.5 77 (hyd)	167			i s^h	s	s	s	s	chl s lig s	**B1**^{3}, 1586
p1639	**—,—,hemihydrate***	$(CH_3)_2COHCCl_3.\frac{1}{2}H_2O$	186.47	wh	80–1	167			i s^h	v	s	s	s	chl s lig s	**B1**^{2}, 415
p1640	**1-Propanol, 2-nitro-***	$CH_3CH(NO_2)CH_2OH$	105.10			100^{12}	1.1841^{25}_{4}	1.4379^{20}	s	s	s				**B1**^{3}, 1429
p1641	**2-Propanol, 3-nitro-1,1,1-trichloro-***	$O_2NCH_2CHOHCCl_3$	208.43	pr or pl	44.7–5.7	105.5–6.5$^{3.5}$			i	s	s				**B1**^{3}, 1477
p1642	**1-Propanol, 2-phenoxy-**	$CH_3CH(OC_6H_5)CH_2OH$	152.19			124–6^{20}									**B6**^{1}, 85

For explanations, symbols and abbreviations see beginning of table.

No.	Name	Synonyms and Formula	Mol. wt.	Crystalline form, color and specific rotation	m.p. °C	b.p. °C	Density	n_D	Solubility w	al	eth	ace	bz	other solvents	Ref.
	1-Propanol														
p1642¹	**—,3-phenoxy-**	$C_6H_5OCH_2CH_2CH_2OH$	152.19	oil		$249–50^{764}$		1.491^{20}	...	...	...	...	...	...	**B6**^{2}, 151
p1642²	**2-Propanol, 1-phenoxy-**	$CH_3CHOHCH_2OC_6H_5$	152.19			134.5^{20}	1.0622^{20}_{4}	1.5232^{20}	...	...	...	...	...	...	**B6**^{1}, 85
p1643	**—,1-phenyl-***	$CH_3CH_2CHOHC_6H_5$	136.20			217–21	0.994^{23}_{0}	1.5208^{20}	i	s	s	...	...	...	**B6, 502**
p1644	**—,3-phenyl-***	$C_6H_5CH_2CH_2CH_2OH$	136.20		<−18	$236–7^{750}$	1.008^{20}_{4}	1.5278^{20}	s	∞	∞	...	...	...	**B6, 503**
p1645	**2-Propanol, 1-phenyl-***	$C_6H_5COH(CH_3)_2$	139.20		35–7	215–20	0.9727^{19}_{4}	1.5314^{19}	i	s	s	...	s	...	**B6, 506**
p1646	**—,1,1,1,3-tetra-chloro-***	$ClCH_2CHOHCCl_3$	197.89			$95–6^{17}$	1.610^{20}_{4}	1.5145^{20}	...	...	s	...	...	...	**B1**^{2}**, 385**
p1647	**—,1,1,3,3,-tetra-chloro-***	$Cl_2CHCHOHCHCl_2$	197.89			$80–90^{14}$	1.612^{20}_{4}	1.5133^{20}	...	...	s	...	...	...	**B1**^{3}**, 1474**
p1648	**—,1,1,1-tri-chloro-***	Isopral. $CH_3CHOHCCl_3$	163.43		50–1	161.8^{773} $53–5^{12}$			δ	v	v	...	...	...	**B1**^{3}**, 1474**
p1649	**2-Propanone***	Acetone. Dimethyl ketone. CH_3COCH_3	58.08		−95.35	56.2	0.7908^{20}	1.3588^{20}	∞	∞	∞	∞	∞	chl ∞	**B1**^{2}**, 692**
p1650	**—,azine***	Acetone azine. $(CH_3)_2C:NN:C(CH_3)_2$	112.17		−12.5	133	0.8422^{20}	1.4535^{20}	∞	∞	∞	...	...	...	**B1**^{2}**, 717**
p1651	**—,diethyl acetal**	2,2-Diethoxypropane*. $(CH_3)_2C(OC_2H_5)_2$	132.21			46^{60}	0.8714^{22}_{25}		...	s	v	...	...	...	**B1**^{2}**, 715**
p1652	**—,2,4-dinitro-phenylhydrazone***	Acetone DNP. $O_2N—C_6H_3(NO_2)—NHN:C(CH_3)_2$	238.20	ye nd or pl (al^{h})	128				i	s^{h}	s^{h}	...	s	AcOEt s chl s	**B15**^{2}**, 216**
p1653	**—,4-nitrophenyl-hydrazone***	$O_2N—C_6H_4—NHN:C(CH_3)_2$	193.20	og-br or og-ye nd (al)	152				$δ^{h}$	...	...	...	...	aa δ	**B15**^{2}**, 184**
p1654	**—,oxime***	Acetoxime. $(CH_3)_2C:NOH$	73.09	pr	61	134.8^{728}	0.9113^{62}_{4}	1.4156^{20}	s	s	s	...	...	lig s	**B1**^{2}**, 716**
p1655	**—,phenyl-hydrazone***	$C_6H_5NHN:C(CH_3)_2$	148.20	rh	26.6	163^{50}			...	s	s	...	...	dil ac s	**B15**^{2}**, 55**
p1656	**—,—,hydrate***	$C_6H_5NHN:C(CH_3)_2.H_2O$	166.23		35				δ	...	...	...	...	...	**B15**^{1}**, 30**
p1657	**—,semicarbazone***	Acetone semicarbazone. $(CH_3)_2C:NNHCONH_2$	115.14	nd (w, ace)	187d				s^{h} δ	v	s	...	...	...	**B3**^{2}**, 81**
p1658	**1-Propanone, 1(4-acetamido-phenyl)-***	$C_{11}H_{13}NO_2$. *See* p1696	191.24	pa ye nd (w)	161				s	v	v	...	...	...	**B14**^{1}**, 375**
p1659	**2-Propanone, 1-amino-***	Acetonylamine. $CH_3COCH_2NH_2$	73.09	nd (al)	189d				v	s	s	...	...	...	**B4, 314**
p1660	**—,1(2-amino-phenyl)-***	$C_9H_{11}NO$. *See* p1696	149.20	pa ye lf (peth), pl (dil al)	46–7				s	s	s	...	...	chl i ac, os s	**B14**^{2}**, 37**
p1661	**—,1(4-amino-phenyl)-***	$C_9H_{11}NO$. *See* p1696	149.20	pl (al, w)	142				s	δ	δ	...	...	chl s	**B14**^{1}**, 375**
p1662	**—,3-amino-1-phenyl-, hydrochloride***	α-Aminopropiophenone hydrochloride. $C_6H_5COCH(CH_3)NH_2.HCl$	185.66	nd (al-eth)	179–80				δ	δ	i	...	...	...	**B14**^{2}**, 37**
p1663	**2-Propanone, 1-bromo-***	Bromoacetone. CH_3COCH_2Br	136.99		−54	136.5^{725}	1.634^{23}		δ	s	s	s	...	...	**B1, 657**
p1664	**1-Propanone, 1(4-bromo-1-hydroxy-2-naphthyl)-***	(structure: naphthalene ring with OH, $COCH_2CH_3$, Br)	279.13	ye nd	98				i	s	s	...	...	...	**B8, 152**
p1665	**—,2-bromo-2-methyl-1(2,4,6-tri-methylphenyl)-***	α-Bromoisobutyryl-mesitylene. $(CH_3)_3C_6H_2—COCBr(CH_3)_2$	269.20		27	$160–70^{24}$			...	...	s	...	...	...	...
p1666	**—,1(4-bromo-phenyl)-***	C_9H_9BrO. *See* p1696	213.08	nd	48	169^{15}			i	s	s	s	...	CS_2 s	**B7, 302**
p1667	**—,2-bromo-1-phenyl-***	α-Bromopropiophenone. $C_6H_5COCHBrCH_3$	213.08	ye		245–50 $110–2^{4}$	1.4298^{20}_{4}	1.572^{20}	...	s	s	s	s	...	**B7**^{2}**, 233**
p1668	**2-Propanone, 1-chloro-***	Acetonyl chloride. Chloroacetone. CH_3COCH_2Cl	92.53		−44.5	119	1.15^{20}		s	s	s	...	...	chl s	**B1, 653**

For explanations, symbols and abbreviations see beginning of table.

No.	Name	Synonyms and Formula	Mol. wt.	Crystalline form, color and specific rotation	m.p. °C	b.p. °C	Density	n_D	Solubility w	al	eth	ace	bz	other solvents	Ref.
	1-Propanone														
p1669	**1-Propanone, 1(4-chloro-phenyl)-***	C_9H_9ClO. *See* p1696	168.63		35–6				i	s					**B7**, 301
p1670	—,—,oxime*	$C_9H_{10}ClNO$. *See* p1696	183.66	pl (al)	62–2.5					s^h					**B7**, 301
p1671	**—,3-chloro-1-phenyl-***	β-Chloropropiophenone. $C_6H_5COCH_2CH_2Cl$	168.63	pl	57										**B7**, 301
p1672	**2-Propanone, 1-chloro-3-phenyl-***	$C_6H_5CH_2COCH_2Cl$	168.63	nd (chl)	72–3	$159–61^{17}$								chl s^h	**B7**^{2}, 235
p1673	**—,1,3-diamino-,** dihydrochloride*	$(H_2NCH_2)_2CO.2HCl$	161.04	(dil al, dil aa), pr (w+1 or 1½)	180d				v	i	i		i	chl δ aa i	**B4**^{2}, 763
p1674	**1-Propanone, 2,3-dibromo-1,3-diphenyl-*** (one form)	Chalcone dibromide. $C_6H_5CHBrCHBrCOC_6H_5$	386.09	nd (al)	122–3				δ	s					**B7**^{2}, 381
p1675	—,—(one form)*	$C_6H_5CHBrCHBrCOC_6H_5$	368.09	pr or nd (al)	159–60					δ v^h					**B7**^{2}, 381
p1676	**2-Propanone, 1,1-dichloro-***	$CH_3COCHCl_2$	126.98			120	1.305^{18}_{15}		δ	s	∞				**B1**, 654
p1677	**—,1,3-dichloro-***	$ClCH_2COCH_2Cl$	126.98	pl or nd	45	173.4	1.3826^{46}_{4}	1.4714^{46}	s	s	s				**B1**, 655
p1678	**—,1(diethyl-amino)-***	$CH_3COCH_2N(C_2H_5)_2$	129.20			155–6δd 64^{16}			∞	∞	∞				**B4**, 316
p1679	**—,1,3-dihydroxy-***	$HOCH_2COCH_2OH$	90.08		ca. 80				s^h	s^h	s^h	s			**B1**, 846
p1680	**1-Propanone, 1(2-4-dihydroxy-phenyl)-***	4-Propionyl resorcinol. $C_9H_{10}O_3$. *See* p1696	166.17		97–5	$176–8^{6}$			δ	v	v		v		**B8**^{2}, 305
p1681	**—,2,2-dimethyl-1-phenyl-***	Pivalophenone. $C_6H_5COC(CH_3)_3$	162.23			219–21 $97–8^{16}$	0.963^{26}	1.5086^{19}							
p1682	**2-Propanone, 1,3-diphenyl-***	Dibenzyl ketone. $C_6H_5CH_2COCH_2C_6H_5$	210.26	(al, peth)	35	331			i	v	v				**B7**, 445
p1683	**1-Propanone, 1(2-furyl)-**	2-Propionylfuran. 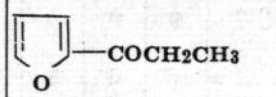 O —COCH2CH3	124.13	cr	28	88^{14}	1.0626^{28}	1.4984^{28}			s				**B17**^{1}, 157
p1684	**2-Propanone, hexachloro-***	Perchloroacetone. $Cl_3CCOCCl_3$	264.77		−3	202–4 110^{40}	1.744^{12}_{12}		δ d^h			s			**B1**, 657
p1685	**—,1-hydroxy-***	Acetol. CH_3COCH_2OH	74.08		−17	145–6d 59^{18}	1.0824^{20}_{20}	1.4295^{20}	∞	∞	∞				**B1**^{2}, 866
p1686	**1-Propanone, 1(1-hydroxy-2-naphthyl)-***	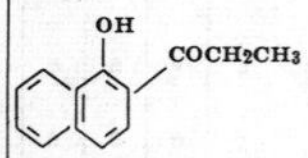 OH COCH2CH3	200.23	grsh ye lf or pl (al)	81				i	s	s				**B8**, 152
p1687	**—,1(2-hydroxy-phenyl)-***	$C_9H_{10}O_2$. *See* p1696	150.18			150^{8}			δ	v	v			alk s	**B8**^{2}, 103
p1688	**—,1(4-hydroxy-phenyl)-**	$C_9H_{10}O_2$. *See* p1696	150.18	wh nd or pr (w)	149				δ v^h	v	v				**B8**^{2}, 104
p1689	**2-Propanone, 1-iodo-***	Iodoacetone. CH_3COCH_2I	183.98	yesh		58.4^{11}	2.17^{15}			s					**B1**^{2}, 719
p1690	**—,—,oxime***	Iodoacetoxime. $CH_3C(:NOH)CH_2I$	198.99	pr (peth)	64.5									peth s^h	**B1**, 660
p1691	**—,1(3-methoxy-phenyl)-***	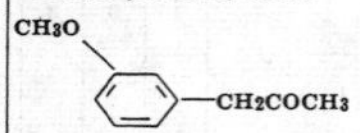 CH3O —CH2COCH3	164.21			258–60	1.0812^{0}			s	s				**B8**, 106
p1692	**—,1(4-methoxy-phenyl)-***	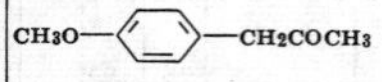 CH3O— —CH2COCH3	164.21		<−15	267–9 $136–7^{14}$	1.0707^{17}_{17}	1.5253^{20}	s	s	s				**B8**^{2}, 105
p1693	**1-Propanone, 2-methyl-1-phenyl-***	Isobutyrophenone. $(CH_3)_2CHCOC_6H_5$	148.21			221	0.9863^{17}_{4}	1.5192^{17}	i	s	s			cyclo-hexanes	**B7**^{2}, 245
p1694	**2-Propanone, pentachloro-***	$Cl_2CHCOCCl_3$	230.31	(w+4)	21 (anh) 15–7 (hyd)	192^{753} 97.5–8.5^{40}	1.69^{15}_{15}		i			v			**B1**, 656
p1695	**—,1-phenoxy-***	$CH_3COCH_2OC_6H_5$	150.18			229–30									**B6**, 151
p1696	**1-Propanone, 1-phenyl-***	Propiophenone. 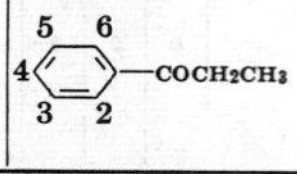 5 6 4 —COCH2CH3 3 2	134.18		18.6	218 $115–20^{21}$	1.0105^{20}	1.5269^{20}	i	s	s				**B7**^{2}, 231

For explanations, symbols and abbreviations see beginning of table.

No	Name	Synonyms and Formula	Mol. wt.	Crystalline form, color and specific rotation	m.p. °C	b.p. °C	Density	n_D	Solubility w	al	eth	ace	bz	other solvents	Ref.
	1-Propanone														
p1697	—,—,oxime*	$C_6H_5C(:NOH)CH_2CH_3$	149.19	lf	53–5	165^{38}			i	s	s				$B7^2$, 232
p1698	**2-Propanone, 1-phenyl-***	Acetonylbenzene. $CH_3COCH_2C_6H_5$	134.18		27	216.5 $104–6^{13}$	1.0157^{20}_{4}	1.5168^{20}	i	v	v		∞	xyl ∞	$B7^2$, 233
p1699	—,**1,1,1,3-tetra-chloro-***	$ClCH_2COCCl_3$	195.86	liq (anh), (w+4)	46 (hyd)	183 $71–2^{13}$	1.624^{15}_{4}	1.497^{18}	δ		v	v			**B1**, 656
p1700	—,**1,1,3,3-tetra-chloro-***	$Cl_2CHCOCHCl_2$	195.86			$180–2^{718}$ cor			v	v	v	v	v		$B1^1$, 345
p1701	—,—,dihydrate*	$Cl_2CHCOCHCl_2.2H_2O$	231.89	tcl pl	48–9										**B1**, 656
p1702	**1-Propanone, 1(3-tolyl)-***	CH_3 —$COCH_2CH_3$	148.21			229–30	1.0059^{0}_{4}		i	s	s	s	s		$B7^2$, 246
p1703	—,**1(4-tolyl)-***	CH_3— —$COCH_2CH_3$	148.21			238–9 106^{8}	0.9926^{20}_{4}	1.5278^{20}	i	s	s	s	s	CS_2 s	$B7^2$, 246
p1704	**2-Propanone, 1,1,1-trichloro-***	CH_3COCCl_3	161.42			134 60^{54}	1.432^{20}_{4}		i	v	v				$B1^2$, 719
p1705	**1-Propanone, 1,3,3-triphenyl-***	$(C_6H_5)_2CHCH_2COC_6H_5$	286.38	nd (al)	96					δ v^h	δ	v	v	chl v lig δ	$B7^2$, 485
p1706	**2-Propanone, 1-ureido-***	Acetonylurea. $CH_3COCH_2NHCONH_2$	116.12	pr	–41	82	0.8018^{4}		s	s	s				
—	**Propargyl alcohol**	*see* **2-Propyn-1-ol***													
—	**Propargyl aldehyde**	*see* **Propynal***													
p1707	**Propenal***	Acrolein. Acrylaldehyde. $CH_2:CHCHO$	56.06		–87.7	52.5–3.5	0.8625^{20}_{0}	1.3998^{20}	v	s	s				$B1^2$, 782
—	—,diacetate	*see* **2-Propene-1,1-diol, diacetate**													
p1708	—,diethyl acetal	3,3-Diethoxy-2-propene*. $CH_2:CHCH(OC_2H_5)_2$	130.18			123.5	0.8543^{15}	1.3983^{25}	δ	∞	∞				$B1^2$, 785
p1709	—,**2-chloro-***	$CH_2:CClCHO$	90.51			40^{30}	1.199^{20}	1.463^{20}			v			CCl_4 v	$B1^2$, 785
p1711	—,**3(2-furyl)-***	Furacrolein. —CH:CHCHO	122.12	ye or wh nd (lig)	51–2	>200d $94–5^{9}$			i s^h	∞	s				$B17^2$, 325
p1712	—,**2-methyl-***	Methacrolein. $CH_2:C(CH_3)CHO$	70.09			68.4	0.830^{20}_{4}	1.4191^{20}	∞	∞	∞				**B1**, 731
—	—,**3-phenyl-***	*see* **Cinnamaldehyde**													
p1713	**Propene***	Propylene. $CH_3CH:CH_2$	42.08	gas	–185.2	–47.8			v	v				aa v	**B1**, 196
p1714	—,**3-amino-***	Allylamine. $H_2NCH_2CH:CH_2$	57.09			58	0.7613^{22}_{4}	1.4194^{22}	∞	∞	∞			chl s	$B4^1$, 622
p1715	—,**1-bromo-***	Propenyl bromide. $CH_3CH:CHBr$	120.99		–116.6	$58–60^{747}$	1.4133^{20}_{4}	1.4519^{20}	i						**B1**, 200
p1716	—,**2-bromo-***	Isopropenyl bromide. $CH_3CBr:CH_2$	120.99		–124.8	48.4	1.362^{20}_{4}	1.4467^{16}							**B1**, 200
p1717	—,**3-bromo-***	Allyl bromide. $BrCH_2CH:CH_2$	120.99		–119.4	70	1.398^{20}_{4}	1.4697^{20}	i	∞	∞			chl, CS_2, CCl_4 s	$B1^3$, 711
p1718	—,**2-bromo-3-cyclohexyl-***	—$CH_2CBr:CH_2$	203.14			84^{10}	1.215^{17}	1.495^{17}			s				$B5^2$, 49
p1719	—,**1-chloro-(*cis*)***	*cis*-Propenyl chloride. $CH_3CH:CHCl$	76.53		–134.8	32.8									**B1**, 198
p1720	—,—(*trans*)*	*trans*-Propenyl chloride. $CH_3CH:CHCl$	76.53		–99	37.4									**B1**, 198
p1721	—,**2-chloro-***	Isopropenyl chloride. $CH_3CCl:CH_2$	76.53		–138.6	23^{738}	0.918^{9}	1.404^{6}							**B1**, 198
p1722	—,**3-chloro-***	Allyl chloride. $ClCH_2CH;CH_2$	76.53		–134.5	45	0.9397^{20}_{4}	1.4154^{20}	i	∞	∞	∞	∞	lig ∞	$B1^3$, 699
p1723	—,**3-chloro-2-(chloromethyl)-***	$(ClCH_2)_2C:CH_2$	125.00		–15, –13	138–8.5 $30–1^{9}$	1.1782^{20}_{4}	1.4754^{20}		v				chl v	$B1^3$, 768
p1724	—,**1-chloro-2-methyl-***	Isocrotyl chloride. $(CH_3)_2C:CHCl$	90.55			$68–9^{775}$	0.925^{16}	1.4221		∞	∞				**B1**, 209
p1725	—,**3-chloro-2-methyl-***	Isobutenyl chloride. $ClCH_2C(CH_3):CH_2$	90.55			71.5–2.5	0.925^{20}	1.427		∞	∞				**B1**, 209

For explanations, symbols and abbreviations see beginning of table.

No.	Name	Synonyms and Formula	Mol. wt.	Crystalline form, color and specific rotation	m.p. °C	b.p. °C	Density	n_D	Solubility w	al	eth	ace	bz	other solvents	Ref.
	Propene														
p1726	**—,1-chloro-1-phenyl-***	$C_6H_5CCl{:}CHCH_3$	152.63			90.5^{9}	1.085^{20}_{4}	1.5635^{15}					s	to s	**B5**^{2}, 372
p1727	**—,1-chloro-3-phenyl-***	$C_6H_5CH_2CH{:}CHCl$	152.63			212–4 cor 76^{13}	1.073^{14}_{4}	1.545^{14}	i		s			to s	**B5**^{2}, 473
p1728	**—,2-chloro-1-phenyl-***	$CH_3CCl{:}CHC_6H_5$	152.63			$118–23^{28}$	1.0738^{19}_{4}	1.5565^{19}					s	chl v	
p1729	**—,3-chloro-1-phenyl-***	Cinnamyl chloride. $ClCH_2CH{:}CHC_6H_5$	152.63		−19	213–5 108^{12}			i	∞	∞				**B5**^{2}, 372
p1730	**—,2,3-dibromo-***	α-Epidibromohydrin. $BrCH_2CBr{:}CH_2$	199.85			139–40 $73–6^{75}$	1.934^{20}_{4}								**B1**, 201
p1731	**—,1,1-dichloro-***	$CH_3CH{:}CCl_2$	110.98			78	$1.1764^{19.5}_{0}$		i					chl s	**B1**, 199
p1732	**—,1,2-dichloro-***	$CH_3CCl{:}CHCl$	110.98			77.0^{757}	1.1818^{20}_{4}	1.4471^{20}	i	v				MeOH, CCl_4 v	**B1**, 199
p1733	**—,1,3-dichloro-** *(cis)***	$ClCH_2CH{:}CHCl$	110.98			112	1.217^{20}_{4}	1.4730^{20}	i		s			chl s	
p1734	—,—*(trans)**	$ClCH_2CH{:}CHCl$	110.98			104.3	1.224^{20}_{4}	1.4682^{20}	i		s			chl s	
p1735	**—,2,3-dichloro-***	$ClCH_2CCl{:}CH_2$	110.98				1.204^{25}_{25}	1.4600^{21}	i	∞	s				**B1**, 199
p1736	**—,3,3-dichloro-***	Acrolein dichloride. $Cl_2CHCH{:}CH_2$	110.98			84.4 cor	1.170^{24}_{24}	1.450^{21}		v	v				**B1**, 199
p1737	**—,1,1-dichloro-2-methyl-***	$(CH_3)_2C{:}CCl_2$	125.01			108.7–9.1 $42–3^{75}$	1.449^{20}_{0}	1.4580^{20}						chl s	**B1**^{3}, 767
p1738	**—,3,3-dichloro-2-methyl-***	$Cl_2CHC(CH_3){:}CH_2$	125.01			$108–12^{762}$ $49–50^{120}$	1.1363^{24}_{4}		i					chl v	
p1739	**—,1,1-dichloro-3-phenyl-***	Cinnamylidene chloride. $C_6H_5CH{:}CHCHCl_2$	187.07	(eth, chl), pl (peth)	58–9	$142–3^{30}$			d		s			chl s	**B5**^{2}, 372
—	**—,3,3-diethoxy-***	*see* **Propenal**, diethyl acetal													
p1741	**—,1,1-diphenyl-***	$CH_3CH{:}C(C_6H_5)_2$	194.26	lf (al)	52	280–1	0.9841	1.593^{23}	i	s			s		**B5**, 644
p1742	**—,1,2-epoxy-***	Allylene oxide. Methyloxirene. $CH_3C{=}CH$ (with O bridging)	56.06			63			δ	∞	∞				**B17**, 20
p1743	**—,3-fluoro-***	Allyl fluoride. $FCH_2CH{:}CH_2$	60.07	gas		−3			δ	v	v				**B1**, 198
p1744	**—,hexafluoro-***	Perfluoropropene. $CF_3CF{:}CF_2$	150.03	gas	−156.2	−29.4	1.583^{-40}_{4}								**B1**^{3}, 697
p1745	**—,3-iodo-***	Allyl iodide. $ICH_2CH{:}CH_2$	167.99	pa ye	−99.3	102–3	1.8454^{22}_{4}	1.5540^{21}	i	s	s			chl s	**B1**^{3}, 114
p1746	**—,3-isocyano-**	Allyl carbylamine. Allyl isocyanide. $CNCH_2CH{:}CH_2$	67.09			98	0.794^{17}_{4}		δ	∞	∞				**B4**, 208
p1747	**—,2-methyl-***	Isobutylene. $(CH_3)_2C{:}CH_2$	56.10	gas		−6.6			i	v	v			sulf s	**B1**, 207
p1748	—,—*(trimer)*	Triisobutylene. $(C_4H_8)_3$	168.31		−76	179–81 56^{10}	0.7590^{20}_{4}	1.4314^{20}							**B1**^{2}, 180
p1749	—,—(tetramer)	Tetraisobutylene. $(C_4H_8)_4$	224.42			109.5^{15}	0.7944^{20}_{4}	1.4482^{20}							**B1**^{2}, 181
p1750	**—,2-methyl-1,1,3-trichloro-***	$ClCH_2C(CH_3){:}CCl_2$	159.45			156 $45–6^{12}$	1.4965^{25}_{4}					v		chl v	**B1**^{3}, 768
p1751	**—,2-methyl-3,3,3-trichloro-***	$Cl_3CC(CH_3){:}CH_2$	159.45			132	1.293^{20}	1.479^{20}	i					chl v	
p1752	**—,2-nitro-1-phenyl-***	$CH_3C(NO_2){:}CHC_6H_5$	163.17	ye nd (peth)	64				i	v	δ			aa s alk i	**B5**, 483
p1753	**—,1,1,2,3,3-pentachloro-***	$Cl_2CHCCl{:}CCl_2$	214.33			183 116^{9}	1.6317^{34}_{4}	1.5313^{20}	i					to v^{h}	**B1**^{2}, 83
p1754	**—,2-phenyl-***	α-Methylstyrene. $C_6H_5C(CH_3){:}CH_2$	118.18			161–2 $60–1^{17}$	0.9165^{10}_{4}	1.5311^{24}	i				s		**B5**^{2}, 374
—	**—,3-phenyl-***	*see* **Benzene, allyl-***													
p1755	**—,1,1,2-trichloro-***	$CH_3CCl{:}CCl_2$	145.43			118 41^{52}	1.387^{14}_{14}	1.4827^{20}						chl s	**B1**^{3}, 707

For explanations, symbols and abbreviations see beginning of table.

No.	Name	Synonyms and Formula	Mol. wt.	Crystalline form, color and specific rotation	m.p. °C	b.p. °C	Density	n_D	Solubility w	al	eth	ace	bz	other solvents	Ref.
	Propene														
p1756	—,**1,2,3-trichloro-***	$ClCH_2CCl:CHCl$	145.43	...	...	142	1.414^{20}_{20}	...	i	v	v	...	...	...	**B1**, 200
p1757	—,**3,3,3-trichloro-***	$Cl_3CCH:CH_2$	145.43	...	−30	114–5 57^{103}	1.369^{20}_{20}	1.4827^{20}	...	...	...	...	...	chl s	$\mathbf{B1^{3}}$, 707
p1758	**2-Propene-1-arsonic acid***	Allylarsonic acid. $CH_2:CHCH_2AsO_3H_2$	166.00	nd (w), pr (al)	129–30	...			v^h	v^h	...	...	...	...	$\mathbf{B4^{2}}$, 998
p1758^{1}	**2-Propene-1,1-diol,** diacetate	Allylidene diacetate. $CH_2:CHCH(O_2CCH_3)_2$	158.15	...	−37.6	180 107^{50}	1.0749^{20}_{20}	1.4193^{20}	δ	∞	∞	...	∞	lig ∞	$\mathbf{B2^{1}}$, 72
p1759	**2-Propene-1-thiol***	Allyl mercaptan. $CH_2:CHCH_2SH$	74.14	...	...	67–8	0.925^{23}_{4}	...	i	∞	∞	...	...	...	**B1**, 440
p1760	**1,2,3-Propenetricarboxylic acid-** *(cis)**	*cis*-Aconitic acid. $HO_2CCH_2C(CO_2H):CHCO_2H$	174.11	nd (w)	125	...			s	...	δ	...	...	...	$\mathbf{B2^{2}}$, 694
p1761	—*(trans)**...	$HO_2CCH_2C(CO_2H):CHCO_2H$.	174.11	lf (w), nd (w, eth)	198–9	...			v	s	s	...	...	...	$\mathbf{B2^{2}}$, 693
p1762	—,triamide*...	Aconitamide. $H_2NOCCH_2C(CONH_2):CHCONH_2$	171.16	ye nd (w)	260d	...			v^h	i	i	...	...	chl i	**B2**, 853
p1763	—,triethyl ester- *(trans)**	$C_2H_5O_2CCH_2C(CO_2C_2H_5):CHCO_2C_2H_5$	258.27		...	275d 159^{9}	1.1064^{20}_{4}	1.4556^{20}	...	v	...	...	...	...	$\mathbf{B2^{2}}$, 694
p1764	—,trimethyl ester- *(trans)**	$CH_3O_2CCH_2C(CO_2CH_3):CHCO_2CH_3$	216.19	...	...	270–1 160^{20}		...	...	s	s	...	...	...	$\mathbf{B2^{2}}$, 694
p1765	—,tripropyl ester*..	$C_3H^n_7O_2CCH_2C(C_3H^n_7):CHCO_2C_3H^n_7$	300.35	...	...	195^{13}		...	i	s	s	...	...	...	**B2**, 852
p1766	**Propenoic acid***..	Acrylic acid. $CH_2:CHCO_2H$..	72.06	...	12–3	141.6 48.5^{15}	1.0511^{20}_{4}	1.4224^{20}	∞	∞	∞	...	...	...	$\mathbf{B2^{2}}$, 383
p1767	—,allyl ester...	Ally acrylate. $CH_2:CHCO_2CH_2CH:CH_2$	112.13	...	...	122 72^{27}	1.0452^{20}_{4}	1.4390^{9}	δ	...	...	...	...	ac s	$\mathbf{B2^{3}}$, 388
p1768	—,amide...	Acrylamide. Propenamide*. $CH_2:CHCONH_2$	71.08	lf (bz),	84–5	...		...	v	...	s	...	...	chl v	$\mathbf{B2^{2}}$, 388
p1769	—,benzyl ester*...	Benzyl acrylate. $CH_2:CHCO_2CH_2C_6H_5$	162.19	...	...	228 94^{6}	1.0573^{20}_{4}	1.5143^{20}	i	s	s	...	...	...	$\mathbf{B6^{2}}$, 418
p1770	—,butyl ester*...	*n*-Butyl acrylate. $CH_2:CHCO_2C_4H^n_9$	128.17	...	...	145–6 39^{10}	0.920^{0}_{9}	1.4185^{20}	i	s	s	...	...	...	$\mathbf{B2^{2}}$, 388
p1771	—,chloride...	Acrylyl chloride. $CH_2:CHCOCl$	90.51	...	...	75–6	1.14^{0}	1.4343^{20}	d	...	...	...	...	chl v	$\mathbf{B2^{2}}$, 388
p1772	—,cyclohexyl ester*	Cyclohexyl acrylate. $CH_2:CHCO_2$—(cyclohexyl ring)	154.21	...	...	182–4 88^{20}	1.0275^{20}_{4}	1.4673^{20}	i	∞	∞	...	...	...	...
p1773	—,ethyl ester*...	Ethyl acrylate. $CH_2:CHCO_2C_2H_5$	100.12	...	...	99.8	0.924^{20}_{4}	1.405^{20}	s	∞	∞	...	...	...	$\mathbf{B2^{2}}$, 386
p1774	—,isobutyl ester*..	Isobutyl acrylate. $CH_2:CHCO_2CH_2CH(CH_3)_2$	128.17	...	...	70^{60}	0.8896^{20}_{4}	1.4150^{20}	...	...	...	...	...	MeOH s	...
p1775	—,methyl ester*...	Methyl acrylate. $CH_2:CHCO_2CH_3$	86.09	...	...	80.5	0.9558^{18}_{4}	1.3984^{20}	δ	s	s	...	...	...	$\mathbf{B2^{2}}$, 385
p1776	—,nitrile...	Acrylonitrile. Vinyl cyanide. $CH_2:CHCN$	53.06	...	−82	77.5–9	0.8060^{20}_{4}	1.393^{20}	s v^h	∞	∞	...	...	...	...
p1777	—,**2-chloro-***...	$CH_2:CClCO_2H$	106.51	nd	65	176–81d	...	...	s	s	s	...	...	...	**B2**, 401
p1778	—,—,ethyl ester*..	$CH_2:CClCO_2C_2H_5$	134.56	...	...	$51–3^{18}$	...	...	...	v	v	...	...	...	...
p1779	—,—,methyl ester*.	$CH_2:CClCO_2CH_3$	120.54	...	...	$57–9^{55}$	1.189^{20}_{4}	1.4420^{20}	...	...	v	...	...	...	...
p1780	—,**3-chloro-***(cis)**	$ClCH:CHCO_2H$	106.51	lf	63–4	...	...	...	...	s	s	...	...	...	$\mathbf{B2^{1}}$, 186
p1781	—,—*(trans)**...	$ClCH:CHCO_2H$	106.51	lf	86	...	...	...	...	s	s	...	...	...	$\mathbf{B2^{1}}$, 186
p1782	—,**2,3-dichloro-***.	$ClCH:CClCO_2H$	140.96	mcl pr (chl), (peth, CS_2)	85–6	...	...	...	v	v	v	...	δ	chl v CS_2 δ	$\mathbf{B2^{1}}$, 186
p1783	—,**3,3-dichloro-***..	$Cl_2C:CHCO_2H$	140.96	nd (peth, sub)	76–7	sub	...	...	δ	...	v	...	...	chl v	**B2**, 401
p1783^{1}	—,**2,3-diphenyl-***.	α-Phenylcinnamic acid. $C_6H_5CH:C(C_6H_5)CO_2H$	224.26	nd (lig, dil al)	172.5–3	sub	...	...	δ^h	s	s	...	...	...	$\mathbf{B9^{2}}$, 482

For explanations, symbols and abbreviations see beginning of table.

No.	Name	Synonyms and Formula	Mol. wt.	Crystalline form, color and specific rotation	m.p. °C	b.p. °C	Density	n_D	Solubility w	al	eth	ace	bz	other solvents	Ref.
	Propenoic acid														
p1784	—,**3(2-furyl)-** (labile form)*	(furan ring, O)—$CH:CHCO_2H$	138.12	wh pr or ta	103–4	sub vac	...	...	δ v^h	...	...	...	s	...	**B18**, 301
p1785	—,—(stable form)*	$C_7H_6O_3$. *See* p1784	138.12	nd (w)	141	286 (sub vac)	...	1.5286^{20}	s^h δ	s	s	...	δ	...	**B18²**, 273
p1786	—,—,benzyl ester	$C_{14}H_{12}O_3$. *See* p1784	228.25	pa ye	42–3	$201–3^{12}$	...	1.5872^{25}	i	s^h	s	s	s	...	**B18²**, 274
p1787	—,—,butyl ester*	$C_{11}H_{14}O_3$. *See* p1784	194.23	...	...	$117–8^{3}$	1.045^{20}	1.5129^{20}	i	s	...	...	...	...	...
p1788	—,—,ethyl ester*	$C_9H_{10}O_3$. *See* p1784	166.18	pa ye	24.5	232, 120^{17}	1.0891_4^{25}	1.5286^{20}	i	∞	∞	...	...	...	**B18¹**, 440
p1789	—,—,methyl ester*	$C_8H_8O_3$. *See* p1784	152.15	...	27	$227–8^{774}$, 112^{15}	...	1.4447^{20}	i	v	v	...	v	...	**B18¹**, 440
p1790	—,—,pentyl ester*	$C_{12}H_{16}O_3$. *See* p1784	208.26	...	...	$116.5–8^{2}$	1.0322_4^{20}	1.5289^{24}	i	...	...	...	...	...	...
p1791	—,—,propyl ester*	$C_{10}H_{12}O_3$. *See* p1784	180.21	...	...	$91–4^{3}$	1.0744_4^{20}	1.5392^{24}	i	s	...	...	...	...	...
p1792	—,**2-methyl-***	Methacrylic acid. $CH_2:C(CH_3)CO_2H$	86.09	...	16	$162–3^{757}$	1.0153_4^{20}	1.4314^{20}	s v^h	∞	∞	...	...	...	**B2²**, 398
p1793	—,—,amide	Methacrylamide. $CH_2:C(CH_3)CONH_2$	85.11	(bz)	109–10	...	...	...	...	s	δ s^h	...	...	...	**B2²**, 399
p1794	—,—,anhydride*	Methacrylic anhydride. $[CH_2:C(CH_3)CO]_2O$	154.17	...	...	89^{5}	...	1.4540^{20}	d	∞	∞	...	...	...	...
p1795	—,—,butyl ester*	Butyl methacrylate. $CH_2:C(CH_3)CO_2(CH_2)_3CH_3$	142.20	...	...	163	0.895_4^{20}	1.426^{16}	i	∞	∞	...	...	...	...
p1796	—,—,ethyl ester*	Ethyl methacrylate. $CH_2:C(CH_3)CO_2C_2H_5$	114.15	...	...	118	0.907_4^{20}	1.414^{20}	δ	∞	∞	...	...	...	**B2²**, 399
p1797	—,—,isobutyl ester*	Isobutyl methacrylate. $CH_2:C(CH_3CO_2)CH_2CH_3(CH_3)_3$	142.20	...	...	155	0.889_{16}^{16}	0.418^{24}	i	∞	∞	...	...	...	...
p1798	—,—,isopropyl ester*	Isopropyl methacrylate. $CH_2:C(CH_3)CO_2CH(CH_3)_2$	128.17	...	...	...	0.890_4^{20}	1.412^{24}	i	∞	∞	...	∞	...	...
p1799	—,—,methyl ester*	Methyl methacrylate. $CH_2:C(CH_3)CO_2CH_3$	100.12	...	ca. −50	100	0.943_4^{20}	1.413^{20}	δ	∞	∞	...	...	...	**B2²**, 398
p1800	—,—,nitrile	Methacrylonitrile. $CH_2:C(CH_3)CN$	67.09	...	−40	90–2	0.7998_4^{20}	1.4013^{29}	i	∞	∞	...	...	...	**B2²**, 399
p1801	—,—,propyl ester*	Propyl methacrylate. $CH_2:C(CH_3)CO_2CH_2CH_2CH_3$	128.17	...	...	141	0.902_{16}^{16}	1.420^{16}	i	∞	∞	...	...	...	...
p1802	—,**3(1-naphthyl)-** (*cis*)*	$C_{10}H_7^{\alpha}CH:CHCO_2H$	198.22	...	154–6	...	...	...	...	s	...	...	...	...	**E12B**, 3287
p1803	—,—(*trans*)*	$C_{10}H_7^{\alpha}CH:CHCO_2H$	198.22	nd (al, w)	211–2	...	...	...	i	s	v	...	...	chl s	**E12B**, 3287
p1804	—,**3(2-naphthyl-***	$C_{10}H_7^{\beta}CH:CHCO_2H$	198.22	nd (al, w)	208–9	...	...	...	s^h	s	...	...	...	...	**E12B**, 3289
p1805	—,**3(2-nitrophenyl)-2-phenyl-(*trans*)***	(NO_2 on benzene ring)—$CH:C(C_6H_5)CO_2H$	269.26	ye cr (al)	196–7	...	...	...	δ^h	s^h	s	...	s	chl δ s^h	**B9²**, 483
p1806	—,**3(3-nitrophenyl)-2-phenyl-(*cis*)***	(O_2N on benzene ring)—$CH:C(C_6H_5)CO_2H$	269.26	nd (al, aa)	195–6	...	...	...	...	δ s^h	...	...	...	...	**B9²**, 483
p1807	—,—(*trans*)	$C_{15}H_{11}NO_4$. *See* p1806	269.26	ye pr (eth), nd (dil al)	182	...	...	...	...	s v^h	s	...	s	CS_2 s, chl s	**B9²**, 483
p1808	—,**3(4-nitrophenyl)-2-phenyl-(*cis*)***	O_2N—(benzene ring)—$CH:C(C_6H_5)CO_2H$	269.26	ye pr (dil al+1w), (abs al+½), lf (bz+½)	144	...	...	...	...	v	...	...	s^h	...	**B9²**, 484
p1809	—,—,(*trans*)*	$C_{15}H_{11}NO_4$. *See* p1809	269.26	ye pr or nd (al)	213–4	...	...	...	δ	s^h	δ	...	δ	chl δ	**B9**, 696
p1810	—,**2-phenyl-***	Atropic acid. $CH_2:C(C_6H_5)CO_2H$	148.16	lf (al), nd (w)	106–7	...	...	267d	δ	s	s	...	s	chl, CS_2 s	**B9²**, 407
—	—,**3-phenyl-***	*see* **Cinnamic acid**													
p1811	—,**trichloro***	$Cl_2C:C(Cl)CO_2H$	175.40	pr (CS_2, eth)	76	221–3, 133^{30} cor	...	...	v^h	s	s	...	...	CCl_4 s	**B2²**, 388
p1812	—,—,chloride	$Cl_2C:C(Cl)COCl$	193.84	...	...	158	...	1.5271^{18}	d	...	...	...	s	CS_2 v	**B2¹**, 187
p1813	—,**triphenyl-**, nitrile	$(C_6H_5)_2C:C(C_6H_5)CN$	281.36	nd (al), pr	166–7	...	...	...	...	s^h	s	...	...	...	**B9²**, 507

For explanations, symbols and abbreviations see beginning of table.

No.	Name	Synonyms and Formula	Mol. wt.	Crystalline form, color and specific rotation	m.p. °C	b.p. °C	Density	n_D	Solubility: w	al	eth	ace	bz	other solvents	Ref.
	2-Propen-1-ol														
p1814	**2-Propen-1-ol***	Allyl alcohol. $CH_2:CHCH_2OH$	58.08		−129	97	0.8540^{20}_{4}	1.4135^{20}	∞	∞	∞				**B1**[2], 474
p1815	**—,2-bromo-***	$CH_2:CBrCH_2OH$	196.98			$153–4^{755}$	1.6^{15}								**B1**, 439
p1816	**—,2-chloro-***	$CH_2:CClCH_2OH$	92.53			136–40 47^{10}	1.1618^{20}_{4}	1.4588^{20}	d^h	s^h					**B1**[3], 1887
p1817	**—,3-chloro-***	$ClCH:CHCH_2OH$	92.53			153	1.162		δ						**B1**, 439
p1818	**—,3(3,5-dimethoxy-4-hydroxyphenyl)-***	Sinapyl alcohol. Syringenin. (3,5-$(CH_3O)_2$-4-HO-C_6H_2)—CH:CHCH₂OH	210.23	pl	66–7				i		s				**B6**[2], 1124
—	**—,3(4-hydroxy-3-methoxyphenyl)-***	*see* **Coniferyl alcohol**													
p1819	**—,3(4-hydroxyphenyl)-***	*p*-Coumaryl alcohol. $HO-C_6H_4-CH:CHCH_2OH$	150.18	pr	124				δ	s	s	s	s	to s MeOH s	
p1820	**—,2-methyl-***	Methallyl alcohol. $CH_2:C(CH_3)CH_2OH$	72.11			114.49	0.8574^{19}	1.4255	v	∞	∞				**B1**, 443
p1821	**—,1-phenyl-***	α-Vinylbenzyl alcohol. $CH_2:CHCH(C_6H_5)OH$	134.18			215–6 $56^{0.05}$	1.0251^{21}_{0}	1.5410^{16}	i	s	s		s	chl s	**B6**[2], 529
p1822	**—,3-phenyl-***	Cinnamic alcohol. Cinnamyl alcohol. $C_6H_5CH:CHCH_2OH$	134.18	wh nd	33	257.5 $127–8^{10}$	1.0440^{20}_{4}	1.5819^{20}	δ	v	v				**B6**[2], 525
p1823	**—,—,acetate**	$C_6H_5CH:CHCH_2O_2CCH_3$	176.22			$145–6^{15}$	1.0566^{20}	1.5425^{20}	i				s		**B6**[2], 527
p1824	**2-Propen-1-one, 3(2-furyl)1-phenyl-***	(2-furyl)—CH:CHCO—C_6H_5	198.22			317 $181–2^{9}$	1.1140^{20}			s	s				**B17**[2], 377
p1825	**—,1(2-hydroxynaphthyl)-***	(2-HO-naphthyl)—COCH:CH—C_6H_5	274.32	og-red lf (al)	129				i	s	s			con sulf s	**B8**[2], 246
—	**Propioin**	*see* **3-Hexanone, 4-hydroxy-***													
—	**Propiolactone**	*see* **Propanoic acid, 3-hydroxy-, β-lactone***													
—	**Propiolaldehyde**	*see* **Propynal***													
—	**Propiolic acid**	*see* **Propyonic acid***													
—	**Propionaldehyde**	*see* **Propanol***													
—	**Propionaldol**	*see* **Pentanal, 3-hydroxy-2-methyl-***													
—	**Propionamide**	*see* **Propanoic acid, amide***													
—	**Propionic acid**	*see* **Propanoic acid***													
—	**Propiophenone**	*see* **1-Propanone, 1-phenyl-***													
—	**Propylal**	*see* **Propanol, diethyl acetal**													
—	**Propyl alcohol**	*see* **1-Propanol***													
—	**Propyl bromide**	*see* **Propane, 1-bromo-***													
—	**Propyl chloride**	*see* **Propane, 1-chloro-***													
—	**Propylene**	*see* **Propene***													
—	**Propylene glycol**	*see* **1,2-Propanediol***													
—	**Propylene oxide**	*see* **Propane, 1,2-epoxy-***													
p1826	**Propylhexedrine**	(cyclohexyl)—$CH_2CH(NHCH_3)CH_3$	155.29			206 84.5^{11}	0.850^{20}_{4}	1.4588^{24}	δ	s				dil ac s	
—	**Propyl iodide**	*see* **Propane, 1-iodo-***													
p1827	**Propyl red**	(2-CO_2H-C_6H_4)—N:N—C_6H_4—$N(CH_2CH_2CH_3)_2$	325.41	vt-bl or purp-red cr (al)						s^h				KOH s	**B16**[2], 165
p1828	**Propynal***	Propargylaldehyde. Propiolaldehyde. $HC\vdots CCHO$	54.05			59–61			v						**B1**, 750
p1829	—,diethyl acetal	3,3-Diethoxypropyne. $CH\vdots CCH(OC_2H_5)_2$	128.17			130.45 37^{11}	0.894^{22}	1.414^{22}			s				**B1**[2], 808 **B4**[2], 1135
p1830	**—,1-phenyl-***	$C_6H_5C\vdots CCHO$	130.15			$127–8^{28}$	1.0639^{16}_{4}	1.6079^{18}							**B7**[2], 317

For explanations, symbols and abbreviations see beginning of table.

No.	Name	Synonyms and Formula	Mol. wt.	Crystalline form, color and specific rotation	m.p. °C	b.p. °C	Density	n_D	Solubility w	al	eth	ace	bz	other solvents	Ref.
	Propyne														
p1831	**Propyne***	Methylacetylene. Propine. $CH_3C{\vdots}CH$	40.07	gas	−102.7	−23.2	0.7062^{-50}_{4}	1.3863^{-40}	...	v	...	...	...	...	B1^2, 222
p1832	—,3-bromo-*	Propargyl bromide. $BrCH_2C{\vdots}CH$	118.97	...	...	88–90	1.579^{19}	1.4942	...	...	...	...	...	...	B1, 223
p1833	—,3-chloro-*	Propargyl chloride. $ClCH_2C{\vdots}CH$	74.51	...	...	65	1.0454^{5}	...	i	∞	∞	...	...	...	B1, 248
p1834	—,3-cyclohexyl-*	(cyclohexyl)—$CH_2C{\vdots}CH$	122.21	...	...	157–8 48^{11}	0.844^{18}	1.4603^{18}	...	...	s	...	...	...	B5^2, 82
p1835	—,1,3-dibromo-*	$BrCH_2C{\vdots}CBr$	197.88	...	...	$73–4^{30}$	2.137	...	i	...	...	...	...	...	B1, 248
—	—,3,3-diethoxy-*	see **Propynal,** diethyl acetal													
p1836	—,1-iodo-*	$CH_3C{\vdots}CI$	165.97	nd (w)	93–4	$110–2^{757}$	2.08^{22}	...	δ	v	v	...	...	...	B1, 248
p1837	—,3-iodo-*	Propargyl iodide. $ICH_2C{\vdots}CH$	165.97	...	...	115	2.0177^{0}	...	...	...	s	...	...	...	B1, 48
p1838	—,3-methoxy-*	Methyl propargyl ether. $CH_3OCH_2C{\vdots}CH$	70.09	...	...	63	0.83^{12}	1.5035^{20}	δ	∞	∞	...	...	...	B1^2, 504
p1839	—,1-phenyl-*	Phenylallylene. $CH_3C{\vdots}CC_6H_5$	116.15	...	...	183	0.942^{15}	1.563^{15}	i	...	s	...	...	...	B1^2, 408
p1840	—,3,3,3-trifluoro-*	$F_3CC{\vdots}CH$	94.04	gas	...	−48.3	...	...	...	...	...	...	...	...	...
p1841	**Propynoic acid***	Propargylic acid. Propiolic acid. $HC{\vdots}CCO_2H$	70.05	...	18	144d	1.1380^{20}_{4}	1.4306^{20}	v	v	s	...	...	chl s	B2, 477
p1842	—,ethyl ester*	$HC{\vdots}CCO_2C_2H_5$	98.10	...	...	119^{745}	0.968^{15}_{15}	1.4433^{15}	i	v	v	...	...	chl v	B2, 477
p1843	—,chloro-*	$ClC{\vdots}CCO_2H$	104.50	(peth)	69–70	...	...	...	...	...	v	...	...	peth δ, v^h	B2^2, 451
p1844	—,(2-chlorophenyl)-*	$C_9H_5ClO_2$. *See* p1849	180.60	(50 % aa, bz)	131–2	...	...	...	i	...	...	...	v^h	aa v	...
p1845	—,(3-chlorophenyl)-*	$C_9H_5ClO_2$. *See* p1849	180.60	(aa)	141–2	...	...	...	...	...	...	...	...	aa v	...
p1846	—,(4-chlorophenyl)-*	$C_9H_5ClO_2$. *See* p1849	180.60	(aa)	147	...	...	...	...	...	...	...	...	aa v^h	...
p1847	—,(2-nitrophenyl)-*	$C_9H_5NO_4$. *See* p1849	191.14	nd or lf (w)	157d	exp	...	...	v^h	s	s	...	...	CS_2 i	B9, 637
p1848	—,(4-nitrophenyl)-*	$C_9H_5NO_4$. *See* p1849	191.14	nd (eth, al)	181d	...	...	...	δ	s^h	s	...	...	CS_2 δ peth i	B9, 637
p1849	—,phenyl-*	(phenyl, positions 2–6)—$C{\vdots}CCO_2H$	146.14	nd (w, CS_2)	137	sub	...	...	δ	v	v	...	...	...	B9, 633
p1850	—,—,ethyl ester*	$C_6H_5C{\vdots}CCO_2C_2H_5$	174.19	...	...	260–70δd 144^{13}	1.0550^{25}_{4}	1.5502^{25}	...	...	...	...	...	...	B9, 634
p1851	**2-Propyn-1-ol***	Propargyl alcohol. Propiolic alcohol. $HC{\vdots}CCH_2OH$	56.06	...	−17	114–5	0.9628^{21}	1.4306^{20}	s	∞	∞	...	...	...	B2^1, 234
p1852	—,acetate*	Propargyl acetate. $HC{\vdots}CCH_2O_2CCH_3$	98.10	...	...	124–5	1.0052^{20}_{4}	1.4204^{20}	δ	s	s	...	...	...	B2, 140
p1853	**Prostigmine bromide**	Neostigmine bromide. $(CH_3)_2NCO_2$—(phenyl)—$N(CH_3)_3Br$	303.20	(al-eth)	167d	...	...	...	v	s	...	...	...	...	...
—	**Protoanemonin**	see **2,4-Pentadienoic acid, 4-hydroxy-, lactone***													
—	**Protocatechaldehyde**	see **Benzaldehyde, 3,4-dihydroxy-**													
—	**Protocatechuic acid**	see **Benzoic acid, 3,4-dihydroxy-**													
p1854	**Protopine**	Fumarine. Macleyine. (structure: N—CH_3)	353.36	mcl pr (al-chl)	206–7	...	...	...	i	δ	...	...	...	chl s	B27^1, 568
p1855	**Protoveratridine**	$C_{31}H_{49}NO_9$	579.74	$[\alpha]_D$ −9 (Py)	272–3	...	...	...	...	s^h	i	...	...	aq ac s	...
p1856	**Protoveratrine**	$C_{32}H_{51}NO_{11}$	625.76	pl	245–50	...	...	...	i	s^h	...	...	...	chl s^h	B9, 951
p1857	**Protoveratrine A**	(structure: CH_3; N; CH_3 OH; OH; CH_3; O; OH; OH; $O_2CCHCH_2CH_3$; $CH_3CH_2CCO_2$; OH; O_2CCH_3; CH_3; CH_3)	793.96	wh pr, $[\alpha]^{24}_D$ −8.3 (chl)	255d	...	...	...	i	...	...	...	...	...	...

For explanations, symbols and abbreviations see beginning of table.

No.	Name	Synonyms and Formula	Mol. wt.	Crystalline form, color and specific rotation	m.p. °C	b.p. °C	Density	n_D	Solubility w	al	eth	ace	bz	other solvents	Ref.
	Protoveratrine B														
p1858	**Protoveratrine B**	Neoprotoveratrine. $C_{41}H_{63}NO_{14}$	809.96	wh pr, $[\alpha]^{24}_D -39$ (Py)	255.4–.8				i						
p1859	**Protoverine**	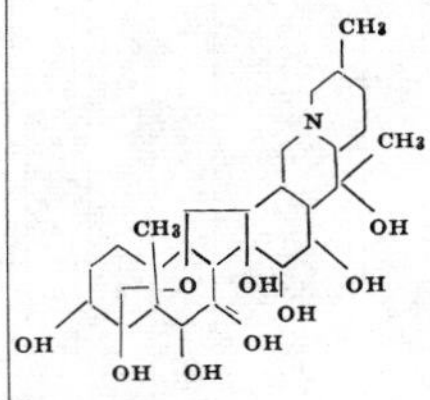	525.65	nd (MeOH) $[\alpha]_D -12$ (Py)	210–6								s	aq ac s MeOH s^h	
—	**Provitamin A**	*see* **β-Carotene**													
—	**Provitamin D₃**	*see* **5,7-Cholestadien-3(β)ol**													
p1860	**Prulaurasin**	*dl*-Mandelonitrile glucoside. 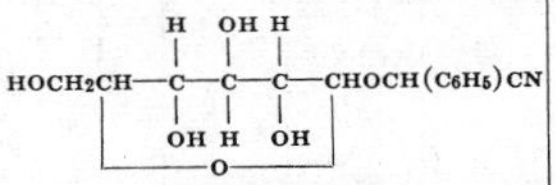	295.29	wh nd $[\alpha]_D -54$	120–2				s	s	i				**B31**, 240
p1861	**Prunasin**	*d*-Mandelonitrile-*β*-*d*-glucoside. 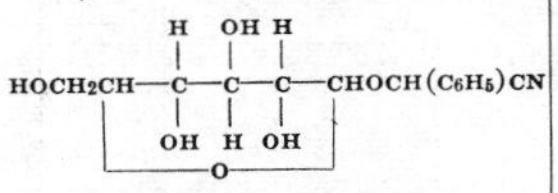	295.29	nd (chl or AcOEt-CCl_4) $[\alpha]^{21}_D -27.0$	149–50				s			s		chl δ^h AcOEt s^h	**B31**, 238
—	**Prunetol**	*see* **Isoflavone, 4′,5,7-trihydroxy**													
—	**Pseudocholestane**	*see* **Coprostane**													
p1862	**Pseudocodeine**	Neoisocodeine. 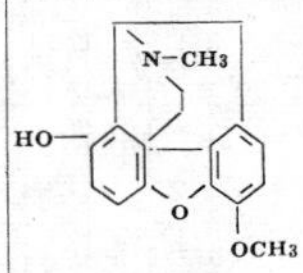	299.36	nd $[\alpha]_D -96.6$ (al)	181–2		1.290^{80}	1.574	δ	s	δ				**B3**, 906
p1863	**Pseudo-conhydrine**	5-Hydroxyconine. $C_8H_{17}NO$.	143.23	nd $[\alpha]^{20}_D +11.0$ (al)	105–6	236			s	s	s				**B21**[1], 191
p1864	**Pseudo-coniceine(*L*)**	γ-Coniceine. 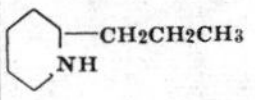	125.21	hyg $[\alpha]^{15}_D$ +122.6		171–2	0.8776^{15}_4	1.4607^{18}	δ	s	s			chl s	**B20**, 146
p1865	**Pseudaconitine**	$C_{36}H_{49}NO_{12}$	687.76	tcl (MeOH-AmOH) $[\alpha]^{15}_D +18.36$ (al)	210–2				i	v	δ				
—	**Pseudocumene**	*see* **Benzene, 1,2,4-trimethyl-***													
—	**Pseudocumenol**	*see* **Benzene, 1-hydroxy-2,4,5-trimethyl-***													
—	**Pseudocumidine**	*see* **Benzene, 1-amino-2,4,5-trimethyl-***													
p1866	**Pseudo-ephedrine(*d*)**	Isoephedrine. ψ-Ephedrine. 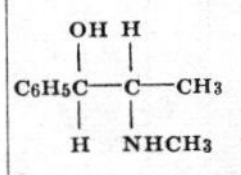	165.23	pr (eth) $[\alpha]^{20}_D +51.9$ (abs al, p=5)	118.7				δ	s	s				**B13**[2], 376
p1867	—(*dl*)	$C_{10}H_{15}NO$. *See* p1866	165.23	nd (eth)	118.2	130^{16}			δ	v	δ		v		**B13**[2], 372
p1868	—(*l*)	$C_{10}H_{15}NO$. *See* p1866	165.23	lf or pr (eth). $[\alpha]^{20}_D -51.9$ (abs al, p=5)	118.7						s				**B13**[2], 377
p1869	—,hydrochloride(*d*)	$C_{10}H_{15}NO.HCl$. *See* p1866	201.69	rh pl or nd $[\alpha]^{20}_D +62$ (c=0.8)	181–2				s	s	s			25% HCl v	**B13**[2], 373
p1870	—,—(*dl*)	$C_{10}H_{15}NO.HCl$. *See* p1866	201.69	nd (abs al)	164				v						**B13**[2], 377

For explanations, symbols and abbreviations see beginning of table.

No.	Name	Synonyms and Formula	Mol. wt.	Crystalline form, color and specific rotation	m.p. °C	b.p. °C	Density	n_D	Solubility w	al	eth	ace	bz	other solvents	Ref.
	Pseudoephedrine														
p1871	—,—(*l*)	$C_{10}H_{15}NO.HCl$. *See* p1866	201.69	nd (abs al or AcOEt) $[\alpha]^{20}_D -62.1$ (w,c=1.8)	182–2.5				v						**B13**², 377
—	**Pseudoergostane**	*see* **Coproergostane**													
p1872	**Pseudo-hyoscyamine**	Norhyoscyamine. 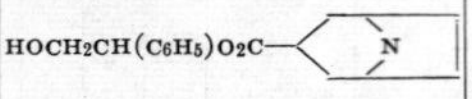 $HOCH_2CH(C_6H_5)O_2C$—	275.35	nd $[\alpha]_D -21$ to -23 (al)	140.5				δ	v	δ			chl s	**B21**¹, 197
p1873	**Pseudoionone**	Citrylideneacetone. $(CH_3)C:CHCH_2CH_2C(CH)_3:CHCH:CHCOCH_3$	192.30	pa ye		143–5^{12}	0.8984^{20}	1.5335^{20}		s	s			chl s MeOH s	**B1**³, 3067
p1874	**Pseudojervine**	$C_{29}H_{43}NO_7$	517.67	wh nd $[\alpha]_D -133$ (al-chl, 1:3)	304–5				i						
p1875	**Pseudomorphine**	1,1′-Dehydrodimorphine. 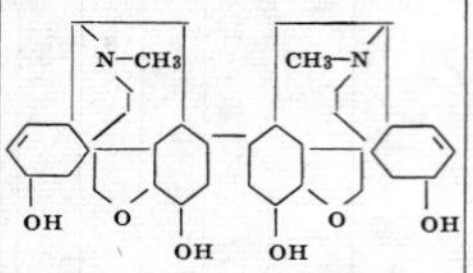	568.68	(aq NH_3+3w) $[\alpha]^{24}_D +44.8$ (1N HCl, c=0.86)	282–3				i	i	i			NH_3 s^h sulf i chl i Py s	**B27**², 886
p1876	—,hydrochloride	$C_{34}H_{36}N_2O_6$. 2HCl. $2H_2O$	677.63	pw $[\alpha]_D -114.76$					δ						**B3**, 910
p1877	**Pseudo-pelletierine**	Pseudopunicine. $C_9H_{15}NO$.	153.23	orh pr	54	246	1.001^{100}	1.4760^{100}	v	s	s			chl s	**B4**²¹, 261
p1878	**Pseudoreserpine**	$C_{32}H_{38}N_2O_9$	594.67	$[\alpha]^{24}_D -65 \pm 2$ (chl)	257–8				i			s			
p1879	**Pseudo-thiohydantoin**	4-Keto-2-iminothiazolidine. 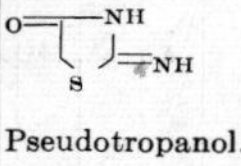	116.14	pr or nd (w)	255–8d				δ v^h	i	i				**B27**², 284
p1880	**Pseudotropine**	Pseudotropanol. 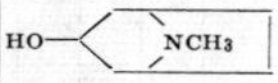	141.21	rh ta or pr (eth)	108	243			v	v	δ			chl s	
p1881	**Psicose**	*d*-Allose. Pseudofructose. 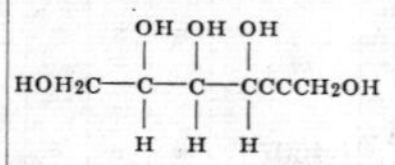	180.16	syr $[\alpha]^{20}_D +3.1$ (w, c=1.6)					s	s		i		MeOH s	**B31**, 349
p1882	**Pteridine**	Pyrimido[4,5-b]pyrazine. 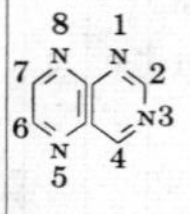	132.13	ye pl	139–40	sub 125–30^{20}			v	s	δ		δ		
p1883	—,**2-amino-4,6-dihydroxy-**	Xantopterin. $C_6H_5N_5O_2$. *See* p1882	179.14	og-ye (w+1)	>410d				i					ac, alk v	
p1884	—,**2-amino-4-hydroxy-**	2-Amino-4-pteridonol. $C_6H_5N_5O$. *See* p1882	163.14	ye	>360										
p1885	**Pukateine**(*l*)	4-Hydroxy-5,6-methylenedioxyaporphin. 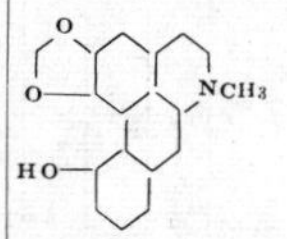	295.34	(al) $[\alpha]^{15}_D -200$ (al, c=6)	200	210–5^2			i	δ s^h	δ s^h			peth δ chl s Py v	**B27**¹, 461
p1886	**Pulegenone**	4-Methyl-1-isopropylcyclopentenone-5. 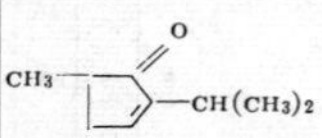	138.22			188–9	0.9144^{20}_0	1.4660^{20}							
p1887	**Pulegone**	4(8)-*p*-Menthen-3-none. $C_{10}H_{16}O$	152.24	$[\alpha]^{20}_D +23.4$		224	0.937^{15}_4	1.4880^{19}	i	∞	∞				**B7**, 81

For explanations, symbols and abbreviations see beginning of table.

No.	Name	Synonyms and Formula	Mol. wt.	Crystalline form, color and specific rotation	m.p. °C	b.p. °C	Density	n_D	Solubility w	al	eth	ace	bz	other solvents	Ref.
	Purine														
p1888	**Purine**	7-Imidazo[4,5-d]pyrimidine. 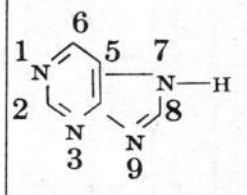	120.11	nd (to or al)	216–7	d			v	v^h	δ	s		chl δ AcOEt s	**B26**, 354
p1889	**—,6-amino-**	Adenine. $C_5H_5N_5$. *See* p1888	135.13	nd or lf (w+3)	360–5	sub			δ v^h	δ	i			chl i ac, alk v	**B26**², 252
—	**—,1,3-dimethyl-2,6-dihydroxy-**	*see* **Theophylline**													
p1890	**—,6-mercapto-**	6-Purinethiol. $C_5H_4N_4S$. *See* p1888	152.18	dk ye hyd	313–4d (−w140)				i $δ^h$						
p1891	**—,2,6,8-trichloro-**	$C_5HCl_3N_4$. *See* p1888	223.45	nd (al)	159–61				$δ^h$	v^h	v	v	v^h	chl v^h	**B26**, 356
—	**—,2,6,8-trihydroxy-**	*see* **Uric acid**													
—	**6(1)Purinone**	*see* **Hypoxanthine**													
—	**Purpurin**	*see* **9,10-Anthraquinone, 1,2,4-trihydroxy-***													
—	**Purpurin bordeaux**	*see* **9,10-Anthraquinone, 1,2,4,5-tetrahydroxy-***													
—	**Purpuroxanthin**	*see* **9,10-Anthraquinone, 1,3-dihydroxy-***													
—	**Putrescine**	*see* **Butane, 1,4-diamino-***													
p1892	**Pyocyanine**	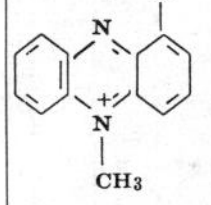	210.24	dk bl nd (w+1)	133d	sub			s^h δ	s^h		s		Py s chl v $PhNO_2$ s	
p1893	**Pyraconitine**	$C_{32}H_{43}O_9N$	585.70	nd	171				δ	s					
p1893¹	**Pyran**	α-Pyran. 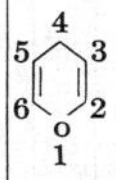	82.10												
p1894	**—,3,4-dihydro-**	C_5H_8O. *See* p1893¹	84.12		86–7		0.922^{19}_{15}	1.4402^{19}	s	s					
—	**—,2,3-dihydro-3,4-dihydroxy-2-hydroxymethyl-**	*see* **Gluconal**													
—	**—,4-oxo-**	*see* **4-Pyrone**													
p1895	**—,tetrahydro-**	Pentamethylene oxide. $C_5H_{10}O$. *See* p1893¹	86.13			88	0.8855^{15}_{4}	1.4195^{18}							
p1896	**Pyranthrone**	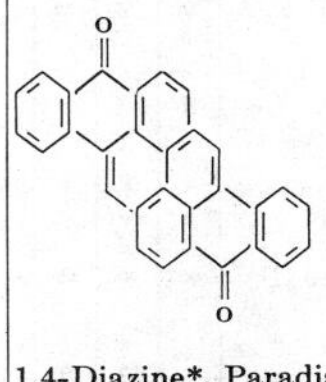	406.44	red-ye or red-br nd ($PhNO_2$)	d, sub vac									$PhNO_2$ $δ^h$ 50% Py δ sulf s (bl)	**E145**, 887
p1897	**Pyrazine**	1,4-Diazine*. Paradiazine. 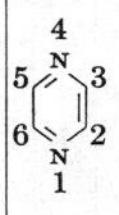	80.09	pr (w)	53	115.5 $-.8^{768}$	1.0311^{61}_{4}	1.4953^{61}	s	s	s				**B23**, 91
p1898	**—,2,5-dimethyl-**	Ketine. $C_6H_8N_2$. *See* p1897	108.14		15	155 cor	0.9887^{20}_{4}	1.4992^{24}	∞	∞	∞				**B23**², 80
p1899	**—,2-methyl-**	$C_5H_6N_2$. *See* p1897	94.12			135	1.0441^{0}_{4}	1.5067^{19}	v	v	v				**B23**, 94
p1900	**Pyrazine-carboxylic acid, amide**	Pyrazinamide. 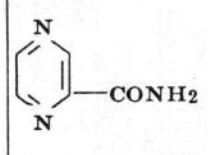	123.12	nd (w^h)	190–1										

For explanations, symbols and abbreviations see beginning of table.

No.	Name	Synonyms and Formula	Mol. wt.	Crystalline form, color and specific rotation	m.p. °C	b.p. °C	Density	n_D	Solubility w	al	eth	ace	bz	other solvents	Ref.
	2,3-Pyrazinedicarboxylic acid														
p1901	**2,3-Pyrazinedi-carboxylic acid**	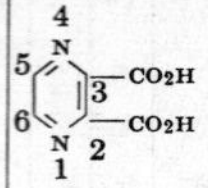	168.11	pr (w+2)	193d				v	δ	δ	s	δ	MeOH s, chl d	**B25**^{2}, 164
p1902	**2,5-Pyrazinedi-carboxylic acid**	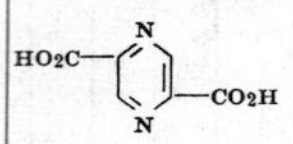	168.11	nd (w+2)	255–6d				δ	δ	δ		δ	chl δ	**B25**^{2}, 168
p1903	**2,6-Pyrazinedi-carboxylic acid**	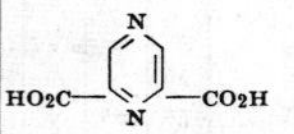	168.11	mcl nd (w+2)	217–8d				δ s^h	s					**B25**, 168
p1904	**2,3-Pyrazinedi-carboxylic acid, 5-methyl-**	$C_7H_6N_2O_4$. *See* p1901	182.14	(w, ace, dil al-eth)	174–5				v^h	s		s^h			**B25**^{2}, 165
p1905	**Pyrazole**	1,2-Diazole. α-Pyrromonazole. 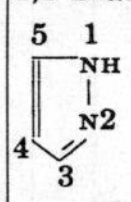	68.08	nd or pr (lig)	69.5–70	186.8		1.4203	s	s	s		s		**B23**, 39
p1906	**—,4-bromo-1,3-dimethyl-**	$C_5H_7BrN_2$. *See* p1905	175.04			$76–7^{10}$	1.4976^{15}_{4}							os v	**B23**^{2}, 54
p1907	**—,4-bromo-1,5-dimethyl-**	$C_5H_7BrN_2$. *See* p1905	175.04		38.5–9.5	85^{10}			δ					peth i, os s dil ac s	**B23**^{2}, 54
p1908	**—,4-bromo-3,5-dimethyl-**	$C_5H_7BrN_2$. *See* p1905	175.04	nd (dil al)	123				s^h					os s	**B23**^{2}, 68
p1909	**—,4-bromo-3-methyl-**	$C_4H_5BrN_2$. *See* p1905	161.07		76–7		1.5638^{100}_{4}								**B23**^{2}, 54
p1910	**—,3-chloro-1,5-dimethyl-**	$C_5H_7ClN_2$. *See* p1905	130.58	pl (w)	46–7	210–2 138^{72}	1.0823^{99}_{4}		δ	v	v	v	v		**B23**^{2}, 49
p1911	**—,4-chloro-3,5-dimethyl-**	$C_5H_7ClN_2$. *See* p1905	130.58	pr (w), (al)	117.5–8.5	220–22								os v	**B23**^{2}, 67
p1912	**—,5-chloro-1,3-dimethyl-**	$C_5H_7ClN_2$. *See* p1905	130.58			157–8 cor	1.1367^{18}_{4}								**B23**^{2}, 49
p1913	**—,5-chloro-3-methyl-**	$C_4H_5ClN_2$. *See* p1905	116.55	(eth, lig)	118–9	258 138^{15}			v^h					os v	**B23**^{2}, 49
—	**—,dihydro-**	*see* **Pyrazoline**													
—	**—,dihydro(oxo)-**	*see* **Pyrazolone**													
p1914	**—,1,3-dimethyl-**	$C_5H_8N_2$. *See* p1905	96.14			136–8	0.9561^{17}_{4}		v						**B23**^{2}, 44
p1915	**—,1,5-dimethyl-**	$C_5H_8N_2$. *See* p1905	96.14			153 cor	0.9813^{17}_{4}		v	v	v				
p1916	**—,3,4-dimethyl-**	$C_5H_8N_2$. *See* p1905	96.14	(peth)	58	111^{10}								os s	**B23**^{3}, 64
p1917	**—,3,5-dimethyl-**	$C_5H_8N_2$. *See* p1905	96.14	(peth, MeOH, al)	107–8	218^{758}			s		v		v	chl, MeOH v	**B23**^{2}, 65
p1918	**—,1-ethyl-**	$C_5H_8N_2$. *See* p1905	96.14			139			v	v	v	v	v	Py v	
p1919	**—,1-isopropyl-**	$C_6H_{10}N_2$. *See* p1905	110.16			143			s	v					
p1920	**—,3-methyl-**	$C_4H_6N_2$. *See* p1905	82.11			204^{752} cor	1.0203^{16}_{4}	1.4941^{14}	∞	∞	∞				**B23**^{11} 44
p1921	**2-Pyrazoline**	4,5-Dihydropyrazole. 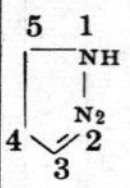	70.10			144	1.017_{20}		∞	∞	s				**B23**, 28
p1922	**—,1-phenyl-**	$C_9H_{10}N_2$. *See* p1921	146.19		52	273			i	s					
p1923	**5-Pyrazolone**	2-Pyrazolin-5-one. 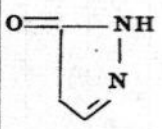	84.08	nd (to, w, xyl)	165	sub									**B24**, 186
—	**3-Pyrazolone, 1,5-dimethyl-2-phenyl-**	*see* **Antipyrine**													

For explanations, symbols and abbreviations see beginning of table.

No.	Name	Synonyms and Formula	Mol. wt.	Crystalline form, color and specific rotation	m.p. °C	b.p. °C	Density	n_D	Solubility w	al	eth	ace	bz	other solvents	Ref.
	5-Pyrazolone														
p1925	**5-Pyrazolone, 3-methyl-**	$C_4H_6N_2O$. *See* p1923	98.11	pr (w), nd (al)	219				v^h	δ^h					**B24**, 19
p1926	**—,3-methyl-1(3-nitrophenyl)-**	$C_{10}H_9N_3O_3$. *See* p1923	219.21	ye amor (aa)	185				δ	δ	δ			alk s aa s^h	**B24**[1], 191
p1927	**—,3-methyl-1-phenyl-**	$C_{10}H_{10}N_2O$. *See* p1923	174.21	pr	127	287^{105}		1.637	s^h	s^h	δ		δ		
p1928	**—,3-methyl-1(4-sulfophenyl)-**	$C_{10}H_{10}N_2O_4S$. *See* p1923	254.27	nd (w+1)	290–320d				s^h	i	i			alk v aa i	**B24**[2], 20
p1929	**5-Pyrazolone-3-carboxylic acid, 1-phenyl-**	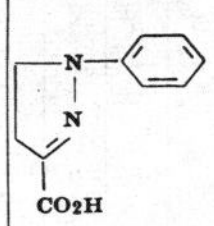	204.19	nd (w)	261				s	s	i				**B25**[2], 219
p1930	**Pyrene***	Benzo(*d,e,f*)phenanthrene. 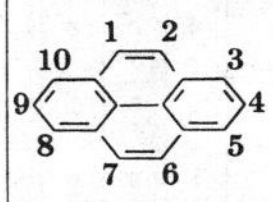	202.06	pa ye	149–50	260^{60}	1.271^{23}_{4}			s				CS_2, to s lig s	**B5**, 693
p1931	**—,3-acetyl-**	$C_{18}H_{12}O$. *See* p1930	244.29	gr-ye lf (al, MeOH)	89–90						s		v	sulf s (og-red) to v	**E14s**, 434
p1932	**—,3-amino-**	$C_{16}H_{11}N$. *See* p1930	217.27	ye nd (hexane), lf (dil al)	117–8					s		s		hexane s ac s	**E14s**, 422
—	**3-Pyrenebutyric acid**	*see* **Butanoic acid, 4(3-pyrenyl)-***													
p1933	**3-Pyrene-carboxylic acid**	$C_{17}H_{10}O_2$. *See* p1930	246.25	ye nd (PhCl, eth-al, sub)	274	sub			δ	δ	s			con sulf s dil alk, chl s	**E14s**, 448
p1934	**4-Pyrene-carboxylic acid**	$C_{17}H_{10}O_2$. *See* p1930	246.25	gr lf or nd ($PhNO_2$)	326									chl, con sulf s	**E14s**, 449
p1935	**Pyrethrin I**	Chrysanthemummono-carboxylic acid, pyrethrolone ester. 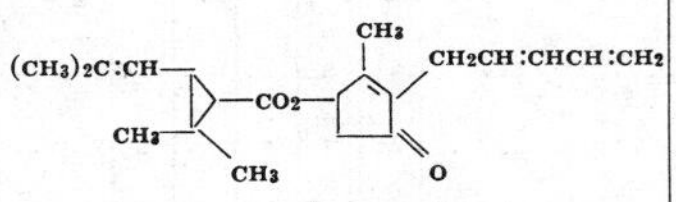	328.43	$[\alpha]^{25}_{D} -32.3$		$170d^{0.1}$		1.5192^{18}	i					peth s CH_3NO_2s CCl_4 s	**B9**[2], 46
p1936	**Pyrethrin II**	Chrysanthemumdicarboxylic acid, methyl ester pryethrolone ester. 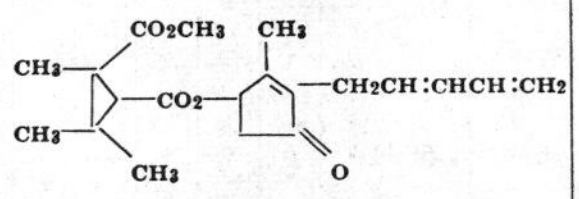	372.47	visc $[\alpha]^{20}_{D} -4.2$		$200d^{0.1}$		1.5290^{21}	i	s				peth s CH_3NO_2 s CCl_4 s	**B9**[2], 565
p1937	**Pyridazine**	1,2-Diazine*. Orthodiazine. 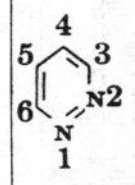	80.09		−8	208	1.1035^{23}_{4}	1.5231^{23}	∞	∞					**B23**, 89
p1938	**—,3,6-dioxo-1,2,3,6-tetra-hydro-**	Maleic acid hydrazide. $C_4H_4N_2O_2$. *See* p1937	112.09	(w)	296–8d				δ s^h	δ^h					
p1939	**Pyridine**	Azine*. 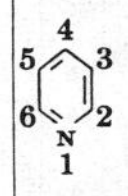	79.10		−42	115.5	0.9819^{20}	1.5095^{20}	∞	∞	∞				**B20**, 181
p1940	—,hydrochloride	Pyridinium hydrochloride. $C_5H_5N.HCl$. *See* p1939	115.56	hyg pl or sc	144.5	218–9			s	s	i		i	chl i	**B20**, 189
p1941	—,nitrate	Pyridinium nitrate. $C_5H_5N.HNO_3$. *See* p1939	142.12	nd (al)	sub				v	s					**B20**[1], 59
p1942	—,picrate	Pyridinium picrate. $C_5H_5N.C_6H_3N_3O_7$. *See* p1939	308.21	ye nd	167–8				s^h	s^h					**B20**[2], 122
p1943	—,sulfate	Pyridinium sulfate. $C_5H_5N.H_2SO_4$. *See* p1939	177.18	(al)					∞	∞	i				**B20**, 190

For explanations, symbols and abbreviations see beginning of table.

No.	Name	Synonyms and Formula	Mol. wt.	Crystalline form, color and specific rotation	m.p. °C	b.p. °C	Density	n_D	Solubility w	al	eth	ace	bz	other solvents	Ref.
	Pyridine														
p1944	**—,2-acetyl-**	Methyl 2-pyridyl ketone. C_7H_7NO. *See* p1939	121.14			192				v	v			min ac v	**B21**^{2}, 247
p1945	**—,3-acetyl-**	Methyl 3-pyridyl ketone. C_7H_7NO. *See* p1939	121.14	ye in air		220 $15–6^{1}$		1.5341^{20}	s	v	v			ac v	**B12**^{2}, 247
p1946	—,—,oxime	$C_7H_8N_2O$. *See* p1939	136.15	(al, bz)	112				s^{h}					min ac s	
p1947	**—,2-allyl-**	C_8H_9N. *See* p1939	119.17			190	0.959		δ	∞	∞				
p1948	**—,2-amino-**	$C_5H_6N_2$. *See* p1939	94.12	lf (lig)	57.5	204				s				os s	**B22**, 428
p1949	**—,3-amino-**	$C_5H_6N_2$. *See* p1939	94.12	lf (lig)	64	252			s	s	s			lig δ	**B22**, 431
p1950	**—,4-amino-**	$C_5H_6N_2$. *See* p1939	94.12	nd (bz)	158				s	s	s		s		**B22**, 433
p1951	**—,2-amino-5-bromo-**	$C_5H_5BrN_2$. *See* p1939	173.02		137					v	i		s	lig i	**B22**, 429
p1952	**—,5-amino-2-butoxy-**	$C_9H_{14}N_2O$. *See* p1939	166.22			$148–50^{12}$	1.037^{25}			s	s			dil ac s	
p1953	**—,2-amino-5-chloro-**	$C_5H_5ClN_2$. *See* p1939	128.56	pl	135–5.5	$127–8^{11}$			s	s				peth i lig i	**B22**^{2}, 332
p1954	**—,2-amino-3,5-dibromo-**	$C_5H_4Br_2N_2$. *See* p1939	251.92	nd	104–4.5				δ	s	s			lig δ	**B22**^{2}, 333
p1955	**—,2-amino-2,5-dichloro-**	$C_5H_4Cl_2N_2$. *See* p1939	163.01	nd or pr	84				δ	s				lig s	**B22**^{2}, 333
p1956	**—,4-amino-2,6-dichloro-**	$C_5H_4Cl_2N_2$. *See* p1939	163.01	nd	176				i	s	s		s		**B22**^{1}, 632
p1957	**—,4-amino-3,5-dinitro-**	$C_5H_4N_4O_4$. *See* p1939	184.11	ye lf	170–1				$δ^{h}$	δ					**B22**^{2}, 341
p1958	**—,2-amino-5-iodo-**	$C_5H_5IN_2$. *See* p1939	220.01	nd or lf	129				δ		v				**B22**^{2}, 334
p1959	**—,2-amino-3-methyl-**	α-Amino-β-picoline. $C_6H_8N_2$. *See* p1939	108.14	hyg	26 − 6.4	22^{748} 95^{8}			v	s				os s lig δ	**B22**^{2}, 342
p1960	**—,2-amino-4-methyl-**	α-Amino-γ-picoline. $C_6H_8N_2$. *See* p1939	108.14	lf or pl (lig)	100–0.5 cor	$115–7^{11}$ cor			v					os s lig i	**B22**^{2}, 342
p1961	**—,2-amino-6-methyl-**	α′-Amino-α-picoline. $C_6H_8N_2$. *See* p1939	108.14	hyg (lig)	41	208–9			v					lig s^{h}	**B22**^{2}, 342
p1962	**—,3-amino-4-methyl-**	β-Amino-γ-picoline. $C_6H_8N_2$. *See* p1939.	108.14	pr	106	254^{735}			v	v	v	v	v	chl v lig i	**B22**^{2}, 343
p1963	**—,3-amino-3-nitro-**	$C_5H_5N_3O_2$. *See* p1939	139.11	ye nd	164				δ						**B22**^{2}, 335
p1964	**—,2-amino-5-nitro-**	$C_5H_5N_3O_2$. *See* p1939	139.11	ye lf	188				δ		δ		δ	lig δ	**B22**^{1}, 631
p1965	**—,4-amino-3-nitro-**	$C_5H_5N_3O_2$. *See* p1939	139.11	ye nd	200				s^{h}	s					**B22**^{2}, 341
p1966	**—,4-benzoyl-**	Phenyl 4-pyridyl ketone. $C_{12}H_9NO$. *See* p1939	183.21	nd (peth)	71–2	$170 − 2^{10}$			$δ^{h}$	s	s		s		**B21**^{2}, 277
p1967	**—,2-benzyl-**	$C_{12}H_{11}N$. *See* p1939	169.23	nd	139	276^{742}	1.067^{0}_{0}		i	s	s				**B20**, 425
p1968	**—,3-benzyl-**	$C_{12}H_{11}N$. *See* p1939	169.23	nd	34	286^{740}	1.061		i	s	s				**B20**, 426
p1969	**—,4-benzyl-**	$C_{12}H_{11}N$. *See* p1939	169.23			287	1.0614^{20}_{0}			s	v				**B20**^{2}, 272
p1970	**—,2-bromo-**	C_5H_4BrN. *See* p1939	158.00			192–4 $74–5^{13}$	1.657^{15}		i					os s	**B20**^{2}, 153
p1971	**—,3-bromo-**	C_5H_4BrN. *See* p1939	158.00	nd (al)	142–3	169–70	1.645^{0}_{4}	1.5694^{20}	s	v	v				**B20**, 233
p1972	**—,5-bromo-2-hydroxy-**	C_5H_4BrNO. *See* p1939	174.00	pr (w, al)	177–8				δ v^{h}	v^{h}	δ			dil alk v con ac v	**B21**^{1}, 202
p1973	**—,2-chloro-**	C_5H_4ClN. *See* p1939	113.55	oil		170	1.205^{15}		δ	s	s				**B20**, 230
p1974	**—,3-chloro-**	C_5H_4ClN. *See* p1939	113.55			148^{744}			s						**B20**, 230
p1975	**—,4-chloro-**	C_5H_4ClN. *See* p1939	113.55			147–8			s	∞					**B20**, 231
p1976	**—,2,3-diamino-** (one form)	$C_5H_7N_3$. *See* p1939	109.13	lf or pl	122	$148–50^{5}$			δ				δ	peth i	**B22**^{1}, 647
p1977	—,—(one form)	$C_5H_7N_3$. *See* p1939	109.13	nd	113				s	s			s^{h}		**B22**^{2}, 394
p1978	**—,2,4-diamino-**	$C_5H_7N_3$. *See* p1939	109.13	hyg lf or nd	107				v				δ	peth i	**B22**^{2}, 394
p1979	**—,2,5-diamino-**	$C_5H_7N_3$. *See* p1939	109.13	nd	109–10	$180–5^{12}$			s	s			δ	peth i	**B22**^{2}, 394
p1980	**—,3,4-diamino-**	$C_5H_7N_3$. *See* p1939	109.13	nd or lf	218–9				i^{h}						**B22**^{2}, 395
p1981	**—,3,5-diamino-**	$C_5H_7N_3$. *See* p1939	109.13	hyg nd	119–20				v				δ	peth i	**B22**^{1}, 648
p1982	**—,3.5-dibromo-**	$C_5H_3Br_2N$. *See* p1939	236.90	nd (al)	112	222			s^{h}	s^{h}	s				**B20**, 233
p1983	**—,1,2-dihydro-1-methyl-**	C_6H_7NO. *See* p1939	109.13	nd	7	250^{740} 121^{10}			∞					peth δ ig δ	**B21**^{2}, 229
p1984	**—,2,4-dihydroxy-**	2,4-Pyridinediol. $C_5H_5NO_2$. *See* p1939	111.10	rh bipym	260–5				s	s	$δ^{h}$				**B21**, 160
p1985	**—,2,6-dihydroxy-**	2,6-Pyridinediol. $C_5H_5NO_2.H_2O$. *See* p1939	129.12	ye pr	202–3				s	s	i			alk δ ac δ	**B21**^{2}, 108
p1986	**—,3,5-diiodo-2-hydroxy-**	$C_5H_3I_2NO$. *See* p1939	346.89	bt br nd	261–2				$δ^{h}$		δ			chl δ AmOH s^{h}	**B21**^{2}, 32
p1987	**—,2,3-dimethyl-**	2,3-Lutidine. C_7H_9N. *See* p1939	107.15			160.7	0.9419^{25}_{4}	1.5057^{20}	i	s	s				**B20**, 243

For explanations, symbols and abbreviations see beginning of table.

No.	Name	Synonyms and Formula	Mol. wt.	Crystalline form, color and specific rotation	m.p. °C	b.p. °C	Density	n_D	Solubility w	al	eth	ace	bz	other solvents	Ref.
	Pyridine														
p1988	**—,2,4-dimethyl-**	2,4-Lutidine. C_7H_9N. *See* p1939	107.15			159–9.5	0.9271$^{25}_{4}$	1.4984^{25}	v i^h	v	v				**B20**^{2}, 160
p1989	**—,2,5-dimethyl-**	2,5-Lutidine. C_7H_9N. *See* p1939	107.15			156.8–.9	0.9261$^{25}_{4}$	1.4982^{25}	v^h δ	v	∞				**B20**, 244
p1990	**—,2,6-dimethyl-**	2,6-Lutidine. C_7H_9N. *See* p1939	107.15			143	0.9200$^{25}_{4}$	1.4953^{25}	∞ s^h	δ	s				**B20**^{2}, 160
p1991	**—,3,4-dimethyl-**	3,4-Lutidine. C_7H_9N. *See* p1939	107.15		178.8^{759}		0.9537$^{25}_{4}$	1.5099^{25}	δ	s	s				**B20**^{2}, 161
p1992	**—,3,5-dimethyl-**	3,5-Lutidine. C_7H_9N. *See* p1939	107.15			171.6	0.9385$^{25}_{4}$	1.5032^{25}	s	s	s				**B20**^{2}, 161
p1993	**—,2(dimethyl-amino)-**	$C_7H_{10}N_2$. *See* p1939	122.18			196 88^{15}	1.0157$^{14}_{14}$							dil ac s	**B22**^{1}, 629
p1994	**—,4(dimethyl-amino)-**	$C_7H_{10}N_2$. *See* p1939	122.18	pl (eth)	114				v	v	s		v	chl v	**B22**^{2}, 341
p1995	**—,2(dimethyl-amino)-6-methyl-**	$C_8H_{12}N_2$. *See* p1939	136.20			198–200			δ	s	s				**B22**^{2}, 342
p1996	**—,2-ethenyl-**	α-Vinylpyridine. C_7H_7N. *See* p1939	105.13			158–9	0.9985^{0}		δ	v	v			chl v	**B20**, 256
p1997	**—,4-ethenyl-**	γ-Vinylpyridine. C_7H_7N. *See* p1939	105.13	red to dk-br		59^{12} par d		1.5490^{25}	s^h	s^h	δ				**B20**^{2}, 170
p1998	**—,2-ethyl-**	C_7H_9N. *See* p1939	107.16			148.6	0.9502^{0}	1.5044	δ	∞	v				**B20**, 241
p1999	**—,3-ethyl-**	C_7H_9N. *See* p1939	107.16			162–5^{762}	0.9539^{0}		δ^h	s	s				**B20**, 242
p2000	**—,4-ethyl-**	C_7H_9N. *See* p1939	107.16			164–5	0.9557^{0}		∞^{h}					dil ac s	**B20**, 243
p2001	**—,2-ethyl-3,5-dimethyl-**	α-Parvoline. $C_9H_{13}N$. *See* p1939	135.21			198–9^{764}	0.9338^{0}		s	v	v				**B20**^{2}, 166
p2002	**—,2-ethyl-4-methyl-**	$C_8H_{11}N$. *See* p1939	121.18			173–5^{748}	0.9239$^{20}_{0}$			s	s				**B20**^{2}, 162
p2003	**—,2-ethyl-6-methyl-**	$C_8H_{11}N$. *See* p1939	121.18			160–1.5	0.9207$^{25}_{4}$	1.4941^{16}	δ	s	s				**B20**^{2}, 162
p2004	**—,3-ethyl-4-methyl-**	β-Collidine. $C_8H_{11}N$. *See* p1939	121.18			198 76^{12}	0.9286$^{17}_{4}$		δ	s	s			chl s	**B20**^{2}, 163
p2005	**—,4-ethyl-2-methyl-**	α-Collidine. $C_8H_{11}N$. *See* p1939	121.18			177–9	0.9227$^{17}_{4}$		s δ^h	s	s		s		**B20**^{2}, 162
p2006	**—,5-ethyl-2-methyl-**	Aldehyde collidine. Aldehydine. $C_8H_{11}N$. *See* p1939	121.18			177.8^{747} 65–6^{17}	0.9219$^{20}_{20}$	1.4971^{20}	δ^h	s	s		s	con sulf s	**B20**, 248
—	**—,hexahydro-**	*see* **Piperidine**													
p2007	**—,2-hydroxy-**	2-Pyridol. α-Pyridone. C_5H_5NO. *See* p1939	95.10	nd (bz)	106–7	280–1			s	s	δ		s	chl s	
p2008	**—,3-hydroxy-**	3-Pyridol. C_5H_5NO. *See* p1939	95.10	nd (bz)	126–9				v	v	δ				**B21**^{2}, 33
p2009	**—,4-hydroxy-**	4-Pyridol. γ-Pyridone. C_5H_5NO. *See* p1939	95.10	pr (w)	148.5	>350 257–60^{10}			v	i	i		i		**B21**^{2}, 34
p2010	**—,2(2-hydroxy-ethyl)-**	2(2-Pyridyl)ethanol. C_7H_9NO. *See* p1939	123.15	hyg		118–21^{15}	1.1111$^{4}_{0}$		v	v	δ			chl v	**B21**, 50
p2011	**—,2(β-hydroxy-isopropyl)-**	2(2-Pyridyl)1-propanol. $C_8H_{11}NO$. *See* p1939	137.18			128–31^{17}			v	v	δ				**B21**, 57
p2012	**—,3(α-hydroxy-isopropyl)-**	2(3-Pyridyl)2-propanol. $C_8H_{11}NO$. *See* p1939	137.18		58	140–1^{12}			s	s				chl s	**B21**^{2}, 37
p2013	**—,2(hydroxy-methyl)-**	2-Pyridylmethanol. C_6H_7NO. *See* p1939	109.12			112–3^{16}			v	v				os v	**B21**, 203
p2014	**—,2(1-hydroxy-propyl)-**	1(2-Pyridyl)1-propanol. $C_8H_{11}NO$. *See* p1939	137.18	pa ye		213–6 112–3^{13}	1.0501$^{20}_{4}$	1.5197^{20}	s	s					**B21**^{2}, 37
p2015	**—,2(2-hydroxy-propyl)-**	1(2-Pyridyl)2-propanol. $C_8H_{11}NO$. *See* p1939	137.18	pr	32	123.5^{20}			v	v				chl v	**B21**^{2}, 37
p2016	**—,2-isopropyl-**	$C_8H_{11}N$. *See* p1939	121.18			158.9	0.9342^{0}		δ	∞	∞				**B20**, 247
p2017	**—,4-isopropyl-**	$C_8H_{11}N$. *See* p1939	121.18			178	0.944		δ	∞	∞				**B20**, 248
p2018	**—,4-methoxy-**	C_6H_7NO. *See* p1939	109.13			191			s						**B21**, 49
p2019	**—,2-methyl**	α-Picoline. C_6H_7N. *See* p1939	93.13			128.8	0.9497$^{15}_{4}$	1.5029^{17}	v	∞	∞				**B20**^{2}, 155
p2020	**—,3-methyl-**	β-Picoline. C_6H_7N. *See* p1939	93.13			143–4	0.9613$^{15}_{4}$	1.5043^{24}	∞	∞	∞				**B20**, 157

For explanations, symbols and abbreviations see beginning of table.

No.	Name	Synonyms and Formula	Mol. wt.	Crystalline form, color and specific rotation	m.p. °C	b.p. °C	Density	n_D	Solubility w	al	eth	ace	bz	other solvents	Ref.
	Pyridine														
p2021	**—,4-methyl-***	γ-Picoline. C_6H_7N. *See* p1939.	93.13	...	...	143.1	0.9571^{15}_{4}	1.5064^{17}	∞	∞	∞	...	...	...	**B20²**, 158
p2022	**—,2(methyl-amino)-**	$C_6H_8N_2$. *See* p1939	108.15	...	15	200–1, 90^{9}	1.052^{29}_{29}	...	s	v	v	...	...	...	**B22²**, 325
p2023	**—,4(methyl-amino)-**	$C_6H_8N_2$. *See* p1939	108.15	pl (eth)	117–8	...	...	...	v	v	v	...	...	aa v	**B22²**, 340
—	**—,2(*N*-methyl-2-pyrrolidyl)-**	*see* **2,2′-Nicotyrine**													
—	**—,3(*N*-methyl-2-pyrrolidyl)-**	*see* **Nicotine**													
—	**—,3(*N*-methyl-2-pyrryl)-**	*see* **β-Nicotyrine**													
p2024	**—,2-phenyl-**	α-Pyridylbenzene. $C_{11}H_9N$. *See* p1939	155.19	pr (al)	115–7	268–9	>1	...	δ	∞	∞	...	...	...	**B20**, 424
p2025	**—,3-phenyl-**	β-Pyridylbenzene. $C_{11}H_9N$. *See* p1939	155.19	pa ye oil or pr (al)	...	$269–70^{749}$	>1	...	δ	s	s	...	...	...	**B20**, 424
p2026	**—,4-phenyl-**	γ-Pyridylbenzene. $C_{11}H_9N$. *See* p1939	155.19	pl (w)	77–8	$280–2^{762}$	...	...	$δ^h$	s	s	...	...	...	**B20**, 424
p2027	**—,2-propyl-**	Conyrine. $C_8H_{11}N$. *See* p1939.	121.18	...	2	166–8, 60^{11}	0.9119^{20}_{4}	1.4925^{20}	δ	∞	∞	...	...	...	**B20²**, 161
p2028	**—,4-propyl-**	$C_8H_{11}N$. *See* p1939	121.18	...	...	184–6	0.9381^{15}	...	δ	v	v	...	...	...	**B20²**, 161
—	**—,3(2-pyrrolidyl)-**	*see* **Nornicotine**													
p2029	**—,tetramethyl-**	β-Parvoline. Parvuline. $C_9H_{13}N$. *See* p1939	135.21	...	...	220	0.916	...	...	...	...	...	...	...	...
p2030	**—,2,4,6-tri-hydroxy-**	2,4,6-Pyridinetriol. $C_5H_5NO_3$. *See* p1939	127.10	nd or pw	230d	...	...	...	δ	s	s	...	...	...	...
p2031	**—,2,3,5-tri-methyl-**	$C_8H_{11}N$. *See* p1939	121.18	...	...	186.7	0.9352^{19}_{4}	1.5057^{25}	δ	s	...	...	s	...	**B20²**, 163
p2032	**—,2,3,6-tri-methyl-**	$C_8H_{11}N$. *See* p1939	121.18	...	...	$172–85^{761}$	0.9225^{25}_{4}	1.5018^{25}	s^h	s	s	...	s	...	**B20²**, 163
p2033	**—,2,4,5-tri-methyl-**	$C_8H_{11}N$. *See* p1939	121.18	...	...	190^{266}	0.9330^{25}_{4}	1.5054^{25}	δ	s	s	...	s	...	**B20²**, 164
p2034	**—,2,4,6-tri-methyl-**	γ-Collidine. *sym*-Collidine. $C_8H_{11}N$. *See* p1939	121.18	...	...	$170–5^{762}$	0.9166^{22}_{4}	1.4959^{25}	s $δ^h$	s	s	...	...	...	**B20²**, 164
p2035	**4-Pyridinecar-boxaldehyde**	Isonicotinaldehyde. N⬡—CHO	107.11	...	...	$77.3–8.1^{12}$	...	1.5352^{25}	s	...	s	...	...	...	**B21**, 287
p2036	**2-Pyridinecar-boxylic acid**	Picolinic acid (structure: 4, 5, 3, 6, N 1, —CO_2H)	123.11	wh to redsh cr pw	136–7	sub	...	...	δ s^h	s	...	...	i	chl i, CS_2 i	**B22²**, 30
p2037	**—,amide**	Picolinamide. $C_6H_6N_2O$. *See* p2035	122.12	mcl pr	107–8	...	...	...	δ	s	...	...	s	lig i	**B22²**, 31
p2038	**—,ethyl ester**	$C_8H_9NO_2$. *See* p2035	151.16	ye in air	0–2	243, 122^{13}	1.1194^{20}	1.5104^{20}	∞	∞	∞	...	...	...	**B22²**, 31
p2039	**—,nitrile**	Picolinonitrile. $C_6H_4N_2$. *See* p2035	104.11	nd or pr (eth)	29	212–5	...	...	s	v	v	...	v	lig δ	**B22**, 36
p2040	**3-Pyridinecar-boxylic acid**	(structure: 4, 5, —CO_2H, 6, 2, N 1)	123.11	nd (w, al)	236–7	sub	1.473	...	δ s^h	s^h	δ	...	...	...	**B22²**, 32
p2041	**—,amide**	Niacin amide. Nicotamide. Nicotine amide. Pellagra preventive vitamin. P. P. factor. $C_6H_6N_2O$. *See* p2040	122.12	wh pw, nd (bz)	124–5	...	1.400	1.466	v	v	...	...	...	glycerol v	**B22²**, 34
p2042	**—,—,*N*,*N*-diethyl-**	$C_{10}H_{14}N_2O$. *See* p2040	178.23	yesh	24–6	269–300 δd, $128–9^{3}$	1.060^{25}_{4}	1.525^{20}	∞	∞	∞	∞	...	chl ∞	**B22²**, 34
p2043	**—,ethyl betaine**	(structure: ⬡—CO_2^-, N^+, C_2H_5)	151.17	hyg pl	84–6	...	...	...	v	δ	i	...	...	os i	**B22**, 43

For explanations, symbols and abbreviations see beginning of table.

No.	Name	Synonyms and Formula	Mol. wt.	Crystalline form, color and specific rotation	m.p. °C	b.p. °C	Density	n_D	Solubility w	al	eth	ace	bz	other solvents	Ref.
	3-Pyridinecarboxylic acid														
p2044	—,ethyl ester	$C_8H_9NO_2$. *See* p2040	151.16	...	...	224 $103–5^{5}$	1.1070^{20}	1.5038^{20}	v	v	v	...	v	...	**B22²**, 33
p2045	—,hydrochloride	$C_6H_5NO_2.HCl$. *See* p2040	159.58	pr or pl, rh bipym	272d	...	...	...	s	s	...	...	...	...	**B22²**, 33
p2046	—,methyl ester	$C_7H_7NO_2$. *See* p2040	137.13	...	38	204	...	...	s	s	...	...	s	...	**B22²**, 33
p2047	—,nitrile	Nicotinonitrile. $C_6H_4N_2$. *See* p2040	104.11	nd (peth-eth)	50–1	201	...	...	v	v	v	...	v	lig δ s^h	**B22²**, 35
p2048	**4-Pyridinecarboxylic acid**	Isonicotinic acid. 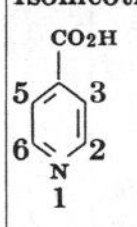	123.11	nd (w)	315–6 cor	sub	...	...	s^h δ	$δ^h$	δ	...	δ	...	**B22²**, 37
p2049	—,anhydride	$C_{12}H_8N_2O_3$. *See* p2048	228.21	nd	302d	...	...	...	...	...	...	...	...	...	**B22²**, 37
p2050	—,ethyl betaine	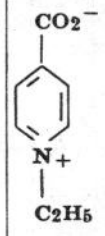	151.17	nd	241d	...	...	...	v	v	...	...	...	...	**B22**, 47
p2051	—,ethyl ester	$C_8H_9NO_2$. *See* p2048	151.17	...	...	219–20 110^{15}	1.0091^{15}	1.4975^{21}	δ	s	v	...	s	chl v	**B22²**, 37
p2052	—,hydrazide	Isoniazid. $C_6H_7N_3O$. *See* p2048	137.14	nd (al)	171	...	...	...	...	...	...	...	...	...	...
p2053	—,methyl ester	$C_7H_7NO_2$. *See* p2048	137.14	...	*ca.* 8.5	207–9d	...	1.5106^{25}	δ	s	s	...	s	...	**B22**, 46
p2054	—,nitrile	Isonicotinonitrile. $C_6H_4N_2$. *See* p2048	104.11	nd (lig-eth)	83	...	...	...	s	s	s	...	s	lig δ	**B22**, 40
p2055	**3-Pyridinecarboxylic acid, 6-amino-**	$C_6H_6N_2O_2$. *See* p2040	138.13	(dil aa+2w)	>300d	...	...	...	...	δ	...	...	...	os δ	...
p2056	**4-Pyridinecarboxylic acid, 2,6-dihydroxy-**	$C_6H_5NO_4$. *See* p2048	155.11	yesh-grsh pw	>300d	...	...	...	i	...	...	...	...	HCl $δ^h$ alk s	...
p2057	**3-Pyridinecarboxylic acid, 2-hydroxy-**	$C_6H_5NO_3$. *See* p2040	139.11	nd (w)	α256d β301–2d	...	...	...	δ s^h	δ	δ	...	...	chl i	**B22²**, 165
p2058	—,**4-hydroxy-**	$C_6H_5NO_3$. *See* p2040	139.11	nd (w+2), (al+2w)	254–5	...	...	...	s^h	δ	...	...	...	chl δ	**B22²**, 165
p2059	—,**6-hydroxy-**	$C_6H_5NO_3$. *See* p2040	139.11	nd (w)	304–d	sub	...	...	i^h	δ	δ	...	δ	chl δ	**B22²**, 165
p2060	**4-Pyridinecarboxylic acid, 3-hydroxy-5(hydroxy-methyl)-**, lactone	$C_8H_7NO_3$. *See* p2048	165.15	...	273–3.5d	...	...	...	s	...	...	...	...	MeOH δ, s^h	...
p2061	**3-Pyridinecarboxylic acid, 5-hydroxy-4(hydroxy-methyl)6-methyl-**, lactone	$C_8H_7NO_3$. *See* p2048	165.15	nd	282–3d	...	...	...	s	δ s^h	...	...	...	...	...
p2062	—,**1-methyl-**	Trigonelline. $C_7H_7NO_2$. *See* p2040	137.14	pr (aq al+1w)	218d (anh) 130 (hyd)	...	...	...	v	v^h δ	δ	...	δ	chl δ	**B22²**, 35
p2063	**4-Pyridinecarboxylic acid, 1,2,3,6-tetrahydro-**, methyl ester	$C_8H_{13}NO_2$. *See* p2048	155.20	...	...	$105–6^{14}$	...	1.4764^{25}	s	s	s	...	...	MeOH s	...
p2064	**2,3-Pyridinedicarboxylic acid.**	Quinolinic acid. 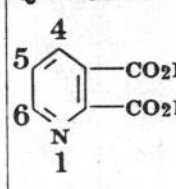	167.12	mcl pr	190d	...	...	...	δ	δ	δ	...	i	...	**B22**, 150
p2065	**2,4-Pyridinedicarboxylic acid**	Lutidinic acid. 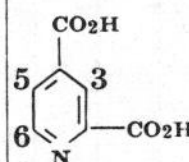	167.12	lf (w+1)	248–50	...	0.942	...	s^h δ	s	δ	...	δ	...	**B22²**, 105

For explanations, symbols and abbreviations see beginning of table.

No.	Name	Synonyms and Formula	Mol. wt.	Crystalline form, color and specific rotation	m.p. °C	b.p. °C	Density	n_D	Solubility w	al	eth	ace	bz	other solvents	Ref.
	2,5-Pyridinedicarboxylic acid														
p2066	**2,5-Pyridinedicarboxylic acid**	Isocinchomeronic acid. 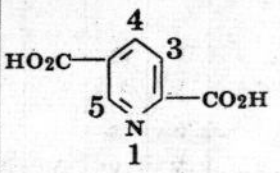	167.12	lf (w+1), (al, dil HCl)	236–7	sub	....	....	s^h	δ^h	i	...	i	HCl s^h	**$B22^2$, 105**
p2067	**2,6-Pyridinedicarboxylic acid**	Dipicolinic acid. 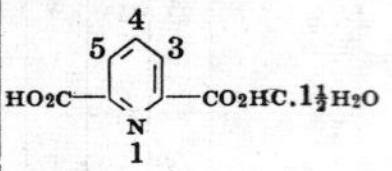 $CO_2HC.1\frac{1}{2}H_2O$	194.14	nd (w+1½)	252 (anh) (228d sealed tube)	....	....	....	δ	s^h δ	...	...	...	aa δ	**$B22^2$, 106**
p2068	**3,4-Pyridinedicarboxylic acid**	Cinchomeronic acid. 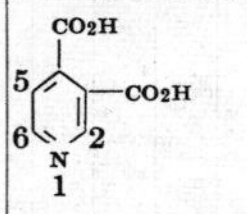	167.12	pr or nd, lf (w)	266d	....	....	....	δ^h	δ	δ	...	δ	chl i	**$B22^2$, 106**
p2069	**3,5-Pyridinedicarboxylic acid**	Dinicotinic acid. 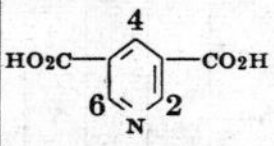	167.12	(ac-w), pr (gl aa)	323d	sub	....	....	δ	...	δ	...	...	HCl s^h aa δ	**$B22^2$, 107**
p2070	**3,4-Pyridinedicarboxylic acid, 4,5-dihydro-2,6-dimethyl-,** diethyl ester	$C_{13}H_{19}NO_4$. *See* p2068	253.30	bl flr, pl (eth)	85	....	....	....	i	v	s^h	...	...	os s lig i	**$B22^2$, 98**
p2071	**3,5-Pyridinedicarboxylic acid, 1,4-dihydro-2,6-dimethyl-,** diethyl ester	$C_{13}H_{19}NO_4$. *See* p2069	253.30	ye nd or lf (al)	183–5	....	....	....	δ^h	s^h	δ	...	...	chl v	**$B22^2$, 98**
p2072	**—,1,4-dihydro-2,4,6-trimethyl-,** diethyl ester	$C_{14}H_{21}NO_4$. *See* p2069	267.33	lt bl flr pl (al)	131	....	....	....	δ	δ v^h	δ	...	...	chl v CS_2 δ	**$B22^2$, 100**
p2073	**3,4-Pyridinedicarboxylic acid, 2,6-dimethyl-,** diethyl ester	$C_{13}H_{17}NO_4$. *See* p2068	251.29	....	16	270d 163^{13}	....	....	...	...	...	...	...	....	**$B22^2$, 109**
p2074	**3,5-Pyridinedicarboxylic acid, 2,6-dimethyl-,** diethyl ester	$C_{13}H_{17}NO_4$. *See* p2069	251.29	nd (al), pr (eth)	75–6	301–2 178^{14}	....	....	i	v	δ	...	v	chl v lig v	**$B22^2$, 110**
p2075	**2,4-Pyridinedicarboxylic acid, 6-methyl-**	Uvitonic acid. $C_8H_7NO_4$. *See* p2065	181.15	(w)	282d	....	....	....	δ	...	...	...	i^h	....	**$B22^2$, 107**
p2076	**3,4-Pyridinedicarboxylic acid, 6-methyl-5-methoxy-**	$C_9H_9NO_5$. *See* p2068	211.17	....	213–5				s^h	...	...	...	...	....	....
—	**Pyridinediol**	*see* **Pyridine, dihydroxy-**													
p2077	**Pyridinepentacarboxylic acid,** dihydrate	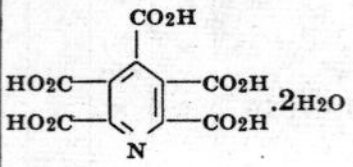 .2H_2O	335.18	(eth, w)	220d	....	....	....	v	δ	i	...	...	....	**B22, 190**
p2078	**3-Pyridinesulfonic acid**	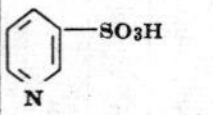	159.16	orh	357d	....	1.718^{25}_{25}	....	v	δ	i	...	...	....	**B22, 387**
p2079	**2,3,4,5-Pyridinetetracarboxylic acid**	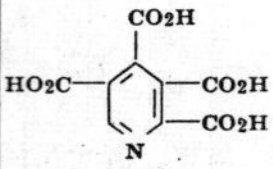	255.14	(w+2 or 3)	160d (−w115)	....	....	....	s	...	...	...	...	....	....
p2080	**2,3,4,6-Pyridinetetracarboxylic acid**	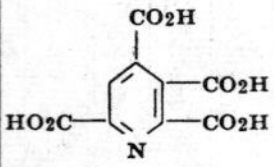	255.14	nd (w+3)	236d	....	....	....	v	δ	δ	...	...	aa s	**$B22^2$, 142**
p2081	**2,3,5,6-Pyridinetetracarboxylic acid**	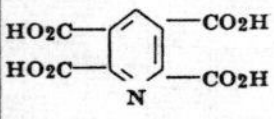	255.14	(w+2)	150d	....	....	....	v	...	...	...	...	....	....

For explanations, symbols and abbreviations see beginning of table.

No.	Name	Synonyms and Formula	Mol. wt.	Crystalline form, color and specific rotation	m.p. °C	b.p. °C	Density	n_D	Solubility w	al	eth	ace	bz	other solvents	Ref.
	2,3,4-Pyridinetricarboxylic acid														
p2082	**2,3,4-Pyridinetri-carboxylic acid**	α-Carbocinchomeronic acid. 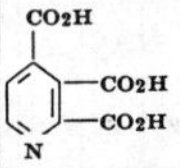 [structure: CO_2H, CO_2H, CO_2H, N]	211.13	lf (dil sulf ±1½w, (w+2)	250				s v^h	δ s^h	i	...	i		**B22**², 136
p2083	**2,3,5-Pyridinetri-carboxylic acid**	Carbodinicotinic acid. 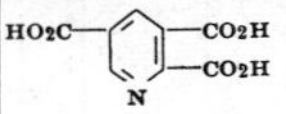 [structure: HO_2C, CO_2H, CO_2H, N]	211.13	pl lf or nd (w+2, dil al +2w)	>100 (130d)				v	δ	δ	...	...	dil al v aa δ	**B22**², 136
p2084	**2,4,5-Pyridinetri-carboxylic acid,** hydrate	Berberonic acid. 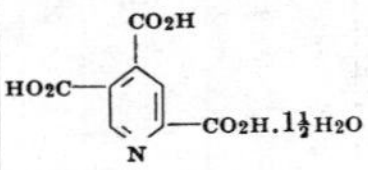 [structure: CO_2H, HO_2C, $CO_2H.1\frac{1}{2}H_2O$, N]	238.16	tcl pr (dil HCl +2w)	243 (anh) 235 (hyd)				δ v^h	$δ^h$	δ	...	δ	dil ac s chl i	**B22**², 136
p2085	**2,4,6-Pyridinetri-carboxylic acid**	Trimesitic acid. 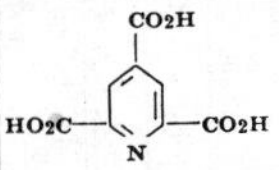 [structure: CO_2H, HO_2C, CO_2H, N]	211.13	nd (w+2)	227d	sub			δ s^h	δ s^h	δ	...	...		**B22**², 136
p2086	**3,4,5-Pyridinetri-carboxylic acid**	β-Carbocinchomeronic acid. 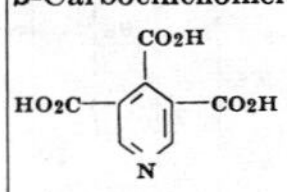 [structure: CO_2H, HO_2C, CO_2H, N]	211.13	lf or pl (w+3), ta (w)	(−w115) 261d				δ s^h	s	...	...	...		**B22**, 186
p2087	**Pyridinium, 1-ethyl-,bromide**	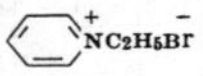 [structure: $\overset{+}{N}C_2H_5\overset{-}{Br}$]	188.07	(al)	111–2				s	s	i	...	...		**B20**, 214
p2088	**—,1-hexadecyl-,** chloride	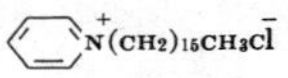 [structure: $\overset{+}{N}(CH_2)_{15}CH_3\overset{-}{Cl}$]	340.00	wh pw	77–83				v	...	δ	...	δ	chl v	
—	**Pyridone**	*see* **Pyridine, hydroxy-**													
p2089	**Pyridoxal,** hydrochloride	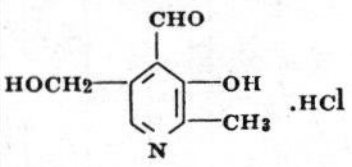 [structure: CHO, $HOCH_2$, OH, CH_3, N] .HCl	203.63	rh	165d				v	δ	...				
p2090	**—,oxime**	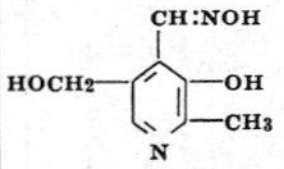 [structure: CH:NOH, $HOCH_2$, OH, CH_3, N]	182.18	(al)	225–6d				...	s^h	...				
p2091	**Pyridoxamine,** dihydrochloride	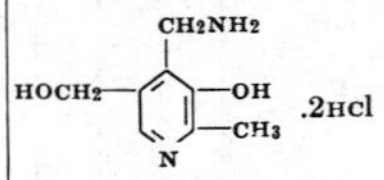 [structure: CH_2NH_2, $HOCH_2$, OH, CH_3, N] .2HCl	241.12	pl	226–7d				v	δ	...				
—	**Pyridoxin**	*see* **Vitamin B_6**													
p2092	**Pyrimidine**	1,3-Diazin*. Niazin. 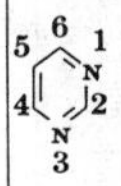 [structure: 5 6 1 N 4 2 N 3]	80.09		22	123.5–4			s	s	...				
p2093	**—,2-amino-**	$C_4H_5N_3$. *See* p2092	95.11		127–8	sub			v	...	...	...	...		**B24**, 80
p2094	**—,4-amino-**	$C_4H_5N_3$. *See* p2092	95.11	lf (AcOEt)	151–2				v	v	...	...	...	AcOEt s^h	**B24**, 81
p2095	**—,2-amino-4,6-dihydroxy-**	$C_4H_5N_3O_2$. *See* p2092	127.11	pl (w+1)	>310				i	i	...	...	...	$PhNO_2$ $δ^h$ ac, alk s	**B24**, 468
p2096	**—,6-amino-2,4-dihydroxy-**	4-Aminouracil. $C_4H_5N_3O_2$. *See* p2092	127.11	(w)	d				...	...	...	...	...	min ac v aa i	**B24**, 469
p2097	**—,2-amino-4,5-dimethyl-**	$C_6H_9N_3$. *See* p2092	123.16	nd (w)	214–5				..	s	δ	s	s	chl s lig δ	**B24**, 91
p2098	**—,4-amino-2,6-dimethyl-**	$C_6H_9N_3$. *See* p2092	123.16	nd (al), pl (bz)	183	sub			s	δ s^h	...	...	δ s^h		**B24**, 89
p2099	**—,6-amino-4,5-dimethyl-**	$C_6H_9N_3$. *See* p2092	123.16	nd (w)	230				...	s^h	δ	s	i	chl s	**B24**, 92
p2100	**—,2-amino-4-methyl-**	$C_5H_7N_3$. *See* p2092	109.14	lf (w), nd (sub)	159–60	sub			s^h	s	...	...	...		**B24**, 84
p2101	**—,2-amino-5-methyl-**	$C_5H_7N_3$. *See* p2092	109.14	pl (sub), pr (w)	193.5				s^h	s	δ	...	δ	lig δ	**B24**, 87

For explanations, symbols and abbreviations see beginning of table.

No.	Name	Synonyms and Formula	Mol. wt.	Crystalline form, color and specific rotation	m.p. °C	b.p. °C	Density	n_D	Solubility w	al	eth	ace	bz	other solvents	Ref.
	Pyrimidine														
p2102	**—,4-amino-2-methyl-**	$C_5H_7N_3$. *See* p2092	109.14	rh (ace)	205				v	s	...	s	s		**B24**, 84
p2103	**—,4-amino-5-methyl-**	$C_5H_7N_3$. *See* p2092	109.14	pl (al, AcOEt)	176				v	s^h	δ	...	s	chl, peth δ lig δ	**B24**, 87
p2104	**—,6-amino-4-methyl-**	$C_5H_7N_3$. *See* p2092	109.14	pr (w), nd, lf (sub)	194–5	sub			s	s	...	...	...		**B24**, 85
p2105	**—,2-amino-5-nitro-**	$C_4H_4N_4O_2$. *See* p2092	140.11	nd (al)	236				δ	s	i	s	i	alk s chl i	**B24**¹, 231
p2106	**—,2-chloro-4(dimethyl-amino)6-methyl-**	$C_7H_{10}ClN_3$. *See* p2092	171.63	br wx so	87	$140–7^{4}$			i	s	...	...	...		
p2107	**—,2,4-dichloro-5-methyl-**	$C_5H_4Cl_2N_2$. *See* p2092	163.02	pl (al)	25–6	235^{759}			δ	v	v	...	v	chl v	**B23**, 93
p2108	**—,2,4-dichloro-6-methyl-**	$C_5H_4Cl_2N_2$. *See* p2092	163.02	nd (lig)	46–7	219			δ	v	v	...	v	chl v lig s^h	**B23**, 92
p2109	**—,2-hydroxy,** hydrochloride	$C_4H_4N_2O.HCl$. *See* p2092	132.56	(al)	203–5				v	s^h	...	...	...		**B24**¹, 231
p2110	**—,2-methyl-**	$C_5H_6N_2$. *See* p2092	94.11		−4 to −5	138			v	...	...	...	...		**B23**, 92
p2111	**—,4(methyl-amino)-**	$C_5H_7N_3$. *See* p2092	109.14	lf (al)	74–5				v	...	s^h	...	...		**B24**², 38
—	**2,4-Pyrimi-dinedione**	*see* **Uracil**													
—	**2(1)Pyrimidone, 4-amino-**	*see* **Cytosine**													
p2112	**Pyrocalciferol**	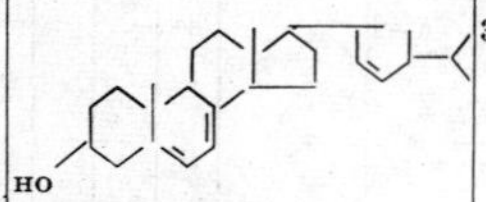	396.63	nd (MeOH), $[\alpha]^{20}_D+502$	93–5				...	s	...	...	...	MeOH s^h	**E14**, 77
—	**Pyrocatechol**	*see* **Benzene, 1,2-dihydroxy-***													
—	***o*-Pyrocatechuic acid**	*see* **Benzoic acid, 2,3-dihydroxy-**													
—	**Pyrocinchonic acid**	*see* **Maleic acid, dimethyl-**													
p2113	**Pyrocoll**	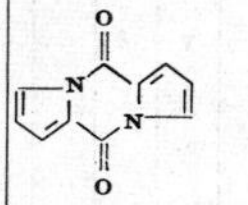	186.17	ye mcl lf (aa)	269	sub			i	s	s	...	δ	chl s aa s^h	**B24**¹, 360
—	**Pyrogallol**	*see* **Benzene, 1,2,3-trihydroxy-***													
—	**Pyrogallo-phthalein**	*see* **Gallein**													
—	**Pyromellitic acid**	*see* **1,2,4,5-Benzenetetra-carboxylic acid***													
—	**Pyromucic acid**	*see* **2-Furancarboxylic acid**													
p2114	**4-Pyrone**	4-Oxo-1,4-pyran. γ-Pyrone. 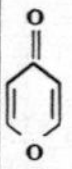	96.08	hyg	32.5	$215–7^{742}$	1.190	1.5238	v	...	...	...	...		**B17**, 271
p2116	**2-Pyrone, 4,6-dimethyl-**	5-Hydroxy-3-methyl-2,4-hexadienoic acid lactone.* 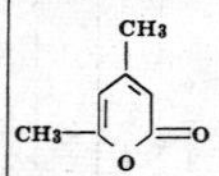	124.14	lf (eth)	51.5	245 126^{11}			v	v	v	...	...	CS_2 δ	**B17**, 291
p2117	**4-Pyrone, 2,6-dimethyl-**	$C_7H_8O_2$. *See* p2114	124.14	pl, nd (sub)	132	248–50	0.9953^{137}		v	v	s	...	...		**B17**², 315
p2118	**2-Pyrone, 3-hydroxy-**	2,5-Dihydroxy-2,4-hexadienoic acid 5-lactone*. Isopyromucic acid. 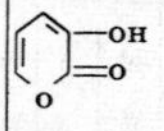	112.08	nd	92	$90–100^{1}$			v	v	s	...	...	lig s^h	**B17**¹, 233
p2119	**4-Pyrone, 5-hydroxy-2(hydroxy-methyl)-**	Kojic acid. $C_6H_6O_4$. *See* p2114	142.11	nd, pr (ace)	152–3	sub			δ s^h	s	δ	...	...	aa δ	**B18**², 57

For explanations, symbols and abbreviations see beginning of table.

No.	Name	Synonyms and Formula	Mol. wt.	Crystalline form, color and specific rotation	m.p. °C	b.p. °C	Density	n_D	Solubility w	al	eth	ace	bz	other solvents	Ref.
	4-Pyrone														
p2120	**—,3-hydroxy-2-methyl-**	Larixinic acid. Maltol. $C_6H_6O_3$. *See* p2114	126.11	red mcl pr (chl) or bipym pr (50 % al)	161–2	sub	...	...	δ v^h	...	δ	...	...	chl v alk, ac s aa δ	**B17**[2], 450
—	**2-Pyrone-5-carboxylic acid**	*see* **Coumalic acid**													
p2121	**4-Pyrone-2-carboxylic acid, 5,6-dihydro-6,6-dimethyl-,** butyl ester	Butopyronoxyl. Indalone. O CH_3— —$CO_2C_4H_9^n$ CH_3 O	226.26	ye or pa red-br	...	256–70	1.052^{25}	1.4745^{25}	i	∞	∞	...	∞	chl ∞	...
p2122	**—,5-hydroxy-**	Comenic acid. O HO— —CO_2H O	156.09	pw	>270	...	...	...	s	i s^h	...	...	...	...	**B18**[2], 352
—	**4-Pyrone-2,6-dicarboxylic acid**	*see* **Chelidonic acid**													
p2123	**Pyrophosphoric acid,** tetraamide, octamethyl-	OMPA. Scharadan. $[(CH_3)_2N]_4P_2O_3$	286.26	...	...	$118–22^{0.3}$	1.1343^{25}	1.462^{25}	s	s	...	...	...	chl s	...
p2124	**—,tetraethyl ester**	$(C_2H_5O)_4P_2O_3$	290.20	...	...	155^5	1.1847^{20}_4	1.4180^{20}	$\infty^{\delta d}$	∞	∞	∞	...	chl, xyl ∞ lig i	**B1**[3], 1330
p2125	**—,dithiono-,** tetraethyl ester	$(C_2H_5O)_4P_2OS_2$	322.33	lt ye	...	$136–9^2$	1.196^{25}_{25}	1.4753^{20}	i	s	...	...	...	...	...
—	**Pyrotartaric acid**	*see* **Succinic acid, methyl-**													
—	**Pyrrocoline, octahydro-**	*see* **Piperolidine**													
p2126	**Pyrrole**	Azole.* 5 1 NH 4 2 3	67.09	...	...	$130–1^{761}$	0.9669^{21}	...	v	δ	δ	...	...	...	**B20**, 159
p2127	**—,1-acetyl-**	C_6H_7NO. *See* p2126	109.12	...	...	181–2	...	...	δ	...	...	...	...	HCl d	**B20**, 165
p2128	**—,2-acetyl-**	C_6H_7NO. *Se* p2126	109.12	mcl nd (w)	90	218–20	...	...	s	s	s	...	...	...	**B21**[2], 236
p2129	**—,1-benzyl-**	$C_{11}H_{11}N$. *See* p2126	157.21	...	...	247 138^{27}	...	...	i	v	v	...	...	...	**B20**[1], 39
p2130	**—,1-butyl-**	$C_8H_{13}N$. *See* p2126	123.20	oil	...	$53–4^{11}$	...	...	...	...	...	...	...	...	**B20**[2], 83
—	**—,dihydro-**	*see* **Pyrroline**													
p2131	**—,2,4-dimethyl-**	C_6H_9N. *See* p2126	95.14	pa bl flr	...	165^{743} cor	0.927	...	δ	v	v	...	...	aa δ	**B20**, 172
p2132	**—,2,5-dimethyl-**	C_6H_9N. *See* p2126	95.14	...	...	$170–2^{765}$	0.9353^{20}_4	1.5025^{20}	...	...	...	...	...	...	**B20**, 173
p2133	**—,2,3-dimethyl-4-ethyl-**	Hemopyrrole. $C_8H_{13}N$. *See* p2126	123.20	...	16–7	198^{725} 88^{11-2}	0.915^{20}_4	...	v^h	...	...	...	...	...	**B20**[2], 91
p2134	**—,2,4-dimethyl-3-ethyl-**	Kryptopyrrole. $C_8H_{13}N$. *See* p2126	123.20	pr	0	197^{710} $84–5^{13}$	0.913^{20}_4	...	δ	s	s	...	s	chl s	**B20**[2], 91
p2135	**—,1,3-diphenyl-**	$C_{16}H_{13}N$. *See* p2126	219.29	pl (al)	122–3	...	...	...	...	s^h	v	...	v	chl s	**B20**[1], 148
p2136	**—,2,5-diphenyl-**	$C_{16}H_{13}N$. *See* p2126	219.29	lf (aa, dil al)	143	...	...	...	i	v	...	...	v	alk i HCl δ (red)	**B20**[2], 313
p2137	**—,1-ethyl-**	C_6H_9N. *See* p2126	95.15	...	...	$129–30^{762}$	0.9009^{20}_4	1.4841^{29}	i	s	...	...	...	...	**B20**, 163
p2138	**—,2-ethyl-**	C_6H_9N. *See* p2126	95.15	...	...	163–5	0.9042^{20}_4	1.4942^{20}	...	...	...	...	...	...	**B20**[2], 85
p2139	**—,3-ethyl-4-methyl-**	Opsopyrrole. $C_7H_{11}N$. *See* p2126	109.17	...	3	$74–5^{11}$	0.9059	1.4913^{20}	...	...	...	...	...	...	**B20**[2], 89
p2140	**—,3-ethyl-2,4,5-trimethyl-**	Phyllopyrrole. $C_9H_{15}N$. *See* p2126	137.23	pl (sub) wh lf (peth),	67.5–8.5	213^{725} 100^{17}	...	...	i	s	s	...	...	lig δ	**B20**[2], 93
p2141	**—,2-isopropyl-**	$C_7H_{11}N$. *See* p2126	109.17	...	...	$171–2^{741}$	0.908^{25}_4	1.491^{25}	...	...	...	...	...	...	**B20**, 176
p2142	**—,1-methyl-**	C_5H_7N. *See* p2126	81.12	...	...	$114–5^{747}$ cor	0.9145^{15}_4	1.4899^{17}	i	∞	∞	...	...	...	**B20**, 163
p2143	**—,2-methyl-**	C_5H_7N. *See* p2126	81.12	...	...	$147–8^{750}$	0.9446^{15}_4	1.5035^{16}	i	∞	∞	...	...	...	**B20**, 170
p2144	**—,3-methyl-**	C_5H_7N. *See* p2126	81.12	...	...	$142–3^{743}$	...	...	...	∞	∞	...	...	...	**B20**[2], 85 **B21**[2], 767
—	**—,oxo(tetrahydro)-**	*see* **Pyrrolidone**													
p2145	**—1-phenyl-**	$C_{10}H_9N$. *See* p2126	143.19	pl (sub) red in air	62	234 140^{38}	...	...	i	s	s	...	s	peth v, chl s	**B20**[2], 83

For explanations, symbols and abbreviations see beginning of table.

No.	Name	Synonyms and Formula	Mol. wt.	Crystalline form, color and specific rotation	m.p. °C	b.p. °C	Density	n_D	Solubility w	al	eth	ace	bz	other solvents	Ref.
	Pyrrole														
p2146	**—,2-phenyl**	$C_{10}H_9N$. *See* p2126	143.19	lf (al), pl (sub)	129	271–2[726]				v	v		v	chl v	**B20²**, 238
p2147	**—,1-propyl-**	$C_7H_{11}N$. *See* p2126	109.17			145.5–6.5									**B20¹**, 44
—	**—,tetrahydro**	*see* **Pyrrolidine**													
p2148	**—,2,3,4,5-tetra-iodo-**	Iodol, C_4HI_4N. *See* p2126	570.68	ye nd (al)	150d				i	s	s	s		chl s	**B20²**, 84
p2149	**—,2,3,4,5-tetra-methyl-**	$C_8H_{13}N$. *See* p2126	123.19	lf	111									peth s[h]	**B20²**, 92
p2150	**2-Pyrrolcarboxal-dehyde**	2-Formylpyrrole. 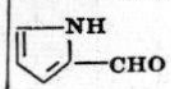	95.10	rh pr (peth)	46–7	217–9 114[15]		1.5939[16]						os v lig δ s[h]	**B21²**, 236
p2151	**2-Pyrrolecar-boxylic acid**	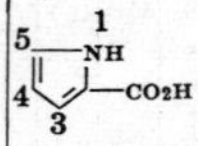	111.10	lf (w)	192d	208d			s	s	s				**B22²**, 15
p2152	**—,4-acetyl-3,5-dimethyl,** ethyl ester	$C_{11}H_{15}NO_3$. *See* p2151	209.25	wh nd (al)	143–4				i	δ					**B27²**, 233
p2152¹	**3-Pyrrolecar-boxylic acid**	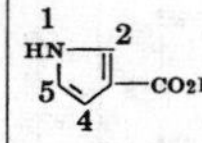	111.10												
p2153	**2-Pyrrolecar-boxylic acid, 3,5-dimethyl-,** ethyl ester	$C_9H_{13}NO_2$. *See* p2151	167.21	(al)	125					s				os s	**B22²**, 21
p2154	**—,2,4-dimethyl-,** ethyl ester	$C_9H_{13}NO_2$. *See* p2152¹	167.21	(eth-lig or peth)	75–6	291 181–2[35]			δ	v	v			peth s[h]	**B22²**, 18
p2155	**3-Pyrrolecar-boxylic acid, 2,5-dimethyl-,** ethyl ester	$C_9H_{13}NO_2$. *See* p2152¹	167.21	rh (al)	117–8	290[731] 130[15]			i	s				dil ac, alk i os s	**B22²**, 22
p2156	**—,4,5-dimethyl-,** ethyl ester	$C_9H_{13}NO_2$. *See* p2152¹	167.21	(dil al)	110–1				δ	v	v			chl v	**B22²**, 20
p2156¹	**2,4-Pyrroledi-carboxylic acid**	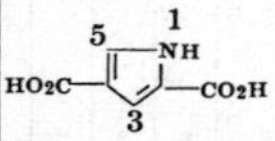	155.11												**B22¹**, 525
p2157	**—,3,5-dimethyl-,** diethyl ester	$C_{14}H_{21}NO_4$. *See* p2156¹	267.33	nd (dil al)	97					s[h]					**B22²**, 102
p2158	**—,3,5-dimethyl-,** diethyl ester	$C_{12}H_{17}NO_4$. *See* p2156¹	239.27	lo nd	136–7				i	δ	s	s	s	chl s lig δ	**B22²**, 94
p2159	**—,5-formyl-3-methyl-,** diethyl ester	$C_{12}H_{15}NO_5$. *See* p2156¹	253.26	nd	124–5				i	s				to s lig i	**B22²**, 278
p2160	**Pyrrolidine**	1-Azacyclopentane. Tetra-hydropyrrole. Tetra-methylenimine. 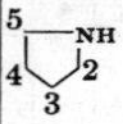	71.12			88.5–9	0.8520^{22}_{4}	1.4270[15]	∞	s	s			chl s	**B20²**, 3
p2161	**—,1-butyl-**	$C_8H_{17}N$. *See* p2160	127.23			154–5[758]		1.437[27]		s				os s	**B20²**, 4
p2162	**—,2-butyl-**	$C_8H_{17}N$. *See* p2160	127.23			154–6[753] 67[18]			δ[h] s	∞				os ∞	**B20²**, 65
p2163	**—,1(chloro-acetyl)-**	$C_6H_{10}ClNO$. *See* p2160	147.61		44–6	112[0.5]									
p2164	**—,1,2-dimethyl-**	$C_6H_{13}N$. *See* p2160	99.18			96[762]	0.7986[15]		∞	v	v				**B20²**, 55
p2165	**—,2,4-dimethyl-**	$C_6H_{13}N$. *See* p2160	99.18			115–7[753]	0.8297^{20}_{4}	1.4325[20]	v	v	v				**B20**, 102
p2166	**—,2,5-dimethyl-**	$C_6H_{13}N$. *See* p2160	99.18			106–8[746]		0.8185^{12}_{4}	∞	∞	∞				**B20²**, 60
p2167	**—,1-methyl-**	$C_5H_{11}N$. *See* p2160	85.15			81–3	0.8188^{20}_{4}	1.4311[20]	∞		s				**B20²**, 4
—	**—,oxo-**	*see* **Pyrrolidone**													
p2168	**—,1-phenyl-**	$C_{10}H_{13}N$. *See* p2160	147.22			110–6[9]	1.0260[25]	1.5803[25]							
—	**2-Pyrrolidinecar-boxylic acid**	*see* **Proline**													
p2169	**2-Pyrrolidone**	γ-Butyrolactam. 2-Oxopyr-rolidine. 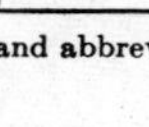	85.11	(peth)	24.6	250.5[742] 133[12]	1.120^{20}_{4}		v	v	v		v	CS_2 v chl v	**B21²**, 213

For explanations, symbols and abbreviations see beginning of table.

No.	Name	Synonyms and Formula	Mol. wt.	Crystalline form, color and specific rotation	m.p. °C	b.p. °C	Density	n_D	Solubility w	al	eth	ace	bz	other solvents	Ref.
	2-Pyrrolidone														
p2170	**—,1,5-dimethyl-**	*N*-Methyl-γ-valerolactam. $C_6H_{11}NO$. *See* p2169	113.16			215–7^{743}			∞		∞				**B21**, 239
p2171	**—,3,3-dimethyl-**	$C_6H_{11}NO$. *See* p2169	113.16	lf (bz)	65–7	237			v	v			v^h		**B21**, 242
p2172	**—,1-methyl-**	C_5H_9NO. *See* p2169	99.13		−17, −16	197–202^{736}	1.0260^{25}_{25}	1.4666^{25}	v					os s	**B21**^{2}, 213
p2173	**Pyrroline**	Dihydropyrrole. (structure: ring with NH)	69.11			90–1^{750}	0.9097^{20}_4	1.4664^{20}	s	∞	∞				**B20**^{2}, 67
—	**Pyruvaldehyde**	*see* **Propanal, 2-oxo-***													
—	**Pyruvic acid**	*see* **Propanoic acid, 2-oxo-***													
—	**Pyruvonitrile**	*see* **Propanoic acid, 2-oxo, nitrile**													

For explanations, symbols and abbreviations see beginning of table.

No.	Name	Synonyms and Formula	Mol. wt.	Crystalline form, color and specific rotation	m.p. °C	b.p. °C	Density	n_D	Solubility w	al	eth	ace	bz	other solvents	Ref.
	Quaterphenyl														
q1	*o,o'*-**Quaterphenyl**		306.41	.....	118	420	.....	.....	i	...	...	s	...	MeOH δ chl v	**$B5^2$**, 669
q1[1]	*p,p'*-**Quaterphenyl**	Tetraphenyl.	306.41	pl (bz)	320	428^{18}	.....	.....	i	i	i	...	s^h	chl i aa s^h	**$B5^2$**, 669
q2	**Quercetrin**		448.39	pa ye nd $[\alpha]_{578}-73.5$ (al)	168.7	250–2	.....	.....	s^h	s	s	...	...	aa s^h	**B31**, 75
—	**Quercitin**	*see* **Flavone, 3,3',4',5,7-pentahydroxy-**													
—	**Quercitol**	*see* **1,2,3,4,5-Cyclohexanepentol***													
q3	**Quillaic acid**	Quillaja sapogenin.	486.70	nd (dil al) $[\alpha]^{20}_D+56.1$	292–3	.....	.....	.....	...	s	s	s	...	Py s aa s	**$B10^2$**, 737
q4	**Quinacrine, dihydrochloride** (*dl*)	Atabrine.	508.92	yesh nd	248–50d	.....	.....	.....	...	δ	i	i	i	MeOH s	
—	**Quilaldic acid**	*see* **2-Quinolinecarboxylic acid**													
—	**Quinaldine**	*see* **Quinoline, 2-methyl-**													
—	**Quinalizarin**	*see* **9,10-Anthraquinone, 1,2,5,8-tetrahydroxy**													
q5	**Quinamine**		312.42	pr (bz), nd $[\alpha]+116$ (al)	185–6	.....	.....	.....	i	v^h	s	...	v	lig v	**$B27^2$**, 667
q6	**Quinazoline**	1,3-Benzodiazine. Benzo(*a*)pyrimidine.	130.15	ye pl (peth)	48.0–.5	241.5^{764}	.....	.....	v	s	...	...	...	oos v	**$B23^2$**, 177
q7	**—,3,4-dihydro-4-oxo-**	$C_8H_6N_2O$. *See* q6	146.15	nd	209–16.5	360	.....	.....	s^h	s^h	...			.....	**B24**, 143
q8	**—,3,4-dihydro-2-phenyl-**	$C_{14}H_{12}N_2$. *See* q6	208.27	pl (bz)	>200	.....	.....	.....	...	v	v	...	...	chl v aa v	**B23**, 239
q9	**—,3,4-dihydro-3-phenyl-**	Orexin. Phenzoline. Cedrarine. $C_{14}H_{12}N_2$. *See* q6	208.27	pl (eth-lig)	95	d	1.290^4	.....	i	v	v	...	v	chl, CS_2 v	**$B23^2$**, 155
q10	**—,3,4-dihydro-4-phenyl-**	$C_{14}H_{12}N_2$. *See* q6	208.27	pl (al, AcOEt)	165–6	.....	.....	.....	...	...	...	...	...	.....	**B23**, 239
q11	**—,2,4-dihydroxy-**	2,4-Quinazolinedione. $C_8H_6N_2O_2$	162.15	nd (w, al) lf (aa)	356 cor	.....	.....	.....	δ^h	δ	δ	...	...	alk s	**$B24^2$**, 197
q12	**—,2,4-dihydroxy-6,8-dinitro-**	$C_8H_4N_4O_6$. *See* q6	252.14	ye gr pr (aa)	274–5d	.....	.....	.....	s^h	δ	δ	...	δ	chl δ aa s^h	**$B24^1$**, 344
q13	**—,2,4-dioxo-1,2,3,4-tetrahydro-**	2,4-Quinazolinedione. Benzoyleneurea. $C_8H_6N_2O_2$	162.15	nd	353–4	.....	.....	.....	δ^h	s	...	...	...	.....	**$B24^2$**, 197
q14	**Quinhydrone**	Benzoquinhydrone.	218.21	red br nd	171	subl	1.401^{20}_4	.....	s^h	v	v	...	...	chl δ lig i	**$B7^2$**, 572

For explanations, symbols and abbreviations see beginning of table.

No.	Name	Synonyms and Formula	Mol. wt.	Crystalline form, color and specific rotation	m.p. °C	b.p. °C	Density	n_D	Solubility w	al	eth	ace	bz	other solvents	Ref.
	Quinic acid														
—	**Quinic acid**	*see* **Cyclohexanecarboxylic acid, 1,2,3,4-tetrahydroxy-***													
q15	**Quinicine**	Quinotoxin. $C_{20}H_{24}N_2O_2$	324.43	red ye amor $[\alpha]_D^{15}+44.1$ (chl)	*ca.* 60				δ	v	v			chl v	B25², 20
q16	—,oxalate(*d*)	$(C_{20}H_{24}N_2O_2)_2.H_2C_2O_4.9H_2O$	901.03	pr (chl), nd (al) $[\alpha]_D^{15}+19.5$ (al-chl)					v^h	s				chl v	B25, 39
q17	**Quinidine**	$C_{20}H_{24}N_2O_2$	324.43	lf (w) $[\alpha]_D^{51}$ +230 (chl), +243.5 (al)	174–5				δ		δ		s	chl v	B23², 414
q18	—,hydrate	Conquinine. $C_{20}H_{24}N_2O_2.2\frac{1}{2}H_2O$. *See* q17	369.46	(dil al)	171.5d				δ	s	s			chl s	B23², 414
q19	—,bisulfate	$C_{20}H_{24}N_2O_2.H_2SO_4.4H_2O$. *See* q17	494.55	pr, nd, flr in sol $[\alpha]_D+184$ (chl)					s	v	δ				
q20	—,hydrochloride	$C_{20}H_{24}N_2O_2.HCl.H_2O$. *See* q17.	378.97		285–9 (anh)				s	s^h				chl δ	
q21	—,sulfate(*d*)	$(C_{20}H_{24}N_2O_2)_2.H_2SO_4.2H_2O$. *See* q17	782.96	pr, nd, sol flr bl, $[\alpha]_D^{25}$ +212 (al)					s	v	δ			chl s	
q22	**Quinine**	$C_{20}H_{24}N_2O_2.3H_2O$	378.47	efflor flaky or micr cr pw $[\alpha]_D^{15}-145.2$ (al)	57			1.625	δ	s	s				
q23	—,bisulfate	$C_{20}H_{24}N_2O_2.H_2SO_4.7H_2O$	548.61	li efflor orh nd $[\alpha]_D^{15}+168.4$ (w)	160d				v	s	δ			chl s	
q24	—,ethyl carbonate	$C_{23}H_{28}N_2O_4$	396.49	nd (dil al)	95				δ	v	v			chl v	B23², 424
q25	—,formate	Quinoform. $C_{20}H_{24}N_2O_2.HCOOH$.	370.45	nd $[\alpha]_D^{20}$ −144.2 (w)	109d				s	s	δ				
q26	—,hydrobromide	$C_{20}H_{24}N_2O_2.HBr.H_2O$.	423.36	silky efflor nd	152–200				s	s	v			chl s	
q27	—,hydrochloride	$C_{20}H_{24}N_2O_2.HCl$.	360.89	silky efflor nd $[\alpha]_D^{15}-145$ (al)	158–60	259d			s	v	δ			chl s alk s	
q28	—,hydrochloride hydrate	$C_{20}H_{24}N_2O_2.HCl.2H_2O$	396.92	silky efflor nd	156–90				v	v	δ			chl v	
q29	—,2-hydroxybenzoate	Quinine salicylate. $C_{20}H_{24}N_2O_2.C_7H_6O_3H_2O$.	480.57	wh	183–7				δ	v	δ			chl v glycerol v	
q30	—,pentanoate	Quinine alerate. $C_{20}H_{24}N_2O_2.CH_3(CH_2)_3COOH.H_2O$	444.61	wh	*ca.* 95				s	v	v				
q31	—,sulfate	$2(C_{20}H_{24}N_2O_2).H_2SO_4$.	746.93	silky nd	235.2				s^h	s	δ				
q32	—,sulfate dihydrate	$2(C_{20}H_{24}O_2N_2).H_2SO_4.2H_2O$.	782.96	silky nd	205				s^h	s	δ			chl δ	
—	**Quininic acid**	*see* **4-Quinolinecarboxylic acid, 6-methoxy-**													
q33	**Quininone**	[structure: N, —CO—, N, CH:CH2, OCH3]	322.41	nd, lf (eth) $[\alpha]_D^{20}+76$	108				i	v	v		v	chl v Peth i	B25², 23
—	**Quinitol**	*see* **1,4-Cyclohexanediol***													
—	**Quinizarin**	*see* **9,10-Anthraquinone, 1,4-dihydro-***													
q34	**Quinoline**	1-Benzazine*. Benzo[*b*]-pyridine. [structure: 8 1 N 2, 7, 6, 3, 5 4]	129.16		fr −15.9	237.10^{760} $112–4^{15}$	1.09376_4^{20}	1.6268^{20}	s	∞	∞			CS_2 ∞	B20², 222

For explanations, symbols and abbreviations see beginning of table.

No.	Name	Synonyms and Formula	Mol. wt.	Crystalline form, color and specific rotation	m.p. °C	b.p. °C	Density	n_D	Solubility w	al	eth	ace	bz	other solvents	Ref.
	Quinoline														
q35	—,hydrochloride	Quinoliniumchloride. $C_9H_7N.HCl$. *See* q34	165.63		134.5				v		s^h		δ^h	chl, s	B20^2, 222
q36	—,hydrogen sulfate	Quinolinium bisulfate. $C_9H_7N.H_2SO_4$. *See* q34	227.15	(al, aa)	163.5–4.5				s	s^h				aa s^h	B20^2, 137
q37	**—,3-acetamido-**	$C_{11}H_{10}N_2O$. *See* q34	186.22	(w)	166–7				δ						B22, 445
q38	**—,4-acetamido-**	$C_{11}H_{10}N_2O$. *See* q34	186.22	nd (w+1)	176	sub			δ						B22, 445
q39	**—,5-acetamido-**	$C_{11}H_{10}N_2O$. *See* q34	186.22	pl (w), pr (dil al)	178				δ	s					B22^2, 353
q40	**—,6-acetamido-**	$C_{11}H_{10}N_2O$. *See* q34	186.22	nd (w)	138				s^h	s	δ		s^h		B22^2, 355
q41	**—,7-acetamido-**	$C_{11}H_{10}N_2O$. *See* q34	186.22	cr (w, al)	167.5				s^h	s^h					B22^1, 640
q42	**—,8-acetamido-**	$C_{11}H_{10}N_2O$. *See* q34	186.22	nd (al)	103					s^h					B22^1, 640
q43	**—,2-amino-**	α-Quinolylamine. $C_9H_8N_2$. *See* q34	144.17	lf (w)	129	sub			v^h	v^h	s	s	δ	chl s	B22^2, 350
q44	—,—(stable form)	β-Quinolylamine. $C_9H_8N_2$. *See* q34	144.17	pl (to), rh bisphenoidic	94				δ	v	v			chl v	B22^2, 352
q45	**—,3-amino-** (unstable form)	$C_9H_8N_2$. *See* q34	144.17	mcl cr (to, w, dil al)	84				δ	v	v			chl v	B22^2, 352
q46	**—,4-amino-**	β-Quinolylamine. $C_9H_8N_2.H_2O$. *See* q34	144.17	nd (w+1H_2O) nd (dil al, bz)	70	100			s	s			i	chl s CS_2 δ	B22^2, 353
q47	**—,5-amino-**	5-Quinolylamine. $C_9H_8N_2$. *See* q34	144.17	ye nd (al), lf (eth)	110	310 sub 184^{10}			δ	s	s		δ	MeOH s	B22^2, 353
q48	**—,6-amino-**	6-Quinolylamine. $C_9H_8N_2$. *See* q34	144.17	(w+2) pr (eth)	118 (anh)	187^{11}			δ	s	δ			NH_3 s	B22^2, 355
q49	**—,7-amino-**	7-Quinolylamine. $C_9H_8N_2$. *See* q34	144.17	nd (+1w)	93.5–94 (anh) 74–75.5 (hyd)					s				ac s	B22^2, 355
q50	**—,8-amino-**	8-Quinolylamine. $C_9H_8N_2$. *See* q34	144.17	pa ye nd (subl) cr (al, lig)	70	157–62^{20-24}			s						B22^2, 356
q51	**—,2-amino-4-hydroxy-**	$C_9H_8N_2O$. *See* q34	160.18	nd (w+1), (al)	303–4					δ			δ	AcOEt δ chl δ alk, ac s	B22^1, 653
q52	**—,5-amino-6-hydroxy-**	$C_9H_8N_2O$. *See* q34	160.18	nd (w+2)	185 (anhy)				s^h	v	δ		δ	chl δ	B22, 501
q53	**—,5-amino-8-hydroxy-**	$C_9H_8N_2O$. *See* q34	160.18	nd (bz)	143				d^h	d^h			δ		B22^1, 653
p54	**—,7-amino-8-hydroxy-**	$C_9H_8N_2O$. *See* q34	160.18	pr (eth)	117–24d				δ	s	s		s	chl s	B22^2, 417
q55	**—,8-amino-6-methoxy-**	$C_{10}H_{10}N_2O$. *See* q34	174.21		41	145–7									B22^2, 413
q56	**—,2-amino-4-methyl-**	2-Aminolepidine. $C_{10}H_{10}N_2$. *See* q34	158.21	pw (bz)	133	320			δ^h	v	v		v	chl v aa v	B22^2, 364
q57	**—,3-amino-2-methyl-**	3-Aminoquinaldine. $C_{10}H_{10}N_2$. *See* q34	158.21	pa ye nd (al), pr (eth)	160–0.5	278^{760} δd 198^{16}			i	v	s		v	chl v	B22^2, 395
q58	**—,4-amino-2-methyl-**	4-Aminoquinaldine. $C_{10}H_{10}N_2$. *See* q34	158.21	nd (bz-lig)	164	333			δ	v	v	v	v^h	lig δ	B22^2, 359
q59	**—,5-amino-2-methyl-**	5-Aminoquinaldine. $C_{10}H_{10}N_2$. *See* q34	158.21	brsh lf or nd (w+1)	117–8 (anh)				v^h	v	δ		v	lig s	B22^2, 360
q60	**—,5-amino-8-methyl-**	$C_{10}H_{10}N_2$. *See* q34	158.21	yesh nd (w or dil al)	143				δ^h	v					B22, 456
q61	**—,6-amino-2-methyl-**	6-Aminoquinaldine. $C_{10}H_{10}N_2$. *See* q34	158.21	br (w or dil al)	187.5				v^h	v				chl v	B22^2, 361
q62	**—,6-amino-4-methyl-**	6-Aminolepidine. $C_{10}H_{10}N_2$. *See* q34	158.21	nd (w)	169–70				s^h	v	s			chl v ac s	B22, 455
q63	**—,7-amino-2-methyl-**	7-Aminoquinaldine. $C_{10}H_{10}N_2$. *See* q34	158.21	nd (w)	148				v^h		s				B22^2, 363
q64	**—,7-amino-8-methyl-**	$C_{10}H_{10}N_2$. *See* q34	158.21	pr (dil al)	128	304			δ	v	v	v	v	lig δ	B22, 456
q65	**—,8-amino-2-methyl-**	8-Aminoquinaldine. $C_{10}H_{10}N_2$. *See* q34	158.21	pr (lig)	56				δ	v	v				
q65[1]	**—,8-amino-6-methyl-**	$C_{10}H_{10}N_2$. *See* q34	158.21	nd	62–4	sub			s	v				oos v	B22^2, 365
—	**—,5-benzamido-8-ethoxy-**	*see* **Analgen**													
q66	**—,2-bromo-**	C_9H_6BrN. *See* q34	208.06	nd (al)	48–9				δ	v^h	v		v	chl v	B20, 362
q67	**—,3-bromo-**	C_9H_6BrN. *See* q34	208.06		12–3	274–6								aa v^h	B20, 363
q68	**—,4-bromo-**	C_9H_6BrN. *See* q34	208.06		29–30	270d			δ					dil ac v	B20, 364
q69	**—,5-bromo-**	C_9H_6BrN. *See* q34	208.06	nd	48	280^{756}								ac s	B20^2, 235
q70	**—,6-bromo-**	C_9H_6BrN. *See* q34	208.06		24	284				s	s			ac s	B20^2, 238
q71	**—,7-bromo-**	C_9H_6BrN. *See* q34	208.06	nd	34	290								ac v	B20, 365
q72	**—,8-bromo-**	C_9H_6BrN. *See* q34	208.06			305–10 δd 165–6^{18}				v				ac s	B20^2, 235
q73	**—,3-bromo-2-hydroxy-**	β-Bromocarbostyril. C_9H_6BrNO. *See* q34	224.06	pr (al)	253	sub				v^h			v^h		B21, 80

For explanations, symbols and abbreviations see beginning of table.

No.	Name	Synonyms and Formula	Mol. wt.	Crystalline form, color and specific rotation	m.p. °C	b.p. °C	Density	n_D	Solubility						Ref.
									w	al	eth	ace	bz	other solvents	
	Quinoline														
q74	**—,4-bromo-2-hydroxy-**	γ-Bromocarbostyril. C_9H_6BrNO. *See* q34	224.06	nd (al)	266–7	sub				v^h					**B21**, 80
q75	**—,5-bromo-2-hydroxy-**	5-Bromocarbostyril. C_9H_6BrNO. *See* q34	224.06	nd (al)	300					v^h					**B21**, 80
q76	**—,5-bromo-6-hydroxy-**	C_9H_6BrNO. *See* q34	224.06	nd (al)	184–5					v^h				dil ac δ	**B21**[1], 221
q77	**—,5-bromo-8-hydroxy-**	C_9H_6BrNO. *See* q34	224.06	nd (al), nd or lf (sub)	124	sub			v^h	v^h			v	chl v	**B21**[1], 222
q78	**—,6-bromo-2-hydroxy-**	C_9H_6BrNO. *See* q34	224.06	ye nd (al)	269				v	v^h	s			chl s ac s alk s	**B21**, 80
q79	**—,7-bromo-2-hydroxy-**	7-Bromocarbostyril. C_9H_6BrNO. *See* q34	224.06	nd (aa), pl (al)	288	sub				v^h	s			chl s	**B21**, 80
q80	**—,8-bromo-5-hydroxy-**	C_9H_6BrNO. *See* q34	224.06	nd	190d					v					**B21**, 85
q81	**—,2-chloro-**	C_9H_6ClN. *See* q34	163.60		37–8	275^{751}	1.2464^{25}_{4}	1.6259^{25}	i	s	s		s	lig s	**B20**[2], 233
q82	**—,3-chloro-**	C_9H_6ClN. *See* q34	163.60	hyg		$256–7^{764}$									**B20**[2], 234
q83	**—,4-chloro-**	C_9H_6ClN. *See* q34	163.60		34	261^{744} 130^{15}	1.251		δ	s	s			dil HCl s	**B20**[2], 234
q84	**—,5-chloro-**	C_9H_6ClN. *See* q34	163.60	(al)	45	256				s^h					**B20**, 360
q85	**—,6-chloro-**	C_9H_6ClN. *See* q34	163.60	pr (eth)	41	$261–2^{740}$									**B20**[2], 234
q86	**—,7-chloro-**	C_9H_6ClN. *See* q34	163.60	nd or pr	31–2	267–8	1.2158^{58}_{4}		δ	v	v		v	chl v	**B20**[2], 234
q87	**—,8-chloro-**	C_9H_6ClN. *See* q34	163.60		fr. −20	288	1.2834^{14}_{4}		s	v	v		v	chl v	**B20**[2], 234
q88	**—,7-chloro-4-hydroxy-**	C_9H_6ClNO. *See* q34	179.61		276–80										**C44**, 2572
q89	**—,5-chloro-8-hydroxy-7-iodo-**	C_9H_5ClINO. *See* q34	305.50	ye br nd	177–8					v^h				aa v^h	**B21**[2], 58
q90	**—,2-chloro-4-methyl-**	2-Chlorolepidine. $C_{10}H_8ClN$. *See* q34	177.64	nd (dil al)	59	296			δ	v	v			chl v	**B20**[2], 245
q91	**—,2-chloro-6-methyl-**	$C_{10}H_8ClN$. *See* q34	177.64	nd (dil al)	116				δ	v	v		v	chl v	**B20**[1], 151
q92	**—,2-chloro-8-methyl-**	$C_{10}H_8ClN$. *See* q34	177.64	nd (eth)	61	286^{734}			δ	v	v		v	chl v	**B20**[1], 152
q93	**—,3-chloro-2-methyl-**	3-Chloroquinaldine. $C_{10}H_8ClN$. *See* q34	177.64	nd (dil al)	71–2					v	v				**B20**, 392
q94	**—,3-chloro-4-methyl-**	3-Chlorolepidine. $C_{10}H_8ClN$. *See* q34	177.64	nd (dil al)	55					s				HCl v	**B20**[1], 150
q95	**—,3-chloro-6-methyl-**	$C_{10}H_8ClN$. *See* q34	177.64	nd (dil MeOH)	85.5				v					MeOH s	**B20**[2], 246
q96	**—,4-chloro-2-methyl-**	4-Chloroquinaldine. $C_{10}H_8ClN$. *See* q34	177.64	nd (+1w)	43 (+1w)	269–70			δ	v	v		v	chl v CS_2 v	**B20**[2], 241
q97	**—,6-chloro-2-methyl-**	5-Chloroquinaldine. $C_{10}H_8ClN$. *See* q34	177.64	lf or nd (dil al)	93–6					s^h	s				**B20**[2], 242
q98	**—,6-chloro-4-methyl-**	6-Chlorolepidine. $C_{10}H_8ClN$. *See* q34	177.64		71–2					v				oos v	**B20**[2], 245
q99	**—,7-chloro-2-methyl-**	7-Chloroquinaldine. $C_{10}H_8ClN$. *See* q34	177.64	nd (eth) (lig)	77–8	$87^{0.5}$					s^h			lig s^h	**B20**[2], 242
q100	**—,8-chloro-5-methyl-**	$C_{10}H_8ClN$. *See* q34	177.64	nd (w)	49				s^h	v	v		v		**B20**[2], 246
q101	**—,8(chloromethyl)-**	$C_{10}H_8ClN$. *See* q34	177.64	nd or pl (peth)	56						s			peth s^h	**B20**, 402
q102	**—,decahydro-(*cis*)**	$C_9H_{17}N$. *See* q34	139.24		−40	$205–6^{735}$ 90^{20}	0.9426^{20}_{4}	1.4926^{20}	δ	s	s				**B20**[2], 73
q103	—,—(*trans, d*)	$C_9H_{17}N$. *See* q34	139.24	$[\alpha]^{25}_{D}$+4.8 (al)	75	200–202				s					
q104	—,—(*trans, dl*)	$C_9H_{17}N$. *See* q34	139.24		48	203^{735}	0.9021^{55}_{4}		s^h	v	v				
q105	—,—(*trans, l*)	$C_9H_{17}N$. *See* q34	139.24	$[\alpha]^{55}_{4}$ −4.5 (al)	74–5	200–201									
q106	**—,5,7-dibromo-8-hydroxy-**	$C_9H_5Br_2NO$. *See* q34	302.96	nd (al)	196	sub			i	s	v	δ	s		**B21**[2], 58
q107	**—,6,8-dibromo-2-hydroxy-**	6,8-Dibromocarbostyril. $C_9H_5Br_2NO$. *See* q34	302.96	nd	230					v^h					**B21**, 80
q108	**—,2,3-dichloro-**	$C_9H_5Cl_2N$. *See* q34	198.05	(dil al)	104–5				i	s	s		s		**B20**, 361
q109	**—,2,4-dichloro-**	$C_9H_5Cl_2N$. *See* q34	198.05	nd (dil al)	67–8	280–2			i	s	s		s	chl s	**B20**[2], 234
q110	**—,2,6-dichloro-**	$C_9H_5Cl_2N$. *See* q34	198.05	nd (eth)	161.5					s	s				**B20**, 361
q111	**—,2,7-dichloro-**	$C_9H_5Cl_2N$. *See* q34	198.05	nd (al)	120 (98–9)	sub 100^{2}			δ	s	s				**B20**[2], 234
q112	**—,3,4-dichloro-**	$C_9H_5Cl_2N$. *See* q34	198.05		69–70										
q113	**—,4,5-dichloro-**	$C_9H_5Cl_2N$. *See* q34	198.05		115.5–6.5										
q114	**—,4,6-dichloro-**	$C_9H_5Cl_2N$. *See* q34	198.05		104										

For explanations, symbols and abbreviations see beginning of table.

No.	Name	Synonyms and Formula	Mol. wt.	Crystalline form, color and specific rotation	m.p. °C	b.p. °C	Density	n_D	Solubility: w	al	eth	ace	bz	other solvents	Ref.
	Quinoline														
q115	**—,4,7-dichloro-**	$C_9H_5Cl_2N$. *See* q34	198.05	(MeOH)	86.4–7.4	148^{10}			...	...	...	...	...		
q116	**—,4,8-dichloro-**	$C_9H_5Cl_2N$. *See* q34	198.05		155–6				...	...	...	...	...		
q117	**—,5,6-dichloro-**	$C_9H_5Cl_2N$. *See* q34	198.05	nd (al)	85				...	s	...	...	...	peth v	**B20**, 361
q118	**—,5,7-dichloro-**	$C_9H_5Cl_2N$. *See* q34	198.05	nd (al)	116–7				...	s	s	...	...		**B20**, 362
q119	**—,5,8-dichloro-**	$C_9H_5Cl_2N$. *See* q34	198.05	nd (al), pl (eth)	97–8	sub			...	s	s	...	...		
q120	**—,6,8-dichloro-**	$C_9H_5Cl_2N$. *See* q34	198.05	nd (al)	103–4				...	s^h	δ	...	...		**B20^{2}**, 234
q121	**—,7,8-dichloro-**	$C_9H_5Cl_2N$. *See* q34	198.05	nd	85.5				...	s	s	...	...		
q122	**—,5,7-dichloro-8-hydroxy**	$C_9H_5Cl_2NO$. *See* q34	214.05	nd (al)	179–80				...	v	...	δ	s		**B21^{2}**, 58
—	**—,8(3-diethyl-amino-propyl-amino)6-methoxy**	*see* **Antimalarine**													
q123	**—,1,2-dihydro-1-methyl-2-oxo-**	*N*-Methyl-2(1)-quinolone. *N*-Methylcarbostyril. *N*-Methyl-*o*-aminocinnamic acid lactam. $C_{10}H_9NO$. *See* q34	159.19	nd (lig)	74	324^{728}			δ	s	...	s	v	chl s	**B21^{1}**, 297
q124	**—,5,7-diiodo-8-hydroxy-**	Diodoquin. $C_9H_5I_2NO$. *See* q34	396.96	yesh nd	210				i	δ	δ	...	...	chl δ	**B21^{2}**, 58
q125	**—,5,8-diiodo-6-hydroxy-**	$C_9H_5I_2NO$. *See* q34	396.96	pa ye	191				i	δ	δ	...	δ	chl δ, alk s	**B21^{2}**, 54
q126	**—,2,3-dimethyl-**	$C_{11}H_{11}N$. *See* q34	157.22	ye rh	68–9	261^{729}	1.1013		i	δ	s	...	...	lig s	**B20**, 406
q127	**—,2,4-dimethyl-**	$C_{11}H_{11}N$. *See* q34	157.22			$264–5^{758}$	1.0611^{15}		δ	v	v	...	...		**B20**, 407
q128	**—,2,6-dimethyl-**	$C_{11}H_{11}N$. *See* q34	157.22	rh pr	60	266–7			δ^h	δ	δ	...	δ		**B20**, 408
q129	**—,2,8-dimethyl-**	*o*-Toluquinaldine. $C_{11}H_{11}N$. *See* q34	157.22		23.2–3.6	255.3^{760} $103–4^{5}$	1.0349^{20}_{4}	1.6022^{20}	δ	s	s	...	...		**B20^{1}**, 154
q130	**—,3,4-dimethyl-**	$C_{11}H_{11}N$. *See* q34	157.22		73–4	290^{737}			i	s	...	...	...		**B20**, 410
q131	**—,5,8-dimethyl-**	$C_{11}H_{11}N$. *See* q34	157.22	nd	4–5	265^{736}	1.070^{21}		δ	s	s	...	...		**B20**, 411
q132	**—,6,8-dimethyl-**	*β*-Cystisolidide. $C_{11}H_{11}N$. *See* q34	157.22			$268–9^{76}$ $133–4^{14}$	1.067^{4}		δ	s	s	...	...		**B20**, 411
q133	**—,6(dimethyl-amino)-2-methyl-**	6(Dimethylamino)quinaldine. $C_{12}H_{14}N_2$. *See* q34	186.26	ye pr (aa)	101	319^{760}			...	v	v	...	v	peth i	**B22^{2}**, 361
q134	**—,2,3-dimethyl-4-hydroxy-**	$C_{11}H_{11}NO$. *See* q34	173.22	pr (w+1)	319–20				δ^h	δ^h	...	...	...	lig δ^h	**B21^{2}**, 68
q135	**—,2,4-dimethyl-6-hydroxy-**	$C_{11}H_{11}NO$. *See* q34	173.22	pr or pl (al)	214				i	v	δ	v	i	ac s alk s	**B21^{2}**, 68
q136	**—,2,4-dimethyl-7-hydroxy-**	$C_{11}H_{11}NO$. *See* q34	173.22	(al)	218				...	s	...	...	...		**B21**, 116
q137	**—,2,4-dimethyl-8-hydroxy-**	$C_{11}H_{11}NO$. *See* q34	173.22	pr (eth)	65	281			i	v	v	v	v	peth δ chl v	**B21**, 116
q138	**—,2,6-dimethyl-4-hydroxy-**	$C_{11}H_{11}NO$. *See* q34	173.22	nd (w+1)	279–80				v^h	...	...	...	...	ac s	**B21^{2}**, 68
q139	**—,2,8-dimethyl-4-hydroxy-**	$C_{11}H_{11}NO$. *See* q34	173.22	lf or pl (w+1)	260–1				δ	v	δ	...	δ	chl δ dil ac v alk v	**B21**, 116
q140	**—,4,6-dimethyl-2-hydroxy-**	$C_{11}H_{11}NO$. *See* q34	173.22	pr (al)	249–50				i	s^h	δ	...	δ	chl δ alk δ dil ac δ	**B31^{2}**, 225
q141	**—,4,7-dimethyl-2-hydroxy-**	$C_{11}H_{11}NO$. *See* q34	173.22	(aa)	220				δ^h	v^h	...	...	...	alk s aa v^h	**B21^{1}**, 225
q142	**—,4,8-dimethyl-2-hydroxy-**	$C_{11}H_{11}NO$. *See* q34	173.22	pl (aa)	217–8				δ	s	...	...	...		**B21^{1}**, 225
q143	**—,6,8-dimethyl-2-hydroxy-**	$C_{11}H_{11}NO$. *See* q34	173.22	nd (al)	201–2				...	s	...	...	...	dil ac i	**B21^{1}**, 225
q144	**—,6,8-dimethyl-5-hydroxy-**	$C_{11}H_{11}NO$. *See* q34	173.22	pl (chl)	197–8	sub			δ	...	...	...	δ	chl δ ac s	**B21**, 117
q145	**—,3-ethyl-2-hydroxy-**	3-Ethylcarbostyril. $C_{11}H_{11}NO$. *See* q34	173.22	(dil HCl)	168				...	...	...	...	...		**B21**, 115
q146	**—,2-hydrazino-**	$C_9H_9N_3$. *See* q34	159.20	(bz)	142–3				...	v	δ	...	...	lig δ	**B22^{1}**, 690
q147	**—,5-hydrazino-**	$C_9H_9N_3$. *See* q34	159.20	ye nd (w)	150–1				s^h	v	...	...	δ	peth i	**B22**, 565
q148	**—,2-hydroxy-**	*o*-Aminocinnamic acid lactam. Carbostyril. 2-Quinolinol. 2(1)Quinolone C_9H_7NO. *See* q34	145.15	pr (al dil al +1w), nd (sub)	199–200 (anh)	sub			δ	s	s	...	...	dil HCl s	**B21^{2}**, 51
q149	**—,3-hydroxy-**	3-Quinolinol. C_9H_7NO. *See* q34	145.15	(to or dil al)	198				i	s	s	...	v^h	chl s	**B21^{2}**, 52
q150	**—,4-hydroxy-**	Kynorine. 4-Quinolinol. 4(1)Quinolone. C_9H_7NO. *See* q34	145.15	nd (w+3)	201 (anh) 100 (hyd)				v^h	v^h	δ	...	δ		**B21^{2}**, 52[?]
q151	**—,5-hydroxy-**	5-Quinolinol. C_9H_7NO. *See* q34	145.15	nd (al)	224	sub			δ^h	v	δ	...	s^h	chl s^h lig i	**B21**, 84

For explanations, symbols and abbreviations see beginning of table.

No.	Name	Synonyms and Formula	Mol. wt.	Crystalline form, color and specific rotation	m.p. °C	b.p. °C	Density	n_D	Solubility						Ref.
									w	al	eth	ace	bz	other solvents	
	Quinoline														
q152	**—,6-hydroxy-**	6-Quinolinol. C_9H_7NO. *See* q34	145.15	pr (al or eth)	193	360	...	...	δ^{h}	δ	δ	...	δ	chl δ ac s alk s	**B21**, 85
q153	**—,7-hydroxy-**	7-Quinolinol. C_9H_7NO. *See* q34	145.15	pr (al), nd (dil al-eth)	238–40	sub	...	...	δ	v	...	...	...	alk s	**B21**, 91
q154	**—,8-hydroxy-**	8-Quinolinol. C_9H_7NO. *See* q34	145.15	pr (al)	75–6	266.6^{752}	...	...	i	v	...	...	v^{h}	chl v^{h}	**B21**, 91
q155	**—,—,sulfate**	Chinosol. $2(C_9H_7NO).H_2SO_4$	388.40	ye pw	175–8	...	...	...	s	δ	i	...	...	...	...
q156	**—,—,sulfate** monohydrate	$2(C_9H_7NO).H_2SO_4.H_2O$	406.42	ye pr	176–9	...	...	...	v	δ^{h}	...	...	...	...	**B21**^{2}, 56
q157	**3-hydroxy-3-methyl-**	3-Methylcarbostyril. $C_{10}H_9NO$. *See* q34	159.18	yesh nd (dil al or ace)	234–5	sub	...	...	...	s	...	s^{h}	...	...	**B21**, 107
q158	**—,2-hydroxy-4-methyl-**	2(1)Lepidone. $C_{10}H_9NO$. *See* q34	159.18	nd (w)	245	270^{17}	...	...	s^{h}	v^{h}	δ	...	δ	chl δ lig δ	**B21**^{2}, 65
q159	**—,2-hydroxy-6-methyl-**	6-Methylcarbostyril. $C_{10}H_9NO$. *See* q34	159.18	nd (al)	233	...	...	...	...	...	...	...	...	oos s	**B21**^{1}, 224
q161	**—,—**(high m.p.)	$C_{10}H_9NO$. *See* q34	159.18	nd (aq ace)	260	...	...	...	δ	v^{h}	δ	v^{h}	δ	lig δ	**B21**^{2}, 59
q162	**—,3-hydroxy-2-methyl-**(low m.p.)	3-Hydroxyquinaldine. $C_{10}H_9NO$. *See* q34	159.18	nd (al)	203–5	...	...	...	δ	s	s	...	...	chl s	...
q163	**—,4-hydroxy-2-methyl-**	4-Hydroxyquinaldine. $C_{10}H_9NO$. *See* q34	159.18	...	230–1	$360d^{760}$	...	...	v^{h}	v	δ	...	δ	...	**B21**, 104
q164	**—,5-hydroxy-2-methyl-**	5-Hydroxyquinaldine. $C_{10}H_9NO$. *See* q34	159.18	lf (al)	232–4	...	...	...	i	δ	v	...	...	alk s Na_2CO_3 i	**B21**, 106
q165	**—,5-hydroxy-6-methyl-**	$C_{10}H_9NO$. *See* q34	159.18	nd (al or sub)	230	sub	...	...	δ^{h}	v	...	...	...	oos v	**B21**, 111
q166	**—,5-hydroxy-8-methyl-**	$C_{10}H_9NO$. *See* q34	159.18	nd (dil al or sub)	262–3	sub	...	...	...	...	...	...	...	NaOH s^{h} chl δ	**B21**, 112
q167	**—,6-hydroxy-2-methyl-**	6-Hydroxyquinaldine. $C_{10}H_9NO$. *See* q34	159.18	(w)	213	$304–5^{760}$ 186^{35}	1.1665^{0}	...	δ	v	v	...	...	ac s alk s	**B21**, 106
q168	**—,6-hydroxy-4-methyl-**	6-Hydroxylepidine. $C_{10}H_9NO$. *See* q34	159.18	nd (w or dil al)	216–8	...	...	...	δ^{h}	v	...	v	...	chl v	**B21**, 109
q169	**—,6-hydroxy-8-methyl-**	$C_{10}H_9NO$. *See* q34	159.18	nd	200	...	...	...	...	...	...	...	...	...	**B21**, 113
q170	**—,7-hydroxy-6-methyl-**	$C_{10}H_9NO$. *See* q34	159.18	nd (al)	244	240^{22} 210^{11} sub	...	...	...	...	...	...	...	...	**B21**, 111
q171	**—,8-hydroxy-2-methyl-**	8-Hydroxyquinaldine. $C_{10}H_9NO$. *See* q34	159.18	pr (al)	74–5	267^{760} sub 100	...	...	i	δ^{h}	δ^{h}	...	δ^{h}	...	**B21**, 106
q172	**—,8-hydroxy-4-methyl-**	8-Hydroxylepidine. $C_{10}H_9NO$. *See* q34	159.18	nd (lig)	141	...	...	...	s^{h}	s^{h}	...	v	v	ac, alk v chl v	**B21**, 109
q173	**—,8-hydroxy-5-methyl-**	$C_{10}H_9NO$. *See* q34	159.18	nd (dil al)	122–4	...	...	...	i	...	...	...	...	ac, alk s	**B21**, 110
q174	**—,8-hydroxy-6-methyl-**	$C_{10}H_9NO$. *See* q34	159.18	nd (chl or bz)	95–6	sub	...	...	δ	v	...	...	...	alk s^{h}	**B21**, 111
q175	**—,8-hydroxy-7-methyl-**	$C_{10}H_9NO$. *See* q34	159.18	nd (dil al)	72–4	100 sub	...	...	...	...	...	...	...	...	**B21**, 111
q176	**—,8-hydroxy-5-nitroso-**	$C_9H_6N_2O_2$. *See* q34	174.16	nd	245d	...	...	...	i	...	δ	...	δ	chl δ	**B21**^{1}, 405
q177	**—,4-hydroxy-2-phenyl-**	$C_{15}H_{11}NO$. *See* q34	221.26	pl or pr	254	...	...	...	δ	δ^{h}	...	...	...	...	**B21**^{2}, 84
q178	**—,2-iodo-**	C_9H_6IN. *See* q34	255.07	nd (dil al)	52–3	...	...	...	δ	...	...	...	...	oos s	**B20**, 370
q179	**—,4-iodo-**	C_9H_6IN. *See* q34	255.07	nd or pr	sub 100	...	...	...	i	v	v	...	...	...	**B20**, 370
q180	**—,5-iodo-**	C_9H_6IN. *See* q34	255.07	nd (al or eth)	sub 101–2	...	...	...	i	s^{h}	s^{h}	...	...	...	**B20**^{1}, 141
q181	**—,6-iodo-**	C_9H_6IN. *See* q34	255.07	lf (w), nd (sub)	91	sub	...	...	s^{h}	v	v	...	...	...	**B20**, 370
q182	**—,8-iodo-**	C_9H_6IN. *See* q34	255.07	nd (al)	36	...	...	...	...	v	v	...	v	...	**B20**^{1}, 141
q183	**—,4-methoxy-**	$C_{10}H_9NO$. *See* q34	159.18	...	41	245 167^{20}	...	...	i	s	s	...	s	...	**B21**^{2}, 53
q184	**—,5-methoxy-**	$C_{10}H_9NO$. *See* q34	159.18	...	...	$275d^{760}$	...	...	...	v	v	...	...	...	**B21**, 85
q185	**—,6-methoxy-**	$C_{10}H_9NO$. *See* q34	159.18	hyg lf	26.5	$304–5d^{760}$ 153^{12}	1.1542^{20}_{20}	...	...	s	s	...	...	dil HCl s	**B21**^{2}, 53
q186	**—,8-methoxy-**	$C_{10}H_9NO$. *See* q34	159.18	nd (peth)	49–50	282^{742} 164^{14}	...	...	...	s	s	...	s	peth s	**B21**^{1}, 222
q187	**—,2-methoxy-6-nitro-**	$C_{10}H_8N_2O_3$. *See* q34	204.19	nd (dil aa, bz or sub)	189–90	...	...	...	δ	δ	s	...	s	con sulf v chl s	**B21**^{1}, 219
q188	**—,2-methoxy-8-nitro-**	$C_{10}H_8N_2O_3$. *See* q34	204.19	nd (dil al)	124–5	...	...	...	...	v	...	...	v	...	**B21**, 82
q189	**—,6-methoxy-5-nitro-**	$C_{10}H_8N_2O_3$. *See* q34	204.19	(al)	104–5	...	...	...	...	s^{h}	...	...	...	...	**B21**, 90
q190	**—,6-methoxy-8-nitro-**	$C_{10}H_8N_2O_3$. *See* q34	204.19	yesh nd	159–60	...	...	...	...	...	...	...	...	chl s, MeOH δ	**B21**^{2}, 54
q191	**—,8-methoxy-5-nitro-**	$C_{10}H_8N_2O_3$. *See* q34	204.19	(al)	151.5	...	...	...	...	v^{h}	...	...	...	...	**B21**, 98

For explanations, symbols and abbreviations see beginning of table.

No.	Name	Synonyms and Formula	Mol. wt.	Crystalline form, color and specific rotation	m.p. °C	b.p. °C	Density	n_D	Solubility w	al	eth	ace	bz	other solvents	Ref.
	Quinoline														
q192	**—,6-methoxy-1,2,3,4-tetrahydro-**	Thalline. $C_{10}H_{13}NO$. *See* q34	163.21	pr	42–3	283^{735}		1.5718^{50}	s^h	v	v	...	v		**B21**, 61
q193	**—,2-methyl-**	Quinaldine. $C_{10}H_9N$. *See* q34	143.18		−2, −1	246.5^{760} 118^{10}	1.0585^{20}_{4}	1.6126^{20}	δ	s	s	...	...	chl s	**B20²**, 238
q194	**—,3-methyl-**	$C_{10}H_9N$. *See* q34	143.18	pr	10–4	259.6^{760}	1.0673^{20}_{4}	1.6171^{20}	...	δ	...	...	...		**B20**, 394
q195	**—,4-methyl-**	Lepidine. $C_{10}H_9N$. *See* q34	143.18	red br	9–10	264.2^{760}	1.0868^{20}_{4}	1.6206^{20}	δ	∞	∞	...	...		**B20²**, 244
q196	**—,5-methyl-**	$C_{10}H_9N$. *See* q34	143.18		19	262.7^{760}	1.0832^{20}_{4}	1.6219^{20}	δ	∞	∞	...	...		**B20²**, 246
q197	**—,6-methyl-**	$C_{10}H_9N$. *See* q34	143.18		ca. −22	258.6^{760}	1.0654^{20}_{4}	1.6157^{20}	δ	s	s	...	...		**B20**, 397
q198	**—,7-methyl**	$C_{10}H_9N$. *See* q34	143.18	ye	39	257.6^{760}	1.0609^{20}_{4}	$1.6149^{20.8}$	δ	s	s	...	...		**B20²**, 246
q199	**—,8-methyl-**	$C_{10}H_9N$. *See* q34	143.18			247.8^{760} 143^{34}	1.0719^{20}_{4}	1.6164^{20}	δ	∞	∞	...	...		**B20²**, 247
q200	**—,2-methyl-5-nitro-**	5-Nitroquinaldine. $C_{10}H_8N_2O_2$. *See* q34	188.18	nd (dil al)	82				δ	v	v	...	...		**B20²**, 243
q201	**—,2-methyl-6-nitro-**	6-Nitroquinaldine. $C_{10}H_8N_2O_2$. *See* q34	188.18	ye nd (w)	173–4				s^h	s	i	...	...	dil HCl v	**B20²**, 243
q202	**—,2-methyl-8-nitro-**	8-Nitroquinaldine. $C_{10}H_8N_2O_2$. *See* q34	188.18	yesh nd (dil al)	137				δ	v	v	...	v		**B20²**, 244
q203	**—,4-methyl-3-nitro-**	3-Nitrolepidine. $C_{10}H_8N_2O_2$. *See* q34	188.18	pr (w)	118				s^h	s	...	...	...		**B20²**, 245
q204	**—,4-methyl-8-nitro-**	8-Nitrolepidine. $C_{10}H_8N_2O_2$. *See* q34	188.18	lf (abs al)	126–7				...	s^h	...	...	...		**B20**, 397
q205	**—,6-methyl-5-nitro-**	$C_{10}H_8N_2O_2$. *See* q34	188.18	pa ye nd (al)	116.7				i	s	...	...	...	oos v	**B20¹**, 151
q206	**—,6-methyl-8-nitro-**	$C_{10}H_8N_2O_2$. *See* q34	188.18	pa ye nd (w)	122				s^h	v	...	...	...	oos v	**B20**, 400
q207	**—,8-methyl-5-nitro-**	$C_{10}H_8N_2O_2$. *See* q34	188.18	pa ye nd (al)	93				...	s	...	...	...	oos v	**B20**, 403
q208	**—,8-methyl-6-nitro-**	$C_{10}H_8N_2O_2$. *See* q34	188.18	(al)	129				δ	v	...	...	...		**B20**, 403
q209	**—,1-methyl-1,2,3,4-tetrahydro-**	Kairoline. $C_{10}H_{13}N$. *See* q34	147.22			$247–50^{758}$	1.022^{20}_{4}	1.5802^{23}	...	v	δ	...	...		**B20²**, 174
q210	**—,3-nitro-**	$C_9H_6N_2O_2$. *See* q34	174.16	nd (dil al)	127–8				$δ^h$	s	...	s	...		**B20²**, 235
q210¹	—,4-nitro-, oxide	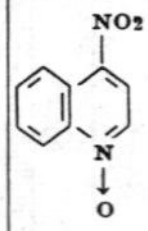	190.16	ye nd	153				...	...	...	...	...		**C50**, 13925
q211	**—,5-nitro-**	$C_9H_6N_2O_2$. *See* q34	174.16	nd (w or al)	72	sub			$δ^h$	s^h	...	...	s		**B20²**, 235
q212	**—,6-nitro-**	$C_9H_6N_2O_2$. *See* q34	174.16	ye pl (HCl-aa), nd (w or dil al)	131	sub 153–4			s^h	s	δ	...	v	dil ac s lig δ	**B20²**, 235
q213	**—,7-nitro-**	$C_9H_6N_2O_2$. *See* q34	174.16	nd (w or al), pl (sub)	132–3				s^h	s^h	s	...	...	chl s	**B20**, 372
q214	**—,8-nitro-**	$C_9H_6N_2O_2$. *See* q34	174.16	mon pr (al)	89–90				δ	s	s	...	s	dil ac s	**B20**, 273
q216	**—,1,2-oxo-1,2,3,4-tetrahydro-** (high m.p.)	3(2-Aminophenyl)propionic acid lactam. Hydrocarbostyril. C_9H_9NO. *See* q34	147.17	pr (al or eth)	163–4	2 01^{15}			i	v	v	...	...	alk $δ^h$	**B21²**, 253
q217	—,—(low m.p.)	Hydroisocarbostyril. C_9H_9NO. *See* q34	147.17	nd (lig)	70–1	>300			δ	v	v	...	...	chl v	**B21**, 290
q218	**—,2-phenyl-**	$C_{15}H_{11}N$. *See* q34	205.25	nd (dil al)	82–3	363^{760}			δ	s^h	s^h	...	v	peth δ	**B20²**, 311
q219	**—,3-phenyl-**	$C_{15}H_{11}N$. *See* q34	205.25	pl (eth)	52	205–7			...	s	s	...	s	chl s	**B20²**, 312
q220	**—,5-phenyl-**	$C_{15}H_{11}N$. *See* q34	205.25	nd (dil al)	82–3				...	...	...	...	...		**J1943**, 441
q221	**—,6-phenyl-**	$C_{15}H_{11}N$. *See* q34	205.25	pl (al, bz or $PhNH_2$)	110–1	260^{77}	1.1945^{20}		δ	v	δ	...	δ	chl v peth δ	**B20**, 483
q222	**—,8-phenyl-**	$C_{15}H_{11}N$. *See* q34	205.25	ye gr		$270–6^{80}$			δ	v	v	...	v	chl v	**B20**, 484
q223	**—,1,2,3,4-tetrahydro-**	$C_9H_{11}N$. *See* q34	133.19	nd	20	$249–50^{755}$	1.0605^{15}_{4}	1.5862^{26}	s	∞	∞	...	...		**B20²**, 173

For explanations, symbols and abbreviations see beginning of table.

No.	Name	Synonyms and Formula	Mol. wt.	Crystalline form, color and specific rotation	m.p. °C	b.p. °C	Density	n_D	Solubility w	al	eth	ace	bz	other solvents	Ref.
	Quinoline														
q224	**—,5,6,7,8-tetra-hydro-**	$C_9H_{11}N$. *See* q34	133.19	...	157–9	222^{760} 92–5^{12}	1.025$^{22}_{4}$	1.5426^{20}	...	...	...	...	...	...	**B20**², 176
q225	**—,2,3,4-tri-methyl-**	$C_{12}H_{13}N$. *See* q34	171.24	...	92	285^{760} 156–8^{12}	...	...	...	...	...	...	...	...	**B20**², 255
q226	**—,2,4,5-or 2,4,7-trimethyl-**	$C_{12}H_{13}N$. *See* q34	171.24	nd (w)	63–4	280–1	1.0337$^{20}_{4}$	1.5973^{20}	...	...	...	...	...	...	**B20**², 257
q227	**—,2,4,6-tri-methyl-**	$C_{12}H_{13}N$. *See* q34	171.24	nd (w or dil al +1w)	43–5 (anh)	281–2^{760} 146–8$^{13.5}$	...	...	δ	v	v	v	v	chl v peth v	**B20**², 256
q228	**—,2,4,8-tri-methyl-**	$C_{12}H_{13}N$. *See* q34	171.24	...	42	269–70	...	1.5855^{50}	...	...	...	...	...	...	**B20**², 257
q229	**—,2,5,6-or-2,6,7-trimethyl-**	$C_{12}H_{13}N$. *See* q34	171.24	nd	69–70	...	...	...	δ	v	v	...	v	...	**B20**, 415
q230	**—,2,5,7-tri-methyl-**	Tetracoline. $C_{12}H_{13}N$. *See* q34	171.24	pr	43	285–7	...	...	...	v	v	...	...	...	...
q231	**—,2,6,8-tri-methyl-**	$C_{12}H_{13}N$. *See* q34	171.24	pr (peth)	44.5–6.5	260^{719}	...	...	i	δ	δ	...	...	peth δ	**B20**, 415
q232	**—,4,5,8-tri-methyl-**	$C_{12}H_{13}N$. *See* q34	171.24	...	73.4	155^{13}	...	...	...	...	...	...	...	...	**Am68**, 644
q233	**2-Quinoline-carboxylic acid**	Quinaldinic acid. 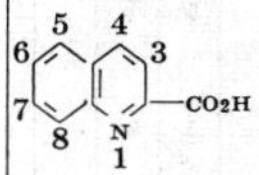	173.17	(bz, w+2)	156 (anh)	...	...	...	v^h	...	...	...	v^h	...	**B22**², 55
q234	—,amide	Quinaldinamide. $C_{10}H_8N_2O$. *See* q233	172.19	nd (dil al, bz-lig)	133	...	...	...	δ^h	v	δ	...	v	chl v dil HCl v	**B22**, 73
q235	—,chloride	Quinaldinyl chloride. $C_{10}H_6ClNO$. *See* q233	191.62	nd (eth or lig)	97–8	...	...	...	δ	d^h	v	...	v	lig δ	**B22**¹, 509
q236	—,methyl ester	Methyl quinaldinate. $C_{11}H_9NO_2$. *See* q233	187.20	nd (lig)	86	...	...	...	...	s	...	...	...	lig s^h	**B22**², 55
q237	—,nitrile	2-Cyanoquinoline. $C_{10}H_6N_2$. *See* q233	154.17	nd (lig)	94	160–70$^{20-23}$			δ	v	v	...	v	chl v lig δ	**B22**¹, 509
q238	**3-Quinoline-carboxylic acid**	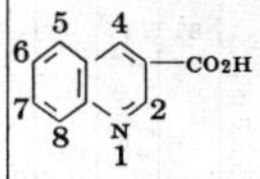	173.17	pl (al or dil al)	275δd	...	...	...	i	δ	...	...	...	dil ac s alk s	**B22**², 56
q239	—,nitrile	3-Cyanoquinoline. $C_{10}H_6N_2$. *See* q238	154.17	(al or sub)	107–8	...	...	...	...	s^h	...	...	...	oos s	**B22**², 57
q240	**4-Quinoline-carboxylic acid**	Cinchoninic acid. 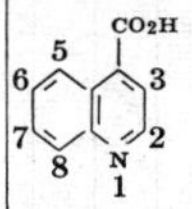	173.17	mcl pr, nd (+1w)	257–8 (anh)	...	...	...	δ	δ	i	...	...	...	**B22**², 57
q241	—,nitrile	4-Cyanoquinoline. Cinchoninonitrile $C_{10}H_6N_2$. *See* q240	154.26	(chl, eth or lig) nd (sub)	103–4	240–5	...	...	...	s	...	...	...	oos v	**B22**², 75
q242	**5-Quinoline-carboxylic acid**	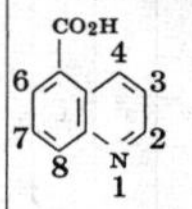	173.17	(ace or sub)	338–40	sub <338	...	...	δ	δ	i	...	i	dil ac v dil alk δ	**B22**¹, 511
q243	—,nitrile	5-Cyanoquinoline. $C_{10}H_6N_2$. *See* q242	154.26	nd (lig, dil al) +1.5w)	87–8 (anh) 70 (hyd)	...	...	...	δ	v	...	...	v	CS_2 v	**B22**, 79
q244	**6-Quinoline-carboxylic acid**	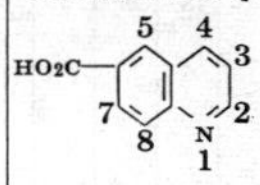	173.17	nd (sub)	291–2	sub <290	...	...	δ	δ^h	...	...	...	dil ac v dil alk v	**B22**¹, 511
q245	**7-Quinoline-carboxylic acid**	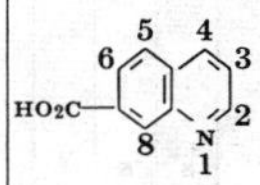	173.17	nd	248–9	sub	...	...	δ^h	s	i	...	...	...	**B22**, 81
q246	**8-Quinoline-carboxylic acid**	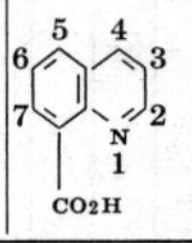	173.17	nd	187	sub	...	...	δ^h	δ	...	...	...	dil ac s dil alk s	**B22**, 81

For explanations, symbols and abbreviations see beginning of table.

No.	Name	Synonyms and Formula	Mol. wt.	Crystalline form, color and specific rotation	m.p. °C	b.p. °C	Density	n_D	Solubility						Ref.
									w	al	eth	ace	bz	other solvents	
	8-Quinolinecarboxylic acid														
q247	—,nitrile	8-Cyanoquinoline. $C_{10}H_6N_2$. *See* q243	154.26	nd (dil al)	84					v					**B22**, 81
q248	**4-Quinoline-carboxylic acid, 2(3-carboxy-4-hydroxyphenyl)-**	Hexophan. $C_{17}H_{11}NO_5$. *See* q240	309.27	yesh pw	283–4				i	δ				chl, alk v lig i	**B22**², 206
q249	**2-Quinoline-carboxylic acid, 4,8-dihydroxy-**	Xanthurenic acid. Xanthuric acid. $C_{10}H_7NO_4$. *See* q233	205.16	ye micr cr (w)	286				i	s				dil HCl s^h alk s Na_2CO_3 s	**Am 73**, 3520
q250	—,4-hydroxy-	Kynurenic acid. 4-Hydroxyquinaldinic acid. $C_{10}H_7NO_3$. *See* q233	189.17	ye nd (dil aa)	287–8				δ^h	s^h	i				**B22**², 174
q251	**4-Quinoline-carboxylic acid, 6-methoxy-**	$C_{11}H_9NO_3$. *See* q240	203.19	pa ye nd	279–80					s^h	δ			chl i alk s	**B22**², 176
q252	—,8-methoxy-2-phenyl-	$C_{17}H_{13}NO_3$. *See* q240	279.28	ye nd (al)	216				i	s	i			chl s alk s ac s	**B22**¹, 559
q253	—,6-methyl-2-phenyl-, ethyl ester	6-Methylcinchophene ethyl ester. $C_{19}H_{17}NO_2$. *See* q240	291.33	(al)	75–6				i	s^h	v			chl v	**B22**¹, 520
q254	**3-Quinoline-carboxylic acid, 2-phenyl-**	$C_{16}H_{11}NO_2$. *See* q238	249.26	nd (al)	230d				δ^h	s					**B22**², 70
q255	**4-Quinoline-carboxylic acid, 2-phenyl-**	Atophan. Cinchophene. $C_{16}H_{11}NO_2$. *See* q240	249.26	nd (MeOH or dil al), ye in air	218				i	s^h		δ^h	δ^h	peth i	**B22**², 70
q256	—,—,allyl ester	Atoquinol. $C_{19}H_{15}NO_2$. *See* q240	289.32	ye nd (dil al)	36	265^{15} $215^{0.8}$			i		v	v	v		**B22**², 71
q257	**5-Quinoline-sulfonic acid, 6-hydroxy-**	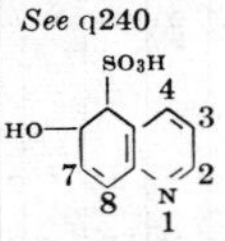	225.23	ye nd (w or al+1w)	270d				s^h		i		i	chl i	**B22**, 407
q258	—,8-hydroxy-	$C_9H_7NO_4S$. *See* q257	225.23	ye (con HCl +2w) lf, nd (dil HCl+1w)	322–3				s^h						**B22**², 313
q259	**7-Quinoline-sulfonic acid, 8-hydroxy-**	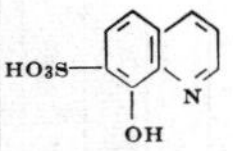	225.23	ye amor	310–2				s^h	δ			i		**B22**, 408
q260	**5-Quinoline-sulfonic acid, 8-hydroxy-7-iodo-**	Loretine. $C_9H_6INO_4S$. *See* q257	351.13	ye pr or lf	260d				δ		δ		δ	chl δ con sulf s	**B22**², 314
q261	**8-Quinoline-sulfonic acid, 5-hydroxy-6-iodo-**	Lorenite. 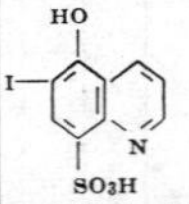	351.13	ye nd or lf	210–30d									con sulf s	**B22**, 406
—	**Quinolinic acid**	*see* **2,3-Pyridinedicarboxylic acid**													
q262	**Quinolinium, *N*-ethyl-, iodide**	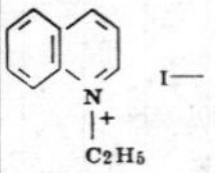	285.13	ye pr (al or MeCN)	158				v	s^h	i			chl δ	**B20**², 231
q263	—,*N*-methyl-, chloride	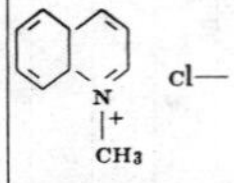	179.65	(al+?w)	126				s	s^h					**B20**¹, 139
—	**Quinolinol**	*see* **Quinoline, hydroxy-**													
—	**Quinone**	*see* **Benzoquinone**													
q264	**Quinovic acid**	$C_{30}H_{46}O_5$	486.70	rh $[\alpha]^{18}_D$+87	298				i	i				alk, Ac_2O s	**E14**, 580
q265	**Quinoxaline**	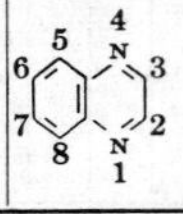 1,4-Benzodiazine. Benzopyrazine. Quinazine.	130.15		29–30	$108–11^{12}$	1.1334^{48}_4	1.6231^{48}	s	s	s			aa s	**B23**², 177

For explanations, symbols and abbreviations see beginning of table.

No.	Name	Synonyms and Formula	Mol. wt.	Crystalline form, color and specific rotation	m.p. °C	b.p. °C	Density	n_D	Solubility w	al	eth	ace	bz	other solvents	Ref.
	Quinoxaline														
q266	**—,6-amino-**	$C_8H_7N_3$. *See* q265	145.17	ye nd (eth)	158–9				v	v	δ			aa δ	**B25**, 326
q267	**—,6-chloro-**	$C_8H_5ClN_2$. *See* q265	164.60	nd (w)	60	$117–9^{10}$			i						**B23**^{2}, 177
q268	**—,2,3-dichloro-**	$C_8H_4Cl_2N_2$. *See* q265	199.04	(al)	148–50				i	v^h			v	chl v	**B23**^{2}, 177
q269	**—,2,3-dihydroxy-**	$C_8H_6N_2O_2$. *See* q265	162.15	nd (w)	410				v	δ	δ		s^h	MeOH s^h	**B24**^{2}, 200
q270	**—,2,3-dimethyl-**	$C_{10}H_{10}N_2$. *See* q265	158.20	nd (w+3), (ace)	106									oos s ac s	**B23**^{2}, 197
q271	**—,2,6-dimethyl-**	$C_{10}H_{10}N_2$. *See* q265	158.20		54	266–8			v	v	v				**B23**, 192
q272	**—,2,3-dioxo-1,2,3,4-tetrahydro-**	2,3-Dihydroxychinoxaline. $C_8H_6N_2O_2$. *See* q265	162.15	nd ($HOCH_2CH_2OH$)	386–90d				i	s^h	δ				**B24**^{2}, 200
q273	**—,2-hydroxy-**	2-Quinoxalinol. $C_8H_6N_2O$. *See* q265	146.15	lf	268–70	$sub200^{0.5}$									**C50**, 10736
q274	**—,5-hydroxy-**	5-Quinoxalinol. $C_8H_6N_2O$. *See* q265	146.15		99.5	sub 90^{25}									**C48**, 8232
q275	**—,6-methoxy-**	$C_9H_8N_2O$. *See* q265	161.18	nd (w)	57.5	128^7									**B23**, 387
q276	**—,2-methyl-**	$C_9H_8N_2$. *See* q265	144.18	ye	180–1	$245–7^{760}$ $125–7^{11}$			∞						**B23**^{2}, 190
q277	**—,6-methyl-**	$C_9H_8N_2$. *See* q265	144.18		218–9	248^{748} 141.5^{29}	1.1164^{20}_{4}	$1.6211^{18.4}$	∞	∞	∞		∞		**B23**, 184
q278	**—,1,2,3,4-tetrahydro-**	$C_8H_{10}N_2$. *See* q265	134.18	pl (w, eth or peth)	99.0–.5	288.5–9.5 $153–4^{14}$			v^h	v			v	CCl_4 v	**B23**^{2}, 106
q279	**Quinuclidine**	1-Azabicyclo[2,2,2]octane. 1,4-Ethylenepiperidine.	111.18	(eth)	158				v	v	v			oos v	**B20**^{2}, 71
q280	**—,3-hydroxy-**	$C_7H_{13}NO$. *See* q279	127.19		221–3										**C48**, 11499
q281	**2-Quinuclidinecarboxylic acid**	$C_8H_{13}NO_2$. *See* q279	155.20	hyg	275–8										

For explanations, symbols and abbreviations see beginning of table.

No.	Name	Synonyms and Formula	Mol. wt.	Crystalline form, color and specific rotation	m.p. °C	b.p. °C	Density	n_D	Solubility w	al	eth	ace	bz	other solvents	Ref.
	Raffinose														
r1	**Raffinose**	Gossypose. Melitose. 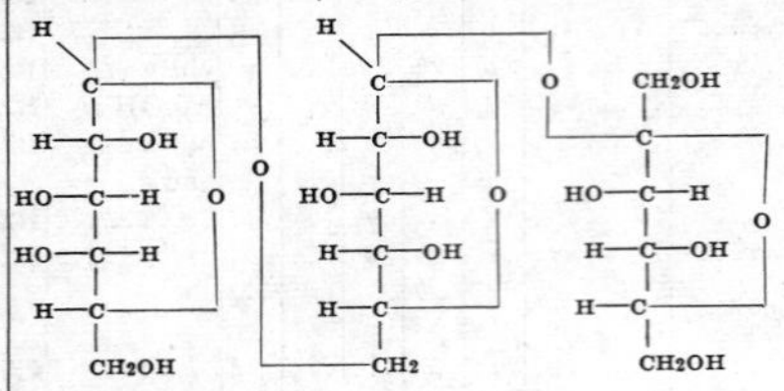 .5H_2O	594.52	wh pw, clusters (dil al) $[\alpha]_D^{20}+105.2$ (c=4)	80 (hyd) 118–9 (anh)	d 130	1.465	......	v	δ	...	...	...	MeOH s Py s	......
r2	**Raunescine, hydrate**	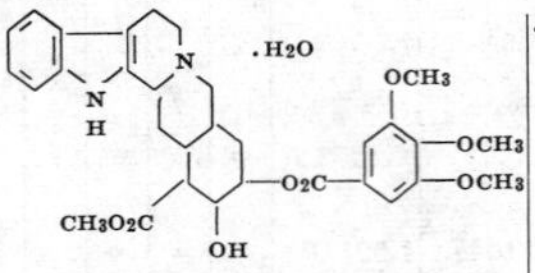 .H_2O	582.62	wh hex pr $[\alpha]_D-74$ (chl)	160–70	......	......	......	i	δ	...	...	...	chl s aa s	......
—	**Reductone**	*see* **Propanal, 3-hydroxy-2-oxo-***													
r3	**Resazurin**	Diazoresorcinol. 3-Isophenoxazinone 10-oxide. 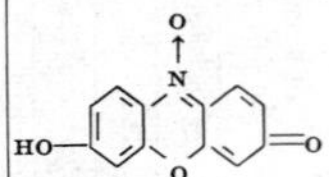	229.18	dk red to grsh pr or pl (aa or AcOEt)	d	sub vac	......	......	i	δ	i	...	...	alk s aa δ	**B27**, 128
r4	**Rescinnamine**	$C_{35}H_{42}N_2O_9$	634.73	nd $[\alpha]_D-98$ (chl)	226	......	......	......	i	δ	...	s	...	chl, AcOEt s	......
r5	**Reserpic acid**	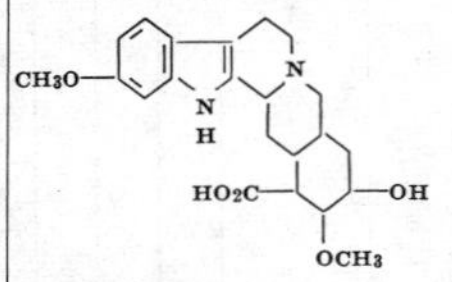	400.46	(MeOH)$[\alpha]_D^{23}$ (of hydrochloride) −81	241–3	......	......	......	...	...	...	...	...	......	......
r6	**Reserpine**	$C_{33}H_{40}N_2O_9$	608.69	lo pr (dil ac) $[\alpha]_D-117.7$ (chl), −164 (Py, c=0.6)	277–8d	......	......	......	i	s^h	...	...	s	chl s	......
—	**—,11-desmethoxy-**	*see* **Canescine**													
r7	**Reserpinine**	$C_{22}H_{26}N_2O_4$	382.43	wh nd, $[\alpha]_D^{23}$ −117±4 (chl)	238–9	......	......	......	i	s	...	...	...	chl s	......
—	**Resodiacetophenone**	*see* **Benzene, 1,5-diacetyl-2,4-dihydroxy-**													
—	**Resorcinol**	*see* **Benzene, 1,3-dihydroxy-***													
—	**Resorcylaldehyde**	*see* **Benzaldehyde, 2,4-dihydroxy-**													
—	**Resorcylic acid**	*see* **Benzoic acid, dihydroxy-**													
r8	**Resorufin**	7-Hydroxy-2-phenoxazone. 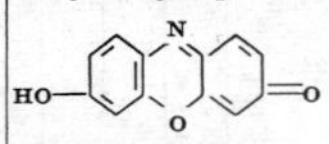	213.20	br or red nd ($PhNH_2$, conc HCl)	......	......	......	......	i	δ	i	...	...	alk v	**B27**[2], 108
—	**Retene**	*see* **Phenanthrene, 7-isopropyl-1-methyl-***													
—	**Retenoquinone**	*see* **9,10-Phenanthraquinone, 7-isopropyl-1-methyl-***													
r10	**Retronecine**	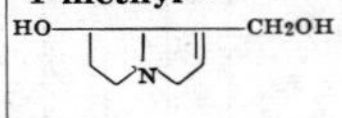	155.20	pr $[\alpha]_D+27.4$ (w), $[\alpha]_D+54.8$ (abs al)	120–1	......	......	......	...	...	...	...	...	......	......
r11	**Rhamnetin**	$C_{16}H_{12}O_7$	316.26	ye nd (al, PhOH)	>300	......	......	......	δ^h	δ^h	...	δ^h	...	dil alk v PhOH s^h	**B18**[2], 237
r12	**Rhamnitol**	Rhamnite. 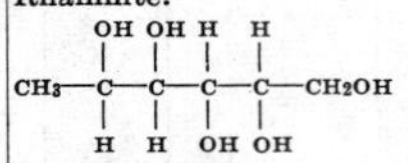 $CH_3-CH(OH)-CH(OH)-CH(OH)-CH(OH)-CH_2OH$	166.18	pr (ace) $[\alpha]_D^{20}$ −12.4 (w)	123	......	......	......	s	s	i	...	...	......	**B1**, 532

For explanations, symbols and abbreviations see beginning of table.

No.	Name	Synonyms and Formula	Mol. wt.	Crystalline form, color and specific rotation	m.p. °C	b.p. °C	Density	n_D	Solubility w	al	eth	ace	bz	other solvents	Ref.
	α-Rhamnose														
r13	**α-Rhamnose**(*d*)	$CH_3-C(H)-C(H)-C(OH)(OH)-C(H)(OH)-CHOH.H_2O$ (ring –O–)	182.18	(w+1), $[\alpha]_D^{20}$ (mut) +3.76 to −0.82 (3 hours), $[\alpha]_s^{16}$ −825 (w)	90–1	....	....	....	s	...	i	...	...	....	**B1**, 439
r14	—(*dl*)	$C_6H_{12}O_5$. *See* r13	164.16	....	151.3–3 (anh)	....	....	....	...	...	...	...	...	....	....
r15	—(*l*)	$C_6H_{12}O_5.H_2O$. *See* r13	182.18	mcl pl (al, w+1) $[\alpha]_D^{20}$ (mut) −11.4 to −9.0 (16 hrs)	92	105^2 sub	....	....	s	...	...	...	...	MeOH s	**B1**[2], 901 **B31**, 66
r16	**β-Rhamnose**(*l*)	$C_6H_{12}O_5$. *See* r13	164.16	nd $[\alpha]_D^{20}$ +31.5	....	....	....	....	s	s	...	...	...	....	**B1**[1], 439 **B31**[2], 901
—	**Rhamnoxanthin**	*see* **Frangulin**													
r17	**Rheadin**	Rhoeadine. $C_{21}H_{21}NO_6$	383.41	nd (chl, eth) (chl) $[\alpha]_D^{19}$ +243	256 –7	....	....	....	δ	δ	δ	...	...	....	....
r18	**Rhizopterin**	$HO_2C-C_6H_4-N(CHO)CH_2-$(pteridine: NH_2, OH)	340.29	lt ye pl (aq $AcONH_4$)	>300	....	....	....	i	...	...	...	...	aq alk s aq NH_3 s Py s	....
r19	**Rhodamine B**	$(C_2H_5)_2N-$xanthene$-N(C_2H_5)_2$, with $C_6H_4CO_2H$	442.54	gr lf (w+4), colorless pr (al, xyl)	165	....	....	....	δ^h s	s	s	...	s	org solv i xyl s^h	**B19**[2], 373
r20	—,hydrochloride	$C_{28}H_{31}N_2O_3.HCl$. *See* r19	479.01	gr or resh vt (w), lf (dil HCl)	....	....	....	....	δ	v	...	...	s	w (red)	**B19**[2], 373
r21	**Rhodanine**	4-Oxo-2-thioxothiazolidine. Rhodanic acid. (ring: O=C–NH(3)–C(=S)–S–C(5))	133.20	lt ye pr (al, w, aa)	170.5–1	....	0.868	....	i s^h	s	s	...	...	MeOH v aa s^h lig i	**B27**[7], 288
r22	—,5[4(dimethyl-amino)-benzylidene]-	$C_{12}H_{12}N_2OS_2$. *See* r21	264.37	red or og nd (al, Py-w)	296	....	....	....	i	δ^h	δ	s	δ	chl δ min ac s lig i	**B27**[2], 484
r23	—,5-ethyl-	$C_5H_7NOS_2$. *See* r21	161.25	ye amor (dil al)	105	....	....	....	i	v	v	v	...	peth i	**B27**[1], 313
r24	—,3-phenyl-	$C_9H_7NOS_2$. *See* r21	209.30	ye pl or nd or pr	194–5	....	....	....	i	δ s^h	δ	s^h	...	chl s	**B27**[2], 288
—	**Rhodinal**	*see* **Citronellal**													
—	**Rhodinol**	*see* **Citronellol**													
r25	**Rhodizonic acid**	5,6-Dihydroxy-5-cyclohexene-1,2,3,4-tetrone*.	170.08	cr	d	....	....	....	d	s	...	...	...	....	**B8**[2], 572
—	**Ribitol**	*see* **Adonitol**													
r26	**Riboflavin**	Lactoflavin. Vitamin B_2. Vitamin G. $C_{17}H_{20}N_4O_6$.	356.78	ye or og-ye nd $[\alpha]_D^{25}$ (mut) −112 to −122 (0.02N NaOH, c=0.5) $[\alpha]_D$ −8.80 (w)	280d	....	....	....	i	i	i	i	...	chl i	....
r27	**Ribose**(*D*)	$H_2C-C(H)(OH)-C(H)(OH)-C(H)(OH)-CHOH$ (ring –O–)	150.13	$[\alpha]_D$ −21.5	86–7	....	....	....	s	δ	...	...	...	....	**B1**[1], 31

For explanations, symbols and abbreviations see beginning of table.

No.	Name	Synonyms and Formula	Mol. wt.	Crystalline form, color and specific rotation	m.p. °C	b.p. °C	Density	n_D	Solubility w	al	eth	ace	bz	other solvents	Ref.
	Ricinidine														
r29	**Ricinidine**	1-Methyl-2-pyridone-3-carbonitrile. 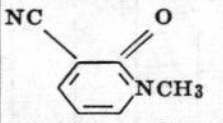	134.14	nd (sub), (chl, al)	140	243^{18}				s^h				chl s^h	**B22**^{2}, 222
r30	**Ricinine**	Ricidine. $C_8H_8N_2O_2$	164.16	pr or nd	204				s^h	δ				chl s^h	**B22**^{1}, 371
—	**Ricinoleic acid**	*see* **9-Octadecenoic acid, 12-hydroxy-(*cis*)***													
—	**Rivanol**	*see* **Acridine, 6,9-diamino-2-ethoxy-**													
—	**Roccelic acid**	*see* **Succinic acid, 2-dodecyl-3-methyl-**													
—	**Rosaniline**	*see* **Methanol, bis(4-aminophenyl)-(4-amino-3-tolyl)-**													
r31	**Rosinduline**	10-Phenyl-3-amino-1,2-benzophenazinium hydroxide. 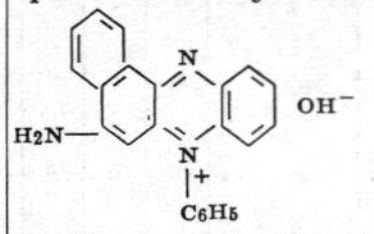	339.40	red-br lf (eth, bz)	198–9				i	v	v		v		**B25**^{2}, 322
—	**Rissic acid**	*see* **Acetic acid, (2-carboxy-4,5-dimethoxyphenoxy)-**													
r32	**Rotenone**	$C_{23}H_{22}O_6$	394.43	nd or lf (al), (CCl_4+1) $[\alpha]_D -132$ (al, c=0.125)	163	$210–20^5$			i	s	δ			chl v os s	**B19**^{2}, 438
r33	**Rubicene**	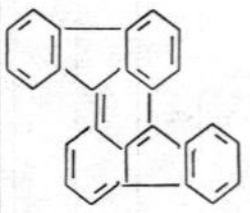	326.40	red nd (xyl)	306					δ	δ			$PhNO_2$ v CS_2 s	**E14**^{5}, 835
r34	**Rubijervine**	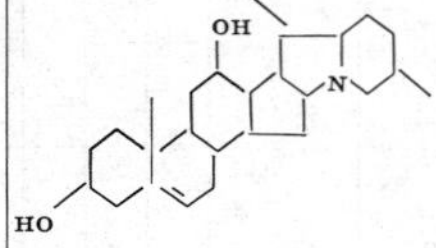	413.65	nd (al+1), $[\alpha]_D^{25}+19.0$ (al, c=1.0)	240–6					s^h	δ		s	chl s aq ac s aq alk i con sulf (red)	
r35	**Rubixanthin**	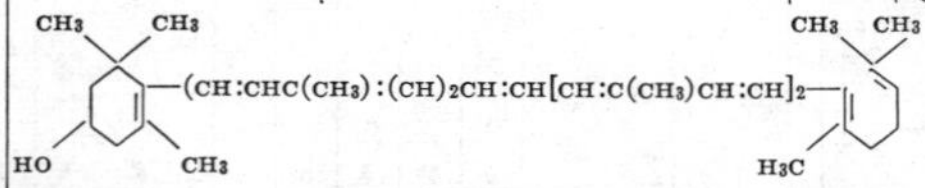	552.85	deep red nd (bz-MeOH), og cr (bz-peth)	160					δ			s	chl s, peth δ	**B30**, 93
r36	**Rubrene**	9,10,11,12-Tetraphenylnaphthacene. 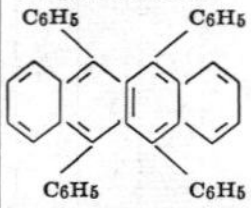	532.69	og red (bz-lig)	334				i	δ	δ		s	CS_2, Py s	**E14**^{5}, 79
—	**Rufigallic acid**	*see* **9,10-Anthraquinone, 1,2,3,5,6,7-hexahydroxy-**													
—	**Rufiopin**	*see* **9,10-Anthraquinone, 1,2,5,6-tetrahydroxy-***													
—	**Rufol**	*see* **Anthracene 1,5-dihydroxy-***													
r37	**Rutaecarpine**	$C_{18}H_{13}N_3O$	287.31	ye pl, nd (AcOEt)	258					δ					**B26**^{2}, 104
r38	**Rutinose**	6(β-1-*L*-Rhamnosido) *D*-glucose. 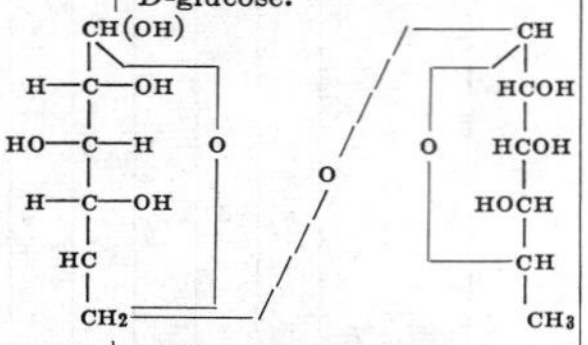	326.30	hyg pw (al, eth) $[\alpha]_D^{20}$ (mut) +3.2 to −0.8	189–2d				v	s					**B31**, 376
—	**Rutonal**	*see* **Barbituric acid, 5-methyl-5-phenyl-**													

Rubixanthin: —(CH:CHC(CH₃):(CH)₂CH:CH[CH:C(CH₃)CH:CH]₂—

For explanations, symbols and abbreviations see beginning of table.

No.	Name	Synonyms and Formula	Mol. wt.	Crystalline form, color and specific rotation	m.p. °C	b.p. °C	Density	n_D	Solubility w	al	eth	ace	bz	other solvents	Ref.
	Sabadine														
s1	**Sabadine**	$C_{29}H_{51}NO_8$	541.71	nd (eth)	238–40				δ	s	δ	s			
—	**Sabinane**	*see* **Thujane**													
s2	**Sabinene**	(structure) $=CH_2$	136.23	$[\alpha]_D$ +89.07		$163\text{–}5^{758}$	0.8430^{25}_{4}	1.4678^{20}	i	∞					**B5²**, 96
s3	—(*l*)	$C_{10}H_{16}$. *See* s2	136.23	$[\alpha]^{15}_D$ −42.5		162–3	0.8468^{20}	1.4674^{17}	i	∞					**B5²**, 96
s4	**Sabinol**(*d*)	$C_{10}H_{16}O$	152.24	$[\alpha]^{18}_D$ +3.94		208–9	0.9488^{19}_{4}	1.4871^{25}			s				**E12A**, 13
s5	**Saccharic acid**(*d*)	$HO_2C—CH(OH)—CH(OH)—CH(OH)—CH(OH)—CO_2H$	210.14	nd (al), $[\alpha]_D$ (mut)+6.9 to +21.4 (w, c=2.5)	125–6				v	v	δ			chl δ	**B3²**, 377
s6	**Saccharin**	2-Sulfobenzoic acid imide. (structure: SO_2, NH, O)	183.18	mcl (ace)	228	sub vac			δ	s	δ				**B27**, 168
s7	**Safrole**	3,4-Methylenedioxy-1-allylbenzene. Shikimole. (structure: $CH_2—O$, O, $—CH_2CH{:}CH_2$)	162.18	mcl	11.2	234.5^{760}	1.0950^{25}_{4}	1.5383^{20}	i	v	v			chl ∞	**B19²**, 39
—	**Salicin**	*see* **Toluene, α,2-dihydroxy-**, glucoside													
—	**Salicylic acid**	*see* **Benzoic acid, 2-hydroxy-**													
—	**Salicylaldehyde**	*see* **Benzaldehyde, 2-hydroxy-**													
—	**Saligenin**	*see* **Toluene, α,2-dihydroxy-**													
—	**Salipyrine**	*see* **Antipyrine,** 2-hydroxybenzoate													
—	**Salol**	*see* **Benzoic acid, 2-hydroxy-**, phenyl ester													
—	**Salophen**	*see* **Benzoic acid, 2-hydroxy-,4-acetamido-**phenyl ester													
s8	**Salsoline**	$C_{11}H_{15}NO_2$	193.24	(al)	221				δ	s^h	i		δ	chl s NaOH s	
s9	**Salvarsan**	606. Arsphenamine. (structure: H_2N, NH_2, HO—, —AS=AS—, $—OH.2HCl.2H_2O$)	475.02	ye to gr powder	185–95d				v^h	δ	i			MeOH s dil HCl s con HCl i	**B16²**, 560
s10	**Sambunigrin**	*d*-Mandelonitrile glucoside. $C_6H_5CH(CN)O.C_6H_{11}O_5$	295.28	nd, $[\alpha]^{18}_D$ −75.1	151–2				s	s				AcOEt s	**B31**, 239
s11	**Samidin**	$C_{21}H_{22}O_7$	386.39	cubic, $[\alpha]$+26 (al)	135–7				i	δ	s			MeOH s	
—	**Sandoptal**	*see* **Barbituric acid, 5-allyl-5-isobutyl-**													
s12	**Sanguinarine**	$C_{20}H_{15}NO_4.H_2O$	351.36	bl nd	213				i	s	s				
s13	**Santalic acid**	Guerbet's acid. (structure: CH_3, CH_3, $CH_2CH_2CH{:}CCO_2H$)	234.33	red		β:202 γ:189		β:1.5136^{20} γ:1.5055^{20}	i	∞					**B9**, 571
s14	**α-Santalol**	Arbeol. $C_{15}H_{24}O$	220.36	$[\alpha]_D$+1		$301\text{–}2^{760}$ 167^{14}	0.977^{25}_{25}	1.4992^{20}	i	s					**B6¹**, 275
s15	**β-Santalol**	$C_{15}H_{24}O$	220.36	$[\alpha]_D$ −41.8		$167\text{–}8^{10}$	0.9717^{20}	1.5091^{20}							**B6²**, 517
s16	**Santene**	C_9H_{14}	122.20			140–2	0.8720^{17}_{4}	1.4657^{17}					s		**E12A**, 540
s17	**Santenic acid** (*cis, d*)	π-Norcamphenic acid. (structure: HO_2C, CH_3, CO_2H, CH_3)	186.20	pr, $[\alpha]^{33}_D$ +38.3 (al)	151–2				s^h	s					**B9²**, 529

For explanations, symbols and abbreviations see beginning of table.

No.	Name	Synonyms and Formula	Mol. wt.	Crystalline form, color and specific rotation	m.p. °C	b.p. °C	Density	n_D	Solubility w	al	eth	ace	bz	other solvents	Ref.
	β-Santenol														
s18	**β-Santenol** *(cis, exo)*	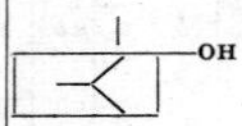	140.22	tab (lig)	97–8	195–9			i		s				**E12A,** 632
s19	**α-Santenone**(*cis*)	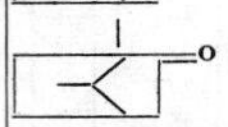	139.20	pl, $[\alpha]^{22}_{D}+11.4$ (al)	56–9					s		s			**E12A,** 271
s20	**Santonin**	Santonic acid. $C_{15}H_{18}O_3$	246.30	orh, $[\alpha]_D$ +173–6 (al)	173 (120 sub)		1.187^{26}_{4}	1.590 (1.640)	δ^h	δ s^h	δ			chl s	**E12B,** 3736
s21	**Sarmentogenin**	$C_{23}H_{34}O_5$	390.50	pr, (85% al or MeOH-eth) $[\alpha]^{19}_{D}+21.1\pm4$	265–6					s	i	δ	i	MeOH s Py s chl i	**E14,** 237
s22	**Sarpagine**	$C_{19}H_{22}N_2O_2$	310.30	nd, $[\alpha]^{20}_{D}+54$ (Py)	>320				i	s^h					
s23	**Sarsasapogenin**	Parigenin. $C_{27}H_{44}O_3$	416.62	lo pr nd (ace), $[\alpha]^{25}_{D}-75$	199–200					s		s	s	chl s	**E14,** 282
s24	**Scarlet red**	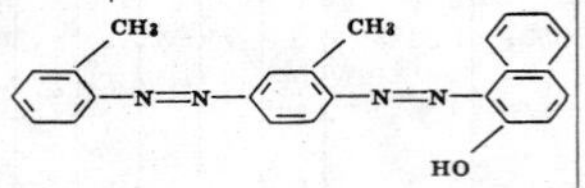	380.43	dark br	181–8δd				i	δ		δ	δ	chl s	
s25	**Scilliroside**	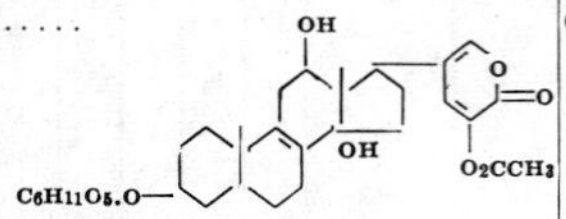	620.67	pr (dil MeOH), $[\alpha]^{20}_{D}-59$	168–70	d			δ	v	i	δ		diox v chl δ AcOEt δ	
—	**Scopolamine**	*see* **Hyoscine**													
—	**Scopoletin**	*see* **Coumarin, 7-hydroxy-6-methoxy-**													
s26	**Scopoline**	Oscin. 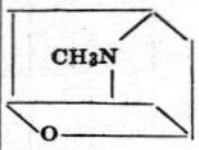	155.19	hyg nd (lig, eth, peth or chl)	108–9	248	1.089^{134}_{4}		s	s	s^h	s		chl δ peth δ	**B27²,** 61
s27	**Scyllitol**	$C_6H_{12}O_6$	180.16	pr (+3w)	353d	1.659^{19}_{4}			δ	δ	i				**B6²,** 1160
—	**Sebacic acid**	*see* **Decanedioic acid***													
s28	**Sedormide**	$H_2NCONHCOCH(CH_2CH{:}CH_2)CH(CH_3)_2$	184.23	nd (al)	194				δ	s^h	δ			chl s	**B3²,** 53
s29	**Selenanthrene**	Diphenyldiselenide. 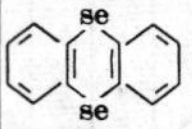	310.10	pr (al, nd (ace)	181	223^{11}				δ s^h	δ			CS_2 s	**B19,** 47
s30	**Selenide, diethyl**	$(C_2H_5)_2Se$	137.08	pa ye		110	1.2300^{20}_{4}	1.4768^{20}	i	v	v		v	chl v	**B1³,** 357
s31	**—,dimethyl**	$(CH_3)_2Se$	109.03			$54–5^{753}$	1.4077^{15}_{4}		i	s	s	δ		chl s to δ MeOH δ	**B1,** 291
s32	**—,—,hexachloro-**	$(Cl_3C)_2Se$	315.72		37									CCl_4 s	**J1947,** 1080
s33	**—,diphenyl-**	$(C_6H_5)_2Se$	233.16	ye nd	2.5	$301–2^{760}$ $136–42^{4}$	1.359^{16}_{4}	1.6478^{16}		∞	∞			xyl s	**B6²,** 318
s34	**—,di(2-tolyl)**	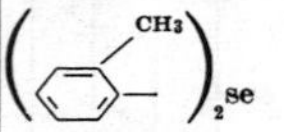	261.23	pl or lf (al)	65	186^{16}				s^h					**B6²,** 343
s35	**—,di(4-tolyl)**	 $(CH_3-C_6H_4-)_2Se$	261.23	rods (al)	69–70	196^{16}				s^h					**B6²,** 402
s36	**Selenonium, diphenyl-,** dichloride	$(C_6H_5)_2SeCl_2$	304.09	pa ye pr (xyl or ace), nd (al-HCl)	183 (142d)				s	s	i	s		chl i CCl_4 i	**B6²,** 318
s37	**—,triphenyl-,** chloride	$(C_6H_5)_3SeCl$	345.74	(AcOEt)	230d				v	v	i	δ		chl s	**B6²,** 318
s38	—,—,fluoride	$(C_6H_5)_3SeF$	329.28	oct deliq	145d				v	v		s		chl v	**J1946,** 1126
s39	—,—,hydrogen fluoride	$(C_6H_5)_3SeHF_2$	349.29	nd (ace)	99							s			**J1946,** 1126

For explanations, symbols and abbreviations see beginning of table.

No.	Name	Synonyms and Formula	Mol. wt.	Crystalline form, color and specific rotation	m.p. °C	b.p. °C	Density	n_D	Solubility w	al	eth	ace	bz	other solvents	Ref.
	Selenophenol														
—	**Selenophenol**	*see* **Phenol, seleno-**													
s39[1]	**Semicarbazide***	Aminourea. Carbamylhydrazine. $NH_2CONHNH_2$	75.07	pr (al)	96				v	s	i	...	i	chl i	
s39[2]	—,hydrochloride	$NH_2CONHNH_2.HCl$	111.54	pr (dil al)	173d				v	i	i	...	...		
s40	—,1,1-diphenyl-*	$H_2NCONHN(C_6H_5)_2$	227.27	nd (al or bz)	195				...	v^h	...	...	v^h	con sulf s	**B15**, 304
s41	—,1,4-diphenyl-*	$C_6H_5NHCONHNHC_6H_5$	227.27	nd or lf (al, bz)	177				δ^h	v	i	...	...		**B15**[2], 106
s42	—,2,4-diphenyl-*	$C_6H_5NHCON(NH_2)C_6H_5$	227.27	lf (al)	165.5				...	s	s	...	s	chl v	**B15**, 277
s43	—,1,1-diphenyl-3-thio-*	$NH_2CSNHN(C_6H_5)_2$	243.33	(al or bz)	202				i	v^h	...	v	v	chl v	**B15**, 304
s44	—,2,4-diphenyl-3-thio-*	$C_6H_5NHCSN(NH_2)C_6H_5$	243.33	lf	139				...	δ, v^h	...	v	v		**B15**[2], 103
s45	—,2-phenyl-*	$NH_2CON(NH_2)C_6H_5$	151.24	nd (bz or al)	120				v^h	v	v	...	δ	chl v	**B15**[2], 103
s46	—,4-phenyl-*	$C_6H_5NHCONHNH_2$	151.24	nd (bz), lf (w)	128				δ^h	v	i	...	...	chl v	**B12**[2], 221
s47	—,1-phenyl-3-thio-*	$NH_2CSNHNHC_6H_5$	167.23	pr (al)	200–1d				δ	s^h	δ	...	δ	chl δ	**B15**[2], 110
s48	—,2-phenyl-3-thio-*	$NH_2CSN(NH_2)C_6H_5$	167.23	(w)	153				δ, s^h	s	...	...	i		**B15**[1], 70
s49	—,4-phenyl-3-thio-*	$C_6H_5NHCSNHNH_2$	167.23	lf (al)	140				...	i	...	...	δ		**B12**[2], 232
s50	—,3-thio-*	Thiosemicarbazide. $NH_2CSNHNH_2$	91.13	lo nd (w)	183				s	s	...	...	...		**B3**[2], 134
s51	—,1(3-tolyl)-	Maretine. CH3 NH2CONHNH—	165.19	lf (w or dil al)	183–4				δ	v^h	i	...	...		
s52	—,1,1,4-triphenyl-	$C_6H_5NHCONHN(C_6H_5)_2$	303.37	nd (al)	206–7				...	δ, v^h	...	...	...		**B15**[2], 115
s53	—,1,4,4-triphenyl-	$(C_6H_5)_2NCONHNHC_6H_5$	303.37	pl (al)	149–50				i	v^h	...	...	...		**B15**[1], 71
s54	—,2,4,4-triphenyl-	$(C_6H_5)_2NCON(NH_2)C_6H_5$	303.37	(dil al)	128				i	v	...	...	...	peth i	**B15**, 277
—	**Seminose**	*see* **Mannose**													
s55	**Serine**(*D*)	2-Amino-3-hydroxypropanoic acid*. $HOCH_2CH(NH_2)CO_2H$	105.09	nd or hex pr, $[\alpha]^{20}_D+6.87$	228	d			v	i	i	...	...		**B4**[2], 919
s56	—(*DL*)	$HOCH_2CH(NH_2)CO_2H$	105.09	mcl pr or lf (w)	246				s	...	...	...	...		**B4**[2], 919
s57	—(*L*)	$HOCH_2CH(NH_2)CO_2H$	105.09	hex pl or pr, $[\alpha]^{20}_D-6.83^0$	228	sub 150^{10}			v	i	i	...	...		**B4**[2], 919
s58	**Serpentine**	$C_{21}H_{20}N_2O_3$	348.39	ye nd, $[\alpha]^{25}_D$ $+292\pm2$ (MeOH)	157–8				i	...	...	...	...		**Am 76**, 2843
s59	**Sesamin**	Asaranin. $C_{19}H_{18}O_6$	242.33		122.5				s	...	...	...	...		
s60	**Shikonine**	d-Alkanin. OH O —CHOHCH2CH:C(CH3)2 OH O	288.30	$[\alpha]^{20}_{644}+135$ (bz, c=1.3)	147				...	...	...	...	...	oos s	**B8**[2], 543
s61	**Silane, butyl-(trichloro)-**	Butylsilicon trichloride. $CH_3(CH_2)_3SiCl_3$	191.56			$148–9^{754}$	1.169^{15}_4	1.4363^{20}	d	d^h	s	...	...		**B4**[1], 582
s62	—,chloro(triethoxy)-	Orthosiliconic acid monochloride, triethyl ester. $(C_2H_5O)_3SiCl$	198.72		−51	155.7^{760}	1.032^{20}_{20}	1.3999^{20}	d	s	...	...	...		
s63	—,dichloro-(diethoxy)-	$(C_2H_5O)_2SiCl_2$	189.12		−130	135.9^{760}	1.130^{20}_{20}		d	s	...	...	...		
s63[1]	—,dichloro-(diethyl)-	$(C_2H_5)_2SiCl_2$	157.12			128–30d			d	d	...	...	...		**B4**, 629
s64	—,dichloro-(diphenyl)-	$(C_6H_5)_2SiCl_2$	253.21			$302–5^{757}$			d	d	...	...	...		**B16**[2], 608
s65	—,diethoxy-(difluoro)-	Diethyl difluorosilicate. $(C_2H_5O)_2SiF_2$	156.21			83^{760}d			d	...	...	...	...		**Am 68**, 76
s67	—,diethyl-(difluoro)-	$(C_2H_5)_2SiF_2$	124.21		−78.7	$61–2^{760}$	>1		i, d^h	...	...	...	...		**J1944**, 454
s68	—,difluoro-(diphenyl)-	$(C_6H_5)SiF_2$	220.30		252^{760} 158^{50}	1.145^{17}_4			i, d^h	...	...	...	s		**J1944**, 454

For explanations, symbols and abbreviations see beginning of table.

No.	Name	Synonyms and Formula	Mol. wt.	Crystalline form, color and specific rotation	m.p. °C	b.p. °C	Density	n_D	Solubility w	al	eth	ace	bz	other solvents	Ref.
	Silane														
s69	**—,dimethyl-**	Dimethylsilicane. $(CH_3)_2SiH_2$.	60.17	gas	−155.2	-19.6^{760}									**B4**[1], 579
s70	**—,ethoxy(trichloro)-**	$(C_2H_5O)SiCl_3$	179.51		−135	101.9^{760}			d	s					
s71	**—,ethoxy(triethyl)-**	$(C_2H_5O)Si(C_2H_5)_3$	160.33			190^{760}	0.8403^{0}		i	∞	∞			sulf s	**B4,** 629
s72	**—,ethoxy(trifluoro)-**	$(C_2H_5O)SiF_3$	130.14	gas	−122	-7^{760}									**J1949,** 1696
s73	**—,ethyl(trichloro)-**	$C_2H_5SiCl_3$	163.51			$97-103^{760}$	1.238^{20}_{4}	1.4257^{20}	d	d					**B4**[1], 580
s74	**—,ethyl(trifluoro)-**	$C_2H_5SiF_3$	114.14		−105	-4.4^{760}			d						**J1944,** 454
s74[1]	**—,ethyl(triethoxy)-**	Triethyl ethaneorthosiliconate. $C_2H_5Si(OC_2H_5)_2$	192.33			158–9	0.9207^{0}		i	∞	∞				**B4,** 630
s74[2]	**—,ethyl(trimethoxy)-**	Trimethyl ethaneorthosiliconate. $C_2H_5Si(OCH_3)_3$	150.25			125–6	0.9747^{0}			s					**B4,** 630
s75	**—,fluoro(triethoxy)-**	$(C_2H_5O)_3SiF$	182.27			134.6^{760}			d						**J1944,** 1696
s76	**—,fluoro(triethyl)-**	$(C_2H_5)_3SiF$	134.27			110^{760}	0.8354^{25}_{4}	1.3900^{25}	i					peth ∞ lig ∞	**Am 58,** 897
—	**—,hydroxy-**	*See* **Silicol**													
s77	**—,methyl-**	Methylsilicane. CH_3SiH_3	46.14		−156.5	−57			i						**B4**[1], 579
s77[1]	**—,methyl-(triethoxy)-**	Trimethyl methaneorthosiliconate. $CH_3Si(OC_2H_5)_3$	178.31			150–1	0.938			s					
s78	**—,phenyl(trifluoro)-**	$C_6H_5SiF_3$	162.19			101.5^{760}	1.212^{17}_{4}		d	s			s		**J1944,** 454
s79	**—,tetraethyl-**	$(C_2H_5)_4Si$	144.34			154.7^{760}	0.7682^{22}_{4}	1.4246^{25}	i						**B4**[2], 1007
s80	**—,tetramethyl-**	$(CH_3)_4Si$	88.20		−91.1	26.5^{760}	0.648^{19}_{4}		i	v	v			sulf i	**B4**[1], 579
s81	**—,triethyl-**	$(C_2H_5)_3SiH$	116.28			107	0.7313^{20}	1.4124^{20}	i					sulf i	**B4,** 625
—	**Silicane**	*see* **Silane**													
s82	**Silicol, triethyl-**	$(C_2H_5)_3SiOH$	132.28			154^{760}	0.8647^{20}_{4}	1.4364^{17}	i	∞	∞				**B4**[1], 581
—	**Sinapic acid**	*see* **Cinnamic acid, 3,5-dimethoxy-4-hydroxy-**													
s83	**Sinapine,** hydrogen sulfate 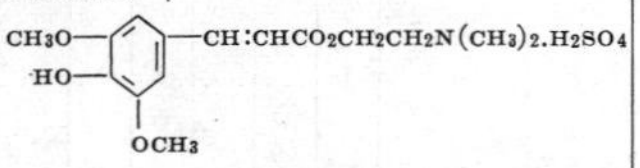	CH_3O—, HO—, OCH_3 (ring) —$CH{:}CHCO_2CH_2CH_2N(CH_3)_2.H_2SO_4$	407.44	lf (al)	186–7d				s	s^h	i				**B10,** 509
s84	—,hydrogen sulfate dihydrate	$C_{16}H_{23}NO_5.H_2SO_4.2H_2O$. *See* s83	443.47	lf (al)	127d				s	s^h	i				**B10,** 509
s85	—,thiocyanate monohydrate	$C_{16}H_{23}NO_5.HSCN.H_2O$. *See* s83	386.46	ye nd	180–1				δ	δ					**B10**[2], 354
—	**Sinapyl alcohol**	*see* **2-Propene-1-ol, 3(3,5-dimethoxy-4-hydroxyphenyl)-***													
s86	**Sinomenine**	Coculine. $C_{19}H_{23}NO_4$.	329.38	nd (bz), $[\alpha]^{26}_{D}-71$	161				δ	s	δ	s	δ	chl s dil alk s	**B21**[2], 470
s87	**α_1-Sitosterol**	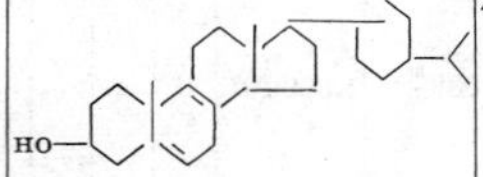	412.70	nd (al), $[\alpha]^{28}_{D}-1.7$ (chl)	164–6					s				chl s	**E14,** 89
s88	**α_2-Sitosterol**	$C_{30}H_{50}O$	426.73	(al-peth), $[\alpha]^{25}_{D}+3.5$ (chl)	156					s				peth δ oos s	**E14,** 89
s89	**α_3-Sitosterol**	$C_{29}H_{48}O$	412.70	pl (al) $(\alpha)^{20}_{D}$ +5.2 (chl)	142–3					s				chl s	**E14,** 90
s90	**β-Sitosterol**	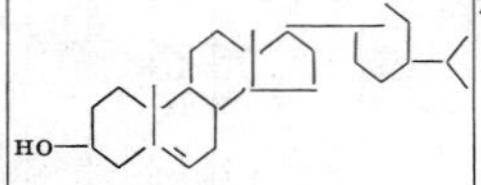	414.72	pl (al), $[\alpha]^{25}_{D}-37$	140					s	s			aa s	**E14,** 90
—	**Skatole**	*see* **Indole, 3-methyl-**													

For explanations, symbols and abbreviations see beginning of table.

No.	Name	Synonyms and Formula	Mol. wt.	Crystalline form, color and specific rotation	m.p. °C	b.p. °C	Density	n_D	Solubility w	al	eth	ace	bz	other solvents	Ref.
	Skimmin														
s91	**Skimmin**	7-Hydroxycoumarin-7-glucoside. Umbelliferone glucoside. 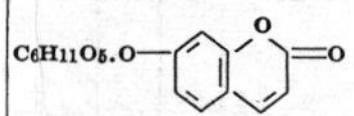 $C_6H_{11}O_5$.O–	324.28	(w+1), $[\alpha]^{18}_D$ −80 (Py)	219–21				s^h	s	i			chl i	
s92	**Smilagenin**	$C_{27}H_{44}O_3$	416.62	silky nd (ace), $[\alpha]^{25}_D$ −69	185					s		s	s	chl s	**E14**, 284
s93	**Solamine**	Solatunine. $C_{45}H_{73}NO_{15}$	868.04	nd (85% al), $[\alpha]^{20}_D$ −60	ca. 285d				i	v^h	i			chl i	
s94	**Solanidine**	Solatubine. $C_{27}H_{43}NO$	397.62	nd (chl-MeOH) $[\alpha]^{21}_D$ −29	218–9	sub δd			i	δ	i		v	chl v MeOH δ	
s95	**Solanine**	Purapurine. Solatunine. $C_{45}H_{73}NO_{16}$. $C_6H_{11}O_4$—O—$C_6H_{10}O_4$—O—$C_6H_{10}O_4$—O—$C_{27}H_{42}N$ (rhamnose galactose glucose solanidine)	884.04	nd (al), flakes (diox) $[\alpha]^{20}_D$ −69 (al)	ca. 190 (284.5d)				i δ^h	s^h	i		i	chl i diox s^h	
s96	**Solasodine**	Purapuridine. Solanidine-S. 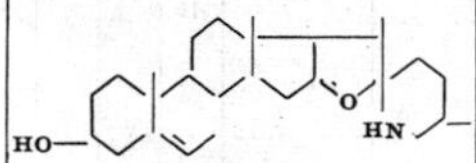	413.62	hex pl, $[\alpha]_D$ −92.4 (bz)	200–2					s^h	δ	s	s	chl v MeOH s diox s aq ac s	**J1942**, 13
—	**Solatubine**	*see* **Solanidine**													
—	**Solatunine**	*see* **Solamine**													
—	**Sorbic acid**	*see* **2,4-Hexadienoic acid***													
s98	**Sorbierite**(*D*,−)	Iditol. 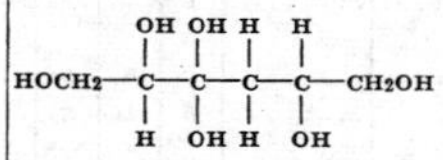 $HOCH_2$—C(H)(OH)—C(H)(OH)—C(OH)(H)—C(OH)(H)—CH_2OH	182.17	pr, $[\alpha]_D$ −3.53 (w, p=2)	73–4				s						**B1**, 544
s99	**Sorbitol**(*D*)	*D*-Glucitol. *D*-Sorbite. 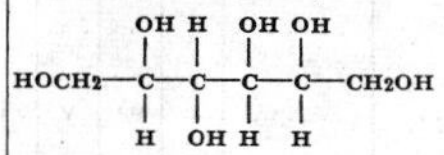	182.17	nd, $[\alpha]^{25}_D$ −1.98 (w)	110	$295^{3.5}$	1.489^{20}_4	1.3330^{20}	s	δ^h	i			Py v^h aa δ	**B1**^{3}, 2385
s100	—(*L*)	Gulitol. $C_6H_{14}O_6$. *See* s99	182.17	[α]+1.7 (w)	89–91										**B1**, 534
s101	—,hexaacetate(*D*)	$C_{18}H_{26}O_{12}$. *See* s99	434.39	pr (w)	98–9				δ^h	v	δ			AcOEt s	**B2**^{2}, 163
s102	**Sorbose**(*D*)	Pseudotagatose. *D*-Sorbinose. 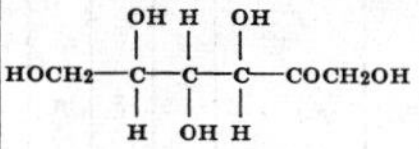 $HOCH_2$—C—C—C—$COCH_2OH$	180.16	rh (al), $[\alpha]^{20}_D$ +42.9 (w)	165		1.612^{17}		s	δ	δ				**B31**, 345
s103	—(*DL*)	β-Acrose. $C_6H_{12}O_6$. *See* s102	180.17	rh	154		1.638^{17}		s	δ^h					**B31**, 348
s104	—(*L*)	$C_6H_{12}O_6$. *See* s102	180.17	rh, $[\alpha]^{20}_D$ −42.8	165		1.612^{17}		s	δ^h				MeOH δ	**B31**, 346
—	**Sorbyl alcohol**	*see* **2,4-Hexadien-1-ol***													
—	**Sparassol**	*see* **Benzoic acid, 2-hydroxy-4-methoxy-6-methyl-**, methyl ester													
s105	**Sparteine**	Lupinidine. $C_{15}H_{26}N_2$	234.37	$[\alpha]^{21}_D$ −16.42 (al)	30–1	325^{754} 173^8	1.0196^{20}_4	1.5312^{20}	δ	v	v			chl v	**B2**^{2}, 621
s106	—,sulfate pentahydrate	$C_{15}H_{26}N_2.H_2SO_4.5H_2O$	422.53	pr, $[\alpha]^{17}_D$ −15.3	242			1.5289	i	v					**B23**^{2}, 99
—	**Spinacane**	*see* **Squalene, perhydro-**													
—	**Spinacene**	*see* **Squalene**													
—	**Spinulosin**	*see* **1,4-Benzoquinone, 2,5-dihydroxy-3-methoxy-6-methyl-***													
s107	**Spiro[5,5]hendecane, 2,4,8,10-tetroxa-**	Pentaerythritol dimethylene ether. Spiro-bi(1,3-dioxane). 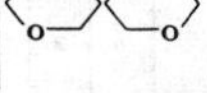	160.17		46.5	147^{53}			v	v	s	s	s	CCl_4 s	**B19**, 423

For explanations, symbols and abbreviations see beginning of table.

No.	Name	Synonyms and Formula	Mol. wt.	Crystalline form, color and specific rotation	m.p. °C	b.p. °C	Density	n_D	Solubility w	al	eth	ace	bz	other solvents	Ref.
	Squalene														
s108	**Squalene**	Spinacene. $C_{30}H_{50}$	410.73			280^{17}	0.8562^{20}_{4}	1.4965^{20}	i	δ	s	s		CCl_4 s	B1², 250
s109	—,**perhydro-**	2,6,10,15,19,23-Hexamethyltetracosane*. Spinacane. $C_{30}H_{62}$. *See* s108	422.80		−80	274^{10}	0.8125^{15}_{4}	1.4525^{20}	i	s^h			∞		B1², 145
s110	**Stachydrine**	Hygric acid methylbetaine. *N*-Methylproline methylbetaine. $C_7H_{13}NO_2$.	143.18	deliq $[\alpha]^{25}_{D}-40.25$	235				s	s	i			chl δ	B22², 4
s111	—,oxalate	$C_7H_{13}NO_2.H_2C_2O_4$	233.22	nd	106					i					
s112	**Starch**	$(C_6H_{10}O_5)_x$		amor powder	d				i	i					
s113	—,triacetate	$C_{216}H_{288}O_{144}$(?)		powder, $[\alpha]^{25}_{D}+72.5$	180 (>270 δd)				δ	δ		s	s	chl i	C46, 8401
—	**Stearic acid**	*see* **Octadecanoic acid***													
—	**Stearolyic acid**	*see* **9-Octadecynoic acid***													
—	**Stearone**	*see* **18-Pentatriacontanone***													
—	**Stearoxylic acid**	*see* **Octadecanoic acid, 9,10-dioxo-***													
s113¹	**Stibine, bis(trifluoromethyl)-bromo-***	$(F_3C)_2SbBr$	339.70			113^{760}									J1957, 3708
s113²	—,**bis(trifluoromethyl)-chloro-***	$(F_3C)_2SbCl$	295.24			ca. 88^{760} 17^{20}			d^h						J1957, 3708
s114	—,**bis(trifluoromethyl)iodo-***	$(F_3C)_2SbI$	386.70		−42	ca. 129 16^{8}									J1957, 3708
s115	—,**dibromo(trifluoromethyl)-***	F_3CSbBr_2	350.61			ca. 157^{760} $34^{2.5}$									J1957, 3708
s116	—,**dichloro-(phenyl)-***	$C_6H_5SbCl_2$	269.77	nd or ta	62	$110–5^{10}$				v				MeOH s aa s	B16², 574
s117	—,**diiodo(trifluoromethyl)-***	F_3CSbI_2	444.61	bt ye	4–8	$>200^{760}$d			i						J1957, 3708
s120	—,**triethyl-***	$(C_2H_5)_3Sb$	208.94		−29	158.5^{750}			i	s	s				B4, 618
s121	—,**trimethyl-***	$(CH_3)_3Sb$	166.86			80.6^{760}	1.523^{15}		δ	i	s				B4², 1004
s122	—,**triphenyl-***	$(C_6H_5)_3Sb$	353.08	tcl pl	51–2	>360 $>220^{1}$	1.4343^{25}_{4}	1.6948^{42}	i	s	v		v	chl, CS_2 v aa v	B16², 573
s123	—,**tris(trifluoromethyl)-**	$(F_3C)_3Sb$	328.79		−58	73^{760}			i						J1957, 3708
s124	—,**tri(2-tolyl)-**	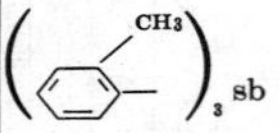 CH₃ ( )₃ sb	395.16	(al)	79–80					s^h	v		v	chl v peth v	B16, 892
s125	—,**tri(3-tolyl)-**	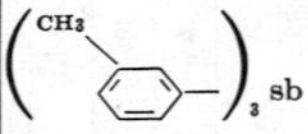 CH₃ ( )₃ sb	395.16	(peth)	67–8		1.3957^{16}_{4}			s	v		v	chl v aa v	B16, 892
s126	—,**tri(4-tolyl)-**	 (CH₃— —)₃ sb	395.16	rh (eth)	127–8		1.3595^{16}			δ	s		s	chl s peth δ	B16², 574
s127	**Stigmastanol**	Fucostanol. β-Sitostanol. 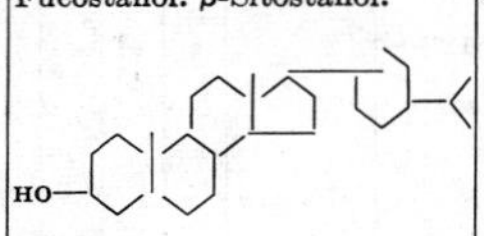 HO—	416.71	(+1w), $[\alpha]^{20}_{D}+25$	144–5 138–9 (+1w)									chl s	E14, 91
s128	**Stigmasterol**	$C_{29}H_{48}O$	412.67	(al+1w), $[\alpha]^{22}_{D}-51$	170				i	s^h				chl s oos s	E14, 88
s129	**Stilbene**(*trans*)	*trans*-1,2-Diphenylethene.* 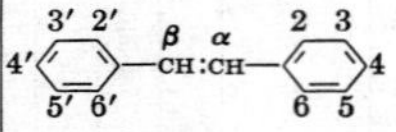 3′ 2′ β α 2 3; 4′ —CH:CH— 4; 5′ 6′ 6 5	180.25		124	306–7	0.9707	1.6264^{17}	i	δ s^h	v		v		B5², 537
s130	—,**2,2′-diamino-** (*cis*)	$C_{14}H_{14}N_2$. *See* s129	210.2	red nd (w)	123										B13¹, 85
s131	—,—(*trans*)	$C_{14}H_{14}N_2$. *See* s129	210.27	purp-br nd	190–1 (176)					s^h					B13¹, 85
s132	—,**4,4′-diamino-**	$C_{14}H_{14}N_2$. *See* s129	210.27	flakes (lig)	90–2										B13, 249
s133	—,**2,2′-dihydroxy-** (α-form)	$C_{14}H_{12}O_2$. *See* s129	212.24	nd (al)	95					v	v				B6², 987
s134	—,—(β-form)	$C_{14}H_{12}O_2$. *See* s129	212.24	flat nd	197					δ	s		s		B6, 1022

For explanations, symbols and abbreviations see beginning of table.

No.	Name	Synonyms and Formula	Mol. wt.	Crystalline form, color and specific rotation	m.p. °C	b.p. °C	Density	n_D	Solubility w	al	eth	ace	bz	other solvents	Ref.
	Stilbene														
s135	**—,3,5-dihydroxy-** *(trans)*	Pinosylvin. $C_{14}H_{12}O_2$. *See* s129	212.24	nd (aa)	156	...	...	...	i	...	...	s	s	chl s aa s	**C33,** 8989
s136	**—,4,4′-dihydroxy-** *(trans)*	Stilbestrol. $C_{14}H_{12}O_2$. *See* s129	212.24	nd (aa), ta (al)	284	...	...	...	...	δ s^h	δ	v	s	aa δ	**B6,** 1022
s137	**—,4,4′-dimethoxy-**	Bianisal. Photoanethole. $C_{16}H_{16}O_2$. *See* s129	240.29	lf (bz or aa)	210	sub	...	...	...	...	...	...	s	aa s	**B6**[2], 988
s138	**—,α,β-dinitro-** (low m.p.)	$C_{14}H_{10}N_2O_4$. *See* s129	270.25	ye pym (al)	104–7	d	...	...	...	s	s	v	v	chl v	**B5**[2], 542
s139	—,—(high m.p.)	$C_{14}H_{10}N_2O_4$. *See* s129	270.25	pa ye nd or pr (al)	186–7	...	...	...	...	δ s^h	...	s	s	chl s aa s	**B5**[2], 541
s140	**—,2,2′-dinitro-** (low m.p.)	$C_{14}H_{10}N_2O_4$. *See* s129	270.25	ye nd (aa)	126	...	...	...	i	...	...	...	...	aa s^h	**B5,** 637
s141	—,—(high m.p.)	$C_{14}H_{10}N_2O_4$. *See* s129	270.25	pa ye nd (chl)	196	420 exp	...	...	...	δ	δ	...	$δ^h$	chl s^h CS_2 δ	**B5**[1], 306
s142	**—,2,4-dinitro-** (low m.p.)	$C_{14}H_{10}N_2O_4$. *See* s129	270.25	ye pl (aa)	127	...	...	...	...	...	...	...	v	chl s aa s^h	**B5**[2], 541
s143	—,—(high m.p.)	$C_{14}H_{10}N_2O_4$. *See* s129	270.25	pa ye (aa)	143–4	412 exp	...	...	...	i	...	...	s	CS_2, chl, xyl s	**B5**[2], 540
s144	**—,2,6-dinitro-**	$C_{14}H_{10}N_2O_4$. *See* s129	270.25	ye nd (bz)	114 (86)	...	...	...	i	...	...	...	s^h	chl s	**B5**[1], 306
s145	**—,3,4′-dinitro-** (low m.p.)	$C_{14}H_{10}N_2O_4$. *See* s129	270.25	ye nd (aa)	155	...	...	...	...	δ	...	s	s	chl s aa s^h	**B5**[2], 541
s146	—,—(high m.p.)	$C_{14}H_{10}N_2O_4$. *See* s129	270.25	ye nd (aa, $PhNO_2$ or Py)	217	...	...	...	...	$δ^h$	...	δ	$δ^h$	chl δ AcOEt δ	**B5**[2], 541
s147	**—,4,4′-dinitro-** (low m.p.)	$C_{14}H_{10}N_2O_4$. *See* s129	270.25	ye nd (aa, chl or $PhNO_2$)	234–5 (210–6)	...	...	...	...	δ	δ	δ	s	chl s lig δ	**B5**[2], 541
s148	—,—(high m.p.)	$C_{14}H_{10}N_2O_4$. *See* s129	270.25	pa ye lf (aa), nd (aa or $PhNO_2$)	293 (286)	...	...	...	...	δ	δ	s	δ	chl δ aa s^h	**B5**[2], 541
—	**Stilbestrol**	*see* **Stilbene, 4,4′-dihydroxy-***(trans)*													
s151	**Strophantidin**	$C_{23}H_{32}O_6$	404.51	(al+½w) $[\alpha]_D^{20}$ +41 (al)	232 (anh) 177–8 (hyd)	...	...	...	i	s	i	s	...	chl δ peth i	**E14,** 252
s152	**δ-Strophanthin**	Onabain. $C_{29}H_{44}O_{12}$	584.67	hyg pl (+9w) $[\alpha]_D^{50}$ −34	200	...	...	...	δ s^h	v	...	...	...	...	**E14,** 251
s153	**Strychnine**	$C_{21}H_{22}N_2O_2$	334.42	orh pr (al), $[\alpha]_D^{18}$ −139.3 (chl, c=1)	268–90	270^5	1.36_4^{20}	...	$δ^h$	δ	i	...	δ	chl s to δ peth i	**B27**[2], 723
s154	—,hydrochloride	$C_{21}H_{22}N_2O_2.HCl.2H_2O$	406.91	$[\alpha]_D$ −28.3 (w, c=0.7)	...	...	...	...	s	δ	i	...	...	chl δ	**B27**[2], 730
s155	—,nitrate	$C_{21}H_{22}N_2O_2.HNO_3$	397.43	nd	280–310	...	1.627	...	s	$δ^h$	i	...	...	chl δ	**B27**[2], 732
s156	—,sulfate	$(C_{21}H_{22}N_2O_2)_2.H_2SO_4.5H_2O$	857.00	mcl $[\alpha]_{589}$ 13.25	200d	...	...	...	s v^h	δ	i	...	...	...	...
—	**Styphnic acid**	*see* **Benzene, 2,4-dihydroxy-1,3,5-trinitro-***													
—	**Styptol**	*see* **Cotarnine,** phthalate													
s157	**Styracin**	Cinnamyl cinnamate. $C_6H_5CH{:}CHCO_2CH_2CH{:}CHC_6H_5$	264.33	nd	44	1.085	...	...	i	s	v	...	...	...	**B9**[2], 388
s158	**Styracitol**	H H OH OH $HOCH_2$—C—C—C—C—CH_2 OH H H —O—	164.16	pr $[\alpha]_D^{17}$ −50 (w)	155	...	...	...	v	i	i	i	i	MeOH δ, s^h	**B17**[2], 235 **B18**[2], 620
s159	**Styrene**	Ethenylbenzene*. Phenylethylene. Vinylbenzene. 5 6 α β 4 ⟨ring⟩ —CH:CH_2 3 2	104.14	...	−33	145–6	0.9090_4^{20}	1.5463^{20}	i	s	s	s	...	MeOH s CS_2 s	**B5**[2], 362
s160	**—,α-bromo-**	$C_6H_5CBr{:}CH_2$	183.06	...	−44	$86–7^{14}$	1.4025^{23}	1.5881^{20}	...	...	...	...	...	...	**B5**[2], 367
s161	**—,β-bromo-**	$C_6H_5CH{:}CHBr$	183.06	...	+6(−8)	107^{22-3}	1.4269^{16}	1.5990^{22}	i	∞	∞	...	...	...	**B5**[2], 368
s162	**—,2-bromo-**	C_8H_7Br. *See* s159	183.06	...	fr −52.8	206.2^{760} $64–5^3$	1.4160_4^{20}	1.5927^{20}	...	...	...	...	...	...	...
s163	**—,3-bromo-**	C_8H_7Br. *See* s159	183.06	...	...	$90–4^{20}$	1.4059_4^{20}	1.5855	...	...	...	...	...	...	**Am59,** 1176
s164	**—,4-bromo-**	C_8H_7Br. *See* s159	183.06	lf	4.5	89^{16}	1.400_4^{20}	1.5933^{20}	i	...	...	...	...	chl v aa s	**B5**[2], 367
s165	**—,α-chloro-**	$C_6H_5CCl{:}CH_2$	138.59	...	−23	199	1.0984_4^{25}	1.5623^{17}	i	s	s	...	...	...	**B5**[2], 367

For explanations, symbols and abbreviations see beginning of table.

No.	Name	Synonyms and Formula	Mol. wt.	Crystalline form, color and specific rotation	m.p. °C	b.p. °C	Density	n_D	Solubility w	al	eth	ace	bz	other solvents	Ref.
	Styrene														
s166	**—,β-chloro-**	$C_6H_5CH{:}CHCl$	138.59			199	1.1095^{18}_{4}	1.5736^{25}	i	s	s				**B5²**, 367
s167	**—,2-chloro-**	C_8H_7Cl. *See* s159	138.59		−63.15	188.7^{760}	1.100^{20}	1.5648^{20}		s	s	s		CCl_4 s aa s	**Am 66**, 1295
s168	**—,3-chloro-**	C_8H_7Cl. *See* s159	138.59			$62\text{–}3^{6}$	1.090^{20}_{4}	1.5619^{20}	i	s	s				**Am 66**, 1296
s169	**—,4-chloro-**	C_8H_7Cl. *See* s159	138.59			74^{12}	1.1554^{20}_{4}	1.5658^{20}	i	s	s				**B5²**, 367
s170	**—,2-fluoro-**	C_8H_7F. *See* s159	122.14			$32\text{–}4^{3}$	1.030^{20}_{4}	1.5197^{20}	i	s	s				**Am 66**, 1295
s171	**—,3-fluoro-**	C_8H_7F. *See* s159	122.14			$30\text{–}1^{4}$	1.025^{20}_{4}	1.5173^{20}	i	s	s				**Am 66**, 1295
s172	**—,4-fluoro-**	C_8H_7F. *See* s159	122.14		−34.5	67.4^{50}	1.024^{20}_{4}	1.5158^{20}	i	s	s				**Am 68**, 1159
s173	**—,2-hydroxy-**	*o*-Vinylphenol. C_8H_8O. *See* s159	120.14	nd	29	108^{15} (77^{15})	1.0468^{36}	1.577^{35}	s	v	v				**B6**, 560
s174	**—,3-hydroxy-**	*m*-Vinylphenol. C_8H_8O. *See* s159	120.14			114.6^{16}	1.0468^{35}_{4}								
s175	**—,3-hydroxy-4-methoxy-**	5-Vinylguaiacol. Hesperetol. $C_9H_{10}O_2$. *See* s159	150.18		57				δ	s	s				**B6**, 954
s176	**—,2-methoxy-**	*o*-Vinylanisole. $C_9H_{10}O$. *See* s159	134.18	nd	29	195–200 $83\text{–}4^{12}$	1.0609^{19}_{4}	1.5388^{20}	i	s	s			oos s	**B6¹**, 277
s177	**—,3-methoxy-**	*m*-Vinylanisole. $C_9H_{10}O$. *See* s159	134.18			$114\text{–}6^{16-7}$	0.999^{16}_{4}	1.555^{16}	i	s	s				**B6**, 561
s178	**—,4-methoxy-**	*p*-Vinylanisole. $C_9H_{10}O$. *See* s159	134.18			$204\text{–}5^{756}$ $95\text{–}6^{16}$	1.0001^{13}_{4}	1.5642^{13}	i-	s	s				**B6²**, 520
s179	**—,β-nitro-**	$C_6H_5CH{:}CHNO_2$	149.15	ye pr (peth or al)	58	250^{760} 150^{14}			i $δ^h$	s	v			chl v CS_2 v peth s	**B5²**, 368
s180	**—,4-nitro-**	$C_8H_7NO_2$. *See* s159	149.15	pr (lig)	29	d				v^h	v			aa s lig s	
—	**Styrene oxide**	*see* **Benzene, (1,2-epoxyethyl)-***													
s181	**β-Styrenesulfonic acid, chloride**	$C_6H_5CH{:}CHSO_2Cl$	202.67	flakes (bz)	85					d			s^h		**B11²**, 87
s182	**Subboric acid, tetramethyl ester**	$(CH_3O)_2BB(OCH_3)_2$	145.78		−24.3	93^{760} extrap 18^{35}			d						
—	**Suberaldehyde**	*see* **Octanedial***													
—	**Suberic acid**	*see* **Octanedioic acid***													
s183	**Succinaldehyde**	Butanedial*. $OCHCH_2CH_2CHO$	86.09			$169\text{–}70^{760}$ δd	1.064^{20}_{4}	1.4262^{18}	s	s	s			os v	**B1²**, 824
—	**Succinamic acid**	*see* **Succinic acid,** monoamide													
—	**Succinamide**	*see* **Succinic acid,** diamide													
—	**Succinanil**	*see* **Succinic acid,** imide, *N*-phenyl-													
—	**Succinanilic acid**	*see* **Succinic acid,** monoamide, *N*-phenyl-													
s185	**Succinic acid**	Butanedioic acid*. $HO_2CCH_2CH_2CO_2H$	118.09	tcl or mcl pr	182	235d	1.572^{25}_{4}	1.450	δ v^h	s	δ		i	MeOH s	**B2²**, 540
s186	**—,anhydride**	Succinic anhydride. (O=C–CH₂–CH₂–C=O ring with O)	100.07	nd (al), rh pym (chl or CCl_4)	119.6	261			i	s	δ			chl s	**B17²**, 428
s187	**—,diamide**	Succinamide. $H_2NCOCH_2CH_2CONH_2$	116.12	orh nd (w)	260	125.5 sub			δ s^h	i	i				**B2²**, 554
s188	**—,dibenzyl ester**	$C_6H_5CH_2O_2CCH_2CH_2CO_2CH_2C_6H_5$	298.33	orh pl or ta (PrOH)	47–8	245^{15}	1.256	1.596	i	s	s		s		**B6²**, 418
s189	**—,dibutyl ester**	$C_4H_9^nO_2CCH_2CH_2CO_2C_4H_9^n$	230.30		−29.25	274.5^{760}	0.9760^{20}_{4}	1.4288^{20}	i	s	s				**B2²**, 551
s190	**—,di-*sec*-butyl ester**	$C_4H_9^sO_2CCH_2CH_2CO_2C_4H_9^s$	230.30			256^{760}	0.9735^{20}_{4}	1.4238^{25}							**B2²**, 559

For explanations, symbols and abbreviations see beginning of table.

No.	Name	Synonyms and Formula	Mol. wt.	Crystalline form, color and specific rotation	m.p. °C	b.p. °C	Density	n_D	Solubility w	al	eth	ace	bz	other solvents	Ref.
	Succinic acid														
s191	—,di(2-carboxyphenyl) ester	Diaspirin. Succinylsalicylic acid. 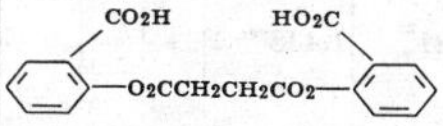 (structure: CO_2H, HO_2C, $—O_2CCH_2CH_2CO_2—$)	358.29	nd (aa or al)	176–8				δ	δ s^h	δ			aa δ s^h	**B10**^{2}, 43
s192	—,dichloride	Succinyl chloride. $ClCOCH_2CH_2COCl$	154.99	pl and lf	18.5	190–2	1.4073$^{20}_{20}$	1.473^{15}	d	d			s		**B2**, 613
s193	—,diethyl ester	Ethyl succinate. $C_2H_5O_2CCH_2CH_2CO_2C_2H_5$	174.19		−22	217.7^{760}	1.0406$^{20}_{4}$	1.4201^{20}	i	∞	∞				**B2**^{2}, 550
s195	—,dimethyl ester	Methyl succinate. $CH_3O_2CCH_2CH_2CO_2CH_3$	146.14		19	196.4^{760}	1.1208$^{20}_{4}$	1.4197^{20}	δ	s					**B2**^{2}, 549
s196	—,dinitrile	Succinonitrile. Ethylene dicyanide. $NCCH_2CH_2CN$	80.09		54.5	265–7^{760}	0.9848$^{63}_{4}$	1.4166^{63}	v	s	δ		s		**B2**, 615
s197	—,diphenyl ester	$C_6H_5O_2CCH_2CH_2CO_2C_6H_5$	270.29	lf (al)	120	222.5^{15}			i	s^h	s		s		**B6**, 155
s198	—,dipropyl ester	$C_3H_7^nO_2CCH_2CH_2CO_2C_3H_7^n$	202.24		−10.4	250.8^{760}	1.0011$^{20}_{4}$	1.4252^{20}						os s	**B2**^{2}, 551
s199	—,di(4-tolyl)ester 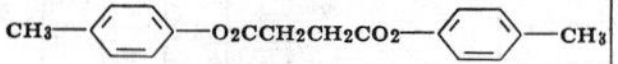	(structure: $CH_3—C_6H_4—O_2CCH_2CH_2CO_2—C_6H_4—CH_3$)	298.32	nd or lf (al)	120–1					s			s	os s aa s lig δ	**B6**^{2}, 379
s200	—,imide	Succinimide. 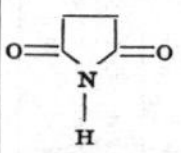 (structure)	99.09	pl (al), rh (ace)	126–7	287–8	1.418		δ v^h	δ s^h	i				**B21**, 369
s201	—,—,*N*-benzyl-	$C_{11}H_{11}NO_2$. *See* s200	189.22	nd (w), pr (al)	103	390–400			v^h	v	s		v^h	chl v CS_2 δ peth i	**B21**^{2}, 306
s202	—,—,*N*-bromo-	NBS. $C_4H_4BrNO_2$. *See* s200	177.99			178 δd	2.098		δ		δ	v		AcOEt v Ac_2O δ	**B21**^{2}, 304
s203	—,—,*N*(2-bromophenyl)-	$C_{10}H_8BrNO_2$. *See* s200	254.09	(dil al)	91				i	s					**B21**^{2}, 304
s204	—,—,*N*(3-bromophenyl)-	$C_{10}H_8BrNO_2$. *See* s200	254.09	nd (al)	118				s^h	v^h					**B21**^{2}, 304
s205	—,—,*N*(4-bromophenyl)-	$C_{10}H_8BrNO_2$. *See* s200	254.09	pr (al)	171				i^h					os v	**B21**^{2}, 304
s206	—,—,*N*-carbethoxy-	$C_7H_9NO_4$. *See* s200	171.15	(eth-lig)	44				v					os v	**B21**^{2}, 305
s207	—,—,*N*-chloro-	NCS. $C_4H_4ClNO_2$. *See* s200	133.53	pl	148–50		1.65		δ	δ				lig δ	**B21**^{2}, 306
s208	—,—,*N*(chloromethyl)-	$C_5H_6ClNO_2$. *See* s200	147.56		58	158–60^{12}			v					oos v	**B21**^{2}, 305
s209	—,—,*N*(4-chlorophenyl)-	$C_{10}H_8ClNO_2$. *See* s200	209.63	nd (dil al)	170					v				os v	**B21**^{2}, 303
s210	—,—,*N*(ethoxymethyl)-	$C_7H_{11}NO_3$. *See* s200	157.17	nd	31–2	262 151–2^{14}			v	v	δ				**B21**^{2}, 304
s211	—,—,*N*(4-ethoxyphenyl)-	$C_{12}H_{13}NO_3$. *See* s200	219.24	pr (al)	155				$δ^h$	s^h	i			aa s^h	**B21**, 377
s212	—,—,*N*-ethyl-	$C_6H_9NO_2$. *See* s200	127.14	(eth)	26	236			v	v	v				**B21**, 373
s213	—,—,*N*(hydroxymethyl)-	$C_5H_7NO_3$. *See* s200	129.12	lf (bz)	66				v					os δ	**B21**^{2}, 304
s214	—,—,*N*-iodo-	$C_4H_4INO_2$. *See* s200	224.99	(ace)	135		2.245		v	s	δ	s			**B21**^{2}, 307
s215	—,—,*N*-methyl-	$C_5H_7NO_2$. *See* s200	113.12	nd (eth-peth, al, ace)	71	234			s	s	v				**B21**^{2}, 303
s216	—,—,*N*(1-naphthyl)-	$C_{14}H_{11}NO_2$. *See* s200	225.25	nd (dil al)	153				i	v^h				MeOH v^h	**B21**, 376
s217	—,—,*N*(2-naphthyl)-	$C_{14}H_{11}NO_2$. *See* s200	225.25	nd (al or AcOEt)	183				i	s			s	AcOEt δ	**B21**, 376
s218	—,—,*N*-phenyl-	Succinanil. $C_{10}H_9NO_2$. *See* s200	175.19	mcl pr or nd (w or al)	155–6	*ca.* 400	1.356		i $δ^h$	s^h	s				**B21**^{2}, 303
s219	—,monoamide	Succinamic acid. $HO_2CCH_2CH_2CONH_2$	117.11	nd (w or ace)	157				s	δ		s^h	i	lig i	**B2**^{2}, 554
s220	—,—,*N*-phenyl-	Succinanilic acid. $HO_2CCH_2CH_2CONHC_6H_5$	193.20	nd (w)	147–8				δ s^h	s	v				**B12**, 295
s221	—,monobenzyl ester	$HO_2CCH_2CH_2CO_2CH_2C_6H_5$	208.21	bz	59				i				δ	lig δ	**B6**^{2}, 418
s222	—,monochloride monomethyl ester	$CH_3O_2CCH_2CH_2COCl$	150.56			93^{18}			d^h	d^h					**B2**^{2}, 553
s223	—,monoethyl ester(*dl*)	$HO_2CCH_2CH_2COC_2H_5$	146.14	pr or nd	108–9	180–3			v	v	v			chl δ, peth i	**B2**^{2}, 584
s224	—,monoethyl ester monomethyl ester	$CH_3O_2CCH_2CH_2CO_2C_2H_5$	160.17		<−20	208.2^{760}	1.076$^{20}_{4}$		i	v	v				**B2**, 609

For explanations, symbols and abbreviations see beginning of table.

No.	Name	Synonyms and Formula	Mol. wt.	Crystalline form, color and specific rotation	m.p. °C	b.p. °C	Density	n_D	Solubility w	al	eth	ace	bz	other solvents	Ref.
	Succinic acid														
s226	—,piperazinium salt	$HO_2CCH_2CH_2CO_2H.C_4H_{10}N_2$	204.23		205–6d				s	s^h	i				**Am56,** 1759
s227	**—,acetyl-,** diethyl ester	$C_2H_5O_2CCH(COCH_3)CH_2CO_2C_2H_5$	216.24			254–6[760] 139[12]	1.081^{20}_{4}	1.438[16]	i	s	s		s	CS_2 s	**B3²,** 486
—	**—,amino-**	*see* **Aspartic acid**													
s228	**—,benzyl-(*dl*)**	$C_6H_5CH_2CH(CO_2H)CH_2CO_2H$	208.22	nd	163–4				s^h	s	s		δ		**B9²,** 628
s229	**—,bromo-(*dl*)**	$HO_2CCH_2CHBrCO_2H$	196.98		160–1		2.073		s	s				aa δ	**B2²,** 560
s230	**—,2,3-diacetyl-,** diethyl ester (diketo form)	Ethyl diacetosuccinate. $C_2H_5O_2CCH(COCH_3)CH(COCH_3)CO_2C_2H_5$	258.27	mcl pr or nd (al)	92				δ	s		s	s	chl v lig δ	**B3²,** 505
s231	**—,2,3-dibromo-** (*d*)	$HO_2CCHBrCHBrCO_2H$	275.89	pl $[\alpha]^{24}_{D}+126.3$ (AcOEt)	157–8										**B2²,** 561
s232	—,—(*dl*)	$HO_2CCHBrCHBrCO_2H$	275.89	(w or AcOEt)	171				s						**B2²,** 561
s233	—,—(*l*)	$HO_2CCHBrCHBrCO_2H$	275.89	nd (bz), $[\alpha]^{13}_{D}-148$ (AcOEt, c=5.79)	157–8				v	v		v		AcOEt, MeOH v, chl, CCl_4, peth δ	**B2¹,** 269
s234	—,—(*meso*)	$HO_2CCHBrCHBrCO_2H$	275.89		269(closed tube)	sub			δ v^h	v	v			chl δ	**B2²,** 561
s235	**—,2,3-dichloro-** (*d*)	$HO_2CCHClCHClCO_2H$	186.98	mcl $[\alpha]^{19}_{D}$ +3.6 (w, c=3)	166–7		1.820[15]								**B2²,** 556
s236	—,—(*dl*)	$HO_2CCHClCHClCO_2H$	186.98	pr (w or eth-peth)	175d		1.844[15]		s^h	s	s	s	δ	chl s	**B2²,** 557
s237	—,—(*meso*)	$HO_2CCHClCHClCO_2H$	186.98	hex pr (w)	220d				s	v	v	v	δ	chl v	**B2²,** 558
s238	—,—,dichloride(*dl*)	$ClCOCHClCHClCOCl$	223.87		39	78.5[7]			d	d	s				**B2²,** 558
s239	—,—,—(*meso*)	$ClCOCHClCHClCOCl$	223.87			105–6[45] 79–80[15]			d	d				CCl_4 s	**B2²,** 558
s240	—,—,diethyl ester(*dl*)	$C_2H_5O_2CCHClCHClCO_2C_2H_5$	243.09			132[15]	1.1963^{77}_{4}	1.4512[20]			v				**B3²,** 556
s241	—,—,—(*meso*)	$C_2H_5O_2CCHClCHClCO_2C_2H_5$	243.09	nd (dil al)	63	125.5[12.5]	1.1490^{99}_{4}	1.4266[65]	i	v	v				**B2²,** 556
s242	**—,2,3-diethyl-,** diphenyl ester	3,4-Dicarbophenoxyhexane. $C_6H_5O_2CCH(C_2H_5)CH(C_2H_5)CO_2C_6H_5$	326.40	nd (lig)	107–8				i	δ		s	s	to s CS_2s	**B6,** 156
—	**—,2,3-dihydroxy-**	*see* **Tartaric acid**													
s243	**—,2,3-dimethoxy-,** dimethyl ester(*d*)	$CH_3O_2CCH(OCH_3)CH(OCH_3)CO_2CH_3$	206.20	pr, $[\alpha]^{18}_{D}+79.9$ (MeOH, c=2)	57	135[12]	1.1317^{60}_{4}	1.4340[20]						MeOH s	**B3²,** 328 **B16²,** 87
—	**—,dioxo-,** dihydrate	*see* **Succinic acid, tetrahydroxy-**													
s244	**—,2-dodecyl-3-methyl-**	2,3-Pentadecanedicarboxylic acid*. Roccelic acid. $CH_3(CH_2)_{11}CH(CO_2H)CH(CH_3)CO_2H$	300.44	lf (aa, bz or al), rods (ace) $[\alpha]^{20}_{D}+14.6$ (al)	132–3				i	s^h	v		s^h	peth i $NaHCO_3$ s	**B2,** 734
s245	**—,ethyl-(*d*)**	$HO_2CCH_2CH(C_2H_5)CO_2H$	146.14	pr or nd, $[\alpha]^{16}_{D}+20.6$ (aa, c=3.7)	84–5	180–3			v	v	v			chl δ peth i	**B2²,** 584
s246	—,—(*l*)	$HO_2CCH_2CH(C_2H_5)CO_2H$	146.14	pr or nd $[\alpha]^{24}_{D}-20.8$ (ace, c=4.6)	84–5	180–3			v	v	v			chl δ peth i	**B2²,** 584
s247	**—,formyl-,** diethyl ester	$C_2H_5O_2CCH_2CH(CHO)CO_2C_2H_5$	202.21			114–9[5]		1.4486[25]	i	∞	∞				**B3²,** 485
—	**—,(hydroxy-methyl)-γ-lactone**	*see* **Paraconic acid**													
s248	**—,2-hydroxy-3-methyl-(*d*)-**	d-Citramalic acid. $HO_2CCHOHCH(CH_3)CO_2H$	148.12	$[\alpha]^{14}_{D}+34.67$ (w, p=74.2)	108–9										**B3,** 443
s249	—,—(*dl*)	$HO_2CCHOHCH(CH_3)CO_2H$	148.12	mcl pr	119	sub			v	s		s	i	peth i	**B3²,** 294
s250	**—,isopropylidene-**	Teraconic acid. $(CH_3)_2C{:}C(CO_2H)CH_2CO_2H$	158.16	tcl	160				s^h	s	s		δ		**B2,** 786
s250[1]	**—,mercapto-(*d*)**	*d*-Thiomalic acid. $HO_2CCH_2CH(SH)CO_2H$	150.15	$[\alpha]^{13}_{D}+64.4$ (al), +76.1 (ace)	152–3				s	s		s			**B3²,** 287

For explanations, symbols and abbreviations see beginning of table.

No.	Name	Synonyms and Formula	Mol. wt.	Crystalline form, color and specific rotation	m.p. °C	b.p. °C	Density	n_D	Solubility						Ref.
									w	al	eth	ace	bz	other solvents	
	Succinic acid														
s250[2]	—,—(*dl*)	$HO_2CCH_2CH(SH)CO_2H$	150.15	amor	149–50	...	...	...	v	v	s	v	i	...	**B3**^{2}, 291
s250[3]	—,—(*l*)	$HO_2CCH_2CH(SH)CO_2H$	150.15	$[\alpha]_D^{17}$ −64.8 (al), −75.8 (ace)	152–3	...	...	...	s	s	δ	s	δ	...	**B3**^{2}, 287
s251	**—,methyl-**	Pyrotartaric acid. $HO_2CCH_2CH(CH_3)CO_2H$	132.12	tcl pr	112.5	d	...	1.4303	...	...	...	...	...	chl δ	**B2**, 636
s252	**—,2-methyl-3-oxo-**, diethyl ester	Diethyl methyloxaloacetate. $C_2H_5O_2CCOCH(CH_3)CO_2C_2H_5$	202.21	...	...	$144-6^{21}$	...	...	i	∞	∞	...	...	...	**B3**^{2}, 484
s253	**—,methylene-**	Itaconic acid. $CH_2{:}C(CO_2H)CH_2CO_2H$	130.10	rh	167–8	d	1.632	...	s	s	δ	...	δ	chl δ CS_2 δ	**B2**^{2}, 650
s254	—,—,anhydride	Itaconic anhydride. (ring structure: =CH_2; O=, =O; O)	112.09	rh bipym pr (eth or chl), scales (aa)	68.5	$114-5^{18}$	...	...	d^h	d^h	δ	...	...	chl v	**B17**^{2}, 449
s255	—,—,dichloride	Itaconyl chloride. $CH_2{:}C(COCl)CH_2(COCl)$	166.99	...	...	89^{17}	...	1.4919^{20}	d^h	d^h	...	...	...	...	**B2**, 762
s256	—,—,diethyl ester	Diethyl itaconate. $CH_2{:}C(CO_2C_2H_5)CH_2CO_2C_2H_5$	186.21	...	...	228 $119-20^{20}$	1.0467_4^{20}	1.4377^{20}	...	∞	s	...	s	...	**B2**^{2}, 651
s257	—,—,dimethyl ester	Dimethyl itaconate. $CH_2{:}C(CO_2CH_3)CH_2CO_2CH_3$	158.16	hyg mcl (MeOH)	38	208^{760} 108^{11}	1.1241_4^{18}	1.4442^{20}	...	s	s	...	...	MeOH s	**B2**, 762
s258	**—,oxo-,** diethyl ester	Diethyl oxaloacetate. $C_2H_5O_2CCH_2COCO_2C_2H_5$	188.18	...	...	$131-2^{24}$	1.13_4^{20}	1.4561^{17}	i	∞	∞	...	∞	...	**B3**^{2}, 479
s259	**—,2-oxo-3-phenyl-**, 1-ethyl ester 4-nitrile	Ethyl phenylcyanopyruvate. $C_6H_5CH(CN)COCO_2C_2H_5$	217.23	(eth-lig)	130	206^{20}	...	...	...	v	δ	...	...	chl v alk s	**B10**^{2}, 607
s260	**—,phenyl-**(*d*)	$HO_2CCH_2CH(C_6H_5)CO_2H$	194.19	pr (w), $[\alpha]_D^{16.5}$ +148.3 (al, c = 1.5)	173–4	...	...	...	δ v^h	s	...	v	δ	MeOH s	**B9**^{1}, 380
s261	—,—(*dl*)	$HO_2CCH_2CH(C_6H_5)CO_2H$	194.19	lf or nd (w)	168	d	...	...	δ v^h	v	v	v	i	chl, CS_2, peth i	**B9**^{2}, 619
s262	—,—(*l*)	$HO_2CCH_2CH(C_6H_5)CO_2H$	194.19	$[\alpha]_D^{15}$ −173.3	173–4	...	...	...	...	...	...	v	...	MeOH s	**B9**^{1}, 381
s263	—,—,anhydride(*d*)	(ring structure: C_6H_5; O=, =O; O)	176.17	nd (bz-peth), $[\alpha]_D^{15}$ +100.9 (al)	84	...	...	...	d^h	s	...	...	v	chl v peth δ CCl_4	**B17**^{1}, 259
s264	—,—,—(*dl*)	$C_{10}H_8O_3$. *See* s263	176.17	mcl pr or nd (eth)	53–4	$204-6^{22}$	...	...	i	v	...	...	...	oos v	**B17**^{2}, 473
s265	—,—,—(*l*)	$C_{10}H_8O_3$. *See* s263	176.17	$[\alpha]_D^{14}$ −100.9 (bz)	84	...	...	...	d^h	...	...	...	v	chl v	**B17**^{1}, 259
s266	—,**(3-phenyl-propenyl)-**	$C_6H_5CH_2CH{:}CHCH(CO_2H)CH_2CO_2H$	234.25	lf (eth), (bz)	112	...	...	...	s	...	v	v	v	...	**B9**, 909
s267	**—,tetrahydroxy-**	$HO_2CC(OH)_2C(OH)_2CO_2H$	182.09	...	114–5	...	...	...	d	...	...	...	...	...	**B3**^{2}, 500
s268	**—,tetramethyl-**	$HO_2CC(CH_3)_2C(CH_3)_2CO_2H$	174.20	tcl (60% MeOH, lig or AcOEt), mcl and tcl (eth or ace)	190–2	sub	1.30	...	δ	δ	...	...	δ	...	**B2**^{2}, 601
s269	—,—,dinitrile	$NCC(CH_3)_2C(CH_3)_2CN$	136.20	mcl pl, lf and pr (dil al)	169	...	1.070	...	...	s^h	...	...	...	...	**B2**^{1}, 290
—	**Succinimide**	*see* **Succinic acid, imide**													
s273	**Sucrose**	Cane sugar. Saccharose. (structure: O; OH H; $HOCH_2$—C—C—C—CCH_2OH; H H OH; O; H H OH H; $HOCH_2C$—C—C—C—CH; OH H OH; O)	342.30	mcl, $[\alpha]_D^{20}$ +65.6	185–6	...	1.588^{15}	1.5376	δ s^h	δ	i	...	...	...	**B31**, 424
s274	—,octaacetate	$C_{28}H_{38}O_{19}$. *See* s273	678.60	nd (al), $[\alpha]_D^{20}$ +59.6 (chl)	72.3	...	1.27^{16}	...	δ^h	...	...	...	...	oos s	**B31**, 453
s275	**Sudan III**	Tetrazobenzene-β-naphthol. (structure: OH; naphthyl—N=N—phenyl—N=N—phenyl)	352.40	br lf with gr lustre (aa)	195	...	...	...	i	s	δ	...	...	xyl s chl s	**B16**^{2}, 75

For explanations, symbols and abbreviations see beginning of table.

No.	Name	Synonyms and Formula	Mol. wt.	Crystalline form, color and specific rotation	m.p. °C	b.p. °C	Density	n_D	Solubility w	al	eth	ace	bz	other solvents	Ref.
	Sudan G														
—	**Sudan G**	*see* **Azobenzene, 2,4-dihydroxy-**													
—	**Sudan yellow**	*see* **Azobenzene 1-naphthalene, 2′-hydroxy-**													
s276	**Sulfadiazine**	2-Sulfanilamidopyrimidine. Sulfapyrimidine. 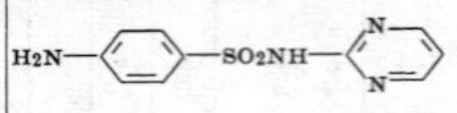	250.28	(w)	255–6				δ	δ		δ			
s277	**Sulfaguanidine**	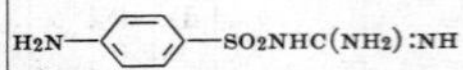 $H_2N-C_6H_4-SO_2NHC(NH_2):NH$	214.25	nd (w)	190–3				δ s^h	δ		δ		dil ac s	
s278	—monohydrate	$C_7H_{10}N_4O_2S.H_2O$	232.27	nd (w)	143 (sealed tube)				δ s^h	δ		δ			
s279	**Sulfamerazine**	Sulfamethyldiazine. 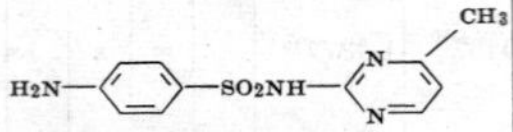	264.30		234–8				δ	δ	i	δ		chl i	
—	**Sulfanilamide**	*see* **Benzenesulfonic acid, 4-amino-, amide**													
—	**Sulfanilic acid**	*see* **Benzenesulfonic acid, 4-amino-***													
s280	**Sulfamethazine**	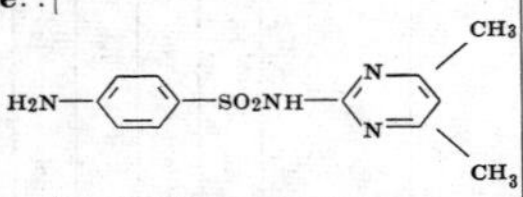	278.33	pa ye (w+½)	197–9				δ					ac, alk s	**Am64,** 567
—	**Sulfamic acid**	*see* **Sulfuric acid, monoamide**													
s281	**Sulfapyrazine**	*N′*(2-Pyrazinyl)sulfanilamide. 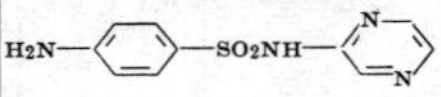	250.28	nd ($PhNO_2$)	251				i	δ	i	δ		chl i diox i	**Am63,** 3153
s282	**Sulfapyridine**	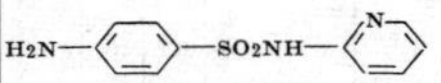	249.29	ye og (al)	278–304 (190–1)				s^h	s^h	...		i	CCl_4 i	**C34,** 1814[2]
s283	**Sulfaquinoxaline**	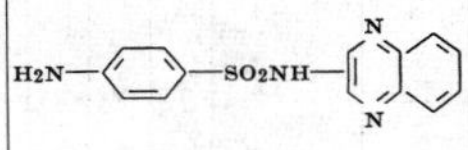	300.33		247–8				i	i	...	δ		aq alk s	
s283[1]	**Sulfathiadiazole**	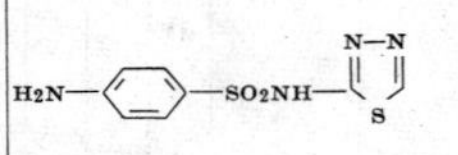	256.32		218				δ	s				Py s	**C40,** 5411[3]
s284	**Sulfathiazole**	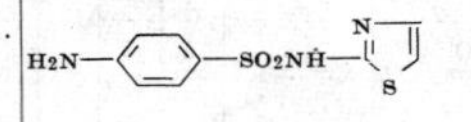	255.31	br pl or rods	200–4 (form I) 175 (form II)				δ	δ					**C40,** 7518[1]
s285	—,**phthalyl-**	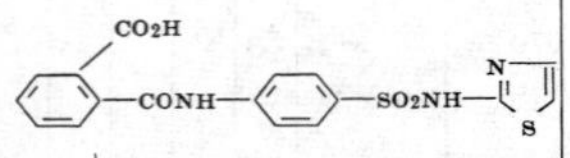	403.43		272–7d				i	δ	i			chl i ac s alk s	
s286	—,**4-nitro-**	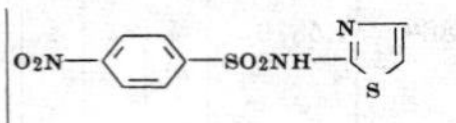	285.30	pa ye pw	225.6				i	δ	i		i	chl i dil alk v	
s287	—,**succinyl-**	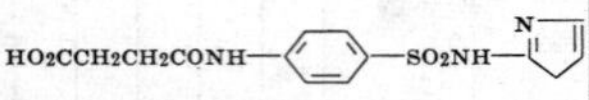	355.39		192–5				i	δ	i	δ		chl i alk s	
s288	**Sulfide, benzyl phenyl**	Benzylthiobenzene. $C_6H_5CH_2SC_6H_5$	200.29	lf (al)	44.5	197^{27}			i	s^h	s			con sulf s	**B6[2]**, 428
—	—,**bis(2-amino-2-carboxyethyl)**	*see* **Lanthionine**													
—	—,**bis(dimethylarsine)**	*see* **Cacodyl sulfide**													
s290	—,**butyl ethyl**	1(Ethylthio)butane*. $CH_3(CH_2)_3SC_2H_5$	118.24		−95.1	144.2^{760}	0.8376^{20}_{4}	1.4491^{20}						chl s	**B1[3]**, 1522
s291	—,**butyl methyl**	1(Methylthio)butane*. $CH_3(CH_2)_3SCH_3$	104.22			123.2^{760}	0.8426^{20}_{4}	1.4477^{20}	...	v				MeOH v^h	**B1[3]**, 1521

For explanations, symbols and abbreviations see beginning of table.

No.	Name	Synonyms and Formula	Mol. wt.	Crystalline form, color and specific rotation	m.p. °C	b.p. °C	Density	n_D	Solubility w	al	eth	ace	bz	other solvents	Ref.
	Sulfide														
s292	—,**diallyl**	Allyl sulfide. $(CH_2{:}CHCH_2)_2S$	114.20	...	−83	139	0.8877^{27}_{4}	1.4877^{27}	δ	∞	∞	...	...	CCl_4, CS_2 s	**B1**^{2}, 478
s293	—,**dibenzoyl**	$C_6H_5COSCOC_6H_5$	242.30	pr (al)	48	d	...	...	i	...	v	...	...	...	**B9**^{2}, 289
s294	—,**dibenzyl**	Benzyl sulfide. $(C_6H_5CH_2)_2S$	214.33	ta (eth or chl)	49–50	d	1.0712^{50}_{50}	...	i	s	s	...	...	...	**B6**^{2}, 429
s295	—,**dibutyl**(α-form)	1(Butylthio)butane*. Butyl sulfide. $CH_3(CH_2)_3S(CH_2)_3CH_3$	146.30	...	−79.7	182	0.8386^{16}_{0}	1.4530^{20}	i	v	v	...	...	...	**B1**^{2}, 399
s296	—,—(β-form)	$CH_3(CH_2)_3S(CH_2)_3CH_3$	146.30	...	...	190–230d	...	...	i	...	...	...	...	oos s	...
s297	—,—,**2,2′-dimethyl-**(*d*)	*d*-Di-*act*-amyl sulfide. $CH_3CH_2CH(CH_3)CH_2SCH_2CH(CH_3)CH_2CH_3$	174.35	$[\alpha]^{20}_{D}+24.5$	98	...	0.8362^{20}_{4}	...	...	...	...	...	...	...	**B1**, 387
s298	—,—,**3,3′-dimethyl-**	Isoamyl sulfide. $(CH_3)_2CHCH_2CH_2SCH_2CH_2CH(CH_3)_2$	174.35	colorless to pa ye	...	216^{760}	0.8323^{20}_{4}	1.4527^{20}	i	∞	v	...	...	...	**B1**^{3}, 1646
s299	—,**di-*sec*-butyl**	*sec*-Butyl sulfide. $CH_3CH_2CH(CH_3)SCH(CH_3)CH_2CH_3$	146.30	...	...	164.5^{739}	0.8348^{20}_{4}	1.4506^{20}	i	v	v	...	...	...	**B1**, 373
s300	—,**di(dimethyl-thiocarbamyl)**	$(CH_3)_2NCSSCSN(CH_3)_2$	208.35	ye	107	...	1.40	...	...	δ	δ	...	...	chl δ	**B4**, 76
s301	—,**diethenyl**	Vinyl sulfide. $CH_2{:}CHSCH{:}CH_2$	86.15	...	...	101	0.9125	...	δ	∞	∞	...	...	...	**B1**^{2}, 474
s302	—,**diethoxy**	Ethoxyl sulfide. $C_2H_5OSOC_2H_5$.	122.19	...	...	117^{733}	0.9940^{20}_{4}	1.4234^{20}	s	v	...	...	...	os v	**B1**^{3}, 1314
s303	—,**diethyl**	Ethyl sulfide. $C_2H_5SC_2H_5$	90.19	...	−103.3	92^{761}	0.8369^{20}_{4}	1.4424^{20}	δ	s	s	...	...	...	**B1**^{2}, 343
s304	—,—,**2,2′-bis-(ethylamino)-**	$C_2H_5NHCH_2CH_2SCH_2CH_2NHC_2H_5$	176.32	...	...	$231\text{–}3^{758}$	...	...	∞	v	v	...	...	...	**B4**, 287
s305	—,—,**2-chloro-**	$ClCH_2CH_2SC_2H_5$	124.64	...	...	156^{760}	1.0663^{20}_{4}	1.4780^{16}	...	...	...	...	...	chl s	**B1**^{2}, 343
s306	—,—,**2,2′-diamino-**	$H_2NCH_2CH_2SCH_2CH_2NH_2$	120.22	ye	...	$231\text{–}3^{755}$	...	...	∞	...	...	...	...	...	**B4**, 287
—	—,—,**2,2′-diamino-2,2′-dicarboxy-**	*see* **Lanthionine**													
s307	—,—,**2,2′-dichloro-**	Mustard gas. Yperite. $ClCH_2CH_2SCH_2CH_2Cl$	159.08	ye	13–4	217^{760}	1.2746^{20}_{4}	1.5313^{15}	i	s	s	...	...	...	**B1**^{2}, 349
s308	—,—,**2,2′-dihydroxy-**	2,2′-Thiodiethanol. $HOCH_2CH_2SCH_2CH_2OH$	122.19	...	−10	153^{8}	1.1819^{20}_{4}	1.5203^{20}	∞	∞	s	...	δ	chl ∞ AcOEt ∞	**B1**^{3}, 2122
s309	—,—,**2,2′-diphenoxy-**	$C_6H_5OCH_2CH_2SCH_2CH_2OC_6H_5$	274.39	nd (al)	54	...	...	...	i	s	s	...	...	...	**B6**^{2}, 150
s310	—,**diheptyl**	Heptyl sulfide. $CH_3(CH_2)_6S(CH_2)_6CH_3$	230.46	...	...	298 164^{20}	0.8416^{20}_{4}	1.4606^{20}	i	...	s	...	...	...	**B1**^{3}, 1685
s311	—,**dihexyl**	Hexyl sulfide. $CH_3(CH_2)_5S(CH_2)_5CH_3$	202.39	...	...	230 113.5^{4}	0.841	1.459	...	...	...	...	...	...	...
s312	—,**diisopropyl**	Isopropyl sulfide. $(CH_3)_2CHSCH(CH_3)_2$	118.24	...	...	120.4	0.8135^{20}_{4}	1.4395^{20}	i	s	s	...	...	...	**B1**^{3}, 1479
s313	—,**dimethyl**	Methyl sulfide. CH_3SCH_3	62.14	...	−83.2	37.3^{760}	0.8458^{21}_{4}	...	i	s	s	...	...	MeOH s	**B1**, 288
s313[1]	—,—,**hexafluoro-**	F_3CSCF_3	170.09	...	...	-22.2^{760}	...	...	...	...	...	...	...	...	**J1952**, 2198
s314	—,**1,1′-dinaphthyl**	α-Naphthyl sulfide. $C_{10}H_7^{\alpha}SC_{10}H_7^{\alpha}$	286.40	nd (al)	110	$289\text{–}90^{15}$	...	...	...	i s^h	...	...	v	CS_2 v	**B6**^{2}, 588
s315	—,**2,2′-dinaphthyl**	β-Naphthyl sulfide. $C_{10}H_7^{\beta}SC_{10}H_7^{\beta}$	286.40	pl (al)	151	$295\text{–}6^{15}$	...	...	...	δ^h	...	...	...	CS_2 v	**B6**^{2}, 611
s316	—,**dipentyl**	Amyl sulfide. $CH_3(CH_2)_4S(CH_2)_4CH_3$	174.35	...	−51.3	230^{760}	0.8409^{20}_{4}	1.4556^{20}	i	...	s	...	...	...	**B1**^{3}, 1608
s317	—,**diphenyl**	Phenyl sulfide. [structure: 3′ 2′ / 4′ ring —S— ring 4 / 5′ 6′; 2 3 / 6 5]	186.27	...	<−40	$108\text{–}9^{15}$ 296	1.1185^{15}_{15}	1.635^{19}	i	s^h	∞	...	∞	CS_2 ∞	**B6**, 299
s318	—,—,**2-amino-**	$C_{12}H_{11}NS$. *See* s317	201.29	pl (al)	35.5	257.5^{100} 212^{25}	...	...	...	s v^h	...	...	...	...	**B13**^{2}, 200

For explanations, symbols and abbreviations see beginning of table.

No.	Name	Synonyms and Formula	Mol. wt.	Crystalline form, color and specific rotation	m.p. °C	b.p. °C	Density	n_D	Solubility w	al	eth	ace	bz	other solvents	Ref.
	Sulfide														
s319	**—,—,4-amino-**	$C_{12}H_{11}NS$. *See* s317	201.29	nd (dil al)	95.8	282.3^{100} 242.5^{29}			δ^h	v	v				**B13**², 298
s320	**—,—,2,4′-diamino-**	$C_{12}H_{12}N_2S$. *See* s317	216.30	nd (w or al)	58				s^h	v	v		v	peth i	**B13**², 299
s321	**—,—,4,4′-diamino-**	*p,p′*-Thiodianiline. $C_{12}H_{12}N_2S$. *See* s317	216.30	nd (w)	108				δ^h	v	v		v^h		**B13**², 299
s322	**—,—,4,4′-dichloro-2,2′-dinitro-**	$C_{12}H_6Cl_2N_2O_4S$. *See* s317	345.13	br-ye nd (90% aa)	149–50					i			v		**B6**², 312
s323	**—,—,4,4′-dihydroxy-**	$C_{12}H_{10}O_2S$. *See* s317	218.27	mcl pr (al)	151				i	v	v				**B6**, 860
s324	**—,—,2,2′-dimethyl-**	o-Anisyl sulfide. $C_{14}H_{14}O_2S$. *See* s317	246.32	pl (al)	73	$252–3^{10}$				v	v		v	peth i	**B6**, 794
s325	**—,—,2,2′-dinitro-**	$C_{12}H_8N_2O_4S$. *See* s317	276.28	ye pl (aa-al)	122–3	sub δd				δ				aa δ	**B6**², 305
s326	**—,—,2,4-dinitro-**	$C_{12}H_8N_2O_4S$. *See* s317	276.28	pa ye nd (ace or bz-al)	117				i	δ			v	aa v	**B6**², 315
s327	**—,—,4,4′-dinitro-**	$C_{12}H_8N_2O_4S$. *See* s317	276.28	pa ye nd	174–5 (157)				i	δ s^h				con sulf s	**B6**², 311
s328	**—,—,2-nitro-**	$C_{12}H_9NO_2S$. *See* s317	231.27	ye-og nd (lig or al-eth)	81	210^{15}				s	s			peth i con sulf s	**B6**², 305
s329	**—,—,4-nitro-**	$C_{12}H_9NO_2S$. *See* s317	231.27	pa ye pr (lig)	55	240^{25}				s	s				**B6**², 311
s330	**—,—,2,2′4,4′-tetranitro-**	$C_{12}H_6N_4O_8S$. *See* s317	366.28	ye nd or pl (aa)	193–4					i			i	CS_2 i con HNO_3 s	**B6**², 315
s332	**—,dipropyl**	Propyl sulfide. $CH_3CH_2CH_2SCH_2CH_2CH_3$	118.23		fr −101.9	142^{770}	0.8385^{20}_{4}		i	s	s				**B1**², 372
s333	**—,—,2,2′-dichloro-**	$CH_3CHClCH_2SCH_2CHClCH_3$	187.14		−40	122^{23}	1.1569^{25}_{4}			s					
s334	**—,—,3,3′-dichloro-**	$Cl(CH_2)_3S(CH_2)_3Cl$	187.14			162^{43} $111–2^{7}$	1.774^{25}_{4}	1.5075^{20}		s	s			to s	**B1**³, 1438
s335	**—,2,2′-ditolyl**	*o*-Tolyl sulfide. 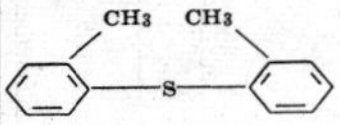	214.32	pl (al)	64	285 174^{15}				s^h	v			chl v CS_2 v	**B6**², 342
s336	**—,2,4′-(ditolyl)**	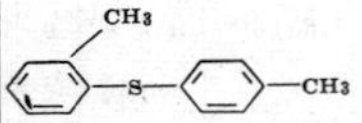	214.32			173^{11}	1.0774^{15}_{4}			s	s				**B6**, 418
s337	**—,3,4′-ditolyl**	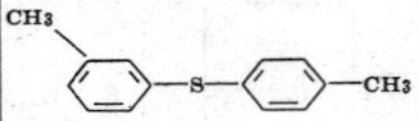	214.32	nd (al)	27.8	179^{11}				s	s				**B6**, 418
s338	**—,4,4′-ditolyl**	*p*-Tolyl sulfide. 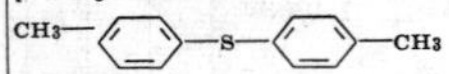	214.32	nd (al)	57.3	179^{11}			i	v^h	v		v^h		**B6**², 395
s340	**—,ethyl isobutyl**	$C_2H_5SCH_2CH(CH_3)_2$	118.24			134–5	0.8321^{20}_{4}	1.1442^{20}						chl s	**B1**³, 1566
s341	**—,ethyl methyl**	$CH_3SC_2H_5$	76.15		−105.91	66^{760}	0.837^{20}	1.4404^{20}	i	∞	s				**B1**, 343
s342	**—,—,2-chloro-**	$CH_3SCH_2CH_2Cl$	110.61			140	1.1245^{20}_{20}	1.4902^{30}			s	s			**B1**², 347
—	**—,ethyl phenyl**	*see* **Benzene, (ethylthio)-***													
s343	**—,ethyl propyl**	$C_2H_5SCH_2CH_2CH_3$	104.22		−117	118.5^{760}	0.8370^{20}_{4}	1.4462^{20}		s					**B1**³, 1432
s345	**—,2-furyl phenyl**	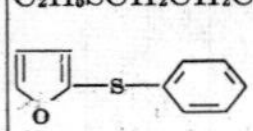	176.23			$119–20^{8}$	1.1341^{26}	1.5977^{26}							
s346	**—,isobutyl methyl**	$CH_3SCH_2CH(CH_3)_2$	104.32			112.5^{760}	0.8335^{20}_{4}	1.4433^{20}		s	s				**B1**³, 1565
s347	**—,isopropyl methyl**	$CH_3SCH(CH_3)_2$	90.19		−101.5	84.8^{760}									**B1**, 367
—	**—,methyl phenyl**	*see* **Benzene, (methylthio)-***													
s348	**—,methyl propyl**	$CH_3SCH_2CH_2CH_3$	90.19		−113	95.5^{760}	0.8424^{20}	1.4442^{20}	s						**B1**³, 1432
s350	**—,pentamethylene**	Tetrahydrothiopyran. Thiacyclohexane. 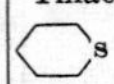	102.20		19(13)	$140–2^{755}$	0.9849^{20}_{4}	1.5067^{20}	i					os s	**B17**², 18
s351	**—,trimethylene**	Thiacyclobutane. 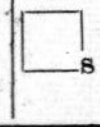	74.14			94^{750}	1.0284^{23}_{4}	1.5059^{23}	i	v				os v	**B17**², 12

For explanations, symbols and abbreviations see beginning of table.

No.	Name	Synonyms and Formula	Mol. wt.	Crystalline form, color and specific rotation	m.p. °C	b.p. °C	Density	n_D	Solubility						Ref.
									w	al	eth	ace	bz	other solvents	
	Sulfonal														
—	**Sulfonal**	*see* **Propane, 2,2-bis(ethylsulfonyl)-***													
s352	**Sulfone, benzyl phenyl**	$C_6H_5CH_2SO_2C_6H_5$	232.31	nd	148	...	1.126^{153}_{4}	...	i s^h	δ	δ	...	δ	...	**B6**^{2}, 428
s353	**—,benzyl (4-tolyl)**	C_6H_5—CH_2SO_2—C_6H_4—CH_3	246.23	nd (al)	144–5	...	...	...	...	v^h	...	...	v	aa v	**B6**, 455
—	**—,diphenylene**	*see* **Dibenzothiophene,** dioxide													
s354	**—,dibenzyl**	Benzyl sulfone. $C_6H_5CH_2SO_2CH_2C_6H_5$	246.33	nd (al-bz)	151	290δd	...	...	i	δ	...	v	s	aa s	**B6**^{2}, 430
s355	**—,dibutyl**	Butyl sulfone. $CH_3(CH_2)_3SO_2(CH_2)_3CH_3$	178.30	pl (w or al)	46	...	...	...	i	s	s	...	...	...	**B1**^{2}, 400
s356	**—,diethyl**	Ethyl sulfone. $C_2H_5SO_2C_2H_5$	122.19	rh pl	73–4	246^{755}	1.357^{20}_{4}	...	s	...	s^h	...	v	peth i	**B1**^{2}, 345
s356[1]	**—,—,2,2′-diphenoxy-**	$C_6H_5OCH_2CH_2SO_2CH_2CH_2OC_6H_5$	306.39	pink lf (al)	108	...	...	...	i	s	...	...	...	...	**B6**^{2}, 150
s357	**—,diisopropyl**	Isopropyl sulfone. $(CH_3)_2CHSO_2CH(CH_3)_2$	150.24	(eth)	36	...	...	...	v	...	s^h	...	...	...	**B1**^{2}, 387
s358	**—,dimethyl**	Methyl sulfone. $CH_3SO_2CH_3$	94.14	pr	109	233.5^{760}	1.1702^{110}_{0}	1.4226	s	s	...	...	s	...	**B1**, 289
s359	**—,diphenyl**	Phenyl sulfone. (4′ 5′ 6′ 3′ 2′)C_6H_5—SO_2—C_6H_5(2 3 4 6 5)	218.26	mcl pr (bz), pl (al), nd (w)	125	377.8 232^{18}	1.252^{20}_{4}	...	i δ^h	s^h	s	...	s	...	...
s360	**—,—,2-amino-**	$C_{12}H_{11}NO_2S$ *See* s359	233.29	lf (dil al)	122	...	...	...	i	v	...	...	v	aa v	**B13**, 399
s361	**—,—,4-amino-**	$C_{12}H_{11}NO_2S$ *See* s359	233.29	nd (al)	176	...	...	...	i	s	...	...	v^h	...	**B13**^{2}, 298
s362	**—,—,4-chloro-**	Sulphenone. $C_{12}H_9ClO_2S$. *See* s359	252.72	...	94(90)	...	...	...	i	δ	...	v	v	...	...
s363	**—,—,4,4′-diacetamido-**	$C_{16}H_{16}N_2O_4S$. *See* s359	332.37	pa ye nd (eth or dil aa), lf (aq al)	282–5	...	...	...	i	s	...	...	...	...	**B13**^{2}, 303
s364	**—,—,2,4-diamino-**	$C_{12}H_{12}N_2O_2S$ *See* s359	248.30	nd (al)	188	...	...	...	i	v^h	i	...	i	...	**B13**, 553
s365	**—,—,3,3′-diamino-**	$C_{12}H_{12}N_2O_2S$. *See* s359	248.30	pr	168	...	...	...	v^h	v^h	...	...	...	...	**B13**^{2}, 219
s366	**—,—,4,4′-diamino-**	$C_{12}H_{12}N_2O_2S$ *See* s359	248.30	lf (dil al)	176.5	...	...	...	...	s	...	...	...	...	**B13**^{2}, 300
s367	**—,—,4,4′-dichloro-**	$C_{12}H_8Cl_2O_2S$. *See* s359	287.16	mcl	148–9	...	...	...	i	δ s^h	...	...	...	...	**B6**^{2}, 297
s368	**—,—,4,4′-diethoxy-**	$C_{16}H_{18}O_4S$ *See* s359	306.39	lf (al or aa)	163	...	...	...	...	v^h	v	...	...	aa s^h	**B6**, 861
s369	**—,—,2,2′-dihydroxy-**	$C_{12}H_{10}O_4S$. *See* s359	250.27	nd (bz)	179 (164)	...	...	...	s	v	v	...	δ	peth i	**B6**^{1}, 396
s370	**—,—,2,5-dihydroxy-**	$C_{12}H_{10}O_4S$ *See* s359	250.27	pr (w), nd (dil al)	196	...	...	...	δ s^h	v	δ	...	...	...	**B6**^{2}, 1072
s371	**—,—,3,3′-dihydroxy-**	$C_{12}H_{10}O_4S$ *See* s359	250.27	...	186–7	...	...	...	δ	s	...	...	...	aa δ	**B6**^{2}, 827
s372	**—,—,4,4′-dihydroxy-**	$C_{12}H_{10}O_4S$ *See* s359	250.27	nd (w), rh bipym	240–1	...	1.3663^{15}	...	i s^h	s	s	...	δ	alk s	**B6**^{2}, 853
s373	**—,—,2,2′-dimethoxy-**	$C_{14}H_{14}O_4S$. *See* s359	278.30	nd (bz)	157–8	...	...	...	...	v^h	...	...	δ^h	xyl δ^h peth i con sulf s^h	**B6**, 794
s374	**—,—,4,4′-dimethoxy-**	Anisyl sulfone. $C_{14}H_{14}O_4S$ *See* s359	278.30	lf or pr (al), nd (al-eth)	129–30	sub	...	...	i	v^h	...	...	...	...	**B6**, 861
s375	**—,dipropyl**	Propyl sulfone. $CH_3CH_2CH_2SO_2CH_2CH_2CH_3$	150.23	scales	29.5	...	1.0278^{50}_{4}	...	δ	s	s	...	...	...	**B1**^{2}, 373
s376	**—,2,2′-ditolyl**	*o*-Tolyl sulfone. (CH_3)C_6H_4—SO_2—C_6H_4(CH_3)	246.33	pl (al)	134.5	...	...	...	i	v	v	...	v	...	**B6**^{2}, 342
s377	**—,3,4′-ditolyl**	(CH_3)C_6H_4—SO_2—C_6H_4—CH_3	246.33	nd (aa)	116	...	...	...	...	...	...	...	...	...	**B6**^{1}, 208
s378	**—,4,4′-ditolyl**	*p*-Tolyl sulfone. CH_3—C_6H_4—SO_2—C_6H_4—CH_3	246.33	pr (bz), nd (w or al)	159	405^{714}	...	...	δ s^h	s^h	δ	...	s	chl s CS_2 s	**B6**^{2}, 395
s381	**—,phenyl (2-tolyl)**	C_6H_5—SO_2—C_6H_4(CH_3)	232.30	pl (al)	81	...	...	...	i	s	v	...	v	lig δ	**B6**, 371

For explanations, symbols and abbreviations see beginning of table.

No.	Name	Synonyms and Formula	Mol. wt.	Crystalline form, color and specific rotation	m.p. °C	b.p. °C	Density	n_D	Solubility						Ref.
									w	al	eth	ace	bz	other solvents	
	Sulfone														
s382	**—,phenyl (4-tolyl)**	(C₆H₅)—SO_2—(C₆H₄)—CH_3	232.30	pl (al)	127–8				i	δ			s	aa s	**B6²**, 395
—	**Sulfoxide, diphenylene**	*see* **Dibenzothiophene,** monoxide													
s383	**—,dibenzyl**	$C_6H_5CH_2SOCH_2C_6H_5$	230.33	pl (al or w)	134–5	210d			i s^h	v	s				**B6²**, 429
s384	**—,dibutyl**	Butyl sulfoxide. $CH_3(CH_2)_3SO(CH_2)_3CH_3$	162.30	nd	32	d	0.8317^{23}_{4}		i	s	s				**B1²**, 400
s385	**—,diethyl**	Ethyl sulfoxide. $C_2H_5SOC_2H_5$	106.19	syr	15(4)	$88–90^{15}$			s	s	s				**B1²**, 345
s386	**—.dimethyl**	Methyl sulfoxide. CH_3SOCH_3	78.13		18.45	189^{760}	1.1014^{20}_{4}		s	s	s	s		AcOEt s	**B1**, 289
s387	**—,diphenyl**	Phenyl sulfoxide. 4′(C₆H₅, 5′ 6′ 3′ 2′)—SO—(C₆H₅, 2 3 6 5)4	202.28	pr (lig)	70	$206–8^{13}$				v	v		v	peth i aa v	**B6²**, 290
s388	**—,—,4-amino-**	$C_{12}H_{11}NOS$. *See* s387	217.29	nd (w)	152				s^h	v	v				**B13**, 534
s389	**—,— 4-amino-4′-nitro-**	$C_{12}H_{10}N_2O_3S$ *See* s387	262.29	ye	132				δ	v					**Am73**, 4356
s390	**—,—,4,4′-diamino-**	*p,p′*-Sulfinyldianiline. $C_{12}H_{12}N_2OS$. *See* s387	232.31	pr (w or al)	175d				s^h	s^h					**B13**, 53
s391	**—,—,4,4′-dihydroxy-**	$C_{12}H_{10}O_3S$. *See* s387	234.28	nd (ace)	195					s		s			**B6**, 860
s392	**—,diproply**	Propyl sulfoxide. $CH_3CH_2CH_2SOCH_2CH_2CH_3$	134.24	nd	18		0.9654^{20}_{4}		δ	s	s				**B1²**, 373
s393	**—,4,4′-ditolyl**	*p*-Tolyl sulfoxide. CH_3—(C₆H₄)—SO—(C₆H₄)—CH_3	230.33	(lig)	92					v	v		v	chl v aa v lig s^h	**B6²**, 395
s394	**Sulfuric acid,** diamide, *N,N,N′,N′*-tetramethyl-	Tetramethylsulfamide. $(CH_3)_2NSO_2N(CH_3)_2$	152.22	pl or nd	73	225^{760} extrap $126^{30.6}$			δ	s					**Am76**, 220
s395	—,dibutyl ester*	Butyl sulfate. $[CH_3(CH_2)_3O]_2SO_2$	210.29			109.5^{4}	1.0616^{20}	1.42126^{20}	i						**J1943**, 16
s396	—,didecyl ester*	1-Decanol sulfate. Decyl sulfate. $[CH_3(CH_2)_9O]_2SO_2$	378.62		37.6										**Am56**, 1204
s397	—,didodecyl ester*	Dodecyl sulfate. $[CH_3(CH_2)_{11}O]_2SO_2$	434.73		48.5										**Am56**, 1204
s398	—,diethyl ester*	Ethyl sulfate. $(C_2H_5O)_2SO_2$	154.19		−24.5	208δd 96^{15}	1.1774^{20}_{4}	1.4004^{20}	i d^h	∞ d^h	∞				**B1²** 327
s399	—,diheptyl ester*	Heptyl sulfate. $[CH_3(CH_2)_6O]_2SO_2$	294.46		13	$146.6^{1.5}$	0.9819^{25}_{25}	1.4362^{25}							**B1³**, 1683
s400	—,dihexadecyl ester*	Cetyl sulfate. $[CH_3(CH_2)_{15}O]_2SO_2$	546.94		66.2				s						
s401	—,dihexyl ester*	Hexyl sulfate. $[CH_3(CH_2)_5O]_2SO_2$	266.40			125.3^{2}	1.0036^{21}_{0}	1.433^{21}							**B1³**, 1656
s402	—,dimethyl ester*	Methyl sulfate. $(CH_3O)_2SO_2$	126.13		−31.75	188.5^{760}d 76^{15}	1.3322^{20}_{4}	1.3874^{20}	s	∞	s		s		**B1**, 283
s403	—,di(3-methylbutyl) ester*	Isoamyl sulfate. $[(CH_3)_2CHCH_2CH_2O]_2SO_2$	238.35		−20	$139–41^{12}$									**B1²**, 434
s404	—,dioctadecyl ester*	Octadecyl sulfate $[CH_3(CH_2)_{17}O]_2SO_2$	603.05		70.5										
s405	—,dipentyl ester*	Amyl sulfate. $[CH_3(CH_2)_4O]_2SO_2$	238.35		14	$128–30^{3}$	1.029^{20}_{0}	1.4290^{20}							**B1²**, 418
s406	—,dipropyl ester*	Propyl sulfate. $(CH_3CH_2CH_2O)_2SO$	182.24			121^{20}	1.1064^{20}_{4}	1.4135^{20}	i δd					peth v	**B1²**, 368
s407	—,ethylene ester	Ethylene sulfate. Glycol sulfate. (cyclic —O—SO_2—O—)	124.12	nd or pr (bz-lig)	99	sub	1.735^{13}_{4}		d				s	os v	**B1³**, 2110
s408	—,monoamide, *N*-cyclohexyl-	Cyclohexylsulfamic acid. (C₆H₁₁)—$NHSO_3H$	179.24		169–70				$δd^h$					alk v	
s409	—,mono(2-aminoethyl) ester*	$H_2NCH_2CH_2OSO_3H$	141.15	mcl pr	230d										**B4²**, 718

For explanations, symbols and abbreviations see beginning of table.

No.	Name	Synonyms and Formula	Mol. wt.	Crystalline form, color and specific rotation	m.p. °C	b.p. °C	Density	n_D	Solubility w	al	eth	ace	bz	other solvents	Ref.
	Sulfuric acid														
s410	—,monobutyl ester*	Butyl hydrogen sulfate. $CH_3(CH_2)_3OSO_3H$	154.19	syr		d			v	s	s				**B1**², 397
s411	—,monochloride monoethyl ester	Ethyl chlorosulfonate*. $C_2H_5OSO_2Cl$	144.58			151–4	1.3502^{25}_{4}	1.416^{20}	d	d	s			chl s	**B1**², 327
s412	—,monochloride monomethyl ester	Methyl chlorosulfonate. CH_3OSO_2Cl	130.55			$133–5^{760}$d 48^{29}	1.4805^{25}_{4}	1.4138^{18}	i	d^h				os s	**B1**³, 1199
s413	—,monoethyl ester*	Ethylsulfuric acid. Ethyl hydrogen sulfate. $C_2H_5OSO_3H$	126.13			280d	1.316^{17}_{4}		d^h v	δ	δ				**B1**², 326
s414	—,monomethyl ester*	Methyl bisulfate. Methyl hydrogen sulfate. CH_3OSO_3H	112.10		<−30	130–40d			v	s	∞				**B1**, 283
s415	**Sulfurous acid,** diamide, *N,N,N′,N′*-tetramethyl-	Tetramethylthionamide. $(CH_3)_2NSON(SH_3)_2$	136.22		31	209^{760} extrap 78^{16}					s			oos s	**Am76,** 220
s416	—,dibutyl ester*	Butyl sulfite. $[CH_3(CH_2)_3O]_2SO$	194.29			230 116^{19}	0.9944^{22}_{4}	1.4310^{20}							**B1**³, 1506
s417	—,diethyl ester*	Ethyl sulfite. $(C_2H_5O)_2SO$	138.19			157^{763}	1.0829^{20}_{4}	1.4144^{20}	d	s	s				**B1**², 326
s418	—,diisobutyl ester*	Isobutyl sulfite. $[(CH_3)_2CHCH_2O]_2SO$	194.29			209^{741}	0.9862^{20}_{4}	1.4268^{20}							**B1**³, 1561
s419	—,dimethyl ester*	Methyl sulfite. $(CH_3O)_2SO$	110.13			126^{760}	1.2183^{15}_{4}		s^h	s	s				**B1**, 282
s420	—,ethylene ester	Ethylene sulfite. Glycol sulfite. (ring: O, SO, O)	108.12		−11	173^{760}	1.4402^{20}_{4}	1.4463^{20}	v d^h	v	v	v	v	AcOEt v	**B1**³, 2110
s421	—,monochloride, monoethyl ester	Ethyl chlorosulfinate. $C_2H_5OS(O)Cl$	128.58			52.5^{44} 32^{16}	1.2766^{25}_{4}	1.4550^{25}	d	d	s				**B1**³, 1316
s422	**Sylvestrene**(*d*)	Carvestrene. *d*-1,8(9)-*m*-menthadiene. $C_{10}H_{16}$.	136.23	$[\alpha]^{18}_{D}+83.18$ (undil), +66.3 (chl, c=4.3)		$175–6^{751}$	0.863^{20}_{4}	1.4780^{18}	i	∞	∞				**B5**², 84
—	**Syntomycein**	*see* Chloromycetin													
—	**Syringaldehyde**	*see* **Benzaldehyde, 4-hydroxy-3,5-dimethoxy-**													
—	**Syringenin**	*see* **2-Propen-1-ol, 3(3,5-dimethoxy-4-hydroxyphenyl)-**													
—	**Syringic acid**	*see* **Benzoic acid, 3,5-dimethoxy-4-hydroxy-**													
—	**Syringidin**	*see* **Malvidin**													
s423	**Syringin**	$HOCH_2CH{:}CH—C_6H_2(OCH_3)_2—O.C_6H_{11}O_5$	372.36	acicular cr (w) nd (al), $[\alpha]_D$ −17.1	192				δ s^h	s	i			con HNO_3 s con sulf s	**B31,** 222

For explanations, symbols and abbreviations see beginning of table.

No.	Name	Synonyms and Formula	Mol. wt.	Crystalline form, color and specific rotation	m.p. °C	b.p. °C	Density	n_D	Solubility w	al	eth	ace	bz	other solvents	Ref.
	Tachysterol														
t1	**Tachysterol**	HO— (structure)	396.66	$[\alpha]_D^{18}-70$ (bz)		220 vac			i					MeOH i os s	E12A, 175
t2	**Tagatose**(*D*)	HOCH₂COC(OH)(H)—C(OH)(H)—C(H)(H)—CH₂OH	180.16	(dil al), $[\alpha]_D^{22}+1$, $[\alpha]_D^{60}-26$	124				v	i				MeOH δ	B31, 348
t3	**Talitol**(*D*)	*D*-Altritol. *D*-Talite. $HOCH_2-C(OH)(H)-C(H)(OH)-C(H)(OH)-C(H)(OH)-CH_2OH$	182.17	$[\alpha]+3.2$ (w)	87–8										B1, 533
t4	**Talonic acid,** hemihydrate(*D*)	$HOCH_2C(H)(OH)-C(OH)(H)-C(OH)(H)-C(OH)(H)-CO_2H.\frac{1}{2}H_2O$	205.17	(aq al+½w) $[\alpha]_D^{25}$ (mut) +16.7 to −21.6 (w, c=4)	125				v						B3², 352
t5	—,γ-lactone(*D*)	$HOCH_2-C(H)(OH)-C(H)-C(OH)(H)-C(OH)(H)-CO$ (—O— ring)	178.14	pr (al), [α] −28.4 (w)	135–7										Am54, 1593
t6	**Talose**(*D*)	$HOCH_2-C(HO)(OH)-C(H)(H)-C(OH)(H)-C(OH)(H)-CHO$	180.16	(al:α form) (MeOH: β form) $(\alpha)_D^{27}$ (mut), +29 to +19.7 (w, c=1, α form) $[\alpha]_D^{20}$ (mut) +11.5 to +21 (w, c=4, β form)	130–5(α) 120–1(β)										B31, 283
t7	**Tannic acid**	Gallotannic acid. Tannin $C_{76}H_{52}O_{46}$	1701.23	pa ye to pa br amor or flakes	210–δd				s	v	i	v	i	chl i CS_2 i	
—	**Tannin**	*see* **Tannic acid**													
t8	**Taraxanthin**	$C_{40}H_{56}O_4$	600.89	ye pr, $[\alpha]_{644}^{22}$ +200 (AcOEt)	185−6					s^h	v		s	MeOH δ CS_2 peth s	B30, 100
t9	**Tartaric acid**(*d*)	*d*-2,3-Dihydroxybutanedioic acid*. *d*-2,3-Dihydroxy-succinic acid. $HO_2CCHOHCHOHCO_2H$	150.09	mcl (anh), rh (w+1), $[\alpha]_D^{20}$ +12 (20% aq)	160–70		1.582 (hyd)	1.4955	v	v	δ	s			B3², 308
t10	—(*dl*)	Racemic acid. $HO_2CCHOHCHOHCO_2H.H_2O$	168.10	mcl pr (w or al+1w)	203–4		1.788		s	s	δ				B3², 335
t11	—(*meso*)	$HO_2CCHOHCHOHCO_2H$	150.09	tcl	140		1.666_4^{20}	1.5–1.6	v	s	δ				B3, 528
t12	—,anhydride diacetate(*d*)	α,α′-Diacetoxysuccinic anhydride. CH_3CO_2—(ring, O=, =O, O)—O_2CCH_3	216.14	nd (bz)	135					v	v	s		aa δ	B18², 143
t13	—,dibenzyl ester(*d*)	$C_6H_5CH_2O_2CCHOHCHOHCO_2CH_2C_6H_5$	330.32	$[\alpha]_D^{15}+19.26$	ca. 50	250–70⁴	1.2036^{72}			s				Py s	B6², 421
t14	—,dibutyl ester(*d*)	$CH_3(CH_2)_3O_2CCHOHCHOHCO_2(CH_2)_3CH_3$	262.30	pr, $[\alpha]_D^{14}+10.1$ (undil)	21.8	320^{760} 175^5	1.0909_4^{20}	1.4451^{20}							B3², 332
t15	—,—(*dl*)	$CH_3(CH_2)_3O_2CCHOHCHOHCO_2(CH_2)_3CH_3$	262.30			320^{765} 185^{12}	1.0879_4^{18}			s				os s	B3², 337

For explanations, symbols and abbreviations see beginning of table.

No.	Name	Synonyms and Formula	Mol. wt.	Crystalline form, color and specific rotation	m.p. °C	b.p. C	Density	n_D	Solubility w	al	eth	ace	bz	other solvents	Ref.
	Tartaric acid														
t16	—,diethyl ester(*d*)	$C_2H_5O_2CCHOHCHOHCO_2C_2H_5$	206.19	$[\alpha]_D^{16}+7.9$ (undil)	18.7	280^{760} 142^{8}	1.2056_4^{20}	1.4468^{20}	s	s				aa ∞	**B3**[2], 329
t17	—,—(*dl*)	Diethyl racemate. $C_2H_5O_2CCHOHCHOHCO_2C_2H_5$	206.19		18.7	281^{765} 158^{14}	1.2046_4^{20}	1.4438^{20}	δ	∞	∞			oos s	**B3**[2], 337
t18	—,—(*l*)	$C_2H_5O_2CCHOHCHOHCO_2C_2H_5$	206.19	$[\alpha]_D^{20}-7.55$		162^{19}	1.2054_4^{20}								**B3**[1], 181
t19	—,—(*meso*)	$C_2H_5O_2CCHOHCHOHCO_2C_2H_5$	206.19		55	157.5^{14}	1.1350_4^{99}	1.4315^{65}							**B3**, 530
t20	—,diethyl ester diacetate(*d*)	$C_2H_5O_2CCH(O_2CCH_3)CH(O_2CCH_3)CO_2C_2H_5$	290.27	mcl, $[\alpha]_D^{100}+6.3$	67–8	$191–2^{727}$ 163^{10}	1.1149_4^{66}		δ s^h	s v^h	v				**B3**[2], 331
t21	—,diisobutyl ester-(*d*)	$(CH_3)_2CHCH_2O_2CCHOHCHOHCO_2CH_2CH(CH_3)_2$	262.29	$[\alpha]_D+11.8$	68	323–5	1.0145^{100}			s					**B3**, 518
t22	—,—(*dl*)	$(CH_3)_2CHCH_2O_2CCHOHCHOHCO_2CH_2CH(CH_3)_2$	262.29		58	311^{768} 195^{13}				v				os s	**B3**[2], 337
t23	—,diisopropyl ester(*d*)	$(CH_3)_2CHO_2CCHOHCHOHCO_2CH(CH_3)_2$	234.24	$[\alpha]_D^{20}+14.9$		275^{765} 152^{12}								os s	**B3**[2], 331
t24	—,—(*dl*)	$(CH_3)_2CHO_2CCHOHCHOHCO_2CH(CH_3)_2$	234.24		34	275^{760} 154^{12}	1.1166_4^{20}							os s	**B3**[2], 337
t25	—,dimethyl ester(*d*)	$CH_3O_2CCHOHCHOHCO_2CH_3$	178.14	$[\alpha]_D^{0}-2.68$ (AcOEt). $[\alpha]_D^{50}+6.72$ (MeOH, c=16)	48 (61)	280^{760} 166^{12}	1.306_4^{45}		s	v				chl v	**B3**[2], 328
t26	—,—(*dl*)	$CH_3O_2CCHOHCHOHCO_2CH_3$	178.14	orh nd (bz), ta (chl)	90 (st) 84 (unst)	282^{760} 169^{20}	1.2604^{90}			v					**B3**[2], 336
t27	—,—(*meso*)	$CH_3O_2CCHOHCHOHCO_2CH_3$	178.14	nd (chl), (MeOH)	114	sub	$98^{0.01}$ sub		s	v				os v	**B3**[2], 339
t28	—,di(2-methyl-butyl) ester (*d*)	$C_2H_5CH(CH_3)CH_2O_2CCHOHCHOHCO_2CH_2CH(CH_3)C_2H_5$	290.36	$[\alpha]_D^{20}+17.73$		208^{20}	1.0636_4^{20}			s					**B3**, 519
t29	—,dinitrate(*d*)	$HO_2CCH(ONO_2)CH(ONO_2)CO_2H$	240.09	nd (eth-bz) $[\alpha]_D^{20}+13.7$ (MeOH, p=9)	d				d	v	v	v	i	peth i	**B3**[2], 328
t30	—,dipropyl ester(*d*)	$CH_3CH_2CH_2O_2CCHOHCHOHCO_2CH_2CH_2CH_3$	234.24	$[\alpha]+12$		303^{760}	1.1390_4^{20}		s	v	v				**B3**, 516
t31	—,—(*dl*)	$CH_3CH_2CH_2O_2CCHOHCHOHCO_2CH_2CH_2CH_3$	234.24		25	286^{765} 167^{11}	1.1256_4^{20}							os s	**B3**[2], 337
t32	—,monoethyl ester (*d*)	$HO_2CCHOHCHOHCO_2C_2H_5$	178.14	rh, $[\alpha]_D+21.8$ (w)	90				s	s	i				**B3**, 527
t33	—,piperazinium salt(*d*)	$HO_2CCHOHCHOHCO_2H.C_4H_{10}N_2$	236.23		248–54				s^h	s	δ				**Am70**, 2758
t34	—,—(*l*)	$HO_2CCHOHCHOHCO_2H.C_4H_{10}N_2$	236.23		140–1				s	s	δ				**Am70**, 2758
—	**Tartronic acid**	*see* **Malonic acid, hydroxy-**													
t35	**Taurine**	2-Aminoethanesulfonic acid. $H_2NCH_2CH_2SO_3H$	125.15	mcl pr	328 (317d)				s	i	i				**B4**, 528
t36	—,*N*,*N*-**dimethyl-**	2(Dimethylamino)ethane-sulfonic acid*. $(CH_3)_2NCH_2CH_2SO_3H$	153.20	pr (MeOH), pl (w+1)	315–6 270–80d				v	i	i			MeOH δ^h	**B4**[2], 951
t37	—,*N*-**methyl-**	2(Methylamino)ethane-sulfonic acid*. $CH_3NHCH_2CH_2SO_3H$	139.18	pr	241–2				s	i	i				**B4**, 529
t38	—,*N*-**methyl-*N*-phenyl-**	$C_6H_5N(CH_3)CH_2CH_2SO_3H$	215.28	pa vt (al)	239–40					v^h					**B12**[2], 285
t39	—,*N*-**phenyl-**	$C_6H_5NHCH_2CH_2SO_3H$	201.25	lf (w), pr (al)	277–80d				v	i	i				**B12**[2], 284
t40	**Taurocholic acid**	$C_{26}H_{45}NO_7S$	515.72	hyg amor$[\alpha]_D^{18}$ +38.8 (dil al)	ca. 125d				v	v^h	δ			AcOEt δ	**E14**, 195
t41	**Telluride, diethyl**	Ethyl telluride. $C_2H_5TeC_2H_5$	185.73	red-ye		$137–8^{760}$	1.599_4^{15}	1.5182^{15}	δ	s					**B1**[2], 359
t42	—,**dimethyl**	Methyl telluride. CH_3TeCH_3	157.68	pa ye		94^{770}	>1		d	v	i				**B1**, 291

For explanations, symbols and abbreviations see beginning of table.

No.	Name	Synonyms and Formula	Mol. wt.	Crystalline form, color and specific rotation	m.p. °C	b.p. °C	Density	n_D	Solubility w	al	eth	ace	bz	other solvents	Ref.
	Tephrosin														
t43	**Tephrosin**	Hydroxydequelin. 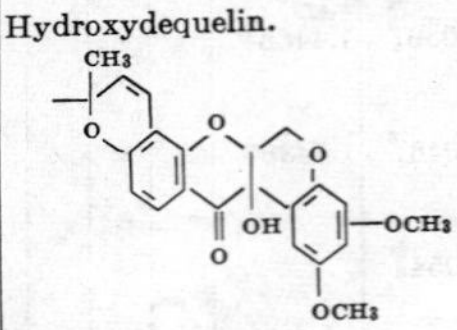	410.41	pr (MeOH-chl)	197–8				i		s	s		MeOH s^h chl s^h	**Am 55,** 759
—	**Teraconic acid**	*see* **Succinic acid, isopropylidene-**													
t44	**Terebic acid**(*dl*)	2,2-Dimethylparaconic acid. Teribinic acid. 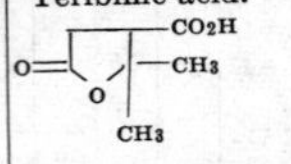	158.15	mcl pr	175		0.8155^{24}_{4}		δ	s^h	δ				**B18,** 377
—	**Terephthalaldehyde**	*see* **1,4-Benzenedicarboxaldehyde**													
—	**Terephthalaldehydic acid**	*see* **Benzoic acid, 4-formyl-**													
—	**Terephthalic acid**	*see* **1,4-Benzenedicarboxylic acid***													
t45	**Terpenolic acid**	Terpenylic acid. 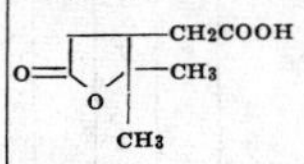	172.18	lf or pr (w+1)	89–90 57 (+1w)	130–40 sub			s v^h						**B18²,** 316
t46	*o*-**Terphenyl**	1,2-Diphenylbenzene. 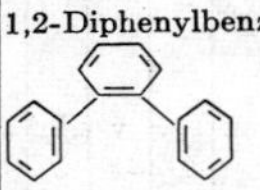	230.31	mcl pr (MeOH)	57	332^{760} 160–70²			i			s	s	MeOH s chl s	**Am 62,** 658
t47	*m*-**Terphenyl**	1,3-Diphenylbenzene. 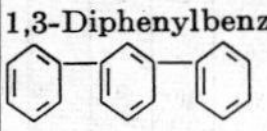	230.31	ye nd (al)	87	365^{760}			i	s	s		s		**B5²,** 611
t48	α-**Terpinene**	*p*-Mentha-1,3-diene. $C_{10}H_{16}$.	136.24			175^{755}	0.8375^{20}_{4}	1.4784^{20}	i	∞	∞				**B5,** 126
t49	α-**Terpineol**(*dl*)	*dl-p*-Menth-1-en-8-ol. $C_{10}H_{18}O$.	154.25		35	220^{760}	0.9337^{20}_{4}	1.4783^{20}	δ	v	v			chl s	**B6¹,** 41
t50	—,acetate	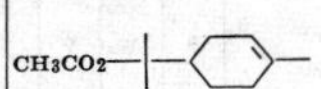 CH_3CO_2–	196.29	*d*:$[\alpha]_D$+52.5 *l*:$[\alpha]^{20}_{D}$−73		140^{40} 104–6¹¹	0.9659^{20}_{4}	1.4689^{21}	i	s			s		**B6²,** 66
t51	**Terpinol,** hydrate (*cis*)	*cis*-Terpin hydrate. 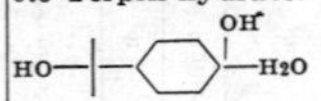	214.31	rh	sub 100 123d			1.51–1.52	δ	s	δ				**B6,** 745
t52	**Terpinolene**	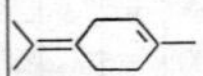	136.24			185^{760}	0.855^{20}	1.4823^{20}	i	∞	∞				**B5,** 133
t53	**3-Terpinolenone**	Piperitenone. 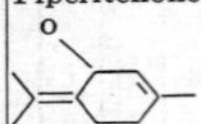	150.22	$[\alpha]_{546}$−0.1		120–2¹⁴	0.9774^{20}_{4}	1.5294^{20}		v	v				**J1939,** 1496
t54	**Terramycin**	Oxytetracycline. $C_{22}H_{30}N_2O_{11}$.	498.49		181–2d	1.634^{20}			δ	δ					
t55	**Testosterone**	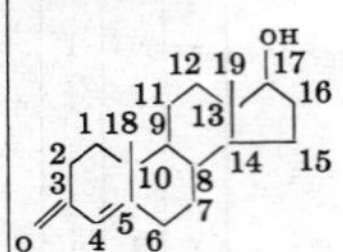	288.43	nd (dil ace) $[\alpha]_D$+109 (al, c=4)	155				i	s	s				**Am 76,** 1962
t55¹	—,4,5-dihydro-17-methyl-	$C_{20}H_{32}O_2$ *See* t55	304.48	(AcOEt)	192–3									AcOEt δ	**E14,** 145
t56	—,17-ethenyl	17-Vinyltestosterone. $C_{21}H_{30}O_2$. *See* t55	314.47	pr nd (pentane-eth) $[\alpha]_D$+87.6	140–1						s	s		MeOH s chl s	**C31,** 2224

For explanations, symbols and abbreviations see beginning of table.

PHYSICAL CONSTANTS OF ORGANIC COMPOUNDS (Continued)

No.	Name	Synonyms and Formula	Mol. wt.	Crystalline form, color and specific rotation	m.p. °C	b.p. °C	Density	n_D	Solubility						Ref.
									w	al	eth	ace	bz	other solvents	
	Testosterone														
t57	**—,17-ethyl-**	$C_{21}H_{32}O_2$. *See* t55	316.49	nd (AcOEt) $[\alpha]^{20}_D+71.2$	143–4	...	...	...	...	...	...	s^h	...	AcOEt s^h	**E14,** 145
t58	**—,17-ethynyl-**	$C_{21}H_{28}O_2$. *See* t55	312.46	$[\alpha]_D+22$	270–2	sub vac	...	...	i	δ	...	...	...	AcOEt s^h diox, Py s	**C34,** 18213
t59	**—,17-methyl-**	$C_{20}H_{30}O_2$. *See* t55	302.46	nd (hexane) $[\alpha]^{25}_D+76.4$ (al)	163–4	...	...	...	i	s	s	...	...	MeOH s os s	**E14s,** 2645
t61	**Tetrabenzotriaza-porphyrin**	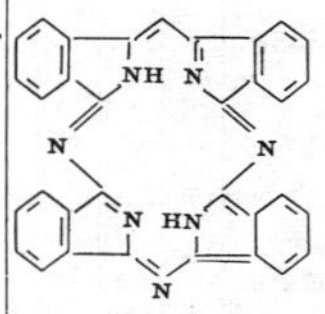	513.57	purple nd and pl (quinoline)	...	...	...	...	...	...	...	...	...	Py δ	**J1939,** 1809
—	**Tetracaine**	*see* **Benzoic acid, 4(butyl-amino)-,** (2-dimethyl-aminoethyl) ester													
—	**Tetracene**	*see* **Naphthacene**													
t62	**Tetracosane***	$CH_3(CH_2)_{22}CH_3$	338.67	...	51	243^{15}	0.7665^{70}_{4}	1.4283^{70}	i	δ	v	...	...	...	**B1,** 175
—	**Tetracyclone**	*see* **Cyclopentadienone, tetraphenyl-***													
t63	**6,8-Tetradeca-diyne***	$CH_3(CH_2)_4C⋮CC⋮C(CH_2)_4CH_3$	190.33	...	...	$118–9^4$	0.8699^{16}_{4}	...	...	...	...	...	...	...	...
t64	**Tetradecanal***	Myristaldehyde. Tetradecyl aldehyde. $CH_3(CH_2)_{12}CHO$	212.38	lf	30	166^{24}	...	...	i	s	s	s	...	oo s	**B1**[2]**,** 770
t65	—,dimethyl acetal	1,1-Dimethoxytetradecane*. $CH_3(CH_2)_{12}CH(OCH_3)_2$	258.45	...	...	$134–6^4$	...	1.4342^{25}	...	v	v	...	...	MeOH v	...
t66	—,oxime	Myristaldoxime. $CH_3(CH_2)_{12}CH{:}NOH$	227.39	flakes or nd(al)	82–3	...	...	...	i	v	...	...	δ	chl δ	**B1,** 716
t67	**Tetradecane***	$CH_3(CH_2)_{12}CH_3$	198.40	...	6	253.5^{760}	0.7627^{20}_{4}	1.4290^{20}	i	v	v	...	...	...	**B1,** 171
t68	**—,1-amino-***	Myristylamine. $CH_3(CH_2)_{13}NH_2$	213.41	...	37–8	291.2^{760} 162^{15}	...	...	i	v	v	s	v	chl v	**B4,** 201
t69	**—,1-bromo-***	Myristyl bromide. $CH_3(CH_2)_{13}Br$	277.30	...	5.7	181^{21}	1.0124^{25}_{4}	1.4605^{20}	i	s	...	∞	∞	chl v	...
t70	**—,1-chloro-***	Myristyl chloride. $CH_3(CH_2)_{13}Cl$	232.84	...	...	$154–5^{15}$	0.8589^{20}_{4}	1.4450^{20}	i	...	...	...	...	os s	**B1**[3]**,** 550
t71	**—,1,14-dibromo-***	Tetradecamethylene bromide. $Br(CH_2)_{14}Br$	356.20	lf (al-eth), (al)	49	$190–2^8$	...	...	i	v^h	s	...	...	chl v	**B1**[3]**,** 551
t72	**—,1-phenyl-***	Tetradecylbenzene*. $CH_3(CH_2)_{13}C_6H_5$	274.49	...	16.1	358.9^{760} 210^{12}	0.8559^{20}_{4}	1.4818^{20}	...	...	...	...	...	...	...
t73	**Tetradecanedioic acid,** dimethyl ester*	$CH_3O_2C(CH_2)_{12}CO_2CH_3$	286.42	nd (MeOH)	42–3	$191–2^{10}$	...	...	i	...	δ	...	...	MeOH i, s^h	**B2**[2]**,** 620
t74	**1,14-Tetrade-canediol***	Tetradecamethylene glycol. $HO(CH_2)_{14}OH$	230.39	nd (bz)	84.5	200^9	...	...	...	s	s	...	...	...	**B1**[3]**,** 2241
t75	**1-Tetradecane-sulfonic acid***	$CH_3(CH_2)_{13}SO_3H$	278.46	...	65.5 (anh) 55–6 (+1w)	...	0.9996^{25}_{4}	...	s	...	...	...	...	...	**Am 57,** 1905
t76	**1-Tetradecane-thiol***	*n*-Tetradecyl mercaptan. $CH_3(CH_2)_{13}SH$	230.45	...	...	$176–80^{22}$	0.8484^{20}_{20}	1.4601^{20}	i	s	s	...	...	...	**Am 55,** 1090
t77	**Tetradecanoic acid***	Myristic acid. $CH_3(CH_2)_{12}CO_2H$	228.36	lf	54.25	149.3^1	0.8439^{80}_{4}	1.4305^{60}	i	s	δ	s	v	chl s	**B2**[2]**,** 324
t78	—,amide	Myristamide. $CH_3(CH_2)_{12}CONH_2$	227.38	lf	103	217^{12}	...	...	i	s	δ	δ	δ	CCl_4 δ	...
t79	—,anhydride*	$[CH_3(CH_2)_{12}CO]_2O$	438.71	lf	53.4	vac distb	0.8502^{70}_{4}	1.4335	i	s	s	...	...	...	**B2**[2]**,** 329
t80	—,benzyl ester	Benzyl myristate. $CH_3(CH_2)_{12}CO_2CH_2C_6H_5$	318.48	...	20.5	229.3^{11}	0.9321^{25}_{25}	...	i	s	v	...	...	...	...
t81	—,chloride	Myristoyl chloride. $CH_3(CH_2)_{12}COCl$	246.83		−1	168^{15}	...	...	d	d	s	...	...	...	**B2**[2]**,** 329
t82	—,ethyl ester*	Ethyl myristate. $CH_3(CH_2)_{12}CO_2C_2H_5$	256.42	...	12.3	295^{760} 195^{30}	...	...	i	s	δ	...	...	...	**B2**[2]**,** 326

For explanations, symbols and abbreviations see beginning of table.

No.	Name	Synonyms and Formula	Mol. wt.	Crystalline form, color and specific rotation	m.p. °C	b.p. °C	Density	n_D	Solubility w	al	eth	ace	bz	other solvents	Ref.
	Tetradecanoic acid														
t83	—,isopropyl ester*	Isopropyl myristate. $CH_3(CH_2)_{12}CO_2CH(CH_3)_2$	270.44			140.2^{2}	0.8532^{20}	1.4325^{25}							
t84	—,methyl ester*	Methyl myristate. $CH_3(CH_2)_{12}CO_2CH_3$	242.39		19	295^{751}, 127^{2}			i	∞	∞	∞	∞	chl CCl_4 ∞ MeOH ∞ AcOEt ∞	**B2²**, 326
t85	—,nitrile	Myristonitrile. $CH_3(CH_2)_{12}CN$	209.36		19.25	226.5^{100}, 119^{1}	0.8281^{19}_{4}	1.4392^{25}	i	∞	∞	∞	∞	chl ∞	**B2²**, 329
t86	—,propyl ester*	Propyl myristate. $CH_3(CH_2)_{12}CO_2CH_2CH_2CH_3$	270.44			147^{2}	0.8592^{20}	1.4356^{25}							
t87	**1-Tetradecanol***	Myristyl alcohol. $CH_3(CH_2)_{13}OH$	214.38	lf	39–40	263.2, 167^{15}	0.8236^{38}_{4}		i	v	v	v	v	chl v	**B1²**, 465
t88	**2-Tetradecanone***	*n*-Dodecyl methyl ketone. $CH_3(CH_2)_{11}COCH_3$	212.36	(dil al)	33–4	$205–6^{100}$			i	s				os s	**B1²**, 770
t89	**3-Tetradecanone***	Ethyl *n*-undecyl ketone. $CH_3(CH_2)_{10}COC_2H_5$	212.36	(MeOH)	34	152^{16}			i	s				os s	**B1**, 716
t90	**1-Tetradecene***	1-Tetradecylene. $CH_3(CH_2)_{11}CH{:}CH_2$	196.36		−12	246, 127^{15}	0.7852^{0}_{4}	1.4932^{22}	i	v	v				**B1**, 226
t91	**5,9-Tetradecadien-7-yne, 6,9-dimethyl-***	$CH_3(CH_2)_3CH{:}C(CH_3)C{\vdots}CC(CH_3){:}CH(CH_2)_3CH_3$	218.37			$95–8^{0.5}$	0.8241^{20}_{4}	1.4866^{20}							**Am 55**, 1655
t92	**2-Tetradecyne***	$CH_3(CH_2)_{10}C{\vdots}CCH_3$	194.35		6.5	252.5, 134^{15}	0.8000^{20}_{4}		i	v	v				**B1**, 262
t93	**7-Tetradecyne***	$CH_3(CH_2)_5C{\vdots}C(CH_2)_5CH_3$	194.35			144^{30}	0.7991^{25}_{4}	1.4330^{25}							**Am 60**, 1717
t94	**Tetraethylene glycol**	$HOCH_2CH_2OCH_2CH_2OCH_2CH_2OCH_2CH_2OH$	194.29		−6.2	328^{760}, 230^{25}	1.1285^{15}_{4}	1.4598^{20}	v					diox s	**B1**, 468
t95	—,dimethyl ether	$CH_3OCH_2CH_2OCH_2CH_2OCH_2CH_2OCH_2CH_2OCH_3$	222.28			275.8	1.0132^{20}_{20}		∞	s					
t96	—,monooctadecanoate	Tetraethylene glycol monostearate. $CH_3(CH_2)_{16}CO_2OCH_2CH_2OCH_2CH_2OCH_2CH_2OCH_2CH_2CH$	460.68		40 (35)	328^{760}	1.1285^{15}_{4}	1.4593^{20}							**Am 58**, 813
t97	**Tetraethylenepentamine**	$H_2NCH_2CH_2NHCH_2CH_2NHCH_2CH_2NHCH_2CH_2NH_2$	189.34			$151–2^{1}$		1.5015^{25}							
t98	**Tetralin**	1,2,3,4-Tetrahydronaphthalene*. 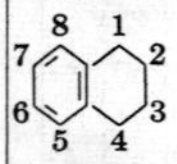	132.20		−31	207.3^{760}	0.9729^{20}_{4}	1.5461^{20}	i	v	v			$PhNH_2$ s	**B5²**, 382
t99	—,6-acetyl-	$C_{12}H_{14}O$. *See* t98	174.24			289–91d, 182^{20}									**B7²**, 305
t100	—,2-amino-	$C_{10}H_{13}N$. *See* t98	147.21		38	278–9	1.028		s^h	v					**B12**, 1200
t101	—,5-amino-	$C_{10}H_{13}N$. *See* t98	147.21			276.8	1.054^{23}	1.5896	δ	s	s			ac s	
t102	—,1,4-dioxo-	2,3-Dihydro-1,4-napthoquinone. $C_{10}H_8O_2$. *See* t98	160.17	lf (hexane), nd (peth)	98–9					s					**E12B**, 2806
t103	—,1,2,3,4-tetrabromo-	Naphthalene tetrabromide. $C_{10}H_8Br_4$. *See* t98	447.83	mcl pr (chl)	111d				i	i, $δ^h$	i		s^h	CS_2 s, chl s^h	**E12B**, 334
t104	**1-Tetralincarboxylic acid**(*dl*)	1,2,3,4-Tetrahydro-1-naphthoic acid. 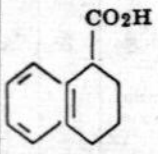	176.22	tcl pr (AcOEt)	85				δ, s^h	v				os v	**E12B**, 4078
t105	**2-Tetralincarboxylic acid**(*dl*)	1,2,3,4-Tetrahydro-2-naphthoic acid. 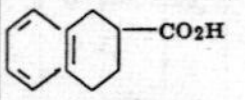	176.22	nd (dil al)	97	$168–70^{15}$			δ				s	CS_2 s, chl s	**E12B**, 4081
t106	**5-Tetralincarboxylic acid**	5,6,7,8-Tetrahydro-1-naphthoic acid. 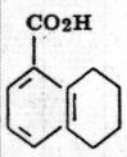	176.22	nd (aa or dil al)	150				$δ^h$	v					**E12B**, 2079

For explanations, symbols and abbreviations see beginning of table.

6-Tetralincarboxylic acid

No.	Name	Synonyms and Formula	Mol. wt.	Crystalline form, color and specific rotation	m.p. °C	b.p. °C	Density	n_D	Solubility: w	al	eth	ace	bz	other solvents	Ref.
t107	**6-Tetralincarboxylic acid**	5,6,7,8-Tetrahydro-2-naphthoic acid. 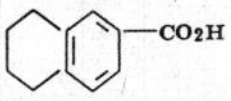	176.22	(aa or bz or dil al)	154	216^{14}				v			s	aa s	**E12B,** 4082
t108	**1,4-Tetralincarboxylic acid, 1-phenyl-(*d,α*)**	*d-α*-Isatropic acid. 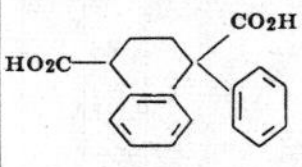	296.33	pr $[\alpha]_D+8.95$ (al, c = 12.6)	196–7										**B9**[1], 417
t109	—,—(*dl,α*)	$C_{18}H_{16}O_4$. *See* t108	296.33		238–9				δ^h	δ	i		i	CS_2 i	**B9**[1], 416
t110	—,—(*dl,β*)	$C_{18}H_{16}O_4$. *See* t108	296.33	ta (w)	208–9				s^h					aa s	**B9**[1], 417
t111	—,—(*l,β*)	$C_{18}H_{16}O_4$. *See* t108	296.33	$[\alpha]_D-8.8$ (al, c = 5)	197										**B9**[1], 417
—	**Tetralindione**	*see* **Tetralin, dioxo-**													
t112	**5-Tetralinsulfonic acid**	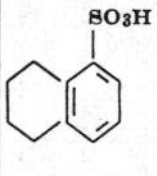	212.26	(chl + 1w), (w or min ac + 2w)	105–10 (+1w)				v^h					chl δ^h	**B11**[2], 87
t113	—,amide	$C_{10}H_{13}NO_2S$. *See* t112	211.31	lf (al or 30 % aa)	139–40					s^h				dil NaOH v	**B11**[2], 87
t114	—,chloride	$C_{10}H_{11}ClO_2S$. *See* t112	230.17	pl (peth)	70.5				d^h	d^h	s			peth δ	**B11**[2], 87
t115	**6-Tetralinsulfonic acid**	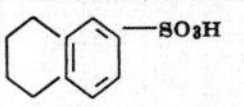	212.26	(chl or dil sulf)	75				s		s			chl v	**B11**[2], 88
t116	—,chloride	$C_{10}H_{11}ClO_2S$. *See* t115	230.71	pl (eth)	58	$197–200^{18}$			d^h		s				**B11**[2], 88
t117	**1-Tetralone**	1-Oxo-1,2,3,4-tetrahydronaphthalene* 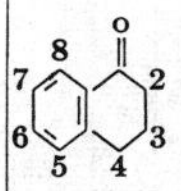	146.19		7	255–7, 129^{12}	1.0988^{16}_{4}	1.571^{16}							**B7**, 370
t118	**2-Tetralone**	2-Oxotetrahydronaphthalene. 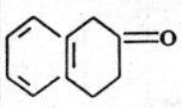	146.19		18	234–40, 138^{16}	1.1055^{17}_{4}				s		s		**B7**[2], 295
t119	**1-Tetralone, 7-ethyl-**	$C_{12}H_{14}O$. *See* t117	174.24			$152–3^{12}$	1.0556^{17}_{4}	1.5599^{17}							**B7**[2], 305
—	**Tetranthrene**	*see* **Phenanthrene, 1,2,3,4-tetrahydro-***													
t120	**Tetraphenylene**	Tetrabenzo-$\Delta^{1,3,5,7}$ cyclooctatetraene. 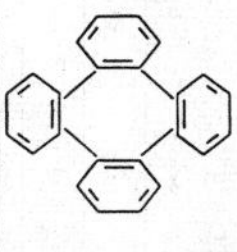	304.39	(al or AcOEt)	233	$200^{0.2}$sub				s^h				oos δ, AcOEt s^h, $PhNO_2$ s^h	**E14s,** 752
t121	**Tetraphosphoric acid, hexaethyl ester**	$[(C_2H_5O)_2P(O)O]_3PO$	506.28	hyg	ca. −40	>150d	1.2917^{27}_{4}	1.4273^{27}	d	∞		∞	∞	oos ∞, peth i	
t122	**Tetrasiloxane, decamethyl-**	$(CH_3)_3SiOSi(CH_3)_2OSi(CH_3)_2OSi(CH_3)_3$	310.58		ca. −70	194	0.8536	1.3895^{20}	i	δ			s	peth s	
t123	**1-Tetratriacontanol***	*n*-Carnatyl alcohol. $CH_3(CH_2)_{33}OH$	494.90	nd	91.9										**B1**[1], 223
t124	**1,2,4,5-Tetrazine**	*s*-Tetrazine. 	82.07	dk red pr	99	sub			s						**B26**, 353
t125	**1,2,3,4-Tetrazole**	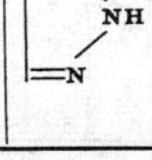	70.06	pl	155	sub			s	s	i				**B26**, 346

For explanations, symbols and abbreviations see beginning of table.

No.	Name	Synonyms and Formula	Mol. wt.	Crystalline form, color and specific rotation	m.p. °C	b.p. °C	Density	n_D	Solubility w	al	eth	ace	bz	other solvents	Ref.
	1,2,3,4-Tetrazolium														
t126	**1,2,3,4-Tetrazolium, 2,3,5-triphenyl-,** chloride	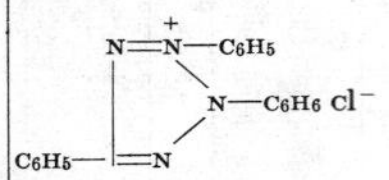	334.80	nd (al or chl)	243d	...	...	...	s	s	i	s	...	...	...
—	**Tetrolic acid**	*see* **2-Butynoic acid***													
—	**Tetronal**	*see* **Pentane, 3,3-bis(ethylsulfonyl)-***													
—	**Tetronic acid**	*see* **Furan, 2,4-dioxo-, tetrahydro-**													
—	**Tetryl**	*see* **Aniline, *N*-methyl-*N*, 2,4,6-tetranitro-**													
—	**Thalline**	*see* **Quinoline, 6-methoxy-1,2,3,4-tetrahydro-**													
—	**Thapsic acid**	*see* **Hexadecanedioic acid***													
t127	**Thebaine**	Paramorphine. $C_{19}H_{21}NO_3$.	311.37	pl (eth), pr (al) $[\alpha]^{25}_D -215.5$	193	...	1.305^{20}_4	...	i	δ	δ	...	...	$PhNH_2$ v	**B27²**, 177
t128	—,hydrochloride monohydrate	$C_{19}H_{21}NO_3.HCl.H_2O$.	365.85	pl $[\alpha]^{20}_D -157$	...	...	...	...	s	s	s	...	...	...	...
t129	**Thebainone**	$C_{18}H_{21}NO_3$. 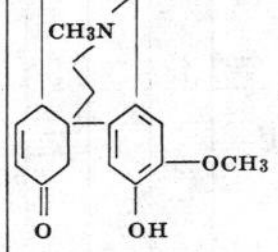	299.36	nd or pr (al, aa, AcOEt) $[\alpha]^{28}_D -47$	151–2	...	...	...	δ	δ s^h	δ	s	s	chl s AcOEt s^h MeOH δ	**B21²**, 448
—	**Theine**	*see* **Caffeine**													
t130	**Theobromine**	3,7-Dimethylxanthine. 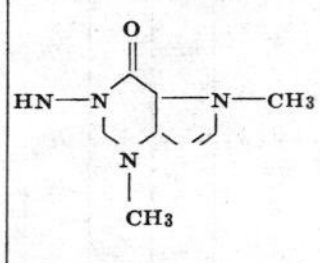	180.17	rh	337 (351)	sub	...	...	δ	...	i	...	i	...	**B26**, 457
t131	**Theophylline**	1,3-Dimethylxanthine. 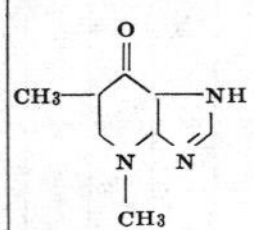	180.17	nd or pl	272	...	...	...	v^h	δ	i	...	...	...	**B26**, 455
t132	**Thevetin**	$C_{42}H_{66}O_{18}$	858.98	pl (*i*-PrOH+3w) $[\alpha]^{26}_D -62.5$	195d	...	...	...	δ	s	i	i	i	MeOH s Py s chl i	**E14**, 231
t134	**Thialdine**	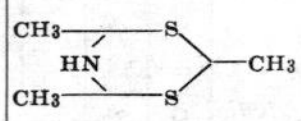	163.31	mcl pr	46	d	1.0632^{50}_{20}	...	δ	δ	v	...	...	...	**B27**, 461
—	**Thiamine hydrochloride**	*see* **Vitamin B_1**													
t135	**Thianthrene**	Diphenylene disulfide. 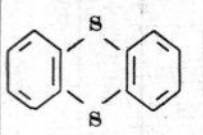	216.33	mcl pr	158–9	366^{760}	...	...	i	s^h	δ	...	s	...	**B19**, 45
t136	**Thiazole**	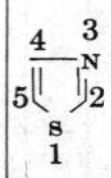	85.13	...	...	116.8	1.1998^{17}_4	1.5969^{20}	...	s	s	...	...	...	**B27**, 15
t137	—,2-amino-	$C_3H_4N_2S$. *See* t136	100.14	ye pl	92	d	...	...	v^h	δ	δ	...	...	dil HCl v	**B27²**, 205
t138	—,2-amino-4-methyl-	$C_4H_6N_2S$. *See* t136	114.17	hyg	42	281–2δd 136^{30-40}	...	...	v	v	v	...	...	...	**B27²**, 206

For explanations, symbols and abbreviations see beginning of table.

PHYSICAL CONSTANTS OF ORGANIC COMPOUNDS (Continued)

No.	Name	Synonyms and Formula	Mol. wt.	Crystalline form, color and specific rotation	m.p. °C	b.p. °C	Density	n_D	Solubility						Ref.
									w	al	eth	ace	bz	other solvents	
	Thiazole														
t139	—,**2-amino-5-methyl-**	$C_4H_6N_2S$. *See* t136	114.17	pl (w)	94–5				δ s^h	v	s				**B27**, 162
t140	—,**2-amino-5-sulfanilyl-**	H_2N—⟨ ⟩—SO_2—N S—NH_2	255.32	nd (al)	219–21δd				i	s	s	v		diox v AcOEt s dil ac s	
—	—,**dihydro-**	*see* **Thiazoline**													
t141	—,**2,4-dimethyl-**	C_5H_7NS. *See* t136	113.18			144–5	1.0562^{20}_{4}		s	s	s				**B27**[2], 10
t142	—,**4,5-dimethyl-**	C_5H_7NS. *See* t136	113.18			158					s				**Am 74**, 5778
t143	—,**2,4-diphenyl-**	$C_{15}H_{11}NS$. *See* t136	237.33	lf (al)	92–3	360	1.1554^{98}_{4}			v	v			ac s	**B24**[2], 43
t144	—,**5(2-hydroxyethyl)-4-methyl-**	C_6H_9NOS. *See* t136	143.21	colorless to pa ye		135^7			v	s	s		s	chl s	**Am 71**, 2931
t145	—,**2-mercapto-4-methyl-**	$C_4H_5NS_2$. *See* t136	131.22	ye	88–9	188^3			δ	s					**B27**, 161
t146	—,**2-methyl-**	C_4H_5NS. *See* t136	99.16			$129–30^{760}$			s	s		s			**B27**, 16
t147	—,**4-methyl-**	C_4H_5NS. *See* t136	99.16			132^{743}	1.112^{25}		s	s	s				**B27**, 16
—	—,**tetrahydro-**	*see* **Thiazolidine**													
t149	**5-Thiazolecarboxylic acid**	N S —CO_2H	129.14	lt ye nd	217–18 (196–7)				s^h		s				**C48**, 2688
t150	—,**2-amino-**, ethyl ester	$C_6H_8N_2O_2S$. *See* t149	172.21		163–4	213–5d			δ						
t151	—,**4-methyl-**	$C_5H_5NO_2S$. *See* t149	143.17	pr or pl (w), nd (al)	280d (257d)				$δ^h$	s^h	δ		i		**B27**[2], 376
t152	—,—,ethyl ester	$C_7H_9NO_2S$. *See* t149	171.22	pr	28	$232–3^{735}$ $110–5^{15}$								oos v	**B27**, 316
t153	—,—,—,hydrochloride	$C_7H_9NO_2S.HCl$. *See* t149	207.68	nd (al)	155d					s^h					**J1939**, 443
t154	—,**4-methyl-2-sulfanilamido-**	H_2N—⟨ ⟩—SO_2NH—N S —CH_3 —CO_2H	313.36		195d				δ						**C38**, 2250
t155	**Thiazolidine**	Tetrahydrothiazole. NH S	89.16			164–5	1.131^{25}_{4}	1.551^{30}	∞						**Am59**, 200
t156	**2,4-Thiazoidinedione**	2,4-Dioxothiazolidine. O NH S =O	117.13	pl (w), pr (al)	128	179^{19}			δ s^h	δ v^h	s				**B27**[2], 284
t157	**2-Thiazolidinethione**	2-Thiothiazolidone. NH S =S	119.21	nd (w or MeOH)	106–7				s^h	δ	i		s^h	chl δ CS_2 i alk v	**B27**[2], 198
—	**4-Thiazolidone, 2-thioxo-**	*see* **Rhodamine**													
t158	**4,5-Thiazoline**	4,5-Dihydrothiazole. $Δ^2$-Thiazoline. 4 3 N 5 2 S 1	87.14			138^{750}					s		s		**B27**[1], 206
t159	—,**2-amino-**	*S,N*-Ethyleneisothiourea. $C_3H_6N_2S$. *See* t158	102.16	nd or flakes (bz)	84–5	d			v	v			v^h	chl v	**B27**[2], 194
t160	—,**2-mercapto-**	2-Thiazolinethiol. $C_3H_5NS_2$ *See* t158	119.21	nd (w or MeOH)	106–7				s^h	s^h	δ		s^h	chl s^h CS_2 i	**B27**[2], 118
—	**Thiazone**	*see* **3-Isophenothiazin-3-one**													
—	**Thienone**	*see* **Ketone, di(2-thienyl)**													
t161	**Thiirane**	Ethylene sulfide. CH_2—CH_2 S	60.12			55–6d	1.0368^{0}_{4}	1.4914^{18}		δ	δ				**B17**[2], 12
—	**Thioacetic acid**	*see* **Acetic acid, thio-**													
—	**Thioanisole**	*see* **Benzene, (methylthio)-**													
—	**Thiocarbazide**	*see* **Cabazide, 3-thio-**													
t162	**Thiochrome**	CH_3— N N N S N —CH_2CH_2OH —CH_3	262.33	pr (chl)	227–8				s	δ	δ	δ		MeOH s chl δ	**C29**, 6242[3]

For explanations, symbols and abbreviations see beginning of table.

No.	Name	Synonyms and Formula	Mol. wt.	Crystalline form, color and specific rotation	m.p. °C	b.p. °C	Density	n_D	Solubility w	al	eth	ace	bz	other solvents	Ref.
	Thiocyanic acid														
t163	**Thiocyanic acid**	Sulfocyanic acid. HSCN	59.09		5	d			∞	∞	∞				**B3**², 107
t164	—,allyl ester	Allyl rhodanide. CH_2:$CHCH_2SCN$	99.15			161	1.056^{15}		δ	v	v				**B3**, 177
t165	—,4-aminophenyl ester	p-Thiocyanatoaniline. $C_7H_6N_2S$. *See* t180	150.20	nd (w), (dil al)	57–8				δ	v	s		s		**B13**², 299
t166	—,benzyl ester	Benzyl rhodanide. Benzyl thiocyanate. $C_6H_5CH_2SCN$	149.22	pr	41	vac distb			i	s	v			CS_2 v	**B6**², 434
t167	—,butyl ester	*n*-Butyl thiocyanate. $CH_3(CH_2)_3SCN$	115.19			185^{743}	0.9563^{25}	1.4636^{22}	i	s	s				**B3**², 122
t168	—,*tert*-butyl ester	$(CH_3)_3CSCN$	115.19		10.5	140^{770}d	0.9187^{10}								**B3**, 175
t169	—,4-chlorophenyl ester	C_7H_4ClNS. *See* t180	169.64	nd (al)	35–6				i	v				os v	**B6**², 298
t170	—,4(dimethyl-amino)-phenyl ester	$C_9H_{10}N_2S$. *See* t180	178.23	nd (lig, w, al)	73–4					δ s^h	s^h				**B13**², 301
t171	—,ethyl ester	Ethyl rhodanate. Ethyl thiocyanate. C_2H_5SCN	87.14		−85.5	145^{758}	0.9964_4^{25}	1.4684^{15}	i	∞	∞				**B3**, 175
t172	—,heptyl ester	*n*-Heptyl rhodanide. $CH_3(CH_2)_6SCN$	157.28			234–6 136^{28}	0.92^{20}								**B3**², 122
t173	—,isobutyl ester	$(CH_3)_2CHCH_2SCN$	115.19		−59	175.4^{760}									**B3**, 177
t174	—,isopropyl ester	$(CH_3)_2CHSCN$	101.16			$152–3^{754}$	0.963^{20}		i	∞	∞				**B3**, 177
t175	—,methyl ester	Methyl rhodanate. Methyl thiocyanate. CH_3SCN	73.11		−51	132.9^{757}	1.0678_4^{25}	1.4669^{25}	δ	∞	∞				**B3**, 175
t176	—,3-methylbutyl ester	Isoamyl rhodanate. Isoamyl thiocyanate. $(CH_3)_2CHCH_2CH_2SCN$	129.22			197^{760}			i	s	s				**B3**, 177
t177	—,1-naphthyl ester	α-Naphthyl thiocyanate. $C_{10}H_7^{\alpha}SCN$	185.25	(peth)	55									peth s^h	**B6**², 588
t178	—,2-naphthyl ester	β-Naphthyl thiocyanate. $C_{10}H_7^{\beta}SCN$	185.25		35										**B6**², 611
t179	—,octyl ester	*n*-Octyl rhodanide. $CH_3(CH_2)_7SCN$	172.30		105	$122–4^{11}$		1.4649^{20}	i				s^h		
t180	—,phenyl ester	Thiocyanatobenzene. Phenyl rhodanide. Phenyl thiocyanate. 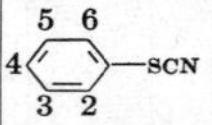 (ring positions 5 6 / 4 / 3 2 —SCN)	135.18			$232–3^{760}$	1.155_{18}^{18}		i	s	s				**B5**, 312
t181	—,propyl ester	Propyl rhodanate. Propyl thiocyanate. $CH_3CH_2CH_2SCN$	101.17			163									**B3**, 177
t182	—,4-tolyl ester	p-Tolyl rhodanide.  CH_3—⟨ ⟩—SCN	149.22			240–5 $155–8^{40,50}$			i	s			s	chl s	**B6**², 398
—	**Thiocyanuric acid**	*see* **Cyanuric acid, thio-**													
t183	**Thiodiglycolic acid**	Methyl sulfide α,α-dicarboxylic acid. Thiodiacetic acid. $S(CH_2CO_2H)_2$	150.16	(w)	129				v	v					**B3**², 178
—	**Triglycolamic acid**	*see* **Acetic acid, nitrilotri-**													
—	**Thioglyolic acid**	*see* **Acetic acid, mercapto-**													
—	**Thiohydantoin**	*see* **Hydantoin, thio-**													
t184	**Thioindigo**	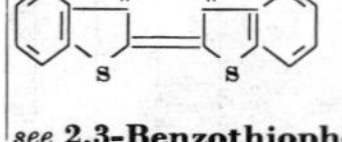 (structure: O, O; S, S)	296.37	br-red or red mcl nd (bz)	359	sub			i	i δ^h			s^h	xyl s $PhNO_2$ v^h CS_2, chl δ^h	**B19**¹, 690
—	**Thioisatin**	*see* **2,3-Benzothiophene-quinone**													
—	**Thiomalic acid**	*see* **Succinic acid, 2-mercapto-**													
t185	**Thiomorpholine**	1,4-Thiazan. 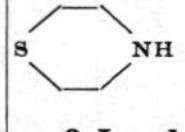 (structure: S, NH)	103.19			169^{758}			∞					oos ∞	**B17**², 4
—	**Thionol**	*see* **3-Isophenothiazin-3-one, 7-hydroxy-**													
—	**Thionaphthene**	*see* **2,3-Benzothiophene**													

For explanations, symbols and abbreviations see beginning of table.

No.	Name	Synonyms and Formula	Mol. wt.	Crystalline form, color and specific rotation	m.p. °C	b.p. °C	Density	n_D	Solubility w	al	eth	ace	bz	other solvents	Ref.
	Thionin														
t186	**Thionin**	7-Amino-2-phenothiazim. Lanth's violet. 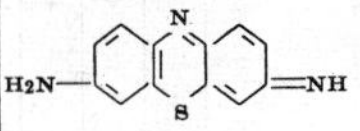 H_2N— N S =NH	227.28	dk br or gr pl or nd					δ^h	δ	δ		s	chl s ac s	**B27**[2]**,** 447
t187	**Thiophene**	Thiofuran. 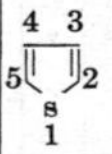 4 3 / 5 2 / S / 1	84.14		−38.4	84.12^{760}	1.0583_4^{25}	1.5256^{25}						oos ∞	**B17,** 29
t188	**—,2-acetamido-**	C_6H_7NOS. *See* t187	141.18	lf (w)	161–2				δ^h	v	δ	v	δ		**B17**[1]**,** 136
t189	**—,2-acetyl-**	Methyl 2-thienyl ketone. C_6H_6OS. *See* t187	126.18		9	213.5	1.1679_4^{22}	1.5667^{20}	δ	∞	∞				**B17**[2]**,** 314
t190	**—,2-acetyl-5-bromo-**	C_6H_5BrOS. *See* t187	205.09	nd (al)	94					δ v^h					**B17,** 288
t191	**—,2-acetyl-5-chloro-**	C_6H_5ClOS. *See* t187	160.63	ta (al or eth)	52				v	v	v				**B17,** 287
t192	**—,2-amino-**	Thiophenine. C_4H_5NS. *See* t187	99.17	pa ye		$77–9^{11}$			v	v	i				**B17,** 248
t192[1]	**—,2-benzoyl-**	Phenyl 2-thienyl ketone. $C_{11}H_8OS$. *See* t187	188.24	nd (dil al or peth)	55–6	300^{760}	1.1890_4^{54}		i	v^h	v^h				**B17**[2]**,** 372
t193	**—,2-bromo-**	C_4H_3BrS. *See* t187	163.05			149–51	1.684_4^{20}	1.5866	i		v	v			**B17,** 39
t194	**—,2-bromo-5-chloro-**	C_4H_2BrClS. *See* t187	197.51		−22, −20	70^{18}	1.803_{25}^{25}	1.5925^{25}							**Am70,** 2379
t195	**—,2-bromo-5-iodo-**	C_4H_2BrIS. *See* t187	288.97			116^{13}									
t196	**—,2-bromo-3-methyl-**	C_5H_5BrS. *See* t187	177.07			175^{729}	1.5844_4^{18}	1.5731			s		s		**B17**[2]**,** 40
t197	**—,2-bromo-5-methyl**	C_5H_5BrS. *See* t187	177.07	colorless to pa ye		177^{740} (117)	1.5529_4^{20}	1.5673^{20}			s		s		**B17**[2]**,** 39
t198	**—,2-chloro-**	C_4H_3ClS. *See* t187	118.59		−71.9	128.3	1.2863_4^{20}	1.5487^{20}	i	∞	∞				**B17,** 32
t199	**—,2-chloro-5-butyl-**	$C_8H_{11}ClS$. *See* t187	174.69			$117–8^{88}$	1.0842_4^{17}	1.5162^{20}							
t200	**—,2-chloro-5-iodo-**	C_4H_2ClIS. *See* t187	244.48		−25	$95–6^{14}$									
t201	**—,2-chloro-5-methyl-**	C_5H_5ClS. *See* t187	132.61			$154–5^{742}$	1.2016_4^{17}	1.5372^{20}		s				os s	**B17,** 37
t202	**—,2,5-dibromo-**	$C_4H_2Br_2S$. *See* t187	241.94		−6	211^{760}	2.147_{23}^{23}	1.6288^{20}	i	v	v				**B17,** 33
t203	**—,2,5-dibromo-3,4-dinitro-**	$C_4Br_2N_2O_4S$. *See* t187	331.94		139–40					δ					**B17,** 36
t204	**—,2,5-dichloro-**	$C_4H_2Cl_2S$. *See* t187	153.03		−40.5	162^{760}	1.4422_4^{20}	1.5626^{20}	i	∞	∞				**B17,** 33
t205	**—,2,5-diiodo-**	$C_4H_2I_2S$. *See* t187	335.93	lf	40.5	$139–40^{15}$			i	v					**B17,** 35
t206	**—,2,3-dimethyl-**	2,3-Thioxene. C_6H_8S. *See* t187	112.19		−49	141^{760}	1.0021_4^{20}	1.5188^{20}	i	v	v				**B17,** 40
t207	**—,2,4-dimethyl-**	2,4-Thioxene. C_6H_8S. *See* t187	112.19			$137–8^{760}$	0.9956_{20}^{20}	1.5130^{20}	i	s	s				**B17,** 41
t208	**—,2,5-dimethyl-**	α,α-Thioxene. C_6H_8S. *See* t187	112.19		−62.6	$136–7^{760}$	0.985_4^{20}	1.5146^{20}	i	s	s				**B17,** 41
t209	**—,2,5-dinitro-**	$C_4H_2N_2O_4S$. *See* t187	174.14	ye mcl pr	80–2	290^{760}			s^h	s	v				**B17,** 35
t210	**—,2-ethyl-**	C_6H_8S. *See* t187	112.19			$132–4^{760}$	0.990_{24}^{24}	1.5127^{20}	i	v	v				**B17,** 39
t211	**—,3-ethyl-**	C_6H_8S. *See* t187	112.19			$135–6^{760}$		1.0012^{16}	i	s	v				**B17,** 40
t212	**—,2-hydroxy-5-methyl-**	2,5-Thiotenol. C_5H_6OS. *See* t187	114.17			85^{40}			δ	v	v				**B17,** 252
t213	**—,2-iodo-**	C_4H_3IS. *See* t187	210.04		−41, −40	$180–2^{760}$					v				**B17,** 34
t214	**—,2-iodo-5-nitro-**	$C_4H_2INO_2S$. *See* t187	255.03	ye pr	74					s					**B17,** 35
t215	**—,2-methyl-**	α-Thiotoluene. C_5H_6S. *See* t187	98.17		−63.5	112.5^{760}	1.0140_4^{25}	1.5174^{25}	i	∞	∞				**B17,** 37
t216	**—,3-methyl-**	β-Thiotoluene. C_5H_6S. *See* t187	98.17		−69	115.4^{760}	1.0162_4^{25}	1.5172^{25}	i	∞	∞				**B17,** 38

For explanations, symbols and abbreviations see beginning of table.

No.	Name	Synonyms and Formula	Mol. wt.	Crystalline form, color and specific rotation	m.p. °C	b.p. °C	Density	n_D	Solubility w	al	eth	ace	bz	other solvents	Ref.
	Thiophene														
t217	**—,2(methyl-amino)-**	*N*-Methylthiophenine. C_5H_7NS. *See* t187	113.18			88–92[15]									**B17**, 136
t218	**—,2-methyl-5-phenyl-**	$C_{11}H_{10}S$. *See* t187	174.27	nd	51	270–2				v	v			lig v	**B17**, 67
t219	**—,2-nitro-**	$C_4H_3NO_2S$. *See* t187	129.14	lt ye mcl nd	46.5	224–5	1.3644_4^{43}		i	v				peth δ	**B17**, 35
t220	**—,2-nitro-3,4,5-trichloro-**	$C_4Cl_3NO_2S$. *See* t187	232.47	red-ye nd	86(70)					s	v		v		**B17**, 33
t222	**—,tetrabromo-**	C_4Br_4S. *See* t187	399.74	nd	117	326			i	s^h	v				**B17**, 34
t223	**—,tetrachloro-**	C_4Cl_4S. *See* t187	221.92	nd	29	233.4[760]	1.7036_4^{30}	1.5915[30]	i	v	∞				**B17**, 33
t224	**—,tetrahydro-**	Tetramethylene sulfide. 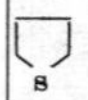 (ring structure, S)	88.17		−96.2	119–20	0.9998_0^{20}	1.5047[20]	i	∞	∞			os ∞	**B13**[2], 15
t225	**—,2,3,5-tribromo-**	C_4HBr_3S. *See* t187	320.84	nd	29	260			i	v^h	v				**B17**, 34
t226	**—,2,3,5-tribromo-4-nitro-**	$C_4Br_3NO_2S$. *See* t187	365.84	red ye nd	106					δ	v				**B17**, 35
t227	**—,2,3,5-trichloro-**	C_4HCl_3S. *See* t187	187.48		−16.1	198.7	1.5856_4^{20}	1.5791[20]	i	∞	∞				**B17**, 33
t228	**—,2,3,5-tri-methyl-**	$C_7H_{10}S$. *See* t187	126.22			163–5[746]	0.9752_4^{20}	1.5131[20]							**B17**[2], 43
	Thiopheneacetic acid	*see* **Acetic acid, thienyl-**													
t229	**2-Thiophene-carboxaldehyde**	2-Formylthiophene. 2-Thiophenealdehyde. 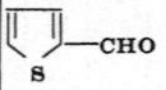 (structure: —CHO, S)	112.15	pa ye		198[760]	1.215_{21}^{21}	1.5920[20]	i	v	s				**B17**, 285
t230	—,oxime	2-Thiophenealdoxime.  (structure: CH:NOH, S)	127.17	nd	133						v				**B17**, 285
t231	—,phenylhydrazone	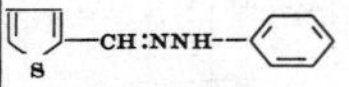 (structure: —CH:NNH—C₆H₅, S)	202.28	ye nd	134.5				i	s					**B17**, 286
t232	**2-Thiophene-carboxylic acid**	α-Thiophenic acid. 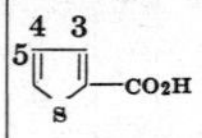 (structure: 4 3 5, —CO2H, S)	128.15	nd (w)	129–30	260d			v^h	v	v				**B18**, 289
t233	—,chloride	C_5H_3ClOS. *See* t232	146.60			206–8			d	d^h					**B18**[2], 269
t234	—,ethyl ester	$C_7H_8O_2S$. *See* t232	156.20			218 96[18]	1.1623_4^{16}			s				os s	**B18**[2], 269
t235	**—,3-methyl-**	$C_6H_6O_2S$. *See* t232	142.18	nd	144				δ v^h	v	v				
t236	**2,3-Thiophenedi-carboxylic acid**	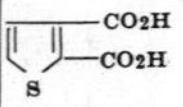 (structure: —CO2H, —CO2H, S)	172.16	pr or nd (w)	270d				$δ^h$		v				**B18**, 327
t237	**2,4-Thiophenedi-carboxylic acid**	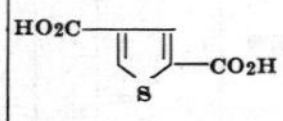 (structure: HO2C—, —CO2H, S)	172.16		280	sub			δ v^h						**B18**, 328
t238	**2,5-Thiophenedi-carboxylic acid**	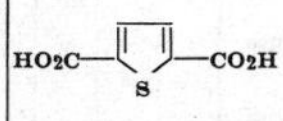 (structure: HO2C—, —CO2H, S)	172.16		>350	sub			δ	s	s				**B18**, 330
t239	—,diethyl ester	$C_{10}H_{12}O_4S$. *See* t238	228.27	nd (al)	51.5					s					**B18**, 331
t240	**2-Thiophene-sulfonic acid,** amide	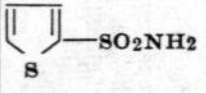 (structure: —SO2NH2, S)	163.22	nd (w)	142				δ s^h						**B18**, 576
t241	—,chloride	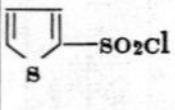 (structure: —SO2Cl, S)	182.65		28	sub or distb			d	d^h	s				**B18**, 567
t242	**3-Thiophene-sulfonic acid,** amide	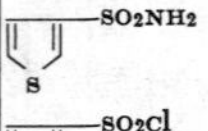 (structure: —SO2NH2, S)	163.22	pl (w)	152–3				δ						**B18**, 568
t243	—,chloride	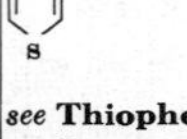 (structure: —SO2Cl, S)	182.66	(eth)	43				d^h	d^h	v			lig i	**B18**, 568
—	**Thiophenine**	*see* **Thiophene, amino-**													
—	**Thiophenol**	*see* **Benzene, mercapto-***													
—	**Thiophosphoric acid**	*see* **Phosphoric acid, thiono-***													

For explanations, symbols and abbreviations see beginning of table.

No.	Name	Synonyms and Formula	Mol. wt.	Crystalline form, color and specific rotation	m.p. °C	b.p. °C	Density	n_D	Solubility						Ref.
									w	al	eth	ace	bz	other solvents	
	Thiophithene														
t244	**Thiophithene**	Thienothiophene. 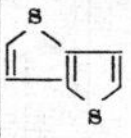	140.23	bipym orh (lig)	56	$221–4^{760}$								lig δ	
t245	**Thiophthene**	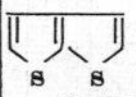	140.23	nd (eth-so CO_2)	6.5	$224–6^{760}$ 106^{16}									**B19**[1], 612
—	**Thiopyran, tetrahydro-**	*see* **Sulfide, pentamethylene**													
t246	**Thiopyrine**	1,5-Dimethyl-2-phenyl-3-thio-3-pyrazolone. 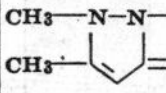	204.29	(w)	166				δ s^h	s	s				**B24**, 56
—	**Thiosemicarbazide**	*see* **Semicarbazide, 3-thio-**													
—	**Thiosinamine**	*see* **Urea, 1-allyl-2-thio-**													
—	**Thiourea**	*see* **Urea, 2-thio-**													
—	**Thiourethan**	*see* **Carbamic acid, thiono-,** ethyl ester													
t248	**Thioxanthene**	Thiaxanthene 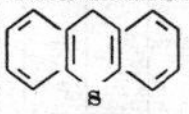	198.28	nd (al-chl)	128	340^{730}			i	δ s^h	δ s^h			chl v peth s	**B17**, 74
t249	**Thioxanthone**	9-Oxothioxanthene. 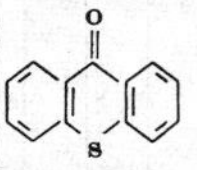	212.27	ye nd	209	$371–3^{715}$ (sub)			i	δ			s^h	CS_2 s chl s^h peth i	**B17**[1], 191
—	**Threite**(*D*)	*see* **Erithrite**													
t250	**Threonic acid**(*D*)	2,3,4-Trihydroxybutyric acid*. 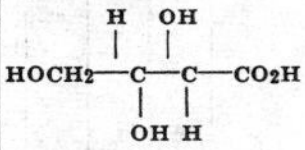 $HOCH_2—CH(OH)—CH(OH)—CO_2H$	136.11	syr, $[\alpha]_D$ ca. −30 (w)					s	s^h		s^h		AcOEt s^h	**B3**[2], 272
t251	—(*DL*)	$C_4H_8O_5$. *See* t250	136.11		99				s	s^h		s^h		AcOEt s^h	**B3**[2], 272
t252	—(*L*)	$C_4H_8O_5$. *See* t250	136.11	syr, $[\alpha]_D$ +9.54 (w)					s	s^h		s		AcOEt s^h	**B3**[2], 272
t253	**Threonine**(*D*)	*D*-2-Amino-3-hydroxybutyric acid. $CH_3CHOHCH(NH_2)CO_2H$	119.12	$[\alpha]^{26}_D$ −28.3 (c=1.1)	255–7d				s	i	i			chl i	
t254	—(*DL*)	$CH_3CHOHCH(NH_2)CO_2H$	119.12	orh	229–30 (+½w)				s v^h	i	i			chl i	**B4**, 514
t255	**Threose**(*D*)	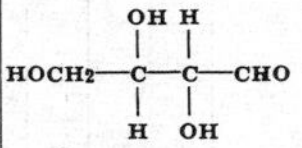 $HOCH_2—CH(OH)—CH(OH)—CHO$	120.10	hyg syr or nd $[\alpha]^{22}_D$ (mut) +29.1 to +19.6 (w)	126–32	1			v	δ	i			peth i	**B1**, 855
t256	**Thujane**	Sabinane. 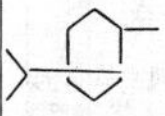	138.25	$[\alpha]_D$ +73.1 +62		158^{756}	0.8157^{20}_0	1.4404^{20}							**E12A**, 6
t257	**α-Thujene**(*d*)	$C_{10}H_{16}$	136.24	$[\alpha]^{20}_D$ +37.7, +38.8		152^{699}		1.4501^{30}				s		chl s	**E12A**, 8
t258	—(*l*)	$C_{10}H_{16}$	136.24	$[\alpha]_D$ −37.2		151^{759}	0.8301^{20}_4	1.4515^{20}				s		chl s	**E12A**, 7
t259	**α-Thujone**	$C_{10}H_{16}O$	152.23	$[\alpha]^{18}_D$ −19.94		200–1	0.9125^{20}	1.4490^{25}	i	∞	∞				**B7**, 92
t260	**Thymidine**	Thymine-2-desoxyriboside. 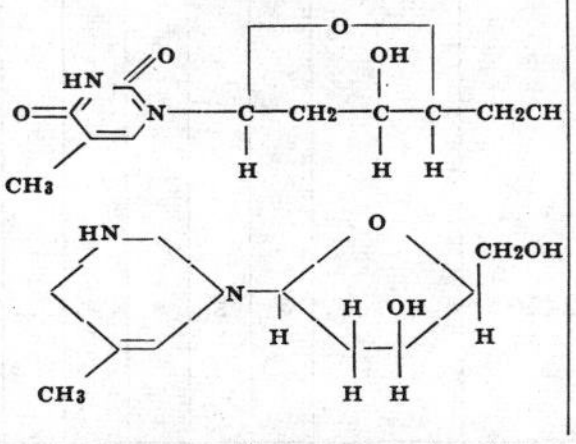	242.23	nd (AcOEt) $[\alpha]^{25}_D$ +30.6	185				s	s^h		s^h		Py, MeOH s chl $δ^h$ AcOEt s^h	

For explanations, symbols and abbreviations see beginning of table.

No.	Name	Synonyms and Formula	Mol. wt.	Crystalline form, color and specific rotation	m.p. °C	b.p. °C	Density	n_D	Solubility w	al	eth	ace	bz	other solvents	Ref.
	Thymine														
—	**Thymine**	*see* **Uracil, 5-methyl**													
—	**Thymohydroquinone**	*see* **Benzene, 1,4-dihydroxy-2-isopropyl-5-methyl-***													
—	**Thymol**	*see* **Benzene, 2-hydroxy-1-isopropyl-4-methyl***													
t263	**Thumol blue**	Thymolsulfonephthalein. 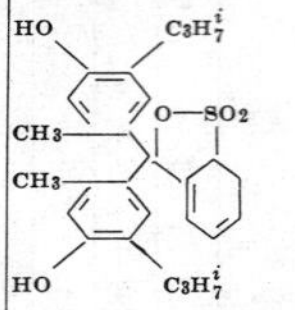	466.58	gr-red (al or eth or aa)	200–20d	...	...	...	δ	s	...	δ	δ	chl, CCl_4, to δ aniline s aa s	**B19**[2], 112
t264	**Thymolphthalein**	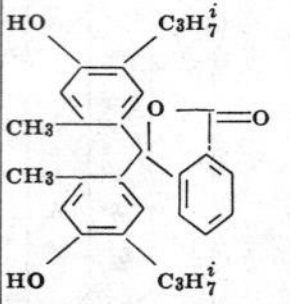	430.52	pr	246–7	...	...	...	...	s^h	s	s	...	alk s	**B8**[1], 381
—	**Thymoquinone**	*see* **1,4-Benzoquinone, 2-isopropyl-5-methyl***													
—	*o*-**Thymotic acid**	*see* **Benzoic acid, 2-hydroxy-3-isopropyl-6-methyl-**													
—	**Thymylamine**	*see* **Benzene, 2-amino-1-isopropyl-4-methyl***													
t266	**Thyroxine**(*d*)	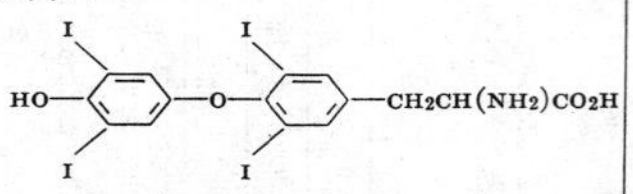 HO—C₆H₂I₂—O—C₆H₂I₂—CH₂CH(NH₂)CO₂H	776.93	nd, $[\alpha]^{25}_{546}$ +2.97	237d	...	...	...	δ	i	...	...	...	...	**B14**[2], 366
t267	−(*l*)	$C_{15}H_{11}I_4NO_4$ *See* t266	776.93	nd, $[\alpha]^{21}_{546}$ −3.8	235–6	...	...	...	δ	i	...	...	...	...	**B14**[2], 366
—	**Tiglaldehyde**	*see* **2-Butenal, 2-methyl-***													
—	**Tiglic acid**	*see* **2-Butenoic acid, 2-methyl-(*trans*)***													
t268	**Tigogenin**	$C_{27}H_{44}O_3$	416.62	lf (al+1w), pr (ace) $[\alpha]^{25}_D$ −49 (Py)	206–10	...	...	...	...	s^h	s	s	...	MeOH s, peth s CCl_4 s^h	**E14**, 280
—	**TNT**	*see* **Toluene, 2,4,6-trinitro-**													
t269	α-**Tocopherol**	Vitamin E. 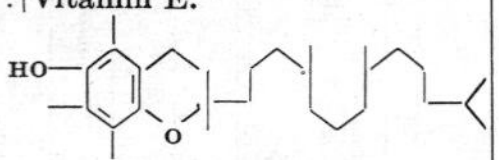	430.69	pa ye	...	350d, $140^{10.0}$	...	...	i	s	s	...		...	...
t270	β-**Tocopherol**	Vitamin E. 5,8-Dimethyltocol. 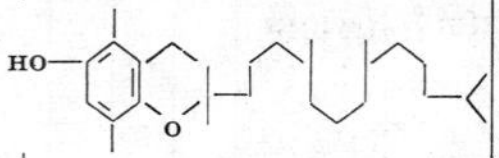	416.66	pa ye, $[\alpha]^{20}_D$ +6.37	...	200–$10^{0.1}$	...	...	i	∞	∞	∞	...	chl ∞	...
t271	γ-**Tocopherol**	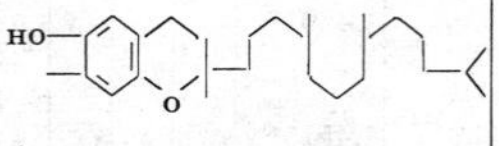	416.66	pa ye, $[\alpha]^{20}_D$ −2.4 (al)	−3	200–$10^{0.1}$	...	...	i	∞	∞	∞	...	chl ∞	...
t272	δ-**Tocopherol**	8-Methyltocol. 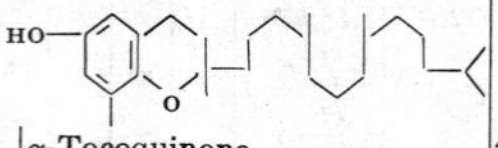	402.64	pa ye, $[\alpha]^{25}_{546}$ +3.4 (al, c=15.5)	...	...	...	...	i	v	v	v	...	chl v	...
t273	α-**Tocopherolquinone**	α-Tocoquinone. 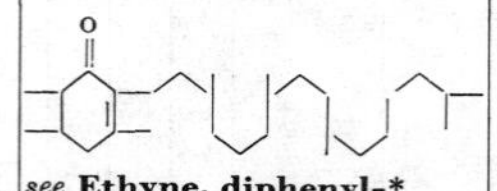	446.71	ye	...	$120^{0.002}$	...	...	...	...	s	...	...	peth s MeOH δ, s^h	**Am73**, 5148
—	**Tolan**	*see* **Ethyne, diphenyl-***													
t274	**Tolbutamide**	1-Butyl-3-(*p*-tolylsulfonyl)urea. 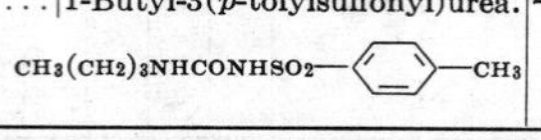 $CH_3(CH_2)_3NHCONHSO_2—C_6H_4—CH_3$	270.34	orh	127–9	...	1.245^{25}	...	$δ^h$	s	s	...	...	chl s	...

For explanations, symbols and abbreviations see beginning of table.

No.	Name	Synonyms and Formula	Mol. wt.	Crystalline form, color and specific rotation	m.p. °C	b.p. °C	Density	n_D	Solubility: w	al	eth	ace	bz	other solvents	Ref.
	Tolidine														
—	**Tolidine**	*see* **Biphenyl, diamino(dimethyl)-**													
—	**α-Tolualdehyde**	*see* **Acetaldehyde, phenyl-**													
—	**Tolualdehyde**	*see* **Benzaldehyde, methyl-**													
t275	**Toluene**	Methylbenzene*. Phenylmethane*. $^{\alpha}CH_3$ (ring positions 1–6)	92.13		−95	110.6	0.8669^{20}	1.4961^{20}	i	∞	∞	s	∞	CS_2 s lig s	**B5**, 280
t276	**—,2-acetamido-3-bromo-**	$C_9H_{10}BrNO$. *See* t275	228.10	nd (bz)	166								s^h		**B12²**, 455
t277	**—,2-acetamido-4-bromo-**	$C_9H_{10}BrNO$. *See* t275	228.10	nd (bz)	165.5								v^h		**B12²**, 456
t278	**—,2-acetamido-5-bromo-**	$C_9H_{10}BrNO$. *See* t275	228.10	nd (dil al or lig)	159–60				s^h	v					**B12²**, 456
t279	**—,4-acetamido-2-bromo-**	$C_9H_{10}BrNO$. *See* t275	228.10	(bz or dil al)	113					v			v^h		**B12¹**, 436
t280	**—,4-acetamido-3-bromo-**	$C_9H_{10}BrNO$. *See* t275	228.10	nd (al)	118					v^h				xyl v^h	**B12²**, 532
t281	**—,α-amino-**	Benzylamine. $C_6H_5CH_2NH_2$	107.15			185^{770} 84^{24}	0.9813^{20}_{4}	1.5402^{20}	∞	∞	∞				**B12²**, 540
t282	**—,—,hydrochloride**	Benzylamine hydrochloride. $C_6H_5CH_2NH_2.HCl$	143.62	pl (al)	255–8				v	s^h				chl i	**B12²**, 540
t283	**—,2-amino-**	*o*-Methylaniline. *o*-Toluidine. C_7H_9N. *See* t275	107.15		−27.7 (−16.25)	199.7^{760} 81^{10}	0.9989^{20}_{4}	1.5688^{20}	δ	∞	∞				**B12²**, 429
t285	**—,—,hydrochloride**	*o*-Toluidine hydrochloride. $C_7H_9N.HCl$. *See* t275	143.62	mcl pr (cold w), rh pym (w^h)	215	242.2^{760}			v	s	i				**B12²**, 432
t286	**—,3-amino-**	3-Methylaniline. *m*-Toluidine. C_7H_9N. *See* t275	107.15		−43.6 (−31.5)	203.2^{760}	0.9916^{18}_{4}	1.5686^{20}	δ	∞	∞				**B12²**, 463
t287	**—,—,hydrochloride**	*m*-Toluidine hydrochloride. $C_7H_9N.HCl$. *See* t275	143.62	lf	228	250^{760}			v	v					**B12**, 856
t288	**—,4-amino-**	4-Methylaniline. *p*-Toluidine. C_7H_9N. *See* t275	107.15	lf	43.5	200.4^{760}	0.9659^{45}_{4}	1.5534^{45}	δ	v				Py v	**B12²**, 482
t289	**—,—,hydrochloride**	*p*-Toluidine hydrochloride. $C_7H_9N.HCl$. *See* t275	143.62	mcl nd	243	257.5^{760}			v	v	i		i	CS_2 i	**B12²**, 487
—	**—,amino-*N*-benzylidene-**	*see* **Benzaldehyde, imine, *N*-tolyl-**													
t291	**—,α-amino-3-bromo-**	*m*-Bromobenzylamine. C_7H_8BrN. *See* t275	186.07			244–5					v			ac s	**B12²**, 575
t292	**—,α-amino-4-bromo-**	*p*-Bromobenzylamine. C_7H_8BrN. *See* t275	186.07		20	250–1 $126–7^{15}$					v			ac s	**B12²**, 575
t293	**—,2-amino-3-bromo-**	C_7H_8BrN. *See* t275	186.07			$105–7^{2-3}$					v			ac s	**B12²**, 455
t294	**—,2-amino-4-bromo-**	C_7H_8BrN. *See* t275	186.07	lf	32	253–7 δd				s				ac s	**B12²**, 456
t295	**—,2-amino-5-bromo-**	C_7H_8BrN. *See* t275	186.07	(al)	59.5	240			δ	v				aa v	**B12²**, 456
t296	**—,3-amino-4-bromo-**	C_7H_8BrN. *See* t275	186.07	pr	46					v				ac s	**B12²**, 474
t297	**—,3-amino-5-bromo-**	C_7H_8BrN. *See* t275	186.07		37–8	255–60 $150–1^{15}$				s				ac s	**B12²**, 474
t298	**—,4-amino-2-bromo-**	C_7H_8BrN. *See* t275	186.07		25–6	254–7					v			con ac s	**B12²**, 532
t299	**—,4-amino-3-bromo-**	C_7H_8BrN. *See* t275	186.07	lf	26	240 $120–2^{30}$	1.510^{20}		i	s	s				**B12²**, 532
t300	**—,5-amino-2-bromo-**	C_7H_8BrN. *See* t275	186.07	pl (50% al), (al)	80	240			i	v					**B12²**, 474
t301	**—,4-amino-5-bromo-2-nitro-**	$C_7H_7BrN_2O_2$. *See* t275	231.06	br to pa ye nd (al or aa)	121										**B12¹**, 441
t302	**—,2-amino-4-chloro-**	C_7H_8ClN. *See* t275	141.60		26	237^{722}									**B12²**, 453
t303	**—,2-amino-5-chloro-**	C_7H_8ClN. *See* t275	141.60	pl (al)	29–30	241				s^h					**B12²**, 454
t304	**—,2-amino-6-chloro-**	C_7H_8ClN. *See* t275	141.60			242–4			s^h	s	i		i		**B12²**, 455
t305	**—,3-amino-4-chloro-**	C_7H_8ClN. *See* t275	141.60	pl	29–30	228–30									**B12²**, 473
t306	**—,4-amino-2-chloro-**	C_7H_8ClN. *See* t275	141.60		26	$242–4^{760}$									**B12²**, 530

For explanations, symbols and abbreviations see beginning of table.

No.	Name	Synonyms and Formula	Mol. wt.	Crystalline form, color and specific rotation	m.p. °C	b.p. °C	Density	n_D	Solubility						Ref.
									w	al	eth	ace	bz	other solvents	
	Toluene														
t307	**—,4-amino-3-chloro-**	C_7H_8ClN. *See* t275	141.60		7	223–4									**B12**, 989
t308	**—,5-amino-2-chloro-**	C_7H_8ClN. *See* t275	141.60		83–4	241			δ	v				lig δ	**B12**[1], 404
t309	**—,4-amino-2,6-dinitro-**	$C_7H_7N_3O_4$. *See* t275	197.16	ye nd (w or aa)	171				δ^h	v			s	chl s	**B12**[1], 442
t310	**—,2-amino-3-hydroxy-**	2-Amino-*m*-cresol. C_7H_9NO. *See* t275	123.16	pl	148–50	sub									**B13**[2], 324
t311	**—,2-amino-4-hydroxy-**	3-Amino-*p*-cresol. C_7H_9NO. *See* t275	123.16	(w or eth), lf (sub)	157–9	sub									**B13**[2], 337
t312	**—,2-amino-5-hydroxy-**	4-Amino-*m*-cresol. C_7H_9NO. *See* t275	123.16	(bz)	179				δ	v	v				**B13**[2], 330
t313	**—,2-amino-6-hydroxy-**	3-Amino-*o*-cresol. C_7H_9NO. *See* t275	123.16	nd (w)	129				δ		δ				**B13**, 579
t314	**—,3-amino-2-hydroxy-**	6-Amino-*o*-cresol. C_7H_9NO. *See* t275	123.16	pl (w)	89				s^h	s	s	s	s	MeOH, chl s CCl_4 s peth i	**B13**[1], 212
t315	**—,3-amino-4-hydroxy-**	2-Amino-*p*-cresol. C_7H_9NO. *See* t275	123.16	rh (bz), sc (eth), lf or nd (sub)	137	sub			δ	v	v		δ	chl v lig i	**B13**[2], 338
t316	**—,3-amino-5-hydroxy-**	5-Amino-*m*-cresol. C_7H_9NO. *See* t275	123.16		79	245									
t318	**—,4-amino-3-hydroxy-**	6-Amino-*m*-cresol. C_7H_9NO. *See* t275 .	123.16	nd (bz)	162d				s^h	s	s	s		MeOH s lig i	**B13**, 590
t319	**—,5-amino-2-hydroxy-**	4-Amino-*o*-cresol. C_7H_9NO. *See* t275-	123.16	nd or lf (bz)	175	sub			δ	s	s		δ		**B13**[2], 319
t320	**—,α-amino-2-hydroxy-5-nitro-**	2-Hydroxy-5-nitrobenzylamine. $C_7H_8N_2O_3$. *See* t275	168.16	ye nd or lf (w or dil NH_3)	253d				s^h	s^h					**B13**, 587
t321	**—,α-amino-4-hydroxy-3-nitro**	4-Hydroxy-3-nitrobenzylamine. $C_7H_8N_2O_3$. *See* t275	168.16	og red nd (w+1)	225d				s^h						**B13**, 610
t322	**—,3-amino-2-hydroxy-5-nitro-**	$C_7H_8N_2O_3$. *See* t275	168.16	red-br nd (bz)	165d					s^h			s^h		**B13**[2], 319
t323	**—,3-amino-4-hydroxy-5-nitro-**	$C_7H_8N_2O_3$. *See* t275	168.16	red-br (al)	119					s^h					**B13**[2], 345
t324	**—,5-amino-2-hydroxy-3-nitro-**	$C_7H_8N_2O_3$. *See* t275	168.16	br-red nd (al)	118					s^h					**B13**, 578
t325	**—,5-amino-4-hydroxy-2-nitro-**	$C_7H_8N_2O_3$. *See* t275	168.16	ye og (al)	*ca.* 200d				s^h	s^h				dil HCl s	**B13**[2], 346
t326	**—,6-amino-3-hydroxy-2-nitro-**	$C_7H_8N_2O_3$. *See* t275	168.16	red-br nd (al)	201					s^h				dil ac v dil alk v	**B13**, 595
t327	**—,2-amino-4-iodo-**	C_7H_8IN. *See* t275	233.05	nd (aq al)	48–9	273d				s				dil ac s aa s	**B12**[1], 391
t328	**—,2-amino-5-iodo-**	C_7H_8IN. *See* t275	233.05	nd (dil al), pr (gasoline)	91–2				s^h	v	v		v	aa v lig v	**B12**[2], 457
t329	**—,3-amino-2-iodo-**	C_7H_8IN. *See* t275	233.05	pr	41–2				i	v	v			os s	**B12**[2], 474
t330	**—,3-amino-4-iodo-**	C_7H_8IN. *See* t275	233.05	nd (dil al), br in air	48 (38)					s	s			chl s	**B12**[2], 475
t331	**—,3-amino-5-iodo-**	C_7H_8IN. *See* t275	233.05	nd (peth)	78					s				oos s	**B12**[1], 406
t332	**—,4-amino-2-iodo-**	C_7H_8IN. *See* t275	233.05	nd (dil al or peth)	39					v				oos v aa v	**B12**[2], 533
t333	**—,4-amino-3-iodo-**	C_7H_8IN. *See* t275	233.05	pr	40	d				v	v		v	chl v peth v	**B12**[2], 533
t334	**—,5-amino-2-iodo-**	C_7H_8IN. *See* t275	233.05	lf or pl (al or peth)	46				δ^h	s	s		v	aa v lig v	**B12**[2], 474
t335	**—,α-amino-3-methoxy-**	*m*-Methoxybenzylamine. $C_8H_{11}NO$. *See* t275	137.18			140^{37}			i	s	s				**B13**[2], 335
t336	**—,α-amino-4-methoxy-**	*p*-Methoxybenzylamine. $C_8H_{11}NO$. *See* t275	137.18		$236–7^{760}$ $133–4^{33}$	1.050^{15}			v	v	v				**B13**[2], 347
t337	**—,2-amino-3-methoxy-**	$C_8H_{11}NO$. *See* t275	137.18	nd (w)	31	$119–21^{16}$			s^h	v				os v	**B13**[2], 324
t338	**—,2-amino-4-methoxy-**	$C_8H_{11}NO$. *See* t275	137.18	nd (w)	47	253 140^{20}			s^h		v				**B13**[2], 337
t339	**—,2-amino-5-methoxy-**	$C_8H_{11}NO$. *See* t275	137.18	(lig)	29–30 (13–4)	248–9 $146–7^{23}$				s				lig s^h	**B13**[2], 330
t340	**—,3-amino-4-methoxy-**	$C_8H_{11}NO$. *See* t275	137.18	nd or lf (al, lig or peth)	48.5	235			s	v	v		v	peth s^h	**B13**[2], 338
t341	**—,4-amino-2-methoxy-**	$C_8H_{11}NO$. *See* t275	137.18		58	250–2			δ	v			v	lig v	**B13**[1], 213
t342	**—,4-amino-3-methoxy-**	$C_8H_{11}NO$. *See* t275	137.18	pa ye		237–9 $179–80^{46}$								oos s	**B13**[2], 326
t343	**—,5-amino-2-methoxy-**	C_8H_{11} NO. *See* t275	137.18		59–60				δ v^h	v	v	v	v		**B13**[2], 320

For explanations, symbols and abbreviations see beginning of table.

No.	Name	Synonyms and Formula	Mol. wt.	Crystalline form, color and specific rotation	m.p. °C	b.p. °C	Density	n_D	Solubility w	al	eth	ace	bz	other solvents	Ref.
	Toluene														
t344	**—,2-amino-3-nitro-**	$C_7H_8N_2O_2$. *See* t275	152.15	og-ye-pr (45% al)	96	...	1.900^{100}_{4}	...	δ	s	s	...	s	chl s	**B12²**, 458
t345	**—,2-amino-4-nitro-**	$C_7H_8N_2O_2$. *See* t275	152.15	ye mcl pr (al)	107 (129)	...	...	...	δ	s	s	s	...	...	**B12²**, 459
t346	**—,2-amino-5-nitro-**	$C_7H_8N_2O_2$. *See* t275	152.15	ye mcl pr or nd (w), (al or lig)	134–5 (107)	1.1586^{140}_{4}	...	...	$δ^h$	v	...	...	s	aa s	**B12²**, 459
t347	**—,2-amino-6-nitro-**	$C_7H_8N_2O_2$. *See* t275	152.15	ye rh nd (w), lf (al)	97 (91)	305d	...	...	$δ^h$	v	v	...	v	...	**B12²**, 460
t348	**—,3-amino-2-nitro-**	$C_7H_8N_2O_2$. *See* t275	152.15	ye-og or og-ye pr or nd (bz-peth)	107–8 (54)	...	...	...	δ	v	s	...	...	...	**B12²**, 476
t349	**—,3-amino-4-nitro-**	$C_7H_8N_2O_2$. *See* t275	152.15	ye lf (w), pl (dil al)	112 (109)	...	...	...	s^h	v	s	...	s	chl s	**B12²**, 476
t350	**—,3-amino-5-nitro-**	$C_7H_8N_2O_2$. *See* t275	152.15	ye-red or red-br nd	98	...	...	...	δ	s	v	...	s	...	**B12²**, 476
t352	**—,4-amino-2-nitro-**	$C_7H_8N_2O_2$. *See* t275	152.15	ye nd (w)	77–8	...	...	...	δ s^h	v^h	s	...	s	CS_2 δ	**B12²**, 534
t353	**—,4-amino-3-nitro-**	$C_7H_8N_2O_2$. *See* t275	152.15	red nd (dil al), mcl pr (al)	117	1.312^{17}	...	...	$δ^h$	s	...	...	...	...	**B12²**, 535
t354	**—,5-amino-2-nitro-**	$C_7H_8N_2O_2$. *See* t275	152.15	lt ye (w or dil al)	138 (133–4)	...	...	...	δ s^h	s	s	...	...	ac v	**B12²**, 476
t355	**—,3-amino-α,α,α-trifluoro-**	*m*-Aminobenzotrifluoride. $C_7H_6F_3N$. *See* t275	161.13	...	...	187.5^{764} $74–5^{10}$	...	...	δ	s	s	...	...	...	**B12²**, 473
t356	**—,α-azido-**	Benzyl azide. $C_6H_5CH_2N_3$	133.16	...	...	108.23	1.0655^{25}	1.5341^{25}	i	∞	∞	...	...	...	**B5²**, 274
t357	**—,α-bromo-**	Benzyl bromide. $C_6H_5CH_2Br$	171.04	pr	−3.9	201^{760} 114^{15}	1.4380^{22}_{0}	...	i	∞	∞	...	...	...	**B5²**, 237
t358	**—,2-bromo-**	*o*-Tolyl bromide. C_7H_7Br. *See* t275	171.04	...	−26	181^{760}	1.4222^{28}_{4}	...	i	v	v	...	v	...	**B5**, 304
t359	**—,3-bromo-**	*m*-Tolyl bromide. C_7H_7Br. *See* t275	171.04	...	−39.8	183.7	1.4019^{20}_{4}	1.551	i	s	∞	...	...	...	**B5**, 305
t360	**—,4-bromo-**	*p*-Tolyl bromide. C_7H_7Br. *See* t275	171.04	(al)	28.5	$184–5^{760}$	1.3898^{20}_{4}	1.5490	i	s	s	...	s	...	**B5**, 305
t361	**—,α-bromo-2-chloro-**	*o*-Chlorobenzyl bromide. C_7H_6BrCl. *See* t275	205.49	...	...	120^{10}	...	...	d^h	...	...	...	s	chl s	**B5²**, 238
t362	**—,2-bromo-α-chloro-**	*o*-Bromobenzyl chloride. C_7H_6BrCl. *See* t275	205.49	...	...	$110–1^{15}$	...	...	i	v	v	...	...	...	**B5²**, 238
t363	**—,3-bromo-α-chloro-**	*m*-Bromobenzyl chloride. C_7H_6BrCl. *See* t275	205.49	...	22–3	119^{18}	...	...	d	s	...	...	...	...	**B5²**, 238
t364	**—,4-bromo-α-chloro-**	*p*-Bromobenzyl chloride. C_7H_6BrCl. *See* t275	205.49	nd (al or peth)	40	236^{760} $110–1^{9}$	...	...	i	v^h	v	...	...	...	**B5²**, 238
t365	**—,5-bromo-2,α-dihydroxy-**	5-Bromosaligenin, Bromosalisol. $C_7H_7BrO_2$. *See* t275	203.04	lf (bz)	113 (109)	...	...	...	δ v^h	v	v	...	s	chl s AcOEt v	**B6²**, 879
t366	**—,2-bromo-α-hydroxy-**	*o*-Bromobenzyl alcohol. C_7H_7BrO. *See* t275	187.05	nd	80	...	...	...	s^h	v	v	...	...	lig s	**B6²**, 423
t367	**—,2-bromo-4-hydroxy-**	3-Bromo-*p*-cresol. C_7H_7BrO. *See* t275	187.05	nd (peth)	55–6	245–7	...	...	δ	v	...	...	v	MeOH v, peth s^h	**B6²**, 384
t368	**—,2-bromo-5-hydroxy-**	4-Bromo-*m*-cresol. C_7H_7BrO. *See* t275	187.05	nd (lig or w)	62	$137–43^{16}$	...	...	δ	...	...	...	...	peth δ Py v	**B6²**, 357
t369	**—,2-bromo-6-hydroxy-**	3-Bromo-*o*-cresol. C_7H_7BrO. *See* t275	187.05	nd	95	...	...	...	δ	s	...	...	s	...	**B6**, 360
t370	**—,3-bromo-4-hydroxy-**	2-Bromo-*p*-cresol. C_7H_7BrO. *See* t275	187.05	...	...	218–9	1.5468^{25}_{25}	...	...	s	...	...	s	...	**B6²**, 384
t371	**—,3-bromo-5-hydroxy-**	5-Bromo-*m*-cresol. C_7H_7BrO. *See* t275	187.05	nd (w)	56–7	$161–2^{28}$	...	...	δ	v	v	...	...	...	**B6²**, 357
t372	**—,4-bromo-α-hydroxy-**	*p*-Bromobenzyl alcohol. C_7H_7BrO. *See* t275	187.05	nd (lig)	76	...	...	...	...	v	v	...	v	CS_2 v	**B6²**, 423
t373	**—,4-bromo-2-hydroxy-**	5-Bromo-*o*-cresol. C_7H_7BrO. *See* t275	187.05	nd (lig or peth)	80	...	...	...	...	...	...	...	...	oos v lig δ s^h	**B6²**, 333
t374	**—,4-bromo-3-hydroxy-**	6-Bromo-*m*-cresol. C_7H_7BrO. *See* t275	187.05	...	38	...	...	...	...	v	...	...	...	os v	**B6²**, 357
t375	**—,5-bromo-2-hydroxy-**	4-Bromo-*o*-cresol. C_7H_7BrO. *See* t275	187.05	nd (al or peth)	63–4	235 $137–43^{18}$	...	...	...	v	...	...	...	os v	**B6²**, 333
t376	**—,α-bromo-2-nitro-**	*o*-Nitrobenzyl bromide. $C_7H_6BrNO_2$. *See* t275	216.04	(dil al)	46–7	...	...	...	i	s	s	...	s	...	**B5¹**, 164
t377	**—,α-bromo-3-nitro-**	*m*-Nitrobenzyl bromide. $C_7H_6BrNO_2$. *See* t275	216.04	nd or pl	59–9	...	...	...	i	s	...	...	...	...	**B5**, 334
t378	**—,α-bromo-4-nitro-**	*p*-Nitrobenzyl bromide. $C_7H_6BrNO_2$. *See* t275	216.04	nd (al)	99–100	...	...	...	δ	v	v	...	...	...	**B5**, 334

For explanations, symbols and abbreviations see beginning of table.

No.	Name	Synonyms and Formula	Mol. wt.	Crystalline form, color and specific rotation	m.p. °C	b.p. °C	Density	n_D	Solubility w	al	eth	ace	bz	other solvents	Ref.
	Toluene														
t379	**—,2-bromo-3-nitro-**	$C_7H_6BrNO_2$. *See* t275	216.04	pr	41–2	157^{22}			i	v					**B5²**, 255
t380	**—,2-bromo-4-nitro-**	$C_7H_6BrNO_2$. *See* t275	216.04	nd (al)	78					δ v^h	v			CS_2 v	**B5²**, 256
t381	**—,2-bromo-5-nitro-**	$C_7H_6BrNO_2$. *See* t275	216.04	(al)	78					δ v^h	v			CS_2 v	**B5²**, 256
t382	**—,2-bromo-6-nitro-**	$C_7H_6BrNO_2$. *See* t275	216.04	pa ye nd (dil al)	42	143^{22}			i	s v^h					**B5²**, 255
t383	**—,3-bromo-2-nitro-**	$C_7H_6BrNO_2$. *See* t275	216.04	pa ye nd	27	$129–30^{10}$			i	δ	δ				**B5²**, 255
t384	**—,3-bromo-4-nitro-**	$C_7H_6BrNO_2$. *See* t275	216.04	pa ye pr or nd (MeOH)	37	$156–8^{19}$			i	v	v				**B5²**, 256
t385	**—,3-bromo-5-nitro-**	$C_7H_6BrNO_2$. *See* t275	216.04	pa ye nd (MeOH)	84	269–70								MeOH δ	**B5²**, 256
t386	**—,4-bromo-2-nitro-**	$C_7H_6BrNO_2$. *See* t275	216.04	pa ye nd (dil al)	46–7	256–7 130^{12}				δ	δ			con sulf s	**B5²**, 255
t387	**—,4-bromo-3-nitro-**	$C_7H_6BrNO_2$. *See* t275	216.04	pa ye nd (MeOH)	35			1.5682^{50}	i					MeOH s^h con sulf s	**B5²**, 256
t388	**—,5-bromo-2-nitro-**	$C_7H_6BrNO_2$. *See* t275	216.04	(al)	56	267			i	δ s^h					**B5²**, 255
t389	**—,3-butoxy-**	Butyl *m*-tolyl ether. $C_{11}H_{16}O$. *See* t275	164.25			229.2	0.9407^{0}_{0}	1.4970^{20}							**B6**, 377
t390	**—,4-butoxy-**	Butyl *p*-tolyl ether. $C_{11}H_{16}O$. *See* t275	164.25			229.5	0.9419^{0}_{0}								**B6**, 393
t391	**—,α-chloro-**	Benzyl chloride. $C_6H_5CH_2Cl$.	126.58		−39	179.3^{760} 66^{11}	1.100^{25}_{4}	1.5391^{20}	i d^h	∞	∞			chl ∞	**B5²**, 227
t392	**—,2-chloro-**	*o*-Tolyl chloride. C_7H_7Cl. *See* t275	126.58		−34	159	1.0817^{20}_{4}	1.5238^{20}	i	s	∞		s	chl s	**B5**, 290
t393	**—,3-chloro-**	*m*-Tolyl chloride. C_7H_7Cl. *See* t275	126.58		−47.8	162	1.0722^{20}_{4}	1.5214^{19}	i	s	∞		s	chl s	**B5**, 291
t394	**—,4-chloro-**	*p*-Tolyl chloride. C_7H_7Cl. *See* t275	126.58		7.5	162	1.0697^{20}_{4}	1.5199^{19}	i	s	∞			chl s aa s	**B5**, 292
t395	**—,α-chloro-α,α-difluoro-**	Benzodifluoro chloride. $C_6H_5CClF_2$	162.57			142.6^{770}	1.254^{13}		i	s	s				**B5**, 295
t397	**—,2-chloro-α-hydroxy-**	*o*-Chlorobenzyl alcohol. C_7H_7ClO. *See* t275	142.58	lf or nd (dil al)	74	230			δ	v	v			lig v^h	**B6²**, 422
t398	**—,2-chloro-3-hydroxy-**	2-Chloro-*m*-cresol. C_7H_7ClO. *See* t275	142.58	pr	55–6	196^{760}			δ					aa v	**B6²**, 355
t399	**—,2-chloro-4-hydroxy-**	3-Chloro-*p*-cresol. C_7H_7ClO. *See* t275	142.58	nd	55–6	228^{760}			δ	v	v			aa v	**B6**, 402
t400	**—,2-chloro-5-hydroxy-**	4-Chloro-*m*-cresol. C_7H_7ClO. *See* t275	142.58		66	235–8			δ						**B6²**, 355
t401	**—,2-chloro-6-hydroxy-**	3-Chloro-*o*-cresol. C_7H_7ClO. *See* t275	142.58	lo nd (w)	86	225			δ s^h	s	s		s		**B6**, 359
t402	**—,3-chloro-2-hydroxy-**	6-Chloro-*o*-cresol. C_7H_7ClO. *See* t275	142.58			$188–9^{740}$					s				**B6²**, 332
t403	**—,3-chloro-4-hydroxy-**	2-Chloro-*p*-cresol. C_7H_7ClO. *See* t275	142.58			$195–6^{760}$	1.2106^{25}_{25}	1.5200^{27}	δ	s	s		s	aa s	**B6²**, 383
t404	**—,4-chloro-α-hydroxy-**	*p*-Chlorobenzyl chloride. C_7H_7ClO. *See* t275	142.58	nd (w), pr (bz or bz-lig)	75	235			$δ^h$	v	v				**B6²**, 423
t405	**—,4-chloro-2-hydroxy-**	5-Chloro-*o*-cresol. C_7H_7ClO. *See* t275	142.58	nd (peth or gasoline)	73–4					v			v	alk v aa v	**B6²**, 332
t406	**—,4-chloro-3-hydroxy-**	6-Chloro-*m*-cresol. C_7H_7ClO. *See* t275	142.58	pr	45–6	196	1.215^{15}		s						**B6²**, 355
t407	**—,5-chloro-2-hydroxy-**	4-Chloro-*o*-cresol. C_7H_7ClO. *See* t275	142.58	nd	48	223			δ						**B6²**, 332
t408	**—,α-chloro-4-hydroxy-3-nitro-**	$C_7H_6ClNO_3$. *See* t275	187.59	ye nd (bz, lig or al), lf (peth)	75					s^h				aa s^h lig s^h	**B6**, 413
t409	**—,α-chloro-4-methoxy-**	*p*-Methoxybenzyl chloride. C_8H_9ClO. *See* t275	156.62			$116–20^{15}$	1.072^{0}								**B6²**, 383
t410	**—,α-chloro-2-nitro-**	*o*-Nitrobenzyl chloride. $C_7H_6ClNO_2$. *See* t275	171.58	(lig)	48–9			1.5557^{62}	i	s	s			aa s	**B5**, 327
t411	**—,α-chloro-3-nitro-**	*m*-Nitrobenzyl chloride. $C_7H_6ClNO_2$. *See* t275	171.58	nd (lig)	45–7	175–$83^{30–5}$		1.5577^{62}	i	s	s			aa s	**B5**, 329
t412	**—,α-chloro-4-nitro-**	*p*-Nitrobenzyl chloride. $C_7H_6ClNO_2$. *See* t275	171.58	pl or nd (al)	71			1.5647^{62}	i	s	s			MeOH s	**B5**, 329
t413	**—,2-chloro-4-nitro-**	$C_7H_6ClNO_2$. *See* t275	171.58	nd	65–5	260^{760}		1.5470^{69}	i^h	s	s				**B5²**, 253

For explanations, symbols and abbreviations see beginning of table.

No.	Name	Synonyms and Formula	Mol. wt.	Crystalline form, color and specific rotation	m.p. C	b.p. °C	Density	n_D	Solubility						Ref.
									w	al	eth	ace	bz	other solvents	
	Toluene														
t414	**—,2-chloro-6-nitro-**	$C_7H_6ClNO_2$. *See* t275	171.58		37	237–41		1.5377^{69}	i						**B5**^{2}, 251
t415	**—,3-chloro-4-nitro-**	$C_7H_6ClNO_2$. *See* t275	171.58	lt ye nd	22	146^{19}			i						**B5**^{2}, 253
t416	**—,3-chloro-5-nitro-**	$C_7H_6ClNO_2$. *See* t275	171.58	nd	58.4			1.5404^{69}	i						**B5**^{1}, 162
t417	**—,4-chloro-2-nitro-**	$C_7H_6ClNO_2$. *See* t275	171.58	mcl nd	37	115.5^{11}	1.2559^{80}		i	s^h	s				**B5**^{2}, 251
t418	**—,4-chloro-3-nitro-**	$C_7H_6ClNO_2$. *See* t275	171.58		7	260^{745} 118^{15}			i						**B5**^{2}, 252
t419	**—,5-chloro-2-nitro-**	$C_7H_6ClNO_2$. *See* t275	171.58	ye	24.9			1.5495^{69}	i						**B5**^{1}, 162
t420	**—,2,3-diamino-**	$C_7H_{10}N_2$. *See* t275	122.17		63–4	255			s	s	s			oos s	**B13**^{2}, 60
t421	**—,2,4-diamino-**	$C_7H_{10}N_2$. *See* t275	122.17	nd (w), pr (al)	99(71)	292 $148–50^{18}$			v^h	v	v				**B13**^{2}, 60
t422	**—,2,5-diamino-**	$C_7H_{10}N_2$. *See* t275	122.17	ta (bz)	64	273–4			v	v	v			aa δ v^h	**B13**^{2}, 62
t423	**—,2,6-diamino-**	$C_7H_{10}N_2$. *See* t275	122.17	pr (bz or w)	106				s	s					**B13**^{2}, 64
t424	**—,3,4-diamino-**	$C_7H_{10}N_2$. *See* t275	122.17	lf (lig)	88.5	sub 265			v						**B13**^{2}, 64
t425	**—,3,5-diamino-**	$C_7H_{10}N_2$. *See* t275	122.17		<0	283–5			v	s	s				**B13**, 164
t426	**—,—,hydrochloride**	$C_7H_{10}N_2.2HCl$. *See* t275	195.09	nd	255–60d				v		i				**B13**, 164
t427	**—,α,α-dibromo-**	Benzal bromide. Benzylidine bromide. $C_6H_5CHBr_2$	249.94			156^{23}	1.51^{15}	1.541^{20}	i	∞	∞				**B5**, 308
t428	**—,α,2-dibromo-**	*o*-Bromobenzyl bromide. $C_7H_6Br_2$. *See* t275	249.94		31	129^{19}			d^h	s	s			CS_2 s aa s	**B5**^{2}, 239
t429	**—,α,3-dibromo-**	*m*-Bromobenzyl bromide. $C_7H_6Br_2$. *See* t275	249.94	nd or lf	40				d^h	δ	v			CS_2 v aa v	**B5**^{2}, 239
t430	**—,α,4-dibromo**	*p*-Bromobenzyl bromide. $C_7H_6Br_2$. *See* t275	249.94	nd (MeOH or al)	63				δ d^h	δ v^h	v		v	CS_2 v aa v	**B5**^{2}, 239
t431	**—,2,5-dibromo-**	$C_7H_6Br_2$. *See* t275	249.94			236	1.8127^{19}		i						**B5**^{1}, 155
t432	**—,3,5-dibromo-**	$C_7H_6Br_2$. *See* t275	249.94	nd	39	246			i						**B5**^{2}, 239
t433	**—,2,4-dibromo-6-hydroxy-**	3,5-Dibromo-*o*-cresol. $C_7H_6Br_2O$. *See* t275	265.94	nd (peth)	98–101	$283–7^{758}$								peth s^h	**B6**^{2}, 334
t434	**—,3,5-dibromo-2-hydroxy-**	4,6-Dibromo-*o*-cresol. $C_7H_6Br_2O$. *See* t275	265.94	nd (peth or 50 % al)	58	$263–6^{745}$d			δ					oos v	**B6**^{2}, 334
t435	**—,3,6-dibromo-2-hydroxy-**	3,6-Dibromo-*o*-cresol. $C_7H_6Br_2O$. *See* t275	265.94		38	255–60								os v	**B6**^{1}, 176
t436	**—,α,α-dibromo-4-nitro-**	*p*-Nitrobenzylidene bromide. $C_7H_5Br_2NO_2$. *See* t275	294.94	nd (al)	84				i	v	v				**B5**, 336
t437	**—,α,α-dichloro-**	Benzal chloride. Benzylidene chloride. $C_6H_5CHCl_2$	161.03		−16.4	205.2^{760}	1.2557^{14}	1.5502^{20}	i	s	s				**B5**^{2}, 232
t438	**—,α,2-dichloro-**	*o*-Chlorobenzyl chloride. $C_7H_6Cl_2$. *See* t275	161.03		−17	207^{760} $94–5^{10}$	1.2699^{0}_{4}								**B5**^{2}, 231
t439	**—,α,3-dichloro**	*m*-Chlorobenzyl chloride. $C_7H_6Cl_2$. *See* t275	161.03			$215–6^{753}$ $110–1^{25}$	1.2695^{15}_{4}		i d^h	v					**B5**^{2}, 231
t440	**—,α,4-dichloro-**	*p*-Chlorobenzyl chloride. $C_7H_6Cl_2$. *See* t275	161.03	nd	31	214^{758} 117^{20}			i	δ s^h	v		v	CS_2 s aa v	**B5**^{2}, 231
t441	**—,2,3-dichloro-**	$C_7H_6Cl_2$. *See* t275	161.03			$207–8^{760}$ $61–2^{3}$		1.5511^{20}	i				v		**B5**, 295
t442	**—,2,4-dichloro-**	$C_7H_6Cl_2$. *See* t275	161.03			196–7	1.2460^{20}_{20}		i						**B5**^{2}, 230
t443	**—,2,5-dichloro-**	$C_7H_6Cl_2$. *See* t275	161.03		5	200^{770}	1.2535^{20}_{20}		i				s		**B5**^{2}, 231
t444	**—,2,6-dichloro-**	$C_7H_6Cl_2$. *See* t275	161.03			198^{760}			i					chl s	**B5**^{2}, 231
t445	**—,3,4-dichloro-**	$C_7H_6Cl_2$. *See* t275	161.03			200^{741}	1.2512^{20}_{20}		i						**B5**^{1}, 152
t446	**—,3,5-dichloro-**	$C_7H_6Cl_2$. *See* t275	161.03		26	$201–2^{760}$			i						**B5**, 296
t447	**—,α,α-dichloro-α-fluoro-**	Benzodichloro fluoride. $C_6H_5CCl_2F$	179.02			178–80	1.3138^{11}	1.5180^{11}	d	s^h					**B5**, 298
t448	**—,2,4-dichloro-3-hydroxy-**	2,6-Dichloro-*m*-cresol. $C_7H_6Cl_2O$. *See* t275	177.03		27	241–2 $80–5^{4}$					s			chl v	**B6**^{2}, 356
t449	**—,2,4-dichloro-5-hydroxy-**	4,6-Dichloro-*m*-cresol. $C_7H_6Cl_2O$. *See* t275	177.03		71–2	235–6								chl v	....
t450	**—,2,6-dichloro-3-hydroxy-**	2,4-Dichloro-*m*-cresol. $C_7H_6Cl_2O$. *See* t275	177.03		58–9	$235–6^{745}$ $75–80^{4}$					s			chl v	**B6**^{2}, 356
t451	**—,3,5-dichloro-2-hydroxy-**	4,6-Dichloro-*o*-cresol. $C_7H_6Cl_2O$. *See* t275	177.03	nd (w or peth)	55	266.5 $73–8^{4}$			δ s^h	v	v			chl v CS_2 v	**B6**^{2}, 332
t452	**—,3,5-dichloro-4-hydroxy-**	2,6-Dichloro-*p*-cresol. $C_7H_6Cl_2O$. *See* t275	177.03	nd (lig)	39	$138–9^{28}$			δ	v	v			aa v	**B6**^{2}, 383
t453	**—,4,5-dichloro-2-hydroxy-**	4,5-Dichloro-*o*-cresol. $C_7H_6Cl_2O$. *See* t275	177.03	nd (peth)	101					v			v	peth δ aa v	**B6**^{2}, 333
t454	**—,α,α-dichloro-3-nitro-**	*m*-Nitrobenzylidene chloride. $C_7H_5Cl_2NO_2$. *See* t275	206.03	mcl (al)	65				i	v^h	v^h				**B5**, 332

For explanations, symbols and abbreviations see beginning of table.

No.	Name	Synonyms and Formula	Mol. wt.	Crystalline form, color and specific rotation	m.p. °C	b.p. °C	Density	n_D	Solubility						Ref.
									w	al	eth	ace	bz	other solvents	
	Toluene														
t455	**—,α,α-dichloro-4-nitro-**	*p*-Nitrobenzylidene chloride. $C_7H_5Cl_2NO_2$. *See* t275	206.03	pr (al)	46				i	v	v	...	...		**B5**, 332
t456	**—,2,5-diethoxy-**	$C_{11}H_{16}O_2$. *See* t275	180.24	nd (gasoline)	24–5	247–9	1.0134^{15}		...	∞	∞	...	∞	chl ∞	**B6**[1], 429
t457	**—,2(diethyl-amino)-**	*N,N*-Diethyl-*o*-toluidine. $C_{11}H_{17}N$. *See* t275	163.26	pr (w)	72–3	$208–9^{755}$			δ	s	s	...	...		**B12**[2], 436
t458	**—,4(diethyl-amino)-**	*N,N*-Diethyl-*p*-toluidine. $C_{11}H_{17}N$. *See* t275	163.26			229^{770}	0.9242^{16}		δ	∞	∞	...	...		**B12**[2], 492
t459	**—,α,α-difluoro-**	Benzal fluoride. Benzylidene fluoride. $C_6H_5CHF_2$	128.12			139.9	1.1357^{20}	1.4577^{20}	i	s	...	...	...		**B5**[2], 224
t460	**—,α,2-dihydroxy-**	Saligenin. $C_7H_8O_2$. *See* t275	124.13	lf (bz), rh (w) lf (sub)	86				s v^h	v	v	...	δ		**B6**[2], 877
t461	**—,—,glucoside**	Salicin. Saligenin-β, *D*-glucoside. HOCH₂–C–C–C–C–CH–O–(C₆H₄, CH₂OH) [ring O; OH H OH / H H OH H; β]	286.28	rh nd or lf $[\alpha]_D^{20}+62.6$	200–1	240d	1.434^{26}		s v^h	s	i	...	...	chl i	**B16**, 214
t461[1]	**—,α,3-dihydroxy-**	$C_7H_8O_2$. *See* t275	124.14	nd (bz)	73	300^{760}d	1.161^{25}		v^h	v	v	...	...	chl δ	**B6**[2], 881
t462	**—,α,4-dihydroxy-**	$C_7H_8O_2$. *See* t275	124.14	pr or nd (w)	124–5	252^{760}			v	v	v	...	δ	chl, peth i	**B6**[2], 882
t463	**—,2,3-dihydroxy-**	$C_7H_8O_2$. *See* t275	124.14	lf (bz)	68	241^{760} $134–6^{15}$			v	v	s	...	v	chl v	**B6**[2], 858
t464	**—,2,4-dihydroxy-**	Cresorcinol. $C_7H_8O_2$. *See* t275	124.14	(bz-peth)	105–7	270–5			s	s	s	...	δ	peth δ	**B6**[2], 859
t465	**—,2,5-dihydroxy-**	Toluhydroquinone. 2-Methyl hydroquinone. $C_7H_8O_2$. *See* t275	124.14	rh pl (bz)	124–5	163^{11}			v	v	v	...	δ		**B6**, 874
t466	**—,—,diacetate**	2,5-Diacetoxytoluene. $C_{11}H_{12}O_4$. *See* t275	208.21	nd (aa), pr (lig)	52				...	v	v	...	δ		**B6**[1], 429
t467	**—,3,4-dihydroxy-**	4-Homopyrocatechol. $C_7H_8O_2$. *See* t275	124.14	lf (bz-lig), pr (bz)	65	251^{766} $143–6^{26}$	1.1287_4^{74}	1.5425^{74}	v	v	v	...	...	os s	**B6**[2], 865
t468	**—,3,5-dihydroxy-**	Orcinol. $C_7H_8O_2$. *See* t275	124.14		107–8	289–90	1.290^4		s	v	v	...	...		
t469	**—,α,4-dihydroxy-3-methoxy-**	Vanillyl alcohol. $C_8H_{10}O_3$. *See* t275	154.17	pr (w), nd (bz)	115	d			s^h	s^h	s^h	...	...		**B6**[2], 1083
t470	**—,2,4-di-mercapto-**	Dithiocresorcinol. $C_7H_8S_2$. *See* t275	156.27		36–7	263^{760}			...	...	...	...	...		**B6**, 873
t471	**—,2,3-di-methoxy-**	$C_9H_{12}O_2$. *See* t275	152.19			$202–3^{760}$			...	s	s	...	...		**B6**[2], 859
t472	**—,2,4-di-methoxy-**	$C_9H_{12}O_2$. *See* t275	152.19			211			δ	s	s	...	...		**B6**[1], 428
t473	**—,3,4-di-methoxy-**	$C_9H_{12}O_2$. *See* t275	152.19	pr (eth)	22	$219–21^{760}$	1.0509_4^{25}	1.5257^{25}	i	...	...	...	...	os v	**B6**[2], 866
t474	**—,3,5-di-methoxy-**	$C_9H_{12}O_2$. *See* t275	152.19			$227–8^{751}$ 102^8	1.0478^{15}	1.5234^{20}	i	v	v	...	v	CS_2 v aa v	**B6**[2], 877
t475	**—,α(dimethyl-amino)-**	*N,N*-Dimethylbenzylamine. $C_6H_5CH_2N(CH_3)_2$	135.21			$180–2^{760}$ $73–4^{15}$	0.915_4^0		δ^h	δ	∞	...	...		**B12**[2], 545
t476	**—,2(dimethyl-amino)-**	*N,N*-Dimethyl-*o*-toluidine. $C_9H_{13}N$. *See* t275	135.21		−60	185.3^{760}			δ	∞	∞	...	...		**B12**[2], 435
t477	**—,3(dimethyl-amino)-**	*N,N*-Dimethyl-*m*-toluidine. $C_9H_{13}N$. *See* t275	135.21			212^{760}	0.9410_4^{20}	1.5492^{20}	...	∞	∞	...	...		**B12**[2], 466
t478	**—,4(dimethyl-amino)-**	*N,N*-Dimethyl-*p*-toluidine. $C_9H_{13}N$. *See* t275	135.21			211^{760}	0.9366_4^{20}	1.5366^{20}	i	∞	∞	...	...		**B12**[2], 491
t479	**—,2,4-dinitro-**	$C_7H_6N_2O_4$. *See* t275	182.14	ye nd or mcl pr	70–1	300 δd	1.321^{70}	1.442 (1.756)	i	s	s	...	s		**B5**, 339
t480	**—,2,5-dinitro-**	$C_7H_6N_2O_4$. *See* t275	182.14	nd (al)	52.5		1.282^{111}		...	v	...	...	v	CS_2 v	**B5**, 341
t481	**—,2,6-dinitro-**	$C_7H_6N_2O_4$. *See* t275	182.14	rh nd	66		1.2833^{111}	1.479 (1.734)	...	s	...	...	...		**B5**, 341
t482	**—,3,4-dinitro-**	$C_7H_6N_2O_4$. *See* t275	182.14	ye nd (CS_2)	61		1.2594^{111}		i	s	...	...	...	CS_2 δ	**B5**, 341
t483	**—,3,5-dinitro-**	$C_7H_6N_2O_4$. *See* t275	182.14	ye rh nd	92.6	sub	1.2772^{111}		δ	s	s	...	...	CS_2 s	**B5**, 341
t484	**—,2,4-dinitro-6-hydroxy-**	3,5-Dinitro-*o*-cresol. Sinox. $C_7H_6N_2O_5$. *See* t275	198.14	ye	85.8				i	s	s	...	...		
t484[1]	**—,3,5-dinitro-2-hydroxy-**	4,6-Dinitro-o-cresol. $C_7H_6N_2O_5$. *See* t275	198.14	ye pr or nd (al)	86.5				δ	s	s	s	...	peth δ	**B6**[2], 341
t485	**—,3,5-dinitro-4-hydroxy-**	2,6-Dinitro-*p*-cresol. $C_7H_6N_2O_5$. *See* t275	198.14	ye nd (eth or peth)	85				i	s	v	...	v		**B6**[2], 391
—	**—,α-ethoxy-**	*see* **Ether, benzyl ethyl**													
t486	**—,2-ethoxy-**	Ethyl *o*-tolyl ether. *o*-Cresol ethyl ether. $C_9H_{12}O$. *See* t275	136.20			184^{760} 70^{12}	0.9592_4^{13}	1.508^{13}	i	s	s	...	...		**B6**[2], 329

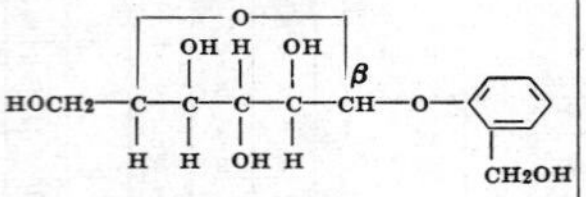

For explanations, symbols and abbreviations see beginning of table.

No.	Name	Synonyms and Formula	Mol. wt.	Crystalline form, color and specific rotation	m.p. °C	b.p. °C	Density	n_D	Solubility: w	al	eth	ace	bz	other solvents	Ref.
	Toluene														
t487	**—,3-ethoxy-**	Ethyl *m*-tolyl ether. *m*-Cresol ethyl ether. $C_9H_{12}O$. *See* t275	136.20			192	0.949^{20}	1.513^{20}	i	s	s				**B6**², 352
t488	**—,4-ethoxy-**	Ethyl *p*-tolyl ether. *p*-Cresol ethyl ether. $C_9H_{12}O$. *See* t275	136.20			188.9^{760}	0.9509^{18}_{4}	1.5058^{18}							**B6**², 376
t489	**—,2(ethylamino)-**	*N*-Ethyl-*o*-toluidine. $C_9H_{13}N$. *See* t275	135.21		<−15	212–4 95.5^{10}	0.948^{25}_{4}			s					**B12**², 435
t490	**—,3(ethylamino)-**	*N*-Ethyl-*m*-toluidine. $C_9H_{13}N$. *See* t275	135.21			215 $111–2^{20}$				s					**B12**², 466
t491	**—,4(ethylamino)-**	*N*-Ethyl-*p*-toluidine. $C_9H_{13}N$. *See* t275	135.21			217	0.9391^{16}			s					**B12**², 492
t492	**—,α-fluoro-**	Benzyl fluoride. $C_6H_5CH_2F$	110.13	nd (fr)	−35	139.8^{753}	1.0228^{25}_{4}	1.4892^{25}	d						**B5**², 224
t493	**—,2-fluoro-**	*o*-Tolyl fluoride. C_7H_7F. *See* t275	110.13		<−80	115	1.0014^{17}	1.4716^{17}	i	v	v				**B5**², 223
t494	**—,3-fluoro-**	*m*-Tolyl fluoride. C_7H_7F. *See* t275	110.13		−110.8	$114–5^{758}$	0.9986^{20}	1.4691^{20}	i	v	v				**B5**², 223
t495	**—,4-fluoro-**	*p*-Tolyl fluoride. C_7H_7F. *See* t275	110.13			116^{756}	1.0007^{15}	1.470	i	v	v				**B5**², 223
t496	**—,α,α,2,3,4,5,6-heptachloro-**	Pentachlorobenzylidene chloride. C_7HCl_7. *See* t275	333.26	lf (al)	119.5	334 199^{13}				δ v^h					**B5**¹, 153
t497	**—,α-hydroxy-**	Benzyl alcohol. $C_6H_5CH_2OH$	108.14		−15.3	205.35^{760} 93^{10}	1.0419^{20}_{4}	1.5396^{20}	s	s	s	s		MeOH, chl s	**46**², 403
t498	**—,2-hydroxy-**	*o*-Cresol. 2-Methylphenol*. C_7H_8O. *See* t275	108.14		30	191–2	1.0465^{20}_{4}	1.4453^{20}	s	v	v			oos s	**B6**², 322
t499	**—,3-hydroxy-**	*m*-Cresol. 3-Methylphenol*. C_7H_8O. *See* t275	108.14		11.1	202.8^{760} 80^{6}	1.0336^{20}_{4}	1.5398^{20}	δ s^h	∞				os s	**B6**², 344
t500	**—,4-hydroxy-**	*p*-Cresol. 4-Methylphenol*. C_7H_8O. *See* t275	108.14	pr	34.8	201.9^{760}	1.0347^{20}_{4}	1.5395^{20} (suc)	δ s^h	∞				os s	**B6**², 368
t502	**—,α-hydroxylamino-**	*N*-Benzylhydroxylamine. $C_6H_5CH_2NHOH$	123.16	nd (peth or lig)	57				δ	s					**B15**², 17
t503	**—,2-hydroxylamino-**	*o*-Tolylhydroxylamine. C_7H_9NO. *See* t275	123.16	nd (bz-eth)	44				i	v	v		v	lig δ	**B15**², 14
t504	**—,3-hydroxylamino-**	*m*-Tolylhydroxylamine. C_7H_9NO. *See* t275	123.16	lf (bz-peth)	68.5				$δ^h$	v	v		v	chl v lig δ	**B15**², 15
t505	**—,4-hydroxylamino-**	*p*-Tolylhydroxylamine. C_7H_9NO. *See* t275	1231.6	lf (bz)	98	115–20d			δ	v	v		δ v^h	chl v lig i	**B15**², 16
t506	**—,2-hydroxylamino-6-nitro-**	$C_7H_8N_2O_3$. *See* t275	168.15	colorless or ye (bz)	120–1				δ	v	v		δ s^h	peth δ	**B15**², 14
t507	**—,4-hydroxylamino-2-nitro-**	$C_7H_8N_2O_3$. *See* t275	168.15	ye (bz)	108–9								s^h		**B15**², 16
t508	**—,α-hydroxy-2-methoxy-**	Saligenin 2-methyl ether. $C_8H_{10}O_2$. *See* t275	138.17			249^{760} 119^{8}	1.0395^{15}_{15}	1.549^{17}	i	s	∞				**B6**², 278
t509	**—,4-hydroxy-3-methoxy-**	Cresol. $C_8H_{10}O_2$. *See* t275	138.17	pr	5.5	221	1.098^{20}_{4}	1.5353^{25}	δ	∞	∞		∞	chl ∞ aa ∞	**B6**², 865
t511	**—,α-hydroxy-4-methoxy-**	Anisyl alcohol. $C_8H_{10}O_2$. *See* t275	138.17	nd	25	259.1^{760} $134–5^{12}$	1.109^{26}_{4}	1.5420^{25}	s	v	v				**B6**², 883
t512	**—,4-hydroxy-2-(methylamino)-**	$C_8H_{11}NO$. *See* t275	137.18	(bz-lig)	*ca.* 108					s	s				**B13**, 599
t513	**—,α-hydroxy-3,4-methylenedioxy-**	Piperonyl alcohol. $C_8H_8O_3$. *See* t275	152.15	nd (peth)	56	157^{16}			δ	s	s		s	chl s lig i MeOH s	**B19**², 77
t514	**—,α-hydroxy-2-nitro-**	*o*-Nitrobenzyl alcohol. $C_7H_7NO_3$. *See* t275	153.14	nd (w)	74	270^{760} 168^{20}			δ	s	s				**B6**², 424
t515	**—,α-hydroxy-3-nitro-**	*m*-Nitrobenzyl alcohol. $C_7H_7NO_3$. *See* t275	153.14	rh nd (w)	30.5	$175–80^{3}$			s	s	s				**B6**², 424
t516	**—,α-hydroxy-4-nitro-**	*p*-Nitrobenzyl alcohol. $C_7H_7NO_3$. *See* t275	153.14	nd (w)	93	250–60d 185^{12}			δ s^h	s	s				**B6**², 424
t517	**—,2-hydroxy-3-nitro-**	$C_7H_7NO_3$. *See* t275	153.14	ye pr (aq al or peth)	69.5	$102–3^{9}$			i	v	v				**B6**², 338
t518	**—,2-hydroxy-4-nitro-**	$C_7H_7NO_3$. *See* t275	153.14	ye nd (lig)	118				δ	s	s		s	CS_2 δ lig δ	**B6**², 339
t519	**—,2-hydroxy-5-nitro-**	$C_7H_7NO_3$. *See* t275	153.14	ye or colorless nd (w or aq al+w), pl (bz)	96(75) 30–40 (hyd)	$186–90^{9}$			δ	v	v		v	aa v	**B6**², 339
t521	**—,2-hydroxy-6-nitro-**	$C_7H_7NO_3$. *See* t275	153.14	lf ye nd (w)	147				$δ^h$	s	s				**B6**¹, 178

For explanations, symbols and abbreviations see beginning of table.

No.	Name	Synonyms and Formula	Mol. wt.	Crystalline form, color and specific rotation	m.p. °C	b.p. °C	Density	n_D	Solubility w	al	eth	ace	bz	other solvents	Ref.
	Toluene														
t522	**—,3-hydroxy-4-nitro-**	$C_7H_7NO_3$. *See* t275	153.14	ye mcl pl (eth or bz)	56				δ	s	s		s		**B6**^{2}, 359
t523	**—,3-hydroxy-5-nitro-**	$C_7H_7NO_3$. *See* t275	153.14	lt ye (bz)	90–1				$δ^h$						**B6**, 386
t524	**—,—,monohydrate.**	$C_7H_7NO_3$. H_2O. *See* t275	171.16	lt ye nd (w)	60–1				$δ^h$	v	v		δ		**B6**^{2}, 385
t525	**—,4-hydroxy-2-nitro-**	$C_7H_7NO_3$. *See* t275	153.14	ye pr (eth)	77				i	v	v		δ	CS_2 δ lig δ	**B6**^{2}, 387
t526	**—,4-hydroxy-3-nitro-**	$C_7H_7NO_3$. *See* t275	153.14	ye nd (al or w)	36.5	125^{22}	1.2399^{39}_{4}	1.574^{40}	$δ^h$	v	v				**B6**^{2}, 388
t527	**—,5-hydroxy-2-nitro-**	$C_7H_7NO_3$. *See* t275	153.14	nd or pr (w)	129				$δ^h$	s	s		s	chl s	**B6**^{2}, 361
t528	**—,2-hydroxy-5-nitroso-**	1,4-Toluquinone 4-oxime. $C_7H_7NO_2$. *See* t275	137.13	nd (w)	134–5d				δ v^h	v	v		s	chl v CCl_4 s alk v	**B7**^{2}, 589
t529	**—,5-hydroxy-2-nitroso-**	1,4-Toluquinone 2-oxime. $C_7H_7NO_2$. *See* t275	137.13	nd (w or bz), pr (aa)	165d				$δ^h$	v	s		v	aa s	**B7**^{2}, 589
t530	**—,2-hydroxy-3,4,5,6-tetrabromo-**	$C_7H_4Br_4O$. *See* t275	423.77	ye nd (chl or aa)	208	d			i	s	s		s	chl s aa δ lig δ	**B6**^{2}, 337
t531	**—,3-hydroxy-2,4,5,6-tetrabromo-**	$C_7H_4Br_4O$. *See* t275	423.77	nd (chl)	194									chl s^h alk s	**B6**^{2}, 358
t532	**—,4-hydroxy-2,3,5,6-tetrabromo-**	$C_7H_4Br_4O$. *See* t275	423.77	nd (al or chl)	209 (198)					s	v			chl v alk v aa v^h	**B6**^{2}, 386
t533	**—,2-hydroxy-3,4,5,6-tetrachloro-**	$C_7H_4Cl_4O$. *See* t275	245.94	nd (lig)	190					v	v		v	aa v lig s^h	**B6**^{2}, 333
t534	**—,3-hydroxy-2,4,5,6-tetrachloro-**	$C_7H_4Cl_4O$. *See* t275	245.94	nd (peth)	189–90					s	s		s	KOH s	**B6**^{1}, 189
t535	**—,4-hydroxy-2,3,5,6-tetrachloro-**	$C_7H_4Cl_4O$. *See* t275	245.94	nd (dil al)	190					s				chl v alk v lig δ	**B6**, 404
t536	**—,2-hydroxy-3,4,5-trinitro-**	$C_7H_5N_3O_7$. *See* t275	243.14	og ye pr (aa)	102				$δ^h$	v	v	v		chl v AcOEt v	**B6**, 369
t537	**—,3-hydroxy-2,4,6-trinitro-**	Methylpicric acid. $C_7H_5N_3O_7$. *See* t275	243.14	lt ye nd (w) or al)	109.5	150 exp			$δ^h$	s	s		s	chl s	**B6**^{2}, 363
t538	**—,α-iodo-**	Benzyl iodide. $C_6H_5CH_2I$	218.05	col or ye nd	24.5	93^{10}	1.734^{25}	1.6334^{25}	i	s	s		s	CS_2 δ	**B5**^{2}, 241
t539	**—,2-iodo-**	*o*-Tolyl iodide. C_7H_7I. *See* t275	218.05			211–2	1.713^{20}_{4}	1.609^{20}	i	∞	∞				**B5**^{2}, 240
t540	**—,3-iodo-**	*m*-Tolyl iodide. C_7H_7I. *See* t275	218.05			213	1.698^{20}		i	∞	∞				**B5**^{2}, 241
t541	**—,4-iodo-**	*p*-Tolyl iodide. C_7H_7I. *See* t275	218.05	lf	35	211	1.678^{40}_{4}		i	v	v				**B5**^{2}, 241
t542	**—,α-isocyano-**	Benzyl isocyanide. Benzyl carbylamine. $C_6H_5CH_2NC$	117.15			198–200δd $93–4^{55}$	0.972^{15}								**B12**, 1041
t543	**—,α-mercapto-**	Benzyl mercaptan. $C_6H_5CH_2SH$	124.21			194–5	1.058^{20}		i	v	v			CS_2 s	**B6**, 453
t544	**—,2-mercapto-**	*o*-Toluenethiol. C_7H_8S. *See* t275	124.21	pl or lf	15	194.3^{760}			i	s	v				**B6**^{2}, 342
t545	**—,3-mercapto-**	*m*-Toluenethiol. *m*-Thiocresol. C_7H_8S. *See* t274	124.21		<−20	195^{760}			i	s	∞				**B6**^{2}, 365
t546	**—,4-mercapto-**	*p*-Toluenethiol. *p*-Thiocresol. C_7H_8S. *See* t275	124.21	lf (eth or dil al)	42–3	195^{760}			i	s	v				**B6**^{2}, 392
—	**—,α-methoxy-**	*see* **Ether, benzyl methyl.**													
t547	**—,2-methoxy-**	*o*-Methylanisole. $C_8H_{10}O$. *See* t275	122.17			171.3	0.9851^{15}_{15}	1.5199^{15}	i	v	v				**B6**^{2}, 328
t548	**—,3-methoxy-**	*m*-Methylanisole. $C_8H_{10}O$. *See* t275	122.17			177.2^{760}	0.9697^{25}_{25}	1.5164^{13}	i	s	s		s		**B6**^{2}, 351
t549	**—,4-methoxy-**	*p*-Methylanisole. $C_8H_{10}O$. *See* t275	122.17			176.5^{760}	0.9689^{25}_{25}	1.5124^{19}	i	s	s				**B6**^{2}, 375
t550	**—,2(methylamino)-**	*N*-Methyl-*o*-toluidine. $C_8H_{11}N$. *See* t275	121.18			206	0.9769^{20}_{4}	1.5649^{20}	i	∞	∞				**B12**^{2}, 435
t551	**—,3(methylamino)-**	*N*-Methyl-*m*-toluidine. $C_8H_{11}N$. *See* t275	121.18			206–7			i	∞	∞				**B12**^{1}, 398
t552	**—,4(methylamino)-**	*N*-Methyl-*p*-toluidine. $C_8H_{11}N$. *See* t275	121.18			$209–11^{761}$ 76^{6}	0.9348^{55}_{4}		i	∞	∞				**B12**^{2}, 491

For explanations, symbols and abbreviations see beginning of table.

No.	Name	Synonyms and Formula	Mol. wt.	Crystalline form, color and specific rotation	m.p. °C	b.p. °C	Density	n_D	Solubility w	al	eth	ace	bz	other solvents	Ref.
	Toluene														
t556	**—,α-nitro-**	Phenylnitromethane*. $C_6H_5CH_2NO_2$	137.14	ye		217–36 118[16]	1.1598_0^{20}	1.5344[20]			s				**B5**, 325
t557	**—,2-nitro-**	$C_7H_7NO_2$. *See* t275	137.14	ye	−2.9 (−9.3)	220.4[760]	1.1629[20]	1.544[25]	i	∞	∞				**B5²**, 243
t558	**—,3-nitro-**	$C_7H_7NO_2$. *See* t275	137.14	pa ye	15	232.6[760]	1.1571_4^{20}	1.5466[20]	i	s	∞		s		**B6²**, 246
t559	**—,4-nitro-**	$C_7H_7NO_2$. *See* t275	137.14	orh	51.7	238.3[760]	1.299_4^0	1.5382[15]	i	v	v		v	CCl_4 v CS_2 v	**B6²**, 247
t560	**—,2-nitroso-**	C_7H_7NO. *See* t275	121.14	nd or pr	72				δ	v	v			chl v	**B6²**, 243
t561	**—,3-nitroso-**	C_7H_7NO. *See* t275	121.14	nd	53				i	δ	s			chl s	**B6**, 318
t562	**—,4-nitroso-**	C_7H_7NO. *See* t275	121.14	nd	48.5				i				v	chl v	**B6²**, 243
t563	**—,3-nitro-4-triazo-**	$C_7H_6N_4O_2$. *See* t275	178.15	ye nd or pl (lig or dil al)	38										**B5¹**, 174
t563¹	**—,2-nitro-α,α,α-trifluoro-**	*o*-Nitrobenzotrifluoride. $C_7H_4F_3NO_2$. *See* t275	191.11	(al)	32.5	216.3[765]				v			v	aa v	**B5²**, 251
t564	**—,3-nitro-α,α,α-trifluoro-**	*m*-Nitrobenzotrifluoride. $C_7H_4F_3NO_2$. *See* t275	191.11		−2.4	201.5[768]	1.4357_4^{15}	1.4742[19]	i	s	s				**B5¹**, 162
t565	**—,α,α,2,3,4-pentachloro-**	$C_7H_3Cl_5$. *See* t275	264.37	(lig)	84	275–85							δ v[h]		**B5¹**, 153
t566	**—,α,α,2,3,6-pentachloro-**	$C_7H_3Cl_5$. *See* t275	264.37	nd (MeOH)	83	145–50[12]							s	MeOH δ	
t567	**—,α,α,2,4,5-pentachloro-**	$C_7H_3Cl_5$. *See* t275	264.37		<0	280−1	1.607_{22}^{22}		i				s		**B5¹**, 153
t568	**—,α,α,2,4,6-pentachloro-**	$C_7H_3Cl_5$. *See* t275	264.37	(MeOH)	27	158[15]								MeOH s[h]	
t569	**—,2,3,4,5,6-pentachloro-**	$C_7H_3Cl_5$. *See* t275	264.37	nd (bz or peth)	24–5	301				δ[h]	δ[h]			to s, CS_2 δ	**B5²**, 234
t570	**—,2-propoxy-**	Propyl *o*-tolyl ether. *o*-Cresol propyl ether. $C_{10}H_{14}O$. *See* t275	150.22			204.1	0.9517_0^0								**B6**, 352
t571	**—,3-propoxy-**	Propyl *m*-tolyl ether. *m*-Cresol propyl ether. $C_{10}H_{14}O$. *See* t275	150.22			210.6	0.9484_0^0								**B6**, 376
t572	**—,4-propoxy-**	Propyl *p*-tolyl ether. *p*-Cresol propyl ether. $C_{10}H_{14}O$. *See* t275	150.22			210.4	0.9497_0^0								**B6**, 393
t573	**—,2,3,5,6-tetra-bromo-**	$C_7H_4Br_4$. *See* t275	407.77	nd	116–7				i	δ s[h]				os s	**B5**, 310
t574	**—,α,α,α,2-tetra-chloro-**	*o*-Chlorobenzo trichloride. $C_7H_4Cl_4$. *See* t275	229.94		30	260 129.5[13]			i d[h]					os s	**B5²**, 234
t575	**—,α,α,α-3-tetra-chloro-**	*m*-Chlorobenzo trichloride. $C_7H_4Cl_4$. *See* t275	229.94			255	1.495[14]		i d[h]					os s	**B5**, 303
t576	**—,α,α,α,4-tetra-chloro-**	*p*-Chlorobenzo trichloride. $C_7H_4Cl_4$. *See* t275	229.94			245			i					os v	**B5**, 303
t577	**—,α,α,2,5-tetra-chloro-**	$C_7H_4Cl_4$. *See* t275	229.94	cubic (chl)	42				i					os v	**B5²**, 234
t578	**—,α,α,3,4-tetra-chloro-**	$C_7H_4Cl_4$. *See* t275	229.94			257	1.518_{22}^{22}		i	s	s		s	aa s	**B5**, 302
t579	**—,α,α,3,5-tetra-chloro-**	$C_7H_4Cl_4$. *See* t275	229.94	(MeOH or dil aa)	36.5									oos v	
t580	**—,α,2,4,5-tetra-chloro-**	$C_7H_4Cl_4$. *See* t275	229.94			273	1.547[20]		i	s				os s	**B5**, 302
t581	**—,2,3,4,5-tetra-chloro-**	$C_7H_4Cl_4$. *See* t275	229.94	nd (MeOH or dil al)	98.1				i	v				os s	**B5²**, 233
t582	**—,2,3,4,6-tetra-chloro-**	$C_7H_4Cl_4$. *See* t275	229.94	nd (al or eth)	96	276.5			i	s			v	CS_2 v	**B5²**, 234
t583	**—,2,3,5,6-tetra-chloro-**	$C_7H_4Cl_4$. *See* t275	229.94	nd (MeOH)	93–4				i	s	v			MeOH s[h]	**B5¹**, 153
—	**—,triacetoxy-**	*see* **Toluene, trihydroxy-,** triacetate													
t586	**—,3,4,5-triamino-**	$C_7H_{11}N_3$. *See* t275	137.18	colorless nd (bz), red in air	105				v	v			s		**B13²**, 147
t587	**—,α-triazo-**	Benzyl azide. $C_6H_5CH_2N_3$	133.15			108[23]	1.0321_4^{25}	1.5233[25]	i	∞	∞				**B5²**, 274
t588	**—,2-triazo-**	*o*-Tolyl azide. $C_7H_7N_3$. *See* t275	133.15	pa ye	<−10	90.5[30]					s		s[h]		**B5²**, 273
t589	**—,3-triazo-**	*m*-Tolyl azide. $C_7H_7N_3$. *See* t275	133.15			92.5[31]					s				**B5²**, 273

For explanations, symbols and abbreviations see beginning of table.

No.	Name	Synonyms and Formula	Mol wt.	Crystalline form, color and specific rotation	m.p. °C	b.p. °C	Density	n_D	Solubility w	al	eth	ace	bz	other solvents	Ref.
	Toluene														
t590	—,4-triazo-	*p*-Tolyl azide. $C_7H_7N_3$. *See* t275	133.15	...	...	93^{32}	1.0527^{22}_{4}	...	i	s	s	...	...	...	**B5**[2], 273
t591	—,α,2,4-tribromo-	$C_7H_5Br_3$. *See* t275	328.86	...	40–1	...	...	...	i	s	...	...	...	oos v	**B5**[2], 240
t592	—,α,3,5-tribromo-	$C_7H_5Br_3$. *See* t275	328.86	pl or nd (al)	96	173^{19}	...	...	i	v	...	...	...	os s	**B5**[2], 240
t593	—,2,3,4-tribromo-	$C_7H_5Br_3$. *See* t275	328.86	rh pl (lig-CS_2, aa)	45–6	...	2.456^{20}	...	i	...	...	...	...	aa s^h	**B5**[1], 156
t594	—,2,3,5-tribromo-	$C_7H_5Br_3$. *See* t275	328.86	mcl pr (eth-to)	53–4	...	2.467^{17}	...	i	...	v	...	...	...	**B5**[1], 156
t595	—,2,3,6-tribromo-	$C_7H_5Br_3$. *See* t275	328.86	mcl pr or pl (lig or chl)	60.5	...	2.471^{17}	...	i	...	...	...	...	chl δ lig s^h	**B5**[1], 156
t596	—,2,4,5-tribromo-	$C_7H_5Br_3$. *See* t275	328.86	mcl pr (eth-al), nd (al)	113.5	...	2.472^{17}	...	i	δ	...	...	...	...	**B5**[2], 240
t597	—,2,4,6-tribromo-	$C_7H_5Br_3$. *See* t275	328.86	lo nd or mcl pr (eth-aa)	68.5	290	2.479^{17}	...	i	δ	s	...	...	...	**B5**[1], 156
t598	—,α,α,α-trichloro-	Benzotrichloride. $C_6H_5CCl_3$	195.48	...	−4.75	220.6^{760}	1.3723^{20}	...	i	s	s	...	s	...	**B5**[2], 233
t599	—,α,α,2-trichloro-	*o*-Chlorobenzylidene chloride. $C_7H_5Cl_3$. *See* t275	195.48	...	...	228.5	1.399^{15}	...	i	...	...	...	...	os s	**B5**, 300
t600	—,α,2,6-trichloro-	$C_7H_5Cl_3$. *See* t275	195.48	(lig, eth, al-eth)	39–40	...	...	...	...	v	v	...	...	...	...
t601	—,α,3,4-trichloro-	$C_7H_5Cl_3$. *See* t275	195.48	...	...	241	...	...	i	s	...	...	...	os s	**B5**, 300
t602	—,α,3,5-trichloro-	$C_7H_5Cl_3$. *See* t275	195.48	(MeOH)	36	...	...	...	...	...	...	...	...	MeOH s^h	...
t603	—,2,3,4-trichloro-	$C_7H_5Cl_3$. *See* t275	195.48	nd (al or MeOH)	41	$231–2^{716}$	...	...	i	s	...	...	...	os v	**B5**[2], 232
t604	—,2,3,5-trichloro-	$C_7H_5Cl_3$. *See* t275	195.48	nd (al)	45–6	$229–31^{757}$	...	...	i	s	...	...	...	os s	**B5**, 299
t605	—,2,3,6-trichloro-	$C_7H_5Cl_3$. *See* t275	195.48	nd (al)	45–6	...	...	...	i	s	...	...	...	os s	**B5**, 299
t606	—,2,4,5-trichloro-	$C_7H_5Cl_3$. *See* t275	195.48	nd or lf (al)	82.4	$229–30^{716}$	...	...	i	s	...	...	...	os s	**B5**[2], 232
t607	—,2,4,6-trichloro-	$C_7H_5Cl_3$. *See* t275	195.48	nd (al)	33–4	...	...	...	i	s	...	...	...	os s	**B5**[2], 232
t608	—,3,4,5-trichloro-	$C_7H_5Cl_3$. *See* t275	195.48	...	45	$246–7^{768}$	...	...	i	s	...	...	...	os s	**B5**, 299
t609	—,α,α,α-trifluoro-	Benzo trifluoride. $C_6H_5CF_3$	146.11	...	−29	103	1.1886^{20}	1.4149^{13}	i	∞	∞	...	...	...	**B5**[2], 224
t610	—,α,α,2-tri-hydroxy-, triacetate	Salicylaldehyde triacetate. $C_{13}H_{14}O_6$. *See* t275	266.26	nd or pl (al or Ac_2O)	104–5	...	...	...	i	δ v^h	s	...	s	chl s CCl_4 s	**B8**[2], 41
t610[1]	—,α,α,3-tri-hydroxy-, triacetate	*m*-Acetoxybenzal acetate. $C_{13}H_{14}O_6$. *See* t275	266.26	lf (w-al)	76	...	...	...	δ	s	s	...	...	...	**B8**, 60
t611[1]	—,α,α,4-tri-hydroxy-, tricetate	*p*-Acetoxybenzal acetate. $C_{13}H_{14}O_6$. *See* t275	266.26	pr (eth or lig)	94	...	...	...	i s^h	δ s^h	v	...	...	lig s^h	**B8**[1], 530
t611	—,α,2,5-tri-hydroxy-	Gentisyl alcohol. $C_7H_8O_3$. *See* t275	140.13	nd (chl)	100	sub vac	...	...	v	v	v	...	δ	chl v lig i	...
t612	—,2,4,5-tri-hydroxy-, triacetate.	2,4,5-Triacetoxytoluene. $C_{13}H_{14}O_6$. *See* t275	266.26	(al)	114–5	...	...	...	s	s	...	...	s	...	**B6**, 1109
t613	—,2,4,6-tri-hydroxy-, triacetate	2,4,6-Triacetoxyto luene. $C_{13}H_{14}O_6$. *See* t275	266.26	nd (lig)	52	...	...	...	$δ^h$	v^h	...	...	v	AcOEt v lig s^h	**B6**, 1111
t614	—,3,4,5-tri-hydroxy-	5-Methylpyrogallol. $C_7H_8O_3$. *See* t275	140.13	pa br nd (bz)	129	...	...	...	...	...	...	...	...	...	**B6**[2], 1081
t615	—,2,3,4-triiodo-	$C_7H_5I_3$. *See* t275	469.84	pa br nd (al)	92	...	...	...	i	δ	...	...	v	...	**B5**[1], 157
t616	—,2,3,5-triiodo-	$C_7H_5I_3$. *See* t275	469.84	og pl (al)	72–3	...	...	...	i	δ	...	...	...	...	**B5**[1], 157
t617	—,2,3,6-triiodo-	$C_7H_5I_3$. *See* t275	469.84	nd (al)	80.5	...	...	...	i	s^h	...	...	...	...	**B5**[1], 158
t618	—,2,4,5-triiodo-	$C_7H_5I_3$. *See* t275	469.84	pa br nd (al)	118–20	...	...	...	i	δ	...	...	...	...	**B5**[1], 158
t619	—,2,4,6-triiodo-	$C_7H_5I_3$. *See* t275	469.84	nd (al)	118–9	300d	...	...	i	δ	...	...	...	...	**B5**, 317
t620	—,3,4,5-triiodo-	$C_7H_5I_3$. *See* t275	469.84	nd (al)	122–3	...	...	...	i	s	...	...	...	...	**B5**, 317
t621	—,2,3,4-trinitro-	$C_7H_5N_3O_6$. *See* t275	227.13	tcl lf (al)	112	exp 290–310	1.62	...	i	δ	v	...	...	...	**B5**[2], 266
t622	—,2,4,5-trinitro-	$C_7H_5N_3O_6$. *See* t275	227.13	pa ye rh bipym	104	288–93	...	...	i	δ	s	s	s	aa s^h	**B5**[2], 268
t623	—,2,4,6-trinitro-	TNT. $C_7H_5N_3O_6$. *See* t275	227.13	orh	82	240 exp	1.654	...	i	δ	s	v	v	to, Py v	**B5**[2], 268
t624	**α-Toluenearsonic acid**	Benzylarsonic acid. Phenylmethanearsonic acid*. $C_6H_5CH_2AsO(OH)_2$	216.06	nd	167–8	...	...	...	δ s^h	δ s^h	...	...	...	...	**B16**[2], 461
t625	**2-Toluenearsonic acid**	*o*-Tolylarsonic acid. (structure: benzene ring with CH_3 and $AsO(OH)_2$ ortho)	216.06	nd (w)	163–4	...	...	...	$δ^h$	v	...	...	...	...	**B16**[2], 460
t626	**3-Toluenearsonic acid**	*m*-Tolylarsonic acid. (structure: benzene ring with CH_3 and $AsO(OH)_2$ meta)	216.05	nd (w)	150	...	...	...	s^h	s	...	...	...	...	**B16**[2], 460
t627	**4-Toluenearsonic acid**	*p*-Tolylarsonic acid. CH_3—⟨benzene⟩—$AsO(OH)_2$	216.05	nd (w)	d	...	...	...	δ v^h	...	...	...	...	aa s^h	**B16**[2], 460

For explanations, symbols and abbreviations see beginning of table.

No.	Name	Synonyms and Formula	Mol. wt	Crystalline form, color and specific rotation	m.p. °C	b.p. °C	Density	n_D	Solubility w	al	eth	ace	bz	other solvents	Ref.
	α-Tolueneboronic acid														
t628	**α-Tolueneboronic acid**	Benzylboric acid. $C_6H_5CH_2B(OH)_2$	135.96	(w or bz)	140				δ	s	...	...	s		**B16**[2], 639
t629	**2-Tolueneboronic acid**	*o*-Tolylboronic acid.	135.96	nd or pl (w)	d	d			δ	s	s	...	s		**B16**, 921
t630	**3-Tolueneboronic acid**	*m*-Tolylboronic acid.	135.96		137–40				δ s^h	v	v	...	...		**B16**, 921
t631	**4-Tolueneboronic acid**	*p*-Tolylboronic acid.	135.96	nd	245				δ v^h	...	s	...	...		**B16**[2], 638
t632	**2-Toluenesulfinic acid**	*o*-Tolylsulfinic acid.	156.20	nd	80	d			s	v	...	...	...	os v	**B11**[2], 6
t633	**4-Toluenesulfinic acid**	*p*-Tolylsulfinic acid.	156.20	rh pl or lo nd (w)	86–7				s^h	v	v	...	δ^h		**B11**[2], 6
t634	—,chloride	*p*-Toluenesulfinyl chloride.	174.65	nd	54–8	115–20[4]			d^h	d^h	...	...	...	chl s	**B11**[2], 9
t635	**α-Toluenesulfonic acid, amide**	Benzylsulfonamide. $C_6H_5CH_2SO_2NH_2$	171.22	pr or nd (w or al)	105				v^h	v^h	...	...	...		**B11**[2], 73
t635[1]	—,—,*N*(2-tolyl)-		261.35	(dil al)	83				...	v	...	...	...	alk s	**B12**[2], 452
t636	—,—,*N*(3-tolyl)-		261.35	(al)	75				...	v	...	...	...	alk s	**B12**[2], 473
t637	—,—,*N*(4-tolyl)-		261.35	pr (al)	113				...	s^h	s	...	...	alk s	**B12**[2], 528
t638	—,chloride	Benzylsulfonyl chloride. $C_6H_5CH_2SO_2Cl$	190.65	pr (eth) nd (bz)	92–3				d^h	d^h	v	...	v		**B11**[2], 73
t639	**2-Toluenesulfonic acid**		172.20	hyg pl	67.5	128.8[25]			v	s	i	...	...		**B11**, 83
t640	—,amide	*o*-Toluenesulfonamide.	171.22	oct (al), pr (w)	156.3				δ	s	δ	...	...		**B11**, 86
t641	—,—,*N*-methyl-	$C_8H_{11}NO_2S$. *See* t640	185.25	pl (lig-bz)	74–5				δ^h	v	...	v	i^h	chl v	**B11**, 87
t642	—,—,*N*-phenyl-	$C_{13}H_{13}NO_2S$. *See* t640	247.32	pr (dil al)	136				i	v	...	...	...		**B12**, 566
t644	—,bromide	*o*-Toluenesulfonyl bromide. $C_7H_7BrO_2S$. *See* t639	235.11		13	138[10]			i d^h	d^h	...	...	...	os ∞	**B11**, 36
t645	—,chloride	*o*-Toluenesulfonyl chloride. $C_7H_7ClO_2S$. *See* t639	190.65		10.2	154[36]	1.3383^{20}_{4}	1.5565^{20}	i d^h	s^h	s	...	s		**B11**[2], 39
t646	—,2-tolyl ester	$C_{14}H_{14}O_3S$. *See* t639	262.33	(al)	50–1				...	s^h	...	...	s		**B11**, 85
t647	—,3-tolyl ester	$C_{14}H_{14}O_3S$. *See* t639	262.33	(al)	60				...	s^h	...	...	s		**B11**, 85
t648	—,4-tolyl ester	$C_{14}H_{14}O_3S$. *See* t639	262.33	(al)	70–1				...	s^h	...	...	s		**B11**, 85
t649	**3-Toluenesulfonic acid**		172.20	nd					v	s	s	...	...		**B11**, 94
t650	—,amide	*m*-Toluenesulfonamide.	171.22	pl (w), mcl pr (w)	108				δ v^h	v					**B11**[1], 23
t651	—,—,*N*-phenyl-	*m*-Toluenesulfonanilide. $C_{13}H_{13}NO_2S$. *See* t650	247.32	pr (al)	96				i	s	s	...	...		**B12**, 566

For explanations, symbols and abbreviations see beginning of table.

No.	Name	Synonyms and Formula	Mol. wt.	Crystalline form, color and specific rotation	m p. °C	b.p. °C	Density	n_D	Solubility w	al	eth	ace	bz	other solvents	Ref.
	4-Toluenesulfonic acid														
t652	**4-Toluenesulfonic acid**	2 3 CH_3— —SO_3H 6 5	172.20	hyg pl (w+?), mcl lf or pr	104–5	140^{20}			v	s	s				**B11**, 97
t653	—,amide	*p*-Toluenesulfonamide. 2 3 CH_3— —SO_2NH_2 6 5	171.22	mcl pl (w+2)	137.5				δ	s	δ				**B11**, 104
t654	—,—,*N*,*N*-dichloro-	Dichloramine T. $C_7H_7Cl_2NO_2S$. *See* t653	240.11	pa ye pr (peth or chl)	183				δ	s	s		s	chl, CCl_4 s peth δ	**B11²**, 63
t655	—,—,*N*-ethyl-	$C_9H_{13}NO_2S$. *See* t653	199.28	pl (dil al or lig)	64					s					**B11²**, 56
t656	—,—,*N*-methyl-	$C_8H_{11}NO_2S$. *See* t653	185.19	rh	78–9		1.340		δ	v	s				**B11²**, 56
t657	—,—,*N*-methyl-*N*-nitroso-	$C_8H_{10}N_2O_3S$. *See* t653	214.25		60				i	v^h	v				**B11¹**, 29
t658	—,—,*N*-methyl-*N*-phenyl-	$C_{14}H_{15}NO_2S$. *See* t653	261.35	pl or mcl pr (AcOEt)	95				i	v	v			AcOEt s^h	**B12²**, 305
t659	—,—,*N*-methyl-*N*(2-tolyl)-	$C_{15}H_{17}NO_2S$. *See* t653	275.38	pr (al)	119–20				i	v^h					**B12¹**, 388
t660	—,—,*N*-methyl-*N*(4-tolyl)-	$C_{15}H_{17}NO_2S$. *See* t653	275.38	pr	60				i	δ				oos δ	**B12²**, 529
t661	—,—,*N*-phenyl-	*p*-Toluenesulfonanilide. $C_{13}H_{13}NO_2S$. *See* t653	247.32	dimorphic, α:tcl. β:mcl pr (dil al or bz)	103–4				i	v			s^h	aa s	**B12²**, 298
t662	—,—,*N*(2-tolyl)-	$C_{14}H_{15}NO_2S$. *See* t653	261.35	rh bipym or (al), nd (dil aa)	110					s	v	v	v	chl v	**B12²**, 452
t663	—,—,*N*(4-tolyl)-	$C_{14}H_{15}NO_2S$. *See* t653	261.35	tcl pr or nd (aa)	118–9					v^h			s	chl s	**B12²**, 528
t664	—,butyl ester	$C_{11}H_{16}O_3S$. *See* t652	228.31			164–6^6	1.1319^{20}_4	1.5046^{20}	i		s				**B11²**, 46
t665	—,chloride	*p*-Toluenesulfonyl chloride. $C_7H_7ClO_2S$. *See* t652	190.65	tcl (eth or peth)	71	145–6^{15}			i	s	s		v		**B11**, 103
t666	—,2-chloroethyl ester	$C_9H_{11}ClO_3S$. *See* t652	234.71			210^{21}			i						**B11²**, 45
t667	—,2-chloropropyl ester	$C_{10}H_{13}ClO_3S$. *See* t652	248.74			216–7^{17}	1.2674^{20}_4	1.5225^{21}							**B11²**, 45
t668	—,ethylene ester	Ethylene *p*-toluenesulfonate. CH_3— —$SO_3CH_2CH_2O_3S$— —CH_3	370.44		126–7										**Am77**, 4899
t669	—,ethyl ester	$C_9H_{12}O_3S$. *See* t652	200.26	mcl pr (aa)	34–5	173^{15}	1.166^{46}_4		i	s	s				**B11²**, 45
t670	—,isopropyl ester	$C_{10}H_{14}O_3S$. *See* t652	214.28		19.87		1.5065^{20}								**Am77**, 4899
t671	—,methyl ester	$C_8H_{10}O_3S$. *See* t652	186.23	mcl lf or pr	28–9	292^{760} 140^{20}			δ	v	v		s	chl v	**B11²**, 44
t672	—,propyl ester	$C_{10}H_{14}O_3S$. *See* t652	214.28		<−20	189^9	1.144^{20}_4	1.4998^{20}							**B11²**, 45
t673	—,2-tolyl ester	$C_{14}H_{14}O_3S$. *See* t652	262.33	nd	54–5					s^h					**B11**, 100
t674	—,2,4,6-tribromo-phenyl ester	$C_{13}H_9Br_3O_3S$. *See* t652	485.03	(al)	113					s^h					**B11²**, 47
t675	**3-Toluenesulfonic acid, 6-aceta-mido-**, chloride	$C_9H_{10}ClNO_3S$. *See* t649	247.71	nd (bz)	159				d	d^h			δ s^h		**B14²**, 448
t676	**2-Toluenesulfonic acid, 4-amino-**	$C_7H_9NO_3S$. *See* t639	187.22	mcl pr	d				δ	i					**B14²**, 446
t677	—,—,amide	$C_7H_{10}N_2O_2S$. *See* t640	186.24	nd or pr	164				δ v^h	v^h					**B14²**, 446
t678	—,5-amino-	$C_7H_9NO_3S$. *See* t639	187.22	pl	>275				δ						**B14²**, 446
t679	—,6-amino-	$C_7H_9NO_3S$. *See* t639	187.22	nd (+½w)	130d				δ	i					**B14**, 722
t680	**3-Toluenesulfonic acid, 4-amino-**	$C_7H_9NO_3S$. *See* t649	187.22	lt ye nd	132d										**B14²**, 447
t681	**4-Toluenesulfonic acid, 2-amino-**	$C_7H_9NO_3S$. *See* t652	187.22	pl, nd or pr (w+1)					δ	i					**B14**, 730
t682	—,3-amino-, amide	$C_7H_{10}N_2O_2S$. *See* t653	186.24	pr	176				δ s^h	δ	i		i		**B14**, 728
t683	**2-Toluenesulfonic acid, 5-iso-propyl-**, amide	$C_{10}H_{15}NO_2S$. *See* t640	213.29	pl (al), pr or lf (dil al)	75				$δ^h$					lig s	**B11**, 140
t684	**3-Toluenesulfonic acid, 4-iso-propyl-**, amide	$C_{10}H_{15}NO_2S$. *See* t650	213.30	flakes (dil al)	149.9				$δ^h$	v^h	v				**B11²**, 84
t685	**4-Toluenesulfonic acid, 3-iso-propyl-**, amide	$C_{10}H_{15}NO_2S$. *See* t653	213.30	nd (w)	162				$δ^h$	v					**B11**, 139

For explanations, symbols and abbreviations see beginning of table.

No.	Name	Synonyms and Formula	Mol. wt.	Crystalline form, color and specific rotation	m.p. °C	b.p. °C	Density	n_D	Solubility w	al	eth	ace	bz	other solvents	Ref.
	2-Toluenesulfonic acid														
t686	**2-Toluenesulfonic acid, 4-nitro-,** dihydrate	$C_7H_7NO_5S.\ 2H_2O$. *See* t639	253.23	pl (w)	133.5				v						**B11**[1], 23
t687	—,**5-nitro-**	$C_7H_7NO_5.S$ *See* t639	217.20	pr or pl (w+2)	133.5 (anh) 130 (+2w)				v	v	v			chl v	**B11**[2], 41
t688	—,**4-propyl-,** amide	$C_{10}H_{15}NO_2S$. *See* t640	213.30	pl (dil al or bz)	101–2				v				δ		**B11**, 138
—	**α-Toluic acid**	*see* **Acetic acid, phenyl-**													
—	**Toluic acid**	*see* **Benzoic acid, methyl-**													
—	**Toluidine**	*see* **Toluene, amino-**													
—	**Toluquinone**	*see* **Benzoquinone, methyl-***													
t689	**Tomatidine**	$C_{27}H_{45}NO_2$	415.67	pl, $[\alpha]_D-8.8$ (MeOH)	210–1					s^h	s				**Am73**, 4018
t690	**Tomatine**	$C_{50}H_{83}NO_{21}$	1034.21	nd	236–7				i	s	i			MeOH s diox s, peth i	
t691	**α-Toxicerol**(*dl*)	Hydroxydequelin. $C_{23}H_{22}O_7$. 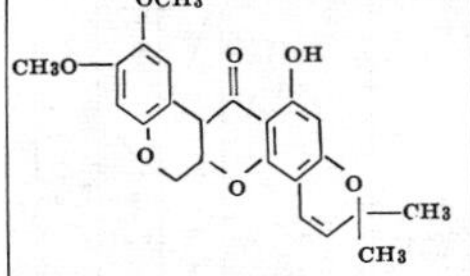	410.43	gr ye pl	218–20			1.580 (1.768)		δ				chl s^h	**J1938**, 513
t692	—(*l*)	$C_{23}H_{22}O_7$. *See* t691	410.43	pa gr ta (al) $[\alpha]_D-70$ (bz), −53 (bz)	125–7										**J1938**, 513
t693	**β-Toxicerol**(*dl*)	$C_{23}H_{22}O_7$. *See* t691	410.43	pa ye	169–70										**J1941**, 878
t694	—,**dihydro-**(*l*)	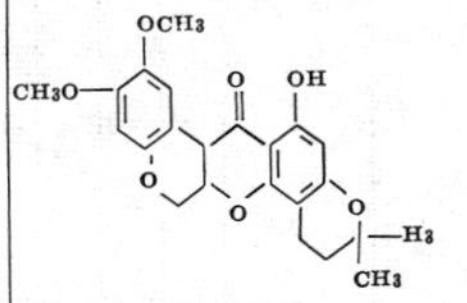	412.44	pa ye pl and rods $[\alpha]^{20}_D-30$	179 (206)										**J1940**, 1178
t695	**Torularhodin**	Torulene. $C_{37}H_{48}O_2$	524.79	red nd (MeOH-eth), vt-bk (bz-MeOH)	201–3d					δ^h		s		chl, CS_2 s Py s peth i	
t696	**Trasentin,** hydrochloride	2(Diethylamino)ethyl diphenylacetate hydrochloride, $(C_6H_5)_2CHCO_2CH_2CH_2N(C_2H_5)_2.HCl$	347.89		114–5				s	δ	i		i		
—	**Traumatic acid**	*see* **2-Dodecenedioic acid***													
t697	**Trehalose**	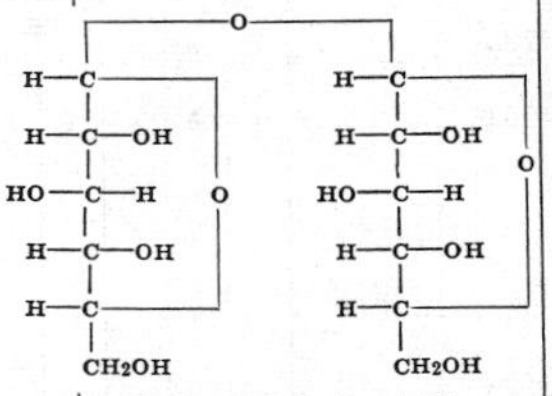	342.30	$[\alpha]^{20}_D+197.1$ (w, c=6)	203				v^h	s^h	i				**B31**, 378
t698	—,dihydrate	$C_{12}H_{22}O_{11}.2H_2O$. *See* t697	378.33	(dil al), $[\alpha]^{20}_D$ +178.3 (w, c=7)	103				v	s^h	i				**B31**, 378
—	**Triacetamide**	*see* **Acetic acid,** amide, *N,N*-diacetyl-													
—	**Triacetin**	*see* **Glycerol,** triacetate													
t699	**Triacontane***	$CH_3(CH_2)_{28}CH_3$	422.83	orh	65.5	304^{15}	0.7750^{78}_4	1.4352^{70}	i	δ^h	s		v^h		**B1**[2], 145
t700	**1-Triacontanol***	Myricyl alcohol. $CH_3(CH_2)_{29}OH$	438.83	nd (eth), flakes (bz)	88		0.777^{95}		i	s	v		v	oos s^h	**B1**, 432
t701	**1,3,5-Triazine***	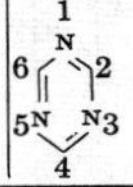	81.08												

For explanations, symbols and abbreviations see beginning of table.

No.	Name	Synonyms and Formula	Mol. wt.	Crystalline form, color and specific rotation	m.p. °C	b.p. °C	Density	n_D	Solubility w	al	eth	ace	bz	other solvents	Ref.
	1,3,5-Triazine														
t701[1]	**1,3,5-Triazine, 2,4-diamino-***	$C_3H_5N_5$. *See* t701	111.11	nd (w)	329d	...	...	...	s^h	δ	...	...	...	...	**B26**^{1}, 65
t702	**—,2,4-diamino-6-phenyl-***	$C_9H_9N_5$. *See* t701	187.21	nd	225	...	...	...	...	s	s	...	...	...	**B26**^{1}, 69
t703	**—,hexahydro-1,3,5-trinitro-***	Cyclonite. Hexogen. $C_3H_6N_6O_6$. *See* t701	222.12	orh (ace)	203.5	...	1.82^{20}	...	...	...	δ	s	...	MeOH δ AcOEt δ CCl_4, CS_2 i	**B26**^{2}, 5
t703[1]	**—,hexahydro-1,3,5-triphenyl-***	$C_{21}H_{21}N_3$. *See* t701	315.42	nd (lig), pr (eth or chl-al)	143	185	...	...	i	δ	s	...	s	chl s	**B26**^{2}, 3
—	**—,trichloro-***	*see* **Cyanuric acid**, trichloride													
—	**—,trihydroxy-***	*see* **Cyanuric acid**													
t704	**1,3,5-Triazine-2,4,6-tricarboxylic acid**	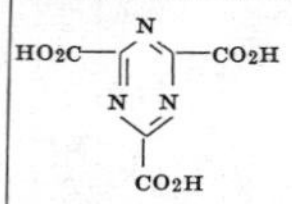 HO_2C— N —CO_2H N N CO_2H	213.11	...	250d	...	...	...	δ	i	i	...	...	...	**B26**^{2}, 168
t705	—,triethyl ester	$C_{12}H_{15}N_3O_6$. *See* t704	297.27	nd (al)	165	d	...	...	i	i $δ^h$	i	...	i	...	**B26**^{2}, 168
t706	—,trinitrile	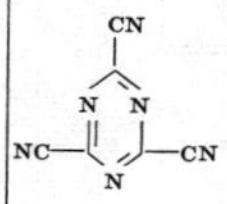 CN N N NC— —CN N	156.11	mcl pr (bz+1)	119	262^{771}	...	...	d^h	...	...	...	s v^h	CCl_4 s	**B26**, 591
t707	**1,2,3-Triazole**	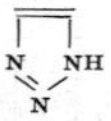 N NH N	69.07	...	23	$208–9^{742}$	1.1861_4^{25}	1.4854^{25}	s	...	...	...	...	...	**B26**^{1}, 5
t708	**1,2,4-Triazole**	 5 1 N HN N 4 2 3	69.07	pr (w), nd (al, eth, chl or bz)	120	260	1.132^{153}	1.4854^{25}	v	v	δ	...	δ	...	**B26**^{2}, 7
t709	**—,3-amino-**	$C_2H_4N_4$. *See* t708	84.08	(w, al or AcOEt)	159	...	...	...	v	v	δ	δ	...	AcOEt δ	**B26**^{2}, 76
t710	**—,4-amino-**	$C_2H_4N_4$. *See* t708	84.08	nd (al or chl)	82–3	...	...	...	s	s	...	...	...	chl δ, peth δ	**B26**^{2}, 7
—	**Tribenzoin**	*see* **Glycerol**, tribenzoate													
—	**Tributyrin**	*see* **Glycerol**, tributanoate													
—	**Tricaproin**	*see* **Glycerol**, trihexanoate													
—	**Tricaprylin**	*see* **Glycerol**, trioctanoate													
—	**Tricarballylic acid**	*see* **1,2,3-Propanetricarboxylic acid***													
—	**Tricarbonimide,** trimethyl ester	*see* **Isocyanuric acid,** trimethyl ester													
—	**Tricetin**	*see* **Flavone, 3′,4′,5,5′,7-pentahydroxy-**													
t711	**Tricosane***	$CH_3(CH_2)_{21}CH_3$	324.64	lf (eth-al)	47.3	243^{15}	0.7785_4^{48}	...	i	δ	s	...	...	...	**B1**^{2}, 141
t712	**12-Tricosanone***	Diundecyl ketone. Laurone. $[CH_3(CH_2)_{10}]_2CO$	338.62	lf (al)	69.3	0.8086_4^{69}	1.4283^{80}	...	i	δ	s	δ	s	chl s	**B1**^{2}, 774
—	**Tricyclodecane**	*see* **α-Dicyclopentadiene, tetrahydro-**													
t713	**1,12-Tridecadiyne***	$HC{:}C(CH_2)_9C{:}CH$	176.30	...	−3	115.5^{12}	0.8262_4^{21}	1.454^{20}	...	...	...	...	...	...	...
t714	**Tridecanal***	$CH_3(CH_2)_{11}CHO$	198.35	...	14	156^{23}	...	...	i	s	...	...	...	os v	**B1**^{2}, 769
t715	—,oxime*	$CH_3(CH_2)_{11}CH{:}NOH$	213.37	nd (dil al)	80.5	...	...	...	i	δ	v	...	δ	chl v peth δ	**B1**^{2}, 769
t716	**Tridecane***	$CH_3(CH_2)_{11}CH_3$	184.37	...	−5.5	243^{760}	0.7559_4^{20}	1.4233^{20}	i	v	v	...	...	...	**B1**^{3}, 134
t717	**—,1-amino-***	$CH_3(CH_2)_{12}NH_2$	199.38	...	27	265^{760} $101–2^1$	...	...	δ	s	s	...	...	...	**B4**^{1}, 388
t718	**—,1-bromo-***	Tridecyl bromide. $CH_3(CH_2)_{12}Br$	263.27	...	6	162^{16}	1.0262_4^{20}	1.4601^{20}	i	...	...	...	...	chl v	**B1**^{2}, 548
t719	**—,1,13-dibromo-***	$Br(CH_2)_{13}Br$	342.17	...	8–10	$188–92^{13}$	...	...	i	...	s	...	...	chl v	**B1**^{3}, 548
t720	**1,12-Tridecanediol***	$CH_3CHOH(CH_2)_{11}OH$	216.36	(dil al)	60–1	$188–90^8$	...	...	...	v^h	...	...	...	peth δ	**B1**^{2}, 563
t721	**1,13-Tridecanediol***	Tridecamethylene glycol. $HO(CH_2)_{13}OH$	216.36	(bz)	76.5	$195–7^{10}$	...	...	...	s	...	...	...	aa v^h	**B1**^{3}, 2240

For explanations, symbols and abbreviations see beginning of table.

No.	Name	Synonyms and Formula	Mol. wt.	Crystalline form, color and specific rotation	m.p. °C	b.p. °C	Density	n_D	Solubility w	al	eth	ace	bz	other solvents	Ref.
	Tridecanoic acid														
t722	**Tridecanoic acid***	$CH_3(CH_2)_{11}CO_2H$	214.35	...	41.8	236^{100} 140.5^{1}	...	...	i	v	v	v	...	aa v	**B2²**, 322
t723	—,amide	$CH_3(CH_2)_{11}CONH_2$	213.37	lf (al)	100	...	...	...	...	v	v	...	...	...	**B2¹**, 159
t724	—,methyl ester*	$CH_3(CH_2)_{11}CO_2CH_3$	228.38	...	fr 6.5	...	...	...	...	∞	...	...	...	...	...
t725	—,nitrile	1-Cyanododecane. $CH_3(CH_2)_{11}CN$	195.35	...	...	275	...	...	...	v	v	...	...	...	**B2**, 364
t726	**1-Tridecanol***	$CH_3(CH_2)_{12}OH$	200.37	...	30.5	152^{14}	0.8223^{31}_{4}	...	i	s	s	...	...	...	**B1²**, 464
t727	**2-Tridecanone***	Hendecyl methyl ketone. $CH_3(CH_2)_{10}COCH_3$	198.35	...	fr 27.5	263^{760} 145^{10}	0.8763^{15}	...	i	v	v	v	...	CCl_4 v	**B1²**, 769
t728	**3-Tridecanone**	Ethyl decyl ketone. $CH_3(CH_2)_9COCH_2CH_3$	198.35	pl	25	140^{17}	...	...	i	...	...	s	...	...	**B1**, 371
t729	**7-Tridecanone***	Dihexyl ketone. Enanthone. $CH_3(CH_2)_5CO(CH_2)_5CH_3$	198.35	lf (al)	29.7	261^{760}	0.825^{30}	...	...	v	v	...	...	chl v lig v	**B1²**, 769
t730	**1-Tridecene***	Tridecylene. $CH_3(CH_2)_{10}CH{:}CH_2$	182.35	...	−13 (−23)	232.8^{760} 104^{11}	0.7658^{20}_{4}	1.4340^{20}	i	v	v	...	...	...	**B1¹**, 98
—	**Tridecylene**	*see* **Tridecene**													
t731	**2-Tridecyne**	$CH_3(CH_2)_9C{\vdots}CCH_3$	180.32	...	30	184^{15}	0.8016^{20}_{4}	...	...	...	...	...	...	...	...
—	**Triethanolamine**	*see* **Amine, triethyl, 2,2′2″-trihydroxy-**													
t732	**Triethylene glycol**	$HOCH_2CH_2OCH_2CH_2OCH_2CH_2OH$	150.18	hyg	−5	276^{760} 165^{14}	1.1274^{15}_{4}	1.4578^{15}	∞	∞	δ	...	...	to s peth i	B1², 521
t733	—,3-aminopropyl ether	$HOCH_2CH_2OCH_2CH_2OCH_2CH_2OCH_2CH_2CH_2NH_2$	207.27	glassy	−50	184^{10}	1.0682^{20}_{20}	1.4668^{20}	∞	∞	...	...	...	...	...
t734	—,diacetate	$CH_3CO_2CH_2CH_2OCH_2CH_2OCH_2CH_2O_2CCH_3$	234.25	...	...	300	...	...	∞	∞	∞	...	...	...	**B2**, 141
t735	—,monobutyl ether	$HOCH_2CH_2OCH_2CH_2OCH_2CH_2O(CH_2)_3CH_3$	206.28	...	...	$274–89^{760}$	0.9890^{20}_{4}	1.4389^{20}	∞	v	...	...	...	MeOH v CCl_4 v	...
t736	**Triethylene-tetramine**	$H_2NCH_2(CH_2NHCH_2)_2CH_2NH_2$	146.24	...	12	266–7 157^{20}	...	...	s^h	s	...	...	...	ac s	**B4²**, 695
t737	**Triglycine**	*N*(*N*-Glycylglycyl)glycine. $H_2NCH_2CONHCH_2CONHCH_2CO_2H$	189.17	nd (w-al)	246	...	...	...	s^h	i	i	...	...	...	**B4²**, 806
t738	—,*N*-phthalyl-	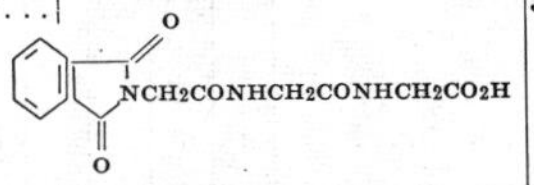	319.28	nd (al)	234–5d	...	...	...	v^h	v^h	i	...	...	chl i peth i	**B21²**, 358
—	**Trigonelline**	*see* **3-Pyridinecarboxylic acid, 1-methyl-**													
—	**Triisovalerin**	*see* **Glycerol**, tri(3-methylbutyrate)													
—	**Trilaurin**	*see* **Glycerol**, tridodecanoate													
—	**Trimellitic acid**	*see* **1,2,4-Benzenetricarboxylic acid***													
—	**Trimesic acid**	*see* **1,3,5-Benzenetricarboxylic acid***													
—	**Trimesitic acid**	*see* **2,4,6-Pyridinetricarboxylic acid**													
t739	**Trimethadione**	3,5,5-Trimethyl-2,4-oxazolidinedione. 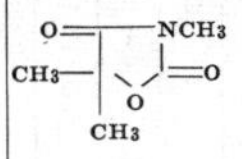	143.14	...	46	$78–80^{5}$	...	...	s	v	v	...	v	chl v peth i	...
—	**Trimethylene glycol**	*see* **1,3-Propanediol***													
—	**Trimethylenimine**	*see* **Azetidine**													
—	**Trimethylethenylammonium hydroxide**	*see* **Neurine**													
—	**Trimyristin**	*see* **Glycerol**, tritetradecanoate													
—	**Triolein**	*see* **Glycerol**, tri(*cis*-9-octadecenoate)													
—	**Trional**	*see* **Butane, 2,2-bis(ethylsulfonyl)-***													

For explanations, symbols and abbreviations see beginning of table.

No.	Name	Synonyms and Formula	Mol. wt.	Crystalline form, color and specific rotation	m.p. °C	b.p. °C	Density	n_D	Solubility w	al	eth	ace	bz	other solvents	Ref.
	1,3,5-Trioxane														
t741	**1,3,5-Trioxane**	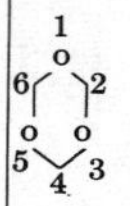 (ring: O1, 2, O3, 4, O5, 6)	90.08	rh nd	64	114.5^{759} 46^{1} sub	1.17^{65}	...	v	s	s	...	s	CCl_4, CS_2 s peth i	**B19**^{2}, 392
—	—,triimine	*see* **Cyamelide**													
—	—,2,4,6-triethenyl-	*see* **Metacrolein**													
—	—,2,4,6-triisopropyl-	*see* **Paraisobutyraldehyde**													
—	—,2,4,6-tripropyl-	*see* **Parabutyraldehyde**													
—	—,2,4,6-trimethyl-	*see* **Paraldehyde**													
—	**Tripalmitin**	*see* **Glycerol**, trihexadecanoate													
—	**Triphenylene**	*see* **9,10-Benzophenanthrene**													
—	**Triphenylmethane**	*see* **Methane, triphenyl-***													
—	**Triphosgene**	*see* **Carbonic acid**, bis(trichloromethyl) ester*													
—	**Tripropionin**	*see* **Glycerol**, tripropanoate													
t743	**Trisiloxane, octamethyl-**	$(CH_3)_3SiOSi(CH_3)_2OSi(CH_3)_3$	236.54	...	*ca.* −80	153	0.8200	1.3848^{20}	...	δ	...	...	s	peth s	...
t744	**—,1,1,3,5,5-pentamethyl-1,3,5-triphenyl-**	$C_6H_5Si(CH_3)_2OSi(CH_3)(C_6H_5)OSi(CH_3)_2C_6H_5$	422.75	...	...	$169^{0.7}$	1.0227^{20}_{4}	1.5280^{20}	i	...	...	...	s	lig s	...
—	**Tristearin**	*see* **Glycerol**, trioctadecanoate													
t745	**Trisulfide, diallyl**	$CH_2{:}CHCH_2S_3CH_2CH{:}CH_2$	178.34	...	...	$112–22^{16}$	1.0845^{15}	...	i	i	∞	...	...	...	**B1**, 441
t746	**—,diethyl**	$CH_3CH_2S_3CH_2CH_3$	154.32	ye	...	$96–7^{26}$	1.114^{20}	1.5689^{13}	...	...	...	...	...	...	**B1**^{3}, 1378
t747	**—,dimethyl, hexabromo-**	$Br_3CS_3CBr_3$	599.67	rh (eth)	125d	d	...	...	i	δ	s	...	s	chl, peth s CS_2 v	...
t748	**—,diphenyl, 2,2′-dinitro-**	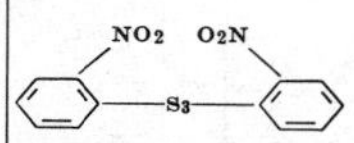 (NO_2 O_2N; C_6H_4—S_3—C_6H_4)	340.40	ye nd (al)	175–6	...	...	...	...	δ	...	...	...	os δ KOH s	**B6**^{2}, 308
—	**Tritan**	*see* **Methane, triphenyl-***													
t749	**1,3,5-Trithiane**	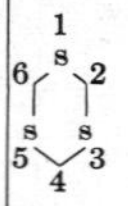 (ring: S1, 2, S3, 4, S5, 6)	138.26	pr	216	sub	...	...	i	δ	δ	...	s	...	**B19**, 382
t750	**—,2,4,6-trimethyl-**(α-form)	Trithioacetaldehyde. $C_6H_{12}S_3$. *See* t749	180.34	mcl	101	245–8	...	...	i	s	s	...	v	chl v CS_2 s	**B19**^{2}, 399
t751	—,—(β-form)	$C_6H_{12}S_3$. *See* t749	180.34	rh nd (ace)	125–6	245–8	...	...	i	s	s	...	v	chl v	**B19**^{2}, 399
t752	—,—(γ-form)	$C_6H_{12}S_3$. *See* t749	180.34	...	81	100	...	...	...	...	...	...	...	...	...
t753	**—,2,4,6-trimethyl-2,4,6-triphenyl-**	Trithioacetophenone. $C_{24}H_{24}S_3$. *See* t749	408.68	nd (al)	122	...	...	...	i	δ s^h	v	v	...	chl v	**B19**, 398
t754	**—,2,4,6-triphenyl-**(low m.p.)	Trithiobenzaldehyde, $C_{21}H_{18}S_3$. *See* t749	366.54	nd (bz-al)	167	d	...	...	i	δ	δ	δ	s	MeOH δ chl s CS_2 s	**B19**^{1}, 808
t755	—,—(high m.p.)	$C_{21}H_{18}S_3$. *See* t749	366.54	(bz or thiophene)	226	...	...	...	...	i	δ	δ	i	chl, CS_2 δ MeOH i	**B19**^{2}, 408
—	**Trithioacetaldehyde**	*see* **1,3,5-Trithiane, 2,4,6-trimethyl-**													
—	**Trithiobenzaldehyde**	*see* **1,3,5-Trithiane, 2,4,6-triphenyl-**													
—	**Trithiocyanuric acid**	*see* **Cyanuric acid, thio-**													
t756	**α-Tritisterol**	$C_{30}H_{50}O$	426.70	nd (MeOH-ace), $[\alpha]^{20}_{D}$+54.3 (al)	114–5	...	...	...	...	s	...	...	...	peth s chl s	...
t757	**β-Tritisterol**	$C_{30}H_{50}O$	426.70	nd (MeOH) $[\alpha]_D$+49.2 (al)	97	...	...	...	...	s	...	...	...	...	...
t758	**Tropacocaine**	Benzoyl-ψ-tropeine. 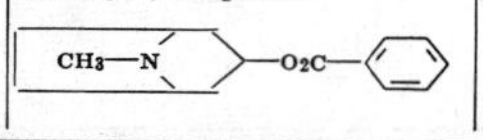 (CH_3—N ring —O_2C—C_6H_5)	245.31	pl or tab	49	d	1.0426^{100}_{4}	1.5080^{100}	δ^h	v	v	...	v	chl v peth v	**B21**^{2}, 23

For explanations, symbols and abbreviations see beginning of table.

No.	Name	Synonyms and Formula	Mol. wt.	Crystalline form, color and specific rotation	m.p. °C	b.p. °C	Density	n_D	Solubility w	al	eth	ace	bz	other solvents	Ref.
	Tropacocaine														
t759	—,hydrochloride	$C_{15}H_{19}NO_2.HCl$. *See* t758	281.78	pr (al)	283d				s	δ	i				**B21**, 39
t760	**Tropane**	2,3-Dihydro-8-methylnortropidine. 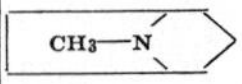 CH_3—N	125.21			163–9	0.9259^{15}_{15}		δ						**B20**[2], 69
t761	**Tropeine, benzoyl-**	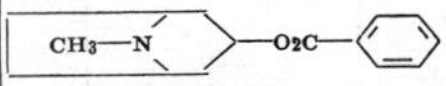 CH_3—N—O_2C—	245.31		41–2 37(+⅝w)				s	s	s				**B21**[2], 17
t763	—,dihydrate	$C_{15}H_{19}NO_2.2H_2O$. *See* t761	281.34	pl	58				s	s	s				**B21**[2], 17
—	**Tropic acid**	*see* **Propanoic acid, 3-hydroxy-2-phenyl-**													
t764	**Tropine**	3-Tropanol. 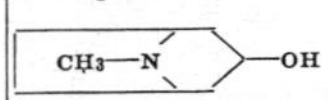 CH_3—N—OH	141.21	hyg pl (eth)	63	233			v	v	s			chl s	**B21**[2], 17
—	**Tropilidene**	*see* **1,3,5-Cycloheptatriene***													
t765	**Tropinic acid**(*d*)	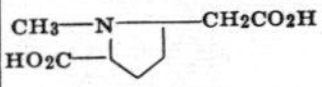 CH_3—N—CH_2CO_2H HO_2C—	187.19	(w or dil al) $[\alpha]_D + 14.8$	253d				s	δ s^h	i			aa i	**B22**, 123
t766	—(*dl*)	$C_8H_{13}NO_4$. *See* t765	187.19	nd (dil al)	251d				v	δ	i			aa i	**B22**, 124
t767	—(*l*)	$C_8H_{13}NO_4$. *See* t765	187.19	(w), $[\alpha]^{20}_D - 14.8$	243				s^h	δ	i			aa i	**B22**, 124
t768	**Tropinone**	3-Tropanone. 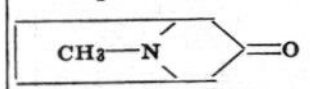 CH_3—N—=O	139.19	lo nd (peth)	42	$224\text{–}5^{714}$ 113^{25}	1.9872^{100}_4	1.4598^{100}			s			oos v	**B21**[2], 225
—	**Tropolone**	*see* **2,4,6-Cycloheptatrien-1-one, 2-hydroxy-**													
—	**Tropone**	*see* **2,4,6-Cycloheptatrien-1-one***													
t769	**Truxane**	Diindene. 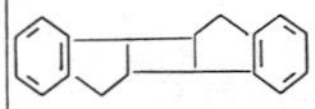	232.33	(peth)	116					v^h	v			aa v	**E14**, 417
t770	—,**tetraphenyl**	Dimer of triphenylallene. 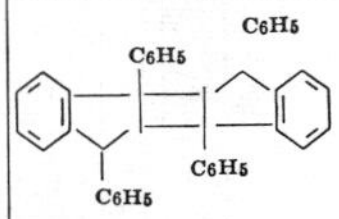 C_6H_5 C_6H_5 C_6H_5 C_6H_5	536.72	lf (bz-al)	210					δ			s^h	chl v	**E14s**, 494
t771	α-**Truxilline**	Cocamine. 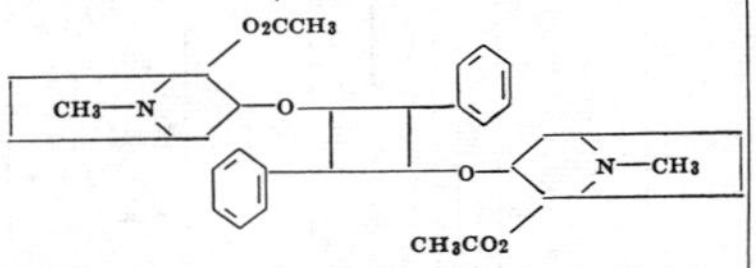 O_2CCH_3 CH_3—N—O— —O— N—CH_3 CH_3CO_2	658.76	amor, $[\alpha]_D < 0$	ca. 80				δ	v	v			chl v peth i	**B22**, 202
t772	**Tryptophan**(*D*)	α-Amino-β-indolepropionic acid. 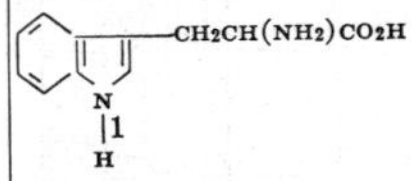 $CH_2CH(NH_2)CO_2H$ N H	204.22		281–2										**C40**, 7244[2]
t773	—(*DL*)	$C_{11}H_{12}N_2O_2$. *See* t772	204.22	pl (50% al)	sub 185–90 289 (264–6)				δ s^h	δ					**B22**[2], 470
t774	—(*L*)	$C_{11}H_{12}N_2O_2$. *See* t772	204.22	lf or pl (dil al), $[\alpha]^{20}_D + 6.1$ (aq NaOH, p = 11), $[\alpha]^{23}_D - 31.5$ (aq HCl, c = 1)	289d				δ s^h	δ	i			chl i	**B22**[2], 466
t775	—,*N*-**acetyl**-(*DL*)	$C_{13}H_{14}N_2O_3$. *See* t772	246.26	pl	206–7				s	s	v			NaOH v	**B22**[2], 469
t776	—,—(*L*)	$C_{13}H_{14}N_2O_3$. *See* t772	246.26	$[\alpha]^{15}_D + 25$ (95% al, c = 1)	185–7				s	s					**Am 66**, 350
t777	—,*N*-**methyl**-(*DL*)	$C_{12}H_{14}N_2O_2$. *See* t772	218.25	nd	245d (297d)				δ	s					**J1938**, 1910
t778	—,—(*L*)	$C_{12}H_{14}N_2O_2$. *See* t772	218.25	$[\alpha]^{21}_D + 4.4$	295d				δ	s					**J1938**, 1910

For explanations, symbols and abbreviations see beginning of table.

No.	Name	Synonyms and Formula	Mol. wt.	Crystalline form, color and specific rotation	m.p. °C	b.p. °C	Density	n_D	Solubility w	al	eth	ace	bz	other solvents	Ref.
	Tuberculostearic acid														
—	**Tuberculostearic acid**	*see* **Octadecanoic acid, 10-methyl-***													
t779	**Tuduranine**		297.34	nd (eth) $[\alpha]_D<0$	ca. 125									os v	
t780	**Turanose**		342.31	pr (w-al) $[\alpha]^{20}_D$ (mut) +27.2 to +75.8	157d				s	δ				MeOH s^h	**B31**, 454
t781	**Tyramine**	$p(\beta$-Aminoethyl)phenyl. HO— —$CH_2CH_2NH_2$	137.18	pr (bz or al), nd (w)	164–5	205–7^{25}			δ s^h	s			δ	to i xyl s	**C29**, 4^7
t782	**Tyrosine**(*D*)	2-Amino-3(4-hydroxyphenyl)-propanoic acid*. HO— —$CH_2CH(NH_2)CO_2H$	181.19	$[\alpha]^{25}_D$ +10.3 (1N HCl, c=4)	310–4d				δ^h					dil HCl s	**B14^2**, 365
t783	—(*DL*)	$C_9H_{11}NO_3$. *See* t782	181.19	nd	316–20d				δ	i δ^h	i				**B14^2**, 379
t784	—(*L*)	$C_9H_{11}NO_3$. *See* t782	181.19	nd (w) $[\alpha]^{22}_D$ −10.6 (1N HCl, c=4)	342–4d	sub			δ^h	i	i				**B14^2**, 366
t785	—,amide(*L*)	$C_9H_{12}N_2O_2$. *See* t782	180.23	pl or pr (al) $[\alpha]^{20}_D$ +19.5 (w)	153–4				s	v				MeOH v	**B14**, 612
t786	—,ethyl ester(*L*)	$C_{11}H_{15}NO_3$. *See* t782	209.25	pr (AcOEt) $[\alpha]^{20}_D$ +20.4 (MeOH)	108–9				δ^h	s	δ		v^h	MeOH v AcOEt s alk s	**B14^2**, 372
t787	—,methyl ester	$C_{10}H_{13}NO_3$. *See* t782	195.12	pr (AcOEt) $[\alpha]^{20}_D$ +25.75 (MeOH)	135–6				δ^h	s	δ		δ	MeOH v AcOEt s alk s	**B14^2**, 372
t788	—,**3-bromo-**(*L*)	$C_9H_{10}BrNO_3$. *See* t782	260.10	(w+1), nd (w+2)	246–9d				δ v^h						**B14^2**, 383
t789	—,**3,5-dibromo-**(*L*)	$C_9H_9Br_2NO_3$. *See* t782	339.01	$[\alpha]^{20}_D$ −5.5 (1N HCl, c=5)	242–5				δ^h	δ s^h	i			ac, alk s	**B14^2**, 377
—	—,**diiodo-**	*see* **Iodogorgoic acid**													
t790	—,*N*-**methyl-**	Andirine. Surinamine. $C_{10}H_{13}NO_3$. *See* t782	195.21	nd, $[\alpha]^{21}_D$ +19.8 (dil al)	280d (246)				δ	i	i			NH_4OH v	**B14^1**, 665

For explanations, symbols and abbreviations see beginning of table.

No.	Name	Synonyms and Formula	Mol. wt.	Crystalline form, color and specific rotation	m.p. °C	b.p. °C	Density	n_D	Solubility w	al	eth	ace	bz	other solvents	Ref.
	Umbellatine														
u1	**Umbellatine**	$C_{21}H_{21}NO_8$	415.39	ye nd	206–7	...	...	...	δ s^h	s	...	δ	...	chl δ	...
—	**Umbellic acid**	*see* **Cinnamic acid, 2,4-dihydroxy-**													
—	**Umbelliferone**	*see* **Coumarin, 7-hydroxy-**													
—	**Undecane***	*see* **Hendecane***													
—	**Undecanoic acid***	*see* **Hendecanoic acid***													
—	**Undecanol***	*see* **Hendecanol***													
—	**Undecanone***	*see* **Hendecanone***													
—	**Undecenoic acid***	*see* **Hendecenoic acid**													
—	**Undecyl alcohol**	*see* **1-Hendecanol***													
—	**Undecylic acid**	*see* **Hendecanoic acid**													
—	**Undecyne***	*see* **Hendecyne***													
u2	**Uracil**	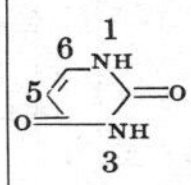	112.09	nd (w)	338	...	...	...	δ v^h	v	v	...	...	dil NH_3 s	**B24**^{2}, 168
u3	**—,5-amino-**	$C_4H_5N_3O_2$. *See* u2	127.11	nd (w)	d	...	...	...	i δ^h	...	...	...	...	alk^h s ac^h s	**B24**^{2}, 266
u4	**—,5,6-dihydro-**	Hydrouracil. β-Lactylurea. $C_4H_6N_2O_2$. *See* u2	114.10	nd (w)	275	...	...	...	v	s	...	...	...	MeOH s chl s	**B24**^{2}, 140
u4[1]	**—,5,6-dihydro-5-methyl-**	Dihydrothymine. $C_5H_8N_2O_2$. *See* u2	128.13	(w or al)	264–5	...	...	...	v^h	δ	...	...	...	...	**B24**^{1}, 306
u5	**—,5-hydroxy-**	Isobarbituric acid. $C_4H_4N_2O_3$. *See* u2	128.09	pr (w)	d>300	...	...	...	...	...	...	...	...	alk s	**B24**^{1}, 408
u6	**—,1-methyl-**	$C_5H_6N_2O_2$. *See* u2	126.12	pr	174–5	...	...	...	v	v	...	...	i	...	**B24**^{2}, 170
u7	**—,3-methyl-**	$C_5H_6N_2O_2$. *See* u2	126.12	pr (al), nd (w)	232	...	...	...	v	v^h	...	...	...	dil alk s	**B24**^{2}, 170
u7[1]	**—,5-methyl-**	Thymine. $C_5H_6N_2O_2$. *See* u2	126.12	nd (al), pl (w)	226	sub	...	...	δ	δ	δ	...	...	...	**B24**, 353
u8	**—,6-methyl-**	$C_5H_6N_2O_2$. *See* u2	126.12	oct, pr or nd (w or al)	270–80d	...	...	...	s	s^h	δ	...	...	NH_3 v	**B24**^{2}, 182
u9	**—,6-methyl-2-thio-**	$C_5H_6N_2OS$. *See* u2	142.19	pl (w)	d>300	...	...	...	i s^h	δ	δ	...	δ	MeOH δ	**B24**^{2}, 183
u10	**—,6-methyl-4-thio-**	$C_5H_6N_2OS$. *See* u2	142.19	ye pr	d>250	...	...	...	δ^h	δ	...	...	...	...	**B24**, 352
u11	**—,5-nitro-**	$C_4H_3N_3O_4$. *See* u2	157.09	nd (al)	exp	...	...	...	δ	s^h	...	...	...	...	**B24**^{2}, 171
u12	**—,6-propyl-2-thio-**	$C_7H_{10}N_2OS$. *See* u2	170.23	pw	219	...	...	...	δ	δ	δ	...	i	chl δ	...
u13	**—,2-thio-**	$C_4H_4N_2OS$. *See* u2	128.15	pr (w or al)	d>280	...	...	...	δ	δ	...	...	...	anh HFs	**B24**^{2}, 171
u14	**—,4-thio-**	$C_4H_4N_2OS$. *See* u2	128.15	yesh nd (w)	328d	...	...	...	s^h	δ	...	...	...	...	**B24**, 323
u15	**6-Uracilcarboxylic acid**	Orotic acid. 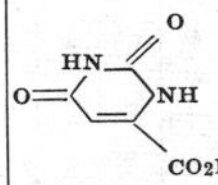	156.10	pr (w+1)	347d	...	...	...	δ	...	...	...	...	os i	**B25**^{2}, 249
u16	—,monohydrate	$C_5H_4N_2O_4.H_2O$. *See* u15	174.11	pr (w)	d125–30	...	...	...	δ s^h	...	...	...	...	os i	**B25**^{2}, 249
—	**Uramil**	*see* **Barbituric acid, 5-amino-**													
u17	**Uranine**	Sodium salt of fluorescein. 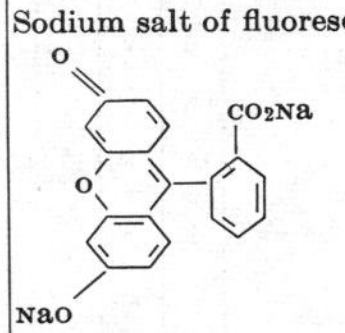	376.29	ye pw	...	...	...	...	s	s	...	...	...	glycerol s dil ace s	**B19**^{2}, 252
—	***p*-Urazine**	*see* **Diurea**													
u18	**Urea***	Carbamide. Carbonyl diamide. H_2NCONH_2	60.06	tetr	132.7	d	1.32_4^{18}	1.484 (1.602)	v	v	...	...	...	chl i MeOH v abs al s	**B3**^{2}, 35
u19	—,nitrate*	$H_2NCONH_2.HNO_3$	123.07	mcl lf	152d	...	1.63	...	δ v^h	s	i	...	i	chl i HNO_3 δ	**B3**^{2}, 45
u20	—,oxalate	$2(H_2NCOH_2).HO_2CCO_2H$	210.15	mcl (w+1)	173d	...	1.585	...	s	δ	i	...	...	...	**B3**^{2}, 48
u21	**—,1-acetyl-**	$CH_3CONHCONH_2$	102.09	nd (al)	216–7	180–90 sub	...	...	δ s^h	s	δ	...	...	...	**B3**^{2}, 49
u22	**—,1-acetyl-3-methyl-**	$CH_3CONHCONHCH_3$	116.12	tcl (w, al or AcOEt)	180	d	...	...	s v^h	δ	δ	...	...	...	**B4**^{2}, 568
u23	**—,1-acetyl-2-thio-**	*N*(Thiocarbamyl)acetamide. $CH_3CONHCSNH_2$	118.16	pr (w^h)	167	...	...	...	δ s^h	s	δ	...	...	dil NaOH s	**B3**^{2}, 131

For explanations, symbols and abbreviations see beginning of table.

No.	Name	Synonyms and Formula	Mol. wt.	Crystalline form, color and specific rotation	m.p. °C	b.p. °C	Density	n_D	Solubility w	al	eth	ace	bz	other solvents	Ref.
	Urea														
u24	—,*S*-**acetyl-2-thio-**, hydrochloride	$H_2NC(:NH)SCOCH_3.HCl$	154.62	pl	109d				v	δ					**B3**^{2}, 133
u25	—,**1-allyl-**	$H_2NCONHCH_2CH:CH_2$	100.12	nd (al)	85				∞	∞	δ			chl, CS_2 δ	**B4**^{2}, 664
u26	—,**1-allyl-3-phenyl-**	$C_6H_5NHCONHCH_2CH:CH_2$	176.22	nd (bz)	115.5				δ	s			s		**B12**, 350
u27	—,**1-allyl-2-thio-**	$H_2NCSNHCH_2CH:CH_2$	116.19	mcl or rh	78.4		1.219^{20}_{20}		s^h	s	δ		i		**B4**, 212
u28	—,**1(4-aminobenzenesulfonyl)-***	1-Sulfanylurea. Sulfacarbamide. $H_2N-C_6H_4-SO_2NHCONH_2$	215.23		143–7 (158–60d)	sub 320			δ	s^h					**Am 64**, 1682
u29	—,**1(4-aminino-benzene-sulfonyl)2-thio-**	1-Sulfamylthiourea. Sulfathiourea. $H_2N-C_6H_4-SO_2NHCSNH_2$	231.30	pw, la pl (w)	182 (200d)				i						**Am 68**, 761
u30	—,**1-benzoyl-**	$H_2NCONHCOC_6H_5$	164.16	flakes	214.5					s					**B9**^{2}, 172
u31	—,**1-benzoyl-2-thio-**	$H_2NCSNHCOC_6H_5$	180.22	pr (dil al)	169–70				δ	s	i				**B9**^{2}, 173
u32	—,**1-benzyl-**	$H_2NCONHCH_2C_6H_5$	150.18	nd (al)	147.5	d 200			δ		i	s	δ		**B12**^{2}, 563
u33	—,**1-benzyl-2-thio-**	$H_2NCSNHCH_2C_6H_5$	166.24	pr (w)	164.5–5.5				i	δ					**B12**^{2}, 564
u33[1]	—,**1,3-bis(2-ethoxyphenyl)-***	$C_{17}H_{20}N_2O_3$. *See* u71	316.36	pr (dil al)	125					s^h					**B13**^{2}, 179
u33[2]	—,**1,3-bis(4-ethoxyphenyl)-**	$C_{17}H_{20}N_2O_3$. *See* u71	316.36	nd (aa), pr (abs al)	225–6				i	s^h				aa s^h	**B13**^{2}, 254
u34	—,**1,3-bis(2,4-dinitrophenyl)-***	$C_{13}H_8N_6O_9$. *See* u71	392.24	lf or pr (al), nd (con HNO_3)	204				i	δ	δ				**B12**^{2}, 410
u35	—,**1,3-bis-(hydroxy-methyl)-***	*N,N′*-Dimethylolurea. $HOCH_2NHCONHCH_2OH$	120.12	pr (abs al), pl (w-al)	137–8	d260			δ	s^h				MeOH s	**B3**^{2}, 49
u36	—,**1,3-bis(1-hydroxy-2,2,2-trichloroethyl)-***	Dichloralurea. EH2. Crag herbicide. $Cl_3CCHOHNHCONHCHOHCCl_3$	354.85		191d				i						
u37	—,**1(2-bromo-2-ethyl-butanoyl)-***	Adalin. Carbromal. Uradal. $H_2NCONHCOCBr(C_2H_5)_2$	237.11	(dil al)	117–8				δ s^h			s	s		**B3**^{2}, 52
u38	—,**1(2-bromo-3-methyl-butanoyl)-***	Bromoisovalum. Bromural. Pivadorm. $H_2NCONHCOCHBrCH(CH_3)_2$	223.08	nd or lf (to)	151 (160)	sub			δ v^h	s v^h	v	s	s	alk v, chl s lig i	**B3**^{2}, 51
u39	—,**1(2-bromo-phenyl)-***	$C_7H_7BrN_2O$. *See* u109	215.07	nd (al)	202				i	s			δ	chl δ	**B12**, 632
u40	—,**1(3-bromo-phenyl)-***	$C_7H_7BrN_2O$. *See* u109	215.07	nd (abs al-eth-lig)	164–5					s				os i	**B12**, 634
u41	—,**1(4-bromo-phenyl)-***	$C_7H_7BrN_2O$. *See* u109	215.07	nd (bz)	ca. 220	d ca. 260 (d 296)			$δ^h$	v^h	v		v	aa v lig δ	**B11**^{2}, 351
u42	—,**1-butyl-***	$H_2NCONH(CH_2)_3CH_3$	116.16	ta (w), nd (bz)	96(86)				v	v					**B4**^{2}, 634
u43	—,**1-*sec*-butyl-**	$H_2NCONHCH_2CH(CH_3)_2$	116.16	nd *d* form: $[\alpha]^{20}_D+24.1$ (al, c=1.5), +27.6 (chl, c=0.3)	169–70 (*dl* form) 166 form (*d*)					s					**B4**, 160
u44	—,**1-*tert*-butyl-**	$H_2NCONHC(CH_3)_3$	116.16	nd (w or dil al)	172 (183)				δ	v			δ		**B4**^{1}, 377
u45	—,**1-butyl-3-phenyl-2-thio-***	$C_6H_5NHCSNH(CH_2)_3CH_3$	208.32	pr	85(62–3)								s	CCl_4 s	**B12**^{2}, 226
u48	—,**1(2-chloro-phenyl)-***	$C_7H_7ClN_2O$. *See* u109	170.60	pr (w)	152				s	v	δ	v	δ		**B12**, 60
u49	—,**3(4-chloro-phenyl)-1,1-dimethyl-***	CMU Weedkiller (Du Pont). $Cl-C_6H_4-NHCON(CH_3)_2$	198.66	gy-wh	170.5–1.5				i	δ		δ			
u50	—,**1(2-chloro-phenyl)2-thio-***	$C_7H_7ClN_2S$. *See* u112	186.67	nd	146–6.5					v			s		**B12**^{2}, 318
u51	—,**1(4-chloro-phenyl)2-thio-***	$C_7H_7ClN_2S$. *See* u112	186.67	pl or nd (al)	178					s^h			s		**B12**^{2}, 329
u52	—,**1,3-diacetyl-**	$CH_3CONHCONHCOCH_3$	144.13	nd (al, 50% aa)	144–5	sub d 179–80			δ	δ					**B3**^{2}, 50
u55	—,**1,1-diethyl-***	$H_2NCON(C_2H_5)_2$	116.16	pl (abs eth), nd (eth)	75				v	v	s		v	lig v	**B4**^{2}, 611
u56	—,**1,3-diethyl-***	$CH_3CH_2NHCONHCH_2CH_3$	116.16	ta (lig), dlq nd (al)	112	263	1.0415	1.4616^{49}	v	v	v				**B4**^{2}, 608
u57	—,**1,3-diethyl-1,3-diphenyl-***	$C_6H_5N(C_2H_5)CON(C_2H_5)C_6H_5$	268.36	(al or w)	79				i	v					**B12**^{2}, 238
u58	—,**1,3-diethyl-1,3-diphenyl-2-thio-***	$C_6H_5N(C_2H_5)CSN(C_2H_5)C_6H_5$	284.43	pl (lig), pr (al)	75.5				i	v^h				lig s^h	**B12**, 624

For explanations, symbols and abbreviations see beginning of table.

No.	Name	Synonyms and Formula	Mol. wt.	Crystalline form, color and specific rotation	m.p. °C	b.p. °C	Density	n_D	Solubility w	al	eth	ace	bz	other solvents	Ref.
	Urea														
u59	**—,1,3-diethyl-2-thio-***	$CH_3CH_2NHCSNHCH_2CH_3$	132.23	...	77	d		...	s	s	v	...	...	...	**B4²**, 610
u60	**—,1,1-dimethyl-***	$NH_2CON(CH_3)_2$	88.11	mcl pr (al or chl)	182	...	1.255	...	s	δ	i	...	...	...	**B4²**, 573
u61	**—,1,3-dimethyl-***	$CH_3NHCONHCH_3$	88.11	rh bipym (chl-eth)	107	268–70	1.142	...	v	v	i	...	...	...	**B4²**, 568
u62	**—,1,3-dimethyl-1,3-diphenyl-***	$C_6H_5N(CH_3)CON(CH_3)C_6H_5$	240.31	pl (al)	121	350	...	...	i	v	v	...	v	...	**B12²**, 236
u64	**—,1,3-dimethyl-2-thio-***	$CH_3NHCSNHCH_3$	104.17	dlq pl	60–2 (51.5)	...	...	...	v	v	δ	v	δ	chl v CS₂, peth i	**B4²**, 573
u65	**—,1,1-di(2-naphthyl)-***	$H_2NCON(C_{10}H_7^{\beta})_2$	312.37	nd (al)	192–3	...	...	...	...	δ	...	...	δ	aa δ	**B12**, 1297
u66	**—,1,3-di(1-naphthyl)-***	$C_{10}H_7^{\alpha}NHCONHC_{10}H_7^{\alpha}$	312.37	nd ($PhNO_2$, Py)	293	...	...	...	δ^h	...	...	...	...	Py s $PhNO_2$ s	**B12²**, 692
u67	**—,1,3-di(2-naphthyl)-***	$C_{10}H_7^{\beta}NHCONHC_{10}H_7^{\beta}$	312.37	nd (ace, $PhNO_2$, aa)	301–2	...	...	...	...	δ^h	δ	...	δ	$PhNO_2$ δ^h *i*-AmOH s	**B12²**, 733
u68	**—,1,3-di(1-naphthyl)2-thio-***	$C_{10}H_7^{\alpha}NHCSNHC_{10}H_7^{\alpha}$	328.44	nd ($PhNO_2$), lf (to)	207.5	...	...	...	...	δ^h	δ	...	δ	to, $PhNO_2$ s^h CS_2 δ	**B12²**, 696
u69	**—,1,3-di(2-naphthyl)2-thio-***	$C_{10}H_7^{\beta}NHCSNHC_{10}H_7^{\beta}$	328.44	lf (aa, $PhNO_2$)	203 (192–3)	...	...	...	...	δ^h	δ^h	...	...	os δ	**B12²**, 723
u70	**—,1,1-diphenyl-***	$H_2NCON(C_6H_5)_2$	212.25	ta	189	d	1.276	...	δ	s	s	...	...	chl s	**B12²**, 241
u71	**—,1,3-diphenyl-***	Carbanilide. 3′ 2′ 2 3 / 4′ ⟨ring⟩—NHCONH—⟨ring⟩ 4 / 5′ 6′ 6 5	212.25	rh bipym, pr (al)	237–7.5	262^{760}	1.239	...	i	s	s	...	i	chl i	**B12²**, 207
u72	**—,1,3-diphenyl-1-methyl-***	*N*-Methylcarbanilide. $C_6H_5NHCON(CH_3)C_6H_5$	226.28	nd (al), (gasoline)	106	$203–5^{760}$	...	...	δ^h	δ	v	...	...	chl s bz-aa v	**B12¹**, 251
u73	**—,1,3-diphenyl-*S*-methyl-2-thio-**	*S*-Methyl-*N,N′*-diphenylisothiourea. $C_6H_5N{:}C(SCH_3)NHC_6H_5$	242.35	nd (al)	109–10	...	...	...	...	s^h	...	...	...	...	**B12²**, 248
u74	**—,1,1-diphenyl-2-thio-***	$H_2NCSN(C_6H_5)_2$	228.32	pr (al)	210d	...	...	...	i	s	...	...	...	...	**B12²**, 242
u75	**—,1,3-diphenyl-2-thio-***	$C_6H_5NHCSNHC_6H_5$	228.32	lf (al)	154 (189)	...	...	...	δ	v	v	...	...	chl v olive oil v	**B12²**, 227
u76	**—,1,1-dipropyl-2-thio-***	$H_2NCSN(CH_2CH_2CH_3)_2$	160.28	...	67	...	...	...	...	...	...	...	...	...	**B4**, 144
u77	**—,1,3-dipropyl-2-thio-***	$CH_3CH_2CH_2NHCSNHCH_2CH_2CH_3$	160.28	lf (w)	71	...	...	...	δ s^h	...	...	...	...	...	**B4**, 143
u78	**—,1,3-diiso-propyl-2-thio-***	$(CH_3)_2CHNHCSNHCH(CH_3)_2$	160.28	nd (w^h)	161	...	...	...	δ^h	...	...	...	...	...	**B4**, 155
u79	**—,1,3-di(2-tolyl)2-thio-**	CH_3 CH_3 ⟨ring⟩—NHCSNH—⟨ring⟩	256.37	nd (al, sub)	165–6	$216–8^{760}$ (sub)	...	...	i	s^h	δ	...	s^h	chl s	**B12²**, 446
u80	**—,1,3-di(4-tolyl)2-thio-**	CH_3—⟨ring⟩—NHCSNH—⟨ring⟩—CH_3	256.37	rh bipym pr	176 cor	...	...	...	i	δ	s	...	...	...	**B12²**, 514
u81	**—,1(3-ethoxy-phenyl)-***	$C_9H_{12}N_2O_2$. *See* u109	180.21	nd	112	...	...	...	δ	δ s^h	...	...	δ	...	**B13**, 418
u82	**—,1(4-ethoxy-phenyl)-***	$C_9H_{12}N_2O_2$. *See* u109	180.21	lf (dil al)	171.5	d	...	...	δ	s	...	...	...	AcOEt v	**B13²**, 253
u83	**—,1-ethyl-***	$H_2NCONHCH_2CH_3$	88.11	nd (al-eth), (bz)	92.1–.4	d	...	...	v	v	...	...	v^h	abs eth, i CS_2 i chl v	**B4²**, 607
u84	**—,1-ethyl-1-phenyl-***	$H_2NCON(C_2H_5)C_6H_5$	164.21	ta (peth)	62.3–.5	...	...	...	s	...	...	...	...	...	**B12²**, 237
u85	**—,1-ethyl-3-phenyl-***	$C_6H_5NHCONHCH_2CH_3$	164.21	nd (dil al)	100	...	...	...	...	s	...	...	...	...	**B12²**, 205
u86	**—,1-ethyl-3-phenyl-2-thio-***	$C_6H_5NHCSNHCH_2CH_3$	180.27	...	107 (102)	...	...	...	...	s^h	...	...	s^h	...	**B12²**, 226
u87	**—,1-ethyl-2-seleno-**	$H_2NCSeNHCH_2CH_3$	151.07	nd (al-peth)	*ca.* 125	...	...	...	v^h	s^h	...	...	...	...	**B4²**, 610
u88	**—,1-ethylidene-2-thio-***	$H_2NCSN{:}CHCH_3$	102.16	(al)	*ca.* 212d	...	...	...	i	δ s^h	δ	...	...	oos i	**B3¹**, 76
u89	**—,1-hydroxy-***	Carbamide oxide. $H_2NCONHOH$	76.06	nd (al^h)	139–40	d	...	...	v	δ v^h	...	...	...	...	**B3²**, 78
u90	**—,1(hydroxy-methyl)-***	Methylolurea. $H_2NCONHCH_2OH$	90.08	pr (al)	110–1	...	...	...	v	s^h	i	...	...	MeOH s	**B3¹**, 27

For explanations, symbols and abbreviations see beginning of table.

No.	Name	Synonyms and Formula	Mol. wt.	Crystalline form, color and specific rotation	m.p. °C	b.p. °C	Density	n_D	Solubility w	al	eth	ace	bz	other solvents	Ref.
	Urea														
u91	**—,1(2-iodo-3-methyl-butanoyl)***	Iodival. α-Iodoisovalerylurea. $H_2NCONHCOCHICH(CH_3)_2$	270.07	lf (al)	180–1				δ s^h	v	i	...	δ	dil NaOH s	**B3**^{1}, 29
u92	**—,isobutyl-**	$H_2NCONHCH_2CH(CH_3)_2$	116.16	pr (w), nd (ace)	141				...	...	i	δ	δ	...	**B4**^{1}, 376
u93	**—,1(2-isopropyl-4-pentenoyl)-***	$H_2NCONHCOCH(C_3H_7^i)CH_2CH{:}CH_2$	184.24	nd (al)	194–4.5				i δ^h	s v^h	δ	...	...	...	**B3**^{2}, 53
u94	**—,1(4-methoxy-phenyl)-***	$C_8H_{10}N_2O_2$. *See* u109	166.18	pl (w)	168				δ	...	s	...	...	...	...
u95	**—,1-methyl-***	$H_2NCONHCH_3$	74.08	rh pr (w or al)	102	d			v	v	i	...	i	CS_2 i	**B4**^{2}, 567
u95¹	**—,1(2-methyl-2-butyl)-***	*tert*-Amylurea. $H_2NCONHC(CH_3)_2CH_2CH_3$	130.19	mcl (w^h)	157 (151)				δ	...	...	...	...	...	**B4**^{1}, 379
u96	**—,1(3-methyl-butyl)-***	Isoamylurea. $H_2NCONHCH_2CH_2CH(CH_3)_2$	130.19	actinomorphic	150				δ	...	...	...	...	...	**B4**^{1}, 383
u97	**—,1-methyl-3(1-naphthyl)-2-thio-***	$C_{10}H_7^{\alpha}NHCSNHCH_3$	216.31	pl (al)	198				...	s^h	...	...	...	...	**B12**^{2}, 696
u98	**—,1-methyl-1-nitroso-***	$H_2NCON(NO)CH_3$	103.09	colorless to yesh pl (eth)	123–4d				i v^h	v	v	v	s	chl s	**B4**^{1}, 342
u99	**—,1-methyl-3-phenyl-2-thio-***	$C_6H_5NHCSNHCH_3$	166.25	ta, pl	112–3				...	s	...	...	...	...	**B12**^{2}, 225
u100	**—,*S*-methyl-2-thio-**	2-Methylthiopseudourea. $H_2NC({:}NH)SCH_3$	90.14	lf (ace)	79				s	s	...	s	...	...	**B3**^{2}, 132
u100¹	—,—,hydroiodide	*S*-Methylisothiouronium iodide. $H_2NC({:}NH)SCH_3.HI$	218.07	pr	117				v	v	...	...	...	...	**B3**, 192
u100²	—,—,nitrate	*S*-Methylthiouronium nitrate. $H_2NC({:}NH)SCH_3.HNO_3$	153.17	(HNO_3)	109–10				s	v	...	...	...	MeOH v HNO_3 s^h	**B3**^{1}, 78
u101	—,—,sulfate	*S*-Methylisothiouronium sulfate. $2[H_2NC({:}NH)SCH_3].H_2SO_4$	278.39	nd (al or w)	244d				δ s^h	i s^h	...	...	...	...	**B3**^{2}, 132
u102	**—,1-methyl-2-thio-***	$H_2NCSNHCH_3$	90.14	pr	119				v	v	δ	s	...	...	**B4**^{2}, 572
u103	**—,1(1-naphthyl)-***	$H_2NCONHC_{10}H_7^{\alpha}$	186.22	nd (al)	213–4				i	v	s	...	...	...	**B12**, 1238
u104	**—,1(2-naphthyl)-***	$H_2NCONHC_{10}H_7^{\beta}$	186.22	nd (al)	213–4				i	v^h	...	...	...	...	**B12**, 1292
u105	**—,1(1-naphthyl)-3-phenyl-2-thio-***	$C_6H_5NHCSNHC_{10}H_7^{\alpha}$	278.38	lf	162–3				...	i	...	...	...	...	**B12**, 500
u106	**—,1(1-naphthyl)-2-thio-***	$H_2NCSNHC_{10}H_7^{\alpha}$	202.27	pr (al)	198				i	δ	δ	δ	...	...	**C42**, 4560
u107	**—,1(2-naphthyl)-2-thio-***	$H_2NCSNHC_{10}H_7^{\beta}$	202.27	lf (al)	186				...	s^h	...	...	...	os δ	**B12**^{2}, 723
u108	**—,1-nitro-***	$H_2NCONHNO_2$	105.06	pl (al-peth)	158.4–.8d	exp			δ d^h	v	v	v	δ	peth δ chl δ	**B3**^{2}, 99
—	**—,1-oxalyl-**	*see* **Parabanic acid**													
u109	**—,1(2-phenoxy-ethyl)-***	$H_2NCONHCH_2CH_2OC_6H_5$	180.21	nd (w+?), (50% al)	120–1				s^h	s	...	...	...	...	**B6**^{1}, 91
u110	**—,1-phenyl-**	H_2NCONH— (ring: 2 3 4 5 6)	136.15	mcl (w, al)	147 (d)	238			δ s^h	s^h	s	...	...	AcOEt s	**B12**^{2}, 204
u111	**—,1(phenyl-acetyl)-**	Phenylcarbamide. $H_2NCONHCOCH_2C_6H_5$	178.19	(al)	209				...	s^h	...	...	...	...	...
u112	**—,1-phenyl-2-thio-***	Phenylthiocarbamide. H_2NCSNH— (ring: 2 3 4 5 6)	152.22	nd (w), pr (al)	154				δ s^h	s	...	...	...	NaOH s	**B12**^{2}, 225
u112¹	**—,*S*-phenyl-2-thio-***	*S*-Phenylisothiourea. $C_6H_5SC({:}NH)NH_2$	152.22	nd (bz)	96–7d				...	...	...	...	s	...	**B6**^{1}, 146
u113	**—,1-propyl-***	$H_2NCONHCH_2CH_2CH_3$	102.14	pr (al)	107				δ	s	...	...	...	...	...
u114	**—,2-seleno-**	Selenourea. $H_2NCSeNH_2$	123.02	pr or nd (w)	200d (213d)				s v^h	δ s^h	...	...	...	abs al s^h	**B3**^{1}, 87
—	**—,1(sulfonamyl-phenyl)-***	*see* **Benzenesulfonic acid, ureido-,** amide													
u115	**—,1,1,3,3-tetra-ethyl-***	$(C_2H_5)_2NCON(C_2H_5)_2$	172.22			210–5			i	...	...	...	...	alk, ac i	**B4**^{2}, 611

For explanations, symbols and abbreviations see beginning of table.

No.	Name	Synonyms and Formula	Mol. wt.	Crystalline form, color and specific rotation	m.p. °C	b.p. °C	Density	n_D	Solubility w	al	eth	ace	bz	other solvents	Ref.
	Urea														
u116	**—,1,1,3,3-tetra-methyl-***	$(CH_3)_2NCON(CH_3)_2$	116.16	...	...	177.5^{766} cor	0.972^{15}	...	...	δ	δ	...	...	...	**B4²**, 574
u117	**—,1,1,3,3-tetra-methyl-2-thio-***	$(CH_3)_2NCSN(CH_3)_2$	132.23	...	78	245	...	...	s v^h	s	δ	...	...	...	**B4²**, 576
u118	**—,1,1,3,3-tetra-phenyl-***	$(C_6H_5)_2NCON(C_6H_5)_2$	364.43	rh (bz)	183	...	1.222	...	i	δ^h	...	...	...	...	**B12²**, 242
u119	**—,1,1,3,3-tetra-phenyl-2-thio-***	$(C_6H_5)_2NCSN(C_6H_5)_2$	380.51	nd (al)	194.5–5.5	...	...	...	i	δ	s	...	v	con sulf s	**B12²**, 242
u120	**—,2-thio-***	Thiourea*. H_2NCSNH_2	76.12	rh (al)	180–2	...	1.405	...	s	s	i	...	...	...	**B3²**, 128
u121	**—,2-thio-1(2-tolyl)-**	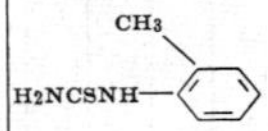 CH_3 / H_2NCSNH— (ring)	166.25	nd (dil al), (w)	162	...	...	...	v^h	v	δ	...	...	...	**B12²**, 445
u122	**—,2-thio-1(3-tolyl)-**	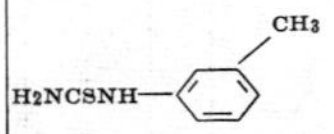 CH_3 / H_2NCSNH— (ring)	166.25	pr (al)	110–1	...	...	...	s^h	δ	δ	...	...	...	**B12²**, 470
u123	**—,2-thio-1(4-tolyl)-**	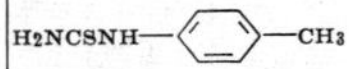 H_2NCSNH— (ring) —CH_3	166.25	pl (al)	188–9	...	...	...	δ s^h	s^h	...	...	...	...	**B12²**, 514
u123¹	**—,2-thio-1,1,3-trimethyl-***	$CH_3NHCSN(CH_3)_2$	118.21	pr (bz-lig)	87–8	...	...	...	v	v	...	...	s	chl s	**B4²**, 576
u124	**—,2-thio-1,1,3-triphenyl-***	$C_6H_5NHCSN(C_6H_5)_2$	304.42	nd (al)	152	...	...	...	...	δ	...	...	...	...	**B12**, 432
u125	**—,1(2-tolyl)-**	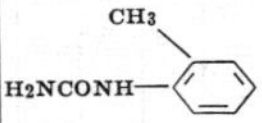 CH_3 / H_2NCONH— (ring)	150.18	lf (al)	190–1	...	...	...	i δ^h	v	v	δ	...	...	**B12²**, 444
u126	**—,1(3-tolyl)-**	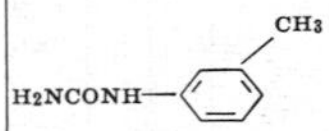 CH_3 / H_2NCONH— (ring)	150.18	lf (w)	142	...	...	...	s	v	δ	...	...	...	**B12²**, 470
u127	**—,1(4-tolyl)-**	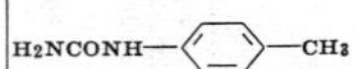 H_2NCONH— (ring) —CH_3	150.18	nd (w), pl (w-aa)	181–2 (187)	...	...	...	δ^h	v	i	s	i	...	**B12²**, 511
u128	**—,1,1,3-tri-methyl-***	$CH_3NHCON(CH_3)_2$	102.14	pr (eth)	75.5	cor 232.5^{764}	...	...	s	s	δ	...	δ	...	**B4¹**, 335
u129	**1-Ureacarboxylic acid,** ethyl ester*	Allophanic acid ethyl ester. $H_2NCONHCO_2C_2H_5$	132.12	nd (w), (bz)	190	d	...	...	i s^h	δ	i	...	δ s^h	...	**B3²**, 56
u130	**—,2-phenyl-,** nitrile	1-Cyano-3-phenylurea. $C_6H_5NHCONHCN$	161.16	nd	125	...	...	...	...	s	...	...	...	...	**B12²**, 210
—	**Urethan**	*see* **Carbamic acid,** ethyl ester													
—	**Urethylan**	*see* **Carbamic acid,** methyl ester													
u131	**Uric acid**	2,6,8(1,3,9)Purinetrione. 2,6,8-Trihydroxypurine. 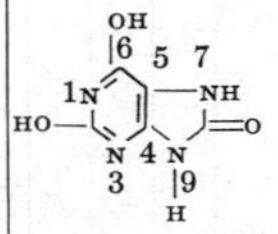 (structure: OH; 6 5 7; 1N —NH; HO— =O; N 4 N; 3 9; H)	168.11	...	d	d	1.89	...	i i^h	i	...	...	...	alk s aq Na_2CO_3 s glycerol s	**B26²**, 293
u132	**—,1-methyl-**	$C_6H_6N_4O_3$. *See* u131	182.14	nd	400	...	...	...	i^h	...	...	...	...	...	**B26²**, 299
u133	**—,3-methyl-**	$C_6H_6N_4O_3$. *See* u131	182.14	pr (w+1)	>350	...	1.6104^{25} (anh) 1.6334^{25} (hyd)	...	δ^h	δ	...	...	...	alk s	**B26²**, 299
u134	**—,7-methyl-**	$C_6H_6N_4O_3$. *See* u131	182.14	pl (w^h)	370–80d	...	1.706^{25}_4	...	δ^h	...	...	...	...	NaOH s	**B26²**, 299
u135	**Uridine**	3(*d*)-Ribofuranosidouracil. 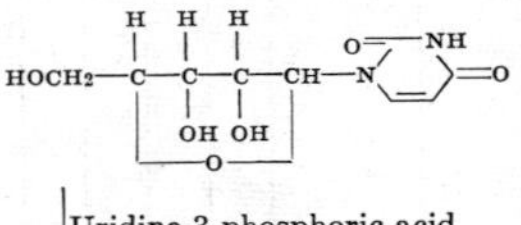 (structure: H H H; $HOCH_2$—C—C—C—CH—N; OH OH; O; O=; NH; =O)	244.21	nd (aq al) $[\alpha]^{20}_D+4$ (w)	164–5	...	...	...	s	s^h	...	...	...	Py s	**J1947**, 338
u136	**Uridylic acid**	Uridine-3-phosphoric acid. 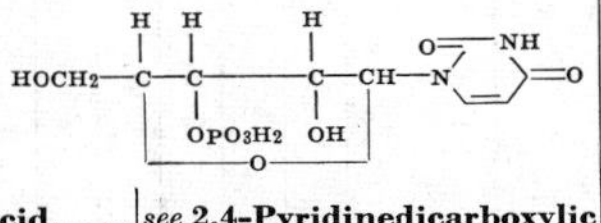 (structure: H H H; $HOCH_2$—C—C—C—CH—N; OPO_3H_2 OH; O; O=; NH; =O)	324.19	pr (MeOH) $[\alpha]^{20}_D+10.5$	202d	...	...	...	v	...	...	...	...	MeOH s^h	...
—	**Uritonic acid**	*see* **2,4-Pyridinedicarboxylic acid, 6-methyl-**													

For explanations, symbols and abbreviations see beginning of table.

No.	Name	Synonyms and Formula	Mol. wt.	Crystalline form, color and specific rotation	m.p. °C	b.p. °C	Density	n_D	Solubility						Ref.
									w	al	eth	ace	bz	other solvents	
	Urocanic acid														
u137	**Urocanic acid**	Iminazolylacrylic acid. Urocaninic acid. 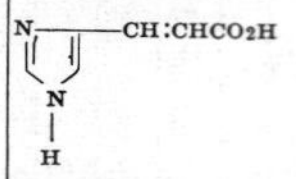	138.13	pr (w+2)	231 (anh) 218–24 (hyd)	...	...	...	s^h	i	i	s^h	...	...	**B25²**, 25, 121
u138	**Urochloralic acid**	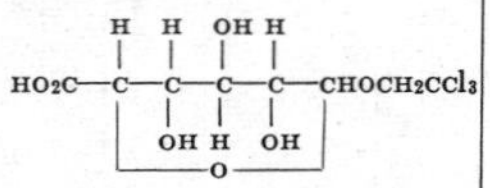 β,β,β-Trichloroethyl-*D*-glucuronide.	325.23	levorotatory nd	142	...	...	...	v	v	δ	...	...	...	**B31**, 266
—	**Uronium, *S*-methyl-2-thio-**, inorganic salt	*see* **Urea, *S*-methyl-2-thio-,** inorganic salt													
—	**Uropterin**	*see* **Xanthopterin**													
—	**Urotropine**	*see* **Hexamethylenetetramine**													
—	**Ursocholanic acid**	*see* **Cholanic acid**													
u139	**Ursodesoxycholic acid**	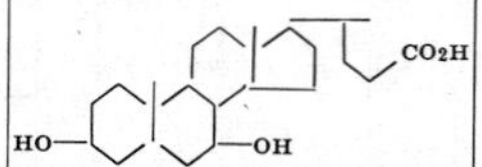 17(β)(1-Methyl-3-carboxypropyl)-etiocholane-3(α),-7(β)-diol.	392.58	pl (al) $[\alpha]^{20}_D+57$	203	...	...	...	...	v	δ	...	...	...	**E14**, 180
u140	**Ursolic acid**	Malol. Prunol. Urson. $C_{30}H_{48}O_3$.	456.72	pr (abs al), nd (dil al) $[\alpha]^{21}_D+67.5$ (1N al KOH)	285–8	...	...	...	i	δ δ^h	δ	s	...	peth, CS_2 i MeOH δ chl δ	**B10²**, 202
u141	**Usnic acid**(*d*)	Usninic acid. 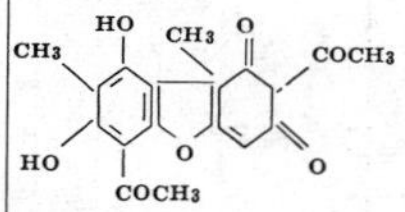	344.32	ye orh pr (ace) $[\alpha]^{20}_D+469$ $[\alpha]^{16}_D+509$ (chl, c=0.7)	204	...	...	...	i	i	...	δ	...	os δ	...
u142	—(*dl*)	$C_{18}H_{16}O_7$. *See* u143	344.32	...	193–4	...	...	...	i					...	**B18²**, 241
u143	—(*l*)	$C_{18}H_{16}O_7$. *See* u143	344.32	$[\alpha]^{20}_D-480$	195–7	...	...	...	...	s	s	...	...	chl v	**B18²**, 241
—	**Uvitic acid**	*see* **1,3-Benzenetricarboxylic acid, 5-methyl-***													
u144	**Uzarin**	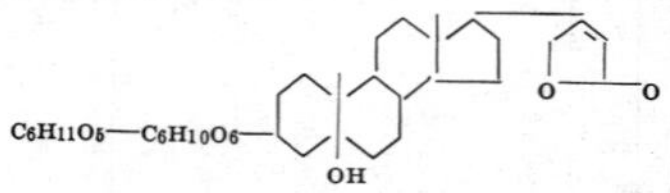 $C_6H_{11}O_5$—$C_6H_{10}O_6$—	698.81	pr $[\alpha]_D-27$	268–70	...	...	...	δ	...	i	i	...	Py v chl i	**E14**, 233

For explanations, symbols and abbreviations see beginning of table.

No.	Name	Synonyms and Formula	Mol. wt.	Crystalline form, color and specific rotation	m.p. °C	b.p. °C	Density	n_D	Solubility w	al	eth	ace	bz	other solvents	Ref.
	Vaccenic acid														
—	**Vaccenic acid**....	*see* **11-Octadecenoic acid**(*trans*)*													
v1	**Vacciniin**........	6-Benzoyl-*D*-glucose. 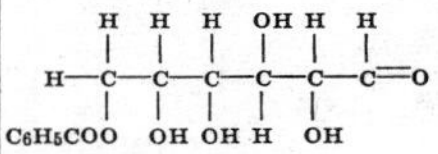	284.26	amor (aq ace +1w) $[\alpha]_D^{21}$+48 (al)	104–6	....	....	....	s	s	i	s	δ	peth i chl δ	**B31**, 123
—	**Valeraldehyde**....	*see* **Pentanal***													
—	**Valeric acid**......	*see* **Pentanoic acid***													
—	**Valerolactone**....	*see* **Pentanoic acid, hydroxy-**, lactone*													
—	**Valeronitrile**.....	*see* **Pentanoic acid**, nitrile													
—	**Valerophenone**...	*see* **1-Pentanone, 1-phenyl-***													
v2	**Valine**(*D*)........	*l*-α-Aminoisovaleric acid. *l*-2-Amino-3-methyl-butanoic acid*. $(CH_3)_2CHCH(NH_2)CO_2H$	117.15	pl (aq al) $[\alpha]_D^{20}$−29.04 (20% al)	293d	....	....	....	s	...	...	...	...	....	**B4²**, 852
v3	—(*DL*)..........	$(CH_3)_2CHCH(NH_2)CO_2H$....	117.15	pl (al)	298d	sub	....	....	s	δ	δ	...	...	....	**B4²**, 854
v4	—(*L*)............	*d*-α-Aminoisovaleric acid. $(CH_3)_2CHCH(NH_2)CO_2H$	117.15	lf (w-al) $[\alpha]_D^{23}$+22.9 (20% al, c=0.8)	315	....	....	1.230	s	δ	δ	...	...	....	....
v5	—,β-**hydroxy**-(*dl*) .	*dl*-2-Amino-3-hydroxy-3-methylbutanoic acid. $(CH_3)_2C(OH)CH(NH_2)CO_2H$	133.15	pl (al)	218d	....	....	....	s	i	i	...	i	....	**B4²**, 942
—	**Valone**..........	*see* **1,3-Indandione, 2(3-methylbutyl)-**													
—	**Valylene**.........	*see* **1-Buten-3-yne, 2-methyl-***													
—	**Vanillic acid**.....	*see* **Benzoic acid, 4-hydroxy-3-methoxy-**													
—	**Vanillyl alcohol**..	*see* **Toluene,**α**,4-dihydroxy-3-methoxy**													
—	**Vanillin**.........	*see* **Benzaldehyde, 4-hydroxy-3-methoxy-**													
v7	**Vasicine**(*dl*)......	Peganine. $C_{11}H_{12}N_2O$.	188.22	nd (al)	210	....	....	....	δ	s	δ	s	δ	chl s	**B23²**, 342
v8	—(*l*).............	$C_{11}H_{12}N_2O$................	188.22	nd (al) $[\alpha]_D^{14}$ −62 (al, c=2.4)	212	....	....	....	δ	s	δ	...	δ	chl s	**B23²**, 342
—	**Veratraldehyde**...	*see* **Benzaldehyde, 2,4-dimethoxy-**													
v9	**Veratramine**.....	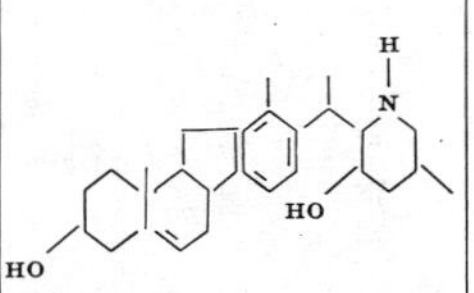	409.62	nd $[\alpha]$−71 (chl)	....	209.0–10.5	....	....	...	s^h	...	...	s	dil ac s dil alk i chl s	....
—	**Veratric acid**.....	*see* **Benzoic acid, 3,4-dimethoxy-**													
v10	**Veratridine**....... 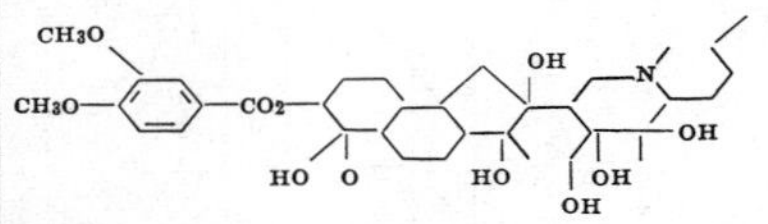		673.80	yesh amor $[\alpha]_D^{22}$+8	180	....	....	....	s^h	s^h	...	...	...	....	....
v11	**Veratrine**........	Cevadine. (A mixture of meta alkaloids). $C_{32}H_{49}NO_9$	591.75	rh (+2 al) $[\alpha]_D^{17}$ +12.5 (al)	205d	....	....	....	δ	s	s	...	...	....	....
—	**Veratrole**.........	*see* **Benzene, 1,2-dimethoxy-***													
—	**Veratrophenone**..	*see* **Benzophenone, 3,3',4,4'-tetramethoxy-**													
v13	**Veratrosine**......	$C_{33}H_{49}NO_7$................	571.76	nd $[\alpha]_D$+53 (al-chl)	242–3	....	....	....	i	...	...	...	...	....	....

For explanations, symbols and abbreviations see beginning of table.

No.	Name	Synonyms and Formula	Mol. wt.	Crystalline form, color and specific rotation	m.p. °C	b.p. °C	Density	n_D	Solubility w	al	eth	ace	bz	other solvents	Ref.
	Verbenone														
v14	**Verbenone,**(*d*)	*d*-4-Piperone. 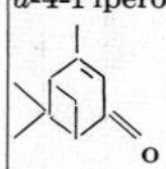	150.22	$[\alpha]_D^{18}+250$		227–8 100^{16}	0.978^{20}	1.4993^{18}	s					oos s	B7^{1}, 104
v15	—(*l*)	$C_{10}H_{14}O$. *See* v14	150.22	$[\alpha]_D-144$	6.5	$227–8^{760}$ 100^{16}	0.982^{15}	1.4994		s			s		**E12A,** 515
—	**Vernine**	*see* **Guanosine**													
—	**Veronal**	*see* **Barbituric acid, 5,5-diethyl-**													
—	**Vetivazulene**	*see* **Azulene, 2,4-dimethyl-2-isopropyl-**													
v16	**α-Vetivone**	α-Vetiverone. 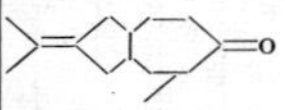	218.34	(pentane) $[\alpha]_D+238.2$ (al, c=6.9)	51.0–1.5	$144.0–4.5^{2}$	1.0035_4^{20}	1.5370^{20}							**E12A,** 436
v17	**β-Vetivone**	β-Vetiverone. $C_{15}H_{22}O$. *See* v16	218.34	(pentane) $[\alpha]_D^{20}-38.9$ (al, c=10.6)	44.0–4.5	$141–2^{2}$	1.002_4^{15}	1.5252^{20}							**E12A,** 436
v18	**Vicianose**	6(β-Arabinoside)*D*-glucose. 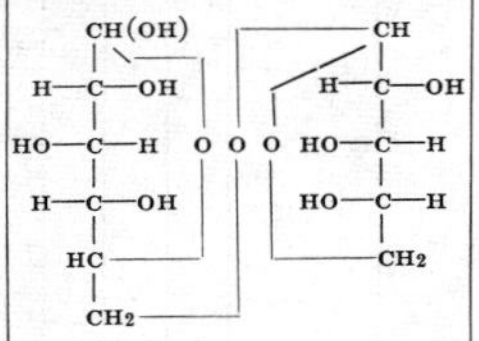	312.28	nd (dil al) $[\alpha]_D^{20}$ (mut) +63 to +40	ca. 210										**B31,** 371
v19	**Vicine**	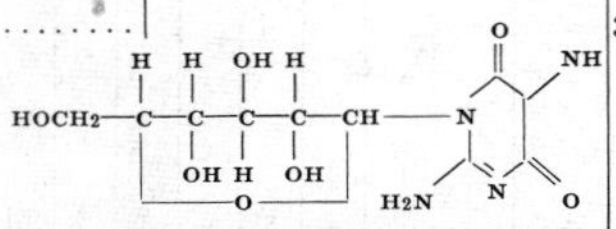	304.26	nd (w or dil al +1w) $[\alpha]_D^{25}-12$ (c=10)	239–42d (anh)				δ	δ				ac v alk v	**B31,** 163
—	**Vinaconic acid**	*see* **1,1-Cyclopropanedicarboxylic acid***													
—	**Vinetine**	*see* **Oxyacanthine**													
—	**Vinyl acetate**	*see* **Acetic acid, ethenyl ester**													
—	**Vinyl amine**	*see* **Ethene, amino-***													
—	**Vinyl bromide**	*see* **Ethene, bromo-***													
—	**Vinyl chloride**	*see* **Ethene, chloro-***													
—	**Vinyl ether**	*see* **Ether, diethenyl-**													
—	**Vinyl fluoride**	*see* **Ethene, fluoro-***													
—	**Vinyl iodide**	*see* **Ethene, iodo-***													
—	**Violanthrone**	*see* **Dibenzanthrone-**													
v20	**Violaxanthin**	Zeaxanthin diepoxide. 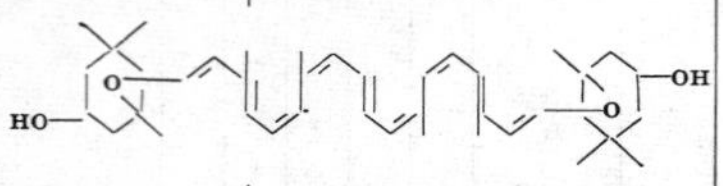	600.89	red pr (MeOH, al-eth or CS_2) (Optically active)	203					s	s				**B30,** 99
v21	**Violuric acid**	Alloxan-5-oxime. 5-Isonitrosobarbituric acid. Oximidomesoxalylurea. 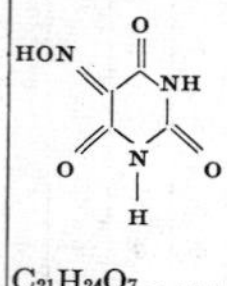	157.09	pa ye orh	203–4d				δ	s					**B24^{2},** 304
v22	**Visnadin**	$C_{21}H_{24}O_7$	388.42	nd $[\alpha]_D+9$ (al)	84–6				i	s	s				**Am 79,** 3534
v23	**Visnagin**	$C_{13}H_{10}O_4$	230.22	nd (w)	142–5				δ	δ				chl v	
v24	**Vitamin A**	Axerophthol. $C_{20}H_{30}O$	286.46	ye pr (peth)	62–4				i					oos v	
v25	**Vitamin B$_1$**	Thiamine hydrochloride. Thiamin chloride. $C_{12}H_{18}Cl_2N_4OS$.	337.27	mcl pl (MeOH, al, dil al)	248d				v	δ	i		i	chl i	
—	**Vitamin B$_2$**	*see* **Riboflavin**													

For explanations, symbols and abbreviations see beginning of table.

No.	Name	Synonyms and Formula	Mol. wt.	Crystalline form, color and specific rotation	m.p. °C	b.p. °C	Density	n_D	Solubility						Ref.
									w	al	eth	ace	bz	other solvents	
	Vitamin B_6														
v26	**Vitamin B_6**.......	Pyridoxin. $C_8H_{11}NO_3$.	169.18	nd (ace)	160	sub	....	....	v	s	δ	s	...	chl δ	....
v27	—,hydrochloride...	Adermine. Pyridoxine hydrochloride. $C_8H_{11}NO_3.HCl$.	205.64	pl (al, ace)	204–6	....	....	....	v	δ	i	δ	...	....	....
—	**Vitamin C**.......	*see* **Ascorbic acid**(*l*)													
—	**Vitamin D**.......	*see* **Calciferol**													
—	**Vitamin E**.......	*see* **Tocopherol**													
—	**Vitamin K**.......	*see* **1,4-Naphthoquinone, 2-hydroxy-3-methyl-***													
v28	**Vitamin K_1**.......	Natural vitamin K. $C_{31}H_{46}O_2$.	450.68	ye $[\alpha]_D^{20}-0.4$ (bz, 57.5%)	−20	$140–45^{10^{-3}}$ >100–120 d	0.967_{25}^{25}	1.525^{25}	i	s	s	s	s	peth s	....
—	**Vitamin K_3**.......	*see* **1,4-Naphthoquinone, 2-methyl-***													
—	**Vitamin K_6**.......	*see* **Naphthalene, 1,4-diamino-2-methyl-,** dihydrochloride*.													
v29	**Vomicine**........	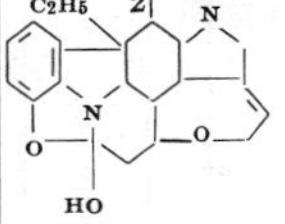	380.43	nd (dil al), pr (ace) $[\alpha]_D^{22}$ +80.4 (al)	282	....	....	....	...	s^h	δ	...	...	chl v	....

For explanations, symbols and abbreviations see beginning of table.

No.	Name	Synonyms and Formula	Mol. wt.	Crystalline form, color and specific rotation	m.p. °C	b.p. °C	Density	n_D	Solubility w	al	eth	ace	bz	other solvents	Ref.
	Xanthaline														
—	**Xanthaline**.......	*see* **Papaveraldine**													
x1	**Xanthene**........	Dibenzo-1,4-pyran. Diphenylmethane oxide. 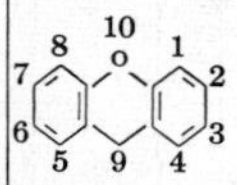	182.22	wh ye lf (al)	100.5	315 sub	...	...	δ	δ	s	...	s	chl s	**B17**², 72
x2	**—9-hydroxy-**.....	5-Hydroxyxanthene. Dibenz-γ-pyranol. Xanthydrol. $C_{13}H_{10}O_2$. *See* x1	198.22	nd (aq al)	*ca.* 125	...	...	...	δ	s	s	...	...	chl s	**B27**¹, 72
x3	**—,9-phenyl-**......	$C_{19}H_{14}O$. *See* x1............	258.32	(al)	145.1–5.5	...	...	...	...	s^h	...	...	s	lig s^h	**B27**¹, 38
x4	**9-Xanthene-carboxylic acid**	Xanthanoic acid 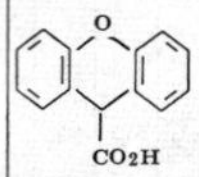	226.23	nd (dil al or MeOH)	223–4	...	...	...	...	s^h	s	...	...	peth s^h	**B12**², 279
x5	*peri*-**Xantheno-xanthene**	1,1-Binaphthylene, 2:8′ 2′:8-dioxide. 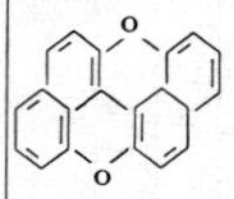	282.30	ye pr (chl)	242	400^{20-25}	...	...	...	...	...	...	s	chl s^h	**B19**², 52
—	**Xantic acid**......	*see* **Formic acid, (ethoxy)dithio-**													
x6	**Xanthine**........	2,6(1,3)Purinedione. 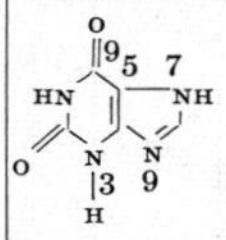	152.11	yesh pl (w)	d	...	...	...	s	i	i	...	...	ac, alk s chl i	**B26**², 260
—	**—,3,7-dimethyl**...	*see* **Theobromine**													
x7	**—,7-methyl-**.....	Heteroxanthine. $C_6H_6N_4O_2$. *See* x6	166.14	nd (w)	380d	...	...	...	δ	i	i	...	...	chl i	**B26**², 263
—	**—,1,3,7-trimethyl-**	*see* **Caffeine**													
x9	**Xanthione**.......	9-Xanthenethione. 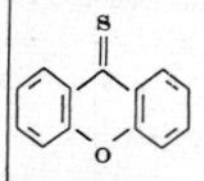	212.27	red nd (al)	156	...	...	...	i	δ	s	...	s	...	**B17**², 382
—	**Xanthogenamide.**	*see* **Carbamic acid, thiono-**, ethyl ester													
x10	**Xanthogen,** diethyl-	Auligen. $C_2H_5OC(:S)SSC(:S)OC_2H_5$	242.40	ye nd or pl	32	112d	1.2604^{25}_{4}	...	i	v^h	v	...	v	peth v	**B3**², 154
x11	**Xanthogenic acid**	Ethyl xanthogenic acid. Thiolothionocarbonic acid *O*-ethyl ester. C_2H_5OCSSH	122.20	...	fr −53	25d	...	...	δ	...	...	...	...	...	**B3**², 151
x12	**Xanthone**........	Diphenylene ketone oxide. Dibenzopyrone. 9-Xanthenone. 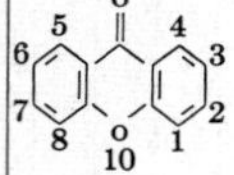	196.19	nd (al)	174	$349\text{–}50^{730}$	...	...	i	δ	δ	...	s	chl s, peth δ	**B17**², 378
x13	**—,1,2-dihydroxy-**.	$C_{13}H_8O_4$. *See* x12...........	228.19	pa ye nd (aq al+3w)	245–8	...	...	...	...	s	...	...	...	Py s	**B18**², 85
x14	**—,1,3-dihydroxy-**.	$C_{13}H_8O_4$. *See* x12...........	228.19	nd (aq al+1w)	259	...	...	...	...	s	δ	...	...	alk v	**B18**², 85
x15	**—,1,6-dihydroxy-**.	Isoeuxanthone. $C_{13}H_8O_4$. *See* x12	228.19	ye nd (dil al)	245–6	...	...	...	i	s	δ	...	...	...	**B18**, 113
x16	**—,1,7-dihydroxy-**.	Euxanthone. $C_{13}H_8O_4$. *See* x12	228.19	nd (to), pl (al)	240	sub	...	...	δ	s^h	δ	...	...	conc alk s	**B18**², 87
x17	**—,1,8-dihydroxy-**.	$C_{13}H_8O_4$. *See* x12...........	228.19	ye lf (bz)	187	...	...	...	...	s^h	δ	...	s^h	alk, chl s	**B18**², 87
x18	**—,2,3-dihydroxy-**.	$C_{13}H_8O_4$. *See* x12...........	228.19	ye nd (al)	294	...	...	...	i	v	δ	v	v	chl δ	**B18**, 116
x19	**—,2,7-dihydroxy-**.	$C_{13}H_8O_4$. *See* x12...........	228.19	ye nd (al or eth)	>330	sub	...	...	...	s	s	...	...	...	**B18**², 88
x20	**—,3,4-dihydroxy-**.	$C_{13}H_8O_4$. *See* x12...........	228.19	red ye nd (dil al+3w)	240 (anh)	...	...	...	δ	v	...	...	...	alk s	**B18**², 88

For explanations, symbols and abbreviations see beginning of table.

No.	Name	Synonyms and Formula	Mol. wt.	Crystalline form, color and specific rotation	m.p. °C	b.p. °C	Density	n_D	Solubility w	al	eth	ace	bz	other solvents	Ref.
	Xanthone														
x21	**—,3,6-dihydroxy-**	$C_{13}H_8O_4$. *See* x12	228.19	nd (dil al), pr (sub)	300–50d	...	...	...	i	v	...	...	i	chl δ	**B18**[1], 357
x22	**—,1,7-dihydroxy-3-methoxy-**	Gentianin. Gentisin. $C_{14}H_{10}O_5$. *See* x12	258.22	ye rh	267	400 sub	...	...	i	v	...	i	...	Py s	**B28**[2], 158
x23	**Xanthophyll**	Lutein. Luteol. 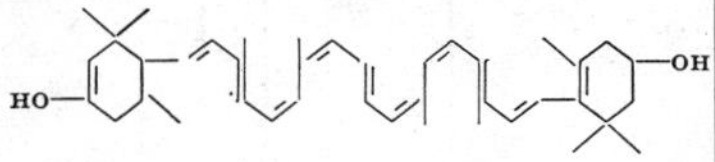	568.85	ye or vt pr (eth-MeOH) (Optically active)	195	...	...	...	i	s	s	...	s	peth s	**B30**, 94
x24	**Xanthopterin**	Uropterin. 2-Amino-4,6-pteridinediol. 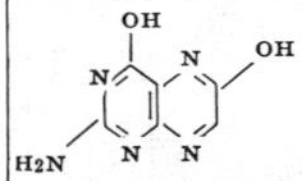	179.14	hyg ye amor	>410d	$98–100^{18}$	1.559	...	i	δ	δ	...	...	...	...
—	**Xanthopurpurin**	*see* **9,10-Anthraquinone, 1,3-dihydroxy-**													
x25	**Xanthotoxin**	Ammoidin. 8-Methoxypsoralen. Zanthotoxin. $C_{12}H_8O_4$.	216.20	pr (dil al), nd (peth or bz-peth)	145–6	...	...	...	δ	v^h	δ	δ	...	peth δ	**B14**[1], 711
x26	**Xanthotoxol**	8-Hydroxy-4′:5′, 6:7-furocoumarin. 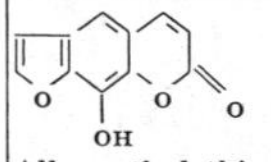	202.17	...	246	...	...	...	i	...	...	...	...	diox s^h	**Am72**, 4826
x27	**Xanthoxyletin**	Alloxanthylethin. Xanthoxylin-N. 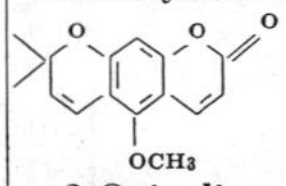	258.28	pr (MeOH)	133	...	...	...	i	s^h	...	...	s	alk v, chl s	**J1936**, 627
—	**Xanthurenic acid**	*see* **2-Quinolinecarboxylic acid, 4,8-dihydroxy-**													
—	**Xanthuric acid**	*see* **2-Quinolinecarboxylic acid, 4,8-dihydroxy-**													
—	**Xanthydrol**	*see* **Xanthene, 9-hydroxy-**													
x28	**Xanthyletin**	2,2-Dimethyl-chromenocoumarin. 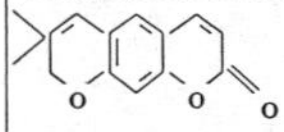	228.25	pr	131.5	$140–5^{0.1}$	...	...	...	s^h	...	...	...	peth s^h	**J1936**, 1828
—	**Xenylamine**	*see* **Biphenyl, 4-amino-**													
—	**Xylene**	*see* **Benzene, dimethyl-***													
—	**Xylenol**	*see* **Benzene, dimethyl(hydroxy)-***													
—	**Xylenyl alcohol**	*see* **Benzene, dimethyl(hydroxy)-(hydroxymethyl)**													
—	**Xylic acid**	*see* **Benzoic acid, 2,4-dimethyl-***													
—	**Xylidine**	*see* **Benzene, amino(dimethyl)-***													
x29	**Xylitol**	Xylit. 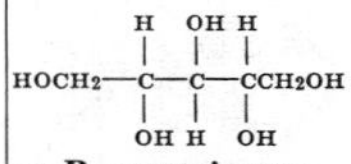	152.15	mcl (al)	93–4.5	216–7	...	...	v	s	...	...	...	Py s	**B1**[2], 604
—	**Xyloquinone**	*see* **Benzoquinone, dimethyl-***													
x30	**Xylose**(*D*)	Wood sugar. 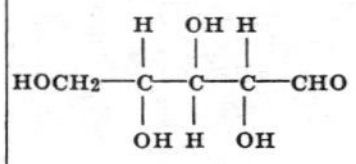	150.13	mcl nd $[\alpha]^{20}_D$ (mut) +69.5 to +19.1 (w, c=3)	144	...	1.525^{20}_4	...	v	s	δ	...	...	...	**B31**, 47
x31	—,osazone	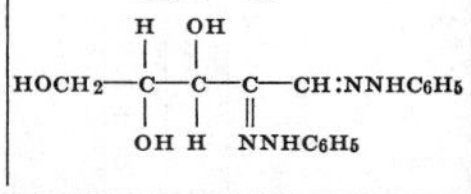	328.36	pa ye nd $[\alpha]_D$ −40.9 (al)	159	...	...	...	δ	s	s	s	...	...	**B31**, 61

For explanations, symbols and abbreviations see beginning of table.

No.	Name	Synonyms and Formula	Mol. wt.	Crystalline form, color and specific rotation	m.p. °C	b.p. °C	Density	n_D	Solubility w	al	eth	ace	bz	other solvents	Ref.
	Yamogenin														
y1	**Yamogenin**	$C_{27}H_{42}O_3$	414.60	pl $[\alpha]_D^{25}-123$	195–200						s^h			MeOH s^h	**Am 77,** 3086
—	**Yellow AB**	*see* **Azo, benzene 1-naphthalene, 2′amino-**													
y3	**Yobyrine**	Yobirine. 1-*o*-Methylbenzoyl-9H-pyrid-[3,4-b]indole.	272.34	nd (dil al)	217	$150^{0.01}$				s	s^h		s^h	chl s	**B23**[2], 263
y4	**Yohimbine**	Corynine. Quebrachine. Hydroergotocin. Aphrodine. $C_{21}H_{26}N_2O_3$.	354.43	orh (dil al) $[\alpha]_D^{20}+108$ (Py, c=1)	235	$159^{0.01}$ sub			δ	δ	δ		$δ^h$	chl δ	**B25**[2], 201
y5	—,hydrochloride	$C_{21}H_{26}N_2O_3.HCl$	390.90	orh nd or pl (w or dil HCl), pr (al) $[\alpha]_D^{22}+106.4$ (w, c=2)	295				s^h	δ	i	i			**B25**[2], 201
y6	—,nitrate	$C_{21}H_{26}N_2O_3.HNO_3$	417.45	pr (w)	269				v						**B25**[2], 201
—	**Yperite**	*see* **Sulfide, diethyl, 2,2′-dichloro-**													
y7	**Yuccagenin**	$C_{27}H_{42}O_4$	430.61	nd (al) $[\alpha]_D-122$	248					s^h	s^h			diox s	**J1956,** 1167

For explanations, symbols and abbreviations see beginning of table.

No.	Name	Synonyms and Formula	Mol. wt.	Crystalline form, color and specific rotation	m.p. °C	b.p. °C	Density	n_D	Solubility						Ref.
									w	al	eth	ace	bz	other solvents	
	Zagadinine														
z1	**Zagadinine**	$C_{27}H_{43}NO_7$	493.65	orh (al) [α] −45 (chl)	201–4	...	...	...	δ	s^h	s	...	s^h	ac, s chl s alk i	**Am35,** 258
z2	**Zeazanthin**	Zeaxanthol. $C_{40}H_{56}O_2$	568.85	ye pr (MeOH) rh (chl-eth)	215.5	$226–9^{0.06}$	...	...	i	...	s	...	s	chl s peth i^h	**B30,** 97
—	**Zymostanone**	*see* **3-Cholestanone**													
z3	**Zymosterol**	$\Delta^{8(14).24(25')}$-Cholestadienol. $C_{27}H_{44}O$.	384.62	pl (MeOH), nd (ace) $[\alpha]^{20}_{D}+50$	109–10	$160^{0.001}$ sub	...	...	...	...	...	s^h	...	MeOH s^h chl s	**E14s,** 1559

For explanations, symbols and abbreviations see beginning of table.

FORMULA INDEX OF ORGANIC COMPOUNDS

C_3

C_4

C_5

C_6

C_7

C_8

C_9

C_{12}

C_{14}

C_{15}

C_{16}

$C_{16}H_{34}S$: h252
$C_{16}H_{35}N$: a860, a860[1], h244
$C_{16}H_{36}IN$: a966
$C_{16}H_{48}O_6Si_7$: h195
$C_{16}H_{48}O_8Si_8$: c927

C_{17}

$C_{17}H_7NO_4$: r3
$C_{17}H_9BrO$: b198
$C_{17}H_9ClO$: b199
$C_{17}H_{10}O$: b197
$C_{17}H_{10}O_2$: b200, b201, p1933, p1934
$C_{17}H_{11}N$: n387
$C_{17}H_{11}NO_5$: q248
$C_{17}H_{12}$: b196, b1229
$C_{17}H_{12}Cl_2O$: p70, p71
$C_{17}H_{12}N_2O_5$: p67, p68, p69
$C_{17}H_{12}O_2$: b1289, b1290, n193
$C_{17}H_{12}O_3$: b1674, b1675, n373
$C_{17}H_{13}Br$: c643
$C_{17}H_{13}N$: h154, n92[1], n92[2]
$C_{17}H_{13}NO$: b1247
$C_{17}H_{13}NO_3$: q252
$C_{17}H_{13}N_3O_5S_2$: s285
$C_{17}H_{14}$: c903, c904, c905, m284, m285
$C_{17}H_{14}N_2$: b1260
$C_{17}H_{14}N_2S$: u105
$C_{17}H_{14}O$: e446, e447, m376, m377, n90, n91, n92, p72, p72[1]
$C_{17}H_{14}O_2$: a253[1], a253[2], a253[3]
$C_{17}H_{14}O_3$: p61, p62
$C_{17}H_{15}ClO_7$: m122
$C_{17}H_{15}N$: a914, a915, a916, a917
$C_{17}H_{16}O$: c246
$C_{17}H_{16}O_3$: p1330
$C_{17}H_{16}O_4$: p163, p1477
$C_{17}H_{16}O_7$: e615
$C_{17}H_{17}ClO_6$: g190
$C_{17}H_{17}N$: b2169
$C_{17}H_{17}NO_2$: a1378
$C_{17}H_{18}ClNO_2$: a1379
$C_{17}H_{18}N_2O_2$: e53
$C_{17}H_{18}O_2$: p1342, p1571
$C_{17}H_{18}O_3$: p1445
$C_{17}H_{18}O_4$: g83
$C_{17}H_{18}O_5$: b2038, b2039, b2040, b2041, b2042, b2043, b2044, b2045, b2046, b2047, b2048
$C_{17}H_{19}ClN_2S$: c268
$C_{17}H_{19}NO_3$: c453, e140, e141, i170, m399, m400, p1040
$C_{17}H_{19}NO_4$: m403
$C_{17}H_{20}$: p136, p137
$C_{17}H_{20}ClNO_3$: m402
$C_{17}H_{20}Cl_2N_2S$: c269
$C_{17}H_{20}N_2O$: b1974, u57
$C_{17}H_{20}N_2O_3$: u33[1], u33[2]
$C_{17}H_{20}N_2O_4$: a1385, a1387, a1389
$C_{17}H_{20}N_2S$: b1975, p719, u58
$C_{17}H_{20}N_4O_3$: x31
$C_{17}H_{20}N_4O_6$: r26
$C_{17}H_{20}O_6$: m423
$C_{17}H_{21}ClN_2S$: p720
$C_{17}H_{21}NO_2$: a1369
$C_{17}H_{21}NO_3$: c456
$C_{17}H_{21}NO_4$: c445, c446, c447, h744, h745, h754, h755
$C_{17}H_{21}N_3$: a1473
$C_{17}H_{22}BrNO_4$: h746, h747, h748, h749, h750, h751, h752
$C_{17}H_{22}ClNO_2$: a1370
$C_{17}H_{22}ClNO_4$: c449, c450, h752, h753
$C_{17}H_{22}ClN_3$: a1474
$C_{17}H_{22}N_2$: m197
$C_{17}H_{22}N_2O$: m356
$C_{17}H_{22}N_2O_7$: h756
$C_{17}H_{23}ClN_2O_2$: n314
$C_{17}H_{23}CrNO_8$: c448
$C_{17}H_{23}NO_3$: a1469, h758, h759, p1872
$C_{17}H_{24}BrNO_3$: h760
$C_{17}H_{24}ClNO_3$: a1470, h761
$C_{17}H_{24}N_2O_5S$: s85
$C_{17}H_{24}O$: b301
$C_{17}H_{24}O_2$: c362, c363, c364
$C_{17}H_{24}O_3$: b1668
$C_{17}H_{24}O_4$: f114
$C_{17}H_{24}O_9$: s423
$C_{17}H_{25}ClN_2O$: d193
$C_{17}H_{25}N_3O$: a1358
$C_{17}H_{26}N_2S_2$: c107
$C_{17}H_{28}$: h64
$C_{17}H_{30}O$: c610
$C_{17}H_{32}O_4$: s244
$C_{17}H_{33}N$: h73
$C_{17}H_{34}$: h80
$C_{17}H_{34}Br_2$: h69
$C_{17}H_{34}N_2S_2$: c103
$C_{17}H_{34}O$: h65, h77, h78
$C_{17}H_{34}O_2$: h70, h264, t83, t86
$C_{17}H_{35}Br$: h68
$C_{17}H_{35}N$: p981
$C_{17}H_{36}$: h66
$C_{17}H_{36}O$: h74, h75, h76
$C_{17}H_{37}N$: h67

C_{18}

$C_{18}H_{10}N_2O_6$: i61
$C_{18}H_{10}O_2$: b194, b195, c318, c319, n6, n7
$C_{18}H_{10}O_4$: b2179, b2202
$C_{18}H_{11}NO_2$: p844, p845
$C_{18}H_{12}$: b181, b1956, b1957, c313, e603, n1
$C_{18}H_{12}N_2$: b2368, b2369, b2370, b2371, b2372, b2373, b2374, b2375, b2376, b2377, b2378
$C_{18}H_{12}N_2O$: a1382
$C_{18}H_{12}O$: b184, p1931
$C_{18}H_{12}O_2$: a586, b2100
$C_{18}H_{12}O_3$: b1834, b1835
$C_{18}H_{12}O_4$: d159, d160
$C_{18}H_{13}N$: c153
$C_{18}H_{13}NO_3$: p850
$C_{18}H_{13}NS$: p721
$C_{18}H_{13}N_3O$: r37
$C_{18}H_{13}N_3O_2$: b1215
$C_{18}H_{14}$: b182, n2, t46, t47
$C_{18}H_{14}N_4O_2$: b306
$C_{18}H_{14}O$: k5, k6, k20, k21
$C_{18}H_{14}O_3$: c353
$C_{18}H_{14}O_8$: s187
$C_{18}H_{15}As$: a1443[1]
$C_{18}H_{15}B$: b2411
$C_{18}H_{15}FSe$: s39
$C_{18}H_{15}N$: a947
$C_{18}H_{15}N_3$: a1602
$C_{18}H_{15}OP$: p757
$C_{18}H_{15}O_3P$: p803
$C_{18}H_{15}O_4P$: p780
$C_{18}H_{15}P$: p756
$C_{18}H_{15}Sb$: s122
$C_{18}H_{16}$: c905, h544, t769
$C_{18}H_{16}F_2Se$: s38
$C_{18}H_{16}N_2$: h710
$C_{18}H_{16}N_2O$: a1489, a1490, a1491, a1492, a1493
$C_{18}H_{16}N_2O_2$: a989
$C_{18}H_{16}N_2O_6S$: q155, q156
$C_{18}H_{16}O_2$: p546, s157
$C_{18}H_{16}O_4$: e93, t108, t109, t111
$C_{18}H_{16}O_7$: u143, u144, u145
$C_{18}H_{17}Cl_4N_3O_8S_2$: h595
$C_{18}H_{17}NO_3$: p1885
$C_{18}H_{17}N_3$: a1494
$C_{18}H_{18}$: p505[1]
$C_{18}H_{18}N_2O_4$: a1361
$C_{18}H_{18}N_6$: b1
$C_{18}H_{18}O_2$: b1833, e94, e95, e96, h415
$C_{18}H_{18}O_3$: b2684
$C_{18}H_{18}O_4$: b2287, b2338, s189, s199
$C_{18}H_{18}O_5$: e483
$C_{18}H_{18}O_6$: t13
$C_{18}H_{19}NO_2$: a844, a845, a846, a1374
$C_{18}H_{19}NO_3$: e144, t779
$C_{18}H_{19}N_3O_2$: a1535
$C_{18}H_{20}$: i27
$C_{18}H_{20}Br_2N_2O_4$: b1416, b1422, b1429
$C_{18}H_{20}Cl_2N_2O_4$: b1472, b1487
$C_{18}H_{20}N_{10}O_{16}$: a1407, a1408
$C_{18}H_{20}O_2$: e97
$C_{18}H_{21}Cl_2NO_3$: b34
$C_{18}H_{21}NO_3$: c455, c460, e138, e139, e143, e145, i130, p1862, t129
$C_{18}H_{21}NO_4$: e146
$C_{18}H_{22}ClNO_3$: c457
$C_{18}H_{22}N_2O_2$: h600
$C_{18}H_{22}N_2O_5$: p916
$C_{18}H_{22}N_2S$: p718
$C_{18}H_{22}N_4O_4$: g48, g49, g220
$C_{18}H_{22}O_2$: b2270, b2271, b2272, b2281, e98, e161, i156
$C_{18}H_{22}O_4$: n569
$C_{18}H_{22}O_8$: b1186
$C_{18}H_{23}ClN_2O_2$: h601
$C_{18}H_{23}NO_3$: c461
$C_{18}H_{23}NO_4$: e5
$C_{18}H_{23}N_3$: a1475
$C_{18}H_{24}NO_7P$: c458
$C_{18}H_{24}O_2$: e159, i155
$C_{18}H_{24}O_3$: e160
$C_{18}H_{26}O_4$: p825
$C_{18}H_{27}NO_3$: c86
$C_{18}H_{27}NO_5$: e610
$C_{18}H_{27}N_3O$: p370
$C_{18}H_{28}N_2O_4S$: a979, a981
$C_{18}H_{28}O$: d314
$C_{18}H_{28}O_2$: d305
$C_{18}H_{28}O_4$: b2095
$C_{18}H_{29}NO$: d297
$C_{18}H_{30}$: b788, o8
$C_{18}H_{30}Br_6O_2$: o45
$C_{18}H_{30}O$: b824
$C_{18}H_{30}O_2$: o64, o65, o66
$C_{18}H_{30}O_4$: a1480
$C_{18}H_{32}Br_4O_2$: o60
$C_{18}H_{32}CaN_2O_{10}$: p5, p6
$C_{18}H_{32}I_2O_2$: o85
$C_{18}H_{32}N_2O_5S$: p3
$C_{18}H_{32}O_2$: c257, c258, o4, o5, o96
$C_{18}H_{32}O_4$: o44
$C_{18}H_{32}O_5$: a1479
$C_{18}H_{32}O_6$: g113
$C_{18}H_{32}O_{16}$: m153, r1
$C_{18}H_{33}N$: o83
$C_{18}H_{34}$: o93, o94, o95
$C_{18}H_{34}Br_2O_2$: o41
$C_{18}H_{34}O$: o68
$C_{18}H_{34}O_2$: o71, o72, o84
$C_{18}H_{34}O_3$: o52, o54, o56, o58, o86
$C_{18}H_{34}O_4$: d33, h386
$C_{18}H_{35}ClO$: o27
$C_{18}H_{35}N$: o35
$C_{18}H_{35}NO$: o73, o74
$C_{18}H_{36}$: o69, o70
$C_{18}H_{36}Br_2$: o17
$C_{18}H_{36}ClNO$: a318
$C_{18}H_{36}N_2S_4$: d225
$C_{18}H_{36}O$: o9, o91, o92

$C_{18}H_{36}O_2$: a207, d55, h72, h260
$C_{18}H_{36}O_3$: h262[1], o46, o47, o48, o49, o50
$C_{18}H_{36}O_4$: o42, o43
$C_{18}H_{37}Br$: o15
$C_{18}H_{37}Cl$: o16
$C_{18}H_{37}I$: o18
$C_{18}H_{37}NO$: o22
$C_{18}H_{38}$: o11
$C_{18}H_{38}O$: o63
$C_{18}H_{38}O_2$: h241, o20
$C_{18}H_{38}S$: o21
$C_{18}H_{39}N$: a937, o12
$C_{18}H_{40}ClN$: o14
$C_{18}H_{54}O_7Si_8$: o202
$C_{18}H_{54}O_9Si_9$: c851

C_{19}

$C_{19}H_{10}Br_4O_5S$: b2427
$C_{19}H_{11}NO_2$: p724
$C_{19}H_{12}O$: b193
$C_{19}H_{12}O_2$: b1224, b1225
$C_{19}H_{12}O_5$: i244
$C_{19}H_{12}O_6$: d161
$C_{19}H_{12}O_7$: c552
$C_{19}H_{13}N$: a740
$C_{19}H_{13}NO$: c143
$C_{19}H_{13}N_3O_6$: m318
$C_{19}H_{13}N_3O_7$: m383[1], m384
$C_{19}H_{14}$: b185, b186, b187, b188, b189, b190, b191, b192, c315, c316, c317, f75
$C_{19}H_{14}N_2O_2$: a1578, a1581, a1584
$C_{19}H_{14}O$: f67[2], x3
$C_{19}H_{14}O_2$: p1579, p1825
$C_{19}H_{14}O_3$: a1476
$C_{18}H_{14}O_5$: p66
$C_{19}H_{14}O_5S$: p714
$C_{19}H_{15}Cl$: m243
$C_{19}H_{15}ClN_4$: t126
$C_{19}H_{15}N$: c144
$C_{19}H_{15}NO$: b1246
$C_{19}H_{15}NO_2$: q256
$C_{19}H_{15}N_3$: a733
$C_{19}H_{16}$: b2234, b2235, m312
$C_{19}H_{16}N_2$: b1257, b1259, b1962, y3
$C_{19}H_{16}N_2S$: u124
$C_{19}H_{16}O$: m381
$C_{19}H_{16}O_3$: m317, m360, o243
$C_{19}H_{16}O_4$: c543
$C_{19}H_{16}O_{10}$: e614
$C_{19}H_{17}ClN_2$: b1258
$C_{19}H_{17}N$: a797, m183, m184
$C_{19}H_{17}NO_2$: q253
$C_{19}H_{17}NO_3$: c569
$C_{19}H_{17}N_3$: g207, g208
$C_{19}H_{17}N_3O$: e616, e617, s52, s53, s54
$C_{19}H_{18}N_2$: m193
$C_{19}H_{18}O_3$: c902, p63, p64, p65
$C_{19}H_{18}O_5$: e15
$C_{19}H_{18}O_6$: s59
$C_{19}H_{19}N$: c492
$C_{19}H_{19}NO_3$: e24
$C_{19}H_{19}NO_4$: b2433, b2434
$C_{19}H_{19}N_3$: m181, m182, m315
$C_{19}H_{19}N_3O$: m382
$C_{19}H_{19}N_7O_6$: f104
$C_{19}H_{20}ClN_3O$: g209
$C_{19}H_{20}O_3$: p232
$C_{19}H_{21}NO_3$: i214, o306, t127
$C_{19}H_{21}NO_5$: c464
$C_{19}H_{21}NO_7S$: e142
$C_{19}H_{22}ClNO_3$: o307, t128
$C_{19}H_{22}N_2O$: a1371, a1372, c321, c322, c326, c327
$C_{19}H_{22}N_2O_2$: a1380, a1381, c563, s22
$C_{19}H_{22}N_2O_6$: o308
$C_{19}H_{22}O_2$: h566
$C_{19}H_{23}ClN_2O$: c324
$C_{19}H_{23}NO_3$: a1376
$C_{19}H_{23}NO_4$: c451, c452, s86
$C_{19}H_{23}NO_5$: m401
$C_{19}H_{23}NO_6$: m117
$C_{19}H_{23}N_3O_2$: e113, e114, e115, p1827
$C_{19}H_{23}N_3O_4$: c331
$C_{19}H_{24}ClNO_3$: m407
$C_{19}H_{24}Cl_2N_2O$: c329
$C_{19}H_{24}N_2O$: c320, c333, h721
$C_{19}H_{24}N_2O_2$: c467, c493, h723, q5
$C_{19}H_{24}N_2O_5S$: c323, c328
$C_{19}H_{24}O_3$: a750
$C_{19}H_{26}ClN_2O$: c330
$C_{19}H_{26}N_2O$: c564
$C_{19}H_{27}ClN_2O_3$: a1373
$C_{19}H_{27}NO_4$: e604
$C_{19}H_{28}ClNO_4$: e606
$C_{19}H_{28}O_2$: t55
$C_{19}H_{30}O_2$: a992, d299, e56
$C_{19}H_{32}$: a991
$C_{19}H_{32}O$: b534[1]
$C_{19}H_{34}Br_4O_2$: o62
$C_{19}H_{34}O_2$: o7, o98
$C_{19}H_{36}$: n515
$C_{19}H_{36}O_2$: o80, o81
$C_{19}H_{36}O_4$: m85
$C_{19}H_{38}O$: n511, n513, n514
$C_{19}H_{38}O_2$: h71, h263, h266, n509, o33, o51
$C_{19}H_{38}O_4$: g93, g94
$C_{19}H_{39}NO$: n512
$C_{19}H_{40}$: n505
$C_{19}H_{40}O$: n510
$C_{19}H_{40}O_3$: g85

C_{20}

$C_{20}H_6I_4Na_2O_5$: e155
$C_{20}H_8Br_4O_5$: e43
$C_{20}H_8I_4O_5$: e154
$C_{20}H_{10}Br_4O_4$: p711
$C_{20}H_{10}Cl_4O_4$: p712
$C_{20}H_{10}I_4O_4$: p713
$C_{20}H_{10}Na_2O_5$: u17
$C_{20}H_{10}O_2$: x5
$C_{20}H_{12}$: b1226, b1227, b1228, b2062, p479
$C_{20}H_{12}O$: b2064, b2065
$C_{20}H_{12}O_3$: f40
$C_{20}H_{12}O_5$: f41, f42, f43, f100, f101, h736
$C_{20}H_{12}O_7$: g9
$C_{20}H_{13}Br$: f57, f58
$C_{20}H_{13}Cl$: f59, f60, f61
$C_{20}H_{13}N$: b2063
$C_{20}H_{14}$: a3, a1234, b2207, b2208, c275, e35, f54
$C_{20}H_{14}N_2$: a1643, a1644, a1645
$C_{20}H_{14}N_2O$: a1656, a1657
$C_{20}H_{14}N_4$: p1050
$C_{20}H_{14}O$: e505, e506, e507
$C_{20}H_{14}O_2$: b2210, b2211, p899
$C_{20}H_{14}O_4$: b572, b576, b579, p710, p827
$C_{20}H_{14}O_5$: f102
$C_{20}H_{14}O_7$: g10
$C_{20}H_{14}S$: s314, s315
$C_{20}H_{14}S_2$: d245, d246
$C_{20}H_{15}Br$: e404
$C_{20}H_{15}Cl$: e407
$C_{20}H_{15}N$: a858
$C_{20}H_{15}NO_4$: s12
$C_{20}H_{15}N_3$: a1646, d102, d103
$C_{20}H_{16}$: b183, c314, e426
$C_{20}H_{16}N_2$: h664, h665
$C_{20}H_{16}N_2O_2$: a1588, a1593, a1596
$C_{20}H_{16}N_4$: n489
$C_{20}H_{16}O_2$: a643, b2855
$C_{20}H_{16}O_4$: b1558, i177, p715
$C_{20}H_{18}$: e279, e280, m272
$C_{20}H_{18}N_2O$: a1531
$C_{20}H_{18}N_2O_7$: b2166
$C_{20}H_{18}N_4O_2$: b591, b592, b593
$C_{20}H_{18}O$: c793, e399, e400
$C_{20}H_{18}O_2$: b510, b511, h196
$C_{20}H_{18}O_8$: a788, e37
$C_{20}H_{19}N$: a819
$C_{20}H_{19}NO_5$: b2163, c262, p8, p1854
$C_{20}H_{19}NO_7$: c521
$C_{20}H_{20}ClNO_5$: c263
$C_{20}H_{20}N_2$: b308
$C_{20}H_{20}O_6$: c477, c559
$C_{20}H_{21}NO_3$: g8
$C_{20}H_{21}NO_4$: b2168, c78, c79, c80, e79, e80, h719, h720, p9
$C_{20}H_{21}N_3O$: m355
$C_{20}H_{22}ClNO_4$: p10
$C_{20}H_{22}ClNO_6$: b2165
$C_{20}H_{22}N_2O_2$: g12, h732, p937, q33
$C_{20}H_{22}N_2O_4$: p937
$C_{20}H_{22}O_3$: h441
$C_{20}H_{22}O_4$: s242
$C_{20}H_{22}O_8$: p1049
$C_{20}H_{23}ClN_2O_2$: g13
$C_{20}H_{23}NO_4$: i133
$C_{20}H_{23}N_7O_7$: f105
$C_{20}H_{24}N_2O_2$: e90, i186, i187, q15, q17, q18, q22
$C_{20}H_{24}N_2O_6$: b1782
$C_{20}H_{24}N_4O_8$: a555
$C_{20}H_{24}O_4$: c555, d191
$C_{20}H_{25}BrN_2O_2$: q26
$C_{20}H_{25}ClN_2O_2$: q20, q27, q28
$C_{20}H_{25}NO_3$: a1375
$C_{20}H_{25}NO_4$: c454, l18, l19, l20
$C_{20}H_{26}Br_4Na_2$: e44
$C_{20}H_{26}ClNO_2$: t696
$C_{20}H_{26}N_2O$: p465
$C_{20}H_{26}N_2O_2$: a752, h733, h734, h735
$C_{20}H_{26}N_2O_3$: e11
$C_{20}H_{26}N_2O_4$: a585, b1808, b1769, b1774
$C_{20}H_{26}N_2O_6S$: q19, q23
$C_{20}H_{26}O_2$: b2269, b2304, h324, h325
$C_{20}H_{26}O_4$: p819
$C_{20}H_{27}NO_{11}$: a982, a983, n454
$C_{20}H_{28}N_2$: b2236
$C_{20}H_{28}O_5$: e31
$C_{20}H_{30}O$: v24
$C_{20}H_{30}O_2$: a1, p919, t59
$C_{20}H_{32}N_2O_6S$: e51, h606
$C_{20}H_{32}O_2$: e27, t55[1]
$C_{20}H_{34}$: t72
$C_{20}H_{34}O_2$: c70, o67
$C_{20}H_{36}Br_4O_2$: o61
$C_{20}H_{36}N_2O_8S$: h607
$C_{20}H_{36}O_2$: o6, o97
$C_{20}H_{36}O_4$: l51
$C_{20}H_{38}$: e29, p905
$C_{20}H_{38}O_2$: o78, o79
$C_{20}H_{38}O_3$: d44, o53, o55, o57, o59, o88, o89
$C_{20}H_{38}O_4$: e17
$C_{20}H_{40}$: e28
$C_{20}H_{40}O$: e23, e24, e26, p906
$C_{20}H_{40}O_2$: a246, d46, e18, h258, o29
$C_{20}H_{40}O_3$: o30[1]
$C_{20}H_{41}NO$: e25

MELTING POINT INDEX OF ORGANIC COMPOUNDS

Temperatures in °C; where values are not precisely known, or where there is a range of melting points, the compound is listed according to the lower temperature.

−215: h768
−196: m291
−192: c184
−190: p1107
−185: b2940, b1713
−183: e162, m297
−182: b848, m175
−181: m241
−180: p376
−173: h373
−169: e401, p210, m308
−162: b2410
−160: e405
−159: p1185
−157: b2983, s77
−155: m253
−154: p148, p412
−151: k1, p1189
−150: s69
−148: b2984, p46
−147: h421
−146: b2461, b2556, h547
−145: m244
−144: e328
−143: c591, e245
−142: c876, e421, m276
−141: b2513
−140: b2645
−139: b2941, b2995, e199, p1721
−138: b2485, c186, c874, e499, p375
−137: b2557, p44, p100, p403
−136: b2435, h201, p377, p410, p1079, p1722
−135: c899, h549, m256, p413, p1719, s70
−134: b2991, p416
−133: a1642, h548
−132: b2454, b2556, b2583, b2987, h579
−131: b2512, p1294
−130: p84, s63
−129: b2557, p184
−127: c907, p1587
−126: c688, h365, o211, p47
−125: h584, p1716
−124: a19, e450, h585, p131
−123: b2511, b2993, b3070, p1133
−122: c878, e411, e528, e539, m261, s71
−121: c817, h358
−120: b2460, h128, h130
−119: c897[1], c910[2], e195, h197, h372, o137, p133, p140, p1717
−118: h371, o213, p732
−117: a939, b2895, b2957, b3078, e336, p1134, p1715, s343
−116: b2643, c820, e477, e534, h116
−115: e458[1], m216, p141
−114: c185
−113: a22[1], c875, h550, p1080, s348
−112: a183, b2500, b2501, b2510, d67, e537, n483, p152
−111: c674, e241, e277, p211, p1293, t494
−110: b3001, p221, p1119
−109: b2436, e4660, h126, h99, p155
−108: c871, e253, n485, o135, p1210
−107: c691, o212, p153, p421
−106: b2944, b2958, e204, h178, p420, s341
−105: b2495, b2487, c877, f111, h378, h581, o214, o223, p101, p456, s74
−104: a813, b2522, b2569, c811, h369
−103: b2568, e501, h119, o224, p1831, s303
−102: o208, s332, s347
−101: d158, o209, p154, p448, p1109, p1192
−100: a197, a200, b2556, h318, h536, h730, p1161
−99: a215, b928, b2471, f141, p99, p1720, p1745
−98: a221, b2669, e471, m349
−97: b2686, c865, e215, f111[2], k3, m228, m252, o304
−96: a174, a850, b771, b847, c825, c873, l41, p229, t224
−95: a258, c694, h320, p90, p1649, s290, t275
−94: a948, b758, c868, e218, e249, e452, f139, p1314
−93: a201, a216, b2408, f146, m176, m213, p219, p419, p1211
−92: b2400, b2661, b2662, b2926, e448, f106, f132, p77, p1326
−91: h364, p142, p222, p1120, s80
−90: b2864, c687, e255, h93, o249, p447, p1193, p1312, p1525, p1588
−89: b2665, c665
−88: b372, e425, h580, o210, p1527, p1707
−87: b2912, c668, h308, h432, p1324
−86: e497, m214, p144, t171
−85: b2563, b2820, c666, e373, h89, h326, p365, p1532
−84: a199, b2403, b3018, e163
−83: b374, h180, h330, m22, p1108, s292, s313
−82: e591, h171, h181, p1776
−81: b3077, f158, h216, o134, p1083, p1530
−80: b2542, b2562, e384, e412, f135, h311, h420, h472, h682, i216, k4, m24, o221, p371, s109, s295, t743
−79: c580, c619, c647, c667, n473, o159, p231, p1281, s67
−78: a178, a226, a1641, b924, d173, h124, h282, n529, p346
−76: b2447, c895, h491, p209, p1332, p1748
−75: a866, b373, b2657, g108, h123, h288, i198
−74: b2523, b2541, b2933, b2985, e467, h443, p1316
−73: a172, a217, a847, b850, b2677, b2682, d68, d77, f145, p121, p1328
−72: p1534, t198
−71: a253, h436, p1322
−70: c854, h170, p474, p1159, t122
−69: b2992, e509, h98, t216
−68: e266, p1114
−67: e213, e350, h212, h376, h433, p151
−66: b960, d66, e600, f109, h149, m282
−65: a639, f220, h158, h490, n560
−64: a640, b247, b889, b3089, h431, m290, m243[1], p139, p1218, t215
−63: a1128, b257, b642, e211, m306, o165, p225, s167, t208
−62: a198, b772
−61: a848, b974, c604, f120
−60: a204, b887, b2887, e478, g112, h295, h387, n536, o222, t476
−59: a1424, b2394[1], c862, p441, t173
−58: b376, b2671, e229, e292, h96, s126
−57: a52, a1157, b35, b641, f167, h517, m413, p362, p1320
−56: b2412, b2681, h152, h316, o112, p218, p776, p1153
−55: h435, o116, p85, p464, p923
−54: b2636, m54, p1522
−53: b403, b975, e408, f168, s162, x11
−52: b925, b1299, b2921, b3015, m246
−51: a173, b843, b2481, b2486, c894, d268, i16, n519, p147, p216, p295, p1319, s62, s316, t175
−50: a241, b2827, c590, c664, e413, i118, m23, p920, p925, p1799, t733
−49: d213, e27, h46, h156, o128, o158, t206
−48: a827, b2976[1], d37, h125, h151, p1136, t393
−47: b661, b2907, c763, h440, h478, p1518
−46: b2596, m230, o172, o176, o273, p224, p1205
−45: b407, d6, e367, h438, i143, n528, o105, p1310, p1663, p1668
−44: a630, a823, b782, b2394, c651, d74, e267, s160
−43: b246, b449, b778, c170, e366, h90, h142, o164, o166, p161, p1466
−42: b2788, c677, d197, e259, p230, p288, p348, p1177, p1327, p1939, s119

−41: c593, e312, h430, p1706
−40: a893, c160, h411, o170, p117
−40: p472, p1475, p1800, q102, s333, t121, t204, t213, t359
−39: a1085, a1428, b779, b2540, b2829, d180, f277, h185, o188, t391
−38: a247, b2802, c898, e403, p1195, b1758, t187
−37: b859, c669, e274, h51, h383, n537, o140
−36: b2468, b2826, e216, e276, e564
−35: b567, b1273, d46, d315, e242, e272, h166, h184, o174, o206, p134, p214, p1163, t492
−34: c581[1], c663, c692, d45, h434, m264, n539, o271, p1154, s172, t392
−33: a374, c704, e273, h55, h150, h186, p291, p1480, s159
−32: b2772, b3071, d7, l74, s402, t386
−31: b309, b828, e285, o168, p268, t98
−30: a852, a1429, a1430, b746, b2191, b2472, b2913, c183, d20, d22, m105, m294, o70, o262, p739, p1757
−29: b2390, e261, p162, s120, s190, t609
−28: a1132, b450, p749, b2672, h439, m286, n476
−27: b780, b2831, c860, e214, e389, h52, p1411, t283
−26: a369, a817, b973, c618, e594, h386, p115, t358
−25: b536, b660, b849, b1796, b2611, c3, d177, e599, h19, h474, m89, n460, s398, t200
−24: a583, b278, b942, b1266, c488, h134, p160, s182
−23: a227, b773, h94, m296, p202, s165
−22: a402, a938, b1267, c524, d36, e293, e420, m32, n34, n217, n463, q197, s193
−21: a1116, b313, b2453, e236, h155, h385, i236, m225, n516, o197, o207, p1178, p1212, p1302
−20: a102, b780, b2180, b2491, b2435, d47, h135, h392, h729, m373, n425, p345, p1129, p1135, p1174, q87, s403, t194, v28
−19: c892, e573, h321, h322, m3[2], p1487, p1729
−18: b1231, c640, c823, d10, d48, f11, h466, n168, n527, o173, p188
−17: a515, a627, b341, b535, b643, b748, b763, d300, e196, e318, h15, h389, n498, o229, p1685, p1851, p2172, t438
−16: b259, b2550, c468, c785, h583, q34, t227, t437
−15: a190, a358, a637, a925, c105, c181, i203, k18, m269, m411, n466, p248, p1017, p1229, p1723, t497
−14: a175, a1043, b370, b1791, b2888, d49, m218, n181
−13: b1291, b1842, b2758, e283, h724, n185, o303, p270, p1650, t730
−12: a1146, b312, b775, b1286, b2896, c611, c681, e500, h541, h717, m345, o263, p1169, t90
−11: b122, h146, o66, s419
−10: a1046, b806, c866, d168, d284, d288, e266, h144, n215, o167, s307
−9: a451, b107, b343, b750, b1671, c759, c765, d64, d290, d320, h543, n154, p952, t556
−8: a892[2], b2746, c339, c880, h413, m414, n461, p1937
−7: b513, b2516, b2651, c583, c620, f276, n182, n552
−6: a1002, b207, b423, b783, b941, b1658, c659, c660, e409, h312, i8, i228, m169, n93, p1208, t94, t202
−5: b248, b277, c1, d18, g117, m408, n554, o4, o139, o145, p1004, t598, t716, t732
−4: a932, b1468, b2464, b2703, b3016, c360, c654, h16, m233, p2110, t357
−3: a1044, b3091, c483, c810, h36, p1684, t271, t557, t713
−2: a1134, b1804, b2590, b2760, b2869, b3088, c594, d9, d58, d301, f227, h53, h427, i49, q193, t564
−1: b1268, e265, h417, n382, n567, o83, p367, t81
0: a296, a420, a566, a825, b328, b800, b2320, e264, e406, h33, p613, p2038, p2134
1: b369, b1662, b2470, b2977, b3082, c173, c335, c656, d35, h42, h391, o163, p1296
2: a1105, a1118, b422, b718, b754, b755, b2186, c616, p855, p2027
3: b154, b349, b784, b1281, d39[1], f117, g106, h395, m1, o95, p40, p2139, s32
4: b135, b338, b784, b2187, b2463, b2551, b2553, b3002, d53, d304, e470, h41, h270, h452, m242, p613, q131, s116
5: b342, b1612, b2321, b2393, b2620, b2889, d303
5: f12, h302, h728, m249, n179, p659, s164, t163, t443
6: a1145, b202, b891, b1666, b2252, b2555, b2763[1], f10, m263, o92, o144, p19, p592, p1094, s161, s386, t67, t69, t245, t509
6: t718, v14, v15
7: b362, b484, b512, b1779, b2152, c637, c653, d57, m13, m236, m303, p1983, t92, t117, t307, t418, t724
8: c356, d8, d299, e207, e416, e588, f115, g118, m239[1], o86, p613, t394, t719
9: a986, b110, b209, b329, b2302, c162, e611, g99, m215, p116, p130, p2053, q195, t188, t554
10: a712, b836, b861, b935, d46, d240, e208, e212, e340, h25, h137, h173, h357, h390, h716, i196, i222, m289, p32, q194
11: a407, b67, b227, b1412, b2929, h80, p788, p818, p1187, s7, t168, t499
12: a613, b478, b563, b934, b1019, b2932, c676, d198, e553, h32, h40, h246, h259, h568, n253, o51, p15, p1766, q67, t736
13: b116, b351, b662, b715, b1317, b2731, g115, m299, p310, s307, s399, t82, t644
14: a1135, b179, b1483, b2139, b3007, c855, c856, d62, e275, h43, h289, m67, m409, n462, p1542, s405, t714
15: a626, a1162, b713, b794, b1097, b2908, b2924, d21, g186, h39, h245, m311, n184, n534, p1435, p1898, p2022, s385, t544, t558
16: a924, a1066, b38, b208, b228, b2727, b2739, h31, h194[1], h258, o71, o86, o89, o161, p660, p1224, p1404, p1792, p2073, p2133
17: a58, b68, b215, b357, b953, b2622, c827, c917, c919, f246, h20, h35, m277, o186, p510, p1513[1]
18: a601, a1017, b1660, b2299, b2380, c471, c912, c899, c930, d175, e398, e511, g70, h243, h252, h297, n542, o69, p38, p256, p261, p762, p1455, p1841, s392, t118
19: a21, a207, a1083, b1265, b1609, b1811, b1821, b2621, b2624, h49, h353, h694, p34, p205, p1696, q196, s192, s195, t16, t18, t84
20: a645, a647, a669, a933, b114, b269, b366, b914, b1885, b2182, b3019, d230, d313, e387, h274, h275, m235, m332, n319, n549, o21, o80, q223, t292, t670
21: b54, b279, b353, b552, b821, b1494,

e513, f269, h66, h296, o16, p1694, t80

22: a487, b380, b391, b677, b785, c361, c755, d61, h139, h249, m86, m247, m251, o40, o157, p2092, t14, t363, t415, t473

23: b275, b387, b647, b864, c790, d311, o27, o34, o93, p127, p512, q129, t707

24: a1076, b396, b473, b1891, b2125, b2506, d305, e587, h37, h50, h695, n155, p30, p1636, p2042, q70, t456

25: a619, a1221, b321, b645, b1319, b1500, b1768, b2625, b3003, c529, c742, c758, d44, e224, h260, i188, m410, o31, p57, p789, p803, p1130, p1454, p1595, p1788, p2107, p2169, t31, t298, t419, t511, t538, t728

26: a584, a987, a1141, b260, b682, b732, b785, b795, b820, b1657, b1959, g80, h635, m271, n222, p35, p1344, p1456, p1633, p1959, s212, t299, t302, t306, t446

27: a148, a612, b398, b432, b435, b553, b638, b1198, b1720, b1784, b2529, d28, d38, e313, e510, e612, f229, h250, h409, h535, o107, p1629, p1655, p1665, p1698, p1789, q185, t383, t448, t568, t717, t727

28: a705, a706, a830, a832, b170, b438, b442, b644, b744, c609, c749, c779, d285, e28, f194, h22, h71, h86, h394, n550, o11, o26, o32, o147, p1683, s337, t152, t241

29: a292, b707, b714, b752, b2318, b2417, b2419, f217, h264, h582, h712, m245, m374, o39, o41, p596, p650, p661, p662, p1186, p2039, q68, q265, s173, s176, s180, t223, t225, t303, t305, t339, t360

30: a355, a400, a841, b291, b361, b1807, b1840, b2221, c549, c859, d199, d309, d312, e586, h72, h161, h303, i139, m25, o15, o36, o94, p461, p468, s105, s375, t64, t287, t498, t574, t729, t731

31: a220, a482, a1660, b829, b1420, b2151, b2817, c698, d42, d293, f143, f284, h165, h265, o192, p266, q86, s210, s415, t337, t428, t440, t515, t726

32: a283, a380, a542, a562, a1016, a1662, b559, b1263, b2145, b2248, b2381, d164, h68, h457, h476, i19, m50, n68, n126, n329, n505, p1179[1], p2015, s384, t294, x10

33: a48, a472, a671, a1024, a1137, a1598, b472, b808, b833, b1806, b2195, b2558, c610, d278, e434[1], e517, h73, h273, n8, n254, n255, o102, p251, p431, p514, p593, p614, p1822, p2114, t88, t563[1], t607

34: a246, a442, a480, a634, b43, b689, b756, b826, b2310, b2311, b2622, b3019, c705, c826, c850, f228, h240, m178, m186, m336, n24, o18, o29, o154, p820, p914, p1627, p1968, q71, q83, t24, t89, t541, t669

35: a432, a453, a1050, a1552, b41, b119, b320, b335, b355, b651, b810, b1631, b1848, b1856, c752, d205, d243, f285, g102, h26, h65, h455, h456, i231, m164, m170, m285, m379[1], n142, o182, p494, p1039, p1645, p1656, p1669, p1682, t49, t169, t178, t387, t500

36: a332, a684, a860, a1648, b265, b438, b836, b1073, b1162, b2294, c355, c748, f287, g151, h29, h257, h745, m190, n180, n236, n507, o92, p642, p815, q182, q256, s318, s357, t470, t510, t602, t639

37: a67, a481, a495, a802, a1176, b152, b160, b232, b396, b438, b1665, b2019, b2323, c359, d281, e16, f271, h367, i18, n57, n218, n515, o53, p33, p299, p1538, q81, s35, t68, t297, t384, t414, t417, t526, t579, t762

38: a285, a596, a701, a1158, b606, b862, b1083, b1149, b1421, b1495, b1839, c26, c727, d55, d212, d263, f13, f258, g117[1], h28, h69, h672, m163, n48, n322, o9, o59, p441[1], p1092, p1370, p1430, p2046, s257, s396, t100, t374, t434, t563

39: a134, a235, a547, a808, a1049, a1102, a1422, b529, b1327, b1479, b1810, b2592, b3057, c354, c470, e556, h271, o33, p62, p1035, p1907, q198, s238, t87, t332, t432, t452, t600

40: a243, a287, a386, a1597, b48, b352, b617, b639, b2715, b3085, c135, c169, c568, c608, c914, c936, e45, e47, e492, f250, g74, h59, i35, i193, j6, n249, p39, p363, p467, p621, p627, p657, p795, p1085, p1395, p1550, t96, t333, t364, t429, t591

41: a642, a1125, b94, b310, b394, b438, b2010, b2013, b2330, b2626, b2922, b3065, c472, c626, c767, c770, c901, c916, d206, d291, e223, h344, i146, o35, o55, 057, o75, o266, p1095, p1961, q55, q85, q183, t166, t205, t329, t379, t761, t603

42: a681, a895, a909, a1054, a1431, a1471, b103, b285, b1018, b1174, b1198, b1262, b1425, b1471, b1485, b1870, b1976, b2338, c112, c342, c347, d298, e102, e572, i135, m165, m279, o298, p611, p816, p1342, p1554, p1610, p1786, q192, q228, t73, t138, t382, t546, t577, t722, t768

43: a384, a1067, a1377, a1556, a1569, b84, b356, b435, b555, b562, b876, b1679, b2780, c572, c722, d683, h1, h58, h115, h406, h408, n328, n475, o84, o152, p204, p438, p496, p555, p615, p656, q96, q227, q230, t243, t289

44: a51, a291, a376, a1175, b133, b162, b473, b526, b916, b949, b958, b1028, b1715, b2014, b2713, b2714

44: b2872, c413, c571, c857, d272, d282[1], d295, h237, h668, i26, i162, o28, o296, p935, p1253, p1363, p2163, s157, s206, s288, t503, v17

45: a322, a363, a950, b37, b761, b815, b1092, b1277, b2194, b2627, b2639, b2846, b3036, b3039, c23, c25, c38, c41, c893, d264, e284, h711, i225, m6, m115, m221, m348, n246, o72, o264, o265, p631, p679, p794, p950, p1626, p1641, p1677, q84, q231, t406, t411, t593, t604, t605, t608

46: a239, a271, a1081, a1144, a1167, a1554, b262, b303, b381, b414, b834, b1602, b1699, b1979, b3038, c109, d314, f68, h244, h662, i134, m117, m185, n533, o25, p787, p886, p1191, p1511, p1660, p1910, p2108, p2150, s355, t134, t296, t334, t376, t386, t545, t739

47: a25, a299, b69, b305, b377, b534[1], b971, b1153, b1623, b1849, b2007, c469, c579, e34, e382, f125, h276, h648, i22, l49, m173, n572, p1105, s107, s189, t219, t338, t711

48: a61, a403, a659, a849, a1171, b205, b502, b605, b931, b1501, b1959, b2482, b2783, b2847, c92, c98, d255, f257, g170,

m358, n46, n210, n390, p306, p1341, p1652, q64, s46, s54, t156, t248

129: a499, a1022, a1164, a1228, b640, b788, b1764, b1920, b1954, b2023, b2264, b2761, b3058, c902, d229, e304, f48, h451, h633, i30, i174, n72, n102, n104, n106, n188, p1, p65, p607, p698, p1065, p1497, p1758, p1825, p1958, p2146, q43, q208, s374, t183, t232, t313, t527, t614

130: a147, a460, a489, a695, a905, a943, a949, a1633, a1637, b21, b42, b91, b130, b137, b511, b602, b650, b1170, b1278, b1880, b1998, b2038, b2200, c358, g33, g52, g206, l55, m2, m314, n397, p570, p840, p1040, p1117, p1499, p2062, s259, t6, t679, t687

131: a171, a1539, b264, b501, b655, b824, b1372, b1693, b1884, b2273, c48, c377, c559, c599, c769, d135, f259, h669, i12, m27, n238, n323, n431, p584, p598, p813, p931, p1345, p1844, p2072, q212, x28

132: a70, a92, a666, a720, a1038, a1559, b64, b384, b1556, b1582, b1610, b2117, b2131, b2133, b2218, b2764, c500, c519, d162, d224, e134, e324, g149, h719, h720, m134, m135, m136, n103, p574, p918, p995, p1557, p2117, q213, s244, s389

133: a288, a470, a477, a540, a887, a1063, a1475, a1500, b245, b487, b614, b961, b1041, b1183, b1244, b1560, b1944, b1946, b2050, b2192, b2499, b2716, e9, e68, f222, h298, l36, m59, n208, n326, p24, p1509, p1519, p1892, q56, q234, t230, t354, u18, x27

134: a151, a157, a473, a739, a1365, a1536, a1601, b146, b155, b253, b494, b629, b957, b1192, b1242, b1653, b1983, b2020, b2265, b2328, b3005, c80, c714, c716, d31, e126, f252, g30, g44, g164, g189, g207, h414, h610, i107, m128, p648, p1540, s376, s383, t346, t528, t686, t687

135: a120, a604, a466, a605, a660, a1241, a1453, a1561, b23, b49, b60, b95, b386, b1057, b1279, b1332, b1474, b1651, b2012, c350, c513, c514, c528, c903, e55, e84, e104, e177, f291, h627, h673, h714, i205, k12, m17, m335, n327, n424, p465, p525, p575, p831, p929, p1369, p1953, q35, s11, s214, t5, t12, t231, t786

136: a97, a793, a1468, a1473, b498, b586, b631, b938, b1090, b1368, b1711, b2122, b2124, b2279, b2287, b2498, b2528, c191, c262, d232, f255, f256, i132, m367, n121, n138, p785, p2036, p2158, t211, t642

137: a1114, a1510, b912 b1131, b1202, b1352, b1613, b1774, b1795, b1945, b1962, b2230, b2262, b2803, c14, c248, c597, l8, l35, n67, n74, n98, n187, n302, n423, n426, p490, p913, p1849, p1951, q202, t315, t630, u35

138: a452, a696, a1499, a1584, b5, b191, b308, b724, b2029, b2129, b2154, b2290, b2353, b2720, b2743, c50, c277, c336, c338, c353, c389, d125, e80, e308, e327, f85, f251, f264, h205, m394, n160, n361, p178, p1149, q40, s127, t351, t653

139: a83, a876, a1149, b4, b74, b933, b1514, b1614, b2268, b3059, c49, c78, c111, c340, c909, e186, d127, d246, h196, i172, m193, p470, p497, p1036, p1882, p1967, s44, t114, t203, u89

140: a95, a418, a1306, b70, b239, b1457, b1518, b1525, b1652, b1844, b2051, c274, c834, c883, c920, f94, g47, h14, h138, h665, l15, l40, m204, m365, n150, n280, o143, p76, p591, p693, p904, p1845, r29, s49, s90, s406, t11, t34, t56, t628

141: a84, a549, a611, a1035, b68, b87, b192, b625, b963, b1029, b1169, b1260, b1564, b1817, b1875, b1877, b1882, b2153, b2882, c346, c420, c429, d86, e76, e127, f65[1], f207, g168, i39, i69, i106, k15, m351, m523, n506, p671, p826, p841, p900, p1037, p1872, q172, u92

142: a674, a1399, b604, b1075, b1111, b1187, b1363, b1465, b1466, b2411, c726, e79, e524, f3, f50, f86, h710, i9, m92, p581, p1262, p1661, p1971, q146, s89, t240, u126, u137, v23

143: a293, a889, a1152, a1603, b1360, b1430, b1845, b1958, b1961, b2095, c18, c307, c601, e269, f130, h708, i97, i163, n77, n81, n84, p485, p489, p495, p843, p2136, p2152, q53, q60, s143, s278, t57, t703[1], t710, u28

144: a109, a966, a1446, a1514, a1549, a1594, a1644, b149, b543, b1143, b1193, b1535, b1541, b1580, b1654, b1931, b1991, b2005, b2106, b2201, b2369, c88, c289, c891, d214, e123, f58, m33, m331, n73, n162, n174, n198, n556, p644, p670, p1808, s127, s353, t141, t235, u52, x30

145: a378, a387, a424, a1097, a1199, a1447, b283, b1085, b1165, b1213, b1252, b1603, b1724, b1988, b2026, b2047, b2123, b2163, b2189, c241, c254, c588, e250, f176, f205, g68, g128, g208, h325, h623, h638, h640, l48, m102, n318, p571, p608, p629, p863, p1038, p1041, p1307, p1477, p1940, s39, t4, x3, x25

146: a98, a434, a1232, a1492, b16, b161, b222, b893, b1259, b1314, b1547, b1753, b1895, b2046, b2240, b2253, b2298, c286, c411, c584, d80, d193, e415, e617, g39, h589[5], h707, i41, m123, u50

147: a428, a585, a677, a1052, b71, b589, b668, b1461, b1517, b1557, b1664, b1709, b1786, b1812, b1836, b1861, b1929, c391, c427, f75, g216, m48, n140, n200, n436, p9, p602, p610, p1269, p1846, s60, s220, t521, u110

148: a129, a501, a509, a557, a565, a650, a788, a1170, a1200, a1213, a1362, a1631, b193, b939, b1141, b1354, b1356, b1570, b1659, b1731, b2267, b2281, b2377, c352, c535, c606, c715, c735, e128, g127, i42, i99, i154, i182, n262, n413, o276, p515, p849, p1362, p1785, q63, q268, s207, s352, s367, t310, u32

149: a773, a1100, a1235, b99, b188, b509, b582, b636, b1008, b1107, b1530, b1982, b2030, b2216, c300, c510, f61, f247, f248, h702, h760, h761, i34, m180, m200, p203, p835, p1057, p1058, p1688, p1861, p1930, p2009, q16, s53, s250[2], s322

150: a86, a641, a778, a1006, a1304, a1491, a1566, b15, b588, b1093, b1410, b1488, b1533, b1536, b1992, b2398, b2833[1], c266, c281, c289, c302, d227, e232, e252, e305, e307, f74, g41, g71, h172, h770, i58, l52, m133, m182, n61, n83, n310, o142, o280, p523, p938, p2081, p2148, q147, t106, t626, t684, u96

151: a12, a131, a441, a602, a758, a760, a958, a1039, a1161, a1361, b1442, b1587, b1917, b1952, b2422, b2543, c138, c265, g201, h637, m132, m198, n15, n352, n400, n458[1], p2094, q191, r14, s10, s17, s315, s323, s354, t129, u38

152: a272, a421, a494, a497, a1099, a1107, a1226, b270, b1850, b1879, b1936, d13, d244, e10, e119, e610, h163, m6[2], n389, p164, p505, p568, p728, p1059, p1118, p1653, p2119, q26, s250[1], s250[3], s338, t242, u19, u48, u124

153: a83[1], a136, a143, a152, a168, a294, a467, a469, a1030, b98, b138, b803, b986, b1109, b1135, b1377, b1431, b1532, b1573, b1722, b1846, b2087, b2449, c428, c633, g34, g144, g160, h380, h664, m153, n408, p565, q210[1], s48, s216, t785

154: a160, a273, a491, b985, b1540, b1571, b1621, b1625, b2283, c136, g179, k24, n306, n317, n395, n432, p185, p563, p578, p585, p1802, s103, t107, u75, u112

155: a389, a575, a1234, a1582, b47, b101, b131, b185, b879, b1127, b1225, b1345, b1417, b1418, b1435, b1473, b1503, b1529, b1801, b1937, b2022, b2166, b2293, c247, c276, c304, c463, c464, c515, c517, c905, e129, f76, h625, m10, m174, m301, n153, n209, p1441, q116, s145, s158, s211, s218, t55, t125, t153

156: a93, a518, a526, a539, a556, a870, a1059, a1237, a1383, a1384, a1481, a1448, b7, b19, b25, b44, b633, b1336, b1388, b1411, b1683, b1712, b2162, b2272, c730, d87, d103, f83, i5, i43, m88, n440, p471, p706, p723, p757, p1539, q28, q233, s88, s135, t640, x9

157: a24, a529, a548, a635, a1403, a1515, b144, b1249, b1408, c157, c401, c455, c587, c595, f72, h611, h622, h692, p3, p241, p502, p869, p1100, p1546, p1848, q224, s58, s219, s231, s233, s373, t311, t780, u95[1]

158: a138, a492, a722, a752, a868, a993, a1033, a1224, b90, b187, b241, b507, b1254, b1489, b1703, b1926, b2105, b2150, b2227, b2285, c21, c67, c70, c370, c415, c681, f67[1], f84, f179, o153, p263, p553, p944, p1950, q27, q262, q266, q279, t135, u108

159: a156, a274, a468, a1388, a1390, a1596, b33, b995, b1220, b1247, b1454, b1490, b1506, b1649, b2370, c376, c378, c543, d126, e113, e164, e333, e334, g6, h294, h653, m209, n34, n163, n211, p61, p1298, p1443, p1675, p1891, p2100, q190, s378, t278, t675, t710, x31

160: a192, a490, a995, a1215, a1458, b20, b59, b918, b956, b1070, b1086, b1441, b1940, b2612, b3010, e158, e515, g46, g157, i77, m120, m319, n315, p493, p566, p569, p1214, p1324, p2079, q23, q57, r2, r35, s229, s250, v26

161: a127, a1585, b32, b147, b539, b1402, b1838, b2090, b2142, b2207, b2254, b2286, b2404, b2767, b2818, c316, c421, c564, f183, h393, i56, l34, n159, p1658, p1986, s86, t189, t317, u78

162: a27, a73, a785, a1010, a1263, a1305, a1540, b181, b590, b1021, b1215, b1689, b1894, b1997, b2128, b2276, b2770, c885, e53, e130, e138, e614, f63, h324, h589[7], m126, m208, n212, n365, p678, p2120, q110, t318, t685, u105, u121

163: a759, a1031, a1204, a1563, b583, b634, b695, b944, b1113, b1222, b1251, b1280, b1455, b1584, b1632, b1746, b1814, b1907, b1908, b1909, b1994, b2119, b2421, c255, c270, c516, c713, h589[2], n26, n403, p1114[1], q216, r32, s228, s368, t59, t150, t625, t760

164: a719, a1110, a1506, a1657, b66, b504, b1201, b1224, b1355, b1374, b1460, b1515, b1538, b1568, b1696, b1910, b1951, b2247, b2266, b2284, b2314, b2544, c69, c706[1], d83, i178, m206, m381, m380, p873, p885, p1279, p1482, p1870, p1963, q36, q58, s87, t677, t781, u40, u134,

165: a5, a113, a140, a965, a1214, a1363, a1386, a1415, a1470, b33, b79, b791, b936, b962, b1331, b1399, b1432, b1448, b1592, b1803, b2053, b2102, b2184, b2327, b2331, b2395, c290, c518, c527, d269, d317, e109, e310, f220[1], g3, g56, h614, h671, m142, m181, m201, n428, p887, p1266, p1442, p1923, p2089, q10, r19, s102, s104, t322, t529, t705, u33, u79

166: a68, a169, a1490, a1616, a1659, b540, b577, b1137, b1227, b1450, b1492, b1932, b2213, b2374, c733, f111[1], f177, h450, l20, n277, p56, p882, p1082, p1297, p1371, p1428, p1813, q37, s235, t246, t276, t277

167: a1320, b635, b913, b1226, b1298, b1504, b1552, b1719, b1756, b1776, e131, m160, m381[1], n308, p577, p1853, p1942, s42, s253, t624, t754, u23

168: a23, a130, a435, a505, a1456, a1477, b195, b1238, b1297, b1537, b1661, b1697, b1762, b1835, b1933, b1943, b1956, b2114, b2232, b2747, c158, c410, e142, f25, f66, g54, g220, h658, h732, l73, m124, m125, n387, n396, p503, p532, p1637, q41, q145, s25, s261, s365, t9, u94

169: a737, a869, a1153, b150, b295, b688, b1256, b1422, b1496, b1511, b1575, b2193, b2413, b2613, c137, c151, c558, d145, h45, n355, p579, p1029, q2, q62, s269, s408, t693, u31, u43

170: a724, a851, a1262, a1309, a1541[1], a1605, b197, b470, b497, b577, b1206, b1216, b1245, b1379, b1404, b1585, b2162, c11, c140, c152, c602, c858, e14, e146, e615, h299, m19, m96, m364, m407, n120, n433, p625, p731, p1957, s128, s209

171: b1240, b1394, b1682, b1717, b2135, c13, c400, c719, d140, d142, d253, e302, f67, g217, h388, h733, i124, i130, k14, m298, m386, m405, n3, n410, n453, p67, p550, p1251, p1893, p2052, q14, r21, s205, s232, t309, u82

172: a1, a103, a858, a963, a1541, b694, b921, b982, b1367, b1509, b1740, b1782, b1876, b2217, c68, c198, c285, c402, c835, d85, h12, h735, m396, n32, n401, n406, p199, p727, p729, p1060, q18, u44

173: a710, a774, a1172, a1646, b145, b996, b1340, b1499, b1607, b1723, b1984, b2295, c437, i126, l71, m203, n33, n176, n415, n420, n434, o150, p851, p866, p1437, p1791[1], q201, s20, s260, s262, u20

174: a1191, a1494, b12, b698, b837, b1005, b1531, b1706, b1728, b1834, b1921, b1974, b2168, b2469, c32, c79, c227, c229, c406, c706[2], c738, d110, e61, e98, e167, h148, i2, p525, p556, p917, p1055, p1904, q17, s327, u6, x12

175: a137, a1036, b25, b700, b802, b992, b1369, b1634, b1758, b1574, b1986, b2155, b2368, c275, e296, c372, c393, c462, c512, c925, c926, c927, c928, d167, e56, g23, g137, h8, h613, h624, h734, i6, m210, m316, n263, o258, p524, p597, p1611, q155, s236, s284, s390, t44, t48, t319, t748

176: a133, a385, a723, a780, a1239, b11, b22, b1039, b1167, b1212, b1348, b1555, b1620, b2085, c19, c31, c64, f59, f91, i153, n157, n336, n407, n445, n450, p1956, p2103, q38, q156, s187, s361, t682, u80

177: a170, a636, a934, b699, b1445, b1590, b1627, b1727, b1837, b1897, b2054, c278, c504, f42, g57, i51, m195, m205, p529, p1972, q89, s41, s151, s366

178: a957, a1032, a1129, a1366, a1402, a1450, b198, b201, b254, b911, b1323, b1491, b1493, b1508, b1721, b2062, b2376, b2428, c33, c197, c379, c934, d144, e141, e143, e159, e309[1], f2, g187, h616, h709, i86, n161, n335, n349, p157, p576, q39, s202, u51

179: a91, a303, a430, a1008, b637, b1481, b1563, b1638, b1800, b1809, b1971, b2132, b2164, c634, c638, c708, c710, f43, f249, g202, n394, p1204, p1215, p1586, p1662, s369, t312, t694

180: a46, a304, a541, a1219, a1509, b29, b132, b189, b491, b696, b1004, b1246, b1451, b1520, b1567, b1742, b2334, c388, c503, c886, d191, d231, e191, e193, f90, g214, h609, i103, m36, m52, n398, n419, o302, p626, p844, p1049, p1579, p1673, q122, q276, s85, s113, u22, u91, u120, v10

181: a1009, a1028, b10, b1395, b2136, b2224, c330, c884, d260, e57, f97, i46, i155, n414, o45, p1848, p1862, p1869, s24, s29, t54, u127

182: a715, a740, a1253, a1316, a1460, a1465, b199, b723, b1114, b1589, b1736, b2581, c63, e60, e107, e121, e262, f40, h589[3], i91, n393, p1054, p1410, p1807, p1871, p1875, s185, u29, u60

183: a1195, a1480, b697, b1330, b1630, b1729, b1872, b2204, c366, c392, c566, c706[3], f52, m140, m211, m363, m392, p2071, p2098, q29, s36, s50, s51, s127, u118

184: a133, a306, a1323, a1639, b970, b1387, b1558, b1739, b2079, b2137, c82, c196, d287, e62, f15, g65, l18, l19, l50, n411, n456, n569, p810, q76

185: a35, a992, a1190, b489, b955, b977, b1304, b1443, b1444, b1775, b2177, c195, c287, c369, c371, c481, c542, e74, e75, e231, f185[1], f245, f266, g155, g215, h747, i133, i211, l38, n22, n324, n541, p66, p549, p583, p850, p903, p1926, q5, q52, s9, s92, s273, t8, t260, t776

186: a984, a1029, a1030[1], a1493, a1516, b984, b991, b1133, b1572, b1644, b2134, c320, c390, d154, g126, i47,m355, h372, o156, p561, p716, p867, p1034, s83, s139, s371, u107

187: a105, a891, a994, b902, b1329, b1437, b1650, b1847, b1906, c58, c149, c201, c402, c449, f46, i108, i131, n19, o289, p856, p1429, p1657, q246, x17

188: a1113, a1258, a1278, b1108, b1140, b1544, b2009, b2208, b2642, c56, c550, c732, e89, e166, g1, h292, n448, p6, p478, p865, p878, p1964, q61, s364, u123

189: b1229, b1710, c728, e49, e100, f21, f114, h418, h738, i177, m382, m384, n289, n489, o269, p864, p1659, q187, r38, s13, t534, u70

190: a301, a538, a875, a1192, a1251, a1380, a1381, a1455, a1457, a1643, b994, b1308, b1424, b1524, b1769, b2289, b2373, b2855, c547, c933, d138, e37, e422, f173, g28, g66, g177, h589[1], h601, i187, k25, m139, n62, n130, n165, n430, n442, o255, p876, p1062, p1900, p2064, q80, s95, s131, s268, s277, t533, t535, u125, u129

191: a110, a877, a1084, b14, b1449, b2157, b2697, c394, f92, f185[2], g62, l9, n350, p684, p860, q125, u36

192: a3, b96, b621, b2205, b2159, b2183, c548, c707, f254, g165, g173, h589[2], i127, n17, n412, p1285, p1593, p2151, q3, s287, s423, t55[1], u65

193: a1292, a1314, b983, b1482, b1576, b1878, b1902, b2031, b2383, b2384, b2693, c73, c369, c398, g59, g143[1], h746, n134, n451, p240, p573, p937, p1063, p1901, q152, s330, t127, u144

194: a1261, a1288, a1472, a1505, a1622, b623, b1403, b1643, b1904, b2199, c75, d121, g134, h544, h666, h667, n251, n332, n333, n429, n570, p2101, p2104, r24, s28, t531, u93

195: a537, a744, a1183, a1378, b89, b286, b1738, b1903, d258, e6, e31, e86, e99, e114, f81, g36, h159, m49, m138, n23, n75, n325, n444, o308, p5, p871, p907, p1353, p1806, s40, s275, s391, t132, t154, u119, u145, x23, y1

196: a1207, a1298, a1407, b408, b1230, b1381, b1702, b1707, b2367, c273, e115, i186, n356, p1805, q106, s141, s370

197: a108, b1188, b1453, b1901, c450, c520, e35, f88, g194, h607, h639, i11, i125, p546, q144, s134, s280, t43, t108, t111, u46

198: a1004, b1700, b2226, b2238, c30, c65, c71, c407, c563, e4, e247, g21, g64, g138, g142, h589[4], i88, i104, m28, n29, n201, n245, p1761, q149, r31, u97

199: a114, a270, a985, a1544, b508, b1250, b1307, b1553, b1616, b1957, b2385, b2433, d122, g163, i213, m29, m135, n283, p877, q148, s23, s166

200: a609, a725, a768, a1379, a1398, a1408, b789, b1002, b1526, b1655, b2032, b2033, b2052, b2233, b2307, b2696, b2856, c309, c944, d81, d194, e606, e613, f51, g20, g129, g159, h11, h752, i63, l14, l76, m401, m402, n49, p1885, p1965, q169, s47, s96, s152, s156, s284, t263, t265, t324, t461, u114

201: a141, a433, a892, b990, b1690, c931, d176, l7, p888, p1265, q143, q150, t326, t695, z1

202: a543, a903, a1220, a1280, a1282, b1463, b1507, b1578, b1591, b1681, b1928, b2034, b2096, b2115, b2255, c57, c74, c233, f190, h445, i15, i69[2], i89, j1, m315, n16, n340, n399, o306, p890, p1985, s13, s43, u39, u135

203: a675, b18, b171, b1217, b1456, b1549, b1808, b2358, b2695, c337, e2, e81, i44, i72, i102, i214, n252, p720, p911, p2109, q162, t10, t697, t703, u69, u141, v20, v21

204: a714, a721, a1240, a1259, a1413, a1488, b578, b1577, b1606, b1975, b2040, c368, c408, c541, e59, e92, e139, i87, n316, n342, p883, p915, p1411, r30, u34, u143, v27

205: a845, a1389, a1501, a1542, b126, b1209, b1551, b1593, b1646, b2198, b2378, c325, c380, c416, c523, d104, e3, e116, h477, h606, i1[1], m187, n63, n338, p1069, p2102, q32, s226, t10, v11

206: a302, a555, a989, a1387, a1504, a1523, b1181, b1497, b1617, b1647, b1752, b1995, c332, c641, g58, g133, h397, h762, l51, n80, n416, n421[1], p812, p881, p1854, s52, t775, u1

207: a429, a1074, a1230, a1385, b740, b1200, b1915, b1924, b2064, c259, c322, c385, e41, n391, p533, p847, t268, u68,

208: a266, a941, a1236, a1277, a1307, a1466, a1645, b999, b1365, b1542, b1743, b1867, b2223, b2414, b2416, b2687, c409, c436, g48, g200, g213, i45, i165, j2, j3, n66, n176[1], n370, p1804, t530

209: a128, a257, a728, a928, a1368, a1401, b976, b1384, b1390, b1725, b2434, d128, d130, d152, h748, i189, i190, i195, p560, p1064, q7, t110, t249, t532, u34, u111, v9

210: a789, a1209, a1502, b527, b1205, b1310, b1351, b1359, b1510, b1527, b1714, b1726, b1854, b1863, b1911, b2036, c247, c284, c310, c430, d32, e200, g29, i29, m149, n353, o278, p8, p1859, p1865, q124, q261, s137, t689, t770, u74, v7, v18

211: a747, a1632, b1366, b1869, b2094, b2415, c486, i208, n337, p919, p1803

212: a1264, a1285, b30, b591, b1200, b1438, b1759, b1996, c228, c537, c680, d252, e111, e135, e270, h9, h631, i114, l66, m62, m318, m326, m383, n2, n278, n314, n454, p580, p1444, u88, v8

213: a960, a1281, b24, b123, b1346, b1440, b1548, b1755, b2209, c741, d112, g61, h756, n65, n378, o233, p1809, p2076, q167, s12, u103, u104

214: a982, a1353, b27, b987, b1446, b2100, b2698, c81, c150, c397, g196, i113, i115, n341, n405, p239, p2097, q135, u30

215: a186, a746, a1148, a1350, b937, b1433, b1437, b1688, b1705, b1939, c399, c488, c635, g148, l64, o232, p55, p1068, p2048, t285, t749, z2

216: a161, a874, a920, a1193, a1478, a1543, b1006, b1104, b1463, b1855, b2065, b2120, b2156, b2341, g24, h604, h630, n357, n374, n568, p845, p1888, q168, q252, u21

217: a14, a873, a942, a1246, a1404, a1410, a1412, b307, b585, b965, b1022, b1228, b1472, b1859, b1918, b2336, b2386, b2426, b2699, c505, c506, f275[1], h603, n175, n384, o260, p1202, p1903, q142, s146, t149, y3

218: a711, a1624, b249, b2210, c85, c511, e48, e50, g22, g190, i82, n164, n261, n409, p207, p884, p1980, p2062, q136, q255, q277, s94, s283[1], t691, v5

219: a1372, b1032, b1106, b1434, b1487, b2340, h444, n346, p814, p1925, q134, s91, t140, u12

220: a586, a750, a1406, a1607, a1622[1], b28, b1172, b1498, b1545, b1645, b2188, b2291, b2350, c61, c453, c458, c489, c509, c557, c867, c929, e118, g185, h381, h612, i105, n18, n427, o284, p1070, p2077, q141, s237

221: a729, b32, b33, b584, c59, c60, c232, c487, c709, g18, q280, s8, s344

222: a761, a871, a1621, a1626, b306, b1030, c444, c737, d187, e159, f89, g136, i112, l5, n135, p466, p1585

223: a828, a844, a846, a1202, a1296, a1411, b1189, b1258, b1757, b2292, b2432, c386, e423, l17, m21, n339, p1048, x4

224: a962, a1400, b27, b1151, b1207, b1429, b1912, e71, g26, g60, l75, n21, n452[1], p10, p558, q151

225: a503, a1217, b143, b838, c200, c231, c539, e54, f270, h696, i31, m45, m344, n382, p567, p712, p1118[1], p2090, s286, t321, t569, t702, u33[2]

226: a1000, a1208, a1257, a1312, a1321, b787, b951, b1362, b1704, e110, p2091, r4, t755, u7[1]

227: a1296[1], b1204, b1416, b1439, b1447, b2037, f18, f98, f99, h231, h626, l70, m177, m398, p2085, t162

228: a1047, b917, b1000, b1067, b1528, b1701, b1898, b2130, b2332, b2335, c534, e108, e117, g19, h736, i84, m188, n343, p542, p910, p2067, s6, s55, s57, t288

229: a765, a1297, b1409, b1733, b1819, b1942, b2288, c374, d84, d147, d148, f53, f93, g193, n452, p559, p715, t254

230: a864, a969, a975, a1342[1], b184, b626, b2127, b2212, b2349, b2379, b2430, c232, c946, d111, e325, f24, g212, h723, m343, m400, n360, p623, p683, p879, p2030, p2099, q107, q163, q165, q254, s37, s409

231: a1495, b592, b988, b1344, c945, h34, h721, m302, m406, p1338, u136

232: a1143, a1174, a1627, b261, b1349, b1534, b1554, d216, g124, g176, n449, p516, p1053, q164, s151, u7

233: a732, a1322, b628, b1732, b1999, h618, p534, q159, t120

234: a731, a1459, b13, b717, b974, b1640, c552, e30, f239, i57, l37, n359, n459, p1337, p1339, p1506, q157, s147, s279, t738

235: a126, a748, a955, a1045, a1338, a1414, a1461, a1498, b31, b1001, b1158, b1358, b1393, b1516, b1559, b1737, b2109, c534, e1, e157, e233, h189, i120, l77, n64, p498, p2084, q31, s110, t267, y4

236: a1198, a1276, a1356, b840, b1550, b2641, e140, f288, i85, i209, l25, n5, n369, p552, p2040, p2066, p2080, p2105

237: a307, a1609, b250, b1554, b1561, c8, d79, e66, e67, h597, h619, h646, p62, p1071, t266, u71

238: a1405, b1, b1067, b1182, b1199, b2308, e37, g203, g209, l74, p68, p830, p1052, p1073, p1278, q153, r7, s1

239: a880, a1370, b1053, b2063, h2, n274, n383, n478, t38, t109, v19

240: a775, a1001, a1003, a1524, b1343, b1391, b1905, b2072, c239, c318, c384, c507, e51, g181, g185, h697, j5, l79, m317, n345, r34, s372, x16, x20

241: a132, a136, b1195, c326, c412, d133, e141, f189, g132, h190, i207, p2050, r5, t37

242: a267, a1308, a1507, b1185, b1480, b1853, b2594, c139, c324, c508, e112, g162, h591, n58, s106, t789, v13, x5

243: a956, a1254, a1347, b1771, b2172, g161, h5, i110, m194, m406, o267, p11, p1067,

BOILING POINT INDEX OF ORGANIC COMPOUNDS

Temperatures in °C; where there is a range of values, the compound is listed according to the lower value.

−205: c184
−161: m175
−129: m297
−112: c161
−95: h768
−89: e162
−84: m291
−83: c185
−82: m308
−81: m241
−79: c159
−78: m276
−77: e249
−76: e421
−75: m260
−60: m226
−57: s77
−56: k1
−51: e417, m259
−50: c186
−48: p1713, p1840
−47: e277
−40: m230
−38: e204, p1081
−37: a853, e245, p468
−35: e225, m232, p1079
−33: c907, m309
−32: a1642, e594
−29: m253
−28: p1831
−27: p754
−26: e268
−25: b3078, e221, e499
−24: f113
−23: m228, m244
−22: m342, s313^{1}
−21: f106, c581^{1}
−20: b2410, m292, s69
−14: e405, p748
−12: a1440, m219, n498, p1189
−9: m237
−7: a852, p1747, s71
−6: a944, b2940, m176
−4: b2436, c593, s74
−3: p1185, p1743
−2: o304
−1: b2485, e256, m330, p1168
0: b2995, c604
1: a1641, b2944, p734, p1190
2: c603
3: a939, m213
4: b2941, c911, e218
5: b2994, b3066, e593
6: b2454, m345
7: e547, p1080
8: b3070, e539, p732
9: m256, p1174
10: b2468
12: b2452, h767
13: c583, c590, e199, e210, e241
15: m274
16: e403
17: e163, n494, p759
18: m222
19: a1416^{1}, b2435
20: b2992
21: a19, a204, b2402, c909, m261, p735
23: e563, p1721
24: m248, p760
25: a1419, m231, p745, p747, p1188, x11
26: h724, p46, s80
27: b3071
28: b2479
29: b3079, p375, p398
31: e222, f141
32: b2564, b3069, e411, e577, f193, m340
33: a1443, p1109, p1719
34: a193, a904, b2461, k3
35: e477, p1134, p1180
36: a1436, e534, m216, p84, p377
37: e328, p742, p1720, s313
38: e195, p376
39: b3077, b2991, b2993, b2462, c592, e475, e581, p447
40: a639, b2460, m252, p464, p755
41: b2984, c865, p443
42: m282, p45
43: e431
44: c899, p44, p1317
45: c897, c910, f111, n497, p1114, p1722
46: a1418, e275
47: e276, p1133, p1181
48: e271, e413, p736, p1716
49: c868, h563, p47, p1083, p1108
50: b2556, e335, e541, g186
51: a183, p1141, p1183
52: a1421, a1441, c814, p463
53: a202, d158, e546, m277, p1707
54: b2653, f135, p411, p412, p417
55: b73, b3067, e402, e460, p73
56: a827, a1483, b2983, e418, e538, p448, p1649
57: a221, c183, c632, e215, h562, n502, p1294
58: a1434, b2557, b2952, e575, p416, p1714, p1715
59: b2440, e501, p442, p475, p1120, p1828
60: a1417, b2562, e412, h281, i140, i143, p148, p418
61: b2561, c90, c582, m306, p456, s67
62: a910, h656, p143, p410
63: a833, a1482, b2487, b2489, b2950, b2953, c631, f213, n492, o263, p149, p1742, p1838
64: a22^{1}, a796, b2953, e557, f220, p1296
65: b2400, b2987, f214, m250, m321, m349, m349^{2}, m385, n485, p1833
66: b2488, e429, h280, p414, p1175, s341
67: b2390, h548, h549, h550, p413, p446, p1293, p1759
68: b2441, b2512, b2513, b2951, e176, e452, e497, f140, h320, h547, n491, p415, p1712, p1724
69: m214
70: a931, b2396, b2986, e197, h574, p1163
71: a197, a1426, b2403, b2957, e458^{1}, h579, m173, p1119, p1717
72: a637, c876, e253, e549, f163, h279, p403, p457, p1099, p1725
73: b2985, c591, p452, p737, p1129, s126
74: e273
75: b2459, b2954, c662 e565, p445, p1094, p1771
76: b2447, b2471, b2956, c677, f215, m296, p407
77: a174, a199, a1429, a1430, b2495, b3073, f200^{1}, p402, p1732
78: b2486, b2511, b3001, e336, d208, h278, h311, h540, n490, p1092, p1731, p1776
79: b3054, e254, e337, e545, f201, h541, p131, p1324
80: a241, a854, b202, b2912, h657, h765, p1314
81: b3072, c628, c637, e451, f146, h581, p133, p401, p1111, p1176, p1775, p2167, b2611, s121
82: d210, e562, h124, h282, p405, p1588, p1633, p1706
83: b2646, c811, e216, e430, e537, e550, e578, f216, p453, p1140, s65
84: a848, b2496, d215, e292, e561, f112, h308, h573, h580 h652, m339, p400, p1135, p1736, t187
85: a30, a630, b779, b2523, b2644, b2645, b2955, e229, i136, p467, p743, s347
86: c629, e459^{1}, f116, h312, n483, p134, p406, p1087, p1894
87: a33, b2457, e425, h682, p454, p1197
88: a589, b2439, f109, p1160, p1295, p1832, p1895, s113^{2}
89: a925, b2632, b2962, b2988, m217, p87, p462, p1125, p1193, p2160
90: a217, b2180, b2502, e435, h371, p132, p391, p1091, p1525, p1800, p2173
91: b2458, b2501, c173, e223, e461, e528, p404
92: a45, a216, a1420, b2480, e450, e551, h369, h372, h373, h374, h584, p86, p1532, p1545, s303
93: b2444, b2481, b2524, e224, f206, f209, h81, h82, h197, h730, p140, p386, p1124, s182
94: a810, a1428, b2963, e448, e469, h586, p408, p1735, s351, t42
95: a182, b2397, b2408, b2451, b2494, b2926, f158, h201, i245, p385, p390, p409
96: a55, a1443^{3}, b2961, e257, h200, h651, p1161, p2164, s348
97: a813, b2643, b2966, b3048, f133, m246, p100, p387, p444, p1326, p1587, p1814, s73
98: a52, a812, b2446, b3038, b3063, e201, e294, f139, h93, h198, h199, p101, p455, p1746
99: a522, b2442, b2443, b2522, b2943, e580, h216, n499, p153, p1316
100: b1082, b1102, b2520, b2521, b2865, b2866, b2867, c4,

c688, d198, e323, e342, e465, f160, h585, h683, m264, m269, m286, n484, p48, p1195, p1196, p1773, p1799, q46, s386, t283, t752

101: b2500, b2979, c691, e454[1], f115, h674, i142, o241, p419, p1128, p1434, s79, s301

102: a22, a258, b492, b2491, b2665, b3091, e243, e455, f196, f197, p77, p346, p1192, p1745, s70

103: a59, b2674, b2896, c820, e366, f161, p85, p348, p392, t609

104: b2389, b2490, b2541, b2933, c818, c819, c874, e502, m225, n501, p388, p1534, p1734

105: a41, a1427, b2944, d197, e576, h218, p211, p1164, p1435

106: a413, b2178, b2394, b2601, b2921, h358, h681, m278, m305, p952, p2166

107: a411, b3090, e196, e481, f132, f153, f210, h83, m239, m258, o235, p379, p393, p1002, p1225, p1432, s81

108: a214, a367, a811, e409, e458, e535, o292[1], p99, p198, p1165, p1738

109: b2450, b2965, c2, c181, e548, e582, h87, h362, h364, m238, p389, p1086, p1375, p1737

110: a36, a893, b2517, b2964, c630, c817, e211, f156, f212, h363, i69[1], n488, p53, p111, p152, p1527, p1836, s30, s76

111: a816, b2456, b2960, h361, p111[1], p381, p473, p802, p1323, t275

112: a179, a180, a416, b2897, b2971, e408, e449, e454, e558, h89, h217, h365, m290, p394, p421, p1733, x10

113: b2898, b3046, b3047, e274, e331, h360, h553, h557, o288, p155, p201, p210, p1625, s113, s346, t215

114: a25, c908, d204, e244, f219, f219[1], p432, p1757, p1820, p1851, p2142, t494

115: a54, a187, b2478, b2515, b2519, b2537, b2799, b2806, c627, c872, e255, e585, f166, f204, h127, h129, h359, m341, m419, p112, p154, p366, p382, p434, p1155, p1837, p2165, t216, t493, t741

116: a181, b2514, b2542, b2809, b2974, c619, e207, e485, f200, h322, p80, p113, p141, p315, p1113, p1897, p1939, t495

117: a215, a836, b2509, b2864, b2915, b2917, b2938, e462, h126, h130, h366, o293, p91, p365, p383, p436, p1003, p1004, p1123, p1136, p1150, p1399, p1416, s302, t136

118: a34, a58, a620, a1423, b2516, b2570, b2577, b2680, c611, e208, e554, h128, h368, m229, p92, p128, p142, p364, p1121, p1198, p1433, p1514, p1632, p1755, p1796

119: b2946, b2647, b3017, b3018, c663, e187, h123, h701, i231, p314, p440, p1668, p1842, s343, t224

120: a1416, b2388, b2569, b2648, b2669, b2833, b2888, b2889, b2916, c669, c428, e472, h564, p461, p1162, p1211, p1676, s312

121: a184, b2510, b3014, b3045, e420, h552, h554, o208, o214, p396, p1531

122: b2508, b3015, b3068, h331, h634, i73, o211, o212, o213, p151, p334, p339, p378, p384, p435, p437, p1139, p1767

123: a375, e219, e289, h310, h332, h518, h555, p380[1], p1084, p1228, p1332, p1383, s291

124: a194, a225, b2538, b2574, b2972, b2980, b3012, c665, c667, c670, e203, f142, h378, k2, n481, p355, p971, p1006, p1708, p1853, p2092

125: a32, b2801, b2970, b2978, c160, c666, c668, e373, e478, f208, h214, h655, o112, o210, o221, p88, p114, p935, p1005, p1303

126: a337, b2467, b3012[1], c170, h517, m299, o209, p1604, s419

127: a178, a445, e479, e500, h538, o192, p103, p209, p221, p228, p969, p1007, p1194, p1524, p1798

128: a177, b2659, b2673, b2975, c875, e350, e536, e560, f167, h316, m408, p15, p122, p751, p970, p976, p1608, p1631, s63[1], t198

129: a53, b2724, b2765, b2892, b2893, b2945, b3089, c816, e266, f159, h313, h321, n496, p338, p2019, p2137, s119, t146

130: a176, a253[1], a972, b450, b2455, b2545, b2568, b2800, b2894, b2976, c664, e463, g130, h226, h323, h718, p90, p336, p441, p1829, p2126, s414

131: a374, a524, b2539, b2810, b2911, c878, c894, e212, e566, h482, i228, o224, p124, p940, p982, p1088, p1210, p1347, p1447, p1630[1]

132: a253[4], a821, b407, b2191, b2395, b2506, b2759, b2973, c671, c674, e457, f145, f210[2], h330, m257, p397, p1085, p1138, p1313, p1380, p1606, p1751, t147, t210

133: a253[2], a1424, b2546, b2939, c614, c879, e467, h117, h551, h572, m242, o223, p323, p460, p1147, p1381, p1601, p1650, s412, t175

134: a420, b2507, b2595, e246, h289, h483, h533, h534, h571, p335, p946, p1537, p1544, p1704, s340

135: b487, b2878, b2895, c640, c727, c913, e359, e363, f211, h118, h119, h340, h342, h484, m247, n480, o106, o286, p104, p354, p1213, p1401, p1899, s75

136: a172, b758, b2762, e552, h481, h529, h576, p1151, p1166, p1816, p1914, t208

137: b2190, b2884, c870, h120, h288, h588, i230, o222, p313, p374, p977, p1322, p1485, p1663, s63, t41

138: a1377, b449, b662, b2505, b2549, b2550, b2774, b2879, c612, c684, c861, g156, h203, h341, m223, m416, p129, p324, p972, p975, p985, p1126, p1723, p2110, t158, t207

139: a847, b661, b914, b2578, b2583, b2914, c696, c992, h421, h530, p202, p433, p1730, p1918, s292

140: a402, a1438, c126, c695, c895, e486, f111[2], h116, h204, h479, h480, i222, p286, p325, p327, p424, p974, p1226, p1630, s16, s342, t168, t459, t492

141: a264, b2877, c892, e370, h95, h122, h221, o138, p139, p229, p294, p1122, p1144, p1153, p1302, p1801, s350, t206

142: a226, a253[3], b778, b2200, b2808, b3031, c110, c651, c815, c860, e217, e471, e544, f151, m283, o129, o137, o234, p94, p983, p984, p1756, p1766, p2144, s332, t395

143: a265, a415, a616, b2610, b2652, b2686, b2737, b2805, b2887, b2924, b3013, b3088, c480, c862, f289, h100, h327, h328, h501, h503, i164, o134, o136, p108, p110, p290, p1386, p1919, p1990, p2020, p2021

144: a219, a281, a369, a523, b444, b660, b2551, b2572, b2775, b2885, b2886, b3052, c896, c897, h101, h186, h531, o135, p145, p1436, p1841, p1921, s290

145: a185, a287, b2547, b2664, b2726, c689, e169, f194[1], h522, h543, h589, h645, o238, p115, p146, p222, p932, p978, p1208, p1472, p1599, p1685, p1770, q55, s159, t171

146: a897, b606, b2423, b2424, b2504, b2890, b2934, e267, e312, e356, m414, p1227, p1312, p1617, p2147

147: a277, a448, b2573, b2936, c172, d7, f221, h505, h524, m224, n486, o204, p96, p123, p337, p395, p1172, p1377, p1975, p2143

148: a202, a236[1], a237, b1678, b2571, b2740, b2863, g114, p118, p330, p345, p1025, p1389, p1974, s61

149: a186, a211, a253, c855, e347, h185, h520, m303, o101, p95, p998, p1159, p1184, p1400, p1530, p1998, t193

150: a282, a372, b2608, b2706, c3, c605, e391, e542, g115, h339, h502, h677, n309, n519, n563, p109, p369, p1485, p1578, s77[1]

151: a462, b2981, c476, c478, c854, e398, e569, f195, h184, h420, h436, h525, p1620, s411, t258

152: b847, b2472, b2662, b2663, b2881, b2913, b3042, c474, c639, h209, i66, i216, p333, p358, p361, p1019, p1218, p1346, p1379, t174, t257

153: a627, a856, b341, b2582, b2923, b2976[1], c877, d238, f120, h432, i237, n562, o217, p150, p331, p986, p1219, p1815, p1817, p1915, q212, t743

154: b781, b859, b1274, c687, c873, f8, f137, f164, h578, p227, p1518, p2161, p2162, s82, t201

155: a918, b309, b2406, b2730, b2959, c697, c758, e338, e540, f7, f152, g97, g98, h90, h94, h131, h194, h202, h326, h337, m147, m346, n495, n561, p144, p249, p295, p351, p356, p945, p947, p987, p1390, p1468, p1480, p1535, p1543, p1678, p1797, p1898, s79, s114

156: a31, a198, a948, b203, b2412, b2811, b2977, c469, c643, c694, c785, f276 h170, h179, h572, p147, p923, p988, p1024, p1229, p1592, p1750, s62, s305

157: a210, a412, b782, b2195, b2672[1], b2927, c871, e372, e474, f9, h577, o218, p319, p322, p482, p1834, p1912, p1989, s417

158: a446, b2529, b2679, c23, c25, c169, c468, c475, c479, c739, h478, h521, m240, m249, o236, p800, p1046, p1132, p1200, p1622, p1812, p1996, t124, t256

159: a279, a536, a820, b928, b3049, e368[1], e377, h98, h715, i219, i220, i221, p328, p1348, p1988, p2016, s120, t392

160: a202, a418, a461, c24, c190, f242, g210, h168, h171, h438, h492, i229, m77, m304, n560, o128, p321, p749, p996, p1131, p1182, p1325, p1591, p2003

161: b771, b2482, b2536, b2540, b3029, c742, h181, m333, o246, p1000, p1603, p1754, p1987, t164

162: a175, a238, a1481, b772, b2792, b3026, c581, c869, e261, f134, f277, h210, h224, h351, i149, m336, o121, p93, p941, p942, p943, p999, p1045, p1209, p1364, p1388, p1648, p1792, p1999, s3, t204, t393, t394

163: a28, a188, a379, b845, b2725, b3087, c1, c642, e424, h180, h193, h318, h470, o108, o115, o199, o270, p1795, p2138, s2, t181, t228

164: a227, a414, b2483, b2651, b2676[1], b2668, c576, c675, c825, c836, f202, h219, h225, h496, h497, i123, p362, p924, p925, p1221, p1365, p1430, p1654, p2000, t155

165: a102, a453, b770, b844, b975, b2609, b2667, b2708, b2781, b2919, b3075, c759, c804, h110[1], h356, h515, i122, e47, o247, p921, p1027, p1542, p2131, s299

166: b2533, c189, c484, c485, e358, g158, h508, i121, m162, m424, o240, p920, p963, p1470, p1515, p1595, p1628, p2027

167: a44, a366, a623, b2661, b2782, c649, c761, c852, e543, h192, h222, h473, i218, o110, o216, p347, p430, p922, p1017, p1154, p1638, p1639

168: a232, a614, b376, b2548, c470, c483, c807, d1, f199, h187[1], h417, h433, h475, h499, m157, m159, m161, o155, o215, p234, p1001, p1310, p1600

169: a21, a208, b2189, b2910, b3004, b3007, b3016, c105, c176, c471, c472, c673, c686, c806, c841, e364, g79, m158, o104, o198, p1328, p1533, p1971, s183, t185

170: a39, a206[1], a252, a371, a487, a510, b854, b974, b2729, b2832, c124, c808, d66, d67, d68, d164, e340, e348, e410, g153, g201, h488, h500, h511, h513, h654, e42, m30, m92, m171, m335, o105, p226, p798, p954, p1018, p1973, p2034, p2132

171: b748, b749, b2807, e319, e371, h510, m369, o122, p483, p1403, p1448, p1864, p2141, t547

172: a228, b536, b746, b2552, b2705, b2869, c188, c797, h152, h348, h493, o100, o149, p771, p1145, p1147, p1476, p1992

173: b75, b373, b374, b843, b1231, b2732, b3037, c334, c473, c492, c765, c766, e473, f194, f227, h212, h559, o283, p125, p205, p320, p1281, p1392, p1589, p1677, p2002, p2032, s419

174: a49, a189, a286, a625, a1132, b537, b2503, b2626, b2718, c763, c764, d20, e290, e444, h141, n500, o197, p964, p1021, p1022, p1621

175: a1484, b848, b925, b2534, b2734, b2738, b3081, c117, c207, c930, d75, d268, h223, h346, h582, h611, m97, p220, p613, p956, p961, p1243, p1378, s422, t173, t196

176: b849, b973, b2796, b2798, b3025, c335, d203, d241, e213, e368, e529, h142, h166, h207, h352, l43, o205, p1618, p1777

177: a205, a417, a838, b846, b850, b2631, b2741, b2797, b2813, b2802, c95, c648, d77, f119, f136, f154, i16, m378, o201, p1268, p2005, t197, t548, t549, u116

178: a284, a508, b35, b2623, b2624, b2676, b2707, c647, c798, c847, e488, h174, h353, l41, m100, o196, o254, p39, p116, p1023, p1393, p1596, p2006, p2017, t447

179: a42, a465, a921, b535, b2391, b2675, b2677, c617, e227, o258, p1220, p1748, p1991, t391

180: a29, a441, a1134, b40, b447, b2766, b2828, c685, e352, f124, g78, h96,

h302, n305, o113, o188, o200, p1287, p1700, p1785[1], s223, s245, s246, t213, t475
181: b566, b2703, c94, c644, c701, c773, f234, h370, m263, p997, p2127, t358
182: a406, b567, b709, b780, b888, b1316, b2596, c912, e189, e553, h175, i49, i232, p121, p555, p761, p1475, p1522, p1615, p1772, s295
183: a290, a365, a463, a730, a1432, b372, b565, b889, b2151, b2580, b2622, b2763, d29, h154, h177, m24, p930, p1419, p1699, p1753, p1839
184: a231, a410, a1002, b887, b2785, b3030, c699, f279, h491, i54, i150, m362, p688, p1096, p1376, p2028, t359, t360, t486
185: a101, a229, a807, b672, b673, b2620, b2634, b2655, b2831, b3009, b3039, c92, c616, c776, d7, d266, e248, e442, e583, f186, h443, n517, p250, p1158, p1415, t52, t167, t281, t476, t703[1]
186: b674, b767, b2682, b2709, b2717, b2733, b2758, c777, e19, f182, i24, o159, o266, p83, p214, p219, p958, p1179, p1236, p1373, p1905
187: a212, a230, a624, a633, a855, a1133, b2812, c106, d209, e172, e272, e498, f286, h48, h149, i152, p156, p423, p1590, p2031
188: a280, a823, b422, b1631, b2409, b2464, c524, e173, e531, h178, n553, n554, o118, p189, p980, p1008, p1513, p1886, t355, t402, t488
189: a309, a310, a311, b671, b828, b884, b927, b2615, b2760, b3080, c842, d170, e353, e456, m295, p252, p1177, p1234, p1290, s167, s402
190: b421, b765, b890, b943, b1236, b2380, b2777, b2778, b2937, c171, c575, c588, c594, e285, e406, e509, h183, n522, o244, p192, p243, p965, p1223, p1320, p1947, p2033, s72, s192, s195, s296
191: a206, a628, a634, a938, b371, b745, b1291, c781, e242, h111, p304, p1205, p1305, p1389, p2018, t498
192: a66, a407, a843, b420, b712, b2614, c133, c661, e433, e496[1], h46, h47, h163, p1694, p1944, p1970, t487
193: a48, b403, b462, b769, c26, c690, c780, d166, d179, d262, f10, f11, h743, i235, n552, o170, p284, p967, p1315, p1404
194: a20, a249, a251, a405, a849, a929, a1105, b423, b711, b2772, c97, c845, e384, h413, m324, m371, o186, p592, t122, t543, t544
195: a65, a806, b39, b457, b469, b710, b942, b2735, b3027, b3083, d6, e87, e175, e214, h19, h91, h532, i13, i151, n189, p14[1], p23, p1318, s18, s176, t403, t545, t546
196: a255, a1157, b313, b314, b768, b1234, b2636, b2906, b3002, c179[1], c700, c782, d201, m74, o165, p255, p960, p1993, t398, t406, t442
197: a368, b122, b157, b400, b446, b468, b766, b885, b1233, b1235, b1268, b2535, c784, e293, f122, f228, h86, m3, p287, p288, p779, p1146, p2134, p2172, t176
198: a610, a793, b401, b463, b568, b784, b884, b2657, e283, e315, f144, f233, h55, h295, h711, i8, l44, p1014, p1216, p1361, p1995, p2001, p2004, p2133, t229, t444, t542
199: a201, a419, a1229, b158, b707, b1286, b2909, b3040, d265, e349, m23, o158, o194, p500, s165, t227, v20
200: a980, b43, b402, b470, b764, b841, b1763, b2160, b2606, b2736, c535, c812, c848, c857, d11[1], e198, e322, f185[1], g112, g118, m34, m106, m233, p291, p991, p1097, p1222, p1553, p2022, q103, q105, t259, t289, t443, t445, u32
201: b399, b2152, b2795, c693, c783, f13, f281, m99, p2047, t357, t446
202: a191, a645, a707, a861, b339, b776, b2630, c76, e388, n520, n521, o116, p247, p768, p1474, p1684, t471, t500, t564
203: a275, a520, a617, a978, a1130, a1140, b406, b642, b777, b941, b2776, b3086, c89, c167, f131, h125, h349, p770, p1344, p1408, p1431, q104, t286, t499, u72
204: a260, a1128, b159, b312, b330, b440, b481, b641, b708, b2617, c33, e387, f192, h344, m167, p231, p431, p1384, p1526, p1920, p1948, p2046, s178, t570
205: a121, b329, b458, b459, b460, b924, b1791, b2618, b2710, b2755, b2968, c440, e220, e436, f257, h427, h454, h587, m3[2], o172, p258, p662, p1009, p1039, q102, q219, t437, t497
206: a444, a562, a837, a899, b647, b759, b785, b2629, b2654, b2788, b2871, b3028, c93, c692, c786, d64, f223, h158, h407, m32, m415, n154, o256, p268, p631, p804, p1260, p1826, s162, s398, t233, t550, t551
207: b2422, b2607, c77, c439, c650, e354, h431, m78, m379, p1333, p1391, p2053, t98, t438, t441
208: a261, b954, b2672, c168, c438, d18, d180, e367, h568, o289, p428, p1206, p2151, s4, s224, s257, t457, t707
209: a112, a1043, a1044, b246, b763, b2688, b3036, c31, c652, d322, h153, h468, m169, o166, p196, p204, p262, p303, p1245, p1452, s415, s418, t552
210: a247, a285, a1083, b350, b370, b439, b643, b1635, b2416, b2721, b2739, b2756, b3019, c898, d58, d62, e260, e432, e476, f203, m15, m168, n14, o132, p1143, p1307, p1319, p1417, p1910, s383, t572, u115
211: a703, b682, b747, b891, b1649, c5, c843, d63, e296, e437, f240, h495, p186, p193, p425, p632, p1329, p1407, t202, t472, t478, t539, t540, t571
212: a262, a263, b354, b441, b678, b680, b794, b2414, b2666, b2859, b3023, e443, g16, i184, i185, n186, n463, n464, n543, p190, p264, p1727, p2039, t477, t489
213: b67, b68, b69, b649, b953, b1273, b1788, b1790, b1799, b2670, c880, d315, e57, e62, f169, m13, n538, p261, p614, p1201, p1254, p1304, p1729, p2014, p2115, p2140, t540
214: a409, b225, b227, b228, b247, b327, b328, b760, b886, b2348, b2603, c311, c921, e438, h410, m95, n461, n462, o273, p161, p1010, p1244, p1484, p1636, t188, t440
215: a512, a1247, b466, b488, b783, b883, b1190, b2094, b2453, b2804, b2903, e532, h466, i180, m380, n185, n536, p218, p776, p1505, p1645, p1821, p2114, p2170, t439, t490
216: a173, a1085, b248, b345, b353, b960, b1313, b2602, b2653, c130, c702, d284,

h408, h474, m164, m165, p159, p342, p659, p679, s298, t563[1], x29

217: a621, a629, a1014, b340, b393, b648, b1806, b2187, c209, h472, m380, p650, p1011,p1643, p1698, p2150, s307, t491

218: a582, a687, b107, b230, b355, b514, b675, b952, b959, b1266, b1283, b1807, b2393, c210, c882, d177, e378, h717, i167, i168, m25, p187, p778, p1696, p1917, p1940, p2128, s193, t234, t370, u79

219: b108, b679, b761, b1501, c927, e389, e394, i226, p270, p633, p990, p1148[1], p1224, p1613, p1614, p1681, p2051, p2108, t473

220: a631, b37, b110, b207, b226, b256, b442, b513, b1361, b1483, b2746, b2773, c87, c96, d21, e439, f278, g88, i17, i36, l45, o239, p615, p1261, p1406, p1481, p1911, p1945, p2023, t49

221: a60, a61, b428, b512, b827, b1789, b1798, b2145, c219, c220, f232, h471, i236, p138, p305, p1582, p1693, p1811, t244, t509, t556, t598

222: a615, a704, a1178, b204, b336, b1288, b1476, c208, d25, m40, m60, m89, m119, n537, p117, p248, p1028, p1042, p1277, p1306, p1635, p1959, p1982, q224, t557

223: a62, b103, b349, b351, b796, b1486, b1671, b2417, b2418, c166[1], c464, e496, h144, m379[1], o26, p344, p685, p1105, p1594, t307, t407, t510

224: a312, a1145, b105, b224, b229, b272, b892, b2838, c180, c441, d48, f188, h42, h437, n466, n539, o237, p1171, p1887, p2044, t219, t245, t768

225: a570, a607, a608, a1042, a1162, b257, b266, b369, b570, b676, b715, b819, b1421, b1804, b2415, b2419, b2671, c156, c204, e47, f181, h147, h716, i118, m3[1], m94, p741, p1572, s394, t401

226: a705, b271, b456, b683, b762, b1797, c467, d235, e291,e419, h376, h440, i135, o176, p225, p301, p953, p1142

227: a580, a662, b855, b2683, c672, h40, h684, m417, p995, p1789, t451, t474, v14, v15

228: a678, b56, b337, b702, b703, b773, b795, b820, b1767, b2842, b2849, b2929, h33, h37, h39, h43, h460, i74, p223, p1327, p1769, s256, t305, t399

229: a80, a1016, b214, b277, b352, b3057, c425, c596, d57, d163, e148[1], f191, g15, h35, h36, h685, m26, o268, p326, p1695, p1702, t385, t458, t599, t604, t606

230: a376, a613, a922, a1046, a1050, b54, b82, b120, b153, b208, b348, b455, b705, b821, b822, b2387, b2587, b2713, b2714, b2803, b2930, c47, c98, c212, c213, d217, e181, f222, m14, m71, m110, n549, o141, p256, p1198[1], p1425, s311, s316, s416, t390, t397

231: a1086, a1139, b1299, b1478, b2116, b2700, c211, c751, d173, e440, p302, p612, s304, s306, s309, t603

232: a256, a403, a679, a1048, a1116, b311, b318, b553, b706, b922, b1254, b1471, b1662, c165, c203, c698, d45, f235, g87, m79, p1043, p1788, t152, t180, u128

233: a408, b213, b338, b347, b368, b551, b797, b2931, c174, c202, d222, f225, m54, p635, p640, p1253, t223, t558, t730, t764

234: a583, a1141, b260, b276, b389, b454, b704, b742, b825, b1470, i203, p195, p1270, p2145, s215, s358, t118, t172, t716

235: a573, a923, a1337, b85, b151, b320, b473, b563, b677, b793, b864, b1069, b1429, b1468, b1888, b2405, b2626, b2658, b2901, b3002, e313, f246, m7, m56, o245, p993, p1044, p1560, p1610, p2107, s7, s185, t340, t375, t404, t449, t450

236: b395, b550, b689, b1680, b1727, b1773, b1805, b2858, e441, h687[1], m109, p1644, s212, t336, t364, t431

237: a664, b179, b321, b525, b681, b832, b1484, b2928, h675, h676, p160, p593, p949, p1575, p2171, q34, t302, t342

238: a611, a829, b212, b255, b258, b367, b484, b644, b753, b754, b800, b2139, b2186, b2850, c129, e343, p300, p594, p1246, p1703, t400, t559, u110

239: a699, b287, b830, b1477, c866, e265, h385, o162, p1427

240: a370, a495, a698, a700, a713, a946, a1115, b125, b215, b289, b392, b569, b877, b1667, b1781, c162, c227, c826, e287, e556, e573, h452, h671, h680, h703, h705, i195, m38, m101, n217, n218, p989, p1235, p1258, q241, t182, t295 t299, t300, t623

241: a633, a1041, b259, b524, b609, b971, d47, h695, q6, t303, t308, t448, t463, t601

242: a292, a400, a601, b319, b474, b774, b1243, b1282, b1543, d213, h20, t285, t304, t306

243: a220, a683, a1147, b152, b316, b933, b1469, d49, f207[2], h31, h694, i35, i188, i217, m348, o262, p665, p752, p1559, p1880, p2038

244: a597, b396, b831, b2681, c37, c46, c846, h411, h672, h704, p664, t291

245: a572, a587, a599, a605, a669, a1004, a1077, b283, b478, b645, b713, b751, b775, b798, b801, b826, b1014, b1412, b1425, b2816, c704, c810, c917, d168, d169, d207, h156, h435, m57, n471, o164, p224, p299, p1255, p1667, p2116, q183, q276, t367, t576, t750, t751, u117

246: a680, a835, b115, b430, b472, b564, b718, b932, b1414, b1766, b3011, c164, c496, d313, n473, p651, p652, p703, p1877, s356, t90, t608

247: a581, a661, a712, b84, b114, b135, b390, b1015, b1267, b2148, d174, d285, h644, i18, p1186, p1573, p2129, q193, q209, t456

248: a1117, b206, b211, b382, b391, b809[1], b812, b2125, d212, h52, p2117, q199, q277, s26, t339

249: a1118, b380, b752, b823, b1239, c36, i225, p1187, p1642[1], q223, t432, t508

250: a484, a571, a647, b154, b288, b387, b552, b818, b935, b2399, b2841, c29, f187, f190, h145, h160, h406, h714, m68, p179, p368, p740, p1983, q2, s179, s198, t288, t292, t341, t516

251: a930, a1017, a1078, a1146, b743, b934, b1415, b1611, c823, p2169, t467

252: a1076, b343, b2932, c601, d206, h134, i197, i212, p782, p1949, s68, t462

253: a243, a481, a1135, n534, p739, t92, t294, t338

254: a1165, b209, b358, b714, b931, b1284, b1413, b1765, b2274, c339, l56, p309, p634, s227, t67, t298

255: a655, a1425, b116, b275, b344, b415, b515, b555, b811, b916, b972, b1270, b1780, b1811, b2320, c27, c477, d311, e611, h428, i21, i23, i65, i196, l2, m421[1], o130, p367, q129, t117, t297, t420, t434, t575

256: a709, a825, b744, b1319, b2214, b2299, b2392, d205, e346, h426, i7, f198, n35, n110, p595, p738, p2121, q82, q84, s191, t386

257: b232, b265, b342, b388, b554, j4, p1015, t578

258: a124, b862, b863, b1071, b1512, b1619, b2839, c355, e494, e510, g106, i201, k10, n181, p1035, p1691, p1822, p1913, q198, t290

259: a1026, b274, b356, b451, b559, b1495, b1622, b1658, b1666, b1785, g74, h434, m420, q27, q197, t511

260: a149, a153[1], a667, a1056, a1080, b2, b221, b299, b1272, b1772, b2381, b2780, b3065, i171, m412, o96, o174, o282, p172, p188, p1850, q94, q231, t225, t232, t413, t418, t574, t708

261: b1419, b1612, b2635, h151, o9, q83, q85, q126, s186, t729

262: a134, a1187, b817, b1287, b1320, b1426, c230, d303, h139, n168, p230, p1253, p1577, s210, t706, u71

263: a937, b117, b160, b418, b1027, b1513, b1792, i238, n109, n258, p641, p807, p1581, p1863, q196, t87, t435, t470, t727

264: a1027, b379, b610, b1615, c62[1], h51, o140, p777, q127, q195

265: a18, a600, a701, b133, b506, b531, b603, b791, b815, b915, c38, c42, c163, c348, h535, i71, i199, i200, k3[1], m271, n169, o242, p643, p1549, p1551, p1556, q131, s196, t424, t717

266: b357, b558, b1810, p1555, q128, q271, t736

267: a73, a574, b494, b816, b1276, b1381, b2322, e570, o273, p1692, q86, q154, q171, t388

268: b231, b560, b1317, b1582, h146, m87, o145, p1214, p2024, q132, u61

269: b1494, b1779, b2496[2], g90, i144, l65, n215, p2025, q228, t385

270: a254, a670, a686, a830, a1024, a1186, a1658, b601, b605, b1322, b1693, b1784, b1822, b2847, c59, c60, d42, e495, h464, h596, i41, n574, p931, p1764, q68, t218, t464, t515

271: a841, a1075, b950, f125, p32, p165, p704, p2146

272: a1079, b1275, b1663, b1874, b2771, c356, c529, e309, i70, p1538

273: a809, a1019, b716, b861[1], b2301, b2321, b2361, d301, g75, g169, h49, n476, p1922, t327, t422, t580

274: a493, b164, b1800, b2248, b2300, b2318, c39, c549, i153, i166, n170, n216, q67, t735

275: a1081, a1445, b636, b755, b842, b949, b1842, b1848, b1583, b1618, b1775, c7, c264, c304, h7, h50, h182, h439, m113, m293, p1763, q81, q184, s190, t23, t24, t565, t725

276: a986, b220, b587, b1018, b1673, b1749, b2310, h150, m425, n467, p638, p1967, t95, t732

277: b583, b919, b920, b1747, c333, e416, o168, p136, t101, t582

278: a4, b93, b270, b486, b575, f127, q57, t100

279: a1066, b1851, c367, m76, p681

280: a1109, b298, b836, b1685, b1686, b2302, b2363, c124[1], e490, f149, h161, i9, i193, j6, n179, n180, n477, o163, o301, p1569, p1741, p2007, q69, q109, q226, s413, t16, t25, t567

281: a842, b103, b574, n93, n94, q137, q227, t18, t138

282: b493, b534, b807, b809, b1020, b1960, e395, h389, o144, p697, p824, p2026, q186, t26

283: a155, a213, b301, b757, c901, h194[1], q192, t425, t433

284: b756, b2249, h274, q70

285: a567, b139, b523, b577, b640, b1271, c62, e237, e588, i134, k18, n147, n150, p843, q225, q230, s335

286: a535, a717, a909, a1087, b1211, b1821, b2312, c351, e236, e390, e508, h243, m294, n144, p162, p829, p1785, p1968, q92, q280, t31

287: b638, b2846, c433, d113, h248, p1969, s200

288: a1088, b835, b906, b1328, c403, c681, e525, f126, g191, n190, n196, q87, t622

289: a648, b837, g124, p775, q278, t99, t468

290: b604, b1244, g70, h155, n330, p897, p2155, q71, q130, s354, t209, t597, t621

291: b1715, b2250, b2364, n8, n126, n143, n172, n375, n527, o167, p215, t17, t68

292: a1185, l26, t421, t671

293: a649, a892[2], a1233, b522, c544, f45, n219, p297, p746, p842, q90

294: a1368, b170, b1219, b1326, b1992, n88, n191

295: a831, a892[1], a1513, b1776, b2303, b2701, d31, e239, h391, n142, p510, p822, p870, t82, t84

296: a688, b517, b732, b926, b1849, b2242, b2362, e240, n24, n322, p821, p2042, s317

297: a860, b278, b872, b1692, b2365, m366, n319

298: b788, e470, m239[1], n248, p496, p1308, s310

299: b727, b2220, h80, i27

300: a697, a803, a1182, b873, b1024, b1300, b1315, b1398, b1931, c350, e602, g175, k23, m46, m186, n36, n315, n324, p1167, p1292, t192[1], t379, t461[1], t619, t734

301: b1540, h726, m375, n25, p2074, s14, s32

302: a863, b1019, b1028, b2222, b3044, c526, p158, s64

303: a145, a167, h66, i111, t30

304: m178, n239, n328, p828, q64, q185

305: a164, a1169, b870, b2311, b2314, b2366, e326, e511, m255, n213, o11, p1114[1], q72, t347

306: a818, b1959, d35, d273, n38, n184, n331, s129

307: a168, b1301, b2293, o173, p33

308: h74, m50, n214, n329

309: b961, n320, p684

310: a1015, b1337, b1566, b1845, b2244, b2404, b2992, c682, c548, d248, e376, g111, m243, q47

311: g120, t22

312: a271, a865, b279, c175, e513

313: a1210, b219, o93, p878

314: b1295, b1297, b1302, b2017, e29, t284

315: a40, a804, a1176, b435, b437, b2267, e165, e503, p950, t287

316: a913, b1303

317: b1837, b2252, e491, n220, n221, p1824

318: p28

319: a10, a805, a901, b725, h77, q133

320: a828, a1211, b280, b331, b2317, b2323, h294, k7, m281, n87, p955, q56, t14, t15

321: a802

322: b787, e586, h244

PHYSICAL CONSTANTS OF ORGANOMETALLIC COMPOUNDS

Metallic salts of organic acids will be found in the preceding table.

No.	Name	Formula	Mol. wt.	Crystalline form, color and index of refraction	Sp. gr. or density	Melting point, °C	Boiling point, °C	Solubility in grams per 100 ml of: Cold water	Hot water	Alcohol, acids, etc.
	Aluminum									
1	Aluminum isopropoxide	$Al[OCH(CH_3)_2]_3$	204.25	col. liq		118	$145–150^{5}$ solidifies			
2	Aluminum isopropylate	$Al(OC_3H_7)_3$	204.25	brittle wh. solid		118	$125–130^{4}$	d.		s. bz.
3	Aluminum magnesium ethoxide	$Mg[Al(OC_2H_5)_4]_2$	438.77	iceblue cr.		129	225^{4}			s. bz.
4	Diethylaluminum chloride	$(C_2H_5)_2AlCl$	120.56	col. liq.		−50	$125–126^{60}$			
5	Diethylaluminum malonate	$Al(C_7H_{11}O_4)_3$	504.48	wh. need. or pr.	1.084^{100}	98		i.		s. org. solv.
6	Dimethylaluminum chloride	$(CH_3)_2AlCl$	92.50	col. liq		−50	$83–84^{200}$			
7	Methylaluminum dichloride	CH_3AlCl_2	112.92		1.00^{22}	72.7	$97–100^{100}$			
8	Triethoxyaluminum	$Al(OC_2H_5)_3$	162.17	sol.	1.142^{20}	150–160	d.			i. al., sl. s. bz., v. sl. s. eth.
9	Triethylaluminum	$Al(C_2H_5)_3$	114.17	col. liq., ign. in air, $1.480^{6.5}$	0.837	<−18 (−50.5)	194	exp.; d. to $Al(OH)_3 + C_2H_6$		
10	Triethylaluminum etherate	$4Al(C_2H_5)_3.3(C_2H_5)_2O$	679.04	col. liq			112^{16}	exp.		d. al; s. bz., eth.
11	Trimethylaluminum	$Al(CH_3)_3$	72.09	col. liq, ign. in air, 1.432^{12}	0.752	0(15)	130 (125^{26})	d. to $Al(OH)_3 + CH_4$		s. eth., al.
12	Triphenylaluminum	$Al(C_6H_5)_3$	258.30	wh. need		196–200		d.		d. al., chl., CCl_4; s. bz.
	Antimony									
1	Antimony ethoxide(ous) (triethyl antimonite)	$Sb(C_2H_5O)_3$	256.93	col. liq	1.524^{17}		95^{11}	d.		s. org. liqs.
2	Pentamethylantimony	$Sb(CH_3)_5$	196.93	(exist?)		96–100		i.	i.	
3	Phenyldimethyl-antimony	$C_6H_5Sb(CH_3)_2$	228.93	col. oil, fumes in air			112^{15-18}			
4	Phenylstibinic acid	$(C_6H_5SbO_2)_3H_2O]2H_20$	746.61	wh. fine cr.				i.		s. chl, al., dil., a.; sl. s. acet.
5	Tetramethyldistibyl	$(C_6H_5)_2Sb.Sb(C_6H_5)_2$	551.93	col.		121–122 (in N_2)				
6	Tributylstibene	$(C_4H_9)_3Sb$	293.10	col. liq.			$133–134^{14}$	i.	i.	s. org. solv.
7	Triethylantimony	$Sb(C_2H_5)_3$	208.94	liq. 15 1.42	1.324^{16}	<−29 (−98)	159.5	i.	i.	s. al., eth.,
8	Triethylantimony chloride	$(C_2H_5)_3SbCl_2$	279.84	col. liq.	1.540^{17}		d.	i.		s. al., eth.; d. conc. H_2SO_4
9	Trimethylantimony	$Sb(CH_3)_3$	166.86	liq. n_D^{15}	1.523^{15}		80.6	sl. s.	sl. s.	s. eth.; i. al.
10	Triphenylantimony	$Sb(C_6H_5)_3$	353.07	col. tricl. pl.	1.4343^{25} (1.4998)	50, (46–53)	$>220^{1}$; $>360^{760}$	i.	i.	s. org. solv.; sl. s. al.
11	Triphenylantimony dichloride	$(C_6H_5)_3SbCl_2$	423.98	wh. sol.		143		i.	i.	s. bz., CS_2, hot al.
12	Triphenylantimony sulfide	$(C_6H_5)_3SbS$	385.13	wh. cr.		108–110		i.	i.	s. org. solv.
	Arsenic									
1	Acetylarsanilic acid	$CH_3CONHC_6H_4AsO(OH)_2$	259.09	cryst.		>200				s. $NaCO_3$; v. sl. s. HCl; i. eth.
2	3-Amino-4-hydroxy-phenylarsonic acid	$H_2O_3AsC_6H_3(OH)(NH_2)$	233.06	col. prism			d.290	sl. s.		s. alk.; min. a.; i. org. solv.
3	2-Aminophenylarsonic acid (Arsanilic acid)	$H_2NC_6H_4AsO(OH)_2$	217.06	need.		153		s.		s. al.; alk., ac.; sl. s. eth.
4	4-Aminophenylarsonic acid (*p*-Arsanilic acid)	$H_2NC_6H_5AsO(OH)_2$	217.06	wh. need.			d. 300	sl. s.	s.	s. alk., dil. a.; sl. s. al.; i. eth., acet.
5	Arsanilic acid (*p*) (*p*-aminophenylarsinic acid)	$H_2NC_6H_4AsO(OH)_2$	217.06	wh. need		232				s. eth., MeOH; sl. s. al., acet; i. bz., chl.
6	Arsenoacetic acid	$(AsCH_2COOH)_2$	267.93	sm. yel. need.		>260, d. 205		i.		s. pyr., alks., alk. carb; i. al., eth., chl.

No.	Name	Formula	Mol. wt.	Crystalline form, color and index of refraction	Sp. gr. or density	Melting point, °C	Boiling point, °C	Solubility in grams per 100 ml of		
								Cold water	Hot water	Alcohol, acids, etc.
	Arsenic									
7	Arsenobenzene	$C_6H_5As: AsC_6H_5$	304.06	wh. need.		212		i.		s. bz., chl., CS_2; sl. s. al.; i. eth.
8	Arsenophenyl-glycinamide	$H_2NCOCH_2NHC_6H_4AsO(OH)_2$	274.11			280			v. s.	v. sl. s. al.
9	Benzophenone-4-arsonic acid	$H_2O_3AsC_6H_4COC_6H_5$	306.15	lustr. pl.		260		i.		s. al., alk; i. bz., eth.
10	Cacodyl oxide(dicacodyl oxide)	$[(CH_3)_2As]_2O$	225.98	col. liq.	1.486^{15}	−25	149–51	sl s.		s. al., eth.
11	Cacodyl sulfide (dicacodyl sulfide)	$[(CH_3)_2As]_2S$	242.05	oil			211	sl. s.		s. al., eth.
12	β-Chlorovinyl dichlorarsine	$C_2H_2Cl_3As$	207.32	liq.	1.8648^{20}_{4}	0.1	76.1^{10}			
13	Dimethylarsine (cacodyl hydride)	$(CH_3)_2AsH$	106.00	col. liq., ign. in air	1.213^{29}		35.6^{747}			s. al., eth., chl., bz., CS_2
14	Dimethylarsinic acid (cacodylic acid)	$As(CH_3)_2O.OH$	138.00	odorl., col. pr.		200		82.9^{22}	v. s.	s. al.; i. eth.
15	Dimethylbromarsine (cacodyl bromide)	$(CH_3)_2AsBr$	184.90	yel. oil			130			
16	Dimethylchlorarsine (cacodyl chloride)	$(CH_3)_2AsCl$	140.44	col. liq., infl.	>1	<−45	106.5–107, (109)	i.		v. s. al.; i. eth.
17	Dimethylcyanoarsine (Cacodylcyanide)	$(CH_3)_2AsCN$	131.01	lustr., col. pr., very pois.		33	140	sl. s.		s. al., eth.
18	Diphenylarsinic acid	$(C_6H_5)_2AsO(OH)$	262.14	wh. need.		174	subl. 190–200	s.		s. al., alk.; i. eth.; bz.
19	Diphenylchloroarsine	$(C_6H_5)_2AsCl$	264.59	pa. yel. liq.	1.4223^{26}		333 (in CO_2)	i.		s. abs. al.; eth.; bz.; NH_4OH.
20	Ethylarsinedisulfide	$C_2H_5AsS_2$	168.11	yel. oil	1.836^{24}			i.		s. bz.; chl.; i. al., eth.
21	Ethylarsonic acid	$C_2H_5AsO(OH)_2$	154.00	need.		99.9		70^{25}		39.4^{25} 95 % al.
22	p-Hydroxybenzene-arsonic acid (Phenol-p-arsonic acid)	$HO(C_6H_4)AsO(OH)_2$	218.04	monocl. pr.		d. 174		s.	s.	s. al.; acet. a.; dil. a.; sl. s. acet. v. sl. s. eth.
23	Methylarsine	CH_3AsH_2	91.97	col. liq.			2	0.00085		s. al., eth.
24	Methyldichloroarsine	CH_3AsCl_2	160.86	col. liq.			133	d.		
25	Phenylarsine	$C_6H_5AsH_2$	154.04	col. oil	1.356^{25}_{25}		148			sl. a., eth., CS_2
26	Phenylarsonic acid (benzenearsonic acid)	$C_6H_5AsO_3H_2$	202.04	col. pr.	1.760	158–62 d.		3.36^{28}	31.6^{84}	18.4^{25} 95 %al
27	Phenylcyclotetra-methylenearsine	$C_4H_8AsC_6H_5$	208.14	col. oil	1.2794^{20}_{4}		128.5^{15-16}	sl. s.		s. al., eth.
28	Phenyldimethylarsine	$(C_6H_5)(CH_3)_2As$	182.10	col. liq.			200	i.		s. al., bz.
29	Sodiumarsenophenyl-glycine	$(AsC_6H_5NHCH_2COONa)_2$	494.12	liq.		d.		s.		
30	Tetraethyldiarsine (ethyl cacodyl)	$[(C_2H_5)_2As]_2$	266.09	oil			185–90	i.		s. al, eth..
31	Tetramethylbiarsine (Tetramethyl-biarsine)	$[As(CH_3)_2]_2$	209.98	col.-yel. oily liq., highly poisonous	1.447^{15}	−5	163	sl. s.		s. al., eth.
32	Tribenzylarsine	$(C_6H_5CH_2)_3As$	348.32	col. monocl.		104		i.		s. eth., bz.; sl. s. al.
33	Tri(β-chlorovinyl)arsine	$(CHClCH)_3As$	259.40	col. liq.	3–4	$151–158^{28}$	$151–158^{25}$	i.		i. al., dil. a.
34	Triethylarsine (arsenic triethyl)	$As(C_2H_5)_3$	162.11	col. liq. n^{20}_{D} 1.467	1.152		140^{736}	i.		
35	Trimethylarsine (arsenic trimethyl)	$As(CH_3)_3$	120.03	col. liq.	1.124		70	sl. s.		s. eth.
36	Triphenylarsine (arsenic triphenyl)	$As(C_6H_5)_3$	306.24	wh. need. or rhomb. pl., 1.6139^{48}	1.2225^{48}	60–60.5(57)	>360 (In CO_2)	i.		v. s. eth., bz.; sl. s. cold. al.
37	Triphenylarsine-dihydroxide	$(C_6H_5)_3As(OH)_2$	340.26	col. hex. pr.		115–116		s.		s. al.; u. sl. s. eth.
38	Triphenylarsinesulfide	$(C_6H_5)_3AsS$	338.31	lustr. need.				i.		sl. s. hot al.; i. eth., a.

No.	Name	Formula	Mol. wt.	Crystalline form, color and index of refraction	Sp. gr. or density	Melting point, °C	Boiling point, °C	Solubility in grams per 100 ml of		
								Cold water	Hot water	Alcohol, acids, etc.
	Beryllium									
1	Di-*n*-butylberyllium	$Be(C_4H_9)_2$	123.24	col. liq.			170^{25}	d.	d.	
2	Diethylberyllium	$Be(C_2H_5)_2$	67.14	col. liq.		12	110^{15}	d. to C_2H_6		
3	Dimethylberyllium	$Be(CH_3)_2$	39.09	wh. need.			subl. 200	d. to CH_4		
4	Dipropylberyllium	$Be(C_3H_7)_2$	95.19	liq.		<-17	245			
	Bismuth									
1	Bismuthethyl-camphorate	$C_{36}H_{57}BiO_{12}$	890.83	wh. amorph. solid		61–67		i.	i.	s. eth., chl.
2	Diphenylbismuthyl iodide	$(C_6H_5)_2BiI$	490.10	yel. flakes		132–134		i.		s. acet.
3	Methylbismuthine	CH_3BiH_2	226.03	liq. (exist ?)	2.30^{18}		110	i.	i.	s. al., eth.
4	Triethylbismuthine (bismuth triethyl)	$Bi(C_2H_5)_3$	296.17	liq.	1.82		107^{79}	i.		s. al., eth.
5	Trimethylbismuthine (bismuth trimethyl)	$Bi(CH_3)_3$	254.09	liq.	2.300^{18}		110	i.	i.	s. al., eth.
6	Trinitrotriphenyl-bismuth dinitrate	*p*-$(NO_2C_6H_4)_3Bi(NO_3)_2$	699.30	pa. yel. cr.		d. 140–147				s. eth. ac., gla. ac. a.
7	Triphenylbismuth diacetate	$(C_6H_5)_3Bi(CH_3COO)_2$	558.39	wh. cr.		152–153				s. acet.
8	Triphenylbismuthine (bismuth triphenyl)	$Bi(C_6H_5)_3$	440.30	monocl., tan. cr.	1.585	78				v. s. chl.; s. eth., acet.; sl. s. al.
9	Tri-*n*-propylbismuth	$(C_3H_7)_3Bi$	338.25	col. liq.	1.621		$86–87^{8}$			s. eth.
10	Tri-*m*-tolylbismuth dichloride	*m*-$(CH_3C_6H_4)_3BiCl_2$	553.29	wh. cr. (from acet.)		132–133				s. acet., al-chl.
	Boron									
1	Aminophenylboric acid (*m*)	$(NH_2C_6H_4)B(OH)_2$	136.95	wh. hex. pl.		d.	d.	sl. s.		s. al.; sl. s. eth.
2	Amylboric acid (*n*)	$(C_5H_{11})B(OH)_2$	115.97	col. fl.		93–4, d.	d.	s.	s.	s. eth., dichloroethane
3	Anisylboric acids, *o,m,p* (methoxyphenyl-boric acids)	$CH_3OC_6H_4B(OH)_2$	151.96	wh. cr.		d.	d.	sl. s.		s. al., eth., bz.
4	Benzeneboronic acid	$C_6H_5B(OH)_2$	121.93			215–216		s.		s. bz., CCl_4; v. s. eth., MeOH
5	Borine carbonyl	BH_3CO	41.85	col. unst. gas		−137.0	−63	d.	d.	
6	Butylboric acid (*n*)	$C_4H_9B(OH)_2$	101.94	col. cr.		92–4	d.	s.	s.	v. s. al., eth., chl., acet., acet. a. and esters; sl. s. bz., CCl_4, pet. eth.
7	Butylboric acid (*tert*)	$C_4H_9B(OH)_2$	101.94	wh. cr.		105 d.	d.	s.		s. eth.
8	*n*-Butylphenylchloro-boronite	*n*-$C_4H_9O(C_6H_5)BCl$	196.49	n_D^{20} 1.4996	1.021^{20}_{4}	−32	$65^{0.4}$	d.		
9	Diethoxyboron chloride	$(C_2H_5O)_2BCl$	136.29	col. liq.			112.3	d.	d.	
10	Diisoamyloxyboron chloride	$(C_5H_{11}O)_2BCl$	220.55	col. liq.			$110–15^{14}$	d.	d.	
11	Dimethoxyborine	$(CH_3O)_2BH$	73.89	col. liq., unst.		−130.6	25.9	d.	d.	
12	Dimethoxyboron chloride	$(CH_3O)_2BCl$	108.33	col. liq.		−87.5	74.7	d.	d.	
13	Dimethylboric acid (dimethylhydroxy-borine)	$(CH_3)_2BOH$	57.89	col. liq.			0^{36}	v. s.		
14	Dimethylboric anhydride	$(CH_3)_2BOB(CH_3)_2$	97.76	col.		−37.3	43	hyd.	hyd.	
15	Dimethylborine trimethylamine	$(CH_3)_3NBH(CH_3)_2$	101.00	col. liq.		−18.0	d. 172	d.	d.	s. eth.
16	Dimethylboron bromide	$(CH_3)_2BBr$	120.79	col. liq. or gas		−123.4	22	d.	d.	
17	Dimethylboron iodide	$(CH_3)_2BI$	167.79	col. liq.		−110.7	65	d.	d.	
18	Dimethyldiborane(1,1) (unsym.)	$B_2H_4(CH_3)_2$	55.72	col. gas		−150.2	−2.6	d.	d.	
19	Dimethyldiborane(1,2) (sym.)	$B_2H_4(CH_3)_2$	55.72	col. unst. gas		−125	4.9	d.	d.	

No.	Name	Formula	Mol. wt.	Crystalline form, color and index of refraction	Sp. gr. or density	Melting point, °C	Boiling point, °C	Solubility in grams per 100 ml of		
								Cold water	Hot water	Alcohol, acids, etc.
	Boron									
20	Dimethyltriborine triamine (*B*)	$(CH_3)_2B_3N_3H_4$	108.55	col. liq.	...	−48	107	hyd.	hyd.	...
21	Dimethyltriborine triamine (*N*)	$(CH_3)_2B_3N_3H_4$	108.55	col. liq.	...	...	108	hyd.	hyd.	...
22	Dimethyltriborine triamine (*N-B*)	$(CH_3)_2B_3N_3H_4$	108.55	col. liq.	...	...	124	hyd.	hyd.	...
23	Diphenylboric acid (diphenylhydroxyborine)	$(C_6H_5)_2BOH$	182.03	col. radiating cr.	...	264–67	$215–35^{17}$	i.	i.	s. eth., al., pet. eth.
24	Diphenylboron bromide	$(C_6H_5)_2BBr$	244.93	col. visc. liq. or cr.	...	25	$150–60^{8}$	d.	d.	s. bz.
25	Diphenylboron chloride	$(C_6H_5)_2BCl$	200.48	col. visc. liq.	...	...	271	d.	d.	s. bz., pet. eth.
26	Di-*p*-tolylboric anhydride	$(C_7H_7)_2BOB(C_7H_7)_2$	402.16	wh. powd.	...	78	...	i.	...	s. al., eth., bz.
27	Ethoxyboron dichloride	$C_2H_5OBCl_2$	126.78	col. liq.	...	...	77.9	d.	d.	...
28	Ethyl boric acid	$(C_2H_5)B(OH)_2$	73.89	wh. cr.	...	subl. 40	...	s.	s.	s. al., eth.
29	Furanylboric acid (*β*)	$(C_4H_3O)B(OH)_2$	111.89	wh. cr.	...	110 d.	d.	s.	...	v. s. eth., al., acet.; sl. s. bz., tol.
30	Hexylboric acid (*n*)	$C_6H_{13}B(OH)_2$	130.00	wh. cr.	...	88–90, d.	d.	sl. s.	...	s. eth.
31	Isobutylboric acid	$C_4H_9B(OH)_2$	101.94	col. cr.	...	106–12, d.	d.	s.	...	s. eth., dichloroethane
32	Isopropylamineborine	$(CH_3)_2CHNH_2BH_3$	72.95	wh. cr.	0.829^{25}_{4}	62	d. 75	sl. s.	sl. s.	s. al., eth.
33	Methoxyboron dichloride	CH_3OBCl_2	112.75	col. liq.	...	−15	58.0	d.	d.	...
34	Methoxyboron difluoride	CH_3OBF_2	79.84	col. liq.	$1.417^{35.5}$; $1.354^{76.5}$	41.9	86	d.	d.	...
35	Methylboric acid	$CH_3B(OH)_2$	59.86	wh. pl.	...	d.	d.	sl. s.	...	s. al., eth.
36	Methylborine trimethylammine	$(CH_3)_3NBH_2CH_3$	86.97	col. liq.	...	0.8	177	d.	d.	s. eth.
37	Methyldiborane	$B_2H_5CH_3$	41.70	col. very unst. gas	...	...	-80^{50}; d. appr. −20	d.	d.	
38	Methyltriborine triamine (*B*)	$CH_3B_3N_3H_5$	94.53	col. liq.	...	−59	87	hyd.	hyd.	...
39	Methyltriborine triamine (*N*)	$CH_3B_3N_3H_5$	94.53	col. liq.	...	...	84	hyd.	hyd.	...
40	Nitrophenylboric acids, *o*, *m*, *p*	$NO_2C_6H_4B(OH)_2$	166.93	yel. need. or pr.	...	d.	d.	sl. s.	s.	s. al., eth.
41	Phenylboron dibromide	$C_6H_5BBr_2$	247.74	col. cr.	...	34	100^{20}	d	d.	s. bz.
42	Phenylboron dichloride	$C_6H_5BCl_2$	158.82	col. liq. n_D^{20} 1.5385	1.194^{20}_{4}	7	175	d.	d.	s. bz.
43	Sodium tri-*α*-naphthyl boride	$Na_2B(C_{10}H_7)_3$	438.29	bl. cr. (purple in dil. soln.)	...	...	d.	d.	...	s. eth.; sl. s. lgr.
44	Tetramethoxydiborine	$(CH_3O)_4B_2$	145.76	col. liq.	...	−24	d. 93; 21^{44}	d.	d.	...
45	Tetramethylammoniumborane	$(CH_3)_4NBH_4$	88.99	wh. cr.	0.813^{25}_{4}	...	>150	48^{20}	61^{40}	0.39; 1.04 abs. al. 95% al.
46	Tetramethyldiborane (1,1,2,2)	$B_2H_2(CH_3)_4$	83.78	col. liq.	...	−72.5	68.6	d.	d.	...
47	Tetramethyltriborine triamine (*N-B-B′-B″*)	$(CH_3)_4B_3N_3H_2$	136.61	col. liq.	...	...	158	hyd.	hyd.	...
48	Thiophenylboric acid (*α*) ("thienylboric" acid)	$(C_4H_3S)B(OH)_2$	127.96	col. star-formed need.	...	134	d.	s.	...	s. eth., al., acet., bz., CCl_4
49	Tribenzylborine	$B(C_6H_5CH_2)_3$	284.21	prismatic need. or col. oily liq.	...	47	230^{13}	i.	...	v. s. al., bz.; sl. s. eth.
50	Tri-*n*-butylborine	$B(C_4H_9)_3$	182.16	col. mobile liq.	...	...	$90–1^{9}$; $108–110^{20}$	i.	i.	v. s. eth., al.
51	Tri-*tert*-butylborine	$B(C_4H_9)_3$	182.16	col. mobile liq.	...	glass at low temp.	71^{12}	i.	...	s. eth.
52	Tri-*n*-butyltriborine trioxane (*n*-butyl boron oxide)	$(C_4H_9)_3B_3O_3$	251.78	col. liq.	...	...	154^{30}	hyd.	hyd.	v. s. eth.

No.	Name	Formula	Mol. wt.	Crystalline form, color and index of refraction	Sp. gr. or density	Melting point, °C	Boiling point, °C	Solubility in grams per 100 ml of: Cold water	Hot water	Alcohol, acids, etc.
	Boron									
53	Tri-*tert*-butyltriborine trioxane (*tert*-butyl boric oxide)	$(C_4H_9)_3B_3O_3$	251.78	col. liq.		20	$66–8^{5}$	hyd.	hyd.	s. eth.
54	Trichloroborine dimethyletherate	$(CH_3)_2OBCl_3$	163.24	col. cr.		d. 76	d.	d.		
55	Trichloroborine trimethylammine	$(CH_3)_3NBCl_3$	176.28	col. cr.		243			s.	s. al.
56	Tricyclohexylborine (boron tricyclohexyl)	$B(C_6H_{11})_3$	260.27	col. interlocking cr.		100	194^{15}	i.		s. eth.
57	Triethyl borate (triethoxyborine)	$B(OC_2H_5)_3$	146.00	col. liq., 1.381	0.8746^{10}; $0.864^{26.5}$	−84.8	117.4, (120)	d.		
58	Triethylboron (triethylborine)	$B(C_2H_5)_3$	98.00	col. liq.	0.6961^{23}	−92.9	$0^{12.5}$	i.	i.	s. al., eth.
59	Tri-*n*-hexyltriborine trioxane (hexylboric oxide)	$(C_6H_{13})_3B_3O_3$	335.94	col. liq., 1.4323^{20}	0.8876		$178–82^{24}$	hyd.	hyd.	s. org. solv.
60	Triisoamyl borate (triisoamyloxyborine)	$B(OC_5H_{11})_3$	272.24	liq., 1.421	0.872^{0}		255			
61	Triisoamylborine	$B(C_5H_{11})_3$	224.24	col. mobile liq., $1.43207^{22.6}$	0.76		119^{14}	i.		s. eth.
62	Tri-*p*-anisylborine	$B(CH_3OC_6H_4)_3$	332.21	wh. need.		128		i.	i.	s. al., eth., bz.
63	Triisobutyl borate (triisobutoxyborine)	$B(OC_4H_9)_3$	230.16	liq., 1.408	0.864^{0}		212			
64	Triisobutylborine	$B(C_4H_9)_3$	182.16	col., mobile liq., $1.41882^{22.8}$	0.74		188.86^{20}	i.	i.	s. eth.
65	Trimethoxyboroxine	$(BO)_3(OCH_3)_3$	173.55	col. liq., 1.3986	1.216^{25}	−30	(flashpt. 37)			v. s. bz., tol., CCl_4
66	Trimethylamminoborine	$(CH_3)_3NBH_3$	72.94	col. hex., columns or need.	0.792^{25}_{4}	94	172	i.	sl. s.	s. eth., al., bs., NH_3
67	Trimethyl borate (trimethoxyborine)	$B(OCH_3)_3$	103.92	col. liq. n_D^{20} 1.3610	0.915; $0.9205^{24.3}$	−29	68.7, (65)	d.		s. al., eth.
68	Trimethylboron (trimethylborine)	$B(CH_3)_3$	55.92	col. gas	1.9108 g/1, 0.625^{-100}	−161.5	−20.2	v. sl. s.		v. s. al., eth.
69	Trimethyldiborane (1,1,2)	$B_2H_3(CH_3)_3$	69.75	col. liq.		−123	45.5	d.		
70	Trimethyltriborine triamine (*B*)	$(CH_3)_3B_3N_3H_3$	122.58	col. cr. or liq.		31.5	129	hyd.	hyd.	
71	Trimethyltriborine triamine (*N*)	$(CH_3)_3B_3N_3H_3$	122.58	col. liq.			134	hyd.	hyd.	
72	Trimethyltriborine triamine (*N-B-B'*)	$(CH_3)_3B_3N_3H_3$	122.58	col. liq.			139	hyd.	hyd.	
73	Trimethyltriborine trioxane (methylboric anhydride)	$(CH_3)_3B_3O_3$	122.54	col. mobile liq.		−37vc,	79.3^{755}	hyd.	hyd.	s. eth.
74	Tri-β-naphthyl borate	$B(C_{10}H_7O)_3$	440.31	col. leaflets		115		d.		s. bz.
75	Tri-α-naphthylborine	$B(C_{10}H_7)_3$	392.31	col. need.		203	d.	i.		sl. s. eth., al.; v. s. bz., CCl_4, chl., CS_2
76	Triphenyl borate (triphenoxyborine)	$(C_6H_5O)_3B$	280.13	col. cr.		*ca.* 35	>360	d.		s. eth., bz.
77	Triphenylborine ammine*	$(C_6H_5)_3BNH_3$	259.16	col. cr.		d. 216	d.			s. al.; sl. s. bz.
78	Triphenylboron	$B(C_6H_5)_3$	242.13	hex. need.		136	203^{15}	d.		d. al.; s. bz.
79	Tripropyl borate (tripropoxyboron)	$B(OC_3H_7)_3$	188.08	liq.	liq. 0.867^{16}		175			
80	Tri-*n*-propylborine	$B(C_3H_7)_3$	140.08	col., mobile liq., $1.41352^{22.8}$	0.725		156, 60^{20}	i.		s. eth.
81	Tri-*sec*-propylborine	$B(C_3H_7)_3$	140.08	col., mobile liq.			148–54, $33–5^{12}$	i.		s. eth.

* This compound is the prototype of numerous stable complex compounds formed from organic amines and tri-aryl-borines.

No.	Name	Formula	Mol. wt.	Crystalline form, color and index of refraction	Sp. gr. or density	Melting point, °C	Boiling point, °C	Solubility in grams per 100 ml of		
								Cold water	Hot water	Alcohol, acids, etc.
	Boron									
82	Tri-*p*-tolylborine	$B(CH_3C_6H_4)_3$	284.21	separate wh. cr.		175	233^{12}	i.		v. s. bz.; sl. s. eth.
83	Tri-*p*-xylylborine	$B(CH_3C_6H_3CH_3)_3$	326.29	col. bushed need.		147	221^{12}	i.		v. s. bz., chl., CCl_4; sl. s. eth.
	Cadmium									
1	Dibutylcadmium	$Cd(C_4H_9)_2$	226.63	oil n_D 1.5155	$1.3056^{19.5}$	−48	$103.5^{12.5}$	d.	d.	
2	Diethylcadmium	$Cd(C_2H_5)_2$	170.52	oil	$1.6564^{18.1}$	−21	64^{19}	d.	d.	v. s. eth.
3	Diisoamylcadmium	$Cd(C_5H_{11})_2$	254.69	oil n_D 1.5039	1.2210^{19}	−115	121.5^{15}			
4	Diisobutylcadmium	$Cd(C_4H_9)_2$	226.63	oil n_D 1.4997	1.2693^{18}	−37	90.5^{20}	d.	d.	
5	Dimethylcadmium	$Cd(CH_3)_2$	142.47	oil	$1.9846^{17.9}$	−4.5	105.5^{758}	d.	d.	
6	Dipropylcadmium	$Cd(C_3H_7)_2$	198.58	oil n_D 1.5291	$1.4201^{17.6}$	−83	$84^{21.5}$	d.	d.	
	Calcium									
1	Dianilinecalcium	$Ca(NHC_6H_5)_2$	224.32	wh. cr.		d.		d.		i. eth., bz., lgr.
2	Ethylcalcium iodide	C_2H_5CaI	196.05	amor. powd.				d.		sl. s.
3	Glycocollcalcium	$(CH_2NHCOO)Ca$	113.13	cr.				s.		
	Chromium									
1	Anilinechromium tricarbonyl	$H_2NC_6H_5Cr(CO)_3$	229.16	yel. cr.		173–5	subl. 110–130			s. org. solv. HCl
2	Anisolechromium tricarbonyl	$CH_3OC_6H_5Cr(CO)_3$	244.17	yel. cr.		80–92	subl. 70–80 (vac.)			
3	Benzoic acid chromium tricarbonyl	$HO_2CC_6H_5Cr(CO)_3$	258.15	or. cr.		201–2	subl. 150 (vac.)			
4	Benzenechromium tricarbonyl	$C_6H_6Cr(CO)_3$	214.14	yel. monocl.		165.5–166.5 (162–3)	subl. 60–90			s. bz., s. org. solv.
5	Bis-cyclopentadienyl-chromium	$(C_5H_5)_2Cr$	182.19	red cr.		170–2	subl. 75–90 (vac.)	d.		d. in CS_2, CCl_4
6	Bis-diphenyl chromium iodide ("Tetraphenyl-chromium" iodide)	$(C_6H_5C_6H_5)_2Cr^+I^-$	487.33	or. yel.		177–8 (178)		s.		s. chl.; i. pet. eth.
7	Chlorobenzene-chromium tricarbonyl	$ClC_6H_5Cr(CO)_3$	248.59	yel.		96–8; 150–5 d.	50–8 (vac.)			
8	Cyclopentadienyl-chromium dinitrosomethyl	$C_5H_5Cr(NO)_2CH_3$	192.14	grn.		83.0				s. org. solv.
9	Dimethylanilinechrom-ium tricarbonyl	$(CH_3)_2NC_6H_5Cr(CO)_3$	257.21	cr.		146–146.5				s. org. solv.
10	Diphenylbenzene-chromium iodide ("Triphenyl chrom-ium" iodide)	$C_6H_5C_6H_5Cr(I)(C_6H_5)$	410.22	or. yel. pl.		110–2 (146–8)		s.		sl. s. et., al., chl., *m*-di-nitro-benzene.
11	Mesitylenechromium tricarbonyl	$(CH_3)_3C_6H_3Cr(CO)_3$	256.22	yel. cr.		177–8 (165, 170 d)	subl. 80–100			s. org. solv.
12	Methylbenzoate chro-mium tricarbonyl	$CH_3O_2CC_6H_5Cr(CO)_3$	272.18	red cr.		93–5	80–90 (vac.)			
13	Phenolchromium tri-carbonyl	$HOC_6H_5Cr(CO)_3$	230.14	yel.		193–240				
14	Thiophenechromium tricarbonyl	$C_4H_4SCr(CO)_3$	220.17	red cr.			subl. 85–95 (vac.)			s. bz., eth, pet. eth.
15	Toluenechromium tricarbonyl	$CH_3C_6H_5Cr(CO)_3$	228.17	yel.		82–3 (80° d.)	subl.			s. org. solv.
16	*m*-Xylenechromium tricarbonyl	$(CH_3)_2C_6H_4Cr(CO)_3$	242.20	yel.		104–5	subl. 65			
17	*o*-Xylenechromium tricarbonyl	$(CH_3)_2C_6H_4Cr(CO)_3$	242.20	yel.		88–90	subl. 65			
18	*p*-Xylenechromium tricarbonyl	$(CH_3)_2C_6H_4Cr(CO)_3$	242.20	yel. cr.		97–8	subl. 65			
	Cesium									
1	Monocesiumacetylide	CsC_2H	157.94	col. cr.		300		d.		s. NH_3; i. bz.
	Cobalt									
1	Bis-dimethylglyoxime cobaltochloride	$HON{:}C(CH_3)C(CH_3){:}NOH.Co$ $HON{:}C(CH_3)C(CH_3){:}NOCl_2$	361.07	lt. grn. cr.				s.		s. al.

No.	Name	Formula	Mol. wt.	Crystalline form, color and index of refraction	Sp. gr. or density	Melting point, °C	Boiling point, °C	Solubility in grams per 100 ml of: Cold water	Hot water	Alcohol, acids, etc.
	Cobalt									
2	Cobalt(ous) hexamethylenetetramine	$CoCl_2C_6H_{12}N_4$	270.03	ultramarine blue				s.		
3	Cobalt(ous) hydroxyquinone	$Co(C_{10}H_5O_3)_2$	405.23	ruby red		d. 210–15				
	Copper									
1	Cyclopentadienyltriethylphosphine copper	$C_5H_5CuP(C_2H_5)_3$	246.80	wh. cr.		127–8		i.		s. dil. HCl, pet. eth., eth., bz, pyr.
2	Diazoaminobenzene(ous)	$CuN_3(C_6H_5)_2$	259.77	or. cr.		d. 270		i.		s. bz.; i. al., lgr.
	Gallium									
1	Dimethylgallium amide	$Ga(CH_3)_2NH_2$	115.81	wh. cr.			subl. 60 vac.			
2	Dimethylgallium chloride monammine	$Ga(CH_3)_2Cl.NH_3$	152.27	wh. cr.		54		d.	d.	v. s. NH_3; s. eth.
3	Dimethylgallium chloride diammine	$Ga(CH_3)_2Cl.2NH_3$	169.30	wh. cr.		112		d.	d.	v. s. NH_3; i. eth.
4	Methylgallium dichloride	$Ga(CH_3)Cl_2$	155.66	wh. cr.		75		d.		v. s. eth.
5	Methylgallium dichloride monammine	$Ga(CH_3)Cl_2.NH_3$	172.69	wh. cr.				d.		i. eth.
6	Methylgallium dichloride pentammine	$Ga(CH_3)Cl_2.5NH_3$	240.81	wh. cr.		d. >80		d.		i. NH_3
7	Triethylgallium	$Ga(C_2H_5)_3$	156.91	col. liq.	1.0576[30]	−82.3	142.6	d.		s. eth.
8	Triethylgallium monammine	$Ga(C_2H_5)_3.NH_3$	173.94	col. liq.				d.		
9	Triethylgallium monoetherate	$Ga(C_2H_5)_3.(C_2H_5)_2O$	231.03	col. liq.				d.		s. eth.
10	Trimethylgallium	$Ga(CH_3)_3$	114.83	col. liq.		−19	55.7 ± 2[762]	d.		s. eth., NH_3
11	Trimethylgallium monammine	$Ga(CH_3)_3.NH_3$	131.86	wh. cr.		31	subl. vac.	d.		s. eth., NH_3; i. pet. eth.
12	Trimethylgallium monoetherate	$Ga(CH_3)_3.(C_2H_5)_2O$	188.95	col. liq.		<−76	99	d.		s. NH_3, eth.
	Germanium									
1	Amyltriphenylgermanium (*n*)	$Ge(C_5H_{11})(C_6H_5)_3$	375.05	col. pl.		42–3	i.	i.	i.	v. s. bz., pet. eth.; sl. s. me. al.
2	Benzyltriphenylgermanium	$Ge(CH_2C_6H_5)(C_6H_5)_3$	395.04	col. pl.		82.5–3.5		i.	i.	v. s. bz., pet. eth., chl.; sl. s. iso-propyl al.; i. me. al.
3	Bis-acetylacetone germanium dibromide	$[CH(CCH_3O)_2]_2GeBr_2$	430.63	col. micr. cr.		226				sl. s. h. acetyl acet.; i. org. solv.
4	Bis-acetylacetone germanium dichloride	$[CH(CCH_3O)_2]_2GeCl_2$	341.72	col. pr.		240d.				sl. s. org. solv.
5	Bis(5-oxy-2,8-dithiooctane) germanium	$[(SCH_2CH_2)_2O]_2Ge$	354.06	col. cr.		159.0–.5				s. bz., abs. al.
6	Bis-propionylacetone germanium dichloride	$[CHC_2(C_2H_5)(CH_3)O_2]_2GeCl_2$	369.77	wh. cr. powd.		128–9				s. c. chl.; i. pet. eth.
7	Bis-tribenzyl germanyl sulfide	$[(C_6H_5CH_2)_3Ge]_2S$	724.05	col. cr.		124				s. al., me. al., bz.; i. alk.
8	Bis-tribiphenylyl germanyl sulfide	$[(C_6H_5C_6H_4)_3Ge]_2S$	1096.48	col. cr.		238				s. org. solv.; i. alk.
9	Bis-trichlorogermanyl methane	$CH_2(GeCl_3)_2$	371.93	col. liq.			110[18]	hyd.	hyd.	s. org. solv.
10	Bis-tricyclohexylgermanium disulfide	$[(C_6H_{11})_3Ge]_2S_2$	708.24	col. cr.		87–8		i.		s. abs. al.
11	Bis-triethylgermanyl sulfide	$[(C_2H_5)_3Ge]_2S$	351.62	col. oily liq.			148–50[12]			s. org. solv.; i. alk.
12	Bis-triphenylgermanyl sulfide	$[(C_6H_5)_3Ge]_2S$	639.88	col. cr.		138				s. org. solv.; i. alk.
13	Bis-tritolylgermanyl sulfide	$[(C_6H_4CH_3)_3Ge]_2S$	724.05	col. cr.		156–7				s. org. solv.; i. alk.

No.	Name	Formula	Mol. wt.	Crystalline form, color and index of refraction	Sp. gr. or density	Melting point, °C	Boiling point, °C	Solubility in grams per 100 ml of		
								Cold water	Hot water	Alcohol, acids, etc.
	Germanium									
14	Butyltriphenyl-germanium (*n*)	$Ge(C_4H_9)(C_6H_5)_3$	361.01	col. need.		84.5–5.5		i.	i.	v. s. pet. eth., bz., chl., eth., sl. s. isopropyl al., i. me. al.
15	Cyclopentamethylene germanium dichloride	$(CH_2)_5GeCl_2$	213.63	col. liq.			$55–60^{12}$			
16	Diethylcyclopenta-methylenegermanium (1,1)	$(CH_2)_5Ge(C_2H_5)_2$	200.85	col. liq.			52^{13}			
17	Diethyldiphenyl-germanium	$Ge(C_2H_5)_2(C_6H_5)_2$	284.93	col. liq.			316	i.	i.	v. s. org. solv.
18	Diethyldinylgermanium	$(C_2H_5)_2Ge)CH{:}CH_2)_2$	184.81	n_D^{20} 1.4575	1.0193^{20}_{4}		149.8^{760}			
19	Diethylgermanium bromide	$(C_2H_5)_2GeBr_2$	290.53	col. liq.		<−33	202	d.	d.	d. liq. NH_3; s. org. solv.
20	Diethylgermanium chloride	$(C_2H_5)_2GeCl_2$	201.62	col. liq.		−39 to −37	175	d.	d.	d. liq. NH_3; s. org. solv.
21	Diethylgermanium imine	$(C_2H_5)_2GeNH$	145.73	col. liq.			$100^{0.01}$			
22	Diethylgermanium iodide	$(C_2H_5)_2GeI_2$	384.52	col. liq.		−2 to −1	252	d.	d.	d. liq. NH_3; s. org. solv.
23	Diethylgermanium oxide(α)	$[(C_2H_5)_2GeO]_x$	146.71	stable wh. amor. sol.		175		i.	i.	i. org. solv., liq. NH_3
24	Diethylgermanium oxide(β)	$[(C_2H_5)_2GeO]_3$	440.14	unst. col. liq.		18		i.	i.	s. org. solv.; i. liq. NH_3
25	Diphenylgermanium	$[(C_6H_5)_2Ge]_4$	907.21	wh. cr.		294–5		i.	i.	sl. s. bz., tol., chl.; i. pet. eth.
26	Diphenylgermanium dibromide	$(C_6H_5)_2GeBr_2$	386.62	col. liq.			$120^{.007}$ $205–7^{512}$	hyd.	hyd.	s. org. solv.
27	Diphenylgermanium dichloride	$(C_6H_5)_2GeCl_2$	297.71	col. liq.	71	9	223^{13}	hyd.	hyd.	s. org. solv.
28	Diphenylgermanium difluoride	$(C_6H_5)_2GeF_2$	264.80	col. liq.			$100^{0.007}$	hyd.	hyd.	s. org. solv.
29	Diphenyl-*sec*-propyl-germanium bromide	$(C_6H_5)_2(C_3H_7)GeBr$	349.80	col. liq.			$215–50^{13}$			
30	Di-*p*-tolylgermanium dibromide	$(CH_3C_6H_4)_2GeBr_2$	414.68	lt. yel. liq.			$230–33^{13}$	hyd.	hyd.	s. org. solv.
31	Di-*p*-tolylphenyl-germanium bromide	$(CH_3C_6H_4)_2(C_6H_5)GeBr$	411.87	col. pr.		119				
32	Di-triphenylgermanyl methane	$[(C_6H_5)_3Ge]_2CH_2$	621.85	lt. col. pr.		132–33		i.		v. s. bz., eth., pet. eth., chl.; i. liq. NH_3, al.
33	Ethyl-*tris*-*p*-biphenylyl-germanium	$Ge(C_2H_5)(C_6H_4C_6H_5)_3$	561.27	col. cr.		154–6				
34	Ethylgermanium oxide	$(C_2H_5GeO)_2O$	251.30	wh. powd.		>300	d.	s.	s.	s. HCl, al.; i. pet. eth.
35	Ethylgermanium tri-bromide bromide	$C_2H_5GeBr_3$	341.38	col. liq.		<−33	200^{763}	d.	d.	d. liq. NH_3; s. bz., eth.
36	Ethylgermanium trichloride	$C_2H_5GeCl_3$	208.01	col. liq.		<−33	144^{762}	d.	d.	d. liq. NH_3; s. bz., eth.
37	Ethylgermanium trifluoride	$C_2H_5GeF_3$	158.65	col. liq.		−16.5 to −15.5	112^{750}	d.	d.	d. liq. NH_3; s. bz., eth.
38	Ethylgermanium triiodide	$C_2H_5GeI_3$	482.37	yel. liq.		−2.5 to −1.5	281^{755} d. >350	d.	d.	d. liq. NH_3; s. bz., eth.
39	Ethylphenyldi-*p*-tolyl-germanium	$Ge(C_2H_5)(C_6H_5)(C_6H_4CH_3)_2$	361.03	wh. cr.		55				
40	Ethyl-*sec*-propyl-diphenylgermanium	$Ge(C_2H_5)(C_3H_7)(C_6H_5)_2$	298.95	liq.		175–90				
41	Ethyltribenzyl-germanium	$Ge(C_2H_5)(CH_2C_6H_5)_3$	375.05	col. cr.		56–7				s. meth. al.
42	Ethyltriphenyl-germanium	$Ge(C_2H_5)(C_6H_5)_3$	332.97	col. sld.		78.0–.5		i.	i.	s. eth., pet. eth., bz., chl., acet.; i. meth. al.
43	Hexabenzyldigermane	$[(C_6H_5CH_2)_3Ge]_2$	691.98	col. cr.		183–4				s. glac. acet. a.

No.	Name	Formula	Mol. wt.	Crystalline form, color and index of refraction	Sp. gr. or density	Melting point, °C	Boiling point, °C	Solubility in grams per 100 ml of: Cold water	Hot water	Alcohol, acids, etc.
	Germanium									
44	Hexaethyldigermane	$[(C_2H_5)_3Ge]_2$	319.55	col. liq.		<−60	265^{758}	i.	i.	s. bz., eth.
45	Hexaphenyldigermane	$[(C_6H_5)_3Ge]_2$	607.82	wh. cr.		340		i.	i.	sl. s. h. bz., h. chl.; i. liq. NH_3, lgr.
46	Hexaphenyldigermane tribenzene	$[(C_6H_5)_3Ge]_2.3C_6H_6$	842.16	col. cr.		d. −C_6H_6		i.		s. bz.
47	Hexa-*p*-tolyldigermane	$[(CH_3C_6H_4)_3Ge]_2$	691.98	col. cr.		226–7				
48	Hexavinylgermanium	$(CH_2{:}CH)_6Ge_2$	307.44	n_D^{25} 1.5217	1.171$^{25}_{4}$		55$^{0.35}$			
49	Methyltriphenyl-germanium	$Ge(CH_3)(C_6H_5)_3$	318.95	trans. col. cr.		70.5−1.0		i.	i.	v. s. bz., eth., acet., chl., pet. eth.; i. c. meth. al., liq. NH_3
50	Octaphenyltrigermane	$(C_6H_5)_8Ge_3$	834.62	wh. cr.		247–8		i.	i.	s. h. bz., h. chl.
51	Phenylethyl-*sec*-propyl-germanium bromide	$(C_6H_5)(C_2H_5)[CH(CH_3)_2]$ GeBr	301.76	col. oil; opt. act., *d. & l.* forms			130–5^{13}			
52	Phenylgermanium tribromide	$C_6H_5GeBr_3$	389.42	col. liq.			120–2^{12}	hyd.	hyd.	d. liq. NH_3; s. org. solv.
53	Phenylgermanium trichloride	$C_6H_5GeCl_3$	256.06	col. liq.			105–6^{13}	hyd.	hyd.	d. liq. NH_3; s. org. solv.
54	Phenylgermanium triiodide	$C_6H_5GeI_3$	530.41	wh. sol., dec. by light		55–6		hyd.	hyd.	d. liq. NH_3; s. glac. acet. a.; org. solv.
55	Phenyltri-*p*-tolyl-germanium	$Ge(C_6H_5)(C_6H_4CH_3)_3$	423.10	wh. pr.		191		i.	i.	s. org. solv.
56	Propyltriphenyl-germanium(*n*)	$Ge(C_3H_7)(C_6H_5)_3$	347.00	col. need.		86.0–.5		i.	i.	v. s. chl., bz., pet. eth.; sl. s. isopropyl al.; i. meth. al., liq. NH_3
57	Tetra-*i*-amylgermanium	$Ge(C_5H_{11})_4$	357.16	col. oily liq., 1.457$^{17.5}$	0.9147$^{20}_{20}$		163–4			
58	Tetra-*n*-amylthio-germanium	$Ge[S(CH_2)_4CH_3]_4$	485.42	col. liq., 1.5336^{25}	1.0697^{25}		240–1$^{3-4}$			s. bz., abs. al.
59	Tetraanhydro-tetrakisdiphenyl-germanediol (*cyclo*)	$[Ge(C_6H_5)_2O]_4$	971.20	monocl. pr. & cubes		218				s. ethyl acetate, pet. eth., eth.
60	Tetrabenzylgermanium	$Ge(CH_2C_6H_5)_4$	437.13	col. sol.		107–8				
61	Tetra-*p*-biphenylyl-germanium	$Ge(C_6H_4C_6H_5)_4$	685.41	wh. need.		270–2		i.	i.	s. bz.
62	Tetra-*p*-bromophenyl-thiogermanium	$Ge(SC_6H_4Br)_4$	824.88	col. cr.		196.0–.5				s. bz., abs. al.
63	Tetra-*n*-butyl-germanium	$Ge(C_4H_9)_4$	301.06	col. oily liq.			178–80^{733}?			
64	Tetra-*p*-*tert*-butyl-phenylthiogermanium	$Ge[SC_6H_4C(CH_3)_3]_4$	733.71	col. tetrag.		155–6				sl. s. al., glac. acet. a.; s. pet. eth., eth., acet.; v. s. chl., bz.
65	Tetra-*n*-butylthio-germanium	$Ge[S(CH_2)_3CH_3]_4$	429.31	liq., 1.5439^{25}	1.1072^{25}		222.5$^{4.5}$			s. abs. al.
66	Tetra-*sec*-butylthio-germanium	$Ge[SCH(CH_3)(C_2H_5)]_4$	429.31	liq., 1.5497^{25}	1.1119^{25}		200.5^{4}			s. bz.
67	Tetra-*tert*-butylthio-germanium	$Ge[SC(CH_3)_3]_4$	429.31	tetrag. columns		172–73	subl. 170^{4}			s. abs. al.
68	Tetracetylthio-germanium	$Ge[SCH_2(CH_2)_{14}CH_3]_4$	1102.61	wh. cr.		50–1				v. s. chl., bz.; s. pet. eth., eth.; sl. s. acet. al., glac. acet. a.
69	Tetracyclohexylthio-germanium	$Ge(SC_6H_{11})_4$	533.46	2 cr. mod. α (stab.) tetrag	α1.270^{15}	84		i.		s. abs. al., pet. eth.
				(metastab.) monocl.	β1.259^{15}	88		i.		s. abs. al., pet. eth.

No.	Name	Formula	Mol. wt.	Crystalline form, color and index of refraction	Sp. gr. or density	Melting point, °C	Boiling point, °C	Solubility in grams per 100 ml of		
								Cold water	Hot water	Alcohol, acids, etc.
	Germanium									
70	Tetraethoxyl-germanium (tetraethyl germanate)	$Ge(OC_2H_5)_4$	252.84	col. liq.		−81	185–7			
71	Tetraethylgermanium	$Ge(C_2H_5)_4$	188.84	col. oil, $1.443^{17.5}$; 1.554^{0}; 1.439^{30}	1.198^{0} $0.991^{24.5}_{24.5}$	−90	162.5–3.0	d.	d.	s. bz., eth., HCl
72	Tetraethylthio-germanium	$Ge(SC_2H_5)_4$	317.09	liq., 1.5886^{25}	1.2547^{25}		$164.5–5.0^{5}$			s. bz.
73	Tetra-iso-butylthio-germanium	$Ge[SCH_2CH(CH_3)_2]_4$	429.31	liq., 1.5381^{25}	1.0984^{25}		$199–200^{4-5}$			s. al.
74	Tetraisopropylthio-germanium	$Ge[SCH(CH_3)_2]_4$	373.20	liq., 1.5535^{25}	1.1478^{25}	15	$162–64^{4}$			s. abs. al.
75	Tetramethylgermanium	$Ge(CH_3)_4$	132.73	col. liq.	1.006^{0}	−88	43.4			s. al., eth., bz.
76	Tetramethylthio-germanium	$Ge(SCH_3)_4$	260.90	liq., 1.6379^{25}	1.4364^{25}	−3	$138–40^{4}$			s. al., bz.
77	Tetraphenoxy-germanium	$Ge(OC_6H_5)_4$	445.01	col. oil			$210–20^{0.3}$			s. bz.
78	Tetraphenylgermanium	$Ge(C_6H_5)_4$	381.02	tetr., col.		235.7	>400	i.	i.	s. chl., bz., tol.; sl. s. eth., acet., lgr.
79	Tetra(2-phenylethyl)-germanium	$Ge(C_6H_5C_2H_4)_4$	493.23	col. cr.		56–7				s. eth., al.
80	Tetraphenylthio-germanium	$Ge(SC_6H_5)_4$	509.27	col., rhomb. cr., 1.7348, 1.782 (H green)		101.5				s. bz., abs. al., meth. al.
81	Tetra-*n*-propyl-germanium	$Ge(C_3H_7)_4$	244.95	col. mob. liq., $1.451^{17.5}$	0.9539^{20}_{20}	−73	225^{746}			
82	Tetrapropylthio-germanium	$Ge(SC_3H_7)_4$	373.20	liq., 1.5612^{25}	1.1662^{25}		$191–92^{5}$			s. abs. al.
83	Tetra-*N*-pyrryl-germanium	$Ge(C_4H_4N)_4$	336.92	lt. yel. cr.		202				s. pet. eth., chl.
84	Tetra-α-thienyl-germanium	$Ge(C_4H_3S)_4$	405.12	wh. need., doubly refract.		149–50		i.	i.	s. bz., tol., acet., chl., CCl_4; sl. s. al., meth. al.; i. pet. eth.
85	Tetra-*o*-tolylgermanium	$Ge(C_6H_4CH_3)_4$	437.13	wh. hex. cr.		175–6		i.	i.	s. CCl_4, bz., xylene; sl. s. h. al.; i. pet. eth., al.
86	Tetra-*m*-tolyl-germanium	$Ge(C_6H_4CH_3)_4$	437.13	wh. need.		146		i.	i.	s. bz., tol., CCl_4; sl. s. meth. al.
87	Tetra-*p*-tolyl-germanium	$Ge(C_6H_4CH_3)_4$	437.13	wh. rhbdr. tab.		227		i.	i.	s. bz.
88	Tetra-*p*-tolylthio-germanium	$Ge(SC_6H_4CH_3)_4$	565.38	col., rhomb. cr., 1.726, 1.771 (H green)		110–11				s. bz., abs. al.
89	Tetravinylgermanium	$(CH_2{:}CH)_4Ge$	180.77	n^{25}_{D} 1.4676	1.040^{25}_{4}		54^{27}			
90	Tolylgermanium tribromide(*p*)	$(CH_3C_6H_4)GeBr_3$	403.45	col. liq.		$155–6^{13}$		hyd.	hyd.	d. liq. NH_3; s. org. solv.
91	Tolylgermanium trichloride(*p*)	$(CH_3C_6H_4)GeCl_3$	270.08	col. liq.			$115–6^{12}$	hyd.	hyd.	d. liq. NH_3; s. org. solv.
92	Tolylgermanium triiodide(*p*)	$(CH_3C_6H_4)GeI_3$	544.44	col. cr., sensit. to light		72		hyd.	hyd.	d. liq. NH_3; s. org. solv.
93	Trianhydrotetrakis-diphenylgermanediol	$[HO\text{-}Ge(C_6H_5)_2\text{-}O\text{-}Ge(C_6H_5)_2]_2O$	989.23			149				s. eth. acetate
94	Tribenzylgermanium bromide	$(C_6H_5CH_2)_3GeBr$	425.90	col. cr.		145				
95	Tribenzylgermanium chloride	$(C_6H_5CH_2)_3GeCl$	381.44	col. cr.		155				

No.	Name	Formula	Mol. wt.	Crystalline form, color and index of refraction	Sp. gr. or density	Melting point, °C	Boiling point, °C	Solubility in grams per 100 ml of Cold water	Hot water	Alcohol, acids, etc.
	Germanium									
96	Tribenzylgermanium fluoride	$(C_6H_5CH_2)_3GeF$	364.99	col. need.		96				
97	Tribenzylgermanium iodide	$(C_6H_5CH_2)_3GeI$	472.90	col. cr.		141				
98	Tribenzylgermanium oxide	$[(C_6H_5CH_2)_3Ge]_2O$	707.99			135				s. pet. eth.
99	Tri-*p*-biphenylyl-germanium bromide	$(C_6H_5C_6H_4)_3GeBr$	612.12	wh. cr.		242				s. bz.
100	Tri-*tert*-butylthio-germanium chloride	$Ge[SC(CH_3)_3]_3Cl$	375.58	col. cr.		66–7	$156–7^{3-4}$			s. al., org. solv.
101	Tri-*n*-butylvinyl-germanium	$(n\text{-}C_4H_9)_3Ge(CH{:}CH_2)$	270.99	n_D^{20} 1.4598	0.9479_4^{20}		$108–109^2$			
102	Tricyclohexyl-germanium bromide	$(C_6H_{11})_3GeBr$	401.96	col. cr.		110		hyd.	hyd.	s. al.
103	Tricyclohexyl-germanium chloride	$(C_6H_{11})_3GeCl$	357.51	col. cr.		102		hyd.	hyd.	s. meth. al.
104	Tricyclohexyl-germanium fluoride	$(C_6H_{11})_3GeF$	341.05	col. need.		92		hyd.	hyd.	s. meth. al.
105	Tricyclohexyl-germanium hydroxide	$(C_6H_{11})_3GeOH$	339.06			176–7				s. al., bz., pet. eth.
106	Tricyclohexyl-germanium iodide	$(C_6H_{11})_3GeI$	448.96	col. cr.		99–100		hyd.	hyd.	s. meth. al.
107	Triethylgermanium bromide	$(C_2H_5)_3GeBr$	239.69	col. liq.		−33	190.9	hyd.	hyd.	s. bz., eth., chl., CCl_4
108	Triethylgermanium chloride	$(C_2H_5)_3GeCl$	195.23	col. liq.		<-50	175.9	hyd.	hyd.	s. bz., eth., chl., CCl_4
109	Triethylgermanium fluoride	$(C_2H_5)_3GeF$	178.77	col. liq.			149.0^{751}	hyd.	hyd.	s. bz., eth., chl., CCl_4
110	Triethylgermanium hydride	$(C_2H_5)_3GeH$	160.78	col. liq.			124.4^{751}	i.	i.	s. bz., eth.; i. liq. NH_3
111	Triethylgermanium-imine	$[(C_2H_5)_3Ge]_2NH$	334.57	col. liq.			$100^{0.1}$	hyd.	hyd.	s. bz., eth., CCl_4, chl.; i. liq. NH_3
112	Triethylgermanium-iodide	$(C_2H_5)_3GeI$	286.68	col. liq.		<-50	212.3	hyd.	hyd.	s. bz., eth., chl., CCl_4; i. liq. NH_3
113	Triethylgermanium oxide	$[(C_2H_5)_3Ge]_2O$	335.55	col. liq.		<-50	253.9	i.	i.	s. $C_2H_5NH_2$, bz., eth. org. solv.; i. liq. NH_3
114	Triethylphenyl-germanium	$Ge(C_2H_5)_3(C_6H_5)$	236.88	col. liq.			$116–7^{13}$	i.		s. org. solv.
115	Triethyl-*p*-tolyl-germanium	$Ge(C_2H_5)_3CH_3C_6H_4$	250.91	col. liq.			$125–6^{12}$	i.		s. org. solv.
116	Triethyl-2,2,2,-tri-phenyldigermane-(1,1,1)	$(C_2H_5)_3GeGe(C_6H_5)_3$	463.69	rhomb. cr.		89.5–90.5		i.		v. s. bz., chl.; s. pet. eth., al.; sl. s. meth. al.
117	Triethylvinyl-germanium	$(C_2H_5)_3Ge(CH{:}CH_2)$	186.82	n_D^{20} 1.4501	1.0048_4^{20}		61^{28}			
118	Trimethylgermanium bromide	$Ge(CH_3)_3Br$	197.60	col. oily liq., 1.4705	1.544_{40}^{18}	−25	113.7	d.	d.	s. org. solv.
119	Trimethylphenyl-germanium	$Ge(CH_3)_3(C_6H_5)$	194.80	col. liq.			182–3	i.		s. org. solv.
120	Trimethylstannyl-tri-phenylgermanium	$(CH_3)_3SnGe(C_6H_5)_3$	467.71	wh. cr.		88		i.		s. pet. eth., CCl_4, chl., bz.; sl. s. al.; i. liq. NH_3
121	Triphenylanisyl-germanium	$Ge(C_6H_5)_3(CH_3OC_6H_4)$	411.04	wh. sld.		158–9				s. al., glac. acet. a.
122	Triphenyldimethyl-aminophenyl-germanium	$Ge(C_6H_5)_3C_6H_4N(CH_3)_2$	424.09	wh. need.		140–1				
123	Triphenylgermanium amide	$(C_6H_5)_3GeNH_2$	319.93	wh. ppt.		d. $-NH_3$		d.	d.	i. liq. NH_3

No.	Name	Formula	Mol. wt.	Crystalline form, color and index of refraction	Sp. gr. or density	Melting point, °C	Boiling point, °C	Solubility in grams per 100 ml of		
								Cold water	Hot water	Alcohol, acids, etc.
	Germanium									
124	Triphenylgermanium bromide	$(C_6H_5)_3GeBr$	383.82	hex. col.		138.7		i.	hyd.	s. bz., chl.; sl. s. lgr.
125	Triphenylgermanium chloride.	$(C_6H_5)_3GeCl$	339.36	wh. cr.		117–8	285^{12}	i.	hyd.	s. bz., eth.
126	Triphenylgermanium fluoride	$(C_6H_5)_3GeF$	322.91	wh. cr.		76.6		i.	hyd.	v. s. bz., eth., lgr., chl.; i. liq. NH_3
127	Triphenylgermanium hydride	$(C_6H_5)_3GeH$	304.92	wh. cr. (two) forms)		α, 47; β, 27		i.	i.	v. s. bz., tol., eth., chl., CCl_4; sl. s. liq. NH_3
128	Triphenylgermanium hydroxide	$(C_6H_5)_3GeOH$	320.92	wh. cr.		134.2		i.	i.	s. bz., chl.; sl. s. lgr.
129	Triphenylgermanium iodide	$(C_6H_5)_3GeI$	430.81	wh. cr.		157		hyd.	hyd.	s. bz., eth.
130	Triphenylgermanium oxide	$[(C_6H_5)_3Ge]_2O$	623.82	col. pl.		183–4		i.	i.	s. bz., lgr., eth.
131	Triphenylgermanium-sodium	$Ge(C_6H_5)_3Na$	326.90	lt. yel.		v. high		d.	d.	v. s. liq. NH_3; sl. s. eth., bz.
132	Triphenylgermanium-sodiumoxide	$Ge(C_6H_5)_3ONa$	342.90	wh. sld.		high		d.	d.	i. liq. NH_3
133	Triphenylgermanium-sodiumtriammine	$Ge(C_6H_5)_3Na.3NH_3$	377.99	yel. sld.		d.		d.	d.	v. s. liq. NH_3
134	Triphenyl-*m*-tolyl germanium	$Ge(C_6H_5)_3(C_6H_4CH_3)$	395.04	wh. anisotropic need.		136.5–8.5		i.	i.	s. bz., h. pet. eth.; i. meth. al.
135	Triphenyl-*p*-tolyl germanium	$Ge(C_6H_5)_3(C_6H_4CH_3)$	395.04	wh. sld.		123–4				s. org. solv.
136	Tri-*o*-tolylgermanium bromide	$(CH_3C_6H_4)_3GeBr$	425.90	col. oil, (blue fluores.)			$205–10^{1}$			
137	Tri-*m*-tolylgermanium bromide	$(CH_3C_6H_4)_3GeBr$	425.90	wh. need., anisotropic		78.0–.9	$222–3^{1}$			s. CCl_4, bz., eth.
138	Tri-*p*-tolylgermanium bromide	$(CH_3C_6H_4)_3GeBr$	425.90	col. cr.		128–9				s. pet. eth.
139	Tri-*o*-tolylgermanium chloride	$(CH_3C_6H_4)_3GeCl$	381.44	col. oil			$216–22^{1}$			
140	Tri-*m*-tolylgermanium chloride	$(CH_3C_6H_4)_3GeCl$	381.44	sm. silky need.; opt. act.		84–5	$221–4^{1}$			s. pet. eth., bz.; i. c. meth. al.
141	Tri-*p*-tolylgermanium chloride	$(CH_3C_6H_4)_3GeCl$	381.44	wh. cr.		121				s. pet. eth.
142	Tri-*o*-tolylgermanium hydroxide	$(CH_3C_6H_4)_3GeOH$	363.00	amor. powd.			$212–4^{1}$			
143	Tri-*m*-tolylgermanium oxide	$[(CH_3C_6H_4)_3Ge]_2O$	707.98	wh. cr.		125.0–.2		i.	i.	s. al., bz., pet. eth.
144	Tri-*p*-tolylgermanium oxide	$[(CH_3C_6H_4)_3Ge]_2O$	707.98	wh. pr. anisotropic		148–50		i.	i.	s. h. lgr., c. bz.; sl. s. eth., al., meth. al.
145	Tri-*m*-tolyl-*p*-tolyl-germanium	$Ge(C_6H_4CH_3)_3(C_6H_4CH_3)$	437.13	wh. sld.		98.5–100.5				s. meth. al.
146	Tri-*p*-tolyl-*o*-tolyl-germanium	$Ge(C_6H_4CH_3)_3(C_6H_4CH_3)$	437.13	wh. cr.		164–6				
147	Tri-triphenylgermanium nitride	$[(C_6H_5)_3Ge]_3N$	925.74	col. need.		163–4		hyd.	hyd.	s. lgr., eth., bz.
148	Tris-acetylacetone-germanium cupribromide	$[(C_5H_7O_2)_3Ge]CuBr_3$	673.19	gr.-blk. cr.		139				i. chl.
149	Tris-acetylacetone-germanium cuprobromide	$[(C_5H_7O_2)_3Ge]CuBr_2$	593.28	col. rect. pr.		165–6				s. chl.
150	Tris-acetylacetone-germanium cuprochloride	$[(C_5H_7O_2)_3Ge]CuCl_2$	504.37	col. pr.		147–8				s. chl.
151	Tris-acetylacetone-germanium dicuprobromide	$[(C_5H_7O_2)_3Ge]Cu_2Br_3$	736.73	col. pr.		195d.				s. h. acet. acet.; i. chl.

No.	Name	Formula	Mol. wt.	Crystalline form, color and index of refraction	Sp. gr. or density	Melting point, °C	Boiling point, °C	Solubility in grams per 100 ml of: Cold water	Hot water	Alcohol, acids, etc.
	Gold									
1	Aminopyridinotribromogold	$H_2NC_5H_4NAuBr_3$	530.81	blk. pr.		160d.		s.		s. chl.
2	α-Auromercaptoacetanilide	C_8H_8AuNOS	363.19	gray-yel. powd.		238–241		i.		i. a., alk., eth., bz.
3	Diethylgold bromide	$[(C_2H_5)_2AuBr]_2$	670.00	col. need.		58; subl. vac.	d. 70, expl.			s. NH_4OH
4	Ethylenediaminodibutylgold bromide	$(CH_2NH_2)_2Au(C_4H_9)_2Br$	451.21	col. need.		190 d.		s.		s. al.
5	Ethylenediaminodipropylgold bromide	$(CH_2NH_2)_2Au(C_3H_7)_2Br$	423.15	col. cr.		volat. 130	d. 190			
6	Pyridinotribromogold	$C_5H_5NAuBr_3$	515.80	red need.		150 d.		s.		s. al., act.
7	Quinolinotribromogold	$C_9H_7NBr_3Au$	565.86	deep red lust. pr.		d. >200				s. chl.
	Indium									
1	Trimethylindium	$In(CH_3)_3$	159.93	col. cr.	1.568^{19}_{19}	89.0–.8	subl.	d.		s. eth.
	Iridium									
1	Cyclopentadienyliridium cyclopentadiene	$(C_5H_5)Ir(C_5H_6)$	323.40	yel. cr.		130–132	subl.			s. bz., acet., pet. eth.; sl. s. meth. alc.; alc.
	Iron									
1	(Acetylcyclopentadienyl) cyclopentadienyl iron (Monoacetylferrocene)	$(C_5H_4COCH_3)FeC_5H_5$	228.08	lng. or. (red) need.		85–6	86–7			sl. s. isooctane; s. H_2SO_4, HCl
2	*Di*(4-aminobutyl)-cyclopentadienyl iron	$[C_5H_4(CH_2)_4NH_2]_2Fe$	328.28	or. cr.		137–8				s. eth. al.; sl. s. *n*-heptane
3	(Aminocyclopentadienyl)cyclopentadienyliron (Aminoferrocene)	$(C_5H_4NH_2)FeC_5H_5$	201.05	yel. cr.		155				s. eth., al.
4	Mono(*p*-anisylisonitrile)tetracarbonyliron	$Fe(CO)_4(NCC_6H_4OCH_3)$	301.04	yel. pr. need.		39–40				s. eth., pet. eth., bz., chl.
5	Azulene-di-iron pentacarbonyl	$C_{10}H_8Fe_2(CO)_5$	379.92	sld.		100d.				
6	(Benzoylcyclopentadienyl)cyclopentadienyliron (Benzoyl ferrocene)	$(C_6H_5COC_5H_4)FeC_5H_5$	290.15	red need.		108.1–108.3 (108–9)				sl. s. meth. al.
7	(Benzylcyclopentadienyl)cyclopentadienyliron. (Phenylferrocenylmethane)	$(C_6H_5CH_2C_5H_4)FeC_5H_5$	276.16	yel. cr.		70–4 73.4				s. al.
8	Benzylcyclopentadienyliron dicarbonyl bromide	$C_6H_5CH_2C_5H_4Fe(CO)_2Br$	347.00	red cr.		82				s. eth.; i. lgr.
9	Bicycloheptadieneiron tricarbonyl. (Bicyclo-[2:2:1] hepta-2:5-dieneiron tricarbonyl)	$C_7H_8Fe(CO)_3$	232.02	or.-red liq. (yel. liq.)			$60.2^{0.2}$			s. cold H_2SO_4
10	Butadieneiron tricarbonyl	$C_4H_6Fe(CO)_3$	193.97	pa. yel.		19				sl. s. meth. al. ligr.
11	(Carboxyazidecyclopentadienyl)cyclopentadienyliron	$(C_5H_4CON_3)Fe(C_5H_5)$	255.06	cr.		74–5				s. eth.
12	(Carboxycyclopentadienyl)cyclopentadienyliron (Ferrocene monocarboxylic acid)	$(C_5H_4CO_2H)FeC_5H_5$	230.05	rdsh. br. need.		225–330 (208.5 d.) (219–225)				sl. s. chl.
13	(Carboxylamidecyclopentadienyl) cyclopentadienyliron	$(C_5H_4CONH_2)FeC_5H_5$	229.06	cr.		168–170				s. chl.

No.	Name	Formula	Mol. wt.	Crystalline form, color and index of refraction	Sp. gr. or density	Melting point, °C	Boiling point, °C	Solubility in grams per 100 ml of Cold water	Hot water	Alcohol, acids, etc.
	Iron									
14	*Di*(3-Carboxypropionyl)cyclopentadienylron	$[C_5H_4CO(CH_2)_2CO_2H]_2Fe$	386.19	cr.		164–6d.				
15	Chloromercuriferrocene	$(C_5H_4HgCl)FeC_5H_5$	421.07	golden-yel. leaf.		193–4d. (194–6)				
16	(*p*-Chlorophenylcyclopentadienyl) cyclopentadienyliron	$(ClC_6H_4C_5H_4)FeC_5H_5$	296.58	yel. cr.		122				s. al., H_2SO_4; sl. s. ac. a.
17	(Cyanocyclopentadienyl)cyclopentadienyliron. (Ferrocenyl cyanide)	$(C_5H_4CN)FeC_5H_5$	211.05			107–8 (103–4)				s. CH_2Cl_2; sl.. s. *n*-heptane
18	*Di*[(3-Cyanopropionyl)-cyclopentadienyl]iron	$[C_5H_4CO(CH_2)_2CN]_2Fe$	348.19	or. cr.		133–4				sl. s. tol., al.
19	Cycloheptatrienyliron dicarbonyl	$C_7H_8Fe(CO)_2$	204.01	yel. liq.		$70^{0.4}$				
20	Cyclohexadieneiron tricarbonyl	$C_6H_8Fe(CO)_3$	220.01	yel. liq.			96^{12}			s. acet.
21	Cyclooctatetraenediiron hexacarbonyl	$C_8H_8Fe_2(CO)_6$	383.91	ye.-or.		d. <185				v. sl. s. common org. solv.; sl. s. bz.
22	Cyclooctatetraeneiron tricarbonyl	$C_8H_8Fe(CO)_3$	244.04	lng. red (deep red) need.		94–5(92) 155 d.				sl. s. hot hexane; s. common org. solv.
23	Cyclopentadienyliron dicarbonylbromide	$C_5H_5FeBr(CO)_2$	256.88			18–102 d.				s. chl.; i. lgr.
24	Cyclopentadienyliron dicarbonyliodide	$C_5H_5FeI(CO)_2$	303.88			117–8 d.				i. lgr.
25	(Cyclopentenylcyclopentadienyl)cyclopentadienyliron	$(C_5H_7C_5H_4)FeC_5H_5$	252.14	cr.		64–5				sl. s. meth. al., bz. gasoline
26	(Cyclopentylcyclopentadienyl)cyclopentadienyliron	$(C_5H_9C_5H_4)FeC_5H_5$	254.16	red liq.		16.3 frz.				s. bz.
27	1,1′-Di(acetylcyclopentadienyl)iron. (Diacetyl ferrocene)	$(C_5H_4COCH_3)_2Fe$	270.12	red cr.		130–1				
28	1-1′Di(benzhydryldicyclopentadienyl)-iron. (1,1′-Dibenzhydrylferrocene)	$[(C_6H_5)_2CHC_5H_4]_2Fe$	518.49	yel. need.		162–3				s. eth., cyclohexane; sl. s. acet.
29	1,1′-Di(benzoylcyclopentadienyl)iron. (1,1′-Dibenzoyl ferrocene)	$(C_6H_5COC_5H_4)_2Fe$	394.26	purple need.		106.5–106.7 (105–106)				s. CH_2Cl_2
30	1,1′-Di(benzylcyclopentadienyl)iron. (1,1′-Dibenzyl ferrocene)	$(C_6H_5CH_2C_5H_4)_2Fe$	366.29	yel. need.		102 (97–8) (105–6)				sl. s. meth. al.; s. *n*-butanol.
31	1,1′-Di(carboxycyclopentadienyl)iron (Ferrocene dicarboxylic acid)	$(C_5H_4CO_2H)_2Fe$	274.06				subl. <230			
32	(1,3-Dicarboxycyclopentadienyl)iron (Ferrocene 1,3-dicarboxylic acid)	$[C_5H_3(CO_2H)_2]FeC_5H_5$	274.06	red need.		206–206.5 d.				
33	1,1′-Di(chloromercuri)-ferrocene	$(C_5H_4HgCl)_2Fe$	656.11	sld. yel. powd.		no melt up to 300				
34	1,1′-Di(*p*-chlorophenyl-cyclopentadienyl)iron. (1,1′-Di′-*p*-chlorophenyl ferrocene).	$(ClC_6H_4C_5H_4)_2Fe$	407.13	cr.		192				s. me. al., H_2SO_4; sl. s. pet. eth.
35	Dicyclopentadienyl-diiron tetracarbonyl	$(C_5H_5)_2Fe_2(CO)_4$	353.92	cr.		192				s. pyr.

No.	Name	Formula	Mol. wt.	Crystalline form, color and index of refraction	Sp. gr. or density	Melting point, °C	Boiling point, °C	Solubility in grams per 100 ml of		
								Cold water	Hot water	Alcohol, acids, etc.
	Iron									
36	Diferrocenyl mercury	$(C_5H_5FeC_5H_4)_2Hg$	570.65	or. cr.		235–6 d. 248–9 233–4				s. bz. sl. s. xylene
37	1,1'-Di(hydroxybenzyl-cyclopentadienyl)-iron. (Ferrocenyl-bis-phenyl methanol)	$[C_6H_5CH(OH)C_5H_4]_2Fe$	398.29	yel. leaf		136–7				sl. s. meth. al., al.
38	(Dimethylamino-methylcyclopenta-dienyl)cyclopenta-dienyl iron. (Di-methylaminomethyl-ferrocene).	$[(CH_3)_2NCH_2C_5H_4]Fe\ C_5H_5$	243.13	amber oil 1.5839^{25}_{4}			$91–2^{0.45}$			s. eth.
39	1,1'-Di(methylcarboxy-cyclopentadienyl)iron	$(C_5H_4CO_2CH_3)_2Fe$	302.11	red cr.			114–5			
40	(1,3-Diphenylcyclo-pentadienyl) cyclo-pentadienyliron. (1,3-Diphenylferrocene)	$[(C_6H_5)_2C_5H_3]Fe(C_5H_5)$	338.24	or. need.		107				s. lgr.
41	1,2-Diphenyl-1,2-di-ferrocenylethanediol	$[C(C_5H_5FeC_5H_4)(OH)(C_6H_5)]_2$	582.31	yel. cr		125–145 d.				s. eth., chl., dioxane,; sl. s. bz.
42	1,1'-Diphenyl-dicyclo-pentadieenyliron. (1,1'-Diphenyl-ferrocene)	$(C_6H_5C_5H_4)_2Fe$	338.24	or.-yel.-br. leaf		154 (140–4)				s. bz.; sl. s. lt. pet.
43	1,1'-Di(trimethylsilyl) ferrocene	$[(CH_3)_3Si\ C_5H_4]_2Fe$	330.40	1.5454^{25}_{D}		16	$87–88^{0.06}$			s. eth.
44	*bis*-(Ethylisonitrile)tri-carbonyl iron	$Fe(CO)_3(CNC_2H_5)_2$	250.04	yel. pl. need.		65.5–66				s. eth., bz., al.
45	Ferrocene. (Dicyclo-pentadienyliron)	$C_5H_5FeC_5H_5$	186.04	yel. need.		172.5–173	subl.	i.	i.	s. al., eth., bz., MeOH
46	Ferrocenyl-acetic acid	$(C_5H_4CH_2CO_2H)FeC_5H_5$	244.08	lt. yel. (yel.) need		150–2 (125–135)				s. eth.; i. pentane; sl. s. meth. al.
47	1,1'-Ferrocenyl-diacetic acid	$(C_5H_4CH_2CO_2H)_2Fe$	302.11	yel.		sintered 140–3(d)				s. chl., eth.; i. pentane
48	β-Ferrocenyl-ethyl alcohol	$(C_5H_4CH_2CH_2OH)Fe(C_5H_5)$	230.09			49–50				s. eth.; i. pentane
49	β-Ferrocenyl-propionic acid	$[C_5H_4CO(CH_2)_2CO_2H]FeC_5H_5$	286.11	or. cr.		166.5–167.5				s. CH_2Cl_2; sl. s. meth. al.
50	(Formylcyclopenta-dienyl)cyclopenta-dienyliron. (Ferrocene monoaldehyde)	$(C_5H_4CHO)FeC_5H_5$	214.05	sld. red (rdsh br.)cr.		120–2 (130–2)				s. eth., chl.; sl. s. al.
51	(Hydroxybenzylcyclo-pentadienyl)cyclo-pentadienyliron. (Phenyl-ferrocenyl methanol)	$[C_6H_5CH(OH)C_5H_4]FeC_5H_5$	292.16	yel. cr.		81–2 (80–80.5)				s. eth.; i. pet. eth.
52	(Hydroxymethylcyclo-pentadienyl)cyclo-pentadienyliron. (Hydroxymethyl-ferrocene)	$(C_5H_4CH_2OH)FeC_5H_5$	216.06	yel. pl. lng. need.		81–2 (74–6)				s. chl.
53	*p*-Hydroxyphenyl-cyclopentadienyl)-cyclopentadienyliron	$(HOC_6H_4C_5H_4)FeC_5H_5$	278.14	cr. yel. grn. fluorescence		165				s. al.
54	(Methylcarboxycyclo-pentadienyl)cyclo-pentadienyliron	$(C_5H_4CO_2CH_3)FeC_5H_5$	244.08			70–1				s. meth. al.
55	*bis*(Methylisonitrile)-tricarbonyliron	$Fe(CO)_3(CNCH_3)_2$	221.98			100–130 d.				s. al., eth., bz., ac. a.
56	Mono(ethylisonitrile)-tetracarbonyliron	$Fe(CO)_4(CNC_2H_5)$	222.97	yel.		−3		i. d.		s. org. solv.

No.	Name	Formula	Mol. wt.	Crystalline form, color and index of refraction	Sp. gr. or density	Melting point, °C	Boiling point, °C	Solubility in grams per 100 ml of: Cold water	Hot water	Alcohol, acids, etc.
	Iron									
57	(*p*-Nitrophenylcyclopentadienyl)cyclopentadienyliron	$(NO_2C_6H_4C_5H_4)FeC_5H_5$	307.13	dk. purple cr.		163				s. al.
58	(Phenylcyclopentadienyl) cyclopentadienyliron (Phenylferrocene)	$(C_6H_5C_5H_5)FeC_5H_5$	263.14	yel. (or.) cr.		110–111 (109–110)				s. al.
59	Phenyl-diferrocenylmethane	$(C_5H_5FeC_5H_4)_2CHC_6H_5$	460.19	cr.		123–4				s. eth., al.
60	1,3,1′,3′-Tetraphenyldicyclopentadienyliron. (1,3,1′,3′-tetraphenylferrocene)	$[(C_6H_5)_2C_5H_3]_2Fe$	490.43	flat or. pr. (deep or. red)		219–220 (220–222)				sl. s. acet.
61	Thiopheniron dicarbonyl	$C_4H_4SFe(CO)_2$	220.03	pale red cr.		51	subl.			s. org. solv.
62	Trimethylsilylferrocene	$[C_5H_4Si(CH_3)_3]FeC_5H_5$	258.22	dk. or.-red liq. 1.5696^{25}_4		23	$64–5^{0.45}$			
	Lanthanum									
1	Hexaantipyrineiodide	$La(COC_{10}H_{12}N_2)_6I_3$	1649.01	yel. cr.		268–9 d.		41.8^{20}		
	Lead									
1	Hexaethyldilead (triethyllead)	$Pb_2(C_2H_5)_6$	588.75	liq.	1.471		d.	i.		
2	Tetraethyllead	$Pb(C_2H_5)_4$	323.44	col. liq.; or. flame, grn. marg. 1.5195^{30}	1.659^{11}	−136.80	200d.; 91^{19}	i.		s. bz., pet., al., eth.
3	Tetra-iso-amyllead	$Pb(C_5H_{11})_4$	491.76	col. oily liq., 1.4946	n_D $1.2332^{20}_{4.5}$			i.		s. al., bz., eth
4	Tetraisobutyllead	$Pb[CH_2CH(CH_3)_2]_4$	435.66	col. liq., 1.5042	1.324	−23				
5	Tetraisopropyllead	$Pb[CH(CH_3)_2]_4$	379.55	col. liq., dec. in air, 1.5223	1.4504	−53.5	120^{14}; $133–8^{27}$	i.		s. bz. pet. eth.
6	Tetramethyllead	$Pb(CH_3)_4$	267.33	liq. 1.5120^{20}.	1.995	−27.5	110	i.		s. bz., pet. eth., al.
7	Tetraphenyllead	$Pb(C_6H_5)_4$	515.62	wh. need.	1.530^{20}	227.7	d. 270	i.		s. bz.
8	Tetra-*n*-propyllead	$Pb(C_3H_7)_4$	379.55	col. liq., 1.5094	1.44		126^{13}			s. bz., pet. eth.; sl. s. al.
9	Tetravinyllead	$(CH_2{:}CH)_4Pb$	315.37	n_D^{20} 1.5462	1.7882^{20}_4		$69–70^{11}$			
10	Tricyclohexyllead	$(C_6H_{11})_3Pb$	456.65	yel. hex. pl., decol. in light		d. 195				
11	Triphenylbenzyllead	$(C_6H_5)_3PbCH_2C_6H_5$	529.64	col. pl.		93	d. 205–210			v. s. bz., eth., chl.
12	Triphenylethyllead	$(C_6H_5)_3Pb(C_2H_5)$	467.57	wh. cr. $n_{H\infty}^{61}$ 1.6263	1.5885^{61}_4 (vac)	42				s. hot al.
13	Triphenylmethyllead	$(C_6H_5)_3PbCH_3$	453.55	col. rhomb. (from bz.)		60	d. 260			sl. sl al.; eth.; v. s. bz., chl.
14	Triphenylleadbromide	$(C_6H_5)_3PbBr$	518.42	need.		166		i.		s. al., bz., eth.
	Lithium									
1	*n*-Butyllithium	n-C_4H_9Li	64.06	inflamm. liq.		subl. vac. 80 100	d. 150	d.	d.	s. bz., eth.
2	Ethyllithium	LiC_2H_5	36.00	hex. transp. pl.		95 (in N_2)	subl. 95	d.		d. eth.; s. bz lgr.
	Magnesium									
1	Dimethylmagnesium	$(CH_3)_2Mg$(polymer?)	54.38	wh. sol., inflamm. in air		stable to 240				v. sl. s. eth.
2	Diphenylmagnesium	$(C_6H_5)_2Mg$	178.53	wh. sol.		d. *ca*, 280				i. bz., eth.
	Magnesiumaluminum ethoxide-*see* Aluminum(3)									

No.	Name	Formula	Mol. wt.	Crystalline form, color and index of refraction	Sp. gr. or density	Melting point, °C	Boiling point, °C	Solubility in grams per 100 ml of: Cold water	Hot water	Alcohol, acids, etc.
	Manganese									
1	Acetylcyclopentadienyl manganese tricarbonyl	$CH_3COC_5H_4Mn(CO)_3$	246.10			41.5–42.5 (39.5–40.0)				s. pet. eth.
2	Acetylmanganese pentacarbonyl	$CH_3COMn(CO)_5$	238.04	wh. cr.		54–5				
3	(Benzoylcyclopentadienyl)manganese tricarbonyl	$C_5H_4COC_6H_5Mn(CO)_3$	308.17	yel. cr.		73.5–74.5				s. bz., hexane; i. pet. eth.
4	Benzoylmanganese pentacarbonyl	$C_6H_5COMn(CO)_5$	300.11	wh.		95.6				
5	Benzylmanganese pentacarbonyl	$C_6H_5CH_2Mn(CO)_5$	286.12	yel.		37.5–38.5				
6	(*tert*-Butylcyclopentadienyl) manganese tricarbonyl	$C_4H_9C_5H_4Mn(CO)_3$	260.17	1.5600^{20}_D			150^{28-29}			
7	(γ-Butyrocyclopentadienyl)manganese tricarbonyl	$[C_5H_4(CH_2)_3CO_2H]Mn(CO)_3$	290.16	yel. cr.		63–5				s. bz., eth., al.
8	(Carboxycyclopentadienyl)manganese tricarbonyl	$(C_5H_4CO_2H)Mn(CO)_3$	248.08	yel. cr.		195–6				s. al.; sl. s. bz.
9	(Di-*tert*-butylcyclopentadienyl) manganese tricarbonyl	$(C_4H_9)_2C_5H_3Mn(CO)_3$	316.28			70–1	$164–8^{10}$			
10	(Dicyclohexyl cyclopentadienyl) manganese tricarbonyl	$(C_6H_{11})_2C_5H_3Mn(CO)_3$	368.36	1.5804^{20}_D			$149–152^{2}$			
11	(Formylcyclopentadienyl)manganese tricarbonyl	$(C_5H_4CHO)Mn(CO)_3$	232.08	red oil.			$105^{0.15}$			s. org. solv.
12	(Hexylcyclopentadienyl)manganese tricarbonyl	$C_6H_{13}C_5H_4Mn(CO)_3$	288.23	1.5686^{20}_D			$110–114^{2}$			s. eth.
13	(Methylcyclopentadienyl)manganese (benzene)	$CH_3C_5H_4MnC_6H_6$	212.18	red cr.		116–8				
14	Methylmanganese pentacarbonyl	$CH_3Mn(CO)_5$	210.03	col. cr.		94.5–95				
15	Phenylmanganese pentacarbonyl	$C_6H_5Mn(CO)_5$	272.10	wh. cr.		52				
16	(Tri-*tert*-butylcyclopentadienyl) manganese tricarbonyl	$(C_4H_9)_3C_5H_2Mn(CO)_3$	372.39	1.5326^{20}_D			$160–3^{3.5}$			
17	(δ-Valerocyclopentadienyl)manganese tricarbonyl	$[C_5H_4(CH_2)_4CO_2H]Mn(CO)_3$	304.18	yel. cr.			69–71			s. bz., eth., al.
	Mercury									
1	Aminophenylmercuric acetate(*p*)	$C_6H_4(NH_2)HgO_2C_2H_3$	351.76	col. pr.		167		i.	i.	s. dil. a.; sl. s. chl., al.; i. eth.
2	Biphenylmercury	$Hg(C_6H_5C_6H_4)_2$	507.00	sm. scales		216		difficultly	soluble in	common solv.
3	*n*-Butylmercuric chloride	C_4H_9ClHg	293.16	wh. need.		130		14×10^{-4}, 18°	3.3×10^{-4}, 100°	0.5^{18} al.; 5.6^{18} $CHCl_3$
4	Chloromercuriphenol(*o*)	$C_6H_4OHHgCl$	329.15			152.5^{0}				s. NaOH
5	Di-*n*-amylmercury	$Hg(C_5H_{11})_2$	342.88	1.4998	1.6369		133^{10}			
6	Di-(*dl*)-amylmercury	$Hg(C_5H_{11})_2$	342.88	1.5014	1.6700		93^{1}			
7	Dibenzylmercury	$Hg(C_7H_7)_2$	382.86	long brittle col. need.						s. al., eth., chl., CS_2, ac. a., bz. eth. acet.; sl. s. lgr.

No.	Name	Formula	Mol. wt.	Crystalline form, color and index of refraction	Sp. gr. or density	Melting point, °C	Boiling point, °C	Solubility in grams per 100 ml of		
								Cold water	Hot water	Alcohol, acids, etc.
	Mercury									
8	Di-*n*-butylmercury	$Hg(C_4H_9)_2$	314.82	liq. 1.5057	1.7779		105^{10}			
9	Diethylmercury	$Hg(C_2H_5)_2$	258.71	col. liq. of hazel odor 1.5399	liq. 2.444		159	i.	i.	v. s. eth.; sl. s. al.
	Diferrocenylmercury- *see* **Iron** (36)									
10	Di-*n*-hexylmercury	$Hg(C_6H_{13})_2$	370.94	1.4973	1.5361		158^{10}			
11	Diisoamylmercury	$Hg(C_5H_{11})_2$	342.89	1.4989	1.6397		125^{10}			
12	Diisobutylmercury	$Hg(C_4H_9)_2$	314.82	col. liq., 1.4965	1.835^{15}; 1.7678	volat. 100	205–7; 86^{10}	v. sl. s.		s. eth. al.
13	Diisopropylmercury	$Hg(C_3H_7)_2$	286.78	1.5263	2.0024		63^{10}			
14	Dimethylaminophenylmercuric acetate(*p*)	$C_6H_4N(CH_3)_2HgO_2C_2H_3$	379.82	long. col. need		165		i.	i.	s. bz., chl., al., dil. a.
15	Dimethylaniline-mercury(*p*)	$Hg[C_6H_4N(CH_3)_2]_2$	440.95	lust. need.		169				s. chl.; sl. s. al., eth., dil. HCl
16	Dimethylmercury	$Hg(CH_3)_2$	230.66	col. liq., sweet odor, 1.5327	3.069		96			s. al., eth.
17	Dinaphthylmercury(α)	$Hg(C_{10}H_7)_2$	454.92	wh. rhomb.	1.929	188 (243)	249	i.	sl. s.	s. h. CS_2, chl.; sl. s. bz., eth. v. sl. s. h. bz
18	Dinaphthylmercury(β)	$Hg(C_{10}H_7)_2$	454.92	cryst. from bz.		247–8		i.		sl. s. al., eth.
19	Diphenylmercury	$Hg(C_6H_5)_2$	354.81	wh. glassy need.	2.318	121.8, subl.	$204^{10.5}$ >306 d.	i.	i.	s. chl., CS_2, bz.; sl. s. eth., h. al.
20	Dipropylmercury	$Hg(C_3H_7)_2$	286.78	col. mobile liq., 1.5170	2.0208		189–91; 73^{10}	i.		v. s. eth.; s. al.
21	Ditolylmercury(*o*)	$Hg(C_7H_7)_2$	382.86	wh. tabl.		107	219^{14}			s. h. bz.
22	Ditolylmercury(*m*)	$Hg(C_7H_7)_2$	382.86	col. or. lt. yel. need.		102		i.		s. bz., chl., acet., eth. acet.
23	Ditolylmercury(*p*)	$Hg(C_7H_7)_2$	382.86	need.		238		i.		s. h. bz., chl., CS_2; sl. s. c. al.
24	Ethane hexamercarbide	$C_2Hg_6O_2(OH)_2$	1293.58	yelsh.-wh. powd.		exp. 230		i.	i.	i.
25	Ethylmercuric chloride	C_2H_5HgCl	265.10	silv. irid. leaf.	3.482	193	subl. 40	i.		v. s. h. al.; sl. s. eth.
26	Ethylmercuric hydroxide	C_2H_4HgOH	245.65	silv. irid. leaf.		37		i.	i.	v. s. h. al.; s. eth.; sl. s. c. al.
27	Ethylmercuric iodide	C_2H_5HgI	356.56	cr. from EtOH		186				s. al.
28	Mercury ethyl-mercaptide(ic)	$Hg(SC_2H_5)_2$	322.85	leaf.		76–7	d.	i.		d. a.
29	Mercuryphenyl-mercaptide(ic)	$Hg(SC_6H_5)_2$	418.93	yelsh. need.		153 d.	d.	i.		sl. s. h. al.; v. s. bz., pyr.
30	Methylmercuric chloride	CH_3HgCl	251.09	wh. cr., disg. odor	4.063	170	volat. 100			
31	Methylmercuric iodide	CH_3HgI	342.53	col. pearly leaf.		143		i.		v. s. meth. al.; s. eth., al.
32	Naphthylmercuric acetate(α)	$C_{10}H_7HgO_2C_2H_3$	386.81	fine need.		154		i.	i.	s. al., ac. a., bz., CS_2, fats; sl. s. eth.
33	Naphthylmercuric chloride(α)	$C_{10}H_7HgCl$	363.22	silk quad. tabl.		188–9		i.	i.	sl. s. bz., al.
34	Phenylmercuric acetate	$C_6H_5HgO_2C_2H_3$	336.75	rhomb. sm. wh. lust. pr.		149		sl. s.	sl. s.	s. glac. ac. a. ac. a., bz., al.
35	Phenylmercuric bromide	C_6H_5HgBr	357.61	rhomb. wh. lust. tabl.		276		i.	i.	s. al., bz., pyr.
36	Phenylmercuric chloride	C_6H_5HgCl	313.15	wh. satiny leaf.		251				sl. s. h. al., bz., pyr., eth.
37	Phenylmercuric cyanide	C_6H_5HgCN	303.73	rhomb. long pr.		204			sl. s.	s. h. al., bz.
38	Phenylmercuric iodide	C_6H_5HgI	404.60	rhomb. satiny tabl.		266		i.		s. chl., CS_2; sl. s. al., eth., bz.
39	Phenylmercuric nitrate	$C_6H_5HgNO_3$	339.70	rhomb. tabl.		176–86		i.	sl. s.	s. h. al., bz.

No.	Name	Formula	Mol. wt.	Crystalline form, color and index of refraction	Sp. gr. or density	Melting point, °C	Boiling point, °C	Solubility in grams per 100 ml of: Cold water	Hot water	Alcohol, acids, etc.
	Mercury									
40	Phenylmercuric nitrate, basic	$C_6H_5HgOH.C_6H_5HgNO_3$	634.41	wh.-grayish powd.		d. 178–184		0.8		sl. s. al., glyc.
41	Tolylmercuric bromide(*p*)	C_7H_7HgBr	371.63	thin lust. gray sc.		228				s. chl., al., bz.; i. c. CS_2
42	Tolylmercuric chloride(*p*)	C_7H_7HgCl	327.19	rhomb. silky tabl.		233		i.	i.	sl. s. h. al., bz , chl., acet., pyr.; i. eth.
	Molybdenum									
1	Benzenemolybdenum tricarbonyl	$C_6H_6Mo(CO)_3$	258.09	yel. monocl.		120–5 d.				
2	Bicycloheptadiene-molybdenum tetra-carbonyl	$C_7H_8Mo(CO)_4$	300.12	yel. pl.		76–7				sl. s. pet. eth.
3	Cycloheptatriene-molybdenum tri-carbonyl	$C_7H_8Mo(CO)_3$	272.11	or.-red pl.		100.5–101.5				s. lt. pet.; chl., bz.
4	*bis*-Cyclopentadienyl-bimolybdenum pentacarbonyl (*bis*-Cyclopentadienyl-μ-pentacarbon monoxide bimolybdenum I)	$(C_5H_5)_2Mo_2(CO)_5$	462.12	purp.-red		215–7 d.				s. chl.; sl. s. CCl_4 al., CS_2, bz.
5	Mesitylenemolybdenum tricarbonyl	$(CH_3)_3C_6H_3Mo(CO)_3$	300.17	yel.		130–40 d.	90 subl.			
6	Tropeniummolybdenum tricarbonyl fluoroborate	$C_7H_7Mo(CO)_3BF_4$	357.91	lt. or. need.		>270				
	Neodymium									
1	Neodymium hexaanti-pyrine iodide	$[Nd(COC_{10}H_{12}N_2)_6I_3]$	1654.34	rose cr.		270–2		12.7^{20}		
	Nickel									
1	Dicyclopentadienyl-nickel	$(C_5H_5)_2Ni$	188.90	drk. grn. cr.			130 subl.			sl. s. lgr.; s. al.
2	*bis*(Cyclopentadienyl-nickel carbonyl)	$(C_5H_5NiCO)_2$	303.63	purple red.		136		i.		s. bz., chl., pet. eth., eth., al.
3	Cyclopentadienylnickel carbonyl iodide	$C_5H_5Ni(CO)I$	278.72	blk. vlt.		20 d.				sl. s. org. solv.
4	Dicyclopentadiene-nickel	$(C_5H_6)_2Ni$	190.92	red need.		41–42 cl. tube				
	Niobium									
1	*bis*-Cyclopentadienylni-obium monohydroxy-dibromide	$(C_5H_5)_2Nb(OH)Br_2$	399.92	or.-red				s.		
2	*bis*-Cyclopentadienylni-obium tribromide	$(C_5H_5)_2NbBr_3$	462.82	reddsh. br. cr.		260 d.		s.		s. chl., al.; i. pet. eth.; sl. s. bz.;
	Palladium									
1	Cycloocta-1,5-dienedi-bromopalladium	$C_8H_{12}PdBr_2$	374.40	or.-red. need.		213 d.		i.		sl. s. glac.; i. acet., ac. a.
2	Cycloocta-1,5-dienedi-chloropalladium	$C_8H_{12}PdCl_2$	285.49	pa. or. need.	d. 205–210					s. h. glac. ac. a.; sl. s. c. al., bx. sl. s. h. bz., chl., acet.
3	Cyclohexenepalladous chloride	$(C_6H_{10}PdCl_2)_2$	519.00	lt. br. cr.						v. i. in common org. solv.
4	Dicyclopentadienedi-chloropalladium	$(C_5H_6)_2PdCl_2$	309.51			165–70 d.		d.		s. bz., eth., chl.

No.	Name	Formula	Mol. wt.	Crystalline form, color and index of refraction	Sp. gr. or density	Melting point, °C	Boiling point, °C	Solubility in grams per 100 ml of: Cold water	Hot water	Alcohol, acids, etc.
	Palladium									
5	*bis*(Dicyclopentadiene-methoxide)dichlorodipalladium	$C_{22}H_{30}O_2Pd_2Cl_2$	610.19	yel. pl.		166–70 d.		d.		i. meth. al.; s. boil. chl.; sl. s. c. chl., bz.
6	*bis*(Dicyclopentadiene-isopropoxide)dichlorodipalladium	$C_{26}H_{38}O_2Pd_2Cl_2$	666.30	yel. pl.		150–60 d.		i.		i. meth. al.; s. boil. chl.; sl. s. cold chl.; bz.
7	*bis*(Dicyclopentadiene-*n*-propoxide) dichlorodipalladium	$C_{26}H_{38}O_2Pd_2Cl_2$	666.30	yel. pl.		150–56 d.		i.		i. meth. al.; s. boil chl.; sl. s. cold chl.; bz.
8	Di(8-methoxycyclooct-4-enyl)dibromopalladium	$(C_8H_{12}OMe)_2Pd_2Br_2$	651.06	pa. yel. cr.		125–35 d.				
9	Di(8-methoxycyclooct-4-enyl)dichloropalladium	$(C_8H_{12}OMe)_2Pd_2Cl_2$	562.14	pa. yel. cr.		130–35 d.				sl. s. meth. al.,
10	Ethylenepalladous chloride	$C_4H_8Pd_2Cl_4$	410.72	lt. canary yel. need.						s. xylene; i. pet. eth.
11	1,5-Hexadienedichloropalladium. (Biallyldichloropalladium)	$C_6H_{10}PdCl_2$	259.46	brnsh. yel. cr.						sl. s. bz.; i. pet. eth.
12	Styrenepalladous chloride	$C_{16}H_{16}Pd_2Cl_4$	562.92	lt. red. br. need.						sl. s. bz.; i. common org. solv.
13	Trimethylethylenepalladous chloride	$C_5H_{10}PdCl_2$	247.45	red. pr.		85–90 d.		s. with d.		s. bz.; i. isopentane, gasoline
	Platinum									
1	Bromochlorothylene amine platinum	$C_2H_4PtNH_3BrCl$	355.55	pa. grn. pr.		d. 163		sl. s.		i. c. HCl
2	*cis*-2-Buteneplatinous chloride	$(C_4H_8PtCl_2)_2$	644.21			d. 165–75				s. liq. butene
3	*trans*-2-Buteneplatinous chloride	$(C_4H_8PtCl_2)_2$	644.21	pink cr.		110–115 (d. 125–135)				s. liq. butene
4	Cyclohexenedibromoplatinum (Cyclohexene-platinous bromide)	$(C_6H_{10}PtBr_2)_2$	874.11	lng. slend. or need.		150–1				s. chl.; v. sl. s. bz., al.; i. eth. ac. a.
5	Cyclohexenedichloroplatinum (Cyclohexeneplatinous chloride)	$(C_6H_{10}PtCl_2)_2$	696.29	lng. sl. or. need.		145–6				v. s. al., acet.; s. chl., eth.; sl. s. bz. i. ac. a.
6	Cycloocta-1,5-dienedichloroplatinum	$C_8H_{12}PtCl_2$	374.18	wh. need.		d. >220				s. boil. chl., ac. a., CH_2Cl_2
7	Cycloocta-1,5-dienediiodoplatinum	$C_8H_{12}PtI_2$	557.08	or.-yel. need.		d. 250				s. boil. CH_2Cl_2
8	Cyclooctatetraenedichloroplatinum	$C_8H_8PtCl_2$	370.15	or.						s. org. solv.
9	Cyclooctatetraenediiodoplatinum	$C_8H_8PtI_2$	553.05	or.-red. cr. powd.		Turns black (no change in cr. form)				s. boil. chl.
10	Cycloocta-1,5-dienedibromoplatinum	$C_8H_{12}PtBr_2$	463.09	pa. yel. need.		d. >200				s. h. ch., ac. ac. CH_2Cl_2
11	Cyclopropaneplatinous chloride	$C_3H_6PtCl_2$	308.08	br. powd.		d. >100		v. sl. s.		s. 10N HCl, al.; sl. s. eth., acet., chl., CCl_4
12	Diallyletherdichloroplatinum	$C_6H_{10}OPtCl_2$	364.14	wh. cr.		180 d.		i.	i.	s. h. chl.
13	*trans*-Dichloroethylenedichloroplatinum (*trans*-Dichloroethyleneplatinous chloride)	$[C_2H_2Cl_2PtCl_2]_2$	725.88	or. cr.		155–60				s. acet., chl., bz., nitrobz.
14	Dicyclopentadienedibromoplatinum	$C_{10}H_{12}PtBr_2$	487.11	yel. pr.		d. 200–225		i.		s. h. chl.; i. eth.

No.	Name	Formula	Mol. wt.	Crystalline form, color and index of refraction	Sp. gr. or density	Melting point, °C	Boiling point, °C	Solubility in grams per 100 ml of: Cold water	Hot water	Alcohol, acids, etc.
	Platinum									
15	Dicyclopentadienedichloroplatinum	$(C_5H_6)_2PtCl_2$	398.21	sm. ivory need.		218–220 d.				i. al.; s. h. chl.
16	Dicyclopentadienediiodoplatinum	$(C_5H_6)_2PtI_2$	581.11			209–211 d.		i.		s. acet.
17	Dicyclopentadienemethoxy chloroplatinum	$C_{11}H_{15}OPtCl$	393.78	wh. cr.						i. eth.; s. chl.
18	Dimethylplatinum diiodide	$(CH_3)_2PtI_2$	478.97	blk.				i.	i.	i. eth.; et. ac. acet.
19	Dipentenedichloroplatinum (Dipenteneplatinous chloride)	$[C_{10}H_{16}PtCl_2]_2$	804.47	pa. yel. biax. pr.		151–2				v. s. chl.; s. acet.; i. eth., al., ac. a., bz.
20	β-Dipenteneplatinous iodide	$C_{10}H_{16}PtI_2$	585.14	red cr.		122–4 d.				i. chl.; s. acet.
21	Dipyridinecyclopropanedichloroplatinum	$C_{13}H_{16}N_2Cl_2Pt$	466.28			d. >100		i.	d.	s. warm 10N HCl
22	Dipyridinotrimethyliodoplatinum	$C_{13}H_{19}N_2PtI$	525.30	pa. yel. cr.		168		sl. s.	s. s.	sl. s. al.; s. bz., chl., acet., h. al.
23	Dipyridinotrimethylplatinic iodide	$(CH_3)_3Pt(Py)_2I$	525.31	col. pr.		168				
24	Ethyl(trimethylplatinic) acetoacetate	$C_6H_9O_3Pt(CH_3)_3$	369.33	wh. hex. pl.		200 d.		i.		s. org. solv.
25	Ethylenedichloroplatinum (Ethyleneplatinous chloride)	$[C_2H_4PtCl_2]_2$	588.10	or. tabl.		170–180 d. 160–165		sl. s.		v. s. acet., al., eth.; s. bz., chl. sl. s. NaCl soln.
26	1,5-Hexadienedichloroplatinum (Biallyldichloroplatinum)	$C_6H_{10}PtCl_2$	348.15	or.-pa. yel.		172–3		i.	s. sl. with d.	s. chl.; v. sl. s. eth.; sl. s. bz.
27	1,5-Hexadienediiodoplatinum (Biallyldiiodoplatinum)	$C_6H_{10}PtI_2$	531.05	or. red cr.		d.		i.		s. chl.
28	Hexamethyldiplatinum	$(CH_3)_6Pt_2$	480.39	col. cr.		d.		i.	i.	v. s. bz., acet.; eth., sl. s. c. pet. eth.
29	Isobutylenedichloroplatinum (Isobutyleneplatinous chloride)	$[C_4H_8PtCl_2]_2$	644.21	or. rhomb. pl.		144–5		i.		v. s., chl., eth., bz., al. acet., NaCl soln.; sl. s. ac. a.; i. pet. eth.
30	8-Methoxycyclooct-4-enyl *p*-toluidinechloroplatinum)	$(C_8H_{12}OCH_3)(CH_3C_6H_4NH_2)PtCl$	476.92	wh. need.		140–2				s. bz., meth. al., nitrobz.; i. lgr.
31	Methoxydicyclopentadiene-*p*-toluidinechloroplatinum	$C_{18}H_{24}ONPtCl$	500.93	wh. need.		160–170				v. s. c. chl.; s. bz., meth. al., acet. eth.; i. nitrobz.
32	Methylplatinum pentaiodide	CH_3PtI_5	844.65	blk. cr.				sl. s.		sl. s. acet., et. ac., s. eth.
33	Methylplatinum triiodide	CH_3PtI_3	590.84	blk. amor.					i.	s. HCl
34	Tetramethyl platinum	$(CH_3)_4Pt$	255.23	lg. col. hex.		d.		i.	i.	v. s. bz., acet., eth., pet. eth.
35	Sesquimethylenediaminotrimethylplatiniciodide	$C_{12}H_{42}N_6I_2Pt_2$	914.50	col. cr.		d. 269		s.		
36	*p*-Toluidineethylenedichloroplatinum	$(C_2H_4)(CH_3C_6H_4NH_2)PtCl_2$	401.20	lem. yel. cr.		d. 125–135		i.		v. s. bz., acet., meth. al., eth., chl.; sl. s. cyclohexane; i. lt. pet.
37	Trimethylplatinicmonoethylenediamineiodide	$(CH_3)_3Pt(NH_2CH_2CH_2NH_2)I$	427.19			204 d.		s.		

No.	Name	Formula	Mol. wt.	Crystalline form, color and index of refraction	Sp. gr. or density	Melting point, °C	Boiling point, °C	Solubility in grams per 100 ml of		
								Cold water	Hot water	Alcohol, acids, etc.
	Platinum									
38	Trimethylplatinum benzoylacetone	$C_{13}H_{18}O_2Pt$	401.37	lng. wh. pr.						v. sl. s. lgr.; v. s. meth. al.
39	Trimethylplatinum chloride	$(CH_3)_3PtCl$	275.64	wh. cr.						sl. s. c. pet. eth.; s. h. pet. eth.
40	Trimethylplatinum dipropionyl methane	$C_7H_{11}O_2Pt(CH_3)_3$	367.36	wh. need.		d. 190		i.		s. org. solv.
41	Trimethylplatinum iodide	C_3H_9PtI	367.09	cr. yel. powd.		d. *ca* 250		i.	i.	sl. s. al. eth. acet., bz., s. warm bz.
42	Trimethylpyridino-platinum iodide	$[(CH_3)_3Pt(C_5H_5N)I]_2$	892.40			d. >150				s. pyr., chl., bz.
43	Pinenedichloroplatinum (Pineneplatinous chloride)	$[C_{10}H_{16}PtCl_2]_2$	804.48	pr.		138–141				v. s. chl.; s. bz., acet.; sl. s. eth., et. ac., ac. a.
44	Stilbenedichloro-platinum (Stilbene-platinous chloride)	$[C_{14}H_{12}PtCl_2]_2$	892.50	sm. hex. or. pr.		191–2				w. s. chl., acet., nitrobz.; v. sl. s. bz., al.; i. ac. a.
45	Styrenedichloro-platinum (Styrene-platinous chloride)	$[C_8H_8PtCl_2]_2$	704.30	or. hex. pr.		169–171		i.		s. chl., eth., acet.; v. sl. s. ac. a., NaCl soln.; sl. s. bz.
46	Styrenedibromo-platinum (Styrene-platinous bromide)	$[C_8H_8PtBr_2]_2$	918.12	pink hex. pr.		153–4				s. chl.; sl. s., bz., acet.; i. eth., al., ac. a.
	Potassium									
1	Potassium saccharate, acid(*d*)	$KHC_6H_8O_8$	248.23	rhomb. need.				1.1^{6}		
2	Potassium-*m*-nitrophenoxide	$KOC_6H_4NO_2.2H_2O$	213.23	flat or. need.	1.691^{20}	$-2H_2O$, 130	d.	16.3^{15}		s. al.
3	Potassium-*p*-nitrophenoxide	$KOC_6H_4NO_2.2H_2O$	213.23	yel. leaf	1.652^{20}	$-2H_2O$, 130	d.	7.5^{15}		sl. s. al.
	Rhenium									
1	Bis-2:2′-dipyridyl rhenichloride	$(C_{10}H_9N_2)_2ReCl_6$	713.31	pa. gr. cr.			sl. s.			
2	Dipyridyl *per*-rhenate(2:2′)	$(C_5H_4N)_2HReO_4$	407.39	col. need.			2.1			
3	Dipyridyl rheni-chloride(2:2′)	$(C_5H_4N)_2ReCl_6$	555.12	yel. need.			sl. s.			
4	Trimethylrhenium	$Re(CH_3)_3$	231.31	col. oil		60				
5	Tripyridyl rheni-chloride(2:2′:2″)	$(C_5H_4N)_3HReCl_6$	634.21	pa. grn. cr.			i.			
	Rubidium									
1	Monorubidium acetylide	$RbHC_2$	110.50	col. cr., very hygr.		d. 300				s. NH_3; i. eth.
	Silicon									
1	Aminotrisilane	$(H_3Si)_3N$	107.34	liq.	0.895^{-106}	−105.6	52	d.		s. eth.
2	Benzyltriethoxysilane	$C_6H_5CH_2Si(OC_2H_5)_3$	254.39	liq.	0.9864^{20}_{4}		253	i.	i.	s. al., acet., eth.
3	Butyltrifluorosilane	$C_4H_9SiF_3$	142.21	col. liq.	1.006^{25}_{4}	−96.6	52.4	d.		s. eth.
4	Carboxyethyl-dimethylsilane	$(CH_3)_2SiCH_2CH_2COOH$	131.23	n_D 1.4279		22				
5	Chloromethylsilicane	SiH_2ClCH_3	80.59		0.935^{-80}	−134.1	7			
6	Chlorotri-isocyanatesilane	$ClSi(NCO)_3$	189.59	solid; n_D^{20} 1.4507	1.437^{20}					s. bz., CS_2, CCl_4

No.	Name	Formula	Mol. wt.	Crystalline form, color and index of refraction	Sp. gr. or density	Melting point, °C	Boiling point, °C	Solubility in grams per 100 ml of: Cold water	Hot water	Alcohol, acids, etc.
	Silicon									
7	Di-*p*-aminoazobenzene fluosilicate	$(NH_2C_6H_4N_2C_6H_5)_2.H_2SiF_6$	538.57	long cinnamon br. need.	......	220 d.	......	......	......	187^{3}, 95% al.
8	Di-*p*-aminobenzoic acid fluosilicate	$(NH_2C_6H_4COOH)_2.H_2SiF_6$	418.37	pr. wh. long, narrow	......	242	......	......	......	0.91^{25}, 95% al.
9	Dianiline fluosilicate	$(C_6H_5NH_2)_2.H_2SiF_6$	330.33	irreg. wh. pl.	......	subl. 230	......	v. s.	......	s. h. al.
10	Di-*p*-biphenylyl diphenylsilane	$(C_6H_5)_2Si(C_6H_4C_6H_5)_2$	488.71	col. cr.	1.14^{20}	170	570	......	......	0.40^{20} al., 36.46^{20} bz., 12.91^{20} pyr.
11	Dichloromethylsilicane	$SiHCl_2CH_3$	115.02	......	0.93^{0}	−93	......	......	......	......
12	Didiphenylamine fluosilicate	$[(C_6H_5)_2NH]_2H_2SiF_6$	482.55	wh. rods forming rosettes	......	169	......	......	......	2.4492^{26}, 95% al.
13	Diethoxydibutoxysilane	$(C_2H_5O)_2Si(C_4H_9O)_2$	264.45	col. liq.; n_D^{20} 1.4010	0.909	......	128^{32}	d.	d.	s. org. solv.
14	Diethylaniline fluosilicate	$(C_6H_5NHC_2H_5)_2H_2SiF_6$	386.46	wh. pointed pr.	......	165.3	......	......	......	$.979^{26}$, 95% al.
15	Diethyldichlorosilane	$(C_2H_5)_2SiCl_2$	157.13	col. liq.; n_D^{20} 1.4809	1.0504_4^{20}	......	129	d.		s. eth.
16	Diethylsilanediol	$(C_2H_5)_2Si(OH)_2$	120.23	col. cr.	......	96	d. 140	s.	......	s. org. solv.
17	Dimethyoxdichlorosilane	$(CH_3O)_2SiCl_2$	161.07	col. liq.	1.2529_{15}^{0}	......	98–103	......	......	s. al., bz., eth.
18	Dimethylaniline fluosilicate	$(C_6H_5NHCH_3)_2H_2SiF_6$	358.38	monocl. wh. need.	......	......	......	......	......	s. h. al.; i. c. al.
19	Dimethyldi(*β*-chloroethoxy)silane	$(CH_3)_2Si(OC_2H_4Cl)_2$	217.17	col. liq. n_D^{20} 1.4420	1.135_4^{25}	......	213	i.	i.	s. al., acet., eth.
20	Dimethylfluorochlorosilane	$(CH_3)_2SiFCl$	112.61	col. liq.	1.18_4^{25}	−85.1	36.4	d.	......	s. eth.
21	Dimethylsilicane	$SiH_2(CH_3)_2$	60.17	......	0.68^{-80}	−150	−20.1	......	......	......
22	Di-*α*-naphthylamine fluosilicate	$(C_{10}H_7NH_2)_2.H_2SiF_6$	430.47	rosettes of wh. need.	......	218	......	......	......	$.1504^{25}$, 95% al.
23	Di-*β*-naphthylamine fluosilicate	$(C_{10}H_7NH_2)_2H_2SiF_6$	430.47	hex. wh. pl.	......	236.3	......	......	......	$.0816^{25}$; $.1248^{35}$, 95% al.
24	Di-*m*-nitraniline fluosilicate	$(C_6H_4NH_2NO_2)_2.H_2SiF_6$	420.50	rhomb. wh. pl.	......	200	......	......	......	$.121^{25}$; $.4736^{35}$, 95% al.
25	Dinitrosodiphenylamine	$[(C_6H_5)_2N=NO]_2.H_2SiF_6$	540.54	butterfly shaped indigo cr.	......	124.5	......	......	......	$.84^{25}$ 95% al.
26	Diphenylarsinophenylene triethylsilane	$(C_2H_5)_3SiC_6H_4As(C_6H_5)_2$	420.51	liq., n_D^{20} 1.61455	1.1661_4^{21}	......	$279–281^{17}$	......	......	......
27	Diphenyldichlorophenoxysilane	$(C_6H_5)_2Si(OC_6H_4Cl)_2$	437.40	col. liq., n_4^{20} 1.5510	1.2027_4^{20}	......	144	i.	i.	s. al., acet., eth.
28	Di-*o*-toluidine fluosilicate	$(C_6H_4NH_2CH_3)_2.H_2SiF_6$	358.41	wh. rhomb.	......	......	......	......	......	s. h. al.; i. c. al.
29	Di-*m*-toluidine fluosilicate	$(C_6H_4NH_2CH_3)_2.H_2SiF_6$	358.41	wh. rect. pr. pl.	......	......	......	......	......	s. h. al.; i. c. al.
30	Di-*p*-toluidine fluosilicate	$(C_6H_4NH_2CH_3)_2.H_2SiF_6$	358.41	wh. need., unst. irreg. outline	......	......	......	......	......	......
31	Docosamethyldecasiloxane	$(CH_3)_{22}O_9Si_{10}$	755.63	n_D^{20} 1.3988	0.925	......	183^{41}	i.	i.	s. bz.; sl. s. al.
32	Dodecamethylcyclohexasiloxane	$(CH_3)_{12}Si_6O_6$	444.93	oily liq., n_D^{20} 1.4015	0.9672	−3	128^{20}	......	......	......
33	Dodecamethylpentasiloxane	$(CH_3)_{12}O_4Si_5$	384.85	liq., n_D^{20} 1.3925	0.8755	*ca.* −80	229^{710}	i.	i.	s. bz.; sl. s. al.
34	Eicosamethylnonasiloxane	$(CH_3)_{20}O_7Si_9$	665.47	liq., n_D^{20} 1.3980	0.918	......	$173^{4.9}$	i.	i.	s. bz.; sl. s. al.

No.	Name	Formula	Mol. wt.	Crystalline form, color and index of refraction	Sp. gr. or density	Melting point, °C	Boiling point, °C	Solubility in grams per 100 ml of		
								Cold water	Hot water	Alcohol, acids, etc.
	Silicon									
35	Ethyldiethoxy-acetoxysilane	$C_2H_5Si(OC_2H_5)_2(OCOCH_3)$	206.31	liq., n_D^{20} 1.404	1.020_4^{20}	...	94^{16}	i.	i.	s. al., acet., eth.
36	Ethyldiethoxychloro-silane	$C_2H_5SiCl(OC_2H_5)_2$	182.72	...	...	...	...	i.	i.	s. al., acet., eth.
37	Ethylisocyanatesilane	$C_2H_5Si(NCO)_3$	183.20	sol., n_D^{25} 1.4468	1.2192^{20}	183.5	...	sl. d.	...	s. bz., CS_2, CCl_4
38	Ethyltriethoxysilane	$C_2H_5Si(OC_2H_5)_3$	192.33	liq., n_D^{20} 1.3853	0.9207_4^{20}	...	158.5	i.	i.	s. al., acet., eth.
39	Ethyltriphenylsilicane	$(C_2H_5)(C_6H_5)_3Si$	288.47	rhomboidal pr.	...	76	...	i.	...	s. chl., bz., eth., acet.; sl. s. al.
40	Hexadecamethylcyclo-octasiloxane	$(CH_3)_{16}Si_8O_8$	593.24	oily liq., n_D^{20} 1.4060	...	*ca.* 30	$175.^{20}$	...	...	...
41	Hexamethylsilicane	$Si_2(CH_3)_6$	146.38	col. liq.	0.723_4^{20}	12.5–14	112.5	i.	i.	s. acet., bz., eth.; col.
42	Hexamethylmethyl-enedisilane	$(CH_3)_3SiCH_2Si(CH_3)_3$	160.41	liq., n_D^{20} 1.4172	0.7520_4^{20}	...	134^{760}	...	...	...
43	Hydroxymethyltri-methylsilane	$OHCH_2Si(CH_3)_3$	104.23	liq., n_D^{25} 1.4169	0.8261_4^{20}	...	121.6	...	...	...
44	Methylsilicane	SiH_3CH_3	46.14	...	0.62^{-67}	−156.4	31	...	...	...
45	Methyltriethoxysilane	$CH_3Si(OC_2H_5)_3$	178.31	liq., n_D^{20} 1.3861	0.9383_4^{20}	...	151	i.	i.	s. al., acet., eth.
46	Methyltriphenyl-silicane	$(CH_3)(C_6H_5)_3Si$	274.44	...	...	67.3	...	i.	...	s. eth., chl., bz.; sl. s. al.
47	Octadecamethyl-octasiloxane	$(CH_3)_{18}O_7Si_8$	607.31	liq., n_D^{20} 1.3970	0.913	$153^{5.1}$	...	...	...	s. bz., sl. s. al.
48	Octamethylcyclo-tetrasiloxane	$(CH_3)_8Si_4O_4$	296.62	oily liq., n_D^{20} 1.3968	0.9558	175.0^{0}	...	...	...	...
49	Octamethyltrisiloxane	$(CH_3)_8O_2Si_3$	236.54	liq., n_D^{20} 1.3848	0.8200	−80	153	...	...	s. bz.; sl. s. al.
50	Phenylenediamine fluosilicate(*m*)	$Si(C_6H_5CH_2)_4$	252.24	choc. br. need.-like pr.	...	243–4	...	...	...	0.065^{26}, 95 % al.
51	Phenylenediamine fluosilicate(*p*)	$C_6H_4(NH_2)_2H_2SiF_6$	252.24	pink irreg. six-sided pl.	...	d.	...	...	...	0.014^{26}, 95 % al.
52	Phenylisocyanatesilane	$C_6H_5Si(NCO)_3$	231.24	solid, n_D^{20} 1.5210	1.273^{20}	251.9	...	sl. d.	...	s. bz., CS_2, CCl_4
53	Phenyltrichlorosilicane	$C_6H_5SiCl_3$	211.55	liq.	1.326_4^{18}	...	201.5	d.	...	s. eth., chl.; d. al.
54	Silicobenzoic acid	C_6H_5SiOOH	138.20	col. flaky resin	...	40–50	215.6–	...	...	s. eth., bz., chl.; sl. s. al., ac. a.
55	Tetra-*m*-aminophenyl-silane	$(NH_2C_6H_4)_4Si$	396.57	solid	...	380				...
56	Tetrabenzylsilicane	$Si(C_6H_5CH_2)_4$	392.62	...	...	...	127.5	i.	...	s. eth., chl., bz.; sl. s. al.
57	Tetra-*p*-Biphenylyl-silane	$Si(C_6H_4C_6H_5)_4$	640.91	col. cr.		283	600	...	...	0.049^{20} al.; 0.27^{20} bz.; 16.69^{20} pyr.
58	Tetradecamethylcyclo-heptasiloxane	$C_{14}H_{42}Si_7O_7$	519.09	oily liq., n_D^{20} 1.4040	0.9730	−26	154^{20}	...	...	...
59	Tetradecamethyl-hexasiloxane	$C_{14}H_{42}O_5Si_6$	459.00	liq., n_D^{20} 1.3948	0.8910	−100	142^{20}	...	...	s. bz.; sl. s. al.
60	Tetraethylsilicane	$Si(C_2H_5)_4$	144.33	liq., 1.4246	0.762^{25}	...	152.8–3.2	...	...	...

No.	Name	Formula	Mol. wt.	Crystalline form, color and index of refraction	Sp. gr. or density	Melting point, °C	Boiling point, °C	Solubility in grams per 100 ml of		
								Cold water	Hot water	Alcohol, acids, etc.
	Silicon									
61	Tetraethylthiosilane	$(C_2H_5S)_4Si$	272.59	liq., n_D^{25} 1.5638	1.0860_4^{25}	−6	170^{12}			
62	Tetrahexyloxysilane	$(C_6H_{13}O)_4Si$	432.77	col. liq., n_D^{20} 1.4300	0.876		$232–234^{13}$	d.	d.	s. org. solv.
63	Tetraisopropyl-mercaptane silicon	$C_{12}H_{28}S_6Si$	392.83	solid, n_D^{35} 1.5350	1.0099_4^{35}	335	$176–178^{13}$			
64	Tetramethoxysilane	$(OCH_3)_4Si$	152.22	col. liq., n_D^{20} 1.3681	1.0523		$121–122^{759}$	d.	d.	s. org. solv.
65	Tetramethyl-mercaptanesilicon	$(CH_3)_4S_4Si$	216.48	solid, n_D^{20} 1.5989	1.1888_4^{25}	31	$144–146^{12}$			
66	Tetramethylsilicane	$Si(CH_3)_4$	88.23	col. liq.	0.651^{15}		26.5			s. eth.; i. conc. H_2SO_4
67	Tetraphenoxysilane	$(C_6H_5O)_4Si$	400.51	col. cr.		47–48	$415–420^{7}$	d.	d.	s. org. solv.
68	Tetraphenylsilicane	$Si(C_6H_5)_4$	336.51	col. flocc. amor. part.						s. acet. anh. and chloro-sulphonic acid
69	Tetrapropoxysilane	$(OC_3H_7)_4Si$	264.44	col. liq., n_D^{20} 1.4015	0.918		$225–227^{769}$	d.	d.	s. org. solv.
70	Tetratriethyl-siloxysilane	$Si[OSi(CH_3)_2C_2H_5]_4$	440.96	col. liq., 7_D^{20} 1.4112	$0.895_{15.6}^{20}$	−45	102^{2}			
71	Thioisocyanatotri-ethylsilane	$(C_2H_5)_3SiNCS$	173.35	liq., n_D^{20} 1.4944	0.934_4^{20}		210.5			
72	Tolidine fluosilicate(*o*)	$(CH_3NH_2C_6H_3)_2H_2SiF_6$	356.39	tiny, micr. wh. pr.		268–9				$.013^{26}$, 0.41^{36} 95 % al.
73	Tri-*p*-Biphenylyl-phenylsilane	$(C_6H_5)_3SiC_6H_4C_6H_5$	412.61	col. cr.	1.10^{20}	174	580			0.238^{20} al.; 17.776^{20} bz.; 2.725^{20} pyr.
74	Trichloromethyl-triethyloxysilane	$CCl_3Si(OC_2H_5)_3$	281.64	col. liq., n_D^{20} 1.4320	1.8660_4^{20}		81^{3}	i.	i.	s. al., acet., eth.
75	Triethylbromosilane	$(C_2H_5)_3SiBr$	195.18	col. liq., n_D^{20} 1.4561	1.777_4^{20}		162	d.		s. eth.
76	Triethylchlorosilane	$(C_2H_5)_3SiCl$	150.73	col. liq., n_D^{20} 1.4314	0.8967_4^{20}		144^{735}	d.		s. eth.
77	Triethylfluorosilane	$(C_2Y_5)_3SiF$	134.27	col. liq., n_D^{20} 1.3915	0.9354_4^{20}		$109–100^{745}$	d.		s. eth.
78	Triethylphenylsilicane	$(C_2H_5)_3(C_6H_5)Si$	192.38			149	230			
79	Trimethylchloro-methylsilane	$(CH_3)_3SiCH_2Cl$	122.67	col. liq., n_D^{20} 1.4180	0.8791_4^{20}		97.1^{734}			
80	Trimethylethoxysilane	$(CH_3)_3Si(OC_2H_5)$	118.25	liq., n_D^{20} 1.3741	0.7573_4^{20}		75^{745}	i.	i.	s. al., acet., eth.
81	Triphenylacetoxysilane	$(C_6H_5)_3SiOCOCH_3$	318.45	col. cr.		91–92				
82	Vinyltriphenoxysilane	$CH_2{:}CHSi(OC_6H_5)_3$	334.45	liq., n_D^{20} 1.5617	1.1300_4^{20}		210.2^{7}	i.	i.	s. al., acet., eth.
	Silver									
1	Acetylene silver nitrate (complex.)	$Ag_2C_2.AgNO_3$	409.64	fine. nd. and crosses		detonating pt. *ca.* 212				
2	Cycloocta-1,3-dienesilver nitrate	$C_8H_{12}.2AgNO_3$	447.94	col. need.		150 d.				s. h. meth. al.
3	Cycloocta-1,4-dienesilver nitrate	$C_8H_{12}.2AgNO_3$	447.94	lg. col. hex. tab.		110–1				sl. s. boil. meth. al.
4	Cycloocta-1,5-dienesilver nitrate	$C_8H_{12}.AgNO_3$	278.06	lng. col. need.		135–6 (128.5–131)				v. s. meth. al.

No.	Name	Formula	Mol. wt.	Crystalline form, color and index of refraction	Sp. gr. or density	Melting point, °C	Boiling point, °C	Solubility in grams per 100 ml of		
								Cold water	Hot water	Alcohol, acids, etc.
	Silver									
5	*cis-trans*-1,3-Cyclooctadienesilver nitrate	$C_8H_{12}.AgNO_3$	278.06	wh. cr.		126–127.5 d.				sl. s. abs. al.
6	Cyclooctatetraene dimer silver nitrate	$C_{16}H_{16}.AgNO_3$	378.18	pr.		153				s. h. al.
7	Cyclooctatetraenesilver nitrate	$C_8H_8.AgNO_3$	274.03	lg. cr.		173 d.		d.		s. aq. $AgNO_3$; s. al.
8	Cyclooctatetraenylphenylketonesilver nitrate	$C_{15}H_{12}O.AgNO_3$	378.13	pa. yel. need.		121.4–122 d.				sl. s. abs. al.
9	Cycloocta-1,3,6-trienesilver nitrate	$C_8H_{10}.3AgNO_3$	615.80	flat need		138–9				sl. s. meth. al.
10	1,3,5-Cyclooctatriene-silver nitrate	$C_8H_{10}.AgNO_3$	276.04	cr.		125–6				sl. s. abs. al.
11	*cis*-Cyclooctenesilver nitrate	$(C_8H_{14})_2.AgNO_3$	390.28	lg. cr.		51				v. s. meth. al.
12	1,2-Dimethylcyclooctatetraenesilver nitrate	$C_{10}H_{12}.2AgNO_3$	471.95	cr.		142.5-144.5				s. h. abs. al.
13	Ethylcyclooctatetraene-silver nitrate	$C_{10}H_{12}.2AgNO_3$	471.95	gray wh. cr.		124–125.5				i. eth.; sl. s. abs. al.
14	Ethylsilver	C_2H_4Ag	135.93			d. >-40				
15	Methylcyclooctatetraenesilver nitrate	$2C_9H_{10}.3AgNO_3$	745.98	lt. yel. pr.		123–124.5				sl. s. abs. al.; i. eth.
16	Methylsilver	$AgCH_3$	122.90	yel.		d. >-40				meth. al. 0.04^{-80}; al. 0.04^{-80}; amyl al. 0.06^{-80}
17	Phenylcyclooctatetraenesilver nitrate	$C_6H_5.C_8H_7.AgNO_3$	350.12	yel. gr. cr.		144.5 d.				i. c. eth.; s. h. abs. al.
18	*n*-Propylcyclooctatetraenesilver nitrate	$C_{11}H_{14}.2AgNO_3$	485.98	pa. yel. cr.		141 d.				s. h. abs. al.
19	Silver acetylide silver nitrate complex	$Ag_2C_2.6AgNO_3$	1259.01	wh. rhomb.		d. >308		d.		i. acet.
	Sodium									
1	Bis(1-methylamyl)-sodium sulfosuccinate (Aerosol MA)	$C_{16}H_{29}NaO_7S$	388.46	wh. waxlike pellets				34.3^{25}	44.7^{70}	s. acet., bz., glyc., CCl_4
2	Hydroxydionsodium (Viadril)	$C_{25}H_{35}NaO_6$	454.54	wh. powd.		d. 193–203		s.	s.	s. acet., chl.
3	Sodium acetamide	$NaNHCOCH_3$	81.05	wh. tabl.		300–50 d.		d.		d. al.; sl. s. bz., liq. NH_3
4	Sodium acetylide (ethinylsodium)	$NaHC_2$	48.02	wh. cr.		d. >210		d.		d. a.; s. liq. NH_3
5	Sodium anilide	$NaNHC_6H_5$	115.11	wh. cr. v. hygr.		d.		d.		d. a., al.; s. liq. NH_3
6	Sodium anthraquinone-β-sulfonate ("Silver salt")	$NaC_{14}H_7O_5S.H_2O$	328.28	silvery leaf		d.		0.84	27	v. sl. s. al.
7	Sodium arsanilate (atoxyl, soamim)	$NaC_6H_7O_3NAs.6H_2O$	347.13	wh. cr. powd., monocl.				$16^{1.7}$		sl. s. al.; s. meth. al.
8	Sodium benzamide	$NaNHCOC_6H_5$	143.12	wh. powd.		d.		d.		d. al.; i. eth., bz., chl.
9	Sodium-*N*-chloro-*p*-toluenesulfonamide	$NaC_7H_7O_2NClS.3H_2O$	281.69	col. pr.		expl. 175–180		s.	v. s.	
10	Sodium ethoxide (sodium ethylate)	$NaOC_2H_5.2C_2H_5OH$	160.19	wh. powd. or need.		$-2C_2H_5OH$, 200	d.	d.	d.	v. s. al.; i. NH_3
11	Sodium, ethyl-	NaC_2H_5	52.05	wh. cr., d. air		d.		d.		d. al., eth.; s. diethylzinc; i. bz., lgr.
12	Sodium-β-naphthoxide	$NaOC_{10}H_7$	166.16	wh. powd., v. hygr.		d.		s.		v. s. al., eth.; i. lgr.
13	Sodium-*p*-nitrobenzene isodiazotate	$NaC_6H_4O_3N_3.2H_2O$	225.14	gold. leaf. or need.		$-H_2O$ over H_2SO_4	exp.	v. s.		

No.	Name	Formula	Mol. wt.	Crystalline form, color and index of refraction	Sp. gr. or density	Melting point, °C	Boiling point, °C	Solubility in grams per 100 ml of: Cold water	Hot water	Alcohol, acids, etc.
	Sodium									
14	Sodium-*p*-nitrophenoxide	$NaOC_6H_4NO_2.4H_2O$	233.15	yel. monocl. pr.	....	$-2H_2O$, 36; $-4H_2O$, 120	d.	5.97[25]	....	sl. s. al.
15	Sodium-*o*-sulfo-benzoicimide (soluble saccharin)	$NaC_7H_4O_3NS.2H_2O$	241.20	wh. tabl.	....	$-H_2O$	....	v. s.	....	sl. s. h. al.
16	Triphenylborylsodium*	$NaB(C_6H_5)_3$	265.12	yel.-or. silky need.	....	....	d.	d.	d.	0.08[18] eth.
17	Triphenylmethylsodium	$NaC(C_6H_5)_3$	266.32	red. cr.	....	....	....	d.	....	s. eth., bz., liq. NH_3
	Tantalum									
1	*bis*-Cyclopentadienyl-tantalum tribromide	$(C_5H_5)_2TaBr_3$	550.87	rust. cr.	....	280 d.	....	s.	....	sl. s. chl.;
2	Tantalumpenta-methoxide	$Ta(OCH_3)_5$	336.12	wh. solid	....	50	189[10]	d.	....	....
	Tellurium									
1	Di-*n*-butyl telluride	$(C_4H_9)_2Te$	241.83	yel. oil	1.334[40]	....	132–5	....	....	....
2	Diethyl telluride	$(C_2H_5)_2Te$	185.72	....	....	....	137–8	sl. s.	....	....
3	Diethyltelluroketone	$C_2H_5.CTe.C_2H_5$	197.74	yelsh., oily liq. n_D^{20} 1.5480	0.8821_4^{15}	....	69–72[8]	i.	i.	v. s. eth., 95% al. s.
4	Dimethyl telluride	$(CH_3)_2Te$	157.67	pa. yel. oil.; garlic-like odor	....	sld. in liq. air	82.97[770]	....	....	....
5	Dimethyltelluroketone (Telluroacetone)	CH_3CTeCH_3	169.68	col. oily liq., n_D^{25} 1.4883	0.8578_4^{18}	....	55–58[10]	i.	i.	s. eth., sl. s. 95% al.
6	Dimethyltelluronium dibromide(α)	$C_2H_6Br_2Te$	317.49	or. leaf-like cr.	....	142 d.	....	....	....	s. al., eth.
7	Dimethyltelluronium dichloride(α)	$C_2H_6Cl_2Te$	228.58	leaf-like cr.	....	92	....	s.	s.	s. al., eth.
8	Dimethyltelluronium dichloride(β)	$C_2H_6Cl_2Te$	228.58	leaf-like cr.	....	134	....	....	....	s. al., eth.
9	Dimethyltelluronium diiodide(α)	$C_2H_6I_2Te$	411.48	red cr.	....	127 d.	....	i.	v. sl. s.	s. chl., bz.
10	Di-*p*-phenetyl ditelluride	$(C_2H_5OC_6H_4)_2Te_2$	497.52	or. brown need.	1.666	108	d. >108	....	....	s. lgr.
11	Ditelluromethane	CH_2Te_2	269.23	dk. red. amor. sld.	....	d. 124	....	i.	i.	i.
12	Ethylmethyl-tellurophetone	$C_2H_5CTeCH_3$	183.71	dark yel. oil, 1.5055[25]	1.8711	....	63–6	....	....	....
	Thallium									
1	Di-*n*-Butylthallium chloride	$(n\text{-}C_4H_9)_2TlCl$	354.06	col. pl. or flakes (from pyr.)	....	d. 240–250 (expl.)	....	sl. s.		s. hot bz.; 0.28[30] al.; 2.98[30] pyr; v. sl. s. eth.
2	Dicyclohexylthallium chloride	$(C_6H_{11})_2TlCl$	406.13	long col. need.	....	d. 210–230	....	v. sl. s.	i.	s. hot al., hot bz.; 1.34[30] pyr.; i. eth.
3	Diethylthallium hydroxide	$Tl(C_2H_5)_2OH$	279.50	....	....	127–128	....	v. s.	....	v. s. al.
4	Di-*n*-hexylthallium chloride	$(n\text{-}C_6H_{13})_2TlCl$	410.16	col. pl.	....	d. 198	....	0.0117[23]	....	s. pyr.; 0.123[23] al.; 0.0304[23] bz.; i. HNO_3
5	Di-*n*-hexylthallium nitrate	$(n\text{-}C_6H_{13})_2TlNO_3$	436.72	shiny flakes	....	d. 271	....	0.008[23]	....	0.1856[32] al.; v. sl. s. bz.
6	Diisoamylthallium chloride	$(i\text{-}C_5H_{11})_2TlCl$	382.11	col. cr.	....	d. 253	....	sl. s.	....	0.31[30] al.; 1.74[30] pyr.
7	Dipropylyhallium chloride	$Tl(C_3H_7)_2Cl$	326.00	silvery leaflets	....	d. 200	....	....	....	s. eth.; i. HCl
8	Triethylthallium	$Tl(C_2H_5)_3$	291.56	yel. liq.	1.957_{23}^{23}	−63	192	....	....	....

* All of the tri-arylborines form analogous addition-salts of the alkali metals.

No.	Name	Formula	Mol. wt.	Crystalline form, color and index of refraction	Sp. gr. or density	Melting point, °C	Boiling point, °C	Solubility in grams per 100 ml of		
								Cold water	Hot water	Alcohol, acids, etc.
	Thallium									
9	Trimethylthallium	$Tl(CH_3)_3$	249.48	col. need.	....	38.5	147; may expl. 90	d.	....	v. s. bz., eth.
	Thorium									
1	Thoriumisotetrapropoxide	$Th(OC_3H_7)_4$	468.39	wh. cr.	....	$80^{0.1}$	subl. 200–$210^{0.1}$	d.	....	s. bz.
	Tin									
1	Amyltetrathioortho-stannate(*n*)	$Sn(SC_5H_{11})_4$	531.52	....	....	....	$162^{.004}$	....	....	....
2	Amyltetrathioortho-stannate(*tert*)	$[CH_3CH_2C(CH_3)_2S]_4Sn$	531.52	....	....	44	....	....	....	....
3	Bis(tributyltin) oxide	$(C_4H_9)_3Sn_2O$	424.73	yel. liq.	1.17^{25}	....	254^{50}	....	0.1 %	....
4	Bis(triphenytin) sulfide	$(C_6H_5)_3Sn_2S$	500.76	solid	....	141–143	....	....	....	....
5	Bromobenzenetetra-thioorthostannate(*p*)	$Sn(SC_6H_4Br)_4$	870.98	....	....	217	....	....	....	....
6	Butylbenzenetetra-thioorthostannate(*p*)	$Sn(SC_6H_4C_4H_9)_4$	779.81	....	....	106	....	....	....	....
7	Butyltetrathioortho-stannate(*n*)	$Sn(SC_4H_9)_4$	475.41	....	....		....	$136^{.001}$	....	
8	Butyltetrathioortho-stannate(*sec*)	$Sn(SC_4H_9)_4$	475.41	....	....	....	$111^{.001}$	....	....	....
9	Butyltin trichloride	$C_4H_9SnCl_3$	282.17	col. liq., 1.5244^{20}	1.7^{20}	−63	98^{10}	....	....	s. org. solv.
10	*n*-Butylvinyltin dichloride	*n*-$C_4H_9(CH_2{:}CH)SnCl_2$	273.78	solid, n_D^{25} 1.5254	1.533_4^{25}	27–28	$99–101^3$	....	....	....
11	Carbomethoxyphenyl-trichlorostannane(*o*)	$(CH_3OCOC_6H_4)SnCl_3$	360.19	....	....	164	....	....	....	....
12	Chlorobenzenetetra-thioorthostannate(*p*)	$Sn(SC_6H_4Cl)_4$	693.15	....	....	189	....	....	....	....
13	Cyclohexyltetrathio-orthostannate	$Sn(SC_6H_{11})_4$	579.56	....	....	53–4	....	....	....	....
14	Diallylsdibutyltin	$(C_4H_9)_2(CH_2{:}CHCH_2)_2Sn$	315.07	liq., 1.4986	1.0999	....	$93^{0.1}$	....	....	....
15	Diallyldiphenyltin	$(C_6H_5)_2(CH_2{:}CHCH_2)_2Sn$	355.05	liq., 1.6013^{25}	1.2688^{25}	....	$173–174^5$	....	....	....
16	Di-*o*-anisyldichloro-stannane	$(CH_3OC_6H_4)_2SnCl_2$	403.86	....	....	113	....	....	....	....
17	Dibenzyldiethyl-stannane	$(C_6H_5CH_2)_2Sn(C_2H_5)_2$	359.08	liq.	1.+	<20	$223–4^{20}$	....	....	s. org. solv.
18	Dibenzylethylpropyl-stannane	$(C_6H_5CH_2)_2(C_2H_5)(C_3H_7)Sn$	373.11	liq.	....	>0	$220–5^{15}$	....	....	misc. all org. solv.
19	Dibenzyltin acetate	$(C_6H_5CH_2)_2Sn(OCOCH_3)_2$	419.05	col. need. f. al.	....	136–7	....	....	....	s. acet. chl., bz.
20	Dibenzyltin dibromide	$(C_6H_5CH_2)_2SnBr_2$	460.78	col. need. f. pet.	....	130	....	....	....	s. acet., al., eth., chl., CCl_4
21	Dibenzyltin dichloride	$(C_6H_5CH_2)_2SnCl_2$	371.86	col. need. f. acet.-HCl	....	163–4	....	....	....	s. acet., al., eth., chl., CCl_4, h. ac. a.
22	Dibenzyltin diiodide	$(C_6H_5CH_2)_2SnI_2$	554.77	col. lng. silky yel. need. f. pet. eth.	....	86–7	....	....	....	s. acet., al., eth., chl., CCl_4
23	Dibutyldiphenyltin	$(C_4H_9)_2(C_6H_5)_2Sn$	387.14	col. liq., 1.5602	1.1882	....	180^2	....	....	....
24	Dibutyldivinyltin	$(C_4H_9)_2(CH_2{:}CH)_2Sn$	287.02	liq., 1.4824	1.1270	....	$78–80^2$	....	....	....
25	Dibutyltin diacetate	$(C_4H_9)_2Sn(OOCH_3)_2$	326.99	col. liq., 1.482^{20}	1.32	10	$142–145^{10}$	i.	i.	s. org. solv.
26	Dibutyltin dibromide	$(C_4H_9)_2SnBr_2$	392.74	sm. need.	....	20	$118–170^{28}$	i.	i.	....
27	Dibutyltin dichloride	$(C_4H_9)_2SnCl_2$	303.83	need. wh. 1.499^{50}	1.36^{50}	142^{10}	d. 113.6^{60}	d.	d.	s. eth., bz., al.
28	Dibutyltin dilaurate	$(C_4H_9)_2Sn(OOC_{11}H_{23})_2$	631.55	oily liq., 1.474^{14}	1.05	27	....	i.	i.	s. ac., bz., eth., CCl_4
29	Dibutyltin oxide	$(C_4H_9)_2SnO$	248.92	wh. amorph. powd.	1.58	d.	....	i.	i.	i. org. solv.

PHYSICAL CONSTANTS OF ORGANOMETALLIC COMPOUNDS (Continued)

No.	Name	Formula	Mol. wt.	Crystalline form, color and index of refraction	Sp. gr. or density	Melting point, °C	Boiling point, °C	Solubility in grams per 100 ml of: Cold water	Hot water	Alcohol, acids, etc.
	Tin									
30	Dibutyltin sulfide	$(C_4H_9)_2SnS$	264.99	col.-yel. lip., 1.58	1.417					
31	Dibutylvinyltin bromide	$(C_4H_9)_2(CH_2{:}CH)SnBr$	339.88	col. liq., 1.4970	1.3913		$96^{0.55}$			
32	Dibutylvinyltin chloride	$(C_4H_9)_2(CH_2{:}CH)SnCl$	295.42	liq. 1.4987^{20}	1.2662^{20}		$112–114^{4}$			
33	Dichlorodi-*m*-tolyl stannane	$(CH_3C_6H_4)_2SnCl_2$	371.86			39–40				
34	Diethyldibromo-dipyridinetin	$(C_2H_5)_2SnBr_2(C_5H_5N)_2$	494.84			140				
35	Diethyldiisoamyltin	$(C_2H_5)_2Sn(C_5H_{11})_2$	319.10		1.0725^{19}		$131^{13.5}$			
36	Diethyldiisobutyltin	$(C_2H_5)_2Sn(C_4H_9)_2$	291.05		1.1030		108.2^{13}			
37	Diethyldiphenyltin	$(C_2H_5)_2(C_6H_5)_2Sn$	331.03				$154–6^{4}$			
38	Diethylisoamyltin bromide	$(C_2H_5)_2(C_5H_{11})SnBr$	327.87		1.4881^{17}		137.5^{17}			
39	Diethylisoamyltin chloride	$(C_2H_5)_2(C_5H_{11})SnCl$	283.41		$1.2994^{18.9}$		125.5^{13}			
40	Diethylisobutyltin bromide	$(C_2H_5)_2(C_4H_9)SnBr$	313.84		1.5108		122^{17}			
41	Diethyl-*n*-propyltin bromide	$(C_2H_5)_2(C_3H_7)SnBr$	299.81		1.5910^{21}		112.2^{16}			
42	Diethyl-*n*-propyltin chloride	$(C_2H_5)_2(C_3H_7)SnCl$	255.36		$1.3848^{15.7}$		108^{17}			
43	Diethyl-*n*-propyltin fluoride	$(C_2H_5)_2(C_3H_7)SnF$	238.90			271				6.93^{31}, meth. al.; 3.78^{12} al.; $.05^{31}$ bz.
44	Diethyltin	$Sn(C_2H_5)_2$	176.81	sl. yel. oily liq.	1.654	<-12	150 d.	i.	i.	s. bz., eth., lgr., chl., CCl_4
45	Diethyltin dibromide	$(C_2H_5)_2SnBr_2$	336.63	col. need.	2.068^{14}	63	232–3	s.	s.	s. eth., org. solv.
46	Diethyltin dichloride	$(C_2H_5)_2SnCl_2$	247.72	wh. need.		84–5	220	s.	s.	s. HCl, org. solv.
47	Diethyltin difluoride	$(C_2H_5)_2SnF_2$	214.81	sp. pl. or long rhomb. tab. f. meth. al.		229				$.45^{31}$, al.; 2.64^{31}, meth. al.; 0.47^{31}, bz.
48	Diethyltin diiodide	$(C_2H_5)_2SnI_2$	430.62	wh. need.		44.5–5.0	240–5 d.	v. sl. s.	sl. s.	s. org. solv.
49	Diethyltin oxide	$(C_2H_5)_2SnO$	192.81	wh. powd.		infus.		i.	i.	s. HCl, dil. al., conc. alk.; i. org. solv.
50	Diisoamyltin dibromide	$(C_5H_{11})_2SnBr_2$	420.79			−25 to −24				
51	Diisoamyltin dichloride	$(C_5H_{11})_2SnCl_2$	331.88			28				
52	Diisoamyltin diiodide	$(C_5H_{11})_2SnI_2$	514.79	oily liq.			$202–5^{8}$			
53	Diisobutyltin dichloride	$(CH_3)_2(CHCH_2)_2SnCl_2$	273.76	liq.	1.5012	9	135.5			
54	Diisobutyltin diiodide	$(C_4H_9)_2SnI_2$	486.73				290–5			
55	Diisopropyltin dibromide	$(C_3H_7)_2SnBr_2$	364.69	pa. yel. hyg. cr.		54		d.	d.	i. org. solv.
56	Diisopropyltin dichloride	$(C_3H_7)_2SnCl_2$	275.77	col. transp. cr.		80–4		s.	s.	s. al., h. bz., glac. ac. a.
57	Diisopropyltin oxide	$(C_3H_7)_2SnO$	220.87			d.		i.	i.	s. h. HCl; i. org. solv., alk.
58	Dimethyldibromo-dipyridinetin	$(CH_3)_2SnBr_2(C_5H_5N)_2$	466.78			172				
59	Dimethyldichloro-dipyridinetin	$(CH_3)_2SnCl_2(C_5H_5N)_2$	377.87			163				
60	Dimethyldiethyltin	$(CH_3)_2Sn(C_2H_5)_2$	206.88	col. liq.	1.2319^{19}	<-13	144–6	i.	i.	s. org. solv.
61	Dimethyldiisobutyltin	$(CH_3)_2Sn(C_4H_9)_2$	262.99		$1.1179^{20.1}$		$85^{16.5}$			
62	Dimethyldioctyltin	$(CH_3)_2(C_8H_{17})_2Sn$	375.21	liq., 1.4659	1.0168		$121–122^{0.2}$			
63	Dimethyldivinyltin	$(CH_3)_2Sn(CH{:}CH_2)_2$	202.85	n_D^{25} 1.4720	1.284_4^{25}		$120–121^{760}$			
64	Dimethylethylpropyltin	$(CH_3)_2(C_2H_5)C_3H_7Sn$	220.91		1.2014_{20}^{20}		149–51			

No.	Name	Formula	Mol. wt.	Crystalline form, color and index of refraction	Sp. gr. or density	Melting point, °C	Boiling point, °C	Solubility in grams per 100 ml of: Cold water	Hot water	Alcohol, acids, etc.
	Tin									
65	Dimethylethyltin iodide	$(CH_3)_2C_2H_5SnI$	304.73	1.5705[18]	2.0264^{20}_{20}		77–8[11]; 185–7[718]			
66	Dimethyltin	$[(CH_3)_2Sn]_x$	$(148.76)_x$	yel. sld.				i.	i.	i. org. solv.
67	Dimethyltin dibromide	$(CH_3)_2SnBr_2$	308.58	col. pr.		74–6	208–13	s.	s.	s. org. solv.
68	Dimethyltin dichloride	$(CH_3)_2SnCl_2$	219.67	col. cr.		90 (107)	188–90	s.		s. org. solv.
69	Dimethyltin difluoride	$(CH_3)_2SnF_2$	186.76	wh. fine pl.			d. <360	4.66[30.7]		.08[31], al., 33[31], meth. al.
70	Dimethyltin diiodide	$(CH_3)_2SnI_2$	402.57	rhomb. wh.	2.872	43 (30)	228	sl. s.	s.	s. org. solv.
71	Dimethyltin oxide	$(CH_3)_2SnO$	164.76	wh. powd.	1.269	infus.	d.	i.	i.	s. a., NaOH; i. org. solv., NH_4OH
72	Dimethyltin sulfide	$(CH_3)_2SnS$	180.82			148				
73	Dimethylvinyltin bromide	$(CH_3)_2(CH_2{:}CH)SnBr$	255.72	solid n^{25}_D 1.5350	1.738[66]	59–61[27]				
74	Dimethylvinyltin iodide	$(CH_3)_2(CH{:}CH_2)SnI$	302.71	n^{25}_D 1.3762	2.033^{25}_4		57.5–59[5.2]			
75	Di-α-naphthyltin	$Sn(C_{10}H_7)_2$	373.02			200	d. 225			
76	Diphenyldivinyltin	$(C_6H_5)_2Sn(CH{:}CH_2)_2$	327.00	n^{25}_D 1.5949	1.3195^{25}_4		153–154[5]			
77	Diphenyldivinyltin	$(C_6H_5)_2Sn(CH{:}CH_2)_2$	326.98	liq., 1.5227[20]	1.3049[20]	143–144[5]				
78	Di(phenylthiol)-diphenylstannane	$Sn(C_6H_5)_2(SC_6H_5)_2$	491.25			65–65.5				
79	Diphenyltin	$Sn(C_6H_5)_2$	272.90	yel. amor. powd.		225.7; (126–30)		i.	i.	s. chl., bz., eth.; i. abs. al.
80	Diphenyltin dibromide	$(C_6H_5)_2SnBr_2$	432.72	col. cr.		38	230[42]			s. al., eth.
81	Diphenyltin dichloride	$(C_6H_5)_2SnCl_2$	343.81	col. cr.		42	333–7 d.	v. sl. s., d.		s. al., eth., lgr.
82	Diphenyltin difluoride	$(C_6H_5)_2SnF_2$	310.90			360				
83	Diphenyltin diiodide	$(C_6H_5)_2SnI_2$	526.71	col. cr.		71–72	176–82[2]	i.	i.	s. org. solv.
84	Diphenyltin hydroxy-chloride	$(C_6H_5)_2Sn(OH)Cl$	325.36	amor. wh. powd.		187		i.	i.	s. conc. a.; i. org. solv.
85	Diphenyltin oxide	$(C_6H_5)_2SnO$	288.90	col. amor. powd.		infus.		i.	i.	s. conc. a.; i. org. solv.
86	Diphenyltin sulfide	$(C_6H_5)_2SnS$	304.97	wh. cr.		171–173				
87	Di-*n*-propyldibromodi-pyridinetin	$(C_3H_7)_2SnBr_2(C_5H_5N)_2$	522.89			128				
88	Dipropyltin dibromide	$(C_3H_7)_2SnBr_2$	364.69	col. need.		49		v. sl. s.		s. org. solv.
89	Dipropyltin dichloride	$(C_3H_7)_2SnCl_2$	275.77	col. cr.		81		v. sl. s.		s. org. solv.
90	Dipropyltin difluoride	$(C_3H_7)_2SnF_2$	242.87	leaf		205		0.22[32]		.93[32], al.. 1.91[32], meth. al.
91	Dipropyltin diiodide	$(C_3H_7)_2SnI_2$	458.68	col. oily liq.		< −15	270–3	i.	i.	s. org. solv.
92	Di-*m*-tolylstannane	$(CH_3C_6H_4)_2SnO$	316.96	wh. amor. infus.				i.	i.	s. min. a.; i. org. solv.
93	Di-*m*-tolyl thiostannane	$(CH_3C_6H_4)_2SnS$	333.02			121.5–2				v. s. chl., bz., eth. acetate, pyr., eth.; s. HCl
94	Di-*p*-tolyltin	$(CH_3C_6H_4)_2Sn$	300.96	or.-yel. amor. powd.		111.5	d. <245			s. bz.
95	Di-*o*-tolyltin dichloride	$(CH_3C_6H_4)_2SnCl_2$	371.86			49–50				
96	Di-*p*-tolyltin dichloride	$(CH_3C_6H_4)_2SnCl_2$	371.86			49–50				
97	Ditriphenylstannyl-methane	$[(C_6H_5)_3Sn]_2CH_2$	714.05	wh. cr. sld.		104.5				v. s. bz. eth., chl.; s. h. pet. eth.
98	Divinylbutyltin chloride	$(C_4H_9)(CH_2{:}CH)_2SnCl$	265.35	liq., 1.4970	1.370		82.84[3]			
99	Divinyltin dichloride	$(CH_2{:}CH)_2SnCl_2$	243.69	col. liq., 1.541	1.762		54–56[3]	i.	i.	s. bz., MeOH
100	Di-*p*-xylyltin	$[(CH_3)_2C_6H_3]_2Sn$	329.01			157	d. 240			
101	Dodecyltetrathioortho-stannate(*n*)	$Sn(SC_{12}H_{25})_4$	924.28			35.5				
102	Ethylchlorostannic acid	$H_2SnC_2H_5Cl_5$	327.03	col. deliq. pr.		d.		d.		
103	Ethyldiisoamyltin bromide	$(C_2H_5)(C_5H_{11})_2SnBr$	369.95		1.3650		154–5[16]			

No.	Name	Formula	Mol. wt.	Crystalline form, color and index of refraction	Sp. gr. or density	Melting point, °C	Boiling point, °C	Solubility in grams per 100 ml of		
								Cold water	Hot water	Alcohol, acids, etc.
	Tin									
104	Ethyldiisobutyltin bromide	$(C_2H_5)(C_4H_9)_2SnBr$	341.90		$1.4089^{19.5}$		130.6^{13}			
105	Ethylmethylpropyltin-iodide	$(CH_3)(C_2H_5)(C_3H_7)SnI$	332.78	1.5548^{17}	1.8182^{20}_{20}		$108–11^{11}$; $226–30^{720}$ sl. d.			
106	Ethyl-*n*-propyldi-isoamyltin	$(C_2H_5)(C_3H_7)Sn(C_5H_{11})_2$	333.13		$1.0654^{21.9}$		141^{17}			
107	Ethylpropyltin dichloride	$(C_2H_5)(C_3H_7)SnCl_2$	261.76	need. f. lt. pet.		57–8		s.		s. eth., al.
108	Ethyl stannic acid	C_2H_5SnOOH	180.77	wh. amor. gel. or powd.		d. below red	heat	i.	i.	s. dil. min. a., KOH; i. al., eth., chl., xylene
109	Ethyltetrathioortho-stannate	$Sn(SC_2H_5)_4$	363.19					$105^{.001}$		
110	Ethyltin tribromide	$C_2H_5SnBr_3$	387.48	col. feath. cr.		310		s.		s. al.
111	Ethyltin triiodide	$C_2H_5SnI_3$	528.47				$181–4^{19}$			
112	Ethyltri-*n*-butyltin	$C_2H_5(C_4H_9)_3Sn$	319.10	1.4732	1.0783		129^{10}			
113	Ethyltri-*n*-propyltin	$C_2H_5(C_3H_7)_3Sn$	277.02				101^{10}			
114	Hexabutylditin	$[(C_4H_9)_3Sn]_2$	580.08	col. liq.	1.1480	198^{10}		i.	i.	s. org. solv.
115	Hexadecyltetrathio-orthostannate(*n*)	$Sn(SC_{16}H_{33})_4$	1148.71			53–54				
116	Hexaethyl disitannane	$[Sn(C_2H_5)_3]_2$	411.75				160^{23}			
117	Hexaethylditin	$[(C_2H_5)_3Sn]_2$	411.76	liq.	1.412^{0}		d. 270			
118	Hexaphenylditin	$[(C_6H_5)_3Sn]_2$	700.02	wh. cr.		232.5	d. <280	i.	i.	.029 eth.; 18.08 chl.; 7.82 bz.
119	Hexa-*p*-tolylditin	$[(C_6H_4CH_3)_3Sn]_2$	784.18	flat tabl. f. bz.		143.5	d. 335			sl. s. bz., eth.; v. sl. s. abs. al.
120	Hexa-*p*-xylylditin	$[(CH_3)_2C_6H_3)_3Sn]_2$	868.36	flat rhomb. tabl. f. bz.-al.		192.5	d. 368			$21^{30.4}$ bz.
121	Isopropylstannic acid	C_3H_7SnOOH	194.79	wh. amor.		d.		i.		s. dil. min. a., KOH; i. org. solv.
122	Isopropyltetrathio-orthostannate	$Sn(SC_3H_7)_4$	419.30				$92^{.001}$			
123	Isopropyltin tribromide	$C_3H_7SnBr_3$	401.51	pa. yel. deliq. pr.		112				s. glac. ac. a.; sl. s. h. bz., chl.; i. dry eth.
124	Isopropyltin trichloride	$C_3H_7SnCl_3$	268.14				75^{16}			
125	Methylstannic acid	$(CH_3)SnOOH$	166.73	wh. amor. powd.		infus.		i.	i.	s. a., alk.; i. org. solv.
126	Methyltetrathio-orthostannate	$Sn(SCH_3)_4$	307.09			31	$81^{.001}$			
127	Methyltin tribromide	CH_3SnBr_3	373.45	wh. need.		53–5	$210–11^{746}$	s.		s. eth., al., bz., lgr., hyd. by alk.
128	Methyltin trichloride	CH_3SnCl_3	240.08	col. cr.		43	171	s.		hyd. by alk.; s. org. solv.
129	Methyltin triiodide	CH_3SnI_3	514.44	lt. yel. need.		86.5		s.	s.	s. eth., al., bz., chl., meth. al.
130	Methyltribromo-dipyridinetin	$CH_3SnBr_3(C_5H_5N)_2$	531.66			203				
131	Methyltri-*n*-butyltin	$CH_3(C_4H_9)_3Sn$	305.07	1.4735	1.0898^{20}_{4}		121^{10}			
132	Methyltri-*n*-propyltin	$CH_3(C_3H_7)_3Sn$	262.99				93^{10}			
133	Phenylbenzyltin dichloride	$(C_6H_5)(C_6H_5CH_2)SnCl_2$	357.84	col. need. f. dil. HCl		83–4	80–100			
134	Phenyltetrathio-orthostannate	$Sn(SC_6H_5)_4$	555.37			67				
135	Phenyltin tribomide	$C_6H_5SnBr_3$	435.52				$182–3^{29}$			
136	Phenyltin trichloride	$C_6H_5SnCl_3$	302.16	col. liq.			$142–3^{25}$	i.	i.	s. bz., MeOH

No.	Name	Formula	Mol. wt.	Crystalline form, color and index of refraction	Sp. gr. or density	Melting point, °C	Boiling point, °C	Solubility in grams per 100 ml of: Cold water	Hot water	Alcohol, acids, etc.
	Tin									
137	Phenyltribenzyltin	$(C_6H_5)Sn(C_6H_5CH_2)_3$	469.20	liq.			290^{5}			s. all ord. org. solv. except al.
138	Propyltetrathioortho-stannate(*n*)	$Sn(SC_3H_7)_4$	419.30				$123^{.001}$			
139	Propyltin triiodide	$C_3H_7SnI_3$	542.49				d. 200^{16}			
140	Propyltri-*n*-amyltin(*n*)	$C_3H_7(C_5H_{11})_3Sn$	375.21	1.4732	1.0368		163^{10}			
141	Stannic bisacetyl-acetone dibromide	$(C_5H_7O_2)_2SnBr_2$	476.73	col. six-sided cr.			187			s. bz., chl., acet.; sl. s. eth., CCl_4
142	Stannic bisacetyl-acetone dichloride	$(C_5H_7O_2)_2SnCl_2$	387.82	col. six-sided cr.			202–3	s.		s. bz., acet.
143	Stannic bisbenzoyl-acetone dibromide	$(C_{10}H_{16}O_2)_2SnBr_2$	614.98	pa. yel. powd.			213–4			sl. s. org. solv.
144	Stannic bisdibenzoyl-methane dibromide	$(C_{15}H_{10}O_2)_2SnBr_2$	723.00	sulfur-yel. cr.			276–8	i.		sl. s. org. solv.
145	Stannic bis-3-ethyl-acetylacetone dibromide	$(C_7H_{11}O_2)_2SnBr_2$	532.84	col. six-sided pr.		164–6				s. c. chl., bz.; sl. s. lt. pet.
146	Tetra-*dl*-amyltin	$Sn(C_5H_{11})_4$	403.26	1.4730	1.0222		174^{10}			
147	Tetra-*n*-amyltin	$Sn(C_5H_{11})_4$	403.26	col. stable liq., 1.4720	1.0206		181^{10}			
148	Tetraaquastannic bisacetylacetone stannibromide	$(C_5H_7O_2)_2Sn(OH_2)_4SnBr_6$	987.12	col. tab. pr.		105–7				s. bz.
149	Tetrabenzyltin	$Sn(C_6H_5CH_2)_4$	483.23	col. pr. f. lt. pet.		42–3		i.	i.	s. common org. solv.; sl. s. lt. pet.
150	Tetra-*n*-butyltin	$Sn(C_4H_9)_4$	347.16	col. stable liq., 1.4730	1.0572	<−70	145^{10}			
151	Tetracyclohexyltin	$Sn(C_6H_{11})_4$	451.31	wh. micr. grains		263–4		i.	i.	6.25^{30} bz.; 0.86^{30} al.; s. chl. CS_2
152	Tetraethyltin	$Sn(C_2H_5)_4$	234.94	col. liq. $n_D^{19.7}$ 1.4724	1.187^{23}	−112	181	i.	i.	s. org. solv.
153	Tetra-*n*-heptyltin	$Sn(C_7H_{15})_4$	515.48	1.4698	0.9748		239^{10}			
154	Tetra-*n*-hexyltin	$Sn(C_6H_{13})_4$	459.37	1.4706	0.9959		209^{10}			
155	Tetraisoamyltin	$Sn(C_5H_{11})_4$	403.26	liq.	$1.035^{19.6}$		188^{24}			
156	Tetraisobutyltin	$Sn(C_4H_9)_4$	347.16	col. liq.	1.054^{23}	−13	267; $143^{16.5}$	i.	i.	s. org. solv.
157	Tetralauryltin	$(C_{12}H_{25})_4Sn$	796.02	1.4736^{20}	0.895	15–16		i.	i.	
158	Tetramethyltin	$Sn(CH_3)_4$	178.83	col. liq. 1.4386	1.314^{0}	−54.8	78	i.	i.	s. org. solv.
159	Tetra-*n*-octyltin	$Sn(C_8H_{17})_4$	571.59	1.4691	0.9605		268^{10}			
160	Tetraphenyltin	$Sn(C_6H_5)_4$	427.12	tetr. col. f. xylene	1.490^{0}	226	>420	i.	i.	s. h. bz., pyr., CCl_4, chl., ac. a.; sl. s. al.
161	Tetrapropyltin	$Sn(C_3H_7)_4$	291.05	col. liq.	$1.1065^{20.2}$		222–5	i.	i.	s. org. solv.
162	Tetra-*o*-tolyltin	$Sn(C_6H_4CH_3)_4$	483.23	wh. cr. powd.		158–9 (215)		i.	i.	s. bz., eth.; i. al.
163	Tetra-*m*-tolyltin	$Sn(C_6H_4CH_3)_4$	483.23	col. need.		128.5		i.	i.	s. bz., h. eth., h. al.
164	Tetra-*p*-tolyltin	$Sn(C_6H_4CH_3)_4$	483.23	col. need.		230–3		i.	i.	s. bz., chl., CS_2, pyr.; sl. s. al., eth.
165	Tetravinyltin	$(CH{:}CH_2)_4Sn$	226.87	col. liq., n_D^{25} 1.4993	1.267_4^{25}		$55–57^{17}$			
166	Tetra-*m*-xylyltin	$[(CH_3)_2C_6H_3]_4Sn$	539.33	rhomb. need. f. bz.-al.		219.5	d. 360			$.314^{30}$ al.; 5.28^{30} eth.; 35.1^{30} bz.; 43.2^{30} chl.
167	Tetra-*p*-xylyltin	$[(CH_3)_2C_6H_3]_4Sn$	539.33	wh. quad. pr.		272–3	d. 360	i.	i.	0.015^{30} al.; 1.73^{30} bz.; 2.80^{30} chl. 0.29^{30} eth.; $.017^{30}$ meth. al.

No.	Name	Formula	Mol. wt.	Crystalline form, color and index of refraction	Sp. gr. or density	Melting point, °C	Boiling point, °C	Solubility in grams per 100 ml of Cold water	Hot water	Alcohol, acids, etc.
	Tin									
168	Tolylstannonic acid(*o*)	$CH_3C_6H_4SnO_2H$	242.83	amor. powd.		d. 295				
169	Tolylstannonic acid(*m*)	$CH_3C_6H_4SnO_2H$	242.83				d. 295	i.	i.	s. c. meth. al., al., eth., chl., eth. acet., pyr., a. and bases; i. pet. eth.
170	Tolyltetrathioortho-stannate(*p*)	$Sn(SCH_3C_6H_4)_4$	611.48			100				
171	Tolyltin trichloride(*o*)	$CH_3C_6H_4SnCl_3$	316.18		1.7619		$154–8^{20}$			
172	Tolyltin trichloride(*p*)	$CH_3C_6H_4SnCl_3$	316.18		1.7522		$156–7^{23}$	sl. d.		
173	Tolyltrichloro-stannane(*m*)	$CH_3C_6H_4SnCl_3$	316.18	col. liq.	1.7516	<−20	$150–1^{23}$			
174	Triallylbutyltin	$(CH_2{:}CHCH_3)_3(C_4H_9)Sn$	302.05	liq., 1.5162	1.3315		$116–119^{10}$			
175	Tri-*n*-amyltin bromide	$(C_5H_{11})_3SnBr$	412.03	1.4963	1.2678					
176	Tribenzylethyltin	$(C_6H_5CH_2)_3(C_2H_5)Sn$	421.15	col. tabl. f. al.-lt. pet.		31–2				s. eth. bz., chl.; sl. s. al.
177	Tribenzyltin chloride	$(C_6H_5CH_2)_3SnCl$	427.54	wh. need.		142–4	d.	i.	i.	s. ac. a., acet., bz., eth., chl. pyr.; i. al.
178	Tribenzyltin hydroxide	$(C_6H_5CH_2)_3SnOH$	409.10	rhomb., col. tabl.		117–21				s. h. al., CS_2, bz.; sl. s. eth., lgr.; i. KOH
179	Tribenzyltiniodide	$(C_6H_5CH_2)_3SnI$	519.00	need. like pr. f. glac. ac. a.		102–3				
180	Tributyltin acetate	$(C_4H_9)_3Sn(OOCH_3)$	349.08	wh., waxy solid	1.27	80–83		i.	i.	s. bz., MeOH
181	Tri-*n*-butyltin bromide	$(C_4H_9)_3SnBr$	369.95	1.5000	1.3365					
182	Tri-*n*-Butylvinyltin	$(n\text{-}C_4H_9)_3Sn(CH{:}CH_2)$	317.08	n_D^{25} 1.4761	1.085_4^{25}		114^3			
183	Triethyl-*n*-amyltin	$(C_2H_5)_3C_5H_{11}Sn$	277.02				102^{10}			
184	Triethyl(*p*-dimethyl-aminophenyl) stannane	$(C_2H_5)_3(CH_3)_2NC_6H_4Sn$	326.05	1.5610^{22}	1.2425		$172–3^3$			
185	Triethyl-*o*-hydroxy-phenyl stannane	$(C_2H_5)_3OHC_6H_4Sn$	298.98	1.5377^{25}	1.3229^{25}		$197–200^3$			
186	Triethylisoamyltin	$(C_2H_5)_3Sn(C_5H_{11})$	277.02		$1.1203^{20.1}$		$111^{18.5}$			
187	Triethylisobutyltin	$(C_2H_5)_3Sn(C_4H_9)$	262.99		$1.139^{20.3}$		96.5^{17}			
188	Triethylphenyltin	$(C_2H_5)_3Sn(C_6H_5)$	282.98	col. liq.	1.2639		254	i.	i.	s. al., eth., org. solv.
189	Triethyl-*n*-propyltin	$(C_2H_5)_3Sn(C_3H_7)$	248.97		$1.1680^{20.6}$		82^{13}			
190	Triethyltin	$(C_2H_5)_3Sn$	205.88	col. liq.	1.3774	<−75	161^{23}	i.	i.	s. al., org. solv.
191	Triethyltin bromide	$(C_2H_5)_3SnBr$	285.79	col. liq.	1.630	−13.5	223–4	v. sl. s.		s. org. solv.
192	Triethyltin chloride	$(C_2H_5)_3SnCl$	241.33	col. liq. $1.5017^{23.3}$	$1.4288^{23.3}$	15.5	208–10	i.		s. org. solv.
193	Triethyltin ethoxide	$(C_2H_5)_3Sn(OC_2H_5)$	250.94	col. liq.	1.2634		190	d.		s. org. solv.
194	Triethyltin hydroxide	$(C_2H_5)_3SnOH$	222.88	col. cr.		43	271	s.	s.	s. org. solv.
195	Triethyltin iodide	$(C_2H_5)_3SnI$	332.78	col. liq.	1.833	−34.5	225 (231)	v. sl. s.		s. org. solv.
196	Triisoamyltin bromide	$(C_5H_{11})_3SnBr$	412.03		$1.2613^{20.7}$	21	177^{15}			
197	Triisoamyltin chloride	$(C_5H_{11})_3SnCl$	367.57		$1.1290^{34.2}$	−30.2	174^{13}			
198	Triisoamyltin fluoride	$(C_5H_{11})_3SnF$	351.12	need.		288				1.03^{31} al.; $.967^{31}$ bz.; 1.22^{31} meth. al.
199	Triisoamyltin iodide	$(C_5H_{11})_3SnI$	459.02		$1.3777^{26.5}$	−22	182^{13}			
200	Triisobutylethyltin	$(C_4H_9)_3Sn(C_2H_5)$	319.10		1.0779^{21}		125^{16}			
201	Triisobutylisoamyltin	$(C_4H_9)_3Sn(C_5H_{11})$	361.18		$1.0356^{26.8}$		$152.9^{16.5}$			
202	Triisobutyltin bromide	$(C_4H_9)_3SnBr$	369.95		1.3523	−26.5	148^{13}			
203	Triisobutyltin chloride	$(C_4H_9)_3SnCl$	325.49		$1.1290^{34.2}$	+30.2	174^{13}			
204	Triisobutyltin fluoride	$(C_4H_9)_3SnF$	309.04	fine long pr.		244				$.414^{32}$ al.; 0.614^{32} meth. al., $.1^{32}$bz.
205	Triisobutyltin iodide	$(C_4H_9)_3SnI$	416.94	col. liq.	$1.378^{26.5}$	−22	284–6			s. eth., org. solv.
206	Triisopropyltin bromide	$(C_3H_7)_3SnBr$	327.87		$1.4263^{25.2}$	−49	133^{12}			s. org. solv.
207	Triisopropyltin idodide	$(C_3H_7)_3SnI$	374.86		$1.4378^{22.2}$		151^{13}			

No.	Name	Formula	Mol. wt.	Crystalline form, color and index of refraction	Sp. gr. or density	Melting point, °C	Boiling point, °C	Solubility in grams per 100 ml of: Cold water	Hot water	Alcohol, acids, etc.
	Tin									
208	Trimethyldecyltin	$(CH_3)_3(C_{10}H_{21})Sn$	305.07	liq., 1.4602	1.0487		$67^{0.05}$			
209	Trimethyldodecyltin	$(CH_3)_3(C_{12}H_{25})Sn$	333.13	liq., 1.4610	1.0285		$93–98^{0.15}$			
210	Trimethylethyltin	$(CH_3)_3(C_2H_5)Sn$	192.86	col. liq.			108.2	i.	i.	s. org. solv.
211	Trimethyltin	$(CH_3)_3Sn$	163.80	col. liq.	1.570^{25}	23	182	i.	i.	s. org. solv.
212	Trimethyltin bromide	$(CH_3)_3SnBr$	243.70	col. cr. or liq.		27	165	s.	s.	s. org. solv.
213	Trimethyltin chloride	$(CH_3)_3SnCl$	195.25	col. cr.		37	154	s.	s.	s. org. solv.
214	Trimethyltin fluoride	$(CH_3)_3SnF$	182.79	wh. short. thick rect. pr.		360 seal. tube	d. <375			2.45^{31} meth. al.; 1.08^{31} al., 0.05^{31} bz.
215	Trimethyltin hydride	$(CH_3)_3SnH$	164.80	col. oily liq.			60	v. sl. s.		s. org. solv.
216	Trimethyltin hydroxide	$(CH_3)_3SnOH$	180.80	col. pr.		118 d.	subl. >80	s.	s.	s. a. al.,, bz., chl., CCl_4, alk.
217	Trimethyltin iodide	$(CH_3)_3SnI$	290.70	col. liq.	2.1432	3.4	170	v. sl. s.		s. bz., al., eth., acet.
218	Trimethyltin oxide	$[(CH_3)_3Sn]_2O$	343.59	wh. amor. powd.		d.		i.	i.	s. a., alk.; i. org. solv.
219	Trimethyltin sulfide	$[(CH_3)_3Sn]_2S$	359.65	lt. yel. oil	1.649^{25}	6	233.5	i.		s. org. solv., HNO_3
220	Triphenylallyltin	$(C_6H_5)_3(CH_2{:}CHCH_2)Sn$	391.08	wh. powd.		73–74		i.	i.	s. org. solv.
221	Triphenylbenzyltin	$(C_6H_5)_3Sn(C_6H_5CH_2)$	441.14	col. pl. f. al.		90	250^{3}			s. org. solv. except al.
222	Triphenylbutyltin	$(C_6H_5)_3(C_4H_9)Sn$	407.13	solid		59 60	222^{3}			sl. s. MeOH
223	Triphenylethyltin	$(C_6H_5)_3SnC_2H_5$	379.07	wh. pr. f. al.	1.2953^{62}	56				
224	Triphenylmethyltin	$(C_6H_5)_3SnCH_3$	365.05	col. tetr. f. eth.	$1.3113^{63.85}$	60–1				s. bz., chl., eth.
225	Triphenyl-α-naphthyltin	$(C_6H_5)_3Sn(C_{10}H_7)$	477.18	col. pr.		125				s. bz., chl., eth.
226	Triphenyltin	$(C_6H_5)_3Sn$	350.01	wh. powd.		232.5	d. 280	i.	i.	$.079^{30}$ al.; 7.82^{30} bz.; 0.92^{30} eth.; 18.1^{30} chl.
227	Triphenyltin bromide	$(C_6H_5)_3SnBr$	429.92	col. cr.		120.5	$249^{13.5}$	i.	i.	s. al., eth. org. solv.
228	Triphenyltin chloride	$(C_6H_5)_3SnCl$	385.46	col. cr.		106	$240^{13.5}$	i.	i.	s. org. solv.
229	Triphenyltin fluoride	$(C_6H_5)_3SnF$	369.01	fine pr.		357		sl. s.		sl. s. c. al., eth.
230	Triphenyltin hydroxide	$(C_6H_5)_3SnOH$	367.02			118				
231	Triphenyltin iodide	$(C_6H_5)_3SnI$	476.91	4-sided monocl. wh.		121	$253^{13.5}$	i.	i.	s. org. solv.
232	Triphenyl-*p*-tolyltin	$(C_6H_5)_3Sn(C_7H_7)$	441.14	need. f. eth.		124				s. bz., chl., eth.
233	Triphenyl-*p*-xylytin	$(C_6H_5)_3Sn[C_6H_3(CH_3)_2]_2$	560.33	col. lng.hex. sheets f. al.		100.5				s. bz., chl. eth.
234	Tri-*n*-propyl-*n*-butyltin	$(C_3H_7)_3SnC_4H_9$	305.07				121^{10}			
235	Tri-*n*-propylethyltin	$(C_3H_7)_3Sn(C_2H_5)$	277.02		$1.1225^{21.8}$		$117.5^{23.3}$			
236	Tri-*n*-propylisobutyltin	$(C_3H_7)_3Sn(C_4H_9)$	305.07		$1.0841^{24.1}$		128^{8}			
237	Tri-*n*-propyltin chloride	$(C_3H_7)_3SnCl$	283.41	col. liq.	1.2678^{28}	−23.5	123^{13}			s. org. solv.
238	Tri-*n*-propyltin fluoride	$(C_3H_7)_3SnF$	266.96	flat pr.		275				4.26^{31} meth. al.; 2.73^{31} al.; 0.118^{31} bz.
239	Tri-*n*-propyltin iodide	$(C_3H_7)_3SnI$	374.86	col. liq.	1.692^{13}	−53	260–2; 141^{13}			s. org. solv.
240	Tri-*o*-tolyltin bromide	$(C_6H_4CH_3)_3SnBr$	472.00	rhomb. tab f. al.		99.5				s. bz., eth.; sl. s. al.
241	Tri-*p*-tolyltin bromide	$(C_6H_4CH_3)_3SnBr$	472.00	rhbdr. f. al.		98.5				s. bz., eth.; sl. s. al.
242	Tri-*o*-tolyltin chloride	$(C_6H_4CH_3)_3SnCl$	427.54	short, thick pr. f. al.		99.5				s. bz., eth.; sl. s. al.
243	Tri-*m*-tolyltin chloride	$(C_6H_4CH_3)_3SnCl$	427.54			108				
244	Tri-*p*-tolytin chloride	$(C_6H_4CH_3)_3SnCl$	427.54	rhomb. pl. f. al.		97.5				sl. s. al. bz., eth.
245	Tri-*p*-tolyltin fluoride	$(C_6H_4CH_3)_3SnF$	411.09	hairlike felted need		305				s. al.
246	Tri-*p*-tolyltin hydroxide	$(C_6H_4CH_3)_3SnOH$	409.10			108–9				
247	Tri-*o*-tolytiniodide	$(C_6H_4CH_3)_3SnI$	519.00	rhomb. cr. fr. al.-eth		119.5				s. bz., eth.; sl. s. al.

No.	Name	Formula	Mol. wt.	Crystalline form, color and index of refraction	Sp. gr. or density	Melting point, °C	Boiling point, °C	Solubility in grams per 100 ml of: Cold water	Hot water	Alcohol, acids, etc.
	Tin									
248	Tri-*p*-tolyltiniodide	$(C_6H_4CH_3)_3SnI$	519.00	rhomb. pl. fr. eth.-al.		120.5				s. bz.,-eth.; sl. s. al.
249	Tritriphenylstannylmethane	$[(C_6H_5)_3Sn]_3CH$	1063.05	wh. cr. sld.		128		i.	i.	v. s. bz., eth. chl.; s. h. pet. eth.; sl. s. c. pet. eth., al.
250	Trivinyldecyltin	$(CH_2{:}CH)_3(C_{10}H_{21})Sn$	341.11	1.4820	1.0672		$90–94^{0.06}$			
251	Trivinylhexyltin	$(CH_2{:}CH)_3(C_6H_{13})Sn$	285.00	1.4851	1.1266		$57–91^{0.03}$			
252	Trivinyloctylitin	$(CH_2{:}CH)_3(C_8H_{17})Sn$	313.05	1.4819	1.0865	$90–93^{02}$				
253	Trivinyltin chloride	$(CH_2{:}CH)_3SnCl$	235.28	1.5235	1.5139		$59–60^{5}$			
254	Tri-*p*-xylyltin bromide	$[(CH_3)_2C_6H_3]_3SnBr$	514.08	lng. hex. cr. f. al.		151				s. bz., chl., eth.; i. c. al.
255	Tri-*p*-xylyltin chloride	$[(CH_3)_2C_6H_3]SnCl$	469.63	6-cornered col. f. al.		141.5				s. bz., chl., eth.; i. c. al.
256	Tri-*m*-xylyltin fluoride	$[(CH_3)_2C_6H_3]_3SnF$	453.17	fine felted need.		205				s. bz., eth., al.
257	Tri-*p*-xylyltin fluoride	$[(CH_3)_2C_6H_3]_3SnF$	453.17	fine lng. need.		247				sl. s. bz., h. eth., al.
258	Tri-*p*-xylyltin iodide	$[(CH_3)_2C_6H_3]_3SnI$	561.08	hex. tabl. f. al.		159.5				s. bz., chl., eth.; i. c. al.
259	Vinyltin trichloride	$(CH_2{:}CH)SnCl_3$	252.10	1.5361	1.9981		$63–65^{10}$			
	Titanium									
1	*bis*-(cyclopentadienyl)-titanium diiodide	$(C_5H_5)_2TiI_2$	431.90	purp.		319 d.				sl. s. tol.; s. org. solv.
2	*bis*-(cyclopentadienyl)-titanium dichloride	$(C_5H_5)_2TiCl_2$	249.00	red-or.		289–291				sl. s. tol., chl., al., eth., bz., CS_2 CCl_4; i. pet. eth.
3	*bis*-(cyclopentadienyl)-titanium dibromide	$(C_5H_5)_2TiBr_2$	337.91	dk. red cr.	1.920	240–3 (314 ± 2)	subl.			s. tol., org. solv.
4	Di(*p*-dimethylaminophenol)[*bis*(cyclopentadienyl)] titanium	$[(CH_3)_2NC_6H_3OH]_2Ti(C_5H_5)_2$	450.44	maroon		unst.				
5	Diphenyl[*bis*(cyclopentadienyl)] titanium	$(C_6H_5)_2Ti(C_5H_5)_2$	332.30	or.-yel.		146–8 d.				s. CH_2Cl_2; i. pet. eth.
6	(Di-*m*-tolyl)[*bis*(cyclopentadienyl)] titanium	$(CH_3C_6H_4)_2Ti(C_5H_5)_2$	360.36	or.-yel.		135–40 d.				
7	(Di-*p*-tolyl)[*bis*(cyclopentadienyl)] titanium	$(CH_3C_6H_4)_2Ti(C_5H_5)_2$	360.36	or.-yel.						i. pet. eth.
8	Methylaminotitanium trichloride	$(CH_3N)_2TiCl_3$	213.34	bluish-grn. cr.	1.33^{25}	subl. 5–20	d. 75	s.		dil. HCl, dil. H_2SO_4
9	Monochlorotriethoxytitanium	$TiCl(OC_2H_5)_3$	218.54	pa. yel. liq.			170^{16}			
10	Phenyltitanium triisopropylate	$(C_6H_5)Ti[(OCH(CH_3)_2)]_3$	302.27	wh. cr.		100–120 d.				s. pet. eth.
11	Tetraethoxytitanium	$Ti(OC_2H_5)_4$	228.15	col. oily liq. supercooled, n_D^{25} 1.5082	1.1066^{25}		$133–135^{5}$			
12	Tetramethoxytitanium	$Ti(OCH_3)_4$	172.04	solid		210	243^{52}	i.	i.	i. bz.
13	Titanium isopropoxide	$Ti(OC_3H_7)_4$	284.25	liq.	0.9550	20	58^{1}			
14	Titaniummonochlorotri-2-ethoxyethoxide	$Ti(OCH_2CH_2OCH_2CH_3)Cl$	350.70	col. liq., n_D^{25} 1.516	1.203^{20}		182^{1}			
15	Titanium tetra-*n*-hexoide	$Ti[O{:}(CH_2)_5{:}CH_3]_4$	452.58	liq., n_D^{25} 1.483.	0.950^{25}		336			
16	Titanium tetra-isobutoxide	$Ti[OCH_2CH(CH_3)_2]_4$	340.36	cr.; liq. n_D^{25} 1.475	liq., 0.960	30	269^{760}			

PHYSICAL CONSTANTS OF ORGANOMETALLIC COMPOUNDS (Continued)

No.	Name	Formula	Mol. wt.	Crystalline form, color and index of refraction	Sp. gr. or density	Melting point, °C	Boiling point, °C	Solubility in grams per 100 ml of: Cold water	Hot water	Alcohol, acids, etc.
	Titanium									
17	Titanium tetra-*n*-pentoxide	$Ti[O:(CH_2)_4:CH_3]_4$	396.47	liq. n_D^{25} 1.485	0.974^{25}	...	314	...	...	...
	Tungsten									
1	*bis*-(Cyclopentadienyl)-hexacarbonmonoxide-bitungsten *bis*-(Cyclopentadienylbi-tungsten hexacarbonyl)	$(C_5H_5)_2W_2(CO)_6$	665.95	purp. red. cr.	...	240–2 d.	...	...	...	s. chl., CCl_4, CS_2; i. ligr.
2	Mesitylenetungsten tricarbonyl	$(CH_3)_3C_6H_3W(CO)_3$	388.08	yel.	...	160–165 d.	subl.	...	...	...
	Uranium									
1	Uranium(IV)-dibenzoylmethane	$U(C_6H_5COCH{:}COC_6H_5)_4$	1131.04	vlt. blk. cr.	...	192–193	subl. vac.	...	...	v. s., bz.
2	Uranium(IV)-diethylamide	$U[N(C_2H_5)_2]_4$	526.55	grn. liq.	...	35.5–36.5	...	d.	...	s. bz., s. eth.
3	Uranium(V)ethoxide	$U(OC_2H_5)_5$	463.33	dk. brn. liq.	1.711^{25}	...	$160^{0.05}$	...	...	s. al., eth. bz.; d. acet.
4	Uranium isipropoxide	$U(OC_4H_9)_5$	603.61	brn. cr.	...	100–104	$192^{0.009}$	...	...	...
5	Uranium(V)methoxide	$U(OCH_3)_5$	393.20	red, cr. solid	...	210	subl. $190–210^{0.01}$	d.	...	s. eth., bz.
6	Uranium(V)2,2,2-tri-fluororethoxide	$U(OCH_2CF_3)_5$	733.19	grn.-brn. solid	...	$130^{0.008}$	...	...	...	s. eth., s. bz.
	Vanadium									
1	*bis*-Cyclopentadienyl-vanadium dichloride	$(C_5H_5)_2VCl_2$	252.04	pa. gr. cr.	...	1.60 g/ml d >250	...	...	...	s. chl., al.; sl. s. eth., CS_2, CCl_4, bz.; i. pet. eth.
2	*bis*-Cyclopentadienyl-vanadium dibromide	$(C_5H_5)_2VBr_2$	340.95	dk. gn.	...	unst.	...	...	...	s. chl., CCl_4, lgr.
	Zinc									
1	Di-*n*-butylzinc	$Zn(CH_2CH_2CH_2CH_3)_2$	179.60	liq.	...	...	$81–2^{9}$	d.	...	...
2	Diethylzinc	$Zn(C_2H_5)_2$	123.49	col. liq. ign. in air or Cl	1.182^{18}	...	118	d.	...	...
3	Dimethylzinc	$Zn(CH_3)_2$	95.44	col. liq.	$1.386^{10.5}$	−42.2	46	d.	...	d. al., a.; s. eth. xylene
4	Diphenylzinc	$Zn(C_6H_5)_2$	219.58	wh. cr.	...	107 (in H_2)	280–285 (in H_2)	d.	...	v. s. bz., eth., s. $CHCl_3$
5	Di-*n*-propylzinc	$Zn(CH_2CH_2CH_3)_2$	151.55	liq. n_4^{48} 1.4845	...	...	146; $39–40^{8}$	d.	...	...
6	Di-*o*-tolylzinc	$Zn(C_6H_4CH_3)_2$	247.64	wh. cr.	...	207–10	...	...	...	s. xylene; v. sl. s. pet. eth.
	Zirconium									
1	*bis*-Cyclopentadienyl-zirconium dibromide	$(C_5H_5)_2ZrBr_2$	381.23	col. cr.	...	260 d.	...	s.	...	...

SUBLIMATION DATA FOR ORGANIC COMPOUNDS

Compiled by Mansel Davies

The tables quote the parameters from what appear to be the best data in the literature expressed in the form

$$\log_{10} p(\text{mm}) = A - B/T$$

and the temperature range for which they apply. The corresponding heats and entropies (taking the standard state of the vapor to be

$$\Delta H \text{ (sublimation)} = 2.303 \text{ R.B. cal/mol.}$$
$$\Delta S \text{ (sublimation)} = 2.303R(A - 2.881) \text{ cal/mole °K.}$$

Compound	Temp. range °C	A	B	Ref.
Acenaphthene	18 to 37	11.758	4290.5	1
Acetamide	25 to 77	11.8468	4050.1	2
Acetic acid	−35 to +10	8.502	2177.4	3
" , *m*-cresyl ester	2 to 44	9.759	3170	14
Acetophenone, 1-chloro-	5 to 50	13.779	4740	14
" , 1-chloro-*o*-nitro	23 to 54	14.24	5413	14
" , 1-chloro-*m*-nitro	26 to 70	14.080	5700	14
" , *p*-methoxy	3 to 27	11.367	4056	1
Acetone, benzoyl	5 to 26	12.317	4375	1
Adipic acid	86 to 133	15.463	6757	4
Anthracene	65 to 80	12.638	5320	5
"	105 to 125	12.002	5102	6
" , 9.10 diphenyl	208 to 229	16.058	8213	22
Anthraquinone	224 to 286	12.305	5747	3
Arachidic acid	63 to 73	25.453	10,424	23
Arsine, diphenylcyano	23 to 53	10.724	4420	14
Azobenzene (*cis*)	30 to 60	9.652	3914	7
" (*trans*)	30 to 60	9.721	3911	7
Behenic acid	71 to 79	23.604	10,100	23
Benzanthrone	—	13.416	6030	8
Benzene	−30 to 5	9.846	2309	3
"	−58 to −30	9.556	2241	3
" , *p*-chloroiodo	30 to 50	9.819	3200	15
" , *p*-dichloro	10 to 50	11.985	3570	17
" , *α*-hexachloro	51 to 71	11.950	4850	14
" , *β*-hexachloro	95 to 117	11.790	5375	14
" , *γ*-hexchloro	60 to 92	15.515	6022	14
" , *λ*-hexachloro	55 to 75	12.635	5100	14
" , -hexamethyl		11.070	4215	37
" , 1,2,3-trichloro	16 to 30	10.662	3440	6
" , 1,2,4-trichloro	6 to 25	10.445	3254	6
" , 1,3,5-trichloro	9 to 28	9.176	2956	6
Benzil	45 to 67	12.708	5140	1
Benzoic acid	70 to 114	12.870	4776	9
" , *p*-hydroxy	125 to 160	13.623	6063	9
" , *o*-methoxy	80 to 95	11.871	4746	9
Benzophenone (stable)	16 to 42	17.46	4966	10
" , (meta stable)	11 to 25	17.19	4818	10
Benzoquinone	—	10.00	3280	18
" , 2.6-dichloro	1 to 42	9.85	3670	18
" , trichloro	28 to 54	12.03	4630	18
" , tetrachloro	60 to 83	12.06	5170	18
" , *p*-xylo	0 to 20	11.53	4030	18
Bibenzyl	13 to 34	12.194	4386	1
Biphenyl	6 to 26	11.168	3959	1
Butyramide	25 to 68	12.739	4546	2
Butyramide	63 to 109	12.594	4513	2
Camphor	0 to 180	8.799	2797	3
Capramide	80 to 97	16.471	6577	2
" *N*-methyl	30 to 52	14.594	5371	11
Capric acid	16 to 28	17.130	6119	23
Caproamide	65 to 95	13.328	4968	2
ϵCaprolactam	21 to 41	11.839	4339	34
Caprylamide	52 to 101	14.920	5783	2
Carbamic acid, *n*-butyl ester	19 to 43	14.582	4919	11
" , ethyl ester	19 to 43	14.090	4646	11
" , *n*-hexyl ester	18 to 41	14.748	5018	11
" , methyl ester	14 to 32	11.966	3883	11
Carbon tetrabromide				
(monoclinic)	22 to 46	9.3867	2841	12
(cubic)	48 to 56	8.5670	2579	12
Carbon tetrachloride	−64 to −48	9.089	2027	13
o-Cresol, 3,5-dinitro	17 to 51	14.140	5400	14
Cyclohexane	−5 to +5	8.594	1953	3
Cyclo-trimethylene-trinitramine	110 to 138	11.870	5850	16
Diphenylamine	25 to 51	12.434	4654	21
Dodecanedioic acid	102 to 123	17.728	8006	4
Eicosanedioic acid	107 to 122	18.185	8644	4
Enanthamide	72 to 93	13.617	5182	2
Ethane, 1.1:pp dichloro diphenyl tri-chloro	66 to 100	14.191	6160	14
" , hexachloro (cubic)	13 to 174	8.731	2677	26
" , hexachloro (triclinic)	13 to 174	9.890	3077	26
Ethylene dibromide	−21 to +8	9.884	2606	13
Ethylene, *trans* di-iodo	−8 to 20	5.86	2130	19
Fluorene	33 to 49	11.325	4324	5
Formic acid	−5 to +8	12.486	3160	36
2 Fuoric acid	44 to 55	14.62	5667	25
Hendecanoic acid	20 to 28	16.432	6037	24
Heneicosanoic acid	68 to 73	22.602	9642	24
Heptadecanoic acid	48 to 58	21.836	8769	24
Hydroquinone tetrachloro	77 to 86	10.08	4650	18
" , *p*-xylo	59 to 88	12.36	5280	18
Lauramide	76 to 95	19.169	7980	2
Lauric acid	22 to 41	19.897	7322	23
Methane	−194 to −184	7.651	5169	3
" , triphenyl	52 to 76	12.661	5228	1
Myristamide	85 to 100	20.940	8746	2
Myristic acid	38 to 52	18.740	7291	23
Naphthalene	6 to 21	11.597	3783	5
1-Naphthol	25 to 39	13.074	4873	1
	39 to 50	11.526	4389	1
2-Naphthol	25 to 39	13.356	5109	1
	39 to 58	11.660	4579	1
Nonadecanoic acid	58 to 64	35.916	13,815	24
n-Octadecane	15 to 25	22.83	7995	27
Oxalic acid, anhyd.	60 to 105	12.223	4727	29
" , anhyd. (*α*)	38 to 52	13.17	5130	28
" , anhyd. (*β*)	38 to 50	12.57	4875	28
Oxamic acid	82 to 90	12.58	5639	30
Oxamide	80 to 96	12.57	5893	30
Palmitamide	91 to 105	22.690	9489	2
Palmitic acid	46 to 60	20.217	8069	23
Pelargonamide	80 to 97	15.249	5997	2
Pentadecanoic acid	38 to 48	23.110	8813	24
Pentaerythritol (tetrag)	106 to 135	16.17	7528	28
" , tetranitrate	97 to 138	17.73	7750	16
Phenanthrene	37 to 50	11.388	4519	5
Phenol	5 to 32	11.421	3540	14
" , *p*-acetyl	47 to 75	12.216	5003	31
" , *p*-benzyl	40 to 62	12.600	5072	31
" , *p*-*tert* butyl	8 to 30	12.332	4402	31
" , 2-*tert* butyl-4-methyl	2 to 20	11.685	4036	31
" , 4-*tert* butyl-2-methyl	3 to 24	11.199	3952	31
Phenol, *p*-formyl	39 to 63	11.795	4762	31
" , *p*-methoxy	5 to 27	13.132	4624	31
" , *o*-phenyl	19 to 40	11.754	4331	31
" , *p*-phenyl	54 to 74	12.056	5068	31
" , 2:4:6-tri*tert* butyl	18 to 40	11.507	4383	31
Phthalic anhydride	30 to 60	12.249	4632	32
Propionamide	45 to 73	12.041	4139	2
Pyrene	72 to 85	11.270	4904	5
Pyrrole 2-carboxylic acid	77 to 81	16.60	6633	25
Rubeanic acid	87 to 105	12.713	5515	30
Salicylic acid	95 to 134	12.859	4969	9
Sebacic acid	102 to 130	18.911	8395	4
Stearamide	94 to 106	24.449	10,230	2
Steraric acid	57 to 67	21.180	8696	23
Suberic acid	106 to 134	16.937	7472	4
Succinic acid	99 to 128	14.068	6132	4
d-Tartaric acid, dimethyl ester	35 to 44	16.610	5903	20
dl- " , "	42 to 85	16.127	5941	20
Thapsic acid	104 to 125	17.165	7885	4
2-Thenoic acid	42 to 50	13.53	5065	25
Thymol	0 to 40	14.201	4766	14
Toluene, 2,4,6-trinitro	50 to 143	15.34	6180	33
Tridecanoic acid	31 to 39	20.939	7764	24
Valeramide	60 to 101	12.846	4666	2

References

1. Aihara, Bull. Chem. Soc., Japan.
2. Davies, Jones and Thomas, Trans. Faraday Soc., **55**, 1100 (1959).
3. This Handbook, 41st Edition, p. 2428 et seq.
4. Davies and Thomas, Trans. Faraday Soc., **56**, 185 (1960).
5. Bradley and Cleasby, J. Chem. Soc., 1690 (1953).
6. Sears and Hopke, J. Amer. Chem. Soc., **71**, 1632 (1949).
7. Bright et al., Research, **3**, 185 (1950).
8. Inokuchi et al., Bull. Ghem. Soc., Japan, **25**, 299 (1952).
9. Davies and Jones, Trans. Farad. Soc., **50**, 1042 (1954).
10. Neumann and Volker, Z. Physik. Chem. **161A**, 33 (1932).
11. Davies and Jones, Trans. Farad. Soc., **55**, 1329 (1959).
12. Bradley and Drury, ibid., **55**, 1844 (1959).
13. Nitta and Seki, J. Chem. Soc., Japan, **69**, 85 (1948).
14. Balson, Trans. Faraday Soc., **43**, 54 (1947).
15. Ewald, ibid., **49**, 1401 (1953).
16. Edwards, ibid., **49**, 152 (1953).
17. Darkis et al., Ind. Eng. Chem., **32**, 946 (1940).
18. Coolidge and Coolidge, J. Amer. Chem. Soc., **49**, 100 (1927).
19. Broadway and Fraser, J. Chem. Soc., 429 (1933).
20. Crowell and Jones, J. Phys. Chem., **58**, 666 (1954).
21. Aihara, J. Chem. Soc., Japan, **74**, 437, 1953.
22. Stevens, J. Chem. Soc., 2973 (1953).
23. Davies, Malpass and Stenhagen, Arkiv for Kemi.
24. Thomas, M.Sc. thesis, Univ. of Wales, 1959.
25. Bradley and Care, J. Chem. Soc., 1688 (1953).
26. Ivin and Dainton, Trans. Farad. Soc., **43**, 32 (1947).
27. Bradley and Shellard, Proc. Roy. Soc., **198A**, 239 (1949).
28. Bradley and Cotson, J. Chem. Soc., 1684 (1953).
29. Noyes and Wobbe, J. Amer. Chem. Soc., **48**, 1882 (1926).
30. Bradley and Cleasby, J. Chem. Soc., 1681 (1953).
31. Aihara, Bull Chem. Soc., Japan.
32. Crooks and Feetham, J. Chem. Soc., 899 (1946).
33. Edwards, Trans. Faraday Soc., **46**, 423 (1950).
34. Aihara, J. Chem. Soc., Japan, **74**, 631 (1953).
35. Seki and Suzuki, Bull. Chem. Soc., Japan, **70**, 387 (1949).
36. Coolidge, J. Amer. Chem. Soc., **53**, 1874 (1930).
37. Nitta et al. J. Chem. Soc., Japan, **70**, 387 (1949).

MISCIBILITY OF ORGANIC SOLVENT PAIRS

Table A

Doctor J. S. Drury

Industrial and Engineering Chemistry Vol. 44, No. 11, Nov. 1952

(Reprinted by permission)

The classifications were made by shaking together 5 ml. of each of the solvents listed in a test tube for 1 minute, then allowing the mixture to settle. If no interfacial meniscus was observed, the solvent pair was considered miscible. If such a meniscus was present, the solvent pair was regarded as immiscible. The classification of immiscible is a qualitative one since solvent pairs may exhibit some degree of partial miscibility while existing as separate phases. Solvent pairs possessing a pronounced degree of partial miscibility are designated by the symbol Is.

Compound number	Compounds	Acetone	Acetyl acetone	2-Amino-2-methyl-1-propanol	Aniline	Benzaldehyde	Benzene	Benzin	Benzyl alcohol	Butyl acetate	Butyl alcohol	*n*-Butyl ether	Capryl alcohol	Carbon tetrachloride	Diacetone alcohol	Diethanolamine	Diethyl cellosolve	Diethyl ether	Dimethylaniline	Ethyl alcohol	Ethyl benzoate	Ethylene glycol	2-Ethylhexanol	Formamide	Furfuryl alcohol	Glycerol	Hydroxyethyl-ethylenediamine	Isoamyl alcohol	Methyl isobutyl ketone	Nitromethane	Dibutoxytetra-ethylene glycol	Pyridine	Triethanolamine	Trimethylene glycol
1	Acetone	..	M	M	..	M	M	M	M	M	M	M	M	M	M	M	M	M	M	M	M	M	M	M	M	I	M	M	M	M	M	M	..	M
2	Acetyl acetone	M	..	R	..	M	M	M	M	M	M	M	M	M	M	R	M	M	M	M	M	M	M	M	M	I	R	M	M	M	M	M	..	M
3	Adiponitrile	M	M	M	M	..	M	..	M	M	M	I	..	I	M	..	M	I	M	M	M	I	I	M	M	I	M	I	M	..	M	M	M	M
4	2-Amino-2-methyl-1-propanol	M	R	..	..	M	M	I	M	M	M	Is	M	M	R	M	M	M	M	M	M	M	M	M	M	M	M	M	M	M	M	M	..	M
5	Benzaldehyde	M	M	M	..	..	M	M	M	M	M	M	M	M	M	I	M	M	M	M	M	Is	M	M	M	Is	R	M	M	M	M	M	..	M
6	Benzene	M	M	M	..	M	..	M	M	M	M	M	M	M	Is	I	M	M	M	M	M	I	M	I	M	I	I	M	M	I	M	M	..	I
7	Benzin	M	M	I	..	M	M	..	I	M	M	M	M	M	I	I	M	M	M	M	M	M	M	I	I	I	Is	M	M	M	M	M	..	I
8	Benzonitrile	M	M	M	M	..	M	..	M	M	M	M	..	M	M	..	M	M	M	M	M	I	M	I	M	I	M	M	M	..	M	M	M	I
9	Benzothiazole	M	M	M	M	..	M	..	M	M	M	M	..	M	M	..	M	M	M	M	M	M	M	I	M	I	M	M	M	..	M	M	M	M
10	Benzyl alcohol	M	M	M	..	M	M	I	..	M	M	M	M	M	M	M	M	M	M	M	M	M	M	M	M	M	M	M	M	M	M	M	..	M
11	Benzyl mercaptan	M	M	I	M	..	M	..	M	M	M	M	..	M	M	..	M	M	M	M	M	I	M	I	M	I	M	M	M	..	M	M	R	I
12	Butyl acetate	M	M	M	..	M	M	M	M	..	M	M	M	M	M	I	M	M	M	M	M	Is	M	I	M	I	I	M	M	M	M	M	..	Is
13	Butyl alcohol	M	M	M	..	M	M	M	M	M	..	M	M	M	M	M	M	M	M	M	M	M	M	M	M	M	M	M	M	M	M	M	..	M
14	*n*-Butyl ether	M	M	Is	..	M	M	M	M	M	M	..	M	M	M	I	M	M	M	M	M	I	M	I	M	I	I	M	M	I	M	M	..	I
15	Capryl alcohol	M	M	M	..	M	M	M	M	M	M	M	..	M	M	M	M	M	M	M	M	M	M	I	M	I	M	M	M	Is	M	M	..	M
16	Carbon tetrachloride	M	M	M	..	M	M	M	M	M	M	M	M	..	Is	I	M	M	M	M	M	I	M	I	M	I	I	M	M	M	M	M	..	I
17	Diacetone alcohol	M	M	R	..	M	Is	I	M	M	M	M	M	Is	..	M	M	M	M	M	M	M	M	M	M	I	R	M	M	M	M	M	..	M
18	Diethanolamine	M	R	M	..	I	I	I	M	I	M	I	M	I	M	..	I	I	Is	M	M	M	M	M	M	M	M	M	I	I	I	M	..	M
19	Diethyl Cellosolve	M	M	M	..	M	M	M	M	M	M	M	M	M	M	I	..	M	M	M	M	M	M	M	M	I	I	M	M	M	M	M	..	M
20	Diethyl ether	M	M	M	..	M	M	M	M	M	M	M	M	M	M	I	M	..	M	M	M	I	M	I	M	I	I	M	M	M	M	M	..	I
21	Dimethylaniline	M	M	M	..	M	M	M	M	M	M	M	M	M	M	Is	M	M	..	M	M	I	M	I	M	I	I	M	M	M	M	M	..	I
22	Di-*N*-propylaniline	M	M	I	M	..	M	..	M	M	M	M	..	M	I	..	M	M	M	M	M	I	M	I	M	I	I	M	M	..	M	M	I	I
23	Ethyl alcohol	M	M	M	..	M	M	M	M	M	M	M	M	M	M	M	M	M	M	..	M	M	M	M	M	M	M	M	M	M	M	M	..	M
24	Ethyl benzoate	M	M	M	..	M	M	M	M	M	M	M	M	M	M	M	M	M	M	M	..	I	M	I	M	I	M	M	M	M	M	M	..	Is
25	Ethyl isothiocyanate	M	M	R	M	..	M	..	M	M	M	M	..	M	M	..	M	M	M	M	M	I	M	I	M	I	R	M	M	..	M	M	M	I
26	Ethyl thiocyanate	M	M	M	M	..	M	..	M	M	M	M	..	M	M	..	M	M	M	M	M	I	M	I	M	I	M	M	M	..	M	M	M	I
27	Ethylene glycol	M	M	M	..	Is	I	M	M	Is	M	I	M	I	M	M	M	I	I	M	I	..	M	M	M	M	M	M	I	I	M	M	..	M
28	2-Ethylhexanol	M	M	M	..	M	M	M	M	M	M	M	M	M	M	M	M	M	M	M	M	M	..	I	M	I	M	M	M	I	M	M	..	M
29	Formamide	M	M	M	..	M	I	I	M	I	M	I	I	I	M	M	M	I	I	M	I	M	I	..	M	M	M	M	Is	M	M	M	..	M
30	Furfuryl alcohol	M	M	M	..	M	M	I	M	M	M	M	M	M	M	M	M	M	M	M	M	M	M	M	..	M	M	M	M	M	M	M	..	M
31	Glycerol	I	I	M	..	I	I	I	M	I	M	I	I	I	I	M	I	I	I	M	I	M	I	M	M	..	M	I	I	I	I	M	..	M
32	Hydroxyethyl-ethylenediamine	M	R	M	..	R	I	Is	M	I	M	I	M	I	R	M	I	I	I	M	M	M	M	M	M	M	..	M	M	M	M	M	..	M
33	Isoamyl alcohol	M	M	M	..	M	M	M	M	M	M	M	M	M	M	M	M	M	M	M	M	M	M	M	M	I	M	..	M	M	M	M	..	M
34	Isoamyl sulfide	M	M	I	M	..	M	..	M	M	M	M	..	M	I	..	M	M	M	M	M	I	M	I	I	I	I	M	M	..	M	M	I	I
35	Isobutyl mercaptan	M	M	M	M	..	M	..	M	M	M	M	..	M	M	..	M	M	M	M	M	I	M	I	M	I	I	M	M	..	M	M	R	R
36	Methyl disulfide	M	M	M	M	..	M	..	M	M	M	M	..	M	M	..	M	M	M	M	M	I	M	I	M	I	I	M	M	..	M	M	I	R
37	Methyl isobutyl ketone	M	M	M	..	M	M	M	M	M	M	M	M	M	M	I	M	M	M	M	M	I	M	Is	M	I	M	M	..	M	M	M	..	I
38	Nitromethane	M	M	M	..	M	I	M	M	M	M	I	Is	M	M	I	M	M	M	M	M	I	I	M	M	I	M	M	M	..	M	M	..	I
39	Dibutoxytetra-ethylene glycol	M	M	M	..	M	M	M	M	M	M	M	M	M	M	I	M	M	M	M	M	M	M	M	M	I	M	M	M	M	..	M	..	M
40	Pyridine	M	M	M	..	M	M	M	M	M	M	M	M	M	M	M	M	M	M	M	M	M	M	M	M	M	M	M	M	M	M	..	..	M
41	Tri-*n*-butylamine	M	M	I	I	..	M	..	M	M	M	M	..	M	I	..	M	M	M	M	M	I	M	I	M	I	I	M	M	..	M	M	I	I
42	Trimethylene glycol	M	M	M	..	M	I	I	M	Is	M	I	M	I	M	M	M	I	I	M	Is	M	M	M	M	M	M	M	I	I	M	M	..	..

MISCIBILITY OF ORGANIC SOLVENT PAIRS (Continued)

Tables B and C

W. M. Jackson and J. S. Drury

The classifications were made at 20°C in the following manner. One-milliliter portions of each solvent comprising a pair were shaken together for approximately a minute. If no interfacial meniscus was observed after the contents of the tube were allowed to settle, the solvent pair was considered to be miscible, M. If a meniscus was observed without apparent change in the volume of either solvent, the pair was regarded as immiscible, I. This classification is a qualitative one, since solvent pairs may exhibit various degrees of partial miscibility while existing as separate phases. If an obvious change occurred in the volume of each solvent, but a meniscus was present, the pair was classified as partially miscible, S. The designation R indicates that the two solvents reacted.

Table B

Compound number	Compounds	Acetone	Isoamyl acetate	*n*-Amyl cyanide	Benzene	Benzyl ether	2-Bromoethyl acetate	Chloroform	Cinnamaldehyde	Di-*n*-amylamine	Di-*n*-butyl carbonate	Diethylacetic acid	Diethylenetriamine	Diethyl formamide	Diisobutyl ketone	Diisopropylamine	Di-*n*-propyl aniline	Ethyl alcohol	Ethyl benzoate	Ethyl ether	Ethyl phenylacetate	Heptadecanol[a]	3-Heptanol	*n*-Heptyl acetate	*n*-Hexyl ether	Methyl isopropyl ketone	4-Methyl-*n*-valeric acid	*o*-Phenetidine	Sulfuric acid (concd.)	Tetradecanol[a]	Tri-*n*-butyl phosphate	Triethylene glycol	Triethylenetetramine	2,6,8-Trimethyl 4-nonanone	Compound number
1	Acetone	..	M	M	M	M	M	M	M	M	M	M	M	M	M	M	M	M	M	M	M	M	M	M	M	M	M	M	R	M	M	M	M	M	1
2	Isoamyl acetate	M	..	M	M	M	M	M	M	M	M	M	M	M	M	M	M	M	M	M	M	M	M	M	M	M	M	M	R	M	M	I	M	M	2
3	*n*-Amyl cyanide	M	M	..	M	M	M	M	M	M	M	M	M	M	M	M	M	M	M	M	M	M	M	M	M	M	M	M	R	M	M	M	M	M	3
4	Benzene	M	M	M	..	M	S	M	M	M	M	M	M	M	M	M	M	M	M	M	M	M	M	M	M	M	M	M	I	M	M	S	M	M	4
5	Benzyl ether	M	M	M	M	..	M	M	M	M	M	M	M	M	M	M	M	M	M	M	M	M	M	M	M	M	M	M	R	M	M	I	M	M	5
6	2-Bromoethyl acetate	M	M	M	S	M	..	M	M	R	M	M	R	R	M	R	M	M	M	M	M	M	M	M	S	M	M	M	R	M	M	M	R	M	6
7	Chloroform	M	M	M	M	M	M	..	M	M	M	M	M	M	M	M	M	M	M	M	M	M	M	M	M	M	M	M	I	M	M	M	M	M	7
8	Cinnamaldehyde	M	M	M	M	M	M	M	..	M	M	M	R	M	M	M	M	M	M	M	M	M	M	M	M	M	M	R	R	M	M	M	R	M	8
9	Di-*n*-amylamine	M	M	M	M	M	R	M	M	..	M	R	I	R	M	M	M	M	M	M	M	M	M	M	M	M	M	M	R	M	M	S	I	M	9
10	Di-*n*-butyl carbonate	M	M	M	M	M	M	M	M	M	..	M	M	M	M	M	M	M	M	M	M	M	M	M	M	M	M	M	R	M	M	I	I	M	10
11	Diethylacetic acid	M	M	M	M	M	M	M	M	R	M	..	R	M	M	R	M	M	M	M	M	M	M	M	M	M	M	M	R	M	M	M	R	M	11
12	Diethylenetriamine	M	M	M	M	M	R	M	R	I	M	R	..	R	M	M	M	M	M	M	M	M	M	R	I	R	R	M	R	M	M	M	M	I	12
13	Diethyl formamide	M	M	M	M	M	R	M	M	R	M	M	R	..	M	R	M	M	M	M	M	M	M	M	I	M	M	M	R	M	M	M	R	M	13
14	Diisobutyl ketone	M	M	M	M	M	M	M	M	M	M	M	M	M	..	M	M	M	M	M	M	M	M	M	M	M	M	M	R	M	M	I	M	M	14
15	Diisopropylamine	M	M	M	M	M	R	M	M	M	M	R	M	R	M	..	M	M	M	M	M	M	M	M	M	M	R	M	R	M	M	M	M	M	15
16	Di-*n*-propylaniline	M	M	M	M	M	M	M	M	M	M	M	M	M	M	M	..	M	M	M	M	M	M	M	M	M	M	M	R	M	M	I	M	M	16
17	Ethyl alcohol	M	M	M	M	M	M	M	M	M	M	M	M	M	M	M	M	..	M	M	M	M	M	M	M	M	M	M	M	M	M	M	M	M	17
18	Ethyl benzoate	M	M	M	M	M	M	M	M	M	M	M	M	M	M	M	M	M	..	M	M	M	M	M	M	M	M	M	M	M	M	M	M	M	18
19	Ethyl ether	M	M	M	M	M	M	M	M	M	M	M	M	M	M	M	M	M	M	..	M	M	M	M	M	M	M	M	R	M	M	I	M	M	19
20	Ethyl phenylacetate	M	M	M	M	M	M	M	M	M	M	M	M	M	M	M	M	M	M	M	..	M	M	M	M	M	M	M	R	M	M	M	M	M	20
21	Heptadecanol[a]	M	M	M	M	M	M	M	M	M	M	M	M	M	M	M	M	M	M	M	M	..	M	M	M	M	M	M	R	M	M	I	M	M	21
22	3-Heptanol	M	M	M	M	M	M	M	M	M	M	M	M	M	M	M	M	M	M	M	M	M	..	M	M	M	M	M	R	M	M	M	M	M	22
23	*n*-Heptyl acetate	M	M	M	M	M	M	M	M	M	M	M	R	M	M	M	M	M	M	M	M	M	M	..	M	M	M	M	R	M	M	I	R	M	23
24	*n*-Hexyl ether	M	M	M	M	M	S	M	M	M	M	M	I	I	M	M	M	M	M	M	M	M	M	M	..	M	M	M	R	M	M	I	I	M	24
25	Methyl isopropyl ketone	M	M	M	M	M	M	M	M	M	M	M	R	M	M	M	M	M	M	M	M	M	M	M	M	..	M	M	R	M	M	M	R	M	25
26	4-Methyl-*n*-valeric acid	M	M	M	M	M	M	M	M	M	M	M	R	M	M	R	M	M	M	M	M	M	M	M	M	M	..	M	R	M	M	M	R	M	26
27	*o*-Phenetidine	M	M	M	M	M	M	M	R	M	M	M	M	M	M	M	M	M	M	M	M	M	M	M	M	M	M	..	R	M	M	M	M	M	27
28	Sulfuric acid (concd.)	R	R	R	I	R	R	I	R	R	R	R	R	R	R	R	R	M	M	R	R	R	R	R	R	R	R	R	..	R	R	R	R	R	28
29	Tetradecanol[a]	M	M	M	M	M	M	M	M	M	M	M	M	M	M	M	M	M	M	M	M	M	M	M	M	M	M	M	R	..	M	I	M	M	29
30	Tri-*n*-butyl phosphate	M	M	M	M	M	M	M	M	M	M	M	M	M	M	M	M	M	M	M	M	M	M	M	M	M	M	M	R	M	..	M	M	M	30
31	Triethylene glycol	M	I	M	S	I	M	M	M	S	I	M	M	M	I	M	I	M	M	I	M	I	M	I	I	M	M	M	R	I	M	..	M	I	31
32	Triethylenetetramine	M	M	M	M	M	R	M	R	I	I	R	M	R	M	M	M	M	M	M	M	M	M	R	I	R	R	M	R	M	M	M	..	I	32
33	2,6,8-Trimethyl 4-nonanone	M	M	M	M	M	M	M	M	M	M	M	I	M	M	M	M	M	M	M	M	M	M	M	M	M	M	M	R	M	M	I	I	..	33

[a] Union Carbide name.

MISCIBILITY OF ORGANIC SOLVENT PAIRS (Continued)

Table C

Compound number	Compounds	Acetone	Isoamyl acetate	n-Amyl cyanide	Anisaldehyde	Benzene	Benzyl ether	Chloroform	o-Cresol	Diisobutyl ketone	Diethylacetic acid	Diethyl formamide	Di-n-propyl aniline	Ethyl alcohol	Ethyl ether	3-Heptanol	n-Heptyl acetate	n-Hexyl ether	α-Methylbenzylamine	α-Methylbenzyldiethanolamine	α-Methylbenzyldimethylamine	α-Methylbenzylethanolamine	2-Methyl-5-ethylpyridine	Methyl isopropyl ketone	4-Methyl-n-valeric acid	o-Phenetidine	2-Phenylethylamine	Isopropanolamine	Pyridine	Salicylaldehyde	Tetradecanol[a]	Tri-n-butyl phosphate	Triethylenetetramine	2,6,8-Trimethyl 4-nonanone
1	1,3-Butylene glycol	M	I	M	I	I	I	M	M	I	M	M	I	M	S	M	I	I	M	M	M	M	M	M	M	M	M	M	M	M	M	M	M	I
2	2,3-Butylene glycol	M	M	M	M	S	I	M	M	M	M	M	I	M	M	M	M	I	M	M	M	M	M	M	M	M	M	M	M	M	M	M	M	I
3	2-Chloroethanol	M	M	M	M	M	M	M	M	M	M	M	M	M	M	M	M	M	R	M	M	M	R	M	M	M	R	R	M	M	M	M	M	M
4	3-Chloro-1,2-propanediol	M	M	M	M	I	M	M	M	M	M	M	I	M	M	M	M	I	R	M	M	M	M	M	M	M	R	R	M	M	S	M	R	S
5	Dibutyl hydrogen phosphite	M	M	M	M	M	M	M	M	M	M	M	M	M	M	M	M	M	M	M	M	M	M	M	M	M	M	M	M	M	M	M	M	M
6	Diethylene glycol dibutyl ether	M	M	M	M	M	M	M	M	M	M	M	M	M	M	M	M	M	R	M	S	M	M	M	M	M	R	R	M	M	M	M	R	M
7	Diethylene glycol diethyl ether	M	M	M	M	M	M	M	M	M	M	M	M	M	M	M	M	M	M	M	M	M	M	M	M	M	M	M	M	M	M	M	M	M
8	Diethylene glycol monobutyl ether	M	M	M	M	M	M	M	M	M	M	M	M	M	M	M	M	M	M	M	M	M	M	M	M	M	M	M	M	M	M	M	M	M
9	Diethylene glycol monoethyl ether	M	M	M	M	M	M	M	M	M	M	M	M	M	M	M	M	I	M	M	M	M	M	M	M	M	M	M	M	M	M	M	M	M
10	Diethylene glycol monomethyl ether	M	M	M	M	M	M	M	M	M	M	M	M	M	M	M	M	I	M	M	M	M	M	M	M	M	M	M	M	M	M	M	M	M
11	Dipropylene glycol	M	M	M	M	M	M	M	M	M	M	M	M	M	M	M	M	I	M	M	M	M	M	M	M	M	M	M	M	M	M	M	M	M
12	Ethylene diacetate	M	M	M	M	M	M	M	M	M	M	M	M	M	M	M	M	I	M	M	M	M	M	M	M	M	M	M	M	M	M	M	M	M
13	Ethylene glycol	M	I	I	I	I	I	S	M	I	M	M	I	M	I	M	I	I	M	M	M	M	M	I	M	M	M	M	M	I	I	S	M	I
14	Ethyl glycol ethylbutyl ether	M	M	M	M	M	M	M	M	M	M	M	M	M	M	M	M	M	M	M	M	M	M	M	M	M	M	M	M	M	M	M	M	M
15	Ethylene glycol monobutyl ether	M	M	M	M	M	M	M	M	M	M	M	M	M	M	M	M	M	M	M	M	M	M	M	M	M	M	M	M	M	M	M	M	M
16	Ethylene glycol monoethyl ether	M	M	M	M	M	M	M	M	M	M	M	M	M	M	M	M	M	M	M	M	M	M	M	M	M	M	M	M	M	M	M	M	M
17	Ethylene glycol monomethyl ether	M	M	M	M	M	M	M	M	M	M	M	M	M	M	M	M	M	M	M	M	M	M	M	M	M	M	M	M	M	M	M	M	M
18	Ethylene glycol monophenyl ether	M	M	M	M	M	M	M	M	M	M	M	M	M	M	M	M	M	M	M	M	M	M	M	M	M	M	M	M	M	M	M	M	M
19	Glycerol	I	I	I	I	I	I	I	M	I	I	M	I	M	I	I	I	I	M	M	I	M	M	I	I	I	M	M	M	I	I	I	M	I
20	1,2-Propanediol	M	M	M	M	I	I	M	M	I	M	M	I	M	S	M	I	I	M	M	M	M	M	M	M	M	M	M	M	M	M	M	M	I
21	1,3-Propanediol	M	I	I	I	I	I	M	M	I	M	M	I	M	I	M	I	I	M	M	M	M	M	M	M	M	M	M	M	I	S	M	M	I
22	Triethylene glycol	M	I	M	M	S	I	M	M	I	M	M	I	M	I	M	I	I	M	M	M	M	M	M	M	M	M	M	M	M	I	M	M	I
23	Triethyl phosphate	M	M	M	M	M	M	M	M	M	M	M	M	M	M	M	M	M	M	M	M	M	M	M	M	M	M	M	M	M	M	M	M	M
24	Trimethylene chlorohydrin	M	M	M	M	M	M	M	M	M	M	M	M	M	M	M	M	M	R	M	M	M	R	M	M	M	R	R	M	M	M	M	R	M

[a] Union Carbide name.

STEROID HORMONES AND OTHER STEROIDAL SYNTHETICS

Compiled by Erwin Di Cyan

The field of steroids has expanded considerably and rapidly in degree and in kind, because synthetic steroids have been synthesized which though resembling the hormones in the body have no natural counterpart, but exert an effect comparable to those of the natural hormones.

In fact, the term steroid hormone thus becomes a misnomer in 1961 when applied to the newer synthetically prepared steroids which do not have a counterpart in the body of man or other animals—as prednisone. (A hormone, by definition, is a material with certain functions and characteristics, *secreted by the ductless glands*. That part of the definition cannot be met by prednisone or by similar steroids as these are not secreted by the ductless, or endocrine glands.)

All the hormones as well as the synthetic analogues have in common the cyclopentanophenanthrene nucleus. Although chemically very similar, a comparatively slight structural change is in many instances productive of substances which have physiologically dissimilar effects, often acting upon different physiologic systems. But in many cases a small change in structure will result merely in an accentuation of certain effects of the unchanged hormone or steroid substance.

The Cyclopentanophenanthrene Nucleus

Classification. Classification becomes a bizarre problem by reason of the (a) overlapping uses to which these substances are put, and (b) the multiple purposes for which the hormones or synthetic substances are used. Indeed, the steroids may be classified by structure; that however would be uninformative to the student as to their use. Classification by origin, as adrenal, would also be unsuitable because, for example, a number of the adrenal corticosteroids are not found in the adrenal cortex at all, but merely resemble the natural hormones found in the adrenal cortex.

For those reasons the hormonal or hormonelike entries in the tables are classified by-and-large, by pharmacologic effects. Even that classification has its disparities as for example, the use of male sex hormones, i.e. the androgens, is neither limited to men, nor to uses which entail their effect upon male sex characteristics. (Attention is invited to the reading list at the end of this article, which lists several excellent texts devoted to, or containing material on steroids.)

Uses. Originally, the use of steroid hormones was largely based upon one or more of the following predicates:

(a) To supplement the progressively declining secretion of a specific hormone due to natural biologic aging of the organism; in the menopause as an example of such declining secretion, a female sex hormone is used for such supplementation;

(b) To make available to the body a specific hormone, the natural secretion of which is inhibited because of a congenital or developmental anomaly; the underdevelopment of male secondary sex characteristics is an example of such an inhibited secretion, in which a male sex hormone is used—and correspondingly, female sex hormones in underdevelopment in females;

(c) To cause a reversal of hormonal balance in the treatment of diseases peculiar to a sex; for example, in the case of cancer of the female breast, a male sex hormone is administered, and in cancer of the prostrate, a female sex hormone is used;

(d) To mimic a natural function, as menstruation, by the administration of estrogens—on withdrawal of which bleeding occurs; or by the alternate use of estrogenic and progestational—both female sex hormones.

Since the finding that cortisone ameliorates the symptoms of rheumatoid arthritis (1949) the adrenal corticosteroid hormones and especially the synthetically prepared steroid analogues which have no natural counterpart in the body, have been successfully employed in the treatment of diseases not related to sex or sex function.

Androgens and Anabolic Agents. The agents listed in the tables under this classification have the effect of male sex hormones (androgens) i.e., to stimulate sexual maturation, in the "male climacteric," etc. But all androgens have in greater or lesser degree the ability to stimulate muscle development, i.e., an anabolic effect. Among the synthetically prepared agents which have no counterpart in the body (Methandrostenolone or Oxymetholone) are those which have a lessened androgenic, but a heightened anabolic effect. These qualities are determined by biologic tests on animals but principally confirmed by clinical use in man. The anabolic effect includes remineralization of bone, which may be partially demineralized (osteoporosis) by age, or by certain drugs, as the adrenal corticosteroids (q.v.).

Anabolic agents are used for muscle and bone nutrition in men as well as women. The reason for the high interest in

STEROID HORMONES AND OTHER STEROIDAL SYNTHETICS (Continued)

synthetic steroidal substances for anabolic use, is based on the need for materials, which within a given effective dose have a greater anabolic-to-androgenic ratio than such androgens as methyl testosterone. Otherwise, the administration of androgens to women produces manifestations of virilism, such as growth of hair on the face, a deepening of the voice, etc. Androgens are also used in the female in the suppression of excessive bleeding and in the treatment of cancer of the breast and cervix. (For other androgen-like agents, see also Progestogens and Progestins.)

Estrogens. Estrogenic agents hasten sexual maturation in the female. Therefore, they are used in underdevelopment in the female. The widest use of estrogens is in the treatment of the menopause, in which they supplement from without, the secretion of natural estrogens by the ovary, which begins to decline at about the 40th year. The menopause is usually a slow process, and the declining secretion gives rise to various symptoms during the time that the secretion declines, until adjustment to the new status takes place. The menopause, a period of physical and psychological stress is made less precipitous by estrogens.

Frequently, a menopause must be quickly induced, as in cancer of the ovary or in uterine hemorrhage. This is done by radiation or by the removal of the uterus. Severe vasomotor symptoms occur when the menopause is thus suddenly induced. Estrogens—among other drugs—are used in the amelioration of these symptoms.

Estrogens (especially diethylstilbestrol which though not a hormone has an estrogenic effect) are also used in the control of cancer of the prostrate in the male. Note the inverse correspondence to the use of male sex hormones in cancer of breast in female.

Progestogens and Progestins (Including 19-Norsteroid Compounds). The agents under that listing include progesterone, a female sex hormone, as well as progestins, i.e., synthetic progesterone-like compounds which have no natural counterpart in the body. Their use includes a variety of conditions: functional uterine bleeding, absence of menstruation (amenorrhea) used at times with estrogens, painful menstruation (dysmenorrhea), infertility, habitual abortion in order to maintain pregnancy, and in fact, to suppress ovulation hence their use as antifertility drugs.

Adrenal Corticosteroids, Including Antiinflammatory, Antiallergic and Antirheumatic Agents. The adrenal cortex secretes a large number of hormones. They usually differ from each other in the accentuation of some phases of their properties. Virtually all of the cortical hormones are catabolic, thus having an effect in this respect, diametrically opposed to the androgens which are anabolic. Nearly all the cortical hormones—differing in degree from each other—cause retention of sodium and water by the body and hasten the excretion of potassium. These effects are utilized in the treatment of adrenal insufficiency or Addison's disease, in which conversely, there is an undue excretion of sodium and a strong retention of potassium. Desoxycorticosterone is used in Addison's disease because it has a particularly strong sodium retaining and potassium excreting effect.

Since the finding in 1949 of the usefulness of cortisone in profoundly reducing the symptoms of rheumatoid arthritis, the adrenal corticosteroids, including hydrocortisone, a natural hormone secreted by the adrenal cortex, and particularly the synthetic analogues not found in the body, as prednisone, have been used in the treatment of a wide variety of inflammatory diseases—especially diseases of collagen tissue. The same antiinflammatory effect is also brought into use in the reduction of inflammations associated with diseases of the skin, allergy, asthma, and in such systemic diseases as disseminated lupus erythematosus, also a collagen disease.

The drawbacks of cortisone, also shared in lesser measure by hydrocortisone, gave the impetus to the synthesis of steroidal substances not native to the body but differing somewhat from cortisone and hydrocortisone, in order to reduce the drawbacks attendant to the use of the latter. The sideeffects—especially those of cortisone—are retention of water and sodium, excretion of potassium, loss of mineral from bone leading to osteoporosis and fractures, hypertention, at times diabetes, personality changes or gastric ulcer. Prednisone and prednisolone among others (see tables) are two such steroidal synthetics which have the effects of cortisone, but fewer or less severe sideeffects. Whereas the synthetic steroidal substances are superior to cortisone with respect to lessened sideeffects, it cannot be said that the sideeffects are absent—they vary in degree from substance to substance.

Diuretic, Antidiuretic and Local Anesthetic Agents. Aldosterone, a natural hormone of the adrenal cortex promotes retention in the body of sodium and water, and facilitates excretion of potassium. Hence its effect is almost diametrically opposed to diuretics—especially the thiazide diuretics. Aldosterone is much more active in this respect than desoxycorticosterone, and is used in the treatment of Addison's disease, a hypofunction of the adrenal glands.

Spironolactone is an antagonist to aldosterone—the latter when elaborated in the body in excessive amounts gives rise to a syndrome called aldosteronism. Spironolactone, a synthetically produced steroid does not have a natural counterpart in the body, is diuretic when mercurial or thiazide diuretics are ineffective; it prevents sodium retention and potassium excretion—effects opposite to aldosterone. Hence spironolactone is used in aldosteronism, against edema, in the treatment of congestive heart failure and in other conditions in which an accumulation of water, and waterretaining salt, is to be corrected.

STEROID HORMONES AND OTHER STEROIDAL SYNTHETICS (Continued)

Doses. The amount of substance which comprises a dose of steroid hormones, or of the steroidal synthetics varies from substance to substance—from 0.1 mg for an estradiol ester, to 50 mg for a 19-norsteroid compound. The dose is conditioned upon the order of activity of the substance, the purpose for which it is administered, as well as the patient's response. However, as additional steroids for hormonal use are synthesized—especially those with adrenocortical activity, their average dose is usually smaller than the previously available steroid. The smaller effective dose of the more recent steroid is cited as an advantage over the previously available steroid.

However, a smaller dose cannot be claimed as an inherent advantage of a new steroid in comparison with an existing one, unless the lower dosage exhibits either greater or more prolonged activity or lesser sideeffects. One cannot meaningfully compare a dose, milligram for milligram, without taking into consideration if a heightened effect of the smaller dose produces fewer sideeffects. For example, it does not make any difference if a given effect and the same accompanying sideeffects are produced by a 50 mg or a 5 mg dose.

Reading List

Clark, F. & Grant, J. K. *The Biosynthesis & Secretion of Adrenocortical Steroids;* Biochemical Society Symposia #18; Cambridge University Press, New York; 1960.

Dorfman, R. I. & Shipley, R. A. *Androgens: Biochemistry, Physiology & Clinical Significance;* John Wiley & Sons, New York; 1956.

Fieser, L. F. & Fieser, M. *Steroids;* Reinhold Publishing Corp., New York; 1959.

Heftmann, E. & Mosettig, E. *Biochemistry of Steroids;* Reinhold Publishing Corp., New York; 1960.

Paschkis, K. E., Rakoff, A. E., & Cantarow, A. *Clinical Endocrinology;* Paul B. Hoeber, Inc., New York; 1958.

Sunderman, F. W. & Sunderman, F. W., Jr. *Lipids & Steroid Hormones in Clinical Medicine;* J. B. Lippincott Co., Philadelphia; 1960.

Stecher, P. G., Editor. *The Merck Index;* Merck & Co., Inc., Rahway, N.J.; 1960.

ADRENAL CORTICOSTEROIDS, INCLUDING ANTIINFLAMMATORY, ANTIALLERGIC AND ANTIRHEUMATIC AGENTS

Names & synonyms:	CHLOROPREDNISONE ACETATE; 6α-chloroprednisone acetate; 6α-chloro-$\Delta^{1,4}$-pregnadien-17β,21-diol-3,11,20-trione 21-acetate.	CORTICOSTERONE; 11,21-dihydroxyprogesterone; Δ^4-pregnene-11β,21-diol-3,20-dione; 11β,21-dihydroxy-4-pregnene-3,20-dione; Kendall compound B; Reichstein substance H.	CORTISONE; 17-hydroxy-11-dehydrocorticosterone; 17α,21-dihydroxy-4-pregnene-3,11,20-trione; Δ^4-pregnene-17α,21-diol-3,11,20-trione; Kendall compound E; Wintersteiner compound F.
Formulae:	CH_2OCOCH_3, C=O, H_3C, OH, O, H_3C, O, Cl	CH_2OH, C=O, HO, H_3C, H_3C, O; $C_{21}H_{30}O_4$	CH_2OH, C=O, H_3C, OH, O, H_3C, O; $C_{21}H_{28}O_5$
Molecular weight	436.6	346.40	360.4
Melting point (°C)	207–213	180–182	220–224
Specific rotation	$(\alpha)\frac{25}{D}$ + 137 – +142 (100 mg. in 10 ml. chloroform)	$(\alpha)\frac{15}{D}$ + 222 (110 mg. in 10 ml. alcohol)	$(\alpha)\frac{25}{D}$ + 209 (120 mg. in 10 ml. alcohol)
Absorption max.		240 mμ	237 mμ

ADRENAL CORTICOSTEROIDS, INCLUDING ANTIINFLAMMATORY, ANTIALLERGIC AND ANTIRHEUMATIC AGENTS (Continued)

Names & synonyms:	DESOXYCORTICOSTERONE; deoxycorticosterone; 11-desoxycorticosterone; 21-hydroxyprogesterone; 4-pregnen-21-ol-3,20-dione; Kendall desoxy compound B; Reichstein substance Q.	DESOXYCORTICOSTERONE ACETATE; DCA; 11-desoxycorticosterone acetate.	DEXAMETHASONE; hexadecadrol; 9α-fluoro-16α-methyl prednisolone; 9α-fluoro-11β,17α-21-trihydroxy-16α-methyl-1,4-pregnadiene-3,20-dione; 16α-methyl-9α-fluoro-1,4-pregnadiene-11β.17α-21-triol-3,20-dione; 16α-methyl-9α-fluoro-Δ¹-hydrocortisone; 1-dehydro-16α-methyl-9α-fluorohydrocortisone.
Formulae:	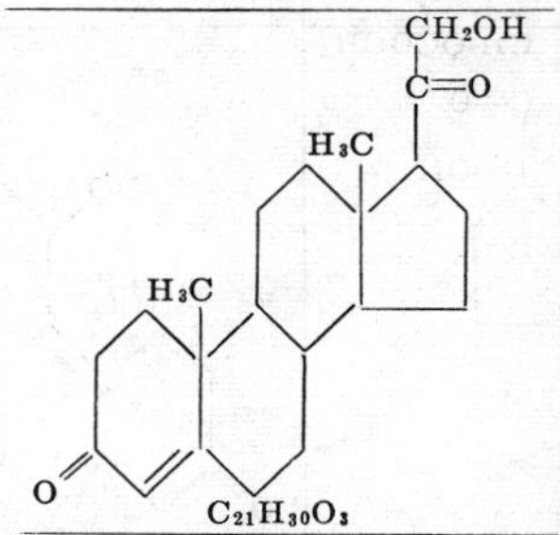 $C_{21}H_{30}O_3$	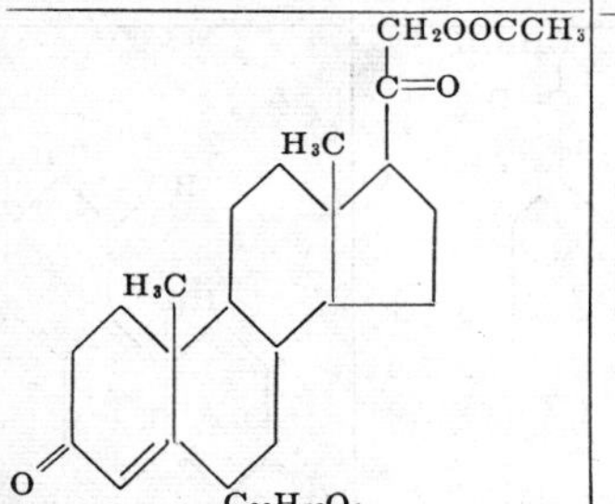 $C_{23}H_{32}O_4$	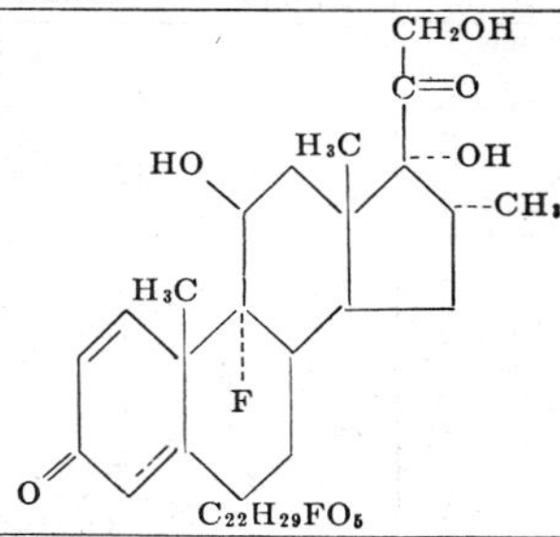 $C_{22}H_{29}FO_5$
Molecular weight	330.2	372.4	392.4
Melting point (°C)	140–142	154–160	262–264
Specific rotation	$(\alpha)\frac{22}{D}$ + 176 – +178 (100 mg. in 10 ml. alcohol)	$(\alpha)\frac{20}{D}$ + 168 – +178 (100 mg. in 10 ml. dioxane)	$(\alpha)\frac{25}{D}$ + 78 (100 mg. in 10 ml. dioxane)
Absorption max.	240 mμ		

Names & synonyms:	DICHLORISONE ACETATE; 9α-11β-dichloro-1,4-pregnadiene-17α,21-diol-3,20-dione-21-acetate.	FLUOROHYDROCORTISONE; fludrocortisone; 9α-fluorohydrocortisone; 9α-fluorocortisol; fluohydrisone; 9α-fluoro-11β,17α,21-trihydroxy-4-pregnene-3,20-dione; 9α-fluoro-17-hydroxycorticosterone.	FLUOROMETHOLONE; 9α-fluoro-11β,17α-dihydroxy-6α-methyl-1,4-pregnadiene-3,20-dione; 21-desoxy-9α-fluoro-6α-methyl-prednisolone.
Formulae:	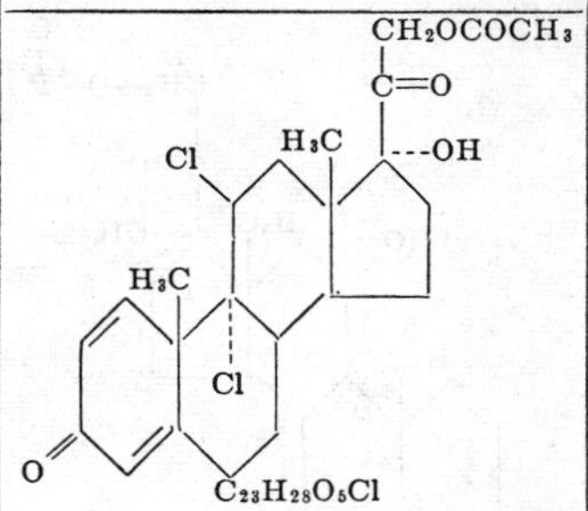 $C_{23}H_{28}O_5Cl$	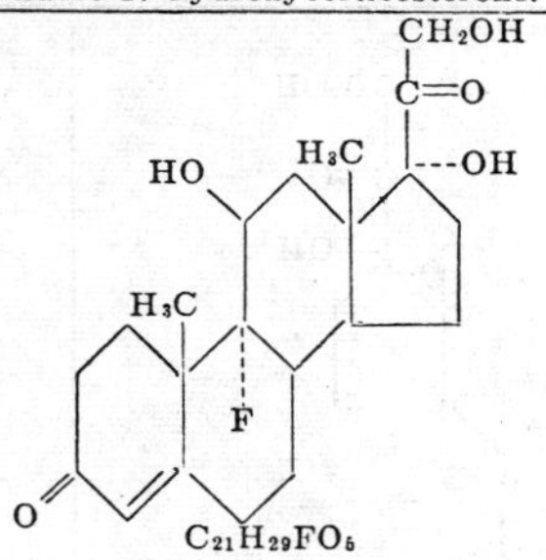 $C_{21}H_{29}FO_5$	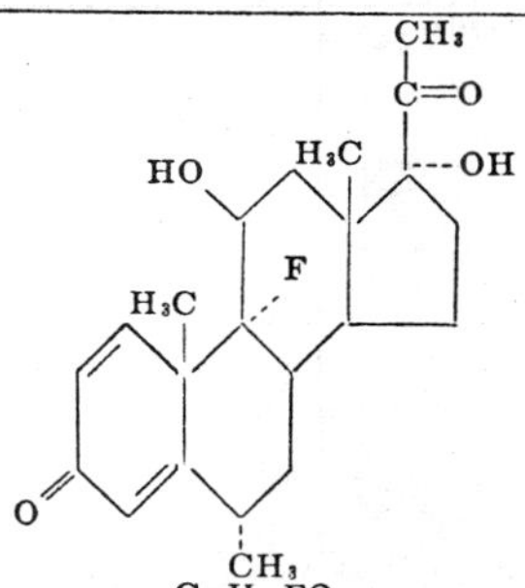 $C_{22}H_{24}FO_4$
Molecular weight	455.3	380.4	376.4
Melting point (°C)	235 (dec.)	260–262 (dec.)	290 (dec.)
Specific rotation	$(\alpha)\frac{25}{D}$ + 160 – 168 (100 mg. in 10 ml. dioxane)	$(\alpha)\frac{23}{D}$ + 139 (55 mg. in 10 ml. alcohol)	$(\alpha)\frac{25}{D}$ + 56 (pyridine)
Absorption max.	237 mμ – 316 – 337 (ϵ_1^1)		239 mu (aM = 15,050) methanol

ADRENAL CORTICOSTEROIDS, INCLUDING ANTIINFLAMMATORY, ANTIALLERGIC AND ANTIRHEUMATIC AGENTS (Continued)

Names & synonyms:	HYDROCORTISONE; cortisol; 17-hydroxycorticosterone; hydrocortisone free alcohol; 11β,17α,21-trihydroxy-4-pregnene-3,20-dione; 4-pregnene-11β,17α,21-triol-3,20-dione; Kendall compound F; Reichstein substance M.	HYDROCORTISONE ACETATE; cortisol acetate; hydrocortisone-21-acetate; 17-hydroxycorticosterone-21-acetate.	METHYLPREDNISOLONE; 6α-methylprednisolone; Δ^1-6α-methylhydrocortisone; 1-dehydro-6α-methylhydrocortisone; 11β,17α,21-trihydroxy-6α-methyl-1,4-pregnadiene-3,20-dione.
Formulae:	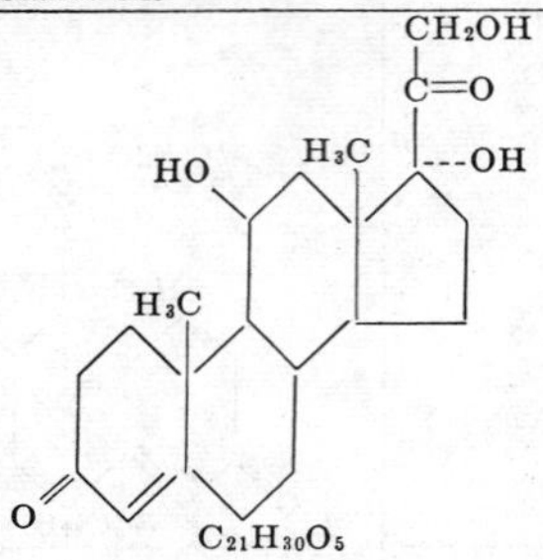 $C_{21}H_{30}O_5$	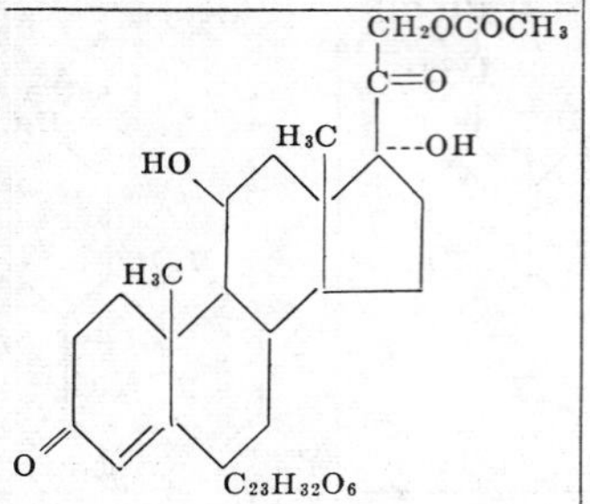 $C_{23}H_{32}O_6$	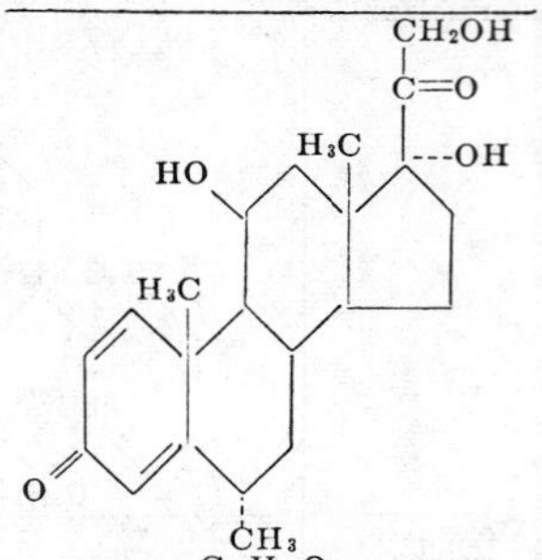 $C_{22}H_{30}O_5$
Molecular weight	362.5	404.5	374.5
Melting point (°C)	215–220 (dec.)	223 (dec.)	230–240 (dec.)
Specific rotation	$(\alpha)\frac{25}{D}$ + 150 – +156 (100 mg. in 10 ml. dioxane)	$(\alpha)\frac{25}{D}$ + 158 – +165 (100 mg. in 10 ml. dioxane)	$(\alpha)\frac{25}{D}$ + 85 (dioxane)
Absorption max.	242 mμ	242 mμ (methanol)	243 mμ

Names & synonyms:	PREDNISOLONE; metacortandralone; Δ^1-dehydrocortisol; delta F; Δ^1-hydrocortisone; Δ^1-dehydrohydrocortisone; 1,4-pregnadiene-3,20-dione-11β,17α,21-triol; 11β,17α,21-trihydroxy-1,4-pregnadiene-3,20-dione.	PREDNISOLONE PHOSPHATE SODIUM; disodium prednisolone 21-phosphate.
Formulae:	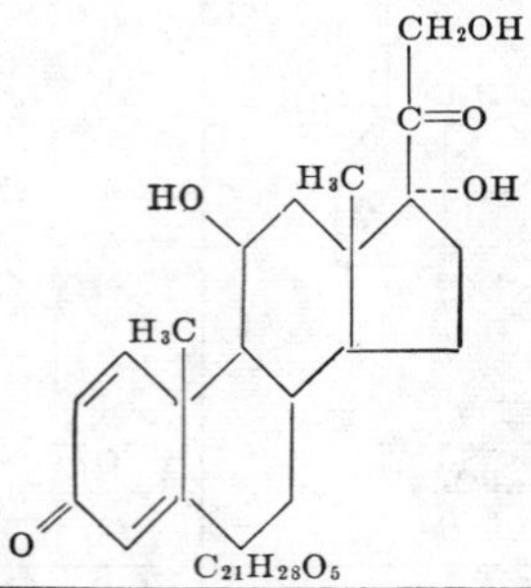 $C_{21}H_{28}O_5$	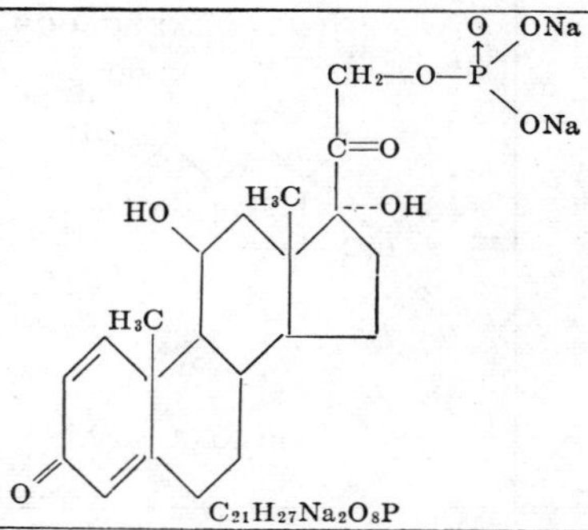 $C_{21}H_{27}Na_2O_8P$
Molecular weight	360.4	484.4
Melting point (°C)	240 (dec.)	
Specific rotation	$(\alpha)\frac{25}{D}$ + 97 – +103 (100 mg. in 10 ml. dioxane)	$(\alpha)\frac{25}{D}$ + 102.5 (100 mg. in 10 ml. H_2O)
Absorption max.	242 mμ (ϵ = 15,000) methanol	243 mμ

ADRENAL CORTICOSTEROIDS, INCLUDING ANTIINFLAMMATORY, ANTIALLERGIC AND ANTIRHEUMATIC AGENTS (Continued)

Names & synonyms:	PREDNISONE; metacortandricin; Δ^1-dehydrocortisone; delta E; Δ^1-cortisone; 1,4-pregnadiene-17α,21-diol-3,11,20-trione; 17α,21-dihydroxy-1,4-pregnadiene-3,11,20-trione.	TRIAMCINOLONE; 9α-fluoro-16α-hydroxyprednisolone; 9α-fluoro-11β,16α,17α,21-tetrahydroxy-1,4-pregnadiene-3,20-dione.
Formulae:	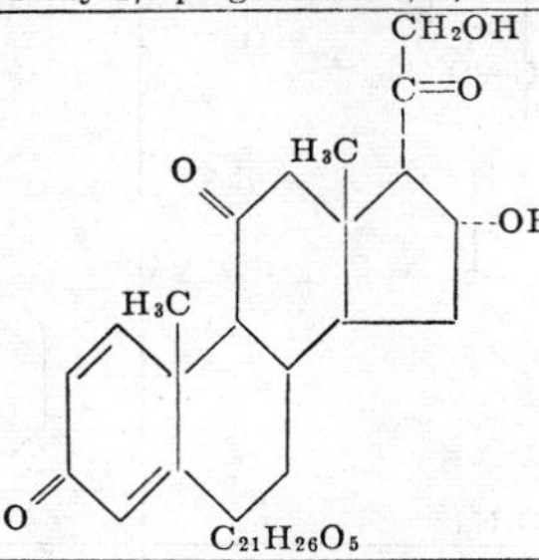 $C_{21}H_{26}O_5$	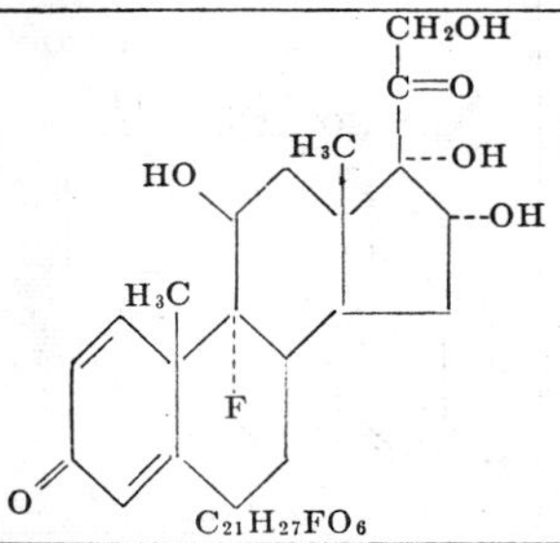 $C_{21}H_{27}FO_6$
Molecular weight	358.4	394.4
Melting point (°C)	230–235 (dec.)	260–262.5 (dec.)
Specific rotation	$(\alpha)\frac{25}{D}$ + 167 – +175 (100 mg. in 10 ml. dioxane)	$(\alpha)\frac{25}{D}$ + 75 (200 mg. in 100 ml. acetone)
Absorption max.	238 mμ (ϵ = 15,500) methanol	238 mμ (ϵ = 15,800)

Names & synonyms:	TRIAMCINOLONE ACETONIDE; 9α-fluoro-11β,21-dihydroxy-16α,17α-isopropylidenedioxy-1,4-pregnadiene-3,20-dione; 9α-fluoro-16α-hydroxyprednisolone 16,17-acetonide.	TRIAMCINOLONE DIACETATE; 16α,21-diacetoxy-9α-fluoro-11β,17α-dihydroxy-1,4-pregnadiene-3,20-dione; 9α-fluoro-16α-hydroxyprednisolone 16,21-diacetate.
Formulae:	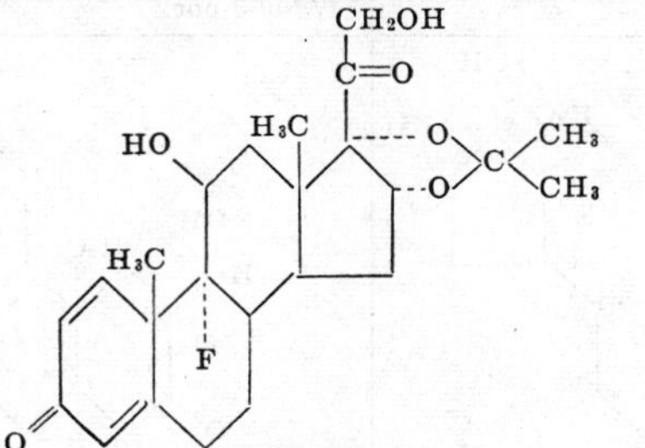 $C_{24}H_{31}FO_6$	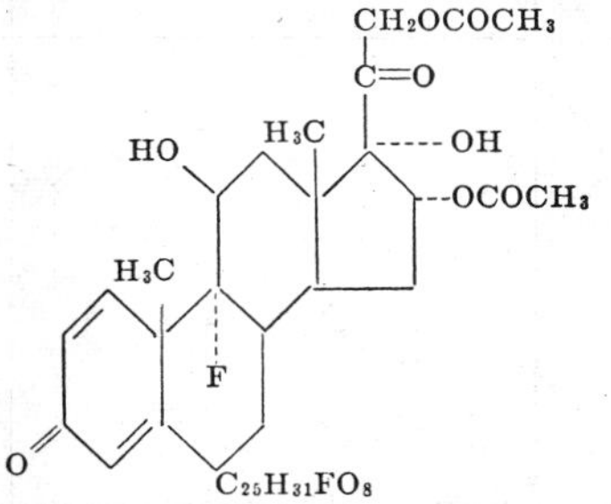 $C_{25}H_{31}FO_8$
Molecular weight	434.4	478.49
Melting point (°C)	274–278 (dec.); 292–294	variable: 158–235
Specific rotation	$(\alpha)\frac{25}{D}$ + 109 – +112 (53.7 mg. in 10 ml. chloroform)	$(\alpha)\frac{25}{D}$ + 22 (78.8 mg. in 10 ml. chloroform)
Absorption max.	238–239 mμ (ϵ = 14,600)	239 mμ (ϵ = 15,200)

DIURETIC, ANTIDIURETIC AND LOCAL ANESTHETIC AGENTS

Names & synonyms:	ALDOSTERONE; electrocortin; 18-oxocorticosterone; 18-formyl-11β,21-dihydroxy-4-pregnene-3,20-dione.	HYDROXYDIONE SODIUM; 21-hydroxypregnane-3,20-dione-21-sodium hemisuccinate.	SPIRONOLACTONE; 3-(3-oxo-7α-acetylthio-17β-hydroxy-4-androsten-17α-yl)-propionic acid γ lactone.
Formulae:	$C_{21}H_{28}O_5$	$C_{25}H_{35}O_6Na$	$C_{24}H_{32}O_4S$
Molecular weight	360.4	454.5	416.5
Melting point (°C)	108–112 (hydrate); 164 (anhydrous)	193–203 (dec.)	135; 202 (dec.)
Specific rotation	$(\alpha)\frac{25}{D}$ + 161 (10 mg. in 10 ml. chloroform)	$(\alpha)\frac{25}{D}$ + 95 (chloroform) for free acid.	$(\alpha)\frac{25}{D}$ − 34 (chloroform)
Absorption max.	240 mμ (log ε = 4.20 monohydr.; ε mol. 15,000 anhydr.)	280 mμ (ε = 93.2)	ϵ^{238} = 20,200

ANDROGENS AND ANABOLIC AGENTS

Names & synonyms:	ANDROSTERONE; cis-androsterone; 3α-hydroxy-17-androstanone; androstane-3α-ol-17-one.	FLUOXYMESTERONE; 9α-fluoro-11β-hydroxy-17α-methyltestosterone; 9α-fluoro-11β,17β-dihydroxy-17α-methyl-4-androsten-3-one.	HYDROXYMETHYLENE METHYL DIHYDROTESTOSTERONE; 2-hydroxymethylene-17α-methyl-dihydrotestosterone; 2-hydroxymethylene-17α-methyl-androstan-17β-ol-3-one.
Formulae:	$C_{19}H_{30}O_2$	$C_{20}H_{29}FO_3$	
Molecular weight	290.4	336.4	332.5
Melting point (°C)	185–185.5	270 (dec.)	174–182
Specific rotation	$(\alpha)\frac{15}{D}$ + 85 − +90 (150 mg. in 10 ml. dioxane)	$(\alpha)\frac{25}{D}$ + 107 − +109 (alcohol)	$(\alpha)\frac{25}{D}$ + 34 − +38 (200 mg. in 10 ml. dioxane)
Absorption max.		240 mμ (ε = 16,700) alcohol	

ANDROGENS AND ANABOLIC AGENTS (Continued)

Names & synonyms:	METHANDROSTENOLONE; 17α-methyl-17β-hydroxy-1,4-androstadien-3-one.	METHYLANDROSTENEDIOL; MAD; methandriol; 17α-methyl-5-androsten-3β,17β-diol.	METHYL TESTOSTERONE; 17-methyl testosterone; 17α-methyl-Δ^4-androsten-17-β-ol-3-one; 17(β)-hydroxy-17(α-methyl-4-androsten-3-one.
Formulae:	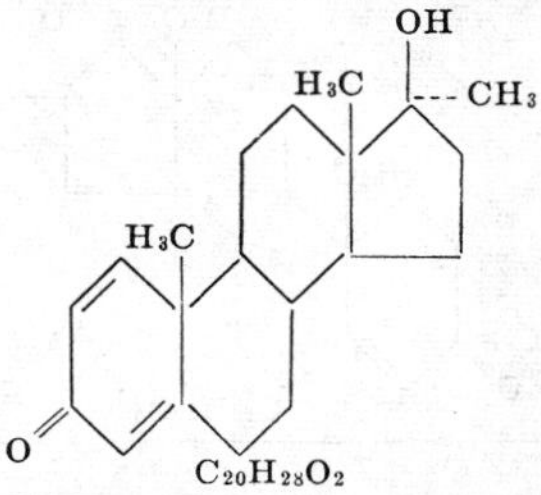 $C_{20}H_{28}O_2$	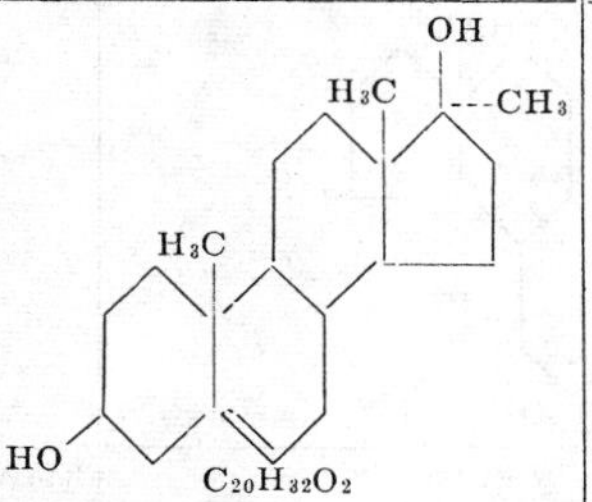 $C_{20}H_{32}O_2$	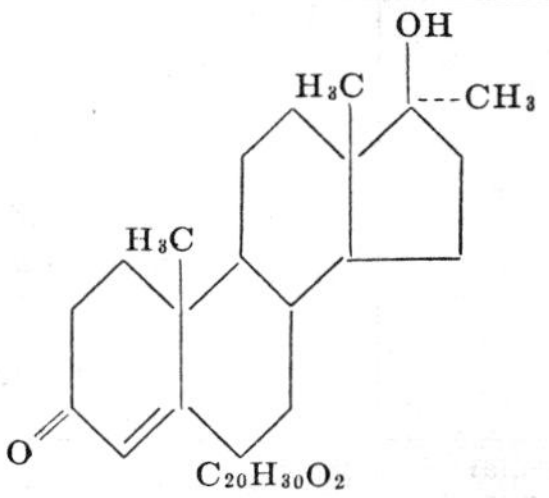 $C_{20}H_{30}O_2$
Molecular weight	300.4	304.4	302.4
Melting point (°C)	166–167	205–207	161–166
Specific rotation	$(\alpha)\frac{20}{D}$ + 9 – +17 (100 mg. in 10 ml. alcohol)	$(\alpha)\frac{20}{D}$ – 73 (100 mg. in 10 ml. alcohol)	$(\alpha)\frac{25}{D}$ + 69 – +75 (100 mg. in 10 ml. dioxane)
Absorption max.			

Names & synonyms:	OXYMETHOLONE; 17β-hydroxy-2-hydroxymethylene-17α-methyl-3-androstanone; 2-hydroxymethylene-17-α-methyl dihydrotestosterone.	PROMETHOLONE; 2α-methyl-dihydro-testosterone propionate; 2α-methyl-5α-androstane-17β-ol-3-one-propionate.
Formulae:	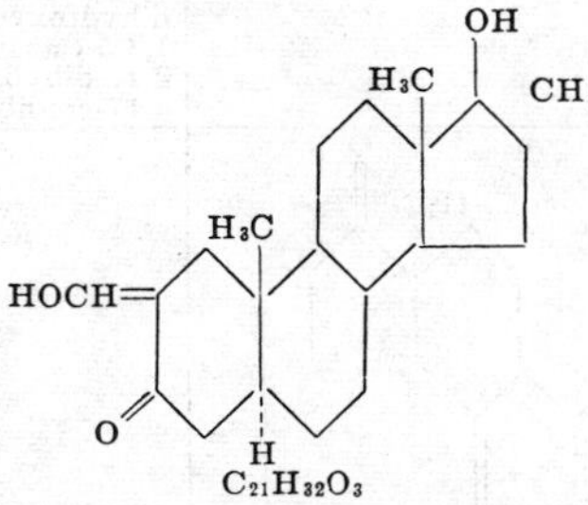 $C_{21}H_{32}O_3$	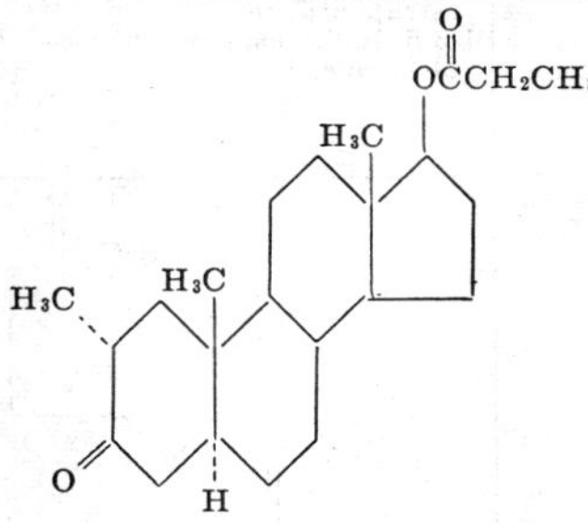
Molecular weight	332.4	360.5
Melting point (°C)	182	124–130
Specific rotation	$(\alpha)\frac{25}{D}$ = +36 (200 mg. in 10 ml. dioxane)	$(\alpha)\frac{25}{D}$ + 22 – +29 (200 mg. in 10 ml. chloroform)
Absorption max.	E_1^1 = 345 at 277 mμ (in abs. ethanol made 0.01 N with conc. HCl).	without significant absorption from 220–300 mμ (methanol)

ANDROGENS AND ANABOLIC AGENTS (Continued)

Names & synonyms:	TESTOSTERONE; trans-testosterone; Δ^4-androsten-17-β-ol-3-one; 17β-hydroxy-4-androsten-3-one.	TESTOSTERONE PROPIONATE; Δ^4-androstene-17-β-propionate-3-one.
Formulae:	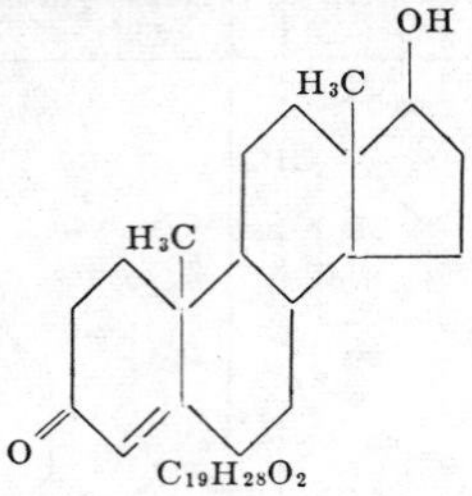 $C_{19}H_{28}O_2$	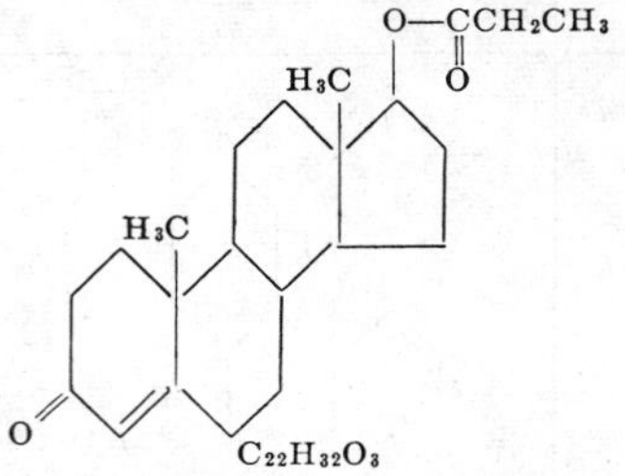 $C_{22}H_{32}O_3$
Molecular weight	288.4	344.4
Melting point (°C)	151–156	118–122
Specific rotation	$(\alpha)\frac{24}{D}$ + 109 (400 mg. in 10 ml. alcohol)	$(\alpha)\frac{25}{D}$ + 83 − +90 (100 mg. in 10 ml. dioxane)
Absorption max.	238 mμ	

ESTROGENS

Names & synonyms:	EQUILENIN; 3-hydroxy-17-keto-$\Delta^{1,3,5-10,6,8}$ estrapentaene; 1,3,5-10,6,8-estrapentaen-3-ol-17-one.	EQUILIN; 3-hydroxy-17-keto-$\Delta^{1,3,5-10,7}$ estratetraene; 1,3,5,7-estratetraen-3-ol-17-one.	ESTRADIOL (formerly called α-estradiol); β-estradiol; dihydrofolliculin; dihydroxyestrin; 1,3,5-estratriene-3,17β-diol; 3,17-dihydroxy-$\Delta^{1,3,5-10}$-estratriene; 3,17-epidihydroxyestratriene.
Formulae:	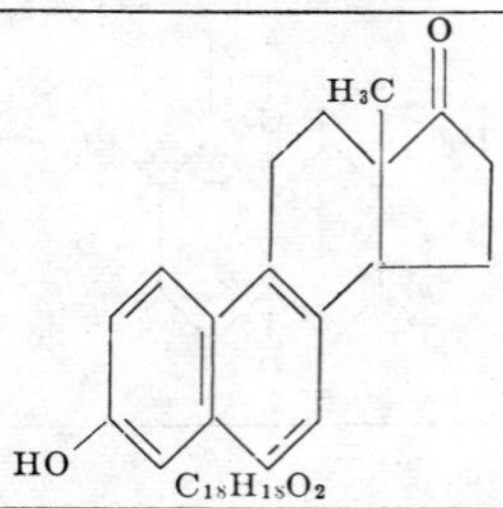 $C_{18}H_{18}O_2$	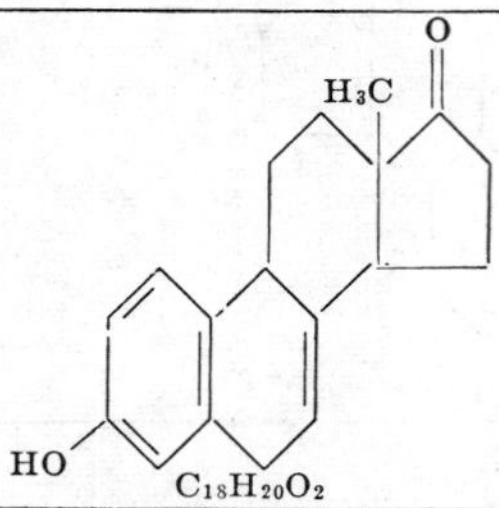 $C_{18}H_{20}O_2$	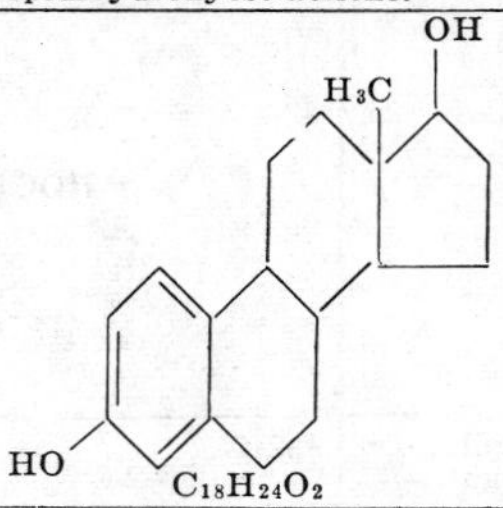 $C_{18}H_{24}O_2$
Molecular weight	266.3	268.3	272.3
Melting point (°C)	258–259	236–240	173–179
Specific rotation	$(\alpha)\frac{25}{D}$ + 89 (dioxane)	$(\alpha)\frac{25}{D}$ + 308 (200 mg. in 10 ml. dioxane); + 325 (200 mg. in 10 ml. alcohol).	$(\alpha)\frac{25}{D}$ + 76 − +83 (100 mg. in 10 ml. dioxane)
Absorption max.	231, 270, 282, 292, 325, 340 mμ	283–285 mμ	225, 280 mμ

ESTROGENS (Continued)

Names & synonyms:	ESTRADIOL BENZOATE; β-estradiol-3-benzoate; estradiol monobenzoate.	ESTRIOL; trihydroxyestrin; $\Delta^{1,3,5-10}$-estratriene-3-16-cis-17-trans-diol; 1,3,5-estratriene-3,16α,17β-triol.	ESTRONE; folliculin; ketohydroxyestrin; 1,3,5-estratrien-3-ol-17-one.
Formulae:	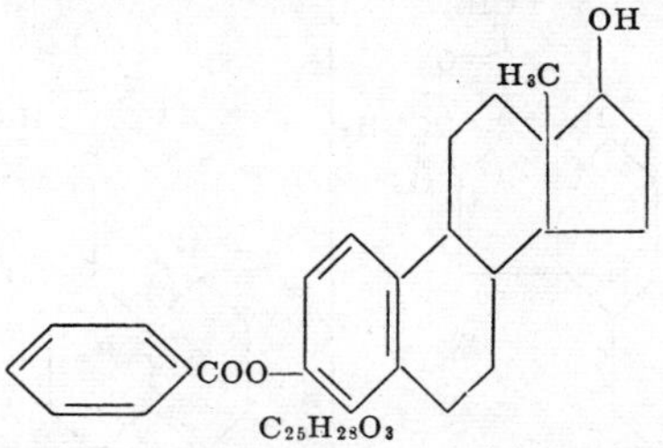 $C_{25}H_{28}O_3$	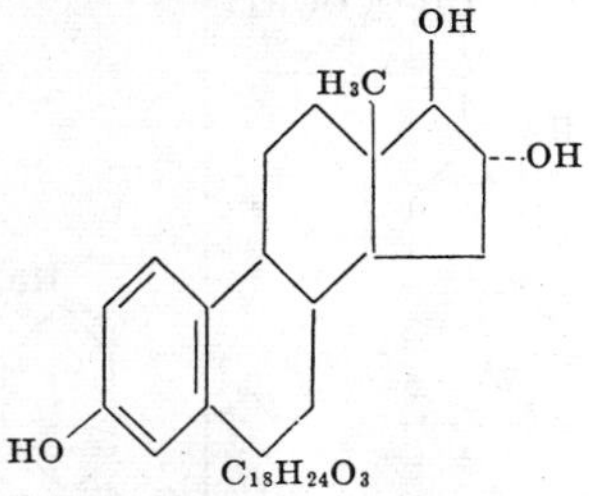 $C_{18}H_{24}O_3$	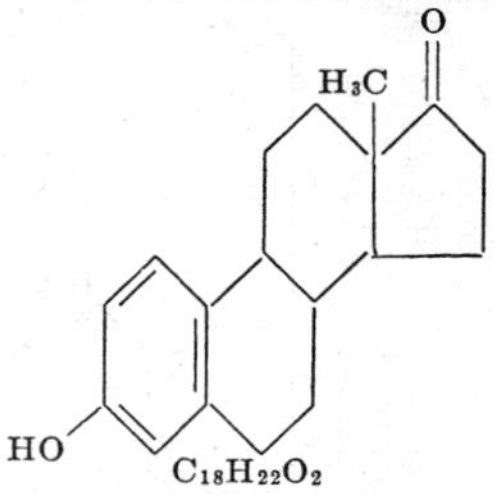 $C_{18}H_{22}O_2$
Molecular weight	376.4	288.3	270.3
Melting point (°C)	191–196	282	258–262
Specific rotation	$(\alpha)\frac{25}{D}$ + 58 – +63 (200 mg. in 10 ml. dioxane)	$(\alpha)\frac{25}{D}$ + 53 – +63 (40 mg. in 1 ml. dioxane)	$(\alpha)\frac{25}{D}$ + 158 – +168 (100 mg. in 10 ml. dioxane)
Absorption max.		280 mμ	283–285 mμ

Names & synonyms:	ESTRONE BENZOATE	ETHYNYL ESTRADIOL; 17-ethinyl estradiol; 17α-ethynyl-1,3,5-estratriene-3,17β-diol.
Formulae:	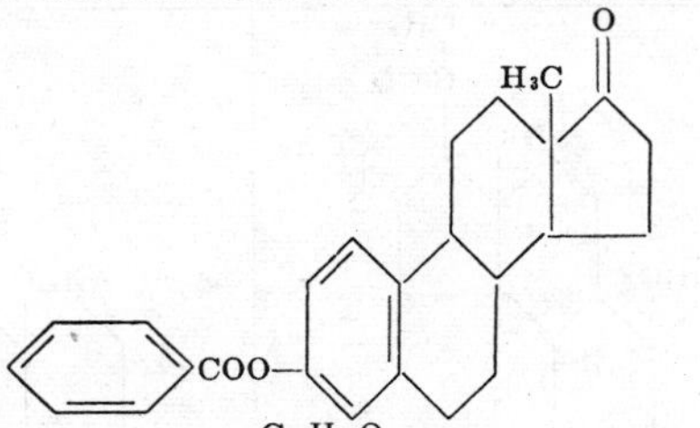 $C_{25}H_{26}O_3$	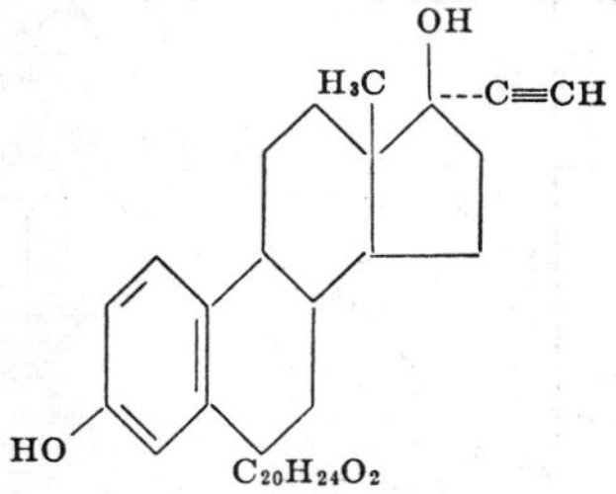 $C_{20}H_{24}O_2$
Molecular weight	374.4	296.4
Melting point (°C)	220	141–146
Specific rotation	$(\alpha)\frac{25}{D}$ + 120 (dioxane)	$(\alpha)\frac{25}{D}$ + 1 – +10 (100 mg. in 10 ml. dioxane)
Absorption max.		248 mμ

PROGESTOGENS AND PROGESTINS (INCLUDING 19-NORSTEROID COMPOUNDS)

Names & synonyms:	ACETOXYPREGNENOLONE; 21-acetoxypregnenolone; prebediolone acetate; Δ^5-pregnene-3β,21-diol-20-one-21-monoacetate; 21-acetoxy-5-pregnene-3-ol-20-one; 3-hydroxy-21-acetoxy-5-pregnen-20-one.	CHLORMADINONE; 6-chloro-Δ^6-dehydro-17α-acetoxyprogesterone; 6-chloro-$\Delta^{4,6}$-pregnadiene-17α-ol-3,20-dioneacetate.	ETHISTERONE; anhydrohydroxyprogesterone; ethinyl testosterone; pregneninolone; 17α ethynyl testosterone; 17α-ethynyl-17β-hydroxy-4-androsten-3-one.
Formulae:	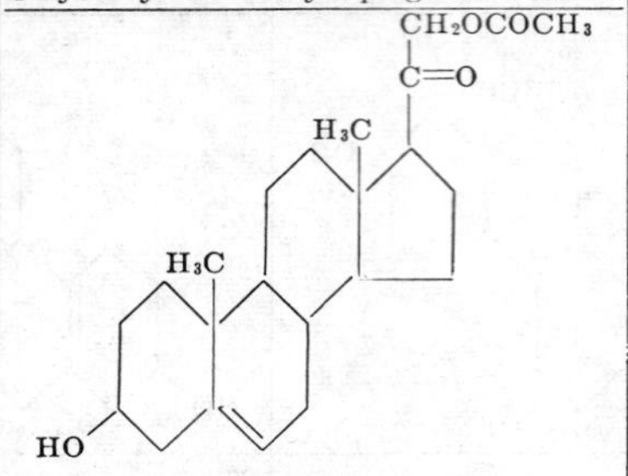 $C_{23}H_{34}O_4$	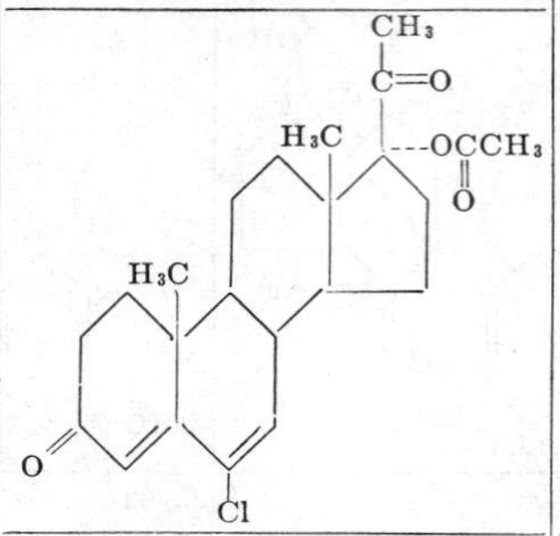	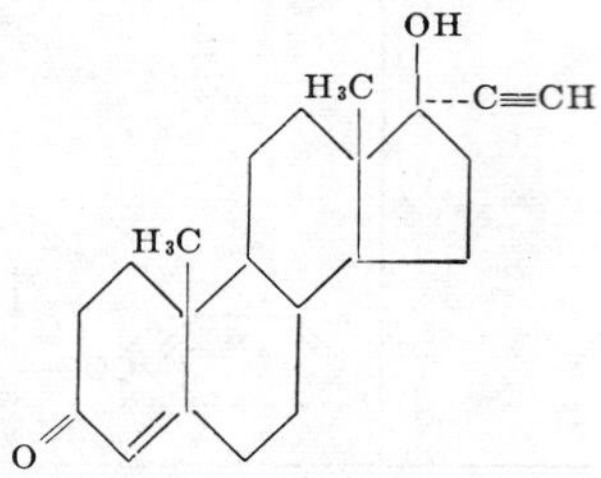 $C_{21}H_{28}O_2$
Molecular weight	374.5	404.9	312.4
Melting point (°C)	184–185	204–212	266–273
Specific rotation	$(\alpha)\frac{20}{D} + 37 - +43$ (dioxane)	$(\alpha)\frac{25}{D}$ 0 to -6 (200 mg. in 10 ml. chloroform)	$(\alpha)\frac{25}{D} - 32°$ (100 mg. in 10 ml. pyridine)
Absorption max.		284 mμ (methanol) Log $\epsilon = 4.34 \pm 0.02$	241 mμ (methanol)

Names & synonyms:	HYDROXYMETHYLPROGESTERONE; medroxyprogesterone; 17α-hydroxy-6α-methylprogesterone; 17α-hydroxy-6α-methyl-4-pregnene-3,20-dione.	HYDROXYMETHYLPROGESTERONE ACETATE; medroxyprogesterone acetate; 17α-hydroxy-6α-methylprogesterone acetate; 17α-hydroxy-6α-methyl-4-pregnene-3,20-dione acetate.	HYDROXYPROGESTERONE; 17α-hydroxyprogesterone; 17α-hydroxy-4-pregnene-3,20 dione; 4-pregnen-17α-ol-3,20-dione.
Formulae:	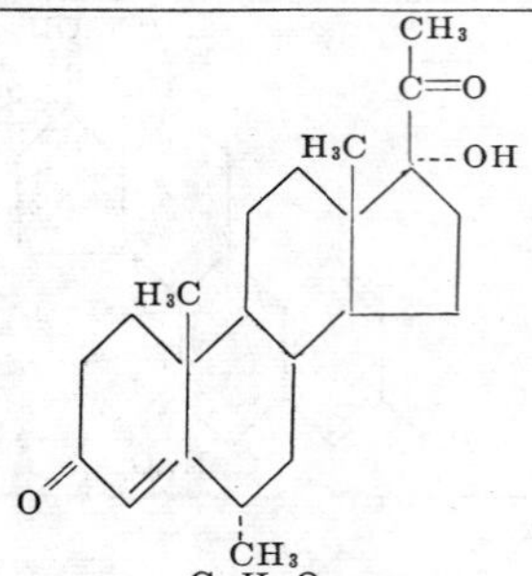 $C_{22}H_{32}O_3$	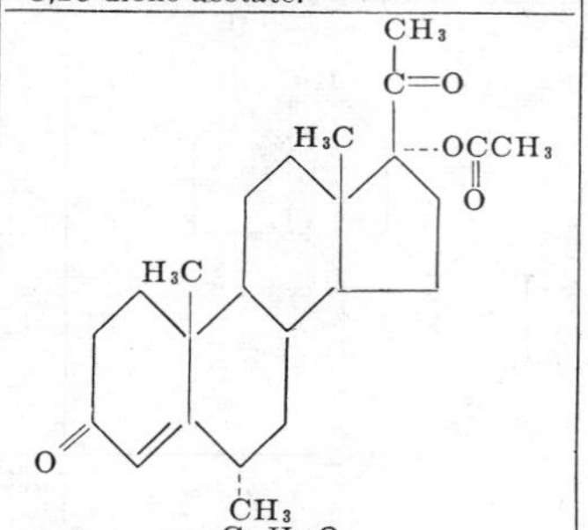 $C_{24}H_{34}O_4$	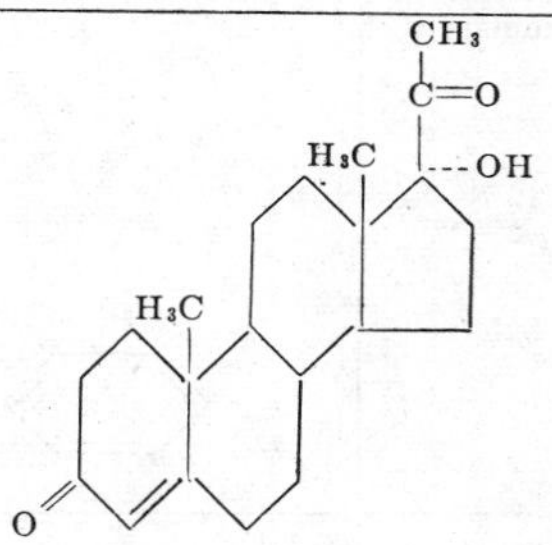 $C_{21}H_{30}O_3$
Molecular weight	344.5	386.5	330.4
Melting point (°C)	220–223.5	202–207	276
Specific rotation	$(\alpha)\frac{25}{D} + 75$	$(\alpha)\frac{25}{D} + 51$ (dioxane)	$(\alpha)\frac{17}{D} + 105$ (104 mg. in 10 ml. chloroform)
Absorption max.	241 mμ (ϵ = 16,150)	241 mμ (α_M = 16,500) ethanol	

PROGESTOGENS AND PROGESTINS (INCLUDING 19-NORSTEROID COMPOUNDS) (Continued)

Names & synonyms:	HYDROXYPROGESTERONE ACETATE; 17α-acetoxyprogesterone; 17α-hydroxyprogesterone acetate; 17α-hydroxy-4-pregnene-3,20 dione acetate.	HYDROXYPROGESTERONE CAPROATE; 17α-hydroxyprogesterone caproate; 17α-hydroxy-4-pregnene-3,20-dione caproate.	NORETHANDROLONE; 17α-ethyl-19-nortestosterone; 17α-ethyl-17-hydroxy-4-norandrosten-3-one; 17α-ethyl-17-hydroxy-19-norandrost-4-en-3-one.
Formulae:	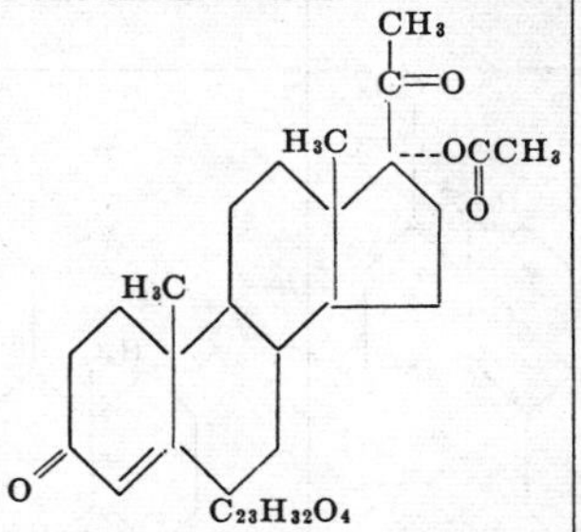 $C_{23}H_{32}O_4$	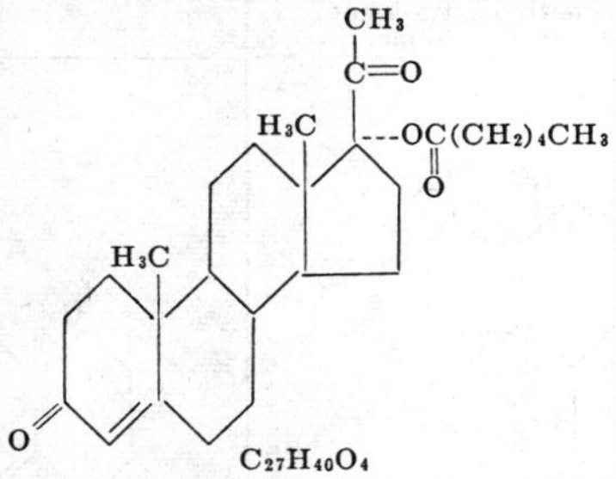 $C_{27}H_{40}O_4$	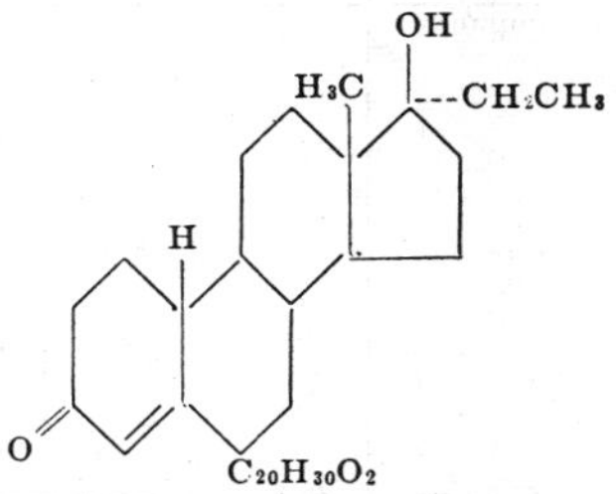 $C_{20}H_{30}O_2$
Molecular weight	372.5	428.6	302.4
Melting point (°C)	249–250	121–123	136–141
Specific rotation	$(\alpha)\frac{25}{D} + 72$ (chloroform)	$(\alpha)\frac{25}{D} + 57$ (chloroform)	$(\alpha)\frac{25}{D} + 21$ (dioxane)
Absorption max.	240 mμ (a_M = 16,875) ethanol		240 mμ (ϵ = 16,500)

Names & synonyms:	NORETHISTERONE; norethindrone; 19-norethisterone; 17α-ethynyl-19-nor-Δ^4-androstan-17β-ol-3-one; 17α-ethynyl-19-nor-testosterone; 17-hydroxy-3-oxo-19-nor-17α-pregn-4-ene-20-yne; 17-hydroxy-19-nor-17α-pregn-4-en-20-yn-3-one.	NORETHYNODREL; 17α-ethynyl-17β-hydroxy-5(10)-estren-3-one.
Formulae:	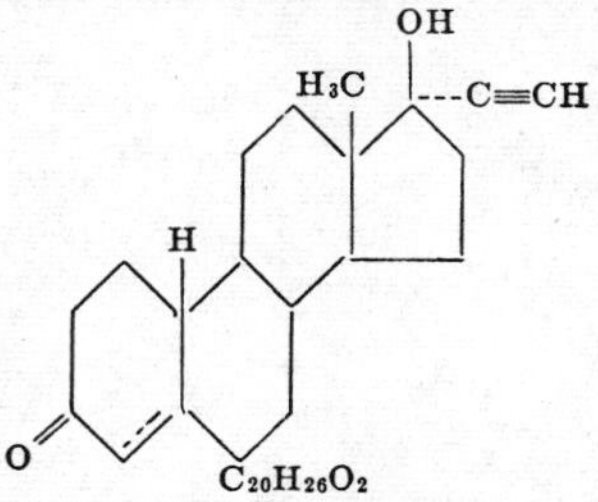 $C_{20}H_{26}O_2$	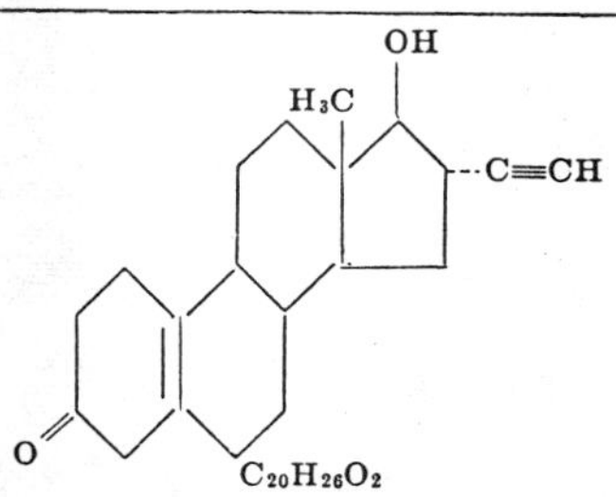 $C_{20}H_{26}O_2$
Molecular weight	298.4	298.4
Melting point (°C)	200–207	178–180
Specific rotation	$(\alpha)\frac{25}{D} - 30 - -36$ (200 mg. in 10 ml. dioxane)	$(\alpha)\frac{25}{D} + 125$ (dioxane)
Absorption max.	240 mμ (ϵ_1^1 = 535)	

PROGESTOGENS AND PROGESTINS (INCLUDING 19-NORSTEROID COMPOUNDS) (Continued)

Names & synonyms:	NORMETHISTERONE; 19-normethisterone; normethandrolone; metalutin; normetandrone; 17α-methyl-19-nor-Δ^4-androsten-17β-ol-3-one; 17α-methyl-19-nor-testosterone; 17β-hydroxy-3-oxo-17α-methyl-estra-4-ene; 17β-hydroxy-17-methyl-estr-4-en-3-one.	PREGNENOLONE; Δ^5-pregnenolone; Δ^5-pregnen-3β-ol-20-one; 17β(1-ketoethyl)-Δ^5-androstene-3β-ol.	PROGESTERONE; progestin; progestone; pregnendione; Δ^4-pregnene-3,20-dione.
Formulae:	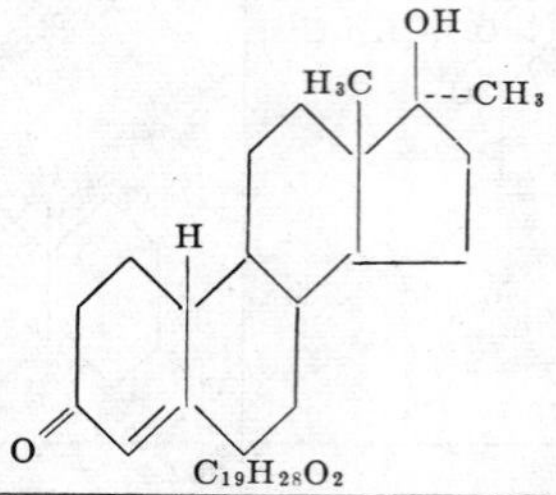 $C_{19}H_{28}O_2$	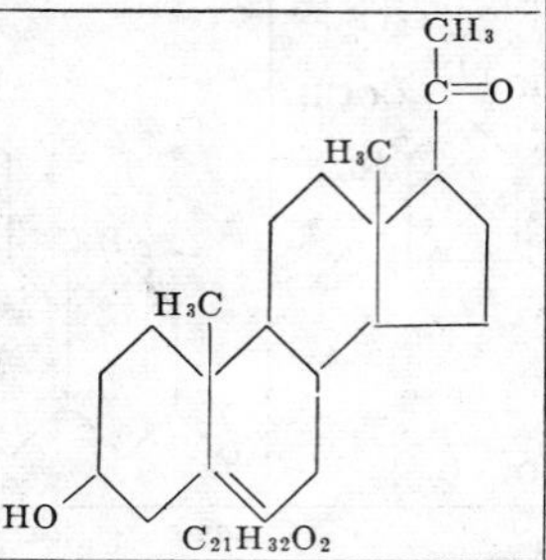 $C_{21}H_{32}O_2$	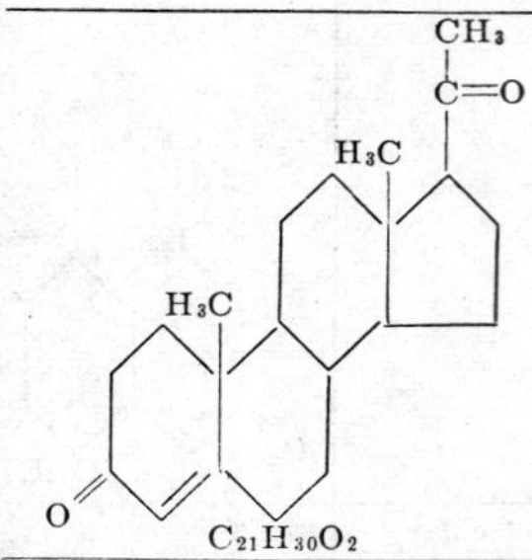 $C_{21}H_{30}O_2$
Molecular weight	288.4	308.4	314.4
Melting point (°C)	150–156	193	(β) isomer 121; (α) isomer 127–131
Specific rotation	$(\alpha)\frac{25}{D}$ + 26.3 (200 mg. in 10 ml. chloroform)	$(\alpha)\frac{20}{D}$ + 28 – +30 (alcohol)	$(\alpha)\frac{20}{D}$ + 172 – +182 (200 mg. in 10 ml. dioxane)
Absorption max.	240 mμ – 570		240 mμ

IONIZATION CONSTANTS AND pH VALUES AT THE ISOELECTRIC POINTS OF THE AMINO ACIDS IN WATER AT 25°C

The majority of the recorded values are true thermodynamic constants calculated from electrometric force measurements of cells without liquid junctions. The values for the constants given in the table were derived from the classical, the zwitterionic (Bjerrum), and the acidic (Brönsted) formulations of ionization and the corresponding mass law expressions. pH values at the isoelectric points were calculated from the expression, $pI = \frac{1}{2}(pk_{a1} + pk_w - pk_{b1})$. The error is approximately 0.5 per cent when this expression is used to calculate pI values for cystine, tyrosine, and diiodotyrosine.

Amino acid	Classical				Zwitterionic				Acidic				pI	Ref. no.
	pk_{a1}	pk_{a2}	pk_{b1}	pk_{b2}	pK_{A1}	pK_{A2}	pK_{B1}	pP_{B2}	pK_1	pK_2	pK_3	pK_4		
DL-Alanine	9.866	.....	11.649	.....	2.348	....	4.131	....	2.348	9.866	.....	.....	6.107	1
L-Arginine	12.48	.....	4.96	11.99	2.01	....	1.52	4.96	2.01	9.04	12.48	....	10.76	2
L-Aspartic acid	3.86	9.82	11.93	.....	2.10	3.86	4.18	....	2.10	3.86	9.82	.....	2.98	3
L-Cystine	8.00	10.25	11.95	12.96	1.04	2.05	3.75	6.00	1.04	2.05	8.00	10.25	5.02	4
Diiodo-L-tyrosine	6.48	7.82	11.88	.....	2.12	6.48	6.18	....	2.12	6.48	7.82	.....	4.29	5,6
L-Glutamic acid	4.07	9.47	11.90	.....	2.10	4.07	4.53	....	2.10	4.07	9.47	.....	3.08	7
Glycine	9.778	.....	11.647	.....	2.350	....	4.219	....	2.350	9.778	.....	.....	6.064	8
L-Histidine	9.18	.....	7.90	12.23	1.77	....	4.82	7.90	1.77	6.10	9.18	.....	7.64	3
Hydroxy-L-proline	9.73	.....	12.08	.....	1.92	....	4.27	....	1.92	9.73	.....	.....	5.82	9
DL-Isoleucine	9.758	.....	11.679	.....	2.318	....	4.239	....	2.318	9.758	.....	.....	6.038	1
DL-Leucine	9.744	.....	11.669	.....	2.328	....	4.253	....	2.328	9.744	.....	.....	6.036	1
L-Lysine	10.53	.....	5.05	11.82	2.18	....	3.47	5.05	2.18	8.95	10.53	.....	9.47	2
DL-Methionine	9.21	.....	11.72	.....	2.28	....	4.79	....	2.28	9.21	.....	.....	5.74	10
DL-Phenylalanine	9.24	.....	11.42	.....	2.58	....	4.76	....	2.58	9.24	.....	.....	5.91	11
L-Proline	10.60	.....	12.0	.....	2.00	....	3.40	....	2.00	10.60	.....	.....	6.3	12
DL-Serine	9.15	.....	11.79	.....	2.21	....	4.85	....	2.21	9.15	.....	.....	5.68	9
L-Tryptophan	9.39	.....	11.62	.....	2.38	....	4.61	....	2.38	9.39	.....	.....	5.88	13
L-Tyrosine	9.11	10.07	11.80	.....	2.20	9.11	3.93	....	2.20	9.11	10.07	.....	5.63	6
DL-Valine	9.719	.....	11.711	.....	2.286	....	4.278	....	2.286	9.719	.....	.....	6.002	1

References

1. Smith, P. K., Taylor, A. C., and Smith, E. R. B., J. Biol. Chem., **122,** 109 (1937–38).
2. Schmidt, C. L. A., Kirk, P. L., and Appleman, W. K., J. Biol. Chem., **88,** 285 (1930).
3. Greenstein, J. P., J. Biol. Chem., **93,** 479 (1931).
4. Borsook, H., Ellis, E. L., and Huffman, H. M., J. Biol. Chem., **117,** 281 (1937).
5. Dalton, J. B., Kirk, P. L., and Schmidt, C. L. A., J. Biol. Chem., **88,** 589 (1930).
6. Winnek, P. S., and Schmidt, C. L. A., J. Gen. Physiol., **18,** 889 (1935).
7. Simms, H. S., J. Gen. Physiol., **11,** 629 (1928); **12,** 231 (1928).
8. Owen, B. B., J. Am. Chem. Soc., **56,** 24 (1934).
9. Kirk, P. L., and Schmidt, C. L. A., J. Biol. Chem., **81,** 237 (1929).
10. Emerson, O. H., Kirk, P. L., and Schmidt, C. L. A., J. Biol. Chem., **92,** 449 (1931).
11. Miyamoto, S., and Schmidt, C. L. A., J. Biol. Chem., **90,** 165 (1931).
12. McCay, C. M., and Schmidt, C. L. A., J. Gen. Physiol., **9,** 333 (1926).
13. Schmidt, C. L. A., Appleman, W. K., and Kirk, P. L., J. Biol. Chem., **85,** 137 (1929–30).

Ionization Constants of the Amino Acids in Aqueous Ethanol Solutions

Amino acid	pK_1	pK_2	pK_3	Volume per cent ethanol	Temperature °C	Ref. No.
Alanine	3.55	10.02	.....	72	25	1
Arginine	3.34	9.40	14.1	72	25	1
Aspartic acid	2.85	5.20	10.51	72	25	1
Glutamic acid	3.16	5.63	10.75	72	25	2
Glycine	2.66	9.82	.....	10	19.5	2
	2.96	9.76	.....	40	19.5	2
	3.46	9.82	.....	72	25	1
	3.79	9.99	.....	90	19.5	2
Histidine	3.00	5.85	9.45	72	25	1
Isoleucine	3.69	9.81	.....	72	25	1
Lysine	2.75	8.95	10.53	48	25	1
	3.56	8.95	10.49	84	25	1
Proline	3.04	10.55	.....	72	25	1
Valine	3.60	9.73	.....	72	25	1

References

1. Jukes, T. H., and Schmidt, C. L. A., J. Biol. Chem., **105,** 359 (1934).
2. Michaelis, L., and Mizutani, M., Z. physik. Chem., **116,** 135 (1925).

Ionization Constants of the Amino Acids in Aqueous Formaldehyde Solution[a]

Amino acid	Mole per cent formaldehyde				
	0.99	3.95	5.60	10.0	17.9
DL-Alanine	8.36	7.42	6.96[b]	6.56	6.10
L-Arginine	....	3.45[c]	3.40[d]	....	....
L-Aspartic acid	....	......	7.21[d]	⋜3.8[e] 6.85[f]	....
L-Glutamic acid	....	......	6.91[d]	⋜4.2[e] 6.8[f]	....
Glycine	7.16	6.08	5.92[b]	5.34	5.04
L-Histidine	....	7.90[c]	7.90[d]	....	....
Hydroxy-L-Proline	....	......	7.19[d]	....	....
L-Leucine	8.44	7.50	6.92[d]	6.62	6.20
DL-Leucine	8.44	7.48	......	6.60	6.20
L-Lysine	....	7.35[c]	7.15[d]	....	....
L-Phenylalanine	....	......	6.62[d]	5.9[e]	....
DL-Phenylalanine	8.09	7.16	6.80[b]	6.35	6.13
L-Proline	....	......	7.78[d]	....	....
DL-Serine	6.66	5.74	5.63[b]	....	4.94
L-Tryptophan	....	......	6.88[d]	....	....
L-Tyrosine	....	......	7.50[d]	6.2[e] >9[f]	....
DL-Valine	8.52	7.65	7.47[b]	....	6.52

[a] Dunn and Weiner (1), pK_2 at 22°.
[b] Dunn and Loshakoff (2), pK_2 at 22°.
[c] Levy (3) pK_2 at 30° for arginine and pK_3 at 30° for histidine and lysine.
[d] Levy and Silberman (4), pK_2 at 30°, pK_3 at 30° for histidine and lysine.
[e] Harris (5), pK_2 at 25° for aspartic acid, glutamic acid, phenylalanine and tyrosine.
[f] Harris (5), pK_3 at 30° for aspartic acid, glutamic acid, and tyrosine.

References

1. Dunn, M. S., and Weiner, J. G., J. Biol. Chem., **117,** 381 (1937).
2. Dunn, M. S., and Loshakoff, A., J. Biol. Chem., **113,** 691 (1936).
3. Levy, M., J. Biol. Chem., **109,** 365 (1935).
4. Levy M., and Silberman, D. E., J. Biol. Chem., **118,** 723 (1937).
5. Harris, L. J., Proc. Roy. Soc. London, Series B, **95,** 440 (1923–24).

Specific Rotations of the Amino Acids Using Sodium Light (5893 Å)

Abbreviations

c—grams of solute per 100 ml. of solution.
d—density of the solution.
p—grams of solute per 100 grams of solution.
l—length of the tube in decimeters.
α—observed rotation in angular degrees.
[α]—specific rotation in angular degrees calculated from

$$[\alpha]^t_\lambda = \frac{\alpha \times 100}{c \times l} = \frac{\alpha \times 100}{p \times d \times l}$$

where *t* is temperature in °C and λ is wave length of the incident light in Ångstroms.

A—prepared from a protein or other naturally occurring material.
B—prepared by resolution of the inactive synthetic form.
C—prepared by resolution of the inactive racemized form.
D—prepared from the inactive synthetic form by a biological method.
E—prepared from the inactive racemized form by a biological method.
?—source not given.

Source	c	Solvent	d	p	Moles acid or base per mole amino acid	l	Temp. °C.	α	[α]	Ref. No.
					L-Alanine					
A	5.790	0.97 N HCl	1.033	5.605	1.5	2	15	+1.70	+14.7	1
A	10.3	Water	1.03	1.00	0	2	22	+0.55	+2.7	2
A	1.781	3 N NaOH	......	......	15	2	20	....	+3.0	3
					D-Alanine					
B	1.344	6 N HCl	......	......	39.4	2	30.4	−0.392	−14.6	4
					L-Arginine					
A	1.653	6.0 N HCl	......	......	63	4.001	23.4	+1.777	+26.9	5
A	3.48	Water	......	......	0	2	20	....	+12.5	6
A	0.87	0.50 N NaOH	......	......	10	2	20	....	+11.8	6
					L-Aspartic acid					
A	2.002	6.0 N HCl	......	......	39	4.001	24.0	+1.972	+24.6	7
A	1.3300	Water	......	......	0	3	18	....	+4.7	3
A	1.3300	3 N NaOH	......	......	30	3	18	....	−1.7	3
					D-Aspartic acid					
C	4.289	0.97 N HCl	1.032	4.156	3	1	20	−1.09	−25.5	8
					L-Cystine					
A	0.9974	1.02 N HCl	1.0181	0.9797	24.6	2	24.35	−4.277	−214.40	9
A	0.400	0.20 N NaOH	......	......	12	2	18.5	....	−70.0	3
					D-Cystine					
C	....	1 N HCl	......	1	24	. ...	20	....	+223	10
					Diiodo-D-tyrosine					
A	5.08	1.1 N HCl	1.05	4.84	9.4	1	20	+0.15	+2.89	11
A	4.41	13.4 N NH_4OH	0.9779	4.51	132	1	20	+0.10	+2.27	11
					L-Glutamic acid					
A	1.002	6.0 N HCl	......	......	87	4.001	22.4	+1.25	+31.2	12
A	1.471	Water	......	......	0	2	18	....	+11.5	3
A	1.471	1 N NaOH	......	......	10	2	18	....	+10.96	3
					D-Glutamic acid					
C	5.425	0.37 N HCl	1.0233	5.3011	1	1	20	−1.63	−30.05	8
					L-Histidine					
A	1.480	6.0 N HCl	......	......	63	4.001	22.7	+0.766	+13.0	7
A	1.128	Water	1.0012	1.127	0	4	25.00	−1.714	−39.01	13
A	0.775	0.50 N NaOH	......	......	10	2	20	....	−10.9	6
					D-Histidine					
?	4.000	1.0 N HCl	......	......	4	1	20	−0.407	−10.2	14
B	2.66	Water	......	......	0	2	23	+2.11	+39.8	14
					Hydroxy-L-proline					
A	1.31	1.0 N HCl	......	......	10	2	20	....	−47.3	6
A	1.001	Water	......	......	0	4.001	22.5	−3.009	−75.2	7
A	0.655	0.50 N NaOH	......	......	10	2	20	....	−70.6	6
					Hydroxy-D-proline					
B	4.48	Water	1.03	4.35	0	1	21	+3.37	+75.2	16
					Allo-Hydroxy-L-proline					
B	2.617	Water	1.014	2.581	0	1	18	−1.52	−58.1	16
					Allo-Hydroxy-D-proline					
B	2.530	Water	1.013	2.998	0	1	17	+1.48	+58.5	16
					L-Isoleucine					
B	5.09	6.1 N HCl	1.098	4.64	15	1	20	+2.07	+40.61	17
B	3.10	Water	1.008	3.08	0	2	20	+0.70	+11.29	17
A	3.34	0.33 N NaOH	1.017	3.28	1.3	2	20	+0.74	+11.09	18
					D-Isoleucine					
B	4.53	6.1 N HCl	1.083	4.18	17	1	20	−1.85	−40.86	17
B	3.12	Water	1.006	3.10	0	2	20	−0.66	−10.55	17
					D-allo-Isoleucine					
D	5.14	6.0 N HCl	1.094	4.70	15.0	2	20	−3.80	−36.95	19
B	2.00	Water	......	......	0	1	20	−0.285	−14.2	20
					L-allo-Isoleucine					
B	3.97	6.0 N HCl	......	......	20	1	20	+1.50	+38.1	20
B	2.00	Water	......	......	0	1	20	+0.28	+14.0	20
					L-Leucine					
A	1.999	6.0 N HCl	......	......	38	4.001	25.9	+1.212	+15.1	5
A	2.001	Water	......	......	0	4.001	24.7	−0.863	−10.8	5
A	1.31	3.00 N NaOH	......	......	30	2	20	....	+7.6	3
					D-Leucine					
?	4.0	6.0 N HCl	1.1	3.664	19	2	20	+1.26	−15.6	21
?	....	Water	......	2.08	0	2	20	+0.43	+10.34	38
					L-Lysine					
A	2.00	6.0 N HCl	......	......	43	4	22.9	+1.652	+25.9	5
A	6.496	Water	......	......	0	2	20	+1.90	+14.6	22
					D-Lysine					
B	2.00	0.27 N HCl	......	......	2	2	20	−0.939	−23.48	23
					L-Methionine					
B	0.80	Water	......	......	0	2	25	−0.13	−8.11	24
					D-Methionine					
B	0.80	0.2001 N HCl	......	......	4	2	25	−0.34	−21.18	24
B	0.80	Water	......	......	0	2	25	+0.13	+8.12	24
B	0.80	0.6 N $NaHCO_3$	......	......	11	2	25	−0.12	−7.47	24
					L-Phenylalanine					
B	1.936	Water	1.0040	1.928	0	2	20	−1.36	−35.14	27
					D-Phenylalanine					
B	3.814	5.4 N HCl	1.0895	3.501	23	2	20	+0.54	+7.07	28
B	2.043	Water	1.0045	2.034	0	2	20	+1.43	+35.0	27
					L-Proline					
A	0.575	0.50 N HCl	......	......	10	2	20	....	−52.6	6
A	1.001	Water	......	......	0	4.001	23.4	−3.402	−85.0	7
B	2.42	0.6 N KOH	1.031	2.35	3	1	20	−2.25	−93.0	29
					D-Proline					
B	3.90	Water	1.01	3.865	0	1	20	+3.18	+81.5	29
					L-Serine					
B	9.344	1 N HCl	1.0465	8.929	1	1	25	+1.35	+14.45	30
B	10.414	Water	1.0414	9.997	0	2	20	−1.42	−6.83	30
					D-Serine					
B	9.359	1 N HCl	1.0465	8.943	1	1	25	−1.34	−14.32	30
B	10.412	Water	1.0414	9.998	0	2	20	+1.43	+6.87	30
					D-Threonine					
B	....	Water	......	1.092	0	2	26	−0.625	−28.3	31
					L-Threonine					
B	....	Water	......	1.331	0	2	26	+0.780	+28.4	31
					D-allo-Threonine*					
B	....	Water	......	1.634	0	2	26	−0.302	−9.1	31
					L-allo-Threonine					
B	....	Water	......	1.643	0	2	26	+0.320	+9.6	31
					L-Thyroxine					
A	....	0.13 N NaOH in 70% EtOH by weight	......	3	3	1	......	−0.147	−4.4	32
					L-Tryptophan					
A	1.02	0.50 N HCl	......	......	10	2	20	....	+2.4	6
A	1.004	Water	......	......	0	4.001	22.7	−1.266	−31.5	7
A	2.426	0.5 N NaOH	1.0243	2.368	4.2	1	20	+0.15	+6.17	33
					D-Tryptophan					
C	0.5024	Water	......	......	0	2	25	+0.326	+32.45	34
					L-Tyrosine					
B	4.40	6.3 N HCl	1.116	3.94	28	2	20	−0.76	−8.64	35
A	0.906	3.0 N NaOH	......	......	60	3	18	....	−13.2	3
					D-Tyrosine					
B	5.1484	6.3 N HCl	1.1175	4.6071	24	2	20	+0.89	+8.64	35
					L-Valine					
B	3.4	6.0 N HCl	1.1	3.05	20	2	20	+1.93	+28.8	36
B	3.58	Water	1.007	3.56	0	2	20	+0.46	+6.42	36
					D-Valine					
B	3.2	6.0 N HCl	1.1	2.91	21	2	20	−1.86	−29.04	36
E	6.24	Water	1.00	6.24	0	1	20	−0.37	−6.06	37

* The levorotatory allothreonine probably belongs to the D family and its enantiomorph to the L family.

Specific Rotations of the Amino Acids Using Sodium Light (5893 Å) (Continued)

References

1. Clough G. W., J. Chem. Soc., **113**, 526 (1918).
2. Fischer, E., and Raske, K., Ber, **40**, 3717 (1907).
3. Lutz, O., and Jirgensons, B., Ber., **63**, 448 (1930).
4. Dunn, M. S., Butler, A. W., and Naiditch, M. J., unpublished data.
5. Dunn, M. S., and Courtney, G., unpublished data.
6. Lutz, O., and Jirgensons, B., Ber., **64**, 1221 (1931).
7. Dunn, M. S., and Stoddard, M. P., unpublished data.
8. Fischer, E., Ber., **32**, 2451 (1899).
9. Toennies, G., and Lavine, T. F., J. Biol. Chem., **89**, 153 (1930).
10. Loring, H. S., and du Vigneaud, V., J. Biol. Chem., **107**, 267 (1934).
11. Abderhalden, E., and Guggenheim, M., Ber., **41**, 1237 (1908).
12. Dunn, M. S., and Sexton, E. L., unpublished data.
13. Dunn, M. S., and Frieden, E. H., unpublished data.
14. Cox, G. J., and Berg, C. P., J. Biol. Chem., **107**, 497 (1934).
15. Dakin, H. D., Biochem. J., **13**, 398 (1919).
16. Leuchs, H., and Bormann, K., Ber., **52**, 2086 (1919).
17. Locquin, R., Bull. Soc. Chim., (4), **1**, 601 (1907).
18. Ehrlich, F., Ber., **37**, 1809 (1904).
19. Ehrlich, F., Ber., **40**, 2538 (1907).
20. Abderhalden, E., and Zeisset, W., Z. physiol. Chem., **196**, 121 (1931)
21. Fischer, E., and Warburg, O., Ber., **38**, 3997 (1905).
22. Vickery, H. B., private communication, April, 1940.
23. Berg, C. P., J. Biol. Chem., **115**, 9 (1936); private communication, June, 1940.
24. Windus, W., and Marvel, C. S., J. Am. Chem. Soc., **53**, 3490 (1931).
27. Fischer, E., and Schoeller, W., Ann., **357**, 1 (1907).
28. Fischer, E., and Mouneyrat, A., Ber., **33**, 2383 (1900).
29. Fischer, E., and Zemplén, G., Ber., **42**, 2989 (1909).
30. Fischer, E., and Jacobs, W. A., Ber., **39**, 2942 (1906).
31. West, H. D., and Carter, H. E., J. Biol. Chem., **119**, 109 (1937); private communication from H. E. Carter, July, 1940.
32. Foster, G. L., Palmer, W. W., and Leland, J. P., J. Biol. Chem., **115**, 467 (1936).
33. Abderhalden, E., and Baumann, L., Z. physiol. Chem., **55**, 412 (1908).
34. Berg, C. P., J. Biol. Chem., **100**, 79 (1933); private communication, July, 1940.
35. Fischer, E., Ber., **32**, 3638 (1899).
36. Fischer, E., Ber., **39**, 2320 (1906).
37. Ehrlich, F., and Wendel, A., Biochem. Z. **8**, 399 (1908).
38. Ehrlich, F., Biochem. Z., **1**, 8 (1906).

Solubilities of the Amino Acids in Grams per 100 Grams of Water

Amino acid	Temperature, °C.					Ref. No.
	0°	25°	50°	75°	100°	
DL-Alanine	12.11	16.72	23.09	31.89	44.04	1
L-Alanine	12.73	16.65	21.79	28.51	37.30	1
DL-Aspartic acid	0.262	0.778	2.000	4.456	8.594	1
L-Aspartic acid	0.209	0.500	1.199	2.875	6.893	1
L-Cystine‡ × 10²	0.502	1.096	2.394	5.229	11.42	2
Diiodo-DL-tyrosine × 10	0.149	0.340	0.773	......	......	3
Diiodo-L-tyrosine × 10	0.204	0.617	1.862	5.62	17.00	1
DL-Glutamic acid	0.855	2.054	4.934	11.86	28.49	1
L-Glutamic acid	0.341	0.864	2.186	5.532	14.00	1
Glycine	14.18	24.99	39.10	54.39	67.17	1
L-Histidine	......	4.19	......	......	......	4
Hydroxy-L-Proline	28.86	36.11	45.18	51.67*	......	5
DL-Isoleucine	1.826	2.229	3.034	4.607	7.802	1
L-Isoleucine	3.791	4.117	4.818	6.076	8.255	2
DL-Leucine	0.797	0.991	1.406	2.276	4.206	1
L-Leucine	2.270	2.426†	2.887†	3.823	5.638	1
DL-Methionine	1.818	3.381	6.070	10.52	17.60	2
DL-Phenylalanine	0.997	1.411	2.187	3.708	6.886	1
L-Phenylalanine	1.983	2.965	4.431	6.624	9.900	2
L-Proline × 10⁻¹	12.74	16.23	20.67	23.90*	......	3
DL-Serine	2.204	5.023	10.34	19.21	32.24	2
L-Tryptophan	0.823	1.136	1.706	2.795	4.987	2
DL-Tyrosine × 10	0.147	0.351	0.836	......	......	3
L-Tyrosine × 10	0.196	0.453	1.052	2.438	5.650	1
D-Tyrosine × 10	0.196	0.453	1.052	......	......	3
DL-Valine	5.98	7.09	9.11	12.61	18.81	1
L-Valine	8.34	8.85	9.62	10.24*	......	6

* Value at 65°.

† Dunn and Stoddard (7) report 2.19 g. at 25° for L-Leucine rendered methionine-free by repeated recrystallization from 6 *N* HCl. Hlynka (8) found 2.20 g. at 25° and 2.66 g. at 50° for L-Leucine rendered methionine-free [by S. W. Fox (9)] by fractional crystallization of the formyl derivative and identical values for D-Leucine obtained by resolution of the DL form.

‡ The following values were found by Loring and du Vigneaud (10): DL-Cystine (0.0049 g), D-Cystine (0.0108 g), and *meso*-cystine (0.0056 g) at 25°.

References

1. Dalton, J. B., and Schmidt, C. L. A., J. Biol. Chem., **103**, 549 (1933).
2. Dalton, J. B., and Schmidt, C. L. A., J. Biol. Chem., **109**, 241 (1935).
3. Winnek, P. S., and Schmidt, C. L. A., J. Gen. Physiol., **18**, 889 (1934–35).
4. Dunn, M. S., Frieden, E. H., and Brown, H. V., unpublished data.
5. Tomiyama, T., and Schmidt, C. L. A., J. Gen. Physiol., **19**, 379 (1935–36).
6. Dalton, J. B., and Schmidt, C. L. A., J. Gen. Physiol., **19**, 767 (1935–36).
7. Dunn, M. S., and Stoddard, M. P., unpublished data.
8. Hlynka, I., Thesis (1939), California Institute of Technology, Pasadena, California.
9. Fox, S. W., Science, **84**, 163 (1936).
10. Loring, H. S., and du Vigneaud, V., J. Biol. Chem., **107**, 270 (1934).

Solubilities of the Amino Acids in Grams per 100 Grams of Water-Ethanol Mixtures

DL-Alanine

Per cent ethanol by volume	Temp. °C	Grams amino acid per 100 grams solvent	Ref. No.
24.93	0.00	3.84	1
50.10	0.00	1.16	1
74.50	0.00	0.305	1
95.14	0.00	0.0167	1
10	25	12.25	2
24.93	24.97	7.09	1
50.10	24.97	2.52	1
74.20	24.97	0.573	1
95.14	25.09	0.0329	1
25.28	45.16	10.6	1
50.10	44.96	4.25	1
74.20	44.98	0.949	1
95.14	45.19	0.0545	1
24.93	64.96	15.9	1
50.10	64.94	6.68	1
74.20	64.94	1.48	1
95.09	65.15	0.0851	1

DL-Aspartic acid

Per cent ethanol by volume	Temp. °C	Grams amino acid per 100 grams solvent	Ref. No.
24.93	0.03	0.0703	1
50.10	0.03	0.0267	1
74.20	0.02	0.0111	1
24.55	25.06	0.266	1
50.25	25.06	0.0992	1
74.28	25.14	0.317	1
95.14	25.07	0.0020	1
24.74	45.25	0.680	1
50.18	45.25	0.255	1
74.28	45.27	0.0608	1
95.14	45.21	0.0042	1
24.93	64.91	1.53	1
50.10	64.91	0.588	1
74.20	65.07	0.132	1
95.14	65.00	0.0129	1

L-Aspartic acid

Per cent ethanol by volume	Temp. °C	Grams amino acid per 100 grams solvent	Ref. No.
20	25	0.204	3
50	25	0.0633	3
70	25	0.0224	3
90	25	0.0034	3

L-Glutamic acid

Per cent ethanol by volume	Temp. °C	Grams amino acid per 100 grams solvent	Ref. No.
24.74	0.01	0.0855	1
50.18	0.01	0.0371	1
74.28	0.03	0.0163	1
24.56	25.05	0.292	1
50.25	25.08	0.131	1
74.35	25.07	0.0370	1
95.14	25.04	0.0044	1
24.55	45.01	0.811	1
50.18	45.27	0.378	1
74.35	44.93	0.0885	1
95.14	45.20	0.0127	1

Glycine

Per cent ethanol by volume	Temp. °C	Grams amino acid per 100 grams solvent	Ref. No.
24.93	0.02	3.95	1
50.10	0.02	1.03	1
74.50	0.02	0.200	1
95.09	0.01	0.0080	1
10	25	17.13	2
24.93	24.97	8.72	1
50.10	24.97	2.47	1
74.20	24.97	0.448	1
95.14	25.09	0.0172	1
24.93	44.98	15.0	1
50.10	44.98	4.62	1
74.20	44.97	0.756	1
95.14	45.19	0.0294	1
24.93	65.11	24.5	1
50.10	65.10	8.03	1
74.20	65.07	1.23	1
95.14	65.00	0.0488	1

Solubilities of the Amino Acids in Grams per 100 Grams of Water-Ethanol Mixtures (Continued)

Per cent ethanol by volume	Temp. °C	Grams amino acid per 100 grams solvent	Ref. No.
L-Isoleucine			
80	20	0.46	4
80	78–80	1.16	4
L-allo-Isoleucine			
80	20	0.81	4
80	78–80	1.97	4
DL-Leucine			
24.93	0.00	0.251	1
50.10	0.00	0.118	1
74.50	0.00	0.0693	1
95.14	0.00	0.0116	1
10	25	0.771	2
24.93	24.97	0.493	1
50.10	24.97	0.318	1
74.20	24.97	0.175	1
95.14	25.09	0.0258	1
24.93	45.24	0.853	1
50.10	45.24	0.633	1
74.50	45.18	0.323	1
95.14	45.18	0.0471	1
24.93	65.16	1.45	1
50.10	65.20	1.16	1
74.20	65.15	0.584	1
95.09	65.07	0.0844	1

Per cent ethanol by volume	Temp. °C	Grams amino acid per 100 grams solvent	Ref. No.
L-Leucine			
20	25	1.33	2
60	25	0.641	2
90	25	0.123	2
L-Proline			
100	19	1.5	5
DL-Serine			
24.93	0.00	0.1530	1
50.10	0.00	0.146	1
74.50	0.00	0.0304	1
95.14	0.00	0.0008	1
24.93	25.14	1.54	1
50.10	25.14	0.461	1
74.50	25.10	0.0840	1
95.14	25.09	0.0028	1
24.93	45.15	3.14	1
50.10	45.04	0.985	1
74.20	45.04	0.185	1
95.14	45.18	0.0058	1
24.93	65.26	5.99	1
50.10	65.25	1.88	1
74.50	65.24	0.318	1
95.14	65.01	0.0152	1

Per cent ethanol by volume	Temp. °C	Grams amino acid per 100 grams solvent	Ref. No.
DL-Threonine			
95	25	0.07*	6
DL-allo-Threonine			
95	25	0.03*	6
L-Tyrosine			
95	17	0.10	7
DL-Tyrosine			
95.09	0.00	0.0031	8
25.28	24.85	0.0285	8
50.99	24.75	0.0226	8
74.63	24.75	0.0117	8
95.09	25.24	0.0032	8
25.28	45.15	0.0630	8
50.99	45.16	0.0513	8
74.63	44.93	0.0230	8
95.09	44.98	0.0035	8
95.09	65.06	0.0067	8

* Grams per 100 ml. of solution.

Per cent ethanol by volume	Temp. °C	Grams amino acid per 100 grams solvent	Ref. No.
DL-Valine			
24.93	0.02	2.10	1
50.10	0.02	0.769	1
74.20	0.02	0.269	1
95.14	0.01	0.0277	1
10	25	5.50	2
25.28	24.85	3.30	1
50.99	24.85	1.53	1
74.35	24.93	0.570	1
95.14	25.04	0.0569	1
24.55	44.91	5.10	1
50.25	44.92	2.74	1
74.35	44.92	0.999	1
95.14	45.21	0.0979	1
24.55	65.07	7.44	1
50.10	64.94	4.49	1
74.20	64.34	1.62	1
95.09	65.15	0.167	1
L-Valine			
20	25	5.11	2
40	25	2.93	2
60	25	1.61	2
80	25	0.52	2

References

1. Dunn, M. S., and Ross, F. J., J. Biol. Chem., **125**, 309 (1938).
2. Cohn, E. J., McMeekin, T. L., Edsall, J. T., and Weare, J. H., J. Am. Chem. Soc., **56**, 2270 (1934).
3. McMeekin, T. L., Cohn, E. J., and Weare, J. H., J. Am. Chem. Soc., **57**, 626 (1935).
4. Abderhalden, E., and Zeisset, W., Z. physiol. Chem., **196**, 121 (1931).
5. Kapfhammer, J., and Eck, R., Z. physiol. Chem., **170**, 294 (1927).
6. West, H. D., and Carter, H. E., J. Biol. Chem., **119**, 109 (1937).
7. Stutzer, A., Z. anal. Chem., **31**, 501 (1892).
8. Dunn, M. S., and Ross, F. J., unpublished data.

Solubilities of the Amino Acids in Grams per 100 Grams of Organic Solvent

Solvent	Grams amino acid per 100 grams solvent	Temp. °C	Ref. No.
DL-Alanine			
Ethanol	0.0087	25	1
L-Aspartic acid			
Ethanol	0.000196	25	2
L-Glutamic acid			
Ethanol	0.000347	25	2
Ethanol	0.0056	44.93	3
Glycine			
Acetone	0.000291	25	4
Butanol	0.000892	25	4
Ethanol	0.0037	25	1
Formamide	0.558	25	4
Methanol	0.0407	25	4

Solvent	Grams amino acid per 100 grams solvent	Temp. °C	Ref. No.
L-Isoleucine			
Ethanol	0.09	20	5
Ethanol	0.13	78–80	5
L-allo-Isoleucine			
Ethanol	0.13	20	5
Ethanol	0.19	78–80	5
L-Leucine			
Ethanol	0.0217	25	1
L-Proline			
Ethanol	1.5	19	6
DL-Valine			
Ethanol	0.0136	0.03	3
Ethanol	0.019	25	1

References

1. Cohn, E. J., McMeekin, T. L., Edsall, J. T., and Weare, J. H., J. Am Chem. Soc., **56**, 2270 (1934).
2. McMeekin, T. L., Cohn, E. J., and Weare, J. H., J. Am. Chem. Soc., **57** 626 (1935).
3. Dunn, M. S., and Ross, F. J., J. Biol. Chem., **125**, 309 (1938).
4. McMeekin, T. L., Cohn, E. J., and Weare, J. H., J. Am. Chem. Soc., **58** 2173 (1936).
5. Abderhalden, E., and Zeisset, W., Z. physiol. Chem., **196**, 121 (1931).
6. Kapfhammer, J., and Eck, R., Z. physiol. Chem., **170**, 294 (1927).

Densities of Crystalline Amino Acids

Amino acid	Density	Ref. No.
DL-Alanine	1.424	1
L-Alanine	1.401	2
β-Alanine	1.404	1
DL-α-Amino-*n*-butyric acid	1.231	1
α-Aminoisobutyric acid	1.278	1
L-Arginine	1.1	3
L-Aspartic acid	1.66	3
DL-Glutamic acid	1.460	4
L-Glutamic acid	1.538	4
Glycine*	1.601	3
	1.607	1

Amino acid	Density	Ref. No.
DL-Leucine	1.191	1
L-Leucine	1.165	1
DL-Methionine	1.340	5
DL-Serine	1.537	5
L-Tyrosine	1.456	1
DL-Valine	1.316	1
L-Valine	1.230	1

* The density of glycine at 50° is 1.5753 according to Houck (6) who concluded that the figure 1.1607, reported by Curtius (7) and reproduced in chemical handbooks, is a typographical error.

References

1. Cohn, E. J., McMeekin, T. L., Edsall, J. T., and Weare, J. H., J. Am Chem. Soc., **56**, 2270 (1934).
2. Dalton, J. B., and Schmidt, C. L. A., J. Biol. Chem., **103**, 549 (1933).
3. Huffman, H. M., Ellis, E. L., and Fox, S. W., J. Am. Chem. Soc., **58**, 172 (1936). Huffman, H. M., Fox, S. W., and Ellis, E. L., J. Am. Chem Soc., **59**, 2144 (1937).
4. Schmidt, C. L. A., Chemistry of the Amino Acids and Proteins, C. C Thomas, Springfield, Illinois, 1938, p. 900.
5. Albrecht, G., and Dunn, M. S., unpublished data.
6. Houck, R. C., J. Am. Chem. Soc., **52**, 2420 (1930).
7. Curtius, T., J. prakt. Chem., **26**, 145 (1882).

CARBOHYDRATES

These data for carbohydrates were compiled originally for the Biology Data Book by M. L. Wolfram, G. G. Maher and R. G. Pagnucco (1964). Data are reproduced here by permission of the copyright owners of the above publication, the Federation of American Societies for Experimental Biology, Washington, D.C. pp. 351–359.

All data are for crystalline substances, unless otherwise specified. Selection of substances was restricted to natural carbohydrates found free (or in chemical combination and released on hydrolysis) and to biological oxidation products of the natural carbohydrates. The nomenclature conforms with that of the British-American report as published in the *Journal of Organic Chemistry,* 28:281 (1963). Substances have been arranged alphabetically under the name of the parent sugar within groups formulated according to increasing carbon content (excluding carbon in substituents), with synonymous common names in parentheses. **Melting Point:** b.p. = boiling point; d. = decomposes; s. = sinters. **Specific Rotation** was determined in water at concentrations of 1–5 g per 100 ml. of solution and at 20°–25°C, unless otherwise specified; other temperatures or wavelengths are shown in brackets; *c* = grams solute per 100 ml of solution.

Part I. NATURAL MONOSACCHARIDES: ALDOSES AND KETOSES

	Substance (Synonym)	Chemical Formula	Melting Point °C	Specific Rotation $[\alpha]_D$
	(A)	(B)	(C)	(D)
	Aldoses			
1	D-Glyceraldehyde	$C_3H_6O_3$	....	+13.5 ± 0.5 (syrup)
2	D-Glyceraldehyde, 3-deoxy-3,3-*C*-bis-(hydroxymethyl)- (Cordycepose)	$C_5H_{10}O_4$	....	−26 (*c* 0.6, C_2H_5OH
3	D-Glyceraldehyde, 3,3-bis(*C*-hydroxy-methyl)- (Apiose)	$C_5H_{10}O_5$	....	+5.6 (*c* 10) [15°] syrup
4	β-D-Arabinose	$C_5H_{10}O_5$	155	−175 → −103
5	D-Arabinose, 2-*O*-methyl-	$C_6H_{12}O_5$	Syrup	−102
6	α-L-Arabinose	$C_5H_{10}O_5$	158 amorphous	+55.4 → +105
7	β-L-Arabinose	$C_5H_{10}O_5$	160	+190.6 → +104.5
8	DL-Arabinose	$C_5H_{10}O_5$	163.5–164.5	None
9	α-L-Lyxose	$C_5H_{10}O_5$	105	+5.8 → +13.5
10	L-Lyxose, 5-deoxy-3-*C*-formyl- (Streptose)	$C_6H_{10}O_5$	....	....
11	L-Lyxose, 3-*C*-formyl- (Hydroxy-streptose)	$C_6H_{10}O_6$	....	....
12	Pentose, 4,5-anhydro-5-deoxy-D-*erythro*-	$C_5H_8O_3$	....	....
13	Pentose, 2-deoxy-D-*erythro*-	$C_5H_{10}O_4$	96–98	−91 → −58
14	D-Ribose	$C_5H_{10}O_5$	87	−23.1 → −23.7
15	D-Ribose, 2-*C*-hydroxymethyl- (Hamamelose)	$C_6H_{12}O_6$	....	−7.1 [λ578]
16	α-D-Xylose	$C_5H_{10}O_5$	145	+93.6 → +18.8
17	D-Xylose, 5-deoxy-	$C_5H_{10}O_4$	....	+16
18	β-D-Xylose, 2-*O*-methyl-	$C_6H_{12}O_5$	137–138	−21 → +34
19	α-D-Xylose, 3-*O*-methyl-	$C_6H_{12}O_5$	95	+45 → +19
20	D-Allose, 6-deoxy-	$C_6H_{12}O_5$	140–143 146–148	+1.6 [18°] (*c* 0.6) −4.7 → 0
21	D-Allose, 6-deoxy-2,3-di-*O*-methyl- (Mycinose)	$C_8H_{16}O_5$	102–106	−46 → −29
22	Amicetose (a trideoxy hexose)	$C_6H_{12}O_3$	Oil, b.p. 65–70	+28.6 ($CHCl_3$)
23	Antiarose	$C_6H_{12}O_5$	....	Levo
24	α-D-Galactose	$C_6H_{12}O_6$	167	+150.7 → +80.2
25	β-D-Galactose	$C_6H_{12}O_6$	143–145	+52.8 → +80.2
26	D-Galactose, 3,6-anhydro-	$C_6H_{10}O_5$	....	+21.3 [10°]
27	α-D-Galactose, 6-deoxy- (D-Fucose; Rhodeose)	$C_6H_{12}O_5$	140–145	+127 → +76.3 (*c* 10)
28	D-Galactose, 6-deoxy-3-*O*-methyl- (Digitalose)	$C_7H_{14}O_5$	106[1], 119[2]	+106

CARBOHYDRATES (Continued)

Part I. NATURAL MONOSACCHARIDES: ALDOSES AND KETOSES (Continued)

	Substance (Synonym) (A)	Chemical Formula (B)	Melting Point °C (C)	Specific Rotation $[\alpha]_D$ (D)
	Aldoses (Con't)			
29	D-Galactose, 6-deoxy-4-*O*-methyl-	$C_7H_{14}O_5$	131–132	+82
30	D-Galactose, 6-deoxy-2,3-di-*O*-methyl-	$C_8H_{16}O_5$	....	+73
31	α-D-Galactose, 3-*O*-methyl-	$C_7H_{14}O_6$	144–147	+150.6 → +108.6
32	α-D-Galactose, 6-*O*-methyl-	$C_7H_{14}O_6$	122–123	+117 → +77.3
33	L-Galactose	$C_6H_{12}O_6$	....	*See* D-Galactose
34	α-L-Galactose, 3,6-anhydro-	$C_6H_{10}O_5$	....	−39.4 → −25.2
35	α-L-Galactose, 6-deoxy- (L-Fucose)	$C_6H_{12}O_5$	145	−124.1 → −76.4
36	L-Galactose, 6-deoxy-2-*O*-methyl-	$C_7H_{14}O_5$	149–150	−75 ± 4 (*c* 0.5)
37	L-Galactose, 6-sulfate	$C_6H_{12}O_9S$	....	−47 (*c* 0.2) (Na salt)
38	DL-Galactose	$C_6H_{12}O_6$	143–144, 163	None (racemic)
39	α-D-Glucose	$C_6H_{12}O_6$	146, 83 (H_2O)	+112 → +52.7
40	β-D-Glucose	$C_6H_{12}O_6$	148–150	+18.7 → +52.7
41	D-Glucose, 6-acetate	$C_7H_{14}O_7$	135	+48
42	D-Glucose, 2,3-di-*O*-methyl-	$C_8H_{16}O_6$	85–86, 121	+50
43	D-Glucose, 6-*O*-benzoyl- (Vaccinin)	$C_{13}H_{16}O_7$	Amorphous	+48 (C_2H_5OH)
44	α-D-Glucose, 6-deoxy- (Chinovose; Epirhamnose; Glucomethylose; Isorhamnose; Isorhodeose; Quinovose)	$C_6H_{12}O_5$	139–140	+73.3 → +29.7 (*c* 8)
45	α-D-Glucose, 6-deoxy-3-*O*-methyl- (D-Thevetose)	$C_7H_{14}O_5$	116	+84 → +33
46	D-Glucose, 6-sulfonic acid, 6-deoxy- (6-Sulfoquinovose)	$C_6H_{12}O_8S$	173–174	+87[3]
47	D-Glucose, 3-*O*-methyl-	$C_7H_{14}O_6$	162–167	+98 → +59.5
48	α-L-Glucose	$C_6H_{12}O_6$	141–143	−95.5 → −51.4
49	L-Glucose, 6-deoxy-3-*O*-methyl- (L-Thevetose)	$C_7H_{14}O_5$	126–129	−36.9 ± 2
50	D-Gulose, 6-deoxy-	$C_6H_{12}O_5$	....	....
51	Hexose, 2-deoxy-D-*arabino*-[4]	$C_6H_{12}O_5$	148	+46.6 [18°]
52	Hexose, 2,6-dideoxy-3-*O*-methyl-D-*arabino*- (D-Oleandrose)	$C_7H_{14}O_4$	....	−11
53	Hexose, 3,6-dideoxy-D-*arabino*- (Tyvelose)	$C_6H_{12}O_4$	....	+24 ± 2
54	Hexose, 2,6-dideoxy-3-*O*-methyl-L-*arabino*- (L-Oleandrose)	$C_7H_{14}O_4$	62–63	+11.9 ± 2.5
55	Hexose, 3,6-dideoxy-L-*arabino*- (Ascarylose)	$C_6H_{12}O_4$	....	−24 ± 2
56	Hexose, 2,6-dideoxy-3-*O*-methyl-D-*lyxo*- (Diginose)	$C_7H_{14}O_4$	90–92	+56 ± 4
57	Hexose, 2,6-dideoxy-L-*lyxo*- (L-Fucose, 2-deoxy-)	$C_6H_{12}O_4$	103–106	−61.6
58	Hexose, 2,6-dideoxy-3-*O*-methyl-L-*lyxo*-	$C_7H_{14}O_4$	78–85	−65
59	Hexose, 2,6-dideoxy-D-*ribo*- (Digitoxose; D-Altrose, 2,6-dideoxy-)	$C_6H_{12}O_4$	110	+46.4
60	Hexose, 2,6-dideoxy-3-*O*-methyl-D-*ribo*- (Cymarose)	$C_7H_{14}O_4$	93	+52
61	Hexose, 3,6-dideoxy-D-*ribo*- (Paratose)	$C_6H_{12}O_4$	....	+10 ± 2 (*c* 0.9)
62	Hexose, 4,6-dideoxy-3-*O*-methyl-D-*ribo*- (D-Gulose, 4,6-dideoxy-3-*O*-methyl-; Chalcose)	$C_7H_{14}O_4$	96–99	+120 → +76
63	Hexose, 2,6-dideoxy-D-*xylo*- (Boivinose)	$C_6H_{12}O_4$	96–98	−3.9 → +3.9

Part I. NATURAL MONOSACCHARIDES: ALDOSES AND KETOSES (Continued)

	Substance (Synonym)	Chemical Formula	Melting Point °C	Specific Rotation $[\alpha]_D$
	(A)	(B)	(C)	(D)
	Aldoses (Con't)			
64	Hexose, 2,6-dideoxy-3-*O*-methyl-D-*xylo*- (Sarmentose)	$C_7H_{14}O_4$	78–79	$+12 \rightarrow +15.8$
65	Hexose, 3,6-dideoxy-D-*xylo*- (Abequose)	$C_6H_{12}O_4$	....	-3.2 ± 0.6
66	Hexose, 2,6-dideoxy-3-*C*-methyl-L-*xylo*- (Mycarose)	$C_7H_{14}O_4$	129–129	−31.1
67	Hexose, 2,6-dideoxy-3-*C*-methyl-3-*O*-methyl-L-*xylo*-(Cladinose)	$C_8H_{16}O_4$	oil, b.p. 120–132 (0.25 mm)	−23.1
68	Hexose, 3,6-dideoxy-L-*xylo*- (Colitose)	$C_6H_{12}O_4$	....	+4 (H_2O); -51 ± 2 (CH_3OH)
69	D-Idose[5]	$C_6H_{12}O_6$	....	....
70	L-Idose, 1,6-anhydro-	$C_6H_{10}O_5$	....	....
71	α-D-Mannose	$C_6H_{12}O_6$	133	$+29.3 \rightarrow +14.5$
72	β-D-Mannose	$C_6H_{12}O_6$	132	$-16.3 \rightarrow +14.5$
73	D-Mannose, 6-deoxy- (D-Rhamnose)	$C_6H_{12}O_5$	86–90	−7.0
74	α-L-Mannose, 6-deoxy-monohydrate (L-Rhamnose)	$C_6H_{14}O_6$	93–94	$-8.6 \rightarrow +8.2$
75	β-L-Mannose, 6-deoxy-	$C_6H_{12}O_5$	123–125	$+38.4 \rightarrow +8.9$
76	L-Mannose, 6-deoxy-2-*O*-methyl-	$C_7H_{14}O_5$	....	....
77	L-Mannose, 6-deoxy-3-*O*-methyl- (L-Acofriose)	$C_7H_{14}O_5$	114–115	+30 [18°]
78	L-Mannose, 6-deoxy-2,4-di-*O*-methyl-	$C_8H_{16}O_5$	82	−19 [16°]
79	L-Mannose, 6-deoxy-5-*C*-methyl-4-*O*-methyl-(Noviose)	$C_8H_{16}O_5$	128–130	+19.9 (50% C_2H_5OH)
80	Rhodinose (a 2,3,6-trideoxyhexose)	$C_6H_{12}O_3$	....	-11 ± 1.6
81	D-Talose	$C_6H_{12}O_6$	128–132	+16.9
82	D-Talose, 6-deoxy- (D-Talomethylose)	$C_6H_{12}O_5$	129–131	+20.6
83	L-Talose, 6-deoxy- (L-Talomethylose)	$C_6H_{12}O_5$	116–118	-19.5 ± 2 [18°]
84	L-Talose, 6-deoxy-2-*O*-methyl- (L-Acovenose)	$C_7H_{14}O_5$	....	−19.4
85	Heptose, D-*glycero*-D-*galacto*-	$C_7H_{14}O_7$	139–140	$+47 \rightarrow +64$ (*c* 0.5)
86	Heptose, D-*glycero*-D-*manno*-	$C_7H_{14}O_7$	....	....
87	Heptose, D-*glycero*-L-*manno*-	$C_7H_{14}O_7$	....	....
	Ketoses			
88	Dihydroxyacetone	$C_3H_6O_3$	80 (dimer)	None
89	Tetrulose, L-*glycero*-[8] (L-Erythrulose; Ketoerythritol; L-Threulose)	$C_4H_8O_4$	Syrup	+12
90	Pentulose, D-*erythro*- (Adonose; D-Ribulose)	$C_5H_{10}O_5$	Syrup	+16.6 [27°]
91	Pentulose, L-*erythro*- (L-Ribulose)	$C_5H_{10}O_5$	....	−16.6
92	Pentulose, D-*threo*- (D-Xylulose)	$C_5H_{10}O_5$	....	−33
93	Pentulose, 5-deoxy-D-*threo*-	$C_5H_{10}O_4$	....	-5 ± 1 (CH_3OH)
94	Pentulose, L-*threo*- (L-Xylulose; L-Lyxulose; Xyloketose)	$C_5H_{10}O_5$	Syrup	+33.1
95	Hexulose, β-D-*arabino*-(β-D-Fructose; Levulose)	$C_6H_{12}O_6$	102–104[7]	$-133.5 \rightarrow -92$
96	Hexulose, 6-deoxy-D-*arabino*- (D-Rhamnulose)	$C_6H_{12}O_5$	....	-13 ± 2
97	Hexulose, D-*lyxo*- (D-Tagatose)	$C_6H_{12}O_6$	131–132	$+2.7 \rightarrow -4, -5$
98	5-Hexulose, D-*lyxo*	$C_6H_{12}O_6$	158	−86.6
99	Hexulose, 6-deoxy-L-*lyxo*- (L-Fuculose)	$C_6H_{12}O_5$	....	....

Part I. NATURAL MONOSACCHARIDES: ALDOSES AND KETOSES (Continued)

	Substance (Synonym)	Chemical Formula	Melting Point °C	Specific Rotation $[\alpha]_D$
	(A)	(B)	(C)	(D)
	Ketoses (Con't)			
100	Hexulose, D-*ribo*- (D-Psicose)	$C_6H_{12}O_6$	Amorphous	+4.7
101	Hexulose, L-*xylo*- (L-Sorbose)	$C_6H_{12}O_6$	159–161	−43.1
102	Hexulose, 6-deoxy-L-*xylo*-	$C_6H_{12}O_5$	88	−25 ± 2 (*c* 0.7)
103	Heptulose, D-*altro*- (Sedoheptulose; Sedoheptose)	$C_7H_{14}O_7$	Amorphous	+2.5 (*c* 10)
104	Heptulose·hemihydrate, L-*galacto*- (Perseulose)	$C_7H_{14}O_7 \cdot \frac{1}{2}H_2O$	110–115	−90 → −80
105	Heptulose, L-*gulo*-	$C_7H_{14}O_7$	....	−28
106	Heptulose, D-*ido*-	$C_7H_{14}O_7$	172	−34 ± 8 (*c* 0.3)
107	Heptulose, D-*manno*- (Mannoketoheptose; D-Mannotagatoheptose)	$C_7H_{14}O_7$	152	+29.4
108	Heptulose, D-*talo*-	$C_7H_{14}O_7$	....	....
109	Octulose, D-*glycero*-L-*galacto*-	$C_8H_{16}O_8$	....	−57, −43.4 → −13.4
110	Octulose, D-*glycero*-D-*manno*-	$C_8H_{16}O_8$	....	+20 (CH_3OH)

[1] Original melting point. [2] Melting point after four-months' storage. [3] As a methyl glycoside cyclohexylamine salt. [4] Included because of speculations concerning it in biological processes. [5] Either D-idose or L-altrose is in the polysaccharide varianose. [6] Early literature refers to this as D-erythrose. [7] The $\cdot\frac{1}{2}H_2O$ and $\cdot 2H_2O$ forms also exist.

Part II. NATURAL MONOSACCHARIDES: AMINO SUGARS

	Substance (Synonym)	Chemical Formula	Melting Point °C	Specific Rotation $[\alpha]_D$
	(A)	(B)	(C)	(D)
	Aldosamines			
1	D-Ribose, 3-amino-3-deoxy-	$C_5H_{11}NO_4$	158–158.5 d.	−24.6 (hydrochloride)
2	D-Galactose, 2-amino-2-deoxy- (Galactosamine; Chondrosamine)	$C_6H_{13}NO_5$	185	+121 → +80 (hydrochloride)
3	α-L-Galactose, 2-amino-2,6-dideoxy- (L-Fucosamine)	$C_6H_{13}NO_4$	192–193 d.	−119 → −92 [27°] (hydrochloride)
4	α-D-Glucose, 2-amino-2-deoxy- (Glucosamine; Chitosamine)	$C_6H_{13}NO_5$	88	+100 → +47.5
5	β-D-Glucose, 2-amino-2-deoxy-	$C_6H_{13}NO_5$	110–111	+28 → +47.5
6	D-Glucose, 3-amino-3-deoxy- (Kanosamine)	$C_6H_{13}NO_5$	128 d.	+19 [14°]
7	D-Glucose, 6-amino-6-deoxy-	$C_6H_{13}NO_5$	161–162 d.	+23 → +50.1 (hydrochloride)
8	D-Glucose, 2,6-diamino-2,6-dideoxy- (Neosamine C)	$C_6H_{14}N_2O_4$	>230	+61.5 (dihydrochloride)
9	D-Glucose, 3,6-dideoxy-3-dimethylamino- (Mycaminose)	$C_8H_{17}NO_4$	115–116	+31 (hydrochloride)
10	D-Glucose, 4,6-dideoxy-4-dimethylamino-	$C_8H_{17}NO_4$	192–193	+45.5 (hydrochloride)
11	L-Glucose, 2-deoxy-2-methylamino-	$C_7H_{15}NO_5$	130–132	−64
12	D-Gulose, 2-amino-1,6-anhydro-2-deoxy-	$C_6H_{11}NO_4$	250–260 d.	+41 ± 2 (hydrochloride)
13	D-Gulose, 2-amino-2-deoxy-	$C_6H_{13}NO_5$	152–162 d.	+5.6 → −18.7 (hydrochloride)

Part II. NATURAL MONOSACCHARIDES: AMINO SUGARS (Continued)

	Substance (Synonym)	Chemical Formula	Melting Point °C	Specific Rotation $[\alpha]_D$
	(A)	(B)	(C)	(D)
	Aldosamines (Con't)			
14	Hexose, 3,4,6-trideoxy-3-dimethylamino-D-*xylo*- (Desosamine; Picrocine)	$C_8H_{17}NO_3$	189–191 d.	+49.5 (*c* 10) (hydrochloride)
15	Hexose, a 4-acetamido-2-amino-2,4,6-trideoxy-	$C_8H_{16}N_2O_4$	216–219	+115 → +94 [26°] (*c* 0.05)
16	Hexose, an amino-deoxy-3-*O*-carboxyethyl-	$C_9H_{17}NO_7$	....	....
17	Hexose, a 2,6-diamino-2,6-dideoxy- (Neosamine B; Paramose)	$C_6H_{14}N_2O_4$	135–150 d.	+17.5 (*c* 0.9 (hydrochloride)
18	Hexose, a 3-dimethylamino-2,3,6-trideoxy- (Rhodosamine)	$C_8H_{17}NO_3$	....	....
19	D-Mannose, 2-amino-2-deoxy- (Mannosamine)	$C_6H_{13}NO_5$	142 d.	−4.3 (*c* 9) (hydrochloride)
20	D-Mannose, 3-amino-3,6-dideoxy- (Mycosamine)	$C_6H_{13}NO_4$	162	−11.5 (hydrochloride)
21	D-Talose, 2-amino-2-deoxy- (Talosamine)	$C_6H_{13}NO_5$	151–153	+3.4 → −5.7 (*c* 0.9) (hydrochloride)
22	L-Talose, 2-amino-2,6-dideoxy- (Pneumosamine)	$C_6H_{13}NO_4$	162–163	+6.9 → +10.4 (hydrochloride)
	Ketosamines			
23	Pentulose, 1-(*o*-carboxyanilino)-1-deoxy-D-*erythro*-	$C_{12}H_{14}NO_6$	....	....
24	Hexulose, 1-(*o*-carboxyanilino)-1-deoxy-D-*arabino*-	$C_{13}H_{16}NO_7$	....	....
25	Hexulose, 5-amino-5-deoxy-L-*xylo*-	$C_6H_{13}NO_5$	174–176	−62
26	Hexulose, 6-deoxy-6-(*N*-methylacetamido)-L-*xylo*-	$C_9H_{17}NO_6$	....	....

Part III. NATURAL ALDITOLS AND INOSITOLS (with Inososes and Inosamines)

	Substance (Synonym)	Chemical Formula	Melting Point °C	Specific Rotation $[\alpha]_D$
	(A)	(B)	(C)	(D)
	Alditols			
1	Glycerol	$C_3H_8O_3$	20	None
2	Glycerol, 1-deoxy- (1,2-Propane-diol)[1]	$C_3H_8O_2$	Oil, b.p. 188–189	None (racemic)
3	Erythritol	$C_4H_{10}O_4$	118–120	None (meso)
4	Erythritol, 1,4-dideoxy- (2,3-Butylene-glycol)	$C_4H_{10}O_2$	25, 34	None (meso)
5	D-Threitol, 1,4-dideoxy-	$C_4H_{10}O_2$	19	−13.0
6	L-Threitol, 1,4-dideoxy-	$C_4H_{10}O_2$	....	+10.2
7	DL-Threitol, 1,4-dideoxy-	$C_4H_{10}O_2$	7.6	None (racemic)
8	D-Arabinitol	$C_5H_{12}O_5$	103	+7.82 (*c* 8, borax solution)
9	L-Arabinitol	$C_5H_{12}O_5$	101–102	−32 (*c* 0.4, 5% molybdate)
10	Ribitol (Adomitol)	$C_5H_{12}O_5$	102	None (meso)
11	Galactitol (Dulcitol)	$C_6H_{14}O_6$	186–188	None (meso)

CARBOHYDRATES (Continued)
Part III. NATURAL ALDITOLS AND INOSITOLS
(with Inososes and Inosamines) (Continued)

	Substance (Synonym)	Chemical Formula	Melting Point °C	Specific Rotation $[\alpha]_D$
	(A)	(B)	(C)	(D)
		Alditols (Con't)		
12	D-Glucitol (Sorbitol)	$C_6H_{14}O_6$	112	−1.8 [15°]
13	D-Glucitol, 1,5-anhydro- (Polygalitol)	$C_6H_{12}O_5$	140–141	+42.4
14	L-Iditol	$C_6H_{14}O_6$	73.5	−3.5 (*c* 10)
15	D-Mannitol	$C_6H_{14}O_6$	166	−0.21
16	D-Mannitol, 1,5-anhydro- (Styracitol)	$C_6H_{12}O_5$	157	−49.9
17	Heptitol, D-*glycero*-D-*galacto*- (Heptitol, L-*glycero*-D-*manno*-; Perseitol)	$C_7H_{16}O_7$	183–185, 188	−1.1
18	Heptitol, D-*glycero*-D-*gluco*- (Heptitol, L-*glycero*-D-*talo*-; β-Sedoheptitol)	$C_7H_{16}O_7$	131–132	+46 (5% NH_4 molybdate)
19	Heptitol, D-*glycero*-D-*manno*- (Heptitol, D-*glycero*-D-*talo*-; Volemitol)	$C_7H_{16}O_7$	153	+2.65
20	Octitol, D-*erythro*-D-*galacto*-	$C_8H_{18}O_8 \cdot H_2O$	169–170	−11 (5% NH_4 molybdate)
		Inositols		
21	Betitol (a dideoxy inositol)	$C_6H_{12}O_4$	224	....
22	Bioinosose (*scyllo*-Inosose; *myo*-Inosose-2; a deoxy keto inositol)	$C_6H_{10}O_6$	198–200	None (meso)
23	*h*-Bornesitol (a *myo*-inositol monomethyl ether)	$C_7H_{14}O_6$	200	+31.6
24	*l*-Bornesitol (a *myo*-inositol monomethyl ether)	$C_7H_{14}O_6$	205–206	−32.1
25	Conduritol (a 2,3-dehydro-2,3-dideoxyinositol)	$C_6H_{10}O_4$	142–143	None (meso)
26	Cordycepic acid (a tetrahydroxycyclohexanecarboxylic acid)[2]	$C_7H_{12}O_6$	....	....
27	Dambonitol (a *myo*-inositol dimethyl ether)	$C_8H_{16}O_6$	206	None (meso)
28	DL-Inositol	$C_6H_{12}O_6$	253	None (racemic)
29	*d*-Inositol	$C_6H_{12}O_6$	....	+60
30	*l*-Inositol	$C_6H_{12}O_6$	240	−65
31	Laminitol (a *C*-methyl *myo*-inositol)	$C_7H_{14}O_6$	266–269	−3
32	Liriodendritol (a *myo*-inositol dimethyl ether)	$C_8H_{16}O_6$	224	−25
33	*muco*-Inositol monomethyl ether	$C_7H_{14}O_6$	322–325	....
34	*myo*-Inositol (*meso*-Inositol)	$C_6H_{12}O_6$	217–218	None (meso)
35	*d*-*myo*-Inosose-1 (a deoxy keto inositol)	$C_6H_{10}O_6$	138–139	+19.6
36	Mytilitol (a *C*-methyl *scyllo*-inositol)	$C_7H_{14}O_6$	259	None (meso)
37	*neo*-Inosamine-2 (a deoxy amino inositol)	$C_6H_{13}O_5N$	239–241 d.	None (meso)
38	*d*-Ononitol (a *myo*-inositol monomethyl ether)	$C_7H_{14}O_6$	172	+6.6
39	*h*-Pinitol (a *dextro*-inositol monomethyl ether)	$C_7H_{14}O_6$	186	+65.5
40	*l*-Pinitol (a *levo*-inositol monomethyl ether)	$C_7H_{14}O_6$	186	−65
41	*l*-Quebrachitol (a *levo*-inositol monomethyl ether)	$C_7H_{14}O_6$	190–191	−80.2 [28°]
42	*d*-Quercitol (a deoxy *dextro*-inositol)	$C_6H_{12}O_5$	235	+24.2
43	*d*-Quinic acid (a trideoxy carboxy *dextro*-inositol)	$C_7H_{12}O_6$	164	+44 (*c* 10)

CARBOHYDRATES (Continued)
Part III. NATURAL ALDITOLS AND INOSITOLS
(with Inososes and Inosamines) (Continued)

	Substance (Synonym) (A)	Chemical Formula (B)	Melting Point °C (C)	Specific Rotation $[\alpha]_D$ (D)
	Inositols (Con't)			
44	*l*-Quinic acid (a trideoxy carboxy *levo*-inositol)	$C_7H_{12}O_6$	162	−42.1
45	Quinic acid, 5-dehydro-	$C_7H_{10}O_6$	140–142 (138 s.)	−82.4 [28°]
46	Scyllitol (*scyllo*-Inositol; Cocositol)	$C_6H_{12}O_6$	352–353	None (meso)
47	Sequoyitol (a *myo*-inositol monomethyl ether)	$C_7H_{14}O_6$	234–235	None (meso)
48	Shikimic acid (a 3,4-anhydro-quinic acid)	$C_7H_{10}O_5$	183–184	−200 [16°]
49	Shikimic acid, 5-dehydro-	$C_7H_8O_5$	150–152	−57.5 [28°] (EtOH)
50	Streptamine (2,4-diaminodideoxy-scyllitol)	$C_6H_{14}O_4N_2$	88, 210–250 d.	None (meso)
51	Streptamine, 2-deoxy-	$C_6H_{14}O_3N_2$	....	None (meso)
52	Streptadine (1,3-Dideoxy-1,3-diguanidino-scyllitol)	$C_8H_{18}N_6O_4$	....	None (meso)
53	Viburnitol (a deoxy *levo*-inositol)[3]	$C_6H_{12}O_5$	174	−73.9

[1] The 1-phosphate ester of this diol is said to occur in brain tissue and sea-urchin eggs. [2] Strong evidence that cordycepic acid is really D-mannitol. [3] Not an enantiomorph of *d*-quercitol; other isomeric relationship is involved.

Part IV. NATURAL ALDONIC, URONIC, AND ALDARIC ACIDS

	Substance (Synonym) (A)	Chemical Formula (B)	Melting Point °C (C)	Specific Rotation $[\alpha]_D$ (D)
	Aldonic Acids			
1	D-Glyceric acid	$C_3H_6O_4$	Gum	Dextro
2	L-Glyceric acid	$C_3H_6O_4$	Gum	Levo
3	D-Arabinonic acid	$C_5H_{10}O_6$	114–116	+10.5 (*c* 6)
4	L-Arabinonic acid	$C_5H_{10}O_6$	118–119	−9.6 → −41.7[1]
5	L-Arabinonic-1,4-lactone	$C_5H_8O_5$	97–99	−72
6	D-Ribonic acid	$C_5H_{10}O_6$	112–113	−17.0
7	D-Xylonic acid	$C_5H_{10}O_6$	....	−2.9 → +20.1[1]
8	L-Xylonic acid	$C_5H_{10}O_6$	....	−91.8[1]
9	D-Altronic acid	$C_6H_{12}O_7$	....	+11.5 → +24.8[1] (Ca salt, *N* HCl)
10	D-Galactonic acid	$C_6H_{12}O_7$	122	−11.2 → +57.6[1]
11	D-Gluconic acid	$C_6H_{12}O_7$	130–132 (110–112 s.)	−6.7 → +11.9[1]
12	L-Gulonic acid	$C_6H_{12}O_7$	Exists only in soln.	[ca. 0°]
13	Hexsonic acid, 2-deoxy-D-*arabino*-	$C_6H_{12}O_6$	93–95	+68 (lactone)
14	2-Hexulosonic acid, D-*arabino*-	$C_6H_{10}O_7$	....	−81.7 (Na salt)
15	2-Hexulosonic acid, 3-deoxy-D-*erythro*-	$C_6H_{10}O_6$	....	−29.2 (*c* 6, Ca salt)
16	2-Hexulosonic acid, D-*lyxo*-	$C_6H_{10}O_7$	169	−5
17	5-Hexulosonic acid, D-*arabino*-	$C_6H_{10}O_7$	108–109	....
18	5-Hexulosonic acid, D-*xylo*-	$C_6H_{10}O_7$	....	−14.5
19	D-Mannonic acid	$C_6H_{12}O_7$	....	−15.6
20	D-Gluconic acid, *O*-β-D-galactopyranosyl- (1 → 4)- (Lactobionic acid)	$C_{12}H_{22}O_{12}$	....	+25.1 (Ca salt)

Part IV. NATURAL ALDONIC, URONIC, AND ALDARIC ACIDS (Continued)

	Substance (Synonym)	Chemical Formula	Melting Point °C	Specific Rotation $[\alpha]_D$
	(A)	(B)	(C)	(D)
	Uronic Acids			
21	L-Lyxuronic acid	$C_5H_8O_6$	....	....
22	β-D-Galacturonic acid	$C_6H_{10}O_7$	160	+27 → +55.6
23	α-D-Galacturonic acid·monohydrate	$C_6H_{12}O_8$	159–160 (110–115 s.)	+97.9 → +50.9
24	D-Galacturonic acid, 2-amino-2-deoxy-	$C_6H_{11}O_6N$	160 d.	+84.5 (pH 2 HCl)
25	β-D-Glucuronic acid	$C_6H_{10}O_7$	156	+11.7 → +36.3
26	D-Glucuronic acid, 2-amino-2-deoxy-	$C_6H_{11}O_6N$	120–172 d.	+55
27	D-Glucuronic acid, 3-*O*-methyl-	$C_7H_{12}O_7$	Syrup	+6
28	L-Guluronic acid	$C_6H_{10}O_7$	....	....
29	L-Iduronic acid	$C_6H_{10}O_7$	....	+30
30	β-D-Mannuronic acid	$C_6H_{10}O_7$	165–167	−47.9 → −23.9
31	α-D-Mannuronic acid·monohydrate	$C_6H_{12}O_8$	110 s., 120–130 d.	+16 → −6.1 (*c* 6.8)
	Aldaric Acids			
32	D-Tartaric acid	$C_4H_6O_6$	170	−15
33	L-Tartaric acid	$C_4H_6O_6$	170	+15 [15°]
34	L-Malic acid	$C_4H_6O_5$	100	−2.3 (*c* 8.4)

[1] Equilibrates with the lactone.

WAXES

These data for waxes were compiled originally for the Biology Data Book by A. H. Warth. Data are reproduced here by permission of the copyright owners of the above publication, the American Societies for Experimental Biology, Washington, D.C. p. 382.

Specific Gravity (column C) was calculated at the specified temperature, degrees centigrade, and referred to water at the same temperature. **Density**, shown in parentheses (column C), and **Refractive Index** (column D) were measured at the specified temperature, degrees centigrade.

	Wax	Melting Point °C	Specific Gravity or (Density)	Refractive Index $n\frac{°C}{D}$	Iodine Value	Acid Value	Saponification Value
	(A)	(B)	(C)	(D)	(E)	(F)	(G)
1	Bamboo leaf	79–80	$(0.961^{25°})$	....	7.8[1]	14.5	43.4
2	Bayberry (myrtle)	46.7–48.8	$(0.985^{15°})$	$1.436^{80°}$	2.9[2]–3.9[3]	3.5	20.5–21.7
3	Beeswax, crude	62–66	$(0.927–0.970^{15°})$	$1.439–1.483^{40°}$	6.8–16.4[2]	16.8–35.8	89.3–149.0
4	Beeswax, white, U.S.P.	61–69	$(0.959–0.975^{15°})$	$1.447–1.465^{65°}$	7–11[3]	17–24	90–96
5	Beeswax, yellow	62–65	$(0.960–0.964^{15°})$	$1.443–1.449^{65°}$	6–11	18–24	90–97
6	Candelilla, refined	67–69	$(0.982–0.986^{15°})$	$1.454–1.463^{85°}$	14.4–20.4	12.7–18.1	35–86
7	Cape berry[4]	40.5–45.0	$(1.004–1.007^{15°})$	$1.450^{45°}$	0.6–2.4	2.5–3.7	211–215
8	Carandá	79.7–84.5	$(0.990^{25°})$	....	8.0–8.9	5.0–9.5	64.5–78.5
9	Carnauba	83–86	$0.990–1.001^{15°}$	$1.467–1.472^{40°}$	7.2–13.5	2.9–9.7	78–95
10	Castor oil, hydrogenated	83–88	$(0.980–0.990^{20°})$	....	2.5–8.5	1.0–5.0	177–181
11	Chinese insect	81.5–84.0	$0.950–0.970^{15°}$	$1.457^{40°}$	1.4	0.2–1.5	73–93
12	Cotton	68–71	$0.959^{15°}$	....	24.5	32	70.6
13	Cranberry	207–218	$(0.970–0.975^{15°})$	....	44.2–53.2[2]	42.2–59.1	131–134
14	Douglas-fir bark	59.0–72.8	$(1.030^{25°})$	$1.468^{80°}$	25.8–62.5	58.6–80.7	112–200
15	Esparto	67.5–78.1	$0.988^{15°}$	....	22–23	22.7–23.9	69.8–79.3
16	Flax	61.5–69.8	$0.908–0.985^{15°}$	....	21.6–28.8	17.5–48.3	77.5–101.5
17	Ghedda, E. Indian beeswax	60.5–66.4	$0.956–0.973^{15°}$	$1.440^{50°}$	5.6–12.6	5.8–7.9	84.5–118.3
18	Indian corn	80–81	....	....	4.2[2]	1.9	120.3
19	Japan wax	48–53	$0.975–0.993^{15°}$	....	4.5–12.5	6–20	206.5–237.5
20	Jojoba	11.2–11.8	$0.864–0.899^{25°}$	$1.465^{25°}$	81.7–88.4[2]	0.2–0.6	92.2–95.0
21	Madagascar	88	....	....	3.2–5.3	17.7–28.0	140.0–159.6
22	Microcrystalline, amber	64–91	$0.913–0.943^{15°}$	$1.424–1.452^{80°}$	0	0	0
23	Microcrystalline, white	71–89	$0.928–0.941^{15°}$	$1.441^{80°}$	0	0	0
24	Montan, crude	76–86	$(1.010–1.020^{25°})$	....	13.9–17.6	22.7–31.0	59.4–92.0
25	Montan, refined	77–84	$(1.010–1.030^{25°})$	....	10–14	24–43	72–103
26	Orange peel	44.0–46.5	$0.985^{15°}$	$1.502^{20°}$	115.7[2]	48.3	120.9
27	Ouricury, refined	79.0–83.8	$1.053^{15°}$	....	6.9–7.8[2]	3.4–21.1	61.8–85.8
28	Ozocerite, refined	74.4–75.0	$0.907–0.920^{15°}$	....	0	0	0
29	Palm	74–86	$(0.991–1.045^{15°})$	....	8.9–16.9[2]	5.0–10.6	64.5–104.0
30	Paraffin, American	49–63	$0.896–0.925^{15°}$	$1.442–1.448^{80°}$	0	0	0
31	Peat wax, natural	73–76	$0.980^{15°}$	....	16–40	60.0–73.3	73.9–136.0
32	Rice bran, refined	75.3–79.9	....	$1.469^{30°}$	11.1–19.4	15–17	56.9–104.4
33	Shellac wax	79–82	$0.971–0.980^{15°}$	....	6.0–8.8[3]	12.1–24.3	63.8–83.0
34	Sisal hemp	74–81	$1.007–1.010^{15°}$	....	28–29[2]	16–19[2]	56–58
35	Sorghum grain	77–82	....	....	15.7–20.9	10.1–16.2	16–44
36	Spanish moss	79–80	....	....	33.0	25.0	120.4
37	Spermaceti	42–50	$0.905–0.945^{15°}$	$1.440^{70°}$	4.8–5.9	2.0–5.2	108–134
38	Sugarcane, crude	52–67	$0.988–0.998^{25°}$	....	32–84	24–57	128–177
39	Sugarcane, double-refined	77–82	$0.961–0.979^{25°}$	$1.510^{25°}$	13–29	8–23	55–95
40	Wool wax, refined	36–43	$0.932–0.945^{15°}$	$1.478–1.482^{40°}$	15.0–46.9	5.6–22.0	80–127

[1] Wijs test. [2] Hanus test. [3] Hubl test. [4] *Myrica cordifolia*.

TRADE NAMES OF DYESTUFF INTERMEDIATES

Trade name	Chemical name
A acid	1,7-Hydroxynaphthalene-3,6-disulfonic acid
Acetyl H acid	N-Acetyl-1-amino-8-naphthol-3,6-disulfonic acid
Alen's acid	1-Naphthylamine-3,6-disulfonic acid (also Freund's ac.)
Alizarin	1,2-Dihydroxyanthraquinone
Amido acid	2-Amino-7-hydroxynaphthalene-5-sulfonic acid
Amido J acid	2-Naphthylamine-5,7-disulfonic acid
Amino G acid	7-Amino-1,3-naphthalene disulfonic acid 2-Naphthylamine-6,8-disulfonic acid
Amino R acid	3-Amino-2,7-naphthalene disulfonic acid 2-Naphthylamine-3,6-disulfonic acid
Aminophenolic acid V	1-Amino-3-oxybenzene-5-sulfonic acid
Aminophenol sulfonic acid III	1-Amino-3-oxybenzene-6-sulfonic acid
Andresen's acid	1-Naphthol-3,8-disulfonic acid
Anisidine	o-Aminophenol methylether
Anthrachrysone	1,3,5,7-Tetrahydroanthraquinone
Anthraflavic acid	2,6-Dihydroxyanthraquinone
Anthranilic acid	o-Amidobenzoic acid
Anthrarufin	1,5-Dihydroxyanthraquinone
Anthranol	9-Hydroxyanthracene
α Anthrol	1-Hydroxyanthracene
Armstrong's acid	Naphthalene-1,5-disulfonic acid
Armstrong & Wynne acid	1-Naphthol-3-sulfonic acid
Armstrong & Wynne acid II	2-Naphthylamino-5,7-disulfonic acid
B acid	8-Amino-1-naphthol-4,6-disulfonic acid
Badische acid	2-Naphthylamino-8-sulfonic acid
Bayer's acid	2-Naphthol-8-sulfonic acid
Benzidine	p,p′-Diaminodiphenyl
Bronner's acid	2-Naphthylamino-6-sulfonic acid
β acid	Anthraquinone-2-sulfonic acid
C acid CLT acid	6-Chloro-m-toluidine-4-sulfonic acid
Casella's acid	2-Naphthol-7-sulfonic acid (F acid)
Chicago acid	1-Amino-8-naphthyl-2,4-disulfonic acid
Chloro H acid	8-Chloro-1-naphthol-3,6-disulfonic acid
Chromogene I	4,5-Dihydroxy-2,7-naphthalene disulfonic acid
Chromotrope acid	1,8-Dihydronaphthalene-3,6-disulfonic acid
Chromotropic acid	4,5-Dihydroxy-2,7-naphthalene disulfonic acid
Chrysazine	1,8-Dihydroxyanthraquinone
Cleve's acid	1-Naphthylamine-3-sulfonic acid
Cleve's acid	1-Naphthylamine-5-sulfonic acid
Cleve's acid	1-Naphthylamine-6-sulfonic acid
Cleve's acid	1-Naphthylamine-7-sulfonic acid
α Coccinic acid	1-Methyl-5-oxybenzyl-2,4-dicarbonic acid
Cresidine	3-Amino-4-methoxytoluene
Cresotic acid	Cresol carboxylic acid
Croceine acid	2-Naphthol-1-sulfonic acid
DS	4,4′-Diamino-2,2′-stilben disulfuric acid
DTS	Dihydrothio-p-toluidine sulfonic acid
Dahl's acid	2-Naphthylamine-5-sulfonic acid
Dahl's acid II	1-Naphthylamine-4,6-disulfonic acid
Dahl's acid III	1-Naphthylamine-4,7-disulfonic acid
Dimethyl-γ-acid	7-Dimethylamino-1-naphthol sulfonic acid
Dioxy G acid	1,7-Dihydroxynaphthalene-3-sulfonic acid
Dioxy J acid	1,6-Dihydroxynaphthalene-3-sulfonic acid
Dioxy S acid	1,8-Dihydroxynaphthalene-4-sulfonic acid
Diphenylblack base	p-Aminodiphenylamine
Disulfo acid S	1-Naphthylamine-4,8-disulfonic acid
δ acid	1-Naphthol-4,8-disulfonic acid 1-Naphthylamine-4,8-disulfonic acid

TRADE NAMES OF DYESTUFF INTERMEDIATES (Continued)

Trade name	Chemical name
Ebert & Merz acid	Naphthalene-2,7-disulfonic acid Naphthalene-2,6-disulfonic acid
Ethyl-γ-acid	7-Ethylamino-1-naphthol-3-sulfonic acid
Ethyl F acid	7-Ethylamino-2-naphthalene sulfonic acid
Ewer & Pick's acid	Naphthalene-1,6-disulfonic acid
ε acid	1-Naphthol-3,8-disulfonic acid 1-Naphthylamine-3,8-disulfonic acid
F acid	2-Naphthol-7-sulfonic acid (Casella's acid)
Fast Black B base	4,4′-Diamino diphenylamine
Fast Blue base	Dianisidine
Fast Blue Red O base	3-Nitro-p-phenetidine
Fast Bordeaux GP	3-Nitro-p-anisidine
Fast Orange GR	o-Nitroaniline
Fast Orange R	m-Nitroaniline
Fast Red base AL	α-Aminoanthraquinone
Fast Red B base	5-Nitro-o-anisidine
Fast Red GG base	p-Nitroaniline
Fast Red GL base	3-Nitro-p-toluidine
Fast Red 3 GL base	4-Chloro-2-nitroaniline
Fast Red RL base	5-Nitro-o-toluidine
Fast Scarlet G base	4-Nitro-α-toluidine
Fast Scarlet R base	4-Nitro-O-anisidine
Forsling's acid I	2-Naphthylamine-8-sulfonic acid
Forsling's acid II	2-Naphthylamine-5-sulfonic acid
Freund's acid	1-Naphthylamine-3,6-disulfonic acid
G acid	2-Naphthol-6,8-disulfonic acid
GR acid	α-Naphthol-3,6-disulfonic acid
Gallic acid	3,4,5-Trihydroxybenzoic acid
γ acid	2-Amino-8-naphthol-6-sulfonic acid
H acid	1-Amino-8-naphthol-3,6-disulfonic acid
Histazarin	2,3-Dihydroxyanthraquinone
Isoanthraflavic acid	2,7-Dihydroxyanthraquinone
J acid	2-Amino-5-naphthol-7-sulfonic acid
K acid	1-Amino-8-naphthol-4,6-disulfonic acid
Kalle's acid	1-Naphthylamine-2,7-disulfonic acid
Ketone base	Tetramethyl aminobenzophenone
Koch's acid	1-Naphthylamine-3,6,8-trisulfonic acid
L acid	1-Naphthol-5-sulfonic acid
Laurent's acid	1-Naphthylamine-5-sulfonic acid
Lepidine	4-Methylquinoline
Leucotrop	Phenyldimethyl benzylammonium chloride
M acid	1-Amino-5-naphthol-7-sulfonic acid
Mesidine	2,4,6-Thimethylaniline
Metanilic acid	Aniline-m-sulfonic acid
Methyl-γ-acid	7-Methyl-8-naphthol disulfonic acid
Michler's hydrol	Tetramethyl diaminobenzohydrol
Michler's ketone	Tetramethyl diaminobenzophenone
Myrbane oil	Nitrobenzene
Naphthacetol	1-Acetylamino-4-naphthol
Naphthazarin	5,8-Dihydroxy-1,4-naphthoquinone
Naphthionic acid	1-Naphthylamine-4-sulfonic acid
o-Naphthionic acid	1-Naphthylamine-2-sulfonic acid
Naphthol AS	Anilide of hydronaphthoic acid
Naphthoresorcine	1,3-Dihydroxynaphthalene
Nekal BX	Na-salt of 1,4-bis,sec-butylnaphthalene-6-sulfonic acid
Nevile and Winther's acid	1-Naphthol-4-sulfonic acid
Nigrotic acid	1,7,3,6-Dihydroxysulfonaphthoic acid
Nitron 1,2,4-acid	1-Amino-8-nitro-7-naphthol-4-sulfonic acid
Nitroso base	p-Nitrodimethyl aniline

Trade name	Chemical name
NW acid	Nevile and Winther s acid
Oxy L acid	1-Naphthol,5-sulfonic acid
Oxy Tobias acid	β-Naphthol-1-sulfonic acid
Peri acid	1-Naphthylamine-8-sulfonic acid
p-Phenetidine	p-Aminophenol ethylether
Phenyl gamma acid	2-Phenylamine-8-naphthol-6-sulfonic acid
Phenyl Peri acid	Phenyl-1-napthylamine-8-sulfonic acid
Phosxgene	Carbonyl chloride
Phthalic acid	O-Benzenedicarbolic acid
Picramic acid	2-Amino-4,6-dinitrophenol
Picric acid	2,4,6-Trinitrophenol
Pirio's acid	4-Amino-1-naphthalene sulfonic acid
Primuline base	p-Toluidine heated with sulfur
Purpurine	1,2,4-Trihydoxyanthraquinone
Pyrogallol	1,2,3-Trihydroxybenzene
Quinaldine	2-Methylquinoline
Quinazarin	1,4-Dihydroxyanthraquinone
R acid	2-Naphthol-3,6-disulfonic acid
2 R acid	2-Amino-8-naphthol-3,6-disulfonic acid
Red acid	1,5-Dihydroxynaphthalene-3,7-disulfonic acid
RG acid	1-Naphthol-3,6-disulfonic acid
Resorcinol	1,3-Dihydroxybenzene
Rumpff acid	2-Naphthol-8-sulfonic acid (Croceine acid)
S acid	1-Amino-8-naphthol-4-sulfonic acid
2 S acid	1-Amino-8-naphthol-2,4-disulfonic acid
Salicylic acid	o-Hydroxybenzoic acid
Schäffer's acid	2-Naphthol-6-sulfonic acid
Schäffer and Baum acid	α-Naphthol-2-sulfonic acid
Schollkopf's acid	1-Naphthol-4,8-disulfonic acid 1-Naphthylamine-4,8-disulfonic acid 1-Naphthylamine-8-sulfonic acid
Sulfanilic acid	Aniline-p-sulfonic acid
Thiocarbanilide	Diphenylthiourea
Tobias acid	2-Naphthylamine-1-sulfonic acid
Tolidine	Di-p-aminoditolyl
Toluidine	Aminotoluene
Violet acid	α-Naphthol-3,6-disulfuric acid
Xylidine	Aminoxylene
Y acid	2-Naphthol-6,8-disulfuric acid
Yellow acid	1,3-Dihydroxynaphthalene-5,7-disulfuric acid
1:2:4 acid	1-Amino-2-naphthol-4-sulfonic acid

TRADE NAMES, COMPOSITION AND MANUFACTURERS OF SOME PLASTICS

Trade name	Composition	Manufacturer
A-100 SM	Acrylics	Resolite Corp.
Abson	Acrylonitrile-butadiene ABS polymers	B. F. Goodrich Chemical Co.
Acco	Polymers, copolymers of acrylics and methacrylate esters	Acco Polymers Company
Acetex	Synthetic resin latices	United States Rubber Co., Naugatuck Chemical Div.
Acrilan	Methyl methacrylate	American Viscose Corp.
Actol	Polyesters	Allied Chemical Corp.
Adipol BCA	Di-butoxyethyl adipate	Food Machinery and Chemical Corp.
Agile	Polyesters, polyethylenes, polypropylenes, polyvinyl chlorides	American Agile Corp.
Alathon	Polyethylene resins	E. I. duPont de Nemours & Co., Inc.
Alkor	Furane resin cement	Atlas Mineral Products
Alpex	Hydrocarbon resins	Alkydol Laboratories, Div. of Reichhold Chemicals, Inc.
Alpha	Nylons, polyethylenes, ABS-polymers, polypropylenes, polyvinyl chlorides	Alpha Plastics, Inc.
Ameripol	High density polyethylenes	Goodrich-Gulf Chemicals, Inc.
Ameripol CB	Polybutadienes	" " " "
Ameripol SBR	Butediene-styrene rubber	" " " "
Amerith	Cellulose nitrate	Celanese Corp. of America
Amphenol	Polystyrene	American Phenolic Co.
Amres	Phenolics, resorcinol, urea	American Marietta Co.
Anaconda	Polyethylenes, polyvinyl chlorides	Anaconda American Brass Co.
Aquaplex	Alkyd-resin emulsions	Rohm & Hass Company
Araldite	Epoxy resins	Ciba Products Corp.
Arcolite	Phenolic	Consolidated Molded Products Co.
Aritemp	Epoxy resins	Aries Laboratories, Inc.
Arociro	Polychlorinated polyphenyls	Monsanto Chemical Company
Arodures	Ureas	National Distillers & Chemical Corp.
Atlac	Polyester resins in solid state	Atlas Chemical Industries, Inc.
Avisco	Urea-formaldehyde compounds	American Viscose Corporation
AviSun	Polypropylene resins	AviSun Corporation
Bakelite	Acrylics, epoxies, phenolics, polyethylenes, copolymers	Union Carbide Corp., Plastics Div.
Bavick-11	Acrylics	J. T. Baker Chemical Co.
Beckacite	Maleic, fumaric, modified phenolic resins	Reichold Chemicals, Inc.
Beckamine	Alcohol-soluble urea-formaldehyde resins for laminates, adhesives, coating	" " "
Beckopol	High melt point modified phenolic resin	" " "
Blacar	Vinyl chloride resin; vinyl compound	Cary Chemicals, Inc.
Boltaflex	Supported and unsupported flexible vinyl sheeting	The General Tire & Rubber Co.
Boltaron	ABS polymers, polyethylenes, polypropylenes, polyvinyl chlorides	Bolta Products Company
Bondstrand	Epoxies, polyesters	Amercoat Corporation
Brea	Polyethylenes	" "
Budene	Polybutadiene	Goodyear Tire & Rubber Company
Butacite	Polyvinyl butyral resins	E. I. duPont de Nemours & Co.
Butakon	Butadiene-acrylonitriles	Imperial Chemical Ind. Ltd.
Buton	Thermosetting butadiene-styrene resins	Enjay Chemical Company, Div. Humble Oil & Refining Company
Butvar	Polyvinyl butyral resins	Shawinigan Resins Corporation

TRADE NAMES, COMPOSITION AND MANUFACTURERS OF SOME PLASTICS (Continued)

Trade name	Composition	Manufacturer
C-8 Epoxy	Epoxy resins	Union Carbide Corporation
Cadco	Acrylics, ABS polymers, nylons, phenolics, polyesters, polyethylenes, polypropylenes	Cadillac Plastic & Chemical Co.
Cadco Penton	Chlorinated polyethers	" " " "
Cadco Epoxy	Epoxies	" " " "
Cadco Teflon	Fluorocarbons	" " " "
Cadco PVC	Polyvinyl chlorides urethanes, elastomers	" " " "
Carbo-Korez	Phenolics	Atlas Mineral Products
Carbomix	Butadiene-styrene rubber	Copolymer Rubber & Chemical Corp.
Carbowax	Polyethylene glycols	Union Carbide Corporation
Cardolite	Phenol-aldehyde resins	Irvington Varnish Co.
Catabond	Phenolic resins	Catalin Corp. of America
Catalin	Acrylics, ABS polymers, nylons, phenolics, polyethylenes, polypropylenes, polystyrene	" " " "
Ceilcrete	Polyesters	The Ceilcote Co.
Centri-Cast	Epoxies, polyesters	Apex Fiberglass Products Co.
Chemigum	Butadiene-acrylonitrile rubber and latices	The Goodyear Tire & Rubber Co.
Chemline	Epoxies	A. O. Smith Corporation
Chempro	Fluorocarbons	Chemical & Power Products
Chemtite	Epoxies, phenolics	Johns-Manville Corp.
Cis-4	Polybutadienes	Phillips Petroleum Company
Cohrlastic	Silicon rubber	Connecticut Hardrubber Co.
Coltrock	Phenolics	Colt's Plastics Co.
Conolite	Polyester resins and laminates	Shellmar-Betner, Div. Continental Can Co.
Copo	Butadiene-styrene rubber	Copolymer Rubber & Chemical Corp.
Coroband	Furanes	The Ceilcote Company
Corocrete	Polyesters	" " "
Corvel	Epoxies, vinyls Fusion bond finishes	The Polymer Co.
Corvic	Polyvinyl chloride resins	Canadian Industries Limited
Cosmalite	Phenol-formaldehyde resins	Colton Chemical Co.
CR	Polyesters	Resolite Corp.
CR-39	Polycarbonates	Pittsburgh Plate Glass Co.
Crystalex	Acrylics	Rohm & Hass Company
Cumar	Paracoumarone-indene resins	Allied Chemical Corporation
Cycolac	ABS polymers Acrylonitrile-butadiene-styrene, copolymer	Borg-Warner Corporation
Cymel	Melamines molding compounds, adhesive and laminating resins	American Cyanamid Company
Dacovin	Polyvinyl chlorides	Diamond Alkali Company
Dapon	Diallyl phthalate	Food Machinery and Chemical Corp.
Delrin	Acetal resins	E. I. duPont de Nemours & Co., Inc.
D.E.R.	Epoxy resins	Dow Chemical Company
Devran	Epoxy resins	Devoe & Reynolds
Diamond PVS	Polyvinyl chloride resins; copolymer resins	Diamond Alkali Company
Dow Corning	Silicones	Dow Chemical Company
Duradene	Butanediene-styrene rubber	Firestone Tire & Rubber Co.
Duragene 1203	Polybutadienes	The General Tire & Rubber Co.
Durcon	Epoxies	The Duriron Co.
Dur-X	Polyethylenes	Johns-Manville Company

TRADE NAMES, COMPOSITION AND MANUFACTURERS OF SOME PLASTICS (Continued)

Trade name	Composition	Manufacturer
Durez	Phenolic resins	Durez Plastic Division, Hooker Chemical Corporation
Dylan	Polyethylene	Koppers Company, Inc.
Dylene	Polystyrene	" " "
Dylite	Expandable polystyrene	" " "
Dyphene	Phenol-formaldehyde resins	The Sherwin-Williams Co.
Ebolene	Fluorocarbons	Chicago Gasket Co.
Enjay Butyl	Butylrubber	Enjay Chemical Company, Div. Humble Oil & Refining Co.
Enjay Butyl latex	Butylrubber in aqueous solution	Enjay Chemical Company, Div. Humble Oil & Refining Co.
Epikote	Epoxy resins	The Shell Chemical Company
Epiphen	Epoxide resins	The Borden Chemical Company
Epi-Rez	Epoxy resins	Jones-Dabney Company
Epi-Tex	Epoxy ester resins	" " "
Epolene	Low molecular weight polyethylene resins	Eastman Kodak Company
Epoxical	Epoxies	United States Gypsum Co.
Epon	Epoxy resins	The Shell Chemical Company
Escon	Polypropylene resins	Enjay Chemical Co., Div. Humble Oil & Refining Company
Estane	Polyurethane elastomers	B. F. Goodrich Chemical Company
Ethafoam	Polyethlene, low density	The Dow Chemical Company
eXtren 200	Polyesters	Universal Molded Fiberglass Co.
Fibercast	Epoxies, polyesters	Fibercast, Div. of Youngstown Sheet & Tube Co.
Fibestos	Cellulose acetate	Monsanto Chemical Co.
Fire-Snuf	Polyesters	Resolite Corp.
Flakoline	Polyesters	The Ceilcote Co.
Flamenol	Polyvinyl chlorides	General Electric Company
Flovic	Polyvinyl chlorides	Canadian Industries, Limited
Fluorogreen	Teflon with glass and ceramic fibers, fluorocarbons	John L. Dore Co.
Fluororay	Fluorocarbons	Raybestos-Manhattan, Inc.
Formica	Melamines	Formica Corp.
Formvar	Polyvinyl formal resins	Shawinigan Resins Corp.
Forticel	Cellulose propionate sheet films, molding powders	Celanese Polymer Co.
Fortiflex-A	Linear polyolefines	Celanese Polymer Company
Fortiflex-B	Linear copolymers	" " "
Fortiflex-C	Medium density polyethylenes	" " "
Fortiflex-D	Low-density polyethylenes	" " "
Fosta-Tuf-Flex	Polystyrene, high impact	Foster-Grant, Inc.
FR-N	Nitrile rubber	The Firestone Tire & Rubber Co., Firestone Plastics Company
FR-S	Butadiene-styrene rubber	The Firestone Tire & Rubber Co., Firestone Plastics Company
Frostwhite	Polystyrene, medium impact	Sheffield Plastics, Inc.,
Furnane	Furanes	Atlas Mineral Products Co.
Gabrite	Urea-formaldehyde	Chemore Corporation
Gaco	Epoxies	Gates Engineering Division, Glidden Company
GE Methylon	Phenolics	General Electric Company, Chemical Materials Dept.
Gelva	Polyvinyl acetate resins	Shawinigan Resins Corp.
	" " copolymers	
Gelvatol	Polyvinyl alcohol resins	" " "

TRADE NAMES, COMPOSITION AND MANUFACTURERS OF SOME PLASTICS (Continued)

Trade name	Composition	Manufacturer
Genco	Acrylics, ABS polymers Polyethylenes, copolymers	General Plastics Corp.
GenEpoxy	Epoxy resins for adhesives, coatings, etc.	General Mills, Inc.
Genetron	Fluorinated hydrocarbons, monomers and polymers	Allied Chemical Corp., General Chemical Div.
Gen-Flo	Styrene-butadiene latex	General Tire & Rubber Co.
Genthanes	Urethane elastomers; prepolymers	" " " "
Geon	Polyvinyl chlorides	B. F. Goodrich Chemical Co.
Gering	Acrylics, ABS polymers, nylons, polyethylenes, polypropylenes polyvinyl chlorides, copolymers	Gering Plastics Company
Ger-Pak	Polyethylene film	Gering Plastics Company
Glykon	Polyester resins	The General Tire & Rubber Co.
GP excel	Polyvinyl chlorides	Glamorgan Plastics, Div. of Glamorgan Pipe & Foundry Co.
Grace	Polystyrenes	W. R. Grace & Company, Polymer Chemical Division
Grex	High density polyethylenes, polypropylenes	W. R. Grace & Company, Polymer Chemical Division
Halon	Fluorohalocarbon resins	Allied Chemical Corp.
Haveg	Phenolic resins, furanes	Haveg Industries, Inc.
Haysite	Reinforced polyester	Haysite Corp.
Hercose AP	Cellulose acetate-propionate	Hercules Powder Company
Herculoid	Cellulose nitrate	" " "
Hetrofoam	Polyurethane fire-retardant rigid foam	Hooker Chemical Corp., Durez Plastics Div.
Hetron	Fire retardant polyester resin	Hooker Chemical Corp., Durez Plastics Div.
Hi-flexible	Polyethylenes	Triangle Conduit & Cable Co.
Hycar	Butadiene-styrene, butadiene-acronitriles, acrylate emulsions	B. F. Goodrich Chemical Co.
Hystran	Epoxies, polyesters	Lamtex Industries of Koppers Co.
Iporca	Urea resin and foam	Badische Aniline & Soda Fabrik
Isothane	Urethane elastomers	Carborundum Company
J-M ABS	ABS polymers	Johns-Manville Company
J-M PVC	Polyvinyl chlorides	" " "
Kaylite	Polyvinyl chlorides	Kaykor Products Corp.
Kayrex	Polyvinyl chlorides	" " "
Kel-F	Chlorotrifluoroethylene, molding resins and dispersions	Minnesota Mining & Manufacturing Co.
Koroseal	Polyvinyl chlorides	B. F. Goodrich Industrial Products Co.
Kralac A-EP	High styrene resins	United States Rubber Co., Naugatuck Chemical Div.
Kralac Latex	Styrene-butadiene copolymers	United States Rubber Co., Naugatuck Chemical Div.
Kralastic	ABS Polymers, copolymers	United States Rubber Co., Naugatuck Chemical Div.
Kynar	Polyvinyldene fluoride	Pennsalt Chemical Corp.
La Favorite	Polyvinyl chlorides	La Favorita Rubber Mfg. Co.
Laminac	Polyester resins	American Cyanamid Company
Lemac	Polyvinyl acetate	Borden Chemical Company
Lexan	Polycarbonate resin	General Electric Company, Chemical Materials Dept:
Lucite	Acrylic resin	E. I. duPont de Nemours & Co.
Lustran	ABS polymers	Monsanto Chemical Company

TRADE NAMES, COMPOSITION AND MANUFACTURERS OF SOME PLASTICS (Continued)

Trade name	Composition	Manufacturer
Lustrex	Styrene molding and extrusion resins	Monsanto Chemical Company
Lytron RJ-100	Styrene molding and extrusion resins	" " "
Madurit	Melamine resins for plastic industry	Casella Farbwerke Mainkur, A.G.
Maplen	Polypropylenes	Chemore Corporation
Maraglas	Crystal clear epoxy resin	The Marblette Corporation
Marawood	Carvable epoxy resin	" " "
Marbon 8000	High styrene resins	Borg-Warner Corporation, Marbon Chemical Division
Marco	Polyester resins	Celanese Corp. of America
Marlex	Polyethylenes, polypropylenes, copolymers	Phillips Petroleum Company
Marvinol	Vinyl chloride resins and compounds	United States Rubber Company, Naugatuck Chemical Division
Melantine	Melamine resins	Ciba Limited
Melmac	Melamine resins	American Cyanamid Company
Melopas	Polyamide formaldehyde	Ciba Limited
Meltiplast 101	Polyethylenes	International Protected Metals
Meltiplast 301	Polyethylenes	" " "
Meltiplast 501	Chlorinated polyesters	" " "
Merlon	Polycarbonate resins	Mobay Chemical Co.
Methacrol	Acrylic emulsions	E. I. duPont de Nemours & Co., Inc.
Micarta	Melamines, phenolics, polyesters	Westinghouse Electric Co.
Microthene	Polyethylenes	U. S. Industrial Chemicals Co.
Monsanto	Polyethylene resins	Monsanto Chemical Company
MR Resin	Polyester resins	Celanese Corporation of America
Multrathane	Urethane elastomers	Mobay Chemical Company
Napcofoam	Polyurethane, flexible	Nopco Chemical Company
Natsyn	Polyisoprene	The Goodyear Tire & Rubber Co.
Naugatex 2700 series	Butadiene-styrene latices	United States Rubber Co., Naugatuck Chemical Division
Neville	Coumarone-indene resins	Neville Chemical Company
Niacet	Vinyl acetate, vinyl chloride	Union Carbide Corp.
Nopcofoam	Polyurethane plastics (flexible)	Nopco Chemical Co., Plastics Div.
Novodur	ABS copolymers	Farbenfabriken Bayer, A.G.
Nu-Klad	Epoxies	Amercoat Corporation
Nypene	Polystyrene resins	Neville Chemical Company
Opalon	Vinyl chloride resins and compounds	Monsanto Chemical Co.
Palmetto	Fluorocarbons	Greene, Tweed & Co.
Paracon	Polyester rubber	Bell Telephone Laboratories
Paracril	Nitrile rubber	United States Rubber Co., Naugatuck Chemical Div.
Paradene	Coumarone-indene resins	Neville Chemical Company
Paraplex	Polyester resins, acrylic modified polyester resin	Rohm & Haas Company
Pentacite	Pentaerythritol resins	Reichold Chemicals, Inc.
Permelite	Melamines	Melamine Plastics, Inc.
Petrothene	Polyethylene resin, polypropylene resin	U. S. Industrial Chemicals Co.
Pfaudlon 201	Epoxies	Pfaudler Co.
Pfaudlon 301	Chlorinated polyethers	" "
Philprene	Butadiene-styrene rubber	Phillips Petroleum Co.
Piccoflex	Styrene copolymer resins	Pennsylvania Industrial Chemical Corp.
Piccolastic	Styrene polymer resins	Pennsylvania Industrial Chemical Corp.
Plaskon	Nylons, melamines, phenolics polyesters	Allied Chemical Corp.
Pleogen	Polyester resins	American Petrochemical Corp.
Plexiglas	Acrylics	Rohm & Haas Company

TRADE NAMES, COMPOSITION AND MANUFACTURERS OF SOME PLASTICS (Continued)

Trade name	Composition	Manufacturer
Plioflex	Polyvinyl chlorides	The Goodyear Tire & Rubber Co.
Pliofoam	Expanded urea resins	" " " " "
Pliolite	Styrene-butadiene resins	" " " " "
Plio-Tuf series	Modified styrene resins	" " " " "
Pliovic	Polyvinyl chlorides	" " " " "
Plyophen	Phenol-formaldehyde resins	Reichhold Chemicals, Inc.
Poly-Eth	Polyethylene resins	Spencer Chemical Co.
Polylite	Polyester resins	Reichhold Chemicals, Inc.
Polypenco	Acrylics, chlorinated polyethers, fluorocarbons, nylons, polycarbonates,	Polymer Corp.
Red Thread	Epoxies, glassfiber filled	A. O. Smith Corp.
Resimene	Urea and melamine resins	Monsanto Chemical Co.
Resinox	Phenolic resins and compounds	" " "
Rezklad	Epoxies	Atlas Mineral Products, Div. Electric Storage Battery Co.
Rhonite	Urea resins	Rohm & Hass Company
Rock Island	Epoxies	Rock Island Fiberglass Pipe Co.
Root-Pruf	Butadiene-styrene	Triangle Conduit & Cable Co.
Roylar	Polyurethanes	United States Rubber Co., Naugatuck Chemical Division
RX	Epoxies, fluorocarbons, phenolics	Rogers Corp.
Ryercite	Phenolics	Joseph T. Ryerson & Son, Inc.
Ryertex	Phenolics	" " "
Ryertex-Omicron	Polyvinyl chlorides	" " "
S-4	Natural latex, liquid	Firestone Tire & Rubber Co.
Saran	Polyvinylidene chloride, vinylidene chloride copolymers	Saran Lined Pipe Co.
Sauereisen	Furanes	Sauereisen Cement Co.
Seilon ETH	Polyethylenes	Seiberling Rubber Co.
" PRO	Polypropylenes	" " "
" PVC	Polyvinyl chlorides	" " "
" S-3	ABS polymers	" " "
" UR	Urethane polymers	" " "
Semi-rigid	ABS polymers	Triangle Conduit & Cable Co.
Silastic	Silicon rubber	The Dow Chemical Company
Solprene	Fluoro elastomers	Phillips Petroleum Company
S-polymers	Butadiene-styrene copolymers	Esso Laboratories
Stauffer	ABS polymers, polyethylenes, polypylenes, polyvinyl chlorides	Stauffer Chemical Company, Molded Products Div.
Stylplast	Urea-formaldehyde compounds	American Viscose Corporation
Styrex	Styrene-acrylonitrile copolymer	The Dow Chemical Company
Super Becamine	Melamine-formaldehyde resins	Reichhold Chemicals, Inc.
Super Dylan	Polyethylene	Koppers Company, Inc.
Supreme	Polyethylenes	Johns-Manville Company
Sylkyd	Silicon alkyd resins	The Dow Chemical Co.
Sylplast	Urea formaldehyde	Food Machinery and Chemical Corp., Organic Chemicals Div.
Synthane	Epoxies, melamines, phenolics, silicones	Synthane Corporation
T/Na-100	Polyvinylfluorides	The Ruberoid Company
Teflon	Fluorocarbons, tetrafluoroethylene (TFE) fluorinated ethylpropylene resins (FEP)	E. I. duPont de Nemours & Co. Inc.
Tempron	Polybutadienes	Ace Molded Products Co.
Temp-R-Tape	Fluorocarbons	Connecticut Hardrubber Co.
Tenite	Cellulose acetate, cellulose-acetate-polyethylene, polypropylenes, urethane elastomers, copolymers	Eastman Chemical Products, Inc.

TRADE NAMES, COMPOSITION AND MANUFACTURERS OF SOME PLASTICS (Continued)

Trade name	Composition	Manufacturer
Tetran	Fluorocarbons	Pennsalt Chemicals Corp.
Texin	Urethane elastomers	Mobay Chemical Company
Thermo-Seal	Polyethylenes, polypropylenes	Cabot Piping Systems, Plastics Division
Thioment	Polyisoprenes	Atlas Mineral Products Co.
Trans-4	Rubberlike materials	Phillips Petroleum Company, Bartlesville, Okla.
TPC	Polyvinyl chlorides	Thermoplastics Corp.
ttP	ABS polymers, chlorinated polyesters, fluorocarbons, polyvinylchlorides	Cabot Piping Systems, Plastics Division
Tyril	Styrene-acrylonitrile, copolymer molding resins	The Dow Chemical Company
Ultrapas	Melamine resins	Dynamite A.G.
Ultrathene	Ethylene vinyl acetate	U. S. Industrial Chemicals Co.
Ultron	Polyvinyl chlorides	Monsanto Chemical Company
Unox	Epoxides	Union Carbide Corporation
Upalon	Polyvinyl chlorides	Monsanto Chemical Company
Vibrathane	Urethane elastomers	United States Rubber Co., Naugatuck Chemical Div.
Vibrin	Polyester resins	United States Rubber Co., Naugatuck Chemical Div.
Vinelle	Vinyl compounds	The General Tire & Rubber Co.
Vipla	Polyvinyl chloride, vinyl acetate, copolymers	Chemore Corp.
Vitalic	Chlorosulfonated polyethylenes, polybutadienes, polyisoprenes, ethylpropylenes, urethane elastomers	Continental Rubber Works
Vitel	Polyesters	Goodyear Chemical, Div., The Goodyear Tire & Rubber Co.
Viton	Synthetic rubbers	E. I. duPont de Nemours & Co., Inc.
Vitroplast	Polyester cement	Atlas Mineral Products Div.
Vygen	Polyvinyl chloride resins	The General Tire & Rubber Co.
Vyron	Polyvinyl chlorides	Monsanto Chemical Co.
X 2 B	Natural latex	The Firestone Tire & Rubber Co.
Yardley	ABS polymers, polyethylenes, polyvinyl chlorides	Celanese Plastics Co.
Zerlon	Methyl methacrylate styrene resins	The Dow Chemical Co.

Properties of Commercial Plastics

Of the many plastics commercially available in each chemical class only one or a very few individuals have been selected for this table as typical of the class. In some cases the range of properties has been expanded to include several grades or types. It is impractical to include a comprehensive list of materials or known properties of these materials in a table of convenient size. Properties vary widely with amount and kind of modifier such as filler and plasticizer. Within any type of thermoplastic resins molecular weight is an important variable. This property is controlled to afford the best physical properties available consistent with economical processing properties.

The information shown refers in all cases, except for "Forms available" and "Fabrication," to material in the fabricated form, which in the case of thermosetting materials means commercially cured. Physical and electrical properties will vary, to a greater or less degree with different materials, with humidity conditioning environment and with orientation. Strength values are quoted on the basis of short time tests at normal room temperature and are not suitable for engineering design purposes for load bearing applications. Maximum continuous service temperature refers to unloaded structures. The user of this table is referred to the specifications and test procedures of the American Society for Testing Materials.

PROPERTIES OF COMMERCIAL PLASTICS

Chemical Class	Cellulose Acetate	Cellulose Acetate	Cellulose Acetate Butyrate
Resin Type	Thermoplastic	Thermoplastic	Thermoplastic
Subclass or Modification	Soft	Hard	Soft
FORMS AVAILABLE Cs—castings, F—film, Fb—fibers, I—impregnants, L—laminations, Lq—lacquers, Mf—monofilaments, P—Powder, pellet, or granules, R—rods, tubes, or other extruded forms, S—sheets.	F, Lq, P, R, S	F, Lq, P, R, S	F, Lq, P, R, S
FABRICATION Cl—calendering, Cs—casting, E—extrusion, F—hot forming or drawing, I—impregnation, MB—blow molding, MC—compression molding, MI—injection molding, S—spreading.	Cs, E, F, MB, MC, MI, S	Cs, E, F, MB, MC, MI, S	Cs, E, F, MB, MC, MI, S
ELECTRICAL PROPERTIES			
D.C. Resistivity, ohm-cm	10^{10}–10^{13}	10^{10}–10^{13}	10^{10}–10^{12}
Dielectric constant, 60 cps	3.5–7.5	3.5–7.5	3.5–6.4
Dielectric constant, 10^6 cps	3.2–7.0	3.2–7.0	3.2–6.2
Dissipation factor, 60 cps	0.01–0.06	0.01–0.06	0.01–0.04
Dissipation factor, 10^6 cps	0.01–0.10	0.01–0.10	0.01–0.04
MECHANICAL PROPERTIES			
Modulus of elasticity, 10^3 psi	86–250	190–400	74–126
Tensile strength, psi	1,900–4,700	4,600–8,500	1,900–3,800
Ultimate elongation, %	32–50	6–40	60–74
Yield stress, psi	2,200–4,200	4,100–7,600	1,200–2,600
Yield strain, %			
Rockwell hardness	R 49–R 103	R 101–R 123	R 59–R 95
Notched Izod impact strength, ft.lb/in	2.0–5.2	0.4–2.7	2.5–5.4
Specific gravity	1.27–1.34	1.27–1.34	1.15–1.22
THERMAL PROPERTIES			
Burning rate	Medium	Medium	Medium
Heat distortion 264 psi, °C	44–57	60–113	49–58
Specific heat, cal/g	0.3–0.42	0.3–0.42	0.3–0.4
Linear thermal expansion coefficient, 10^{-5}, °C	8–16	8–16	11–17
Maximum continuous service temperature, °C			
CHEMICAL RESISTANCE			
Mineral acids, weak	Fair to good	Fair to good	Good
Mineral acids, strong	Poor	Poor	Fair to good
Oxidizing acids, concentrated	Very poor	Very poor	
Alkalies, weak	Poor	Poor	Good
Alkalies, strong	Very poor	Very poor	Poor
Alcohols	Poor	Poor	Poor
Ketones	Poor	Poor	Poor
Esters	Poor	Poor	Poor
Hydrocarbons, aliphatic	Fair to good	Fair to good	Fair to good
Hydrocarbons, aromatic	Poor to fair	Poor to fair	Poor
Oils, vegetable, animal, mineral	Fair to good	Fair to good	Good
MISCELLANEOUS PROPERTIES			
Clarity	Excellent	Excellent	Good to excellent
Color	Pale to colorless	Pale to colorless	Pale to colorless
Refractive index, n_D	1.46–1.50	1.46–1.50	1.46–1.49
Applicable ASTM specifications and test methods	D786, D706, D257, D150, D638, D785, D256, D792, D648, D696, D543, D542	D786, D706, D257, D150, D638, D785, D256, D792, D648, D696, D543, D542	D707, D257, D150, D638, D785, D256, D792, D648, D696

PROPERTIES OF COMMERCIAL PLASTICS

Chemical Class	Cellulose Acetate Butyrate	Nylon	Polycarbonates
Resin Type	Thermoplastic	Thermoplastic	Thermoplastic
Subclass or Modification	Hard	6/6	Unfilled
FORMS AVAILABLE Cs—castings, F—film, Fb—fibers, I—impregnants, L—laminations, Lq—lacquers, Mf—monofilaments, P—Powder, pellet, or granules, R—rods, tubes, or other extruded forms, S—sheets.	F, Lq, P, R, S	F, Fb, Mf, P, R, S	F, Fb, Mf, P, R, S
FABRICATION Cl—calendering, Cs—casting, E—extrusion, F—hot forming or drawing, I—impregnation, MB—blow molding, MC—compression molding, MI—injection molding, S—spreading.	Cs, E, F, MB, MC, MI, S	E, F, MB, MC, MI	Cs, E, F, MB, MC, MI
ELECTRICAL PROPERTIES			
D.C. Resistivity, ohm-cm	10^{10}–10^{12}		2 x 10^{16}
Dielectric constant, 60 cps	3.5–6.4	4.0–4.6	3.17
Dielectric constant, 10^6 cps	3.2–6.2	3.4–3.6	2.96
Dissipation factor, 60 cps	0.01–0.04	0.014–0.04	0.0009
Dissipation factor, 10^6 cps	0.01–0.04	0.04	0.01
MECHANICAL PROPERTIES			
Modulus of elasticity, 10^3 psi	150–200		290–325
Tensile strength, psi	5,000–6,800	9,000–12,000	8,000–9,500
Ultimate elongation, %	38–54	60–300	20–100
Yield stress, psi	3,600–6,100		8,000–10,000
Yield strain, %			
Rockwell hardness	R 108–R 117	R 108–R 120	M 70–M 180
Notched Izod impact strength, ft.lb/in.	0.7–2.4	1.0–2.0	8–16
Specific gravity	1.19–1.25	1.13–1.15	1.2
THERMAL PROPERTIES			
Burning rate	Medium	Self extinguishing	Self extinguishing
Heat distortion 264 psi, °C	70–99		135–145
Specific heat, cal/g	0.3–0.4	0.4	0.3
Linear thermal expansion coefficient, 10^{-5}, °C	11–17	8.0	6.6
Maximum continuous service temperature, °C		80–150	138–143
CHEMICAL RESISTANCE			
Mineral acids, weak	Good	Very good	Excellent
Mineral acids, strong	Fair to good	Poor	Fair
Oxidizing acids, concentrated		Poor	
Alkalies, weak	Good	No effect	Poor
Alkalies, strong	Poor	No effect	Poor
Alcohols	Poor	Good	Poor
Ketones	Poor	Good	Poor
Esters	Poor	Good	Poor
Hydrocarbons, aliphatic	Fair to good	Very good	Poor
Hydrocarbons, aromatic	Poor	Fair to good	Poor
Oils, vegetable, animal, mineral	Good	Good	Poor
MISCELLANEOUS PROPERTIES			
Clarity	Good to excellent	Clear	Clear
Color	Pale to colorless	Pale amber to colorless	Colorless
Refractive index, n_D	1.46–1.49	1.53	1.60
Applicable ASTM specifications and test methods	D707, D257, D150, D638, D785, D256, D792, D648, D696, D543, D542	D257, D150, D638, D785, D256, D792, D648, D696, D543, D542	D257, D150, D638, D785, D256, D792, D648, D696, D543, D542

Chemical Class	Polyethylene	Polyethylene	Polyethylene
Resin Type	Thermoplastic	Thermoplastic	Thermoplastic
Subclass or Modification	Low density	Medium density	High density
FORMS AVAILABLE Cs—castings, F—film, Fb—fibers, I—impregnants, L—laminations, Lq—lacquers, Mf—monofilaments, P—Powder, pellet or granules, R—rods, tubes, or other extruded forms, S—sheets.	F, Mf, P, R, S	F, Mf, P, R, S	F, Fb, Mf, P, R, S
FABRICATION Cl—calendering, Cs—casting, E—extrusion, F—hot forming or drawing, I—impregnation, MB—blow molding, MC—compression molding, MI—injection molding, S—spreading.	Cl, E, F, MB, MC, MI	Cl, E, F, MB, MC, MI	Cl, E, F, MB, MC, MI
ELECTRICAL PROPERTIES			
D.C. Resistivity, ohm-cm	$>10^{15}$	$>10^{15}$	$>10^{15}$
Dielectric constant, 60 cps	2.3–2.35	2.3	2.3–2.35
Dielectric constant, 10^6 cps	2.3–2.35	2.3	2.3–2.35
Dissipation factor, 60 cps	<0.0005	<0.0005	<0.0005
Dissipation factor, 10^6 cps	<0.0005	<0.0005	<0.0005
MECHANICAL PROPERTIES			
Modulus of elasticity, 10^3 psi	14–38	35–90	85–160
Tensile strength, psi	1,000–1,400	1,200–3,500	3,100–5,500
Ultimate elongation, %	400–700	50–600	15–100
Yield stress, psi	1,100–1,700	1,500–2,600	2,400–5,000
Yield strain, %	20–40	10–20	5–10
Rockwell hardness			R 30–R 50
Notched Izod impact strength, ft.lb/in.	no break	0.5–>16	1.5–20
Specific gravity	0.91–0.925	0.926–0.941	0.941–0.965
THERMAL PROPERTIES			
Burning rate	Very slow	Slow	Slow
Heat distortion 264 psi, °C			
Specific heat, cal/g	0.55	0.55	0.55
Linear thermal expansion coefficient, 10^{-5}, °C	10–20	14–16	11–13
Maximum continuous service temperature, °C	60–77	71–93	92–200
CHEMICAL RESISTANCE			
Mineral acids, weak	Good	Excellent	Excellent
Mineral acids, strong	Good	Excellent	Excellent
Oxidizing acids, concentrated	Good to poor	Good to poor	Good to poor
Alkalies, weak	Good	Excellent	Excellent
Alkalies, strong	Good	Excellent	Excellent
Alcohols	Excellent to poor	Excellent to poor	Excellent to poor
Ketones	Excellent to poor	Excellent to poor	Excellent to poor
Esters	Excellent to poor	Excellent to poor	Excellent to poor
Hydrocarbons, aliphatic	Fair	Fair	Fair
Hydrocarbons, aromatic	Fair	Good	Fair
Oils, vegetable, animal, mineral	Good	Excellent	Good
MISCELLANEOUS PROPERTIES			
Clarity	Translucent	Translucent	Translucent
Color	Colorless	Colorless	Colorless
Refractive index, n_D	1.50–1.54	1.52–1.54	1.54
Applicable ASTM specifications and test methods	D702, D788, D257, D150, D412, D638, D696, D543, D542, D1248	D257, D150, D412, D638, D785, D256, D696, D543, D542, D1248	D257, D150, D412, D638, D785, D256, D696, D543, D542, D1248

PROPERTIES OF COMMERCIAL PLASTICS

PROPERTIES OF COMMERCIAL PLASTICS

Chemical Class	Methylmethacrylate	Polypropylene	Polypropylene
Resin Type	Thermoplastic	Thermoplastic	Thermoplastic
Subclass or Modification	Unmodified	Unmodified	Copolymer
FORMS AVAILABLE Cs—castings, F—film, Fb—fibers, I—impregnants, L—laminations, Lq—lacquers, Mf—monofilaments, P—Powder, pellet, or granules, R—rods, tubes, or other extruded forms, S—sheets.	Cs, P, R, S	F, Fb, Mf, P, R, S	F, Fb, Mf, P, R, S
FABRICATION Cl—calendering, Cs—casting, E—extrusion, F—hot forming or drawing, I—impregnation, MB—blow molding, MC—compression molding, MI—injection molding, S—spreading	Cs, E, F, Lq, MB, MC, MI	Cl, E, F, MB, MC, MI	Cl, E, F, MB, MC, MI
ELECTRICAL PROPERTIES			
D.C. Resistivity, ohm-cm	$>10^{14}$	$>10^{15}$	10^{17}
Dielectric constant, 60 cps	3.5–4.5	2.2–2.6	2.3
Dielectric constant, 10^6 cps	3.0–3.5	2.2–2.6	2.3
Dissipation factor, 60 cps	0.04–0.06	<0.0005	0.0001–0.0005
Dissipation factor, 10^6 cps	0.02–0.03	0.0005–0.002	0.0001–0.002
MECHANICAL PROPERTIES			
Modulus of elasticity, 10^3 psi	350–500	1.4–1.7	
Tensile strength, psi	7,000–11,000	4,300–5,500	2,900–4,500
Ultimate elongation, %	2.0–10	>220	200–700
Yield stress, psi		4,900	
Yield strain, %		15	
Rockwell hardness	M 80–M 105	93	R 50–R 96
Notched Izod impact strength, ft.lb/in	0.3–0.6	1.0	1.1–12
Specific gravity	1.18–1.20	0.90	0.90
THERMAL PROPERTIES			
Burning rate	Slow	Medium	Medium
Heat distortion 264 psi, °C	66–99		
Specific heat, cal/g	0.35	0.5	0.5
Linear thermal expansion coefficient, 10^{-5}, °C	5.0–9.0	5.8–10	8–10
Maximum continuous service temperature, °C	60–93		190–240
CHEMICAL RESISTANCE			
Mineral acids, weak	Good	Excellent	Excellent
Mineral acids, strong	Fair to poor	Excellent	Excellent
Oxidizing acids, concentrated	Attacked	Good to poor	Poor
Alkalies, weak	Good	Excellent to good	Excellent
Alkalies, strong	Poor	Excellent to good	Good
Alcohols		Excellent to good	Good below 80°C
Ketones	Dissolves	Excellent to good	Good below 80°C
Esters	Dissolves	Excellent to good	Good below 80°C
Hydrocarbons, aliphatic	Good	Good to fair	Good below 80°C
Hydrocarbons, aromatic	Softens	Good to fair	Good below 80°C
Oils, vegetable, animal, mineral	Good	Good	
MISCELLANEOUS PROPERTIES			
Clarity	Excellent	Transparent	Transparent
Color	Colorless	Colorless to sl. yellow	Colorless to sl. yellow
Refractive index, n_D	1.48–1.50	1.49	
Applicable ASTM specifications and test methods	D257, D150, D638, D785, D256, D792, D648, D696, D543, D542	D257, D150, D412, D638, D785, D256, D648, D543, D542	D257, D150, D412, D638, D785, D256, D648, D543, D542

PROPERTIES OF COMMERCIAL PLASTICS

Chemical Class	Polystyrene	Polystyrene-acrylonitrile	Polytetrafluoro ethylene
Resin Type	Thermoplastic	Thermoplastic	Thermoplastic
Subclass or Modification	Unmodified	Unmodified	Unmodified
FORMS AVAILABLE Cs—castings, F—film, Fb—fibers, I—impregnants, L—laminations, Lq—lacquers, Mf—monofilaments, P—Powder, pellet, or granules, R—rods, tubes, or other extruded forms, S—sheets.	F, Fb, Mf, P, R, S	F, Mf, P, R, S	F, L, P, R, S
FABRICATION Cl—calendering, Cs—casting, E—extrusion, F—hot forming or drawing, I—impregnation, MB—blow molding, MC—compression molding, MI—injection molding, S—spreading.	E, F, MB, MC, MI	Cl, E, F, MB, MC, MI	E, F, MC, MI
ELECTRICAL PROPERTIES			
D.C. Resistivity, ohm-cm	$>10^{16}$	10^{13}–10^{17}	10^{18}
Dielectric constant, 60 cps	2.5–2.65	2.6–3.4	2.
Dielectric constant, 10^6 cps	2.5–2.65	2.5–3.1	2.
Dissipation factor, 60 cps	0.0001–0.0003	0.006–0.008	0.0002
Dissipation factor, 10^6 cps	0.0001–0.0004	0.007–0.01	0.0002
MECHANICAL PROPERTIES			
Modulus of elasticity, 10^3 psi	400–600	$>10^{16}$	33–65
Tensile strength, psi	5,000–10,000	9,000–12,000	2,000–4,500
Ultimate elongation, %	1.0–2.5	1.0–2.5	200–400
Yield stress, psi			1,600–2,000
Yield strain, %			50–75
Rockwell hardness	M 65–M 85	M 75–M 90	D 50–D 65
Notched Izod impact strength, ft.lb/in	0.25–0.60	0.3–0.6	2.5–4.0
Specific gravity	1.04–1.08	1.05–1.1	2.1–2.3
THERMAL PROPERTIES			
Burning rate	Medium to slow	Slow	Self extinguishing
Heat distortion 264 psi, °C		91–104	60
Specific heat, cal/g	0.32–0.35	0.32–0.35	0.25
Linear thermal expansion coefficient, 10^{-5}, °C	6.0–8.0	3.6–3.8	10
Maximum continuous service temperature, °C	66–82	77–88	260
CHEMICAL RESISTANCE			
Mineral acids, weak	Excellent	Excellent	Excellent
Mineral acids, strong	Excellent	Good to excellent	Excellent
Oxidizing acids, concentrated	Poor	Poor	Excellent
Alkalies, weak	Excellent	Excellent	Excellent
Alkalies, strong	Excellent	Good to excellent	Excellent
Alcohols	Excellent	Good to excellent	Excellent
Ketones	Dissolves	Dissolves	Excellent
Esters	Poor	Dissolves	Excellent
Hydrocarbons, aliphatic	Poor	Good	Excellent
Hydrocarbons, aromatic	Dissolves	Fair to good	Excellent
Oils, vegetable, animal, mineral	Fair to poor	Good to excellent	Excellent
MISCELLANEOUS PROPERTIES			
Clarity	Transparent	Transparent	Translucent
Color	Colorless	Colorless to amber	Colorless to grey
Refractive index, n_D	1.59–1.60	1.56–1.57	1.30–1.40
Applicable ASTM specifications and test methods	D257, D150, D638, D785, D256, D792, D648, D696, D543, D542	D257, D150, D638, D785, D256, D792, D648, D696, D543, D542	

PROPERTIES OF COMMERCIAL PLASTICS

Chemical Class	Polytrifluorochloro ethylene	Polyvinylchloride and Vinylchloride acetate	Polyvinylchloride and Vinylchloride acetate
Resin Type	Thermoplastic	Thermoplastic	Thermoplastic
Subclass or Modification	Unmodified	Unmodified, rigid	Plasticized (non rigid)
FORMS AVAILABLE Cs—castings, F—film, Fb—fibers, I—impregnants, L—laminations, Lq—lacquers, Mf—monofilaments, P—Powder, pellet, or granules, R—rods, tubes, or other extruded forms, S—sheets.	F, Mf, P, R, S	F, Fb, I, Lq, Mf, P, R, S	F, L, P, R, S
FABRICATION Cl—calendering, Cs—casting, E—extrusion, F—hot forming or drawing, I—impregnation, MB—blow molding, MC—compression molding, MI—injection molding, S—spreading.	Cs, E, F, I, MC, MI, S	Cl, Cs, E, F, I, MB, MC, MI, S	Cl, Cs, E, MB, MC, MI, S
ELECTRICAL PROPERTIES			
D.C. Resistivity, ohm-cm	10^{18}	10^{12}–10^{16}	10^{11}–10^{14}
Dielectric constant, 60 cps	2.2–2.8	3.2–4.0	5.0–9.0
Dielectric constant, 10^6 cps	2.3–2.5	3.0–4.0	3.0–4.0
Dissipation factor, 60 cps	0.001	0.01–0.02	0.03–0.05
Dissipation factor, 10^6 cps	0.005	0.006–0.02	0.06–0.1
MECHANICAL PROPERTIES			
Modulus of elasticity, 10^3 psi	150	200–600	
Tensile strength, psi	4,500–6,000	5,000–9,000	1,500–3,000
Ultimate elongation, %	250	2.0–40	200–400
Yield stress, psi	4,200		
Yield strain, %	10	1.0–5.0	
Rockwell hardness	J 75–J 95	R 110–R 120	
Notched Izod impact strength, ft.lb/in	2.5–4.0	0.4–2.0	
Specific gravity	2.1–2.3	1.36–1.4	1.15–1.35
THERMAL PROPERTIES			
Burning rate	Self extinguishing	Self extinguishing	Slow to self extinguishing
Heat distortion 264 psi, °C		60–80	
Specific heat, cal/g	0.22	0.2–0.28	0.36–0.5
Linear thermal expansion coefficient, 10^{-5}, °C	7.0	5.0–18	7.0–25
Maximum continuous service temperature, °C	200	70–74	80–105
CHEMICAL RESISTANCE			
Mineral acids, weak	Excellent	Excellent	Fair to good
Mineral acids, strong	Excellent	Good to excellent	Fair to good
Oxidizing acids, concentrated	Excellent	Fair to good	Poor to fair
Alkalies, weak	Excellent	Excellent	Fair to good
Alkalies, strong	Excellent	Good	Fair to good
Alcohols	Excellent	Excellent	Fair
Ketones	Excellent	Poor	Poor
Esters	Excellent	Poor	Poor
Hydrocarbons, aliphatic	Excellent	Excellent	Poor
Hydrocarbons, aromatic	Excellent	Poor	Poor
Oils, vegetable, animal, mineral	Excellent	Excellent	Poor
MISCELLANEOUS PROPERTIES			
Clarity	Transparent	Transparent	Transparent
Color	Colorless to pale	Colorless to amber	Colorless to amber
Refractive index, n_D	1.43	1.54	1.50–1.55
Applicable ASTM specifications and test methods	D1430, D257, D150, D638, D785, D256, D792, D648, D696, D543,	D708, D728, D257, D150, D638, D256, D792, D648, D696, D543,	D1432, D257, D150, D412, D792, D543, D542

Chemical Class	Epoxy	Melamine-Formaldehyde	Melamine-Formaldehyde
Resin Type	Thermosetting	Thermosetting	Thermosetting
Subclass or Modification	Unfilled	α-Cellulose filled	Mineral filled, (electrical)
FORMS AVAILABLE Cs—castings, F—film, Fb—fibers, I—impregnants, L—laminations, Lq—lacquers, Mf—monofilaments, P—Powder, pellet, or granules, R—rods, tubes, or other extruded forms, S—sheets.	Cs, Lq	P, R, S	P, R, S
FABRICATION Cl—calendering, Cs—casting, E—extrusion, F—hot forming or drawing, I—impregnation, MB—blow molding, MC—compression molding, MI—injection molding, S—spreading.	Cs, I, S	MC	MC
ELECTRICAL PROPERTIES			
D.C. Resistivity, ohm-cm	10^{12}–10^{14}	10^{12}–10^{14}	10^{13}–10^{14}
Dielectric constant, 60 cps	3.5–5.0	7.9–9.4	10.2
Dielectric constant, 10^6 cps	3.4–4.4	7.2–8.4	6.1
Dissipation factor, 60 cps	0.001–0.005	0.03–0.08	0.10
Dissipation factor, 10^6 cps	0.03–0.05	0.03–0.043	0.051
MECHANICAL PROPERTIES			
Modulus of elasticity, 10^3 psi	>300	1,300	1,950
Tensile strength, psi	4,000–13,000	7,000–13,000	5,500–6,500
Ultimate elongation, %	2.0–6.0	0.6–0.9	
Yield stress, psi			
Yield strain, %			
Rockwell hardness	M 75–M 110	M 110–M 124	E 90
Notched Izod impact strength, ft.lb/in	0.2–1.0	0.24–0.35	0.3–0.4
Specific gravity	1.115	1.47–1.52	1.78
THERMAL PROPERTIES			
Burning rate	Slow	Self extinguishing	Self extinguishing
Heat distortion 264 psi, °C	Up to 120	204	130
Specific heat, cal/g	0.25–0.4	0.4	
Linear thermal expansion coefficient, 10^{-5}, °C	4.5–9.0	2.0–5.7	2.1–4.3
Maximum continuous service temperature, °C	80	99.0	149
CHEMICAL RESISTANCE			
Mineral acids, weak	Excellent	Good	Fair
Mineral acids, strong	Fair to good	Poor	Poor
Oxidizing acids, concentrated	Poor	Poor	Poor
Alkalies, weak	Excellent	Good	Fair
Alkalies, strong	Excellent	Poor	Poor
Alcohols	Excellent	Good	Good
Ketones	Poor	Good	Good
Esters		Good	Good
Hydrocarbons, aliphatic	Excellent	Good	Good
Hydrocarbons, aromatic	Excellent	Good	Good
Oils, vegetable, animal, mineral	Excellent	Good	Good
MISCELLANEOUS PROPERTIES			
Clarity	Transparent	Translucent	Opaque
Color	Colorless	Colorless	Dark
Refractive index, n_D	1.58		
Applicable ASTM specifications and test methods	D257, D150, D651, D785, D256, D792, D648, D696, D543	D704, D257, D150, D638, D785, D256, D792, D648, D696, D543	D704, D257, D150, D638, D785, D256, D792, D648, D696, D543

PROPERTIES OF COMMERCIAL PLASTICS

Chemical Class	Phenol-Formaldehyde	Phenol-Formaldehyde	Phenol-Formaldehyde
Resin Type	Thermosetting	Thermosetting	Thermosetting
Subclass or Modification	Cordfilled	Cellulose filled	Unfilled Cast Phenolic, mechanical and chemical grade
FORMS AVAILABLE Cs—castings, F—film, Fb—fibers, I—impregnants, L—laminations, Lq—lacquers, Mf—monofilaments, P—Powder, pellet, or granules, R—rods, tubes, or other extruded forms, S—sheets.	L, P, S	L, P, S	Cs, R, S
FABRICATION Cl—calendering, Cs—casting, E—extrusion, F—hot forming or drawing, I—impregnation, MB—blow molding, MC—compression molding, MI—injection molding, S—spreading.	MC	MC	Cs, F
ELECTRICAL PROPERTIES			
D.C. Resistivity, ohm-cm	10^{11}–10^{12}	10^{11}–10^{13}	1.0-7.0 x 10^{12}
Dielectric constant, 60 cps	7.0–10.0	5.0–9.0	6.5–7.5
Dielectric constant, 10^6 cps	5.0–6.0	4.0–7.0	4.0–5.5
Dissipation factor, 60 cps	0.1–0.3	0.04–0.3	0.10–0.15
Dissipation factor, 10^6 cps	0.04–0.09	0.03–0.07	0.04–0.05
MECHANICAL PROPERTIES			
Modulus of elasticity, 10^3 psi	900–1,300	800–1,200	4.0–5.0
Tensile strength, psi	6,000–9,000	6,500–8,500	6,000–9,000
Ultimate elongation, %	0.5–1.0	0.6–1.0	1.5–2.0
Yield stress, psi			
Yield strain, %			
Rockwell hardness		M 110–M 120	M 93–M 120
Notched Izod impact strength, ft.lb/in	4.0–8.0	0.24–0.34	0.25–0.4
Specific gravity	1.36–1.43	1.32–1.55	1.307–1.318
THERMAL PROPERTIES			
Burning rate	Self extinguishing	Self extinguishing	Self extinguishing
Heat distortion 264 psi, °C	121–127	143–171	74–80
Specific heat, cal/g		0.35–0.40	
Linear thermal expansion coefficient, 10^{-5}, °C		3.0–4.5	6.0–8.0
Maximum continuous service temperature, °C	121	149–177	
CHEMICAL RESISTANCE			
Mineral acids, weak	Variable	Variable	Fair to good
Mineral acids, strong	Poor	Poor	Poor to good
Oxidizing acids, concentrated	Poor	Poor	Poor
Alkalies, weak	Variable	Variable	Poor to good
Alkalies, strong	Poor	Poor	Poor
Alcohols	Good	Good to excellent	Good to excellent
Ketones	Poor to fair	Fair	Fair
Esters	Fair to good	Fair to good	Fair to good
Hydrocarbons, aliphatic	Fair to good	Excellent	Good to excellent
Hydrocarbons, aromatic	Fair to good	Excellent	Good
Oils, vegetable, animal, mineral	Good	Excellent	Excellent
MISCELLANEOUS PROPERTIES			
Clarity	Opaque	Opaque	Clear
Color			Colorless to amber
Refractive index, n_D			
Applicable ASTM specifications and test methods	D700, D257, D150, D638, D651, D785, D256, D792, D648, D543	D700, D257, D150, D638, D651, D785, D256, D792, D648, D696,	D257, D150, D638, D785, D256, D792, D648, D696, D543

Chemical Class	Polyester (Styrene-Alkyd)	Silicones	Urea Formaldehyde
Resin Type	Thermosetting	Thermosetting	Thermosetting
Subclass or Modification	Glassfiber mat reinforced	Mineral filled	*a*-Cellulose filled
FORMS AVAILABLE Cs—castings, F—film, Fb—fibers, I—impregnants, L—laminations, Lq—lacquers, Mf—monofilaments, P—Powder, pellet, or granules, R—rods, tubes, or other extruded forms, S—sheets.	L, S	P	P, R, S
FABRICATION Cl—calendering, Cs—casting, E—extrusion, F—hot forming or drawing, I—impregnation, MB—blow molding, MC—compression molding, MI—injection molding, S—spreading.	I	MC	MC
ELECTRICAL PROPERTIES			
D.C. Resistivity, ohm-cm	10^{11}	$>10^{12}$	0.5–5.0
Dielectric constant, 60 cps	4.0–5.5	3.5–3.6	7.7–9.5
Dielectric constant, 10^6 cps	4.0–5.5	3.4–3.6	6.7–8.0
Dissipation factor, 60 cps	0.01–0.04	0.004	0.036–0.043
Dissipation factor, 10^6 cps	0.01–0.06	0.005–0.007	0.025–0.035
MECHANICAL PROPERTIES			
Modulus of elasticity, 10^3 psi	500–1,500		1,300–1,400
Tensile strength, psi	30,000–50,000	3,000–4,000	5,500–13,000
Ultimate elongation, %	0.5–1.5		0.6
Yield stress, psi			
Yield strain, %			
Rockwell hardness	M 80–M 120	M 85–M 95	E 94–E 97
Notched Izod impact strength, ft.lb/in	7.0–30	0.25–0.35	0.24–0.40
Specific gravity	1.5–2.1	1.8–2.8	1.47–1.52
THERMAL PROPERTIES			
Burning rate	Self extinguishing	Self extinguishing	Self extinguishing
Heat distortion 264 psi, °C	93–288	>260	130
Specific heat, cal/g	0.2–0.4	0.2–0.3	0.6
Linear thermal expansion coefficient, 10^{-5}, °C	1.8–3.0	2.0–4.0	2.2–3.6
Maximum continuous service temperature, °C	121–204	288	77
CHEMICAL RESISTANCE			
Mineral acids, weak	Good	Fair to good	Poor
Mineral acids, strong	Poor	Poor to good	Poor
Oxidizing acids, concentrated	Poor		Poor
Alkalies, weak	Good	Fair	Fair
Alkalies, strong	Poor	Poor	Poor
Alcohols	Good	Poor	Good
Ketones	Poor	Poor	Good
Esters	Good		Good
Hydrocarbons, aliphatic	Good	Fair to good	Good
Hydrocarbons, aromatic	Poor to fair	Poor	Good
Oils, vegetable, animal, mineral	Good	Good	
MISCELLANEOUS PROPERTIES			
Clarity	Translucent	Opaque	Translucent
Color	Colorless	Pale to dark	Colorless
Refractive index, n_D			1.54–1.56
Applicable ASTM specifications and test methods	D257, D150, D638, D785, D256, D792, D648, D696, D543	D257, D150, D785, D256, D792, D648, D696, D543	D705, D257, D150, D638, D785, D256, D792, D648

PROPERTIES OF COMMERCIAL PLASTICS

Chemical Class	Acrylonitrile-Butadiene-Styrene (ABS)	Acetal	Alkyd resins
Resin Type	Thermoplastic	Thermoplastic	Thermosetting
Subclass or Modification	High Heat resistant	Homopolymer	Synthetic fiber filled
FORMS AVAILABLE Cs—castings, F—film, Fb—fibers, I—impregnants, L—laminations, Lq—lacquers, Mf—monofilaments, P—Powder, pellet, or granules, R—rods, tubes, or other extruded forms, S—sheets.	P, S, L, R	C, R	P,
FABRICATION Cl—calendering, Cs—casting, E—extrusion, F—hot forming or drawing, I—impregnation, MB—blow molding, MC—compression molding, MI—injection molding, S—spreading.	Cl, E, MB, MI	MI, E	Cs, MC, MI
ELECTRICAL PROPERTIES			
D.C. Resistivity, ohm-cm			
Dielectric constant, 60 cps	2.4–5.0		3.8–5.0
Dielectric constant, 10^6 cps	2.4–3.8	3.7	3.6–4.7
Dissipation factor, 60 cps	0.003–0.008		0.012–0.026
Dissipation factor, 10^6 cps	0.007–0.015	0.004	0.01–0.016
MECHANICAL PROPERTIES			
Modulus of elasticity, 10^3 psi			
Tensile strength, psi	7,000–8,000	10,000–12,000	4,500–6,500
Ultimate elongation, %	1.0–20	15–75	
Yield stress, psi	4,000–9,000		10,000–13,000
Yield strain, %			
Rockwell hardness	R 110–115	M 94, R 120	E 76
Notched Izod impact strength, ft.lb/in	2.0–4.0	1.4–2.3	0.50–4.5
Specific gravity	1.06–1.08	1.43	1.24–2.6
THERMAL PROPERTIES			
Burning rate	Slow	Slow	Self extinguishing
Heat distortion 264 psi, °C	115–118		
Specific heat, cal/g	0.3–0.4	0.35	
Linear thermal expansion coefficient, 10^{-5}, °C	6.0–6.5	8.1	4.0–5.5
Maximum continuous service temperature, °C	88–110	84	149–220
CHEMICAL RESISTANCE			
Mineral acids, weak	Good	Fair	Good
Mineral acids, strong	Good	Poor	Fair
Oxidizing acids, concentrated	Poor	Poor	
Alkalies, weak	Good	Poor	Good
Alkalies, strong	Good	Poor	Fair
Alcohols	Good	Good	Fair to good
Ketones	Poor	Good	Fair to good
Esters	Poor	Good	Fair to good
Hydrocarbons, aliphatic	Fair	Good	Fair to good
Hydrocarbons, aromatic	Fair	Good	Fair to good
Oils, vegetable, animal, mineral	Good	Good	
MISCELLANEOUS PROPERTIES			
Clarity	Translucent to opaque	Translucent to opaque	Opaque
Color	Colorless	Colorless	Colorless
Refractive index, n_D		1.48	
Applicable ASTM specifications and test methods	D638, D150, D792, D651, D648, D256, D758, D696, D543	D638, D150, D792, D651, D648, D256, D758, D696, D543	D638, D150, D792, D651, D648, D256, D758, D543

AZEOTROPES

The following two tables contain data for some binary and ternary systems which exhibit constant boiling points. The boiling points of the compounds in the systems as well as those of the azeotropes are given.

BINARY SYSTEM

No.	Components: Compounds	Components: B.P. °C	Azeotrope: B.P. °C	Percent composition: In Azeotrope	Percent composition: Upper	Percent composition: Lower	Relative vol. of layers at 20°C	Spec. grav. of Azeotrope or layers
1	a. Acetaldehyde b. Ethyl ether	21.0 34.6	20.5	76.0 24.0				0.762
2	a. Acetaldehyde 1551 mm b. Ethyl ether	55.0 70.0	52.6	77.5 22.5				0.763
3	a. Acetamide b. Benzaldehyde	222.0 179.5	178.6	6.5 93.5				
4	a. Acetamide b. o-Bromophenol	222.0 195.0	223.0	50.0 50.0				
5	a. Acetamide b. p-Chloronitrobenzene	222.0 242.0	213.6	55.0 45.0				
6	a. Acetamide b. o-Chlorotoluene	222.0 159.0	157.8	8.0 92.0				
7	a. Acetamide b. p-Dibromobenzene	222.0 219.0	199.35	18.0 82.0				
8	a. Acetamide b. Glycolmonoacetate	222.0 182.0	190.7	5.0 95.0				
9	a. Acetamide b. 2-Methyl-5-ethyl pyridine	222.0 174.0	176.9	5.4 94.6				0.926
10	a. Acetamide b. o-Nitrotoluene	222.0 222.3	206.5	32.5 67.5				
11	a. Acetamide b. Nitrobenzene	222.0 210.9	202.0	24.0 76.0				
12	a. Acetamide b. Octyl alcohol	222.0 195.0	194.5	9.5 90.5				
13	a. Acetamide b. o-Toluidine	222.0 199.84	198.6	12.0 88.0				
14	a. Acetamide b. o-Xylene	222.0 144.1	142.6	11.0 89.0				
15	a. Acetic acid b. Benzene	118.1 80.1	80.1	2.0 98.0				0.882
16	a. Acetic acid b. Bromobenzene	118.1 156.0	118.4	95.0 5.0				
17	a. Acetic acid b. Butyl ether	118.1 142.0	116.7	81.0 19.0				0.983
18	a. Acetic acid b. Chlorobenzene	118.1 132.0	114.7	58.5 41.5				
19	a. Acetic acid b. Cyclohexane	118.1 81.4	79.7	2.0 98.0				
20	a. Acetic acid b. Cyclohexane	118.1 83.0	81.8	6.5 93.5				
21	a. Acetic acid b. 1,1-Dibromoethane	118.1 109.5	103.7	25.0 75.0				
22	a. Acetic acid b. 1,2-Dibromoethane	118.1 131.6	114.4	55.0 45.0				
23	a. Acetic acid b. Dibromomethane	118.1 98.2	94.8	16.0 84.0				
24	a. Acetic acid b. Dimethylformamide	118.1 153.0	159.0	26.0 74.0				1.004
25	a. Acetic acid b. 1,4-Dioxane	118.1 101.5	119.5	77.0 23.0				1.05
26	a. Acetic acid b. Epichlorohydrin	118.1 117.0	115.1	34.5 65.5				
27	a. Acetic acid b. Ethyl benzene	118.1 136.15	114.7	66.0 34.0				0.882
28	a. Acetic acid b. Pyridine	118.1 115.3	139.7	35.0 65.0				1.024
29	a. Acetic acid b. Tetrachloroethylene	118.1 121.02	107.4	38.5 61.5				
30	a. Acetic acid b. Trichloroethylene	118.1 87.0	86.5	3.8 96.2				
31	a. Acetic acid b. Toluene	118.1 110.6	105.4	28.0 72.0				0.905
32	a. Acetic acid b. Triethylamine	118.1 89.5	163.0	69.0 31.0				1.023
33	a. Acetic acid b. m-Xylene	118.1 139.1	115.4	27.0 73.0				0.908
34	a. Acetic acid b. Water	118.1 100.0	76.6	3.0 97.0				
35	a. Acetone b. 2-Bromopropane	56.5 59.6	54.12	42.0 58.0				

AZEOTROPES

The following two tables contain data for some binary and ternary systems which exhibit constant boiling points. The boiling points of the compounds in the systems as well as those of the azeotropes are given.

BINARY SYSTEM

AZEOTROPES (Continued)

BINARY SYSTEM (Continued)

No.	Components: Compounds	Components: B.P. °C	Azeotrope: B.P. °C	Percent composition: In Azeo-trope	Percent composition: Upper	Percent composition: Lower	Relative vol. of layers at 20°C	Spec. grav. of Azeo-trope or layers
36	a. Acetone b. Carbon disulfide	56.5 46.3	39.3	33.0 67.0				1.04
37	a. Acetone b. Chloroform	56.5 61.2	64.7	20.0 80.0				1.268
38	a. Acetone b. 1-Chloropropane	56.5 47.2	45.8	15.0 85.0				
39	a. Acetone b. Cyclohexane	56.5 81.4	53.0	67.0 33.0				
40	a. Acetone b. Cyclopentane	56.5 49.5	41.0	36.0 64.0				
41	a. Acetone b. Diethyl amine	56.5 55.5	51.3	38.2 61.8				0.732
42	a. Acetone b. Hexane	56.5 69.0	49.8	59.0 41.0				
43	a. Acetone b. Iodoethane	56.5 72.2	55.0	60.0 40.0				
44	a. Acetone b. Isobutyl amine	56.5 68.0	<56.0	<96.0 >4.0				
45	a. Acetone b. Isobutyl chloride	56.5 68.9	55.8	73.0 27.0				
46	a. Acetone b. Isopropyl ether	56.5 67.5	53.3	56.5 43.5				0.764
47	a. Acetone 775.5 mm b. Isopropyl ether	78.3 91.9	75.0	55.0 45.0				0.769
48	a. Acetone b. Methanol	56.5 64.65	55.7	88.0 12.0				0.795
49	a. Acetone 11.6 atm. b. Methanol	151.0 143.0	140.0	44.0 56.0				0.796
50	a. Acetone 4.56 atm. b. Methanol	108.0 109.0	102.0	68.0 32.0				0.796
51	a. Acetone b. Methyl acetate	56.5 57.1	55.6	48.0 52.0				0.854
52	a. Acetone b. Water	56.5 100.0	56.08	88.5 11.5				
53	a. Acetone 1034 mm b. Water	84.0 126.0	81.4	98.7 1.3				0.795
54	a. Acetone cyanohydrin b. Water	dec. 100.0	99.9	15.0 85.0				1.001
55	a. Acetonitrile b. Benzene	82.0 80.1	73.0	34.0 66.0				
56	a. Acetonitrile b. Diethyl amine	82.0 55.5	54.5	8.0 92.0				0.705$^{25/20}$
57	a. Acetonitrile b. Ethanol	82.0 78.5	72.9	43.0 57.0				0.788
58	a. Acetonitrile b. Ethyl acetate	82.0 77.15	74.8	23.0 77.0				
59	a. Acetonitrile b. Isopropyl ether	82.0 67.5	61.7	17.0 83.0				0.743
60	a. Acetonitrile b. Methanol	82.0 64.65	63.45	81.0 19.0				
61	a. Acetophenone b. Octyl alcohol	202.3 195.0	195.0	12.5 87.5				
62	a. Acetonitrile 1240 mm b. Pentane	118.0 65.0	58.0	13.0 87.0	3.1 96.9	85.3 14.7	U 90 L 10	U 0.616 L 0.779
63	a. Acetonitrile 152 mm b. Triethyl amine	36.5 43.4	29.0	36.8 63.2				0.749
64	a. Acetonitrile b. Triethyl amine	82.0 89.7	70.3	37.2 62.8				0.749
65	a. Acetonitrile b. Water	82.0 100.0	76.5	83.7 16.3				0.818
66	a. Acetylene b. Ethane	−83.6 −88.3	−94.5	40.7 59.3				
67	a. Acrolein b. Water	52.5 100.0	52.4	97.4 2.6				homog.
68	a. Acrylonitrile b. Benzene	78.0 80.1	73.3	47.0 53.0				
69	a. Acrylonitrile b. Water	78.0 100.0	70.6	85.7 14.3	96.8 3.2	7.3 92.7	U 89.5 L 10.5	U 0.813 L 0.99
70	a. Allyl acetate b. Water	104.1 100.0	83.0	83.3 16.7	98.4 1.6	3.0 97.0	U 84.0 L 16.0	U 0.930 L 0.998
71	a. Allyl acetone b. Water	129.5 100.0	92.1	64.7 35.3	97.75 2.25	2.2 97.8	U 69.0 L 31.0	U 0.849 L 0.997

AZEOTROPES (Continued)
BINARY SYSTEM (Continued)

No	Components: Compounds	Components: B.P. °C	Azeotrope: B.P. °C	Percent composition: In Azeotrope	Percent composition: Upper	Percent composition: Lower	Relative vol. of layers at 20°C	Spec. grav. of Azeotrope or layers
72	a. Allyl alcohol b. Benzene	97.1 80.1	76.8	17.4 82.6				0.874
73	a. Allyl alcohol b. Carbon tetrachloride	97.1 76.8	72.3	11.5 88.5				1.450
74	a. Allyl alcohol b. Hexane	97.0 69.0	65.5	4.5 95.5				
75	a. Allyl alcohol b. Propylacetate	97.0 101.6	94.6	52.0 48.0				
76	a. Allyl alcohol b. Toluene	97.1 110.6	91.5	50.0 50.0				homog.
77	a. Allyl alcohol b. Trichloroethylene	97.1 87.0	81.0	16.0 84.0				1.313
78	a. Allyl alcohol b Water	97.1 100.0	88.2	72.9 27.1				0.905
79	a. Allyl cyanide b. Water	118.9 100.0	89.4	66.0 34.0	97.49 2.51	3.55 96.45	U 70.0 L 30.0	U 0.837 L 0.994
80	a. 2-Aminoethanol b. Chlorobenzene	172.2 132.0	128.6	13.5 86.5				
81	a. 2-Aminoethanol b. o-Chlorotoluene	172.2 159.0	146.5	26.0 74.0				
82	a. Amyl acetate b. Glycol (1–2 ethanediol)	148.7 197.2	147.6	94.0 6.0				
83	a. *tert* Amyl alcohol b. Cyclohexane	101.8 81.4	78.5	16.0 84.0				
84	a. Aniline b. o-Cresol	184.4 191.5	191.3	8.0 92.0				
85	a. Aniline b. Glycol (1–2 ethanediol)	184.4 197.2	180.55	76.0 24.0				
86	a. Aniline b. Octyl alcohol	184.4 195.0	184.0	83.0 17.0				
87	a. *tert* Amyl alcohol b. Toluene	101.8 110.63	100.5	56.0 44.0				
88	a. Benzaldehyde b. α-Chlorotoluene	179.5 179.0	177.9	50.0 50.0				
89	a. Benzaldehyde b. Phenol	179.5 182.0	185.6	49.0 51.0				

No.	Components: Compounds	Components: B.P. °C	Azeotrope: B.P. °C	Percent composition: In Azeotrope	Percent composition: Upper	Percent composition: Lower	Relative vol. of layers at 20°C	Spec. grav. of Azeotrope or layers
90	a. Benzene b. Cyclohexane	80.1 81.4	77.8	55.0 45.0				0.834
91	a. Benzene b. Ethanol	80.1 78.5	67.8	67.6 32.4				0.848
92	a. Benzene b. Ethyl nitrate	80.1 88.7	80.03	88.0 12.0				
93	a. Benzene b. Formic acid	80.1 100.7	71.05	69.0 31.0				
94	a. Benzene b. Heptane	80.1 98.43	80.1	99.3 0.7				
95	a. Benzene b. Isobutanol	80.1 108.3	79.8	90.7 9.3				0.870
96	a. Benzene b. Isopropanol	80.1 82.3	71.5	66.7 33.3				0.838
97	a. Benzene b. Methanol	80.1 64.65	58.3	60.5 39.5				0.844
98	a. Benzene b. Methylethyl ketone	80.1 79.6	78.4	62.5 37.5				0.853
99	a. Benzene b. 2-Methyl-2-propanol	80.1 82.8	74.0	63.4 36.6				0.842
100	a. Benzene b. Nitromethane	80.1 101.0	79.15	86.0 14.0				
101	a. Benzene b. Propanol	80.1 97.2	77.1	83.1 16.9				0.865
102	a. Benzene b. Water	80.1 100.0	69.4	91.1 8.9	99.94 0.06	0.07 99.93	U 92.0 L 8.0	U 0.880 L 0.999
103	a. Benzonitrile b. p-Bromotoluene	190.7 185.0	184.3	15.0 85.0				
104	a. Benzonitrile b. o-Cresol	190.7 191.5	196.0	49.0 51.0				
105	a. Benzyl alcohol b. p-Cresol	205.2 202.5	206.8	62.0 38.0				
106	a. Benzyl alcohol b. Nitrobenzene	205.2 210.9	204.2	62.0 38.0				
107	a. Benzyl alcohol b. Water	205.2 100.0	99.9	9.0 91.0				

AZEOTROPES (Continued)
BINARY SYSTEM (Continued)

AZEOTROPES (Continued)
BINARY SYSTEM (Continued)

No.	Components: Compounds	Components: B.P. °C	Azeotrope: B.P. °C	Percent composition: In Azeotrope	Percent composition: Upper	Percent composition: Lower	Relative vol. of layers at 20°C	Spec. grav. of Azeotrope or layers
108	a. Bromobenzene b. Chloroacetic acid	156.0 189.0	154.3	89.0 11.0				
109	a. 1-Bromobutane b. Ethanol	101.6 78.5	75.0	57.0 43.0				
110	a. 1-Bromobutane b. Glycol	101.6 197.2	101.3	98.3 1.7				
111	a. 1-Bromobutane b. Propylacetate	101.6 101.6	99.9	52.0 48.0				
112	a. Bromodichloromethane b. Ethanol	90.2 78.5	75.5	72.0 28.0				
113	a. Bromoethane b. Ethanol	38.0 78.5	37.0	97.0 3.0				
114	a. o-Bromophenol b. o-Cresol	195.0 191.5	189.8	25.0 75.0				
115	a. o-Bromophenol b. Octyl alcohol	195.0 195.0	204.0	50.0 50.0				
116	a. 1-Bromopropane b. Hexane	70.9 69.0	67.2	50.0 50.0				
117	a. 2-Bromopropene b. Ethanol	48.4 78.5	46.2	94.0 6.0				
118	a. o-Bromotoluene b. Caproic acid	181.75 205.0	180.8	94.0 6.0				
119	a. o-Bromotoluene b. Chloroacetic acid	181.75 189.0	173.0	68.0 32.0				
120	a. o-Bromotoluene b. Ethylacetoacetate	181.75 180.0	174.7	49.0 51.0				
121	a. o-Bromotoluene b. Glycol	181.75 197.2	166.8	75.0 25.0				
122	a. m-Bromotoluene b. Octyl alcohol	183.7 195.0	184.1	91.0 9.0				
123	a. p-Bromotoluene b. Octyl alcohol	185.0 195.0	184.6	90.0 10.0				
124	a. 2,3-Butanedione b. Ethanol	88.0 78.5	73.9	53.0 47.0				homog.
125	a. 1-Butanol b. Butyl acetate	117.7 126.5	117.6	67.2 32.8				0.832
126	a. 1-Butanol b. Butyl ether	117.7 142.0	117.6	82.5 17.5				0.804
127	a. 1-Butanol b. Butyronitrile	117.7 118.0	113.0	50.0 50.0				
128	a. 1-Butanol b. Chlorobenzene	117.7 132.0	115.3	56.0 44.0				
129	a. 1-Butanol b. 1-Chloro-2-propanone	117.7 119.0	112.5	43.0 57.0				
130	a. 1-Butanol b. Cyclohexane	117.7 81.4	79.8	10.0 90.0				homog.
131	a. 1-Butanol b. 1,1-Dibromoethane	117.7 109.5	104.5	20.0 80.0				
132	a. 1-Butanol b. 1,2-Dibromoethane	117.7 131.6	114.8	44.0 56.0				
133	a. 1-Butanol b. Epichlorohydrin	117.7 117.0	112.0	43.0 57.0				
134	a. 1-Butanol b. Ethylenediamine	117.7 116.9	124.7	64.3 35.7				0.849
135	a. 1-Butanol b. Ethyl nitrate	117.7 88.7	87.45	4.0 96.0				
136	a. 1-Butanol b. Heptane	117.7 98.4	93.3	18.0 82.0				0.701
137	a. 1-Butanol b. Hexaldehyde	117.7 128.5	116.8	77.1 22.9				0.829
138	a. 1-Butanol b. Hexane	117.7 69.0	67.0	3.0 97.0				0.69
139	a. 1-Butanol b. 1-Iodopropane	117.7 102.4	99.5	13.5 86.5				
140	a. 1-Butanol b. 3-Iodopropene	117.7 103.1	98.7	13.0 87.0				
141	a. 1-Butanol b. Methylchloroacetate	117.7 130.0	116.3	74.0 26.0				
142	a. 1-Butanol b. Nitroethane	117.7 114.8	107.7	45.0 55.0				
143	a. 1-Butanol b. Nitromethane	117.7 101.0	97.8	30.0 70.0				

AZEOTROPES (Continued)
BINARY SYSTEM (Continued)

No.	Components: Compounds	Components: B.P. °C	Azeotrope: B.P. °C	Percent composition: In Azeotrope	Percent composition: Upper	Percent composition: Lower	Relative vol. of layers at 20°C	Spec. grav. of Azeotrope or layers
144	a. 1-Butanol	117.7	110.2	50.0				
	b. Octane	125.8		50.0				
145	a. 1-Butanol	117.7	126.5	51.0				0.843
	b. Propylenediamine	120.9		49.0				
146	a. 1-Butanol	117.7	118.7	71.0				
	b. Pyridine	115.3		29.0				
147	a. 1-Butanol	117.7	110.0	32.0				
	b. Tetrachloroethylene	121.2		68.0				
148	a. 1-Butanol	117.7	105.6	27.0				0.846
	b. Toluene	110.6		73.0				
149	a. 1-Butanol	117.7	86.65	3.0				
	b. Trichloroethylene	87.0		97.0				
150	a. 1-Butanol	117.7	93.0	55.5	79.9	7.7	U 71.5	U 0.849
	b. Water	100.0		44.5	20.1	92.3	L 28.5	L 0.990
151	a. 1-Butanol 30 mm	48.0	29.0	47.6	79.9	7.7	U 59.0	U 0.849
	b. Water	29.0		52.4	20.1	92.3	L 41.0	L 0.989
152	a. 1-Butanol	117.7	116.8	75.0				
	b. o-Xylene	144.41		25.0				
153	a. 2-Butanol	99.5	76.0	18.0				
	b. Cyclohexane	81.4		82.0				
154	a. 2-Butanol	99.5	67.2	8.0				
	b. Hexane	69.0		92.0				
155	a. 2-Butanol	99.5	95.3	55.0				
	b. Toluene	110.63		45.0				
156	a. 2-Butanol	99.5	88.5	68.0				0.863
	b. Water	100.0		32.0				
157	a. *tert* Butanol	82.8	74.0	36.6				
	b. Benzene	80.1		63.4				
158	a. *tert* Butanol	82.8	71.3	37.0				
	b. Cyclohexane	81.4		63.0				
159	a. *tert* Butanol	82.8	63.7	22.0				
	b. Hexane	69.0		78.0				
160	a. 2-Butanone	79.6	71.8	40.0				
	b. Cyclohexane	81.4		60.0				
161	a. 2-Butanone	79.6	77.0	18.0				
	b. Ethylacetate	77.15		82.0				

No.	Components: Compounds	Components: B.P. °C	Azeotrope: B.P. °C	Percent composition: In Azeotrope	Percent composition: Upper	Percent composition: Lower	Relative vol. of layers at 20°C	Spec. grav. of Azeotrope or layers
162	a. 1-Butenylmethyl ether, cis-	72.0	64.0	93.9	99.79	0.32	U 95.2	U 0.777
	b. Water	100.0		6.1	0.21	99.68	L 4.8	L 1.000
163	a. 1-Butenylmethyl ether, trans-	76.7	67.0	92.8	99.80	0.01	U 94.2	U 0.787
	b. Water	100.0		7.2	0.20	99.99	L 5.8	L 1.000
164	a. 1-Butoxy-2propanol	170.1	98.6	28.0	80.8	6.4	U 31.0	U 0.910
	b. Water	100.0		72.0	19.2	93.6	L 69.0	L 0.997
165	a. Butyl acetate	126.5	125.9	95.0				0.879
	b. Butyl ether	142.1		5.0				
166	a. Butyl acetate	126.5	125.6	69.0				
	b. 2-Chloroethanol	128.0		31.0				
167	a. Butyl acetate	126.5	125.8	64.3				0.896
	b. 2-Ethoxyethanol	135.6		35.7				
168	a. Butyl acetate	126.5	126.0	82.5				
	b. Isoamyl alcohol	130.5		17.5				
169	a. Butyl acetate	126.5	119.0	52.0				
	b. Octane	125.8		48.0				
170	a. Butyl acetate	126.5	90.7	72.9	98.8	0.68	U 75.8	U 0.882
	b. Water	100.0		27.1	1.2	99.32	L 24.2	L 0.998
171	a. Butyl acetoacetate	213.9	99.4	15.9	97.2	0.4	U 16.0	U 0.970
	b. Water	100.0		84.1	2.8	99.6	L 84.0	L 1.000
172	a. Butylamine	77.1	76.5	60.0				
	b. Cyclohexane	81.4		40.0				
173	a. Butylamine	77.1	82.2	51.0				$0.766^{30/20}$
	b. Ethanol	78.5		49.0				
174	a. Butylamine	77.1	84.7	40.0				0.775
	b. Isopropanol	82.3		60.0				
175	a. Butylamine 575 mm	69.0	69.0	98.7				homog.
	b. Water	92.4		1.3				
176	a. n-Butyl aniline	240.9	99.8	5.6	99.76	0.01	U 6.0	U 0.929
	b. Water	100.0		94.4	0.24	99.99	L 94.0	L 1.000
177	a. Butyl benzoate	250.2	99.9	5.0	0.01	99.68	U 95.0	U 1.000
	b. Water	100.0		95.0	99.99	0.32	L 5.0	L 1.007
178	a. Butyl butyrate	166.4	97.9	47.0	99.52	0.06	U 50.6	U 0.871
	b. Water	100.0		53.0	0.48	99.94	L 49.4	L 0.998
179	a. Butyl 2-ethoxyethanol	171.1	170.6	42.0				0.887
	b. Dibutyl acetal	189.4		58.0				

AZEOTROPES (Continued)
BINARY SYSTEM (Continued)

No.	Components: Compounds	Components: B.P. °C	Azeotrope: B.P. °C	Percent composition: In Azeotrope	Percent composition: Upper	Percent composition: Lower	Relative vol. of layers at 20°C	Spec. grav. of Azeotrope or layers
180	a. Butyl 2-ethoxyethanol b. Water	171.1 100.0	98.8	20.8 79.2				0.989
181	a. Butyl chloride b. Water	78.0 100.0	68.0	93.0 7.0	99.92 0.08	0.11 99.89	U 94.0 L 6.0	U 0.887 L 1.000
182	a. Butyl ether b. 2-Ethoxy ethanol	142.0 135.1	127.0	50.0 50.0				homog.
183	a. Butyl ether b. Ethylene glycol	142.0 197.2	139.5	93.6 6.4	98.0 2.0	1.0 99.0	U 95.0 L 5.0	U 0.777 L 1.114
184	a. Butyl ether b. Morpholine	142.0 128.3	126.7	27.0 73.0				0.924
185	a. Butyl ether b. Propionic acid	142.0 141.1	136.0	55.0 45.0				
186	a. Butyl ether b. Water	142.0 100.0	94.1	66.6 33.4	99.97 0.03	0.19 99.81	U 72.0 L 28.0	U 0.769 L 0.998
187	a. Butylisopropenyl ether b. Water	114.8 100.0	86.3	81.2 18.8	0.01 99.99	99.99 0.01	U 84.7 L 15.3	U 0.790 L 1.000
188	a. 2-Butyl octanol b. Water	253.4 100.0	99.9	2.5 97.5	98.9 1.1	0.01 99.99	U 3.0 L 97.0	U 0.85 L 1.00
189	a. Butyl salicylate b. Water	268.2 100.0	99.9	4.2 95.8	0.02 99.98	99.87 0.13	U 96.0 L 4.0	U 1.000 L 1.075
190	a. Butyraldehyde b. Ethanol	75.7 78.5	70.7	39.4 60.6				0.835
191	a. Butyraldehyde b. Water	75.7 100.0	68.0	90.3 9.7	96.8 3.2	7.1 92.9	U 94.0 L 6.0	U 0.815
192	a. Butyric acid b. o-Dichlorobenzene	163.5 180.0	163.0	65.0 35.0				
193	a. Butyric acid b. Water	163.5 100.0	99.4	18.4 81.6				1.007
194	a. Butyric acid b. o-Xylene	163.5 144.41	143.0	10.0 90.0				
195	a. Butyric acid b. m-Xylene	163.5 139.1	138.1	6.6 93.4				
196	a. Butyric acid b. p-Xylene	163.5 138.4	137.5	5.4 94.6				
197	a. Butyronitrile b. Toluene	118.0 110.63	107.0	27.0 73.0				

AZEOTROPES (Continued)
BINARY SYSTEM (Continued)

No.	Components: Compounds	Components: B.P. °C	Azeotrope: B.P. °C	Percent composition: In Azeotrope	Percent composition: Upper	Percent composition: Lower	Relative vol. of layers at 20°C	Spec. grav. of Azeotrope or layers
198	a. Butyronitrile b. Water	118.0 100.0	88.7	67.5 32.5	97.45 2.55	3.53 96.47	U 72.9 L 27.1	U 0.795 L 0.994
199	a. Caproic acid b. m-Cresol	205.0 202.8	201.9	13.0 87.0				
200	a. Carbon disulfide b. 1,1-Dichloroethane	46.2 57.2	46.0	94.0 6.0				homog.
201	a. Carbon disulfide b. Ethyl acetate	46.2 77.06	46.1	97.0 3.0				1.249
202	a. Carbon disulfide b. Ethyl ether	46.2 34.6	34.4	1.0 99.0				0.719
203	a. Carbon disulfide b. Ethyl formate	46.2 54.3	39.4	63.0 37.0				
204	a. Carbon disulfide b. Isopropanol	46.2 82.3	44.6	92.0 8.0				1.202
205	a. Carbon disulfide b. Methanol	46.2 64.65	37.7	86.0 14.0	50.8 49.2	97.2 2.8	U 26.3 L 73.7	U 0.979 L 1.261
206	a. Carbon disulfide b. Methyl acetate	46.2 57.0	40.2	70.0 30.0				1.126
207	a. Carbon disulfide b. Methylethyl ketone	46.2 79.6	45.9	84.7 15.3				1.157
208	a. Carbon disulfide b. 2-Methyl-2-propanol	46.2 82.2	45.3	94.0 6.0				1.219
209	a. Carbon disulfide b. Propyl chloride	46.2 46.6	45.2	55.0 45.0				homog.
210	a. Carbon disulfide b. Water	46.2 100.0	43.6	98.0 2.0	0.29 99.71	99.99 0.01	U 2.3 L 97.7	U 1.001 L 1.265
211	a. Carbon tetrachloride b. Ethanol	76.8 78.5	65.0	84.2 15.8				1.377
212	a. Carbon tetrachloride b. Ethyl acetate	76.8 77.1	74.8	57.0 43.0				1.202
213	a. Carbon tetrachloride 286 mm b. Ethyl acetate	47.5 50.5	47.4	82.2 17.8				homog.
214	a. Carbon tetrachloride b. Ethylene dichloride	76.8 84.0	75.3	78.0 22.0				1.500
215	a. Carbon tetrachloride b. Isobutanol	76.8 108.3	75.8	94.5 5.5				1.515

AZEOTROPES (Continued)
BINARY SYSTEM (Continued)

AZEOTROPES (Continued)
BINARY SYSTEM (Continued)

No.	Components: Compounds	Components: B.P. °C	Azeotrope: B.P. °C	Percent composition: In Azeotrope	Percent composition: Upper	Percent composition: Lower	Relative vol. of layers at 20°C	Spec. grav. of Azeotrope or layers
216	a. Carbon tetrachloride b. Isopropanol	76.8 82.3	67.0	82.0 18.0				1.344
217	a. Carbon tetrachloride b. Methanol	76.8 64.65	55.7	79.4 20.6				1.322
218	a. Carbon tetrachloride b. Methylethyl ketone	76.8 79.6	73.8	71.0 29.0				1.247
219	a. Carbon tetrachloride b. 2-Methyl-2-Propanol	76.8 82.2	69.5	83.0 17.0				1.358
220	a. Carbon tetrachloride b. Propanol	76.8 97.2	72.8	88.5 11.5				1.437
221	a. Carbon tetrachloride b. Water	76.8 100.0	66.8	95.9 4.1	0.03 99.97	99.97 0.03	U 6.4 L 93.6	U 1.000 L 1.597
222	a. 2-Ethoxyethanol b. Toluene	135.6 110.6	110.0	10.0 90.0				0.874
223	a. 2-Ethoxyethanol b. Water	135.6 100.0	99.4	28.8 71.2				1.003
224	a. 2-Ethoxyethanol 200 mm b. Water	96.5 66.3	66.4	15.0 85.0				
225	a. Chloral b. Heptane	98.0 98.43	93.0	53.0 47.0				
226	a. Chloral b. Water	97.75 100.0	95.0	93.0 7.0				homog.
227	a. Chloroacetic acid b. o-Chlorotoluene	189.0 159.0	156.8	12.0 88.0				
228	a. Chloroacetic acid b. Naphthalene	189.0 217.9	187.1	78.0 22.0				
229	a. Chloroacetic acid b. Valeric acid	189.0 187.0	186.3	3.0 97.0				
230	a. Chloroacetic acid b. o-Xylene	189.0 144.4	143.5	12.0 88.0				
231	a. Chlorobenzene b. 2-Chloroethanol	132.0 128.0	120.0	58.0 42.0				
232	a. Chlorobenzene b. 2-Ethoxyethanol	132.0 135.1	127.2	68.0 32.0				
233	a. Chlorobenzene b. Isoamyl alcohol	132.0 130.5	124.4	66.0 34.0				
234	a. Chlorobenzene b. Isobutyl alcohol	132.0 108.39	107.1	37.0 63.0				
235	a. Chlorobenzene b. Isobutyric acid	132.0 154.4	131.2	92.0 8.0				
236	a. Chlorobenzene b. Propanol	132.0 97.2	96.5	20.0 80.0				
237	a. Chlorobenzene b. Propionic acid	132.0 141.1	128.9	82.0 18.0				
238	a. Chlorobenzene b. Pyrrole	132.0 131.0	124.5	57.0 43.0				
239	a. Chlorobenzene b. Water	132.0 100.0	90.2	71.6 28.4				
240	a. 1-Chlorobutane b. Ethanol	78.0 78.5	65.7	79.7 20.3				
241	a. 1-Chlorobutane b. Isopropyl alcohol	78.0 82.3	70.8	77.0 23.0				
242	a. 2-Chloroethanol b. Ethyl benzene	128.0 136.15	121.0	62.0 38.0				
243	a. 2-Chloroethanol b. Isoamyl alcohol	128.8 130.5	127.8	75.0 25.0				
244	a. 2-Chloroethyl ether b. Water	179.2 100.0	98.0	34.5 65.5	1.07 98.93	99.72 0.28	U 70.4 L 29.6	U 1.002 L 1.220
245	a. Chloroform b. Ethanol	61.2 78.5	59.4	93.0 7.0				1.403
246	a. Chloroform b. Hexane	61.2 68.7	60.0	72.0 28.0				1.101
247	a. Chloroform b. Methanol	61.2 64.65	53.5	87.0 13.0				1.342
248	a. Chloroform b. Methylethyl ketone	61.2 79.6	79.9	17.0 83.0				0.877
249	a. Chloroform b. Water	61.2 100.0	56.3	97.0 3.0	0.8 99.2	99.8 0.2	U 4.4 L 95.6	U 1.004 L 1.491
250	a. Chloroisopropyl ether b. Water	187.0 100.0	98.5	37.4 62.6	0.17 99.83	99.86 0.14	U 65.0 L 35.0	U 1.00 L 1.11
251	a. o-Chloronitrobenzene b. Diethylene glycol	245.7 244.5	233.5	59.0 41.0				

AZEOTROPES (Continued)
BINARY SYSTEM (Continued)

No.	Components: Compounds	Components: B.P. °C	Azeotrope: B.P. °C	Percent composition: In Azeotrope	Percent composition: Upper	Percent composition: Lower	Relative vol. of layers at 20°C	Spec. grav. of Azeotrope or layers
252	a. Chloronitromethane b. Tetrachloroethylene	122.5 121.02	115.2	45.0 55.0				
253	a. o-Chlorophenol b. o-Dichlorobenzene	175.6 180.0	173.6	48.0 52.0				
254	a. o-Chlorophenol b. Phenol	175.6 182.0	174.5	25.0 75.0				
255	a. p-Chlorophenol b. Naphthalene	217.0 217.9	216.3	36.5 63.5				
256	a. p-Chlorophenol b. Nitrobenzene	217.0 210.9	219.9	8.0 92.0				
257	a. 1-Chloro-2-propanone b. Toluene	119.0 110.63	109.2	28.5 71.5				
258	a. o-Chlorotoluene b. Cyclohexanol	159.0 161.5	155.5	62.0 38.0				
259	a. o-Chlorotoluene b. 2-Furaldehyde	159.0 161.7	155.4	65.0 35.0				
260	a. o-Chlorotoluene b. Glycol	159.0 197.2	152.5	87.0 13.0				
261	a. o-Chlorotoluene b. Hexyl alcohol	159.0 157.2	153.5	56.0 44.0				
262	a. o-Chlorotoluene b. Isovaleric acid	159.0 176.7	157.5	88.0 12.0				
263	a. o-Chlorotoluene b. Phenol	159.0 182.0	159.2	97.0 3.0				
264	a. o-Chlorotoluene b. Valeric acid	159.0 187.0	158.5	95.0 5.0				
265	a. o-Cresol b. Octyl alcohol	191.5 195.0	196.9	38.0 62.0				
266	a. m-Cresol b. Octyl alcohol	202.8 195.0	203.3	62.0 38.0				
267	a. p-Cresol b. Octyl alcohol	202.5 195.0	202.3	70.0 30.0				
268	a. Croton aldehyde b. Water	104.0 100.0	84.0	75.2 24.8	90.4 9.6	15.6 84.4	U 81.4 L 18.6	U 0.876 L 0.987
269	a. 1,3-Cyclohexadiene b. Cyclohexane	80.5 81.4	79.0	45.0 55.0				
270	a. Cyclohexane b. Ethanol	81.4 78.5	64.9	69.5 30.5				
271	a. Cyclohexane b. Ethyl acetate	81.4 77.15	72.8	46.0 54.0				
272	a. Cyclohexane b. Ethyl nitrate	81.4 88.7	74.5	64.0 36.0				
273	a. Cyclohexane b. Isobutanol	81.0 108.3	78.1	86.0 14.0				
274	a. Cyclohexane b. Isopropanol	81.0 82.3	68.6	67.0 33.0				0.777
275	a. Cyclohexane b. Isopropyl acetate	81.4 89.0	78.9	75.0 25.0				
276	a. Cyclohexane b. Methanol	81.0 64 65	54.2	63.0 37.0	97.0 3.0	39.0 61.0	U 43.0 L 57.0	homog.
277	a. Cyclohexane b. Nitromethane	81.4 101.0	70.2	72.0 28.0				
278	a. Cyclohexane b. Propanol	81.0 97.2	74.3	80.0 20.0				
279	a. Cyclohexane b. Vinyl acetate	81.0 72.0	67.4	38.7 61.3				0.850
280	a. Cyclohexane b. Water	81.0 100.0	69.8	91.5 8.5	99.99 0.01	0.01 99.99	U 93.2 L 6.8	U 0.780 L 1.000
281	a. Cyclohexanol b. 2-Furaldehyde	161.5 161.7	156.5	94.5 5.5				
282	a. Cyclohexanol b. Phenol	161.5 182.0	183.0	13.0 87.0				
283	a. Cyclohexanol b. Propyl chloroacetate	161.5 163.5	159.0	53.0 47.0				
284	a. Cyclohexanol b. Water	161.5 99.1	~97.8	~20 ~80				
285	a. Cyclohexanol b. o-Xylene	161.5 144.41	143.0	14.0 86.0				
286	a. Cyclohexanone b. Water	155.65 100.0	95.0	38.4 61.6	92.0 8.0	2.3 97.7	U 41.5 L 58.5	U 0.953 L 1.000
287	a. Cyclopentanol b. Water	140.85 100.0	96.25	42.0 58.0				

No.	Components: Compounds	Components: B.P. °C	Azeotrope: B.P. °C	Percent composition: In Azeotrope	Percent composition: Upper	Percent composition: Lower	Relative vol. of layers at 20°C	Spec. grav. of Azeotrope or layers
288	a. Cyclopentanol b. m-Xylene	140.0 139.10	132.8	40.0 60.0				
289	a. Cyclopentanone b. Water	130.65 100.0	93.5	58.0 42.0	86.2 13.8	29.3 70.7	U 51.0 L 49.0	U 0.965 L 0.998
290	a. Diacetone alcohol b. Water	169.2 100.0	99.6	13.0 87.0				1.002
291	a. Diallyl acetal b. Water	150.9 100.0	95.3	59.0 41.0	99.3 0.7	0.7 99.3	U 62.0 L 38.0	U 0.876 L 0.998
292	a. Diallyl amine b. Water	110.5 100.0	87.2	76.0 24.0	71.1 28.9	5.09 94.91		0.852
293	a. p-Dibromobenzene b. Nitrobenzene	219.0 210.9	210.5	22.5 77.5				
294	a. p-Dibromobenzene b. o-Nitrotoluene	219.0 222.3	218.0	73.0 27.0				
295	a. 1,2-Dibromobutane b. Glycol	140.5 197.2	139.0	94.0 6.0				
296	a. 1,2-Dibromoethane b. Glycol	131.6 197.2	130.85	96.5 3.5				
297	a. Dibromomethane b. Ethanol	98.2 78.5	76.0	62.0 38.0				
298	a. Dibutyl acetal b. Water	189.4 100.0	98.7	33.7 66.3	99.8 0.2	0.03 99.97	U 33.0 L 67.0	U 0.83 L 1.000
299	a. Dibutyl amine b. Water	159.6 100.0	97.0	49.5 50.5	93.8 6.2	0.47 99.53	U 58.0 L 42.0	U 0.780 L 0.990
300	a. Dibutyl ethanolamine b. Water	228.7 100.0	99.9	9.0 91.0	93.2 6.8	0.39 99.61	U 10.0 L 90.0	U 0.877 L 0.999
301	a. Dibutyl formal b. Water	181.8 100.0	98.2	38.0 62.0				heterog.
302	a. Dibutyl fumarate b. Water	285.1 100.0	99.9	1.5 98.5	99.62 0.38	0.1 99.9	U 1.5 L 98.5	U 0.987 L 1.000
303	a. Dibutyl maleate b. Water	280.6 100.0	99.9	1.6 98.4	99.36 0.64	0.01 99.99	U 1.7 L 98.3	U 0.996 L 1.000
304	a. o-Dichlorobenzene b. 2-Furaldehyde	180.0 161.7	161.0	22.0 78.0				
305	a. o-Dichlorobenzene b. Glycol	180.0 197.2	165.8	80.0 20.0				
306	a. o-Dichlorobenzene b. Phenol	180.0 182.0	173.7	65.0 35.0				
307	a. o-Dichlorobenzene b. Valeric acid	180.0 187.0	175.8	78.0 22.0				
308	a. p-Dichlorobenzene b. Furfuryl alcohol	173.4 171.0	172.5	30.0 70.0				
309	a. p-Dichlorobenzene b. Phenol	173.4 182.0	171.1	74.8 25.2				
310	a. 1,2-Dichloroethane b. Ethanol	83.5 78.5	70.5	63.0 37.0				
311	a. 1,2-Dichloroethane b. Isobutyl alcohol	83.5 108.39	83.45	93.5 6.5				
312	a. Di(2-chloroethyl)formal b. Water	218.1 100.0	99.4	13.2 86.8	0.78 99.22	99.56 0.44	U 89.0 L 11.0	U 1.002 L 1.233
313	a. Dichloromethane b. Ethanol	40.1 78.5	<39.85	>95.0 <5.0				
314	a. Dichloromethane b. Glycol	40.1 197.2	168.7	86.0 14.0				
315	a. 2,2-Dichloropropane b. Isopropyl alcohol	69.7 82.3	66.8	83.0 17.0				
316	a. 2,3-Dichloropropanol b. Water	182.0 100.0	99.4	13.0 87.0	11.3 88.7	89.7 10.3	U 99.0 L 1.0	U 1.04 L 1.33
317	a. Dicyclopentadiene b. Water	170.0 100.0	98.0	32.3 67.7	99.99 0.01	0.05 99.95	U 33.0 L 67.0	U 0.978 L 1.000
318	a. Diethylacetal b. Water	102.1 100.0	82.6	85.7 14.3	98.8 1.2	5.5 94.5	U 88.0 L 12.0	U 0.826 L 0.99
319	a. Diethylaminoethyl amine b. Water	144.9 100.0	99.8	20.5 79.5				0.988
320	a. Diethyl butyral b. Water	146.3 100.0	94.2	65.5 34.5	99.52 0.48	0.46 99.54	U 30.0 L 70.0	U 0.830 L 0.999
321	a. Diethylene glycol b. Naphthalene	244.5 217.9	212.6	22.0 78.0				
322	a. Diethylene glycol b. Nitrobenzene	244.5 210.9	210.0	10.0 90.0				
323	a. Diethylene glycol b. o-Nitrotoluene	244.5 222.3	218.2	17.5 82.5				

AZEOTROPES (Continued)
BINARY SYSTEM (Continued)

AZEOTROPES (Continued)
BINARY SYSTEM (Continued)

No.	Components: Compounds	Components: B.P. °C	Azeotrope: B.P. °C	Percent composition: In Azeotrope	Percent composition: Upper	Percent composition: Lower	Relative vol. of layers at 20°C	Spec. grav. of Azeotrope or layers
324	a. Diethyleneglycol dibutyl ether b. Water	254.6 100.0	99.8	5.3 94.7	98.6 1.4	0.3 99.7	U 6.0 L 94.0	U 0.887 L 1.000
325	a. Diethyl ethanolamine b. Water	162.1 100.0	98.9	25.6 74.4				0.993
326	a. Diethyl formal b. Ethanol	87.5 78.5	74.2	43.0 57.0				homog.
327	a. Diethyl fumarate b. Water	218.1 100.0	99.5	12.5 87.5	0.40 99.60	98.87 1.13	U 88.0 L 12.0	U 1.000 L 1.052
328	a. Di(2-ethylhexyl)acetal b. Water	dec. 100.0	99.9	1.0 99.0	99.9 0.1	0.1 99.9	U 1.2 L 98.8	U 0.845 L 1.000
329	a. Diethylisopropanol amine b. Water	159.5 100.0	97.2	45.0 55.0				0.962
330	a. Diethyl ketone b. Propanol	102.7 97.2	94.9	43.0 57.0				homog.
331	a. Diethyl maleate b. Water	225.0 100.0	99.6	11.9 88.1	1.50 98.50	98.60 1.40	U 89.0 L 11.0	U 1.001 L 1.069
332	a. Diethyl phthalate b. Water	296.0 100.0	99.9	1.6 98.4	0.02 99.98	99.5 0.5	U 98.7 L 1.3	U 1.000 L 1.118
333	a. Diethyl pimelate b. Water	268.1 100.0	100.0	1.7 98.3	98.7 1.3	0.14 99.86	U 1.7 L 98.3	U 0.995 L 1.000
334	a. Diethyl succinate b. Water	217.7 100.0	99.9	9.0 91.0	2.05 97.95	98.2 1.8	U 93.0 L 7.0	U 1.005 L 1.04
335	a. Dihexyl amine b. Water	239.8 100.0	99.8	7.2 92.8	97.5 2.5	1.1 98.9	U 7.7 L 92.3	U 0.801 L 0.999
336	a. Diisobutylene b. Water	102.6 100.0	82.0	88.0 12.0	99.99 0.01	0.01 99.99	U 91.0 L 9.0	U 0.721 L 1.000
337	a. Diisobutyl ketone b. Morpholine	168.0 128.3	128.0	2.0 98.0				0.996
338	a. Diisobutyl ketone b. Water	168.0 100.0	97.0	48.1 51.9	99.25 0.75	0.05 99.95	U 53.4 L 46.6	U 0.810 L 1.000
339	a. Diisopropyl amine b. Water	84.1 100.0	74.1	91.0 9.0				homog.
340	a. Diisopropylethanol amine b. Water	190.9 100.0	99.2	15.0 85.0	81.8 18.2	1.28 98.72	U 18.0 L 82.0	U 0.90 L 0.998
341	a. Diisopropyl maleate b. Water	228.7 100.0	99.9	7.0 93.0	0.23 99.77	99.19 0.81	U 93.0 L 7.0	U 1.000 L 1.012
342	a. Dimethyl acetal b. Methanol	64.5 64.65	57.5	75.8 24.2				0.841
343	a. Dimethyl acetal b. Water	64.5 100.0	61.3	96.4 3.6				heterog.
344	a. 1,3-Dimethylbutyl amine b. Water	108.5 100.0	89.5	71.4 28.6				0.83
345	a. Dimethyl butyral b. Water	114.0 100.0	87.3	79.7 20.3	99.1 0.9	2.9 97.1	U 82.3 L 17.7	U 0.851 L 0.997
346	a. Dimethyl formal b. Methanol	42.3 64.65	41.8	92.1 7.9				0.860
347	a. 2,5-Dimethyl furan b. Methanol	93.3 64.65	61.5	49.0 51.0				0.841
348	a. 2,5-Dimethyl furan b. Water	93.3 100.00	77.0	88.3 11.7	99.87 0.13	0.13 99.87	U 89.0 L 11.0	U 0.902 L 1.000
349	a. 2,6-Dimethyl-4-heptanol b. Water	178.1 100.0	98.5	29.6 70.4	99.01 0.99	0.06 99.94	U 34.4 L 65.6	U 0.814 L 1.000
350	a. Dimethyl isobutyral b. Water	104.7 100.0	83.9	85.7 14.3	99.23 0.77	3.48 96.52	U 87.7 L 12.3	U 0.847 L 0.995
351	a. Dimethyl phthalate b. Water	282.9 100.0	100.0	1.1 98.9	0.43 99.57	98.5 1.5	U 99.0 L 1.0	U 1.001 L 1.191
352	a. Dimethyl pimelate b. Water	248.9 100.0	99.9	3.2 96.8	0.99 99.01	98.0 2.0	U 97.8 L 2.2	U 1.005 L 1.04
353	a. 1,4-Dioxane b. Water	101.3 100.0	87.8	81.6 18.4				1.040
354	a. 1,3-Dioxolane b. Heptane	75.6 98.4	72.3	81.0 19.0				0.957
355	a. Dipropyl acetal b. Water	146.6 100.0	95.0	63.4 36.6	99.66 0.34	0.41 99.59	U 67.5 L 32.5	U 0.837 L 0.999
356	a. Dipropyl ketone b. Water	144.0 100.0	94.3	59.5 40.5	99.1 0.9	0.4 99.6	U 64.6 L 35.4	U 0.820 L 1.000
357	a. Epichlorohydrin b. Propanol	117.0 97.2	96.0	23.0 77.0				
358	a. Epichlorohydrin b. Toluene	117.0 110.6	108.3	26 0 74.0				

AZEOTROPES (Continued)
BINARY SYSTEM (Continued)

No.	Components: Compounds	Components: B.P. °C	Azeotrope: B.P. °C	Percent composition: In Azeotrope	Percent composition: Upper	Percent composition: Lower	Relative vol. of layers at 20°C	Spec. grav. of Azeotrope or layers
359	a. Epichlorohydrin b. Water	117.0 100.0	88.5	74.0 26.0	5.9 94.1	98.8 1.2	U 30.0 L 70.0	U 1.010 L 1.175
360	a. Ethanol b. Ethyl acetate	78.5 77.1	71.8	31.0 69.0				0.863
361	a. Ethanol b. Ethylbutyl ether	78.5 92.2	73.8	49.3 50.7				0.776
362	a. Ethanol b. Ethylene dichloride	78.5 83.5	71.0	33.5 66.5				1.049
363	a. Ethanol b. Ethyl nitrate	78.5 88.7	71.85	44.0 56.0				
364	a. Ethanol b. Heptane	78.5 98.4	72.0	48.0 52.0				0.729
365	a. Ethanol b. Hexane	78.5 69.0	58.7	21.0 79.0				0.687
366	a. Ethanol b. Iodoethane	78.5 72.2	63.0	14.0 86.0				
367	a. Ethanol b. Iodomethane	78.5 42.5	41.2	3.2 96.8				
368	a. Ethanol b. Isobutyl chloride	78.5 68.9	61.3	16.5 83.7				homog.
369	a. Ethanol b. Isoprene	78.5 34.0	32.65	3.0 97.0				
370	a. Ethanol b. Isopropyl acetate	78.5 89.0	76.8	53.0 47.0				
371	a. Ethanol b. Isopropyl ether	78.5 67.5	64.0	17.1 82.9				0.741
372	a. Ethanol b. Methylethyl ketone	78.5 79.6	74.8	34.0 66.0				0.802
373	a. Ethanol b. Nitromethane	78.5 101.0	75.95	73.2 26.8				
374	a. Ethanol b. Tetrachloroethylene	78.5 121.02	76.75	~63.0 ~37.0				
375	a. Ethanol b. Thiophene	78.5 84.12	70.0	45.0 55.0				
376	a. Ethanol b. Toluene	78.5 110.6	76.7	68.0 32.0				0.815
377	a. Ethanol b. Trichloroethylene	78.5 87.1	70.9	27.0 73.0				1.197
378	a. Ethanol b. 1,1,2-Trichlorotrifluoro-ethane	78.5 47.7	43.8	3.8 96.2				1.517
379	a. Ethanol b. Triethyl amine	78.5 89.5	76.9	51.0 49.0				0.775
380	a. Ethanol b. Vinylbutyl ether	78.5 94.2	73.0	48.0 52.0				0.786
381	a. Ethanol b. Vinylisobutyl ether	78.5 83.4	69.2	33.0 67.0				0.778
382	a. Ethanol b. Water	78.5 100.0	78.2	95.6 4.4				0.804
383	a. Ethanol 3 atm. b. Water	109.0 134.0	109.0	95.2 4.8				0.805
384	a. Ethanol 95 mm b. Water	33.5 51.0	33.4	99.5 0.5				0.792
385	a. Ethoxy diglycol b. Ethylene glycol	202.8 197.5	192.0	54.5 45.5				1.050
386	a. 2-Ethoxyethanol b. Toluene	135.1 110.63	110.2	10.8 89.2				
387	a. 2-Ethoxyethanol b. o-Xylene	135.1 144.41	130.8	55.0 45.0				
388	a. Ethyl acetate b. Iodoethane	77.15 72.2	70.0	22.0 78.0				
389	a. Ethyl acetate b. Isopropanol	77.1 82.3	74.8	77.0 23.0				0.869
390	a. Ethyl acetate b. Methanol	77.1 64.65	62.1	51.4 48.6				0.846
391	a. Ethyl acetate b. Water	77.1 100.0	70.4	91.9 8.1	96.7 3.3	8.7 91.3	U 95.0 L 5.0	U 0.907 L 0.999
392	a. Ethyl acrylate b. Ethanol	99.8 78.5	77.5	27.3 72.7				homog.
393	a. Ethyl acrylate b. Water	99.8 100.0	81.0	84.9 15.1	98.5 1.5	2.0 98.0	U 87.0 L 13.0	U 0.910$^{30/20}$ L 0.998$^{30/20}$
394	a. n-Ethyl aniline b. Water	204.8 100.0	99.2	16.1 83.9	99.3 0.7	0.2 99.8	U 16.6 L 83.4	U 0.963 L 1.000

AZEOTROPES (Continued)
BINARY SYSTEM (Continued)

AZEOTROPES (Continued)

BINARY SYSTEM (Continued)

No.	Components: Compounds	Components: B.P. °C	Azeotrope: B.P. °C	Percent composition: In Azeotrope	Percent composition: Upper	Percent composition: Lower	Relative vol. of layers at 20°C	Spec. grav. of Azeotrope or layers
395	a. Ethyl benzene b. Water	136.0 100.0	92.0	67.0 33.0	99.95 0.05	0.02 99.98	U 70.0 L 30.0	U 0.870 L 1.000
396	a. 2-Ethyl butanol b. Water	146.0 100.0	96.7	42.0 58.0	95.44 4.56	0.43 99.57	U 48.0 L 52.0	U 0.841 L 0.999
397	a. 2-Ethylbutyl acetate b. Water	162.3 100.0	97.0	47.6 52.4	99.43 0.57	0.06 99.94	U 51.3 L 48.7	U 0.871 L 1.000
398	a. Ethylbutyl ether b. Methanol	92.2 64.65	62.6	44.0 56.0				0.770
399	a. Ethylbutyl ether b. Water	92.2 100.0	76.6	88.1 11.9	99.6 0.4	0.44 99.56	U 91.0 L 9.0	U 0.753 L 0.998
400	a. Ethylbutyl ketone b. Water	148.5 100.0	94.6	57.8 42.2	99.2 0.8	1.4 98.6	U 62.5 L 37.5	U 0.822 L 0.997
401	a. 2-Ethylbutyraldehyde b. Water	116.9 100.0	87.5	76.3 23.7	99.2 0.8	0.4 99.6	U 80.2 L 19.8	U 0.816 L 0.997
402	a. 2-Ethyl butyric acid b. Water	194.2 100.0	99.7	13.0 87.0	96.7 3.3	1.55 98.45	U 12.0 L 88.0	U 0.929 L 1.000
403	a. Ethyl carbamate b. Octyl alcohol	180.0 195.0	183.5	72.5 27.5				
404	a. Ethyl carbamate b. Phenol	180.0 182.0	190.8	53.5 46.5				
405	a. Ethyl crotonate b. Water	137.8 100.0	93.5	62.0 38.0	98.48 1.52	0.63 99.37	U 65.0 L 35.0	U 0.921 L 1.000
406	a. n-Ethylcyclohexyl amine b. Water	164.0 100.0	97.1	42.0 58.0	77.0 23.0	1.5 98.5	U 56.0 L 44.0	U 0.895 L 0.998
407	a. Ethylene chlorohydrin b. Water	128.7 100.0	97.8	42.3 57.7				1.093
408	a. Ethylene diamine b. Isobutanol	116.5 108.3	120.5	50.0 50.0				0.856
409	a. Ethylene diamine b. Toluene	116.5 110.6	103.0	30.0 70.0				homog.
410	a. Ethylene diamine b. Water	116.5 100.0	119.0	81.6 18.4				0.953
411	a. Ethylene diamine 300 mm b. Water	90.5 75.9	94.5	77.1 22.9				0.963
412	a. Ethylene diamine 50 mm b. Water	49.0 38.0	55.0	71.0 29.0				
413	a. Ethylene dichloride b. Isopropanol	84.0 82.3	72.7	60.8 39.2				1.012
414	a. Ethylene dichloride b. Methanol	84.0 64.65	59.5	65.0 35.0				1.045
415	a. Ethylene dichloride b. Trichloroethylene	84.0 87.1	82.3	59.1 40.9				1.330
416	a. Ethylene dichloride b. Water	84.0 100.0	71.6	91.8 8.2	0.8 99.2	99.8 0.2	U 10.0 L 90.0	U 1.002 L 1.254
417	a. Ethylene glycol 10 mm b. 2-Ethylhexylether	91.0 135.0	87.0				U 50.0 L 50.0	heterog.
418	a. Ethylene glycol 50 mm b. Hexylether	123.0 137.0	112.8	35.6 64.4	0.1 99.9	99.9 0.1	U 71.8 L 28.2	U 0.795 L 1.115
419	a. Ethylene glycol b. Phenylether	197.5 257.4	192.3	64.5 35.5	0.22 99.78	98.28 1.72	U 35.3 L 64.7	U 1.068[30/20] L 1.108[30/20]
420	a. Ethyl ether b. Isoprene	34.6 34.0	33.2	48.0 52.0				
421	a. Ethyl ether b. Methyl formate	34.6 31.50	28.2	44.0 56.0				
422	a. Ethyl ether b. Methyl sulfide	34.6 37.5	34.0	80.0 20.0				
423	a. Ethyl ether b. Water	34.6 100.0	34.2	98.8 1.2				0.720
424	a. Ethyl formate b. Methanol	54.2 64.65	51.0	84.0 16.0				
425	a. Ethyl formate b. Water	54.2 100.0	52.6	95.0 5.0	95.5 4.5	13.6 86.4	U >99.5 L <0.5	U 0.920[26/20] L 0.995[26/20]
426	a. 2-Ethylhexanoic acid b. Water	228.0 100.0	99.9	3.6 96.4	98.77 1.23	0.25 99.75	U 3.7 L 96.3	U 0.906 L 1.000
427	a. 2-Ethyl hexanol 25 mm b. Phenol	95.6 89.6	95.0	66.0 34.0				0.900
428	a. 2-Ethyl hexanol b. Water	185.0 100.0	99.1	20.0 80.0	97.4 2.6	0.10 99.90	U 23.0 L 77.0	U 0.838 L 1.000
429	a. Ethylhexyl acetate b. Water	199.0 100.0	99.0	26.5 73.5	99.45 0.55	0.03 99.97	U 25.0 L 75.0	U 0.873 L 1.000
430	a. 2-Ethylhexyl amine b. Water	169.1 100.0	98.2	36.0 64.0	74.7 25.3	0.25 99.75	U 52.0 L 48.0	U 0.848 L 1.000

AZEOTROPES (Continued)
BINARY SYSTEM (Continued)

AZEOTROPES (Continued)
BINARY SYSTEM (Continued)

No.	Components: Compounds	Components: B.P. °C	Azeotrope: B.P. °C	Percent composition: In Azeotrope	Percent composition: Upper	Percent composition: Lower	Relative vol. of layers at 20°C	Spec. grav. of Azeotrope or layers
431	a. 2-Ethylhexyl chloride b. Water	173.0 100.0	97.3	45.0 55.0	99.9 0.1	0.1 99.9	U 48.0 L 52.0	U 0.883 L 1.000
432	a. 2-Ethylhexyl ether b. Water	269.8 100.0	99.8	3.6 96.4	99.97 0.03	0.01 99.99	U 4.0 L 96.0	U 0.911 L 0.998
433	a. 2-Ethylhexyl hexanoate b. Water	267.2 100.0	99.9	3.6 96.4	99.81 0.19	0.01 99.99	U 4.1 L 95.9	U 0.865 L 1.000
434	a. Ethylidene acetone b. Water	123.5 100.0	92.0	71.4 28.6	82.8 17.2	38.0 62.0	U 76.3 L 23.7	U 0.892 L 0.975
435	a. Ethyl propionate b. Propanol	99.0 97.2	93.4	49.0 51.0				
436	a. Ethyl propionate 350 mm b. Water	76.0 79.7	61.0	86.7 13.3	98.6 1.4	1.7 98.3	U 89.0 L 11.0	U 0.886 L 0.998
437	a. Formic acid b. Toluene	100.7 110.63	85.8	50.0 50.0				
438	a. Formic acid b. Water	100.7 100.0	107.1	77.5 22.5				
439	a. 2-Furaldehyde b. o-Xylene	161.7 144.41	140.5	13.0 87.0				
440	a. Glycol b. Naphthalene	197.2 217.9	183.9	51.0 49.0				
441	a. Glycol b. Nitrobenzene	197.2 210.9	185.9	59.0 41.0				
442	a. Glycol b. Octyl alcohol	197.2 195.0	184.4	36.5 63.5				
443	a. Glycol b. Quinoline	197.2 237.7	196.4	79.5 20.5				
444	a. Glycol b. Styrene	197.2 146.0	139.5	16.5 83.5				
445	a. Glycol b. Toluene	197.2 110.63	110.2	6.5 93.5				
446	a. Glycol b. o-Toluidine	197.2 199.84	186.5	42.5 57.5				
447	a. Glycol b. o-Xylene	197.2 144.4	139.6	16.0 84.0				
448	a. Glycol diacetate b. Water	190.8 100.0	99.7	15.4 84.6				1.024
449	a. Heptane b. Methanol	98.43 64.65	59.1	48.5 51.5				
450	a. Heptane b. Vinyl acetate	98.4 72.7	72.0	16.5 83.5				0.880
451	a. Heptane b. Water	98.4 100.0	79.2	87.1 12.9	99.98 0.02	0.01 99.99	U 90.8 L 9.2	U 0.685 L 1.000
452	a. 2-Heptyl acetate b. Water	176.4 100.0	97.8	41.1 58.9	99.49 0.51	0.03 99.97	U 45.0 L 55.0	U 0.864 L 1.000
453	a. 3-Heptyl acetate b. Water	173.8 100.0	97.5	42.4 57.6	99.56 0.44	0.03 99.97	U 46.0 L 54.0	U 0.863 L 1.000
454	a. Hexaldehyde b. Water	128.5 100.0	91.0	68.7 31.3	98.8 1.2	0.6 99.4	U 73.4 L 26.6	U 0.817 L 0.997
455	a. Hexane b. Isobutyl chloride	69.0 68.9	66.3	45.0 55.0				homog.
456	a. Hexane b. Isopropanol	69.0 82.3	61.0	78.0 22.0				0.686
457	a. Hexane b. Isopropyl ether	69.0 67.5	67.5	47.0 53.0				
458	a. Hexane b. Methanol	69.0 64.65	50.0	73.1 26.9	85.0 15.0	42.0 58.0	U 67.8 L 32.2	U 0.675 L 0.724
459	a. Hexane b. Methylethyl ketone	69.0 79.6	64.3	71.7 28.3				0.698
460	a. Hexane b. Nitromethane	69.0 101.0	62.0	79.0 21.0				
461	a. Hexane b. Propanol	69.0 97.2	65.7	96.0 4.0				0.67
462	a. Hexane b. Propionitrile	69.0 97.1	63.5	91.0 9.0				
463	a. Hexane b. Water	69.0 100.0	61.6	94.4 5.6			U 96.2 L 3.8	U 0.660 L 1.000
464	a. Hexanoic acid b. Water	205.0 100.0	99.8	7.9 92.1	94.6 5.4	1.1 98.9	U 8.5 L 91.5	U 0.934 L 0.999
465	a. Hexanol b. Water	158.0 100.0	97.8	32.8 67.2	92.8 7.2	0.58 99.42	U 39.0 L 61.0	U 0.835 L 0.999
466	a. 2-Hexenal b. Water	149.0 100.0	95.1	51.4 48.6	98.3 1.7	0.2 99.8	U 56.3 L 43.7	U 0.848 L 0.999

AZEOTROPES (Continued)
BINARY SYSTEM (Continued)

No.	Components: Compounds	Components: B.P. °C	Azeotrope: B.P. °C	Percent composition: In Azeotrope	Percent composition: Upper	Percent composition: Lower	Relative vol. of layers at 20°C	Spec. grav. of Azeotrope or layers
467	a. Hexyl acetate b. Water	169.2 100.0	97.4	39.0 61.0	99.44 0.56	0.05 99.95	U 42.3 L 57.7	U 0.873 L 1.000
468	a. Hexyl amine b. Water	132.7 100.0	95.5	51.0 49.0				0.881
469	a. Hexyl chloride b. Water	133.9 100.0	91.8	70.3 29.7	99.83 0.17	0.01 99.99	U 73.0 L 27.0	U 0.873 L 1.000
470	a. Hydrogen bromide b. Water	−67.0 100.0	126.0	47.5 52.5				1.481
471	a. Hydrogen chloride b. Water	−83.7 100.0	108.6	20.2 79.8				1.102
472	a. Hydrogen chloride 6.8 atm. b. Water	−31.0 169.0	177.0	14.8 85.2				
473	a. Hydrogen fluoride b. Water	19.4 100.0	111.4	35.6 64.4				
474	a. Hydrogen iodide b. Water	−35.5 100.0	127.0	57.0 43.0				
475	a. Hydrogen nitrate b. Water	86.0 100.0	121.0	68.5 31.5				1.405
476	a. Iodoethane b. Propanol	72.2 97.19	70.0	93.0 7.0				
477	a. Iodomethane b. Isopropyl alcohol	42.5 82.3	42.4	98.2 1.8				
478	a. Iodomethane b. Methanol	42.5 64.65	37.8	95.5 4.5				
479	a. 1-Iodopropane b. Isopropyl alcohol	102.4 82.3	79.8	58.0 42.0				
480	a. Isoamyl acetate b. Isoamyl alcohol	142.5 130.5	129.1	2.6 97.4				
481	a. Isoamyl alcohol b. Water	132.05 100.0	95.15	50.40 49.60				
482	a. Isoamyl alcohol b. o-Xylene	130.5 144.41	127.0	>52.0 <48.0				
483	a. Isobutyl acetate b. Isobutyl alcohol	116.5 108.39	107.4	45.0 55.0				
484	a. Isobutyl alcohol b. Propyl acetate	108.39 101.6	101.0	17.0 83.0				
485	a. Isobutyl alcohol b. Toluene	108.3 110.6	101.2	44.5 55.5				0.836
486	a. Isobutyl alcohol b. Trichloroethylene	108.3 87.1	85.4	9.0 91.0				1.368
487	a. Isobutyl alcohol b. Water	108.3 100.0	89.7	70.0 30.0	85.0 15.0	8.7 91.3	U 82.3 L 17.7	U 0.839 L 0.988
488	a. Isobutyl chloride b. Isopropanol	68.9 82.3	63.8	81.0 19.0				homog.
489	a. Isobutyl chloride b. Methanol	68.9 64.65	53.1	79.4 20.6				homog.
490	a. Isophorone b. Water	215.2 100.0	99.5	16.1 83.9	95.7 4.3	1.2 98.8	U 16.0 L 84.0	U 0.929 L 0.999
491	a. Isoprene b. 2-Methyl-2-butene	34.0 38.4	34.0	86.0 14.0				
492	a. Isopropyl acetate b. Methanol	89.0 64.65	64.0	29.8 70.2				0.816
493	a. Isopropyl acetate b. Water	89.0 100.0	75.9	88.9 11.1	98.2 1.8	2.9 97.1	U 91.4 L 8.6	U 0.870 L 0.995
494	a. Isopropyl alcohol b Isopropyl acetate	82.3 93.0	80.1	52.6 47.4				0.822
495	a. Isopropyl alcohol b. Methylethyl ketone	82.3 79.6	77.3	30.0 70.0				0.800
496	a. Isopropyl alcohol b. Nitromethane	82.3 101.0	79.3	71.8 28.2				
497	a. Isopropyl alcohol b. Tetrachloroethylene	82.3 121.2	81.7	81.0 19.0				
498	a. Isopropyl alcohol b. Toluene	82.3 110.63	80.6	58.0 42.0				
499	a. Isopropyl alcohol b. Trichloroethylene	82.3 87.1	74.0	28.0 72.0				1.182
500	a. Isopropyl alcohol b. Vinylacetate	82.3 72.7	70.8	22.4 77.6				0.890
501	a. Isopropyl alcohol b. Water	82.3 100.0	80.4	87.8 12.2				0.818
502	a. Isopropyl benzene b. Water	152.4 100.0	95.0	56.2 43.8	99.95 0.05	0.01 99.99	U 60.0 L 40.0	U 0.864 L 1.000

AZEOTROPES (Continued)
BINARY SYSTEM (Continued)

No.	Components: Compounds	Components: B.P. °C	Azeotrope: B.P. °C	Percent composition: In Azeotrope	Percent composition: Upper	Percent composition: Lower	Relative vol. of layers at 20°C	Spec. grav. of Azeotrope or layers
503	a. Isopropyl chloride b. Water	36.5 100.0	35.0	99.0 1.0	99.67 0.33	0.31 99.69	U 99.0 L 1.0	U 0.862 L 0.999
504	a. Isopropyl ether b. Water	67.5 100.0	62.2	95.4 4.6	99.43 0.57	0.90 99.10	U 97.0 L 3.0	U 0.727 L 0.998
505	a. Isovaleraldehyde b. Water	92.5 100.0	77.0	88.0 12.0				
506	a. Mesityl oxide b. Water	128.7 100.0	91.8	65.3 34.7	96.6 3.4	2.8 97.2	U 69.8 L 30.2	U 0.860 L 0.995
507	a. Methacryl aldehyde b. Water	67.5 100.0	63.6	92.3 7.7	98.02 1.98	2.65 97.35	U 94.9 L 5.1	U 0.845 L 0.996
508	a. Methanol b. Methyl acetate	64.65 57.0	54.0	18.7 81.3				0.908
509	a. Methanol 4.4 atm. b. Methyl acetate	107.0 107.0	99.0	29.0 71.0				0.890
510	a. Methanol b. Methyl acrylate	64.65 80.5	62.5	54.0 46.0				homog.
511	a. Methanol b. Nitromethane	64.65 101.0	64.5	92.0 8.0				
512	a. Methanol b. Octane	64.65 125.8	63.0	72.0 28.0				
513	a. Methanol b. Toluene	64.65 110.6	63.7	72.4 27.6				0.813
514	a. Methanol b. Trichloroethylene	64.65 87.1	60.2	36.0 64.0				1.126
515	a. Methanol b. 1,1,2-trichloro-1,2,2,-trifluoro ethane	64.65 47.5	39.9	6.0 94.0				1.476
516	a. Methanol b. Trimethyl borate	64.65 65.0	54.0	27.0 73.0				0.892
517	a. Methanol b. Vinyl acetate	64.65 72.7	58.5	36.6 63.4				0.880
518	a. Methanol b. 1-Methoxy-1,3-butadiene	64.65 90.9	62.0	57.5 42.5				
519	a. 1-Methoxy-1,3-butadiene b. Water	90.9 100.0	76.2	87.3 12.7	99.74 0.26	0.31 99.69	U 89.4 L 10.6	U 0.832 L 0.999

AZEOTROPES (Continued)
BINARY SYSTEM (Continued)

No.	Components: Compounds	Components: B.P. °C	Azeotrope: B.P. °C	Percent composition: In Azeotrope	Percent composition: Upper	Percent composition: Lower	Relative vol. of layers at 20°C	Spec. grav. of Azeotrope or layers
520	a. 3-Methoxy butyl acetate b. Water	171.3 100.0	96.5	34.6 65.4	95.9 4.1	6.2 93.8	U 33.0 L 67.0	U 0.960 L 1.005
521	a. 2-Methoxy ethanol b. Toluene	124.6 110.6	105.9	25.0 75.0				0.887
522	a. 2-Methoxy ethanol b. Water	124.6 100.0	99.9	15.3 84.7				$1.005^{30/20}$
523	a. 1-Methoxy-2-propanol b. Water	118.5 100.0	97.5	51.0 49.0				0.994
524	a. Methyl acetate b. Water	57.0 100.0	56.1	95.0 5.0				0.940
525	a. Methylamyl ketone b. Water	150.5 100.0	95.2	54.6 45.4	98.5 1.5	0.43 99.57	U 59.5 L 40.5	U 0.819 L 0.999
526	a. α-Methylbenzyl amine b. Water	187.4 100.0	99.4	16.2 83.8	52.0 48.0	4.8 95.2	U 26.0 L 74.0	U 0.994 L 1.000
527	a. α-Methylbenzyl ether b. Water	286.7 100.0	100.0	1.3 98.7	0.01 99.99	99.78 0.22	U 98.7 L 1.3	U 1.000 L 1.003
528	a. n-Methylbutyl amine b. Water	91.1 100.0	82.7	85.0 15.0				0.802
529	a. n-Methyldibutyl amine b. Water	162.9 100.0	96.5	52.0 48.0	99.6 0.4	0.07 99.93	U 58.9 L 41.1	U 0.761 L 1.000
530	a. Methylene chloride b. Water	40.0 100.0	38.8	99.0 1.0	2.0 98.0	99.9 <0.1	U 1.6 L 98.4	U 1.009 L 1.328
531	a. Methylethyl ketone b. Water	79.6 100.0	73.4	88.0 12.0				0.834
532	a. 2-Methyl-5-ethyl pyridine b. Water	177.8 100.0	98.4	28.0 72.0	80.6 19.4	1.22 98.78	U 35.0 L 65.0	U 0.960 L 1.002
533	a. 5-Methyl-2-hexanone b. Water	144.0 100.0	94.7	56.0 44.0	98.6 1.4	0.55 99.45	U 63.0 L 37.0	U 0.814 L 1.004
534	a. Methylisobutyl ketone b. Water	115.1 100.0	87.9	76.0 24.0	98.4 1.6	2.0 98.0	U 80.4 L 19.6	U 0.806 L 0.999
535	a. Methylisopropenyl ketone b. Water	97.9 100.0	81.5	81.6 18.4	97.0 3.0	0.91 99.09	U 86.0 L 14.0	U 0.853 L 0.997
536	a. 2-Methyl pentanal b. Propanol	118.3 97.2	95.0	14.0 86.0				0.814
537	a. 2-Methyl pentanal b. Water	118.3 100.0	88.5	77.0 23.0	99.17 0.83	0.42 99.58	U 81.0 L 19.0	U 0.811 L 0.999

AZEOTROPES (Continued)
BINARY SYSTEM (Continued)

No.	Components: Compounds	Components: B.P. °C	Azeotrope: B.P. °C	Percent composition: In Azeo-trope	Percent composition: Upper	Percent composition: Lower	Relative vol. of layers at 20°C	Spec. grav. of Azeo-trope or layers
538	a. 2-Methyl pentanoic acid b. Water	196.4 100.0	99.4	12.1 87.9	97.1 2.9	1.3 98.7	U 13.0 L 87.0	U 0.924 L 1.000
539	a. 2-Methyl pentanol b. Water	148.0 100.0	97.2	40.0 60.0	94.6 5.4	0.3 99.7	U 47.0 L 53.0	U 0.826 L 1.003
540	a. 4-Methyl-2-pentanol b. Water	131.0 100.0	94.3	56.7 43.3	94.2 5.8	1.7 98.3	U 64.0 L 36.0	U 0.820 L 0.999
541	a. 4-Methyl-2-pentyl acetate b. Water	146.1 100.0	94.8	63.3 36.7	99.42 0.58	0.13 99.87	U 67.0 L 33.0	U 0.858 L 0.998
542	a. Methylphenyl carbinol b. Water	205.0 100.0	99.7	11.0 89.0	2.3 97.7	94.1 5.9	U 91.0 L 9.0	U 1.00 L 1.01
543	a. Methylphenyl ketone b. Water	202.3 100.0	99.1	18.5 81.5	0.55 99.45	98.35 1.65	U 81.0 L 19.0	U 1.00 L 1.03
544	a. 2-Methyl-2-propanol b. Water	82.6 100.0	79.9	88.3 11.7				0.814
545	a. 2-Methylpropyl acetate b. Water	117.3 100.0	88.4	78.0 22.0	98.98 1.02	0.63 99.37	U 81.0 L 19.0	U 0.874 L 1.000
546	a. Methylpropyl ketone b. Water	101.7 100.0	83.8	80.4 19.6	96.7 3.3	4.3 95.7	U 85.0 L 15.0	U 0.812 L 0.99
547	a. Nitric acid b. Water	86.0 100.0	120.5	68.0 32.0				
548	a. Nitroethane b. Toluene	114.8 110.63	106.2	25.0 75.0				
549	a. Nitromethane b. Toluene	101.0 110.63	96.5	55.0 45.0				
550	a. Nitromethane b. Trichloroethylene	101.0 87.0	81.4	20.0 80.0				
551	a. o-Nitrophenol b. Propionamide	217.25 213.0	211.2	75.2 24.8				
525	a. Nonane b. Water	150.8 100.0	95.0	60.2 39.8	100.0	100.0	U 68.0 L 32.0	U 0.719 L 1.000
553	a. Octane b. Water	125.7 100.0	89.6	74.5 25.5	99.99 0.01	0.02 99.98	U 80.0 L 20.0	U 0.720 L 1.000
554	a. 1-Octanol b. Phenol	195.0 182.0	195.4	87.0 13.0				
555	a. 1-Octanol b. Water	195.0 100.0	99.4	10.0 90.0				
556	a. Paraldehyde b. Water	124.5 100.0	90.8	74.8 25.2	98.9 1.1	10.5 89.5	U 73.0 L 27.0	U 0.998 L 1.01
557	a. Pentachloroethane b. Phenol	162.0 182.0	160.9	90.5 9.5				
558	a. Pentane b. Water	36.1 100.0	34.6	98.6 1.4	99.95 0.05	0.04 99.96	U 99.1 L 0.9	U 0.627 L 1.000
559	a. 2,4-Pentanedione b. Water	140.6 100.0	94.4	59.0 41.0	95.5 4.5	16.6 83.4	U 55.0 L 45.0	U 0.981 L 1.011
560	a. Pentanol b. Water	138.0 100.0	95.4	45.0 55.0				heterog.
561	a. 3-Pentanol b. Water	115.6 100.0	91.5	65.0 35.0	90.1 9.9	5.5 94.5	U 74.0 L 26.0	U 0.833 L 0.992
562	a. Phenol b. Water	182.0 100.0	99.52	9.21 90.79				
563	a. Phenyl ether b. Water	259.0 100.0	99.8	4.3 95.7	0.02 99.98	99.97 0.03	U 96.0 L 4.0	U $0.997^{30/20}$ L $1.067^{30/20}$
564	a. Propanol b. Propyl acetate	97.2 101.6	94.0	63.0 37.0				0.833
565	a. Propanol b. Tetrachloroethylene	97.19 121.02	94.05	48.0 52.0				
566	a. Propanol b. Toluene	97.2 110.6	92.6	49.0 51.0				0.836
567	a. Propanol b. Trichloroethylene	97.2 87.1	81.8	17.0 83.0				1.287
568	a. Propanol b. Water	97.2 100.0	88.1	71.8 28.2				0.866
569	a. Propanol b. m-Xylene	97.19 139.10	97.08	94.0 6.0				
570	a. Propionaldehyde b. Water	48.8 100.0	47.5	98.0 2.0				0.81
571	a. Propionamide b. o-Xylene	213.0 144.41	144.0	2.0 98.0				
572	a. Propionic acid b. Water	141.6 100.0	99.9	17.7 82.3				1.016
573	a. Propionic acid b. o-Xylene	141.1 144.4	135.4	43.0 57.0				

AZEOTROPES (Continued)
BINARY SYSTEM (Continued)

AZEOTROPES (Continued)

BINARY SYSTEM (Continued)

No.	Components: Compounds	Components: B.P. °C	Azeotrope: B.P. °C	Percent composition: In Azeotrope	Percent composition: Upper	Percent composition: Lower	Relative vol. of layers at 20°C	Spec. grav. of Azeotrope or layers
574	a. Propionitrile b. Water	97.2 100.0	82.2	76.0 24.0	95.0 5.0	9.4 90.6	U 83.3 L 16.7	U 0.793 L 0.983
575	a. Propyl acetate b. Water	101.6 100.0	82.4	86.0 14.0				
576	a. Propyl chloride b. Water	46.6 100.0	44.0	97.8 2.2	99.7 0.3	0.3 99.7	U 98.0 L 2.0	U 0.892 L 1.000
577	a. Propylene chlorohydrin b. Water	127.0 100.0	95.4	54.2 45.8				0.076
578	a. Propylene diamine b. Toluene	120.9 110.6	105.0	32.0 68.0				0.865
579	a. Propylene dichloride b. Water	96.8 100.0	78.4	89.4 10.6	0.26 99.74	99.94 0.06	U 12.0 L 88.0	U 1.000 L 1.159
580	a. Pyridine b. 3-Pentanol	115.5 115.6	117.4	45.0 55.0				
581	a. Pyridine b. Toluene	115.3 110.63	110.2	22.0 78.0				
582	a. Pyridine b. Water	115.5 100.0	92.6	57.0 43.0				1.010
583	a. Styrene b. Water	145.2 100.0	93.9	59.1 40.9	99.95 0.05	0.03 99.97	U 61.0 L 39.0	U 0.91 L 1.00
584	a. Styrene oxide b. Water	194.2 100.0	99.2	22.4 77.6	0.3 99.7	99.5 0.5	U 79.1 L 20.9	U 1.000 L 1.054
585	a. Tetrachloroethylene b. Water	121.0 100.0	88.5	82.8 17.2	0.02 99.98	99.99 <0.01	U 25.0 L 75.0	U 1.000 L 1.625
586	a. Tetrahydrobenzonitrile b. Water	195.1 100.0	98.8	21.7 78.3	99.46 0.54	0.63 99.37	U 22.3 L 77.7	U 0.958 L 0.999
587	a. 1,4-Thioxane b. Water	149.2 100.0	95.6	52.0 48.0	6.85 93.15	98.38 1.62	U 53.0 L 47.0	U 1.014 L 1.120
588	a. Toluene b. Water	110.6 100.0	85.0	79.8 20.2	99.95 0.05	0.06 99.94	U 82.0 L 18.0	U 0.868 L 1.000
589	a. Triallyl amine b. Water	151.1 100.0	95.0	62.0 38.0	99.57 0.43	0.13 99.87	U 67.0 L 33.0	U 0.802 L 1.000
590	a. Tributyl amine b. Water	213.9 100.0	99.8	18.0 82.0	99.7 0.3	0.01 99.99	U 22.0 L 78.0	U 0.781 L 1.000
591	a. 1,1,2-Trichloroethane b. Water	113.7 100.0	86.0	83.6 16.4	0.45 99.55	99.95 0.05	U 22.0 L 78.0	U 1.000 L 1.443
592	a. 1,1,2-Trichlorotrifluoro-ethane b. Water	47.5 100.0	44.5	99.0 1.0	1.0 99.0	99.9 0.1	U 2.0 L 98.0	U 1.00 L 1.57
593	a. Tridecanol b. Water	244.0 100.0	100.0	2.2 97.8	98.74 1.26	0.01 99.99	U 2.6 L 97.4	U 0.843 L 1.000
594	a. Triethyl amine b. Water	89.5 100.0	75.8	90.0 10.0				0.769
595	a. Triglycol trichloride b. Water	240.9 100.0	99.7	6.0 94.0	1.9 98.1	99.2 0.8	U 96.5 L 3.5	U 1.003 L 1.188
596	a. Trimethyltetrahydro benzaldehyde b. Water	204.5 100.0	99.0	23.0 77.0	99.72 0.28	0.01 99.99	U 24.5 L 75.5	U 0.919 L 1.000
597	a. Valeraldehyde b. Water	103.3 100.0	83.0	81.0 19.0	98.7 1.3	1.35 98.65	U 85.0 L 15.0	U 0.810 L 0.998
598	a. Valeric acid b. Water	186.2 100.0	99.8	11.0 89.0	87.0 13.0	2.4 97.6	U 10.0 L 90.0	U 0.957 L 1.000
599	a. Vinyl acetate b. Water	72.7 100.0	66.0	92.7 7.3	98.97 1.03	2.0 98.0	U 94.0 L 6.0	U 0.933 L 0.999
600	a. Vinylallyl ether b. Water	67.4 100.0	60.0	94.6 5.4	99.73 0.27	0.4 99.6	U 95.8 L 4.2	U 0.806 L 0.999
601	a. Vinyl benzoate b. Water	dec. 100.0	99.3	17.4 82.6	0.01 99.99	99.68 0.32	U 83.5 L 16.5	U 1.000 L 1.070
602	a. Vinylbutyl ether b. Water	94.2 100.0	77.5	88.4 11.6	99.91 0.09	0.3 99.7	U 91.0 L 9.0	U 0.780 L 1.000
603	a. Vinyl butyrate b. Water	116.7 100.0	87.2	79.60 20.40	99.7 0.3	0.3 99.7	U 81.0 L 19.0	U 0.902 L 0.999
604	a. Vinyl-2-chloroethyl ether b. Water	109.1 100.0	84.0	83.0 17.0	0.61 99.39	99.63 0.37	U 17.0 L 83.0	U 1.000 L 1.049
605	a. Vinyl crotonate b. Water	133.9 100.0	92.0	69.0 31.0	98.9 1.1	0.3 99.7	U 71.0 L 29.0	U 0.944 L 1.000
606	a. Vinylethyl ether b. Water	35.5 100.0	34.6	98.5 1.5	99.8 0.2	0.9 99.1	U 99.3 L 0.7	U 0.754 L 0.999
607	a. Vinyl-2-ethyl hexanoate b. Water	185.2 100.0	98.6	32.0 68.0	99.80 0.20	0.01 99.99	U 35.0 L 65.0	U 0.875 L 1.000
608	a. Vinyl-2-ethylhexyl ether b. Water	177.7 100.0	97.8	40.9 59.1	99.95 0.05	0.01 99.99	U 46.0 L 54.0	U 0.810 L 1.000

AZEOTROPES (Continued)
BINARY SYSTEM (Continued)

AZEOTROPES (Continued)
BINARY SYSTEM (Continued)

No.	Components		Azeotrope					
	Compounds	B.P. °C	B.P. °C	Percent composition: In Azeotrope	Percent composition: Upper	Percent composition: Lower	Relative vol. of layers at 20°C	Spec. grav. of Azeotrope or layers
609	a. 2-Vinyl-5-ethyl pyridine b. Water	dec. 100.0	99.4	15.0 85.0	94.64 5.36	0.01 99.99	U 16.0 L 84.0	U 0.95 L 1.00
610	a. Vinyl isobutyl ether b. Water	83.4 100.0	70.5	92.2 7.8	99.92 0.08	0.2 99.8	U 94.0 L 6.0	U 0.771 L 1.000
611	a. Vinyl isobutyrate b. Water	105.4 100.0	83.5	83.0 17.0	99.70 0.30	0.36 99.64	U 85.0 L 15.0	U 0.891 L 1.000
612	a. Vinylisopropyl ether b. Water	55.7 100.0	51.8	97.3 2.7	99.81 0.19	0.64 99.36	U 98.0 L 2.0	U 0.760 L 0.991
613	a. Vinyl-2-methyl pentanoate b. Water	148.8 100.0	95.0	62.0 38.0	99.81 0.19	0.03 99.97	U 65.0 L 35.0	U 0.881 L 1.000
614	a. Vinyl propionate b. Water	94.9 100.0	79.0	87.0 13.0	99.40 0.60	0.82 99.18	U 88.0 L 12.0	U 0.918 L 0.999
615	a. Vinylpropyl ether b. Water	65.1 100.0	59.0	95.0 5.0	99.8 0.2	0.4 99.6	U 96.0 L 4.0	U 0.770 L 0.999
616	a. Vinyltrimethyl bromyl ether b. Water	223.4 100.0	99.6	15.7 84.3	99.86 0.14	0.01 99.99	U 18.7 L 81.3	U 0.808 L 1.000
617	a. m-Xylene b. Water	139.1 100.0	94.5	60.0 40.0	99.95 0.05	0.05 99.95	U 63.4 L 36.6	U 0.868 L 1.000

AZEOTROPES (Continued)
TERNARY SYSTEM

No.	Components		Azeotrope					
	Compounds	B.P. °C	B.P. °C	Percent composition: In Azeotrope	Percent composition: Upper	Percent composition: Lower	Relative vol. of layers at 20°C	Spec. grav. of Azeotrope or layers
1	a. Acetone b. Chloroform c. Methanol	56.5 61.2 64.65	57.5	30.0 47.0 23.0				homog.
2	a. Acetone b. Chloroform c. Water	56.5 61.26 100.0	60.4	38.4 57.6 4.0				
3	a. Acetone b. Cyclohexane c. Methanol	56.5 81.4 64.65	51.5	43.5 40.5 16.0				
4	a. Acetone 775.5 mm b. Isopropyl ether c. Water	78.5 91.9 120.9	75.0	49.0 48.0 3.0				0.769
5	a. Acetone b. Methanol c. Chloroform	56.5 64.65 61.2	57.5	30.0 23.0 47.0				homog.
6	a. Acetone b. Methanol c. Methyl acetate	56.5 64.65 57.0	53.7	5.8 17.4 76.8				0.898
7	a. Acetonitrile b. Benzene c. Water	82.0 80.1 100.0	66.0	23.3 68.5 8.2				
8	a. Acetonitrile b. Ethanol c. Triethyl amine	82.0 78.5 89.5	70.1	34.0 8.0 58.0				0.757
9	a. Acetonitrile b. Ethanol c. Water	82.0 78.5 100.0	72.9	44.0 55.0 1.0				0.793
10	a. Acetonitrile b. Isopropyl ether c. Water	82.0 68.3 100.0	59.0	13.0 82.0 5.0	13.0 85.5 1.5	13.0 1.0 86.0	U 97.0 L 3.0	U 0.742 L 0.976
11	a. Acetonitrile b. Trichloroethylene c. Water	82.0 87.0 100.0	67.0	20.5 73.1 6.4				
12	a. Acetonitrile b. Triethyl amine c. Water	82.0 89.5 100.0	68.6	31.0 63.0 6.0				0.77
13	a. Acrylonitrile b. Ethanol c. Water	78.0 78.5 100.0	69.5	71.0 20.3 8.7				
14	a. Allyl alcohol b. Benzene c. Water	97.1 80.1 100.0	68.2	9.2 82.2 8.6	8.7 90.7 0.6	17.7 0.4 80.9	U 91.2 L 8.8	U 0.877 L 0.985

No.	Components: Compounds	Components: B.P. °C	Azeotrope: B.P. °C	Percent composition: In Azeotrope	Percent composition: Upper	Percent composition: Lower	Relative vol. of layers at 20°C	Spec. grav. of Azeotrope or layers
15	a. Allyl alcohol	97.1	65.2	11.0	25.6	10.1	U 8.6	U 0.777
	b. Carbon tetrachloride	76.8		84.0	2.7	89.1	L 91.4	L 1.464
	c. Water	100.0		5.0	71.7	0.8		
16	a. Allyl alcohol	97.0	66.2	81.0				
	b. Cyclohexane	81.4		11.0				
	c. Water	100.0		8.0				
17	a. Allyl alcohol	97.0	68.0	11.0				
	b. Cyclohexene	83.0		80.5				
	c. Water	100.0		8.5				
18	a. Allyl alcohol	97.1	59.7	5.0	3.6	34.8	U 98.2	U 0.668
	b. Hexane	69.0		90.0	95.9	0.8	L 1.8	L 0.964
	c. Water	100.0		5.0	0.5	64.4		
19	a. Allyl alcohol	97.1	80.6	31.4	32.4	27.4	U 83.0	U 0.856
	b. Toluene	110.6		53.4	64.9	2.2	L 17.0	L 0.95
	c. Water	100.0		15.2	2.7	70.4		
20	a. Allyl alcohol	97.1	71.4	12.5	21.6	12.5	U 11.8	U 0.983
	b. Trichloroethylene	87.1		80.0	4.7	86.5	L 88.2	L 1.345
	c. Water	100.0		7.5	73.7	1.0		
21	a. Benzene	80.1	64.6	74.1	86.0	4.8	U 85.5	U 0.866
	b. Ethanol	78.5		18.5	12.7	52.1	L 14.2	L 0.892
	c. Water	100.0		7.4	1.3	43.1		
22	a. Benzene	80.1	65.7	72.0	77.5	0.5	U 93.6	U 0.855
	b. Isopropanol	82.3		19.8	20.2	14.4	L 6.4	L 0.966
	c. Water	100.0		8.2	2.3	85.1		
23	a. Benzene	80.1	68.2	65.1	71.3	0.1	U 92.5	U 0.858
	b. Methylethyl ketone	79.6		26.1	28.1	5.2	L 7.5	L 0.992
	c. Water	100.0		8.8	0.6	94.7		
24	a. Benzene	80.1	67.3	70.5	76.7	0.1	U 92.4	U 0.857
	b. 2-Methyl-2-propanol	82.6		21.4	21.6	3.2	L 7.6	L 0.979
	c. Water	100.0		8.1	1.7	96.7		
25	a. Benzene	80.1	68.5	82.4	92.2	0.1	U 92.1	U 0.873
	b. Propanol	97.2		9.0	7.7	15.3	L 7.9	L 0.979
	c. Water	100.0		8.6	0.1	84.6		
26	a. Benzene	80.1	67.0	82.3				
	b. Propyl alcohol	97.19		10.1				
	c. Water	100.0		7.6				
27	a. Bromodichloromethane	90.2	72.0	>70.0				
	b. Ethanol	78.5		<22.5				
	c. Water	100.0		7.5				
28	a. 1-Butanol	117.7	90.7	8.0	11.0	2.0	U 75.5	U 0.874
	b. Butylacetate	126.5		63.0	86.0	1.0	L 24.5	L 0.997
	c. Water	100.0		29.0	3.0	97.0		

No.	Components: Compounds	Components: B.P. °C	Azeotrope: B.P. °C	Percent composition: In Azeotrope	Percent composition: Upper	Percent composition: Lower	Relative vol. of layers at 20°C	Spec. grav. of Azeotrope or layers
29	a. 1-Butanol 100 mm	69.77	46.0	26.0	41.0	2.6	U 65.0	U 0.862$^{25/20}$
	b. Butylacrylate			33.0	53.0	0.4	L 35.0	L 0.990$^{25/20}$
	c. Water	51.6		41.0	6.0	97.0		
30	a. 1-Butanol	117.7	90.6	34.6	46.0	4.9	U 76.5	U 0.796
	b. Butyl ether	142.0		34.5	49.0	0.3	L 23.5	L 0.995
	c. Water	100.0		29.9	5.0	94.8		
31	a. 1-Butanol	117.7	77.4	2.0	2.0	2.0	U 92.0	U 0.78
	b. Vinylbutyl ether	94.2		88.0	98.0	0.3	L 8.0	L 1.00
	c. Water	100.0		10.0	0.1	98.0		
32	a. 2-Butanol	99.5	85.5	27.4	31.7	4.6	U 86.0	U 0.858
	b. 2-Butyl acetate	112.2		52.4	62.3	0.6	L 14.0	L 0.994
	c. Water	100.0		20.2	6.0	94.8		
33	a. 2-Butanol	99.5	86.6	56.1	65.0	10.0	U 86.0	U 0.816$^{25/20}$
	b. Butyl ether	142.0		19.2	23.0	0.2	L 14.0	L 0.981$^{25/20}$
	c. Water	100.0		24.7	12.0	89.8		
34	a. 2-Butanol	99.5	67.0					heterog.
	b. Cyclohexane	81.0						
	c. Water	100.0						
35	a. 2-Butanol	99.5	77.5	19.0	20.0	9.0 ± 1	U 92.0	U 0.736
	b. Diisobutylene	102.6		70.0	78.8	0.5	L 8.0	L 0.987
	c. Water	100.0		11.0	1.2	91.0 ± 1		
36	a. 2-Butanol	99.5	61.1					heterog.
	b. Hexane	69.0						
	c. Water	100.0						
37	a. 1-Butenyl methyl ether		61.4	78.9	84.8	2.1	U 94.0	U 0.787
	b. Ethanol	78.5		14.3	12.6	37.1	L 6.0	L 0.996
	c. Water	100.0		6.8	2.6	60.8		
38	a. Butyl amine	77.1	81.8	50.0				0.795
	b. Ethanol	78.5		42.5				
	c. Water	100.0		7.5				
39	a. Butyl amine	77.1	83.0	47.0				homog.
	b. Isopropanol	82.3		40.5				
	c. Water	100.0		12.5				
40	a. Butyraldehyde	75.7	67.2	80.0	82.0		U 97.8	
	b. Ethanol	78.5		11.0	11.0		L 2.2	L 0.838
	c. Water	100.0		9.0	7.0			
41	a. Carbon disulfide	46.2	41.3	93.4				homog.
	b. Ethanol	78.5		5.0				
	c. Water	100.0		1.6				
42	a. Carbon disulfide	46.2	37.0					homog.
	b. Methanol	64.65						
	c. Methyl acetate	57.0						

AZEOTROPES (Continued)
TERNARY SYSTEM (Continued)

AZEOTROPES (Continued)

TERNARY SYSTEM (Continued)

No.	Components: Compounds	Components: B.P. °C	Azeotrope: B.P. °C	Percent composition: In Azeotrope	Percent composition: Upper	Percent composition: Lower	Relative vol. of layers at 20°C	Spec. grav. of Azeotrope or layers
43	a. Carbon tetrachloride	76.8	61.8	86.3	7.0	94.8	U 15.2	U 0.935
	b. Ethanol	78.5		10.3	48.5	5.2	L 84.8	L 1.519
	c. Water	100.0		3.4	44.5	<0.1		
44	a. Carbon tetrachloride	76.8	65.7	74.8	0.1	77.3	U 3.9	U 0.993
	b. Methylethyl ketone	79.6		22.2	5.5	22.6	L 96.1	L 1.313
	c. Water	100.0		3.0	94.4	0.1		
45	a. Carbon tetrachloride	76.8	65.4	84.0	0.1	88.0	U 7.0	U 0.979
	b. Propanol	97.2		11.0	15.0	11.0	L 93.0	L 1.436
	c. Water	100.0		5.0	84.9	1.0		
46	a. Chloroform	61.2	55.5	92.5	1.0	95.8	U 6.2	U 0.976
	b. Ethanol	78.5		4.0	18.2	3.7	L 93.8	L 1.441
	c. Water	100.0		3.5	80.8	0.5		
47	a. Chloroform	61.2	52.6	81.0	32.0	83.0	U 3.0	U 1.022
	b. Methanol	64.65		15.0	41.0	14.0	L 97.0	L 1.399
	c. Water	100.0		4.0	27.0	3.0		
48	a. Croton aldehyde	104.0	78.0	7.3				0.810
	b. Ethanol	78.5		87.9				
	c. Water	100.0		4.8				
49	a. Cyclohexane	81.0	62.1	76.0				heterog.
	b. Ethanol	78.5		17.0				
	c. Water	100.0		7.0				
50	a. Cyclohexane	81.0	68.3					homog.
	b. Ethylacetate	77.1						
	c. Isopropanol	82.3						
51	a. Cyclohexane	81.4	50.8	33.6				
	b. Methanol	64.65		17.8				
	c. Methylacetate	57.1		48.6				
52	a. Cyclohexane	81.0	63.6	35.0	62.4	0.1	U 94.5	U 0.769
	b. Methylethyl ketone	79.6		60.0	37.0	10.0	L 5.5	L 0.98
	c. Water	100.0		5.0	0.6	89.9		
53	a. Cyclohexane	81.0	65.0	71.0				heterog.
	b. 2-Methyl-2-propanol	82.6		21.0				
	c. Water	100.0		8.0				
54	a. Cyclohexane	81.0	66.6	81.5				heterog.
	b. Propanol	97.2		10.0				
	c. Water	100.0		8.5				
55	a. Cyclohexane	83.0	66.1	71.0				
	b. Isopropanol	82.3		21.5				
	c. Water	100.0		7.5				
56	a. 1,2-Dichloroethane	83.7	66.7	78.0				
	b. Ethanol	78.5		17.0				
	c. Water	100.0		5.0				

AZEOTROPES (Continued)

TERNARY SYSTEM (Continued)

No.	Components: Compounds	Components: B.P. °C	Azeotrope: B.P. °C	Percent composition: In Azeotrope	Percent composition: Upper	Percent composition: Lower	Relative vol. of layers at 20°C	Spec. grav. of Azeotrope or layers
57	a. Diethylformal	87.5	73.2	69.5				heterog.
	b. Ethanol	78.5		18.4				
	c. Water	100.0		12.1				
58	a. Diethyl ketone	102.7	81.2	60.0				heterog.
	b. Propanol	97.2		20.0				
	c. Water	100.0		20.0				
59	a. Diisobutylene	102.6	72.3	59.1	70.0	5.4	U 83.0	
	b. Isopropanol	82.3		31.6	26.9	55.2	L 17.0	
	c. Water	100.0		9.3	3.1	39.4		
60	a. Dipropyl acetal	146.6	87.6	21.0				heterog.
	b. Propanol	97.2		51.6				
	c. Water	100.0		27.4				
61	a. Dipropyl formal	137.4	86.4	47.2				heterog.
	b. Propanol	97.2		44.5				
	c. Water	100.0		8.0				
62	a. Ethanol	78.5	70.2	8.4				heterog.
	b. Ethyl acetate	77.1		82.6				$0.901^{25/20}$
	c. Water	100.0		9.0				
63	a. Ethanol	78.5	77.1	48.3				0.867
	b. Ethyl acrylate	99.8		41.6				
	c. Water	100.0		10.1				
64	a. Ethanol	78.5	71.6	4.2	2.6	33.4	U 95.5	U 0.774
	b. Ethylbutyl ether	92.2		86.5	91.2	1.4	L 4.5	L 0.941
	c. Water	100.0		9.3	6.2	65.2		
65	a. Ethanol	78.5	67.8	15.7	41.8	12.5	U 13.3	U 0.941
	b. Ethylenedichloride	84.0		77.1	11.6	85.2	L 86.7	L 1.167
	c. Water	100.0		7.2	46.6	2.3		
66	a. Ethanol	78.5	66.0	15.8	14.6	33.0	U 94.5	U 0.747
	b. Ethylisobutyl ether			77.7	82.6	7.8	L 5.5	L 0.90
	c. Water	100.0		6.5	2.8	59.2		
67	a. Ethanol	78.5	68.8	33.0	5.0	75.9	U 64.6	U 0.686
	b. Heptane	98.4		60.9	94.8	9.1	L 35.4	L 0.801
	c. Water	100.0		6.1	0.2	15.0		
68	a. Ethanol	78.5	56.0	12.0	3.0	75.0	U 90.0	U 0.672
	b. Hexane	69.0		85.0	96.5	6.0	L 10.0	L 0.833
	c. Water	100.0		3.0	0.5	19.0		
69	a. Ethanol	78.5	58.6	13.0				heterog.
	b. Isobutyl chloride	68.9		82.5				
	c. Water	100.0		4.5				
70	a. Ethanol	78.5	74.8	19.4				0.874
	b. Isopropyl acetate	93.0		70.8				
	c. Water	100.0		9.8				

AZEOTROPES (Continued)

TERNARY SYSTEM (Continued)

No.	Compounds	B.P. °C	B.P. °C	Percent composition In Azeo-trope	Upper	Lower	Relative vol. of layers at 20°C	Spec. grav. of Azeo-trope or layers
71	a. Ethanol	78.5	61.0	6.5	5.9	20.2	U 97.1	U 0.737
	b. Isopropyl ether	67.5		89.5	92.9	1.8	L 2.9	L 0.967
	c. Water	100.0		4.0	1.2	78.0		
72	a. Ethanol	78.5	62.0	8.6	8.1	17.0	U 96.0	U 0.743
	b. Methylbutyl ether			85.1	89.6	2.2	L 4.0	L 0.95
	c. Water	100.0		6.3	2.3	80.8		
73	a. Ethanol	78.5	73.2	14.0				0.832
	b. Methylethyl ketone	79.6		75.0				
	c. Water	100.0		11.0				
74	a. Ethanol	78.5	66.0	14.7	13.7	30.5	U 95.0	U 0.745
	b. Propylisopropyl ether			78.3	83.3	0.7	L 5.0	L 0.925
	c. Water	100.0		7.0	3.0	68.8		
75	a. Ethanol	78.5	74.4	37.0	15.6	54.8	U 46.5	U 0.849
	b. Toluene	110.6		51.0	81.3	24.5	L 53.5	L 0.855
	c. Water	100.0		12.0	3.1	20.7		
76	a. Ethanol	78.5	74.7	15.0				0.774
	b. Triethyl amine	89.5		75.0				
	c. Water	100.0		10.0				
77	a. Ethanol	78.5	60.0	22.0	20.3	38.7	U 91.0	U 0.777
	b. Vinyl-isobutyl ether	83.4		70.0	77.6	3.6	L 9.0	L 0.90
	c. Water	100.0		8.0	2.1	57.7		
78	a. Ethanol	78.5	57.0	21.2	20.6	32.0	U 95.5	U 0.722
	b. Vinylpropyl ether	65.1		73.7	77.8	0.2	L 4.5	L 0.923
	c. Water	100.0		5.1	1.6	67.8		
79	a. 2-Ethoxy ethanol	135.6	97.7	11.0	0.5	17.0	U 43.0	U 0.81
	b. Vinyl-2-ethylhexyl ether	177.7		38.0	99.4	0.1	L 57.0	L 1.00
	c. Water	100.0		51.0	0.1	82.9		
80	a. Ethyl acetate	77.1	68.3					homog.
	b. Isopropanol	82.3						
	c. Cyxlohexane	81.0						
81	a. Ethylbutyl ether	92.2	73.4	67.7	73.1	0.5	U 94.0	U 0.762
	b. Isopropanol	82.3		21.9	22.5	14.5	L 6.0	L 0.97
	c. Water	100.0		10.4	4.4	85.0		
82	a. Ethylene dichloride	84.0	69.7	75.3	0.9	78.1	U 7.0	U 0.968
	b. Isopropanol	82.3		19.0	20.7	18.8	L 93.0	L 1.117
	c. Water	100.0		7.7	78.4	31.0		
83	a. Hexane	69.0	45.0	59.0				0.73
	b. Methanol	64.65		14.0				
	c. Methyl acetate	57.0		27.0				
84	a. Isoamyl acetate	142.5	93.6	24.0				
	b. Isoamy alcohol	130.5		31.2				
	c. Water	100.0		44.8				
85	a. Isobutyl acetate	116.5	86.8	46.5				
	b. Isobutyl alcohol	108.39		23.1				
	c. Water	100.0		30.4				
86	a. Isopropanol	82.3	75.5	13.0	13.0	11.5	U 94.0	U 0.870
	b. Isopropyl acetate	89.0		76.0	81.4	2.9	L 6.0	L 0.981
	c. Water	100.0		11.0	5.6	85.6		
87	a. Isopropanol	82.3	61.8	4.0	4.0	5.0	U 97.2	U 0.732
	b. Isopropyl ether	67.5		91.0	94.7	1.0	L 2.8	L 0.990
	c. Water	100.0		5.0	1.3	94.0		
88	a. Isopropanol 776 mm	100.7	81.0	7.0	7.0	9.0	U 96.1	U 0.737
	b. Isopropyl ether	91.9		87.0	91.5	1.0	L 3.9	L 0.984
	c. Water	120.9		6.0	1.5	90.0		
89	a. Isopropanol	82.3	73.4	1.0				0.834
	b. Methylethyl ketone	79.6		88.0				
	c. Water	100.0		11.0				
90	a. Isopropanol	82.3	76.3	38.2	38.2	38.0	U 92.0	U 0.845
	b. Toluene	110.6		48.7	53.3	1.0	L 8.0	L 0.930
	c. Water	100.0		13.1	8.5	61.0		
91	a. Isopropyl ether	67.5	61.0	92.4	97.4	3.0	U 97.0	U 0.724
	b. Propanol	97.2		2.0	2.0	3.0	L 3.0	L 0.990
	c. Water	100.0		5.6	0.6	94.0		
92	a. Methanol	64.65	67.9	81.20				
	b. Methylchloro acetate	131.5		13.54				
	c. Water	100.0		5.26				
93	a. 2-Methoxy ethanol	124.6	97.7	4.0	0.2	6.0	U 45.0	U 0.81
	b. Vinyl-2-ethylhexyl ether	177.7		39.0	99.7	0.1	L 55.0	L 1.00
	c. Water	100.0		57.0	0.1	93.9		
94	a. Propanol	97.2	82.2	19.5				
	b. Propyl acetate	101.6		59.5				
	c. Water	100.0		21.0				
95	a. Propanol	97.2	71.6	12.0				heterog.
	b. Trichloroethylene	87.1		81.0				
	c. Water	100.0		7.0				
96	a. Propyl alcohol	97.19	74.8	20.2				
	b. Propyl ether	91.0		68.1				
	c. Water	100.0		11.7				

AZEOTROPES (Continued)
TERNARY SYSTEM (Continued)

HEAT OF FORMATION OF INORGANIC OXIDES

The ΔH_o values are given in gram calories per mole. The a, b, and I values listed here make it possible for one to calculate the ΔF and ΔS values by use of the following equations:

$$\Delta F_t = \Delta H_o + 2.303aT \log T + b \times 10^{-3}T^2 + c \times 10^5T^{-1} + IT$$
$$\Delta S_t = -a - 2.303a \log T - 2b \times 10^{-3}T + c \times 10^5T^{-2} - I$$

Ref: Bulletin 542, U. S. Bureau of Mines, 1954.

Coefficients in Free-Energy Equations

Reaction and temperature range of validity		ΔH_o	2.303a	b	c	I
$2\,Ac(c) + 3/2\,O_2(g) = Ac_2O_3(c)$	(298.16°–1,000° K.)	−446,090	−16.12	.....	.....	+109.89
$2\,Al(c) + 1/2\,O_2(g) = Al_2O(g)$	(298.16°–931.7° K.)	−31,660	+14.97	.....	.....	−72.74
$2\,Al(l) + 1/2\,O_2(g) = Al_2O(g)$	(931.7°–2,000° K.)	−38,670	+10.36	.....	.....	−51.53
$Al(c) + 1/2\,O_2(g) = AlO(g)$	(298.16°–931.7° K.)	+10,740	+5.76	.....	.....	−37.61
$Al(l) + 1/2\,O_2(g) = AlO(g)$	(931.7°–2,000° K.)	+8,170	+5.76	.....	.....	−34.85
$2\,Al(c) + 3/2\,O_2(g) = Al_2O_3$ (*corundum*).	(298.16°–931.7° K.)	−404,080	−15.68	+2.18	+3.935	+123.64
$2\,Al(l) + 3/2\,O_2(g) = Al_2O_3$ (*corundum*).	(931.7°–2,000° K.)	−407,950	−6.19	−.78	+3.935	+102.37
$2\,Am(c) + 3/2\,O_2(g) = Am_2O_3(c)$	(298.16°–1,000° K.)	−422,090	−16.12	.....	.....	+107.89
$Am(c) + O_2(g) = AmO_2(c)$	(298.16°–1,000° K.)	−240,600	−4.61	.....	.....	+55.91
$2\,Sb(c) + 3/2\,O_2(g) = Sb_2O_3$ (*cubic*)	(298.16°–842° K.)	−169,450	+6.12	−6.01	−.30	+52.21
$2\,Sb(c) + 3/2\,O_2(g) = Sb_2O_3$ (*orthorhombic*).	(298.16°–903° K.)	−168,060	+6.12	−6.01	−.30	+50.56
$2\,Sb(l) + 3/2\,O_2(g) = Sb_2O_3$ (*orthorhombic*).	(903°–928° K.)	−175,370	+15.29	−7.75	−.30	+33.12
$2\,Sb(l) + 3/2\,O_2(g) = Sb_2O_3(l)$	(928°–1,698° K.)	−173,940	−32.84	+.75	−.30	+166.52
$2\,Sb(l) + 3/2\,O_2(g) = 1/2\,Sb_4O_6(g)$	(1,698°–1,713° K.)	−132,760	+10.91	+.75	−.30	+.96
$2\,Sb(g) + 3/2\,O_2(g) = 1/2\,Sb_4O_6(g)$	(1,713°–2,000° K.)	−234,760	−.74	+.75	−.30	+98.17
$2\,Sb(c) + 2\,O_2(g) = Sb_2O_4(c)$	(298.16°–903° K.)	−208,310	+6.31	−5.36	−.40	+73.02
$2\,Sb(l) + 2\,O_2(g) = Sb_2O_4(c)$	(903°–1,500° K.)	−215,610	+15.47	−7.10	−.40	+55.61
$6\,Sb(c) + 13/2\,O_2(g) = Sb_6O_{13}(c)$	(298.16°–903° K.)	−649,160	+38.46	−25.13	−1.30	+192.54
$6\,Sb(c) + 13/2\,O_2(g) = Sb_6O_{13}(c)$	(903°–1,500° K.)	−691,370	+14.13	−30.35	−1.30	+315.93
$2\,Sb(c) + 5/2\,O_2(g) = Sb_2O_5(c)$	(298.16°–903° K.)	−226,060	+37.12	−22.66	−.50	+18.61
$2\,Sb(l) + 5/2\,O_2(g) = Sb_2O_5(c)$	(903°–1,500° K.)	−240,130	+29.01	−24.40	−.50	+59.74
$2\,As(c) + 3/2\,O_2(g) = As_2O_3$ (*orthorhombic*).	(298.16°–542° K.)	−154,870	+29.54	−21.33	−.30	−8.83
$2\,As(c) + 3/2\,O_2(g) = As_2O_3$ (*monoclinic*).	(298.16°–586° K.)	−150,760	+29.54	−21.33	−.30	−16.95
$2\,As(c) + 3/2\,O_2(g) = As_2O_3(l)$	(542°–730.3° K.)	−156,260	−43.29	+2.97	−.30	+180.95
$2\,As(c) + 3/2\,O_2(g) = 1/2\,As_4O_6(g)$	(730.3°–883° K.)	−135,930	+.46	+2.97	−.30	+26.88
$1/2\,As_4(g) + 3/2\,O_2(g) = 1/2\,As_4O_6(g)$	(883°–2,000° K.)	−154,450	−2.90	+.75	−.30	+59.71
$2\,As(c) + 2\,O_2(g) = As_2O_4(c)$	(298.16°–883° K.)	−173,690	+21.52	−13.42	−.40	+34.38
$1/2\,As_4(g) + 2\,O_2(g) = As_2O_4(c)$	(883°–1,500° K.)	−192,210	+18.15	−15.64	−.40	+67.22
$2\,As(c) + 5/2\,O_2(g) = As_2O_5(c)$	(298.16°–883° K.)	−217,080	+12.32	−4.65	−.50	+80.50
$1/2\,As_4(g) + 5/2\,O_2(g) = As_2O_5(c)$	(883°–2,000° K.)	−235,600	+8.96	−6.87	−.50	+113.33
$Ba(\alpha) + 1/2\,O_2(g) = BaO(c)$	(298.16°–648° K.)	−134,590	−7.60	+.87	+.42	+45.76
$Ba(\beta) + 1/2\,O_2(g) = BaO(c)$	(648°–977° K.)	−134,140	−3.34	−.56	+.42	+34.01
$Ba(l) + 1/2\,O_2(g) = BaO(c)$	(977°–1,911° K.)	−135,900	−2.19	−.56	+.42	+32.37
$Ba(g) + 1/2\,O_2(g) = BaO(c)$	(1,911°–2,000° K.)	−176,400	−8.01	−0.56	+0.42	+72.66
$Ba(\alpha) + O_2(g) = BaO_2(c)$	(298.16°–648° K.)	−154,830	−11.05	+.87	+.42	+74.48
$Ba(\beta) + O_2(g) = BaO_2(c)$	(648°–977° K.)	−154,380	−6.79	−.56	+.42	+62.73
$Ba(l) + O_2(g) = BaO_2(c)$	(977°–1,500° K.)	−156,140	−5.64	−.56	+.42	+61.09
$Be(c) + 1/2\,O_2(g) = BeO(c)$	(298.16°–1,556° K.)	−144,220	−1.91	−.46	+1.24	+30.64
$Be(l) + 1/2\,O_2(g) = BeO(c)$	(1,556°–2,000° K.)	−144,300	+6.06	−1.75	+1.485	+7.25
$Bi(c) + 1/2\,O_2(g) = BiO(c)$	(298.16°–544° K.)	−50,450	−4.61	.....	.....	+35.51
$Bi(l) + 1/2\,O_2(g) = BiO(c)$	(544°–1,600° K.)	−52,920	−4.61	.....	.....	+40.05
$2\,Bi(c) + 3/2\,O_2(g) = Bi_2O_3(c)$	(298.16°–544° K.)	−139,000	−11.56	+2.15	−.30	+96.52
$2\,Bi(l) + 3/2\,O_2(g) = Bi_2O_3(c)$	(544°–1,090° K.)	−142,270	+2.30	−3.25	−.30	+67.55
$2\,Bi(l) + 3/2\,O_2(g) = Bi_2O_3(l)$	(1,090°–1,600° K.)	−147,350	−32.84	+.75	−.30	+174.59
$2\,B(c) + 3/2\,O_2(g) = B_2O(c)$	(298.16°–723° K.)	−304,690	+11.72	−7.55	+.355	+34.25
$2\,B(c) + 3/2\,O_2(g) = B_2O_3(gl)$	(298.16°–723° K.)	−298,670	+26.57	−15.90	−.30	−10.40
$2\,B(c) + 3/2\,O_2(g) = B_2O_3(l)$	(723°–2,000° K.)	−308,100	−38.41	+5.15	−.30	+173.24
$Cd(c) + 1/2\,O_2(g) = CdO(c)$	(298.16°–594° K.)	−62,330	−2.05	+.71	−.10	+29.17
$Cd(l) + 1/2\,O_2(g) = CdO(c)$	(594°–1,038° K.)	−63,240	+2.07	−.76	−.10	+20.14
$Cd(g) + 1/2\,O_2(g) = CdO(c)$	(1,038°–2,000° K.)	−89,320	−2.83	−.76	−.10	+60.05
$Ca(\alpha) + 1/2\,O_2(g) = CaO(c)$	(298.16°–673° K.)	−151,850	−6.56	+1.46	+.68	+43.93
$Ca(\beta) + 1/2\,O_2(g) = CaO(c)$	(673°–1,124° K.)	−151,730	−4.14	+.41	+.68	+37.63
$Ca(l) + 1/2\,O_2(g) = CaO(c)$	(1,124°–1,760° K.)	−153,480	−1.36	−.29	+.68	+31.49
$Ca(g) + 1/2\,O_2(g) = CaO(c)$	(1,760°–2,000° K.)	−194,670	−7.18	−.29	+.68	+73.84
$Ca(\alpha) + O_2(g) = CaO_2(c)$	(298.16°–500° K.)	−158,230	−12.32	+1.46	+.68	+78.28
C(*graphite*) $+ 1/2\,O_2(g) = CO(g)$	(298.16°–2,000° K.)	−25,400	+2.05	+.27	−1.095	−28.79

HEAT OF FORMATION OF INORGANIC OXIDES (Continued)
Coefficients in Free-Energy Equations (Continued)

Reaction and temperature range of validity		ΔH_o	2.303a	b	c	I
C(*graphite*) + $O_2(g)$ = $CO_2(g)$	(298.16°–2,000° K.)	−93,690	+1.63	−.07	−.23	−5.64
2 Ce(c) + 3/2 $O_2(g)$ = $Ce_2O_3(c)$	(298.16°–1,048° K.)	−435,600	−4.60	.....	.....	+92.84
2 Ce(l) + 3/2 $O_2(g)$ = $Ce_2O_3(c)$	(1,048°–1,900° K.)	−440,400	−4.60	.....	.....	+97.42
Ce(c) + $O_2(g)$ = $CeO_2(c)$	(298.16°–1,048° K.)	−245,490	−6.42	+2.34	−.20	+67.79
Ce(l) + $O_2(g)$ = $CeO_2(c)$	(1,048°–2,000° K.)	−247,930	+.71	−.66	−.20	+51.73
2 Cs(c) + 1/2 $O_2(g)$ = $Cs_2O(c)$	(298.16°–301.5°K.)	−75,900	.....	.....	.....	+36.60
2 Cs(l) + 1/2 $O_2(g)$ = $Cs_2O(c)$	(301.5°–763° K.)	−76,900	.....	.....	.....	+39.92
2 Cs(l) + 1/2 $O_2(g)$ = $Cs_2O(l)$	(763°–963° K.)	−75,370	−9.21	.....	.....	+64.47
2 Cs(g) + 1/2 $O_2(g)$ = $Cs_2O(l)$	(963°–1,500° K.)	−113,790	−23.03	.....	.....	+145.60
2 Cs(c) + $O_2(g)$ = $Cs_2O_2(c)$	(298.16°–301.5° K.)	−96,500	−2.30	.....	.....	+62.30
2 Cs(l) + $O_2(g)$ = $Cs_2O_2(c)$	(301.5°–870° K.)	−97,800	−4.61	.....	.....	+72.34
2 Cs(l) + $O_2(g)$ = $Cs_2O_2(l)$	(870°–963° K.)	−96,060	−18.42	.....	.....	+110.94
2 Cs(g) + $O_2(g)$ = $Cs_2O_2(l)$	(963°–1,500° K.)	−134,000	−31.08	.....	.....	+188.11
2 Cs(c) + 3/2 $O_2(g)$ = $Cs_2O_3(c)$	(298.16°–301.5° K.)	−112,690	−11.51	.....	.....	+110.10
2 Cs(l) + 3/2 $O_2(g)$ = $Cs_2O_3(c)$	(301.5°–775° K.)	−113,840	−12.66	.....	.....	+116.77
2 Cs(l) + 3/2 $O_2(g)$ = $Cs_2O_3(l)$	(775°–963° K.)	−110,740	−26.48	.....	.....	+152.70
2 Cs(g) + 3/2 $O_2(g)$ = $Cs_2O_3(l)$	(963°–1,500° K.)	−148,680	−39.14	.....	.....	+229.87
Cs(c) + $O_2(g)$ = $CsO_2(c)$	(298.16°–301.5° K.)	−63,590	−11.51	.....	.....	+72.29
Cs(l) + $O_2(g)$ = $CsO_2(c)$	(301.5°–705° K.)	−64,240	−12.66	.....	.....	+77.30
Cs(l) + $O_2(g)$ = $CsO_2(l)$	(705°–963° K.)	−61,770	−18.42	.....	.....	+90.20
Cs(g) + $O_2(g)$ = $CsO_2(l)$	(963°–1,500° K.)	−80,500	−24.18	.....	.....	+126.83
$Cl_2(g)$ + 1/2 $O_2(g)$ = $Cl_2O(g)$	(298.16°–2,000° K.)	+17,770	−.71	−.12	+.49	+16.81
1/2 $Cl_2(g)$ + 1/2 $O_2(g)$ = $ClO(g)$	(298.16°–1,000° K.)	+33,000	.....	.....	.....	−.24
1/2 $Cl_2(g)$ + $O_2(g)$ = $ClO_2(g)$	(298.16°–2,000° K.)	+24,150	−.76	−.105	−.665	+19.08
1/2 $Cl_2(g)$ + 3/2 $O_2(g)$ = $ClO_3(g)$	(298.16°–500° K.)	+37,740	+5.76	.....	.....	+21.42
$Cl_2(g)$ + 7/2 $O_2(g)$ = $Cl_2O_7(g)$	(298.16°–500° K.)	+65,040	+12.66	.....	.....	+78.01
2 Cr(c) + 3/2 $O_2(g)$ = $Cr_2O_3(\beta)$	(298.16°–1,823° K.)	−274,670	−14.07	+2.01	+.69	+105.65
2 Cr(l) + 3/2 $O_2(g)$ = $Cr_2O_3(\beta)$	(1,823°–2,000° K.)	−278,030	+2.33	−.35	+1.57	+58.29
Cr(c) + $O_2(g)$ = $CrO_2(c)$	(298.16°–1,000° K.)	−142,500	.....	.....	.....	+42.00
Cr(c) + 3/2 $O_2(g)$ = $CrO_3(c)$	(298.16°–471° K.)	−141,590	−13.82	.....	.....	+103.90
Cr(c) + 3/2 $O_2(g)$ = $CrO_3(l)$	(471°–600° K.)	−141,580	−32.24	.....	.....	+153.14
Co(α, β) + 1/2 $O_2(g)$ = CoO(c)	(298.16°–1,400° K.)	−56,910	+.69	.....	.....	+16.03
Co(γ) + 1/2 $O_2(g)$ = CoO(c)	(1,400°–1,763° K.)	−58,160	−1.15	.....	.....	+22.71
Co(l) + 1/2 $O_2(g)$ = CoO(c)	(1,763°–2,000° K.)	−65,680	−6.22	.....	.....	+43.43
3 Co(α, β, γ) + 2 $O_2(g)$ = $Co_3O_4(c)$	(298.16°–1,500° K.)	−207,300	−2.30	.....	.....	+90.56
2 Cu(c) + 1/2 $O_2(g)$ = $Cu_2O(c)$	(298.16°–1,357° K.)	−40,550	−1.15	−1.10	−.10	+21.92
2 Cu(l) + 1/2 $O_2(g)$ = $Cu_2O(c)$	(1,357°–1,502° K.)	−43,880	+8.47	−2.60	−.10	−3.72
2 Cu(l) + 1/2 $O_2(g)$ = $Cu_2O(l)$	(1,502°–2,000° K.)	−37,710	−12.48	+.25	−.10	+54.44
Cu(c) + 1/2 $O_2(g)$ = CuO(c)	(298.16°–1,357° K.)	−37,740	−.64	−1.40	−.10	+24.87
Cu(l) + 1/2 $O_2(g)$ = CuO(c)	(1,357°–1,720° K.)	−39,410	+4.17	−2.15	−.10	+12.05
Cu(l) + 1/2 $O_2(g)$ = CuO(l)	(1,720°–2,000° K.)	−41,060	−11.35	+.25	−.10	+59.09
$F_2(g)$ + 1/2 $O_2(g)$ = $F_2O(g)$	(298.16°–2,000° K.)	+5,070	−.41	−.15	+.535	+16.04
2 Ga(c) + 1/2 $O_2(g)$ = $Ga_2O(c)$	(298.16°–302.7° K.)	−81,110	+10.32	−5.75	−.10	−3.66
2 Ga(l) + 1/2 $O_2(g)$ = $Ga_2O(c)$	(302.7°–1,000° K.)	−83,360	+13.49	−5.75	−.10	−4.08
2 Ga(c) + 3/2 $O_2(g)$ = $Ga_2O_3(c)$	(298.16°–302.7° K.)	−256,240	+14.64	−3.75	−.30	+32.23
2 Ga(l) + 3/2 $O_2(g)$ = $Ga_2O_3(c)$	(302.7°–2,000° K.)	−258,490	+17.82	−3.75	−.30	+31.79
Ge(c) + 1/2 $O_2(g)$ = GeO(c)	(298.16°–1,200° K.)	−60,900	+1.27	−1.49	−.10	+17.19
Ge(c) + 1/2 $O_2(g)$ = GeO(g)	(298.16°–1,200° K.)	−21,870	+6.72	−.075	−.10	−41.25
Ge(c) + $O_2(g)$ = $GeO_2(gl)$	(298.16°–1,200° K.)	−127,830	+4.28	−2.52	−0.20	+30.54
2 Au(c) + 3/2 $O_2(g)$ = $Au_2O_3(c)$	(298.16°–500° K.)	−2,160	−10.36	.....	.....	+95.14
Hf(c) + $O_2(g)$ = HfO_2(*monocl.*)	(298.16°–2,000° K.)	−268,380	−9.74	−.28	+1.54	+78.16
$H_2(g)$ + 1/2 $O_2(g)$ = $H_2O(l)$	(298.16°–373.16° K.)	−70,600	−18.26	+.64	−.04	+91.67
$H_2(g)$ + 1/2 $O_2(g)$ = $H_2O(g)$	(298.16°–2,000° K.)	−56,930	+6.75	−.64	−.08	−8.74
$D_2(g)$ + 1/2 $O_2(g)$ = $D_2O(l)$	(298.16°–374.5° K.)	−72,760	−18.10	.....	.....	+93.59
$D_2(g)$ + 1/2 $O_2(g)$ = $D_2O(g)$	(298.16°–2,000° K.)	−58,970	+5.50	−.75	+.085	−3.74
1/2 $H_2(g)$ + 1/2 $O_2(g)$ = OH(g)	(298.16°–2,000° K.)	+10,350	+.90	+.005	−.26	−6.69
$H_2(g)$ + $O_2(g)$ = $H_2O_2(l)$	(298.16°–425° K.)	−47,140	−13.52	−7.13	.....	+99.39
$H_2(g)$ + $O_2(g)$ = $H_2O_2(g)$	(298.16°–1,500° K.)	−32,570	+4.77	−.96	+.97	+13.84
2 In(c) + 3/2 $O_2(g)$ = $In_2O_3(c)$	(298.16°–429.6° K.)	−220,410	+5.43	−.50	−.30	+59.49
2 In(l) + 3/2 $O_2(g)$ = $In_2O_3(c)$	(429.6°–2,000° K.)	−220,970	+13.22	−3.00	−.30	+41.36
$I_2(c)$ + 5/2 $O_2(g)$ = $I_2O_5(c)$	(298.16°–386.8° K.)	−42,040	+2.30	.....	.....	+113.71
$I_2(l)$ + 5/2 $O_2(g)$ = $I_2O_5(c)$	(386.8°–456° K.)	−43,490	+16.12	.....	.....	+81.70
$I_2(g)$ + 5/2 $O_2(g)$ = $I_2O_5(c)$	(456°–500° K.)	−58,020	−6.91	.....	.....	+174.79
Ir(c) + $O_2(g)$ = $IrO_2(c)$	(298.16°–1,300° K.)	−39,480	+8.17	−6.39	−.20	+20.33
0.947 Fe(α) + 1/2 $O_2(g)$ = $Fe_{0.947}O(c)$	(298.16°–1,033° K.)	−65,320	−11.26	+2.61	+0.44	+48.60
0.947 Fe(β) + 1/2 $O_2(g)$ = $Fe_{0.947}O(c)$	(1,033°–1,179° K.)	−62,380	+4.08	−.75	+.235	+3.00
0.947 Fe(γ) + 1/2 $O_2(g)$ = $Fe_{0.947}O(c)$	(1,179°–1,650° K.)	−66,750	−8.04	+.67	−.10	+42.28
0.947 Fe(γ) + 1/2 $O_2(g)$ = $Fe_{0.947}O(l)$	(1,650°–1,674° K.)	−64,200	−18.72	+1.67	−.10	+73.45
0.947 Fe(δ) + 1/2 $O_2(g)$ = $Fe_{0.947}O(l)$	(1,674°–1,803° K.)	−59,650	−6.84	+.25	−.10	+34.81

HEAT OF FORMATION OF INORGANIC OXIDES (Continued)

Coefficients in Free-Energy Equations (Continued)

Reaction and temperature range of validity	ΔH_0	2.303*a*	*b*	*c*	*I*
$0.947\ Fe(l) + 1/2\ O_2(g) = Fe_{0.947}O(l)$ (1,803°–2,000° K.)	−63,660	−7.48	+.25	−.10	+39.12
$3\ Fe(\alpha) + 2\ O_2(g) = Fe_3O_4$ (*magnetite*) (298.16°–900° K.)	−268,310	+5.87	−12.45	+.245	+73.11
$3\ Fe(\alpha) + 2\ O_2(g) = Fe_3O_4(\beta)$ (900°–1,033° K.)	−272,300	−54.27	+11.65	+.245	+233.52
$3\ Fe(\beta) + 2\ O_2(g) = Fe_3O_4(\beta)$ (1,033°–1,179° K.)	−262,990	−5.71	+1.00	−.40	+89.19
$3\ Fe(\gamma) + 2\ O_2(g) = Fe_3O_4(\beta)$ (1,179°–1,674° K.)	−276,990	−44.05	+5.50	−.40	+213.52
$3\ Fe(\delta) + 2\ O_2(g) = Fe_3O_4(\beta)$ (1,674°–1,803° K.)	−262,560	−6.40	+1.00	−.40	+91.05
$3\ Fe(l) + 2\ O_2(g) = Fe_3O_4(\beta)$ (1,803°–1,874° K.)	−275,280	−8.74	+1.00	−.40	+104.84
$3\ Fe(l) + 2\ O_2(g) = Fe_3O_4(l)$ (1,874°–2,000° K.)	−257,240	−26.89	+1.00	−.40	+155.46
$2\ Fe(\alpha) + 3/2\ O_2(g) = Fe_2O_3$ (*hematite*) (298.16°–950° K.)	−200,000	−13.84	−1.45	+1.905	+108.26
$2\ Fe(\alpha) + 3/2\ O_2(g) = Fe_2O_3(\beta)$ (950°–1,033° K.)	−202,960	−42.64	+7.85	+.13	+188.48
$2\ Fe(\beta) + 3/2\ O_2(g) = Fe_2O_3(\beta)$ (1,033°–1,050° K.)	−196,740	−10.27	+.75	−.30	+92.26
$2\ Fe(\beta) + 3/2\ O_2(g) = Fe_2O_3(\gamma)$ (1,050°–1,179° K.)	−193,200	−.39	−.13	−.30	+59.96
$2\ Fe(\gamma) + 3/2\ O_2(g) = Fe_2O_3(\gamma)$ (1,179°–1,674° K.)	−202,540	−25.95	+2.87	−.30	+142.85
$2\ Fe(\delta) + 3/2\ O_2(g) = Fe_2O_3(\gamma)$ (1,674°–1,800° K.)	−192,920	−.85	−.13	−.30	+61.21
$2\ La(c) + 3/2\ O_2(g) = La_2O_3(c)$ (298.16°–1,153° K.)	−431,120	−13.31	+.80	+1.34	+112.36
$2\ La(l) + 3/2\ O_2(g) = La_2O_3(c)$ (1,153°–2,000° K.)	−434,330	−4.88	−.80	+1.34	+91.17
$Pb(c) + 1/2\ O_2(g) = PbO$ (*red*) (298.16°–600.5° K.)	−52,800	−2.76	−.80	−.10	+32.49
$Pb(l) + 1/2\ O_2(g) = PbO$ (*red*) (600.5°–762° K.)	−53,780	−.51	−1.75	−.10	+28.44
$Pb(c) + 1/2\ O_2(g) = PbO$ (*yellow*) (298.16°–600.5° K.)	−52,040	+.81	−2.00	−.10	+22.13
$Pb(l) + 1/2\ O_2(g) = PbO$ (*yellow*) (600.5°–1,159° K.)	−53,020	+3.06	−2.95	−.10	+18.08
$Pb(l) + 1/2\ O_2(g) = PbO(l)$ (1,159°–1,745° K.)	−53,980	−12.94	+.25	−.10	+64.22
$Pb(c) + 1/2\ O_2(g) = PbO(g)$ (298.16°–600.5° K.)	+10,270	+1.91	+1.08	+.295	−23.21
$Pb(l) + 1/2\ O_2(g) = PbO(g)$ (600.5°–2,000° K.)	+9,300	+4.17	+.13	+.295	−27.29
$3\ Pb(c) + 2\ O_2(g) = Pb_3O_4(c)$ (298.16°–600.5° K.)	−174,920	+8.82	−8.20	−.40	+72.78
$3\ Pb(l) + 2\ O_2(g) = Pb_3O_4(c)$ (600.5°–1,000° K.)	−177,860	+15.59	−11.05	−.40	+60.57
$Pb(c) + O_2(g) = PbO_2(c)$ (298.16°–600.5° K.)	−66,120	+.64	−2.45	−.20	+45.58
$Pb(l) + O_2(g) = PbO_2(c)$ (600.5°–1,100° K.)	−67,100	+2.90	−3.40	−.20	+41.50
$2\ Li(c) + 1/2\ O_2(g) = Li_2O(c)$ (298.16°–452° K.)	−142,220	−3.06	+5.77	−.10	+34.19
$2\ Li(l) + 1/2\ O_2(g) = Li_2O(c)$ (452°–1,600° K.)	−141,380	+16.97	−2.63	−.10	−17.05
$2\ Li(c) + O_2(g) = Li_2O_2(c)$ (298.16°–452° K.)	−151,880	−1.38	.....	.....	+54.83
$2\ Li(l) + O_2(g) = Li_2O_2(c)$ (452°–500° K.)	−153,260	−1.38	.....	.....	+57.88
$Mg(c) + 1/2\ O_2(g) = MgO$ (*periclase*) (298.16°–923° K.)	−144,090	−1.06	+.13	+.25	+29.16
$Mg(l) + 1/2\ O_2(g) = MgO$ (*periclase*) (923°–1,393° K.)	−145,810	+1.84	−.62	+.64	+23.07
$Mg(g) + 1/2\ O_2(g) = MgO$ (*periclase*) (1,393°–2,000° K.)	−180,700	−3.75	−.62	+.64	+65.69
$Mg(c) + O_2(g) = MgO_2(c)$ (298.16°–500° K.)	−150,230	−9.12	+.13	+.25	+70.84
$Mn(\alpha) + 1/2\ O_2(g) = MnO(c)$ (298.16°–1,000° K.)	−92,600	−4.21	+.97	+.155	+29.66
$Mn(\beta) + 1/2\ O_2(g) = MnO(c)$ (1,000°–1,374° K.)	−91,900	+1.84	−.39	+.34	+12.15
$Mn(\gamma) + 1/2\ O_2(g) = MnO(c)$ (1,374°–1,410° K.)	−89,810	+7.30	−.72	+.34	−6.05
$Mn(\delta) + 1/2\ O_2(g) = MnO(c)$ (1,410°–1,517° K.)	−89,390	+8.68	−.72	+.34	−10.70
$Mn(l) + 1/2\ O_2(g) = MnO(c)$ (1,517°–2,000° K.)	−93,350	+7.99	−.72	+.34	−5.90
$3\ Mn(\alpha) + 2\ O_2(g) = Mn_3O_4(\alpha)$ (298.16°–1,000° K.)	−332,400	−7.41	+.66	+.145	+106.62
$3\ Mn(\beta) + 2\ O_2(g) = Mn_3O_4(\alpha)$ (1,000°–1,374° K.)	−330,310	+10.75	−3.42	+.70	+54.07
$3\ Mn(\gamma) + 2\ O_2(g) = Mn_3O_4(\alpha)$ (1,374°–1,410° K.)	−324,050	+27.12	−4.41	+.70	−.50
$3\ Mn(\delta) + 2\ O_2(g) = Mn_3O_4(\alpha)$ (1,410°–1,445° K.)	−322,800	+31.27	−4.41	+.70	−14.46
$3\ Mn(\delta) + 2\ O_2(g) = Mn_3O_4(\beta)$ (1,445°–1,517° K.)	−328,870	−4.56	+1.00	−.40	+95.20
$3\ Mn(l) + 2\ O_2(g) = Mn_3O_4(\beta)$ (1,517°–1,800° K.)	−340,730	−6.63	+1.00	−.40	+109.60
$2\ Mn(\alpha) + 3/2\ O_2(g) = Mn_2O_3(c)$ (298.16°–1,000° K.)	−230,610	−5.96	−.06	+.945	+80.74
$2\ Mn(\beta) + 3/2\ O_2(g) = Mn_2O_3(c)$ (1,000°–1,374° K.)	−229,210	+6.15	−2.78	+1.315	+45.70
$2\ Mn(\gamma) + 3/2\ O_2(g) = Mn_2O_3(c)$ (1,374°–1,410° K.)	−225,030	+17.06	−3.44	+1.315	+9.33
$2\ Mn(\delta) + 3/2\ O_2(g) = Mn_2O_3(c)$ (1,410°–1,517° K.)	−224,200	+19.82	−3.44	+1.315	+0.05
$2\ Mn(l) + 3/2\ O_2(g) = Mn_2O_3(c)$ (1,517°–1,700° K.)	−232,110	+18.44	−3.44	+1.315	+9.65
$Mn(\alpha) + O_2(g) = MnO_2(c)$ (298.16°–1,000° K.)	−126,400	−8.61	+.97	+1.555	+70.14
$2\ Hg(l) + 1/2\ O_2(g) = Hg_2O(c)$ (298.16°–629.88° K.)	−22,400	−4.61	.....	.....	+43.29
$2\ Hg(g) + 1/2\ O_2(g) = Hg_2O(c)$ (629.88°–1,000° K.)	−53,800	−16.12	.....	.....	+125.36
$Hg(l) + 1/2\ O_2(g) = HgO$ (*red*) (298.16°–629.88° K.)	−21,760	+.85	−2.47	−.10	+24.81
$Hg(g) + 1/2\ O_2(g) = HgO$ (*red*) (629.88°–1,500° K.)	−36,920	−2.92	−2.47	−.10	+59.42
$Mo(c) + O_2(g) = MoO_2(c)$ (298.16°–2,000° K.)	−132,910	−3.91	.....	.....	+47.42
$Mo(c) + 3/2\ O_2(g) = MoO_3(c)$ (298.16°–1,068° K.)	−182,650	−8.86	−1.55	+1.54	+90.07
$Mo(c) + 3/2\ O_2(g) = MoO_3(l)$ (1,068°–1,500° K.)	−179,770	−36.34	+1.40	−.30	+167.61
$2\ Nd(c) + 3/2\ O_2(g) = Nd_2O_3$ (*hexagonal*). (298.16°–1,113° K.)	−435,150	−16.19	+3.21	+1.78	+125.68
$2\ Nd(l) + 3/2\ O_2(g) = Nd_2O_3$ (*hexagonal*). (1,113°–1,500° K.)	−437,090	+4.03	−2.13	+1.78	+71.77
$Np(c) + O_2(g) = NpO_2(c)$ (298.16°–913° K.)	−246,450	−3.45	.....	.....	+52.44
$Np(l) + O_2(g) = NpO_2(c)$ (913°–1,500° K.)	−249,010	−4.61	.....	.....	+58.68
$Ni(\alpha) + 1/2\ O_2(g) = NiO(c)$ (298.16°–633° K.)	−57,640	−4.61	+2.16	−.10	+34.41
$Ni(\beta) + 1/2\ O_2(g) = NiO(c)$ (633°–1,725° K.)	−57,460	−.14	−.46	−.10	+23.27
$Ni(l) + 1/2\ O_2(g) = NiO(c)$ (1,725°–2,000° K.)	−58,830	+7.23	−1.36	−.10	+1.76
$2\ Nb(c) + 2\ O_2(g) = Nb_2O_4(c)$ (298.16°–2,000° K.)	−382,050	−9.67	.....	.....	+116.23

HEAT OF FORMATION OF INORGANIC OXIDES (Continued)
Coefficients in Free-Energy Equations (Continued)

Reaction	Temperature range of validity	ΔH_0	2.303*a*	*b*	*c*	*I*
$2\,Nb(c) + 5/2\,O_2(g) = Nb_2O_5(c)$	(298.16°–1,785° K.)	−458,640	−16.14	−.56	+1.94	+157.66
$2\,Nb(c) + 5/2\,O_2(g) = Nb_2O_5(l)$	(1,785°–2,000° K.)	−463,630	−66.04	+2.21	−.50	+317.84
$N_2(g) + 1/2\,O_2(g) = N_2O(g)$	(298.16°–2,000° K.)	+18,650	−1.57	−.27	+.92	+23.47
$1/2\,N_2(g) + 1/2\,O_2(g) = NO(g)$	(298.16°–2,000° K.)	+21,590	−.28	+.45	−.03	−2.20
$N_2(g) + 3/2\,O_2(g) = N_2O_3(g)$	(298.16°–500° K.)	+17,390	−.35	......	.....	+54.30
$1/2\,N_2(g) + O_2(g) = NO_2(g)$	(298.16°–2,000° K.)	+7,730	+.53	−.265	+.605	+13.74
$N_2(g) + 2\,O_2(g) = N_2O_4(g)$	(298.16°–1,000° K.)	+1,370	+2.14	−3.24	+1.38	+68.34
$N_2(g) + 5/2\,O_2(g) = N_2O_5(c)$	(298.16°–305° K.)	−10,200	.....	......	.....	+131.70
$N_2(g) + 5/2\,O_2(g) = N_2O_5(g)$	(298.16°–500° K.)	+3,600	.....	......	.....	+86.50
$Os(c) + 2\,O_2(g) = OsO_4$ (*yellow*)	(298.16°–329° K.)	−92,260	+14.97	......	.....	+32.45
$Os(c) + 2\,O_2(g) = OsO_4$ (*white*)	(298.16°–315° K.)	−90,560	+14.97	......	.....	+27.60
$Os(c) + 2\,O_2(g) = OsO_4(l)$	(315°–403° K.)	−88,970	+9.67	......	.....	+35.78
$Os(c) + 2\,O_2(g) = OsO_4(g)$	(298.16°–1,000° K.)	−81,200	+8.17	−2.86	+1.94	+20.34
$3/2\,O_2(g) = O_3(g)$	(298.16°–2,000° K.)	+33,980	+2.03	−.48	+.36	+11.45
P (*white*) $+ 1/2\,O_2(g) = PO(g)$	(298.16°–317.4° K.)	−9,370	+2.53	......	.....	−25.40
$P(l) + 1/2\,O_2(g) = PO(g)$	(317.4°–553° K.)	−9,390	+3.45	......	.....	−27.63
$1/4\,P_4(g) + 1/2\,O_2(g) = PO(g)$	(553°–1,500° K.)	−12,640	+2.30	......	.....	−18.61
4 P (*white*) $+ 5\,O_2(g) =$ P_4O_{10} (*hexagonal*).	(298.16°–317.4° K.)	−711,520	+95.67	−51.50	−1.00	−28.24
$4\,P(l) + 5\,O_2(g) = P_4O_{10}$ (*hex.*)	(317.4°–553° K.)	−711,800	+97.98	−51.50	−1.00	−33.13
$P_4(g) + 5\,O_2(g) = P_4O_{10}$ (*hex.*)	(553°–631° K.)	−725,560	+87.45	−51.07	−2.405	+20.87
$P_4(g) + 5\,O_2(g) = P_4O_{10}(g)$	(631°–1,500° K.)	−722,330	−43.45	+2.93	−2.405	+348.20
$Pu(c) + O_2(g) = PuO_2(c)$	(298.16°–1,500° K.)	−246,450	−3.45	......	.....	+52.48
$Po(c) + O_2(g) = PoO_2(c)$	(298.16°–900° K.)	−61,510	−9.21	......	.....	+72.80
$2\,K(c) + 1/2\,O_2(g) = K_2O(c)$	(298.16°–336.4° K.)	−86,400	.....	......	.....	+33.90
$2\,K(l) + 1/2\,O_2(g) = K_2O(c)$	(336.4°–1,049° K.)	−87,380	+1.15	......	.....	+33.90
$2\,K(g) + 1/2\,O_2(g) = K_2O(c)$	(1,049°–1,500° K.)	−133,090	−16.12	......	.....	+129.64
$2\,K(c) + O_2(g) = K_2O_2(c)$	(298.16°–336.4° K.)	−118,300	−2.30	......	.....	+59.60
$2\,K(l) + O_2(g) = K_2O_2(c)$	(336.4°–763° K.)	−119,780	−4.61	......	.....	+69.85
$2\,K(l) + O_2(g) = K_2O_2(l)$	(763°–1,049° K.)	−118,250	−18.42	......	.....	+107.66
$2\,K(g) + O_2(g) = K_2O_2(l)$	(1,049°–1,500° K.)	−161,870	−31.08	......	.....	+187.49
$2\,K(c) + 3/2\,O_2(g) = K_2O_3(c)$	(298.16°–336.4° K.)	−126,640	−12.66	......	.....	+111.75
$2\,K(l) + 3/2\,O_2(g) = K_2O_3(c)$	(336.4°–703° K.)	−127,790	−12.66	......	.....	+115.16
$2\,K(l) + 3/2\,O_2(g) = K_2O_3(l)$	(703°–1,000° K.)	−125,330	−27.63	......	.....	+154.28
$K(c) + O_2(g) = KO_2(c)$	(298.16°–336.4° K.)	−68,940	−10.36	......	.....	+66.45
$K(l) + O_2(g) = KO_2(c)$	(336.4°–653° K.)	−69,510	−10.36	......	.....	+68.15
$K(l) + O_2(g) = KO_2(l)$	(653°–1,000° K.)	−67,880	−18.42	......	.....	+88.34
$K(c) + 3/2\,O_2(g) = KO_3(c)$	(298.16°–336.4° K.)	−63,340	−10.36	......	.....	+85.85
$K(l) + 3/2\,O_2(g) = KO_3(c)$	(336.4°–500° K.)	−63,910	−10.36	......	.....	+87.55
$2\,Pr(c) + 3/2\,O_2(g) = Pr_2O_3$(*c, C-type*)	(298.16°–1,205° K.)	−440,600	−4.60	......	.....	+78.38
$2\,Pr(l) + 3/2\,O_2(g) = Pr_2O_3$(*c, C-type*)	(1,205°–2,000° K)	−446,100	−4.60	......	.....	+82.94
$6\,Pr(c) + 11/2\,O_2(g) = Pr_6O_{11}(c)$	(298.16°–1,205° K.)	−1,374,000	....	......	.....	+241.04
$6\,Pr(l) + 11/2\,O_2(g) = Pr_6O_{11}(c)$	(1,205°–1,500° K.)	−1,390,500	....	......	.....	+254.73
$Pr(c) + O_2(g) = PrO_2(c)$	(298.16°–1,200° K.)	−230,990	−6.42	+2.34	−.20	+61.07
$Ra(c) + 1/2\,O_2(g) = RaO(c)$	(298.16°–1,000° K.)	−130,000	....	......	.....	+23.50
$Re(c) + 3/2\,O_2(g) = ReO_3(c)$	(298.16°–433° K.)	−149,090	−16.12	......	.....	+110.49
$Re(c) + 3/2\,O_2(g) = ReO_3(l)$	(433°–1,000° K.)	−146,750	−31.32	......	.....	+145.16
$2Re(c) + 7/2\,O_7(g) = Re_2O_7(c)$	(298.16°–569° K.)	−301,470	−34.54	......	.....	+250.57
$2\,Re(c) + 7/2\,O_7(g) = Re_2O_7(l)$	(569°–635.5° K.)	−295,810	−73.68	......	.....	+348.45
$2\,Re(c) + 7/2\,O_2(g) = Re_2O_7(g)$	(635.5°–1,500° K.)	−256,460	+3.45	......	.....	+70.33
$2\,Re(c) + 4\,O_2(g) = Re_2O_8(c)$	(298.16°–420° K.)	−313,870	−41.45	......	.....	+293.57
$2\,Re(c) + 4\,O_2(g) = Re_2O_8(l)$	(420°–600° K.)	−318,470	−87.50	......	.....	+425.32
$2\,Rh(c) + 1/2\,O_2(g) = Rh_2O(c)$	(298.16°–2,000° K.)	−23,740	−8.06	......	.....	+35.64
$Rh(c) + 1/2\,O_2(g) = RhO(c)$	(298.16°–1,500° K.)	−22,650	−7.37	......	.....	+40.54
$2\,Rh(c) + 3/2\,O_2(g) = Rh_2O_3(c)$	(298.16°–1,500° K.)	−70,060	−13.58	......	.....	+101.72
$2\,Rb(c) + 1/2\,O_2(g) = Rb_2O(c)$	(298.16°–312.2° K.)	−78,900	....	......	.....	+32.20
$2\,Rb(l) + 1/2\,O_2(g) = Rb_2O(c)$	(312.2°–750° K.)	−79,950	....	......	.....	+35.56
$2\,Rb(l) + 1/2\,O_2(g) = Rb_2O(l)$	(750°–952° K.)	−78,830	−10.36	......	.....	+63.85
$2\,Rb(g) + 1/2\,O_2(g) = Rb_2O(l)$	(952°–1,500° K.)	−120,290	−23.03	......	.....	+145.14
$2\,Rb(c) + O_2(g) = Rb_2O_2(c)$	(298.16°–312.2° K.)	−102,000	−2.30	......	.....	+57.40
$2\,Rb(l) + O_2(g) = Rb_2O_2(c)$	(312.2°–840° K.)	−103,360	−4.61	......	.....	+67.52
$2\,Rb(l) + O_2(g) = Rb_2O_2(l)$	(840°–952° K.)	−101,680	−18.42	......	.....	+105.91
$2\,Rb(g) + O_2(g) = Rb_2O_2(l)$	(952°–1,500 K.)	−143,130	−31.08	......	.....	+187.16
$2\,Rb(c) + 3/2\,O_2(g) = Rb_2O_3(c)$	(298.16°–312.2° K.)	−118,190	−11.51	......	.....	+104.70
$2\,Rb(l) + 3/2\,O_2(g) = Rb_2O_3(c)$	(312.2°–760° K.)	−119,400	−12.66	......	.....	+111.43
$2\,Rb(l) + 3/2\,O_2(g) = Rb_2O_3(l)$	(760°–952° K.)	−116,740	−27.63	......	.....	+151.06
$2\,Rb(g) + 3/2\,O_2(g) = Rb_2O_3(l)$	(952°–1,500° K.)	−157,720	−39.14	......	.....	+228.39
$Rb(c) + O_2(g) = RbO_2(c)$	(298.16°–312.2° K.)	−52,330	−10.36	......	.....	+66.25
$Rb(l) + O_2(g) = RbO_2(c)$	(312.2°–685° K.)	−65,120	−11.51	......	.....	+71.30

HEAT OF FORMATION OF INORGANIC OXIDES (Continued)
Coefficients in Free-Energy Equations (Continued)

Reaction and temperature range of validity	ΔH_o	2.303a	b	c	I
$Rb(l) + O_2(g) = RbO_2(l)$ (685°–952° K.)	−63,070	−18.42	.....	.....	+87.89
$Rb(g) + O_2(g) = RbO_2(l)$ (952°–1,500° K.)	−83,560	−24.18	.....	.....	+126.57
$Ru(\alpha, \beta, \gamma) + O_2(g) = RuO_2(c)$ (298.16°–1,500° K.)	−57,290	−6.91	.....	.....	+62.01
$2\ Sm(c) + 3/2\ O_2(g) = Sm_2O_3(c)$ (298.16°–1,623° K.)	−430,600	−4.60	.....	.....	+78.38
$2\ Sm(l) + 3/2\ O_2(g) = Sm_2O_3(c)$ (1,623°–2,000° K.)	−438,000	−4.60	.....	.....	+82.94
$2\ Sc(c) + 3/2\ O_2(g) = Sc_2O_3(c)$ (298.16°–1,673° K.)	−409,960	+7.78	−1.84	−0.30	+52.73
$2\ Sc(l) + 3/2\ O_2(g) = Sc_2O_3(c)$ (1,673°–2,000° K.)	−412,950	+18.47	−2.93	−.30	+21.88
$Se(c) + 1/2\ O_2(g) = SeO(g)$ (298.16°–490° K.)	+9,280	−3.04	+4.40	+.30	−14.78
$Se(l) + 1/2\ O_2(g) = SeO(g)$ (490°–1,027° K.)	+9,420	+8.70	.....	+.30	−44.50
$1/2\ Se_2(g) + 1/2\ O_2(g) = SeO(g)$ (1,027°–2,000° K.)	−7,400	−.37	.....	+.19	−.80
$Se(c) + O_2(g) = SeO_2(c)$ (298.16°–490° K.)	−53,770	+14.94	−9.41	−.20	+6.94
$Se(l) + O_2(g) = SeO_2(c)$ (490°–595° K.)	−53,640	+27.59	−14.31	.....	−25.05
$Se(l) + O_2(g) = SeO_2(g)$ (595°–1,027° K.)	−32,840	+6.79	.....	.....	−10.80
$1/2\ Se_2(g) + O_2(g) = SeO_2(g)$ (1,027°–2,000° K.)	−49,000	−.74	.....	.....	+27.61
$Si(c) + 1/2\ O_2(g) = SiO(g)$ (298.16°–1,683° K.)	−21,090	+3.84	+.16	−.295	−33.14
$Si(l) + 1/2\ O_2(g) = SiO(g)$ (1,683°–2,000° K.)	−30,170	−7.78	−.12	+.25	−40.01
$Si(c) + O_2(g) = SiO_2(\alpha\text{-}quartz)$ (298.16°–848° K.)	−210,070	+3.98	−3.32	+.605	+34.59
$Si(c) + O_2(g) = SiO_2(\beta\text{-}quartz)$ (848°–1,683° K.)	−209,920	−3.36	−.19	+.745	+53.44
$Si(l) + O_2(g) = SiO_2(\beta\text{-}quartz)$ (1,683°–1,883° K.)	−219,000	+.58	−.47	−.20	+46.58
$Si(l) + O_2(g) = SiO_2(l)$ (1,883°–2,000° K.)	−228,590	−15.66	.....	.....	+103.97
$Si(c) + O_2(g) = SiO_2(\alpha\text{-}crist.)$ (298.16°–523° K.)	−207,330	+19.96	−9.75	−.745	−9.78
$Si(c) + O_2(g) = SiO_2(\beta\text{-}crist.)$ (523°–1,683° K.)	−209,820	−3.34	−.24	−.745	+53.35
$Si(l) + O_2(g) = SiO_2(\beta\text{-}crist.)$ (1,683°–2,000° K.)	−218,900	+.60	−.52	−.20	+46.49
$Si(c) + O_2(g) = SiO_2(\alpha\text{-}trid.)$ (298.16°–390° K.)	−207,030	+22.29	−11.62	−.745	−15.64
$Si(c) + O_2(g) = SiO_2(\beta\text{-}trid.)$ (390°–1,683° K.)	−209,350	−1.59	−.54	−.745	+47.86
$Si(l) + O_2(g) = SiO_2(\beta\text{-}trid.)$ (1,683°–1,953° K.)	−218,430	+2.35	−.82	−.20	+41.00
$2\ Ag(c) + 1/2\ O_2(g) = Ag_2O(c)$ (298.16°–1,000° K.)	−7,740	−4.14	.....	.....	+27.84
$2\ Ag(c) + O_2(g) = Ag_2O_2(c)$ (298.16°–500° K.)	−6,620	−3.22	.....	.....	+52.17
$2\ Na(c) + 1/2\ O_2(g) = Na_2O(c)$ (298.16°–371° K.)	−99,820	−7.51	+5.47	−.10	+50.43
$2\ Na(l) + 1/2\ O_2(g) = Na_2O(c)$ (371°–1,187° K.)	−100,150	+4.97	−2.45	−.10	+22.19
$2\ Na(g) + 1/2\ O_2(g) = Na_2O(c)$ (1,187°–1,190° K.)	−156,200	−20.72	.....	.....	+145.48
$2\ Na(g) + 1/2\ O_2(g) = Na_2O(l)$ (1,190°–2,000° K.)	−150,250	−23.03	.....	.....	+147.58
$2\ Na(c) + O_2(g) = Na_2O_2(c)$ (298.16°–371° K.)	−122,500	−2.30	.....	.....	+57.51
$2\ Na(l) + O_2(g) = Na_2O_2(c)$ (371°–733° K.)	−124,320	−5.76	.....	.....	+71.30
$2\ Na(l) + O_2(g) = Na_2O_2(l)$ (733°–1,187° K.)	−123,220	−20.72	.....	.....	+112.66
$2\ Na(g) + O_2(g) = Na_2O_2(l)$ (1,187°–1,500° K.)	−174,800	−31.08	.....	.....	+187.97
$Na(c) + O_2(g) = NaO_2(c)$ (298.16°–371° K.)	−63,040	−8.06	.....	.....	+56.98
$Na(l) + O_2(g) = NaO_2(c)$ (371°–1,000° K.)	−64,220	−11.51	.....	.....	+69.04
$Sr(c) + 1/2\ O_2(g) = SrO(c)$ (298.16°–1,043° K.)	−142,410	−6.79	+.305	+.675	+44.33
$Sr(l) + 1/2\ O_2(g) = SrO(c)$ (1,043°–1,657° K.)	−143,370	−2.42	−.38	+.675	+32.77
$Sr(g) + 1/2\ O_2(g) = SrO(c)$ (1,657°–2,000° K.)	−181,180	−8.24	−.38	+.675	+74.32
$Sr(c) + O_2(g) = SrO_2(c)$ (298.16°–1,000° K.)	−155,510	−11.40	+.305	+.675	+75.44
$S(rh) + 1/2\ O_2(g) = SO(g)$ (298.16°–368.6° K.)	+19,250	−1.24	+2.95	+.225	−18.84
$S(mon) + 1/2\ O_2(g) = SO(g)$ (368.6°–392° K.)	+19,200	−1.29	+3.31	+.225	−18.72
$S(l\lambda,\mu) + 1/2\ O_2(g) = SO(g)$ (392°–718° K.)	+20,320	+10.22	−.17	+.225	−50.05
$1/2\ S_2(g) + 1/2\ O_2(g) = SO(g)$ (298.16°–2,000° K.)	+3,890	+.07	.....	.....	−1.50
$S(rh) + O_2(g) = SO_2(g)$ (298.16°–368.6° K.)	−70,980	+.83	+2.35	+.51	−5.85
$S(mon) + O_2(g) = SO_2(g)$ (368.6°–392° K.)	−71,020	+.78	+2.71	+.51	−5.74
$S(l\lambda,\mu) + O_2(g) = SO_2(g)$ (392°–718° K.)	−69,900	+12.30	−.77	+.51	−37.10
$1/2\ S_2(g) + O_2(g) = SO_2(g)$ (298.16°–2,000° K.)	−86,330	+2.42	−.70	+.31	+10.71
$S(rh) + 3/2\ O_2(g) = SO_3(c\text{—}I)$ (298.16°–335.4° K.)	−111,370	−6.45	.....	.....	+88.32
$S(rh) + 3/2\ O_2(g) = SO_3(c\text{—}II)$ (298.16°–305.7° K.)	−108,680	−11.97	.....	.....	+94.95
$S(rh) + 3/2\ O_2(g) = SO_3(l)$ (298.16°–335.4° K.)	−107,430	−21.18	.....	.....	+113.76
$S(rh) + 3/2\ O_3(g) = SO_3(g)$ (298.16°–368.6° K.)	−95,070	+1.43	+0.66	+1.26	+16.81
$S(mon) + 3/2\ O_2(g) = SO_3(g)$ (368.6°–392° K.)	−95,120	+1.38	+1.02	+1.26	+16.93
$S(l\lambda,\mu) + 3/2\ O_2(g) = SO_3(g)$ (392°–718° K.)	−94,010	+12.89	−2.46	+1.26	−14.40
$1/2\ S_2(g) + 3/2\ O_2(g) = SO_3(g)$ (298.16°–1,500° K.)	−110,420	+3.02	−2.39	+1.06	+33.41
$2\ Ta(c) + 5/2\ O_2(g) = Ta_2O_5(c)$ (298.16°–2,000° K.)	−492,790	−17.18	−1.25	+2.46	+161.68
$Tc(c) + O_2(g) = TcO_2(c)$ (298.16°–500° K.)	−103,400	.....	.....	.....	+41.00
$Tc(c) + 3/2\ O_2(g) = TcO_3(c)$ (298.16°–500° K.)	−129,000	.....	.....	.....	+64.50
$2\ Tc(c) + 7/2\ O_2(g) = Tc_2O_7(c)$ (298.16°–392.7° K.)	−266,000	.....	.....	.....	+147.00
$2\ Tc(c) + 7/2\ O_2(g) = Tc_2O_7(l)$ (392.7°–500° K.)	−258,930	.....	.....	.....	+129.00
$Te(c) + 1/2\ O_2(g) = TeO(g)$ (298.16°–723° K.)	+43,110	+1.91	+.84	+.315	−27.22
$Te(l) + 1/2\ O_2(g) = TeO(g)$ (723°–1,360° K.)	+39,750	+6.08	+.09	+.315	−33.94
$1/2\ Te_2(g) + 1/2\ O_2(g) = TeO(g)$ (1,360°–2,000° K.)	+23,730	−.90	+.09	+.315	−.29
$Te(c) + O_2(g) = TeO_2(c)$ (298.16°–723° K.)	−78,090	−2.10	−2.35	−.20	+51.27
$Te(l) + O_2(g) = TeO_2(c)$ (723°–1,006° K.)	−81,530	+1.84	−3.10	−.20	+45.30
$Te(l) + O_2(g) = TeO_2(l)$ (1,006°–1,300° K.)	−82,090	−21.74	+.50	−.20	+113.04
$2\ Tl(\alpha) + O_2(g) = Tl_2O(c)$ (298.16°–505.5° K.)	−44,110	−6.91	.....	.....	+42.30

HEAT OF FORMATION OF INORGANIC OXIDES (Continued)
Coefficients in Free-Energy Equations (Continued)

Reaction and temperature range of validity	ΔH_0	2.303*a*	*b*	*c*	*I*
$2\,Tl(\beta) + O_2(g) = Tl_2O(c)$ (505.5°–573° K.)	−44,260	−6.91	.....	.....	+42.60
$2\,Tl(\beta) + 1/2\,O_2(g) = Tl_2O(l)$ (573°–576° K.)	−40,880	−13.82	.....	.....	+55.76
$2\,Tl(l) + 1/2\,O_2(g) = Tl_2O(l)$ (576°–773° K.)	−42,320	−11.51	.....	.....	+51.89
$2\,Tl(l) + 1/2\,O_2(g) = Tl_2O(g)$ (773°–1,730° K.)	−18,400	+11.51	.....	.....	−45.55
$2\,Tl(g) + 1/2\,O_2(g) = Tl_2O(g)$ (1,730°–2,000° K.)	−104,670	.....	.....	.....	+41.59
$2\,Tl(\alpha) + 3/2\,O_2(g) = Tl_2O_3(c)$ (298.16°–505.5° K.)	−99,410	−16.12	.....	.....	+119.09
$2\,Tl(\beta) + 3/2\,O_2(g) = Tl_2O_3(c)$ (505.5°–576° K.)	−99,560	−16.12	.....	.....	+119.39
$2\,Tl(l) + 3/2\,O_2(g) = Tl_2O_3(c)$ (576°–990° K.)	−101,010	−13.82	.....	.....	+115.55
$2\,Tl(l) + 3/2\,O_2(g) = Tl_2O_3(l)$ (990°–1,500° K.)	−94,550	−27.63	.....	.....	+150.39
$2\,Tl(\alpha) + 2\,O_2(g) = Tl_2O_4(c)$ (298.16°–505.5° K.)	−117,680	−23.03	.....	.....	+161.19
$2\,Tl(\beta) + 2\,O_2(g) = Tl_2O_4(c)$ (505.5°–576° K.)	−117,830	−23.03	.....	.....	+161.49
$2\,Tl(l) + 2\,O_2(g) = Tl_2O_4(c)$ (576°–1,000° K.)	−119,270	−20.72	.....	.....	+157.63
$Th(c) + O_2(g) = ThO_2(c)$ (298.16°–2,000° K.)	−294,350	−5.25	+.59	+.775	+62.81
$Sn(c) + 1/2\,O_2(g) = SnO(c)$ (298.16°–505° K.)	−68,600	−3.57	+1.65	−.10	+32.59
$Sn(l) + 1/2\,O_2(g) = SnO(c)$ (505°–1,300° K.)	−69,670	+3.06	−1.50	−.10	+18.39
$Sn(c) + 1/2\,O_2(g) = SnO(g)$ (298.16°–505° K.)	−1,000	−.97	+3.24	+.32	−17.41
$Sn(l) + 1/2\,O_2(g) = SnO(g)$ (505°–2,000° K.)	−2,070	+5.66	+.09	+.32	−31.62
$Sn(c) + O_2(g) = SnO_2(c)$ (298.16°–505° K.)	−142,010	−14.00	+2.45	+2.38	+90.74
$Sn(l) + O_2(g) = SnO_2(c)$ (505°–1,898° K.)	−143,080	−7.37	−.70	+2.38	+76.53
$Sn(l) + O_2(g) = SnO_2(l)$ (1,898°–2,000° K.)	−139,130	−21.97	+.50	−.20	+120.11
$Ti(\alpha) + 1/2\,O_2(g) = TiO(\alpha)$ (298.16°–1,150° K.)	−125,010	−4.01	−.29	+.83	+36.28
$Ti(\beta) + 1/2\,O_2(g) = TiO(\alpha)$ (1,150°–1,264° K.)	−125,040	+1.17	−1.55	+.83	+21.90
$Ti(\beta) + 1/2\,O_2(g) = TiO(\beta)$ (1,264°–2,000° K.)	−125,210	−1.77	−1.25	−.10	+30.83
$Ti(\alpha) + 1/2\,O_2(g) = TiO(g)$ (298.16°–1,150° K.)	+11,710	+3.71	+1.07	−.10	−35.50
$Ti(\beta) + 1/2\,O_2(g) = TiO(g)$ (1,150°–2,000° K.)	+11,680	+8.89	−.19	−.10	−49.88
$2\,Ti(\alpha) + 3/2\,O_2(g) = Ti_2O_3(\alpha)$ (298.16°–473° K.)	−360,660	+32.08	−23.49	−.30	−10.66
$2\,Ti(\alpha) + 3/2\,O_2(g) = Ti_2O_3(\beta)$ (473°–1,150° K.)	−369,710	−30.95	+2.62	+4.80	+162.79
$2\,Ti(\beta) + 3/2\,O_2(g) = Ti_2O_3(\beta)$ (1,150°–2,000° K.)	−369,760	−20.59	+.10	+4.80	+134.03
$3\,Ti(\alpha) + 5/2\,O_2(g) = Ti_3O_5(\alpha)$ (298.16°–450° K.)	−587,980	−4.19	−9.72	−.50	+131.05
$3\,Ti(\alpha) + 5/2\,O_2(g) = Ti_3O_5(\beta)$ (450°–1,150° K.)	−586,330	−18.31	+1.03	−.50	+159.98
$3\,Ti(\beta) + 5/2\,O_2(g) = Ti_3O_5(\beta)$ (1,150°–2,000° K.)	−586,420	−2.76	−2.75	−.50	+116.81
$Ti(\alpha) + O_2(g) = TiO_2$ (*rutile*) (298.16°–1,150° K.)	−228,360	−12.80	+1.62	+1.975	+82.81
$Ti(\beta) + O_2(g) = TiO_2$ (*rutile*) (1,150°–2,000° K.)	−228,380	−7.62	+.36	+1.975	+68.43
$W(c) + O_2(g) = WO_2(c)$ (298.16°–1,500° K.)	−137,180	−1.38	.....	.....	+45.56
$4\,W(c) + 11/2\,O_2(g) = W_4O_{11}(c)$ (298.16°–1,700° K.)	−745,730	−32.70	.....	.....	+321.84
$W(c) + 3/2\,O_2(g) = WO_3(c)$ (298.16°–1,743° K.)	−201,180	−2.92	−1.81	−.30	+70.89
$W(c) + 3/2\,O_2(g) = WO_3(l)$ (1,743°–2,000° K.)	−203,140	−35.74	+1.13	−.30	+173.27
$U(\alpha) + O_2(g) = UO_2(c)$ (298.16°–935° K.)	−262,880	−19.92	+3.70	+2.13	+100.54
$U(\beta) + O_2(g) = UO_2(c)$ (935°–1,045° K.)	−260,660	−4.28	−.31	+1.78	+55.50
$U(\gamma) + O_2(g) = UO_2(c)$ (1,045°–1,405° K.)	−262,830	−6.54	−.31	+1.78	+64.41
$U(l) + O_2(g) = UO_2(c)$ (1,405°–1,500° K.)	−264,790	−5.92	.....	.....	+63.50
$3\,U(\alpha) + 4\,O_2(g) = U_3O_8(c)$ (298.16°–935° K.)	−863,370	−56.57	+10.68	+5.20	+330.19
$3\,U(\beta) + 4\,O_2(g) = U_3O_8(c)$ (935°–1,045° K.)	−856,720	−9.67	−1.35	+4.15	+195.12
$3\,U(\gamma) + 4\,O_2(g) = U_3O_8(c)$ (1,045°–1,405° K.)	−863,230	−16.44	−1.35	+4.15	+221.79
$3\,U(l) + 4\,O_2(g) = U_3O_8(c)$ (1,405°–1,500° K.)	−869,460	−10.91	−1.35	+4.15	+208.82
$U(\alpha) + 3/2\,O_2(g) = UO_3$ (*hex*) (298.16°–935° K.)	−294,090	−18.33	+3.49	+1.535	+114.94
$U(\beta) + 3/2\,O_2(g) = UO_3$ (*hex*) (935°–1,045° K.)	−291,870	−2.69	−.52	+1.185	+69.90
$U(\gamma) + 3/2\,O_2(g) = UO_3$ (*hex*) (1,045°–1,400° K.)	−294,040	−4.95	−.52	+1.185	+78.80
$V(c) + 1/2\,O_2(g) = VO(c)$ (298.16°–2,000° K.)	−101,090	−5.39	−.36	+.53	+38.69
$V(c) + 1/2\,O_2(g) = VO(g)$ (298.16°–2,000° K.)	+52,090	+1.80	+1.04	+.35	−28.42
$2\,V(c) + 3/2\,O_2(g) = V_2O_3(c)$ (298.16°–2,000° K.)	−299,910	−17.98	+.37	+2.41	+118.83
$2\,V(c) + 2\,O_2(g) = V_2O_4(\alpha)$ (298.16°–345° K.)	−342,890	−11.03	+3.00	−.40	+117.38
$2\,V(c) + 2\,O_2(g) = V_2O_4(\beta)$ (345°–1,818° K.)	−345,330	−24.36	+1.30	+3.545	+155.55
$2\,V(c) + 2\,O_2(g) = V_2O_4(l)$ (1,818°–2,000° K.)	−339,880	−59.59	+3.00	−0.40	+264.42
$6\,V(c) + 13/2\,O_2(g) = V_6O_{13}(c)$ (298.16°–1,000° K.)	−1,076,340	−95.33	.....	.....	+557.61
$2\,V(c) + 5/2\,O_2(g) = V_2O_5(c)$ (298.16°–943° K.)	−381,960	−41.08	+5.20	+6.11	+228.50
$2\,V(c) + 5/2\,O_2(g) = V_2O_5(l)$ (943°–2,000° K.)	−365,840	−38.91	+3.25	−.50	+207.54
$2\,Y(c) + 3/2\,O_2(g) = Y_2O_3(c)$ (298.16°–1,773° K.)	−419,600	+2.76	−1.73	−.30	+66.36
$2\,Y(l) + 3/2\,O_2(g) = Y_2O_3(c)$ (1,773°–2,000° K.)	−422,850	+13.36	−2.75	−.30	+35.56
$Zn(c) + 1/2\,O_2(g) = ZnO(c)$ (298.16°–692.7° K.)	−84,670	−6.40	+.84	+.99	+43.25
$Zn(l) + 1/2\,O_2(g) = ZnO(c)$ (692.7°–1,180° K.)	−85,520	−1.45	−.36	+.99	+31.25
$Zn(g) + 1/2\,O_2(g) = ZnO(c)$ (1,180°–2,000° K.)	−115,940	−7.28	−.36	+.99	+74.94
$Zr(\alpha) + O_2(g) = ZrO_2(\alpha)$ (298.16°–1,135° K.)	−262,980	−6.10	+.16	+1.045	+65.00
$Zr(\beta) + O_2(g) = ZrO_2(\alpha)$ (1,135°–1,478° K.)	−264,190	−5.09	−.40	+1.48	+63.58
$Zr(\beta) + O_2(g) = ZrO_2(\beta)$ (1,478°–2,000° K.)	−262,290	−7.76	+.50	−.20	+69.50

REFRACTORY MATERIALS

Compiled by Albert P. Levitt and Aram Tarpinian

Borides

Name	Formula	Molecular Weight	Melting Point (°C)	Crystalline Form	Lattice Parameter Å	X-Ray Density gm/cm³
Titanium Boride	TiB_2	69.54	2980[6]	Hexagonal AlB_2-type [C 32]	a = 3.028[28] c = 3.228	4.52
Zirconium Boride	ZrB_2	112.86	3040 ± 50[30]	Hexagonal AlB_2-type [C 32]	a = 3.169[28] c = 3.530	6.09
	ZrB_{12}	221.06	2680[30]	Face-centered cubic	a = 7.408[33]	
Hafnium Boride	HfB_2	200.14	3100[31]	Hexagonal AlB_2-type [C 32]	a = 3.14[6] c = 3.47	10.5
Vanadium Boride	VB_2	72.59	2100[29]	Hexagonal AlB_2-type [C 32]	a = 2.998[28] c = 3.057	5.10
Niobium Boride	NbB	103.73	>.2000[3]	Orthorhombic	a = 3.298 b = 8.724 c = 3.137[35]	......
	NbB_2	114.55	2900[36] (decomposes)	Hexagonal AlB_2-type [C 32]	a = 3.086[28] c = 3.306	7.21
Tantalum Boride	TaB	191.77	>.2000[3]	Orthorhombic	a = 3.276 b = 8.669 c = 3.157[37]	14.29
	TaB_2	202.59	3000[6]	Hexagonal AlB_2-type [C 32]	a = 3.088[28] c = 3.241	12.60
Chromium Boride	CrB_2	73.65	1850[29]	Hexagonal AlB_2-type [C 32]	a = 2.969 c = 3.066[39]	5.16
Molybdenum Boride	MoB	106.77	2180[40]	Tetragonal	a = 3.110[41] c = 16.95	8.77
	MoB_2	117.59	2100[40]	Hexagonal AlB_2-type [C 32]	a = 3.05 c = 3.113[42]	7.78
	Mo_2B	202.72	2000[40] (decomposes)	Tetragonal $CuAl_2$-type [C 16]	a = 5.543[41] c = 4.735	9.31
Tungsten Boride	WB	194.68	2860[29]	Tetragonal	a = 3.115[6], c = 16.92	16.0
	W_2B	378.54	2770[29]	Tetragonal $CuAl_2$-type [C 16]	a = 5.564[41] c = 4.740	16.72
Thorium Boride	ThB_4	275.53	>.2500[6]	Tetragonal D_{4k}^{5}-P4/mbm	a = 7.256[43] c = 4.113	8.45
Uranium Boride	UB_{12}	367.91	>.1500[6] (decomposes)	Face-centered cubic	a = 7.473[44]	5.82

Borides

Name	Thermal Conductivity (cal-sec⁻¹-cm⁻²-cm-°C⁻¹)	Electrical Resistivity (Microhm-cm)	Ductility[a] Relative Scale	Resistance[b] to Oxidation	Hardness[c]
Titanium Boride	0.0624 @ 200°C[28] (15% porosity)	28.4[28]	3[3]	2–3[3]	3400 kg/mm²[29]
Zirconium Boride	0.0550 @ 200°C[28]	9.2 @ 20°C[31]	2[3]	2–3[3]	8 mohs[32]
	0.029 @ Rm. Temp.[33]	60–80 @ Rm. Temp.[30]	......	......	......
Hafnium Boride	......	10[31]	3[3]	......	......
Vanadium Boride	16 @ 20°C[34]	......	......	......	8–9 mohs[6]
Niobium Boride	......	6.45 @ Rm. Temp.[36]	......	......	......
	0.040 @ 20°C[36]	32.0 @ Rm. Temp.[36]	......	......	>.8 mohs[36]
Tantalum Boride	......	100 @ Rm. Temp.[38]	3[3]	3[3]	......
	0.026 @ 20°C[6]	68 @ Rm. Temp.[38]	......	......	......
Chromium Boride	......	21 @ Rm. Temp.[38]	......	......	1800 kg/mm²[29]
Molybdenum Boride	......	α-MoB = 45 @ Rm. Temp.[40]	3[3]	3[3]	8 mohs[20]
		β-MoB = 25 @ Rm. Temp.			1570 kg/mm²[29]
	......	45 @ Rm. Temp.[40]	......	......	1280 kg/mm²[29]
	......	40 @ Rm. Temp.[40]	......	......	8–9 mohs[20]
Tungsten Boride	......	......	3[3]	3[3]	9 mohs[6]
Thorium Boride	......	......	......	......	......
Uranium Boride	......	......	......	......	......

Carbides

Name	Formula	Molecular Weight	Melting Point (°C)	Crystalline Form	Lattice Parameter Å	X-Ray Density gm/cm³
Titanium Carbide	TiC	59.91	3160[4]	Cubic NaCl-type (B1)	a = 4.32[6]	4.938
Zirconium Carbide	ZrC	103.23	3030[4]	Cubic NaCl-type (B1)	a = 4.689[5]	6.44
Hafnium Carbide	HfC	190.51	3890[9]	Cubic NaCl-type (B1)	a = 4.46[10]	12.7
Vanadium Carbide	VC	62.96	2830[1]	Cubic NaCl-type (B1)	a = 4.16[6]	5.8
Niobium Carbide	NbC	104.92	3500[9]	Cubic NaCl-type (B1)	a = 4.461	7.85
Tantalum Carbide	TaC	192.96	3880[9], 4730[4]	Cubic NaCl-type (B1)	a = 4.455[14]	14.5
Chromium Carbide	Cr_3C_2	180.05	1895[15]	Orthorhombic ($D5_{10}$)	a = 2.82, b = 5.53 c = 11.47[16]	6.7
	Cr_7C_3	400.01	1780[15]	Hexagonal [$C_{3v}{}^{4}$]	a = 14.01 c = 4.532[17]	6.92
Molybdenum Carbide	Mo_2C	203.91	2690[6]	Hexagonal (L′3)	a = 3.002 c = 4.724[18]	9.2
	MoC	107.96	2695[9]	Face-centered cubic	a = 4.28[19]	
Tungsten Carbide	W_2C	379.73	2730[21]	Hexagonal (L′3)	a = 2.98[6] c = 4.71	17.34
	WC	195.87	2630[21] (decomposes)	Hexagonal (L′3)	a = 2.900[22] c = 2.831	15.77
Thorium Carbide	ThC	244.06	2625[23]	Cubic NaCl-type (B1)	a = 5.34[9]	10.67
	ThC_2	256.07	2655[29]	Monoclinic [C 2/c]	a = 6.53[24] b = 4.24	9.6
Uranium Carbide	UC	250.08	2450–2500[25]	Cubic NaCl-type (B1)	a = 4.955[26]	13.6
	UC_2	262.09	2350–2400[25]	Body-centered tetragonal $CaCl_2$-type	a = 3.517[27] c = 5.987	11.68
Graphite	C	12.01	3700 ± 100[3]	Hexagonal	ortho-hexagonal axes a_0 = 2.46 b_0 = 4.28 c_0 = 6.71	2.25

REFRACTORY MATERIALS (Continued)

Carbides (Continued)

Name	Thermal Conductivity (cal-sec^{-1}-cm^{-2}-cm-°C^{-1})	Electrical Resistivity (Microhm-cm)	Ductility[a] Relative Scale	Resistance[b] to Oxidation	Hardness[c]
Titanium Carbide	0.049 @ Rm. Temp.[6]	180–250[4]	3[3]	3[4]	3200 kg/mm^2[8]
Zirconium Carbide	0.049 @ Rm. Temp.[6]	70 @ Rm. Temp.[4]	3[3]	3[3]	2600 kg/mm^2[6]
Hafnium Carbide	....	109 @ Rm. Temp.[1]	3[3]	3[3]	....
Vanadium Carbide	....	150[6]	3[3]	....	2800 kg/mm^2[11]
					2100 kg/mm^2[12]
Niobium Carbide	0.034 @ Rm. Temp.[7]	74[6]	2–3[3]	3[3]	2470 kg/mm^2[13]
Tantalum Carbide	0.053 @ Rm. Temp.[6]	30[6]	2–3[3]	3[3]	1800 kg/mm^2[8]
Chromium Carbide	....	....	3[3]	3[3]	1300 kg/mm^2[8]
	....	....	3[3]	3[3]	....
Molybdenum Carbide	....	97[6]	3[3]	5[3]	1800 kg/mm^2[6]
	....	....	3[3]	5[3]	>.7 mohs[20]
Tungsten Carbide	....	80[6]	3[3]	5[3]	3000 kg/mm^2[6]
	....	53[6]	3[3]	5[3]	2400 kg/mm^2[6]
Thorium Carbide	....	....	3[3]	3[3]	....
Uranium Carbide	....	....	3[3]	....	....
	....	....	....	5[3]	....
Graphite	0.268–0.451 @ Rm. Temp.	65 @ Rm. Temp.[71]	3[3]	4[3]	....

Nitrides

Name	Formula	Molecular Weight	Melting Point (°C)	Crystalline Form	Lattice Parameter Å	X-Ray Density gm/cm^3
Titanium Nitride	TiN	61.91	2930-N_2 liberated on melting[46]	Cubic NaCl-type (B1)	a = 4.23[53]	5.43
Zirconium Nitride	ZrN	105.22	2980[47]	Cubic NaCl-type (B1)	a = 4.567[53]	7.349
Hafnium Nitride	HfN	192.60	3310[6]	....	....	......
Vanadium Nitride	VN	64.96	2050[46]	Cubic NaCl-type (B1)	a = 4.129[6]	6.102
Niobium Nitride	NbN	106.92	2050[46]	Cubic NaCl-type (B1)	a = 4.41 to 4.375[6]	7.28
Tantalum Nitride	Ta_2N	375.77	3090[46]	Hexagonal	a = 3.05[48] c = 4.95[53]	14.1
Chromium Nitride	CrN	66.02	1500 (decomposes)	Cubic NaCl-type (B1)	a = 4.140[6]	6.14
Thorium Nitride	ThN	246.13	2630[48]	Cubic NaCl-type (B1)	a = 5.2[48]	......
Uranium Nitride	UN	....	2650	Cubic NaCl-type (B1)	a = 4.880[58]	14.32

Nitrides

Name	Thermal Conductivity (cal-sec^{-1}-cm^{-2}-cm-°C^{-1})	Electrical Resistivity (Microhm-cm)	Ductility[a] Relative Scale	Resistance[b] to Oxidation	Hardness[c]
Titanium Nitride	....	21.7 @ Rm. Temp.	3[3]	3[3]	Between 9 & 10 mohs[46]
Zirconium Nitride	....	13.6 @ Rm. Temp.[46]	3[3]	3[3]	+8 mohs[6]
Hafnium Nitride	....	....	....	....	....
Vanadium Nitride	....	200 @ Rm. Temp.[46]	3[3]	....	....
Niobium Nitride	....	200 @ Rm. Temp.[46]	5[3]	3[3]	+8 mohs[46]
Tantalum Nitride	....	135 @ Rm Temp[47]	3[3]	5[3]	+8
Chromium Nitride	....	....	....	....	....
Thorium Nitride	....	....	....	....	....
Uranium Nitride	....	....	....	....	....

Oxides

Name	Formula	Molecular Weight	Melting Point (°C)	Crystalline Form	Lattice Parameter Å	X-Ray Density gm/cm^3
Aluminum Oxide	Al_2O_3	101.92	2015[1]	Rhombohedral	a = 5.13 axial angle = 55° 6′	3.965
Beryllium Oxide	BeO	25.02	2550[1]	Hexagonal	a = 2.70 c = 4.39	3.03
Cerium Oxide	CeO_2	172.3	2600[1]	Face-centered cubic	a = 5.41	7.13
Chromic Oxide	Cr_2O_3	152.02	2265[1]	Rhombohedral	a = 5.38 axial angle = 54° 50′	5.21
Hafnium Oxide	HfO_2	210.6	2777[1]	Face-centered cubic	a = 5.11	9.68
Magnesium Oxide	MgO	40.32	2800[1]	Face-centered cubic	a = 4.20	3.58
Silicon Oxide	SiO_2	60.06	1728[1]	Hexagonal	a = 4.90 c = 5.39	2.32 (low cristobalite)
Thorium Oxide	ThO_2	264.12	3300[1]	Face-centered cubic	a = 5.59	9.69
Titanium Oxide	TiO_2	79.90	1840[1]	Tetragonal	a = 4.58 c = 2.98	4.24
Uranium Oxide	UO_2	270.07	2280[1]	Face-centered cubic	a = 5.47	10.96
Zirconium Oxide	ZrO_2	123.22	2677[1]	Monoclinic	a = 5.21 c = 5.37	5.56

REFRACTORY MATERIALS (Continued) Oxides (Continued)

Name	Thermal Conductivity (cal-sec^{-1}-cm^{-2}-cm-°C^{-1})	Electrical Resistivity (Microhm-cm)	Ductility[a] Relative Scale	Resistance[b] to Oxidation	Hardness[c]
Aluminum Oxide	.0723 @ 100°C[2]	1×10^{22} @ 14°C[1] 3×10^{19} @ 300°C[1] 3.5×10^{14} @ 800°C[1]	3[3]	1[3]	9 mohs[1]
Beryllium Oxide	.525 @ 100°C[2]	4×10^{14} @ 600°C[1] 5×10^{12} @ 1100°C[1] 8×10^{8} @ 2100°C[1]	....	....	9 mohs[1]
Cerium Oxide	....	6.5×10^{10} @ 800°C[1] 3.4×10^{8} @ 1200°C[1]	....	....	6 mohs[1]
Chromium Oxide	....	1.3×10^{9} @ 350°C[1] 2.3×10^{7} @ 1200°C[1] 6.8×10^{3} @ 600°C[1] 4.5×10^{3} @ 1100°C[1]	3[3]	1[3]	....
Hafnium Oxide	....	5×10^{15} @ 400°C[1] 1×10^{3} @ 1500°C[1]	....	....	....
Magnesium Oxide	.0860 @ 100°C[2]	2×10^{14} @ 850°C[1] 3×10^{13} @ 980°C[1] 4.5×10^{8} @ 2100°C[1]	....	....	6 mohs[1]
Silicon Oxide	....	1×10^{21} @ 20°C[1] 7×10^{12} @ 600°C[1] 4×10^{5} @ 1300°C[1] (vitreous)	3[4] (vitreous)	1[4] (vitreous)	6–7 mohs[1] (cristobalite)
Thorium Oxide	.0245 @ 100°C[2]	2.6×10^{13} @ 550[1] 8×10^{11} @ 800°C[1] 1.5×10^{10} @ 1200°C[1]	....	....	6.5 mohs[1]
Titanium Oxide	.0156 @ 100°C[2]	1.2×10^{10} @ 800°C[1] 8.5×10^{6} @ 1200°C[1]	....	....	5.5–6.0 mohs[1]
Uranium Oxide	.0234 @ 100°C[2]	3.8×10^{10} @ 20°C[1] 5×10^{8} @ 500°C[1]	....	....	....
Zirconium Oxide	.00466 @ 100°C[2]	1×10^{12} @ 385°C[1] 2.2×10^{10} @ 700°C[1] 3.6×10^{a} @ 1200°C[1]	3[3]	1[3]	6.5 mohs[1]

Silicides

Name	Formula	Molecular Weight	Melting Point (°C)	Crystalline Form	Lattice Parameter Å	X-Ray Density gm/cm^3
Titanium Silicide	$TiSi_2$	104.02	1540[49]	Orthorhombic	a = 8.24 b = 4.79 c = 8.52[59]	4.13
	Ti_5Si_3	323.68	2120[49]	Hexagonal	a = 7.465 c = 5.162[60]	4.32
Vanadium Silicide	VSi_2	107.07	1750[50]	Hexagonal $CrSi_2$-type	a = 4.562 c = 6.359[51]	4.71
Niobium Silicide	$NbSi_2$	....	1950[50]	Hexagonal $CrSi_2$-type	a = 4.785 c = 6.576[51]	5.29
Tantalum Silicide	$TaSi_2$	237.00	2400[50]	Hexagonal $CrSi_2$-type	a = 4.773 c = 6.552[51]	8.83
Chromium Silicide	$CrSi_2$	108.13	1570[51]	Hexagonal	a = 4.42 c = 6.35[51]	......
	CrSi	80.07	1870 decomposes in presence of C	Tetragonal (C 11b)	a = 3.20 c = 7.86[6]	......
Molybdenum Silicide	$MoSi_2$	152.07	1870 decomposes in presence of C	Tetragonal (C 11b)	a = 3.20 c = 7.86[64][51]	6.24
Tungsten Silicide	WSi_2	240.04	2050[50]	Tetragonal $MoSi_2$ Structure (C 11b)	a = 3.21 c = 7.83[66]	9.3
Uranium Silicide	βUSi_2	294.19	1700[6]	Hexagonal	a = 3.85[6] c = 4.06	9.25
	U_3Si_2	770.33	1665[6]	Tetragonal	a = 7.3151[6] c = 3.8925	12.20
Boron Carbide	B_4C	55.29	2450	Hexagonal	a_0 = 5.60[6] c_0 = 12.12	2.52
Boron Nitride	BN	24.83	2730[6]	Hexagonal (B 12)	a = 2.51 ± .02 c = 6.70 ± .04[52]	2.25
Silicon Carbide (β)	SiC	40.07	Trans. to α at 2100	Face-centered cubic	a_0 = 4.3590[6]	3.22
(α)	SiC	....	2700[6]	Hexagonal (Wurtzite)	a_0 = 3.081[6] c_0 = 5.0394	......

Name	Thermal Conductivity (cal-sec^{-1}-cm^{-2}-cm-°C^{-1})	Electrical Resistivity (Microhm-cm)	Ductility[a] Relative Scale	Resistance[b] to Oxidation	Hardness[c]
Titanium Silicide	....	123 (hot pressed)[1]	3[3]	4[3]	870 kg/mm^2[70]
	....	....	3[3]	4[3]	986 kg/mm^2[49]
Vanadium Silicide	....	9.5[68]	....	....	1090 kg/mm^2[50]
Niobium Silicide	....	6.3[68]	2[3]	4[3]	1050 kg/mm^2[50]
Tantalum Silicide	....	8.5[68]	2[3]	3[3]	1560 kg/mm^2[50]
Chromium Silicide	....	....	....	....	1150 kg/mm^2[50]
Molybdenum Silicide	0.075 @ Rm Temp. to 200°C[6]	21.5 @ Rm. Temp. 18.9 @ −80°C[69]	2[3]	1[3]	1290 kg/mm^2[50]
Tungsten Silicide	....	33.4[68]	....	1–2[3]	1310 kg/mm^2[68] 1090 kg/mm^2[68]
Uranium Silicide	....	....	....	....	....
Boron Carbide	0.05 @ 20–425°C[1]	0.30–0.80[1]	3[3]	3[3]	9.3 mohs
Boron Nitride	....	1900 @ 2000°C[1]	3[3]	2[3]	2.0 mohs
Silicon Carbide (β)	0.10 @ 20–425°C[1]	107–200 ohm cm[1] @ Rm. Temp.	3[3]	2[3]	9.2 mohs
(α)	....	....	....	....	....

a. *Ductility:* (1) Capable of being severely drawn, rolled, or otherwise worked without failure. (2) Capable of withstanding slight deformation, or consisting of individually ductile crystals fragily bound together. (3) Incapable of being worked; of glass-like brittleness.

b. *Oxidation resistance:* Classed according to the temperature range in which the rate of attack by air would cause severe erosion or failure of the coated specimen within a few hours: (1) above 1700°C; (2) 1400–1700°C; (3) 1100–1400°C; (4) 800–1100°C; and (5) 500–800°C.

c. *Microhardness* values taken with 100 gram load.

d. () Figures in parentheses refer to references at end of this table.

References

1. I. E. Campbell, High Temperature Technology, Electrochemical Society, John Wiley & Sons, (1956).
2. W. D. Kingery; J. Franch; R. L. Coble; and T. Vasilos; "Thermal Conductivity: X, Data for Several Pure Oxide Materials Corrected to Zero Porosity," Journal American Ceramic Society, 37, (2) 107–110 (1954).
3. C. F. Powel; I. E. Campbell; and B. W. Gonser; "Vapor Plating," John Wiley & Sons, (1955).
4. E. Friederich and L. Sittig, Z. anorg. allg. Chem., 144, 169 (1925).
5. J. T. Norton and A. L. Morory, Trans. AIME 185, 133 (1949).
6. P. Schwarzkopf and R. Kieffer, "Refractory Hard Metals," The Macmillan Co., New York, (1953).
7. P. Schwarzkopf and S. J. Sindeband, Electrochemical Society, 97th Meeting, Cleveland, Ohio (1950).
8. R. Kieffer and F. Kölbl, Powder Met. Bull. 4, 4 (1949).
9. C. Agte and H. Alterthum, Z. techn. Physik 11, 182 (1930).
10. K. Becker and F. Ebert, Z. Physik 31, 268 (1925).
11. O. Ruff and W. Martin, Z. angew. chem. 25, 49 (1912).
12. J. Hinnuber, Z. VDI 92, 111 (1950).
13. L. S. Foster; L. W. Forbes, Jr.; L. B. Friar; L. S. Moody; and W. H. Smith; Journal American Ceramic Society, 33, 27 (1950).
14. F. H. Ellinger, Trans. ASM 31, 89 (1943).
15. D. S. Bloom and N. J. Grant, Trans. AIME 188, 41 (1950).
16. K. Hellstrom and A. Westgren, Svensk kem Tidskr 45, 141 (1933).
17. A. Westgren, Jernkont, Ann. 119, 231 (1935).
18. K. Kuo and G. Hägg, Nature 170, 245 (1952).
19. H. Nowotny and R. Kieffer, Z. anorg. allg. Chem. 267, 261 (1952).
20. G. Weiss, Ann. chem. 1, 446 (1946).
21. L. Brewer; L. A. Bromely; P. W. Gilles; and N. L. Lofgren; "The Chemistry and Metallurgy of Miscellaneous Materials: Thermodynamics," McGraw-Hill Book Co., Inc. (1950).
22. K. Becker, Z. Physik 51, 481 (1928).
23. H. A. Wilhelm and P. Chiotti, Trans. ASM 42, 1295 (1950).
24. E. B. Hunt and R. E. Rundle, Journal American Chem. Soc. 73, 4777 (1951).
25. W. Mallet; A. F. Gerds; and H. R. Nelson: J. Electrochem. Soc. 99, 197 (1952).
26. L. M. Litz; A. B. Gurrett; and F. C. Croxton; J. Am. Chem. Soc. 70, 1718 (1948).
27. R. E. Rundle; N. C. Baenziger; A. S. Wilson; and R. A. McDonald; J. Am. Chem. Soc. 70, 99 (1948).
28. J. T. Norton; H. Blumenthal; and S. J. Sindeband; Trans. AIME 185, 749 (1949).
29. E. R. Honak, Thesis, Tech, Hochsch. Graz (1951).
30. F. W. Glaser and B. Post, J. Metals (1953).
31. K. Moers, Z. anorg. allg. Chem. 198, 262 (1931).
32. L. Andrieux, Rev. mét. 45, 49 (1948).
33. B. Post and F. W. Glaser, J. Metals 4, 631 (1952).
34. K. Moers, Z. anorg. allg. Chem. 198, 243 (1931).
35. L. H. Anderson and R. Kiessling, Acta Chem. Scand. 4, 160, 209 (1950).
36. F. W. Glaser, J. Metals 4, 391 (1952).
37. R. Kiessling, Acta Chem. Scand. 3, 603 (1949).
38. K. Moers, Z. anorg. u. allgem. Chem. 198, 262–275 (1931).
39. R. Kiessling, Acta Chem. Scand. 3, 595 (1949).
40. R. Steintz, J. Metals 4, 148 (1952).
41. R. Kiessling, Acta Chem. Scand. 1, 893 (1947).
42. F. Bertaut and P. Blum, Acta Cryst. 4, 72 (1951).
43. A. Zalkin and D. H. Templeton, J. Chem. Physics 18, 391 (1950).
44. F. Bertaut and P. Blum, Compt. rend. 229, 666 (1949).
45. H. N. Baumann, Jr., 99th Meeting of Electrochem. Soc., Washington, D. C. April (1951).
46. E. Friederich and L. Sittig, Z. anorg. allg. Chem. 143, 293 (1925).
47. C. Agte and K. Moers, Z. anorg. Chem. 198, 233 (1931).
48. P. Chiotti, J. Am. Ceramic Soc. 35, 123 (1952).
49. M. Hansen; H. D. Klasler; and D. J. McPherson; Trans. ASM 44, 518 (1952).
50. E. Cerwenka, Thesis, Tech. Hochsch. Graz (1951).
51. H. J. Wallbaum, Z. Metallkunde 33, 378 (1941).
52. R. S. Pease, Acta Cryst, 5, 356 (1952).
53. A. E. van Arkel, Physica 4, 296 (1924).
54. F. H. Horn, and W. T. Ziegler, J. Am. Chem. Soc. 69, 2762 (1947).
55. L. Pauling; D. H. Killeffer; and A. Linz; Molybdenum Compounds, Interscience Publishers Co., Inc., N. Y. (1952).
56. G. Hägg, Z. phys. Chem. (B) 7, 339 (1930).
57. R. Kiessling and Y. H. Lier, J. Metals 3, 639 (1951).
58. R. E. Rundle; N. C. Baenziger; A. S. Wilson; and R. A. McDonald; J. Am. Chem. Soc. 70, 99 (1941).
59. F. Laues and H. J. Wallbaum, Z. Krist. (A) 101, 78 (1939).
60. P. Pietrokowsky and P. Duwez, J. Metals 3, 772 (1951).
61. S. V. Naray Szako, Z. Krist. A97, 223 (1937).
62. C. E. Lundin; D. J. McPherson; and M. Hansen; Am. Soc. Metals, 34th Ann. Convention, Preprint No. 41, (1952).
63. H. J. Wallbaum, Z. Metallkunde 31, 362 (1939).
64. W. H. Zachariasen, Z. Physik. Chem. 128, 39 (1927).
65. D. H. Templeton and C. H. Dauben, Acta Cryst. 3, 261 (1950).
66. H. Nowotny; R. Kieffer; and H. Schachner; Mh. Chem. 83, 1243 (1952).
67. G. Brauer and A. Mitices, Z. anorg. allg. Chem. 249, 325 (1942).
68. E. Gallistl, Thesis, Tech. Hochsch. Graz (1951).
69. F. W. Glaser, J. Appl. Physics 22, 103 (1951).
70. E. Cerwenka, Thesis, Tech. Hochsch. Graz (1951).
71. L. M. Currie; V. C. Hamister; and H. G. MacPherson; A Paper presented at the United Nations International Conference on The Peaceful Uses of Atomic Energy, Geneva, Switzerland (1955).

THERMODYNAMIC PROPERTIES OF ELEMENTS AND OXIDES

Thermodynamic calculations over a wide range of temperatures are generally made with the aid of algebraic equations representing the characteristic properties of the substances being considered. The necessary integrations and differentiations, or other mathematical manipulations, are then most easily effected.

The most convenient starting point in making such calculations for a given substance is the heat capacity at constant pressure. From this quantity and a knowledge of the properties of any phase transitions, the other thermodynamic properties may be computed by the well-known equations given in standard texts on thermodynamics.

Empirical heat capacity equations are generally of the form of a power series with the absolute temperature T as the independent variable:

$$C_p = a' + (b' \times 10^{-3})T + (c' \times 10^{-6})T_2$$

or

$$C_p = a'' + (b'' \times 10^{-3})T + \frac{d \times 10^5}{T^2}.$$

Since both forms are used in the ensuing, let

(1) $$C_p = a + (b \times 10^{-3})T + (c \times 10^{-6})T^2 + \frac{d \times 10^5}{T^2}.$$

The constants a, b, c, and d are to be determined either experimentally or by some theoretical or semi-empirical approach.

The heat content or enthalpy H is determined from the heat capacity by a simple integration over the range of temperatures for which (1) is applicable. Thus, if 298° K is taken as a reference temperature,

(2) $$H_T - H_{298} = \int_{298}^{T} C_p dT$$

$$= a(T - 298) + \tfrac{1}{2}(b \times 10^{-3})(T_2 - 298^2) + \frac{1}{3}(c \times 10^{-6})(T^3 - 298^3) - (d \times 10^5)\left(\frac{1}{T} - \frac{1}{298}\right)$$

$$= aT + \tfrac{1}{2}(b \times 10^{-3})T^2 + \frac{1}{3}(c \times 10^{-6})T^3 - \frac{d \times 10^5}{T} - A,$$

where all the constants on the right hand side of the equation have been incorporated in the term $-A$.

In general, the enthalpy is given by a sum of terms such as (2) for each phase of the substance involved in the temperature range considered plus terms which represent the heats of transitions:

$$H_T - H_{298} = \sum \int_{T_1}^{T_2} C_p dT + \sum \Delta H_{tr}.$$

In a similar manner, the entropy S is obtained from (1) by performing the integration

(3) $$S_T - S_{298} = \int_{298}^{T} (C_p/T)dt$$

$$= a\ln(T/298) + (b \times 10^{-3})(T - 298) + \tfrac{1}{2}(c \times 10^{-6})(T^2 - 298^2) - \tfrac{1}{2}(d \times 10^5)\left(\frac{1}{T^2} - \frac{1}{298^2}\right)$$

$$= a\ln T + (b \times 10^{-3})T + \tfrac{1}{2}(c \times 10^{-6})T^2 \frac{-\tfrac{1}{2}(d \times 10^5)}{T^2} - B'$$

or

(4) $$S_T = 2.303\, a \log T + (b \times 10^{-3})T + \tfrac{1}{2}(c \times 10^{-6})T^2 - \frac{\tfrac{1}{2}(d \times 10^5)}{T^2} - B$$

where

(5) $$B = B' - S_{298}.$$

From the definition of free energy F:

$$F = H - TS$$

the quantity

$$F_T - H_{298} = (H_T - H_{298}) - TS_T$$

is obtained from (2) and (4):

(6) $$F_T - H_{298} = -2.303aT \log T - \tfrac{1}{2}(b \times 10^3)T^2 - \frac{1}{6}(c \times 10^{-6})T^3 - \frac{\tfrac{1}{2}(d \times 10^5)}{T} + (B+a)T + A$$

and also the free energy function

(7) $$\frac{F_T - H_{298}}{T} = -2.303a \log T - \tfrac{1}{2}(b \times 10^3)T - \frac{1}{6}(c \times 10^{-6})T^2 - \frac{\tfrac{1}{2}(d \times 10^5)}{T^2} + (B+a) - \frac{A}{T}$$

In the following two tables there has been collected values of the constants. The first column lists the element or the oxide. The second column gives the phase to which they are applicable. The third, fourth, and fifth columns specify the thermodynamic properties for the transition to the succeding phase. In column 6, the value of the entropy at 298.15° K, the reference temperature, is given. The remaining columns, except for the last, give the values of the constants a, b, c, d, A, and B required in the thermodynamic equations.

All values throughout the table which represent estimates have been enclosed in parentheses.

The heat capacities at temperatures beyond the range of experimental determination were estimated by extrapolation. Where no experimental values were found, use of analogy with compounds of neighboring elements in the Periodic Table was employed.

THERMODYNAMIC PROPERTIES OF THE ELEMENTS

Ref.: U. S. Atomic Energy Commission Report ANL-5750.

Element	Phase	Temperature of Transition (°K)	Heat of Transition (kcal/g atom)	Entropy of Transition (e.u.)	Entropy at 298°K (e.u.)	Element	a	b	c	d	A (kcal/g atom)	B (e.u.)
Ac	solid	(1090)	(2.5)	(2.3)	(13)	Ac	(5.4)	(3.0)	—	—	(1.743)	(18.7)
	liquid	(2750)	(70)	(25)	—		(8)	—	—	—	(0.295)	(31.3)
Ag	solid	1234	2.855	2.313	10.20	Ag	5.09	1.02	—	0.36	1.488	19.21
	liquid	2485	60.72	24.43	—		7.30	—	—	—	0.164	30.12
	gas	—	—	—	—		(4.97)	—	—	—	(−66.34)	(−12.52)
Al	solid	931.7	2.57	2.76	6.769	Al	4.94	2.96	—	—	1.604	22.26
	liquid	2600	67.9	26	—		7.0	—	—	—	0.33	30.83
Am	solid	(1200)	(2.4)	(2.0)	(13)	Am	(4.9)	(4.4)	—	—	(1.657)	(16.2)
	liquid	2733	51.7	18.9	—		(8.5)	—	—	—	(0.409)	(34.5)
As	solid	883	31/4	35.1/4	8.4	As	5.17	2.34	—	—	1.646	21.8
Au	solid	1336.16	3.03	2.27	11.32	Au	6.14	−0.175	0.92	—	1.831	23.65
	liquid	2933	74.21	25.30	—		7.00	—	—	—	−0.631	26.99
B	solid	2313	(3.8)	(1.6)	1.42	B	1.54	4.40	—	—	0.655	8.67
	liquid	2800	75	27	—		(6.0)	—	—	—	(−4.599)	(31.4)
Ba	solid, α	648	0.14	0.22	16	Ba	5.55	1.50	—	—	1.722	16.1
	solid, β	977	1.83	1.87	—		5.55	1.50	—	—	1.582	15.9
	liquid	1911	35.665	18.63	—		(7.4)	—	—	—	(0.843)	(25.3)
	gas	—	—	—	—		(4.97)	—	—	—	(−39.65)	(−11.7)
Be	solid	1556	2.335	1.501	2.28	Be	5.07	1.21	—	−1.15	1.951	27.62
	liquid	—	—	—	—		5.27	—	—	—	−1.611	25.68
Bi	solid	544.2	2.63	4.83	13.6	Bi	5.38	2.60	—	—	1.720	17.8
	liquid	1900	41.1	21.6	—		7.60	—	—	—	−0.087	25.6
	gas	—	—	—	—		(4.97)	—	—	—	(−46.19)	(−15.9)
C	solid	—	—	—	1.3609	C	4.10	1.02	—	−2.10	1.972	23.484
Ca	solid, α	723	0.24	0.33	9.95	Ca	5.24	3.50	—	—	1.718	20.95
	solid, β	1123	2.2	1.96	—		6.29	1.40	—	—	1.689	26.01
	liquid	1755	38.6	22.0	—		7.4	—	—	—	−0.147	30.28
	gas	—	—	—	—		(4.97)	—	—	—	(−43.015)	(−9.88)
Cd	solid	594.1	1.46	2.46	12.3	Cd	5.31	2.94	—	—	1.714	18.8
	liquid	1040	23.86	22.94	—		7.10	—	—	—	0.798	26.1
	gas	—	—	—	—		(4.97)	—	—	—	(−25.28)	(−11.7)
Ce	solid	1048	2.1	2.0	13.8	Ce	4.40	6.0	—	—	1.579	13.1
	liquid	2800	73	26	—		(7.9)	—	—	—	(−0.148)	(29.1)
Cl_2	gas	—	—	—	53.286	Cl_2	8.76	0.27	—	−0.65	2.845	−2.929
Co	solid, α	723	0.005	0.007	6.8	Co	4.72	4.30	—	—	1.598	21.4
	solid, β	1398	0.095	0.068	—		3.30	5.86	—	—	0.974	13.1
	solid, γ	1766	3.7	2.1	—		9.60	—	—	—	3.961	50.5
	liquid	3370	93	28	—		8.30	—	—	—	−2.034	38.7
Cr	solid	2173	3.5	1.6	5.68	Cr	5.35	2.36	—	−0.44	1.848	25.75
	liquid	2495	72.97	29.25	—		9.40	—	—	—	1.556	50.13
	gas	—	—	—	—		(4.97)	—	—	—	(−82.47)	(−13.8)
Cs	solid	301.9	0.50	1.7	19.8	Cs	7.42	—	—	—	2.212	22.5
	liquid	963	16.32	17.0	—		8.00	—	—	—	1.887	24.1
	gas	—	—	—	—		(4.97)	—	—	—	(−17.35)	(−13.6)
Cu	solid	1356.2	3.11	2.29	7.97	Cu	5.41	1.50	—	—	1.680	23.30
	liquid	2868	72.8	25.4	—		7.50	—	—	—	0.024	34.05
F_2	gas	—	—	—	48.58	F_2	8.29	0.44	—	−0.80	2.760	−0.76
Fe	solid, α	1033	0.410	0.397	6.491	Fe	3.37	7.10	—	0.43	1.176	14.59
	solid, β	1180	0.217	0.184	—		10.40	—	—	—	4.281	55.66
	solid, γ	1673	0.15	0.084	—		4.85	3.00	—	—	0.396	19.76
	solid, δ	1808	3.86	2.14	—		10.30	—	—	—	4.382	55.11
	liquid	3008	84.62	28.1	—		10.00	—	—	—	−0.021	50.73
Ga	solid	302.94	1.335	4.407	9.82	Ga	5.237	3.33	—	—	1.710	21.01
	liquid	2700	—	—	—		(6.645)	—	—	—	(0.648)	(23.64)
Ge	solid	1232	8.3	6.7	10.1	Ge	5.90	1.13	—	—	1.764	23.8
	liquid	2980	68	23	—		(7.3)	—	—	—	(−5.668)	(25.7)
H_2	gas	—	—	—	31.211	H_2	6.62	0.81	—	—	2.010	6.75
Hf	solid	(2600)	(6.0)	(2.3)	13.1	Hf	(6.00)	(0.52)	—	—	(1.812)	(21.2)
Hg	liquid	629.73	13.985	22.208	18.46	Hg	6.61	—	—	—	1.971	19.20
	gas	—	—	—	—		4.969	—	—	—	−13.048	−13.54
In	solid	430	0.775	1.80	13.88	In	5.81	2.50	—	—	1.844	19.97
	liquid	2440	53.8	22.0	—		7.50	—	—	—	1.564	27.34
	gas	—	—	—	—		(4.97)	—	—	—	(−58.42)	(−14.46)
Ir	solid	2727	6.6	2.4	8.7	Ir	5.56	1.42	—	—	1.721	23.4
K	solid	336.4	0.5575	1.657	15.2	K	1.3264	19.405	—	—	1.258	−1.86
	liquid	1052	18.88	17.95	—		8.8825	−4.565	2.9369	—	1.923	32.55
	gas	—	—	—	—		(4.97)	—	—	—	(−19.689)	(−9.46)
La	solid	1153	(2.3)	(2.0)	13.7	La	6.17	1.60	—	—	1.911	21.9
	liquid	3000	80	27	—		(7.3)	—	—	—	(−0.15)	(26.0)
Li	solid	459	0.69	1.5	6.70	Li	3.05	8.60	—	—	1.292	12.92
	liquid	1640	32.48	19.81	—		7.0	—	—	—	1.509	32.00
	gas	—	—	—	—		(4.97)	—	—	—	(−34.30)	(−2.84)
Mg	solid	923	2.2	2.4	7.77	Mg	5.33	2.45	—	−0.103	1.733	23.39
	liquid	1393	31.5	22.6	—		(8.0)	—	—	—	0.942	36.97
	gas	—	—	—	—		(4.97)	—	—	—	(−34.78)	(−7.60)
Mn	solid, α	1000	0.535	0.535	7.59	Mn	5.70	3.38	—	−0.37	1.974	26.11
	solid, β	1374	0.545	0.397	—		8.33	0.66	—	—	2.672	41.02
	solid, γ	1410	0.430	0.305	—		10.70	—	—	—	4.760	56.84
	solid, δ	1517	3.5	2.31	—		11.30	—	—	—	5.176	60.88
	liquid	2368	53.7	22.7	—		11.00	—	—	—	1.221	56.38
	gas	—	—	—	—		6.26	—	—	—	−63.704	−3.13
Mo	solid	2883	(5.8)	(2.0)	6.83	Mo	5.48	1.30	—	—	1.692	24.78
N_2	gas	—	—	—	45.767	N_2	6.76	0.606	0.13	—	2.044	−7.064
Na	solid	371	0.63	1.7	12.31	Na	5.657	3.252	0.5785	—	1.836	20.92
	liquid	1187	23.4	20.1	—		8.954	−4.577	2.540	—	1.924	36.0
	gas	—	—	—	—		(4.97)	—	—	—	(−24.40)	(−8.7)
Nb	solid	2760	(5.8)	(2.1)	8.3	Nb	5.66	0.96	—	—	1.730	24.24
Nd	solid	1297	(2.55)	(1.97)	13.9	Nd	5.61	5.34	—	—	1.910	19.7
	liquid	(2750)	(61)	(22)	—		(9.1)	—	—	—	−0.606	35.8
Ni	solid, α	626	0.092	0.15	7.137	Ni	4.06	7.04	—	—	1.523	18.095
	solid, β	1728	4.21	2.44	—		6.00	1.80	—	—	1.619	27.16
	liquid	3110	90.48	29.0	—		9.20	—	—	—	0.251	45.47

THERMODYNAMIC PROPERTIES OF THE ELEMENTS

Element	Phase	Temperature of Transition (°K)	Heat of Transition (kcal/g atom)	Entropy of Transition (e.u.)	Entropy at 298°K (e.u.)	Element	a	b	c	d	A (kcal/g atom)	B (e.u.)
Np	solid	913	(2.3)	(2.5)	(14)	Np	(5.3)	(3.4)	—	—	(1.731)	(17.9)
	liquid	(2525)	(55)	(22)	—		(9.0)	—	—	—	(1.392)	(37.5)
O_2	gas	—	—	—	49.003	O_2	8.27	0.258	—	−1.877	3.007	−0.750
Os	solid	2970	(6.4)	(2.2)	7.8	Os	5.69	0.88	—	—	1.736	24.9
P_4	solid, white	317.4	0.601	1.89	42.4	P_4	13.62	28.72	—	—	5.338	43.8
	liquid	553	11.9	21.5	—		19.23	0.51	—	−2.98	6.035	66.7
	gas	—	—	—	—		(19.5)	(−0.4)	(1.3)	—	(−6.32)	(46.1)
Pa	solid	(1825)	(4.0)	(2.2)	(13.5)	Pa	(5.2)	(4.0)	—	—	(1.728)	(17.3)
	liquid	(4500)	(115)	(26)	—		(8.0)	—	—	—	(−3.823)	(28.8)
Pb	solid	600.6	1.141	1.900	15.49	Pb	5.64	2.30	—	—	1.784	17.33
	liquid	2023	42.5	21.0	—		7.75	−0.73	—	—	1.362	27.11
	gas	—	—	—	—		(4.97)	—	—	—	(−45.25)	(−13.6)
Pd	solid	1828	4.12	2.25	8.9	Pd	5.80	1.38	—	—	1.791	24.6
	liquid	3440	89	26	—		(9.0)	—	—	—	(1.215)	(43.8)
Po	solid	525	(2.4)	(4.6)	13	Po	(5.2)	(3.2)	—	—	(1.693)	(17.6)
	liquid	(1235)	(24.6)	(19.9)	—		(9.0)	—	—	—	(0.847)	(35.2)
	gas	—	—	—	—		(4.97)	—	—	—	(−28.73)	(−13.5)
Pr	solid	1205	(2.5)	(2.1)	(13.5)	Pr	(5.0)	(4.6)	—	—	(1.705)	(16.4)
	liquid	3563	—	—	—		(8.0)	—	—	—	(−0.519)	(30.0)
Pt	solid	2042.5	5.2	2.5	10.0	Pt	5.74	1.34	—	0.10	1.737	23.0
	liquid	4100	122	29.8	—		(9.0)	—	—	—	(0.406)	(42.6)
Pu	solid	913	(2.26)	(2.48)	(13.0)	Pu	(5.2)	(3.6)	—	—	(1.710)	(17.7)
	liquid		—	—	—		(8.0)	—	—	—	(0.506)	(31.0)
Ra	solid	1233	(2.3)	(1.9)	(17)	Ra	(5.8)	(1.2)	—	—	(1.783)	(16.4)
	liquid	(1700)	(35)	(21)	—		(8.0)	—	—	—	(1.284)	(28.6)
	gas	—	—	—	—		(4.97)	—	—	—	(−38.87)	(−14.5)
Rb	solid	312.0	0.525	1.68	16.6	Rb	3.27	13.1	—	—	1.557	5.9
	liquid	952	18.11	19.0	—		7.85	—	—	—	1.814	26.5
	gas	—	—	—	—		(4.97)	—	—	—	(−19.04)	(−12.3)
Re	solid	3440	(7.9)	(2.3)	(8.89)	Re	(5.85)	(0.8)	—	—	(1.780)	(24.7)
Rh	solid	2240	(5.2)	(2.3)	7.6	Rh	5.40	2.19	—	—	1.707	23.8
	liquid	4150	127	30.7	—		(9.0)	—	—	—	(−0.923)	(44.4)
Ru	solid, α	1308	0.034	0.026	6.9	Ru	5.25	1.50	—	—	1.632	23.5
	solid, β	1473	0	0	—		7.20	—	—	—	2.867	35.5
	solid, γ	1773	0.23	0.13	—		7.20	—	—	—	2.867	35.5
	solid, δ	2700	(6.1)	(2.3)	—		7.50	—	—	—	3.169	37.6
S	solid, α	368.6	0.088	0.24	7.62	S	3.58	6.24	—	—	1.345	14.64
	solid, β	392	0.293	0.747	—		3.56	6.95	—	—	1.298	14.54
	liquid	717.76	2.5	3.5	—		5.4	5.0	—	—	1.576	24.02
$\frac{1}{2}S_2$	gas	—	—	—	—	$\frac{1}{2}S_2$	(4.25)	(0.15)	—	(−1.0)	(−2.859)	(9.57)
Sb	solid (α,β,γ	903.7	4.8	5.3	10.5	Sb	5.51	1.74	—	—	1.720	21.4
	liquid	1713	46.665	27.3	—		7.50	—	—	—	−1.992	28.1
$\frac{1}{2}Sb_2$	gas	—	—	—	—	$\frac{1}{2}Sb_2$	4.47	—	—	−0.11	−53.876	−21.7
Sc	solid	1670	(4.0)	(2.4)	(9.0)	Sc	(5.13)	(3.0)	—	—	1.663	21.1
	liquid	3000	80	27	—		(7.50)	—	—	—	(−2.563)	31.3
Se	solid	490.6	1.25	2.55	10.144	Se	3.30	8.80	—	—	1.375	11.28
	liquid	1000	14.27	14.27	—		7.0	—	—	—	0.881	27.34
Si	solid	1683	11.1	6.60	4.50	Si	5.70	1.02	—	−1.06	2.100	28.88
	liquid	2750	71	26	—		7.4	—	—	—	−7.646	33.17
Sm	solid	1623	3.7	2.3	(15)	Sm	(6.7)	(3.4)	—	—	(2.149)	(24.2)
	liquid	(2800)	(70)	(25)	—		(9.0)	—	—	—	(−2.296)	(33.4)
Sn	solid (α,β)	505.1	1.69	3.35	12.3	Sn	4.42	6.30	—	—	1.598	14.8
	liquid	2473	(55)	(22)	—		7.30	—	—	—	0.559	26.2
	gas	—	—	—	—		(4.97)	—	—	—	(−60.21)	(−14.3)
Sr	solid	1043	2.2	2.1	13.0	Sr	(5.60)	(1.37)	—	—	(1.731)	(19.3)
	liquid	1657	33.61	20.28	—		(7.7)	—	—	—	(0.976)	(30.4)
	gas	—	—	—	—		(4.97)	—	—	—	(−37.16)	(−10.2)
Ta	solid	3250	7.5	2.3	9.9	Ta	5.82	0.78	—	—	1.770	23.4
Tc	solid	(2400)	(5.5)	(2.3)	(8.0)	Tc	(5.6)	(2.0)	—	—	(1.759)	(24.5)
	liquid	(3800)	(120)	(32)	—		(11)	—	—	—	(3.459)	(59.4)
Te	solid, α	621	0.13	0.21	11.88	Te	4.58	5.25	—	—	1.599	15.78
	solid, β	723	4.28	5.92	—		4.58	5.25	—	—	1.469	15.57
	liquid	1360	11.9	8.75	—		9.0	—	—	—	−0.988	34.96
$\frac{1}{2}Te_2$	gas	—	—	—	—	$\frac{1}{2}Te_2$	4.47	—	—	−0.10	−19.048	−6.47
Th	solid	2173	(4.6)	(2.1)	12.76	Th	8.2	−0.77	2.04	—	2.591	33.64
	liquid	4500	(130)	(29)	—		(8.0)	—	—	—	(−7.602)	(26.84)
Ti	solid, α	1155	0.950	0.822	7.334	Ti	5.25	2.52	—	—	1.677	23.33
	solid, β	2000	(4.6)	(2.3)	—		7.50	—	—	—	1.645	35.46
	liquid	3550	(101)	(28)	—		(7.8)	—	—	—	(−2.355)	(35.45)
Tl	solid, α	508.3	0.082	0.16	15.4	Tl	5.26	3.46	—	—	1.722	15.6
	solid, β	576.8	1.03	1.79	—		7.30	—	—	—	2.230	26.4
	liquid	1730	38.81	22.4	—		7.50	—	—	—	1.315	25.9
	gas	—	—	—	—		(4.97)	—	—	—	(−41.88)	(−15.4)
U	solid, α	938	0.665	0.709	12.02	U	3.25	8.15	—	0.80	1.063	8.47
	solid, β	1049	1.165	1.111	—		10.28	—	—	—	3.493	48.27
	solid, γ	1405	(3.0)	(2.1)	—		9.12	—	—	—	1.110	39.09
	liquid	3800	—	—	—		(8.99)	—	—	—	(−2.073)	36.01
V	solid	2003	(4.0)	(2.0)	7.05	V	5.57	0.97	—	—	1.704	24.97
	liquid	3800	—	—	—		(8.6)	—	—	—	1.827	44.06
W	solid	3650	8.42	2.3	8.0	W	5.74	0.76	—	—	1.745	24.9
Y	solid	1750	(4.0)	(2.3)	(11)	Y	(5.6)	(2.2)	—	—	(1.767)	(21.6)
	liquid	3500	(90)	(26)	—		(7.5)	—	—	—	(−2.277)	(29.6)
Zn	solid	692.7	1.595	2.303	9.95	Zn	5.35	2.40	—	—	1.702	21.25
	liquid	1180	27.43	23.24	—		7.50	—	—	—	1.020	31.35
	gas	—	—	—	—		(4.97)	—	—	—	(−29.407)	(−9.81)
Zr	solid, α	1135	0.920	0.811	9.29	Zr	6.83	1.12	—	−0.87	2.378	30.45
	solid, β	2125	(4.9)	(2.3)	—		7.27	—	—	—	1.159	31.43
	liquid	(3900)	(100)	(26)	—		(8.0)	—	—	—	(−2.190)	(34.7)

THERMODYNAMIC PROPERTIES OF THE OXIDES

For description of headings and symbols see preceding table.
Ref.: U. S. Atomic Energy Report ANL-5750.

Oxide	Phase	Temperature of Transition (°K)	Heat of Transition kcal/mole	Entropy of Transition (e.u.)	Entropy at 298°K (e.u.)
Ac_2O_3	solid	(2250)	(20)	(8.9)	(36.5)
	liquid	—	—	—	—
Ag_2O	solid	dec. 460	—	—	29.09
Ag_2O_2	solid	dec	—	—	(20.4)
Al_2O_3	solid	2300	26	11	12.186
	liquid	dec.	—	—	—
Am_2O_3	solid	(2225)	(17)	(7.6)	(37)
	liquid	(3400)	(85)	(25)	—
AmO_2	solid	dec.	—	—	(20)
As_2O_3	solid, α	503	4.1	8.2	25.6
	solid, β	586	4.4	7.5	—
	liquid	730	7.15	9.79	—
	gas	—	—	—	—
AsO_2	solid	(1200)	(9.0)	(7.5)	(13)
	liquid	(dec.)	—	—	—
As_2O_5	solid	dec. >1100	—	—	25.2
Au_2O_3	solid	dec.	—	—	30
B_2O_3	solid	723	5.27	7.29	12.91
	liquid	2520	(55)	(22)	—
Ba_2O	solid	(880)	(5.2)	(5.9)	(23.5)
	liquid	(1040)	(20)	(19)	—
	gas	—	—	—	—
BaO	solid	2196	13.8	6.28	16.8
	liquid	3000	(62)	(21)	—
BaO_2	solid	723	(5.7)	(7.9)	(18.5)
	liquid	dec. 1110	—	—	—
BeO	solid	dec.	—	—	3.37
BiO	solid	(1175)	(3.7)	(3.1)	(15)
	liquid	(1920)	(54)	(28)	—
	gas	—	—	—	—
Bi_2O_3	solid	1090	6.8	6.2	36.2
	liquid	(dec.)	—	—	—
CO	gas	—	—	—	47.30
CO_2	gas	—	—	—	51.06
CaO	solid	2860	(18)	(6.3)	9.5
CdO	solid	dec.	—	—	13.1
Ce_2O_3	solid	1960	(20)	(10)	(33.5)
	liquid	(3500)	(80)	(23)	—
CeO_2	solid	3000	(19)	(6.3)	17.7
CoO	solid	2078	(12)	(5.8)	10.5
	liquid	(2900)	(61)	(21)	—
Co_3O_4	solid	dec., 1240	—	—	(35.5)
Cr_2O_3	solid	2538	(25)	(10)	19.4
CrO_2	solid	dec. 700	—	—	(11.5)
CrO_3	solid	460	(6.1)	(13)	(17.5)
	liquid	(1000)	(25)	(25)	—
	gas	—	—	—	—
Cs_2O	solid	763	(4.58)	(6.0)	(23)
	liquid	dec.	—	—	—
Cs_2O_2	solid	867	(5.5)	(6.3)	(40)
	liquid	dec.	—	—	—
Cs_2O_3	solid	775	(7.75)	(10)	(47)
	liquid	dec.	—	—	—
Cu_2O	solid	1503	13.4	8.92	22.44
	liquid	dec.	—	—	—
CuO	solid	1609	(8.9)	(5.5)	10.4
	liquid	dec.	—	—	—
FeO	solid	1641	7.5	4.6	12.9
	liquid	(2700)	(55)	(20)	—
Fe_3O_4	solid, α	900	(0)	(0)	35.0
	solid, β	dec.	—	—	—
Fe_2O_3	solid, α	950	0.16	0.17	21.5
	solid, β	1050	0	0	—
	solid, γ	dec.	—	—	—
Ga_2O	solid	(925)	(8.5)	(9.2)	(22.5)
	liquid	(1000)	(20)	(20)	—
	gas	—	—	—	—
Ga_2O_3	solid	2013	(22)	(11)	20.23
	liquid	(2900)	(75)	(26)	—
GeO	solid	983	(50)	(51)	(12.5)
	gas	—	—	—	—
GeO_2	solid (α,β)	1389	10.5	7.56	(12.5)
	liquid	(2625)	(61)	(23)	—
H_2O	liquid	373.16	9.770	26.18	16.716
	gas	—	—	—	—
HfO_2	solid	3063	(17)	(5.6)	14.18
Hg_2O	solid	dec.	—	—	(30)
HgO	solid	dec.	—	—	16.839
In_2O	solid	(600)	(4.5)	(7.5)	(28)
	liquid	(800)	(16)	(20)	—
	gas	—	—	—	—
InO	solid	(1325)	(4.0)	(3.0)	(14.5)
	liquid	(2000)	(60)	(30)	—
	gas	—	—	—	—
In_2O_3	solid	(2000)	(20)	(10)	30.1
	liquid	(3600)	(85)	(24)	—

Oxide	a	b	c	d	A (kcal/mole)	B (e.u.)
Ac_2O_3	(20.0)	(20.4)	—	—	(6.870)	(80.9)
	(40)	—	—	—	(−19.767)	(180.5)
Ag_2O	13.26	7.04	—	—	4.266	48.56
Ag_2O_2	(16.4)	(12.2)	—	—	(5.432)	(76.7)
Al_2O_3	26.12	4.388	—	−7.269	10.422	142.03
	(33)	—	—	—	(−11.655)	(174.1)
Am_2O_3	(20.0)	(15.6)	—	—	(6.657)	(81.6)
	(38.5)	—	—	—	(−7.796)	(181.8)
AmO_2	(14.0)	(6.8)	—	—	(4.477)	(61.8)
As_2O_3	8.37	48.6	—	—	4.656	36.6
	8.37	48.6	—	—	0.556	28.4
	(39)	—	—	—	(5.760)	(187.6)
	(21.5)	—	—	—	(−14.164)	(62.5)
AsO_2	(8.5)	(9.4)	—	—	(2.952)	(38.2)
	(21)	—	—	—	(2.184)	(108.0)
As_2O_5	(31.1)	(16.4)	—	(−5.4)	(11.813)	(159.9)
Au_2O_3	(23.5)	(4.8)	—	—	(7.220)	(105.3)
B_2O_3	8.73	25.40	—	−1.31	4.171	45.04
	30.50	—	—	—	7.822	161.59
Ba_2O	(20.0)	(2.2)	—	—	(6.061)	(91.1)
	(22)	—	—	—	(1.769)	(96.8)
	(15)	—	—	—	(−25.51)	(29.0)
BaO	12.74	1.040	—	−1.984	4.510	57.2
	(13.9)	—	—	—	(− 9.341)	(57.5)
BaO_2	(13.6)	(2.0)	—	—	(4.144)	(59.6)
	(21)	—	—	—	(3.241)	(99.0)
BeO	8.69	3.65	—	−3.13	3.803	48.99
BiO	(9.7)	(3.0)	—	—	(3.025)	(41.2)
	(14)	—	—	—	(2.306)	(64.9)
	(8.9)	—	—	—	(−61.49)	(−1.8)
Bi_2O_3	23.27	11.05	—	—	7.429	99.7
	(35.7)	—	—	—	(7.614)	(168.3)
CO	6.60	1.2	—	—	2.021	−9.34
CO_2	7.70	5.3	−0.83	—	2.490	−5.64
CaO	10.00	4.84	—	−1.08	3.559	49.5
CdO	9.65	2.08	—	—	2.970	42.5
Ce_2O_3	(23.0)	(9.0)	—	—	(7.258)	(100.2)
	(37)	—	—	—	(−2.591)	(178.5)
CeO_2	15.0	2.5	—	—	4.579	68.5
CoO	(9.8)	(2.2)	—	—	(3.020)	(46.0)
	(15.5)	—	—	—	(−1.886)	(79.2)
Co_3O_4	(29.5)	(17.0)	—	—	(9.551)	(137.6)
Cr_2O_3	28.53	2.20	—	−3.736	9.857	145.9
CrO_2	(16.1)	(3.0)	—	(−3.0)	(5.946)	(82.8)
CrO_3	(18.1)	(4.0)	—	(−2.0)	(6.245)	(87.9)
	(27)	—	—	—	(3.381)	(127.0)
	(20)	—	—	—	(−28.62)	(53.6)
Cs_2O	(16.5)	(5.4)	—	—	(5.160)	(72.6)
	(22)	—	—	—	(3.205)	(99.0)
Cs_2O_2	(21.4)	(11.4)	—	—	(6.887)	(85.3)
	(29.5)	—	—	—	(4.125)	(123.8)
Cs_2O_3	(24.0)	(22.6)	—	—	(8.160)	(96.5)
	(35)	—	—	—	(2.148)	(142.2)
Cu_2O	(13.4)	(8.6)	—	—	(4.378)	(54.9)
	(21.5)	—	—	—	(3.721)	(96.0)
CuO	14.34	6.2	—	—	4.551	61.11
	(22)	—	—	—	(−4.339)	(98.91)
FeO	9.27	4.80	—	—	(2.977)	(43.8)
	(14.5)	—	—	—	(−3.721)	(69.2)
Fe_3O_4	12.38	1.62	—	−0.38	3.826	58.3
	(14.5)	—	—	—	(−2.399)	(66.7)
Fe_2O_3	21.88	48.20	—	—	8.666	104.0
	48.00	—	—	—	12.652	238.3
	23.49	18.6	—	−3.55	9.021	119.9
Ga_2O	36.00	—	—	—	11.979	187.6
	31.71	1.8	—	—	8.467	159.7
	(13.8)	(8.6)	—	—	(4.497)	(58.7)
Ga_2O_3	(21.5)	—	—	—	(−0.559)	(94.1)
	(14)	—	—	—	(−28.06)	(22.3)
GeO	11.77	25.2	—	—	(4.630)	(54.35)
	(35.5)	—	—	—	(−20.66)	(173.2)
GeO_2	(10.4)	(2.6)	—	(−0.5)	(3.384)	(47.8)
	(8.2)	(0.4)	—	(−0.2)	(−49.67)	(−20.2)
H_2O	11.2	7.17	—	—	(3.658)	(53.5)
	(21.7)	—	—	—	(1.149)	(111.9)
HfO_2	18.03	—	—	—	5.376	86.01
Hg_2O	7.17	2.56	—	−0.08	−8.290	−3.56
HgO	17.39	2.08	—	−3.48	6.445	87.48
	(15.6)	(28)	—	—	(4.776)	(59.7)
	(8.5)	(7.0)	—	—	(2.846)	(33.7)
In_2O	(14.7)	(7.8)	—	—	(4.730)	(58.1)
	(22)	—	—	—	(3.206)	(92.6)
	(15)	—	—	—	(−18.39)	(25.8)
InO	(10.0)	(3.2)	—	—	(3.124)	(43.4)
	(14)	—	—	—	(1.615)	(64.9)
	(9.0)	—	—	—	(−68.38)	(−3.1)
In_2O_3	(22.6)	(6.0)	—	—	(7.005)	(100.5)
	(35)	—	—	—	(−0.195)	(172.8)

THERMODYNAMIC PROPERTIES OF THE OXIDES

Oxide	Phase	Temperature of Transition (°K)	Heat of Transition kcal/mole	Entropy of Transition (e.u.)	Entropy at 298°K (e.u.)	Oxide	a	b	c	d	A (kcal/mole)	B (e.u.)
Ir_2O_3	solid	(1450)	(10)	(6.8)	(26.5)	Ir_2O_3	(21.8)	(14.4)	—	—	(7.140)	(102.0)
	liquid	(2250)	(50)	(22)	—		(35)	—	—	—	(0.706)	(170.3)
	gas	—	—	—	—		(20)	(10)	—	—	(−57.73)	(54.8)
IrO_2	solid	dec. 1373	—	—	(15.9)	IrO_2	9.17	15.20	—	—	3.410	40.9
K_2O	solid	(980)	(6.8)	(6.9)	(23)	K_2O	(15.9)	(6.4)	—	—	(5.025)	(69.5)
	liquid	dec.	—	—	—		(22)	—	—	—	(1.130)	(98.3)
K_2O_2	solid	763	(7.0)	(9.2)	(27)	K_2O_2	(20.8)	(5.4)	—	—	(6.442)	(93.1)
	liquid	(1800)	(45)	(25)	—		(29)	—	—	—	(4.127)	(134.2)
	gas	—	—	—	—		(20)	—	—	—	(−57.07)	(41.7)
K_2O_3	solid	703	(6.1)	(8.7)	(33.5)	K_2O_3	(19.1)	(23.2)	—	—	(6.750)	(82.2)
	liquid	(975)	(25)	(26)	—		(35.5)	—	—	—	(6.447)	(164.7)
	gas	—	—	—	—		(20)	(5.0)	—	—	(−31.29)	(37.3)
KO_2	solid	653	(4.9)	(7.5)	27.9	KO_2	(15.0)	(12.0)	—	—	(5.006)	(61.1)
	liquid	dec.	—	—	—		(24)	—	—	—	(3.424)	(105.5)
La_2O_3	solid	2590	(18)	(7)	(36.5)	La_2O_3	28.86	3.076	—	−3.275	9.840	(130.7)
Li_2O	solid	2000	(14)	(7)	9.06	Li_2O	(11.4)	(5.4)	—	—	(3.639)	(57.5)
	liquid	2600	(56)	(22)	—		(21)	—	—	—	(−1.961)	(112.7)
Li_2O_2	solid	dec. 470	—	—	(16.5)	Li_2O_2	(17.0)	(5.4)	—	—	(5.309)	(82.0)
MgO	solid	3075	18.5	5.8	6.4	MgO	10.86	1.197	—	−2.087	3.991	57.0
MgO_2	solid	dec. 361	—	—	(20.5)	MgO_2	(12.1)	(2.4)	—	—	(3.714)	(49.2)
MnO	solid	2058	13.0	6.32	14.27	MnO	11.11	1.94	—	−0.88	3.689	50.10
	liquid	dec.	—	—	—		(13.5)	—	—	—	(−8.543)	(58.02)
Mn_3O_4	solid, α	1445	4.97	3.44	35.5	Mn_3O_4	34.64	10.82	—	−2.20	11.312	166.3
	solid, β	1863	(33)	(18)	—		50.20	—	—	—	17.376	260.4
	liquid	(2900)	(75)	(26)	—		(49)	—	—	—	(−17.86)	(233.4)
Mn_2O_3	solid	dec. 1620	—	—	26.4	Mn_2O_3	24.73	8.38	—	−3.23	8.829	118.8
MnO_2	solid	dec. 1120	—	—	12.7	MnO_2	16.60	2.44	—	−3.88	6.359	84.8
MoO_2	solid	(2200)	(16)	(7.3)	(14.5)	MoO_2	(16.2)	(3.0)	—	(−3.0)	(5.973)	(80.4)
	liquid	dec. 2250	—	—	—		(23)	—	—	—	(−2.463)	(118.4)
MoO_3	solid	1068	12.54	11.74	18.68	MoO_3	13.6	13.5	—	—	4.655	62.83
	liquid	1530	33	22	—		(28.4)	—	—	—	(0.222)	(139.88)
	gas	—	—	—	—		(18.1)	—	—	—	(−48.54)	(42.8)
N_2O	gas	—	—	—	52.58	N_2O	10.92	2.06	—	−2.04	4.032	11.40
Na_2O	solid	1193	(7.1)	(6.0)	17.4	Na_2O	15.70	5.40	—	—	4.921	73.7
	liquid	dec.	—	—	—		(22)	—	—	—	(1.494)	(105.9)
Na_2O_2	solid	dec. 919	—	—	22.6	Na_2O_2	(20.2)	(3.8)	—	—	(6.192)	(93.6)
NaO_2	solid	(825)	(6.2)	(7.5)	27.7	NaO_2	(16.2)	(3.6)	—	—	(4.990)	(65.7)
	liquid	(1300)	(28)	(22)	—		(23)	—	—	—	(3.175)	(100.9)
	gas	—	—	—	—		(15)	—	—	—	(−35.22)	(22.0)
NbO	solid	(2650)	(16)	(6.0)	(12)	NbO	(9.6)	(4.4)	—	—	(3.058)	(44.0)
NbO_2	solid	(2275)	(16)	(7.0)	(12.7)	NbO_2	(17.1)	(1.6)	—	(−2.8)	(6.109)	(84.6)
	liquid	(3800)	(85)	(22)	—		(24)	—	—	—	(1.033)	(127.2)
Nb_2O_5	solid	1733	(28)	(16)	32.8	Nb_2O_5	21.88	28.2	—	—	7.776	100.3
	liquid	(3200)	(80)	(25)	—		(44.2)	—	—	—	(−24.09)	(201.6)
Nd_2O_3	solid	2545	(22)	(8.8)	(35.3)	Nd_2O_3	28.99	5.760	—	(−4.159)	10.295	(133.9)
NiO	solid	2230	(12.1)	(5.43)	9.22	NiO	13.69	0.83	—	−2.915	5.097	70.67
	liquid	dec.	—	—	—		(14.3)	—	—	—	(−7.861)	(67.91)
NpO_2	solid	(2600)	(15)	(5.7)	19.19	NpO_2	(17.7)	(3.2)	—	(−2.6)	(6.292)	(84.08)
Np_2O_5	solid	dec. 800–900°K	—	—	(43)	Np_2O_5	(32.4)	(12.6)	—	—	(10.22)	(145.4)
OsO_2	solid	dec. 923	—	—	(14.5)	OsO_2	(11.5)	(6.0)	—	—	(3.696)	(52.8)
OsO_4	solid	313.3	3.41	10.9	34.7	OsO_4	(16.4)	(23.1)	—	(−2.4)	(6.726)	(67.0)
	liquid	403	9.45	23.4	—		(33)	—	—	—	(6.612)	(143.0)
	gas	—	—	—	—		16.46	8.60	—	−4.6	(−7.644)	(25.3)
P_2O_3	liquid	448.5	4.5	10	(34)	P_2O_3	(34.5)	—	—	—	(10.287)	(162.6)
	gas	—	—	—	—		(15)	(10)	—	—	(−1.953)	(38.0)
PO_2	solid	(350)	(2.7)	(7.7)	(11.5)	PO_2	(11.3)	(5.0)	—	—	(3.591)	(54.4)
	liquid	(dec)	—	—	—		(20)	—	—	—	(3.640)	(95.9)
P_2O_5	solid	631	8.8	13.9	33.5	P_2O_5	8.375	5.40	—	—	4.897	30.3
	gas	—	—	—	—		36.80	—	—	—	3.284	165.6
PaO_2	solid	(2560)	(20)	(7.8)	(17.8)	PaO_2	(14.4)	(2.6)	—	—	(4.409)	(65.0)
Pa_2O_5	solid	(2050)	(26)	(13)	(37.5)	Pa_2O_5	(28.4)	(11.4)	—	—	(8.975)	(127.7)
	liquid	(3350)	(95)	(28)	—		(48)	—	—	—	(−0.800)	(241.1)
PbO	solid, red	762	(0.4)	(0.5)	16.2	PbO	10.60	4.00	—	—	3.338	45.4
	solid, yellow	1159	2.8	2.4	—		9.05	6.40	—	—	2.454	36.4
	liquid	1745	51	29	—		(14.6)	—	—	—	1.788	65.7
	gas	—	—	—	—		(8.1)	(0.4)	—	—	(−59.94)	(−11.0
Pb_3O_4	solid	dec.	—	—	50.5	Pb_3O_4	(31.1)	(17.6)	—	—	(10.055)	(132.0)
PbO_2	solid	dec.	—	—	18.3	PbO_2	12.7	7.80	—	—	4.133	56.4
PdO	solid	dec. 1150	—	—	(9.1)	PdO	3.30	14.2	—	—	1.615	(13.9)
PoO_2	solid	(825)	(5.5)	(6.7)	(17)	PoO_2	(14.3)	(5.6)	—	—	(4.513)	(66.1)
	liquid	(dec.)	—	—	—		(22)	—	—	—	(3.460)	(106.5)
Pr_2O_3	solid	(2200)	(22)	(10)	(35.5)	Pr_2O_3	(29.0)	(4.0)	—	(−4.0)	(10.166)	(133.2)
	liquid	(4000)	(90)	(23)	—		(36)	—	—	—	(−6.298)	(168.3)
PrO_2	solid	dec. 700	—	—	(17)	PrO_2	(17.6)	(3.4)	—	(−2.8)	(6.338)	(85.9)
PtO	solid	dec. 780	—	—	(13.5)	PtO	(9.0)	(6.4)	—	—	(2.968)	(39.7)
Pt_3O_4	solid	(dec.)	—	—	(41)	Pt_3O_4	(30.8)	(17.4)	—	—	(9.957)	(139.7)
PtO_2	solid	723	(4.6)	(6.4)	(16.5)	PtO_2	(11.1)	(9.6)	—	—	(3.736)	(49.6)
	liquid	dec. 750	—	—	—		(21)	—	—	—	(3.785)	(101.5)
PuO	solid	(1290)	(7.2)	(5.6)	(20)	PuO	(12.0)	(2.4)	—	—	(3.685)	(49.1)
	liquid	(2325)	(47)	(20)	—		(14.5)	—	—	—	(−2.287)	(58.3)
	gas	—	—	—	—		(8.9)	—	—	—	(−62.307)	(−5.3)
Pu_2O_3	solid	(1880)	(16)	(8.5)	(38)	Pu_2O_3	(21.2)	(18.2)	—	—	(7.130)	(88.2)
	liquid	(3250)	(75)	(23)	—		(40)	—	—	—	(−5.691)	(187.2)
PuO_2	solid	(2400)	(15)	(6.2)	(19.7)	PuO_2	(17.1)	(3.4)	—	(−2.6)	(6.122)	(80.2)
	liquid	(3500)	(90)	(26)	—		(20.5)	—	—	—	(−10.62)	(92.2)
RaO	solid	(>2500)	—	—	(17)	RaO	(10.5)	(2.0)	—	—	(3.220)	(43.4)
Rb_2O	solid	(910)	(5.7)	(6.3)	(27)	Rb_2O	(15.4)	(5.8)	—	—	(4.850)	(62.5)
	liquid	dec.	—	—	—		(22)	—	—	—	(2.754)	(95.9)
Rb_2O_2	solid	843	(7.3)	(8.7)	(27.5)	Rb_2O_2	(20.9)	(8.0)	—	—	(6.587)	(94.0)
	liquid	(dec.)	—	—	—		(29)	—	—	—	(3.273)	(133.2)
Rb_2O_3	solid	762	(7.6)	(10)	(32.5)	Rb_2O_3	(20.5)	(13.0)	—	—	(6.690)	(88.2)
	liquid	dec.	—	—	—		(34)	—	—	—	(5.603)	(157.8)
RbO_2	solid	685	(4.1)	(6.0)	(21.5)	RbO_2	(13.8)	(6.4)	—	—	(4.399)	(59.0)
	liquid	dec.	—	—	—		(21)	—	—	—	(3.720)	(95.7)
ReO_2	solid	(1475)	(12)	(8.1)	(15)	ReO_2	(10.8)	(9.8)	—	—	(3.656)	(49.5)

THERMODYNAMIC PROPERTIES OF THE OXIDES

Oxide	Phase	Temperature of Transition (°K)	Heat of Transition kcal/mole	Entropy of Transition (e.u.)	Entropy at 298°K (e.u.)	Oxide	a	b	c	d	A (kcal/mole)	B (e.u.)
ReO_2	liquid	(3250)	(80)	(25)	—	ReO_2	(24.5)	—	—	—	(1.204)	(127.0)
ReO_3	solid	433	5.2	12	19.8	ReO_3	(18.0)	(5.8)	—	—	(5.625)	(84.5)
	liquid	dec.	—	—	—		29	—	—	—	(4.644)	(136.8)
Re_2O_7	solid	569	15.8	27.8	44	Re_2O_7	(41.8)	(14.8)	—	(−3.0)	(14.127)	(200.3)
	liquid	635.5	17.7	27.9	—		(65.7)	—	—	—	(9.203)	(314.7)
	gas	—	—	—	—		(38.2)	—	—	—	(−25.97)	(109.3)
ReO_4	solid	420	(4.2)	(10)	(34.5)	ReO_4	(21.4)	(10.8)	—	(−2.0)	(7.531)	(91.8)
	liquid	(460)	(9.3)	(20)	—		(33)	—	—	—	(6.775)	(146.7)
	gas	—	—	—	—		(16.5)	(8.6)	—	(−5.0)	(−8.118)	(30.6)
Rh_2O	solid	dec. 1400	—	—	(25.5)	Rh_2O	15.59	6.47	—	—	4.936	(65.3)
RhO	solid	dec. 1394	—	—	(12)	RhO	(9.84)	(5.53)	—	—	(3.179)	(45.7)
Rh_2O_3	solid	dec. 1388	—	—	(23)	Rh_2O_3	20.73	13.80	—	—	6.794	(99.2)
RuO_2	solid	dec. 1400	—	—	(12.5)	RuO_2	(11.4)	(6.0)	—	—	3.666	(54.2)
RuO_4	solid	300	(3.2)	(11)	(32.5)	RuO_4	(20)	—	—	—	(5.963)	(81.5)
	liquid	dec.	—	—	—		(33)	—	—	—	(6.663)	(144.9)
SO_2	gas	—	—	—	59.40	SO_2	11.4	1.414	—	−2.045	4.148	7.12
Sb_2O_3	solid	928	14.74	15.88	29.4	Sb_2O_3	19.10	17.1	—	—	6.455	84.5
	liquid	1698	8.92	5.25	—		(36)	—	—	—	(0.035)	(168.2)
	gas	—	—	—	—		(20.8)	—	—	—	(−34.70)	(49.9)
SbO_2	solid	dec.	—	—	15.2	SbO_2	11.30	8.1	—	—	3.725	51.6
Sb_2O_5	solid	dec.	—	—	29.9	Sb_2O_5	(22.4)	(23.6)	—	—	(7.728)	(104.8)
Sc_2O_3	solid	(2500)	(23)	(9.3)	24.8	Sc_2O_3	23.17	5.64	—	—	7.159	108.9
SeO	solid	(1375)	(7.6)	(5.5)	(11)	SeO	(9.1)	(3.8)	—	—	(2.882)	(42.0)
	liquid	(2075)	(45)	(22)	—		(15.5)	—	—	—	(0.490)	(77.5)
	gas	—	—	—	—		8.20	0.50	—	−0.80	(−58.54)	(0.7)
SeO_2	solid	603	(24.5)	(40.6)	(15)	SeO_2	(12.8)	(6.1)	—	(−0.2)	(4.150)	(59.9)
	gas	—	—	—	—		(14.5)	—	—	—	(−20.45)	(26.4)
SiO	solid	(2550)	(12)	(4.7)	(6.5)	SiO	(7.3)	(2.4)	—	—	(2.283)	(35.8)
SiO_2	solid, β	856	0.15	0.18	10.06	SiO_2	11.22	8.20	—	−2.70	4.615	57.83
	solid, α	1883	2.04	1.08	—		14.41	1.94	—	—	4.602	73.67
	liquid	dec. 2250	—	—	—		(20)	—	—	—	(9.649)	(111.08)
Sm_2O_3	solid	(2150)	(20)	(9.3)	(36.5)	Sm_2O_3	(25.9)	(7.0)	—	—	(8.033)	(113.2)
	liquid	(3800)	(80)	(21)	—		(36)	—	—	—	(−6.431)	(166.3)
SnO	solid	(1315)	(6.4)	(4.9)	13.5	SnO	9.40	3.62	—	—	2.964	41.1
	liquid	(1800)	(60)	(33)	—		(14.5)	—	—	—	(0.141)	(68.1)
	gas	—	—	—	—		(9.0)	—	—	—	(−69.76)	(−6.4
SnO_2	solid	1898	(11.39)	(5.95)	12.5	SnO_2	17.66	2.40	—	−5.16	7.103	91.7
	liquid	(3200)	(75)	(23)	—		(22.5)	—	—	—	(0.304)	(117.7)
SrO	solid	2703	16.7	6.2	13.0	SrO	12.34	1.120	—	−1.806	4.335	58.7
SrO_2	solid	dec. 488	—	—	(14.8)	SrO_2	(16.8)	(2.2)	—	(−3.0)	(6.113)	(83.3)
Ta_2O_5	solid	2150	(16)	(7.4)	34.2	Ta_2O_5	29.2	10.0	—	—	9.151	135.2
	liquid	—	—	—	—		(46)	—	—	—	(6.158)	(235.1)
TcO_2	solid	(2400)	(18)	(7.5)	(13.5)	TcO_2	(10.4)	(9.2)	—	—	(3.510)	(48.6)
	liquid	(4000)	(105)	(26)	—		(25)	—	—	—	(−5.946)	(132.7)
TcO_3	solid	(dec. <1200)	—	—	(19.5)	TcO_3	(19.4)	(5.2)	—	(−2.0)	(6.686)	(93.7)
Tc_2O_7	solid	392.7	(11)	(28)	(42.5)	Tc_2O_7	(39.1)	(18.6)	—	(−2.4)	(13.29)	(187.2)
	liquid	583.8	(14)	(24)	—		(64)	—	—	—	(10.02)	(299.8)
	gas	—	—	—	—		(25)	(28)	—	—	(−21.98)	(43.8)
TeO	solid	(1020)	(7.1)	(7.0)	(13)	TeO	(8.6)	(6.2)	—	—	(2.840)	(37.8)
	liquid	(1775)	(50)	(28)	—		(15.5)	—	—	—	(−0.448)	(72.3)
	gas	—	—	—	—		(8.9)	—	—	—	(−62.16)	(−5.2)
TeO_2	solid	1006	3.2	3.2	16.99	TeO_2	13.85	6.87	—	—	4.435	63.97
	liquid	dec.	—	—	—		(20)	—	—	—	(3.940)	(96.4)
ThO	solid	(2150)	(13)	(6.0)	(16)	ThO	(11.0)	(2.4)	—	—	(3.386)	(47.4)
	liquid	(3250)	(65)	(20)	—		(15)	—	—	—	(−6.561)	(66.9)
ThO_2	solid	3225	(18)	(5.6)	15.59	ThO_2	16.45	2.346	—	−2.124	5.721	80.03
TiO	solid, α	1264	0.82	0.65	8.31	TiO	10.57	3.60	—	−1.86	3.935	54.03
	solid, β	dec. 2010	—	—	—		11.85	3.00	—	—	4.108	61.71
Ti_2O_3	solid, α	473	0.215	0.455	18.83	Ti_2O_3	7.31	53.52	—	—	4.559	38.78
	solid, β	2400	(24)	(10)	—		34.68	1.30	—	−10.20	13.605	184.48
	liquid	3300	—	—	—		(37.5)	—	—	—	(−7.796)	(193.2)
Ti_3O_5	solid, α	450	2.24	4.98	30.92	Ti_3O_5	35.47	29.50	—	—	11.887	179.98
	solid, β	(2450)	(50)	(20)	—		41.60	8.00	—	—	10.230	202.80
	liquid	(3600)	(85)	(24)	—		(60)	—	—	—	(−18.701)	(306.4)
TiO_2	solid	2128	(16)	(7.5)	12.01	TiO_2	17.97	0.28	—	−4.35	6.829	92.92
	liquid	dec. 3200	—	—	—		(21.4)	—	—	—	(−2.610)	(111.08)
Tl_2O	solid	573	(5.0)	(8.7)	23.8	Tl_2O	(15.8)	(6.0)	—	(−0.3)	(5.078)	(68.2)
	liquid	773	(17)	(22)	—		(22.1)	—	—	—	(2.651)	(96.0)
	gas	—	—	—	—		(13.7)	—	—	—	(−20.94)	(18.0)
Tl_2O_3	solid	990	(12.4)	(13)	(33.5)	Tl_2O_3	(23.0)	(5.0)	—	—	(7.080)	(99.0)
	liquid	(dec)	—	—	—		(35.5)	—	—	—	(4.604)	(167.8)
UO	solid	(2750)	(14)	(5.1)	(16)	UO	(10.6)	(2.0)	—	—	(3.249)	(45.0)
UO_2	solid	3000	—	—	18.63	UO_2	19.20	1.62	—	−3.957	7.124	93.37
U_3O_8	solid	dec.	—	—	(66)	U_3O_8	(65)	(7.5)	—	(−10.9)	(23.37)	(312.7)
UO_3	solid	dec. 925	—	—	23.57	UO_3	22.09	2.54	—	−2.973	7.696	104.72
VO	solid	(2350)	(15)	(6.4)	9.3	VO	11.32	1.61	—	−1.26	3.869	56.4
	liquid	(3400)	(70)	(21)	—		(14.5)	—	—	—	(−8.157)	(70.9)
V_2O_3	solid	2240	(24)	(11)	23.58	V_2O_3	29.35	4.76	—	−5.42	10.780	148.12
	liquid	dec. 3300	—	—	—		(38)	—	—	—	(−6.028)	(193.4)
V_3O_5	solid	(2100)	(42)	(20)	(32)	V_3O_5	(36)	(30)	—	—	(12.07)	(182.1)
	liquid	(dec.)	—	—	—		(55.6)	—	—	—	(−54.72)	(249.1)
VO_2	solid, α	345	1.02	2.96	12.32	VO_2	14.96	—	—	—	4.460	72.92
	solid, β	1818	13.60	7.48	—		17.85	1.70	—	−3.94	5.680	89.09
	liquid	dec. 3300	—	—	—		25.50	—	—	—	2.962	135.87
V_2O_5	solid	943	15.56	16.50	313	V_2O_5	46.54	−3.90	—	−13.22	18.136	240.2
	liquid	(2325)	(63)	(27)	—		45.60	—	—	—	2.122	220.1
	gas	—	—	—	—		(40)	—	—	—	(−73.90)	(149.6)
WO_2	solid	(1543)	(11.5)	(7.45)	(15)	WO_2	(17.6)	(4.2)	—	(−4.0)	(6.772)	(88.8)
	liquid	dec. 2125	—	—	—		(24)	—	—	—	(−0.112)	(121.8)
WO_3	solid	1743	(17)	(9.8)	19.90	WO_3	17.33	7.74	—	—	5.511	81.15
	liquid	(2100)	(43)	(20)	—		(30)	—	—	—	(−1.162)	(152.5)
	gas	—	—	—	—		(18)	—	—	—	(−69.36)	(40.2)
Y_2O_3	solid	(2500)	(25)	(10)	(29.5)	Y_2O_3	(26.0)	(8.2)	—	(−2.2)	(8.846)	(122.3)
ZnO	solid	dec.	—	—	10.4	ZnO	11.71	1.22	—	−2.18	4.277	57.88
ZrO_2	solid, α	1478	1.420	0.961	12.03	ZrO_2	16.64	1.80	—	−3.36	6.168	85.21
	solid, β	2950	20.8	7.0	—		17.80	—	—	—	4.270	89.96

VALUES OF CHEMICAL THERMODYNAMIC PROPERTIES

All values of energy given in these tables are expressed, insofar as possible, in terms of the thermochemical calorie, now defined in terms of the absolute joule. 1 thermochemical calorie = 1 calorie = 4.1840 absolute joule = 4.1833 international joule. The notations used in these tables are as follows:

$\Delta Hf°$ = the standard heat of formation of a given substance from its elements at 25°C.
$\Delta Ff°$ = the standard free energy of formation of a given substance from its elements at 25°C.
$\log_{10} Kf$ = the logarithm of the equilibrium constant for the reaction for forming a given substance from its elements at 25°C.
S° = the entropy of the given substance in its thermodynamic reference state at the reference temperature at 25°C.
c = crystalline; in certain cases where a substance exists in more than one crystal-line form there is an indication as to which form is concerned.
g = gaseous.
am = amorphous.
aq = aqueous; unless otherwise indicated the aqueous solution is taken as the hypothetical ideal state of unit molality.
gls = glass.
lq = liquid.
ppt = precipitate.

The values in these tables were taken from Circular of the National Bureau of Standards 500, "Selected Values of Chemical Thermodynamic Properties, issued February 1, 1952. **

Substance	State	ΔHf°	ΔFf°	Log₁₀ Kf	S°
Aluminum					
Al	*g*	75.00	65.3	−47.86	39.303
	c	0.00	0.00	0.000	6.77
Al^{+++}	*aq*	1307.44	....	....	....
$AlBr_3$	*c*	−125.8	−120.7	88.47	44
	aq	−211.9	....	....	....
Al_4C_3	*c*	−30.9	−29.0	21.26	25
$Al(CH_3)_3$	*lq*	−26.9	....	....	....
Al_2Cl_6	*g*	−303.6	....	....	....
$AlCl_3$	*c*	−166.2	−152.2	111.56	40
	aq, 600	−245.5	....	....	....
$AlCl_3 \cdot 6H_2O$	*c*	−641.1	−542 4	397.57	90
AlF_3	*c*	−311	−294	215.5	23
$AlF_3 \cdot 3H_2O$	*c*	549.1	−490.4	359.46	50
	aq	−361 4	....	....	....
$AlF_3 \cdot \frac{1}{2}H_2O$	*c*	−357.4	−333.6	244.52	28
AlI_3	*c*	−75.2	−75.0	54.97	48
	aq	−165.8	....	....	....
AlN	*c*	−57.7	−50.1	36.72	5
$Al(NO_2)_3$	*aq*	−273.65	....	....	....
$Al(NO_3)_3 \cdot 6H_2O$	*c*	−680.65	−525.82	385.419	111.8
$Al(NO_3)_3 \cdot 9H_2O$	*c*	−897.34	−700.2	513.24	136
$Al_2 \cdot O_3(\alpha)$*	*c*	−399.09	−376.77	276.167	12.19
(γ)	*c*	−384.84	....	....	....
$Al_2O_3 \cdot H_2O$	*c*	−471	−435	318.8	23.15
$Al_2O_3 \cdot 3H_2O$†	*c*	−613.7	−547.9	401.60	33.51
$Al(OH)_3$	*am*	−304.2	....	....	....
Al_2S_3	*c*	−121.6	−117.7	86.27	23
$Al_2(SO_4)_3$	*c*	−820.98	−738.99	541.670	57.2
	aq	−897.1	....	....	....
$Al_2(SO_4)_3 \cdot 6H_2O$	*c*	−1268.14	−1105.14	810.054	112.1
$Al_2(SO_4)_3 18H_2O$	*c*	−2118.5	....	....	....
$AlNH_4(SO_4)_2$	*c*	−561.24	−485.95	356.195	51.7
	aq	−591.74	....	....	....
$AlNH_4(SO_4)_2 \cdot 12H_2O$	*c*	−1419.40	−1179.02	864.207	166.6
Ammonium					
NH_3	*g*	−11.04	−3.976	2.914	46.01
	aq	−19.32	−6.37	−4.669	26.3
	aq ∞	−19.32	....	....	....
NH_4^+	*aq*	−31.74	−19.00	13 927	26.97
NH_4OH	*aq*	−87.64	....	....	....
$NH_4H_2AsO_4$	*c*	−251.47	−197.24	144.574	41.12
	aq	247.9	....	....	....
$(NH_4)_2HAsO_4$	*c*	−280.24	....	....	....
	aq	−278.3	....	....	....
$(NH_4)_3AsO_4$	*c*	−306.11	....	....	....
	aq	−303	....	....	....
$(NH_4)_3AsO_4 \cdot 3H_2O$	*c*	−516.6	....	....	....
NH_4BO_2	*aq*	−215.2	....	....	....
NH_4BO_3	*aq*	−193.7	....	....	....
$NH_4BO_3 \cdot H_2O$	*c*	−270.8	....	....	....
$(NH_4)_2HBO_3$	*aq*	−305.5	....	....	....
NH_4Br	*c*	−64.61	....	....	....
	aq, ∞	−60.74	....	....	....
$(NH_4)_2CO_3$	*aq*	−225.11	−164.22	120.371	41.2
NH_4HCO_3	*c*	−203.7	....	....	....
	aq	−196.92	−159.31	116.772	49.7
$NH_4CO_2NH_2$	*c*	−154.21	−109.47	80.240	39.70
	aq	−150.4	....	....	....
NH_4CN	*c*	0.0	....	....	....
	aq	4.6	20.4	....	....
NH_4CNO	*c*	−74.7	....	....	....
	aq	−68.5	....	....	....
NH_4CNS	*c*	−20.0	....	....	....
	aq	−14.5	4.4	....	....
$NH_4C_2H_3O_2$	*c*	−147.8	....	....	....
	aq, 400	−148.1	....	....	....
$(NH_4)_2C_2O_4$	*c*	−268.72	....	....	....
	aq, 2100	−260.6	....	....	....
Ammonium					
$(NH_4)_2C_2O_4 \cdot H_2O$	*c*	−340.62	....	....	....
$NH_4HC_2O_4$	*aq*	−227.02	....	....	....
NH_4Cl	*c*	−75.38	−48.73	35.718	22.6
	aq, ∞	−71.76	....	....	....
NH_4ClO_4	*c*	−69.42	....	....	....
NH_4F	*c*	−111.6	....	....	....
	aq, ∞	−110.40	....	....	....
NH_4I	*c*	−48.30	....	....	....
	aq, ∞	−45.11	....	....	....
NH_4NO_2	*c*	−63.1	....	....	....
	aq	−57.1	....	....	....
NH_4NO_3	*c*	−87.27	....	....	....
	aq	−81.11	....	....	....
$NH_4H_2PO_4$	*c*	−346.75	−290.46	212.903	36.32
	aq, 500	−342.91	....	....	....
$(NH_4)_2HPO_4$	*c*	−376.12	....	....	....
	aq, 500	−373.04	....	....	....
$(NH_4)_3PO_4$	*c*	−401 8	....	....	....
	aq, 660	−394.0	....	....	....
$(NH_4)_3PO_4 \cdot 3H_2O$	*c*	−612.8	....	....	....
$(NH_4)_2PtCl_4$	*c*	−195.3	....	....	....
	aq	−186.9	....	....	....
$(NH_4)_2S$	*aq*	−54.5	....	....	....
NH_4HS	*c*	−38.10	....	....	....
	aq	−35.1	....	....	....
NH_4S_4	*c*	−34.0	....	....	....
	aq	−29.7	....	....	....
$(NH_4)_2S_5$	*c*	−69.4	....	....	....
	aq	−65.4	....	....	....
$(NH_4)_2SO_3$	*c*	−212.0	....	....	....
	aq	−211.3	....	....	....
$(NH_4)_2SO_3 \cdot H_2O$	*c*	−284.22	....	....	....
NH_4HSO_3	*c*	−183.8	....	....	....
	aq	−181.5	....	....	....
$(NH_4)_2SO_4$	*c*	−281.86	−215.19	157.732	52.65
	aq, ∞	−280.38	....	....	....
NH_4HSO_4	*c*	−244.83	....	....	....
	aq	−245.6	....	....	....
$(NH_4)_2S_2O_8$	*c*	−396.4	....	....	....
	aq	−387.8	....	....	....
$(NH_4)_2Se$	*aq*	−26.2	....	....	....
NH_4HSe	*aq*	−5.6	....	....	....
$(NH_4)_2SiF_6$	*c*	−629.7	....	....	....
	aq	−622.0	....	....	....
NH_4VO_3	*c*	−251.2	−211.8	155.25	33.6
NH_2OH	*c*	−25.5	....	....	....
	aq	−21.7	....	....	....
$NH_2OH \cdot HCl$	*c*	−74.0	....	....	....
	aq	−70.7	....	....	....
$NH_2OH \cdot HNO_3$	*c*	−86.3	....	....	....
	aq	−80.5	....	....	....
$(NH_2OH)_2 \cdot H_2SO_4$	*c*	−282.5	....	....	....
	aq	−276.7	....	....	....
$NH_2OH \cdot H_2SO_4$	*c*	−245.1	....	....	....
	aq	−244.4	....	....	....
N_2H_4	*lq*	12.05	....	....	....
	aq, 300	8.16	....	....	....
$N_2H_4 \cdot H_2O$	*lq*	−57.95	....	....	....
$N_2H_4 \cdot HCl$	*c*	−46.93	....	....	....
	aq	−41.7	....	....	....
$N_2H_4 \cdot 2HCl$	*c*	−90.0	....	....	....
	aq	−84.0	....	....	....
$(N_2H_4)_2 \cdot H_2SO_4$	*aq*	−217.8	....	....	....
$N_2H_4 \cdot H_2SO_4$	*c*	−231.6	....	....	....
	aq	−223.4	....	....	....
HN_3	*g*	70.3	78.5	−57.54	56.74
	aq, 100	61.51	....	....	....

* Corundum.
† Hydrargillite.
** All values are expressed as kilo-cals per gram mole

VALUES OF CHEMICAL THERMODYNAMIC PROPERTIES (Continued)

Substance	State	ΔHf°	ΔFf°	Log_{10} Kf	S°
Antimony					
Sb	*g*	60.8	51.1	−37.46	43.06
	c, III	0.00	0.00	0.000	10.5
SbO^+	*aq*	....	−42.0	30.79	......
Sb_2	*g*	52.	40	−29.32	60.9
$SbBr_3$	*c*	−62.1	....	....	......
$SbCl_3$	*g*	−75.2	−72.3	52.99	80.8
	c	−91.34	−77.62	56.894	44.5
$SbCl_5$	*g*	−93.9	....	....	......
	lq	−104.8	....	....	......
SbOCl	*c*	−90.8	....	....	......
SbF_3	*c*	−217.2	....	....	......
	aq	−216.1	....	....	......
SbI_3	*c*	−23.0	....	....	......
	aq	−22.6	....	....	......
Sb_2O_3	*aq*	−166.5	....	....	......
Sb_2O_4	*c*	−214	−188	137.8	30.3
Sb_2O_5	*c*	−234.4	−200.5	146.96	29.9
	aq	−226.4	....	....	......
Sb_2O_6	*c*	−336.8	−298.0	218.43	59.8
Sb_2S_3 (black)	*c*	−43.5	....	....	......
(orange)	*am*	−36.0	....	....	......
$Sb_2(SO_4)_3$	*c*	575.3	....	....	......
H_3SbO_4	*aq*	−215.7	....	....	......
Argon					
A	*g*	0.00	0.00	0.000	36.983
$A \cdot 5H_2O$	*c*	−357.2	....	....	......
Arsenic					
As (gray)	*c*	0.00	0.00	0.000	8.4
(β)	*am*	1.0	....	....	......
(yellow)	*c*	3.53	....	....	......
AsO^+	*aq*	....	−39.1	28.69	......
As_2	*g*	29.6	17.5	−12.83	57.3
As_4	*g*	35.7	25.2	−18.47	69
$AsBr_3$	*c*	−46.61	....	....	......
$AsCl_3$	*g*	−71.5	−68.5	50.21	78.2
	lq	−80.2	−70.5	51.68	55.8
AsF_3	*g*	−218.3	−214.7	157.37	69.08
	lg	−226.8	−215.5	157.96	43.31
AsH_3	*g*	41.0	....	....	......
$AsH_3 \cdot 6H_2O$	*c*	−386.7	....	....	......
AsI_3	*c*	−13.7	....	....	......
As_2O_5	*c*	−218.6	−184.6	135.41	25.2
	aq	−224.6	....	....	......
$As_2O_5 \cdot 4H_2O$	*c*	−500.3	....	....	......
$3As_2O_5 \cdot 5H_2O$	*c*	−1007.5	....	....	......
$As_2O_3 \cdot As_2O_5$	*c*	−351.1	....	....	......
As_4O_6 (oct)	*c*	−313.94	−275.36	201.835	51.2
(mon)	*c*	−312.8	....	....	......
	aq	−299.4	....	....	......
As_2S_3	*c*	−35	....	....	......
H_3AsO_3	*aq*	−177.3	−152.9	112	47.0
H_3AsO_4	*c*	−215.2	....	....	......
	aq	−214.8	−183.8	134.72	49.3
Barium					
Ba	*g*	41.96	34.60	−25.361	40.699
	c	0.00	0.00	0.000	16
Ba^{++}	*aq*	−128.67	−134.0	98.22	3
$Ba_3(AsO_4)_2$	*c*	−817.8	....	....	......
$BaHAsO_4 \cdot H_2O$	*c*	−411.5	....	....	......
$Ba(H_2AsO_4)_2 \cdot 2H_2O$	*c*	−694.7	....	....	......
$Ba(C_2H_3O_2)_2$	*c*	−355.1	....	....	......
$Ba(C_2H_3O_2)_2 3H_2O$	*c*	−567.3	....	....	......
$BaBr_2$	*c*	−180.4	....	....	......
	aq	−186.47	−183.1	134.21	42
$BaBr_2 \cdot H_2O$	*c*	−254.9	....	....	......
$BaBr_2 \cdot 2H_2O$	*c*	−326.3	....	....	......
$BaOBr_2$	*aq*	−181.6	....	....	......
$Ba(HCO_2)_2$	*c*	−326.5	....	....	......
$Ba(CN)_2$	*c*	−47.9	....	....	......
	aq	−50.7	....	....	......
$Ba(CN)_2 \cdot H_2O$	*c*	−120.1	....	....	......
$Ba(CN)_2 \cdot 2H_2O$	*c*	−191.1	....	....	......
$BaCN_2$	*c*	−63.8	....	....	......
$Ba(CNO)_2$	*c*	−209.6	....	....	......
$BaCO_3$	*c*	−291.3	−272.2	199.52	26.8
	aq	−290.30	−260.2	190.72	−10
$Ba(HCO_3)_2$	*aq*	−459.0	−414.6	303.90	48
$BaC_2O_4 \cdot \frac{1}{2}H_2O$	*c*	−363.7	....	....	......
$BaC_2O_4 \cdot 2H_2O$	*c*	−470.1	....	....	......
$BaC_2O_4 \cdot 3\frac{1}{2}H_2O$	*c*	−575.3	....	....	......
$BaCl_2$	*c*	−205.56	−193.8	142.05	30
	aq	−208.72	−196.7	144.18	29
$BaCl_2 \cdot H_2O$	*c*	−278.4	−253.1	185.52	40
$BaCl_2 \cdot 2H_2O$	*c*	−349.35	−309.7	227.01	48.5
$BaOCl_2$	*aq*	−194.0	....	....	......
$Ba(OCl)_2$	*aq*	−176.1	....	....	......
$Ba(ClO_2)_2$	*c*	−158.2	....	....	......
$Ba(ClO_3)_2$	*c*	−181.7	....	....	......
	aq	−175.6	....	....	......
Barium					
$Ba(ClO_3)_2 \cdot H_2O$	*c*	−254.9	....	....	......
$Ba(ClO_4)_2$	*c*	−192.8	....	....	......
$Ba(ClO_4)_2 \cdot 3H_2O$	*c*	−405.4	....	....	......
$BaPtCl_6$	*c*	−286.8	....	....	......
	aq	−296.1	....	....	......
$BaPtCl_6 \cdot 6H_2O$	*c*	−707.2	....	....	......
$BaCrO_4$	*c*	−341.3	....	....	......
BaF_2	*c*	−286.9	....	....	......
	aq	−286.0	−265.3	....	......
$BaSiF_6$	*c*	−691.6	....	....	......
BaH	*g*	52	46	−33.7	52.97
BaH_2	*c*	−40.9	....	....	......
BaI_2	*c*	−144.0	....	....	......
	aq	−155.41	−158.7	116.32	55
$BaI_2 \cdot H_2O$	*c*	−219.8	....	....	......
$BaI_2 \cdot 2H_2O$	*c*	−290.9	....	....	......
$BaI_2 \cdot 2\frac{1}{2}H_2O$	*c*	−326.0	....	....	......
$BaI_2 \cdot 7H_2O$	*c*	−640.1	....	....	......
$Ba(IO_3)_2$	*aq*	−238.4	....	....	......
$Ba(IO_3)_2 \cdot H_2O$	*c*	−319.6	....	....	......
$BaMoO_4$	*c*	−373.8	....	....	......
$Ba(N_3)_2$	*c*	−8.0	....	....	......
	aq	0.2	....	....	......
Ba_3N_2	*c*	−86.9	....	....	......
$Ba(NO_2)_2$	*c*	−174.0	....	....	......
$Ba(NO_2)_2 \cdot H_2O$	*c*	−254.5	....	....	......
$Ba(NO_3)_2$	*c*	−237.06	−190.0	139.27	51.1
	aq	−227.41	−186.8	136.92	73
BaO	*c*	−133.4	−126.3	92.58	16.8
BaO_2	*c*	−150.5	....	....	......
$BaO_2 \cdot H_2O$	*c*	−223.5	....	....	......
$BaO_2 \cdot 8H_2O$	*c*	−719.3	....	....	......
$Ba(OH)_2$	*c*	−226.2	....	....	......
	aq	−238.58	−209.2	153.34	−2
$Ba(OH)_2 \cdot H_2O$	*c*	−299.0	....	....	......
$Ba(OH)_2 \cdot 8H_2O$	*c*	−799.5	....	....	......
$Ba_3(PO_4)_2$	*c*	−998.0	....	....	......
$BaHPO_4$	*c*	−465.8	....	....	......
$Ba(H_2PO_4)_2$	*c*	−749.6	....	....	......
BaS	*g*	41	....	....	......
	c	−106.0	....	....	......
	aq	−118.4	....	....	......
$Ba(HS)_2$	*aq*	−134.8	....	....	......
$Ba(HSO_3)_2$	*aq*	−430.7	....	....	......
$BaSO_3$	*c*	−282.6	....	....	......
$BaSO_4$	*c*	−350.2	−323.4	237.05	31.6
	aq, ∞	−345.57	−311.3	228.18	7
BaS_2O_6	*aq*	−409.3	....	....	......
$BaS_2O_6 \cdot 2H_2O$	*c*	−552.5	....	....	......
BaS_2O_8	*aq*	−454.0	....	....	......
$BaS_2O_8 \cdot 4H_2O$	*c*	−738.7	....	....	......
BaS_4O_6	*aq*	−401.3	....	....	......
$BaS_4O_6 \cdot 2H_2O$	*c*	−554.5	....	....	......
BaSe	*c*	−142.0	....	....	......
$BaSeO_4$	*c*	−280.0	....	....	......
$BaSiO_3$	*c*	−359.5	....	....	......
Ba_2SiO_4	*c*	−496.8	....	....	......
$BaWO_4$	*c*	−407.7	....	....	......
Beryllium					
Be	*g*	76.63	67.60	−49.550	32.55
	c	0.00	0.00	0.000	2.28
Be^{++} (in acid solution)	*aq*	....	−93	....	......
$BeBr_2$	*c*	−88.4	....	....	......
$BeBr_2$ (in HCl)	*aq*	151	....	....	......
$BeCl_2$	*c*	−122.3	....	....	......
$BeCl_2$ (in HCl)	*aq*	173.4	....	....	......
$BeCl_2 \cdot 4H_2O$	*c*	−436.8	....	....	......
BeF_2	*aq*	−251.4	....	....	......
BeH	*g*	78.1	71.3	−52.26	40.84
BeI_2	*c*	−50.6	....	....	......
BeI_2 (in HCl)	*aq*	−120	....	....	......
$BeMoO_4$	*c*	−330	....	....	......
Be_3N_2	*c*	−135.7	−122.4	....	......
$Be(NO_3)_2$	*aq*	−188.3	....	....	......
BeO	*g*	11.8	5.7	−4.18	47.18
	c	−146.0	−139.0	101.88	3.37
$Be(OH)_2$	*c* (α)	−216.8	....	....	......
	c (β)	−216.1	....	....	......
BeS	*c*	−55.9	....	....	......
$BeSO_4$	*c*	−286.0	....	....	......
$BeSO_4 \cdot H_2O$	*c*	−361	....	....	......
$BeSO_4 \cdot 2H_2O$	*c*	−433.2	....	....	......
$BeSO_4 \cdot 4H_2O$	*c*	−576.3	....	....	......
$BeSO_4 \cdot 4BeO$	*c*	−871.4	....	....	......
Bismuth					
Bi	*g*	49.7	40.4	−29.61	44.67
	c	0.00	0.00	0.000	13.6
BiO^+	*aq*	....	−35.54	25.317	......
$BiCl_3$	*g*	−64.7	−62.2	45.59	85 3
	c	−90.61	−76.23	55.876	45.3

VALUES OF CHEMICAL THERMODYNAMIC PROPERTIES (Continued)

Substance	State	ΔHf°	ΔFf°	Log₁₀ Kf	S°
Bismuth					
$BiCl_3$ (in HCl)	aq	−101.6	......	......	......
$BiOCl$	c	−87.3	−77.0	56.44	20.6
Bi_2O_3	c	−137.9	−118.7	87.01	36.2
$Bi(OH)_3$	c	−169.6	......	......	......
Bi_2S_3	c	−43.8	−39.4	28.88	35.3
Boron					
B	g	97.2	86.7	−63.55	36.649
	c	0.00	0.00	0.000	1.56
B	am	0.4	......	......	......
BBr_3	g	−44.6	−51.0	37.38	77.49
	lq	−52.8	−52.4	38.41	54.7
BCl_3	g	−94.5	−90.9	66.63	69.29
	lq	−100.0	−90.6	66.41	50.0
B_4C	c	......	......	......	6.47
$B(CH_3)_3$	lq	−31.4	......	......	......
BF_3	g	−265.4	−261.3	191.53	60.70
	aq	−289.8	......	......	......
BF_4^-	aq	−365	−347	251.4	40
HBF_4	aq	−365	......	......	......
BH	g	73.8	67.1	−49.18	39.62
B_2H_6	g	7.5	19.8	−14.51	55.66
B_5H_9	g	15.0	39.6	−29.03	65.88
$B_{10}H_{14}$	c	8	......	......	......
BN	g	90.6	......	......	......
	c	−32.1	−27.2	19.94	8
BO	g	−5.3	−11.6	8.50	47.22
B_2O_3	c	−302.0	−283.0	207.44	12.91
	gls	−297.6	−280.4	205.53	18.8
B_2S_3	c	−57.0	......	......	......
HBO_2	c	−186.9	−170.5	124.97	11
	aq	−186.9	......	......	......
HBO_3	aq	−251.8	−217.63	159.520	7.3
$H_2B_4O_7$	c	−676.5	......	......	......
H_3BO_3	c	−260.2	−230.2	168.73	21.41
	aq	−255.2	−230.24	168.762	38.2
Bromine					
Br_2	g	7.34	0.75	−0.551	58.64
	lq	0.00	0.00	0.000	36.4
	aq	−1.1	......	......	......
$Br_2 \cdot 10H_2O$	c	−700	......	......	......
BrCl	g	3.51	−0.21	0.154	57.34
HBr	g	−8.66	−12.72	9.327	47.44
	aq, 1	−18.56	......	......	......
......	aq, 3	−24.36	......	......	......
	aq	−28.90	−24.57	18.012	19.29
$HBrO_3$	aq	−11.63	5.00	......	......
Cadmium					
Cd	g	26.97	18.69	−13.700	40.07
(α)	c	0.00	0.00	0.000	12.3
(γ)	c	......	0.14	−0.103	......
$CdBr_2$	c	−75.15	−70.14	51.412	31.9
	aq	−75.822	−67.6	......	......
$CdBr_2 \cdot 4H_2O$	c	−356.32	−297.64	218.166	74.7
$CdCl_2$	c	−93.00	−81.88	60.017	28.3
	aq	−97.39	−81.28	59.577	11.6
$CdCl_2 \cdot H_2O$	c	−164.13	−140.13	102.713	40.8
$CdCl_2 \cdot 2\frac{1}{2}H_2O$	c	−269.97	−225.47	165.266	55.6
$Cd(CH_3)_2$	lq	16.2	......	......	......
$Cd(C_2H_5)_2$	lq	14.7	......	......	......
$Cd(CN)_2$	c	39.0	......	......	......
	aq	30.3	......	......	......
$Cd(ONC)_2$	c	37.7	......	......	......
$CdCO_3$	c	−178.7	−160.2	117.42	25.2
CdF_2	c	−164.9	−154.8	113.47	27
	aq	−173.6	......	......	......
CdH	g	62.54	55.73	−40.849	50.76
CdI_2	c	−48.0	−48.00	35.183	40.2
	aq, ∞	−44.66	......	......	......
$Cd(N_3)_2$	c	107.8	......	......	......
Cd_3N_2	c	38.6	......	......	......
$Cd(NO_3)_2$	c	−107.98	......	......	......
	aq	−116.04	−71.41	52.342	20.4
$Cd(NO_3)_2 \cdot 2H_2O$	c	−251.19	......	......	......
$Cd(NO_3)_2 \cdot 4H_2O$	c	−394.02	......	......	......
CdO	c	−60.86	−53.79	39.427	13.1
$Cd(OH)_2$	c	−133.26	−112.46	82.432	22.8
CdS	c	−34.5	−33.6	24.63	17
$CdSO_4$	c	−221.36	−195.99	143.657	32.8
	aq	−234.20	−195.92	143.607	−10.47
$CdSO_4 \cdot H_2O$	c	−294.37	−254.84	186.794	41.1
Cd_3Sb_2	c	7.83	1.59	−1.165	78.8
CdSe	g	1.6	......	......	......
CdTe	c	−24.30	−23.82	17.460	22.6
Calcium					
Ca	g	46.04	37.98	−27.839	36.99
	c	0.00	0.00	0.000	9.95
Ca^{++}	aq	−129.77	−132.18	96.886	−13.2
$2CaO \cdot Al_2O_3$	c	−704	......	......	......
$2CaOAl_2O_3 \cdot 5H_2O$	c	−1078	......	......	......
$3CaO \cdot Al_2O_3$	c	−861	......	......	......
$3CaO \cdot Al_2O_3 \cdot 6H_2O$	c	−1329	......	......	......
$4CaO \cdot Al_2O_3$	c	−1026	......	......	......
$Ca_3(AsO_4)_2$	c	−796	......	......	......
$CaHAsO_4$	aq	−344.1	......	......	......
$CaHAsO_4 \cdot H_2O$	c	−410	−363	266.1	35

Substance	State	ΔHf°	ΔFf°	Log₁₀ Kf	S°
Calcium					
$Ca(H_2AsO_4)_2$	aq	−560.8	......	......	......
$CaBr_2$	c	−161.3	−156.8	114.93	31
	aq	−187.57	−181.33	132.912	25.4
$CaBr_2 \cdot 6H_2O$	c	−597.2	......	......	......
$CaO \cdot B_2O_3$	c	−483.3	−457.7	335.49	25.1
$CaO \cdot 2B_2O_3$	c	−798.8	−752.4	551.50	32.2
$2CaO \cdot B_2O_3$	c	−651.6	−618.6	453.42	34.7
$3CaO \cdot B_2O_3$	c	−817.7	−777.1	569.60	43.9
CaC_2	c	−15.0	−16.2	11.87	16.8
$CaCO_3$ (calcite)	c	−288.45	−269.78	197.745	22.2
(aragonite)	c	−288.49	−269.53	197.562	21.2
$Ca(HCO_3)_2$	aq	−460.13	−412.80	302.577	32.2
$Ca(HCO_2)_2$	c	−323.5	......	......	......
CaC_2O_4	c	−332.2	......	......	......
$CaC_2O_4 \cdot H_2O$	ppt	−399.1	−360.6	264.31	37.28
$CaC_2O_4 \cdot 2H_2O$	c	−469.1	−416.9	305.58	47
$Ca(C_2H_3O_2)_2$	c	−355.0	......	......	......
	aq	−362.5	......	......	......
$Ca(C_2H_3O_2)_2H_2O$	c	−425.1	......	......	......
$Ca(CN)_2$	c	−44.2	......	......	......
	aq	......	−54	......	......
$CaCN_2$	c	−84.0	......	......	......
$CaCl_2$	c	−190.0	−179.3	131.42	27.2
$CaCl_2$	aq	−209.82	−194.88	142.884	13.1
$CaCl_2 \cdot H_2O$	c	−265.1	......	......	......
$CaCl_2 \cdot 2H_2O$	c	−335.5	......	......	......
$CaCl_2 \cdot 4H_2O$	c	−480.2	......	......	......
$CaCl_2 \cdot 6H_2O$	c	−623.15	......	......	......
$CaCl_2 \cdot 2CaO$	c	−505	......	......	......
$CaCl_2 \cdot 3CaO$	c	−654	......	......	......
$CaCl_2 \cdot 3CaO \cdot 3H_2O$	c	−910.6	......	......	......
$CaCl_2 \cdot 3CaO \cdot 16H_2O$	c	−1833	......	......	......
$CaCl_2 \cdot 3C_2H_5OH$	c	−399.8	......	......	......
$CaCl_2 \cdot 4C_2H_5OH$	c	−467.4	......	......	......
$CaOCl_2$	c	−178.6	......	......	......
	aq	−189.1	......	......	......
$CaOCl_2 \cdot H_2O$	c	−249.2	......	......	......
$Ca(OCl)_2$	aq	−180.0	......	......	......
$CaCrO_4$	c	−329.6	−305.3	223.78	32
	aq	−336.0	......	......	......
CaF_2	c	−290.3	−277.7	203.55	16.46
	aq	−287.09	−264.34	193.758	−17.8
CaH	g	58.7	......	......	......
CaH_2	c	−45.1	−35.8	26.24	10
CaI_2	c	−127.8	−126.6	92.80	34
	aq	−156.51	−156.88	114.991	39.1
$CaI_2 \cdot 8H_2O$	c	−700.7	......	......	......
CaN_6	c	75.8	......	......	......
Ca_3N_2	c	−103.2	−88.1	64.57	25
$Ca(NO_2)_2$	c	−178.3	......	......	......
	aq	−180.5	......	......	......
$Ca(NO_3)_2$	c	−224.0	−177.34	129.988	46.2
	aq	−228.51	−185.00	135.602	56.8
$Ca(NO_3)_2 \cdot 2H_2O$	c	−368.00	−293.51	215.139	64.3
$Ca(NO_3)_2 \cdot 3H_2O$	c	−437.18	−349.0	255.81	74
$Ca(NO_3)_2 \cdot 4H_2O$	c	−509.37	−406.5	297.96	81
CaO	c	−151.9	−144.4	105.84	9.5
CaO_2	c	−157.5	......	......	......
$Ca(OH)_2$	c	−235.80	−214.33	157.101	18.2
	aq	−239.68	−207.37	152.000	−18.2
Ca_3P_2	c	−120.5	......	......	......
$Ca_3(PO_4)_2$	c (α)	−986.2	−929.7	681.46	57.6
	c (β)	−988.9	−932.0	683.14	56.4
$CaHPO_4$	c	−435.2	−401.5	294.29	21
$CaHPO_4 \cdot 2H_2O$	c	−576.0	−514.6	377.19	40
$Ca(H_2PO_4)_2$	ppt	−744.4	......	......	......
CaS	c	−115.3	−114.1	83.63	13.5
	aq	−119.8	......	......	......
$CaSO_3 \cdot 2H_2O$	c	−421.2	−374.1	274.21	44
$CaSO_4$ (anhydrite)	c	−342.42	−315.56	231.301	25.5
(soluble α)	c	−340.27	−313.52	229.806	25.9
(soluble β)	c	−339.21	−312.46	229.029	25.9
	aq	−346.67	−309.52	226.874	−9.1
$CaSO_4 \cdot \frac{1}{2}H_2O$	c (α)	−376.47	−343.02	251.429	31.2
	c (β)	−375.97	−342.78	251.253	32.1
$CaSO_4 \cdot 2H_2O$	c	−483.06	−429.19	314.590	46.36
CaS_2O_3	aq, 1000	−283.4	......	......	......
$CaS_2O_3 \cdot 6H_2O$	c	......	−602.2	441.40	......
CaSe	c	−74.7	−73.5	53.87	16
$CaSi_2$	c	−36	......	......	......
Ca_2Si	c	−50	......	......	......
Ca_2Si_2	c	−72	......	......	......
$CaSiO_3$*	c (α)	−377.4	−357.4	261.97	20.9
	c (β)	−378.6	−358.2	262.55	19.6
Ca_2SiO_4	c (β)	−538.0	......	......	......
	c (γ)	−539.0	......	......	......
Ca_3SiO_5	c	−688.4	......	......	......
$CaWO_4$	c	−392.5	......	......	......
Carbon**					
C	g	171.70	160.85	−117.897	37.76
C (diamond)	c	0.45	0.69	−0.502	0.58
(graphite)	c	0.00	0.00	0.000	1.36
CO_2	g	−94.05	−94.26	69.092	51.06
	aq	−98.69	−92.31	67.662	29.0
CO	g	−26.42	−32.81	24.048	47.30

* (α) Pseudowollastonite; (β) wollast----

** For the values of organic compounds see table following zirconium compounds.

VALUES OF CHEMICAL THERMODYNAMIC PROPERTIES (Continued)

Substance	State	ΔHf°	ΔFf°	Log₁₀ Kf	S°
Cerium					
Ce	c	0.00	0.00	0.000	13.8
Ce^{+++}	aq	−173.7	−170.5	124.97	−44
$CeCl_3$	c (α)	−260.3			
	aq	−293.9	−264 5	193.87	−5
Ce_3H_8	c	−170			
CeI_3	c (α)	−164.4			
	aq	−213.9	−207.5	152.09	34
CeO_2	c	−233			
$CeO_2·2H_2O$	c	−389			
CeS_2	c	−153.9			
Ce_2S_3	c	−298.7			
$Ce(SO_4)_2$	c	−560			
$Ce_2(SO_4)_3$	aq	−998.3	−873.0	639.90	−76
$Ce_2(SO_4)_3·5H_2O$	c	−1308			
$Ce_2(SO_4)_3·8H_2O$	c		−1340.2	982.35	
$Ce_2(SO_4)_3·9H_2O$	c		−1396.8	1023.83	
Cesium					
Cs	c	0.00	0.00	0.000	19.8
Cs^+	aq	−59.2	−67.41	49.111	31.8
CsBr	c	−94.3	−91.6	67.142	29
	aq	−88.1	−91.98	67.420	51
Cs_2CO_3	c	−267.4			
$CsHCO_3$	c	−228.4			
	aq	−224.4			
CsCl	c	−103.5			
	aq	−99.2	−88.76	65.060	45.0
$CsClO_4$	c	−103.86	−73.28	53.713	41.89
	aq	−90.6	−69.98	51.294	75.3
CsF	c	−126.9			
	aq	−135.9	−133.49	97.846	29.5
Cs_2SiF_6	c	−669.5			
CsH	g	29.0	24.3	−17.81	51.25
CsI	c	−80.5	−79.7	58.419	31
	aq	−72.6	−79.76	58.463	57.9
$CsNH_2$	c	−25.4			
$CsNO_3$	c	−118.11			
	aq	−108.6	−93.82	68.769	86.8
Cs_2O	c	−75.9			
Cs_2O_2	c	−96.2			
CsOH	c	−97.2			
	aq	−114.2	−105.00	76.964	29.3
$CsOH·H_2O$	c	−186.9			
$CsReO_4$	c	−257.2			
	aq	−249.5			
Cs_2S	c	−81.1			
	aq	−108.4			
CsHS	c	−62.9			
	aq	−63.2			
Cs_2SO_4	c	−339.38			
	aq	−335.3	−312.16	228.809	67.7
$CsHSO_4$	c	−274.0			
CsHSe	c	−36.7			
	aq	−34.6			
Chlorine					
Cl_2	g	0.00	0.00	0.000	53.29
Cl^-	aq	−40.023	−31.350	22.98	13.17
ClF	g	−13.3	−13.6	9.97	52.05
ClO	g	33			
ClO_2	g	24.7	29.5	−21.623	59.6
ClO_3	g	37.0			
Cl_2O	g	18.20	22.40	−16.419	63.70
Cl_2O_7	g	63.4			
HCl	g	−22.06	−22.77	16.690	44.62
	aq	−40.02	−31.35	22.979	13.17
HClO	aq, 400	−28.18			
	aq, 1000	−27.83			
$HClO_2$	aq	−14.0			
$HClO_3$	aq	−23.50			
$HClO_4$	lq (lq)	−11.1			
	aq	−31.41			
$HClO_4·H_2O$	c	−92.1			
$HClO·2H_2O$	lq	−162.8			
Chromium					
Cr	g	80.5	69.8	−51.16	41.64
	c	0.00	0.00	0.000	5.68
Cr^{+++}	aq	1310			
Cr_3C_2	c	−21.0	−21.2	15.54	20.4
Cr_4C	c	−16.4	−16.8	12.31	25.3
Cr_7C_3	c	−42.5	−43.8	32.10	48.0
$CrCl_2$	c	−94.56	−85.15	62.414	27.4
	aq	−113.2			
$CrCl_2·2H_2O$	c	−237			
$CrCl_2·3H_2O$	c	−309.0			
$CrCl_2·4H_2O$	c	−384.5			
$CrCl_3$	c	−134.6	−118.0	86.492	30.0
$CrCl_4$	g	−104.			
CrO_2Cl_2	lq	−135.7			
CrF_2	c	−181.0			
CrF_3	c	−265.2			
Cr_7H_2	c	−3.7			
CrI_2	c	−54.2			
Chromium	aq	−59.9			
CrN	c	−29.8			
Cr_2N	c	−23.4			
Cr_2O_3	c	−269.7	−250.2	183.39	19.4
$Cr_2O_3·H_2O$	c	−358			
$Cr_2O_3·2H_2O$	c	−439			
$Cr_2O_3·3H_2O$	c	−517			
$Cr(OH)_3$	c	−247.1			
H_2CrO_4	aq	−213.3	−178.5	130.84	17.5
Cobalt					
Co	g	105	94	−68.9	42.88
	c	0.00	0.00	0.000	6.8
Co^{++}	aq	−16.1	−12.3	9.016	−37.1
$CoBr_2$	c	−55.5			
	aq	−73 9			
$CoBr_2·6H_2O$	c	−485.1			
Co_3C	c	9.5	7.1	−5.20	29.8
$CoCO_3$	c	−172.7	−155.36		
$CoCl_2$	c	−77.8	−67.5	49.48	25.4
	aq	−96.1			
$CoCl_2·2H_2O$	c	−222 9			
$CoCl_2·4H_2O$	c	−367 2			
$CoCl_2·6H_2O$	c	−508.9			
CoF_2	c	−159			
	aq	−173.6			
$CoF_2·4H_2O$	c		−380.9	279.19	
CoF_3	c	−187			
CoH	c	−4.1			
CoH_2	c	−10.2			
CoI_2	c	−24.4			
	aq	−42.8			
$Co(IO_3)_2$	c	−124 3			
	aq	−125.9			
$Co(IO_3)_2·2H_2O$	c	−264.9			
$Co(IO_3)_2·4H_2O$	c	−401 9			
$Co(NO_3)_2$	c	−102.9			
	aq	−114 8			
$Co(NO_3)_2·6H_2O$	c	−529.7			
CoO	c	−57.2	−51.0	37.38	10.5
Co_3O_4	c	−210			
$Co(OH)_2$	c	−131.2	−108.9		
$Co(OH)_3$	c	−176 6	−142.0		
CoP	c	−35			
CoP_3	c	−65			
Co_2P	c	−47 3			
CoS	c	−20 2	−19.8		
	ppt	−21.4			
Co_2S_3	c	−51			
$CoSO_4$	c	−207.5	−182.1	133.48	27.1
	aq	−232 0			
$CoSO_4·6H_2O$	c	−643.2			
$CoSO_4·7H_2O$	c	−713.8			
Columbium (See Niobium)					
Copper					
Cu	g	81.52	72.04	−52.804	39.74
	c	0.00	0.00	0.000	7.96
Cu^+	aq	12.4	12.0	−8.796	−6.3
Cu^{++}	aq	15.39	15.53	−11.383	−23.6
CuBr	g	38	28	−20.5	59.22
	c	−25.1	−23.81	17.452	21.9
$CuBr_2$	c	−33.8			
$CuBr_2·4H_2O$	c	316.4			
$CuCO_3$	c	−142.2	−123.8	90.744	21
$Cu(C_2H_3O_2)_2$	c	−213.2			
$Cu(C_2H_3O_2)_2·H_2C$	c	−284.2			
CuONC	c	26.3			
CuCl	g	32	25	−18.3	56.50
	c	−32.5	−28.2	20.67	20.2
$CuCl_2$	c	−49.2			
CuF	g	44			
CuF_2	c	−126.9			
$CuF_2·2H_2O$	c	−274.5	−235.2	172.40	36.2
CuH	g	71	64	−46.9	46.89
CuI	g	62	50	−36.6	61.06
	c	−16.2	−16.62	12.182	23.1
CuI_2	c	−1.7			
CuN_3	c	60.5			
Cu_3N	c	17.8			
$Cu(NO_3)_2$	c	−73.4			
CuO	g	35			
	c	−37.1	−30.4	22.28	10.4
Cu_2O	c	−39.84	−34.98	25.640	24.1
$Cu(OH)_2$	c	107.2			
CuS	c	−11.6	−11.7	8.576	15.9
Cu_2S	c	−19.0	−20.6	15.10	28 9
$CuSO_4$	c	−184.00	−158.2	115.96	27.1
	aq	−201.51	−161 81	118.604	19.5
$CuSO_4·H_2O$	c	−259.00	−219.2	160.67	35.8
$CuSO_4·3H_2O$	c	−402.27	−334 6	245.26	53.8
$CuSO_4·5H_2O$	c	−544 45	−449.3	329 33	73.0
Cu_2SO_4	c	−179.2			
	aq	−190.8			
CuSe	c	−15.1			

* (α) Pseudowollastonite; (β) wollastonite.
** For the values of organic compounds see table following zirconium compounds.

Substance	State	ΔHf°	ΔFf°	Log_{10} Kf	S°
Dysprosium					
Dy^{+++}	aq	−166.0	−161.2	118.16	......
$DyCl_3$	c (β)	−237.8	......	......	......
	c (γ)	−234.8	......	......	......
	aq	−286.1	−255.2	187.06	......
DyI_3	c (β)	−144.5	......	......	......
	aq	−206.1	−198.2	145.28	......
$Dy(OH)_3$	c	......	−305.8	224.15	......
$Dy_2(SO_4)_3$	aq	−982.7	−854.4	626.26	......
$Dy_2(SO_4)_3{\cdot}8H_2O$	c	......	−1322.0	969.01	......
Erbium					
Er^{+++}	aq	−162.3	−157.5	115.45	......
$ErCl_3$	c (γ)	−231.8	......	......	......
	aq	−282.4	−251.5	184.35	......
ErI_3	c (β)	−140.0	......	......	......
ErI_3	aq	−202.4	−194.5	142.57	......
$Er(OH)_3$	c	−340.5	......	......	......
$Er_2(SO_4)_3$	aq	−975.3	−847.0	620.84	......
$Er_2(SO_4)_3{\cdot}8H_2O$	c	......	−1313.9	963.07	......
Europium					
Eu^{+++}	aq	−169.3	−165.1	121.02	......
$EuCl_3$	c (α)	−247.1	......	......	......
	aq	−289.4	−259.2	189.99	......
$Eu_2(SO_4)_3$	aq	−989.3	−862.2	631.98	......
$Eu_2(SO_4)_3{\cdot}8H_2O$	c	......	−1331.0	975.60	......
Fluorine					
F_2	g	0.00	0.00	0.000	48.6
F^-	aq	−78.66	66.08	48.435	−2.3
F_2O	g	5.5	9.7	−7.126	58.95
HF	g	−64.2	−64.7	47.402	41.47
	aq	−78.66	−66.08	48.435	−2.3
Gadolinium					
Gd	g	87	77	−56.4	46.41
Gd^{+++}	aq	−168.8	−164.6	120.65	−47.1
$GdCl_3$	c (α)	−245.5	......	......	......
	aq	−288.9	−258.6	189.55	−7.6
GdI_3	c (β)	−147.6	......	......	......
	aq	−208.9	−201.6	147.77	−31.3
$Gd(OH)_3$	c	......	−308.1	225.83	......
$Gd_2(SO_4)_3$	aq	−988.3	−861.2	631.25	−81.9
$Gd_2(SO_4)_3{\cdot}8H_2O$	c	−1518.9	−1329.8	974.72	155.8
Gallium					
Ga	g	66.0	57.0	−41.78	40.38
Ga^{+++}	aq	50.4	36.6	26.83	−83
$GaBr_3$	c	−92.4	......	......	......
$GaCl_3$	c	−125.4	......	......	......
	aq	−170.5	−130.7	95.801	−43
GaI_3	c	−51.2	......	......	......
GaN	c	−25	......	......	......
GaO	g	66	......	......	......
Ga_2O	c	−82	......	......	......
Ga_2O_3	c	−258	......	......	......
$Ga(OH)_3$	c	......	−199	145.9	......
Germanium					
Ge	g	78.44	69.50	−50.943	40.11
	c	0.00	0.00	0.000	10.14
GeBr	g	34.8	......	......	......
GeCl	g	32.6	32.45	......	......
$GeCl_4$	lq	−130	......	......	......
GeH_4	g	......	......	......	51.21
Ge_3N_4	c	−14.8	......	......	......
GeO	g	−22.8	−28.2	20.63	52.56
GeO_2 (gls)	am	−128.3	......	......	......
(tetr.)	c	−128.3	......	......	......
GeS	g	1.35	......	......	......
H_2GeO_3	aq	−199.3	......	......	......
Gold					
Au	g	82.29	72.83	−53.383	43.12
AuO_3^{---}	aq	......	−5.8	4.25	......
$AuBr_3$	c	−13.0	......	......	......
	aq	−9.2	......	......	......
$HAuBr_4$	aq	−45.5	−38.1	27.93	75
$HAuBr_4{\cdot}5H_2O$	c	−398.5	......	......	......
$AuCl_3$	c	−28.3	......	......	......
	aq	−32.8	......	......	......
$AuCl_3{\cdot}2H_2O$	c	−167.7	......	......	......
$HAuCl_4$	aq	−77.8	−56.2	41.19	61
$HAuCl_4{\cdot}3H_2O$	c	−279.2	......	......	......
$HAuCl_4{\cdot}4H_2O$	c	−356.9	......	......	......
AuI	c	0.2	......	......	......
Au_2O_3	c	19.3	39.0	−28.6	30
$Au(OH)_3$	c	−100.0	−69.3	50.80	29
Hafnium					
Hf	g	......	......	......	44.65
	c	0.00	0.00	0.000	13.1
HfO_2	c	−271.5	......	......	......
Helium					
He	g	0.00	0.00	0.000	30.13
Holmium					
Ho^{+++}	aq	−163.7	−159.2	116.69	......
$HoCl_3$	c (γ)	−232.8	......	......	......
	aq	−283.8	−253.2	185.59	......
HoI_3	c (β)	−141.7	......	......	......
	aq	−203.8	−196.2	143.81	......
$Ho_2(SO_4)_3$	aq	−978.1	−850.4	623.33	......
$Ho_2(SO_4)_3{\cdot}8H_2O$	c	......	−1318.0	966.08	......

Substance	State	ΔHf°	ΔFf°	Log_{10} Kf	S°
Hydrogen					
H_2 (mass 1)*	g	0.00	0.00	0.000	31.21
(mass 2)**	g	0.00	0.00	0.000	34.602
H*	g	52.09	48.58	−35.60	27.39
H**	g	52.98	49.36	−36.18	29.46
H_2O*	g	−57.80	−54.64	40.047	45.11
	lq	−68.32	−56.69	41.553	16.72
H_2O**	g	−59.56	−56.07	41.096	47.38
	lq	−70.41	−58.21	42.664	18.16
H*H**O	g	−58.74	−55.83	40.921	47.66
	lq	−69.39	−57.93	42.459	18.85
H_2O_2	g	−31.83	......	......	......
	lq	−44.84	−28.2	......	......
	aq	−45.68	......	......	......
Indium					
In	g	58.2	49.6	−36.36	41.51
In^{+++}	aq	27.3	−32.0	23.46	−62
$InBr_3$	c	−96.5	......	......	......
InCl	g	−18	−23	16.9	59.3
	c	−44.5	......	......	......
$InCl_2$	c*	−86.8	......	......	......
$InCl_3$	c	−128.4	......	......	......
InH	g	51	45	−33.0	49.6
InI	g	20	9	−6.6	63.8
InI_3	c	−55.0	......	......	......
InN	c	−4.8	......	......	......
InO	g	91	......	......	......
In_2O_3	c	−222.5	......	......	......
$In(OH)_3$	c	−214	−182	133.4	25
$In_2(SO_4)_3$	c	−695	......	......	......
Iodine					
I_2	g	14.88	4.63	−3.394	62.28
	c	0.00	0.00	0.000	27.9
I^-	aq	−13.37	−12.35	9.252	26.14
IBr	g	9.75	0.91	0.667	61.80
ICl	g	4.20	−1.32	0.968	59.12
ICl_3	c	−21.1	−5.36	3.929	41.1
HI	g	6.20	0.31	−0.227	49.31
	aq	−13.37	−12.35	9.052	26.14
I_2O_5	c	−42.34	......	......	......
HIO_3	c	−57.03	......	......	......
Iridium					
Ir	g	165	154	−112.9	46.25
	c	0.00	0.00	0.000	8.7
IrCl	c	−22.3	......	......	......
$IrCl_2$	c	−42.8	......	......	......
$IrCl_3$	c	−61.5	......	......	......
IrF_6	lq	−130	......	......	......
IrO_2	c	−40.1	......	......	......
IrS_2	c	−30	......	......	......
Ir_2S_3	c	−51	......	......	......
Iron					
Fe	g	96.68	85.76	−62.861	43.11
	c	0.00	0.00	0.000	6.49
Fe^{++}	aq	−21.0	−20.30	14.88	−27.1
Fe^{+++}	aq	−11.4	−2.52	1.847	−70.1
$FeBr_2$	c	−60.02	......	......	......
$FeBr_3$	aq	−98.1	......	......	......
Fe_3C (cementite)	c	5.0	3.5	−2.56	25.7
$Fe(CO)_5$	lq	−187.8	......	......	......
$FeCO_3$ (siderite)	c	−178.70	−161.06	118.055	22.2
$H_3Fe(CN)_6$	aq	153.	......	......	......
$H_4Fe(CN)_6$	c	127.8	......	......	......
	aq	127.4	......	......	......
$FeCl_2$	c	−81.5	−72.2	52.92	28.6
	aq	−101.0	......	......	......
$FeCl_2{\cdot}2H_2O$	c	−228.2	......	......	......
$FeCl_2{\cdot}4H_2O$	c	−370.7	......	......	......
$FeCl_3$	c	−96.8	......	......	......
	aq	−127.9	−96.5	......	......
$FeCl_3{\cdot}6H_2O$	c	−532.0	......	......	......
$FeCr_2O_4$	c	−341.9	−317.7	232.87	34.9
FeF_2	aq	−177.8	......	......	......
FeF_3	aq	−243.1	......	......	......
FeI_2	c	−29.98	......	......	......
	aq	−49.03	......	......	......
Fe_2N	c	−0.9	2.6	−1.91	24.2
Fe_4N	c	−2.55	0.89	−0.652	37.3
$Fe(NO_3)_2$	aq	−118.9	......	......	......
$Fe_{0.95}O$ "FeO" (wüstite)	c	−63.7	−58.4	42.81	12.9
Fe_2O_3 (hematite)	c	−196.5	−177.1	129.81	21.5
Fe_3O_4 (magnetite)	c	−267.0	−242.4	177.68	35.0
$Fe(OH)_2$	c	−135.8	−115.57	84.711	19
$Fe(OH)_3$	c	−197.0	......	......	......
FeP	c	−28	......	......	......
FeP_2	c	−42	......	......	......
Fe_2P	c	−36	......	......	......
Fe_3P	c	−40	......	......	......
$FePO_4$	c	−299.6	......	......	......
$FePO_4{\cdot}2H_2O$	c	−440.8	......	......	......
$FePO_4{\cdot}4H_2O$	c	−578.8	......	......	......
FeS	c (α)	−22.72	−23.32	17.093	16.1
	c (β)	−21.35	......	......	......
FeS_2 (pyrites)	c	−42.52	−39.84	29.202	12.7
(markasite)	c	−36.88	......	......	......

* Atomic weight 1.0078.
** Atomic weight 2.0142.

VALUES OF CHEMICAL THERMODYNAMIC PROPERTIES (Continued)

Substance	State	ΔHf°	ΔFf°	Log_{10} Kf	S°
Krypton					
Kr	g	0.00	0.00	0.000	39.19
$Kr \cdot 5H_2O$	c	−357.1	....	....	......
Lanthanum					
La	g	88	79	−57.9	43.57
	c	0.00	0.00	0.000	13.7
La^{+++}	aq	−176.2	−172.9	126.73	−44
$La_2(CN_2)_3$	c	−229	....	....	....
$LaCl_3$	c (α)	−263.6	....	....	....
	aq	−296.3	−266.9	195.63	−5
LaI_3	c (α)	−167.4	....	....	....
	aq	−216.3	−209.9	153.85	34
LaN	c	−72.1	....	....	....
La_2O_3	c	−458.	....	....	....
$La(OH)_3$	c	....	−312.8	229.28	....
La_2S_3	c	−306.8	....	....	....
$La_2(SO_4)_3$	aq	−1003.1	−877.8	643.41	−76
Lanthanum					
$La_2(SO_4)_3 \cdot 9H_2O$	c	....	−1403.1	1028.45	....
Lead					
pb	g	46.34	38.47	−28.198	41.89
	c	0.00	0.00	0.000	15.51
Pb^{++}	aq	0.39	−5.81	4.259	5.1
$PbBr_2$	c	−66.21	−62.24	45.621	38.6
	aq	−57.41	−54.95	40.278	43.7
$PbCO_3$	c	−167.3	−149.7	109.73	31.3
$PbCO_3 \cdot PbO$	c	−220.0	−195.6	143.37	48.5
$PbCO_3 \cdot 2PbO$	c	−273	−242	177.4	65
PbC_2O_4	c	−205.1	....	....	....
$Pb(C_2H_3O_2)_2$	c	230.5	....	....	....
	aq	−232.6	....	....	....
$Pb(C_2H_3O_2)_2 \cdot 3H_2O$	c	−443.1	....	....	....
$Pb(C_2H_5)_4$	lq	52	....	....	....
$PbCl_2$	c	−85.85	−75.04	55.003	32.6
	aq	−79.65	−68.51	50.217	31.4
PbF_2	c	−158.5	−148.1	108.55	29
PbF_4	c	−222.3	....	....	....
PbI_2	c	−41.85	−41.53	30.441	42.3
	aq	−26.35	−30.51	22.363	57.8
$Pb(N_3)_2$	c	104.3	....	....	....
$Pb(NO_3)_2$	c	−107.35	....	....	....
	aq	−98.35	−58.64	42.982	75.1
PbO	g	....	....	....	57.4
(red)	c	−52.40	−45.25	33.168	16.2
(yellow)	c	−52.07	−45.05	33.021	16.6
PbO_2	c	−66.12	−52.34	38.364	18.3
Pb_2O	c	−51.2	....	....	....
Pb_3O_4	c	−175.6	−147 6	108.19	50.5
$Pb(OH)_2$	c	−123.0	−100.6	73.738	21
$Pb_3(PO_4)_2$	c	−620.3	−581.4	426.16	84.45
PbS	c	−22.54	−22.15	16.236	21.8
$PbSO_4$	c	−219 50	−193.89	142.119	35.2
PbS_2O_3	c	−150.1	....	....	....
PbSe	c	−18	....	....	....
$PbSeO_4$	c	−148	....	....	....
$PbSiO_3$	c	−258 8	−239.0	175.18	27
Pb_2SiO_4	c	−312.7	−285.7	209.41	43
Lithium					
Li	g	37.07	29.19	−21.396	33.14
	c	0.00	0.00	0 000	6.70
Li^+	aq	−66.54	−70.22	51.470	3.4
LiBr	g	−41	−50	36.6	53.78
	c	−83.72	....	....	....
	aq	−95.45	−94.69	69.483	22.7
$LiBr \cdot H_2O$	c	−158.34	....	....	....
$LiBr \cdot 2H_2O$	c	−229.94	....	....	....
$LiBr \cdot 3H_2O$	c	−301.9	....	....	....
Li_2CO_3	c	−290.54	−270.66	198.390	21.60
	aq	−294.74	−266.66	195.458	−5.9
$LiHCO_3$	aq	−231.73	−210.53	154.316	29.5
LiCl	g	−53	−58	42.5	51.01
	c	−97.70	....	....	....
	aq	−106.58	−101.57	74.449	16.6
$LiCl \cdot H_2O$	c	−170.31	−151.2	110.83	24.8
$LiCl \cdot 2H_2O$	c	−242.1	....	....	....
$LiCl \cdot 3H_2O$	c	−313.5	....	....	....
LiF	c	−146.3	−139.5	102.32	8.57
	aq	−145.21	−136.30	99.906	1.1
LiH	g	30.7	25.2	−18.47	40.77
	c	−21.61	−16.72	12.255	5.9
LiI	g	−16	−26	19.1	55.68
	c	−64.79	....	....	....
	aq	−79.92	−82.57	−60.523	29.5
$LiI \cdot \frac{1}{2}H_2O$	c	−103.8	....	....	....
$LiI \cdot H_2O$	c	−141.16	....	....	....
$LiI \cdot 2H_2O$	c	−213.03	....	....	....
$LiI \cdot 3H_2O$	c	−285.02	....	....	....
Li_3N	c	−47.2	....	....	....
$LiNO_2$	c	−96.6	....	....	....
$LiNO_3$	c	−115.28	....	....	....
	aq	−115.93	−96.63	70.828	38.4
$LiNO_3 \cdot 3H_2O$	c	−328.6	....	....	....
Li_2O	c	−142.4	....	....	....
Li_2O_2	c	−151.7	....	....	....
	aq	−159.0	....	....	....
Lithium					
LiOH	c	−116.45	−106.1	77.77	12
	aq	−121.51	−107.82	79.031	0.9
$LiOH \cdot H_2O$	c	−188.77	−164.8	120.80	22
Li_2SO_4	c	−342.83	....	....	....
	aq	−350.01	−317.78	232.928	10.9
$Li_2SO_4 \cdot H_2O$	c	−414.20	....	....	....
Lutetium					
Lu	g	87	....	....	44.14
	c	0.00	0.00	0.000	....
Lu^{++}	aq	−160.1	−155.0	113.61	....
$LuCl_3$	c (γ)	−227.9	....	....	....
	aq	−280.2	−249.0	182.51	....
LuI_3	c (β)	−133.2	....	....	....
	aq	−200.2	−192.0	140.73	....
$Lu_2(SO_4)_3$	aq, ∞	−970.9	−842.0	617.17	....
$Lu_2(SO_4)_3 \cdot 8H_2O$	c	....	−1308.1	958.82	....
Magnesium					
Mg	g	35.9	27.6	−20.23	35.50
	c	0.00	0.00	0.000	7.77
Mg^{++}	aq	−110.41	−108.99	79.89	−28.2
$Mg_3(AsO_4)_2$	c	−731.3	....	....	....
$MgHAsO_4$	aq	−349.2	....	....	....
$Mg(H_2AsO_4)_2$	aq	−541.0	....	....	....
$Mg(NH_4)AsO_4 \cdot 6H_2O$	c	−800.7	....	....	....
Mg_3Bi_2	c	−36.5	....	....	....
$MgBr_2$	c	−123.7	....	....	....
	aq	−168.21	−158.14	115.914	10.4
$MgBr_2 \cdot 6H_2O$	c	−575.4	−491.0	359.90	95
$Mg(CN)_2$	aq	−31.9	....	....	....
$MgCN_2$	c	−60.3	....	....	....
$MgCO_3$	c	−266	−246	180.3	15.7
$MgCl_2$	c	−153.40	−141.57	103.769	21.4
	aq	−190.46	−171.69	125.846	−1.9
$MgCl_2 \cdot H_2O$	c	−231.15	−206.11	151.076	32.8
$MgCl_2 \cdot 2H_2O$	c	−305.99	−267.32	195.942	43.0
$MgCl_2 \cdot 4H_2O$	c	−454.00	−390.49	286.224	63.1
$MgCl_2 \cdot 6H_2O$	c	−597.42	−305.65	370.635	87.5
Mg(OH)Cl	c	−191.3	−175.0	128.27	19.8
$Mg(ClO_4)_2$	c	−140.6	....	....	....
$Mg(ClO_4)_2 \cdot 2H_2O$	c	−290.7	....	....	....
$Mg(ClO_4)_2 \cdot 4H_2O$	c	−438.6	....	....	....
$Mg(ClO_4)_2 \cdot 6H_2O$	c	−583.2	....	....	....
$MgCrO_4$	c	−318.3	....	....	....
	aq	−321.2	....	....	....
MgF_2	c	−263.5	−250.8	183.83	13.68
MgH	g	41	34	−24.9	47.61
MgI_2	c	-86.0	....	....	....
	aq	−137.15	−133.69	97.993	24.1
Mg_3N_2	c	−110.24	....	....	....
$Mg(NO_3)_2$	c	−188.72	−140.63	103.080	39.2
	aq	−209.15	−161.81	118.605	41.8
$Mg(NO_3)_2 \cdot 6H_2O$	c	−624.36	....	....	....
MgO	c	−143.84	−136.13	99.781	6.4
MgO_2	c	−148.9	....	....	....
$Mg(OH)_2$	c	−221.00	−199.27	146.062	15.09
$Mg_3(PO_4)_2$	c	−961.5	....	....	....
$Mg(NH_4)PO_4 \cdot 6H_2O$	c	−881.0	....	....	....
MgS	c	−83.0	....	....	....
$MgSO_4$	c	−305.5	−280.5	152.83	21.9
	aq	−327.31	−286.33	209.876	−24.0
$MgSO_4 \cdot 2H_2O$	c	−381.9	....	....	....
$MgSO_4 \cdot 4H_2O$	c	−595.5	....	....	....
$MgSO_4 \cdot 6H_2O$	c	−736.6	....	....	....
$MgSO_4 \cdot 7H_2O$	c	−808.7	....	....	....
Mg_3Sb_2	c	−68.1	....	....	....
$MgSiO_3$	c	−357.9	−337.2	247.16	16.2
Mg_2SiO_4 (forsterite)	c	−488.2	−459.8	337.03	22.7
Mg_2Sn	c	−17.0	....	....	....
$MgWO_4$	c	−345.2	....	....	....
Manganese					
Mn	g	68.34	58.23	−42.682	41.49
	c (α)	0.00	0.00	0.000	7.59
	c (γ)	0.37	0.33	−0.242	7.72
Mn^{++}	aq	−52.3	−53.4	39.14	−20
$MnBr_2$	c	−90.7	....	....	....
	aq	−110.1	....	....	....
$MnBr_2 \cdot H_2O$	c	−164.4	....	....	....
$MnBr_2 \cdot 4H_2O$	c	−367.5	....	....	....
$MnBr_3$	aq	−111	....	....	....
Mn_3C	c	−1	−1	0.73	23.6
$MnCO_3$	c	−213.9	−195.4	143.23	20.5
	ppt	−211.0	....	....	....
	aq	−213.9	−179.6	131.64	−32.7
MnC_2O_4	c	−258.2	....	....	....
$MnC_2O_4 \cdot 2H_2O$	c	−388.6	....	....	....
$MnC_2O_4 \cdot 3H_2O$	c	−455.4	....	....	....
$Mn(C_2H_3O_2)_2$	c	−273.0	....	....	....
	aq	−285.2	....	....	....
$Mn(C_2H_3O_2)_2 \cdot 4H_2O$	c	−556.9	....	....	....
$MnCl_2$	c	−115.3	−105.5	77.330	28.0
$MnCl_2 \cdot H_2O$	c	−188.5	....	....	....
$MnCl_2 \cdot 2H_2O$	c	−276.7	....	....	....

Substance	State	ΔHf°	ΔFf°	Log₁₀ Kf	S°
Manganese					
$MnCl_2\cdot 4H_2O$	c	−407.0			
MnF_2	c	−189	−179	131.2	22.2
	aq	−209.2			
MnI_2	c	−59.3			
MnI_2	aq	−79.0			
$MnI_2\cdot H_2O$	c	−127.9			
$MnI_2\cdot 2H_2O$	c	−194.5			
$MnI_2\cdot 4H_2O$	c	−327.5			
$MnI_2\cdot 6H_2O$	c	−451			
$Mn(N_3)_2$	c	92.2			
Mn_5N_2	c	−57.8			
Mn_3N_2	c	−81			
$Mn(NO_3)_2$	c	−166.32			
$Mn(NO_3)_2\cdot 3H_2O$	c	−355.1			
$Mn(NO_3)_2\cdot 6H_2O$	c	−566.50			
MnO	g	34.6			
	c	−92.0	−86.8	63.62	14.4
MnO_2	c	−124.5	−111.4	81.655	12.7
Mn_2O_3	c	−232.1			
Mn_3O_4	c	−331.4	−306.0	224.29	35.5
$Mn(OH)_2$	am	−165.8	−145.9	106.94	21.1
$Mn(OH)_3$	am	−212			
$Mn_3(PO_4)_2$	c	−771			
MnS (green)	c	−48.8	−49.9	36.58	18.7
(red)	c	−47.6			
$MnSO_4$	c	−254.24	−228.48	167.473	26.8
$MnSO_4$	aq	−269.2	−230.7	169.10	−16
$MnSO_4\cdot H_2O$	c (I)	−328.5			
	c (II)	−322.4			
$MnSO_4\cdot 4H_2O$	c	−539.3			
$MnSO_4\cdot 5H_2O$	c	−609.6			
$MnSO_4\cdot 7H_2O$	c	−750.0			
$Mn_2(SO_4)_3$	c	−666.9			
	aq	−699			
$MnSiO_3$	c	−302.5	−283.3	207.65	21.3
	gls	−294.0			
Mercury					
Hg	g	14.54	7.59	−5.563	41.8
	lq	0.00	0.00	0.00	18.5
Hg^{++}	aq		39.38	−28.865	
Hg_2^{++}	aq		36.79	−26.967	
$HgBr$	g	23	18	−13.2	65.0
Hg_2Br_2	c	−49.42	−42.714	31.309	50.9
$HgBr_2$	c	−40.5	−38.8		
$Hg(CN)_2$	c	62.5			
$Hg(ONC)_2$	c	64			
$Hg(CNS)_2$	c	48.0			
HgC_2O_4	c	−161.8			
$Hg(C_2H_3O_2)_2$	c	−199.4			
$Hg_2(C_2H_3O_2)_2$	c	−201.1			
$Hg(CH_3)_2$	lq	18.			
$Hg(C_2H_5)_2$	lq	15			
$HgCl$	g	19	14	−10.3	62.2
Hg_2Cl_2	c	−63.32	−50.35	36.906	46.8
$HgCl_2$	c	−55.0	−42.2		
HgF	g	14			
HgH	g	58.06	52.60	−38.555	52.42
HgI	g	33	23	−16.9	67.1
Hg_2I_2 (yellow)	c	−28.91	−26.60	19.497	57.2
HgI_2 (red)	c	−25.2			
(yellow)	c	−24.55			
$Hg_2(N_3)_2$	c	133			
$Hg(NO_3)_2\cdot \frac{1}{2}H_2O$	c	−93.0			
$Hg_2(NO_3)_2\cdot 2H_2O$	c	−206.9			
HgO (red)	c	−21.68	−13.990	10.255	17.2
(yellow)	c	−21.56	−13.959	10.232	17.5
Hg_2O	c	−21.8	−12.80		
$Hg(OH)_2$	aq		−66.0	43.38	
HgS (red)	c	−13.90	−11.67	8.554	18.6
(black)	c	−12.90	−11.05	8.100	19.9
$HgSO_4$	c	−168.3			
Hg_2SO_4	c	−177.34	−149.12	109.303	47.98
Molybdenum					
Mo	g	155.5	144.2	−105.70	43.46
	c	0.00	0.00	0.000	6.83
$MoBr_2$	c	−29			
$MoBr_3$	c	−41			
$MoBr_4$	c	−45			
$MoBr_5$	c	−51			
Mo_2C	c	4.3	2.9	−2.13	19.7
$MoCl_2$	c	−44			
$MoCl_3$	c	−65			
$MoCl_4$	c	−79			
$MoCl_5$	c	−90.8			
$MoCl_6$	c	−90			
MoI_2	c	−12			
MoI_3	c	−15			
MoI_4	c	−18			
MoI_5	c	−18			
Mo_2N	c	−8 3			
MoO_2	c	−130			
MoO_3	c	−180.33	−161.95	118.707	18.68
	aq	−188.1			
Molybdenum					
MoO_4	aq	−173.5			
MoO_5	aq	−155.1			
MoS_2	c	−55.5	−53.8	39.43	15.1
MoS_3	c	−61.2			
H_2MoO_4 (white)	c	−256.9			
$H_2MoO_4\cdot H_2O$ (yellow)	c	−331.4			
Neodymium					
Nd	g	87			
	c	0.00	0.00	0.000	
Nd^{+++}	aq	−171.2	−167.6	122.85	
$NdCl_3$	c (α)	−254.3			
	aq	−291.3	−261.6	191.75	
$NdCl_3\cdot 6H_2O$	c	−692.3			
$Nd(OH)_3$	c		−309.6	226.93	
NdI_3	c (α)	−158.9			
	aq	−211.3	−204.6	149.97	
Nd_2O_3	c	−442.0			
Nd_2S_3	c	−281.8			
$Nd_2(SO_4)_3$	c	−948.1			
	aq	−993.1	−867.2	635.65	
$Nd_2(SO_4)_3\cdot 5H_2O$	c	−1318.4			
$Nd_2(SO_4)_3\cdot 8H_2O$	c	−1524.7	−1334.5	978.17	
Neon					
Ne	g	0.00	0.00	0.000	34.95
Neptunium					
Np	c	0.00	0.00	0.000	
$NpBr_3$	c	−174			
$NpBr_4$	c	−183			
$NpCl_3$	c	−216			
$NpCl_4$	c	−237			
$NpCl_5$	c	−246			
NpF_3	c	−360			
NpF_4	c	−428			
NpI_3	c	−120			
Nickel					
Ni	g	101.61	90.77	−66.533	43.59
	c	0.00	0.00	0.000	7.20
Ni^{++}	aq	−15.3	−11.1	8.136	−38.1
$NiBr_2$	c	−54.2			
$NiBr_2\cdot 3H_2O$	c	−277.8			
Ni_3C	c	11.0			
$Ni(CN)_2$	c	27.1			
$NiCO_3$	c		−146.7	107.53	
$NiCl_2$	c	−75.5	−65.1	47.72	25.6
$NiCl_2\cdot 2H_2O$	c	−220.8			
$NiCl_2\cdot 4H_2O$	c	−364.7			
Nickel					
$NiCl_2\cdot 6H_2O$	c	−505.8	−410.5	300.89	75.2
NiF_2	c	−159.5			
$NiF_2\cdot 4H_2O$	c		−379.9	278.46	
NiH	g	93			
	c	−2.7			
NiH_2	c	−6.2			
NiI_2	c	−20.5			
$Ni(IO_3)_2$	c	−124.5			
$Ni(IO_3)_2\cdot 2H_2O$	c	−264.9			
	c (I)	−263.9			
	c (II)	−401.7			
$Ni(IO_3)_2\cdot 4H_2O$	c	−102.2			
$Ni(NO_3)_2$	c	−531.4			
$Ni(NO_3)_2\cdot 6H_2O$	g	59.3	51.8	−37.97	57
NiO	c	−58.4	−51.7	37.90	9.22
	c	−128.6	−108.3	79.382	19
$Ni(OH)_2$	c	−162.1			
$Ni(OH)_3$	c	−17.5			
NiS	c	−213.0	−184.9	135.53	18.6
$NiSO_4$	aq	−232.2	−188.4	138.09	−34.0
	c	−644.98			
$NiSO_4\cdot 6H_2O$ (green)	c	−642.5	−531.0	389.2	73.1
(blue)	c	−712.9			
$NiSO_4\cdot 7H_2O$					
Niobium					
Nb	g	184.5	173.7	−127.32	44.49
	c	0.00	0.00	0.000	8.3
	c	−387.8			
Nb_2O_4	c	−463.2			
Nb_2O_5					
Nitrogen (NH_4OH; NH_2OH; N_2H_4; and N_3H; see under ammonia)					
N_2	g	0.00	0.00	0.000	45.77
$NOBr$	g	19.56	19.70	−14.44	65.16
$NOCl$	g	12.57	15.86	−11.625	63.0
NF_3	g	−27.2			
NO	g	21.60	20.72	−15.187	50.34
NO_2	g	8.09	12.39	−9.082	57.47
N_2O_4	g	2.31	23.49	−17.219	72.73
N_2O	g	19.49	24.76	−18.149	52.58
N_2O_5	g	3.6			
	c	−10.0			
HNO_3	lq	−41.40	−19.10	14.000	37.19
	aq	−49.37	−26.41	19.36	35.0
$HNO_3\cdot H_2O$	lq	−112.96	−78.41	57.474	51.83
$HNO_3\cdot 3H_2O$	lq	−252.20	−193.70	141.980	82.92
Osmium					
Os	g	174	163	−119.5	45.97
	c	0.00	0.00	0.000	7.8
OsO_4	g	−79.9	−67.9	49.77	65.6

VALUES OF CHEMICAL THERMODYNAMIC PROPERTIES (Continued)

Substance	State	ΔHf°	ΔFf°	Log_{10} Kf	S°
Osmium					
OsO_4	g	−79.9	−67.9	49.77	65.6
(white)	c	−91.7	−70.5	51.68	34.7
(yellow)	c	−93.4	−70.7	51.82	29.7
	aq	....	−68.59	50.276	....
OsS_2	c	−35	....	....	....
H_2OsO_5	aq	....	−125.28	91.829	....
Oxygen					
O_2	g	0.00	0.00	0.000	49.003
O_3	g	34.0	39.06	−28.631	56.8
OH^-	aq	−54.957	−37.595	27.56	−2.519
Palladium					
Pd	g	93	84	−61.6	39.91
	c	0.00	0.00	0.000	8.9
$PdBr_2$	c	−24.9	....	....	....
	aq	−205.6	....	....	....
$Pd(CN)_2$	c	52.1	....	....	....
$PdCl_2$	c	−45.4	....	....	....
$PdCl_4^{--}$	aq	−128.3	−96.7	70.88	41.
H_2PdCl_4	aq	−129.3	....	....	....
Pd_2H	c	−8.9	....	....	....
PdO	c	−20.4	....	....	....
$Pd(OH)_2$	c	−92.1	....	....	....
$Pd(OH)_4$	c	−169.4	....	....	....
Phosphorus					
P	g	75.18	66.71	−48.897	38.98
(white)	c	0.00	0.00	0.000	10.6
(red)	c	−4.4	....	....	....
(black)	c	−10.3	....	....	....
P_2	g	33.82	24.60	−18.031	52.13
P_4	g	13.12	5.82	−4.266	66.90
PBr_3	g	−35.9	−41.2	30.19	83.11
PBr_5	c	−66	....	....	....
$POBr_3$	c	−114.6	....	....	....
PCl_3	g	−73.22	−68.42	50.151	74.49
PCl_5	g	−95.35	−77.59	56.872	84.3
$POCl_3$	g	−141.5	−130.3	95.508	77.59
PH_3	g	2.21	4.36	−3.196	50.2
PI_3	c	−10.9	....	....	....
PN	g	−20.2	−25.3	18.54	50.45
HPO_3	c	−228.2	....	....	....
	aq	−234.9	....	....	....
H_3PO_2	c	−145.5	....	....	....
	aq	−145.6	....	....	....
H_3PO_3	c	−232.2	....	....	....
	aq	−232.2	....	....	....
H_3PO_4	c	−306.2	....	....	....
$H_4P_2O_7$	c	−538.0	....	....	....
Platinum					
Pt	g	121.6	110.9	−81.288	45.96
	c	0.00	0.00	0.000	10.0
$PtBr_4$	c	−41.3	....	....	....
$H_2PtBr_6{\cdot}9H_2O$	c	−734.7	....	....	....
$PtCl_2$	c	−35.5	....	....	....
$PtCl_4$	c	−62.9	....	....	....
$PtCl_4^{--}$	aq	−123.4	−91.9	67.36	42.
$PtCl_4{\cdot}5H_2O$	c	−425.8	....	....	....
$PtCl_6^{--}$	aq	−167.4	123.1	90.231	52.6
H_2PtCl_6	aq	−167.3	....	....	....
$H_2PtCl_6{\cdot}6H_2O$	c	−572.9	....	....	....
PtI_4	c	−21.6	....	....	....
$Pt(OH)_2$	c	−87.2	−68.2	49.99	26.5
PtS	c	−20.8	....	....	....
PtS_2	c	−27.8	....	....	....
Plutonium					
$PuBr_3$	c	−187.2	....	....	....
$PuCl_3$	c	−230.0	....	....	....
PuOCl	c	−222.8	....	....	....
PuF_3	aq	−374.6	....	....	....
PuH_2	c	−31.7	....	....	....
PuI_3	c	−133	....	....	....
PuO_2	c	−251	....	....	....
Polonium					
PoO_2	c	....	−46.3	33.7	....
Potassium					
K	g	21.51	14.62	−10.716	38.30
	c	0.00	0.00	0.000	15.2
K^+	ao	−60.04	−67.466	49.452	24.5
$KAl(SO_4)_7$	c	−589.24	−534.29	391.627	48.9
$KAl(SO_4)_2{\cdot}12H_2O$	c	−1447.74	--1227.8	899.96	164.3
KH_2AsO_4	c	−271.5	−237.0	173.72	37.08
KBr	c	−93.73	−90.63	66.430	23.05
	aq	−88.94	−92.04	67.464	43.8
KBrO	aq	−82.0	....	....	....
$KBrO_3$	c	−79.4	−58.2	42.66	35.65
	aq	69.6	−56.6	41.49	63.4
K_2CO_3	c	−273.93	....	....	....
$K_2CO_3{\cdot}\frac{1}{2}H_2O$	c	−210.43	....	....	....
$K_2CO_3{\cdot}1\frac{1}{2}H_2O$	c	−283.40	....	....	....
$KHCO_3$	c	−229.3	....	....	....
$KC_2H_3O_2$	c	−173.2	....	....	....
$K_2C_2O_4$	c	−320.8	....	....	....
$K_2C_2O_4{\cdot}H_2O$	c	−392.17	....	....	....
KCN	c	−26.90	....	....	....
KCNO	c	−98.5	....	....	....
Potassium					
KCNS	c	−48.62	....	....	....
KCl	g	−51.6	−56.2	41.19	57.24
	c	−104.18	−97.592	71.534	19.76
	aq	−100.06	−98.82	72.431	37.7
KClO	aq	−85.4	....	....	....
$KClO_3$	e	−93.50	−69.29	50.789	34.17
	aq, 100	−84.09	....	....	....
	aq	−83.54	−68.09	49.909	63.5
$KClO_4$	c	−103.6	−72.7	53.29	36.1
	aq	−91.45	−70.04	51.338	68.0
K_2CrO_4	c	−330.49	....	....	....
$K_2Cr_2O_7$	c	−485.90	....	....	....
KF	c	−134.46	−127.42	93.397	15.91
$KF{\cdot}2H_2O$	c	−277.0	−242.7	177.90	36
$KF{\cdot}4H_2O$	c	−418.0	....	....	....
KHF_2	c	−219.98	−203.73	149.331	24.92
KI	c	−78.31	−77.03	56.462	24.94
	aq	−73.41	−79.82	58.504	50.6
KIO_3	c	−121.5	−101.7	74.54	36.20
	aq	−115.0	−99.9	73.22	52.2
KIO_4	aq, 1200	−97.6	....	....	....
$KMnO_4$	c	−194.4	−170.6	125.05	41.04
KNH_2	c	−28.3	....	....	....
KNO_2	c	−88.5	....	....	....
KNO_3	c	−117.76	−93.96	68.871	31.77
	aq	−109.41	−93.88	68.813	69.5
K_2O	c	−86.4	....	....	....
K_2O_2	c	−118	....	....	....
K_2O_4	c	−134	....	....	....
KOH	c	−101.78	....	....	....
	aq	−115.00	−105.06	77.008	22.0
K_2S	c	−100	....	....	....
K_2S_4	c	−113.0	....	....	....
K_2SO_3	c	−266.9	....	....	....
K_2SO_4	c	−342.66	−314.62	230.612	42.0
$KHSO_4$	c	−276.8	....	....	....
$K_2S_2O_6$	c	−413.6	....	....	....
$K_2S_2O_8$	c	−458.3	....	....	....
$K_2S_4O_6$	c	−422	....	....	....
K_2Se	c	−79.3	....	....	....
Praseodymium					
Pr	g	87	....	....	....
	c	0.00	0.00	0.000	....
Pr^{+++}	aq	−172.7	−169.1	123.95	....
$PrCl_3$	c (α)	−257.8	....	....	....
	aq	−292.8	−263.1	192.85	....
$PrCl_3{\cdot}H_2O$	c	−330.8	....	....	....
$PrCl_3{\cdot}7H_2O$	c	−764.5	....	....	....
PrI_3	c (α)	−162.0	....	....	....
	aq	−212.8	−206.1	151.07	....
PrO_2	c	−234.0	....	....	....
Pr_2O_3	c	−444.5	....	....	....
$Pr(OH)_3$	c	....	−310.7	227.74	....
$Pr_2(SO_4)_3$	aq	−996.1	−870.2	637.84	....
$Pr_2(SO_4)_3{\cdot}8H_2O$	c	....	−1337.0	980.00	....
Promethium					
Pm^{+++}	aq	−170.4	....	....	....
$PmCl_3$	c	−251.9	....	....	....
	aq	−290.5	....	....	....
Radium					
Ra	g	31	23	−16.8	42.15
	c	0.00	0.00	0.000	17
Ra^{++}	aq	−126.	−134.5	98.58	13
$RaCl_2{\cdot}2H_2O$	c	−351	−311.7	228.47	50
$Ra(NO_3)_2$	c	−237	−190.3	139.49	52
RaO	c	−125	....	....	....
$RaSO_4$	c	−352	−326.0	238.95	34
Radon					
Rn	g	0.00	0.00	0.000	42.10
Rhenium					
Re	g	189	179	−131.2	45.13
	c	0.00	0.00	0.000	10
ReF_6	g	−273	....	....	....
Re_2O_7	c	−297.5	....	....	....
Re_2O_8	c	−148.3	....	....	....
ReS_2	c	−44.3	....	....	....
ReO_4^-	aq	−190.3	....	....	....
Rhodium					
Rh	g	138	127	−93.09	44.39
	c	0.00	0.00	0.000	7.6
RhCl	c	−16	....	....	....
$RhCl_2$	c	−36	....	....	....
$RhCl_3$	c	−56	....	....	....
RhO	c	−21.7	....	....	....
RH_2O	c	−22.7	....	....	....
Rh_2O_3	c	−68.3	....	....	....
Rubidium					
Rb	g	20.51	13.35	−9.785	40.63
	c	0.00	0.00	0.000	16.6
Rb^{+++}	aq	−58.9	−67.45	47.974	29.7
RbBr	c	−93.03	−90.38	66.247	25.88
	aq	−87.8	−92.02	67.449	49.0
Rb_2CO_3	c	−269.6	....	....	....
$Rb_2CO_3{\cdot}H_2O$	c	−344.2	....	....	....

VALUES OF CHEMICAL THERMODYNAMIC PROPERTIES (Continued)

Substance	State	ΔHf°	ΔFf°	Log_{10} Kf	S°
Rubidium					
$RbHCO_3$	c	−228.5	......	......	......
$RbCNS$	c	−54	......	......	......
	aq	−41.7	......	......	......
$RbCl$	c	−102.9	−98.48	......	......
	aq	−98.9	−98.80	72.419	42.9
$RbClO_3$	c	−93.8	−69.8	51.16	36.3
	aq	−82.4	−68.07	49.894	68.7
$RbClO_4$	c	−103.87	−73.19	53.647	38.4
	aq	−90.3	−70.02	51.324	73.2
RbF	c	−131.28	......	......	......
	aq	−137.6	−133.53	97.876	27.4
$RbHF_2$	c	−217.3	......	......	......
RbI	c	−78.5	−77.8	57.03	28.21
	aq	−72.3	−79.80	58.492	55.5
$RbNO_3$	c	−117.04	......	......	......
	aq	−108.5	−93.86	68.798	64.7
$RbNH_2$	c	−25.7	......	......	......
Rb_2O	c	−78.9	......	......	......
Rb_2O_2	c	−101.7	......	......	......
$RbOH$	c	−98.9	......	......	......
	aq	−113.9	−105.05	77.000	27.2
$RbOH\cdot H_2O$	c	−177.8	......	......	......
$RbOH\cdot 2H_2O$	c	−250.8	......	......	......
Rb_2S	c	−83.2	......	......	......
$RbHS$	c	−62.4	......	......	......
$RbHSO_4$	c	−273.7	......	......	......
Rb_2SO_4	c	−340.50	......	......	......
	aq	−334.7	−312.24	228.868	33.8
Ruthenium					
Ru	g	160	149	−109.2	44.57
	c	0.00	0.00	0.000	6.9
$RuCl_3$	c	−63	......	......	......
RuO_2	c	−52.5	......	......	......
RuS_2	c	−48.1	−44.1	......	......
Samarium					
Sm	g	87	......	......	43.74
Sm^{+++}	aq	−169.8	−165.9	121.60	......
$SmCl_3$	c (α)	−249.8	......	......	......
	aq	−289.9	−259.9	190.50	......
SmI_3	c (β)	−153.4	......	......	......
	aq	−209.9	−202.9	148.72	......
$Sm(OH)_3$	c	......	−308.8	226.35	......
$Sm_2(SO_4)_3$	aq	−990.3	−863.8	633.15	......
$Sm_2(SO_4)_3\cdot 8H_2O$	c	......	−1332.6	976.78	......
Scandium					
Sc	g	93	......	......	41.76
Sc^{+++}	aq	−148.8	−143.7	105.33	......
$ScBr_3$	c	−179.4	......	......	......
	aq	−235.5	−217.4	159.35	......
$ScCl_3$	c	−220.8	......	......	......
	aq	−268.9	−237.7	174.23	......
Selenium					
Se	g	48.37	38.77	−28.42	42.21
Se (gray, hex.)	c	0.00	0.00	0.000	10.0
(red, mon.)	c	0.2	......	......	......
	gls	1.05	......	......	......
	ppt	1.05	......	......	......
H_2Se	g	20.5	17	−12.468	52.9
	aq	18.1	18.4	−13.49	39.9
SeO_2	c	−55.00	......	......	......
H_2SeO_3	c	−126.5	......	......	......
	aq	−122.39	−101.8	74.622	45.7
H_2SeO_4	c	−128.6	......	......	......
	aq	−145.3	−105.42	77.276	5.7
Silicon					
Si	g	88.04	77.41	−56.740	40.12
	c	0.00	0.00	0.000	4.47
$SiBr_4$	lq	−95.1	......	......	......
SiC	c	−26.7	−26.1	19.13	3.94
$SiCl_4$	g	−145.7	−136.2	99.83	79.2
	lq	−153.0	−136.9	100.34	57.2
SiF_4	g	−370	−360	263.9	68.0
SiH_4	g	−14.8	−9.4	6.89	48.7
SiI_4	c	−31.6	......	......	......
Si_3N_4	c	−179.3	−154.7	113.39	22.4
SiO_2 (quartz)	c	−205.4	−192.4	141.03	10.00
(cristobalite)	c	−205.0	−192.1	140.81	10.19
(tridymite)	c	−204.8	−191.9	140.66	10.36
	gls	−202.5	−190.9	139.92	11.2
	aq	−201.0	......	......	......
SiS_2 (white)	c	−34.7	......	......	......
H_2SiF_6	aq	−557.2	......	......	......
H_2SiO_3	c	−270.7	......	......	......
H_2SiO_4	c	−340.6	......	......	......
H_2SiO_5	c	−472.4	......	......	......
$H_6Si_2O_7$	c	−611.2	......	......	......
Silver					
Ag	g	69.12	59.84	−43.862	41.32
	c	0.00	0.00	0.000	10.21
Ag^+	aq	25.31	18.43	−13.51	17.67
$AgBr$	c	−23.78	−22.39	16.807	25.60
Ag_2CO_3	c	−120.97	−104.48	76.582	40.0
$AgC_2H_3O_2$	c	−93.41	......	......	......
$AgCN$	c	34.94	39.20	−28.733	20.0

Substance	State	ΔHf°	ΔFf°	Log_{10} Kf	S°
Silver					
$AgCNO$	c	−21.1	......	......	......
$AgONC$	c	43.2	......	......	......
$AgCNS$	c	21.0	......	......	......
$AgCl$	g	23.23	16.79	−12.307	58.5
	c	−30.36	−26.22	19.222	22.97
$AgClO_2$	c	0.00	16.0	−11.73	32.16
$AgClO_3$	c	−5.73	......	......	......
$AgClO_4$	c	−7.75	......	......	......
Ag_2CrO_4	c	−170.15	−148.57	108.90	51.8
AgF	c	−48.5	−44.2	32.40	20
	aq	−53.35	−47.65	34.927	15.4
AgH	g	67.7	60.8	−44.56	48.86
AgI	c	−14.91	−15.85	11.618	27.3
$AgIO_3$	c	−42.02	−24.08	......	35.7
AgN_3	c	66.8	......	......	......
$AgNO_2$	c	−10.61	4.74	−3.477	30.62
	aq	−0.09	9.79	−7.176	49.0
$AgNO_3$	c	−29.43	−7.69	5.637	33.68
	aq	−24.06	−7.98	5.849	52.67
Ag_2O	c	−7.31	−2.59	1.896	29.09
Ag_2O_2	c	−6.3	......	......	......
Ag_2S	c (α)	−7.60	−9.62	7.051	34.8
	c (β)	−7.01	−9.36	6.860	35.9
Ag_2SO_4	c	−170.50	−147.17	107.873	47.8
Sodium					
Na	g	25.98	18.67	−13.685	36.72
	c	0.00	0.00	0.000	12.2
Na^+	aq	−57.28	−62.59	45.88	14.4
Na_3AsO_4	c	−365	......	......	......
$Na_3AsO_4\cdot 12H_2O$	c	−1213.9	......	......	......
$NaBO_2$	c	−253	......	......	......
$NaBO_3\cdot 4H_2O$	c	−504.8	......	......	......
$Na_2B_4O_7$	c	−777.7	......	......	......
$Na_2B_4O_7\cdot 4H_2O$	c	−1072.9	......	......	......
$Na_2B_4O_7\cdot 5H_2O$	c	−1143.5	......	......	......
$Na_2B_4O_7\cdot 10H_2O$	c	−1497.2	......	......	......
Na_3BiO_4	c	−288.	......	......	......
$NaBr$	c	−86.03	......	......	......
	aq	−86.18	−87.16	63.889	33.7
$NaBr\cdot 2H_2O$	c	−227.25	......	......	......
$NaCN$	c	−21.46	......	......	......
$NaCN\cdot \frac{1}{2}H_2O$	c	−56.19	......	......	......
$NaCN\cdot 2H_2O$	c	−162.25	......	......	......
$NaCNO$	c	−95.6	......	......	......
$NaCNS$	c	−41.73	......	......	......
	aq	−40.1	......	......	......
Na_2CO_3	c	−270.3	−250.4	183.53	32.5
$Na_2CO_3\cdot H_2O$	c	−341.8	......	......	......
$Na_2CO_3\cdot 7H_2O$	c	−765.1	......	......	......
$Na_2CO_3\cdot 10H_2O$	c	−975.6	......	......	......
$NaHCO_3$	c	−226.5	−203.6	149.24	24.4
$NaC_2H_3O_2$	c	−169.8	......	......	......
$NaC_2H_3O_2\cdot 3H_2O$	c	−383.50	......	......	......
$Na_2C_2O_4$	c	−314.3	......	......	......
$NaCl$	g	−43.50	......	......	......
	c	−98.23	−91.79	67.277	17.30
	aq	−97.302	−93.94	68.856	27.6
$NaClO$	aq	−82.7	......	......	......
$NaClO_2$	c	−72.65	......	......	......
$NaClO_3$	c	−85.73	......	......	......
	aq, 400	−80.74	−62.8	......	......
	aq, 1000	−80.73	......	......	......
	aq, 5000	−80.74	......	......	......
	aq	−80.78	−63.21	46.332	53.4
$NaClO_4$	c	−92.18	......	......	......
	aq	−88.69	−65.16	47.688	57.9
Na_2CrO_4	c	−317.6	......	......	......
$Na_2CrO_4\cdot 4H_2O$	c	−601.3	......	......	......
$Na_2Cr_2O_7$	aq, 600	−463.4	......	......	......
NaF	c	−136.0	−129.3	94.78	14.0
	aq	−135.94	−128.67	94.313	12.1
$NaHF_2$	c	−216.6	......	......	......
NaH	g	29.88	24.78	−18.163	44.93
	c	−13.7	......	......	......
NaI	g	−20.94	......	......	......
	c	−68.84	......	......	......
	aq	−70.65	−70.94	51.998	40.5
$NaI\cdot 2H_2O$	c	−211.05	......	......	......
	aq, ∞	−112.3	......	......	......
$NaNH_2$	c	−28.4	......	......	......
$NaNO_2$	c	−85.9	......	......	......
$NaNO_3$	c	−101.54	−87.45	64.100	27.8
	aq	−106.65	−89.00	65.236	49.4
Na_2O	c	−99.4	−90.0	65.97	17.4
Na_2O_2	c	−120.6	......	......	......
$NaOH$	c	−101.99	......	......	......
	aq	−112.24	−100.184	73.434	11.9
$NaOH\cdot H_2O$	c	−175.17	−149.00	109.215	20.2
$NaPO_3$	c	−288.6	......	......	......
NaH_2PO_3	c	−289.4	......	......	......
$NaH_2PO_3\cdot 2\frac{1}{2}H_2O$	c	−454.8	......	......	......
Na_2HPO_3	c	−338.0	......	......	......
$Na_2HPO_3\cdot 5H_2O$	c	−684.2	......	......	......
NaH_2PO_4	aq, 300	−367.7	......	......	......

VALUES OF CHEMICAL THERMODYNAMIC PROPERTIES (Continued)

Substance	State	ΔHf°	ΔFf°	Log₁₀ Kf	S°
Sodium					
Na_2HPO_4	c	−417.4	....	....	....
$Na_2HPO_4 \cdot 2H_2O$	c	−560.2	....	....	....
$Na_2HPO_4 \cdot 7H_2O$	c	−913.3	....	....	....
$Na_2HPO_4 \cdot 12H_2O$	c	−1266.4	....	....	....
Na_3PO_4	c	−460	....	....	....
$Na_3PO_4 \cdot 12H_2O$	c	−1309.0	....	....	....
$NaNH_4HPO_4 \cdot 4H_2O$	c	−682.7	....	....	....
$NaH_3P_2O_7$	c	−602.7	....	....	....
$NaH_3P_2 \cdot O_7H_2O$	c	−670.6	....	....	....
$Na_2H_2P_2O_7$	c	−663.4	....	....	....
$Na_2H_2P_2O_7 \cdot 6H_2O$	c	−1085.5	....	....	....
$Na_3HP_2O_7$	c	−711.4	....	....	....
$Na_3HP_2O_7 \cdot H_2O$	c	−788.2	....	....	....
$Na_3HP_2O_7 \cdot 6H_2O$	c	−1135.7	....	....	....
$Na_4P_2O_7$	c	−760.8	....	....	....
$Na_4P_2O_7 \cdot 10H_2O$	c	−1468.2	....	....	....
Na_2PbO_3	c	−205.	....	....	....
Na_2PtCl_6	c	−273.6	....	....	....
$Na_2PtCl_6 \cdot 2H_2O$	c	−418.5	....	....	....
$Na_2PtCl_6 \cdot 6H_2O$	c	−702.5	....	....	....
Na_2S	c	−89.2	....	....	....
$Na_2S \cdot 4\frac{1}{2}H_2O$	c	−416.9	....	....	....
$Na_2S \cdot 5H_2O$	c	−452.7	....	....	....
$Na_2S \cdot 9H_2O$	c	−736.7	....	....	....
Na_2S_4	c	−98.4	....	....	....
Na_2SO_3	c	−260.6	−239.5	175.55	34.9
$Na_2SO_3 \cdot 7H_2O$	c	−753.4	....	....	....
Na_2SO_4	c	−330.90	−302.78	221.934	35.73
	aq	−331.46	−302.52	221.743	32.9
$Na_2SO_4 \cdot 10H_2O$	c	−1033.48	−870.93	638.380	141.7
$NaHSO_4$	c	−269.2	....	....	....
$NaHSO_4 \cdot H_2O$	c	−339.2	....	....	....
$Na_2S_2O_3$	c	−267.0	....	....	....
$Na_2S_2O_3 \cdot 5H_2O$	c (I)	−621.89	....	....	....
	c (II)	−620.60	....	....	....
$Na_2S_2O_5$	c	−349.1	....	....	....
$Na_2S_2O_6$	c	−399.9	....	....	....
$Na_2S_2O_6 \cdot 2H_2O$	c	−542.5	....	....	....
$Na_2S_3O_6 \cdot 3H_2O$	c	−623.0	....	....	....
$Na_2S_4O_6 \cdot 2H_2O$	c	−550.0	....	....	....
Na_2Se	c	−63.0	....	....	....
$Na_2Se \cdot 4\frac{1}{2}H_2O$	c	−398.2	....	....	....
$Na_2Se \cdot 9H_2O$	c	−709.1	....	....	....
$Na_2Se \cdot 16H_2O$	c	−1199.4	....	....	....
$NaHSe$	c	−27.8	....	....	....
Na_2SeO_4	c	−258.	....	....	....
Na_2SiO_3	c	−363	−341	249.9	27.2
$Na_2SiO_3 \cdot 5H_2O$	c	−720.0	....	....	....
$Na_2SiO_3 \cdot 9H_2O$	c	−1002.0	....	....	....
Na_2SiF_6	c	−677	....	....	....
Na_2SnO_3	c	−276	....	....	....
Na_4SnO_4	aq, 1200	−455.5	....	....	....
Na_2UO_4	c	−501	....	....	....
$Na_2U_2O_7 \cdot 1\frac{1}{2}H_2O$	c	−880	....	....	....
Na_3VO_4	c	−420	....	....	....
Na_2WO_4	c	−395	....	....	....
Na_2ZnO_2	c	−188	....	....	....
Strontium					
Sr	g	39.2	26.3	−19.28	39.33
	c	0.00	0.00	0.000	13.0
Sr^{++}	aq	−130.38	−133.2	97.49	−9.4
$Sr_3(AsO_4)_2$	c	−800.7	....	....	....
$SrBr_2$	c	−171.1	....	....	....
	aq	−188.2	−182.1	133.48	29.2
$SrBr_2 \cdot H_2O$	c	−246.2	....	....	....
$SrBr_2 \cdot 6H_2O$	c	−604.4	....	....	....
$SrCO_3$	c	−291.2	−271.9	199.30	23.2
$Sr(HCO_3)_2$	aq	−460.7	−413.8	303.31	36.0
$Sr(C_2H_3O_2)_2$	c	−356.7	....	....	....
	aq	−362.7	−311.8	....	....
$Sr(C_2H_3O_2)_2 \cdot \frac{1}{2}H_2O$	c	−391.2	....	....	....
SrC_2O_4	aq	−327.4	−294.3	215.72	3
$SrC_2O_4 \cdot H_2O$	c	....	360.8	264.46	....
$SrC_2O_4 \cdot 2\frac{1}{2}H_2O$	c	−504.2	....	....	....
$Sr(CN)_2 \cdot 4H_2O$	c	−335.2	....	....	....
$SrCl_2$	c	−198.0	−186.7	136.85	28
	aq	−210.43	−195.7	143.45	16.9
$SrCl_2 \cdot H_2O$	c	−271.7	....	....	....
$SrCl_2 \cdot 2H_2O$	c	−343.7	....	....	....
$SrCl_2 \cdot 6H_2O$	c	−627.1	....	....	....
SrF_2	c	−290.3	....	....	....
SrH	g	52.4	45.8	−33.57	49.43
SrH_2	c	−42.3	....	....	....
SrI_2	c	−135.5	....	....	....
	aq	−157.1	−157.7	115.59	42.9
$SrI_2 \cdot H_2O$	c	−212.2	....	....	....
$SrI_2 \cdot 2H_2O$	c	−282.8	....	....	....
$SrI_2 \cdot 6H_2O$	c	−571.2	....	....	....
$Sr(N_3)_2$	c	48.9	....	....	....
Sr_3N_2	c	−93.4	....	....	....
$Sr(NO_2)_2$	c	−179.3	....	....	....
$Sr(NO_3)_2$	c	−233.25	....	....	....
	aq	−229.02	−185.8	136.19	60.06
$Sr(NO_3)_2 \cdot 4H_2O$	c	−514.5	....	....	....

Substance	State	ΔHf°	ΔFf°	Log₁₀ Kf	S°
Strontium					
SrO	c	−141 1	−133.8	98.07	13.0
SrO_2	c	−153.6	....	....	....
$SrO_2 \cdot 8H_2O$	c	−722.6	....	....	....
$Sr(OH)_2$	c	−229.3	....	....	....
	aq	−240.29	−208.2	152.61	−14.4
$Sr(OH)_2 \cdot H_2O$	c	−302.3	....	....	....
$Sr(OH)_2 \cdot 8H_2O$	c	−801.2	....	....	....
$Sr_3(PO_4)_2$	c	−987.3	....	....	....
$SrHPO_4$	c	−431.3	....	....	....
$Sr(H_2PO_4)_2 \cdot H_2O$	c	−819.4	....	....	....
SrS	g	19	....	....	....
	c	−108.1	....	....	....
$SrSO_4$	c	−345.3	−318.9	233.75	29.1
	aq	−347.38	−310.3	227.45	−5.3
Sulfur					
S	g	53.25	43.57	−31.9362	40.085
(rhb)	c	0.00	0.00	0.000	7.62
(mon)	c	0.071	0.023	−0.017	7.78
S_2Br_2	lq	−3.6	....	....	....
S_2Cl_2	lq	−14.4	....	....	....
S_2Cl_4	lq	−24.0	....	....	....
$SOCl_2$	lq	−49.2	....	....	....
SO_2Cl_2	lq	−93 0	−75	....	....
SF_6	g	−262	−237	173.7	69.5
H_2S	g	−4.815	−7.892	5.785	49.15
	aq	−9.4	−6.54	4.797	29.2
$H_2S \cdot 6H_2O$	c	−431.2	....	....	....
SO_2	g	−70.96	−71.79	52.621	59.40
SO_3	g	−94.45	−88.52	64.884	61.24
	lq	−104.67	....	....	....
S_2O_7	c	−194.3	....	....	....
H_2SO_3	aq, 200	−146.82	−128.54	....	....
	aq, 500	−147.17	....	....	....
	aq, 1000	−147.59	....	....	....
	aq, 5000	−148.73	....	....	....
H_2SO_4	lq	−193.91	....	....	....
	aq	−216.90	−177.34	129.99	4.1
S^{--}	aq	10	20	−14.660	5.3
SO_3^{--}	aq	−149.2	−118.8	87.08	10.4
SO_4^{--}	aq	−216.90	−177.34	129.99	4.1
$S_2O_3^{--}$	aq	−154.	−127.2	93.24	29.
HS^-	aq	−4.22	3.01	−2.206	14.6
HSO_3^-	aq	−150.09	−126.03	92.378	31.64
HSO_4^-	aq	−211.70	−179.94	131.893	30.32
Tantalum					
Ta	g	185	175	−128.3	44.24
	c	0.00	0.00	0.000	9.9
TaN	c	−58.2	....	....	....
Ta_2O_5	c	−499.9	−470.6	344.94	34.2
Technetium					
Tc	c	0.00	0.00	0.000	9
Tellurium					
Te	g	47.6	38.1	−27.93	43.64
	c	0.00	0.00	0.000	11.88
	(am. ppt)	2.7	....	....	....
Te_2	g	41.0	29.0	−21.26	64.07
$TeBr_4$	c	−49.8	....	....	....
$TeCl_4$	c	−77.2	....	....	....
TeF_6	g	−315	−292	214.0	80.67
H_2Te	g	36.9	33.1	−24.26	56
TeO_2	c	−77.69	−64.60	47.353	16.99
H_2TeO_3	c	−144.7	−115.7	84.81	47.7
	aq	−144.7	....	....	....
H_2TeO_4	aq	−166.7	....	....	....
$H_2TeO_4 \cdot 2H_2O$	c	−306.6	−245.3	179.80	47
Terbium					
Tb	g	87	....	....	....
	c	0.00	0.00	0.000	....
Tb^{+++}	aq	−168.4	−163.9	120.14	....
$TbCl_3$	c (β)	−241.6	....	....	....
	aq	−288.5	−257.9	189.04	....
$Tb_2(SO_4)_3$	aq	−987.5	−859.8	630.22	....
$Tb_2(SO_4)_3 \cdot 8H_2O$	c	....	−1328.2	973.55	....
Thallium					
Tl	g	44.5	36.2	−26.53	43.23
	c	0.00	0.00	0.000	15.4
Tl^+	aq	1.38	−7.755	5.68	30.4
Tl^{+++}	aq	27.7	50.0	−36.65	−106
TlBr	g	−5.	−14	10.3	63.8
	c	−41.2	−39.7	29.10	26.6
	aq	−27.5	−32.3	23.68	49.7
$TlBr_3$	c	−59.0	....	....	....
$TlBr_3 \cdot 4H_2O$	c	−334.6	....	....	....
$TlBrO_3$	c	−24.0	....	....	....
	aq	−10.2	1.3	−0.953	68.6
TlCl	g	−16.	−22	16.1	61.1
	c	−48.99	−44.19	32.391	25.9
	aq	−38.64	−39.105	28.663	43.6
$TlCl_3$	c	−83.9	....	....	....
$TlCl_3 \cdot 4H_2O$	c	−367.7	....	....	....
TlF	g	−33	....	....	....
TlH	g	48	42	−30.8	51.39
TlI	c (II)	−29.7	−29.7	21.77	29.4
	aq	−12.0	−19.9	14.59	56.5

VALUES OF CHEMICAL THERMODYNAMIC PROPERTIES (Continued)

Substance	State	ΔHf°	ΔFf°	Log_{10} Kf	S°
Thallium					
$TlIO_3$	c	....	−47.6	34.89	......
	aq	−53.5	−40.1	29.39	58.1
$TlNO_3$	c	−58.01	−36.07	26.439	38.2
	aq	−47.99	−34.22	25.083	65.6
Tl_2O	c	−41.9	−32.5	23.82	23.8
$TlOH$	c	−56.9	−45.5	33.35	17.3
$Tl(OH)_3$	c	−122.6	....	....	......
Tl_2S	c	−20.8	....	....	......
Tl_2SO_4	c	−221.7	....	....	......
	aq	−214.14	−192.85	141.356	65.1
Thorium					
Th	c	0.00	0.00	0.000	13.6
Th^{++++}	aq	−183.0	....	....	......
$ThBr_4$	c	−227.1	....	....	......
ThC_2	c	−45	....	....	......
$ThCl_4$	c	−285	....	....	......
ThF_4	c	−477	....	....	......
ThH_4	c	−43	....	....	......
ThI_4	c	−131	....	....	......
$Th(NO_3)_4$	aq, 100	−380.48	....	....	......
ThO_2	c	−292	....	....	......
$Th(OH)_4$	c*	−421.5	....	....	......
Th_2S_3	c	−262.0	....	....	......
$Th(SO_4)_2$	c	−602	....	....	......
Thulium					
Tm	c	0.00	0.00	0.000	......
Tm^{+++}	aq	−161.3	−156.5	114.71	......
$TmCl_3$	c (γ)	−229.5	....	....	......
	aq	−281.4	−250.5	183.61	......
TmI_3	c (β)	−137.8	....	....	......
	aq	−201.4	−193.5	141.83	......
Tin					
Sn	g	72	64	−47.6	40.25
(gray), cubic	c (III)	0.6	1.1	−0.806	10.7
(tet), white	c (II)	0.00	0.00	0.000	12.3
$SnBr_2$	c	−63.6	....	....	......
$SnBr_4$	c (I)	−97.1	....	....	......
$SnBr_4 \cdot 8H_2O$	lq	−656.6	....	....	......
$Sn(C_2H_5)_4$	lq	−49	....	....	......
$SnCl_2$	c	−83.6	....	....	......
$SnCl_2 \cdot 2H_2O$	c	−225.9	....	....	......
$SnCl_4$	lq	−130.3	−113.3	83.047	61.8
$SnOCl_2$	aq	−147.5	....	....	......
SnI_2	c	−34.4	....	....	......
SnO	c	−68.4	−61.5	45.08	13.5
SnO_2	c	−138.8	−124.2	91.037	12.5
$Sn(OH)_2$	c	−138.3	−117.6	86.199	23.1
$Sn(OH)_4$	c	−270.5	....	....	......
SnS	c	−18.6	−19.7	14.44	23.6
$Sn(SO_4)_2$	c	−393.4	....	....	......
Titanium					
Ti	g	112	101	−74	43.07
	c	0.00	0.00	0.000	7.24
$TiBr_2$	c	−95	....	....	......
$TiBr_3$	c	−132	....	....	......
$TiBr_4$	c	−155	....	....	......
TiC	c	−54	−53	38.8	5.8
TiCl	g	122	....	....	......
$TiCl_2$	c	−114	....	....	......
$TiCl_3$	c	−165	....	....	......
$TiCl_4$	g	....	....	....	84.4
	lq	−179.3	−161.2	118.16	60.4
TiF_2	c	−198	....	....	......
TiF_3	c	−315	....	....	......
TiF_4	c	−370	....	....	......
H_2TiF_6	lq	−555.1	....	....	......
TiI_2	c	−61	....	....	......
TiI_3	c	−80	....	....	......
TiI_4	c	−102	....	....	......
TiN	c	−73.0	−66.1	48.45	7.2
TiO	g	43	....	....	......
	c (II)	....	....	....	8.31
TiO_2 (rutile)	c (III)	−218.0	−203.8	149.38	12.01
(hydrated)	(*am. ppt*)	−207.	....	....	......
Ti_2O_3	c (II)	−367	−346	253.6	18.83
Ti_3O_5	c	−584	−550	403.1	30.92
Tungsten					
W	g	201.6	191.6	−140.44	41.55
	c	0.00	0.00	0.000	8.0
WC	c	−9.09	....	....	......
WO_2	c	−136.3	....	....	......
WO_3 (yellow)	c	−200.84	−182.47	133.748	19.90
$WO_3 \cdot H_2O$ (H_2WO_4)	c	−279.6	....	....	......
W_2O_5	c	−337.9	....	....	......
WS_2	c	−46.3	−46.2	33.86	23
Uranium					
U	g	125	....	....	......
	c (III)	0.00	0.00	0.000	12.03
U^{+++}	aq	−123.0	−124.4	91.18	−30
U^{++++}	aq	−146.7	−138.4	101.44	−78
UBr_3	c	−170.1	−164.7	120.72	49
UBr_4	c	−196.6	−188.5	138.17	58
UC_2	c	−42	−42	30.8	14
UCl_3	c	−213.0	−196.9	144.32	37.99
UCl_4	c	−251.2	−230.0	168.59	47.4
Uranium					
UCl_5	c	−262.1	−237.4	174.01	62
UCl_6	c	−272.4	−241.5	177.02	68.3
UF_3	c	−357	−339	248.5	26
UF_4	c	−443	−421	308.6	36.1
UF_5	c	−488	−461	337.9	43
UF_6	g	−505	−485	355.5	90.76
UH_3	c	−30.4	....	....	......
UI_3	c	−114.7	−115.3	84.51	56
UI_4	c	−127.0	−126.1	92.43	65
$UIBr_3$	c	−177.1	....	....	......
$UICl_3$	c	−219.9	−204.4	149.82	54
UN	c	−80	−75	55	18
U_2N_3	c	−213	−194	142.2	29
UO_2	c	−270	−257	188.4	18.6
UO_2^+	aq	−247.4	−237.6	174.16	12
UO_2^{++}	aq	−250.4	−236.4	173.28	−17
UO_3	c	−302	−283	207.4	23.57
$UO_3 \cdot H_2O$	c	−375.4	....	....	......
$UO_3 \cdot 2H_2O$	c	−446.2	....	....	......
$UO_4 2H_2O$	c	−436	....	....	......
U_3O_8	c	−898	....	....	......
$U(OH)_3^{+++}$	aq	−204.1	−193.5	141.83	−30
$U(SO_4)_2$	c	−563	....	....	......
UO_2Br_2	aq (dil)	−308.2	....	....	......
$UO_2(C_2H_3O_2)_2$	aq (dil)	−484.0	....	....	......
$UO_2(C_2H_3O_2)_2 \cdot 2H_2O$	c	−624.9	....	....	......
$UO_2(C_2H_3O_2)_2 \cdot NH_4C_2H_3O_2 \cdot 6H_2O$	c	−1045.8	....	....	......
UO_2Cl_2	aq (dil)	−331	....	....	......
UO_2CrO_4	aq (dil)	−456.7	....	....	......
$UO_2CrO_4 \cdot 5\frac{1}{2}H_2O$	c	−838.8	....	....	......
$UO_2(NO_3)_2$	c	−329.2	−273.1	200.18	66
	aq	−349.1	−289.2	211.98	53
$UO_2(NO_3)_2 \cdot H_2O$	c	−404.8	−335.3	245.77	76
$UO_2(NO_3)_2 \cdot 2H_2O$	c	−480.0	−396.6	290.70	85
$UO_2(NO_3)_2 \cdot 3H_2O$	c	−552.2	−454.7	333.29	94
$UO_2(NO_3) \cdot 6H_2O$	c	−764.3	−625.0	458.12	120.85
UO_2SO_4	aq	−467.3	−413.7	303.24	−13
$UO_2SO_4 \cdot 3H_2O$	c	−666.8	−586.0	429.53	63
Vanadium					
V	g	120	109	−79.90	43.55
	c	0.00	0.00	0.000	7.05
VCl_2	c	−108	−97	71.1	23.2
VCl_3	c	−137	−120	87.96	31.3
VCl_4	lq	−138	....	....	......
VN	c	−41	−35	25.7	8.91
VO	g	52	....	....	......
V_2O_2	c	−200	....	....	......
V_2O_3	c	−290	−271	198.6	23.58
V_2O_4	c (II)	−344	−318	233.1	24.65
V_2O_5	c	−373	−344	252.1	31.3
$VOCl_3$	c	−172	....	....	......
$VOSO_4$	c	−312.5	....	....	......
Wolfram (See Tungsten)					
Xenon					
Xe	g	0.00	0.00	0.000	40.53
$Xe \cdot 6H_2O$	c	−428	....	....	......
Ytterbium					
Yb	g	87	....	....	41.30
	c	0.00	0.00	0.000	......
Yb^{++}	aq	....	−129.0	94.56	......
Yb^{+++}	aq	−160.6	−155.5	113.98	......
$YbCl_3$	c (γ)	−228.7	....	....	......
	aq	−280.7	−249.5	182.88	......
$Yb_2(SO_4)_3$	aq	−971.9	−843.0	617.91	......
$Yb_2(SO_4)_3 \cdot 8H_2O$	c	....	−1308.8	959.33	......
Yttrium					
Y	g	103	....	....	42.87
	c	0.00	0.00	0.000	......
Y^{+++}	aq	−168.0	−164.1	120.28	......
YCl_3	c (γ)	−234.8	....	....	......
	aq	−288.1	−258.1	189.18	......
YI_3	c (β)	−143.2	....	....	......
	aq	−208.1	−201.1	147.40	......
$Y(OH)_3$	c	−337.6	−308.3	225.98	......
$Y_2(SO_4)_3$	aq	−986.7	−860.2	630.51	......
$Y_2(SO_4)_3 \cdot 8H_2O$	c	−1512.6	−1327.3	972.89	......
Zinc					
Zn	g	31.19	22.69	−16.631	38.45
	c	0.00	0.00	0.000	9.95
Zn^{++}	aq	−36.43	−35.184	25.79	−25.45
$ZnBr_2$	c	−78.17	−74.142	54.345	32.84
	aq	−94.23	−84.33	61.814	13.13
$ZnBr_2 \cdot 2H_2O$	c	−220.9	−190.6	139.71	56.2
$Zn(CN)_2$	c	18.4	....	....	......
$ZnCO_3$	c	−194.2	−174.8	128.13	19.7
$Zn(C_2H_3O_2)_2$	c	−258.1	....	....	......
$Zn(C_2H_3O_2)_2 \cdot H_2O$	c	−329.2	....	....	......
$Zn(C_2H_3O_2)_2 \cdot 2H_2O$	c	−398.8	....	....	......
$ZnC_2O_4 \cdot 2H_2O$	c	−373.8	....	....	......
$Zn(CH_3)_2$	lq	6.0	....	....	......
$Zn(C_2H_5)_2$	lq	5.0	....	....	......
$ZnCl_2$	c	−99.40	−88.26	64.690	25.9
	aq	−116.48	−97.88	71.748	0.89

* Soluble.

VALUES OF CHEMICAL THERMODYNAMIC PROPERTIES (Continued)

Substance	State	ΔHf°	ΔFf°	Log_{10} Kf	S°
Zinc					
ZnF_2	*aq*	−187.9	−166	....	....
ZnH	*g*	54.4	47.5	−34.82	48.70
ZnI_2	*c*	−49.98	−50.01	36.657	38.0
	aq	−63.17	−59.88	43.894	26.83
$Zn(NO_3)_2$	*c*	−115.12	....	....	....
$Zn(NO_3)_2 \cdot H_2O$	*c*	−191.99	....	....	....
$Zn(NO_3)_2 \cdot 2H_2O$	*c*	−264.94	....	....	....
$Zn(NO_3)_2 \cdot 4H_2O$	*c*	−405.76			
$Zn(NO_3)_2 \cdot 6H_2O$	*c*	−550.92	....	....	....
ZnO	*c*	−83.17	−76.05	55.744	10.5
$Zn(OH)_2$ (stable)	*c* (I)	−153.5	....	....	....
(unstable)	*c* (II)	−153.1	....	....	....
ZnS	*g*	−14.0	....	....	....
(sphalerite)	*c* (II)	−48.5	−47.4	34.74	13.8
(würtzite)	*c* (I)	−45.3	....	....	....
$ZnSO_4$	*c*	−233.88	−208.31	152.688	29.8
	aq	−253.33	−212.52	155 774	−21.3
$ZnSO_4 \cdot H_2O$	*c*	−310.6	−269.9	197 83	34.9
$ZnSO_4 \cdot 6H_2O$	*c*	−663.3	−555.0	406.81	86.8
$ZnSO_4 \cdot 7H_2O$	*c*	−735.1	−611.9	448.51	92.4
Zirconium					
Zr	*g*	125	115	−84.29	43.31
	c	0.00	0.00	0.000	9.18
$ZrBr_2$	*c*	−120	....	....	....
$ZrBr_3$	*c*	−174	....	....	....
$ZrBr_4$	*c*	−192	....	....	....
ZrC	*c*	−45	....	....	....
$ZrCl_2$	*c*	−145	....	....	....
$ZrCl_3$	*c*	−208	....	....	....
$ZrCl_4$	*c*	−230	−209	153.2	44.5
ZrF_2	*c*	−230	....	....	....
ZrF_3	*c*	−350	....	....	....
ZrF_4	*c*	−445	....	....	....
ZrI_2	*c*	−90	....	....	....
ZrI_3	*c*	−128	....	....	....
ZrI_4	*c*	−130	....	....	....
ZrN	*c*	−82.2	−75.4	55.27	9.23
$ZrO(NO_3)_2 \cdot 2H_2O$	*c*	−456.3	....	....	....
$ZrO(NO_3)_2 \cdot 3H_2O$	*c*	−527.3	....	....	....
$ZrO(NO_3)_2 \cdot 3\frac{1}{2}H_2O$	*c*	−562.8	....	....	....
$ZrO(NO_3)_2 \cdot 6H_2O$	*c*	−737.6	....	....	....
ZrO_2	*c*	−258.2	−244.4	179.14	12.03
$Zr(OH)_4$	*c*	−411.2	....	....	....
$Zr(OH)_4 \cdot H_2O$	*c*	−482.9	....	....	....
$Zr(OH)_4 \cdot 2H_2O$	*c*	−554.2	....	....	....
$Zr(SO_4)_2$	*c*	−597.4	....	....	....
$Zr(SO_4)_2 \cdot H_2O$	*c*	−677.9	....	....	....
$Zr(SO_4)_2 \cdot 4H_2O$	*c*	−893.1	....	....	....

Formula	Name	State	ΔHf°	ΔFf°	Log_{10} Kf	S°
C	Carbon	*g*	171.698	160.845	−117.897	37.761
	(diamond)	*c*	0.4532	0.6850	−0.5021	0.5829
	(graphite)	*c*	0.000	0.000	0.000	1.3609
CO	Carbon monoxide	*g*	−26.4157	−32.8079	24.048	47.301
CO_2	Carbon dioxide	*g*	−94.052	−94.260	69.091	51.061
		aq	−98.69	−92.31	67.662	29.0
CO_3^{--}	Carbonate ion	*aq*	−161.63	−126.22	92.517	−12.7
CH_3	Methyl	*g*	32.0	....	....	....
CH_4	Methane	*g*	−17.889	−12.140	8.8985	44.50
$HCOO^-$	Formate ion	*aq*	−98.0	−80.0	58.64	21.9
HCO_3^-	Bicarbonate ion	*aq*	−165.18	−140.31	102.845	22.7
CH_2O	Formaldehyde	*g*	−27.7	−26.3	19.28	52.26
CH_2O_2	Formic acid, monomer	*g*	−86.67	−80.24	58.815	60.0
	Formic acid	*lq*	−97.8	−82.7	60.62	30.82
		aq	−98.0	−85.1	62.38	39.1
H_2CO_3	Carbonic acid	*aq*	−167.0	−149.00	109.215	45.7
CH_3OH	Methanol	*g*	−48.08	−38.69	28.359	56.8
		lq	−57.02	−39.73	29.122	30.3
CF_4	Tetrafluoromethane	*g*	−162.5	−151.8	111.27	62.7
CCl_4	Tetrachloromethane	*g*	−25.5	−15.3	11.21	73.95
		lq	−33.3	−16.4	12.02	51.25
$COCl_2$	Carbonyl chloride	*g*	−53.30	−50.31	36.877	69.13
CH_3Cl	Chloromethane	*g*	−19.6	−14.0	10.26	55.97
CH_2Cl_2	Dichloromethane	*g*	−21.	−14.	10.3	64.68
		lq	−28	−15.1	11.07	42.7
$CHCl_3$	Trichloromethane	*g*	−24	−16	11.7	70.86
		lq	−31.5	−17.1	12.53	48.5
CBr_4	Tetrabromomethane	*g*	12.	8.6	−6.30	85.6
$CHBr_3$	Bromomethane	*g*	−8.5	−6.2	4.54	58.74
CH_2Br_2	Dibromomethane	*g*	−1.	−1.4	1.03	70.16
$CHBr_3$	Tribromomethane	*g*	6.	3.8	−2.78	79.18
		lq	−4.8	0.7	−0.51	53.0
CH_3I	Iodomethane	*g*	4.9	5.3	−3.88	60.85
		lq	−2.0	4.9	−3.59	38.9

VALUES OF CHEMICAL THERMODYNAMIC PROPERTIES (Continued)

Formula	Name	State	ΔHf°	ΔFf°	Log₁₀ Kf	S°
CH_2I_2	Diodomethane	*lq*	15.9	....	....	....
CHI_3	Triodomethane	*c*	33.6	....	....	....
CS_2	Carbon disulfide	*g*	27.55	15.55	−11.398	56.84
		lq	21.0	15.2	−11.14	36.10
COS	Carbon oxysulfide	*g*	−32.80	−40.45	29.649	55.34
CN^-	Cyanide ion	*aq*	36.1	39.6	−29.03	28.2
CNO^-	Cyanate ion	*aq*	−33.5	−23.6	17.30	31.1
$C(NO_2)_4$	Tetranitromethane	*lq*	8.8	....	....	....
HCN	Hydrogen cyanide	*g*	31.2	28.7	−21.04	48.23
		lq	25.2	29.0	−21.26	26.97
	Hydrocyanic acid	*aq*	25.2	26.8	−19.64	30.8
CH_5N	Methyl amine	*g*	−6.7	6.6	−4.84	57.73
CH_2N_2	Cyanamide	*c*	9.2	....	....	....
CH_5N_3	Guanidine	*c*	−17.0	....	....	....
HCNO	Cyanic acid	*aq*	−35.1	−28.9	21.18	43.6
CH_3O_2N	Nitromethane	*lq*	−21.28	2.26	−1.656	41.1
CH_4ON_2	Urea	*c*	−79.634	−47.120	34.538	25.00
		aq	−76.30	−48.72	35.711	41.55
$CH_6O_2N_2$	Ammonium carbamate	*c*	−154.21	−109.47	80.240	39.70
CNCl	Cyanogen chloride	*g*	34.5	32.9	−24.12	56.31
CH_6NCl	Methylamine hydrochloride	*c*	−68.31	−35.09	25.720	33.13
CNI	Cyanogen iodide	*g*	54.6	47.7	−34.96	61.26
		c	40.4	42.6	−31.23	30.8
		aq	43.2	43.3	−31.74	37.9
CNS^-	Thiocyanate ion	*aq* (very dilute)	17.2	....	....	....
CH_4N_2S	Thiourea	*c*	−22.1	....	....	....
$C_2O_4^{--}$	Oxalate ion	*aq*	−197.0	−161.3	118.23	12.2
C_2H_2	Ethyne (acetylene)	*g*	54.194	50.000	−36.649	47.997
$C_2H_2 \cdot 6H_2O$	Ethyne hexahydrate	*c*	−371.1	....	....	....
C_2H_4	Ethene (ethylene)	*g*	12.496	16.282	−11.935	52.45
C_2H_6	Ethane	*g*	−20.236	−7.860	5.7613	54.85
$HC_2O_4^-$	Bioxalate ion	*aq*	−195.5	−167.1	122.48	36.7
C_2H_2O	Ketene	*g*	−14.6	....	....	....
$C_2H_2O_2$	Glyoxal	*c*	−83.7	....	....	....
$C_2H_2O_4$	Oxalic acid	*c*	−197.6	−166.8	122.28	28.7
$C_2H_3O_2^-$	Acetate ion	*aq*	−116.843	....	....	....
C_2H_4O	Ethylene oxide	*g*	−12.19	−2.79	2.045	58.1
C_2H_4O	Acetaldehyde	*g*	−39.76	−31.96	23.426	63.5
$C_2H_4O_2$	Acetic acid	*lq*	−116.4	−93.8	68.75	38.2
$C_2H_4O_4$	Formic acid dimer	*g*	−187.7	−163.8	120.09	83.1
C_2H_6O	Ethanol	*g*	−56.24	−40.30	29.536	67.4
		lq	−66.356	−41.77	30.617	38.4
	Dimethyl ether	*g*	−44.3	−27.3	19.98	63.72
$C_2H_6O_2$	1-2 Ethanediol	*lq*	−108.58	−77.12	56.528	39.9
C_2H_5Cl	Chloroethane	*g*	−25.1	−12.7	9.331	65.90
$C_2H_4Cl_2$	1-2 Dichloroethane	*lq*	−39.7	−19.2	14.11	49.84
C_2HOCl_3	Chloral trichloroacetaldehyde	*lq*	−51.1	....	....	....
$C_2HO_2Cl_3$	Trichloroacetic acid	*c*	−122.8	....	....	....
C_2H_5Br	Bromoethane	*g*	−13.0	....	....	....
		lq	−20.4	....	....	....
$C_2H_4Br_2$	1-2 Dibromoethane	*lq*	−19.30	−4.94	3.621	53.37
C_2H_5I	Iodoethane	*lq*	−7.4	....	....	....
$C_2H_4I_2$	1-2 Diiodoethane	*c*	0.1	....	....	....
C_2H_6S	Ethanethiol	*lq*	−16.	....	....	....
	Dimethyl sulfide	*g*	−6.9	3.7	−2.74	68.28
		lq	−13.6	3.4	−2.50	46.94
C_2N_2	Cyanogen	*g*	73.60	70.81	−51.903	57.86
C_2H_3N	Acetonitrile	*g*	21.0	25.2	−18.47	58.18
		lq	12.7	24.0	−17.59	34.5
	Methyl isocyanide	*g*	35.9	40.0	−29.32	58.78
C_2H_7N	Ethylamine	*g*	−11.6	....	....	....
$C_2H_5O_2N$	Aminoacetic acid (glycine)	*c*	−126.33	−88.61	64.950	26.1
	Nitroethane	*lq*	−30.	....	....	....
C_2H_3SN	Methyl thiocyanate	*lq*	25.	....	....	....
	Methyl isothiocyanate	*c*	18.	....	....	....
$C_2H_7O_3SN$	2-Aminoethylsulfonic acid (taurine)	*c*	−181.66	....	....	....

VALUES OF CHEMICAL THERMODYNAMIC PROPERTIES OF HYDROCARBONS

The values in this table are from the American Petroleum Institute Research Project 44. The values are for the ideal gas state at 298.16°K and are expressed as kilo-cal per mole. It is frequently possible to calculate values for compounds not listed since the following increments are known for an addition of a methylene group, CH_2, to the following types of compounds:

Normal alkyl cyclohexanes
Normal paraffins
Normal alkyl benzenes
Normal alkyl cyclopentanes
Normal monoolefins (1-alkenes)
Normal acetylenes (1-alkynes)

For each of the above types of compounds the increments per CH_2 group are:

ΔHf°	−4.926	Kcal/gm mole
ΔFf°	+2.048	"
Log_{10} Kf	−1.5012	"
S°	+9.183	"

	Compounds	ΔHf°	ΔFf°	Log_{10} Kf	S°
CH_4	Methane	−17.889	−12.140	8.8985	44.50
C_2H_2	Ethyne (Acetylene)	54.194	50.000	−36.6490	47.997
C_2H_4	Ethene (Ethylene)	12.496	16.282	−11.9345	52.54
C_2H_6	Ethane	−20.236	− 7.860	5.7613	54.85
C_3H_4	Propadiene (Allene)	45.92	48.37	−35.4519	58.30
C_3H_4	Propyne (Methyl-acetylene)	44.319	46.313	−33.9469	59.30
C_3H_6	Propene (Propylene)	4.879	14.990	−10.9875	63.80
C_3H_8	Propane	−24.820	− 5.614	4.1150	64.51
C_4H_6	1,2-Butadiene	39.55	48.21	−35.3377	70.03
C_4H_6	1,3-Butadiene	26.75	36.43	−26.7004	66.62
C_4H_6	1-Butyne (Ethyl Acetylene)	39.70	48.52	−35.5616	69.51
C_4H_6	2-Butyne (Dimethylacetylene)	35.374	44.725	−32.7823	67.71
C_4H_8	1-Butene	0.280	17.217	−12.6199	73.48
C_4H_8	cis-2-Butene	− 1.362	16.046	−11.7618	71.90
C_4H_8	trans-2-Butene	− 2.405	15.315	−11.2255	70.86
C_4H_8	2-Methylpropane (Isobutene)	− 3.343	14.582	−10.6888	70.17
C_4H_{10}	*n*-Butane	−29.812	− 3.754	2.7516	74.10
C_4H_{10}	2-Methylpropane (Isobutane)	−31.452	− 4.296	3.1489	70.42
C_5H_8	1-Pentyne	34.50	50.17	−36.7712	79.10
C_5H_8	2-Pentyne	30.80	46.41	−34.0177	79.30
C_5H_8	3-Methyl-1-butyne	32.60	49.12	−36.0061	76.23
C_5H_8	1,2-Pentadiene	34.80	50.29	−36.861	79.7
C_5H_8	cis-1,3-Pentadiene (cis-piperylene)	18.70	34.88	−25.563	77.5
C_5H_8	trans-1,3-Pentadiene (trans-piperylene)	18.60	35.07	−25.707	76.4
C_5H_8	1,4-Pentadiene	25.20	40.69	−29.824	79.7
C_5H_8	2,3-Pentadiene	33.10	49.22	−36.074	77.6
C_5H_8	3-Methyl-1,2-butadiene	31.00	47.47	−34.657	76.4
C_5H_8	2-Methyl-1,3-butadiene (Isoprene)	18.10	34.87	−25.560	75.44
C_5H_{10}	1-Pentene	− 5.000	18.787	−13.7704	83.08
C_5H_{10}	cis-2-Pentene	− 6.710	17.173	−12.5874	82.76
C_5H_{10}	trans-2-Pentene	− 7.590	16.575	−12.1495	81.81
C_5H_{10}	2-Methyl-1-butene	− 8.680	15.509	−11.3680	81.73
C_5H_{10}	3-Methyl-1-butene	− 6.920	17.874	−13.1017	79.70
C_5H_{10}	2-Methyl-2-butene	−10.170	14.267	−10.4572	80.90
C_5H_{10}	Cyclopentane	−18.46	9.23	− 6.7643	70.00
C_5H_{12}	*n*-Pentane	−35.00	− 1.96	1.4366	83.27
C_6H_6	Benzene	19.820	30.989	−22.7143	64.34
C_6H_{10}	1-Hexyne	29.55	52.19	−38.258	88.27
C_6H_{12}	1-Hexene	− 9.96	20.80	−15.2491	92.25
C_6H_{12}	cis-2-Hexene	−11.56	19.18	−14.0549	92.35
C_6H_{12}	trans-2-Hexene	−12.56	18.46	−13.5291	91.40
C_6H_{12}	cis-3-Hexene	−11.56	19.66	−14.4094	90.73
C_6H_{12}	trans-3-Hexene	−12.56	18.86	−13.8262	90.04
C_6H_{12}	2-Methyl-1-pentene	−13.56	17.48	−12.8135	91.32
C_6H_{12}	3-Methyl-1-pentene	−11.02	20.28	−14.8655	90.45
C_6H_{12}	4-Methyl-1-pentene	−11.66	19.90	−14.5865	89.58
C_6H_{12}	2-Methyl-2-pentene	−14.96	16.34	−11.9780	90.45
C_6H_{12}	cis-3-Methyl-2-pentene	−14.32	16.98	−12.4471	90.45
C_6H_{12}	trans-3-Methyl-2-pentene	−14.32	16.74	−12.2697	91.26
C_6H_{12}	cis-4-Methyl-2-pentene	−13.26	18.40	−13.4903	89.23
C_6H_{12}	trans-4-Methyl-2-pentene	−14.26	17.77	−13.0216	88.02
C_6H_{12}	2-Ethyl-1-butene	−12.92	18.51	−13.5690	90.01
C_6H_{12}	2,3-Dimethyl-1-butene	−14.78	17.43	−12.7782	89.39
C_6H_{12}	3,3-Dimethyl-1-butene	−14.25	19.04	−13.9578	83.79
C_6H_{12}	2,3-Dimethyl-2-butene	−15.91	16.52	−12.1073	86.67
C_6H_{12}	Methylcyclopentane	−25.50	8.55	− 6.2649	81.24
C_6H_{12}	Cyclohexane	−29.43	7.59	− 5.5605	71.28
C_6H_{14}	*n*-Hexane	−39.96	0.05	0.037	92.45
C_7H_8	Methylbenzene (Toluene)	11.950	29.228	−21.4236	76.42
C_7H_{12}	1-Heptyne	24.62	54.24	−39.759	97.25
C_7H_{12}	1-Heptene	−14.85	22.84	−16.742	101.43
C_7H_{14}	Ethylcyclopentane	−30.37	10.59	− 7.7632	90.62
C_7H_{14}	1,1-Dimethylcyclopentane	−33.05	9.33	− 6.8372	85.87
C_7H_{14}	1,cis-2-Dimethylcyclopentane	−30.96	10.93	− 8.0107	87.51
C_7H_{14}	1,trans-2-Dimethylcyclopentane	−32.67	9.17	− 6.7224	87.67
C_7H_{14}	1,cis-3-Dimethylcyclopentane	−31.93	9.91	− 7.2648	87.67
C_7H_{14}	1,trans-3-Dimethylcyclopentane	−32.47	9.37	− 6.8690	87.67
C_7H_{14}	Methylcyclohexane	−36.99	6.52	− 4.7819	82.06
C_7H_{16}	*n*-Heptane	−44.89	2.09	− 1.532	101.64
C_8H_8	Ethenylbenzene (Styrene)	35.32	51.10	−37.4532	82.48
C_8H_{10}	Ethylbenzene	7.120	31.208	−22.8750	86.15
C_8H_{10}	1,2-Dimethylbenzene (*o*-Xylene)	4.540	29.177	−21.3860	84.31
C_8H_{10}	1,3-Dimethylbenzene (*m*-Xylene)	4.120	28.405	−20.8202	85.49
C_8H_{10}	1,4-Dimethylbenzene (*p*-Xylene)	4.290	28.952	−21.2214	84.23
C_8H_{14}	1-Octyne	19.70	56.29	−41.260	106.63
C_8H_{16}	1-Octene	−19.82	24.89	−18.244	110.61
C_8H_{16}	*n*-propylcyclopentane	−35.39	12.54	− 9.195	99.80
C_8H_{16}	1,1-Dimethylcyclohexane	−43.26	8.42	− 6.174	87.24
C_8H_{16}	cis-1,2-Dimethylcyclohexane	−41.15	9.85	− 7.225	89.51
C_8H_{16}	trans-1,2-Dimethylcyclohexane	−43.02	8.24	− 6.038	88.65
C_8H_{16}	cis-1,3-Dimethylcyclohexane	−44.16	7.13	− 5.228	88.54
C_8H_{16}	trans-1,3-Dimethylcyclohexane	−42.20	8.68	− 6.363	89.92
C_8H_{16}	cis-1,4-Dimethylcyclohexane	−42.22	9.07	− 6.650	88.54
C_8H_{16}	trans-1,4-Dimethylcyclohexane	−44.12	7.58	− 5.552	87.19
C_8H_{18}	*n*-Octane	−49.82	4.14	− 3.035	110.82
C_8H_{18}	2-Methylheptane	−51.50	3.06	− 2.243	108.81
C_8H_{18}	3-Methylheptane	−50.82	3.29	− 2.412	110.32
C_8H_{18}	4-Methylheptane	−50.69	4.00	− 2.932	108.35
C_8H_{18}	3-Ethylhexane	−50.40	3.95	− 2.895	109.51
C_8H_{18}	2,2-Dimethylhexane	−53.71	2.56	− 1.876	103.06
C_8H_{18}	2,3-Dimethylhexane	−51.13	4.23	− 3.101	106.11
C_8H_{18}	2,4-Dimethylhexane	−52.44	2.80	− 2.052	106.51
C_8H_{18}	2,5-Dimethylhexane	−53.21	2.50	− 1.832	104.93
C_8H_{18}	3,3-Dimethylhexane	−52.61	3.17	− 2.324	104.70
C_8H_{18}	3,4-Dimethylhexane	−50.91	4.14	− 3.035	107.15
C_8H_{18}	2-Methyl-3-ethylpentane	−50.48	5.08	− 3.724	105.43
C_8H_{18}	3-Methyl-3-ethylpentane	−51.38	4.76	− 3.489	103.48
C_8H_{18}	2,2,3-Trimethylpentane	−52.61	4.09	− 2.998	101.62
C_8H_{18}	2,2,4-Trimethylpentane	−53.57	3.13	− 2.294	101.62
C_8H_{18}	2,3,3-Trimethylpentane	−51.73	4.52	− 3.313	103.14
C_8H_{18}	2,3,4-Trimethylpentane	−51.97	4.32	− 3.167	102.99
C_8H_{18}	2,2,3,3-tetramethylbutane	−53.99	4.88	− 3.577	94.34
C_9H_{10}	Isopropenylbenzene (*α*-Methylstyrene, 2-phenyl-1-propene)	27.00	49.84	−36.531	91.70
C_9H_{10}	1-Methyl-2-ethenylbenzene (*o*-Methylstyrene)	28.30	51.14	−37.484	91.70
C_9H_{10}	1-Methyl-3-ethenylbenzene (*m*-Methylstyrene)	27.60	50.02	−36.665	93.1
C_9H_{10}	1-Methyl-4-ethenylbenzene (*p*-Methylstyrene)	27.40	50.24	−36.825	91.7
C_9H_{12}	*n*-Propylbenzene	1.870	32.810	−24.049	95.74
C_9H_{12}	isopropylbenzene (Cumene)	0.940	32.738	−23.996	92.87
C_9H_{12}	1,3,5-trimethylbenzene (Mesitylene)	− 3.840	28.172	−20.6497	92.15
C_9H_{16}	1-Nonyne	14.77	58.34	−42.761	115.82
C_9H_{18}	1-Nonene	−24.74	26.94	−19.747	119.80
C_9H_{18}	*n*-Butylcyclopentane	−40.22	14.69	−10.768	108.99
C_9H_{20}	*n*-Nonane	−54.74	6.18	− 4.536	120.00
$C_{10}H_{14}$	*n*-Butylbenzene	− 3.30	34.62	−25.374	104.91
$C_{10}H_{18}$	1-Decyne	9.85	60.39	−44.262	125.00
$C_{10}H_{20}$	1-Decene	−29.67	28.99	−21.249	128.98
$C_{10}H_{22}$	*n*-Decane	−59.67	8.23	6.037	129.19
$C_{11}H_{22}$	1-Undecene	−34.60	31.03	−22.745	138.16
$C_{11}H_{24}$	*n*-Undecane	−64.60	10.28	− 7.539	138.37
$C_{12}H_{24}$	1-Dodecene	−39.52	33.08	−24.297	147.34

HEAT OF DILUTION OF ACIDS

From National Standards Reference Data Systems NSRDS-NBS 2

Vivian B. Parker

ΔH_{diln}, the integral heat of dilution, is the change in enthalpy, per mole of solute, when a solution of concentration m_1 is diluted to a final finite concentration m_2. When the dilution is carried out by addition of an infinite amount of solvent, so the final solution is infinitely dilute, the enthalpy change is the integral heat of dilution to infinite dilution. Since Φ_L, the relative apparent molal enthalpy, is equal to and opposite in sign to this, only Φ_L is referred to here.

Φ_L, cal/mole, at 25°C

n	m	HF	HCl	$HClO_4$	HBr	HI	HNO_3	CH_2O_2	$C_2H_4O_2$
∞	0.00	0	0	0	0	0	0	0	0
500,000	.000111	300	5	5	5	5	5	9	40
100,000	.000555	900	10	10	9	9	11	13	50
50,000	.00111	1,300	16	14	13	12	15	20	53
20,000	.00278	1,800	25	22	22	20	23	23	55
10,000	.00555	2,130	34	30	31	29	31	25	58
7,000	.00793	2,250	40	35	37	34	36	26	59
5,000	.01110	2,360	47	40	44	41	42	26	61
4,000	.01388	2,450	54	43	49	46	46	27	62
3,000	.01850	2,550	60	47	56	52	51	28	62
2,000	.02775	2,700	74	54	68	63	59	28	63
1,500	.03700	2,812	85	58	77	71	65	29	64
1,110	.05000	2,927	97	62	89	81	73	29	65
1,000	.05551	2,969	102	62	92	84	76	29	65
900	.0617	2,989	107	63	97	88	78	30	66
800	.0694	3,015	113	64	102	92	81	31	67
700	.0793	3,037	120	65	108	96	84	32	68
600	.0925	3,057	129	65	115	102	88	32	68
555.1	.1000	3,060	133	65	119	105	89	32	69
500	.1110	3,077	140	65	124	108	92	32	70
400	.1388	3,097	156	64	135	116	97	33	72
300	.1850	3,126	176	61	150	125	103	34	76
277.5	.2000	3,129	182	59	155	128	105	35	79
200	.2775	3,142	212	50	176	140	117	36	82
150	.3700	3,148	242	36	197	154	118	39	88
111.0	.5000	3,156	280	18	225	170	119	42	97
100	.5551	3,160	295	+12	235	176	120	44	101
75	.7401	3,167	343	−14	270	194	121	49	113
55.51	1.0000	3,179	405	−48	314	223	121	54	130
50	1.1101	3,184	431	−61	331	234	121	56	147
40	1.3877	3,192	493	−91	379	260	121	60	155
37.00	1.5000	3,194	518	−103	398	269	121	62	162
30	1.8502	3,200	595	−138	455	301	124	65	183
27.75	2.0000	3,203	627	−149	477	315	126	66	192
25	2.2202	3,208	674	−162	510	336	130	67	204
22.20	2.5000	3,211	732	−173	550	365	139	68	218
20	2.7753	3,214	792	−182	590	396	149	69	233
18.50	3.0000	3,216	838	−187	624	427	159	69	245
15.86	3.500	3,221	946	−196	709	503	189	69	268
15	3.7004	3,227	988	−195	743	536	203	69	277
13.88	4.0000	3,234	1,052	−188	796	588	229	69	291
12.33	4.5000	3,246	1,171	−175	887	676	265	69	313
12	4.6255	3,249	1,190	−170	911	700	277	69	318
11.10	5.0000	3,256	1,271	−150	983	764	313	69	333
10	5.5506	3,265	1,396	−117	1,097	855	368	68	353
9.5	5.8427	3,269	1,462	−97	1,156	920	400	68	363
9.251	6.0000	3,272	1,498	−84	1,196	950	418	67	368
9.0	6.1674	3,274	1,535	−72	1,230	980	437	67	373
8.5	6.5301	3,278	1,618	−40	1,313	1,050	480	66	383
8.0	6.9383	3,282	1,710	+4	1,401	1,115	530	65	392
7.929	7.0000	3,283	1,725	11	1,416	1,130	538	65	394

HEAT OF DILUTION OF ACIDS (Continued)

n	m	HF	HCl	$HClO_4$	HBr	HI	HNO_3	CH_2O_2	$C_2H_4O_2$
7.5	7.4008	3,286	1,820	61	1,497	1,210	595	63	402
7.0	7.9295	3,290	1,942	135	1,608	1,325	661	61	411
6.938	8.0000	3,291	1,960	146	1,622	1,340	667	61	412
6.5	8.5394	3,296	2,090	229	1,738	1,450	745	58	420
6.167	9.0000	3,302	2,202	306	1,845	1,570	805	55	426
6.0	9.2510	3,305	2,265	348	1,903	1,630	840	53	429
5.551	10.0000	3,316	2,447	481	2,078	1,820	940	49	436
5.5	10.0920	3,317	2,472	499	2,102	1,850	950	49	437
5.0	11.1012	3,335	2,721	730	2,344	2,100	1,098	43	445
4.5	12.3346	3,362	3,025	1,144	2,655	2,460	1,270	37	453
4.0	13.8765	3,400	3,404	1,574	3,089	2,960	1,495	29	462
3.700	15.0000	3,428	3,680	1,893	3,415	3,350	1,645	26	469
3.5	15.8589	3,450	3,882	2,150	3,668	3,660	1,770	21	473
3.25	17.0788	3,483	4,160	2,460	4,005	4,110	1,920	17	481
3.0	18.5020	3,520	4,460	2,880	4,370	4,630	2,101	13	488
2.775	20.0000	3,557	4,750	3,300	4,760	5,190	2,270	9	496
2.5	22.2024	3,607	5,180	4,000	5,300	6,000	2,520	+4	506
2.0	27.7530	3,712	6,260	5,500	6,650	.....	3,060	−5	528
1.5	37.0040	.....	8,240	.....	8,530	.....	3,770	−13	532
1.0	55.506	.....	10,900	.....	11,670	.....	4,715	+11	518
0.5	111.012	.....	.....	.....	.....	.....	.....	77	495
0.25	222.02	.....	.....	.....	.....	.....	.....	129	.....

HEATS OF SOLUTION

From National Standards Reference Data Systems NSRDS-NBS 2

Vivian B. Parker

ΔH°_∞ 25°C for uni-univalent electrolytes in H_2O

Substance	State	ΔH°_∞	Substance	State	ΔH°_∞	Substance	State	ΔH°_∞
		cal/mole			*cal/mole*			*cal/mole*
HF	g	−14,700	$LiBr\cdot 2H_2O$	c	−2,250	KCl	c	4,115
HCl	g	−17,888	$LiBrO_3$	c	340	$KClO_3$	c	9,890
$HClO_4$	l	−21,215	LiI	c	−15,130	$KClO_4$	c	12,200
$HClO_4\cdot H_2O$	c	−7,875	$LiI\cdot H_2O$	c	−7,090	KBr	c	4,750
HBr	g	−20,350	$LiI\cdot 2H_2O$	c	−3,530	$KBrO_3$	c	9,830
HI	g	−19,520	$LiI\cdot 3H_2O$	c	140	KI	c	4,860
HIO_3	c	2,100	$LiNO_2$	c	−2,630	KIO_3	c	6,630
HNO_3	l	−7,954	$LiNO_2\cdot H_2O$	c	1,680	KNO_2	c	3,190
HCOOH	l	−205	$LiNO_3$	c	−600	KNO_3	c	8,340
CH_3COOH	l	−360				$KC_2H_3O_2$	c	−3,665
			NaOH	c	−10,637	KCN	c	2,800
NH_3	g	−7,290	$NaOH\cdot H_2O$	c	−5,118	KCNO	c	4,840
NH_4Cl	c	3,533	NaF	c	218	KCNS	c	5,790
NH_4ClO_4	c	8,000	NaCl	c	928	$KMnO_4$	c	10,410
NH_4Br	c	4,010	$NaClO_2$	c	80			
NH_4I	c	3,280	$NaClO_2\cdot 3H_2O$	c	6,830	RbOH	c	−14,900
NH_4IO_3	c	7,600	$NaClO_3$	c	5,191	$RbOH\cdot H_2O$	c	−4,310
NH_4NO_2	c	4,600	$NaClO_4$	c	3,317	$RbOH\cdot 2H_2O$	c	210
NH_4NO_3	c	6,140	$NaClO_4\cdot H_2O$	c	5,380	RbF	c	−6,240
$NH_4C_2H_3O_2$	c	−570	NaBr	c	−144	$RbF\cdot H_2O$	c	−100
NH_4CN	c	4,200	$NaBr\cdot 2H_2O$	c	4,454	$RbF\cdot 1\frac{1}{2}H_2O$	c	320
NH_4CNS	c	5,400	$NaBrO_3$	c	6,430	RbCl	c	4,130
CH_3NH_3Cl	c	1,378	NaI	c	−1,800	$RbClO_3$	c	11,410
$(CH_3)_3NHCl$	c	350	$NaI\cdot 2H_2O$	c	3,855	$RbClO_4$	c	13,560
$N(CH_3)_4Cl$	c	975	$NaIO_3$	c	4,850	RbBr	c	5,230
$N(CH_3)_4Br$	c	5,800	$NaNO_2$	c	3,320	$RbBrO_3$	c	11,700
$N(CH_3)_4I$	c	10,055	$NaNO_3$	c	4,900	RbI	c	6,000
			$NaC_2H_3O_2$	c	−4,140	$RbNO_3$	c	8,720
$AgClO_4$	c	1,760	$NaC_2H_3O_2\cdot 3H_2O$	c	4,700			
$AgNO_2$	c	8,830	NaCN	c	290	CsOH	c	−17,100
$AgNO_3$	c	5,400	$NaCN\cdot \frac{1}{2}H_2O$	c	790	$CsOH\cdot H_2O$	c	−4,900
			$NaCN\cdot 2H_2O$	c	4,440	CsF	c	−8,810
LiOH	c	−5,632	NaCNO	c	4,590	$CsF\cdot H_2O$	c	−2,500
$LiOH\cdot H_2O$	c	−1,600	NaCNS	c	1,632	$CsF\cdot 1\frac{1}{2}H_2O$	c	−1,300
LiF	c	1,130				CsCl	c	4,250
LiCl	c	−8,850	KOH	c	−13,769	$CsClO_4$	c	13,250
$LiCl\cdot H_2O$	c	−4,560	$KOH\cdot H_2O$	c	−3,500	CsBr	c	6,210
$LiClO_4$	c	−6,345	$KOH\cdot 1\frac{1}{2}H_2O$	c	−2,500	$CsBrO_3$	c	12,060
$LiClO_4\cdot 3H_2O$	c	7,795	KF	c	−4,238	CsI	c	7,970
LiBr	c	−11,670	$KF\cdot 2H_2O$	c	1,666	$CsNO_3$	c	9,560
$LiBr\cdot H_2O$	c	−5,560						

HEAT CAPACITY OF AQUEOUS SOLUTIONS OF VARIOUS ACIDS

From National Standards Reference Data Systems NSRDS-NBS 2

Vivian B. Parker

Φ_C is the apparent molal heat capacity of the solute, equal to $[(1000 + mM_2)C - 1000C°]/m$ where C and C° are the specific heats (per unit mass) of the solution and pure solvent, respectively, m is the molality, and M_2 is the molecular weight of the solute.

Φ_C, cal/deg mole, at 25°C

n	*m*	HF	HCl	HBr	HI	HIO_3	HNO_3	CH_2O_2	$C_2H_4O_2$	$C_3H_6O_2$
∞	0.00	−25.5	−32.6	−33.9	−34.0	−29.6	−20.7	−21.0	−1.5	+26.7
500,000	.000111	−23.0	.....	.....	.....	.....	.....	−9.8	+25.8	38.0
100,000	.000555	−18.8	−32.4	−33.8	−33.9	−29.4	−20.6	−1.2	32.6	45.1
50,000	.00111	−16.6	−32.4	−33.7	−33.8	−29.3	−20.5	+3.1	34.0	48.3
20,000	.00278	−12.4	−32.3	−33.6	−33.7	−29.1	−20.4	10.2	35.7	52.2
10,000	.00555	−8.6	−32.2	−33.4	−33.5	−28.5	−20.3	12.7	36.9	54.1
7,000	.00793	−6.7	−32.1	−33.4	−33.4	−28.4	−20.2	13.7	37.4	54.8
5,000	.01110	−4.9	−32.0	−33.3	−33.3	−28.1	−20.1	14.3	37.8	55.3
4,000	.01388	−3.6	−31.9	−33.2	−33.2	−27.7	−20.1	14.7	37.9	55.6
3,000	.01850	−2.2	−31.8	−33.1	−33.1	−27.2	−20.0	15.1	38.1	56.1
2,000	.02775	−0.5	−31.6	−32.9	−32.9	−25.8	−19.8	15.8	38.5	56.7
1,500	.03700	+0.4	−31.4	−32.7	−32.7	−24.9	−19.6	16.4	38.7	57.1
1,000	.05551	1.6	−31.2	−32.5	−32.5	−23.0	−19.3	16.8	39.0	57.7
900	.0617	1.8	−31.2	−32.4	−32.4	−22.3	−19.2	17.0	39.0	57.8
800	.0694	2.0	−31.1	−32.3	−32.3	−21.5	−19.1	17.2	39.1	57.9
700	.0793	2.2	−30.9	−32.2	−32.1	−20.2	−19.0	17.3	39.2	58.0
600	.0925	2.4	−30.8	−32.1	−32.0	−18.7	−18.9	17.5	39.3	58.3
500	.1110	2.7	−30.6	−31.9	−31.7	−16.7	−18.7	17.7	39.4	58.4
400	.1388	2.8	−30.3	−31.6	−31.4	−13.9	−18.4	18.0	39.4	58.6
300	.1850	3.2	−30.0	−31.2	−30.9	−10.1	−17.9	18.3	39.4	58.8
200	.2775	3.3	−29.3	−30.6	−30.2	−4.7	−17.2	18.7	39.3	58.9
150	.3700	3.5	−28.8	−30.1	−29.6	−0.7	−16.4	18.9	39.2	58.8
100	.5551	3.8	−27.8	−29.1	−28.4	+4.8	−15.0	19.0	39.1	58.7
75	.7401	4.1	−27.0	−28.5	−27.5	9.2	−13.7	19.2	38.9	58.6
50	1.1101	4.5	−25.6	−26.8	−25.9	15.7	−11.5	19.8	38.6	58.3
40	1.3877	4.9	−24.8	−25.9	−24.8	19.0	−9.9	19.8	38.4	58.0
30	1.8502	5.3	−23.6	−24.5	−23.3	23.1	−7.3	20.0	38.0	57.1
25	2.2202	5.6	−22.7	−23.6	−22.2	25.7	−5.4	20.1	37.7	56.2
20	2.7753	5.7	−21.5	−22.2	−20.8	.....	−2.7	20.2	37.1	55.0
15	3.7004	6.0	−19.8	−20.3	.....	.....	+1.4	20.4	36.3	53.1
12	4.6255	6.2	−18.2	−18.5	.....	.....	5.0	20.4	35.4	51.3
10	5.5506	6.3	−16.8	−16.8	.....	.....	8.5	20.6	34.8	49.7
9.5	5.8427	6.4	−16.4	−16.3	.....	.....	9.2	20.6	34.7	.....
9.0	6.1674	6.4	−15.8	−15.7	.....	.....	10.3	20.6	34.5	.....
8.5	6.5301	6.5	−15.4	−15.1	.....	.....	11.4	20.7	34.4	.....
8.0	6.9383	6.6	−14.8	−14.4	.....	.....	12.5	20.7	34.2	.....
7.5	7.4008	6.7	−14.2	−13.2	.....	.....	13.7	20.8	34.0	.....
7.0	7.9295	6.9	−13.5	−12.7	.....	.....	14.9	20.8	33.8	.....
6.5	8.5394	7.0	−12.7	−11.9	.....	.....	16.1	20.8	33.5	.....
6.0	9.2510	7.1	−11.8	−10.8	.....	.....	17.1	20.9	33.3	.....
5.5	10.0920	7.2	−10.8	−9.6	.....	.....	18.3	20.9	33.0	.....
5.0	11.1012	7.3	−9.6	−8.2	.....	.....	19.3	21.0	32.8	.....
4.5	12.3346	7.4	−8.7	−6.8	.....	.....	20.4	21.1	32.5	.....
4.0	13.8765	7.5	−6.6	−5.5	.....	.....	21.3	21.2	32.2	.....
3.5	15.8589	7.6	−4.7	−4.0	.....	.....	22.1	21.3	31.8	.....
3.25	17.0788	7.6	.....	−3.2	.....	.....	22.6	21.4	31.7	.....
3.0	18.5020	7.7	.....	−2.3	.....	.....	23.0	21.5	31.5	.....
2.5	22.2024	7.8	.....	−0.4	.....	.....	23.8	21.6	31.2	.....
2.0	27.7530	7.9	.....	.....	.....	.....	24.6	21.7	30.8	.....
1.5	37.0040	.....	.....	.....	.....	.....	25.2	21.8	30.4	.....
1.0	55.506	.....	.....	.....	.....	.....	25.7	22.0	30.1	.....

THERMODYNAMIC FORMULAS

Compiled by Doctor E. A. Coomes

Legend: p = Pressure
V = Volume
T = Temperature
n = Number of mols
S = Entropy
U = Internal energy (some books use E)
C_p = Molal specific heat at constant pressure
β = Coefficient volume expansion
K = Compressibility
H = U + pV = Total heat or enthalpy
A = U − TS = Helmholtz free energy
G = H − TS = Gibbs' free energy (some books use F).

Use of Table.—Partial derivatives of the first order for the eight fundamental thermodynamic variables, namely, p, V, T, U, S, H, A, G, may be obtained in terms of $(\partial V/\partial T)_p$, $(\partial V/\partial p)_T$, and $(\partial H/\partial T)_p$; the latter three are connected to measurable quantities as follows:

$$\frac{1}{V}\left(\frac{\partial V}{\partial T}\right)_p = \beta;\ -\frac{1}{V}\left(\frac{\partial V}{\partial p}\right)_T = K;\ \left(\frac{\partial H}{\partial T}\right)_p = nC_p$$

Computation by the Table.—The method of using the table will become apparent in working several examples. Suppose it is desired to know $(\partial H/\partial p)_S$ in terms of p, V, and T. Under caption "Constant" move horizontally to column marked "S"; across from "H" beside caption "differential" find "$-VnC_p/T$." Across from "p" in column "S" find "$-nC_p/T$"; $(\partial H/\partial p)_S$ is found by taking the ratio of the two:

$$(\partial H/\partial p)_S = (-VnC_p/T)/(-nC_p/T) = V$$

To find $(\partial S/\partial V)_T$ in terms of p, V, T, move to column "T" under "constant." Opposite "S" beside "differential" find "$(\partial V/\partial T)_p$"; opposite "V" find "$-(\partial V/\partial p)_T$."
Taking the ratio:

$$(\partial S/\partial V)_T = (\partial V/\partial T)_p/-(\partial V/\partial p)_T = (\partial p/\partial T)_V.$$

Differential	Constant T	Constant p	Constant V	Constant S
T	0	1	$\left(\frac{\partial V}{\partial p}\right)_T$	$-\left(\frac{\partial V}{\partial T}\right)_p$
p	-1	0	$-\left(\frac{\partial V}{\partial T}\right)_p$	$-\frac{nC_p}{T}$
V	$-\left(\frac{\partial V}{\partial p}\right)_T$	$\left(\frac{\partial V}{\partial T}\right)_p$	0	$\left(-\frac{1}{T}\right)\left[nC_p\left(\frac{\partial V}{\partial p}\right)_T + T\left(\frac{\partial V}{\partial T}\right)_p^2\right]$
S	$\left(\frac{\partial V}{\partial T}\right)_p$	$\frac{nC_p}{T}$	$\left(\frac{1}{T}\right)\left[nC_p\left(\frac{\partial V}{\partial p}\right)_T + T\left(\frac{\partial V}{\partial T}\right)_p^2\right]$	0
U	$T\left(\frac{\partial V}{\partial T}\right)_p + p\left(\frac{\partial V}{\partial p}\right)_T$	$nC_p - p\left(\frac{\partial V}{\partial T}\right)_p$	$nC_p\left(\frac{\partial V}{\partial p}\right)_T + T\left(\frac{\partial V}{\partial T}\right)_p^2$	$\left(\frac{p}{T}\right)\left[nC_p\left(\frac{\partial V}{\partial p}\right)_T + T\left(\frac{\partial V}{\partial T}\right)_p^2\right]$
H	$-V + T\left(\frac{\partial V}{\partial T}\right)_p$	nC_p	$nC_p\left(\frac{\partial V}{\partial p}\right)_T + T\left(\frac{\partial V}{\partial T}\right)_p^2 - V\left(\frac{\partial V}{\partial T}\right)$	$-\frac{VnC_p}{T}$
A	$p\left(\frac{\partial V}{\partial p}\right)_T$	$-S - p\left(\frac{\partial V}{\partial T}\right)_p$	$-S\left(\frac{\partial V}{\partial p}\right)_T$	$\left(\frac{1}{T}\right)\left[pnC_p\left(\frac{\partial V}{\partial p}\right)_T + pT\left(\frac{\partial V}{\partial T}\right)_p^2 + TS\left(\frac{\partial V}{\partial T}\right)_p\right]$
G	$-V$	$-S$	$-V\left(\frac{\partial V}{\partial T}\right)_p - S\left(\frac{\partial V}{\partial p}\right)_T$	$\left(-\frac{1}{T}\right)\left[nC_pV - TS\left(\frac{\partial V}{\partial T}\right)_p\right]$

Differential	Constant U	Constant H	Constant A	Constant G
T	$-T\left(\frac{\partial V}{\partial T}\right)_p - p\left(\frac{\partial V}{\partial p}\right)_T$	$V - T\left(\frac{\partial V}{\partial T}\right)_p$	$-p\left(\frac{\partial V}{\partial p}\right)_T$	V
p	$-nC_p + p\left(\frac{\partial V}{\partial T}\right)_p$	$-nC_p$	$S + p\left(\frac{\partial V}{\partial T}\right)_p$	S
V	$-nC_p\left(\frac{\partial V}{\partial p}\right)_T - T\left(\frac{\partial V}{\partial T}\right)_p^2$	$-nC_p\left(\frac{\partial V}{\partial p}\right)_T - T\left(\frac{\partial V}{\partial T}\right)_p^2 + V\left(\frac{\partial V}{\partial T}\right)_p$	$S\left(\frac{\partial V}{\partial p}\right)_T$	$V\left(\frac{\partial V}{\partial T}\right)_p + S\left(\frac{\partial V}{\partial p}\right)_T$
S	$\left(-\frac{p}{T}\right)\left[nC_p\left(\frac{\partial V}{\partial p}\right)_T + T\left(\frac{\partial V}{\partial T}\right)_p^2\right]$	$\frac{VnC_p}{T}$	$\left(-\frac{1}{T}\right)\left[pnC_p\left(\frac{\partial V}{\partial p}\right)_T + pT\left(\frac{\partial V}{\partial T}\right)_p^2 + TS\left(\frac{\partial V}{\partial T}\right)_p\right]$	$\left(\frac{1}{T}\right)\left[nC_pV - TS\left(\frac{\partial V}{\partial T}\right)_p\right]$
U	0	$V\left[nC_p - p\left(\frac{\partial V}{\partial T}\right)_p\right] + p\left[nC_p\left(\frac{\partial V}{\partial p}\right)_T + T\left(\frac{\partial V}{\partial T}\right)_p^2\right]$	$-p\left[nC_p\left(\frac{\partial V}{\partial p}\right)_T + T\left(\frac{\partial V}{\partial T}\right)_p^2\right] - S\left[T\left(\frac{\partial V}{\partial T}\right)_p + p\left(\frac{\partial V}{\partial p}\right)_T\right]$	$V\left[nC_p - p\left(\frac{\partial V}{\partial T}\right)_p\right] - S\left[T\left(\frac{\partial V}{\partial T}\right)_p + p\left(\frac{\partial V}{\partial p}\right)_T\right]$
H	$-V\left[nC_p - p\left(\frac{\partial V}{\partial T}\right)_p\right] - p\left[nC_p\left(\frac{\partial V}{\partial p}\right)_T + T\left(\frac{\partial V}{\partial T}\right)_p^2\right]$	0	$\left[S + p\left(\frac{\partial V}{\partial T}\right)_p\right] \times \left[V - T\left(\frac{\partial V}{\partial T}\right)_p - pnC_p\left(\frac{\partial V}{\partial p}\right)_T\right]$	$VnC_p + VS - TS\left(\frac{\partial V}{\partial T}\right)_p$
A	$p\left[nC_p\left(\frac{\partial V}{\partial p}\right)_T + T\left(\frac{\partial V}{\partial T}\right)_p^2\right] + S\left[T\left(\frac{\partial V}{\partial T}\right)_p + p\left(\frac{\partial V}{\partial p}\right)_T\right]$	$-\left[S + p\left(\frac{\partial V}{\partial T}\right)_p\right] \times \left[V - T\left(\frac{\partial V}{\partial T}\right)_p - pnC_p\left(\frac{\partial V}{\partial p}\right)_T\right]$	0	$-S\left[V + p\left(\frac{\partial V}{\partial p}\right)_T\right] + pV\left(\frac{\partial V}{\partial T}\right)_p$
G	$-V\left[nC_p - p\left(\frac{\partial V}{\partial T}\right)_p\right] + S\left[T\left(\frac{\partial V}{\partial T}\right)_p + p\left(\frac{\partial V}{\partial p}\right)_T\right]$	$-VnC_p - VS + TS\left(\frac{\partial V}{\partial T}\right)_p$	$S\left[V + p\left(\frac{\partial V}{\partial p}\right)_T\right] + pV\left(\frac{\partial V}{\partial T}\right)_p$	0

LIMITS OF INFLAMMABILITY

Reprinted from "Combustion Flame and Explosions of Gases," B. Lewis and G. von Elbe, authors, Academic Press (1951), publishers, by special permission.

The limits of inflammability given in the following tables were all determined at atmospheric pressure and room* temperature for upward propagation in a tube or bomb 2 inches or more in diameter. Values are on a percentage-by-volume basis.

LIMITS OF INFLAMMABILITY OF GASES AND VAPORS IN AIR

Compound	Empirical formula	Limits of inflammability	
		Lower	Upper
Paraffin hydrocarbons			
Methane	CH_4	5.00	15.00
Ethane	C_2H_6	3.00	12.50
Propane	C_3H_8	2.12	9.35
Butane	C_4H_{10}	1.86	8.41
Isobutane	C_4H_{10}	1.80	8.44
Pentane	C_5H_{12}	1.40	7.80
Isopentane	C_5H_{12}	1.32	.....
2,2-Dimethylpropane	C_5H_{12}	1.38	7.50
Hexane	C_6H_{14}	1.18	7.40
Heptane	C_7H_{16}	1.10	6.70
2,3-Dimethylpentane	C_7H_{16}	1.12	6.75
Octane	C_8H_{18}	0.95	.....
Nonane	C_9H_{20}	0.83	.....
Decane	$C_{10}H_{22}$	0.77	5.35
Olefins			
Ethylene	C_2H_4	2.75	28.60
Propylene	C_3H_6	2.00	11.10
Butene-1	C_4H_8	1.65	9.95
Butene-2	C_4H_8	1.75	9.70
Amylene	C_5H_{10}	1.42	8.70
Acetylenes			
Acetylene	C_2H_2	2.50	80.00
Aromatics			
Benzene	C_6H_6	1.40	7.10
Toluene	C_7H_8	1.27	6.75
o-Xylene	C_8H_{10}	1.00	6.00
Cyclic hydrocarbons			
Cyclopropane	C_3H_6	2.40	10.40
Cyclohexane	C_6H_{12}	1.26	7.75
Methylcyclohexane	C_7H_{14}	1.15	.....
Terpenes			
Turpentine	$C_{10}H_{16}$	0.80	.....
Alcohols			
Methyl alcohol	CH_4O	6.72	36.50
Ethyl alcohol	C_2H_6O	3.28	18.95
Allyl alcohol	C_3H_6O	2.50	18.00
n-Propyl alcohol	C_3H_8O	2.15	13.50
Isopropyl alcohol	C_3H_8O	2.02	11.80
n-Butyl alcohol	$C_4H_{10}O$	1.45	11.25
Isobutyl alcohol	$C_4H_{10}O$	1.68	.....
n-Amyl alcohol	$C_5H_{12}O$	1.19	.....
Isoamyl alcohol	$C_5H_{12}O$	1.20	.....
Aldehydes			
Acetaldehyde	C_2H_4O	3.97	57.00
Crotonic aldehyde	C_4H_6O	2.12	15.50
Furfural	$C_5H_4O_2$	2.10	.....
Paraldehyde	$C_6H_{12}O_3$	1.30	.....
Ethers			
Methylethyl ether	C_3H_8O	2.00	10.00
Diethyl ether	$C_4H_{10}O$	1.85	36.50
Divinyl ether	C_4H_6O	1.70	27.00
Ketones			
Acetone	C_3H_6O	2.55	12.80
Methylethyl ketone	C_4H_8O	1.81	9.50
Methylpropyl ketone	$C_5H_{10}O$	1.55	8.15
Methylbutyl ketone	$C_6H_{12}O$	1.35	7.60

Compound	Empirical formula	Limits of inflammability	
		Lower	Upper
Acids			
Acetic acid	$C_2H_4O_2$	5.40	.....
Hydrocyanic acid	HCN	5.60	40.00
Esters			
Methyl formate	$C_2H_4O_2$	5.05	22.70
Ethyl formate	$C_3H_6O_2$	2.75	16.40
Methyl acetate	$C_3H_6O_2$	3.15	15.60
Ethyl acetate	$C_4H_8O_2$	2.18	11.40
Propyl acetate	$C_5H_{10}O_2$	1.77	8.00
Isopropyl acetate	$C_5H_{10}O_2$	1.78	7.80
Butyl acetate	$C_6H_{12}O_2$	1.39	7.55
Amyl acetate	$C_7H_{14}O_2$	1.10	.....
Hydrogen			
Hydrogen	H_2	4.00	74.20
Nitrogen compounds			
Ammonia	NH_3	15.50	27.00
Cyanogen	C_2N_2	6.60	42.60
Pyridine	C_5H_5N	1.81	12.40
Ethyl nitrate	$C_2H_5NO_3$	3.80	.....
Ethyl nitrite	$C_2H_5NO_2$	3.01	50.00
Oxides			
Carbon monoxide	CO	12.50	74.20
Ethylene oxide	C_2H_4O	3.00	80.00
Propylene oxide	C_3H_6O	2.00	22.00
Dioxan	$C_4H_8O_2$	1.97	22.25
Diethyl peroxide	$C_4H_{10}O_2$	2.34	.....
Sulfides			
Carbon disulfide	CS_2	1.25	50.00
Hydrogen sulfide	H_2S	4.30	45.50
Carbon oxysulfide	COS	11.90	28.50
Chlorides			
Methyl chloride	CH_3Cl	8.25	18.70
Ethyl chloride	C_2H_5Cl	4.00	14.80
Propyl chloride	C_3H_7Cl	2.60	11.10
Butyl chloride	C_4H_9Cl	1.85	10.10
Isobutyl chloride	C_4H_9Cl	2.05	8.75
Allyl chloride	C_3H_5Cl	3.28	11.15
Amyl chloride	$C_5H_{11}Cl$	1.60	8.63
Vinyl chloride	C_2H_3Cl	4.00	21.70
Ethylene dichloride	$C_2H_4Cl_2$	6.20	15.90
Propylene dichloride	$C_3H_6Cl_2$	3.40	14.50
Bromides			
Methyl bromide	CH_3Br	13.50	14.50
Ethyl bromide	C_2H_5Br	6.75	11.25
Allyl bromide	C_3H_5Br	4.36	7.25
Amines			
Methyl amine	CH_5N	4.95	20.75
Ethyl amine	C_2H_7N	3.55	13.95
Dimethyl amine	C_2H_7N	2.80	14.40
Propyl amine	C_3H_9N	2.01	10.35
Diethyl amine	$C_4H_{11}N$	1.77	10.10
Trimethyl amine	C_3H_9N	2.00	11.60
Triethyl amine	$C_6H_{15}N$	1.25	7.90

* The upper limits of some vapors were determined at somewhat higher temperatures because of their low vapor pressures.

LIMITS OF INFLAMMABILITY OF GASES AND VAPORS IN OXYGEN

Compound	Formula	Limits of inflammability	
		Lower	Upper
Hydrogen	H_2	4.65	93.9
Deuterium	D_2	5.00	95.0
Carbon monoxide	CO	15.50	93.9
Methane	CH_4	5.40	59.2
Ethane	C_2H_6	4.10	50.5
Ethylene	C_2H_4	2.90	79.9
Propylene	C_3H_6	2.10	52.8
Cyclopropane	C_3H_6	2.45	63.1
Ammonia	NH_3	13.50	79.0
Diethyl ether	$C_4H_{10}O$	2.10	82.0
Divinyl ether	C_4H_6O	1.85	85.5

FLAME AND BEAD TESTS

Flame Colorations

VIOLET

Potassium compounds. Purple red through blue glass. Easily obscured by sodium flame. Bluish green through green glass. Rubidium and Caesium compounds impart same flame as potassium compounds

BLUES

Azure.—Copper chloride. Copper bromide gives azure blue followed by green. Other copper compounds give same coloration when moistened with hydrochloric acid.

Light Blue.—Lead, Arsenic, Selenium.

GREENS

Emerald.—Copper compounds except the halides, and when not moistened with hydrochloric acid.

Pure Green.—Compounds of thallium and tellurium.

Yellowish.—Barium compounds. Some molybdenum compounds. Borates, especially when treated with sulphuric acid or when burned with alcohol.

Bluish.—Phosphates with sulphuric acid.

Feeble.—Antimony compounds. Ammonium compounds.

Whitish.—Zinc.

REDS

Carmine.—Lithium compounds. Violet through blue glass. Invisible through green glass. Masked by barium flame.

Scarlet.—Strontium compounds. Violet through blue glass. Yellowish through green glass. Masked by barium flame.

Yellowish.—Calcium compounds. Greenish through blue glass. Green through green glass. Masked by barium flame.

YELLOW

Yellow.—All sodium compounds. Invisible with blue glass.

Borax Beads

Abbreviations employed: s., saturated; s.s., supersaturated; n.s; saturated; h., hot; c., cold.

Substance	Oxidizing flame	Reducing flame
Aluminum	Colorless (h.c., n.s.); opaque (s.s.)	Colorless; opaque (s.)
Antimony	Colorless; yellow or brownish (h., s.s.)	Gray and opaque
Barium	Colorless (n.s.)	....
Bismuth	Colorless; yellow or brownish (h., s.s.)	Gray and opaque
Cadmium	Colorless	Gray and opaque
Calcium	Colorless (n.s.)	....
Cerium	Red (h.)	Colorless (h.c.)
Chromium	Green (c.)	Green
Cobalt	Blue (h.c.)	Blue (h.c.)
Copper	Green (h.); blue (c.)	Red (c.): opaque (s.s.): colorless (h.)
Iron	Yellow or brownish red (h., n.s.)	Green (s.s.)
Lead	Colorless; yellow or brownish (h., s.s.)	Gray and opaque
Magnesium	Colorless (n.s.)	....
Manganese	Violet (h.c.)	Colorless (h.c.)
Molybdenum	Colorless	Yellow or brown (h.)
Nickel	Brown; red (c.)	Gray and opaque
Silicon	Colorless (h.c.); opaque (s.s.)	Colorless; opaque (s.)
Silver	Colorless (n.s.)	Gray and opaque
Strontium	Colorless (n.s.)	....
Tin	Colorless (h.c.); opaque (s.s.)	Colorless; opaque (s.)
Titanium	Colorless	Yellow (h.); violet (c.)
Tungsten	Colorless	Brown
Uranium	Yellow or brownish (h., n.s.)	Green
Vanadium	Colorless	Green

Beads of Microcosmic Salt

$NaNH_4HPO_4$

Substance	Oxidizing flame	Reducing flame
Aluminum	Colorless; opaque (s.)	Colorless; not clear (s.s.)
Antimony	Colorless (n.s.)	Gray and opaque
Barium	Colorless; opaque (s.)	Colorless; not clear (s.s.)
Bismuth	Colorless (n.s.)	Gray and opaque
Cadmium	Colorless (n.s.)	Gray and opaque
Calcium	Colorless; opaque (s.)	Colorless; not clear (s.s.)
Cerium	Yellow or brownish red (h., s.)	Colorless
Chromium	Red (h., s.); green (c.)	Green (c.)
Cobalt	Blue (h.c.)	Blue (h.c.)
Copper	Blue (c.); green (h.)	Red and opaque (c.)
Iron	Yellow or brown (h., s.)	Colorless; yellow or brownish (h.)
Lead	Colorless (n.s.)	Gray and opaque
Magnesium	Colorless; opaque (s.)	Colorless; not clear (s.s.)
Manganese	Violet (h.c.)	Colorless
Molybdenum	Colorless; green (h.)	Green (h.)
Nickel	Yellow (c.); red (h., s.)	Yellow (c.); red (h.); gray and opaque
Silicon	(Swims undissolved)	(Swims undissolved)
Silver	....	Gray and opaque
Strontium	Colorless; opaque (s.)	Colorless; not clear (s.s.)
Tin	Colorless; opaque (s.)	Colorless
Titanium	Colorless (n.s.)	Violet (c.); yellow or brownish (h.)
Uranium	Green; yellow or brownish (h., s.)	Green (h.)
Vanadium	Yellow	Green
Zinc	Colorless (n.s.)	Gray and opaque

Sodium Carbonate Bead

Substance	Oxidizing flame	Reducing flame
Manganese	Green	Colorless

PREPARATION OF REAGENTS

The following pages present directions for the preparation of various reagents. The collection has been prepared with the active collaboration of W. D. Bonner, R. K. Carleton, L. L. Carrick, Giles B. Cooke, E. J. Cragoe, Thos. De Vries, James L. Kassner, Thos. W. Mason, F. C. Mathers, M. G. Mellon, W. C. Pierce, J. H. Reedy, Arthur A. Vernon and S. R. Wood. Many others have contributed valuable suggestions.

Volumes have been stated in milliliters (ml) and liters (l). One milliliter is equivalent to 1.000027 cubic centimeters (cm^3 or cc.). Masses are indicated in grams (g).

The relation to molar solution (*M*) or normal solution (*N*) is indicated in many cases.

Distilled water should be used.

LABORATORY REAGENTS FOR GENERAL USE

DILUTE ACIDS, 3 molar. Use the amount of concentrated acid indicated and dilute to one liter.

Acetic acid, 3 *N*. Use 172 ml of 17.4 *M* acid (99–100 %).

Hydrochloric acid, 3 *N*. Use 258 ml of 11.6 *M* acid (36 % HCl).

Nitric acid, 3 *N*. Use 183 ml of 16.4 *M* acid (69% HNO_3).

Phosphoric acid, 9 *N*. Use 205 ml of 14.6 *M* acid (85% H_3PO_4).

Sulfuric acid, 6 *N*. Use 168 ml of 17.8 *M* acid (95% H_2SO_4).

DILUTE BASES.

Ammonium hydroxide, 3 *M*, 3 *N*. Dilute 200 ml of concentrated solution (14.8 *M*, 28 % NH_3) to 1 liter.

Barium hydroxide, 0.2 *M*, 0.4 *N*. Saturated solution, 63 g per liter of $Ba(OH)_2.8H_2O$. Use some excess, filter off $BaCO_3$ and protect from CO_2 of the air with soda lime or ascarite in a guard tube.

Calcium hydroxide, 0.02 *M*, 0.04 *N*. Saturated solution, 1.5 g per liter of $Ca(OH)_2$. Use some excess, filter off $CaCO_3$ and protect from CO_2 of the air.

Potassium hydroxide, 3 *M*, 3 *N*. Dissolve 176 g of the sticks (95 %) in water and dilute to 1 liter.

Sodium hydroxide, 3 *M*, 3 *N*. Dissolve 126 g of the sticks (95 %) in water and dilute to 1 liter.

GENERAL REAGENTS. (See also *Standard Solutions for Volumetric Analysis*, and *Decinormal Solutions of Salts and Other Reagents.*)

Aluminum chloride, 0.167 *M*, 0.5 *N*. Dissolve 22 g of $AlCl_3$ in 1 liter of water.

Aluminum nitrate, 0.167 *M*, 0.5 *N*. Dissolve 58 g of $Al(NO_3)_3.7.5H_2O$ in 1 liter of water.

Aluminum sulfate, 0.083 *M*, 0.5 *N*. Dissolve 56 g of $Al_2(SO_4)_3.18H_2O$ in 1 liter of water.

Ammonium acetate, 3 *M*, 3 *N*. Dissolve 230 g of $NH_4C_2H_3O_2$ in water and dilute to 1 liter.

Ammonium carbonate, 1.5 *M*. Dissolve 144 g of the commercial salt (mixture of $(NH_4)_2CO_3.H_2O$ and $NH_4CO_2NH_2$) in 500 ml of 3 *N* NH_4OH and dilute to 1 liter.

Ammonium chloride, 3 *M*, 3 *N*. Dissolve 160 g of NH_4Cl in water. Dilute to 1 liter.

Ammonium molybdate.

1. 0.5 *M*, 1 *N*. Mix well 72 g of pure MoO_3 (or 81 g of H_2MoO_4) with 200 ml of water, and add 60 ml of conc. ammonium hydroxide. When solution is complete, filter and pour filtrate, *very slowly* and with *rapid stirring*, into a mixture of 270 ml of conc. HNO_3 and 400 ml of water. Allow to stand over night, filter and dilute to 1 liter.

2. The reagent is prepared as two solutions which are mixed as needed, thus always providing fresh reagent of proper strength and composition. Since ammonium molybdate is an expensive reagent, and since an acid solution of this reagent as usually prepared keeps for only a few days, the method proposed will avoid loss of reagent and provide more certain results for quantitative work.

Solution 1. Dissolve 100 g of ammonium molybdate (C.P. grade) in 400 ml of water and 80 ml of 15 *M* NH_4OH. Filter if necessary, though this seldom has to be done.

Solution 2. Mix 400 ml of 16 *M* nitric acid with 600 ml of water.

For use, mix the calculated amount of solution 1 with twice its volume of solution 2, adding solution 1 to solution 2 slowly with vigorous stirring. Thus, for amounts of phosphorus up to 20 mg, 10 ml of solution 1 to 20 ml of solution 2 is adequate. Increase amount as needed.

Ammonium nitrate, 1 *M*, 1 *N*. Dissolve 80 g of NH_4NO_3 in 1 liter of water.

Ammonium oxalate, 0.25 *M*, 0.5 *N*. Dissolve 35.5 g of $(NH_4)_2C_2O_4.H_2O$ in water. Dilute to 1 liter.

Ammonium sulfate, 0.25 *M*, 0.5 *N*. Dissolve 33 g of $(NH_4)_2SO_4$ in 1 liter of water.

Ammonium sulfide, colorless.

1. 3 *M*. Treat 200 ml of conc. NH_4OH with H_2S until saturated, keeping the solution cold. Add 200 ml of conc. NH_4OH and dilute of 1 liter.

2. 6 *N*. Saturate 6 *N* ammonium hydroxide (40 ml conc. ammonia solution + 60 ml H_2O) with washed H_2S gas. The ammonium hydroxide bottle must be completely full and must be kept surrounded by ice while being saturated (about 48 hours for two liters). The reagent is best preserved in brown, completely filled, glass-stoppered bottles.

LABORATORY REAGENTS (Continued)

Ammonium sulfide, yellow. Treat 150 ml of conc. NH_4OH with H_2S until saturated, keeping the solution cool. Add 250 ml of conc. NH_4OH and 10 g of powdered sulfur. Shake the mixture until the sulfur is dissolved and dilute to 1 liter with water. In the solution the concentration of $(NH_4)_2S_2$, $(NH_4)_2S$ and NH_4OH are 0.625, 0.4 and 1.5 normal respectively. On standing, the concentration of $(NH_4)_2S_2$ increases and that of $(NH_4)_2S$ and NH_4OH decreases.

Antimony pentachloride, 0.1 *M*, 0.5 *N*. Dissolve 30 g of $SbCl_5$ in 1 liter of water.

Antimony trichloride, 0.167 *M*, 0.5 *N*. Dissolve 38 g of $SbCl_3$ in 1 liter of water.

Aqua regia. Mix 1 part concentrated HNO_3 with 3 parts of concentrated HCl. This formula should include one volume of water if the aqua regia is to be stored for any length of time. Without water, objectionable quantities of chlorine and other gases are evolved.

Barium chloride, 0.25 *M*, 0.5 *N*. Dissolve 61 g of $BaCl_2 \cdot 2H_2O$ in water. Dilute to 1 liter.

Barium hydroxide, 0.1 *M*, about 0.2 *N*. Dissolve 32 g of $Ba(OH)_2 \cdot 8H_2O$ in 1 liter of water.

Barium nitrate, 0.25 *M*, 0.5 *N*. Dissolve 65 g of $Ba(NO_3)_2$ in 1 liter of water.

Bismuth chloride, 0.167 *M*, 0.5 *N*. Dissolve 53 g of $BiCl_3$ in 1 liter of dilute HCl. Use 1 part HCl to 5 parts water.

Bismuth nitrate, 0.083 *M*, 0.25 *N*. Dissolve 40 g of $Bi(NO_3)_3 \cdot 5H_2O$ in 1 liter of dilute HNO_3. Use 1 part of HNO_3 to 5 parts of water.

Cadmium chloride, 0.25 *M*, 0.5 *N*. Dissolve 46 g of $CdCl_2$ in 1 liter of water.

Cadmium nitrate, 0.25 *M*, 0.5 *N*. Dissolve 77 g of $Cd(NO_3)_2 \cdot 4H_2O$ in 1 liter of water.

Cadmium sulfate, 0.25 *M*, 0.5 *N*. Dissolve 70 g of $CdSO_4 \cdot 4H_2O$ in 1 liter of water.

Calcium chloride, 0.25 *M*, 0.5 *N*. Dissolve 55 g of $CaCl_2 \cdot 6H_2O$ in water. Dilute to 1 liter.

Calcium nitrate, 0.25 *M*, 0.5 *N*. Dissolve 41 g of $Ca(NO_3)_2$ in 1 liter of water.

Chloroplatinic acid.

1. 0.0512 *M*, 0.102 *N*. Dissolve 26.53 g of $H_2PtCl_6 \cdot 6H_2O$ in water. Dilute to 100 ml. Contains 0.100 g Pt per ml.

2. Make a 10% solution by dissolving 1 g of $H_2PtCl_6 \cdot 6H_2O$ in 9 ml of water. Shake thoroughly to insure complete mixing. Keep in a dropping bottle.

Chromic chloride, 0.167 *M*, 0.5 *N*. Dissolve 26 g of $CrCl_3$ in 1 liter of water.

Chromic nitrate, 0.167 *M*, 0.5 *N*. Dissolve 40 g of $Cr(NO_3)_3$ in 1 liter of water.

Chromic sulfate, 0.083 *M*, 0.5 *N*. Dissolve 60 g of $Cr_2(SO_4)_3 \cdot 18H_2O$ in 1 liter of water.

Cobaltous nitrate, 0.25 *M*, 0.5 *N*. Dissolve 73 g of $Co(NO_3)_2 \cdot 6H_2O$ in 1 liter of water.

Cobaltous sulfate, 0.25 *M*, 0.5 *N*. Dissolve 70 g of $CoSO_4 \cdot 7H_2O$ in 1 liter of water.

Cupric chloride, 0.25 *M*, 0.5 *N*. Dissolve 43 g of $CuCl_2 \cdot 2H_2O$ in 1 liter of water.

Cupric nitrate, 0.25 *M*, 0.5 *N*. Dissolve 74 g of $Cu(NO_3)_2 \cdot 6H_2O$ in 1 liter of water.

Cupric sulfate, 0.5 *M*, 1 *N*. Dissolve 124.8 g of $CuSO_4 \cdot 5H_2O$ in water to which 5 ml of H_2SO_4 has been added. Dilute to 1 liter.

Ferric chloride, 0.5 *M*, 1.5 *N*. Dissolve 135.2 g of $FeCl_3 \cdot 6H_2O$ in water containing 20 ml of conc. HCl. Dilute to 1 liter.

Ferric nitrate, 0.167 *M*, 0.5 *N*. Dissolve 67 g of $Fe(NO_3)_3 \cdot 9H_2O$ in 1 liter of water.

Ferric sulfate, 0.25 *M*, 0.5 *N*. Dissolve 140.5 g of $Fe_2(SO_4)_3 \cdot 9H_2O$ in water containing 100 ml of conc. H_2SO_4. Dilute to 1 liter.

Ferrous ammonium sulfate, 0.5 *M*, 1 *N*. Dissolve 196 g of $Fe(NH_4SO_4)_2 \cdot 6H_2O$ in water containing 10 ml of conc. H_2SO_4. Dilute to 1 liter. Prepare fresh solutions for best results.

Ferrous sulfate, 0.5 *M*, 1 *N*. Dissolve 139 g of $FeSO_4 \cdot 7H_2O$ in water containing 10 ml of conc. H_2SO_4. Dilute to 1 liter. Solution does not keep well.

Lead acetate, 0.5 *M*, 1 *N*. Dissolve 190 g of $Pb(C_2H_3O_2)_2 \cdot 3H_2O$ in water. Dilute to 1 liter.

Lead nitrate, 0.25 *M*, 0.5 *N*. Dissolve 83 g of $Pb(NO_3)_2$ in 1 liter of water.

Lime water. See *Calcium hydroxide.*

Magnesium chloride, 0.25 *M*, 0.5 *N*. Dissolve 51 g of $MgCl_2 \cdot 6H_2O$ in 1 liter of water.

Magnesium chloride reagent. Dissolve 50 g of $MgCl_2 \cdot 6H_2O$ and 100 g of NH_4Cl in 500 ml of water. Add 10 ml of conc. NH_4OH, allow to stand over night and filter if a precipitate has formed. Make acid to methyl red with dilute HCl. Dilute to 1 liter. Solution contains 0.25 *M* $MgCl_2$ and 2 *M* NH_4Cl. Solution may also be diluted with 133 ml of conc. NH_4OH and water to make 1 liter. Such a solution will contain 2 *M* NH_4OH.

Magnesium nitrate, 0.25 *M*, 0.5 *N*. Dissolve 64 g of $Mg(NO_3)_2 \cdot 6H_2O$ in 1 liter of water.

Magnesium sulfate, 0.25 *M*, 0.5 *N*. Dissolve 62 g of $MgSO_4 \cdot 7H_2O$ in 1 liter of water.

Manganous chloride, 0.25 *M*, 0.5 *N*. Dissolve 50 g of $MnCl_2 \cdot 4H_2O$ in 1 liter of water.

Manganous nitrate, 0.25 *M*, 0.5 *N*. Dissolve 72 g of $Mn(NO_3)_2 \cdot 6H_2O$ in 1 liter of water.

Manganous sulfate, 0.25 *M*, 0.5 *N*. Dissolve 69 g of $MnSO_4 \cdot 7H_2O$ in 1 liter of water.

Mercuric chloride, 0.25 *M*, 0.5 *N*. Dissolve 68 g of $HgCl_2$ in water. Dilute to 1 liter.

Mercuric nitrate, 0.25 *M*, 0.5 *N*. Dissolve 81 g of $Hg(NO_3)_2$ in 1 liter of water.

Mercuric sulfate, 0.25 *M*, 0.5 *N*. Dissolve 74 g of $HgSO_4$ in 1 liter of water.

Mercurous nitrate. Use 1 part $HgNO_3$, 20 parts water and 1 part HNO_3.

Nickel chloride, 0.25 *M*, 0.5 *N*. Dissolve 59 g of $NiCl_2.6H_2O$ in 1 liter of water.

Nickel nitrate, 0.25 *M*, 0.5 *N*. Dissolve 73 g of $Ni(NO_3)_2.6H_2O$ in 1 liter of water.

Nickel sulfate, 0.25 *M*, 0.5 *N*. Dissolve 66 g of $NiSO_4.6H_2O$ in 1 liter of water.

Potassium bromide, 0.5 *M*, 0.5 *N*. Dissolve 60 g of KBr in 1 liter of water.

Potassium carbonate, 1.5 *M*, 3 *N*. Dissolve 207 g of K_2CO_3 in 1 liter of water.

Potassium chloride, 0.5 *M*, 0.5 *N*. Dissolve 37 g of KCl in 1 liter of water.

Potassium chromate, 0.25 *M*, 0.5 *N*. Dissolve 49 g of K_2CrO_4 in 1 liter of water.

Potassium cyanide, 0.5 *M*, 0.5 *N*. Dissolve 33 g of KCN in 1 liter of water.

Potassium dichromate, 0.125 *M*. Dissolve 37 g of $K_2Cr_2O_7$ in 1 liter of water.

Potassium ferricyanide, 0.167 *M*, 0.5 *N*. Dissolve 55 g of $K_3Fe(CN)_6$ in 1 liter of water.

Potassium ferrocyanide, 0.5 *M*, 2 *N*. Dissolve 211 g of $K_4Fe(CN)_6.3H_2O$ in water. Dilute to 1 liter.

Potassium iodide, 0.5 *M*, 0.5 *N*. Dissolve 83 g of KI in 1 liter of water.

Potassium nitrate, 0.5 *M*, 0.5 *N*. Dissolve 51 g of KNO_3 in 1 liter of water.

Potassium sulfate, 0.25 *M*, 0.5 *N*. Dissolve 44 g of K_2SO_4 in 1 liter of water.

Silver nitrate, 0.5 *M*, 0.5 *N*. Dissolve 85 g of $AgNO_3$ in water. Dilute to 1 liter.

Sodium acetate, 3 *M*, 3 *N*. Dissolve 408 g of $NaC_2H_3O_2.3H_2O$ in water. Dilute to 1 liter.

Sodium carbonate, 1.5 *M*, 3 *N*. Dissolve 159 g of Na_2CO_3, or 430 g of $Na_2CO_3.10H_2O$ in water. Dilute to 1 liter.

Sodium chloride, 0.5 *M*, 0.5 *N*. Dissolve 29 g of NaCl in 1 liter of water.

Sodium cobaltinitrite, 0.08 *M* (reagent for potassium). Dissolve 25 g of $NaNO_2$ in 75 ml of water, add 2 ml of glacial acetic acid and then 2.5 g of $Co(NO_3)_2.6H_2O$. Allow to stand for several days, filter and dilute to 100 ml. Reagent is somewhat unstable.

Sodium hydrogen phosphate, 0.167 *M*, 0.5 *N*. Dissolve 60 g of $Na_2HPO_4.12H_2O$ in 1 liter of water.

Sodium nitrate, 0.5 *M*, 0.5 *N*. Dissolve 43 g of $NaNO_3$ in 1 liter of water.

Sodium sulfate, 0.25 *M*, 0.5 *N*. Dissolve 36 g of Na_2SO_4 in 1 liter of water.

Sodium sulfide, 0.5 *M*, 1 *N*. Dissolve 120 g of $Na_2S.9H_2O$ in water and dilute to 1 liter. Or, saturate 500 ml of 1 *M* NaOH (21 g of 95% NaOH sticks) with H_2S, keeping the solution cool, and dilute with 500 ml of 1 *M* NaOH.

Stannic chloride, 0.125 *M*, 0.5 *N*. Dissolve 33 g of $SnCl_4$ in 1 liter of water.

Stannous chloride, 0.5 *M*, 1 *N*. Dissolve 113 g of $SnCl_2.2H_2O$ in 170 ml of conc. HCl, using heat if necessary. Dilute with water to 1 liter. Add a few pieces of tin foil. Prepare solution fresh at frequent intervals.

Stannous chloride (for Bettendorf test). Dissolve 113 g of $SnCl_2.2H_2O$ in 75 ml of conc. HCl. Add a few pieces of tin foil.

Strontium chloride, 0.25 *M*, 0.5 *N*. Dissolve 67 g of $SrCl_2.6H_2O$ in 1 liter of water.

Zinc nitrate, 0.25 *M*, 0.5 *N*. Dissolve 74 g of $Zn(NO_3)_2.6H_2O$ in 1 liter of water.

Zinc sulfate, 0.25 *M*, 0.5 *N*. Dissolve 72 g of $ZnSO_4.7H_2O$ in 1 liter of water.

SPECIAL SOLUTIONS AND REAGENTS

Aluminon (qualitative test for aluminum). Aluminon is a trade name for the ammonium salt of aurin tricarboxylic acid. Dissolve 1 g of the salt in 1 liter of distilled water. Shake the solution well to insure thorough mixing.

Bang's reagent (for glucose estimation). Dissolve 100 g of K_2CO_3, 66 g of KCl and 160 g of $KHCO_3$ in the order given in about 700 ml of water at 30° C. Add 4.4 g of $CuSO_4$ and dilute to 1 liter after the CO_2 is evolved. This solution should be shaken only in such a manner as not to allow entry of air. After 24 hours 300 ml are diluted to 1 liter with saturated KCl solution, shaken gently and used after 24 hours; 50 ml equivalent to 10 mg glucose.

Barfoed's reagent (test for glucose). See *Cupric acetate.*

Baudisch's reagent. See *Cupferron.*

Benedict's solution (qualitative reagent for glucose). With the aid of heat, dissolve 173 g of sodium citrate and 100 g of Na_2CO_3 in 800 ml of water. Filter, if necessary, and dilute to 850 ml. Dissolve 17.3 g of $CuSO_4.5H_2O$ in 100 ml of water. Pour the latter solution, with constant stirring, into the arbonate-citrate solution, and make up to 1 liter.

Benzidine hydrochloride solution (for sulfate determination). Make a paste of 8 g of benzidine hydrochloride ($C_{12}H_8(NH_2)_2.2HCl$) and 20 ml of water, add 20 ml of HCl (sp. gr. 1.12) and dilute to 1 liter with water. Each ml of this solution is equivalent to 0.00357 g of H_2SO_4.

Bertrand's reagent (glucose estimation). Consists of the following solutions:

(a) Dissolve 200 g of Rochelle salts and 150 g of NaOH in sufficient water to make 1 liter of solution.

(b) Dissolve 40 g of $CuSO_4$ in enough water to make 1 liter of solution.

(c) Dissolve 50 g of $Fe_2(SO_4)_3$ and 200 g of H_2SO_4 (sp. gr. 1.84) in sufficient water to make 1 liter of solution.

(d) Dissolve 5 g of $KMnO_4$ in sufficient water to make 1 liter of solution.

Bial's reagent (for pentose). Dissolve 1 g of orcinol ($CH_3.C_6H_3(OH)_2$) in 500 ml of 30% HCl to which 30 drops of a 10% solution of $FeCl_3$ has been added.

Boutron-Boudet soap solution.

(a) Dissolve 100 g of pure castile soap in about 2500 ml of 56% ethyl alcohol.

(b) Dissolve 0.59 g of $Ba(NO_3)_2$ in 1 liter of water.

Adjust the castile soap solution so that 2.4 ml of it will give a permanent lather with 40 ml of solution (b). When adjusted, 2.4 ml of soap solution is equivalent to 220 parts per million of hardness (as $CaCO_3$) for a 40 ml sample.

See also *Soap solution.*

Brucke's reagent (protein precipitation). See *Potassium iodide-mercuric iodide.*

Clarke's soap solution (or A.P.H.A. standard method). Estimation of hardness in water.

(a) Dissolve 100 g of pure powdered castile soap in 1 liter of 80% ethyl alcohol and allow to stand over night.

(b) Prepare a standard solution of $CaCl_2$ by dissolving 0.5 g of $CaCO_3$ in HCl (sp. gr. 1.19), neutralize with NH_4OH and make slightly alkaline to litmus, and dilute to 500 ml. One ml is equivalent to 1 mg of $CaCO_3$.

Titrate (a) against (b) and dilute (a) with 80% ethyl alcohol until 1 ml of the resulting solution is equivalent to 1 ml of (b) after making allowance for the lather factor (the amount of standard soap solution required to produce a permanent lather in 50 ml of distilled water). One ml of the adjusted solution after subtracting the lather factor is equivalent to 1 mg of $CaCO_3$.

See also *Soap solution.*

Cobalticyanide paper (Rinnmann's test for Zn). Dissolve 4 g of $K_3Co(CN)_6$ and 1 g of $KClO_3$ in 100 ml of water. Soak filter paper in solution and dry at 100° C. Apply drop of zinc solution and burn in an evaporating dish. A green disk is obtained if zinc is present.

Cochineal. Extract 1 g of cochineal for four days with 20 ml of alcohol and 60 ml of distilled water. Filter.

Congo red. Dissolve 0.5 g of congo red in 90 ml of distilled water and 10 ml of alcohol.

Cupferron (Baudisch's reagent for iron analysis). Dissolve 6 g of the ammonium salt of nitroso-phenyl-hydroxylamine (cupferron) in 100 ml of H_2O. Reagent good for one week only and must be kept in the dark.

Cupric acetate (Barfoed's reagent for reducing monosaccharides). Dissolve 66 g of cupric acetate and 10 ml of glacial acetic acid in water and dilute to 1 liter.

Cupric oxide, ammoniacal; Schweitzer's reagent (dissolves cotton, linen and silk, but not wool).

1. Dissolve 5 g of cupric sulfate in 100 ml of boiling water, and add sodium hydroxide until precipitation is complete. Wash the precipitate well, and dissolve it in a minimum quantity of ammonium hydroxide.

2. Bubble a slow stream of air through 300 ml of strong ammonium hydroxide containing 50 g of fine copper turnings. Continue for one hour.

Cupric sulfate in glycerin-potassium hydroxide (reagent for silk). Dissolve 10 g of cupric sulfate, $CuSO_4.5H_2O$, in 100 ml of water and add 5 g of glycerin. Add KOH solution slowly until a deep blue solution is obtained.

Cupron (benzoin oxime). Dissolve 5 g in 100 ml of 95% alcohol.

Cuprous chloride, acidic (reagent for CO in gas analysis).

1. Cover the bottom of a two-liter flask with a layer of cupric oxide about one-half inch deep, suspend a bunch of copper wire so as to reach from the bottom to the top of the solution, and fill the flask with hydrochloric acid (sp. gr. 1.10). Shake occasionally. When the solution becomes nearly colorless, transfer to reagent bottles, which should also contain copper wire. The stock bottle may be refilled with dilute hydrochloric acid until either the cupric oxide or the copper wire is used up.

Copper sulfate may be substituted for copper oxide in the above procedure.

2. Dissolve 340 g of $CuCl_2.2H_2O$ in 600 ml of conc. HCl and reduce the cupric chloride by adding 190 ml of a saturated solution of stannous chloride or until the solution is colorless. The stannous chloride is prepared by treating 300 g of metallic tin in a 500 ml flask with conc. HCl until no more tin goes into solution.

3. (Winkler method). Add a mixture of 86 g of CuO and 17 g of finely divided metallic Cu, made by the reduction of CuO with hydrogen, to a solution of HCl, made by diluting 650 ml of conc. HCl with 325 ml of water. After the mixture has been added slowly and with frequent stirring, a spiral of copper wire is suspended in the bottle, reaching all the way to the bottom. Shake occasionally, and when the solution becomes colorless, it is ready for use.

Cuprous chloride, ammoniacal (reagent for CO in gas analysis).

1. The acid solution of cuprous chloride as prepared above is neutralized with ammonium hydroxide until an ammonia odor persists. An excess of metallic copper must be kept in the solution.

2. Pour 800 ml of acidic cuprous chloride, prepared by the Winkler method, into about 4 liters of water. Transfer the

SPECIAL SOLUTIONS AND REAGENTS (Continued)

precipitate to a 250 ml graduate. After several hours, siphon off the liquid above the 50 ml mark and refill with 7.5% NH_4OH solution which may be prepared by diluting 50 ml of conc. NH_4OH with 150 ml of water. The solution is well shaken and allowed to stand for several hours. It should have a faint odor of ammonia.

Dichlorofluorescein indicator. Dissolve 1 g in 1 liter of 70% alcohol or 1 g of the sodium salt in 1 liter of water.

Dimethylglyoxime (diacetyl dioxime), 0.01 *N*. Dissolve 0.6 g of dimethylglyoxime, $(CH_3CNOH)_2$, in 500 ml of 95% ethyl alcohol. This is an especially sensitive test for nickel, a very definite crimson color being produced.

Diphenylamine (reagent for rayon). Dissolve 0.2 g in 100 ml of concentrated sulfuric acid.

Diphenylamine sulfonate (for titration of iron with $K_2Cr_2O_7$). Dissolve 0.32 g of the barium salt of diphenylamine sulfonic acid in 100 ml of water, add 0.5 g of sodium sulfate and filter off the precipitate of $BaSO_4$.

Diphenylcarbazide. Dissolve 0.2 g of diphenylcarbazide in 10 ml of glacial acetic acid and dilute to 100 ml with 95% ethyl alcohol.

Esbach's reagent (estimation of protein). To a water solution of 10 g of picric acid and 20 g of citric acid, add sufficient water to make one liter of solution.

Eschka's compound. Two parts of calcined ("light") magnesia are thoroughly mixed with one part of anhydrous sodium carbonate.

Fehling's solution (reagent for reducing sugars).

(a) Copper sulfate solution. Dissolve 34.66 g of $CuSO_4 \cdot 5H_2O$ in water and dilute to 500 ml.

(b) Alkaline tartrate solution. Dissolve 173 g of potassium sodium tartrate (Rochelle salts, $KNaC_4H_4O_6.4H_2O$) and 50 g of NaOH in water and dilute when cold to 500 ml.

For use, mix equal volumes of the two solutions at the time of using.

Ferric-alum indicator. Dissolve 140 g of ferric-ammonium sulfate crystals in 400 ml of hot water. When cool, filter, and make up to a volume of 500 ml with dilute (6 *N*) nitric acid.

Folin's mixture (for uric acid). To 650 ml of water add 500 g of $(NH_4)_2SO_4$, 5 g of uranium acetate and 6 g of glacial acetic acid. Dilute to 1 liter.

Formaldehyde-sulfuric acid (Marquis' reagent for alkaloids). Add 10 ml of formaldehyde solution to 50 ml of sulfuric acid.

Froehde's reagent. See *Sulfomolybdic acid.*

Fuchsin (reagent for linen). Dissolve 1 g of fuchsin in 100 ml of alcohol.

Fuchsin-sulfurous acid (Schiff's reagent for aldehydes). Dissolve 0.5 g of fuchsin and 9 g of sodium bisulfite in 500 ml of water, and add 10 ml of HCl. Keep in well-stoppered bottles and protect from light.

Gunzberg's reagent (detection of HCl in gastric juice). Prepare as needed a solution containing 4 g of phloroglucinol and 2 g of vanillin in 100 ml of absolute ethyl alcohol.

Hager's reagent. See *Picric acid.*

Hanus solution (for iodine number). Dissolve 13.2 g of resublimed iodine in one liter of glacial acetic acid which will pass the dichromate test for reducible matter. Add sufficient bromine to double the halogen content, determined by titration (3 ml is about the proper amount). The iodine may be dissolved by the aid of heat, but the solution should be cold when the bromine is added.

Iodine, tincture of. To 50 ml of water add 70 g of I_2 and 50 g of KI. Dilute to 1 liter with alcohol.

Iodo-potassium iodide (Wagner's reagent for alkaloids). Dissolve 2 g of iodine and 6 g of KI in 100 ml of water.

Litmus (indicator). Extract litmus powder three times with boiling alcohol, each treatment consuming an hour. Reject the alcoholic extract. Treat residue with an equal weight of cold water and filter; then exhaust with five times its weight of boiling water, cool and filter. Combine the aqueous extracts.

Magnesia mixture (reagent for phosphates and arsenates). Dissolve 55 g of magnesium chloride and 105 g of ammonium chloride in water, barely acidify with hydrochloric acid, and dilute to 1 liter. The ammonium hydroxide may be omitted until just previous to use. The reagent, if completely mixed and stored for any period of time, becomes turbid.

Magnesium reagent. See *S and O reagent.*

Magnesium uranyl acetate. Dissolve 100 g of $UO_2(C_2H_3O_2)_2.2H_2O$ in 60 ml of glacial acetic acid and dilute to 500 ml. Dissolve 330 g of $Mg(C_2H_3O_2)_2.4H_2O$ in 60 ml of glacial acetic acid and dilute to 200 ml. Heat solutions to the boiling point until clear, pour the magnesium solution into the uranyl solution, cool and dilute to 1 liter. Let stand over night and filter if necessary.

Marme's reagent. See *Potassium-cadmium iodide.*

Marquis' reagent. See *Formaldehyde-sulfuric acid.*

Mayer's reagent (white precipitate with most alkaloids in slightly acid solutions). Dissolve 1.358 g of $HgCl_2$ in 60 ml of water and pour into a solution of 5 g of KI in 10 ml of H_2O. Add sufficient water to make 100 ml.

Methyl orange indicator. Dissolve 1 g of methyl orange in 1 liter of water. Filter, if necessary.

Methyl orange, modified. Dissolve 2 g of methyl orange and 2.8 g of xylene cyanole FF in 1 liter of 50% alcohol.

Methyl red indicator. Dissolve 1 g of methyl red in 600 ml of alcohol and dilute with 400 ml of water.

Methyl red, modified. Dissolve 0.50 g of methyl red and 1.25 g of xylene cyanole FF in 1 liter of 90% alcohol. Or, dissolve 1.25 g of methyl red and 0.825 g of methylene blue in 1 liter of 90% alcohol.

SPECIAL SOLUTIONS AND REAGENTS (Continued)

Millon's reagent (for albumins and phenols). Dissolve 1 part of mercury in 1 part of cold fuming nitric acid. Dilute with twice the volume of water and decant the clear solution after several hours.

Mixed indicator. Prepared by adding about 1.4 g of xylene cyanole FF to 1 g of methyl orange. The dye is seldom pure enough for these proportions to be satisfactory. Each new lot of dye should be tested by adding additional amounts of the dye until a test portion gives the proper color change. The acid color of this indicator is like that of permanganate; the neutral color is gray; and the alkaline color is green. Described by Hickman and Linstead, J. Chem. Soc. (Lon.), **121,** 2502 (1922).

Molisch's reagent. See *α-Naphthol.*

α-Naphthol (Molisch's reagent for wool). Dissolve 15 g of α-naphthol in 100 ml of alcohol or chloroform.

Nessler's reagent (for ammonia). Dissolve 50 g of KI in the smallest possible quantity of cold water (50 ml). Add a saturated solution of mercuric chloride (about 22 g in 350 ml of water will be needed) until an excess is indicated by the formation of a precipitate. Then add 200 ml of 5 *N* NaOH and dilute to 1 liter. Let settle, and draw off the clear liquid.

Nickel oxide, ammoniacal (reagent for silk). Dissolve 5 g of nickel sulfate in 100 ml of water, and add sodium hydroxide solution until nickel hydroxide is completely precipitated Wash the precipitate well and dissolve in 25 ml of concentrated ammonium hydroxide and 25 ml of water.

***p*-Nitrobenzene-azo-resorcinol** (reagent for magnesium). Dissolve 1 g of the dye in 10 ml of *N* NaOH and dilute to 1 liter.

Nitron (detection of nitrate radical). Dissolve 10 g of nitron ($C_{20}H_{16}N_4$, 4, 5-dihydro-1, 4-diphenyl-3, 5-phenylimino-1, 2, 4-triazole) in 5 ml of glacial acetic acid and 95 ml of water. The solution may be filtered with slight suction through an alundum crucible and kept in a dark bottle.

α-Nitroso-β-naphthol. Make a saturated solution in 50% acetic acid (1 part of glacial acetic acid with 1 part of water). Does not keep well.

Nylander's solution (carbohydrates). Dissolve 20 g of bismuth subnitrate and 40 g of Rochelle salts in 1 liter of 8% NaOH solution. Cool and filter.

Obermayer's reagent (for indoxyl in urine). Dissolve 4 g of $FeCl_3$ in one liter of HCl (sp. gr. 1.19).

Oxine. Dissolve 14 g of HC_9H_6ON in 30 ml of glacial acetic acid. Warm slightly, if necessary. Dilute to 1 liter.

Oxygen absorbent. Dissolve 300 g of ammonium chloride in one liter of water and add one liter of concentrated ammonium hydroxide solution. Shake the solution thoroughly. For use as an oxygen absorbent, a bottle half full of copper turnings is filled nearly full with the NH_4Cl-NH_4OH solution and the gas passed through.

Pasteur's salt solution. To one liter of distilled water add 2.5 g of potassium phosphate, 0.25 g of calcium phosphate, 0.25 g of magnesium sulfate and 12.00 g of ammonium tartrate.

Pavy's solution (glucose reagent). To 120 ml of Fehling's solution, add 300 ml of NH_4OH (sp. gr. 0.88) and dilute to 1 liter with water.

Phenanthroline ferrous ion indicator. Dissolve 1.485 g of phenanthroline monohydrate in 100 ml of 0.025 *M* ferrous sulfate solution.

Phenolphthalein. Dissolve 1 g of phenolphthalein in 50 ml of alcohol and add 50 ml of water.

Phenolsulfonic acid (determination of nitrogen as nitrate). Dissolve 25 g of phenol in 150 ml of conc. H_2SO_4, add 75 ml of fuming H_2SO_4 (15% SO_3), stir well and heat for two hours at 100° C.

Phloroglucinol solution (pentosans). Make a 3% phloroglucinol solution in alcohol. Keep in a dark bottle.

Phosphomolybdic acid (Sonnenschein's reagent for alkaloids).

1. Prepare ammonium phosphomolybdate and after washing with water, boil with nitric acid and expel NH_3; evaporate to dryness and dissolve in 2 *N* nitric acid.

2. Dissolve ammonium molybdate in HNO_3 and treat with phosphoric acid. Filter, wash the precipitate, and boil with aqua regia until the ammonium salt is decomposed. Evaporate to dryness. The residue dissolved in 10% HNO_3 constitutes Sonnenschein's reagent.

Phosphoric acid—sulfuric acid mixture. Dilute 150 ml of conc. H_2SO_4 and 100 ml of conc. H_3PO_4 (85%) with water to a volume of 1 liter.

Phosphotungstic acid (Scheibler's reagent for alkaloids).

1. Dissolve 20 g of sodium tungstate and 15 g of sodium phosphate in 100 ml of water containing a little nitric acid.

2. The reagent is a 10% solution of phosphotungstic acid in water. The phosphotungstic acid is prepared by evaporating a mixture of 10 g of sodium tungstate dissolved in 5 g of phosphoric acid (sp. gr. 1.13) and enough boiling water to effect solution. Crystals of phosphotungstic acid separate.

Picric acid (Hager's reagent for alkaloids, wool and silk). Dissolve 1 g of picric acid in 100 ml of water.

Potassium antimonate (reagent for sodium). Boil 22 g of potassium antimonate with 1 liter of water until nearly all of the salt has dissolved, cool quickly, and add 35 ml of 10% potassium hydroxide. Filter after standing over night.

Potassium-cadmium iodide (Marme's reagent for alkaloids). Add 2 g of CdI_2 to a boiling solution of 4 g of KI in 12 ml of water, and then mix with 12 ml of saturated KI solution.

Potassium hydroxide (for CO_2 absorption). Dissolve 360 g of KOH in water and dilute to 1 liter.

Potassium iodide-mercuric iodide (Brucke's reagent for proteins). Dissolve 50 g of KI in 500 ml of water, and saturate with mercuric iodide (about 120 g). Dilute to 1 liter.

SPECIAL SOLUTIONS AND REAGENTS (Continued)

SPECIAL SOLUTIONS AND REAGENTS (Continued)

Potassium pyrogallate (for oxygen absorption). For mixtures of gases containing less than 28% oxygen, add 100 ml of KOH solution (50 g of KOH to 100 ml of water) to 5 g of pyrogallol. For mixtures containing more than 28% oxygen the KOH solution should contain 120 g of KOH to 100 ml of water.

Pyrogallol, alkaline.

(a) Dissolve 75 g of pyrogallic acid in 75 ml of water.

(b) Dissolve 500 g of KOH in 250 ml of water. When cool, adjust until sp. gr. is 1.55.

For use, add 270 ml of solution (b) to 30 ml of solution (a).

Rosolic acid (indicator). Dissolve 1 g of rosolic acid in 10 ml of alcohol and add 100 ml of water.

S and O reagent (Suitsu and Okuma's test for Mg). Dissolve 0.5 g of the dye (*o-p*-dihydroxy-monazo-*p*-nitrobenzene) in 100 ml of 0.25 *N* NaOH.

Scheibler's reagent. See *Phosphotungstic acid.*

Schiff's reagent. See *Fuchsin-sulfurous acid.*

Schweitzer's reagent. See *Cupric oxide, ammoniacal.*

Soap solution (reagent for hardness in water). Dissolve 100 g of dry castile soap in 1 liter of 80% alcohol (5 parts alcohol to 1 part water). Allow to stand several days and dilute with 70% to 80% alcohol until 6.4 ml produces a permanent lather with 20 ml of standard calcium solution. The latter solution is made by dissolving 0.2 g of $CaCO_3$ in a small amount of dilute HCl, evaporating to dryness and making up to 1 liter.

Sodium bismuthate (oxidation of manganese). Heat 20 parts of NaOH nearly to redness in an iron or nickel crucible and add slowly 10 parts of basic bismuth nitrate which has been previously dried. Add two parts of sodium peroxide, and pour the brownish-yellow fused mass on an iron plate to cool. When cold, break up in a mortar, extract with water, and collect on an asbestos filter.

Sodium hydroxide (for CO_2 absorption). Dissolve 330 g of NaOH in water and dilute to 1 liter.

Sodium nitroprusside (reagent for hydrogen sulfide and wool). Use a freshly prepared solution of 1 g of sodium nitroprusside in 10 ml of water.

Sodium oxalate, according to Sörensen (primary standard). Dissolve 30 g of the commercial salt in 1 liter of water, make slightly alkaline with sodium hydroxide, and let stand until perfectly clear. Filter and evaporate the filtrate to 100 ml. Cool and filter. Pulverize the residue and wash it several times with small volumes of water. The procedure is repeated until the mother liquor is free from sulfate and is neutral to phenolphthalein.

Sodium plumbite (reagent for wool). Dissolve 5 g of sodium hydroxide in 100 ml of water. Add 5 g of litharge and boil until dissolved.

Sodium polysulfide. Dissolve 480 g of $Na_2S.9H_2O$ in 500 ml of water, add 40 g of NaOH and 18 g of sulfur. Stir thoroughly and dilute to 1 liter with water.

Sonnenschein's reagent. See *Phosphomolybdic acid.*

Starch solution.

1. Make a paste with 2 g of soluble starch and 0.01 g of HgI_2 with a small amount of water. Add the mixture slowly to 1 liter of boiling water and boil for a few minutes. Keep in a glass stoppered bottle. If other than soluble starch is used, the solution will not clear on boiling; it should be allowed to stand and the clear liquid decanted.

2. A solution of starch which keeps indefinitely is made as follows: Mix 500 ml of saturated NaCl solution (filtered), 80 ml of glacial acetic acid, 20 ml of water and 3 g of starch. Bring slowly to a boil and boil for two minutes.

3. Make a paste with 1 g of soluble starch and 5 mg of HgI_2, using as little cold water as possible. Then pour about 200 ml of boiling water on the paste and stir immediately. This will give a clear solution if the paste is prepared correctly and the water actually boiling. Cool and add 4 g of KI. Starch solution decomposes on standing due to bacterial action, but this solution will keep a long time if stored under a layer of toluene.

Stoke's reagent. Dissolve 30 g of $FeSO_4$ and 20 g of tartaric acid in water and dilute to 1 liter. Just before using, add concentrated NH_4OH until the precipitate first formed is redissolved.

Sulfanilic acid (reagent for nitrites). Dissolve 0.5 g of sulfanilic acid in a mixture of 15 ml of glacial acetic acid and 135 ml of recently boiled water.

Sulfomolybdic acid (Froehde's reagent for alkaloids and glucosides). Dissolve 10 g of molybdic acid or sodium molybdate in 100 ml of conc. H_2SO_4.

Tannic acid (reagent for albumen, alkaloids and gelatin). Dissolve 10 g of tannic acid in 10 ml of alcohol and dilute with water to 100 ml.

Titration mixture. See *Zimmermann-Reinhardt reagent.*

***o*-Tolidine solution** (residual chlorine in water analysis). Prepare 1 liter of dilute HCl (100 ml of HCl (sp. gr. 1.19) in sufficient water to make 1 liter). Dissolve 1 g of *o*-tolidine in 100 ml of the dilute HCl and dilute to 1 liter with dilute HCl solution.

Trinitrophenol solution. See *Picric acid.*

Turmeric paper. Impregnate white, unsized paper with the tincture, and dry.

Turmeric tincture (reagent for borates). Digest ground turmeric root with several quantities of water which are discarded. Dry the residue and digest it several days with six times its weight of alcohol. Filter.

Uffelmann's reagent (turns yellow in presence of a lactic acid). To a 2% solution of pure phenol in water, add a water solution of $FeCl_3$ until the phenol solution becomes violet in color.

Wagner's reagent. See *Iodo-potassium iodide.*

Wagner's solution (used in phosphate rock analysis to prevent precipitation of iron and aluminum). Dissolve 25 g of citric acid and 1 g of salicylic acid in water and dilute to 1 liter. Use 50 ml of the reagent.

Wij's iodine monochloride solution (for iodine number). Dissolve 13 g of resublimed iodine in 1 liter of glacial acetic acid which will pass the dichromate test for reducible matter. Set aside 25 ml of this solution. Pass into the remainder of the solution dry chlorine gas (dried and washed by passing through H_2SO_4 (sp. gr. 1.84)) until the characteristic color of free iodine has been discharged. Now add the iodine solution which was reserved, until all free chlorine has been destroyed. A slight excess of iodine does little or no harm, but an excess of chlorine must be avoided. Preserve in well stoppered, amber colored bottles. Avoid use of solutions which have been prepared for more than 30 days.

Wij's special solution (for iodine number—Analyst **58,** 523–7, 1933). To 200 ml of glacial acetic acid that will pass the dichromate test for reducible matter, add 12 g of dichloroamine T (paratoluene-sulfonedichloroamide), and 16.6 g of dry KI (in small quantities with continual shaking until all the KI has dissolved). Make up to 1 liter with the same quality of acetic acid used above and preserve in a dark colored bottle.

Zimmermann-Reinhardt reagent (determination of iron). Dissolve 70 g of $MnSO_4.4H_2O$ in 500 ml of water, add 125 ml of conc. H_2SO_4 and 125 ml of 85 % H_3PO_4, and dilute to 1 liter.

Zinc chloride solution, basic (reagent for silk). Dissolve 1000 g of zinc chloride in 850 ml of water, and add 40 g of zinc oxide. Heat until solution is complete.

Zinc uranyl acetate (reagent for sodium). Dissolve 10 g of $UO_2(C_2H_3O_2)_2.2H_2O$ in 6 g of 30 % acetic acid with heat, if necessary, and dilute to 50 ml. Dissolve 30 g of $Zn(C_2H_3O_2)_2.2H_2O$ in 3 g of 30 % acetic acid and dilute to 50 ml. Mix the two solutions, add 50 mg of NaCl, allow to stand over night and filter.

DECI-NORMAL SOLUTIONS OF SALTS AND OTHER REAGENTS

Atomic and molecular weights in the following table are based upon the 1965 atomic weight scale and the isotope C-12. The weight in grams of the compound in 1 cc of the following deci-normal solutions is found by dividing the H equivalent in the last column by 1000.

Name	Formula	Atomic or molecular weight	Hydrogen equivalent	0.1 Hydrogen equivalent in g
Acetic acid	$HC_2H_3O_2$	60.0530	$HC_2H_3O_2$	6.0053
Ammonia	NH_3	17.0306	NH_3	1.7031
Ammonium ion	NH_4^+	18.0386	NH_4	1.8039
Ammonium chloride	NH_4Cl	53.4916	NH_4Cl	5.3492
Ammonium sulfate	$(NH_4)_2SO_4$	132.1388	$\frac{1}{2}(NH_4)_2SO_4$	6.6069
Ammonium thiocyanate	NH_4CNS	76.1204	NH_4CNS	7.6120
Barium	Ba	137.34	$\frac{1}{2}Ba$	6.867
Barium carbonate	$BaCO_3$	197.3494	$\frac{1}{2}BaCO_3$	9.8675
Barium chloride hydrate	$BaCl_2·2H_2O$	244.2767	$\frac{1}{2}BaCl_2·2H_2O$	12.2138
Barium hydroxide	$Ba(OH)_2$	171.3547	$\frac{1}{2}Ba(OH)_2$	8.5677
Barium oxide	BaO	153.3394	$\frac{1}{2}BaO$	7.6670
Bromine	Br	79.909	Br	7.9909
Calcium	Ca	40.08	$\frac{1}{2}Ca$	2.004
Calcium carbonate	$CaCO_3$	100.0894	$\frac{1}{2}CaCO_3$	5.0045
Calcium chloride	$CaCl_2$	110.9860	$\frac{1}{2}CaCl_2$	5.5493
Calcium chloride hydrate	$CaCl_2.6H_2O$	219.0150	$\frac{1}{2}CaCl_2.6H_2O$	10.9508
Calcium hydroxide	$Ca(OH)_2$	74.0947	$\frac{1}{2}Ca(OH)_2$	3.7047
Calcium oxide	CaO	56.0794	$\frac{1}{2}CaO$	2.8040
Chlorine	Cl	35.453	Cl	3.5453
Citric acid	$C_6H_8O_7·H_2O$	210.1418	$\frac{1}{3}C_6H_8O_7·H_2O$	7.0047
Cobalt	Co	58.9332	$\frac{1}{2}Co$	2.9466
Copper	Cu	63.54	$\frac{1}{2}Cu$	3.177
Copper oxide (cupric)	CuO	79.5394	$\frac{1}{2}CuO$	3.9770
Copper sulfate hydrate	$CuSO_4·5H_2O$	249.6783	$\frac{1}{2}CuSO_4·5H_2O$	12.4839
Cyanogen	$(CN)_2$	26.0179	CN	2.6018
Hydrochloric acid	HCl	36.4610	HCl	3.6461
Hydrocyanic acid	HCN	27.0258	HCN	2.7026
Iodine	I	126.9044	I	12.6904
Lactic acid	$C_3H_6O_3$	90.0795	$C_3H_6O_3$	9.0080
Malic acid	$C_4H_6O_5$	134.0894	$\frac{1}{2}C_4H_6O_5$	6.7045
Magnesium	Mg	24.312	$\frac{1}{2}Mg$	1.2156
Magnesium carbonate	$MgCO_3$	84.3214	$\frac{1}{2}MgCO_3$	4.2161
Magnesium chloride	$MgCl_2$	95.2180	$\frac{1}{2}MgCl_2$	4.7609
Magnesium chloride hydrate	$MgCl_2·6H_2O$	203.2370	$\frac{1}{2}MgCl_2·6H_2O$	10.1623
Magnesium oxide	MgO	40.3114	$\frac{1}{2}MgO$	2.0156
Manganese	Mn	54.938	$\frac{1}{2}Mn$	2.7469
Manganese sulfate	$MnSO_4$	150.9996	$\frac{1}{2}MnSO_4$	7.5500
Mercuric chloride	$HgCl_2$	271.4960	$\frac{1}{2}HgCl_2$	13.5748
Nickel	Ni	58.71	$\frac{1}{2}Ni$	2.9356
Nitric acid	HNO_3	63.0129	HNO_3	6.3013
Nitrogen	N	14.0067	N	1.4007
Nitrogen pentoxide	N_2O_5	108.0104	$\frac{1}{2}N_2O_5$	5.4005
Oxalic acid	$H_2C_2O_4$	90.0358	$\frac{1}{2}H_2C_2O_4$	4.5018
Oxalic acid hydrate	$H_2C_2O_4·2H_2O$	126.0665	$\frac{1}{2}H_2C_2O_4·2H_2O$	6.3033
Oxalic acid anhydride	C_2O_3	72.0205	$\frac{1}{2}C_2O_3$	3.6010
Phosphoric acid	H_3PO_4	97.9953	$\frac{1}{3}H_3PO_4$	3.2665
Potassium	K	39.102	K	3.9102
Potassium bicarbonate	$KHCO_3$	100.1193	$KHCO_3$	10.0119
Potassium carbonate	K_2CO_3	138.2134	$\frac{1}{2}K_2CO_3$	6.9106
Potassium chloride	KCl	74.5550	KCl	7.4555
Potassium cyanide	KCN	65.1199	KCN	6.5120
Potassium hydroxide	KOH	56.1094	KOH	5.6109
Potassium oxide	K_2O	94.2034	$\frac{1}{2}K_2O$	4.7102

DECI-NORMAL SOLUTIONS OF SALTS AND OTHER REAGENTS (Continued)

Name	Formula	Atomic or molecular weight	Hydrogen equivalent	0.1 Hydrogen equivalent in g
Potassium permanganate for Co estimation	$KMnO_4$	158.0376	$\frac{1}{6}KMnO_4$	2.6339
Potassium permanganate for Mn estimation	$KMnO_4$	158.0376	$\frac{1}{3}KMnO_4$	5.2678
Potassium tartrate	$K_2H_4C_4O_6$	226.2769	$\frac{1}{2}K_2H_4C_4O_6$	11.3139
Silver	Ag	107.87	Ag	10.787
Silver nitrate	$AgNO_3$	169.8749	$AgNO_3$	16.9875
Sodium	Na	22.9898	Na	2.2990
Sodium bicarbonate	$NaHCO_3$	84.0071	$NaHCO_3$	8.4007
Sodium carbonate	Na_2CO_3	105.9890	$\frac{1}{2}Na_2CO_3$	5.2995
Sodium chloride	NaCl	58.4428	NaCl	5.8443
Sodium hydroxide	NaOH	39.9972	NaOH	3.9997
Sodium oxide	Na_2O	61.9790	$\frac{1}{2}Na_2O$	3.0990
Sodium sulfide	Na_2S	78.0436	$\frac{1}{2}Na_2S$	3.9022
Succinic acid	$H_2C_4H_4O_4$	118.0900	$\frac{1}{2}H_2C_4H_4O_4$	5.9045
Sulfuric acid	H_2SO_4	98.0775	$\frac{1}{2}H_2SO_4$	4.9039
Sulfur trioxide	SO_3	80.0622	$\frac{1}{2}SO_3$	4.0031
Tartaric acid	$C_4H_6O_6$	150.0888	$\frac{1}{2}C_4H_6O_6$	7.5044
Zinc	Zn	65.37	$\frac{1}{2}Zn$	3.269
Zinc sulfate	$ZnSO_4 \cdot 7H_2O$	287.5390	$\frac{1}{2}ZnSO_47 \cdot H_2O$	14.3769

DECI–NORMAL SOLUTIONS OF OXIDATION AND REDUCTION REAGENTS

Atomic and molecular weights in the following table are based upon the 1965 atomic weight scale and the isotope C-12. The weight in grams of the compound in 1 cc of the following deci-normal solutions is found by dividing the H equivalent in the last column by 1000.

Name	Formula	Atomic or molecular weight	Hydrogen equivalent	0.1 Hydrogen equivalent in g
Antimony	Sb	121.75	$\frac{1}{2}Sb$	6.0875
Arsenic	As	74.9216	$\frac{1}{2}As$	3.7461
Arsenic trisulfide	As_2S_3	246.0352	$\frac{1}{4}As_2S_3$	6.1509
Arsenous oxide	As_2O_3	197.8414	$\frac{1}{4}As_2O_3$	4.9460
Barium peroxide	BaO_2	169.3388	$\frac{1}{2}BaO_2$	8.4669
Barium peroxide hydrate	$BaO_2{\cdot}8H_2O$	313.4615	$\frac{1}{2}BaO_2{\cdot}8H_2O$	15.6730
Calcium	Ca	40.08	$\frac{1}{2}Ca$	2.004
Calcium carbonate	$CaCO_3$	100.0894	$\frac{1}{2}CaCO_3$	5.0045
Calcium hypochlorite	$Ca(OCl)_2$	142.9848	$\frac{1}{4}Ca(OCl)_2$	3.5746
Calcium oxide	CaO	56.0794	$\frac{1}{2}CaO$	2.8040
Chlorine	Cl	35.453	Cl	3.5453
Chromium trioxide	CrO_3	99.9942	$\frac{1}{3}CrO_3$	3.3331
Ferrous ammonium sulfate	$FeSO_4(NH_4)SO_4{\cdot}6H_2O$	392.0764	$FeSO_4(NH_4)_2SO_4{\cdot}6H_2O$	39.2076
Hydroferrocyanic acid	$H_4Fe(CN)_6$	215.9860	$H_4Fe(CN)_6$	21.5986
Hydrogen peroxide	H_2O_2	34.0147	$\frac{1}{2}H_2O_2$	1.7007
Hydrogen sulfide	H_2S	34.0799	$\frac{1}{2}H_2S$	1.7040
Iodine	I	126.9044	I	12.6904
Iron	Fe	55.847	Fe	5.5847
Iron oxide (ferrous)	FeO	71.8464	FeO	7.1846
Iron oxide (ferric)	Fe_2O_3	159.6922	$\frac{1}{2}Fe_2O_3$	7.9846
Lead peroxide	PbO_2	239.1888	$\frac{1}{2}PbO_2$	11.9594
Manganese dioxide	MnO_2	86.9368	$\frac{1}{2}MnO_2$	4.3468
Nitric acid	HNO_3	63.0129	$\frac{1}{3}HNO_3$	2.1004
Nitrogen trioxide	N_2O_3	76.0116	$\frac{1}{4}N_2O_3$	1.9002
Nitrogen pentoxide	N_2O_5	108.0104	$\frac{1}{5}N_2O_5$	1.8001
Oxalic acid	$C_2H_2O_4$	90.0358	$\frac{1}{2}C_2H_2O_4$	4.5018
Oxalic acid hydrate	$C_2H_2O_4{\cdot}2H_2O$	126.0665	$\frac{1}{2}C_2H_2O_4{\cdot}2H_2O$	6.3033
Oxygen	O	15.9994	$\frac{1}{2}O$	0.8000
Potassium dichromate	$K_2Cr_2O_7$	294.1918	$\frac{1}{6}K_2Cr_2O_7$	4.9032
Potassium chlorate	$KClO_3$	122.5532	$\frac{1}{6}KClO_3$	2.0425
Potassium chromate	K_2CrO_4	194.1076	$\frac{1}{3}K_2CrO_4$	6.4733
Potassium ferrocyanide	$K_4Fe(CN)_6$	368.3621	$K_4Fe(CN)_6$	36.8362
Potassium ferrocyanide	$K_4Fe(CN)_6{\cdot}3H_2O$	422.4081	$K_4Fe(CN)_6{\cdot}3H_2O$	42.2408
Potassium iodide	KI	166.0064	KI	16.6006
Potassium nitrate	KNO_3	101.1069	$\frac{1}{3}KNO_3$	3.3702
Potassium perchlorate	$KClO_4$	138.5526	$\frac{1}{8}KClO_4$	1.7319
Potassium permanganate	$KMnO_4$	158.0376	$\frac{1}{5}KMnO_4$	3.1608
Sodium chlorate	$NaClO_3$	106.4410	$\frac{1}{6}NaClO_3$	1.7740
Sodium nitrate	$NaNO_3$	84.9947	$\frac{1}{3}NaNO_3$	2.8332
Sodium thiosulfate	$Na_2S_2O_3{\cdot}5H_2O$	248.1825	$Na_2S_2O_3{\cdot}5H_2O$	24.8183
Stannous chloride	$SnCl_2$	189.5960	$\frac{1}{2}SnCl_2$	9.4798
Stannous oxide	SnO	134.6894	$\frac{1}{2}SnO$	6.7345
Sulfur dioxide	SO_2	64.0628	$\frac{1}{2}SO_2$	3.2031
Tin	Sn	118.69	$\frac{1}{2}Sn$	5.935

ORGANIC ANALYTICAL REAGENTS

Compiled by John H. Yoe

Determination	Reagent	Reference
Acetate	*o*-Nitrobenzaldehyde	Feigl, p. 342 (3)
Aldehydes	Dimethyl-dihydro-resorcin (Dimedon)	Ind. Eng. Chem., Anal. Ed. **3**, 365 (1931)
Aluminum	Alizarin S	J. Am. Chem. Soc. **50**, 748 (1928)
	Ammonium salt of aurin tricarboxylic acid ("Aluminon")	J. Am. Chem. Soc. **49**, 2395 (1927); J. Am. Chem. Soc. **55**, 2437 (1933)
	Ammonium salt of nitrosophenyl hydroxylamine ("Cupferron")	Bull. soc. chim. Belg. **36**, 288 (1927)
	Eriochrome cyanine	Z. anal. Chem. **96**, 91 (1934)
	Hematoxylin	Ind. Eng. Chem. **16**, 233 (1924)
	8-Hydroxyquinoline	J. Am. Chem. Soc. **50**, 1900 (1928)
	Morin	Feigl, p. 182 (2); Ind. Eng. Chem., Anal. Ed. **12**, 229 (1940)
	Quinalizarine	J. Am. Pharm. Assoc. **17**, 260 (1928)
	Sodium or zinc salt of 4-sulfo-2 hydroxy-α-naphthalene-azo-β-naphthol (Pontachrome Blue Black R)	Sandell, p. 241 (6)
	Urea	Ind. Eng. Chem., Anal. Ed. **9**, 357 (1937)
Ammonia	Hematoxylin	Helv. Chim. Acta **12**, 730 (1929)
	p-Nitrobenzenediazonium chloride	Feigl, p. 235 (2)
	Phenol and sodium hypochloride	Snell, Vol. II, p. 818 (8)
	Tannin—$AgNO_3$	Snell, Vol. II, p. 819 (8)
	Thymol	J. Biol. Chem. **131**, 309 (1939)
	Zinc oxinate	Feigl, p. 674 (9)
Antimony	Hexamethylenetetramine	Z. anal. Chem. **67**, 298 (1925)
	9-Methyl-2,3,7-tryhydroxyfluorone	Helv. Chim. Acta **20**, 1427 (1937)
	Phenylthiohydantoic acid	Compt. rend. **176**, 1221 (1923)
	Pyridine	Analyst **53**, 373 (1928)
	Pyrogallol	Z. anal. Chem. **64**, 44 (1924)
	Rhodamine B	Z. anal. Chem. **70**, 400 (1927)
Arsenic	Cocaine-molybdate	Biochem. Z. **185**, 14 (1927)
	N-Ethyl-8-hydroxy-tetrahydroquinoline hydrochloride	Z. anal. Chem. **99**, 180 (1934)
	Quinine arsenomolybdate	Analyst **47**, 317 (1922)
	Strychnine-molybdate	Ann. Chim. applicata **23**, 517 (1933)
Barium	Sodium rhodizonate	Feigl, p. 216 (2)
	Tetrahydroxyquinone	Welcher, Vol. I, p. 225 (9)
Beryllium	1-Amino-4-hydroxy-anthraquinone	Ind. Eng. Chem., Anal. **13**, 809 (1941)
	Aurin trycarboxylic	J. Am. Chem. Soc. **48**, 2125 (1926); ibid., **50**, 353 (1928)
	Curcumin	J. Am. Chem. Soc. **50**, 393 (1928)
	1,4-Dihydroxyanthraquinone (Quinizarin)	Ind. Eng. Chem. Anal. Ed., **18**, 179 (1946)
	1,4-Dihydroxyanthraquinone-2-sulfonic acid (quinizarine-2-sulfonic acid)	Sandell, p. 318 (6)
	8-Hydroxyquinoline	Bur. Standards J. Research **3**, 91 (1929)
	Morin	Ind. Eng. Chem., Anal. Ed. **12**, 674, 762 (1940)
	Naphthochrome Azurine 2B	Sandell, p. 318 (6)
	Naphthochrome Green G	Sandell, p. 318 (6)
	p-Nitrobenzeneazo-orcinol	Mikrochemie **14**, 315 (1934)
	1,2,5,8-Tetrahydroxy-anthraquinone (Quinalizarin)	Siemens-Konzerns, Beryllium, p. 25 (1932)
Bismuth	Caffeine (as sulfate or nitrate)	Ind. Eng. Chem., Anal. Ed. **14**, 43 (1942)
	Cinchonine	Scott, p. 158 (7)
	Diethyldithiocarbamate	Sandell, p. 338 (6)
	Dimercaptothiodiazole	Z. anal. Chem. **98**, 184 (1934); ibid., **100**, 408 (1935)
	Dimethylglyoxime	Z. anal. Chem. **72**, 11 (1927)
	Diphenylthiocarbazone	Z. Angew. Chem. **47**, 685 (1934); Ind. Eng. Chem., Anal. Ed. **7**, 285 (1935)
	8-Hydroxyquinoline	Z. anal. Chem. **72**, 177 (1927)
	Phenyldithiobiazolonethiol	J. Indian Chem. Soc., **21**, 240, 347 (1944)
	Pyrogallol	Z. anal. Chem. **65**, 448 (1925)
	Thiourea	Z. anal. Chem. **94**, 161 (1933)
Boron	Curcumin	Chem. News **87**, 27 (1903)
	1,1-Dianthramide	Boltz, p. 346 (1)
	Diaminochrysazin	Anal. Chem. **29**, 1251 (1957)
	Diaminoanthrarufin	Ibid.
	Tribromoanthrarufin	Ibid.
	Mannitol	Scott, p. 168 (7)
	Methyl alcohol	J. Am. Chem. Soc. **50**, 1385 (1928)
	p-Nitrobenzeneazo-chromotropic acid (Chromotrope 2B)	Feigl, p. 341 (2)
	Quinalizarin	Ind. Eng. Chem., Anal. Ed. **11**, 540 (1939)
	Titan Yellow (Clayton Yellow)	Compt. rend. **138**, 1046 (1904)
	Turmeric	Ind. Eng. Chem., Anal. Ed. **4**, 180 (1932)
Bromine	Fluorescein	Snell, Vol. II, p. 725 (8)
	Fuchsin	Snell, Vol. II, p. 724 (8)
	Phenol red	Snell, Vol. II, p. 725 (8)
Cadmium	Allythiourea	Helvetica Chim. Acta. **12**, 718 (1929)
	Di-β-naphthylthiocarbazone	Ind. Eng. Chem., Anal. Ed., **16**, 333, (1944)
	Dinitrodiphenylcarbazide	Feigl, p. 96 (2)
	Diphenylcarbazide	Feigl, p. 99 (2)
	Diphenylthiocarbazone	Z. angew. Chem. **47**, 685 (1934); Ind. Eng. Chem., Anal. Ed. **11**, 364 (1939)
	Ethylenediamine	Z. anal. Chem. **77**, 340 (1929)
	Hexamethylenetetramine alliodide	C. A., **24**, 311 (1930)
	β-Naphthoquinoline	Analyst, **58**, 667 (1933)
	Nitrophenolarsinic acid	Mikrochemie **8**, 277 (1930)
	4-Nitrophthalene-diazo-aminobenzene-4-azobenzene (Cadion 2B)	C. A. **32**, 2871 (1938)
	Phenyl-trimethyl-ammonium iodide	Analyst **58**, 667 (1933)
	Pyridine	Z. Anal. Chem. **73**, 279 (1928)
Calcium	Alizarin	Biochem. J. **16**, 494 (1922); Yoe, Vol. I, p. 139 (2)
	1-amino-2-naphthol-4-sulfonic acid	J. Biol. Chem. **81**, 1 (1929)
	Ammonium oxalate	Snell, Vol. II, p. 592 (8)
	Ammonium stearate	J. Biol. Chem. **29**, 169 (1917); Yoe, Vol. II, p. 119 (3)
	2,5-Dichloro-3,6-dihydroxyquinone (Chloranilic acid)	Anal. Chem., **20**, 76 (1948)
	Dihydroxytartaric acid osazone (sodium salt)	Feigl, p. 221 (2)
	Picrolonic acid	Biochem. Z. **265**, 85 (1933)
	Potassium oleate	Sandell, p. 377 (6)
	Pyrogallol carboxylic acid	Sandell, p. 380 (6)
	Sodium sulforicinate	Biochem. Z. **137**, 157 (1923); Yoe, Vol. II, p. 125 (3)
Cerium	Benzidine	Feigl, p. 211 (2)
	Brucine	Sandell, p. 386 (6)
	Gallic acid	Snell, Vol. II, p. 608 (8)
	8-Hydroxyquinoline	Analyst, **75**, 275 (1948)
	Malachite Green	C. A. **30**, 5143 (1936), **31**, 626 (1937)
	Morphine	Sandell, p. 386 (6)
	Sulfanilic acid	Sandell, p. 386 (6)
Cesium	Dipicrylamine (Hexanitrodiphenylamine)	Mikrochemie 18, 175 (1935)
Chlorate	Aniline hydrochloride	Snell, Vol. II, p. 717 (8)
Chlorine	Benzidine hydrochloride	Ind. Eng. Chem., Anal. Ed. **4**, 2 (1932)
	Dimethyl-*p*-phenylene-diamine	Chem. Weekblad. **23**, 203 (1926)
	Oleic acid	J. Soc. Chem. Ind. **42**, 427A (1923)
	Sodium sulforicinate	Biochem. Z., **137**, 157 (1923); Yoe, Vol. II, p. 125 (3)
	Thymolphthalein	Ind. Eng. Chem. **19**, 112 (1927)
	o-Tolidine	Yoe, Vol. I, p. 157 (2)
Columbium	Benzidine	Feigl, p. 171 (2)
	1,8-Dihydroxynaphthalene-3,6-Disulfonate	Ind. Eng. Chem. **5**, 298 (1913)
	s-Diphenylcarbazide	J. Am. Chem. Soc. **50**, 2363 (1928)
	Pyrogallol dimethyl ether	C. A. **4**, 3178 (1910)
	Serichrome Blue R	Ind. Eng. Chem., Amal. Ed., **4**, 245 (1932)

Determination	Reagent	Reference
Cobalt	Anthranilic acid (*o*-aminobenzoic acid)	Z. anal. Chem. **93**, 241 (1933)
	Cysteine hydrochloride	J. Biol. Chem. **83**, 367 (1929)
	Dimethylglyoxime	J. Am. Chem. Soc. **43**, 482 (1921)
	3,5-Dimethylpyrazole	Ind. Eng. Chem., Anal. Ed. **2**, 38 (1930)
	Dinitrosoresorcinol	J. Am. Chem. Soc. **45**, 1439 (1923)
	Formaldoxime reagent	C. A. **27**, 927 (1933)
	o-Nitrosocresol	Sandell, p. 422 (6)
	α-Nitroso-β-naphthol	Chem. Zeit. **46**, 430 (1922)
	Nitrose-R-salt	J. Am. Chem. Soc. **43**, 746 (1921)
	β-Nitroso-α-naphthol	Ind. Eng. Chem., Anal. Ed. **12**, 405 (1940)
	o-Nitrosoresorcinol	Ind. Eng. Chem., Anal. Ed. **15**, 310 (1943)
	o-Nitrosophenol	Sandell, p. 422 (6)
	Phenylthiohydantoic acid	J. Am. Chem. Soc. **44**, 2219 (1922)
	Rubeanic acid (Dithio-oxamide)	Anal. Chim. Acta **20**, 332, 435 (1959)
	2,2′,2″-Terpyridyl	Ind. Eng. Chem., Anal. Ed., **15**, 74 (1943)
	Niobium	
Columbium	See **Nobium**	
Copper	Ammonium salt of nitrosophenyl-hydroxylamine ("Cupferron")	Ind. Eng. Chem. **3**, 629 (1911)
	m-Benzamino-semicarbazide	Snell, Vol. II, p. 129 (6)
	Benzidine	Z. anal. Chem. **67**, 31 (1925)
	α-Benzionoxime (Cupron)	Ber. **56**, 2083 (1923)
	Benzotriazole	Ind. Eng. Chem., Anal. Ed. **13**, 349 (1941)
	2-Carboxy-a′-hydroxy-5′ sulfoformazylbenzene (Zincon)	Anal. Chem. **26**, 1345 (1954)
	Diacetyl-dioxime	Analyst **54**, 333 (1929)
	Dibenzyldithiocarbamate	Sandell, p. 443 (6)
	Dihydroxyethyldithiocarbamic acid	Sandell, p. 443 (6)
	p-Dimethylaminobenzalrhodanine	J. Am. Chem. Soc. **52**, 2222 (1930) J. Am. Chem. Soc. **47**, 1268 (1925)
	s-Diphenylcarbazide	Chem. Weekblad. **21**, 20 (1924)
	Diphenylthiocarbazone (Dithizone)	J. Assoc. Official Agr. Chem. **18**, 192 (1935)
	Hydroquinone	Bull. soc. chim. **31**, 1176 (1922)
	Isatin	Rec trav chim. **42**, 199 (1923)
	Mercaptobenzothiazole	Z. anal. Chem. **102**, 24, 108 (1935)
	α-Naphthol	Bull. soc. chim. **31**, 1176 (1922)
	β-Naphthol	Am. J. Pharm. **105**, 62 (1933)
	Phenolphthalein	Compt. rend. **173**, 1082 (1921)
	Phenylthiohydantoic acid	J. Am. Chem. Soc. **44**, 225 (1922)
	Piperidinium piperidyl-dithioformate	Analyst **56**, 736 (1931)
	Potassium ethyl xanthate	Yoe, Vol. I, p. 184 (2)
	Pyridine	Z. anal. Chem. **67**, 27 (1925)
	Rubeanic acid (Dithio-oxamide)	Anal. Chim. Acta **20**, 332, 435 (1959)
	Salicylaldoxime	J. Chem. Soc. (1933) 314; Feigl, p. 40 (8)
	Salicylic acid	Yoe, Vol. I, p. 183 (2)
	Sodium diethyldithiocarbamate	Analyst **54**, 650 (1929)
	o-Toluidine	Z. anal. Chem. **67**, 31 (1925)
	Urobilin	Chem. Weekblad. **27**, 552 (1930)
Cyanide	Benzidine	Feigl, p. 276 (2)
	Phenolphthalin	Analyst **60**, 294 (1935)
	Picric acid	Helv. Chim. Acta **12**, 713 (1929); J. Am. Chem. Soc. **51**, 1171 (1929)
Fluoride	Acetylacetone	Ind. Eng. Chem., Anal. Ed. **5**, 300 (1933)
	Alizarin sodium sulfonate—$Zr(NO_3)_4$	Ind. Eng. Chem., Anal. Ed. **7**, 23 (1935)
	p-Dimethylaminoazophenylarsonic acid	Feigl, p. 271 (2)
	Triphenyltin chloride	J. Am. Chem. Soc. **54**, 4625 (1932)
	Quinalizarine—$Zr(NO_3)_4$	Ind. Eng. Chem., Anal. Ed. **6**, 61 (1934)
Gallium	Camphoric acid	Welcher, Vol. II, p. 31 (9)
	8-Hydroxyquinoline	Ind. Eng. Chem., Anal. Ed. **13**, 844 (1941)
	Morin	Mikrochemie **20**, 194 (1936)
	Quinalizarin	J. Am. Chem. Soc. **59**, 40 (1937)
Germanium	Benzidine	Mikrochemie **18**, 66 (1935)
	Quinalizarin	Mikrochemie **18**, 48 (1935)
	Tannin	Welcher, Vol. II, p. 160 (9)
Gold	Benzidine	Bull. Chim. Farm. **52**, 461 (1912)
	o-Dianisidine	Sandell, p. 507 (6)
	Dimethylaminobenzylidene rhodanine	Feigl, p. 127 (2)
	Formaldehyde	Bull. soc. chim. **31**, 717 (1922)
	p-Fuchsine	Ind. Eng. Chem., Anal. Ed., **18**, 400 (1946)
	Malachite Green	Sandell, p. 506 (6)
	m-Phenylenediamine sulfate	Chem. Zeit. **36**, 934 (1912)
	Phenylhydrazine	Ann. chim. anal. **12**, 90 (1907)
	o-Toluidine	Analyst **44**, 94 (1919)
Hafnium	2-Hydroxy-5-methylazobenzene-4-sulfonic acid	Feigl, p. 288 (4)
Hydrogen sulfide	*p*-phenylenedimethyldiamine sulfate	Yoe, Vol. I, p. 375 (2)
Indium	Diphenylthiocarbazone (Dithizone)	Sandell, p. 516 (6)
	8-Hydroxyquinoline	Z. anorg. allgem. Chem. **209**, 12[illegible] (1932)
	Morin	Mikrochim. Acta **2**, 287 (1937)
Iodine	Starch	Snell, Vol. II, p. 740 (8)
	o-Toluidine	J. Am. Chem. Soc. **47**, 1000 (1925)
Iridium	Benzidine	Snell, Vol. II, p. 526 (8)
	Malachite Green, leuco base	C. A. **24**, 2689 (1930)
Iron	Acetylacetone	J. Am. Chem. Soc. **26**, 967 (1904)
	Alloxantin	Compt. rend. **180**, 519 (1925)
	Ammonium salt of nitrosophenyl hydroxylamine ("Cupferon")	Ind. Eng. Chem. **3**, 629 (1911)
	Bis-*p*-chlorophenylphosphoric acid	Feigl, p. 294 (4)
	Cysteine	Biochem. Z. **187**, 255 (1927)
	Dimethyl glyoxime	Z. anorg. Chemie **89**, 401 (1914)
	Dinitrosoresorcinol	J. Am. Chem. Soc. **47**, 1268 (1925)
	Diisonitrosoacetone	Chem. Listy **23**, 496 (1929); C. A. **24**, 801 (1930)
	Dioximes (various)	Anal. Chem. **19**, 1017 (1947)
	Diphenylamine	J. Am. Chem. Soc. **46**, 263 (1924)
	2,2-Bipyridyl	Snell, Vol. II, p. 316 (8)
	Disodium-1,2-dihydroxybenzene-3,5-disulfonate ("Tiron")	Ind. Eng. Chem., Anal. Ed. **16**, 111 (1944)
	Hexamethylenetetramine	Bull soc. chim. Rom. **2**, 89 (1921)
	4 Hydroxybiphenyl-3-carboxylic acid	J. Am. Chem. Soc. **70**, 648 (1948)
	8-Hydroxyquinoline	Bull. soc. chim. biol., **17**, 432 (1935)
	4-Hydroxybiphenyl-3-carboxylic acid	J. Am. Chem. Soc. **70**, 648 (1948)
	7-Iodo-8-hydroxyquinoline-5-sulfonic acid (Ferron)	J. Am. Chem. Soc. **59**, 872 (1937)
	Isonitrosoacetophenone	Ber. **60**, 527 (1927)
	Isonitrosodimethyldihydroresorcinol	Anal. Chem., **20**, 1205 (1948)
	Kojic acid	Ind. Eng. Chem., Anal. Ed., **13**, 612 (1941)
	α-Nitrose-β-naphthol	Bull. soc. chim. **35**, 641 (1924)
	Nitroso-R salt	Ind. Chem., Anal. Ed., **14**, 756 (1942); **16**, 276 (1944)
	o-Phenanthroline	Ind. Eng. Chem., Anal. Ed. **9**, 67 (1937)
	Protacatechvic acid	J. Biol. Chem., **137**, 417 (1941)
	Pyramidone	Pharm.Weekblad, **63**, 1121 (1926)
	Pyrocatechol	Helv. chim. Acta **9**, 835 (1926)
	Salicylaldoxime	Ind Eng. Chem., Anal. Ed., **12**, 448 (1940)
	Salicylic acid	J. Chem. Soc. **93**, 93 (1908)
	Salicylsulfonic acid	Snell, Vol. II, p. 321 (8)
	Sulfosalicylic acid	Biochem. Z. **181**, 391 (1927)
	Thioglycollic acid	J. Am. Chem. Soc. **49**, 1916 (1927)
Lanthanum	8-Hydroxyquinoline	Z. anal. Chem. **107**, 191 (1936)
	Sodium alizarinesulfonate	J. Am. Chem. Soc., **53**, 1217 (1931)
Lead	Ammonium thiocyanate and pyridine	Z. anal. Chem. **72**, 289 (1927)
	Aniline	Ind. Eng. Chem. **11**, 1055 (1919); Yoe, Vol. I, p. 257 (2)
	Anthranilic acid	Z. anal. Chem. **101**, 85 (1935)
	Carminic acid	Mikrochemie **7**, 301 (1929)
	s-Diphenylcarbazide	Yoe, Vol. I, p. 255 (2)
	Diphenylthiocarbazone (Dithizone)	Snell, Vol. IIA, p. 10 (8)
	Hematein	Yoe, Vol. I, p. 257 (2)
	Salicylaldoxime	Ind. Eng. Chem., Anal. Ed. **14**, 359 (1942)
	Tetramethyldiamidodiphenylmethane	Snell, Vol. II, p. 43 (8)
	Thiourea	Z. anorg. Chem., **234**, 224 (1937)

ORGANIC ANALYTICAL REAGENTS (Continued)

Determination	Reagent	Reference
Lithium	Ammonium stearate	J. Am. Chem. Soc. **52**, 2754 (1930)
	Hexamethylenetetramine	Welcher, Vol. III, p. 131 (9)
	Pyridine	Welcher, Vol. III, p. 33 (9)
Magnesium	Brilliant yellow	Sandell, p. 598 (6)
	Curcumin	Ind. Eng. Chem., Anal. Ed. **4**, 426 (1932)
	Dimethylamine	Z. anorg. Chem. **26**, 347 (1901)
	Hydroquinone	Yoe, Vol. I, p. 264 (2)
	8-Hydroxyquinoline	Z. anal. Chem. **71**, 122 (1927)
	p-Nitrobenzeneazo-α-naphthol	Feigl, p. 225 (2)
	p-Nitrobenzeneazo-resorcinol	J. Am. Chem. Soc. **51**, 1456 (1929)
	Oleic acid	Yoe, Vol. I, p. 270 (2)
	Quinalizarin	Feigl, p. 224 (2)
	Sodium 1-azo-2-hydroxy-3-(2,4-dimethylcarboxaniliodonaphthalene-1'(2-hydroxybenzene-5-sulfonate)	Anal. Chem. **28**, 202 (1956); cf. Anal. Chim. Acta **16**, 155 (1957)
	Thiazole yellow	Sandell, p. 598 (6)
	Titan yellow	Sandell, p. 591 (6)
	Tropeoline OO	Rec. trav. chim., **61**, 849 (1942)
Manganese	Benzidine	Snell, Vol. II, p. 397 (8)
	Formaldoxime	Ind. Eng. Chem., Anal. Ed. **9**, 445 (1937); **12**, 307 (1940)
	Tetramethyldiaminodiphenylmethane	Feigl, p. 174 (2); Snell, Vol. IIA, p. 311 (8)
	4,4-Tetramethyldiaminotriphenylmethane	J. Biol. Chem., **168**, 537 (1947)
Mercury	Anthranilic acid	Z. anal. Chem. **101**, 88 (1935)
	p-Dimethylaminobenzalrhodamine	J. Am. Chem. Soc. **52**, 2222 (1930)
	Di-β-naphthylthiocarbazone	Sandell, p. 630 (6)
	s-Diphenylcarbazide	Z. angew. Chem. **39**, 791 (1926)
	Diphenylthiocarbazone (Dithizone)	Z. anal. Chem. **103**, 241 (1935)
	Potassium *s*-diphenylcarbazone	Snell, Vol. II 78 (8)
	Strychnine	" II 76 (8)
Molybdenum	α-Benzoin-oxime (Cupron)	B. S. J. Research **9**, 1 (1932)
	Disodium-1,2-dihydroxy-benzene-3,5-disulfonate (Tiron)	Anal. Chim. Acta **8**, 546 (1953)
	4-Methyl-1,2-dimercaptobenzene (Dithiol)	J. Am. Pharm. Assoc., **37**, 255 (1948)
	Phenylhydrazine	Ber. **36**, 512 (1903)
	Potassium ethyl xanthate and chloroform	J. Am. Chem. Soc. **44**, 1462 (1922)
	Tannic acid	Chem. Eng. Mining Rev. **11**, 258 (1919)
Nickel	α-Benzil-dioxime	Analyst **38**, 316 (1913)
	Cyclohexanedione-dioxime	Ann. **437**, 148 (1925)
	Dicyandiamidine sulfate	Chem. Zeit. **31**, 335, 911 (1907)
	Diethyldithiocarbamate	Ind. Eng. Chem., Anal. Ed., **18**, 206 (1946)
	Dimethylglyoxime	Chem. Weekblad. **21**, 358 (1924)
	Formaldoxime	Snell, Vol. IIA, p. 265 (8)
	α-Furildioxime	J. Am. Chem. Soc. **47**, 918 (1925)
	Potassium dithiooxalate	J. Am. Chem. Soc. **54**, 1866 (1932)
	Rubianic acid (Dithio-oxamide)	Anal. Chim. Acta **20**, 332 435 (1959)
Niobium (Columbium)	Ammonium salt of nitrosophenyl hydroxylamine ("Cupferron")	Hillebrand et al., p. 120 (5)
	8 Hydroxyquinolin	Wechler, Vol. I, p. 302 (2)
Nitrate	Brucine	Yoe, Vol. I, p. 318 (2)
	Diphenylamine sulfonic acid	J. Am. Chem. Soc. **55**, 1448 (1933)
	Diphenylbenzidine	Yoe, Vol. I, p. 316 (2)
	Diphenyl-endo-anilo-hydrotriazole ("Nitron")	Fales and Kenny, Inorg. Quant. Anal., p. 347 (1938)
	Phenoldisulfonic acid	Yoe, Vol. I, p. 313 (2)
	Pyrogallol	" p. 319 (2)
	Strychnine sulfate	" p. 320 (2)
	2:4-Xylenol	J. Assoc. Off. Agri. Chem. **18**, 459 (1935)

Determination	Reagent	Reference
Nitrite	Antipyrin	Yoe, Vol. I, p. 311 (2)
	Dimethylaniline	" p. 311 (2)
	Dimethyl-α-Naphthylamine	Ind. Eng. Chem., Anal. Ed. **1**, 28 (1929)
	Diphenylamine sulfate	Yoe, Vol. I, p. 654 (2)
	α-Naphthylamine and β-Naphthylamine-6,8-Disulfonic acid	J. Pharmacol. **51**, 398 (1934)
	α-Naphthylamine hydrochloride	Yoe, Vol. I, p. 309 (2)
	m-Phenylenediamine	Yoe, Vol. I, p. 310 (2)
	Sulfanilic acid and α-naphthylamine	Yoe, Vol. I, p. 308 (2)
Osmium	*s*-Diphenylthiourea	Yoe and Sarver, p. 155 (12)
	1-Naphthylamine-4,6,8-trisulfonic acid	Anal. Chim. Acta **20**, 205 (1959)
	Strychnine	Welcher, Vol. IV, p. 272 (9)
	Thiourea	Compt. rend. **167**, 235 (1918)
Oxygen	Indigo carmine	Snell, Vol. IIA, p. 726 (8)
	Pyrogallol	Dennis, Gas Analysis, p. 174 (1929)
Palladium	Dimethylglyoxime	J. Am. Chem. Soc. **57**, 2565 (1935)
	p-Fuchsine	Ind. Eng. Chem., Anal. Ed., **18**, 400 (1946)
	β-Furfuraldoxime	Ind. Eng. Chem., Anal. Ed. **14**, 491 (1942)
	6-Nitroquinoline	J. Am. Chem. Soc. **50**, 3018 (1928)
	p-Nitrosodiphenylamine	J. Am. Chem. Soc. **61**, 2058 (1939)
	p-Nitrosodimethylaniline	J. Am. Chem. Soc. **63**, 3224 (1941)
	p-Nitrosodiethylaniline	Ibid.
Phosphate	Thiomalic acid	Talanta **2**, 223 (1959)
	1,2,4-Aminonaphthosulfonic acid	Yoe, Vol. I, p. 348 (2)
	Hydroquinone	Yoe, Vol. I, pp. 346 and 353 (2)
	Quinine-molybdate	Yoe, Vol. I, p. 343 (2)
	Strychnine-molybdate	Yoe, Vol. II, p. 142 (3)
Phosphorus	Hydrazine sulfate	Yoe, Vol. I, p. 341 (2)
Platinum	*p*-Nitrosodimethylaniline	Anal. Chem. **26**, 1335, 1340 (1954)
Potassium	6-Chloro-5-nitrotoluene-3-sulfonic acid	Mikrochem. **14**, 368 (1934)
	Dipicrylamine	Z. angew. Chem. **49**, 827 (1936)
	Picric acid	J. Am. Chem. Soc. **53**, 539 (1931)
Rhodium	Thiomalic acid	Talanta **2**, 239 (1959)
Ruthenium	*s*-Diphenylthiourea	Yoe and Sarver, p. 155 (12)
	5-Hydroxyquinoline-8-carboxylic acid	Canadian J. Research B25, **49**, (1945)
	1-Naphthylamine-3,5,7-trisulfonic acid	Anal. Chim Acta **20**, 211 (1959)
	Rubianic acid (Dithio-oxamide)	Mikrochemie **15**, 295 (1934)
	Thioglycolyl-β-amidonaphthalide (Thionalid)	Ind. Eng. Chem., Anal. Ed. **12**, 5611 (1940)
	Thiourea	Sandell, p. 781 (6)
Scandium	Morin	Mikrochem. Acta **2**, 9, 287 (1937)
Selenium	Codeine phosphate	Arch Pharm. **252**, 161 (1914)
	Hydrazine	Boltz, p. 321 (1)
	Hydroquinone	Am. J. Sci. **15**, 253 (1928)
	Hydroxylamine hydrochloride	J. Am. Chem. Soc. **47**, 2456 (1925)
	Pyrrol	Snell, Vol. II, p. 779 (8)
	Thiourea	Ann. chim. applicata **17**, 357 (1927); Feigl, p. 231 (8)
Silver	Chromotropic acid	Helvetica Chim. Acta **12**, 714 (1929)
	Dichlorofluorescein	J. Am. Chem. Soc. **51**, 3273 (1929)
	p-Dimethylaminobenzalrhodamine	J. Am. Chem. Soc. **52**, 2222 (1930)
	Diphenylthiocarbazone (Dithizone)	Z. anal. Chem. **101**, 1 (1935)
	Formazylcarboxylic acid	J. Anal. Chem. Russ., **2**, 131 (1934); abs. in Analyst, **73**, 352 (1948)
	Methylamine	Mikrochemie **7**, 233 (1929)
	2-Thio-5-keto-4-carbethoxy-1,3-dihydropyrimidine	Ind. Eng. Chem., Anal. Ed. **14**, 148 (1942)
Sodium	6,8-Dichlorobenzoylurea	J. Org. Chem. **3**, 414 (1938)
	Dihydroxy-tartaric acid	J. Russ. Phys. Chem. Soc. **60**, 661 (1928)
	Uranyl zinc acetate	J. Am. Chem. Soc. **51**, 1664 (1929)
Strontium	Sodium rhodizonate	Mikrochemie **2**, 187 (1924)
Sulfide	*p*-Aminodimethylaniline	Snell, Vol. IIA, p. 658 (8)
Sulfur	*p*-Phenylenedimethyldiamine-hydrochloride	Yoe, Vol. I, p. 373 (2)

ORGANIC ANALYTICAL REAGENTS

Determination	Reagent	Reference
Tantalum	Ammonium salt of nitrosophenyl hydroxylamine ("Cupferron")	Hillebrand et. al., p. 120 (5)
	Pyrogallol	Sandell, p. 695 (6)
Tellurium	Hydrazine hydrochloride	J. Am. Chem. Soc. **47**, 2456 (1925)
	Hydroquinone	Am. J. Sci. **15**, 253 (1928)
	Thiourea	Ann. chim. applicata **17**, 359 (1927)
Thallium	Diphenylthiocarbazone (Dithizone)	Analyst **60**, 394 (1935)
	Thionalid (Thioglycolyl-β-amidonaphthalide)	Z. angew. Chem. **48**, 430, 597 (1935)
Thorium	1-Amino-4-hydroxyanthraquinone	Ind. Eng. Chem., Anal. Ed. **13**, 809 (1941)
	Cupferron (Ammonium nitrosophenylhydroxylamine)	Chem. Ztg. **33**, 1298 (1908)
	8-Hydroxyquinoline	Z. anal. Chem. **100**, 98 (1935)
	Phenylarsonic acid	J. Am. Chem. Soc. **48**, 895 (1926)
Tin	Ammonium salt of nitrosophenyl hydroxylamine ("Cupferron")	Hillebrand et al., p. 120 (5)
	Cacotheline	Ind. Eng. Chem., Anal. Ed. **7**, 26 (1935)
	4-Chloro-1,2-dimercaptobenzene (1-chloro-benzene-3,4-dithiol)	J. Chem. Soc. **149**, 175 (1936)
	Hematoxylin	Sandell, p. 866 (6)
	Quinalizarin	Sandell, p. 866 (6)
	Toluene-3,4-dithiol (4-methyl-1,2-dimercaptobenzene)	Analyst **61**, 242 (1936); Ibid., **62**, 661 (1937)
Titanium	Ammonium salt of nitrosophenyl hydroxylamine ("Cupferron")	Hillebrand et al., p. 119 (5); Z. anal. Chem. **83**, 345 (1931)
	Chromotropic acid	Feigl, p. 197 (2)
	5,7-Dibromo-8-hydroxyquinoline	Z. anorg. Chem. **204**, 215 (1932)
	Dihydroxymaleic acid	Snell, Vol. II, p. 445 (8)
	Disodium 1,2-dihydroxybenzene-3,5-disulfonate ("Tiron")	Ind. Eng. Chem., Anal. Ed. **19**, 100 (1947)
	Gallic acid	Snell, Vol. II, p. 444 (8)
	p-Hydroxyphenylarsonic acid	Ind. Eng. Chem., Anal. Ed. **10**, 642 (1938)
	8-Hydroxyquinoline	Z. anal. Chem. **81**, 1 (1930)
	Resoflavine (Color Index 1015)	Anal. Chim. Acta, **1**, 244 (1947)
	Tannic acid	Analyst **55**, 605 (1930)
	Thymol	Yoe, Vol. I, p. 381 (2)
Tungsten (Wolfram)	Anti-1,5-di-(*p*-methoxyphenyl)-1-hydroxylamino-3-oximino-4-pentene ("Wolfron")	Ind. Eng. Chem., Anal. Ed. **16**, 45 (1944)
	Benzidine	Ber. **38**, 783 (1905)
	α-Benzoinoxime ("Cupron")	Bur. Std. J. Research **9**, 1 (1932)
	Cinchonine	Hillebrand et al., p. 689 (5)
	Hydroquinone	Z. angew. Chem. **44**, 237 (1931)
	Phenylhydrazine	Bull. soc. chim. Belg. **38**, 385 (1929)
	Rhodamine B.	Snell, Vol. II, p. 469 (8)
Tungsten (Wolfram) (Cont.)	Toluene-3,4-dithiol	Analyst, **69**, 109 (1944); J. Am. Pharm. Assoc., **37**, 255 (1948)
	Uric acid	Ann. chim. anal. **9**, 371 (1904)
	α-Benzoinoxime ("Cupron")	Bur. Std. J. Research **9**, 1 (1932)
	Cinchonine	Hillebrand et al., p. 689 (5)
	Hydroquinone	Z. angew. Chem. **44**, 237 (1931)
	Phenylhydrazine	Bull. soc. chim. Belg. **38**, 385 (1929)
	Rhodamine B.	Snell, Vol. II, p. 469 (8)
	Toluene-3,4-dithiol	Analyst, **69**, 109 (1944); J. Am. Pharm. Assoc., **37**, 255 (1948)
	Uric acid	Ann. chim. anal. **9**, 371 (1904)
Uranium	Dibenzoylmethane	Anal. Chem. **25**, 1200 (1953)
	o-Hydroxybenzoic acid	Snell, Vol. II, p. 492 (8)
	α-Quinaldinic acid	Z. anal. Chem. **95**, 400 (1933)
	Sodium diethyldithiocarbamate	Sandell, p. 921 (6)
	Sodium salicylate	Chem. Zeit. **43**, 739 (1919)
	Thioglycolic acid (Mercapto acetic acid)	Anl. Chem., **21**, 1093 (1949)!
Urea	Xanthydrol	Mikrochem. **14**, 132 (1934)
Vanadium	Ammonium benzoate	C. A. **29**, 2880 (1935)
	Aniline	C. A. **24**, 567 (1930)
	Benzidine	J. Applied Chem. (U.S.S.R.) **17**, 83 (1944); C. A. **39**, 1115 (1945)
	Diphenylamine	Yoe, Vol. I, p. 715 (2)
	Diphenylbenzidine	Ind. Eng. Chem. **20**, 764 (1928)
	8-Hydroxyquinoline	Feigl, p. 125 (2)
	Safranine	Vol. Anal., Vol. II, p. 326 (6)
	Strychnine	Yoe, Vol. I, p. 393 (2)
Wolfram	See Tungsten	
Zinc	Anthranilic acid	Z. anal. Chem. **91**, 332 (1933)
	2 Carboxy-2'hydroxy-5'-sulfoformazybenzene (Zincon)	Anal. Chem. **26**, 1345 (1954)
	Di-β-napthylthiocarbazone	Sandell, pp. 945, 962 (6)
	Diphenylamine	J. Am. Chem. Soc. **49**, 2214 (1927)
	Diphenylbenzidine	J. Am. Chem. Soc. **49**, 356 (1927)
	Diphenylthiocarbazone (Dithizone)	Ind. Eng. Chem., Anal. Ed. **9**, 127 (1937)
	8-Hydroxyquinoline	Z. anal. Chem. **71**, 171 (1927)
	5-Nitroquinaldic acid	Ind. Eng. Chem., Anal. Ed. **10**, 335 (1938)
	Pyridine	Z. anal. Chem. **73**, 356 (1928)
	α-Quinaldinic acid	Z. anal. Chem. **100**, 324 (1935)
	Resorcinol	Yoe, Vol. I, p. 396 (2)
	Urobilin	J. Ind. Hyg. **7**, 273 (1925)
Zirconium	Ammonium salt of nitrosophenyl hydroxylamine ("Cupferron")	Hillebrand et al., p. 572 (5)
	5-Chlorobromamine acid	Ind. Eng. Chem., Anal. Ed **15**, 73 (1943)
	p-Dimethylaminoazophenylarsenic acid	Ind. Eng. Chem., Anal. Ed., **13**, 603 (1941).
	2-Hydroxy-5-methylazobenzene-4-sulfonic acid	Feigl, p. 288 (4)
	Morin	Sandell, p. 971 (6)
	Phenylarsenic acid	J. Am. Chem. Soc. **48**, 895 (1926)
	Sodium alizarin sulfonate (Alizarin S)	Chem. Weekblad **20**, 404 (1924); Feigl, p. 200 (2)

(1) Colorimetric Determination of Nonmetals, Boltz, Edition 1958.

(2) Feigl, Spot Tests in Inorganic Analysis, 5th English Ed., translated by Oesper, 1958.

(3) Feigl, Spot Tests in Organic Analysis, 5th English Ed., translated by Oesper, 1956.

(4) Feigl, Chemistry of Specific, Selective and Sensitive Reactions translated by Oesper, 1949.

(5) Hillebrand, Lundell, Bright and Hoffman, Applied Inorganic Analysis, 2nd Edition, 1953.

(6) Sandell, Colorimetric Determination of Traces of Metals, 3rd Edition, 1959.

(7) Scott's Standard Methods of Chemical Analysis, Furman, 5th Edition 1939.

(8) Snell, Snell and Snell, Colorimetric Methods of Analysis, Volume II (1949), II A (1959).

(9) Welcher, Organic Analytical Reagents, Vols. I–IV, 1947–1948.

(10) Yoe, Photometric Chemical Analysis, Vol. I, Colorimetry, 1928.

(11) Yoe, Photometric Chemical Analysis Vol. II, Nephelometry, 1929

(12) Yoe and Sarver, Organic Analytical Reagents, 1941

TRUE CAPACITY OF GLASS VESSELS FROM THE WEIGHT OF THE CONTAINED WATER OR MERCURY WHEN WEIGHED IN AIR WITH BRASS WEIGHTS*

A glass vessel containing G grams of water at a temperature of t°C has, at the same temperature, a capacity $V = W_t \times G$ cubic centimeters. Similarly when filled with G grams of mercury at a temperature of t°C the capacity at the same temperature is given by $V = M_t \times G$ cubic centimeters.

A glass vessel containing G grams of water at a temperature of t°C has a capacity at a temperature of 18°C given by $V = W_{18°} \times G$ cubic centimeters. Similarly when filled with G grams of mercury at a temperature of t°C the capacity at a temperature of 18°C is given by $V = M_{18°} \times G$ cubic centimeters. The true volume at temperature of 25°C when the weighing is made at t° is similarly obtained by use of the values under $W_{25°}$ and $M_{25°}$ for water and mercury respectively.

t°C	W_t	M_t	$W_{18°}$	$M_{18°}$	$W_{25°}$	$M_{25°}$
0	1.001193	0.0735501	1.001643	0.0735832	1.001818	0.0735960
1	1.001133	0.0735636	1.001559	0.0735949	1.001734	0.0736077
2	1.001092	0.0735771	1.001492	0.0736066	1.001668	0.0736194
3	1.001068	0.0735907	1.001443	0.0736183	1.001618	0.0736311
4	1.001060	0.0736037	1.001410	0.0736294	1.001586	0.0736423
5	1.001068	0.0736172	1.001394	0.0736411	1.001569	0.0736540
6	1.001092	0.0736308	1.001392	0.0736529	1.001568	0.0736657
7	1.001131	0.0736492	1.001406	0.0736695	1.001581	0.0736824
8	1.001184	0.0736628	1.001435	0.0736812	1.001610	0.0736941
9	1.001252	0.0736763	1.001477	0.0736929	1.001652	0.0737058
10	1.001333	0.0736894	1.001534	0.0737042	1.001709	0.0737171
11	1.001428	0.0736975	1.001603	0.0737104	1.001779	0.0737233
12	1.001536	0.0737111	1.001686	0.0737222	1.001862	0.0737351
13	1.001657	0.0737241	1.001782	0.0737333	1.001957	0.0737463
14	1.001790	0.0737377	1.001890	0.0737451	1.002066	0.0737581
15	1.001935	0.0737513	1.002010	0.0737569	1.002186	0.0737698
16	1.002092	0.0737644	1.002143	0.0737681	1.002318	0.0737810
17	1.002261	0.0737780	1.002286	0.0737798	1.002462	0.0737927
18	1.002441	0.0737911	1.002441	0.0737911	1.002617	0.0738039
19	1.002633	0.0738047	1.002608	0.0738028	1.002783	0.0738157
20	1.002835	0.0738183	1.002785	0.0738146	1.002960	0.0738275
21	1.003047	0.0738314	1.002972	0.0738258	1.003148	0.0738398
22	1.003271	0.0738450	1.003170	0.0738376	1.003346	0.0738505
23	1.003504	0.0738581	1.003379	0.0738489	1.003554	0.0738618
24	1.003748	0.0738717	1.003597	0.0738607	1.003773	0.0738736
25	1.004001	0.0738848	1.003825	0.0738719	1.004001	0.0738848
26	1.004264	0.0738985	1.004063	0.0738837	1.004239	0.0738966
27	1.004537	0.0739116	1.004310	0.0738950	1.004486	0.0739079
28	1.004819	0.0739253	1.004567	0.0739068	1.004743	0.0739197
29	1.005110	0.0739384	1.004833	0.0739181	1.005009	0.0739310
30	1.005410	0.0739520	1.005109	0.0739299	1.005284	0.0739428

* Assuming 25×10^{-6} as the coefficient of cubic expansion for glass.

Reduction of Weighings to Vacuo

If the apparent mass of a body is m, its density d_m, the density of the weights d_w and the density of the air d_a the true mass in vacuo is,

$$M = m + md_a\left(\frac{1}{d_m} - \frac{1}{d_w}\right)$$

REDUCTIONS OF WEIGHINGS IN AIR TO VACUO

When the weight M in grams of a body is determined in air, a correction is necessary for the buoyancy of the air. The following table is computed for an air density of 0.0012. The corrected weight=M+kM/1000, values of k being found in the table.

Density of body weighed	Correction factor, k.			Density of body weighed	Correction factor, k.		
	Pt Ir weights	Brass weights	Quartz or Al weights		Pt Ir weights	Brass weights	Quartz or Al weights
.5	+2.34	+2.26	+1.95	1.6	+ .69	+ .61	+ .30
.6	+1.94	+1.86	+1.55	1.7	+ .65	+ .56	+ .25
.7	+1.66	+1.57	+1.26	1.8	+ .62	+ .52	+ .21
.75	+1.55	+1.46	+1.15	1.9	+ .58	+ .49	+ .18
.80	+1.44	+1.36	+1.05	2.0	+ .54	+ .46	+ .15
.85	+1.36	+1.27	+0.96	2.5	+ .43	+ .34	+ .03
.90	+1.28	+1.19	+ .88	3.0	+ .34	+ .26	— .05
.95	+1.21	+1.12	+ .81	4.0	+ .24	+ .16	— .15
1.00	+1.14	+1.06	+ .75	6.0	+ .14	+ .06	— .25
1.1	+1.04	+0.95	+ .64	8.0	+ .09	+ .01	— .30
1.2	+0.94	+ .86	+ .55	10.0	+ .06	— .02	— .33
1.3	+ .87	+ .78	+ .47	15.0	+ .03	— .06	— .37
1.4	+ .80	+ .71	+ .40	20.0	+ .004	— .08	— .39
1.5	+ .75	+ .66	+ .35	22.0	— .001	— .09	— .40

SOLUBILITY CHART

Abbreviations: W, soluble in water; A, insoluble in water but soluble in acids; w, sparingly soluble in water but soluble in acids; a, insoluble in water and only sparingly soluble in acids; I, insoluble in both water and acids; d, decomposes in water. * Certain salts occur in two modifications.

No.		Al	NH_4	Sb	Ba	Bi	Cd	Ca	Cr	Co	Cu	Au′	Au‴	H	Fe″	Fe‴
1	Acetates $-(C_2H_3O_2)$	W $Al(-)_3$	W $NH_4(-)$	....	W $Ba(-)_2$	W $Bi(-)_3$	W $Cd(-)_2$	W $Ca(-)_2$	W $Cr(-)_3$	W $Co(-)_2$	W $Cu(-)_2$	....	....	W $C_2H_4O_2$	W $Fe(-)_2$	W $Fe_2(-)_6$
2	Arsenate $-(AsO_4)$	a $Al(-)$	W $(NH_4)_3(-)$	A $Sb(-)$	w $Ba_3(-)_2$	A $Bi(-)$	A $Cd_3(-)_2$	w $Ca_3(-)_2$	....	A $Co_3(-)_2$	A $Cu_3(-)_2$	....	....	W H_3AsO_4	A $Fe_3(-)_2$	A $Fe(-)$
3	Arsenite $-(AsO_3)$	....	W NH_4AsO_2	A $Sb(-)$	....	....	....	w $Ca_3(-)_2$	....	A $Co_3H_6(-)_4$	A $CuH(-)$	....	....	....	....	....
4	Benzoate $-(C_7H_5O_2)$	....	W $NH_4(-)$	....	W $Ba(-)_2$	A $Bi(-)_3$	W $Cd(-)_2$	W $Ca(-)_2$	....	W $Co(-)_2$	w $Cu(-)_2$	....	....	W $C_7H_6O_2$	W $Fe(-)_2$	A $Fe_2(-)_6$
5	Bromide	W $AlBr_3$	W NH_4Br	d $SbBr_3$	W $BaBr_2$	d $BiBr_3$	W $CdBr_2$	W $CaBr_2$	W(I)* $CrBr_3$	W $CoBr_2$	W $CuBr_2$	w $AuBr$	W $AuBr_3$	W HBr	W $FeBr_2$	W $FeBr_3$
6	Carbonate	....	W $(NH_4)_2CO_3$	....	w $BaCO_3$	....	A $CdCO_3$	w $CaCO_3$	W $CrCO_3$	A $CoCO_3$	....	....	....	....	w $FeCO_3$	....
7	Chlorate $-(ClO_3)$	W $Al(-)_3$	W $NH_4(-)$	....	W $Ba(-)_2$	W $Bi(-)_3$	W $Cd(-)_2$	W $Ca(-)_2$	....	W $Co(-)_2$	W $Cu(-)_2$	....	....	W $HClO_3$	W $Fe(-)_2$	W $Fe(-)_3$
8	Chloride	W $AlCl_3$	W NH_4Cl	W $SbCl_3$	W $BaCl_2$	d $BiCl_3$	W $CdCl_2$	W $CaCl_2$	I $CrCl_3$	W $CoCl_2$	W $CuCl_2$	w $AuCl$	W $AuCl_3$	W HCl	W $FeCl_2$	W $FeCl_3$
9	Chromate $-(CrO_4)$	....	W $(NH_4)_2(-)$	....	A $Ba(-)$	....	A $Cd(-)$	W $Ca(-)$	....	A $Co(-)$	....	....	....	....	....	A $Fe_2(-)_3$
10	Citrate $-(C_6H_5O_7)$	W $Al(-)$	W $(NH_4)_3(-)$	....	w $Ba_3(-)_2$	A $Bi(-)$	A $Cd_3(-)_2$	w $Ca_3(-)_2$	....	w $Co_3(-)_2$	....	....	....	W $C_6H_8O_7$	....	W $Fe(-)$
11	Cyanide	....	W NH_4CN	....	W $Ba(CN)_2$	w $Bi(CN)_3$	W $Cd(CN)_2$	W $Ca(CN)_2$	A $Cr(CN)_2$	A $Co(CN)_2$	A $Cu(CN)_2$	w $AuCN$	W $Au(CN)_3$	W HCN	a $Fe(CN)_2$	....
12	Ferricy'de $-(Fe(CN)_6)$	....	W $(NH_4)_3(-)$	....	w $Ba_3(-)_2$	....	A $Cd_3(-)_2$	W $Ca_3(-)_2$	....	I $Co_3(-)_2$	I $Cu_3(-)_2$	....	....	W $H_3(-)$	I $Fe_3(-)_2$	....
13	Ferrocy'de $-(Fe(CN)_6)$	w $Al_4(-)_3$	W $(NH_4)_4(-)$	....	W $Ba_2(-)$	....	A $Cd_2(-)$	W $Ca_2(-)$	....	I $Co_2(-)$	I $Cu_2(-)$	....	....	W $H_4(-)$	I $Fe_2(-)$	a $Fe_4(-)_3$
14	Fluoride	W AlF_3	W NH_4F	W SbF_3	w BaF_2	W BiF_3	W CdF_2	w CaF_2	W(a)* CrF_3	W CoF_2	w CuF_2	....	....	W HF	w FeF_2	w FeF_3
15	Formate $-(CHO_2)$	W $Al(-)_3$	W $NH_4(-)$	....	W $Ba(-)_2$	W $Bi(-)_3$	W $Cd(-)_2$	W $Ca(-)_2$	....	W $Co(-)_2$	W $Cu(-)_2$	....	....	W CH_2O_2	W $Fe(-)_2$	W $Fe(-)_3$
16	Hydroxide	A $Al(OH)_3$	W NH_4OH	....	W $Ba(OH)_2$	A $Bi(OH)_3$	A $Cd(OH)_2$	W $Ca(OH)_2$	A $Cr(OH)_3$	A $Co(OH)_2$	A $Cu(OH)_2$	W $AuOH$	A $Au(OH)_3$	....	A $Fe(OH)_2$	A $Fe(OH)_3$
17	Iodide	W AlI_3	W NH_4I	d SbI_3	W BaI_2	A BiI_3	W CdI_2	W CaI_2	W CrI_2	W CoI_2	a CuI	a AuI	a AuI_3	W HI	W FeI_2	W FeI_3
18	Nitrate	W $Al(NO_3)_3$	W NH_4NO_3	....	W $Ba(NO_3)_2$	d $Bi(NO_3)_3$	W $Cd(NO_3)_2$	W $Ca(NO_3)_2$	W $Cr(NO_3)_3$	W $Co(NO_3)_2$	W $Cu(NO_3)_2$	....	....	W HNO_3	W $Fe(NO_3)_2$	W $Fe(NO_3)_3$
19	Oxalate $-(C_2O_4)$	A $Al_2(-)_3$	W $(NH_4)_2(-)$	....	w $Ba(-)$	A $Bi_2(-)_3$	w $Cd(-)$	A $Ca(-)$	W $Cr(-)$	A $Co(-)$	A $Cu(-)$	....	....	W $C_2H_2O_4$	A $Fe(-)$	W $Fe_2(-)_3$
20	Oxide	a Al_2O_3	....	w Sb_2O_3	W BaO	A Bi_2O_3	A CdO	w CaO	a Cr_2O_3	A CoO	A CuO	Au_2O	A Au_2O_3	W H_2O_2	A FeO	A Fe_2O_3
21	Phosphate	A $AlPO_4$	W $NH_4H_2PO_4$	....	A $Ba_3(PO_4)_2$	A $BiPO_4$	A $Cd_3(PO_4)_2$	w $Ca_3(PO_4)_2$	w $Cr_2(PO_4)_2$	A $Co_3(PO_4)_2$	A $Cu_3(PO_4)_2$	....	....	W H_3PO_4	A $Fe_3(PO_4)_2$	w $FePO_4$
22	Silicate, $-(SiO_3)$	I $Al_2(-)_3$	....	....	W $Ba(-)$	....	A $Cd(-)$	w $Ca(-)$	....	A Co_2SiO_4	A $Cu(-)$	....	....	I H_2SiO_3	....	....
23	Sulfate	W $Al_2(SO_4)_3$	W $(NH_4)_2SO_4$	A $Sb_2(SO_4)_3$	a $BaSO_4$	d $Bi_2(SO_4)_3$	W $CdSO_4$	w $CaSO_4$	W(I)* $Cr_2(SO_4)_3$	W $CoSO_4$	W $CuSO_4$	....	....	W H_2SO_4	W $FeSO_4$	w $Fe_2(SO_4)_3$
24	Sulfide	d Al_2S_3	W $(NH_4)_2S$	A Sb_2S_3	d BaS	A Bi_2S_3	A CdS	w CaS	d Cr_2S_3	A CoS	A CuS	I Au_2S	I Au_2S_3	W H_2S	A FeS	d Fe_2S_3
25	Tartrate $-(C_4H_4O_6)$	w $Al_2(-)_3$	W $(NH_4)_2(-)$	W $Sb_2(-)_3$	w $Ba(-)$	A $Bi_2(-)_3$	A $Cd(-)$	w $Ca(-)$	....	w $Co(-)$	w $Cu(-)$	....	....	W $C_4H_6O_6$	w $Fe(-)$	W $Fe_2(-)_3$
26	Thiocy'te	....	W NH_4CNS	....	W $Ba(CNS)_2$	....	....	W $Ca(CNS)_2$	....	W $Co(CNS)_2$	d $CuCNS$	....	....	W $CNSH$	W $Fe(CNS)_2$	W $Fe(CNS)_3$

No.		Pb	Mg	Mn	Hg′	Hg″	Ni	K	Ag	Na	Sn⁗	Sn″	Sr	Zn	Pt
1	Acetate $-(C_2H_3O_2)$	W $Pb(-)_2$	W $Mg(-)_2$	W $Mn(-)_2$	w $Hg(-)$	W $Hg(-)_2$	W $Ni(-)_2$	W $K(-)$	w $Ag(-)$	W $Na(-)$	W $Sn(-)_4$	d $Sn(-)_2$	W $Sr(-)_2$	W $Zn(-)_2$	....
2	Arsenate $-(AsO_4)$	A $PbH(-)$	A $Mg_3(-)$	w $MnH(-)$	A $Hg_3(-)$	w $Hg_3(-)_2$	A $Ni_3(-)_2$	W $K_3(-)$	A $Ag_3(-)$	W $Na_3(-)$	....	....	w $SrH(-)$	A $Zn_3(-)_2$	....
3	Arsenite $-(AsO_3)$	....	W $Mg_3(-)_2$	A $Mn_3H_6(-)_4$	A $Hg_3(-)$	A $Hg_3(-)$	A $Ni_3H_6(-)_4$	W K_3AsO_3	A $Ag_3(-)$	W $Na_2H(-)$	....	A $Sn_3(-)_2$	w $Sr_3(-)_2$	....	....
4	Benzoate $-(C_7H_5O_2)$	w $Pb(-)_2$	W $Mg(-)_2$	W $Mn(-)_2$	A $Hg_2(-)_2$	w $Hg(-)_2$	w $Ni(-)_2$	W $K(-)$	w $Ag(-)$	W $Na(-)$	....	....	....	W $Zn(-)_2$	....
5	Bromide	W $PbBr_2$	W $MgBr_2$	W $MnBr_2$	A $HgBr$	W $HgBr_2$	W $NiBr_2$	W KBr	a $AgBr$	W $NaBr$	W $SnBr_4$	W $SnBr_2$	W $SrBr_2$	W $ZnBr_2$	w $PtBr_4$
6	Carbonate	A $PbCO_3$	w $MgCO_3$	w $MnCO_3$	A Hg_2CO_3	....	w $NiCO_3$	W K_2CO_3	A Ag_2CO_3	W Na_2CO_3	....	....	w $SrCO_3$	w $ZnCO_3$	....
7	Chlorate $-(ClO_3)$	W $Pb(-)_2$	W $Mg(-)_2$	W $Mn(-)_2$	W $Hg(-)$	W $Hg(-)_2$	W $Ni(-)_2$	W $K(-)$	W $Ag(-)$	W $Na(-)$	....	W $Sn(-)_2$	W $Sr(-)_2$	W $Zn(-)_2$	....
8	Chloride	W $PbCl_2$	W $MgCl_2$	W $MnCl_2$	a $HgCl$	W $HgCl_2$	W $NiCl_2$	W KCl	a $AgCl$	W $NaCl$	W $SnCl_4$	W $SnCl_2$	W $SrCl_2$	W $ZnCl_2$	W $PtCl_4$
9	Chromate $-(CrO_4)$	A $Pb(-)$	W $Mg(-)$	....	w $Hg_2(-)$	w $Hg(-)$	A $Ni(-)$	W $K_2(-)$	w $Ag_2(-)$	W $Na_2(-)$	W $Sn(-)_2$	A $Sn(-)$	w $Sr(-)$	w $Zn(-)$	....
10	Citrate $-(C_6H_5O_7)$	W $Pb_3(-)_2$	W $Mg_3(-)_2$	w $MnH(-)$	w $Hg_3(-)$	....	W $Ni_3(-)_2$	W $K_3(-)$	w $Ag_3(-)$	W $Na_3(-)$	....	....	A $SrH(-)$	w $Zn_3(-)_2$	....

SOLUBILITY CHART (Continued)

No.		Pb	Mg	Mn	Hg′	Hg″	Ni	K	Ag	Na	Sn''''	Sn″	Sr	Zn	Pt
11	Cyanide	w $Pb(CN)_2$	W $Mg(CN)_2$	....	A $HgCN$	W $Hg(CN)_2$	a $Ni(CN)_2$	W KCN	a $AgCN$	W $NaCN$	....	....	W $Sr(CN)_2$	A $Zn(CN)_2$	I $Pt(CN)_2$
12	Ferricy'de $-Fe(CN)_6$	w $Pb_3(-)_2$	W $Mg_3(-)_2$	....	......	A $Hg_3(-)_2$	I $Ni_3(-)_2$	W $K_3(-)$	I $Ag_3(-)$	W $Na_3(-)$	....	A $Sn_3(-)_2$	W $Sr_3(-)_2$	A $Zn_3(-)_2$	....
13	Ferrocy'de $-Fe(CN)_6$	a $Pb_2(-)$	W $Mg_2(-)$	A $Mn_2(-)$	......	I $Hg_2(-)$	I $Ni_2(-)$	W $K_4(-)$	I $Ag_4(-)$	W $Na_4(-)$	....	a $Sn_2(-)$	W $Sr_2(-)$	I $Zn_2(-)$	....
14	Fluoride	w PbF_2	w MgF_2	A MnF_2	d HgF	d HgF_2	w NiF_2	W KF	W AgF	W NaF	W SnF_4	W SnF_2	w SrF_2	w ZnF_2	W PtF_4
15	Formate $-(CHO_2)$	W $Pb(-)_2$	W $Mg(-)_2$	W $Mn(-)_2$	w $Hg(-)$	W $Hg(-)_2$	W $Ni(-)_2$	W $K(-)$	W $Ag(-)$	W $Na(-)$	....	....	W $Sr(-)_2$	W $Zn(-)_2$	....
16	Hydroxide	w $Pb(OH)_2$	A $Mg(OH)_2$	A $Mn(OH)_2$	......	A $Hg(OH)_2$	w $Ni(OH)_2$	W KOH	....	W $NaOH$	w $Sn(OH)_4$	A $Sn(OH)_2$	W $Sr(OH)_2$	A $Zn(OH)_2$	A $Pt(OH)_4$
17	Iodide	w PbI_2	W MgI_2	W MnI_2	A HgI	w HgI_2	W NiI_2	W KI	I AgI	W NaI	d SnI_4	W SnI_2	W SrI_2	W ZnI_2	I PtI_2
18	Nitrate	W $Pb(NO_3)_2$	W $Mg(NO_3)_2$	W $Mn(NO_3)_2$	W $HgNO_3$	W $Hg(NO_3)_2$	W $Ni(NO_3)_2$	W KNO_3	W $AgNO_3$	W $NaNO_3$	....	d $Sn(NO_3)_2$	W $Sr(NO_3)_2$	W $Zn(NO_3)_2$	W $Pt(NO_3)_4$
19	Oxalate $-(C_2O_4)$	A $Pb(-)$	w $Mg(-)$	w $Mn(-)$	a $Hg_2(-)$	A $Hg(-)$	A $Ni(-)$	W $K_2(-)$	a $Ag_2(-)$	W $Na_2(-)$	....	A $Sn(-)$	w $Sr(-)$	A $Zn(-)$	....
20	Oxide	w PbO	A MgO	A MnO	A Hg_2O	w HgO	A NiO	W K_2O	w Ag_2O	d Na_2O	A SnO_2	A SnO	W SrO	w ZnO	A PtO
21	Phosphate	A $Pb_3(PO_4)_2$	w $Mg_3(PO_4)_2$	w $Mn_3(PO_4)_2$	A Hg_3PO_4	A $Hg_3(PO_4)_2$	A $Ni_3(PO_4)_2$	W K_3PO_4	A Ag_3PO_4	W Na_3PO_4	....	A $Sn_3(PO_4)_2$	A $Sr_3(PO_4)_2$	A $Zn_3(PO_4)_2$	....
22	Silicate $-(SiO_3)$	A $Pb(-)$	A $Mg(-)$	I $Mn(-)$	......	....	....	W $K_2(-)$	....	W $Na_2(-)$	....	....	A $Sr(-)$	A $Zn(-)$	....
23	Sulfate	w $PbSO_4$	W $MgSO_4$	W $MnSO_4$	w Hg_2SO_4	d $HgSO_4$	W $NiSO_4$	W K_2SO_4	w Ag_2SO_4	W Na_2SO_4	W $Sn(SO_4)_2$	W $SnSO_4$	w $SrSO_4$	W $ZnSO_4$	W $Pt(SO_4)_2$
24	Sulfide	A PbS	d MgS	A MnS	I Hg_2S	I HgS	A NiS	W K_2S	A Ag_2S	W Na_2S	A SnS_2	A SnS	W SrS	A ZnS	I PtS
25	Tartrate $-(C_4H_4O_6)$	A $Pb(-)$	w $Mg(-)$	w $Mn(-)$	I $Hg_2(-)$	....	A $Ni(-)$	W $K_2(-)$	w $Ag_2(-)$	W $Na_2(-)$	....	W $Sn(-)$	w $Sr(-)$	w $Zn(-)$	....
26	Thiocy'te	w $Pb(CNS)_2$	W $Mg(CNS)_2$	W $Mn(CNS)_2$	A $HgCNS$	w $Hg(CNS)_2$	....	W $KCNS$	I $AgCNS$	W $NaCNS$	....	....	W $Sr(CNS)_2$	W $Zn(CNS)_2$	....

BUFFER SOLUTIONS

OPERATIONAL DEFINITIONS OF pH

Prepared by R. A. Robinson

The operational definition of pH is:

$$pH = pH(s) + E/k$$

where E is the e.m.f. of the cell:

$$H_2|\text{Solution, pH}|\text{Saturated KCl}|\text{Solution, pH(s)}|H_2$$

the half cell on the left containing the solution whose pH is being measured and that on the right a standard buffer mixture of known pH; $k = 2.303RT/F$, where R is the gas constant, T the temperature in degrees Kelvin and F the value of the faraday.

Alternatively, the cell:

Glass electrode|Solution, pH|Saturated calomel electrode

can be used, the glass electrode being calibrated using a standard buffer mixture or, if possible, two standard buffer mixtures whose pH values lie on either side of that of the solution which is being measured. Suitable standard buffer mixtures are:

0.05 M potassium hydrogen phthalate (pH = 4.008 at 25°C)
0.025 M potassium dihydrogen phosphate
0.025 M disodium hydrogen phosphate (pH = 6.865 at 25°C)
0.01 M borax (pH = 9.180 at 25°C)

For most purposes pH can be equated to $-\log_{10} \gamma_{H^+} m_{H^+}$, i.e., to the negative logarithm of the hydrogen ion activity. There is a small difference between those two quantities if pH > 9.2 or pH < 4.0, given by:

$$-\log \gamma_{H^+} m_{H^+} = pH + 0.014(pH - 9.2) \text{ for } pH > 9.2$$
$$= pH + 0.009(4.0 - pH) \text{ for } pH < 4.0$$

It should be noted that in the table titled "Solutions giving Round Values of pH at 25°C" it is $-\log \gamma_{H^+} m_{H^+}$ and not pH which is quoted when there is a difference between them.

References:

R. G. Bates, "Electrometric pH Determinations: Theory and Practice" Wiley, New York, 1954.
R. A. Robinson and R. H. Stokes, "Electrolyte Solutions," 2nd edition, Butterworths, London; Academic Press, Inc. New York, 1959. R. C. Bates, J. Res. of N.B.S. 66 A, 179 (1962).

National Bureau of Standards
R. G. Bates and S. F. Acree, Res. **34,** 373 (1945); W. J. Hammer, C. D. Pinching and S. F. Acree, ibid. **36,** 47 (1946); G. G. Manor, N. J. DeLollis, P. W. Lindwall and S. F. Acree, ibid., **36,** 543 (1946); R. G. Bates, ibid., **39,** 411 (1947); R. G. Bates, V. E. Bower, R. G. Miller and E. R. Smith, ibid., **47,** 433 (1951); V. E. Bower, R. G. Bates and E. R. Smith, ibid., **51,** 189 (1953); V. E. Bower and R. G. Bates, ibid., **55,** 197 (1955); R. G. Bates, V. E. Bower and E. R. Smith, ibid., **56,** 305 (1956); V. E. Bower and R. G. Bates, ibid., **59,** 261; R. G. Bates and V. E. Bower, Anal. Chem., **28,** 1322 (1956).

PROPERTIES OF STANDARD AQUEOUS BUFFER SOLUTIONS AT 25°C

Solution	Buffer substance	Molality m	Weight of salt in air per liter solution	Density g/ml	Molarity M	Dilution value $\Delta pH_{\frac{1}{2}}$	ΔpH_s^a	Buffer value, equiv. per pH	Temp coeff., dpH_s/dt. Units per °C
Tetroxalate	$KH_3(C_2O_4)_2{\cdot}2H_2O$	0.05	12.61	1.0032	0.04962	+0.186	−0.0028	0.070	+0.001
Tartrate	$KHC_4H_4O_6$, sat. sol'n. at 25°C	0.0341	......	1.0036	0.034	+0.049	−0.0003	0.027	−0.0014
Phthalate	$KHC_8O_4H_4$	0.05	10.12	1.0017	0.04958	+0.052	−0.0009	0.016	+0.0012
Phosphate	KH_2PO_4 + Na_2HPO_4	0.025[b]	3.39 3.53	1.0028	0.0249[b]	+0.080	−0.0006	0.029	−0.0028
Phosphate	KH_2PO_4 + Na_2HPO_4	0.008695[c] 0.03043[d]	1.179 4.30	1.0020	0.008665[c] 0.03032[d]	+0.07[e]	−0.0005	0.016	−0.0028
Borax	$Na_2B_4O_7{\cdot}10H_2O$	0.01	3.80	0.9996	0.009971	+0.01	−0.0001	0.020	−0.0082
Calcium hydroxide	$Ca(OH)_2$, sat. sol'n. at 25°C	0.0203	......	0.9991	0.02025	−0.28	+0.0014	0.09	−0.033

[a] $\Delta pH_s = pH_s$ (M Molar solution) − pH_s (m molal solution).
[b] Concentration of each phosphate salt.
[c] KH_2PO_4.
[d] Na_2HPO_4.
[e] Calculated value.

BUFFER SOLUTIONS (Continued)

SOLUTIONS GIVING ROUND VALUES OF pH AT 25°C

Reproduced from "Electrolyte Solutions" by permission from Robinson and Stokes, authors, and Butterworth's Scientific Publications.

A*		B*		C*		D*		E*	
pH	x	pH	x	pH	x	pH	x	pH	x
1.00	67.0	2.20	49.5	4.10	1.3	5.80	3.6	7.00	46.6
1.10	52.8	2.30	45.8	4.20	3.0	5.90	4.6	7.10	45.7
1.20	42.5	2.40	42.2	4.30	4.7	6.00	5.6	7.20	44.7
1.30	33.6	2.50	38.8	4.40	6.6	6.10	6.8	7.30	43.4
1.40	26.6	2.60	35.4	4.50	8.7	6.20	8.1	7.40	42.0
1.50	20.7	2.70	32.1	4.60	11.1	6.30	9.7	7.50	40.3
1.60	16.2	2.80	28.9	4.70	13.6	6.40	11.6	7.60	38.5
1.70	13.0	2.90	25.7	4.80	16.5	6.50	13.9	7.70	36.6
1.80	10.2	3.00	22.3	4.90	19.4	6.60	16.4	7.80	34.5
1.90	8.1	3.10	18.8	5.00	22.6	6.70	19.3	7.90	32.0
2.00	6.5	3.20	15.7	5.10	25.5	6.80	22.4	8.00	29.2
2.10	5.1	3.30	12.9	5.20	28.8	6.90	25.9	8.10	26.2
2.20	3.9	3.40	10.4	5.30	31.6	7.00	29.1	8.20	22.9
		3.50	8.2	5.40	34.1	7.10	32.1	8.30	19.9
		3.60	6.3	5.50	36.6	7.20	34.7	8.40	17.2
		3.70	4.5	5.60	38.8	7.30	37.0	8.50	14.7
		3.80	2.9	5.70	40.6	7.40	39.1	8.60	12.2
		3.90	1.4	5.80	42.3	7.50	41.1	8.70	10.3
		4.00	0.1	5.90	43.7	7.60	42.8	8.80	8.5
						7.70	44.2	8.90	7.0
						7.80	45.3	9.00	5.7
						7.90	46.1		
						8.00	46.7		

F*		G*		H*		I*		J*	
pH	x	pH	x	pH	x	pH	x	pH	x
8.00	20.5	9.20	0.9	9.60	5.0	10.90	3.3	12.00	6.0
8.10	19.7	9.30	3.6	9.70	6.2	11.00	4.1	12.10	8.0
8.20	18.8	9.40	6.2	9.80	7.6	11.10	5.1	12.20	10.2
8.30	17.7	9.50	8.8	9.90	9.1	11.20	6.3	12.30	12.8
8.40	16.6	9.60	11.1	10.00	10.7	11.30	7.6	12.40	16.2
8.50	15.2	9.70	13.1	10.10	12.2	11.40	9.1	12.50	20.4
8.60	13.5	9.80	15.0	10.20	13.8	11.50	11.1	12.60	25.6
8.70	11.6	9.90	16.7	10.30	15.2	11.60	13.5	12.70	32.2
8.80	9.6	10.00	18.3	10.40	16.5	11.70	16.2	12.80	41.2
8.90	7.1	10.10	19.5	10.50	17.8	11.80	19.4	12.90	53.0
9.00	4.6	10.20	20.5	10.60	19.1	11.90	23.0	13.00	66.0
9.10	2.0	10.30	21.3	10.70	20.2	12.00	26.9		
		10.40	22.1	10.80	21.2				
		10.50	22.7	10.90	22.0				
		10.60	23.3	11.00	22.7				
		10.70	23.8						
		10.80	24.25						

*A. 25 ml of 0.2 molar KCl + x ml of 0.2 molar HCl.

*B. 50 ml of 0.1 molar potassium hydrogen phthalate + x ml of 0.1 molar HCl.

*C. 50 ml of 0.1 molar potassium hydrogen phthalate + x ml of 0.1 molar NaOH.

*D. 50 ml of 0.1 molar potassium dihydrogen phosphate + x ml 0.1 molar NaOH.

*E. 50 ml of 0.1 molar tris(hydroxymethyl) aminomethane + x ml of 0.1 M HCl.

*F. 50 ml of 0.025 molar borax + x ml of 0.1 molar HCl.

*G. 50 ml of 0.025 molar borax + x ml of 0.1 molar NaOH.

*H. 50 ml of 0.05 molar sodium bicarbonate + x ml of 0.1 molar NaOH.

*I. 50 ml of 0.05 molar disodium hydrogen phosphate + x ml of 0.1 molar NaOH.

*J. 25 ml of 0.2 molar KCl + x ml of 0.2 molar NaOH.

STANDARD VALUES OF pH AT TEMPERATURE 0–95°C

Temperature	Tetroxalate 0.05 molal	Tartrate 0.0341 molal (sat'd at 25°C)	Phthalate 0.05 molal	Phosphate[a]	Phosphate[b]	Borax 0.01 molal	Calcium hydroxide (sat'd at 25°C)
0	1.666	.....	4.003	6.984	7.534	9.464	13.423
5	1.668	.....	3.999	6.951	7.500	9.395	13.207
10	1.670	.....	3.998	6.923	7.472	9.332	13.003
15	1.672	.....	3.999	6.900	7.448	9.276	12.810
20	1.675	.....	4.002	6.881	7.429	9.225	12.627
25	1.679	3.557	4.008	6.865	7.413	9.180	12.454
30	1.683	3.552	4.015	6.853	7.400	9.139	12.289
35	1.688	3.549	4.024	6.844	7.389	9.102	12.133
38	1.691	3.548	4.030	6.840	7.384	9.081	12.043
40	1.694	3.547	4.035	6.838	7.380	9.068	11.984
45	1.700	3.547	4.047	6.834	7.373	9.038	11.841
50	1.707	3.549	4.060	6.833	7.367	9.011	11.705
55	1.715	3.554	4.075	6.834	.....	8.985	11.574
60	1.723	3.560	4.091	6.836	.....	8.962	11.449
70	1.743	3.580	4.126	6.845	.....	8.921	......
80	1.766	3.609	4.164	6.859	.....	8.885	......
90	1.792	3.650	4.205	6.877	.....	8.850	......
95	1.806	3.674	4.227	6.886	.....	8.833	......

[a] Solution 0.025 m KH_2PO_4 and 0.025 m Na_2HPO_4.
[b] Solution 0.008695 m KH_2PO_4 and 0.03043 m Na_2HPO_4.

APPROXIMATE pH VALUES

The following tables give approximate pH values for a number of substances such as acids, bases, foods, biological fluids, etc. All values are rounded off to the nearest tenth and are based on measurements made at 25° C. A few buffer systems with their pH values are also given.

From Modern pH and Chlorine Control, W. A. Taylor & Co., by permission

ACIDS

Substance	pH
Hydrochloric, N	0.1
Hydrochloric, 0.1N	1.1
Hydrochloric, 0.01N	2.0
Sulfuric, N	0.3
Sulfuric, 0.1N	1.2
Sulfuric, 0.01N	2.1
Orthophosphoric, 0.1N	1.5
Sulfurous, 0.1N	1.5
Oxalic, 0.1N	1.6
Tartaric, 0.1N	2.2
Malic, 0.1N	2.2
Citric, 0.1N	2.2
Formic, 0.1N	2.3
Lactic, 0.1N	2.4
Acetic, N	2.4
Acetic, 0.1N	2.9
Acetic, 0.01N	3.4
Benzoic, 0.01N	3.1
Alum, 0.1N	3.2
Carbonic (saturated)	3.8
Hydrogen sulfide, 0.1N	4.1
Arsenious (saturated)	5.0
Hydrocyanic, 0.1N	5.1
Boric, 0.1N	5.2

BASES

Substance	pH
Sodium hydroxide, N	14.0
Sodium hydroxide, 0.1N	13.0
Sodium hydroxide, 0.01N	12.0
Potassium hydroxide, N	14.0
Potassium hydroxide, 0.1N	13.0
Potassium hydroxide, 0.01N	12.0
Sodium metasilicate, 0.1N	12.6
Lime (saturated)	12.4
Trisodium phosphate, 0.1N	12.0
Sodium carbonate, 0.1N	11.6
Ammonia, N	11.6
Ammonia, 0.1N	11.1
Ammonia, 0.01N	10.6
Potassium cyanide, 0.1N	11.0
Magnesia (saturated)	10.5
Sodium sesquicarbonate, 0.1M	10.1
Ferrous hydroxide (saturated)	9.5
Calcium carbonate (saturated)	9.4
Borax, 0.1N	9.2
Sodium bicarbonate, 0.1N	8.4

BIOLOGIC MATERIALS

Substance	pH
Blood, plasma, human	7.3–7.5
Spinal fluid, human	7.3–7.5
Blood, whole, dog	6.9–7.2
Saliva, human	6.5–7.5
Gastric contents, human	1.0–3.0
Duodenal contents, human	4.8–8.2
Feces, human	4.6–8.4
Urine, human	4.8–8.4
Milk, human	6.6–7.6
Bile, human	6.8–7.0

FOODS

Substance	pH
Apples	2.9–3.3
Apricots	3.6–4.0
Asparagus	5.4–5.8
Bananas	4.5–4.7
Beans	5.0–6.0
Beers	4.0–5.0
Beets	4.9–5.5
Blackberries	3.2–3.6
Bread, white	5.0–6.0
Butter	6.1–6.4
Cabbage	5.2–5.4
Carrots	4.9–5.3
Cheese	4.8–6.4
Cherries	3.2–4.0
Cider	2.9–3.3
Corn	6.0–6.5
Crackers	6.5–8.5
Dates	6.2–6.4
Eggs, fresh white	7.6–8.0
Flour, wheat	5.5–6.5
Gooseberries	2.8–3.0
Grapefruit	3.0–3.3
Grapes	3.5–4.5
Hominy (lye)	6.8–8.0
Jams, fruit	3.5–4.0
Jellies, fruit	2.8–3.4
Lemons	2.2–2.4
Limes	1.8–2.0
Maple syrup	6.5–7.0
Milk, cows	6.3–6.6
Olives	3.6–3.8
Oranges	3.0–4.0
Oysters	6.1–6.6
Peaches	3.4–3.6
Pears	3.6–4.0
Peas	5.8–6.4
Pickles, dill	3.2–3.6
Pickles, sour	3.0–3.4
Pimento	4.6–5.2
Plums	2.8–3.0
Potatoes	5.6–6.0
Pumpkin	4.8–5.2
Raspberries	3.2–3.6
Rhubarb	3.1–3.2
Salmon	6.1–6.3
Sauerkraut	3.4–3.6
Shrimp	6.8–7.0
Soft drinks	2.0–4.0
Spinach	5.1–5.7
Squash	5.0–5.4
Strawberries	3.0–3.5
Sweet potatoes	5.3–5.6
Tomatoes	4.0–4.4
Tuna	5.9–6.1
Turnips	5.2–5.6
Vinegar	2.4–3.4
Water, drinking	6.5–8.0
Wines	2.8–3.8

ACID BASE INDICATORS

Indicator	Approximate pH range	Color-change	Preparation
Methyl Violet	0.0–1.6	yel to bl	0.01–0.05% in water
Crystal Violet	0.0–1.8	yel to bl	0.02% in water
Ethyl Violet	0.0–2.4	yel to bl	0.1 g in 50 ml of MeOH + 50 ml of water
Malachite Green	0.2–1.8	yel to bl grn	water
Methyl Green	0.2–1.8	yel to bl	0.1% in water
2-(p-dimethylaminophenylazo)pyridine	0.2–1.8 4.4–5.6	yel to bl red to yel	0.1% in EtOH
o-Cresolsulfonephthalein (Cresol Red)	0.4–1.8 7.0–8.8	yel to red yel to red	0.1 g in 26.2 ml 0.01N NaOH + 223.8 ml water
Quinaldine Red	1.0–2.2	col to red	1% in EtOH
p-(p-dimethylaminophenylazo)-benzoic acid, Na-salt (Paramethyl Red)	1.0–3.0	red to yel	EtOH
m-(p-anilnophenylazo)benzene sulfonic acid, Na-salt (Metanil Yellow)	1.2–2.4	red to yel	0.01% in water
4-Phenylazodiphenylamine	1.2–2.6	red to yel	0.01 g in 1 ml 1N HCl + 50 ml EtOH + 49 ml water
Thymolsulfonephthalein (Thymol Blue)	1.2–2.8 8.0–9.6	red to yel yel to bl	0.1 g in 21.5 ml 0.01N NaOH + 229.5 ml water
m-Cresolsulfonephthalein (Metacresol Purple)	1.2–2.8 7.4–9.0	red to yel yel to purp	0.1 g in 26.2 ml 0.01N NaOH + 223.8 ml water
p-(p-anilinophenylazo)benzenesulfonic acid, Na-salt (Orange IV)	1.4–2.8	red to yel	0.01% in water
4-o-Tolylazo-o-toluidine	1.4–2.8	or to yel	water
Erythrosine, disodium salt	2.2–3.6	or to red	0.1% in water
Benzopurpurine 48	2.2–4.2	vt to red	0.1% in water
N,N-dimethyl-p-(m-tolylazo)aniline	2.6–4.8	red to yel	0.1% in water
4,4′-Bix(2-amino-1-naphthylazo)2,2′-stilbenedisulfonic acid	3.0–4.0	purp to red	0.1 g in 5.9 ml 0.05N NaOH + 94.1 ml water
Tetrabromophenolphthaleinethyl ester, K-salt	3.0–4.2	yel to bl	0.1% in EtOH
3′,3″,5′,5″-tetrabromophenolsulfonephthalein (Bromophenol Blue)	3.0–4.6	yel to bl	0.1 g in 14.9 ml 0.01N NaOH + 235.1 ml water
2,4-Dinitrophenol	2.8–4.0	col to yel	saturated water solution
N,N-Dimethyl-p-phenylazoaniline (p-Dimethylaminoazobenzene)	2.8–4.4	red to yel	0.1 g in 90 ml in EtOH + 10 ml water
Congo Red	3.0–5.0	blue to red	0.1% in water
Methyl Orange-Xylene Cyanole solution	3.2–4.2	purp to grn	ready solution
Methyl Orange	3.2–4.4	red to yel	0.01% in water
Ethyl Orange	3.4–4.8	red to yel	0.05–0.2% in water or aqueous EtOH
4-(4-Dimethylamino-1-naphthylazo)-3-methoxybenzenesulfonic acid	3.5–4.8	vt to yel	0.1% in 60% EtOH
3′,3″,5′,5″-Tetrabromo-m-cresolsulfonephthalein (Bromocresol Green)	3.8–5.4	yel to blue	0.1 g in 14.3 ml 0.01N NaOH + 235.7 ml water
Resazurin	3.8–6.4	or to vt	water
4-Phenylazo-1-naphthylamine	4.0–5.6	red to yel	0.1% in EtOH
Ethyl Red	4.0–5.8	col to red	0.1 g in 50 ml MeOH + 50 ml water
2-(p-Dimethylaminophenylazo)-pyridine	0.2–1.8 4.4–5.6	yel to red red to yel	0.1% in EtOH
4-(p-ethoxyphenylazo)-m-phenylenediamine monohydrochloride	4.4–5.8	or to yel	0.1% in water
Lacmoid	4.4–6.2	red to bl	0.2% in EtOH
Alizarin Red S	4.6–6.0	yel to red	dilute solution in water
Methyl Red	4.8–6.0	red to yel	0.02 g in 60 ml EtOH + 40 ml water

Indicator	Approximate pH range	Color-change	Preparation
Propyl Red	4.8–6.6	red to yel	EtOH
5′,5″-Dibromo-o-cresolsulfone-phthalein (Bromocresol Purple)	5.2–6.8	yel to purp	0.1 g in 18.5 ml 0.01N NaOH + 231.5 ml water
3′,3″-Dichlorophenolsulfonephthalein (Chlorophenol Red)	5.2–6.8	yel to red	0 1 g in 23.6 ml 0.01N NaOH + 226.4 ml water
p-Nitrophenol	5.4–6.6	col to yel	0.1% in water
Alizarin	5.6–7.2	yel to red	0.1% in MeOH
	11.0–12.4	red to purp	
2-(2,4-Dinitrophenylazo)-1-naphthol-3, 6-disulfonic acid, di-Na salt	6.0–7.0	yel to bl	0.1% in water
3′,3″-Dibromothymolsulfonephthalein (Bromothymol Blue)	6.0–7.6	yel to bl	0.1 g in 16 ml 0.01N NaOH + 234 ml water
6,8-Dinitro-2,4-(1H)quinazolinedione (m-Dinitrobenzoylene urea)	6.4–8.0	col to yel	25 g in 115 ml M NaOH + 50 ml boiling water 0.292 g of NaCl in 100 ml water
Brilliant Yellow	6.6–7.8	yel to or	1% in water
Phenolsulfonephthalein (Phenol Red)	6.6–8.0	yel to red	0.1 g in 28.2 ml 0.01N NaOH + 221.8 ml water
Neutral Red	6.8–8.0	red to amb	0.01 g in 50 ml EtOH + 50 ml water
m-Nitrophenol	6.8–8.6	col to yel	0.3% in water
o-Cresolsulfonephthalein (Cresol Red)	0.0–1.0	red to yel	0.1 g in 26.2 ml 0.01N
	7.0–8.8	yel to red	NaOH + 223.8 ml water
Curcumin	7.4–8.6	yel to red	EtOH
	10.2–11.8		
m-Cresolsulfonephthalein (Metacresol	1.2–2.8	red to yel	0.1 g in 26.2 ml 0.01N
Purple)	7.4–9.0	yel to purp	NaOH + 223.8 ml water
4,4′-Bis(4-amino-1-naphthylazo) 2,2′stilbene disulfonic acid	8.0–9.0	bl to red	0.1 g in 5.9 ml 0.05N NaOH + 94.1 ml water
Thymolsulfonephthalein (Thymol Blue)	1.2–2.8	red to yel	0.1 g in 21.5 ml 0.01N
	8.0–9.6		NaOH + 228.5 ml water
o-Cresolphthalein	8.2–9.8	col to red	0.04% in EtOH
p-Naphtholbenzene	8.2–10.0	or to bl	1% in dil. alkali
Phenolphthalein	8.2–10.0	col to pink	0.05 g in 50 ml EtOH + 50 ml water
Ethyl-bis(2,4-dimethylphenyl)acetate	8.4–9.6	col to bl	saturated solution in 50% acetone alcohol
Thymolphthalein	9.4–10.6	col to bl	0.04 g in 50 ml EtOH + 50 ml water
5-(p-Nitrophenylazo)salicylic acid, Na-salt (Alizarin Yellow R)	10.1–12.0	yel to red	0.01% in water
p-(2,4-Dihydroxyphenylazo)benzenesulfonic acid, Na-salt	11.4–12.6	yel to or	0.1% in water
5,5′-Indigodisulfonic acid, di-Na-salt	11.4–13.0	bl to yel	water
2,4,6-Trinitrotoluene	11.5–13.0	col to or	0.1–0.5% in EtOH
1,3,5-Trinitrobenzene	12.0–14.0	col to or	0.1–0.5% in EtOH
Clayton Yellow	12.2–13.2	yel to amb	0.1% in water

FLUORESCENT INDICATORS

Jack DeMent

Fluorescent indicators are substances which show definite changes in fluorescence with change in pH. Some fluorescent materials are not suitable for indicators since their change in fluorescence is too gradual. Fluorescent indicators find greatest utility in the titration of opaque, highly turbid or deeply colored solutions. A long wavelength ultraviolet ("black light") lamp in a dimly lighted room provides the best environment for titrations involving fluorescent indicators, although bright daylight is sometimes sufficient to evoke a response in the bright green, yellow and orange fluorescent indicators. Titrations are carried out in non-fluorescent glassware. One should check the glassware prior to use to make certain that it does not fluoresce due to the wavelengths of light involved in the titration. The meniscus of the liquid in the burette can be followed when a few particles of an insoluble fluorescent solid are dropped onto its surface.

In this table the indicators are arranged by approximate pH range covered. In the case of some of the dyestuffs the end point may vary slightly with the source or manufacturer.

pH 0 to 2

Indicator	C.I.	From pH	To pH
Benzoflavine	—	0.3, yellow fl.	1.7, green fl.
3,6-Dioxyphthalimide	—	0, blue fl.	2.4, green fl.
Eosine YS	768	0, yellow colored	3.0, yellow fl.
Erythrosine	772	0, yellow colored	3.6, yellow fl.
Esculin	—	1.5, colorless	2, blue fl.
4-Ethoxyacridone	—	1.2, green fl.	3.2, blue fl.
3,6-Tetramethyldiaminooxanthone	—	1.2, green fl.	3.4, blue fl.

pH 2 to 4

Indicator	C.I.	From pH	To pH
Chromotropic acid	—	3.5, colorless	4.5, blue fl.
Fluorescein	766	4, colorless	4.5, green fl.
Magdala Red	—	3.0, purple colored	4.0, fl.
α-Naphthylamine	—	3.4, colorless	4.8, blue fl.
β-Naphthylamine	—	2.8, colorless	4.4, violet fl.
Phloxine	774	3.4, colorless	5.0, bright yellow fl.
Salicylic acid	—	2.5, colorless	3.5, blue fl.

pH 4 to 6

Indicator	C.I.	From pH	To pH
Acridine	788	4.9, green fl.	5.1, violet colored
Dichlorofluorescein	—	4.0, colorless	5.0, green fl.
3,6-Dioxyxanthone	—	5.4, colorless	7.6, blue-violet fl.
Erythrosine	772	4.0, colorless	4.5, yellow-green fl.
β-Methylesculetin	—	4.0, colorless	6.2, blue fl.
Neville-Winther acid	—	6.0, colorless	6.5, blue fl.
Resorufin	—	4.4, yellow fl.	6.4, weak orange fl.
Quininic acid	—	4.0, yellow colored	5.0, blue fl.
Quinine [first end point]	—	5.0, blue fl.	6.1, violet fl.

pH 6 to 8

Indicator	C.I.	From pH	To pH
Acid R Phosphine	—	(claimed for range pH 6.0–7.0)	
Brilliant Diazol Yellow	—	6.5, colorless	7.5, violet fl.
Cleves acid	—	6.5, colorless	7.5, green fl.
Coumaric acid	—	7.2, colorless	9.0, green fl.
3,6-Dioxyphthalic dinitrile	—	5.8, blue fl.	8.2, green fl.
Magnesium 8-hydroxyquinolinate	—	6.5, colorless	7.5, golden fl.
β-Methylumbelliferone	—	7.0, colorless	7.5, blue fl.
1-Naphthol-4-sulfonic acid	—	6.0, colorless	6.5, blue fl.
Orcinaurine	—	6.5, colorless	8.0, green fl.
Patent Phosphine	789	(for the range pH 6.0–7.0, green-yellow fl.)	
Thioflavine	816	(for the region pH 6.5–7.0, yellow fl.)	
Umbelliferone	—	6.5, colorless	7.6, blue fl.

FLUORESCENT INDICATORS (Continued)

pH 8 to 10

Indicator	C.I.	From pH	To pH
Acridine Orange	788	8.4, orange colored	10.4, green fl.
Ethoxyphenylnaphthostilbazonium chloride	—	9, green fl.	11, non-fl.
G Salt	—	9.0, dull blue fl.	9.5, bright blue fl.
Naphthazol derivatives	—	8.2, colorless	10.0, yellow or green fl.
α-Naphthionic acid	—	9, blue fl.	11, green fl.
2-Naphthol-3,6-disulfonic acid	—	9.5, dark blue fl.	Light blue fl. at higher pH
β-Naphthol	—	8.6, colorless	Blue fl. at higher pH
α-Naphtholsulfonic acid	—	8.0, dark blue fl.	9.0, bright violet fl.
1,4-Naphtholsulfonic acid	—	8.2, dark blue fl.	Light blue fl. at higher pH
Orcinsulfonphthalein	—	8.6, yellow colored	10.0 fl.
Quinine [second end point]	—	9.5, violet fl.	10.0, colorless
R-Salt	—	9.0, dull blue fl.	9.5, bright blue fl.
Sodium 1-naphthol-2-sulfonate	—	9.0, dark blue fl.	10.0, bright violet fl.

pH 10 to 12

Indicator	C.I.	From pH	To pH
Coumarin	—	9.8, deep green fl.	12, light green fl.
Eosine BN	771	10.5, colorless	14.0, yellow fl.
Papaverine (permanganate oxidized)	—	9.5, yellow fl.	11.0, blue fl.
Schaffers Salt	—	5.0, violet fl.	11.0, green-blue fl.
SS-Acid (sodium salt)	—	10.0, violet fl.	12.0, yellow colored

pH 12 to 14

Indicator	C.I.	From pH	To pH
Cotarnine	—	12.0, yellow fl.	13.0, white fl.
α-Naphthionic acid	—	12, blue fl.	13, green fl.
β-Naphthionic acid	—	12, blue fl.	13, violet fl.

FORMULAS FOR CALCULATING TITRATION DATA, *p*H VS. ML. OF REAGENT

A Substance Titrated V_0 ml. of solution M_0 its molarity	B Initial $[H^+]$ or $[OH^-]$	C Intermediate Points 10, 50, 99, etc., per cent neutralized, V_1 ml. of reagent of M_1 molarity added	D Equivalence Point	E Excess of Reagent V_1 volume reagent M_1 its molarity V_T total volume
(1) Strong Acid	$[H^+] = M_0$	$[H^+] = \frac{V_0M_0 - V_1M_1}{V_0 + V_1}$	$\sqrt{K_w}$	$[OH^-] = \frac{V_1M_1}{V_T}$
(2) Strong Base	$[OH^-] = M_0$	$[OH^-] = \frac{V_0M_0 - V_1M_1}{V_0 + V_1}$	$\sqrt{K_w}$	$[H^+] = \frac{V_1M_1}{V_T}$
(3) Weak Acid ($K_a = 10^{-5}$ to 10^{-8})	$[H^+] = \sqrt{M_0K_a}$	$[H^+] = \frac{[Acid]}{[Salt]} K_a$	$[OH^-] = \sqrt{\frac{K_w}{K_a} c}$	$[OH^-] = \frac{V_1M_1}{V_T}$ (Value in column D to be added)
(4) Weak Base ($K_b = 10^{-5}$ to 10^{-8})	$[OH^-] = \sqrt{M_0K_b}$	$[OH^-] = \frac{[Base]}{[Salt]} K_b$	$[H^+] = \sqrt{\frac{K_w}{K_b} c}$	$[H^+] = \frac{V_1M_1}{V_T}$ (Value in column D to be added)
(5) Salt of a Very Weak Acid (e.g. KCN)	$[OH^-] = \sqrt{\frac{K_w}{K_a} c}$	$[H^+] = \frac{[Acid]}{[Salt]} K_a$	$[H^+] = \sqrt{[Acid]K_a}$	$[H^+] = \frac{V_1M_1}{V_T}$ (Correct for value in column D)
(6) Salt of a Very Weak Base	$[H^+] = \sqrt{\frac{K_w}{K_b} c}$	$[OH^-] = \frac{[Base]}{[Salt]} K_b$	$[OH^-] = \sqrt{[Base]K_b}$	$[OH^-] = \frac{V_1M_1}{V_T}$ (Add to $[OH^-]$ found in column D)

CONVERSION FORMULAE FOR SOLUTIONS HAVING CONCENTRATIONS EXPRESSED IN VARIOUS WAYS

A = Weight per cent of solute
B = Molecular weight of solvent
E = Molecular weight of solute
F = Grams of solute per liter of solution
G = Molality
M = Molarity
N = Mole fraction
R = Density of solution grams per cc

Concentration of solute— SOUGHT	Concentration of solute—GIVEN A	B	G	M	F
A	—	$\frac{100N \times E}{N \times E + (1 - N)B}$	$\frac{100G \times E}{1000 + G \times E}$	$\frac{M \times E}{10R}$	$\frac{F}{10R}$
B	$\frac{\frac{A}{E}}{\frac{A}{E} + \frac{100 - A}{B}}$	—	$\frac{B \times G}{B \times G + 1000}$	$\frac{B \times M}{M(B - E) + 1000R}$	$\frac{B \times F}{F(B - E) + 1000R \times E}$
G	$\frac{1000A}{E(100 - A)}$	$\frac{100N}{B - N \times B}$	—	$\frac{1000M}{1000R - (M \times E)}$	$\frac{1000F}{E(1000R - F)}$
M	$\frac{10R \times A}{E}$	$\frac{10R \times N}{N \times E + (1 - N)B}$	$\frac{1000R \times G}{1000 + E \times G}$	—	$\frac{F}{E}$
F	10AR	$\frac{10R \times N \times E}{N \times E + (1 - N)B}$	$\frac{1000R \times G \times E}{1000 + G \times E}$	$M \times E$	—

POTENTIALS OF ELECTROCHEMICAL REACTIONS AT 25° C

Compiled by Thomas De Vries

The reactions are alphabetically arranged according to the element being reduced and the signs of the electrode potentials are given according to the Stockholm Convention (IUPAC, 1935). They can also be called reduction potentials.

Reaction	Potential, volts
$Ag^{+} + e = Ag$	+0.7996
$Ag^{++} + e = Ag^{+}$(4f $HClO_4$)	+1.987
$AgBr + e = Ag + Br^{-}$	+0.0713
$AgBrO_3 + e = Ag + BrO_3^{-}$	+0.680
$Ag_2CO_3 + 2e = 2Ag + CO_3^{--}$	+0.4769
$AgC_2H_3O_2 + e = Ag + C_2H_3O_2^{-}$	+0.64
$AgC_2O_4 + 2e = Ag + C_2O_4^{--}$	+0.4776
$AgCl + e = Ag + Cl^{-}$	+0.2223
$AgCN + e = Ag + CN^{-}$	−0.02
$Ag_2CrO_4 + 2e = 2Ag + CrO_4^{--}$	+0.4463
$Ag_4Fe(CN)_6 + 4e = 4Ag + Fe(CN)_6^{----}$	+0.1943
$AgI + e = Ag + I^{-}$	−0.1519
$AgIO_3 + e = Ag + IO_3^{-}$	+0.3551
$Ag_2MoO_4 + 2e = 2Ag + MoO_4^{--}$	+0.49
$AgNO_2 + e = Ag + NO_2^{-}$	+0.59
$Ag_2O + H_2O + 2e = 2Ag + 2OH^{-}$	+0.342
$2AgO + H_2O + 2e = Ag_2O + 2OH^{-}$	+0.599
$AgOCN + e = Ag + OCN^{-}$	+0.41
$Ag_2S + 2e = 2Ag + S^{--}$	−0.7051
$Ag_2S + 2H^{+} + 2e = 2Ag + H_2S$	−0.0366
$AgSCN + e = Ag + SCN^{-}$	+0.0895
$Ag_2SO_4 + 2e = 2Ag + SO_4^{--}$	+0.653
$Ag_2SeO_3 + 2e = 2Ag + SeO_3^{--}$	+0.3629
$Ag_2WO_4 + 2e = 2Ag + WO_4^{--}$	+0.466
$Al^{+++} + 3e = Al$(0.1f NaOH)	−1.706
$H_3AsO_4 + 2H^{+} + 2e = HAsO_2 + 2H_2O$(1f HCl)	+0.58
$AsO_4^{---} + 2H_2O + 2e = AsO_2^{-} + 4OH^{-}$(1f NaOH)	−0.08
$Au^{+++} + 3e = Au$	+1.42
$AuBr_2^{-} + e = Au + 2Br^{-}$	+0.963
$AuBr_4^{-} + 3e = Au + 4Br^{-}$	+0.858
$AuCl_4^{-} + 3e = Au + 4Cl^{-}$	+0.994
$H_2BO_3^{-} + 5H_2O + 8e = BH_4^{-} + 8OH^{-}$	−1.24
$Ba^{++} + 2e = Ba$	−2.90
$Ba^{++} + 2e = Ba(Hg)$	−1.570
$BiCl_4^{-} + 3e = Bi + 4Cl^{-}$	+0.168
$BiOCl + 2H^{+} + 3e = Bi + Cl^{-} + 2H_2O$	+0.1583
$Br_2(aq) + 2e = 2Br^{-}$	+1.087
$Br_2(l) + 2e = 2Br^{-}$	+1.065
$BrO^{-} + H_2O + 2e = Br^{-} + 2OH^{-}$(1f NaOH)	+0.70
$BrO_3^{-} + 6H^{+} + 6e = Br^{-} + 3H_2O$	+1.44
$BrO_3^{-} + 3H_2O + 6e = Br^{-} + 6OH^{-}$	+0.61
$C_6H_4O_2 + 2H^{+} + 2e = C_6H_4(OH)_2$	+0.6992
$Ca^{++} + 2e = Ca$	−2.76
Calomel electrode, satd. KCl	+0.2415
Calomel electrode, satd. NaCl	+0.2360
Calomel electrode, molal KCl	+0.2800
Calomel electrode, normal KCl	+0.2807
Calomel electrode, 0.1n KCl	+0.3337
Calomel electrodes used with buffers: add +0.0028 to above values for liquid junction.	
$Cd^{++} + 2e = Cd$	−0.4026
$Cd^{++} + 2e = Cd(Hg)$	−0.3521
$Cd(OH)_2 + 2e = Cd(Hg) + 2OH^{-}$	−0.761
$CdSO_4 \cdot \frac{8}{3}H_2O + 2e = Cd(Hg) + CdSO_4$(sat'd-aq)	−0.4346
$Ce^{+++} + 3e = Ce$	−2.335
$Ce^{+++} + 3e = Ce(Hg)$	−1.4373
$Ce^{++++} + e = Ce^{+++}$	+1.4430
$CeOH^{+++} + H^{+} + e = Ce^{+++} + H_2O$	+1.7134
Ce(IV) + e = Ce(III)(0.5f H_2SO_4)	+1.4587
$Cl_2(g) + 2e = 2Cl^{-}$	+1.3583
$HClO + H^{+} + 2e = Cl^{-} + H_2O$	+1.49
$ClO^{-} + H_2O + 2e = Cl^{-} + 2OH^{-}$	+0.90
$ClO_3^{-} + 6H^{+} + 6e = Cl^{-} + 3H_2O$	+1.45
$ClO_4^{-} + 8H^{+} + 8e = Cl^{-} + 4H_2O$	+1.37
$ClO_2(aq) + e = ClO_2^{-}$	+0.954
$Co^{++} + 2e = Co$	−0.28
$Co^{+++} + e = Co^{++}$(3f HNO_3)	+1.842
$Cr^{+++} + e = Cr^{++}$	−0.41
$Cr^{+++} + 3e = Cr$	−0.557
$HCrO_4^{-} + 7H^{+} + 3e = Cr^{+++} + 4H_2O$	+1.195
Cr(VI) + 3e = Cr(III)(2f H_2SO_4)	+1.10
Cr(VI) + 3e = Cr(III)(1f NaOH)	−0.12
$Cs^{+} + e = Cs$	−2.923
$Cu^{++} + e = Cu^{+}$	+0.158
$Cu^{+} + e = Cu$	+0.522
$Cu^{++} + 2e = Cu$	+0.3402
$Cu^{++} + 2e = Cu(Hg)$	+0.345
$2D^{+} + 2e = D_2$	−0.044
$F_2 + 2e = 2F^{-}$	+2.87
$Fe^{+++} + e = Fe^{++}$	+0.770
$Fe^{+++} + e = Fe^{++}$(1f HCl)	+0.700
$Fe^{+++} + e = Fe^{++}$(1f $HClO_4$)	+0.747
$Fe^{+++} + e = Fe^{++}$(1f H_3PO_4)	+0.438
$Fe^{+++} + e = Fe^{++}$(0.5f H_2SO_4)	+0.679
$Fe^{++} + 2e = Fe$	−0.409
$Fe(CN)_6^{---} + e = Fe(CN)_6^{----}$(1f H_2SO_4)	+0.69
$Fe(CN)_6^{---} + e = Fe(CN)_6^{----}$(0.01f NaOH)	+0.46
Fe(phenanthroline)$_3^{+++}$ + e = Fe(ph)$_3^{++}$	+1.14
Fe(phenathroline)$_3^{+++}$ + e = Fe(ph)$_3^{++}$(2f H_2SO_4)	+1.056
$Ga^{+++} + 3e = Ga$	−0.560
$2H^{+} + 2e = H_2$	0.0000
$2H_2O + 2e = H_2 + 2OH^{-}$	−0.8277
$2Hg^{++} + 2e = Hg_2^{++}$	+0.905
$Hg_2^{++} + 2e = 2Hg$	+0.7961
$Hg^{++} + 2e = Hg$	+0.851
$Hg_2(AcO)_2 + 2e = 2Hg + 2AcO^{-}$	+0.5113
$Hg_2Br_2 + 2e = 2Hg + 2Br^{-}$	+0.1396
$Hg_2Cl_2 + 2e = 2Hg + 2Cl^{-}$	+0.2682
$Hg_2Cl_2 + 2e = 2Hg + 2Cl^{-}$(0.1f NaCl)	+0.3419
$Hg_2I_2 + 2e = 2Hg + 2I^{-}$	−0.0405
$HgO + 2H_2O + 2e = Hg + 2OH^{-}$	+0.097
$Hg_2HPO_4 + H^{+} + 2e = 2Hg + H_2PO_4^{-}$	+0.639
$Hg_2SO_4 + 2e = 2Hg + SO_4^{--}$	+0.6158
$I_2 + 2e = 2I^{-}$	+0.535
$I_3^{-} + 2e = 3I^{-}$	+0.5338
$IO_3^{-} + 6H^{+} + 5e = \frac{1}{2}I_2 + 3H_2O$	+1.195
$In^{+++} + e = In^{++}$	−0.49
$In^{++} + e = In^{+}$	−0.40
$In^{+++} + 3e = In$	−0.338
$K^{+} + e = K$	−2.924
$Li^{+} + e = Li$	−3.045
$Mg^{++} + 2e = Mg$	−2.375
$Mn^{++} + 2e = Mn$	−1.029
$MnO_2 + 4H^{+} + 2e = Mn^{++} + 2H_2O$	+1.208
$MnO_4^{-} + e = MnO_4^{--}$	+0.564
$MnO_4^{-} + 4H^{+} + 3e = MnO_2 + 2H_2O$	+1.679
$MnO_4^{-} + 2H_2O + 3e = MnO_2 + 4OH^{-}$	+0.588
$MnO_4^{-} + 8H^{+} + 5e = Mn^{++} + 4H_2O$	+1.491
$Na^{+} + e = Na$	−2.7109
Nb(V) + 2e = Nb(III)(2f HCl)	+0.344
$Nd^{+++} + 3e = Nd$	−2.246
$NO_3^{-} + H_2O + 2e = NO_2^{-} + 2OH^{-}$	0.0
$Ni^{++} + 2e = Ni$	−0.23
$NiO_2 + 4H^{+} + 2e = Ni^{++} + 2H_2O$	+1.93
$NiO_2 + 2H_2O + 2e = Ni(OH)_2 + 2OH^{-}$	+0.76
Np(VI) + e = Np(V)(1f $HClO_4$)	+1.137
Np(V) + e = Np(IV)(1f $HClO_4$)	+0.739
Np(IV) + e = Np(III)(1f $HClO_4$)	+0.155
$Np^{+++} + 3e = Np$	−1.9
$O_2 + 2H^{+} + 2e = H_2O_2$	+0.682
$O_2 + 2H_2O + 2e = H_2O_2 + 2OH^{-}$	−0.146
$O_2 + 4H^{+} + 4e = 2H_2O$	+1.229
$O_2 + 2H_2O + 4e = 4OH^{-}$	+0.401
$H_2O_2 + 2H^{+} + 2e = 2H_2O$	+1.776
$Pb^{++} + 2e = Pb$	−0.1263
$Pb^{++} + 2e = Pb(Hg)$	−0.1205
$PbBr_2 + 2e = Pb(Hg) + 2Br^{-}$	−0.275
$PbCl_2 + 2e = Pb(Hg) + 2Cl^{-}$	−0.262
$PbF_2 + 2e = Pb(Hg) + 2F^{-}$	−0.3444
$PbI_2 + 2e = Pb(Hg) + 2I^{-}$	−0.358
$PbO_2 + 4H^{+} + 2e = Pb^{++} + 2H_2O$	+1.46
$PbO + H_2O + 2e = Pb + 2OH^{-}$	−0.576
$PbO_2 + SO_4^{--} + 4H^{+} + 2e = PbSO_4 + 2H_2O$	+1.685
$PbHPO_4 + H^{+} + 2e = Pb(Hg) + HPO_4^{-}$	−0.2448
$PbSO_4 + 2e = Pb(Hg) + SO_4^{--}$	−0.3505
$Pd^{++} + 2e = Pd$(4f $HClO_4$)	+0.987
$Pd^{++} + 2e = Pd$(1f HCl)	+0.623
Pu(VI) + 2e = Pu(IV)(1f HCl)	+1.052
Pu(VI) + e = Pu(V)(1f $HClO_4$)	+0.9184
Pu(V) + e = Pu(IV)(0.5f HCl)	+1.099
Pu(IV) + e = Pu(III)(1f $HClO_4$)	+0.982
Quinhydrone electrode: H^{+}, a = 1	+0.6995
$Rb^{+} + e = Rb$	−2.925
$ReO_4^{-} + 8H^{+} + 7e = Re + 4H_2O$	+0.367
$ReO_4^{-} + 2H^{+} + e = ReO_3(c) + 2H_2O$	+0.768
$RuO_4(c) + e = RuO_4^{-}$	+1.00
$RuO_4^{-} + e = RuO_4^{--}$	+0.59
Ru(IV) + e = Ru(III)(2f HCl)	+0.858
Ru(IV) + e = Ru(III)(0.1f $HClO_4$)	+0.49
Ru(III) + e = Ru(II)(0.1f $HClO_4$)	−0.11
Ru(III) + e = Ru(II)(1–6f HCl)	−0.084
$S_4O_6^{--} + 2e = 2S_2O_3^{--}$	+0.10
$SO_4^{--} + H_2O + 2e = SO_3^{--} + 2OH^{-}$	−0.92
$Sb_2O_3 + 6H^{+} + 6e = 2Sb + 3H_2O$	+0.1445
$SbO^{+} + 2H^{+} + 3e = Sb + 2H_2O$	+0.212
Sb(V) + 2e = Sb(III)(3.5f HCl)	+0.75
$H_2SeO_3 + 4H^{+} + 4e = Se + 3H_2O$	+0.74
$Sn^{++} + 2e = Sn$	−0.1364
Sn(IV) + 2e = Sn(II)(0.1f HCl)	+0.070
Sn(IV) + 2e = Sn(II)(1f HCl)	+0.139
$Sr^{++} + 2e = Sr$	−2.89
$Sr^{++} + 2e = Sr(Hg)$	−1.793
$TcO_4^{-} + 4H^{+} + 3e = TcO_2(c) + 2H_2O$	+0.738
$TcO_4^{-} + 8H^{+} + 7e = Tc + 4H_2O$	+0.472
Te(IV) + 4e = Te(2.5f HCl)	+0.63
$TeO_2 + 4H^{+} + 4e = Te + 2H_2O$	+0.593
$Ti^{+++} + e = Ti^{++}$	−2.
$Ti(OH)^{+++} + H^{+} + e = Ti^{+++} + H_2O$	+0.06
$Tl^{+++} + 2e = Tl^{+}$	+1.247
Tl(III) + 2e = Tl(I)(1f HCl)	+0.783
$Tl^{+} + e = Tl$	−0.3363
$Tl^{+} + e = Tl(Hg)$	−0.3338
$TlBr + e = Tl(Hg) + Br^{-}$	−0.606
$TlCl + e = Tl(Hg) + Cl^{-}$	−0.555
$TlI + e = Tl(Hg) + I^{-}$	−0.769
$Tl_2SO_4 + 2e = Tl(Hg) + SO_4^{--}$	−0.4360
U(VI) + e = U(V)(1f $HClO_4$)	+0.063
$UO_2^{+} + 4H^{+} + e = U^{++++} + 2H_2O$	+0.62
U(IV) + e = U(III)(1f $HClO_4$)	−0.631
$UO_2^{++} + 4H^{+} + 2e = U^{++++} + 2H_2O$	+0.334
U(V) + e = U(IV)(1f HCl)	+1.02
$V(OH)_4^{+} + 2H^{+} + e = VO^{++} + 3H_2O$	+1.000
V(V) + e = V(IV)(1f NaOH)	−0.74
$VO^{++} + 2H^{+} + e = V^{+++} + H_2O$	+0.337
$V^{+++} + e = V^{++}$	−0.255
$Zn^{++} + 2e = Zn$	−0.7628
$Zn^{++} + 2e = Zn(Hg)$	−0.7628
$ZnSO_4 \cdot 7H_2O + 2e = Zn(Hg) + SO_4^{--}$(in sat'd $ZnSO_4$)	−0.7993

DISSOCIATION CONSTANTS OF ORGANIC BASES IN AQUEOUS SOLUTION

No.	Compound	Temp. °C	Step	pK_a	K_a
a1	Acetamide	25		0.63	2.34×10^{-1}
a2	Acridine	20		5.58	2.63×10^{-6}
a3	α-Alanine	25		2.345	4.52×10^{-3}
a4	Alanine, glycyl-	25		3.153	7.03×10^{-4}
a5	Alanine, methoxy- (DL)	25		2.037	9.18×10^{-3}
a6	Alanine, phenyl	25	2	9.18	6.61×10^{-10}
a7	Allothreonine	25	1	2.108	7.80×10^{-3}
	Allothreonine	25	2	9.096	8.02×10^{-10}
a8	*n*-Amylamine	25		10.63	2.34×10^{-11}
a9	Aniline	25		4.63	2.34×10^{-5}
a10	Aniline, *n*-allyl	25		4.17	6.76×10^{-5}
a11	Aniline, 4-(p-aminobenzoyl)	25	1	2.932	1.17×10^{-3}
a12	Aniline, 4-benzyl	25		2.17	6.76×10^{-3}
a13	Aniline, 2-bromo	25		2.53	2.95×10^{-3}
a14	Aniline, 3-bromo	25		3.58	2.63×10^{-4}
a15	Aniline, 4-bromo	25		3.86	1.38×10^{-4}
a16	Aniline, 4-bromo-N,N-dimethyl	25		4.232	5.86×10^{-5}
a17	Aniline, o-chloro	25		2.65	2.24×10^{-3}
a18	Aniline, m-chloro	25		3.46	3.47×10^{-4}
a19	Aniline, p-chloro	25		4.15	7.08×10^{-5}
a20	Aniline, 3-chloro-N,N-dimethyl	20		3.837	1.46×10^{-4}
a21	Aniline, 4-chloro-N,N-dimethyl	20		4.395	4.03×10^{-5}
a22	Aniline, 3,5-dibromo	25		2.34	4.57×10^{-3}
a23	Aniline, 2,4-dichloro	22		2.05	8.91×10^{-3}
a24	Aniline, N,N-diethyl	22		6.61	2.46×10^{-7}
a25	Aniline, N,N-dimethyl	25		5.15	7.08×10^{-6}
a26	Aniline, N,N-dimethyl-3-nitro	25		2.626	2.37×10^{-3}
a27	Aniline, N-ethyl	24		5.12	7.59×10^{-6}
a28	Aniline, 2-fluoro	25		3.20	6.31×10^{-4}
a29	Aniline, 3-fluoro	25		3.50	3.16×10^{-4}
a30	Aniline, 4-fluoro	25		4.65	2.24×10^{-5}
a31	Aniline, 2-iodo	25		2.60	2.51×10^{-3}
a32	Aniline, N-methyl	25		4.848	1.41×10^{-5}
a33	Aniline, 4-methylthio	25		4.35	4.46×10^{-5}
a34	Aniline, 3-nitro	25		2.466	3.42×10^{-3}
a35	Aniline, 4-nitro	25		1.0	1.00×10^{-1}
a36	Aniline, 2-sulfonic acid	25	2	2.459	3.47×10^{-3}
a37	Aniline, 3-sulfonic acid	25	2	3.738	1.82×10^{-4}
a38	Aniline, 4-sulfonic acid	25	2	3.227	5.92×10^{-4}
a39	*o*-Anisidine	25		4.52	3.02×10^{-5}
a40	*m*-Anisidine	25		4.23	5.89×10^{-5}
a41	*p*-Anisidine	25		5.34	4.57×10^{-6}
a42	Arginine	25	1	1.8217	1.51×10^{-2}
		25	2	8.9936	1.01×10^{-9}
a43	Asparagine	20	1	2.213	6.12×10^{-3}
		20	2	8.85	1.41×10^{-9}
a44	Asparagine, glycyl	25	1	2.942	1.14×10^{-3}
		18	2	8.44	3.63×10^{-9}
a45	DL-Aspartic acid	1	1	2.122	7.55×10^{-3}
		1	2	4.006	1.00×10^{-4}
a46	Azetidine (Trimethylimidine)	25		11.29	5.12×10^{-12}
a47	Aziridine	25		8.01	9.77×10^{-9}
b1	Benzene, 4-aminoazo	25		2.82	1.51×10^{-3}
b2	Benzene, 2-aminoethyl (β-Phenylamine)	25		9.84	1.45×10^{-10}
b3	Benzene, 4-dimethylaminoazo	25		3.226	5.94×10^{-4}
b4	Benzidine	30	1	4.66	2.19×10^{-5}
		30	2	3.57	2.69×10^{-4}
b5	Benzimidazole	25		5.532	2.94×10^{-6}
b6	Benzimidazole, 2-ethyl	25		6.18	6.61×10^{-7}
b7	Benzimidazole, 2-methyl	25		6.19	6.46×10^{-7}
b8	Benzimidazole, 2-phenyl	25	1	5.23	5.89×10^{-6}
		25	2	11.91	1.23×10^{-12}
b9	Benzoic acid, 2-amino (Anthranilic acid)	25	1	2.108	7.80×10^{-3}
		25	2	4.946	1.13×10^{-5}
b10	Benzoic acid, 4-amino	25	1	2.501	3.15×10^{-3}
		25	2	4.874	1.33×10^{-5}
b11	Benzylamine	25		9.33	4.67×10^{-10}
b12	Betaine	0		1.83	1.48×10^{-2}
b13	Biphenyl, 2-amino	22		3.82	1.51×10^{-4}
b14	Bornylamine(trans-)	25		10.17	6.76×10^{-11}

No.	Compound	Temp. °C	Step	pK_a	K_a
b15	Brucine	25	1	8.28	5.24×10^{-9}
b16	Butane, 1-amino-3-methyl	25		10.60	2.51×10^{-11}
b17	Butane, 2-amino-2-methyl	19		10.85	1.41×10^{-11}
b18	Butane, 1,4-diamino (Putrescine)	10	1	11.15	7.08×10^{-12}
		10	2	9.71	1.95×10^{-10}
b19	*n*-Butylamine	20		10.77	1.69×10^{-11}
b20	*t*-Butylamine	18		10.83	1.48×10^{-11}
b21	Butyric acid, 4-amino	25	1	4.0312	9.31×10^{-5}
		25	2	10.5557	2.78×10^{-11}
b22	*n*-Butyric acid, glycyl-2-amino	25	1	3.1546	7.01×10^{-4}
c1	Cacodylic acid	25	1	1.57	2.69×10^{-2}
			2	6.27	5.37×10^{-7}
c2	β-Chlortriethylammonium	25		8.80	1.59×10^{-9}
c3	Cinnoline	20		2.37	4.27×10^{-3}
c4	Codeine	25		8.21	6.15×10^{-9}
c5	Cyclohexaneamine, *n*-butyl	25		11.23	5.89×10^{-12}
c6	Cyclohexylamine	24		10.66	2.19×10^{-11}
c7	Cystine	30	1	1.90	1.25×10^{-2}
		30	2	8.24	5.76×10^{-9}
d1	*n*-Decylamine	25		10.64	2.29×10^{-11}
d2	Diethylamine	40		10.489	3.24×10^{-11}
d3	Diisobutylamine	21		10.91	1.23×10^{-11}
d4	Diisopropylamine	28.5		10.96	1.09×10^{-11}
d5	Dimethylamine	25		10.732	1.85×10^{-11}
d6	*n*-Diphenylamine	25		0.79	1.62×10^{-1}
d7	*n*-Dodecaneamine (Laurylamine)	25		10.63	2.35×10^{-11}
e1	*d*-Ephedrine	10		10.139	7.26×10^{-11}
e2	*l*-Ephedrine	10		9.958	1.10×10^{-10}
e3	Ethane, 1-amino-3-methoxy	10		9.89	1.29×10^{-10}
e4	Ethane, 1,2-bismethylamino	25	1	10.40	3.98×10^{-11}
		25	2	8.26	5.50×10^{-9}
e5	Ethanol, 2-amino	25		9.50	3.16×10^{-10}
e6	Ethylamine	20		10.807	1.56×10^{-11}
e7	Ethylenediamine	0	1	10.712	1.94×10^{-11}
		0	2	7.564	2.73×10^{-8}
g1	*l*-Glutamic acid	25	1	2.13	7.41×10^{-3}
		25	2	4.31	4.90×10^{-5}
e2	Glutamic acid, α-monoethyl	25	1	3.846	1.42×10^{-4}
		25	2	7.838	1.45×10^{-8}
e3	*l*-Glutamine	—		9.28	5.25×10^{-10}
e4	*l*-Glutathione	25	2	3.59	2.57×10^{-4}
e5	Glycine	25	1	2.3503	4.46×10^{-3}
		25	2	9.7796	1.68×10^{-10}
e6	Glycine, *n*-acetyl	25		3.6698	2.14×10^{-4}
e7	Glycine, dimethyl	5		10.3371	4.60×10^{-11}
e8	Glycine, glycyl	25		3.1397	7.25×10^{-4}
e9	Glycine, glycylglycyl	25	1	3.225	5.96×10^{-4}
		25	2	8.090	8.13×10^{-9}
e10	Glycine, leucyl	25	1	3.25	5.62×10^{-4}
		25	2	8.28	5.25×10^{-9}
e11	Glycine, methyl (Sarcosine)	25	1	2.21	6.16×10^{-3}
		25	2	10.12	7.58×10^{-11}
g12	Glycine, phenyl	25	1	1.83	1.48×10^{-2}
		25	2	4.39	4.07×10^{-5}
g13	Glycine, N,*n*-propyl	25	1	2.35	4.46×10^{-3}
		25	2	10.19	6.46×10^{-11}
g14	Glycine, tetraglycyl	20	1	3.10	7.94×10^{-4}
		20	2	8.02	9.55×10^{-9}
g15	Glycylserine	25	1	2.9808	1.04×10^{-3}
		25	2	8.38	4.17×10^{-9}
h1	Hexadecaneamine	25		10.63	2.35×10^{-11}
h2	Heptane, 1-amino	25		10.66	2.19×10^{-11}
h3	Heptane, 2-amino	19		10.88	1.58×10^{-11}
h4	Heptane, 2-methylamino	17		10.99	1.02×10^{-11}
h5	Hexadecaneamine	25		10.61	2.46×10^{-11}

DISSOCIATION CONSTANTS OF ORGANIC BASES IN AQUEOUS SOLUTION (*Continued*)

No.	Compound	Temp. °C	Step	pK_a	K_a
h6	Hexamethylenediamine	0	1	111.857	1.39×10^{-12}
		0	2	0.762	1.73×10^{-11}
h7	Hexanoic acid, 6-amino.	25	1	4.373	4.23×10^{-5}
		25	2	10.804	1.57×10^{-11}
h8	*n*-Hexylamine	25		10.56	2.75×10^{-11}
h9	*dl*-Histidine	25	1	1.80	1.58×10^{-2}
		25	2	6.04	9.12×10^{-7}
		25	3	9.33	4.67×10^{-10}
h10	Histidine, β-alanyl (Carnosine)	20	1	2.73	1.86×10^{-3}
		20	2	6.87	1.35×10^{-7}
		20	3	9.73	1.48×10^{-10}
i1	Imidazol	25		6.953	1.11×10^{-7}
i2	Imidazol, 2,4-dimethyl.	25		8.359	5.50×10^{-9}
i3	Imidazol, 1-methyl (Oxalmethyline)	25		6.95	1.12×10^{-7}
i4	Indane, 1-amino (*d*-1-Hydrindamine	22.5		9.21	6.17×10^{-10}
i5	Isobutyric acid, 2-amino	25	1	2.357	4.30×10^{-3}
		25	2	10.205	6.23×10^{-11}
i6	Isoleucine	25	1	2.318	4.81×10^{-3}
		25	2	9.758	1.74×10^{-10}
i7	Isoquinoline (Leucoline)	20		5.42	3.80×10^{-6}
i8	Isoquinoline, 1-amino	20		7.59	2.57×10^{-8}
i9	Isoquinoline, 7-hydroxy	20	1	5.68	2.09×10^{-6}
		20	2	8.90	1.26×10^{-9}
l1	*L*-Leucine	25	1	2.328	4.70×10^{-3}
		25	2	9.744	1.80×10^{-10}
l2	Leucine, glycyl	25		3.18	6.61×10^{-4}
m1	Methionine	25	1	2.22	6.02×10^{-3}
		25	2	9.27	5.37×10^{-10}
m2	Methylamine	25		10.657	2.70×10^{-11}
m3	Morphine	25		8.21	6.16×10^{-9}
m4	Morpholine	25		8.33	4.67×10^{-9}
n1	Naphthalene, 1-amino-6-hydroxy	25		3.97	1.07×10^{-4}
n2	Naphthalene, dimethylamino	25		4.566	2.72×10^{-5}
n3	α-Naphthylamine	25		3.92	1.20×10^{-4}
n4	β-Naphthylamine	25		4.16	6.92×10^{-5}
n5	α-Naphthylamine, *n*-methyl	27		3.67	2.13×10^{-4}
n6	Neobornylamine(cis-)	25		10.01	9.77×10^{-11}
n7	Nicotine	25	1	8.02	9.55×10^{-9}
		25	2	3.12	7.59×10^{-4}
n8	*n*-Nonylamine	25		10.64	2.29×10^{-11}
n9	Norleucine	25		2.335	4.62×10^{-3}
o1	Octadecaneamine	25		10.60	2.51×10^{-11}
o2	Octylamine	25		10.65	2.24×10^{-11}
o3	Ornithine	25	1	1.705	1.97×10^{-2}
		25	2	8.690	2.04×10^{-9}
p1	Papaverine	25		6.40	3.98×10^{-7}
p2	Pentane, 3-amino	17		10.59	2.57×10^{-11}
p3	Pentane, 3-amino-3-methyl	16		11.01	9.77×10^{-12}
p4	*n*-Pentadecylamine	25		10.61	2.46×10^{-11}
p5	Pentanoic acid, 5-amino(Valeric acid)	25	1	4.270	5.37×10^{-5}
		25	2	10.766	1.71×10^{-11}
p6	Perimidine	20		6.35	4.47×10^{-7}
p7	Phenanthridine	20		5.58	2.63×10^{-6}
p8	1,10-Phenanthroline	25		4.84	1.44×10^{-5}
p9	*o*-Phenetidine (2-Ethoxyaniline)	28		4.43	3.72×10^{-5}
p10	*m*-Phenetidine (3-Ethoxyaniline)	25		4.18	6.60×10^{-5}
p11	*p*-Phenetidine (4-Ethoxyaniline)	28		5.20	6.31×10^{-6}
p12	α-Picoline	20		5.97	1.07×10^{-6}
p13	β-Picoline	20		5.68	2.09×10^{-6}
p14	γ-Picoline	20		6.02	9.55×10^{-7}
p15	Pilocarpine	30		6.87	1.35×10^{-7}

No.	Compound	Temp. °C	Step	pK_a	K_a
p16	Piperazine	23.5	1	9.83	1.48×10^{-10}
		23.5	2	5.56	2.76×10^{-6}
p17	Piperazine, 2,5-dimethyl(trans-)	25	1	9.66	2.19×10^{-10}
		25	2	5.20	6.31×10^{-6}
p18	Piperidine	25		11.123	7.53×10^{-12}
p19	Piperidine, 3-acetyl	25		3.18	6.61×10^{-4}
p20	Piperidine, 1-*n*-butyl	23		10.47	3.39×10^{-11}
p21	Piperidine, 1,2-dimethyl	25		10.22	6.03×10^{-11}
p22	Piperidine, 1-ethyl	23		10.45	3.55×10^{-11}
p23	Piperidine, 1-methyl	25		10.08	8.32×10^{-11}
p24	Piperidine, 2,2,6,6-tetramethyl	25		11.07	8.51×10^{-12}
p25	Piperidine, 2,2,4-trimethyl	30		11.04	9.12×10^{-12}
p26	Proline	25	1	1.952	1.11×10^{-2}
		25	2	10.640	2.29×10^{-11}
p27	Proline, hydroxy	25	1	1.818	1.52×10^{-2}
		25	2	9.662	2.18×10^{-10}
p28	Propane, 1-amino-2,2-dimethyl	25		10.15	7.08×10^{-11}
p29	Propane, 1,2-diamino	25	1	9.82	1.52×10^{-10}
		25	2	6.61	2.46×10^{-7}
p30	Propane, 1,3-diamino	10	1	10.94	1.15×10^{-11}
		10	2	9.03	9.33×10^{-10}
p31	Propane, 1,2,3-triamino	20	1	9.59	2.57×10^{-10}
		20	2	7.95	1.12×10^{-8}
p32	Propanoic acid, 3-amino (β-Alanine)	25	1	3.551	2.81×10^{-4}
		25	2	10.238	5.78×10^{-11}
p33	Propylamine	20		10.708	1.96×10^{-11}
p34	Pteridine	20		4.05	8.91×10^{-5}
p35	Pteridine, 2-amino-4,6-dihydroxy	20	2	6.59	2.57×10^{-7}
		20	3	9.31	4.90×10^{-10}
p36	Pteridine, 2-amino-4-hydroxy	20	1	2.27	5.37×10^{-3}
		20	2	7.96	1.10×10^{-8}
p37	Pteridine, 6-chloro	20		3.68	2.09×10^{-4}
p38	Pteridine, 6-hydroxy-4-methyl	20	1	4.08	8.32×10^{-5}
			2	6.41	3.89×10^{-7}
p39	Purine	20	1	2.30	5.01×10^{-3}
		20	2	8.96	1.10×10^{-9}
p40	Purine, 6-amino (Adenine)	25	1	4.12	7.59×10^{-5}
		25	2	9.83	1.48×10^{-10}
p41	Purine, 2-dimethylamino	20	1	4.00	1.00×10^{-4}
		20	2	10.24	5.75×10^{-11}
p42	Purine, 8-hydroxy	20	1	2.56	2.75×10^{-3}
		20	2	8.26	9.49×10^{-9}
p43	Pyrazine	27		0.65	2.24×10^{-1}
p44	Pyrazine, 2-methyl	27		1.45	3.54×10^{-2}
p45	Pyrazine, methylamino.	25		3.39	4.07×10^{-4}
p46	Pyridazine	20		2.24	5.76×10^{-3}
p47	Pyrimidine, 2-amino	20		3.45	3.54×10^{-4}
p48	Pyrimidine, 2-amino-4,6-dimethyl	20		4.82	1.51×10^{-5}
p49	Pyrimidine, 2-amino-5-nitro	20		0.35	4.46×10^{-1}
p50	Pyridine	25		5.25	5.62×10^{-6}
p51	Pyridine, 2-aldoxime	20	1	3.59	2.57×10^{-4}
		20	2	10.18	6.61×10^{-11}
p52	Pyridine, 2-amino	20		6.82	1.51×10^{-7}
p53	Pyridine, 4-amino	25		9.1141	7.69×10^{-10}
p54	Pyridine, 2-benzyl	25		5.13	7.41×10^{-6}
p55	Pyridine, 3-bromo	25		2.84	1.45×10^{-3}
p56	Pyridine, 3-chloro	25		2.84	1.45×10^{-3}
p57	Pyridine, 2,5-diamino	20		6.48	3.31×10^{-7}
p58	Pyridine, 2,3-dimethyl (2,3-Lutidine)	25		6.57	2.69×10^{-7}
p59	Pyridine, 2,4-dimethyl (2,4-Lutidine)	25		6.99	1.02×10^{-7}
p60	Pyridine, 3,5-dimethyl (3,5-Lutidine)	25		6.15	7.08×10^{-7}
p61	Pyridine, 2-ethyl	25		5.89	1.28×10^{-6}
p62	Pyridine, 2-formyl	20		3.80	1.59×10^{-4}
p63	Pyridine, 2-hydroxy (2-Pyridol)	20	1	0.75	9.82×10^{-1}
		20	2	11.65	2.24×10^{-12}

DISSOCIATION CONSTANTS OF ORGANIC BASES IN AQUEOUS SOLUTION (*Continued*)

No.	Compound	Temp. °C	Step	pK_a	K_a
p64	Pyridine, 4-hydroxy	20	1	3.20	6.31×10^{-4}
		20	2	11.12	7.59×10^{-12}
p65	Pyridine, methoxy	25		6.47	3.30×10^{-7}
p66	Pyridine, 4-methylamino	20		9.65	2.24×10^{-10}
p67	Pyridine, 2,4,6-trimethyl	25		7.43	3.72×10^{-8}
p68	Pyrrolidine	25		11.27	5.37×10^{-12}
p69	Pyrrolidine, 1,2-dimethyl	26		10.20	6.31×10^{-11}
p70	Pyrrolidine, *n*-methyl	25		10.32	4.79×10^{-11}
q1	Quinazoline	20		3.43	3.72×10^{-4}
q2	Quinazoline, 5-hydroxy	20	1	3.62	2.40×10^{-4}
		20	2	7.41	3.89×10^{-8}
q3	Quinine	25	1	8.52	3.02×10^{-9}
		25	2	4.13	7.41×10^{-5}
q4	Quinoline	20		4.90	1.25×10^{-5}
q5	Quinoline, 3-amino	20		4.91	1.23×10^{-5}
q6	Quinoline, 3-bromo	25		2.69	2.04×10^{-3}
q7	Quinoline, 8-carboxy	25		1.82	1.51×10^{-2}
q8	Quinoline, 3-hydroxy (3-Quinolinol)	20	1	4.28	5.25×10^{-5}
		20	2	8.08	8.32×10^{-9}
q9	Quinoline, 8-hydroxy (8-Quinolinol)	20	1	5.017	1.21×10^{-6}
		25	2	9.812	1.54×10^{-10}
q10	Quinoline, 8-hydroxy-5-sulfo	25	1	4.112	7.73×10^{-5}
		25	2	8.757	1.75×10^{-9}
q11	Quinoline, 6-methoxy	20		5.03	9.33×10^{-6}
q12	Quinoline, 2-methyl (Quinaldine)	20		5.83	1.48×10^{-6}
q13	Quinoline, 4-methyl (Lepidine)	20		5.67	2.14×10^{-6}
q14	Quinoline, 5-methyl	20		5.20	6.31×10^{-6}
q15	Quinoxaline (Quinazine)	20		0.56	3.63×10^{-1}
s1	Serine (2-amino-3-hydroxypropanoic acid)	25	1	2.186	5.49×10^{-3}
		25	2	9.208	6.19×10^{-10}
s2	Strychnine	25		8.26	5.49×10^{-9}
t1	Taurine (2-Aminoethane sulfonic acid)	25	2	9.0614	8.69×10^{-10}
t2	Tetradecaneamine (Myristilamine)	25		10.62	2.40×10^{-11}
t3	Thiazole	20		2.44	3.63×10^{-3}
t4	Thiazole, 2-amino	20		5.36	4.36×10^{-5}
t5	Threonine	25	1	2.088	8.16×10^{-3}
		25	2	9.10	7.94×10^{-10}
t6	*o*-Toluidine	25		4.44	3.63×10^{-5}
t7	*m*-Toluidine	25		4.73	1.86×10^{-5}
t8	*p*-Toluidine	25		5.08	8.32×10^{-6}
t9	1,3,5-Triazine, 2,4,6-triamino	25		5.00	1.00×10^{-5}
t10	Tridecaneamine	25		10.63	2.35×10^{-11}
t11	Triethylamine	18		11.01	9.77×10^{-12}
t12	Trimethylamine	25		9.81	1.55×10^{-10}
t13	Tryptophan	25	1	2.43	3.72×10^{-3}
		25	2	9.44	3.63×10^{-10}
t14	Tyrosine	25	2	9.11	7.76×10^{-10}
		25	3	10.13	7.41×10^{-11}
t15	Tyrosineamide	25		7.33	4.68×10^{-8}
u1	Urea	21		0.10	7.94×10^{-1}
v1	Valine	25	1	2.286	5.17×10^{-3}
		25	2	9.719	1.91×10^{-10}

DISSOCIATION CONSTANTS OF INORGANIC BASES IN AQUEOUS SOLUTIONS

(Approximately 0.1–0.01 N)

Compound	T°C	Step	K_b	pK_b
Ammonium hydroxide	25		1.79×10^{-5}	4.75
Arsenous oxide	25		1.1×10^{-4}	3.96
Beryllium hydroxide	25	2	5×10^{-11}	10.30
Calcium hydroxide	25	1	3.74×10^{-3}	2.43
Calcium hydroxide	30	2	4.0×10^{-2}	1.40
Deuteroammonium hydroxide	25		1.1×10^{-5}	4.96
Hydrazine	20		1.7×10^{-6}	5.77
Hydroxylamine	20		1.07×10^{-8}	7.97
Lead Hydroxide	25		9.6×10^{-4}	3.02
Silver Hydroxide	25		1.1×10^{-4}	3.96
Zinc Hydroxide	25		9.6×10^{-4}	3.02

DISSOCIATION CONSTANTS OF ORGANIC ACIDS IN AQUEOUS SOLUTIONS

Compound	T°C	Step	K	pK
Acetic	25		1.76×10^{-5}	4.75
Acetoacetic	18		2.62×10^{-4}	3.58
Acrylic	25		5.6×10^{-5}	4.25
Adipamic	25		2.35×10^{-5}	4.63
Adipic	25	1	3.71×10^{-5}	4.43
Adipic	25	2	3.87×10^{-5}	4.41
d-Alanine	25		1.35×10^{-10}	9.87
Allantoin	25		1.10×10^{-9}	8.96
Alloxanic	25		2.3×10^{-7}	6.64
α-Aminoacetic (glycine)	25		1.67×10^{-10}	9.78
o-Aminobenzoic	25		1.07×10^{-7}	6.97
m-Aminobenzoic	25		1.67×10^{-5}	4.78
p-Aminobenzoic	25		1.2×10^{-5}	4.92
o-Aminobenzosulfonic	25		3.3×10^{-3}	2.48
m-Aminobenzosulfonic	25		1.85×10^{-4}	3.73
p-Aminobenzosulfonic	25		5.81×10^{-4}	3.24
Anisic	25		3.38×10^{-5}	4.47
o-β-Anisylpropionic	25		1.59×10^{-5}	4.80
m-β-Anisylpropionic	25		2.24×10^{-5}	4.65
p-β-Anisylpropionic	25		2.04×10^{-5}	4.69
Ascorbic	24	1	7.94×10^{-5}	4.10
Ascorbic	16	2	1.62×10^{-12}	11.79
DL-Aspartic	25	1	1.38×10^{-4}	3.86
DL-Aspartic	25	2	1.51×10^{-10}	9.82
Barbituric	25		9.8×10^{-5}	4.01
Benzoic	25		6.46×10^{-5}	4.19
Benzosulfonic	25		2×10^{-1}	0.70
Bromoacetic	25		2.05×10^{-3}	2.69
o-Bromobenzoic	25		1.45×10^{-3}	2.84
m-Bromobenzoic	25		1.37×10^{-4}	3.86
n-Butyric	20		1.54×10^{-5}	4.81
iso-Butyric	18		1.44×10^{-5}	4.84
Cacodylic	25		6.4×10^{-7}	6.19
n-Caproic	18		1.43×10^{-5}	4.83
iso-Caproic	18		1.46×10^{-5}	4.84
Chloroacetic	25		1.40×10^{-3}	2.85
o-Chlorobenzoic	25		1.20×10^{-3}	2.92
m-Chlorobenzoic	25		1.51×10^{-4}	3.82
p-Chlorobenzoic	25		1.04×10^{-4}	3.98
α-Chlorobutyric	R.T.		1.39×10^{-3}	2.86
β-Chlorobutyric	R.T.		8.9×10^{-5}	4.05
γ-Chlorobutyric	R.T.		3.0×10^{-5}	4.52
o-Chlorocinnamic	25		5.89×10^{-5}	4.23
m-Chlorocinnamic	25		5.13×10^{-5}	4.29
p-Chlorocinnamic	25		3.89×10^{-5}	4.41
o-Chlorophenoxyacetic	25		8.91×10^{-4}	3.05
m-Chlorophenoxyacetic	25		8.51×10^{-4}	3.07
p-Chlorophenoxyacetic	25		7.94×10^{-4}	3.10
o-Chlorophenylacetic	25		1.18×10^{-5}	4.07
m-Chlorophenylacetic	25		7.25×10^{-5}	4.14
p-Chlorophenylacetic	25		6.46×10^{-5}	4.19
β-(*o*-Chlorophenyl) propionic	25		2.63×10^{-4}	4.58
β-(*m*-Chlorophenyl) propionic	25		2.57×10^{-5}	4.59
β-(*p*-Chlorophenyl) propionic	25		2.46×10^{-5}	4.61
α-Chloropropionic	25		1.47×10^{-3}	2.83
β-Chloropropionic	25		1.04×10^{-4}	3.98
cis-Cinnamic	25		1.3×10^{-4}	3.89
trans-Cinnamic	25		3.65×10^{-5}	4.44
Citric	18	1	8.4×10^{-4}	3.08
Citric	18	2	1.8×10^{-5}	4.74
Citric	18	3	4.0×10^{-6}	5.40
o-Cresol	25		6.3×10^{-11}	10.20
m-Cresol	25		9.8×10^{-11}	10.01
p-Cresol	25		6.7×10^{-11}	10.17
Crotonic (*trans*-)	25		2.03×10^{-5}	4.69
Cyanoacetic	25		3.65×10^{-3}	2.45
γ-Cyanobutyric	25		3.80×10^{-3}	2.42
o-Cyanophenoxyacetic	25		1.05×10^{-3}	2.98
m-Cyanophenoxyacetic	25		9.33×10^{-4}	3.03
p-Cyanophenoxyacetic	25		1.18×10^{-3}	2.93
Cyanopropionic	25		3.6×10^{-3}	2.44
Cyclohexane-1:1-dicarboxylic	25	1	3.55×10^{-4}	3.45
Cyclohexane-1:1-dicarboxylic	25	2	7.76×10^{-7}	6.11
Cyclopropane-1:1-dicarboxylic	25	1	1.51×10^{-2}	1.82
Cyclopropane-1:1-dicarboxylic	25	2	3.72×10^{-8}	7.43
DL-Cysteine	30	1	7.25×10^{-9}	8.14
DL-Cysteine	30	2	4.6×10^{-11}	10.34
L-Cystine	25	1	1.4×10^{-8}	7.85
L-Cystine	25	2	1.4×10^{-10}	9.85
Deuteroacetic (in D_2O)	25		5.5×10^{-6}	5.25
Dichloroacetic	25		3.32×10^{-2}	1.48
Dichloroacetylacetic	?		7.8×10^{-3}	2.11
Dichlorophenol (2,3-)	25		3.6×10^{-8}	7.44
Dihydroxybenzoic (2,2-)	25		1.14×10^{-3}	2.94
Dihydroxybenzoic (2,5-)	25		1.08×10^{-3}	2.97
Dihydroxybenzoic (3,4-)	25		3.3×10^{-5}	4.48
Dihydroxybenzoic (3,5-)	25		9.1×10^{-5}	4.04
Dihydroxymalic	25		1.2×10^{-2}	1.92
Dihydroxytartaric	25		1.2×10^{-2}	1.92
Dimethylglycine	25		1.3×10^{-10}	9.89
Dimethylmalic	25	1	6.83×10^{-4}	3.17
Dimethylmalic	25	2	8.72×10^{-7}	6.06
Dimethylmalonic	25		7.08×10^{-4}	3.15
Dinicotinic	25		1.6×10^{-3}	2.80
Dinitrophenol (2,4-)	15		1.1×10^{-4}	3.96
Dinitrophenol (3,6-)	15		7.1×10^{-6}	5.15
Diphenylacetic	25		1.15×10^{-4}	3.94
Ethylbenzoic	25		4.47×10^{-5}	4.35
Ethylphenylacetic	25		4.27×10^{-5}	4.37
Fluorobenzoic	17		1.25×10^{-3}	2.90
Formic	20		1.77×10^{-4}	3.75
Fumaric (*trans*-)	18	1	9.30×10^{-4}	3.03
Fumaric (*trans*-)	18	2	3.62×10^{-5}	4.44
Furancarboxylic	25		7.1×10^{-4}	3.15
Furoic	25		6.76×10^{-4}	3.17
Gallic	25		3.9×10^{-5}	4.41
Glutaramic	25		3.98×10^{-5}	4.60
Glutaric	25	1	4.58×10^{-5}	4.34
Glutaric	25	2	3.89×10^{-6}	5.41
Glycerol	25		7×10^{-15}	14.15
Glycine	25		1.67×10^{-10}	9.87
Glycol	25		6×10^{-15}	14.22
Glycolic	25		1.48×10^{-4}	3.83
Heptanoic	25		1.28×10^{-5}	4.89
Hexahydrobenzoic	25		1.26×10^{-5}	4.90
Hexanoic	25		1.31×10^{-5}	4.88
Hippuric	25		1.57×10^{-4}	3.80
Histidine	25		6.7×10^{-10}	9.17
Hydroquinone	20		4.5×10^{-11}	10.35
o-Hydroxybenzoic	19	1	1.07×10^{-3}	2.97
o-Hydroxybenzoic	18	2	4×10^{-14}	13.40
m-Hydroxybenzoic	19	1	8.7×10^{-5}	4.06
m-Hydroxybenzoic	19	2	1.2×10^{-10}	9.92
p-Hydroxybenzoic	19	1	3.3×10^{-5}	4.48
p-Hydroxybenzoic	19	2	4.8×10^{-10}	9.32
β-Hydroxybutyric	25		2×10^{-5}	4.70
γ-Hydroxybutyric	25		1.9×10^{-5}	4.72
β-Hydroxypropionic	25		3.1×10^{-5}	4.51
γ-Hydroxyquinoline	20		3.1×10^{-10}	9.51
Iodoacetic	25		7.5×10^{-4}	3.12
o-Iodobenzoic	25		1.4×10^{-3}	2.85
m-Iodobenzoic	25		1.6×10^{-4}	3.80
Itaconic	25	1	1.40×10^{-4}	3.85
Itaconic	25	2	3.56×10^{-6}	5.45
Lactic	100		8.4×10^{-4}	3.08
Lutidinic	25		7.0×10^{-3}	2.15
Lysine	25		2.95×10^{-11}	10.53
Maleic	25	1	1.42×10^{-2}	1.83
Maleic	25	2	8.57×10^{-7}	6.07
Malic	25	1	3.9×10^{-4}	3.40
Malic	25	2	7.8×10^{-6}	5.11
Malonic	25	1	1.49×10^{-3}	2.83
Malonic	25	2	2.03×10^{-6}	5.69
DL-Mandelic	25		1.4×10^{-4}	3.85
Mesaconic	25	1	8.22×10^{-4}	3.09
Mesaconic	25	2	1.78×10^{-5}	4.75
Mesitylenic	25		4.8×10^{-5}	4.32
Methyl-*o*-aminobenzoic	25		4.6×10^{-6}	5.34
Methyl-*m*-aminobenzoic	25		8×10^{-6}	5.10
Methyl-*p*-aminobenzoic	25		9.2×10^{-6}	5.04
o-Methylcinnamic	25		3.16×10^{-5}	4.50
m-Methylcinnamic	25		3.63×10^{-5}	4.44
p-Methylcinnamic	25		2.76×10^{-5}	4.56
β-Methylglutaric	25		5.75×10^{-5}	4.24
n-Methylglycine	18		1.2×10^{-10}	9.92
Methylmalonic	25		1.17×10^{-4}	3.07
Methylsuccinic	25	1	7.4×10^{-5}	4.13
Methylsuccinic	25	2	2.3×10^{-6}	5.64
o-Monochlorophenol	25		3.2×10^{-9}	8.49
m-Monochlorophenol	25		1.4×10^{-9}	8.85
p-Monochlorophenol	25		6.6×10^{-10}	9.18
Naphthalenesulfonic	25		2.7×10^{-1}	0.57
α-Naphthoic	25		2×10^{-4}	3.70
β-Naphthoic	25		6.8×10^{-5}	4.17
α-Naphthol	25		4.6×10^{-10}	9.34
β-Naphthol	25		3.1×10^{-10}	9.51
Nitrobenzene	0		1.05×10^{-4}	3.98
o-Nitrobenzoic	18		6.95×10^{-3}	2.16
m-Nitrobenzoic	25		3.4×10^{-4}	3.47
p-Nitrobenzoic	25		3.93×10^{-4}	3.41
o-Nitrophenol	25		6.8×10^{-8}	7.17
m-Nitrophenol	25		5.3×10^{-9}	8.28
p-Nitrophenol	25		7×10^{-8}	7.15
o-Nitrophenylacetic	25		1.00×10^{-5}	4.00
m-Nitrophenylacetic	25		1.07×10^{-4}	3.97
p-Nitrophenylacetic	25		1.41×10^{-4}	3.85
o-β-Nitrophenylpropionic	25		3.16×10^{-5}	4.50
p-β-Nitrophenylpropionic	25		3.39×10^{-5}	4.47
Nonanic	25		1.09×10^{-5}	4.96
Octanoic	25		1.28×10^{-5}	4.89
Oxalic	25	1	5.90×10^{-2}	1.23
Oxalic	25	2	6.40×10^{-5}	4.19

DISSOCIATION CONSTANTS OF ORGANIC ACIDS IN AQUEOUS SOLUTIONS (Continued)

Compound	T°C	Step	K	pK
Phenol	20		1.28×10^{-10}	9.89
Phenylacetic	18		5.2×10^{-5}	4.28
o-Phenylbenzoic	25		3.47×10^{-4}	3.46
γ-Phenylbutyric	25		1.74×10^{-5}	4.76
α-Phenylpropionic	25		2.27×10^{-5}	4.64
β-Phenylpropionic	25		4.25×10^{-5}	4.37
o-Phthalic	25	1	1.3×10^{-3}	2.89
o-Phthalic	25	2	3.9×10^{-6}	5.51
m-Phthalic	25	1	2.9×10^{-4}	3.54
m-Phthalic	18	2	2.5×10^{-5}	4.60
p-Phthalic	25	1	3.1×10^{-4}	3.51
p-Phthalic	16	2	1.5×10^{-5}	4.82
Picric	25		4.2×10^{-1}	0.38
Pimelic	25		3.09×10^{-5}	4.71
Propionic	25		1.34×10^{-5}	4.87
iso-Propylbenzoic	25		3.98×10^{-5}	4.40
2-Pyridinecarboxylic	25		3×10^{-6}	5.52
3-Pyridinecarboxylic	25		1.4×10^{-5}	4.85
4-Pyridinecarboxylic	25		1.1×10^{-5}	4.96
Pyrocatechol	20		1.4×10^{-10}	9.85
Quinolinic	25		3×10^{-3}	2.52
Resorcinol	25		1.55×10^{-10}	9.81
Saccharin	18		2.1×10^{-12}	11.68
Suberic	25		2.99×10^{-5}	4.52
Succinic	25	1	6.89×10^{-5}	4.16
Succinic	25	2	2.47×10^{-6}	5.61
Sulfanilic	25		5.9×10^{-4}	3.23
α-Tartaric	25	1	1.04×10^{-3}	2.98
α-Tartaric	25	2	4.55×10^{-5}	4.34
meso-Tartaric	25	1	6×10^{-4}	3.22
meso-Tartaric	25	2	1.53×10^{-5}	4.82
Theobromine	18		1.3×10^{-8}	7.89
Terephthalic	25		3.1×10^{-4}	3.51
Thioacetic	25		4.7×10^{-4}	3.33
Thiophenecarboxylic	25		3.3×10^{-4}	3.48
o-Toluic	25		1.22×10^{-4}	3.91
m-Toluic	25		5.32×10^{-5}	4.27
p-Toluic	25		4.33×10^{-5}	4.36
Trichloroacetic	25		2×10^{-1}	0.70
Trichlorophenol	25		1×10^{-6}	6.00
Trihydroxybenzoic (2,4,6-)	25		2.1×10^{-2}	1.68
Trimethylacetic	18		9.4×10^{-6}	5.03
Trinitrophenol (2,4,6-)	25		4.2×10^{-1}	0.38
Tryptophan	25		4.2×10^{-10}	9.38
Tyrosine	17		3.98×10^{-9}	8.40
Uric	12		1.3×10^{-4}	3.89
n-Valeric	18		1.51×10^{-5}	4.82
iso-Valeric	25		1.7×10^{-5}	4.77
Veronal	25		3.7×10^{-8}	7.43
Vinylacetic	25		4.57×10^{-5}	4.34
Xanthine	40		1.24×10^{-10}	9.91

DISSOCIATION CONSTANTS OF INORGANIC ACIDS IN AQUEOUS SOLUTIONS

(Approximately 0.1–0.01 N)

Compound	T°C	Step	K	pK
Arsenic	18	1	5.62×10^{-3}	2.25
Arsenic	18	2	1.70×10^{-7}	6.77
Arsenic	18	3	3.95×10^{-12}	11.60
Arsenious	25		6×10^{-10}	9.23
o-Boric	20	1	7.3×10^{-10}	9.14
o-Boric	20	2	1.8×10^{-13}	12.74
o-Boric	20	3	1.6×10^{-14}	13.80
Carbonic	25	1	4.30×10^{-7}	6.37
Carbonic	25	2	5.61×10^{-11}	10.25
Chromic	25	1	1.8×10^{-1}	0.74
Chromic	25	2	3.20×10^{-7}	6.49
Germanic	25	1	2.6×10^{-9}	8.59
Germanic	25	2	1.9×10^{-13}	12.72
Hydrocyanic	25		4.93×10^{-10}	9.31
Hydrofluoric	25		3.53×10^{-4}	3.45
Hydrogen sulfide	18	1	9.1×10^{-8}	7.04
Hydrogen sulfide	18	2	1.1×10^{-12}	11.96
Hydrogen peroxide	25		2.4×10^{-12}	11.62
Hypobromous	25		2.06×10^{-9}	8.69
Hypochlorous	18		2.95×10^{-8}	7.53
Hypoiodous	25		2.3×10^{-11}	10.64
Iodic	25		1.69×10^{-1}	0.77
Nitrous	12.5		4.6×10^{-4}	3.37
Periodic	25		2.3×10^{-2}	1.64
o-Phosphoric	25	1	7.52×10^{-3}	2.12
o-Phosphoric	25	2	6.23×10^{-8}	7.21
o-Phosphoric	18	3	2.2×10^{-13}	12.67
Phosphorous	18	1	1.0×10^{-2}	2.00
Phosphorous	18	2	2.6×10^{-7}	6.59
Pyrophosphoric	18	1	1.4×10^{-1}	0.85
Pyrophosphoric	18	2	3.2×10^{-2}	1.49
Pyrophosphoric	18	3	1.7×10^{-6}	5.77
Pyrophosphoric	18	4	6×10^{-9}	8.22
Selenic	25	2	1.2×10^{-2}	1.92
Selenious	25	1	3.5×10^{-3}	2.46
Selenious	25	2	5×10^{-8}	7.31
m-Silicic	R.T.	1	2×10^{-10}	9.70
m-Silicic	R.T.	2	1×10^{-12}	12.00
o-Silicic	30	1	2.2×10^{-10}	9.66
o-Silicic	30	2	2×10^{-12}	11.70
o-Silicic	30	3	1×10^{-12}	12.00
o-Silicic	30	4	1×10^{-12}	12.00
Sulfuric	25	2	1.20×10^{-2}	1.92
Sulfurous	18	1	1.54×10^{-2}	1.81
Sulfurous	18	2	1.02×10^{-7}	6.91
Telluric	18	1	2.09×10^{-8}	7.68
Telluric	18	2	6.46×10^{-12}	11.29
Tellurous	25	1	3×10^{-3}	2.48
Tellurous	25	2	2×10^{-8}	7.70
Tetraboric	25	1	$\sim 10^{-4}$	~4.00
Tetraboric	25	2	$\sim 10^{-9}$	~9.00

DISSOCIATION CONSTANTS (K_b) OF AQUEOUS AMMONIA FROM 0 TO 50°C.

Temperature °C.	pK_b	K_b
0	4.862	1.374×10^{-5}
5	4.830	1.479×10^{-5}
10	4.804	1.570×10^{-5}
15	4.782	1.652×10^{-5}
20	4.767	1.710×10^{-5}
25	4.751	1.774×10^{-5}
30	4.740	1.820×10^{-5}
35	4.733	1.849×10^{-5}
40	4.730	1.862×10^{-5}
45	4.726	1.879×10^{-5}
50	4.723	1.892×10^{-5}

Values of K_b accurate to ±0.005; determined by e.m.f. method by: R. G. Bates and G. D. Pinching, J. Am. Chem. Soc., 1950, **72**, 1393.

IONIZATION CONSTANT FOR WATER (K_w)

$-\log_{10} K_w$	Temperature °C.	$-\log_{10} K_w$	Temperature °C.
14.9435	0	13.8330	30
14.7338	5	13.6801	35
14.5346	10	13.5348	40
14.3463	15	13.3960	45
14.1669	20	13.2617	50
14.0000	24	13.1369	55
13.9965	25	13.0171	60

IONIZATION CONSTANTS OF ACIDS IN WATER AT VARIOUS TEMPERATURES

Acids		0°	5°	10°	15°	20°	25°	30°	35°	40°	45°	50°
Formic	$K_A \cdot 10^4$	1.638	1.691	1.728	1.749	1.765	1.772	1.768	1.747	1.716	1.685	1.650
Acetic	$K_A \cdot 10^5$	1.657	1.700	1.729	1.745	1.753	1.754	1.750	1.728	1.703	1.670	1.633
Propionic	$K_A \cdot 10^5$	1.274	1.305	1.326	1.336	1.338	1.336	1.326	1.310	1.280	1.257	1.229
n-Butyric	$K_A \cdot 10^5$	1.563	1.574	1.576	1.569	1.542	1.515	1.484	1.439	1.395	1.347	1.302
Chloracetic	$K_A \cdot 10^3$	1.528	....	1.488	....	....	1.379	....	....	1.230	....	....
Lactic	$K_A \cdot 10^4$	1.287	....	....	....	....	1.374	....	....	....	....	1.270
Glycollic	$K_A \cdot 10^4$	1.334	....	....	....	....	1.475	....	....	....	....	1.415
Oxalic	$K_{2A} \cdot 10^5$	5.91	5.82	5.70	5.55	5.40	5.18	4.92	4.67	4.41	4.09	3.83
Malonic	$K_{2A} \cdot 10^6$	2.140	2.165	2.152	2.124	2.076	2.014	1.948	1.863	1.768	1.670	1.575
Phosphoric	$K_A \cdot 10^3$	8.968	....	....	....	....	7.516	....	....	....	....	5.495
Phosphoric	$K_{2A} \cdot 10^8$	4.85	5.24	5.57	5.89	6.12	6.34	6.46	6.53	6.58	6.59	6.55
Boric	$K_A \cdot 10^{10}$	....	3.63	4.17	4.72	5.26	5.79	6.34	6.86	7.38	....	8.32
Carbonic	$K_{1A} \cdot 10^7$	2.64	3.04	3.44	3.81	4.16	4.45	4.71	4.90	5.04	5.13	5.19
Phenol-sulfonic	$K_{2A} \cdot 10^{10}$	4.45	5.20	6.03	6.92	7.85	8.85	9.89	10.94	12.00	13.09	14.16
Glycine	$K_{1A} \cdot 10^7$	....	3.82	3.99	4.17	4.32	4.46	4.57	4.66	4.73	4.77	4.79
Citric	$K_{1A} \cdot 10^4$	6.03	6.31	6.69	6.92	7.21	7.45	7.66	7.78	7.96	7.99	8.04
	$K_{2A} \cdot 10^5$	1.45	1.54	1.60	1.65	1.70	1.73	1.76	1.77	1.78	1.76	1.75
	$K_{3A} \cdot 10^7$	4.05	4.11	4.14	4.13	4.09	4.02	3.99	3.78	3.69	3.45	3.28

Reproducibility between various workers is about ± $(0.01 - 0.02) \cdot 10^5$.
All values are on the m-scale.

COMPOSITION OF SOME INORGANIC ACIDS AND BASES

The following acids and bases are frequently supplied as concentrated aqueous solutions. This table presents certain data concerning these solutions.

	Formula weight	Molarity	Specific gravity	Weight percent	
Acetic acid	60.05	17.5	1.05	99–100	CH_3COOH
Ammonium hydroxide	35.05	7.4	0.90	28–30	NH_4OH
Hydriodic acid	127.91	5.5	1.5	47–47.5	HI
Hydrobromic acid	80.93	9.0	1.5	47–49	HBr
Hydrochloric acid	36.46	12.0	1.18	36.5–38	HCl
Hydrofluoric acid	20.01	28.9	1.17	48–51	HF
Phosphoric acid	98.00	14.7	1.7	85	H_3PO_4
Sulfuric acid	98.08	18.0	1.84	95–98	H_2SO_4

IONIZATION CONSTANTS FOR DEUTERIUM OXIDE FROM 10 TO 50°C.

From NBS Technical Note 400

The subscript *m* indicates values on the molal scale, whereas the subscript *c* indicates values on the molar scale.

t, °C	pK_m	pK_c
10	15.526	15.439
20	15.136	15.049
25	14.955	14.869
30	14.784	14.699
40	14.468	14.385
50	14.182	14.103

EQUIVALENT CONDUCTANCES OF SOME ELECTROLYTES IN AQUEOUS SOLUTIONS AT 25°C

Compound	Concentration in gram equivalents per 1000 cm³							
	Infinite dilution	0.0005	0.001	0.005	0.01	0.02	0.05	0.1
$AgNO_3$	133.36	131.36	130.51	127.20	124.76	121.41	115.24	109.14
$BaCl_2$	139.98	135.96	134.34	128.02	123.94	119.09	111.48	105.19
$CaCl_2$	135.84	131.93	130.36	124.25	120.36	115.65	108.47	102.4
$Ca(OH)_2$	257.9			232.9	225.9	213.9		
$CuSO_4$	133.6	121.6	115.26	94.07	83.12	72.20	59.05	50.58
HCl	426.16	422.74	421.36	415.80	412.00	407.24	399.09	391.32
KBr	151.9			146.09	143.43	140.48	135.68	131.39
KCl	149.86	147.81	146.95	143.35	141.27	138.34	133.37	128.96
$KClO_4$	140.04	138.76	137.87	134.16	131.46	127.92	121.62	115.20
$K_3Fe(CN)_6$	174.5	166.4	163.1	150.7				
$K_4Fe(CN)_6$	184.5		167.24	146.09	134.83	122.82	107.70	97.87
$KHCO_3$	118.0	116.10	115.34	112.24	110.08	107.22		
KI	150.38			144.37	142.18	139.45	134.97	131.11
KIO_4	127.92	125.80	124.94	121.24	118.51	114.14	106.72	98.12
KNO_3	144.96	142.77	141.84	138.48	132.82	132.41	126.31	120.40
$KReO_4$	128.20	126.03	125.12	121.31	118.49	114.49	106.40	97.40

Compound	Concentration in gram equivalents per 1000 cm³							
	Infinite dilution	0.0005	0.001	0.005	0.01	0.02	0.05	0.1
$LaCl_3$	145.8	139.6	137.0	127.5	121.8	115.3	106.2	99.1
LiCl	115.03	113.15	112.40	109.40	107.40	104.65	100.11	95.86
$LiClO_4$	105.98	104.18	103.44	100.57	98.61	96.18	92.20	88.56
$MgCl_2$	129.40	125.61	124.11	118.31	114.55	110.04	103.08	97.10
NH_4Cl	149.7		146.8	143.5	141.28	138.33	133.29	128.75
NaCl	126.45	124.50	123.74	120.65	118.51	115.51	111.06	106.74
$NaClO_4$	117.48	115.64	114.87	111.75	109.59	106.96	102.40	98.43
NaI	126.94	125.36	124.25	121.25	119.24	116.70	112.79	108.78
$NaOOCCH_3$	91.0	89.2	88.5	85.72	83.76	81.24	76.92	72.80
$NaOOCC_2H_5$	85.9		83.5	80.9	79.1	76.6		
$NaOOCC_3H_7$	82.70	81.04	80.31	77.58	75.76	73.39	69.32	65.27
NaOH	247.8	245.6	244.7	240.8	238.0			
Na_2SO_4	129.9	125.74	124.15	117.15	112.44	106.78	97.75	89.98
$SrCl_2$	135.80	131.90	130.33	124.24	120.24	115.54	108.25	102.19
$ZnSO_4$	132.8	121.4	114.53	95.49	84.91	74.24	61.20	52.64

THE EQUIVALENT CONDUCTANCE OF THE SEPARATE IONS

(From Smithsonian Physical Tables)

Ion.	0°	18°	25°	50°	75°	100°	128°	156°
K	40.4	64.6	74.5	115	159	206	263	317
Na	26.	43.5	50.9	82	116	155	203	249
NH_4	40.2	64.5	74.5	115	159	207	264	319
Ag	32.9	54.3	63.5	101	143	188	245	299
½Ba	33.	55.	65.	104	149	200	262	322
½Ca	30.	51.	60.	98	142	191	252	312
⅓La	35.	61.	72.	119	173	235	312	388
Cl	41.1	65.5	75.5	116	160	207	264	318
NO_3	40.4	61.7	70.6	104	140	178	222	263
$C_2H_3O_2$	20.3	34.6	40.8	67	96	130	171	211

Ion.	0°	18°	25°	50°	75°	100°	128°	156°
½SO_4	41.	68.	79.	125	177	234	303	370
½C_2O_4	39.	63.	73.	115	163	213	275	336
⅓$C_6H_5O_7$	36.	60.	70.	113	161	214		
¼$Fe(CN)_6$	58.	95.	111.	173	244	321		
H	240.	314.	350.	465	565	644	722	777
OH	105.	172.	192.	284	360	439	525	592

ACTIVITY COEFFICIENTS OF ACIDS, BASES AND SALTS

The following coefficients are valid only at 25°C. The concentrations are expressed as molalities.

Name	0.1	0.2	0.3	0.4	0.5	0.6	0.7	0.8	0.9	1.0
$AgNO_3$	0.734	0.657	0.606	0.567	0.536	0.509	0.485	0.464	0.446	0.429
$AlCl_3$	(0.337)	0.305	0.302	0.313	0.331	0.356	0.388	0.429	0.479	0.539
$Al_2(SO_4)_3$	(0.0350)	0.0225	0.0176	0.0153	0.0143	0.0140	0.0142	0.0149	0.0159	0.0175
$CdSO_4$	(0.150)	0.102	0.082	0.069	0.061	0.055	0.050	0.046	0.043	0.041
$CrCl_3$	(0.331)	0.298	0.294	0.300	0.314	0.335	0.362	0.397	0.436	0.481
$Cr(NO_3)_3$	(0.319)	0.285	0.279	0.281	0.291	0.304	0.322	0.344	0.371	0.401
$Cr_2(SO_4)_3$	(0.0458)	0.0300	0.0238	0.0207	0.0190	0.0182	0.0181	0.0185	0.0194	0.0208
CsBr	0.754	0.694	0.654	0.626	0.603	0.586	0.571	0.558	0.547	0.538
CsCl	0.756	0.694	0.656	0.628	0.606	0.589	0.575	0.563	0.553	0.544
CsI	0.754	0.692	0.651	0.621	0.599	0.581	0.567	0.554	0.543	0.533
$CsNO_3$	0.733	0.655	0.602	0.561	0.528	0.501	0.478	0.458	0.439	0.422
CsOAc	0.799	0.771	0.761	0.759	0.762	0.768	0.776	0.783	0.792	0.802
$CuSO_4$	(0.150)	0.104	0.083	0.071	0.062	0.056	0.052	0.048	0.045	0.043
HBr	0.805	0.782	0.777	0.781	0.789	0.801	0.815	0.832	0.850	0.871
HCl	0.796	0.767	0.756	0.755	0.757	0.763	0.772	0.783	0.795	0.809
$HClO_4$	0.803	0.778	0.768	0.766	0.769	0.776	0.785	0.795	0.808	0.823
HI	0.818	0.807	0.811	0.823	0.839	0.860	0.883	0.908	0.935	0.963
HNO_3	0.791	0.754	0.735	0.725	0.720	0.717	0.717	0.718	0.721	0.724
KBr	0.772	0.722	0.693	0.673	0.657	0.646	0.636	0.629	0.622	0.617
KCl	0.770	0.718	0.688	0.666	0.649	0.637	0.626	0.618	0.610	0.604
KCNS	0.769	0.716	0.685	0.663	0.646	0.633	0.623	0.614	0.606	0.599
KF	0.775	0.727	0.700	0.682	0.670	0.661	0.654	0.650	0.646	0.645
KI	0.778	0.733	0.707	0.689	0.676	0.667	0.660	0.654	0.649	0.645
KNO_3	0.739	0.663	0.614	0.576	0.545	0.519	0.496	0.476	0.459	0.443
KOAc	0.796	0.766	0.754	0.750	0.751	0.754	0.759	0.766	0.774	0.783

Name	0.1	0.2	0.3	0.4	0.5	0.6	0.7	0.8	0.9	1.0
KOH	0.798	0.760	0.742	0.734	0.732	0.733	0.736	0.742	0.749	0.756
LiBr	0.796	0.766	0.756	0.752	0.753	0.758	0.767	0.777	0.789	0.803
LiCl	0.790	0.757	0.744	0.740	0.739	0.743	0.748	0.755	0.764	0.774
$LiClO_4$	0.812	0.794	0.792	0.798	0.808	0.820	0.834	0.852	0.869	0.887
LiI	0.815	0.802	0.804	0.813	0.824	0.838	0.852	0.870	0.888	0.910
$LiNO_3$	0.788	0.752	0.736	0.728	0.726	0.727	0.729	0.733	0.737	0.743
LiOAc	0.784	0.742	0.721	0.709	0.700	0.691	0.689	0.688	0.688	0.689
$MgSO_4$	(0.150)	0.108	0.088	0.076	0.068	0.062	0.057	0.054	0.051	0.049
$MnSO_4$	(0.150)	0.106	0.085	0.073	0.064	0.058	0.053	0.049	0.046	0.044
NaBr	0.782	0.741	0.719	0.704	0.697	0.692	0.689	0.687	0.687	0.687
NaCl	0.778	0.735	0.710	0.693	0.681	0.673	0.667	0.662	0.659	0.657
$NaClO_4$	0.775	0.729	0.701	0.683	0.668	0.656	0.648	0.641	0.635	0.629
NaCNS	0.787	0.750	0.731	0.720	0.715	0.712	0.710	0.710	0.711	0.712
NaF	0.765	0.710	0.676	0.651	0.632	0.616	0.603	0.592	0.582	0.573
NaH_2PO_4	0.744	0.675	0.629	0.593	0.563	0.539	0.517	0.499	0.483	0,468
NaI	0.787	0.751	0.735	0.727	0.723	0.723	0.724	0.727	0.731	0.736
$NaNO_3$	0.762	0.703	0.666	0.638	0.617	0.599	0.583	0.570	0.558	0.548
NaOAc	0.791	0.757	0.744	0.737	0.735	0.736	0.740	0.745	0.752	0.757
NaOH	0.766	0.727	0.708	0.697	0.690	0.685	0.681	0.679	0.678	0.678
$NiSO_4$	(0.150)	0.105	0.084	0.071	0.063	0.056	0.052	0.047	0.044	0.042
RbBr	0.763	0.706	0.673	0.650	0.632	0.617	0.605	0.595	0.586	0.578
RbCl	0.764	0.709	0.675	0.652	0.634	0.620	0.608	0.599	0.590	0.583
RbI	0.762	0.705	0.671	0.647	0.629	0.614	0.602	0.591	0.583	0.575
$RbNO_3$	0.734	0.658	0.606	0.565	0.534	0.508	0.485	0.465	0.446	0.430
RbOAc	0.796	0.767	0.756	0.753	0.755	0.759	0.766	0.773	0.782	0.792
$TlNO_3$	0.702	0.606	0.545	0.500	....	....	....	....	....	....
$ZnSO_4$	(0.150)	0.104	0.083	0.071	0.063	0.057	0.052	0.048	0.046	0.043

SPECIFIC HEAT OF WATER

Heat Capacity of Air-free Water 0°–100°C at 1 Atmosphere Pressure

The heat capacity of air-free water is given in international steam table calories per gram and in absolute joules per gram. (1 absolute joule—0.238846 I.T. Cal.).

The enthalpy or heat content is given for air-free water in I.T. Cal. per gram and in absolute joules per gram.

From Osborne, Stimson and Ginnings; B. of S. Jour. Res. **23.** 238, 1939.

Temp. °C.	Thermal Capacity		Enthalpy		Temp. °C	Thermal Capacity		Enthalpy	
	Cal./g/°C	Joules/g/°C	Cal./g	Joules/g		Cal/g/°C	Joules/g/°C	Cal/g	Joules/g
0	1.00738	4.2177	0.0245	0.1026	50	.99854	4.1807	50.0079	209.3729
1	1.00652	4.2141	1.0314	4.3184	51	.99862	4.1810	51.0065	213.5538
2	1.00571	4.2107	2.0376	8.5308	52	.99871	4.1814	52.0051	217.7350
3	1.00499	4.2077	3.0429	12.7400	53	.99878	4.1817	53.0039	221.9166
4	1.00430	4.2048	4.0475	16.9462	54	.99885	4.1820	54.0027	226.0984
5	1.00368	4.2022	5.0515	21.1498	55	.99895	4.1824	55.0016	230.2806
6	1.00313	4.1999	6.0549	25.3508	56	.99905	4.1828	56.0006	234.4632
7	1.00260	4.1977	7.0578	29.5496	57	.99914	4.1832	56.9997	238.6462
8	1.00213	4.1957	8.0602	33.7463	58	.99924	4.1836	57.9989	242.8296
9	1.00170	4.1939	9.0621	37.9410	59	.99933	4.1840	58.9982	247.0134
10	1.00129	4.1922	10.0636	42.1341	60	.99943	4.1844	59.9975	251.1976
11	1.00093	4.1907	11.0647	46.3255	61	.99955	4.1849	60.9970	255.3822
12	1.00060	4.1893	12.0654	50.5155	62	.99964	4.1853	61.9966	259.5673
13	1.00029	4.1880	13.0659	54.7041	63	.99976	4.1858	62.9963	263.7529
14	1.00002	4.1869	14.0660	58.8916	64	.99988	4.1863	63.9962	267.9390
15	.99976	4.1858	15.0659	63 0779	65	1.00000	4.1868	64.9961	272.1256
16	.99955	4.1849	16.0655	67.2632	66	1.00014	4.1874	65.9962	276.3127
17	.99933	4.1840	17.0650	71.4476	67	1.00026	4.1879	66.9964	280.5003
18	.99914	4.1832	18.0642	75.6312	68	1.00041	4.1885	67.9967	284.6885
19	.99897	4.1825	19.0633	79.8141	69	1.00053	4.1890	68.9972	288.8772
20	.99883	4.1819	20.0622	83.9963	70	1.00067	4.1896	69.9977	293.0665
21	.99869	4.1813	21.0609	88.1778	71	1.00081	4.1902	70.9985	297.2564
22	.99857	4.1808	22.0596	92.3589	72	1.00096	4.1908	71.9994	301.4469
23	.99847	4.1804	23.0581	96.5395	73	1.00112	4.1915	73.0004	305.6381
24	.99838	4.1800	24.0565	100.7196	74	1.00127	4.1921	74.0016	309.8299
25	.99828	4.1796	25.0548	104.8994	75	1.00143	4.1928	75.0030	314.0224
26	.99821	4.1793	26.0530	109.0788	76	1.00160	4.1935	76.0045	318.2155
27	.99814	4.1790	27.0512	113.2580	77	1.00177	4.1942	77.0062	322.4094
28	.99809	4.1788	28.0493	117.4369	78	1.00194	4.1949	78.0080	326.6039
29	.99804	4.1786	29.0474	121.6157	79	1.00213	4.1957	79.0101	330.7992
30	.99802	4 1785	30.0455	125 7943	80	1.00229	4.1964	80.0123	334.9952
31	.99799	4.1784	31.0435	129.9727	81	1.00248	4.1972	81.0147	339.1920
32	.99797	4.1783	32.0414	134.1510	82	1.00268	4.1980	82.0172	343.3897
33	.99797	4.1783	33.0394	138.3293	83	1.00287	4.1988	83.0200	347.5881
34	.99795	4.1782	34.0374	142.5076	84	1.00308	4.1997	84.0230	351.7873
35	.99795	4.1782	35.0353	146.6858	85	1.00327	4.2005	85.0262	355.9874
36	.99797	4.1783	36.0333	150.8641	86	1.00349	4.2014	86.0295	360.1883
37	.99797	4.1783	37.0312	155.0423	87	1.00370	4.2023	87.0331	364.3902
38	.99799	4.1784	38.0292	159.2207	88	1.00392	4.2032	88.0369	368.5929
39	.99802	4.1785	39.0272	163.3991	89	1.00416	4.2042	89.0410	372.7966
40	.99804	4.1786	40.0253	167.5777	90	1.00437	4.2051	90.0452	377.0012
41	.99807	4.1787	41.0233	171.7563	91	1.00461	4.2061	91.0497	381.2068
42	.99811	4.1789	42.0214	175.9351	92	1.00485	4.2071	92.0545	385.4135
43	.99816	4.1791	43.0195	180.1141	93	1.00509	4.2081	93.0594	389.6211
44	.99819	4.1792	44.0177	184.2933	94	1.00535	4.2092	94.0647	393.8297
45	.99826	4.1795	45.0159	188.4726	95	1.00561	4.2103	95.0701	398.0395
46	.99830	4.1797	46.0142	192.6522	96	1.00588	4.2114	96.0759	402.2503
47	.99835	4.1799	47.0125	196.8320	97	1.00614	4.2125	97.0819	406.4622
48	.99842	4.1802	48.0109	201.0120	98	1.00640	4.2136	98.0882	410.6753
49	.99847	4.1804	49.0094	205.1923	99	1.00669	4.2148	99.0947	414.8895
					100	1.00697	4.2160	100.1015	419.1049

SPECIFIC HEAT OF WATER (Continued)

Enthalpy of Air-saturated Water
1 Atmosphere Pressure 0–100°C

Temp. °C	Enthalpy Cal/g	Enthalpy Joules/g	Temp. °C	Enthalpy Cal/g	Enthalpy Joules/g
0	0	0	75	74.9907	313.9712
5	5.0276	21.0496	80	80.0019	334.9519
10	10.0402	42.0363	85	85.0180	355.9532
15	15.0431	62.9826	90	90.0395	376.9773
20	20.0400	83.9034	95	95.0671	398.0270
25	25.0332	104.8089	100	100.1016	419.1053
30	30.0244	125.7063			
35	35.0149	146.6003			
40	40.0055	167.4949			
45	44.9968	188.3928			
50	49.9896	209.2964			
55	54.9842	230.2077			
60	59.9811	251.1289			
65	64.9808	272.0619			
70	69.9839	293.0087			

Specific Heat of Super-heated Steam

Specific heat of steam under constant pressure given in atmospheres and at temperatures above saturation in Cal./g/°C

Temp. °C	Pressure in atmospheres 1	2	4	6	8	10	12
110	0.481						
120	0.477	0.498					
130	0.475	0.494					
140	0.473	0.489					
150	0.472	0.486	0.519				
160	0.471	0.483	0.512	0.549			
170	0.470	0.481	0.507	0.538			
180	0.469	0.479	0.502	0.528	0.561	0.602	
190	0.469	0.478	0.498	0.522	0.549	0.583	0.625
200	0.469	0.478	0.495	0.515	0.539	0.567	0.601
210	0.470	0.477	0.493	0.510	0.531	0.555	0.584
220	0.470	0.477	0.491	0.506	0.524	0.545	0.569
230	0.471	0.477	0.489	0.504	0.519	0.537	0.557
240	0.472	0.477	0.488	0.501	0.515	0.530	0.548
250	0.473	0.477	0.488	0.499	0.512	0.525	0.540
260	0.474	0.478	0.487	0.498	0.509	0.521	0.534
270	0.474	0.478	0.487	0.497	0.507	0.518	0.529
280	0.475	0.479	0.487	0.496	0.505	0.515	0.525
290	0.476	0.480	0.487	0.495	0.504	0.513	0.523
300	0.477	0.481	0.488	0.495	0.503	0.511	0.519
310	0.478	0.482	0.488	0.495	0.502	0.510	0.518
320	0.480	0.483	0.489	0.496	0.502	0.509	0.516
330	0.482	0.484	0.490	0.496	0.502	0.508	0.515
340	0.483	0.485	0.491	0.496	0.502	0.507	0.513
350	0.484	0.486	0.492	0.497	0.502	0.507	0.512
360	0.485	0.487	0.492	0.497	0.502	0.507	0.511
370	0.486	0.488	0.493	0.498	0.503	0.507	0.511
380	0.488	0.490	0.494	0.498	0.503	0.507	0.511
390	0.489	0.491	0.495	0.499	0.503		
400	0.490	0.492	0.496	0.500	0.504		
410	0.492	0.494	0.497	0.501	0.505		
420	0.494	0.496	0.498	0.502	0.506		
430	0.495	0.497	0.500	0.504	0.507		
440	0.497	0.499	0.501	0.505	0.508		
450	0.498	0.500	0.503	0.506	0.509		
460	0.500	0.501	0.505	0.507	0.510		
470	0.502	0.503	0.506	0.508	0.512		
480	0.504	0.505	0.507	0.509	0.513		
490	0.505	0.506	0.509	0.511	0.514		
500	0.506	0.508	0.510	0.512	0.515		

Specific Heat of Water Above 100°C

Mean specific heat of water in 15°C calories between 0°C and the temperature stated.

Heat content (Enthalpy) in joules per gram between 0°C and the temperature stated.

From data by Osborne, Stimson and Fiock, B of S Jour. Res. 5, 411, 1930.

Temp. °C	Specific heat mean 0–t°C	Heat content 0–t joules/g	Temp. °C	Specific heat mean 0–t°C	Heat content 0–t joules/g
100	1.0008	418.75	190	1.0153	807.15
110	1.0015	460.97	200	1.0181	852.02
120	1.0025	503.36	210	1.0212	897.35
130	1.0037	545.93	220	1.0247	943.24
140	1.0050	588.71	230	1.0285	989.75
150	1.0067	631.75	240	1.0326	1036.97
160	1.0083	675.06	250	1.0376	1084.97
170	1.0103	718.66	260	1.0423	1133.87
180	1.0127	762.72	270	1.0483	1184.32

Specific Heat of Ice—Cal./g/°C

Temp. °C	Specific heat	Observer	Temp. °C.	Specific heat	Observer
−252 to −188	.146	Dieterici, 1903	−31.8	.4454	Dickinson-Osborne, 1915
−250	.0361		−23.7	.4599	Dickinson-Osborne, 1915
−200	.162	Mean	−24.5	.4605	Dickinson-Osborne, 1915
−188 to −78	.285	Dieterici, 1903	−20.8	.4668	Dickinson-Osborne, 1915
−180	.199	Nernst, 1910	−14.8	.4782	Dickinson-Osborne, 1915
−160	.230	Nernst, 1910	−14.6	.4779	Dickinson-Osborne, 1915
−150	.246		−11.0	.4861	Dickinson-Osborne, 1915
−140	.262	Nernst, 1910	− 8.1	.4896	Dickinson-Osborne, 1915
−100	.329	Mean	− 4.3	.4989	Dickinson-Osborne, 1915
− 78 to −18	.463	Dieterici, 1903	− 4.5	.4984	Dickinson-Osborne, 1915
− 60	.392		− 4.9	.4932	Dickinson-Osborne, 1915
− 38.3	.4346	Dickinson-Osborne, 1915	− 2.6	.5003	Dickinson-Osborne, 1915
− 34.3	.4411	Dickinson-Osborne, 1915	− 2.2	.5018	Dickinson-Osborne, 1915
− 30.6	.4488	Dickinson-Osborne, 1915			
		Water Below 0°C			
− 6	1.0119	Martinetti, 1890	− 3	1.0102	Martinetti, 1890
− 5	1.0155	Barnes, 1902	− 2	1.0097	Martinetti, 1890
− 5	1.0113	Martinetti, 1890	− 1	1.0092	Martinetti, 1890
− 4	1.0105	Martinetti, 1890			

HEAT CAPACITY OF MERCURY

The specific heat of solid mercury is given in relation to water at 15°C The values are from Carpenter and Stoodley, Phil. Mag. **10,** 249, 1930. Heat capacity is given in calories per gram and in calories per gram atom (1 cal = 4.1840 absolute joules and the atomic weight of mercury 200.61). Values for the liquid and vapor are from Douglas, Ball and Ginnings, Jour of Res. Bureau of Standards **46,** 334, 1951.

Temp. °C	Specific heat	Heat capacity cal/g-atom
	Solid	Solid
−75.6	.0319	6.3995
−72.9	.0324	6.4998
−65.4	.0324	6.4998
−59.5	.0324	6.4998
−44.9	.0336	6.7405
−42.2	.0336	6.7405
−40.0	.0337	6.7606
Temp. °C	**Heat capacity cal/g**	**Heat capacity cal/g-atom**
	Liquid	Liquid
−38.88	.033686	6.7578
−20	.033534	6.7272
0	.033382	6.6967
20	.033240	6.6683
25	.033206	6.6615
40	.033109	6.6419
60	.032987	6.6176
80	.032877	6.5954
100	.032776	6.5752
120	.032686	6.5571
140	.032606	6.5410
160	.032535	6.5270
180	.032476	6.5150

Temp. °C	Heat capacity cal/g	Heat capacity cal/g-atom
	Liquid	Liquid
200	.032426	6.5050
220	.032386	6.4970
240	.032356	6.4910
260	.032336	6.4860
280	.032325	6.4847
300	.032322	6.4843
320	.032330	6.4858
340	.032346	6.4890
356.58	.032366	6.4930
360	.032371	6.4940
380	.032404	6.5005
400	.032445	6.5087
420	.032494	6.5186
440	.032550	6.5298
460	.032614	6.5426
480	.032684	6.5567
500	.032762	6.5723
	Vapor	Vapor
0	.02476	4.968
100	.02476	4.968
200	.02477	4.969
300	.02480	4.975
400	.02489	4.993
500	.02507	5.030

HEAT CAPACITY (C_p) OF SOME COMMON ORGANIC COMPOUNDS

The C_p values quoted are for 25°C and are in joule (g mole)$^{-1}$
Joules × 0.2390 = calories
s = solid: 1 = liquid: g = gas.

Name	Formula	State	C_p	Name	Formula	State	C_p
Acetic acid	CH_3COOH	1	123.4	1, 4-Dioxane	$C_4H_8O_2$	1	153
Acetone	CH_3COCH_3	1	125	Ethanol	C_2H_5OH	1	114.4
Aniline	$C_6H_5NH_2$	1	191			g	73.6
Benzene	C_6H_6	1	136.1	Ethyl acetate	$CH_3COOC_2H_5$	1	170
		g	81.7	Formic acid	$HCOOH$	1	99.0
Carbon dioxide	CO_2	g	37.1	Glucose	$C_6H_{12}O_6$	s	219
Carbon disulfide	CS_2	1	75.6	Glycine	NH_2CH_2COOH	s	100
		g	45.6	n-Hexane	C_6H_{14}	1	195
Carbon tetrachloride	CCl_4	1	131.7			g	147
		g	83.4	Methane	CH_4	g	35.8
Chlorobenzene	C_6H_5Cl	1	150.1	Methanol	CH_3OH	1	81.6
Chloroform	$CHCl_3$	1	116.3	Nitrobenzene	$C_6H_5NO_2$	1	187.3
1, 4-Dichlorobenzene	$C_6H_4Cl_2$	s	143	Phenol	C_6H_5OH	s	135
Diethyl ether	CH_3OCH_3	1	171	Toluene	$C_6H_5CH_3$	1	166
n-Dotriacontane	$C_{32}H_{66}$	s	877				

SPECIFIC HEAT OF THE ELEMENTS AT 25°C

C_p = cal/g

Element	Kelly: Bureau of Mines Bulletin 592 (1961)	Hultgren: Selected values of Thermodynamic properties of Metals and Alloys (1963)	N.B.S. Circular #500 Part 1 (1952)
Aluminum	0.215	0.215	0.2154
Antimony	0.049	0.0495	0.0501
ARgon	0.124	......	0.124
Arsenic	0.0785	......	0.0796
Barium	0.046	0.0362	0.0458
Beryllium	0.436	0.436	0.4733
Bismuth	0.0296	0.0238	0.0292
Boron	0.245	......	0.2463
Bromine (Br_2)	0.113	0.0537	
Cadmium	0.0555	0.0552	0.0554
Calcium	0.156	0.155	0.1566
Carbon (Diamond)	0.124	......	0.120
" (Graphite)	0.170	......	0.172
Cerium	0.049	0.0459	0.0442
Cesium	0.057	0.0575	0.0558
Chlorine (Cl_2)	0.114	......	0.114
Chromium	......	0.107	0.1073
Cobalt	0.109	0.107	0.1037
Columbium	See Niobium		
Copper	0.092	0.0924	0.0920
Dysprosium	0.0414	0.0414	
Erbium	0.0401	0.0401	
Europium	0.0421	0.0326	
Fluorine (F_2)	0.197	0.197	
Gadolinium	0.055	0.056	
Gallium	0.089	0.088	0.0911
Germanium	0.077		
Gold	0.0308	0.0308	0.0305
Hafnium	0.035	0.028	0.0344
Helium	1.24	......	1.242
Hydrogen (H_2)	3.41	......	3.42
Holmium	0.0393	0.0394	
Indium	0.056	0.0556	0.0570
Iodine (I_2)	0.102	......	0.034
Iridium	0.0317	0.0312	0.0305
Iron (α)	0.106	0.1075	0.1078
Krypton	0.059	......	0.059
Lanthanum	0.047	0.0479	0.0475
Lead	0.038	0.0305	0.0308
Lithium	0.85	0.834	0.814
Lutetium	0.037	0.0285	
Magnesium	0.243	0.245	0.235
Manganese (α)	0.114	0.114	0.1147
" (β)	0.119	......	0.1120
Mercury	0.0331	0.0333	0.0331
Molybdenum	0.0599	0.0597	0.0584
Neodymium	0.049	0.0453	0.0499
Neon	0.246	......	0.246
Nickel	0.106	0.1061	0.1057
Niobium	0.064	0.0633	
Nitrogen (N_2)	0.249	......	0.249
Osmium	0.03127	0.0310	0.0310
Oxygen (O_2)	0.219	0.219	
Palladium	0.0584	0.0583	0.0590

SPECIFIC HEAT OF THE ELEMENTS AT 25°C (Continued)

Element	Kelly: Bureau of Mines Bulletin 592 (1961)	Hultgren: Selected values of Thermodynamic properties of Metals and Alloys (1963)	N.B.S. Circular #500 Part 1 (1952)
Phosphorus, white	0.181	......	0.178
" red, triclinic	0.160		
Platinum	0.0317	0.0317	0.0325
Polonium	0.030		
Potassium	0.180	0.180	0.1787
Praseodymium	0.046	(0.0467)	0.0482
Promethium	0.0442		
Protactinium	0.029		
Radium	0.0288		
Radon	0.0224	......	0.0224
Rhenium	0.0329	0.0330	0.0327
Rhodium	0.0583	0.0580	0.0592
Rubidium	0.0861	0.0860	0.0850
Ruthenium	0.057	0.0569	
Samarium	0.043	0.0469	
Scandium	0.133	0.1173	
Selenium (Se_2)	0.0767	......	0.0535
Silicon	0.168	......	0.169
Silver	0.0566	0.0562	0.0564
Sodium	0.293	0.292	0.2952
Strontium	0.0719	0.0719	0.0684
Sulfur, yellow	0.175	......	0.177
Tantalum	0.0334	0.0334	0.0335
Technetium	0.058		
Tellurium	0.0481	......	0.482
Terbium	0.0437	0.0435	
Thallium	0.0307	0.0307	0.0310
Thorium	0.0271	0.0281	0.0331
Thulium	0.0382		
Tin (α)	0.0510	0.0519	0.0518
Tin (β)	0.0530	0.0543	0.0530
Titanium	0.125	0.1248	0.1231
Tungsten	0.0317	0.0322	0.0324
Uranium	0.0276	0.0278	0.0276
Vanadium	0.116	0.116	0.1147
Xenon	0.0378	......	0.0379
Ytterbium	0.0346	0.0287	
Yttrium	0.068	0.0713	
Zinc	0.0928	0.0922	0.0916
Zirconium	0.0671	0.0660	

SPECIFIC HEAT AND ENTHALPY OF SOME SOLIDS AT LOW TEMPERATURES

R. J. Corruccini and J. J. Gniewek

For a more extensive listing of data one is referred to N.B.S. Monograph 21 (1960)

Joules/gm × 453.6 = joules/lb × 0.239 = cal/gm × 0.4299 = Btu/lb

METALS

T	Aluminum		Beryllium		Bismuth		Cadmium	
	C_p	$H - H_0$	C_p	$H - H_0$	C_p	$H - H_0$	C_p	$H - H_0$
°K	jg^{-1} deg^{-1} K	jg^{-1}	jg^{-1} deg^{-1} K	jg^{-1}	jg^{-1} deg^{-1} K	jg^{-1}	jg^{-1} deg^{-1} K	jg^{-1}
1	0.000 10[a]	....	....	....			....	....
1	.000 051	0.000 025	0.000 025	0.000 013	0.000 00598	0.000 00158	0.000 008	0.000 003
2	.000 108	.000 105	.000 051	.000 051	.000 0461	.000 0233	.000 033	.000 022
3	.000 176	.000 246	.000 079	.000 116	.000 170	.000 123	.000 090	.000 082
4	.000 261	.000 463	.000 109	.000 209	.000 493	.000 432	.000 21	.000 22
6	.000 50	.001 21	.000 180	.000 496	.002 14	.002 88	.001 30	.001 5
8	.000 88	.002 6	.000 271	.000 944	.005 47	.010 2	.004 3	.007 0
10	.001 4	.004 9	.000 389	.001 60	.010 4	.025 9	.008 0	.019
15	.004 0	.018	.000 842	.004 57	.023 8	.111	.025	.102
20	.008 9	.048	.001 61	.010 5	.036 3	.262	.046	.28
25	.017 5	.112	.002 79	.021 2	.047 7	.472	.066	.56
30	.031 5	.232	.004 50	.039 2	.057 2	.734	.086	.94
35	.051 5	.436	....	....	....	....	....	....
40	.077 5	.755	.009 96	.109	.072 7	1.38	.117	1.96
50	.142	1.85	.019 2	.253	.084 6	2.17	.141	3.26
60	.214	3.64	.034 1	.523	.093 5	3.06	.159	4.76
70	.287	6.15	.056 2	.971	.100	4.03	.172	6.43
80	.357	9.37	.090 6	1.69	.105	5.05	.182	8.20
90	.422	13.25	.139	2.82	.108	6.12	.190	10.1
100	.481	17.76	.199	4.51	.111	7.21	.196	12.0

T	Chromium		Copper		Germanium[b]		Gold	
	C_p	$H - H_0$	C_p	$H - H_0$	C_p	$H - H_0$	C_p	$H - H_0$
°K	jg^{-1} deg^{-1} K	jg^{-1}	jg^{-1} deg^{-1} K	jg^{-1}	g^{-1} deg^{-1} K	jg^{-1}	jg^{-1} deg^{-1} K	jg^{-1}
1	0.000 0285	0.000 0142	0.000 012	0.000 006	0.000 000 528	0.000 000 132	0.000 006	0.000 002
2	.000 058	.000 0573	.000 028	.000 025	.000 004 23	.000 002 11	.000 025	.000 016
3	.000 089	.000 131	.000 053	.000 064	.000 014 4	.000 010 7	.000 070	.000 061
4	.000 124	.000 237	.000 091	.000 13	.000 034 4	.000 034 3	.000 16	.000 17
6	.000 206	.000 567	.000 23	.000 44	.000 125	.000 179	.000 50	.000 78
8	.000 312	.001 07	.000 47	.001 12	.000 335	.000 612	.001 2	.002 4
10	.000 451	.001 82	.000 86	.002 4	.000 813	.001 69	.002 2	.005 6
15	.001 02	.005 28	.002 7	.010 7	.004 45	.013 6	.007 4	.028
20	.002 10	.012 8	.007 7	.034	.012 5	.054 0	.015 9	.086
25	.003 92	.027 4	.016	.090	.024 0	.145	.026 3	.191
30	.006 83	.053 2	.027	.195	.036 6	.296	.037 1	.349
40	.017 1	.163	.060	.61	.061 7	.786	.057 2	.821
50	.035 8	.421	.099	1.40	.085 8	1.52	.072 6	1.47
60	.062 1	.904	.137	2.58	.108	2.50	.084 2	2.25
70	.093	1.68	.173	4.13	.131	3.70	.092 8	3.14
80	.127	2.77	.205	6.02	.153	5.12	.099 2	4.10
90	.161	4.21	.232	8.22	.173	6.74	.104 3	5.12
100	.193	5.98	.254	10.6	.191	8.55	.108 3	6.18

T	Indium		α-Iron[c]		γ-Iron[d]		Lead	
	C_p	$H - H_0$	C_p	$H - H_0$	C_p	$H - H_0$	C_p	$H - H_0$
°K	jg^{-1} deg^{-1} K	jg^{-1}	jg^{-1} deg^{-1} K	jg^{-1}	jg^{-1} deg^{-1} K	jg^{-1}	jg^{-1} deg^{-1} K	jg^{-1}
1	0.000 029	0.000 011	0.000 090	0.000 045	....	....	0.000 026	0.000 010
1	[a].000 019	[a].000 006	....	....	....	....	[a].000 012	[a].000 003
2	.000 138	.000 085	.000 183	.000 181	....	....	.000 12	.000 07
2	[a].000 141	[a].000 073	....	....	....	....	[a].000 09	[a].000 05
								.000 28
3	.000 410	.000 341	.000 279	.000 412	....	....	.000 33	[a].000 23
3	[a].000 464	[a].000 357	....	....	....	....	[a].000 31	
[e]3.40	.000 584	.000 537	....	....	....	....	....	....
3.40	[a].000 669	[a].000 581	....	....	....	....	....	....
4	.000 95	.000 99	.000 382	.000 742	....	....	.000 7	.000 8
4	....	....	....	....	....	....	[a].000 7	[a].000 7
5	....	....	....	....	....	....	.001 5	.001 8
5	....	....	....	....	....	....	[a].001 5	[a].001 8
6	.003 59	.005 20	.000 615	.001 73	....	....	.002 9	.003 9
6	....	....	....	....	....	....	[a].003 0	[a].004 0
7	....	....	....	....	....	....	.004 8	.008
7	....	....	....	....	....	....	[a].005 0	[a].008
8	.008 55	.017 0	.000 90	.003 23	....	....	.007 3	.014
10	.015 5	.040 8	.001 24	.005 37	....	....	.013 7	.034
15	.036 7	.170	.002 49	.014 5	....	....	.033 5	.150
20	.060 8	.413	.004 5	.031 6	0.007	0	.053 1	.368
25	.085 7	.778	.007 5	.061	....	....	.068 1	.672
30	.108	1.265	.012 4	.110	.016	.11	.079 6	1.042
40	.141	2.52	.029	.31	.041	.39	.094 4	1.920
50	.162	4.04	.055	.73	.090	1.0_2	.103	2.91
60	.176	5.73	.087	1.43	$.13_7$	2.1_6	.108	3.97
70	.186	7.53	.121	2.46	$.18_0$	3.7_5	.112	5.07
80	.193	9.42	.154	3.84	$.21_8$	5.7_4	.114	6.20
90	.198	11.38	.186	5.55	$.25_5$	8.1_1	.116	7.35
100	.203	13.39	.216	7.56	$.28_8$	$10._8$	.118	8.53

METALS (*Continued*)

T	Molybdenum		Nickel		Palladium	
	C_p	$H - H_0$	C_p	$H - H_0$	C_p	$H - H_0$
°K	$jg^{-1}\,deg^{-1}\,K$	jg^{-1}	$jg^{-1}\,deg^{-1}\,K$	jg^{-1}	$jg^{-1}\,deg^{-1}\,K$	jg^{-1}
1	0.000 0229	0.000 0105	0.000 120	0.000 060	0.000 099	0.000 0493
2	.000 0472	.000 0445	.000 242	.000 241	.000 203	.000 200
2	....	....	....	....	....	....
3	.000 0745	.000 105	.000 369	.000 546	.000 318	.000 459
3	....	....	....	....	....	....
4	.000 106	.000 194	.000 503	.000 98	.000 447	.000 840
4	....	....	....	....	....	....
5	....	....	....	....	....	....
5	....	....	....	....	....	....
6	.000 191	.000 484	.000 82	.002 28	.000 891	.002 31
6	....	....	....	....	....	....
7	....	....	....	....	....	....
7	....	....	....	....	....	....
8	.000 317	.000 981	.001 19	.004 28	.001 41	.004 60
8	....	....	....	....	....	....
9	....	....	....	....	....	....
9	....	....	....	....	....	....
10	.000 498	.001 78	.001 62	.007 1	.002 10	.008 07
15	.001 31	.006 10	.003 1	.018 5	.004 71	.024 5
20	.002 87	.016 1	.005 8	.041	.009 22	.058 6
25	.005 77	.037 4	.010 1	.079	.016 0	.120
30	.009 60	.072 9	.016 7	.145	.025 8	.223
40	.023 6	.232	.038 1	.413	.050 7	.600
50	.041 0	.554	.068 2	.937	.077 7	1.24
60	.061 9	1.07	.103	1.79	.101	2.14
70	.083 8	1.80	.139	3.00	.122	3.26
80	.104	2.74	.173	4.56	.139	4.56
90	.123	3.88	.204	6.45	.154	6.03
100	.139	5.20	.232	8.63	.167	7.63

T	Platinum		Rhodium		Silicon[l]		Silver	
	C_p	$H - H_0$	C_p	$H - H_0$	C_p	$H - H_0$	C_p	$H - H_0$
°K	$jg^{-1}\,deg^{-1}\,K$	jg^{-1}	$jg^{-1}\,deg^{-1}\,K$	jg^{-1}	$jg^{-1}\,deg^{-1}\,K$	jg^{-1}	$jg^{-1}\,deg^{-1}\,K$	jg^{-1}
1	0.000 035	0.000 0175	0.000 048	0.000 024	0.000 000 263	0.000 000 0658	0.000 0072	0.000 0032
2	.000 074	.000 071	.000 097	.000 096	.000 002 10	.000 001 05	.000 0239	.000 0176
3	.000 122	.000 168	.000 147	.000 218	.000 007 09	.000 005 32	.000 0595	.000 0574
4	.000 186	.000 320	.000 201	.000 392	.000 016 8	.000 016 8	.000 124	.000 146
6	.000 37	.000 85	.000 32	.000 91	.000 059 6	.000 085 3	.000 39	.000 62
8	.000 67	.001 88	.000 47	.001 70	.000 140	.000 279	.000 91	.001 87
10	.001 12	.003 65	.000 65	.002 81	.000 275	.000 679	.001 8	.004 52
15	.003 3	.013 5	.001 35	.007 65	.001 09	.003 74	.006 4	.023 3
20	.007 4	.039 5	.002 71	.017 4	.003 37	.013 8	.015 5	.076
25	.013 7	.092	.005 61	.037 3	.008 49	.042 3	.028 7	.185
30	.021 2	.182	.010 6	.077 1	.017 1	.105	.044 2	.368
40	.038	.48	.026 6	.256	.044 0	.400	.078	.979
50	.055	.95	.048 9	.633	.078 5	1.00	.108	1.91
60	.068	1.56	.072 4	1.238	.115	1.97	.133	3.12
70	.079	2.29	.094	2.07	.152	3.31	.151	4.54
80	.088	3.12	.114	3.11	.188	5.01	.166	6.13
90	.094	4.02	.132	4.34	.224	7.06	.177	7.85
100	.100	5.01	.147	5.74	.259	9.47	.187	9.67

T	Sodium[m]		Tantalum		Tin (white)		Titanium	
	C_p	$H - H_0$	C_p	$H - H_0$	C_p	$H - H_0$	C_p	$H - H_0$
°K	$jg^{-1}\,deg^{-1}\,K$	jg^{-1}	$jg^{-1}\,deg^{-1}\,K$	jg^{-1}	$jg^{-1}\,deg^{-1}\,K$	jg^{-1}	$jg^{-1}\,deg^{-1}\,K$	jg^{-1}
1	0.000 081	0.000 035	0.000 032	0.000 016	0.000 0170	0.000 0079	0.000 071	0.000 035
1	....	....	[a].000 0063	[a].000 0021	[a].000 0041	[a].000 0009	....	....
2	.000 289	.000 204	.000 068	.000 065	.000 047	.000 0383	.000 146	.000 143
2	....	....	[a].000 054	[a].000 026	[a].000 048	[a].000 0228	....	....
3	.000 76	.000 70	.000 112	.000 155	.000 109	.000 113	.000 226	.000 329
3	....	....	[a].000 178	[a].000 138	[a].000 151	[a].000 116	....	....
[n]3.72	....	....	....	....	.000 198	.000 221	....	....
3.72	....	....	....	....	[a].000 285	[a].000 270	....	....
4	.001 60	.001 84	.000 171	.000 295	.000 245	.000 283	.000 317	.000 599
4	....	....	[a].000 352	[a].000 400	....	....	....	....
[o]4.39	....	....	.000 201	.000 368	....	....	....	....
4.39	....	....	[a].000 433	[a].000 553	....	....	....	....
5	.002 98	.004 08	....	....	.000 54	.000 65	....	....
6	.005 1	.008 1	.000 333	.000 776	.001 27	.001 51	.000 54	.001 45
8	.012 2	.024 7	.000 648	.001 73	.004 2	.006 8	.000 84	.002 81
10	.023 8	.060 2	.001 17	.003 52	.008 1	.019 0	.001 26	.004 89
12	.039 7	.123	....	....	....	....	....	....
14	.063	.225	....	....	....	....	....	....
15	....	....	.003 60	.014 5	.022 6	.093	.003 3	.015 6
16	.093	.380	....	....	....	....	....	....
18	.124	.597	....	....	....	....	....	....
20	.155	.875	.008 23	.043 2	.040	.251	.007 0	.040
25	.259	1.90	.015 3	.102	.058	.498	.013 4	.090
30	.364	3.45	.024 0	.202	.076	.834	.024 5	.182
40	.544	8.03	.043 0	.540	.106	1.75	.057 1	.581
50	.695	14.2	.060 4	1.06	.130	2.93	.099 2	1.358
60	.793	21.7	.075 4	1.74	.148	4.33	.146 7	2.592
70	.86	30.0	.087 9	2.56	.162	5.88	.189	4.27
80	.91	38.9	.097 6	3.49	.173	7.55	.230	6.37
90	.95	48.2	.105	4.50	.182	9.33	.267	8.86
100	.98	57.9	.111	5.58	.189	11.18	.300	11.69

SPECIFIC HEAT AND ENTHALPY OF SOME SOLIDS AT LOW TEMPERATURES (Continued)

METALS (*Continued*)

T	Tungsten		Zinc	
	C_p	$H - H_0$	C_p	$H - H_0$
°K	jg^{-1} deg^{-1} K	jg^{-1}	jg^{-1} deg^{-1} K	jg^{-1}
1	0.000 0074	0.000 0037	0.000 011	0.000 005
2	.000 0158	.000 0152	.000 028	.000 023
3	.000 0262	.000 0360	.000 058	.000 065
4	.000 0393	.000 0685	.000 11	.000 14
6	.000 0783	.000 182	.000 29	.000 53
8	.000 141	.000 396	.000 96	.001 6
10	.000 234	.000 765	.002 5	.005 0
15	.000 725	.002 97	.011	.034
20	.001 89	.009 27	.026	.125
25	.004 21	.023 7	.049	.31
30	.007 83	.053 4	.076	.62
40	.018 4	.181	.125	1.62
50	.033 2	.436	.171	3.11
60	.048 3	.843	.208	5.01
70	.060 5	1.39	.236	7.23
80	.071 5	2.05	.258	9.70
90	.081 0	2.81	.277	12.38
100	.088 8	3.66	.293	15.24

[a] Superconducting.

[b] In germanium the electronic specific heat is markedly dependent on impurities. The values given are for pure germanium (negligible electronic specific heat).

[c] α-Iron is the form that is thermodynamically stable at low temperatures. It has the body-centered cubic lartice which is the basis of the ferritic steels.

[d] γ-Iron is stable between 910° and 1,400°C. It has the face-centered cubic structure which is the basis of the austenitic steels. Since pure γ-iron is not stable at low temperatures the above values were calculated by application of the Kopp-Neumann rule to experimental data on two austenitic Fe-Mn alloys and are of uncertain accuracy.

[e] Superconducting transition temperature.

[i] Superconducting transition temperature of mercury.

[j] Melting temperature of mercury.

[l] In silicon the electronic specific heat, γT, is markedly dependent on impurities. Values of the coefficient, γ, from zero to 2.4×10^{-6} jg^{-1} deg^{-2} K have been reported. The values in the above table are for pure silicon ($\gamma = 0$).

[m] It has been shown (Barrett 1956, Hull & Rosenberg 1959) that sodium partially transforms at low temperatures from the normal body-centered cubic structure to close-packed hexagonal. The transformation is of the martensitic type and is promoted by cold-working at the low temperatures. Inasmuch as none of the calorimetric measurements on sodium were accompanied by crystallographic analysis, the tabulated data below 100°K are to some degree ambiguous.

[n] Superconducting transition temperature of tin.

[o] Superconducting transition temperature of tantalum.

CONSTANTS OF DEBYE-SOMMERFELD EQUATION

$C_v = \gamma T + \alpha T^3$; $\alpha = 12\pi^4 R/5\theta_0^3$; $0 \lesssim T \lesssim T_{max}$; T_{max} = maximum temperature to which the equation can be used with the limiting value of θ.

Substance	$10^6\gamma$	γ	$10^6\alpha$	θ_0	T_{max}
Metals:	jg^{-1} deg^{-2} K	mjg-atom^{-1} deg^{-2} K	jg^{-1} deg^{-4} K	deg K	deg K
Aluminum	50.4	1.36	0.93	426	4
Beryllium	25	0.226	.138	1160	20
Bismuth	0.32	.067	5.66	118	2
Cadmium	5.6	.63	2.69	186	3
Chromium	28.3	1.47	0.165	610	4
Copper	10.81	0.687	.746	344.5	10
Germanium	(a)	(a)	.528	370	2
Gold	3.75	0.74	2.19	165	15
Indium	15.8	1.81	13.1	109	2
α-Iron	90	5.0	0.349	464	10
Lead	15.1	3.1	10.6	96	4
Magnesium	54	1.32	1.19	406	4
α-Manganese	251	13.8	0.328	476	12
Molybdenum	23	2.18	.238	440	4
Nickel	120	7.0	.39	440	4
Niobium	85	7.9	.64	320	1
Palladium	98	10.5	.89	274	4
Platinum	34.1	6.7	.72	240	3
Rhodium	48	4.9	.173	478	4
Silicon	(a)	(a)	.263	640	4
Silver	5.65	0.610	1.58	225	4
Sodium[a]	60	1.37	21.4	158	4
Tantalum	31.5	5.7	0.69	250	4
Tin (white)	14.7	1.75	2.21	195	2
Titanium	71	3.4	0.54	420	10
Tungsten	7	1.3	.16	405	4
Zinc	9.6	0.63	1.10	300	4
Alloys:					
Constantan[a]	113	6.9	0.56	384	15
Monel[a]	108	6.5	.62	374	20
Other inorg. subs.:					
Diamond	....	....	0.0152	2200	50
Ice	....	....	15.2	192	10
Pyrex	....	....	3.14	......	5
Organic subs.:					
Glyptal	....	....	27	......	4
Lucite	....	....	35	......	4
Polystyrene	....	....	63	......	4

(a) Superconducting.

BOILING POINT OF WATER

(Hydrogen Scale)

Pressure mm	Tenths of millimeters .0	.1	.2	.3	.4	.5	.6	.7	.8	.9
700	97.714	718	722	725	729	733	737	741	745	749
701	753	757	761	765	769	773	777	781	785	789
702	792	796	800	804	808	812	816	820	824	828
703	832	836	840	844	847	851	855	859	863	867
704	871	875	879	883	887	891	895	899	902	906
705	97.910	914	918	922	926	930	934	938	942	946
706	949	953	957	961	965	969	973	977	981	985
707	989	993	996	*000	*004	*008	*012	*016	*020	*024
708	98.028	032	036	040	043	047	051	055	059	063
709	067	071	075	079	082	086	090	094	098	102
710	98.106	110	114	118	121	125	129	133	137	141
711	145	149	153	157	160	164	168	172	176	180
712	184	188	192	195	199	203	207	211	215	219
713	223	227	230	234	238	242	246	250	254	258
714	261	265	269	273	277	281	285	289	292	296
715	98.300	304	308	312	316	320	323	327	331	335
716	339	343	347	351	355	358	362	366	370	374
717	378	382	385	389	393	397	401	405	409	412
718	416	420	424	428	432	436	440	443	447	451
719	455	459	463	467	470	474	478	482	486	490
720	98.493	497	501	505	509	513	517	520	524	528
721	532	536	540	544	547	551	555	559	563	567
722	570	574	578	582	586	590	593	597	601	605
723	609	613	617	620	624	628	632	636	640	643
724	647	651	655	659	662	666	670	674	678	682
725	98.686	689	693	697	701	705	709	712	716	720
726	724	728	732	735	739	743	747	751	755	758
727	762	766	770	774	777	781	785	789	793	797
728	800	804	808	812	816	819	823	827	831	835
729	838	842	846	850	854	858	861	865	869	873
730	98.877	880	884	888	892	896	899	903	907	911
731	915	918	922	926	930	934	937	941	945	949
732	953	956	960	964	968	972	975	979	983	987
733	991	994	998	*002	*006	*010	*013	*017	*021	*025
734	99.029	032	036	040	044	048	051	055	059	063
735	99.067	070	074	078	082	085	089	093	097	101
736	104	108	112	116	119	123	127	131	135	138
737	142	146	150	153	157	161	165	169	172	176
738	180	184	187	191	195	199	203	206	210	214
739	218	221	225	229	233	236	240	244	248	252
740	99.255	259	263	267	270	274	278	282	285	289
741	293	297	300	304	308	312	316	319	323	327
742	331	334	338	342	346	349	353	357	361	364
743	368	372	376	379	383	387	391	394	398	402
744	406	409	413	417	421	424	428	432	436	439
745	99.443	447	451	454	458	462	466	469	473	477
746	481	484	488	492	495	499	503	507	510	514
747	518	522	525	529	533	537	540	544	548	551
748	555	559	563	566	570	574	578	581	585	589
749	592	596	600	604	607	611	615	619	622	626
750	99.630	633	637	641	645	648	652	656	659	663
751	667	671	674	678	682	686	689	693	697	700
752	704	708	712	715	719	723	726	730	734	738
753	741	745	749	752	756	760	764	767	771	775
754	778	782	786	790	793	797	801	804	808	812
755	99.815	819	823	827	830	834	838	841	845	849
756	852	856	860	863	867	871	875	878	882	886
757	889	893	897	900	904	908	911	915	919	923
758	926	930	934	937	941	945	948	952	956	959
759	963	967	970	974	978	982	985	989	993	996
760	100.000	004	007	011	015	018	022	026	029	033
761	037	040	044	048	052	055	059	063	066	070
762	074	077	081	085	088	092	096	099	103	107
763	110	114	118	121	125	129	132	136	140	143
764	147	151	154	158	162	165	169	173	176	180
765	100.184	187	191	195	198	202	206	209	213	216
766	220	224	227	231	235	238	242	246	249	253
767	257	260	264	268	271	275	279	283	286	290
768	293	297	300	304	308	311	315	319	322	326
769	330	333	337	341	344	348	352	355	359	363
770	100.366	370	373	377	381	384	388	392	395	399
771	403	406	410	414	417	421	424	428	432	435
772	439	442	446	450	453	457	461	464	468	472
773	475	479	483	486	490	493	497	501	504	508
774	511	515	519	522	526	530	533	537	540	544
775	100.548	551	555	559	562	566	569	573	577	580
776	584	588	591	595	598	602	606	609	613	616
777	620	624	627	631	634	638	642	645	649	653
778	656	660	663	667	671	674	678	681	685	689
779	692	696	689	703	707	710	714	718	721	725
780	100.728	732	735	739	743	746	750	753	757	761
781	764	768	772	775	779	782	786	789	793	797
782	800	804	807	811	815	818	822	825	829	833
783	836	840	843	847	851	854	858	861	865	869
784	872	876	879	883	886	890	894	897	901	904
785	100.908	912	915	919	922	926	929	933	937	940
786	944	947	951	954	958	962	965	969	972	976
787	979	983	987	990	994	997	*001	*005	*008	*012
788	101.015	019	022	026	029	033	037	040	044	047
789	051	054	058	062	065	069	072	076	079	083
790	101.087	090	094	097	101	104	108	112	115	119
791	122	126	129	133	136	140	144	147	151	154
792	158	161	165	168	172	176	179	183	186	190
793	193	197	200	204	207	211	215	218	222	225
794	229	232	236	239	243	246	250	254	257	261
795	101.264	268	271	275	278	282	286	289	293	296
796	300	303	307	310	314	317	321	324	328	332
797	335	339	342	346	349	353	356	360	363	367
798	370	374	377	381	385	388	392	395	399	402
799	406	409	413	416	420	423	427	430	434	437
800	101.441	...	...	...	...	...	...	...	...	...

For lower pressures see under Vapor Tension of Water

MELTING POINTS OF MIXTURES OF METALS

(Smithsonian Physical Tables)

Melting-points, °C.

Metals	Percentage of metal in second column. 0%	10%	20%	30%	40%	50%	60%	70%	80%	90%	100%
Pb. Sn.	326*	295	276	262	240	220	190	185	200	216	232
Bi.	322	290	...	...	179	145	126	168	205	...	268
Te.	322	710	790	880	917	760	600	480	410	425	446
Ag.	328	460	545	590	620	650	705	775	840	905	959
Na.	...	360	420	400	370	330	290	250	200	130	96
Cu.	326	870	920	925	945	950	955	985	1005	1020	1084
Sb.	326	250	275	330	395	440	490	525	560	600	632
Al. Sb.	650	750	840	925	945	950	970	1000	1040	1010	632
Cu.	650	630	600	560	540	580	610	755	930	1055	1084
Au.	655	675	740	800	855	915	970	1025	1055	675	1062
Ag.	650	625	615	600	590	580	575	570	650	750	954
Zn.	654	640	620	600	580	560	530	510	475	425	419
Fe.	653	860	1015	1110	1145	1145	1220	1315	1425	1500	1515
Sn.	650	645	635	625	620	605	590	570	560	540	232
Sb. Bi.	632	610	590	575	555	540	520	470	405	330	268
Ag.	630	595	570	545	520	500	505	545	680	850	959
Sn.	622	600	570	525	480	430	395	350	310	255	232
Zn.	632	555	510	540	570	565	540	525	510	470	419
Ni. Sn.	1455*	1380	1290	1200	1235	1290	1305	1230	1060	800	232
Na. Bi.	96	425	520	590	645	690	720	730	715	570	268
Cd.	96	125	185	245	285	325	330	340	360	390	322
Cd. Ag.	322	420	520	610	700	760	805	850	895	940	954
Tl.	321	300	285	270	262	258	245	230	210	235	302
Zn.	322	280	270	295	313	327	340	355	370	390	419
Au. Cu.	1063	910	890	895	905	925	975	1000	1025	1060	1084
Ag.	1064	1062	1061	1058	1054	1049	1039	1025	1006	982	963
Pt.	1075	1125	1190	1250	1320	1380	1455	1530	1610	1685	1775
K. Na.	62	17.5	−10	−3.5	5	11	26	41	58	77	97.5
Hg.	...	...	...	...	...	90	110	135	162	265	...
Tl.	62.5	133	165	188	205	215	220	240	280	305	301
Cu. Ni.	1080	1180	1240	1290	1320	1355	1380	1410	1430	1440	1455
Ag.	1082	1035	990	945	910	870	830	788	814	875	960
Sn.	1084	1005	890	755	725	680	630	580	530	440	232
Zn.	1084	1040	995	930	900	880	820	780	700	580	419
Ag. Zn.	959	850	755	705	690	660	630	610	570	505	419
Sn.	959	870	750	630	550	495	450	420	375	300	232
Na. Hg.	96.5	90	80	70	60	45	22	55	95	215	...

* The data in this table are compiled from various sources,— hence variations in the melting point of the metals as shown in this column.

The triple point of water, 0.01°C (273.16°K) is the thermodynamic point for temperature measurements.

MELTING AND BOILING POINTS OF THE ELEMENTS

Element	Melting Point °C	Boiling Point °C
Actinium	1050	3200 ± 300
Aluminum	660.1	2467
Americium	>850	...
Antimony	630.5	1380
Argon	−189.2	−185.7
Arsenic (gray)	817 (28 atm.)	613 (sub.)
Astatine	...	...
Barium	725	1638
Berkelium	...	...
Beryllium	1278 ± 5	2970^{760}
Bismuth	271.3	1560 ± 5^{760}
Boron	2300	2550 (sub.)
Bromine	−7.2	58.78
Cadmium	321.03	765
Calcium	842–8	1487
Californium	...	...
Carbon	>3550	4827
Cerium	795	3468
Cesium	28.5	690
Chlorine	−100.98	−34.6
Chromium	1890	2482
Cobalt	1492	2900
Columbium (See *Niobium*)		
Copper	1083 ± 0.1	2595
Curium	...	...
Deuterium (See *Hydrogen*)		
Dysprosium	1407	2600
Einsteinium	...	...
Element 102	...	...
Erbium	1497	2900
Europium	826	1439
Fermium	...	...
Fluorine	$-219.62^{1\ atm.}$	$-188.14^{1\ atm.}$
Francium	...	...
Gadolinium	1312	~3000
Gallium	29.78	2403
Germanium	937.4	2830
Gold	1063.0	2966
Hafnium	2150	5400
Helium	$-272.2^{26\ atm.}$	−268.6
Holmium	1461	2600
Hydrogen	−259.14	−252.5
Indium	156.61	2000 ± 10
Iodine	113.5	$184.35^{atm.}$
Iridium	2443	4527 ± 100
Iron	1535	3000
Krypton	−156.6	−152.30 ± 0.10
Lanthanum	920	3469
Lawrencium	...	...
Lead	327.3	1744
Lithium	179	1317
Lutetium	1652	3327
Magnesium	651	1107

Element	Melting Point °C	Boiling Point °C
Manganese	1244 ± 3	2097
Mendelevium	...	...
Mercury	−38.87	356.58
Molybdenum	2610	5560
Neodymium	1024	3027
Neon	−248.67	−245.92
Neptunium	640 ± 1	...
Nickel	1453	2732
Niobium	2468 ± 10	4927
Nitrogen	−209.86	−195.8
Osmium	3000 ± 10	~5000
Oxygen	−218.4	−183.0
Ozone	−192.7 ± 2	−111.9
Palladium	1552	2927
Phosphorus (white)	44.1	280
Platinum	1769	3827 ± 100
Plutonium	639.5 ± 2	3235 ± 19
Polonium	254	962
Potassium	63.65	774
Praseodymium	935	3127
Promethium	1035	2730
Protactinium	~1230?	...
Radium	700	<1737
Radon	−71	−61.8
Rhenium	3180	5627
Rhodium	1960	3727 ± 100
Rubidium	38.89	688
Ruthenium	2250	3900
Samarium	1072	1900
Scandium	1539	2727
Selenium (gray)	217	684.9 ± 1.0
Silicon	1410	2355
Silver	960.8	2212
Sodium	97.81 ± 0.03	892
Strontium	769	1384
Sulfur (rhombic)	112.8	444.6
(monoclinic)	119.0	444.6
Tantalum	2996	5425 ± 100
Technetium	2200 ± 50	...
Tellurium	449.5 ± 0.3	989.8 ± 3.8
Terbium	1356	2800
Thallium	303.5	1457 ± 10
Thorium	~1700	~4000
Thulium	1545	1727
Tin	231.91	2270
Titanium	1675	3260
Tungsten	3380	5927
Uranium	1132.3 ± 0.8	3818
Vanadium	1890 ± 10	~3000
Wolfram (See *Tungsten*)		
Xenon	−111.9	−107.1 ± 3
Ytterbium	824 ± 5	1427
Yttrium	1495 ± 5	2927
Zinc	419.4	907
Zirconium	1852 ± 2	3578

BOILING POINTS AND TRIPLE POINTS OF SOME LOW BOILING ELEMENTS

(Compiled 1960)

Temperature relative to ice point of 273.15°K.

Element	Boiling point °K.	Triple point °K.	λ point °K
He	4.215		2.174
p-H_2	20.27	13.81	
n-H_2	20.379	13.95	
N_2	77.35		
O_2	90.18	54.34	
A	83.81		

MOLECULAR ELEVATION OF THE BOILING POINT

(Most values from Hoyt, C.S. and Fink, C.K., Journal of Physical Chemistry, Vol. 41. No. 3., March, 1937.)

Molecular elevation of the boiling point showing the elevation of the boiling point in degrees C due to the addition of one gram molecular weight of the dissolved substance to 1000 grams of any one of the solvents below. The correction in the last column gives the number of degrees to be subtracted for each mm. of difference between the barometric reading and 760 mm.

Solvent	K_B	Barometric Correction per mm.
Acetic acid	3.07	0.0008
Acetone	1.71	0.0004
Aniline	3.52	0.0009
Benzene	2.53	0.0007
Bromobenzene	6.26	0.0016
Carbon bisulfide	2.34	0.0006
Carbon tetrachloride	5.03	0.0013
Chloroform	3.63	0.0009
Cyclohexane	2.79	0.0007
Ethanol (ethyl alcohol)	1.22	0.0003
Ethyl acetate	2.77	0.0007
Ethyl ether	2.02	0.0005
n-Hexane	2.75	0.0007
Methanol (methyl alcohol)	0.83	0.0002
Methyl acetate	2.15	0.0005
Nitrobenzene	5.24	0.0013
n-Octane	4.02	0.0010
Phenol	3.56	0.0009
Toluene	3.33	0.0008
Water	0.512	0.0001

MOLECULAR DEPRESSION OF THE FREEZING POINT

Showing the depression of the freezing point due to the addition of one gram molecular weight of dissolved substance, for various solvents.

Solvent	Depression for one gram molecular weight dissolved in 100 gms. °C
Acetic acid	39.0
Benzene	49.0
Benzophenone	98.0
Diphenyl	80.0
Diphenylamine	86.0
Ethylene dibromide	118.0
Formic acid	27.7
Naphthalene	68–69
Nitrobenzene	70.0
Phenol	74.0
Stearic acid	45.0
Triphenyl methane	124.5
Urethane	51.4
Water	18.5–18.7

CORRECTION OF BOILING POINTS TO STANDARD PRESSURE

By H. B. Hass and R. F. Newton

This correction may be made by using the equation:

$$\Delta t = \frac{(273.1 + t)(2.8808 - \log p)}{\phi + .15(2.8808 - \log p)} \quad (1)$$

where Δt = degrees C to be added to the observed boiling point.
t = the observed boiling point.
$\log p$ = the logarithm of the observed pressure in millimeters of mercury.
ϕ = the entropy of vaporization at 760 mm.

The value of ϕ may be estimated from the graph and the table. Substances not included in the table may be classified by grouping them with compounds which bear a close physical or structural resemblance to them.

Example 1. Benzene boils at 20°C. at 75 mm pressure. What is its normal boiling point? We do not find benzene in the table but we find hydrocarbons in group 2, and a group 2 compound with a boiling point of 20° has a ϕ of 4.6.

Substituting in the equation

$$\Delta t = \frac{(273.1 + 20)(2.8808 - 1.8751)}{4.60 + .15(2.8808 - 1.8751)} = 62°$$

Adding this to 20° gives 82° as a first approximation.

The graph shows that the ϕ for a compound of group 2 boiling at 82° is 4.72 instead of 4.60 which we originally used. Since ϕ is in the denominator, this increase will lower our Δt by the ratio, 4.60/4.72, or the corrected Δt is $62 \times 4.60/4.72 = 60.4$. Adding Δt to t, gives 80.4° as a second approximation.

The formula can best be used in a slightly different form when the reverse calculation is desired, *i.e.*, when one calculates the vapor pressure at a given temperature, lower than the normal boiling point.

$$2.8808 - \log p = \frac{\phi \Delta t}{273.1 + t - .15\Delta t} \quad (2)$$

Example 2. Alcohol boils at 78.4°C. What is its vapor pressure at 20°C.? Substituting in equation 2:

$$2.8808 - \log p = \frac{6.06 \times 58.4}{293.1 - (.15 \times 58.4)} = 1.245$$
$$\log p = 2.8808 - 1.245 = 1.6358$$
$$p = 43.2 \text{ mm.}$$

Here no second approximation is necessary, since the correct value of ϕ was taken immediately, the normal boiling point having been known.

Compound	Group	Compound	Group
Acetaldehyde	3	Amines	3
Acetic acid	4	*n*-Amyl alcohol	8
Acetic anhydride	6	Anthracene	1
Acetone	3	Anthraquinone	1
Acetophenone	4	Benzaldehyde	2

CORRECTION OF BOILING POINTS TO STANDARD PRESSURE(Continued)

Compound	Group	Compound	Group
Benzoic acid	5	Hydrogen cyanide	3
Benzonitrile	2	Isoamyl alcohol	7
Benzophenone	2	Isobutyl alcohol	8
Benzyl alcohol	5	Isobutyric acid	6
Butylethylene	1	Isocaproic acid	7
Butyric acid	7	Methane	1
Camphor	2	Methanol	7
Carbon monoxide	1	Methyl amine	5
Carbon oxysulfide	2	Methyl benzoate	3
Carbon suboxide	2	Methyl ether	3
Carbon sulfoselenide	2	Methyl ethyl ether	3
m.p. Chloroanilines	3	Methyl ethyl ketone	2
Chlorinated derivatives	Same group as though Cl were H	Methyl fluoride	3
		Methyl formate	4
o.m.p. Cresols	4	Methyl salicylate	2
Cyanogen	4	Methyl silicane	1
Cyanogen chloride	3	α, β Naphthols	3
Dibenzyl ketone	2	Nitrobenzene	3
Dimethyl amine	4	Nitromethane	3
Dimethyl oxalate	4	o.m.p. Nitrotoluenes	2
Dimethyl silicane	2	o.m.p. Nitrotoluidines	2
Esters	3	Phenanthrene	1
Ethanol	8	Phenol	5
Ethers	2	Phosgene	2
Ethylamine	4	Phthalic anhydride	2
Ethylene glycol	7	Propionic acid	5
Ethylene oxide	3	*n*-Propyl alcohol	8
Formic acid	3	Quinoline	2
Glycol diacetate	4	Sulfides	2
Halogen derivatives	Same group as though halogen were hydrogen.	Tetranitromethane	3
		Trichloroethylene	1
		Valeric acid	7
Heptylic acid	7	Water	6
Hydrocarbons	2		

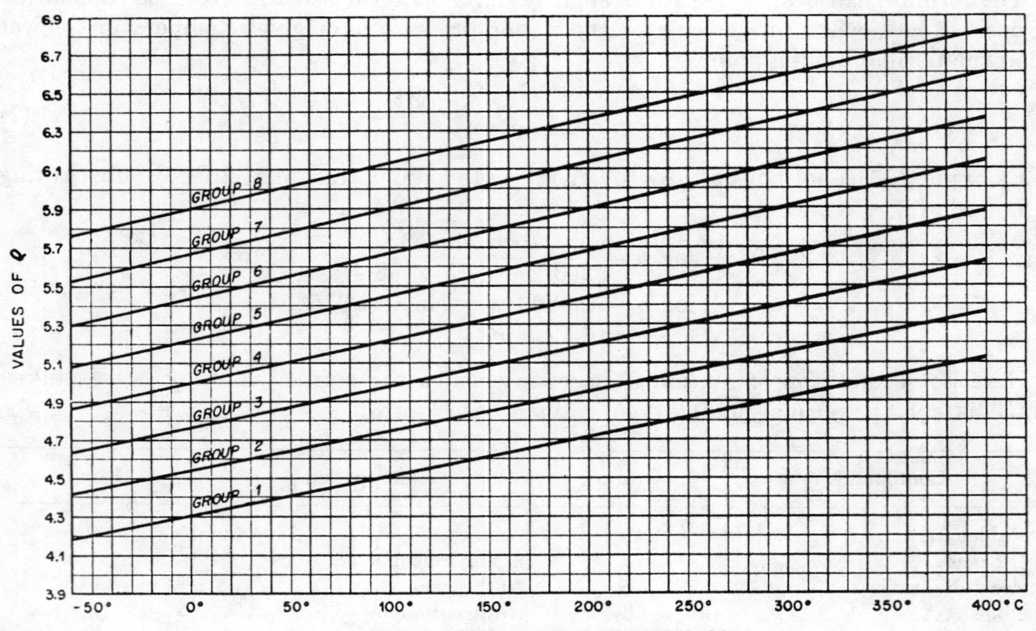

VAN DER WAALS' CONSTANTS FOR GASES

(Calculated from Amagat units in Landolt-Bornstein Physical Chemical Tables)

Van der Waals' equation is an equation of state for real gases. It may be written

$$\left(P + \frac{a}{V^2}\right)(V - b) = RT \text{ for one mole.} \qquad \text{or} \qquad \left(P + \frac{n^2a}{V^2}\right)(V - nb) = nRT \text{ for } n \text{ moles.}$$

The term a is a measure of the attractive force between the molecules. The term b is due to the finite volume of the molecules and to their general incompressibility. It is known that a and b vary to some extent with temperature.

The values for a and b in the following table are those to be used when the pressure is in atmospheres and the volume is in liters. Thus R in the above equation will be 0.08206 liter atmospheres per mole per degree. T is degrees Kelvin.

Name	Formula	a $\frac{(\text{liters})^2 \times \text{atm.}}{(\text{mole})^2}$	b $\frac{\text{liters}}{\text{mole}}$
Acetic acid	CH_3CO_2H	17.59	0.1068
Acetic anhydride	$(CH_3CO)_2O$	19.90	0.1263
Acetone	$(CH_3)_2CO$	13.91	0.0994
Acetonitrile	CH_3CN	17.58	0.1168
Acetylene	C_2H_2	4.390	0.05136
Ammonia	NH_3	4.170	0.03707
Amyl formate	$HCO_2C_5H_{11}$	27.58	0.1730
Amylene	C_5H_{10}	15.90	0.1207
*Iso*amylene	C_5H_{10}	18.08	0.1405
Aniline	$C_6H_5NH_2$	26.50	0.1369
Argon	A	1.345	0.03219
Benzene	C_6H_6	18.00	0.1154
Benzonitrile	C_6H_5CN	33.39	0.1724
Bromobenzene	C_6H_5Br	28.56	0.1539
n-Butane	C_4H_{10}	14.47	0.1226
iso-Butane	C_4H_{10}	12.87	0.1142
iso-Butyl acetate	$CH_3CO_2C_4H_9$	28.50	0.1833
iso-Butyl alcohol	C_4H_9OH	17.03	0.1143
iso-Butyl benzene	$C_6H_5C_4H_9$	38.59	0.2144
iso-Butyl formate	$HCO_2C_4H_9$	22.54	0.1476
Butyronitrile	C_3H_7CN	25.72	0.1596
Capronitrile	$C_5H_{11}CN$	34.16	0.1984
Carbon dioxide	CO_2	3.592	0.04267
Carbon disulfide	CS_2	11.62	0.07685
Carbon monoxide	CO	1.485	0.03985
Carbon oxysulfide	COS	3.933	0.05817
Carbon tetrachloride	CCl_4	20.39	0.1383
Chlorine	Cl_2	6.493	0.05622
Chlorobenzene	C_6H_5Cl	25.43	0.1453
Chloroform	$CHCl_3$	15.17	0.1022
m-Cresol	$CH_3C_6H_4OH$	31.38	0.1607
Cyanogen	C_2N_2	7.667	0.06901
Cyclohexane	C_6H_{12}	22.81	0.1424
Cymene	$C_{10}H_{14}$	42.16	0.2336
Decane	$C_{10}H_{12}$	48.55	0.2905
Di-isobutyl	C_8H_{18}	34.97	0.2296
Diethylamine	$(C_2H_5)_2NH$	19.15	0.1392
Dimethylamine	$(CH_3)_2NH$	10.38	0.08570
Dimethylaniline	$C_6H_5N(CH_3)_2$	37.49	0.1970
Diphenyl	$(C_6H_5)_2$	52.79	0.2480
Diphenyl methane	$(C_6H_5)_2CH_2$	38.20	0.2240
Dipropylamine	$(C_2H_7)_2NH$	27.72	0.1820
Di-isopropyl	$(C_3H_7)_2$	23.13	0.1669
Durene	$C_{10}H_{14}$	45.32	0.2424
Ethane	C_2H_6	5.489	0.06380
Ethyl acetate	$CH_3CO_2C_2H_5$	20.45	0.1412
Ethyl alcohol	C_2H_5OH	12.02	0.08407
Ethylamine	$C_2H_5NH_2$	10.60	0.08409
Ethyl benzene	$C_2H_5C_6H_5$	28.60	0.1667
Ethyl butyrate	$C_3H_7CO_2C_2H_5$	30.07	0.1919
Ethyl isobutyrate	$C_3H_7CO_2C_2H_5$	28.87	0.1994
Ethyl chloride	C_2H_5Cl	10.91	0.08651
Ethyl ether	$(C_2H_5)_2O$	17.38	0.1344
Ethyl formate	$HCO_2C_2H_5$	14.80	0.1056
Ethyl mercaptan	C_2H_5SH	11.24	0.08098
Ethyl propionate	$C_2H_5CO_2C_2H_5$	24.39	0.1615
Ethyl sulfide	$(C_2H_5)_2S$	18.75	0.1214
Ethylene	C_2H_4	4.471	0.05714
Ethylene bromide	$(CH_2Br)_2$	13.98	0.08664
Ethylene chloride	$(CH_2Cl)_2$	16.91	0.1086
Ethylidene chloride	CH_3CHCl_2	15.50	0.1073
Fluorobenzene	C_6H_5F	19.93	0.1286
Germanium tetrachloride	$GeCl_4$	22.60	0.1485
Helium	He	0.03412	0.02370
n-Heptane	C_7H_{16}	31.51	0.2065
n-Hexane	C_6H_{14}	24.39	0.1735
Hydrogen	H_2	0.2444	0.02661
Hydrogen bromide	HBr	4.451	0.04431
Hydrogen chloride	HCl	3.667	0.04081
Hydrogen selenide	H_2Se	5.268	0.04637
Hydrogen sulfide	H_2S	4.431	0.04287
Iodobenzene	C_6H_5I	33.08	0.1656
Krypton	Kr	2.318	0.03978
Mercury	Hg	8.093	0.01696
Mesitylene	$(CH_3)_3C_6H_3$	34.32	0.1979
Methane	CH_4	2.253	0.04278
Methyl acetate	$CH_3CO_2CH_3$	15.29	0.1091
Methyl alcohol	CH_3OH	9.523	0.06702
Methylamine	CH_3NH_2	7.130	0.05992
Methyl butyrate	$C_3H_7CO_2CH_3$	23.94	0.1569
Methyl isobutyrate	$C_3H_7CO_2CH_3$	24.50	0.1637
Methyl chloride	CH_3Cl	7.471	0.06483
Methyl ether	$(CH_3)_2O$	8.073	0.07246
Methyl ethyl ether	$CH_3OC_2H_5$	11.95	0.09775
Methyl ethyl sulfide	$CH_3SC_2H_5$	19.23	0.1304
Methyl fluoride	CH_3F	4.631	0.05264
Methyl formate	HCO_2CH_3	10.84	0.08068
Methyl propionate	$C_2H_5CO_2CH_3$	19.91	0.1360
Methyl sulfide	$(CH_3)_2S$	12.87	0.09213
Methyl valerate	$C_4H_9CO_2CH_3$	28.96	0.1845
Naphthalene	$C_{10}H_8$	39.74	0.1937
Neon	Ne	0.2107	0.01709
Nitric oxide	NO	1.340	0.02789
Nitrogen	N_2	1.390	0.03913
Nitrogen dioxide	NO_2	5.284	0.04424
Nitrous oxide	N_2O	3.782	0.04415
n-Octane	C_8H_{18}	37.32	0.2368
Oxygen	O_2	1.360	0.03183
n-Pentane	C_5H_{12}	19.01	0.1460
iso-Pentane	C_5H_{12}	18.05	0.1417
Phenetole	$C_6H_5OC_2H_5$	35.16	0.1963
Phosphine	PH_3	4.631	0.05156
Phosphonium chloride	PH_4Cl	4.054	0.04545
Phosphorus	P	52.94	0.1566
Propane	C_3H_8	8.664	0.08445
Propionic acid	$C_2H_5CO_2H$	20.11	0.1187
Propionitrile	C_2H_5CN	16.44	0.1064
Propyl acetate	$CH_3CO_2C_3H_7$	24.63	0.1619
Propyl alcohol	C_3H_7OH	14.92	0.1019
Propylamine	$C_3H_7NH_2$	14.99	0.1090
Propyl benzene	$C_6H_5C_3H_7$	35.85	0.2028
iso-Propyl benzene	$C_6H_5C_3H_7$	35.64	0.2025
Propyl chloride	C_3H_7Cl	15.91	0.1141
Propyl formate	$HCO_2C_3H_7$	18.95	0.1280
Propylene	C_2H_6	8.379	0.08272
Pseudo-cumene	$C_6H_3(CH_3)_3$	36.61	0.2021
Silicon fluoride	SiF_4	4.195	0.05571
Silicon tetrahydride	SiH_4	4.320	0.05786
Stannic chloride	$SnCl_4$	26.91	0.1642
Sulfur dioxide	SO_2	6.714	0.05636
Thiophene	C_4H_4S	20.72	0.1270
Toluene	$C_6H_5CH_3$	24.06	0.1463
Triethylamine	$(C_2H_5)_3N$	27.17	0.1831
Trimethylamine	$(CH_3)_3N$	13.02	0.1084
Xenon	Xe	4.194	0.05105
m-Xylene	$C_6H_4(CH_3)_2$	30.36	0.1772
o-Xylene	$C_6H_4(CH_3)_2$	29.98	0.1755
p-Xylene	$C_6H_4(CH_3)_2$	30.93	0.1809
Water	H_2O	5.464	0.03049

VAN DER WAALS' RADII IN Å

		H 1.2
N 1.5	O 1.40	F 1.35
P 1.9	S 1.85	Cl 1.80
As 2.0	Se 2.00	Br 1.95
Sb 2.2	Te 2.20	I 2.15

Methyl group CH_3 and methylene CH_2: 2.0. Half thickness of aromatic nucleus 1.85.

EMERGENT STEM CORRECTION FOR LIQUID-IN-GLASS THERMOMETERS

Accurate thermometers are calibrated with the entire stem immersed in the bath which determines the temperature of the thermometer bulb. However, for reasons of convenience it is common practice when using a thermometer to permit its stem to extend out of the apparatus. Under these conditions both the stem and the mercury in the exposed stem are at a temperature different from that of the bulb. This introduces an error into the observed temperature. Since the coefficient of thermal expansion of glass is less than that of mercury, the observed temperature will be less than the true temperature if the bulb is hotter than the stem and greater than the true temperature, providing the thermal gradient is reversed. For exact work the magnitude of this error can only be determined by experiment. However, for most purposes it is sufficiently accurate to apply the following equation which takes into account the difference of the thermal expansion of glass and mercury:

$$T_c = T_o + F \times L(T_o - T_m)$$

Where

T_c = corrected temperature

T_o = observed temperature

T_m = mean temperature of exposed stem. The mean temperature of the exposed stem may be determined by fastening the bulb of a second thermometer against the midpoint of the exposed liquid column.

L = the length of the exposed column in degrees above the surface of the substance whose temperature is being determined.

F = correction factor. For approximate work and when the liquid in the thermometer is mercury a value for F of 0.00016 is generally used. For more accurate work with mercury filled thermometers values as given in the following table are used. For thermometers filled with organic liquids it is customary to use 0.001 for the value of F.

Values of F for various glasses

Tm°C.	Corning 0041	Corning 8800	Corning 8810	Jena 16 III	Jena 59 III
50	0.000157	0.000166	0.000156	0.000158	0.000164
150	0.000159	0.000167	0.000157	0.000158	0.000165
250	0.000163	0.000168	0.000161	0.000161	0.000170
350	0.000168	0.000173	0.000166	....	0.000177

PRESSURE OF AQUEOUS VAPOR

Vapor Pressure of Ice

Pressure of aqueous vapor over ice in mm of Hg for temperatures from −98 to 0°C.

Temp. °C	0	2	4	6	8
−90	.000070	.000048	.000033	.000022	.000015
−80	.00040	.00029	.00020	.00014	.00010
−70	.00194	.00143	.00105	.00077	.00056
−60	.00808	.00614	.00464	.00349	.00261
−50	.02955	.0230	.0178	.0138	.0106
−40	.0966	.0768	.0609	.0481	.0378
−30	.2859	.2318	.1873	.1507	.1209

Temp. °C	0.0	0.2	0.4	0.6	0.8
−29	0.317	0.311	0.304	0.298	0.292
−28	0.351	0.344	0.337	0.330	0.324
−27	0.389	0.381	0.374	0.366	0.359
−26	0.430	0.422	0.414	0.405	0.397
−25	0.476	0.467	0.457	0.448	0.439
−24	0.526	0.515	0.505	0.495	0.486
−23	0.580	0.569	0.558	0.547	0.536
−22	0.640	0.627	0.615	0.603	0.592
−21	0.705	0.691	0.678	0.665	0.652
−20	0.776	0.761	0.747	0.733	0.719
−19	0.854	0.838	0.822	0.806	0.791
−18	0.939	0.921	0.904	0.887	0.870
−17	1.031	1.012	0.993	0.975	0.956
−16	1.132	1.111	1.091	1.070	1.051
−15	1.241	1.219	1.196	1.175	1.153
−14	1.361	1.336	1.312	1.288	1.264
−13	1.490	1.464	1.437	1.411	1.386
−12	1.632	1.602	1.574	1.546	1.518
−11	1.785	1.753	1.722	1.691	1.661
−10	1.950	1.916	1.883	1.849	1.817
− 9	2.131	2.093	2.057	2.021	1.985
− 8	2.326	2.285	2.246	2.207	2.168
− 7	2.537	2.493	2.450	2.408	2.367
− 6	2.765	2.718	2.672	2.626	2.581
− 5	3.013	2.962	2.912	2.862	2.813
− 4	3.280	3.225	3.171	3.117	3.065
− 3	3.568	3.509	3.451	3.393	3.336
− 2	3.880	3.816	3.753	3.691	3.630
− 1	4.217	4.147	4.079	4.012	3.946
− 0	4.579	4.504	4.431	4.359	4.287

VAPOR PRESSURE OF WATER BELOW 100°C

Pressure of aqueous vapor over water in mm of Hg for temperatures from −15.8 to 100°C. Values for fractional degrees between 50 and 89 were obtained by interpolation.

Temp. °C	0.0	0.2	0.4	0.6	0.8
−15	1.436	1.414	1.390	1.368	1.345
−14	1.560	1.534	1.511	1.485	1.460
−13	1.691	1.665	1.637	1.611	1.585
−12	1.834	1.804	1.776	1.748	1.720
−11	1.987	1.955	1.924	1.893	1.863
−10	2.149	2.116	2.084	2.050	2.018
− 9	2.326	2.289	2.254	2.219	2.184
− 8	2.514	2.475	2.437	2.399	2.362
− 7	2.715	2.674	2.633	2.593	2.553
− 6	2.931	2.887	2.843	2.800	2.757
− 5	3.163	3.115	3.069	3.022	2.976
− 4	3.410	3.359	3.309	3.259	3.211
− 3	3.673	3.620	3.567	3.514	3.461
− 2	3.956	3.898	3.841	3.785	3.730
− 1	4.258	4.196	4.135	4.075	4.016
− 0	4.579	4.513	4.448	4.385	4.320
0	4.579	4.647	4.715	4.785	4.855
1	4.926	4.998	5.070	5.144	5.219
2	5.294	5.370	5.447	5.525	5.605
3	5.685	5.766	5.848	5.931	6.015
4	6.101	6.187	6.274	6.363	6.453
5	6.543	6.635	6.728	6.822	6.917
6	7.013	7.111	7.209	7.309	7.411
7	7.513	7.617	7.722	7.828	7.936
8	8.045	8.155	8.267	8.380	8.494
9	8.609	8.727	8.845	8.965	9.086
10	9.209	9.333	9.458	9.585	9.714
11	9.844	9.976	10.109	10.244	10.380
12	10.518	10.658	10.799	10.941	11.085
13	11.231	11.379	11.528	11.680	11.833
14	11.987	12.144	12.302	12.462	12.624
15	12.788	12.953	13.121	13.290	13.461
16	13.634	13.809	13.987	14.166	14.347
17	14.530	14.715	14.903	15.092	15.284
18	15.477	15.673	15.871	16.071	16.272
19	16.477	16.685	16.894	17.105	17.319
20	17.535	17.753	17.974	18.197	18.422
21	18.650	18.880	19.113	19.349	19.587
22	19.827	20.070	20.316	20.565	20.815
23	21.068	21.324	21.583	21.845	22.110
24	22.377	22.648	22.922	23.198	23.476
25	23.756	24.039	24.326	24.617	24.912
26	25.209	25.509	25.812	26.117	26.426
27	26.739	27.055	27.374	27.696	28.021
28	28.349	28.680	29.015	29.354	29.697
29	30.043	30.392	30.745	31.102	31.461
30	31.824	32.191	32.561	32.934	33.312
31	33.695	34.082	34.471	34.864	35.261
32	35.663	36.068	36.477	36.891	37.308
33	37.729	38.155	38.584	39.018	39.457
34	39.898	40.344	40.796	41.251	41.710
35	42.175	42.644	43.117	43.595	44.078
36	44.563	45.054	45.549	46.050	46.556
37	47.067	47.582	48.102	48.627	49.157
38	49.692	50.231	50.774	51.323	51.879
39	52.442	53.009	53.580	54.156	54.737
40	55.324	55.91	56.51	57.11	57.72
41	58.34	58.96	59.58	60.22	60.86
42	61.50	62.14	62.80	63.46	64.12
43	64.80	65.48	66.16	66.86	67.56
44	68.26	68.97	69.69	70.41	71.14
45	71.88	72.62	73.36	74.12	74.88
46	75.65	76.43	77.21	78.00	78.80
47	79.60	80.41	81.23	82.05	82.87
48	83.71	84.56	85.42	86.28	87.14
49	88.02	88.90	89.79	90.69	91.59
50	92.51	93.5	94.4	95.3	96.3
51	97.20	98.2	99.1	100.1	101.1
52	102.09	103.1	104.1	105.1	106.2
53	107.20	108.2	109.3	110.4	111.4
54	112.51	113.6	114.7	115.8	116.9
55	118.04	119.1	120.3	121.5	122.6
56	123.80	125.0	126.2	127.4	128.6
57	129.82	131.0	132.3	133.5	134.7
58	136.08	137.3	138.5	139.9	141.2
59	142.60	143.9	145.2	146.6	148.0
60	149.38	150.7	152.1	153.5	155.0
61	156.43	157.8	159.3	160.8	162.3
62	163.77	165.2	166.8	168.3	169.8
63	171.38	172.9	174.5	176.1	177.7
64	179.31	180.9	182.5	184.2	185.8
65	187.54	189.2	190.9	192.6	194.3
66	196.09	197.8	199.5	201.3	203.1
67	204.96	206.8	208.6	210.5	212.3
68	214.17	216.0	218.0	219.9	221.8
69	223.73	225.7	227.7	229.7	231.7
70	233.7	235.7	237.7	239.7	241.8
71	243.9	246.0	248.2	250.3	252.4
72	254.6	256.8	259.0	261.2	263.4
73	265.7	268.0	270.2	272.6	274.8
74	277.2	279.4	281.8	284.2	286.6
75	289.1	291.5	294.0	296.4	298.8
76	301.4	303.8	306.4	308.9	311.4
77	314.1	316.6	319.2	322.0	324.6
78	327.3	330.0	332.8	335.6	338.2
79	341.0	343.8	346.6	349.4	352.2
80	355.1	358.0	361.0	363.8	366.8
81	369.7	372.6	375.6	378.8	381.8
82	384.9	388.0	391.2	394.4	397.4
83	400.6	403.8	407.0	410.2	413.6
84	416.8	420.2	423.6	426.8	430.2
85	433.6	437.0	440.4	444.0	447.5
86	450.9	454.4	458.0	461.6	465.2
87	468.7	472.4	476.0	479.8	483.4
88	487.1	491.0	494.7	498.5	502.2
89	506.1	510.0	513.9	517.8	521.8
90	525.76	529.77	533.80	537.86	541.95
91	546.05	550.18	554.35	558.53	562.75
92	566.99	571.26	575.55	579.87	584.22
93	588.60	593.00	597.43	601.89	606.38
94	610.90	615.44	620.01	624.61	629.24
95	633.90	638.59	643.30	648.05	652.82
96	657.62	662.45	667.31	672.20	677.12
97	682.07	687.04	692.05	697.10	702.17
98	707.27	712.40	717.56	722.75	727.98
99	733.24	738.53	743.85	749.20	754.58
100	760.00	765.45	770.93	776.44	782.00
101	787.57	793.18	798.82	804.50	810.21

VAPOR PRESSURE OF WATER ABOVE 100° C.

Based on values given by Keyes in the International Critical Tables.

Temp. °C	Pressure mm	Pressure Pounds per sq. in.	Temp. °F	Temp. °C	Pressure mm	Pressure Pounds per sq. in.	Temp. °F	Temp °C	Pressure mm	Pressure Pounds per sq. in.	Temp. °F	Temp. °C	Pressure mm	Pressure Pounds per sq. in.	Temp. °F
100	760.	14.696	212.0	170	5940.92	114.879	338.0	240	25100.52	485.365	464.0	310	74024.00	1431.390	590.0
101	787.51	15.228	213.8	171	6085.32	117.671	339.8	241	25543.60	493.933	465.8	311	75042.40	1451.083	591.8
102	815.86	15.776	215.6	172	6233.52	120.537	341.6	242	25994.28	502.647	467.6	312	76076.00	1471.070	593.6
103	845.12	16.342	217.4	173	6383.24	123.432	343.4	243	26449.52	511.450	469.4	313	77117.20	1491.203	595.4
104	875.06	16.921	219.2	174	6538.28	126.430	345.2	244	26912.36	520.400	471.2	314	78166.00	1511.484	597.2
105	906.07	17.521	221.0	175	6694.08	129.442	347.0	245	27381.28	529.467	473.0	315	79230.00	1532.058	599.0
106	937.92	18.136	222.8	176	6852.92	132.514	348.8	246	27855.52	538.638	474.8	316	80294.00	1552.632	600.8
107	970.60	18.768	224.6	177	7015.56	135.659	350.6	247	28335.84	547.926	476.6	317	81373.20	1573.501	602.6
108	1004.42	19.422	226.4	178	7180.48	138.848	352.4	248	28823.76	557.360	478.4	318	82467.60	1594.663	604.4
109	1038.92	20.089	228.2	179	7349.20	142.110	354.2	249	29317.00	566.898	480.2	319	83569.60	1615.972	606.2
110	1074.56	20.779	230.0	180	7520.20	145.417	356.0	250	29817.84	576.583	482.0	320	84686.80	1637.575	608.0
111	1111.20	21.487	231.8	181	7694.24	148.782	357.8	251	30324.00	586.370	483.8	321	85819.20	1659.472	609.8
112	1148.74	22.213	233.6	182	7872.08	152.221	359.6	252	30837.76	596.305	485.6	322	86959.20	1681.516	611.6
113	1187.42	22.961	235.4	183	8052.96	155.719	361.4	253	31356.84	606.342	487.4	323	88114.40	1703.854	613.4
114	1227.25	23.731	237.2	184	8236.88	159.275	363.2	254	31885.04	616.556	489.2	324	89277.20	1726.339	615.2
115	1267.98	24.519	239.0	185	8423.84	162.890	365.0	255	32417.80	626.858	491.0	325	90447.60	1748.971	617.0
116	1309.94	25.330	240.8	186	8616.12	166.609	366.8	256	32957.40	637.292	492.8	326	91633.20	1771.897	618.8
117	1352.95	26.162	242.6	187	8809.92	170.356	368.6	257	33505.36	647.888	494.6	327	92826.40	1794.969	620.6
118	1397.18	27.017	244.4	188	9007.52	174.177	370.4	258	34059.40	658.601	496.4	328	94042.40	1818.483	622.4
119	1442.63	27.896	246.2	189	9208.16	178.057	372.2	259	34618.76	669.417	498.2	329	95273.60	1842.291	624.2
120	1489.14	28.795	248.0	190	9413.36	182.025	374.0	260	35188.00	680.425	500.0	330	96512.40	1866.245	626.0
121	1536.80	29.717	249.8	191	9620.08	186.022	375.8	261	35761.80	691.520	501.8	331	97758.80	1890.346	627.8
122	1586.04	30.669	251.6	192	9831.36	190.107	377.6	262	36343.20	702.763	503.6	332	99020.40	1914.742	629.6
123	1636.36	31.642	253.4	193	10047.20	194.281	379.4	263	36932.20	714.152	505.4	333	100297.20	1939.431	631.4
124	1687.81	32.637	255.2	194	10265.32	198.499	381.2	264	37529.56	725.703	507.2	334	101581.60	1964.267	633.2
125	1740.93	33.664	257.0	195	10488.76	202.819	383.0	265	38133.00	737.372	509.0	335	102881.20	1989.398	635.0
126	1795.12	34.712	258.8	196	10715.24	207.199	384.8	266	38742.52	749.158	510.8	336	104196.00	2014.822	636.8
127	1850.83	35.789	260.6	197	10944.76	211.637	386.6	267	39361.92	761.135	512.6	337	105526.00	2040.540	638.6
128	1907.83	36.891	262.4	198	11179.60	216.178	388.4	268	39986.64	773.215	514.4	338	106871.20	2066.552	640.4
129	1966.35	38.023	264.2	199	11417.48	220.778	390.2	269	40619.72	785.457	516.2	339	108224.00	2092.710	642.2
130	2026.16	39.180	266.0	200	11659.16	225.451	392.0	270	41261.16	797.861	518.0	340	109592.00	2119.163	644.0
131	2087.42	40.364	267.8	201	11905.40	230.213	393.8	271	41910.20	810.411	519.8	341	110967.60	2145.763	645.8
132	2150.42	41.582	269.6	202	12155.44	235.048	395.6	272	42566.08	823.094	521.6	342	112358.40	2172.657	647.6
133	2214.64	42.824	271.4	203	12408.52	239.942	397.4	273	43229.56	835.923	523.4	343	113749.20	2199.550	649.4
134	2280.76	44.103	273.2	204	12666.16	244.924	399.2	274	43902.16	848.929	525.2	344	115178.00	2227.179	651.2
135	2347.26	45.389	275.0	205	12929.12	250.008	401.0	275	44580.84	862.053	527.0	345	116614.40	2254.954	653.0
136	2416.34	46.724	276.8	206	13197.40	255.196	402.8	276	45269.40	875.367	528.8	346	118073.60	2283.171	654.8
137	2488.16	48.113	278.6	207	13467.96	260.428	404.6	277	45964.04	888.799	530.6	347	119532.80	2311.387	656.6
138	2560.67	49.515	280.4	208	13742.32	265.733	406.4	278	46669.32	902.437	532.4	348	121014.80	2340.044	658.4
139	2634.84	50.950	282.2	209	14022.76	271.156	408.2	279	47382.20	916.222	534.2	349	122504.40	2368.848	660.2
140	2710.92	52.421	284.0	210	14305.48	276.623	410.0	280	48104.20	930.183	536.0	350	124001.60	2397.799	662.0
141	2788.44	53.920	285.8	211	14595.04	282.222	411.8	281	48833.80	944.291	537.8	351	125521.60	2427.191	663.8
142	2867.48	55.448	287.6	212	14888.40	287.895	413.6	282	49570.24	958.532	539.6	352	127049.20	2456.730	665.6
143	2948.80	57.020	289.4	213	15184.80	293.626	415.4	283	50316.56	972.963	541.4	353	128599.60	2486.710	667.4
144	3031.64	58.622	291.2	214	15488.04	299.490	417.2	284	51072.76	987.586	543.2	354	130157.60	2516.837	669.2
145	3116.76	60.268	293.0	215	15792.80	305.383	419.0	285	51838.08	1002.385	545.0	355	131730.80	2547.258	671.0
146	3203.40	61.944	294.8	216	16104.40	311.408	420.8	286	52611.76	1017.345	546.8	356	133326.80	2578.119	672.8
147	3292.32	63.663	296.6	217	16420.56	317.522	422.6	287	53395.32	1032.497	548.6	357	134945.60	2609.422	674.6
148	3382.76	65.412	298.4	218	16742.04	323.738	424.4	288	54187.24	1047.810	550.4	358	136579.60	2641.018	676.4
149	3476.24	67.220	300.2	219	17067.32	330.028	426.2	289	54989.04	1063.314	552.2	359	138228.80	2672.908	678.2
150	3570.48	69.042	302.0	220	17395.64	336.377	428.0	290	55799.20	1078.980	554.0	360	139893.20	2705.093	680.0
151	3667.00	70.908	303.8	221	17731.56	342.872	429.8	291	56612.40	1094.705	555.8	361	141572.80	2737.571	681.8
152	3766.56	72.833	305.6	222	18072.80	349.471	431.6	292	57448.40	1110.871	557.6	362	143275.20	2770.490	683.6
153	3866.88	74.773	307.4	223	18417.84	356.143	433.4	293	58284.40	1127.036	559.4	363	144992.80	2803.703	685.4
154	3970.24	76.772	309.2	224	18766.68	362.888	435.2	294	59135.60	1143.496	561.2	364	146733.20	2837.357	687.2
155	4075.88	78.815	311.0	225	19123.12	369.781	437.0	295	59994.40	1160.102	563.0	365	148519.20	2871.892	689.0
156	4183.80	80.901	312.8	226	19482.60	376.732	438.8	296	60860.80	1176.856	564.8	366	150320.40	2906.722	690.8
157	4293.24	83.018	314.6	227	19848.92	383.815	440.6	297	61742.40	1193.903	566.6	367	152129.20	2941.698	692.6
158	4404.96	85.178	316.4	228	20219.80	390.987	442.4	298	62624.00	1210.950	568.4	368	153960.80	2977.116	694.4
159	4519.72	87.397	318.2	229	20596.76	398.276	444.2	299	63528.40	1228.439	570.2	369	155815.20	3012.974	696.2
160	4636.00	89.646	320.0	230	20978.28	405.654	446.0	300	64432.80	1245.927	572.0	370	157692.40	3049.273	698.0
161	4755.32	91.953	321.8	231	21365.12	413.134	447.8	301	65352.40	1263.709	573.8	371	159584.80	3085.866	699.8
162	4876.92	94.304	323.6	232	21757.28	420.717	449.6	302	66279.60	1281.638	575.6	372	161507.60	3123.047	701.6
163	5000.04	96.685	325.4	233	22154.00	428.388	451.4	303	67214.40	1299.714	577.4	373	163468.40	3160.963	703.4
164	5126.96	99.139	327.2	234	22558.32	436.207	453.2	304	68156.80	1317.937	579.2	374	165467.20	3199.613	705.2
165	5256.16	101.638	329.0	235	22967.96	444.128	455.0	305	69114.40	1336.454	581.0				
166	5386.88	104.165	330.8	236	23382.92	452.152	456.8	306	70072.00	1354.971	582.8				
167	5521.40	106.766	332.6	237	23802.44	460.264	458.6	307	71052.40	1373.929	584.6				
168	5658.20	109.412	334.4	238	24229.56	468.523	460.4	308	72048.00	1393.181	586.4				
169	5798.04	112.116	336.2	239	24661.24	476.871	462.2	309	73028.40	1412.139	588.2				

VAPOR PRESSURE OF MERCURY

Vapor pressure of mercury in mm. of Hg for temperatures from −38 to 400°C. Note that the values for the first four lines only, are to be multiplied by 10^{-6}

Temp. °C	0	2	4	6	8
	10^{-6}	10^{-6}	10^{-6}	10^{-6}	10^{-6}
−30	4.78	3.59	2.66	1.97	1.45
−20	18.1	14.0	10.8	8.28	6.30
−10	60.6	48.1	38.0	29.8	23.2
− 0	185.	149.	119.	95.4	76.2
+ 0	.000185	.000228	.000276	.000335	.000406
+10	.000490	.000588	.000706	.000846	.001009
20	.001201	.001426	.001691	.002000	.002359
30	.002777	.003261	.003823	.004471	.005219
40	.006079	.007067	.008200	.009497	.01098
50	.01267	.01459	.01677	.01925	.02206
60	.02524	.02883	.03287	.03740	.04251
70	.04825	.05469	.06189	.06993	.07889
80	.08880	.1000	.1124	.1261	.1413
90	.1582	.1769	.1976	.2202	.2453
100	.2729	.3032	.3366	.3731	.4132
110	.4572	.5052	.5576	.6150	.6776
120	.7457	.8198	.9004	.9882	1.084
130	1.186	1.298	1.419	1.551	1.692
140	1.845	2.010	2.188	2.379	2.585
150	2.807	3.046	3.303	3.578	3.873
160	4.189	4.528	4.890	5.277	5.689
170	6.128	6.596	7.095	7.626	8.193
180	8.796	9.436	10.116	10.839	11.607
190	12.423	13.287	14.203	15.173	16.200

Temp. °C	0	2	4	6	8
200	17.287	18.437	19.652	20.936	22.292
210	23.723	25.233	26.826	28.504	30.271
220	32.133	34.092	36.153	38.318	40.595
230	42.989	45.503	48.141	50.909	53.812
240	56.855	60.044	63.384	66.882	70.543
250	74.375	78.381	82.568	86.944	91.518
260	96.296	101.28	106.48	111.91	117.57
270	123.47	129.62	136.02	142.69	149.64
280	156.87	164.39	172.21	180.34	188.79
290	197.57	206.70	216.17	226.00	236.21
300	246.80	257.78	269.17	280.98	293.21
310	305.89	319.02	332.62	346.70	361.26
320	376.33	391.92	408.04	424.71	441.94
330	459.74	478.13	497.12	516.74	537.00
340	557.90	579.45	601.69	624.64	648.30
350	672.69	697.83	723.73	750.43	777.92
360	806.23	835.38	865.36	896.23	928.02
370	960.66	994.34	1028.9	1064.4	1100.9
380	1138.4	1177.0	1216.6	1257.3	1299.1
390	1341.9	1386.1	1431.3	1477.7	1525.2
400	1574.1	....	....	....	....

VAPOR PRESSURES OF SOME OF THE ELEMENTS* AT VARIOUS TEMPERATURES °C

Element	Melting point (°C)	Pressure (mm Hg)					
		10^{-5}	10^{-4}	10^{-3}	10^{-2}	10^{-1}	1
Ag	961	767	848	936	1047	1184	1353
Al	660	724	808	889	996	1123	1279
Au	1063	1083	1190	1316	1465	1646	1867
Ba	717	418	476	546	629	730	858
Be	1284	942	1029	1130	1246	1395	1582
Bi	271	474	536	609	698	802	934
C	....	2129	2288	2471	2681	2926	3214
Cd	321	148	180	220	264	321	....
Co	1478	1249	1362	1494	1649	1833	2056
Cr	1900	907	992	1090	1205	1342	1504
Cu	1083	946	1035	1141	1273	1432	1628
Fe	1535	1094	1195	1310	1447	1602	1783
Hg	−38.9	−23.9	−5.5	18.0	48.0	82.0	126
In	157	667	746	840	952	1088	1260
Ir	2454	1993	2154	2340	2556	2811	3118
Mg	651	287	331	383	443	515	605
Mn	1244	717	791	878	980	1103	1251
Mo	2622	1923	2095	2295	2533	....	....
Ni	1455	1157	1257	1371	1510	1679	1884
Os	2697	2101	2264	2451	2667	2920	3221
Pb	328	483	548	625	718	832	975
Pd	1555	1156	1271	1405	1566	1759	2000
Pt	1774	1606	1744	1904	2090	2313	2582
Sb	630	466	525	595	678	779	904
Si	1410	1024	1116	1223	1343	1485	1670
Sn	232	823	922	1042	1189	1373	1609
Ta	2996	2407	2599	2820	....	....	....
W	3382	2554	2767	3016	3309	....	....
Zn	419	211	248	292	343	405	....
Zr	2127	1527	1660	1816	2001	2212	2459

* The values given in this table are from a variety of sources, not all of which are in agreement. The table is to be used, therefore, only as a general guide. For a detailed discussion of vapor pressure data, see Dr. Dushman's book, pages 752–754.

VAPOR PRESSURE OF CARBON DIOXIDE

SOLID

From Bureau of Standards Journal of Research

(Mercury column, density = 13.5951 g/cm³, g = 980.665)

Pressure in microns of mercury

°C	0	1	2	3	4	5	6	7	8	9
−180	0.013	0.008	0.006	0.004	0.003	0.0017	0.0011	0.0007	0.0005	0.0003
−170	.37	.27	.20	.14	.10	.074	.052	.037	.026	.018
−160	5.9	4.6	3.6	2.7	2.1	1.58	1.19	.90	.67	.50
−150	60.5	48.8	39.2	31.4	25.1	19.9	15.8	12.4	9.8	7.6
−140	431	359	298	247	204	168	138	113	92	75

Pressure in mm of mercury

°C	0	1	2	3	4	5	6	7	8	9
−130	2.31	1.97	1.68	1.43	1.22	1.03	0.87	0.73	0.61	0.51
−120	9.81	8.57	7.46	6.49	5.63	4.88	4.22	3.64	3.13	2.69
−110	34.63	30.76	27.27	24.14	21.34	18.83	16.58	14.58	12.80	11.22
−100	104.81	94.40	84.91	76.27	68.43	61.30	54.84	48.99	43.71	38.94
− 90	279.5	254.7	231.8	210.8	191.4	173.6	157.3	142.4	128.7	116.2
− 80	672.2	618.3	568.2	521.7	478.5	438.6	401.6	367.4	335.7	306.5
− 70	1486.1	1377.3	1275.6	1180.5	1091.7	1008.9	931.7	859.7	792.7	730.3
− 60	3073.1	2865.1	2669.7	2486.3	2314.2	2152.8	2001.5	1859.7	1726.9	1602.5
− 50	......	......	......	......	......	......	......	3780.9	3530.2	3294.6

LIQUID

°C	0	1	2	3	4	5	6	7	8	9
−50	5127.8	4922.7	4723.9	4531.1	4344.3	4163.2	3987.9	3818.2*	3653.9*	3495.0*
−40	7545	7271	7005	6746	6494	6250	6012	5781	5557	5339
−30	10718	10363	10017	9679	9350	9029	8716	8412	8115	7826
−20	14781	14331	13891	13461	13040	12630	12229	11838	11455	11082
−10	19872	19312	18764	18228	17703	17189	16686	16194	15712	15241
− 0	26142	25457	24786	24127	23482	22849	22229	21622	21026	20443
0	26142	26840	27552	28277	29017	29771	30539	31323	32121	32934
10	33763	34607	35467	36343	37236	38146	39073	40017	40980	41960
20	42959	43977	45014	46072	47150	48250	49370	50514	51680	52871
30	54086	55327	......	......	......	......	......	......	......	......

* Undercooled liquid.

Critical temperature = 31.0°C. Triple point, −56.602 ± 0.005°C; 3885.2 ± 0.4 mm.

VAPOR PRESSURE

The following table is an abridged form of the very extensive compilation by Daniel R. Stull and published in Industrial and Engineering Chemistry **39,** 517 (1947).

The table gives the temperatures in degrees centigrade at which the vapor of the compound listed at the left has the pressure indicated at the top of the column. Organic compounds are listed in the order of their empirical formulae and inorganic compounds in the alphabetic order of their names. Pressures greater than one atmosphere are listed in separate tables.

Abbreviations:

d = decomposes
d = dextrorotatory
dl = inactive (50% *d* and 50% *l*)
e = explodes
l = levorotatory
M.P. = melting point
p = polymerizes
s = solid

INORGANIC COMPOUNDS
Pressures Less than One Atmosphere

Name	Formula	Temperature, °C						M.P.	Citation No.
		1 mm	10 mm	40 mm	100 mm	400 mm	760 mm		
Aluminum	Al	1284	1487	1635	1749	1947	2056	660	(153, 229, 469)
Aluminum borohydride	AlB_3H_{12}	s	−42.9	−20.9	−3.9	+28.1	45.9	−64.5	(386)
Aluminum bromide	$AlBr_3$	81.3$_s$	118.0	150.6	176.1	227.0	256.3	97.5	(9, 127, 229)
Aluminum chloride	$AlCl_3$	100.0$_s$	123.8$_s$	139.9$_s$	152.0$_s$	171.6$_s$	180.2$_s$	192.4	(127, 130, 229, 264, 409, 410, 457)
Aluminum fluoride	AlF_3	1238	1324	1378	1422	1496	1537	1040	(362)
Aluminum iodide	AlI_3	178.0$_s$	225.8	265.0	294.5	354.0	385.5	......	(127, 229)
Aluminum oxide	Al_2O_3	2148	2385	2549	2665	2874	2977	2050	(229, 359, 369)
Ammonia	NH_3	−109.1$_s$	−91.9$_s$	−79.2$_s$	−68.4	−45.4	−33.6	−77.7	(23, 25, 33, 49, 62, 82, 142, 175, 192, 223, 231, 279, 305, 334, 420)
Deutero ammonia	ND_3	s	s	s	−67.4	−45.4	−33.4	−74.0	(443)
Ammonium azide	NH_4N_3	29.2$_s$	59.2$_s$	80.1$_s$	95.2$_s$	120.4$_s$	133.8$_s$	......	(131)
Ammonium bromide	NH_4Br	198.3$_s$	252.0$_s$	290.0$_s$	320.0$_s$	370.9$_s$	396.0$_s$	......	(403)
Ammonium carbamate	$NH_4CO_2NH_2$	−26.1$_s$	−2.9$_s$	+14.0$_s$	26.7$_s$	48.0$_s$	58.3$_s$	......	(48, 107)
Ammonium chloride	NH_4Cl	160.4$_s$	209.8$_s$	245.0$_s$	271.5$_s$	316.5$_s$	337.8$_s$	520	(333, 403)
Ammonium hydrogen sulfide	NH_4HS	−51.1	−28.7	−12.3	0.0	+21.8	33.3	......	(200)
Ammonium iodide	NH_4I	210.9$_s$	263.5$_s$	302.8$_s$	331.8$_s$	381.0$_s$	404.9$_s$	......	(403)
Ammonium cyanide	NH_4CN	−50.6$_s$	−28.6$_s$	−12.6$_s$	−0.5$_s$	+20.5$_s$	31.7$_s$	36	(201)
Antimony	Sb	886	1033	1141	1223	1364	1440	630.5	(153, 229, 255, 348, 485)
Antimony tribromide	$SbBr_3$	93.9	142.7	177.4	203.5	250.2	275.0	96.6	(9, 102)
Antimony trichloride	$SbCl_3$	49.2$_s$	85.2	117.8	143.3	192.2	219.0	73.4	(8, 9, 43, 102, 229, 264)
Antimony pentachloride	$SbCl_5$	22.7	61.8	91.0	114.1	....	....	2.8	(8, 9, 43, 229)
Antimony triiodide	SbI_3	163.6$_s$	223.5	267.8	303.5	368.5	401.0	167	(9, 102)
Antimony trioxide	Sb_2O_3	574$_s$	666	812	957	1242	1425	656	(183, 229)
Argon	A	−218.2$_s$	−210.9$_s$	−204.9$_s$	−200.5$_s$	−190.6$_s$	−185.6	−189.2	(41, 83, 84, 85, 193, 229, 295, 329, 466)
Arsenic (metallic)	As	372$_s$	437$_s$	483$_s$	518$_s$	579$_s$	610$_s$	814	(143, 194, 229, 322, 348, 367)
Arsenic tribromide	$AsBr_3$	41.8	85.2	118.7	145.2	193.6	220.0	......	(9)
Arsenic trichloride	$AsCl_3$	−11.4	+23.5	50.0	70.9	109.7	130.4	−18	(9, 20, 229, 264)
Arsenic trifluoride	AsF_3	s	s	−2.5	+13.2	41.5	56.3	−5.9	(376)
Arsenic pentafluoride	AsF_5	−117.9$_s$	−103.1$_s$	−92.4$_s$	−84.3$_s$	−64.0	−52.8	−79.8	(229, 352)
Arsenic hydride (arsine)	AsH_3	−142.6$_s$	−124.7$_s$	−110.2	−98.0	−75.2	−62.1	−116.3	(215)
Arsenic trioxide	As_2O_3	212.5$_s$	259.7$_s$	299.2$_s$	332.5	412.2	457.2	312.8	(229, 287, 373, 402, 405, 418, 479)
Barium	Ba	s	1049	1195	1301	1518	1638	850	(167, 229, 355)
Beryllium borohydride	BeB_2H_8	+1.0$_s$	28.1$_s$	46.2$_s$	58.6$_s$	79.7$_s$	90.0$_s$	123	(60)
Beryllium bromide	$BeBr_2$	289$_s$	342$_s$	379$_s$	405$_s$	451$_s$	474$_s$	490	(229, 326)
Beryllium chloride	$BeCl_2$	291$_s$	346$_s$	384$_s$	411	461	487	405	(229, 326)
Beryllium iodide	BeI_2	283$_s$	341$_s$	382$_s$	411$_s$	461$_s$	487$_s$	488	(229, 326)
Bismuth	Bi	1021	1136	1217	1271	1370	1420	271	(17, 153, 154, 155, 229, 255, 348)
Bismuth tribromide	$BiBr_3$	s	282	327	360	425	461	218	(116, 229)
Bismuth trichloride	$BiCl_3$	s	264	311	343	405	441	230	(116, 229, 264)
Borine carbonyl	BH_3CO	−139.2	−121.1	−106.6	−95.3	−74.8	−64.0	−137.0	(59)
Boron tribromide	BBr_3	−41.4	−10.1	+14.0	33.5	70.0	91.7	−45	(229, 421)
Boron trichloride	BCl_3	−91.5	−66.9	−47.8	−32.4	−3.6	+12.7	−107	(229, 308, 334, 430)
Boron trifluoride	BF_3	−154.6	−141.3$_s$	−131.0$_s$	−123.0	−108.3	−110.7	−126.8	(39, 118, 229. 319, 352)
Dihydrodiborane	B_2H_6	−159.7	−144.3	−131.6	−120.9	−99.6	−86.5	−169	(229, 419, 423)
Diborane hydrobromide	B_2BrH_5	−93.3	−66.3	−45.4	−29.0	0.0	+16.3	−104.2	(229, 424)
Triborine triamine	$B_3H_6N_3$	−63.0$_s$	−35.3	−13.2	+4.0	34.3	50.6	−58.2	(229, 427)
Tetrahydrotetraborane	B_4H_{10}	−90.9	−64.3	−44.3	−28.1	+0.8	16.1	−119.9	(229, 423, 425, 426)
Dihydropentaborane	B_5H_9	s	−30.7	−8.0	+9.6	40.8	58.1	−47.0	(229, 423)
Tetrahydropentaborane	B_5H_{11}	−50.2	−19.9	+2.7	20.1	51.2	67.0	......	(229, 426)
Dihydrodecaborane	$B_{10}H_{14}$	60.0$_s$	90.2$_s$	117.4	142.3	d	....	99.6	(229, 428)
Bromine	Br_2	−48.7$_s$	−25.0$_s$	−8.0$_s$	+9.3	41.0	58.2	−7.3	(88, 102, 171, 203, 217, 229, 281, 331, 345, 381, 458, 487)
Bromine pentafluoride	BrF_5	−69.3$_s$	−41.9	−21.0	−4.5	+25.7	40.0	−61.4	(229, 365)
Cadmium	Cd	394	484	553	611	711	765	320.9	(17, 42, 61, 109, 111, 129, 163, 181, 210, 229, 243, 348)

INORGANIC COMPOUNDS (Continued)
Pressures Less than One Atmosphere (Continued)

Name	Formula	Temperature, °C						M.P.	Citation No.
		1 mm	10 mm	40 mm	100 mm	400 mm	760 mm		
Cadmium chloride	$CdCl_2$	s	656	736	797	908	967	568	(156, 229, 264)
Cadmium fluoride	CdF_2	1112	1286	1400	1486	1651	1751	520	(362)
Cadmium iodide	CdI_2	416	512	584	640	742	796	385	(229, 387)
Cadmium oxide	CdO	1000_s	1149_s	1257_s	1341_s	1484_s	1559_s	....	(119, 184, 229)
Calcium	Ca	s	983	1111	1207	1388	1487	851	(167, 229, 318, 346, 355)
Carbon	C	3586_s	3946_s	4196_s	4373_s	4660_s	4827_s	.....	(4, 176, 177, 229, 241, 242, 266, 377, 445, 480)
Carbon tetrabromide	CBr_4	s	s	96.3	119.7	163.5	189.5	90.1	(38)
Carbon tetrachloride	CCl_4	-50.0_s	−19.6	+4.3	23.0	57.8	76.7	−22.6	(104, 179, 279, 334, 335, 413, 495)
Carbon tetrafluoride	CF_4	-184.6_s	−169.3	−158.8	−150.7	−135.5	−127.7	−183.7	(229, 270)
Carbon dioxide	CO_2	-134.3_s	-119.5_s	-108.6_s	-100.2_s	-85.7_s	-78.2_s	−57.5	(5, 46, 117, 172, 175, 211, 229, 272, 292, 302, 334, 420, 464, 467, 477, 497)
Carbon suboxide	C_3O_2	−94.8	−71.0	−52.0	−36.9	−8.9	+6.3	−107	(237, 435)
Carbon disulfide	CS_2	−73.8	−44.7	−22.5	−5.1	+28.0	46.5	−110.8	(175, 229, 265, 328, 334, 335, 420, 467, 490)
Carbon subsulfide	C_3S_2	14.0	54.9	85.6	109.9	p	....	+0.4	(229, 429)
Carbon selenosulfide	$CSSe$	−47.3	−16.0	+8.6	28.3	65.2	85.6	−75.2	(229, 436)
Carbon monoxide	CO	-222.0_s	-215.0_s	-210.0_s	-205.7_s	−196.3	−191.3	−205.0	(16, 67, 77, 78, 87, 229, 293, 463)
Carbonyl chloride	$COCl_2$	−92.9	−69.3	−50.3	−35.6	−7.6	+8.3	−104	(14, 135, 229, 288, 309)
Carbonyl selenide	$COSe$	−117.1	−95.0	−76.4	−61.7	−35.6	−21.9	....	(325)
Carbonyl sulfide	COS	−132.4	−113.3	−98.3	−85.9	−62.7	−49.9	−138.8	(196, 229, 230, 422)
Chloropicrin	CCl_3NO_2	−25.5	+7.8	33.8	53.8	91.8	111.9	−64	(20, 31)
Chlorotrifluoromethane	$CClF_3$	−149.5	−134.1	−121.9	−111.7	−92.7	−81.2	....	(102, 447)
Cyanogen	C_2N_2	-95.8_s	-76.8_s	-62.7_s	-51.8_s	−33.0	−21.0	−34.4	(74, 83, 102, 118, 313, 444)
Cyanogen bromide	$CBrN$	-35.7_s	-10.0_s	$+8.6_s$	22.6_s	46.0_s	61.5	58	(20, 229)
Cyanogen chloride	$CClN$	-76.7_s	-53.8_s	-37.5_s	-24.9_s	−2.3	+13.1	−6.5	(229, 334)
Cyanogen fluoride	CFN	-134.4_s	-118.5_s	-106.4_s	-97.0_s	-80.5_s	-72.6_s	....	(80, 229)
Cyanogen iodide	CIN	25.2_s	57.7_s	80.3_s	97.6_s	126.1_s	141.1_s	....	(229, 493)
Deuterocyanic acid	CDN	-68.9_s	-46.7_s	-30.1_s	-17.5_s	+10.0	26.2	−12	(258)
Dichlorodifluoromethane	CCl_2F_2	−118.5	−97.8	−81.6	−68.6	−43.9	−29.8	....	(102, 144)
Dichlorofluoromethane	$CHCl_2F$	−91.3	−67.5	−48.8	−33.9	−6.2	+8.9	−135	(233)
Chlorodifluoromethane	$CHClF_2$	−122.8	−103.7	−88.6	−76.4	−53.6	−40.8	−160	(40, 233)
Trichlorofluoromethane	CCl_3F	−84.3	−59.0	−39.0	−23.0	+6.8	23.7	....	(233)
Cesium	Cs	279	375	449	509	624	690	28.5	(35, 132, 161, 229, 246, 251, 357, 394, 462)
Cesium bromide	$CsBr$	748	887	993	1072	1221	1300	636	(229, 367, 475)
Cesium chloride	$CsCl$	744	884	989	1069	1217	1300	646	(122, 229, 367, 475)
Cesium fluoride	CsF	712	844	947	1025	1170	1251	683	(229, 370, 371, 475)
Cesium iodide	CsI	738	873	976	1055	1200	1280	621	(229, 367, 475)
Chlorine	Cl_2	-118.0_s	-101.6_s	−84.5	−71.7	−47.3	−33.8	−100.7	(139, 166, 171, 214, 229, 239, 312, 452)
Chlorine fluoride	ClF	s	−139.0	−128.8	−120.8	−107.0	−100.5	−145	(229, 361)
Chlorine trifluoride	ClF_3	s	−71.8	−51.3	−34.7	−4.9	+11.5	−83	(229, 360)
Chlorine monoxide	Cl_2O	−98.5	−73.1	−54.3	−39.4	−12.5	+2.2	−116	(147, 229)
Chlorine dioxide	ClO_2	s	−59.0	−42.8	−29.4	−4.0	+11.1	−59	(229, 234)
Dichlorine hexoxide	Cl_2O_6	+7.5	42.0	68.0	87.7	123.8	142.0	3.5	(149)
Dichlorine heptoxide	Cl_2O_7	−45.3	−13.2	+10.3	29.1	62.2	78.8	−91	(148, 229)
Chlorosulfonic acid	HSO_3Cl	32.0	64.0	87.6	105.3	136.1	151.0_d	−80	(9)
Chromium	Cr	1616	1845	2013	2139	2361	2482	1615	(153, 229)
Chromium carbonyl	$Cr(CO)_6$	36.0	68.3	91.2	108.0	137.2	151.0	....	(484)
Chromyl chloride	CrO_2Cl_2	−18.4	+13.8	38.5	58.0	95.2	117.1	....	(229, 276)
Cobaltous chloride	$COCl_2$	s	s	770	843	974	1050	735	(229, 264)
Cobalt nitrosyl tricarbonyl	$Co(CO)_2NO$	s	s	+11.0	29.0	62.0	80.0	−11	(30)
Columbium pentafluoride	CbF_5	s	86.3	121.5	148.5	198.0	225.0	75.5	(229, 368)
Copper	Cu	1628	1879	2067	2207	2465	2595	1083	(153, 154, 155, 165, 218, 229, 262, 348, 359, 469)
Cuprous bromide	Cu_2Br_2	572	718	844	951	1189	1355	504	(209, 229, 474)
Cuprous chloride	Cu_2Cl_2	546	702	838	960	1249	1490	422	(229, 264, 474)
Cuprous iodide	Cu_2I_2	s	656	786	907	1158	1336	605	(156, 209, 229, 474)
Ferric chloride	$FeCl_2$	194.0_s	235.5_s	256.8_s	272.5_s	298.0_s	319.0	304	(208, 229, 264)
Ferrous chloride	$FeCl_2$	....	700	779	842	961	1026	....	(229, 264)
Fluorine	F_2	−223.0	−214.1	−207.7	−202.7	−193.2	−187.9	−223	(65, 229)
Fluorine monoxide	F_2O	−196.1	−182.3	−173.0	−165.8	−151.9	−144.6	−223.9	(229, 363, 364)
Gallium	Ga	1349	1541	1680	1784	1974	2071	30	(165, 229)
Gallium trichloride	$GaCl_3$	48.0_s	76.5_s	107.5	132.0	176.3	200.0	77.0	(126, 253)
Germanium hydride	GeH_4	−163.0	−145.3	−131.6	−120.3	−100.2	−88.9	−165	(79, 229, 307, 383)
Germanium bromide	$GeBr_4$	s	56.8	88.1	113.2	161.6	189.0	26.1	(45, 229)
Germanium chloride	$GeCl_4$	−45.0	−15.0	+8.0	27.5	63.8	84.0	−49.5	(229, 254, 289)
Trichlorogermane	$GeHCl_3$	−41.3	−13.0	+8.8	26.5	58.3_d	75.0_d	−71.1	(94, 229)
Tetramethylgermanium	$Ge(CH_3)_4$	−73.2	−45.2	−23.4	−6.3	+26.0	44.0	−88	(93, 229)
Digermane	Ge_2H_6	−88.7	−60.1	−38.2	−20.3	+13.3	31.5	−109	(92, 229)
Trigermane	Ge_3H_8	−36.9	−0.9	+26.3	47.9	88.6	110.8	−105.6	(92, 229)

INORGANIC COMPOUNDS (Continued)
Pressures Less than One Atmosphere (Continued)

Name	Formula	Temperature, °C						M.P.	Citation No.
		1 mm	10 mm	40 mm	100 mm	400 mm	760 mm		
Gold	Au	1869	2154	2363	2521	2807	2966	1063	(165, 229, 348, 359, 469)
Helium	He	−271.7	−271.3	−270.7	−270.3	−269.3	−268.6	......	(227, 228, 229, 300, 301)
Hydrogen	H_2	-263.3_s	-261.3_s	-259.6_s	−257.9	−254.5	−252.5	−259.1	(58, 71, 173, 225, 229, 298, 299, 396, 454, 455)
Hydrogen deuteride	HD	......	−259.8	−258.2	−256.6	−253.0	−251.0	......	(395)
Hydrogen bromide	HBr	-138.8_s	-121.8_s	-108.3_s	-97.7_s	−78.0	−66.5	−87.0	(18, 103, 169, 229, 415, 416)
Hydrogen chloride	HCl	-150.8_s	-135.6_s	-123.8_s	−114.0	−95.3	−84.8	−114.3	(50, 69, 103, 118, 141, 169, 175, 223, 229, 416, 420)
Hydrogen cyanide	HCN	-71.0_s	-47.7_s	-30.9_s	-17.8_s	+10.2	25.9	−13.2	(44, 164, 229, 258, 314, 397, 401)
Hydrogen fluoride	HF	s	−65.8	−45.0	−28.2	+2.5	19.7	−83.7	(76, 229, 398, 400)
Hydrogen iodide	HI	-123.3_s	-102.3_s	$-85\ 6_s$	-72.1_s	−48.3	−35.1	−50.9	(103, 169, 229, 415, 416)
Hydrogen peroxide	H_2O_2	15.3	50.4	77.0	97.9	137.4_d	158.0_d	−0.9	(229, 260)
Hydrogen selenide	H_2Se	-115.3_s	-97.9_s	-84.7_s	-74.2_s	−53.6	−41.1	−64	(57, 90, 91, 229, 292, 417)
Hydrogen sulfide	H_2S	-134.3_s	-116.3_s	-102.3_s	-91.6_s	−71.8	−60.4	−85.5	(118, 137, 229, 235, 236, 292, 334, 415, 416)
Hydrogen disulfide	H_2S_2	−43.2	−15.2	+6.0	22.0	49.6	64.0	−89.7	(64, 229)
Hydrogen teluride	H_2Te	-96.4_s	-75.4_s	-59.1_s	−45.7	−17.2	−2.0	−49.0	(56, 229, 417)
Hydroxylamine	NH_2OH	s	47.2	64.6	77.5	99.2	110.0	34.0	(9, 55, 259)
Iodine	I_2	38.7_s	73.2_s	97.5_s	116.5	159.8	183.0	112.9	(10, 21, 22, 96, 136, 145, 159, 229, 282, 331, 336, 440, 483, 486)
Iodine pentafluoride	IF_5	−15.2	+8.5	32.2	50.0	81.2	97.0	8.0	(351)
Iodine heptafluoride	IF_7	-87.0_s	-63.0_s	-45.3_s	-31.9_s	$-8\ 3_s$	$+4.0_s$	5.5	(229, 358)
Iron	Fe	1787	2039	2224	2360	2605	2735	1535	(153, 218, 229, 350)
Iron pentacarbonyl	$Fe(CO)_5$	......	+4.6	30.3	50.3	86.1	105.0	−21	(229, 450)
Krypton	Kr	-199.3_s	-187.2_s	-178.4_s	-171.8_s	-159.0_s	−152.0	−156.7	(3, 221, 229, 315, 317, 329)
Lead	Pb	973	1162	1309	1421	1630	1744	327.5	(110, 153, 154, 155, 165, 197, 229, 255, 341, 342, 348, 469, 471)
Lead bromide	$PbBr_2$	513	610	686	745	856	914	373	(156, 229, 465, 474)
Lead chloride	$PbCl_2$	547	648	725	784	893	954	501	(105, 156, 207, 229, 465, 474)
Lead fluoride	PbF_2	s	904	1003	1080	1219	1293	855	(229, 474)
Lead iodide	PbI	479	571	644	701	807	872	402	(156, 209, 229)
Lead oxide	PbO	943	1085	1189	1265	1402	1472	890	(119, 229)
Lead sulfide	PbS	852_s	975_s	1048_s	1108_s	1221	1281	1114	(229, 382)
Lithium	Li	723	881	1003	1097	1273	1372	186	(36, 37, 167, 229, 257)
Lithium bromide	LiBr	748	888	994	1076	1226	1310	547	(229, 367, 475)
Lithium chloride	LiCl	783	932	1045	1129	1290	1382	614	(229, 264, 367, 475)
Lithium fluoride	LiF	1047	1211	1333	1425	1591	1681	870	(229, 371, 475)
Lithium iodide	LiI	723	841	927	993	1110	1171	446	(229, 367, 475)
Magnesium	Mg	621_s	743	838	909	1034	1107	651	(153, 167, 229, 355, 469)
Magnesium chloride	$MgCl_2$	778	930	1050	1142	1316	1418	712	(229, 264)
Manganese	Mn	1292	1505	1666	1792	2029	2151	1260	(133, 153, 229, 349)
Manganous chloride	$MnCl_2$	s	778	879	960	1108	1190	650	(229, 264)
Mercury	Hg	126.2	184.0	228.8	261.7	323.0	357.0	−38.9	(26, 34, 109, 134, 159, 178, 181, 182, 185, 210, 222, 229, 240, 243, 271, 273, 277, 285, 316, 320, 330, 334, 342, 348, 439, 460, 494)
Mercuric bromide	$HgBr_2$	136.5_s	179.8_s	211.5_s	237.8	290.0	319.0	237	(212, 220, 229, 324, 465, 483)
Mercuric chloride	$HgCl_2$	136.2_s	180.2_s	212.5_s	237.0_s	275.5_s	304.0	277	(124, 212, 229, 324, 336, 387, 438, 483)
Mercuric iodide	HgI_2	157.5_s	204.5_s	238.2_s	261.8	324.2	354.0	259	(98, 212, 229, 324, 339, 438, 483)
Molybdenum	Mo	3102	3535	3859	4109	4553	5560	2622	(218, 229, 252)
Molybdenum hexafluoride	MoF_6	-65.5_s	-40.8_s	-22.1_s	-8.0_s	+17.2	36.0	17	(229, 347)
Molybdenum trioxide	MoO_3	734_s	814	892	955	1082	1151	795	(120, 205, 229)
Neon	Ne	-257.3_s	-254.6_s	-252.6_s	-251.0_s	−248.1	−246.0	−248.7	(71, 229, 297, 455)
Nickel	Ni	1810	2057	2234	2364	2603	2732	1452	(218, 229, 349)
Nickel chloride	$NiCl_2$	671_s	759_s	821_s	866_s	945_s	987_s	1001	(125, 229, 264)

INORGANIC COMPOUNDS (Continued)
Pressures Less than One Atmosphere (Continued)

Name	Formula	Temperature, °C						M.P.	Citation No.
		1 mm	10 mm	40 mm	100 mm	400 mm	760 mm		
Nickel carbonyl	$Ni(CO)_4$	s	s	-23.0	-6.0	+25.8	42.5	-25	(7, 96, 97, 229, 274)
Nitrogen	N_2	-226.1$_s$	-219.1$_s$	-214.0$_s$	-209.7	-200.9	-195.8	-210.0	(15, 70, 86, 100, 123, 138, 173, 174, 193, 229, 292, 321, 463, 467, 489)
Nitrogen trifluoride	NF_3	s	-170.7	-160.2	-152.3	-137.4	-129.0	-183.7	(229, 270)
Nitric oxide	NO	-184.5$_s$	-178.2$_s$	-171.7$_s$	-166.0$_s$	-156.8$_s$	-151.7	-161	(1, 146, 170, 216, 229, 279, 292, 294)
Nitrous oxide	N_2O	-143.4$_s$	-128.7$_s$	-118.3$_s$	-110.3$_s$	-96.2$_s$	-88.5	-90.9	(29, 32, 53, 62, 115, 188, 229, 247, 327, 464)
Nitrogen tetroxide	N_2O_4	-55.6$_s$	-36.7$_s$	-23.9$_s$	-14.7$_s$	+8.0	21.0	-9.3	(19, 108, 158, 229, 275, 317, 332, 374, 380, 385)
Nitrogen pentoxide	N_2O_5	-36.8$_s$	-16.7$_s$	-2.9$_s$	+7.4$_s$	24.4$_s$	32.4	30	(89, 229, 375)
Nitrosyl chloride	$NOCl$	s	s	-60.2	-46.3	-20.3	-6.4	-64.5	(52, 229, 451)
Nitrosyl fluoride	NOF	-132.0	-114.3	-100.3	-88.8	-68.2	-56.0	-134	(366)
Nitroxyl fluoride	NO_2F	-143.7$_s$	-126.2	-112.8	-102.3	-83.2	-72.0	-139	(366)
Osmium tetroxide (white)	O_sO_4	-5.6$_s$	+26.0$_s$	50.5	71.5	109.3	130.0	42	(229, 245, 290, 372, 468)
Osmium tetroxide (yellow)	OsO_4	3.2$_s$	31.3$_s$	51.7$_s$	71.5	109.3	130.0	56	(229, 245, 290, 372, 468)
Oxygen	O_2	-219.1$_s$	-210.6	-204.1	-198.8	-188.8	-182.96	-218.4	(15, 28, 58, 70, 86, 100, 114, 172, 174, 226, 229, 292, 296, 420, 456, 467, 488)
Ozone	O_3	-180.4	-163.2	-150.7	-141.0	-122.5	-111.9	-192.1 ±.1	(229, 337, 338, 414)
Phosphorus (yellow)	P	76.6	128.0	166.7	197.3	251.0	280.0	44.1	(72, 186, 219, 229, 263, 322)
Phosphorus (violet)	P	237$_s$	287$_s$	323$_s$	349$_s$	391$_s$	417$_s$	590	(186, 229, 286, 406, 408)
Phosphorus (black)	P	290$_s$	338$_s$	371$_s$	393$_s$	432$_s$	453$_s$	......	(47, 186, 229, 286, 408)
Phosphorus tribromide	PBr_3	7.8	47.8	79.0	103.6	149.7	175.3	-40	(9)
Phosphorus trichloride	PCl_3	-51.6	-21.3	+2.3	21.0	56.9	74.2	-111.8	(229, 334)
Phosphorus pentachloride	PCl_5	55.5$_s$	83.2$_s$	102.5$_s$	117.0$_s$	147.2$_s$	162.0$_s$	......	(403)
Phosphorus hydride (Phosphene)	PH_3	s	s	-129.4	-118.8	-98.3	-87.5	-132.5	(50, 175, 229, 416, 420)
Phosphonium bromide	PH_4Br	-43.7$_s$	-21.2$_s$	-5.0$_s$	+7.4$_s$	28.0$_s$	38.3$_d$	......	(202, 213)
Phosphonium chloride	PH_4Cl	-91.0$_s$	-74.0$_s$	-61.5$_s$	-52.0$_s$	-35.4$_s$	-27.0$_s$	-28.5	(50, 442)
Phosphonium iodide	PH_4I	-25.2$_s$	-1.1$_s$	16.1$_s$	29.3$_s$	51.6$_s$	62.3$_s$	......	(213, 403)
Phosphorus trioxide	P_2O_3	s	53.0	84.0	108.3	150.3	173.1	22.5	(229, 384, 461)
Phosphorus oxychloride	$POCl_3$	s	2.0	27.3	47.4	84.3	105.1	2	(12, 13, 229)
Phosphorus pentoxide (stable form)	P_2O_5	384$_s$	442$_s$	481$_s$	510$_s$	556$_s$	591	569	(187, 229, 407)
Phosphorus pentoxide (Metastable) form)	P_2O_5	189	236	270	294	336	358	......	(187, 229, 407)
Phosphorus thiobromide	$PSBr_3$	50.0	83.6	108.0	126.3	157.8	175.0$_d$	38	(9)
Phosphorus thiochloride	$PSCl_3$	-18.3	+16.1	42.7	63.8	102.3	124.0	-36.2	(9)
Platinum	Pt	2730	3146	3469	3714	4169	4407	1755	(218, 229, 252)
Potassium	K	341	443	524	586	708	774	62.3	(106, 122, 161, 163, 181, 229, 246, 257, 285, 357, 478)
Potassium bromide	KBr	795	982	1050	1137	1297	1383	730	(122, 229, 367, 473)
Potassium chloride	KCl	821	968	1078	1164	1322	1407	790	(122, 156, 162, 195, 204, 229, 243, 367, 473)
Potassium fluoride	KF	885	1039	1156	1245	1411	1502	880	(229, 371, 475)
Potassium hydroxide	KOH	719	863	976	1064	1233	1327	380	(204, 229, 475)
Potassium iodide	KI	745	887	995	1080	1238	1324	723	(122, 156, 229, 365, 475)
Radon	Rn	-144.2$_s$	-126.3$_s$	-111.3$_s$	-99.0$_s$	-75.0$_s$	-61.8	-71	(152, 229, 244)
Rhenium heptoxide	Re_2O_7	212.5$_s$	248.0$_s$	272.0$_s$	289.0$_s$	336.0	362.4	296	(229, 291)
Rubidium	Rb	297	389	459	514	620	679	38.5	(161, 229, 357, 394)
Rubidium bromide	RbBr	781	923	1031	1114	1267	1352	682	(229, 367, 475)
Rubidium chloride	RbCl	792	937	1047	1133	1294	1381	715	(229, 367, 475)
Rubidium fluoride	RbF	921	1016	1096	1168	1322	1408	760	(229, 371, 475)
Rubidium iodide	RbI	748	884	991	1072	1223	1304	642	(229, 367, 475)
Selenium	Se	356	442	506	554	637	680	217	(99, 206, 229, 322)
Selenium dioxide	SeO_2	157.0$_s$	202.5$_s$	234.1$_s$	258.0$_s$	297.7$_s$	317.0$_s$	340	(6, 206, 229)
Selenium hexafluoride	SeF_6	-118.6$_s$	-98.9$_s$	-84.7$_s$	-73.9$_s$	-55.2$_s$	-45.8$_s$	-34.7	(229, 238, 491)
Selenium oxychloride	$SeOCl_2$	34.8	71.9	98.0	118.0	151.7	168.0	8.5	(229, 256)
Selenium tetrachloride	$SeCl_4$	74.0$_s$	107.4$_s$	130.1$_s$	147.5$_s$	176.4$_s$	191.5$_d$	......	(492)
Silane	SiH_4	-179.3	-163.0	-150.3	-140.5	-122.0	-111.5	-185	(2, 229, 431)
Silicon	Si	**1724**	**1888**	**2000**	**2083**	**2220**	2287	1420	(229, 359, 470)
Silicon dioxide	SiO_2	s	1732	1867	1969	2141	2227	1710	(229, 369)
Silicon tetrachloride	$SiCl_4$	-63.4	-34.4	-12.1	+5.4	38.4	56.8	-68.8	(24, 224, 229, 334, 434)
Silicon tetrafluoride	SiF_4	-144.0$_s$	-130.4$_s$	-120.8$_s$	-113.3$_s$	-100.7$_s$	-94.8$_s$	-90	(40, 229, 310, 347)
Bromosilane	SiH_3Br	s	-77.3	-57.8	-42.3	-13.3	+2.4	-93.9	(229, 431)
Chlorosilane	SiH_3Cl	-117.8	-97.7	-81.8	-68.5	-44.5	-30.4	......	(431)
Fluorosilane	SiH_3F	-153.0	-141.2	-130.8	-122.4	-106.8	-98.0	......	(112)
Iodosilane	SiH_3I	s	-43.7	-21.8	-4.4	+27.9	45.4	-57.0	(112)

VAPOR PRESSURE (Continued)

INORGANIC COMPOUNDS (Continued)

Pressures Less than One Atmosphere (Continued)

Name	Formula	Temperature, °C						M.P.	Citation No.
		1 mm	10 mm	40 mm	100 mm	400 mm	760 mm		
Bromodichlorofluorosilane	$SiBrCl_2F$	−86.5	−59.0	−37.0	−19.5	+15.4	35.4	−112.3	(389)
Bromotrifluorosilane	$SiBrF_3$	s	s	s	s	−55.9	−41.7	−70.5	(388)
Chlorotrifluorosilane	$SiClF_3$	−144.0$_s$	−127.0	−112.8	−101.7	−81.0	−70.0	−142	(40, 229, 392)
Dibromochlorofluorosilane	$SiBr_2ClF$	−65.2	−35.6	−12.0	+6.3	43.0	59.5	−99.3	(389)
Dibromodifluorosilane	$SiBr_2F_2$	s	−66.8	−47.4	−31.9	−2.6	+13.7	−66.9	(389)
Dibromosilane	SiH_2Br_2	−60.9	−29.4	−5.2	+14.1	50.7	70.5	−70.2	(229, 432)
Dichlorodifluorosilane	$SiCl_2F_2$	−124.7	−102.9	−85.0	−70.3	−45.0	−31.8	−139.7	(40, 229, 392)
Difluorosilane	SiH_2F_2	−146.7	−130.4	−117.6	−107.3	−87.6	−77.8	......	(112)
Diiodosilane	SiH_2I_2	s	18.0	52.6	79.4	125.5	149.5	−1.0	(113)
Disilane	Si_2H_6	−114.8	−91.4	−72.8	−57.5	−29.0	−14.3	−132.6	(229, 431, 434)
Disiloxane	$(SiH_3)_2O$	−112.5	−88.2	−70.4	−55.9	−29.3	−15.4	−144.2	(229, 434)
Fluorotrichlorosilane	$SiCl_3F$	−92.6	−68.3	−48.8	−33.2	−4.0	+12.2	−120.8	(40)
Hexachlorodisilane	Si_2Cl_6	+4.0	38.8	65.3	85.4	120.6	139.0	−1.2	(229, 267)
Hexachlorodisiloxane	$(SiCl_3)_2O$	−5.0	+29.4	55.2	75.4	113.6	135.6	−33.2	(229, 434)
Hexafluorodisilane	Si_2F_6	−81.0$_s$	−63.1$_s$	−50.6$_s$	−41.7$_s$	−26.4$_s$	−18.9$_s$	−18.6	(229, 391, 392)
Octachlorotrisilane	Si_3Cl_8	46.3	89.3	121.5	146.0	189.5	211.4	......	(229, 267)
Tetrasilane	Si_4H_{10}	−27.7	+4.3	28.4	47.4	81.7	100.0	−93.6	(229, 431)
Tribromofluorosilane	$SiBr_3F$	−46.1	−15.1	+9.2	28.6	64.6	83.8	−82.5	(388)
Tribromosilane	$SiHBr_3$	−30.5	+3.4	30.0	51.6	90.2	111.8	−73.5	(390)
Trichlorosilane	$SiHCl_3$	−80.7	−53.4	−32.9	−16.4	+14.5	31.8	−126.6	(229, 437)
Trifluorosilane	$SiHF_3$	−152.0$_s$	−138.2$_s$	−127.3	−118.7	−102.8	−95.0	−131.4	(112)
Trisilane	Si_3H_8	−68.9	−40.0	−16.9	+1.6	35.5	53.1	−117.2	(229, 431)
Disilazane	$(SiH_3)_3N$	−68.7	−40.4	−18.5	−1.1	+31.0	48.7	−105.7	(229, 433)
Silver	Ag	1357	1575	1743	1865	2090	2212	960.5	(153, 154, 155, 163, 165, 218, 229, 348, 469, 471)
Silver chloride	$AgCl$	912	1074	1200	1297	1467	1564	455	(229, 264, 474)
Silver iodide	AgI	820	983	1111	1210	1400	1506	552	(209, 229)
Sodium	Na	439	549	633	701	823	892	97.5	(106, 134, 150, 160, 161, 163, 181, 229, 248, 257, 340, 343, 344, 357, 446, 472, 478)
Sodium bromide	$NaBr$	806	952	1063	1148	1304	1392	755	(229, 367, 473)
Sodium chloride	$NaCl$	865	1017	1131	1220	1379	1465	800	(122, 156, 162, 195, 229, 243, 264, 362, 367, 473)
Sodium cyanide	$NaCN$	817	983	1115	1214	1401	1497	564	(198, 229)
Sodium fluoride	NaF	1077	1240	1363	1455	1617	1704	992	(229, 371, 475)
Sodium hydroxide	$NaOH$	739	897	1017	1111	1286	1378	318	(229, 473)
Sodium iodide	NaI	767	903	1005	1083	1225	1304	651	(156, 229, 367, 475)
Stannic bromide	$SnBr_4$	s	72.7	105.5	131.0	177.7	204.7	31.0	(9)
Stannic chloride	$SnCl_4$	−22.7	+10.0	35.2	54.7	92.1	113.0	−30.2	(229, 279, 481, 494, 496)
Stannic hydride	SnH_4	−140.0	−118.5	−102.3	−89.2	−65.2	−52.3	−149.9	(229, 306)
Stannic iodide	SnI_4	s	175.8	218.8	254.2	315.5	348.0	144.5	(283)
Stannous chloride	$SnCl_2$	316	391	450	493	577	623	246.8	(229, 264)
Strontium	Sr	s	898	1018	1111	1285	1384	800	(167, 229, 355)
Strontium oxide	SrO	2068$_s$	2262$_s$	2410$_s$	....	....	....	2430	(75, 229)
Sulfur	S	183.8	243.8	288.3	327.2	399.6	444.6	112.8	(17, 34, 54, 66, 73, 157, 191, 229, 268, 278, 284, 323, 334, 353, 354, 482)
Sulfur hexafluoride	SF_6	−132.7$_s$	−114.7$_s$	−101.5$_s$	−90.9$_s$	−72.6$_s$	−63.5$_s$	−50.2	(229, 238)
Sulfur dioxide	SO_2	−95.5$_s$	−76.8$_s$	−60.5	−46.9	−23.0	−10.0	−73.2	(25, 33, 51, 63, 68, 140, 142, 175, 229, 261, 280, 334, 385, 415, 420)
Sulfur monochloride	S_2Cl_2	−7.4	+27.5	54.1	75.3	115.4	138.0	−80	(168, 229, 453)
Sulfuryl chloride	SO_2Cl_2	s	−24.8	−1.0	+17.8	51.3	69.2	−54.1	(229, 448, 449)
Sulfur trioxide (α)	SO_3	−39.0$_s$	−16.5$_s$	−1.0$_s$	+10.5$_s$	32.6	44.8	16.8	(27, 151, 229, 411, 412)
Sulfur trioxide (β)	SO_3	−34.0$_s$	−12.3$_s$	+3.2$_s$	14.3$_s$	32.6	44.8	32.3	(27, 151, 229, 411, 412)
Sulfur trioxide (γ)	SO_3	−15.3$_s$	+4.3$_s$	17.9$_s$	28.0$_s$	44.0$_s$	51.6$_s$	62.1	(27, 151, 229, 411, 412)
Sulfuric acid	H_2SO_4	145.8	194.2	229.7	257.0	305.0	330.0d	10.5	(9)
Thionyl bromide	$SOBr_2$	−6.7	+31.0	58.8	80.6	119.2	139.5	−52.2	(229, 269)
Thionyl chloride	$SOCl_2$	−52.9	−21.9	+2.2	21.4	56.5	75.4	−104.5	(11, 13, 229)
Tantalum pentafluoride	TaF_5	s	s	s	130.0	194.0	230.0	96.8	(229, 368)
Tellurium	Te	520	650	753	838	997	1087	452	(101, 229)
Tellurium tetrachloride	$TeCl_4$	s	233	273	304	360	392	224	(229, 399)
Tellurium hexafluoride	TeF_6	−111.3$_s$	−92.4$_s$	−78.4$_s$	−67.9$_s$	−48.2$_s$	−38.6$_s$	−37.8	(229, 238, 491)
Thallium	Tl	825	983	1103	1196	1364	1457	303.5	(143, 229, 255, 469)
Thallium bromide	$TlBr$	s	522	598	653	759	819	460	(229, 465, 474)
Thallium chloride	$TlCl$	s	517	589	645	748	807	430	(229, 465, 474)
Thallium iodide	TlI	440	531	607	663	763	823	440	(229, 465, 474)
Tin	Sn	1492	1703	1855	1968	2169	2270	231.9	(153, 154, 155, 165, 229, 348, 469, 471)
Titanium tetrachloride	$TiCl_4$	−13.9	+21.3	48.4	71.0	112.7	136.0	−30	(11, 229)
Tungsten	W	3990	4507	4886	5168	5666	5927	3370	(218, 229, 250, 498, 499)
Tungsten hexafluoride	WF_6	−71.4$_s$	−49.2$_s$	−33.0$_s$	−20.3$_s$	+1.2	17.3	−0.5	(229, 347)
Uranium hexafluoride	UF_6	−38.8$_s$	−13.8$_s$	+4.4$_s$	18.2$_s$	42.7$_s$	55.7$_s$	69.2	(229, 356)
Vanadyl trichloride	$VOCl_3$	−23.2	+12.2	40.0	62.5	103.5	127.2	......	(128)

INORGANIC COMPOUNDS (Continued)
Pressures Less than One Atmosphere (Continued)

Name	Formula	Temperature, °C						M.P.	Citation No.
		1 mm	10 mm	40 mm	100 mm	400 mm	760 mm		
Water	H_2O	− 17.3$_s$	+ 11.3	34.1	51.6	83.0	100.0	0.0	(81, 95, 104, 121, 189, 190, 217, 222, 229, 232, 303, 304, 378, 379, 404, 476, 481)
Xenon	Xe	−168.5$_s$	−152.8$_s$	−141.2$_s$	−132.8$_s$	−117.1$_s$	−108.0	−111.6	(3, 180, 229, 311, 315, 329)
Zinc	Zn	487	593	673	736	844	907	419.4	(17, 42, 61, 109, 154, 155, 163, 181, 210, 229, 255, 342, 348)
Zinc chloride	$ZnCl_2$	428	508	566	610	689	732	365	(208, 229, 264)
Zinc fluoride	ZnF_2	970	1086	1175	1254	1417	1497	872	(362)
Zirconium tetrabromide	$ZrBr_4$	207$_s$	250$_s$	281$_s$	301$_s$	337$_s$	357$_s$	450	(229, 326)
Zirconium tetrachloride	$ZrCl_4$	190$_s$	230$_s$	259$_s$	279$_s$	312$_s$	331$_s$	437	(229, 326)
Zirconium tetraiodide	ZrI_4	264$_s$	311$_s$	344$_s$	369$_s$	409$_s$	431$_s$	499	(229, 326)

Pressures Greater than One Atmosphere

Name	Formula	Temperature, °C							Citation No.
		1 atm.	2 atm.	5 atm.	10 atm.	20 atm.	40 atm.	60 atm.	
Ammonia	NH_3	− 33.6	− 18.7	+ 4.7	25.7	50.1	78.9	98.3	(23, 25, 33, 49, 62, 82, 142, 175, 192, 223, 231, 279, 305, 334, 420)
Argon	A	−185.6	−179.0	−166.7	−154.9	−141.3	−124.9	......	(41, 83, 193, 229, 295, 329, 466)
Boron trichloride	BCl_3	12.7	33.2	66.0	96.7	135.4	......	......	(229, 308, 334, 430)
Boron trifluoride	BF_3	−100.7	− 89.4	− 72.6	− 57.7	− 40.0	− 19.0	......	(39, 118, 229, 319, 352)
Bromine	Br_2	58.2	78.8	110.3	139.8	174.0	216.0	243.5	(88, 102, 171, 203, 217, 229, 281, 331, 345, 381, 458, 487)
Carbontetrachloride	CCl_4	76.7	102.0	141.7	178.0	222.0	276.0	......	(104, 179, 279, 334, 335, 413, 495)
Carbon dioxide	CO_2	− 78.2$_s$	− 69.1$_s$	− 56.7	− 39.5	− 18.9	+ 5.9	22.4	(5, 46, 117, 172, 175, 211, 229, 272, 292, 302, 334, 420, 464, 467, 477, 497)
Carbon disulfide	CS_2	46.5	69.1	104.8	136.3	175.5	222.8	256.0	(175, 229, 265, 328, 334, 335, 420, 467, 490)
Carbon monoxide	CO	−191.3	−183.5	−170.7	−161.0	−149.7	......	......	(16, 67, 77, 78, 87, 229, 293, 463)
Carbonyl chloride	$COCl_2$	+ 8.3	27.3	57.2	85.0	119.0	159.8	......	(14, 135, 229, 288, 309)
Chlorotrifluoromethane	$CClF_3$	− 81.2	− 66.7	− 42.7	− 18.5	+ 12.0	52.8	......	(102, 447)
Cyanogen	C_2N_2	− 21.0	− 4.4	+ 21.4	44.6	72.6	106.5	......	(74, 83, 102, 118, 313, 444)
Dichlorodifluoromethane	CCl_2F_2	− 29.8	− 12.2	+ 16.1	42.4	74.0	......	......	(102, 144)
Dichlorofluoromethane	$CHCl_2F$	8.9	28.4	59.0	87.0	121.2	162.6	......	(233)
Chlorodifluoromethane	$CHClF_2$	− 40.8	− 24.7	+ 0.3	24.0	52.0	85.3	......	(40, 233)
Trichlorofluoromethane	CCl_3F	23.7	44.1	77.3	108.2	146.7	194.0	......	(233)
Chlorine	Cl_2	− 33.8	− 16.9	+ 10.3	35.6	65.0	101.6	127.1	(139, 166, 171, 214, 229, 239, 312, 452)
Helium	He	−268.6	−268.0	......	......	......	......	......	(227, 228, 229, 300, 301)
Hydrogen	H_2	−252.5	−250.2	−246.0	−241.8	......	......	......	(58, 71, 173, 225, 229, 298, 299, 396, 454, 455)
Hydrogen bromide	HBr	− 66.5	− 51.5	− 29.1	− 8.4	+ 16.8	48.1	70.6	(18, 103, 169, 229, 415, 416)
Hydrogen chloride	HCl	− 84.8	− 71.4	− 50.5	− 31.7	− 8.8	+ 17.8	36.2	(50, 69, 103, 118, 141, 169, 175, 223, 229, 416, 420)
Hydrogen cyanide	HCN	25.9	45.8	75.8	102.7	135.0	169.9	......	(44, 164, 229, 258, 314, 397, 401)
Hydrogen iodide	HI	− 35.1	− 18.9	+ 7.3	32.0	62.2	100.7	127.5	(103, 169, 229, 415, 416)
Hydrogen sulfide	H_2S	− 60.4	− 45.9	− 22.3	− 0.4	+ 25.5	55.8	76.3	(118, 137, 229, 235, 236, 292, 334, 415, 416)
Hydrogen selenide	H_2Se	− 41.1	− 25.2	0.0	+ 23.4	50.8	84.6	108.7	(57, 90, 91, 229, 292, 417)
Krypton	Kr	−152.0	−143.5	−130.0	−118.0	−101.7	− 78.4	......	(3, 221, 229, 315, 317, 329)
Neon	Ne	−246.0	−243.8	−239.9	−236.0	−230.8	......	......	(71, 229, 297, 455)
Nitrogen	N_2	−195.8	−189.2	−179.1	−169.8	−157.6	......	......	(15, 70, 86, 100, 123, 138, 173, 174, 193, 229, 292, 321, 463, 467, 489)
Nitric oxide	NO	−151.7	−145.1	−135.7	−127.3	−116.8	−103.2	− 94.8	(1, 146, 170, 216, 229, 279, 292, 294)
Nitrous oxide	N_2O	− 88.5	− 76.8	− 58.0	− 40.7	− 18.8	+ 8.0	27.4	(29, 32, 53, 62, 115, 188, 229, 247, 327, 464)
Nitrogen tetroxide	N_2O_4	21.0	37.3	59.8	79.4	100.3	121.4	132.2	(19, 108, 158, 229, 275, 317, 332, 374, 380, 385)
Oxygen	O_2	−183.1	−176.0	−164.5	−153.2	−140.0	−124.1	......	(15, 28, 58, 70, 86, 100, 114, 172, 174, 226, 229, 292, 296, 420, 456, 467, 488)
Silicon tetrafluoride	SiF_4	− 94.8$_s$	− 84.4	− 67.9	− 52.6	− 33.4	......	......	(40, 229, 310, 347)
Chlorotrifluorosilane	$SiClF_3$	− 70.0	− 57.3	− 37.2	− 18.6	+ 4.1	......	......	(40, 229, 392)
Dichlorodifluorosilane	$SiCl_2F_2$	− 31.8	− 15.1	+ 11.6	36.6	66.2	......	......	(40, 229, 392)
Fluorotrichlorosilane	$SiCl_3F$	12.2	32.4	64.6	94.2	131.8	......	......	(40)
Stannic chloride	$SnCl_4$	113.0	141.3	184.3	223.0	270.0	......	......	(229, 279, 481, 494, 496)
Sulfur dioxide	SO_2	− 10.0	+ 6.3	32.1	55.5	83.8	118.0	141.7	(25, 33, 51, 63, 68, 140, 142, 175, 229, 261, 280, 334, 385, 415, 420)
Sulfur trioxide	SO_3	44.8	60.0	82.5	104.0	138.0	175.0	198.0	(27, 151, 229, 411, 412)
Water	H_2O	100.0	120.1	152.4	180.5	213.1	251.1	276.5	(81, 95, 104, 121, 189, 190, 217, 222, 229, 232, 303, 304, 378, 379, 404, 476, 481)

VAPOR PRESSURE—CITATIONS
Inorganic Compounds

(1) Adwentowski, Bull. intern. acad. sci., Cracovie, [II] **1909**, 742.
(2) Adwentowski and Drozdowski, Ibid., [A] **1911**, 330–44.
(3) Allen and Moore, J. Am. Chem. Soc., **53**, 2522–7 (1931).
(4) Alterthum and Koref, Z. Elektrochem., **31**, 658–62 (1925).
(5) Amagat, Compt. rend., **114**, 1093–8 (1892).
(6) Amelin and Belyakov, J. Phys. Chem. (U.S.S.R.), **18**, 446–8, (1944).
(7) Anderson, J. Chem. Soc., **1930**, 1653–6.
(8) Anschütz and Evans, Ber., **19**, 1994–5 (1886).
(9) Anschütz and Reitter, "Die Destillation unter vermindertem Druck im Laboratorium," 2nd ed., Bonn, Cohen, 1895.
(10) Arctowski, Z. anorg. allgem. Chem., **12**, 427–30 (1896).
(11) Arii, Bull. Inst. Phys. Chem. Research (Tokyo), **8**, 714 (1929).
(12) Ibid., **8**, 545–51 (1929).
(13) Arii, Science Reports Tôhoku Imp. Univ., First Ser., **22**, 182–99 (1933).
(14) Atkinson, Heycock, and Pope, J. Chem. Soc., **117**, 1410–26 (1920).
(15) Baly, Phil. Mag., [5] **49**, 517 (1900).
(16) Baly and Donnan, J. Chem. Soc., **81**, 907–23 (1902).
(17) Barus, Phil. Mag., [5] **29**, 141–57 (1890).
(18) Bates, Halford, and Anderson, J. Chem. Phys., **3**, 531–4 (1935).
(19) Baume and Robert, Compt. rend., **168**, 1199–201 (1919).
(20) Baxter, Bezzenberger, and Wilson, J. Am. Chem. Soc., **42**, 1386–93 (1920).
(21) Baxter and Grose, Ibid., **37**, 1061–72 (1915).
(22) Baxter, Hickey and Holmes, Ibid., **29**, 127–36 (1907).
(23) Beattie and Lawrence, Ibid., **52**, 6–14 (1930).
(24) Becker and Meyer, Z. anorg. allgem. Chem., **43**, 251–66 (1907).
(25) Bergstrom, J. Phys. Chem., **26**, 358–76 (1922).
(26) Bernhardt, Physik Z., **26**, 265–75 (1925).
(27) Berthoud, Helv. Chim. Acta, **5**, 513–32 (1922).
(28) Bestelmeyer, Ann. Physik., [4] **14**, 87–98 (1904).
(29) Black, Van Praagh, and Topley, Trans. Faraday Soc., **26**, 196–7 (1930).
(30) Blanchard, Rafter, and Adams, J. Am. Chem. Soc., **56**, 16–7 (1934).
(31) Blaszkowska and Zakrzewska, Roczniki Chem., **8**, 210–8 (1928).
(32) Blue and Giauque, J. Am. Chem. Soc., **57**, 991–7 (1935).
(33) Blümcke, Ann. [2] **34**, 10–21 (1888).
(34) Bodenstein. Z. physik. Chem., **30**, 113–39 (1899).
(35) Boer and Dippel, Z. Physik, Chem., **B21**, 273–7 (1933).
(36) Bogros, Ann. phys., [10] **17**, 199–282 (1932).
(37) Bogros, Compt. rend., **191**, 560–2 (1930).
(38) Bolas, J. Chem. Soc., **24**, 780 (1871).
(39) Booth and Carter, J. Phys. Chem., **36**, 1359–63 (1932).
(40) Booth and Swinehart, J. Am. Chem. Soc., **57**, 1337–42 (1935).
(41) Born, Ann. Physik., [4] **69**, 473–504 (1922).
(42) Braune, Z. Anorg. allgem. Chem., **111**, 109–47 (1920).
(43) Braune and Tiedje, Ibid., **152**, 39–51 (1926).
(44) Bredig and Teichmann, Z. Electrochem., **31**, 449–54 (1925).
(45) Brewer and Dennis, J. Phys. Chem., **31**, 1101–5 (1927).
(46) Bridgman, J. Am. Chem. Soc., **49**, 1174–83 (1927).
(47) Bridgman, Ibid., **36**, 1344–63 (1914).
(48) Briggs and Migrdichian, J. Phys. Chem., **28**, 1121–35 (1924).
(49) Brill, Ann. Physik., [4] **21**, 170–180 (1906).
(50) Briner, J. chim. phys., **4**, 476 (1906).
(51) Briner and Cardoso, Ibid., **6**, 665 (1908).
(52) Briner and Pylkoff, Ibid., **10**, 640–79 (1912).
(53) Britton, Trans. Faraday Soc., **25**, 520–25 (1929).
(54) Brown and Muir, Phys. Rev., **41**, 111 (1932).
(55) Brühl, Ber., **26**, 2508–20 (1893).
(56) Bruylants, Bull. sci. acad. roy. belg., **1920**, 472–8.
(57) Bruylants, and Dondeyne, Ibid., [V] **8**, 387–405 (1922).
(58) Bulle, Physik, Z., **14**, 860–2 (1913).
(59) Burg and Schlesinger, J. Am. Chem. Soc., **59**, 780–7 (1937).
(60) Ibid., **62**, 3425–9 (1940).
(61) Burmeister and Jellinek, Z. physik. Chem., **A165**, 121–32 (1933).
(62) Burrell and Robertson, J. Am. Chem. Soc., **37**, 2482–6 (1915).
(63) Burrell and Robertson, U. S. Bur. Mine, Tech. Paper **142** (1916).
(64) Butler and Maass, J. Am. Chem. Soc., **52**, 2184–98 (1930).
(65) Cady and Hildebrand, Ibid., **52**, 3839–43 (1930).
(66) Callender and Griffiths, Chem. News, **63**, 1–2 (1891).
(67) Cardoso, J. chim. phys., **13**, 312–50 (1915).
(68) Cardoso and Fiorentino, Ibid., **23**, 841–7 (1926).
(69) Cardoso and Germann, Ibid., **11**, 632 (1913).
(70) Cath, Verslag. Akad. Wetenschappen, **27**, 553–60 (1918).
(71) Cath and Onnes, Ibid., **26**, 490 (1917).
(72) Centnerszwer, Z. physik. Chem., **85**, 99–112 (1913).
(73) Chappuis, Trav. bur. intern. poids measures, **16**, (1914).
(74) Chappuis and Rivière, Compt. rend., **104**, 1504–5 (1887).
(75) Claassen and Veenemans, Z. Physik., **80**, 342–51 (1933).
(76) Claussen and Hildebrand, J. Am. Chem. Soc., **56**, 1820 (1934).
(77) Clayton and Giauque, Ibid., **54**, 2610–26 (1932).
(78) Clusius and Teske, Z. physik. Chem., **6B**, 135–51 (1929).
(79) Corey, Laubengayer, and Dennis, J. Am. Chem. Soc., **47**, 112–7 (1925).
(80) Cosslett, Z. anorg. allgem. Chem., **201**, 75–80 (1931).
(81) Crafts, J. chim. phys., **13**, 105–61 (1915).
(82) Cragoe, Refrig. Eng., **12**, 131–42 (1925).
(83) Crommelin, Verslag. Akad. Wetenschappen, **22**, 510–20 (1913).
(84) Ibid., **16**, 477–85 (1913).
(85) Ibid., **22**, 1212–5 (1913).
(86) Ibid., **23**, 991–4 (1914).
(87) Crommelin, Bijleved, and Brown, Proc. Acad. Sci. Amsterdam, **34**, 1314–7 (1931).
(88) Cuthbertson and Cuthbertson, Proc. Roy. Soc. (London), **85**, 306–8 (1911).
(89) Daniels and Bright, J. Am. Chem. Soc., **42**, 1131–41 (1920).
(90) De Forcrand and Fonezes-Diacon, Compt. rend., **134**, 171–3 (1902)
(91) Ibid., **134**, 229–31 (1902).
(92) Dennis, Corey, and Moore, J. Am. Chem. Soc., **46**, 657–74 (1924).
(93) Dennis and Hance, J. Phys. Chem. **30**, 1055–9 (1926).
(94) Dennis, Orndorff, and Tabern, J. Phys. Chem., **30**, 1049–54 (1926).
(95) Derby, Daniels and Gutsche, J. Am. Chem. Soc., **36**, 793–804 (1914)
(96) Dewer, Proc. Chem. Soc. (London), **14**, 241–8 (1898).
(97) Dewer and Jones, Proc. Roy. Soc. (London), **A71**, 427–39 (1903).
(98) Ditte, Compt. rend., **140**, 1162–7 (1905).
(99) Dodd, J. Am. Chem. Soc., **42**, 1579–94 (1920).
(100) Dodge and Davis, Ibid., **49**, 610–20 (1927).
(101) Doolan and Partington, Trans. Faraday Soc., **20**, 342–4 (1924).
(102) Dow Chemical Co. files.
(103) Drozdowski and Pietrzak, Bull. intern. acad. sci. Cracovie, **A1913**, 219–40.
(104) Drucker, Jimeno, and Kangro, Z. physik. Chem., **90**, 513–52 (1915).
(105) Eastman and Duschak, U. S. Bur. Mines, Tech. Paper **225**, (1919).
(106) Edmondson and Egerton, Proc. Roy. Soc. (London), **A113**, 520–42 (1927).
(107) Egan, Potts, and Potts, Ind. Eng. Chem., **38**, 454–6 (1946).
(108) Egerton, J. Chem. Soc., **105**, 647–57 (1914).
(109) Egerton, Phil. Mag., [6] **33**, 33 (1917).
(110) Egerton, Proc. Roy. Soc. (London), **103A**, 469–86 (1923).
(111) Egerton and Raleigh, J. Chem. Soc., **123**, 3024–32 (1923).
(112) Emeléus and Maddock, Ibid., **1944**, 293–6.
(113) Emeléus, Maddock, and Reid, Ibid., **1941**, 353–8.
(114) Estreicher and Olszewski, Phil. Mag., [5] **40**, 454 (1895).
(115) Eucken and Donath, Z. physik. Chem., **124**, 181–203 (1926).
(116) Evnevitsch and Suchodski, J. Russ. Phys.-Chem. Soc., **61**, 1503–12 (1929).
(117) Falck, Physik. Z., **9**, 433–40 (1908).
(118) Faraday, Phil. Trans., **135A**, 155 (1845).
(119) Feiser, Metall u. Erz, **26**, 269–84 (1929).
(120) Ibid., **28**, 297–302 (1931).
(121) Fenby, Chem. Age (London), **2**, 434–5 (1920).
(122) Fiock and Rodebush, J. Am. Chem. Soc., **48**, 2522–8 (1926).
(123) Fischer and Alt, Ann. Physik, [4] **9**, 1149 (1902).
(124) Fischer and Biltz, Z. anorg. allgem. Chem., **176**, 81–111 (1928).
(125) Fischer and Gewehr, Ibid., **222**, 303–11 (1935).
(126) Fischer and Jübermann, Ibid., **227**, 227–36 (1936).
(127) Fischer, Rahlfs, and Benze, Ibid., **205**, 1–41 (1932).
(128) Flood, Gorrissen, and Veimo, J. Am. Chem. Soc., **59**, 2494–5 (1937).
(129) Folger and Rodebush, Ibid., **45**, 2080–90 (1923).
(130) Friedel and Crafts, Compt. rend., **106**, 1764–70 (1888).
(131) Frost, Cothran, and Browne, J. Am. Chem. Soc., **55**, 3516–8 (1933).
(132) Füchtbauer and Bartels, Z. Physik., **4**, 337–42 (1921).
(133) Gayler, Metallwirtschaft, **9**, 677–9 (1930).
(134) Gebhardt, Ber. physik. Ges., **3**, 184–8 (1905).
(135) German and Taylor, J. Am. Chem. Soc., **48**, 1154–9 (1926).
(136) Gerry and Gillespie, Phys. Rev., [2] **40**, 269–80 (1932).
(137) Giauque and Blue, J. Am. Chem. Soc., **58**, 831–7 (1936).
(138) Giauque and Clayton, Ibid., **55**, 4875–89 (1933).
(139) Giauque and Powell, Ibid., **61**, 1970–4 (1939).
(140) Giauque and Stephenson, Ibid., **60**, 1389–94 (1938).
(141) Giauque and Wiebe, Ibid., **50**, 101–22 (1928).
(142) Gibbs, Ibid., **27**, 851–65 (1905).
(143) Gibson, dissertation, Breslau (1911).
(144) Gilkey, Gerard, and Bixler, Ind. Eng. Chem., **23**, 364–7 (1931).
(145) Gillespie and Fraser, J. Am. Chem. Soc., **58**, 2260–3 (1936).
(146) Goldschmidt, Z. Physik., **20**, 159–65 (1923).
(147) Goodeve, J. Chem. Soc., **1930**, 2733–7).
(148) Goodeve and Powney, Ibid., **1932**, 2078–81.
(149) Goodeve and Richardson, Ibid., **1937**, 294–300.
(150) Gordon, J. Chem. Phys., **4**, 100–2 (1936).
(151) Grau and Roth, Z. anorg. allgem. Chem. **188**, 173–185 (1930).
(152) Gray and Ramsay, J. Chem. Soc., **95**, 1073–85 (1909).
(153) Greenwood, Proc. Roy. Soc. (London), **A82**, 396–408 (1909).
(154) Ibid., **A83**, 483–91 (1910).
(155) Greenwood, Z. physik. Chem., **76**, 484–90 (1911).
(156) Greiner and Jellinek, Ibid., **A165**, 97–120 (1933).
(157) Gruener, J. Am. Chem. Soc., **29**, 1396–402 (1907).
(158) Guye and Drouguinine, J. chim. phys., **8**, 473 (1910).
(159) Haber and Kirschbaum, Z. Elektrochem., **20**, 296–305 (1914).
(160) Haber and Zisch, Z. Physik., **9**, 302–26 (1922).
(161) Hackspill, Ann. chim. phys., [8] **28**, 613–96 (1913).
(162) Hackspill and Grandadam, Ann. chimie, [10] **5**, 218–50 (1926).
(163) Hansen, Ber., **42**, 210–4 (1909).
(164) Hara and Shinozaki, Tech. Reports. Tôhoku Imp. Univ., **4**, 145–52 (1924).
(165) Harteck, Z. physik. Chem., **134**, 1–20 (1928).
(166) Ibid., **134**, 21–5 (1928).
(167) Hartman and Schneider, Z. Anorg. allgem. Chem., **180**, 275–83 (1929).
(168) Harvey and Schuette, J. Am. Chem. Soc., **48**, 2065–8 (1926).
(169) Henglein, Z. Physik., **18**, 64–9 (1923).
(170) Henglein and Krüger, Z. anorg. allgem. Chem., **130**, 181–7 (1923).
(171) Henglein, Von Rosenberg, and Muchlinski, Z. Physik., **11**, 1–11 (1922).
(172) Henning, Ann. Physik., [4] **43**, 282–94 (1914).
(173) Henning, Z. Physik., **40**, 775–85 (1927).
(174) Henning and Hense, Ibid., **23**, 105–16 (1924).
(175) Henning and Stock, Ibid., **4**, 226–40 (1921).
(176) Herbst, Physik. Z., **27**, 366–71 (1926).
(177) Herbst, Z. tech., Physik., **7**, 467–8 (1926).
(178) Hertz, Ann., [2] **17**, 177–200 (1882).
(179) Herz and Rathmann, Chem.-Ztg., **36**, 1417–8 (1912).
(180) Heuse and Otto, Z. tech. Physik., **13**, 277–8 (1932).
(181) Heycock and Lamplough, Proc. Chem. Soc. (London), **28**, 3–4 (1912).
(182) Hill, Phys. Rev., **20**, 259–66 (1922).
(183) Hincke, J. Am. Chem. Soc., **52**, 3869–77 (1930).
(184) Ibid., **55**, 1751–3 (1933).
(185) Hirst and Olson, J. Am. Chem. Soc., **51**, 2398–403 (1929).
(186) Hittorf, Ann., [2] **126**, 193–228 (1865).
(187) Hoeflake and Scheffer, Rec. trav. chim., **45**, 191–200 (1926).
(188) Hoge, J. Research Natl. Bur. Standards, **34**, 281–93 (1945).
(189) Holborn and Baumann, Ann. Physik., [4] **31**, 945–70 (1910).
(190) Holborn and Henning, Ibid., **26**, 833–83 (1908).
(191) Ibid., **35**, 761–4 (1911).
(192) Holst, thesis, Zurich, 1914.
(193) Holst and Hamburger, Z. physik. Chem., **91**, 513–47 (1916).
(194) Horiba, Proc. Acad. Sci. Amsterdam, **25**, 387 (1923).
(195) Horiba and Baba, Bull. Chem. Soc. Japan, **3**, 11–7 (1928).

VAPOR PRESSURE—CITATIONS (Continued)
Inorganic Compounds

(196) Ilosvay, Bull. Soc. Chem., [2] **37**, 294 (1882).
(197) Ingold, J. Chem. Soc., **121**, 2419–32 (1922).
(198) Ibid., **123**, 885–91 (1923).
(199) International Critical Tables, **III**, 201–49 New York, McGraw-Hill Book Co., Inc., 1928.
(200) Isambert, Compt. rend., **92**, 919–22 (1881).
(201) Ibid., **94**, 958–60 (1882).
(202) Ibid., **96**, 643–8 (1883).
(203) Isnardi, Ann. Physik, [4] **61**, 264–72 (1920).
(204) Jackson and Morgan, Ind. Eng. Chem., **13**, 110–8 (1921).
(205) Jaeger and Germs, Z. anorg. allgem. Chem., **119**, 145–73 (1921).
(206) Jannek and Meyer, Ibid., **83**, 51–96 (1913).
(207) Jellinek and Golubowski, Z. physik. Chem., **A147**, 461–9 (1930).
(208) Jellinek and Koop, Ibid., **145A**, 305–29 (1929).
(209) Jellinek and Rudat, Ibid., **A143**, 55–61 (1929).
(210) Jenkens, Proc. Roy. Soc. (London), **110A**, 456–63 (1926).
(211) Jenkins and Pye, Phil. Trans. Roy. Soc., **A213**, 67 (1913).
(212) Johnson, J. Am. Chem. Soc., **33**, 777–81 (1911).
(213) Ibid., **34**, 877–80 (1912).
(214) Johnson and McIntosh, J. Am. Chem. Soc., **31**, 1138–44 (1909).
(215) Johnson and Pechukas, Ibid., **59**, 2065–8 (1937).
(216) Johnston and Giauque, Ibid., **51**, 3194–214 (1929).
(217) Jolly and Briscoe, J. Chem. Soc., **129**, 2154–9 (1926).
(218) Jones, Langmuir, and MacKay, Phys. Rev., [2] **30**, 201–214 (1927).
(219) Joubert, Compt. rend., **78**, 1853–5 (1874).
(220) Jung and Ziegler, Z. physik. Chem., **150**, 139–44 (1930).
(221) Justi, Physik. Z., **36**, 571–4 (1935).
(222) Kahlbaum, Z. physik, Chem., **13**, 14–55 (1894).
(223) Karwat, Ibid., **112**, 486–90 (1924).
(224) Kearby, J. Am. Chem. Soc., **58**, 374–5 (1936).
(225) Keesom, Byl, and Van der Horst, Proc. Acad. Sci., Amsterdam, **34**, 1223 (1931).
(226) Keesom, Van der Horst, and Jansen, Ibid., **32**, 1167–70 (1929).
(227) Keesom, Weber, and Norgaard, Ibid., **32**, 864 (1929).
(228) Keesom, Weber, and Schmidt, Ibid., **32**, 1314 (1929).
(229) Kelley, U. S. Bur. Mines, Bull. **383** (1935).
(230) Kemp and Giauque, J. Am. Chem. Soc., **59**, 79–84 (1937).
(231) Keyes and Brownlee, Ibid., **40**, 25–45 (1918).
(232) Keyes and Smith, Mech. Eng., **53**, 132–5 (1931).
(233) Kinetic Chemical Corporation bulletins.
(234) King and Partington, J. Chem. Soc., **129**, 925–9 (1926).
(235) Klemenc, Z. Elektrochem., **38**, 592–5 (1932).
(236) Klemenc and Bankowski, Z. anorg. allgem. Chem., **208**, 348–66 (1932).
(237) Klemens and Wagner, Ber., **70B**, 1880–2 (1937).
(238) Klemm and Henkel, Z. Anorg. allgem. Chem., **207**, 73–86 (1932).
(239) Knietsch, Ann., [2] **259**, 100–24 (1890).
(240) Knudsen, Ann. Physik, [4] **29**, 179–93 (1909).
(241) Kohn, Z. Physik., **3**, 143–56 (1920).
(242) Kohn and Guckel, Ibid, **27**, 305–57 (1924).
(243) Kordes and Raaz, Z. anorg. allgem. Chem., **181**, 225–36 (1929).
(244) Kovarik, Phil. Mag., [VII] **4**, 1262–75 (1927).
(245) Krauss and Wilken, Z. anorg. allgem. Chem., **145**, 151–68 (1925).
(246) Kröner, Ann. Physik, [4] **40**, 438–52 (1913).
(247) Kuenen, Phil. Mag., [5] **40**, 173 (1895).
(248) Landenberg and Thiele, Z. Physik. Chem., **B7**, 161–87 (1930).
(249) Landolt-Börnstein, "Physikalisch-Chemische Tabellen," Hauptwerk, 1332–77; **I** Ergänzungsband, 721–42; **II** Ergänzungsband, 1290–1310; **III** Ergänzungsband, 2430–62; Berlin, Julius Springer, 1923–36.
(250) Langmuir, Phys. Rev., [2] **2**, 329–42 (1913).
(251) Langmuir and Kingdon, Science, **57**, 58–60 (1923).
(252) Langmuir and MacKay, Phys. Rev., [2] **4**, 377–86 (1914).
(253) Laubengayer and Schirmer, J. Am. Chem. Soc., **62**, 1578–83 (1940).
(254) Laubengayer and Tabern, J. Phys. Chem., **30**, 1947–8 (1926).
(255) Leitgebel, Z. anorg. allgem. Chem., **202**, 305–24 (1931).
(256) Lenher, Smith, and Town, J. Phys. Chem., **26**, 156–60 (1922).
(257) Lewis, Z. Physik, **69**, 786–809 (1931).
(258) Lewis and Schutz, J. Am. Chem. Soc., **56**, 1002 (1934).
(259) Lobry de Bruyn, Rec. trav. chim., **10**, 100–12 (1891).
(260) Maass and Hiebert, J. Am. Chem. Soc., **46**, 2693–700 (1924).
(261) Maass and Maass, Ibid., **50**, 1352–68 (1928).
(262) Mack, Osterhof, and Kraner, Ibid., **45**, 617–23 (1923).
(263) Mac Rae and Van Voorhis, Ibid., **43**, 547–53 (1921).
(264) Maier, U. S. Bur. Mines, Tech. Paper **360**, (1925).
(265) Mali, Z. Anorg. Chem., **149**, 150–6 (1925).
(266) Marshall and Norton, J. Am. Chem. Soc., **55**, 431–2 (1933).
(267) Martin, J. Chem. Soc., **105**, 2836–60 (1914).
(268) Matthies, Physik, Z., **7**, 395–7 (1906).
(269) Mayes and Partington, J. Chem. Soc., **129**, 2594–605 (1926).
(270) Menzel and Mohry, Z. anorg. allgem. Chem. **210**, 257–63 (1933).
(271) Menzies, Z. physik. Chem., **130**, 90–8 (1927).
(272) Meyers and Van Dusen, J. Research Natl. Bur. Standards, **10**, 381–412 (1933).
(273) Millar, J. Am. Chem. Soc., **49**, 3003–10 (1927).
(274) Mittasch, Z. physik. Chem., **40**, 1–83 (1902).
(275) Mittasch, Kusz, and Schlenter, Z. Anorg. allgem. Chem., **159**, 1–36 (1926).
(276) Moles and Gómez, Z. physik. Chem., **80**, 513–30 (1912).
(277) Morley, Ibid., **49**, 95–100 (1904).
(278) Mueller and Burgess, J. Franklin Inst., **188**, 264–5 (1919).
(279) Mündel, Z. phys. Chem., **85**, 435–65 (1913).
(280) Mund, Bull. sci. acad. roy. belg., [V] **5**, 529 (1919).
(281) Nadejdine, Bull. Acad. imp. Sci. St. Petersburg, **30**, 327–30 (1886).
(282) Naumann, dissertation, Berlin (1907).
(283) Negishi, J. Am. Chem. Soc., **58**, 2293–6 (1936).
(284) Neumann, Z. physik. Chem., **A171**, 416–20 (1934).
(285) Neumann and Völker, Ibid., **161A**, 33–45 (1932).
(286) Nicolaieff, Compt. rend., **186**, 1621–4 (1928).
(287) Niederschulte, dissertation, Erlangen (1903).
(288) Nikitin, J. Russ. Phys. Chem. Soc., **52**, 235–49 (1920).
(289) Nilson and Pettersson, Z. physik. Chem., **1**, 27–38 (1887).
(290) Ogawa, Bull. Chem. Soc. Japan, **6**, 302–17 (1931).
(291) Ibid., **7**, 265–73 (1932.)
(292) Olszewski, Ann., [2] **31**, 58–74 (1887).
(293) Olszewski, Compt. rend., **99**, 706–7 (1884).
(294) Ibid., **100**, 940–3 (1885).
(295) Olszewski, Z. physik. Chem., **16**, 380–4 (1895).
(296) Onnes and Braak, Commun. Kamerlingh Onnes **Lab. Univ.** Leiden, No. **107a** (1908).
(297) Onnes, Crommelin, and Cath, Ibid., No. **151**, 15 (1917).
(298) Onnes, Crommelin, and Cath, Verslag. Akad. Wetenschappen, **26**, 124 (1917).
(299) Onnes and Keesom, Ibid., **22**, 389 (1913).
(300) Onnes and Weber, Comm. Phys. Lab. Leiden, No. **147b** (1915).
(301) Onnes and Weber, Proc. Acad. Sci. Amsterdam. **18**, 493 (1915).
(302) Onnes and Weber, Verslag. Akad. Wetenschappen, **22**, 226 (1913).
(303) Osborne, Stimson, Fiock, and Ginnings, J. Research Natl. Bur. Standards, **10**, 155–88 (1933).
(304) Osborne, Stimson, and Ginnings, Ibid., **23**, 197–270 (1939).
(305) Overstreet and Giauque, J. Am. Chem. Soc., 254–9 (1937).
(306) Paneth, Haken, and Rabinowitsch, Ber., **57**, 1891–903 (1924).
(307) Paneth and Rabinowitsch, Ber., **58**, 1138–63 (1925).
(308) Parker and Robinson, J. Chem. Soc., **130**, 2977–81 (1927).
(309) Paterno and Mazzuchelli, Gazz. chim. ital., **50**, **I**, 3052 (1920).
(310) Patnode and Papish, J. Phys. Chem., **34**, 1494–6 (1930).
(311) Patterson, Cripps, and Whytlow-Gray, Proc. Roy. Soc. (London), **86**, 579–90 (1912).
(312) Pellaton, J. chim. phys., **13**, 426–64 (1915).
(313) Perry and Bardwell, J. Am. Chem. Soc., **47**, 2629–32 (1925).
(314) Perry and Porter, Ibid., **48**, 299–302 (1926).
(315) Peters and Weil, Z. physik. Chem., **148A**, 27–35 (1930).
(316) Pfaundler, Ann., [2] **63**, 36–43 (1897).
(317) Pickering, U. S. Natl. Bur. Standards, Sci. Tech. Papers, **21**, 597–629 (1926).
(318) Pilling, Phys. Rev., [2] **18**, 362–8 (1921).
(319) Pohland and Harlos, Z. anorg. allgem. Chem., **207**, 242–5 (1932).
(320) Poindexter, Phys. Rev., **26**, 859–68 (1925).
(321) Porter and Perry, J. Am. Chem. Soc., **48**, 2059–60 (1926).
(322) Preuner and Brockmöller, Z. physik. Chem., **81**, 129–70 (1913).
(323) Preuner and Schupp, Ibid., **68**, 129–68 (1910).
(324) Prideaux, J. Chem. Soc., **97**, 2032–44 (1910).
(325) Purcell and Zahoorbux, J. Chem. Soc., **140**, 1029–35 (1937).
(326) Rahlfs and Fisher, Z. anorg. allgem. Chem., **211**, 349–67 (1933).
(327) Ramsay and Shields, J. Chem. Soc., **63**, 833–7 (1893).
(328) Ramsay and Shields, Z. physik. Chem., **12**, 433–76 (1893).
(329) Ramsay and Travers, Ibid., **38**, 641–89 (1901).
(330) Ramsay and Young, J. Chem. Soc., **49**, 37–50 (1886).
(331) Ibid., **49**, 453–62 (1886).
(332) Ramsay and Young, Trans. Roy. Soc. (London), **A177**, 86 (1886).
(333) Rassow, Z. anarg. allgem. Chem., **114**, 117–50 (1920).
(334) Regnault, Mém. Paris, **26**, 339 (1862).
(335) Rex, Z. physik. Chem., **55**, 355–70 (1916).
(336) Richter, Ber., **19**, 1057–60 (1886).
(337) Riesenfeld and Beja, Z. anorg. allgem. Chem., **132**, 179–200 (1923).
(338) Riesenfeld and Schwab, Z. Physik., **11**, 12–21 (1922).
(339) Rinse, Rec. trav. chim., **47**, 33–6 (1928).
(340) Rodebush and DeVries, J. Am. Chem. Soc., **47**, 2488–93 (1925).
(341) Rodebush and Dixon, Ibid., **47**, 1036–43 (1925).
(342) Rodebush and Dixon, Phys. Rev., [2] **26**, 851–8 (1925).
(343) Rodebush and Henry, J. Am. Chem. Soc., **52**, 3159–61 (1930).
(344) Rodebush and Walters, Ibid., **52**, 2654–65 (1930).
(345) Roozeboom, Rec. trav. chim., **3**, 67–76 (1884).
(346) Rudeberg, Phys. Rev., **46**, 763–7 (1934).
(347) Ruff and Ascher, Z. anorg. allgem. Chem., **196**, 413–20 (1931).
(348) Ruff and Bergdahl, Ibid., **106**, 76–94 (1919).
(349) Ruff and Bormann, Ibid., **88**, 365–85 (1914).
(350) Ibid., **88**, 397–409 (1914).
(351) Ruff and Braida, Ibid., **220**, 43–8 (1934).
(352) Ruff, Braida, Bretschneider, Menzel and Plaut, Ibid., 206, 59–64 (1932).
(353) Ruff and Graf, Ber., **40**, 4199–205 (1907).
(354) Ruff and Graf, Z. anorg. allgem. Chem., **58**, 209–12 (1908).
(355) Ruff and Hartmann, Ibid., **133**, 29–45 (1924).
(356) Ruff and Heinzelmann, Ibid., **72**, 63–84 (1911).
(357) Ruff and Johannsen, Ber., **38**, 3601–4 (1905).
(358) Ruff and Keim, Z. anorg. Chem., **193**, 176–86 (1930).
(359) Ruff and Konschak, Z. Elektrochem., **32**, 515–25 (1926).
(360) Ruff and Krug, Z. anorg. allgem. Chem., **190**, 270–6 (1930).
(361) Ruff and Laass, Ibid., **183**, 214–22 (1929).
(362) Ruff and LeBoucher, Ibid., **219**, 376–81 (1934).
(363) Ruff and Menzel, Ibid., **190**, 257–66 (1930).
(364) Ibid., **198**, 39–52 (1931).
(365) Ibid., **202**, 49–61 (1931).
(366) Ruff, Menzel, and Neumann, Z. Anorg. allgem. Chem., **208**, 293–303 (1932).
(367) Ruff and Mugdan, Ibid., **117**, 147–71 (1921).
(368) Ruff and Schiller, Ibid., **72**, 329–57 (1911).
(369) Ruff and Schmidt, Ibid., **117**, 172–90 (1921).
(370) Ibid., **123**, 83–8 (1922).
(371) Ruff, Schmidt, and Mugdan, Z. anorg. allgem. Chem., **123**, 83–8 (1922).
(372) Ruff and Tschirch, Ber., **46**, 929–49 (1913).
(373) Rushton and Daniels, J. Am. Chem. Soc., **48**, 384–9 (1926).
(374) Russ, Z. physik. Chem., **82**, 217–22 (1913).
(375) Russ and Pokorny, Monatsh., **34**, 1027–60 (1913).
(376) Russell, Rundle, and Yost, J. Am. Chem. Soc., **63**, 2825–8 (1941).
(377) Ryschkewitsch, Z. Elektrochem., **31**, 54–63 (1925).
(378) Scheel and Heuse, Ann. Physik, [4] **29**, 723–37 (1909).
(379) Ibid., [4] **31**, 715–36 (1910).
(380) Scheffer and Treub, Z. physik. Chem., **81**, 308–32 (1913).
(381) Scheffer and Voogd, Rec. trav. chim., **45**, 214–23 (1926).
(382) Schenck and Albers, Z. anorg. allgem. Chem., **105**, 145–66 (1919).
(383) Schenck and Imker, Ber., **58B**, 271–2 (1925).
(384) Schenck, Mihr, and Banthien, Ibid., **39**, 1506–21 (1906).

VAPOR PRESSURE—CITATIONS (Continued)
Inorganic Compounds

(385) Scheuer, Anz. Akad. Wiss. Wien, **48,** 304 (1911).
(386) Schlesinger, Sanderson and Burg, J. Am. Chem. Soc., **62,** 3421–5 (1940).
(387) Schmidt and Walter, Ann. Physik, **72,** 565 (1923).
(388) Schumb and Anderson, J. Am. Chem. Soc., **58,** 994–6 (1936).
(389) Ibid., **59,** 651–3 (1937).
(390) Schumb and Bickford, Ibid., **56,** 852–4 (1934).
(391) Schumb and Gamble, Ibid., **53,** 3191–2 (1931).
(392) Ibid., **54,** 583–90 (1932).
(393) Ibid., **54,** 3943–9 (1932).
(394) Scott, Phil. Mag., [6] **47,** 32–50 (1924).
(395) Scott and Brickwedde, Phys. Rev., [2] **48,** 483 (1935).
(396) Scott et al., J. Chem. Phys., **2,** 454–64 (1934).
(397) Shirado, Bull. Chem. Soc. Japan, **2,** 85–95 (1927).
(398) Simons, J. Am. Chem. Soc., **46,** 2179–83 (1924).
(399) Ibid., **52,** 3488–93 (1930).
(400) Simons and Hildebrand, Ibid., **46,** 2183–91 (1924).
(401) Sinozaki, Hara, and Mitsukuri, Tech. Repts. Tôhoku Imp. Univ., **6,** 157 (1926).
(402) Smellie, J. Soc. Chem. Ind., **42,** 466–8 (1923).
(403) Smith and Calvert, J. Am. Chem. Soc., **36,** 1363–82 (1914).
(404) Smith and Menzies, Ibid., **32,** 907–14 (1910).
(405) Smits and Beljaars, Proc. Acad. Sci. Amsterdam, **34,** 1141 (1931).
(406) Smits and Bokhorst, Z. physik. Chem., **91,** 249–312 (1916).
(407) Smits and Dienum, Proc. Acad. Sci. Amsterdam, **33,** 514–25 (1930).
(408) Smits, Meyer, and Beck, Verslag Akad. Wetenschappen, **24,** 939 (1915).
(409) Smits, Meyering, and Kamermans, Proc. Acad. Sci. Amsterdam, **34,** 1327–39 (1931).
(410) Ibid., **35,** 193–6 (1932).
(411) Smits and Moerman, Z. physik. Chem., **B35,** 69–81 (1937).
(412) Smits and Shoenmaker, J. Chem. Soc., **125,** 2554–73 (1924).
(413) Smythe and Engel, J. Am. Chem. Soc., **51,** 2646–70 (1929).
(414) Spangenberg, Z. physik. Chem., **119,** 419–38 (1926).
(415) Steele and Bagster, J. Chem. Soc., **97,** 2607–20 (1911).
(416) Steele, McIntosh, and Archibald, Z. physik. Chem., **55,** 129–99 (1906).
(417) Stein, J. Chem. Soc., **134,** 2134–8 (1931).
(418) Stelzner, dissertation, Erlangen (1901).
(419) Stock and Friederici, Ber., **46,** 1959–71 (1913).
(420) Stock, Henning and Kuss, Ibid., **54,** 1119–29 (1921).
(421) Stock and Kuss, Ibid., **47,** 3113–5 (1914).
(422) Ibid., **50,** 159–64 (1917).
(423) Ibid., **56B,** 789–808 (1923).
(424) Stock, Kuss, and Priess, Ibid., **47,** 3115–49 (1914).
(425) Stock and Massenez, Ibid., **45,** 3539–68 (1912).
(426) Ibid., **59,** 2210–5 (1926).
(427) Ibid., **59,** 2215–23 (1926).
(428) Ibid., **62,** 90–9 (1929).
(429) Stock and Praetorius, Ibid., **45,** 3568 (1912).
(430) Stock and Priess, Ibid., **47,** 3109–13 (1914).
(431) Stock and Somieski, Ibid., **49,** 111–57 (1916).
(432) Ibid., **50,** 1739–54 (1917).
(433) Ibid., **54,** 740–58 (1921).
(434) Stock, Somieski, and Wintgen, Ibid., **50,** 1754–64 (1917).
(435) Stock and Stoltzenberg, Ibid., **50,** 498–502 (1917).
(436) Stock and Willfroth, Ibid., **47,** 144–54 (1914).
(437) Stock and Zeidler, Ibid., **56,** 986–97 (1923).
(438) Stock and Zimmermann, Monatsh., **53–54,** 786–90 (1929).
(439) Ibid., **55,** 1–2 (1930).
(440) Strassmann, dissertation, Hanover (1929).
(441) Stull, Ind. Eng. Chem. **39,** 517 (1947).
(442) Tamman, "Kristallisieren und Schmelzen; ein Bertrag zur Lehre der Anderungen des Aggregatzustandes", p. 289, Leipzig, Barth, 1903.
(443) Taylor and Jungers, J. Am. Chem. Soc., **55,** 5057–8 (1933).
(444) Terwen, Z. physik. Chem., **91,** 469–99 (1916).
(445) Thiel and Ritter, Z. anorg. allgem. Chem., **132,** 125–52 (1923).
(446) Thiele, Ann. Physik., [5] **14,** 937–70 (1932).
(447) Thornton, Burg, and Schlesinger, J. Am. Chem. Soc., **55,** 3177–82 (1933).
(448) Trautz, Z. Elektrochem., **14,** 271–2 (1908).
(449) Ibid., **14,** 534–44 (1908).
(450) Trautz and Badstübner, Ibid., **35,** 799–802 (1929).
(451) Trautz and Gerwig, Z. anorg. allgem. Chem., **134,** 409–16 (1924).
(452) Ibid., **134,** 417–20 (1924).
(453) Trautz, Rick, and Acker, Z. Elektrochem., **35,** 122–4 (1929).
(454) Travers and Jacquerod, Trans. Roy. Soc. (London), **200A,** 155 (1903).
(455) Travers and Jacquerod, Z. physik. Chem., **45,** 435–60 (1903).
(456) Travers, Senter, and Jacquerod, Ibid., **45,** 416–34 (1903).
(457) Treadwell and Terebesi, Helv. Chim. Acta, **15,** 1053–66 (1932).
(458) Tsuruta, Physik. Z., **1,** 417–9 (1900).
(459) Ultee, Ber., **39,** 1856–8 (1906).
(460) Van der Plaats, Rec. trav. chim., **5,** 149–83 (1886).
(461) Van Doormaal and Scheffer, Ibid., **50,** 1100–4 (1931).
(462) Van Leimpt, Ibid., **55,** 157–60 (1936).
(463) Verschoyle, Trans. Roy. Soc. (London), **A230,** 189 (1931).
(464) Villard, Ann. chim. phys., [7] **10,** 387 (1897).
(465) Volmer, Physik. Z., **30,** 590–6 (1929).
(466) Von Rechenberg, "Destillation," Leipzig (1923).
(467) Von Siemens, Ann. Physik, [4] **42,** 871–88 (1913).
(468) Von Wartenberg, Ann., [2] **440,** 97–110 (1924).
(469) Von Wartenberg, Z. anorg. allgem. Chem., **56,** 320–36 (1908).
(470) Ibid., **79,** 71–87 (1913).
(471) Von Wartenberg, Z. Elektrochem., **19,** 482–9 (1913).
(472) Ibid., **20,** 443–9 (1914).
(473) Von Wartenberg and Albrecht, Ibid., **27,** 162–7 (1921).
(474) Von Wartenberg and Bosse, Ibid., **28,** 384–7 (1922).
(475) Von Wartenberg and Schulz, Ibid., **27,** 568–73 (1921).
(476) Washburn, Monthly Weather Rev., **52,** 488–90 (1924).
(477) Weber, Verslag. Akad. Wetenschappen, **22,** 380 (1914).
(478) Weiler, Ann. Physik., [5] **1,** 361–99 (1929).
(479) Welch and Duschak, U. S. Bur. Mines, Tech. Paper **81,** (1915).
(480) Wertenstein and Jedrzejewski, Compt. rend., **177,** 316–22 (1923).
(481) Wertheimer, Ber. deut. physik. Ges., **21,** 435–53 (1919).
(482) West and Menzies, J. Phys. Chem., **33,** 1880–92 (1929).
(483) Wiedmann, Stelzner, and Niederschulte, Ber. deut. physik. Ges., **3,** 159–62 (1905).
(484) Windsor and Blanchard, J. Am. Chem. Soc., **56,** 823–5 (1934).
(485) Winkler, dissertation, Müchen (1917).
(486) Wright, J. Chem. Soc., **107,** 1527–31 (1915).
(487) Ibid., **109,** 1134–9 (1916).
(488) Wroblewski, Compt. rend., **98,** 982–5 (1884).
(489) Ibid., **102,** 1010–2 (1886).
(490) Wüllener and Grotrain, Ann., [2] **11,** 545–604 (1880).
(491) Yost and Claussen, J. Am. Chem. Soc., **55,** 885–91 (1933).
(492) Yost and Kircher, Ibid., **52,** 4680–5 (1930).
(493) Yost and Stone, Ibid., **55,** 1889–95 (1933).
(494) Young, J. Chem. Soc., **59,** 626–34 (1891).
(495) Ibid., **59,** 911–36 (1891).
(496) Young, Sci. Proc. Roy. Dublin Soc., **12,** 374–443 (1909–1910).
(497) Zeleny and Smith, Physik, Z., **7,** 667–71 (1906).
(498) Zwikker, Physica, **5,** 249–60 (1925).
(499) Ibid., **5,** 319 (1925).

VAPOR PRESSURE (Continued) ORGANIC COMPOUNDS Pressures Less than One Atmosphere

Name	Formula	Temperature °C						M.P.	Citation No.
		1 mm	10 mm	40 mm	100 mm	400 mm	760 mm		
Cyanogen bromide	CBrN	-35.7_s	-10.0_s	$+8.6_s$	22.6_s	46.0_s	61.5	58	(26, 208)
Carbon tetrabromide	CBr_4	s	s	96.3	119.7	163.5	189.5	90.1	(48)
Chlorotrifluoromethane	$CClF_3$	−149.5	−134.1	−121.9	−111.7	− 92.7	− 81.2	......	(113, 430)
Cyanogen chloride	CClN	-76.7_s	-53.8_s	-37.5_s	-24.9_s	− 2.3	+ 13.1	− 6.5	(208, 344)
Dichlorodifluoromethane	CCl_2F_2	−118.5	− 97.8	− 81.6	− 68.6	− 43.9	− 29.8	......	(113, 146)
Carbonyl chloride	CCl_2O	− 92.9	− 69.3	− 50.3	− 35.6	− 7.6	+ 8.3	−104	(20, 145, 208, 305, 311)
Trichlorofluoromethane	CCl_3F	− 84.3	− 59.0	− 39.0	− 23.0	+ 6.8	23.7	......	(212)
Trichloronitromethane	CCl_3NO_2	− 25.5	+ 7.8	33.8	53.8	91.8	111.9	− 64	(26, 46, 47)
Carbontetrachloride	CCl_4	-50.0_s	− 19.6	+ 4.3	23.0	57.8	76.7	− 22.6	(115, 175, 289, 343, 352, 394, 481)
Cyanogen fluoride	CFN	-134.4_s	-118.5_s	-106.4_s	-97.0_s	-80.5_s	-72.6_s	......	(94, 208)
Carbontetrafluoride	CF_4	-184.6_s	−169.3	−158.8	−150.7	−135.5	−127.7	......	(208, 271)
Tribromomethane	$CHBr_3$	s	34.0	63.6	85.9	127.9	150.5	8.5	(113, 201)
Chlorodifluoromethane	$CHClF_2$	−122.8	−103.7	− 88.6	− 76.4	− 53.6	− 40.8	−160	(50, 212)
Dichlorofluoromethane	$CHCl_2F$	− 91.3	− 67.5	− 48.8	− 33.9	− 6.2	+ 8.9	−135	(212)
Trichloromethane	$CHCl_3$	− 58.0	− 29.7	− 7.1	+ 10.4	42.7	61.3	− 63.5	(32, 115, 175, 345, 352, 365)
Hydrocyanic acid	CHN	-70.8_s	-48.2_s	-31.3_s	-18.8_s	+ 9.8	25.8	− 14	(252)
Dibromomethane	CH_2Br_2	− 35.1	− 2.4	+ 23.3	42.3	79.0	98.6	− 52.8	(113, 352)
Dichloromethane	CH_2Cl_2	− 70.0	− 43.3	− 22.3	− 6.3	24.1	40.7	− 96.7	(317, 352)
Formaldehyde	CH_2O	s	− 88.0	− 70.6	− 57.3	− 33.0	− 19.5	− 92	(264, 399)
Formic acid	CH_2O_2	-20.0_s	$+2.1_s$	24.0	43.8	80.3	100.6	8.2	(7, 91, 203, 219, 246, 353, 373)
Dichloromethylarsine	CH_3AsCl_2	− 11.1	+ 24.3	51.5	73.0	112.7	134.5	− 59	(26)
Borine carbonyl	CH_3BO	−139.2	−121.1	−106.6	− 95.3	− 74.8	− 64.0	......	(64)
Methyl bromide	CH_3Br	-96.3_s	− 72.8	− 54.2	− 39.4	− 11.9	+ 3.6	− 93	(117, 323)
Methyl chloride	CH_3Cl	s	− 92.4	− 76.0	− 63.0	− 38.0	− 24.0	− 97.7	(54, 182, 241, 350, 451)

ORGANIC COMPOUNDS (Continued)

Pressures Less than One Atmosphere (Continued)

Name	Formula	Temperature °C						M.P.	Citation No.
		1 mm	10 mm	40 mm	100 mm	400 mm	760 mm		
Trichloromethylsilane	CH_3Cl_3Si	− 60.8	− 30.7	− 7.0	+ 12.1	47.0	66.4	− 90	(113)
Methyl fluoride	CH_3F	−147.3	−131.6	−119.1	−109.0	− 89.5	78.2	....	(89, 113, 281)
Methyl iodide	CH_3I	s	− 45.8	− 24.2	− 7.0	+ 25.3	42.4	− 64.4	(351, 352, 363)
Formamide	CH_3NO	70.5	109.5	137.5	157.5	193.5	210.5$_d$	....	(11, 113)
Nitromethane	CH_3NO_2	− 29.0	+ 2.8	27.5	46.6	82.0	101.2	− 29	(154, 470)
Methane	CH_4	−205.9$_s$	−195.5$_s$	−187.7$_s$	−181.4	−168.8	−161.5	−182.5	(6, 72, 73, 93, 123, 206, 359, 407)
Dichloromethylsilane	CH_4Cl_2Si	− 75.0	− 47.8	− 26.2	− 9.0	+ 23.7	41.9	....	(113, 114, 410)
Methanol	CH_4O	− 44.0	− 16.2	+ 5.0	21.2	49.9	64.7	− 97.8	(111, 289, 319, 338, 353)
Methanethiol	CH_4S	− 90.7	− 67.5	− 49.2	− 34.8	− 7.9	+ 6.8	−121	(39)
Chloromethylsilane	CH_5ClSi	− 95.0	− 71.0	− 51.7	− 36.4	− 7.8	+ 8.7	....	(410)
Methylamine	CH_5N	− 95.8$_s$	− 73.8	− 56.9	− 43.7	− 19.7	− 6.3	− 93.5	(19, 38, 130, 186, 324)
Methylsilane	CH_6Si	−138.5	−120.0	−104.8	− 93.0	− 70.3	− 56.9	....	(410)
2-Methyldisilazane	CH_9NSi_2	− 76.3	− 50.1	− 29.6	− 13.1	+ 17.2	34.0	....	(118)
Cyanogen iodide	CIN	25.2$_s$	57.7$_s$	80.3$_s$	97.6$_s$	126.1$_s$	141.1$_s$	....	(208, 476)
Tetranitromethane	CN_4O_8	s	22.7	48.4	68.9	105.9	125.7$_d$	13	(272)
Carbon monoxide	CO	−222.0$_s$	−215.0$_s$	−210.0$_s$	−205.7$_s$	−196.3	−191.3	−205.0	(23, 72, 73, 82, 83, 102, 208, 306, 447)
Carbonyl sulfide	COS	−132.4	−113.3	− 98.3	− 85.9	− 62.7	− 49.9	−138.8	(190, 208, 209, 408)
Carbonyl selenide	$COSe$	−117.1	− 95.0	− 76.4	− 61.7	− 35.6	− 21.9	....	(330)
Carbon dioxide	CO_2	−134.3$_s$	−119.5$_s$	−108.6$_s$	−100.2$_s$	− 85.7$_s$	− 78.2$_s$	− 57.5	(4, 5, 53, 124, 167, 168, 196, 208, 242, 275, 307, 308, 343, 407, 448, 454, 463, 493)
Carbon Selenosulfide	$CSSe$	− 47.3	− 16.0	+ 8.6	28.3	65.2	85.6	− 75.2	(208, 412)
Carbon disulfide	CS_2	− 73.8	− 44.7	− 22.5	− 5.1	+ 28.0	46.5	−110.8	(168, 208, 263, 331, 343, 352, 407, 454, 475)
Trichloroacetyl bromide	C_2BrCl_3O	− 7.4	+ 29.3	57.2	79.5	120.2	143.0	....	(11)
1-Chloro-1,2,2-trifluoroethylene	C_2ClF_3	−116.0	− 95.9	− 79.7	− 66.7	− 41.7	− 27.9	−157.5	(49, 50, 113)
1,2-Dichloro-1,2-difluoroethylene	$C_2Cl_2F_2$	− 82.0	− 57.3	− 38.2	− 23.0	+ 5.0	20.9	−112	(49, 113)
1,2-Dichloro-1,1,2,2-tetrafluoroethane	$C_2Cl_2F_4$	− 95.4	− 72.3	− 53.7	− 39.1	− 12.0	+ 3.5	− 94	(212)
1,1,2-Trichloro-1,2,2-trifluoroethane	$C_2Cl_3F_3$	− 68.0$_s$	− 40.3$_s$	− 18.5	− 1.7	+ 30.2	47.6	− 35	(212)
Tetrachloroethylene	C_2Cl_4	− 20.6$_s$	+ 13.8	40.1	61.3	100.0	120.8	− 19.0	(113, 175, 176)
1,1,2,2-Tetrachloro-1,2-difluoroethane	$C_2Cl_4F_2$	− 37.5$_s$	− 5.0$_s$	+ 19.8$_s$	38.6	73.1	92.0	26.5	(113, 184)
Hexachloroethane	C_2Cl_6	32.7$_s$	73.5$_s$	102.3$_s$	124.2$_s$	163.8$_s$	185.6$_s$	186.6	(296, 402)
Tribromoacetaldehyde	C_2HBr_3O	18.5	58.0	87.8	110.2	151.6	174.0$_d$	....	(113, 201)
Trichloroethylene	C_2HCl_3	− 43.8	− 12.4	+ 11.9	31.4	67.0	86.7	− 73	(113, 175)
Trichloroacetaldehyde	C_2HCl_3O	− 37.8	− 5.0	20.2	40.2	77.5	97.7	− 57	(113, 280)
Trichloroacetic acid	$C_2HCl_3O_2$	51.0$_s$	88.2	116.3	137.8	175.2	195.6	57	(11)
Pentachloroethane	C_2HCl_5	+ 1.0	39.8	69.9	93.5	137.2	160.5	− 22	(175, 296, 402)
Acetylene	C_2H_2	−142.9$_s$	−128.2$_s$	−116.7$_s$	−107.9$_s$	− 92.0$_s$	− 84.0$_s$	− 81.5	(12, 66, 67, 69, 217, 449)
1,1,1,2-Tetrabromoethane	$C_2H_2Br_4$	58.0	95.7	123.2	144.0	181.0	200.0$_d$	....	(11)
1,1,2,2-Tetrabromoethane	$C_2H_2Br_4$	65.0	110.0	144.0	170.0	217.5	243.5	....	(11, 113)
cis-1,2-Dichloroethylene	$C_2H_2Cl_2$	− 58.4	− 29.9	− 7.9	+ 9.5	41.0	59.0	− 80.5	(113, 175, 176)
trans-1,2-Dichloroethylene	$C_2H_2Cl_2$	− 65.4$_s$	− 38.0	− 17.0	− 0.2	+ 30.8	47.8	− 50.0	(113, 175, 176)
1,1-Dichloroethene	$C_2H_2Cl_2$	− 77.2	− 51.2	− 31.1	− 15.0	+ 14.8	31.7	−122.5	(113)
Dichloroacetic acid	$C_2H_2Cl_2O_2$	44.0	82.6	111.8	134.0	173.7	194.4	9.7	(11, 313)
1,1,1,2-Tetrachloroethane	$C_2H_2Cl_4$	− 16.3	+ 19.3	46.7	68.0	108.2	130.5	− 68.7	(113, 402)
1,1,2,2-Tetrachloroethane	$C_2H_2Cl_4$	− 3.8	+ 33.0	60.8	83.2	124.0	145.9	− 36	(113, 175, 296, 402)
1-Bromoethylene	C_2H_3Br	− 95.4	− 68.8	− 48.1	− 31.9	− 1.1	+ 15.8	−138	(378)
Bromoacetic acid	$C_2H_3BrO_2$	54.7	94.1	124.0	146.3	186.7	208.0	49.5	(11)
1,1,2-Tribromoethane	$C_2H_3Br_3$	32.6	70.6	100.0	123.5	165.4	188.4	− 26	(9, 11)
1-Chloroethylene	C_2H_3Cl	−105.6	− 83.7	− 66.8	− 53.2	− 28.0	− 13.8	−153.7	(104, 113)
Chloroacetic acid	$C_2H_3ClO_2$	43.0$_s$	81.0	109.2	130.7	169.0	189.5	61.2	(203, 204, 313)
1,1,1-Trichloroethane	$C_2H_3Cl_3$	− 52.0	− 21.9	+ 1.6	20.0	54.6	74.1	− 30.6	(113, 361, 402)
1,1,2-Trichloroethane	$C_2H_3Cl_3$	− 24.0	+ 8.3	35.2	55.7	93.0	113.9	− 36.7	(113, 402)
Trichloroacetaldehyde hydrate	$C_2H_3Cl_3O_2$	− 9.8$_s$	+ 19.5$_s$	39.7$_s$	55.0	82.1	96.2$_d$	51.7	(113)
1-Fluoroethylene	C_2H_3F	−149.3	−132.2	−118.0	−106.2	− 84.0	− 72.2	−160.5	(75)
Acetonitrile	C_2H_3N	− 47.0$_s$	− 16.3	+ 7.7	27.0	62.5	81.8	− 41	(159)
Methyl thiocyanate	C_2H_3NS	− 14.0	+ 21.6	49.0	70.4	110.8	132.9	− 51	[illegible]
Methyl isothiocyanate	C_2H_3NS	− 34.7$_s$	+ 5.4$_s$	38.2	59.3	97.8	119.0	35.5	(25)
Ethylene	C_2H_4	−168.3	−153.2	−141.3	−131.8	−113.9	−103.7	−169	(6, 57, 66, 93, 116, 244, 359, 407, 448, 453)
1-Bromo-1-chloroethane	C_2H_4BrCl	− 36.0$_s$	− 9.4$_s$	+ 10.4$_s$	28.0	63.4	82.7	16.6	(113, 402)
1-Bromo-2-chloroethane	C_2H_4BrCl	− 28.8$_s$	+ 4.1	29.7	49.5	86.0	106.7	− 16.6	(113, 402)
1,2-Dibromoethane	$C_2H_4Br_2$	− 27.0$_s$	+ 18.6	48.0	70.4	110.1	131.5	10	(11, 31, 113, 349, 351, 438)
1,1-Dichloroethane	$C_2H_4Cl_2$	− 60.7	− 32.3	− 10.2	+ 7.2	39.8	57.4	− 96.7	(113, 352, 402)
1,2-Dichloroethane	$C_2H_4Cl_2$	− 44.5$_s$	− 13.6	+ 10.0	29.4	64.0	82.4	− 35.3	(113, 133, 315, 352, 402)
1,1-Difluoroethane	$C_2H_4F_2$	−112.5	− 91.7	− 75.8	− 63.2	− 39.5	− 26.5	−117	(75)
Acetaldehyde	C_2H_4O	− 81.5	− 56.8	− 37.8	− 22.6	+ 4.9	20.2	−123.5	(113, 147, 248)
Ethylene oxide	C_2H_4O	− 89.7	− 65.7	− 46.9	− 32.1	− 4.9	+ 10.7	−111.3	(260)
Acetic acid	$C_2H_4O_2$	− 17.2$_s$	+ 17.5	43.0	63.0	99.0	118.1	16.7	(113, 203, 246, 258, 289, 335, 353, 373, 480)
Methyl formate	$C_2H_4O_2$	− 74.2	− 48.6	− 28.7	− 12.9	16.0	32.0	− 99.8	(297, 491)
Mercaptoacetic acid	$C_2H_4O_2S$	60.0	101.5	131.8	154.0$_d$	....	....	− 16.5	(113)
Ethyl bromide	C_2H_5Br	− 74.3	− 47.5	− 26.7	− 10.0	+ 21.0	38.4	−117.8	(113, 347, 391, 438)

ORGANIC COMPOUNDS (Continued)
Pressures Less than One Atmosphere (Continued)

Name	Formula	Temperature °C						M.P.	Citation No.
		1 mm	10 mm	40 mm	100 mm	400 mm	760 mm		
Ethyl chloride	C_2H_5Cl	− 89.8	− 65.8	− 47.0	− 32.0	− 3.9	+ 12.3	−139	(38, 113, 195, 346, 402)
2-Chloroethanol	C_2H_5ClO	− 4.0	+ 30.3	56.0	75.0	110.0	128.8	− 69	(11, 113)
Trichloroethylsilane	$C_2H_5Cl_3Si$	− 27.9	+ 3.6	27.9	46.3	80.3	99.5	− 40	(114)
Trichloroethoxysilane	$C_2H_5Cl_3OSi$	− 32.4	0.0	+ 25.3	45.2	82.2	102.4	....	(114)
Ethyl fluoride	C_2H_5F	−117.0	− 97.7	− 81.8	− 69.3	− 45.5	− 32.0	....	(50)
Ethyltrifluorosilane	$C_2H_5F_3Si$	− 95.4	− 73.7	− 56.8	− 43.6	− 19.1	− 5.4	....	(119)
Ethyl Iodide	C_2H_5I	− 54.4	− 24.3	− 0.9	+ 18.0	52.3	72.4	−105	(113, 348, 351, 352, 394)
Acetamide	C_2H_5NO	65.0$_s$	105.0	135.8	158.0	200.0	222.0	81	(11)
Acetaldoxime	C_2H_5NO	− 5.8$_s$	+ 25.8	48.6	66.2	98.0	115.0	47	(11)
Nitroethane	$C_2H_5NO_2$	− 21.0	+ 12.5	38.0	57.8	94.0	114.0	− 90	(31, 90)
Di(nitrosomethyl)amine	$C_2H_5N_3O_2$	+ 3.2	40.0	68.2	90.3	131.3	153.0	....	(11)
Ethane	C_2H_6	−159.5	−142.9	−129.8	−119.3	− 99.7	− 88.6	−183.2	(6, 30, 65, 84, 93, 153, 261, 262, 359, 453)
Dichlorodimethylsilane	$C_2H_6Cl_2Si$	− 53.5	− 23.8	− 0.4	+ 17.5	51.9	70.3	− 86.0	(113)
Ethanol	C_2H_6O	− 31.3	− 2.3	+ 19.0	34.9	63.5	78.4	−112	(32, 113, 115, 202, 273, 289, 329, 337, 343, 365, 374, 394)
Dimethyl ether	C_2H_6O	−115.7	− 93.3	− 76.2	− 62.7	− 37.8	− 23.7	−138.5	(55, 74, 211, 260, 350)
1,2-Ethanediol	$C_2H_6O_2$	53.0	92.1	120.0	141.8	178.5	197.3	− 15.6	(11, 71, 113, 136, 141)
Dimethyl sulfide	C_2H_6S	− 75.6	− 49.2	− 28.4	− 12.0	+ 18.7	36.0	− 83.2	(39)
Ethanethiol	C_2H_6S	− 76.7	− 50.2	− 29.8	− 13.0	+ 17.7	35.5	−121	(39)
Dimethylantimony	C_2H_6Sb	44.0	86.0	118.3	143.5	187.2	211.0	....	(309)
Ethylamine	C_2H_7N	− 82.3$_s$	− 58.3	− 39.8	− 25.1	+ 2.0	16.6	− 80.6	(38)
Dimethylamine	C_2H_7N	− 87.7	− 64.6	− 46.7	− 32.6	− 7.1	+ 7.4	− 96	(38, 390, 467)
1,2-Ethanediamine	$C_2H_8N_2$	− 11.0$_s$	+ 21.5	45.8	62.5	99.0	117.2	8.5	(181)
Dimethylsilane	C_2H_8Si	−115.0	− 93.1	− 75.7	− 61.4	− 35.0	− 20.1	....	(410)
Dimethyldiborane	$C_2H_{10}B_2$	−106.5	− 82.1	− 62.4	− 47.0	− 18.8	− 2.6	−150.2	(371)
2-Ethyldisilazane	$C_2H_{11}NSi_2$	− 62.0	− 32.2	− 8.3	+ 10.4	45.9	65.9	−127	(118)
Cyanogen	C_2N_2	− 95.8$_s$	− 76.8$_s$	− 62.7$_s$	− 51.8$_s$	− 33.0	− 21.0	− 34.4	(79, 101, 113, 125, 318, 426)
Acrylonitrile	C_3H_3N	− 51.0	− 20.3	+ 3.8	22.8	58.3	78.5	− 82	(113)
Propadiene	C_3H_4	−120.6	−101.0	− 85.2	− 72.5	− 48.5	− 35.0	−136	(113, 244, 255)
Propyne	C_3H_4	−111.0$_s$	− 90.5	− 74.3	− 61.3	− 37.2	− 23.3	−102.7	(113, 165, 262, 286)
2,3-Dibromopropene	$C_3H_4Br_2$	− 6.0	+ 30.0	57.8	79.5	119.5	141.2	....	(113, 204)
Methyl dichloroacetate	$C_3H_4Cl_2O_2$	3.2	38.1	64.7	85.4	122.6	143.0	....	(11)
2-Propenal	C_3H_4O	− 64.5	− 36.7	− 15.0	+ 2.5	34.5	52.5	− 87.7	(283)
Acrylic acid	$C_3H_4O_2$	+ 3.5$_s$	39.0	66.2	86.1	122.0	141.0	14	(342)
Pyruvic acid	$C_3H_4O_3$	21.4	57.9	85.3	106.5	144.7	165.0$_d$	13.6	(11)
1,2,3-Tribromopropene	$C_3H_5Br_3$	47.5	90.0	122.8	148.0	195.0	220.0	16.5	(113)
1-Chloropropene	C_3H_5Cl	− 81.3	− 54.1	− 32.7	− 15.1	+ 18.0	37.0	− 99.0	(113)
3-Chloropropene	C_3H_5Cl	− 70.0	− 42.9	− 21.2	− 4.5	+ 27.5	44.6	−136.4	(113, 192)
Epichlorohydrine	C_3H_5ClO	− 16.5	+ 16.6	42.0	62.0	98.0	117.9	− 25.6	(11)
Methyl chloroacetate	$C_3H_5ClO_2$	− 2.9	+ 30.0	54.5	73.5	109.5	130.3	− 31.9	(113, 297)
1,1,1-Trichloropropane	$C_3H_5Cl_3$	− 28.8	+ 4.2	29.9	50.0	87.5	108.2	− 77.7	(113)
1,2,3-Trichloropropane	$C_3H_5Cl_3$	+ 9.0	46.0	74.0	96.1	137.0	158.0	− 14.7	(11)
Allyltrichlorosilane	$C_3H_5Cl_3Si$	− 20.7	+ 13.2	39.2	59.3	97.1	118.0	....	(114)
Propionitrile	C_3H_5N	− 35.0	− 3.0	+ 22.0	41.4	77.7	97.1	− 91.9	(159)
3-Hydroxypropionitrile	C_3H_5NO	58.7	102.0	134.1	157.7	200.0	221.0	....	(113, 210)
Ethylisothiocyanate	C_3H_5NS	− 13.2$_s$	+ 22.8	50.8	71.9	110.1	131.0	− 5.9	(25)
Nitroglycerine	$C_3H_5N_3O_9$	127	188	235	e	....	....	11	(100, 268)
Propylene	C_3H_6	−131.9	−112.1	− 96.5	− 84.1	− 60.9	− 47.7	−185	(6, 66, 67, 137, 244, 262, 359, 382)
Cyclopropane	C_3H_6	−116.8	− 97.5	− 82.3	− 70.0	− 46.9	− 33.5	−126.6	(113)
2-Bromo-2-nitrosopropane	C_3H_6BrNO	− 33.5	− 4.3	+ 17.9	35.2	66.2	83.0	....	(321)
1,2-Dibromopropane	$C_3H_6Br_2$	− 7.0	+ 29.4	57.2	78.7	118.5	141.6	− 55.5	(11, 113, 204)
1,3-Dibromopropane	$C_3H_6Br_2$	+ 9.7	48.0	77.8	101.3	144.1	167.5	− 34.4	(113, 204)
2,3-Dibromo-1-propanol	$C_3H_6Br_2O$	57.0	98.2	129.8	153.0	196.0	219.0	....	(11)
1,2-Dichloropropane	$C_3H_6Cl_2$	− 38.5	− 6.1	+ 19.4	39.4	76.0	96.8	....	(301)
1,3-Dichloro-2-propanol	$C_3H_6Cl_2O$	28.0	64.7	93.0	114.8	153.5	174.3	....	(11, 199, 201)
Acetone	C_3H_6O	− 59.4	− 31.1	− 9.4	+ 7.7	39.5	56.5	− 94.6	(29, 115, 129, 343, 364)
Allyl alcohol	C_3H_6O	− 20.0	+ 10.5	33.4	50.0	80.2	96.6	−129	(25, 291)
Propylene oxide	C_3H_6O	− 75.0	− 49.0	− 28.4	− 12.0	+ 17.8	34.5	−112.1	(113)
Propionic acid	$C_3H_6O_2$	4.6	39.7	65.8	85.8	122.0	141.1	− 22	(203, 204, 353, 373)
Methyl acetate	$C_3H_6O_2$	− 57.2	− 29.3	− 7.9	+ 9.4	40.0	57.8	− 98.7	(113, 491)
Ethyl formate	$C_3H_6O_2$	− 60.5	− 33.0	− 11.5	+ 5.4	37.1	54.3	− 79	(297, 491)
Methyl glycolate	$C_3H_6O_3$	+ 9.6	45.3	72.3	93.7	131.7	151.5	....	(113)
Methoxyacetic acid	$C_3H_6O_3$	52.5	92.0	122.0	144.5	184.2	204.0	....	(113)
1-Bromopropane	C_3H_7Br	− 53.0	− 23.3	− 0.3	+ 18.0	52 0	71.0	−109.9	(113, 352)
2-Bromopropane	C_3H_7Br	− 61.8	− 32.8	− 10.1	+ 8 0	41.5	60 0	− 89 0	(113, 352)
1-Chloropropane	C_3H_7Cl	− 68.3	− 41.0	− 19.5	− 2.5	+ 29.4	46.4	−122.8	(38, 113, 160, 352)
2-Chloropropane	C_3H_7Cl	− 78.8	− 52 0	− 31.0	− 13.7	+ 18.0	36.5	−117	(113, 352)
Trichloroisopropylsilane	$C_3H_7Cl_3Si$	− 24.3	+ 9.9	36.5	57.8	96.8	118.5	....	(114)
1-Iodopropane	C_3H_7I	− 36.0	− 2.4	+ 23.6	43.8	81 8	102.5	− 98 8	(352)
2-Iodopropane	C_3H_7I	− 43.3	− 11.7	+ 13.2	32.8	69 5	89.5	− 90	(352)
Propionamide	C_3H_7NO	65.0$_s$	105.0	134.8	156.0	194.0	213.0	79	(11, 113)
1-Nitropropane	$C_3H_7NO_2$	− 9.6	+ 25.3	51.8	72.3	110.6	131 6	−108	(90)
2-Nitropropane	$C_3H_7NO_2$	− 18.8	+ 15.8	41 8	62.0	99.8	120.3	− 93	(90)
Ethyl Carbamate	$C_3H_7NO_2$	s	77.8	105.6	126 2	164.0	184 0	49	(11)
Propane	C_3H_8	−128.9	−108.5	− 92.4	− 79.6	− 55.6	− 42.1	−187.1	(6, 30, 66, 93, 137, 153, 262, 359)
Dichloroethoxymethylsilane	$C_3H_8Cl_2OSi$	− 33.8	− 1.3	+ 24.4	44.1	80.3	100 6	....	(114)
1-Propanol	C_3H_8O	− 15.0	+ 14.7	36.4	52.8	82.0	97.8	−127	(113, 289, 340, 374)

VAPOR PRESSURE (Continued)
ORGANIC COMPOUNDS (Continued)
Pressures Less than One Atmosphere (Continued)

Name	Formula	Temperature °C						M.P.	Citation No.
		1 mm	10 mm	40 mm	100 mm	400 mm	760 mm		
2-Propanol	C_3H_8O	−26.1	+2.4	23.8	39.5	67.8	82.5	−85.8	(310)
Ethyl methyl ether	C_3H_8O	−91.0	−67.8	−49.4	−34.8	−7.8	+7.5	......	(39)
1,2-Propanediol	$C_3H_8O_2$	45.5	83.2	111.2	132.0	168.1	188.2	......	(71, 113, 368)
1,3-Propanediol	$C_3H_8O_2$	59.4	100.6	131.0	153.4	193.8	214.2	......	(113, 141, 368)
2-Methoxyethanol	$C_3H_8O_2$	−13.5	+22.0	47.8	68.0	104.3	124.4	......	(141)
Glycerol	$C_3H_8O_3$	125.5	167.2	198.0	220.1	263.0	290.0	17.9	(11, 71, 134, 205, 353, 403)
1-Propanethiol	C_3H_8S	−56.0	−26.3	−3.2	+15.3	49.2	67.4	−112	(422)
Trimethylborine	C_3H_9B	−118.0	−92.4	−74.7	−60.8	−34.7	−20.1	......	(413)
Chlorotrimethylsilane	C_3H_9ClSi	−62.8	−34.0	−11.4	+6.0	39.4	57.9	......	(113, 424)
Trimethylgallium	C_3H_9Ga	−62.3$_s$	−31.7$_s$	−9.0	+8.0	39.0	55.6	−19	(240)
Propylamine	C_3H_9N	−64.4	−37.2	−16.0	+0.5	31.5	48.5	−83	(38)
Trimethylamine	C_3H_9N	−97.1	−73.8	−55.2	−40.3	−12.5	+2.9	−117.1	(18, 390, 418, 467)
Trimethyl phosphate	$C_3H_9O_4P$	26.0	67.8	100.0	124.0	167.8	192.7	......	(77)
Trimethyldiborane	$C_2H_{12}B_2$	−74.0	−44.8	−22.0	−4.4	+27.8	45.5	−122.9	(371)
Carbon suboxide	C_3O_2	−94.8	−71.0	−52.0	−36.9	−8.9	+6.3	−107	(411)
Carbon subsulfide	C_3S_2	14.0	54.9	85.6	109.9	......	......	+0.4	(208, 409)
Trichloroacetic anhydride	$C_4Cl_6O_3$	56.2	99.6	131.2	155.2	199.8	223.0	......	(11)
1,3-Butadiyne	C_4H_2	−82.5$_s$	−61.2$_s$	−45.9$_s$	−34.0	−6.1	+9.7	−34.9	(414, 421)
α,β-Dibromomaleic anhydride	$C_4H_2Br_2O_3$	50.0	92.0	123.5	147.7	192.0	215.0	......	(11)
trans-Fumaryl chloride	$C_4H_2Cl_2O_2$	+15.0	51.8	79.5	101.0	140.0	160.0	......	(11)
Maleic anhydride	$C_4H_2O_3$	44.0$_s$	78.7	111.8	135.8	179.5	202.0	58	(11, 189)
2-Nitrothiophene	$C_4H_3NO_2S$	48.2	92.0	125.8	151.5	199.6	224.5	46	(22)
Butenyne	C_4H_4	−93.2	−70.0	−51.7	−37.1	−10.1	+5.3	......	(304)
Succinyl chloride	$C_4H_4Cl_2O_2$	39.0	78.0	107.5	130.0	170.0	192.5	17	(11)
Chloroacetic anhydride	$C_4H_4Cl_2O_3$	67.2	108.0	138.2	159.8	197.0	217.0	46	(313)
Succinic anhydride	$C_4H_4O_3$	92.0$_s$	128.2	163.0	189.0	237.0	261.0	119.6	(11, 234)
1,4-Dioxane-2,6-dione	$C_4H_4O_4$	s	116.6	148.6	173.2	217.0	240.0	97	(11)
Thiophene	C_4H_4S	−40.7$_s$	−10.9	+12.5	30.5	64.7	84.4	−38.3	(113, 127)
Selenophene	C_4H_4Se	−39.0	−4.0	+24.1	47.0	89.8	114.3	......	(56)
α-Chlorocrotonic acid	$C_4H_5ClO_2$	70.0	108.0	135.6	155.9	193.2	212.0	......	(11)
Ethyl chloroglyoxylate	$C_4H_5ClO_3$	−5.1	+29.9	56.0	76.6	114.7	135.0	......	(11)
Ethyl trichloroacetate	$C_4H_5Cl_3O_2$	20.7	57.7	85.5	107.4	146.0	167.0	......	(11)
3-Butenenitrile	C_4H_5N	−19.6	+14.1	40.0	60.2	98.0	119.0	......	(159)
Methacrylonitrile	C_4H_5N	−44.5	−12.5	+12.8	32.8	32.8	90.3	......	(113)
cis-Crotononitrile	C_4H_5N	−29.0	+4.0	30.0	50.1	88.0	108.0	......	(159)
trans-Crotononitrile	C_4H_5N	−19.5	+15.0	41.8	62.8	101.5	122.8	......	(159)
Succinimide	$C_4H_5NO_2$	115.0$_s$	157.0	192.0	217.4	263.5	287.5	125.5	(11)
Allylisothiocyanate	C_4H_5NS	−2.0	+38.3	67.4	89.5	129.8	150.7	−80	(11, 25, 299, 201)
1,2-Butadiene	C_4H_6	−89.0	−64.2	−44.3	−28.3	+1.8	18.5	......	(113)
1,3-Butadiene	C_4H_6	−102.8	−79.7	−61.3	−46.8	−19.3	−4.5	−108.9	(52, 162, 189, 445)
Cyclobutene	C_4H_6	−99.1	−75.4	−56.4	−41.2	−12.2	+2.4	......	(163)
1 Butyne	C_4H_6	−92.5	−68.7	−50.0	−34.9	−6.9	+8.7	−130	(286)
2-Butyne	C_4H_6	−73.0$_s$	−50.5$_s$	−33.9$_s$	−18.8	+10.6	27.2	−32.5	(164, 287)
Ethyl dichloroacetate	$C_4H_6Cl_2O_2$	9.6	46.3	74.0	96.1	135.9	156.5	......	(11)
2-Chloroethyl chloroacetate	$C_4H_6Cl_2O_2$	46.0	86.0	116.0	140.0	182.2	205.0	......	(113)
cis-Crotonic acid	$C_4H_6O_2$	33.5	69.0	96.0	116.3	152.2	171.9$_d$	15.5	(11)
trans-Crotonic acid	$C_4H_6O_2$	s	80.0	107.8	128.0	165.5	185.0	72	(11)
Methyl acrylate	$C_4H_6O_2$	−43.7	−13.5	+9.2	28.0	61.8	80.2$_p$	......	(113)
Methacrylic acid	$C_4H_6O_2$	25.5	60.0	86.4	106.6	142.5	161.0	15	(342)
Vinyl acetate	$C_4H_6O_2$	−48.0	−18.0	+5.3	23.3	55.5	72.5	......	(267)
Acetic anhydride	$C_4H_6O_3$	1.7	36.0	62.1	82.2	119.8	139.6	−73	(33, 113, 201)
Dimethyl oxalate	$C_4H_6O_4$	20.0	56.0	83.6	104.8	143.3	163.3	......	(11, 97, 144)
cis-1 Bromo-1-butene	C_4H_7Br	−44.0	−12.8	+11.5	30.8	66.8	86.2	......	(250)
trans-1-Bromo-1-butene	C_4H_7Br	−38.4	−6.4	+18.4	38.1	75.0	94.7	−100.3	(250)
2-Bromo-1-butene	C_4H_7Br	−47.3	−16.8	+7.2	26.3	61.9	81.0	−133.4	(250)
cis-2-Bromo-2-butene	C_4H_7Br	−39.0	−7.2	+17.7	37.5	74.0	93.0	−111.2	(250)
trans-2-Bromo-2-butene	C_4H_7Br	−45.0	−13.8	+10.5	29.9	66.0	85.5	−114.6	(250)
1-Bromo-2-Butanone	C_4H_7BrO	+6.2	41.8	68.2	89.2	126.3	147.0	......	(285)
2-Methylpropionyl bromide	C_4H_7BrO	13.5	50.6	79.4	101.6	141.7	163.0	......	(11)
1,1,2-Tribromobutane	$C_4H_7Br_3$	45.0	87.8	120.2	146.0	192.0	216.2	......	(250)
1,2,2-Tribromobutane	$C_4H_7Br_3$	41.0	83.2	116.0	141.8	188.0	213.8	......	(250)
2,2,3-Tribromobutane	$C_4H_7Br_3$	38.2	79.8	111.8	136.3	182.2	206.5	......	(250)
Ethyl chloroacetate	$C_4H_7ClO_2$	+1.0	37.5	65.2	86.0	123.8	144.2	−26	(11, 298)
1,2,3-Trichlorobutane	$C_4H_7Cl_3$	+0.5	40.0	71.5	96.2	143.0	169.0	......	(285)
Butyronitrile	C_4H_7N	−20.0	+13.4	38.4	59.0	96.8	117.5	......	(159)
Diacetamide	$C_4H_7NO_2$	70.0$_s$	108.0	138.2	160.6	202.0	223.0	78.5	(11, 171)
1-Butene	C_4H_8	−104.8	−81.6	−63.4	−48.9	−21.7	−6.3	−130	(6, 84, 112, 244, 359)
cis-2-Butene	C_4H_8	−96.4	−73.4	−54.7	−39.8	−12.0	+3.7	−138.9	(6, 84, 213, 244, 359)
trans-2-Butene	C_4H_8	−99.4	−76.3	−57.6	−42.7	−14.8	+0.9	−105.4	(6, 213, 244, 359)
2-Methylpropene	C_4H_8	−105.1	−81.9	−63.8	−49.3	−22.2	−6.9	−140.3	(6, 84, 85, 86, 113, 244, 359)
Cyclobutane	C_4H_8	−92.0$_s$	−67.9$_s$	−48.4	−32.8	−3.4	+12.9	−50	(113, 163)
Methylcyclopropane	C_4H_8	−96.0	−72.8	−54.2	−39.3	−11.3	+4.5	......	(110)
2-Bromoethyl 2-chloroethyl ether	C_4H_8BrClO	36.5	76.3	106.6	129.8	172.3	195.8	......	(113)
1,2-Dibromobutane	$C_4H_8Br_2$	7.5	46.1	76.0	99.8	143.5	166.3	−64.5	(250)
dl-2,3-Dibromobutane	$C_4H_8Br_2$	+5.0	41.6	72.0	95.3	138.0	160.5	......	(250)
meso-2,3-Dibromobutane	$C_4H_8Br_2$	+1.5	39.3	68.2	91.7	134.2	157.3	−34.5	(250)
1,4-Dibromobutane	$C_4H_8Br_2$	32.0	72.4	104.0	128.7	173.8	197.5	−20	(113)
1,2-Dibromo-2-methylpropane	$C_4H_8Br_2$	−28.8	+10.5	42.3	68.8	119.8	149.0	−70.3	(11, 86)
1,3-Dibromo-2-methylpropane	$C_4H_8Br_2$	14.0	53.0	83.5	107.4	150.6	174.6	......	(113)
Di(2-bromoethyl) ether	$C_4H_8Br_2O$	47.7	88.5	119.8	144.0	188.0	212.5	......	(113)
1,2-Dichlorobutane	$C_4H_8Cl_2$	−23.6	+11.5	37.7	60.2	100.8	123.5	......	(285)
2,3-Dichlorobutane	$C_4H_8Cl_2$	−25.2	+8.5	35.0	56.0	94.2	116.0	−80.4	(437)
1,1-Dichloro-2-methylpropane	$C_4H_8Cl_2$	−31.0	+2.6	28.2	48.2	85.4	−106.0$_d$	......	(113)
1,2-Dichloro-2-methylpropane	$C_4H_8Cl_2$	−25.8	+6.7	32.0	51.7	87.8	108.0	......	(113)
1,3-Dichloro-2-methylpropane	$C_4H_8Cl_2$	−3.0	+32.0	58.6	78.8	115.4	135.0	......	(113)
di(2-chloroethyl) ether	$C_4H_8Cl_2O$	23.5	62.0	91.5	114.5	155.4	178.5	......	(113, 133, 141)

VAPOR PRESSURE (Continued)
ORGANIC COMPOUNDS (Continued)
Pressures Less than One Atmosphere (Continued)

Name	Formula	Temperature °C						M.P	Citation No.
		1 mm	10 mm	40 mm	100 mm	400 mm	760 mm		
1,2-Epoxy-2-methylpropane	C_4H_8O	−69.0	−40.3	−17.3	+1.2	36.0	55.5	......	(113)
2-Butanone	C_4H_8O	−48.3	−17.7	+6.0	25.0	60.0	79.6	−85.9	(113)
1,4-Dioxane	$C_4H_8O_2$	−35.8s	−1.2s	+25.2	45.1	81.8	101.1	10	(141, 173)
Butyric acid	$C_4H_8O_2$	25.5	61.5	88.0	108.0	144.5	163.5	−4.7	(204, 332, 373, 374)
Isobutyric acid	$C_4H_8O_2$	14.7	51.2	77.8	98.0	134.5	154.5	−47	(203, 204, 353, 373, 374)
Ethyl acetate	$C_4H_8O_2$	−43.4	−13.5	+9.1	27.0	59.3	77.1	−82.4	(113, 455, 491)
Methyl propionate	$C_4H_8O_2$	−42.0	−11.8	+11.0	29.0	61.8	79.8	−87.5	(491)
Propyl formate	$C_4H_8O_2$	−43.0	−12.6	+10.8	29.5	62.6	81.3	−92.9	(297, 491)
Isopropyl formate	$C_4H_8O_2$	−52.0	−22.7	−0.2	+17.8	50.5	68.3	......	(113, 297, 491)
α-Hydroxyisobutyric acid	$C_4H_8O_3$	73.5s	110.5	138.0	157.7	193.8	212.0	79	(11)
Ethyl glycolate	$C_4H_8O_3$	14.3	50.5	78.1	99.8	138.0	158.2	......	(113)
1-Bromobutane	C_4H_9Br	−33.0	−0.3	+24.8	44.7	81.7	101.6	−112.4	(113)
1-Bromo-2-butanol	C_4H_9BrO	23.7	55.8	79.5	97.6	128.3	145.0	......	(285)
1-Chlorobutane	C_4H_9Cl	−49.0	−18.6	+5.0	24.0	58.8	77.8	−123.1	(113, 249, 333, 394)
sec-Butyl chloride	C_4H_9Cl	−60.2	−29.2	−5.0	+14.2	50.0	68.0	−131.3	(113, 439)
Isobutyl chloride	C_4H_9Cl	−53.8	−24.5	−1.9	+16.0	50.0	68.9	−131.2	(113, 438)
tert-Butyl chloride	C_4H_9Cl	s	s	−19.0	−1.0	+32.6	51.0	−26.5	(87, 113)
2-(2-Chloroethoxy) ethanol	$C_4H_9ClO_2$	53.0	90.7	118.4	139.5	176.5	196.0	......	(113)
1-Iodo-2-methylpropane	C_4H_9I	−17.0	+17.0	42.8	63.5	100.3	120.4	−90.7	(11)
Ethyl methylcarbamate	$C_4H_9NO_2$	26.5	63.2	91.0	112.0	149.8	170.0	......	(375)
Propyl carbamate	$C_4H_9NO_2$	52.4	90.0	117.7	138.3	175.8	195.0	......	(11)
Di(nitrosoethyl)amine	$C_4H_9N_2O_2$	18.5	57.7	87.6	111.0	153.5	176.9	......	(11)
Butane	C_4H_{10}	−101.5	−77.8	−59.1	−44.2	−16.3	−0.5	−135	(6, 66, 93, 359, 382)
2-Methylpropane	C_4H_{10}	−109.2	−86.4	−68.4	−54.1	−27.1	−11.7	−145	(6, 16, 93, 187, 359, 382)
Dichlorodiethylsilane	$C_4H_{10}Cl_2Si$	−9.2	+25.4	51.6	71.8	110.0	130.4	......	(114)
Diethyldifluorosilane	$C_4H_{10}F_2Si$	−56.8	−28.8	−7.3	+9.8	40.5	58.0	......	(119)
Butyl alcohol	$C_4H_{10}O$	−1.2	+30.2	53.4	70.1	100.8	117.5	−79.9	(68, 174, 203, 204, 314)
sec-Butyl alcohol	$C_4H_{10}O$	−12.2	+16.9	38.1	54.1	83.9	99.5	−114.7	(68)
Isobutyl alcohol	$C_4H_{10}O$	−9.0	+21.7	44.1	61.5	91.4	108.0	−108	(68, 203, 219, 291, 374)
tert-Butyl alcohol	$C_4H_{10}O$	−20.4s	+5.5s	24.5s	39.8	68.0	82.9	25.3	(68, 310)
Diethyl ether	$C_4H_{10}O$	−74.3	−48.1	−27.7	−11.5	+17.9	34.6	−116.3	(31, 113, 115, 289, 338, 423a)
Methyl propyl ether	$C_4H_{10}O$	−72.2	−45.4	−24.3	−8.1	+22.5	39.1	......	(113)
1,3-Butanediol	$C_4H_{10}O_2$	22.2s	85.3	117.4	141.2	183.8	206.5	77	(189)
2,3-Butanediol	$C_4H_{10}O_2$	44.0	80.3	107.8	127.8	164.0	182.0	22.5	(368)
1,2-Dimethoxyethane	$C_4H_{10}O_2$	−48.0	−15.3	+10.7	31.8	70.8	93.0	......	(254)
2,2′-Thiodiethanol	$C_4H_{10}O_2S$	42.0	128.0	210.0d	285d	......	......	......	(25)
Diethylene glycol	$C_4H_{10}O_3$	91.8	133.8	164.3	187.5	226.5	244.8	......	(71, 141, 355)
1,2,3-Butanetriol	$C_4H_{10}O_3$	102.0	146.0	178.0	202.5	243.5	264.0	......	(113)
Diethyl sulfite	$C_4H_{10}O_3S$	10.0	46.4	74.2	96.3	137.0	159.0	......	(11, 13, 14)
Diethyl sulfate	$C_4H_{10}O_4S$	47.0	87.7	118.0	142.5	185.5	209.5d	−25.0	(11, 450)
Diethyl sulfide	$C_4H_{10}S$	−39.6	−8.0	+16.1	35.0	69.7	88.0	−99.5	(25, 39)
Diethyl selenide	$C_4H_{10}Se$	−25.7	+7.0	31.2	51.8	88.0	108.0	......	(420)
Diethylzinc	$C_4H_{10}Zn$	−22.4	+11.7	38.0	59.1	97.3	118.0	−28	(161)
Diethylamine	$C_4H_{11}N$	s	−33.0	−11.3	+6.0	38.0	55.5	−38.9	(38)
Isobutylamine	$C_4H_{11}N$	−50.0	−21.0	+1.3	18.8	50.7	68.6	−85.0	(390)
1,3-Dichlorotetramethyldisiloxane	$C_4H_{12}Cl_2Si_2$	−7.4	+28.3	55.7	76.9	116.3	138.0	−37	(312)
Tetramethyllead	$C_4H_{12}Pb$	−29.0s	+4.4	30.3	50.8	89.0	110.0	−27.5	(420)
Tetramethylsilane	$C_4H_{12}Si$	−83.8	−58.0	−37.4	−20.9	+10.0	27.0	−102.1	(15, 114)
Tetramethyltin	$C_4H_{12}Sn$	−51.3	−20.6	+3.5	22.8	58.5	78.0	......	(62, 420)
Tetramethyldiborane	$C_4H_{14}B_2$	−59.6	−27.4	−3.4	+15.3	49.8	68.6	−72.5	(371)
3-Bromopyridine	C_5H_4BrN	16.8	55.2	84.1	107.8	150.0	173.4	......	(113)
2-Chloropyridine	C_5H_4ClN	13.3	51.7	81.7	104.6	147.7	170.2	......	(113)
2-Furaldehyde	$C_5H_4O_2$	18.5	54.8	82.1	103.4	141.8	161.8	−36.5	(193, 369, 459)
Citraconic anhydride	$C_5H_4O_3$	47.1	88.9	120.3	145.4	189.8	213.5	......	(11)
Pyridine	C_5H_5N	−18.9	+13.2	38.0	57.8	95.6	115.4	−42	(11, 181, 199, 203, 204, 443)
Glutaryl chloride	$C_5H_6Cl_2O_2$	56.1	97.8	128.3	151.8	195.3	217.0	......	(11)
Glutaronitrile	$C_5H_6N_2$	91.3	140.0	176.4	205.5	257.3	286.2	......	(11, 234)
Furfuryl alcohol	$C_5H_6O_2$	31.8	68.0	95.7	115.9	151.8	170.0	......	(113, 121)
Glutaric anhydride	$C_5H_6O_3$	100.8	149.5	185.5	212.5	261.0	287.0	......	(11, 234)
Pyrotartaric anhydride	$C_5H_6O_3$	69.7	114.2	147.8	173.8	221.0	247.4	......	(11)
2-Methylthiophene	C_5H_6S	−27.4	+6.0	32.3	53.1	91.8	112.5	−63.5	(126)
3-Methylthiophene	C_5H_6S	−24.5	+9.1	35.4	55.8	93.8	115.4	−68.9	(126)
Propyl chloroglyoxylate	$C_5H_7ClO_3$	9.7	43.5	68.8	88.0	123.0	150.0	......	(11)
Tiglonitrile	C_5H_7N	−25.5	+9.2	36.7	58.2	99.7	122.0	......	(113)
Angelonitrile	C_5H_7N	−8.0	+28.0	55.8	77.5	117.7	140.0	......	(113)
α-Ethylacrylonitrile	C_5H_7N	−29.0	+5.0	31.8	53.0	92.2	114.0	......	(113)
Ethyl cyanoacetate	$C_5H_7NO_2$	67.8	106.0	133.8	152.8	187.8	206.0	......	(113)
Isoprene	C_5H_8	−79.8	−53.3	−32.6	−16.0	+15.4	32.6	−146.7	(113, 189)
1,3-Pentadiene	C_5H_8	−71.8	−45.0	−23.4	−6.7	+24.7	42.1	......	(113)
1,4-Pentadiene	C_5H_8	−83.5	−57.1	−37.0	−20.6	+8.3	26.1	......	(244)
Tiglaldehyde	C_5H_8O	−25.0	+10.0	37.0	57.7	95.5	116.4	......	(113)
Levulinaldehyde	$C_5H_8O_2$	28.1	68.0	98.3	121.8	164.0	187.0	......	(158)
Tiglic acid	$C_5H_8O_2$	52.0s	90.2	119.0	140.5	179.2	198.5	64.5	(11, 113, 220, 257)
α-Valerolactone	$C_5H_8O_2$	37.5	79.8	101.9	136.5	182.3	207.5	......	(11, 377)
α-Ethylacrylic acid	$C_5H_8O_2$	47.0	82.0	108.1	127.5	160.7	179.2	......	(45, 113, 401)
Ethyl acrylate	$C_5H_8O_2$	−29.5	+2.0	26.0	44.5	80.0	99.5	−71.2	(113)
Methyl methacrylate	$C_5H_8O_2$	−30.5	+1.0	25.5	47.0	82.0	101.0	......	(113)
Levulinic acid	$C_5H_8O_3$	102.0	141.8	169.5	190.2	227.4	245.8d	33.5	(11)
Glutaric acid	$C_5H_8O_4$	155.5	196.0	226.3	247.0	283.5	303.0	97.5	(11, 234)
Dimethyl malonate	$C_5H_8O_4$	35.0	72.0	100.0	121.9	159.8	180.7	−62	(11)
Ethyl *α*-chloropropionate	$C_5H_9ClO_2$	+6.6	41.9	68.2	89.3	126.2	146.5	......	(11)
Isopropyl chloroacetate	$C_5H_9ClO_2$	+3.8	40.2	68.7	90.3	128.0	148.6	......	(297)
Valeronitrile	C_5H_9N	−6.0	+30.0	57.8	78.6	118.7	140.8	......	(159)
α-Hydroxybutyronitrile	C_5H_9NO	41.0	77.8	104.8	125.0	159.8	178.8	......	(113, 169, 170, 442)

ORGANIC COMPOUNDS (Continued)
Pressures Less than One Atmosphere (Continued)

Name	Formula	Temperature °C 1 mm	10 mm	40 mm	100 mm	400 mm	760 mm	M.P.	Citation No.
1-Pentene	C_5H_{10}	−80.4	−54.5	−34.1	−17.7	+12.8	30.1	......	(6, 113, 359)
2-Methyl-2-butene	C_5H_{10}	−75.4	−47.9	−26.7	−9.9	+21.6	38.5	−133	(244)
2-Methyl-1-butene	C_5H_{10}	−89.1	−64.3	−44.1	−28.0	+2.5	20.2	−135	(290)
Cyclopentane	C_5H_{10}	−68.0	−40.4	−18.6	−1.3	+31.0	49.3	−93.7	(6, 359, 472)
1,2-Dibromopentane	$C_5H_{10}Br_2$	19.8	58.0	87.4	110.1	151.8	175.0	......	(11)
2-Chloroethyl 2-chloroisopropyl ether	$C_5H_{10}Cl_2O$	24.7	63.0	92.4	115.8	156.5	180.0	......	(113)
2-Chloroethyl 2-chloropropyl ether	$C_5H_{10}Cl_2O$	29.8	70.0	101.5	125.6	169.8	194.1	......	(113)
Di(2-chloroethoxy)methane	$C_5H_{10}Cl_2O_2$	53.0	94.0	125.5	149.6	192.0	215.0	......	(113)
Allyldichloroethylsilane	$C_5H_{10}Cl_2Si$	−3.0	34.2	62.7	85.2	127.0	150.3	......	(114)
3-Pentanone	$C_5H_{10}O$	−12.7	+17.2	39.4	56.2	86.3	102.7	−42	(113)
2-Pentanone	$C_5H_{10}O$	−12.0	+17.9	39.8	56.8	86.8	103.3	−77.8	(113, 491)
3-Methyl-2-butanone	$C_5H_{10}O$	−19.9	+8.3	29.6	45.5	73.8	88.9	−92	(11, 199, 201)
4-Hydroxy-3-methyl-2-butanone	$C_5H_{10}O_2$	44.6	81.0	108.2	129.0	165.5	185.0	......	(113)
Valeric acid	$C_5H_{10}O_2$	42.2	79.8	107.8	128.3	165.0	184.4	−34.5	(11, 198, 199, 200, 203, 204)
Isovaleric acid	$C_5H_{10}O_2$	34.5	71.3	98.0	118.9	155.2	175.1	−37.6	(11, 198, 199, 200, 203)
Ethyl propionate	$C_5H_{10}O_2$	−28.0	+3.4	27.2	45.2	79.8	99.1	−72.6	(491)
Propyl acetate	$C_5H_{10}O_2$	−26.7	+5.0	28.8	47.8	82.0	101.8	−92.5	(11, 113, 199, 491)
Isopropyl acetate	$C_5H_{10}O_2$	−38.3	−7.2	+17.0	35.7	69.8	89.0	......	(152)
Methyl butyrate	$C_5H_{10}O_2$	−26.8	+5.0	29.6	48.0	83.1	102.3	......	(11, 491)
Methyl isobutyrate	$C_5H_{10}O_2$	−34.1	−2.9	+21.0	39.6	73.6	92.6	−84.7	(491)
Butyl formate	$C_5H_{10}O_2$	−26.4	+6.1	31.6	51.0	86.2	106.0	......	(297)
Isobutyl formate	$C_5H_{10}O_2$	−32.7	−0.8	+24.1	43.4	79.0	98.2	−95.3	(292, 380)
sec-Butyl formate	$C_5H_{10}O_2$	−34.4	−3.1	+21.3	40.2	75.2	93.6	......	(297)
Diethyl carbonate	$C_5H_{10}O_3$	−10.1	+23.8	49.5	69.7	105.8	125.8	−43	(11)
1-Bromo-3-methylbutane	$C_5H_{11}Br$	−20.4	+13.6	39.8	60.4	99.4	120.4	......	(199, 201)
1-Iodo-3-methylbutane	$C_5H_{11}I$	−2.5	+34.1	62.3	84.4	125.8	148.2	......	(11)
Piperidine	$C_5H_{11}N$	s	+3.9	29.2	49.0	85.7	106.0	−9	(11, 199, 201)
Isobutyl carbamate	$C_5H_{11}NO_2$	s	96.4	125.3	147.2	186.0	206.5	65	(11)
Isoamyl nitrate	$C_5H_{11}NO_3$	+5.2	40.3	67.6	88.6	126.5	147.5	......	(11)
Pentane	C_5H_{12}	−76.6	−50.1	−29.2	−12.6	+18.5	36.1	−129.7	(6, 28, 93, 274, 359, 482, 485)
2-Methylbutane	C_5H_{12}	−82.9	−57.0	−36.5	−20.2	+10.5	27.8	−159.7	(6, 93, 359, 485, 486, 492)
2,2-Dimethylpropane	C_5H_{12}	−102.0	−76.7s	−56.1s	−39.1s	−7.1	+9.5	−16.6	(6, 17, 113, 253, 359, 465)
Amyl alcohol	$C_5H_{12}O$	+13.6	44.9	68.0	85.8	119.8	137.8	......	(68, 148)
Isoamyl alcohol	$C_5H_{12}O$	+10.0	40.8	63.4	80.7	113.7	130.6	−117.2	(11, 68, 199, 3)
2-Pentanol	$C_5H_{12}O$	+1.5	32.2	54.1	70.7	102.3	119.7	......	(68)
tert-Amyl alcohol	$C_5H_{12}O$	−12.9s	+17.2	38.8	55.3	85.7	101.7	−11.9	(11, 68)
Ethyl propyl ether	$C_5H_{12}O$	−64.3	−35.0	−12.0	+6.8	41.6	61.7	......	(39)
2,3,4-Pentanetriol	$C_5H_{12}O_3$	155.0	204.5	239.6	263.5	307.0	327.2	......	(113)
Ethoxytrimethylsilane	$C_5H_{14}OSi$	−50.9	−20.7	+3.7	22.1	56.3	75.7	......	(114)
Ethyltrimethylsilane	$C_5H_{14}Si$	−60.6	−31.8	−9.0	+9.2	42.8	62.0	......	(466)
Ethyltrimethyltin	$C_5H_{14}Sn$	−30.0	+3.8	30.0	50.0	87.6	108.8	......	(62)
Chloranil	$C_6Cl_4O_2$	70.7s	97.8s	116.1s	129.5s	151.3s	162.6s	290	(92)
Hexachlorobenzene	C_6Cl_6	114.4s	166.4s	206.0s	235.5	283.5	309.4	230	(113)
Pentachlorobenzene	C_6HCl_5	98.6	144.3	178.5	205.5	251.6	276.0	85.5	(113)
Pentachlorophenol	C_6HCl_5O	s	s	211.2	239.6	285.0	309.3d	188.5	(113)
3-Bromo-2,4,6-trichlorophenol	$C_6H_2BrCl_3O$	112.4	163.2	200.5	229.3	278.0	305.8	......	(113)
1,2,3,4-Tetrachlorobenzene	$C_6H_2Cl_4$	68.5	114.7	149.2	175.7	225.5	254.0	46.5	(113)
1,2,3,5-Tetrachlorobenzene	$C_6H_2Cl_4$	58.2	104.1	140.0	168.0	220.0	246.0	54.5	(113)
1,2,4,5-Tetrachlorobenzene	$C_6H_2Cl_4$	s	s	146.0	173.5	220.5	245.0	139	(113)
2,3,4,6-Tetrachlorophenol	$C_6H_2Cl_4O$	100.0	145.3	179.1	205.2	250.4	275.0	69.5	(113)
2-Bromo-4,6-dichlorophenol	$C_6H_3BrCl_2O$	84.0	130.8	165.8	193.2	242.0	268.0	68	(113)
1,2,3-Trichlorobenzene	$C_6H_3Cl_3$	40.0s	85.6	119.8	146.0	193.5	218.5	52.5	(113)
1,2,4-Trichlorobenzene	$C_6H_3Cl_3$	38.4	81.7	114.8	140.0	187.7	213.0	17	(113)
1,3,5-Trichlorobenzene	$C_6H_3Cl_3$	s	78.0	110.8	136.0	183.0	208.4	63.5	(113)
2,4,5-Trichlorophenol	$C_6H_3Cl_3O$	72.0	117.3	151.5	178.0	226.5	251.8	62	(113)
2,4,6-Trichlorophenol	$C_6H_3Cl_3O$	76.5	120.2	152.2	177.8	222.5	246.0	68.5	(113)
1,4-Dibromobenzene	$C_6H_4Br_2$	61.0s	87.7	120.8	146.5	192.5	218.6	87.5	(11, 113, 243)
1,4-Bromochlorobenzene	C_6H_4BrCl	32.0	72.7	103.8	128.0	172.6	196.9	......	(113)
1,2-Dichlorobenzene	$C_6H_4Cl_2$	20.0	59.1	89.4	112.9	155.8	179.0	−17.6	(113)
1,3-Dichlorobenzene	$C_6H_4Cl_2$	12.1	52.0	82.0	105.0	149.0	173.0	−24.2	(113)
1,4-Dichlorobenzene	$C_6H_4Cl_2$	s	54.8	84.8	108.4	150.2	173.9	53.0	(113)
2,4-Dichlorophenol	$C_6H_4Cl_2O$	53.0	92.8	123.4	146.0	187.5	210.0	45.0	(113)
2,6-Dichlorophenol	$C_6H_4Cl_2O$	59.5	101.0	131.6	154.6	197.7	220.0	......	(113)
2,4,6-Trichloroaniline	$C_6H_4Cl_3N$	134.0	170.0	195.8	214.6	246.4	262.0	78	(11)
Dichlorophenylarsine	$C_6H_5AsCl_2$	61.8	116.0	151.0	178.9	228.8	256.5	......	(113)
Bromobenzene	C_6H_5Br	+2.9	40.0	68.6	90.8	132.3	156.2	−30.7	(11, 199, 204, 289, 358, 478)
Chlorobenzene	C_6H_5Cl	−13.0	+22.2	49.7	70.7	110.0	132.2	−45.2	(11, 199, 289, 478)
2-Chlorophenol	C_6H_5ClO	12.1	51.2	82.0	106.0	149.8	174.5	7.0	(113)
3-Chlorophenol	C_6H_5ClO	44.2	86.1	118.0	143.0	188.7	214.0	32.5	(113)
4-Chlorophenol	C_6H_5ClO	49.8	92.2	125.0	150.0	196.0	220.0	42	(113)
Benzenesulfonylchloride	$C_6H_5ClO_2S$	65.9	112.0	147.7	174.5	224.0	251.5d	14.5	(11, 235)
Phenyl dichlorophosphate	$C_6H_5Cl_2O_2P$	66.7	110.0	143.4	168.0	213.0	239.5	......	(11, 113)
Trichlorophenylsilane	$C_6H_5Cl_3Si$	33.0	74.2	105.8	130.5	175.7	201.0	......	(114)
Fluorobenzene	C_6H_5F	−43.4s	−12.4	+11.5	30.4	65.7	84.7	−42.1	(113, 416, 478)
Trifluorophenylsilane	$C_6H_5F_3Si$	−31.0	+0.8	25.4	44.2	78.7	98.3	......	(119)
Iodobenzene	C_6H_5I	24.1	64.0	94.4	118.3	163.9	188.6	−28.5	(115, 289, 358, 478)
Nitrobenzene	$C_6H_5NO_2$	44.4	84.9	115.4	139.9	185.8	210.6	+5.7	(11, 58, 199, 204)
2-Nitrophenol	$C_6H_5NO_3$	49.3	90.4	122.1	146.4	191.0	214.5	45	(11)
1,5-Hexadiene-3-yne	C_6H_6	−45.1	−14.0	+10.0	29.5	64.4	84.0	......	(304)
Benzene	C_6H_6	−36.7	−11.5s	+7.6	26.1	60.6	80.1	+5.5	(3, 6, 11, 32, 109, 131, 197, 204, 289, 343, 359, 366, 367, 392, 415, 478)

VAPOR PRESSURE (Continued)
ORGANIC COMPOUNDS (Continued)
Pressures Less than One Atmosphere (Continued)

Name	Formula	Temperature °C						M.P.	Citation No.
		1 mm	10 mm	40 mm	100 mm	400 mm	760 mm		
2-Chloroaniline	C_6H_6ClN	46.3	84.8	115.6	139.5	183.7	208.8	......	(204)
3-Chloroaniline	C_6H_6ClN	63.5	102.0	133.6	158.0	203.5	228.5	− 10.4	(204)
4-Chloroaniline	C_6H_6ClN	59.3_s	102.1	135.0	159.9	206.6	230.5	70.5	(11)
4-Chlorophenol	C_6H_6ClO	54.3	95.8	126.8	150 6	194.3	217.0	42	(11)
2-Nitroaniline	$C_6H_6N_2O_2$	104.0	150.4	186.0	213.0	260.0	284.5_d	71.5	(35)
3-Nitroaniline	$C_6H_6N_2O_2$	119.3	167.8	204.2	232.1	280.2	305.7_d	114	(35)
4-Nitroaniline	$C_6H_6N_2O_2$	142.4_s	194.4	234.2	261.8	310.2_d	336.0_d	146.5	(35)
Phenol	C_6H_6O	40.1_s	73.8	100.1	121.4	160.0	181.9	40.6	(11, 34, 113, 151, 174, 203, 204)
Pyrocatechol	$C_6H_6O_2$	s	118.3	150.6	176.0	221.5	245.5	105	(11)
Resorcinol	$C_6H_6O_2$	108.4_s	152.1	185.3	209.8	253.4	276.5	110.7	(11)
Hydroquinone	$C_6H_6O_2$	132.4_s	163.5_s	192.0	216.5	262.5	286.2	170.3	(404)
Pyrogallol	$C_6H_6O_3$	s	167.7	204 2	232.0	281.5	309.0_d	133	(11)
Benzenethiol	C_6H_6S	18.6	56.0	84.2	106.6	146.7	168.0	......	(11)
Aniline	C_6H_7N	34.8	69.4	96.7	119.9	161.9	184.4	− 6.2	(11, 34, 143, 151, 204, 341)
2-Picoline	C_6H_7N	− 11.1	+ 24.4	51.2	71.4	108.4	128.8	− 70	(11, 199)
Ethylene-bis-(chloroacetate)	$C_6H_8Cl_2O_4$	112.0	158.0	191.0	215.0	259.5	283.5	......	(113)
1,3-Phenylenediamine	$C_6H_8N_2$	99.8	147.0	182.5	209.9	259.0	285.5	62.8	(11)
Phenylhydrazine	$C_6H_8N_2$	71.8	115.8	148.2	173.5	218.2	243.5_d	19.5	(11, 113)
α-Methylglutaric anhydride	C_6H[illegible]O_3	93.8	141.8	177.5	205.0	255.5	282.5	......	(11)
α,α-Dimethylsuccinic anhydride	$C_6H_8O_3$	61.4	102.0	132.3	155.3	197.5	219.5	......	(11)
Dimethyl maleate	$C_6H_8O_4$	45.7	86.4	117.2	140.3	182.2	205.0	......	(11)
Isobutyl dichloroacetate	$C_6H_{10}Cl_2O_2$	28.6	67.5	96.7	119.8	160.0	183.0	......	(11)
Diallyldichlorosilane	$C_6H_{10}Cl_2Si$	+ 9.5	47.4	76.4	99.7	142.0	165.3	......	(114)
Cyclohexanone	$C_6H_{10}O$	+ 1.4	38.7	67.8	90.4	132.5	155.6	− 45.0	(440)
Mesityl oxide	$C_6H_{10}O$	− 8.7	+ 26.0	51.7	72.1	109.8	130.0	− 59	(11, 199)
Isocaprolactone	$C_6H_{10}O_2$	38.3	80.3	112 3	137.2	182 1	207.0	......	(11)
Propionic anhydride	$C_6H_{10}O_3$	20.6	57.7	85.6	107.2	146.0	167.0	− 45	(11, 198, 199)
Ethyl acetoacetate	$C_6H_{10}O_3$	28.5	67.3	96.2	118.5	158.2	180.8	− 45	(11, 199)
Methyl levulinate	$C_6H_{10}O_3$	39 8	79.7	109.5	133 0	175 8	197.7	......	(376)
Adipic acid	$C_6H_{10}O_4$	159.5	205.5	240 5	265.0	312.5	337.5	152	(11, 234)
Diethyl oxalate	$C_6H_{10}O_4$	47.4	83 8	110.6	130 8	166.2	185.7	− 40.6	(11, 199, 201)
Glycol diacetate	$C_6H_{10}O_4$	38.3	77.1	106.1	128.0	168.3	190.5	− 31	(423)
Dimethyl-*l*-malate	$C_6H_{10}O_5$	75.4	118.3	150.1	175.1	219.5	242.6	......	(11)
Dimethyl-*d*-tartrate	$C_6H_{10}O_6$	102.1	148.2	182.4	208.8	255.0	280.0	61.5	(11, 149)
Dimethyl-*dl*-tartrate	$C_6H_{10}O_6$	100.4	147.5	182.4	209.5	257.4	282.0	**89**	**(11, 149)**
Diallyl sulfide	$C_6H_{10}S$	− 9.5	+ 26.6	54.2	75.8	116.1	138.6	**− 83**	**(11, 25)**
Ethyl α-bromoisobutyrate	$C_6H_{11}BrO_2$	10.6	48.0	77.0	99.8	141.2	163.6	......	**(11)**
***sec*-Butyl chloroacetate**	$C_6H_{11}ClO_2$	17.0	54.6	83.6	105.5	146.0	167.8	......	**(297)**
Capronitrile	$C_6H_{11}N$	+ 9.2	47.5	76.9	99.8	141.0	163.7	......	**(159)**
1-Hexene	C_6H_{12}	− 57.5	− 28.1	− 5.0	+ 13.0	46.8	66.0	**− 98.5**	**(474)**
Cyclohexane	C_6H_{12}	$- 45.3_s$	$- 15.9_s$	+ 6.7	25.5	60.8	80.7	**+ 6.6**	**(6, 109, 293, 359, 367, 487)**
Methylcyclopentane	C_6H_{12}	− 53.7	− 23.7	− 0.6	+ 17.9	52.3	71.8	**−142.4**	**(6, 359, 471)**
Dichlorodiisopropyl ether	$C_6H_{12}Cl_2O$	29.6	68.2	97.3	119.7	159.8	182.7	......	**(113)**
Bis-2-chloroethyl acetal	$C_6H_{12}Cl_2O_2$	56.2	97.6	127.8	150.7	190.5	212.6	......	**(113)**
2-Hexanone	$C_6H_{12}O$	+ 7.7	38.8	62.0	79.8	111.0	127.5	**− 56.9**	**(113)**
4-Methyl-2-pentanone	$C_6H_{12}O$	− 1.4	+ 30.0	52.8	70.4	102.0	119.0	**− 84.7**	**(113)**
Allyl propyl ether	$C_6H_{12}O$	− 39.0	− 7.9	+ 16.4	35.8	71.4	90.5	......	**(254)**
Allyl isopropyl ether	$C_6H_{12}O$	− 43.7	− 12.9	+ 10.9	29.0	61.7	79.5	......	**(254)**
Cyclohexanol	$C_6H_{12}O$	21.0_s	56.0	83.0	103.7	141.4	161.0	**23.9**	**(113)**
Caproic acid	$C_6H_{12}O_2$	71.4	99.5	125.0	144.0	181.0	202.0	**− 1.5**	**(11, 203)**
Isocaproic acid	$C_6H_{12}O_2$	66.2	94.0	120.4	141.4	181.0	207.7	**− 35**	**(11, 203, 204)**
4-Hydroxy-4-methyl-2-pentanone	$C_6H_{12}O_2$	22.0	58.8	86.7	108.2	147.5	167.9	**− 47**	**(113)**
Methyl isovalerate	$C_6H_{12}O_2$	− 19.2	+ 14.0	39.8	59.8	96.7	116.7	......	**(380)**
Ethyl butyrate	$C_6H_{12}O_2$	− 18.4	+ 15.3	41.5	62.0	100.0	121.0	**− 93.3**	**(380)**
Ethyl isobutyrate	$C_6H_{12}O_2$	− 24.3	+ 8.4	33.8	53.5	90.0	110.1	**− 88.2**	**(292, 380)**
Propyl propionate	$C_6H_{12}O_2$	− 14.2	+ 19.4	45.0	65.2	102.0	122.4	**− 76**	**(11, 380)**
Isobutyl acetate	$C_6H_{12}O_2$	− 21.2	+ 12.8	39.2	59.7	97.5	118.0	**− 98.9**	**(11, 199, 292, 380)**
Isoamyl formate	$C_6H_{12}O_2$	− 17.5	+ 17.1	44.0	65.4	102.7	123.3	......	**(380)**
Paraformaldehyde	$C_6H_{12}O_3$	$- 9.4_s$	$+ 24.1_s$	49.5_s	69.0_s	104.3_s	124.0_s	**155 ± 5**	**(11)**
***sec*-Butyl glycolate**	$C_6H_{12}O_3$	28.3	66.0	94.2	116.4	155.6	177.5	......	**(113)**
Hexane	C_6H_{14}	− 53.9	− 25.0	− 2.3	+ 15.8	49.6	68.7	**− 95.3**	**(6, 115, 251, 289, 359, 394, 427, 474)**
2-Methylpentane	C_6H_{14}	− 60.9	− 32.1	− 9.7	+ 8.1	41.6	60.3	**−154**	**(6, 21, 359, 471)**
3-Methylpentane	C_6H_{14}	− 59.0	− 30.1	− 7.3	+ 10.5	44.2	63.3	**−118**	**(6, 359, 471)**
2,2-Dimethylbutane	C_6H_{14}	− 69.3	− 41.5	− 19.5	− 2.0	+ 31.0	49.7	**− 99.8**	**(6, 21, 359, 471)**
2,3-Dimethylbutane	C_6H_{14}	− 63.6	− 34.9	− 12.4	+ 5.4	39.0	58.0	**−128.2**	**(6, 359, 471, 488)**
1-Hexanol	$C_6H_{14}O$	24.4	58.2	83.7	102.8	138.8	157.0	**− 51 6**	**(68, 185, 380)**
2-Hexanol	$C_6H_{14}O$	14.6	45.0	67.9	87.3	121.8	139.9	......	**(185)**
3-Hexanol	$C_6H_{14}O$	+ 2.5	36.7	62.2	81.8	117.0	135.5	......	**(185)**
2-Methyl-1-pentanol	$C_6H_{14}O$	15.4	49 6	74.7	94.2	129.8	147.9	......	**(184a, 185)**
2-Methyl-2-pentanol	$C_6H_{14}O$	− 4.5	+ 27.6	51.3	69.2	102.6	121 1	**−103**	**(185)**
2-Methyl-4-pentanol	$C_6H_{14}O$	− 0.3	+ 33.3	58.2	78.0	113.5	131.7	......	**(185)**
Dipropyl ether	$C_6H_{14}O$	− 43.3	− 11.8	+ 13.2	33.0	69.5	89.5	**−122**	**(113)**
Diisopropyl ether	$C_6H_{14}O$	− 57.0	− 27.4	− 4.5	+ 13.7	48.2	67.5	**− 60**	**(133)**
Acetal	$C_6H_{14}O_2$	− 23.0	+ 8.0	31.9	50.1	84.0	102.0	......	**(11, 199)**
1,2-Diethoxyethane	$C_6H_{14}O_2$	− 33.5	+ 1.6	29.7	51.8	94.1	119.5	......	**(254)**
Di(2-methoxyethyl) ether	$C_6H_{14}O_3$	13.0	50.0	77.5	99.5	138.5	159.8	......	**(141)**
Diethylene glycol, ethyl ether	$C_6H_{14}O_3$	45.3	85.8	116.7	140.3	180.3	201.9	......	**(113)**
Dipropyleneglycol	$C_6H_{14}O_3$	73.8	116.2	147.4	169.9	210.5	231.8	......	**(71, 113)**
Triethyleneglycol	$C_6H_{14}O_4$	114.0	158.1	191.3	214.6	256.6	278.3	......	**(71, 141)**
Triethylboron	$C_6H_{15}B$	......	−148.0	−131.4	−116.0	− 81.0	− 56.2	......	**(413)**
Chlorotriethylsilane	$C_6H_{15}ClSi$	− 4.9	+ 32.0	60.2	82.3	123.6	146.3	......	**(114)**
Triethyl phosphate	$C_6H_{15}O_4P$	39.6	82.1	115.7	141.6	187.0	211.0	......	**(76)**
Triethylthallium	$C_6H_{15}Tl$	+ 9.3	51.7	85.4	112.1	163.5	192.1_d	**− 63.0**	**(357)**

VAPOR PRESSURE (Continued)
ORGANIC COMPOUNDS (Continued)
Pressures Less than One Atmosphere (Continued)

Name	Formula	Temperature °C						M.P.	Citation No.
		1 mm	10 mm	40 mm	100 mm	400 mm	760 mm		
Diethoxydimethylsilane	$C_6H_{16}O_2Si$	− 19.1	+ 13.3	38.0	57.6	93.2	113.5	......	(114)
Trimethylpropylsilane	$C_6H_{16}Si$	− 46.0	− 13.9	+ 11.3	31.6	69.2	90.0	......	(466)
Trimethylpropyltin	$C_6H_{16}Sn$	− 12.0	+ 21.8	48.5	69.8	109.6	131.7	......	(62)
1,5-Dichlorohexamethyltrisiloxane	$C_6H_{18}Cl_2O_2Si_3$	26.0	65.1	94.8	118.2	160.2	184.0	− 53	(312)
Hexamethylcyclotrisiloxane	$C_6H_{18}O_3Si_3$	s	s	s	78.7	114.7	134.0	64	(188, 312)
Hexamethyldisiloxane	$C_6H_{18}OSi_2$	− 29.0	+ 2.8	26.7	45.6	80.0	99.2	......	(114)
3,4-Dichloro-α,α,α-trifluorotoluene	$C_7H_3Cl_2F_3$	11.0	52.2	84.0	109.2	150.5	172.8	− 12.1	(113)
2-Chloro-α,α,α-trifluorotoluene	$C_7H_4ClF_3$	0.0	37.1	65.9	88.3	130.0	152.2	− 6.0	(113)
2-α,α,α-tetrachlorotoluene	$C_7H_4Cl_4$	69.0	117.9	155.0	185.0	233.0	262.1	28.7	(113)
Benzoyl bromide	C_7H_5BrO	47.0	89.8	122.6	147.7	193.7	218.5	0	(11)
Benzoyl chloride	C_7H_5ClO	32.1	73.0	103.8	128.0	172.8	197.2	− 0.5	(11, 203, 204)
α,α,α-Trichlorotoluene	$C_7H_5Cl_3$	45.8	87.6	119.8	144.3	189.2	213.5	− 21.2	(11)
α,α,α-Trifluorotoluene	$C_7H_5F_3$	− 32.0s	+ 0.4	25.7	45.3	82.0	102.2	− 29.3	(113)
Benzonitrile	C_7H_5N	28.2	69.2	99.6	123.5	166.7	190.6	− 12.9	(113, 204)
Phenyl isocyanide	C_7H_5N	12.0	49.7	78.3	101.0	142.3	165.0d	......	(11, 235)
Phenyl isocyanate	C_7H_5NO	10.6	48.5	77.7	100.6	142.7	165.6	......	(113)
2-Nitrobenzaldehyde	$C_7H_5NO_3$	85.8	133.4	168.8	196.2	246.8	273.5	40.9	(11)
3-Nitrobenzaldehyde	$C_7H_5NO_3$	96.2	142.8	177.7	204.3	252.1	278.3	58	(11)
Phenyl isothiocyanate	C_7H_5NS	47.2	89.8	122.5	147.7	194.0	218.5	− 21.0	(11, 25, 198, 199)
α,α-dichlorotoluene	$C_7H_6Cl_2$	35.4	78.7	112.1	138.3	187.0	214.0	− 16.1	(11)
Benzaldehyde	C_7H_6O	26.2	62.0	90.1	112.5	154.1	179.0	− 26	(11, 199, 204)
Benzoic acid	$C_7H_6O_2$	96.0s	132.1	162.6	186.2	227.0	249.2	121.7	(11, 189, 204, 218, 303)
Salicylaldehyde	$C_7H_6O_2$	33.0	73.8	105.2	129.4	173.7	196.5	− 7	(113)
4-Hydroxybenzaldehyde	$C_7H_6O_2$	121.2	169.7	206.0	233.5	282.6	310.0	115.5	(11)
Salicylic acid	$C_7H_6O_3$	113.7s	146.2s	172.2	193.4	230.5	256.0	159	(404)
α-Bromotoluene	C_7H_7Br	32.2	73.4	104.8	129.8	175.2	198.5	− 4	(11)
2-Bromotoluene	C_7H_7Br	24.4	62.3	91.0	112.0	157.3	181.8	− 28	(128, 201, 415)
3-Bromotoluene	C_7H_7Br	14.8	64.0	93.9	117.8	160.0	183.7	− 39.8	(128, 415)
4-Bromotoluene	C_7H_7Br	10.3	61.1	91.8	116.4	160.2	184.5	28.5	(11, 113, 199, 128)
4-Bromoanisole	C_7H_7BrO	48.8	91.9	125.0	150.1	197.5	223.0	12.5	(11)
α-Chlorotoluene	C_7H_7Cl	22.0	60.8	90.7	114.2	155.8	179.4	− 39	(11, 199, 201)
2-Chlorotoluene	C_7H_7Cl	+ 5.4	43.2	72.0	94.7	137.1	159.3	......	(11, 113, 128 415)
3-Chlorotoluene	C_7H_7Cl	+ 4.8	43.2	73.0	96.3	139.7	162.3	......	(128, 415)
4-Chlorotoluene	C_7H_7Cl	+ 5.5	43.8	73.5	96.6	139.8	162.3	+ 7.3	(11, 113, 128, 199, 415)
2-Fluorotoluene	C_7H_7F	− 24.2	+ 8.9	34.7	55.3	92.8	114.0	− 80	(113)
3-Fluorotoluene	C_7H_7F	− 22.4	+ 11.0	37.0	57.5	95.4	116.0	−110.8	(113)
4-Fluorotoluene	C_7H_7F	− 21.8	+ 11.8	37.8	58.1	96.1	117.0	......	(113)
2-Iodotoluene	C_7H_7I	37.2	79.8	112.4	138.1	185.7	211.0	......	(11)
2-Nitrotoluene	$C_7H_7NO_2$	50.0	93.8	126.3	151.5	197.7	222.3	− 4.1	(11, 36, 204, 302)
3-Nitrotoluene	$C_7H_7NO_2$	50.2	96.0	130.7	156.9	206.8	231.9	15.1	(36, 302)
4-Nitrotoluene	$C_7H_7NO_2$	53.7	100.5	136.0	163.0	212.5	238.3	51.9	(11, 36, 204)
Toluene	C_7H_8	− 26.7	+ 6.4	31.8	51.9	89.5	110.6	− 95.0	(6, 11, 24, 107, 115, 199, 204, 239, 253, 359)
Benzyldichlorosilane	$C_7H_8Cl_2Si$	45.3	83.2	111.8	133.5	173.0	194.3	......	(114)
Dichloromethylphenylsilane	$C_7H_8Cl_2Si$	35.7	77.4	109.5	134.2	180.2	205.5	......	(114)
Dichloro-4-tolysilane	$C_7H_8Cl_2Si$	46.2	84.2	113.2	135.5	175.2	196.3	......	(114)
Anisole	C_7H_8O	+ 5.4	42.2	70.7	93.0	133.8	155.5	− 37.3	(11)
Benzyl alcohol	C_7H_8O	58.0	92.6	119.8	141.7	183.0	204.7	− 15.3	(11, 204)
2-Cresol	C_7H_8O	38.2	76.7	105.8	127.4	168.4	190.8	30.8	(11, 204, 406)
3-Cresol	C_7H_8O	52.0	87.8	116.0	138.0	179.0	202.8	10.9	(11, 204, 406)
4-Cresol	C_7H_8O	53.0	88.6	117.7	140.0	179.4	201.8	35.5	(11, 204, 406)
3,5-Dimethyl-1,2-pyrone	$C_7H_8O_2$	78.6	122.0	152.7	177.5	221.0	245.0	51.5	(11)
2-Methoxyphenol	$C_7H_8O_2$	52.4	92.0	121.6	144.0	184.1	205.0	28.3	(11)
Ethyl 2-furoate	$C_7H_8O_3$	37.6	77.1	107.5	130.4	172.5	195.0	34	(11)
Benzylamine	C_7H_9N	29.0	67.7	97.3	120.0	161.3	184.5	......	(11)
N-Methylaniline	C_7H_9N	36.0	76.2	106.0	129.8	172.0	195.5	− 57	(11, 204, 300)
2-Toluidine	C_7H_9N	44.0	81.4	110.0	133.0	176.2	199.7	− 16.3	(11, 37, 204)
3-Toluidine	C_7H_9N	41.0	82.0	113.5	136.7	180.6	203.3	− 31.5	(11, 37, 204)
4-Toluidine	C_7H_9N	42.0	81.8	111.5	133.7	176.9	200.4	44.5	(11, 37, 204)
2-Methoxyaniline	C_7H_9NO	61.0	101.7	132.0	155.2	197.3	218.5	5.2	(11)
Toluene-2,4-diamine	$C_7H_{10}N_2$	106.5	151.7	185.7	211.5	256.0	280.0	99	(11)
4-Tolylhydrazine	$C_7H_{10}N_2$	82.0	123.8	154.1	178.0	219.5	242.0d	65.5	(11)
Trimethysuccinic anhydride	$C_7H_{10}O_3$	53.5	97.4	131.0	156.5	205.5	231.0	......	(11)
Dimethyl citraconate	$C_7H_{10}O_4$	50.8	91.8	122.6	145.8	188.0	210.5	......	(11)
Dimethyl itaconate	$C_7H_{10}O_4$	69.3	106.6	133.7	153.7	189.8	208.0	38	(11)
trans-Dimethyl mesaconate	$C_7H_{10}O_4$	46.8	87.8	118.0	141.5	183.5	206.0	......	(11)
2-Cyano-2-butyl acetate	$C_7H_{11}NO_2$	42.0	82.0	111.8	133.8	173.4	195.2	......	(113, 170)
Butyl acrylate	$C_7H_{12}O_2$	− 0.5	+ 35.5	63.4	85.1	125.2	147.4	− 64.6	(113)
Ethyl levulinate	$C_7H_{12}O_3$	47.3	87.3	117.7	141.3	183.0	206.2	......	(376)
Pimelic acid	$C_7H_{12}O_4$	163.4	212.0	247.0	272.0	318.5	342.1	103	(11, 234)
Diethyl malonate	$C_7H_{12}O_4$	40.0	81.3	113.3	136.2	176.8	198.9	− 49.8	(11, 180)
Enanthyl chloride	$C_7H_{13}ClO$	34.2	64.6	86.4	102.7	130.7	145.0	......	(11, 229)
Enanthonitrile	$C_7H_{13}N$	21.0	61.6	92.6	116.8	160.0	184.6	......	(159)
Ethylcyclopentane	C_7H_{14}	− 32.2	− 0.1	+ 25.0	45.0	82.3	103.4	−138.6	(6, 359, 471)
Methylcyclohexane	C_7H_{14}	− 35.9	− 3.2	+ 22.0	42.1	79.6	100.9	−126.4	(6, 253, 359, 415, 471)
2-Heptene	C_7H_{14}	− 35.8	− 3.5	+ 21.5	41.3	78.1	98.5	......	(446)
Enanthaldehyde	$C_7H_{14}O$	12.0	43.0	66.3	84.0	125.5	155.0	− 42	(11, 199, 203,
2-Heptanone	$C_7H_{14}O$	19.3	55.5	81.2	100.0	133.2	150.2	......	(415)
4-Heptanone	$C_7H_{14}O$	23.0	55.0	78.1	96.0	127.3	143.7	− 32.6	(11)
2,4-Dimethyl-3-pentanone	$C_7H_{14}O$	+ 5.2	36.7	59.6	77.0	108.0	123.7	......	(113)
Enanthic acid	$C_7H_{14}O_2$	78.0	113.2	139.5	160.0	199.6	221.5	− 10	(11, 203, 204)
Methyl caproate	$C_7H_{14}O_2$	+ 5.0	42.0	70.0	91.4	129.8	150.0d	......	(2)
Ethyl isovalerate	$C_7H_{14}O_2$	− 6.1	+ 28.7	55.2	75.9	114.0	134.3	− 99.3	(11, 380)
Propyl butyrate	$C_7H_{14}O_2$	− 1.6	+ 34.0	61.5	82.6	121.7	142.7	− 95.2	(11, 380)
Propyl isobutyrate	$C_7H_{14}O_2$	− 6.2	+ 28.3	54.3	73.9	112.0	133.9	......	(380)
Isopropyl isobutyrate	$C_7H_{14}O_2$	− 16.3	+ 17.0	42.4	62.3	100.0	120.5	......	(489)
Isobutyl propionate	$C_7H_{14}O_2$	− 2.3	+ 32.3	58.5	79.5	116.4	136.8	− 71	(380)

ORGANIC COMPOUNDS (Continued)
Pressures Less than One Atmosphere (Continued)

Name	Formula	Temperature °C						M.P.	Citation No.
		1 mm	10 mm	40 mm	100 mm	400 mm	760 mm		
Isoamyl acetate	$C_7H_{14}O_2$	0.0	+ 35.2	62.1	83.2	121.5	142.0	......	(11)
Heptane	C_7H_{16}	− 34.0	− 2.1	+ 22.3	41.8	78.0	98.4	− 90.6	(6, 251, 269, 289, 359, 394, 483)
2-Methylhexane	C_7H_{16}	− 40.4	− 9.1	+ 14.9	34.1	69.8	90.0	−118.2	(6, 359, 471)
3-Methylhexane	C_7H_{16}	− 39.0	− 7.8	+ 16.4	35.6	71.6	91.9	−119.4	(6, 359, 471)
3-Ethylpentane	C_7H_{16}	− 37.8	− 6.8	+ 17.5	36.9	73.0	93.5	−118.6	(6, 359, 471)
2,2-Dimethylpentane	C_7H_{16}	− 49.0	− 18.7	+ 5.0	23.9	59.2	79.2	−123.7	(6, 359, 471)
2,3-Dimethylpentane	C_7H_{16}	− 42.0	− 10.3	+ 13.9	33.3	69.4	89.8	−135	(6, 359, 471)
2,4-Dimethylpentane	C_7H_{16}	− 48.0	− 17.1	+ 6.5	25.4	60.6	80.5	−119.5	(6, 359, 471)
3,3-Dimethylpentane	C_7H_{16}	− 45.9	− 14.4	+ 9.9	29.3	65.5	86.1	−135.0	(6, 359, 471)
2,2,3-Trimethylbutane	C_7H_{16}	s	− 18.8	+ 5.2	24.4	60.4	80.9	− 25.0	(6, 359, 471)
1-Heptanol	$C_7H_{16}O$	42.4	74.7	99.8	119.5	155.6	175.8	34.6	(11, 68)
Triethyl orthoformate	$C_7H_{16}O_3$	+ 5.5	40.5	67.5	88.0	125.7	146.0	......	(11)
Triethoxymethylsilane	$C_7H_{18}O_3Si$	− 1.5	+ 34.6	61.7	82.7	121.8	143.5	......	(114)
Butyltrimethylsilane	$C_7H_{18}Si$	− 23.4	+ 9.9	35.9	56.3	93.8	115.0	......	(466)
Triethylmethylsilane	$C_7H_{18}Si$	− 18.2	+ 16.6	44.0	65.6	105.3	127.0	......	(466)
Phthaloyl chloride	$C_8H_4Cl_2O_2$	86.3_s	134.2	170.0	197.8	248.3	275.8	88.5	(11)
Phthalic anhydride	$C_8H_4O_3$	96.5_s	134.0	172.0	202.3	256.8	284.5	130.8	(97, 189, 284)
α,α-Dichlorophenylacetonitrile	$C_8H_5Cl_2N$	56.0	98.1	130.0	154.5	199.5	223.5	......	(11)
Pentachloroethylbenzene	$C_8H_5Cl_5$	96.2	148.0	186.2	216.0	269.3	299.0	......	(113)
Phenylglyoxylonitrile	C_8H_5NO	44.5	85.5	116.6	141.0	185.0	208.0	33.5	(11)
2,3-Dichlorostyrene	$C_8H_6Cl_2$	61.0	104.6	137.8	163.5	210.0_p	235.0_p	......	(113, 276)
2,4-Dichlorostyrene	$C_8H_6Cl_2$	53.5	97.4	129.2	153.8	200.0_p	225.0_p	......	(113, 276)
2,5-Dichlorostyrene	$C_8H_6Cl_2$	55.5	98.2	131.0	155.8	202.5_p	227.0_p	......	(113, 276)
2,6-Dichlorostyrene	$C_8H_6Cl_2$	47.8	90.0	122.4	147.6	193.5_p	217.0_p	......	(113, 276)
3,4-Dichlorostyrene	$C_8H_6Cl_2$	57.2	100.4	133.7	158.2	205.7_p	230.0_p	......	(113, 276)
3,5-Dichlorostyrene	$C_8H_6Cl_2$	53.5	97.4	129.2	153.8	200.0_p	225.0_p	......	(113, 276)
3,4,5,6-Tetrachloro-1,2-xylene	$C_8H_6Cl_4$	94.4	140.3	174.2	200.5	248.3	273.5	......	(11)
1,2,3,5-Tetrachloro-4-ethylbenzene	$C_8H_6Cl_4$	77.0	126.0	162.1	191.6	243.0	270.0	......	(113)
Phenylglyoxal	$C_8H_6O_2$	s	87.8	115.5	136.2	173.5	193.5	73	(354)
Phthalide	$C_8H_6O_2$	95.5	144.0	181.0	210.0	261.8	290.0	73	(11)
Piperonal	$C_8H_6O_3$	87.0	132.0	165.7	191.7	238.5	263.0	37	(11)
3-Chlorostyrene	C_8H_7Cl	25.3	65.2	96.5	121.2	165.7_p	190.0_p	......	(113)
4-Chlorostyrene	C_8H_7Cl	28.0	67.5	98.0	122.0	166.0_p	191.0_p	− 15.0	(113)
Phenylacetyl chloride	C_8H_7ClO	48.0	89.0	119.8	143.5	186.0	210.0	......	(11)
2-Tolunitrile	C_8H_7N	36.7	77.9	110.0	135.0	180.0	205.2	− 13	(11)
4-Tolunitrile	C_8H_7N	42.5	85.8	109.5	145.2	193.0	217.6	29.5	(11)
Phenylacetonitrile	C_8H_7N	60.0	103.5	136.3	161.8	208.5	233.5	− 23.8	(11)
2-Tolyl isocyanide	C_8H_7N	25.2	64.0	94.0	117.7	159.9	183.5	......	(11, 294)
2-Nitrophenyl acetate	$C_8H_7NO_4$	100.0	142.0	172.8	194.1	233.5	253.0_d	......	(11)
2-Methylbenzothiazole	C_8H_7NS	70.0	111.2	141.2	163.9	204.5	225.5	15.4	(95)
Benzylisothiocyanate	C_8H_7NS	79.5	121.8	153.0	177.7	220.4	243.0	......	(277)
Styrene	C_8H_8	− 7.0	+ 30.8	59.8	82.0_p	122.5_p	145.2_p	− 30.6	(11, 44, 63, 113, 253)
(1,2-Dibromoethyl) benzene	$C_8H_8Br_2$	86.0	129.8	161.8	186.3	230.0	254.0	......	(11)
1,2-Dichloro-3-ethylbenzene	$C_8H_8Cl_2$	46.0	90.0	123.8	149.8	197.0	222.1	− 40.8	(113)
1,2-Dichloro-4-ethylbenzene	$C_8H_8Cl_2$	47.0	92.3	127.5	153.3	201.7	226.6	− 76.4	(113)
1,4-Dichloro-2-ethylbenzene	$C_8H_8Cl_2$	38.5	83.2	118.0	144.0	191.5	216.3	− 61.2	(113)
Acetophenone	C_8H_8O	37.1	78.0	109.4	133.6	178.0	202.4	20.5	(11, 156, 203, 204)
Phenyl acetate	$C_8H_8O_2$	38.2	78.0	108.1	131.6	173.5	195.9	......	(11)
Phenylacetic acid	$C_8H_8O_2$	97.0	141.3	173.6	198.2	243.0	265.5	76.5	(11)
Anisaldehyde	$C_8H_8O_2$	73.2	117.8	150.5	176.7	223.0	248.0	2.5	(11, 316)
Methyl benzoate	$C_8H_8O_2$	39.0	77.3	107.8	130.8	174.7	199.5	− 12.5	(11, 203, 204)
Methyl salicylate	$C_8H_8O_3$	54.0	95.3	126.2	150.0	197.5	223.2	− 8.3	(11, 334, 341)
Vanillin	$C_8H_8O_3$	107.0	154.0	188.7	214.5	260.0	285.0	81.5	(11)
Dehydroacetic acid	$C_8H_8O_4$	91.7	137.3	171.0	197.5	244.5	269.0	......	(11)
2-Bromo-1,4-xylene	C_8H_9Br	37.5	78.8	110.6	135.7	181.0	206.7	+ 9.5	(113, 166)
1-Bromo-4-ethylbenzene	C_8H_9Br	30.4	74.0	108.5	135.5	182.0	206.0	− 45.0	(11, 113)
(2-Bromoethyl) benzene	C_8H_9Br	48.0	90.5	123.2	148.2	194.0	219.0	......	(113)
1-Chloro-2-ethylbenzene	C_8H_9Cl	17.2	56.1	86.2	110.0	152.2	177.6	− 80.2	(113)
1-Chloro-3-ethylbenzene	C_8H_9Cl	18.6	58.2	89.2	113.6	156.7	181.1	− 53.3	(113)
1-Chloro-4-ethylbenzene	C_8H_9Cl	19.2	60.0	91.8	116.0	159.8	184.3	− 62.6	(113)
1-Chloro-2-ethoxybenzene	C_8H_9ClO	45.8	86.5	117.8	141.8	185.5	208.0	......	(11)
4-Chlorophenethyl alcohol	C_8H_9ClO	84.0	129.0	162.0	188.1	234.5	259.3	......	(113)
Acetanilide	C_8H_9NO	114.0	162.0	199.6	227.2	277.0	303.8	113.5	(11)
Methyl anthranilate	$C_8H_9NO_2$	77.6	124.2	159.7	187.8	238.5	266.5	24	(383)
4-Nitro-1,3-xylene	$C_8H_9NO_2$	65.6	109.8	143.3	168.5	217.5	244.0	+ 2	(302)
Ethylbenzene	C_8H_{10}	− 9.8	+ 25.9	52.8	74.1	113.8	136.2	− 94.9	(6, 11, 265, 359, 415, 471, 474)
2-Xylene	C_8H_{10}	− 3.8	+ 32.1	59.5	81.3	121.7	144.4	− 25.2	(6, 207, 253, 265, 302, 359, 415, 474)
3-Xylene	C_8H_{10}	− 6.9	+ 28.3	55.3	76.8	116.7	139.1	− 47.9	(6, 11, 207, 253, 265, 302, 359, 415, 474)
4-Xylene	C_8H_{10}	− 8.1	+ 27.3	54.4	75.9	115.9	138.3	+ 13.3	(6, 207, 253, 265, 302, 359, 415, 474)
Dichloroethoxyphenylsilane	$C_8H_{10}Cl_2OSi$	52.4	94.6	126.2	151.4	197.2	222.2	......	(114)
Dichloroethylphenylsilane	$C_8H_{10}Cl_2Si$	48.5	92.4	126.7	153.3	203.5	230.0	......	(114)
2-Ethylphenol	$C_8H_{10}O$	46.2	87.0	117.9	141.8	184.5	207.5	− 45	(406)
3-Ethylphenol	$C_8H_{10}O$	60.0	100.2	130.0	152.0	193.3	214.0	− 4	(406)
4-Ethylphenol	$C_8H_{10}O$	59.3	100.2	131.3	154.2	197.4	219.0	46.5	(406)
2,3-Xylenol	$C_8H_{10}O$	56.0_s	97.6	129.2	152.2	196.0	218.0	75	(406)
2,4-Xylenol	$C_8H_{10}O$	51.8	91.3	121.5	143.0	184.2	211.5	25.5	(406)
2,5-Xylenol	$C_8H_{10}O$	51.8_s	91.3	121.5	143.0	184.2	211.5	74.5	(406)
3,4-Xylenol	$C_8H_{10}O$	66.2	107.7	138.0	161.0	203.6	225.2	62.5	(406)
3,5-Xylenol	$C_8H_{10}O$	62.0_s	102.4	133.3	156.0	197.8	219.5	68	(406)
Phenetole	$C_8H_{10}O$	18.1	56.4	86.6	108.4	149.8	172.0	− 30.2	(11, 199)
α-Methyl benzyl alcohol	$C_8H_{10}O$	49.0	88.0	117.8	140.3	180.7	204.0	......	(113)
Phenethylalcohol	$C_8H_{10}O$	58.2	100.0	130.5	154.0	197.5	219.5	......	(113, 395, 396, 456)
4,6-Dimethylresorcinol	$C_8H_{10}O_2$	49.0	90.7	122.5	147.3	192.0	215.0	......	(11)

ORGANIC COMPOUNDS (Continued)

Pressures Less than One Atmosphere (Continued)

Name	Formula	Temperature °C 1 mm	10 mm	40 mm	100 mm	400 mm	760 mm	M.P.	Citation No.
2-Phenoxyethanol	$C_8H_{10}O_2$	78.0	121.2	152.2	176.5	221.0	245.3	11.6	(113)
Diethyl dioxosuccinate	$C_8H_{10}O_6$	70.0	112.0	143.8	167.7	210.8	233.5	....	(11)
Chlorodimethylphenylsilane	$C_8H_{11}ClSi$	29.8	70.0	101.2	124.7	168.6	193.5	....	(114)
N-Ethylaniline	$C_8H_{11}N$	38.5	80.6	113.2	137.3	180.8	204.0	−63.5	(11, 203, 204, 300)
N,N-Dimethylaniline	$C_8H_{11}N$	29.5	70.0	101.6	125.8	169.2	193.1	+2.5	(11, 203, 204, 300)
4-Ethylaniline	$C_8H_{11}N$	52.0	93.8	125.7	149.8	194.2	217.4	−4	(113)
2,4-Xylidine	$C_8H_{11}N$	52.6	93.0	123.8	146.8	188.3	211.5	....	(11, 199)
2,6-Xylidine	$C_8H_{11}N$	44.0	87.0	120.2	146.0	193.7	217.9	....	(302)
2-Phenetidine	$C_8H_{11}NO$	67.0	108.6	139.9	163.5	207.0	228.0	....	(11)
2-Anilinoethanol	$C_8H_{11}NO$	104.0	149.6	183.7	209.5	254.5	279.6	....	(113)
Dimethyl arsanilate	$C_8H_{12}AsNO_2$	15.0	51.8	79.7	101.0	140.3	160.5	....	(11)
Diethyleneglycol-bis-chloroacetate	$C_8H_{12}Cl_2O_5$	148.3	195.8	229.0	252.0	291.8	313.0	....	(113)
Diethyl maleate	$C_8H_{12}O_4$	57.3	100.0	131.8	156.0	201.7	225.0	....	(11)
Diethyl fumarate	$C_8H_{12}O_4$	53.2	95.3	126.7	151.1	195.8	218.5	+0.6	(11)
Dimethylphenylsilane	$C_8H_{12}Si$	+5.3	42.6	71.4	94.2	136.4	159.3	....	(114)
Ethyl-α-ethylacetoacetate	$C_8H_{14}O_3$	40.5	80.2	110.3	133.8	175.6	198.0	....	(11)
Propyl levulinate	$C_8H_{14}O_3$	59.7	99.9	130.1	154.0	198.0	221.2	....	(376)
Isopropyl levulinate	$C_8H_{14}O_3$	48.0	88.0	118.1	141.8	185.2	208.2	....	(376)
Dipropyl oxalate	$C_8H_{14}O_4$	53.4	93.9	124.6	148.1	190.3	213.5	....	(11)
Diisopropyl oxalate	$C_8H_{14}O_4$	43.2	81.9	110.5	132.6	171.8	193.5	....	(11)
Diethyl succinate	$C_8H_{14}O_4$	54.6	96.6	127.8	151.1	193.8	216.5	−20.8	(11, 180)
Diethyl isosuccinate	$C_8H_{14}O_4$	39.8	80.0	111.0	134.8	177.7	201.3	....	(11)
Suberic acid	$C_8H_{14}O_4$	172.8	219.5	254.6	279.0	322.8	345.5	142	(234)
Diethyl malate	$C_8H_{14}O_5$	80.7	125.6	157.8	183.9	229.5	253.4	....	(11)
Diethyl-*dl*-tartrate	$C_8H_{14}O_6$	100.0	147.2	181.7	208.0	254.3	280.0	....	(11)
Diethyl-*d*-tartrate	$C_8H_{14}O_6$	102.0	148.0	182.3	208.5	254.8	280.0	....	(11)
(2-Bromoethyl) cyclohexane	$C_8H_{15}Br$	38.7	80.5	113.0	138.0	186.2	213.0	....	(113)
Caprylonitrile	$C_8H_{15}N$	43.0	80.4	110.6	134.8	179.5	204.5	....	(159)
Ethyl *N,N*-diethyloxamate	$C_8H_{15}NO_3$	76.0	121.7	154.4	180.3	226.5	252.0	....	(11)
2-Methyl-2-heptene	C_8H_{16}	−16.1	+17.8	44.0	64.6	102.2	122.5	....	(11)
1,1-Dimethylcyclohexane	C_8H_{16}	−24.4	+10.3	37.3	57.9	97.2	119.5	−34	(6, 359, 471)
cis-1,2-Dimethylcyclohexane	C_8H_{16}	−15.9	+18.4	45.3	66.8	107.0	129.7	−50.0	(6, 359, 471)
trans-1,2-Dimethylcyclohexane	C_8H_{16}	−21.1	+13.0	39.7	61.0	100.9	123.4	−88.0	(6, 359, 471)
cis-1,3-Dimethylcyclohexane	C_8H_{16}	−22.7	+11.2	37.5	58.5	97.8	120.1	−76.2	(6, 359, 471)
trans-1,3-Dimethylcyclohexane	C_8H_{16}	−19.4	+14.9	41.4	62.5	102.1	124.4	−92.0	(6, 359, 471)
cis-1,4-Dimethylcyclohexane	C_8H_{16}	−20.0	+14.5	41.1	62.3	101.9	124.3	−87.4	(6, 359, 471)
trans-1,4-Dimethylcyclohexane	C_8H_{16}	−24.3	+10.1	36.5	57.6	97.0	119.3	−36.9	(6, 359, 471)
Ethylcyclohexane	C_8H_{16}	−14.5	+20.6	47.6	69.0	109.1	131.8	−111.3	(6, 359, 471)
Caprylaldehyde	$C_8H_{16}O$	73.4	101.2	120.0	133.9	156.5	168.5	....	(11, 225)
Cyclohexaneethanol	$C_8H_{16}O$	50.4	90.0	119.8	142.7	183.5	205.4	....	(113)
6-Methyl-3-hepten-2-ol	$C_8H_{16}O$	41.6	76.7	102.7	122.6	156.6	175.5	....	(431, 434, 460)
6-Methyl-5-hepten-2-ol	$C_8H_{16}O$	41.9	77.8	104.0	123.8	156.6	174.3	....	(113)
2-Octanone	$C_8H_{16}O$	23.6	60.9	89.8	111.7	151.0	172.9	−16	(11)
2,2,4-Trimethyl-3-pentanone	$C_8H_{16}O$	14.7	46.4	69.8	87.6	118.4	135.0	....	(113)
Caprylic acid	$C_8H_{16}O_2$	92.3	124.0	150.6	172.2	213.9	237.5	16	(11, 203, 204)
Ethyl isocaproate	$C_8H_{16}O_2$	11.0	48.0	76.3	98.4	139.2	160.4	....	(11)
Propyl isovalerate	$C_8H_{16}O_2$	+8.0	45.1	72.8	95.0	135.0	155.9	....	(380)
Isobutyl butyrate	$C_8H_{16}O$	+4.6	42.2	71.7	94.0	135.7	156.9	....	(11, 380)
Isobutyl isobutyrate	$C_8H_{16}O_2$	+4.1	39.9	67.2	88.0	126.3	147.5	−80.7	(11, 203, 380)
Amyl propionate	$C_8H_{16}O_2$	+8.5	46.3	75.5	97.6	138.4	160.2	....	(380)
Tetraethyleneglycol chlorohydrin	$C_8H_{16}ClO_4$	110.1	156.1	190.0	214.7	258.2	281.5	....	(141)
1-Iodoöctane	$C_8H_{17}I$	45.8	90.0	123.8	150.0	199.3	225.5	−45.9	(11, 228)
Ethyl-*l*-leucinate	$C_8H_{17}NO_2$	27.8	72.1	106.0	131.8	167.3	184.0	....	(419)
Octane	C_8H_{18}	−14.0	+19.2	45.1	65.7	104.0	125.6	−56.8	(6, 251, 253, 359, 471, 474, 483a)
2-Methylheptane	C_8H_{18}	−21.0	+12.3	37.9	58.3	96.2	117.6	−109.5	(6, 359, 471)
3-Methylheptane	C_8H_{18}	−19.8	+13.3	38.9	59.4	97.4	118.9	−120.8	(6, 359, 471)
4-Methylheptane	C_8H_{18}	−20.4	+12.4	38.0	58.3	96.3	117.7	−121.1	(6, 359, 471)
2,2-Dimethylhexane	C_8H_{18}	−29.7	+3.1	28.2	48.2	85.6	106.8	....	(6, 359, 471)
2,3-Dimethylhexane	C_8H_{18}	−23.0	+9.9	35.6	56.0	94.1	115.6	....	(6, 359, 471)
2,4-Dimethylhexane	C_8H_{18}	−26.9	+5.2	30.5	50.6	88.2	109.4	....	(6, 359, 471)
2,5-Dimethylhexane	C_8H_{18}	−26.7	+5.3	30.4	50.5	87.9	109.1	−90.7	(6, 359, 471, 488)
3,3-Dimethylhexane	C_8H_{18}	−25.8	+6.1	31.7	52.5	90.4	112.0	....	(6, 359, 471)
3,4-Dimethylhexane	C_8H_{18}	−22.1	+11.3	37.1	57.7	96.0	117.7	....	(6, 359, 471)
3-Ethylhexane	C_8H_{18}	−20.0	+12.8	38.5	58.9	97.0	118.5	....	(6, 359, 471)
2,2,3-Trimethylpentane	C_8H_{18}	−29.0	+3.9	29.5	49.9	88.2	109.8	−112.3	(6, 359, 471)
2,2,4-Trimethylpentane	C_8H_{18}	−36.5	−4.3	+20.7	40.7	78.0	99.2	−107.3	(6, 132, 253, 359, 471)
2,3,3-Trimethylpentane	C_8H_{18}	−25.8	+6.9	33.0	53.8	92.7	114.8	−101.5	(6, 359, 471)
2,3,4-Trimethylpentane	C_8H_{18}	−26.3	+7.1	32.9	53.4	91.8	113.5	−109.2	(6, 359, 471)
2-Methyl-3-ethylpentane	C_8H_{18}	−24.0	+9.5	35.2	55.7	94.0	115.6	−114.5	(6, 359, 471)
3-Methyl-3-ethylpentane	C_8H_{18}	−23.9	+9.9	36.2	57.1	96.2	118.3	−90	(6, 359, 471)
2,2,3,3-Tetramethylbutane	C_8H_{18}	−17.4	+13.5	36.8	54.8	87.4	106.3	+100.7	(6, 132, 253, 359, 471, 495)
Tetramethylpiperazine	$C_8H_{18}N_2$	23.7	61.7	90.0	113.8	157.8	183.5	....	(113)
1-Octanol	$C_8H_{18}O$	54.0	88.3	115.2	135.2	173.8	195.2	−15.4	(68, 113, 278, 279, 494)
2-Octanol	$C_8H_{18}O$	32.8	70.0	98.0	119.8	157.5	178.5	−38.6	(11, 199)
1,2-Dipropoxy ethane	$C_8H_{18}O_2$	−38.8	+5.0	42.3	74.2	140.0	180.0	....	(254)
Diethylene glycol butyl ether	$C_8H_{18}O_3$	70.0	107.8	135.5	159.8	205.0	231.2	....	(113)
Tetraethylene glycol	$C_8H_{18}O_5$	153.9	197.1	228.0	250.0	228.0	307.8	....	(71, 141)
Dibutyl sulfide	$C_8H_{18}S$	+21.7	66.4	96.0	118.6	159.0	182.0	−79.7	(25)
Dibutyl disulfide	$C_8H_{18}S_2$	+34.6	94.0	145.1	188.0	275.5	330.5	....	(25)
Diisobutylamine	$C_8H_{19}N$	−5.1	+30.6	57.8	79.2	118.0	139.5	−70	(11)
Tetraethoxysilane	$C_8H_{20}O_4Si$	16.0	52.6	81.1	103.6	146.2	168.5	....	(114, 398)
Tetraethyllead	$C_8H_{20}Pb$	38.4	74.8	102.4	123.8	161.8	183.0	−136	(61)
Amyltrimethylsilane	$C_8H_{20}Si$	−9.2	+26.7	54.4	76.2	116.6	139.0	....	(466)
Tetraethylsilane	$C_8H_{20}Si$	−1.0	+36.3	65.3	88.0	130.2	153.0	....	(466)

VAPOR PRESSURE (Continued)
ORGANIC COMPOUNDS (Continued)
Pressures Less than One Atmosphere Continued

Name	Formula	Temperature °C						M.P.	Citation No.
		1 mm	10 mm	40 mm	100 mm	400 mm	760 mm		
Tetraethyl bistibine	$C_8H_{20}Sb_2$	97.0	151.2	193.2	225.6	286.2	320.3	......	(309)
1,3-Diethoxytetramethyldisiloxane	$C_8H_{22}O_3Si_2$	14.8	51.2	78.7	100.3	139.8	160.7	......	(114)
1,7-Dichlorooctamethyltetrasiloxane	$C_8H_{24}Cl_2O_3Si_4$	53.3	95.8	127.8	152.7	197.8	222.0	− 62	(312)
Octamethyltrisiloxane	$C_8H_{24}O_2Si_3$	7.4	43.1	70.0	91.1	129.4	150.2	......	(114, 469)
Octamethylcyclotetrasiloxane	$C_8H_{24}O_4Si_4$	21.7	59.0	87.4	110.0	149.6	171.2	17.4	(188, 312, 469)
Coumarin	$C_9H_6O_2$	106.0	153.4	189.0	216.5	264.7	291.0	70	(11)
Quinoline	C_9H_7N	59.7	103.8	136.7	163.2	212.3	237.7	− 15	(11, 199, 477)
Isoquinoline	C_9H_7N	63.5	107.8	141.6	167.6	214.5	240.5	24.6	(11)
Indene	C_9H_8	16.4	58.5	90.7	114.7	157.8	181.6	− 2	(63)
Cinnamylaldehyde	C_9H_8O	76.1	120.0	152.2	177.7	222.4	246.0	− 7.5	(11)
trans-Cinnamic acid	$C_9H_8O_2$	127.5$_s$	173.0	207.1	232.4	276.7	300.0	133	(11)
Skatole	C_9H_9N	95.0	139.6	171.9	197.4	242.5	266.2	95	(78)
Ethyl 3-nitrobenzoate	$C_9H_9NO_4$	108.1	155.0	192.6	220.3	270.6	298.0	47	(11)
α-Methyl styrene	C_9H_{10}	7.4	47.1	77.8	102.2	143.0	165.4$_p$	− 23.2	(113)
β-Methyl styrene	C_9H_{10}	17.5	57.0	87.7	111.7	154.7	179.0	− 30.1	(113, 436)
4-Methyl styrene	C_9H_{10}	16.0	55.1	85.0	108.6	151.2	175.0$_p$	......	(216)
Propenylbenzene	C_9H_{10}	17.5	57.0	87.7	111.7	154.7	179.0	− 30.1	(113)
2,4-Xylaldehyde	$C_9H_{10}O$	59.0$_s$	99.0	129.7	152.2	194.1	215.5	75	(51, 138, 139)
Cinnamyl alcohol	$C_9H_{10}O$	72.6	117.8	151.0	177.8	224.6	250.0	33	(11)
Propiophenone	$C_9H_{10}O$	50.0	92.2	124.3	149.3	194.2	218.0	21	(11)
2-Vinylanisole	$C_9H_{10}O$	41.9	81.0	110.0	132.3	172.1	194.0$_p$	......	(215)
3-Vinylanisole	$C_9H_{10}O$	43.4	83.0	112.5	135.3	175.8	197.5$_p$	......	(215)
4-Vinylanisole	$C_9H_{10}O$	45.2	85.7	116.0	139.7	182.0	204.5$_p$	......	(215)
Benzyl acetate	$C_9H_{10}O_2$	45.0	87.6	119.6	144.0	189.0	213.5	− 51.5	(179)
Ethyl Benzoate	$C_9H_{10}O_2$	44.0	86.0	118.2	143.2	188.4	213.4	− 34.6	(11)
Hydrocinnamic acid	$C_9H_{10}O_2$	102.2	148.7	183.3	209.0	255.0	279.8	48.5	(11)
Ethyl salicylate	$C_9H_{10}O_3$	61.2	104.2	136.7	161.5	207.0	231.5	1.3	(11, 199)
N-Methylacetanilide	$C_9H_{11}NO$	s	118.6	152.2	179.8	227.4	253.0	102	(11)
Ethyl carbanilate	$C_9H_{11}NO_2$	107.8	143.7	168.8	187.9	220.0	237.0	52.5	(11)
1,2,3-Trimethylbenzene	C_9H_{12}	16.8	55.9	85.4	108.8	152.0	176.1	− 25.5	(6, 359, 471)
1,2,4-Trimethylbenzene	C_9H_{12}	13.6	50.7	79.8	102.8	145.4	169.2	− 44.1	(6, 359, 471, 474)
1,3,5-Trimethylbenzene	C_9H_{12}	9.6	47.4	76.1	98.9	141.0	164.7	− 44.8	(6, 11, 207, 253, 415, 471, 474)
2-Ethyltoluene	C_9H_{12}	9.4	47.6	76.4	99.0	141.4	165.1	......	(6,113,359,471)
3-Ethyltoluene	C_9H_{12}	7.2	44.7	73.3	95.9	137.8	161.3	− 95.5	(6,113,359,471, 495)
4-Ethyltoluene	C_9H_{12}	7.6	44.9	73.6	96.3	136.4	162.0	......	(6, 216, 359, 471)
Cumene	C_9H_{12}	2.9	38.3	66.1	88.1	129.2	152.4	− 96.0	(6, 253, 359, 471, 474)
Propylbenzene	C_9H_{12}	6.3	43.4	71.6	94.0	135.7	159.2	− 99.5	(6, 359, 471)
2-Ethylanisole	$C_9H_{12}O$	29.7	69.0	98.8	122.3	164.2	187.1	......	(215)
3-Ethylanisole	$C_9H_{12}O$	33.7	73.9	104.8	129.2	172.8	196.5	......	(215)
4-Ethylanisole	$C_9H_{12}O$	33.5	73.9	104.7	128.4	172.3	196.5	......	(215)
3-Phenyl-1-propanol	$C_9H_{12}O$	74.7	116.0	147.4	170.3	212.8	235.0	......	(429)
2-Isopropylphenol	$C_9H_{12}O$	56.6	97.0	127.5	150.3	192.6	214.5	15.5	(113)
3-Isopropylphenol	$C_9H_{12}O$	62.0	104.1	136.2	160.2	205.0	228.0	26	(113)
4-Isopropylphenol	$C_9H_{12}O$	67.0	108.0	139.8	163.3	206.1	228.2	61	(113)
Benzyl ethyl ether	$C_9H_{12}O$	26.0	65.0	95.4	118.9	161.5	185.0	......	(11)
Chloroethoxymethylphenylsilane	$C_9H_{13}ClOSi$	44.8	94.6	117.8	142.6	187.7	212.0	......	(114)
2,4,5-Trimethylaniline	$C_9H_{13}N$	68.4	109.0	139.8	162.0	203.7	234.5	67	(11)
N,N-Dimethyl-2-toluidine	$C_9H_{13}N$	28.8	66.2	95.0	118.1	161.5	184.8	− 61	(11, 204)
N,N-Dimethyl-4-toluidine	$C_9H_{13}N$	50.1	86.7	116.3	140.3	185.4	209.5	......	(11, 204)
4-Cumidine	$C_9H_{13}N$	60.0	102.2	134.2	158.0	203.2	227.0	......	(113)
Phorone	$C_9H_{14}O$	42.0	81.5	111.3	134.0	175.3	197.2	28	(11)
Isophorone	$C_9H_{14}O$	38.0	81.2	114.5	140.6	188.7	215.2	......	(393)
cis-Diethyl citraconate	$C_9H_{14}O_4$	59.8	103.0	135.7	160.0	206.5	230.3	......	(11)
Diethyl itaconate	$C_9H_{14}O_4$	51.3	95.2	128.2	154.3	203.1	227.9	......	(11)
Diethyl mesaconate	$C_9H_{14}O_4$	62.8	105.3	137.3	161.6	205.8	229.0	......	(11)
Trimethyl citrate	$C_9H_{14}O_7$	106.2	160.4	194.2	219.6	264.2	287.0$_d$	78.5	(11)
Isobutyl levulinate	$C_9H_{16}O_3$	65.0	105.9	136.2	160.2	205.5	229.9	......	(376)
Azelaic acid	$C_9H_{16}O_4$	178.3	225.5	260.0	286.5	332.8	356.5$_d$	106.5	(234)
Diethyl ethylmalonate	$C_9H_{16}O_4$	50.8	91.6	122.4	146.0	188.7	211.5	......	(11)
Diethyl glutarate	$C_9H_{16}O_4$	65.6	109.7	142.8	167.8	212.8	237.0	......	(180)
2-Nonanone	$C_9H_{18}O$	32.1	72.3	103.4	127.4	171.2	195.0	− 19	(183)
Azelaldehyde	$C_9H_{18}O$	33.3	71.6	100.2	123.0	163.4	185.0	......	(458)
Pelargonic acid	$C_9H_{18}O_2$	108.2	137.4	163.7	184.4	227.5	253.5	12.5	(11, 203, 223)
Methyl caprylate	$C_9H_{18}O_2$	34.2	74.9	105.3	128.0	170.0	193.0$_d$	− 40	(2)
Isobutyl isovalerate	$C_9H_{18}O_2$	16.0	53.8	82.7	105.2	146.4	168.7	......	(380)
Isoamyl butyrate	$C_9H_{18}O_2$	21.2	59.9	90.0	113.1	155.3	178.6	......	(380)
Isoamyl isobutyrate	$C_9H_{18}O_2$	14.8	52.8	81.8	104.4	146.0	168.8	......	(380)
Iodononane	$C_9H_{19}I$	70.0	109.0	138.1	159.8	199.3	219.5	......	(225, 228)
Nonane	C_9H_{20}	+ 1.4	38.0	66.0	88.1	128.2	150.8	− 53.7	(11, 223, 484, 495)
1-Nonanol	$C_9H_{20}O$	59.5	99.7	129.0	151.3	192.1	213.5	− 5	(11, 228)
Dipropyleneglycol, isopropyl ether	$C_9H_{20}O_3$	46.0	86.2	117.0	140.3	183.1	205.6	......	(113)
Tripropyleneglycol	$C_9H_{20}O_4$	96.0	140.5	173.7	199.0	244.3	267.2	......	(113)
Hexyltrimethylsilane	$C_9H_{22}Si$	+ 6.7	44.8	74.0	97.2	139.9	163.0	......	(466)
Triethylpropylsilane	$C_9H_{22}Si$	15.2	54.0	83.7	107.4	149.8	173.0	......	(466)
1-Bromonaphthalene	$C_{10}H_7Br$	84.2	133.6	170.2	198.8	252.0	281.1	5.5	(11, 203, 204, 334)
1-Chloronaphthalene	$C_{10}H_7Cl$	80.6	118.6	153.2	180.4	230.8	259.3	− 20	(203, 204)
Dicyclopentadiene	$C_{10}H_{12}$	s	47.6	77.9	101.7	144.2	166.6$_d$	32.9	(63)
Naphthalene	$C_{10}H_8$	52.6$_s$	85.8	119.3	145.5	193.2	217.9	80.2	(1, 8, 11, 24, 34, 98, 140, 194, 288, 299, 303, 372, 400, 404, 417)
Dichloro-1-naphthylsilane	$C_{10}H_8Cl_2Si$	106.2	149.2	181.7	205.9	249.7	273.3	......	(114)
1-Naphthol	$C_{10}H_8O$	94.0$_s$	142.0	177.8	206.0	255.8	282.5	96	(11, 270)
2-Naphthol	$C_{10}H_8O$	s	145.5	181.7	209.8	260.6	288.0	122.5	(11, 270, 400)
1-Naphthylamine	$C_{10}H_9N$	104.3	153.8	191.5	220.0	272.2	300.8	50	(11)

ORGANIC COMPOUNDS (Continued)

Pressures Less than One Atmosphere (Continued)

Name	Formula	Temperature °C						M.P.	Citation No.
		1 mm	10 mm	40 mm	100 mm	400 mm	760 mm		
2-Naphthylamine	$C_{10}H_9N$	108.0	157.6	195.7	224.3	277.4	306.1	111.5	(11)
2-Methylquinoline	$C_{10}H_9N$	75.3	119.0	150.8	176.2	211.7	246.5	− 1	(11)
1,3-Divinylbenzene	$C_{10}H_{10}$	32.7	73.8	105.5	130.0	175.2	199.5_p	− 66.9	(113)
4-Phenyl-3-buten-2-one	$C_{10}H_{10}O$	81.7	127.4	161.3	187.8	235.4	261.0	41.5	(11)
α-Methylcinnamic acid	$C_{10}H_{10}O_2$	125.7	169.8	201.8	224.8	266.8	288.0	......	(11)
Methyl cinnamate	$C_{10}H_{10}O_2$	77.4	123.0	157.9	185.8	235.0	263.0	33.4	(370)
Safrole	$C_{10}H_{10}O_2$	63.8	107.6	140.1	165.1	210.0	233.0	11.2	(383)
1,2-Phenylene diacetate	$C_{10}H_{10}O_4$	98.0	145.7	179.8	206.5	253.3	278.0	......	(11)
Dimethyl phthalate	$C_{10}H_{10}O_4$	100.3	147.6	182.8	210.0	257.8	283.7	......	(142)
2,4-Dimethylstyrene	$C_{10}H_{12}$	34.2	75.8	107.7	132.3	177.5	202.0_p	......	(216)
2,5-Dimethylstyrene	$C_{10}H_{12}$	29.0	69.0	100.2	124.7	168.7	193.0_p	......	(216)
3-Ethylstyrene	$C_{10}H_{12}$	28.3	68.3	99.2	123.2	167.2	191.5_p	......	(113)
4-Ethylstyrene	$C_{10}H_{12}$	26.0	66.3	97.3	121.5	165.0	189.0_p	......	(216)
Tetralin	$C_{10}H_{12}$	38.0	79.0	110.4	135.3	181.8	207.2	− 31.0	(177, 253)
Anethole	$C_{10}H_{12}O$	62.6	106.0	139.3	164.2	210.5	235.3	22.5	(11)
4-Methylpropiophenone	$C_{10}H_{12}O$	59.6	103.8	138.0	164.2	212.7	238.5	......	(214)
Estragole	$C_{10}H_{12}O$	52.6	93.7	124.6	148.5	192.0	215.0	......	(436)
Cuminal	$C_{10}H_{12}O$	58.0	102.0	135.2	160.0	206.7	232.0	......	(11, 199, 203, 204)
4-Vinylphenetole	$C_{10}H_{12}O$	64.0	105.6	136.3	159.8	202.8	225.0_p	......	(215)
Eugenol	$C_{10}H_{12}O_2$	78.4	123.0	155.8	182.2	228.3	253.5	......	(11, 40)
Isoeugenol	$C_{10}H_{12}O_2$	86.3	132.4	167.0	194.0	242.3	267.5	− 10	(383)
Chavibetol	$C_{10}H_{12}O_2$	83.6	127.0	159.8	185.5	229.8	254.0	......	(113)
Propyl benzoate	$C_{10}H_{12}O_2$	54.6	98.0	131.8	157.4	205.2	231.0	− 51.6	(11)
2-Phenoxyethyl acetate	$C_{10}H_{12}O_3$	82.6	128.0	162.3	189.2	235.0	259.7	− 6.7	(113)
2-Chloroethyl α-methylbenzyl ether	$C_{10}H_{13}ClO$	62.3	106.0	139.6	164.8	210.8	235.0	......	(113)
4-*tert*-Butylphenyl dichlorophosphate	$C_{10}H_{13}Cl_2O_2P$	96.0	146.0	184.3	214.3	268.2	299.0	......	(113)
1,2,3,4-Tetramethylbenzene	$C_{10}H_{14}$	42.6	81.8	111.5	135.7	180.0	204.4	− 6.2	(259, 495)
1,2,3,5-Tetramethylbenzene	$C_{10}H_{14}$	40.6	77.8	105.8	128.3	173.7	197.9	− 24.0	(259)
1,2,4,5-Tetramethylbenzene	$C_{10}H_{14}$	45.0_s	74.6_s	104.2	128.1	172.1	195.9	79.5	(253, 259)
4-Ethyl-1,3-xylene	$C_{10}H_{14}$	26.3	66.4	97.2	121.2	164.4	188.4	......	(216, 495)
5-Ethyl-1,3-xylene	$C_{10}H_{14}$	22.1	62.1	92.6	116.5	159.6	183.7	......	(113, 495)
2-Ethyl-1,4-xylene	$C_{10}H_{14}$	25.7	65.6	96.0	120.0	163.1	186.9	......	(216, 495)
1,2-Diethylbenzene	$C_{10}H_{14}$	22.3	62.0	92.5	116.2	159.0	183.5	− 31.4	(140, 495, 501)
1,3-Diethylbenzene	$C_{10}H_{14}$	20.7	59.9	90.4	114.4	156.9	181.1	− 83.9	(113, 253, 495)
1,4-Diethylbenzene	$C_{10}H_{14}$	20.7	60.3	91.1	115.3	159.0	183.8	− 43.2	(113, 216, 253, 495)
Cymene	$C_{10}H_{14}$	17.3	57.0	87.0	110.8	153.5	177.2	− 68.2	(11, 253, **474**, 498)
Butylbenzene	$C_{10}H_{14}$	22.7	62.0	92.4	116.2	159.2	183.1	− 88.0	(253, 495)
Isobutylbenzene	$C_{10}H_{14}$	14.1	53.7	83.3	107.0	149.6	172.8	− 51.5	(474, 495)
sec-Butylbenzene	$C_{10}H_{14}$	18.6	57.0	86.2	109.5	150.3	173.5	− 75.5	(253, 495)
tert-Butylbenzene	$C_{10}H_{14}$	13.0	51.7	80.8	103.8	145.8	168.5	− 58	(253)
Carvacrol	$C_{10}H_{14}O$	70.0	113.2	145.2	169.7	213.8	237.0	+ 0.5	(11, 59, 113, 385)
Carbone	$C_{10}H_{14}O$	57.4	100.4	133.0	157.3	203.5	227.5	......	(11, 59, 362)
Cuminyl alcohol	$C_{10}H_{14}O$	74.2	118.0	150.3	176.2	221.7	246.6	......	(11)
4-Ethylphenetole	$C_{10}H_{14}O$	48.5	89.5	119.8	143.5	185.7	208.0	......	(215)
Thymol	$C_{10}H_{14}O$	64.3	107.4	139.8	164.1	209.2	231.8	51.5	(11, 113, 322)
4-Isobutylphenol	$C_{10}H_{14}O$	72.1	115.5	147.2	171.2	214.7	237.0	......	(11)
4-*sec*-Butylphenol	$C_{10}H_{14}O$	71.4	114.8	147.8	172.4	217.6	242.1	......	(113)
2-*sec*-Butylphenol	$C_{10}H_{14}O$	57.4	100.8	133.4	157.3	203.8	228.0	......	(113)
2-*tert*-Butylphenol	$C_{10}H_{14}O$	56.6	98.1	129.2	153.5	196.3	219.5	......	(113)
4-*tert*-Butylphenol	$C_{10}H_{14}O$	70.0	114.0	146.0	170.2	214.0	238.0	99	(113, 406)
Nicotine	$C_{10}H_{14}N_2$	61.8	107.2	142.1	169.5	219.8	247.3	......	(490)
N-Diethylaniline	$C_{10}N_{15}N$	49.7	91.9	123.6	147.3	192.4	215.5	− 34.4	(11, 199, 204, 300)
N-Phenyliminodiethanol	$C_{10}H_{15}NO_2$	145.0	195.8	233.0	260.6	311.3	337.8	......	(113)
Camphene	$C_{10}H_{16}$	s	47.2_s	75.7	97.9	138.7	160.5	50	(383)
Dipentene	$C_{10}H_{16}$	14.0	53.8	84.3	108.3	150.5	174.6	......	(11, 253, 320)
d-Limonene	$C_{10}H_{16}$	14.0	53.8	84.3	108.3	151.4	175.0	− 96.9	(11, 253)
Myrcene	$C_{10}H_{16}$	14.5	53.2	82.6	106.0	148.3	171.5	......	(327, 387)
α-Phellandrene	$C_{10}H_{16}$	20.0	58.0	87.8	110.6	152.0	175.0	......	(461)
α-Pinene	$C_{10}H_{16}$	− 1.0	+ 37.3	66.8	90.1	132.3	155.0	− 55	(253, 320, 351)
β-Pinene	$C_{10}H_{16}$	+ 4.2	42.3	71.5	94.0	136.1	158.3	......	(320)
Terpenoline	$C_{10}H_{16}$	32.3	70.6	100.0	122.7	163.5	185.0	......	(320, 383)
Diethyl arsanilate	$C_{10}H_{16}AsNO_3$	38.0	74.8	102.6	123.8	161.0	181.0	......	(11)
d-Camphor	$C_{10}H_{16}O$	41.5_s	82.3_s	114.0_s	138.0_s	182.0	209.2	178.5	(1, 105, 303, 336, 404, 444)
l-Dihydrocarvone	$C_{10}H_{16}O$	46.6	90.0	123.7	149.7	197.0	223.0	......	(383)
α-Citral	$C_{10}H_{16}O$	61.7	103.9	135.9	160.0	205.0	228.0_d	......	(383)
d-Fenchone	$C_{10}H_{16}O$	28.0	68.3	99.5	123.6	166.8	191.0	5	(42)
Pulegone	$C_{10}H_{16}O$	58.3	94.0	121.7	143.1	189.8	221.0	......	(383)
α-Thyjone	$C_{10}H_{16}O$	38.3	79.3	110.0	134.0	177.8	201..	......	(265)
Ethoxydimethylphenylsilane	$C_{10}H_{16}OSi$	36.3	76.2	107.2	131.4	175.0	199.5	......	(114)
Campholenic acid	$C_{10}H_{16}O_2$	97.6	139.8	170.0	193.7	234.0	256.0	......	(11)
Diosphenol	$C_{10}H_{16}O_2$	66.7	109.0	141.2	165.6	209.5	232.0	......	(383)
Fencholic acid	$C_{10}H_{16}O_2$	101.7	142.3	171.8	194.0	237.8	264.1	19	(113)
cis-Decalin	$C_{10}H_{18}$	22.5	64.2	97.2	123.2	169.9	194.6	− 43.3	(389)
trans-Decalin	$C_{10}H_{18}$	− 0.8	+ 47.2	85.7	114.6	160.1	186.7	− 30.7	(389)
d-Citronellal	$C_{10}H_{18}O$	44.0	84.8	116.1	140.1	183.8	206.5	......	(432)
Cineol	$C_{10}H_{18}O$	15.0	54.1	84.2	108.2	151.6	176.0	− 1	(11)
Dihydrocarveol	$C_{10}H_{18}O$	63.9	105.0	136.1	159.8	202.8	225.0	......	(362, 441)
dl-Fenchyl alcohol	$C_{10}H_{18}O$	45.8	82.1	110.8	132.3	173.2	201.0	35	(320)
Geraniol	$C_{10}H_{18}O$	69.2	110.0	141.8	165.3	207.8	230.0	......	(383)
d-Linalool	$C_{10}H_{18}O$	40.0	79.8	109.9	133.8	175.6	198.0	......	(383)
Nerol	$C_{10}H_{18}O$	61.7	104.0	136.1	159.8	203.5	226.0	......	(179, 397)
α-Terpineol	$C_{10}H_{18}O$	52.8	94.3	126.0	150.1	194.3	217.5	35	(320)
Citronellic acid	$C_{10}H_{18}O_2$	99.5	141.4	171.9	195.4	236.6	257.0	......	(386, 432, 433, 435)
Amyl levulinate	$C_{10}H_{18}O_3$	81.3	124.0	155.8	180.5	227.4	253.2	......	(376)
Isoamyl levulinate	$C_{10}H_{18}O_3$	75.6	118.8	151.7	177.0	222.7	247.9	......	(376)
Diethyl ethylmethylmalonate	$C_{10}H_{18}O_4$	44.7	85.7	116.7	140.8	184.1	207.5	......	(11)

ORGANIC COMPOUNDS (Continued)
Pressures Less than One Atmosphere (Continued)

Name	Formula	Temperature °C						M.P.	Citation No.
		1 mm	10 mm	40 mm	100 mm	400 mm	760 mm		
Diethyl adipate	$C_{10}H_{18}O_4$	74.0	123.0	154.6	179.0	219.1	240.0	− 21	(180)
Diisobutyl oxalate	$C_{10}H_{18}O_4$	63.2	105.3	137.5	161.8	205.8	229.5	....	(11)
Dipropyl succinate	$C_{10}H_{18}O_4$	77.5	122.2	154.8	180.3	226.5	250.8	....	(11)
Sebacic acid	$C_{10}H_{18}O_4$	183.0	232.0	268.2	294.5	332.8	352.3d	134.5	(11, 234)
Dipropyl-*d*-tartrate	$C_{10}H_{18}O_6$	115.6	163.5	199.7	227.0	275.6	303.0	....	(11)
Diisopropyl-*d*-tartrate	$C_{10}H_{18}O_6$	103.7	148.2	181.8	207.3	251.8	275.0	....	(11)
Camphylamine	$C_{10}H_{19}N$	45.3	83.7	112.5	134.6	173.8	195.0	....	(11)
Menthane	$C_{10}H_{20}$	+ 9.7	48.3	78.3	102.1	146.0	169.5	....	(253)
1-Decene	$C_{10}H_{20}$	14.7	53.7	83.3	106.5	149.2	172.0	....	(11, 150)
1,2-Dibromodecane	$C_{10}H_{20}Br_2$	95.7	137.3	167.4	190.2	229.8	250.4	....	(11, 150)
Citronellol	$C_{10}H_{20}O$	66.4	107.0	137.2	159.8	201.0	221.5	....	(432)
Capraldehyde	$C_{10}H_{20}O$	51.9	92.0	122.2	145.3	186.3	208.5	....	(405)
l-Menthol	$C_{10}H_{20}O$	56.0	96.0	126.1	149.4	190.2	212.0	42.5	(11, 31, 325)
Decan-2-one	$C_{10}H_{20}O$	44.2	85.8	117.1	142.0	186.7	211.0	+ 3.5	(11, 183, 224)
Capric acid	$C_{10}H_{20}O_2$	125.0	152.2	179.9	200.0	240.3	268.4	31.5	(11, 203, 223, 224, 225, 231)
Isoamyl isovalerate	$C_{10}H_{20}O_2$	27.0	68.6	100.6	125.1	169.5	194.0	....	(11, 203)
Decane	$C_{10}H_{22}$	16.5	55.7	85.5	108.6	150.6	174.1	− 29.7	(11, 223, 224, 253, 484, 496)
2,7-Dimethyloctane	$C_{10}H_{22}$	+ 6.3	42.3	71.2	93.9	136.0	159.7	− 52.8	(253, 474)
Decyl alcohol	$C_{10}H_{22}O$	69.5	111.3	142.1	165.8	208.8	231.0	+ 7	(11, 225, 228)
Diisoamyl ether	$C_{10}H_{22}O$	18.6	57.0	86.3	109.6	150.3	173.4	....	(11, 247)
2-Butyl-2-ethylbutane-1,3-diol	$C_{10}H_{22}O_2$	94.1	136.8	167.8	191.9	233.5	255.0	....	(11, 150)
Dihydrocitronellol	$C_{10}H_{22}O$	68.0	103.0	127.6	145.9	176.8	193.5	....	(155)
Dipropylene glycol monobutyl ether	$C_{10}H_{22}O_3$	64.7	106.0	136.3	159.8	203.8	227.0	....	(113)
Diisoamyl sulfide	$C_{10}H_{22}S$	43.0	87.6	120.0	145.3	191.0	216.0	....	(25)
Heptyltrimethylsilane	$C_{10}H_{24}Si$	22.3	62.1	92.4	116.5	159.8	184.0	....	(466)
Butyltriethylsilane	$C_{10}H_{24}Si$	27.1	67.5	98.3	123.2	167.5	192.0	....	(466)
1,5-Diethoxyhexamethyltrisiloxane	$C_{10}H_{28}O_4Si_3$	41.8	80.7	110.0	133.2	174.0	196.6	....	(114)
Decamethyltetrasiloxane	$C_{10}H_{30}O_3Si_4$	35.3	74.3	104.0	127.3	169.8	193.5	....	(114, 469)
Decamethylcyclopentasiloxane	$C_{10}H_{30}O_5Si_5$	45.2	86.2	117.7	142.0	186.0	210.0	− 38.0	(188, 312, 469)
1-Naphthoic acid	$C_{11}H_8O_2$	156.0	196.8	225.0	245.8	281.4	300.0	160.5	(11)
2-Naphthoic acid	$C_{11}H_8O_2$	160.8	202.8	231.5	252.7	289.5	308.5	184	(11)
Ethyl-*trans*-cinnamate	$C_{11}H_{12}O_2$	87.6	134.0	169.2	196.0	245.0	271.0	12	(11)
1-Phenyl-1,3-pentanedione	$C_{11}H_{12}O_2$	98.0	144.0	178.0	204.5	251.2	276.5	....	(43)
Ethyl benzoylacetate	$C_{11}H_{12}O_3$	107.6	150.3	181.8	205.0	244.7	265.0d	....	(11)
Myristicine	$C_{11}H_{12}O_3$	95.2	142.0	177.7	205.0	253.5	280.0	....	(384, 428)
2,4,5-Trimethylstyrene	$C_{11}H_{14}$	48.1	91.6	124.2	149.8	196.1	221.2p	....	(216)
2,4,6-Trimethylstyrene	$C_{11}H_{14}$	37.5	79.7	111.8	136.8	182.3	207.0p	....	(216)
4-Isopropylstyrene	$C_{11}H_{14}$	34.7	76.0	108.0	132.8	178.0	202.5p	....	(216)
Isovalerophenone	$C_{11}H_{14}O$	58.3	101.4	133.8	158.0	204.2	228.0	....	(11)
Pivalophenone	$C_{11}H_{14}O$	57.8	99.0	130.4	154.0	197.7	220.0	....	(295)
2,3,5-Trimethylacetophenone	$C_{11}H_{14}O$	79.0	122.3	154.2	179.7	224.3	247.5	....	(383)
Isobutyl benzoate	$C_{11}H_{14}O_2$	64.0	108.6	141.8	166.4	212.8	237.0	....	(11, 199)
4-Allylveratrole	$C_{11}H_{14}O_2$	85.0	127.0	158.3	183.7	226.2	248.0	....	(40)
3,5-Diethyltoluene	$C_{11}H_{16}$	34.0	75.3	107.0	131.7	176.5	200.7	....	(113, 498)
1,2,4-Trimethyl-5-ethylbenzene	$C_{11}H_{16}$	43.7	84.6	106.0	140.3	184.5	208.1	....	(216)
1,3,5-Trimethyl-2-ethylbenzene	$C_{11}H_{16}$	38.8	80.5	113.2	137.9	183.5	208.0	....	(216)
3-Ethylcumene	$C_{11}H_{16}$	28.3	68.8	99.9	124.3	168.2	193.0	....	(113)
4-Ethylcumene	$C_{11}H_{16}$	31.5	72.0	103.3	127.2	171.8	195.8	....	(216)
sec-Amylbenzene	$C_{11}H_{16}$	29.0	69.2	100.0	124.1	168.0	193.0	....	(11, 503)
4-*tert*-Butyl-2-cresol	$C_{11}H_{16}O$	74.3	118.0	150.8	176.2	221.8	247.0	....	(113)
2-*tert*-Butyl-4-cresol	$C_{11}H_{16}O$	70.0	112.0	143.9	167.0	210.0	232.6	....	(406)
4-*tert*-Amylphenol	$C_{11}H_{16}O$	s	125.5	160.3	189.0	239.5	266.0	93	(11)
Ethylcamphoronic anhydride	$C_{11}H_{16}O_5$	118.2	165.0	199.8	226.6	272.8	298.0	....	(11)
Bornyl formate	$C_{11}H_{18}O_2$	47.0	89.3	121.2	145.8	190.2	214.0	....	(383)
Geranyl formate	$C_{11}H_{18}O_2$	61.8	104.3	136.2	160.7	205.8	230.0	....	(432)
Neryl formate	$C_{11}H_{18}O_2$	57.3	99.7	131.5	155.6	200.0	224.5	....	(178, 397)
Diethoxymethylphenylsilane	$C_{11}H_{18}O_2Si$	56.5	97.2	127.5	151.2	193.8	216.5	....	(114)
Diethyl-γ-oxoazelate	$C_{11}H_{18}O_5$	121.0	165.7	197.7	221.6	264.5	286.0	....	(11)
10-Hendecenoic acid	$C_{11}H_{20}O_2$	114.0	156.3	188.7	213.5	254.0	275.0	24.5	(11, 60)
Menthyl formate	$C_{11}H_{20}O_2$	47.3	90.0	123.0	148.0	194.2	219.0	....	(425)
2-Ethylhexyl acrylate	$C_{11}H_{20}O_2$	50.0	91.8	123.7	147.9	192.2	216.0	....	(113)
Octyl acrylate	$C_{11}H_{20}O_2$	58.5	102.0	135.6	159.1	204.0	227.0	....	(113)
Hexyl levulinate	$C_{11}H_{20}O_3$	90.0	134.7	167.8	193.6	241.0	266.8	....	(376)
Hendecan-2-one	$C_{11}H_{22}O$	68.2	108.9	139.0	161.0	202.3	224.0	15	(11, 223, 224, 328)
Methyl caprate	$C_{11}H_{22}O_2$	63.7	108.0	139.0	161.5	202.9	224.0d	− 18	(2)
Hendecanoic acid	$C_{11}H_{22}O_2$	101.4	149.0	185.6	212.5	262.8	290.0	29.5	(11, 221, 223)
Hendecane	$C_{11}H_{24}$	32.7	73.9	104.4	128.1	171.9	195.8	− 25.6	(11, 223, 224, 484, 504)
Hendecan-2-ol	$C_{11}H_{24}O$	71.1	112.8	143.7	167.2	209.8	232.0	....	(183, 266)
Trimethyloctylsilane	$C_{11}H_{26}Si$	41.8	82.3	113.0	136.5	179.5	202.0	....	(466)
Amyltriethylsilane	$C_{11}H_{26}Si$	41.8	83.8	116.0	141.2	186.3	211.0	....	(466)
1-Bromobiphenyl	$C_{12}H_9Br$	98.0	150.6	190.8	221.8	277.7	310.0	90.5	(113)
2-Bromo-4-Phenylphenol	$C_{12}H_9BrO$	100.0	152.3	193.8	224.5	280.2	311.0	95	(113)
2-Chlorobiphenyl	$C_{12}H_9Cl$	89.3	134.7	169.9	197.0	243.8	267.5	34	(113)
4-Chlorobiphenyl	$C_{12}H_9Cl$	96.4	146.0	183.8	212.5	264.5	292.9	75.5	(113)
2-Chloro-3-phenylphenol	$C_{12}H_9ClO$	118.0	169.7	207.4	237.0	289.4	317.5	+ 6	(113)
2-Chloro-6-phenylphenol	$C_{12}H_9ClO$	119.8	170.7	208.2	237.1	289.5	317.0	....	(113)
2-Xenyl dichlorophosphate	$C_{12}H_9Cl_2PO$	138.2	187.0	223.8	251.5	301.5	328.5	....	(113)
Carbazole	$C_{12}H_9N$	s	s	s	265.0	323.0	354.8	244.8	(388)
Acenaphthene	$C_{12}H_{10}$	s	131.2	168.2	197.5	250.0	277.5	95	(11, 253, 288)
Biphenyl	$C_{12}H_{10}$	70.6	117.0	152.5	180.7	229.4	254.9	69.5	(11, 34, 80, 103, 143, 194, 379)
Diphenyl chlorophosphate	$C_{12}H_{10}ClPO_3$	121.5	182.0	227.9	265.0	337.2	378.0	....	(113)
Dichlorodiphenylsilane	$C_{12}H_{10}Cl_2Si$	109.6	158.0	195.5	223.8	275.5	304.0	....	(114)
Difluorodiphenylsilane	$C_{12}H_{10}F_2Si$	68.4	115.5	149.8	176.3	225.4	252.5	....	(119)
Azobenzene	$C_{12}H_{10}N_2$	103.5	151.5	187.9	216.0	266.1	293.0	68	(11)
1-Acetonaphthone	$C_{12}H_{10}O$	115.6	161.5	196.8	223.8	270.5	295.5	....	(360)
2-Acetonaphthone	$C_{12}H_{10}O$	120.2	168.5	203.8	229.8	275.8	301.0	55.5	(360)
Diphenyl ether	$C_{12}H_{10}O$	66.1	114.0	150.0	178.8	230.7	258.5	27	(113)

VAPOR PRESSURE (Continued)

ORGANIC COMPOUNDS (Continued)

Pressures Less than One Atmosphere (Continued)

Name	Formula	Temperature °C						M.P.	Citation No.
		1 mm	10 mm	40 mm	100 mm	400 mm	760 mm		
2-Phenylphenol	$C_{12}H_{10}O$	100.0	146.2	180.3	205.9	251.8	275.0	56.5	(113)
4-Phenylphenol	$C_{12}H_{10}O$	s	176.2	213.0	240.9	285.5	308.0	164.5	(113)
Diphenyl sulfide	$C_{12}H_{10}S$	96.1	145.0	182.8	211.8	263.9	292.5	......	(11, 232, 237)
Diphenyl disulfide	$C_{12}H_{10}S$	131.6	180.0	214.8	241.3	285.8	310.0	61	(11, 232, 237)
Diphenyl selenide	$C_{12}H_{10}Se$	105.7	154.4	192.2	220.8	273.2	301.5	+ 2.5	(11, 232, 237)
Diphenylamine	$C_{12}H_{11}N$	108.3	157.0	194.3	222.8	274.1	302.0	52.9	(11, 159)
1-Ethylnaphthalene	$C_{12}H_{12}$	70.0	116.8	152.0	180.0	230.8	258.1d	− 27	(113)
1,1-Diphenylhydrazine	$C_{12}H_{12}N_2$	126.0	176.1	213.5	242.5	294.0	322.2	44	(11)
2-Cyclohexyl-4,6-dinitrophenol	$C_{12}H_{14}N_2O_5$	132.8	175.9	206.7	229.0	269.8	291.5	......	(113)
Eugenyl acetate	$C_{12}H_{14}O_3$	101.6	148.0	183.0	209.7	257.4	282.0	29.5	(122)
Apiole	$C_{12}H_{14}O_4$	116.0	160.2	193.7	218.0	262.1	285.0	30	(81, 88)
Diethyl phthalate	$C_{12}H_{14}O_4$	108.8	156.0	192.1	219.5	267.5	294.0	......	(11)
2,5-Diethylstyrene	$C_{12}H_{16}$	49.7	92.6	125.8	151.0	198.0	223.0p	......	(216)
Phenylcyclohexane	$C_{12}H_{16}$	67.5	111.3	144.0	169.3	214.6	240.0	+ 7.5	(113, 253)
Isoamyl benzoate	$C_{12}H_{16}O_2$	72.0	121.6	158.3	186.8	235.8	262.0	......	(11, 199)
1,2,4-Triethylbenzene	$C_{12}H_{18}$	46.0	88.5	121.7	146.8	193.7	218.0	......	(113)
1,4-Diisopropylbenzene	$C_{12}H_{18}$	40.0	81.8	114.0	138.7	184.3	209.0	......	(113, 500)
1,3-Diisopropylbenzene	$C_{12}H_{18}$	34.7	76.0	107.9	132.3	177.6	202.0	−105	(113)
2-*tert*-Butyl-4-ethylphenol	$C_{12}H_{18}O$	76.3	121.0	154.0	179.0	223.8	247.8	......	(406)
4-*tert*-Butyl-2,5-xylenol	$C_{12}H_{18}O$	88.2	135.0	169.8	195.0	241.3	265.3	......	(406)
4-*tert*-Butyl-2,6-xylenol	$C_{12}H_{18}O$	74.0	119.0	152.2	176.0	217.8	239.8	......	(406)
6-*tert*-Butyl-2,4-xylenol	$C_{12}H_{18}O$	70.3	115.0	148.5	172.0	214.2	236.5	......	(406)
6-*tert*-Butyl-3,4-xylenol	$C_{12}H_{18}O$	83.9	127.0	159.7	184.0	226.7	249.5	......	(406)
d-Bornyl acetate	$C_{12}H_{20}O_2$	46.9	90.2	123.7	149.8	197.5	223.0	29	(383)
Geranyl acetate	$C_{12}H_{20}O_2$	73.5	117.9	150.0	175.2	219.8	243.3d	......	(41)
Linalyl acetate	$C_{12}H_{20}O_2$	55.4	96.0	127.7	151.8	196.2	220.0d	......	(383)
Triethoxyphenylsilane	$C_{12}H_{20}O_3Si$	71.0	112.6	143.5	167.5	210.5	233.5	......	(114)
Triethyl citrate	$C_{12}H_{20}O_7$	107.0	144.0	190.4	217.8	267.5	294.0d	......	(11)
Trimethallyl phosphate	$C_{12}H_{21}PO_4$	93.7	149.8	192.0	225.7	288.5	324.0	......	(113)
Citronellyl acetate	$C_{12}H_{22}O_2$	74.7	113.0	140.5	161.0	197.8	217.0	......	(432)
Menthyl acetate	$C_{12}H_{22}O_2$	57.4	100.0	132.1	156.7	202.8	227.0	......	(383)
Dimethyl sebacate	$C_{12}H_{22}O_4$	104.0	156.2	196.0	222.6	269.6	293.5	38	(180)
Diisoamyl oxalate	$C_{12}H_{22}O_4$	85.4	131.4	165.7	192.2	240.0	265.0	......	(11)
Diisobutyl-*d*-tartrate	$C_{12}H_{22}O_6$	117.8	169.0	208.5	239.5	294.0	324.0	73.5	(11)
1-Dodecene	$C_{12}H_{24}$	47.2	87.8	118.6	142.3	185.5	208.0	− 31.5	(11, 226)
Triisobutylene	$C_{12}H_{24}$	18.0	56.5	86.7	110.0	153.0	179.0	......	(406)
Dodecan-2-one	$C_{12}H_{24}O$	77.1	120.4	152.4	177.5	222.5	246.5	......	(11, 223, 224)
Lauraldehyde	$C_{12}H_{24}O$	77.7	123.7	157.8	184.5	231.8	257.0	44.5	(11, 222, 225, 230)
Lauric acid	$C_{12}H_{24}O_2$	121.0	166.0	201.4	227.5	273.8	299.2	48	(11, 157, 221, 223, 224, 225, 226)
Dodecane	$C_{12}H_{26}$	47.8	90.0	121.7	146.2	191.0	216.2	− 9.6	(11,27,223,224, 225,497,502)
Dodecyl alcohol	$C_{12}H_{26}O$	91.0	134.7	167.2	192.0	235.7	259.0	24	(11, 225)
Tripropylene glycol monoisopropyl ether	$C_{12}H_{26}O_4$	82.4	127.3	161.4	187.8	232.8	256.6	......	(113)
Triisobutylamine	$C_{12}H_{27}N$	32.3	69.8	97.8	119.7	157.8	179.0	− 22	(11)
Dodecylamine	$C_{12}H_{27}N$	82.8	127.8	157.4	182.1	225.0	248.0	......	(11, 230)
Triethylhexylsilane	$C_{12}H_{28}Si$	52.4	96.4	130.0	156.0	204.6	230.0	......	(466)
1,7-Diethoxyoctamethyltetrasiloxane	$C_{12}H_{34}O_5Si_4$	67.7	108.6	319.0	162.0	204.0	227.5	......	(114)
Dodecamethylpentasiloxane	$C_{12}H_{36}O_4Si_5$	56.6	98.0	128.8	162.8	196.5	220.5	......	(114, 469)
Dodecamethylcyclohexasiloxane	$C_{12}H_{36}O_6Si_6$	67.3	110.0	141.8	166.3	210.6	236.0	− 3.0	(188, 312, 469)
Acridine	$C_{13}H_9N$	129.4	184.0	224.2	256.0	314.3	346.0	110.5	(11)
Fluorene	$C_{13}H_{10}$	s	146.0	185.2	214.7	268.6	295.0	113	(11, 135, 288)
Benzophenone	$C_{13}H_{10}O$	108.2	157.6	195.7	224.4	276.8	305.4	48.5	(11, 70, 194, 452)
Phenyl benzoate	$C_{13}H_{10}O_2$	106.8	157.8	197.6	227.8	283.5	314.0	70.5	(11)
Salol	$C_{13}H_{10}O_3$	117.8	167.0	205.0	233.8	284.8	313.0	42.5	(11)
Diphenylmethane	$C_{13}H_{12}$	76.0	122.8	157.8	186.3	237.5	264.5	26.5	(97, 99)
Benzhydrol	$C_{13}H_{12}O$	110.0	162.0	200.0	227.5	275.6	301.0	68.5	(11)
Benzyl phenyl ether	$C_{13}H_{12}O$	95.4	144.0	180.1	209.2	259.8	287.0	......	(11)
1-Propionaphthone	$C_{13}H_{12}O$	124.0	171.0	206.9	233.5	280.2	306.0	......	(360)
Chloromethyldiphenylsilane	$C_{13}H_{13}ClSi$	105.0	152.7	189.2	216.0	266.5	295.5	......	(114)
Methyldiphenylamine	$C_{13}H_{13}N$	103.5	149.7	184.0	210.1	257.0	282.0	− 7.6	(11)
2-Isopropylnaphthalene	$C_{13}H_{14}$	76.0	123.4	159.0	187.6	238.5	266.0	......	(113)
Methyldiphenylsilane	$C_{13}H_{14}Si$	88.0	132.8	166.4	193.7	241.5	266.8	......	(114)
Enanthophenone	$C_{13}H_{18}O$	100.0	145.5	178.9	204.2	248.3	271.3	......	(11, 229)
Heptylbenzene	$C_{13}H_{20}$	64.0	110.0	144.0	170.2	217.8	244.0	......	(11, 229, 499)
α-Ionone	$C_{13}H_{20}O$	79.5	123.0	155.6	181.2	225.2	250.0	......	(256)
Bornyl propionate	$C_{13}H_{22}O_2$	64.6	108.0	140.4	165.7	211.2	235.0	......	(383)
2-Tridecanone	$C_{13}H_{26}O$	86.8	131.8	165.7	191.5	238.3	262.5	28.5	(11, 223, 224)
Methyl laurate	$C_{13}H_{26}O_2$	87.8	133.2	166.0	190.8	d	d	5	(2)
Tridecanoic acid	$C_{13}H_{26}O_2$	137.8	181.0	212.4	236.0	276.5	299.0	41	(11, 221)
Tridecane	$C_{13}H_{28}$	59.4	104.0	137.7	162.5	209.4	234.0	− 6.2	(11, 223, 484)
Tripropyleneglycol, monobutyl ether	$C_{13}H_{28}O_4$	101.5	147.0	179.8	204.4	247.0	269.5	......	(113)
Decyltrimethylsilane	$C_{13}H_{30}Si$	67.4	111.0	144.0	169.5	215.5	240.0	......	(466)
Triethylheptylsilane	$C_{13}H_{30}Si$	70.0	114.6	148.0	174.0	221.0	247.0	......	(466)
Anthraquinone	$C_{14}H_8O_2$	190.0s	234.2s	264.3s	285.0s	346.2	379.9	286	(299, 404)
1,4-Dihydroxyanthraquinone	$C_{14}H_8O_4$	196.7	259.8	307.4	344.5	413.0d	450.0d	194	(464)
Anthracene	$C_{14}H_{10}$	145.0s	187.2s	217.5s	250.0	310.2	342.0	217.5	(11, 97, 288, 299, 303, 381)
Phenanthrene	$C_{14}H_{10}$	118.2	173.0	215.8	249.0	308.0	340.2	99.5	(11, 288, 299)
Benzil	$C_{14}H_{10}O_2$	128.4	183.0	224.5	255.8	314.3	347.0	95	(11, 238, 473)
Benzoic anhydride	$C_{14}H_{10}O_3$	143.8	198.0	239.8	270.4	328.8	360.0	42	(11)
1,1-Diphenylethylene	$C_{14}H_{12}$	87.4	135.0	170.8	198.6	249.8	277.0	......	(11)
trans-Diphenylethylene	$C_{14}H_{12}$	113.2	161.0	199.0	227.4	287.3	306.5	124	(11)
Desoxybenzoin	$C_{14}H_{12}O$	123.3	173.5	212.0	241.3	293.0	321.0	60	(10, 11)
Benzoin	$C_{14}H_{12}O$	135.6	188.0	227.6	258.0	313.5	343.0	132	(11)
Dibenzyl	$C_{14}H_{14}$	86.8	136.0	173.7	202.8	255.0	284.0	51.5	(11)
2-Isobutyronaphthone	$C_{14}H_{14}O$	133.2	181.0	215.6	242.3	288.2	313.0	......	(360)
Dibenzylamine	$C_{14}H_{15}O$	118.3	165.6	200.2	227.3	274.3	300.0	− 26	(11)

Name	Formula	Temperature °C						M.P.	Citation No.
		1 mm	10 mm	40 mm	100 mm	400 mm	760 mm		
Ethyldiphenylamine	$C_{14}H_{15}N$	98.3	146.0	182.0	209.8	258.8	286.0	......	(11)
1,2-Dichlorotetraethylbenzene	$C_{14}H_{20}Cl_2$	105.6	155.0	192.2	220.7	272.8	302.0	......	(113)
1,4-Dichlorotetraethylbenzene	$C_{14}H_{20}Cl_2$	91.7	143.8	183.2	212.0	265.8	296.5	......	(113)
2-(4-*tert*-Butylphenoxy) ethyl acetate	$C_{14}H_{20}O_3$	118.0	165.8	201.5	228.0	277.6	304.4	......	(113)
1,2,4,5-Tetraethylbenzene	$C_{14}H_{22}$	65.7	111.6	145.8	172.4	221.4	248.0	11.6	(505)
2,4-Di-*tert*-butylphenol	$C_{14}H_{22}O$	84.5	130.0	164.3	190.0	237.0	260.8	......	(406)
Bornyl butyrate	$C_{14}H_{24}O_2$	74.0	118.0	150.7	176.4	222.2	247.0	......	(383)
Bornyl isobutyrate	$C_{14}H_{24}O_2$	70.0	114.0	147.2	172.2	218.2	243.0	......	(277)
Geranyl butyrate	$C_{14}H_{24}O_2$	96.8	139.0	170.1	193.8	235.0	257.4	......	(120)
Geranyl isobutyrate	$C_{14}H_{24}O_2$	90.7	133.0	164.0	187.7	228.5	251.0	......	(120)
Diethyl sebacate	$C_{14}H_{26}O_4$	125.3	172.1	207.5	234.4	280.3	305.5	1.3	(11)
2-Tetradecanone	$C_{14}H_{28}O$	99.3	145.5	179.8	206.0	253.0	278.0	......	(11, 223)
Myristaldehyde	$C_{14}H_{28}O$	99.0	148.3	186.0	214.5	267.9	297.8	23.5	(11, 222, 225, 230)
Myristic acid	$C_{14}H_{28}O_2$	142.0	190.8	223.5	250.5	294.6	318.0	57.5	(11, 157, 221, 223, 225)
1-Chlorotetradecane	$C_{14}H_{29}Cl$	98.5	148.2	187.0	215.5	267.5	296.0	+ 0.9	(113)
Tetradecane	$C_{14}H_{30}$	76.4	120.7	152.7	178.5	226.8	252.5	5.5	(11, 223, 224, 253, 484)
Tetradecylamine	$C_{14}H_{31}N$	102.6	152.0	189.0	215.7	264.6	291.2	......	(11, 230)
Triethyloctylsilane	$C_{14}H_{32}Si$	73.7	120.6	155.7	184.3	235.0	262.0	......	(466)
1,9-Diethoxydecamethylpentasiloxane	$C_{14}H_{40}O_6Si_5$	89.0	131.5	162.2	187.0	230.0	253.3	......	(114)
Tetradecamethylhexasiloxane	$C_{14}H_{42}O_5Si_6$	73.7	117.6	149.8	175.2	220.5	245.5	......	(114, 469)
Tetradecamethylcycloheptasiloxane	$C_{14}H_{42}O_7Si_7$	86.3	131.5	165.3	191.8	239.2	264.0	− 32	(188, 312, 469)
1,3-Diphenyl-2-propanone	$C_{15}H_{14}O$	125.5	177.6	216.6	246.6	301.7	330.5	34.5	(479)
1-Biphenyloxy-2,3-epoxypropane	$C_{15}H_{14}O_2$	135.3	187.2	226.3	255.0	309.8	340.0	......	(113)
1-Isovaleronaphthone	$C_{15}H_{16}O$	136.0	184.0	219.7	246.7	294.0	320.0	......	(360)
4,4-Isopropylidenebisphenol	$C_{15}H_{16}O_2$	193.0	240.8	273.0	297.0	339.0	360.5	......	(113)
Ethoxymethyldiphenylsilane	$C_{15}H_{18}OSi$	109.0	152.7	186.0	211.8	256.8	282.0	......	(114)
Helenin	$C_{15}H_{20}O_2$	157.7	192.1	215.2	232.6	260.6	275.0	76	(11)
Cadinene	$C_{15}H_{24}$	101.3	146.0	179.8	205.6	250.7	275.0	......	(383)
2,6-Di-*tert*-butyl-4-cresol	$C_{15}H_{24}O$	85.8	131.0	164.1	190.0	237.6	262.5	......	(406)
4,6-Di-*tert*-butyl-2-cresol	$C_{15}H_{24}O$	86.2	132.4	167.4	194.0	243.4	269.3	......	(406)
4,6-Di-*tert*-butyl-3-cresol	$C_{15}H_{24}O$	103.7	150.0	185.3	211.0	257.1	282.0	......	(406)
Champacol	$C_{15}H_{26}O$	100.0	148.0	184.0	211.9	261.2	288.0	91	(462)
Triethyl camphoronate	$C_{15}H_{26}O_6$	s	166.0	201.8	228.6	276.0	301.0	135	(11)
Methyl myristate	$C_{15}H_{30}O_2$	115.0	160.8	195.8	222.6	269.8	295.8	18.5	(2)
Pentadecane	$C_{15}H_{32}$	91.6	135.4	167.7	194.0	242.8	270.5	10	(11, 223, 224, 484)
Tetrapropylene glycol monoisopropyl ether	$C_{15}H_{32}O_5$	116.6	163.0	197.7	223.3	268.3	292.7	......	(113)
Dodecyltrimethylsilane	$C_{15}H_{34}Si$	91.2	137.7	172.1	199.5	248.0	273.0	......	(466)
Benzyl cinnamate	$C_{16}H_{14}O_2$	173.8	221.5	255.8	281.5	326.7	350.0	39	(172)
Di(α-methylbenzyl) ether	$C_{16}H_{18}O$	96.7	144.0	179.6	206.8	254.8	281.0	......	(113)
Diethoxydiphenylsilane	$C_{16}H_{20}O_2Si$	111.5	157.6	193.2	220.0	259.7	296.0	......	(114)
Dibutyl phthalate	$C_{16}H_{22}O_4$	148.2	198.2	235.8	263.7	313.5	340.0	......	(142)
Pentaethylchlorobenzene	$C_{16}H_{25}Cl$	90.0	140.7	178.2	208.0	257.2	285.0	......	(113)
Pentaethylbenzene	$C_{16}H_{26}$	86.0	135.8	171.9	200.0	250.2	277.0	......	(113)
2,6-Di-*tert*-butyl-4-ethylphenol	$C_{16}H_{26}O$	89.1	137.0	172.1	198.0	244.0	268.6	......	(406)
4,6-Di-*tert*-butyl-3-ethylphenol	$C_{16}H_{26}O$	111.5	157.4	192.3	218.0	264.6	290.0	......	(406)
Muscone	$C_{16}H_{30}O$	118.0	170.0	210.0	241.5	297.2	328.0	......	(457)
Palmitonitrile	$C_{16}H_{31}N$	134.3	185.8	223.8	251.5	304.5	332.0	31	(11, 233, 236)
1-Hexadecene	$C_{16}H_{32}$	101.6	146.2	178.8	205.3	250.0	274.0	4	(11, 226)
Tetraisobutylene	$C_{16}H_{32}$	63.8	108.5	142.2	167.5	214.6	240.0	......	(406)
2-Hexadecanone	$C_{16}H_{32}O$	109.8	167.3	203.7	230.5	279.8	307.0	......	(11, 223, 224)
Palmitaldehyde	$C_{16}H_{32}O$	121.6	171.8	210.0	239.5	292.3	321.0	34	(11, 222, 225, 230)
Palmitic acid	$C_{16}H_{32}O_2$	153.6	205.8	244.4	271.5	326.0	353.8	64.0	(11, 150, 157, 221, 223, 224, 225)
Hexadecane	$C_{16}H_{34}$	105.3	149.8	181.3	208.5	258.3	287.5	18.5	(11, 223, 224, 228, 484)
Cetyl alcohol	$C_{16}H_{34}O$	122.7	177.8	219.8	251.7	312.7	344.0	49.3	(11, 225)
Cetylamine	$C_{16}H_{35}N$	123.6	176.0	215.7	245.8	300.4	330.0	......	(11, 230, 233)
Decyltriethylsilane	$C_{16}H_{36}Si$	108.5	155.6	191.7	218.3	267.5	293.0	......	(466)
1,1,1-Diethoxydodecamethylhexasiloxane	$C_{16}H_{46}O_7Si_6$	103.6	147.5	180.0	205.5	250.0	273.5	......	(114)
Hexadecamethylheptasiloxane	$C_{16}H_{48}O_6Si_7$	93.2	138.5	171.8	198.0	244.7	270.0	......	(114, 469)
Hexadecamethylcyclooctasiloxane	$C_{16}H_{48}O_8Si_8$	103.5	150.5	186.3	213.8	263.0	290.0	31.5	(188, 469)
Benzanthrone	$C_{17}H_{10}O$	225.0	297.2	350.0	390.0	......	......	174	(464)
4-*tert*-Butylphenyl salicylate	$C_{17}H_{18}O_3$	166.2	225.0	270.7	305.8	370.6	404.0d	......	(113)
Menthyl benzoate	$C_{17}H_{24}O_2$	123.2	170.0	204.3	230.4	277.1	301.0	54.5	(383)
2-Heptadecanone	$C_{17}H_{34}O$	129.6	178.0	214.3	242.0	291.7	319.5	......	(11, 223, 224)
Methyl palmitate	$C_{17}H_{34}O_2$	134.3	184.3	d	......	......	......	30	(2)
Heptadecane	$C_{17}H_{36}$	115.0	160.0	195.8	223.0	274.5	303.0	22.5	(11, 223, 224, 484)
Tetradecyltrimethylsilane	$C_{17}H_{38}Si$	120.0	166.2	201.5	227.8	275.0	300.0	......	(466)
Tri-2-chlorophenylthiophosphate	$C_{18}H_{12}Cl_3O_3PS$	188.2	231.2	261.7	283.8	322.0	341.3	......	(113)
Triphenyl phosphate	$C_{18}H_{15}O_4P$	193.5	249.8	290.3	322.5	379.2	413.5	49.4	(113)
Hexaethylbenzene	$C_{18}H_{30}$	s	150.3	187.7	216.0	268.5	298.3	130	(113)
2,4,6-Tri-*tert*-butylphenol	$C_{18}H_{30}O$	95.2	142.0	177.4	203.0	250.6	276.3	......	(406)
Oleic Acid	$C_{18}H_{34}O_2$	176.5	223.0	257.2	286.0	334.7	360.0d	14	(11, 122, 134, 234)
Elaidic acid	$C_{18}H_{34}O_2$	171.3	223.5	260.8	288.0	337.0	362.0	51.5	(11, 230, 234)
Stearaldehyde	$C_{18}H_{36}O$	140.0	192.1	230.8	260.0	313.8	342.5	63.5	(11, 222, 225, 230)
Stearic acid	$C_{18}H_{36}O_2$	173.7	225.0	263.3	291.0	343.0	370.0d	69.3	(11, 150, 221, 222, 225, 227, 230, 234)
Octadecane	$C_{18}H_{38}$	119.6	169.6	207.4	236.0	288.0	317.0	28	(223, 225, 484)
2-Methylheptadecane	$C_{18}H_{38}$	119.8	168.7	204.8	231.5	279.8	306.5	......	(245)
1-Octadecanol	$C_{18}H_{38}O$	150.3	202.0	240.4	269.4	320.3	349.5	58.5	(11, 225)

VAPOR PRESSURE (Continued)

ORGANIC COMPOUNDS (Continued)

Pressures Less than One Atmosphere (Continued)

Name	Formula	Temperature °C 1 mm	10 mm	40 mm	100 mm	400 mm	760 mm	M.P.	Citation No.
Ethylcetylamine	$C_{18}H_{39}N$	133.2	186.0	226.5	256.8	313.0	342.0$_d$	......	(11, 233)
1,1,3-Diethoxytetradecamethylheptasiloxane	$C_{18}H_{52}O_8Si_7$	119.0	163.5	197.0	223.2	268.3	293.5	......	(114)
Octadecamethyloctasiloxane	$C_{18}H_{54}O_7Si_8$	105.8	152.3	187.5	214.5	263.5	290.0	......	(114, 469)
Triphenylmethane	$C_{19}H_{16}$	169.7	197.0	215.5	228.4	249.8	259.2	93.4	(11, 70, 383)
Diphenyl-2-tolyl thiophosphate	$C_{19}H_{17}O_3PS$	159.7	201.6	230.6	252.5	290.0	310.0	......	(113)
Nonadecane	$C_{19}H_{40}$	133.2	183.5	220.0	248.0	299.8	330.0	32	(11, 223, 224, 484)
Ethoxytriphenylsilane	$C_{20}H_{20}OSi$	167.0	213.5	247.0	273.5	319.5	344.0	......	(114)
Diethylhexadecylamine	$C_{20}H_{43}N$	139.8	194.0	235.0	265.5	324.6	355.0	......	(11, 233)
1,1,5-Diethoxyhexadecamethyloctasiloxane	$C_{20}H_{58}O_9Si_8$	133.7	179.7	213.8	240.0	286.0	311.5	......	(114)
Eicosamethylnonasiloxane	$C_{20}H_{60}O_8Si_9$	144.0	189.0	220.5	244.3	286.0	307.5	......	(469)
Tritolyl phosphate	$C_{21}H_{21}O_4P$	154.6	198.0	229.7	252.2	292.7	313.0	......	(113)
Heneicosane	$C_{21}H_{44}$	152.6	205.4	243.4	272.0	323.8	350.5	40.4	(11, 223, 224, 484)
Erucic acid	$C_{22}H_{42}O_2$	206.7	254.5	289.1	314.4	358.8	381.5$_d$	33.5	(11, 234)
Brassidic acid	$C_{22}H_{42}O_2$	209.6	256.0	290.0	316.2	359.6	382.5$_d$	61.5	(11, 234)
Docosane	$C_{22}H_{46}$	157.8	213.0	254.5	286.0	343.5	376.0	44.5	(11, 223, 224, 468, 484)
Docosamethyldecasiloxane	$C_{22}H_{66}O_9Si_{10}$	160.3	202.8	233.8	255.0	293.8	314.0	......	(469)
Tricosane	$C_{23}H_{48}$	170.0	223.0	261.3	289.8	339.8	366.5	47.7	(11, 223, 224, 484)
Tetracosane	$C_{24}H_{50}$	183.8	237.6	276.3	305.2	358.0	386.4	51.1	(11, 223, 224, 468, 484)
Tetracosamethylhendecasiloxane	$C_{24}H_{72}O_{10}Si_{11}$	175.2	216.7	246.2	266.3	303.7	322.8	......	(469)
Pentacosane	$C_{25}H_{52}$	194.2	248.2	285.6	314.0	365.4	390.3	53.3	(484)
Hexacosane	$C_{26}H_{54}$	204.0	257.4	295.2	323.2	374.6	399.8	56.6	(11, 468, 484)
Dicarvacryl-2-tolyl phosphate	$C_{27}H_{33}O_4P$	180.2	221.8	251.5	272.5	309.8	330.0	......	(113)
Heptacosane	$C_{27}H_{56}$	211.7	266.8	305.7	333.5	385.0	410.6	59.5	(11, 223, 484)
Octacosane	$C_{28}H_{58}$	226.5	277.4	314.2	341.8	388.9	412.5	61.6	(11, 468, 484)
Nonacosane	$C_{29}H_{60}$	234.2	286.4	323.2	350.0	397.2	421.8	63.8	(484)
Dicarvacryl-mono-(6-chloro-2-xenyl) phosphate	$C_{32}H_{34}ClO_4P$	204.2	249.3	280.5	304.9	342.0	361.0	......	(113)

Pressures Greater than One Atmosphere

Name	Formula	Temperature °C 1 atm.	2 atm.	5 atm.	10 atm.	20 atm.	40 atm.	60 atm.	Citation No.
Chlorotrifluoromethane	$CClF_3$	− 81.2	− 66.7	− 42.7	− 18.5	+ 12.0	52.8	......	(113, 430)
Dichlorodifluoromethane	CCl_2F_2	− 29.8	− 12.2	+ 16.1	42.4	74.0	.....	....	(113, 146)
Carbonyl chloride	CCl_2O	8.3	27.3	57.2	85.0	119.0	159.8	....	(20, 145, 208, 305, 311)
Trichlorofluoromethane	CCl_3F	23.7	44.1	77.3	108.2	146.7	194.0	....	(212)
Carbontetrachloride	CCl_4	76.7	102.0	141.7	178.0	222.0	276.0	....	(115, 175, 289, 343, 352, 394, 481)
Chlorodifluoromethane	$CHClF_2$	− 40.8	− 24.7	+ 0.3	24.0	52.0	85.3	....	(50, 212)
Dichlorofluoromethane	$CHCl_2F$	8.9	28.4	59.0	87.0	121.2	162.6	....	(212)
Trichloromethane	$CHCl_3$	61.3	83.9	120.0	152.3	191.8	237.5	......	(32, 115, 175, 345, 352, 365)
Hydrocyanic acid	CHN	25.8	45.5	75.5	103.5	134.2	170.2	....	(252)
Methyl bromide	CH_3Br	3.6	23.3	54.8	84.0	121.7	170.2	....	(117, 323)
Methyl chloride	CH_3Cl	− 24.0	− 6.4	+ 22.0	47.3	77.3	113.8	137.5	(54, 182, 241, 350, 451)
Methyl fluoride	CH_3F	− 78.2	− 64.5	− 42.0	− 21.0	+ 2.6	26.5	43.5	(89, 113, 281)
Methyl iodide	CH_3I	42.4	65.5	101.8	138.0	176.5	228.5	....	(351, 352, 363)
Methane	CH_4	−161.5	−152.3	−138.3	−124.8	−108.5	− 86.3	....	(6, 72, 93, 123, 206, 359, 407)
Methanol	CH_4O	64.7	84.0	112.5	138.0	167.8	203.5	224.0	(111, 289, 319, 338, 353)
Methanethiol	CH_4S	6.8	26.1	55.9	83.4	117.5	157.7	185.0	(39)
Methylamine	CH_5N	− 6.3	+ 10.1	36.0	59.5	87.8	121.8	144.6	(19, 38, 130, 186, 324)
Carbon monoxide	CO	−191.3	−183.5	−170.7	−161.0	−149.7	......	....	(23, 72, 82, 83, 102, 208, 306, 447)
Carbon dioxide	CO_2	− 78.2$_s$	− 69.1$_s$	− 56.7	− 39.5	− 18.9	+ 5.9	22.4	(4, 5, 53, 124, 167, 168, 196, 208, 242, 275, 307, 308, 343, 407, 448, 454, 463, 493)
Carbon disulfide	CS_2	46.5	69.1	104.8	136.3	175.5	222.8	256.0	(168, 208, 263, 331, 343, 352, 407, 454, 475)
1-Chloro-1,2,2-trifluoroethylene	C_2ClF_3	− 27.9	− 11.1	+ 15.5	40.0	71.1	......	....	(49, 50, 113)
1,2-Dichloro-1,1,2,2-tetrafluoroethane	$C_2Cl_2F_4$	3.5	22.8	54.0	82.3	117.5	......	....	(212)
1,1,2-Trichloro-1,2,2-trifluoroethane	$C_2Cl_3F_3$	47.6	70.0	105.5	138.0	177.7	......	....	(212)
Acetylene	C_2H_2	− 84.0$_s$	− 71.6	− 50.2	− 32.7	− 10.0	+ 16.8	34.8	(12, 66, 67, 69, 217, 449)
cis-1,2-Dichloroethylene	$C_2H_2Cl_2$	59.0	82.1	119.3	152.3	194.0	244.5	....	(113, 175, 176)
trans-1,2-Dichloroethylene	$C_2H_2Cl_2$	47.8	69.8	104.0	135.7	174.0	220.0	....	(113, 175, 176)
Ethylene	C_2H_4	−103.7	− 90.8	− 71.1	− 52.8	− 29.1	− 1.5	....	(6, 66, 93, 57, 116, 244, 359, 407, 448, 453)
1,2-Dibromoethane	$C_2H_4Br_2$	131.5	157.7	200.0	237.0	269.0	295.0	304.5	(11, 31, 113, 349, 351, 438)
1,1-Dichloroethane	$C_2H_4Cl_2$	57.3	80.2	117.3	150.3	192.7	243.0	....	(113, 352, 402)
1,2-Dichloroethane	$C_2H_4Cl_2$	83.7	108.1	147.8	183.5	226.5	272.0	....	(113, 133, 315, 352, 402)
Acetic acid	$C_2H_4O_2$	118.1	143.5	180.3	214.0	252.0	297.0	....	(113, 203, 246, 258, 289, 335, 353, 373, 480)
Methyl formate	$C_2H_4O_2$	32.0	51.9	83.5	112.0	147.2	188.5	....	(297, 491)
Ethyl bromide	C_2H_5Br	38.4	60.2	95.0	126.8	164.3	206.5	229.5	(113, 347, 391, 438)
Ethyl chloride	C_2H_5Cl	12.3	32.5	64.0	92.6	127.3	167.0	....	(38, 113, 195, 346, 402)
Ethyl fluoride	C_2H_5F	− 32.0	− 16.7	+ 7.7	30.2	57.5	90.0	....	(50)
Ethane	C_2H_6	− 88.6	− 75.0	− 52.8	− 32.0	− 6.4	+ 23.6	....	(6, 30, 65, 84, 93, 153, 261, 262, 359, 453)

VAPOR PRESSURE (Continued)
ORGANIC COMPOUNDS (Continued)
Pressures Greater than One Atmosphere (Continued)

Name	Formula	Temperature °C							Citation No.
		1 atm.	2 atm.	5 atm.	10 atm.	20 atm.	40 atm.	60 atm.	
Ethanol	C_2H_6O	78.4	97.5	126.0	151.8	183.0	218.0	242.0	(32, 113, 115, 202, 273, 289, 329, 337, 343, 365, 374, 394)
Dimethyl ether	C_2H_6O	− 23.7	− 6.4	+ 20.8	45.5	75.7	112.1	....	(55, 74, 211, 260, 350)
Ethanethiol	C_2H_6S	35.0	56.6	90.7	121.9	159.5	204.7	....	(39)
Dimethyl sulfide	C_2H_6S	36.0	57.8	92.3	124.5	163.8	209.0	....	(39)
Ethylamine	C_2H_7N	16.6	35.7	65.3	91.8	124.0	163.0	....	(38)
Dimethylamine	C_2H_7N	7.4	25.0	53.9	80.0	111.7	149.8	....	(38, 390, 467)
Cyanogen	C_2N_2	− 21.0	− 4.4	+ 21.4	44.6	72.6	106.5	....	(79, 101, 113, 125, 318, 426)
Propadiene	C_3H_4	− 35.0	− 18.4	+ 8.0	33.2	64.5	103.5	....	(113, 244, 255)
Propyne	C_3H_4	− 23.3	− 7.1	+ 19.5	43.8	74.0	111.5	....	(113, 165, 262, 286)
Propylene	C_3H_6	− 47.7	− 31.4	− 4.8	+ 19.8	49.5	85.0	....	(6, 66, 67, 137, 244, 262, 359, 382)
Acetone	C_3H_6O	56.5	78.6	113.0	144.5	181.0	214.5	....	(29, 115, 129, 343, 364)
Propionic acid	$C_3H_6O_2$	141.1	160.0	186.0	203.5	220.0	233.0	....	(203, 204, 353, 373)
Methyl acetate	$C_3H_6O_2$	57.8	79.5	113.1	144.2	181.0	225.0	....	(113, 491)
Ethyl formate	$C_3H_6O_2$	54.3	76.0	110.5	142.2	180.0	225.0	....	(297, 491)
Propane	C_3H_8	− 42.1	− 25.6	+ 1.4	26.9	58.1	94.8	....	(6, 30, 66, 93, 137, 153, 262, 359)
1-Propanol	C_3H_8O	97.8	117.0	149.0	177.0	210.8	250.0	....	(113, 289, 340, 374)
2-Propanol	C_3H_8O	82.5	101.3	130.2	155.7	186.0	220.2	....	(310)
Ethyl methyl ether	C_3H_8O	7.5	26.5	56.4	84.0	108.0	160.0	....	(39)
Propylamine	C_3H_9N	48.5	69.8	102.8	133 4	170.0	214.5	....	(38)
1,3-Butadiene	C_4H_6	− 4 5	+ 15.3	47.0	76.0	114.0	158.0	....	(52, 162, 189, 445)
Acetic anhydride	$C_4H_6O_3$	139.6	162.0	194.0	221.5	253.0	288.5	....	(33, 113, 201)
Dimethyl oxalate	$C_4H_6O_4$	163 3	189.6	228.7	......	......	......	....	(11, 97, 144)
Butyric acid	$C_4H_8O_2$	163.5	188.3	225.0	257.0	295.0	338.0	....	(204, 332, 373, 374)
Isobutyric acid	$C_4H_8O_2$	154.5	179.8	217.0	250.0	289.0	336.0	....	(203, 204, 353, 373, 374)
Ethyl acetate	$C_4H_8O_2$	77.1	100.6	136.6	169.7	209.5	......	....	(113, 455, 491)
Methyl propionate	$C_4H_8O_2$	79.8	103.0	139.8	172.6	212.5	......	....	(491)
Propyl formate	$C_4H_8O_2$	81.3	104.3	142.0	176.4	217.5	......	....	(297, 491)
Butane	C_4H_{10}	− 0.5	+ 18.8	50.0	79.5	116.0	......	....	(6, 66, 93, 359, 382)
2-Methylpropane	C_4H_{10}	− 11.7	+ 7.5	39.0	66.8	99.5	......	....	(6, 16, 93, 187, 359, 382)
Butyl alcohol	$C_4H_{10}O$	117.5	139.8	172.5	203.0	237.0	277.0	....	(68, 174, 203, 204, 314)
sec-Butyl alcohol	$C_4H_{10}O$	99.5	118.2	147.5	172.0	204.0	251.0	....	(68)
Isobutyl alcohol	$C_4H_{10}O$	108.0	127.3	156.2	182.0	212.5	251.0	....	(68, 203, 219, 374)
tert-Butyl alcohol	$C_4H_{10}O$	82.9	102.0	130.0	154.2	184.5	222.5	....	(68, 310)
Diethyl ether	$C_4H_{10}O$	34.6	56.0	90.0	122.0	159.0	......	....	(31, 113, 115, 289, 338, 423a)
Diethyl sulfide	$C_4H_{10}S$	88.0	112.0	153.8	190.2	234.0	......	....	(25, 39)
Diethylamine	$C_4H_{11}N$	55.5	77.8	113.0	145.3	184.5	......	....	(38)
Tetramethylsilane	$C_4H_{12}Si$	27.0	48.0	82.0	113.0	152.0	......	....	(15, 114)
Ethyl propionate	$C_5H_{10}O_2$	99.1	123.8	162.7	197.8	240.0	......	....	(491)
Propyl acetate	$C_5H_{10}O_2$	101.8	126.8	165.7	200.5	242.8	......	....	(11, 113, 199, 491)
Isobutyl formate	$C_5H_{10}O_2$	98.2	121.8	157.8	192.4	234.0	......	....	(292, 380)
Methyl butyrate	$C_5H_{10}O_2$	102.3	127.5	166.7	203.0	244.5	......	....	(11, 491)
Methyl isobutyrate	$C_5H_{10}O_2$	92.6	116.7	155.2	190.2	232.0	......	....	(491)
Pentane	C_5H_{12}	36.1	58.0	92.4	124.7	164.3	......	....	(6, 28, 93, 274, 359, 482, 485)
2-Methylbutane	C_5H_{12}	27.8	48.8	82.8	114.5	154.0	......	....	(6, 93, 359, 485, 486, 492)
2,2-Dimethylpropane	C_5H_{12}	+ 9.5	29.5	61.1	90 7	127.6	......	....	(6, 17, 113, 253, 359, 465)
Ethyl propyl ether	$C_5H_{12}O$	61.7	85.3	123.1	156.2	197.2	......	....	(39)
Bromobenzene	C_6H_5Br	156.2	186.2	232.5	274.5	327.0	387.5	....	(11, 199, 204, 289, 358, 478)
Chlorobenzene	C_6H_5Cl	132.2	160.2	205.0	245.3	292.8	349.8	....	(11, 199, 289, 478)
Fluorobenzene	C_6H_5F	84 7	109.9	148.5	184.4	227.6	279.3	....	(113, 416, 478)
Iodobenzene	C_6H_5I	188.6	220.0	270.0	315.7	371.5	437.2	....	(115, 289, 358, 478)
Benzene	C_6H_6	80.1	103.8	142.5	178.8	221.5	272.3	....	(3, 6, 11, 32, 109, 131, 197, 204, 289, 343, 359, 366, 392, 415, 478)
Phenol	C_6H_6O	181.9	208.0	248.2	283.8	328.7	382.1	418.7	(11, 34, 113, 151, 174, 203, 204)
Aniline	C_6H_7N	184.4	212.8	254.8	292.7	342.0	400.0	....	(11, 34, 143, 151, 204, 341)
Cyclohexane	C_6H_{12}	80.7	106.0	146.4	184.0	228.4	......	....	(6, 109, 293, 359, 487)
Ethyl isobutyrate	$C_6H_{12}O_2$	110.1	135.5	174.2	210.0	253.0	......	....	(292, 380)
Hexane	C_6H_{14}	68.7	93.0	131.7	166.6	209.4	......	....	(6, 115, 251, 289, 359, 394, 427, 474)
2,3-Dimethylbutane	C_6H_{14}	58.0	82.0	120.3	155.7	198.7	......	....	(6, 359, 471, 488)
Toluene	C_7H_8	110.6	136.5	178.0	215.8	262.5	319.0	....	(6, 11, 24, 107, 115, 199, 204, 239, 253, 359)
Heptane	C_7H_{16}	98.4	124.8	165.7	202.8	247.5	......	....	(6, 251, 269, 289, 359, 394, 483)
Ethylbenzene	C_8H_{10}	136.2	163.5	207.5	246.3	294.5	......	....	(6, 11, 265, 359, 415, 471, 474)
Octane	C_8H_{18}	125.6	152.7	196.2	235.8	181.4	......	....	(6, 251, 253, 359, 471, 474, 492)
Dodecane	$C_{12}H_{26}$	216.2	249.2	300.0	345.8	......	......	....	(11, 27, 223, 224, 225)

VAPOR PRESSURE—CITATIONS
Organic Compounds

(1) Allen, J. Chem. Soc., **77**, 400–16 (1900).
(2) Althouse and Triebold, Ind. Eng. Chem. Anal. Ed., **16**, 605–6 (1944).
(3) Altschul, Z. Physik. Chem., **11**, 577–97 (1893).
(4) Amagat, Compt. rend., **114**, 1093–8 (1892).
(5) Amagat, J. Phys., (3) **1**, 297 (1892).
(6) Am. Petroleum Inst. Research Project, **44**, Natl. Bur. Standards, tables ending with letter K (June 30, 1946).
(7) André, Compt. rend., **126**, 1105–7 (1898).
(8) Andrews, J. Phys. Chem., **30**, 1497–1500 (1926).
(9) Anschütz, Ann. **221**, 133–57 (1883).
(10) Anschütz, and Berns, Ber., **20**, 1389–93 (1887).
(11) Anschütz and Reitter, "Die Destillation unter vermindertem Druck im Laboratorium," 2nd ed., Bonn, Cohen, 1895.
(12) Ansdell, Chem. News, **40**, 136–8 (1879).
(13) Arbusow, Chem. Zentr., **1909, II**, 684–5.
(14) Arbusow, J. Russ. Phys.-Chem. Soc., **41**, 429–46 (1909).
(15) Aston, Kennedy, and Messerly, J. Am. Chem. Soc., **63**, 2343–8 (1941)
(16) Aston, Kennedy, and Schumann, Ibid., **62**, 2059–63 (1940).
(17) Aston and Messerly, Ibid., **58**, 2354–61 (1936).
(18) Aston, Sagenkahn, Szasz, Moessen, and Zuhr, Ibid., **66**, 1171–7 (1944).
(19) Aston, Siller, and Messerly, Ibid., **59**, 1743–51, (1937).
(20) Atkinson, Heycock, and Pope, J. Chem. Soc., **117**, 1410–26 (1920)

Organic Compounds

(21)Auwers and Lange, Ann., **409,** 149–82 (1915).
(22)Babasinian and Jackson, J. Am. Chem. Soc., **51,** 2147–51 (1929)
(23)Baly and Donnan, J. Chem. Soc., **81,** 907–23 (1902).
(24)Barker, Z. Physik Chem., **71,** 235–53 (1910).
(25)Bauer and Burschkies, Ber., **68B,** 1238–43 (1935).
(26)Baxter, Bezzenberger, and Wilson, J. Am. Chem. Soc., **42,** 1386–93 (1920).
(27)Beale and Docksey, J. Inst. Petroleum Tech., **21,** 860–70 (1935).
(28)Beall, Refiner Natural Gasoline Mfr., **14,** 588–9 (1935).
(29)Beare, McVicar, and Ferguson, J. Phys. Chem., **34,** 1310–8 (1930).
(30)Beattie, Hadlock, and Poffenberger, J. Chem. Phys., **3,** 93–6 (1935).
(31)Beckman, Fuchs, and Gernhardt, Z. Physik. Chem., **18,** 473–513 (1895).
(32)Beckmann and Liesche, Ibid., **88,** 13–34 (1914).
(33)Ibid., **88,** 419–27 (1914).
(34)Ibid., **89,** 111–24, (1915).
(35)Berliner and May, J. Am. Chem. Soc., **47,** 2350–6 (1925).
(36)Ibid., **48,** 2630–4 (1926).
(37)Ibid., **49,** 1007–11 (1927).
(38)Berthoud, J. chim. phys., **15,** 3–29 (1917).
(39)Berthoud and Brum, Ibid., **21,** 143–60 (1924).
(40)Bertram and Gildemeister, J. prakt. Chem., **II, 39,** 349–55 (1889).
(41)Ibid., 49, 185–96 (1894).
(42)Bertram and Helle, Ibid., **II, 61,** 293–306 (1900).
(43)Beyer and Claisen, Ber., **20,** 2178–88 (1887).
(44)Biltz, Ann., **297,** 263–78 (1897).
(45)Blaise and Luttringer, Bull. soc. chim. Mém., **III, 33,** 761–83 (1905).
(46)Blaszkowska and Zakrzewska, Chem. Zentr., **1929, II,** 3076.
(47)Blaszkowska and Zakrzewska, Roczniki Chem., **8,** 210–18 (1928).
(48)Bolas and Groves, J. Chem. Soc., **24,** 773–85 (1871).
(49)Booth, Burchfield, Bixby, and McKelvey, J. Am. Chem. Soc., **55,** 2231–5 (1933).
(50)Booth and Swinehart, Ibid., **57,** 1337–42 (1935).
(51)Bouveault, Bull. soc. chim. **III, 17,** 363–72 (1897).
(52)Brickwedde, Bekkedahl, Meyers, Rands, and Scott, J. Research Natl. Bur. Standards, **35,** 39–87 (1945).
(53)Bridgeman, J. Am. Chem. Soc., **49,** 1174–83 (1927).
(54)Brinkman, Thesis, Amsterdam.
(55)Briner and Cardoso, J. chim. physic., **6,** 663 (1908).
(56)Briscoe and Peel, J. Chem. Soc., **131,** 1741–7 (1928).
(57)Britton, Trans. Faraday Soc., **55,** 520–5 (1929).
(58)Brückner, Z. anorg. allgem. Chem., **199,** 91–2 (1931).
(59)Brühl, Ber., **32,** 1222–36 (1899).
(60)Brunner, Ibid., **19,** 2224–8 (1886).
(61)Buckler and Norrish, J. Chem. Soc., **139,** 1567–9 (1936).
(62)Bullard and Haussmann, J. Phys. Chem., **34,** 743–7 (1930).
(63)Burchfield, J. Am. Chem. Soc., 64, 2501 (1942).
(64)Burg and Schlesinger, Ibid., 59, 780–7 (1937).
(65)Burrell and Robertson, Ibid., 37, 1893–1902 (1915).
(66)Ibid., 37, 2188–93 (1915).
(67)Burrell and Robertson, U S. Bur. Mines, Tech. Paper 142 (1916).
(68)Butler et al., J. Chem. Soc., **138,** 280–5 (1935).
(69)Cailletet, Compt. rend., **85,** 851–4 (1877).
(70)Callendar and Griffiths, Chem. News, **63,** 1–2 (1891).
(71)Carbide and Carbon Chemicals Corp. bulletins (1945).
(72)Cardoso, J. chim. phys., **13,** 312–50 (1915).
(73)Ibid., **13,** 332 (1915)
(74)Cardoso and Bruno, Ibid., 20, 347–51 (1923).
(75)Carpmael, British Patent 469,421 (1937).
(76)Cavalier, Ann. chim. phys., **18,** 449–507 (1899).
(77)Cavalier, Bull. soc. chim., **19,** 883–7 (1898).
(78)Chamuleau and Noyous, Acta Brevia Neerland, Physiol. Pharmacol. Microbiol., **2,** 94–5 (1932).
(79)Chappuis and Rivière, Compt. rend., **104,** 1504–5 (1887).
(80)Chipman and Peltier, Ind. Eng. Chem., **21,** 1106–8 (1929).
(81)Ciamician and Silber, Ber , **29,** 1799–1811 (1896).
(82)Clayton and Giauque, J. Am. Chem. Soc., **54,** 2610–26 (1932).
(83)Clusius and Teske, Z. physik Chem., **6B,** 135–51 (1929).
(84)Coffin and Maass, J. Am. Chem Soc., **50,** 1427–37 (1926).
(85)Coffin and Maass, Chem. Zentr., **1928, I,** 1643.
(86)Coffin and Maass, Trans. Roy. Soc. Can., [3] **21,** 33–40 (1927).
(87)Coffin, Sutherland, and Maass, Can. J Research, **2,** 267–78 (1930).
(88)Cole, Philippine J. Sci., **40,** 499–502 (1929).
(89)Collie, J. Chem. Soc., **55,** 110–13 (1889).
(90)Commercial Solvents Corp. bull. (1944).
(91)Coolidge, J. Am. Chem. Soc., **52,** 1874–87 (1930).
(92)Coolidge and Coolidge, Ibid., **49,** 100–4 (1927).
(93)Copson and Frolich, Ind. Eng. Chem., **21,** 1116–7 (1929).
(94)Cosslett, Z. anorg. allgem. Chem., **201,** 75–80 (1931).
(95)Courtot and Metzger, Compt. rend., **219,** 487–90 (1944).
(96)Cox, Ind. Eng. Chem., **15,** 592 (1923).
(97)Crafts, Ber , **20,** 709–16 (1887).
(98)Crafts, J. chim. phys., **11,** 429–77 (1913).
(99)Ibid., **13,** 105–61 (1915).
(100)Crater, Ind. Eng Chem., **21,** 674–6 (1929).
(101)Crommelin, Proc. "Konink." Akad. Wetenschappen Amsterdam (1913).
(102)Crommelin, Bijleved, and Brown, Proc. Acad. Sci. Amsterdam, **34,** 1314 (1931).
(103)Cunningham, Power, **72,** 374–7 (1930).
(104)Dana, Burdick, and Jenkins, J. Am. Chem. Soc., **49,** 2801–6 (1927).
(105)Datin, Ann. Physik., [9] **5,** 218–40 (1916).
(106)Davis, Ind. Eng. Chem , **17,** 735 (1925).
(107)Ibid., **22,** 380–1 (1930).
(108)Davis and Calingaert, Ibid., **17,** 1287 (1925).
(109)Déjardin, Ann. phys., [9] **11,** 253–5 (1919).
(110)Demjanoff, Ber., **28,** 22–4 (1895).
(111)Dittmar and Fawsitt, Edinburgh Trans., **33, II,** 509 (1886–1887).
(112)Doss, "Physical Constants of Principle Hydrocarbons," 4th ed., p. 20, New York, Texas Co., 1943.
(113)Dow Chemical Co. files.
(114)Dow Corning Corp. files.
(115)Drücker, Jimeno, and Kangro, Z. physik. Chem., **90,** 513–52 (1915).
(116)Egan and Kemp, J. Am. Chem. Soc., **59,** 1264–8 (1937).
(117)Ibid., **60,** 2097–101 (1938).
(118)Emeléus and Miller, J. Chem. Soc., **142,** 819–23 (1939).
(119)Emeléus and Wilkins, Ibid., **146,** 454–6 (1944).
(120)Erdmann, Ber., **31,** 356–60 (1898).
(121)Ibid., **35,** 1855–62 (1902).
(122)Erdmann, J. prakt. Chem., **II, 56,** 143–56 (1897).
(123)Eucken and Berger, Z. ges. Kälte-Ind., **41,** 147 (1934).
(124)Falck, Physik. Z., **9,** 433–40 (1908).
(125)Faraday, Phil. Trans., **1351,** 155 (1845).
(126)Fawcett, J. Am. Chem. Soc., **68,** 1420–2 (1946).
(127)Fawcett and Rasmussen, Ibid., 67, 1705–9 (1945).
(128)Feitler, Z. physik. Chem. **4,** 66–88 (1889).
(129)Felsing and Durban, J. Am. Chem. Soc. **48,** 2885–93 (1926).
(130)Felsing and Thomas, Ind. Eng. Chem., **21,** 1269–72 (1929).
(131)Ferche, Ann. Physik., **44,** 265–87 (1891).
(132)Ferguson, Freed, and Morris, J. Phys. Chem., **37,** 87–91 (1933).
(133)Fife and Reid, Ind. Eng. Chem., **22,** 513–15 (1930).
(134)Fischer and Harries, Ber., **35,** 2158–63 (1902).
(135)Fittig and Schmitz, Ann., **193,** 134–42 (1878).
(136)deForcand, Compt. rend., **132,** 688–92 (1901).
(137)Francis and Robbins, J. Am. Chem. Soc., **55,** 4339–42 (1933).
(138)Friedr. Bayer and Co., Chem. Zentr., **1898, II,** 951–2.
(139)Ibid., 1899, **I,** 461–2.
(140)Fries and Bestian, Ber., **69,** 715–22 (1936).
(141)Gallaugher and Hibbert, J. Am. Chem. Soc., **59,** 2521–5 (1937).
(142)Gardner and Brewer, Ind. Eng. Chem., **29,** 179–81 (1937).
(143)Garrick, Trans. Faraday Soc., **23,** 560–3 (1927).
(144)Germann and Pickering, International Critical Tables, **III,** 248 (1928).
(145)Germann and Taylor, J. Am. Chem. Soc., **48,** 1154–9 (1926).
(146)Gilkey, Gerard, Bixler, Ind. Eng. Chem., **23,** 364–7 (1931).
(147)Gilmore, J. Soc. Chem. Ind. (Trans.), **41,** 293–4 (1922).
(148)Grassi, Nuovo cimento, [3] **24,** 109 (1888).
(149)Gróh, Ber., **45,** 1441–7 (1912).
(150)Grosjean, Ibid., **25,** 478–81 (1892).
(151)Guye and Mallet, Arch. sci. phys. nat., **13,** 30, 129, 274, 462 (1902).
(152)Haggerty and Weiler, J. Am. Chem. Soc., **51,** 1621–6 (1929).
(153)Hainlen, Ann. **282,** 229–45 (1894).
(154)Halban, Z. physik. Chem., **84,** 129–59 (1913).
(155)Haller and Martine, Compt. rend., **140,** 1298–1303 (1905).
(156)Ham, Churchill, and Ryder, J. Franklin Inst., **186,** 13–28 (1918).
(157)Hansen, Z. physik. Chem., **74,** 65–114 (1910).
(158)Harries, Ber., **31,** 37–47 (1898).
(159)Heim, Bull. soc. chim. Belg., **42,** 467–82 (1933).
(160)Hein. Z. physik. Chem., **86,** 385–426 (1914).
(161)Hein and Schramm, Ibid., **A149,** 408–16 (1930).
(162)Heisig, J. Am. Chem. Soc., **55,** 2304–11 (1933).
(163)Ibid., **63,** 1698–9 (1941).
(164)Heisig and Davis, Ibid., **57,** 339–40 (1935).
(165)Heisig and Hurd, Ibid., **55,** 3485–7 (1933).
(166)Henglein, Z. Physik, **18,** 64–9 (1923).
(167)Henning, Ann. Physik, [4] **43,** 282–94 (1914).
(168)Henning and Stock, Z. Physik, **4,** 226–40 (1921).
(169)Henry, Bull. acad. roy. méd. Belg., [3] **36,** 241–62 (1898).
(170)Henry, Chem. Zentr., **1899, I,** 194–5.
(171)Hentschel, Ber., **23,** 2394–401 (1890).
(172)Ibid., **31,** 246–9 (1898).
(173)Herz and Lorentz, Z. physik. Chem., **A140,** 406–22 (1929).
(174)Herz and Neukirch, Ibid., **104,** 433–50 (1923).
(175)Herz and Rathmann, Chem.-Ztg., **36,** 1417–8 (1912).
(176)Ibid., **37,** 621–2 (1913).
(177)Herz and Schuftan, Z. physik. Chem., **101,** 269–85 (1922).
(178)Hess and Zeitschel, J. prakt. Chem. **II, 66,** 481–516 (1902).
(179)Hesse and Muller, Ber., **32,** 565–74 (1899).
(180)Hieber and Reindl, Z. Elektrochem., **46,** 559–70 (1940).
(181)Hieber and Woerner, Ibid., **40,** 252–6 (1934).
(182)Holst, Thesis, Zurich (1914).
(183)Houben, Ber., **35,** 3587–92 (1902).
(184)Hovorka and Geiger, J. Am. Chem. Soc., **55,** 4759–61 (1933).
(184a) Hovorka, Lankelma, and Nanjoks, Ibid., **55,** 4820–2 (1933).
(185)Hovorka, Lankelma, and Stanford, Ibid., **60,** 820–7 (1938).
(186)Hsia, Z. tech. Phys., **12,** 550–1 (1931).
(187)Hückel and Rassmann, J. prakt. Chem. [N. F.], **136,** 30–40 (1933).
(188)Hunter, Hyde, Warrick and Fletcher, J. Am. Chem. Soc. **68,** 667–72 (1946).
(189)I. G. Oppau and Ludwigshafen, unpublished report.
(190)Ilosvay, Bull. Soc. Chem., [2] **37,** 294 (1882).
(191)International Critical Tables, **III,** 201–49, New York, McGraw-Hill Book Co., Inc., (1928).
(192)Ioffe and Yampol'skaya, J. Applied Chem. (U.S.S.R.), **17,** 527 8 (1944).
(193)Jaeger, Z. Anorg. allgem. Chem., **101,** 155 (1917).
(194)Jaquerod and Wassmer, Ber., **37,** 2531–4 (1904).
(195)Jenkin and Shorthose, Dept. Sci. Ind. Research (Brit.), Special Report 14 (1923).
(196)Jenkins and Pye, Phil. Trans. Roy. Soc. (London), **A213,** 67 (1913).
(197)Jolly and Birscoe, J. Chem. Soc., **129,** 2154–9 (1926).
(198)Kahlbaum, Ber., **16,** 2476–84 (1883).
(199)Ibid., **17,** 1245–62 (1884).
(200)Ibid., **19,** 2863–5 (1886).
(201)Kahlbaum, "Sedetemperatur und Druck", Leipzig (1885).
(202)Kahlbaum, "Studien uber Dampfspannkraft", p. 116, Basel, 1897.
(203)Kahlbaum, Z. physik. Chem., **13,** 14–55 (1894).
(204)Ibid., **26,** 577–658 (1898).
(205)Kailan, Z. anal. Chem. **51,** 81–101 (1912).
(206)Karwat, Z. physik, Chem., **112,** 486–90 (1924).
(207)Kassel, J. Am. Chem. Soc., **58,** 670 (1936).
(208)Kelley, U. S. Bur. Mines, Bull. 383 (1935).
(209)Kemp and Giauque, J. Am. Chem. Soc., **59,** 79–84 (1937).
(210)Kendall and McKenzie, "Organic Syntheses", Vol. III, pp. 57–9, New York, John Wiley & Sons, Inc., 1923.

VAPOR PRESSURE—CITATIONS (Continued)
Organic Compounds

(211)Kennedy, Sagenkahn, and Aston, J. Am. Chem. Soc., **63**, 2267–72 (1941).
(212)Kinetic Chemicals Corp. bulletins (1933–1942).
(213)Kistiakowski, Ruoff, Smith, and Vaughan, J. Am. Chem. Soc., **57**, 876–82 (1935).
(214)Klages, Ber., **35**, 2245–62 (1902).
(215)Ibid., **36**, 3584–97 (1903).
(216)Klages and Keil, Ibid., **36**, 1632–45 (1903).
(217)Klemenc, Bankowski, and Von Frugnoni, Naturwissenschaften, **22**, 465 (1934).
(218)Klosky, Woo, and Flanigan J. Am. Chem. Soc., **49**, 1280–4 (1927).
(219)Konowalow, Ann. Physik., **14**, 34–52 (1881).
(220)Kopp, Ann., **195**, 81–92 (1879).
(221)Krafft, Ber., **12**, 1664–8 (1879).
(222)Ibid., **13**, 1413–21 (1880).
(223)Ibid., **15**, 1687–711 (1882).
(224)Ibid., **15**, 1711–28 (1882).
(225)Ibid., **16**, 1714–26 (1883).
(226)Ibid., **16**, 3018–24 (1883).
(227)Ibid., **17**, 1627–31 (1884).
(228)Ibid., **19**, 2218–23 (1886).
(229)Ibid., **19**, 2982–8 (1886).
(230)Ibid., **23**, 2360–4 (1890).
(231)Krafft and Bürger, Ibid., **17**, 1378–80 (1884).
(232)Krafft and Lyons, Ibid., **27**, 1761–8 (1894).
(233)Krafft and Moye, Ibid., **22**, 811–5 (1889).
(234)Krafft and Noerdlinger, Ibid., **22**, 816–20 (1889).
(235)Krafft and Roos, Ibid., **25**, 2255–62 (1892).
(236)Krafft and Stauffer, Ibid., **15**, 1728–31 (1882).
(237)Krafft and Vorster, Ibid., **26**, 2813–22 (1893).
(238)Krafft and Weilandt, Ibid., **29**, 1316–28 (1896).
(239)Krase and Goodman, Ind. Eng. Chem., **22**, 13 (1930).
(240)Kraus and Toonder, Proc. Natl. Acad. Sci. U. S., **19**, 292–8 (1933).
(241)Kuenen, Arch. néerland, sci., **26**, 354 (1898).
(242)Kuenen and Robson, Phil. Mag., [6] **3**, 149 (1902).
(243)Küster, Z. physik. Chem. **51**, 222–42 (1905).
(244)Lamb and Roper, J. Am. Chem. Soc., **62**, 806–14 (1940).
(245)Landa and Riedl, Collection Czechoslov. Chem. Commun., **2**, 520–30 (1930).
(246)Landolt, Ann. Suppl., **6**, 129–81 (1868).
(246a)Landolt-Börnstein, "Physikalisch-Chemische Tabellen", Hauptwerk, 1353–77; **I** Ergänzungsband, 729–42; **II** Ergänzungsband, 1302–10; **III** Ergänzungsband, 2450–62, Berlin, Julius Springer, 1923–36.
(247)Lecat, Ann. soc. sci. Bruxelles, **49B**, 17–27 (1929).
(248)deLeeuw, Z. phys. Chem., **77**, 284–314 (1911).
(249)Lenth, J. Am. Chem. Soc., **55**, 3283 (1933).
(250)Lépingle, Bull. soc. chim. Mém., **39**, 741–62 (1926).
(251)Leslie and Carr, Ind. Eng. Chem., **17**, 810–17 (1925).
(252)Lewis and Schutz, J. Am. Chem. Soc., **56**, 1002 (1934).
(253)Linder, J. Phys. Chem. **35**, 531–5 (1931).
(254)Lippert, Ann., **276**, 148–99 (1893).
(255)Livingston and Heisig, J. Am. Chem. Soc., **52**, 2409–10 (1930).
(256)Lobry and Bruyn, Ber., **26**, 266–7 (1893).
(257)Lucas and Prater, J. Am. Chem. Soc., **59**, 1682–6 (1939).
(258)MacDougall, Ibid., **58**, 2585–91 (1936).
(259)MacDougall and Smith, Ibid., **52**, 1998–2001 (1930).
(260)Maass and Boomer, Ibid., **44**, 1709–28 (1922).
(261)Maass and McIntosh, Ibid., **36**, 737–42 (1914).
(262)Maass and Wright, Ibid., **43**, 1098–111 (1921).
(263)Mali, Z. anorg. Chem., **149**, 150–6 (1925).
(264)Mali and Ghosh, Quart. J. Indian Chem. Soc., **1**, 37–43 (1924).
(265)Mangold, Wien, Akad. Ber., **102**, **IIA**, 1071 (1893).
(266)Mannich, Ber., **35**, 2144–6 (1902).
(267)Marsden and Cuthberton, Can. J. Research, **9**, 419–23 (1933).
(268)Marshall, J. Chem. Soc., **89**, 1371–86 (1906).
(269)Mathews, J. Am. Chem. Soc., **48**, 562–76 (1926).
(270)May, Berliner, and Lynch, Ibid., **49**, 1012–6 (1927).
(271)Menzel and Mohry, Z. Anorg. allgem. Chem., **210**, 257–63 (1933).
(272)Menzies, J. Am. Chem. Soc., **41**, 1336–7 (1919).
(273)Merriman, J. Chem. Soc., **103**, 628–36 (1913).
(274)Messerly and Kennedy, J. Am. Chem. Soc., **62**, 2988–91 (1940).
(275)Meyers and Van Dusen, J. Research Natl. Bur. Standards, **10**, 381–412 (1933).
(276)Michalek and Clark, Chem. Eng. News, **22**, 1559–63 (1944).
(277)Minguin, Gregoire, and Bollemont, Compt. rend., **134**, 608–10 (1902).
(278)Möslinger, Ann. **185**, 26–74 (1887).
(279)Möslinger, Ber., **9**, 998–1008 (1876).
(280)Moitessier and Engel, Jahresber. Chemie, **1880**, 142.
(281)Moles and Batuecas, J. chim. phys., **17**, 537–88 (1919).
(282)Monhaupt, Chem.-Ztg., **32**, 573 (1908).
(283)Monrew, Bontaric, and Dufraisse, J. chim. phys. **18**, 333–47 (1920).
(284)Monroe, Ind. Eng. Chem., **12**, 969–71 (1920).
(285)Montmollin and Matile, Helv. Chim. Acta, **7**, 106–11 (1924).
(286)Moorehouse and Maass, Can. J. Research, **5**, 306–12 (1931).
(287)Ibid., **11**, 637–43 (1934).
(288)Mortimer and Murphy, Ind. Eng. Chem., **15**, 1140–2 (1923).
(289)Mündel, Z. phys. chem., **85**, 435–65 (1913).
(290)Nadejdine, J. Russ. Phys. Chem. Soc., **14**, 157 (1882).
(291)Ibid., **15**, **25** (1883).
(292)Nadejdine, Rep. phys., **23**, 759 (1887).
(293)Nagornow and Rolinjanz, Ann. inst. anal. phys. chim. **(U.S.S.R.) 2**, 371–400 (1924).
(294)Nef, Ann., **270**, 267–335 (1892).
(295)Ibid., **310**, 316–35 (1900).
(296)Nelson, Ind. Eng. Chem., **22**, 971–2 (1930).
(297)Ibid., **20**, 1380–2 (1928).
(298)Ibid., **20**, 1382–4 (1928).
(299)Nelson and Senseman, Ibid., **14**, 58–62 (1922).
(300)Nelson and Wales, J. Am. Chem. Soc., **47**, 867–72 (1925).
(301)Nelson and Young, Ibid., **55**, 2429–31 (1933).
(302)Neubeck, Z. physik. Chem., **1**, 649–66 (1887).
(303)Niederschulte, Dissertation Erlangen, 1903.
(304)Nieuwland, Calcott, Downing, and Carter, J. Am. Chem. Soc., **53**, 4197–202 (1931).
(305)Nikitin, J. Russ. Phys.-Chem. Soc., **52**, 235–49 (1920).
(306)Olszewski, Compt. rend., **99**, 706–7 (1884).
(307)Olszewski, Ann. Physik., **31**, 58–74 (1887).
(308)Onnes and Weber, Verslag Akad. Wetenschappen, **22**, 226 (1913)
(309)Paneth and Loleit, J. Chem. Soc., **138**, 366–71 (1935).
(310)Parks and Barton, J. Am. Chem. Soc., **50**, 24–6 (1928).
(311)Paterno and Mazzuchelli, Gazz. chim. ital., **50**, **I**, 30–52 (1920).
(312)Patnode and Wilcock, J. Am. Chem. Soc., **68**, 358–63 (1946).
(313)Patterson, Ber., **38**, 210–3 (1905).
(314)Pawlewski, Ibid., **16**, 2633–6 (1883).
(315)Pearce and Peters, J. Phys. Chem., **33**, 871–8 (1929).
(316)Perkin, J. Chem. Soc., **55**, 549–51 (1881).
(317)Perry, J. Phys. Chem., **31**, 1737–41 (1927).
(318)Perry and Bardwell, J. Am. Chem. Soc., **47**, 2629–32 (1925).
(319)Pesce and Evdokimoff, Gazz. chim. ital., **70**, 712–6 (1940).
(320)Pickett and Peterson, Ind. Eng. Chem., **21**, 325–6 (1929).
(321)Piloty, Ber., **31**, 454–7 (1898).
(322)Pinette, Ann. **243**, 32–63 (1888).
(323)Plank and Hsia, Z. ges. Kälte-Ind., **38**, 97 (1931).
(324)Plank and Vahl, Forch. Gebiete Ingenieurw., Ausgabe **2**, 11–8 (1931).
(325)Power and Kleber, Arch. Pharm. **232**, 634–48 (1894).
(326)Ibid., 649–59 (1894).
(327)Power and Kleber, Pharm. Rev., **13**, 60 (1895).
(328)Power and Lees, J. Chem. Soc., **81**, 1585–94 (1902).
(329)Price, Ibid , **107**, 188–98 (1915).
(330)Purcell and Zahoorbux, Ibid., **140**, 1029–35 (1937).
(331)Ramsay and Shields, Z. Physik, Chem., **12**, 433–76 (1893).
(332)Ramsay and Young, Ber., **19**, 2107–14 (1886).
(333)Ramsay and Young, J. Chem. Soc. **47**, 42–5 (1885).
(334)Ibid., **47**, 640–57 (1885).
(335)Ibid., **49**, 790–812 (1886).
(336)Ramsay and Young, Phil. Trans., **A175**, 37 (1884).
(337)Ibid., **1771**, 123 (1886).
(338)Ibid., **A178**, 57 (1887).
(339)Ibid., **A178**, 313 (1887).
(340)Ibid., **180**, 137 (1889).
(341)Ramsay and Young, Z. physik. Chem., **1**, 237–57 (1887).
(342)Ratchford, Rehberg, and Fisher, J. Am. Chem. Soc., **66**, 1864–7 (1944).
(343)Regnault, Mém. Paris, **26**, 339 (1862).
(344)Ibid., **26**, 375 (1862).
(345)Ibid., **26**, 403 (1862).
(346)Ibid., **26**, 440 (1862).
(347)Ibid., **26**, 448 (1862).
(348)Ibid., **26**, 455 (1862).
(349)Ibid., **26**, 462 (1862).
(350)Ibid., **26**, 535 (1862).
(351)Regnault, Rel. des exp. **2**, 440–62.
(352)Rex, Z. physik. Chem. **55**, 355–70 (1916).
(353)Richardson, J. Chem. Soc., **49**, 761–76 (1886).
(354)Riley and Gray, "Organic Syntheses," Vol. **XV**, pp. 67–9, New York, John Wiley & Sons, Inc., 1935.
(355)Rinkenbach, Ind. Eng. Chem., **19**, 474–6 (1927).
(356)Robinson "Elements of Fractional Distillation", 1st ed., New York, McGraw-Hill Book Co., Inc., 1922.
(357)Rochow and Dennis, J. Am. Chem. Soc. **57**, 486–7 (1935).
(358)Rolla, Atti accad Lincei, **18**, 365–73 (1909).
(359)Rossini, Am. Petroleum Inst. Research Project 44; Willingham et al., Ibid.
(360)Rousset, Bull. soc. chim. Mém., **III**, **15**, 58–72 (1896).
(361)Rubin, Levedahl, and Yost, J. Am. Chem. Soc., **66**, 279–82 (1944).
(362)Rupe and Schlochoff, Ber., **38**, 1719–25 (1905).
(363)Sajotschewsky, Beibl. Ann. Physik., **3**, 741 (1879).
(364)Sameshima, J. Am. Chem. Soc., **40**, 1482–508 (1918).
(365)Scatchard and Raymond, Ibid., **60**, 1278–87 (1938).
(366)Scatchard, Wood, and Mochel, Ibid., **61**, 3206–10 (1939).
(367)Scatchard, Wood, and Mochel, J. Phys. Chem., **43**, 119–130 (1939).
(368)Schierholtz and Staples, J. Am. Chem. Soc., **57**, 2709–11 (1935).
(369)Schiff, Ann., **220**, 71–113 (1883).
(370)Schimmel and Co., bulletins.
(371)Schlesinger and Walker, J. Am. Chem. Soc., **57**, 621–5 (1935).
(372)Schlumberger, J. Gasbeleucht., **55**, 1257–60 (1912).
(373)Schmidt, Z. physik. Chem., **7**, 433–67 (1891).
(374)Ibid., **8**, 628–46 (1891).
(375)Ibid., **58**, 513–40 (1907).
(376)Schuette and Cowley, J. Am. Chem. Soc., **53**, 3485–9 (1931).
(377)Schuette and Thomas, Ibid., **52**, 2028–30 (1930).
(378)Schütze and Widenmann, Helv. Chim. Acta, **20**, 936–49 (1937).
(379)Schultz, Ann., **174**, 201–35 (1874).
(380)Schumann, Ann. Physik., **12**, 40–65 (1881).
(381)Schweitzer, Ann., **264**, 193–6 (1891).
(382)Seibert and Burrell, J. Am. Chem. Soc., **37**, 2683–94 (1915).
(383)Semmler, "Die Ätherischen Öle", Vol. **I–IV**, Leipzig, Veit & Co., 1906–7.
(384)Semmler, Ber., **23**, 1803–10 (1890).
(385)Ibid., **25**, 3352–4 (1892).
(386)Ibid., **26**, 2254–8 (1893).
(387)Ibid., **34**, 3122–30 (1901).
(388)Senseman and Nelson, Ind. Eng. Chem., **15**, 382–3 (1923).
(389)Seyer and Mann, J. Am. Chem. Soc., **67**, 328–9 (1945).
(390)Simon and Huter, Z. Elektrochem., **41**, 28–33 (1935).
(391)Smith, J. Chem. Soc., **134** (2), 2573–83 (1931).
(392)Smith and Menzies, J. Am. Chem. Soc., **32**, 1448–59 (1910).
(393)Smyth and Seaton, J. Ind. Hyg. Toxicol., **22**, 477–83 (1940).
(394)Smythe and Engel, J. Am Chem. Soc., **51**, 2646–70 (1929).
(395)Soden and Rojahn, Ber. **33**, 1720–4 (1900).
(396)Ibid., **33**, 3063–5 (1900).
(397)Soden and Treff, Chem.-Ztg., **27**, 897 (1903).
(398)Solana and Moles, Anales soc. españ. fís. quím., **30**, 886–917 (1932).

VAPOR PRESSURE—CITATIONS (Continued)
Organic Compounds

(399)Spence and Wild, J. Chem. Soc., **138,** 506–9 (1935).
(400)Speranski, Z. physik. Chem. **46,** 70–8 (1903).
(401)Ssemenow, J. Russ. Phys.-Chem. Soc., **31,** 115–35 (1889).
(402)Staedel, Ber., **15,** 2559–72 (1882).
(403)Stedman, Trans. Faraday Soc., **24,** 289–98 (1928).
(404)Stelzner, Dissertation Erlangen (1901).
(405)Stephan, J. prakt. Chem., **II, 62,** 523–35 (1900).
(406)Stevens, Ind. Eng. Chem., **35,** 655–60 (1943).
(407)Stock, Henning, and Kuss, Ber., **54,** 1119–29 (1921).
(408)Stock and Kuss, Ibid., **50,** 159–64 (1917).
(409)Stock and Praetorius, Ibid., **45,** 3568 (1912).
(410)Stock and Somieski, Ibid., **52,** 695–724 (1919).
(411)Stock and Stoltzenberg, Ibid., **50,** 498–502 (1917).
(412)Stock and Willfroth, Ibid., **47,** 144–54 (1914).
(413)Stock and Zeidler, Ibid., **54,** 531–41 (1921).
(414)Straus and Kollek, Ibid., **59,** 1664–81 (1926).
(415)Stuckey and Saylor, J. Am. Chem. Soc., **62,** 2922–5 (1940).
(416)Stull, Ibid., **59,** 2726–33 (1937).
(417)Swan and Mack, Ibid., **47,** 2112–6 (1925).
(418)Swift and Hochanadel, Ibid., **67,** 880–1 (1945).
(419)Takahashi and Yaginuma, Jap. J. Chem., **4,** 19 (1920).
(420)Tanaka and Nagai, Proc. Imp. Acad. (Tokyo), **5,** 78–9 (1929).
(421)Tannenberger, Ber., **66,** 484–6 (1933).
(422)Taylor and Layng, J. Chem. Phys., **1,** 798–808 (1933).
(423)Taylor and Rinkenbach, J. Am. Chem. Soc., **48,** 1305–9 (1926).
(423a Taylor and Smith, Ibid., **44,** 2450–63 (1922).
(424)Taylor and Walden, Ibid., **66,** 842–3 (1944).
(425)Tchugaeff, Ber., **31,** 360–8 (1898).
(426)Terwen, Z. physik. Chem., **91,** 469–99 (1916).
(427)Thomas and Young, J. Chem. Soc., **67,** 1071–5 (1895).
(428)Thoms, Ber., **36,** 3446–51 (1903).
(429)Thoms and Biltz, Z. allgem. österr. Apoth-Ver., **42,** 943–7 (1904).
(430)Thornton, Burg, and Schlesinger, J. Am. Chem. Soc., **55,** 3177–82 (1933).
(431)Tiemann, Ber., **31,** 2989–92 (1898).
(432)Tiemann and Schmidt, Ibid., **29,** 903–26 (1896).
(433)Ibid., **30,** 33–8 (1897).
(434)Tiemann and Semmler, Ber., **26,** 2708–29 (1893).
(435)Ibid., **31,** 2889–99 (1898).
(436)Tiffeneau, Compt. rend, **139,** 481–6 (1904).
(437)Timmermans, Bull. soc. chim. Belg., **36,** 502–8 (1927).
(438)Timmermans and Martin, J. Chim. phys., **23,** 733–46 (1926).
(439)Ibid., **25,** 411–51 (1928).
(440)Treon, Crutchfield, and Kitzmiller, J. Ind. Hyg. Toxicol., **25,** 199–214 (1943).
(441)Tschugaeff, Ber., **33,** 735–6 (1900).
(442)Ultee, Ibid., **39,** 1856–8 (1906).
(443)Van der Meulen and Mann, J. Am. Chem. Soc., **53,** 451–3 (1931).
(444)Vanstone, J. Chem. Soc., **97,** 429–43 (1910).
(445)Vaughan, J. Am. Chem. Soc., **54,** 3863–76 (1932).
(446)Venable, Ber., **13** (2), 1649–52 (1880).
(447)Verschoyle, Trans. Roy. Soc. (London), **A230,** 189 (1931).
(448)Villard, Ann. chim. phys., [7] **10,** 387 (1897).
(449)Villard, Compt. rend., **120,** 1262–5 (1895).
(450)Villiers, Bull. soc. chim. **34,** 25–7 (1880).
(451)Vincent and Chappuis, Compt. rend., **100,** 1216–8 (1885).
(452)Volmer and Kirchoff, Z. physik. Chem., **115,** 233–8 (1925).
(453)Von Rechenberg, "Destillation", Leipzig, 1923.
(454)Von Siemens, Ann. Physik., [4] 42, 871–88 (1913).
(455)Wade and Merriman, J. Chem. Soc., **101,** 2438–43 (1912).
(456)Walbaum, Ber., **33,** 2299–302 (1900).
(457)Walbaum, J. prakt. Chem., **II, 73,** 488–93 (1906).
(458)Walbaum and Stephan, Ber., **33,** 2302–8 (1900).
(459)Walden, Z. physik. Chem., **70,** 569–619 (1910).
(460)Wallach, Ann. **275,** 171 (1893).
(461)Ibid., **336,** 1–46 (1904).
(462)Wallach et al., Ibid., **279,** 366–97 (1894).
(463)Weber, Verslag Akad. Wetenschappen, **22,** 380 (1914).
(464)Wenzel and Pirak, Collection Czechoslov. Chem. Commun., **6,** 54–9 (1934).
(465)Whitmore and Fleming, J. Am. Chem. Soc., **55,** 3803–6 (1933).
(466)Whitmore et al., Ibid., **68,** 475–81 (1946).
(467)Wiberg and Sütterlin, Z. Eleketrochem., **41,** 151–3 (1935).
(468)Wilbrand and Beilstein, Z. Chemie, **7,** 153–60 (1864).
(469)Wilcock, J. Am. Chem. Soc., **68,** 691–6 (1946).
(470)Williams, Ibid., **47,** 2644–52 (1925).
(471)Willingham et al., J. Research Natl. Bur. Standards, **35,** 219 (1945).
(472)Wislicenus, Ann., **275,** 309–82 (1893).
(473)Wittenberg and Meyer, Ber., **16,** 500–8 (1883).
(474)Woringer, Z. physik. Chem., **34,** 257–89 (1900).
(475)Wüllener and Grotrian, Wied Ann., **11,** 545–604 (1880).
(476)Yost and Stone, J. Am. Chem. Soc., **55,** 1889–95 (1933).
(477)Young, J. Chem. Soc. **55,** 483–5 (1889).
(478)Ibid., **55,** 486–521 (1889).
(479)Ibid., **59,** 621–6 (1891).
(480)Ibid., **59,** 903–11 (1891).
(481)Ibid., **59,** 911–36 (1891).
(482)Ibid., **71,** 446–57 (1897).
(483)Ibid., **73,** 675–81 (1898).
(483a Ibid., **77,** 1145–51 (1900).
(484)Young, Proc. Roy. Irish Acad., **38B,** 65–92 (1928).
(485)Young, Sci. Proc. Roy. Dublin Soc., **12,** 374–432 (1910).
(486)Young, Z. physik. Chem., **29,** 193–241 (1899).
(487)Young and Fortey, J. Chem. Soc., **75,** 873–83 (1899).
(488)Ibid., **77,** 1126–51 (1900).
(489)Ibid., **81,** 783–6 (1902).
(490)Young and Nelson, Ind. Eng. Chem., **21,** 321–2 (1929).
(491)Young and Thomas, J. Chem. Soc., **63,** 1191–262 (1893).
(492)Ibid., **71,** 440–6 (1897).
(493)Zeleny and Smith, Physik. Z, **7,** 667–71 (1906).
(494)Zincke, Ann. **152,** 1–21 (1869).
(495)American Petroleum Institute Research Project 44, Nat. Bur. Standards, "Selected Values of Properties of Hydrocarbons."
(496)Boord, et al, Ind. Eng. Chem., to be published.
(497)Deanesly and Carleton, J. Phys. Chem., **45,** 1104–23 (1941).
(498)Gibbons and co-workers, J. Am. Chem. Soc. **68,** 1130–1 (1946).
(499)Gilman and Meals, J. Org. Chem., **8,** 126–46 (1943).
(500)Heise and Töhl, Ann., **270,** 155–71 (1892).
(501)Karabinos and co-workers, J Am. Chem. Soc., **68,** 2107–8 (1946).
(502)Mair and Streiff, J. Research Natl. Bur. Standards, **24,** 395–414 (1940).
(503)O'Connor and Sowa, J. Am. Chem. Soc., **60,** 125–7 (1938).
(504)Shephard and co-workers, Ibid., **53,** 1948–58 (1931).
(505)Smith and Guss, Ibid., **62,** 2630–1 (1940).

VAPOR PRESSURE
Variation with Temperature

The following table gives the value of the constants *a* and *b* in the following equation:

$$\log_{10} p = -\frac{0.05223a}{T} + b$$

where *p* is the pressure in mm of mercury of the saturated vapor at the absolute temperature *T*. ($T = t°C + 273.1$). The values obtained by the use of the equation given above are valid within the temperature ranges indicated for each of the compounds.

Elements and Inorganic Compounds

Compound	Formula	Temp. range °C	*a*	*b*
Aluminum oxide	Al_2O_3	1840 to 2200 liq.	540,000	14.22
Ammonia	NH_3	−127 to −78 sol.	31,211	9.9974
Ammonium bromide	NH_4Br	250 to 400 sol.	90,208	9.9404
Ammonium chloride	NH_4Cl	100 to 400 sol.	83,486	10.0164
Ammonium cyanide	NH_4CN	7 to 17 sol.	41,484	9.978
Ammonium iodide	NH_4I	300 to 400 sol.	95,730	10.2700
Ammonium sulfhydrate	NH_4HS	6 to 40 sol.	46,025	10.7500
Antimony	Sb	1070 to 1325 liq.	189,000	9.051
Argon	A	−208 to −189 sol.	7,814.5	7.5741
		−189 to −183 liq.	6,826.0	6.9605
Arsenic	As	800 to 860 liq.	47,100	6.692
		440 to 815 sol.	133,000	10.800
Arsenous oxide	As_2O_3	100 to 310 sol.	111,350	12.127
		315 to 490 liq.	52,120	6.513
Barium	Ba	930 to 1130 liq.	350,000	15.765
Bismuth	Bi	1210 to 1420 liq.	200,000	8.876
Bismuth trichloride	$BiCl_3$	91 to 213 sol.	13,125	2.681
Cadmium	Cd	150 to 320.9 sol.	109,000	8.564
		500 to 840 liq.	99,900	7.897
Cadmium iodide	CdI_2	385 to 450 liq.	122,200	9.269
Cesium	Cs	200 to 350 liq	73,400	6.949
Cesium chloride	CsCl	986 to 1295 liq	163,200	8.340

VAPOR PRESSURE (Continued)

Compound	Formula	Temp. range °C	*a*	*b*
Calcium	Ca	960 to 1110 liq.	370,000	16.240
Carbon	C	3880 to 4430 liq.	540,000*	9.596*
Carbon dioxide	CO_2	−135 to −56.7 sol.	26,179.3	9.9082
Carbon monoxide	CO	−220 to −206 liq.	6,354	6.976
Chlorine	Cl	−154 to −103 sol.	29,293	9.950
Cobalt	Co	2375 liq.	309,000	7.571
Copper	Cu	2100 to 2310 liq.	468,000	12.344
Cuprous chloride	Cu_2Cl_2	878 to 1369 liq.	80,700	5.454
Cyanogen	$(CN)_2$	−72 to −28 sol.	32,437	9.6539
		−32 to −6 liq.	23,750	7.808
Ferrous chloride	$FeCl_2$	700 to 930 sol.	135,200	8.33
Gold	Au	2315 to 2500 liq.	385,000	9.853
Hydriodic acid	HI	−97 to −51 sol.	24,160	8.259
		−50 to −34 liq.	21,580	7.630
Hydrobromic acid	HBr	−114 to −86 sol.	22,420	8.734
		−86 to −66 liq.	17,960	7.427
Hydrochloric acid	HCl	−158 to −110 sol.	19,588	8.4430
Hydrocyanic acid	HCN	−8 to +27 liq.	27,830	7.7446
Hydrofluoric acid	HF	−83 to +48 liq.	25,180	7.370
Hydrogen peroxide	H_2O_2	10 to 90 liq.	48,530	8.853
Hydrogen sulfide	H_2S	−110 to −83 sol.	20,690	7.880
Iron	Fe	2220 to 2450 liq.	309,000	7.482
Krypton	Kr	−189 to −169 sol.	10,065	7.1770
		−169 to −150 liq.	9,377.0	6.92387
Lead	Pb	525 to 1325 liq.	188,500	7.827
Lead bromide	$PbBr_2$	735 to 918 liq.	118,000	8.064
Lead chloride	$PbCl_2$	500 to 950 liq.	141,900	8.961
Lithium bromide	LiBr	1010 to 1265 liq.	152,700	8.068
Lithium chloride	LiCl	1045 to 1325 liq.	155,900	7.939
Lithium fluoride	LiF	1398 to 1666 liq.	218,400	8.753
Lithium iodide	LiI	940 to 1140 liq.	143,600	8.011
Magnesium	Mg	900 to 1070 liq.	260,000	12.993
Manganese	Mn	1510 to 1900 liq.	267,000	9.300
Mercuric bromide	$HgBr_2$	111 to 235 sol.	79,800	10.181
		238 to 331 liq.	61,250	8.284
Mercuric chloride	$HgCl_2$	60 to 130 sol.	85,030	10.888
		130 to 270 sol.	78,850	10.094
		275 to 309 liq.	61,020	8.409
Mercuric iodide	HgI_2	100 to 250 sol.	82,340	10.057
		266 to 360 liq.	62,770	8.115
Mercury	Hg	−80 to −38.87 sol.	73,000	10.383
		400 to 1300 liq.	58,700	7.752
Molybdenum	Mo	1800 to 2240 sol.	680,000	10.844
Nitrogen	N_2	−215 to −210 sol.	6,881.3	7.66558
Nitrogen dioxide	NO	−200 to −161 sol.	16,423	10.048
		−163.7 to −148 liq.	13,040	8.440
Nitrogen monoxide	N_2O	−144 to −90 sol.	23,590	9.579
		−90.1 to −88.7 liq.	16,440	7.535
Nitrogen pentoxide	N_2O_5	−30 to +30 sol.	57,180	12.647
Nitrogen tetroxide	N_2O_4	−100 to −40 sol.	55,160	13.400
		−40 to −10 sol.	45,440	11.214
		−8 to +43.2 liq.	33,430	8.814
Nitrogen trioxide	N_2O_3	−25 to 0 liq.	39,400	10.30
Phosphorus (white)	P	20 to 44.1 sol.	63,123	9.6511
Phosphorus (violet)	P	380 to 590 sol.	108,510	11.0842
Platinum	Pt	1425 to 1765 sol.	486,000	7.786
Potassium	K	260 to 760 liq.	84,900	7.183
Potassium bromide	KBr	906 to 1063 liq.	168,100	8.2470
		1095 to 1375 liq.	163,800	7.936
Potassium chloride	KCl	906 to 1105 liq.	174,500	8.3526
		1116 to 1418 liq.	169,700	8.130
Potassium fluoride	KF	1278 to 1500 liq.	207,500	9.000
Potassium hydroxide	KOH	1170 to 1327 liq.	136,000	7.330
Potassium iodide	KI	843 to 1028 liq.	157,600	8.0957
		1063 to 1333 liq.	155,700	7.949
Rubidium	Rb	250 to 370 liq.	76,000	6.976
Rubidium chloride	RbCl	1142 to 1395 liq.	198,600	9.111
Silicon	Si	1200 to 1320 sol.	170,000	5.950
Silicon dioxide	SiO_2	1860 to 2230 liq.	506,000	13.43
Silver	Ag	1650 to 1950 liq.	250,000	8.762
Silver chloride	AgCl	1255 to 1442 liq.	185,500	8.179
Sodium	Na	180 to 883 liq.	103,300	7.553
Sodium bromide	NaBr	1138 to 1394 liq.	161,600	7.948
Sodium chloride	NaCl	976 to 1155 liq.	180,300	8.3297
		1156 to 1430 liq.	185,800	8.548
Sodium cyanide	NaCN	800 to 1360 liq.	155,520	7.472
Sodium fluoride	NaF	1562 to 1701 liq.	218,200	8.640
Sodium hydroxide	NaOH	1010 to 1402 liq.	132,000	7.030
Sodium iodide	NaI	1063 to 1307 liq.	165,100	8.371
Stannic chloride	$SnCl_4$	−52 to −38 sol.	46,740	9.824
Strontium	Sr	940 to 1140 liq.	360,000	16.056
Sulfur dioxide	SO_2	−95 to −75 sol.	35,827	10.5916
Sulfur trioxide	SO_3	24 to 48 liq.	43,450	10.022
Thallium	Tl	950 to 1200 liq.	120,000	6.140
Thallium chloride	TlCl	665 to 807 liq.	105,200	7.974
Tin	Sn	1950 to 2270 liq.	328,000	9.643
Tungsten	W	2230 to 2770 sol.	897,000	9.920
Zinc	Zn	250 to 419.4 sol.	133,000	9.200
		600 to 985 liq.	118,000	8.108

* Based on boiling point of 3927° C or 4200° absolute.

Organic Compounds

Compound	Formula	Temp. range °C	*a*	*b*
Acetaldehyde	C_2H_4O	−24.3 to +27.5 liq.	27,707	7.8206
Acetic acid	$C_2H_4O_2$	−35 to 10 sol.	41,689	8.502
Acetic anhydride	$C_4H_6O_3$	100 to 140 liq.	45,585	8.688

VAPOR PRESSURE (Continued)

Compound	Formula	Temp. range °C	a	b
Acetylene	C_2H_2	−140 to −82 sol.	21,914	8.933
Aniline	C_6H_7N	145 to185 liq.	45,951.6	8.1278
Anthracene	$C_{14}H_{10}$	100 to 600 sol.	70,390	8.706
		100 to 160 liq.	72,000	8.91
		223 to 342 liq.	59,219	7.910
Anthraquinone	$C_{14}H_8O_2$	224 to 286 sol.	110,040	12.305
Benzene	C_6H_6	−58 to −30 sol.	42,904	9.556
		−30 to +5 sol.	44,222	9.846
		0 to 42 liq.	34,172	7.9622
		42 to 100 liq.	32,295	7.6546
Benzoic acid	$C_7H_6O_2$	60 to 110 sol.	63,820	9.033
Benzophenone	$C_{13}H_{10}O$	260 to 308 liq.	58,221	8.137
Benzoyl chloride	C_7H_5ClO	140 to 200 liq.	45,416	7.9245
Benzyl alcohol	C_7H_8O	100 to 135 liq.	59,491	9.5152
		135 to 205 liq.	53,118	8.6977
Butane	C_4H_{10}	−100 to +12 liq.	23,450	7.395
iso-Butane	C_4H_{10}	−115 to −34 liq.	21,273	7.25
n-Butyl alcohol	$C_4H_{10}O$	75 to 117.5 liq.	46,774	9.1362
Butyric acid	$C_4H_8O_2$	80 to 165 liq.	51,103	9.010
Bromobenzene	C_6H_5Br	−26 to −15 liq.	42,500	8.075
p-Bromochlorobenzene	C_6H_4BrCl	23 to 63 sol.	69,755	11.629
Camphor	$C_{10}H_{16}O$	0 to 180 sol.	53,559	8.799
Carbon tetrachloride	CCl_4	−70 to −50 sol.	34,608	8.05
		−19 to +20 liq.	33,914	8.004
Chlorobenzene	C_6H_5Cl	−35 to −15 liq.	42,250	8.500
Cyclohexane	C_6H_{12}	−5 to +5 sol.	37,394	8.594
p-Dichlorobenzene	$C_6H_4Cl_2$	30 to 50 sol.	72,218	12.480
Dichloroethane-1,1	$C_2H_4Cl_2$	0 to 30 liq.	31,706	7.909
Dichloroethane-1,2	$C_2H_4Cl_2$	0 to 30 liq.	35,598	8.126
Diphenylamine	$C_{12}H_{11}N$	278 to 284 liq.	57,350	8.088
Ethyl chloride	C_2H_5Cl	−30 to +30 liq.	26,319	7.691
Ethylene	C_2H_4	−160 to −104 liq.	14,396	7.330
Ethylene bromide	$C_2H_4Br_2$	10 to 150 liq.	38,082	7.792
Heptane	C_7H_{16}	−63 to −40 liq.	37,358	8.2585
Hexane	C_6H_{14}	−10 to +90 liq.	31,679	7.724
Iodobenzene	C_6H_5I	−30 to +18 liq.	43,000	7.500
Methane	CH_4	−194 to −184 sol.	9,896.2	7.6509
		−174 to −163 liq.	8,516.9	6.8626
Methyl alcohol	CH_4O	−62 to −44 liq.	39,234	8.9547
		−10 to +80 liq.	38,324	8.8017
Methyl chloride	CH_3Cl	−47 to −10 liq.	21,988	7.481
Methyl ether	C_2H_6O	−70 to −20 liq.	23,025	7.720
Methyl fluoride	CH_3F	−102 to −76 liq.	17,053	7.445
Methyl salicylate	$C_8H_8O_3$	175 to 215 liq.	48,670	8.008
Naphthalene	$C_{10}H_8$	0 to 80 sol.	71,401	11.450
		120 to 200 liq.	47,362	7.927
o-Nitroaniline	$C_6H_6N_2O_2$	150 to 260 liq.	63,881	8.8684
m-Nitroaniline	$C_6H_6N_2O_2$	170 to 260 liq.	65,880	8.8188
p-Nitroaniline	$C_6H_6N_2O_2$	190 to 260 liq.	77,345	9.5595
Nitrobenzene	$C_6H_5NO_2$	112 to 209 liq.	48,955	8.192
Nitromethane	CH_3NO_2	47 to 100 liq.	36,914	8.033
Oxalic acid	$C_2H_2O_4$	55 to 105 sol.	90,502.6	12.2229
n-Pentane	C_5H_{12}	−20 to +50 liq.	27,691	7.558
Phenol	C_6H_6O	116 to 180 liq.	49,644	8.587
Phthalic anhydride	$C_8H_4O_3$	160 to 285 liq.	54,920	8.022
Propane	C_3H_8	−136 to −40 liq.	19,037	7.217
Propionic acid	$C_3H_6O_2$	20 to 140 liq.	46,150	8.715
n-Propyl alcohol	C_3H_8O	−45 to −10 liq.	47,274	9.5180
Propyl bromide	C_3H_7Br	0 to 30 liq.	32,430	7.821
Propyl chloride	C_3H_7Cl	0 to 50 liq.	28,894	7.593
Propylene	C_3H_6	−95 to −48 liq.	19,693	7.4463
Quinoline	C_9H_7N	180 to 240 liq.	49,720	7.969
Tetrachloroethane-1, 1, 1, 2	$C_2H_2Cl_4$	105 to 145 liq.	36,508	7.605
Tetrachloroethane-1, 1, 2, 2	$C_2H_2Cl_4$	26 to 145 liq.	39,729	7.846
Toluene	C_7H_8	−92 to +15 liq.	39,198	8.330

VAPOR PRESSURE OF NITRIC ACID

Temperature °C	Vapor Pressure, mm. of Hg	
	100% HNO_3	90% of HNO_3
0	14.4	5.5
10	26.6	11.
20	47.9	20.
30	81.3	37.3
40	133.	64.4
50	208.	107.
70	467.	242.
80	670.	352.
90	937.	504.
100	1282.	710.

VAPOR PRESSURE OF THE ELEMENTS

Rudolf Loebel

This table lists the temperature in degrees Celsius at which an element has a vapor pressure indicated by the headings of the columns. For pressures of one atmosphere and lower, the pressures are given in millimeters of mercury; for pressures above one atmosphere, the pressures are given in atmospheres.

Element		mm Hg					atm.				
		1	10	100	400	760	2	5	10	20	40
Aluminum	Al	1540	1780	2080	2320	2467	2610	2850	3050	3270	3530
Antimony	Sb		960	1280	1570	1750	1960	2490			
Arsenic	As	380	440	510	580	610					
Barium	Ba	860	1050	1300	1520	1640	1790	2030	2230		
Beryllium	Be	1520	1860	2300	2770	2970	3240	3730	4110	4720	5610
Bismuth	Bi		1060	1280	1450	1560	1660	1850	2000	2180	
Boron	B	2660	3030	3460	3810	4000					
Bromine	Br	−60	−30	+9	39	59	78	110			
Cadmium	Cd	393	486	610	710	765	830	930	1030	1120	1240
Calcium	Ca	800	970	1200	1390	1490	1630	1850	2020	2290	
Cesium	Cs		373	513	624	690					
Chlorine	Cl	−123	−101	−71	−46	−34	−17	+9	30	55	97
Chromium	Cr	1610	1840	2140	2360	2480	2630	2850	3010	3180	
Cobalt	Co	1910	2170	2500	2760	2870	3040	3270			
Copper	Cu		1870	2190	2440	2600	2760	3010	3500	3460	3740
Fluorine	F			−203	−193	−188	−180.7	−169.1	−159.6		
Gallium	Ga	1350	1570	1850	2060	2180	2320	2560	2730		
Germanium	Ge		2080	2440	2710	2830	2970	3200	3430		
Gold	Au	1880	2160	2520	2800	2940	3120	3490	3630	3890	
Indium	In				1960	2080	2230	2440	2600		
Iridium	Ir	2830	3170	3630	3960	4130	4310	4650			
Iron	Fe	1780	2040	2370	2620	2750	2900	3150	3360	3570	
Iodine	I	40	72	115	160	185	216	265			
Lanthanum	La				3230	3420	3620	3960	4270		
Lead	Pb	970	1160	1420	1630	1740	1880	2140	2320	2620	
Lithium	Li	750	890	1080	1240	1310	1420	1518			
Magnesium	Mg	620	740	900	1040	1110	1190	1330	1430	1560	
Manganese	Mn		1510	1810	2050	2100	2360	2580	2850		
Mercury	Hg			260	330	356.9	398	465	517	581	657
Molybdenum	Mo	3300	3770	4200	4580	4830	5050	5340	5680	5980	
Neodymium	Nd				2870	3100	3300	3680	3990		
Nickel	Ni	1800	2090	2370	2620	2730	2880	3120	3300	3310	
Palladium	Pd	1470	2290	2670	2950	3140	3270	3560	3840		
Phosphorus	P		127	199	253	283	319				
Platinum	Pt	2600	2940	3360	3650	3830	4000	4310	4570	4860	
Polonium	Po	472	587	752	890	960	1060	1200	1340		
Potassium	K			590	710	770	850	950	1110	1240	1420
Rhodium	Rh	2530	2850	3260	3590	3760	3930	4230	4440		
Rubidium	Rb		390	527	640	700					
Selenium	Se		429	547	640	685	750	850	920	1010	1120
Silver	Ag	1310	1540	1850	2060	2210	2360	2600	2850	3050	3300
Sodium	Na	440	546	700	830	890	980	1120	1230	1370	
Strontium	Sr	740	900	1100	1280	1380	1480	1670	1850	2030	
Sulfur	S		246	333	407	445	493	574	640	720	
Tellurium	Te	520	633	792	900	962	1030	1160	1250		
Thallium	Tl		1000	1210	1370	1470	1560	1750	1900	2050	2260
Tin	Sn	1610	1890	2270	2580	2750	2950	3270	3540	3890	
Titanium	Ti	2180	2480	2860	3100	3260	3400	3650	3800		
Tungsten	W	3980	4490	5160	5470	5940	6260	6670	7250	7670	
Uranium	U	2450	2800	3270	3620	3800	4040	4420			
Vanadium	V	2290	2570	2950	3220	3380	3540	3800			
Zinc	Zn		590	730	840	907	970	1090	1180	1290	

CONCENTRATIVE PROPERTIES OF AQUEOUS SOLUTIONS: CONVERSION TABLES

A. V. Wolf and Morden G. Brown

The table columns are:

A% = anhydrous compound weight percent, e.g., grams solute per 100 grams solution.
H% = hydrated compound weight percent.
D_{20}^{20} = specific gravity of solution at 20°C.
C_s = anhydrous solute concentration, grams/liter.
M = molar concentration, gram-mols/liter.
C_w = total water concentration, grams/liter.
$C_0 - C_w$ = water displaced by anhydrous solute, grams/liter.
$(n - n_0) \times 10^4$ = index of refraction increment above index of refraction of pure water × 10^4; "refraction."
n = index of refraction at 20°C. relative to air for sodium yellow light.
Δ = freezing point depression, °C.
S = osmosity, molar concentration of NaCl with same freezing point or osmotic pressure as solution.

Some related measures are:

Density, kg/1 = $.99823 \times D_{20}^{20} = D_4^{20}$
$(C_s + C_w)/1000$ = Density, kg/1
Molality, g-mol/(kg water), = $1000 \times M/C_w$
Osmolality = $\Delta/1.86$
Concentration relative to water, g/(kg water) = $1000 \times C_s/C_w$

Relative specific refractivity = contribution of unit weight of solute relative to unit weight of water to the imaginary concentration of water with same refractive index as solution, equivalent concentration of water = $3694.1788 \times n - 300 \times n^2 - 3394.1788$. The refractive index for this equation is relative to a vacuum whereas the tabulated values are relative to air.

The above equation for the equivalent concentration of water was found to provide better constancy of relative specific refractivity than the equation from the Lorentz and Lorenz relation, ECW = $4848.3431 - 14545.029/(n^2 + 2)$.

References of data sources at end of each table are respectively for specific gravity, index of refraction and freezing point depression. Letters indicate:

A, original data by A. V. Wolf.
H, The Chemical Rubber Co. Handbook of Chemistry and Physics, 44th Edition (1961).
C, International Critical Tables (1926–1930).
W, B. Wagner's *Tabellen zum Eintauchrefraktometer*, Private Publication (1928).
L, Lange's Handbook of Chemistry (1961).
S, Scatchard et al., papers in J.A.C.S.

CONCENTRATIVE PROPERTIES OF AQUEOUS SOLUTIONS (Continued)

1 ACETIC ACID, CH_3COOH

MOLECULAR WEIGHT = 60.05

RELATIVE SPECIFIC REFRACTIVITY = 1.068

A % by wt.	D^{20}_{20}	C_s g/l	M g-mol/l	C_w g/l	$(C_o - C_w)$ g/l	$(n - n_o) \times 10^4$	n	Δ °C	S g-mol/l
.00	1.0000	.0	.000	998.2	.0	0	1.3330	.00	.000
1.00	1.0015	10.0	.166	989.7	8.5	7	1.3337	.31	.091
2.00	1.0029	20.0	.333	981.1	17.1	15	1.3345	.62	.181
3.00	1.0044	30.1	.501	972.5	25.7	22	1.3352	.93	.272
4.00	1.0058	40.2	.669	963.9	34.4	30	1.3359	1.25	.364
5.00	1.0072	50.3	.837	955.2	43.0	37	1.3367	1.56	.457
6.00	1.0087	60.4	1.006	946.5	51.7	44	1.3374	1.88	.551
7.00	1.0101	70.6	1.175	937.8	60.5	52	1.3382	2.21	.645
8.00	1.0115	80.8	1.345	929.0	69.3	59	1.3389	2.54	.740
9.00	1.0130	91.0	1.516	920.2	78.1	66	1.3396	2.87	.835
10.00	1.0144	101.3	1.686	911.3	86.9	74	1.3403	3.21	.930
11.00	1.0158	111.5	1.857	902.5	95.8	81	1.3411	3.55	1.025
12.00	1.0172	121.8	2.029	893.5	104.7	88	1.3418	3.89	1.121
13.00	1.0186	132.2	2.201	884.6	113.6	95	1.3425	4.24	1.224
14.00	1.0200	142.5	2.374	875.6	122.6	102	1.3432	4.60	1.321
15.00	1.0213	152.9	2.547	866.6	131.6	110	1.3439	4.96	1.418
16.00	1.0227	163.3	2.720	857.5	140.7	117	1.3447	5.32	1.515
17.00	1.0240	173.8	2.894	848.4	149.8	124	1.3454	5.69	1.611
18.00	1.0254	184.2	3.068	839.3	158.9	131	1.3461	6.06	1.708
19.00	1.0267	194.7	3.243	830.1	168.1	138	1.3468	6.43	1.804
20.00	1.0280	205.2	3.418	820.9	177.3	145	1.3475	6.81	1.899
22.00	1.0306	226.3	3.769	802.4	195.8	158	1.3488	7.58	2.090
24.00	1.0331	247.5	4.122	783.8	214.5	172	1.3502	8.37	2.278
26.00	1.0356	268.8	4.476	765.0	233.3	185	1.3515	9.17	2.465
28.00	1.0379	290.1	4.831	746.0	252.2	198	1.3528	9.98	2.649
30.00	1.0403	311.5	5.188	726.9	271.3	211	1.3541	10.81	2.831
32.00	1.0425	333.0	5.546	707.7	290.6	224	1.3554	11.65	3.009
34.00	1.0447	354.6	5.905	688.3	310.0	236	1.3566	12.50	3.183
36.00	1.0468	376.2	6.264	668.8	329.5	249	1.3578	13.35	3.352
38.00	1.0488	397.8	6.625	649.1	349.1	260	1.3590	14.20	3.516
40.00	1.0507	419.6	6.987	629.3	368.9	272	1.3602		
42.00	1.0526	441.3	7.349	609.4	388.8	283	1.3613		
44.00	1.0544	463.1	7.712	589.4	408.8	294	1.3624		
46.00	1.0561	484.9	8.076	569.3	429.0	305	1.3635		
48.00	1.0577	506.8	8.440	549.0	449.2	316	1.3645		
50.00	1.0593	528.7	8.804	528.7	469.5	326	1.3656		
52.00	1.0608	550.6	9.169	508.3	490.0	336	1.3666		
54.00	1.0622	572.6	9.535	487.8	510.5	346	1.3676		
56.00	1.0636	594.6	9.901	467.2	531.1	355	1.3685		
58.00	1.0650	616.6	10.268	446.5	551.7	365	1.3695		
60.00	1.0663	638.6	10.635	425.8	572.5	374	1.3704		

H W C

2 ACETONE, CH_3COCH_3

MOLECULAR WEIGHT = 58.05

RELATIVE SPECIFIC REFRACTIVITY = 1.350

A % by wt.	D^{20}_{20}	C_s g/l	M g-mol/l	C_w g/l	$(C_o - C_w)$ g/l	$(n - n_o) \times 10^4$	n	Δ °C	S g-mol/l
.00	1.0000	.0	.000	998.2	.0	0	1.3330	.00	.000
.50	.9993	5.0	.086	992.5	5.7	4	1.3333	.16	.046
1.00	.9985	10.0	.172	986.8	11.4	7	1.3337	.32	.093
1.50	.9978	14.9	.257	981.1	17.1	11	1.3340	.48	.140
2.00	.9971	19.9	.343	975.5	22.8	14	1.3344	.64	.187
2.50	.9964	24.9	.428	969.8	28.4	18	1.3348	.81	.235
3.00	.9957	29.8	.514	964.1	34.1	21	1.3351	.97	.283
3.50	.9950	34.8	.599	958.5	39.7	25	1.3355	1.13	.331
4.00	.9943	39.7	.684	952.9	45.4	28	1.3358	1.30	.379
4.50	.9937	44.6	.769	947.3	51.0	32	1.3362	1.46	.427
5.00	.9930	49.6	.854	941.7	56.6	36	1.3366	1.62	.475
5.50	.9923	54.5	.939	936.1	62.1	39	1.3369	1.79	.523
6.00	.9917	59.4	1.023	930.5	67.7	43	1.3373	1.95	.571
6.50	.9910	64.3	1.108	925.0	73.3	47	1.3377	2.12	.620
7.00	.9904	69.2	1.192	919.4	78.8	50	1.3380	2.29	.667
7.50	.9897	74.1	1.276	913.9	84.4	54	1.3384	2.45	.715
8.00	.9891	79.0	1.361	908.4	89.9	58	1.3388	2.62	.763
8.50	.9885	83.9	1.445	902.9	95.4	62	1.3392	2.79	.811
9.00	.9879	88.8	1.529	897.4	100.9	65	1.3395	2.95	.858
9.50	.9873	93.6	1.613	891.9	106.3	69	1.3399	3.12	.905
10.00	.9867	98.5	1.697	886.4	111.8	73	1.3403	3.29	.952

A A A

3 AMMONIUM CHLORIDE, NH_4Cl

MOLECULAR WEIGHT = 53.50
RELATIVE SPECIFIC REFRACTIVITY = 1.244

A % by wt.	D^{20}_{20}	C_s g/l	M g-mol/l	C_w g/l	$(C_o - C_w)$ g/l	$(n - n_o) \times 10^4$	n	Δ °C	S g-mol/l
.00	1.0000	.0	.000	998.2	.0	0	1.3330	.00	.000
.50	1.0016	5.0	.093	994.8	3.4	10	1.3340	.32	.092
1.00	1.0031	10.0	.187	991.3	6.9	19	1.3349	.64	.185
1.50	1.0046	15.0	.281	987.8	10.4	29	1.3359	.95	.278
2.00	1.0063	20.1	.376	984.4	13.8	39	1.3369	1.27	.371
2.50	1.0078	25.2	.470	980.9	17.3	49	1.3379	1.59	.465
3.00	1.0094	30.2	.565	977.4	20.9	58	1.3388	1.91	.559
3.50	1.0109	35.3	.660	973.8	24.4	68	1.3398	2.24	.654
4.00	1.0125	40.4	.756	970.3	28.0	78	1.3408	2.57	.749
4.50	1.0140	45.5	.851	966.7	31.6	87	1.3417	2.91	.845
5.00	1.0155	50.7	.947	963.1	35.2	97	1.3427	3.25	.941
5.50	1.0170	55.8	1.044	959.4	38.8	106	1.3436	3.59	1.038
6.00	1.0186	61.0	1.140	955.8	42.5	116	1.3446	3.94	1.135
6.50	1.0201	66.2	1.237	952.1	46.2	126	1.3455	4.30	1.239
7.00	1.0216	71.4	1.334	948.4	49.9	135	1.3465	4.66	1.338
7.50	1.0231	76.6	1.432	944.7	53.6	145	1.3474	5.03	1.437
8.00	1.0245	81.8	1.529	940.9	57.3	154	1.3484	5.40	1.536
8.50	1.0260	87.1	1.627	937.2	61.1	163	1.3493	5.78	1.635
9.00	1.0275	92.3	1.725	933.4	64.9	173	1.3503	6.16	1.734
9.50	1.0290	97.6	1.824	929.6	68.7	182	1.3512	6.55	1.834
10.00	1.0304	102.9	1.923	925.7	72.5	192	1.3522	6.94	1.933
11.00	1.0333	113.5	2.121	918.0	80.2	210	1.3540		
12.00	1.0362	124.1	2.320	910.3	88.0	229	1.3559		
13.00	1.0391	134.8	2.520	902.4	95.8	248	1.3578		
14.00	1.0419	145.6	2.722	894.5	103.8	266	1.3596		
15.00	1.0447	156.4	2.924	886.5	111.8	285	1.3615		
16.00	1.0475	167.3	3.127	878.4	119.8	303	1.3633		
17.00	1.0503	178.2	3.332	870.2	128.0	322	1.3652		
18.00	1.0531	189.2	3.537	862.0	136.2	340	1.3670		
19.00	1.0558	200.3	3.743	853.7	144.5	358	1.3688		
20.00	1.0586	211.3	3.950	845.4	152.9	376	1.3706		
22.00	1.0640	233.7	4.367	828.4	169.8	412	1.3742		
24.00	1.0693	256.2	4.788	811.2	187.0	448	1.3778		

H W (S + C)

4 AMMONIUM HYDROXIDE, NH_4OH

MOLECULAR WEIGHT = 35.05
RELATIVE SPECIFIC REFRACTIVITY = 1.283

NH_4OH %	NH_3 %	D^{20}_{20}	C_s g/l	M g-mol/l	C_w g/l	$(C_o - C_w)$ g/l	$(n - n_o) \times 10^4$	n	Δ °C	S g-mol/l
.00	.00	1.0000	.0	.000	998.2	.0	0	1.3330	.00	.000
1.00	.49	.9979	10.0	.284	986.2	12.1	2	1.3332	.57	.166
2.00	.97	.9957	19.9	.567	974.1	24.1	5	1.3335	1.15	.334
3.00	1.46	.9936	29.8	.849	962.1	36.1	7	1.3337	1.73	.506
4.00	1.94	.9915	39.6	1.130	950.2	48.1	9	1.3339	2.32	.678
5.00	2.43	.9894	49.4	1.409	938.3	59.9	12	1.3342	3.23	.937
6.00	2.92	.9874	59.1	1.687	926.5	71.7	14	1.3344	3.73	1.076
7.00	3.40	.9853	68.9	1.964	914.7	83.5	17	1.3346	4.26	1.228
8.00	3.89	.9833	78.5	2.240	903.3	95.2	19	1.3349	4.81	1.379
9.00	4.37	.9813	88.2	2.515	891.4	106.8	21	1.3351	5.40	1.536
10.00	4.86	.9793	97.8	2.789	879.8	118.4	24	1.3354	6.02	1.698
11.00	5.34	.9773	107.3	3.062	868.3	130.0	27	1.3356	6.67	1.864
12.00	5.83	.9754	116.8	3.333	856.8	141.4	29	1.3359	7.36	2.035
13.00	6.32	.9734	126.3	3.604	845.4	152.9	32	1.3362	8.08	2.211
14.00	6.80	.9715	135.8	3.874	834.0	164.2	34	1.3364	8.85	2.391
15.00	7.29	.9696	145.2	4.142	822.7	175.5	37	1.3367	9.66	2.576
16.00	7.77	.9677	154.6	4.410	811.4	186.8	39	1.3369	10.51	2.765
17.00	8.26	.9658	163.9	4.676	800.2	198.0	42	1.3372	11.41	2.957
18.00	8.75	.9639	173.2	4.942	789.0	209.2	45	1.3375	12.36	3.154
19.00	9.23	.9621	182.5	5.206	777.9	220.3	47	1.3377	13.36	3.354

4 AMMONIUM HYDROXIDE, NH_4OH

NH_4OH %	NH_3 %	D^{20}_{20}	C_s g/l	M g-mol/l	C_w g/l	$(C_o - C_w)$ g/l	$(n - n_o) \times 10^4$	n	Δ °C	S g-mol/l
20.00	9.72	.9603	191.7	5.470	766.9	231.4	50	1.3380	14.41	3.556
22.00	10.69	.9567	210.1	5.994	744.9	253.4	56	1.3386	16.71	3.971
24.00	11.66	.9531	228.3	6.515	723.1	275.2	61	1.3391	19.27	4.394
26.00	12.63	.9496	246.5	7.031	701.4	296.8	67	1.3397	22.13	
28.00	13.60	.9461	264.4	7.545	680.0	318.2	72	1.3402	25.33	
30.00	14.58	.9427	282.3	8.055	658.7	339.5	78	1.3408	28.91	
32.00	15.55	.9393	300.1	8.561	637.6	360.6	84	1.3414	32.92	
34.00	16.52	.9360	317.7	9.064	616.7	381.5	90	1.3420	37.43	
36.00	17.49	.9328	335.2	9.563	595.9	402.3	95	1.3425	42.50	
38.00	18.46	.9295	352.6	10.060	575.3	423.0	101	1.3431	48.22	
40.00	19.44	.9263	369.9	10.553	554.8	443.4	107	1.3437	54.69	
42.00	20.41	.9232	387.0	11.043	534.5	463.7	113	1.3443	62.02	
44.00	21.38	.9200	404.1	11.529	514.3	483.9	119	1.3449	70.35	
46.00	22.35	.9170	421.1	12.013	494.3	504.0	125	1.3455	79.85	
48.00	23.32	.9139	437.9	12.493	474.4	523.8	131	1.3461	90.72	
50.00	24.29	.9109	454.6	12.971	454.6	543.6	137	1.3467		
52.00	25.27	.9079	471.3	13.445	435.0	563.2	143	1.3473		
54.00	26.24	.9049	487.8	13.917	415.5	582.7	149	1.3479		
56.00	27.21	.9019	504.2	14.385	396.2	602.1	154	1.3484		
58.00	28.18	.8990	520.5	14.850	376.9	621.3	160	1.3490		
60.00	29.15	.8961	536.7	15.313	357.8	640.4	166	1.3496		

H W C

5 AMMONIUM SULFATE, $(NH_4)_2SO_4$

MOLECULAR WEIGHT = 132.15

RELATIVE SPECIFIC REFRACTIVITY = .885

A % by wt.	D^{20}_{20}	C_s g/l	M g-mol/l	C_w g/l	$(C_o - C_w)$ g/l	$(n - n_o) \times 10^4$	n	Δ °C	S g-mol/l
.00	1.0000	.0	.000	998.2	.0	0	1.3330	.00	.000
.50	1.0030	5.0	.038	996.2	2.0	8	1.3338	.17	.048
1.00	1.0059	10.0	.076	994.1	4.1	17	1.3346	.33	.094
1.50	1.0089	15.1	.114	992.0	6.2	25	1.3355	.48	.139
2.00	1.0119	20.2	.153	989.9	8.3	33	1.3363	.63	.183
2.50	1.0149	25.3	.192	987.7	10.5	41	1.3371	.78	.226
3.00	1.0178	30.5	.231	985.5	12.7	49	1.3379	.92	.269
3.50	1.0208	35.7	.270	983.3	14.9	58	1.3388	1.07	.312
4.00	1.0238	40.9	.309	981.1	17.2	66	1.3396	1.21	.353
4.50	1.0267	46.1	.349	978.8	19.4	74	1.3404	1.35	.395
5.00	1.0297	51.4	.389	976.5	21.7	82	1.3412	1.49	.436
5.50	1.0327	56.7	.429	974.1	24.1	90	1.3420	1.63	.477
6.00	1.0356	62.0	.469	971.8	26.5	98	1.3428	1.77	.517
6.50	1.0386	67.4	.510	969.4	28.9	107	1.3436	1.91	.558
7.00	1.0416	72.8	.551	966.9	31.3	115	1.3445	2.05	.598
7.50	1.0445	78.2	.592	964.5	33.8	123	1.3453	2.19	.638
8.00	1.0475	83.6	.633	962.0	36.3	131	1.3461	2.32	.679
8.50	1.0504	89.1	.674	959.4	38.8	139	1.3469	2.46	.719
9.00	1.0534	94.6	.716	956.9	41.4	147	1.3477	2.61	.759
9.50	1.0563	100.2	.758	954.3	44.0	155	1.3485	2.75	.800
10.00	1.0593	105.7	.800	951.7	46.6	163	1.3493	2.89	.841
11.00	1.0651	117.0	.885	946.3	51.9	178	1.3508	3.18	.922
12.00	1.0710	128.3	.971	940.8	57.4	194	1.3524	3.47	1.005
13.00	1.0769	139.7	1.058	935.2	63.0	210	1.3540	3.77	1.087
14.00	1.0827	151.3	1.145	929.5	68.8	225	1.3555	4.07	1.176
15.00	1.0885	163.0	1.233	923.6	74.6	241	1.3571		
16.00	1.0943	174.8	1.323	917.6	80.6	256	1.3586		
17.00	1.1001	186.7	1.413	911.5	86.8	271	1.3601		
18.00	1.1059	198.7	1.504	905.2	93.0	286	1.3616		
19.00	1.1116	210.8	1.596	898.8	99.4	301	1.3631		
20.00	1.1174	223.1	1.688	892.3	105.9	316	1.3646		

H W S

6 BARIUM CHLORIDE, $BaCl_2 \cdot 2H_2O$

MOLECULAR WEIGHT = 208.27 FORMULA WEIGHT, HYDRATE = 244.31
RELATIVE SPECIFIC REFRACTIVITY = .555

A % by wt.	H % by wt.	D^{20}_{20}	C_s g/l	M g-mol/l	C_w g/l	$(C_o - C_w)$ g/l	$(n - n_o) \times 10^4$	n	Δ °C	S g-mol/l
.00	.00	1.0000	.0	.000	998.2	.0	0	1.3330	.00	.000
.50	.59	1.0044	5.0	.024	997.6	.6	7	1.3337	.12	.033
1.00	1.17	1.0088	10.1	.048	996.9	1.3	15	1.3345	.23	.066
1.50	1.76	1.0132	15.2	.073	996.3	2.0	23	1.3352	.35	.100
2.00	2.35	1.0177	20.3	.098	995.6	2.6	30	1.3360	.46	.134
2.50	2.93	1.0222	25.5	.122	994.9	3.3	38	1.3368	.58	.167
3.00	3.52	1.0267	30.7	.148	994.2	4.1	45	1.3375	.69	.201
3.50	4.11	1.0313	36.0	.173	993.5	4.8	53	1.3383	.81	.235
4.00	4.69	1.0359	41.4	.199	992.7	5.5	61	1.3391	.93	.270
4.50	5.28	1.0405	46.7	.224	992.0	6.3	68	1.3398	1.05	.306
5.00	5.87	1.0452	52.2	.250	991.2	7.0	76	1.3406	1.18	.343
5.50	6.45	1.0499	57.6	.277	990.4	7.8	84	1.3414	1.30	.381
6.00	7.04	1.0547	63.2	.303	989.6	8.6	92	1.3422	1.43	.419
6.50	7.62	1.0594	68.7	.330	988.8	9.4	100	1.3430	1.57	.458
7.00	8.21	1.0643	74.4	.357	988.0	10.2	108	1.3438	1.70	.497
7.50	8.80	1.0691	80.0	.384	987.2	11.0	116	1.3446	1.84	.538
8.00	9.38	1.0740	85.8	.412	986.4	11.9	124	1.3454	1.98	.579
8.50	9.97	1.0790	91.6	.440	985.5	12.7	132	1.3462	2.12	.621
9.00	10.56	1.0839	97.4	.468	984.7	13.6	140	1.3470	2.27	.663
9.50	11.14	1.0890	103.3	.496	983.8	14.5	148	1.3478	2.42	.706
10.00	11.73	1.0940	109.2	.524	982.9	15.3	157	1.3487	2.57	.750
11.00	12.90	1.1043	121.3	.582	981.1	17.1	173	1.3503	2.89	.841
12.00	14.08	1.1148	133.5	.641	979.3	19.0	190	1.3520	3.22	.934
13.00	15.25	1.1254	146.0	.701	977.4	20.9	208	1.3538	3.56	1.030
14.00	16.42	1.1362	158.8	.762	975.4	22.8	225	1.3555	3.92	1.130
15.00	17.60	1.1472	171.8	.825	973.4	24.8	243	1.3573	4.30	1.239
16.00	18.77	1.1584	185.0	.888	971.4	26.9	261	1.3591	4.69	1.346
17.00	19.94	1.1698	198.5	.953	969.2	29.0	279	1.3609	5.10	1.456
18.00	21.11	1.1814	212.3	1.019	967.1	31.2	297	1.3627	5.53	1.570
19.00	22.29	1.1932	226.3	1.087	964.8	33.4	316	1.3646		
20.00	23.46	1.2052	240.6	1.155	962.5	35.7	335	1.3665		
22.00	25.81	1.2299	270.1	1.297	957.6	40.6	373	1.3703		
24.00	28.15	1.2553	300.7	1.444	952.4	45.9	412	1.3742		
26.00	30.50	1.2816	332.6	1.597	946.7	51.6	452	1.3782		

H W C

7 CALCIUM CHLORIDE, $CaCl_2 \cdot 2H_2O$

MOLECULAR WEIGHT = 110.99 FORMULA WEIGHT, HYDRATE = 147.03
RELATIVE SPECIFIC REFRACTIVITY = .865

A % by wt.	H % by wt.	D^{20}_{20}	C_s g/l	M g-mol/l	C_w g/l	$(C_o - C_w)$ g/l	$(n - n_o) \times 10^4$	n	Δ °C	S g-mol/l
.00	.00	1.0000	.0	.000	998.2	.0	0	1.3330	.00	.000
.50	.66	1.0041	5.0	.045	997.3	.9	12	1.3342	.22	.063
1.00	1.32	1.0083	10.1	.091	996.4	1.8	24	1.3354	.44	.127
1.50	1.99	1.0124	15.2	.137	995.5	2.7	36	1.3366	.66	.192
2.00	2.65	1.0166	20.3	.183	994.5	3.7	48	1.3378	.88	.257
2.50	3.31	1.0208	25.5	.230	993.5	4.7	60	1.3390	1.10	.320
3.00	3.97	1.0250	30.7	.277	992.5	5.8	72	1.3402	1.33	.387
3.50	4.64	1.0292	36.0	.324	991.4	6.8	84	1.3414	1.57	.458
4.00	5.30	1.0334	41.3	.372	990.3	7.9	96	1.3426	1.82	.531
4.50	5.96	1.0376	46.6	.420	989.2	9.0	108	1.3438	2.08	.608
5.00	6.62	1.0419	52.0	.469	988.1	10.2	120	1.3450	2.36	.688
5.50	7.29	1.0462	57.4	.518	986.9	11.3	133	1.3463	2.64	.770
6.00	7.95	1.0505	62.9	.567	985.7	12.5	145	1.3475	2.94	.856
6.50	8.61	1.0548	68.4	.617	984.5	13.8	157	1.3487	3.26	.944
7.00	9.27	1.0591	74.0	.667	983.2	15.0	170	1.3499	3.58	1.035
7.50	9.94	1.0634	79.6	.717	981.9	16.3	182	1.3512	3.93	1.130
8.00	10.60	1.0678	85.3	.768	980.6	17.6	194	1.3524	4.28	1.234
8.50	11.26	1.0722	91.0	.820	979.3	18.9	207	1.3537	4.65	1.335
9.00	11.92	1.0766	96.7	.871	977.9	20.3	219	1.3549	5.04	1.439
9.50	12.58	1.0810	102.5	.924	976.6	21.7	232	1.3562	5.44	1.546
10.00	13.25	1.0854	108.4	.976	975.2	23.1	245	1.3575	5.85	1.655
11.00	14.57	1.0944	120.2	1.083	972.3	25.9	270	1.3600	6.73	1.880
12.00	15.90	1.1034	132.2	1.191	969.3	28.9	296	1.3626	7.69	2.115
13.00	17.22	1.1126	144.4	1.301	966.2	32.0	322	1.3652	8.71	2.359
14.00	18.55	1.1218	156.8	1.412	963.0	35.2	348	1.3678	9.81	2.611

7 CALCIUM CHLORIDE, $CaCl_2{\cdot}2H_2O$

A % by wt.	H % by wt.	D_{20}^{20}	C_s g/l	M g-mol/l	C_w g/l	$(C_o - C_w)$ g/l	$(n - n_o) \times 10^4$	n	Δ °C	S g-mol/l
15.00	19.87	1.1311	169.4	1.526	959.8	38.5	374	1.3704	11.00	2.871
16.00	21.20	1.1406	182.2	1.641	956.4	41.8	401	1.3731	12.28	3.137
17.00	22.52	1.1502	195.2	1.759	953.0	45.3	428	1.3758	13.64	3.409
18.00	23.84	1.1599	208.4	1.878	949.4	48.8	456	1.3786	15.11	3.685
19.00	25.17	1.1697	221.8	1.999	945.8	52.5	483	1.3813		
20.00	26.49	1.1796	235.5	2.122	942.0	56.2	511	1.3841		
22.00	29.14	1.1997	263.5	2.374	934.1	64.1				
24.00	31.79	1.2202	292.3	2.634	925.7	72.5				

H W C

7.1 CESIUM CHLORIDE, CsCl

MOLECULAR WEIGHT = 168.37

RELATIVE SPECIFIC REFRACTIVITY = .465

A % by wt.	D_{20}^{20}	C_s g/l	M g-mol/l	C_w g/l	$(C_o - C_w)$ g/l	$(n - n_o) \times 10^4$	n	Δ °C	S g-mol/l
.00	1.0000	.0	.000	998.2	.0	0	1.3330	.00	.000
1.00	1.0076	10.1	.060	995.8	2.5	8	1.3338	.20	.057
2.00	1.0153	20.3	.120	993.3	5.0	15	1.3345	.40	.114
3.00	1.0232	30.6	.182	990.7	7.5	23	1.3353	.59	.172
4.00	1.0311	41.2	.245	988.2	10.1	31	1.3361	.80	.231
5.00	1.0392	51.9	.308	985.5	12.7	39	1.3369	1.00	.292
6.00	1.0475	62.7	.373	982.9	15.3	48	1.3378	1.21	.353
7.00	1.0558	73.8	.438	980.2	18.0	56	1.3386	1.42	.414
8.00	1.0643	85.0	.505	977.5	20.8	65	1.3395	1.63	.477
9.00	1.0730	96.4	.573	974.7	23.5	73	1.3403	1.85	.541
10.00	1.0818	108.0	.641	971.9	26.4	82	1.3412	2.07	.605
11.00	1.0907	119.8	.711	969.0	29.3	91	1.3421	2.29	.669
12.00	1.0997	131.7	.782	966.1	32.2	100	1.3430	2.52	.735
13.00	1.1089	143.9	.855	963.1	35.2	110	1.3439	2.75	.801
14.00	1.1183	156.3	.928	960.0	38.2	119	1.3449	2.99	.867
15.00	1.1278	168.9	1.003	956.9	41.3	128	1.3458	3.22	.934
16.00	1.1375	181.7	1.079	953.8	44.5	138	1.3468	3.46	1.002
17.00	1.1473	194.7	1.156	950.6	47.7	148	1.3478	3.71	1.070
18.00	1.1573	207.9	1.235	947.3	51.0	158	1.3488	3.96	1.139
19.00	1.1674	221.4	1.315	943.9	54.3	168	1.3498	4.21	1.215
20.00	1.1777	235.1	1.396	940.5	57.7	178	1.3508	4.47	1.286
22.00	1.1989	263.3	1.564	933.5	64.8	199	1.3529		
24.00	1.2207	292.4	1.737	926.1	72.1	220	1.3550		
26.00	1.2433	322.7	1.916	918.4	79.8	242	1.3572		
28.00	1.2666	354.0	2.103	910.4	87.9	265	1.3595		
30.00	1.2908	386.6	2.296	902.0	96.3	288	1.3618		
32.00	1.3158	420.3	2.496	893.1	105.1	312	1.3642		
34.00	1.3417	455.4	2.705	883.9	114.3	337	1.3666		
36.00	1.3685	491.8	2.921	874.3	124.0	362	1.3692		
38.00	1.3963	529.6	3.146	864.2	134.1	388	1.3718		
40.00	1.4251	569.0	3.380	853.5	144.7	414	1.3744		
42.00	1.4550	610.0	3.623	842.4	155.8	442	1.3772		
44.00	1.4861	652.7	3.877	830.8	167.5	470	1.3800		
46.00	1.5185	697.3	4.141	818.5	179.7	500	1.3829		
48.00	1.5522	743.7	4.417	805.7	192.5	530	1.3860		
50.00	1.5874	792.3	4.706	792.3	206.0	562	1.3891		
52.00	1.6241	843.0	5.007	778.2	220.0	595	1.3924		
54.00	1.6625	896.2	5.323	763.4	234.8	629	1.3959		
56.00	1.7029	951.9	5.654	747.9	250.3	665	1.3995		
58.00	1.7453	1010.5	6.001	731.7	266.5	703	1.4033		
60.00	1.7900	1072.1	6.367	714.7	283.5	744	1.4074		
62.00	1.8373	1137.1	6.754	696.9	301.3	787	1.4117		
64.00	1.8875	1205.9	7.162	678.3	319.9	833	1.4163		

A A A

8 CITRIC ACID, $(COOH)CH_2C(OH)(COOH)CH_2COOH \cdot 1H_2O$

MOLECULAR WEIGHT = 192.12 FORMULA WEIGHT, HYDRATE = 210.14
RELATIVE SPECIFIC REFRACTIVITY = .956

A % by wt.	H % by wt.	D^{20}_{20}	C_s g/l	M g-mol/l	C_w g/l	$(C_o - C_w)$ g/l	$(n - n_o)$ $\times 10^4$	n	Δ °C	S g-mol/l
.00	.00	1.0000	.0	.000	998.2	.0	0	1.3330	.00	.000
.50	.55	1.0020	5.0	.026	995.2	3.0	6	1.3336	.06	.016
1.00	1.09	1.0040	10.0	.052	992.2	6.0	13	1.3343	.11	.033
1.50	1.64	1.0061	15.1	.078	989.2	9.0	19	1.3349	.17	.049
2.00	2.19	1.0081	20.1	.105	986.2	12.0	25	1.3355	.23	.066
2.50	2.73	1.0102	25.2	.131	983.2	15.0	32	1.3362	.29	.083
3.00	3.28	1.0123	30.3	.158	980.2	18.1	38	1.3368	.35	.101
3.50	3.83	1.0144	35.4	.184	977.1	21.1	45	1.3375	.41	.118
4.00	4.38	1.0165	40.6	.211	974.1	24.2	51	1.3381	.47	.136
4.50	4.92	1.0186	45.8	.238	971.0	27.2	58	1.3388	.53	.154
5.00	5.47	1.0207	50.9	.265	968.0	30.3	64	1.3394	.59	.172
5.50	6.02	1.0228	56.2	.292	964.9	33.4	71	1.3401	.65	.190
6.00	6.56	1.0250	61.4	.320	961.8	36.5	77	1.3407	.72	.208
6.50	7.11	1.0271	66.6	.347	958.7	39.6	84	1.3414	.78	.227
7.00	7.66	1.0293	71.9	.374	955.6	42.7	91	1.3421	.84	.245
7.50	8.20	1.0315	77.2	.402	952.4	45.8	97	1.3427	.91	.264
8.00	8.75	1.0336	82.5	.430	949.3	49.0	104	1.3434	.97	.283
8.50	9.30	1.0358	87.9	.457	946.1	52.1	111	1.3441	1.04	.302
9.00	9.84	1.0380	93.3	.485	943.0	55.3	117	1.3447	1.10	.321
9.50	10.39	1.0403	98.6	.513	939.8	58.5	124	1.3454	1.17	.341
10.00	10.94	1.0425	104.1	.542	936.6	61.7	131	1.3461	1.23	.360
11.00	12.03	1.0469	115.0	.598	930.1	68.1	144	1.3474		
12.00	13.13	1.0514	125.9	.656	923.6	74.6	158	1.3488		
13.00	14.22	1.0559	137.0	.713	917.0	81.2	172	1.3502		
14.00	15.31	1.0605	148.2	.771	910.4	87.8	185	1.3515		
15.00	16.41	1.0651	159.5	.830	903.7	94.5	199	1.3529		
16.00	17.50	1.0697	170.8	.889	896.9	101.3	213	1.3543		
17.00	18.59	1.0743	182.3	.949	890.1	108.1	226	1.3556		
18.00	19.69	1.0790	193.9	1.009	883.2	115.1	240	1.3570		
19.00	20.78	1.0836	205.5	1.070	876.2	122.0	254	1.3584		
20.00	21.88	1.0883	217.3	1.131	869.1	129.1	268	1.3598		
22.00	24.06	1.0978	241.1	1.255	854.8	143.5	296	1.3625		
24.00	26.25	1.1074	265.3	1.381	840.1	158.1	323	1.3653		
26.00	28.44	1.1171	289.9	1.509	825.2	173.1	351	1.3681		
28.00	30.63	1.1269	315.0	1.639	809.9	188.3	379	1.3709		
30.00	32.81	1.1368	340.4	1.772	794.4	203.9	407	1.3737		

H W A

9 COBALT CHLORIDE, $CoCl_2 \cdot 6H_2O$

MOLECULAR WEIGHT = 129.85 FORMULA WEIGHT, HYDRATE = 237.95
RELATIVE SPECIFIC REFRACTIVITY = .757

A % by wt.	H % by wt.	D^{20}_{20}	C_s g/l	M g-mol/l	C_w g/l	$(C_o - C_w)$ g/l	$(n - n_o)$ $\times 10^4$	n	Δ °C	S g-mol/l
.00	.00	1.0000	.0	.000	998.2	.0	0	1.3330	.00	.000
.50	.92	1.0045	5.0	.039	997.7	.5	11	1.3341	.19	.055
1.00	1.83	1.0090	10.1	.078	997.2	1.1	23	1.3352	.38	.110
1.50	2.75	1.0136	15.2	.117	996.6	1.6	34	1.3364	.58	.167
2.00	3.66	1.0182	20.3	.157	996.0	2.2	45	1.3375	.77	.224
2.50	4.58	1.0228	25.5	.197	995.5	2.8	57	1.3387	.97	.283
3.00	5.50	1.0274	30.8	.237	994.8	3.4	69	1.3398	1.18	.345
3.50	6.41	1.0321	36.1	.278	994.2	4.0	80	1.3410	1.40	.409
4.00	7.33	1.0368	41.4	.319	993.6	4.7	92	1.3422	1.63	.476
4.50	8.25	1.0415	46.8	.360	992.9	5.3	104	1.3434	1.86	.544
5.00	9.16	1.0463	52.2	.402	992.2	6.0	116	1.3445	2.10	.615
5.50	10.08	1.0511	57.7	.444	991.6	6.7	127	1.3457	2.36	.688
6.00	10.99	1.0560	63.2	.487	990.8	7.4	139	1.3469	2.62	.763
6.50	11.91	1.0608	68.8	.530	990.1	8.1	152	1.3481	2.89	.840
7.00	12.83	1.0657	74.5	.574	989.4	8.9	164	1.3494	3.17	.920
7.50	13.74	1.0707	80.2	.617	988.6	9.6	176	1.3506	3.46	1.001
8.00	14.66	1.0756	85.9	.662	987.8	10.4	188	1.3518	3.76	1.085
8.50	15.58	1.0807	91.7	.706	987.0	11.2				
9.00	16.49	1.0857	97.5	.751	986.2	12.0				
9.50	17.41	1.0908	103.4	.797	985.4	12.8				

CONCENTRATIVE PROPERTIES OF AQUEOUS SOLUTIONS (Continued)

9 COBALT CHLORIDE, $CoCl_2 \cdot 6H_2O$

A % by wt.	H % by wt.	D_{20}^{20}	C_s g/l	M g-mol/l	C_w g/l	$(C_o - C_w)$ g/l	$(n - n_o) \times 10^4$	n	Δ °C	S g-mol/l
10.00	18.32	1.0959	109.4	.842	984.6	13.7				
11.00	20.16	1.1063	121.5	.935	982.8	15.4				
12.00	21.99	1.1168	133.8	1.030	981.0	17.2				
13.00	23.82	1.1275	146.3	1.127	979.2	19.1				
14.00	25.65	1.1383	159.1	1.225	977.2	21.0				
15.00	27.49	1.1493	172.1	1.325	975.2	23.0				
16.00	29.32	1.1605	185.4	1.427	973.1	25.1				
17.00	31.15	1.1719	198.9	1.532	971.0	27.2				
18.00	32.98	1.1835	212.7	1.638	968.8	29.5				
19.00	34.82	1.1953	226.7	1.746	966.5	31.7				
20.00	36.65	1.2073	241.0	1.856	964.2	34.1				

H A C

10 CUPRIC SULFATE, $CuSO_4 \cdot 5H_2O$

MOLECULAR WEIGHT = 159.61 FORMULA WEIGHT, HYDRATE = 249.69
RELATIVE SPECIFIC REFRACTIVITY = .511

A % by wt.	H % by wt.	D_{20}^{20}	C_s g/l	M g-mol/l	C_w g/l	$(C_o - C_w)$ g/l	$(n - n_o) \times 10^4$	n	Δ °C	S g-mol/l
.00	.00	1.0000	.0	.000	998.2	.0	0	1.3330	.00	.000
.50	.78	1.0051	5.0	.031	998.3	−.1	9	1.3339	.07	.020
1.00	1.56	1.0103	10.1	.063	998.5	−.2	18	1.3348	.14	.039
1.50	2.35	1.0155	15.2	.095	998.5	−.3	28	1.3358	.20	.057
2.00	3.13	1.0208	20.4	.128	998.6	−.4	37	1.3367	.26	.075
2.50	3.91	1.0261	25.6	.160	998.7	−.4	46	1.3376	.32	.092
3.00	4.69	1.0314	30.9	.194	998.7	−.4	56	1.3386	.37	.108
3.50	5.48	1.0367	36.2	.227	998.7	−.4	65	1.3395	.43	.124
4.00	6.26	1.0421	41.6	.261	998.7	−.4	75	1.3405	.49	.140
4.50	7.04	1.0475	47.1	.295	998.6	−.4	84	1.3414	.54	.156
5.00	7.82	1.0529	52.6	.329	998.5	−.3	94	1.3424	.59	.172
5.50	8.60	1.0584	58.1	.364	998.4	−.2	103	1.3433	.65	.188
6.00	9.39	1.0639	63.7	.399	998.3	−.1	113	1.3443	.70	.204
6.50	10.17	1.0694	69.4	.435	998.1	.1	122	1.3452	.76	.220
7.00	10.95	1.0750	75.1	.471	997.9	.3	132	1.3462	.81	.237
7.50	11.73	1.0805	80.9	.507	997.7	.5	142	1.3472	.87	.254
8.00	12.52	1.0862	86.7	.543	997.5	.7	151	1.3481	.93	.271
8.50	13.30	1.0918	92.6	.580	997.2	1.0	161	1.3491	.99	.289
9.00	14.08	1.0975	98.6	.618	996.9	1.3	171	1.3501	1.05	.307
9.50	14.86	1.1032	104.6	.655	996.6	1.6	181	1.3510	1.11	.325
10.00	15.64	1.1090	110.7	.694	996.3	1.9	190	1.3520	1.18	.344
11.00	17.21	1.1206	123.0	.771	995.6	2.7	210	1.3540	1.31	.382
12.00	18.77	1.1324	135.6	.850	994.7	3.5	230	1.3560	1.45	.423
13.00	20.34	1.1444	148.5	.930	993.8	4.4	251	1.3581	1.60	.466
14.00	21.90	1.1566	161.6	1.013	992.9	5.4	271	1.3601	1.75	.513
15.00	23.47	1.1690	175.0	1.097	991.9	6.3	293	1.3623		
16.00	25.03	1.1817	188.7	1.182	990.9	7.4	314	1.3644		
17.00	26.59	1.1947	202.7	1.270	989.8	8.4	337	1.3667		
18.00	28.16	1.2080	217.1	1.360	988.8	9.4	360	1.3690		

H W C

11 CREATININE, $CH_3NC(:NH)NHCOCH_2$

MOLECULAR WEIGHT = 113.12
RELATIVE SPECIFIC REFRACTIVITY = 1.302

A % by wt.	D_{20}^{20}	C_s g/l	M g-mol/l	C_w g/l	$(C_o - C_w)$ g/l	$(n - n_o) \times 10^4$	n	Δ °C	S g-mol/l
.00	1.0000	.0	.000	998.2	.0	0	1.3330	.00	.000
.50	1.0012	5.0	.044	994.5	3.8	9	1.3339	.07	.021
1.00	1.0025	10.0	.088	990.7	7.5	19	1.3349	.15	.042
1.50	1.0038	15.0	.133	986.9	11.3	29	1.3358	.22	.064
2.00	1.0050	20.1	.177	983.2	15.1	38	1.3368	.30	.087
2.50	1.0063	25.1	.222	979.4	18.8	48	1.3378	.38	.110
3.00	1.0076	30.2	.267	975.6	22.6	58	1.3387	.46	.133
3.50	1.0089	35.2	.312	971.9	26.4	67	1.3397		
4.00	1.0102	40.3	.357	968.1	30.2	77	1.3407		
4.50	1.0115	45.4	.402	964.3	34.0	87	1.3417		

11 CREATININE, $CH_3NC(:NH)NHCOCH_2$

A % by wt.	D^{20}_{20}	C_s g/l	M g-mol/l	C_w g/l	$(C_o - C_w)$ g/l	$(n - n_o)$ × 10^4	n	Δ °C	S g-mol/l
5.00	1.0128	50.6	.447	960.5	37.8	97	1.3427		
5.50	1.0142	55.7	.492	956.7	41.6	107	1.3437		
6.00	1.0155	60.8	.538	952.9	45.3	117	1.3447		
6.50	1.0169	66.0	.583	949.1	49.1	127	1.3457		
7.00	1.0183	71.2	.629	945.3	52.9	137	1.3467		
7.50	1.0197	76.3	.675	941.6	56.7	148	1.3477		
8.00	1.0212	81.6	.721	937.8	60.4	158	1.3488		

A A A

12 DEXTRAN, $(C_6H_{10}O_5)_x$

RELATIVE SPECIFIC REFRACTIVITY = 1.040

A % by wt.	D^{20}_{20}	C_s g/l	M g-mol/l	C_w g/l	$(C_o - C_w)$ g/l	$(n - n_o)$ × 10^4	n	Δ °C	S g-mol/l
.00	1.0000	.0	.000	998.2	.0	0	1.3330	.00	.000
.50	1.0020	5.0	.000	995.2	3.0	8	1.3337		
1.00	1.0040	10.0	.000	992.2	6.0	15	1.3345		
1.50	1.0060	15.1	.000	989.2	9.1	23	1.3353		
2.00	1.0080	20.1	.000	986.1	12.1	30	1.3360		
2.50	1.0100	25.2	.000	983.1	15.2	38	1.3368		
3.00	1.0121	30.3	.000	980.0	18.2	46	1.3376		
3.50	1.0141	35.4	.000	976.9	21.3	54	1.3384		
4.00	1.0162	40.6	.001	973.8	24.4	61	1.3391		
4.50	1.0182	45.7	.001	970.7	27.5	69	1.3399		
5.00	1.0203	50.9	.001	967.6	30.7	77	1.3407		
5.50	1.0224	56.1	.001	964.4	33.8	85	1.3415		
6.00	1.0244	61.4	.001	961.3	37.0	93	1.3423		
6.50	1.0265	66.6	.001	958.1	40.1	101	1.3431		
7.00	1.0286	71.9	.001	954.9	43.3	109	1.3439		
7.50	1.0307	77.2	.001	951.7	46.5	117	1.3447		
8.00	1.0328	82.5	.001	948.5	49.7	125	1.3455		
8.50	1.0350	87.8	.001	945.3	52.9	133	1.3463		
9.00	1.0371	93.2	.001	942.1	56.1	141	1.3471		
9.50	1.0392	98.6	.001	938.9	59.4	149	1.3479		
10.00	1.0414	104.0	.001	935.6	62.6	157	1.3487		
11.00	1.0457	114.8	.002	929.1	69.2	174	1.3504		
12.00	1.0501	125.8	.002	922.5	75.8	190	1.3520		

Note: The molar concentration, M corresponds to an average molecular weight of 72,000.
(Snyder, et al, J. Res. N.B.S. 53:131) (ibid) —

13 ETHANOL, CH_3CH_2OH

MOLECULAR WEIGHT = 46.07
RELATIVE SPECIFIC REFRACTIVITY = 1.361

A % by wt.	D^{20}_{20}	C_s g/l	M g-mol/l	C_w g/l	$(C_o - C_w)$ g/l	$(n - n_o)$ × 10^4	n	Δ °C	S g-mol/l
.00	1.0000	.0	.000	998.2	.0	0	1.3330	.00	.000
1.00	.9980	10.0	.216	986.3	11.9	6	1.3335	.40	.116
2.00	.9962	19.9	.432	974.6	23.7	12	1.3342	.81	.236
3.00	.9945	29.8	.646	962.9	35.3	18	1.3348	1.23	.358
4.00	.9928	39.6	.860	951.4	46.8	25	1.3354	1.65	.484
5.00	.9912	49.5	1.074	940.0	58.3	31	1.3361	2.09	.611
6.00	.9896	59.3	1.287	928.6	69.6	38	1.3368	2.54	.741
7.00	.9881	69.0	1.499	917.3	80.9	45	1.3375	3.00	.872
8.00	.9866	78.8	1.710	906.1	92.2	52	1.3382	3.47	1.004
9.00	.9851	88.5	1.921	894.9	103.4	59	1.3389	3.97	1.142
10.00	.9837	98.2	2.131	883.8	114.5	66	1.3396	4.48	1.288
11.00	.9823	107.9	2.341	872.7	125.5	73	1.3403	5.00	1.431
12.00	.9809	117.5	2.551	861.7	136.5	81	1.3411	5.55	1.576
13.00	.9796	127.1	2.759	850.7	147.5	88	1.3418	6.12	1.724
14.00	.9782	136.7	2.967	839.8	158.5	96	1.3425	6.72	1.876

13 ETHANOL, CH_3CH_2OH

A % by wt.	D_{20}^{20}	C_s g/l	M g-mol/l	C_w g/l	$(C_o - C_w)$ g/l	$(n - n_o) \times 10^4$	n	Δ °C	S g-mol/l
15.00	.9769	146.3	3.175	828.9	169.4	103	1.3433	7.34	2.031
16.00	.9755	155.8	3.382	818.0	180.2	110	1.3440	7.99	2.189
17.00	.9742	165.3	3.589	807.2	191.1	118	1.3448	8.68	2.351
18.00	.9729	174.8	3.794	796.4	201.9	125	1.3455	9.40	2.517
19.00	.9716	184.3	4.000	785.6	212.7	132	1.3462	10.15	2.686
20.00	.9702	193.7	4.204	774.8	223.4	140	1.3469	10.94	2.857
22.00	.9675	212.5	4.612	753.3	244.9	154	1.3484	12.63	3.208
24.00	.9647	231.1	5.017	731.9	266.4	168	1.3497	14.46	3.565
26.00	.9618	249.6	5.418	710.4	287.8	181	1.3511	16.41	3.919
28.00	.9587	268.0	5.817	689.1	309.2	194	1.3523	18.45	4.263
30.00	.9556	286.2	6.211	667.7	330.5	206	1.3535	20.50	
32.00	.9522	304.2	6.602	646.4	351.9	217	1.3547	22.42	
34.00	.9487	322.0	6.989	625.0	373.2	227	1.3557		
36.00	.9450	339.6	7.372	603.7	394.5	237	1.3567		
38.00	.9412	357.0	7.749	582.5	415.8	245	1.3575		
40.00	.9371	374.2	8.122	561.3	437.0	253	1.3583		
42.00	.9329	391.1	8.489	540.1	458.1	260	1.3590		
44.00	.9285	407.8	8.852	519.0	479.2	266	1.3596		
46.00	.9240	424.3	9.209	498.1	500.2	271	1.3601		
48.00	.9193	440.5	9.562	477.2	521.0	276	1.3606		
50.00	.9147	456.5	9.909	456.5	541.7	280	1.3610		
52.00	.9100	472.4	10.253	436.0	562.2	284	1.3614		
54.00	.9055	488.1	10.595	415.8	582.4	288	1.3618		
56.00	.9011	503.7	10.934	395.8	602.4	293	1.3623		
58.00	.8972	519.4	11.275	376.1	622.1	299	1.3629		
60.00	.8937	535.3	11.619	356.8	641.4	307	1.3637		

H W C

14 ETHYLENE GLYCOL, CH_2OHCH_2OH

MOLECULAR WEIGHT = 62.07

RELATIVE SPECIFIC REFRACTIVITY = 1.149

A % by wt.	D_{20}^{20}	C_s g/l	M g-mol/l	C_w g/l	$(C_o - C_w)$ g/l	$(n - n_o) \times 10^4$	n	Δ °C	S g-mol/l
.00	1.0000	.0	.000	998.2	.0	0	1.3330	.00	.000
.50	1.0006	5.0	.080	993.8	4.4	5	1.3335	.15	.042
1.00	1.0012	10.0	.161	989.5	8.8	9	1.3339	.30	.085
1.50	1.0018	15.0	.242	985.1	13.2	14	1.3344	.45	.130
2.00	1.0025	20.0	.322	980.7	17.6	19	1.3349	.60	.175
2.50	1.0031	25.0	.403	976.3	22.0	23	1.3353	.76	.220
3.00	1.0037	30.1	.484	971.9	26.4	28	1.3358	.92	.267
3.50	1.0043	35.1	.565	967.4	30.8	33	1.3363	1.08	.314
4.00	1.0049	40.1	.646	963.0	35.2	37	1.3367	1.24	.362
4.50	1.0056	45.2	.728	958.6	39.6	42	1.3372	1.41	.411
5.00	1.0062	50.2	.809	954.2	44.1	47	1.3377	1.57	.460
5.50	1.0068	55.3	.891	949.7	48.5	52	1.3381	1.75	.510
6.00	1.0074	60.3	.972	945.3	52.9	56	1.3386	1.92	.561
6.50	1.0081	65.4	1.054	940.9	57.4	61	1.3391	2.10	.612
7.00	1.0087	70.5	1.136	936.4	61.8	66	1.3396	2.27	.664
7.50	1.0093	75.6	1.217	932.0	66.3	71	1.3400	2.46	.717
8.00	1.0100	80.7	1.299	927.5	70.7	75	1.3405	2.64	.769
8.50	1.0106	85.7	1.381	923.1	75.2	80	1.3410	2.83	.823
9.00	1.0112	90.8	1.464	918.6	79.6	85	1.3415	3.02	.877
9.50	1.0119	96.0	1.546	914.1	84.1	90	1.3420	3.21	.932
10.00	1.0125	101.1	1.628	909.6	88.6	95	1.3424	3.41	.987

A A A

15 FERRIC CHLORIDE, $FeCl_3 \cdot 6H_2O$

MOLECULAR WEIGHT = 162.22 FORMULA WEIGHT, HYDRATE = 270.32
RELATIVE SPECIFIC REFRACTIVITY = .947

A % by wt.	H % by wt.	D_{20}^{20}	C_s g/l	M g-mol/l	C_w g/l	$(C_o - C_w)$ g/l	$(n - n_o) \times 10^4$	n	Δ °C	S g-mol/l
.00	.00	1.0000	.0	.000	998.2	.0	0	1.3330	.00	.000
.50	.83	1.0043	5.0	.031	997.5	.7	14	1.3344	.19	.056
1.00	1.67	1.0086	10.1	.062	996.8	1.5	28	1.3358	.38	.110
1.50	2.50	1.0129	15.2	.093	995.9	2.3	42	1.3372	.57	.166
2.00	3.33	1.0171	20.3	.125	995.0	3.2	55	1.3385	.81	.235
2.50	4.17	1.0214	25.5	.157	994.1	4.2	69	1.3399	.96	.280
3.00	5.00	1.0256	30.7	.189	993.0	5.2	83	1.3413	1.13	.329
3.50	5.83	1.0298	36.0	.222	992.0	6.2	96	1.3426	1.30	.381
4.00	6.67	1.0340	41.3	.255	990.9	7.3	110	1.3440	1.49	.436
4.50	7.50	1.0383	46.6	.288	989.8	8.4	124	1.3453	1.69	.494
5.00	8.33	1.0425	52.0	.321	988.7	9.6	137	1.3467	1.90	.556
5.50	9.17	1.0468	57.5	.354	987.5	10.7	151	1.3481	2.13	.621
6.00	10.00	1.0511	63.0	.388	986.3	11.9	165	1.3495	2.36	.690
6.50	10.83	1.0555	68.5	.422	985.1	13.1	179	1.3509	2.61	.761
7.00	11.66	1.0599	74.1	.457	984.0	14.3	193	1.3523	2.88	.837
7.50	12.50	1.0643	79.7	.491	982.8	15.5	208	1.3538	3.16	.915
8.00	13.33	1.0688	85.4	.526	981.6	16.7	222	1.3552	3.45	.997
8.50	14.16	1.0733	91.1	.561	980.3	17.9	237	1.3567	3.76	1.083
9.00	15.00	1.0779	96.8	.597	979.1	19.1	252	1.3581	4.08	1.178
9.50	15.83	1.0825	102.7	.633	977.9	20.3	266	1.3596	4.42	1.272
10.00	16.66	1.0871	108.5	.669	976.6	21.6	281	1.3611	4.77	1.368
11.00	18.33	1.0964	120.4	.742	974.1	24.1	312	1.3642	5.53	1.571
12.00	20.00	1.1059	132.5	.817	971.4	26.8	342	1.3672	6.36	1.787
13.00	21.66	1.1153	144.7	.892	968.6	29.6	373	1.3703	7.27	2.014
14.00	23.33	1.1246	157.2	.969	965.4	32.8	403	1.3733	8.26	2.253
15.00	25.00	1.1343	169.8	1.047	962.5	35.8	434	1.3764	9.33	2.502
16.00	26.66	1.1441	182.7	1.126	959.3	38.9				
17.00	28.33	1.1539	195.8	1.207	956.1	42.2				
18.00	29.99	1.1638	209.1	1.289	952.6	45.6				
19.00	31.66	1.1738	222.6	1.372	949.1	49.2				
20.00	33.33	1.1838	236.3	1.457	945.4	52.9				
22.00	36.66	1.2043	264.5	1.630	937.7	60.5				
24.00	39.99	1.2254	293.6	1.810	929.7	68.6				
26.00	43.33	1.2473	323.7	1.996	921.4	76.9				
28.00	46.66	1.2699	354.9	2.188	912.7	85.5				
30.00	49.99	1.2934	387.3	2.388	903.8	94.5				
32.00	53.32	1.3176	420.9	2.595	894.4	103.8				
34.00	56.66	1.3426	455.7	2.809	884.5	113.7				
36.00	59.99	1.3681	491.7	3.031	874.1	124.2				
38.00	63.32	1.3941	528.8	3.260	862.8	135.4				
40.00	66.66	1.4200	567.0	3.495	850.5	147.7				

H W C

16 FORMIC ACID, HCOOH

MOLECULAR WEIGHT = 46.03
RELATIVE SPECIFIC REFRACTIVITY = .909

A % by wt.	D_{20}^{20}	C_s g/l	M g-mol/l	C_w g/l	$(C_o - C_w)$ g/l	$(n - n_o) \times 10^4$	n	Δ °C	S g-mol/l
.00	1.0000	.0	.000	998.2	.0	0	1.3330	.00	.000
1.00	1.0027	10.0	.217	990.9	7.3	6	1.3336	.39	.113
2.00	1.0054	20.1	.436	983.6	14.7	12	1.3342	.81	.237
3.00	1.0081	30.2	.656	976.1	22.1	18	1.3348	1.24	.362
4.00	1.0107	40.4	.877	968.6	29.6	24	1.3354	1.67	.487
5.00	1.0134	50.6	1.099	961.0	37.2	30	1.3360	2.10	.613
6.00	1.0160	60.9	1.322	953.3	44.9	35	1.3365	2.53	.738
7.00	1.0186	71.2	1.546	945.6	52.6	41	1.3371	2.97	.862
8.00	1.0211	81.5	1.772	937.8	60.4	47	1.3377	3.41	.986
9.00	1.0237	92.0	1.998	929.9	68.3	52	1.3382	3.85	1.110
10.00	1.0263	102.4	2.226	922.0	76.2	58	1.3388	4.30	1.240
11.00	1.0288	113.0	2.454	914.0	84.2	63	1.3393	4.75	1.363
12.00	1.0313	123.5	2.684	905.9	92.3	68	1.3398	5.21	1.486
13.00	1.0338	134.2	2.915	897.8	100.4	74	1.3404	5.67	1.607
14.00	1.0363	144.8	3.146	889.6	108.6	79	1.3409	6.13	1.727
15.00	1.0388	155.5	3.379	881.4	116.8	84	1.3414	6.60	1.846
16.00	1.0412	166.3	3.613	873.1	125.1	89	1.3419	7.07	1.964
17.00	1.0437	177.1	3.848	864.7	133.5	94	1.3424	7.54	2.081
18.00	1.0461	188.0	4.084	856.3	141.9	99	1.3429	8.02	2.197
19.00	1.0486	198.9	4.321	847.8	150.4	104	1.3434	8.51	2.312

16 FORMIC ACID, HCOOH

A % by wt.	D^{20}_{20}	C_s g/l	M g-mol/l	C_w g/l	$(C_o - C_w)$ g/l	$(n - n_o) \times 10^4$	n	Δ °C	S g-mol/l
20.00	1.0510	209.8	4.558	839.3	158.9	109	1.3439	9.00	2.425
22.00	1.0558	231.9	5.037	822.1	176.2	119	1.3449	9.99	2.650
24.00	1.0606	254.1	5.520	804.6	193.6	128	1.3458	11.00	2.871
26.00	1.0653	276.5	6.007	787.0	211.3	138	1.3468	12.03	3.088
28.00	1.0701	299.1	6.498	769.1	229.1	147	1.3477	13.09	3.302
30.00	1.0748	321.9	6.993	751.0	247.2	156	1.3486	14.18	3.512
32.00	1.0795	344.8	7.492	732.8	265.5	166	1.3495	15.30	3.720
34.00	1.0842	368.0	7.995	714.3	283.9	175	1.3505	16.45	3.925
36.00	1.0889	391.3	8.501	695.7	302.5	184	1.3514	17.64	4.129
38.00	1.0936	414.8	9.013	676.9	321.4	193	1.3523	18.88	4.332
40.00	1.0983	438.6	9.528	657.8	340.4	202	1.3532	20.17	
42.00	1.1030	462.5	10.047	638.6	359.6	211	1.3541	21.52	
44.00	1.1078	486.5	10.570	619.2	379.0	220	1.3549	22.94	
46.00	1.1125	510.8	11.098	599.7	398.6	229	1.3558	24.45	
48.00	1.1172	535.3	11.630	579.9	418.3	237	1.3567	26.05	
50.00	1.1219	560.0	12.165	560.0	438.3	246	1.3576	27.77	
52.00	1.1267	584.8	12.705	539.8	458.4	255	1.3585	29.61	
54.00	1.1314	609.9	13.250	519.5	478.7	264	1.3594	31.59	
56.00	1.1361	635.1	13.798	499.0	499.2	273	1.3603	33.72	
58.00	1.1409	660.5	14.350	478.3	519.9	282	1.3611	36.00	
60.00	1.1456	686.1	14.906	457.4	540.8	290	1.3620	38.40	

H W C

17 D-FRUCTOSE (LEVULOSE), $C_6H_{12}O_6$

MOLECULAR WEIGHT = 180.16

RELATIVE SPECIFIC REFRACTIVITY = 1.020

A % by wt.	D^{20}_{20}	C_s g/l	M g-mol/l	C_w g/l	$(C_o - C_w)$ g/l	$(n - n_o) \times 10^4$	n	Δ °C	S g-mol/l
.00	1.0000	.0	.000	998.2	.0	0	1.3330	.00	.000
1.00	1.0039	10.0	.056	992.1	6.1	14	1.3344	.10	.030
2.00	1.0079	20.1	.112	986.0	12.3	29	1.3358	.21	.061
3.00	1.0118	30.3	.168	979.8	18.5	43	1.3373	.32	.093
4.00	1.0158	40.6	.225	973.5	24.8	57	1.3387	.43	.126
5.00	1.0199	50.9	.283	967.2	31.1	72	1.3402	.55	.159
6.00	1.0239	61.3	.340	960.8	37.5	87	1.3417	.67	.194
7.00	1.0280	71.8	.399	954.3	43.9	102	1.3431	.79	.229
8.00	1.0321	82.4	.457	947.8	50.4	116	1.3446	.91	.266
9.00	1.0362	93.1	.517	941.3	56.9	131	1.3461	1.04	.304
10.00	1.0404	103.9	.576	934.7	63.5	147	1.3477	1.17	.342
11.00	1.0446	114.7	.637	928.0	70.2	162	1.3492	1.31	.381
12.00	1.0488	125.6	.697	921.3	76.9	177	1.3507	1.44	.422
13.00	1.0530	136.7	.758	914.5	83.7	193	1.3523	1.59	.463
14.00	1.0573	147.8	.820	907.7	90.6	208	1.3538	1.73	.506
15.00	1.0616	159.0	.882	900.8	97.5	224	1.3554	1.88	.550
16.00	1.0659	170.2	.945	893.8	104.5	240	1.3570		
17.00	1.0703	181.6	1.008	886.8	111.5	256	1.3586		
18.00	1.0747	193.1	1.072	879.7	118.6	272	1.3602		
19.00	1.0791	204.7	1.136	872.5	125.7	288	1.3618		
20.00	1.0835	216.3	1.201	865.3	133.0	304	1.3634		
22.00	1.0925	239.9	1.332	850.6	147.6	337	1.3667		
24.00	1.1016	263.9	1.465	835.7	162.5	370	1.3700		
26.00	1.1108	288.3	1.600	820.5	177.7	404	1.3734		
28.00	1.1202	313.1	1.738	805.1	193.1	439	1.3768		
30.00	1.1296	338.3	1.878	789.3	208.9	473	1.3803		
32.00	1.1392	363.9	2.020	773.3	224.9	509	1.3839		
34.00	1.1490	390.0	2.165	757.0	241.3	545	1.3875		
36.00	1.1588	416.4	2.312	740.3	257.9	581	1.3911		
38.00	1.1688	443.4	2.461	723.4	274.8	618	1.3948		
40.00	1.1790	470.8	2.613	706.1	292.1	656	1.3985		
42.00	1.1892	498.6	2.767	688.5	309.7	694	1.4023		
44.00	1.1996	526.9	2.925	670.6	327.6	732	1.4062		
46.00	1.2102	555.7	3.084	652.3	345.9	771	1.4101		
48.00	1.2208	585.0	3.247	633.7	364.5	811	1.4141		
50.00	1.2316	614.7	3.412	614.7	383.5	851	1.4181		
52.00	1.2426	645.0	3.580	595.4	402.9	892	1.4222		
54.00	1.2536	675.8	3.751	575.7	422.6	934	1.4264		
56.00	1.2649	707.1	3.925	555.6	442.7	976	1.4306		
58.00	1.2762	738.9	4.101	535.1	463.2	1018	1.4348		
60.00	1.2877	771.2	4.281	514.2	484.1	1061	1.4391		

(Jackson and Mathews, J. Res. N.B.S. 8:403) (ibid) C

18 D-GLUCOSE (DEXTROSE), $C_6H_{12}O_6 \cdot 1H_2O$

MOLECULAR WEIGHT = 180.16 FORMULA WEIGHT, HYDRATE = 198.17
RELATIVE SPECIFIC REFRACTIVITY = 1.027

A % by wt.	H % by wt.	D^{20}_{20}	C_s g/l	M g-mol/l	C_w g/l	$(C_o - C_w)$ g/l	$(n - n_o)$ × 10^4	n	Δ °C	S g-mol/l
.00	.00	1.0000	.0	.000	998.2	.0	0	1.3330	.00	.000
.50	.55	1.0019	5.0	.028	995.1	3.1	7	1.3337	.05	.015
1.00	1.10	1.0038	10.0	.056	992.0	6.2	14	1.3344	.10	.030
1.50	1.65	1.0057	15.1	.084	988.9	9.4	21	1.3351	.16	.045
2.00	2.20	1.0076	20.1	.112	985.7	12.5	28	1.3358	.21	.061
2.50	2.75	1.0096	25.2	.140	982.6	15.6	35	1.3365	.27	.077
3.00	3.30	1.0115	30.3	.168	979.4	18.8	42	1.3372	.32	.093
3.50	3.85	1.0134	35.4	.197	976.2	22.0	50	1.3380	.38	.109
4.00	4.40	1.0154	40.5	.225	973.1	25.2	57	1.3387	.43	.126
4.50	4.95	1.0173	45.7	.254	969.8	28.4	64	1.3394	.49	.142
5.00	5.50	1.0193	50.9	.282	966.6	31.6	71	1.3401	.55	.159
5.50	6.05	1.0213	56.1	.311	963.4	34.8	79	1.3409	.61	.176
6.00	6.60	1.0233	61.3	.340	960.2	38.1	86	1.3416	.67	.194
6.50	7.15	1.0252	66.5	.369	956.9	41.3	93	1.3423	.73	.212
7.00	7.70	1.0272	71.8	.398	953.6	44.6	101	1.3431	.79	.229
7.50	8.25	1.0292	77.1	.428	950.4	47.9	108	1.3438	.85	.248
8.00	8.80	1.0313	82.4	.457	947.1	51.1	116	1.3445	.91	.266
8.50	9.35	1.0333	87.7	.487	943.8	54.5	123	1.3453	.98	.285
9.00	9.90	1.0353	93.0	.516	940.5	57.8	131	1.3460	1.04	.304
9.50	10.45	1.0373	98.4	.546	937.1	61.1	138	1.3468	1.11	.323
10.00	11.00	1.0394	103.8	.576	933.8	64.4	146	1.3476	1.17	.342

A A C

19 GLYCEROL, $CH_2OHCHOHCH_2OH$

MOLECULAR WEIGHT = 92.09
RELATIVE SPECIFIC REFRACTIVITY = 1.100

A % by wt.	D^{20}_{20}	C_s g/l	M g-mol/l	C_w g/l	$(C_o - C_w)$ g/l	$(n - n_o)$ × 10^4	n	Δ °C	S g-mol/l
.00	1.0000	.0	.000	998.2	.0	0	1.3330	.00	.000
1.00	1.0024	10.0	.109	990.6	7.7	12	1.3341	.20	.058
2.00	1.0047	20.1	.218	982.9	15.4	23	1.3353	.41	.119
3.00	1.0071	30.2	.327	975.2	23.1	35	1.3365	.63	.182
4.00	1.0095	40.3	.438	967.4	30.9	46	1.3376	.85	.248
5.00	1.0119	50.5	.548	959.6	38.7	58	1.3388	1.08	.315
6.00	1.0143	60.7	.660	951.7	46.5	70	1.3400	1.31	.384
7.00	1.0167	71.0	.771	943.8	54.4	82	1.3412	1.56	.455
8.00	1.0191	81.4	.884	935.9	62.3	94	1.3424	1.81	.528
9.00	1.0215	91.8	.997	927.9	70.3	106	1.3436	2.06	.603
10.00	1.0240	102.2	1.110	919.9	78.3	118	1.3448	2.33	.679
11.00	1.0264	112.7	1.224	911.9	86.3	130	1.3460	2.60	.757
12.00	1.0289	123.2	1.338	903.8	94.4	142	1.3472	2.88	.837
13.00	1.0313	133.8	1.453	895.7	102.6	155	1.3485	3.17	.919
14.00	1.0338	144.5	1.569	887.5	110.7	167	1.3497	3.47	1.003
15.00	1.0363	155.2	1.685	879.3	118.9	179	1.3509	3.77	1.088
16.00	1.0388	165.9	1.802	871.0	127.2	192	1.3522	4.09	1.182
17.00	1.0413	176.7	1.919	862.8	135.5	204	1.3534	4.42	1.272
18.00	1.0438	187.6	2.037	854.4	143.8	217	1.3547	4.75	1.364
19.00	1.0463	198.5	2.155	846.0	152.2	230	1.3560	5.10	1.457
20.00	1.0489	209.4	2.274	837.6	160.6	242	1.3572	5.46	1.552
22.00	1.0540	231.5	2.513	820.6	177.6	268	1.3598	6.21	1.747
24.00	1.0591	253.7	2.755	803.5	194.8	294	1.3624	7.01	1.949
26.00	1.0642	276.2	2.999	786.1	212.1	320	1.3650	7.86	2.157
28.00	1.0694	298.9	3.246	768.6	229.6	346	1.3676	8.76	2.371
30.00	1.0747	321.8	3.495	750.9	247.3	373	1.3703	9.72	2.591
32.00	1.0799	345.0	3.746	733.0	265.2	400	1.3730	10.75	2.816
34.00	1.0852	368.3	4.000	715.0	283.3	427	1.3757	11.83	3.046
36.00	1.0905	391.9	4.256	696.7	301.5	454	1.3784	12.99	3.280
38.00	1.0959	415.7	4.514	678.2	320.0	482	1.3812	14.21	3.517
40.00	1.1012	439.7	4.775	659.6	338.7	510	1.3840	15.50	3.757
42.00	1.1066	464.0	5.038	640.7	357.5	538	1.3868		
44.00	1.1120	488.4	5.304	621.6	376.6	566	1.3896		
46.00	1.1175	513.1	5.572	602.4	395.9	595	1.3924		
48.00	1.1229	538.1	5.843	582.9	415.3	623	1.3953		
50.00	1.1284	563.2	6.116	563.2	435.0	652	1.3982		
52.00	1.1339	588.6	6.391	543.3	454.9	681	1.4011		
54.00	1.1394	614.2	6.669	523.2	475.1	711	1.4040		
56.00	1.1448	640.0	6.949	502.8	495.4	740	1.4070		
58.00	1.1503	666.0	7.232	482.3	516.0	770	1.4099		
60.00	1.1558	692.2	7.517	461.5	536.7	799	1.4129		

H (Hoyt, J. Eng. Chem. 26:329) C

20 HYDROCHLORIC ACID, HCl

MOLECULAR WEIGHT = 36.47
RELATIVE SPECIFIC REFRACTIVITY = 1.168

A % by wt.	D^{20}_{20}	C_s g/l	M g-mol/l	C_w g/l	$(C_o - C_w)$ g/l	$(n - n_o) \times 10^4$	n	Δ °C	S g-mol/l
.00	1.0000	.0	.000	998.2	.0	0	1.3330	.00	.000
.50	1.0025	5.0	.137	995.8	2.5	12	1.3342	.50	.144
1.00	1.0050	10.0	.275	993.2	5.0	23	1.3353	.98	.287
1.50	1.0075	15.1	.414	990.6	7.6	34	1.3364	1.50	.438
2.00	1.0099	20.2	.553	988.0	10.3	46	1.3376	2.04	.596
2.50	1.0124	25.3	.693	985.3	12.9	57	1.3387	2.61	.762
3.00	1.0148	30.4	.833	982.7	15.6	69	1.3399	3.22	.933
3.50	1.0173	35.5	.975	980.0	18.3	80	1.3410	3.85	1.110
4.00	1.0198	40.7	1.116	977.2	21.0	92	1.3421	4.52	1.301
4.50	1.0222	45.9	1.259	974.5	23.7	103	1.3433	5.23	1.490
5.00	1.0247	51.1	1.402	971.7	26.5	115	1.3445	5.96	1.683
5.50	1.0272	56.4	1.546	969.0	29.3	126	1.3456	6.74	1.880
6.00	1.0296	61.7	1.691	966.1	32.1	138	1.3468	7.54	2.081
6.50	1.0321	67.0	1.836	963.3	34.9	149	1.3479	8.39	2.284
7.00	1.0346	72.3	1.982	960.5	37.8	161	1.3491	9.28	2.490
7.50	1.0370	77.6	2.129	957.6	40.7	173	1.3502	10.20	2.698
8.00	1.0395	83.0	2.276	954.7	43.6	184	1.3514	11.17	2.907
8.50	1.0420	88.4	2.424	951.7	46.5	196	1.3526	12.18	3.118
9.00	1.0445	93.8	2.573	948.8	49.5	208	1.3537	13.23	3.329
9.50	1.0469	99.3	2.722	945.8	52.4	219	1.3549	14.33	3.540
10.00	1.0488	104.7	2.871	942.3	55.9	229	1.3559	15.47	3.752
11.00	1.0540	115.7	3.173	936.4	61.8	253	1.3583	17.90	4.173
12.00	1.0592	126.9	3.479	930.4	67.8	278	1.3607	20.53	
13.00	1.0643	138.1	3.787	924.3	73.9	302	1.3632	23.38	
14.00	1.0695	149.5	4.098	918.1	80.1				
15.00	1.0746	160.9	4.412	911.8	86.4				
16.00	1.0798	172.5	4.729	905.4	92.8				
17.00	1.0849	184.1	5.048	898.9	99.3				
18.00	1.0901	195.9	5.371	892.3	106.0				
19.00	1.0952	207.7	5.696	885.5	112.7				
20.00	1.1003	219.7	6.023	878.7	119.6				
22.00	1.1105	243.9	6.687	864.7	133.6				
24.00	1.1207	268.5	7.362	850.2	148.0				
26.00	1.1308	293.5	8.048	835.3	162.9				
28.00	1.1409	318.9	8.744	820.0	178.2				
30.00	1.1510	344.7	9.451	804.3	194.0				
32.00	1.1610	370.9	10.169	788.1	210.1				
34.00	1.1710	397.4	10.898	771.5	226.7				
36.00	1.1809	424.4	11.636	754.4	243.8				
38.00	1.1908	451.7	12.385	737.0	261.3				
40.00	1.2006	479.4	13.145	719.1	279.2				

H W C

21 INULIN, $(C_6H_{10}O_5)_x$

RELATIVE SPECIFIC REFRACTIVITY = 1.039

A % by wt.	D^{20}_{20}	C_s g/l	M g-mol/l	C_w g/l	$(C_o - C_w)$ g/l	$(n - n_o) \times 10^4$	n	Δ °C	S g-mol/l
.00	1.0000	.0	.000	998.2	.0	0	1.3330	.00	.000
.50	1.0019	5.0	.001	995.1	3.1	7	1.3337	.00	.000
1.00	1.0038	10.0	.002	992.0	6.2	14	1.3344	.00	.001
1.50	1.0057	15.1	.003	988.9	9.4	21	1.3351	.01	.002
2.00	1.0076	20.1	.004	985.7	12.5	29	1.3358	.01	.002
2.50	1.0095	25.2	.005	982.6	15.7	36	1.3366	.01	.003
3.00	1.0115	30.3	.006	979.4	18.8	43	1.3373	.02	.004
3.50	1.0134	35.4	.007	976.2	22.0	50	1.3380	.02	.005
4.00	1.0154	40.5	.008	973.0	25.2	58	1.3388	.02	.006
4.50	1.0173	45.7	.009	969.8	28.4	65	1.3395	.02	.006
5.00	1.0193	50.9	.010	966.6	31.6	72	1.3402	.03	.007
5.50	1.0213	56.1	.011	963.4	34.9	80	1.3410	.03	.008
6.00	1.0232	61.3	.012	960.1	38.1	87	1.3417	.03	.009
6.50	1.0252	66.5	.013	956.9	41.3	95	1.3425	.04	.010
7.00	1.0272	71.8	.014	953.6	44.6	102	1.3432	.04	.012
7.50	1.0292	77.1	.015	950.4	47.9	110	1.3440	.05	.013
8.00	1.0312	82.4	.016	947.1	51.2	117	1.3447	.05	.014
8.50	1.0333	87.7	.017	943.8	54.5	125	1.3455	.05	.015
9.00	1.0353	93.0	.018	940.5	57.8	133	1.3463	.06	.016
9.50	1.0373	98.4	.019	937.1	61.1	140	1.3470	.06	.018
10.00	1.0394	103.8	.020	933.8	64.4	148	1.3478	.07	.019

Notes: Supersaturated solutions. The molar concentrations correspond to an average molecular weight of 5200.
A A A

22 LACTOSE, $C_{12}H_{22}O_{11}\cdot 1H_2O$

MOLECULAR WEIGHT = 342.30 FORMULA WEIGHT, HYDRATE = 360.31
RELATIVE SPECIFIC REFRACTIVITY = 1.036

A % by wt.	H % by wt.	D^{20}_{20}	C_s g/l	M g-mol/l	C_w g/l	$(C_o - C_w)$ g/l	$(n - n_o)$ $\times 10^4$	n	Δ °C	S g-mol/l
.00	.00	1.0000	.0	.000	998.2	.0	0	1.3330	.00	.000
.50	.53	1.0020	5.0	.015	995.2	3.0	7	1.3337	.03	.007
1.00	1.05	1.0040	10.0	.029	992.2	6.1	15	1.3345	.05	.015
1.50	1.58	1.0060	15.1	.044	989.1	9.1	22	1.3352	.08	.023
2.00	2.11	1.0080	20.1	.059	986.1	12.2	30	1.3360	.11	.031
2.50	2.63	1.0100	25.2	.074	983.0	15.3	38	1.3367	.14	.040
3.00	3.16	1.0120	30.3	.089	979.9	18.3	45	1.3375	.17	.048
3.50	3.68	1.0140	35.4	.103	976.8	21.5	53	1.3383	.20	.057
4.00	4.21	1.0160	40.6	.119	973.7	24.6	60	1.3390	.23	.065
4.50	4.74	1.0181	45.7	.134	970.5	27.7	68	1.3398	.26	.074
5.00	5.26	1.0201	50.9	.149	967.4	30.9	76	1.3406	.29	.083
5.50	5.79	1.0221	56.1	.164	964.2	34.0	83	1.3413	.32	.093
6.00	6.32	1.0242	61.3	.179	961.0	37.2	91	1.3421	.35	.102
6.50	6.84	1.0262	66.6	.195	957.8	40.4	99	1.3429	.39	.112
7.00	7.37	1.0283	71.9	.210	954.6	43.6	107	1.3437	.42	.121
7.50	7.89	1.0304	77.1	.225	951.4	46.8	114	1.3444	.45	.131
8.00	8.42	1.0324	82.4	.241	948.2	50.1	122	1.3452	.49	.142
8.50	8.95	1.0345	87.8	.256	944.9	53.3	130	1.3460		
9.00	9.47	1.0366	93.1	.272	941.7	56.6	138	1.3468		
9.50	10.00	1.0387	98.5	.288	938.4	59.9	146	1.3476		
10.00	10.53	1.0408	103.9	.304	935.1	63.1	154	1.3484		
11.00	11.58	1.0451	114.8	.335	928.5	69.8	170	1.3500		
12.00	12.63	1.0493	125.7	.367	921.8	76.5	186	1.3516		
13.00	13.68	1.0536	136.7	.399	915.0	83.2	202	1.3532		
14.00	14.74	1.0579	147.9	.432	908.2	90.0	218	1.3548		
15.00	15.79	1.0623	159.1	.465	901.4	96.9	235	1.3565		
16.00	16.84	1.0667	170.4	.498	894.5	103.8	251	1.3581		
17.00	17.89	1.0711	181.8	.531	887.5	110.8	268	1.3598		
18.00	18.95	1.0756	193.3	.565	880.5	117.8	285	1.3615		

(McDonald and Turcotte, J. Res. N.B.S. 4:63) (ibid) C

23 LEAD NITRATE, $Pb(NO_3)_2$

MOLECULAR WEIGHT = 331.23
RELATIVE SPECIFIC REFRACTIVITY = .473

A % by wt.	D^{20}_{20}	C_s g/l	M g-mol/l	C_w g/l	$(C_o - C_w)$ g/l	$(n - n_o)$ $\times 10^4$	n	Δ °C	S g-mol/l
.00	1.0000	.0	.000	998.2	.0	0	1.3330	.00	.000
.50	1.0044	5.0	.015	997.6	.6	6	1.3336	.07	.019
1.00	1.0088	10.1	.030	996.9	1.3	12	1.3342	.13	.038
1.50	1.0132	15.2	.046	996.3	2.0	18	1.3348	.20	.056
2.00	1.0177	20.3	.061	995.6	2.7	24	1.3354	.26	.074
2.50	1.0221	25.5	.077	994.8	3.4	30	1.3360	.32	.091
3.00	1.0266	30.7	.093	994.1	4.1	36	1.3366	.37	.108
3.50	1.0312	36.0	.109	993.3	4.9	42	1.3372	.43	.124
4.00	1.0357	41.4	.125	992.6	5.7	48	1.3378	.48	.140
4.50	1.0403	46.7	.141	991.8	6.5	54	1.3384	.54	.156
5.00	1.0449	52.2	.157	990.9	7.3	60	1.3390	.59	.171
5.50	1.0496	57.6	.174	990.1	8.1	66	1.3396	.64	.186
6.00	1.0543	63.1	.191	989.3	9.0	72	1.3402	.69	.201
6.50	1.0590	68.7	.207	988.4	9.8	78	1.3408	.74	.215
7.00	1.0638	74.3	.224	987.6	10.7	85	1.3415	.79	.230
7.50	1.0686	80.0	.242	986.7	11.6	91	1.3421	.84	.244
8.00	1.0734	85.7	.259	985.8	12.5	97	1.3427	.88	.258
8.50	1.0783	91.5	.276	984.9	13.4	103	1.3433	.93	.271
9.00	1.0832	97.3	.294	984.0	14.3	110	1.3440	.98	.285
9.50	1.0881	103.2	.312	983.0	15.2	116	1.3446	1.02	.299
10.00	1.0931	109.1	.329	982.1	16.1	123	1.3453	1.07	.312
11.00	1.1033	121.1	.366	980.2	18.0	136	1.3466	1.16	.340
12.00	1.1137	133.4	.403	978.3	19.9	149	1.3479	1.26	.367
13.00	1.1241	145.9	.440	976.2	22.0	163	1.3493	1.35	.395
14.00	1.1348	158.6	.479	974.2	24.0	177	1.3506	1.45	.424
15.00	1.1458	171.6	.518	972.2	26.0	191	1.3521	1.55	.453
16.00	1.1570	184.8	.558	970.2	28.0	206	1.3536	1.65	.482
17.00	1.1685	198.3	.599	968.2	30.1	221	1.3551	1.75	.513
18.00	1.1803	212.1	.640	966.1	32.1	236	1.3566	1.86	.543
19.00	1.1922	226.1	.683	964.0	34.3	252	1.3582	1.96	.573
20.00	1.2043	240.4	.726	961.8	36.5	268	1.3598	2.06	.603
22.00	1.2290	269.9	.815	957.0	41.3	299	1.3629	2.26	.659
24.00	1.2542	300.5	.907	951.5	46.7	331	1.3661	2.40	.700
26.00	1.2797	332.1	1.003	945.3	53.0				
28.00	1.3051	364.8	1.101	938.0	60.2				
30.00	1.3302	398.4	1.203	929.5	68.7				

C A C

24 LITHIUM CHLORIDE, LiCl

MOLECULAR WEIGHT = 42.40
RELATIVE SPECIFIC REFRACTIVITY = 1.005

A % by wt.	D^{20}_{20}	C_s g/l	M g-mol/l	C_w g/l	$(C_o - C_w)$ g/l	$(n - n_o) \times 10^4$	n	Δ °C	S g-mol/l
.00	1.0000	.0	.000	998.2	.0	0	1.3330	.00	.000
.50	1.0029	5.0	.118	996.2	2.1	10	1.3340	.41	.119
1.00	1.0059	10.0	.237	994.1	4.2	20	1.3350	.83	.242
1.50	1.0088	15.1	.356	991.9	6.3	31	1.3361	1.27	.371
2.00	1.0117	20.2	.476	989.7	8.5	41	1.3371	1.72	.503
2.50	1.0146	25.3	.597	987.5	10.7	51	1.3381	2.19	.640
3.00	1.0175	30.5	.719	985.2	13.0	61	1.3391	2.68	.781
3.50	1.0204	35.7	.841	982.9	15.3	71	1.3401	3.19	.926
4.00	1.0233	40.9	.964	980.6	17.6	81	1.3411	3.73	1.076
4.50	1.0262	46.1	1.087	978.2	20.0	91	1.3421	4.29	1.238
5.00	1.0290	51.4	1.211	975.9	22.4	102	1.3431	4.88	1.399
5.50	1.0319	56.7	1.336	973.4	24.8	112	1.3441	5.50	1.564
6.00	1.0348	62.0	1.462	971.0	27.2	122	1.3452	6.15	1.733
6.50	1.0377	67.3	1.588	968.5	29.7	132	1.3462	6.84	1.906
7.00	1.0405	72.7	1.715	966.0	32.2	142	1.3472	7.55	2.083
7.50	1.0434	78.1	1.842	963.5	34.8	152	1.3482	8.30	2.263
8.00	1.0463	83.6	1.971	960.9	37.3	162	1.3492	9.09	2.447
8.50	1.0492	89.0	2.100	958.3	39.9	172	1.3502	9.91	2.633
9.00	1.0521	94.5	2.229	955.7	42.6	182	1.3512	10.77	2.821
9.50	1.0549	100.0	2.359	953.0	45.2	192	1.3522	11.66	3.010
10.00	1.0578	105.6	2.490	950.4	47.9	202	1.3532	12.59	3.201
11.00	1.0636	116.8	2.754	944.9	53.3	223	1.3553	14.55	3.581
12.00	1.0694	128.1	3.021	939.4	58.8	243	1.3573	16.63	3.957
13.00	1.0752	139.5	3.291	933.8	64.4	264	1.3593	18.82	4.322
14.00	1.0811	151.1	3.563	928.1	70.1	284	1.3614	21.09	
15.00	1.0870	162.8	3.839	922.3	75.9	305	1.3635	23.39	
16.00	1.0929	174.6	4.117	916.4	81.8	326	1.3656		
17.00	1.0988	186.5	4.398	910.4	87.8	347	1.3677		
18.00	1.1049	198.5	4.682	904.4	93.9	368	1.3698		
19.00	1.1109	210.7	4.969	898.2	100.0	389	1.3719		
20.00	1.1170	223.0	5.260	892.0	106.2	411	1.3741		
22.00	1.1294	248.0	5.850	879.3	118.9				
24.00	1.1419	273.6	6.452	866.3	131.9				
26.00	1.1548	299.7	7.069	853.0	145.2				
28.00	1.1678	326.4	7.699	839.4	158.9				
30.00	1.1812	353.7	8.343	825.4	172.9				

H A C

25 MAGNESIUM CHLORIDE, $MgCl_2 \cdot 6H_2O$

MOLECULAR WEIGHT = 95.23 FORMULA WEIGHT, HYDRATE = 203.33
RELATIVE SPECIFIC REFRACTIVITY = .916

A % by wt.	H % by wt.	D^{20}_{20}	C_s g/l	M g-mol/l	C_w g/l	$(C_o - C_w)$ g/l	$(n - n_o) \times 10^4$	n	Δ °C	S g-mol/l
.00	.00	1.0000	.0	.000	998.2	.0	0	1.3330	.00	.000
.50	1.07	1.0041	5.0	.053	997.3	.9	13	1.3343	.26	.076
1.00	2.14	1.0082	10.1	.106	996.3	1.9	26	1.3355	.55	.161
1.50	3.20	1.0123	15.2	.159	995.3	2.9	38	1.3368	.79	.231
2.00	4.27	1.0164	20.3	.213	994.3	3.9	51	1.3381	1.05	.306
2.50	5.34	1.0205	25.5	.267	993.3	5.0	64	1.3394	1.32	.387
3.00	6.41	1.0247	30.7	.322	992.2	6.1	77	1.3407	1.62	.473
3.50	7.47	1.0288	35.9	.377	991.0	7.2	90	1.3419	1.93	.563
4.00	8.54	1.0329	41.2	.433	989.9	8.4	102	1.3432	2.26	.659
4.50	9.61	1.0371	46.6	.489	988.7	9.6	115	1.3445	2.61	.759
5.00	10.68	1.0413	52.0	.546	987.5	10.8	128	1.3458	2.97	.864
5.50	11.74	1.0454	57.4	.603	986.2	12.0	141	1.3471	3.36	.974
6.00	12.81	1.0496	62.9	.660	984.9	13.3	154	1.3483	3.78	1.089
6.50	13.88	1.0538	68.4	.718	983.6	14.7	166	1.3496	4.21	1.214
7.00	14.95	1.0580	73.9	.776	982.2	16.0	179	1.3509	4.67	1.339
7.50	16.01	1.0623	79.5	.835	980.8	17.4	192	1.3522	5.14	1.468
8.00	17.08	1.0665	85.2	.894	979.4	18.8	205	1.3535	5.65	1.601
8.50	18.15	1.0707	90.9	.954	978.0	20.2	218	1.3548	6.17	1.738
9.00	19.22	1.0750	96.6	1.014	976.5	21.7	231	1.3561	6.73	1.879
9.50	20.28	1.0793	102.3	1.075	975.0	23.2	244	1.3574	7.31	2.023

25 MAGNESIUM CHLORIDE, $MgCl_2 \cdot 6H_2O$

A % by wt.	H % by wt.	D^{20}_{20}	C_s g/l	M g-mol/l	C_w g/l	$(C_o - C_w)$ g/l	$(n - n_o) \times 10^4$	n	Δ °C	S g-mol/l
10.00	21.35	1.0835	108.2	1.136	973.5	24.8	257	1.3586	7.91	2.170
11.00	23.49	1.0922	119.9	1.259	970.3	27.9	282	1.3612	9.21	2.474
12.00	25.62	1.1008	131.9	1.385	967.0	31.2	308	1.3638	10.62	2.789
13.00	27.76	1.1096	144.0	1.512	963.6	34.6	334	1.3664	12.16	3.113
14.00	29.89	1.1184	156.3	1.641	960.1	38.1	360	1.3690	13.83	3.445
15.00	32.03	1.1272	168.8	1.772	956.5	41.8	387	1.3717	15.64	3.782
16.00	34.16	1.1362	181.5	1.906	952.7	45.5	413	1.3743	17.60	4.122
17.00	36.30	1.1452	194.3	2.041	948.8	49.4	439	1.3769		
18.00	38.43	1.1543	207.4	2.178	944.9	53.4	466	1.3796		
19.00	40.57	1.1635	220.7	2.317	940.8	57.5	493	1.3823		
20.00	42.70	1.1728	234.1	2.459	936.5	61.7	519	1.3849		
22.00	46.97	1.1916	261.7	2.748	927.8	70.5	573	1.3903		
24.00	51.24	1.2108	290.1	3.046	918.6	79.7	628	1.3958		
26.00	55.51	1.2304	319.3	3.353	908.9	89.4				
28.00	59.78	1.2505	349.5	3.670	898.8	99.5				
30.00	64.05	1.2711	380.6	3.997	888.2	110.1				

H W C

26 MAGNESIUM SULFATE, $MgSO_4 \cdot 7H_2O$

MOLECULAR WEIGHT = 120.39 FORMULA WEIGHT, HYDRATE = 246.50
RELATIVE SPECIFIC REFRACTIVITY = .575

A % by wt.	H % by wt.	D^{20}_{20}	C_s g/l	M g-mol/l	C_w g/l	$(C_o - C_w)$ g/l	$(n - n_o) \times 10^4$	n	Δ °C	S g-mol/l
.00	.00	1.0000	.0	.000	998.2	.0	0	1.3330	.00	.000
.50	1.02	1.0051	5.0	.042	998.3	−.1	10	1.3340	.09	.026
1.00	2.05	1.0102	10.1	.084	998.3	−.1	20	1.3350	.18	.051
1.50	3.07	1.0153	15.2	.126	998.3	.0	30	1.3360	.26	.076
2.00	4.10	1.0204	20.4	.169	998.2	.0	40	1.3370	.35	.101
2.50	5.12	1.0255	25.6	.213	998.1	.1	51	1.3380	.44	.127
3.00	6.14	1.0307	30.9	.256	998.0	.2	61	1.3390	.53	.152
3.50	7.17	1.0359	36.2	.301	997.8	.4	71	1.3401	.61	.178
4.00	8.19	1.0411	41.6	.345	997.6	.6	81	1.3411	.70	.204
4.50	9.21	1.0463	47.0	.390	997.4	.8	91	1.3421	.79	.230
5.00	10.24	1.0515	52.5	.436	997.2	1.1	101	1.3431	.88	.257
5.50	11.26	1.0568	58.0	.482	996.9	1.3	111	1.3441	.97	.283
6.00	12.29	1.0621	63.6	.528	996.6	1.6	121	1.3451	1.03	.301
6.50	13.31	1.0674	69.3	.575	996.3	2.0	131	1.3461	1.11	.325
7.00	14.33	1.0727	75.0	.623	995.9	2.4	141	1.3471	1.20	.351
7.50	15.36	1.0781	80.7	.670	995.5	2.8	151	1.3481	1.29	.377
8.00	16.38	1.0835	86.5	.719	995.1	3.2	161	1.3491	1.39	.406
8.50	17.40	1.0889	92.4	.767	994.6	3.6	171	1.3501	1.49	.435
9.00	18.43	1.0944	98.3	.817	994.1	4.1	182	1.3511	1.59	.466
9.50	19.45	1.0998	104.3	.866	993.6	4.6	192	1.3522	1.70	.498
10.00	20.48	1.1053	110.3	.917	993.0	5.2	202	1.3532	1.82	.532
11.00	22.52	1.1164	122.6	1.018	991.9	6.4	222	1.3552	2.07	.604
12.00	24.57	1.1276	135.1	1.122	990.6	7.7	243	1.3572	2.34	.682
13.00	26.62	1.1390	147.8	1.228	989.2	9.1	263	1.3593	2.63	.766
14.00	28.67	1.1504	160.8	1.335	987.6	10.6	284	1.3614	2.95	.857
15.00	30.71	1.1620	174.0	1.445	986.0	12.3	304	1.3634	3.29	.954
16.00	32.76	1.1737	187.5	1.557	984.2	14.0	325	1.3655	3.67	1.058
17.00	34.81	1.1856	201.2	1.671	982.3	15.9	346	1.3676		
18.00	36.86	1.1976	215.2	1.787	980.3	17.9	367	1.3697		
19.00	38.90	1.2097	229.4	1.906	978.1	20.1	388	1.3718		
20.00	40.95	1.2220	244.0	2.026	975.9	22.4	409	1.3739		
22.00	45.05	1.2469	273.8	2.275	970.9	27.4	451	1.3781		
24.00	49.14	1.2724	304.8	2.532	965.3	32.9	494	1.3824		
26.00	53.24	1.2984	337.0	2.799	959.1	39.1	537	1.3867		

H W C

27 MALTOSE, $C_{12}H_{22}O_{11} \cdot 1H_2O$

MOLECULAR WEIGHT = 342.29 FORMULA WEIGHT, HYDRATE = 360.31
RELATIVE SPECIFIC REFRACTIVITY = 1.030

A % by wt.	H % by wt.	D^{20}_{20}	C_s g/l	M g-mol/l	C_w g/l	$(C_o - C_w)$ g/l	$(n - n_o) \times 10^4$	n	Δ °C	S g-mol/l
.00	.00	1.0000	.0	.000	998.2	.0	0	1.3330	.00	.000
.50	.53	1.0020	5.0	.015	995.3	3.0	7	1.3337	.03	.008
1.00	1.05	1.0041	10.0	.029	992.3	6.0	15	1.3345	.06	.016
1.50	1.58	1.0061	15.1	.044	989.2	9.0	22	1.3352	.08	.024
2.00	2.11	1.0081	20.1	.059	986.2	12.0	29	1.3359	.11	.032
2.50	2.63	1.0101	25.2	.074	983.1	15.1	37	1.3367	.14	.040
3.00	3.16	1.0121	30.3	.089	980.0	18.2	44	1.3374	.17	.049
3.50	3.68	1.0142	35.4	.104	976.9	21.3	52	1.3382	.20	.057
4.00	4.21	1.0162	40.6	.119	973.8	24.4	59	1.3389	.23	.066
4.50	4.74	1.0182	45.7	.134	970.7	27.6	67	1.3397	.26	.075
5.00	5.26	1.0202	50.9	.149	967.5	30.7	75	1.3404	.29	.083
5.50	5.79	1.0222	56.1	.164	964.3	33.9	82	1.3412	.32	.092
6.00	6.32	1.0243	61.3	.179	961.1	37.1	90	1.3420	.35	.101
6.50	6.84	1.0263	66.6	.195	957.9	40.4	97	1.3427	.38	.111
7.00	7.37	1.0283	71.9	.210	954.6	43.6	105	1.3435	.42	.120
7.50	7.89	1.0303	77.1	.225	951.3	46.9	113	1.3443	.45	.129
8.00	8.42	1.0323	82.4	.241	948.0	50.2	120	1.3450	.48	.139
8.50	8.95	1.0343	87.8	.256	944.7	53.5	128	1.3458	.51	.149
9.00	9.47	1.0363	93.1	.272	941.4	56.8	136	1.3466	.55	.159
9.50	10.00	1.0383	98.5	.288	938.0	60.2	144	1.3474	.58	.169
10.00	10.53	1.0403	103.8	.303	934.6	63.6	152	1.3482	.62	.180
11.00	11.58	1.0443	114.7	.335	927.8	70.4	167	1.3497	.69	.201
12.00	12.63	1.0483	125.6	.367	920.9	77.3	183	1.3513	.77	.224
13.00	13.68	1.0523	136.6	.399	913.9	84.3	199	1.3529	.85	.247
14.00	14.74	1.0563	147.6	.431	906.8	91.4	216	1.3545	.93	.272
15.00	15.79	1.0603	158.8	.464	899.7	98.6	232	1.3562	1.02	.297

(Browne and Zerban, Sugar Chemistry) (McDonald, J. Res. N.B.S. 46:165) (A + C)

28 MANGANESE SULFATE, $MnSO_4 \cdot 1H_2O$

MOLECULAR WEIGHT = 151.00 FORMULA WEIGHT, HYDRATE = 169.01
RELATIVE SPECIFIC REFRACTIVITY = .545

A % by wt.	H % by wt.	D^{20}_{20}	C_s g/l	M g-mol/l	C_w g/l	$(C_o - C_w)$ g/l	$(n - n_o) \times 10^4$	n	Δ °C	S g-mol/l
.00	.00	1.0000	.0	.000	998.2	.0	0	1.3330	.00	.000
.50	.56	1.0049	5.0	.033	998.1	.1	9	1.3339	.08	.023
1.00	1.12	1.0098	10.1	.067	997.9	.3	18	1.3348	.15	.043
1.50	1.68	1.0147	15.2	.101	997.7	.6	27	1.3357	.22	.063
2.00	2.24	1.0196	20.4	.135	997.4	.8	36	1.3366	.29	.084
2.50	2.80	1.0246	25.6	.169	997.2	1.0	45	1.3375	.36	.104
3.00	3.36	1.0296	30.8	.204	996.9	1.3	54	1.3384	.43	.124
3.50	3.92	1.0346	36.1	.239	996.6	1.6	63	1.3393	.50	.145
4.00	4.48	1.0396	41.5	.275	996.3	1.9	72	1.3402	.57	.166
4.50	5.04	1.0447	46.9	.311	995.9	2.3	81	1.3411	.64	.186
5.00	5.60	1.0498	52.4	.347	995.6	2.6	90	1.3420	.78	.227
5.50	6.16	1.0550	57.9	.384	995.2	3.0	100	1.3430	.83	.242
6.00	6.72	1.0602	63.5	.421	994.8	3.4	109	1.3439	.88	.258
6.50	7.28	1.0654	69.1	.458	994.4	3.8	118	1.3448	.94	.274
7.00	7.83	1.0707	74.8	.495	994.0	4.3	127	1.3457	1.00	.291
7.50	8.39	1.0760	80.6	.533	993.5	4.7	136	1.3466	1.06	.309
8.00	8.95	1.0813	86.4	.572	993.1	5.2	146	1.3476	1.12	.328
8.50	9.51	1.0867	92.2	.611	992.6	5.6	155	1.3485	1.19	.348
9.00	10.07	1.0921	98.1	.650	992.1	6.1	164	1.3494	1.26	.369
9.50	10.63	1.0976	104.1	.689	991.6	6.7	174	1.3503	1.33	.390
10.00	11.19	1.1031	110.1	.729	991.1	7.2	183	1.3513	1.41	.412
11.00	12.31	1.1143	122.4	.810	989.9	8.3	202	1.3532	1.57	.460
12.00	13.43	1.1256	134.8	.893	988.8	9.5	221	1.3551	1.75	.512
13.00	14.55	1.1371	147.6	.977	987.5	10.7	240	1.3570	1.94	.567
14.00	15.67	1.1487	160.5	1.063	986.2	12.1	259	1.3589	2.15	.627
15.00	16.79	1.1606	173.8	1.151	984.7	13.5	279	1.3609	2.37	.692
16.00	17.91	1.1726	187.3	1.240	983.2	15.0	298	1.3628	2.61	.761
17.00	19.03	1.1848	201.1	1.331	981.6	16.6	318	1.3648	2.87	.834
18.00	20.15	1.1971	215.1	1.424	979.9	18.3	337	1.3667	3.15	.913
19.00	21.27	1.2096	229.4	1.519	978.0	20.2	357	1.3687	3.45	.997
20.00	22.39	1.2222	244.0	1.616	976.1	22.2	376	1.3706	3.77	1.087

H W C

29 D-MANNITOL, $CH_2OH(CHOH)_4CH_2OH$

MOLECULAR WEIGHT = 182.17
RELATIVE SPECIFIC REFRACTIVITY = 1.068

A % by wt.	D^{20}_{20}	C_s g/l	M g-mol/l	C_w g/l	$(C_o - C_w)$ g/l	$(n - n_o) \times 10^4$	n	Δ °C	S g-mol/l
.00	1.0000	.0	.000	998.2	.0	0	1.3330	.00	.000
.50	1.0018	5.0	.027	995.0	3.2	7	1.3337	.05	.014
1.00	1.0035	10.0	.055	991.7	6.5	14	1.3344	.10	.029
1.50	1.0053	15.1	.083	988.5	9.8	22	1.3352	.15	.044
2.00	1.0070	20.1	.110	985.2	13.1	29	1.3359	.21	.059
2.50	1.0088	25.2	.138	981.9	16.4	36	1.3366	.26	.074
3.00	1.0106	30.3	.166	978.5	19.7	44	1.3373	.31	.090
3.50	1.0124	35.4	.194	975.2	23.0	51	1.3381	.37	.106
4.00	1.0141	40.5	.222	971.8	26.4	58	1.3388	.42	.122
4.50	1.0159	45.6	.251	968.5	29.8	66	1.3395	.48	.138
5.00	1.0177	50.8	.279	965.1	33.1	73	1.3403	.53	.155
5.50	1.0195	56.0	.307	961.7	36.5	80	1.3410	.59	.171
6.00	1.0213	61.2	.336	958.3	39.9	88	1.3418	.65	.188
6.50	1.0230	66.4	.364	954.9	43.4	95	1.3425	.71	.206
7.00	1.0248	71.6	.393	951.4	46.8	102	1.3432	.77	.223
7.50	1.0266	76.9	.422	948.0	50.3	110	1.3440	.83	.241
8.00	1.0284	82.1	.451	944.5	53.7	117	1.3447	.89	.259
8.50	1.0302	87.4	.480	941.0	57.2	125	1.3455	.95	.277
9.00	1.0320	92.7	.509	937.5	60.7	132	1.3462	1.01	.295
9.50	1.0338	98.0	.538	934.0	64.3	140	1.3470	1.08	.314
10.00	1.0357	103.4	.568	930.4	67.8	147	1.3477	1.14	.333

A A (C + A)

30 METHANOL, CH_3OH

MOLECULAR WEIGHT = 32.04
RELATIVE SPECIFIC REFRACTIVITY = 1.244

A % by wt.	D^{20}_{20}	C_s g/l	M g-mol/l	C_w g/l	$(C_o - C_w)$ g/l	$(n - n_o) \times 10^4$	n	Δ °C	S g-mol/l
.00	1.0000	.0	.000	998.2	.0	0	1.3330	.00	.000
1.00	.9982	10.0	.311	986.5	11.7	2	1.3332	.56	.163
2.00	.9965	19.9	.621	974.8	23.4	4	1.3334	1.14	.333
3.00	.9948	29.8	.930	963.2	35.0	7	1.3337	1.74	.510
4.00	.9931	39.7	1.238	951.7	46.6	9	1.3339	2.37	.692
5.00	.9914	49.5	1.545	940.2	58.1	11	1.3341	3.03	.879
6.00	.9898	59.3	1.851	928.7	69.5	14	1.3344	3.71	1.069
7.00	.9881	69.0	2.156	917.3	80.9	16	1.3346	4.41	1.270
8.00	.9865	78.8	2.460	906.0	92.2	19	1.3349	5.13	1.464
9.00	.9849	88.5	2.763	894.7	103.5	22	1.3352	5.86	1.657
10.00	.9834	98.2	3.065	883.5	114.8	24	1.3354	6.57	1.840
11.00	.9818	107.8	3.366	872.3	126.0	27	1.3357	7.31	2.024
12.00	.9803	117.4	3.666	861.1	137.1	30	1.3360	8.07	2.208
13.00	.9788	127.0	3.965	850.0	148.2	33	1.3363	8.86	2.394
14.00	.9772	136.6	4.264	838.9	159.3	35	1.3365	9.67	2.579
15.00	.9757	146.1	4.561	827.9	170.3	38	1.3368	10.51	2.764
16.00	.9742	155.6	4.858	816.9	181.3	41	1.3371	11.37	2.949
17.00	.9727	165.1	5.154	805.9	192.3	44	1.3374	12.25	3.133
18.00	.9712	174.5	5.448	795.0	203.2	47	1.3376	13.16	3.315
19.00	.9697	183.9	5.742	784.1	214.1	49	1.3379	14.10	3.496
20.00	.9682	193.3	6.035	773.2	225.0	52	1.3382	15.05	3.675
22.00	.9653	212.0	6.618	751.6	246.7	58	1.3388	17.04	4.027
24.00	.9623	230.5	7.197	730.0	268.2	63	1.3393	19.11	4.368
26.00	.9592	249.0	7.773	708.6	289.7	68	1.3398	21.26	
28.00	.9562	267.3	8.344	687.2	311.0	73	1.3403	23.49	
30.00	.9531	285.4	8.911	666.0	332.3	78	1.3408	25.79	
32.00	.9499	303.4	9.473	644.8	353.4	82	1.3412	28.15	
34.00	.9467	321.3	10.031	623.7	374.5	86	1.3416	30.57	
36.00	.9433	339.0	10.584	602.7	395.6	90	1.3420	33.05	
38.00	.9399	356.5	11.131	581.7	416.5	93	1.3423	35.58	
40.00	.9364	373.9	11.673	560.8	437.4	96	1.3426		
44.00	.9290	408.0	12.739	519.3	478.9	99	1.3429		
48.00	.9210	441.3	13.778	478.1	520.2	100	1.3430		
52.00	.9131	474.0	14.797	437.5	560.7	101	1.3431		
56.00	.9048	505.8	15.792	397.4	600.8	100	1.3430		
60.00	.8962	536.7	16.757	357.8	640.4	96	1.3426		
64.00	.8871	566.8	17.694	318.8	679.4	91	1.3421		
68.00	.8778	595.8	18.602	280.4	717.8	84	1.3414		
72.00	.8682	624.0	19.481	242.7	755.6	75	1.3405		
76.00	.8584	651.2	20.331	205.6	792.6	66	1.3396		
80.00	.8483	677.5	21.151	169.4	828.9	54	1.3384		
84.00	.8381	702.7	21.940	133.9	864.4	42	1.3372		
88.00	.8274	726.8	22.691	99.1	899.1	27	1.3357		

L W (A + L)

31 NICKEL SULFATE, $NiSO_4 \cdot 6H_2O$

MOLECULAR WEIGHT = 154.76 FORMULA WEIGHT, HYDRATE = 262.85
RELATIVE SPECIFIC REFRACTIVITY = .527

A % by wt.	H % by wt.	D^{20}_{20}	C_s g/l	M g-mol/l	C_w g/l	$(C_o - C_w)$ g/l	$(n - n_o) \times 10^4$	n	Δ °C	S g-mol/l
.00	.00	1.0000	.0	.000	998.2	.0	0	1.3330	.00	.000
.50	.85	1.0054	5.0	.032	998.6	−.3	10	1.3340	.08	.021
1.00	1.70	1.0107	10.1	.065	998.8	−.6	20	1.3350	.14	.041
1.50	2.55	1.0161	15.2	.098	999.1	−.8	30	1.3360	.21	.061
2.00	3.40	1.0215	20.4	.132	999.3	−1.0	40	1.3370	.28	.081
2.50	4.25	1.0268	25.6	.166	999.4	−1.2	50	1.3380	.35	.101
3.00	5.10	1.0322	30.9	.200	999.5	−1.3	60	1.3390	.42	.120
3.50	5.94	1.0376	36.3	.234	999.6	−1.3	70	1.3400	.48	.139
4.00	6.79	1.0431	41.6	.269	999.6	−1.3	80	1.3410	.55	.158
4.50	7.64	1.0485	47.1	.304	999.5	−1.3	90	1.3420	.61	.177
5.00	8.49	1.0539	52.6	.340	999.5	−1.2	100	1.3430	.67	.196
5.50	9.34	1.0594	58.2	.376	999.3	−1.1	110	1.3440	.74	.214
6.00	10.19	1.0648	63.8	.412	999.2	−.9	120	1.3450	.80	.233

(H + A) W C

32 NITRIC ACID, HNO_3

MOLECULAR WEIGHT = 63.02
RELATIVE SPECIFIC REFRACTIVITY = .823

A % by wt.	D^{20}_{20}	C_s g/l	M g-mol/l	C_w g/l	$(C_o - C_w)$ g/l	$(n - n_o) \times 10^4$	n	Δ °C	S g-mol/l
.00	1.0000	.0	.000	998.2	.0	0	1.3330	.00	.000
.50	1.0027	5.0	.079	995.9	2.3	6	1.3336	.28	.081
1.00	1.0054	10.0	.159	993.6	4.6	13	1.3342	.54	.157
1.50	1.0081	15.1	.240	991.3	7.0	19	1.3349	.83	.240
2.00	1.0109	20.2	.320	988.9	9.3	25	1.3355	1.11	.325
2.50	1.0136	25.3	.401	986.5	11.7	32	1.3362	1.41	.412
3.00	1.0164	30.4	.483	984.1	14.1	38	1.3368	1.71	.500
3.50	1.0191	35.6	.565	981.7	16.5	44	1.3374	2.02	.589
4.00	1.0219	40.8	.647	979.3	18.9	51	1.3381	2.33	.680
4.50	1.0247	46.0	.730	976.8	21.4	57	1.3387	2.65	.771
5.00	1.0274	51.3	.814	974.3	23.9	63	1.3393	2.97	.863
5.50	1.0302	56.6	.898	971.8	26.4	70	1.3400	3.30	.956
6.00	1.0329	61.9	.982	969.2	29.0	76	1.3406	3.64	1.051
6.50	1.0358	67.2	1.066	966.8	31.5	83	1.3413	3.98	1.146
7.00	1.0387	72.6	1.152	964.3	33.9	89	1.3419	4.33	1.249
7.50	1.0416	78.0	1.237	961.8	36.4	96	1.3426	4.69	1.347
8.00	1.0446	83.4	1.324	959.3	38.9	103	1.3433	5.06	1.445
8.50	1.0475	88.9	1.410	956.8	41.5	110	1.3440	5.43	1.544
9.00	1.0504	94.4	1.497	954.2	44.1	116	1.3446	5.81	1.643
9.50	1.0533	99.9	1.585	951.6	46.6	123	1.3453	6.20	1.743
10.00	1.0563	105.4	1.673	949.0	49.3	130	1.3460	6.59	1.844
11.00	1.0622	116.6	1.851	943.7	54.6	143	1.3473	7.41	2.047
12.00	1.0681	127.9	2.030	938.3	60.0	157	1.3487	8.25	2.251
13.00	1.0740	139.4	2.212	932.8	65.5	170	1.3500	9.14	2.458
14.00	1.0800	150.9	2.395	927.2	71.0	184	1.3514	10.06	2.666
15.00	1.0861	162.6	2.580	921.5	76.7	197	1.3527	11.02	2.875
16.00	1.0921	174.4	2.768	915.8	82.5	211	1.3541	12.02	3.085
17.00	1.0983	186.4	2.957	909.9	88.3	225	1.3555	13.06	3.296
18.00	1.1044	198.4	3.149	904.0	94.2	238	1.3568	14.15	3.507
19.00	1.1106	210.6	3.343	898.0	100.2	252	1.3582	15.29	3.719
20.00	1.1169	223.0	3.538	891.9	106.3	266	1.3596	16.48	3.931
22.00	1.1296	248.1	3.936	879.5	118.7	294	1.3624		
24.00	1.1425	273.7	4.343	866.7	131.5	323	1.3653		
26.00	1.1555	299.9	4.759	853.6	144.6	352	1.3681		
28.00	1.1688	326.7	5.184	840.0	158.2	380	1.3710		
30.00	1.1822	354.0	5.618	826.0	172.2	409	1.3739		
32.00	1.1956	381.9	6.060	811.6	186.7				
34.00	1.2091	410.4	6.512	796.6	201.6				
36.00	1.2225	439.3	6.971	781.0	217.2				
38.00	1.2357	468.7	7.438	764.8	233.4				
40.00	1.2486	498.6	7.911	747.8	250.4				

H W C

33 PHOSPHORIC ACID, H_3PO_4

MOLECULAR WEIGHT = 98.00

RELATIVE SPECIFIC REFRACTIVITY = .729

A % by wt.	D^{20}_{20}	C_s g/l	M g-mol/l	C_w g/l	$(C_o - C_w)$ g/l	$(n - n_o) \times 10^4$	n	Δ °C	S g-mol/l
.00	1.0000	.0	.000	998.2	.0	0	1.3330	.00	.000
.50	1.0028	5.0	.051	996.0	2.2	5	1.3335	.13	.036
1.00	1.0055	10.0	.102	993.7	4.5	10	1.3340	.24	.068
1.50	1.0082	15.1	.154	991.4	6.9	15	1.3344	.35	.100
2.00	1.0110	20.2	.206	989.0	9.2	19	1.3349	.46	.133
2.50	1.0137	25.3	.258	986.6	11.6	24	1.3354	.58	.168
3.00	1.0164	30.4	.311	984.2	14.1	29	1.3358	.70	.202
3.50	1.0191	35.6	.363	981.7	16.5	33	1.3363	.82	.238
4.00	1.0218	40.8	.416	979.2	19.0	38	1.3368	.94	.274
4.50	1.0246	46.0	.470	976.7	21.5	42	1.3372	1.07	.311
5.00	1.0273	51.3	.523	974.2	24.0	47	1.3377	1.13	.329
5.50	1.0300	56.6	.577	971.7	26.6	51	1.3381	1.28	.374
6.00	1.0328	61.9	.631	969.1	29.1	56	1.3386	1.44	.420
6.50	1.0355	67.2	.686	966.5	31.7	60	1.3390	1.59	.466
7.00	1.0383	72.6	.740	963.9	34.3	65	1.3395	1.75	.513
7.50	1.0410	77.9	.795	961.3	37.0	70	1.3399	1.92	.561
8.00	1.0438	83.4	.851	958.6	39.6	74	1.3404	2.08	.609
8.50	1.0466	88.8	.906	956.0	42.3	79	1.3409	2.25	.657
9.00	1.0494	94.3	.962	953.3	45.0	83	1.3413	2.42	.706
9.50	1.0522	99.8	1.018	950.6	47.6	88	1.3418	2.59	.756
10.00	1.0551	105.3	1.075	947.9	50.4	92	1.3422	2.77	.806
11.00	1.0608	116.5	1.189	942.4	55.8	102	1.3432	3.13	.908
12.00	1.0665	127.8	1.304	936.9	61.3	111	1.3441	3.50	1.013
13.00	1.0724	139.2	1.420	931.3	66.9	120	1.3450	3.89	1.120
14.00	1.0783	150.7	1.538	935.7	72.5	130	1.3460	4.29	1.236
15.00	1.0843	162.4	1.657	920.0	78.2	140	1.3470	4.70	1.349
16.00	1.0903	174.1	1.777	914.3	84.0	149	1.3479	5.13	1.464
17.00	1.0965	186.1	1.899	908.5	89.8	159	1.3489	5.57	1.581
18.00	1.1027	198.1	2.022	902.6	95.6	169	1.3499	6.03	1.700
19.00	1.1090	210.3	2.146	896.7	101.5	180	1.3510	6.50	1.821
20.00	1.1154	222.7	2.272	890.7	107.5	190	1.3520	6.99	1.944
22.00	1.1283	247.8	2.528	878.5	119.7	211	1.3540	8.03	2.198
24.00	1.1415	273.5	2.791	866.0	132.2	231	1.3561	9.15	2.460
26.00	1.1549	299.7	3.059	853.1	145.1	252	1.3582	10.35	2.731
28.00	1.1686	326.6	3.333	839.9	158.3	274	1.3604	11.66	3.011
30.00	1.1825	354.1	3.614	826.3	171.9	295	1.3625	13.08	3.299
32.00	1.1967	382.3	3.901	812.3	185.9	317	1.3647		
34.00	1.2111	411.0	4.194	797.9	200.3	338	1.3668		
36.00	1.2258	440.5	4.495	783.1	215.1	360	1.3690		
38.00	1.2407	470.6	4.802	767.9	230.4	382	1.3712		
40.00	1.2559	501.5	5.117	752.2	246.0	404	1.3734		

H W C

34 POTASSIUM BROMIDE, KBr

MOLECULAR WEIGHT = 119.01

RELATIVE SPECIFIC REFRACTIVITY = .629

A % by wt.	D^{20}_{20}	C_s g/l	M g-mol/l	C_w g/l	$(C_o - C_w)$ g/l	$(n - n_o) \times 10^4$	n	Δ °C	S g-mol/l
.00	1.0000	.0	.000	998.2	.0	0	1.3330	.00	.000
.50	1.0036	5.0	.042	996.8	1.4	6	1.3336	.15	.042
1.00	1.0072	10.1	.084	995.4	2.9	12	1.3342	.29	.084
1.50	1.0108	15.1	.127	993.9	4.3	18	1.3348	.44	.126
2.00	1.0145	20.3	.170	992.4	5.8	24	1.3354	.58	.169
2.50	1.0182	25.4	.214	990.9	7.3	30	1.3360	.73	.212
3.00	1.0218	30.6	.257	989.4	8.8	36	1.3366	.88	.256
3.50	1.0256	35.8	.301	987.9	10.3	42	1.3372	1.03	.299
4.00	1.0293	41.1	.345	986.4	11.9	49	1.3378	1.18	.343
4.50	1.0330	46.4	.390	984.8	13.4	55	1.3385	1.33	.388
5.00	1.0368	51.7	.435	983.2	15.0	61	1.3391	1.48	.432
5.50	1.0406	57.1	.480	981.7	16.6	67	1.3397	1.63	.477
6.00	1.0444	62.6	.526	980.0	18.2	73	1.3403	1.79	.522
6.50	1.0483	68.0	.572	978.4	19.8	80	1.3410	1.94	.568
7.00	1.0522	73.5	.618	976.8	21.4	86	1.3416	2.10	.613
7.50	1.0561	79.1	.664	975.1	23.1	92	1.3422	2.26	.659
8.00	1.0600	84.6	.711	973.5	24.8	99	1.3429	2.42	.705
8.50	1.0639	90.3	.759	971.8	26.5	105	1.3435	2.58	.752
9.00	1.0679	95.9	.806	970.1	28.2	111	1.3441	2.74	.798
9.50	1.0719	101.6	.854	968.3	29.9	118	1.3448	2.91	.845

34 POTASSIUM BROMIDE, KBr

A % by wt.	D^{20}_{20}	C_s g/l	M g-mol/l	C_w g/l	$(C_o - C_w)$ g/l	$(n - n_o)$ × 10^4	n	Δ °C	S g-mol/l
10.00	1.0759	107.4	.902	966.6	31.6	124	1.3454	3.07	.892
11.00	1.0840	119.0	1.000	963.1	35.2	137	1.3467	3.41	.988
12.00	1.0922	130.8	1.099	959.5	38.8	151	1.3480	3.76	1.084
13.00	1.1005	142.8	1.200	955.8	42.4	164	1.3494	4.12	1.189
14.00	1.1090	155.0	1.302	952.0	46.2	177	1.3507	4.48	1.289
15.00	1.1175	167.3	1.406	948.2	50.0	191	1.3521	4.85	1.391
16.00	1.1262	179.9	1.511	944.3	53.9	205	1.3534	5.24	1.493
17.00	1.1350	192.6	1.618	940.4	57.8	218	1.3548	5.63	1.598
18.00	1.1439	205.5	1.727	936.4	61.9	232	1.3562	6.04	1.703
19.00	1.1530	218.7	1.837	932.2	66.0	247	1.3577	6.45	1.810
20.00	1.1621	232.0	1.950	928.1	70.2	261	1.3591	6.88	1.917
22.00	1.1809	259.3	2.179	919.4	78.8	290	1.3620	7.77	2.136
24.00	1.2001	287.5	2.416	910.5	87.7	320	1.3650	8.71	2.358
26.00	1.2200	316.6	2.661	901.2	97.0	351	1.3681	9.68	2.582
28.00	1.2404	346.7	2.913	891.5	106.7	382	1.3712		
30.00	1.2615	377.8	3.174	881.5	116.7	414	1.3744		
32.00	1.2832	409.9	3.444	871.1	127.2	447	1.3777		
34.00	1.3056	443.1	3.723	860.2	138.1	481	1.3811		
36.00	1.3287	477.5	4.012	848.9	149.4	515	1.3845		
38.00	1.3525	513.0	4.311	837.1	161.2	550	1.3880		
40.00	1.3770	549.8	4.620	824.8	173.5	587	1.3916		

H W C

35 POTASSIUM CARBONATE, $K_2CO_3 \cdot 1\frac{1}{2}H_2O$

MOLECULAR WEIGHT = 138.20 FORMULA WEIGHT, HYDRATE = 165.23

RELATIVE SPECIFIC REFRACTIVITY = .609

A % by wt.	H % by wt.	D^{20}_{20}	C_s g/l	M g-mol/l	C_w g/l	$(C_o - C_w)$ g/l	$(n - n_o)$ × 10^4	n	Δ °C	S g-mol/l
.00	.00	1.0000	.0	.000	998.2	.0	0	1.3330	.00	.000
.50	.60	1.0045	5.0	.036	997.7	.5	9	1.3339	.17	.049
1.00	1.20	1.0090	10.1	.073	997.1	1.1	17	1.3347	.34	.098
1.50	1.79	1.0135	15.2	.110	996.5	1.7	26	1.3356	.50	.146
2.00	2.39	1.0180	20.3	.147	995.9	2.3	35	1.3365	.66	.193
2.50	2.99	1.0226	25.5	.185	995.2	3.0	43	1.3373	1.07	.311
3.00	3.59	1.0271	30.8	.223	994.5	3.7	52	1.3382	1.18	.346
3.50	4.18	1.0317	36.0	.261	993.8	4.4	61	1.3390	1.31	.382
4.00	4.78	1.0362	41.4	.299	993.0	5.2	69	1.3399	1.43	.419
4.50	5.38	1.0408	46.8	.338	992.2	6.0	78	1.3408	1.56	.457
5.00	5.98	1.0454	52.2	.378	991.4	6.8	86	1.3416	1.70	.497
5.50	6.58	1.0501	57.7	.417	990.6	7.7	95	1.3425	1.84	.538
6.00	7.17	1.0547	63.2	.457	989.7	8.6	103	1.3433	1.99	.580
6.50	7.77	1.0593	68.7	.497	988.7	9.5	112	1.3442	2.14	.624
7.00	8.37	1.0640	74.3	.538	987.8	10.4	120	1.3450	2.29	.669
7.50	8.97	1.0687	80.0	.579	986.8	11.4	129	1.3459	2.45	.715
8.00	9.56	1.0734	85.7	.620	985.8	12.5	138	1.3467	2.62	.763
8.50	10.16	1.0781	91.5	.662	984.7	13.5	146	1.3476	2.79	.812
9.00	10.76	1.0828	97.3	.704	983.6	14.6	155	1.3484	2.97	.863
9.50	11.36	1.0876	103.1	.746	982.5	15.7	163	1.3493	3.15	.914
10.00	11.96	1.0924	109.0	.789	981.4	16.8	172	1.3501	3.34	.968
11.00	13.15	1.1020	121.0	.876	979.0	19.2	189	1.3518	3.74	1.079
12.00	14.35	1.1116	133.2	.964	976.5	21.7	206	1.3535	4.17	1.203
13.00	15.54	1.1214	145.5	1.053	973.9	24.4	223	1.3552	4.62	1.327
14.00	16.74	1.1312	158.1	1.144	971.1	27.1	240	1.3569	5.10	1.458
15.00	17.93	1.1411	170.9	1.236	968.2	30.0	257	1.3587	5.62	1.594
16.00	19.13	1.1511	183.8	1.330	965.2	33.0	274	1.3604	6.17	1.736
17.00	20.32	1.1611	197.0	1.426	962.1	36.2	291	1.3621	6.75	1.884
18.00	21.52	1.1713	210.5	1.523	958.8	39.5	308	1.3638	7.37	2.039
19.00	22.72	1.1815	224.1	1.622	955.4	42.9	325	1.3655	8.03	2.199
20.00	23.91	1.1919	238.0	1.722	951.8	46.4	342	1.3672	8.74	2.365
22.00	26.30	1.2128	266.3	1.927	944.3	53.9	376	1.3706	10.28	2.714
24.00	28.69	1.2341	295.7	2.139	936.3	62.0	410	1.3740	12.02	3.085
26.00	31.09	1.2558	325.9	2.358	927.6	70.6	445	1.3774	13.99	3.477
28.00	33.48	1.2778	357.1	2.584	918.4	79.9	479	1.3809	16.23	3.886
30.00	35.87	1.3001	389.4	2.817	908.5	89.7	513	1.3843	18.75	4.311
32.00	38.26	1.3228	422.6	3.058	897.9	100.3				
34.00	40.65	1.3459	456.8	3.305	886.7	111.5				
36.00	43.04	1.3692	492.0	3.560	874.7	123.5				
38.00	45.43	1.3927	528.3	3.823	862.0	136.3				
40.00	47.82	1.4165	565.6	4.093	848.4	149.8				

H W C

36 POTASSIUM CHLORIDE, KCl

MOLECULAR WEIGHT = 74.55

RELATIVE SPECIFIC REFRACTIVITY = .757

A % by wt.	D^{20}_{20}	C_s g/l	M g-mol/l	C_w g/l	$(C_o - C_w)$ g/l	$(n - n_o) \times 10^4$	n	Δ °C	S g-mol/l
.00	1.0000	.0	.000	998.2	.0	0	1.3330	.00	.000
.50	1.0032	5.0	.067	996.4	1.8	7	1.3337	.23	.067
1.00	1.0064	10.0	.135	994.6	3.7	14	1.3343	.46	.133
1.50	1.0096	15.1	.203	992.7	5.5	20	1.3350	.69	.200
2.00	1.0128	20.2	.271	990.8	7.4	27	1.3357	.92	.267
2.50	1.0160	25.4	.340	988.9	9.4	34	1.3364	1.15	.335
3.00	1.0192	30.5	.409	986.9	11.3	41	1.3371	1.38	.404
3.50	1.0225	35.7	.479	984.9	13.3	47	1.3377	1.61	.472
4.00	1.0257	41.0	.549	982.9	15.3	54	1.3384	1.85	.541
4.50	1.0289	46.2	.620	980.9	17.3	61	1.3391	2.09	.610
5.00	1.0322	51.5	.691	978.8	19.4	68	1.3398	2.32	.678
5.50	1.0354	56.8	.763	976.8	21.5	75	1.3404	2.56	.747
6.00	1.0387	62.2	.835	974.7	23.6	81	1.3411	2.80	.816
6.50	1.0420	67.6	.907	972.5	25.7	88	1.3418	3.05	.885
7.00	1.0453	73.0	.980	970.4	27.8	95	1.3425	3.29	.954
7.50	1.0486	78.5	1.053	968.2	30.0	102	1.3432	3.54	1.023
8.00	1.0519	84.0	1.127	966.0	32.2	109	1.3438	3.79	1.092
8.50	1.0552	89.5	1.201	963.8	34.4	115	1.3445	4.04	1.167
9.00	1.0585	95.1	1.276	961.6	36.7	122	1.3452	4.29	1.238
9.50	1.0619	100.7	1.351	959.3	39.0	129	1.3459	4.55	1.308
10.00	1.0652	106.3	1.426	957.0	41.2	136	1.3466	4.81	1.379
11.00	1.0719	117.7	1.579	952.3	45.9	150	1.3479	5.34	1.520
12.00	1.0787	129.2	1.733	947.6	50.7	163	1.3493	5.89	1.664
13.00	1.0855	140.9	1.890	942.7	55.5	177	1.3507	6.45	1.809
14.00	1.0924	152.7	2.048	937.8	60.4	191	1.3521		
15.00	1.0993	164.6	2.208	932.8	65.5	205	1.3535		
16.00	1.1063	176.7	2.370	927.7	70.6	219	1.3549		
17.00	1.1133	188.9	2.534	922.4	75.8	233	1.3563		
18.00	1.1204	201.3	2.701	917.1	81.1	247	1.3577		
19.00	1.1276	213.9	2.869	911.7	86.5	262	1.3592		
20.00	1.1348	226.6	3.039	906.2	92.0	276	1.3606		
22.00	1.1494	252.4	3.386	895.0	103.2	305	1.3635		
24.00	1.1643	278.9	3.742	883.3	114.9	335	1.3665		

H W (S + C)

37 POTASSIUM CHROMATE, K_2CrO_4

MOLECULAR WEIGHT = 194.20

RELATIVE SPECIFIC REFRACTIVITY = .836

A % by wt.	D^{20}_{20}	C_s g/l	M g-mol/l	C_w g/l	$(C_o - C_w)$ g/l	$(n - n_o) \times 10^4$	n	Δ °C	S g-mol/l
.00	1.0000	.0	.000	998.2	.0	0	1.3330	.00	.000
.50	1.0040	5.0	.026	997.2	1.0	11	1.3341	.08	.022
1.00	1.0080	10.1	.052	996.2	2.0	22	1.3352	.16	.045
1.50	1.0121	15.2	.078	995.1	3.1	33	1.3363	.24	.068
2.00	1.0161	20.3	.104	994.0	4.2	44	1.3374	.32	.091
2.50	1.0202	25.5	.131	992.9	5.3	55	1.3385	.40	.114
3.00	1.0243	30.7	.158	991.8	6.4	67	1.3397	.48	.138
3.50	1.0284	35.9	.185	990.7	7.6	78	1.3408	.56	.162
4.00	1.0325	41.2	.212	989.5	8.8	89	1.3419	.64	.187
4.50	1.0367	46.6	.240	988.3	10.0	100	1.3430	.73	.212
5.00	1.0408	51.9	.268	987.0	11.2	112	1.3441	.81	.237
5.50	1.0450	57.4	.295	985.8	12.4	123	1.3453	.90	.262
6.00	1.0492	62.8	.324	984.5	13.7	134	1.3464	.99	.287
6.50	1.0535	68.4	.352	983.2	15.0	146	1.3476	1.07	.313
7.00	1.0577	73.9	.381	981.9	16.3	157	1.3487	1.16	.339
7.50	1.0620	79.5	.409	980.6	17.7	168	1.3498	1.21	.354
8.00	1.0663	85.1	.438	979.2	19.0	180	1.3510	1.32	.385
8.50	1.0706	90.8	.468	977.8	20.4	191	1.3521	1.42	.416
9.00	1.0749	96.6	.497	976.4	21.8	203	1.3533	1.53	.448
9.50	1.0793	102.3	.527	975.0	23.2	215	1.3545	1.64	.480
10.00	1.0836	108.2	.557	973.5	24.7	226	1.3556	1.75	.512
11.00	1.0925	120.0	.618	970.6	27.7	250	1.3580	1.98	.579
12.00	1.1014	131.9	.679	967.5	30.7	273	1.3603	2.21	.647
13.00	1.1104	144.1	.742	964.3	33.9	297	1.3627	2.46	.716
14.00	1.1195	156.5	.806	961.1	37.1	321	1.3651	2.71	.788

37 POTASSIUM CHROMATE, K_2CrO_4

A % by wt.	D^{20}_{20}	C_s g/l	M g-mol/l	C_w g/l	$(C_o - C_w)$ g/l	$(n - n_o) \times 10^4$	n	Δ °C	S g-mol/l
15.00	1.1287	169.0	.870	957.7	40.5	345	1.3675	2.96	.861
16.00	1.1381	181.8	.936	954.3	43.9	369	1.3699	3.23	.936
17.00	1.1475	194.7	1.003	950.8	47.5	393	1.3723	3.50	1.012
18.00	1.1571	207.9	1.071	947.1	51.1	418	1.3748	3.79	1.091
19.00	1.1667	221.3	1.139	943.4	54.9	443	1.3773	4.08	1.178
20.00	1.1765	234.9	1.210	939.5	58.7	468	1.3798	4.38	1.262
22.00	1.1964	262.8	1.353	931.6	66.7				
24.00	1.2168	291.5	1.501	923.2	75.1				
26.00	1.2378	321.3	1.654	914.3	83.9				
28.00	1.2592	352.0	1.812	905.0	93.2				
30.00	1.2812	383.7	1.976	895.2	103.0				
32.00	1.3037	416.4	2.144	884.9	113.3				
34.00	1.3267	450.3	2.319	874.1	124.1				
36.00	1.3503	485.3	2.499	862.7	135.5				
38.00	1.3745	521.4	2.685	850.7	147.5				
40.00	1.3992	558.7	2.877	838.0	160.2				

H W C

38 POTASSIUM DICHROMATE, $K_2Cr_2O_7$

MOLECULAR WEIGHT = 294.21

RELATIVE SPECIFIC REFRACTIVITY = .833

A % by wt.	D^{20}_{20}	C_s g/l	M g-mol/l	C_w g/l	$(C_o - C_w)$ g/l	$(n - n_o) \times 10^4$	n	Δ °C	S g-mol/l
.00	1.0000	.0	.000	998.2	.0	0	1.3330	.00	.000
.50	1.0035	5.0	.017	996.7	1.5	9	1.3339	.10	.028
1.00	1.0070	10.1	.034	995.1	3.1	18	1.3348	.19	.055
1.50	1.0105	15.1	.051	993.6	4.7	27	1.3357	.28	.080
2.00	1.0140	20.2	.069	992.0	6.3	37	1.3367	.36	.104
2.50	1.0175	25.4	.086	990.3	7.9	46	1.3376	.44	.126
3.00	1.0211	30.6	.104	988.7	9.5	55	1.3385	.51	.146
3.50	1.0246	35.8	.122	987.0	11.2	64	1.3394		
4.00	1.0282	41.1	.140	985.3	12.9	74	1.3404		
4.50	1.0318	46.3	.158	983.6	14.6	83	1.3413		
5.00	1.0354	51.7	.176	981.9	16.3	92	1.3422		
5.50	1.0390	57.0	.194	980.1	18.1	102	1.3432		
6.00	1.0426	62.4	.212	978.3	19.9	111	1.3441		
6.50	1.0463	67.9	.231	976.5	21.7	121	1.3451		
7.00	1.0499	73.4	.249	974.7	23.5	130	1.3460		
7.50	1.0536	78.9	.268	972.9	25.4	140	1.3469		
8.00	1.0573	84.4	.287	971.0	27.2	149	1.3479		
8.50	1.0610	90.0	.306	969.1	29.1	159	1.3489		
9.00	1.0647	95.7	.325	967.2	31.1	168	1.3498		
9.50	1.0684	101.3	.344	965.2	33.0	178	1.3508		
10.00	1.0722	107.0	.364	963.3	35.0	188	1.3518		

H W C

39 POTASSIUM IODIDE, KI

MOLECULAR WEIGHT = 166.021

RELATIVE SPECIFIC REFRACTIVITY = .649

A % by wt.	D^{20}_{20}	C_s g/l	M g-mol/l	C_w g/l	$(C_o - C_w)$ g/l	$(n - n_o) \times 10^4$	n	Δ °C	S g-mol/l
.00	1.0000	.0	.000	998.2	.0	0	1.3330	.00	.001
.50	1.0036	5.0	.030	996.9	1.4	6	1.3336	.11	.031
1.00	1.0073	10.1	.061	995.5	2.7	13	1.3343	.22	.062
1.50	1.0110	15.1	.091	994.1	4.1	20	1.3350	.32	.093
2.00	1.0148	20.3	.122	992.7	5.5	26	1.3356	.43	.124
2.50	1.0185	25.4	.153	991.3	6.9	33	1.3363	.54	.156
3.00	1.0223	30.6	.184	989.9	8.3	40	1.3370	.65	.188
3.50	1.0261	35.9	.216	988.5	9.8	47	1.3376	.76	.220
4.00	1.0299	41.1	.248	987.0	11.2	53	1.3383	.87	.252
4.50	1.0338	46.4	.280	985.5	12.7	60	1.3390	.98	.284

39 POTASSIUM IODIDE, KI

A % by wt.	D^{20}_{20}	C_s g/l	M g-mol/l	C_w g/l	$(C_o - C_w)$ g/l	$(n - n_o) \times 10^4$	n	Δ °C	S g-mol/l
5.00	1.0377	51.8	.312	984.1	14.2	67	1.3397	1.09	.317
5.50	1.0416	57.2	.344	982.6	15.7	74	1.3404	1.20	.350
6.00	1.0455	62.6	.377	981.1	17.2	81	1.3411	1.31	.383
6.50	1.0495	68.1	.410	979.5	18.7	88	1.3418	1.42	.416
7.00	1.0535	73.6	.443	978.0	20.2	95	1.3425	1.54	.450
7.50	1.0575	79.2	.477	976.5	21.8	102	1.3432	1.65	.483
8.00	1.0615	84.8	.511	974.9	23.3	110	1.3439	1.76	.514
8.50	1.0656	90.4	.545	973.3	24.9	117	1.3447	1.88	.550
9.00	1.0697	96.1	.579	971.7	26.5	124	1.3454	2.00	.585
9.50	1.0739	101.8	.613	970.1	28.1	131	1.3461	2.13	.621
10.00	1.0780	107.6	.648	968.5	29.7	139	1.3469	2.25	.658
11.00	1.0864	119.3	.719	965.2	33.0	153	1.3483	2.51	.731
12.00	1.0950	131.2	.790	961.9	36.4	169	1.3498	2.77	.806
13.00	1.1036	143.2	.863	958.4	39.8	184	1.3514	3.04	.882
14.00	1.1124	155.5	.936	955.0	43.3	199	1.3529	3.31	.959
15.00	1.1213	167.9	1.011	951.4	46.8	215	1.3545	3.59	1.038
16.00	1.1304	180.5	1.087	947.8	50.4	231	1.3561	3.88	1.118
17.00	1.1396	193.4	1.165	944.2	54.1	247	1.3577	4.18	1.206
18.00	1.1489	206.4	1.243	940.5	57.8	264	1.3593	4.48	1.289
19.00	1.1584	219.7	1.323	936.7	61.6	280	1.3610	4.79	1.374
20.00	1.1681	233.2	1.405	932.8	65.4	297	1.3627	5.11	1.460
22.00	1.1878	260.9	1.571	924.8	73.4	332	1.3662	5.78	1.636
24.00	1.2081	289.4	1.743	916.6	81.7	367	1.3697	6.48	1.817
26.00	1.2292	319.0	1.921	908.0	90.2	404	1.3734	7.23	2.003
28.00	1.2509	349.6	2.106	899.1	99.2	442	1.3772	8.02	2.195
30.00	1.2734	381.4	2.297	889.8	108.4	481	1.3811	8.85	2.392
32.00	1.2967	414.2	2.495	880.2	118.0	522	1.3852	9.74	2.595
34.00	1.3208	448.3	2.700	870.2	128.0	564	1.3894	10.69	2.804
36.00	1.3458	483.6	2.913	859.8	138.5	607	1.3937	11.70	3.019
38.00	1.3716	520.3	3.134	848.9	149.3	652	1.3982	12.78	3.240
40.00	1.3984	558.4	3.363	837.5	160.7	698	1.4028	13.94	3.467

H W C

40 POTASSIUM NITRATE, KNO_3

MOLECULAR WEIGHT = 101.10

RELATIVE SPECIFIC REFRACTIVITY = .651

A % by wt.	D^{20}_{20}	C_s g/l	M g-mol/l	C_w g/l	$(C_o - C_w)$ g/l	$(n - n_o) \times 10^4$	n	Δ °C	S g-mol/l
.00	1.0000	.0	.000	998.2	.0	0	1.3330	.00	.000
.50	1.0031	5.0	.050	996.3	1.9	5	1.3335	.17	.048
1.00	1.0062	10.0	.099	994.4	3.8	9	1.3339	.33	.096
1.50	1.0094	15.1	.149	992.5	5.7	14	1.3344	.49	.142
2.00	1.0125	20.2	.200	990.5	7.7	19	1.3349	.64	.187
2.50	1.0157	25.3	.251	988.6	9.7	24	1.3354	.79	.231
3.00	1.0189	30.5	.302	986.6	11.7	28	1.3358	.94	.274
3.50	1.0221	35.7	.353	984.5	13.7	33	1.3363	1.08	.316
4.00	1.0253	40.9	.405	982.5	15.7	38	1.3368	1.22	.357
4.50	1.0285	46.2	.457	980.4	17.8	43	1.3373	1.36	.398
5.00	1.0317	51.5	.509	978.4	19.9	47	1.3377	1.50	.438
5.50	1.0349	56.8	.562	976.3	22.0	52	1.3382	1.63	.477
6.00	1.0382	62.2	.615	974.1	24.1	57	1.3387	1.76	.515
6.50	1.0414	67.6	.668	972.0	26.2	62	1.3391	1.89	.554
7.00	1.0447	73.0	.722	969.8	28.4	66	1.3396	2.02	.591
7.50	1.0480	78.5	.776	967.6	30.6	71	1.3401	2.15	.628
8.00	1.0513	84.0	.830	965.4	32.8	76	1.3406	2.27	.664
8.50	1.0546	89.5	.885	963.2	35.0	81	1.3410	2.40	.699
9.00	1.0579	95.0	.940	961.0	37.3	85	1.3415	2.52	.734
9.50	1.0612	100.6	.995	958.7	39.5	90	1.3420	2.64	.768
10.00	1.0646	106.3	1.051	956.4	41.8	95	1.3425	2.75	.800
11.00	1.0713	117.6	1.164	951.8	46.5	104	1.3434		
12.00	1.0781	129.1	1.277	947.0	51.2	114	1.3444		
13.00	1.0849	140.8	1.393	942.2	56.0	123	1.3453		
14.00	1.0918	152.6	1.509	937.3	60.9	133	1.3463		
15.00	1.0988	164.5	1.627	932.3	65.9	143	1.3472		
16.00	1.1058	176.6	1.747	927.3	71.0	152	1.3482		
17.00	1.1129	188.9	1.868	922.1	76.1	162	1.3492		
18.00	1.1201	201.3	1.991	916.9	81.4	172	1.3501		
19.00	1.1273	213.8	2.115	911.5	86.7	181	1.3511		
20.00	1.1346	226.5	2.241	906.1	92.2	191	1.3521		
22.00	1.1493	252.4	2.497	894.9	103.3	211	1.3540		
24.00	1.1644	279.0	2.759	883.3	114.9	230	1.3560		

H W S

41 POTASSIUM OXALATE, $K_2C_2O_4 \cdot 1H_2O$

MOLECULAR WEIGHT = 166.21   FORMULA WEIGHT, HYDRATE = 184.23
RELATIVE SPECIFIC REFRACTIVITY = .682

A % by wt.	H % by wt.	D^{20}_{20}	C_s g/l	M g-mol/l	C_w g/l	$(C_o - C_w)$ g/l	$(n - n_o)$ × 10^4	n	Δ °C	S g-mol/l
.00	.00	1.0000	.0	.000	998.2	.0	0	1.3330	.00	.000
.50	.55	1.0036	5.0	.030	996.9	1.4	7	1.3337	.14	.040
1.00	1.11	1.0072	10.1	.060	995.4	2.8	14	1.3344	.27	.078
1.50	1.66	1.0109	15.1	.091	993.9	4.3	21	1.3351	.41	.117
2.00	2.22	1.0145	20.3	.122	992.4	5.8	28	1.3358	.54	.156
2.50	2.77	1.0181	25.4	.153	990.9	7.3	35	1.3364	.67	.195
3.00	3.33	1.0218	30.6	.184	989.4	8.8	41	1.3371	.81	.234
3.50	3.88	1.0255	35.8	.216	987.8	10.4	48	1.3378	.94	.274
4.00	4.43	1.0292	41.1	.247	986.2	12.0	55	1.3385	1.08	.314
4.50	4.99	1.0329	46.4	.279	984.6	13.6	62	1.3392	1.21	.354
5.00	5.54	1.0366	51.7	.311	983.0	15.2	69	1.3399	1.35	.394
5.50	6.10	1.0403	57.1	.344	981.3	16.9	76	1.3406	1.49	.434
6.00	6.65	1.0440	62.5	.376	979.7	18.6	83	1.3413	1.62	.475
6.50	7.20	1.0478	68.0	.409	978.0	20.3	90	1.3420	1.76	.515
7.00	7.76	1.0516	73.5	.442	976.2	22.0	97	1.3427	1.90	.556
7.50	8.31	1.0554	79.0	.475	974.5	23.7	104	1.3434	2.04	.596
8.00	8.87	1.0592	84.6	.509	972.7	25.5	111	1.3441	2.18	.637
8.50	9.42	1.0630	90.2	.543	970.9	27.3	118	1.3448		
9.00	9.98	1.0668	95.8	.577	969.1	29.1	125	1.3455		
9.50	10.53	1.0707	101.5	.611	967.3	31.0	132	1.3462		
10.00	11.08	1.0746	107.3	.645	965.4	32.8	139	1.3469		

A W C

42 POTASSIUM PHOSPHATE, DIHYDROGEN (MONOBASIC), KH_2PO_4

MOLECULAR WEIGHT = 136.09
RELATIVE SPECIFIC REFRACTIVITY = .627

A % by wt.	D^{20}_{20}	C_s g/l	M g-mol/l	C_w g/l	$(C_o - C_w)$ g/l	$(n - n_o)$ × 10^4	n	Δ °C	S g-mol/l
.00	1.0000	.0	.000	998.2	.0	0	1.3330	.00	.000
.50	1.0036	5.0	.037	996.8	1.4	6	1.3336	.13	.036
1.00	1.0072	10.1	.074	995.3	2.9	12	1.3342	.25	.072
1.50	1.0108	15.1	.111	993.8	4.4	18	1.3348	.37	.108
2.00	1.0144	20.3	.149	992.3	5.9	24	1.3354	.50	.144
2.50	1.0180	25.4	.187	990.8	7.5	29	1.3359	.62	.179
3.00	1.0216	30.6	.225	989.2	9.1	35	1.3365	.74	.214
3.50	1.0252	35.8	.263	987.5	10.7	41	1.3371	.85	.249
4.00	1.0288	41.1	.302	985.9	12.4	47	1.3377	.97	.283
4.50	1.0324	46.4	.341	984.2	14.1	52	1.3382	1.09	.317
5.00	1.0360	51.7	.380	982.4	15.8	58	1.3388	1.20	.350
5.50	1.0396	57.1	.419	980.7	17.6	64	1.3393	1.31	.383
6.00	1.0432	62.5	.459	978.9	19.4	69	1.3399	1.42	.415
6.50	1.0468	67.9	.499	977.0	21.2	75	1.3405	1.53	.447
00	1.0504	73.4	.539	975.2	23.1	80	1.3410	1.64	.478
7.50	1.0541	78.9	.580	973.3	25.0	86	1.3416	1.74	.509
8.00	1.0577	84.5	.620	971.3	26.9	91	1.3421	1.84	.539
8.50	1.0613	90.1	.662	969.4	28.8	96	1.3426	1.94	.568
9.00	1.0649	95.7	.703	967.4	30.8	102	1.3432	2.04	.596
9.50	1.0686	101.3	.744	965.4	32.9	107	1.3437	2.14	.624
10.00	1.0722	107.0	.786	963.3	34.9	112	1.3442	2.23	.651

A A (A + C)

43 POTASSIUM PHOSPHATE, MONOHYDROGEN (DIBASIC), $K_2HPO_4 \cdot 3H_2O$

MOLECULAR WEIGHT = 174.18 FORMULA WEIGHT, HYDRATE = 228.23
RELATIVE SPECIFIC REFRACTIVITY = .588

A % by wt.	H % by wt.	D^{20}_{20}	C_s g/l	M g-mol/l	C_w g/l	$(C_o - C_w)$ g/l	$(n - n_o)$ × 10^4	n	Δ °C	S g-mol/l
.00	.00	1.0000	.0	.000	998.2	.0	0	1.3330	.00	.000
.50	.66	1.0043	5.0	.029	997.5	.7	8	1.3338	.12	.035
1.00	1.31	1.0085	10.1	.058	996.7	1.6	15	1.3345	.25	.071
1.50	1.97	1.0128	15.2	.087	995.8	2.4	23	1.3352	.37	.107
2.00	2.62	1.0171	20.3	.117	994.9	3.3	30	1.3360	.49	.143
2.50	3.28	1.0213	25.5	.146	994.0	4.2	37	1.3367	.62	.179
3.00	3.94	1.0256	30.7	.176	993.1	5.1	45	1.3375	.73	.213
3.50	4.59	1.0300	36.0	.207	992.2	6.1	52	1.3382	.85	.248
4.00	5.25	1.0342	41.3	.237	991.1	7.1	59	1.3389	.97	.282
4.50	5.90	1.0386	46.7	.268	990.1	8.1	67	1.3397	1.08	.315
5.00	6.56	1.0430	52.1	.299	989.1	9.1	74	1.3404	1.19	.347
5.50	7.22	1.0475	57.5	.330	988.1	10.1	82	1.3412		
6.00	7.87	1.0519	63.0	.362	987.1	11.2	90	1.3419		
6.50	8.53	1.0564	68.5	.393	986.0	12.2	97	1.3427		
7.00	9.18	1.0609	74.1	.425	984.9	13.3	105	1.3435		
7.50	9.84	1.0654	79.8	.458	983.8	14.5	112	1.3442		
8.00	10.49	1.0699	85.4	.490	982.6	15.6	120	1.3450		

A A A

44 POTASSIUM SULFATE, K_2SO_4

MOLECULAR WEIGHT = 174.26
RELATIVE SPECIFIC REFRACTIVITY = .567

A % by wt.	D^{20}_{20}	C_s g/l	M g-mol/l	C_w g/l	$(C_o - C_w)$ g/l	$(n - n_o)$ × 10^4	n	Δ °C	S g-mol/l
.00	1.0000	.0	.000	998.2	.0	0	1.3330	.00	.000
.50	1.0040	5.0	.029	997.2	1.0	6	1.3336	.14	.039
1.00	1.0080	10.1	.058	996.2	2.1	13	1.3343	.26	.075
1.50	1.0121	15.2	.087	995.1	3.1	19	1.3349	.38	.110
2.00	1.0161	20.3	.116	994.0	4.2	25	1.3355	.50	.144
2.50	1.0201	25.5	.146	992.9	5.3	31	1.3361	.61	.177
3.00	1.0242	30.7	.176	991.7	6.5	38	1.3368	.73	.211
3.50	1.0283	35.9	.206	990.6	7.7	44	1.3374	.84	.244
4.00	1.0324	41.2	.237	989.3	8.9	50	1.3380	.95	.277
4.50	1.0365	46.6	.267	988.1	10.1	56	1.3386	1.06	.310
5.00	1.0406	51.9	.298	986.8	11.4	62	1.3392	1.17	.342
5.50	1.0447	57.4	.329	985.5	12.7	68	1.3398		
6.00	1.0489	62.8	.361	984.2	14.0	75	1.3405		
6.50	1.0530	68.3	.392	982.8	15.4	81	1.3411		
7.00	1.0572	73.9	.424	981.4	16.8	87	1.3417		
7.50	1.0614	79.5	.456	980.0	18.2	93	1.3423		
8.00	1.0656	85.1	.488	978.6	19.6	99	1.3429		
8.50	1.0698	90.8	.521	977.1	21.1	105	1.3435		
9.00	1.0740	96.5	.554	975.6	22.6	111	1.3441		
9.50	1.0782	102.3	.587	974.1	24.2	117	1.3447		
10.00	1.0825	108.1	.620	972.5	25.7	123	1.3453		

A W C

45 PROPYLENE GLYCOL, $CH_2OHCHOHCH_3$

MOLECULAR WEIGHT = 76.09
RELATIVE SPECIFIC REFRACTIVITY = 1.224

A % by wt.	D^{20}_{20}	C_s g/l	M g-mol/l	C_w g/l	$(C_o - C_w)$ g/l	$(n - n_o)$ × 10^4	n	Δ °C	S g-mol/l
.00	1.0000	.0	.000	998.2	.0	0	1.3330	.00	.000
.50	1.0004	5.0	.066	993.6	4.6	5	1.3335	.12	.034
1.00	1.0008	10.0	.131	989.0	9.2	11	1.3340	.24	.069
1.50	1.0012	15.0	.197	984.4	13.8	16	1.3346	.36	.105
2.00	1.0016	20.0	.263	979.8	18.4	21	1.3351	.49	.142
2.50	1.0020	25.0	.329	975.2	23.0	26	1.3356	.62	.179
3.00	1.0024	30.0	.395	970.6	27.7	32	1.3361	.75	.217
3.50	1.0028	35.0	.460	965.9	32.3	37	1.3367	.88	.256
4.00	1.0031	40.1	.526	961.3	36.9	42	1.3372	1.02	.296
4.50	1.0035	45.1	.592	956.7	41.6	47	1.3377	1.15	.337
5.00	1.0039	50.1	.659	952.0	46.2	53	1.3382	1.29	.378
5.50	1.0043	55.1	.725	947.4	50.9	58	1.3388	1.44	.420
6.00	1.0046	60.2	.791	942.7	55.5	63	1.3393	1.58	.463
6.50	1.0050	65.2	.857	938.0	60.2	68	1.3398		
7.00	1.0054	70.3	.923	933.3	64.9	73	1.3403		
7.50	1.0057	75.3	.990	928.7	69.6	79	1.3409		
8.00	1.0061	80.3	1.056	924.0	74.3	84	1.3414		
8.50	1.0065	85.4	1.122	919.3	79.0	89	1.3419		
9.00	1.0068	90.5	1.189	914.6	83.7	94	1.3424		
9.50	1.0072	95.5	1.255	909.9	88.4	99	1.3429		
10.00	1.0075	100.6	1.322	905.1	93.1	104	1.3434		

A A A

46 PROTEIN IN HUMAN PLASMA OR SERUM

RELATIVE SPECIFIC REFRACTIVITY = 1.275

Pr % by wt.	TS % by wt.	D^{20}_{20}	C_{Pr} g/l	C_{NPr} g/l	C_w g/l	$(C_o - C_w)$ g/l	$(n - n_o)$ × 10^4	n
.0	1.7	1.0079	0	16.9	989	9	27	1.3357
.5	2.2	1.0092	5	16.9	986	13	37	1.3366
1.0	2.7	1.0105	10	16.8	982	16	46	1.3376
1.5	3.2	1.0119	15	16.7	978	20	55	1.3385
2.0	3.6	1.0132	20	16.7	975	24	65	1.3395
2.5	4.1	1.0145	25	16.6	971	27	74	1.3404
3.0	4.6	1.0158	30	16.5	967	31	84	1.3413
3.5	5.1	1.0172	36	16.5	963	35	93	1.3423
4.0	5.6	1.0185	41	16.4	960	39	102	1.3432
4.5	6.1	1.0197	46	16.3	956	42	112	1.3442
5.0	6.6	1.0210	51	16.3	952	46	121	1.3451
5.5	7.1	1.0223	56	16.2	948	50	130	1.3460
6.0	7.6	1.0236	61	16.2	944	54	140	1.3470
6.5	8.1	1.0249	66	16.1	940	58	149	1.3479
7.0	8.6	1.0261	72	16.0	937	62	158	1.3488
7.5	9.1	1.0274	77	16.0	933	66	167	1.3497
8.0	9.5	1.0286	82	15.9	929	69	177	1.3507
8.5	10.0	1.0299	87	15.8	925	73	186	1.3516
9.0	10.5	1.0311	93	15.8	921	77	195	1.3525
9.5	11.0	1.0323	98	15.7	917	81	204	1.3534
10.0	11.5	1.0335	103	15.6	913	85	214	1.3543
11.0	12.5	1.0359	114	15.5	905	93	232	1.3562
12.0	13.5	1.0383	124	15.3	897	101	250	1.3580
13.0	14.5	1.0407	135	15.2	889	110	268	1.3598
14.0	15.4	1.0430	146	15.1	880	118	286	1.3616
15.0	16.4	1.0452	157	14.9	872	126	304	1.3634

Notes: TS = Total solids, Pr = Protein, NPr = Non-protein solids. Protein concentrations have been determined as differences between nonprotein solids and total solids. These protein concentrations are consistent with a Pr/N ratio of 6.54 rather than the conventional 6.25.

A A —

47 SEA WATER

RELATIVE SPECIFIC REFRACTIVITY = .780

Salinity ‰	Chlorinity ‰	TS % by wt.	D_{20}^{20}	C_{TS} g/l	C_w g/l	$(C_o - C_w)$ g/l	$(n - n_o) \times 10^4$	n	Δ °C	S g-mol/l
.00	.00	.00	1.0000	.0	998.2	.0	0	1.3330	.00	.000
5.00	2.76	.50	1.0037	5.0	996.9	1.4	9	1.3339	.27	.077
10.00	5.52	1.00	1.0073	10.1	995.5	2.7	18	1.3348	.53	.155
15.00	8.28	1.50	1.0110	15.1	994.1	4.1	27	1.3357	.80	.233
20.00	11.04	2.00	1.0148	20.3	992.7	5.5	36	1.3366	1.07	.313
25.00	13.78	2.50	1.0185	25.4	991.3	7.0	45	1.3375	1.34	.392
30.00	16.56	3.01	1.0222	30.7	989.8	8.4	54	1.3384	1.61	.471
35.00	19.32	3.51	1.0260	35.9	988.3	9.9	63	1.3393	1.88	.550
40.00	22.08	4.01	1.0299	41.2	986.8	11.4	72	1.3402	2.15	.629
45.00	24.84	4.51	1.0337	46.5	985.3	12.9	81	1.3411	2.42	.707
50.00	27.60	5.01	1.0375	51.9	983.8	14.5	90	1.3420	2.69	.784
55.00	30.36	5.51	1.0413	57.3	982.2	16.0	99	1.3429	2.97	.861

Notes: ‰ = Parts/(1000 parts); TS = Total solids. Salinity and Chlorinity as defined in Sverdrup et al (1942), The Oceans, Prentice-Hall, p. 92. Experimental data from artificial sea water of composition given by Lyman and Fleming (1940) J. Mar. Res. 3:134.

A A A

48 SILVER NITRATE, $AgNO_3$

MOLECULAR WEIGHT = 169.89

RELATIVE SPECIFIC REFRACTIVITY = .474

A % by wt.	D_{20}^{20}	C_s g/l	M g-mol/l	C_w g/l	$(C_o - C_w)$ g/l	$(n - n_o) \times 10^4$	n	Δ °C	S g-mol/l
.00	1.0000	.0	.000	998.2	.0	0	1.3330	.00	.000
.50	1.0044	5.0	.030	997.6	.6	6	1.3336	.10	.029
1.00	1.0087	10.1	.059	996.8	1.4	12	1.3342	.20	.057
1.50	1.0130	15.2	.089	996.0	2.2	17	1.3347	.29	.085
2.00	1.0173	20.3	.120	995.2	3.1	23	1.3352	.39	.113
2.50	1.0216	25.5	.150	994.3	4.0	28	1.3358	.49	.141
3.00	1.0259	30.7	.181	993.3	4.9	33	1.3363	.58	.169
3.50	1.0302	36.0	.212	992.4	5.8	39	1.3369	.68	.197
4.00	1.0346	41.3	.243	991.4	6.8	44	1.3374	.77	.225
4.50	1.0390	46.7	.275	990.5	7.8	49	1.3379	.87	.252
5.00	1.0434	52.1	.307	989.5	8.8	55	1.3385	.96	.280
5.50	1.0479	57.5	.339	988.5	9.7	60	1.3390	1.05	.308
6.00	1.0524	63.0	.371	987.5	10.7	66	1.3396	1.15	.335
6.50	1.0570	68.6	.404	986.5	11.7	72	1.3402	1.24	.362
7.00	1.0616	74.2	.437	985.5	12.7	77	1.3407	1.33	.389
7.50	1.0662	79.8	.470	984.5	13.7	83	1.3413	1.42	.416
8.00	1.0709	85.5	.503	983.5	14.7	89	1.3419	1.52	.443
8.50	1.0757	91.3	.537	982.5	15.7	95	1.3425	1.61	.469
9.00	1.0804	97.1	.571	981.4	16.8	101	1.3431	1.70	.496
9.50	1.0852	102.9	.606	980.4	17.8	107	1.3437	1.78	.522
10.00	1.0901	108.8	.641	979.3	18.9	113	1.3443	1.87	.547
11.00	1.0999	120.8	.711	977.2	21.1	125	1.3455	2.05	.598
12.00	1.1099	133.0	.783	975.0	23.3	137	1.3467	2.22	.648
13.00	1.1201	145.4	.856	972.7	25.5	150	1.3480	2.38	.696
14.00	1.1304	158.0	.930	970.4	27.8	163	1.3493	2.55	.742
15.00	1.1409	170.8	1.006	968.1	30.2	176	1.3506	2.70	.787
16.00	1.1516	183.9	1.083	965.7	32.6	189	1.3519		
17.00	1.1625	197.3	1.161	963.2	35.1	203	1.3532		
18.00	1.1736	210.9	1.241	960.6	37.6	216	1.3546		
19.00	1.1849	224.7	1.323	958.0	40.2	230	1.3560		
20.00	1.1963	238.8	1.406	955.4	42.9	244	1.3574		
22.00	1.2198	267.9	1.577	949.8	48.4	273	1.3603		
24.00	1.2442	298.1	1.755	943.9	54.3	302	1.3632		
26.00	1.2694	329.5	1.939	937.7	60.5	333	1.3663		
28.00	1.2956	362.1	2.132	931.2	67.1	364	1.3694		
30.00	1.3228	396.1	2.332	924.3	73.9	397	1.3727		
32.00	1.3510	431.6	2.540	917.1	81.2	430	1.3760		
34.00	1.3805	468.5	2.758	909.5	88.7	466	1.3795		
36.00	1.4112	507.1	2.985	901.5	96.7	502	1.3832		
38.00	1.4433	547.5	3.222	893.2	105.0	541	1.3871		
40.00	1.4769	589.7	3.471	884.6	113.7	581	1.3911		

H W C

49 SODIUM ACETATE, CH_3COONa

MOLECULAR WEIGHT = 82.04
RELATIVE SPECIFIC REFRACTIVITY = .893

A % by wt.	D^{20}_{20}	C_s g/l	M g-mol/l	C_w g/l	$(C_o - C_w)$ g/l	$(n - n_o) \times 10^4$	n	Δ °C	S g-mol/l
.00	1.0000	.0	.000	998.2	.0	0	1.3330	.00	.000
.50	1.0026	5.0	.061	995.8	2.4	7	1.3337	.22	.062
1.00	1.0051	10.0	.122	993.3	4.9	14	1.3344	.43	.125
1.50	1.0077	15.1	.184	990.8	7.4	21	1.3351	.65	.189
2.00	1.0102	20.2	.246	988.3	9.9	28	1.3358	.88	.256
2.50	1.0128	25.3	.308	985.7	12.5	35	1.3365	1.11	.323
3.00	1.0154	30.4	.371	983.2	15.1	42	1.3372	1.34	.391
3.50	1.0179	35.6	.433	980.5	17.7	49	1.3378	1.58	.461
4.00	1.0205	40.7	.497	977.9	20.3	55	1.3385	1.82	.532
4.50	1.0230	46.0	.560	975.3	23.0	62	1.3392	2.07	.604
5.00	1.0256	51.2	.624	972.6	25.7	69	1.3399	2.32	.678
5.50	1.0281	56.4	.688	969.9	28.4	76	1.3406	2.58	.752
6.00	1.0307	61.7	.752	967.2	31.1	83	1.3413	2.85	.828
6.50	1.0333	67.0	.817	964.4	33.8	90	1.3420	3.12	.906
7.00	1.0358	72.4	.882	961.6	36.6	97	1.3427	3.40	.984
7.50	1.0384	77.7	.948	958.8	39.4	104	1.3434	3.68	1.063
8.00	1.0410	83.1	1.013	956.0	42.2	111	1.3441	3.97	1.143
8.50	1.0435	88.5	1.079	953.2	45.1	118	1.3447	4.27	1.231
9.00	1.0461	94.0	1.146	950.3	47.9	124	1.3454	4.57	1.312
9.50	1.0487	99.5	1.212	947.4	50.8	131	1.3461	4.87	1.394
10.00	1.0513	104.9	1.279	944.5	53.7	138	1.3468	5.17	1.474
11.00	1.0565	116.0	1.414	938.6	59.6	152	1.3482		
12.00	1.0617	127.2	1.550	932.6	65.6	166	1.3496		
13.00	1.0669	138.4	1.688	926.5	71.7	180	1.3510		
14.00	1.0721	149.8	1.826	920.4	77.8	194	1.3524		
15.00	1.0774	161.3	1.966	914.2	84.1	208	1.3538		
16.00	1.0827	172.9	2.108	907.8	90.4	222	1.3552		
17.00	1.0880	184.6	2.250	901.4	96.8	236	1.3566		
18.00	1.0933	196.4	2.395	894.9	103.3	250	1.3580		
19.00	1.0987	208.4	2.540	888.3	109.9	264	1.3594		
20.00	1.1041	220.4	2.687	881.7	116.5	279	1.3609		
22.00	1.1149	244.9	2.985	868.1	130.1	307	1.3637		
24.00	1.1259	269.7	3.288	854.2	144.0	336	1.3666		
26.00	1.1370	295.1	3.597	839.9	158.3	366	1.3696		
28.00	1.1483	320.9	3.912	825.3	172.9	395	1.3725		
30.00	1.1596	347.3	4.233	810.3	187.9	425	1.3755		

H W S

50 SODIUM BICARBONATE, $NaHCO_3$

MOLECULAR WEIGHT = 84.02
RELATIVE SPECIFIC REFRACTIVITY = .676

A % by wt.	D^{20}_{20}	C_s g/l	M g-mol/l	C_w g/l	$(C_o - C_w)$ g/l	$(n - n_o) \times 10^4$	n	Δ °C	S g-mol/l
.00	1.0000	.0	.000	998.2	.0	0	1.3330	.00	.000
.50	1.0036	5.0	.060	996.8	1.4	7	1.3337	.22	.063
1.00	1.0072	10.1	.120	995.4	2.8	14	1.3344	.43	.124
1.50	1.0108	15.1	.180	993.9	4.3	20	1.3350	.63	.184
2.00	1.0145	20.3	.241	992.4	5.8	27	1.3357	.83	.242
2.50	1.0181	25.4	.302	990.9	7.4	34	1.3364	1.03	.300
3.00	1.0217	30.6	.364	989.3	9.0	41	1.3370	1.22	.356
3.50	1.0253	35.8	.426	987.6	10.6	47	1.3377	1.41	.411
4.00	1.0289	41.1	.489	986.0	12.3	54	1.3384	1.59	.464
4.50	1.0325	46.4	.552	984.3	14.0	60	1.3390	1.76	.515
5.00	1.0361	51.7	.616	982.5	15.7	67	1.3396	1.93	.564
5.50	1.0397	57.1	.679	980.7	17.5	73	1.3403	2.09	.612
6.00	1.0432	62.5	.744	978.9	19.3	79	1.3409	2.25	.658

A (A + W) C

51 SODIUM BROMIDE, NaBr

MOLECULAR WEIGHT = 102.91
RELATIVE SPECIFIC REFRACTIVITY = .629

A % by wt.	D_{20}^{20}	C_s g/l	M g-mol/l	C_w g/l	$(C_o - C_w)$ g/l	$(n - n_o) \times 10^4$	n	Δ °C	S g-mol/l
.00	1.0000	.0	.000	998.2	.0	0	1.3330	.00	.000
.50	1.0039	5.0	.049	997.1	1.1	7	1.3337	.17	.049
1.00	1.0078	10.1	.098	995.9	2.3	14	1.3344	.34	.099
1.50	1.0117	15.1	.147	994.8	3.5	21	1.3351	.51	.149
2.00	1.0156	20.3	.197	993.6	4.7	28	1.3358	.69	.199
2.50	1.0196	25.4	.247	992.3	5.9	35	1.3365	.86	.250
3.00	1.0236	30.7	.298	991.1	7.1	42	1.3372	1.04	.302
3.50	1.0276	35.9	.349	989.9	8.3	49	1.3379	1.21	.355
4.00	1.0316	41.2	.400	988.6	9.6	57	1.3386	1.39	.408
4.50	1.0357	46.5	.452	987.4	10.9	64	1.3394	1.58	.461
5.00	1.0398	51.9	.504	986.1	12.2	71	1.3401	1.76	.515
5.50	1.0439	57.3	.557	984.8	13.5	78	1.3408	1.95	.570
6.00	1.0481	62.8	.610	983.5	14.8	86	1.3416	2.14	.625
6.50	1.0522	68.3	.663	982.1	16.1	93	1.3423	2.33	.681
7.00	1.0565	73.8	.717	980.8	17.5	100	1.3430	2.53	.737
7.50	1.0607	79.4	.772	979.4	18.8	108	1.3438	2.73	.794
8.00	1.0649	85.0	.826	978.0	20.2	115	1.3445	2.93	.851
8.50	1.0692	90.7	.882	976.6	21.6	123	1.3453	3.13	.909
9.00	1.0735	96.4	.937	975.2	23.0	130	1.3460	3.34	.968
9.50	1.0779	102.2	.993	973.8	24.5	138	1.3468	3.55	1.027
10.00	1.0823	108.0	1.050	972.3	25.9	146	1.3475	3.77	1.086
11.00	1.0911	119.8	1.164	969.4	28.9	161	1.3491	4.21	1.214
12.00	1.1001	131.8	1.280	966.3	31.9	176	1.3506	4.66	1.338
13.00	1.1091	143.9	1.399	963.3	35.0	192	1.3522	5.14	1.466
14.00	1.1184	156.3	1.519	960.1	38.1	208	1.3538	5.63	1.598
15.00	1.1277	168.9	1.641	956.9	41.4	224	1.3554	6.16	1.734
16.00	1.1372	181.6	1.765	953.6	44.7	241	1.3570	6.72	1.876
17.00	1.1468	194.6	1.891	950.2	48.0	257	1.3587	7.32	2.025
18.00	1.1566	207.8	2.019	946.7	51.5	274	1.3604	7.97	2.184
19.00	1.1665	221.2	2.150	943.2	55.0	291	1.3621		
20.00	1.1766	234.9	2.283	939.6	58.6	308	1.3638		
22.00	1.1972	262.9	2.555	932.2	66.1	343	1.3673		
24.00	1.2184	291.9	2.837	924.4	73.8	378	1.3708		
26.00	1.2404	321.9	3.128	916.3	82.0	415	1.3745		
28.00	1.2630	353.0	3.430	907.8	90.5	453	1.3783		
30.00	1.2864	385.3	3.744	898.9	99.3	492	1.3822		
32.00	1.3106	418.7	4.068	889.7	108.6				
34.00	1.3357	453.3	4.405	880.0	118.2				
36.00	1.3616	489.3	4.755	869.9	128.3				
38.00	1.3885	526.7	5.118	859.3	138.9				
40.00	1.4163	565.5	5.495	848.3	149.9				

H W C

52 SODIUM CARBONATE, $Na_2CO_3 \cdot 10H_2O$

MOLECULAR WEIGHT = 106.00 FORMULA WEIGHT, HYDRATE = 286.16
RELATIVE SPECIFIC REFRACTIVITY = .618

A % by wt.	H % by wt.	D_{20}^{20}	C_s g/l	M g-mol/l	C_w g/l	$(C_o - C_w)$ g/l	$(n - n_o) \times 10^4$	n	Δ °C	S g-mol/l
.00	.00	1.0000	.0	.000	998.2	.0	0	1.3330	.00	.000
.50	1.35	1.0052	5.0	.047	998.4	−.2	11	1.3341	.22	.062
1.00	2.70	1.0104	10.1	.095	998.5	−.3	23	1.3352	.42	.122
1.50	4.05	1.0156	15.2	.143	998.6	−.4	34	1.3364	.62	.180
2.00	5.40	1.0208	20.4	.192	998.6	−.4	45	1.3375	.82	.237
2.50	6.75	1.0260	25.6	.242	998.6	−.4	56	1.3386	1.00	.292
3.00	8.10	1.0312	30.9	.291	998.5	−.3	67	1.3397	1.18	.345
3.50	9.45	1.0364	36.2	.342	998.4	−.1	78	1.3408	1.36	.397
4.00	10.80	1.0416	41.6	.392	998.2	.1	89	1.3419	1.52	.446
4.50	12.15	1.0468	47.0	.444	997.9	.3	99	1.3429	1.68	.493
5.00	13.50	1.0520	52.5	.495	997.7	.6	110	1.3440	1.84	.537
5.50	14.85	1.0573	58.0	.548	997.3	.9	121	1.3451	1.98	.580
6.00	16.20	1.0625	63.6	.600	997.0	1.3	132	1.3462	2.12	.619
6.50	17.55	1.0677	69.3	.654	996.6	1.7	142	1.3472		
7.00	18.90	1.0730	75.0	.707	996.1	2.1	153	1.3483		
7.50	20.25	1.0783	80.7	.762	995.6	2.6	164	1.3494		
8.00	21.60	1.0835	86.5	.816	995.1	3.1	174	1.3504		
8.50	22.95	1.0888	92.4	.872	994.5	3.7	185	1.3515		
9.00	24.30	1.0941	98.3	.927	993.9	4.3	195	1.3525		
9.50	25.65	1.0995	104.3	.984	993.3	5.0	206	1.3536		
10.00	27.00	1.1048	110.3	1.040	992.6	5.7	217	1.3546		
11.00	29.70	1.1156	122.5	1.156	991.1	7.1	238	1.3568		
12.00	32.40	1.1264	134.9	1.273	989.5	8.8	259	1.3589		
13.00	35.10	1.1373	147.6	1.392	987.7	10.5	280	1.3610		
14.00	37.79	1.1483	160.5	1.514	985.8	12.4	301	1.3631		
15.00	40.49	1.1594	173.6	1.638	983.8	14.5	322	1.3652		

H W C

53 SODIUM CHLORIDE, NaCl

MOLECULAR WEIGHT = 58.45
RELATIVE SPECIFIC REFRACTIVITY = .797

A % by wt.	D^{20}_{20}	C_s g/l	M g-mol/l	C_w g/l	$(C_o - C_w)$ g/l	$(n - n_o) \times 10^4$	n	Δ °C	S g-mol/l
.00	1.0000	.0	.000	998.2	.0	0	1.3330	.00	.000
.50	1.0035	5.0	.086	996.8	1.5	9	1.3339	.30	.086
1.00	1.0071	10.1	.172	995.3	3.0	18	1.3347	.59	.172
1.50	1.0107	15.1	.259	993.8	4.5	26	1.3356	.89	.259
2.00	1.0143	20.2	.346	992.2	6.0	35	1.3365	1.19	.346
2.50	1.0178	25.4	.435	990.6	7.6	44	1.3374	1.49	.435
3.00	1.0214	30.6	.523	989.0	9.2	53	1.3382	1.79	.523
3.50	1.0250	35.8	.613	987.4	10.8	61	1.3391	2.10	.613
4.00	1.0286	41.1	.703	985.7	12.5	70	1.3400	2.41	.703
4.50	1.0322	46.4	.793	984.1	14.2	79	1.3409	2.72	.793
5.00	1.0359	51.7	.885	982.3	15.9	88	1.3417	3.05	.885
5.50	1.0395	57.1	.976	980.6	17.6	96	1.3426	3.36	.976
6.00	1.0431	62.5	1.069	978.8	19.4	105	1.3435	3.69	1.069
6.50	1.0468	67.9	1.162	977.0	21.2	114	1.3444	4.02	1.162
7.00	1.0504	73.4	1.256	975.2	23.1	123	1.3453	4.36	1.256
7.50	1.0541	78.9	1.350	973.3	24.9	131	1.3461	4.70	1.350
8.00	1.0578	84.5	1.445	971.4	26.8	140	1.3470	5.05	1.445
8.50	1.0615	90.1	1.541	969.5	28.7	149	1.3479	5.41	1.541
9.00	1.0652	95.7	1.637	967.6	30.6	158	1.3488	5.78	1.637
9.50	1.0689	101.4	1.734	965.6	32.6	167	1.3496	6.16	1.734
10.00	1.0726	107.1	1.832	963.6	34.6	175	1.3505	6.54	1.832
11.00	1.0801	118.6	2.029	959.6	38.7	193	1.3523	7.33	2.029
12.00	1.0876	130.3	2.229	955.4	42.8	211	1.3541	8.16	2.229
13.00	1.0952	142.1	2.432	951.1	47.1	228	1.3558	9.03	2.432
14.00	1.1028	154.1	2.637	946.7	51.5	246	1.3576	9.93	2.637
15.00	1.1105	166.3	2.845	942.3	56.0	264	1.3594	10.88	2.845
16.00	1.1182	178.6	3.056	937.7	60.6	282	1.3612	11.88	3.056
17.00	1.1260	191.1	3.269	933.0	65.3	300	1.3630	12.93	3.269
18.00	1.1339	203.7	3.486	928.2	70.1	318	1.3648	14.04	3.486
19.00	1.1418	216.6	3.705	923.3	75.0	336	1.3666	15.21	3.705
20.00	1.1498	229.6	3.927	918.2	80.0	354	1.3684	16.45	3.927
22.00	1.1660	256.1	4.381	907.9	90.3	391	1.3721	19.18	4.381
24.00	1.1825	283.3	4.847	897.1	101.1	428	1.3758		4.847
26.00	1.1993	311.3	5.325	885.9	112.3	466	1.3796		5.325

H W (S + L)

54 SODIUM HYDROXIDE, NaOH

MOLECULAR WEIGHT = 40.01
RELATIVE SPECIFIC REFRACTIVITY = .701

A % by wt.	D^{20}_{20}	C_s g/l	M g-mol/l	C_w g/l	$(C_o - C_w)$ g/l	$(n - n_o) \times 10^4$	n	Δ °C	S g-mol/l
.00	1.0000	.0	.000	998.2	.0	0	1.3330	.00	.000
.50	1.0056	5.0	.125	998.8	−.6	14	1.3344	.43	.124
1.00	1.0112	10.1	.252	999.3	−1.0	28	1.3358	.88	.257
1.50	1.0167	15.2	.381	999.7	−1.5	42	1.3372	1.29	.378
2.00	1.0223	20.4	.510	1000.1	−1.8	56	1.3386	1.71	.501
2.50	1.0279	25.7	.641	1000.4	−2.2	70	1.3400	2.15	.628
3.00	1.0334	30.9	.773	1000.6	−2.4	84	1.3413	2.60	.757
3.50	1.0390	36.3	.907	1000.8	−2.6	97	1.3427	3.06	.888
4.00	1.0445	41.7	1.042	1001.0	−2.7	111	1.3441	3.54	1.022
4.50	1.0501	47.2	1.179	1001.0	−2.8	124	1.3454	4.03	1.164
5.00	1.0556	52.7	1.317	1001.1	−2.8	138	1.3468	4.53	1.304
5.50	1.0612	58.3	1.456	1001.0	−2.8	151	1.3481	5.06	1.445
6.00	1.0667	63.9	1.597	1000.9	−2.7	164	1.3494	5.59	1.587
6.50	1.0722	69.6	1.739	1000.8	−2.5	178	1.3507	6.15	1.731
7.00	1.0778	75.3	1.882	1000.6	−2.3	191	1.3521	6.72	1.877
7.50	1.0833	81.1	2.027	1000.3	−2.1	204	1.3534	7.31	2.023
8.00	1.0888	87.0	2.173	1000.0	−1.7	217	1.3547	7.91	2.170
8.50	1.0944	92.9	2.321	999.6	−1.3	230	1.3560	8.54	2.318
9.00	1.0999	98.8	2.470	999.1	−.9	243	1.3572	9.18	2.468
9.50	1.1054	104.8	2.620	998.6	−.4	255	1.3585	9.84	2.617
10.00	1.1110	110.9	2.772	998.1	.1	268	1.3598	10.52	2.768
11.00	1.1220	123.2	3.079	996.8	1.4	293	1.3623	11.95	3.070
12.00	1.1331	135.7	3.392	995.3	2.9	318	1.3648	13.45	3.372
13.00	1.1441	148.5	3.711	993.6	4.6	343	1.3673	15.05	3.675
14.00	1.1551	161.4	4.035	991.7	6.6	367	1.3697	16.75	3.977

54 SODIUM HYDROXIDE, NaOH

A % by wt.	D_{20}^{20}	C_s g/l	M g-mol/l	C_w g/l	$(C_o - C_w)$ g/l	$(n - n_o) \times 10^4$	n	Δ °C	S g-mol/l
15.00	1.1662	174.6	4.364	989.5	8.7	392	1.3721		
16.00	1.1772	188.0	4.699	987.1	11.1	415	1.3745		
17.00	1.1882	201.6	5.040	984.5	13.7	439	1.3769		
18.00	1.1993	215.5	5.386	981.6	16.6	462	1.3792		
19.00	1.2103	229.5	5.737	978.6	19.7	485	1.3815		
20.00	1.2213	243.8	6.094	975.3	22.9	508	1.3838		
22.00	1.2432	273.0	6.824	968.0	30.2				
24.00	1.2651	303.1	7.576	959.8	38.4				
26.00	1.2869	334.0	8.348	950.7	47.6				
28.00	1.3086	365.8	9.142	940.5	57.7				
30.00	1.3301	398.3	9.956	929.4	68.8				
32.00	1.3513	431.7	10.789	917.3	81.0				
34.00	1.3722	465.7	11.641	904.1	94.2				
36.00	1.3928	500.5	12.510	889.8	108.4				
38.00	1.4128	535.9	13.394	874.4	123.9				
40.00	1.4322	571.9	14.293	857.8	140.4				

H W C

55 SODIUM MOLYBDATE, $Na_2MoO_4 \cdot 2H_2O$

MOLECULAR WEIGHT = 205.94 FORMULA WEIGHT, HYDRATE = 241.98
RELATIVE SPECIFIC REFRACTIVITY = .641

A % by wt.	H % by wt.	D_{20}^{20}	C_s g/l	M g-mol/l	C_w g/l	$(C_o - C_w)$ g/l	$(n - n_o) \times 10^4$	n	Δ °C	S g-mol/l
.00	.00	1.0000	.0	.000	998.2	.0	0	1.3330	.00	.000
.50	.59	1.0043	5.0	.024	997.5	.7	9	1.3338	.10	.029
1.00	1.18	1.0086	10.1	.049	996.7	1.5	17	1.3347	.22	.062
1.50	1.76	1.0129	15.2	.074	995.9	2.3	26	1.3356	.33	.095
2.00	2.35	1.0173	20.3	.099	995.1	3.1	34	1.3364	.44	.128
2.50	2.94	1.0216	25.5	.124	994.3	3.9	43	1.3373	.55	.161
3.00	3.53	1.0261	30.7	.149	993.5	4.7	52	1.3382	.67	.194
3.50	4.11	1.0305	36.0	.175	992.7	5.6	61	1.3390	.78	.227
4.00	4.70	1.0350	41.3	.201	991.8	6.4	69	1.3399	.90	.261
4.50	5.29	1.0395	46.7	.227	991.0	7.3	78	1.3408	1.01	.295
5.00	5.88	1.0440	52.1	.253	990.1	8.2	87	1.3417	1.12	.328
5.50	6.46	1.0486	57.6	.280	989.2	9.1	96	1.3426	1.24	.362
6.00	7.05	1.0532	63.1	.306	988.3	10.0	105	1.3435	1.35	.396
6.50	7.64	1.0578	68.6	.333	987.3	10.9	115	1.3444		
7.00	8.23	1.0625	74.2	.361	986.4	11.9	124	1.3454		
7.50	8.81	1.0672	79.9	.388	985.4	12.8	133	1.3463		
8.00	9.40	1.0719	85.6	.416	984.4	13.8	142	1.3472		
8.50	9.99	1.0767	91.4	.444	983.5	14.8	152	1.3482		
9.00	10.58	1.0815	97.2	.472	982.5	15.8	161	1.3491		

A A A

56 SODIUM NITRATE, $NaNO_3$

MOLECULAR WEIGHT = 85.01
RELATIVE SPECIFIC REFRACTIVITY = .658

A % by wt.	D^{20}_{20}	C_s g/l	M g-mol/l	C_w g/l	$(C_o - C_w)$ g/l	$(n - n_o) \times 10^4$	n	Δ °C	S g-mol/l
.00	1.0000	.0	.000	998.2	.0	0	1.3330	.00	.000
.50	1.0034	5.0	.059	996.6	1.6	6	1.3336	.21	.061
1.00	1.0067	10.0	.118	994.9	3.3	11	1.3341	.40	.114
1.50	1.0101	15.1	.178	993.2	5.1	17	1.3347	.58	.168
2.00	1.0135	20.2	.238	991.4	6.8	23	1.3353	.77	.223
2.50	1.0168	25.4	.299	989.7	8.6	28	1.3358	.95	.277
3.00	1.0202	30.6	.359	987.9	10.3	34	1.3364	1.14	.332
3.50	1.0237	35.8	.421	986.1	12.1	40	1.3369	1.33	.387
4.00	1.0271	41.0	.482	984.3	14.0	45	1.3375	1.51	.443
4.50	1.0305	46.3	.545	982.4	15.8	51	1.3381	1.70	.498
5.00	1.0340	51.6	.607	980.5	17.7	57	1.3386	1.89	.554
5.50	1.0374	57.0	.670	978.7	19.6	62	1.3392	2.08	.609
6.00	1.0409	62.3	.733	976.7	21.5	68	1.3398	2.28	.665
6.50	1.0444	67.8	.797	974.8	23.4	74	1.3403	2.47	.720
7.00	1.0479	73.2	.861	972.9	25.4	79	1.3409	2.66	.775
7.50	1.0515	78.7	.926	970.9	27.3	85	1.3415	2.86	.830
8.00	1.0550	84.3	.991	968.9	29.3	91	1.3420	3.05	.886
8.50	1.0586	89.8	1.057	966.9	31.3	96	1.3426	3.25	.941
9.00	1.0622	95.4	1.123	964.8	33.4	102	1.3432	3.44	.995
9.50	1.0657	101.1	1.189	962.8	35.4	108	1.3438	3.64	1.050
10.00	1.0693	106.7	1.256	960.7	37.5	113	1.3443	3.84	1.105
11.00	1.0766	118.2	1.391	956.5	41.8	125	1.3455	4.23	1.221
12.00	1.0839	129.8	1.527	952.2	46.1	136	1.3466	4.63	1.330
13.00	1.0913	141.6	1.666	947.8	50.5	148	1.3478	5.03	1.438
14.00	1.0988	153.6	1.806	943.3	55.0	159	1.3489	5.44	1.546
15.00	1.1063	165.6	1.949	938.7	59.6	171	1.3501	5.84	1.652
16.00	1.1139	177.9	2.093	934.0	64.2	183	1.3513	6.25	1.757
17.00	1.1215	190.3	2.239	929.2	69.0	194	1.3524	6.66	1.861
18.00	1.1292	202.9	2.387	924.3	73.9	206	1.3536	7.07	1.964
19.00	1.1370	215.7	2.537	919.4	78.9	218	1.3548	7.48	2.065
20.00	1.1449	228.6	2.689	914.3	83.9	229	1.3559	7.89	2.165
22.00	1.1609	254.9	2.999	903.9	94.4	253	1.3583	8.72	2.361
24.00	1.1771	282.0	3.317	893.0	105.2	277	1.3607	9.54	2.550
26.00	1.1937	309.8	3.644	881.8	116.5	301	1.3631		
28.00	1.2106	338.4	3.980	870.1	128.1	325	1.3655		
30.00	1.2279	367.7	4.325	858.0	140.3	349	1.3679		
32.00	1.2455	397.8	4.680	845.4	152.8	374	1.3704		
34.00	1.2634	428.8	5.044	832.4	165.8	398	1.3728		
36.00	1.2818	460.6	5.419	818.9	179.3	423	1.3753		
38.00	1.3005	493.3	5.803	804.9	193.3	449	1.3779		
40.00	1.3197	527.0	6.199	790.4	207.8	474	1.3804		

H W C

57 SODIUM PHOSPHATE, MONOHYDROGEN (DIBASIC), $Na_2HPO_4 \cdot 7H_2O$

MOLECULAR WEIGHT = 141.98 FORMULA WEIGHT, HYDRATE = 268.09
RELATIVE SPECIFIC REFRACTIVITY = .576

A % by wt.	H % by wt.	D^{20}_{20}	C_s g/l	M g-mol/l	C_w g/l	$(C_o - C_w)$ g/l	$(n - n_o) \times 10^4$	n	Δ °C	S g-mol/l
.00	.00	1.0000	.0	.000	998.2	.0	0	1.3330	.00	.000
.50	.94	1.0050	5.0	.035	998.2	.1	10	1.3340	.17	.048
1.00	1.89	1.0099	10.1	.071	998.0	.2	19	1.3349	.32	.092
1.50	2.83	1.0149	15.2	.107	997.9	.4	29	1.3359	.46	.133
2.00	3.78	1.0198	20.4	.143	997.7	.6	39	1.3368		
2.50	4.72	1.0248	25.6	.180	997.4	.8	48	1.3378		
3.00	5.67	1.0297	30.8	.217	997.1	1.1	57	1.3387		
3.50	6.61	1.0347	36.2	.255	996.7	1.5	67	1.3397		
4.00	7.55	1.0397	41.5	.292	996.3	1.9	76	1.3406		
4.50	8.50	1.0446	46.9	.331	995.9	2.4	85	1.3415		
5.00	9.44	1.0496	52.4	.369	995.4	2.9	94	1.3424		
5.50	10.39	1.0546	57.9	.408	994.8	3.4	103	1.3433		

A W C

58 SODIUM SULFATE, $Na_2SO_4 \cdot 10H_2O$

MOLECULAR WEIGHT = 142.06 FORMULA WEIGHT, HYDRATE = 322.22
RELATIVE SPECIFIC REFRACTIVITY = .543

A % by wt.	H % by wt.	D^{20}_{20}	C_s g/l	M g-mol/l	C_w g/l	$(C_o - C_w)$ g/l	$(n - n_o)$ $\times 10^4$	n	Δ °C	S g-mol/l
.00	.00	1.0000	.0	.000	998.2	.0	0	1.3330	.00	.000
.50	1.13	1.0045	5.0	.035	997.7	.5	8	1.3338	.17	.047
1.00	2.27	1.0091	10.1	.071	997.2	1.0	15	1.3345	.32	.092
1.50	3.40	1.0136	15.2	.107	996.7	1.6	23	1.3353	.47	.135
2.00	4.54	1.0182	20.3	.143	996.1	2.2	31	1.3361	.61	.176
2.50	5.67	1.0228	25.5	.180	995.5	2.8	38	1.3368	.74	.216
3.00	6.80	1.0274	30.8	.217	994.8	3.4	46	1.3376	.87	.254
3.50	7.94	1.0320	36.1	.254	994.1	4.1	53	1.3383	1.00	.292
4.00	9.07	1.0367	41.4	.291	993.4	4.8	61	1.3391	1.13	.328
4.50	10.21	1.0413	46.8	.329	992.7	5.5	69	1.3398	1.24	.364
5.00	11.34	1.0460	52.2	.367	991.9	6.3	76	1.3406	1.36	.397
5.50	12.48	1.0506	57.7	.406	991.1	7.1	84	1.3413	1.46	.428
6.00	13.61	1.0553	63.2	.445	990.3	8.0	91	1.3421	1.56	.456
6.50	14.74	1.0601	68.8	.484	989.4	8.8	98	1.3428		
7.00	15.88	1.0648	74.4	.524	988.5	9.7	106	1.3436		
7.50	17.01	1.0695	80.1	.564	987.6	10.7	113	1.3443		
8.00	18.15	1.0743	85.8	.604	986.6	11.6	121	1.3451		
8.50	19.28	1.0790	91.6	.644	985.6	12.7	128	1.3458		
9.00	20.41	1.0838	97.4	.685	984.5	13.7	135	1.3465		
9.50	21.55	1.0886	103.2	.727	983.5	14.8	143	1.3472		
10.00	22.68	1.0934	109.1	.768	982.3	15.9	150	1.3480		
11.00	24.95	1.1031	121.1	.853	980.0	18.2	164	1.3494		
12.00	27.22	1.1129	133.3	.938	977.6	20.6	179	1.3509		
13.00	29.49	1.1227	145.7	1.026	975.0	23.2	193	1.3523		
14.00	31.75	1.1326	158.3	1.114	972.3	25.9	207	1.3537		
15.00	34.02	1.1426	171.1	1.204	969.5	28.8	222	1.3552		
16.00	36.29	1.1526	184.1	1.296	966.5	31.7	236	1.3566		
17.00	38.56	1.1628	197.3	1.389	963.4	34.8	250	1.3580		
18.00	40.83	1.1730	210.8	1.484	960.1	38.1	264	1.3594		
19.00	43.10	1.1833	224.4	1.580	956.7	41.5	278	1.3608		
20.00	45.36	1.1936	238.3	1.677	953.2	45.0	292	1.3622		
22.00	49.90	1.2145	266.7	1.878	945.7	52.6				
24.00	54.44	1.2358	296.1	2.084	937.5	60.7				

H W C

59 SODIUM TARTRATE, $NaOOC(CHOH)_2COONa \cdot 2H_2O$

MOLECULAR WEIGHT = 194.07 FORMULA WEIGHT, HYDRATE = 230.10
RELATIVE SPECIFIC REFRACTIVITY = .780

A % by wt.	H % by wt.	D^{20}_{20}	C_s g/l	M g-mol/l	C_w g/l	$(C_o - C_w)$ g/l	$(n - n_o)$ $\times 10^4$	n	Δ °C	S g-mol/l
.00	.00	1.0000	.0	.000	998.2	.0	0	1.3330	.00	.000
.50	.59	1.0035	5.0	.026	996.7	1.5	8	1.3338	.11	.032
1.00	1.19	1.0070	10.1	.052	995.2	3.0	17	1.3346	.23	.065
1.50	1.78	1.0106	15.1	.078	993.7	4.6	25	1.3355	.34	.099
2.00	2.37	1.0141	20.2	.104	992.1	6.2	33	1.3363	.46	.132
2.50	2.96	1.0177	25.4	.131	990.5	7.8	42	1.3371	.57	.164
3.00	3.56	1.0212	30.6	.158	988.8	9.4	50	1.3380	.68	.196
3.50	4.15	1.0248	35.8	.185	987.2	11.1	58	1.3388	.78	.228
4.00	4.74	1.0284	41.1	.212	985.5	12.7	67	1.3396	.89	.259
4.50	5.34	1.0320	46.4	.239	983.8	14.5	75	1.3405	.99	.289
5.00	5.93	1.0356	51.7	.266	982.0	16.2	83	1.3413	1.09	.319
5.50	6.52	1.0392	57.1	.294	980.3	18.0	92	1.3422		
6.00	7.11	1.0428	62.5	.322	978.5	19.7	100	1.3430		
6.50	7.71	1.0464	67.9	.350	976.7	21.6	108	1.3438		
7.00	8.30	1.0501	73.4	.378	974.8	23.4	117	1.3447		
7.50	8.89	1.0537	78.9	.407	973.0	25.3	125	1.3455		
8.00	9.49	1.0574	84.4	.435	971.1	27.2	134	1.3463		
8.50	10.08	1.0610	90.0	.464	969.1	29.1	142	1.3472		
9.00	10.67	1.0647	95.7	.493	967.2	31.0	150	1.3480		
9.50	11.26	1.0684	101.3	.522	965.2	33.0	159	1.3489		
10.00	11.86	1.0721	107.0	.551	963.2	35.0	167	1.3497		
11.00	13.04	1.0796	118.5	.611	959.1	39.1	184	1.3514		
12.00	14.23	1.0871	130.2	.671	954.9	43.3	201	1.3531		
13.00	15.41	1.0946	142.1	.732	950.7	47.6	218	1.3548		
14.00	16.60	1.1022	154.0	.794	946.3	52.0	236	1.3565		
15.00	17.79	1.1099	166.2	.856	941.8	56.5	253	1.3583		
16.00	18.97	1.1176	178.5	.920	937.1	61.1	270	1.3600		
17.00	20.16	1.1254	191.0	.984	932.4	65.8				
18.00	21.34	1.1332	203.6	1.049	927.6	70.6				
19.00	22.53	1.1411	216.4	1.115	922.7	75.5				
20.00	23.71	1.1491	229.4	1.182	917.7	80.6				
22.00	26.09	1.1652	255.9	1.319	907.3	91.0				
24.00	28.46	1.1817	283.1	1.459	896.5	101.8				
26.00	30.83	1.1984	311.0	1.603	885.3	113.0				
28.00	33.20	1.2155	339.7	1.751	873.6	124.6				

H W A

60 SODIUM THIOSULFATE, $Na_2S_2O_3{\cdot}5H_2O$

MOLECULAR WEIGHT = 158.13 FORMULA WEIGHT, HYDRATE = 248.21
RELATIVE SPECIFIC REFRACTIVITY = .768

A % by wt.	H % by wt.	D^{20}_{20}	C_s g/l	M g-mol/l	C_w g/l	$(C_o - C_w)$ g/l	$(n - n_o)$ $\times 10^4$	n	Δ °C	S g-mol/l
.00	.00	1.0000	.0	.000	998.2	.0	0	1.3330	.00	.000
.50	.78	1.0041	5.0	.032	997.3	.9	10	1.3340		
1.00	1.57	1.0083	10.1	.064	996.4	1.8	21	1.3350		
1.50	2.35	1.0124	15.2	.096	995.4	2.8	31	1.3361		
2.00	3.14	1.0165	20.3	.128	994.5	3.8	41	1.3371		
2.50	3.92	1.0207	25.5	.161	993.4	4.8	51	1.3381		
3.00	4.71	1.0249	30.7	.194	992.4	5.9	62	1.3392		
3.50	5.49	1.0291	36.0	.227	991.3	6.9	72	1.3402		
4.00	6.28	1.0333	41.3	.261	990.2	8.1	83	1.3412		
4.50	7.06	1.0375	46.6	.295	989.0	9.2	93	1.3423		
5.00	7.85	1.0417	52.0	.329	987.8	10.4	103	1.3433		
5.50	8.63	1.0459	57.4	.363	986.6	11.6	114	1.3444		
6.00	9.42	1.0502	62.9	.398	985.4	12.8	124	1.3454		
6.50	10.20	1.0544	68.4	.433	984.1	14.1	134	1.3464		
7.00	10.99	1.0587	74.0	.468	982.8	15.4	145	1.3475		
7.50	11.77	1.0630	79.6	.503	981.5	16.7	155	1.3485		
8.00	12.56	1.0673	85.2	.539	980.2	18.1	166	1.3496		
8.50	13.34	1.0716	90.9	.575	978.8	19.4	176	1.3506		
9.00	14.13	1.0759	96.7	.611	977.4	20.9	187	1.3517		
9.50	14.91	1.0803	102.4	.648	975.9	22.3	198	1.3527		
10.00	15.70	1.0847	108.3	.685	974.5	23.8	208	1.3538		
11.00	17.27	1.0934	120.1	.759	971.5	26.8	229	1.3559		
12.00	18.84	1.1023	132.0	.835	968.3	29.9	251	1.3581		
13.00	20.41	1.1112	144.2	.912	965.1	33.2	272	1.3602		
14.00	21.98	1.1202	156.6	.990	961.7	36.5	294	1.3624		
15.00	23.54	1.1293	169.1	1.069	958.2	40.0	316	1.3646		
16.00	25.11	1.1385	181.8	1.150	954.7	43.6	338	1.3668		
17.00	26.68	1.1477	194.8	1.232	950.9	47.3	360	1.3690		
18.00	28.25	1.1571	207.9	1.315	947.1	51.1	383	1.3712		
19.00	29.82	1.1665	221.2	1.399	943.2	55.0	405	1.3735		
20.00	31.39	1.1760	234.8	1.485	939.1	59.1	428	1.3758		
22.00	34.53	1.1953	262.5	1.660	930.7	67.5				
24.00	37.67	1.2150	291.1	1.841	921.7	76.5				
26.00	40.81	1.2350	320.5	2.027	912.3	86.0				
28.00	43.95	1.2554	350.9	2.219	902.3	95.9				
30.00	47.09	1.2762	382.2	2.417	891.7	106.5				
32.00	50.23	1.2973	414.4	2.621	880.6	117.6				
34.00	53.37	1.3188	447.6	2.831	868.9	129.4				
36.00	56.51	1.3406	481.8	3.047	856.5	141.7				
38.00	59.65	1.3628	516.9	3.269	843.4	154.8				
40.00	62.79	1.3851	553.1	3.498	829.6	168.6				

H W —

61 SODIUM TUNGSTATE, $Na_2WO_4{\cdot}2H_2O$

MOLECULAR WEIGHT = 293.91 FORMULA WEIGHT, HYDRATE = 329.95
RELATIVE SPECIFIC REFRACTIVITY = .414

A % by wt.	H % by wt.	D^{20}_{20}	C_s g/l	M g-mol/l	C_w g/l	$(C_o - C_w)$ g/l	$(n - n_o)$ $\times 10^4$	n	Δ °C	S g-mol/l
.00	.00	1.0000	.0	.000	998.2	.0	0	1.3330	.00	.000
.50	.56	1.0045	5.0	.017	997.7	.5	5	1.3335	.08	.022
1.00	1.12	1.0091	10.1	.034	997.3	.9	11	1.3341	.16	.045
1.50	1.68	1.0138	15.2	.052	996.8	1.4	17	1.3347	.24	.068
2.00	2.25	1.0184	20.3	.069	996.3	1.9	22	1.3352	.32	.091
2.50	2.81	1.0231	25.5	.087	995.8	2.4	28	1.3358	.40	.114
3.00	3.37	1.0279	30.8	.105	995.3	3.0	34	1.3364	.48	.138
3.50	3.93	1.0326	36.1	.123	994.7	3.5	39	1.3369	.56	.161
4.00	4.49	1.0374	41.4	.141	994.1	4.1	45	1.3375	.64	.185
4.50	5.05	1.0422	46.8	.159	993.5	4.7	51	1.3381	.72	.209
5.00	5.61	1.0470	52.3	.178	992.9	5.3	56	1.3386	.80	.232
5.50	6.17	1.0519	57.8	.197	992.3	5.9	62	1.3392	.88	.256
6.00	6.74	1.0568	63.3	.215	991.7	6.6	68	1.3398	.96	.280
6.50	7.30	1.0618	68.9	.234	991.0	7.2	74	1.3403	1.04	.304
7.00	7.86	1.0667	74.5	.254	990.3	7.9	79	1.3409	1.12	.328
7.50	8.42	1.0718	80.2	.273	989.6	8.6	85	1.3415	1.21	.352
8.00	8.98	1.0768	86.0	.293	988.9	9.3	91	1.3421	1.29	.376
8.50	9.54	1.0819	91.8	.312	988.2	10.1	97	1.3426	1.37	.401
9.00	10.10	1.0870	97.7	.332	987.4	10.8	102	1.3432	1.45	.425

A A A —

62 STRONTIUM CHLORIDE, $SrCl_2 \cdot 6H_2O$

MOLECULAR WEIGHT = 158.54 FORMULA WEIGHT, HYDRATE = 266.64
RELATIVE SPECIFIC REFRACTIVITY = .635

A % by wt.	H % by wt.	D^{20}_{20}	C_s g/l	M g-mol/l	C_w g/l	$(C_o - C_w)$ g/l	$(n - n_o)$ $\times 10^4$	n	Δ °C	S g-mol/l
.00	.00	1.0000	.0	.000	998.2	.0	0	1.3330	.00	.000
.50	.84	1.0044	5.0	.032	997.6	.6	9	1.3339	.22	.064
1.00	1.68	1.0089	10.1	.064	997.0	1.2	18	1.3348	.35	.102
1.50	2.52	1.0134	15.2	.096	996.4	1.8	27	1.3357	.49	.142
2.00	3.36	1.0179	20.3	.128	995.8	2.5	36	1.3366	.63	.184
2.50	4.20	1.0224	25.5	.161	995.1	3.1	45	1.3375	.78	.228
3.00	5.05	1.0270	30.8	.194	994.4	3.8	55	1.3385	.94	.273
3.50	5.89	1.0316	36.0	.227	993.7	4.5	64	1.3394	1.09	.319
4.00	6.73	1.0362	41.4	.261	993.0	5.2	73	1.3403	1.26	.368
4.50	7.57	1.0409	46.8	.295	992.3	5.9	83	1.3412	1.43	.418
5.00	8.41	1.0456	52.2	.329	991.6	6.7	92	1.3422	1.61	.470
5.50	9.25	1.0503	57.7	.364	990.8	7.4	101	1.3431	1.79	.523
6.00	10.09	1.0551	63.2	.399	990.0	8.2	111	1.3441	1.98	.578
6.50	10.93	1.0599	68.8	.434	989.2	9.0	120	1.3450	2.17	.635
7.00	11.77	1.0647	74.4	.469	988.4	9.8	130	1.3460	2.38	.693
7.50	12.61	1.0696	80.1	.505	987.6	10.6	139	1.3469	2.59	.753
8.00	13.45	1.0745	85.8	.541	986.8	11.5	149	1.3479	2.80	.815
8.50	14.30	1.0794	91.6	.578	985.9	12.3	159	1.3489	3.03	.879
9.00	15.14	1.0844	97.4	.614	985.0	13.2	168	1.3498	3.26	.944
9.50	15.98	1.0894	103.3	.652	984.2	14.1	178	1.3508	3.50	1.010
10.00	16.82	1.0944	109.3	.689	983.3	15.0	188	1.3518	3.74	1.079
11.00	18.50	1.1046	121.3	.765	981.4	16.8	208	1.3538	4.26	1.228
12.00	20.18	1.1149	133.6	.842	979.4	18.9	228	1.3558	4.81	1.379
13.00	21.86	1.1255	146.1	.921	977.5	20.8	248	1.3578	5.40	1.536
14.00	23.55	1.1362	158.8	1.002	975.4	22.8	269	1.3599	6.03	1.700
15.00	25.23	1.1470	171.8	1.083	973.3	25.0	290	1.3620	6.70	1.871
16.00	26.91	1.1579	184.9	1.167	971.0	27.3	310	1.3640	7.41	2.047
17.00	28.59	1.1690	198.4	1.251	968.5	29.7	331	1.3661	8.16	2.230
18.00	30.27	1.1801	212.1	1.338	966.0	32.2	352	1.3682	8.97	2.419
19.00	31.96	1.1914	226.0	1.425	963.4	34.9	373	1.3703	9.82	2.614
20.00	33.64	1.2029	240.1	1.515	960.6	37.6	394	1.3724	10.73	2.814
22.00	37.00	1.2262	269.3	1.699	954.7	43.5	437	1.3767	12.73	3.230
24.00	40.36	1.2502	299.5	1.889	948.5	49.7	480	1.3810	14.99	3.664
26.00	43.73	1.2750	330.9	2.087	941.8	56.4	524	1.3854		
28.00	47.09	1.3006	363.5	2.293	934.8	63.4	570	1.3900		
30.00	50.46	1.3272	397.5	2.507	927.4	70.8	617	1.3946		
32.00	53.82	1.3548	432.8	2.730	919.7	78.6				
34.00	57.18	1.3836	469.6	2.962	911.6	86.7				

H W C

63 SUCROSE, $C_{12}H_{22}O_{11}$

MOLECULAR WEIGHT = 342.30
RELATIVE SPECIFIC REFRACTIVITY = 1.029

A % by wt.	D^{20}_{20}	C_s g/l	M g-mol/l	C_w g/l	$(C_o - C_w)$ g/l	$(n - n_o)$ $\times 10^4$	n	Δ °C	S g-mol/l
.00	1.0000	.0	.000	998.2	.0	0	1.3330	.00	.000
1.00	1.0039	10.0	.029	992.1	6.1	14	1.3344	.06	.015
2.00	1.0078	20.1	.059	985.9	12.3	29	1.3359	.11	.032
3.00	1.0117	30.3	.089	979.6	18.6	43	1.3373	.17	.049
4.00	1.0157	40.6	.118	973.3	24.9	58	1.3388	.23	.066
5.00	1.0197	50.9	.149	967.0	31.3	73	1.3403	.29	.084
6.00	1.0237	61.3	.179	960.5	37.7	88	1.3418	.35	.102
7.00	1.0277	71.8	.210	954.1	44.2	103	1.3433	.42	.121
8.00	1.0318	82.4	.241	947.5	50.7	118	1.3448	.49	.141
9.00	1.0359	93.1	.272	941.0	57.3	133	1.3463	.55	.161
10.00	1.0400	103.8	.303	934.3	63.9	148	1.3478	.63	.181
11.00	1.0441	114.7	.335	927.6	70.6	164	1.3494	.70	.203
12.00	1.0483	125.6	.367	920.9	77.3	179	1.3509	.77	.225
13.00	1.0525	136.6	.399	914.1	84.2	195	1.3525	.85	.248
14.00	1.0568	147.7	.431	907.2	91.0	211	1.3541	.93	.271
15.00	1.0610	158.9	.464	900.3	97.9	227	1.3557	1.01	.295
16.00	1.0653	170.2	.497	893.3	104.9	243	1.3573	1.10	.320
17.00	1.0697	181.5	.530	886.3	112.0	259	1.3589	1.18	.346
18.00	1.0740	193.0	.564	879.2	119.1	275	1.3605	1.27	.372
19.00	1.0784	204.5	.598	872.0	126.2	292	1.3622	1.37	.400

63 SUCROSE, $C_{12}H_{22}O_{11}$

A % by wt.	D^{20}_{20}	C_s g/l	M g-mol/l	C_w g/l	$(C_o - C_w)$ g/l	$(n - n_o)$ $\times 10^4$	n	Δ °C	S g-mol/l
20.00	1.0829	216.2	.632	864.8	133.5	308	1.3638	1.46	.428
22.00	1.0918	239.8	.700	850.1	148.1	342	1.3672	1.67	.488
24.00	1.1009	263.8	.771	835.2	163.0	376	1.3706	1.89	.551
26.00	1.1101	288.1	.842	820.0	178.2	410	1.2740	2.12	.619
28.00	1.1195	312.9	.914	804.6	193.6	445	1.3775	2.37	.692
30.00	1.1290	338.1	.988	788.9	209.3	481	1.3811	2.65	.771
32.00	1.1386	363.7	1.063	772.9	225.3	517	1.3847	2.94	.855
34.00	1.1484	389.8	1.139	756.6	241.6	554	1.3884	3.27	.947
36.00	1.1583	416.2	1.216	740.0	258.2	591	1.3921	3.62	1.046
38.00	1.1683	443.2	1.295	723.1	275.1	629	1.3959	4.01	1.161
40.00	1.1785	470.6	1.375	705.9	292.4	668	1.3998	4.45	1.281
42.00	1.1889	498.4	1.456	688.3	309.9	707	1.4037		
44.00	1.1994	526.8	1.539	670.5	327.8	747	1.4076		
46.00	1.2100	555.6	1.623	652.2	346.0	787	1.4117		
48.00	1.2208	584.9	1.709	633.7	364.5	828	1.4158		
50.00	1.2317	614.8	1.796	614.8	383.5	869	1.4199		
52.00	1.2428	645.1	1.885	595.5	402.7	912	1.4242		
54.00	1.2541	676.0	1.975	575.9	422.4	955	1.4284		
56.00	1.2655	707.4	2.067	555.8	442.4	998	1.4328		
58.00	1.2770	739.4	2.160	535.4	462.8	1042	1.4372		
60.00	1.2887	771.9	2.255	514.6	483.6	1087	1.4417		

(N.B.S. Circular C440) (ibid) C

64 SULFURIC ACID, H_2SO_4

MOLECULAR WEIGHT = 98.08

RELATIVE SPECIFIC REFRACTIVITY = .693

A % by wt.	D^{20}_{20}	C_s g/l	M g-mol/l	C_w g/l	$(C_o - C_w)$ g/l	$(n - n_o)$ $\times 10^4$	n	Δ °C	S g-mol/l
.00	1.0000	.0	.000	998.2	.0	0	1.3330	.00	.000
1.00	1.0068	10.1	.102	995.0	3.3	13	1.3343	.41	.118
2.00	1.0135	20.2	.206	991.5	6.8	25	1.3355	.78	.228
3.00	1.0202	30.6	.311	987.8	10.4	37	1.3367	1.14	.333
4.00	1.0269	41.0	.418	984.1	14.2	49	1.3379	1.52	.444
5.00	1.0336	51.6	.526	980.2	18.0	61	1.3391	2.00	.584
6.00	1.0404	62.3	.635	976.2	22.0	73	1.3403	2.49	.727
7.00	1.0472	73.2	.746	972.1	26.1	85	1.3415	3.01	.873
8.00	1.0540	84.2	.858	968.0	30.2	97	1.3427	3.55	1.024
9.00	1.0609	95.3	.972	963.8	34.5	109	1.3439	4.11	1.187
10.00	1.0679	106.6	1.087	959.4	38.8	121	1.3451	4.70	1.350
11.00	1.0750	118.0	1.204	955.1	43.2	134	1.3463	5.33	1.519
12.00	1.0821	129.6	1.322	950.6	47.6	146	1.3476	6.00	1.694
13.00	1.0893	141.4	1.441	946.1	52.2	158	1.3488	6.71	1.875
14.00	1.0966	153.3	1.563	941.4	56.8	171	1.3501	7.48	2.065
15.00	1.1039	165.3	1.685	936.6	61.6	183	1.3513	8.30	2.262
16.00	1.1113	177.5	1.810	931.8	66.4	196	1.3526	9.19	2.469
17.00	1.1188	189.9	1.936	927.0	71.3	209	1.3539	10.15	2.685
18.00	1.1263	202.4	2.063	922.0	76.3	222	1.3551	11.19	2.912
19.00	1.1339	215.1	2.193	916.9	81.4	234	1.3564	12.33	3.149
20.00	1.1416	227.9	2.324	911.6	86.6	247	1.3577	13.58	3.397
22.00	1.1570	254.1	2.591	900.8	97.4	273	1.3603	16.46	3.927
24.00	1.1725	280.9	2.864	889.6	108.7	298	1.3628	19.97	
26.00	1.1883	308.4	3.145	877.8	120.4	324	1.3654	24.28	
28.00	1.2043	336.6	3.432	865.6	132.7	349	1.3679	29.62	
30.00	1.2205	365.5	3.727	852.9	145.4	375	1.3705	36.29	
32.00	1.2370	395.1	4.029	839.6	158.6	400	1.3730	44.66	
34.00	1.2536	425.5	4.338	825.9	172.3	426	1.3756	55.21	
36.00	1.2705	456.6	4.655	811.7	186.5	452	1.3782	68.58	
38.00	1.2877	488.5	4.980	797.0	201.3	477	1.3807		
40.00	1.3052	521.2	5.314	781.7	216.5	503	1.3833		
42.00	1.3230	554.7	5.655	766.0	232.3				
44.00	1.3411	589.0	6.006	749.7	248.6				
46.00	1.3595	624.3	6.365	732.8	265.4				
48.00	1.3783	660.4	6.733	715.5	282.8				
50.00	1.3975	697.5	7.112	697.5	300.7				
52.00	1.4172	735.6	7.500	679.0	319.2				
54.00	1.4373	774.8	7.900	660.0	338.2				
56.00	1.4580	815.1	8.310	640.4	357.8				
58.00	1.4793	856.5	8.733	620.2	378.0				
60.00	1.5013	899.2	9.168	599.5	398.8				

H W (A + L)

65 TRICHLOROACETIC ACID, CCl_3COOH

MOLECULAR WEIGHT = 163.40

RELATIVE SPECIFIC REFRACTIVITY = .864

A % by wt.	D^{20}_{20}	C_s g/l	M g-mol/l	C_w g/l	$(C_o - C_w)$ g/l	$(n - n_o) \times 10^4$	n	Δ °C	S g-mol/l
.00	1.0000	.0	.000	998.2	.0	0	1.3330	.00	.000
.50	1.0025	5.0	.031	995.7	2.5	6	1.3336	.10	.030
1.00	1.0050	10.0	.061	993.2	5.0	13	1.3343	.21	.060
1.50	1.0076	15.1	.092	990.7	7.5	19	1.3349	.32	.091
2.00	1.0101	20.2	.123	988.1	10.1	26	1.3355	.42	.122
2.50	1.0126	25.3	.155	985.5	12.7	32	1.3362	.53	.152
3.00	1.0151	30.4	.186	982.9	15.3	38	1.3368	.63	.183
3.50	1.0177	35.6	.218	980.3	17.9	45	1.3375	.74	.214
4.00	1.0202	40.7	.249	977.7	20.6	51	1.3381	.84	.245
4.50	1.0227	45.9	.281	975.0	23.2	58	1.3387	.95	.277
5.00	1.0253	51.2	.313	972.3	25.9	64	1.3394	1.06	.308
5.50	1.0278	56.4	.345	969.6	28.7	70	1.3400	1.16	.339
6.00	1.0304	61.7	.378	966.8	31.4	77	1.3407	1.27	.370
6.50	1.0329	67.0	.410	964.1	34.2	83	1.3413	1.37	.401
7.00	1.0355	72.4	.443	961.3	37.0	89	1.3419	1.48	.432
7.50	1.0380	77.7	.476	958.5	39.8	96	1.3426	1.59	.464
8.00	1.0406	83.1	.509	955.6	42.6	102	1.3432	1.69	.495
8.50	1.0431	88.5	.542	952.8	45.4	109	1.3438	1.80	.525
9.00	1.0457	93.9	.575	949.9	48.3	115	1.3445	1.90	.556
9.50	1.0483	99.4	.608	947.0	51.2	121	1.3451	2.01	.587
10.00	1.0508	104.9	.642	944.1	54.1	128	1.3458	2.11	.618

A A A

66 UREA, NH_2CONH_2

MOLECULAR WEIGHT = 60.06

RELATIVE SPECIFIC REFRACTIVITY = 1.147

A % by wt.	D^{20}_{20}	C_s g/l	M g-mol/l	C_w g/l	$(C_o - C_w)$ g/l	$(n - n_o) \times 10^4$	n	Δ °C	S g-mol/l
.00	1.0000	.0	.000	998.2	.0	0	1.3330	.00	.000
.50	1.0014	5.0	.083	994.6	3.6	7	1.3337	.15	.043
1.00	1.0028	10.0	.167	991.0	7.3	15	1.3344	.30	.087
1.50	1.0041	15.0	.250	987.3	10.9	22	1.3352	.45	.131
2.00	1.0055	20.1	.334	983.6	14.6	29	1.3359	.61	.176
2.50	1.0068	25.1	.418	979.9	18.3	36	1.3366	.76	.221
3.00	1.0081	30.2	.503	976.2	22.1	43	1.3373	.92	.267
3.50	1.0095	35.3	.587	972.4	25.8	51	1.3381	1.08	.314
4.00	1.0109	40.4	.672	968.7	29.5	58	1.3388	1.23	.360
4.50	1.0122	45.5	.757	964.9	33.3	65	1.3395	1.39	.407
5.00	1.0136	50.6	.842	961.2	37.1	72	1.3402	1.56	.455
5.50	1.0149	55.7	.928	957.4	40.8	80	1.3410	1.72	.502
6.00	1.0163	60.9	1.013	953.6	44.6	87	1.3417	1.88	.550
6.50	1.0176	66.0	1.099	949.8	48.4	94	1.3424	2.05	.599
7.00	1.0190	71.2	1.186	946.0	52.2	102	1.3432	2.21	.647
7.50	1.0204	76.4	1.272	942.2	56.1	109	1.3439	2.38	.696
8.00	1.0217	81.6	1.359	938.3	59.9	116	1.3446	2.55	.744
8.50	1.0231	86.8	1.445	934.5	63.8	124	1.3454	2.72	.793
9.00	1.0245	92.0	1.532	930.6	67.6	131	1.3461	2.90	.842
9.50	1.0258	97.3	1.620	926.7	71.5	139	1.3469	3.07	.892
10.00	1.0272	102.5	1.707	922.9	75.4	146	1.3476	3.25	.941

A A A

67 URINE SOLIDS, CAT

RELATIVE SPECIFIC REFRACTIVITY = 1.08

A % by wt.	D^{20}_{20}	C_s g/l	M g-mol/l	C_w g/l	$(C_o - C_w)$ g/l	$(n - n_o) \times 10^4$	n	Δ °C	S g-mol/l
.00	1.000	0	.000	998	0	0	1.3330	.00	.000
.50	1.002	5		995	3	8	1.3338		
1.00	1.004	10		992	6	16	1.3346		
1.50	1.006	15		989	9	24	1.3354		
2.00	1.008	20		986	13	33	1.3363		
2.50	1.010	25		983	16	41	1.3371		
3.00	1.012	30		979	19	49	1.3379		
3.50	1.013	35		976	22	57	1.3387		
4.00	1.015	41		973	25	65	1.3395		
4.50	1.017	46		970	29	73	1.3403		
5.00	1.019	51		966	32	81	1.3411		
5.50	1.021	56		963	35	89	1.3419		
6.00	1.023	61		960	39	97	1.3427		
6.50	1.025	67		956	42	105	1.3435		
7.00	1.027	72		953	45	113	1.3443		

67 URINE SOLIDS, CAT

A % by wt.	D^{20}_{20}	C_s g/l	M g-mol/l	C_w g/l	$(C_o - C_w)$ g/l	$(n - n_o)$ × 10^4	n	Δ °C	S g-mol/l
7.50	1.028	77		950	49	121	1.3451		
8.00	1.030	82		946	52	129	1.3459		
8.50	1.032	88		943	56	137	1.3466		
9.00	1.034	93		939	59	144	1.3474		
9.50	1.036	98		936	62	152	1.3482		
10.00	1.038	104		932	66	160	1.3490		
11.00	1.041	114		925	73	176	1.3505		
12.00	1.045	125		918	80	191	1.3521		
13.00	1.049	136		911	88	206	1.3536		
14.00	1.052	147		903	95	222	1.3552		
15.00	1.056	158		896	102	237	1.3567		
16.00	1.060	169		888	110	252	1.3582		
17.00	1.063	180		881	117	267	1.3597		
18.00	1.067	192		873	125	282	1.3612		
19.00	1.070	203		865	133	297	1.3626		
20.00	1.074	214		857	141	311	1.3641		

A A

68 URINE SOLIDS, HUMAN

RELATIVE SPECIFIC REFRACTIVITY = 1.02

A % by wt.	D^{20}_{20}	C_s g/l	M g-mol/l	C_w g/l	$(C_o - C_w)$ g/l	$(n - n_o)$ × 10^4	n	Δ °C	S g-mol/l
.00	1.000	0	.000	998	0	0	1.3330	.00	.000
.50	1.002	5		996	3	8	1.3338		
1.00	1.005	10		993	5	16	1.3346		
1.50	1.007	15		990	8	24	1.3354		
2.00	1.009	20		987	11	32	1.3362		
2.50	1.011	25		984	14	40	1.3370		
3.00	1.014	30		981	17	48	1.3377		
3.50	1.016	36		978	20	55	1.3385		
4.00	1.018	41		975	23	63	1.3393		
4.50	1.020	46		972	26	71	1.3401		
5.00	1.022	51		969	29	79	1.3408		
5.50	1.024	56		966	33	86	1.3416		
6.00	1.026	61		963	36	94	1.3424		
6.50	1.028	67		959	39	101	1.3431		
7.00	1.029	72		956	43	109	1.3439		
7.50	1.031	77		952	46	116	1.3446		
8.00	1.033	83		949	50	123	1.3453		
8.50	1.035	88		945	53	131	1.3461		
9.00	1.036	93		941	57	138	1.3468		
9.50	1.038	98		938	61	145	1.3475		
10.00	1.039	104		934	65	152	1.3482		

A A

69 ZINC SULFATE, $ZnSO_4 \cdot 7H_2O$

MOLECULAR WEIGHT = 161.44 FORMULA WEIGHT, HYDRATE = 287.56

RELATIVE SPECIFIC REFRACTIVITY = .494

A % by wt.	H % by wt.	D^{20}_{20}	C_s g/l	M g-mol/l	C_w g/l	$(C_o - C_w)$ g/l	$(n - n_o)$ × 10^4	n	Δ °C	S g-mol/l
.00	.00	1.0000	.0	.000	998.2	.0	0	1.3330	.00	.000
.50	.89	1.0052	5.0	.031	998.4	−.2	9	1.3339	.08	.022
1.00	1.78	1.0104	10.1	.062	998.5	−.3	18	1.3348	.15	.042
1.50	2.67	1.0156	15.2	.094	998.6	−.4	27	1.3357	.22	.062
2.00	3.56	1.0209	20.4	.126	998.7	−.4	36	1.3366	.28	.082
2.50	4.45	1.0261	25.6	.159	998.7	−.5	45	1.3375	.35	.100
3.00	5.34	1.0314	30.9	.191	998.7	−.5	54	1.3384	.41	.119
3.50	6.23	1.0368	36.2	.224	998.7	−.5	63	1.3393	.47	.137
4.00	7.12	1.0421	41.6	.258	998.7	−.4	72	1.3402	.53	.155
4.50	8.02	1.0475	47.1	.291	998.6	−.4	81	1.3411	.59	.172
5.00	8.91	1.0529	52.6	.326	998.5	−.3	90	1.3420	.65	.190
5.50	9.80	1.0584	58.1	.360	998.4	−.1	99	1.3429	.71	.207
6.00	10.69	1.0638	63.7	.395	998.2	−.0	109	1.3438	.77	.224
6.50	11.58	1.0694	69.4	.430	998.1	.2	118	1.3448	.83	.241
7.00	12.47	1.0749	75.1	.465	997.9	.3	127	1.3457	.89	.259
7.50	13.36	1.0805	80.9	.501	997.7	.5	136	1.3466	.95	.276
8.00	14.25	1.0861	86.7	.537	997.5	.8	145	1.3475	1.01	.294
8.50	15.14	1.0918	92.6	.574	997.2	1.0	155	1.3485	1.07	.313
9.00	16.03	1.0975	98.6	.611	997.0	1.2	164	1.3494	1.14	.331
9.50	16.92	1.1033	104.6	.648	996.7	1.5	173	1.3503	1.20	.351
10.00	17.81	1.1091	110.7	.686	996.4	1.8	183	1.3513	1.27	.370
11.00	19.59	1.1209	123.1	.762	995.8	2.4	202	1.3532	1.41	.412
12.00	21.37	1.1328	135.7	.841	995.1	3.1	221	1.3551	1.56	.455
13.00	23.16	1.1449	148.6	.920	994.3	3.9	241	1.3571	1.72	.502
14.00	24.94	1.1573	161.7	1.002	993.5	4.7	261	1.3590	1.89	.552
15.00	26.72	1.1699	175.2	1.085	992.7	5.6	281	1.3611	2.07	.605
16.00	28.50	1.1827	188.9	1.170	991.7	6.5	301	1.3631	2.27	.662

H W C

SPECIFIC GRAVITY OF AQUEOUS INVERT SUGAR SOLUTIONS

t=20 C

Invert sugar is the equimolar mixture of fructose and glucose obtained by hydrolyzing sucrose.

Wt % (Vacuum)	Specific Gravity	G solute per liter of solution	lb solute per cu ft of solution	lb solute per gallon (U.S.) of solution	Wt % (Vacuum)	Specific Gravity	G solute per liter of solution	lb solute per cu ft of solution	lb solute per gallon (U.S.) of solution
1	1.00211	10.021	0.62561	0.08363	44	1.19539	525.972	32.83643	4.38924
2	1.00601	20.120	1.25609	0.16790	45	1.20057	540.257	33.72824	4.50844
3	1.00994	30.298	1.89150	0.25284	46	1.20577	554.654	34.62705	4.62859
4	1.01389	40.556	2.53191	0.33844	47	1.21101	569.175	35.53360	4.74977
5	1.01786	50.893	3.17725	0.42470	48	1.21629	583.819	36.44782	4.87197
6	1.02186	61.312	3.82771	0.51165	49	1.22159	598.579	37.36929	4.99514
7	1.02589	71.812	4.48322	0.59927	50	1.22693	613.465	38.29862	5.11937
8	1.02995	82.396	5.14398	0.68759	51	1.23230	628.473	39.23557	5.24461
9	1.03403	93.063	5.80992	0.77661	52	1.23771	643.609	40.18051	5.37092
10	1.03814	103.814	6.48111	0.86633	53	1.24315	658.870	41.13325	5.49827
11	1.04227	114.649	7.15754	0.95675	54	1.24861	674.249	42.09337	5.62661
12	1.04644	125.573	7.83952	1.04791	55	1.25412	689.766	43.06209	5.75610
13	1.05063	136.582	8.52681	1.13978	56	1.25965	705.404	44.03837	5.88660
14	1.05484	147.678	9.21954	1.23237	57	1.26522	721.175	45.02296	6.01821
15	1.05909	158.864	9.91788	1.32572	58	1.27082	737.076	46.01565	6.15090
16	1.06337	170.139	10.62178	1.41981	59	1.27646	753.111	47.01672	6.28471
17	1.06767	181.504	11.33129	1.51465	60	1.28112	768.672	47.98819	6.41457
18	1.07200	192.960	12.04649	1.61025	61	1.28782	785.570	49.04310	6.55558
19	1.07637	204.510	12.76756	1.70664	62	1.29355	802.001	50.06892	6.69270
20	1.08076	216.152	13.49437	1.80379	63	1.29932	818.572	51.10345	6.83098
21	1.08518	227.888	14.22705	1.90173	64	1.30511	835.270	52.14591	6.97033
22	1.08963	239.719	14.96566	2.00046	65	1.31094	852.111	53.19729	7.11087
23	1.09411	251.645	15.71020	2.09998	66	1.31680	869.088	54.25716	7.25254
24	1.09862	263.669	16.46086	2.20032	67	1.32269	886.202	55.32559	7.39536
25	1.10316	275.790	17.21757	2.30147	68	1.32862	903.462	56.40313	7.53939
26	1.10773	288.010	17.98046	2.40344	69	1.33458	920.860	57.48929	7.68458
27	1.11233	300.329	18.74954	2.50625	70	1.34057	938.399	58.58425	7.83094
28	1.11697	312.752	19.52511	2.60992	71	1.34659	956.079	59.68801	7.79848
29	1.12163	325.273	20.30679	2.71440	72	1.35264	973.901	60.80064	8.12720
30	1.12633	337.899	21.09503	2.81977	73	1.35872	991.866	61.92219	8.27712
31	1.13105	350.626	21.88958	2.92597	74	1.36484	1009.982	63.05318	8.42830
32	1.13581	363.459	22.69075	3.03307	75	1.37099	1028.243	64.19321	8.58069
33	1.14060	376.398	23.49853	3.14104	76	1.37717	1046.649	65.34230	8.73429
34	1.14542	389.443	24.31293	3.24990	77	1.38337	1065.195	66.50012	8.88905
35	1.15027	402.595	25.13401	3.35966	78	1.38962	1083.904	67.66813	9.04518
36	1.15516	415.858	25.96201	3.47034	79	1.39589	1102.753	68.84487	9.20247
37	1.16007	429.226	26.79658	3.58199	80	1.40219	1121.752	70.03098	9.36102
38	1.16502	442.708	27.63826	3.69440	81	1.40852	1140.901	71.22645	9.52082
39	1.17000	456.300	28.48681	3.80728	82	1.41488	1160.202	72.43141	9.68189
40	1.17502	470.008	29.34260	3.92222	83	1.42128	1179.662	73.64630	9.84428
41	1.18006	483.825	30.20519	4.03752	84	1.42770	1199.268	74.87030	10.00790
42	1.18514	497.759	31.07509	4.15380	85	1.43415	1219.028	76.10392	10.17279
43	1.19025	511.808	31.95217	4.27104					

HEAT OF COMBUSTION

For Organic Compounds

The heat of combustion is given in kilogram calories per gram molecular weight of the substance when combustion takes place at atmospheric pressure and 20° C. The final products of combustion are gaseous carbon dioxide, liquid water and nitrogen gas for C, H, N compounds. For method of computing heats of formation see statement following this table.

Name	Formula	Physical state	Heat of combustion, kg. calories
Acetaldehyde	CH_3CHO	liquid	279.0
Acetamide	CH_3CONH_2	solid	282.6
Acetanilide	$CH_3CONHC_6H_5$	solid	1,010.4
Acetic acid	CH_3CO_2H	liquid	209.4
Acetic anhydride	$(CH_3CO)_2O$	liquid	431.9
Acetone	$(CH_3)_2CO$	liquid	426.8
Acetonitrile	CH_3CN	liquid	302.4
Acetophenone	$C_6H_5COCH_3$	solid	988.9
Acetylacetone	$CH_3COCH_2COCH_3$	liquid	615.9
Acetylene	$(CH)_2$	gas	312.0
Acrolein	$CH_2:CHCHO$	liquid	389.6
Acrylic acid	$CH_2:CHCO_2H$	liquid	327.5
Adipic acid	$(CH_2)_4(CO_2H)_2$	solid	669.0
Alanine	$CH_3CH(NH_2)CO_2H$	solid	387.7
Aldol, *see β-hydroxybutyr-aldehyde*			
Alizarin, *see Dihydroxyanthraquinone*			
Allyl alcohol	$CH_2:CHCH_2OH$	liquid	442.4
Allylene	$CH_3C:CH$	gas	465.1
p-Aminoazobenzene	$H_2NC_6H_4N_2C_6H_5$	solid	1,574.0
p-Aminophenol	$HOC_6H_4NH_2$	solid	760.0
Amygdalin	$C_{20}H_{27}O_{11}N$	solid	2,348.4
Amyl acetate	$C_4H_9CO_2C_2H_5$	liquid	1,042.5
Amyl alcohol (ferm.)	$(CH_3)_2CH_2CH_2CH_2OH$	liquid	793.7
Amylene	C_5H_{10}	liquid	803.4
Anethole	$C_{10}H_{12}O$	solid	1,324.4
Aniline	$C_6H_5NH_2$	liquid	811.7
p-Anisidine	$CH_3OC_6H_4NH_2$	solid	924.0
Anisole	$C_6H_5OCH_3$	liquid	905.1
Anthracene	$C_6H_4:(CH)_2:C_6H_4$	solid	1,700.4
Anthraquinone	$C_{14}H_8O_2$	solid	1,544.5
Arabinose	$C_5H_{10}O_5$	solid	559.9
Arabitol	$C_5H_{12}O_5$	solid	661.2
Arachidic acid	$C_{20}H_{40}O_2$	solid	3,025.9
Azelaic acid	$(CH_2)_7(CO_2H)_2$	solid	1,141.7
Azobenzene	$(C_6H_5N)_2$	solid	1,545.9
Azoxybenzene	$(C_6H_5N)_2O$	solid	1,534.5
Behenic acid	$C_{22}H_{44}O_2$	solid	3,338.4
Benzalacetone	$C_6H_5CH:CHCOCH_3$	solid	1,257.4
Benzaldehyde	C_6H_5CHO	liquid	841.3
Benzamide	$C_6H_5CONH_2$	solid	847.6
Benzanilide	$C_6H_5CONHC_6H_5$	solid	1,575.5
Benzene	C_6H_6	liquid	782.3
Benzenediazonium nitrate	$C_6H_5N_2NO_3$	solid	782.6
Benzidine	$(C_6H_4NH_2)_2$	solid	1,560.9
Benzil	$(C_6H_5CO)_2$	solid	1,624.6
* Benzoic acid	$C_6H_5CO_2H$	solid	771.2
Benzoic anhydride	$(C_6H_5CO)_2O$	solid	1,555.1
Benzoin	$C_6H_5.CHOH.COC_6H_5$	solid	1,671.4
Benzonitrile	C_6H_5CN	liquid	865.5
Benzophenone	$(C_6H_5)_2CO$	solid	1,556.5
Benzoyl chloride	C_6H_5COCl	liquid	782.8
Benzoyl peroxide	$(C_6H_5CO)_2O_2$	solid	1,551.7
Benzyl alcohol	$C_6H_5CH_2OH$	liquid	894.3
Benzylamine	$C_6H_5CH_2NH_2$	liquid	969.4
Benzyl carbylamine	$C_6H_5CH_2NC$	liquid	1,046.5
Benzyl chloride	$C_6H_5CH_2Cl$	liquid	886.4
Benzyl cyanide	$C_6H_5CH_2CN$	liquid	1,023.5
Borneol	$C_{10}H_{18}O$	liquid	1,469.6
Brucine	$C_{23}H_{26}O_4N_2$	solid	2,933.0
n-Butyl alcohol	C_4H_9OH	liquid	638.6
tert.-Butyl alcohol, see *Trimethyl carbinol*			
n-Butylamine	$C_4H_9NH_2$	liquid	710.6
sec.-Butylamine	$(CH_3)(C_2H_5):CHNH_2$	liquid	713.0
tert.-Butylamine	$(CH_3)_3CNH_2$	liquid	716.0
tert.-Butylbenzene	$C_6H_5C(CH_3)_3$	liquid	1,400.4
n-Butyramide	$C_3H_7CONH_2$	solid	596.0
n-Butyric acid	$C_3H_7CO_2H$	liquid	524.3
n-Butyronitrile	C_3H_7CN	liquid	613.3
Caffeine	$C_8H_{10}O_2N_4$	solid	1,014.2
Camphene	$C_{10}H_{16}$	solid	1,468.8
Camphor	$C_{10}H_{16}O$	solid	1,411.0
Cane sugar, *see Sucrose*			
Capric acid	$C_9H_{18}O_2$	solid	1,458.1
Caproic acid	$C_5H_{11}CO_2H$	liquid	831.0
Carbon disulfide	CS_2	liquid	246.6
Carbon subnitride	$(C.CN)_2$	solid	514.8
Carbon tetrachloride	CCl_4	liquid	37.3
Carbonyl sulfide	COS	gas	130.5
Carvacrol	$C_{10}H_{14}O$	liquid	1,354.5
Cetyl alcohol	$C_{16}H_{34}O$	solid	2,504.4
Cetyl palmitate	$C_{32}H_{64}O$	solid	4,872.8
Chloracetic acid	$ClCH_2CO_2H$	solid	171.0
o-Chlorobenzoic acid	$ClC_6H_4CO_2H$	solid	734.5
Chloroform	$CHCl_3$	liquid	89.2
Chrysene	$C_{14}H_{10}$	solid	2,139.1
Cinnamic acid (*trans*)	$C_6H_5CH:CHCO_2H$	solid	1,040.2
Cinnamic aldehyde	$C_6H_5CH:CHCHO$	liquid	1,112.3

* Accepted value by Int. Union of Pure and Appld. Chem., Lyons, 1923.

HEAT OF COMBUSTION (Continued)

For Organic Compounds

Name	Formula	Physical state	Heat of combustion, kg. calories
Cinnamic anhydride	$C_{18}H_{14}O_3$	solid	2,091.3
d-Citrene	$C_{10}H_{16}$	liquid	1,473.0
Citric acid (anhydr.)	$C_6H_8O_7$	solid	474.5
Codeine	$C_{18}H_{21}O_3N.H_2O$	solid	2,327.6
Coniine	$C_8H_{17}N$	liquid	1,275.5
Creatine (anhydr.)	$C_4H_9O_2N_3$	solid	559.8
Creatinine	$C_4H_7ON_3$	solid	563.4
o-Cresol	$CH_3C_6H_4OH$	liquid	882.6
o-Cresol	$CH_3C_6H_4OH$	solid	879.5
m-Cresol	$CH_3C_6H_4OH$	liquid	880.5
p-Cresol	$CH_3C_6H_4OH$	liquid	882.5
p-Cresol	$CH_3C_6H_4OH$	solid	880.0
m-Cresolmethyl ether	$CH_3C_6H_4OCH_3$	liquid	1,057.0
Crotonaldehyde	C_3H_5CHO	liquid	542.1
Cyanoacetic acid	NC CH_2CO_2H	solid	298.8
Cyanogen	$(CN)_2$	gas	258.3
Cycloheptanol	$CH_2(CH_2)_5CHOH$	liquid	1,050.2
Cyclohexanol	$CH_2(CH_2)_4CHOH$	liquid	890.7
Cycloheptene	C_7H_{12}	liquid	1,049.9
Cycloheptane	$(CH_2)_7$	liquid	1,087.3
Cyclohexane	$(CH_2)_6$	liquid	937.8
Cyclohexene, *see Tetrahydrobenzene*			
Cyclopentane	$(CH_2)_5$	liquid	783.6
Cyclopropane, *see Trimethylene*			
Cymene	$C_6H_4(CH_3)(CH_3CHCH_3)$—(1, 4)	liquid	1,402.8
Decahydronaphthalene (*cis*)	$C_{10}H_{18}$	liquid	1,502.5
Decahydronaphthalene (*trons*)	$C_{10}H_{18}$	liquid	1,499.5
Decane	$C_{10}H_{22}$	liquid	1,610.2
Dextrose, *see Glucose*			
Diallyl	$(CH_3CH:CH_2)_2$	vapor	903.4
Diamyl ether	$(C_5H_{11})_2O$	liquid	1,609.3
Diamylene	$C_{10}H_{20}$	liquid	1,582.2
Dibenzyl	$(C_6H_5CH_2)_2$	solid	1,810.6
Dibenzyl amine	$(C_6H_5CH_2)_2NH$	solid	1,853.0
o-Dichlorobenzene	$C_6H_4Cl_2$	liquid	671.8
Diethylacetic acid	$(C_2H_5)_2CHCO_2H$	liquid	830.8
Diethyl amine	$(C_2H_5)_2NH$	liquid	716.9
Diethylaniline	$C_6H_5N(C_2H_5)_2$	liquid	1,451.6
Diethyl carbonate	$CO(OC_2H_5)_2$	liquid	647.9
Diethyl ether	$(C_2H_5)_2O$	liquid	651.7
Diethyl ketone	$(C_2H_5)_2CO$	liquid	735.6
Diethyl malonate	$CH_2(CO_2C_2H_5)_2$	liquid	860.4
Diethyl oxalate	$(CO_2C_2H_5)_2$	liquid	716.0
Diethyl succinate	$(CH_2CO_2C_2H_5)_2$	liquid	1,007.3
Dihydrobenzene	C_6H_8	liquid	847.8
Δ_1-Dihydronaphthalene	$C_{10}H_{10}$	liquid	1,296.3
Δ_1-Dihydronaphthalene	$C_{10}H_{10}$	solid	1,298.3
Dihydroxyanthraquinone	$C_{14}H_6O_2(OH)_2$—(1, 2)	solid	1,448.9
Diisoamyl	$[(CH_3)_2CHCH_2CH_2]_2$	liquid	1,615.8
Diisobutylene	$[(CH_3)_2CHCH_2]_2$	liquid	1,252.4
Diisopropyl	$[(CH_3)_2CH]_2$	vapor	993.9
Diisopropyl ketone	$[(CH_3)_2CH]_2CO$	liquid	1,045.5
Dimethyl amine	$(CH_3)_2NH$	liquid	416.7
Dimethylaniline	$C_6H_5N(CH_3)_2$	liquid	1,142.7
Dimethyl carbonate	$CO(OCH_3)_2$	liquid	340.8
Dimethyl ether	$(CH_3)_2O$	gas	347.6
Dimethylethyl carbinol	$C_2H_5(CH_3)_2CHOH$	liquid	784.6
Dimethyl fumarate	$(CHCO_2CH_3)_2$	solid	664.3
2, 5-Dimethylhexane	$(CH_3)_2CH.C_2H_4CH(CH_3)_2$	liquid	1,303.3
3, 4-Dimethylhexane	$[(C_2H_5)(CH_3)CH]_2$	liquid	1,303.7
Dimethyl maleate	$(CHCO_2CH_3)_2$	solid	669.2
Dimethyl oxalate	$(CO_2CH_3)_2$	solid	401.9
2, 2-Dimethylpentane	$(CH_3)_3C.C_3H_7$	liquid	1,148.9
2, 3-Dimethylpentane	$(CH_3)_2CHCH(CH_3)C_2H_5$	liquid	1,148.9
2, 4-Dimethylpentane	$(CH_3)_2CHCH_2CH(CH_3)_2$	liquid	1,148.9
3, 3-Dimethylpentane	$(CH_3)_2C(C_2H_5)_2$	liquid	1,147.9
Dimethyl phthalate	$C_6H_4(CO_2CH_3)_2$	liquid	1,119.7
Dimethyl succinate	$(CH_2CO_2CH_3)_2$	solid	703.3
m-Dinitrobenzene	$C_6H_4(NO_2)_2$	solid	696.8
Dinitrophenol	$C_6H_3(OH)(NO_2)_2$—(1, 2, 4)	solid	648.0
Dinitrotoluene	$C_6H_3(CH_3)(NO_2)_2$—(1, 2, 4)	solid	852.8
Diphenyl	$(C_6H_5)_2$	solid	1,493.6
Diphenyl amine	$(C_6H_5)_2NH$	solid	1,536.2
Diphenyl carbinol	$(C_6H_5)_2CHOH$	solid	1,615.4
Diphenylmethane	$(C_6H_5)_2CH_2$	solid	1,655.0
Diphenylnitrosamine	$(C_6H_5)_2N.NO$	solid	1,532.6
Dipropargyl	$(CH:C.CH_2)_2$	vapor	882.9
Dipropyl ketone	$(C_3H_7)_2CO$	liquid	1,050.5
Dulcitol	$C_6H_{14}O_6$	solid	729.1
Durene	$C_6H_2(CH_3)_4$—(1, 2, 4, 5)	solid	1,393.6
Eicosane	$C_{20}H_{42}$	solid	3,183.1
Erythritol	$C_4H_{10}O_4$	solid	504.1
Ethane	C_2H_6	gas	368.4
Ethine, see *Acetylene*			
Ethyl acetate	$CH_3CO_2C_2H_5$	liquid	536.9
Ethyl acetoacetate	$CH_3COCH_2CO_2C_2H_5$	liquid	690.8
Ethyl alcohol	C_2H_5OH	liquid	327.6
Ethyl amine	$C_2H_5NH_2$	liquid	408.5
Ethylaniline	$C_6H_5NHC_2H_5$	liquid	1,121.5
Ethylbenzene	$C_2H_5C_6H_5$	liquid	1,091.2

HEAT OF COMBUSTION (Continued)

For Organic Compounds

Name	Formula	Physical state	Heat of combustion, kg. calories
Ethyl benzoate	$C_6H_5CO_2C_2H_5$	liquid	1,098.7
Ethyl bromide	C_2H_5Br	vapor	340.5
Ethyl *n*-butyrate	$C_3H_7CO_2C_2H_5$	liquid	851.2
Ethyl carbylamine	C_2H_5NC	liquid	477.1
Ethyl chloride	C_2H_5Cl	vapor	316.7
Ethylcycloheptane	$C_2H_5C_7H_{13}$	liquid	1,406.8
Ethyl formate	$HCO_2C_2H_5$	liquid	391.7
3-Ethylhexane	$(C_2H_5)_2CH.C_3H_7$	liquid	1,302.3
Ethyl iodide	C_2H_5I	liquid	356.0
Ethyl isobutyrate	$(CH_3)_2CHCH_2CO_2C_2H_5$	liquid	845.7
Ethyl isocyanate	C_2H_5NCO	liquid	424.5
Ethyl nitrate	$C_2H_5ONO_2$	vapor	322.4
Ethyl nitrite	C_2H_5ONO	vapor	332.6
3-Ethylpentane	$(C_2H_5)_3CH$	liquid	1,149.9
Ethyl propionate	$C_2H_5CO_2C_2H_5$	liquid	690.8
Ethyl salicylate	$HOC_6H_4CO_2C_2H_5$	liquid	1,051.2
Ethyl valerate	$C_4H_9CO_2C_2H_5$	liquid	1,017.5
Ethylene	$CH_2{:}CH_2$	gas	31.6
Ethylene chloride	$(CH_2Cl)_2$	vapor	271.0
Ethylene diamine	$(CH_2NH_2)_2$	liquid	452.6
Ethylene glycol	$(CH_2OH)_2$	liquid	281.9
Ethylene iodide	$(CH_2I)_2$	solid	324.8
Ethylene oxide	CH_2CH_2O (ring)	liquid	302.1
Ethylidene chloride	CH_3CHCl_2	liquid	267.1
Eugenol	$C_{10}H_{12}O_2$	liquid	1,286.6
Fenchane	$C_{10}H_{18}$	liquid	1,502.6
Fluorene	$(C_6H_4)_2{:}CH_2$	solid	1,584.9
Fluorobenzene	C_6H_5F	liquid	747.2
Formaldehyde	CH_2O	gas	134.1
Formamide	$HCONH_2$	solid	134.9
Formic acid	HCO_2H	liquid	62.8
l-Fructose	$C_6H_{12}O_6$	solid	675.6
Fumaric acid (*trans*)	$(CHCO_2H)_2$	solid	320.0
Furfural	C_4H_3OCHO	liquid	559.5
Galactose	$C_6H_{12}O_6$	solid	670.7
Gallic acid	$C_6H_2(OH)_3CO_2H$—(1, 3, 5, 6)	solid	633.7
d-Glucose	$C_6H_{12}O_6$	solid	673.0
Glutaric acid	$(CH_2)_3(CO_2H)_2$	solid	514.9
Glycerol	$(CH_2OH)_2CHOH$	liquid	397.0
Glyceryl tributyrate	$C_{15}H_{26}O_6$	liquid	1,941.1
Glycine	$H_2NCH_2CO_2H$	solid	234.5
Glycogen	$(C_6H_{10}O_5)x$ per kg	solid	4,186.8
Glycollic acid	CH_2OHCO_2H	solid	166.6
Glycylglycine	$C_4H_8O_3N_2$	solid	470.7
n-Heptaldehyde	$CH_3(CH_2)_5CHO$	liquid	1,062.4
n-Heptane	C_7H_{16}	liquid	1,149.9
Heptine-1	$CH{:}C(CH_2)_4CH_3$	liquid	1,091.2
n-Heptyl alcohol	$CH_3(CH_2)_5CH_2OH$	liquid	1,104.9
Heptyl amine	$C_7H_{15}NH_2$	liquid	1,178.9
Heptylic acid	$C_7H_{14}O_2$	liquid	986.1
n-Hexane	C_6H_{14}	liquid	989.8
Hexachlorbenzene	C_6Cl_6	solid	509.0
Hexachlorethane	C_2Cl_6	solid	110.0
Hexadecane	$C_{16}H_{34}$	solid	2,559.1
Hexahydronaphthalene	$C_{10}H_{14}$	liquid	1,419.3
Hexamethylbenzene	$C_6(CH_3)_6$	solid	1,711.9
Hexamethylenetetramine	$(CH_2)_6N_4$	solid	1,006.7
Hexamethylethane	$[(CH_3)_3C]_2$	solid	1,301.8
Hexyl amine	$C_6H_{13}NH_2$	liquid	1,022.2
Hexylene	C_6H_{12}	liquid	952.6
Hippuric acid	$C_6H_5CONHCH_2CO_2H$	solid	1,012.4
Hydantoic acid	$C_3H_6O_3N_2$	solid	308.6
Hydrazobenzene	$(C_6H_5NH)_2$	solid	1,597.3
Hydroquinol	$C_6H_4(OH)_2$	solid	683.7
Hydroquinoldimethyl ether	$(CH_3O)_2C_6H_4$	solid	1,014.7
p-Hydroxyazobenzene	$HOC_6H_4N_2C_6H_5$	solid	1,502.0
o-Hydroxybenzaldehyde	$C_6H_4(OH)CHO$	liquid	796.0
m-Hydroxybenzaldehyde	$C_6H_4(OH)CHO$	solid	788.7
p-Hydroxybenzaldehyde	$C_6H_4(OH)CHO$	solid	792.7
m-Hydroxybenzoic acid	$HOC_6H_4CO_2H$	solid	726.1
p-Hydroxybenzoic acid	$HOC_6H_4CO_2H$	solid	725.4
β-Hydroxybutyraldehyde	$CH_3CHOHCH_2CHO$	liquid	546.6
Indigo	$C_{16}H_{10}O_2N_2$	solid	1,815.0
Indole	C_8H_7N	solid	1,022.2
Inositol	$C_6H_{12}O_6$	solid	662.1
Iodoform	CHI_3	solid	161.9
Isoamyl amine	$(CH_3)_2CHC_2H_4NH_2$	liquid	866.8
Isobutane	$(CH_3)_3CH$	gas	683.4
Isobutyl alcohol	$(CH_3)_2CH_2CH_2OH$	liquid	638.2
Isobutyl amine	$C_4H_9NH_2$	liquid	713.6
Isobutylene	$(CH_3)_2C{:}CH_2$	gas	647.2
Isobutyraldehyde	$(CH_3)_2CHCHO$	vapor	596.8
Isobutyramide	$(CH_3)_2CHCONH_2$	solid	595.9
Isobutyric acid	$(CH_3)_2CHCO_2H$	liquid	517.4
Isoeugenol	$C_{10}H_{12}O_2$	liquid	1,277.6
Isopentane	C_5H_{12}	gas	843.5(?)
Isopentane	C_5H_{12}	liquid	838.3(?)
Isophthalic acid	$C_6H_4(CO_2H)_2$	solid	768.3
Isopropyl alcohol	$(CH_3)_2CHOH$	liquid	474.8
Isopropylbenzene	$(CH_3)_2CHC_6H_5$	liquid	1,247.3
Isopropyltoluene	$C_6H_4(CH_3)(CH_3CHC_3)$—(1, 3)	liquid	1,409.5
Isopropyltoluene, *see Cymene*			

HEAT OF COMBUSTION (Continued)

For Organic Compounds

Name	Formula	Physical state	Heat of combustion, kg. calories
Isosafrole	$C_{10}H_{10}O_2$	liquid	1,233.9
Lactic acid	$CH_3CHOHCO_2H$	liquid	326.0
Lactose (anhydr.)	$C_{12}H_{22}O_{11}$	solid	1,350.8
Lauric acid	$C_{12}H_{24}O_2$	solid	1,771.7
Leucine	$C_6H_{13}O_2N$	solid	855.6
d-Limonene	$C_{10}H_{16}$	liquid	1,471.2
Maleic acid (*cis*)	$(CHCO_2H)_2$	solid	326.1
Maleic anhydride	$(CHCO)_2O$	solid	333.9
l-Malic acid	$(CHOHCH_2):(CO_2H)_2$	solid	320.1
Malonic acid	$CH_2(CO_2H)_2$	solid	207.2
Maltose	$C_{12}H_{22}O_{11}$	solid	1,350.2
Mandelic acid	$C_6H_5CHOHCO_2H$	solid	890.3
d-Mannitol	$C_6H_{14}O_6$	solid	727.6
Menthene	$C_{10}H_{18}$	liquid	1,523.2
Menthol	$C_{10}H_{20}O$	solid	1,508.8
Mesitylene	$(CH_3)_3C_6H_3$—(1, 3, 5)	liquid	1,243.6
Mesityl oxide	$(CH_3)_2C:CHCOCH_3$	liquid	846.7
Mesotartaric acid	$(CHOH)_2(CO_2H)_2$	solid	276.0
Methane	CH_4	gas	210.8
Methyl acetate	$CH_3CO_2CH_3$	liquid	381.2
Methyl alcohol	CH_3OH	liquid	170.9
Methyl amine	CH_3NH_2	liquid	256.1
Methylaniline	$C_6H_5NHCH_3$	liquid	973.5
Methyl benzoate	$C_6H_5CO_2CH_3$	liquid	943.5
Methyl bromide	CH_3Br	vapor	184.0
Methyl butyl ketone	$CH_3COC_4H_9$	liquid	895.2
Methyl *tert*-butyl ketone, *see Pinacoline*			
Methyl butyrate	$C_3H_7COC_2H_3$	liquid	692.8
Methyl carbylamine	CH_3NC	liquid	320.1
Methyl chloride	CH_3Cl	gas	164.2
Methyl cinnamate	$C_{10}H_{10}O_2$	solid	1,213.0
Methylcyclobutane	$CH_3CHCH_2CH_2CH_2$	liquid	784.2
Methylcycloheptane	$CH_3C_7H_{13}$	liquid	1,244.5
Methylcyclohexane	$CH_3C_6H_{11}$	liquid	1,091.8
Methylcyclopentane	$CH_3CH.C_3H_6CH_2$	liquid	937.9
Methyldiethyl carbinol	$CH_3(C_2H_5)_2CHOH$	liquid	927.0
Methylene chloride	CH_2Cl_2	vapor	106.8
Methylene iodide	CH_2I_2	liquid	178.4
Methylethyl ether	$CH_3OC_2H_5$	vapor	503.4
Methylethyl ketone	$CH_3COC_2H_5$	liquid	582.3
Methyl formate	HCO_2CH_3	liquid	233.1
2-Methylheptane	$(CH_3)_2CH.C_5H_{11}$	liquid	1,306.1
2-Methylhexane	$(CH_3)_2CHC_4H_9$	liquid	1,148.9
3-Methylhexane	$(C_2H_5)(CH_3)CHC_3H_7$	liquid	1,148.9
Methylhexyl ketone	$CH_3COC_6H_{13}$	liquid	1,205.1
Methyl iodide	CH_3I	liquid	194.7
Methyl isobutyrate	$(CH_3)_2CHCO_2CH_3$	liquid	694.2
Methyl isocyanate	CH_3NCO	liquid	269.4
Methylisopropyl ketone	$CH_3COCH(CH_3)_2$	liquid	733.9
Methyl lactate	$CH_3CHOHCO_2CH_3$	liquid	497.2
Methyl propionate	$C_2H_5CO_2CH_3$	vapor	552.3
Methylpropyl ketone	$CH_3COC_3H_7$	liquid	735.6
Methyl salicylate	$HOC_6H_4CO_2CH_3$	liquid	898.3
Milk sugar, *see Lactose*			
Morphine	$C_{17}H_{19}O_3N.H_2O$	solid	2,146.3
Mucic acid	$C_6H_{10}O_8$	solid	483.6
Myristic acid	$C_{14}H_{28}O_2$	solid	2,085.8
Naphthalene	$C_{10}H_8$	solid	1,232.5
α-Naphthoic acid	$C_{10}H_7CO_2H$	solid	1,231.8
β-Naphthoic acid	$C_{10}H_7CO_2H$	solid	1,227.6
α-Naphthol	$C_{10}H_7OH$	solid	1,185.4
β-Naphthol	$C_{10}H_7OH$	solid	1,187.2
α-Naphthonitrile	$C_{10}H_7CN$	solid	1,326.2
β-Naphthonitrile	$C_{10}H_7CN$	solid	1,321.0
α-Naphthoquinone	$C_{10}H_6O_2$	solid	1,100.8
β-Naphthoquinone	$C_{10}H_6O_2$	solid	1,106.4
α-Naphthyl amine	$C_{10}H_7NH_2$	solid	1,263.5
β-Naphthyl amine	$C_{10}H_7NH_2$	solid	1,261.0
Narceine	$C_{23}H_{27}O_8N.2H_2O$	solid	2,802.9
Narcotine	$C_{22}H_{23}O_7N$	solid	2,644.5
Nicotine	$C_{10}H_{14}N_2$	liquid	1,427.7
o-Nitraniline	$C_6H_4(NH_2)(NO_2)$	solid	765.8
m-Nitraniline	$C_6H_4(NH_2)(NO_2)$	solid	765.2
p-Nitraniline	$C_6H_4(NH_2)(NO_2)$	solid	761.0
m-Nitrobenzaldehyde	$O_2NC_6H_4CHO$	solid	800.4
Nitrobenzene	$C_6H_5NO_2$	liquid	739.2
m-Nitrobenzoic acid	$O_2NC_6H_4CO_2H$	solid	729.1
Nitroethane	$C_2H_5NO_2$	liquid	322.2
Nitroglycerine, *see Trinitroglycerol*			
Nitromethane	CH_3NO_2	liquid	169.4
o-Nitrophenol	$HOC_6H_6NO_2$	solid	689.1
m-Nitrophenol	$HOC_5H_4NO_2$	solid	684.4
p-Nitrophenol	$HOC_6H_4NO_2$	solid	688.8
Nitropropane	$C_3H_7NO_2$	liquid	477.9
o-Nitrotoluene	$CH_3C_6H_4NO_2$	liquid	897.0
p-Nitrotoluene	$CH_3C_6H_6NO_2$	solid	888.6
Octahydronaphthalene	$C_{10}H_{16}$	liquid	1,461.7
n-Octane	C_8H_{18}	liquid	1,302.7
Octyl alcohol	$C_8H_{18}O$	liquid	1,262.0
Oleic acid	$C_{18}H_{34}O_2$	liquid	2,657.0
Oxalic acid	$(CO_2H)_2$	solid	60.2
Oxamide	$(CONH_2)_2$	solid	203.2
Palmitic acid	$C_{16}H_{32}O_2$	solid	2,398.4

HEAT OF COMBUSTION (Continued)

For Organic Compounds

Name	Formula	Physical state	Heat of combustion, kg. calories
Papaverine	$C_{20}H_{21}O_4N$	solid	2,478.1
Pentamethylbenzene	$C_6H(CH_3)_5$	solid	1,554.0
n-Pentane	C_5H_{12}	gas	838.3
n-Pentane	C_5H_{12}	liquid	833.4
Phenacetin	$C_{10}H_{13}O_2N$	solid	1,285.2
Phenanthraquinone	$C_{14}H_8O_2$	solid	1,544.0
Phenanthrene	$C_{14}H_{10}$	solid	1,692.5
Phenetole	$C_6H_5OC_2H_5$	liquid	1,060.3
Phenol	C_6H_5OH	solid	732.2
Phenylacetic acid	$C_6H_5CH_2CO_2H$	solid	930.2
Phenylacetylene	$C_6H_5C{:}CH$	liquid	1,024.2
Phenylalanine	$C_9H_{11}O_2N$	solid	1,111.3
p-Phenylenediamine	$C_6H_4(NH_2)_2$	solid	843.4
Phenylethylene, *see Styrene*			
Phenylglycine	$C_2H_5NHCH_2CO_2H$	solid	955.1
Phenylhydrazine	$C_6H_5N_2H_3$	solid	875.4
Phenylhydroxylamine	C_6H_5NHOH	liquid	803.7
Phenyl iodide	C_6H_5I	liquid	770.7
Phloroglucinol	$C_6H_3(OH)_3$	solid	635.7
Phthalic acid	$C_6H_4(CO_2H)_2$	solid	771.0
Phthalic anhydride	$C_6H_4(CO)_2O$	solid	783.4
Phthalimide	$C_8H_5O_2N$	solid	849.5
Picric acid	$C_6H_2(OH)(NO_2)_3$—(1, 2, 4, 6)	solid	611.8
Pinacoline	$CH_3COC(CH_3)_3$	solid	891.8
Piperidine	$C_5H_{11}N$	liquid	826.6
Piperonal	$C_8H_6O_3$	solid	870.7
Propane	C_3H_8	gas	526.3
Propine, *see Allylene*			
Propionaldehyde	C_2H_5CHO	liquid	434.2
Propionamide	$C_2H_5CONH_2$	solid	439.9
Propionic acid	$C_2H_5CO_2H$	liquid	367.2
Propionic anhydride	$(C_2H_5CO)_2O$	liquid	746.6
Propionitrile	C_2H_5CN	liquid	456.4
n-Propyl alcohol	C_3H_7OH	liquid	480.5
Propyl amine	$C_3H_7NH_2$	liquid	558.3
n-Propylbenzene	$C_3H_7C_6H_5$	liquid	1,246.4
Propyl bromide	C_3H_7Br	vapor	497.3
Propyl carbylamine	C_3H_7NC	liquid	639.6
Propyl chloride	C_3H_7Cl	vapor	478.3
Propylene	$CH_3CH{:}CH_2$	gas	490.2
Propylene glycol	$CH_3CHOHCH_2OH$	liquid	431.0
n-Propyl iodide	C_3H_7I	liquid	514.3
n-Propyltoluene	$C_6H_4(CH_3)(C_3H_7)$—(1, 3)	liquid	1,405.4
Pseudocumene	$C_6H_3(CH_3)_3$—(1, 2, 4)	liquid	1,241.7
Pyridine	C_5H_5N	liquid	658.5
Pyrocatechol	$C_6H_4(OH)_2$	solid	684.8
Pyrogallol	$C_6H_3(OH)_3$	solid	638.7
Pyrrole	C_4H_5N	liquid	567.7
Quercitol	$C_6H_{12}O_5$	solid	704.2
Quinoline	C_9H_7N	liquid	1,123.5
Quinone	$O{:}C_6H_4{:}O$	solid	656.6
Raffinose	$C_{18}H_{32}O_{16}$	solid	2,025.5
Retene	$C_{18}H_{18}$	solid	2,306.8
Resorcinol	$C_6H_4(OH)_2$	solid	683.0
Resorcinoldimethyl ether	$(CH_3O)_2C_6H_4$	liquid	1,022.6
Rhamnose	$C_6H_{12}O_6$	solid	718.3
Safrole	$C_{10}H_{10}O_2$	liquid	1,244.1
Salicylaldehyde, *see o-Hydroxybenzaldehyde*			
*Salicylic acid	$HOC_6H_4CO_2H$—(1, 2)	solid	723.1
Sarcosine	$CH_3NHCH_2CO_2H$	solid	401.1
Sebacic acid	$(CH_2)_8(CO_2H)_2$	solid	1,297.3
Skatole	C_9H_9N	liquid	1,170.5
d-Sorbose	$C_6H_{12}O_6$	solid	668.3
Starch	$(C_6H_{10}O_5)x$ per kg	solid	4,178.8
Stearic acid	$C_{18}H_{36}O_2$	solid	2,711.8
Strychnine	$C_{21}H_{22}O_2N_2$	solid	2,685.7
Styrene	$C_6H_5CH{:}CH_2$	liquid	1,047.1
Suberic acid	$(CH_2)_6(CO_2H)_2$	solid	985.2
Succinic acid	$(CH_2CO_2H)_2$	solid	357.1
Succinic acid nitrile	$(CH_2CN)_2$	liquid	545.7
Succinic anhydride	$(CH_2CO)_2O$	solid	369.6
Succinimide	$C_4H_5O_2N$	solid	437.9
Sucrose	$C_{12}H_{22}O_{11}$	solid	1,349.6
Sylvestrene	$C_{10}H_{16}$	liquid	1,464.7
d-Tartaric acid	$(CHOH)_2(CO_2H)_2$	solid	275.1
d, l-Tartaric acid (anhydr.)	$(CHOH)_2(CO_2H)_2$	solid	278.4
Terephthalic acid	$C_6H_4(CO_2H)_2$	solid	770.4
Terpin hydrate	$C_{10}H_{22}O_3$	solid	1,451.0
Terpineol	$C_{10}H_{18}O$	solid	1,469.5
Tetrahydrobenzene	C_6H_{10}	liquid	891.9
Tetrahydronaphthalene	$C_{10}H_{12}$	liquid	1,352.4
Tetramethylmethane	$(CH_3)_4C$	gas	842.6
Tetraphenylmethane	$(C_6H_5)_4C$	solid	3,102.4
Tetryl	$C_7H_5N_5O_8$	solid	842.3
Thebaine	$C_{19}H_{21}O_3N$	solid	2,441.3
Thiophene	C_4H_4S	liquid	670.5
Thujane	$C_{10}H_{18}$	liquid	1,506.4
Thymol	$C_{10}H_{14}O$	liquid	1,353.4
Thymol	$C_{10}H_{14}O$	solid	1,349.7
Thymoquinone	$C_{10}H_{12}O_2$	solid	1,271.3
Toluene	$CH_3C_6H_5$	liquid	934.2
o-Toluic acid	$CH_3C_6H_4CO_2H$	solid	928.9

* Recommended as a secondary thermochemical standard.

HEAT OF COMBUSTION (Continued)

FOR ORGANIC COMPOUNDS

Name	Formula	Physical state	Heat of combustion, kg. calories
m-Toluic acid	$CH_3C_6H_4CO_2H$	solid	928.6
p-Toluic acid	$CH_3C_6H_4CO_2H$	solid	926.9
o-Toluidine	$CH_3C_6H_4NH_2$	liquid	964.3
m-Toluidine	$CH_3C_6H_4NH_2$	liquid	965.3
p-Toluidine	$CH_3C_6H_4NH_2$	solid	958.4
o-Tolunitrile	$CH_3C_6H_4CN$	liquid	1,030.3
Toluquinone	$C_7H_6O_2$	solid	803.2
Triaminotriphenyl carbinol	$(C_6H_4NH_2)_3COH$	solid	2,483.5
Tribenzyl amine	$(C_6H_5CH_2)_3N$	solid	2,762.1
Trichloracetic acid	$Cl_3C.CO_2H$	solid	92.8
Triethyl amine	$(C_2H_5)_3N$	liquid	1,036.8
Triethyl carbinol	$(C_2H_5)_3CHOH$	liquid	1,080.0
Triisoamyl amine	$[(CH_3)_2CHCH_2CH_2]_3N$	liquid	2,459.3
Triisobutyl amine	$[(CH_3)_2CHCH_2]_3N$	liquid	1,973.6
Trimethyl amine	$(CH_3)_3N$	liquid	578.6
2, 2, 3-Trimethylbutane	$(CH_3)_3C.CH(CH_3)_2$	liquid	1,147.9
Trimethyl carbinol	$(CH_3)_3COH$	liquid	629.3
Trimethylene	$CH_2CH_2CH_2$ (ring)	gas	496.8
Trimethylethylene	$(CH_3)_2C{:}CHCH_3$	liquid	796.0
Trimethylethylene	$(CH_3)_2C{:}CHCH_3$	vapor	803.6
2, 2, 4-Trimethylpentane	$(CH_3)_3C.CH_2CH(CH_3)_2$	liquid	1,303.9
Trinitrobenzene	$C_6H_3(NO_2)_3$—(1, 3, 5)	solid	663.7
Trinitroglycerol	$C_3H_5(NO_3)_3$	liquid	368.4
Trinitrotoluene	$C_6H_2(CH_3)(NO_2)_3$—(1, 2, 4, 6)	solid	820.7
Triphenyl amine	$(C_6H_5)_3N$	solid	2,267.8
Triphenylbenzene	$C_6H_3(C_6H_5)_3$—(1, 3, 5)	solid	2,936.7
Triphenyl carbinol	$(C_6H_5)_3CHOH$	solid	2,340.8
Triphenylmethane	$(C_6H_5)_3CH$	solid	2,388.7
Triphenyl methyl	$(C_6H_5)_3C$	solid	2,378.5
Tyrosine	$C_9H_{11}O_3N$	solid	1,070.2
Undecylic acid	$C_{11}H_{22}O_2$	solid	1,615.9
Urea	$(NH_2)_2CO$	solid	151.6
Urethane	$NH_2CO_2C_2H_5$	solid	397.2
Uric acid	$C_5H_4O_3N_4$	solid	460.2
n-Valeric acid	$C_4H_9CO_2H$	liquid	681.6
Vanillin	$C_6H_3(OH)(OCH_3)CHO$—(1, 2, 4)	solid	914.1
o-Xylene	$(CH_3)_2C_6H_4$	liquid	1,091.7
m-Xylene	$(CH_3)_2C_6H_4$	liquid	1,088.4
p-Xylene	$(CH_3)_2C_6H_4$	liquid	1,089.1
Xylose	$C_5H_{10}O_5$	solid	561.5

HEAT OF FORMATION

FOR ORGANIC COMPOUNDS

The heat of formation of a compound "A" is equal to the sum of the heats of formation of the products of combustion minus the heat of combustion (see preceding table) of the compound "A." The heat of formation of:

Free elements	0	kg-cal
CO_2 (gas)	94.38	"
$\frac{1}{2}H_2O$ (liquid from 1 H)	34.19	"
HF (dilute aqueous solution)	75.6	"
SO_2 (gas)	69.3	"
HBr (aqueous solution)	28.54	"
HCl (aqueous solution)	39.46	"
HNO_3 (aqueous solution)	49.80	"
H_2SO_4 (aqueous solution)	207.5	"

Example I

To calculate the heat of formation of methane (CH_4) where

Heat of combustion of methane = 210.8
Heat of formation of CO_2 = 94.38
Heat of formation of $\frac{1}{2}H_2O$ = 34.19

and where the combustion occurs according to the equation:

$$CH_4 + 2O_2 = CO_2\ (gas) + 2H_2O\ (liquid)$$

Then the heat of formation of $CH_4 = 94.38 + 4(34.19) - 210.8 = +20.34$ kg-cal. per gram molecular weight.

Example II

To calculate the heat of formation of ethylene (C_2H_4) where

Heat of combustion of ethylene = 331.6

and the combustion occurs according to the equation:

$$C_2H_4 + 3O_2 = 2CO_2 + 2H_2O$$

The heat of formation of $C_2H_4 = 2(94.38) + 4(34.19) - 331.6 = -6.08$ kg-cal. per gram molecular weight.

Example III

To calculate the heat of formation of ethylamine ($C_2H_5NH_2$) where

Heat of combustion of ethylamine = 408.5

and the combustion occurs according to the equation:

$$C_2H_5NH_2 + 3.75O_2 = 2CO_2 + 3.5(H_2O) + 0.5N_2$$

The heat of formation of $C_2H_5NH_2 = 2(94.38) + 7(34.19) + O(N_2) - 408.5 =$ $+19.59$ kg-cal. per gram molecular weight.

FATS AND OILS

These data for fats and oils were compiled originally for the Biology Data Book by H. J. Harwood, and R. P. Geyer. 1964. Data are reproduced here by permission of the copyright owners of the above publication, the Federation of American Societies for Experimental Biology, Washington, D.C. pp. 380–382.

Values are typical rather than average, and frequently were derived from specific analyses for particular samples (especially the constituent fatty acids). Extreme variations may occur, depending on a number

	Fat or Oil	Source	Constants				
			Melting (or Solidification) Point, °C	Specific Gravity (or Density)	Refractive Index $n_D^{40°C}$	Iodine Value	Saponification Value
	(A)	(B)	(C)	(D)	(E)	(F)	(G)
	Land Animals						
1	Butterfat	*Bos taurus*	32.2	0.911$^{40°/15°}$	1.4548	36.1	227
2	Depot fat	*Homo sapiens*	(15)	0.918$^{15°}$	1.4602	67.6	196.2
3	Lard oil	*Sus scrofa*	(30.5)	0.919$^{15°}$	1.4615	58.6	194.6
4	Neat's-foot oil	*B. taurus*	....	0.910$^{25°}$	1.464$^{25°}$	69–76	190–199
5	Tallow, beef	*B. taurus*	....	....	....	49.5	197
6	Tallow, mutton	*Ovis aries*	(42.0)	0.945$^{15°}$	1.4565	40	194
	Marine Animals						
7	Cod-liver oil	*Gadus morhua*	....	0.925^{25}	1.481$^{25°}$	165	186
8	Herring oil	*Clupea harengus*	....	0.900$^{60°}$	1.4610$^{60°}$	140	192
9	Menhaden oil	*Brevoortia tyrannus*	....	0.903$^{60°}$	1.4645$^{60°}$	170	191
10	Sardine oil	*Sardinops caerulea*	....	0.905$^{60°}$	1.4660$^{60°}$	185	191
11	Sperm oil, body	*Physeter macrocephalus*	....	....	....	76–88	122–130
12	Sperm oil, head	*P. macrocephalus*	....	....	....	70	140–144
13	Whale oil	*Balaena mysticetus*	....	0.892^{60}	1.460$^{60°}$	120	195
	Plants						
14	Babassu oil	*Attalea funifera*	22–26	(0.893$^{60°}$)	1.443$^{60°}$	15.5	247
15	Castor oil	*Ricinus communis*	(−18.0)	0.961$^{15°}$	1.4770	85.5	180.3
16	Cocoa butter	*Theobroma cacao*	34.1	0.964$^{15°}$	1.4568	36.5	193.8
17	Coconut oil	*Cocos nucifera*	25.1	0.924$^{15°}$	1.4493	10.4	268
18	Corn oil	*Zea mays*	(−20.0)	0.922$^{15°}$	1.4734	122.6	192.0
19	Cotton seed oil	*Gossypium hirsutum*	(−1.0)	0.917$^{25°}$	1.4735	105.7	194.3
20	Linseed oil	*Linum usitatissimum*	(−24.0)	0.938$^{15°}$	1.4782$^{25°}$	178.7	190.3
21	Mustard oil	*Brassica hirta*	....	0.9145$^{15°}$	1.475	102	174
22	Neem oil	*Melia azadirachta*	−3	0.917$^{15°}$	1.4615	71	194.5
23	Niger-seed oil	*Guizotia abyssinica*	....	0.925$^{15°}$	1.471	128.5	190
24	Oiticica oil	*Licania rigida*	....	0.974$^{25°}$	....	140–180	......
25	Olive oil	*Olea europaea sativa*	(−6.0)	0.918$^{15°}$	1.4679	81.1	189.7
26	Palm oil	*Elaeis guineensis*	35.0	0.915$^{15°}$	1.4578	54.2	199.1
27	Palm-kernel oil	*E. guineensis*	24.1	0.923$^{15°}$	1.4569	37.0	219.9
28	Peanut oil	*Arachis hypogaea*	(3.0)	0.914$^{15°}$	1.4691	93.4	192.1
29	Perilla oil	*Perilla frutescens*	....	(0.935$^{15°}$)	1.481$^{25°}$	195	192
30	Poppy-seed oil	*Papaver somniferum*	(−15)	0.925$^{15°}$	1.4685	135	194
31	Rapeseed oil	*Brassica campestris*	(−10)	0.915$^{15°}$	1.4706	98.6	174.7
32	Safflower oil	*Carthamus tinctorius*	....	(0.900$^{60°}$)	1.462$^{60°}$	145	192
33	Sesame oil	*Sesamum indicum*	(−6.0)	0.919$^{25°}$	1.4646	106.6	187.9
34	Soybean oil	*Glycine soja*	(−16.0)	0.927$^{15°}$	1.4729	130.0	190.6
35	Sunflower-seed oil	*Helianthus annuus*	(−17.0)	0.923$^{15°}$	1.4694	125.5	188.7
36	Tung oil	*Aleurites fordi*	(−2.5)	0.934$^{15°}$	1.5174$^{25°}$	168.2	193.1
37	Wheat-germ oil	*Triticum aestivum*	....	....	....	125	......

[1] Caproic. [2] Capryli. [3] Capric. [4] Butyric. [5] Decenoic. [6] C_{12} monoethenoic. [7] C_{14} monoethenoic. [8] Gadoleic plus erucic. [9] C_{12} n-pentadecanoic. [10] C_{17} margaric. [11] 12-Methyl tetradecanoic. [12] C_{20} polyethenoic.

of variables such as source, treatment, and age of a fat or oil. **Specific Gravity** (column D) was calculated at the specified temperature (degrees centigrade) and referred to water at the same temperature, unless otherwise specified. **Density,** shown in parentheses (column D), was measured at the specified temperature (degrees centigrade). **Refractive Index** (column E) was measured at 50°C, unless otherwise specified.

Constituent Fatty Acids, g/100 g total fatty acids

	Saturated						Unsaturated				
	Lauric	Myristic	Palmitic	Stearic	Arachidic	Other	Palmitoleic	Oleic	Linoleic	Linolenic	Other
	(H)	(I)	(J)	(K)	(L)	(M)	(N)	(O)	(P)	(Q)	(R)
1	2.5	11.1	29.0	9.2	2.4	2.0[1]; 0.5[2]; 2.3[3]	4.6	26.7	3.6	....	3.6[4]; 0.1[5]; 0.1[6]; 0.9[7] 1.4[8]; 1.0[9]; 1.0[10]; 0.4[11]
2	....	2.7	24.0	8.4	....	....	5	46.9	10.2	....	2.5[8]
3	....	1.3	28.3	11.9	....	....	2.7	47.5	6	....	0.2[7]; 2.1[8]
4	....	.....	17–18	2–3	....	....	....	74–76	....	....	....
5	....	6.3	27.4	14.1	....	....	....	49.6	2.5	....	....
6	....	4.6	24.6	30.5	....	....	....	36.0	4.3	....	....
7	....	5.8	8.4	0.6	....	....	20.0	←29.1→		....	25.4[12]; 9.6[13]
8	....	7.3	13.0	Trace	....	....	4.9	....	....	20.7	30.1[12]; 23.2[13]
9	....	5.9	16.3	0.6	0.6	....	15.5	....	....	29.6	19.0[12]; 11.7[13]; 0.8[14]
10	....	5.1	14.6	3.2	....	....	11.8	←17.8→		....	18.1[12]; 14.0[13]; trace[7]; 15.4[15]
11	1	5	6.5	.....	....	....	26.5	37	19	....	1[13]; 4[7]; 19[16]
12	16	14	8	2	....	3.5[3]	15	17	6.5	....	4[6]; 14[7]; 6.5[16]
13	0.2	9.3	15.6	2.8	....	....	14.4	35.2	....	....	13.6[12]; 5.9[13]; 2.5[7]; 0.2[17]
14	44.1	15.4	8.5	2.7	0.2	0.2[1]; 4.8[2]; 6.6[3]	....	16.1	1.4	....	....
15	←2.4→				....	....	....	7.4	3.1	....	87[18]
16	....	.....	24.4	35.4	....	....	....	38.1	2.1	....	....
17	45.4	18.0	10.5	2.3	0.4[19]	0.8[1]; 5.4[2]; 8.4[3]	0.4	7.5	Trace	....	....
18	....	1.4	10.2	3.0	....	....	1.5	49.6	34.3	....	....
19	....	1.4	23.4	1.1	1.3	....	2.0	22.9	47.8	....	....
20	....	.....	6.3	2.5	0.5	....	....	19.0	24.1	47.4	0.2[14]
21	....	1.3[20]	.....	.....	....	....	....	27.2[20]	16.6[20]	1.8[20]	1.1[14]; 1.0[21]; 51.0[22]
22	....	2.6[20]	14.1[20]	24.0[20]	0.8[20]	....	....	58.5[20]	....	....	....
23	....	3.3[20]	8.2[20]	4.8[20]	0.5[20]	....	....	30.3[20]	57.3[20]	....	....
24	←11.3[23]→					....	....	6.2	....	....	82.5[24]
25	....	Trace	6.9	2.3	0.1	....	....	84.4	4.6	....	....
26	....	1.4	40.1	5.5	....	....	....	42.7	10.3	....	....
27	46.9	14.1	8.8	1.3	....	2.7[2]; 7.0[3]	....	18.5	0.7	....	....
28	....	.....	8.3	3.1	2.4	....	......	56.0	26.0	....	3.1[14]; 1.1[21]
29	←9.6[23]→					....	....	17.8	....	17.5	....
30	....	.....	4.8[20]	2.9[20]	....	....	....	30.1[20]	62.2[20]	....	....
31	....	.....	1	.....	....	....	....	32	15	1	50[22]
32	←6.8[23]→					....	....	18.6	70.1	3.4	....
33	....	.....	9.1	4.3	0.8	....	....	45.4	40.4	....	....
34	0.2	0.1	9.8	2.4	0.9	....	0.4	28.9	50.7	6.5	0.1[7]
35	....	.....	5.6	2.2	0.9	....	....	25.1	66.2	....	....
36	←4.6[23]→					....	....	4.1	0.6	....	90.7[25]
37	←16.0[23]→					....	....	28.1	52.3	3.6	....

[13] C_{22} polyethenoic. [14] Behenic. [15] C_{14} polyethenoic. [16] Gadoleic. [17] C_{24} polyethenoic. [18] Ricinoleic. [19] Includes behenic and lignoceric. [20] Percent by weight. [21] Lignoceric. [22] Erucic. [23] Includes behenic. [24] Licanic. [25] Eleostearic.

LOWERING OF VAPOR PRESSURE BY SALTS IN AQUEOUS SOLUTIONS

The table gives the reduction of the vapor pressure in millimeters due to the presence of the number of grammolecules of salt per liter of water given at the head of the columns, at the temperature 100° C, at which temperature the vapor pressure of pure water is 760 millimeters.

(From Smithsonian Tables.)

Substance	0.5	1.0	2.0	3.0	4.0	5.0	6.0	8.0	10.0
$Al_2(SO_4)_3$	12.8	36.5							
$AlCl_3$	22.5	61.0	179.0	318.0					
BaS_2O_6	6.6	15.4	34.4						
$Ba(OH)_2$	12.3	22.5	39.0						
$Ba(NO_3)_2$	13.5	27.0							
$Ba(ClO_3)_2$	15.8	33.3	70.5	108.2					
$BaCl_2$	16.4	36.7	77.6						
$BaBr_2$	16.8	38.8	91.4	150.0	204.7				
CaS_2O_8	9.9	23.0	56.0	106.0					
$Ca(NO_3)_2$	16.4	34.8	74.6	139.3	161.7	205.4			
$CaCl_2$	17.0	39.8	95.3	166.6	241.5	319.5			
$CaBr_2$	17.7	44.2	105.8	191.0	283.3	368.5			
$CdSO_4$	4.1	8.9	18.1						
CdI_2	7.6	14.8	33.5	52.7					
$CdBr_2$	8.6	17.8	36.7	55.7	80.0				
$CdCl_2$	9.6	18.8	36.7	57.0	77.3	99.0			
$Cd(NO_3)_2$	15.9	36.1	78.0	122.2					
$Cd(ClO_3)_2$	17.5								
$CoSO_4$	5.5	10.7	22.9	45.5					
$CoCl_2$	15.0	34.8	83.0	136.0	186.4				
$Co(NO_3)_2$	17.3	39.2	89.0	152.0	218.7	282.0	332.0		
$FeSO_4$	5.8	10.7	24.0	42.4					
H_3BO_3	6.0	12.3	25.1	38.0	51.0				
H_3PO_4	6.6	14.0	28.6	45.2	62.0	81.5	103.0	146.9	189.5
H_3AsO_4	7.3	15.0	30.2	46.4	64.9				
H_2SO_4	12.9	26.5	62.8	104.0	148.0	198.4	247.0	343.2	
KH_2PO_4	10.2	19.5	33.3	47.8	60.5	73.1	85.2		
KNO_3	10.3	21.1	40.1	57.6	74.5	88.2	102.1	126.3	148.0
$KClO_3$	10.6	21.6	42.8	62.1	80.0				
$KBrO_3$	10.9	22.4	45.0						
$KHSO_4$	10.9	21.9	43.3	65.3	85.5	107.8	129.9	170.0	
KNO_2	11.1	22.8	44.8	67.0	90.0	110.5	130.7	167.0	198.8
$KClO_4$	11.5	22.3							
KCl	12.2	24.4	48.8	74.1	100.9	128.5	152.2		
$KHCO_3$	11.6	23.6	59.0	77.6	104.2	132.0	160.0	210.0	255.0
KI	12.5	25.3	52.2	82.6	112.2	141.5	171.8	225.5	278.5
$K_2C_2O_4$	13.9	28.3	59.8	94.2	131.0				
K_2WO_4	13.9	33.0	75.0	123.8	175.4	226.4			
K_2CO_3	14.4	31.0	68.3	105.5	152.0	209.0	258.5	350.0	
KOH	15.0	29.5	64.0	99.2	140.0	181.8	223.0	309.5	387.8
K_2CrO_4	16.2	29.5	60.0						
$LiNO_3$	12.2	25.9	55.7	88.9	122.2	155.1	188.0	253.4	309.2
$LiCl$	12.1	25.5	57.1	95.0	132.5	175.5	219.5	311.5	393.5
$LiBr$	12.2	26.2	60.0	97.0	140.0	186.3	241.5	341.5	438.0
Li_2SO_4	13.3	28.1	56.8	89.0					
$LiHSO_4$	12.8	27.0	57.0	93.0	130.0	168.0			
LiI	13.6	28.6	64.7	105.2	154.5	206.0	264.0	357.0	445.0
Li_2SiFl_6	15.4	34.0	70.0	106.0					

Substance	0.5	1.0	2.0	3.0	4.0	5.0	6.0	8.0	10.0
$LiOH$	15.9	37.4	78.1						
Li_2CrO_4	16.4	32.6	74.0	120.0	171.0				
$MgSO_4$	6.5	12.0	24.5	47.5					
$MgCl_2$	16.8	39.0	100.5	183.3	277.0	377.0			
$Mg(NO_3)_2$	17.6	42.0	101.0	174.8					
$MgBr_2$	17.9	44.0	115.8	205.3	298.5				
$MgH_2(SO_4)_2$	18.3	46.0	116.0						
$MnSO_4$	6.0	10.5	21.0						
$MnCl_2$	15.0	34.0	76.0	122.3	167.0	209.0			
NaH_2PO_4	10.5	20.0	36.5	51.7	66.8	82.0	96.5	126.7	157.1
$NaHSO_4$	10.9	22.1	47.3	75.0	100.2	126.1	148.5	189.7	231.4
$NaNO_3$	10.6	22.5	46.2	68.1	90.3	111.5	131.7	167.8	198.8
$NaClO_3$	10.5	23.0	48.4	73.5	98.5	123.3	147.5	196.5	223.5
$(NaPO_3)_6$	11.6								
$NaOH$	11.8	22.8	48.2	77.3	107.5	139.1	172.5	243.3	314.0
$NaNO_2$	11.6	24.4	50.0	75.0	98.2	122.5	146.5	189.0	226.2
Na_2HPO_4	12.1	23.5	43.0	60.0	78.7	99.8	122.1		
$NaHCO_3$	12.9	24.1	48.2	77.6	102.2	127.8	152.0	198.0	239.4
Na_2SO_4	12.6	25.0	48.9	74.2					
$NaCl$	12.3	25.2	52.1	80.0	111.0	143.0	176.5		
$NaBrO_3$	12.1	25.0	54.1	81.3	108.8	136.0			
$NaBr$	12.6	25.9	57.0	89.2	124.2	159.5	197.5	268.0	
NaI	12.1	25.6	60.2	99.5	136.7	177.5	221.0	301.5	370.0
$Na_4P_2O_7$	13.2	22.0							
Na_2CO_3	14.3	27.3	53.5	80.2	111.0				
$Na_2C_2O_4$	14.5	30.0	65.8	105.8	146.0				
Na_2WO_4	14.8	33.6	71.6	115.7	162.6				
Na_3PO_4	16.5	30.0	52.5						
$(NaPO_3)_3$	17.1	36.5							
NH_4NO_3	12.8	22.0	42.1	62.7	82.9	103.8	121.0	152.2	180.0
$(NH_4)_2SiFl_6$	11.5	25.0	44.5						
NH_4Cl	12.0	23.7	45.1	69.3	94.2	118.5	138.2	179.0	213.8
NH_4HSO_4	11.5	22.0	46.8	71.0	94.5	118.	139.0	181.2	218.0
$(NH_4)_2SO_4$	11.0	24.0	46.5	69.5	93.0	117.0	141.8		
NH_4Br	11.9	23.9	48.8	74.1	99.4	121.5	145.5	190.2	228.5
NH_4I	12.9	25.1	49.8	78.5	104.5	132.3	156.0	200.0	243.5
$NiSO_4$	5.0	10.2	21.5						
$NiCl_2$	16.1	37.0	86.7	147.0	212.8				
$Ni(NO_3)_2$	16.1	37.3	91.3	156.2	235.0				
$Pb(NO_3)_2$	12.3	23.5	45.0	63.0					
$Sr(SO_3)_2$	7.2	20.3	47.0						
$Sr(NO_3)_2$	15.8	31.0	64.0	97.4	131.4				
$SrCl_2$	16.8	38.8	91.4	156.8	223.3	281.5			
$SrBr_2$	17.8	42.0	101.1	179.0	267.0				
$ZnSO_4$	4.9	10.4	21.5	42.1	66.2				
$ZnCl_2$	9.2	18.7	46.2	75.0	107.0	153.0	195.0		
$Zn(NO_3)_2$	16.6	39.0	93.5	157.5	223.8				

THERMAL CONDUCTIVITY OF GASES

The values in this table are given as cal/(sec)(cm²)(°C/cm) × 10^{-6}. To convert these values to Btu/(hr)(ft²)(°F/ft) × 10^{-6} multiply by 241.909.

Gas °F	−400	−300	−200	−100	−40	−20	0	20	40	60	80	100	120	200
Gas °C	−240	−184.4	−128.9	−73.3	−40	−28.9	−17.8	−6.7	4.4	15.6	26.7	37.8	48.9	93.3
Acetylene	....	....	....	28.10	34.71	37.19	39.67	42.15	45.04	47.94	50.83	53.72	56.62	69.43
Air	....	....	....	....	50.09	52.15	54.22	56.24	58.31	60.34	62.20	64.22	66.04	....
Ammonia	....	....	....	....	43.39	45.87	48.35	50.83	53.31	55.79	58.68	61.58	64.47	....
Argon	....	....	....	....	34.30	35.95	37.19	38.85	40.09	41.33	42.57	44.22	45.46	....
Bromine	....	....	....	....	....	....	9.09	....	....	....	....	11.57	....	....
n-Butane	....	....	....	....	....	....	....	30.99	33.06	35.54	38.02	40.91	43.39	54.14
i-Butane	....	....	....	....	....	....	....	32.65	33.89	36.37	38.85	41.74	44.22	55.79
Carbon dioxide	....	....	....	....	27.90	29.75	31.70	33.68	35.62	37.61	39.67	41.74	43.81	....
Carbon disulfide	....	....	....	....	....	....	14.05	15.29	16.53	17.77	19.01	19.84	....	....
Carbon monoxide	....	....	....	....	47.94	50.00	51.95	53.85	55.87	57.86	59.92	61.99	63.89	....
Chlorine	....	....	....	....	15.29	16.53	17.36	18.18	19.01	20.25	21.08	21.90	23.14	....
Deuterium	....	....	....	....	274.82	285.15	295.07	305.81	309.95	322.34	334.74	343.01	355.40	....
Ethane	....	....	....	23.97	32.65	35.54	38.43	41.33	44.63	47.94	51.24	54.55	58.27	74.39
Ethanol	....	....	....	....	....	....	....	29.34	30.99	32.65	34.71	36.78	....	....
Ethylamine	....	....	....	....	....	....	....	31.41	33.47	35.54	37.61	39.67	42.15	....
Ethylene	....	....	....	26.86	33.06	35.54	38.02	40.50	43.39	46.29	49.18	52.07	54.96	68.19
Fluorine	....	18.18	30.58	43.39	50.83	52.90	55.38	57.86	59.92	61.99	64.06	66.12	68.19	76.04
Helium	84.31	163.24	221.51	274.8	304.99	314.49	324.00	333.50	343.42	352.10	360.36	368.63	376.07	....
Hydrogen	59.92	142.57	227.29	308.7	357.47	371.93	388.46	405.00	417.39	433.92	446.32	458.72	471.11	....
Hydrogen bromide	....	....	....	....	15.29	16.11	16.49	17.77	18.60	19.84	20.66	21.49	....	....
Hydrogen chloride	....	....	....	....	25.62	26.86	28.51	29.75	30.99	32.23	33.89	35.12	....	....
Hydrogen cyanide	....	....	....	....	....	....	23.97	25.62	26.86	28.10	29.75	30.99	32.65	....
Hydrogen sulfide	....	....	....	....	....	....	28.10	29.75	31.41	33.47	....	36.78	....	....
Krypton	....	....	....	....	....	....	19.84	....	....	....	....	23.56	....	....
Methane	....	22.32	36.86	52.07	61.37	64.55	67.86	71.08	74.39	78.11	81.83	85.54	89.26	106.62
Neon	....	....	....	....	97.94	100.84	104.14	107.03	109.93	112.82	115.71	118.19	121.09	....
Nitric oxide	....	....	30.91	42.40	49.01	51.24	53.39	55.54	57.65	59.76	61.99	64.06	66.12	74.39
Nitrogen	....	20.25	33.06	44.22	50.42	52.48	54.55	56.20	58.27	60.34	62.40	64.06	65.71	....
Nitrous oxide	....	....	....	....	28.93	30.91	32.90	35.04	37.15	39.30	41.45	43.81	46.08	....
Oxygen	....	18.84	31.66	43.72	50.54	52.81	54.96	57.24	59.43	61.58	63.64	65.91	68.19	76.87
n-Propane	....	....	....	....	27.69	29.75	32.23	34.71	37.19	39.67	42.47	45.46	48.35	60.75
R-11(CCl_3F)	....	....	....	....	....	12.81	13.64	14.88	15.70	16.53	17.77	18.60	....	....
R-12(CCl_2F_2)	....	....	....	....	....	17.36	18.60	19.42	20.66	21.49	22.73	23.56	....	....
R-21($CHCl_2F$)	....	....	....	....	....	....	21.90	22.32	22.73	23.14	23.56	23.97	....	....
R-22($CHClF_2$)	....	....	....	....	....	....	24.80	25.62	26.45	27.28	28.10	28.93	....	....
Water	....	....	....	....	....	....	34.71	36.78	38.85	40.50	42.57	44.63	46.70	54.96

THERMAL CONDUCTIVITY OF GASEOUS HELIUM, NITROGEN AND WATER

From NSRDS-NBS 8
R. W. Powell, C. Y. Ho, and P. E. Liley

The thermal conductivity, k, is given in the units Milliwatt cm^{-1} °K^{-1}. To convert to Cal(*gm*) hr^{-1} cm^{-1} °K^{-1} multiply the values listed in the table by 0.860421

T (K)	k He	k N_2	T (K)	k He	k N_2	k H_2O	T (K)	k He	k N_2	k H_2O	T (K)	k He	k N_2
0.08	0.00044		150	0.950	0.1385		650	2.64	0.467	0.518	1750	5.57	0.981
0.09	0.00053		160	0.992	0.1474		660	2.67	0.472	0.529	1800	5.70	1.013
0.10	0.00064		170	1.033	0.1562		670	2.69	0.478	0.540	1850	5.83	1.046
0.15	0.00130		180	1.072	0.1651		680	2.72	0.483	0.551	1900	5.96	1.080
0.20	0.00231		190	1.112	0.1739		690	2.75	0.488	0.562	1950	6.08	1.113
0.25	0.0039		200	1.151	0.1826		700	2.78	0.493	0.572	2000	6.20	1.146
0.30	0.0062		210	1.190	0.1908		710	2.81	0.498	0.58	2100	6.44	1.207
0.35	0.0089		220	1.228	0.1989		720	2.84	0.503	0.59	2200	6.69	1.263
0.40	0.0120		230	1.266	0.2067		730	2.87	0.508	0.60	2300	6.93	1.314
0.45	0.0154		240	1.304	0.2145		740	2.90	0.513	0.62	2400	7.16	1.361
0.5	0.0187		250	1.338	0.2222	(0.140)*	750	2.92	0.517	0.63	2500	7.39	1.406
0.6	0.0231		260	1.372	0.2298	(0.148)*	760	2.95	0.522	0.64	2600	7.62	1.449
0.7	0.0252		270	1.405	0.2374	(0.156)*	770	2.98	0.526	0.65	2700	7.85	1.494
0.8	0.0262		280	1.437	0.2449	0.164	780	3.01	0.531	0.66	2800	8.07	1.542
0.9	0.0266		290	1.468	0.2524	0.172	790	3.04	0.536	0.67	2900	8.29	1.590
1.0	0.0269		300	1.499	0.2598	0.181	800	3.07	0.541	0.68	3000	8.51	1.640
1.25	0.0281		310	1.530	0.2671	0.189	810	3.09	0.546	0.69	3100	8.72	1.691
1.5	0.0306		320	1.560	0.2741	0.197	820	3.12	0.551	0.70	3200	8.95	1.743
2.0	0.0393		330	1.590	0.2808	0.205	830	3.15	0.555	0.71	3300	9.16	1.795
2.5	0.0502		340	1.619	0.2874	0.214	840	3.18	0.559	0.72	3400	9.37	1.853
3.0	0.0607		350	1.649	0.2939	0.222	850	3.21	0.564	0.73	3500	9.58	1.915
3.5	0.0732		360	1.678	0.3002	0.231	860	3.23	0.569	0.74	3600	9.79	
4.0	0.0803		370	1.708	0.3065	0.239	870	3.26	0.574	0.75	3700	10.00	
4.5	0.0879		380	1.737	0.3127	0.248	880	3.29	0.578	0.76	3800	10.22	
5.0	0.0962		390	1.766	0.3189	0.256	890	3.32	0.583	0.77	3900	10.43	
6	0.1113		400	1.795	0.3252	0.264	900	3.35	0.587	0.78	4000	10.64	
7	0.1247		410	1.824	0.3314	0.273	910	3.37	0.592		4100	10.85	
8	0.1393		420	1.853	0.3376	0.282	920	3.40	0.596		4200	11.06	
9	0.1523		430	1.882	0.3438	0.291	930	3.43	0.600		4300	11.27	
10	0.1640		440	1.914	0.3501	0.300	940	3.46	0.605		4400	11.48	
12	0.1866		450	1.947	0.3564	0.307	950	3.49	0.609		4500	11.69	
14	0.2067		460	1.980	0.3626	0.317	960	3.52	0.613		4600	11.90	
16	0.2259		470	2.013	0.3688	0.327	970	3.54	0.618		4700	12.11	
18	0.2435		480	2.046	0.3749	0.337	980	3.57	0.622		4800	12.31	
20	0.2582		490	2.080	0.3808	0.347	990	3.60	0.626		4900	12.51	
25	0.2962		500	2.114	0.3864	0.357	1000	3.63	0.631		5000	12.71	
30	0.3330		510	2.15	0.392	0.368	1050	3.76	0.651				
35	0.3669		520	2.18	0.398	0.378	1100	3.89	0.672				
40	0.4000		530	2.22	0.403	0.389	1150	4.03	0.693				
45	0.4314		540	2.25	0.408	0.400	1200	4.16	0.713				
50	0.4623	(0.0485)*	550	2.29	0.414	0.411	1250	4.29	0.733				
60	0.521	(0.0578)*	560	2.33	0.420	0.422	1300	4.43	0.754				
70	0.578	(0.0670)*	570	2.36	0.425	0.432	1350	4.55	0.775				
80	0.631	0.0762	580	2.40	0.431	0.443	1400	4.69	0.797				
90	0.679	0.0852	590	2.43	0.436	0.454	1450	4.82	0.819				
100	0.730	0.0941	600	2.47	0.441	0.464	1500	4.94	0.842				
110	0.776	0.1030	610	2.51	0.446	0.475	1550	5.07	0.867				
120	0.819	0.1119	620	2.54	0.452	0.486	1600	5.21	0.893				
130	0.863	0.1208	630	2.58	0.457	0.497	1650	5.33	0.921				
140	0.907	0.1296	640	2.61	0.462	0.508	1700	5.45	0.950				

THERMAL CONDUCTIVITY

THERMAL CONDUCTIVITY OF DIELECTRIC CRYSTALS

Name	Remarks	Conductivity mw/cm deg K	
		83° K	273° K
Marble	Small crystals, 99.9 % $CaCO_3$	42	33
Do	99.99 % $CaCO_3$	54	38
Do	Large crystals	50	33
Calcite	Main crystal axis perpendicular to rod axis.	180	46
Do	Main crystal axis parallel to rod axis.	293	54
Sylvite	Natural crystal	159	75
KCl	Pressed at 8,000 atm	314	88
KCl	From a melt	402	92
NaCl	do	343	92
NaCl	Pressed at 8,000 atm	251	71
Rock salt	do	180	63
Sylvite	do	343	84
KCl	Pressed at 1,250 atm	243	75
KCl	Pressed at 2,500 atm	368	92
KCl	Pressed at 8,900 atm	402	96
KBr	Pressed at 8,000 atm	92	38
NaBr	do	50	25
KI	do	121	29
KF	do	234	71
NaF	do	519	105
RbI	do	59	33
RbCl	do	29	21
90 % KBr, 10 % KCl	do	50	29
75 % KBr, 25 % KCl	do	29	21
50 % KBr, 50 % KCl	do	25	25
25 % KBr, 75 % KCl	Pressed at 8,000 atm	46	33
10 % KBr, 90 % KCl	do	80	50
50 % KCl, 50 % NaCl	do	188	71
KNO_3	do	17	21
Mercuric chloride	do	17	13
NH_4Cl	do	109	25
NH_4Br	do	67	25
$Ba(NO_3)_2$	do	33	13
Copper sulfate		29	21
Magnesium sulfate		25	25
$K_4Fe(CN_6)$		17	17
Chrom alum		13	21
Potassium alum		13	21
Potassium bichromate	Main crystal axis perpendicular to rod axis.	17	21
Do	Main crystal axis parallel to rod axis.	17	17
Topaz	Mineral		234
Zincblend	do	63	264
Beryll	do	88	84
Tourmaline	do	38	46

THERMAL CONDUCTIVITY OF ORGANIC COMPOUNDS

The values in this table are given as cal/(sec)(cm²)(°C/cm). To convert these values to Btu/(hr)(ft²)(°F/ft) multiply by 242.08.

Substance	k	t, °C	t, °F
Acetaldehyde	0.0004089	21	69.8
Acetic acid	0.0004109	20	68
Acetic anhydride	0.0005286	21	69.8
Acetone	0.0004750	−80	−112
	0.0004543	16	61
	0.0004031	75	167
Allyl alcohol	0.0004295	30	86
Amyl acetate (n)	0.0003085	20	68
(iso)	0.000310	20	68
Amyl alcohol (n)	0.0003874	30–100	86–212
(iso)	0.0003531	30	86
Amyl bromide (n)	0.0002350	18	64.4
Aniline	0.0004237	16.5	61.5
Benzene	0.0003780	22.5	72.5
	0.0003275	50	122
	0.0003630	60	140
	0.0002870	140	284
Bromobenzene	0.0002664	20	68
Butyl acetate (n)	0.000327	20	68
Butyl alcohol (n)	0.0003663	20	68
Carbon tetrachloride	0.0002470	20	68
	0.0002333	50	122
Chlorobenzene	0.0003457	30–100	86–212
Chlorotoluene (p)	0.000310	20	68
Chloroform	0.0002891	16	61
	0.000246	20	68
Cresol (m)	0.0003581	20	68
(p)	0.000345	20.1	68.2
Cumene	0.000298	20	68
Cymene (p)	0.0003217	30	86
Decane	0.0003349	30	86

THERMAL CONDUCTIVITY OF ORGANIC COMPOUNDS (Continued)

Substance	k	t, °C	t, °F
Diethyl ether	0.0003283	30	86
Dichloroethane, 1–2	0.000302	20	68
Di-isopropyl ether	0.000262	20	68
Ethyl acetate	0.0003560	16	60.8
Ethyl alcohol	0.0003995	20	68
Ethyl benzene	0.0003160	20	68
Ethyl bromide	0.0002862	30	86
Ethyl ether	0.0003283	30	86
Ethyl iodide	0.0002651	30	86
Ethylene glycol	0.0006236	20	68
	0.0006323	15	122
	0.0006443	80	176
Freon-12 (CCl_2F_2)	0.0002310	0–75	32–167
Freon-21 ($CHCl_2F$)	0.0003180	0–75	32–167
Freon-22 ($CHClF_2$)	0.0002309	40	104
Freon-113 (CCl_2FCCl_2F)	0.0002379	0–80	32–176
Freon-114 ($C_2H_2F_4$)	0.0002127	0–75	32–167
Glycerol	0.000703	20	68
Heptane (n)	0.0003354	30	86
Heptyl alcohol	0.0003882	70–100	86–212
Hexane (n)	0.0003287	30–100	86–212
Hexyl alcohol (n)	0.0003857	30–100	86–212
Iodobenzene	0.0002874	30–100	86–212
Mesitylene	0.0003246	20	68
Methyl alcohol	0.0004832	20	68
Methyl aniline	0.0004419	21.5	70.5
Methyl chloride	0.0004597	−15 (−) +30	5–86
Methyl cyclohexane	0.0003052	30	86
Methylene chloride	0.0002908	0	32
Nitrobenzene	0.0003907	30–100	86–212
Nitromethane	0.0005142	30	86
Nonane (n)	0.0003374	30–100	86–212
Nonyl alcohol (n)	0.0004014	30–100	86–212
Octane (n)	0.0003469	30	86
Octyl alcohol (n)	0.0003973	30–100	86–212
Oleic acid	0.0005514	26.5	79.7
Palmitic acid	0.0004097	72.5	162.5
Pentachloroethane	0.0002994	20	68
Pentane (n)	0.0003221	30	86
Phenetole	0.0003577	−20	−4
Phenyl hydrazine	0.0004121	25	69.8
Propyl acetate (iso)	0.000321	20	68
Propyl alcohol (iso)	0.0003362	20	68
Propylene chloride	0.0002994	20–50	68–122
Propylene glycol (1–2)	0.0004799	20–80	68–176
Stearic acid	0.0003824	72.5	162.5
Tetrachloroethane (sym)	0.000272	20	68
Tetrachloroethylene	0.0003866	20	68
Toluene	0.0003804	−80	−112
	0.0003221	20	68
	0.0002808	80	176
Trichloroethylene	0.0003246	−60	−76
	0.0002775	20	68
Triethylamine	0.0003498	−80	−112
	0.0002891	20	68
	0.0002664	44.4	112
Xylene (o)	0.0003411	−20 (−) +80	(−4)–176
Xylene (m)	0.0003767	25	77

THERMAL CONDUCTIVITY OF INORGANIC COMPOUNDS

Substance	k	t, °C	t, °F
Ammonia	0.0001198	−15 (−) +30	5–86
Argon	0.0002895	−183	−297
	0.0001677	−133	−207
	0.0000553	−105	−157.5
	0.0000409	−75	−102.5
Carbon dioxide	0.0002040	−50	−58
	0.0002412	−40	−40
	0.0002664	−30	−22
	0.0002746	−20	−4
	0.0002495	0	32
	0.0001677	30	86
Nitrogen	0.0003400	−196	−321.5
	0.0002961	−189	−308
	0.0002028	−158	−253
	0.0000640	−105	−155
Oxygen	0.0000500	(−207)–(−191)	(−340)–(−312)
	0.0000504	(−178)–(−182)	(−288)–(−295)
Water	0.001348	0	72
	0.001429	20	68
	0.001499	40	104
	0.001557	60	140
	0.001598	100	212
	0.001631	149	300
	0.001553	216	420
	0.001404	271	520
	0.001136	327	620

Thermal Conductivity of Miscellaneous Substances

Chlorinated diphenyl 1242	0.0002936	30–100	86–212
Chlorinated diphenyl 1248	0.0002808	30–100	86–212
Kerosene	0.0003572	30	86
Light heat transfer oil	0.0003159	30–100	86–212
Petroleum ether	0.0003118	30	86
Red oil	0.0003366	30	86
Transformer oil	0.0004242	70–100	86–212

THERMAL CONDUCTIVITY OF MATERIALS

(Bureau of Standards Letter Circular No. 227)

D = Density in pounds per cubic foot.
K = Thermal conductivity in B.T.U. per hour, square foot, and temperature gradient of 1 degree Fahrenheit per inch thickness. The lower the conductivity, the greater the insulating values.

Soft Flexible Materials in Sheet Form

		D	K
Dry Zero	Kapok between burlap or paper	1.0	0.24
		2.0	0.25
Cabots Quilt	Eel grass between kraft paper	3.4	0.25
		4.6	0.26
Hair Felt	Felted cattle hair	11.0	0.26
		13.0	0.26
Balsam Wool	Chemically treated wood fibre	2.2	0.27
Hairinsul	75% hair 25% jute	6.3	0.27
	50% hair 50% jute	6.1	0.26
Linofelt	Flax fibres between paper	4.9	0.28
Thermofelt	Jute and asbestos fibres, felted	10.0	0.37
	Hair and asbestos fibres. felted	7.8	0.28

Loose Materials

		D	K
Rock Wool	Fibrous material made from rock, also made in sheet form, felted and confined with wire netting	6.0	0.26
		10.0	0.27
		14.0	0.28
		18.0	0.29
Glass Wool	Pyrex glass, curled	4.0	0.29
		10.0	0.29
Sil-O-Cel	Powdered diatomaceous earth	10.6	0.31
Regranulated Cork	Fine particles	9.4	0.30
	about 3/16 inch particles	8.1	0.31
Thermofill	Gypsum in powdered form	26.	0.52
		34.	0.60
Sawdust	Various	12.0	0.41
	redwood	10.9	0.42
Shavings	Various, from planer	8.8	0.41
Charcoal	From maple, beech and birch, coarse	13.2	0.36
	6 mesh	15.2	0.37
	20 mesh	19.2	0.39

Semi-flexible Materials in Sheet Form

		D	K
Flaxlinum	Flax fibre	13.0	0.31
Fibrofelt	Flax and rye fibre	13.6	0.32

Semi-rigid Materials in Board Form

		D	K
Corkboard	No added binder; very low density	5.4	0.25
Corkboard	No added binder; low density	7.0	0.27
Corkboard	No added binder; medium density	10.6	0.30
Corkboard	No added binder; high density	14.0	0.34
Eureka	Corkboard with asphaltic binder	14.5	0.32
Rock Cork	Rock wool block with binder	14.5	0.326
	Also called "Tucork"		
Lith	Board containing rock wool, flax and straw pulp	14.3	0.40

Stiff Fibrous Materials in Sheet Form

		D	K
Insulite	Wood pulp	16.2	0.34
		16.9	0.34
Celotex	Sugar cane fibre	13.2	0.34
		14.8	0.34

	K
*Masonite	K = 0.33
*Inso-board	0.33
*Maizewood	0.33 to 0.39
*Cornstalk Pith Board	0.24 to 0.30
*Maftex	0.34

THERMAL CONDUCTIVITY OF MATERIALS

(Continued)

Cellular Gypsum

	D	K
Insulex or Pyrocell	8	0.35
	12	0.44
	18	0.59
	24	0.77
	30	1.00

Woods (Across Grain)

	D	K
Balsa	7.3	0.33
	8.8	0.38
	20	0.58
Cypress	29	0.67
White pine	32	0.78
Mahogany	34	0.90
Virginia pine	34	0.98
Oak	38	1.02
Maple	44	1.10

Miscellaneous Building Materials

(Data taken from various sources)

	K		K
Cinder concrete	2 to 3	Limestone	4 to 9
Building gypsum	About 3	Concrete	6 to 9
Plaster	2 to 5	Sandstone	8 to 16
Building brick	3 to 6	Marble	14 to 20
Glass	5 to 6	Granite	13 to 28

* From various commercial laboratories and the work of O. R. Sweeney at Iowa State College.

THERMAL CONDUCTIVITY DATA ON CERAMIC MATERIALS

Description[a]	Class.[b]	Water Abs. %	Bulk Density gms/cc	Thermal Conductivity[c]		
				100°F	200°F	300°F
Single Crystals						
Silicon carbide	5	—	—	52.0	50.0	49.0
Periclase	5	—	—	26.7	22.5	19.5
Sapphire, c-axis	5	—	—	20.2	16.0	14.0
Sapphire, a-axis	5	—	—	18.7	15.0	12.9
Topaz, a-axis	5	—	—	10.8	9.4	7.9
Kyanite, c-axis	5	—	—	10.00	8.6	7.4
Kyanite, b-axis	5	—	—	9.6	8.3	7.1
Spinel, $MgO \cdot Al_2O_3$	5	—	—	6.80	6.20	5.50
Quartz, c-axis	4	—	—	6.40	5.40	5.02
Quartz, a-axis	4	—	—	3.40	3.00	2.60
Rutile, c-axis	5	—	—	5.60	4.80	4.40
Rutile, a-axis	5	—	—	3.20	3.20	3.20
Fluorite	5	—	—	5.30	4.37	3.45
Beryl, aquamarine, c-axis	4	—	—	3.18	3.15	3.12
Beryl, aquamarine, a-axis	4	—	—	2.52	2.52	2.52
Zircon, a-axis	4	—	—	2.45	2.45	2.45
Zircon, c-axis	4	—	—	2.34	2.34	2.35
Polycrystalline Single Oxide Ceramics						
Pure BeO, hot pressed	2	0.03	2.97	125.0	104.0	92.0
MgO (spec. pure)	1	0.83	3.21	21.2	18.4	16.0
SnO_2 98%	1	0.03	6.62	17.5	15.0	12.7
ZnO (yellow)	1	0.00	5.28	16.8	14.6	12.5
ZnO (gray)	1	0.03	5.20	13.6	11.8	10.2
CuO (100%)	1	0.04	6.76	10.2	9.00	7.80
ThO_2, hot pressed	2	—	9.58	8.00	7.02	6.50
CeO_2	1	0.00	6.20	6.63	6.29	5.20
Mn_3O_4	1	0.02	4.21	4.18	3.80	3.41
PbO (100%)	1	0.38	7.98	1.6	1.25	0.98

[a] Composition: 90% MgO, 10% Al_2O_3 designates weight percent. $Li_2O{:}4B_2O_3$ designates mole composition, does not indicate compound formation.

[b] Classification: 1 = research body; 2 = industrial research body; 3 = commercial body; 4 = natural mineral; 5 = synthetic mineral.

[c] Thermal conductivity: Units in Btu/(hr) (sq ft)(°F/ft); to convert to cal/(sec) (sq cm)(°C/cm) multiply by 0.00413. (I) = determination made with high vacuum apparatus, inconel thermodes. No letters following value, determination made with high vacuum thermal conductivity apparatus, copper thermodes.

By permission from Engineering Research Bulletin No. 40 Rutgers University (1958).

THERMAL CONDUCTIVITIES OF GLASSES BETWEEN −150 AND +100°C

E. H. Ratcliffe

Type of glass	Approximate silica contents (wt. %)	Approximate contents other oxides normally present in quantity (wt. %)		Estimated approximate thermal conductivity at various temperatures	
				Temperature (°C)	Thermal conductivity $\left(\frac{\text{cal cm}}{\text{cm}^2\text{ s deg C}}\right) \times 10^4$
(a) Vitreous silica	100	....		−150	20.0
				−100	25.0
				− 50	28.8
				0	31.5
				50	33.7
				100	35.4
(b) 'Vycor' glass	96	B_2O_3	3	−100	24
				0	30
				100	34
(c) General information					
'Crown' glasses	50–75	Various		−100	12–20.5
				30	19–26
				100	21–29
'Flint' glasses	20–55	Various		−100	9–15
				30	13–21
				100	15–23
(d) Pyrex type chemically-resistant borosilicate glasses	80–81	B_2O_3	12–13	−100	21
		Na_2O	4	0	26
		Al	2	100	30
(e) Borosilicate crown glasses	60–65	B_2O_3	15–20	−100	16–17.5
				0	21–22.5
				100	24–25.5
	65–70	B_2O_3	10–15	−100	17.5–19
				0	22.5–24
				100	25.5–27
	70–75	B_2O_3	5–10	−100	19–20.5
				0	24.5–26
				100	27.5–29
(f) (i) Zinc crown glasses	55–65	ZnO	5–15	−100	21–22
		Remainder		0	26–27
		B_2O_3, Al_2O_3		100	28–30
		ZnO	5–15	−100	14–17
		Remainder		0	17–21
		Na_2O, K_2O		100	20–23
		ZnO	15–25	−100	21–22
		Remainder		0	26–27
		B_2O_3, Al_2O_3		100	27–29
		ZnO	15–25	−100	16–19
		Remainder		0	20–23
		Na_2O, K_2O		100	22–25
(f) (ii) Zinc crown glasses	65–75	ZnO	5–15	−100	21–22
		Remainder		0	27–28
		B_2O_3, Al_2O_3		100	29–31
		ZnO	5–15	−100	17–20
		Remainder		0	21–25
		Na_2O, K_2O		100	24–27
		ZnO	15–25	−100	21–23
		Remainder		0	27–28
		B_2O_3, Al_2O_3		100	29–30
		ZnO	15–25	−100	16–20
		Remainder		0	20–24
		Na_2O, K_2O		100	25–29

Type of glass	Approximate silica contents (wt. %)	Approximate contents other oxides normally present in quantity (wt. %)		Estimated approximate thermal conductivity at various temperatures: Temperature (°C)	Thermal conductivity $\left(\frac{\text{cal cm}}{\text{cm}^2\text{ s deg C}}\right) \times 10^4$
(g) Barium crown glasses	31	B_2O_3	12	−100	13
		Al_2O_3	8	0	17
		BaO	48	100	19
	41	B_2O_3	6	−100	14
		Al_2O_3	2	0	18
		ZnO	8	100	20
		BaO	43		
	47	B_2O_3	4	−100	15
		Na_2O	1	0	18
		K_2O	7	100	21
		ZnO	8		
		BaO	32		
	65	B_2O_3	2	−100	17
		Na_2O	5	0	21
		K_2O	15	100	24
		ZnO	2		
		BaO	10		
(h) Borate glasses					
Borate flint glass	9	B_2O_3	36	−100	13
		Na_2O	1	0	16
		K_2O	2	100	19
		PbO	36		
		Al_2O_3	10		
		ZnO	6		
Borate flint glass	.....	B_2O_3	56	−100	12
		Al_2O_3	12	0	16
		PbO	32	100	20
Borate flint glass	.....	B_2O_3	43	−100	9
		Al_2O_3	5	0	13
		PbO	52	100	17
Borate glass	4	B_2O_3	55	−100	15
		Al_2O_3	14	0	19
		PbO	11	100	21
		K_2O	4		
		ZnO	12		
Borate crown glass	.....	B_2O_3	64	−100	12
		Na_2O	8	0	16
		K_2O	3	100	20
		BaO	4		
		PbO	3		
		Al_2O_3	18		
Light borate crown glass	.....	B_2O_3	69	−100	13
		Na_2O	8	0	17
		BaO	5	100	21
		Al_2O_3	18		
Zinc borate glass	.....	B_2O_3	40	−100	16
		ZnO	60	0	18
				100	20
(i) Phosphate crown glasses					
Potash phosphate glass	.....	P_2O_5	70	0	18
		B_2O_3	3	100	20
		K_2O	12		

Type of glass	Approximate silica contents (wt. %)	Approximate contents other oxides normally present in quantity (wt. %)		Estimated approximate thermal conductivity at various temperatures: Temperature (°C)	Thermal conductivity $\left(\frac{\text{cal cm}}{\text{cm}^2\text{ s deg C}}\right) \times 10^4$
(i) Potash phosphate glass (*Continued*)	.....	Al_2O_3	10		
		MgO	4		
Baryta phosphate glass	.....	P_2O_5	60	45	18
		B_2O_3	3		
		Al_2O_3	8		
		BaO	28		
(j) Soda-lime glasses	75	Na_2O	17	−100	18
		CaO	8	0	23
				100	26
	75	Na_2O	12	−100	21
		CaO	13	0	26
				100	28
	72	Na_2O	15	−100	19
		CaO	11	0	24
		Al_2O_3	2	100	27
	65	Na_2O	25	−100	16
		CaO	10	0	20
				100	23
	65	Na_2O	15	−100	20
		CaO	20	0	24
				100	26
	60	Na_2O	20	−100	18
		CaO	20	0	22
				100	24
(k) Other crown glasses					
Crown glass	75	Na_2O	9	−100	19
		K_2O	11	0	24
		CaO	5	100	26
High dispersion crown glass	68	Na_2O	16	−100	16
		ZnO	3	0	20
		PbO	13	100	24
(l) Miscellaneous flint glasses					
(i) Silicate flint glasses					
Light flint glasses	65	PbO	25	−100	16–17
		Others	10	0	21–22
				100	24–25
	55	PbO	35	−100	14–16
		Others	10	0	18–20
				100	21–22
Ordinary flint glass	45	PbO	45	−100	12–14
		Others	10	0	16–18
				100	19–20
Heavy flint glass	35	PbO	60	−100	11–12
		Others	5	0	14–15
				100	17–18
Very heavy flint glasses	25	PbO	73	−100	10–11
		Others	2	0	13–14
				100	15–16
	20	PbO	80	−100	10
				0	12
				100	14

Type of glass	Approximate silica contents (wt. %)	Approximate contents other oxides normally present in quantity (wt. %)		Estimated approximate thermal conductivity at various temperatures: Temperature (°C)	Thermal conductivity $\left(\frac{\text{cal cm}}{\text{cm}^2\text{ s deg C}}\right) \times 10^4$
(ii) Borosilicate flint glass	33	B_2O_3	31	−100	15
		PbO	25	0	20
		Al_2O_3	7	100	23
		K_2O	3		
		Na_2O	1		
(iii) Barium flint glass	50	BaO	24	−100	14
		PbO	6	0	17
		K_2O	8	100	20
		Na_2O	3		
		ZnO	8		
		Sb_2O_3	1		
(m) Other glasses					
(i) Potassium glass	59	K_2O	33	50	21–22
		CaO	8		
(ii) Iron glasses	63	Fe_2O_3	10	−100	19
		Na_2O	17	0	23
		MgO	4	100	25
		CaO	3		
		Al_2O_3	2		
	67	Fe_2O_3	15	0	21–22
		Na_2O_3	18	100	24–25
	62	Fe_2O_3	20	0	20.5–21.5
		Na_2O	18	100	23–24
(ii) Rock glasses					
Obsidian				0	32
				100	35
Artificial diabase				0	27
				100	30

THERMAL CONDUCTIVITY OF CERTAIN METALS

From NSRDS-NBS 8

R. W. Powell, C. Y. Ho, and P. E. Liley

The thermal conductivity, k, is given in the units Watt cm^{-1} $°K^{-1}$.
To convert to Cal(*gm*) hr^{-1} cm^{-1} $°C^{-1}$ multiply the values listed in the tables by 860.421.
To convert to Btu hr^{-1} ft^{-1} $°F^{-1}$ multiply the values listed in the tables by 57.818.
ρ_0 is the residual electrical resistivity and the value of ρ at 4.2°K is used approximately as ρ_0.

T,K	Aluminum 99.996+% $\rho_0 = 0.00315$ μohm cm	Copper 99.999+% $\rho_0 = 0.000851$ μohm cm	Gold 99.999+% $\rho_0 = 0.0055$ μohm cm	Iron 99.998+% $\rho_0 = 0.0327$ μohm cm	Manganin	Platinum 99.999% $\rho_0 = 0.0106$ μohm cm	Silver 99.999+% $\rho_0 = 0.00062$ μohm cm	Tungsten 99.99+% $\rho_0 = 0.0017$ μohm cm
0	0	0	0	0	0	0	0	0
1	7.8	28.7	4.4	0.75	0.0007	2.31	39.4	14.4
2	15.5	57.3	8.9	1.49	0.0018	4.60	78.3	28.7
3	23.2	85.5	13.1	2.24	0.0031	6.79	115	42.6
4	30.8	113	17.1	2.97	0.0046	8.8	147	55.6
5	38.1	138	20.7	3.71	0.0062	10.5	172	67.1
6	45.1	159	23.7	4.42	0.0078	11.8	187	76.2
7	51.5	177	26.0	5.13	0.0095	12.6	193	82.4
8	57.3	189	27.5	5.80	0.0111	12.9	190	85.3
9	62.2	195	28.2	6.45	0.0128	12.8	181	85.1
10	66.1	196	28.2	7.05	0.0145	12.3	168	82.4
11	69.0	193	27.7	7.62	0.0162	11.7	154	77.9
12	70.8	185	26.7	8.13	0.0180	10.9	139	72.4
13	71.5	176	25.5	8.58	0.0197	10.1	124	66.4
14	71.3	166	24.1	8.97	0.0215	9.3	109	60.4
15	70.2	156	22.6	9.30	0.0232	8.4	96	54.8
16	68.4	145	20.9	9.56	0.0250	7.6	85	49.3
18	63.5	124	17.7	9.88	0.0285	6.1	66	40.0
20	56.5	105	15.0	9.97	0.0322	4.9	51	32.6
25	40.0	68	10.2	9.36	0.0410	3.15	29.5	20.4
30	28.5	43	7.6	8.14	0.0497	2.28	19.3	13.1
35	21.0	29	6.1	6.81	0.0583	1.80	13.7	8.9
40	16.0	20.5	5.2	5.55	0.067	1.51	10.5	6.5
45	12.5	15.3	4.6	4.50	0.075	1.32	8.4	5.07
50	10.0	12.2	4.2	3.72	0.082	1.18	7.0	4.17
60	6.7	8.5	3.8	2.65	0.097	1.01	5.5	3.18
70	5.0	6.7	3.58	2.04	0.110	0.90	4.97	2.76
80	4.0	5.7	3.52	1.68	0.120	0.84	4.71	2.56
90	3.4	5.14	3.48	1.46	0.127	0.81	4.60	2.44
100	3.0	4.83	3.45	1.32	0.133	0.79	4.50	2.35
150	2.47	4.28	3.35	1.04	0.156	0.762	4.32	2.10
200	2.37	4.13	3.27	0.94	0.172	0.748	4.30	1.97
250	2.35	4.04	3.20	0.865	0.193	0.737	4.28	1.86
273	2.36	4.01	3.18	0.835	0.206	0.734	4.28	1.82
300	2.37	3.98	3.15	0.803	0.222	0.730	4.27	1.78
350	2.40	3.94	3.13	0.744	0.250	0.726	4.24	1.70
400	2.40	3.92	3.12	0.694	(0.279)	0.722	4.20	1.62
500	2.37	3.88	3.09	0.613	(0.338)	0.719	4.13	1.49
600	2.32	3.83	3.04	0.547	(0.397)	0.720	4.05	1.39
700	2.26	3.77	2.98	0.487		0.723	3.97	1.33
800	2.20	3.71	2.92	0.433		0.729	3.89	1.28
900	2.13	3.64	2.85	0.380		0.737	3.82	1.24
1000	[0.93]**	3.57	(2.78)	0.326		0.748	(3.74)	1.21
1100	[0.96]	3.50	(2.71)	0.297		0.760	(3.66)	1.18
1200	[0.99]	3.42	(2.62)	0.282		0.775	(3.58)	1.15
1300	[1.02]	(3.34)†	(2.51)	0.299		0.791		1.13
1400				0.309		0.807		1.11
1500				0.318		0.824		1.09
1600				(0.327)		0.842		1.07
						0.860		1.05
						0.877		1.03
						(0.895)		1.02
						(0.913)		1.00
								0.98
								0.96
								0.94
								0.925
								0.915
								0.905
								0.900
								(0.895)

* In the table the third significant figure is given only for the purpose of comparison and for smoothness and is not indicative of the degree of accuracy.

** Values in square brackets are for liquid state.

† Values in parentheses are extrapolated.

‡ Estimated.

THERMAL CONDUCTIVITY OF CERTAIN LIQUIDS

From NSRDS-NBS 8
R. W. Powell, C. Y. Ho, and P. E. Liley

The thermal conductivity, k, is given in the units Milliwatt cm^{-1} $°K^{-1}$. To convert to Cal(*gm*) hr^{-1} cm^{-1} $°K^{-1}$ multiply the values listed in the table by 0.860421

T (K)	Helium	Nitrogen	Argon	Carbon tetra-chloride	Diphenyl	m-Terphenyl	Toluene	Water
2.4	0.192							
2.6	0.193							
2.8	0.197							
3.0	0.204							
3.2	0.214							
3.4	0.227							
3.6	0.241							
3.8	0.260							
4.0	0.282							
4.2	0.307							
4.4	(0.335)‡							
4.6	(0.366)‡							
4.8	(0.400)‡							
5.0	(0.437)‡							
5.2	(0.477)‡							
60		1.692†						
65		1.598						
70		1.504						
75		1.411						
80		1.320‡	1.315†					
85		1.229‡	1.258					
90		1.140‡	1.200‡					
95		1.051‡	1.141‡					
100		0.965‡	1.082‡					
105		0.879‡	1.023‡					
110		0.794‡	0.963‡					
115		0.710‡	0.903‡					
120		0.627‡	0.842‡					
125		0.544‡	0.780‡					
130			0.717‡					
135			0.654‡					
140			0.591‡					
145			0.527‡					
150			0.463‡				(1.719)†	
160							(1.694)†	
170							(1.669)†	
180							1.644	
190							1.619	
200							1.594	
210							1.569	
220							1.543	
230				(1.169)†			1.518	
240				(1.150)†			1.492	
250				1.131			1.467	5.22†
260				1.112			1.442	5.39†
270				1.093			1.416	5.55†
280				1.074			1.391	5.74
290				1.055			1.365	5.92
300				1.036			1.340	6.09
310				1.017			1.315	6.23
320				0.997			1.289	6.37
330				0.978	(1.402)†		1.264	6.48
340				0.959	(1.387)†		1.238	6.59
350				0.940	1.373	(1.361)†	1.213	6.68
360				(0.921)	1.359	(1.356)†	1.188	6.75
370				(0.902)	1.345	1.351	1.162	6.80
380				(0.882)	1.331	1.346	1.137	6.84‡
390				(0.863)	1.316	1.341	(1.112)‡	6.86‡
400				(0.844)	1.302	1.335	(1.086)‡	6.86‡
410				(0.825)	1.288	1.329	(1.061)‡	6.86‡
420				(0.806)	1.274	1.323	(1.036)‡	6.84‡
430				(0.787)	1.259	1.317	(1.013)‡	6.81‡
440				(0.768)	1.245	1.310	(0.985)‡	6.78‡
450				(0.749)	1.231	1.304	(0.959)‡	6.73‡
460					1.217	1.297	(0.933)‡	6.67‡
470					1.202	1.290	(0.908)‡	6.61‡
480					1.188	1.283	(0.885)‡	6.53‡
490					1.174	1.276	(0.862)‡	6.45‡
500					1.160	1.268	(0.839)‡	6.35‡
510					1.146	1.261		6.24‡
520					1.131	1.254		6.12‡
530					1.117‡	1.246		5.99‡
540					1.103‡	1.238		5.86‡
550					1.089‡	1.230		5.71‡
560					1.074‡	1.222		5.55‡
570					1.060‡	1.213		5.39‡
580					1.046‡	1.205		5.20‡
590					1.032‡	1.197		5.01‡
600					1.018‡	1.188		4.81‡
610						1.180		4.60‡
620						1.172		4.40‡
630						1.163		(4.20)‡
640						1.155‡		(4.01)‡
650						1.146‡		

† Extrapolated for the supercooled liquid. [Approximate n.m.p. in K: N_2, 63; A, 84; CCl_4, 250; $C_{12}H_{10}$, 342; m-$C_{18}H_{14}$, 361; p-$C_{18}H_{14}$, 486; C_7H_{10}, 178; H_2O, 273.1].

‡Under saturation vapor pressure [Approximate n.b.p. in K: He, 4.3; N_2, 77; A, 88; CCl_4, 350; $C_{12}H_{10}$, 528; m-$C_{18}H_{14}$, 637; p-$C_{18}H_{14}$, 658; C_7H_{10}, 384; H_2O, 373].

STEAM TABLES

Reproduced by permission of the publishers and copyright owners of the 1967 ASME Steam Tables. Further data and information on the thermodynamic and transport properties of steam and water are contained in the above ASME publication. It is obtainable from The American Society of Mechanical Engineers, United Engineering Center, 345 East 47th Street, New York, New York 10017.

Properties of Saturated Steam and Saturated Water (Temperature)

Temp. F	Press. psia	Volume, ft³/lbm Water v_f	Evap. v_{fg}	Steam v_g	Enthalpy, Btu/lbm Water h_f	Evap. h_{fg}	Steam h_g	Entropy, Btu/lbm × F Water s_f	Evap. s_{fg}	Steam s_g	Temp. F
705.47	3208.2	0.05078	0.00000	0.05078	906.0	0.0	906.0	1.0612	0.0000	1.0612	705.47
705.0	3198.3	0.04427	0.01304	0.05730	873.0	61.4	934.4	1.0329	0.0527	1.0856	705.0
704.5	3187.8	0.04233	0.01822	0.06055	861.9	85.3	947.2	1.0234	0.0732	1.0967	704.5
704.0	3177.2	0.04108	0.02192	0.06300	854.2	102.0	956.2	1.0169	0.0876	1.1046	704.0
703.5	3166.8	0.04015	0.02489	0.06504	848.2	115.2	963.5	1.0118	0.0991	1.1109	703.5
703.0	3156.3	0.03940	0.02744	0.06684	843.2	126.4	969.6	1.0076	0.1087	1.1163	703.0
702.5	3145.9	0.03878	0.02969	0.06847	838.9	136.1	974.9	1.0039	0.1171	1.1210	702.5
702.0	3135.5	0.03824	0.03173	0.06997	835.0	144.7	979.7	1.0006	0.1246	1.1252	702.0
701.5	3125.2	0.03777	0.03361	0.07138	831.5	152.6	984.0	0.9977	0.1314	1.1291	701.5
701.0	3114.9	0.03735	0.03536	0.07271	828.2	159.8	988.0	0.9949	0.1377	1.1326	701.0
700.5	3104.6	0.03697	0.03701	0.07397	825.2	166.5	991.7	0.9924	0.1435	1.1359	700.5
700.0	3094.3	0.03662	0.03857	0.07519	822.4	172.7	995.2	0.9901	0.1490	1.1390	700.0
699.0	3073.9	0.03600	0.04149	0.07749	817.3	184.2	1001.5	0.9858	0.1590	1.1447	699.0
698.0	3053.6	0.03546	0.04420	0.07966	812.6	194.6	1007.2	0.9818	0.1681	1.1499	698.0
697.0	3033.5	0.03498	0.04674	0.08172	808.4	204.0	1012.4	0.9783	0.1764	1.1547	697.0
696.0	3013.4	0.03455	0.04916	0.08371	804.4	212.8	1017.2	0.9749	0.1841	1.1591	696.0
695.0	2993.5	0.03415	0.05147	0.08563	800.6	221.0	1021.7	0.9718	0.1914	1.1632	695.0
694.0	2973.7	0.03379	0.05370	0.08749	797.1	228.8	1025.9	0.9689	0.1983	1.1671	694.0
693.0	2954.0	0.03345	0.05587	0.08931	793.8	236.1	1029.9	0.9660	0.2048	1.1708	693.0
692.0	2934.5	0.03313	0.05797	0.09110	790.5	243.1	1033.6	0.9634	0.2110	1.1744	692.0
690.0	2895.7	0.03256	0.06203	0.09459	784.5	256.1	1040.6	0.9583	0.2227	1.1810	690.0
688.0	2857.4	0.03204	0.06595	0.09799	778.8	268.2	1047.0	0.9535	0.2337	1.1872	688.0
686.0	2819.5	0.03157	0.06976	0.10133	773.4	279.5	1052.9	0.9490	0.2439	1.1930	686.0
684.0	2782.1	0.03114	0.07349	0.10463	768.2	290.2	1058.4	0.9447	0.2537	1.1984	684.0
682.0	2745.1	0.03074	0.07716	0.10790	763.3	300.4	1063.6	0.9406	0.2631	1.2036	682.0
680.0	2708.6	0.03037	0.08080	0.11117	758.5	310.1	1068.5	0.9365	0.2720	1.2086	680.0
678.0	2672.5	0.03002	0.08440	0.11442	753.8	319.4	1073.2	0.9326	0.2807	1.2133	678.0
676.0	2636.8	0.02970	0.08799	0.11769	749.2	328.5	1077.6	0.9287	0.2892	1.2179	676.0
674.0	2601.5	0.02939	0.09156	0.12096	744.7	337.2	1081.9	0.9249	0.2974	1.2223	674.0
672.0	2566.6	0.02911	0.09514	0.12424	740.2	345.7	1085.9	0.9212	0.3054	1.2266	672.0
670.0	2532.2	0.02884	0.09871	0.12755	735.8	354.0	1089.8	0.9174	0.3133	1.2307	670.0
668.0	2498.1	0.02858	0.10229	0.13087	731.5	362.1	1093.5	0.9137	0.3210	1.2347	668.0
666.0	2464.4	0.02834	0.10588	0.13421	727.1	370.0	1097.1	0.9100	0.3286	1.2387	666.0
664.0	2431.1	0.02811	1.10947	0.13757	722.9	377.7	1100.6	0.9064	0.3361	1.2425	664.0
662.0	2398.2	0.02789	0.11306	0.14095	718.8	385.1	1103.9	0.9028	0.3434	1.2462	662.0
660.0	2365.7	0.02768	0.11663	0.14431	714.9	392.1	1107.0	0.8995	0.3502	1.2498	660.0
658.0	2333.5	0.02748	0.12023	0.14771	711.1	399.0	1110.1	0.8963	0.3570	1.2533	658.0
656.0	2301.7	0.02728	0.12387	0.15115	707.4	405.7	1113.1	0.8931	0.3637	1.2567	656.0
654.0	2270.3	0.02709	0.12754	0.15463	703.7	412.2	1115.9	0.8899	0.3702	1.2601	654.0
652.0	2239.2	0.02691	0.13124	0.15816	700.0	418.7	1118.7	0.8868	0.3767	1.2634	652.0
650.0	2208.4	0.02674	0.13499	0.16173	696.4	425.0	1121.4	0.8837	0.3830	1.2667	650.0
648.0	2178.1	0.02657	0.13876	0.16534	692.9	431.1	1124.0	0.8806	0.3893	1.2699	648.0
646.0	2148.0	0.02641	0.14258	0.16899	689.4	437.2	1126.6	0.8776	0.3954	1.2730	646.0
644.0	2118.3	0.02625	0.14644	0.17269	685.9	443.1	1129.0	0.8746	0.4015	1.2761	644.0
642.0	2088.9	0.02610	0.15033	0.17643	682.5	448.9	1131.4	0.8716	0.4075	1.2791	642.0
640.0	2059.9	0.02595	0.15427	0.18021	679.1	454.6	1133.7	0.8686	0.4134	1.2821	640.0
638.0	2031.2	0.02580	0.15824	0.18405	675.8	460.2	1136.0	0.8657	0.4193	1.2850	638.0
636.0	2002.8	0.02566	0.16226	0.18792	672.4	465.7	1138.1	0.8628	0.4251	1.2879	636.0
634.0	1974.7	0.02553	0.16633	0.19185	669.1	471.1	1140.2	0.8599	0.4307	1.2907	634.0
632.0	1947.0	0.02539	0.17044	0.19583	665.9	476.4	1142.2	0.8571	0.4364	1.2934	632.0
630.0	1919.5	0.02526	0.17459	0.19986	662.7	481.6	1144.2	0.8542	0.4419	1.2962	630.0
628.0	1892.4	0.02514	0.17880	0.20394	659.5	486.7	1146.1	0,8514	0.4474	1.2988	628.0
626.0	1865.6	0.02501	0.18306	0.20807	656.3	491.7	1148.0	0.8486	0.4529	1.3015	626.0
624.0	1839.0	0.02489	0.18737	0.21226	653.1	496.6	1149.8	0.8458	0.4583	1.3041	624.0
622.0	1812.8	0.02477	0.19173	0.21650	650.0	501.5	1151.5	0.8430	0.4636	1.3066	622.0
620.0	1786.9	0.02466	0.19615	0.22081	646.9	506.3	1153.2	0.8403	0.4689	1.3092	620.0
618.0	1761.2	0.02455	0.20063	0.22517	643.8	511.0	1154.8	0.8375	0.4742	1.3117	618.0
616.0	1735.9	0.02444	0.20516	0.22960	640.8	515.6	1156.4	0.8348	0.4794	1.3141	616.0
614.0	1710.8	0.02433	0.20976	0.23409	637.8	520.2	1158.0	0.8321	0.4845	1.3166	614.0
612.0	1686.1	0.02422	0.21442	0.23865	634.8	524.7	1159.5	0.8294	0.4896	1.3190	612.0
610.0	1661.6	0.02412	0.21915	0.24327	631.8	529.2	1160.9	0.8267	0.4947	1.3214	610.0
608.0	1637.3	0.02402	0.22394	0.24796	628.8	533.6	1162.4	0.8240	0.4997	1.3238	608.0
606.0	1613.4	0.02392	0.22881	0.25273	625.9	537.9	1163.8	0.8214	0.5048	1.3261	606.0
604.0	1589.7	0.02382	0.23374	0.25757	622.9	542.2	1165.1	0.8187	0.5097	1.3284	604.0
602.0	1566.3	0.02373	0.23875	0.26248	620.0	546.4	1166.4	0.8161	0.5147	1.3307	602.0
600.0	1543.2	0.02364	0.24384	0.26747	617.1	550.6	1167.7	0.8134	0.5196	1.3330	600.0
598.0	1520.4	0.02354	0.24900	0.27255	614.3	554.7	1169.0	0.8108	0.5245	1.3353	598.0
596.0	1497.8	0.02345	0.25425	0.27770	611.4	558.8	1170.2	0.8082	0.5293	1.3375	596.0
594.0	1475.4	0.02337	0.25958	0.28294	608.6	562.8	1171.4	0.8056	0.5342	1.3398	594.0
592.0	1453.3	0.02328	0.26499	0.28827	605.7	566.8	1172.6	0.8030	0.5390	1.3420	592.0
590.0	1431.5	0.02319	0.27049	0.29368	602.9	570.8	1173.7	0.8004	0.5437	1.3442	590.0
588.0	1410.0	0.02311	0.27608	0.29919	600.1	574.7	1174.8	0.7978	0.5485	1.3464	588.0
586.0	1388.6	0.02303	0.28176	0.30478	597.3	578.5	1175.9	0.7953	0.5532	1.3485	586.0
584.0	1367.6	0.02295	0.28753	0.31048	594.6	582.4	1176.9	0.7927	0.5580	1.3507	584.0
582.0	1346.7	0.02287	0.29340	0.31627	591.8	586.1	1178.0	0.7902	0.5627	1.3528	582.0
580.0	1326.2	0.02279	0.29937	0.32216	589.1	589.9	1179.0	0.7876	0.5673	1.3550	580.0

Properties of Saturated Steam and Saturated Water (Temperature)

Temp. F	Press. psia	Volume, ft³/lbm Water v_f	Evap. v_{fg}	Steam v_g	Enthalpy, Btu/lbm Water h_f	Evap. h_{fg}	Steam h_g	Entropy, Btu/lbm × F Water s_f	Evap. s_{fg}	Steam s_g	Temp. F
580.0	1326.17	0.02279	0.29937	0.32216	589.1	589.9	1179.0	0.7876	0.5673	1.3550	580.0
578.0	1305.84	0.02271	0.30544	0.32816	586.4	593.6	1179.9	0.7851	0.5720	1.3571	578.0
576.0	1285.74	0.02264	0.31162	0.33426	583.7	597.2	1180.9	0.7825	0.5766	1.3592	576.0
574.0	1265.89	0.02256	0.31790	0.34046	581.0	600.9	1181.8	0.7800	0.5813	1.3613	574.0
572.0	1246.26	0.02249	0.32429	0.34678	578.3	604.5	1182.7	0.7775	0.5859	1.3634	572.0
570.0	1226.88	0.02242	0.33079	0.35321	575.6	608.0	1183.6	0.7750	0.5905	1.3654	570.0
568.0	1207.72	0.02235	0.33741	0.35975	572.9	611.5	1184.5	0.7725	0.5950	1.3675	568.0
566.0	1188.80	0.02228	0.34414	0.36642	570.3	615.0	1185.3	0.7699	0.5996	1.3696	566.0
564.0	1170.10	0.02221	0.35099	0.37320	567.6	618.5	1186.1	0.7674	0.6041	1.3716	564.0
562.0	1151.63	0.02214	0.35797	0.38011	565.0	621.9	1186.9	0.7650	0.6087	1.3736	562.0
560.0	1133.38	0.02207	0.36507	0.38714	562.4	625.3	1187.7	0.7625	0.6132	1.3757	560.0
558.0	1115.36	0.02201	0.37230	0.39431	559.8	628.6	1188.4	0.7600	0.6177	1.3777	558.0
556.0	1097.55	0.02194	0.37966	0.40160	557.2	632.0	1189.2	0.7575	0.6222	1.3797	556.0
554.0	1079.96	0.02188	0.38715	0.40903	554.6	635.3	1189.9	0.7550	0.6267	1.3817	554.0
552.0	1062.59	0.02182	0.39479	0.41660	552.0	638.5	1190.6	0.7525	0.6311	1.3837	552.0
550.0	1045.43	0.02176	0.40256	0.42432	549.5	641.8	1191.2	0.7501	0.6356	1.3856	550.0
548.0	1028.49	0.02169	0.41048	0.43217	546.9	645.0	1191.9	0.7476	0.6400	1.3876	548.0
546.0	1011.75	0.02163	0.41855	0.44018	544.4	648.1	1192.5	0.7451	0.6445	1.3896	546.0
544.0	995.22	0.02157	0.42677	0.44834	541.8	651.3	1193.1	0.7427	0.6489	1.3915	544.0
542.0	978.90	0.02151	0.43514	0.45665	539.3	654.4	1193.7	0.7402	0.6533	1.3935	542.0
540.0	962.79	0.02146	0.44367	0.46513	536.8	657.5	1194.3	0.7378	0.6577	1.3954	540.0
538.0	946.88	0.02140	0.45237	0.47377	534.2	660.6	1194.8	0.7353	0.6621	1.3974	538.0
536.0	931.17	0.02134	0.46123	0.48257	531.7	663.6	1195.4	0.7329	0.6665	1.3993	536.0
534.0	915.66	0.02129	0.47026	0.49155	529.2	666.6	1195.9	0.7304	0.6708	1.4012	534.0
532.0	900.34	0.02123	0.47947	0.50070	526.8	669.6	1196.4	0.7280	0.6752	1.4032	532.0
530.0	885.23	0.02118	0.48886	0.51004	524.3	672.6	1196.9	0.7255	0.6796	1.4051	530.0
528.0	870.31	0.02112	0.49843	0.51955	521.8	675.5	1197.3	0.7231	0.6839	1.4070	528.0
526.0	855.58	0.02107	0.50819	0.52926	519.3	678.4	1197.8	0.7206	0.6883	1.4089	526.0
524.0	841.04	0.02102	0.51814	0.53916	516.9	681.3	1198.2	0.7182	0.6926	1.4108	524.0
522.0	826.69	0.02097	0.52829	0.54926	514.4	684.2	1198.6	0.7158	0.6969	1.4127	522.0
520.0	812.53	0.02091	0.53864	0.55956	512.0	687.0	1199.0	0.7133	0.7013	1.4146	520.0
518.0	798.55	0.02086	0.54920	0.57006	509.6	689.9	1199.4	0.7109	0.7056	1.4165	518.0
516.0	784.76	0.02081	0.55997	0.58079	507.1	692.7	1199.8	0.7085	0.7099	1.4183	516.0
514.0	771.15	0.02076	0.57096	0.59173	504.7	695.4	1200.2	0.7060	0.7142	1.4202	514.0
512.0	757.72	0.02072	0.58218	0.60289	502.3	698.2	1200.5	0.7036	0.7185	1.4221	512.0
510.0	744.47	0.02067	0.59362	0.61429	499.9	700.9	1200.8	0.7012	0.7228	1.4240	510.0
508.0	731.40	0.02062	0.60530	0.62592	497.5	703.7	1201.1	0.6987	0.7271	1.4258	508.0
506.0	718.50	0.02057	0.61722	0.63779	495.1	706.3	1201.4	0.6963	0.7314	1.4277	506.0
504.0	705.78	0.02053	0.62938	0.64991	492.7	709.0	1201.7	0.6939	0.7357	1.4296	504.0
502.0	693.23	0.02048	0.64180	0.66228	490.3	711.7	1202.0	0.6915	0.7400	1.4314	502.0
500.0	680.86	0.02043	0.65448	0.67492	487.9	714.3	1202.2	0.6890	0.7443	1.4333	500.0
498.0	668.65	0.02039	0.66743	0.68782	485.6	716.9	1202.5	0.6866	0.7486	1.4352	498.0
496.0	656.61	0.02034	0.68065	0.70100	483.2	719.5	1202.7	0.6842	0.7528	1.4370	496.0
494.0	644.73	0.02030	0.69415	0.71445	480.8	722.1	1202.9	0.6818	0.7571	1.4389	494.0
492.0	633.03	0.02026	0.70794	0.72820	478.5	724.6	1203.1	0.6793	0.7614	1.4407	492.0
490.0	621.48	0.02021	0.72203	0.74224	476.1	727.2	1203.3	0.6769	0.7657	1.4426	490.0
488.0	610.10	0.02017	0.73641	0.75658	473.8	729.7	1203.5	0.6745	0.7700	1.4444	488.0
486.0	598.87	0.02013	0.75111	0.77124	471.5	732.2	1203.7	0.6721	0.7742	1.4463	486.0
484.0	587.81	0.02009	0.76613	0.78622	469.1	734.7	1203.8	0.6696	0.7785	1.4481	484.0
482.0	576.90	0.02004	0.78148	0.80152	466.8	737.2	1204.0	0.6672	0.7828	1.4500	482.0
480.0	566.15	0.02000	0.79716	0.81717	464.5	739.6	1204.1	0.6648	0.7871	1.4518	480.0
478.0	555.55	0.01996	0.81319	0.83315	462.2	742.1	1204.2	0.6624	0.7913	1.4537	478.0
476.0	545.11	0.01992	0.82958	0.84950	459.9	744.5	1204.3	0.6599	0.7956	1.4555	476.0
474.0	534.81	0.01988	0.84632	0.86621	457.5	746.9	1204.4	0.6575	0.7999	1.4574	474.0
472.0	524.67	0.01984	0.86345	0.88329	455.2	749.3	1204.5	0.6551	0.8042	1.4592	472.0
470.0	514.67	0.01980	0.88095	0.90076	452.9	751.6	1204.6	0.6527	0.8084	1.4611	470.0
468.0	504.83	0.01976	0.89885	0.91862	450.7	754.0	1204.6	0.6502	0.8127	1.4629	468.0
466.0	495.12	0.01973	0.91716	0.93689	448.4	756.3	1204.7	0.6478	0.8170	1.4648	466.0
464.0	485.56	0.01969	0.93588	0.95557	446.1	758.6	1204.7	0.6454	0.8213	1.4667	464.0
462.0	476.14	0.01965	0.95504	0.97469	443.8	761.0	1204.8	0.6429	0.8256	1.4685	462.0
460.0	466.87	0.01961	0.97463	0.99424	441.5	763.2	1204.8	0.6405	0.8299	1.4704	460.0
458.0	457.73	0.01958	0.99467	1.01425	439.3	765.5	1204.8	0.6381	0.8342	1.4722	458.0
456.0	448.73	0.01954	1.01518	1.03472	437.0	767.8	1204.8	0.6356	0.8385	1.4741	456.0
454.0	439.87	0.01950	1.03616	1.05567	434.7	770.0	1204.8	0.6332	0.8428	1.4759	454.0
452.0	431.14	0.01947	1.05764	1.07711	432.5	772.3	1204.8	0.6308	0.8471	1.4778	452.0
450.0	422.55	0.01943	1.07962	1.09905	430.2	774.5	1204.7	0.6283	0.8514	1.4797	450.0
448.0	414.09	0.01940	1.10212	1.12152	428.0	776.7	1204.7	0.6259	0.8557	1.4815	448.0
446.0	405.76	0.01936	1.12515	1.14452	425.7	778.9	1204.6	0.6234	0.8600	1.4834	446.0
444.0	397.56	0.01933	1.14874	1.16806	423.5	781.1	1204.6	0.6210	0.8643	1.4853	444.0
442.0	389.49	0.01929	1.17288	1.19217	421.3	783.2	1204.5	0.6185	0.8686	1.4872	442.0
440.0	381.54	0.01926	1.19761	1.21687	419.0	785.4	1204.4	0.6161	0.8729	1.4890	440.0
438.0	373.72	0.01923	1.22293	1.24216	416.8	787.5	1204.3	0.6136	0.8773	1.4909	438.0
436.0	366.03	0.01919	1.24887	1.26806	414.6	789.7	1204.2	0.6112	0.8816	1.4928	436.0
434.0	358.46	0.01916	1.27544	1.29460	412.4	791.8	1204.1	0.6087	0.8859	1.4947	434.0
432.0	351.00	0.01913	1.30266	1.32179	410.1	793.9	1204.0	0.6063	0.8903	1.4966	432.0
430.0	343.67	0.01909	1.33055	1.34965	407.9	796.0	1203.9	0.6038	0.8946	1.4985	430.0

STEAM TABLES (Continued)

Properties of Saturated Steam and Saturated Water (Temperature)

Temp. F	Press. psia	Volume, ft³/lbm Water v_f	Evap. v_{fg}	Steam v_g	Enthalpy, Btu/lbm Water h_f	Evap. h_{fg}	Steam h_g	Entropy, Btu/lbm ×F Water s_f	Evap. s_{fg}	Steam s_g	Temp. F
430.0	343.674	0.01909	1.3306	1.3496	407.9	796.0	1203.9	0.6038	0.8946	1.4985	430.0
428.0	336.463	0.01906	1.3591	1.3782	405.7	798.0	1203.7	0.6014	0.8990	1.5004	428.0
426.0	329.369	0.01903	1.3884	1.4075	403.5	800.1	1203.6	0.5989	0.9034	1.5023	426.0
424.0	322.391	0.01900	1.4184	1.4374	401.3	802.2	1203.5	0.5964	0.9077	1.5042	424.0
422.0	315.529	0.01897	1.4492	1.4682	399.1	804.2	1203.3	0.5940	0.9121	1.5061	422.0
420.0	308.780	0.01894	1.4808	1.4997	396.9	806.2	1203.1	0.5915	0.9165	1.5080	420.0
418.0	302.143	0.01890	1.5131	1.5320	394.7	808.2	1202.9	0.5890	0.9209	1.5099	418.0
416.0	295.617	0.01887	1.5463	1.5651	392.5	810.2	1202.8	0.5866	0.9253	1.5118	416.0
414.0	289.201	0.01884	1.5803	1.5991	390.3	812.2	1202.6	0.5841	0.9297	1.5137	414.0
412.0	282.894	0.01881	1.6152	1.6340	388.1	814.2	1202.4	0.5816	0.9341	1.5157	412.0
410.0	276.694	0.01878	1.6510	1.6697	386.0	816.2	1202.1	0.5791	0.9385	1.5176	410.0
408.0	270.600	0.01875	1.6877	1.7064	383.8	818.2	1201.9	0.5766	0.9429	1.5195	408.0
406.0	264.611	0.01872	1.7253	1.7441	381.6	820.1	1201.7	0.5742	0.9473	1.5215	406.0
404.0	258.725	0.01870	1.7640	1.7827	379.4	822.0	1201.5	0.5717	0.9518	1.5234	404.0
402.0	252.942	0.01867	1.8037	1.8223	377.3	824.0	1201.2	0.5692	0.9562	1.5254	402.0
400.0	247.259	0.01864	1.8444	1.8630	375.1	825.9	1201.0	0.5667	0.9607	1.5274	400.0
398.0	241.677	0.01861	1.8862	1.9048	372.9	827.8	1200.7	0.5642	0.9651	1.5293	398.0
396.0	236.193	0.01858	1.9291	1.9477	370.8	829.7	1200.4	0.5617	0.9696	1.5313	396.0
394.0	230.807	0.01855	1.9731	1.9917	368.6	831.6	1200.2	0.5592	0.9741	1.5333	394.0
392.0	225.516	0.01853	2.0184	2.0369	366.5	833.4	1199.9	0.5567	0.9786	1.5352	392.0
390.0	220.321	0.01850	2.0649	2.0833	364.3	835.3	1199.6	0.5542	0.9831	1.5372	390.0
388.0	215.220	0.01847	2.1126	2.1311	362.2	837.2	1199.3	0.5516	0.9876	1.5392	388.0
386.0	210.211	0.01844	2.1616	2.1801	360.0	839.0	1199.0	0.5491	0.9921	1.5412	386.0
384.0	205.294	0.01842	2.2120	2.2304	357.9	840.8	1198.7	0.5466	0.9966	1.5432	384.0
382.0	200.467	0.01839	2.2638	2.2821	355.7	842.7	1198.4	0.5441	1.0012	1.5452	382.0
380.0	195.729	0.01836	2.3170	2.3353	353.6	844.5	1198.0	0.5416	1.0057	1.5473	380.0
378.0	191.080	0.01834	2.3716	2.3900	351.4	846.3	1197.7	0.5390	1.0103	1.5493	378.0
376.0	186.517	0.01831	2.4279	2.4462	349.3	848.1	1197.4	0.5365	1.0148	1.5513	376.0
374.0	182.040	0.01829	2.4857	2.5039	347.2	849.8	1197.0	0.5340	1.0194	1.5534	374.0
372.0	177.648	0.01826	2.5451	2.5633	345.0	851.6	1196.7	0.5314	1.0240	1.5554	372.0
370.0	173.339	0.01823	2.6062	2.6244	342.9	853.4	1196.3	0.5289	1.0286	1.5575	370.0
368.0	169.113	0.01821	2.6691	2.6873	340.8	855.1	1195.9	0.5263	1.0332	1.5595	368.0
366.0	164.968	0.01818	2.7337	2.7519	338.7	856.9	1195.6	0.5238	1.0378	1.5616	366.0
364.0	160.903	0.01816	2.8002	2.8184	336.5	858.6	1195.2	0.5212	1.0424	1.5637	364.0
362.0	156.917	0.01813	2.8687	2.8868	334.4	860.4	1194.8	0.5187	1.0471	1.5658	362.0
360.0	153.010	0.01811	2.9392	2.9573	332.3	862.1	1194.4	0.5161	1.0517	1.5678	360.0
358.0	149.179	0.01809	3.0117	3.0298	330.2	863.8	1194.0	0.5135	1.0564	1.5699	358.0
356.0	145.424	0.01806	3.0863	3.1044	328.1	865.5	1193.6	0.5110	1.0611	1.5721	356.0
354.0	141.744	0.01804	3.1632	3.1812	326.0	867.2	1193.2	0.5084	1.0658	1.5742	354.0
352.0	138.138	0.01801	3.2423	3.2603	323.9	868.9	1192.7	0.5058	1.0705	1.5763	352.0
350.0	134.604	0.01799	3.3238	3.3418	321.8	870.6	1192.3	0.5032	1.0752	1.5784	350.0
348.0	131.142	0.01797	3.4078	3.4258	319.7	872.2	1191.9	0.5006	1.0799	1.5806	348.0
346.0	127.751	0.01794	3.4943	3.5122	317.6	873.9	1191.4	0.4980	1.0847	1.5827	346.0
344.0	124.430	0.01792	3.5834	3.6013	315.5	875.5	1191.0	0.4954	1.0894	1.5849	344.0
342.0	121.177	0.01790	3.6752	3.6931	313.4	877.2	1190.5	0.4928	1.0942	1.5871	342.0
340.0	117.992	0.01787	3.7699	3.7878	311.3	878.8	1190.1	0.4902	1.0990	1.5892	340.0
338.0	114.873	0.01785	3.8675	3.8853	309.2	880.5	1189.6	0.4876	1.1038	1.5914	338.0
336.0	111.820	0.01783	3.9681	3.9859	307.1	882.1	1189.1	0.4850	1.1086	1.5936	336.0
334.0	108.832	0.01781	4.0718	4.0896	305.0	883.7	1188.7	0.4824	1.1134	1.5958	334.0
332.0	105.907	0.01779	4.1788	4.1966	302.9	885.3	1188.2	0.4798	1.1183	1.5981	332.0
330.0	103.045	0.01776	4.2892	4.3069	300.8	886.9	1187.7	0.4772	1.1231	1.6003	330.0
328.0	100.245	0.01774	4.4030	4.4208	298.7	888.5	1187.2	0.4745	1.1280	1.6025	328.0
326.0	97.506	0.01772	4.5205	4.5382	296.6	890.1	1186.7	0.4719	1.1329	1.6048	326.0
324.0	94.826	0.01770	4.6418	4.6595	294.6	891.6	1186.2	0.4692	1.1378	1.6071	324.0
322.0	92.205	0.01768	4.7669	4.7846	292.5	893.2	1185.7	0.4666	1.1427	1.6093	322.0
320.0	89.643	0.01766	4.8961	4.9138	290.4	894.8	1185.2	0.4640	1.1477	1.6116	320.0
318.0	87.137	0.01764	5.0295	5.0471	288.3	896.3	1184.7	0.4613	1.1526	1.6139	318.0
316.0	84.688	0.01761	5.1673	5.1849	286.3	897.9	1184.1	0.4586	1.1576	1.6162	316.0
314.0	82.293	0.01759	5.3096	5.3272	284.2	899.4	1183.6	0.4560	1.1626	1.6185	314.0
312.0	79.953	0.01757	5.4566	5.4742	282.1	901.0	1183.1	0.4533	1.1676	1.6209	312.0
310.0	77.667	0.01755	5.6085	5.6260	280.0	902.5	1182.5	0.4506	1.1726	1.6232	310.0
308.0	75.433	0.01753	5.7655	5.7830	278.0	904.0	1182.0	0.4479	1.1776	1.6256	308.0
306.0	73.251	0.01751	5.9277	5.9452	275.9	905.5	1181.4	0.4453	1.1827	1.6279	306.0
304.0	71.119	0.01749	6.0955	6.1130	273.8	907.0	1180.9	0.4426	1.1877	1.6303	304.0
302.0	69.038	0.01747	6.2689	6.2864	271.8	908.5	1180.3	0.4399	1.1928	1.6327	302.0
300.0	67.005	0.01745	6.4483	6.4658	269.7	910.0	1179.7	0.4372	1.1979	1.6351	300.0
298.0	65.021	0.01743	6.6339	6.6513	267.7	911.5	1179.2	0.4345	1.2031	1.6375	298.0
296.0	63.084	0.01741	6.8259	6.8433	265.6	913.0	1178.6	0.4317	1.2082	1.6400	296.0
294.0	61.194	0.01739	7.0245	7.0419	263.5	914.5	1178.0	0.4290	1.2134	1.6424	294.0
292.0	59.350	0.01738	7.2301	7.2475	261.5	915.9	1177.4	0.4263	1.2186	1.6449	292.0
290.0	57.550	0.01736	7.4430	7.4603	259.4	917.4	1176.8	0.4236	1.2238	1.6473	290.0
288.0	55.795	0.01734	7.6634	7.6807	257.4	918.8	1176.2	0.4208	1.2290	1.6498	288.0
286.0	54.083	0.01732	7.8916	7.9089	255.3	920.3	1175.6	0.4181	1.2342	1.6523	286.0
284.0	52.414	0.01730	8.1280	8.1453	253.3	921.7	1175.0	0.4154	1.2395	1.6548	284.0
282.0	50.786	0.01728	8.3729	8.3902	251.2	923.2	1174.4	0.4126	1.2448	1.6574	282.0
280.0	49.200	0.01726	8.6267	8.6439	249.2	924.6	1173.8	0.4098	1.2501	1.6599	280.0

Properties of Saturated Steam and Saturated Water (Temperature)

Temp. F	Press. psia	Volume, ft³/lbm Water v_f	Evap. v_{fg}	Steam v_g	Enthalpy, Btu/lbm Water h_f	Evap. h_{fg}	Steam h_g	Entropy, Btu/lbm ×F Water s_f	Evap. s_{fg}	Steam s_g	Temp. F
280.0	49.200	0.017264	8.627	8.644	249.17	924.6	1173.8	0.4098	1.2501	1.6599	280.0
278.0	47.653	0.017246	8.890	8.907	247.13	926.0	1173.2	0.4071	1.2554	1.6625	278.0
276.0	46.147	0.017228	9.162	9.180	245.08	927.5	1172.5	0.4043	1.2607	1.6650	276.0
274.0	44.678	0.017210	9.445	9.462	243.03	928.9	1171.9	0.4015	1.2661	1.6676	274.0
272.0	43.249	0.017193	9.738	9.755	240.99	930.3	1171.3	0.3987	1.2715	1.6702	272.0
270.0	41.856	0.017175	10.042	10.060	238.95	931.7	1170.6	0.3960	1.2769	1.6729	270.0
268.0	40.500	0.017157	10.358	10.375	236.91	933.1	1170.0	0.3932	1.2823	1.6755	268.0
266.0	39.179	0.017140	10.685	10.703	234.87	934.5	1169.3	0.3904	1.2878	1.6781	266.0
264.0	37.894	0.017123	11.025	11.042	232.83	935.9	1168.7	0.3876	1.2933	1.6808	264.0
262.0	36.644	0.017106	11.378	11.395	230.79	937.3	1168.0	0.3847	1.2988	1.6835	262.0
260.0	35.427	0.017089	11.745	11.762	228.76	938.6	1167.4	0.3819	1.3043	1.6862	260.0
258.0	34.243	0.017072	12.125	12.142	226.72	940.0	1166.7	0.3791	1.3098	1.6889	258.0
256.0	33.091	0.017055	12.520	12.538	224.69	941.4	1166.1	0.3763	1.3154	1.6917	256.0
254.0	31.972	0.017039	12.931	12.948	222.65	942.7	1165.4	0.3734	1.3210	1.6944	254.0
252.0	30.883	0.017022	13.358	13.375	220.62	944.1	1164.7	0.3706	1.3266	1.6972	252.0
250.0	29.825	0.017006	13.802	13.819	218.59	945.4	1164.0	0.3677	1.3323	1.7000	250.0
248.0	28.796	0.016990	14.264	14.281	216.56	946.8	1163.4	0.3649	1.3379	1.7028	248.0
246.0	27.797	0.016974	14.744	14.761	214.53	948.1	1162.7	0.3620	1.3436	1.7056	246.0
244.0	26.826	0.016958	15.243	15.260	212.50	949.5	1162.0	0.3591	1.3494	1.7085	244.0
242.0	25.883	0.016942	15.763	15.780	210.48	950.8	1161.3	0.3562	1.3551	1.7113	242.0
240.0	24.968	0.016926	16.304	16.321	208.45	952.1	1160.6	0.3533	1.3609	1.7142	240.0
238.0	24.079	0.016910	16.867	16.884	206.42	953.5	1159.9	0.3505	1.3667	1.7171	238.0
236.0	23.216	0.016895	17.454	17.471	204.40	954.8	1159.2	0.3476	1.3725	1.7201	236.0
234.0	22.379	0.016880	18.065	18.082	202.38	956.1	1158.5	0.3446	1.3784	1.7230	234.0
232.0	21.567	0.016864	18.701	18.718	200.35	957.4	1157.8	0.3417	1.3842	1.7260	232.0
230.0	20.779	0.016849	19.364	19.381	198.33	958.7	1157.1	0.3388	1.3902	1.7290	230.0
229.0	20.394	0.016842	19.707	19.723	197.32	959.4	1156.7	0.3373	1.3931	1.7305	229.0
228.0	20.015	0.016834	20.056	20.073	196.31	960.0	1156.3	0.3359	1.3961	1.7320	228.0
227.0	19.642	0.016827	20.413	20.429	195.30	960.7	1156.0	0.3344	1.3991	1.7335	227.0
226.0	19.274	0.016819	20.777	20.794	194.29	961.3	1155.6	0.3329	1.4021	1.7350	226.0
225.0	18.912	0.016812	21.149	21.166	193.28	962.0	1155.3	0.3315	1.4051	1.7365	225.0
224.0	18.556	0.016805	21.529	21.545	192.27	962.6	1154.9	0.3300	1.4081	1.7380	224.0
223.0	18.206	0.016797	21.917	21.933	191.26	963.3	1154.5	0.3285	1.4111	1.7396	223.0
222.0	17.860	0.016790	22.313	22.330	190.25	963.9	1154.2	0.3270	1.4141	1.7411	222.0
221.0	17.521	0.016783	22.718	22.735	189.24	964.6	1153.8	0.3255	1.4171	1.7427	221.0
220.0	17.186	0.016775	23.131	23.148	188.23	965.2	1153.4	0.3241	1.4201	1.7442	220.0
219.0	16.857	0.016768	23.554	23.571	187.22	965.8	1153.1	0.3226	1.4232	1.7458	219.0
218.0	16.533	0.016761	23.986	24.002	186.21	966.5	1152.7	0.3211	1.4262	1.7473	218.0
217.0	16.214	0.016754	24.427	24.444	185.21	967.1	1152.3	0.3196	1.4293	1.7489	217.0
216.0	15.901	0.016747	24.878	24.894	184.20	967.8	1152.0	0.3181	1.4323	1.7505	216.0
215.0	15.592	0.016740	25.338	25.355	183.19	968.4	1151.6	0.3166	1.4354	1.7520	215.0
214.0	15.289	0.016733	25.809	25.826	182.18	969.0	1151.2	0.3151	1.4385	1.7536	214.0
213.0	14.990	0.016726	26.290	26.307	181.17	969.7	1150.8	0.3136	1.4416	1.7552	213.0
212.0	14.696	0.016719	26.782	26.799	180.17	970.3	1150.5	0.3121	1.4447	1.7568	212.0
211.0	14.407	0.016712	27.285	27.302	179.16	970.9	1150.1	0.3106	1.4478	1.7584	211.0
210.0	14.123	0.016705	27.799	27.816	178.15	971.6	1149.7	0.3091	1.4509	1.7600	210.0
209.0	13.843	0.016698	28.324	28.341	177.14	972.2	1149.4	0.3076	1.4540	1.7616	209.0
208.0	13.568	0.016691	28.862	28.878	176.14	972.8	1149.0	0.3061	1.4571	1.7632	208.0
207.0	13.297	0.016684	29.411	29.428	175.13	973.5	1148.6	0.3046	1.4602	1.7649	207.0
206.0	13.031	0.016677	29.973	29.989	174.12	974.1	1148.2	0.3031	1.4634	1.7665	206.0
205.0	12.770	0.016670	30.547	30.564	173.12	974.7	1147.9	0.3016	1.4665	1.7681	205.0
204.0	12.512	0.016664	31.135	31.151	172.11	975.4	1147.5	0.3001	1.4697	1.7698	204.0
203.0	12.259	0.016657	31.736	31.752	171.10	976.0	1147.1	0.2986	1.4728	1.7714	203.0
202.0	12.011	0.016650	32.350	32.367	170.10	976.6	1146.7	0.2971	1.4760	1.7731	202.0
201.0	11.766	0.016643	32.979	32.996	169.09	977.2	1146.3	0.2955	1.4792	1.7747	201.0
200.0	11.526	0.016637	33.622	33.639	168.09	977.9	1146.0	0.2940	1.4824	1.7764	200.0
199.0	11.290	0.016630	34.280	34.297	167.08	978.5	1145.6	0.2925	1.4856	1.7781	199.0
198.0	11.058	0.016624	34.954	34.970	166.08	979.1	1145.2	0.2910	1.4888	1.7798	198.0
197.0	10.830	0.016617	35.643	35.659	165.07	979.7	1144.8	0.2894	1.4920	1.7814	197.0
196.0	10.605	0.016611	36.348	36.364	164.06	980.4	1144.4	0.2879	1.4952	1.7831	196.0
195.0	10.385	0.016604	37.069	37.086	163.06	981.0	1144.0	0.2864	1.4985	1.7848	195.0
194.0	10.168	0.016598	37.808	37.824	162.05	981.6	1143.7	0.2848	1.5017	1.7865	194.0
193.0	9.956	0.016591	38.564	38.580	161.05	982.2	1143.3	0.2833	1.5050	1.7882	193.0
192.0	9.747	0.016585	39.337	39.354	160.05	982.8	1142.9	0.2818	1.5082	1.7900	192.0
191.0	9.541	0.016578	40.130	40.146	159.04	983.5	1142.5	0.2802	1.5115	1.7917	191.0
190.0	9.340	0.016572	40.941	40.957	158.04	984.1	1142.1	0.2787	1.5148	1.7934	190.0
189.0	9.141	0.016566	41.771	41.787	157.03	984.7	1141.7	0.2771	1.5180	1.7952	189.0
188.0	8.947	0.016559	42.621	42.638	156.03	985.3	1141.3	0.2756	1.5213	1.7969	188.0
187.0	8.756	0.016553	43.492	43.508	155.02	985.9	1140.9	0.2740	1.5246	1.7987	187.0
186.0	8.568	0.016547	44.383	44.400	154.02	986.5	1140.5	0.2725	1.5279	1.8004	186.0
185.0	8.384	0.016541	45.297	45.313	153.02	987.1	1140.2	0.2709	1.5313	1.8022	185.0
184.0	8.203	0.016534	46.232	46.249	152.01	987.8	1139.8	0.2694	1.5346	1.8040	184.0
183.0	8.025	0.016528	47.190	47.207	151.01	988.4	1139.4	0.2678	1.5379	1.8057	183.0
182.0	7.850	0.016522	48.172	48.189	150.01	989.0	1139.0	0.2662	1.5413	1.8075	182.0
181.0	7.679	0.016516	49.178	49.194	149.00	989.6	1138.6	0.2647	1.5446	1.8093	181.0
180.0	7.511	0.016510	50.208	50.225	148.00	990.2	1138.2	0.2631	1.5480	1.8111	180.0

STEAM TABLES (Continued)

Properties of Saturated Steam and Saturated Water (Temperature)

Temp. F	Press. psia	Volume, ft³/lbm Water v_f	Evap. v_{fg}	Steam v_g	Enthalpy, Btu/lbm Water h_f	Evap. h_{fg}	Steam h_g	Entropy, Btu/lbm×F Water s_f	Evap. s_{fg}	Steam s_g	Temp. F
180.0	7.5110	0.016510	50.21	50.22	148.00	990.2	1138.2	0.2631	1.5480	1.8111	180.0
179.0	7.3460	0.016504	51.26	51.28	147.00	990.8	1137.8	0.2615	1.5514	1.8129	179.0
178.0	7.1840	0.016498	52.35	52.36	145.99	991.4	1137.4	0.2600	1.5548	1.8147	178.0
177.0	7.0250	0.016492	53.46	53.47	144.99	992.0	1137.0	0.2584	1.5582	1.8166	177.0
176.0	6.8690	0.016486	54.59	54.61	143.99	992.6	1136.6	0.2568	1.5616	1.8184	176.0
175.0	6.7159	0.016480	55.76	55.77	142.99	993.2	1136.2	0.2552	1.5650	1.8202	175.0
174.0	6.5656	0.016474	56.95	56.97	141.98	993.8	1135.8	0.2537	1.5684	1.8221	174.0
173.0	6.4182	0.016468	58.18	58.19	140.98	994.4	1135.4	0.2521	1.5718	1.8239	173.0
172.0	6.2736	0.016463	59.43	59.45	139.98	995.0	1135.0	0.2505	1.5753	1.8258	172.0
171.0	6.1318	0.016457	60.72	60.74	138.98	995.6	1134.6	0.2489	1.5787	1.8276	171.0
170.0	5.9926	0.016451	62.04	62.06	137.97	996.2	1134.2	0.2473	1.5822	1.8295	170.0
169.0	5.8562	0.016445	63.39	63.41	136.97	996.8	1133.8	0.2457	1.5857	1.8314	169.0
168.0	5.7223	0.016440	64.78	64.80	135.97	997.4	1133.4	0.2441	1.5892	1.8333	168.0
167.0	5.5911	0.016434	66.21	66.22	134.97	998.0	1133.0	0.2425	1.5926	1.8352	167.0
166.0	5.4623	0.016428	67.67	67.68	133.97	998.6	1132.6	0.2409	1.5961	1.8371	166.0
165.0	5.3361	0.016423	69.17	69.18	132.96	999.2	1132.2	0.2393	1.5997	1.8390	165.0
164.0	5.2124	0.016417	70.70	70.72	131.96	999.8	1131.8	0.2377	1.6032	1.8409	164.0
163.0	5.0911	0.016412	72.28	72.30	130.96	1000.4	1131.4	0.2361	1.6067	1.8428	163.0
162.0	4.9722	0.016406	73.90	73.92	129.96	1001.0	1131.0	0.2345	1.6103	1.8448	162.0
161.0	4.8556	0.016401	75.56	75.58	128.96	1001.6	1130.6	0.2329	1.6138	1.8467	161.0
160.0	4.7414	0.016395	77.27	77.29	127.96	1002.2	1130.2	0.2313	1.6174	1.8487	160.0
159.0	4.6294	0.016390	79.02	79.04	126.96	1002.8	1129.8	0.2297	1.6210	1.8506	159.0
158.0	4.5197	0.016384	80.82	80.83	125.96	1003.4	1129.4	0.2281	1.6245	1.8526	158.0
157.0	4.4122	0.016379	82.66	82.68	124.95	1004.0	1129.0	0.2264	1.6281	1.8546	157.0
156.0	4.3068	0.016374	84.56	84.57	123.95	1004.6	1128.6	0.2248	1.6318	1.8566	156.0
155.0	4.2036	0.016369	86.50	86.52	122.95	1005.2	1128.2	0.2232	1.6354	1.8586	155.0
154.0	4.1025	0.016363	88.50	88.52	121.95	1005.8	1127.7	0.2216	1.6390	1.8606	154.0
153.0	4.0035	0.016358	90.55	90.57	120.95	1006.4	1127.3	0.2199	1.6426	1.8626	153.0
152.0	3.9065	0.016353	92.66	92.68	119.95	1007.0	1126.9	0.2183	1.6463	1.8646	152.0
151.0	3.8114	0.016348	94.83	94.84	118.95	1007.6	1126.5	0.2167	1.6500	1.8666	151.0
150.0	3.7184	0.016343	97.05	97.07	117.95	1008.2	1126.1	0.2150	1.6536	1.8686	150.0
149.0	3.6273	0.016337	99.33	99.35	116.95	1008.7	1125.7	0.2134	1.6573	1.8707	149.0
148.0	3.5381	0.016332	101.68	101.70	115.95	1009.3	1125.3	0.2117	1.6610	1.8727	148.0
147.0	3.4508	0.016327	104.10	104.11	114.95	1009.9	1124.9	0.2101	1.6647	1.8748	147.0
146.0	3.3653	0.016322	106.58	106.59	113.95	1010.5	1124.5	0.2084	1.6684	1.8769	146.0
145.0	3.2816	0.016317	109.12	109.14	112.95	1011.1	1124.0	0.2068	1.6722	1.8789	145.0
144.0	3.1997	0.016312	111.74	111.76	111.95	1011.7	1123.6	0.2051	1.6759	1.8810	144.0
143.0	3.1195	0.016308	114.44	114.45	110.95	1012.3	1123.2	0.2035	1.6797	1.8831	143.0
142.0	3.0411	0.016303	117.21	117.22	109.95	1012.9	1122.8	0.2018	1.6834	1.8852	142.0
141.0	2.9643	0.016298	120.05	120.07	108.95	1013.4	1122.4	0.2001	1.6872	1.8873	141.0
140.0	2.8892	0.016293	122.98	123.00	107.95	1014.0	1122.0	0.1985	1.6910	1.8895	140.0
139.0	2.8157	0.016288	125.99	126.01	106.95	1014.6	1121.6	0.1968	1.6948	1.8916	139.0
138.0	2.7438	0.016284	129.09	129.11	105.95	1015.2	1121.1	0.1951	1.6986	1.8937	138.0
137.0	2.6735	0.016279	132.28	132.29	104.95	1015.8	1120.7	0.1935	1.7024	1.8959	137.0
136.0	2.6047	0.016274	135.55	135.57	103.95	1016.4	1120.3	0.1918	1.7063	1.8980	136.0
135.0	2.5375	0.016270	138.93	138.94	102.95	1016.9	1119.9	0.1901	1.7101	1.9002	135.0
134.0	2.4717	0.016265	142.40	142.41	101.95	1017.5	1119.5	0.1884	1.7140	1.9024	134.0
133.0	2.4074	0.016260	145.97	145.98	100.95	1018.1	1119.1	0.1867	1.7178	1.9046	133.0
132.0	2.3445	0.016256	149.64	149.66	99.95	1018.7	1118.6	0.1851	1.7217	1.9068	132.0
131.0	2.2830	0.016251	153.42	153.44	98.95	1019.3	1118.2	0.1834	1.7256	1.9090	131.0
130.0	2.2230	0.016247	157.32	157.33	97.96	1019.8	1117.8	0.1817	1.7295	1.9112	130.0
129.0	2.1642	0.016243	161.32	161.34	96.96	1020.4	1117.4	0.1800	1.7335	1.9134	129.0
128.0	2.1068	0.016238	165.45	165.47	95.96	1021.0	1117.0	0.1783	1.7374	1.9157	128.0
127.0	2.0507	0.016234	169.70	169.72	94.96	1021.6	1116.5	0.1766	1.7413	1.9179	127.0
126.0	1.9959	0.016229	174.08	174.09	93.96	1022.2	1116.1	0.1749	1.7453	1.9202	126.0
125.0	1.9424	0.016225	178.58	178.60	92.96	1022.7	1115.7	0.1732	1.7493	1.9224	125.0
124.0	1.8901	0.016221	183.23	183.24	91.96	1023.3	1115.3	0.1715	1.7533	1.9247	124.0
123.0	1.8390	0.016217	188.01	188.03	90.96	1023.9	1114.9	0.1697	1.7573	1.9270	123.0
122.0	1.7891	0.016213	192.94	192.95	89.96	1024.5	1114.4	0.1680	1.7613	1.9293	122.0
121.0	1.7403	0.016208	198.01	198.03	88.96	1025.0	1114.0	0.1663	1.7653	1.9316	121.0
120.0	1.6927	0.016204	203.25	203.26	87.97	1025.6	1113.6	0.1646	1.7693	1.9339	120.0
119.0	1.6463	0.016200	208.64	208.66	86.97	1026.2	1113.2	0.1629	1.7734	1.9362	119.0
118.0	1.6009	0.016196	214.20	214.21	85.97	1026.8	1112.7	0.1611	1.7774	1.9386	118.0
117.0	1.5566	0.016192	219.93	219.94	84.97	1027.3	1112.3	0.1594	1.7815	1.9409	117.0
116.0	1.5133	0.016188	225.84	225.85	83.97	1027.9	1111.9	0.1577	1.7856	1.9433	116.0
115.0	1.4711	0.016184	231.93	231.94	82.97	1028.5	1111.5	0.1559	1.7897	1.9457	115.0
114.0	1.4299	0.016180	238.21	238.22	81.97	1029.1	1111.0	0.1542	1.7938	1.9480	114.0
113.0	1.3898	0.016177	244.69	244.70	80.98	1029.6	1110.6	0.1525	1.7980	1.9504	113.0
112.0	1.3505	0.016173	251.37	251.38	79.98	1030.2	1110.2	0.1507	1.8021	1.9528	112.0
111.0	1.3123	0.016169	258.26	258.28	78.98	1030.8	1109.8	0.1490	1.8063	1.9552	111.0
110.0	1.2750	0.016165	265.37	265.39	77.98	1031.4	1109.3	0.1472	1.8105	1.9577	110.0
109.0	1.2385	0.016162	272.71	272.72	76.98	1031.9	1108.9	0.1455	1.8146	1.9601	109.0
108.0	1.2030	0.016158	280.28	280.30	75.98	1032.5	1108.5	0.1437	1.8188	1.9626	108.0
107.0	1.1684	0.016154	288.09	288.11	74.99	1033.1	1108.1	0.1419	1.8231	1.9650	107.0
106.0	1.1347	0.016151	296.16	296.18	73.99	1033.6	1107.6	0.1402	1.8273	1.9675	106.0
105.0	1.1017	0.016147	304.49	304.50	72.99	1034.2	1107.2	0.1384	1.8315	1.9700	105.0

Properties of Saturated Steam and Saturated Water (Temperature)

Temp. F	Press. psia	Volume, ft^3/lbm Water v_f	Evap. v_{fg}	Steam v_g	Enthalpy, Btu/lbm Water h_f	Evap. h_{fg}	Steam h_g	Entropy, Btu/lbm×F Water s_f	Evap. s_{fg}	Steam s_g	Temp. F
105.0	1.10174	0.016147	304.5	304.5	72.990	1034.2	1107.2	0.1384	1.8315	1.9700	105.0
104.0	1.06965	0.016144	313.1	313.1	71.992	1034.8	1106.8	0.1366	1.8358	1.9725	104.0
103.0	1.03838	0.016140	322.0	322.0	70.993	1035.4	1106.3	0.1349	1.8401	1.9750	103.0
102.0	1.00789	0.016137	331.1	331.1	69.995	1035.9	1105.9	0.1331	1.8444	1.9775	102.0
101.0	0.97818	0.016133	340.6	340.6	68.997	1036.5	1105.5	0.1313	1.8487	1.9800	101.0
100.0	0.94924	0.016130	350.4	350.4	67.999	1037.1	1105.1	0.1295	1.8530	1.9825	100.0
99.0	0.92103	0.016127	360.5	360.5	67.001	1037.6	1104.6	0.1278	1.8573	1.9851	99.0
98.0	0.89356	0.016123	370.9	370.9	66.003	1038.2	1104.2	0.1260	1.8617	1.9876	98.0
97.0	0.86679	0.016120	381.7	381.7	65.005	1038.8	1103.8	0.1242	1.8660	1.9902	97.0
96.0	0.84072	0.016117	392.8	392.9	64.006	1039.3	1103.3	0.1224	1.8704	1.9928	96.0
95.0	0.81534	0.016114	404.4	404.4	63.008	1039.9	1102.9	0.1206	1.8748	1.9954	95.0
94.0	0.79062	0.016111	416.3	416.3	62.010	1040.5	1102.5	0.1188	1.8792	1.9980	94.0
93.0	0.76655	0.016108	428.6	428.6	61.012	1041.0	1102.1	0.1170	1.8837	2.0006	93.0
92.0	0.74313	0.016105	441.3	441.3	60.014	1041.6	1101.6	0.1152	1.8881	2.0033	92.0
91.0	0.72032	0.016102	454.5	454.5	59.016	1042.2	1101.2	0.1134	1.8926	2.0059	91.0
90.0	0.69813	0.016099	468.1	468.1	58.018	1042.7	1100.8	0.1115	1.8970	2.0086	90.0
89.0	0.67653	0.016096	482.2	482.2	57.020	1043.3	1100.3	0.1097	1.9015	2.0112	89.0
88.0	0.65551	0.016093	496.8	496.8	56.022	1043.9	1099.9	0.1079	1.9060	2.0139	88.0
87.0	0.63507	0.016090	511.9	511.9	55.024	1044.4	1099.5	0.1061	1.9105	2.0166	87.0
86.0	0.61518	0.016087	527.5	527.5	54.026	1045.0	1099.0	0.1043	1.9151	2.0193	86.0
85.0	0.59583	0.016085	543.6	543.6	53.027	1045.6	1098.6	0.1024	1.9196	2.0221	85.0
84.0	0.57702	0.016082	560.3	560.3	52.029	1046.1	1098.2	0.1006	1.9242	2.0248	84.0
83.0	0.55872	0.016079	577.6	577.6	51.031	1046.7	1097.7	0.0988	1.9288	2.0275	83.0
82.0	0.54093	0.016077	595.5	595.6	50.033	1047.3	1097.3	0.0969	1.9334	2.0303	82.0
81.0	0.52364	0.016074	614.1	614.1	49.035	1047.8	1096.9	0.0951	1.9380	2.0331	81.0
80.0	0.05683	0.016072	633.3	633.3	48.037	1048.4	1096.4	0.0932	1.9426	2.0359	80.0
79.0	0.49049	0.016070	653.2	653.2	47.038	1049.0	1096.0	0.0914	1.9473	2.0387	79.0
78.0	0.47461	0.016067	673.8	673.9	46.040	1049.5	1095.6	0.0895	1.9520	2.0415	78.0
77.0	0.45919	0.016065	695.2	695.2	45.042	1050.1	1095.1	0.0877	1.9567	2.0443	77.0
76.0	0.44420	0.016063	717.4	717.4	44.043	1050.7	1094.7	0.0858	1.9614	2.0472	76.0
75.0	0.42964	0.016060	740.3	740.3	43.045	1051.2	1094.3	0.0839	1.9661	2.0500	75.0
74.0	0.41550	0.016058	764.1	764.1	42.046	1051.8	1093.8	0.0821	1.9708	2.0529	74.0
73.0	0.40177	0.016056	788.8	788.8	41.048	1052.4	1093.4	0.0802	1.9756	2.0558	73.0
72.0	0.38844	0.016054	814.3	814.3	40.049	1052.9	1093.0	0.0783	1.9804	2.0587	72.0
71.0	0.37549	0.016052	840.8	840.9	39.050	1053.5	1092.5	0.0764	1.9852	2.0616	71.0
70.0	0.36292	0.016050	868.3	868.4	38.052	1054.0	1092.1	0.0745	1.9900	2.0645	70.0
69.0	0.35073	0.016048	896.9	896.9	37.053	1054.6	1091.7	0.0727	1.9948	2.0675	69.0
68.0	0.33889	0.016046	926.5	926.5	36.054	1055.2	1091.2	0.0708	1.9996	2.0704	68.0
67.0	0.32740	0.016044	957.2	957.2	35.055	1055.7	1090.8	0.0689	2.0045	2.0734	67.0
66.0	0.31626	0.016043	989.0	989.1	34.056	1056.3	1090.4	0.0670	2.0094	2.0764	66.0
65.0	0.30545	0.016041	1022.1	1022.1	33.057	1056.9	1089.9	0.0651	2.0143	2.0794	65.0
64.0	0.29497	0.016039	1056.5	1056.5	32.058	1057.4	1089.5	0.0632	2.0192	2.0824	64.0
63.0	0.28480	0.016038	1092.1	1092.1	31.058	1058.0	1089.0	0.0613	2.0242	2.0854	63.0
62.0	0.27494	0.016036	1129.2	1129.2	30.059	1058.5	1088.6	0.0593	2.0291	2.0885	62.0
61.0	0.26538	0.016035	1167.6	1167.6	29.059	1059.1	1088.2	0.0574	2.0341	2.0915	61.0
60.0	0.25611	0.016033	1207.6	1207.6	28.060	1059.7	1087.7	0.0555	2.0391	2.0946	60.0
59.0	0.24713	0.016032	1249.1	1249.1	27.060	1060.2	1087.3	0.0536	2.0441	2.0977	59.0
58.0	0.23843	0.016031	1292.2	1292.2	26.060	1060.8	1086.9	0.0516	2.0491	2.1008	58.0
57.0	0.23000	0.016029	1337.0	1337.0	25.060	1061.4	1086.4	0.0497	2.0542	2.1039	57.0
56.0	0.22183	0.016028	1383.6	1383.6	24.059	1061.9	1086.0	0.0478	2.0593	2.1070	56.0
55.0	0.21392	0.016027	1432.0	1432.0	23.059	1062.5	1085.6	0.0458	2.0644	2.1102	55.0
54.0	0.20625	0.016026	1482.4	1482.4	22.058	1063.1	1085.1	0.0439	2.0695	2.1134	54.0
53.0	0.19883	0.016025	1534.7	1534.8	21.058	1063.6	1084.7	0.0419	2.0746	2.1165	53.0
52.0	0.19165	0.016024	1589.2	1589.2	20.057	1064.2	1084.2	0.0400	2.0798	2.1197	52.0
51.0	0.18469	0.016023	1645.9	1645.9	19.056	1064.7	1083.8	0.0380	2.0849	2.1230	51.0
50.0	0.17796	0.016023	1704.8	1704.8	18.054	1065.3	1083.4	0.0361	2.0901	2.1262	50.0
49.0	0.17144	0.016022	1766.2	1766.2	17.053	1065.9	1082.9	0.0341	2.0953	2.1294	49.0
48.0	0.16514	0.016021	1830.0	1830.0	16.051	1066.4	1082.5	0.0321	2.1006	2.1327	48.0
47.0	0.15904	0.016021	1896.5	1896.5	15.049	1067.0	1082.1	0.0301	2.1058	2.1360	47.0
46.0	0.15314	0.016020	1965.7	1965.7	14.047	1067.6	1081.6	0.0282	2.1111	2.1393	46.0
45.0	0.14744	0.016020	2037.7	2037.8	13.044	1068.1	1081.2	0.0262	2.1164	2.1426	45.0
44.0	0.14192	0.016019	2112.8	2112.8	12.041	1068.7	1080.7	0.0242	2.1217	2.1459	44.0
43.0	0.13659	0.016019	2191.0	2191.0	11.038	1069.3	1080.3	0.0222	2.1271	2.1493	43.0
42.0	0.13143	0.016019	2272.4	2272.4	10.035	1069.8	1079.9	0.0202	2.1325	2.1527	42.0
41.0	0.12645	0.016019	2357.3	2357.3	9.031	1070.4	1079.4	0.0182	2.1378	2.1560	41.0
40.0	0.12163	0.016019	2445.8	2445.8	8.027	1071.0	1079.0	0.0162	2.1432	2.1594	40.0
39.0	0.11698	0.016019	2538.0	2538.0	7.023	1071.5	1078.5	0.0142	2.1487	2.1629	39.0
38.0	0.11249	0.016019	2634.1	2634.2	6.018	1072.1	1078.1	0.0122	2.1541	2.1663	38.0
37.0	0.10815	0.016019	2734.4	2734.4	5.013	1072.7	1077.7	0.0101	2.1596	2.1697	37.0
36.0	0.10395	0.016020	2839.0	2839.0	4.008	1073.2	1077.2	0.0081	2.1651	2.1732	36.0
35.0	0.09991	0.016020	2948.1	2948.1	3.002	1073.8	1076.8	0.0061	2.1706	2.1767	35.0
34.0	0.09600	0.016021	3061.9	3061.9	1.996	1074.4	1076.4	0.0041	2.1762	2.1802	34.0
33.0	0.09223	0.016021	3180.7	3180.7	0.989	1074.9	1075.9	0.0020	2.1817	2.1837	33.0
32.018	0.08865	0.016022	3302.4	3302.4	0.0003	1075.5	1075.5	0.0000	2.1872	2.1872	32.018
*32.0	0.08859	0.016022	3304.7	3304.7	−0.0179	1075.5	1075.5	−0.0000	2.1873	2.1873	32.0

*The states here shown are metastable

STEAM TABLES (Continued)

Specific Heat at constant pressure of Steam and of Water

Temp. F	c_p, Btu/lbm × F															Temp. F
Press., psia	1	1.5	2	3	4	6	8	10	15	20	30	40	60	80	100	Press., psia
Sat. Water	0.998	0.998	0.999	1.000	1.000	1.002	1.003	1.004	1.007	1.010	1.014	1.019	1.026	1.033	1.039	Sat. Water
Sat. Steam	0.450	0.452	0.454	0.458	0.461	0.466	0.471	0.475	0.485	0.493	0.508	0.521	0.543	0.564	0.582	Sat. Steam
1500	0.559	0.559	0.559	0.559	0.559	0.559	0.559	0.559	0.559	0.559	0.560	0.560	0.560	0.561	0.561	1500
1480	0.557	0.557	0.557	0.557	0.557	0.557	0.557	0.558	0.558	0.558	0.558	0.558	0.559	0.559	0.559	1480
1460	0.556	0.556	0.556	0.556	0.556	0.556	0.556	0.556	0.556	0.556	0.556	0.557	0.557	0.557	0.558	1460
1440	0.554	0.554	0.554	0.554	0.554	0.554	0.554	0.554	0.554	0.554	0.555	0.555	0.555	0.556	0.556	1440
1420	0.552	0.552	0.552	0.552	0.552	0.552	0.553	0.553	0.553	0.553	0.553	0.553	0.554	0.554	0.555	1420
1400	0.551	0.551	0.551	0.551	0.551	0.551	0.551	0.551	0.551	0.551	0.551	0.552	0.552	0.553	0.553	1400
1380	0.549	0.549	0.549	0.549	0.549	0.549	0.549	0.549	0.549	0.549	0.550	0.550	0.550	0.551	0.551	1380
1360	0.547	0.547	0.547	0.547	0.547	0.547	0.547	0.547	0.548	0.548	0.548	0.548	0.549	0.549	0.550	1360
1340	0.546	0.546	0.546	0.546	0.546	0.546	0.546	0.546	0.546	0.546	0.546	0.547	0.547	0.548	0.548	1340
1320	0.544	0.544	0.544	0.544	0.544	0.544	0.544	0.544	0.544	0.544	0.545	0.545	0.545	0.546	0.546	1320
1300	0.542	0.542	0.542	0.542	0.542	0.542	0.542	0.542	0.542	0.543	0.543	0.543	0.544	0.544	0.545	1300
1280	0.540	0.540	0.540	0.540	0.540	0.540	0.540	0.541	0.541	0.541	0.541	0.541	0.542	0.543	0.543	1280
1260	0.538	0.539	0.539	0.539	0.539	0.539	0.539	0.539	0.539	0.539	0.539	0.540	0.540	0.541	0.541	1260
1240	0.537	0.537	0.537	0.537	0.537	0.537	0.537	0.537	0.537	0.537	0.538	0.538	0.539	0.539	0.540	1240
1220	0.535	0.535	0.535	0.535	0.535	0.535	0.535	0.535	0.535	0.536	0.536	0.536	0.537	0.537	0.538	1220
1200	0.533	0.533	0.533	0.533	0.533	0.533	0.533	0.533	0.534	0.534	0.534	0.534	0.535	0.536	0.536	1200
1180	0.531	0.531	0.531	0.531	0.531	0.531	0.532	0.532	0.532	0.532	0.532	0.533	0.533	0.534	0.535	1180
1160	0.529	0.529	0.530	0.530	0.530	0.530	0.530	0.530	0.530	0.530	0.530	0.531	0.532	0.532	0.533	1160
1140	0.528	0.528	0.528	0.528	0.528	0.528	0.528	0.528	0.528	0.528	0.529	0.529	0.530	0.531	0.531	1140
1120	0.526	0.526	0.526	0.526	0.526	0.526	0.526	0.526	0.526	0.527	0.527	0.527	0.528	0.529	0.530	1120
1100	0.524	0.524	0.524	0.524	0.524	0.524	0.524	0.524	0.525	0.525	0.525	0.526	0.526	0.527	0.528	1100
1080	0.522	0.522	0.522	0.522	0.522	0.522	0.522	0.523	0.523	0.523	0.523	0.524	0.525	0.525	0.526	1080
1060	0.520	0.520	0.520	0.520	0.520	0.521	0.521	0.521	0.521	0.521	0.522	0.522	0.523	0.524	0.524	1060
1040	0.518	0.519	0.519	0.519	0.519	0.519	0.519	0.519	0.519	0.519	0.520	0.520	0.521	0.522	0.523	1040
1020	0.517	0.517	0.517	0.517	0.517	0.517	0.517	0.517	0.517	0.518	0.518	0.518	0.519	0.520	0.521	1020
1000	0.515	0.515	0.515	0.515	0.515	0.515	0.515	0.515	0.515	0.516	0.516	0.517	0.518	0.519	0.519	1000
980	0.513	0.513	0.513	0.513	0.513	0.513	0.513	0.513	0.514	0.514	0.514	0.515	0.516	0.517	0.518	980
960	0.511	0.511	0.511	0.511	0.511	0.511	0.512	0.512	0.512	0.512	0.513	0.513	0.514	0.515	0.516	960
940	0.509	0.509	0.509	0.509	0.509	0.510	0.510	0.510	0.510	0.510	0.511	0.511	0.512	0.514	0.515	940
920	0.507	0.508	0.508	0.508	0.508	0.508	0.508	0.508	0.508	0.509	0.509	0.510	0.511	0.512	0.513	920
900	0.506	0.506	0.506	0.506	0.506	0.506	0.506	0.506	0.506	0.507	0.507	0.508	0.509	0.510	0.512	900
880	0.504	0.504	0.504	0.504	0.504	0.504	0.504	0.504	0.505	0.505	0.506	0.506	0.508	0.509	0.510	880
860	0.502	0.502	0.502	0.502	0.502	0.502	0.503	0.503	0.503	0.503	0.504	0.505	0.506	0.507	0.509	860
840	0.500	0.500	0.500	0.500	0.500	0.501	0.501	0.501	0.501	0.502	0.502	0.503	0.504	0.506	0.507	840
820	0.498	0.498	0.499	0.499	0.499	0.499	0.499	0.499	0.499	0.500	0.501	0.501	0.503	0.504	0.506	820
800	0.497	0.497	0.497	0.497	0.497	0.497	0.497	0.497	0.498	0.498	0.499	0.500	0.501	0.503	0.505	800
780	0.495	0.495	0.495	0.495	0.495	0.495	0.495	0.496	0.496	0.496	0.497	0.498	0.500	0.502	0.503	780
760	0.493	0.493	0.493	0.493	0.493	0.494	0.494	0.494	0.494	0.495	0.496	0.497	0.499	0.500	0.502	760
740	0.491	0.491	0.491	0.492	0.492	0.492	0.492	0.492	0.493	0.493	0.494	0.495	0.497	0.499	0.501	740
720	0.490	0.490	0.490	0.490	0.490	0.490	0.490	0.491	0.491	0.492	0.493	0.494	0.496	0.498	0.500	720
700	0.488	0.488	0.488	0.488	0.488	0.488	0.489	0.489	0.490	0.490	0.491	0.492	0.495	0.497	0.500	700
680	0.486	0.486	0.486	0.486	0.487	0.487	0.487	0.487	0.488	0.489	0.490	0.491	0.494	0.496	0.499	680
660	0.484	0.485	0.485	0.485	0.485	0.485	0.485	0.486	0.486	0.487	0.489	0.490	0.493	0.496	0.499	660
640	0.483	0.483	0.483	0.483	0.483	0.484	0.484	0.484	0.485	0.486	0.487	0.489	0.492	0.495	0.499	640
620	0.481	0.481	0.481	0.481	0.482	0.482	0.482	0.483	0.483	0.484	0.486	0.488	0.491	0.495	0.499	620
600	0.479	0.480	0.480	0.480	0.480	0.480	0.481	0.481	0.482	0.483	0.485	0.487	0.491	0.495	0.499	600
580	0.478	0.478	0.478	0.478	0.478	0.479	0.479	0.480	0.481	0.482	0.484	0.486	0.491	0.495	0.500	580
560	0.476	0.476	0.476	0.477	0.477	0.477	0.478	0.478	0.479	0.481	0.483	0.485	0.490	0.496	0.501	560
540	0.475	0.475	0.475	0.475	0.475	0.476	0.476	0.477	0.478	0.480	0.482	0.485	0.491	0.497	0.503	540
520	0.473	0.473	0.473	0.474	0.474	0.475	0.475	0.476	0.477	0.479	0.482	0.485	0.491	0.498	0.505	520
500	0.472	0.472	0.472	0.472	0.473	0.473	0.474	0.475	0.476	0.478	0.481	0.485	0.492	0.500	0.508	500
480	0.470	0.470	0.470	0.471	0.471	0.472	0.473	0.473	0.475	0.477	0.481	0.485	0.493	0.502	0.511	480
460	0.469	0.469	0.469	0.469	0.470	0.471	0.472	0.472	0.475	0.477	0.481	0.486	0.495	0.505	0.516	460
440	0.467	0.467	0.468	0.468	0.469	0.470	0.470	0.471	0.474	0.476	0.481	0.487	0.498	0.509	0.522	440
420	0.466	0.466	0.466	0.467	0.467	0.468	0.470	0.471	0.473	0.476	0.482	0.488	0.501	0.514	0.528	420
400	0.464	0.465	0.465	0.466	0.466	0.467	0.469	0.470	0.473	0.476	0.483	0.490	0.504	0.520	0.536	400
380	0.463	0.463	0.464	0.464	0.465	0.466	0.468	0.469	0.473	0.477	0.484	0.492	0.509	0.527	0.546	380
360	0.462	0.462	0.462	0.463	0.464	0.466	0.467	0.469	0.473	0.477	0.486	0.495	0.515	0.536	0.558	360
340	0.460	0.461	0.461	0.462	0.463	0.465	0.467	0.469	0.473	0.478	0.488	0.499	0.521	0.546	0.572	340
320	0.459	0.460	0.460	0.461	0.462	0.464	0.467	0.469	0.474	0.480	0.491	0.504	0.530	0.558	1.036	320
300	0.458	0.459	0.459	0.460	0.462	0.464	0.466	0.469	0.475	0.482	0.495	0.509	0.539	1.029	1.029	300
280	0.457	0.458	0.458	0.460	0.461	0.464	0.467	0.469	0.477	0.484	0.500	0.516	1.022	1.022	1.022	280
260	0.456	0.457	0.457	0.459	0.461	0.464	0.467	0.470	0.478	0.487	0.505	1.017	1.017	1.017	1.016	260
240	0.455	0.456	0.457	0.458	0.460	0.464	0.468	0.471	0.481	0.491	1.012	1.012	1.012	1.012	1.012	240
220	0.454	0.455	0.456	0.458	0.460	0.464	0.468	0.473	0.484	1.008	1.008	1.008	1.008	1.008	1.008	220
200	0.453	0.454	0.455	0.458	0.460	0.465	0.470	0.475	1.005	1.005	1.005	1.005	1.005	1.005	1.005	200
180	0.452	0.454	0.455	0.458	0.460	0.466	1.003	1.003	1.003	1.003	1.003	1.003	1.003	1.002	1.002	180
160	0.451	0.453	0.455	0.458	0.461	1.001	1.001	1.001	1.001	1.001	1.001	1.001	1.001	1.001	1.001	160
140	0.451	0.453	0.454	1.000	1.000	1.000	1.000	1.000	0.999	0.999	0.999	0.999	0.999	0.999	0.999	140
120	0.450	0.452	0.999	0.999	0.999	0.999	0.999	0.999	0.999	0.999	0.998	0.998	0.998	0.998	0.998	120
100	0.998	0.998	0.998	0.998	0.998	0.998	0.998	0.998	0.998	0.998	0.998	0.998	0.998	0.998	0.998	100
80	0.998	0.998	0.998	0.998	0.998	0.998	0.998	0.998	0.998	0.998	0.998	0.998	0.998	0.998	0.998	80
60	1.000	1.000	1.000	1.000	1.000	1.000	1.000	1.000	1.000	1.000	1.000	1.000	0.999	0.999	0.999	60
40	1.004	1.004	1.004	1.004	1.004	1.004	1.004	1.004	1.004	1.004	1.004	1.004	1.004	1.004	1.003	40
32	1.007	1.007	1.007	1.007	1.007	1.007	1.007	1.007	1.007	1.007	1.007	1.007	1.007	1.007	1.006	32

STEAM TABLES (Continued)

Specific Heat at constant pressure of Steam and of Water

Temp. F / Press., psia	150	200	300	400	600	800	1000	1500	2000	3000	4000	6000	8000	10000	15000	Temp. F / Press., psia
	c_p, Btu/lbm × F															
Sat. Water	1.054	1.067	1.093	1.118	1.168	1.224	1.286	1.492	1.841	7.646	—	—	—	—	—	Sat. Water
Sat. Steam	0.624	0.661	0.729	0.792	0.915	1.046	1.191	1.667	2.557	13.66	—	—	—	—	—	Sat. Steam
1500	0.562	0.563	0.565	0.567	0.571	0.576	0.580	0.590	0.601	0.623	0.645	0.691	0.737	0.780	0.868	1500
1480	0.561	0.562	0.564	0.566	0.570	0.575	0.579	0.590	0.601	0.623	0.647	0.694	0.742	0.786	0.878	1480
1460	0.559	0.560	0.562	0.565	0.569	0.573	0.578	0.589	0.601	0.624	0.648	0.698	0.747	0.793	0.888	1460
1440	0.557	0.559	0.561	0.563	0.568	0.572	0.577	0.589	0.600	0.625	0.650	0.701	0.753	0.800	0.900	1440
1420	0.556	0.557	0.559	0.562	0.566	0.571	0.576	0.588	0.600	0.625	0.651	0.705	0.759	0.808	0.909	1420
1400	0.554	0.555	0.558	0.560	0.565	0.570	0.575	0.587	0.600	0.626	0.653	0.709	0.765	0.817	0.926	1400
1380	0.553	0.554	0.556	0.559	0.564	0.569	0.574	0.587	0.600	0.627	0.655	0.714	0.773	0.827	0.939	1380
1360	0.551	0.552	0.555	0.558	0.563	0.568	0.573	0.586	0.600	0.628	0.657	0.719	0.781	0.838	0.953	1360
1340	0.549	0.551	0.553	0.556	0.561	0.567	0.572	0.586	0.600	0.629	0.660	0.725	0.790	0.850	0.968	1340
1320	0.548	0.549	0.552	0.555	0.560	0.566	0.571	0.585	0.600	0.630	0.663	0.731	0.800	0.864	0.983	1320
1300	0.546	0.548	0.550	0.553	0.559	0.565	0.570	0.585	0.600	0.632	0.666	0.738	0.811	0.879	0.998	1300
1280	0.545	0.546	0.549	0.552	0.558	0.564	0.570	0.585	0.600	0.634	0.669	0.746	0.824	0.897	1.014	1280
1260	0.543	0.544	0.547	0.550	0.556	0.563	0.569	0.585	0.601	0.636	0.673	0.755	0.838	0.918	1.033	1260
1240	0.541	0.543	0.546	0.549	0.555	0.562	0.568	0.584	0.601	0.638	0.678	0.765	0.855	0.942	1.053	1240
1220	0.540	0.541	0.544	0.548	0.554	0.561	0.567	0.584	0.602	0.641	0.683	0.777	0.875	0.969	1.072	1220
1200	0.538	0.540	0.543	0.546	0.553	0.560	0.567	0.584	0.603	0.644	0.689	0.790	0.897	1.000	1.095	1200
1180	0.536	0.538	0.541	0.545	0.552	0.559	0.566	0.584	0.604	0.647	0.696	0.805	0.922	1.033	1.117	1180
1160	0.535	0.536	0.540	0.544	0.551	0.558	0.565	0.585	0.606	0.652	0.704	0.823	0.952	1.070	1.143	1160
1140	0.533	0.535	0.539	0.542	0.550	0.557	0.565	0.585	0.607	0.656	0.713	0.843	0.986	1.107	1.167	1140
1120	0.531	0.533	0.537	0.541	0.549	0.557	0.565	0.586	0.609	0.662	0.723	0.866	1.025	1.149	1.190	1120
1100	0.530	0.532	0.536	0.540	0.548	0.556	0.564	0.587	0.612	0.668	0.735	0.893	1.070	1.193	1.220	1100
1080	0.528	0.530	0.534	0.538	0.547	0.555	0.564	0.588	0.615	0.676	0.749	0.924	1.120	1.242	1.240	1080
1060	0.527	0.529	0.533	0.537	0.546	0.555	0.564	0.590	0.618	0.685	0.765	0.960	1.176	1.295	1.260	1060
1040	0.525	0.527	0.532	0.536	0.545	0.555	0.565	0.592	0.622	0.695	0.783	1.002	1.238	1.351	1.282	1040
1020	0.523	0.526	0.530	0.535	0.545	0.555	0.565	0.594	0.627	0.707	0.804	1.051	1.306	1.399	1.298	1020
1000	0.522	0.524	0.529	0.534	0.544	0.555	0.566	0.597	0.633	0.721	0.829	1.110	1.382	1.471	1.306	1000
980	0.520	0.523	0.528	0.533	0.544	0.555	0.567	0.601	0.640	0.737	0.858	1.180	1.475	1.531	1.312	980
960	0.519	0.521	0.527	0.532	0.543	0.556	0.568	0.605	0.648	0.756	0.893	1.267	1.598	1.595	1.310	960
940	0.517	0.520	0.526	0.531	0.543	0.556	0.570	0.610	0.658	0.778	0.934	1.376	1.708	1.639	1.299	940
920	0.516	0.519	0.525	0.531	0.544	0.558	0.573	0.617	0.669	0.803	0.984	1.520	1.819	1.667	1.281	920
900	0.515	0.518	0.524	0.530	0.544	0.559	0.576	0.624	0.683	0.834	1.048	1.716	1.932	1.660	1.259	900
880	0.513	0.516	0.523	0.530	0.545	0.561	0.580	0.633	0.699	0.872	1.130	1.993	2.000	1.633	1.232	880
860	0.512	0.515	0.523	0.530	0.546	0.564	0.584	0.644	0.718	0.918	1.240	2.316	2.019	1.593	1.212	860
840	0.511	0.514	0.522	0.530	0.548	0.568	0.590	0.657	0.740	0.977	1.395	2.653	1.978	1.547	1.192	840
820	0.510	0.514	0.522	0.531	0.550	0.572	0.597	0.672	0.767	1.054	1.620	2.886	1.888	1.503	1.175	820
800	0.509	0.513	0.522	0.532	0.553	0.577	0.605	0.690	0.800	1.160	1.967	2.872	1.768	1.459	1.157	800
780	0.508	0.513	0.522	0.533	0.557	0.584	0.615	0.712	0.840	1.312	2.550	2.547	1.670	1.416	1.142	780
760	0.507	0.512	0.523	0.535	0.561	0.592	0.628	0.738	0.892	1.542	4.462	2.156	1.576	1.370	1.126	760
740	0.507	0.512	0.524	0.537	0.567	0.602	0.642	0.770	0.960	1.913	8.119	1.886	1.493	1.332	1.114	740
720	0.506	0.512	0.525	0.540	0.574	0.613	0.660	0.811	1.052	2.584	3.458	1.696	1.421	1.290	1.100	720
700	0.506	0.513	0.528	0.544	0.582	0.627	0.681	0.861	1.181	6.145°	2.237	1.557	1.358	1.250	1.089	700
680	0.506	0.514	0.530	0.549	0.592	0.644	0.707	0.927	1.365	2.469	1.789	1.450	1.303	1.217	1.079	680
660	0.507	0.515	0.534	0.555	0.604	0.665	0.738	1.015	1.639	1.851	1.587	1.369	1.256	1.187	1.071	660
640	0.507	0.517	0.538	0.562	0.619	0.690	0.777	1.135	2.219	1.601	1.454	1.303	1.216	1.157	1.063	640
620	0.509	0.519	0.543	0.571	0.637	0.720	0.826	1.308	1.614	1.455	1.362	1.252	1.184	1.136	1.056	620
600	0.510	0.522	0.550	0.582	0.659	0.757	0.888	1.586	1.453	1.358	1.295	1.211	1.157	1.118	1.052	600
580	0.513	0.526	0.558	0.595	0.685	0.804	0.969	1.393	1.351	1.289	1.243	1.178	1.134	1.102	1.046	580
560	0.516	0.531	0.568	0.611	0.717	0.862	1.079	1.309	1.281	1.237	1.202	1.151	1.115	1.087	1.039	560
540	0.519	0.538	0.580	0.630	0.756	0.937	1.272	1.249	1.229	1.196	1.169	1.128	1.098	1.074	1.031	540
520	0.524	0.545	0.594	0.653	0.804	1.035	1.221	1.204	1.189	1.164	1.142	1.109	1.083	1.062	1.024	520
500	0.530	0.554	0.611	0.680	0.865	1.187	1.181	1.169	1.157	1.137	1.120	1.092	1.069	1.051	1.017	500
480	0.537	0.565	0.632	0.714	1.159	1.154	1.150	1.140	1.131	1.115	1.101	1.077	1.057	1.041	1.010	480
460	0.545	0.578	0.657	0.755	1.132	1.128	1.125	1.117	1.110	1.096	1.084	1.064	1.047	1.033	1.004	460
440	0.556	0.594	0.687	1.113	1.110	1.107	1.104	1.098	1.092	1.080	1.070	1.052	1.038	1.025	0.999	440
420	0.568	0.614	0.724	1.094	1.091	1.089	1.087	1.081	1.076	1.067	1.058	1.042	1.029	1.018	0.994	420
400	0.583	0.636	1.079	1.078	1.076	1.074	1.072	1.067	1.063	1.055	1.047	1.034	1.022	1.011	0.990	400
380	0.601	1.066	1.065	1.065	1.063	1.061	1.059	1.056	1.052	1.044	1.038	1.026	1.015	1.006	0.986	380
360	0.622	1.054	1.054	1.053	1.052	1.050	1.049	1.045	1.042	1.036	1.030	1.019	1.009	1.001	0.982	360
340	1.045	1.044	1.044	1.043	1.042	1.040	1.039	1.036	1.033	1.028	1.022	1.013	1.004	0.996	0.979	340
320	1.036	1.036	1.035	1.034	1.033	1.032	1.031	1.028	1.026	1.021	1.016	1.007	0.999	0.992	0.976	320
300	1.028	1.028	1.028	1.027	1.026	1.025	1.024	1.022	1.019	1.015	1.010	1.002	0.995	0.988	0.973	300
280	1.022	1.022	1.021	1.021	1.020	1.019	1.018	1.016	1.014	1.009	1.005	0.998	0.991	0.985	0.971	280
260	1.016	1.016	1.016	1.015	1.014	1.013	1.013	1.011	1.009	1.005	1.001	0.994	0.988	0.982	0.968	260
240	1.012	1.011	1.011	1.011	1.010	1.009	1.008	1.006	1.004	1.001	0.997	0.991	0.985	0.979	0.966	240
220	1.008	1.008	1.007	1.007	1.006	1.005	1.005	1.003	1.001	0.998	0.994	0.988	0.982	0.977	0.964	220
200	1.005	1.004	1.004	1.004	1.003	1.002	1.002	1.000	0.998	0.995	0.992	0.986	0.980	0.975	0.963	200
180	1.002	1.002	1.002	1.001	1.001	1.000	0.999	0.998	0.996	0.993	0.989	0.983	0.978	0.973	0.961	180
160	1.000	1.000	1.000	0.999	0.999	0.998	0.997	0.996	0.994	0.991	0.987	0.981	0.976	0.971	0.959	160
140	0.999	0.999	0.998	0.998	0.997	0.997	0.996	0.994	0.992	0.989	0.986	0.980	0.974	0.969	0.958	140
120	0.998	0.998	0.997	0.997	0.996	0.996	0.995	0.993	0.991	0.988	0.984	0.978	0.972	0.967	0.957	120
100	0.997	0.997	0.997	0.996	0.996	0.995	0.994	0.992	0.990	0.986	0.983	0.976	0.970	0.965	0.955	100
80	0.998	0.997	0.997	0.996	0.995	0.994	0.994	0.991	0.989	0.985	0.981	0.974	0.968	0.962	0.951	80
60	0.999	0.999	0.998	0.997	0.996	0.995	0.994	0.991	0.989	0.984	0.979	0.970	0.963	0.956	0.942	60
40	1.003	1.003	1.002	1.001	1.000	0.998	0.997	0.993	0.989	0.983	0.976	0.965	0.954	0.945	0.920	40
32	1.006	1.006	1.005	1.004	1.002	1.000	0.999	0.994	0.990	0.983	0.975	0.962	0.949	0.937	0.904	32

°Critical point.

STEAM TABLES (Continued)

Thermal Conductivity of Steam and Water

Temp. F	k, (Btu/hr × ft. × F) × 10^3												
Press., psia	1	2	5	10	20	50	100	200	500	1000	2000	5000	7500
Sat. Water	364.0	373.1	383.8	390.4	395.2	397.4	394.7	386.2	361.7	327.6	271.8	—	—
Sat. Steam	11.6	12.2	13.0	13.8	14.8	16.6	18.4	21.1	27.2	36.5	61.3	—	—
1500	63.7	63.7	63.7	63.7	63.7	63.8	64.0	64.3	65.4	67.1	70.7	82.0	92.2
1450	61.4	61.4	61.5	61.5	61.5	61.6	61.8	62.1	63.2	64.9	68.5	80.1	90.6
1400	59.2	59.2	59.2	59.2	59.3	59.4	59.6	59.9	60.9	62.7	66.3	78.2	89.2
1350	57.0	57.0	57.0	57.0	57.1	57.2	57.3	57.7	58.7	60.5	64.2	76.3	87.9
1300	54.8	54.8	54.8	54.8	54.8	54.9	55.1	55.5	56.5	58.3	62.0	74.6	86.9
1250	52.6	52.6	52.6	52.6	52.6	52.7	52.9	53.2	54.3	56.1	59.9	73.0	86.3
1200	50.4	50.4	50.4	50.4	50.4	50.5	50.7	51.0	52.1	53.9	57.8	71.6	86.2
1150	48.2	48.2	48.2	48.2	48.2	48.3	48.5	48.9	49.9	51.8	55.7	70.5	87.0
1100	46.0	46.0	46.0	46.0	46.1	46.2	46.3	46.7	47.8	49.6	53.7	69.8	89.0
1050	43.9	43.9	43.9	43.9	43.9	44.0	44.2	44.6	45.6	47.5	51.8	69.7	93.4
1000	41.7	41.7	41.8	41.8	41.8	41.9	42.1	42.4	43.5	45.5	50.0	70.7	102.9
950	39.6	39.6	39.7	39.7	39.7	39.8	40.0	40.3	41.4	43.5	48.3	73.5	115.5
900	37.6	37.6	37.6	37.6	37.6	37.7	37.9	38.3	39.4	41.5	46.8	80.2	138.7
850	35.5	35.6	35.6	35.6	35.6	35.7	35.9	36.3	37.4	39.7	45.6	96.7	178.8
800	33.6	33.6	33.6	33.6	33.6	33.7	33.9	34.3	35.5	37.9	44.9	129.6	223.2
750	31.6	31.6	31.6	31.6	31.7	31.8	32.0	32.3	33.6	36.3	45.2	202.5	258.3
700	29.7	29.7	29.7	29.7	29.8	29.9	30.1	30.4	31.8	35.0	47.5◎	262.8	295.1
650	27.8	27.8	27.9	27.9	27.9	28.0	28.2	28.6	30.1	34.1	55.7	304.3	326.7
600	26.0	26.0	26.1	26.1	26.1	26.2	26.4	26.9	28.7	34.1	301.9	333.7	349.3
550	24.3	24.3	24.3	24.3	24.4	24.5	24.7	25.2	27.5	36.1	333.7	356.1	368.0
500	22.6	22.6	22.6	22.6	22.7	22.8	23.0	23.6	26.9	350.8	357.4	373.8	383.6
450	21.0	21.0	21.0	21.0	21.0	21.2	21.4	22.3	368.1	370.6	375.3	387.9	396.5
400	19.4	19.4	19.4	19.4	19.5	19.6	20.0	21.3	383.0	384.9	388.5	398.6	406.4
350	17.9	17.9	17.9	17.9	18.0	18.2	18.8	392.0	392.9	394.4	397.4	406.1	413.2
300	16.5	16.5	16.5	16.5	16.6	16.9	396.9	397.2	398.0	399.3	402.0	409.9	416.4
250	15.1	15.1	15.1	15.2	15.3	396.9	397.0	397.3	398.1	399.4	402.1	409.7	415.8
200	13.8	13.8	13.9	14.0	391.6	391.6	391.8	392.1	393.0	394.4	397.2	404.9	410.6
150	12.7	12.7	380.5	380.5	380.6	380.7	380.8	381.1	382.1	383.7	386.7	394.7	400.3
100	363.3	363.3	363.3	363.3	363.3	363.4	363.6	363.9	365.0	366.6	369.8	378.3	384.1
50	339.1	339.1	339.1	339.1	339.2	339.3	339.4	339.8	340.8	342.5	345.7	354.6	361.0
32	328.6	328.6	328.6	328.6	328.6	328.7	328.9	329.2	330.3	331.9	335.1	344.1	350.8

◎ Critical point.

Ammonia, NH_3

Temp. °F	Abs. press. sat. vap.		Heat content abv. −40°F BTU/lb.		Ht. of vaporiz. BTU/lb.	Heat content abv. −40°C g-cal./g		Ht. of vaporiz. g-cal./g	Temp. °C
	lb./in.²	kg/cm²	Liq.	Vap.		Liq.	Vap.		
−60	5.55	0.390	−21.2	589.6	610.8	−11.8	327.6	339.3	−51.11
−58	5.93	.417	−19.1	590.4	609.5	−10.6	328.0	338.6	−50.00
−56	6.33	.445	−17.0	591.2	608.2	− 9.44	328.4	337.9	−48.89
−54	6.75	.475	−14.8	592.1	606.9	− 8.22	328.9	337.2	−47.78
−52	7.20	.506	−12.7	592.9	605.6	− 7.06	329.4	336.4	−46.67
−50	7.67	0.539	−10.6	593.7	604.3	− 5.89	329.8	335.7	−45.56
−48	8.16	.574	− 8.5	594.4	602.9	− 4.7	330.2	334.9	−44.44
−46	8.68	.610	− 6.4	595.2	601.6	− 3.6	330.7	334.2	−43.33
−44	9.23	.649	− 4.3	596.0	600.3	− 2.4	331.1	333.5	−42.22
−42	9.81	.690	− 2.1	596.8	598.9	− 1.2	331.6	332.7	−41.11
−40	10.41	0.7319	0.0	597.6	597.6	0.0	332.0	332.0	−40.00
−38	11.04	.7762	+ 2.1	598.3	596.2	+ 1.2	332.4	331.2	−38.89
−36	11.71	.8233	4.3	599.1	594.8	2.4	332.8	330.4	−37.78
−34	12.41	.8725	6.4	599.9	593.5	3.6	333.3	329.7	−36.67
−32	13.14	.9238	8.5	600.6	592.1	4.7	333.7	328.9	−35.56
−30	13.90	0.9773	10.7	601.4	590.7	5.94	334.1	328.2	−34.44
−28	14.71	1.034	12.8	602.1	589.3	7.11	334.5	327.4	−33.33
−26	15.55	1.093	14.9	602.8	587.9	8.28	334.9	326.6	−32.22
−24	16.42	1.154	17.1	603.6	586.5	9.50	335.3	325.8	−31.11
−22	17.34	1.219	19.2	604.3	585.1	10.7	335.7	325.1	−30.00
−20	18.30	1.287	21.4	605.0	583.6	11.9	336.1	324.2	−28.89
−18	19.30	1.357	23.5	605.7	582.2	13.1	336.5	323.4	−27.78
−16	20.34	1.430	25.6	606.4	580.8	14.2	336.9	322.7	−26.67
−14	21.43	1.507	27.8	607.1	579.3	15.4	337.3	321.8	−25.56
−12	22.56	1.586	30.0	607.8	577.8	16.7	337.7	321.0	−24.44
−10	23.74	1.669	32.1	608.5	576.4	17.8	338.1	320.2	−23.33
− 8	24.97	1.756	34.3	609.2	574.9	19.1	338.4	319.4	−22.22
− 6	26.26	1.846	36.4	609.8	573.4	20.2	338.8	318.6	−21.11
− 4	27.59	1.940	38.6	610.5	571.9	21.4	339.2	317.7	−20.00
− 2	28.98	2.037	40.7	611.1	570.4	22.6	339.5	316.9	−18.89
0	30.42	2.139	42.9	611.8	568.9	23.8	339.9	316.1	−17.78
2	31.92	2.244	45.1	612.4	567.3	25.1	340.2	315.2	−16.67
4	33.47	2.353	47.2	613.0	565.8	26.2	340.6	314.3	−15.56
6	35.09	2.467	49.4	613.6	564.2	27.4	340.9	313.4	−14.44
8	36.77	2.585	51.6	614.3	562.7	28.7	341.3	312.6	−13.33
10	38.51	2.708	53.8	614.9	561.1	29.9	341.6	311.7	−12.22
12	40.31	2.834	56.0	615.5	559.5	31.1	341.9	310.8	−11.11
14	42.18	2.966	58.2	616.1	557.9	32.3	342.3	309.9	−10.00
16	44.12	3.102	60.3	616.6	556.3	33.5	342.6	309.1	− 8.89
18	46.13	3.243	62.5	617.2	554.7	34.7	342.9	308.2	− 7.78
20	48.21	3.390	64.7	617.8	553.1	35.9	343.2	307.3	− 6.67
22	50.36	3.541	66.9	618.3	551.4	37.2	343.5	306.3	− 5.56
24	52.59	3.697	69.1	618.9	549.8	38.4	343.8	305.4	− 4.44
26	54.90	3.860	71.3	619.4	548.1	39.6	344.1	304.5	− 3.33
28	57.28	4.027	73.5	619.9	546.4	40.8	344.4	303.6	− 2.22

Ammonia, NH_3

Temp. °F	Spec. vol. sat. vap.		Density sat. vap.		Dens. liq. lb./ft.³	Entropy from −40°F BTU/lb./°F		Temp. °C
	ft.³/lb.	m³/kg	lb./ft.³	kg/m³		Liq.	Vap.	
−60	44.73	2.792	0.02235	0.3580	43.91	−0.0517	1.4769	−51.11
−58	42.05	2.625	.02378	.3809		− .0464	1.4713	−50.00
−56	39.56	2.470	.02528	.4049		− .0412	1.4658	−48.89
−54	37.24	2.325	.02685	.4301		− .0360	1.4604	−47.78
−52	35.09	2.191	.02850	.4565		− .0307	1.4551	−46.67
−50	33.08	2.065	0.03023	0.4842	43.49	−0.0256	1.4497	−45.56
−48	31.20	1.948	.03205	.5134		− .0204	1.4445	−44.44
−46	29.45	1.839	.03395	.5438		− .0153	1.4393	−43.33
−44	27.82	1.737	.03595	.5758		− .0102	1.4342	−42.22
−42	26.29	1.641	.03804	.6093		− .0051	1.4292	−41.11
−40	24.86	1.552	0.04022	0.6442	43.08	0.0000	1.4242	−40.00
−38	23.53	1.469	.04251	.6809		.0051	1.4193	−38.89
−36	22.27	1.390	.04489	.7190		.0101	1.4144	−37.78
−34	21.10	1.317	.04739	.7591		.0151	1.4096	−36.67
−32	20.00	1.249	.04999	.8007		.0201	1.4048	−35.56
−30	18.97	1.184	0.05271	0.8443	42.65	0.0250	1.4001	−34.44
−28	18.00	1.124	.05555	.8898		.0300	1.3955	−33.33
−26	17.09	1.067	.05850	.9371		.0350	1.3909	−32.22
−24	16.24	1.014	.06158	.9864		.0399	1.3863	−31.11
−22	15.43	0.9633	.06479	1.038		.0448	1.3818	−30.00
−20	14.68	0.9164	0.06813	1.091	42.22	0.0497	1.3774	−28.89
−18	13.97	.8721	.07161	1.147		.0545	1.3729	−27.78
−16	13.29	.8297	.07522	1.205		.0594	1.3686	−26.67
−14	12.66	.7903	.07898	1.265		.0642	1.3643	−25.56
−12	12.06	.7529	.08289	1.328		.0690	1.3600	−24.44
−10	11.50	0.7179	0.08695	1.393	41.78	0.0738	1.3558	−23.33
− 8	10.97	.6848	.09117	1.460		.0786	1.3516	−22.22
− 6	10.47	.6536	.09555	1.531		.0833	1.3474	−21.11
− 4	9.991	.6237	.1001	1.603		.0880	1.3433	−20.00
− 2	9.541	.5956	.1048	1.679		.0928	1.3393	−18.89
0	9.116	0.5691	0.1097	1.757	41.34	0.0975	1.3352	−17.78
2	8.714	.5440	.1148	1.839		.1022	1.3312	−16.67
4	8.333	.5202	.1200	1.922		.1069	1.3273	−15.56
6	7.971	.4976	.1254	2.009		.1115	1.3234	−14.44
8	7.629	.4763	.1311	2.100		.1162	1.3195	−13.33
10	7.304	0.4560	0.1369	2.193	40.89	0.1208	1.3157	−12.22
12	6.996	.4367	.1429	2.289		.1254	1.3118	−11.11
14	6.703	.4185	.1492	2.390		.1300	1.3081	−10.00
16	6.425	.4011	.1556	2.492		.1346	1.3043	− 8.89
18	6.161	.3846	.1623	2.600		.1392	1.3006	− 7.78
20	5.910	0.3690	0.1692	2.710	40.43	0.1437	1.2969	− 6.67
22	5.671	.3540	.1763	2.824		.1483	1.2933	− 5.56
24	5.443	.3398	.1837	2.943		.1528	1.2897	− 4.44
26	5.227	.3263	.1913	3.064		.1573	1.2861	− 3.33
28	5.021	.3135	.1992	3.191		.1618	1.2825	− 2.22

Ammonia, NH_3 (Continued)

Temp. °F	Abs. press. sat. vap. lb./in.²	Abs. press. sat. vap. kg/cm²	Heat content abv. −40°F BTU/lb. Liq.	Heat content abv. −40°F BTU/lb. Vap.	Ht. of vaporiz. BTU/lb.	Heat content abv. −40°C g-cal./g Liq.	Heat content abv. −40°C g-cal./g Vap.	Ht. of vaporiz. g-cal./g	Temp. °C
30	59.74	4.200	75.7	620.5	544.8	42.1	344.7	302.7	− 1.11
32	62.29	4.379	77.9	621.0	543.1	43.3	345.0	301.7	0.00
34	64.91	4.564	80.1	621.5	541.4	44.5	345.3	300.8	+ 1.11
36	67.63	4.755	82.3	622.0	539.7	45.7	345.6	299.8	2.22
38	70.43	4.952	84.6	622.5	537.9	47.0	345.8	298.8	3.33
40	73.32	5.155	86.8	623.0	536.2	48.2	346.1	297.9	4.44
42	76.31	5.365	89.0	623.4	534.4	49.4	346.3	296.9	5.56
44	79.38	5.581	91.2	623.9	532.7	50.7	346.6	295.9	6.67
46	82.55	5.804	93.5	624.4	530.9	51.9	346.9	294.9	7.78
48	85.82	6.034	95.7	624.8	529.1	53.2	347.1	293.9	8.89
50	89.19	6.271	97.9	625.2	527.3	54.4	347.3	292.9	10.00
52	92.66	6.515	100.2	625.7	525.5	55.67	347.6	291.9	11.11
54	96.23	6.766	102.4	626.1	523.7	56.89	347.8	290.9	12.22
56	99.91	7.024	104.7	626.5	521.8	58.17	348.1	289.9	13.33
58	103.7	7.291	106.9	626.9	520.0	59.39	348.3	288.9	14.44
60	107.6	7.565	109.2	627.3	518.1	60.67	348.5	287.8	15.56
62	111.6	7.846	111.5	627.7	516.2	61.94	348.7	286.8	16.67
64	115.7	8.135	113.7	628.0	514.3	63.17	348.9	285.7	17.78
66	120.0	8.437	116.0	628.4	512.4	64.44	349.1	284.7	18.89
68	124.3	8.739	118.3	628.8	510.5	65.72	349.3	283.6	20.00
70	128.8	9.056	120.5	629.1	508.6	66.94	349.5	282.6	21.11
72	133.4	9.379	122.8	629.4	506.6	68.22	349.7	281.4	22.22
74	138.1	9.709	125.1	629.8	504.7	69.50	349.9	280.4	23.33
76	143.0	10.05	127.4	630.1	502.7	70.78	350.1	279.3	24.44
78	147.9	10.40	129.7	630.4	500.7	72.06	350.2	278.2	25.56
80	153.0	10.76	132.0	630.7	498.7	73.33	350.4	277.1	26.67
82	158.3	11.13	134.3	631.0	496.7	74.61	350.6	275.9	27.78
84	163.7	11.51	136.6	631.3	494.7	75.89	350.7	274.8	28.89
86	169.2	11.90	138.9	631.5	492.6	77.17	350.8	273.7	30.00
88	174.8	12.29	141.2	631.8	490.6	78.44	351.0	272.6	31.11
90	180.6	12.70	143.5	632.0	488.5	79.72	351.1	271.4	32.22
92	186.6	13.12	145.8	632.2	486.4	81.00	351.2	270.2	33.33
94	192.7	13.55	148.2	632.5	484.3	82.33	351.4	269.1	34.44
96	198.9	13.98	150.5	632.6	482.1	83.61	351.4	267.8	35.56
98	205.3	14.43	152.9	632.9	480.0	84.94	351.6	266.7	36.67
100	211.9	14.90	155.2	633.0	477.8	86.22	351.7	265.4	37.78
102	218.6	15.37	157.6	633.2	475.6	87.56	351.8	264.2	38.89
104	225.4	15.85	159.9	633.4	473.5	88.83	351.9	263.1	40.00
106	232.5	16.35	162.3	633.5	471.2	90.17	351.9	261.8	41.11
108	239.7	16.85	164.6	633.6	469.0	91.44	352.0	260.6	42.22
110	247.0	17.37	167.0	633.7	466.7	92.78	352.1	259.3	43.33
112	254.5	17.89	169.4	633.8	464.4	94.11	352.1	258.0	44.44
114	262.2	18.43	171.8	633.9	462.1	95.44	352.2	256.7	45.56
116	270.1	18.99	174.2	634.0	459.8	96.78	352.2	255.4	46.67
118	278.2	19.56	176.6	634.0	457.4	98.11	352.2	254.1	47.78
120	286.4	20.14	179.0	634.0	455.0	99.45	352.2	252.8	48.89
122	294.8	20.73	181.4	634.0	452.6	100.8	352.2	251.4	50.00
124	303.4	21.33	183.9	634.0	450.1	102.2	352.2	250.1	51.11

Ammonia, NH_3 (Continued)

Temp. °F	Spec. vol. sat. vap. ft.³/lb.	Spec. vol. sat. vap. m³/kg	Density sat. vap. lb./ft.³	Density sat. vap. kg/m³	Dens. liq. lb./ft.³	Entropy from −40°F BTU/lb./°F Liq.	Entropy from −40°F BTU/lb./°F Vap.	Temp. °C
30	4.825	0.3012	0.2073	3.321	39.96	0.1663	1.2790	− 1.11
32	4.637	.2895	.2156	3.453		.1708	1.2755	0.00
34	4.459	.2784	.2243	3.593		.1753	1.2721	+ 1.11
36	4.289	.2678	.2332	3.735		.1797	1.2686	2.22
38	4.126	.2576	.2423	3.881		.1841	1.2652	3.33
40	3.971	0.2479	0.2518	4.033	39.49	0.1885	1.2618	4.44
42	3.823	.2387	.2616	4.190		.1930	1.2585	5.56
44	3.682	.2299	.2716	4.350		.1974	1.2552	6.67
46	3.547	.2214	.2819	4.515		.2018	1.2519	7.78
48	3.418	.2134	.2926	4.687		.2062	1.2486	8.89
50	3.294	0.2056	0.3036	4.863	39.00	0.2105	1.2453	10.00
52	3.176	.1983	.3149	5.044		.2149	1.2421	11.11
54	3.063	.1912	.3265	5.230		.2192	1.2389	12.22
56	2.954	.1844	.3385	5.422		.2236	1.2357	13.33
58	2.851	.1780	.3508	5.619		.2279	1.2325	14.44
60	2.751	0.1717	0.3635	5.823	38.50	0.2322	1.2294	15.56
62	2.656	.1658	.3765	6.031		.2365	1.2262	16.67
64	2.565	.1601	.3899	6.245		.2408	1.2231	17.78
66	2.477	.1546	.4037	6.466		.2451	1.2201	18.89
68	2.393	.1494	.4179	6.694		.2494	1.2170	20.00
70	2.312	0.1443	0.4325	6.928	38.00	0.2537	1.2140	21.11
72	2.235	.1395	.4474	7.166		.2579	1.2110	22.22
74	2.161	.1349	.4628	7.413		.2622	1.2080	23.33
76	2.089	.1304	.4786	7.666		.2664	1.2050	24.44
78	2.021	.1262	.4949	7.927		.2706	1.2020	25.56
80	1.955	0.1220	0.5115	8.193	37.48	0.2749	1.1991	26.67
82	1.892	.1181	.5287	8.469		.2791	1.1962	27.78
84	1.831	.1143	.5462	8.749		.2833	1.1933	28.89
86	1.772	.1106	.5643	9.039		.2875	1.1904	30.00
88	1.716	.1071	.5828	9.335		.2917	1.1875	31.11
90	1.661	0.1037	0.6019	9.641	36.95	0.2958	1.1846	32.22
92	1.609	.1004	.6214	9.954		.3000	1.1818	33.33
94	1.559	.09733	.6415	10.28		.3041	1.1789	34.44
96	1.510	.09427	.6620	10.60		.3083	1.1761	35.56
98	1.464	.09140	.6832	10.94		.3125	1.1733	36.67
100	1.419	0.08859	0.7048	11.29	36.40	0.3166	1.1705	37.78
102	1.375	.08584	.7270	11.65		.3207	1.1677	38.89
104	1.334	.08328	.7498	12.01		.3248	1.1649	40.00
106	1.293	.08072	.7732	12.39		.3289	1.1621	41.11
108	1.254	.07829	.7972	12.77		.3330	1.1593	42.22
110	1.217	0.07598	0.8219	13.17	35.84	0.3372	1.1566	43.33
112	1.180	.07367	.8471	13.57		.3413	1.1538	44.44
114	1.145	.07148	.8730	13.98		.3453	1.1510	45.56
116	1.112	.06942	.8996	14.41		.3495	1.1483	46.67
118	1.079	.06736	.9269	14.85		.3535	1.1455	47.78
120	1.047	0.06536	0.9549	15.30	35.26	0.3576	1.1427	48.89
122	1.017	.06349	.9837	15.76		.3618	1.1400	50.00
124	0.987	.0616	1.0132	16.229		.3659	1.1372	51.11

THERMODYNAMIC PROPERTIES (Continued)

THERMODYNAMIC PROPERTIES (Continued)

Carbon Dioxide, CO_2

Temp. °F	Abs. press sat. vap lb./in.²	Abs. press sat. vap kg/cm²	Heat content abv. 32°F BTU/lb. Liq.	Heat content abv. 32°F BTU/lb. Vap.	Heat of vaporiz. BTU/lb.	Heat content abv 0°C g-cal./g Liq.	Heat content abv 0°C g-cal./g Vap.	Heat of vaporiz. g-cal./g	Temp. °C
−20	220.6	15.51	−23.96	102.0	126.0	−13.31	56.67	70.00	−28.89
−18	228.4	16.06	−23.13	102.1	125.2	−12.85	56.72	69.56	−27.78
−16	236.4	16.62	−22.30	102.2	124.5	−12.39	56.78	69.17	−26.67
−14	244.6	17.20	−21.46	102.2	123.7	−11.92	56.78	68.72	−25.56
−12	253.0	17.79	−20.61	102.3	122.9	−11.45	56.83	68.28	−24.44
−10	261.7	18.40	−19.76	102.3	122.0	−10.98	56.83	67.78	−23.33
−8	270.6	19.03	−18.90	102.3	121.2	−10.50	56.83	67.33	−22.22
−6	279.7	19.66	−18.04	102.3	120.3	−10.02	56.83	66.83	−21.11
−4	289.1	20.33	−17.17	102.3	119.5	−9.539	56.83	66.39	−20.00
−2	298.7	21.00	−16.29	102.3	118.6	−9.050	56.83	65.89	−18.89
0	308.6	21.70	−15.41	102.2	117.7	−8.561	56.78	65.39	−17.78
2	318.7	22.41	−14.51	102.2	116.7	−8.061	56.78	64.83	−16.67
4	329.1	23.14	−13.61	102.1	115.8	−7.561	56.72	64.33	−15.56
6	339.8	23.89	−12.71	102.1	114.8	−7.061	56.72	63.78	−14.44
8	350.7	24.66	−11.79	102.0	113.8	−6.550	56.67	63.22	−13.33
10	361.8	25.44	−10.87	101.9	112.8	−6.039	56.61	62.67	−12.22
12	373.3	26.25	−9.934	101.8	111.7	−5.519	56.56	62.06	−11.11
14	385.0	27.07	−8.992	101.7	110.7	−4.996	56.50	61.50	−10.00
16	397.1	27.92	−8.038	101.5	109.6	−4.466	56.39	60.89	−8.89
18	409.4	28.78	−7.076	101.4	108.5	−3.931	56.33	60.28	−7.78
20	422.0	29.67	−6.102	101.2	107.3	−3.390	56.22	59.61	−6.67
22	434.9	30.58	−5.117	101.0	106.1	−2.843	56.11	58.94	−5.56
24	448.1	31.50	−4.121	100.8	104.9	−2.289	56.00	58.28	−4.44
25	454.8	31.98	−3.618	100.7	104.3	−2.010	55.94	57.94	−3.89
27	468.5	32.94	−2.601	100.5	103.1	−1.445	55.83	57.28	−2.78
29	482.5	33.92	−1.570	100.2	101.8	−0.8722	55.67	56.56	−1.67
31	496.8	34.93	−0.525	99.98	100.5	−0.292	55.54	55.83	−0.56
33	511.4	35.95	+0.531	99.69	99.16	+0.295	55.38	55.09	+0.56
35	526.4	37.01	1.604	99.38	97.77	0.8911	55.21	54.32	1.67
37	541.7	38.09	2.697	99.05	96.35	1.498	55.03	53.53	2.78
39	557.4	39.19	3.806	98.69	94.88	2.114	54.83	52.71	3.89
41	573.4	40.31	4.932	98.31	93.37	2.740	54.62	51.87	5.00
43	589.8	41.47	6.080	97.90	91.82	3.378	54.39	51.01	6.11
45	606.5	42.64	7.251	97.46	90.21	4.028	54.14	50.12	7.22
47	623.6	43.84	8.443	96.99	88.55	4.691	53.88	49.19	8.33
49	641.1	45.07	9.664	96.50	86.83	5.369	53.61	48.24	9.44
51	659.0	46.33	10.91	95.97	85.06	6.061	53.32	47.26	10.56
53	677.3	47.62	12.19	95.40	83.21	6.772	53.00	46.23	11.67
55	695.9	48.93	13.49	94.78	81.29	7.494	52.66	45.16	12.78
57	714.9	50.26	14.84	94.13	79.30	8.244	52.29	44.06	13.89
59	734.3	51.63	16.22	93.44	77.22	9.011	51.91	42.90	15.00
61	754.2	53.03	17.65	92.69	75.04	9.806	51.49	41.69	16.11
63	774.5	54.45	19.13	91.88	72.75	10.63	51.04	40.42	17.22
65	795.1	55.90	20.66	91.01	70.35	11.48	50.56	39.08	18.33
67	816.2	57.38	22.25	90.07	67.81	12.36	50.04	37.67	19.44
69	837.8	58.90	23.92	89.04	65.12	13.29	49.47	36.18	20.56
71	859.8	60.45	25.67	87.92	62.25	14.26	48.84	34.58	21.67
73	882.2	62.02	27.52	86.69	59.17	15.29	48.16	32.87	22.78
75	905.1	63.63	29.50	85.33	55.83	16.39	47.41	31.02	23.89
77	928.4	65.27	31.62	83.80	52.17	17.57	46.56	28.98	25.00
79	952.2	66.95	33.95	82.06	48.11	18.86	45.59	26.73	26.11
81	976.5	68.65	36.54	80.03	43.49	20.30	44.46	24.16	27.22
83	1001.0	70.377	39.53	77.60	38.07	21.96	43.11	21.15	28.33
85	1027.0	72.205	43.18	74.47	31.29	23.99	41.37	17.38	29.44
86	1039.0	73.049	45.45	72.46	27.00	25.25	40.26	15.00	30.00
87	1052.0	73.963	48.32	69.84	21.52	26.84	38.80	11.96	30.56
88	1065.0	74.877	52.78	65.62	12.84	29.32	36.46	7.133	31.11

Carbon Dioxide, CO_2

Temp. °F	Spec. vol. sat. vap. ft.³/lb.	Spec. vol. sat. vap. m³/kg	Density sat. vap. lb./ft.³	Density sat. vap. kg/m³	Dens. liq. lb./ft.³	Entropy from 32°F BTU/lb./°F Liq.	Entropy from 32°F BTU/lb./°F Vap.	Temp. °C
−20	0.4166	0.02601	2.401	38.46	64.34	−0.0514	0.2353	−28.89
−18	.4018	.02508	2.489	39.87	64.15	−.0495	.2342	−27.78
−16	.3876	.02420	2.580	41.33	63.94	−.0476	.2331	−26.67
−14	.3739	.02334	2.674	42.83	63.73	−.0458	.2319	−25.56
−12	.3608	.02252	2.772	44.40	63.49	−.0439	.2307	−24.44
−10	0.3482	0.02174	2.872	46.00	63.25	−0.0420	0.2296	−23.33
−8	.3360	.02098	2.976	47.67	63.01	−.0401	.2284	−22.22
−6	.3243	.02025	3.083	49.38	62.76	−.0382	.2273	−21.11
−4	.3131	.01955	3.194	51.16	62.50	−.0362	.2261	−20.00
−2	.3022	.01887	3.309	53.00	62.23	−.0343	.2249	−18.89
0	0.2918	0.01822	3.427	54.89	61.95	−0.0324	0.2237	−17.78
2	.2817	.01759	3.550	56.86	61.65	−.0304	.2225	−16.67
4	.2720	.01698	3.676	58.88	61.36	−.0285	.2213	−15.56
6	.2627	.01640	3.807	60.98	61.07	−.0266	.2201	−14.44
8	.2537	.01584	3.942	63.14	60.77	−.0246	.2189	−13.33
10	0.2450	0.01529	4.082	65.39	60.48	−0.0226	0.2176	−12.22
12	.2366	.01477	4.227	67.71	60.18	−.0206	.2164	−11.11
14	.2285	.01426	4.377	70.11	59.88	−.0186	.2151	−10.00
16	.2207	.01378	4.532	72.59	59.58	−.0166	.2139	−8.89
18	.2131	.01330	4.692	75.16	59.27	−.0146	.2126	−7.78
20	0.2058	0.01285	4.859	77.83	58.95	−0.0126	0.2113	−6.67
22	.1987	.01240	5.031	80.59	58.64	−.0105	.2100	−5.56
24	.1919	.01198	5.211	83.47	58.31	−.0085	.2037	−4.44
25	.1886	.01177	5.303	84.94	58.14	−.0074	.2080	−3.89
27	.1821	.01137	5.492	87.97	57.81	−.0053	.2066	−2.78
29	0.1758	0.01097	5.688	91.11	57.47	−0.0032	0.2053	−1.67
31	.1697	.01059	5.892	94.38	57.12	−.0011	.2039	−0.56
33	.1639	.01023	6.103	97.76	56.77	+.0011	.2025	+0.56
35	.1581	.009870	6.323	101.3	56.41	.0033	.2010	1.67
37	.1526	.009527	6.553	105.0	56.03	.0055	.1996	2.78
39	0.1472	0.009189	6.792	108.8	55.65	0.0077	0.1981	3.89
41	.1420	.008865	7.040	112.8	55.25	.0099	.1965	5.00
43	.1370	.008553	7.300	116.9	54.84	.0122	.1950	6.11
45	.1321	.008247	7.571	121.3	54.41	.0146	.1934	7.22
47	.1273	.007947	7.854	125.8	53.97	.0169	.1918	8.33
49	0.1227	0.007660	8.151	130.6	53.51	0.0193	0.1901	9.44
51	.1182	.007379	8.461	135.5	53.04	.0218	.1884	10.56
53	.1138	.007104	8.787	140.8	52.55	.0243	.1867	11.67
55	.1095	.006836	9.132	146.3	52.05	.0268	.1849	12.78
57	.1053	.006574	9.497	152.1	51.53	.0294	.1830	13.89
59	0.1012	0.006318	9.880	158.3	50.99	0.0321	0.1811	15.00
61	.0972	.00607	10.29	164.8	50.42	.0348	.1790	16.11
63	.0933	.00582	10.72	171.7	49.80	.0377	.1770	17.22
65	.0894	.00558	11.18	179.1	49.14	.0406	.1748	18.33
67	.0856	.00534	11.67	186.9	48.44	.0436	.1725	19.44
69	0.0819	0.00511	12.21	195.6	47.69	0.0468	0.1701	20.56
71	.0782	.00488	12.82	205.4	46.87	.0501	.1675	21.67
73	.0745	.00465	13.43	215.1	45.99	.0536	.1647	22.78
75	.0708	.00442	14.13	226.3	45.05	.0573	.1618	23.89
77	.0671	.00419	14.90	238.7	44.06	.0613	.1585	25.00
79	0.0633	0.00395	15.81	253.2	43.04	0.0656	0.1550	26.11
81	.0592	.00370	16.90	270.7	41.95	.0704	.1509	27.22
83	.0548	.00342	18.25	292.3	40.62	.0759	.1461	28.33
85	.0500	.00312	20.00	320.4	38.76	.0826	.1401	29.44
86	.0474	.00296	21.09	337.8	37.41	.0868	.1363	30.00
87	0.0446	0.00278	22.42	359.1	35.34	0.0921	0.1314	30.56
88	.0401	.00250	24.95	399.6	32.79	.1002	.1237	31.11

THERMODYNAMIC PROPERTIES (Continued)

Sulfur Dioxide, SO_2

Temp. °F	Abs. press. sat. vap. lb./in.²	Abs. press. sat. vap. kg/cm²	Heat content abv. −40°F BTU/lb. Liq.	Heat content abv. −40°F BTU/lb. Vap.	Ht. of vaporiz BTU/lb.	Heat content abv. −40°C g-cal./g Liq.	Heat content abv. −40°C g-cal./g Vap.	Ht. of vaporiz. g-cal./g	Temp. °C
−40	3.136	0.2205	0.00	178.61	178.61	0.00	99.228	99.228	−40.00
−30	4.331	.3045	2.93	179.90	176.97	1.63	99.945	98.317	−34.44
−20	5.883	.4136	5.98	181.07	175.09	3.32	100.59	97.272	−28.89
−10	7.863	.5528	9.16	182.13	172.97	5.09	101.18	96.095	−23.33
0	10.35	.7277	12.44	183.07	170.63	6.911	101.71	94.795	−17.78
2	10.91	0.7670	13.12	183.25	170.13	7.289	101.81	94.517	−16.67
4	11.50	.8085	13.78	183.41	169.63	7.656	101.89	94.239	−15.56
5	11.81	.8303	14.11	183.49	169.38	7.839	101.94	94.100	−15.00
6	12.12	.8521	14.45	183.57	169.12	8.028	101.98	93.956	−14.44
8	12.75	.8964	15.13	183.73	168.60	8.406	102.07	93.667	−13.33
10	13.42	0.9435	15.80	183.87	168.07	8.778	102.15	93.372	−12.22
12	14.12	.9927	16.48	184.01	167.53	9.156	102.23	93.072	−11.11
14	14.84	1.043	17.15	184.14	166.97	9.528	102.30	92.761	−10.00
16	15.59	1.096	17.84	184.28	166.44	9.911	102.38	92.467	− 8.89
18	16.37	1.1509	18.52	184.40	165.88	10.29	102.44	92.156	− 7.78
20	17.18	1.208	19.20	184.52	165.32	10.67	102.51	91.845	− 6.67
22	18.03	1.268	19.90	184.64	164.74	11.06	102.58	91.522	− 5.56
24	18.89	1.328	20.58	184.74	164.16	11.43	102.63	91.200	− 4.44
26	19.80	1.392	21.26	184.84	163.58	11.81	102.69	90.878	− 3.33
28	20.73	1.457	21.96	184.94	162.98	12.20	102.74	90.545	− 2.22
30	21.70	1.526	22.64	185.02	162.38	12.58	102.79	90.211	− 1.11
32	22.71	1.597	23.33	185.10	161.77	12.96	102.83	89.872	0.00
34	23.75	1.670	24.03	185.18	161.15	13.35	102.88	89.528	+ 1.11
36	24.82	1.745	24.72	185.25	160.53	13.73	102.92	89.183	2.22
38	25.95	1.824	25.42	185.31	159.89	14.12	102.95	88.828	3.33
40	27.10	1.905	26.12	185.37	159.25	14.51	102.98	88.472	4.44
42	28.29	1.989	26.81	185.42	158.61	14.89	103.01	88.117	5.56
44	29.52	2.075	27.51	185.46	157.95	15.28	103.03	87.750	6.67
46	30.79	2.165	28.21	185.50	157.29	15.67	103.06	87.383	7.78
48	32.10	2.257	28.92	185.54	156.62	16.07	103.08	87.011	8.89
50	33.45	2.352	29.61	185.56	155.95	16.45	103.09	86.639	10.00
52	34.86	2.451	30.31	185.58	155.27	16.84	103.10	86.261	11.11
54	36.31	2.553	31.00	185.59	154.59	17.22	103.11	85.883	12.22
56	37.80	2.658	31.72	185.61	153.89	17.62	103.12	85.495	13.33
58	39.33	2.765	32.42	185.61	153.19	18.01	103.12	85.106	14.44
60	40.93	2.878	33.10	185.59	152.49	18.39	103.11	84.717	15.56
62	42.58	2.994	33.79	185.57	151.78	18.77	103.09	84.322	16.67
64	44.27	3.112	34.49	185.55	151.06	19.16	103.08	83.922	17.78
66	46.00	3.234	35.19	185.53	150.34	19.55	103.07	83.522	18.89
68	47.78	3.359	35.88	185.50	149.62	19.93	103.06	83.122	20.00
70	49.62	3.489	36.58	185.46	148.88	20.32	103.03	82.711	21.11
72	51.54	3.624	37.28	185.42	148.14	20.71	103.01	82.300	22.22
74	53.48	3.760	37.97	185.37	147.40	21.09	102.98	81.889	23.33
76	55.48	3.901	38.67	185.31	146.64	21.48	102.95	81.467	24.44
78	57.56	4.047	39.36	185.24	145.88	21.87	102.91	81.045	25.56
80	59.68	4.196	40.05	185.17	145.12	22.25	102.87	80.622	26.67
82	61.88	4.351	40.73	185.09	144.36	22.63	102.83	80.200	27.78
84	64.14	4.509	41.43	185.01	143.58	23.02	102.78	79.767	28.89
86	66.45	4.672	42.12	184.92	142.80	23.40	102.73	79.333	30.00
88	68.84	4.840	42.80	184.82	142.02	23.78	102.68	78.900	31.11
90	71.25	5.009	43.50	184.72	141.22	24.17	102.62	78.456	32.22
92	73.70	5.182	44.19	184.61	140.42	24.55	102.56	78.011	33.33
94	76.30	5.364	44.86	184.49	139.62	24.92	102.49	77.567	34.44
96	79.03	5.556	45.54	184.37	138.83	25.30	102.43	77.128	35.56
98	81.77	5.749	46.22	184.25	138.03	25.68	102.36	76.683	36.67
100	84.52	5.942	46.90	184.10	137.20	26.06	102.28	76.222	37.78

Sulfur Dioxide, SO_2

Temp. °F	Spec. vol. sat. vap. ft.³/lb.	Spec. vol. sat. vap. m³/kg	Density sat. vap. lb./ft.³	Density sat. vap. kg/m³	Dens. liq. lb./ft.³	Entropy from −40°F BTU/lb./°F Liq.	Entropy from −40°F BTU/lb./°F Vap.	Temp. °C
−40	22.42	1.400	0.04460	0.7144	95.79	0.00000	0.42562	−40.00
−30	16.56	1.034	.06039	0.9673	94.94	.00674	.41864	−34.44
−20	12.42	0.7754	.08119	1.301	94.10	.01366	.41192	−28.89
−10	9.44	.5893	.1025	1.642	93.27	.02075	.40544	−23.33
0	7.280	.4545	.1374	2.201	92.42	.02795	.39917	−17.78
2	6.923	0.4322	0.1444	2.313	92.25	0.02941	0.39794	−16.67
4	6.584	.4110	.1501	2.404	92.08	.03084	.39670	−15.56
5	6.421	.4009	.1558	2.496	92.00	.03155	.39609	−15.00
6	6.266	.3912	.1596	2.556	91.91	.03228	.39547	−14.44
8	5.967	.3725	.1676	2.685	91.74	.03373	.39426	−13.33
10	5.682	0.3547	0.1760	2.819	91.58	0.03519	0.39306	−12.22
12	5.417	.3382	.1846	2.957	91.41	.03664	.39185	−11.11
14	5.164	.3224	.1936	3.101	91.24	.03808	.39065	−10.00
16	4.926	.3075	.2030	3.252	91.07	.03953	.38946	− 8.89
18	4.701	.2935	.2127	3.407	90.89	.04098	.38827	− 7.78
20	4.487	0.2801	0.2228	3.569	90.71	0.04241	0.38707	− 6.67
22	4.287	.2676	.2332	3.735	90.53	.04385	.38589	− 5.56
24	4.096	.2557	.2441	3.910	90.33	.04528	.38471	− 4.44
26	3.915	.2444	.2559	4.099	90.15	.04671	.38354	− 3.33
28	3.744	.2337	.2671	4.278	89.96	.04814	.38237	− 2.22
30	3.581	0.2236	0.2800	4.485	89.76	0.04956	0.38119	− 1.11
32	3.437	.2146	.2909	4.660	89.58	.05099	.38003	0.00
34	3.283	.2050	.3046	4.879	89.39	.05242	.37887	+ 1.11
36	3.144	.1963	.3181	5.095	89.18	.05384	.37772	2.22
38	3.013	.1881	.3319	5.316	89.00	.05527	.37657	3.33
40	2.887	0.1802	0.3464	5.549	88.81	0.05668	0.37541	4.44
42	2.769	.1729	.3611	5.784	88.62	.05809	.37425	5.56
44	2.656	.1658	.3765	6.031	88.43	.05949	.37311	6.67
46	2.548	.1591	.3925	6.287	88.24	.06090	.37197	7.78
48	2.446	.1527	.4088	6.548	88.05	.06231	.37083	8.89
50	2.348	0.1466	0.4259	6.822	87.87	0.06370	0.36969	10.00
52	2.256	.1408	.4433	7.101	87.67	.06509	.36857	11.11
54	2.167	.1353	.4615	7.392	87.51	.06646	.36743	12.22
56	2.083	.1300	.4801	7.690	87.31	.06785	.36629	13.33
58	2.003	.1250	.4992	7.996	87.13	.06923	.36517	14.44
60	1.926	0.1202	0.5194	8.320	86.95	0.07066	0.36405	15.56
62	1.853	.1157	.5396	8.643	86.77	.07196	.36293	16.67
64	1.783	.1113	.5609	8.984	86.59	.07333	.36181	17.78
66	1.716	.1071	.5827	9.334	86.41	.07469	.36070	18.89
68	1.652	.1031	.6054	9.697	86.22	.07602	.35958	20.00
70	1.590	0.09926	0.6290	10.08	86.02	0.07736	0.35846	21.11
72	1.532	.09564	.6527	10.45	85.82	.07871	.35736	22.22
74	1.476	.09214	.6777	10.86	85.62	.08003	.35624	23.33
76	1.422	.08877	.7030	11.26	85.42	.08135	.35512	24.44
78	1.371	.08559	.7295	11.69	85.23	.08268	.35401	25.56
80	1.321	0.08247	0.7570	12.13	85.03	0.08399	0.35291	26.67
82	1.274	.07953	.7850	12.57	84.84	.08525	.35177	27.78
84	1.229	.07672	.8140	13.04	84.64	.08653	.35065	28.89
86	1.185	.07398	.8440	13.52	84.44	.08783	.34954	30.00
88	1.144	.07142	.8740	14.00	84.25	.08910	.34843	31.11
90	1.104	0.06892	0.9058	14.51	84.05	0.09038	0.34731	32.22
92	1.065	.06649	.9390	15.04	83.86	.09165	.34620	33.33
94	1.028	.06418	.9730	15.59	83.67	.09389	.34508	34.44
96	0.9931	.06200	1.007	16.18	83.47	.09411	.34397	35.56
98	0.9591	.05987	1.043	16.71	83.27	.09532	.34285	36.67
100	0.9262	0.05782	1.080	17.30	83.07	0.09657	0.34173	37.78

THERMODYNAMIC PROPERTIES (Continued)

Butane, $CH_3(CH_2)_2CH_3$

Temp. °F	Abs. press. sat. vap. lb./in.²	Abs. press. sat. vap. kg/cm²	Heat content abv. 32°F BTU/lb. Liq.	Heat content abv. 32°F BTU/lb. Vap.	Ht. of vaporiz. BTU/lb.	Heat content abv. 0°C g-cal./g Liq.	Heat content abv. 0°C g-cal./g Vap.	Ht. of vaporiz. g-cal./g	Temp. °C
0	7.3	0.51	−17.2	153.3	170.5	− 9.56	85.17	94.72	−17.78
10	9.2	0.65	−11.7	156.8	168.5	− 6.50	87.11	93.61	−12.22
20	11.6	0.816	− 6.7	160.3	167.0	− 3.7	89.06	92.78	− 6.67
30	14.4	1.01	− 1.2	164.3	165.5	− 0.67	91.28	91.94	− 1.11
40	17.7	1.24	+ 4.3	167.8	163.5	+ 2.4	93.22	90.83	+ 4.44
50	21.6	1.52	9.8	171.3	161.5	5.4	95.17	89.72	10.00
60	26.3	1.85	15.8	175.3	159.5	8.78	97.39	88.61	15.56
70	31.6	2.22	21.3	178.8	157.5	11.8	99.33	87.50	21.11
80	37.6	2.64	27.3	182.3	155.0	15.2	101.3	86.11	26.67
90	44.5	3.13	33.8	185.8	152.0	18.8	103.2	84.44	32.22
100	52.2	3.67	39.8	189.3	149.5	22.1	105.2	83.06	37.78
110	60.8	4.27	46.3	193.3	147.0	25.7	107.4	81.67	43.33
120	70.8	4.98	52.8	196.3	143.5	29.3	109.1	79.72	48.89
130	81.4	5.72	59.3	199.8	140.5	32.9	111.0	78.06	54.44
140	92.6	6.51	66.3	203.8	137.5	36.8	113.2	76.39	60.00

Isobutane, $(CH_3)_3CH$

Temp. °F	Abs. press. sat. vap. lb./in.²	Abs. press. sat. vap. kg/cm²	Heat content abv. 32°F BTU/lb. Liq.	Heat content abv. 32°F BTU/lb. Vap.	Ht. of vaporiz. BTU/lb.	Heat content abv. 0°C g-cal./g Liq.	Heat content abv. 0°C g-cal./g Vap.	Ht. of vaporiz. g-cal./g	Temp. °C
−20	7.50	0.527	−25.5	140.0	165.5	−14.2	77.78	91.94	−28.89
−10	9.28	0.652	−21.0	142.0	163.0	−11.7	78.89	90.56	−23.33
0	11.6	0.816	−16.5	144.0	160.5	− 9.17	80.00	89.17	−17.78
+10	14.6	1.03	−11.5	147.0	158.5	− 6.39	81.67	88.06	−12.22
20	18.2	1.28	− 6.5	149.5	156.0	− 3.6	83.06	86.67	− 6.67
30	22.3	1.57	− 1.0	152.5	153.5	− 0.56	84.72	85.28	− 1.11
40	26.9	1.89	+ 4.5	155.5	151.0	+ 2.5	86.39	83.89	+ 4.44
50	32.5	2.28	10.5	159.0	148.5	5.83	88.33	82.50	10.00
60	38.7	2.72	16.5	162.5	146.0	9.17	90.28	81.11	15.56
70	45.8	3.22	23.0	166.5	143.5	12.8	92.50	79.72	21.11
80	53.9	3.79	30.0	170.5	140.5	16.7	94.72	78.06	26.67
90	63.3	4.45	37.0	174.5	137.5	20.6	96.94	76.39	32.22
100	73.7	5.18	44.5	179.0	134.5	24.7	99.44	74.72	37.78
110	85.1	5.98	52.5	183.5	131.0	29.2	101.9	72.78	43.33
120	98.0	6.89	60.5	188.0	127.5	33.6	104.4	70.83	48.89
130	112.0	7.87	69.5	193.5	124.0	38.6	107.5	68.89	54.44
140	126.8	8.915	78.5	199.0	120.5	43.6	110.6	66.94	60.00

Propane, C_3H_8

Temp. °F	Abs. press. sat. vap. lb./in.²	Abs. press. sat. vap. kg/cm²	Heat content abv. 32°F BTU/lb. Liq.	Heat content abv. 32°F BTU/lb. Vap.	Ht. of vaporiz. BTU/lb.	Heat content abv. 0°C g-cal./g Liq.	Heat content abv. 0°C g-cal./g Vap.	Ht. of vaporiz. g-cal./g	Temp. °C
−70	7.37	0.518	−55.2	134.3	189.5	−30.7	74.61	105.3	−56.67
−60	9.72	0.683	−50.2	136.8	187.0	−27.9	76.00	103.9	−51.11
−50	12.6	0.886	−44.7	139.8	184.5	−24.8	77.67	102.5	−45.56
−40	16.2	1.14	−39.7	141.8	181.5	−22.1	78.78	100.8	−40.00
−30	20.3	1.43	−34.2	144.8	179.0	−19.0	80.44	99.44	−34.44
−20	25.4	1.79	−29.2	146.8	176.0	−16.2	81.56	97.78	−28.89
−10	31.4	2.21	−23.7	149.8	173.5	−13.2	83.22	96.39	−23.33
0	38.2	2.69	−18.2	152.3	170.5	−10.1	84.61	94.72	−17.78
+10	46.0	3.23	−12.7	155.3	168.0	− 7.06	86.28	93.33	−12.22
20	55.5	3.90	− 7.2	157.8	165.0	− 4.0	87.67	91.67	− 6.67
30	66.3	4.66	− 1.2	160.8	162.0	− 0.67	89.33	90.00	− 1.11
40	78.0	5.48	+ 4.8	163.8	159.0	+ 2.7	91.00	88.33	+ 4.44
50	91.8	6.45	10.8	166.8	156.0	6.00	92.67	86.67	10.00
60	107.1	7.530	16.8	169.8	153.0	9.33	94.33	85.00	15.56
70	124.0	8.718	22.8	172.3	149.5	12.7	95.72	83.06	21.11
80	142.8	10.04	29.3	175.3	146.0	16.3	97.39	81.11	26.67
90	164.0	11.53	35.8	178.3	142.5	19.9	99.06	79.17	32.22
100	187.0	13.15	42.3	180.8	138.5	23.5	100.4	76.94	37.78
110	213.0	14.98	48.8	182.8	134.0	27.1	101.6	74.44	43.33
120	240.0	16.87	55.3	184.3	129.0	30.7	102.4	71.67	48.89

Butane, $CH_3(CH_2)_2CH_3$

Temp. °F	Spec. vol. sat. vap. ft.³/lb.	Spec. vol. sat. vap. m³/kg	Density of sat. vap. lb./ft.³	Density of sat. vap. kg/m³	Density of liq. lb./ft.³	Density of liq. kg/m³	Temp. °C
0	11.1	0.693	0.0901	1.44	38.59	618.1	−17.78
10	8.95	.559	.112	1.79	38.24	612.5	−12.22
20	7.23	.451	.138	2.21	37.89	606.9	− 6.67
30	5.90	.368	.169	2.71	37.54	601.3	− 1.11
40	4.88	.305	.205	3.28	37.19	595.7	+ 4.44
50	4.07	.254	.246	3.94	36.82	589.8	10.00
60	3.40	.212	.294	4.71	36.45	583.9	15.56
70	2.88	.180	.347	5.56	36.06	577.6	21.11
80	2.46	.154	.407	6.52	35.65	571.0	26.67
90	2.10	.131	.476	7.62	35.24	564.5	32.22
100	1.81	.113	.552	8.84	34.84	558.1	37.78
110	1.58	.0986	.633	10.1	34.41	551.2	43.33
120	1.38	.0862	.725	11.6	33.96	544.0	48.89
130	1.21	.0755	.826	13.2	33.49	536.4	54.44
140	1.07	.0668	.934	15.0	32.98	528.3	60.00

Isobutane, $(CH_3)_3CH$

Temp. °F	Spec. vol. sat. vap. ft.³/lb.	Spec. vol. sat. vap. m³/kg	Density of sat. vap. lb./ft.³	Density of sat. vap. kg/m³	Density of liq. lb./ft.³	Density of liq. kg/m³	Temp. °C
−20	10.5	0.655	0.0952	1.52	38.35	614.3	−28.89
−10	8.91	.556	.112	1.79	37.95	607.9	−23.33
0	7.17	.448	.139	2.23	37.60	602.3	−17.78
+10	5.75	.359	.174	2.79	37.20	595.9	−12.22
20	4.68	.292	.214	3.43	36.80	589.5	− 6.67
30	3.86	.241	.259	4.15	36.40	583.1	− 1.11
40	3.22	.201	.311	4.98	36.00	576.6	+ 4.44
50	2.71	.169	.369	5.91	35.60	570.2	10.00
60	2.28	.142	.439	7.03	35.20	563.8	15.56
70	1.94	.121	.515	8.25	34.80	557.4	21.11
80	1.66	.104	.602	9.64	34.35	550.2	26.67
90	1.42	.0886	.704	11.3	33.90	543.0	32.22
100	1.23	.0768	.813	13.0	33.45	535.8	37.78
110	1.07	.0668	.935	15.0	33.00	528.6	43.33
120	0.926	.0578	1.08	17.3	32.50	520.6	48.89
130	0.811	.0506	1.23	19.7	32.00	512.6	54.44
140	0.716	.0447	1.32	21.1	31.80	509.4	60.00

Propane, C_3H_8

Temp. °F	Spec. vol. sat. vap. ft.³/lb.	Spec. vol. sat. vap. m³/kg	Density of sat. vap. lb./ft.³	Density of sat. vap. kg/m³	Density of liq. lb./ft.³	Density of liq. kg/m³	Temp. °C
−70	12.9	0.805	0.0775	1.24	37.40	599.1	−56.67
−60	9.93	.620	.111	1.78	37.00	592.7	−51.11
−50	7.74	.483	.129	2.07	36.60	586.3	−45.56
−40	6.13	.383	.163	2.61	36.19	579.7	−40.00
−30	4.93	.308	.203	3.25	35.78	573.1	−34.44
−20	4.00	.250	.250	4.00	35.37	566.6	−28.89
−10	3.26	.204	.307	4.92	34.96	560.0	−23.33
0	2.71	.169	.369	5.91	34.54	553.3	−17.78
+10	2.27	.142	.441	7.06	34.12	546.5	−12.22
20	1.90	.119	.526	8.43	33.67	539.3	− 6.67
30	1.60	.0999	.625	10.0	33.20	531.8	− 1.11
40	1.37	.0855	.730	11.7	32.73	524.3	+ 4.44
50	1.18	.0737	.847	13.6	32.24	516.4	10.00
60	1.01	.0631	.990	15.9	31.75	508.6	15.56
70	0.883	.0551	1.13	18.1	31.24	500.4	21.11
80	0.770	.0481	1.30	20.8	30.70	491.8	26.67
90	0.673	.0420	1.49	23.9	30.15	482.9	32.22
100	0.591	.0369	1.69	27.1	29.58	473.8	37.78
110	0.519	.0324	1.96	31.4	28.85	462.1	43.33
120	0.459	.0287	2.18	34.9	28.30	453.3	48.89

THERMODYNAMIC PROPERTIES (Continued)

Difluorodichloromethane, CCl_2F_2 ("F-12")

Temp. °F	Abs. press. sat. vap. lb./in.²	Abs. press. sat. vap. kg/cm²	Heat content abv. −40°F BTU/lb. Liq.	Heat content abv. −40°F BTU/lb. Vap.	Ht. of vaporiz. BTU/lb.	Heat content abv. −40°C g-cal./g Liq.	Heat content abv. −40°C g-cal./g Vap.	Ht. of vaporiz. g-cal./g	Temp. °C
−40	9.32	0.655	0	73.50	73.50	0	40.83	40.83	−40.00
−30	12.02	0.845	2.03	74.70	72.67	1.13	41.50	40.37	−34.44
−20	15.28	1.074	4.07	75.87	71.80	2.26	42.15	39.89	−28.89
−10	19.20	1.350	6.14	77.05	70.91	3.41	42.81	39.39	−23.33
0	23.87	1.678	8.25	78.21	69.96	4.58	43.45	38.87	−17.78
+10	29.35	2.064	10.39	79.36	68.97	5.772	44.09	38.32	−12.22
20	35.75	2.513	12.55	80.49	67.94	6.972	44.72	37.74	− 6.67
30	43.16	3.034	14.76	81.61	66.85	8.200	45.34	37.14	− 1.11
40	51.68	3.633	17.00	82.71	65.71	9.444	45.95	36.51	+ 4.44
50	61.39	4.316	19.27	83.78	64.51	10.71	46.54	35.84	10.00
60	72.41	5.091	21.57	84.82	63.25	11.98	47.12	35.14	15.56
70	84.82	5.963	23.90	85.82	61.92	13.28	47.68	34.40	21.11
80	98.76	6.944	26.28	86.80	60.52	14.60	48.22	33.62	26.67
90	114.3	8.036	28.70	87.74	59.04	15.94	48.74	32.80	32.22
100	131.6	9.252	31.16	88.62	57.46	17.31	49.23	31.92	37.78
110	150.7	10.60	33.65	89.43	55.78	18.69	49.68	30.99	43.33
120	171.8	12.08	36.16	90.15	53.99	20.09	50.08	29.99	48.89

Difluorodichloromethane, CCl_2F_2 ("F-12")

Temp. °F	Spec. vol sat. vap. ft.³/lb.	Spec. vol sat. vap. m³/kg	Density of vap. lb./ft.³	Density of vap. kg/m³	Dens. liq. lb./ft.³	Entropy from −40°F BTU/lb./°F Liq.	Entropy from −40°F BTU/lb./°F Vap.	Temp. °C
−40	3.911	0.2442	0.2557	4.096	94.58	0	0.17517	−40.00
−30	3.088	.1928	.3238	5.187	93.59	0.00471	.17387	−34.44
−20	2.474	.1544	.4042	6.474	92.58	.00940	.17275	−28.89
−10	2.003	.1250	.4993	7.998	91.57	.01403	.17175	−23.33
0	1.637	.1022	.6109	9.785	90.52	.01869	.17091	−17.78
+10	1.351	.08434	.7402	11.86	89.45	.02328	.17015	−12.22
20	1.121	.06998	.8921	14.29	88.37	.02783	.16949	− 6.67
30	0.939	.0586	1.065	17.06	87.24	.03233	.16887	− 1.11
40	.792	.0494	1.263	20.23	86.10	.03680	.16833	+ 4.44
50	.673	.0420	1.485	23.79	84.94	.04126	.16785	10.00
60	.575	.0359	1.740	27.87	83.78	.04568	.16741	15.56
70	.493	.0308	2.028	32.48	82.60	.05009	.16701	21.11
80	.425	.0265	2.353	37.69	81.39	.05446	.16662	26.67
90	.368	.0230	2.721	43.58	80.11	.05882	.16624	32.22
100	.319	.0199	3.135	50.22	78.80	.06316	.16584	37.78
110	.277	.0173	3.610	57.82	77.46	.06749	.16542	43.33
120	.240	.0150	4.167	66.75	76.02	.07180	.16495	48.89

Carbon Disulfide, CS_2

Temp. °F	Abs. press. sat. vap. lb./in.²	Abs. press. sat. vap. kg/cm²	Heat content abv. 32°F BTU/lb. Liq.	Heat content abv. 32°F BTU/lb. Vap.	Ht. of vaporiz. BTU/lb.	Heat content abv. 0°C g-cal./g Liq.	Heat content abv. 0°C g-cal./g Vap.	Ht. of vaporiz. g-cal./g	Temp. °C
0	1.10	0.0773	− 8.60	156.90	165.5	− 4.78	87.167	91.94	−17.78
10	1.46	.103	− 5.60	158.90	164.5	− 3.11	88.278	91.39	−12.22
20	1.89	.133	− 3.00	160.20	163.2	− 1.67	89.000	90.67	− 6.67
30	2.36	.166	− 0.50	161.70	162.2	− 0.28	89.833	90.11	− 1.11
40	3.03	.213	+ 2.05	163.25	161.2	+ 1.14	90.695	89.56	+ 4.44
50	3.90	274	4.24	164.24	160.0	2.36	91.245	88.89	10.00
60	4.95	.348	7 20	166.40	159.2	4.00	92.445	88.44	15.56
70	5.85	.411	9.80	167.90	158.1	5.44	93.278	87.83	21.11
80	7.30	.513	11.70	168.60	156.9	6.500	93.667	87.17	26.67
90	9.15	.643	13.80	169.40	155.6	7.667	94.111	86.44	32.22
100	11.08	.7790	16.15	170.55	154.4	8.972	94.750	85.78	37.78
110	13.50	.9491	18.30	171.50	153.2	10.17	95.278	85.11	43.33
120	16.10	1.132	20.01	172.01	152.0	11.12	95.561	84.44	48.89

Carbon Disulfide, CS_2

Temp. °F	Spec. vol. sat. vap. ft.³/lb.	Spec. vol. sat. vap. m³/kg	Density sat. vap. lb./ft.³	Density sat. vap. kg/m³	Temp. °C
0	53.76	3.356	0.0186	0.2979	−17.78
10	43.47	2.714	.0230	.3684	−12.22
20	34.84	2.175	.0287	.4597	− 6.67
30	29.49	1.841	.0339	.5430	− 1.11
40	23.52	1.468	.0425	.6808	+ 4.44
50	20.60	1.286	.0482	.7721	10.00
60	18.00	1.124	.0555	.8890	15.56
70	13.20	0.824	.0758	1.214	21.11
80	10.40	0.649	.0961	1.539	26.67
90	8.30	0.518	.1204	1.929	32.22
100	7.03	0.439	.1369	2.193	37.78
110	5.80	0.362	.1724	2.762	43.33
120	5.10	0.318	.1960	3.140	48.89

Carbon Tetrachloride, CCl_4

Temp. °F	Abs. press. sat. vap. lb./in.²	Abs. press. sat. vap. kg/cm²	Heat content Liq. (BTU/lb.)	Heat content Vap. (BTU/lb.)	Ht. of vaporiz. BTU/lb.	Heat content Liq. (g-cal./g)	Heat content Vap. (g-cal./g)	Ht. of vaporiz. g-cal./g	Temp. °C
20	0.40	0.028	− 2.00	92.45	94.45	− 1.11	51.36	52.47	− 6.67
30	0.60	.042	− 0.25	93.45	93.70	− 0.14	51.92	52.06	− 1.11
40	0.84	.059	+ 1.60	94.80	93.20	+ 0.889	52.67	51.78	+ 4.44
60	1.42	.100	5.95	98.15	92.20	3.31	54.53	51.22	15.56
70	1.85	.130	8.20	99.53	91.40	4.56	55.29	50.78	21.11
80	2.40	.169	9.80	99.87	90.07	5.44	55.48	50.04	26.67
90	3.12	.219	11.60	101.62	90.02	6.444	56.46	50.01	32.22
100	4.00	.281	13.40	102.80	89.40	7.444	57.11	49.67	37.78
110	4.89	.344	15.80	104.50	88.70	8.778	58.06	49.28	43.33
120	5.95	.418	18.06	105.90	87.90	10.03	58.83	48.83	48.89

Carbon Tetrachloride, CCl_4

Temp. °F	Spec. vol. ft.³/lb.	Spec. vol. m³/kg	Density lb./ft.³	Density kg/m³	Temp. °C
20	69.5	4.34	0.01438	0.2303	− 6.67
30	53.0	3.31	.01886	.3021	− 1.11
40	40.0	2.50	.02500	.4005	+ 4.44
60	24.0	1.50	.04166	.6673	15.56
70	19.5	1.22	.05128	.8214	21.11
80	16.0	0.999	.06345	1.016	26.67
90	13.0	0.812	.07692	1.232	32.22
100	10.0	0.624	.1000	1.602	37.78
110	8.5	0.53	.1176	1.884	43.33
120	7.5	0.47	.1333	2.135	48.89

Ethyl Ether, $(C_2H_5)_2O$

Temp. °F	Abs. press. sat. vap. lb./in.²	Abs. press. sat. vap. kg/cm²	Heat content Liq. (BTU/lb.)	Heat content Vap. (BTU/lb.)	Ht. of vaporiz. BTU/lb.	Heat content Liq. (g-cal./g)	Heat content Vap. (g-cal./g)	Ht. of vaporiz. g-cal./g	Temp. °C
0	1.3	0.091	−18.00	153.00	171.0	−10.00	85.000	95.00	−17.78
10	1.8	.13	−12.0	158.43	170.4	− 6.67	88.017	94.67	−12.22
20	2.5	.18	− 6.50	163.50	170.0	− 3.61	90.833	94.44	− 6.67
30	3.4	.24	− 1.50	167.90	169.4	− 0.833	93.278	94.11	− 1.11
40	4.4	.31	+ 4.00	172.40	168.4	+ 2.22	95.778	93.56	+ 4.44
50	5.5	.39	9.57	177.17	167.6	5.32	98.428	93.11	10.00
70	8.8	.62	20.04	185.44	165.4	11.13	103.02	91.89	21.11
80	10.9	.766	26.40	190.60	164.2	14.67	105.89	91.22	26.67
90	13.4	.942	31.50	194.50	163.0	17.50	108.06	90.56	32.22
100	16.0	1.12	36.50	197.50	161.5	20.28	109.72	89.72	37.78

Ethyl Ether, $(C_2H_5)_2O$

Temp. °F	Spec. vol. ft.³/lb.	Spec. vol. m³/kg	Density lb./ft.³	Density kg/m³	Temp. °C
0	38.0	2.37	0.0263	0.4213	−17.78
10	32.5	2.03	.0332	.5318	−12.22
20	27.0	1.69	.0372	.5959	− 6.67
30	21.4	1.34	.0468	.7496	− 1.11
40	17.0	1.06	.0588	.9419	+ 4.44
50	13.2	0.824	.0757	1.213	10.00
70	7.8	0.49	.1280	2.050	21.11
80	6.2	0.39	.1620	2.595	26.67
90	5.1	0.32	.1960	3.140	32.22
100	4.5	0.28	.2220	3.556	37.78

Methyl Chloride, CH_3Cl

Temp. °F	Abs. press. sat. vap.		Heat content abv. 32°F BTU/lb.		Ht. of vaporiz. BTU/lb.	Heat content abv. 0°C g-cal./g		Ht. of vaporiz. g-cal./g	Temp. °C
	lb./in.²	kg/cm²	Liq.	Vap.		Liq.	Vap.		
−20	11.75	0.8261	−19.0	167.36	186.36	−10.6	92.978	103.53	−28.89
−10	15.0	1.055	−15.38	168.83	184.21	− 8.544	93.795	102.34	−23.33
− 5	16.79	1.180	−13.58	169.54	183.12	− 7.544	94.189	101.73	−20.56
0	18.8	1.32	−11.75	170.23	181.98	− 6.528	94.572	101.10	−17.78
+ 5	21.0	1.48	− 9.93	170.96	180.84	− 5.517	94.978	100.47	−15.00
10	23.3	1.64	− 8.06	171.58	179.65	− 4.478	95.322	99.806	−12.22
15	25.9	1.82	− 6.74	172.24	178.47	− 3.744	95.689	99.150	− 9.44
20	28.8	2.02	− 4.32	172.95	177.27	− 2.400	96.083	98.483	− 6.67
25	31.8	2.24	− 2.48	173.63	176.10	− 1.378	96.461	97.833	− 3.89
30	35.2	2.47	− 0.62	174.28	174.90	− 0.344	96.822	97.167	− 1.11
35	38.7	2.72	+ 1.75	174.92	173.77	+ 0.972	97.178	96.539	+ 1.67
40	42.6	3.00	3.15	175.57	172.42	1.75	97.539	95.789	4.44
45	46.9	3.30	5.04	176.20	171.16	2.80	97.889	95.089	7.22
50	51.5	3.62	6.88	176.78	169.90	3.82	98.211	94.389	10.00
55	56.4	3.97	8.80	177.45	168.65	4.89	98.583	93.695	12.78
60	61.6	4.33	10.70	178.05	167.35	5.944	98.917	92.972	15.56
65	67.3	4.73	12.62	178.64	166.02	7.011	99.245	92.233	18.33
70	73.3	5.15	14.52	179.17	164.65	8.067	99.539	91.472	21.11
75	79.2	5.57	16.46	179.78	163.28	9.144	99.878	90.711	23.89
80	85.3	6.00	18.36	180.24	161.88	10.20	100.13	89.933	26.67
85	94.1	6.62	20.12	180.74	160.48	11.18	100.41	89.156	29.44
90	102.1	7.178	22.13	181.22	159.09	12.29	100.68	88.383	32.22
95	110.3	7.755	24.07	181.76	157.70	13.37	100.98	87.611	35.00
100	118.8	8.352	26.06	182.36	156.30	14.48	101.31	86.833	37.78

Methyl Chloride, CH_3Cl

Temp. °F	Spec. vol. sat. vap.		Spec. vol. liq.		Density sat. vap.		Density of liq.		Temp. °C
	ft.³/lb.	m³/kg	ft.³/lb.	m³/kg	lb./ft.³	kg/m³	lb./ft.³	kg/m³	
−20	8.09	.505	.015827	.0009880	0.124	1.98	63.185	1012.1	−28.89
−10	6.46	.403	.015985	.0009979	.155	2.48	62.560	1002.1	−23.33
− 5	5.80	.362	.016013	.0009997	.172	2.76	62.450	1000.3	−20.56
0	5.18	.323	.016146	.001008	.193	3.09	61.936	992.09	−17.78
+ 5	4.68	.292	.016228	.001013	.214	3.42	61.623	987.08	−15.00
10	4.18	.261	.016310	.001018	.239	3.83	61.311	982.08	−12.22
15	3.88	.242	.016388	.001023	.258	4.13	61.022	977.45	− 9.44
20	3.41	.213	.016474	.001028	.293	4.70	60.702	972.32	− 6.67
25	3.09	.193	.016552	.001033	.324	5.18	60.415	967.73	− 3.89
30	2.81	.175	.016645	.001039	.356	5.70	60.077	962.31	− 1.11
35	2.50	.156	.016746	.001045	.400	6.41	59.715	956.51	+ 1.67
40	2.31	.144	.016809	.001049	.433	6.93	59.492	952.94	4.44
45	2.10	.131	.016929	.001057	.476	7.63	59.069	946.17	7.22
50	1.93	.120	.017023	.001063	.518	8.30	58.745	940.98	10.00
55	1.75	.109	.017118	.001069	.571	9.16	58.419	935.76	12.78
60	1.61	.101	.017219	.001075	.621	9.95	58.077	930.28	15.56
65	1.47	.0918	.017318	.001081	.680	10.9	57.742	924.91	18.33
70	1.34	.0837	.017421	.001088	.746	12.0	57.403	919.48	21.11
75	1.24	.0774	.017526	.001094	.806	12.9	57.058	913.96	23.89
80	1.14	.0712	.017632	.001101	.877	14.1	56.714	908.44	26.67
85	1.05	.0655	.017740	.001108	.952	15.3	56.369	902.92	29.44
90	0.98	.061	.017850	.001114	1.02	16.3	56.022	897.36	32.22
95	0.91	.057	.017961	.001121	1.10	17.6	55.675	891.80	35.00
100	0.85	.053	.018074	.001128	1.18	18.8	55.327	886.23	37.78

Mercury, Hg

Temp. °F	Abs. press. sat. vap.		Heat content abv. 32°F BTU/lb.		Ht. of vaporiz. BTU/lb.	Heat content abv. 0°C g-cal./g		Ht. of vaporiz. g-cal./g	Temp. °C
	lb./in.²	kg/cm²	Liq.	Vap.		Liq.	Vap		
402	0.4	0.03	13.81	141.96	128.15	7.672	78.867	71.195	205.56
444	0.8	0.06	15.36	142.60	127.24	8.533	79.222	70.689	228.89
458	1.0	0.07	15.89	142.81	126.92	8.828	79.339	70.511	236.67
485	1.5	0.11	16.90	143.23	126.33	9.389	79.572	70.183	251.67
505	2.0	0.14	17.65	143.54	125.89	9.806	79.745	69.939	262.78
558	4.0	0.28	19.62	144.34	124.72	10.90	80.189	69.289	292.22
591	6.0	0.42	20.87	144.86	123.99	11.59	80.478	68.883	310.56
617	8.0	0.56	21.81	145.24	123.43	12.12	80.689	68.572	325.00
637	10.0	0.703	22.58	145.56	122.98	12.54	80.867	68.322	336.11
676	15.0	1.05	24.04	146.16	122.12	13.36	81.200	67.844	357.78
706	20.0	1.41	25.15	146.61	121.46	13.97	81.450	67.478	374.44
730	25.0	1.76	26.05	146.98	120.93	14.47	81.656	67.183	387.78
751	30.0	2.11	26.81	147.29	120.48	14.89	81.828	66.933	399.44
769	35.0	2.46	27.49	147.57	120.08	15.27	81.983	66.711	409.44
785	40.0	2.81	28.08	147.81	119.73	15.60	82.117	66.517	418.33
799	45.0	3.16	28.62	148.04	119.42	15.90	82.245	66.344	426.11
812	50	3.5	29.11	148.24	119.13	16.17	82.356	66.183	433.33
836	60	4.2	29.99	148.60	118.61	16.66	82.556	65.894	446.67
857	70	4.9	30.75	148.90	118.15	17.08	82.722	65.639	458.33
875	80	5.6	31.44	149.19	117.75	17.47	82.883	65.417	468.33
892	90	6.3	32.06	149.44	117.38	17.81	83.022	65.211	477.78
907	100	7.03	32.63	149.68	117.05	18.13	83.156	65.028	486.11
921	110	7.73	33.16	149.90	116.74	18.42	83.278	64.856	493.89
934	120	8.44	33.66	150.10	116.44	18.70	83.389	64.689	501.11
947	130	9.14	34.12	150.29	116.17	18.96	83.495	64.539	508.33
958	140	9.84	34.55	150.47	115.92	19.19	83.595	64.400	514.44
969	150	10.5	34.96	150.63	115.67	19.42	83.683	64.261	520.56
1000	180	12.7	36.09	151.10	115.01	20.05	83.945	63.894	537.78

Mercury, Hg

Temp. °F	Spec. vol. sat. vap.		Density of sat. vap.		Entropy above 32°F BTU/lb./°F			Temp °C
	ft.³/lb.	m³/kg	lb./ft.³	kg/m³	Liq.	Vap.	Evap.	
402	114.50	7.1480	0.008733	0.1399	.0209	.1696	.1487	205.56
444	59.72	3.728	.016745	0.26822	.0227	.1635	.1408	228.89
458	48.45	3.025	.02064	0.3306	.0233	.1616	.1383	236.67
485	33.14	2.069	.03017	0.4833	.0244	.1581	.1337	251.67
505	25.32	1.581	.03948	0.6324	.0251	.1556	.1305	262.78
558	13.26	0.8278	.07540	1.208	.0271	.1497	.1226	292.22
591	9.096	.5678	.10993	1.7609	.0283	.1462	.1179	310.56
617	6.9630	.43469	.14361	2.3003	.0292	.1439	.1147	325.00
637	5.6610	.35341	.17664	2.8294	.0299	.1420	.1121	336.11
676	3.8923	.24299	.25691	4.1152	.0312	.1387	.1075	357.78
706	2.983	.1862	.3352	5.369	.0322	.1364	.1042	374.44
730	2.429	.1516	.4117	6.595	.0330	.1346	.1016	387.78
751	2.053	.1282	.4871	7.802	.0336	.1331	.0995	399.44
769	1.7815	.11122	.5613	8.991	.0342	.1319	.0977	409.44
785	1.5762	.098399	.6344	10.16	.0346	.1308	.0962	418.33
799	1.4147	.088317	.7069	11.32	.0351	.1300	.0949	426.11
812	1.284	.08016	.7788	12.47	.0355	.1291	.0936	433.33
836	1.086	.06780	.9204	14.74	.0361	.1276	.0915	446.67
857	0.9436	.05891	1.0597	16.974	.0367	.1265	.0898	458.33
875	.8349	.05212	1.1977	19.185	.0372	.1254	.0882	468.33
892	.7497	.04680	1.3338	21.365	.0377	.1247	.0870	477.78
907	.6811	.04252	1.4682	23.518	.0381	.1237	.0856	486.11
921	.6242	.03897	1.6020	25.661	.0385	.1230	.0845	493.89
934	.5767	.03600	1.7340	27.775	.0389	.1224	.0835	501.11
947	.5360	.03346	1.8656	29.883	.0392	.1218	.0826	508.33
958	.5012	.03129	1.9952	31.959	.0395	.1213	.0818	514.44
969	.4706	.02938	2.125	34.04	.0398	.1207	.0809	520.56
1000	.3990	.02491	2.506	40.14	.0406	.1194	.0788	537.78

THERMODYNAMIC PROPERTIES (Concluded)

MISCELLANEOUS PROPERTIES OF COMMON REFRIGERANTS

Refrigerant	Formula	Flash Point °F.	Ignition Temp. °F.	Explosive Limits % by Volume		Vapor Density (Air = 1)	Boiling Point °F	Threshold Limit Value* Parts per Million in Air	Water Soluble	Odor	Refrigerant
				Lower	Upper						
Ammonia	NH_3	....	1204	16	25	0.59	−28	100	yes	yes	Ammonia
Bromotrifluoromethane (Kulene-131)	CF_3Br		nonflammable			5.25	−73.6	....	no	yes	Bromotrifluoromethane (Kulene-131)
Butane	C_4H_{10}	−76	806	1.8	8.4	2.04	33	....	no	no	Butane
Carbon dioxide	CO_2		nonflammable			1.53	−108	5000	yes	no	Carbon dioxide
Carbon tetrachloride	CCl_4		nonflammable			5.32	170	25	no	yes	Carbon tetrachloride
Dichlorodifluoromethane (Freon-12)	CCl_2F_2		nonflammable			4.17	−21.6	....	no	no	Dichlorodifluoromethane (Freon-12)
Dichlorodifluoromethane, 73.8% Ethylidene fluoride, 26.2% (Carrene-7)	CCl_2F_2 CH_3CHF_2		nonflammable			3.24	−28.0	....	no	yes	Dichlorodifluoromethane, 73.8% Ethylidene fluoride, 26.2% (Carrene-7)
Dichloromonofluoromethane (Freon-21)	$CHCl_2F$	practically nonflammable				3.55	48	....	no	yes	Dichloromonofluoromethane (Freon-21)
Dichlorotetrafluoroethane (Freon-114)	$C_2Cl_2F_4$	practically nonflammable				5.89	38	....	no	no	Dichlorotetrafluoroethane (Freon-114)
Ethane	C_2H_6	<20	950	3.0	12.5	1.04	−128	....	no	no	Ethane
Ethylene	C_2H_4	<20	842	3.1	32	0.972	−155	....	yes	yes	Ethylene
Isobutane	$(CH_3)_3CH$	<20	1010	1.8	8.4	2.01	14	....	no	no	Isobutane
Methyl chloride	CH_3Cl	632	1170	10.7	11.4	1.78	−11	100	yes	yes	Methyl chloride
Monochlorodifluoromethane (Freon-22)	$CHClF_2$	practically nonflammable				2.9	−41	....	yes	no	Monochlorodifluoromethane (Freon-22)
Monochlorotrifluoromethane (Freon-13)	$CClF_3$		nonflammable			3.6	−112	....	.....	no	Monochlorotrifluoromethane (Freon-13)
Propane	C_3H_8	<20	871	2.2	9.5	1.56	−45	....	no	no	Propane
Propylene	C_3H_6	<20	927	2.4	10.3	1.49	−53	....	yes	yes	Propylene
Sulfur dioxide	SO_2		nonflammable			2.2	14	10	yes	yes	Sulfur dioxide
Tetrafluoromethane (Freon-14)	CF_4		nonflammable			3.0	−198	....	no	no	Tetrafluoromethane (Freon-14)
Trichloroethylene	C_2HCl_3	nonflammable at normal temperature				4.53	189	200	no	yes	Trichloroethylene
Trichloromonofluoromethane (Freon-11) (Carrene-2)	CCl_3F		nonflammable			4.7	75.3	....	no	no	Trichloromonofluoromethane (Freon-11) (Carrene-2)
Trichlorotrifluoroethane (Freon-113)	$C_2Cl_3F_3$	practically nonflammable				6.4	118	....	no	no	Trichlorotrifluoroethane (Freon-113)

* Maximum average atmospheric concentration of contaminants to which workers may be exposed for an eight-hour work day without injury to health. (American Conference of Governmental Industrial Hygienists: "Threshold Limit Values for 1954.) Where blanks appear in this column no published information was available on threshold limit values.

MISCELLANEOUS PROPERTIES OF COMMON REFRIGERANTS

HYGROMETRIC AND BAROMETRIC TABLES

CONVERSION TABLE FOR BAROMETRIC READINGS

U. S. inches to cm.

Inches.	.00	.01	.02	.03	.04	.05	.06	.07	.08	.09
27.0	68.580	.606	.631	.656	.682	.707	.733	.758	.783	.809
27.1	.834	.860	.885	.910	.936	.961	.987	*.012	*.037	*.063
27.2	69.088	.114	.139	.164	.190	.215	.241	.266	.291	.317
27.3	.342	.368	.393	.418	.444	.469	.495	.520	.545	.571
27.4	.596	.622	.647	.672	.698	.723	.749	.774	.799	.825
27.5	.850	.876	.901	.926	.952	.977	*.002	*.028	*.053	*.079
27.6	70.104	.130	.155	.180	.206	.231	.257	.282	.307	333
27.7	.358	.384	.409	.434	.460	.485	.511	.536	.561	.587
27.8	.612	.638	.663	.688	.714	.739	.765	.790	.815	.841
27.9	.866	.892	.917	.942	.968	.993	*.018	*.044	*.069	*.095
28.0	71.120	.146	.171	.196	.222	.247	.273	.298	.323	.349
28.1	.374	.400	.425	.450	.476	.501	.527	.552	.577	.603
28.2	.628	.654	.679	.704	.730	.755	.781	.806	.831	.857
28.3	.882	.908	.933	.958	.984	*.009	*.035	*.060	*.085	*.111
28.4	72.136	.162	.187	.212	.238	.263	.289	.314	.339	.365
28.5	390	.416	.441	.466	.492	.517	.543	.568	.593	.619
28.6	.644	.670	.695	.720	.746	.771	.797	.822	.847	.873
28.7	.898	.924	.949	.974	*.000	*.025	*.051	*.076	*.101	*.127
28.8	73.152	.178	.203	.228	.254	.279	.305	.330	.355	.381
28.9	.406	.432	.457	.482	.508	.533	.559	.584	.609	.635
29.0	.660	.686	.711	.736	.762	.787	.813	.838	.863	.889
29.1	.914	.940	.965	.990	*.016	*.041	*.067	*.092	* 117	*.143
29.2	74.168	.194	.219	.244	.270	.295	.321	.346	.371	.397
29.3	.422	.448	.473	.498	.524	.549	.575	.600	.625	.651
29.4	.676	.702	.727	.752	.778	.803	.829	.854	.879	.905
29.5	.930	.956	.981	*.006	*.032	*.057	*.083	*.108	*.133	*.159
29.6	75.184	.210	.235	.260	.286	.311	.337	.362	.387	.413
29.7	.438	.464	.489	.514	.540	.565	.591	.616	.641	.667
29.8	.692	.718	.743	.768	.794	.819	.845	.870	.895	.921
29.9	.946	.972	.997	*.022	*.048	*.073	*.099	*.124	*.149	*.175
30.0	76.200	.226	.251	.277	.302	.327	.353	.378	.404	.429
30.1	.454	.480	.505	.531	.556	.581	.607	.632	.658	.683
30.2	.708	.734	.759	.785	.810	.835	.861	.886	.912	.937
30.3	.962	.988	*.013	*.039	*.064	*.089	*.115	*.140	*.166	*.191
30.4	77.216	.242	.267	.293	.318	.343	.369	.394	.420	.445
30.5	.470	.496	.521	.547	.572	.597	.623	.648	.674	.699
30.6	.724	.750	.775	.801	.826	.851	.877	.902	.928	.953
30.7	.978	*.004	*.029	*.055	*.080	*.105	*.131	*.156	*.182	*.207
30.8	78.232	.258	.283	.309	.334	.359	.385	410	.436	.461
30.9	.486	.512	.537	.563	.588	.613	.639	.664	.690	.715

U. S. Inches to Millibars

Based on the relation 1 inch of mercury at 32°F represents a pressure of 33.8639 millibars.

Note:
Figures in last nine columns to be preceded by 7, 8, 9 or 10 as indicated in column 2.

Inches	.00	.01	.02	.03	.04	.05	.06	.07	.08	.09
23.0	7 78.87	79.21	79.55	79.89	80.22	80.56	80.90	81.24	81.58	81.92
23.1	7 82.26	82.59	82.93	83.27	83.61	83.95	84.29	84.63	84.97	85.30
23.2	7 85.64	85.98	86.32	86.66	87.00	87.34	87.67	88.01	88.35	88.69
23.3	7 89.03	89.37	89.71	90.04	90.38	90.72	91.06	91.40	91.74	92.08
23.4	7 92.42	92.75	93.09	93.43	93.77	94.11	94.45	94.79	95.12	95.46
23.5	7 95.80	96.14	96.48	96.82	97.16	97.49	97.83	98.17	98.51	98.85
23.6	7 99.19	99.53	99.87	*00.20	*00.54	*00.88	*01.22	*01.56	*01.90	*02.24
23.7	8 02.57	02.91	03.25	03.59	03.93	04.27	04.61	04.94	05.28	05.62
23.8	8 05.96	06.30	06.64	06.98	07.32	07.65	07.99	08.33	08.67	09.01
23.9	8 09.35	09.69	10.02	10.36	10.70	11.04	11.38	11.72	12.06	12.39
24.0	8 12.73	13.07	13.41	13.75	14.09	14.43	14.77	15.10	15.44	15.78
24.1	8 16.12	16.46	16.80	17.14	17.47	17.81	18.15	18.49	18.83	19.17
24.2	8 19.51	19.85	20.18	20.52	20.86	21.20	21.54	21.88	22.22	22.55
24.3	8 22.89	23.23	23.57	23.91	24.25	24.59	24.92	25.26	25.60	25.94
24.4	8 26.28	26.62	26.96	27.30	27.63	27.97	28.31	28.65	28.99	29.33
24.5	8 29.67	30.00	30.34	30.68	31.02	31.36	31.70	32.04	32.37	32.71
24.6	8 33.05	33.39	33.73	34.07	34.41	34.75	35.08	35.42	35.76	36.10
24.7	8 36.44	36.78	37.12	37.45	37.79	38.13	38.47	38.81	39.15	39.49
24.8	8 39.82	40.16	40.50	40.84	41.18	41.52	41.86	42.20	42.53	42.87
24.9	8 43.21	43.55	43.89	44.23	44.57	44.90	45.24	45.58	45.92	46.26
25.0	8 46.60	46.94	47.27	47.61	47.95	48.29	48.63	48.97	49.31	49.65
25.1	8 49.98	50.32	50.66	51.00	51.34	51.68	52.02	52.35	52.69	53.03
25.2	8 53.37	53.71	54.05	54.39	54.72	55.06	55.40	55.74	56.08	56.42
25.3	8 56.76	57.10	57.43	57.77	58.11	58.45	58.79	59.13	59.47	59.80
25.4	8 60.14	60.48	60.82	61.16	61.50	61.84	62.17	62.51	62.85	63.19

U. S. Inches to Millibars (Continued)

Inches	.00	.01	.02	.03	.04	.05	.06	.07	.08	.09
25.5	8 63.53	63.87	64.21	64.55	64.88	65.22	65.56	65.90	66.24	66.58
25.6	8 66.92	67.25	67.59	67.93	68.27	68.61	68.95	69.29	69.62	69.96
25.7	8 70.30	70.64	70.98	71.32	71.66	72.00	72.33	72.67	73.01	73.35
25.8	8 73.69	74.03	74.37	74.70	75.04	75.38	75.72	76.06	76.40	76.74
25.9	8 77.08	77.41	77.75	78.09	78.43	78.77	79.11	79.45	79.78	80.12
26.0	8 80.46	80.80	81.14	81.48	81.82	82.15	82.49	82.83	83.17	83.51
26.1	8 83.85	84.19	84.53	84.86	85.20	85.54	85.88	86.22	86.56	86.90
26.2	8 87.23	87.57	87.91	88.25	88.59	88.93	89.27	89.60	89.94	90.28
26.3	8 90.62	90.96	91.30	91.64	91.98	92.31	92.65	92.99	93.33	93.67
26.4	8 94.01	94.35	94.68	95.02	95.36	95.70	96.04	96.38	96.72	97.05
26.5	8 97.39	97.73	98.07	98.41	98.75	99.09	99.43	99.76	*00.10	*00.44
26.6	9 00.78	01.12	01.46	01.80	02.13	02.47	02.81	03.15	03.49	03.83
26.7	9 04.17	04.50	04.84	05.18	05.52	05.86	06.20	06.54	06.88	07.21
26.8	9 07.55	07.89	08.23	08.57	08.91	09.25	09.58	09.92	10.26	10.60
26.9	9 10.94	11.28	11.62	11.95	12.29	12.63	12.97	13.31	13.65	13.99
27.0	9 14.33	14.66	15.00	15.34	15.68	16.02	16.36	16.70	17.03	17.37
27.1	9 17.71	18.05	18.39	18.73	19.07	19.40	19.74	20.08	20.42	20.76
27.2	9 21.10	21.44	21.78	22.11	22.45	22.79	23.13	23.47	23.81	24.15
27.3	9 24.48	24.82	25.16	25 50	25.84	26.18	26.52	26.85	27.19	27.53
27.4	9 27.87	28 21	28.55	28.89	29.23	29.56	29.90	30.24	30.58	30.[illegible]
27.5	9 31.26	31.60	31.93	32.27	32.61	32.95	33.29	33.63	33.97	34.31
27.6	9 34.64	34.98	35.32	35.66	36.00	36.34	36.68	37.01	37.35	37.69
27.7	9 38.03	38.37	38.71	39.05	39.38	39.72	40.06	40.40	40.74	41.08
27.8	9 41.42	41.76	42.09	42.43	42.77	43.11	43.45	43.79	44.13	44.46
27.9	9 44.80	45.14	45.48	45.82	46.16	46.50	46.83	47.17	47.51	47.85
28.0	9 48.19	48.53	48.87	49.21	49.54	49.88	50.22	50.56	50.90	51.24
28.1	9 51.58	51.91	52.25	52.59	52.93	53.27	53.61	53.95	54.28	54.62
28.2	9 54.96	55.30	55.64	55.98	56.32	56.66	56.99	57.33	57.67	58.01
28.3	9 58.35	58.69	59 03	59.36	59.70	60.04	60.38	60.72	61.06	61.40
28.4	9 61.73	62.07	62.41	62.75	63.09	63.43	63.77	64.11	64.44	64.78
28.5	9 65.12	65.46	65.80	66.14	66.48	66.81	67.15	67.49	67.83	68.17
28.6	9 68.51	68.85	69.18	69.52	69.86	70.20	70.54	70.88	71.22	71.56
28.7	9 71.89	72.23	72.57	72.91	73.25	73.59	73.93	74.26	74.60	74.94
28.8	9 75.28	75.62	75.96	76.30	76.63	76.97	77.31	77.65	77.99	78.33
28.9	9 78.67	79.01	79.34	79.68	80.02	80.36	80.70	81.04	81.38	81.71
29.0	9 82.05	82.39	82.73	83.07	83.41	83.75	84.08	84.42	84.76	85.10
29.1	9 85.44	85.78	86.12	86.46	86.79	87.13	87.47	87.81	88.15	88.49
29.2	9 88.83	89.16	89.50	89.84	90.18	90.52	90.86	91.20	91.53	91.87
29.3	9 92.21	92.55	92.89	93.23	93.57	93.91	94.24	94.58	94.92	95.26
29.4	9 95.60	95.94	96.28	96.61	96.95	97.29	97.63	97.97	98.31	98.65
29.5	9 98.99	99.32	99.66	*00.00	*00.34	*00.68	*01.02	*01.36	*01.69	*02.03
29.6	10 02.37	02.71	03.05	03.39	03.73	04.06	04.40	04.74	05.08	05.42
29.7	10 05.76	06.10	06.44	06.77	07.11	07.45	07.79	08.13	08.47	08.81
29.8	10 09.14	09.48	09.82	10.16	10.50	10.84	11.18	11.51	11.85	12.19
29.9	10 12.53	12.87	13.21	13.55	13.89	14.22	14.56	14.90	15.24	15.58
30.0	10 15.92	16.26	16.59	16.93	17.27	17.61	17.95	18.29	18.63	18.96
30.1	10 19.30	19.64	19.98	20.32	20.66	21.00	21.34	21.67	22.01	22.35
30.2	10 22.69	23.03	23.37	23.71	24.04	24.38	24.72	25.06	25.40	25.74
30.3	10 26.08	26.41	26.75	27.09	27.43	27.77	28.11	28.45	28.79	29.12
30.4	10 29.46	29.80	30.14	30.48	30.82	31.16	31.49	31.83	32.17	32.51
30.5	10 32.85	33.19	33.53	33.86	34.20	34.54	34.88	35.22	35.56	35.90
30.6	10 36.24	36.57	36.91	37.25	37.59	37.93	38.27	38.61	38.94	39.28
30.7	10 39.62	39.96	40.30	40.64	40.98	41.31	41.65	41.99	42.33	42.67
30.8	10 43.01	43.35	43.69	44.02	44.36	44.70	45.04	45.38	45.72	46.06
30.9	10 46.39	46.73	47.07	47.41	47.75	48.09	48.43	48.76	49.10	49.44
31.0	10 49.78	50.12	50.46	50.80	51.14	51.47	51.81	52.15	52.49	52.83
31.1	10 53.17	53.51	53.84	54.18	54.52	54.86	55.20	55.54	55.88	56.22
31.2	10 56.55	56.89	57.23	57.57	57.91	58.25	58.59	58.92	59.26	59.60
31.3	10 59.94	60.28	60.62	60.96	61.29	61.63	61.97	62.31	62.65	62.99
31.4	10 63.33	63.67	64.00	64.34	64.68	65.02	65.36	65.70	66.04	66.37
31.5	10 66.71	67.05	67.39	67.73	68.07	68.41	68.74	69.08	69.42	69.76
31.6	10 70.10	70.44	70.78	71.12	71.45	71.79	72.13	72.47	72.81	73.15
31.7	10 73.49	73.82	74.16	74.50	74.84	75.18	75.52	75.86	76.19	76.53
31.8	10 76.87	77.21	77.55	77.89	78.23	78.57	78.90	79.24	79.58	79.92
31.9	10 80.26	80.60	80.94	81.27	81.61	81.95	82.29	82.63	82.97	83.31

Note: Figures in last nine columns to be preceded by 9 or 10 as indicated.

CONVERSION TABLE FOR BAROMETRIC READINGS (Continued)

Centimeters to Millibars

Based on the relation 1 centimeter of mercury at 0°C represents a pressure of 13.3322 millibars. Note: Figures in last nine columns to be preceded by 9.

Centimeters	.00	.01	.02	.03	.04	.05	.06	.07	.08	.09
68.0	9 06.59	06.72	06.86	06.99	07.12	07.26	07.39	07.52	07.66	07.79
68.1	9 07.92	08.06	08.19	08.32	08.46	08.59	08.72	08.86	08.99	09.12
68.2	9 09.26	09.39	09.52	09.66	09.79	09.92	10.06	10.19	10.32	10.46
68.3	9 10.59	10.72	10.86	10.99	11.12	11.26	11.39	11.52	11.66	11.79
68.4	9 11.92	12.06	12.19	12.32	12.46	12.59	12.72	12.86	12.99	13.12
68.5	9 13.26	13.39	13.52	13.66	13.79	13.92	14.06	14.19	14.32	14.46
68.6	9 14.59	14.72	14.86	14.99	15.12	15.26	15.39	15.52	15.66	15.79
68.7	9 15.92	16.06	16.19	16.32	16.46	16.59	16.72	16.86	16.99	17.12
68.8	9 17.26	17.39	17.52	17.66	17.79	17.92	18.06	18.19	18.32	18.46
68.9	9 18.59	18.72	18.86	18.99	19.12	19.26	19.39	19.52	19.66	19.79
69.0	9 19.92	20.06	20.19	20.32	20.46	20.59	20.72	20.86	20.99	21.12
69.1	9 21.26	21.39	21.52	21.65	21.79	21.92	22.05	22.19	22.32	22.45
69.2	9 22.59	22.72	22.85	22.99	23.12	23.25	23.39	23.52	23.65	23.79
69.3	9 23.92	24.05	24.19	24.32	24.45	24.59	24.72	24.85	24.99	25.12
69.4	9 25.25	25.39	25.52	25.65	25.79	25.92	26.05	26.19	26.32	26.45
69.5	9 26.59	26.72	26.85	26.99	27.12	27.25	27.39	27.52	27.65	27.79
69.6	9 27.92	28.05	28.19	28.32	28.45	28.59	28.72	28.85	28.99	29.12
69.7	9 29.25	29.39	29.52	29.65	29.79	29.92	30.05	30.19	30.32	30.45
69.8	9 30.59	30.72	30.85	30.99	31.12	31.25	31.39	31.52	31.65	31.79
69.9	9 31.92	32.05	32.19	32.32	32.45	32.59	32.72	32.85	32.99	33.12
70.0	9 33.25	33.39	33.52	33.65	33.79	33.92	34.05	34.19	34.32	34.45
70.1	9 34.59	34.72	34.85	34.99	35.12	35.25	35.39	35.52	35.65	35.79
70.2	9 35.92	36.05	36.19	36.32	36.45	36.59	36.72	36.85	36.99	37.12
70.3	9 37.25	37.39	37.52	37.65	37.79	37.92	38.05	38.19	38.32	38.45
70.4	9 38.59	38.72	38.85	38.99	39.12	39.25	39.39	39.52	39.65	39.79
70.5	9 39.92	40.05	40.19	40.32	40.45	40.59	40.72	40.85	40.99	41.12
70.6	9 41.25	41.39	41.52	41.65	41.79	41.92	42.05	42.19	42.32	42.45
70.7	9 42.59	42.72	42.85	42.99	43.12	43.25	43.39	43.52	43.65	43.79
70.8	9 43.92	44.05	44.19	44.32	44.45	44.59	44.72	44.85	44.99	45.12
70.9	9 45.25	45.39	45.52	45.65	45.79	45.92	46.05	46.19	46.32	46.45
71.0	9 46.59	46.72	46.85	46.99	47.12	47.25	47.39	47.52	47.65	47.79
71.1	9 47.92	48.05	48.19	48.32	48.45	48.59	48.72	48.85	48.99	49.12
71.2	9 49.25	49.39	49.52	49.65	49.79	49.92	50.05	50.19	50.32	50.45
71.3	9 50.59	50.72	50.85	50.99	51.12	51.25	51.39	51.52	51.65	51.79
71.4	9 51.92	52.05	52.19	52.32	52.45	52.59	52.72	52.85	52.99	53.12
71.5	9 53.25	53.39	53.52	53.65	53.79	53.92	54.05	54.19	54.32	54.45
71.6	9 54.59	54.72	54.85	54.99	55.12	55.25	55.39	55.52	55.65	55.79
71.7	9 55.92	56.05	56.19	56.32	56.45	56.59	56.72	56.85	56.99	57.12
71.8	9 57.25	57.39	57.52	57.65	57.79	57.92	58.05	58.19	58.32	58.45
71.9	9 58.59	58.72	58.85	58.99	59.12	59.25	59.39	59.52	59.65	59.79
72.0	9 59.92	60.05	60.19	60.32	60.45	60.59	60.72	60.85	60.98	61.12
72.1	9 61.25	61.38	61.52	61.65	61.78	61.92	62.05	62.18	62.32	62.45
72.2	9 62.58	62.72	62.85	62.98	63.12	63.25	63.38	63.52	63.65	63.78
72.3	9 63.92	64.05	64.18	64.32	64.45	64.58	64.72	64.85	64.98	65.12
72.4	9 65.25	65.38	65.52	65.65	65.78	65.92	66.05	66.18	66.32	66.45
72.5	9 66.58	66.72	66.85	66.98	67.12	67.25	67.38	67.52	67.65	67.78
72.6	9 67.92	68.05	68.18	68.32	68.45	68.58	68.72	68.85	68.98	69.12
72.7	9 69.25	69.38	69.52	69.65	69.78	69.92	70.05	70.18	70.32	70.45
72.8	9 70.58	70.72	70.85	70.98	71.12	71.25	71.38	71.52	71.65	71.78
72.9	9 71.92	72.05	72.18	72.32	72.45	72.58	72.72	72.85	72.98	73.12

Centimeters to Millibars (Continued)

Note: Figures in last nine columns to be preceded by 9 or 10.

Centimeters	.00	.01	.02	.03	.04	.05	.06	.07	.08	.09
73.0	9 73.25	73.38	73.52	73.65	73.78	73.92	74.05	74.18	74.32	74.45
73.1	9 74.58	74.72	74.85	74.98	75.12	75.25	75.38	75.52	75.65	75.78
73.2	9 75.92	76.05	76.18	76.32	76.45	76.58	76.72	76.85	76.98	77.12
73.3	9 77.25	77.38	77.52	77.65	77.78	77.92	78.05	78.18	78.32	78.45
73.4	9 78.58	78.72	78.85	78.98	79.12	79.25	79.38	79.52	79.65	79.78
73.5	9 79.92	80.05	80.18	80.32	80.45	80.58	80.72	80.85	80.98	81.12
73.6	9 81.25	81.38	81.52	81.65	81.78	81.92	82.05	82.18	82.32	82.45
73.7	9 82.58	82.72	82.85	82.98	83.12	83.25	83.38	83.52	83.65	83.78
73.8	9 83.92	84.05	84.18	84.32	84.45	84.58	84.72	84.85	84.98	85.12
73.9	9 85.25	85.38	85.52	85.65	85.78	85.92	86.05	86.18	86.32	86.45
74.0	9 86.58	86.72	86.85	86.98	87.12	87.25	87.38	87.52	87.65	87.78
74.1	9 87.92	88.05	88.18	88.32	88.45	88.58	88.72	88.85	88.98	89.12
74.2	9 89.25	89.38	89.52	89.65	89.78	89.92	90.05	90.18	90.32	90.45
74.3	9 90.58	90.72	90.85	90.98	91.12	91.25	91.38	91.52	91.65	91.78
74.4	9 91.92	92.05	92.18	92.32	92.45	92.58	92.72	92.85	92.98	93.12
74.5	9 93.25	93.38	93.52	93.65	93.78	93.92	94.05	94.18	94.32	94.45
74.6	9 94.58	94.72	94.85	94.98	95.12	95.25	95.38	95.52	95.65	95.78
74.7	9 95.92	96.05	96.18	96.32	96.45	96.58	96.72	96.85	96.98	97.12
74.8	9 97.25	97.38	97.52	97.65	97.78	97.92	98.05	98.18	98.32	98.45
74.9	9 98.58	98.72	98.85	98.98	99.12	99.25	99.38	99.52	99.65	99.78
75.0	9 99.92	*00.05	*00.18	*00.31	*00.45	*00.58	*00.71	*00.85	*00.98	*01.11
75.1	10 01.25	01.38	01.51	01.65	01.78	01.91	02.05	02.18	02.31	02.45
75.2	10 02.58	02.71	02.85	02.98	03.11	03.25	03.38	03.51	03.65	03.78
75.3	10 03.91	04.05	04.18	04.31	04.45	04.58	04.71	04.85	04.98	05.11
75.4	10 05.25	05.38	05.51	05.65	05.78	05.91	06.05	06.18	06.31	06.45
75.5	10 06.58	06.71	06.85	06.98	07.11	07.25	07.38	07.51	07.65	07.78
75.6	10 07.91	08.05	08.18	08.31	08.45	08.58	08.71	08.85	08.98	09.11
75.7	10 09.25	09.38	09.51	09.65	09.78	09.91	10.05	10.18	10.31	10.45
75.8	10 10.58	10.71	10.85	10.98	11.11	11.25	11.38	11.51	11.65	11.78
75.9	10 11.91	12.05	12.18	12.31	12.45	12.58	12.71	12.85	12.98	13.11
76.0	10 13.25	13.38	13.51	13.65	13.78	13.91	14.05	14.18	14.31	14.45
76.1	10 14.58	14.71	14.85	14.98	15.11	15.25	15.38	15.51	15.65	15.78
76.2	10 15.91	16.05	16.18	16.31	16.45	16.58	16.71	16.85	16.98	17.11
76.3	10 17.25	17.38	17.51	17.65	17.78	17.91	18.05	18.18	18.31	18.45
76.4	10 18.58	18.71	18.85	18.98	19.11	19.25	19.38	19.51	19.65	19.78
76.5	10 19.91	20.05	20.18	20.31	20.45	20.58	20.71	20.85	20.98	21.11
76.6	10 21.25	21.38	21.51	21.65	21.78	21.91	22.05	22.18	22.31	22.45
76.7	10 22.58	22.71	22.85	22.98	23.11	23.25	23.38	23.51	23.65	23.78
76.8	10 23.91	24.05	24.18	24.31	24.45	24.58	24.71	24.85	24.98	25.11
76.9	10 25.25	25.38	25.51	25.65	25.78	25.91	26.05	26.18	26.31	26.45
77.0	10 26.58	26.71	26.85	26.98	27.11	27.25	27.38	27.51	27.65	27.78
77.1	10 27.91	28.05	28.18	28.31	28.45	28.58	28.71	28.85	28.98	29.11
77.2	10 29.25	29.38	29.51	29.65	29.78	29.91	30.05	30.18	30.31	30.45
77.3	10 30.58	30.71	30.85	30.98	31.11	31.25	31.38	31.51	31.65	31.78
77.4	10 31.91	32.05	32.18	32.31	32.45	32.58	32.71	32.85	32.98	33.11
77.5	10 33.25	33.38	33.51	33.65	33.78	33.91	34.05	34.18	34.31	34.45
77.6	10 34.58	34.71	34.85	34.98	35.11	35.25	35.38	35.51	35.65	35.78
77.7	10 35.91	36.05	36.18	36.31	36.45	36.58	36.71	36.85	36.98	37.11
77.8	10 37.25	37.38	37.51	37.65	37.78	37.91	38.05	38.18	38.31	38.45
77.9	10 38.58	38.71	38.85	38.98	39.11	39.24	39.38	39.51	39.64	39.78

TEMPERATURE CORRECTION FOR BAROMETER READINGS

Brass Scale—Metric Units

To reduce readings of a mercurial barometer with a brass scale to 0°C subtract the appropriate quantity as found in the table. These values are based on the coefficient of expansion of mercury $(181792 + 0.175t + 0.035116t^2) \times 10^{-9}$, and of brass 0.0000184 per °C. Corrections are in millimeters.

Temp. °C	Observed height in millimeters																	
	620	630	640	650	660	670	680	690	700	710	720	730	740	750	760	770	780	790
0	0.00	0.00	0.00	0.00	0.00	0.00	0.00	0.00	0.00	0.00	0.00	0.00	0.00	0.00	0.00	0.00	0.00	0.00
1	.10	.10	.10	.11	.11	.11	.11	.11	.11	.12	.12	.12	.12	.12	.12	.13	.13	.13
2	.20	.21	.21	.21	.22	.22	.22	.23	.23	.23	.24	.24	.24	.25	.25	.25	.25	.26
3	.30	.31	.31	.32	.32	.33	.33	.34	.34	.35	.35	.36	.36	.37	.37	.38	.38	.39
4	.40	.41	.42	.42	.43	.44	.44	.45	.46	.46	.47	.48	.48	.49	.50	.50	.51	.52
5	0.51	0.51	0.52	0.53	0.54	0.55	0.56	0.56	0.57	0.58	0.59	0.60	0.60	0.61	0.62	0.63	0.64	0.64
6	.61	.62	.63	.64	.65	.66	.67	.68	.69	.70	.71	.71	.72	.73	.74	.75	.76	.77
7	.71	.72	.73	.74	.75	.77	.78	.79	.80	.81	.82	.83	.85	.86	.87	.88	.89	.90
8	.81	.82	.84	.85	.86	.87	.89	.90	.91	.93	.94	.95	.97	.98	.99	1.01	1.02	1.03
9	.91	.92	.94	.95	.97	.98	1.00	1.01	1.03	1.04	1.06	1.07	1.09	1.10	1.12	1.13	1.15	1.16
10	1.01	1.03	1.04	1.06	1.08	1.09	1.11	1.13	1.14	1.16	1.17	1.19	1.21	1.22	1.24	1.26	1.27	1.29
11	1.11	1.13	1.15	1.17	1.18	1.20	1.22	1.24	1.26	1.27	1.29	1.31	1.33	1.35	1.36	1.38	1.40	1.42
12	1.21	1.23	1.25	1.27	1.29	1.31	1.33	1.35	1.37	1.39	1.41	1.43	1.45	1.47	1.49	1.51	1.53	1.55
13	1.31	1.34	1.36	1.38	1.40	1.42	1.44	1.46	1.48	1.50	1.53	1.55	1.57	1.59	1.61	1.63	1.65	1.67
14	1.41	1.44	1.46	1.48	1.51	1.53	1.55	1.57	1.60	1.62	1.64	1.67	1.69	1.71	1.73	1.76	1.78	1.80

CORRECTION FOR BAROMETER (Continued)

Brass Scale—Metric Units (Continued)

Temp. °C	Observed height in millimeters																	
	620	630	640	650	660	670	680	690	700	710	720	730	740	750	760	770	780	790
15	1.52	1.54	1.56	1.59	1.61	1.64	1.66	1.69	1.71	1.74	1.76	1.78	1.81	1.83	1.86	1.88	1.91	1.93
16	1.62	1.64	1.67	1.69	1.72	1.75	1.77	1.80	1.82	1.85	1.88	1.90	1.93	1.96	1.98	2.01	2.03	2.06
17	1.72	1.74	1.77	1.80	1.83	1.86	1.88	1.91	1.94	1.97	1.99	2.02	2.05	2.08	2.10	2.13	2.16	2.19
18	1.82	1.85	1.88	1.91	1.93	1.96	1.99	2.02	2.05	2.08	2.11	2.14	2.17	2.20	2.23	2.26	2.29	2.32
19	1.92	1.95	1.98	2.01	2.04	2.07	2.10	2.13	2.17	2.20	2.23	2.26	2.29	2.32	2.35	2.38	2.41	2.44
20	2.02	2.05	2.08	2.12	2.15	2.18	2.21	2.25	2.28	2.31	2.34	2.38	2.41	2.44	2.47	2.51	2.54	2.57
21	2.12	2.15	2.19	2.22	2.26	2.29	2.32	2.36	2.39	2.43	2.46	2.50	2.53	2.56	2.60	2.63	2.67	2.70
22	2.22	2.26	2.29	2.33	2.36	2.40	2.43	2.47	2.51	2.54	2.58	2.61	2.65	2.69	2.72	2.76	2.79	2.83
23	2.32	2.36	2.40	2.43	2.47	2.51	2.54	2.58	2.62	2.66	2.69	2.73	2.77	2.81	2.84	2.88	2.92	2.96
24	2.42	2.46	2.50	2.54	2.58	2.62	2.66	2.69	2.73	2.77	2.81	2.85	2.89	2.93	2.97	3.01	3.05	3.08
25	2.52	2.56	2.60	2.64	2.68	2.72	2.77	2.81	2.85	2.89	2.93	2.97	3.01	3.05	3.09	3.13	3.17	3.21
26	2.62	2.66	2.71	2.75	2.79	2.83	2.88	2.92	2.96	3.00	3.04	3.09	3.13	3.17	3.21	3.26	3.30	3.34
27	2.72	2.77	2.81	2.85	2.90	2.94	2.99	3.03	3.07	3.12	3.16	3.20	3.25	3.29	3.34	3.38	3.42	3.47
28	2.82	2.87	2.91	2.96	3.00	3.05	3.10	3.14	3.19	3.23	3.28	3.32	3.37	3.41	3.46	3.51	3.55	3.60
29	2.92	2.97	3.02	3.06	3.11	3.16	3.21	3.25	3.30	3.35	3.39	3.44	3.49	3.54	3.58	3.63	3.68	3.72
30	3.02	3.07	3.12	3.17	3.22	3.27	3.32	3.36	3.41	3.46	3.51	3.56	3.61	3.66	3.71	3.75	3.80	3.85
31	3.12	3.17	3.22	3.27	3.32	3.37	3.43	3.48	3.53	3.58	3.63	3.68	3.73	3.78	3.83	3.88	3.93	3.98
32	3.22	3.28	3.33	3.38	3.43	3.48	3.54	3.59	3.64	3.69	3.74	3.79	3.85	3.90	3.95	4.00	4.05	4.11
33	3.32	3.38	3.43	3.48	3.54	3.59	3.64	3.70	3.75	3.81	3.86	3.91	3.97	4.02	4.07	4.13	4.18	4.23
34	3.42	3.48	3.53	3.59	3.64	3.70	3.75	3.81	3.87	3.92	3.98	4.03	4.09	4.14	4.20	4.25	4.31	4.36
35	3.52	3.58	3.64	3.69	3.75	3.81	3.86	3.92	3.98	4.03	4.09	4.15	4.21	4.26	4.32	4.38	4.43	4.49

Brass Scale—English Units

Standard Temperature of scale 62° F; of mercury, 32° F. Zero correction at 28.5° F; subtract corrections above, add below. Owing to the difference in the standard temperature of English and metric scales, readings taken in inches to be reduced to centimeters should *first* be corrected for temperature.

Temp. °F	Observed height in inches																	
	23.0 in.	23.5 in.	24.0 in.	24.5 in.	25.0 in.	25.5 in.	26.0 in.	26.5 in.	27.0 in.	27.5 in.	28.0 in.	28.5 in.	29.0 in.	29.5 in.	30.0 in.	30.5 in.	31.0 in.	31.5 in.
0	+.060	+.061	+.063	+.064	+.065	+.067	+.068	+.069	+.070	.072	.073	.075	.076	.077	.078	.080	.081	.082
2	.056	.057	.058	.060	.061	.062	.063	.065	.065	.067	.068	.069	.070	.072	.073	.074	.075	.077
4	.052	.053	.054	.055	.056	.057	.058	.060	.061	.062	.063	.064	.065	.066	.067	.069	.070	.071
6	.047	.048	.049	.051	.052	.053	.054	.055	.056	.057	.058	.059	.060	.061	.062	.063	.064	.065
8	.043	.044	.045	.046	.047	.048	.049	.050	.051	.052	.053	.054	.054	.056	.056	.057	.058	.059
10	.039	.040	.041	.042	.042	.043	.044	.045	.046	.047	.047	.048	.049	.050	.051	.052	.053	.054
12	.035	.036	.036	.037	.038	.039	.039	.040	.041	.042	.042	.043	.044	.045	.045	.046	.047	.048
14	.031	.031	.032	.033	.033	.034	.035	.035	.036	.037	.037	.038	.039	.039	.040	.041	.041	.042
16	.026	.027	.028	.028	.029	.029	.030	.031	.031	.032	.032	.033	.033	.034	.034	.035	.036	.036
18	.022	.023	.023	.024	.024	.025	.025	.026	.026	.027	.027	.028	.028	.029	.029	.030	.030	.031
20	.018	.018	.019	.019	.020	.020	.020	.021	.021	.022	.022	.022	.023	.023	.024	.024	.024	.025
22	.014	.014	.014	.015	.015	.015	.016	.016	.016	.017	.017	.017	.017	.018	.018	.018	.019	.019
24	.010	.010	.010	.010	.011	.011	.011	.011	.011	.012	.012	.012	.012	.012	.013	.013	.013	.013
26	.005	.006	.006	.006	.006	.006	.006	.006	.006	.007	.007	.007	.007	.007	.007	.007	.007	.008
28	+.001	+.001	+.001	+.001	+.001	+.001	+.001	+.002	+.002	+.002	+.002	+.002	+.002	+.002	+.002	+.002	+.002	+.002
30	−.003	−.003	−.003	−.003	−.003	−.003	−.003	−.003	−.003	−.003	−.003	−.004	−.004	−.004	−.004	−.004	−.004	−.004
32	.007	.007	.007	.008	.008	.008	.008	.008	.008	.008	.009	.009	.009	.009	.009	.009	.009	.010
34	.011	.011	.012	.012	.012	.012	.013	.013	.013	.013	.014	.014	.014	.014	.015	.015	.015	.015
36	.015	.016	.016	.016	.017	.017	.017	.018	.018	.018	.019	.019	.019	.020	.020	.020	.021	.021
38	.020	.020	.020	.021	.021	.022	.022	.023	.023	.023	.024	.024	.025	.025	.026	.026	.026	.027
40	.024	.024	.025	.025	.026	.026	.027	.027	.028	.028	.029	.030	.030	.031	.031	.032	.032	.033
42	.028	.029	.029	.030	.030	.031	.032	.032	.033	.033	.034	.035	.035	.036	.036	.037	.038	.038
44	.032	.033	.033	.034	.035	.036	.036	.037	.038	.038	.039	.040	.040	.041	.042	.043	.043	.044
46	.036	.037	.038	.039	.039	.040	.041	.042	.043	.043	.044	.045	.046	.047	.047	.048	.049	.050
48	.040	.041	.042	.043	.044	.045	.046	.047	.047	.048	.049	.050	.051	.052	.053	.054	.054	.055
50	.045	.046	.046	.048	.048	.050	.050	.052	.052	.053	.054	.055	.056	.057	.058	.059	.060	.061
52	.049	.050	.051	.052	.053	.054	.055	.056	.057	.058	.059	.061	.061	.063	.064	.065	.066	.067
54	.053	.054	.055	.057	.057	.059	.060	.061	.062	.063	.064	.066	.067	.068	.069	.070	.071	.073
56	.057	.058	.060	.061	.062	.063	.064	.066	.067	.068	.069	.071	.072	.073	.074	.076	.077	.078
58	.061	.063	.064	.065	.066	.068	.069	.071	.072	.073	.074	.076	.077	.079	.080	.081	.082	.084
60	.065	.067	.068	.070	.071	.073	.074	.076	.077	.078	.080	.081	.082	.084	.085	.087	.088	.090
62	.069	.071	.073	.074	.076	.077	.079	.080	.082	.083	.085	.086	.088	.089	.091	.092	.094	.095
64	.074	.075	.077	.079	.080	.082	.083	.085	.086	.088	.090	.092	.093	.095	.096	.098	.099	.101
66	.078	.079	.081	.083	.085	.087	.088	.090	.091	.093	.095	.097	.098	.100	.101	.103	.105	.107
68	.082	.084	.085	.088	.089	.091	.093	.095	.096	.098	.100	.102	.103	.105	.107	.109	.110	.113
70	.086	.088	.090	.092	.094	.096	.097	.100	.101	.103	.105	.107	.109	.111	.112	.115	.116	.118
72	.090	.092	.094	.096	.098	.100	.102	.104	.106	.108	.110	.112	.114	.116	.118	.120	.122	.124
74	.094	.096	.098	.101	.103	.105	.107	.109	.111	.113	.115	.117	.119	.121	.123	.126	.127	.130
76	.098	.101	.103	.105	.107	.110	.111	.114	.116	.118	.120	.122	.124	.127	.128	.131	.133	.135
78	.103	.105	.107	.110	.112	.114	.116	.119	.120	.123	.125	.128	.129	.132	.134	.137	.138	.141
80	.107	.109	.111	.114	.116	.119	.121	.123	.125	.128	.130	.133	.135	.137	.139	.142	.144	.147
82	.111	.113	.116	.119	.121	.123	.125	.128	.130	.133	.135	.138	.140	.143	.145	.148	.149	.152
84	.115	.118	.120	.123	.125	.128	.130	.133	.185	.138	.140	.143	.145	.148	.150	.153	.155	.158
86	.119	.122	.124	.127	.130	.133	.135	.138	.140	.143	.145	.148	.150	.153	.155	.159	.161	.164
88	.123	.126	.129	.132	.134	.137	.139	.143	.145	.148	.150	.153	.155	.159	.161	.164	.166	.169
90	.127	.130	.133	.136	.138	.142	.144	.147	.150	.153	.155	.158	.161	.164	.166	.170	.172	.175
92	.132	.134	.137	.141	.143	.146	.149	.152	.154	.158	.160	.163	.166	.169	.172	.175	.177	.181
94	.136	.139	.142	.145	.147	.151	.153	.157	.159	.163	.165	.169	.171	.175	.177	.180	.183	.186
96	.140	.143	.146	.150	.152	.155	.158	.161	.134	.168	.170	.174	.176	.180	.182	.186	.188	.192
98	.144	.147	.150	.154	.156	.160	.163	.166	.169	.172	.175	.179	.181	.185	.188	.191	.194	.197
100	.148	.151	.154	.158	.161	.164	.167	.171	.174	.177	.180	.184	.187	.190	.193	.197	.200	.203

TEMPERATURE CORRECTION, GLASS SCALE

Metric

To reduce readings of a mercurial barometer with a glass scale to 0° C. subtract the appropriate quantity as found in table.

Temp. °C.	Observed height in centimeters.									Temp. °C.	Observed height in centimeters.								
	70 cm.	71 cm.	72 cm.	73 cm.	74 cm.	75 cm.	76 cm.	77 cm.	78 cm.		70 cm.	71 cm.	72 cm.	73 cm.	74 cm.	75 cm.	76 cm.	77 cm.	78 cm.
0	0.000	0.000	0.000	0.000	0.000	0.000	0.000	0.000	0.000	15	0.181	0.184	0.186	0.189	0.191	0.193	0.196	0.198	0.201
1	.012	.012	.013	.013	.013	.013	.013	.013	.014	16	.194	.196	.199	.201	.204	.207	.209	.212	.214
2	.025	.025	.025	.026	.026	.026	.026	.027	.027	17	.205	.208	.210	.213	.216	.219	.221	.224	.227
3	.036	.036	.037	.037	.038	.038	.039	.039	.040	18	.217	.220	.223	.226	.229	.232	.235	.238	.241
4	.048	.049	.049	.050	.051	.051	.052	.053	.053	19	.230	.233	.236	.239	.242	.245	.248	.251	.254
5	0.060	0.061	0.062	0.063	0.064	0.064	0.065	0.066	0.067	20	0.242	0.245	0.248	0.252	0.255	0.258	0.261	0.264	0.268
6	.073	.074	.074	.076	.077	.077	.078	.079	.080	21	.254	.258	.261	.264	.268	.271	.275	.278	.281
7	.085	.086	.087	.088	.089	.091	.092	.093	.094	22	.266	.269	.273	.276	.280	.283	.287	.290	.294
8	.096	.098	.099	.100	.101	.103	.104	.105	.107	23	.278	.282	.285	.289	.293	.296	.300	.304	.308
9	.109	.110	.111	.113	.114	.116	.117	.119	.120	24	.290	.294	.298	.302	.306	.310	.313	.317	.321
10	0.121	0.122	0.124	0.126	0.127	0.129	0.130	0.132	0.134	25	0.303	0.307	0.311	0.315	0.319	0.323	0.327	0.331	0.335
11	.133	.135	.137	.138	.140	.142	.144	.146	.147	26	.315	.319	.323	.327	.332	.336	.340	.344	.348
12	.144	.146	.148	.150	.152	.154	.156	.158	.160	27	.326	.331	.335	.339	.344	.348	.352	.357	.361
13	.157	.159	.161	.163	.165	.167	.169	.171	.174	28	.339	.343	.348	.352	.357	.361	.366	.370	.375
14	.169	.171	.174	.176	.178	.180	.183	.185	.187	29	.351	.356	.360	.365	.370	.374	.379	.384	.388
										30	0.363	0.368	0.373	0.378	0.383	0.387	0.392	0.397	0.402

WEIGHT IN GRAMS OF A CUBIC METER OF SATURATED AQUEOUS VAPOR

(From Smithsonian Tables)

Mass in grams per cubic meter.

Temp. °C	0.0	1.0	2.0	3.0	4.0	5.0	6.0	7.0	8.0	9.0
−20	1.074	.988	.909	.836	.768	.705	.646	.592	.542	.496
−10	2.358	2.186	2.026	1.876	1.736	1.605	1.483	1.369	1.264	1.165
− 0	4.847	4.523	4.217	3.930	3.660	3.407	3.169	2.946	2.737	2.541
+ 0	4.847	5.192	5.559	5.947	6.360	6.797	7.260	7.750	8.270	8.819
+10	9.399	10.01	10.66	11.35	12.07	12.83	13.63	14.84	15.37	16.21
+20	17.30	18.34	19.43	20.58	21.78	23.05	24.38	25.78	27.24	28.78
+30	30.38	32.07	33.83	35.68	37.61	39.63	41.75	43.96	46.26	48.67

EFFICIENCY OF DRYING AGENTS

Compiled by John H. Yoe

A. Drying agents depending upon chemical action (absorption) for their efficiency:*

Substance	Residual water, mg per liter of dry air**	Reference
P_2O_5	<1 mg in 40,000 l,	Morley, Am. J. Sci., **34,** 199 (1887); J.A.C.S., **26,** 1171 (1904).
$Mg(ClO_4)_2$ anhyd.	"Unweighable" in 210 l,	Willard and Smith, J.A.C.S., **44,** 2255 (1922).
BaO	0.00065	Bower, Bur. Std. J. Res., **12,** 241 (1934).
KOH fused	0.002	Baxter and Starkweather, J.A.C.S., **38,** 2038 (1916).
CaO	0.003	Bower, loc. cit.
H_2SO_4	0.003	Baxter and Starkweather, loc. cit.
$CaSO_4$ anhyd.	0.005	Bower, loc. cit.
Al_2O_3	0.005	Ibid.
KOH sticks	0.014	Ibid.
NaOH fused	0.16	Baxter and Starkweather, loc. cit.
$CaBr_2$	0.18	Baxter and Warren, J.A.C.S., **33,** 340 (1911).
$CaCl_2$ fused	0.34	Baxter and Starkweather, loc. cit.
NaOH sticks	0.80	Bower, loc. cit.
$Ba(ClO_4)_2$	0.82	Ibid.
$ZnCl_2$	0.85	Baxter and Warren, loc. cit.
$ZnBr_2$	1.16	Ibid.
$CaCl_2$ granular	1.5	Bower, loc. cit.
$CuSO_4$ anhyd.	2.8	Ibid.

B. Drying agents depending upon physical action (adsorption) for their efficiency:* Alumina (low temperature fired), asbestos, charcoal, clay and porcelain (low temperature fired), glass wool, kieselguhr, silica gel, refrigeration.

* It should be noted that the efficiency of some drying agents (e.g. Al_2O_3 and anhydrous $CaCl_2$, and probably also BaO, anhydrous $Mg(ClO_4)_2$, $Mg(ClO_4)_2 \cdot 3H_2O$, anhydrous $Ba(ClO_4)_2$, and $CaSO_4$) depends upon both adsorption and absorption.

** 30°C. for Bower's values; others 25°C. or room temp.

REDUCTION OF BAROMETER TO SEA LEVEL

The correction to be added to reduce barometric readings to "sea level" values depends principally on three factors: The temperature of the air column (assumed) from the station to sea level, the altitude of the station, and the value of the reading itself. Two tables are provided. Table I is entered with the altitude and assumed temperature and a factor "2000 m" taken out. Table II is entered with the above factor and the approximate barometer reading and the final correction taken out.

The correction is to be added. If B_0 is the corrected or sea level value; B the barometer reading at the station; C the correction,—

$$C = B_0 - B = B\,(10^m - 1)$$

The actual barometer reading at the station should be corrected for temperature of the mercury column by the usual methods before entering the tables or applying the sea level correction.

A complete explanation of the theory of the corrections and a more extended set of tables will be found in the Smithsonian Meteorological Tables.

Latitude Factor

The influence of the latitude on the value of the correction is usually negligible, being overshadowed by uncertainties in the assumed temperature of the air column. For cases where this correction is desirable the table below is provided. The value of the temperature-altitude factor "2000 m" obtained in Table I is corrected for latitude by subtracting for latitudes 0-45° and adding for latitudes from 45-90° the values found. With this corrected value of "2000 m" Table II is entered for the value of the correction.

LATITUDE FACTOR

To be used in connection with Tables I and II, either English or metric units, to obtain latitude corrections to temperature-altitude factor. For latitudes 0-45° subtract the correction. For latitudes 45-90° add the correction.

Temp.—Alt. from Table I	Latitude			
	0°	15°	30°	45°
100	0.3	0.2	0.1	0.0
200	0.5	0.5	0.3	0.0
300	0.8	0.7	0.4	0.0
	90°	75°	60°	45°

METRIC UNITS—TABLE I

Values of the temperature-altitude factor (2000 m.) for entering table II.

Altitude in meters	Assumed temperature of air column °C									
	−16°	−8°	0°	+4°	+8°	+12°	+16°	+20°	+24°	+28°
10	1.2	1.1	1.1	1.1	1.0	1.0	1.0	1.0	1.0	1.0
50	5.8	5.6	5.4	5.3	5.2	5.2	5.1	5.0	4.9	4.9
100	11.5	11.2	10.8	10.7	10.5	10.3	10.2	10.0	9.9	9.7
150	17.3	16.7	16.2	16.0	15.7	15.5	15.3	15.0	14.8	14.6
200	23.0	22.3	21.6	21.3	21.0	20.7	20.3	20.0	19.7	19.5
250	28.8	27.9	27.0	26.6	26.2	25.8	25.4	25.0	24.7	24.3
300	34.5	33.5	32.5	32.0	31.5	31.0	30.5	30.1	29.6	29.2
350	40.3	39.0	37.9	37.3	36.7	36.2	35.6	35.1	34.6	34.0
400	46.0	44.6	43.3	42.6	42.0	41.3	40.7	40.1	39.5	38.9
450	51.8	50.2	48.7	47.9	47.2	46.5	45.8	45.1	44.4	43.8
500	57.5	55.8	54.1	53.3	52.4	51.6	50.9	50.1	49.4	48.6
550	63.3	61.4	59.5	58.6	57.7	56.8	55.9	55.1	54.3	53.5
600	69.0	66.9	64.9	63.9	62.9	62.0	61.0	60.1	59.2	58.3
650	74.8	72.5	70.3	69.2	68.2	67.1	66.1	65.1	64.2	63.2
700	80.6	78.1	75.7	74.6	73.4	72.3	71.2	70.1	69.1	68.1
750	86.3	83.7	81.1	79.9	78.7	77.5	76.3	75.1	74.0	72.9
800	92.1	89.2	86.5	85.2	83.9	82.6	81.4	80.1	79.0	77.8
850	97.8	94.8	92.0	90.5	89.2	87.8	86.4	85.2	83.9	82.7
900	103.6	100.4	97.4	95.9	94.4	93.0	91.5	90.2	88.8	87.5
950	109.3	106.0	102.8	101.2	99.6	98.1	96.6	95.2	93.8	92.4
1000	115.1	111.5	108.2	106.5	104.9	103.3	101.7	100.2	98.7	97.3
1050	120.8	117.1	113.6	111.8	110.1	108.4	106.8	105.2	103.6	102.1
1100	126.6	122.7	119.0	117.2	115.4	113.6	111.9	110.2	108.6	107.0
1150	132.3	128.3	124.4	122.5	120.6	118.8	117.0	115.2	113.5	111.8
1200	138.1	133.8	129.8	127.8	125.9	123.9	122.0	120.2	118.4	116.7
1250	143.8	139.4	135.2	133.1	131.1	129.1	127.1	125.2	123.4	121.6
1300	149.6	145.0	140.6	138.5	136.3	134.3	132.2	130.2	128.3	126.4
1350	155.3	150.6	146.0	143.8	141.6	139.4	137.3	135.2	133.2	131.3
1400	161.1	156.2	151.4	149.1	146.8	144.6	142.4	140.2	138.2	136.2
1450	166.8	161.7	156.8	154.5	152.1	149.7	147.5	145.3	143.1	141.0

METRIC UNITS—TABLE I (Continued)

Altitude in meters	Assumed temperature of air column °C									
	−16°	−8°	0°	+4°	+8°	+12°	+16°	+20°	+24°	+28°
1500	172.6	167.3	162.3	159.8	157.3	154.9	152.5	150.3	148.0	145.9
1550	178.3	172.9	167.7	165.1	162.6	160.1	157.6	155.3	153.0	150.7
1600	184.1	178.5	173.1	170.4	167.8	165.2	162.7	160.3	157.9	155.6
1650	189.8	184.0	178.5	175.7	173.0	170.4	167.8	165.3	162.8	160.5
1700	195.6	189.6	183.9	181.1	178.3	175.6	172.9	170.3	167.8	165.3
1750	201.4	195.2	189.3	186.4	183.5	180.7	178.0	175.3	172.7	170.2
1800	207.1	200.8	194.7	191.7	188.8	185.9	183.1	180.3	177.6	175.0
1850	212.9	206.3	200.1	197.0	194.0	191.0	188.1	185.3	182.6	179.9
1900	218.6	211.9	205.5	202.4	199.3	196.2	193.2	190.3	187.5	184.8
1950	224.4	217.5	210.9	207.7	204.5	201.4	198.3	195.3	192.4	189.6
2000	230.1	223.0	216.3	213.0	209.7	206.5	203.4	200.3	197.4	194.5
2050	235.9	228.6	221.7	218.3	215.0	211.7	208.5	205.3	202.3	199.3
2100	241.6	234.2	227.1	223.7	220.2	216.8	213.5	210.4	207.2	204.2
2150	247.4	239.8	232.5	229.0	225.5	222.0	218.6	215.4	212.2	209.1
2200	253.1	245.4	237.9	234.3	230.7	227.2	223.7	220.4	217.1	213.9
2250	258.9	250.9	243.4	239.6	235.9	232.3	228.8	225.4	222.0	218.8
2300	264.6	256.5	248.8	245.0	241.2	237.5	233.9	230.4	227.0	223.6
2350	270.4	262.1	254.2	250.3	246.4	242.7	239.0	235.4	231.9	228.5
2400	276.1	267.7	259.6	255.6	251.7	247.8	244.0	240.4	236.8	233.4
2450	281.9	273.2	265.0	260.9	256.9	253.0	249.1	245.4	241.8	238.2
2500	287.6	278.8	270.4	266.2	262.2	258.1	254.2	250.4	246.7	243.1
2550	293.4	284.4	275.8	271.6	267.4	263.3	259.3	255.4	251.6	247.9
2600	299.1	290.0	281.2	276.9	272.6	268.5	264.4	260.4	256.6	252.8
2650	304.9	295.5	286.6	282.2	277.9	273.6	269.5	265.4	261.5	257.7
2700	310.6	301.1	292.0	287.5	283.1	278.8	274.5	270.4	266.4	262.5
2750	316.4	306.7	297.4	292.9	288.4	283.9	279.6	275.4	271.4	267.4
2800	322.1	312.3	302.8	298.2	293.6	289.1	284.7	280.4	276.3	272.2
2850	327.9	317.8	308.2	303.5	298.8	294.3	289.8	285.4	281.2	277.1
2900	333.6	323.4	313.6	308.8	304.1	299.4	294.9	290.4	286.2	282.0
2950	339.4	329.0	319.0	314.2	309.3	304.6	299.9	295.5	291.1	286.8
3000	345.1	334.5	324.4	319.5	314.6	309.7	305.0	300.5	296.0	291.7

METRIC UNITS—TABLE II

Values of Correction to be Added

Temp.—alt.—factor	Barometer reading					
	780 mm	760 mm	740 mm	720 mm	700 mm	
1	0.9	0.9	0.9	0.8	0.8	
5	4.5	4.4	4.3	4.2	4.0	
10	9.0	8.8	8.6	8.3	8.1	
15	13.6	13.2	12.9	12.5	12.2	
20	18.2	17.7	17.2	16.8	16.3	
25	22.8	22.2	21.6	21.0	20.4	
30	27.4	26.7	26.0	25.3	24.6	
35	.....	31.2	30.4	29.6	28.8	
	760 mm	740 mm	720 mm	700 mm	680 mm	660 mm
40	35.8	34.9	33.9	33.0	32.0	31.1
45	40.4	39.3	38.3	37.2	36.2	35.1
50	45.0	43.8	42.7	41.5	40.3	39.1
55	49.7	48.4	47.1	45.8	44.5	43.1
60	.....	52.9	51.5	50.1	48.6	47.2
65	.....	57.5	55.9	54.4	52.8	51.3
70	.....	62.1	60.4	58.7	57.1	55.4
75	.....	66.7	64.9	63.1	61.3	59.5
	720 mm	700 mm	680 mm	660 mm	640 mm	
80	69.5	67.5	65.6	63.7	61.7	
85	74.0	72.0	69.9	67.9	65.8	
90	78.6	76.4	74.2	72.1	69.9	
95	83.2	80.9	78.6	76.3	74.0	
100	87.9	85.4	83.0	80.5	78.1	
105	.....	89.9	87.4	84.8	82.2	
110	.....	94.5	91.8	89.1	86.4	
115	.....	99.1	96.3	93.4	90.6	
120	.....	103.7	100.7	97.8	94.8	
125	.....	108.3	105.3	102.2	99.1	
	680 mm	660 mm	640 mm	620 mm	600 mm	
125	105.3	102.2	99.1	96.0	92.9	
130	109.8	106.6	103.3	100.1	96.9	
135	114.3	111.0	107.6	104.3	100.9	
140	118.9	115.4	111.9	108.4	104.9	
145	123.5	119.9	116.3	112.6	109.0	
150	128.2	124.4	120.6	116.9	113.1	
155	.....	128.9	125.0	121.1	117.2	
160	.....	133.5	129.4	125.4	121.4	
165	.....	138.1	133.9	129.7	125.5	
170	.....	142.7	138.4	134.0	129.7	

Temp.—alt.—factor	Barometer reading				
	640 mm	620 mm	600 mm	580 mm	560 mm
170	138.4	134.0	129.7	125.4	121.1
175	142.9	138.4	133.9	129.5	125.0
180	147.4	142.8	138.2	133.6	129.0
185	151.9	147.2	142.4	137.7	132.9
190	156.5	151.6	146.7	141.8	136.9
195	161.1	156.1	151.0	146.0	141.0
200	165.7	160.5	155.4	150.2	145.0
205	170.4	165.0	159.7	154.4	149.1
210	.....	169.6	164.1	158.6	153.2
215	.....	174.1	168.5	162.9	157.3
	620 mm	600 mm	580 mm	560 mm	540 mm
215	174.1	168.5	162.9	157.3	151.7
220	178.7	172.9	167.2	161.4	155.7
225	183.3	177.4	171.5	165.6	159.7
230	188.0	181.9	175.8	169.8	163.7
235	192.6	186.4	180.2	174.0	167.8
240	.....	191.0	184.6	178.2	171.9
245	.....	195.5	189.0	182.5	176.0
250	.....	200.1	193.4	186.8	180.1
255	.....	204.7	197.9	191.1	184.3
260	.....	209.4	202.4	195.4	188.4
	580 mm	560 mm	540 mm	520 mm	
260	202.4	195.4	188.4	181.5	
265	206.9	199.8	192.6	185.5	
270	211.5	204.2	196.9	189.6	
275	216.0	208.6	201.1	193.7	
280	220.6	213.0	205.4	197.8	
285	225.2	217.5	209.7	201.9	
290	229.9	222.0	214.0	206.1	
295	.....	226.5	218.4	210.3	
300	.....	231.0	222.8	214.5	
	560 mm	540 mm	520 mm	500 mm	480 mm
305	235.6	227.2	218.8	210.3	201.9
310	240.2	231.6	223.0	214.4	205.9
315	244.8	236.0	227.3	218.6	209.8
320	249.4	240.5	231.6	222.7	213.8
325	254.1	245.0	236.0	226.9	217.8
330	.....	249.6	240.3	231.1	221.8
335	.....	254.1	244.7	235.3	225.9
340	.....	258.7	249.1	239.6	230.0
345	.....	263.3	253.6	243.8	234.1

REDUCTION OF BAROMETER TO SEA LEVEL
(Continued)

ENGLISH UNITS—TABLE I

Values of the temperature-altitude factor (2000 m.) for entering table II.

Altitude feet	Assumed temperature of air column °F									
	−20	0	+10	+20	+30	+40	+50	+60	+70	+80
200	7.4	7.1	6.9	6.8	6.6	6.5	6.3	6.2	6.1	6.0
400	14.8	14.1	13.8	13.5	13.2	13.0	12.7	12.4	12.2	11.9
600	22.2	21.2	20.7	20.3	19.9	19.5	19.0	18.6	18.2	17.9
800	29.6	28.3	27.7	27.1	26.5	25.9	25.4	24.8	24.3	23.8
1000	37.0	35.3	34.6	33.8	33.1	32.4	31.7	31.1	30.4	29.8
1200	44.3	42.4	41.5	40.6	39.7	38.9	38.1	37.3	36.5	35.8
1400	51.7	49.5	48.4	47.4	46.4	45.4	44.4	43.5	42.6	41.7
1600	59.1	56.5	55.3	54.1	53.0	51.9	50.8	49.7	48.7	47.7
1800	66.5	63.6	62.2	60.9	59.6	58.4	57.1	55.9	54.7	53.6
2000	73.9	70.6	69.1	67.7	66.2	64.8	63.4	62.1	60.8	59.6
2200	81.3	77.7	76.0	74.4	72.9	71.3	69.8	68.3	66.9	65.5
2400	88.7	84.8	82.9	81.2	79.5	77.8	76.1	74.5	73.0	71.5
2600	96.1	91.8	89.9	87.9	86.1	84.3	82.5	80.7	79.1	77.5
2800	103.5	98.9	96.8	94.7	92.7	90.8	88.8	87.0	85.1	83.4
3000	110.9	106.0	103.7	101.5	99.3	97.2	95.2	93.2	91.2	89.4
3200	118.2	113.0	110.6	108.2	106.0	103.7	101.5	99.4	97.3	95.3
3400	125.6	120.1	117.5	115.0	112.6	110.2	107.9	105.6	103.4	101.3
3600	133.0	127.2	124.4	121.8	119.2	116.7	114.2	111.8	109.5	107.2
3800	140.4	134.2	131.3	128.5	125.8	123.2	120.5	118.0	115.5	113.2
4000	147.8	141.3	138.2	135.3	132.4	129.6	126.9	124.2	121.6	119.2
4200	155.2	148.3	145.1	142.1	139.1	136.1	133.2	130.4	127.7	125.1
4400	162.6	155.4	152.0	148.8	145.7	142.6	139.6	136.6	133.8	131.1
4600	170.0	162.5	159.0	155.6	152.3	149.1	145.9	142.8	139.9	137.0
4800	177.3	169.5	165.9	162.3	158.9	155.6	152.2	149.0	145.9	143.0
5000	184.7	176.6	172.8	169.1	165.6	162.0	158.6	155.2	152.0	148.9
5200	192.1	183.7	179.7	175.9	172.2	168.5	164.9	161.5	158.1	154.9
5400	199.5	190.7	186.6	182.6	178.8	175.0	171.3	167.7	164.2	160.8
5600	206.9	197.8	193.5	189.4	185.4	181.5	177.6	173.9	170.3	166.8
5800	214.3	204.8	200.4	196.2	192.0	188.0	184.0	180.1	176.3	172.8
6000	221.7	211.9	207.3	202.9	198.7	194.4	190.3	186.3	182.4	178.7
6200	229.1	219.0	214.2	209.7	205.3	200.9	196.6	192.5	188.5	184.7
6400	236.4	226.0	221.1	216.4	211.9	207.4	203.0	198.7	194.6	190.6
6600	243.8	233.1	228.0	223.2	218.5	213.0	209.3	204.9	200.7	196.6
6800	251.2	240.1	235.0	230.0	225.1	220.4	215.7	211.1	206.7	202.5
7000	258.6	247.2	241.9	236.7	231.8	226.8	222.0	217.3	212.8	208.5
7200	266.0	254.3	248.8	243.5	238.4	233.3	228.4	223.5	218.9	214.4
7400	273.4	261.3	255.7	250.2	245.0	239.8	234.7	229.7	225.0	220.4
7600	280.8	268.4	262.6	257.0	251.6	246.3	241.0	235.9	231.1	226.4
7800	288.1	275.4	269.5	263.8	258.2	252.8	247.4	242.2	237.1	232.3
8000	295.5	282.5	276.4	270.5	264.8	259.2	253.7	248.4	243.2	238.3
8200	302.9	289.6	283.3	277.3	271.5	265.7	260.1	254.6	249.3	244.2
8400	310.3	296.6	290.2	284.0	278.1	272.2	266.4	260.8	255.4	250.2
8600	317.7	303.7	297.1	290.8	284.7	278.7	272.7	267.0	261.4	256.1
8800	325.1	310.7	304.0	297.6	291.3	285.2	279.1	273.2	267.5	262.1
9000	332.5	317.8	310.9	304.3	297.9	291.6	285.4	279.4	273.6	268.0

ENGLISH UNITS—TABLE II

Value of Correction to be Added.

Temp alt. factor	Barometer reading					Temp. alt. factor	Barometer reading				
	31	30	29	28	27		26	25	24	23	22
	in.	in.	in.	in.	in.		in.	in.	in.	in.	in.
1	0.04	0.03	0.03	.....	.....	165	5.44	5.23	5.02		
5	0.18	0.17	0.17	.....	.....	170	5.62	5.40	5.19		
10	0.36	0.35	0.34	0.32	.....	175	.....	5.58	5.36		
15	0.54	0.52	0.51	0.49	.....	180	.....	5.76	5.53	5.30	
20	0.72	0.70	0.68	0.65	.....	185	.....	5.93	5.70	5.46	
25	.....	0.88	0.85	0.82	.....	190	.....	6.11	5.87	5.62	
30	.....	1.05	1.02	0.98	.....	195	.....	6.29	6.04	5.79	
35	.....	1.23	1.19	1.15	.....	200	.....	6.47	6.21	5.96	
40	.....	1.41	1.37	1.32	1.27	205	.....	.....	6.39	6.12	
45	.....	1.60	1.54	1.49	1.44	210	.....	.....	6.56	6.29	
50	.....	.....	1.72	1.66	1.60	215	.....	.....	6.74	6.46	
55	.....	.....	1.90	1.83	1.76	220	.....	.....	6.92	6.63	6.34
60	.....	.....	2.07	2.00	1.93	225	.....	.....	7.10	6.80	6.51
65	.....	.....	2.25	2.18	2.10	230	.....	.....	7.28	6.97	6.67
70	.....	.....	2.43	2.35	2.27	235	.....	.....	7.46	7.15	6.84
75	.....	.....	.....	2.53	2.43	240	.....	.....	.....	7.32	7.00
80	.....	.....	.....	2.70	2.60	245	.....	.....	.....	7.49	7 17
	28	27	26	25	24		23	22	21	20	
	in.	in.	in.	in.	in.		in.	in.	in.	in.	
75	2.53	2.43	2.34	.....	.....	250	7.67	7.34			
80	2.70	2.60	2.51	.....	.....	255	7.85	7.51			
85	2.88	2.78	2.67	.....	.....	260	8.03	7.68	7.33		
90	3.06	2.95	2.84	.....	.....	265	8.21	7.85	7.49		
95	3.24	3.12	3.01	.....	.....	270	8.39	8.02	7.66		
100	3.42	3.29	3.17	.....	....	275	8.57	8.19	7.82		
105	3.60	3.47	3.34	3.21	...	280	.....	8.37	7.99		

ENGLISH UNITS—TABLE II (Cont)

Value of Correction to be Added.

	28	27	26	25	24		23	22	21	20
110	.....	3.85	3.51	3.38	.....	285	.....	8.54	8.16	
115	.....	3.82	3.68	3.54	.....	290	.....	8.72	8.32	
120	.....	4.00	3.85	3.70	.....	295	.....	8.90	8.49	8.09
125	.....	4.18	4.02	3.87	.....	300	.....	9.08	8.66	8.25
130	.....	4.36	4.20	4.04	.....	305	.....	9.26	8.83	8.41
135	.....	4.54	4.37	4.20	.....	310	.....	9.44	9.01	8.58
140	.....	.....	4.55	4.37	4.20	315	.....	9.62	9.18	8.74
145			4.72	4.54	4.36	320	.....	9.80	9.35	8.91
150			4.90	4.71	4.52	325	.....	.....	9.53	9.08
155	.....	.....	5.08	4.88	4.69	330	.....	.....	9.71	9.24
160	.....	.....	5.26	5.06	4.85					

REDUCTION OF BAROMETER TO GRAVITY AT SEA LEVEL

Metric Units

Correction to be subtracted given in millimeters

(From Smithsonian Physical Tables)

Height above sea level in meters	Observed Height of Barometer in Millimeters						
	500	550	600	650	700	750	800
100	.....	.....	.....	.....	.02	.02	.02
200	.....	.....	.....	.....	.04	.05	.05
300	.....	.....	.....	.....	.07	.07	.07
400	.....	.....	.....	.....	.09	.10	.10
500	.....	.....	.....	.....	.11	.12	.13
600	.....	.....	.....	.12	.13	.14	
700	.....	.....	.....	.14	.15	.16	
800	.....	.....	.....	.16	.18	.19	
900	.....	.....	.....	.18	.20	.22	
1000	.....	.18	.19	.20	.22	.24	
1100	.....	.19	.21	.22	.24		
1200	.....	.21	.23	.24	.26		
1300	.....	.22	.24	.26	.29		
1400	.....	.24	.26	.28	.31		
1500	.24	.26	.28	.30	.33		
1600	.25	.28	.30	.32			
1700	.27	.30	.32	.34			
1800	.28	.31	.34	.36			
1900	.30	.33	.36	.39			
2000	.31	.34	.38	.41			
2100	.33	.36	.40				
2200	.35	.38	.41				
2300	.36	.40	.43				
2400	.38	.42	.45				
2500	.39	.43	.47				

English Units

Height above sea level in feet	Observed Height in Inches						
	18	20	22	24	26	28	30
1000	.....	.....	.....	.....	.003	.003	.003
2000	.....	.....	.....	.004	.005	.005	.006
3000	.....	.....	.007	.007	.008	.008	
4000	.....	.....	.009	.009	.010		
4500	.....	.....	.010	.010	.011		
5000	.....	.010	.011	.011	.012		
5500	.....	.011	.012	.013			
6000	.....	.011	.013	.014			
6500	.011	.012	.014	.015			
7000	.012	.013	.015	.016			
7500	.013	.014	.016	.017			
8000	.014	.015	.017				
8500	.015	.016	.018				
9000	.016	.017	.019				
9500	.016	.018	.020				

REDUCTION OF BAROMETER TO LATITUDE 45°

METRIC SCALE

For latitudes below 45°, subtract the correction; for latitudes greater than 45° it is to be added. Corrections in cm.

(From Smithsonian Meteorological Tables.)

Latitude		Observed Height of Barometer in Centimeters					
		68	70	72	74	76	78
25°	65°	0.116	0.120	0.123	0.127	0.130	0.133
26	64	.111	.115	.118	.121	.125	.128
27	63	.106	.110	.113	.116	.119	.122
28	62	.101	.104	.107	.110	.113	.116
29	61	.096	.099	.102	.104	.107	.110
30	60	0.091	0.094	0.096	0.098	0.101	0.104
31	59	.085	.087	.090	.092	.095	.097
32	58	.079	.082	.084	.086	.089	.091
33	57	.074	.076	.078	.080	.082	.084
34	56	.068	.070	.072	.074	.076	.078
35	55	0.062	0.064	0.066	0.067	0.069	0.071
36	54	.056	.058	.059	.061	.063	.064
37	53	.050	.051	.053	.054	.056	.057
38	52	.044	.045	.046	.048	.049	.050
39	51	.038	.039	.040	.041	.042	.043
40	50	0.031	0.032	0.033	0.034	0.035	0.036
41	49	.025	.026	.027	.027	.028	.029
42	48	.019	.019	.020	.021	.021	.022
43	47	.013	.013	.013	.014	.014	.014
44	46	.006	.007	.007	.007	.007	.007

ENGLISH SCALE

Corrections in inches.

Latitude		Observed Height in Inches					
		25	26	27	28	29	30
25°	65°	0.043	0.044	0.046	0.048	0.050	0.051
26	64	.041	.043	.044	.046	.048	.049
27	63	.039	.041	.042	.044	.045	.047
28	62	.037	.039	.040	.042	.043	.045
29	61	.035	.037	.038	.039	.041	.042
30	60	0.033	0.035	0.036	0.037	0.039	0.040
31	59	.031	.032	.034	.035	.036	.037
32	58	.029	.030	.032	.033	.034	.035
33	57	.027	.028	.029	.030	.031	.032
34	56	.025	.026	.027	.028	.029	.030
35	55	0.023	0.024	0.025	0.025	0.026	0.027
36	54	.021	.021	.022	.023	.024	.025
37	53	.018	.019	.020	.021	.021	.022
38	52	.016	.017	.017	.018	.019	.019
39	51	.014	.014	.015	.015	.016	.017
40	50	0.012	0.012	0.012	0.013	0.013	0.014
41	49	.009	.010	.010	.010	.011	.011
42	48	.007	.007	.008	.008	.008	.008
43	47	.005	.005	.005	.005	.005	.006
44	46	.002	.002	.003	.003	.003	.003

RELATIVE HUMIDITY—DEW-POINT

The table gives the relative humidity of the air for temperature *t* and dewpoint *d*.

(From Smithsonian Meteorological Tables.)

RELATIVE HUMIDITY—DEW-POINT (Continued)

Depression of dew-point $t-d$° C.	Dew-point (*d*).				
	−10	0	+10	+20	+30
0.0	100%	100%	100%	100%	100%
0.2	98	99	99	99	99
0.4	97	97	97	98	98
0.6	95	96	96	96	97
0.8	94	94	95	95	96
1.0	92	93	94	94	94
1.2	91	92	92	93	93
1.4	90	90	91	92	92
1.6	88	89	90	91	91
1.8	87	88	89	90	90
2.0	86	87	88	88	89
2.2	84	85	86	87	88
2.4	83	84	85	86	87
2.6	82	83	84	85	86
2.8	80	82	83	84	85
3.0	79	81	82	83	84
3.2	78	80	81	82	83
3.4	77	79	80	81	82
3.6	76	77	79	80	82
3.8	75	76	78	79	81
4.0	73	75	77	78	80
4.2	72	74	76	77	79
4.4	71	73	75	77	78
4.6	70	72	74	76	77
4.8	69	71	73	75	76
5.0	68	70	72	74	75
5.2	67	69	71	73	75
5.4	66	68	70	72	74
5.6	65	67	69	71	73
5.8	64	66	69	70	72
6.0	63	66	68	70	71
6.2	62	65	67	69	71
6.4	61	64	66	68	70
6.6	60	63	65	67	69
6.8	60	62	64	66	68
7.0	59	61	63	66	68
7.2	58	60	63	65	67
7.4	57	60	62	64	66
7.6	56	59	61	63	65
7.8	55	58	60	63	65
8.0	54	57	60	62	64
8.2	54	56	59	61	63
8.4	53	56	58	60	63
8.6	52	55	57	60	62
8.8	51	54	57	59	61
9.0	51	53	56	58	61
9.2	50	53	55	58	60
9.4	49	52	55	57	59
9.6	48	51	54	56	59
9.8	48	51	53	56	58
10.0	47	50	53	55	57
10.5	45	48	51	54	
11.0	44	47	49	52	
11.5	42	45	48	51	
12.0	41	44	47	49	
12.5	39	42	45	48	
13.0	38	41	44	46	
13.5	37	40	43	45	
14.0	35	38	41	44	
14.5	34	37	40	43	
15.0	33	36	39	42	
15.5	32	35	38	40	
16.0	31	34	37	39	
16.5	30	33	36	38	
17.0	29	32	35	37	
17.5	28	31	34	36	
18.0	27	30	33	35	
18.5	26	29	32	34	
19.0	25	28	31	33	
19.5	24	27	30	33	
20.0	24	26	29	32	
21.0	22	25	27		
22.0	21	23	26		
23.0	19	22	24		
24.0	18	21	23		
25.0	17	19	22		
26.0	16	18	21		
27.0	15	17	20		
28.0	14	16	19		
29.0	13	15	18		
30.0	12	14	17		

RELATIVE HUMIDITY FROM WET AND DRY BULB THERMOMETER (CENT. SCALE)

This table gives the approximate relative humidity directly from the reading of the air temperature (dry bulb) (t°C) and the wet bulb (t'°C). It is computed for a barom pressure of 74.27 cm Hg. Errors resulting from the use of this table for air temperatures above −10°C and between 77.5 and 71 cm Hg will usually be within the errors of observa

Condensed from Bulletin of the U. S. Weather Bureau No. 1071

t \ $t-t'$	0.2	0.4	0.6	0.8	1.0	1.2	1.4	1.6	1.8	2.0	2.2	2.4	2.6	2.8	3.0	3.2	3.4	3.6	3.8	4.0	4.5	5.0	5.5	6.0	6.5	7.0	7.5	8.0	8.5	9.0	9.5	10.0	10.5	11.0
−10	93	87	80	74	67	61	54	48	41	35	28	22	16	9	..	..	..	..	..	..	..	..	..	..	..	..	..	..	..	..	..	..	..	..
−9	94	88	81	75	69	63	57	51	45	39	33	27	21	15	9	..	..	..	..	..	..	..	..	..	..	..	..	..	..	..	..	..	..	..
−8	94	88	83	77	71	65	60	54	48	43	37	32	26	20	15	10	..	..	..	..	..	..	..	..	..	..	..	..	..	..	..	..	..	..
−7	95	89	84	78	73	67	62	57	52	46	41	36	31	25	20	15	10	5	..	..	..	..	..	..	..	..	..	..	..	..	..	..	..	..
−6	95	90	85	79	74	69	64	59	54	49	45	40	35	30	25	20	15	11	6	..	..	..	..	..	..	..	..	..	..	..	..	..	..	..
−5	95	90	86	81	76	71	66	62	57	52	48	43	39	34	29	25	20	16	11	7	..	..	..	..	..	..	..	..	..	..	..	..	..	..
−4	95	91	86	82	77	73	68	64	59	55	51	46	42	38	33	29	25	21	17	12	..	..	..	..	..	..	..	..	..	..	..	..	..	..
−3	96	91	87	82	78	74	70	66	62	57	53	49	45	41	37	33	29	25	21	17	8	..	..	..	..	..	..	..	..	..	..	..	..	..
−2	96	92	88	84	79	75	71	68	64	60	56	52	48	44	40	37	33	29	25	22	12	..	..	..	..	..	..	..	..	..	..	..	..	..
−1	96	92	88	84	81	77	73	69	66	62	58	54	51	47	43	40	36	33	29	26	17	8	..	..	..	..	..	..	..	..	..	..	..	..
0	96	93	89	85	81	78	74	71	67	64	60	57	53	50	46	43	40	36	33	29	21	13	5	..	..	..	..	..	..	..	..	..	..	..
1	97	93	90	86	83	80	76	73	70	66	63	59	56	53	49	46	43	40	36	33	25	17	10	..	..	..	..	..	..	..	..	..	..	..
2	97	93	90	87	84	81	78	74	71	68	65	62	59	55	52	49	46	43	40	37	29	22	14	7	..	..	..	..	..	..	..	..	..	..
3	97	94	91	88	84	82	78	76	72	70	67	64	61	58	55	52	49	46	43	40	33	26	19	12	5	..	..	..	..	..	..	..	..	..
4	97	94	91	88	85	82	79	77	74	71	68	65	62	60	57	54	51	48	46	43	36	29	22	16	9	..	..	..	..	..	..	..	..	..
5	97	94	91	88	86	83	80	77	75	72	69	67	64	61	58	56	53	51	48	45	39	33	26	20	13	7	..	..	..	..	..	..	..	..
6	97	94	92	89	86	84	81	78	76	73	70	68	65	63	60	58	55	53	50	48	41	35	29	24	17	11	5	..	..	..	..	..	..	..
7	97	95	92	89	87	84	82	79	77	74	72	69	67	64	62	59	57	54	52	50	44	38	32	26	21	15	10	..	..	..	..	..	..	..
8	97	95	92	90	87	85	82	80	77	75	73	70	68	65	63	61	58	56	54	51	46	40	35	29	24	19	14	8	..	..	..	..	..	..
9	98	95	93	90	88	85	83	81	78	76	74	71	69	67	64	62	60	58	55	53	48	42	37	32	27	22	17	12	7	..	..	..	..	..
10	98	95	93	90	88	86	83	81	79	77	74	72	70	68	66	63	61	59	57	55	50	44	39	34	29	24	20	15	10	6	..	..	..	..
11	98	95	93	91	89	86	84	82	80	78	75	73	71	69	67	65	62	60	58	56	51	46	41	36	32	27	22	18	13	9	5	..	..	..
12	98	96	93	91	89	87	85	82	80	78	76	74	72	70	68	66	64	62	60	58	53	48	43	39	34	29	25	21	16	12	8	..	..	..
13	98	96	93	91	89	87	85	83	81	79	77	75	73	71	69	67	65	63	61	59	54	50	45	41	36	32	28	23	19	15	11	7	..	..
14	98	96	94	92	90	88	86	84	82	79	78	76	74	72	70	68	66	64	62	60	56	51	47	42	38	34	30	26	22	18	14	10	6	..
15	98	96	94	92	90	88	86	84	82	80	78	76	74	73	71	69	67	65	63	61	57	53	48	44	40	36	32	27	24	20	16	13	9	6

t \ $t-t'$	0.5	1.0	1.5	2.0	2.5	3.0	3.5	4.0	4.5	5.0	5.5	6.0	6.5	7 0	7.5	8.0	8.5	9.0	9.5	0.0	10.5	11.0	11.5	12.0	12.5	13.0	13.5	14.0	14.5	15.0	16.0	17.0	18.0	19.0	20.0
16	95	90	85	81	76	71	67	63	58	54	50	46	42	38	34	30	26	23	19	15	12	8	5	..	..	..	..	..	..	..	..	..	..	..	..
17	95	90	86	81	76	72	68	64	60	55	51	47	43	40	36	32	28	25	21	18	14	11	8	..	..	..	..	..	..	..	..	..	..	..	..
18	95	91	86	82	77	73	69	65	61	57	53	49	45	41	38	34	30	27	23	20	17	14	10	7	..	..	..	..	..	..	..	..	..	..	..
19	95	91	87	82	78	74	70	65	62	58	54	50	46	43	39	36	32	29	26	22	19	16	13	10	7	..	..	..	..	..	..	..	..	..	..
20	96	91	87	83	78	74	70	66	63	59	55	51	48	44	41	37	34	31	28	24	21	18	15	12	9	6	..	..	..	..	..	..	..	..	..
21	96	91	87	83	79	75	71	67	64	60	56	53	49	46	42	39	36	32	29	26	23	20	17	14	12	9	6	..	..	..	..	..	..	..	..
22	96	92	87	83	80	76	72	68	64	61	57	54	50	47	44	40	37	34	31	28	25	22	19	17	14	11	8	6	..	..	..	..	..	..	..
23	96	92	88	84	80	76	72	69	65	62	58	55	52	48	45	42	39	36	33	30	27	24	21	19	16	13	11	8	6	..	..	..	..	..	..
24	96	92	88	84	80	77	73	69	66	62	59	56	53	49	46	43	40	37	34	31	29	26	23	20	18	15	13	10	8	5	..	..	..	..	..
25	96	92	88	84	81	77	74	70	67	63	60	57	54	50	47	44	41	39	36	33	30	28	25	22	20	17	15	12	10	8	..	..	..	..	..
26	96	92	88	85	81	78	74	71	67	64	61	58	54	51	49	46	43	40	37	34	32	29	26	24	21	19	17	14	12	10	5	..	..	..	..
27	96	92	89	85	82	78	75	71	68	65	62	58	56	52	50	47	44	41	38	36	33	31	28	26	23	21	18	16	14	12	7	..	..	..	..
28	96	93	89	85	82	78	75	72	69	65	62	59	56	53	51	48	45	42	40	37	34	32	29	27	25	22	20	18	16	13	9	5	..	..	..
29	96	93	89	86	82	79	76	72	69	66	63	60	57	54	52	49	46	43	41	38	36	33	31	28	26	24	22	19	17	15	11	7	..	..	..
30	96	93	89	86	83	79	76	73	70	67	64	61	58	55	52	50	47	44	42	39	37	35	32	30	28	25	23	21	19	17	13	9	5	..	..
31	96	93	90	86	83	80	77	73	70	67	64	61	59	56	53	51	48	45	43	40	38	36	33	31	29	27	25	22	20	18	14	11	7	..	..
32	96	93	90	86	83	80	77	74	71	68	65	62	60	57	54	51	49	46	44	41	39	37	35	32	30	28	26	24	22	20	16	12	9	5	..
33	97	93	90	87	83	80	77	74	71	68	66	63	60	57	55	52	50	47	45	42	40	38	36	33	31	29	27	25	23	21	17	14	10	7	..
34	97	93	90	87	84	81	78	75	72	69	66	63	61	58	56	53	51	48	46	43	41	39	37	35	32	30	28	26	24	23	19	15	12	8	5
35	97	94	90	87	84	81	78	75	72	69	67	64	61	59	56	54	51	49	47	44	42	40	38	36	34	32	30	28	26	24	20	17	13	10	7
36	97	94	90	87	84	81	78	75	73	70	67	64	62	59	57	54	52	50	48	45	43	41	39	37	35	33	31	29	27	25	21	18	15	11	8
37	97	94	91	87	84	82	79	76	73	70	68	65	63	60	58	55	53	51	48	46	44	42	40	38	36	34	32	30	28	26	23	19	16	13	10
38	97	94	91	88	84	82	79	76	74	71	68	66	63	61	58	56	54	51	49	47	45	43	41	39	37	35	33	31	29	27	24	20	17	14	11
39	97	94	91	88	85	82	79	77	74	71	69	66	64	61	59	57	54	52	50	48	46	43	42	39	38	36	34	32	30	28	25	22	18	15	12
40	97	94	91	88	85	82	80	77	74	72	69	67	64	62	59	57	54	53	51	48	46	44	42	40	38	36	35	33	31	29	26	23	20	16	14

REDUCTION OF PSYCHROMETRIC OBSERVATION

For the reduction of observations with the wet and dry bulb thermometer Assuming the relative velocity of the air to the thermometer bulbs is at least three meters per second; if t is the temperature of the air as indicated by the dry bulb, t_w, the temperature of the wet bulb, B, the barometric pressure, and E_w, the vapor tension of water corresponding to t_w, then the actual vapor tension is

$$E = E_w - 0.00066B(t - t_w)(1 + 0.00115t_w)$$

The value of the term

$$0.00066B(t - t_w)(1 + 0.00115t_w)$$

is given in the following table.

(From Miller's Laboratory Physics, Ginn & Co., publishers, by permission.)

$t-t_w$	Barometric Pressure B in Centimeters							
	70.0	71.0	72.0	73.0	74.0	75.0	76.0	77.(
°C	cm	cm	cm	cm	cm	cm	cm	cm
1	0.047	0.048	0.048	0.049	0.050	0.050	0.051	0.05
2	.093	.094	.096	.097	.098	.100	.101	.10
3	.139	.141	.143	.145	.147	.149	.152	.15
4	.186	.189	.191	.194	.197	.199	.202	.20
5	0.232	0.236	0.239	0.243	0.246	0.249	0.252	0.25
6	.279	.283	.287	.291	.295	.299	.303	.30
7	.326	.331	.336	.340	.345	.350	.354	.35
8	.373	.379	.384	.389	.395	.400	.405	.41
9	.421	.427	.432	.438	.444	.450	.456	.46
10	0.468	0.474	0.481	0.488	0.494	0.501	0.508	0.51
11	.515	.522	.530	.537	.544	.551	.559	.56
12	.562	.570	.578	.586	.594	.602	.611	.61
13	.610	.618	.627	.636	.645	.653	.662	.67
14	.658	.667	.676	.686	.695	.705	.714	.72
15	0.706	0.716	0.726	0.736	0.746	0.756	0.766	0.77
16	.754	.764	.775	.786	.796	.807	.818	.82
17	.802	.813	.824	.836	.847	.859	.870	.88
18	.850	.862	.874	.886	.898	.910	.922	.93
19	.898	.911	.923	.936	.949	.962	.975	.98
20	0.946	0.960	0.973	0.987	1.000	1.014	1.027	1.04

CONSTANT HUMIDITY

The following table shows the % humidity and the aqueous tension at the given temperature within a closed space when an excess of the substance indicated is in contact with a saturated aqueous solution of the given solid phase.

Solid phase	t°C.	% humidity	Aq. tension mm Hg
$H_3PO_4.\frac{1}{2}H_2O$	24	9	1.99
$KC_2H_3O_2$	168	13	738
$LiCl.H_2O$	20	15	2.60
$KC_2H_3O_2$	20	20	3.47
KF	100	22.9	174
NaBr	100	22.9	174
NaCl, KNO_3 and $NaNO_3$	16.39	30.49	4.23
$CaCl_2.6H_2O$	24.5	31	7.08
$CaCl_2.6H_2O$	20	32.3	5.61
$CaCl_2.6H_2O$	18.5	35	5.54
CrO_3	20	35	6.08
$CaCl_2.6H_2O$	10	38	3.47
$CaCl_2.6H_2O$	5	39.8	2.59
$Zn(NO_3)_2.6H_2O$	20	42	7.29
$K_2CO_3.2H_2O$	24.5	43	9.82
$K_2CO_3.2H_2O$	18.5	44	6.96
KNO_2	20	45	7.81
KCNS	20	47	8.16
NaI	100	50.4	383
$Ca(NO_3)_2.4H_2O$	24.5	51	11.6
$NaHSO_4.H_2O$	20	52	9.03
$Na_2Cr_2O_7.2H_2O$	20	52	9.03
$Mg(NO_3)_2.6H_2O$	24.5	52	11.9
$NaClO_3$	100	54	410
$Ca(NO_3)_2.4H_2O$	18.5	56	8.86
$Mg(NO_3)_2.6H_2O$	18.5	56	8.86
KI	100	56.2	427
$NaBr.2H_2O$	20	58	10.1
$Mg(C_2H_3O_2)_2.4H_2O$	20	65	11.3
$NaNO_2$	20	66	11.5
NH_4Cl and KNO_3	30	68.6	21.6
KBr	100	69.2	526
NH_4Cl and KNO_3	25	71.2	16.7
NH_4Cl and KNO_3	20	72.6	12.6
$NaClO_3$	20	75	13.0
$(NH_4)_2SO_4$	108	75	754
$NaC_2H_3O_2.3H_2O$	20	76	13.2
$H_2C_2O_4.2H_2O$	20	76	13.2
$Na_2S_2O_3.5H_2O$	20	78	13.5
NH_4Cl	20	79.5	13.8
NH_4Cl	25	79.3	18.6
NH_4Cl	30	77.5	24.4
$(NH_4)_2SO_4$	20	81	14.1
$(NH_4)_2SO_4$	25	81.1	19.1
$(NH_4)_2SO_4$	30	81.1	25.6
KBr	20	84	14.6
Tl_2SO_4	104.7	84.8	768
$KHSO_4$	20	86	14.9
$Na_2CO_3.10H_2O$	24.5	87	20.9
$BaCl_2.2H_2O$	24.5	88	20.1
K_2CrO_4	20	88	15.3
$Pb(NO_3)_2$	103.5	88.4	760
$ZnSO_4.7H_2O$	20	90	15.6
$Na_2CO_3.10H_2O$	18.5	92	14.6
$NaBrO_3$	20	92	16.0
K_2HPO_4	20	92	16.0
$NH_4H_2PO_4$	30	92.9	29.3
$NH_4H_2PO_4$	25	93	21.9
$Na_2SO_4.10H_2O$	20	93	16.1
$NH_4H_2PO_4$	20	93.1	16.2
$ZnSO_4.7H_2O$	5	94.7	6.10
$Na_2SO_3.7H_2O$	20	95	16.5
$Na_2HPO_4.12H_2O$	20	95	16.5
NaF	100	96.6	734
$Pb(NO_3)_2$	20	98	17.0
$CuSO_4.5H_2O$	20	98	17.0
$TlNO_3$	100.3	98.7	759
TlCl	100.1	99.7	761

CONSTANT HUMIDITY WITH SULFURIC ACID SOLUTIONS

The relative humidity and pressure of aqueous vapor of air in equilibrium conditions above aqueous solutions of sulfuric acid are given below.

Density of acid solution	Relative humidity	Vapor pressure at 20°C	Density of acid solution	Relative humidity	Vapor pressure at 20°C
1.00	100.0	17.4	1.30	58.3	10.1
1.05	97.5	17.0	1.35	47.2	8.3
1.10	93.9	16.3	1.40	37.1	6.5
1.15	88.8	15.4	1.50	18.8	3.3
1.20	80.5	14.0	1.60	8.5	1.5
1.25	70.4	12.2	1.70	3.2	0.6

For concentration of sulfuric acid solution refer to tables relating density to percent composition.

VELOCITY OF SOUND

Compiled by Gordon E. Becker, Bell Telephone Laboratories

The data for the Velocity of Sound in Various Materials were compiled from a variety of sources. For more extensive tables one is referred to the following books:

AIP Handbook, Smithsonian Tables.
Mason: Physical Acoustics and the Properties of Solids (1958).
Chalmers and Quarrell: Physical Examination of Metals (1960).
Mason: Piezoelectric Crystals and their Application to Ultrasonics (1950).
Bergmann: Der Ultraschall (Hirzel, 1954).

Definition of Terms: V_l = Velocity of plane longitudinal wave in bulk material
V_s = Velocity of plane transverse (shear) wave
V_{ext} = Velocity of longitudinal wave (extensional wave) in thin rods.

Solids

Substance	Density gm/cc	V_l m/sec	V_s m/sec	V_{ext} m/sec
Metals:				
Aluminum, rolled	2.7	6420	3040	5000
Beryllium	1.87	12890	8880	12870
Brass (70 Cu, 30 Zn)	8.6	4700	2110	3480
Copper, annealed	8.93	4760	2325	3810
Copper, rolled	8.93	5010	2270	3750
Duralumin 17S	2.79	6320	3130	5150
Gold, hard-drawn	19.7	3240	1200	2030
Iron, electrolytic	7.9	5950	3240	5120
Iron, Armco	7.85	5960	3240	5200
Lead, annealed	11.4	2160	700	1190
Lead, rolled	11.4	1960	690	1210
Magnesium, drawn, annealed	1.74	5770	3050	4940
Molybdenum	10.1	6250	3350	5400
Monel metal	8.90	5350	2720	4400
Nickel (unmagnetized)	8.85	5480	2990	4800
Nickel	8.9	6040	3000	4900
Platinum	21.4	3260	1730	2800
Silver	10.4	3650	1610	2680
Steel, mild	7.85	5960	3235	5200
Steel, 347 Stainless	7.9	5790	3100	5000
Steel (1% C)	7.84	5940	3220	5180
Steel (1% C, hardened)	7.84	5854	3150	5070
Tin, rolled	7.3	3320	1670	2730
Titanium	4.5	6070	3125	5080
Tungsten, annealed	19.3	5220	2890	4620
Tungsten, drawn	19.3	5410	2640	4320
Tungsten Carbide	13.8	6655	3980	6220
Zinc, rolled	7.1	4210	2440	3850
Various:				
Fused silica	2.2	5968	3764	5760
Glass, pyrex	2.32	5640	3280	5170
Glass, heavy silicate flint	3.88	3980	2380	3720
Glass, light borate crown	2.24	5100	2840	4540
Lucite	1.18	2680	1100	1840
Nylon 6-6	1.11	2620	1070	1800
Polyethylene	0.90	1950	540	920
Polystyrene	1.06	2350	1120	2240
Rubber, butyl	1.07	1830		
Rubber, gum	0.95	1550		
Rubber, neoprene	1.33	1600		
Brick	1.8			3650
Clay rock	2.2			3480
Cork	0.25			500
Marble	2.6			3810
Paraffin	0.9			1300
Tallow				390
Woods:				
Ash, along the fiber				4670
Ash, across the rings				1390
Ash, along the rings				1260
Beech, along the fiber				3340
Elm, along the fiber				4120
Maple, along the fiber				4110
Oak, along the fiber				3850

Liquids

Substance	Formula	Density gm/cc	Velocity at 25°C m/sec	$-\Delta v/\Delta t$ m/sec °C
Acetone	C_3H_6O	0.79	1174	4.5
Benzene	C_6H_6	0.870	1295	4.65
Carbon disulphide	CS_2	1.26	1149	—
Carbon tetrachloride	CCl_4	1.595	926	2.7
Castor oil	$C_{11}H_{10}O_{10}$	0.969	1477	3.6
Chloroform	$CHCl_3$	1.49	987	3.4
Ethanol	C_2H_6O	0.79	1207	4.0
Ethanol amide	C_2H_7NO	1.018	1724	3.4
Ethyl ether	$C_4H_{10}O$	0.713	985	4.87
Ethylene glycol	$C_2H_6O_2$	1.113	1658	2.1
Glycerol	$C_3H_8O_3$	1.26	1904	2.2
Kerosene	—	0.81	1324	3.6
Mercury	Hg	13.5	1450	—

Liquids (Continued)

Substance	Formula	Density gm/cc	Velocity at 25°C m/sec	$-\Delta v/\Delta t$ m/sec °C
Methanol	CH_4O	0.791	1103	3.2
Nitrobenzene	$C_6H_5NO_2$	1.20	1463	3.6
Turpentine	—	0.88	1255	—
Water (distilled)	H_2O	0.998	1498	−2.4
Water (sea)	—	1.025	1531	−2.4
Xylene hexafluoride	$C_8H_4F_6$	1.37	879	—

Gases and Vapors

Substance	Formula	Density gm/liter	Velocity m/sec	$\Delta v/\Delta t$ m/sec °C
Gases (0°C)				
Air, dry		1.293	331.45	0.59
Ammonia	NH_3	0.771	415	
Argon	A	1.783	319	0.56
Carbon monoxide	CO	1.25	338	0.6
Carbon dioxide	CO_2	1.977	259	0.4
Chlorine	Cl_2	3.214	206	
Deuterium	D_2		890	1.6
Ethane (10°C)	C_2H_6	1.356	308	
Ethylene	C_2H_4	1.260	317	
Helium	He	0.178	965	0.8
Hydrogen	H_2	0.0899	1284	2.2
Hydrogen bromide	HBr	3.50	200	
Hydrogen chloride	HCl	1.639	296	
Hydrogen iodide	HI	5.66	157	
Hydrogen sulfide	H_2S	1.539	289	
Illuminating (Coal gas)			453	
Methane	CH_4	0.7168	430	
Neon	Ne	0.900	435	0.8
Nitric oxide (10°C)	NO	1.34	324	
Nitrogen	N_2	1.251	334	0.6
Nitrous oxide	N_2O	1.977	263	0.5
Oxygen	O_2	1.429	316	0.56
Sulfur dioxide	SO_2	2.927	213	0.47
Vapors (97.1°C)				
Acetone	C_3H_6O		239	0.32
Benzene	C_6H_6		202	0.3
Carbon tetrachloride	CCl_4		145	
Chloroform	$CHCl_3$		171	0.24
Ethanol	C_2H_6O		269	0.4
Ethyl ether	$C_4H_{10}O$		206	0.3
Methanol	CH_4O		335	0.46
Water vapor (134°C)	H_2O		494	

SOUND VELOCITY IN WATER ABOVE 212°F

By permission from the Acoustical Society of America, Volume 31 (1959) and J. C. McDade, D. R. Pardue, A. L. Gedrich and F. Vrataric.

Temperature °F	Velocity m/sec.	Velocity ft/sec.
186.8	1552	5092
200	1548	5079
210	1544	5066
220	1538	5046
230	1532	5026
240	1524	5000
250	1516	4974
260	1507	4944
270	1497	4911
280	1487	4879
290	1476	4843
300	1465	4806
310	1453	4767
320	1440	4724
330	1426	4678
340	1412	4633
350	1398	4587
360	1383	4537
370	1368	4488
380	1353	4439
390	1337	4386
400	1320	4331
410	1302	4272
420	1283	4209
430	1264	4147
440	1244	4081
450	1220	4010
460	1200	3940
470	1180	3880

SOUND VELOCITY IN WATER ABOVE 212°F (Continued)

Temperature °F	Velocity m/sec.	Velocity ft/sec.
480	1160	3800
490	1140	3730
500	1110	3650
510	1090	3570
520	1070	3500
530	1040	3410
540	1010	3320
550	980	3230

MUSICAL SCALES

Equal Tempered Chromatic Scale
$A_4 = 440$

American Standard pitch. Adopted by the American Standards Association in 1936

Note	Frequency	Note	Frequency	Note	Frequency	Note	Frequency
C_0	16.35	C_2	65.41	C_4	261.63	C_6	1046.50
$C\#_0$	17.32	$C\#_2$	69.30	$C\#_4$	277.18	$C\#_6$	1108.73
D_0	18.35	D_2	73.42	D_4	293.66	D_6	1174.66
$D\#_0$	19.45	$D\#_2$	77.78	$D\#_4$	311.13	$D\#_6$	1244.51
E_0	20.60	E_2	82.41	E_4	329.63	E_6	1318.51
F_0	21.83	F_2	87.31	F_4	349.23	F_6	1396.91
$F\#_0$	23.12	$F\#_2$	92.50	$F\#_4$	369.99	$F\#_6$	1479.98
G_0	24.50	G_2	98.00	G_4	392.00	G_6	1567.98
$G\#_0$	25.96	$G\#_2$	103.83	$G\#_4$	415.30	$G\#_6$	1661.22
A_0	27.50	A_2	110.00	**A_4**	**440.00**	A_6	1760.00
$A\#_0$	29.14	$A\#_2$	116.54	$A\#_4$	466.16	$A\#_6$	1864.66
B_0	30.87	B_2	123.47	B_4	493.88	B_6	1975.53
C_1	32.70	C_3	130.81	C_5	523.25	C_7	2093.00
$C\#_1$	34.65	$C\#_3$	138.59	$C\#_5$	554.37	$C\#_7$	2217.46
D_1	36.71	D_3	146.83	D_5	587.33	D_7	2349.32
$D\#_1$	38.89	$D\#_3$	155.56	$D\#_5$	622.25	$D\#_7$	2489.02
E_1	41.20	E_3	164.81	E_5	659.26	E_7	2637.02
F_1	43.65	F_3	174.61	F_5	698.46	F_7	2793.83
$F\#_1$	46.25	$F\#_3$	185.00	$F\#_5$	739.99	$F\#_7$	2959.96
G_1	49.00	G_3	196.00	G_5	783.99	G_7	3135.96
$G\#_1$	51.91	$G\#_3$	207.65	$G\#_5$	830.61	$G\#_7$	3322.44
A_1	55.00	A_3	220.00	A_5	880.00	A_7	3520.00
$A\#_1$	58.27	$A\#_3$	233.08	$A\#_5$	932.33	$A\#_7$	3729.31
B_1	61.74	B_3	246.94	B_5	987.77	B_7	3951.07
						C_8	4186.01

Equal Tempered Chromatic Scale
$A_4 = 435$

International Pitch, adopted 1891

Note	Frequency	Note	Frequency	Note	Frequency	Note	Frequency
C_0	16.17	C_2	64.66	C_4	258.65	C_6	1034.61
$C\#_0$	17.13	$C\#_2$	68.51	$C\#_4$	274.03	$C\#_6$	1096.13
D_0	18.15	D_2	72.58	D_4	290.33	D_6	1161.31
$D\#_0$	19.22	$D\#_2$	76.90	$D\#_4$	307.59	$D\#_6$	1230.37
E_0	20.37	E_2	81.47	E_4	325.88	E_6	1303.53
F_0	21.58	F_2	86.31	F_4	345.26	F_6	1381.04
$F\#_0$	22.86	$F\#_2$	91.45	$F\#_4$	365.79	$F\#_6$	1463.16
G_0	24.22	G_2	96.89	G_4	387.54	G_6	1550.16
$G\#_0$	25.66	$G\#_2$	102.65	$G\#_4$	410.59	$G\#_6$	1642.34
A_0	27.19	A_2	108.75	**A_4**	**435.00**	A_6	1740.00
$A\#_0$	28.80	$A\#_2$	115.22	$A\#_4$	460.87	$A\#_6$	1843.47
B_0	30.52	B_2	122.07	B_4	488.27	B_6	1953.08
C_1	32.33	C_3	129.33	C_5	517.31	C_7	2069.22
$C\#_1$	34.25	$C\#_3$	137.02	$C\#_5$	548.07	$C\#_7$	2192.26
D_1	36.29	D_3	145.16	D_5	580.66	D_7	2322.62
$D\#_1$	38.45	$D\#_3$	153.80	$D\#_5$	615.18	$D\#_7$	2460.73
E_1	40.74	E_3	162.94	E_5	651.76	E_7	2607.05
F_1	43.16	F_3	172.63	F_5	690.52	F_7	2762.08
$F\#_1$	45.72	$F\#_3$	182.89	$F\#_5$	731.58	$F\#_7$	2926.32
G_1	48.44	G_3	193.77	G_5	775.08	G_7	3100.33
$G\#_1$	51.32	$G\#_3$	205.29	$G\#_5$	821.17	$G\#_7$	3284.68
A_1	54.38	A_3	217.50	A_5	870.00	A_7	3480.00
$A\#_1$	57.61	$A\#_3$	230.43	$A\#_5$	921.73	$A\#_7$	3686.93
B_1	61.03	B_3	244.14	B_5	976.54	B_7	3906.17
						C_8	4138.44

Scientific or Just Scale
$C_4 = 256$

Note	Frequency	Note	Frequency	Note	Frequency	Note	Frequency
C_0	16	C_2	64	**C_4**	**256**	C_6	1024
D_0	18	D_2	72	D_4	288	D_6	1152
E_0	20	E_2	80	E_4	320	E_6	1280
F_0	21.33	F_2	85.33	F_4	341.33	F_6	1365.33
G_0	24	G_2	96	G_4	384	G_6	1536
A_0	26.67	A_2	106.67	A_4	426.67	A_6	1706.67
B_0	30	B_2	120	B_4	480	B_6	1920
C_1	32	C_3	128	C_5	512	C_7	2048
D_1	36	D_3	144	D_5	576	D_7	2304
E_1	40	E_3	160	E_5	640	E_7	2560
F_1	42.67	F_3	170.67	F_5	682.67	F_7	2730.67
G_1	48	G_3	192	G_5	768	G_7	3072
A_1	53.33	A_3	213.33	A_5	853.33	A_7	3413.33
B_1	60	B_3	240	B_5	960	B_7	3840
						C_8	4096

Sound Absorption

Data in the following tables have been selected from the official Bulletin of the Acoustical Materials Association and are reproduced by permission of the Association.

TABLE I. GENERAL BUILDING MATERIALS AND FURNISHINGS

The following table gives the sound absorption coefficients of various materials. The reverberation time is defined as the time required for the reverberant sound intensity to decrease to one millionth of its initial intensity, or 60 decibels. The time is given in seconds by the relation, $T_r = 0.049 \left(\frac{V}{\alpha}\right)$, where V is the volume of the room in cubic feet and α is the absorption. The absorption α in Sabins is computed by multiplying the area in square feet of each surface by its absorption coefficient and taking the sum of the products. The absorption of seats and audience is also computed on an area basis using the coefficients given below.

Complete tables of coefficients of the various materials that normally constitute the interior finish of rooms may be found in the various books on architectural acoustics. The following short list will be useful in making simple calculations of the reverberation in rooms.

Materials	Coefficients					
	125 cps	250 cps	500 cps	1000 cps	2000 cps	4000 cps
Brick, unglazed	.03	.03	.03	.04	.05	.07
Brick, unglazed, painted	.01	.01	.02	.02	.02	.03
Carpet, heavy, on concrete	.02	.06	.14	.37	.60	.65
Same, on 40 oz. hairfelt or foam rubber	.08	.24	.57	.69	.71	.73
Same, with impermeable latex backing on 40 oz. hairfelt or foam rubber	.08	.27	.39	.34	.48	.63
Concrete Block, coarse	.36	.44	.31	.29	.39	.25
Concrete Block, painted	.10	.05	.06	.07	.09	.08
Fabrics						
Light velour, 10 oz. per sq. yd., hung straight, in contact with wall	.03	.04	.11	.17	.24	.35
Medium velour, 14 oz. per sq. yd., draped to half area	.07	.31	.49	.75	.70	.60
Heavy velour, 18 oz. per sq. yd., draped to half area	.14	.35	.55	.72	.70	.65
Floors						
Concrete or terrazzo	.01	.01	.015	.02	.02	.02
Linoleum, asphalt, rubber or cork tile on concrete	.02	.03	.03	.03	.03	.02
Wood	.15	.11	.10	.07	.06	.07
Wood parquet in asphalt on concrete	.04	.04	.07	.06	.06	.07
Glass						
Large panes of heavy plate glass	.18	.06	.04	.03	.02	.02
Ordinary window glass	.35	.25	.18	.12	.07	.04
Gypsum Board, ½″ nailed to 2 × 4's 16″ o.c.	.29	.10	.05	.04	.07	.09
Marble or Glazed Tile	.01	.01	.01	.01	.02	.02
Openings						
Stage, depending on furnishings	.25 – .75					
Deep balcony, upholstered seats	.50 – 1.00					
Grills, ventilating	.15 – .50					
Plaster, gypsum or lime, smooth finish on tile or brick	.013	.015	.02	.03	.04	.05

TABLE I. GENERAL BUILDING MATERIALS AND FURNISHINGS (Continued)

Materials	Coefficients					
	125 cps	250 cps	500 cps	1000 cps	2000 cps	4000 cps
Plaster, gypsum or lime, rough finish on lath	.02	.03	.04	.05	.04	.03
Same, with smooth finish	.02	.02	.03	.04	.04	.03
Plywood Panelling, ⅜″ thick	.28	.22	.17	.09	.10	.11
Water Surface, as in a swimming pool	.008	.008	.013	.015	.020	.025
Air, Sabins per 1000 cubic feet					2.3	7.2

ABSORPTION OF SEATS AND AUDIENCE

Values given are in Sabins per square foot of seating area or per unit

	125 cps	250 cps	500 cps	1000 cps	2000 cps	4000 cps
Audience, seated in upholstered seats, per sq. ft. of floor area	.60	.74	.88	.96	.93	.85
Unoccupied cloth-covered upholstered seats, per sq. ft. of floor area	.49	.66	.80	.88	.82	.70
Unoccupied leather-covered upholstered seats, per sq. ft. of floor area	.44	.54	.60	.62	.58	.50
Wooden Pews, occupied, per sq. ft. of floor area	.57	.61	.75	.86	.91	.86
Chairs, metal or wood seats, each, unoccupied	.15	.19	.22	.39	.38	.30

TABLE II. SPECIAL ACOUSTICAL MATERIALS

This table is divided into two parts; one part listing coefficients and the other part listing attenuation factors. *Italicized numbers in that portion of this table concerned with attenuation factors indicate ceiling attenuation greater than that listed;* the actual performance could not be determined because the ceiling-plenum-ceiling-attenuation approached or exceeded that of the partition separating the rooms. This is a result of the partition employed in the tests exceeding in performance most commonly available partitions. Attention is called to the different types of panel mountings which are used during the conduct of these Sound Absorption tests.

TYPES OF MOUNTING

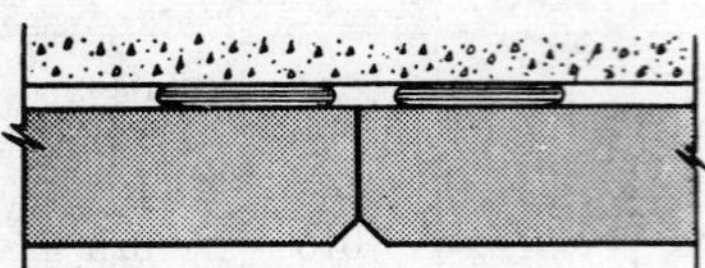

1. Cemented to plaster board with ⅛″ air space. Considered equivalent to cementing to plaster or concrete ceiling.

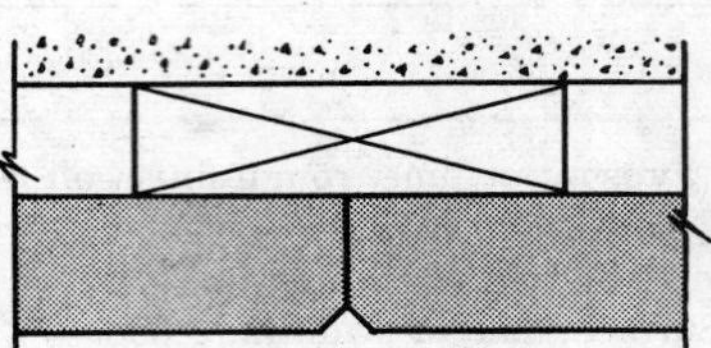

2. Nailed to nominal 1″ × 3″ wood furring 12″ o.c. unless otherwise indicated.

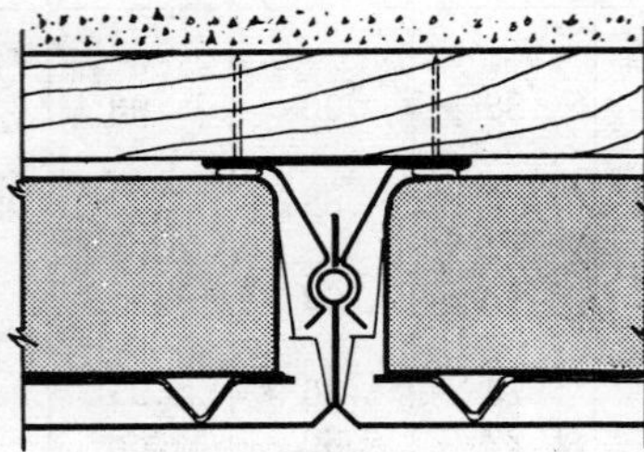

3. Attached to metal supports applied to nominal 1″ × 3″ wood furring.

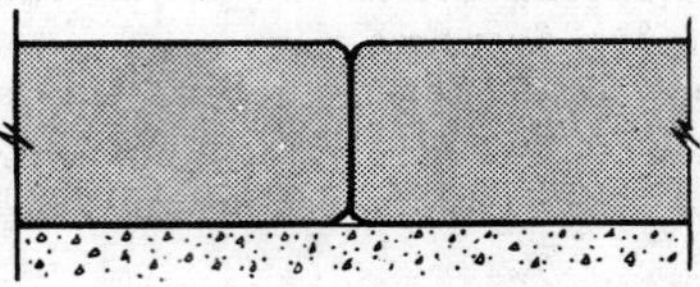

4. Applied directly to ceiling or wall.

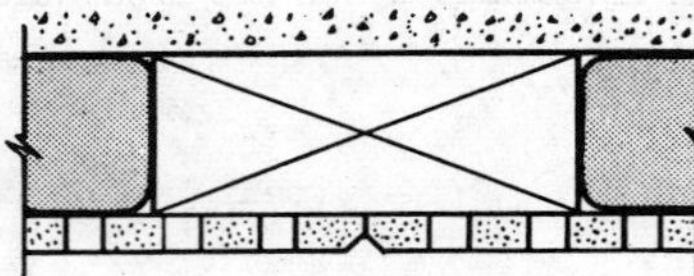

5. Furred 1″, furring 24″ o.c. 1″ Mineral wool between furring.

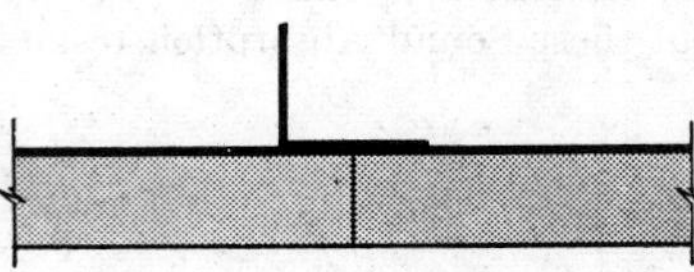

6. Attached to 24 ga. sheet iron, supported by metal angles.

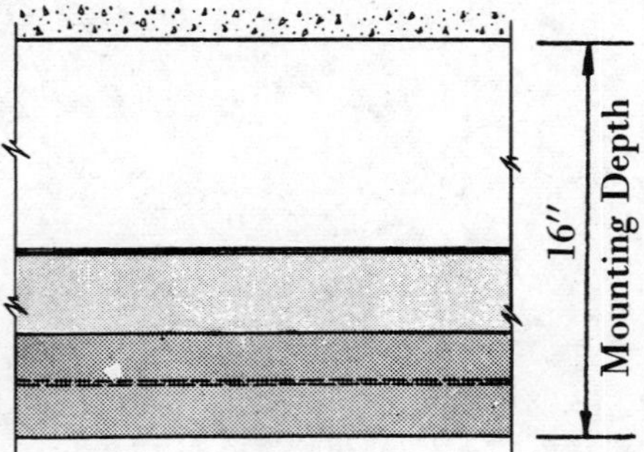

7. Mechanically mounted on special metal supports.

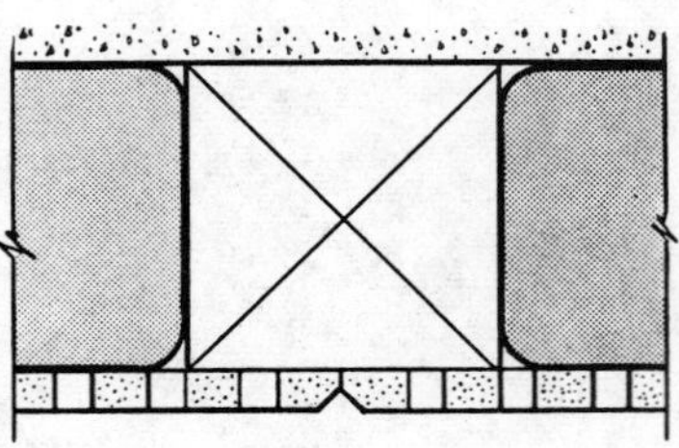

8. Furred 2″, furring 24″ o.c. 2″ Mineral wool between furring.

TABLE II. SPECIAL ACOUSTICAL MATERIALS

* The Federal Specification establishes specific criteria by which materials may be classified from "A" to "D," depending on their performance in the test. No specific terms are given to describe these classes but materials classified "A" are usually considered as "incombustible" and those classified "D" as "combustible." Classes "B" and "C" represent materials of intermediate flame resistance.

For the flame resistance tests for which ratings are shown in this bulletin, materials were mounted by bolting them directly to an asbestos cement board panel.

ARMSTRONG CORK COMPANY

Material	Thickness	Flame* Resistance	Surface	Mounting	Coefficients 125 cps	250 cps	500 cps	1000 cps	2000 cps	4000 cps
Acoustical Fire Guard[7]	⅝″	A	Perforated,[2] painted	X-2	.43	.54	.67	.96	.79	.59
Full Random	¾″	A	Same as above	X-2¼	.44	.60	.83	.99	.66	.56
Acoustical Fire Guard[7]	⅝″	A	Fissured, painted	X-2½	.48	.60	.67	.71	.75	.73
Fissured	¾″	A	Same as above	X-2¾	.42	.68	.74	.74	.71	.64
Travertone Fire Guard[8]	¾″	A	Fissured, painted	7	.75	.68	.69	.85	.93	.90
Travertone	¾″	A	Fissured, painted	1	.08	.32	.79	.93	.87	.80
				X-2¼	.36	.62	.81	.84	.87	.91
Ventilating Travertone	¾″	A	Fissured, painted		Write to manufacturer for sound absorption values					
Travertone Embossed	¾″	A	Embossed, painted	1	.08	.23	.79	.93	.88	.86
				X-10¾	.65	.56	.63	.76	.88	.91
Travertone, Golden and Silver	¾″	A	Fissured, painted	1	.13	.41	.86	.84	.76	.62
			Inlaid silver or gold colored metallic flecks	X-2¼	.37	.68	.77	.68	.71	.64
Minatone Classic	½″	A	Perforated,[6] painted	1	.11	.14	.68	.87	.68	.45
	⅝″	A	Same as above	1	.12	.25	.83	.87	.64	.52
				X-10½	.46	.44	.68	.80	.72	.56
Minaboard Fissured	⅝″	A	Fissured, painted	7	.46	.44	.58	.78	.80	.74
Minaboard Classic	⅝″	A	Perforated,[6] painted	X-2½	.39	.58	.72	.90	.72	.59
Crestone	⅝″	A	Striated, painted	1	.11	.37	.61	.63	.68	.70
				X-2	.39	.63	.57	.56	.65	.72
Cushiontone Straight Row	½″	C, D	Perforated,[1] painted	1	.07	.17	.65	.73	.73	.67
				2	.16	.57	.58	.69	.71	.70
	¾″	C, D	Same as above	1	.12	.31	.83	.93	.75	.65
				2	.18	.66	.69	.94	.77	.63
				X-2¼	.32	.65	.75	.89	.72	.64
Cushiontone Textured	9/16″	C, D	Fissured, perforated, painted	1	.14	.23	.60	.73	.83	.64
				2	.17	.60	.52	.71	.81	.71
	¾″	C, D	Same as above	1	.21	.33	.77	.78	.82	.70
				2	.16	.67	.60	.76	.81	.69
				X-2¼	.39	.64	.64	.72	.79	.68

TABLE II. SPECIAL ACOUSTICAL MATERIALS (Continued)

ARMSTRONG CORK COMPANY

Material	Thick-ness	Flame* Resist-ance	Surface	Mount-ing	Coefficients					
					125 cps	250 cps	500 cps	1000 cps	2000 cps	4000 cps
Cushiontone Classic	½″	C, D	Perforated,[6] painted	1	.10	.19	.64	.78	.72	.52
				2	.11	.57	.54	.66	.71	.54
	¾″	C, D	Same as above	1	.20	.32	.84	.81	.74	.55
				2	.22	.69	.72	.76	.71	.54
				X-2¼	.35	.62	.71	.71	.68	.52
Cushiontone Silver (Classic Pattern)	½″	D	Perforated,[6] painted	1	.20	.25	.57	.62	.56	.39
			Inlaid silver colored metallic flecks	2	.18	.59	.45	.53	.51	.40
Cushiontone Golden (Textured Pattern)	9⁄16″	D	Fissured, perforated	1	.14	.33	.55	.52	.56	.53
			painted. Inlaid gold colored metallic flecks	2	.21	.64	.39	.51	.56	.52
Cushiontone Centennial										
Rhapsody	½″	D	Perforated,[6] painted	1	.22	.30	.62	.54	.49	.43
			with design, tongue and groove	2	.28	.67	.46	.51	.54	.44
Tahiti	½″	D	Same as above	1	.22	.32	.62	.52	.48	.36
				2	.22	.67	.46	.50	.46	.39
Autumn Leaves	½″	D	Same as above	1	.20	.33	.64	.52	.51	.49
				2	.27	.67	.48	.52	.52	.45
Arrestone										
Diagonal	1 9⁄16″	A	Perforated, enameled metal[3]	3	.27	.57	.99	.99	.75	.64
Full Random		A	Perforated, enameled metal[4]	3	.25	.58	.99	.99	.76	.61
Gridtone	¾″	A	Perforated, embossed, painted metal[5]	X-8¾	.64	.84	.85	.78	.68	.40

TABLE II. SPECIAL ACOUSTICAL MATERIALS (Continued)

ARMSTRONG CORK COMPANY

Material	Thickness	Ceiling Detail at Partition	Attenuation Factors—Decibels				
			125 cps	250 cps	500 cps	1000 cps	4000 cps
Acoustical Fire Guard, Full Random	5/8″	Interrupted	31	29	36	44	59
Acoustical Fire Guard, Classic	3/4″	Interrupted	27	29	39	*50*	54
Acoustical Fire Guard, Lay-in Units Classic	5/8″	Interrupted	29	32	37	44	*61*
Travertone	3/4″	Interrupted	29	25	28	32	55
Travertone Vinyl Face	3/4″	Continuous	26	32	40	46	*57*
Minaboard Classic	5/8″	Interrupted	24	28	35	43	56
Cushiontone Classic	3/4″	Interrupted	29	29	40	*50*	49
Ventilating Travertone	3/4″	Interrupted	21	21	23	26	52

[1] Straight row perforated 529 holes per sq. ft., 3/16″ diameter.
[2] Full random perforated 353 holes per sq. ft.; 250 of 3/16″ and 103 of 1/4″ diameters.
[3] Diagonal perforated, enameled metal pan with mineral wool sound-absorbing pad. Pans perforated 1105 holes per sq. ft., 3/32″ diameter; bevels and flanges unperforated. Thickness includes tee bar.
[4] Full random perforated, enameled metal pan with mineral wool sound-absorbing pad. Pan perforated 717 holes per sq. ft., 498 of 7/64″ and 219 of 5/32″ diameters. Thickness includes tee bar.
[5] Random perforations in two sizes and a field of directional bars set in relief, painted metal facing unit backed with 3/4″ glass fiber sound-absorbing pad.
[6] Hundreds of small perforations scattered in lace-like fashion.
[7] Acoustical Ceiling Constructions Classified by Underwriters' Laboratories, Incorporated.

No. 4177-3—5/8″ Acoustical Fire Guard—Design 8—2 Hours
No. 4177-4—5/8″ Acoustical Fire Guard—Design 9—1 Hour
No. 4177-5—3/4″ Acoustical Fire Guard—Design 31—4 Hours
No. 4177-6—3/4″ Acoustical Fire Guard—Design 21—4 Hours

[8] Acoustical Ceiling Constructions Classified by Underwriters' Laboratories, Incorporated.

No. 4177-9—3/4″ Travertone Fire Guard—Design 7—1½ Hours

BALDWIN-EHRET-HILL INC.

Material	Thickness	Flame* Resistance	Surface	Mounting	Coefficients					
					125 cps	250 cps	500 cps	1000 cps	2000 cps	4000 cps
Hiltonia, Regular Pattern, Perforated Mineral Tile	5/8″	A	Perforated,[1] painted	1	.03	.23	.71	.92	.69	.58
				2	.11	.32	.73	.99	.91	.70
				X-10½	.68	.67	.74	.96	.94	.75
Hiltonia, Random Pattern, Perforated Mineral Tile	1/2″	A	Perforated,[2] painted	1	.05	.16	.48	.99	.86	.58
				2	.10	.40	.68	.97	.86	.62
	5/8″	A	Same as above	1	.06	.17	.61	.99	.78	.47
				2	.11	.36	.84	.99	.85	.56
				7	.61	.57	.69	.99	.83	.51
	3/4″	A	Same as above	1	.08	.21	.70	.99	.77	.52
				7	.70	.69	.80	.99	.82	.56

TABLE II. SPECIAL ACOUSTICAL MATERIALS (Continued)

BALDWIN-EHRET-HILL INC.

Material	Thickness	Flame* Resistance	Surface	Mounting	Coefficients 125 cps	250 cps	500 cps	1000 cps	2000 cps	4000 cps
Hiltonia, Pin Point Pattern, Perforated Mineral Tile	½″	A	Perforated,[3] painted	1	.05	.13	.50	.99	.84	.58
	⅝″	A	Same as above	1	.03	.16	.71	.99	.85	.74
				7	.63	.52	.69	.92	.82	.65
	¾″	A	Same as above	1	.10	.21	.90	.99	.85	.61
				X-10¾	.53	.62	.73	.96	.91	.74
Panatone with B-E-H Mineral Fiber Pads	1 9/16″	A	Perforated, enameled metal[4]	3	.26	.60	.99	.98	.78	.66
Claritone Regular Drilled	½″	C, D	Perforated,[1] painted[5]	1	.03	.16	.58	.82	.87	.71
				2	.13	.57	.56	.74	.84	.71
	¾″	C, D	Same as above	1	.04	.29	.77	.88	.81	.55
				2	.08	.65	.65	.87	.87	.65
				X-10¾	.51	.37	.59	.86	.90	.64
B-E-H Ceiling Board	¾″	A	Textured, painted	7	.85	.67	.74	.91	.93	.92

Material	Thickness	Surface	Ceiling Detail at Partition	Attenuation Factors—Decibels 125 cps	250 cps	500 cps	1000 cps	4000 cps
Styltone Fissured Mineral Tile	¾″	See preceding table	Interrupted	28	33	33	37	*58*

[1] Perforated 484 holes per sq. ft., 3/16″ diameter, uniformly spaced.
[2] Perforated 320 holes per sq. ft., 111 of ¼″ and 209 of 3/16″ diameters, randomly spaced.
[3] Perforated 1596 holes per sq. ft., 3/32″, 5/64″, 1/16″, 3/64″ diameters, randomly spaced.
[4] Perforated, enameled metal pan backed with sound-absorbing mineral fiber pad. Pan perforated with 1024 holes per square foot, 0.109″ diameter. Thickness includes tee bar.
[5] Available with standard, factory-applied paint finish or with factory-applied flame-resistant paint finish which gives flame resistance rating of Class "C" and light reflection of .80.

THE CELOTEX CORP.

Material	Thickness	Flame* Resistance	Surface	Mounting	Coefficients 125 cps	250 cps	500 cps	1000 cps	2000 cps	4000 cps
Acousti-Celotex Cane Tile, Standard Pattern	½″	C, D	Perforated,[1] painted[10]	1	.03	.18	.64	.82	.79	.71
				2	.15	.55	.55	.69	.75	.73
				X-2½	.39	.62	.59	.69	.73	.71
	¾″	C, D	Perforated,[1] painted[2]	1	.12	.21	.80	.99	.80	.60
				2	.15	.53	.80	.92	.77	.51
				X-2¾	.31	.71	.80	.90	.76	.63

TABLE II. SPECIAL ACOUSTICAL MATERIALS (Continued)

THE CELOTEX CORP.

Material	Thick-ness	Flame* Resist-ance	Surface	Mount-ing	Coefficients					
					125 cps	250 cps	500 cps	1000 cps	2000 cps	4000 cps
Hush-Tone Tile, Standard Pattern	½"	D	Perforated,[1] painted	2	.15	.55	.55	.69	.75	.73
	¾"	C, D	Same as above	1	.18	.41	.79	.80	.69	.51
				2	.26	.61	.64	.76	.69	.57
				X-2¾	.33	.68	.71	.67	.72	.55
Celotone Standard	¾"	A	Fissured, painted	1	.08	.23	.79	.98	.87	.80
				X-2¾	.28	.58	.96	.84	.89	.78
Acousteel	1 9/16"	A	Perforated, enameled metal[3]	3	.32	.67	.99	.96	.72	.56
Steelacoustic Standard	1¼"	A	Slotted, enameled metal[4]	X-9¼	.25	.63	.78	.70	.81	.62
Random		A	Same as above	X-9¼	.54	.75	.72	.69	.63	.38
Supracoustic	¾"	A	Fissured, painted	7	.89	.87	.75	.95	.86	.60
	1"	A	Same as above	7	.93	.88	.81	.99	.89	.68

Material	Thick-ness	Ceiling Detail at Partition	Attenuation Factors—Decibels				
			125 cps	250 cps	500 cps	1000 cps	4000 cps
Acousti-Celotex Cane Tile, Standard Pattern	¾"	Interrupted	23	25	33	39	50
Perforated Mineral Fiber Tile MP-1, Standard	⅝"	Interrupted	27	29	32	34	57
MP-6, Standard	1"	Interrupted	28	32	39	*49*	59
Celotone Standard	¾"	Interrupted	23	24	27	30	56
Chase Celotone	¾"	Interrupted	28	33	36	44	*64*
Protectone Mineral Fiber Tile PSP-1 Serene	⅝"	Interrupted	30	33	34	39	*60*
PF-4 Natural Fissured	¾"	Interrupted	27	30	29	34	*59*

[1] Perforated 484 holes per sq. ft., 3/16" diameter, 17/32" on centers.

[2] Available either with standard No. 232 two-coat, hot-rolled finish, or with No. 9 flame-resisting factory-applied finish which provides a Class "C" flame resistance rating according to Federal Specification SS-A-118b.

[3] Perforated, enameled metal pan with mineral wool sound-absorbing pad. Perforated in diagonal pattern, 1105 holes per sq. ft., 3/32" diameter. Flanges and bevels unperforated. Thickness includes tee bar. Also available in standard and random perforating patterns.

[4] Slotted enameled metal units with 1¼" mineral fiber absorbent pads laminated.

TABLE II. SPECIAL ACOUSTICAL MATERIALS (Continued)

ELOF HANSSON, INC.

Material	Thickness	Flame* Resistance	Surface	Mounting	Coefficients 125 cps	250 cps	500 cps	1000 cps	2000 cps	4000 cps
Hansonite Regular Perforated Acoustical Tile	½″	C, D	Perforated,[1] painted[2]	1	.09	.19	.58	.73	.78	.73
				2	.13	.59	.49	.66	.76	.75
	¾″	C, D	Same as above	1	.21	.32	.68	.87	.86	.71
				2	.29	.65	.59	.80	.91	.71
Hansonite Random Perforated Mineral Tile	½″	A	Perforated,[3] painted[6]	1	.05	.16	.48	.99	.86	.58
				2	.10	.40	.68	.97	.86	.62
	⅝″	A	Same as above	1	.06	.17	.61	.99	.78	.47
				2	.11	.36	.84	.99	.85	.56
				7	.61	.57	.69	.99	.83	.51
	¾″	A	Same as above	1	.08	.21	.70	.99	.77	.52
				7	.70	.69	.80	.99	.82	.56
Hansoguard,[7] Hansostar Perforated Fire Protective Acoustical Tile	⅝″	A	Perforated,[4] painted[6]	7	.41	.38	.66	.89	.77	.60
Hansoguard[7] Fissured Fire Protective Acoustical Tile	⅝″	A	Fissured, painted	7	.53	.43	.62	.73	.72	.68
Hansoboard Gold Mist Mineral Ceiling Board	⅝″	A	Perforated,[4] painted[5]	7	.56	.51	.69	.89	.76	.52
Hansonite Perforated Asbestos Board	3³⁄₁₆″	A	Perforated,[8] unpainted	X-13¼	.83	.93	.88	.87	.69	.48

TABLE II. SPECIAL ACOUSTICAL MATERIALS (Continued)

ELOF HANSSON, INC.

Material	Thickness	Ceiling Detail at Partition	Attenuation Factors—Decibels				
			125 cps	250 cps	500 cps	1000 cps	4000 cps
Hansonite Fissured Mineral Tile[9]	¾″	Interrupted	23	26	26	29	52
Note 9		Interrupted	25	36	*46*	*54*	*61*
Hansoguard Random Perforated Fire Protective Acoustical Tile	⅝″	Interrupted	26	30	35	40	*61*
Hansopan Regular Perforated Metal Pan with 1¼″ paper wrapped pad	1⁹⁄₁₆″	Interrupted	23	23	25	29	52
Hansoboard Perforated Mineral Ceiling Board	⅝″	Interrupted	26	31	31	37	*61*

[1] Perforated 484 holes per sq. ft., ³⁄₁₆″ diameter, uniformly spaced.
[2] Factory painted white. Also available factory painted white with flame resistant finish which meets Federal Specification SS-A-118b Class "C" requirements.
[3] Perforated 320 holes per sq. ft., 111 of ¼″ and 209 of ³⁄₁₆″ diameters, randomly spaced.
[4] Perforated 1,596 holes per sq. ft. ³⁄₆₄″, ¹⁄₁₆″, ⁵⁄₆₄″ and ³⁄₃₂″ diameters, randomly spaced.
[5] Factory painted white base with textured gold overlay; flame resistant finish meets Federal Specification SS-A-118b Class "C" requirements.
[6] Factory painted white, face and bevels.
[7] Acoustical Ceiling Constructions Classified by Underwriters' Laboratories, Inc. No. 4355-1—⅝″ Hansoguard—Design 18—2 Hours.
[8] Perforated ³⁄₁₆″ diameter holes on ½″ straight centers.
[9] Tile cemented to ⅝″ gypsum board suspended ceiling.

E. F. HAUSERMAN COMPANY

Material	Thickness	Flame* Resistance	Surface	Mounting	Coefficients					
					125 cps	250 cps	500 cps	1000 cps	2000 cps	4000 cps
Acoustic Ceiling	3″	A	Perforated metal[1,2]	4	.71	.60	.81	.99	.69	.50
Acousti-Wall	2¾″	A	Perforated metal[1,3]	4	.70	.47	.77	.99	.69	.44
Signature Wall Panel	2¼″		Perforated metal[1,3]	4	.41	.49	.78	.99	.67	.42

[1] Color as selected, factory baked enamel.
[2] Perforated steel face, airspace with 1″ glass fibre board, ⅜″ gypsum board, rock wool and unperforated back.
[3] Construction similar to Note 2 except ¼″ gypsum board.

TABLE II. SPECIAL ACOUSTICAL MATERIALS (Continued)

JOHNS-MANVILLE SALES CORPORATION

Material	Thickness	Flame* Resistance	Surface	Mounting	Coefficients 125 cps	250 cps	500 cps	1000 cps	2000 cps	4000 cps
Sanacoustic "W," J-M Mineral Wool Pad, Paper Wrapped, 1¼" Thick	1 9/16"	A	Perforated,[1] enameled metal	3 7	.39 .71	.73 .68	.99 .87	.99 .98	.90 .83	.72 .70
Sanacoustic 50/50 Pattern, J-M Mineral Wool Pad, Paper Wrapped, 1¼" thick	1 9/16"	A	Perforated and unperforated, enameled metal[2]	3	.50	.59	.66	.71	.56	.32
Sanacoustic "W," J-M Microlite Blanket; 1½" Thick, 0.75 Lb. Density	3"	A	Perforated,[1] enameled metal[3]	X-12½	.70	.77	.79	.84	.74	.54
Perforated Transite Panels, J-M Microlite Blanket; 1" Thick, 1.0 Lb. Density	1 3/16"	A	Perforated,[4] painted	5	.20	.30	.69	.98	.68	.25
Perforated Transite Panels, J-M Microlite Blanket; 2" Thick, 1.0 Lb. Density	2 3/16"	A	Perforated,[4] painted	8	.32	.62	.99	.86	.63	.37
Spintone Tile, Uniform Drilled Mineral Tile	½"	A	Perforated,[5] painted	1	.09	.23	.62	.75	.77	.77
	⅝"	A	Same as above	1 7	.18 .44	.28 .50	.68 .65	.95 .90	.84 .92	.66 .67
	¾"	A	Same as above	1 7	.20 .54	.31 .55	.78 .70	.95 .92	.78 .85	.67 .71
Spintone Tile, Random Drilled Mineral Tile	½"	A	Perforated,[6] painted	1	.14	.26	.64	.65	.64	.58
	⅝"	A	Same as above	1 7	.20 .58	.35 .50	.68 .67	.71 .92	.80 .91	.78 .72
	¾"	A	Same as above	7	.45	.46	.64	.89	.85	.65
Firedike Tile,[8] Uniform Drilled Mineral Tile	¾"	A	Perforated,[5] painted	X-10¾ 7	.44 .54	.47 .55	.59 .70	.83 .92	.87 .85	.68 .71
Spanglas, Random Pebbled	1¼"	A	Random pebbled, painted	X-11¼ 7	.81 .84	.85 .87	.87 .89	.96 .99	.99 .96	.86 .86
Acousti-Shell	2"		Dome shaped,[7] glass fiber fabric facing	7	.78	.76	.68	.79	.81	.81

TABLE II. SPECIAL ACOUSTICAL MATERIALS (Continued)

JOHNS-MANVILLE SALES CORPORATION

Material	Thickness	Flame* Resistance	Surface	Mounting	Coefficients 125 cps	250 cps	500 cps	1000 cps	2000 cps	4000 cps
Fibretone, Uniform Drilled	½″	C, D	Perforated,[5] painted	1	.03	.16	.58	.82	.87	.71
				2	.13	.57	.56	.74	.84	.71
	¾″	C, D	Same as above	1	.04	.29	.77	.88	.81	.55
				2	.08	.65	.65	.87	.87	.65
				X-10¾	.51	.37	.59	.86	.90	.64
Solo Tile, Random Perforated	⅞″		Perforated,[9] painted	1	.21	.45	.59	.58	.58	.35
Airacoustic	½″	A	Unpainted	6	.35	.43	.40	.74	.85	.78
	1″	A	Same as above	6	.60	.56	.69	.75	.82	.82

Material	Thickness	Ceiling Detail at Partition	Attenuation Factors—Decibels 125 cps	250 cps	500 cps	1000 cps	4000 cps
Firedike Tile, Uniform Drilled Mineral Tile	¾″	Interrupted	33	34	39	46	*65*
Sanacoustic "W," J-M Mineral Wool Pad, Paper Wrapped, 1¼″ Thick	1⁹⁄₁₆″	Interrupted	23	23	25	29	52
Sanacoustic "W," J-M Mineral Wool Pad, Paper Wrapped, 1¼″ Thick. Backed with ⅛″ J-M Flexboard Attenuation Baffles	1⁹⁄₁₆″	Interrupted	30	34	*48*	*54*	*58*
Spintone Tile, Random Drilled Mineral Tile	⅝″	Interrupted	27	31	36	42	59

[1] Perforated enameled metal pan with sound-absorptive element as stated under Material. Pan perforated 925 holes per sq. ft., 0.108″ diameter. Thickness includes tee bar.

[2] One half perforated metal pan backed with sound-absorptive element as stated under Material; one half unperforated metal pan unbacked. Thickness includes tee bar. Pan perforated 1625 holes per sq. ft., 0.10″ diameter.

[3] Blanket supported by tee bars about 1½″ above face of metal pan.

[4] Transite perforated 600 holes per sq. ft., ³⁄₁₆″ diameter.

[5] Perforated 484 holes per sq. ft., ³⁄₁₆″ diameter.

[6] Perforated 312 holes per sq. ft., 132 of ¼″ and 180 of ³⁄₁₆″ diameters.

[7] Three dimensional glass fiber lay-in panel, approximately ⁵⁄₃₂″ thick, with glass fiber fabric finish available in beige, green and blue.

[8] Acoustical Ceiling Constructions Classified by Underwriters' Laboratories, Incorporated.

No. 4400-1—¾″ Firedike Mineral Tile—Design 29—2 Hours

[9] Perforated 350 hexagonal shaped holes per sq. ft., 86 holes ³⁄₁₆″ diameter across the flats, and 264 holes ⁵⁄₃₂″ diameter across the flats.

TABLE II. SPECIAL ACOUSTICAL MATERIALS (Continued)

KAISER GYPSUM COMPANY, INC.

Material	Thick-ness	Flame* Resist-ance	Surface	Mount-ing	Coefficients					
					125 cps	250 cps	500 cps	1000 cps	2000 cps	4000 cps
Kaiser Fir-Tex Perforated Acoustical Tile, Regular Drilled	½″	C, D	Perforated,[1] painted[2]	1	.12	.18	.58	.74	.84	.78
				2	.14	.60	.64	.68	.83	.77
				7	.51	.38	.48	.79	.94	.91
	¾″	C, D	Same as above	1	.19	.24	.71	.97	.88	.65
				2	.27	.58	.63	.94	.92	.71
				X-10¾	.61	.39	.57	.90	.90	.72

[1] Perforated 529 holes per sq. ft., 13⁄64″ diameter.

[2] Factory-painted two coats face and bevels. Also available factory-painted two coats face and bevels with slow-burning finish which meets Federal Specification SS-A-118b, Class "C" flame resistance requirements.

NATIONAL GYPSUM COMPANY

Material	Thick-ness	Flame* Resist-ance	Surface	Mount-ing	Coefficients					
					125 cps	250 cps	500 cps	1000 cps	2000 cps	4000 cps
Acoustifibre, Uniform Pattern	½″	C, D	Perforated,[1] painted[2]	1	.07	.22	.61	.83	.80	.72
				2	.12	.53	.54	.77	.85	.78
				7	.35	.40	.48	.75	.89	.84
	¾″	C, D	Same as above	1	.14	.32	.78	.92	.88	.66
				2	.15	.71	.63	.91	.88	.72
				7	.44	.42	.58	.94	.89	.67
Acoustiroc Fissured Pattern	¾″	A	Fissured, painted	1	.11	.32	.77	.88	.87	.84
				7	.62	.58	.61	.79	.90	.83
Travacoustic	¾″	A	Fissured, painted	1	.09	.23	.67	.99	.92	.89
				7	.69	.71	.73	.89	.88	.90
Travacoustic Sculptured Pattern	¾″	A	Striated, painted	1	.04	.16	.62	.98	.95	.91
				X-10¾	.71	.76	.78	.82	.93	.90
Gold Bond Perforated Asbestos Panels	3⁄16	A	Perforated, autoclaved[4]	7	.82	.88	.66	.73	.66	.51
	15⁄16″	A	Perforated, autoclaved[3]	7	.77	.64	.60	.78	.68	.42
	13⁄16″	A	Perforated, autoclaved[3,5]	5	.08	.20	.57	.96	.69	.24
			Perforated, autoclaved[3]	5	.09	.31	.56	.93	.68	.23
				7	.75	.66	.62	.75	.65	.44
	23⁄16″	A	Same as above	8	.18	.55	.98	.98	.58	.44
				7	.93	.81	.86	.96	.65	.45

TABLE II. SPECIAL ACOUSTICAL MATERIALS (Continued)

NATIONAL GYPSUM COMPANY

Material	Thickness	Ceiling Detail at Partition	Attenuation Factors—Decibels				
			125 cps	250 cps	500 cps	1000 cps	4000 cps
Acoustifibre, Full Random Pattern[6]	¾″	Interrupted	*22*	25	37	45	46
		Interrupted	*31*	*34*	*48*	*53*	*61*
Fire-Shield Solitude							
Full Random Pattern	¾″	Interrupted	33	35	43	*51*	*65*
Needlepoint Pattern	⅝″	Interrupted	31	34	42	*50*	*61*
Fire-Shield Acoustiroc							
Textured Pattern	¾″	Interrupted	28	34	37	45	*63*
Full Random Pattern	¾″	Interrupted	25	26	24	26	53
Fissured Pattern	¾″	Interrupted	27	31	31	36	*60*
Travacoustic[6]	¾″	Interrupted	*24*	25	31	37	50
		Interrupted	*31*	*34*	*47*	*53*	*58*
Acoustimetal Uniform Pattern with RIS Blanket[7]	3⅝″	Interrupted	*22*	26	33	41	49
Fire-Shield Solitude Grid Panels, Fissured	⅝″	Interrupted	29	33	41	*49*	55

[1] Perforated 484 holes per sq. ft., 3⁄16″ diameter, uniformly spaced.
[2] Factory painted two coats, which includes flame resisting paint giving Class "C" flame resistance rating. Also available with Class "D" finish.
[3] Asbestos board perforated 550 holes per sq. ft., 3⁄16″ diameter; autoclaved white finish.
[4] Asbestos board backed with flame resistant paper sound absorbing element, perforated 550 holes per sq. ft., 3⁄16″ diameter; autoclaved white finish.
[5] Pad thickness ¾″.
[6] Tile cemented to ½″ gypsum board suspended ceiling.
[7] Semi-thick Gold Bond RIS Blankets installed on top of pans and pads, vapor barrier side toward the pan.

OWENS-CORNING FIBERGLAS CORPORATION

Material	Thickness	Flame* Resistance	Surface	Mounting	Coefficients					
					125 cps	250 cps	500 cps	1000 cps	2000 cps	4000 cps
Fiberglas Acoustical Tile	½″	A	Textured, painted	1	.14	.19	.66	.78	.80	.64
Type TXW	¾″	A	Same as above	1	.13	.33	.78	.86	.78	.69
				X-10¾	.68	.74	.74	.81	.87	.76
			Same as above	7	.80	.83	.69	.85	.84	.72
Type-TXW-TL	¾″	A	Same as above	7	.70	.54	.68	.81	.81	.59
Fiberglas Frescor Acoustical Tile	¾″	A	Stippled,[1] painted	1	.17	.29	.76	.91	.92	.76
				X-10¾	.70	.81	.77	.83	.88	.77
			Same as above	X-10¾	.63	.74	.77	.82	.87	.73
Type TL	¾″	A	Same as above	7	.60	.49	.72	.89	.85	.63

TABLE II. SPECIAL ACOUSTICAL MATERIALS (Continued)

OWENS-CORNING FIBERGLAS CORPORATION

Material	Thick-ness	Flame* Resist-ance	Surface	Mount-ing	Coefficients					
					125 cps	250 cps	500 cps	1000 cps	2000 cps	4000 cps
Fiberglas Ceiling Board	¾″	A	Textured, painted	X-10¾	.69	.76	.92	.88	.88	.88
	1″	A	Same as above	X-11	.65	.89	.93	.98	.92	.91
	1¼″	A	Same as above	X-11¼	.65	.78	.91	.95	.97	.93
Type TL	¾″	A	Same as above	7	.51	.53	.75	.95	.97	.80
Fiberglas Sonocor Ceiling Board	1″	C	Embossed membrane faced[2]	7	.77	.57	.84	.93	.75	.51
	2″		Same as above	7	.53	.86	.94	.91	.63	.39
	3″		Same as above	7	.84	.87	.99	.85	.58	.30

Material	Thick-ness	Ceiling Detail at Partition	Attenuation Factors—Decibels				
			125 cps	250 cps	500 cps	1000 cps	4000 cps
Fiberglas Acoustical Tile Type TXW-TL	¾″	Interrupted	29	26	33	38	50
Fiberglas Frescor Acoustical Tile Type TL	¾″	Interrupted	29	23	31	36	49
Fiberglas Ceiling Board, Type TL	¾″	Interrupted	24	23	31	38	52
Fiberglas Frescor Ceiling Board, Type TL	¾″	Interrupted	26	24	32	39	54
Fiberglas Cloth Faced Ceiling Board, Type TL	¾″	Interrupted	23	28	32	31	48

[1] Stippled, random white on white.
[2] Thin embossed plastic membrane cemented to board face.

SIMPSON TIMBER COMPANY

Material	Thick-ness	Flame* Resist-ance	Surface	Mount-ing	Coefficients					
					125 cps	250 cps	500 cps	1000 cps	2000 cps	4000 cps
Simpson Perforated Acoustical Tile PCP, Random Perforated	½″		Perforated,[3] painted	1	.14	.27	.63	.76	.71	.49
				2	.08	.66	.55	.68	.62	.50
	¾″		Same as above	1	.15	.42	.77	.78	.62	.45
				2	.19	.65	.67	.76	.58	.42
				7	.39	.46	.65	.77	.61	.49
Simpson Perforated Acoustical Tile, Standard Drilled	½″	C, D	Perforated,[1] painted[2]	1	.13	.19	.61	.80	.83	.74
				2	.07	.65	.52	.62	.75	.77
	¾″	C, D	Same as above	1	.29	.37	.69	.81	.85	.77
				2	.22	.62	.66	.88	.86	.58
				X-10¾	.34	.31	.61	.83	.81	.61

TABLE II. SPECIAL ACOUSTICAL MATERIALS (Continued)

SIMPSON TIMBER COMPANY

Material	Thickness	Flame* Resistance	Surface	Mounting	Coefficients 125 cps	250 cps	500 cps	1000 cps	2000 cps	4000 cps
Simpson Forestone Ceiling Board	1″	C	Fissured, painted	X-10¾	.39	.37	.48	.62	.77	.80
Simpson Fissured Mineral Tile	¾″	A	Fissured, painted	1 X-10¾	.06 .81	.19 .74	.67 .70	.99 .80	.91 .94	.90 .89
Simpson Metal Acoustical Units	1 9/16″	A	Perforated, enameled metal[4]	3	.31	.68	.99	.98	.79	.62

[1] Perforated 484 holes per sq. ft., 3/16″ diameter.

[2] Painted by manufacturer two coats face and bevels. Also available with factory applied finish which meets Federal Specification SS-A-118b, Class "C" flame resistance.

[3] Random drilled 457 holes per sq. ft., 63 holes 13/16 and 92 holes 11/16″ diameters, 302 holes punch perforated, approximately 5/16″ diameter.

[4] Perforated, enameled metal pan with sound-absorbing pad. Pan perforated 1013 holes per sq. ft., 0.109″ diameter. Thickness includes tee bar.

UNITED STATES GYPSUM COMPANY

Material	Thickness	Flame* Resistance	Surface	Mounting	Coefficients 125 cps	250 cps	500 cps	1000 cps	2000 cps	4000 cps
Acoustone "F"	¾″	A	Fissured, painted	1 7	.03 .68	.27 .67	.83 .65	.99 .84	.82 .87	.71 .74
Acoustone "db"	¾″	A	Fissured, painted	7	.58	.48	.67	.99	.92	.85
Auditone Regular Perforated	½″	C, D	Perforated,[1] painted[2]	1 2	.14 .18	.23 .61	.66 .51	.68 .62	.78 .78	.76 .74
	¾″	C, D	Same as above	1 2 7	.24 .23 .36	.42 .62 .35	.64 .56 .62	.69 .72 .88	.81 .83 .83	.79 .85 .74
Perfatone	1 9/16″	A	Perforated,[3] enameled metal	7	.74	.79	.88	.99	.77	.56
Ceiling Board	¾″	A	Fissured, painted	7	.89	.87	.75	.95	.86	.60
	1″	A	Same as above	7	.93	.88	.81	.99	.89	.68

TABLE II. ACOUSTICAL SPECIAL MATERIALS (Continued)

UNITED STATES GYPSUM COMPANY

Material	Thick-ness	Ceiling Detail at Partition	Attenuation Factors—Decibels				
			125 cps	250 cps	500 cps	1000 cps	4000 cps
Acoustone "F"[4]	¾″	Interrupted	23	26	26	29	52
		Interrupted	25	36	*46*	*54*	*61*
Acoustone "db"	¾″	Interrupted	31	33	41	*49*	*59*
Auditone Random Perforated	¾″	Interrupted	26	33	*43*	*50*	49
Perfatone	1 9/16″	Interrupted	21	21	21	24	44
Motif'd Acoustone "db," Galaxy Design No. 33	¾″	Interrupted	27	32	35	41	*61*
Acoustone "120"	¾″	Interrupted	29	32	36	41	*60*

[1] Perforated 529 holes per sq. ft., 3/16″ diameter.

[2] Factory painted face and bevels. Also furnished, factory painted, with special paint finish giving Class "C" flame resistance rating.

[3] Perforated enameled metal, 1105 holes per sq. ft., 3/32″ diameter. Thickness includes tee bar.

[4] Tile cemented to ⅝″ Sheetrock Firecode Gypsum Board suspended ceiling.

WOOD CONVERSION COMPANY

Material	Thick-ness	Flame* Resist-ance	Surface	Mount-ing	Coefficients					
					125 cps	250 cps	500 cps	1000 cps	2000 cps	4000 cps
Nu-Wood, Regular Pattern, Regularly Perforated Cellulose Fiber Tile	½″	C, D	Perforated,[1] painted[6]	1 2	.09 .13	.19 .59	.58 .49	.73 .66	.78 .76	.73 .75
	¾″	C, D	Same as above	1 2	.21 .29	.32 .65	.68 .59	.87 .80	.86 .91	.71 .71
Lo-Tone, Regular Pattern, Regularly Perforated Mineral Tile	⅝″	A	Perforated,[1] painted	1 2 X-10¾	.03 .11 .68	.23 .32 .67	.71 .73 .74	.92 .99 .96	.69 .91 .94	.58 .70 .75
Lo-Tone FR[7] Random Pattern Mineral Tile	⅝″	A	Perforated,[2] painted	7	.48	.41	.57	.96	.89	.59
Lo-Tone Metal Pan Regular Pattern	1 9/16″	A	Perforated,[3] enameled metal	3	.39	.73	.99	.99	.90	.72
Random Pattern		A	Perforated,[4] enameled metal	3	.39	.76	.99	.92	.61	.38
Lo-Tone Asbestos Board, Regularly Perforated	1 3/16″	A	Perforated,[5] painted	5	.20	.32	.62	.99	.70	.30
	2 3/16″	A	Same as above	8	.36	.51	.99	.88	.63	.46

TABLE II. SPECIAL ACOUSTICAL MATERIALS (Continued)

WOOD CONVERSION COMPANY

Material	Thick-ness	Ceiling Detail at Partition	Attenuation Factors—Decibels				
			125 cps	250 cps	500 cps	1000 cps	4000 cps
Lo-Tone, Random Pattern, Random Perforated Mineral Tile	5/8″	Interrupted	26	24	26	28	54
Lo-Tone FR, Random Pattern Mineral Tile	5/8″	Interrupted	26	30	35	40	*61*
Lo-Tone, Metal Pan, Regular Pattern	1 9/16″	Interrupted	23	23	25	29	52

[1]Perforated 484 holes per sq. ft., 3/16″ diameter, uniformly spaced.

[2] Perforated 320 holes per sq. ft., 1/4″ and 3/16″ diameters, randomly spaced.

[3] Perforated, enameled metal pan with sound-absorbing mineral pad. Pan perforated 925 holes per sq. ft., 0.108″ diameter. Thickness includes tee bar.

[4] Perforated, enameled metal pan with sound-absorbing mineral pad. Pan random perforated 293 holes per sq ft., 166 of 9/64″ and 127 of 3/16″ diameters. Thickness includes tee bar.

[5] Perforated 550 holes per sq. ft., 3/16″ diameter.

[6] Factory-painted face and bevels. Also available with factory applied paint finish providing Class "C" flame resistance rating.

[7] Acoustical Ceiling Constructions Classified by Underwriters' Laboratories, Incorporated

No. 4355-1—5/8″ Lo-Tone FR—Design 18—2 Hours
No. 4355-3—5/8″ Lo-Tone FR—Design 24—2 Hours

SPARK-GAP VOLTAGES

**Based on results of the American Institute of Electric Engineers
Air at 760 mm, 25° C.**

Peak voltage, kilovolts	Diameter of spherical electrodes, cm				Needle points
	2.5	5	10	25	
	Length of spark gap cm				
5	0.13	0.15	0.15	0.16	0.42
10	0.27	0.29	0.30	0.32	0.85
15	0.42	0.44	0.46	0.48	1.30
20	0.58	0.60	0.62	0.64	1.75
25	0.76	0.77	0.78	0.81	2.20
30	0.95	0.94	0.95	0.98	2.69
35	1.17	1.12	1.12	1.15	3.20
40	1.41	1.30	1.29	1.32	3.81
45	1.68	1.50	1.47	1.49	4.49
50	2.00	1.71	1.65	1.66	5.20
60	2.82	2.17	2.02	2.01	6.81
70	4.05	2.68	2.42	2.37	8.81
80	.....	3.26	2.84	2.74	11.1
90	.....	3.94	3.28	3.11	13.3
100	.....	4.77	3.75	3.49	15.5
110	.....	5.79	4.25	3.88	17.7
120	.....	7.07	4.78	4.28	19.8
130	.....	.....	5.35	4.69	22.0
140	.....	.....	5.97	5.10	24.1
150	.....	.....	6.64	5.52	26.1
160	.....	.....	7.37	5.95	28.1
170	.....	.....	8.16	6.39	30.1
180	.....	.....	9.03	6.84	32.0
190	.....	.....	10.0	7.30	33.9
200	.....	.....	11.1	7.76	35.7
210	.....	.....	12.3	8.24	37.6
220	.....	.....	13.7	8.73	39.5
230	.....	.....	15.3	9.24	41.4
240	.....	.....	.....	9.76	43.3
250	.....	.....	.....	10.3	45.2
300	.....	.....	.....	13.3	54.7

CORRECTIONS FOR TEMPERATURE AND PRESSURE

Values found in the above table may be corrected for temperature and pressure by multiplying the values given by the appropriate correction factor found below:

Temp. °C.	Pressure mm			
	720	740	760	780
0	1.04	1.06	1.09	1.12
10	1.00	1.02	1.05	1.08
20	0.96	0.99	1.02	1.04
30	0.93	0.96	0.98	1.01

DIELECTRIC CONSTANTS

Dielectric Constants of Pure Liquids

The values listed in the following table were obtained from National Bureau of Standards Circular 514.

The dielectric constants are intended to be the limiting values at low frequencies. the so-called static values. Temperature is the only variable considered explicitly. Usually pressure is atmospheric or insignificantly different with respect to its effect on the dielectric constant. Where data are listed above the normal boiling point, the pressure corresponds to the vapor pressure of the liquid unless otherwise noted in the footnote.

Symbols

ϵ = dielectric constant ($\epsilon_{vacuum=1}$)
t = temperature, Centigrade (°C)
T = temperature, Absolute (°K)
$a = -d\epsilon/dt$
$\alpha = -d \log_{10} \epsilon/dt$
t_1, t_2 = the limits of temperature between which a or α is considered applicable
mp = melting point
bp = boiling point
f = frequency of alternating current in cycles per second

Standard Liquids

		$\epsilon_{20°C}$	$\epsilon_{25°C}$	a (or α)*
C_6H_{12}	Cyclohexane	2.023	2.015	0.0016
CCl_4	Carbon tetrachloride	2.238	2.228	.0020
C_6H_6	Benzene	2.284	2.274	.0020
C_6H_5Cl	Chlorobenzene	5.708	5.621	.00133(α)
$C_2H_4Cl_2$	1,2-Dichloroethane	10.65	10.36	.00240(α)
CH_4O	Methanol	33.62	32.63	.00260(α)

DIELECTRIC CONSTANTS

Dielectric Constants of Pure Liquids (Continued)

Standard Liquids (Continued)

		$\epsilon_{20°C}$	$\epsilon_{25°C}$	a (or α)*
$C_6H_5NO_2$	Nitrobenzene	35.74	34.82	.00225(α)
H_2O	Water	80.37	78.54	.00200(α)
H_2	Hydrogen	1.228 at 20.4°K		.0034
O_2	Oxygen	1.507 at 80.0°K		.0024

* The values of a or α given in this table are derived from data in the vicinity of room temperature and are not necessarily identical with the values listed in the following sections of this table. They may be used to calculate values of dielectric constant between 15° and 30°C without introducing significant error.

Inorganic Liquids

	Substance	ϵ	t°C	a (or α) × 10^2	Range t_1, t_2
A	Argon	1.53_8	−191	0.34	−191, −184
$AlBr_3$	Aluminum bromide	3.38	100	0.33	100,240
AsH_3	Arsine	2.50	−100	0.43	−116, −72
BBr_3	Boron bromide	2.58	0	0.28	−70,80
Br_2	Bromine	3.09	20	0.7	0,50
CO_2	Carbon dioxide	1.60^c	20	....	....
Cl_2	Chlorine	2.10_1	−50	0.31	−65, −33
		1.91	14	0.32	−22,14
		1.7_3	77		
		1.5_4	142		
D_2	Deuterium	1.277	20°K	0.4	18.8,21.2°K
D_2O	Deuterium oxide	78.25	25	(^{d})	0.4,98
F_2	Fluorine	1.54	−202	0.19	−216, −190
$GeCl_4$	Germanium tetrachloride	2.43_0	25	0.240	0,55
HBr	Hydrogen bromide	7.00	−85	0.26(α)	−85, −70
HCl	Hydrogen chloride	6.35	−15	0.288(α)	−85, −15
		12.	−113	....	....
		4.6	28	....	....
HF	Hydrogen fluoride	17_5.	−73	....	....
		13_4.	−42		
		11_1.	−27		
		84.	0		
HI	Hydrogen iodide	3.39	−50	0.8	−51, −37
H_2	Hydrogen	1.228	20.4°K	0.34	14,21°K
H_2O	Water	78.54	25	(^{e})	0,100
		34.5_9	200	(^{y})	100,370
H_2O_2	Hydrogen peroxide	84.2	0	(^{z})	−30,20
H_2S	Hydrogen sulfide	9.26	−85.5	....	....
		9.05	−78.5	....	....
He	Helium	1.055_5	2.06°K	....	....
		1.055_9	2.30^f		
		1.055_3	2.63		
		1.053_9	3.09		
		1.051_8	3.58		
		1.048	4.19		
I_2	Iodine	$11._1$	118	....	....
		$11._7$	140		
		$13._0$	168		
NH_3	Ammonia	25.	−77.7	....	....
		22.4	−33.4	....	....
		18.9	5	....	....
		17.8	15		
		16.9	25		
		16.3	35		
NOBr	Nitrosyl bromide	$13._4$	15	....	....
NOCl	Nitrosyl chloride	$18._2$	12	....	....
N_2	Nitrogen	1.454	−203	0.29	−210, −195
N_2H_4	Hydrazine	$52._9$	20	0.21(α)	0,25
N_2O	Dinitrogen oxide	1.97	−90	....	....
		1.61	0	0.6	−6,14
N_2O_4	Dinitrogen tetroxide	2.5_6^b	15	....	....
O_2	Oxygen	1.507	−193	0.24	−218, −183
P	Phosphorus	4.10	34	....	....
		4.06	46		
		3.86	85		
PCl_3	Phosphorus trichloride	3.43	25	0.84	17,60
PCl_5	Phosphorus pentachloride	2.8_5	160	....	....
$POCl_3$	Phosphoryl chloride	$13._3$	22	....	....
$PSCl_3$	Thiophosphoryl chloride	5.8	22	....	....
$PbCl_4$	Lead tetrachloride	2.78	20	....	....
S	Sulfur	3.52	118	(^{g})	....
		3.48	231		
$SOBr_2$	Thionyl bromide	9.06	20	3.0	at 20

[c] At pressure of 50 atmospheres.

[d] $\epsilon = 78.25[1 - 4.617(10^{-3})(t-25) + 1.22(10^{-5})(t-25)^2 - 2.7(10^{-8})(t-25)^3]$; av. dev. ±0.04%.

[b] $f = 3.6 \times 10^8$ cycles/sec.

[e] $\epsilon = 78.54[1 - 4.579(10^{-3})(t-25) + 1.19(10^{-5})(t-25)^2 - 2.8(10^{-8})(t-25)^3]$; av. dev. ±0.03%.

[f] Liquid transition and discontinuity in variation of dielectric constant with temperature at 2.295°K.

[g] Graphical data in the range 118°–350°C show a minimum near 160° and a broad maximum near 200°.

[y] $\epsilon = 5321/T + 233.76 - 0.9297T + 0.001417T^2 - 0.0000008292T^3$.

[z] $\epsilon = 84.2 - 0.62t + 0.0032t^2$.

Dielectric Constants of Pure Liquids (Continued)

Inorganic Liquids (Continued)

	Substance	ε	t°C	a (or α) × 10²	Range t_1,t_2
$SOCl_2$	Thionyl chloride	9.25	20	3.9	at 20
SO_2	Sulfur dioxide	17.6	−20	0.287(α)	−65,−15
		15.0_8	0	....	....
		$14._1$	20	7.7	14,140
		2.1_0	154[h]		
SO_3	Sulfur trioxide	3.11	18	....	....
S_2Cl_2	Sulfur monochloride	4.79	15	0.146(α)	−41,15
SO_2Cl_2	Sulfuryl chloride	$10._0$	22	....	....
$SbCl_5$	Antimony pentachloride	3.22	20	0.46	2,47
Se	Se lenium	5.40	250	0.25	237,301
$SiCl_4$	Silicon tetrachloride	2.4_0	16	....	....
$SnCl_4$	Tin tetrachloride	2.87	20	0.30	−30,20
$TiCl_4$	Titanium tetrachloride	2.80	20	0.20	−20,20

Organic Liquids

	Substance	ε	t°C	a (or α) × 10²	Range t_1,t_2
	C_1				
CCl_4	Carbon tetrachloride	2.238	20	0.200	−10,60
CN_4O_8	Tetranitromethane	2.52_1	25	....	....
CO_2	Carbon dioxide	1.60_4[c]	0	....	....
CS_2	Carbon disulfide	2.641	20	0.268	−90,130
		3.001	−110		
		2.19	180		
$CHBr_3$	Bromoform	4.39	20	0.105(α)	10,70
$CHCl_3$	Chloroform	4.806	20	0.160(α)	0,50
		6.76	−60	....	....
		6.12	−40		
		5.61	−20		
		3.7_1	100	....	....
		3.3_3	140		
		2.9_3	180		
CHN	Hydrocyanic acid	$158._1$	0	(i)	−13,18
		$114._9$	20	0.63(α)	18,26
CH_2Br_2	Dibromomethane	7.77	10	....	....
		6.68	40		
CH_2Cl_2	Dichloromethane	9.08	20	(j)	−80,25
CH_2I_2	Diiodomethane	5.32	25	....	....
CH_2O_2	Formic acid	$58._5$[a]	16	....	....
CH_3Br	Bromomethane	9.82	0	(k)	−80,0
CH_3Cl	Chloromethane	12.6	−20	(l)	−70,−20
CH_3I	Iodomethane	7.00	20	(m)	−70,40
CH_3NO	Formamide	109.	20	72.	18,25
CH_4	Methane	1.70	−173	0.2	−181,−159
CH_4O	Methanol	32.63	25	0.264(α)	5,55
		64.	−113	....	....
		54.	−80		
		40.	−20		
CH_5N	Methylamine	11.4	−10	0.26(α)	−30,−10
		9.4	25	....	....
	C_2				
C_2HCl_3	Trichloroethylene	3.4_2	ca 16	....	....
C_2HCl_3O	Chloral	4.9_4	20	0.17(α)	15,45
		7.6	−40		
		4.2	62		
$C_2HCl_3O_2$	Trichloroacetic acid	4.6	60	....	....
$C_2H_2Br_2$	*cis*-1,2-Dibromoethylene	7.7_2	0	....	....
		7.0_8	25		
	trans-1,2-Dibromoethylene	2.9_7	0	....	....
		2.8_8	25		
$C_2H_2Br_4$	1,1,2,2-Tetrabromoethane	8.6	3	....	....
		7.0	22		
$C_2H_2Cl_2$	1,1-Dichloroethylene	4.6_7	16	....	....
	cis-1,2-Dichloroethylene	9.20	25	....	....
	trans-1,2-Dichloroethylene	2.14	25	....	....
$C_2H_2Cl_2O_2$	Dichloroacetic acid	8.2	22	....	....
		7.8	61		
C_2H_3ClO	Acetyl chloride	$16._9$	2	....	....
		$15._8$	22		
$C_2H_3ClO_2$	Chloroacetic acid	12.3	60	2.	60,80
C_2H_4O	Ethylene oxide	$13._9$	−1	....	....
	Acetaldehyde	$21._8$[a]	10	....	....
$C_2H_4O_2$	Acetic acid	6.15	20	....	....
		6.29	40		
		6.62	70		
	Methyl formate	8.5	20	5.	0,20
C_2H_5ClO	2-Chloroethanol (Ethylene chlorohydrin)	$25._8$	25	....	....
		$13._2$	132		
C_2H_5NO	Acetamide	59.[a]	83	....	....
$C_2H_5NO_2$	Nitroethane	28.0_6	30	11.4	30,35
C_2H_6O	Ethanol	24.30	25	....	....
		24.3[z]	25	0.270(α)	−5,70
		41.0[z]	−60	0.297(α)	−110,−20
	Methyl ether	5.02	25	2.38	25,100
		2.97	110		
		2.64	120		
		2.37	125		
		2.26	126.1		
		1.90	127.6[p]		

Organic Liquids (Continued)

	Substance	ε	t°C	a (or α) × 10²	Range t_1,t_2
$(C_2H_6OSi)_n$					
n = 4	Octamethylcyclotetrasiloxane	2.39	20	....	....
n = 5	Decamethylcyclopentasiloxane	2.50	20	....	....
n = 6	Dodecamethylcyclohexasiloxane	2.59	20	....	....
n = 7	Tetradecamethylcycloheptasiloxane	2.68	20	....	....
n = 8	Hexadecamethylcyclooctasiloxane	2.74	20	....	....
$C_2H_6O_2$	Glycol	$37._7$	25	0.224(α)	20,100
C_2H_7N	Dimethylamine	6.32	0	....	....
		5.26	25		
	C_3				
C_3H_6	Propene	1.87_5	20	....	....
		1.79_5	45		
		1.69_0	65		
		1.53_0	85		
		1.44_1	90		
		1.33_1	91.9[h]		
C_3H_6O	2-Propen-1-ol (Allyl alcohol)	$21._6$	15	....	....
	Acetone	20.7_0	25	0.205(α)	−60,40
		17.7	56	....	....
	Propionaldehyde	$18._5$[a]	17	....	....
$C_3H_6O_2$	Propionic acid	3.30	10	....	....
		3.44	40		
	Ethyl formate	7.1_6	25	....	....
	Methyl acetate	6.68	25	2.2	25,40
$C_3H_6O_3$	*dl*-Lactic acid	22.	17	....	....
$C_3H_7NO_2$	Ethyl carbamate (Urethan)	14.2	50	5.2	50,70
C_3H_8	Propane	1.61	0	0.20	−90,15
C_3H_8O	1-Propanol	20.1	25	0.293(α)	20,90
		38.	−80	....	....
		29.	−34		
	2-Propanol	18.3	25	0.310(α)	20,70
$C_3H_8O_2$	1,2-Propanediol	$32._0$	20	0.27(α)	at 20
	1,3-Propanediol	$35._0$	20	0.23(α)	at 20
$C_3H_8O_3$	Glycerol	42.5	25	0.208(α)	0,100
C_3H_9N	Isopropylamine	5.5[b]	20	....	....
	Trimethylamine	2.44	25	0.52	0,25
	C_4				
$C_4H_2O_3$	Maleic anhydride	50.[a]	60	....	....
C_4H_4O	Furan	2.95	25	....	....
C_4H_4S	Thiophene	2.76	16	....	....
C_4H_5N	Pyrrole	7.48	18	....	....
C_4H_5NS	Allyl isothiocyanate	$17._2$[b]	18	....	....
C_4H_6O	Vinyl ether	3.94	20	....	....
$C_4H_6O_3$	Acetic anhydride	$22._4$	1	....	....
		$20._7$	19		
C_4H_8O	2-Butanone	18.5_1	20	0.207(α)	−60,60
	Butyraldehyde	13.4	26	....	....
		10.8	77		
$C_4H_8O_2$	Butyric acid	2.97	20	−0.23	10,70
	Isobutyric acid	2.71	10	....	....
		2.73	40		
	Propyl formate	7.7_2[a]	19	....	....
	Ethyl acetate	6.02	25	1.5	at 25
		5.3_0	77	....	....
	Methyl propionate	5.5[n]	19	....	....
	1,4-Dioxane	2.209	25	0.170	20,50
C_4H_9NO	Morpholine	7.33	25	....	....
$C_4H_{10}O$	1-Butanol	17.8	20	0.300(α)	−40,20
		17.1	25	0.335(α)	25,70
		8.2	118	....	....
	2-Methyl-1-propanol	17.7	25	0.377(α)	20,90
		34.	−80	....	....
		26.	−34		
	2-Butanol	15.8	25	....	....
	2-Methyl-2-propanol	10.9	30	....	....
		8.49	50		
		6.89	70		
	Ethyl ether	4.335	20	2.0	at 20
		4.34[z]	20	0.217(α)	−40,30
		10.4	−116	....	....
		3.97	40	0.170(α)	40,140
		2.1_2	180		
		1.8_9	190		
		1.5_3	193.3[h]		

[a] $f = 4 \times 10^8$ cycles/sec.
[b] $f = 3.6 \times 10^8$ cycles/sec.
[c] At pressure of 50 atmospheres.
[h] Critical temperature.
[i] $\log_{10} \epsilon = 2.199 - 0.0079t + 0.00005t^2$.
[j] $\epsilon = (3320/T) - 2.24$.
[k] $\epsilon = (3320/T) - 2.34$.
[l] $\epsilon = 12.6 - 0.061(t + 20) + 0.0005(t + 20)^2$.
[m] $\epsilon = (2160/T) - 0.39$.
[n] $f = 5 \times 10^8$ cycles/sec.
[p] Critical temperature = 126.9°C.
[t] Mixture of *cis-trans* isomers.
[z] Value chosen to conform with the remainder of the tabulated data for this substance.

Dielectric Constants of Pure Liquids (Continued)

ORGANIC LIQUIDS (Continued)

	Substance	ϵ	t°C	a (or α) $\times 10^2$	Range t_1,t_2
$C_4H_{10}Zn$	Diethyl zinc	2.5_5	20	...	...
	C_5				
$C_5H_4O_2$	Furfural	$46._9$	1	...	...
		$41._9$	20		
		$34._9$	50		
C_5H_5N	Pyridine	12.3	25	...	...
		9.4	116	...	...
C_5H_8	1,3-Pentadiene[t]	2.32	25	...	...
	2-Methyl-1,3-butadiene (Isoprene)	2.10	25	0.24	−75,25
$C_5H_8O_2$	2,4-Pentanedione (Acetylacetone)	$25._7$[a]	20	...	...
C_5H_{10}	1-Pentene	2.100	20	...	...
	2-Methyl-1-butene	2.197	20	...	...
	Cyclopentane	1.965	20	...	...
	Ethylcyclopropane	1.933	20	...	...
$C_5H_{10}O$	2-Pentanone	15.4_5	20	0.195(α)	−40,80
		22.0	−60		
	3-Pentanone	17.0_0	20	0.225(α)	0,80
		19.4	−20		
		19.8	−40		
$C_5H_{10}O_2$	Valeric acid	2.6_6	20	...	...
	Isovaleric acid	2.6_4	20	...	...
	Methyl butyrate	5.6[n]	20	...	...
$C_5H_{11}N$	Piperidine	5.8[b]	22	...	...
C_5H_{12}	n-Pentane	1.844	20	0.160	−50,30
		2.011	−90		
		1.984	−70		
	2-Methylbutane	1.843	20	...	...
$C_5H_{12}O$	1-Pentanol	13.9	25	0.23(α)	15,35
	3-Methyl-1-butanol	14.7	25	...	...
		5.8_2	132		
	2-Methyl-2-butanol	5.82	25	...	...
	C_6				
$C_6H_4Cl_2$	o-Dichlorobenzene	9.93	25	0.194(α)	0,50
	m-Dichlorobenzene	5.04	25	0.120(α)	0,50
	p-Dichlorobenzene	2.41	50	0.18	50,80
C_6H_5Br	Bromobenzene	5.40	25	0.115(α)	0,70
C_6H_5Cl	Chlorobenzene	5.708	20	...	...
		5.621	25		
		5.71	20	0.130(α)	0,80
		7.28	−50		
		6.30	−20		
		4.21	130		
C_6H_5ClO	o-Chlorophenol	6.31	25	2.7	25,58
	p-Chlorophenol	9.47	55	3.7	55,65
C_6H_5I	Iodobenzene	4.63	20	...	...
$C_6H_5NO_2$	Nitrobenzene	34.82	25	0.225(α)	10,80
		20.8	130	0.164(α)	130,211
		24.9	90		
		22.7	110		
$C_6H_5NO_3$	o-Nitrophenol	$17._3$	50	6.4	50,60
C_6H_6	Benzene	2.284	20	0.200	10,60
		2.073	129	...	...
		1.966	182		
C_6H_6BrN	m-Bromoaniline	$13._0$[n]	19	...	...
C_6H_6ClN	m-Chloroaniline	$13._4$[n]	19	...	...
$C_6H_6N_2O_2$	o-Nitroaniline	$34._5$	90	3.	90,110
	p-Nitroaniline	$56._3$	160	6.	160,180
C_6H_6O	Phenol	9.78	60	0.32(α)	40,70
C_6H_7N	Aniline	6.89	20	0.148(α)	0,50
		5.93	70	...	...
		4.54	184.6	...	...
	2-Methylpyridine (α-Picoline)	9.8[b]	20	...	...
$C_6H_8N_2$	Phenylhydrazine	7.2	23	...	...
C_6H_{10}	Cyclohexene	2.220	25	...	...
		2.6_0	−105	...	...
$C_6H_{10}O$	Cyclohexanone	18.3	20	...	...
		$19._9$	−40	...	...
$C_6H_{10}O_3$	Ethyl acetoacetate	$15._7$[a]	22	...	...
C_6H_{12}	Cyclohexane	2.023	20	0.160	10,60
	Methylcyclopentane	1.985	20	...	...
	Ethylcyclobutane	1.965	20	...	...
$C_6H_{12}O$	Cyclohexanol	15.0	25	0.437(α)	20,66
		7.2_4	100		
		4.8_8	150		
$C_6H_{12}O_2$	Butyl acetate	5.01	20	1.4	20,40
		6.8_5	−73		
	Ethyl butyrate	5.10	18	1.0	at 20
$C_6H_{12}O_3$	Paraldehyde	13.9	25	...	...
		6.29	128		
C_6H_{14}	n-Hexane	1.890	20	0.155	−10,50
		2.044	−90		
		1.990	−50		
$C_6H_{14}O$	1-Hexanol	13.3	25	0.35(α)	15,35
		8.5_5	75		
	Propyl ether	3.3_9	26	...	...
	Isopropyl ether	3.88	25	1.8	0,25
$C_6H_{15}Al$	Triethyl aluminum	2.9	20	...	...
$C_6H_{15}N$	Dipropylamine	2.9[b]	21	...	...
	Triethylamine	2.42	25	...	...

ORGANIC LIQUIDS (Continued)

	Substance	ϵ	t°C	a (or α) $\times 10^2$	Range t_1,t_2
$C_6H_{15}N$	Dipropylamine	2.9[b]	21	...	...
	Triethylamine	2.42	25	...	...
$C_6H_{18}OSi_2$	$(CH_3)_3Si[OSi(CH_3)_2]_nCH_3$				
$n = 1$	Hexamethyldisiloxane	2.17	20	...	...
$n = 2$	Octamethyltrisiloxane	2.30	20	...	...
$n = 3$	Decamethyltetrasiloxane	2.39	20	...	...
$n = 4$	Dodecamethylpentasiloxane	2.46	20	...	...
$n = 5$	Tetradecamethylhexa-siloxane	2.50	20	...	...
$n = 66$[w]		2.72	20	...	...
	C_7				
C_7H_5ClO	Benzoyl chloride	29.	0	...	...
		23.	20		
C_7H_5NO	Phenyl isocyanate	8.8[b]	20	...	...
C_7H_5NS	Phenyl isothiocyanate	$10._4$[a]	20	...	...
C_7H_6O	Benzaldehyde	$19._7$	0	...	...
		$17._8$	20		
$C_7H_6O_2$	Salicylaldehyde	$17._1$	30	7.	30,40.
C_7H_7Br	o-Bromotoluene	4.28	58	...	...
	m-Bromotoluene	5.36	58	...	...
	p-Bromotoluene	5.49	58	...	...
C_7H_7Cl	o-Chlorotoluene	4.45	20	...	...
		4.16	58	...	...
	m-Chlorotoluene	5.55	20	...	...
		5.04	58	...	...
	p-Chlorotoluene	6.08	20	...	...
		5.55	58	...	...
	α-Chlorotoluene	7.0	13	...	...
$C_7H_7NO_2$	o-Nitrotoluene	27.4	20	15.	at 20
		$21._6$	58	...	...
		11.8	222	...	...
	m-Nitrotoluene	$23._8$	20	...	...
		$21._9$	58	...	...
	p-Nitrotoluene	$22._2$	58	...	...
C_7H_8	Toluene	2.438	0	0.0455(α)	−90,0
		2.379	25	0.243	0,90
		2.15_7	127		
		2.04_2	181		
C_7H_8O	Benzyl alcohol	13.1	20	...	...
		9.47	70	...	...
		6.6	132	...	...
	o-Cresol	11.5	25	11.	25,30
	m-Cresol	11.8	25	0.41(α)	15,50
	p-Cresol	9.9_1	58	...	...
C_7H_9N	o-Toluidine	6.34	18	...	...
		5.71	58	...	...
		4.00	200	...	...
	m-Toluidine	5.95	18	...	...
		5.45	58	...	...
	p-Toluidine	4.98	54	...	...
	N-Methylaniline	5.97	22	...	...
	1-Heptene	2.05	20	...	...
	C_8				
C_8H_8	Styrene (Phenylethylene)	2.43	25	...	...
		2.32	75		
	Acetophenone	17.39	25	4.	at 25
		8.64	202	...	...
$C_8H_8O_2$	Phenyl acetate	5.23	20	0.7	at 20
	Methyl benzoate	6.59	20	0.14(α)	20,50
$C_8H_8O_3$	Methyl salicylate	9.41	30	3.1	30,40
C_8H_{10}	Ethylbenzene	2.412	20	...	...
	o-Xylene	2.568	20	0.266	−20,130
	m-Xylene	2.374	20	0.195	−40,180
	p-Xylene	2.270	20	0.160	20,130
C_8H_{18}	n-Octane	1.948	20	0.130	−50,50
		1.879	70		
		1.817	110		
	2,2,3-Trimethylpentane	1.96	20	...	...
	2,2,4-Trimethylpentane	1.940	20	0.142	−100,100
$C_8H_{18}O$	1-Octanol	10.3_4	20	0.410(α)	20,60
$C_8H_{20}O_4Si$	Tetraethyl silicate	4.1[b]	ca 20	...	...
	C_9				
C_9H_7N	Quinoline	9.00	25	...	...
		5.05	238		
	Isoquinoline	10.7	25	...	...
C_9H_8O	Cinnamaldehyde	16.9	24	...	...
$C_9H_{10}O_2$	Benzyl acetate	5.1[n]	21	...	...
	Ethyl benzoate	6.02	20	2.1	20,40
$C_9H_{10}O_3$	Ethyl salicylate	7.99	30	2.	30,40
C_9H_{12}	Isopropylbenzene (Cumene)	2.38_0	20	...	...
	1,3,5-Trimethylbenzene (Mesitylene)	2.27_9	20	...	...

[a] $f = 4 \times 10^8$ cycles/sec.
[b] $f = 3.6 \times 10^8$ cycles/sec.
[h] Critical temperature.
[n] $f = 5 \times 10^8$ cycles/sec.
[t] Mixture of *cis-trans* isomers.
[x] Value chosen to conform with the remainder of the tabulated data for this substance.
[w] Silicone oil of average molecular weight corresponding to this formula.

ORGANIC LIQUIDS (Continued)

Formula	Substance	ε	t°C	a (or α) × 10²	Range t_1,t_2
C_9H_{20}	n-Nonane	1.972	20	0.135	−10,90
		2.059	−50		
		1.847	110		
		1.787	150		
	C_{10}				
$C_{10}H_8$	Naphthalene	2.54	85	....	....
$C_{10}H_{10}O_4$	Dimethyl phthalate	8.5	24	....	....
$C_{10}H_{16}$	d-Camphene	2.33	ca 40	....	....
	d-Pinene	2.64	25	....	....
	l-Pinene	2.76	20	....	....
	Terpinene	2.7[b]	21	....	....
	d-Limonene	2.3_6	20	....	....
	dl-Limonene (Dipentene)	2.3_0	20	....	....
$C_{10}H_{22}$	n-Decane	1.991	20	0.130	10,110
		2.050	−30		
		1.844	130		
		1.783	170		
$C_{10}H_{22}O$	1-Decanol	8.1	20	....	....
$C_{11}H_{24}$	n-Undecane	2.005	20	0.125	10,130
		2.039	−10		
		1.838	150		
		1.781	190		
	C_{12}				
$C_{12}H_{10}$	Diphenyl	2.53	75	0.18	75,155
$C_{12}H_{10}O$	Phenyl ether	3.65	30	0.7	30,50
$C_{12}H_{11}N$	Diphenylamine	3.3	52	....	....
$C_{12}H_{26}$	n-Dodecane	2.014	20	0.120	10,150
		2.047	−10		
		1.776	210		
	C_{13}				
$C_{13}H_{10}O$	Benzophenone	11.4	50	....	....
	C_{14}				
$C_{14}H_{15}N$	Dibenzylamine	3.6[b]	20	....	....
	C_{16}				
$C_{16}H_{32}O_2$	Palmitic acid	2.30	71	....	....
	C_{18}				
$C_{18}H_{32}O_2$	Linoleic acid	2.61	0	....	....
		2.71	20		
		2.70	70		
		2.60	120		
$C_{18}H_{34}O_2$	Oleic acid	2.46	20	....	....
		2.45	60		
		2.41	100		
$C_{18}H_{36}O_2$	Stearic acid	2.29	70	....	....
		2.26	100	....	....
	Ethyl palmitate	3.20	20	0.4	20,40
		2.71	104	....	....
		2.46	182		
	C_{19}				
$C_{19}H_{16}$	Triphenylmethane	2.45	100	0.14	94,175
$C_{19}H_{38}O_4$	Monopalmitin	5.34	67	....	....
		5.09	80		
	C_{20}				
$C_{20}H_{38}O_2$	Ethyl Oleate	3.17	28	0.48	28,122
$C_{20}H_{40}O_2$	Ethyl Stearate	2.98	40	0.6	32,50
		2.69	100		
		2.48	167		
	C_{21}				
$C_{21}H_{21}O_4P$	Tricresyl phosphate	6.9	40	....	....
	C_{22}				
$C_{22}H_{42}O_2$	Butyl oleate	4.0	25	....	....
$C_{22}H_{44}O_2$	Butyl stearate	3.11[a]	30	0.53	30,35

[b] $f = 3.6 \times 10^8$ cycles/sec.
[a] $f = 5 \times 10^8$ cycles/sec.

DIELECTRIC CONSTANTS

Dielectric Constants of Solids

Compiled by Earle C. Gregg, Jr.

SOLIDS* (17 to 22°C)

Material	Frequency	Dielectric constant	Material	Frequency	Dielectric constant
Acetamide	4×10^8	4.0	Mercuric chloride	10^6	3.2
Acetanilide	....	2.9	Mercurous chloride	10^6	9.4
Acetic acid (2°C)	4×10^8	4.1	Naphthalene	4×10^8	2.52
Aluminum oleate	4×10^8	2.40	Phenanthrene	4×10^8	2.80
Ammonium bromide	10^6	7.1	Phenol (10°C)	4×10^8	4.3
Ammonium chloride	10^6	7.0	Phosphorus, red	10^8	4.1
Antimony trichloride	10^8	5.34	Phosphorus, yellow	10^8	3.6
Apatite ⊥ optic axis	3×10^8	9.50	Potassium aluminum		
" ∥ optic axis	3×10^8	7.41	sulfate	10^6	3.8
Asphalt	$<3 \times 10^6$	2.68	Potassium carbonate		
Barium chloride (anhyd.)	6×10^7	11.4	(15°C)	10^8	5.6
" " ($2H_2O$)	6×10^7	9.4	Potassium chlorate	6×10^7	5.1
Barium nitrate	6×10^7	5.9	Potassium chloride	10^4	5.03

SOLIDS* (17 to 22°C) (Continued)

Material	Frequency	Dielectric constant	Material	Frequency	Dielectric constant
Barium sulfate (15°C)	10^8	11.4	Potassium chromate	6×10^7	7.3
Beryl ⊥ optic axis	10^4	7.02	Potassium iodide	6×10^7	5.6
" ∥ optic axis	10^4	6.08	Potassium nitrate	6×10^7	5.0
Calcite ⊥ optic axis	10^4	8.5	Potassium sulfate	6×10^7	5.9
" ∥ optic axis	10^4	8.0	Quartz ⊥ optic axis	3×10^7	4.34
Calcium carbonate	10^6	6.14	" ∥ optic axis	3×10^7	4.27
Calcium fluoride	10^4	7.36	Resorcinol	4×10^8	3.2
Calcium sulfate ($2H_2O$)	10^4	5.66	Ruby ⊥ optic axis	10^4	13.27
Cassiterite ⊥ optic axis	10^{12}	23.4	" ∥ optic axis	10^4	11.28
Cassiterite ∥ optic axis	10^{12}	24.	Rutile ⊥ optic axis	10^8	86
d-Cocaine	5×10^8	3.10	" ∥ optic axis	10^8	170
Cupric oleate	4×10^8	2.80	Selenium	10^8	6.6
Cupric oxide (15°C)	10^8	18.1	Silver bromide	10^6	12.2
Cupric sulfate (anhyd.)	6×10^7	10.3	Silver chloride	10^6	11.2
Cupric sulfate ($5H_2O$)	6×10^7	7.8	Silver cyanide	10^6	5.6
Diamond			Smithsonite ⊥ optic axis	10^{12}	9.3
"	10^8	5.5	" ∥ optic axis	10^{10}	9.4
Diphenylmethane	4×10^8	2.7	Sodium carbonate (anhyd.)	6×10^7	8.4
Dolomite ⊥ optic axis	10^8	8.0	Sodium carbonate ($10H_2O$)	6×10^7	5.3
" ∥	10^8	6.8	Sodium chloride	10^4	6.12
Ferrous oxide (15°C)	10^8	14.2	Sodium nitrate	....	5.2
Iodine	10^8	4	Sodium oleate	4×10^8	2.75
Lead acetate	10^6	2.6	Sodium perchlorate	6×10^7	5.4
Lead carbonate (15°C)	10^8	18.6	Sucrose (mean)	3×10^8	3.32
Lead chloride	10^6	4.2	Sulfur (mean)	....	4.0
Lead monoxide (15°C)	10^8	25.9	Thallium chloride	10^6	46.9
Lead nitrate	6×10^7	37.7	p-Toluidine	4×10^8	3.0
Lead oleate	4×10^8	3.27	Tourmaline ⊥ optic axis	10^4	7.10
Lead sulfate	10^6	14.3	" ∥ optic axis	10^4	6.3
Lead sulfide (15°)	10^6	17.9	Urea	4×10^8	3.5
Malachite (mean)	10^{12}	7.2	Zircon ⊥, ∥	10^8	12

* For plastics and other insulating materials, refer to table on Properties of Dielectrics.

DIELECTRIC CONSTANTS

Table of Dielectric Constants of Reference Gases at 20°C and 1 Atmosphere

The listed values (ε − 1) refer to the gas at a temperature of 20°C and pressure of 1 atmosphere. The values can be adjusted to slightly different conditions without introducing more than 0.1% error by use of the following equation:

$$\frac{(\epsilon - 1)_{t,p}}{(\epsilon - 1)_{20°,1\,atm}} = \frac{p}{760[1 + 0.003411(t - 20)]}$$

where p = pressure in mm Hg
t = degrees C

t should be between 10 and 30°C and p between 700 and 800 mm Hg. From National Bureau of Standards Circular 537

Substance	$(\epsilon - 1) \cdot 10^6$	Reference
Helium	Radio frequency	
	67.8	Watson
	63.7	Hector
	64.5	Jelatis
	Microwave	
	65.6	Birnbaum
	65.2	Essen
	Optical	
	64.6	Koch
	64.5	Cuthbertson
Hydrogen	Radio frequency	
	$254._0$	Watson
	Microwave	
	253.4	Essen
	Optical	
	254.1	Cuthbertson
	253.6	Koch
	253.7	Kirn
	254.3	Tausz
Oxygen	Radio frequency	
	$494._3$	Watson
	496.2	Jelatis
	Microwave[a]	
	494.9	Birnbaum
	495.0	Essen
	494.9	Essen
	Optical	
	494.5	Cuthbertson
	493.5	Lowery
	494.7	Tausz
	494.4	Ladenberg
Air (dry, CO_2 free)	Radio frequency	
	$537._0$	Watson
	Microwave[a]	
	536.5	Birnbaum
	536.6	Essen
	536.6	Essen
	Optical	
	536.9	Koch
	535.8	Meggers
	536.0	Traub
	536.7	Quarder
	536.4	Lowery
	536.5	Tausz
	536.1	Koster
	536.3	Perard
	535.8	Barrell
Nitrogen	Radio frequency	
	$547._2$	Watson
	Microwave	
	547.3	Birnbaum
	548.0	Essen
	548.0	Essen
	Optical	
	548.9	Cuthbertson
	548.7	Koch
	547.2	Tausz
Carbon dioxide	Radio frequency	
	$921._5$	Watson
	Microwave	
	922.4	Birnbaum
	$920._6$	Essen

[a] These values were derived from measurements of the refractive index after making allowance for the magnetic permeability of oxygen.

DIELECTRIC CONSTANTS (Continued)

Table of Dielectric Constants of Reference Gases at 20°C and 1 Atmosphere

Substance	$(\epsilon - 1) \cdot 10^6$	Reference	Substance	$(\epsilon - 1) \cdot 10^6$	Reference
Argon.......	Radio frequency				
	513.0	Watson			
	516.4	Jelatis			
	Microwave				
	517.7	Essen			
	Optical				
	516.8	Cuthbertson			
	517.8	Quarder			
	517.0	Tausz			
	516.7	Damköhler			

[a] These values were derived from measurements of the refractive index after making allowance for the magnetic permeability of oxygen.

Dielectric Constants of Gases at 760 MM Pressure

Compiled by Earle C. Gregg, Jr.

GASES AT 760 MM PRESSURE

Material	Temperature °C	Frequency cycles/sec.	Dielectric Constant	Material	Temperature °C	Frequency cycles/sec.	Dielectric Constant
Acetalde-	100	$<3 \times 10^6$		*n*-Heptane..	20	$<3 \times 10^6$	1.0035
hyde.....	0	2×10^6	1.0213	Hydrogen..	100	$<3 \times 10^6$	1.000264
Acetone.....			1.0159	Hydrogen			
Acetyl chlo-	0	$<3 \times 10^6$		bromide..	20	$<3 \times 10^6$	1.00313
ride......			1.0217	Hydrogen	0	$<3 \times 10^6$	
Acetylene...	0	$<3 \times 10^6$	1.00134	chloride...	0	$<3 \times 10^6$	1.0046
Air.........			1.000590	Hydrogen	0	$<10^6$	
Ammonia...	0	$<3 \times 10^6$	1.0072	iodide....	100	$<3 \times 10^6$	1.00234
β-Amylene..			1.0028	Hydrogen	0	$<10^6$	
Argon......	23	10^{10}	1.000545	sulfide....	100	$<3 \times 10^6$	1.00030
Benzene....	400	3×10^8	1.0028	Mercury....	180	$<3 \times 10^6$	1.00074
Bromine....	0	$<3 \times 10^6$	1.0128	Methane....	0	$<3 \times 10^6$	1.000944
Butylene....			1.00319	Methyl alco-			
Carbon diox-	100	$<3 \times 10^6$		hol.......	0	$<10^6$	1.0057
ide.......			1.000985	Methyl bro-			
Carbon di-	23	10^{10}		mide.....	0	$<3 \times 10^6$	1.00095
sulfide....			1.0029	Methyl chlo-			
Carbon	23	10^{10}		ride......	0	$<3 \times 10^6$	1.00094
monoxide.			1.00070	Methyl io-			
Carbon tet-	100	$<3 \times 10^6$		dide......	110	$<3 \times 10^6$	1.0063
rachloride.			1.0030	Methyl-	120	$<3 \times 10^6$	
Chloroform .	100	$<3 \times 10^6$	1.0042	amine....			1.0038
Dichlorodi-				Methylene			
fluorome-	100	$<3 \times 10^6$		chloride...	23	10^{10}	1.0065
thane.....	0	$<3 \times 10^6$	1.00029	Neon.......			1.000127
Dichloro-	0	$<3 \times 10^6$		Nitrogen....			1.000580
fluorome-				Nitrome-	23	10^{10}	
thane.....	100	$<3 \times 10^6$	1.00049	thane.....			1.0247
Dimethyla-				Nitrous ox-	23	2.5×10^{10}	
mine.....	0	2×10^6	1.00040	ide (N_2O).	0	$<3 \times 10^6$	1.00113
Ethane.....	0	$<3 \times 10^6$	1.00150	Oxygen.....	100	$<3 \times 10^6$	1.000523
Ethyl alcohol	100	$<3 \times 10^6$	1.0061	*n*-Pentane...	23	10^{10}	1.0025
Ethylamine.			1.00053	*n*-Propyl			
Ethyl bro-	20	$<3 \times 10^6$		chloride...	20	$<3 \times 10^6$	1.0143
mide.....			1.0139	*iso*-Propyl			
Ethyl chlo-	20	$<3 \times 10^6$		chloride...	20	$<3 \times 10^6$	1.0152
ride......			1.0132	Sulfur diox-	100	$<3 \times 10^6$	
Ethyl ether.	23	10^{10}	1.0049	ide.......			1.00075
Ethyl for-	126	$<3 \times 10^6$		Toluene.....	100	$<3 \times 10^6$	1.0043
mate.....			1.0083	Vinyl bro-	20	$<3 \times 10^6$	
Ethyl iodide.	20	$<3 \times 10^6$	1.0140	mide.....	100	$<3 \times 10^6$	1.0081
" " .			1.0089	Water	0	$<3 \times 10^6$	
Ethylene....	110	$<3 \times 10^6$	1.00144	(steam)...	0	$<3 \times 10^6$	1.0126
Helium.....	140	$<3 \times 10^6$	1.0000684	" ..			1.00785

PROPERTIES OF DIELECTRICS

Compiled by Earle G. Gregg, Jr.

In most cases properties have been determined by A S T M (American Society for Testing Materials) test methods at room temperature under standard conditions. Values will in general change considerably with temperature.

Plastics

Material	Dielectric Constant 10^6 Cycles	Di-electric Strength volts/mil	Volume Resistivity Ohms-cm 23°C	Loss Factor*
Allyl resin, cast	3.6–4.5	380	$>4 \times 10^{14}$	0.028–0.06
Aniline formaldehyde resin, no filler	3.5–3.6	600–650	10^{16}–10^{17}	0.006–0.008
Casein	6.1–6.8	400–700	...	0.052
Cellulose acetate, molding	3.2–7.0	250–365	10^{10}–10^{13}	0.01–1.0
sheet	4.0–5.5	250–300	10^{11}–10^{13}	0.04–0.06
Cellulose acetate butyrate	3.2–6.2	250–400	10^{10}–10^{12}	0.01–0.04
Cellulose nitrate (proxylin)	6.4	300–600	$(10\text{–}15) \times 10^{10}$	0.06–0.09
Cold molded compound, inorganic, refractory	...	45	...	...
Cold molded compound, organic nonrefractory	6.0	85–115	1.3×10^{12}	0.07
Ethyl cellulose	2.8–3.9	350–500	10^{12}–10^{14}	0.01–0.06
Epoxy cast resin	3.62	400	10^{16}–10^{17}	0.019
Glycerol phthalate, cast, alkyd	3.7–4.0	300–350	$>10^{14}$	0.025–0.035
Melamine formaldehyde resins, molding				
Alpha cellulose filler	7.2–8.2	300–400	$10^{12} \times 10^{14}$	0.027–0.045
Asbestos filler	6.1–6.7	350–400	2.4×10^{11}	0.041–0.050
Cellulose filler	4.7–7.0	350–400	...	0.032–0.060
Flock filler	...	300–330	...	...
Macerated fabric filler	6.5–6.9	250–350	10^{9}–10^{10}	0.036
Methyl methacrylate, cast	2.7–3.2	450–500	$>10^{15}$	0.02–0.03
molding	2.7–3.2	450–500	$>10^{14}$	0.02–0.03
Mica, glass bonded, compression	7.4–7.85	...	10^{14}–10^{15}	0.0015–0.002
injection	6.9–9.2	...	10^{14}–10^{17}	0.0015–0.012
Nylon (F.M. 3001)	3.5	470	4×10^{14}	0.03
Phenol formaldehyde resins, molding				
Asbestos filler	5.0–7.0	100–350	10^{10}–10^{12}	0.10–0.50
Glass fiber	6.6	140–370	7×10^{12}	0.02
Mica filler	4.2–5.2	300–460	10^{12}–10^{14}	0.005–0.04
No filler	4.5–5.0	300–400	10^{11}–10^{12}	0.015–0.03
Sisal felt	3–5	250–400	10^{11}–10^{12}	0.3–0.5
Wood flour filler	4.–7.	200–425	10^{9}–10^{13}	0.03–0.07
Phenol formaldehyde resins, cast				
Mineral filler	9–15	100–250	10^{9}–10^{12}	0.07–0.2
No filler	4.0–5.5	350–400	10^{12}–10^{13}	0.04–0.05
Polyacrylic ester (filled and vulcanized)	...	400–700	2×10^{11} at 70°C	...
Polyester, cast resin, rigid	2.8–4.1	380–500	10^{14}	0.006–0.026
flexible	4.1–5.2	250–500	...	0.023–0.052
Polyester molding materials, (glass fiber filler)	4.0–4.5	150–400	10^{12}–10^{14}	0.015–0.020
Polyethylene	2.3	460	1.6×10^{13}	<0.0005
Polymonochlorodifluoroethylene	2.5	400	1.2×10^{18}	0.010
Polystyrene molding	2.4–2.65	500–700	10^{17}–10^{19}	0.0001–0.0004
Polytetrafluoroethylene	2.0	480	$>10^{15}$	<0.0002–0.0005
Rubber, hard	2.8	470	2×10^{15}	0.06
chlorinated	3 approx.	...	1.5×10^{13}	0.006
modified, isomerized	2.4–2.7	620	5×10^{16}	0.0008–0.002
Shellac	...	200–600	1.8×10^{9}	...
Silicone molding compound (glass fiber filled)	3.7	185	$10^{11} \times 10^{13}$	0.0017
Styrene, modified, molding (shock resistant type)	2.4–3.8	300–600	10^{12}–10^{17}	0.0004–0.02
Urea formaldehyde resin, alpha cellulose filler	6.4–6.9	300–400	10^{12}–10^{13}	0.028–0.032

Plastics (Continued)

Material	Dielectric Constant 10^6 Cycles	Di-electric Strength volts/mil	Volume Resistivity Ohms-cm 23°C	Loss Factor*
Vinyl butyral, flexible, unfilled	3.92	350	5×10^{10}	0.061
rigid	3.33	400	$>10^{14}$	0.0065
Vinyl chloride rigid	2.8–3.0	700–1300	$>10^{16}$	0.006–0.014
flexible filled	3.5–4.5	600–800	5×10^{14}	0.09–0.10
flexible unfilled	3.5–4.5	800–1000	5×10^{12}	0.09–0.10
Vinyl chloride, acetate rigid	3.0–3.1	425	10^{16}	0.018–0.019
flexible, unfilled	3.3–4.3	300–400	10^{11}–10^{13}	0.04–0.14
flexible, filled	3.3–4.3	250–350	10^{11}–10^{13}	0.04–0.14
Vinyl formal molding compound	3.0	490	...	0.023
Vinylidene chloride	3.0–4.0	350	10^{14}–10^{16}	0.05–0.08

Ceramics

Material	Dielectric Constant 10^6 Cycles	Di-electric Strength volts/mil	Volume Resistivity Ohms-cm 23°C	Loss Factor*
Alumina	4.5–8.4	40–160	10^{11}–10^{14}	0.0002–0.01
Corderite	4.5–5.4	40–250	10^{12}–10^{14}	0.004–0.012
Forsterite	6.2	240	10^{14}	0.0004
Porcelain (Dry Process)	6.0–8.0	40–240	10^{12}–10^{14}	0.003–0.02
Porcelain (Wet Process)	6.0–7.0	90–400	10^{12}–10^{14}	0.006–0.01
Porcelain, Zircon	7.1–10.5	250–400	10^{13}–10^{15}	0.0002–0.008
Steatite	5.5–7.5	200–400	10^{13}–10^{15}	0.0002–0.004
Titanates (Ba, Sr, Ca, Mg, and Pb)	15–12,000	50–300	10^{8}–10^{15}	0.0001–0.02
Titanium Dioxide	14–110	100–210	10^{13}–10^{18}	0.0002–0.005

Waxes

Material	Dielectric Constant 10^6 Cycles	Di-electric Strength volts/mil	Volume Resistivity Ohms-cm 23°C	Loss Factor*
Acrawax C	2.4	...	...	0.005
Beeswax, white	2.75–3.0	...	5×10^{14}	0.025
Beeswax, yellow	2.9	...	8×10^{14}	0.029
Candelilla	2.25–2.50			
Carnauba	2.75–3.0			
Cerese, brown G	2.27	...	...	0.0025
Ceresine	2.25–2.50	...	$>5 \times 10^{18}$	0.0011
Halowax 1001	−4.10	...	2×10^{13}	0.014
Halowax 1013	−4.75	...	...	0.036
Halowax 1014	−4.40	...	...	0.035
Halowax 11–314	2.94	...	...	0.00094
Microcrystalline waxes	2.2–2.5			
Opalwax	3.1	...	...	0.34
Ozokerite wax	2.3	100–150	5×10^{14}	0.0018
Paraffin	2.0–2.5	250	10^{15}–10^{19}	0.003 (900 cps)
Parawax	2.25	...	10^{16}	0.00045
135 A.M.P. wax	2.25	...	...	0.00023

PROPERTIES OF DIELECTRICS

PROPERTIES OF DIELECTRICS (Continued)

GLASSES

Type	Dielectric Constant at 100 mc 20°C	Volume Resistivity 350°C megohm-cm	Loss Factor*
Corning 0010	6.32	10	0.015
Corning 0080	6.75	0.13	0.058
Corning 0120	6.65	100	0.012
Pyrex 1710	6.00	2,500	0.025
Pyrex 3320	4.71	....	0.019
Pyrex 7040	4.65	80	0.013
Pyrex 7050	4.77	16	0.017
Pyrex 7052	5.07	25	0.019
Pyrex 7060	4.70	13	0.018
Pyrex 7070	4.00	1,300	0.0048
Vycor 7230	3.83	....	0.0061
Pyrex 7720	4.50	16	0.014
Pyrex 7740	5.00	4	0.040
Pyrex 7750	4.28	50	0.011
Pyrex 7760	4.50	50	0.0081
Vycor 7900	3.9	130	0.0023
Vycor 7910	3.8	1,600	0.00091
Vycor 7911	3.8	4,000	0.00072
Corning 8870	9.5	5,000	0.0085
G. E. Clear (Silica Glass)	3.81	4,000–30,000	0.00038
Quartz (Fused)	3.75–4.1 (1 mc)	....	0.0002 (1 mc)

* Power factor × dielectric constant equals loss factor.

Dielectric Constant of Water

t°C	ϵ	t°C	ϵ
0	88.00	40	73.28
5	86.04	45	71.59
10	84.11	50	69.94
15	82.22	60	66.74
20	80.36	70	63.68
25	78.54	80	60.76
30	76.75	90	57.98
35	75.00	100	55.33

DIELECTRIC CONSTANT OF DEUTERIUM OXIDE

t	ϵ	$-\frac{d\epsilon}{dt}$	$-\frac{1}{\epsilon}\frac{d\epsilon}{dt}$
°C			
4	85.877	0.3974	4.627 × 10^{-3}
5	85.480	.3956	4.628
10	83.526	.3862	4.624
15	81.618	.3771	4.620
20	79.755	.3681	4.615
25	77.936	.3594	4.611
30	76.161	.3509	4.607
35	74.427	.3425	4.602
40	72.735	.3344	4.597
45	71.083	.3265	4.593
50	69.470	.3187	4.587
55	67.896	.3112	4.583
60	66.358	.3038	4.578
65	64.857	.2967	4.575
70	63.391	.2898	4.571
75	61.959	.2830	4.567
80	60.561	.2765	4.565
85	59.194	.2701	4.563
90	57.859	.2640	4.563
95	56.554	.2581	4.564
100	55.278	.2523	4.564

DIELECTRIC CONSTANTS (Continued)

Dielectric Constant of Liquid Parahydrogen vs. Temperature (°K) and Pressure (atm)

R. J. Corruccini

P atm \ T, °K	20	21	22	23	24	25	26	27	28	29	30	31	32
1	1.2297												
2	1.2302	1.2260	1.2216										
3	1.2306	1.2265	1.2221	1.2174	1.2122								
4	1.2311	1.2270	1.2227	1.2180	1.2129	1.2073	1.2010						
5	1.2315	1.2275	1.2233	1.2186	1.2136	1.2081	1.2020	1.1950					
6	1.2320	1.2280	1.2238	1.2192	1.2143	1.2089	1.2029	1.1962	1.1883				
7	1.2324	1.2285	1.2243	1.2198	1.2150	1.2097	1.2039	1.1973	1.1897	1.1805			
8	1.2329	1.2290	1.2249	1.2204	1.2157	1.2105	1.2048	1.1984	1.1911	1.1824			
9	1.2333	1.2295	1.2254	1.2210	1.2163	1.2112	1.2056	1.1994	1.1924	1.1842	1.1734		
10	1.2337	1.2300	1.2259	1.2216	1.2169	1.2119	1.2065	1.2004	1.1936	1.1857	1.1758	1.1621	
15	1.2358	1.2322	1.2284	1.2243	1.2200	1.2153	1.2103	1.2049	1.1990	1.1924	1.1847	1.1758	1.1645
20	1.2378	1.2343	1.2307	1.2268	1.2227	1.2184	1.2137	1.2088	1.2034	1.1976	1.1913	1.1839	1.1757
25	1.2396	1.2363	1.2328	1.2291	1.2253	1.2211	1.2168	1.2122	1.2073	1.2021	1.1964	1.1903	1.1832
30	1.2414	1.2382	1.2349	1.2313	1.2276	1.2237	1.2196	1.2153	1.2107	1.2059	1.2008	1.1952	1.1891
35	1.2431	1.2400	1.2368	1.2334	1.2298	1.2261	1.2222	1.2181	1.2138	1.2093	1.2046	1.1995	1.1942
40	1.2448	1.2418	1.2386	1.2354	1.2319	1.2284	1.2246	1.2208	1.2167	1.2124	1.2080	1.2033	1.1984
45	1.2464	1.2434	1.2404	1.2372	1.2339	1.2305	1.2269	1.2232	1.2193	1.2153	1.2111	1.2067	1.2021
50	1.2479	1.2450	1.2421	1.2390	1.2358	1.2325	1.2291	1.2255	1.2218	1.2179	1.2139	1.2098	1.2055
60	1.2508	1.2481	1.2453	1.2424	1.2394	1.2363	1.2331	1.2297	1.2263	1.2227	1.2191	1.2153	1.2114
70	1.2535	1.2510	1.2483	1.2455	1.2427	1.2397	1.2367	1.2336	1.2303	1.2270	1.2236	1.2201	1.2165
80	1.2561	1.2536	1.2511	1.2484	1.2457	1.2429	1.2400	1.2371	1.2340	1.2309	1.2277	1.2244	1.2211
90	1.2585	1.2561	1.2537	1.2512	1.2486	1.2459	1.2431	1.2403	1.2374	1.2345	1.2315	1.2284	1.2252
100	1.2608	1.2586	1.2562	1.2538	1.2513	1.2487	1.2461	1.2434	1.2406	1.2378	1.2349	1.2320	1.2290
120	1.2652	1.2631	1.2609	1.2586	1.2563	1.2539	1.2514	1.2489	1.2464	1.2438	1.2412	1.2385	1.2357
140	1.2693	1.2672	1.2651	1.2630	1.2608	1.2586	1.2563	1.2540	1.2516	1.2492	1.2467	1.2442	1.2417
160	1.2730	1.2711	1.2691	1.2671	1.2650	1.2629	1.2607	1.2585	1.2563	1.2540	1.2517	1.2494	1.2470
180	1.2766	1.2747	1.2728	1.2709	1.2689	1.2669	1.2649	1.2628	1.2606	1.2585	1.2563	1.2541	1.2518
200	1.2799	1.2781	1.2763	1.2745	1.2726	1.2707	1.2687	1.2667	1.2647	1.2626	1.2605	1.2584	1.2563
220	1.2831	1.2814	1.2796	1.2779	1.2760	1.2742	1.2723	1.2704	1.2685	1.2665	1.2645	1.2625	1.2605
240		1.2845	1.2828	1.2811	1.2793	1.2775	1.2757	1.2739	1.2720	1.2701	1.2682	1.2663	1.2643
260		1.2874	1.2858	1.2841	1.2824	1.2807	1.2790	1.2772	1.2754	1.2736	1.2717	1.2699	1.2680
280			1.2886	1.2870	1.2853	1.2837	1.2821	1.2803	1.2786	1.2768	1.2751	1.2733	1.2714
300			1.2914	1.2898	1.2882	1.2866	1.2850	1.2833	1.2817	1.2800	1.2782	1.2765	1.2747
320				1.2925	1.2910	1.2894	1.2878	1.2862	1.2846	1.2829	1.2813	1.2796	1.2779
340				1.2951	1.2936	1.2921	1.2905	1.2890	1.2874	1.2858	1.2842	1.2825	1.2809

Note: Values below the stepped line represent an extrapolation of p with density.

Selected Values of Electric Dipole Moments for Molecules in the Gas Phase

Ralph D. Nelson, Jr., David R. Lide, Jr., and Arthur A. Maryott

The following table was abstracted from the publication, "Selected Values of Electric Dipole Moments for Molecules in the Gas Phase" compiled by Nelson, Lide and Maryott and published as part of the National Reference Data Series—National Bureau of Standards (NSRDS—NBS 10). The publication is available from the Superintendent of Documents, U.S. Government Printing Office, Washington, D.C., 20402. Those desiring a complete listing of all compounds in the NSRDS—NBS 10, discussion of the bibliographic procedure and the principal methods of dipole moment measurement should obtain the publication.

Values of the dipole moment, μ, are expressed in the cgs system of units, since this is the system universally used by workers in the field. The numerical values are in debye units, D, ($1\ D = 10^{-18}$ electrostatic units of charge$\times$centimeters). The conversion factor to the Système International is $1\ D = 3.33564 \times 10^{-24}$ coulomb-meter.

Code symbol	Estimated accuracy of value
A	$\pm$ 1% or, for $\mu < 1.0$ D, $\pm$ 0.01 D
B	$\pm$ 2% or, for $\mu < 1.0$ D, $\pm$ 0.02 D
C	$\pm$ 5% or, for $\mu < 1.0$ D, $\pm$ 0.05 D
S	$\mu \equiv 0$ on grounds of molecular symmetry
i	The significance of these values may involve some ambiguity because of the possibility of different conformations or spatial isomers.

Compounds not containing carbon

Formula	Compound name	Selected moment (debyes)	
AgCl	Silver chloride	5.73	C
$AsCl_3$	Arsenic trichloride	1.59	C
AsF_3	Arsenic trifluoride	2.59	B
AsH_3	Arsine	0.20	C
BCl_3	Boron trichloride	0	S
BF_3	Boron trifluoride	0	S
B_2H_6	Diborane	0	S
$B_3H_6N_3$	Triborotriazine (Borazine)	0	S
B_5H_9	Pentaborane	2.13	B
BaO	Barium oxide	7.95	A
BrH	Hydrogen bromide	0.82	B
BrH_3Si	Bromosilane	1.33	B
BrK	Potassium bromide	10.41	B
BrLi	Lithium bromide	7.27	A
Br_2Hg	Mercury dibromide	0	S
Br_4Sn	Tin tetrabromide	0	S
ClCs	Cesium chloride	10.42	A
ClF	Chlorine fluoride	0.88	C
$ClFO_3$	Perchloryl fluoride	0.023	A
$ClGeH_3$	Chlorogermane	2.13	A
ClH	Hydrogen chloride	1.08	B
ClH_3Si	Chlorosilane	1.31	A
ClK	Potassium chloride	10.27	A
ClLi	Lithium chloride	7.13	A
ClNa	Sodium chloride	9.00	A
$ClNO_2$	Nitryl chloride	0.53	A
ClTl	Thallium chloride	4.44	B
Cl_2F_3P	Dichlorotrifluorophosphorus	0.68	C
Cl_2H_2Si	Dichlorosilane	1.17	B
Cl_2Hg	Mercury dichloride	0	S
Cl_2OS	Thionyl chloride	1.45	B
Cl_2O_2S	Sulfuryl chloride	1.81	B
Cl_3F_2P	Trichlorodifluorophosphorus	0	S
Cl_3HSi	Trichlorosilane	0.86	B
Cl_3P	Phosphorus trichloride	0.78	C
Cl_4FP	Tetrachlorofluorophosphorus	0.21	B
Cl_4Ge	Germanium tetrachloride	0	S
Cl_4Si	Silicon tetrachloride	0	S
Cl_4Sn	Tin tetrachloride	0	S
Cl_4Ti	Titanium tetrachloride	0	S
CsF	Cesium fluoride	7.88	A
FH	Hydrogen fluoride	1.82	A
FH_3Si	Fluorosilane	1.27	B
FH_5Si_2	Fluorodisilane	1.26	A
FK	Potassium fluoride	8.60	A
FLi	Lithium fluoride	6.33	A
FNO	Nitrosyl fluoride	1.81	B
FNa	Sodium fluoride	8.16	A
FRb	Rubidium fluoride	8.55	A
FTl	Thallium fluoride	4.23	A
F_2HN	Difluoramine	1.92	A
F_2H_2Si	Difluorosilane	1.55	A
F_2N_2	*cis*-Difluorodiazine	0.16	A
F_2O	Oxygen difluoride	0.297	A
F_2OS	Thionyl fluoride	1.63	A
F_2O_2	Dioxygen difluoride	1.44	C
F_2O_2S	Sulfuryl fluoride	1.12	B
F_2S_2	Sulfur monofluoride ($S = SF_2$ isomer)	1.03	C
F_2S_2	Sulfur monofluoride (FSSF isomer)	1.45	B
F_2Si	Silicon difluoride	1.23	B
F_3HSi	Trifluorosilane	1.27	B
F_3N	Nitrogen trifluoride	0.235	A
F_3NS	Nitridotrifluorosulfur	1.91	B
F_3OP	Phosphoryl fluoride	1.76	B
F_3P	Phosphorus trifluoride	1.03	A
F_3PS	Thiophosphoryl fluoride	0.64	B
F_4N_2	Tetrafluorohydrazine, *gauche* conformation	0.26	B
F_4S	Sulfur tetrafluoride	0.632	A
F_4Si	Silicon tetrafluoride	0	S
F_5P	Phosphorus pentafluoride	0	S
F_5I	Iodine pentafluoride	2.18	C
F_6S	Sulfur hexafluoride	0	S
F_6Se	Selenium hexafluoride	0	S
F_6Te	Tellurium hexafluoride	0	S
F_6U	Uranium hexafluoride	0	S
HI	Hydrogen iodide	0.44	B
HLi	Lithium hydride	5.88	A
HN	Imidyl radical		
HNO_3	Nitric acid	2.17	A
HO	Hydroxyl radical	1.66	A
H_2O	Water	1.85	A
H_2O_2	Hydrogen peroxide	2.2	D
H_2S	Hydrogen sulfide	0.97	A
H_3N	Ammonia	1.47	A
H_3P	Phosphine	0.58	A
H_3Sb	Stibine	0.12	C
H_4N_2	Hydrazine	1.75	C
H_4Si	Silane	0	S
H_6OSi_2	Disilyl ether (disiloxane)	0.24	B
H_6Si_2	Disilane	0	S
HgI_2	Mercury diiodide	0	S
ILi	Lithium iodide	7.43	A
I_4Sn	Tin tetraiodide	0	S
NO	Nitrogen monoxide (nitric oxide)	0.153	A
NO_2	Nitrogen dioxide	0.316	A
N_2O	Dinitrogen oxide (nitrous oxide)	0.167	A
OS	Sulfur monoxide	1.55	A
OS_2	Disulfur monoxide	1.47	B
OSr	Strontium oxide	8.90	A
O_2S	Sulfur dioxide	1.63	A
O_3	Ozone	0.53	B
O_3S	Sulfur trioxide	0	S
O_4Os	Osmium tetroxide	0	S

Compounds containing carbon

Formula	Compound name	Selected moment (debyes)	
$CBrF_3$	Bromotrifluoromethane	0.65	C
CBr_2F_2	Dibromodifluoromethane	0.66	C
$CClF_3$	Chlorotrifluoromethane	0.50	A
CClN	Cyanogen chloride	2.82	B
CCl_2F_2	Dichlorodifluoromethane	0.51	C
CCl_2O	Carbonyl chloride (phosgene)	1.17	A
CCl_2S	Thiocarbonyl chloride	0.29	C
CCl_3F	Trichlorofluoromethane	0.45	C
CCl_3NO_2	Trichloronitromethane	1.89	C
CCl_4	Carbon tetrachloride	0	S
CFN	Cyanogen fluoride	2.17	C
CF_2	Carbon difluoride	0.46	B
CF_2O	Carbonyl fluoride	0.95	A
CF_3I	Iodotrifluoromethane	0.92	C
CF_3NO_2	Trifluoronitromethane	1.44	C
CF_4	Carbon tetrafluoride	0	S
CN_4O_8	Tetranitromethane	0	S
CO	Carbon monoxide	0.112	A
COS	Carbonyl sulfide	0.712	A
COSe	Carbonyl selenide	0.73	B
CO_2	Carbon dioxide	0	S
CS	Carbon monosulfide	1.98	A
CSTe	Thiocarbonyl telluride	0.17	A
CS_2	Carbon disulfide	0	S
$CHBr_3$	Tribromomethane	0.99	B
$CHClF_2$	Chlorodifluoromethane	1.42	B

Selected Values of Electric Dipole Moments for Molecules in the Gas Phase (Continued)

Compounds containing carbon—Continued

Formula	Compound name	Selected moment (debyes)	
$CHCl_2F$	Dichlorofluoromethane	1.29	B
$CHCl_3$	Trichloromethane (chloroform)	1.01	B
$CHFO$	Formyl fluoride	2.02	A
CHF_3	Trifluoromethane	1.65	A
CHN	Hydrogen cyanide	2.98	A
CHP	Methylidyne phosphide (methinophosphide)	0.390	A
CH_2Br_2	Dibromomethane	1.43	B
CH_2ClF	Chlorofluoromethane	1.82	B
CH_2ClNO_2	Chloronitromethane	2.91	B
CH_2Cl_2	Dichloromethane	1.60	B
CH_2F_2	Difluoromethane	1.97	A
CH_2N_2	Cyanogen amide (cyanamide)	4.27	C
CH_2N_2	Diazomethane	1.50	A
CH_2N_2	Diazirine	1.59	C
CH_2O	Methanal (formaldehyde)	2.33	A
CH_2O_2	Methanoic acid (formic acid)	1.41	A
CH_3BF_2	Methyl difluoroborane	1.66	B
CH_3BO	Carbonyl borane	1.80	B
CH_3Br	Bromomethane	1.81	A
CH_3Cl	Chloromethane	1.87	A
CH_3F	Fluoromethane	1.85	A
CH_3I	Iodomethane	1.62	B
CH_3NO	Hydroxyliminomethane (formaldoxime)	0.44	A
CH_3NO	Formyl amide (formamide)	3.73	B
CH_3NOS	Methyl sulfinylamine	1.70	B
CH_3NO_2	Nitromethane	3.46	A
CH_3NO_3	Methyl nitrate	3.12	B
CH_3N_3	Methyl azide	2.17	B
CH_4	Methane	0	S
CH_4F_2Si	Methyl difluorosilane	2.11	A
CH_4O	Methanol	1.70	A
CH_4S	Methanethiol (methyl mercaptan)	1.52	C
CH_5FSi	Methyl monofluorosilane	1.71	A
CH_5N	Methyl amine	1.31	B
CH_5P	Methyl phosphine	1.10	A
CH_6Ge	Methyl germane	0.643	A
CH_6OSi	Methoxysilane	1.17	B
CH_6Si	Methyl silane	0.735	A
CH_6Sn	Methyl stannane	0.68	C
C_2ClF_5	Chloropentafluorethane	0.52	C
C_2F_6	Hexafluorethane	0	S
C_2N_2	Dicyanogen (cyanogen)	0	S
C_2N_2S	Dicyano sulfide	3.02	A
C_2HCl	Chloroacetylene	0.44	A
C_2HCl_5	Pentachloroethane	0.92	C
C_2HF	Fluoroacetylene	0.73	C
C_2HF_3	Trifluoroethylene	1.40	C
C_2HF_5	Pentafluoroethane	1.54	C
C_2H_2	Acetylene	0	S
$C_2H_2Cl_2$	1, 1-Dichloroethylene	1.34	A
$C_2H_2Cl_2$	*cis*-1, 2-Dichloroethylene	1.90	B
$C_2H_2Cl_2O$	Chloroacetyl chloride	2.23	Ci
$C_2H_2Cl_4$	1,1,2,2-Tetrachloroethane	1.32	Ci
C_2H_2FN	Fluorocyanomethane	3.43	C
$C_2H_2F_2$	1,1-Difluoroethylene	1.38	A
$C_2H_2F_2$	*cis*-1,2-Difluoroethylene	2.42	A
$C_2H_2N_2O$	1,2,5-Oxadiazole	3.38	A
$C_2H_2N_2O$	1,3,4-Oxadiazole	3.04	B
$C_2H_2N_2S$	1,2,5-Thiadiazole	1.56	A
$C_2H_2N_2S$	1.3.4-Thiadiazole	3.29	B
C_2H_2O	Methylene carbonyl (ketene)	1.42	B
C_2H_3Br	Bromoethylene	1.42	B
C_2H_3Cl	Chloroethylene	1.45	B
$C_2H_3ClF_2$	1-Chloro-1,1-difluoroethane	2.14	B
C_2H_3ClO	Acetyl chloride	2.72	C
$C_2H_3Cl_3$	1,1,1-Trichloroethane	1.78	B
C_2H_3F	Fluoroethylene	1.43	A
C_2H_3FO	Acetyl fluoride	2.96	A
$C_2H_3F_3$	1,1,1-Trifluoroethane	2.32	B
$C_2H_3F_3$	1,1,2-Trifluoroethane	1.58	B
C_2H_3N	Cyanomethane (acetonitrile)	3.92	A
C_2H_3N	Isocyanomethane	3.85	B
C_2H_4	Ethylene	0	S
C_2H_4ClF	1-Chloro-2-fluoroethane, *gauche* conformation	2.72	C
$C_2H_4ClNO_2$	1-Chloro-1-nitroethane	3.27	B
$C_2H_4Cl_2$	1,1-Dichloroethane	2.06	B
$C_2H_4F_2$	1,1-Difluoroethane	2.27	B
C_2H_4Ge	Germyl acetylene	0.136	A
C_2H_4O	Oxirane (ethylene oxide)	1.89	A
C_2H_4O	Ethanal (acetaldehyde)	2.69	B
$C_2H_4O_2$	Ethanoic acid (acetic acid)	1.74	C
$C_2H_4O_2$	Methyl methanoate (methly formate)	1.77	B
C_2H_4S	Thiirane (ethylene sulfide)	1.85	A
C_2H_4Si	Silyl acetylene	0.316	A
C_2H_5Br	Bromoethane	2.03	A
C_2H_5BrO	Bromomethoxymethane	2.05	Ci
C_2H_5Cl	Chloroethane	2.05	A
C_2H_5ClO	2-Chloroethanol	1.78	Ci
C_2H_5F	Fluoroethane	1.94	B
C_2H_5I	Iodoethane	1.91	B
C_2H_5N	Iminoethane (ethyleneimine)	1.90	A
C_2H_5N	Methyliminomethane ($CH_3N = CH_2$)	1.53	B

Compounds containing carbon—Continued

Formula	Compound name	Selected moment (debyes)	
C_2H_5NO	Acetyl amine (acetamide)	3.76	Bi
C_2H_5NO	Methylaminomethanal (*N*-methylformamide)	3.83	Bi
$C_2H_5NO_2$	Nitritoethane (ethyl nitrite)	2.40	Ci
$C_2H_5NO_2$	Nitroethane	3.65	B
C_2H_6	Ethane	0	S
C_2H_6BF	Dimethyl fluoroborane	1.32	C
C_2H_6O	Ethanol	1.69	Bi
C_2H_6O	Dimethyl ether	1.30	A
C_2H_6OS	Dimethylsulfoxide	3.96	A
$C_2H_6O_2$	1,2-Ethanediol (ethylene glycol)	2.28	Ci
$C_2H_6O_2S$	Dimethyl sulfoxylate (dimethyl sulfone)	4.49	B
C_2H_6S	Ethanethiol	1.58	Bi
C_2H_6S	Dimethyl sulfide	1.50	A
C_2H_6Si	Silyl ethylene	0.66	A
$C_2H_7B_5$	2,4-Dicarbaheptaborane	1.32	B
C_2H_7N	Aminoethane (ethyl amine)	1.22	Ci
C_2H_7N	Dimethyl amine	1.03	B
C_2H_7P	Ethyl phosphine	1.17	Bi
C_2H_7P	Dimethyl phosphine	1.23	A
$C_2H_8N_2$	1,2-Diaminoethane	1.99	Ci
C_2H_8Si	Dimethyl silane	0.75	A
C_2H_8Si	Ethyl silane	0.81	B
C_3O_2	Dicarbonyl carbon (carbon suboxide)	0	S
C_3HF_3	3,3,3-Trifluoropropyne	2.36	B
C_3HN	Cyanoacetylene	3.72	A
$C_3H_2N_2$	Dicyanomethane	3.73	A
C_3H_2O	Propynal	2.47	B
$C_3H_2O_3$	Vinylene carbonate	4.55	A
C_3H_3Br	3-Bromopropyne	1.54	C
C_3H_3Cl	3-Chloropropyne	1.68	C
$C_3H_3F_3$	3,3,3-Trifluoropropene	2.45	B
C_3H_3N	Cyanoethylene	3.87	B
C_3H_3NO	Acetyl cyanide	3.45	B
C_3H_3NS	Thiazole	1.62	B
C_3H_4	Cyclopropene	0.45	A
C_3H_4	Propyne	0.781	A
C_3H_4	Propadiene (allene)	0	S
$C_3H_4Cl_2$	1,1-Dichlorocyclopropane	1.58	B
C_3H_4O	Ethylidene carbonyl (methyl ketene)	1.79	B
C_3H_4O	Propenal, *trans* conformation (acrolein)	3.12	B
$C_3H_4O_2$	2-Oxoöxetane (β-propiolactone)	4.18	A
$C_3H_4O_2$	Vinyl formate	1.49	A
C_3H_5Cl	2-Chloropropene	1.66	B
C_3H_5Cl	*cis*-1-Chloropropene	1.67	C
C_3H_5Cl	*trans*-1-Chloropropene	1.97	C
C_3H_5Cl	3-Chloropropene	1.94	Ci
C_3H_5F	*cis*-1-Fluoropropene	1.46	B
C_3H_5F	2-Fluoropropene	1.61	B
C_3H_5F	3-Fluoropropene, *cis* conformation	1.76	A
C_3H_5F	3-Fluoropropene, *gauche* conformation.	1.94	A
C_3H_5N	Cyanoethane (propionitrile)	4.02	A
C_3H_6	Cyclopropane	0	S
C_3H_6	Propene	0.366	A
$C_3H_6ClNO_2$	1-Chloro-1-nitropropane	3.48	Bi
$C_3H_6Cl_2$	1,2-Dichloropropane		
$C_3H_6Cl_2$	1,3-Dichloropropane	2.08	Bi
$C_3H_6Cl_2$	2,2-Dichloropropane	2.27	C
C_3H_6O	Oxetane (trimethylene oxide)	1.94	A
C_3H_6O	Methyl oxirane (propylene oxide)	2.01	A
C_3H_6O	Propanone (acetone)	2.88	A
C_3H_6O	2-Propen-1-ol (allyl alcohol)	1.60	C
C_3H_6O	Propanal, *cis* conformation (propionaldehyde)	2.52	B
$C_3H_6O_2$	Propanoic acid	1.75	Ci
$C_3H_6O_2$	Methyl acetate	1.72	Ci
$C_3H_6O_2$	Ethyl formate	1.93	Ci
$C_3H_6O_3$	1,3,5-Trioxane	2.08	A
C_3H_6S	Thietane (trimethylene sulfide)	1.85	C
C_3H_6S	Methyl thiirane (propylene sulfide)	1.95	A
C_3H_7Br	1-Bromopropane	2.18	Ci
C_3H_7Br	2-Bromopropane	2.21	C
C_3H_7Cl	1-Chloropropane	2.05	Bi
C_3H_7Cl	2-Chloropropane	2.17	C
C_3H_7F	1-Fluoropropane, *gauche* conformation	1.90	C
C_3H_7F	1-Fluoropropane, *trans* conformation	2.05	B
C_3H_7I	1-Iodopropane	2.04	Ci
C_3H_7NO	N,N-Dimethylformamide	3.82	Bi
C_3H_7NO	Acetyl methylamine (*N*-Methylacetamide)	3.73	Bi
$C_3H_7NO_2$	1-Nitropropane	3.66	Bi
$C_3H_7NO_2$	2-Nitropropane	3.73	B
C_3H_8	Propane	0.084	A
C_3H_8O	1-Propanol	1.68	Bi
C_3H_8O	2-Propanol	1.66	Bi
C_3H_8O	Methoxyethane (methyl ethyl ether)	1.23	Ci
C_3H_9As	Trimethyl arsine	0.86	B
C_3H_9N	Trimethyl amine	0.612	A
C_3H_9N	1-Aminopropane (*n*-propylamine)	1.17	Ci
C_3H_9P	Trimethyl phosphine	1.19	A

Selected Values of Electric Dipole Moments for Molecules in the Gas Phase (Continued)

Compounds containing carbon—Continued

Formula	Compound name	Selected moment (debyes)	
$C_3H_{10}Si$	Trimethyl silane	0.525	A
C_4F_8	Perfluorocyclobutane	0	S
$C_4H_2N_2$	*trans*-1,2-Dicyanoethylene	0	S
$C_4H_4Cl_2$	1,4-Dichloro-2-butyne	2.10	Bi
$C_4H_4F_2$	1,1-Difluoro-1,3-butadiene (*trans* conformation)	1.29	A
C_4H_4O	Furan	0.66	A
$C_4H_4O_2$	Diketene	3.53	B
C_4H_4S	Thiophene	0.55	C
C_4H_5Cl	4-Chloro-1,2-butadiene	2.02	Ci
C_4H_5Cl	1-Chloro-2-butyne	2.19	C
C_4H_5F	2-Fluoro-1,3-butadiene (*trans* conformation)	1.42	A
C_4H_5N	Pyrrole	1.84	C
C_4H_5N	*cis*-1-Cyanopropene	4.08	B
C_4H_5N	*trans*-1-Cyanopropene	4.50	B
C_4H_5N	2-Cyanopropene (methacrylonitrile)	3.69	C
C_4H_6	Cyclobutene	0.132	A
C_4H_6	1-Butyne	0.80	C
C_4H_6	1,2-Butadiene	0.403	A
C_4H_6	1,3-Butadiene	0	S
C_4H_6O	Cyclobutanone	2.99	B
C_4H_6O	*trans*-2-Butenal (crotonaldehyde)	3.67	Bi
C_4H_6O	2-Methylpropenal (methacrolein)	2.68	Ci
C_4H_6O	3-Butene-2-one	3.16	B
C_4H_7Cl	1-Chloro-2-methylpropene	1.95	Bi
$C_4H_7Cl_3$	1,1,2-Trichloro-2-methylpropane	1.86	Ci
C_4H_7F	Fluorocyclobutane	1.94	A
C_4H_7N	1-Cyanopropane	4.07	Bi
C_4H_8	1-Butene	0.34	Ci
C_4H_8	*trans*-2-Butene	0	S
C_4H_8	2-Methylpropene	0.50	A
$C_4H_8Cl_2$	1,4-Dichlorobutane	2.22	Ci
C_4H_8O	Tetrahydrofuran	1.63	C
C_4H_8O	*cis*-2,3-Dimethyloxirane	2.03	A
C_4H_8O	Butanal	2.72	Bi
$C_4H_8O_2$	1,4-Dioxane	0	S
$C_4H_8O_2$	Ethyl acetate	1.78	Ci
C_4H_9Br	1-Bromobutane	2.08	Ci
C_4H_9Br	2-Bromobutane	2.23	Ci
C_4H_9Cl	1-Chlorobutane	2.05	Bi
C_4H_9Cl	2-Chlorobutane	2.04	Ci
C_4H_9Cl	1-Chloro-2-methylpropane	2.00	Ci
C_4H_9Cl	2-Chloro-2-methylpropane	2.13	B
C_4H_9F	2-Fluoro-2-methylpropane	1.96	A
C_4H_9I	1-Iodobutane	2.12	Ci
C_4H_9NO	Propanoyl methylamine (N-methylpropionamide)	3.61	Bi
C_4H_9NO	Acetyl dimethylamine (N,N-dimethylacetamide)	3.81	Bi
$C_4H_9NO_2$	2-Nitrito-2-methylpropane (t-butyl nitrite)	2.74	Ci
$C_4H_9NO_2$	1-Nitrobutane	3.59	Bi
$C_4H_9NO_2$	2-Nitro-2-methylpropane	3.71	B
C_4H_{10}	Butane	$\leqslant$0.05	Ci
C_4H_{10}	2-Methylpropane	0.132	A
$C_4H_{10}O$	1-Butanol	1.66	Bi
$C_4H_{10}O$	2-Methylpropan-1-ol (isobutanol)	1.64	C
$C_4H_{10}O$	Diethyl ether	1.15	Bi
$C_4H_{10}S$	Diethyl sulfide	1.54	Ci
$C_4H_{11}N$	Diethyl amine	0.92	Ci
C_5H_5N	Pyridine	2.19	B
C_5H_5N	1-Cyano-1,3-butadiene	3.90	Ci
C_5H_6	1,3-Cyclopentadiene	0.419	A
C_5H_8	Cyclopentene	0.20	B
C_5H_8	1-Pentyne	0.81	Ci
C_5H_8	2-Methyl-1,3-butadiene (*trans* conformation)	0.25	A
$C_5H_8O_2$	Acetylacetone		Ci
C_5H_9N	1-Cyanobutane	4.12	Bi
C_5H_9N	2-Cyano-2-methylpropane	3.95	A
$C_5H_{10}O_3$	Diethyl carbonate	1.10	Ci
$C_5H_{11}Br$	1-Bromopentane	2.20	Ci
$C_5H_{11}Cl$	1-Chloropentane	2.16	Ci
C_5H_{12}	2-Methylbutane	0.13	C
C_5H_{12}	2,2-Dimethylpropane	0	S
$C_6H_2Cl_2O_2$	2,5-Dichloro-1,4-cyclohexadienedione	0	S
$C_6H_4ClNO_2$	*o*-Chloronitrobenzene	4.64	B
$C_6H_4ClNO_2$	*m*-Chloronitrobenzene	3.73	B
$C_6H_4ClNO_2$	*p*-Chloronitrobenzene	2.83	B
$C_6H_4Cl_2$	*o*-Dichlorobenzene	2.50	B
$C_6H_4Cl_2$	*m*-Dichlorobenzene	1.72	C
$C_6H_4Cl_2$	*p*-Dichlorobenzene	0	S
$C_6H_4FNO_2$	*p*-Fluoronitrobenzene	2.87	B
$C_6H_4F_2$	*m*-Difluorobenzene	1.58	B
$C_6H_4N_2O_4$	*p*-Dinitrobenzene	0	S
$C_6H_4O_2$	1,4-Cyclohexadienedione (*p*-benzoquinone)	0	S
C_6H_5Br	Bromobenzene	1.70	B
C_6H_5Cl	Chlorobenzene	1.69	B
C_6H_5ClO	*p*-Chlorophenol	2.11	C
C_6H_5F	Fluorobenzene	1.60	C
C_6H_5I	Iodobenzene	1.70	C
$C_6H_5NO_2$	Nitrobenzene	4.22	B
C_6H_6	Benzene	0	S
C_6H_6O	Phenol	1.45	C
C_6H_7N	Aminobenzene (aniline)	1.53	C
C_6H_8	1,3-Cyclohexadiene	0.44	B
C_6H_{10}	1-Hexyne	0.83	Ci
C_6H_{10}	3, 3-Dimethyl-1-butyne	0.66	A
$C_6H_{10}Cl_2$	*cis*-le,2a-Dichlorocyclohexane	3.11	C
$C_6H_{12}N_2$	Diisopropylidene hydrazine (dimethyl ketazine)	1.53	Bi
$C_6H_{12}O_2$	Pentyl formate (n-amyl formate)	1.90	Ci
$C_6H_{12}O_3$	2,4,6-Trimethyl-1,3,5-trioxane (paraldehyde)	1.43	C
$C_6H_{14}O$	Dipropyl ether	1.21	Ci
$C_6H_{14}O_2$	1,1-Diethoxyethane		i
$C_6H_{15}N$	Triethyl amine	0.66	Ci
$C_7H_4ClF_3$	*o*-Chloro(trifluoromethyl)benzene	3.46	B
$C_7H_4ClF_3$	*p*-Chloro(trifluoromethyl)benzene	1.58	C
$C_7H_5F_3$	(Trifluoromethyl)benzene	2.86	B
C_7H_5N	Cyanobenzene (benzonitrile)	4.18	B
C_7H_7Cl	*o*-Chlorotoluene	1.56	C
C_7H_7Cl	*p*-Chlorotoluene	2.21	B
C_7H_7F	*o*-Fluorotoluene	1.37	C
C_7H_7F	*m*-Fluorotoluene	1.86	C
C_7H_7F	*p*-Fluorotoluene	2.00	C
$C_7H_7NO_3$	*o*-Nitro(methoxy)benzene	4.83	Bi
$C_7H_7NO_3$	*m*-Nitro(methoxy)benzene	4.55	Bi
$C_7H_7NO_3$	*p*-Nitro(methoxy)benzene	5.26	B
C_7H_8	1,3,5-Cycloheptatriene	0.25	C
C_7H_8	Toluene	0.36	C
C_7H_8O	Phenylmethanol (benzyl alcohol)	1.71	C
C_7H_8O	Methoxybenzene (anisole)	1.38	C
C_7H_9NO	*o*-Amino(methoxy)benzene	1.61	Ci
$C_7H_{14}O_2$	Pentyl acetate (*n*-amyl acetate)	1.75	Ci
$C_7H_{15}Br$	1-Bromoheptane	2.16	Ci
$C_8H_4N_2$	*p*-Dicyanobenzene	0	S
C_8H_8O	Acetylbenzene (acetophenone)	3.02	B
$C_8H_8O_2$	2,5-Dimethyl-1,-4-cyclohexadienedione	0	S
C_8H_{10}	Ethylbenzene	0.59	C
C_8H_{10}	*o*-Xylene	0.62	C
C_8H_{10}	*p*-Xylene	0	S
$C_8H_{12}O_2$	Tetramethylcyclobutane-1,3-dione	0	S
$C_8H_{18}O$	Dibutyl ether	1.17	Ci
C_9H_7N	Quinoline	2.29	C
C_9H_7N	Isoquinoline	2.73	C
$C_9H_{10}O_2$	Ethyl phenylformate (ethyl benzoate)	2.00	Ci
$C_{10}H_8$	Azulene	0.80	B
$C_{10}H_{14}BeO_4$	Bis(2,4-pentanedionato) beryllium	0	S
$C_{12}H_9BrO$	*p*-Bromophenoxybenzene	1.98	C
$C_{12}H_9NO_3$	*p*-Nitrophenoxybenzene	4.54	B
$C_{12}H_{10}$	Phenylbenzene (diphenyl)	0	S
$C_{13}H_{11}BrO$	*p*-Bromophenoxy-*p*-toluene	2.45	C
$C_{14}H_{14}O$	Bis(*p*-tolyl) ether	1.54	C
$C_{15}H_{21}AlO_6$	Tris(2,4-pentanedionato) aluminum	0	S
$C_{15}H_{21}CrO_6$	Tris(2,4-pentanedionato) chromium (III)	0	S
$C_{15}H_{21}FeO_6$	Tris(2,4-pentanedionato) iron (III)	0	S
$C_{20}H_{28}O_8Th$	Tetrakis(2,4-pentanedionato) thorium	0	S

DIPOLE MOMENTS

The method of measurement of the dipole moments is indicated in the following **two tables** by the symbols:

B benzene solution
C carbon tetrachloride solution
D 1,4-dioxane solution
H n-heptane solution
St measurement of Stark effect in microwave spectrum of gas

Dipole Moments for Some Inorganic Compounds

Compound	Dipole Moment $\times 10^{-18}$ e. s. u.	Method
$AlBr_3$	5.14	B
AlI_3	2.48	B
CsCl	10.42	St
CsF	7.875	St
HF	1.92 ± 0.02	..
HCl	1.084 ± 0.003–0.007	..
NBr	0.78	..
HDSe	0.62	St
HI	0.38	..
DCl	1.084 ± 0.003–0.007	..
HNO_3	2.16	St
$HgBr_2$	0.95	B
$HgCl_2$	1.23	B
H_2O	1.87	..
H_2O_2	2.13 ± 0.05	..
H_2S	1.10	..
SO_2	1.60	..
SO_3	0.00	..
SO_2F_2	1.110	St
NH_3	1.3	..
N_2H_4	1.84	..
NO	0.16	..
NO_2	0.29	..
N_2O_4	0.37	..
NOCl	1.83	..
NOBr	1.87	..
PCl_3	0.90–1.16	..
PCl_5	0.0	..
CO	0.10	..
CO_2	0.0	..
SiD_2F_2	1.53	St
SiH_2F_2	1.54	St
$SnCl_4$	0.95	B
SnI_4	0	B
$TiCl_4$	0	C

Dipole Moments for Some Organo-metallic Compounds

Compound	Dipole Moment $\times 10^{-18}$ e. s. u.	Method
Beryllium diethyl	1.0	H
Cadmium diethyl	0.3	H
Mercury diethyl	0.0	H
Magnesium diethyl	4.8	D
Zinc diethyl	0.0	H
Beryllium diphenyl	1.6	B
Cadmium diphenyl	0.6	B
Mercury diphenyl	0.2	B
Magnesium diphenyl	4.9	D
Zinc diphenyl	0.8	B
Chromium (0), diphenyl	0	B
Chromium, ditolyl	0	B
Cobalt, mononitrosyl tricarbonyl	0.72	B
Cyclopentadienyl, chromium dicarbonyl mono nitrosyl	3.23	B
Cyclopentadienyl, manganese tricarbonyl	3.30	B
Cyclopentadienyl, cobalt ducarbonyl	2.87	B
Cyclopentadienyl, vanadium tetracarbonyl	3.17	B
Penta cyclopentadienyl, dicobalt	0	B
Dicyclopentadienyl, iron (II)	0	B
Dicyclopentadienyl, lead (II)	1.63	B
Dicyclopentadienyl, tin (II)	1.02	B
Ethyl lithium	0.87	B
Iron, dinitrosyl dicarbonyl	0.95	B
Iron, tetracarbonyl-diiodide	3.68	B
Iron, tetracarbonyl mono-(methyl isonitrile)	5.07	B
Iron, pentacarbonyl	0.63	B
Iron, bis(p-chlorophenyl cyclopentadienyl	3.12	B
Ruthenium (II), di-indenyl	0	B

Dipole Moments

Dipole Moments of Amino Acid Esters

Substance	$\mu \cdot 10^{18}$ e. s. u.
Glycine ethyl ester	2.11
α-Alanine ethyl ester	2.09
α-Aminobutyric acid ethyl ester	2.13
α-Aminovaleric acid ethyl ester	2.13
Valine ethyl ester	2.11
α-Aminocaproic acid ethyl ester	2.13
β-Alanine ethyl ester	2.14
β-Aminobutyric acid ethyl ester	2.11

Accurate to ±0.01 . 10^{-18} e. s. u.
J. Wyman, Chem. Rev., 1936, **19**, 213.

Dipole Moments of Amides

Substance	
Urea	4.56
Thiourea	4.89
Symm.-dimethylurea	4.8
Tetraethylurea	3.3
Propylurea	4.1
Acetamide	3.6
Sulfamide	3.9
Benzamide	3.6
Valeramide	3.7
Caproamide	3.9

For comprehensive list of dipole moments see Trans. Faraday Soc., 1934, **30**, General Discussion.

Dipole Moments of Somè Hormones and Related Compounds in Dioxan

Compound	
Cholestane-3(β) : 7(α)-diol	2.31
Cholestane-3(β) : 7(β)-diol	2.55
Cholestane	2.98
Δ^5-Cholestane-3(β)ol-7 one	3.79
Androsterone	3.70
β-Androsterone	2.95
Δ^5-Androstene-3(β) : 17(α)-diol	2.89
Δ^5-Androstene-3(β) : 17(β)-diol	2.69
Δ^5-Androstene-3(β)ol-17 one	2.46
Testosterone	4.32
cis-Testosterone	5.17
Δ^4-Androstene-3 : 17 dione	3.32
Isophorone	3.96

Ethylenic >C=C< in a six membered ring and conjugated with >C=0 increases the dipole moment approximately by 1 Debye. Non-conjugated >C=C< in sterols decreases the dipole moment by approximately 0.49. Biological activity is not correlated with dipole moment. W. D. Kumler and G. M. Fohlen, J. Am. Chem. Soc., 1945, **67**, 437.

ELECTRON AFFINITIES

H. O. Pritchard

The tabulated values are given in k cal/gm ion and may be converted into e.v. using the conversion 1 e.v. ≡ 23.06 kcal. All values are experimental determinations, unless marked with a ‡, and refer to the gas phase at 0°K. For more extensive tables see L. M. Branscomb, in "Atomic and Molecular Processes" Ed. D. R. Bates, Academic Press, New York, (1962); H. O. Pritchard, Chemical Reviews, **52,** 529 (1953); F. M. Page, Technical Report DA-91-591-EUC-6-3206 (1965). The compiler is indebted to Drs. L. M. Branscomb and F. M. Page for communicating unpublished data.

Atoms		Molecules		Radicals			
Ion formed	E.A.	Ion formed	E.A.	Ion formed	E.A.	Ion formed	E.A.
H^- ($1s^2$)	17.2‡	NO^-	21	OH^-	42.2	$C_6H_5^-$	51
H^- ($2s^2$) from 2s	10.0‡	NO_2^-	92	SH^-	53	$C_6H_5CH_2^-$	21
H^- (2s2p) from 2s	10.6‡	O_2^-	10.1	NH_2^-	28	CCl_3^-	33
C^-	28.8	O_3^-	45‡	CN^-	59	CH_3O^-	10
O^-	33.8	BF_3^-	73	NF_2^-	75	$C_2H_5O^-$	14
F^-	79.5	CCl_4^-	50	SCN^-	50	n-$C_3H_7O^-$	18
S^-	48	CCl_3H^-	40	CH_3^-	26	CH_3S^-	32
Cl^-	83.3	TCNE*⁻	75	$C_2H_5^-$	22	$C_2H_5S^-$	32
Br^-	77.5	TCNB†⁻	60	n-$C_3H_7^-$	16	$C_6H_5NH^-$	33
I^-	70.6	SF_6^-	35	n-$C_4H_9^-$	15	SF_5^-	84

* TCNE = Tetracyanoethylene.
† TCNB = Tetracyanobenzene.

IONIZATION POTENTIALS OF THE ELEMENTS

Different methods have been employed to measure ionization potentials. Abbreviations of the methods used for data listed in the following table are:

S; Vacuum ultraviolet spectroscopy
SI; Surface ionization, mass spectrometric
EI; Electron impact with mass analysis

El.	At. No.	Ionization potential in volts I	II	III	IV	V	VI	VII	VIII	Meth.
Ar	18	15.755	27.62	40.9	59.79	75	91.3	124	143.46	S
Ac	89	6.9	12.1	20	......	......	......	......	......	S
Ag	47	7.574	21.48	34.82	......	......	......	......	......	S
Al	13	5.984	18.823	28.44	119.96	153.77	190.42	241.38	284.53	S
As	33	9.81	18.63	28.34	50.1	62.6	127.5	......	......	S
At	85	9.5	......	......	......	......	......	......	......	S
Au	79	9.22	20.5	......	......	......	......	......	......	S
B	5	8.296	25.149	37.92	259.298	340.127	......	......	......	S
Ba	56	5.21	10.001	35.5	......	......	......	......	......	S
Be	4	9.32	18.206	153.85	217.657	......	......	......	......	S
Bi	83	7.287	16.68	25.56	45.3	56	88.3	......	......	S
Br	35	11.84	21.6	35.9	47.3	59.7	88.6	103	193	S
C	6	11.256	24.376	47.871	64.476	391.986	489.84	......	......	S
Ca	20	6.111	11.868	51.21	67	84.39	109	128	143.3	S
Cb(Nb)	41	6.88	14.32	25.04	38.3	50	103	125	......	S
Cd	48	8.991	16.904	37.47	......	......	......	......	......	S
Ce	58	5.6	12.3	20	33.3	......	......	......	......	SI
Cl	17	13.01	23.8	39.9	53.5	67.8	96.7	114.27	348.3	S
Co	27	7.86	17.05	33.49	83.1	......	......	......	......	S
Cr	24	6.764	16.49	30.95	50	73	91	161	185	S
Cs	55	3.893	25.1	35	......	......	......	......	......	S
Cu	29	7.724	20.29	36.83	......	......	......	......	......	S
Dy	66	6.8	......	......	......	......	......	......	......	S
Er	68	6.08	......	......	......	......	......	......	......	SI
Eu	63	5.67	11.24	......	......	......	......	......	......	S
F	9	17.418	34.98	62.646	87.14	114.214	157.117	185.139	953.6	S
Fe	26	7.87	16.18	30.643	56.8	......	......	......	151	S
Fr	87	4	......	......	......	......	......	......	......	S
Ga	31	6	20.57	30.7	64.2	......	......	......	......	S
Gd	64	6.16	12	......	......	......	......	......	......	S
Ge	32	7.88	15.93	34.21	44.7	93.4	......	......	......	S
H	1	13.595	......	......	......	......	......	......	......	S
He	2	24.481	54.403	......	......	......	......	......	......	S
Hf	72	7	14.9	23.2	33.3	......	......	......	......	S
Hg	80	10.43	18.751	34.2	49.5**	......	67**	......	......	S
I	53	10.454	19.13	......	......	......	......	......	170	S
In	49	5.785	18.86	28.03	54.4	......	......	......	......	S
Ir	77	9	......	......	......	......	......	......	......	S
K	19	4.339	31.81	46	60.9	82.6	99.7	118	155	S
Kr	36	13.996	24.56	36.9	43.5**	63**	94**	......	......	S
La	57	5.61	11.43	19.17	......	......	......	......	......	S
Li	3	5.39	75.619	122.419	......	......	......	......	......	S
Lu	71	......	14.7	......	......	......	......	......	......	S
Mg	12	7.644	15.031	80.14	109.29	141.23	186.49	224.9	265.957	S
Mn	25	7.432	15.636	33.69	52	76	......	119	196	S
Mo	42	7.10	16.15	27.13	46.4	61.2	68	126	153	S
N	7	14.53	29.593	47.426	77.45	97.863	551.925	666.83	......	S
Na	11	5.138	47.29	71.715	98.88	138.37	172.09	208.444	264.155	S
Nb(Cb)	41	6.88	14.32	25.04	38.3	50	103	125	......	S
Nd	60	5.51	......	......	......	......	......	......	......	SI
Ne	10	21.559	41.07	63.5	97.02	126.3	157.91	......	......	S
Ni	28	7.633	18.15	35.16	......	......	......	......	......	S
O	8	13.614	35.108	54.886	77.394	113.873	138.08	739.114	871.12	S
Os	76	8.5	17	......	......	......	......	......	......	S
P	15	10.484	19.72	30.156	51.354	65.007	220.414	263.31	309.26	S
Pb	82	7.415	15.028	31.93	42.31	68.8	......	......	......	S
Pd	46	8.33	19.42	32.92	......	......	......	......	......	S
Po	84	8.43	......	......	......	......	......	......	......	S
Pr	59	5.46	......	......	......	......	......	......	......	SI
Pt	78	9.0	18.56	......	......	......	......	......	......	S
Pu	94	5.1	......	......	......	......	......	......	......	S
Ra	88	5.277	10.144	......	......	......	......	......	......	S
Rb	37	4.176	27.5	40	......	......	......	......	......	S
Re	75	7.87	16.6	......	......	......	......	......	......	S
Rh	45	7.46	18.07	31.05	......	......	......	......	......	S
Rn	86	10.746	......	......	......	......	......	......	......	S
Ru	44	7.364	16.76	28.46	......	......	......	......	......	S
S	16	10.357	23.4	35	47.29	72.5	88.029	280.99	328.8	S
Sb	51	8.639	16.5	25.3	44.1	56	108	......	......	S
Sc	21	6.54	12.8	24.75	73.9	92	111	139	159	S
Se	34	9.75	21.5	32	43	68	82	155	......	S
Si	14	8.149	16.34	33.488	45.13	166.73	205.11	246.41	303.07	S
Sm	62	5.6	11.2	......	......	......	......	......	......	S
Sn	50	7.342	14.628	30.49	40.72	72.3	......	......	......	S
Sr	38	5.692	11.027	......	57	......	......	......	......	S
Ta	73	7.88	16.2	......	......	......	......	......	......	S
Tb	65	5.98	......	......	......	......	......	......	......	SI
Tc	43	7.28	15.26	29.54	......	......	......	......	......	S
Te	52	9.01	18.6	31	38	60	72	12.7	......	S
Th	90	6.95	......	......	29.38	......	......	......	......	SI
Ti	22	6.82	13.57	27.47	43.24	99.8	120	141	172	S
Tl	81	6.106	20.42	29.8	50.7	......	......	......	......	S
Tm	69	5.81	......	......	......	......	......	......	......	S
U	92	6.08	......	......	......	......	......	......	......	SI
V	23	6.74	14.65	29.31	48	65	129	151	170	S
W	74	7.98	17.7	......	......	......	......	......	......	S
Xe	54	12.127*	21.2	31.3	42	53	58	135	......	EI
Y	39	6.38	12.23	20.5	......	77	......	......	......	S
Yb	70	6.2	12.10	......	......	......	......	......	......	S
Zn	30	9.391	17.96	39.7	......	......	......	......	......	S
Zr	40	6.84	13.13	22.98	34.33	......	99	......	......	S

*These steps by S method.

**These steps by EI method.

IONIZATION POTENTIALS

Compounds

The first ionization potential of the molecules indicated is given in volts.

Compound	Ionization potential I volts	Compound	Ionization potential I volts
Br_2	12.8	F_2	17.8 (calc.)
BrCl	12.9 (calc.)	H_2	15.6
C_2	12	HBr	13.2
CH_2O, formaldehyde	11.3	HCN	14.8
CH_3Br, methyl bromide	10.0	HCl	13.8
CH_3Cl, methyl chloride	10.7	HF	17.7 (calc.)
CH_3I, methyl iodide	9.1	HI	12.8
CH_4, methane	14.5	H_2O	12.56
CN	14	H_2S	10.42
CO	14.1	I_2	9.7
CO_2	14.4	IBr	11.6 (calc.)
CS	10.6	ICl	11.9 (calc.)
CS_2	10.4	N_2	15.51
C_2H_2, acetylene	11.6	NH_3	11.2
C_2H_4, ethylene	12.2	NO	9.5
C_2H_6, ethane	12.8	NO_2	11.0
C_6H_6, benzene	9.6	N_2O	12.9
C_7H_8, toluene	8.5	O_2	12.5
Cl_2	13.2	S_2	10.7
		SO_2	13.1

NUCLEAR SPINS, MOMENTS, AND MAGNETIC RESONANCE FREQUENCIES

Kenneth Lee and Weston A. Anderson
1967

This table contains the published values for the nuclear spins, magnetic moments, and quadrupole moments, and the calculated values for the nuclear magnetic resonance (NMR) frequency and for the relative sensitivities. Only those isotopes with both published spin and magnetic moment values are tabulated. The magnetic and quadrupole moment values were selected from results published during the period from January, 1955 to June, 1967. Earlier references were obtained from H. E. Walchli, A Table of Nuclear Moment Data, U.S. Atomic Energy Commission Report ORNL—1469, Supplement I (1953) and Supplement II (1955), and D. Streminger, J. M. Hollander, and G. T. Seaborg, Table of Isotopes, Rev. Mod. Phys. **30**, 585 (1958). A table containing the known (1963) spin and electromagnetic moment values of nuclear ground and excited states has been compiled by I. Lindgren, Perturbed Angular Correlations; E. Karlsson, E. Mathias, and K. Siegbahn, editors; North-Holland Publishing Co. (1964). The magnetic moments given in this latter table are corrected for the diamagnetic effect. A more complete list of spin and moment results for nuclei in excited states are included in Lindgren's table.

In general, the results chosen for this table were selected with an inclination to NMR measurements and to the precision of the measurement. Only six significant figures are used in this table. Therefore, the number of figures may be less than those published. The experimental methods employed in determining the moments are indicated by the following symbols:

Ab = atomic beam magnetic resonance (hyperfine structure, double or triple resonance or other method)
E = electron spin resonance or electron-nuclear double resonance
M = microwave absorption in gases
Mb = molecular (or diamagnetic) beam magnetic resonance
Mc = miscellaneous
Mo = Mössbauer effect
N = nuclear magnetic resonance
No = nuclear orientation
O = optical spectroscopy (hyperfine structure, band structure, double resonance, or optical pumping)
Qr = quadrupole resonance

Other symbols used in the table are:

A = atomic weight (mass number)
El = element
I = nuclear spin in units of $h/2\pi$
μ = magnetic moment in units of the nuclear magneton $eh/4\pi Mc$
m = metastable excited state
Q = quadrupole moment in units of barns (10^{-24} cm²)
Z = atomic number
* = radioactive isotope
• = magnetic moment observed by NMR
( ) = assumed or estimated values

Assuming a nuclear magneton value of 5.0505×10^{-24} erg/gauss, the NMR frequency was calculated for a total field of 10^4 gauss. The sensitivities, relative to the proton, are calculated from the following expressions:

$$\text{Sensitivity at constant field} = 7.652 \times 10^{-3}\mu^3 (I+1)/I^2$$
$$\text{Sensitivity at constant frequency} = 0.2387\, \mu(I+1).$$

These expressions assume an equal number of nuclei, a constant temperature, and $T_1 = T_2$ (the longitudinal relaxation time equals the transverse relaxation time). These sensitivities represent the ideal induced voltage in the receiver coil at saturation and with a constant noise source. The calculated values are therefore determined under complete optimum conditions and should be regarded as such.

Isotope						Relative Sensitivity for Equal Number of Nuclei		Magnetic Moment μ			Electric Quadrupole Moment Q		
Z	El	A	Spin I	NMR Frequency in MHz for a 10 kilogauss field	Natural Abundance %	At constant field	At constant frequency	In multiples of the nuclear magneton (eh/4πMc)	Reference	Method	In multiples of barns (10^{-24}cm²)	Reference	Method
0	n	1*	1/2	29.167	—	0.322	0.685	−1.91315	1	Ab	—		
1	•H	1	1/2	42.5759	99.985	1.00	1.00	2.79268	2	N	—		
1	•H	2	1	6.53566	1.5×10^{-2}	9.65×10^{-3}	0.409	0.857387	3	N	2.73×10^{-3}	4	Ab
1	•H	3*	1/2	45.4129	—	1.21	1.07	2.97877	5	N	—		
2	•He	3	1/2	32.433	1.3×10^{-4}	0.442	0.762	−2.1274	6, 7	N	—		
3	•Li	6	1	6.2653	7.42	8.50×10^{-3}	0.392	0.82192	8	N	6.9×10^{-4}	9	N
3	•Li	7	3/2	16.546	92.58	0.293	1.94	3.2560	8	N	-3×10^{-2}	10	O
3	Li	8*	2	6.300	—	2.59×10^{-2}	1.184	1.653	11	Ab			
4	•Be	9	3/2	5.9834	100	1.39×10^{-2}	0.703	−1.1774	12, 13	N	5.2×10^{-2}	14	Ab
5	•B	10	3	4.5754	19.58	1.99×10^{-2}	1.72	1.8007	15	N	7.4×10^{-2}	16	Ab
5	•B	11	3/2	13.660	80.42	0.165	1.60	2.6880	17	N	3.55×10^{-2}	16	Ab
6	•C	13	1/2	10.7054	1.108	1.59×10^{-2}	0.251	0.702199	18	N	—		
7	N	13*	1/2	4.91	—	1.53×10^{-3}	0.115	(−)0.322	19	Ab	—		
7	•N	14	1	3.0756	99.63	1.01×10^{-3}	0.193	0.40347	20, 21	N	7.1×10^{-2}	22	Mc
7	•N	15	1/2	4.3142	0.37	1.04×10^{-3}	0.101	−0.28298	20, 21	N	—		
8	O	15*	1/2	11.0	—	1.70×10^{-2}	0.257	0.719	23	Ab	—		
8	•O	17	5/2	5.772	3.7×10^{-2}	2.91×10^{-2}	1.58	−1.8930	24	N	-2.6×10^{-2}	25	M
9	•F	17*	5/2	14.40	—	0.451	3.94	4.720	26	N			
9	•F	19	1/2	40.0541	100	0.833	0.941	2.62727	17, 21	N	—		
9	•F	20*	2	7.977	—	5.26×10^{-2}	1.50	2.093	27	N			
10	Ne	19*	(1/2)	28.75	—	0.308	0.675	−1.886	28	Ab			
10	Ne	21	3/2	3.3611	0.257	2.50×10^{-3}	0.395	−0.66140	29	Ab			
11	Na	21*	3/2	12.126	—	0.116	1.42	2.3861	30	Ab			
11	Na	22*	3	4.436	—	1.81×10^{-2}	1.67	1.746	31	Ab			
11	•Na	23	3/2	11.262	100	9.25×10^{-2}	1.32	2.2161	17	N	0.14-0.15	32, 33	O
11	Na	24*	4	3.221	—	1.15×10^{-2}	2.02	1.690	34	Ab			
12	•Mg	25	5/2	2.6054	10.13	2.67×10^{-3}	0.714	−0.85449	13	N			
13	•Al	27	5/2	11.094	100	0.206	3.04	3.6385	35	N	0.149	36	Ab
14	•Si	29	1/2	8.4578	4.70	7.84×10^{-3}	0.199	−0.55477	37	N	—		
15	•P	31	1/2	17.235	100	6.63×10^{-2}	0.405	1.1305	38	N	—		
15	P	32*	1	1.923	—	2.46×10^{-4}	0.120	−0.2523	39	E			
16	•S	33	3/2	3.2654	0.76	2.26×10^{-3}	0.383	0.64257	37	N	-6.4×10^{-2}	40	M
16	S	35*	3/2	5.08	—	8.50×10^{-3}	0.597	1.00	41	M	4.5×10^{-2}	40	M
17	•Cl	35	3/2	4.1717	75.53	4.70×10^{-3}	0.490	0.82091	8	N	-7.89×10^{-2}	42	Ab
17	•Cl	36*	2	4.8931	—	1.21×10^{-2}	0.919	1.2838	43	N	-1.72×10^{-2}	44	M
17	•Cl	37	3/2	3.472	24.47	2.71×10^{-3}	0.408	0.6833	45	N	-6.21×10^{-2}	42	Ab
18	Ar	37*	3/2	5.08	—	8.50×10^{-3}	0.597	1.0	240	O			
19	K	38*	3	3.491	—	8.82×10^{-3}	1.31	1.374	46	Ab			
19	•K	39	3/2	1.9868	93.10	5.08×10^{-4}	0.233	0.39097	47, 48	N	0.11	49, 50	O
19	K	40*	4	2.470	1.18×10^{-2}	5.21×10^{-3}	1.55	−1.296	51	Mb			
19	•K	41	3/2	1.0905	6.88	8.40×10^{-5}	0.128	0.21459	48	N			
19	K	42*	2	4.345	—	8.50×10^{-3}	0.816	−1.140	52	Ab			
19	K	43*	3/2	0.828	—	3.68×10^{-5}	9.73×10^{-2}	0.163	53	Ab			
20	•Ca	41*	7/2	3.4681	—	1.14×10^{-2}	1.71	−1.5924	54	N			
20	•Ca	43	7/2	2.8646	0.145	6.40×10^{-3}	1.41	−1.3153	55	N			
21	Sc	43*	7/2	10.04	—	0.275	4.95	4.61	56	Ab	−0.26	56	Ab
21	Sc	44*	2	9.76	—	9.63×10^{-2}	1.83	2.56	57	Ab	0.14	57	Ab
21	Sc	44^*_m	6	5.03	—	9.24×10^{-2}	6.62	3.96	57	Ab	0.37	57	Ab
21	•Sc	45	7/2	10.343	100	0.301	5.10	4.7492	58	N	−0.22	59	Ab
21	Sc	46*	4	5.77	—	6.65×10^{-2}	3.62	3.03	60	Ab	0.12	60	Ab
21	Sc	47*	7/2	11.6	—	0.426	5.73	5.33	56	Ab	−0.22	56	Ab
22	Ti	45*	7/2	0.207	—	2.40×10^{-6}	0.102	0.095	61	Ab	1.5×10^{-2}	61	Ab
22	•Ti	47	5/2	2.4000	7.28	2.09×10^{-3}	0.658	−0.78710	62	N			
22	•Ti	49	7/2	2.4005	5.51	3.76×10^{-3}	1.18	−1.1022	62	N			
23	V	49*	7/2	9.71	—	0.249	4.79	4.46	63	E			
23	•V	50	6	4.2450	0.24	5.55×10^{-2}	5.58	3.3413	8	N			
23	•V	51	7/2	11.19	99.76	0.382	5.52	5.139	64	N	-4×10^{-2}	65	O
24	•Cr	53	3/2	2.4065	9.55	9.03×10^{-4}	0.283	−0.47354	66	N			
25	Mn	52*	6	3.907	—	4.33×10^{-2}	5.14	3.075	67	Ab			
25	Mn	52^*_m	2	0.030	—	2.9×10^{-9}	5.73×10^{-3}	0.008	68	Ab			
25	Mn	53*	7/2	11.0	—	0.362	5.42	5.05	69	E			

NUCLEAR SPINS, MOMENTS, AND MAGNETIC RESONANCE FREQUENCIES (Continued)

Isotope			Spin I	NMR Frequency in MHz for a 10 kilogauss field	Natural Abundance %	Relative Sensitivity for Equal Number of Nuclei		Magnetic Moment μ			Electric Quadrupole Moment Q		
Z	El	A				At constant field	At constant frequency	In multiples of the nuclear magneton (eh/4πMc)	Reference	Method	In multiples of barns (10^{-24}cm²)	Reference	Method
25	Mn	54*	(2) (3)	8.4 6.6	—	6.11x10^{-2} 5.98x10^{-2}	1.58 2.48	(2.2) (2.6)	70	No			
25	•Mn	55	5/2	10.501	100	0.175	2.88	3.444	71	E	0.55	72	M
25	Mn	56*	3	8.233	—	0.116	3.09	3.240	73	Ab			
26	•Fe	57	1/2	1.3758	2.19	3.37x10^{-5}	3.23x10^{-2}	0.09024	74	E	—		
27	Co	55*	7/2	10.0	—	0.274	4.94	4.6	75	No			
27	Co	56*	4	7.34	—	0.136	4.60	3.85	76	E			
27	Co	57*	7/2	10.1	—	0.283	4.99	4.65	76	E			
27	Co	58*	2	15.4	—	0.381	2.90	4.05	77	E			
27	•Co	59	7/2	10.054	100	0.277	4.96	4.6163	241 242	N	0.40	78	Ab
27	Co	60*	5	5.793	—	0.101	5.44	3.800	79	E			
28	•Ni	61	3/2	3.8047	1.19	3.57x10^{-3}	0.447	−0.74868	80	N			
29	Cu	61*	3/2	10.8	—	8.22x10^{-2}	1.27	2.13	81	Ab			
29	•Cu	63	3/2	11.285	69.09	9.31x10^{-2}	1.33	2.2206	8	N	−0.16	82	E
29	Cu	64*	1	3.1	—	9.79x10^{-4}	0.191	0.40	83	Ab			
29	•Cu	65	3/2	12.089	30.91	0.114	1.42	2.3789	8	N	−0.15	82	E
29	Cu	66*	1	1.65	—	1.54x10^{-4}	0.103	−0.216	81	Ab			
30	Zn	65*	5/2	2.345	—	1.95x10^{-3}	0.643	0.7692	83	O	−2.4x10^{-2}	83	O
30	•Zn	67	5/2	2.663	4.11	2.85x10^{-3}	0.730	0.8733	37	N	0.15	83	O
31	Ga	68*	1	0.0892	—	2.45x10^{-8}	5.59x10^{-3}	0.0117	84	Ab	3.1x10^{-2}	84	Ab
31	Ga	69	3/2	10.22	60.4	6.91x10^{-2}	1.20	2.011	8	N	0.178	85	Ab
31	Ga	71	3/2	12.984	39.6	0.142	1.52	2.5549	8	N	0.112	85	Ab
31	Ga	72*	3	0.33591	—	7.80x10^{-6}	0.126	−0.13220	86	Ab	0.72	86	Ab
32	Ge	71*	1/2	8.4	—	7.64x10^{-3}	0.197	0.55	87	Ab	—		
32	•Ge	73	9/2	1.4852	7.76	1.40x10^{-3}	1.15	−0.87679	62	N	−0.2	88	M
33	•As	75	3/2	7.2919	100	2.51x10^{-2}	0.856	1.4349	15	N	0.3	89	M
33	As	76*	2	3.45	—	4.27x10^{-3}	0.649	−0.906	90	No			
34	•Se	77	1/2	8.118	7.58	6.93x10^{-3}	0.191	0.5325	8	N	—		
34	Se	79	7/2	2.22	—	2.98x10^{-3}	1.10	−1.02	91	M	0.9	40	M
35	Br	76*	1	4.18	—	2.52x10^{-3}	0.262	(−)0.548	92	Ab	0.27	92	Ab
35	•Br	79	3/2	10.667	50.54	7.86x10^{-2}	1.25	2.0990	8	N	0.33	93	Ab
35	Br	80*	1	3.92	—	2.08x10^{-3}	0.245	0.514	94	Ab	0.20	94	Ab
35	Br	80m*	5	2.008	—	4.20x10^{-3}	1.89	1.317	94	Ab	0.76	94	Ab
35	•Br	81	3/2	11.498	49.46	9.85x10^{-2}	1.35	2.2626	8	N	0.28	93	Ab
35	Br	82*	5	2.479	—	7.90x10^{-3}	2.33	(+)1.626	95	Ab	(+)0.76	95	Ab
36	•Kr	83	9/2	1.638	11.55	1.88x10^{-3}	1.27	−0.9671	96	N	0.15	97	O
36	Kr	85*	9/2	1.6956	—	2.08x10^{-3}	1.31	−1.001	98	O	0.25	98	O
37	Rb	81*	3/2	10.4	—	7.32x10^{-2}	1.22	2.05	99	Ab			
37	Rb	82m*	5	2.29	—	6.20x10^{-3}	2.15	1.50	99	Ab			
37	Rb	83*	5/2	4.33	—	1.23x10^{-2}	1.19	1.42	99	Ab			
37	Rb	84*	2	5.03	—	1.32x10^{-2}	0.945	−1.32	99	Ab			
37	•Rb	85	5/2	4.1108	72.15	1.05x10^{-2}	1.13	1.3482	100	N	0.27	101	Ab
37	Rb	86*	2	6.44	—	2.77x10^{-2}	1.21	−1.69	102	Ab			
37	•Rb	87	3/2	13.931	27.85	0.175	1.64	2.7414	103	N	0.13	101	Ab
38	•Sr	87	9/2	1.8452	7.02	2.69x10^{-3}	1.43	−1.0893	66	N	0.2	104	E
39	•Y	89	1/2	2.0859	100	1.18x10^{-4}	4.90x10^{-2}	−0.13682	47	N	—		
39	Y	90*	2	6.17	—	2.44x10^{-2}	1.16	−1.62	105	Ab	−0.16	105	Ab
39	Y	91*	1/2	2.49	—	1.99x10^{-4}	5.84x10^{-2}	0.163	60	Ab	—		
40	•Zr	91	5/2	3.97249	11.23	9.48x10^{-3}	1.09	−1.30284	106	N			
41	•Nb	93	9/2	10.407	100	0.482	8.07	6.1435	107	N	−0.2	108	O
42	•Mo	95	5/2	2.774	15.72	3.23x10^{-3}	0.760	0.9097	45	N	0.12	109	N
42	•Mo	97	5/2	2.832	9.46	3.43x10^{-3}	0.776	−0.9289	45	N	1.1	109	N
43	•Tc	99*	9/2	9.5830	—	0.376	7.43	5.6572	110	N	0.3	111	O
44	Ru	99	3/2	1.44	12.72	1.95x10^{-4}	0.169	−0.284	112	Mo			
44	Ru	101	5/2	2.1	17.07	1.41x10^{-3}	0.576	−0.69	113	O			
45	•Rh	103	1/2	1.3401	100	3.11x10^{-5}	3.15x10^{-2}	−0.08790	43 114	N	—		
46	•Pd	105	5/2	1.95	22.23	1.12x10^{-3}	0.534	−0.639	115	N			
47	Ag	104*	5	6.1	—	0.118	5.73	4.0	116	Ab			
47	Ag	104m*	2	14.0	—	0.291	2.65	3.7	116	Ab			
47	Ag	105*	1/2	1.54	—	4.73x10^{-5}	3.62x10^{-2}	0.101	117	Ab	—		
47	•Ag	107	1/2	1.7229	51.82	6.62x10^{-5}	4.05x10^{-2}	−0.11301	47 118	N	—		
47	Ag	108*	1	32.0	—	1.13	2.01	4.2	119	Ab			
47	•Ag	109	1/2	1.9807	48.18	1.01x10^{-4}	4.65x10^{-2}	−0.12992	47 118	N	—		
47	Ag	110m*	6	4.557	—	6.87x10^{-2}	5.99	3.587	120	Ab			
47	Ag	111*	1/2	2.21	—	1.40x10^{-4}	5.19x10^{-2}	−0.145	121	Ab	—		
47	Ag	112*	2	0.2077	—	9.00x10^{-7}	3.90x10^{-2}	0.0545	122	Ab			
47	Ag	113*	1/2	2.41	—	1.81x10^{-4}	5.66x10^{-2}	0.158	122	Ab	—		
48	Cd	107*	5/2	1.879	—	1.00x10^{-3}	0.515	−0.6162	123	O	0.8	123	O
48	Cd	109*	5/2	2.529	—	2.44x10^{-3}	0.693	−0.8293	124	O	0.8	124	O
48	•Cd	111	1/2	9.028	12.75	9.54x10^{-3}	0.212	−0.5922	125 126	N	—		
48	•Cd	113	1/2	9.445	12.26	1.09x10^{-2}	0.222	−0.6195	125 126	N	—		
48	Cd	113m*	11/2	1.51	—	2.13x10^{-3}	1.69	−1.09	239	O	−0.79	239	O
48	Cd	115*	1/2	9.862	—	1.24x10^{-2}	0.232	−0.6469	127	O	—		
48	Cd	115m*	11/2	1.447	—	1.87x10^{-3}	1.62	−1.044	127	O	−0.61	127	O
49	•In	113	9/2	9.3099	4.28	0.345	7.22	5.4960	45	N	1.14	97	Ab
49	In	113m*	1/2	3.209	—	4.28x10^{-4}	7.54x10^{-2}	−0.2105	128	Ab	—		
49	In	114m*	5	7.2	—	0.191	6.73	4.7	129	Ab			
49	•In	115*	9/2	9.3301	95.72	0.347	7.23	5.5079	130	N	1.16	97	Ab
49	In	115m*	1/2	3.715	—	6.64x10^{-4}	8.73x10^{-2}	−0.2437	131	Ab	—		
49	In	116*	5	6.42	—	0.137	6.03	4.21	132	Ab			
49	In	116m*	5	6.7	—	0.156	6.30	4.4	129	Ab			
50	•Sn	115	1/2	13.922	0.35	3.50x10^{-2}	0.327	−0.91320	125	N	—		
50	•Sn	117	1/2	15.168	7.61	4.52x10^{-2}	0.356	−0.99490	125	N	—		
50	•Sn	119	1/2	15.869	8.58	5.18x10^{-2}	0.373	−1.0409	125	N	—		
51	•Sb	121	5/2	10.189	57.25	0.160	2.79	3.3415	133	N	−0.5	134	O
51	Sb	122*	2	7.24	—	3.94x10^{-2}	1.36	−1.90	135 136	No			
51	•Sb	123	7/2	5.5176	42.75	4.57x10^{-2}	2.72	2.5334	45	N	−0.7	134	O
52	Te	119*	1/2	4.12	—	9.04x10^{-4}	9.67x10^{-2}	0.27	137	Ab	—		
52	•Te	123	1/2	11.16	0.87	1.8x10^{-2}	0.262	−0.7319	37	N	—		
52	•Te	125	1/2	13.45	6.99	3.15x10^{-2}	0.316	−0.8824	37	N	—		
53	I	125*	5/2	9.0	—	0.116	2.51	3	138	M	−0.66	138	M
53	•I	127	5/2	8.5183	100	9.34x10^{-2}	2.33	2.7937	139	N	−0.69	134	O
53	•I	129*	7/2	5.6694	—	4.96x10^{-2}	2.80	2.6031	139	N	−0.48	140	Qr
53	I	131*	7/2	5.963	—	5.77x10^{-2}	2.94	2.738	141	Ab	−0.41	141	Ab
54	•Xe	129	1/2	11.777	26.44	2.12x10^{-2}	0.277	−0.77247	142	N	—		
54	•Xe	131	3/2	3.4911	21.18	2.76x10^{-3}	0.410	0.68697	142	N	−0.12	143	O
55	Cs	127*	1/2	21.8	—	0.134	0.512	1.43	144	Ab	—		
55	Cs	129*	1/2	22.4	—	0.146	0.526	1.47	144	Ab	—		
55	Cs	130*	1	10.7	—	4.20x10^{-2}	0.668	1.4	144	Ab			
55	Cs	131*	5/2	10.72	—	0.186	2.94	3.517	145	Ab			
55	Cs	132*	2	8.46	—	6.28x10^{-2}	1.59	2.22	144	Ab			
55	•Cs	133	7/2	5.58469	100	4.74x10^{-2}	2.75	2.56422	8	N	−3x10^{-3}	146 147	Ab
55	Cs	134*	4	5.666	—	6.28x10^{-2}	3.55	2.973	148	Ab	0.43	149	O
55	Cs	134m*	8	1.0447	—	1.42x10^{-3}	2.36	1.0964	150	Ab			
55	Cs	135*	7/2	5.9096	—	5.62x10^{-2}	2.91	2.7134	148	Ab			
55	Cs	137*	7/2	6.1459	—	6.32x10^{-2}	3.03	2.8219	148	Ab			
56	•Ba	135	3/2	4.2296	6.59	4.90x10^{-3}	0.497	0.83229	151 152	N	0.25	153	O
56	•Ba	137	3/2	4.7315	11.32	6.86x10^{-3}	0.556	0.93107	151 152	N	0.2	153	O
57	•La	138	5*	5.6171	0.089	9.19x10^{-2}	5.28	3.6844	154	N	2.7	154	N
57	•La	139	7/2	6.0144	99.911	5.92x10^{-2}	2.97	2.7615	107	N	0.21	155	Ab
58	Ce	137*	3/2	4.6	—	6.20x10^{-3}	0.537	0.9	156	No			
58	Ce	137m*	11/2	0.96	—	5.40x10^{-4}	1.07	0.69	157	No			
58	Ce	139*	3/2	5.1	—	8.50x10^{-3}	0.597	1.0	156	No			
58	Ce	141*	7/2	2.1	—	2.57x10^{-3}	1.04	0.97	158	E			

NUCLEAR SPINS, MOMENTS, AND MAGNETIC RESONANCE FREQUENCIES (Continued)

Isotope			Spin I	NMR Frequency in MHz for a 10 kilogauss field	Natural Abundance %	Relative Sensitivity for Equal Number of Nuclei		Magnetic Moment μ			Electric Quadrupole Moment Q		
Z	El	A				At constant field	At constant frequency	In multiples of the nuclear magneton (eh/4πMc)	Reference	Method	In multiples of barns ($10^{-24}cm^2$)	Reference	Method
58	Ce	143*	7/2	2.2	—	2.81×10^{-3}	1.07	1.0	156	No			
59	•Pr	141	5/2	12.5	100	0.293	3.42	4.09	159 160	O	-5.9×10^{-2}	158	Ab
59	Pr	142*	2	1.1	—	1.55×10^{-4}	0.215	0.30	161	Ab	4×10^{-2}	161	Ab
60	Nd	143	7/2	2.315	12.17	3.38×10^{-3}	1.14	−1.063	162	Ab	−0.48	162	Ab
60	Nd	145	7/2	1.42	8.30	7.86×10^{-4}	0.703	−0.654	162	Ab	−0.25	162	Ab
60	Nd	147*	5/2	1.77	—	8.32×10^{-4}	0.484	0.579	158	E			
61	Pm	143*	(5/2) (7/2)	11.6 8.5	—	0.235 0.167	3.17 4.19	(3.8) (3.9)	163	No			
61	Pm	144*	(5) (6)	2.6 2.3	—	9.02×10^{-3} 8.68×10^{-3}	2.43 3.01	(1.7) (1.8)	164	No			
61	Pm	147*	7/2	5.62	—	4.83×10^{-2}	2.77	2.58	165	O	0.7	165	O
61	Pm	148*	1	16	—	0.142	1.00	2.1	166	Ab	0.2	166	Ab
61	Pm	148^{*}_{m}	6	2.3	—	8.68×10^{-3}	3.01	1.8	163	No			
61	Pm	149*	7/2	7.2	—	0.101	3.54	3.3	163	No			
61	Pm	151*	5/2	5.5	—	2.50×10^{-2}	1.50	1.8	167	Ab	1.9	167	Ab
62	Sm	147	7/2	1.76	14.97	1.48×10^{-3}	0.867	−0.807	168	Ab	−0.208	158	Ab
62	Sm	149	7/2	1.40	13.83	7.47×10^{-4}	0.691	−0.643	158	Ab	6.0×10^{-2}	158	Ab
63	•Eu	151	5/2	10.559	47.82	0.178	2.89	3.4630	169	Ab	1.16	170	O
63	Eu	152*	3	4.858	—	2.38×10^{-2}	1.83	1.912	171	Ab			
63	•Eu	153	5/2	4.6627	52.18	1.53×10^{-2}	1.28	1.5292	169 172	Ab	2.9	170	O
63	Eu	154*	3	5.084	—	2.72×10^{-2}	1.91	2.001	158	E			
64	•Gd	155	3/2	1.6	14.73	2.79×10^{-4}	0.191	−0.32	173 174	O	1.6	173	O
64	•Gd	157	3/2	2.0	15.68	5.44×10^{-4}	0.239	−0.40	173 174	O	2	173	O
65	Tb	156*	3	3.8	—	1.15×10^{-2}	1.43	1.5	175	No	1.4	175	No
65	Tb	159	3/2	9.66	100	5.83×10^{-2}	1.13	1.90	158	E	1.3	176	O
65	Tb	160*	3	4.1	—	1.39×10^{-2}	1.53	1.6	177	No	1.9	177	No
66	Dy	155*	(3/2)	1.1	—	7.87×10^{-5}	0.125	0.21	178	No			
66	Dy	157*	(3/2)	1.6	—	2.79×10^{-4}	0.191	0.32	178	No			
66	Dy	161	5/2	1.4	18.88	4.17×10^{-4}	0.384	−0.46	158	E	1.4	158	E
66	Dy	163	5/2	2.0	24.97	1.12×10^{-3}	0.535	0.64	158	E	1.6	158	E
67	Ho	165	7/2	8.73	100	0.181	4.31	4.01	158	Ab	2.82	158	Ab
68	Er	165*	5/2	2.0	—	1.18×10^{-3}	0.543	0.65	166	Ab	2.2	166	Ab
68	Er	167	7/2	1.23	22.94	5.07×10^{-4}	0.607	−0.565	162	Ab	2.83	162	Ab
68	Er	169*	1/2	7.8	—	6.09×10^{-3}	0.183	0.51	179	Ab	—		
68	Er	171*	5/2	2.1	—	1.47×10^{-3}	0.585	0.70	180	Ab			
69	Tm	166*	2	0.19	—	7.17×10^{-7}	3.58×10^{-2}	0.05	181	Ab	4.6	181	Ab
69	•Tm	169	1/2	3.52	100	5.66×10^{-4}	8.27×10^{-2}	−0.231	182 160	Ab	—		
69	Tm	170*	1	2.0	—	2.69×10^{-4}	0.124	0.26	183	Ab	0.61	183	Ab
69	Tm	171*	1/2	3.46	—	5.37×10^{-4}	8.13×10^{-2}	0.227	180	Ab	—		
70	•Yb	171	1/2	7.4990	14.31	5.46×10^{-3}	0.176	0.49188	184 185	O	—		
70	Yb	173	5/2	2.0659	16.13	1.33×10^{-3}	0.566	−0.67755	184 185	O	2.8	186	O
70	Yb	175*	(7/2)	0.33	—	9.40×10^{-6}	0.161	−0.15	187	No			
71	•Lu	175	7/2	4.86	97.41	3.12×10^{-2}	2.40	2.23	188	N	5.68	189	Ab
71	Lu	176*	7	3.4	2.59	3.72×10^{-2}	5.92	3.1	190	Ab	8.0	190	Ab
71	Lu	177*	7/2	4.84	—	3.08×10^{-2}	2.38	2.22	191	Ab	5.51	191	Ab
72	Hf	177	7/2	1.3	18.50	6.38×10^{-4}	0.655	0.61	192 193	O	3	192 193	O
72	Hf	179	9/2	0.80	13.75	2.16×10^{-4}	0.617	−0.47	193	O	3	193	O
73	•Ta	181	7/2	5.096	99.988	3.60×10^{-2}	2.51	2.340	194	N	3	195	O
74	•W	183	1/2	1.7716	14.40	7.20×10^{-5}	4.16×10^{-2}	0.116205	196	N	—		
75	•Re	185	5/2	9.5855	37.07	0.133	2.63	3.1437	13	N	2.8	97 241	O
75	Re	186*	1	13.17	—	7.90×10^{-2}	0.825	1.728	197	Ab			
75	•Re	187*	5/2	9.6837	62.93	0.137	2.65	3.1759	13	N	2.6	97 241	O
75	Re	188*	1	13.55	—	8.59×10^{-2}	0.848	1.777	197	Ab			
76	•Os	187	1/2	0.98059	1.64	1.22×10^{-5}	2.30×10^{-2}	0.06432	198	N	—		
76	•Os	189	3/2	3.3034	16.1	2.34×10^{-3}	0.388	0.65004	199	N	0.8	195	O
77	•Ir	191	3/2	0.7318	37.3	2.53×10^{-5}	8.59×10^{-2}	0.1440	200 201	N	1.5	202	O
77	•Ir	193	3/2	0.7968	62.7	3.27×10^{-5}	9.36×10^{-2}	0.1568	200 201	N	1.5	202	O
78	•Pt	195	1/2	9.153	33.8	9.94×10^{-3}	0.215	0.6004	45	N	—		
79	Au	190*	1	0.496	—	4.20×10^{-6}	3.10×10^{-2}	0.065	203	Ab			
79	Au	194*	1	0.56	—	5.95×10^{-6}	3.49×10^{-2}	0.073	204	Ab			
79	Au	195*	3/2	0.742	—	2.65×10^{-5}	8.71×10^{-2}	0.146	204	Ab			
79	Au	196*	2	2.3	—	1.24×10^{-3}	0.430	0.6	204	Ab			
79	•Au	197	3/2	0.729188	100	2.51×10^{-5}	8.56×10^{-2}	0.143489	205 201	Ab	0.59	206	Ab
79	Au	198*	2	2.227	—	1.14×10^{-3}	0.418	0.5842	207	Ab			
79	Au	199*	3/2	1.358	—	1.62×10^{-4}	0.160	0.2673	207	Ab			
80	Hg	193*	3/2	3.1	—	1.93×10^{-3}	0.364	−0.61	208	O			
80	Hg	193^{*}_{m}	13/2	1.23	—	1.57×10^{-3}	1.88	−1.05	209	O	1.37	209	O
80	Hg	195*	1/2	8.1	—	6.84×10^{-3}	0.190	0.53	210 211	O	—		
80	Hg	195^{*}_{m}	13/2	1.22	—	1.53×10^{-3}	1.86	−1.04	209	O	1.41	209	O
80	Hg	197*	1/2	7.9	—	6.46×10^{-3}	0.186	0.52	212	O	—		
80	•Hg	199	1/2	7.59012	16.84	5.67×10^{-3}	0.178	0.497859	213 45	No			
80	•Hg	201	3/2	2.8099	13.22	1.44×10^{-3}	0.330	−0.55293	214 215	O	0.50	216	O
80	Hg	203*	5/2	2.5	—	2.45×10^{-3}	0.693	0.83	217	O	0.5	217	O
81	Tl	197*	1/2	23.6	—	0.171	0.555	1.55	218	O	—		
81	Tl	199*	1/2	23.9	—	0.178	0.562	1.57	219	O	—		
81	Tl	200*	2	0.57	—	1.94×10^{-5}	0.107	(0.15)	219	O			
81	Tl	201*	1/2	24.1	—	0.181	0.566	1.58	219	O	—		
81	Tl	202*	2	0.57	—	1.94×10^{-5}	0.107	(0.15)	219	O			
81	•Tl	203	1/2	24.332	29.50	0.187	0.571	1.5960	220	N	—		
81	Tl	204*	2	0.34	—	4.05×10^{-6}	6.37×10^{-2}	0.089	221	Ab			
81	•Tl	205	1/2	24.570	70.50	0.192	0.577	1.6116	8	N	—		
82	•Pb	207	1/2	8.90771	22.6	9.16×10^{-3}	0.209	0.584284	222	N	—		
83	Bi	203*	9/2	7.78	—	0.201	6.03	4.59	223	Ab	−0.64	223	Ab
83	Bi	204*	6	5.40	—	0.114	7.10	4.25	223	Ab	−0.41	223	Ab
83	Bi	205*	9/2	9.3	—	0.346	7.22	(5.5)	223	Ab			
83	Bi	206*	6	5.79	—	0.141	7.62	4.56	223	Ab	−0.19	223	Ab
83	•Bi	209*	9/2	6.84178	100	0.137	5.30	4.03896	224	N	−0.4	97	O
83	Bi	210*	1	0.337	—	1.32×10^{-6}	2.11×10^{-2}	0.0442	225	Ab	0.13	225	Ab
84	Po	205*	5/2	0.79	—	7.55×10^{-5}	0.217	0.26	226	Ab	0.17	226	Ab
84	Po	207*	5/2	0.82	—	8.43×10^{-5}	0.226	0.27	226	Ab	0.28	226	Ab
89	Ac	227*	3/2	5.6	—	1.13×10^{-2}	0.656	1.1	227	O	−1.7	227	O
90	Th	229*	5/2	1.2	—	2.74×10^{-4}	0.334	0.4	228	O	4.6	228	O
91	Pa	231*	3/2	9.96	—	6.40×10^{-2}	1.17	1.96	229	E			
91	Pa	233*	3/2	17	—	0.334	2.03	3.4	230	Ab	−3.0	230	Ab
92	U	233*	5/2	1.6	—	6.75×10^{-4}	0.451	0.54	231	E	3.5	231	E
92	U	235*	7/2	0.76	0.72	1.21×10^{-4}	0.376	0.35	231	E	4.1	231	E
93	Np	237*	5/2	18	—	0.926	5.01	(6)	232	E			
94	Pu	239*	1/2	3.05	—	3.67×10^{-4}	7.16×10^{-2}	0.200	233	Ab	—		
94	Pu	241*	5/2	2.09	—	1.38×10^{-3}	0.573	−0.686	234	O			
95	Am	241*	5/2	4.82	—	1.69×10^{-2}	1.32	1.58	235	Ab	4.9	236	O
95	Am	242*	1	2.90	—	8.46×10^{-4}	0.182	0.381	235	Ab	−2.8	237	Ab
95	Am	243*	5/2	4.79	—	1.66×10^{-2}	1.31	1.57	238	O	4.9	236	O
Free electron with g=2.00232			1/2	2.80246×10^{4}	—	2.84×10^{-8}	657	−1836.09	—		—		

NUCLEAR SPINS, MOMENTS, AND MAGNETIC RESONANCE FREQUENCIES (Continued)

REFERENCES

(1) V. W. Cohen, et al., Phys. Rev. **104**, 283 (1956).
(2) H. Sommer, et al., Phys. Rev. **82**, 697 (1951).
(3) T. F. Wimett, Phys. Rev. **91**, 499A (1953).
(4) H. Kopfermann, et al., Z. Physik **144**, 9 (1956).
(5) F. Bloch, et al., Phys. Rev. **71**, 551 (1947).
(6) H. L. Anderson, Phys. Rev. **76**, 1460 (1949).
(7) M. P. Klein, et al., Phys. Rev. **106**, 837 (1957).
(8) H. E. Walchli, Thesis, M.S., U. of Tenn. (1954). AEC Report ORNL-1775.
(9) N. A. Schuster, et al., Phys. Rev. **81**, 157 (1951).
(10) K. C. Brog, et al., Phys. Rev. **153**, 91 (1967).
(11) D. Connor, Phys. Rev. Letters **3**, 429 (1959).
(12) L. C. Brown, et al., J. Chem. Phys. **24**, 751 (1956).
(13) F. Alder, et al., Phys. Rev. **82**, 105 (1951).
(14) A. G. Blachman, et al., Bull Am. Phys. Soc. **11**, 343 (1966).
(15) Y. Ting, et al., Phys. Rev. **89**, 595 (1953).
(16) G. Wessel, Phys. Rev. **92**, 1581 (1953).
(17) G. Lindström, Arkiv Fysik **4**, 1 (1951).
(18) V. Royden, Phys. Rev. **96**, 543 (1954).
(19) A. M. Bernstein, et al., Phys. Rev. **136**, B27 (1964).
(20) L. W. Anderson, et al., Phys. Rev. **116**, 87 (1959).
(21) M. R. Baker, et al., Phys. Rev. **133**, A1533 (1964).
(22) A. Bassompiere, Compt. Rend. **240**, 285 (1955).
(23) E. D. Commins, et al., Phys. Rev. **131**, 700 (1963).
(24) F. Alder, et al., Phys. Rev. **81**, 1067 (1951).
(25) M. J. Stevenson, et al., Phys. Rev. **107**, 635 (1957).
(26) K. Sugimoto, et al., J. Phys. Soc. Japan **21**, 213 (1966).
(27) T. Tsang, et al., Phys. Rev. **132**, 1141 (1963).
(28) E. D. Commins, et al., Phys. Rev. Letters **10**, 347 (1963).
(29) J. T. LaTourette, et al., Phys. Rev. **107**, 1202 (1957).
(30) O. Ames, et al., Phys. Rev. **137**, B1157 (1965).
(31) L. Davis, et al., Phys. Rev. **76**, 1068 (1949).
(32) H. Ackermann, Z. Physik. **194**, 253 (1966).
(33) M. Baumann, et al., Z. Physik **194**, 270 (1966).
(34) Y. W. Chan, et al., Phys. Rev. **150**, 933 (1966).
(35) L. C. Brown, et al., J. Chem. Phys. **24**, 751 (1956).
(36) H. Lew, et al., Phys. Rev. **90**, 1 (1953).
(37) H. E. Weaver, Phys. Rev. **89**, 923 (1953).
(38) T. Kanda, et al., Phys. Rev. **85**, 938 (1952).
(39) G. Feher, et al., Phys. Rev. **107**, 1462 (1957).
(40) G. R. Bird, et al., Phys. Rev. **94**, 1203 (1954).
(41) B. F. Burke, et al., Phys. Rev. **93**, 193 (1954).
(42) V. Jaccarino, et al., Phys. Rev. **83**, 471 (1951).
(43) P. B. Sogo, et al., Phys. Rev. **98**, 1316 (1955).
(44) C. H. Townes, et al., Phys. Rev. **76**, 691 (1949).
(45) W. G. Proctor, et al., Phys. Rev. **81**, 20 (1951).
(46) E. A. Phillips, et al., Phys. Rev. **138**, B773 (1965).
(47) E. Brun, et al., Phys. Rev. **93**, 172 (1954).
(48) O. Lutz, et al., Phys. Letters **24A**, 122 (1967).
(49) G. W. Series, Phys. Rev. **105**, 1128 (1957).
(50) G. J. Ritter, et al., Proc. Roy. Soc. (London) **238**, 473 (1957).
(51) J. T. Eisinger, et al., Phys. Rev. **86**, 73 (1952).
(52) J. M. Kahn, et al., Phys. Rev. **134**, A45 (1964).
(53) F. R. Petersen, et al., Phys. Rev. **116**, 734 (1959).
(54) E. Brun, et al., Phys. Rev. Letters **9**, 166 (1962).
(55) C. D. Jeffries, Phys. Rev. **90**, 1130 (1953).
(56) R. G. Cornwall, et al., Phys. Rev. **141**, 1106 (1966).
(57) D. L. Harris, et al., Phys. Rev. **132**, 310 (1963).
(58) D. M. Hunten, Can. J. Phys. **29**, 463 (1951).
(59) G. Fricke, et al., Z. Physik, **156**, 416 (1959).
(60) F. R. Petersen, et al., Phys. Rev. **128**, 1740 (1962).
(61) R. G. Cornwall, et al., Phys. Rev. **148**, 1157 (1966).
(62) C. D. Jeffries, Phys. Rev. **92**, 1262 (1953).
(63) M. M. Weiss, et al., Bull. Am. Phys. Soc. **2**, 31 (1957).
(64) H. E. Walchli, et al., Phys. Rev. **87**, 541 (1952).
(65) K. Murakawa, et al., J. Phys. Soc. Japan **21**, 1466 (1966).
(66) C. D. Jeffries, et al., Phys. Rev. **91**, 1286 (1953).
(67) K. E. Adelroth, et al., Arkiv Fysik **31**, 549 (1966).
(68) E. A. Phillips, et al., Phys. Rev. **140**, B555 (1965).
(69) W. Dobrowolski, et al., Phys. Rev. **104**, 1378 (1956).
(70) R W. Bauer, et al., Phys. Rev. **120**, 946 (1960).
(71) W. B. Mims, et al., Phys. Letters **24A**, 481 (1967).
(72) A. Javan, et al., Phys. Rev. **96**, 649 (1954).
(73) W. J. Childs, et al., Phys. Rev. **122**, 891 (1961).
(74) P. R. Locher, et al., Phys. Rev. **139**, A991 (1965).
(75) R. W. Bauer, et al., Nucl. Phys. **16**, 264 (1960).
(76) J. M. Baker, et al., Proc. Phys. Soc. (London), **A69**, 354 (1956).
(77) W. Dobrov, et al., Phys. Rev. **108**, 60 (1957).
(78) D. V. Ehrenstein, et al., Z. Physik **159**, 230 (1960).
(79) W. Dobrowolski, et al., Phys. Rev. **101**, 1001 (1956).
(80) L. E. Drain, Phys. Letters **11**, 114 (1964).
(81) B. M. Dodsworth, et al., Phys. Rev. **142**, 638 (1966).
(82) B. Bleaney, et al., Proc. Roy. Soc. (London) **228A**, 166 (1955).
(83) A Lemonick, et al., Phys. Rev. **95**, 1356 (1954).
(84) V. J. Ehlers, et al., Phys. Rev. **127**, 529 (1962).
(85) R. T. Daly, et al., Phys. Rev. **96**, 539 (1954).
(86) W. J. Childs, et al., Phys. Rev. **120**, 2138 (1960).
(87) W. J. Childs, et al., Phys. Rev. **141**, 15 (1966).
(88) J. M. Mays, et al., Phys. Rev. **81**, 940 (1951).
(89) B. P. Dailey, et al., Phys. Rev. **74**, 1245 (1948).
(90) F. M. Pipkin, et al., Phys. Rev. **106**, 1102 (1957).
(91) W. A. Hardy, et al., Phys. Rev. **92**, 1532 (1953).
(92) E. Lipworth, et al., Phys. Rev. **119**, 1053 (1960).
(93) J. G. King, et al., Phys. Rev. **94**, 1610 (1954).
(94) M. B. White, et al., Phys. Rev. **136**, B584 (1964).
(95) H. L. Garvin, et al., Phys. Rev. **116**, 393 (1959).
(96) E. Brun, et al., Helv. Phys. Acta **27**, 173A (1954).
(97) J. E. Mack, Rev. Mod. Phys. **22**, 64 (1950).
(98) E. Rasmussen, et al., Z. Physik **141**, 160 (1955).
(99) J. C. Hubbs, et al., Phys. Rev. **107**, 723 (1957).
(100) H. Walchli, et al., Phys. Rev. **85**, 922 (1952).
(101) B. Senitzky, et al., Phys. Rev. **103**, 315 (1956).
(102) E. H. Bellamy, et al., Phil. Mag. **44**, 33 (1953).
(103) E. Yasaitis, et al., Phys. Rev. **82**, 750 (1951).
(104) J. W. Culvahouse, et al., Phys. Rev. **140**, A1181 (1965).
(105) F. R. Petersen, et al., Phys. Rev. **125**, 284 (1962).
(106) E. Brun, et al., Phys. Rev. **105**, 1929 (1957).
(107) R. E. Sheriff, et al., Phys. Rev. **82**, 651 (1951).
(108) K. Murakawa, Phys. Rev. **98**, 1285 (1955).
(109) A. Narath, et al., Phys. Rev. **143**, 328 (1966).
(110) H. Walchli, et al., Phys. Rev. **85**, 479 (1952).
(111) K. G. Kessler, et al., Phys. Rev. **92**, 303 (1953).
(112) E. Matthias, et al., Phys. Rev. **139**, B532 (1965).
(113) K. Murakawa, J. Phys. Soc. Japan **10**, 919 (1955).
(114) J. A. Seitchik, et al., Phys. Rev. **138**, A148 (1965).
(115) J. A. Seitchik, et al., Phys. Rev. **136**, A1119 (1964).
(116) O. Ames, et al., Phys. Rev. **123**, 1793 (1960).
(117) W. B. Ewbank, et al., Phys. Rev. **129**, 1617 (1963).
(118) P. B. Sogo, et al., Phys. Rev. **93**, 174 (1954).
(119) G. K. Rochester, et al., Phys. Letters 8, **266** (1964).
(120) S. G. Schmelling, et al., Phys. Rev. **154**, 1142 (1967).
(121) G. K. Woodgate, et al., Proc. Phys. Soc. (London) **69**, 581 (1956).
(122) Y. W. Chan, et al., Phys. Rev. **133**, B1138 (1964).
(123) F. W. Byron, et al., Phys. Rev. **132**, 1181 (1963).
(124) P. Thaddeus, et al., Phys. Rev. **132**, 1186 (1963).
(125) W. G. Proctor, Phys. Rev. **79**, 35 (1950).
(126) M. Leduc, et al., Compt. Rend. **B262**, 736 (1966).
(127) M. N. McDermott, et al., Phys. Rev. **134**, B25 (1964).
(128) W. J. Childs, et al., Phys. Rev. **118**, 1578 (1960).
(129) L. S. Goodman, et al., Phys. Rev. **108**, 1524 (1957).
(130) M. Rice, et al., Phys. Rev. **106**, 953 (1957).
(131) J. A. Cameron, et al., Can. J. Phys. **40**, 931 (1962).
(132) P. B. Nutter, Phil. Mag. **1**, 587 (1956).
(133) V. W. Cohen, et al., Phys. Rev. **79**, 191 (1950).
(134) K. Murakawa, Phys. Rev. **100**, 1369 (1955).
(135) F. M. Pipkin, Bull. Am. Phys. Soc. **3**, 8 (1958).
(136) P. C. B. Fernando, et al., Phil. Mag. **5**, 1309 (1960).
(137) K. E. Adelroth, et al., Arkiv Fysik **30**, 111 (1965).
(138) P. C. Fletcher, et al., Phys. Rev. **110**, 536 (1958).
(139) H. Walchli, et al., Phys. Rev. **82**, 97 (1951).
(140) R. Livingston, et al., Phys. Rev. **90**, 609 (1953).
(141) E. Lipworth, et al., Phys. Rev. **119**, 2022 (1960).
(142) D. Brinkmann, Helv. Phys. Acta **36**, 413 (1963).
(143) A. Bohr, et al., Arkiv Fysik **4**, 455 (1952).
(144) W. A. Nierenberg, et al., Phys. Rev. **112**, 186 (1958).
(145) R. D. Worley, et al., Phys. Rev. **140**, B1483 (1965).
(146) P. Buck, et al., Phys. Rev. **104**, 553 (1956).
(147) K. H. Althoff, et al., Naturwiss. **41**, 368 (1954).
(148) H. H. Stroke, et al., Phys. Rev. **105**, 590 (1957).
(149) G. Heinzelmann, et al., Phys. Letters 21, 162 (1966).
(150) V. W. Cohen, et al., Phys. Rev. **127**, 517 (1962).
(151) H. E. Walchli, et al., Phys. Rev. **102**, 1334 (1956).
(152) L. Olschewski, et al., Z. Physik **196**, 77 (1966).
(153) N. I. Kaliteevskii, et al., Soviet Physics—JETP **12**, 661 (1961).
(154) P. B. Sogo, et al., Phys. Rev. **99**, 613 (1955).
(155) K. Murakawa, et al., J. Phys. Soc. Japan **16**, 2533 (1961).
(156) J. N. Haag, et al., Phys. Rev. **129**, 1601 (1963).
(157) J. Blok, et al., Phys. Rev. **143**, 78 (1966).
(158) B. Bleaney, Quantum Electronics Conf., Paris, 1963; P. Grivet and N. Bloembergen, editors; Columbia U. Press (1964).
(159) J. Reader, et al., Phys. Rev. **137**, B784 (1965).
(160) E. D. Jones, Phys. Rev. Letters **19**, 432 (1967).
(161) A. Y. Cabezas, et al., Phys. Rev. **126**, 1004 (1962).
(162) K. F. Smith, et al., Proc. Phys. Soc. (London) **86**, 1249 (1965).
(163) R. W. Grant, et al., Phys. Rev. **130**, 1100 (1963).
(164) D. A. Shirley, et al., Phys. Rev. **121**, 558 (1961).
(165) J. Reader, Phys. Rev. **141**, 1123 (1966).
(166) D. Ali, et al., Phys. Rev. **138**, B1356 (1965).
(167) B. Budick, et al., Phys. Rev. **132**, 723 (1963).
(168) G. K. Woodgate, Proc. Roy. Soc. (London) **A293**, 117 (1966).
(169) L. Evans, et al., Proc. Roy. Soc. (London) **A289**, 114 (1965).
(170) W. Müller, et al., Z. Physik **183**, 303 (1965).
(171) S. S. Alpert, Phys. Rev. **129**, 1344 (1963).
(172) E. L. Boyd, Phys. Rev. **145**, 174 (1966).
(173) N. I. Kaliteevskii, et al., Soviet Phys.—JETP **37**, 629 (1960).
(174) E. L. Boyd, et al., Phys. Rev. Letters **12**, 20 (1964).
(175) C. A. Lovejoy, et al., Nucl. Phys. **30**, 452 (1962).
(176) C. Arnoult, et al., J. Opt. Soc. Am. **56**, 177 (1966).
(177) C. E. Johnson, et al., Phys. Rev. **120**, 2108 (1960).
(178) Q. O. Navarro, et al., Phys. Rev. **123**, 186 (1961).
(179) W. M. Doyle, et al., Phys. Rev. **131**, 1586 (1963).
(180) B. Budick, et al., Phys. Rev. **135**, B1281 (1964).
(181) J. C. Walker, Phys. Rev. **127**, 1739 (1962).
(182) D. Giglberger, et al., Z. Physik **199**, 244 (1967).
(183) A. Y. Cabezas, et al., Phys. Rev. **120**, 920 (1960).
(184) L. Olschewski, et al., Z. Physik **200**, 224 (1967).
(185) A. C. Gossard, et al., Phys. Rev. **133**, A881 (1964).
(186) J. S. Ross, et al., Phys. Rev. **128**, 1159 (1962).
(187) M. A. Grace, et al., Phil. Mag. **2**, 1079 (1957).
(188) A. H. Reddoch, et al., Phys. Rev. **126**, 1493 (1962).
(189) G. J. Ritter, Phys. Rev. **126**, 240 (1962).
(190) I. J. Spalding, et al., Proc. Phys. Soc. (London) **79**, 787 (1962).
(191) F. R. Petersen, et al., Phys. Rev. **126**, 252 (1962).
(192) D. R. Speck, Bull. Am. Phys. Soc. **1**, 282 (1956).
(193) D. R. Speck, et al., Phys. Rev. **101**, 1831 (1956).
(194) L. H. Bennett, et al., Phys. Rev. **120**, 1812 (1960).
(195) K. Murakawa, et al., Phys. Rev. **105**, 671 (1957).
(196) M. P. Klein, et al., Bull. Am. Phys. Soc. **6**, 104 (1961).
(197) L. Armstrong, et al., Phys. Rev. **138**, B310 (1965).
(198) J. Kaufmann, et al., Phys. Letters **24A**, 115 (1967).
(199) H. R. Loeliger, et al., Phys. Rev. **95**, 291 (1954).

REFERENCES (Continued)

(200) A. Narath, (to be published in Phys. Rev.).
(201) A. Narath, et al., Bull. Am. Phys. Soc. **12**, 314 (1967).
(202) W. von Siemens, Ann. Physik **13**, 136 (1953).
(203) Y. W. Chan, et al., Phys. Rev. **144**, 1020 (1966).
(204) Y. W. Chan, et al., Phys. Rev. **137**, B1129 (1965).
(205) H. Dahmen, et al., Z. Physik **200**, 456 (1967).
(206) W. J. Childs, et al., Phys. Rev. **141**, 176 (1966).
(207) P. A. Vanden Bout, et al., Phys. Rev. **158**, 1078 (1967).
(208) H. Kleiman, et al., Phys. Letters **13**, 212 (1964).
(209) W. J. Tomlinson, et al., Nucl. Phys. **60**, 614 (1964).
(210) H. Kleiman, et al., J. Opt. Soc. Am. **53**, 822 (1963).
(211) W. J. Tomlinson, et al., J. Opt. Soc. Am. **53**, 828 (1963).
(212) F. Bitter, et al., Phys. Rev. **96**, 1531 (1954).
(213) B. Cagnac, et al., Compt. Rend. **249**, 77 (1959).
(214) B. Cagnac, Ann. Phys. **6**, 467 (1961).
(215) J. C. Lehmann, et al., Compt. Rend. **257**, 3152 (1963).
(216) J. Blaise, et al., J. Phys. Radium **18**, 193 (1957).
(217) O. Redi, et al., Phys. Letters **8**, 257 (1964).
(218) S. P. Davis, et al., J. Opt. Soc. Am. **56**, 1604 (1966).
(219) R. J. Hull, et al., Phys. Rev. **122**, 1574 (1961).
(220) H. L. Poss, Phys. Rev. **75**, 600 (1949).
(221) G. O. Brink, et al., Phys. Rev. **107**, 189 (1957).
(222) E. B. Baker, J. Chem. Phys. **26**, 960 (1957).
(223) I. Lindgren, et al., Arkiv Fysik **15**, 445 (1959).
(224) Y. Ting, et al., Phys. Rev. **89**, 595 (1953).
(225) S. S. Alpert, et al., Phys. Rev. **125**, 256 (1962).
(226) C. M. Olsmats, et al., Arkiv Fysik **19**, 469 (1961).
(227) M. Fred, et al., Phys. Rev. **98**, 1514 (1955).
(228) V. N. Egorov, Opt. Spectry. (USSR) **16**, 301 (1964).
(229) J. D. Axe, et al., Phys. Rev. **121**, 1630 (1961).
(230) R. Marrus, et al., Nucl. Phys. **23**, 90 (1961).
(231) P. B. Dorain, et al., Phys. Rev. **105**, 1307 (1957).
(232) B. Bleaney, et al., Phil. Mag. **45**, 992 (1954).
(233) J. Faust, et al., Phys. Letters **16**, 71 (1965).
(234) R. J. Champeau, et al., Compt. Rend. **257**, 1238 (1963).
(235) L. Armstrong, et al., Phys. Rev. **144**, 994 (1966).
(236) T. E. Manning, et al., Phys. Rev. **102**, 1108 (1956).
(237) R. Marrus, et al., Phys. Rev. **124**, 1904 (1961).
(238) M. Fred, et al., J. Opt. Soc. Am. **47**, 1076 (1957).
(239) B. Perry, et al., Bull. Am. Phys. Soc. **8**, 345 (1963).
(240) M. M. Robertson, et al., Phys. Rev. **140**, B820 (1965).
(241) R. E. Walstedt, et al., Phys. Rev. **162**, 301 (1967).
(242) R. Freeman, et al., Proc. Roy Soc. (London) **242A**, 455 (1957).

ELECTRON WORK FUNCTIONS OF THE ELEMENTS

From data published up to 1954

Compiled by Herbert B. Michaelson

Preferred values for well-outgassed polycrystalline materials are indicated by an asterisk (*). Because of the anisotropy and allotropy of work function, however, the designation "preferred value" has a rather limited meaning.

Values are expressed in electron-volts.

Each value is followed by a reference to the literature.

Element	Thermionic Work Function	Photoelectric Work Function	Work Function by Contact Potential Method
Ag	3.09 (1) 3.56 (2) 4.08 (3) 4.31 (114)	3.67 (4) 4.1 to 4.75 (5) 4.50 to 4.52 (117) 4.56 (600°)* (6) 4.73 (20°)* (6) 4.75 (111) face (7) 4.81 (100) face (7)	4.21 (116) 4.33 (8) 4.35 (115) 4.44 (9) 4.47 (10)
Al	. . .	2.98 (4) 3.43 (11) 4.08* (12) 4.20 (113) 4.36 (13)	3.38 (9) 4.25 (115)
As		4.72 (118) 5.11 (14)	
Au	4.0 to 4.58 (2) 4.25 (114) 4.32 (3)	4.73 (740°C)* (15) 4.82 (20°C)* (15) 4.86 to 4.92 (5)	4.46 (9)
B	. . .	4.4 to 4.6 (16)	
Ba	2.11 (17)	2.48* (19) 2.49 (20) 2.52 (21)	1.73 (9) 2.39 (18)
Be	. . .	3.17 (14) 3.30 (22) 3.92* (23)	(9)
Bi	. . .	4.14 (11) 4.22 to 4.25* (24) 4.31 (25) 4.34 (118) 4.44 (22) 4.46 (47)	4.17 (9)
C	4.00 (26) 4.34 (27) 4.35 (119) 4.39 (28) 4.60* (120) 4.83 (121) 4.84 (122)	4.81 (29)	
Ca	2.24 (30)	2.42 (14) 2.706* (21) 2.76 (31) 3.20 (32) 3.21 (33)	3.33 (9)
Cd	3.96 (34) 4.01* (35)	3.68 (14) 3.73 (4) 3.94 (11) 4.07* (36) 4.099 (113)	4.00 (9) 4.49 (123)
Ce	2.6 (37)	2.84 (32)	
	4.40* 4.41 (38) (124)	3.90 (31) 4.12 to 4.25 (39)	4.21 (9)
Cr	4.58 4.60* 4.7 (124) (40) (41)	4.37 (25)	4.38 (9)
Cs	1.81* (42)	1.9 (43) 1.96 (44)	
	3.85 (1) 4.26 (2) 4.38 (3) 4.50 (114) 4.55 (a) (45)	4.07 (4) 4.18 (31) 4.70 (125) 4.86 (111) face (46) 5.61 (100) face (46)	4.46* (9, 126) 4.61 (115) 4.86 (123)
Fe	4.04 (48) γ4.21 (38) γ4.31 (124) 4.47 (49) β4.48 (38) 4.76 (127)	3.91 (31) 3.92 (53) β4.62 (128) γ4.68 (128) α4.70 (128) 4.72 (50) 4.77 (51)	4.40 (9)
Ga	4.12 (14)	. . .	3.80 (9)
Ge	. . .	4.29 (31) 4.5 (52) 4.73 (14) 4.80 (16)	4.50 (9)
Hf	3.53 (54)		
Hg	. . .	4.50 (55) 4.52 (56) 4.53* (57, 58)	4.50 (9)
I	. . .	2.8 (rhombic) (129) 5.4 (monoclinic) (129) 6.8 (amorphous) (129)	
Ir	5.3 (130)	. . .	4.57 (9)
K	. . .	2.0 (59) 2.12 (22) 2.24* (44, 43) 2.26 (60)	1.60 (9)
La	3.3 (37)		
Li	. . .	2.28 (43) 2.42 (14)	1.40 (9) 2.49* (61)
Mg	. . .	2.74 (14) <3.0 (62) 3.59 (63) 3.62 (64) 3.68* (23)	3.58 (9) 3.78 (66)
Mn	3.83 (124)	3.76 (14)	4.14 (9)
Mo	4.15 (67) 4.17 (68) 4.19 (69) 4.20* (70) 4.23 (121) 4.32 (71) 4.33 (72) 4.38 (73) 4.44 (74)	4.15 (67) 4.34 (25)	4.08 (75) 4.48 (9)
Na	. . .	2.06 (14) 2.25 (76) 2.28* (19) 2.29 (23) 2.47 (43)	1.60 (9) 2.26 (77)

(a) By field current method.

ELECTRON WORK FUNCTIONS OF THE ELEMENTS (Continued)

Element	Thermionic Work Function	Photoelectric Work Function	Work Function by Contact Potential Method
Nd	3.3 (37)		
Ni	4.50 (124) 4.61 (38) 4.63 (48) 5.03* (78) 5.1 (a) (45) 5.24 (1150°K) (131)	3.67 (4) 4.06 (31) 4.87 (25) 5.01* (51) 5.05 (623°K) (131) 5.20 (1108°K) (131)	4.32 (9) 4.96 (300°K) (79)
Os			4.55 (9)
Pb		3.97 (4) 4.14 (11)	3.94 (9)
Pd	(80)	4.97* (80)	4.49 (9)
Pr	(37)		
Pt	(132) (26) (81) (82) (83)	4.09 (4) 6.35 (84)	4.52 (9) 5.36* (85)
Rb		2.09* (44) 2.16 (43)	
Re	4.74 (133) 5.1* (86)	∽5.0 (87)	
Rh	4.58 (88) 4.80* (89) 4.9 (130)	4.57 (88)	4.52 (9)
Ru	. . .	. . .	4.52 (9)
Sb	. . .	4.01 (90) 4.60 (118)	4.14 (9)
Se	. . .	4.62 (11) 5.11 (14)	4.42 (9)
Si	3.59 (28) 4.02 (134)	4.37 to 4.67 (117)	4.2 (91)
Sm	3.2 (37)		
Sn	. . .	3.62 (4) 3.87 (11) (liq.) 4.21 (92) γ4.38* (92) β4.50* (92)	4.09 (9) 4.64 (123)

Element	Thermionic Work Function	Photoelectric Work Function	Work Function by Contact Potential Method
Sr	. . .	2.06 (93) 2.24 (14) 2.74* (64)	
Ta	4.03 (135) 4.07 (74) 4.10 (94) 4.19* (95)	4.05 (63) 4.12 (32) 4.16 (96)	3.96 (9) 4.25 (97)
Te	. . .	4.04 (117) 4.76 (16)	4.70 (9)
Th	3.35* (73)	3.38 (32) 3.47 (98) 3.57 (11)	3.46 (9)
Ti	3.95 (124)	3.95 (14) 4.17 (98) 4.45 (136)	4.14 (9
Tl	. . .	3.68 (107)	3.84 (9
U	3.27 (99)	3.63 (32)	4.32 (9
V	4.12 (124)	3.77 (14)	4.44 (9
W	4.25 (100) 4.29 (116) face (142) 4.38 (111) face (142) 4.39 (111) face (140) 4.39 (116) face (140) 4.45 (doped) (138) 4.46 (101) 4.52* (001) face (26, 102, 142) 4.53 (103, 138) 4.56 (001) face (140) 4.58 (110) face (71, 73, 142) 4.59 (001) face (139) 4.65 (112) face (142) 4.68 (110) face (140) 4.69 (112) face (140) 5.01 (137)	4.35 (310) face (104) 4.49 (143) 4.50 (211) face (104) 4.54 (105) 4.60 (32)	4.38 (9
Zn	. . .	3.08 (4) 3.28 (106) 3.32 and 3.57 (108) 3.60 (11) 3.89 (31) 4.24 (109) 4.26 (0001) face (110) 4.307 (113)	3.40 3.66 4.28 4.65
Zr	4.12 (54) 4.21* (141)	3.73 (32) 4.33 (136)	3.60 (

(a) By field current method.

REFERENCES

(1) A. Wehnelt and S. Seliger, Zeits. f. Physik **38,** 443 (1926).
(2) I. Ameiser, Zeits. f. Physik **69,** 111 (1931).
(3) A. Goetz, Zeits. f. Physik **43,** 531 (1927).
(4) P. Lukirsky and S. Prilesaev, Zeits. f. Physik **49,** 236 (1928).
(5) R. H. Fowler, Phys. Rev. **38,** 45 (1931).
(6) R. P. Winch, Phys. Rev. **37,** 1269 (1931).
(7) H. E. Farnsworth and R. P. Winch, Phys. Rev. **58,** 812 (1940).
(8) P. A. Anderson, Phys. Rev. **50,** 320 (1936).
(9) O. Klein and E. Lange, Zeits. f. Elektrochemie **44,** 542 (1938).
(10) P. A. Anderson, Phys. Rev. **59,** 1034 (1941).
(11) R. Hamer, Jour. Opt. Soc. Amer. **9,** 251 (1924).
(12) J. Brady and P. Jakobsmeyer, Phys. Rev. **49,** 670 (1936).
(13) E. Gaviola and J. Strong, Phys. Rev. **49,** 441 (1936).
(14) R. Schulze, Zeits, f Physik **92,** 212 (1934).
(15) L. W. Morris, Phys. Rev. **37,** 1263 (1931).
(16) L. Apker, E. Taft, and J. Dickey, Phys. Rev. **74,** 1462 (1948).
(17) A. L. Reimann, *Thermionic Emission* (Chapman and Hall, London, 1934), p. 37.
(18) P. A. Anderson, Phys. Rev. **47,** 958 (1935).
(19) R. J. Maurer, Phys. Rev. **57,** 653 (1940).
(20) R. J. Cashman and E. Bassoe, Phys. Rev. **55,** 63 (1939).
(21) N. C. Jamison and R. J. Cashman, Phys. Rev. **50,** 624 (1936).
(22) R. Suhrmann and A. Schallamach, Zeits. f. Physik **91,** 775 (1934).
(23) M. M. Mann, Jr. and L. A. DuBridge, Phys. Rev. **51,** 120 (1937).
(24) H. Jupnik, Phys. Rev. **60,** 884 (1941).
(25) H. C. Rentschler and D. E. Henry, Jour. Opt. Soc. Amer. **26,** (1936).
(26) S. Dushman, "Thermal Emission of Electrons," International Critic Tables (1929) Vol. VI, pp. 53–4.
(27) A. L. Reiman, Proc. Phys. Soc. **50,** 496 (1938).
(28) A. Braun and G. Busch, Helv. Phys. Acta **20,** #1, 33 (1947).
(29) S. C. Roy, Proc. Roy. Soc. **A112,** 599 (1926).
(30) S. Dushman, Phys. Rev. **21,** 623 (1923).
(31) G. B. Welch, Phys. Rev. **32,** 657 (1928).
(32) C. Rentschler, D. E. Henry, and K. O. Smith, Rev. Sci. Instr. 794 (1932).
(33) I. Liben, Phys. Rev. **51,** 642 (1937).
(34) H. B. Wahlin and L. O. Sordahl, Phys. Rev. **45,** 886 (1934).
(35) A. L. Reimann and C. Kerr Grant, Phil. Mag. **22,** 34 (1936).
(36) H. Bomke, Ann. d. Physik **10,** 579 (1931).

REFERENCES (Continued)

(37) E. E. Schumaker, J. E. Harris, J. Amer. Chem. Soc. **48**, 3108 (1926).
(38) H. B. Wahlin, Phys. Rev. **61**, 509 (1942).
(39) A. B. Cardwell, Phys. Rev. **38**, 2033 (1931).
(40) H. B. Wahlin, Phys. Rev. **73**, 1458 (1948).
(41) H. Koesters, Zeits f. Physik **66**, 807 (1930).
(42) K. H. Kingdon, Rev. **25**, 892 (1925).
(43) A. R. Olpin, quoted by Hughes and DuBridge, *Photoelectric Phenomena* (1932).
(44) J. J. Brady, Phys. Rev. **41**, 613 (1932).
(45) W. P. Dyke, Thesis, University of Washington, (1946).
(46) N. Underwood, Phys. Rev. **47**, 502 (1935).
(47) A. H. Weber and C. J. Eisele, Phys. Rev. **59**, 473A (1941).
(48) W. Distler and G. Monch, Zeits, f. Physik **84**, 271 (1933).
(49) G. Siljeholm, Ann. d. Physik **10**, 178 (1931).
(50) A. B. Cardwell, Proc. Nat. Acad. Sci. **14**, 439 (1928).
(51) G. N. Glasoe, Phys. Rev. **38**, 1490 (1931).
(52) A. H. Smith, Phys. Rev. **75**, 953 (1949).
(53) G. B. Welch, Phys. Rev. **31**, 709A (1928).
(54) C. Zwikker, Phys. Zeits. **30**, 578 (1929).
(55) H. Cassel and A. Schneider, Naturwiss **22**, 464 (1934).
(56) D. Roller, W. H. Jordan, and C. S. Woodward, Phys. Rev. **38**, (1931).
(57) C. B. Kazda, Phys. Rev. **26**, 643 (1925).
(58) W. B. Hales, Phys. Rev. **32**, 950 (1928).
(59) H. E. Ives, J. Opt. Soc. Am. **8**, 551 (1924).
(60) H. Mayer, Ann. d. Phys. **29**, 129 (1937).
(61) P. A. Anderson, Phys. Rev. **75**, 1205 (1949).
(62) C. Kenty, Phys. Rev. **43**, 776(A), (1933).
(63) J. Cashman and S. Huxford, Phys. Rev. **48**, 734 (1935).
(64) R. J. Cashman and E. Bassoe, Phys. Rev. **53**, 919(A), (1938).
(65) R. J. Cashman, Phys. Rev. **54**, 971 (1938).
(66) P. A. Anderson, Phys. Rev. **54**, 753 (1938).
(67) L. A. DuBridge and W. W. Roehr, Phys. Rev. **42**, 52 (1932).
(68) H. B. Wahlin and J. A. Reynolds, Phys. Rev. **48**, 751 (1935).
(69) H. Grover, Phys. Rev. **52**, 982 (1937).
(70) R. W. Wright, Phys. Rev. **60**, 465 (1941).
(71) A. J. Ahearn, Phys. Rev. **44**, 277 (1933).
(72) H. Freitag and F. Kruger, Ann. d. Physik **21**, 697 (1934).
(73) C. Zwikker, Proc. Roy. Acad. Amsterdam **29**, 792 (1926).
(74) S. Dushman, H. N. Rowe, J. Ewald, C. A. Kidner, Phys. Rev. **25**, 338 (1925).
(75) C. W. Oatley, Proc. Roy. Soc. **A155**, 218 (1936).
(76) Z. Berkes, Math. Phys. Lapok **41**, 131 (1934).
(77) E. Patai, Zeits, f. Physik. **59**, 697 (1930).
(78) G. W. Fox and R. M. Bowie, Phys. Rev. **44**, 345 (1933).
(79) R. C. L. Bosworth, Trans. Far. Soc. **35**, 397 (1939).
(80) L. A. DuBridge and W. W. Roehr, Phys. Rev. **39**, 99 (1932).
(81) H. L. Van Velzer, Phys. Rev. **44**, 831 (1933).
(82) L. V. Whitney, Phys. Rev. **50**, 1154 (1936).
(83) L. A. DuBridge, Phys. Rev. **32**, 961 (1928).
(84) L. A. DuBridge, Phys. Rev. **31**, 236 (1928).
(85) C. W. Oatley, Proc. Phys. Soc. **51**, 318 (1939).
(86) C. Agte, H. Alterthum, K. Becker, G. Heyne, and K. Moers. Naturwiss **19**, 108 (1931).
(87) A. Englemann, Ann. d. Physik **17**, 185 (1933).
(88) E. H. Dixon, Phys. Rev. **37**, 60 (1931).
(89) H. B. Wahlin and L. V. Whitney, J. Chem. Phys. **6**, 594 (1938).
(90) V. Middel, Zeits. f. Physik **105**, 358 (1937).
(91) W. E. Meyerhof, Phys. Rev. **71**, 727 (1947).
(92) A. Goetz, Phys. Rev. **33**, 373 (1929).
(93) R. Doepel, Zeits. f. Physik **33**, 237 (1925).
(94) A. B. Cardwell, Phys. Rev. **47**, 628 (1935).
(95) M. D. Fiske, Phys. Rev. **61**, 513 (1942).
(96) A. B. Cardwell, Phys. Rev. **38**, 2041 (1931).
(97) W. Heinze, Zeits. f. Phys. **109**, 459 (1938).
(98) H. C. Rentschler and D. E. Henry, Trans. Electrochem. Soc. **87**, 289 (1945–6).
(99) W. L. Hole and R. W. Wright, Phys. Rev. **56**, 785 (1939).
(100) A. H. Warner, Proc. Nat. Acad. Sci. **13**, 56 (1927).
(101) G. M. Fleming and J. E. Henderson, Phys. Rev. **58**, 887 (1940).
(102) W. B. Nottingham, Phys. Rev. **47**, 806A (1935).
(103) H. Freitag and F. Krueger, Ann. d. Physik **21**, 697 (1934).
(104) C. E. Mendenhall and C. F. DeVoe, Phys. Rev. **51**, 346 (1937).
(105) A. H. Warner, Phys. Rev. **38**, 1871 (1931).
(106) A. Nitsche, Ann. d. Physik **14**, 463 (1932).
(107) R. Suhrmann and H. Csesch, Zeits. Chem. **B28**, 215 (1935).
(108) J. H. Dillon, Phys. Rev. **38**, 408 (1931).
(109) C. F. DeVoe, Phys. Rev. **50**, 481 (1936).
(110) W. Klug and H. Steyskal, Zeits. f. Physik **116**, 415 (1940).
(111) C. W. Oatley, Proc. Roy. Soc. **A155**, 218 (1936).
(112) P. A. Anderson, Phys. Rev. **57**, 122 (1940).
(113) R. Suhrmann and J. Pietrzyk, Zeits. f. Physik **122**, 600 (1944).
(114) S. C. Jain and K. S. Krishman, Proc. Roy. Soc. **A217**, 451 (1953).
(115) E. W. J. Mitchell and J. W. Mitchell, Proc. Roy. Soc. **A210**, 70 (1952).
(116) G. L. Weissler and T. N. Wilson, J. Appl. Phys. **24**, 472 (1953).
(117) S. M. Fainshtein, Zavodskaya Lab. **14**, 64 (1948).
(118) L. Apker, E. Taft and J. Dickey, Phys. Rev. **76**, 270 (1949).
(119) G. Glockler and J. W. Sausville, J. Electrochem. Soc. **95**, 292 (1949).
(120) H. F. Ivey, Phys. Rev. **76**, 567 (1949).
(121) S. B. L. Mathur, Natl. Inst. Sci. India **19**, 153 (1953).
(122) A. S. Bhatnager, Proc. Nat. Acad. Sci. India **A14**, 5 (1944).
(123) R. Hirschberg and E. Lange, Naturwiss. **39**, 131 (1952).
(124) K. S. Krishman and S. C. Jain, Nature **170**, 759 (1952).
(125) R. Ito, J. Phys. Soc. Japan **6**, 188 (1951).
(126) P. A. Anderson, Phys. Rev. **76**, 388 (1949).
(127) S. B. L. Mathur, Proc. Natl. Inst. Sci. India **19**, 165 (1953).
(128) A. B. Cardwell, Phys. Rev. **92**, 554 (1953).
(129) D. C. West, Can. J. Phys. **31**, 691 (1953).
(130) O. A. Weinreich, Phys. Rev. **82**, 573 (1951).
(131) A. B. Cardwell, Phys. Rev. **76**, 125 (1949).
(132) K. Kondo, Rept. Inst. Sci. Tech., Tokyo Univ. **1**, 24 (1947).
(133) R. Levi and G. A. Espersen, Phys. Rev. **78**, 231 (1950).
(134) L. Esaki, J. Phys. Soc. Japan **8**, 347 (1953).
(135) R. J. Munick, W. B. LaBerge and E. A. Coomes, Phys. Rev. **80**, 887 (1950).
(136) H. Malamud and A. D. Krumbein, J. Appl. Phys. **25**, 591 (1954).
(137) G. M. Fleming and J. E. Henderson, Phys. Rev. **58**, 887 (1940).
(138) M. H. Nichols, Phys. Rev. **78**, 158 (1950).
(139) A. A. Brown, L. J. Neelands and H. E. Farnsworth, J. Appl. Phys. **21**, 1 (1950).
(140) C. Herring and M. H. Nichols, Rev. Mod. Phys. **21**, 185 (1949).
(141) A. Wahl, Phys. Rev. **82**, 574 (1951).
(142) G. F. Smith, Phys. Rev. **94**, 295 (1954).
(143) L. Apker, E. Taft and J. Dickey, Phys. Rev. **73**, 46 (1948).

PROPERTIES OF METALS AS CONDUCTORS

Metal.	Resistivity microhm-centimeters 20° C.	Temp. coefficient 20° C.	Specific gravity.	Tensile strength, lbs./in.	Melting point ° C.
*Advance. See *constantan*					
Aluminum	2.824	0.0039	2.70	30,000	659
Antimony	41.7	.0036	6.6	....	630
Arsenic	33.3	.0042	5.73	....	....
Bismuth	120	.004	9.8	....	271
Brass	7	.002	8.6	70,000	900
Cadmium	7.6	.0038	8.6	....	321
*Calido. See *nichrome*					
Climax	87	.0007	8.1	150,000	1250
Cobalt	9.8	.0033	8.71	....	1480
Constantan	49	.00001	8.9	120,000	1190
Copper: annealed	1.7241	.00393	8.89	30,000	1083
hard-drawn	1.771	.00382	8.89	60,000	....
Eureka. See *constantan*					
Excello	92	.00016	8.9	95,000	1500
Gas Carbon	5000	−.0005	....	....	3500
German silver, 18% Ni	33	.0004	8.4	150,000	1100
Gold	2.44	.0034	19.3	20,000	1063
Ideal. See *constantan*					
Iron, 99.98% pure	10	.005	7.8	....	1530
Lead	22	.0039	11.4	3,000	327
Magnesium	4.6	.004	1.74	33,000	651
Manganin	44	.00001	8.4	150,000	910
Mercury	95.783	.00089	13.546	0	−38.9
Molybdenum, drawn	5.7	.004	9.0	....	2500
Monel metal	42	.0020	8.9	160,000	1300
*Nichrome	100	.0004	8.2	150,000	1500
Nickel	7.8	.006	8.9	120,000	1452
Palladium	11	.0033	12.2	39,000	1550
Phosphor bronze	7.8	.0018	8.9	25,000	750
Platinum	10	.003	21.4	50,000	1755
Silver	1.59	.0038	10.5	42,000	960
Steel, E. B. B.	10.4	.005	7.7	53,000	1510
Steel, B. B.	11.9	.004	7.7	58,000	1510
Steel, Siemens-Martin	18	.003	7.7	100,000	1510
Steel, manganese	70	.001	7.5	230,000	1260
Tantalum	15.5	.0031	16.6	....	2850
*Therlo	47	.00001	8.2	....	....
Tin	11.5	.0042	7.3	4,000	232
Tungsten, drawn	5.6	.0045	19	500,000	3400
Zinc	5.8	.0037	7.1	10,000	419

* Trade mark.

Superconductivity

COMPILED BY B. W. ROBERTS

General Electric Research Laboratory, Schenectady, New York

The sensitivity of superconductive properties to the state of the material is most pronounced. Annealing, cold-working, homogeneity, dislocation densities, and the presence of impurity atoms and other scattering centers are all capable of controlling the critical temperature and magnetic properties of a specific material.

Table I lists the elements and some of their superconductive properties.

Table II contains a collection of some superconductive materials. All compositions are reported on an atomic basis; i.e., AB, AB_2, or AB_3 for compounds, unless noted. Solid solutions or odd compositions may be denoted as A_zB_{1-z} or A_zB. A series of three or more alloys is indicated by A_xB_y or by actual indication of the atomic fraction range such as $A_{0-.6}B_{1-.4}$. The critical temperatures of such a series of alloys is denoted by a range of values or possibly the maximum value.

A more extensive listing of superconductive materials, together with references to the original literature may be found in *Progress in Cryogenics*, Vol. IV.

TABLE I. SUPERCONDUCTIVE ELEMENTS

Element	T_c(°K)		H_0(Oe)		θD(°K)	γ $\left[\frac{\text{millijoules}}{\text{mole(°K)}^2}\right]$
	Cal.	Mag.	Cal.	Mag.		
Al	1.183	1.196	104	99	420	1.36
Cd	0.54	0.56	29	30	300	0.71
Ga	1.087	1.091	59.4	51	317	0.601
Hg(α)		4.153		412	87	1.91
Hg(β)		3.949		339	93	1.37
In	3.396	3.4035	278	293	109	1.81
Ir		0.14		19		3.1
La(α)	4.80	~5.0				
La(β)	5.91	6.3		1600	132	6.7
Mo		0.92		98	470	1.91
Nb	9.17	9.13	1944	1980	238	7.53
Os		0.655		65		2.35
Pb	7.23	7.193		803	96.3	3.0
Re	1.699	1.698	188	198	210	2.35
Ru		0.49		66		2.4
Sn	3.722	3.722	303	309	195	1.75
Ta	4.39	4.483	780	830	255	6.4
Tc		8.22				
Th		1.368	131	162	168	4.65
Ti		0.39		100	430	3.3–3.6
Tl	2.36	2.39		171	88	2.8?
U(α)		0.68			206	10.6

TABLE I. SUPERCONDUCTIVE ELEMENTS (Continued)

Element	T_c(°K)		H_0(Oe)		θD(°K)	γ $\left[\frac{\text{millijoules}}{\text{mole(°K)}^2}\right]$
	Cal.	Mag.	Cal.	Mag.		
U(pseudo-γ)		1.80(extrapolated value)				
V	5.03	5.30	1310	1020	338	6.4
Zn	0.852	0.875	51.8	53	309	0.66
Zr		0.546		47	265	2.9

Thin films formed below ~10°K

Be	~6, ~8.4
Bi	~6
Ga	8.4
In	3.95–4.25
Sn	4.6–4.7

Under high pressure

Bi	See Table II

TABLE II. SUPERCONDUCTIVE MATERIALS

Symbols Used:

*	Eutectic alloy	°	Impure material
Δ	Uncertain composition	C	Calorimetric determination
R	Resistance measurements	VA	Valence electron/atom

M Denotes maximum T_c in specimen series.

∞ All cell edges are intended to be quoted in angstrom units.

$T_c'(—)$ Denotes incremental changes in T_c from pure metal.

KEY TO CRYSTAL STRUCTURE TYPES FOUND IN TABLE II

"Struck-turbericht" Type	Example	Class	"Struck-turbericht" Type	Example	Class
A1	Cu	Cubic, FC	$C11_b$	$MoSi_2$	Tetragonal, BC
A2	W	Cubic, BC	C14	$MgZn_2$	Hexagonal
A3	Mg	Hexagonal, close packed	C15	Cu_2Mg	Cubic, FC
			$C15_b$	$AuBe_5$	Cubic
A4	Diamond	Cubic, FC	C16	$CuAl_2$	Tetragonal, BC
A5	White Sn	Tetragonal, BC	C18	FeS_2	Orthorhombic
A6	In	Tetragonal, BC (FC cell usually used)	C32	AlB_2	Hexagonal
			C36	$MgNi_2$	Hexagonal
A7	As	Rhombohedral	C49	$ZrSi_2$	Orthorhombic
A8	Se	Trigonal	C54	$TiSi_2$	Orthorhombic
A10	Hg	Rhombohedral	C_c	Si_2Th	Tetragonal, BC
A12	α-Mn	Cubic, BC	DO_3	BiF_3	Cubic, FC
A13	β-Mn	Cubic	DO_{18}	Na_3As	Hexagonal
A15	β-W	Cubic	DO_{19}	Ni_3Sn	Hexagonal
B1	NaCl	Cubic, FC	DO_{22}	$TiAl_3$	Tetragonal
B2	CsCl	Cubic	DO_e	Ni_3P	Tetragonal, BC
$B8_1$	NiAs	Hexagonal	Dl_3	Al_4Ba	Tetragonal, BC
B10	PbO	Tetragonal	Dl_c	$PtSn_4$	Orthorhombic
B11	γ-CuTi	Tetragonal	$D2_c$	MnU_6	Tetragonal, BC
B18	CuS	Hexagonal	$D5_2$	La_2O_3	Trigonal
B20	FeSi	Cubic	$D5_8$	Sb_2S_3	Orthorhombic
B31	MnP	Orthorhombic	$D7_b$	Ta_3B_4	Orthorhombic
B32	NaTl	Cubic, FC	$D8_2$	Cu_5Zn_8	Cubic, BC
B_f	ζ-CrB	Orthorhombic	$D8_8$	Mn_5Si_3	Hexagonal
B_g	MoB	Tetragonal, BC	$D8_b$	CrFe	Tetragonal
B_h	WC	Hexagonal	$D8_i$	Mo_2B_5	Rhombohedral
B_i	γ′-MoC	Hexagonal	$D10_2$	Fe_3Th_7	Hexagonal
Cl	CaF_2	Cubic, FC	$E9_3$	Fe_3W_3C	Cubic, FC
Cl_b	MgAgAs	Cubic, FC	$L1_0$	CuAu	Tetragonal
C2	FeS_2	Cubic	$L1_2$	Cu_3Au	Cubic
C6	CdI_2	Trigonal	L'_{2b}	ThH_2	Tetragonal, BC

TABLE II. SUPERCONDUCTIVE MATERIALS

Material	T_c(°K)	H_0(Oe)	Crystal Structure ∞
$Ag_{0-.002}Al_y$	$T'_c(-.0543)$		
$AgIn_2$	2.30–2.46		C16
Ag_xPb_y	7.2M		
AgPb*	7.20		
AgSn	3.70R		
Ag_xSn_y	1.5–3.7R		
$AgTh_2$	2.26		C16, a = 7.56 c = 5.84
$Ag_{.94}Tl_{.06}$	2.32R		
$Ag_{.03}Tl_{.97}$	2.68R?		
AgTl*	2.67R		
Al	1.196	99	Al, a = 4.0496
Al	1.183	104	
Al(6063)alloy	1.14	94	
$Al_xGe_{0-.002}$	$T'_c(-.003)$ +.002)		
$Al_{.5}Ge_{.5}Nb_3$	12.6		A13, a = 5.175
Al_3Mg_2	0.84		Cub., FC, a = 28.28
$Al_xMg_{0-.011}$	$T'_c(-.0478)$		
$AlMo_3$	0.58		A15, a = 4.950
$AlNb_3$	17.5, 18.0		A15, a = 5.187
$AlNb_3$ (sintered)	17.1		A15, a = 5.186
$Al_{.22-.31}Nb_y$ (sintered)	16.0–18.0		A15
$Al_{.17-.35}Nb_y$ (melted)	15.2–17.6		A15
$Al_{>.25}Nb_{<.75}$	7–12		σ-phase, $D8_b$
$Al_{1-x}Nb_3Sb_x$	<4.2–15.7		A15
$Al_{.5}Nb_3Sn_{.5}$	16.3		A15
$Al_{0-.8}Nb_3Sn_{1-.2}$	14.6–18.1		A15, a = 5.262–5.292
Al_3Os	5.9		
AlOs	0.39		B2
$Al_xSi_{0-.0505}$	$T'_c(-.0048)$		
Al_3Th	0.75		DO_{19}, a = 6.500, c = 4.626
$Al_xZn_{0-.01}$	$T'_c(-.0444)$		
$AlZr_3$	0.73		$L1_2$, a = 4.37
Al_2Zr	<0.35		C14, a = 5.282 c = 8.748
AsBiPbΔ	9.0R		
AsBiPbSb	9.0R		
$As_xNb_3Sn_{1-x}$	17.91–18.0		A15
$AsNi_{.25}Pd_{.75}$	1.34		C2
$AsNi_{.12}Pd_{.88}$	1.39		C2
$As_{.50}Ni_xPd_y$	1.60M		$B8_1$
$AsNi_{.25}Pd_{.75}$	1.6		$B8_1$, a = 3.66, c = 5.03
AsPb*	8.4		
$As_{.03}Pb_{.97}$	8.4R		
$AsPd_2$	1.7		
As_3Pd_5	1.9		
AsSn*	4.10		
As?Sn?	4.2R		
AuBe	2.64		B20
Au_2Bi	1.84		C15, a = 7.958
$AuNb_3$	11.5		A15, a = 5.21
Au_xPb_y	7.20MR		
AuPb*	7.0R		
$AuPb_2$	7.0R		
$AuPb_2$	4.42		C16
$AuSb_2$	0.58		C2, a = 6.658
$AuSn_4$	2.38		$D1_c$
$AuSn_2$	2.48R		
Au_xSn_y	3.71MR		
AuSn	3.70R		
$AuTh_2$	3.08		C16, a = 7.42, c = 5.95
AuTl*	1.92R		
$Au_{.27}Tl_{.73}$(w/o)	2.04R?		
AuV_3	0.74		A15, a = 4.883
$AuZn_3$	1.21		P. cubic a = 7.893
$Au_{.05}Zr_{.95}$	1.65		A3
$Au_{.07}Zr_{.93}$ (3.75VA)	2.52		A3
$Au_{.12}Zr_{.88}$ (3.64VA)	2.74		A3
$Au_{.10}Zr_{.90}$ (3.7VA)	2.79		A3
$AuZr_3$	0.92		A15
BMo_2	4.74		C16
BNb	8.25		B_f
BRe_2	2.80		
B_3Ru_7	2.58		$D10_2$, a = 7.465, c = 4.715
BTa_2Δ	3.12		C16
BW_2	3.10		C16
BZrΔ	2.8–3.2		Cub., F.C.
$BaBi_3$	5.69	740	Tet., a = 5.188, c = 5.157
$BaRh_2$	6.0		C15, a = 7.852
Be (thin films)	7–9R		
BeII (thin film)	~6		
BeI (thin film)	~8.4		
Bi (films deposited at low temp.)	~6		
BiIII 27,000–28,400 atm.	7.25		

TABLE II. SUPERCONDUCTIVE MATERIALS (Continued)

Material	T_c(°K)	H_0(Oe)	Crystal Structure ∞
BiII 25,000 atm.	3.916		
25,200 atm.	3.90		
26,800 atm.	3.86		
Bi_3Ca	2.00		
$Bi_{.5}Cd_{.12}Pb_{.25}Sn_{.12}$ (w/o)	8.2		
Bi_2Cs	4.75		C15, a = 9.746
BiCu (unstable)	2.20		
$BiIn_2$	5.6		Hex. (disordered)
$Bi_{0-.003}In_y$	T_c'(−.0129 + .0119)		
Bi_2K	3.58		C15, a = 9.501
BiK	3.60?		
BiLi(α)	2.47		$L1_0$, a = 3.361, c = 4.24
BiNa	2.25		$L1_0$, a = 3.46, c = 4.80
$Bi_xNb_3Sn_{1-x}$	18.0–18.2		
Bi_3Ni	4.06		Orthorhombic a = 8.875, b = 4.112, c = 11.477
BiNi	4.25		$B8_1$, a = 4.070, c = 5.35
BiPb*	8.8		
$Bi_{.5}Pb_{.5}$	8.4		a = 9.746
BiPbSb	8.9R		
$Bi_{.50}Pb_{.31}Sn_{.19}$ (w/o)	8.5R		
Bi_2PbSn	8.5R		
Bi_2Pd(α)	1.70		Monoclinic a = 12.74, b = 4.25 c = 5.665 β = 102°35′
BiPd	3.7		Orthorhombic a = 7.203, b = 8.707, c = 10.662
$BiPd_2$	4.00		
Bi_2Pd(β)	4.25		Tet., a = 3.362, b = 12.983
BiPt	1.21, 2.4		$B8_1$, a = 4.315, c = 5.490
Bi_2Pt(β)	0.155	10	Hex., a = 6.44, c = 6.25
$Bi_{0.1-1}Pt Sb_{0.9-0}$	1.21–2.05		$B8_1$, a = 4.13–4.32, c = 5.47–5.48

Material	T_c(°K)	H_0(Oe)	Crystal Structure ∞
Bi_2Rb	4.25		C15, a = 9.609
Bi_4Rh(γ)	2.70		Hex. complex
BiRh	2.06		$B8_1$, a = 4.094, c = 5.663
Bi_3Rh	3.2		Like $NiBi_3$, Ortho., a = 11.522, b = 9.027, c = 4.24
BiSn*	3.48R	130	
BiSnΔ	3.81R		
Bi_xSn_y	3.85–4.18		
$Bi_{0-.01}Sn_y$	3.730–3.734		
Bi_3Sr	5.62	530	$L1_2$, a = 5.042
Bi_5Tl_3	6.4R		
$Bi_{.26}Tl_{.74}$	4.4		$L1_2$, disordered?
$Bi_{.26}Tl_{.74}$	4.15		$L1_2$, ordered?
Bi_2Y_3	2.25		
$Bi_{.3}Zr_{.7}$	1.51		
$CCo_{.01}Ta_{.99}$	Broad, 2–8		
CMo_2	2.78		L_3'
CMo	7.7		Hex.
CMo	9.26		
$C_{.50}Mo_xNb_y$	12.5M		B1
$C_{.50}Mo_xTi_y$	10.2M		B1
$C_{.50}Mo_xV_y$	9.3M		B1
$C_{.50}Mo_xZr_y$	9.5M		B1
$C_xN_yNb_{.50}$	17.8M		
CNb	6.0		B1
CNb	10.3		B1, a = 4.40
$C_{.829-.977}Nb_y$	1.05–11.1		
$C_{.977}Nb$	11.1		
CNb	~14 (Linear extrapolation)		
CRuΔ	2.00R		
CTa_2	3.3		D_6^{4h}, a = 3.09, c = 4.93
$C_{.848-.987}Ta_y$	2.04–9.7		
$C_{.987}Ta$	9.7		
CTa	~11 (Linear extrapolation)		
CTa	2–7		B1, a = 4.455
$C_{.50}Ta_xW_y$	10.5M		
CW_2	2.74		L_3'
CW_2	5.2		Cub. FC, a = 4.25
$CaIr_2$	4–6.15		C15, a = 7.545
CaPb	7.0R		

TABLE II. SUPERCONDUCTIVE MATERIALS (Continued)

Material	T_c(°K)	H_0(Oe)	Crystal Structure ∞
$CaRh_2$	6.40		C15,
			a = 7.525
Cd	0.54C	29	A3,
			a = 2.9788,
			c = 5.6167
Cd	0.56	30	
$Cd_{.3}Hg_{.7}$	1.70		
$Cd_{.4}Hg_{.6}$	1.92		
CdHg	1.77		Tet., B.C.,
			a = 3.940,
			c = 2.916
$Cd_{0-.009}In_y$	$T'_c(-.0535)$		
$Cd_{.97}Pb_{.03}$	~4.20R		
CdSn*	3.65	266	
$Cd_{0-.01}Sn_y$	3.725–3.734		
$Cd_{.17}Tl_{.83}$*	2.30R		
$Cd_{.18}Tl_{.82}$	2.54R		
$Ce_xGd_yRu_{.66}$	3.20–5.20		C15
$Ce_{.96}Gd_{.04}$	5.20		C15
Ru_2			
$Ce_{.92}Gd_{.08}$	5.20		C15
Ru_2			
Ce_xLa_y	1.3–6.30		
$Ce_{.01}La_{.99}$	3.60		
(Ce_xPr_{1-x}) Ru_2	1.4–5.3		C15
$CeRu_2$	4.90		C15,
			a = 7.535
$Co(Fe)Si_2$	1.40M		C1
$CoHf_2$	0.56		$E9_3$,
			a = 12.067
$Co_{.03}In_{.97}$	3.95		
(Thin films			
formed at 6°K)			
$Co_{0-.01}Mo_{.8}Re_{.2}$	~2–10		
$Co(Ni)Si_2$	1.40M		C1
$CoRhSi_4$	2.0–3.0		
$Co_xRh_ySi_{.66}$	3.65M		
$CoSi_2$	1.22	105	C1
Co_3Th_7	1.83		$D10_2$,
			a = 9.833,
			c = 6.200
Co_xTi_y	2.8M		Co in α-Ti
Co_xTi_y	3.8M		Co in β-Ti
CoTi	0.71		A2, SS,
			a = 2.99
$CoTi_2$	3.44		$E9_3$, 11.30
CoU	1.70		Distorted B2
CoU_6	2.29		$D2_c$
$Co_{0-.03}V_y$	1.5–5.3		
$Co_{.10}Zr_{.90}$	3.90		A3, SS
$CoZr_2$	6.30		C16
$Cr_{0-.0004}In_y$	1.7–4.2		
(Deposited at			
low temp)			
$Cr_{.6}Ir_{.4}$	0.4		HCP
$Cr_{.65}Ir_{.35}$	0.59		HCP
$Cr_{.7}Ir_{.3}$	0.76		HCP

Material	T_c(°K)	H_0(Oe)	Crystal Structure ∞
$Cr_{.72}Ir_{.28}$	0.83		
Cr_3Ir	0.45		A15
$Cr_{.8}Ir_{.2}$	0.50		A15
$Cr_{.85}Ir_{.51}$	0.77		A15
$Cr_{.875}Ir_{.125}$	0.77		A15
$Cr_{.9}Ir_{.1}$	0.78		A15
$Cr_{0-.02}Mo_{.8}$	~9–10		
$Re_{.2}$			
$Cr_{0-.1}Nb_y$	4.6–9.2		BCC, SS
$Cr_{.8-.6}Rh_{.2-.4}$	0.5–1.10		A3, SS
$Cr_{.875}Ru_{.125}$	1.20		A15, plus
$Cr_{.85}Ru_{.15}$	1.13		A15, SS
$Cr_{.8}Ru_{.2}$	1.92		A15, SS
Cr_3Ru	2.15		A15, as cast
Cr_3Ru	3.3		A15, annealed
$Cr_{.7}Ru_{.3}$	2.00		A15, SS
Cr_2Ru	2.02		$D8_b$
$Cr_{.6}Ru_{.4}$	2.10		$D8_b$
$Cr_{.6}Ru_{.4}$	1.60		A15, SS
$Cr_{.6}Ru_{.4}$	1.90		A3, SS
$Cr_{.5}Ru_{.5}$	1.30		$D8_b$
$Cr_{.1-.5}Ru_{.9-.5}$	0.34–1.65		A3, SS
Cr_xTi_y	3.6M		Cr in α-Ti
Cr_xTi_y	4.2M		Cr in β-Ti
$Cr_{.25}Ti_{.75}$	3.70		Cr in β = Ti
$Cr_{0-.005}V_y$	2.0–5.3	700–	A2, SS
		1250	
$Cr_{.01}V_{.99}$	4.8	980	A2, SS
$Cr_{0-.1}V_y$	1.9–5.2		A2, SS
$Cu_{.15}In_{.85}^{\Delta}$	3.75		
(Formed at			
6°K)			
$Cu_{.08}In_{.92}^{\Delta}$	4.4		
$Cu_{.04}In_{.96}$	4.4		
Cu_xPb_y	5.7 = 7.7R		
CuS	1.62R		B18
Cu_xSn_y	3.17–3.71		
$Cu_xSn_y^{\Delta}$	3.6–7		
Formed at			
10°K)			
$Cu_xSn_y^{\Delta}$	2.8–3.7		
(Formed at			
300°K)			
$CuTh_2$	3.49		C16,
			a = 7.28,
			c = 5.75
$Cu_{0-.027}V_y$	3.9–5.3		A2, SS
$D_{.11-.13}Nb$	8.76–9.13		
$Dy_{.01}La_{.99}$	3.8		
$Er_{.01}La_{.99}$	5.3		
$Er_{0-.08}La_y$	1.4–6.3		
$Eu_{.01}La_{.99}$	3.80		
$Eu_{0.0} = {}_{15}La_y$	1.5–6.3		
$Fe_{0-.015}In_y^{\Delta}$	0.8–4.2		
(Deposited			
4.2°K)			
$Fe_{0-.01}In_y^{\Delta}$	2.5–4.2		

TABLE II. SUPERCONDUCTIVE MATERIALS (Continued)

Material	T_c(°K)	H_0(Oe)	Crystal Structure ∞
$Fe_{.01}Mo_{0-.3}$ $Nb_{1-.7}$	1–8		
$Fe_{0-.04}Mo_{.8}Re_{.2}$	1–10		
$Fe_{.05}Ni_{.05}Zr_{.90}$	~3.9		
Fe_3Th_7	1.86		$D10_2$, a = 9.823, c = 6.211
Fe_xTi_y	3.2M		Fe in α-Ti
Fe_xTi_y	3.7M		Fe in β-Ti
$Fe_xTI_{.6}Vi_{.4}$	6.8M		
FeU_6	3.86		$D2_c$
$Fe_{.1}Zr_{.9}$	1		A3, SS
Ga	1.091	51	Ortho.
Ga	1.087C	59.4	a = 4.5198, b = 7.6602, c = 4.5258
Ga (thin film) (Formed at low temp)	8.40		?
$Ga_{.5}Ge_{.5}Nb_3$	7.3		A15, a = 5.175
$Ga_{0-.009}In_y$	T'_c(+.0325)		
Ga_2Mo	9.5		
Ga_4Mo	9.8		
$GaMo_3$	0.76		A15, a = 4.943
$GaNb_3$	14.5		A15, a = 5.171
$GaNb_3$ (sintered)	12.5–13.2		A15
$Ga_{0-.3}Nb_3Sn_{1-.7}$	18.03–18.37R		A15
$Ga_0 = {}_3Nb_3$ $Sn_{1-.7}$	18.0–18.17		A15
$Ga_0 = {}_6Nb_3$ $Sn_{1-.4}$	14.0–18.1		A15, a = 5.230a 5288
$Ga_{.8}Nb_3Sn_2$	13.1		Not single phase, A15
$Ga_xNb_3Sn_{1-x}$ 18.0–18.15 R	18.0–18.35		A15
$(Ga_{.5}Si_{.5})V_3$	8.6–11.9		A15
GaV_3	16.8		A15, a = 4.816
GaV_4	10.1		A15
$GaV_{3.34}$	13.0		A15
$GaV_{2.95}$	14.8		A15
$GaV_{2.61}$	10.2		A15
$GaV_{2.48}$	9.6		A15
GaV_2	7.6		A15
$GaV_{1.95}$	5.3		A15
$Gd_{.005}La_{.995}$	3.7		
$Gd_{.006-\ 01}$ La_y	1.0–2.8		
$Gd_{0-.008}$ La_y	1.3–5.7		
$Gd_{.01}La_{.99}$	0.6		

Material	T_c(°K)	H_0(Oe)	Crystal Structure ∞
$Gd_{.09}La_{.91}$ Os_2	≈6 (Superconducting and Ferro.)		
$Gd_xOs_{.66}Y_y$	1.4–4.7 2.2–7.2		
$Gd_{0-.02}Pb^{\Delta}_{1-.98}$ (Deposited at low temp)			
Gd_xRu_2 Th_{1-x}	3.60M		C15
Gelr	4.70		B31
Ge_2La	1.49		Ortho., distorted $ThSi_2$ type
$Ge_{.7}Mo$	1.20		
$GeMo_3$	1.43		A15
$GeNb_3$	6.90		A15, a = 5.166
$GeNb_2\Delta$	1.90		
$Ge_{.22}Nb$	5.3		A15, a = 5.168
$Ge_{.55}Nb_{3.45}$	4.9		A15
$Ge_{.72}Nb_{3.28}$	5.5		A15
Ge_xNb_3 Sn_{1-x}	17.6–18 0		A15
$Ge_{.5}Nb_3Sn_{.5}$	12.6		A15, a = 5.236
Ge_2Pr	Ferro.		C_c
GePt	0.40		B31, a = 6.088, b = 5.733, c = 3.701
Ge_3Rh_5	2.12		Related to $InNi_2$, ortho., a = 5.42, b = 10.32, c = 3.96
GeRh	0.96		B31, a = 5.70, b = 6.48, c = 3.25
Ge_2Sc	1.30–1.31		
$GeTa_2$°	1.60?		
GeV_3	6.01		A15
Ge_2Y	3.80		C_c
$H_{.33}Nb$	7.28		Cub., BC, a = 3.323
$H_{.1}Nb$	7.38		Cub., BC, a = 3.327
$H_{.05}Nb$	7.83		Cub., BC, a = 3.311
$H_{.12}Ta$	2.81		Cub., BC, a = 33.31
$H_{.08}Ta$	3.26		Cub., BC, a = 3.316
$H_{.04}Ta$	3.62		Cub., BC, a = 3.314

TABLE II. SUPERCONDUCTIVE MATERIALS (Continued)

Material	T_c(°K)	H_0(Oe)	Crystal Structure ∞
$Hf_{.9}Mo_{.1}$	2.5		A2 plus A3, SS's
HfN	6.2		B1
Hf_xNZr_y	6.2–10.7		
$Hf_{0-.5}Nb_y$	8.3–9.5		A2, SS
Hf_xNb_y	>1.2		
$Hf_{.65-.85}Os_y$	2.3–2.4		A2, SS
$HfOs_2$	2.69		C14, a = 5.184 c = 8.468
$HfRe_2$	4.80		C14, a = 5.239, c = 8.584
$HfRe_2$	5.61		Hex.
$Hf_{.14}Re_{.86}$	5.86		A12
$Hf_{.025}Re_{.975}$	7.3		A3, SS
$Hf_{.875}Re_{.125}$	1.70		A2, SS
$Hf_{.99}Rh_{.01}$	0.85		C14, a = 5.239, c = 8.584
$Hf_{.98}Rh_{.02}$	1.17		
$Hf_{.97}Rh_{.03}$	1.37		
$Hf_{.96}Rh_{.04}$	1.51		
$Hf_{0-.55}Ta_y$	4.4–6.5		A2, SS
$Hf_{.99}Zr_{.01}$	0.37		A3, SS
Hg(β)	3.949	339	Tet., B.C., a = 3.995 c = 2.825
Hg(α)	4.153	412	A10, a = 3.005, α = 70°31.7
$Hg_{.995-1.0}In_y$	41.3–4.15		Rhomb.
$Hg_{.8-.98}In_y$	3.15–4.07		
$Hg_{.08-.20}In_y$	3.25–4.55		Cub. FC
$Hg_{0-.08}In_y$	3.32–3.62		Tet.
HgIn	3.81		
$Hg_{0-.07}In_y$	3.41–3.34–3.43		
Hg_2K	1.20		Ortho.
Hg_3K	3.18		
Hg_4K	3.27		
Hg_8K	3.42		
Hg_3Li	1.7		Hex.
Hg_2Na	1.62		Hex.
Hg_4Na	3.05		
Hg_xPb	4.14–7.26		
$Hg_{.12-.20}Pt_y$	3.75–3.98		
HgSn	4.20		
$Hg_{0-.01}Sn_y$	3.726–3.734		Tet., SS
$Hg_{.026}Tl_{.974}$	2.30		
$Hg_{.104}Tl_{.896}$	2.96		
$Hg_{.131}Tl_{.896}$	3.25		
$Hg_{.698}Tl_{.302}$	3.888		
$Hg_{.714}Tl_{.286}$	3.875		

Material	T_c(°K)	H_0(Oe)	Crystal Structure ∞
$Hg_{.734}Tl_{.266}$	3.690		
$Hg_{.847}Tl_{.153}$	4.100		
$Hg_{.935}Tl_{.065}$	4.100		
$Hg_{.97}Tl_{.03}$	4.109		
$Hg_{.72}Tl_{.28}$	3.84R		
$Ho_{.01}La_{.99}$	5.20		
$Ho_{0-.042}La_y$	1.3–6.3		
In	3.396C	278	A6, a = 4.5979, c = 4.9467
In	3.4035	293	
In^{Δ} (Deposited at low temp)	3.95–4.25		
$InLa_3$	10.40		$L1_2$, a = 5.07
$In_xMn^{\Delta}_{0-.0004}$ (Deposited at 4.2°K)	2.2–4.2		
$InNb_3$	9.2		A15, a = 5.303
$In_{0-.3}Nb_3Sn_{1-.7}$	18.0–18.19		A15
InPb	6.65		
In_xPb_y	3.39–7.26		
In_xPb_y	4.36	~2000	
$In_xPb_{0-.003}$	$T_c'(-.0147)$		
$In_{.97}Pd^{\Delta}_{.03}$ (Formed at 6°K)	4.45		
$(InSb)_{0-.07}Sn_y$	3.67–3.74		
$InSc_3$			D019, a = 6.421, c = 5.183
$In_xSn_{0-.02}$	$T_c'(-.01 + .125)$		
$In_xSn_{0-.005}$	$T_c'(-.0097)$		
$In_{0-.012}Sn_y$	3.724–3.734		Tet., SS
In_xSn_y	3.40–6.40		
In_xSn_y	3.4–7.3		
$In_{.7}Sn_{.3}$	7.30		
$In_{.5}Sn_{.5}$	6.9		
$In_{.75}Sn_{.25}$	6.5, 5.5		
$In_{0-.03}Sn_y$	3.63–3.73		
$In_xTh_{0-.008}$	T_c' (−.0252).		
$In_{.005}Tl_{.995}$	2.418		
InTl	2.7		
In_xTl_y	2.7–3.374	252 = 284	
$In_{.95-.63}Tl_{.05-.37}$	2.4–3.30		
$In_{.92}Zn^{\Delta}_{.08}$ (Formed at 6°K)	4.55		
$In_{.85}Zn^{\Delta}_{.15}B$	4.6		
Ir	0.140	19	A1, a = 3.8389
IrMo	<1.0		A3

TABLE II. SUPERCONDUCTIVE MATERIALS (Continued)

Material	T_c(°K)	H_0(Oe)	Crystal Structure ∞
$IrMo_3$	8.8		A15, a = 4.974
$IrMo_3$	6.8		$D8_b$, a = 9.631, c = 4.956
$Ir_{.26}Mo_{.74}$	6.7		$D8_b$, a = 9.63, c = 4.96
$Ir_{.26}Mo_{.74}$	9.6		A15, a = 4.972
$Ir_{.37}Nb_{.63}$	2.40		$D8_b$, a = 9.86, c = 5.06
$IrNb\Delta$	7.9		$D8_b$,
Ir_2Nb_3	9.8		$D8_b$, a = 9.834, c = 5.052
$IrNb_3$	1.7		A15, a = 5.131
$Ir_{.287}O_{.14}Ti_{.573}$	5.5		$E9_3$, a = 11.620
$Ir_{.265}O_{.085}Ti_{.65}$	2.30		$E9_3$, a = 12.430
$Ir_{.95}Os_{.05}$	0.6		A3, SS
$Ir_{.85}Os_{.15}$	0.55		A3
$Ir_{.7}Os_{.3}$	0.46		Cub., FC
$Ir_{.7}Os_{.3}$	0.97		A3, Cub., FC
$Ir_{.6}Os_{.4}$	0.98		A3, Cub., FC
$Ir_{.5}Os_{.5}$	0.98		Cub., FC
$Ir_{.4}Os_{.6}$	0.74		Cub., FC
$Ir_{.3}OS_{,7}$	0.49		Cub., FC
$Ir_{.25}Os_{.75}$	0.38		Cub., FC
$Ir_{.2}Os_{.8}$	0.30		Cub., FC
$IrOsY$	2.60		C15
$Ir_{.5}Os_{.5}Y$	2.40		C14
Ir_2Sc	1.03		C15, a = 7.348
Ir_2Sr	5.7		C15, a = 7.700
Ir_2Th	6.50		C15, a = 7.664
Ir_3Th_7	1.52		$D10_2$, a = 10.06, c = 6.290
$IrTi_3$	5.40		A15, a = 5.009
$Ir(Ir_{.33}V_{2.67})$	1.39		A15, a = 4.794
$Ir_{.28}W_{.72}$	4.46		$D8_b$, a = 9.67, c = 5.00
$Ir_{.25}W_{.75}$	2.1–3.82		
Ir_2Y	2.18		C15, a = 7.500–7.520
Ir_2Zr	4.10		C15, a = 7.359
$Ir_{.1}Zr_{.9}$	5.50		A3
$La(\alpha)$	4.80		Hex., a = 3.770, c = 12.159
$La(\beta)$	5.91		A1, a = 5.296
La_xLu_y	2.0–3.4		
$La_{.55}Lu_{.45}$	2.20		La type, Hex.
$La_{.8}Lu_{2.}$	3.40		La type, Hex.
$La_{.99}Lu_{.01}$	5.60		La type, Hex.
$La_{0-.044}Nd_y$	1.4–6.3		
$La_{.99}Nd_{.01}$	4.70		
$LaOs_2$	6.5		C15, a = 7.737
$La_{.99}Pr_{.01}$	5.30		
$LaRu_2$	1.63		C15, a = 7.702
$LaSi_2$	~2.5		C_c
La_5Si_3	1.6		$D8_8$(?)
$La_{.99}Sm_{.01}$	4.50		
$La_xSm_{0-.02}$	1.3–6.3		
$La_xTb_{0-.013}$	1.4–6.3		
$La_{.99}Tb_{.01}$	2.55		
La_xY_y	5.40M		Cub. and Hex.
$La_{.6}Y_{.4}$	1.70		La type, Hex.
$La_{.75}Y_{.25}$	2.50		La type, Hex
$La_{.85}Y_{.15}$	3.20		La type, Hex.
$La_{.95}Y_{.05}$	5.40, 4.4		La type, Hex. + Cub. F. C.
$La_{.99}Yb_{.01}$	5.50		
LiPb	7.20R		
$LuOs_2$	3.49		C14, a = 5.254, c = 8.661
$LuRu_2$	0.86		C14, a = 5.204, c = 8.725
$Mg_{\sim .47}Tl_{\sim .53}$	2.75	220	B2, a = 3.628
$Mn_{0-.003}Pb_y^{\Delta}$ (Deposited at 7°K)	2.3–7.2		
Mn_xTi_y	0.6–2.3		Mn in α = Ti
Mn_xTi_y	1.1–3.0		Mn in β = Ti
MnU_6	2.32		$D2_c$
Mo	0.92	98	A2, a = 3.1468
Mo_2N	5.0		Cub., FC
MoN	12.0		Hex., ordered
$Mo_{0-.03}Nb_y$ (See Fc, Mo, Nb)	1–8.7		
$Mo_{0-.3}Nb_y$	2.6–9.2		A2, SS
$Mo_{0-.42}Nb_y$	0.25–9.2		
$Mo_{.4}Nb_{.6}$	0.60		
$Mo_{.43-.51}Nb_y$	~.07–.181		A2, SS
$Mo_{.51}Nb_{.49}$	~.07		A2, SS
$Mo_{.8}Ni_{0-.02}Re_{.2}$	~8–10		
Mo_3Os	7.20		A15

TABLE II. SUPERCONDUCTIVE MATERIALS (Continued)

Material	T_c(°K)	H_0(Oe)	Crystal Structure ∞
$Mo_{.62}Os_{.38}$	5.65		$D8_b$,a = 9.60, c = 4.93
$MoOs\Delta$	5.20		$D8_b$
$Mo_3P\Delta$	7.00		
Mo_3P	5.31		DO_e
$Mo_{.5}Pd_{.5}$	3.52		A3, SS
Mo_3Re	9.80	530	
$Mo_{.5}Re_{.5}$	6.4, 7.3		$D8_b$, a = 9.61, c = 4.98
Mo_2Re	10.8		
$Mo_{0-.12}Re_y$	1.6–7.9		A3, SS
$Mo_{.23}Re_{.77}$	9.3		χ-phase
$Mo_{.27}Re_{.73}$	6.5, 9.1		σ and χ-phases
$Mo_{.28}Re_{.72}$	6.5		σ and χ-phases
$Mo_{.5}Re_{.5}$	5.7, 11.6		A2, SS and σ-phase
$Mo_{.33-.45}Re_y$	5.7–6.0		σ-phase
$Mo_{.42}Re_{.58}$	6.5		σ-phase
$Mo_{.42}Re_{.58}$	8.4		$D8_b$, a = 9.59, c = 4.97
$Mo_{.35}Re_{.65}$	8.6		$D8_b$, a = 9.57, c = 4.97
$Mo_{.62-.95}Re_y$	1.2–12.2		A2, SS
$MoRe\Delta$	6.0		$D8_b$
$MoRe_3$	9.89, 9.26		A12
$Mo_{.8}Re_{.2}Rh_x$	~10–10+		
$Mo_{.8}Re_{.2}Ru_x$	~10–10+		
$Mo_{.8}Rh_{.2}$	8.20		Cub., B. C., SS
$Mo_{.97}Rh_{.03}$	1.50		Cub., B. C., SS
$MoRh$	1.97		A3. SS
$Mo_{.95}Rh_{.05}$	2.30		Cub., B. C., SS
$Mo_{.92}Rh_{.08}$	3.90		Cub., B. C., SS
$Mo_{.84}Rh_{.16}$	7.70		Cub., B. C., SS
$Mo_{.2}Ru_{.8}\Delta$	1.66		A3, SS
$MoRu$	9.5–10.5		A3, SS
$MoRu\Delta$	6.90		$D8_b$
$Mo_{.025}Ru_{.975}$	0.59		
$Mo_{.05}Ru_{.95}$	0.76		A3, SS
$Mo_{.10}Ru_{.90}$	1.0		A3, SS
$Mo_{.25}Ru_{.75}$	2.6		A3, SS
$Mo_{.6}Ru_{.4}$	7.0		$D8_b$, a = 9.55, c = 4.95
Mo_3Sb_4	2.10		
Mo_3Si	1.30		A15
$MoSi_{.7}$	1.34		
$Mo_{.15}Si_{.25}V_{.60}$	4.54		A15, a = 4.758
$Mo_{.12}Si_{.25}V_{.63}$	5.1		A15
$Mo_{.05}Si_{.25}V_{.70}$	5.59		A15, a = 4.736
$Mo_{.02}Si_{.25}V_{.73}$	10.4		A15
$Mo_{.009}Si_{.248}V_{.743}$	14.0		A15
$Mo_{.005}Si_{.25}V_{.745}$	16.0		A15
$Mo_{.05}Tc_{.95}$	10.8		
$Mo_{.1}Tc_{.9}$	13.4		
$Mo_{.15}Tc_{.85}$	14.0		
$Mo_{.1}Tc_{.8}$	14.0		
$Mo_{.25}Tc_{.75}$	15.8		
$Mo_{.3}Tc_{.7}$	12.0		Annealed, σ = phase, a = 9.5091, c = 4.9448
$Mo_{.35.}Tc_{65}$	13.3		
$Mo_{.4}Tc_{.6}$	14.7, 13.4		
$Mo_{.45}Tc_{.55}$	14.0		
$Mo_{.5}Tc_{.5}\Delta$	12.6		
$Mo_{.5}Tc_{.5}$	14.3		
$Mo_{.6}Tc_{.4}$	12.8		
$Mo_{.5}Tc_{.5}$	14.0		
$Mo_{.065}Ti_{.935}$	2.23		BCC, metastable
$Mo_{.06}Ti_{.94}$	2.04		BCC, metastable
$Mo_{.08}Ti_{.92}$	2.60		BCC, metastable
$Mo_{.09}Ti_{.91}$	3.09		BCC, metastable
Mo_xTi_y	2.0–>4.20		BCC, a = 3.223–3.264
$Mo_{.0625-.086}Ti_y$	2.04–3.09		
$Mo_{.06-.6}Ti_{.94-.4}(\beta)$	1.3–3.7		A2, SS
$Mo_{0-.05}Ti_{1-.95}(\alpha')$	1.9–2.5		A3, SS
$Mo_{.116}U_{.874}(\gamma)$	1.85		A2, SS, a = 3.44
$Mo_{.217}U_{.883}(\gamma)$	2.06		A2, SS, a = 3.40
$Mo_{.305}U_{.695}$ (γ)	1.97		A2, SS, a = 3.39
$Mo_{.14}U_{.86}$	2.02		γ-U
$Mo_{.18}U_{.82}$	2.07		γ-U
$Mo_{.22}U_{.78}$	2.03		γ-U
$Mo_{.25}U_{.75}$	1.94		γ-U
$Mo_{.3}U_{.7}$	1.84		γ-U
Mo_xU_y	1.85–2.06		γ-U, plus Mo
$Mo_{0-.3}U_y$	2.05M		α-U, γ-U plus Mo
Mo_xV_y	0–5.3		
$N_{.62}Nb$	3.80		NbN Tet. + Nb_2N, a = 3.042, c = 4.985
NNb	<5.10		Hex. NbN (NbNI)

TABLE II. SUPERCONDUCTIVE MATERIALS (Continued)

Material	T_c(°K)	H_0(Oe)	Crystal Structure ∞
$N_{.48}Nb$	<5.50		Hex. Nb_2N
$N_{.19}Nb$	5.72		Nb + Nb_8N, a = 3.050, c = 4.946
$N_{.62}Nb$	6.10		Tet. NbN + Hex Nb_2N
$N_{.80}Nb$	5.2–9.7		TbN III + NbN Tet., a = 4.377, c = 4.323
$N_{.19}Nb$	7.20		Hex Nb_2N + Cub. Nb
N_3Nb_4	7.20		Tet., FC
$N_{.82}Nb$	8.66		NbN III + NbN Tet., a = 4.378, c = 4.321
$N_{.8}Nb$	8.90		Tet., NbN
$N_{.82}Nb$	10.60		Cub. + Tet. NbN
$N_{.94}Nb$	8.0–13.1		NbN III, Cub. a = 4.380
$N_{.94}Nb$	7–14		Cub. NbN III
N_xNb_y	11–16		B1, a = 4.377–4.381
NNb	15.60		B1, a = 4.375
$N_{.003}Nb_{.997}$	9.120		A2
$N_{.016}Nb_{.984}$	9.245		A2
N(12.8w/o)Nb	15.2		B1
$N_{.07}NbO_{.07}$	6.2		Cub., BC, a = 3.304
$N_{.77}NkO_{.07}$	11.0		NbN III + NbN Tet., a = 4.377, c = 4.320
$N_{.58}NbO_{.02}$	6.0		NbN Tet. + Nb_2N, a = 3.044, c = 4.929
$N_{.14}NbO_{.03}$	7.0		Nb_2N + Nb
$N_{.66}NbO_{.10}$	8.1		NbN Tet.
$N_{.88}NbO_{.08}$	9.9		NbN III, cub.
$N_{.47}O_{.03}Ti_{.50}$	2.9		Cub., a = 4.230
$N_{.47}O_{.03}Ti_{.50}$	4.86		Cub., a = 4.240
$NO_{.49}O_{.01}Ti_{.50}$	5.58		Cub., a = 4.240
$N_{.48}O_{.02}Ti_{.50}$	5.58		Cub., a = 4.240
$N_{12}O_2V_{86}$(w/o)	5.8		Cub., a = 4.130
$N_{19}O_2V_{79}$(w/o)	6.7		Cub., a = 4.132
$N_{14}O_8V_{78}$(w/o)	8.2		Cub., a = 4.135

Material	T_c(°K)	H_0(Ie)	Crystal Structure ∞
$N_{0.34}Re$	4–5		Cub., FC
$N_{0-.018}Ta_y$	3.63–4.483		
NTi	4.86		B1
NTi	5.60		B1
NV	7.50		B1
NV	8.20		B1
NZr	8.90		B1
NZr	9.05		B1
NZr	10.7		B1
$Na_{.28}Pb_{.72}$	7.20R		
Nb	9.17C	1944	A2, a = 3.3004
Nb	9.13	1980	
$Nb_{.962}O_{.038}$	5.840		A2
$Nb_{.986}O_{.014}$	8.200		A2
$Nb_{.948}O_{.052}$	6.650		A2
$Nb_{.936}O_{.064}$	9.023		A2
Nb_3Os_2	1.78, 1.85		$D8_b$, a = 9.844, c = 5.056
$Nb_{.5}Os_{.5}$	2.86		A12, a = 9.760
$NbOs_2$	2.52		A12, a = 9.655
Nb_3Os	1.05		A15, a = 5.121
NbOsΔ	1.40		$D8_b$
$Nb_3Pb_xSn_{1-x}$	18.0–18.15		
$Nb_{.6}Pd_{.4}$	2.04 – 2.47		A12, a = 9.77
$Nb_{.6}Pd_{.4}$	1.7		A12
NbPdΔ	2.0		$D8_b$
$Nb_{.625}Pt_{.375}$	3.73		$D8_b$, a = 9.91, c = 5.12
$Nb_{.625}Pt_{.375}$	4.2		$D8_b$
$Nb_{.62}Pt_{.38}$	4.01		$D8_b$, a = 9.91, c = 5.13
NbPtΔ	2.40		$D8_b$
Nb_3Pt	9.20		A15, a = 5.153
$Nb_{.18}Re_{.82}$	9.7, 8.89		A12, a = 9.641
NbReΔ	2.0		$D8_b$
$Nb_{.4}Re_{.6}$	2.36		A12, a = 9.781
$Nb_{.4}Re_{.6}$	2.5		$D8_b$, a = 9.77, c = 5.14, plus A12, a = 9.773
$Nb_{.5}Re_{.5}$	2.0–3.8		$D8_b$, a = 9.79, c = 5.10, plus cub., a = 3.189
$Nb_{.38}Re_{.62}$	2.45		A12
$Nb_{.6}Rh_{.4}$	4.04		$D8_b$, a = 9.80. c = 5.07
NbRhΔ	4.10		$D8_b$

TABLE II. SUPERCONDUCTIVE MATERIALS (Continued)

Material	T_c(°K)	H_0(Oe)	Crystal Structure ∞
Nb_3Rh	2.5		A15, a = 5.115
$Nb_{.6}Ru_{.4}$	1.2		
$Nb_{.4}Ru_{.6}$	2.5		
$Nb_{.9}Ru_{.1}$	2.8		
$Nb_{.925}Ru_{.075}$	4.20		
$Nb_3Sb_{0-.3}Sn_{1-.7}$	14.7–18.0		A15
$Nb_3Sb_{0-.7}Sn_{1-.3}$	6.8–18.0		A15, a = 5.292–5.270
Nb_3SiSnV_3	4.0		
$Nb_3Si_{.5}Sn_{.5}$	8.3		A15
$Nb_3Si_{.6}Sn_{.4}$	6.5		A15
Nb_3Sn	18.05		A15, a = 5.289
Nb_3Sn (Strain)	17.5–17.9		A15
Nb_3Sn (Torsion)	16.95–17.9		A15
$Nb_xSn_y^{\Delta}$	8.2–17.9		
$NbSn_2$	2.60	620	Ortho., a = 9.852, b = 5.645, c = 19.126
Nb_xSn_y	17.25–18.18		
$Nb_{.5}Sn_{.5}$	17.91		
Nb_3Sn_2	16.6		Tet., a = 6.901, c = 9.533
$Nb_{.92}Sn_{.08}$	5.6		SS
$Nb_{.8}Sn_{.2}$	18.18		A15
$Nb_{.8}Sn_{.2}$	18.0		A15
$Nb_{.84}Sn_{.16}$	5.6		A15, a = 5.282
$Nb_{.8}Sn_{.2}$	5.6–6.6		A15, a = 5.283–5.282
$Nb_{.76}Sn_{.24}$	18.2		A15, a = 5.288
$Nb_{.72}Sn_{.28}$	16.0		A15, a = 5.286
$Nb_{.8}Sn_{.2}$	7.2		A15, a = 5.287
$Nb_{.76}Sn_{.24}$	17.5		
$Nb_{.8}Sn_{.2}$	18.5		A15, a = 5.290
$Nb_{.72}Sn_{.28}$	18.2		A15, a = 5.290
$Nb_{.76}Sn_{.24}$	18.1		A15, a = 5.291
$Nb_{3(1-x)}SnTa_{3x}$	14.1–18.0		
$Nb_{3x}SnTa_{3(1-x)}$	6.0–18.0		
$Nb_{2.75}SnTa_{.25}$	17.8		A15
$Nb_{2.5}SnTa_{.5}$	17.6		A15
$NbSnTa_2$	10.8		A15, a = 5.287

Material	T_c(°K)	H_0(Oe)	Crystal Structure ∞
Nb_2SnTa	16.4		A15, a = 5.280
$Nb_2SnTa_{.5}V_{.5}$	12.2		A15
$NbSnTaV$	6.2		A15, a = 5.175
$Nb_3Sn_{1-x}Tl_x$	18.0–18.16		
$Nb_3Sn_2V_3$	7.4		A15
$Nb_{2.5}SnV_{0.5}$	14.2		A15
Nb_2SnV	9.8		A15, a = 5.171
$NbSnV_2$	5.5		A15, a = 5.115
$Nb_{.64}Ta_{.36}$	6.8		A2, SS
$Nb_{.47}Ta_{.53}$	6.2		A2, SS
$Nb_{.29}Ta_{.71}$	5.4		A2, SS
$Nb_{.16}Ta_{.84}$	4.85		A2, SS
$Nb_{.06}Ta_{.94}$	4.6		A2, SS
Nb_xTa_y	4.4–9.2		A2, SS
$Nb_{.73}Ta_{.02}Zr_{.25}$	>4.2		
$Nb_{.70}Ta_{.05}Zr_{.25}$	>4.2		
$Nb_{.65}Ta_{.10}Zr_{.25}$	>4.2		
$NbTc_3$	10.5		A12, a = 9.625
$Nb_{0-.15}Ti_{1.-85}$	0.6–5.5 (7.9, annealed)		A3, SS
Nb_2Ti	9.3		
$Nb_{0.2-.36}Ti_y$	1.7–7.5 (quenched)		
$Nb_{.02-.5}Ti_y$	5.8–9.5 (slow cooled)		
$Nb_{.25-1}Ti_y$	6.7–9.6		A2, SS
$Nb_{0-.25}Ti_y$	1.6–6.3		Hex
$Nb_{.5}Ti_{.5}$	9.5		
$Nb_{.222}U_{.778}(\gamma)$	1.98		A2, SS, a = 3.45
$Nb_{.26}U_{.74}$	1.85		γ-U
Nb_xU_y	0.9–1.0		Nb in α-U
Nb_xU_y	1.8–2.0		Nb in γ-U
Nb_xV_y	4.1–9.2		A2, SS
$Nb_{.65-1}W_y$	1.5–9.2		A2, SS
$Nb_{.75}Zr_{.25}$	10.8		
$Nb_{0.4-1}Zr_y$	8.8–10.8		A2, SS
Nb_2Zr	10.80		
$Ni_{.05}Pd_{.95}Te_2$	1.40		C6
$Ni_{.1}Pd_{.9}Te_2$	1.30		C6
Ni_3Th_7	1.98		$D10_2$, a = 9.885, c = 6.225
$Ni_{0.03}V_y$	2.1–5.3		A2, SS
$Ni_{1.5}V_{.5}Zr$	0.43		C15, a = 7.068
$Ni_{.1}Zr_{.9}$	1.50		A3, SS
$NiZr_2$	1.52		
$O_{.105}Pd_{.285}Zr_{.61}$	2.09		$E9_3$, a = 12.470

TABLE II. SUPERCONDUCTIVE MATERIALS (Continued)

Material	T_c(°K)	H_0(Oe)	Crystal Structure ∞
OReTi	5.74		
$O_{.14}Rh_{,.287}Ti_{.573}$	3.37		$E9_3$, a = 11.588
$O_{.105}Rh_{.285}Zr_{.61}$	11.8		$E9_3$, a = 12.408
$O_{.006}Ta_{.994}$	4.185		A2
$O_{.017}Ta_{.983}$	3.78		A2
$O_{.028}Ta_{.972}$	3.48		A2
$O_{.03}V_{.97}$	1.8–2.4		
OV_3Zr_3	7.5		$E9_3$, a = 12.160
Os	0.655	65	A3, a = 2.7353, c = 4.3191
Os	0.58C		
OsReY	2.00		C14
Os_2Sc	4.60		C14, a = 5.179, c = 8.484
OsTa	1.95		A12, a = 9.773
Os_3Th_7	1.51		$D10_2$, a = 10.02, c = 6.285
OsTi	0.46		B2, a = 3.077
$Os_{.34}W_{.66}$	3.81		$D8_b$, a = 9.63, c = 4.98
$Os_{.52-.7}W_y$	0.9–3.7		A3, SS
$Os_{.37-.45}W_y$	3.7–4.1		A3, SS plus $D8_b$
$Os_{.23-33}W_y$	2.5–3.6		$D8_b$
$Os_{.15}W_{.85}$	2.2		BCC, SS plus $D8_b$
$Os_{.1}W_{.9}$	1–2		A2, SS
$Os_{.23}W_{.925}$	0.9		A2, SS
OsW_3	2.21–3.02		
OsWΔ	4.40		$D8_b$
Os_2Y	4.7		C14, a = 5.307, c = 8.786
Os_2Zr	3.0		C14, a = 5.219, c = 8.538
OsZr (4.04 VA)	1.5		A3, SS
OsZr (4.1 VA)	3.0		A3, SS
$Os_{.1}Zr_{.9}$	5.2		A3, SS
OsZr (4.4 VA)	5.2		?
OsZr (4.2 VA)	5.4		A3, SS
OsZr (4.28 VA)	5.6		
OsZr (4.24 VA)	5.6		
PPb	7.8R		Mixture of phases
PPb*	7.8R		
P_4Rh_5Δ	1.22		
PW_3	2.26		
Pb	7.23C		A1, a = 4.9402
Pb	7.193	803	

Material	T_c(°K)	H_0(Oe)	Crystal Structure ∞
Pb_2Pd	2.95		C16
Pb_2Rh	2.66		C16
Pb_4Pt	2.80		Related to C16
PbSb*	6.6		
$Pb_{0-.01}Sn_y$	3.731–3.734		Tet., SS
$Pb_{0-.003}Tl_y$	T'_c(+.045)		
Pb_xTl_y	>4.2	<185–555	
$PbTl_2$	3.75, 4.09		
$Pb_{.352}Tl_{.648}$	3.75		
Pb_3Zr_5	4.60		$D8_8$, a = 8.529, c = 5.864
$PbZr_3$	0.76		A15
$Pd_{.9}Pt_{.1}Te_2$	1.65		C6
$Pd_{.95}Pt_{.05}Te_2$	1.71		C6
$Pd_{.95}Rh_{.05}Te_2$	1.65		C6
$Pd_{.05}Ru_{.05}Zr_{.90}$	~9		
PdSb	1.25		C2, a = 6.459
PdSbl	1.50		$B8_1$
Pd_2Se	2.20		$Pd_{2.8}Se$ + $Pd_{1.1}Se$ phases
Pd_5Se_2	2.30		$Pd_{2.8}Se$ phase
Pd_xSe_y	2.50M		
PdSi	0.93		B31, a = 6.133, b = 5.59. c = 3.381
$Pd_{1.75}Te_2$	2.25; 1.93 (unannealed)		C6
$Pd_{1.5}Te_2$	2.21; 1.87 (unannealed)		C6
$Pd_{1.25}Te_2$	2.20; 1.90 (unannealed)		C6
$Pd_{1.05}Te_2$	1.77; 1.74 (unannealed)		C6
$PdTe_2$(single crystal)	1.53; 1.46 (unannealed)		C6
PdTe	2.30		$B8_1$
$Pd_{.1}Zr_{.9}$	7.5		A3, SS
PtSb	2.10		$B8_1$
PtSi	0.88		B31, a = 5.932, b = 5.595, c = 3.603
Pt_3Ta_7	<1.2–1.5		$D8_b$, a = 9.93, c = 5.16
PtTaΔ	1.0		$D8_b$
$PtTi_3$	0.58		A15, a = 5.032
PtV_3	2.83		A15, a = 4.814
$Pt_{.5}W_{.5}$	1.45		Al, SS
$Pt_{.6-.3}W_{.4-.7}$	0.4–2.15		Al, SS

Material	T_c(°K)	H_0(Oe)	Crystal Structure ∞
$Pt_{.9-.63}W_{.1-.37}$	2.55–2.7		A1 and A2, SS
$Pt_{.98-.95}W_{.02-.05}$	1.1–2.2		A2, SS
Pt_2Y	1.57		C15, a = 7.590
PtZr	3.0		A3, SS
Re	1.699	198	hcp, a = 2.760, c = 4.458
$Re_{.64}Ta_{.36}$	1.46		A12, a = 9.765
Re_3Ta	6.78		A12
ReTaΔ	1.3		$D8_b$
$Re_{.65}Ta_{.35}$	1.58		A12, a = 9.762
$Re_{6.}Ta_{.4}$	1.4		$D8_b$, a = 9.77, c = 5.09 plus A12, a = 9.783
$Re_{24}Ti_5$	6.6		A12, a = 9.587
Re_xTi_y	1.2–2.7		
$Re_{.83}Ti_{.17}$	5.1		A12, a = 9.595
$Re_{.9}V_{.1}$	9.4		
$Re_{.52}W_{.48}$	5.2		σ-phase
$Re_{.84-1}W_y$	1.6–8.0		A3, SS
$Re_{.7}W_{.3}$	4.9		σ-phase
$Re_{.7}W_{.3}$	8.6		χ-phase
$Re_{.15-.4}W_y$	2.3–4.0		A2, SS
$Re_{.5-.7}W_y$	4.8–5.2		σ-phase
$Re_{.6}W_{.4}$	4.9		σ-phase
$Re_{.4}W_{.6}$	5.1		A2, SS and σ-phase
$Re_{.25}W_{.75}$	4.6		A2, SS
$Re_{.15}W_{.85}$	2.4		A2, SS
$Re_{.5}W_{.5}$	5.03		$D8_b$, a = 9.63, c = 5.01
ReWΔ	5.20		$D8_b$
Re_3W	9.0		A12
Re_3W_2	6.0		
Re_2Y	1.83		C14, a = 5.396, c = 8.819
Re_6Zr	7.40		A12, a = 9.698
Re_2Zr	6.0		
Re_2Zr	6.8		C14, a = 5.262, c = 8.593
$Rh_{17}S_{15}$	5.8		Cub. a = 9.911
$RhSe_{1.75}$	6.0		C2, a = 6.015
Rh_xSe_y	6.0M		
$Rh_{.53}Se_{.47}$	6.0		C2
Rh_2Sr	6.2		C15, a = 7.706

Material	T_c(°K)	H_0(Oe)	Crystal Structure ∞
RhTaΔ	2.0		$D8_b$.
$Rh_{.4}Ta_{.6}$	2.35		$D8_b$, a = 9.80, c = 5.09
$RhTe_2$(low temp. form)	1.51		C2, a = 6.441
Rh_3Th_7	2.15		$D10_2$, a = 10.031, c = 6.287
Rh_xTi_y	2.25–3.95		
$Rh_{.05-.15}Ti_y$	2.25–3.95		
RhV_3	0.38		A15, a = 4.784
RhW	2.64–3.37	17.0	A3, SS
$Rh_{.02}Zr_{.98}$	1.5		A3, SS
$Rh_{.03}Zr_{.97}$	6.0		A3, SS
$Rh_{.05}Zr_{.95}$	8.6		A3, SS
$Rh_{.09}Zr_{.91}$	8.7		
$Rh_{.2}Zr_{.8}$	9.0		hcp
$Rh_{.13}Zr_{.87}$	8.6		A3, SS
$Rh_{.14}Zr_{.86}$	9.5		A3, SS
$Rh_{.17}Zr_{.83}$	9.6		
$Rh_{.2}Zr_{.8}$	9.0		
$Rh_{.23}Zr_{.77}$	9.0		
$Rh_{.27}Zr_{.73}$	7.9		
$Rh_{.34}Zr_{.66}$	8.2		
$Rh_{.4}Zr_{.6}$	6.4		
RhZr	2.7		
Ru	0.49	66	A3, a = 2.7058, c = 4.2816
$Ru_2Sc_{1.2}$	1.67		C14, a = 5.119, c = 8.542
Ru_2Th	3.56		C115, a = 7.651
$Ru_{.05}Ti_{.95}$	2.5		
$Ru_{.1}Ti_{.9}$	3.5		
RuTi	1.07		B2, a = 3.067
$Ru_xTi_{.6}V_{.4}$	6.6M		
RuW	7.5		A3, SS
$Ru_{.58}W_{.42}$	5.20		$D8_b$
$Ru_{.4}W_{.6}$	4.67		$D8_b$, a = 9.57, c = 4.96
Ru_2Y	1.52		C14, a = 5.256, c = 8.792
$Ru_{.1}Zr_{.9}$	5.7		A3, SS
Ru_2Zr	1.84		C14, a = 5.144, c = 8.504
Sb_2Sn_3	3.80R		
$Sb_{0-.04}Sn_y$	3.730–3.739		Tet., SS
$Sb_{.02-.08}Sn_y$	2.64 = 3.89	~304–345	
$SbTi_3$	5.80		A15, a = 5.217
Sb_2TlΔ	5.20		
Sb_2Tl_7	5.20R		

TABLE II. SUPERCONDUCTIVE MATERIALS (Continued)

Material	T_c(°K)	H_0(Oe)	Crystal Structure ∞
SbV_3	0.80		A15, a = 4.941
Sb_2Zr_3	1.74		
SiTa	4.25–4.38		Hex.
$Si_2Th(\beta)$	2.41		C32
$Si_2Th(\alpha)$	3.16		C_c
SiV_3	17.0		A15, a = 4.722
$Si_{.263}V_{.737}$	15.8		A15, a = 4.726
$Si_{.206}V_{.794}$	14.5		A15, a = 4.729
SiV_3	17.0		A15, a = 4.728
$Si(V_{.9}Ru_{.1})_3$	2.9		A15, a = 4.707
$Si(V_{.9}Ti_{.1})_3$	10.9		A15, a = 4.736
$Si(V_{.9}Cr_{.1})_3$	11.3		A15, a = 4.697
$Si(V_9Mo_{.1})_3$	11.7		A15, a = 4.732
$Si(V_{.9}Nb_{.1})_3$	12.8		A15, a = 4.756
$Si(V_{.9}Zr_{.1})_3$	13.2		A15, a = 4.724
$Si_9V_3Ge_{.1}$	14.0		A15, a = 4.731
$Si_{.9}V_3Al_{.1}$	14.05		A15, a = 4.727
SiV_3(1.3% Fe, Mn)	14.4		A15, a = 4.720
$Si_{.9}V_3B_{.1}$	15.8		A15, a = 4720
SiV_3(0.4% Fe, Mn)	16.3		A15, a = 4.723
$Si_{.9}V_3C_{.1}$	16.4		A15, a = 4.723
SiV_3(0.25% Fe, Mn)	17.0		A15, a = 5.722
$Si_{.236}V_{.755}(Al_{.009})$	16.12		A15, a = 4.727
$Si_{.255}V_{.692}(Al_{.053})$	<14		A15, a = 4.733
$Si_{.218}V_{.694}(B_{.088})$	16.2		A15, a = 4.722
$Si_{.244}V_{.606}(Be_{.06})$	15.6		A15, a = 4.725_5
$Si_{.233}V_{.761}(Be_{.002})$	16.6		A15, a = 4.725_5
$Si_{.238}V_{.747}(C_{.016})$	16.5		A15, a = 4.724
$Si_{.246}V_{.72}(Ce_{.034})$	15.32		A15, a = 4.729
$Si_{.25}V_{.70}(Cr_{.05})$	<14		A15, a = 4.709
$Si_{.238}V_{.760}(La_{.002})$	16.48		A15, a = 4.727
$Si_{.24}V_{.753}(Mn_{.007})$	16.25		A15, a = 4.721
$Si_{.233}V_{.751}(O_{.016})$	15.94		A15, a = 4.726
$Si_{.255}V_{.697}(Re_{.048})$	<14		A15, a = 4.727
$Si_{.7}W$	2.84		
Si_2W_3	2.87		
Sn	3.722C	303	A5, a = 5.8314, c = 3.1814
Sn	3.722	309	
$SnTa_3$	6.0		A15, a = 5.276
$SnTa_{13}$	6.4		A15, a = 5.278
$SnTa_2V$	3.7		A15, a = 5.174
$SnTaV_2$	2.8		A15, a = 5.041
Sn_xTl_y	2.38–4.2R		
Sn_xTl_y	2.37 = 5.2		
SnV_3	7.0		A15, a = 4.94
SnV_3	3.8		A15, a = 4.96
$SnZn_{0-.01}$	3.731 = 3.734		Tet., SS
Ta	4.39C	780	A2, a = 3.298
Ta	4.483	830	
$Ta_{.19}Ti_{.81}$	>1.2		
$Ta_{1-.7}Ti_{0-.3}$	4.3–6.5		
$Ta_{0-.65}Ti_{1-.35}$	4.4–7.8		
$Ta_{0.8-1}W_y$	1.2–4.4		A2, SS
Tc	8.22		A3, a = 2.743, c = 4.400
Tc_6Zr	9.7		A12, a = 9.636
Th	1.368	162	A1, a = 5.0843
$Th_{.5}Y_{.5}$	1.25		
$Th_{.25}Y_{.75}$	1.8		
$Th_{0-.55}Y_{1-.45}$	1.2–1.8		
Ti	0.39	100	A3, a = 2.9504 c = 4.6833
$Ti_{.5}V_{.5}$	6.09		
$Ti_{1-.85}V_{0-.15}$	0.6–4.4 6.6 (annealed)		a = 2.94, c/a = 1.57
$Ti_{.9}V_{.1}$	6.3		
$Ti_{0-.85}V_{1-.15}$	2.3–7.5		A2, SS
Ti_xV_y	5.3–7.25		
$Ti_{.15-.8}V_{.85-.2}$	3.5–7.30		A2, SS
$Ti_{0-.5}V_{1-.5}$	5.3–6.7	1050–1250	A2, SS

TABLE II. SUPERCONDUCTIVE MATERIALS (Continued)

Material	T_c(°K)	H_0(Oe)	Crystal Structure ∞
Ti_xV_y	>1.2		
$Ti_{.67}Zr_{.33}$	1.36		A3, SS
$Ti_{.5}Zr_{.5}$	1.57		A3, SS
$Ti_{.33}Zr_{.67}$	1.35		A3, SS
$Ti_{.18}Zr_{.82}$	1.03		A3, SS
Tl(α)	2.36C		A3, a = 3.4566, c = 5.5248
Tl	2.39	171	
U(α)	0.68		ortho, a = 2.858, β = 5.877, c = 4.955
U(Pseudo γ)	1.8 (extrap. value)		A2
V	5.03C	1310	A2, a = 3.0282.
V	5.3	1020	
VΔ	1.8–4.8		
V_2Zr	8.8		C15, a = 7.439
W_2Zr	2.16		C15, a = 7.621
Zn	0.852C	51.8	A3, a = 2.6649, c = 4.9468
Zn	0.855	53	
Zr	0.546	47	A3, a = 3.2312, c = 5.1477

TABLES OF PROPERTIES OF SEMICONDUCTORS

Compiled by Dr. Brian Randall Pamplin,
School of Physics,
Bath University of Technology,
Ashley Down,
Bristol, 7, England.

The term "semiconductor" is applied to a material in which electric current is carried by electrons or holes and whose electrical conductivity when extremely pure rises exponentially with temperature and may be increased from this low "intrinsic" value by many orders of magnitude by "doping" with electrically active impurities.

Semiconductors are characterised by an energy gap in the allowed energies of electrons in the material which separates the normally filled energy levels of the *valence band* (where "missing" electrons behave like positively charged current carriers "holes") and the *conduction band* (where electrons behave rather like a gas of free negatively charged carriers with an effective mass dependent on the material and the direction of the electrons' motion). This energy gap depends on the nature of the material and varies with direction in anisotropic crystals. It is slightly dependent on temperature and pressure, and this dependence is usually almost linear at normal temperatures and pressures.

The data is presented in three tables. Table I "General Properties of Semiconductors" lists the main crystallographic and semiconducting properties of a large number of semiconducting materials in three main categories; "Tetrahedral Semiconductors" in which every atom is tetrahedrally co-ordinated to four nearest neighbour atoms (or atomic sites) as for example in the diamond structure; "Octahedral Semiconductors" in which every atom is octahedrally co-ordinated to six nearest neighbour atoms—as for example in the halite structure; and "Other Semiconductors".

Table II gives more detailed information about some better known semiconductors, while Table III gives some information about the electronic energy band structure parameters of the best known materials.

TABLE I

GENERAL PROPERTIES OF SEMICONDUCTORS

(listed by Crystal Structure)

Substance	Lattice Parameters (A° Room temperature)	Density (gm/cc)	Melting Point (°K)	Minimum Room Temperature Energy Gap (eV)	Thermal Conductivity	Heat of Formation k cal/mole	Mobility (Room Temperature) (cm²/V.s) Electrons	Holes	Remarks
PART A TETRAHEDRAL SEMICONDUCTORS									
§A1 Diamond Structure Elements (Strukturbericht symbol A4, Space Group Fd3m - O_h^7)									
C	3.5597	3.51	4300	5.4	2000	161	1800	1400	
Si	5.43072	2.3283	1685	1.107	1240	77.5	1900	500	
Ge	5.65754	5.3234	1231	0.67	640	69.5	3800	1820	
α - Sn	6.4912	5.765	503	0.08		64	2500	2400	
§A2 Sphalerite (Zinc Blende) Structure Compounds (Strukturbericht symbol B3 Space Group F $\bar{4}$ 3m - T_d^2)									
I VII Compounds									
CuF	4.255								
CuCl	5.4057	3.53	695						
CuBr	5.6905	4.72	770	2.94		115			
CuI	6.0427	5.63	878			105			
AgBr						102	4000		
AgI	6.473	5.67				93	30		
II VI Compounds									
BeS	4.865	2.36							
BeSe	5.139	4.315							
BeTe	5.626	5.090							
BePo	8.38	7.3							
Zn0	4.63								see § A3
ZnS	5.4093	4.079	1920	3.54	140	114	180	5(400°C)	see also § A3
ZnSe	5.6676	5.42	1790	2.58	140	101	540	28	
ZnTe	6.101	5.72	1510	2.26	140	90	340	100	
ZnPo									
CdS	5.5818								see also § A3
CdSe	6.05								see § A3
CdTe	6.477	5.86	1370	1.44	55	81	1200	50	
CdPo									
HgS	5.8517	7.73	~2020						
HgSe	6.084	8.25	1070	0.30	10	59	20000		
HgTe	6.429	8.17	943	0.15	20	58	25000	350	
III V Compounds									
BN	3.615	3.49	3000	~4	200	195			
BP(L.T.)	4.538	2.9		~6			500	70	
BAs	4.777								
AlP	5.451	2.85	1770	2.5					
AlAs	5.6622	3.81	1870	2.16		150	1200	420	
AlSb	6.1355	4.218	1330	1.60	600	140	200–400	550	
GaP	5.4505	4.13	1750	2.24	1100	152	300	100	
GaAs	5.65315	5.316	1510	1.35	370	128	8800	400	
GaSb	6.0954	5.619	980	0.67	270	118	4000	1400	
InP	5.86875	4.787	1330	1.27	800	134	4600	150	
InAs	6.05838	5.66	1215	0.36	290	114	33000	460	
InSb	6.47877	5.775	798	0.165	160	107	78000	750	
Other Sphalerite Structure Compounds									
β - SiC	4.348	3.21	3070	2.3			4000		
Ga_2Te_3	5.899	5.75	1063	~1.0	~14	65			
In_2Te_3(H.T.)	6.150	5.8	940	~1.0	~8	47.4	~10		
$MgGeP_2$	5.652								
$ZnSnP_2$	5.65			2.1					
$ZnSnAs_2$(H.T.)	5.851	5.53	1050	~0.7	70				
§ A3 Wurtzite (Zincite) Structure Compounds (Strukturbericht symbol B4, Space Group P 6_3mc - C_{6v}^4)									
I VII Compounds									
CuCl	3.91 6.42		T_c 680°K						
CuBr	4.06 6.66		T_c 658°K						
CuI	4.31 7.09								
AgI	4.580 7.494			2.63					

TABLE 1

GENERAL PROPERTIES OF SEMICONDUCTORS (Continued)

Substance	Lattice Parameters (A° Room temperature)		Density (gm/cc)	Melting Point (°K)	Minimum Room Temperature Energy Gap (eV)	Thermal Conductivity	Heat of Formation k cal/mole	Mobility (Room Temperature) (cm²/V.s) Electrons	Holes	Remarks
II VI Compounds										
BeO	2.698	4.380		2800						
MgTe	4.54	7.39	3.85	~2800						
ZnO	3.24950	5.2069	5.66	2250	3.2	6	154	180		
ZnS	3.8140	6.2576	4.1	2100	3.67		110			
ZnSe	3.996	6.626								
ZnTe	4.27	6.99								
CdS	4.1348	6.7490	4.82	2020	2.42		96	400		
CdSe	4.299	7.010	5.66	1530	1.74		90	650		
CdTe	4.57	7.47			1.50					
III V Compounds										
BP(H.T.)	3.562	5.900								
AlN	3.111	4.978	3.26	~2500	6.02		197			
GaN	3.180	5.166	6.10	1500	3.34		157			
InN	3.533	5.693	6.88	1200	2.0		133			
Other Wurtzite Structure Compounds										
SiC	3.076	5.048								
MnTe	4.078	6.701			~1.0					
Al_2S_3	3.579	5.829	2.55		4.1		426			
Al_2Se_3	3.890	6.30	3.91		3.1		367			
§ A$\bar{4}$ Chalcopyrite Structure Compounds (Strukturbericht symbol $E1_1$, Space Group $I\bar{4}2d - D^{12}_{2d}$)										
I III VI_2 Compounds										
$CuAlS_2$	5.323	10.44	3.47		2.5					
$CuAlSe_2$	5.617	10.92	4.70	1270	1.1					
$CuAlTe_2$	5.976	11.80	5.50	1160	0.88					
$CuGaS_2$	5.360	10.49	4.35							
$CuGaSe_2$	5.618	11.01	5.56	1310	0.96, 1.63					
$CuGaTe_2$	6.013	11.93	5.99	1150	0.82, 1.0					
$CuInS_2$	5.528	11.08	4.75		1.2					
$CuInSe_2$	5.785	11.57	5.77	1250	0.86, 0.92	37				
$CuInTe_2$	6.179	12.365	6.10	970	0.95	49				
$CuTlS_2$	5.580	11.17	6.32							
$CuTlSe_2$(L.T.)	5.844	11.65	7.11	680	1.07					
$CuFeS_2$	5.25	10.32		1150	0.53					
$CuFeSe_2$				850	0.16					
$CuLaS_2$	5.65	10.86								
$AgAlS_2$	5.707	10.28	3.94							
$AgAlSe_2$	5.968	10.77	5.07	1220	0.7					
$AgAlTe_2$	6.309	11.85	6.18	1000	0.56					
$AgGaS_2$	5.755	10.28	4.72							
$AgGaSe_2$	5.985	10.90	5.84	1120	1.66					
$AgGaTe_2$	6.301	11.96	6.05	990	1.1	10				
$AgInS_2$(L.T.)	5.828	11.19	5.00		1.9					
$AgInSe_2$	6.102	11.69	5.81	1053	1.18	30				
$AgInTe_2$	6.42	12.59	6.12	965	0.96, 0.52					
$AgFeS_2$	5.66	10.30	4.53							
II IV V_2 Compounds										
$ZnSiP_2$	5.400	10.441	3.39	1640	2.3			1000		
$ZnGeP_2$	5.465	10.771	4.17	1295	2.2					
$CdSiP_2$	5.678	10.431	4.00	~1470	2.2			1000		
$CdGeP_2$	5.741	10.775	4.48	~1060	1.8					
$CdSnP_2$	5.900	11.518			1.5					
$ZnSiAs_2$	5.61	10.88	4.70	~1350	1.7				50	
$ZnGeAs_2$	5.672	11.153	5.32	~1150	0.85	110				
$ZnSnAs_2$	5.8515	11.704	5.53	~ 910	0.65	150			300	disorders at 910°K
$CdSiAs_2$	5.884	10.882								
$CdGeAs_2$	5.9427	11.2172	5.60	~ 903	0.53	40		70	25	disorders at 903°K
$CdSnAs_2$	6.0944	11.9182	5.72	880	0.26	70		22000	250	

TABLE I.

GENERAL PROPERTIES OF SEMICONDUCTORS (Continued)

Substance	Lattice Parameters (A° Room temperature)	Density (gm/cc)	Melting Point (°K)	Minimum Room Temperature Energy Gap (eV)	Thermal Conductivity	Heat of Formation k cal/mole	Mobility (Room Temperature) Electrons (cm²/V.s)	Holes	Remarks
§ A5 "Defect Chalcopyrite" Structure Compounds (Strukturbericht symbol E3, Space Group $I\,\overline{4} - S_4^2$)									
$ZnAl_2Se_4$	5.503 10.90	4.37							
$ZnAl_2Te_4$(?)	5.104 12.05	4.95							
$ZnGa_2S_4$(?)	5.274 10.44	3.80							
$ZnGa_2Se_4$(?)	5.496 10.99	5.21							
$ZnGa_2Te_4$(?)	5.937 11.87	5.67		1.35					
$ZnIn_2Se_4$	5.711 11.42	5.44	1250	2.6					
$ZnIn_2Te_4$	6.122 12.24	5.83	1075	1.2					
$CdAl_2S_4$	5.564 10.32	3.06							
$CdAl_2Se_4$	5.747 10.68	4.54							
$CdAl_2Te_4$(?)	6.011 12.21	5.10							
$CdGa_2S_4$	5.577 10.08	4.03							
$CdGa_2Se_4$	5.743 10.73	5.32							
$CdGa_2Te_4$	6.093 11.81	5.77							
$CdIn_2Te_4$	6.205 12.41	5.9	1060	(1.26 or 0.9)	4000				
$HgAl_2S_4$	5.488 10.26	4.11							
$HgAl_2Se_4$	5.708 10.74	5.05							
$HgAl_2Te_4$(?)	6.004 12.11	5.81							
$HgGa_2S_4$	5.507 10.23	5.00							
$HgGa_2Se_4$	5.715 10.78	6.18							
$HgIn_2Se_4$	5.764 11.80	6.3	1100	0.6					
$HgIn_2Te_4$(?)	6.186 12.37	6.3	980	0.86		200			
§ A6 Other Tetrahedral Compounds									
aSiC	3.0817 15.1183	3.21	3070	2.86			400		6H structure
$Hg_5Ga_2Te_8$	6.235								B3 with super lattice
$Hg_5In_2Te_8$	6.328			0.7			2000		B3 with super lattice
PART B OCTAHEDRAL SEMICONDUCTORS									
Halite Structure Semiconductors (Strukturbericht symbol B1, Space Group Fm3m $- 0_h^5$)									
SnSe	6.020		1133						
SnTe	6.313		1080(max)	0.5	91				
PbS	5.9362	7.61	1390	0.37	23	104	600	600	
PbSe	6.1243	8.15	1340	0.26	17	94	1000	900	
PbTe	6.454	8.16	1180	0.25	23	94	1600	600	
Selected Other Binary Chalcides									
BiSe	5.99	7.98	880	0.4					
BiTe	6.47								
EuSe	6.191		2300	1.8	2.4				
GdSe	5.771		2400						
NiD	4.1684	6.6	2260	2.0 or 3.7			4	–	
CdO	4.6953		1700	2.5	7	127	100		
SrS	6.0199	3.643	3000	4.1					
Selected Ternary Compounds									
$AgSbSe_2$	5.786	6.60	910	0.58	10.5				
$AgSbTe_2$ (or $Ag_{19}Sb_{29}Te_{52}$)	6.078	7.12	830	0.7, 0.27	8.6, 0.3				
$AgBiS_2$(H.T.)	5.648								
$AgBiSe_2$(H.T.)	5.82								
$AgBiTe_2$(H.T.)	6.155								

TABLE II

SEMICONDUCTING PROPERTIES OF SELECTED MATERIALS

Substance	Minimum Energy Gap (eV) R.T.	0°K	$\frac{dE_g}{dT}$ x 10^4 eV/°C	$\frac{dE_g}{dP}$ x 10^6 eV.cm²/kg	Density of States Electron Effective Mass m_{d_n} (m_o)	Electron Mobility and Temperature Dependence μ_n cm²/V.s	−x	Density of States Hole Effective Mass m_{d_p} (m_o)	Hole Mobility and Temperature Dependence μ_p cm²/V.s	−x
Si	1.107	1.153	−2.3	−2.0	0.58	1,900	2.6	1.06	500	2.3
Ge	0.67	0.744	−3.7	+7.3	0.35	3,800	1.66	0.56	1,820	2.33
αSn	0.08	0.094	−0.5		0.02	2,500	1.65	0.3	2,400	2.0
Te	0.33				0.68	1,100		0.19	560	
III–V Compounds										
AlAs	2.2	2.3				1,200			420	
AlSb	1.6	1.7	−3.5	−1.6	0.09	200	1.5	0.4	500	1.8
GaP	2.24	2.40	−5.4	−1.7	0.35	300	1.5		150	1.5
GaAs	1.35	1.53	−5.0	+9.4	0.068	9,000	1.0	0.5	500	2.1
GaSb	0.67	0.78	−3.5	+12	0.050	5,000	2.0	0.23	1,400	0.9
InP	1.27	1.41	−4.6	+4.6	0.067	5,000	2.0		200	2.4
InAs	0.36	0.43	−2.8	+8	0.022	33,000	1.2	0.41	460	2.3
InSb	0.165	0.23	−2.8	+15	0.014	78,000	1.6	0.4	750	2.1
II–VI Compounds										
ZnO	3.2		−9.5	+0.6	0.38	180	1.5			
ZnS	3.54		−5.3	+5.7		180			5(400°C)	
ZnSe	2.58	2.80	−7.2	+6		540			28	
ZnTe	2.26			+6		340			100	
CdO	2.5 ± .1		−6		0.1	120				
CdS	2.42		−5	+3.3	0.165	400		0.8		
CdSe	1.74	1.85	−4.6		0.13	650	1.0	0.6		
CdTe	1.44	1.56	−4.1	+8	0.14	1,200		0.35	50	
HgSe	0.30				0.030	20,000	2.0			
HgTe	0.15		−1		0.017	25,000		0.5	350	
Halite Structure Compounds										
PbS	0.37	0.28	+4		0.16	800		0.1	1,000	2.2
PbSe	0.26	0.16	+4		0.3	1,500		0.34	1,500	2.2
PbTe	0.25	0.19	+4	−7	0.21	1,600		0.14	750	2.2
Others										
ZnSb	0.50	0.56			0.15	10				1.5
CdSb	0.45	0.57	−5.4		0.15	300			2,000	1.5
Bi_2S_3	1.3					200			1,100	
Bi_2Se_3	0.27					600			675	
Bi_2Te_3	0.13		−0.95		0.58	1,200	1.68	1.07	510	1.95
Mg_2Si		0.77	−6.4		0.46	400	2.5		70	
Mg_2Ge		0.74	−9			280	2		110	
Mg_2Sn	0.21	0.33	−3.5		0.37	320			260	
Mg_3Sb_2		0.32				20			82	
Zn_3As_2	0.93					10	1.1		10	
Cd_3As_2	0.13				0.046	15,000	0.88			
GaSe	2.05		3.8						20	
GaTe	1.66	1.80	−3.6			14	−5			
InSe	1.8					900				
TlSe	0.57		−3.9		0.3	30		0.6	20	1.5
$CdSnAs_2$	0.23				0.05	25,000	1.7			
Ga_2Te_3	1.1	1.55	−4.8							
α-In_2Te_3	1.1	1.2			0.7				50	1.1
β-In_2Te_3	1.0								5	
$Hg_5In_2Te_8$	0.5								11,000	
SnO_2									78	

TABLE III

PART A. DATA ON VALENCE BANDS OF SEMICONDUCTORS (Room Temperature data)

Substance	Band Curvature Effective Mass (Expressed as fraction of free electron mass): Heavy Holes	Light Holes	"Split-off" Band Holes	Energy Separation of "Split-off" Band (eV)	Measured (Light) Hole Mobility $cm^2/V.s$
Semiconductors with Valence Band Maximum at Centre of Brillouin Zone ("Γ")					
Si	0.52	0.16	0.25	0.044	500
Ge	0.34	0.043	0.08	0.3	1,820
Sn	0.3				2,400
AlAs					
AlSb	0.4			0.7	550
GaP				0.13	100
GaAs	0.8	0.12	0.20	0.34	400
GaSb	0.23	0.06		0.7	1,400
InP				0.21	150
InAs	0.41	0.025	0.083	0.43	460
InSb	0.4	0.015		0.85	750
CdTe	0.35				50
HgTe	0.5				350

Semiconductors with Multiple Valence Band Maxima

Substance	Number of Equivalent Valleys Direction	Curvature Effective Masses: Longitudinal m_L	Transverse m_T	Anisotropy $K = \frac{m_L}{m_T}$	Measured (Light) Hole Mobility $cm^2/V.s$
PbSe	4 "L" [111]	0.095	0.047	2.0	1,500
PbTe	4 "L" [111]	0.27	0.02	10	750
Bi_2Te_3	6	0.207	~0.045	4.5	515

PART B. DATA ON CONDUCTION BANDS OF SEMICONDUCTORS (Room Temperature Data)

Single Valley Semiconductors

Substance	Energy Gap (eV)	Effective Mass (m_o)	Mobility ($cm^2/V.s$)	Comments
GaAs	1.35	0.067	8,500	3(or 6?) equivalent [100] valleys 0.36 eV above this maximum with a mobility of ~50
InP	1.27	0.067	5,000	3(or 6?) equivalent [100] valleys 0.4 eV above this minimum.
InAs	0.36	0.022	33,000	equivalent valleys ~1.0 eV above this minimum.
InSb	0.165	0.014	78,000	
CdTe	1.44	0.11	1,000	4(or 8?) equivalent [111] valleys 0.51 eV above this minimum.

Multivalley Semiconductors

Substance	Energy Gap	Number of Equivalent Valleys and Direction	Band Curvature Effective Mass: Longitudinal m_L	Transverse m_T	Anisotropy $K = \frac{m_L}{m_T}$	Comments
Si	1.107	6 in [100] "Δ"	0.90	0.192	4.7	
Ge	0.67	4 in [111] at "L"	1.588	0.0815	19.5	
GaSb	0.67	as Ge (?)	~1.0	~0.2	~5	
PbSe	0.26	4 in [111] at "L"	0.085	0.05	1.7	
PbTe	0.25	4 in [111] at "L"	0.21	0.029	5.5	
Bi_2Te_3	0.13	6			~0.05	

STANDARD CALIBRATION TABLES FOR THERMOCOUPLES

The following tables which represent the Temperature-E. M. F. functions of various thermocouples should be used with appropriate correction curves if precise results are desired. These curves must be determined for each individual couple by plotting ΔE, the difference between the observed and the standard E. M. F., against the standard E. M. F. at three or more fixed temperature points. The value ΔE as shown by such a correction curve is then subtracted algebraically from the observed E. M. F. to give the true E. M. F. reading.

In the following tables the fixed or "cold junction" is at 0° C.; when the cold junction is not maintained at 0° C. the readings of the E. M. F. must be corrected as follows: $Et = E(_{t-tc}) + Etc$ where $E(_{t-tc})$ is the observed reading, Etc is the E. M. F. for the temperature corresponding to the cold junction temperature as read from the standard table and Et is the E. M. F. produced by the hot junction corrected to the value which would be obtained with the cold junction at 0° C. The temperature corresponding to Et is then obtained by reference to the standard table.

Since the E. M. F.-temperature function is not linear the cold junction should be maintained at a temperature very close to that at which the thermocouple was calibrated. Otherwise considerable error will result despite the above correction.

PLATINUM VERSUS PLATINUM-10-PERCENT RHODIUM THERMOCOUPLES

(Electromotive Force in Absolute Millivolts. Temperatures in Degrees C (Int. 1948). Reference Junctions at 0° C.)

°C	0	10	20	30	40	50	60	70	80	90
	Millivolts									
0	0	0.06	0.11	0.17	0.24	0.30	0.36	0.43	0.50	0.57
100	0.64	0.72	0.79	0.87	0.95	1.03	1.11	1.19	1.27	1.35
200	1.44	1.52	1.61	1.69	1.78	1.87	1.96	2.05	2.14	2.23
300	2.32	2.41	2.50	2.59	2.69	2.78	2.87	2.97	3.06	3.16
400	3.25	3.35	3.44	3.54	3.64	3.73	3.83	3.93	4.02	4.12
500	4.22	4.32	4.42	4.52	4.62	4.72	4.82	4.92	5.02	5.12
600	5.22	5.33	5.43	5.53	5.64	5.74	5.84	5.95	6.05	6.16
700	6.26	6.37	6.47	6.58	6.68	6.79	6.90	7.01	7.11	7.22
800	7.33	7.44	7.55	7.66	7.77	7.88	7.99	8.10	8.21	8.32
900	8.43	8.55	8.66	8.77	8.88	9.00	9.11	9.23	9.34	9.46
1000	9.57	9.69	9.80	9.92	10.04	10.15	10.27	10.39	10.51	10.62
1100	10.74	10.86	10.98	11.10	11.22	11.34	11.46	11.58	11.70	11.82
1200	11.94	12.06	12.18	12.30	12.42	12.54	12.66	12.78	12.90	13.02
1300	13.14	13.26	13.38	13.50	13.62	13.74	13.86	13.98	14.10	14.22
1400	14.34	14.46	14.58	14.70	14.82	14.94	15.05	15.17	15.29	15.41
1500	15.53	15.65	15.77	15.89	16.01	16.12	16.24	16.36	16.48	16.60
1600	16.72	16.83	16.95	17.07	17.19	17.31	17.42	17.54	17.66	17.77
1700	17.89	18.01	18.12	18.24	18.36	18.47	18.59	...	...	...

CALIBRATION TABLES FOR THERMOCOUPLES (Continued)
PLATINUM VERSUS PLATINUM-10-PERCENT RHODIUM THERMOCOUPLES

(Electromotive Force in Absolute Millivolts. Temperatures in Degrees F.* Reference Junctions at 32° F.)

°F	0	10	20	30	40	50	60	70	80	90
	Millivolts									
0	...	...	...	...	0.02	0.06	0 09	0.12	0.15	0.19
100	0.22	0.26	0.29	0.33	0.36	0.40	0.44	0.48	0.52	0.56
200	0.60	0.64	0.68	0.72	0.76	0.80	0.84	0.89	0.93	0.97
300	1.02	1.06	1.11	1.15	1.20	1.24	1.29	1.33	1.38	1.43
400	1.47	1.52	1.57	1.62	1.66	1.71	1.76	1.81	1.86	1.91
500	1.96	2.01	2.06	2.11	2.16	2.21	2.26	2.31	2.36	2.41
600	2.46	2.51	2.56	2.61	2.66	2.72	2.77	2.82	2.87	2.92
700	2.98	3.03	3.08	3.14	3.19	3.24	3.29	3.35	3.40	3.45
800	3.51	3.56	3.61	3.67	3.72	3.78	3.83	3.88	3.94	3.99
900	4.05	4.10	4.16	4.21	4.26	4.32	4.37	4.43	4.49	4.54
1000	4.60	4.65	4.71	4.76	4.82	4.87	4.93	4.99	5.04	5.10
1100	5.16	5.21	5.27	5.33	5.38	5.44	5.50	5.56	5.61	5.67
1200	5.73	5.78	5.84	5.90	5.96	6.02	6.07	6.13	6.19	6.25
1300	6.31	6.37	6.42	6.48	6.54	6.60	6.66	6.72	6.78	6.84
1400	6.90	6.96	7.02	7.08	7.14	7.20	7.26	7.32	7.38	7.44
1500	7.50	7.56	7.62	7.68	7.74	7.80	7.86	7.93	7.99	8.05
1600	8.11	8.17	8.23	8.30	8.36	8.42	8.48	8.55	8.61	8.67
1700	8.73	8.80	8.86	8.92	8.98	9.05	9.11	9.17	9.24	9.30
1800	9.37	9.43	9.49	9.56	9.62	9.69	9.75	9.82	9.88	9.94
1900	10.01	10.07	10.14	10.20	10.27	10.33	10.40	10.47	10.53	10.60
2000	10.66	10.73	10.79	10.86	10.93	10.99	11.06	11.12	11.19	11.26
2100	11.32	11.39	11.46	11.52	11.59	11.66	11.72	11.79	11.86	11.92
2200	11.99	12.06	12.12	12.19	12.26	12.32	12.39	12.46	12.52	12.59
2300	12.66	12.72	12.79	12.86	12.92	12.99	13.06	13.12	13.19	13.26
2400	13.33	13.39	13.46	13.53	13.59	13.66	13.73	13.79	13.86	13.92
2500	13.99	14.06	14.12	14.19	14.26	14.32	14.39	14.46	14.52	14.59
2600	14.66	14.72	14.79	14.86	14.92	14.99	15.05	15.12	15.19	15.25
2700	15.32	15.39	15.45	15.52	15.58	15.65	15.72	15.78	15.85	15.91
2800	15.98	16.05	16.11	16.18	16.24	16.31	16.37	16.44	16.51	16.57
2900	16.64	16.70	16.77	16.83	16.90	16.97	17.03	17.10	17.16	17.23
3000	17.29	17.36	17.42	17.49	17.55	17.62	17.68	17.75	17.81	17.88
3100	17.94	18.01	18.07	18.14	18.20	18.27	18.33	18.40	18.46	18.53
3200	18.59	18.66	...	...	...	...	...	...	...	...

* Based on the International Temperature Scale of 1948.

CALIBRATION TABLES FOR THERMOCOUPLES (Continued)

PLATINUM VERSUS PLATINUM-13-PERCENT RHODIUM THERMOCOUPLES

(Electromotive Force in Absolute Millivolts. Temperatures in Degrees C (Int. 1948) Reference Junctions at 0° C.)

°C	0	10	20	30	40	50	60	70	80	90
	Millivolts									
0	0.00	0.06	0.11	0.17	0.23	0.30	0.36	0.43	0.50	0.57
100	0.65	0.72	0.80	0.88	0.96	1.04	1.12	1.21	1.29	1.38
200	1.47	1.55	1.64	1.73	1.83	1.92	2.01	2.11	2.20	2.30
300	2.40	2.49	2.59	2.69	2.79	2.89	2.99	3.09	3.19	3.30
400	3.40	3.50	3.61	3.71	3.82	3.92	4.03	4.13	4.24	4.35
500	4.46	4.56	4.67	4.78	4.89	5.00	5.12	5.23	5.34	5.45
600	5.56	5.68	5.79	5.91	6.02	6.14	6.25	6.37	6.49	6.60
700	6.72	6.84	6.96	7.08	7.20	7.32	7.44	7.56	7.68	7.80
800	7.92	8.05	8.17	8.29	8.42	8.54	8.67	8.80	8.92	9.05
900	9.18	9.30	9.43	9.56	9.69	9.82	9.95	10.08	10.21	10.34
1000	10.47	10.60	10.74	10.87	11.00	11.14	11.27	11.41	11.54	11.68
1100	11.82	11.95	12.09	12.23	12.37	12.50	12.64	12.78	12.92	13.06
1200	13.19	13.33	13.47	13.61	13.75	13.89	14.03	14.17	14.30	14.44
1300	14.58	14.72	14.86	15.00	15.14	15.28	15.42	15.55	15.69	15.83
1400	15.97	16.11	16.25	16.39	16.52	16.66	16.80	16.94	17.08	17.22
1500	17.36	17.49	17.63	17.77	17.91	18.04	18.18	18.32	18.45	18.59
1600	18.73	18.86	19.00	19.14	19.27	19.41	19.55	19.68	19.82	19.95
1700	20.09	...	...	...	...	...	...	...	...	...

(Electromotive Force in Absolute Millivolts. Temperatures in Degrees F.* Reference Junctions at 32° F.)

°F	0	10	20	30	40	50	60	70	80	90
	Millivolts									
0	...	...	...	...	0.02	0.06	0.09	0.12	0.15	0.19
100	0.22	0.26	0.29	0.33	0.36	0.40	0.44	0.48	0.52	0.56
200	0.60	0.64	0.68	0.72	0.76	0.81	0.85	0.89	0.94	0.98
300	1.03	1.08	1.12	1.17	1.21	1.26	1.31	1.36	1.41	1.46
400	1.50	1.55	1.60	1.65	1.70	1.75	1.81	1.86	1.91	1.96
500	2.01	2.07	2.12	2.17	2.22	2.28	2.33	2.38	2.44	2.49
600	2.55	2.60	2.66	2.71	2.77	2.82	2.88	2.94	2.99	3.05
700	3.10	3.16	3.22	3.27	3.33	3.39	3.45	3.50	3.56	3.62
800	3.68	3.74	3.79	3.85	3.91	3.97	4.03	4.09	4.15	4.21
900	4.26	4.32	4.38	4.44	4.50	4.56	4.62	4.69	4.75	4.81
1000	4.87	4.93	4.99	5.05	5.12	5.18	5.24	5.30	5.36	5.43
1100	5.49	5.55	5.61	5.68	5.74	5.81	5.87	5.93	6.00	6.06
1200	6.13	6.19	6.25	6.32	6.38	6.45	6.51	6.58	6.64	6.71
1300	6.77	6.84	6.90	6.97	7.04	7.10	7.17	7.24	7.30	7.37
1400	7.44	7.50	7.57	7.64	7.71	7.77	7.84	7.91	7.98	8.05
1500	8.12	8.18	8.25	8.32	8.39	8.46	8.53	8.60	8.67	8.74
1600	8.81	8.88	8.95	9.02	9.09	9.16	9.23	9.30	9.37	9.45
1700	9.52	9.59	9.66	9.73	9.80	9.87	9.95	10.02	10.09	10.16
1800	10.24	10.31	10.38	10.46	10.53	10.60	10.68	10.75	10.82	10.90
1900	10.97	11.05	11.12	11.20	11.27	11.35	11.42	11.50	11.58	11.65
2000	11.73	11.80	11.88	11.95	12.03	12.11	12.18	12.26	12.34	12.41
2100	12.49	12.56	12.64	12.72	12.80	12.87	12.95	13.03	13.10	13.18
2200	13.26	13.33	13.41	13.49	13.56	13.64	13.72	13.80	13.87	13.95
2300	14.03	14.10	14.18	14.26	14.34	14.41	14.49	14.57	14.64	14.72
2400	14.80	14.88	14.95	15.03	15.11	15.18	15.26	15.34	15.42	15.49
2500	15.57	15.65	15.72	15.80	15.88	15.95	16.03	16.11	16.19	16.26
2600	16.34	16.42	16.49	16.57	16.65	16.73	16.80	16.88	16.96	17.03
2700	17.11	17.19	17.26	17.34	17.42	17.49	17.57	17.65	17.72	17.80
2800	17.88	17.95	18.03	18.10	18.18	18.26	18.33	18.41	18.48	18.56
2900	18.64	18.71	18.79	18.86	18.94	19.02	19.09	19.17	19.24	19.32
3000	19.39	19.47	19.55	19.62	19.70	19.77	19.85	19.92	20.00	20.08
3100	20.15	...	...	...	...	...	...	...	...	...

CALIBRATION TABLES FOR THERMOCOUPLES (Continued)

CHROMEL-ALUMEL THERMOCOUPLES

(Electromotive Force in Absolute Millivolts. Temperatures in Degrees C (Int. 1948) Reference Junctions at 0° C.)

°C	0	1	2	3	4	5	6	7	8	9
−190	−5.60	−5.62	−5.63	−5.65	−5.67	−5.68	−5.70	−5.71	−5.73	−5.74
−180	−5.43	−5.45	−5.46	−5.48	−5.50	−5.52	−5.53	−5.55	−5.57	−5.58
−170	−5.24	−5.26	−5.28	−5.30	−5.32	−5.34	−5.35	−5.37	−5.39	−5.41
−160	−5.03	−5.05	−5.08	−5.10	−5.12	−5.14	−5.16	−5.18	−5.20	−5.22
−150	−4.81	−4.84	−4.86	−4.88	−4.90	−4.92	−4.95	−4.97	−4.99	−5.01
−140	−4.58	−4.60	−4.62	−4.65	−4.67	−4.70	−4.72	−4.74	−4.77	−4.79
−130	−4.32	−4.35	−4.37	−4.40	−4.42	−4.45	−4.48	−4.50	−4.52	−4.55
−120	−4.06	−4.08	−4.11	−4.14	−4.16	−4.19	−4.22	−4.24	−4.27	−4.30
−110	−3.78	−3.81	−3.84	−3.86	−3.89	−3.92	−3.95	−3.98	−4.00	−4.03
−100	−3.49	−3.52	−3.55	−3.58	−3.61	−3.64	−3.66	−3.69	−3.72	−3.75
−90	−3.19	−3.22	−3.25	−3.28	−3.31	−3.34	−3.37	−3.40	−3.43	−3.46
−80	−2.87	−2.90	−2.93	−2.96	−3.00	−3.03	−3.06	−3.09	−3.12	−3.16
−70	−2.54	−2.57	−2.61	−2.64	−2.67	−2.71	−2.74	−2.77	−2.80	−2.84
−60	−2.20	−2.24	−2.27	−2.30	−2.34	−2.37	−2.41	−2.44	−2.47	−2.51
−50	−1.86	−1.89	−1.93	−1.96	−2.00	−2.03	−2.07	−2.10	−2.13	−2.17
−40	−1.50	−1.54	−1.57	−1.61	−1.64	−1.68	−1.72	−1.75	−1.79	−1.82
−30	−1.14	−1.17	−1.21	−1.25	−1.28	−1.32	−1.36	−1.39	−1.43	−1.47
−20	−0.77	−0.80	−0.84	−0.88	−0.92	−0.95	−0.99	−1.03	−1.06	−1.10
−10	−0.39	−0.42	−0.46	−0.50	−0.54	−0.58	−0.62	−0.66	−0.69	−0.73
(−)0	−0.00	−0.04	−0.08	−0.12	−0.16	−0.19	−0.23	−0.27	−0.31	−0.35
(+)0	0.00	0.04	0.08	0.12	0.16	0.20	0.24	0.28	0.32	0.36
10	0.40	0.44	0.48	0.52	0.56	0.60	0.64	0.68	0.72	0.76
20	0.80	0.84	0.88	0.92	0.96	1.00	1.04	1.08	1.12	1.16
30	1.20	1.24	1.28	1.32	1.36	1.40	1.44	1.49	1.53	1.57
40	1.61	1.65	1.69	1.73	1.77	1.81	1.85	1.90	1.94	1.98
50	2.02	2.06	2.10	2.14	2.18	2.23	2.27	2.31	2.35	2.39
60	2.43	2.47	2.51	2.56	2.60	2.64	2.68	2.72	2.76	2.80
70	2.85	2.89	2.93	2.97	3.01	3.05	3.10	3.14	3.18	3.22
80	3.26	3.30	3.35	3.39	3.43	3.47	3.51	3.56	3.60	3.64
90	3.68	3.72	3.76	3.81	3.85	3.89	3.93	3.97	4.01	4.06
100	4.10	4.14	4.18	4.22	4.26	4.31	4.35	4.39	4.43	4.47
110	4.51	4.55	4.60	4.64	4.68	4.72	4.76	4.80	4.84	4.88
120	4.92	4.96	5.01	5.05	5.09	5.13	5.17	5.21	5.25	5.29
130	5.33	5.37	5.41	5.45	5.49	5.53	5.57	5.61	5.65	5.69
140	5.73	5.77	5.81	5.85	5.89	5.93	5.97	6.01	6.05	6.09
150	6.13	6.17	6.21	6.25	6.29	6.33	6.37	6.41	6.45	6.49
160	6.53	6.57	6.61	6.65	6.69	6.73	6.77	6.81	6.85	6.89
170	6.93	6.97	7.01	7.05	7.09	7.13	7.17	7.21	7.25	7.29
180	7.33	7.37	7.41	7.45	7.49	7.53	7.57	7.61	7.65	7.69
190	7.73	7.77	7.81	7.85	7.89	7.93	7.97	8.01	8.05	8.09
200	8.13	8.17	8.21	8.25	8.29	8.33	8.37	8.41	8.46	8.50
210	8.54	8.58	8.62	8.66	8.70	8.74	8.78	8.82	8.86	8.90
220	8.94	8.98	9.02	9.06	9.10	9.14	9.18	9.22	9.26	9.30
230	9.34	9.38	9.42	9.46	9.50	9.54	9.59	9.63	9.67	9.71
240	9.75	9.79	9.83	9.87	9.91	9.95	9.99	10.03	10.07	10.11
250	10.16	10.20	10.24	10.28	10.32	10.36	10.40	10.44	10.48	10.52
260	10.57	10.61	10.65	10.69	10.73	10.77	10.81	10.85	10.89	10.93
270	10.98	11.02	11.06	11.10	11.14	11.18	11.22	11.26	11.30	11.34
280	11.39	11.43	11.47	11.51	11.55	11.59	11.63	11.67	11.72	11.76
290	11.80	11.84	11.88	11.92	11.96	12.01	12.05	12.09	12.13	12.17
300	12.21	12.25	12.29	12.34	12.38	12.42	12.46	12.50	12.54	12.58
310	12.63	12.67	12.71	12.75	12.79	12.83	12.88	12.92	12.96	13.00
320	13.04	13.08	13.12	13.17	13.21	13.25	13.29	13.33	13.37	13.42
330	13.46	13.50	13.54	13.58	13.62	13.67	13.71	13.75	13.79	13.83
340	13.88	13.92	13.96	14.00	14.04	14.09	14.13	14.17	14.21	14.25
350	14.29	14.34	14.38	14.42	14.46	14.50	14.55	14.59	14.63	14.67
360	14.71	14.76	14.80	14.84	14.88	14.92	14.97	15.01	15.05	15.09
370	15.13	15.18	15.22	15.26	15.30	15.34	15.39	15.43	15.47	15.51
380	15.55	15.60	15.64	15.68	15.72	15.76	15.81	15.85	15.89	15.93
390	15.98	16.02	16.06	16.10	16.14	16.19	16.23	16.27	16.31	16.36
400	16.40	16.44	16.48	16.52	16.57	16.61	16.65	16.69	16.74	16.78
410	16.82	16.86	16.91	16.95	16.99	17.03	17.07	17.12	17.16	17.20
420	17.24	17.29	17.33	17.37	17.41	17.46	17.50	17.54	17.58	17.62
430	17.67	17.71	17.75	17.79	17.84	17.88	17.92	17.96	18.01	18.05
440	18.09	18.13	18.17	18.22	18.26	18.30	18.34	18.39	18.43	18.47
450	18.51	18.56	18.60	18.64	18.68	18.73	18.77	18.81	18.85	18.90
460	18.94	18.98	19.02	19.07	19.11	19.15	19.19	19.24	19.28	19.32
470	19.36	19.41	19.45	19.49	19.54	19.58	19.62	19.66	19.71	19.75
480	19.79	19.84	19.88	19.92	19.96	20.01	20.05	20.09	20.13	20.18
490	20.22	20.26	20.31	20.35	20.39	20.43	20.48	20.52	20.56	20.60
500	20.65	20.69	20.73	20.77	20.82	20.86	20.90	20.94	20.99	21.03
510	21.07	21.11	21.16	21.20	21.24	21.28	21.32	21.37	21.41	21.45
520	21.50	21.54	21.58	21.63	21.67	21.71	21.75	21.80	21.84	21.88
530	21.92	21.97	22.01	22.05	22.09	22.14	22.18	22.22	22.26	22.31
540	22.35	22.39	22.43	22.48	22.52	22.56	22.61	22.65	22.69	22.73
550	22.78	22.82	22.86	22.90	22.95	22.99	23.03	23.07	23.12	23.16

* Based on the International Temperature Scale of 1948.

CALIBRATION TABLES FOR THERMOCOUPLES (Continued)

CHROMEL-ALUMEL THERMOCOUPLES

Electromotive Force in Absolute Millivolts. Temperatures in Degrees C (Int. 1948). Reference Junctions at 0° C.)

°C	0	1	2	3	4	5	6	7	8	9
	Millivolts									
560	23.20	23.25	23.29	23.33	23.38	23.42	23.46	23.50	23.54	23.59
570	23.63	23.67	23.72	23.76	23.80	23.84	23.89	23.93	23.97	24.01
580	24.06	24.10	24.14	24.18	24.23	24.27	24.31	24.36	24.40	24.44
590	24.49	24.53	24.57	24.61	24.65	24.70	24.74	24.78	24.83	24.87
600	24.91	24.95	25.00	25.04	25.08	25.12	25.17	25.21	25.25	25.29
610	25.34	25.38	25.42	25.47	25.51	25.55	25.59	25.64	25.68	25.72
620	25.76	25.81	25.85	25.89	25.93	25.98	26.02	26.06	26.10	26.15
630	26.19	26.23	26.27	26.32	26.36	26.40	26.44	26.48	26.53	26.57
640	26.61	26.65	26.70	26.74	26.78	26.82	26.86	26.91	26.95	26.99
650	27.03	27.07	27.12	27.16	27.20	27.24	27.28	27.33	27.37	27.41
660	27.45	27.49	27.54	27.58	27.62	27.66	27.71	27.75	27.79	27.83
670	27.87	27.92	27.96	28.00	28.04	28.08	28.13	28.17	28.21	28.25
680	28.29	28.34	28.38	28.42	28.46	28.50	28.55	28.59	28.63	29.67
690	28.72	28.76	28.80	28.84	28.88	28.93	28.97	29.01	29.05	29.10
700	29.14	29.18	29.22	29.26	29.30	29.35	29.39	29.43	29.47	29.52
710	29.56	29.60	29.64	29.68	29.72	29.77	29.81	29.85	29.89	29.93
720	29.97	30.02	30.06	30.10	30.14	30.18	30.23	30.27	30.31	30.35
730	30.39	30.44	30.48	30.52	30.56	30.60	30.65	30.69	30.73	30.77
740	30.81	30.85	30.90	30.94	30.98	31.02	31.06	31.10	31.15	31.19
750	31.23	31.27	31.31	31.35	31.40	31.44	31.48	31.52	31.56	31.60
760	31.65	31.69	31.73	31.77	31.81	31.85	31.90	31.94	31.98	32.02
770	32.06	32.10	32.15	32.19	32.23	32.27	32.31	32.35	32.39	32.43
780	32.48	32.52	32.56	32.60	32.64	32.68	32.72	32.76	32.81	32.85
790	32.89	32.93	32.97	33.01	33.05	33.09	33.13	33.18	33.22	33.26
800	33.30	33.34	33.38	33.42	33.46	33.50	33.54	33.59	33.63	33.67
810	33.71	33.75	33.79	33.83	33.87	33.91	33.95	33.99	34.04	34.08
820	34.12	34.16	34.20	34.24	34.28	34.32	34.36	34.40	34.44	34.48
830	34.53	34.57	34.61	34.65	34.69	34.73	34.77	34.81	34.85	34.89
840	34.93	34.97	35.02	35.06	35.10	35.14	35.18	35.22	35.26	35.30
850	35.34	35.38	35.42	35.46	35.50	35.54	35.58	35.63	35.67	35.71
860	35.75	35.79	35.83	35.87	35.91	35.95	35.99	36.03	36.07	36.11
870	36.15	36.19	36.23	36.27	36.31	36.35	36.39	36.43	36.47	36.51
880	36.55	36.59	36.63	36.67	36.72	36.76	36.80	36.84	36.88	36.92
890	36.96	37.00	37.04	37.08	37.12	37.16	37.20	37.24	37.28	37.32
900	37.36	37.40	37.44	37.48	37.52	37.56	37.60	37.64	37.68	37.72
910	37.76	37.80	37.84	37.88	37.92	37.96	38.00	38.04	38.08	38.12
920	38.16	38.20	38.24	38.28	38.32	38.36	38.40	38.44	38.48	38.52
930	38.56	38.60	38.64	38.68	38.72	38.76	38.80	38.84	38.88	38.92
940	38.95	38.99	39.03	39.07	39.11	39.15	39.19	39.23	39.27	39.31
950	39.35	39.39	39.43	39.47	39.51	39.55	39.59	39.63	39.67	39.71
960	39.75	39.79	39.83	39.86	39.90	39.94	39.98	40.02	40.06	40.10
970	40.14	40.18	40.22	40.26	40.30	40.34	40.38	40.41	40.45	40.49
980	40.53	40.57	40.61	40.65	40.69	40.73	40.77	40.81	40.85	40.89
990	40.92	40.96	41.00	41.04	41.08	41.12	41.16	41.20	41.24	41.28
1000	41.31	41.35	41.39	41.43	41.47	41.51	41.55	41.59	41.63	41.67
1010	41.70	41.74	41.78	41.82	41.86	41.90	41.94	41.98	42.02	42.05
1020	42.09	42.13	42.17	42.21	42.25	42.29	42.33	42.36	42.40	42.44
1030	42.48	42.52	42.56	42.60	42.63	42.67	42.71	42.75	42.79	42.83
1040	42.87	42.90	42.94	42.98	43.02	43.06	43.10	43.14	43.17	43.21
1050	43.25	43.29	43.33	43.37	43.41	43.44	43.48	43.52	43.56	43.60
1060	43.63	43.67	43.71	43.75	43.79	43.83	43.87	43.90	43.94	43.98
1070	44.02	44.06	44.10	44.13	44.17	44.21	44.25	44.29	44.33	44.36
1080	44.40	44.44	44.48	44.52	44.55	44.59	44.63	44.67	44.71	44.74
1090	44.78	44.82	44.86	44.90	44.93	44.97	45.01	45.05	45.09	45.12
1100	45.16	45.20	45.24	45.27	45.31	45.35	45.39	45.43	45.46	45.50
1110	45.54	45.58	45.62	45.65	45.69	45.73	45.77	45.80	45.84	45.88
1120	45.92	45.96	45.99	46.03	46.07	46.11	46.14	46.18	46.22	46.26
1130	46.29	46.33	46.37	46.41	46.44	46.48	46.52	46.56	46.59	46.63
1140	46.67	46.70	46.74	46.78	46.82	46.85	46.89	46.93	46.97	47.00
1150	47.04	47.08	47.12	47.15	47.19	47.23	47.26	47.30	47.34	47.38
1160	47.41	47.45	47.49	47.52	47.56	47.60	47.63	47.67	47.71	47.75
1170	47.78	47.82	47.86	47.89	47.93	47.97	48.00	48.04	48.08	48.12
1180	48.15	48.19	48.23	48.26	48.30	48.34	48.37	48.41	48.45	48.48
1190	48.52	48.56	48.59	48.63	48.67	48.70	48.74	48.78	48.81	48.85
1200	48.89	48.92	48.96	49.00	49.03	49.07	49.11	49.14	49.18	49.22
1210	49.25	49.29	49.32	49.36	49.40	49.43	49.47	49.51	49.54	49.58
1220	49.62	49.65	49.69	49.72	49.76	49.80	49.83	49.87	49.90	49.94
1230	49.98	50.01	50.05	50.08	50.12	50.16	50.19	50.23	50.26	50.30
1240	50.34	50.37	50.41	50.44	50.48	50.52	50.55	50.59	50.62	50.66
1250	50.69	50.73	50.77	50.80	50.84	50.87	50.91	50.94	50.98	51.02
1260	51.05	51.09	51.12	51.16	51.19	51.23	51.27	51.30	51.34	51.37
1270	51.41	51.44	51.48	51.51	51.55	51.58	51.62	51.66	51.69	51.73
1280	51.76	51.80	51.83	51.87	51.90	51.94	51.97	52.01	52.04	52.08
1290	52.11	52.15	52.18	52.22	52.25	52.29	52.32	52.36	52.39	52.43
1300	52.46	52.50	52.53	52.57	52.60	52.64	52.67	52.71	52.74	52.78

* Based on the International Temperature Scale of 1948.

CALIBRATION TABLES FOR THERMOCOUPLES (Continued)

CHROMEL-ALUMEL THERMOCOUPLES

(Electromotive Force in Absolute Millivolts. Temperatures in Degrees C (Int. 1948). Reference Junctions at 0° C.)

°C	0	1	2	3	4	5	6	7	8	9
	Millivolts									
1300	52.46	52.50	52.53	52.57	52.60	52.64	52.67	52.71	52.74	52.78
1310	52.81	52.85	52.88	52.92	52.95	52.99	53.02	53.06	53.09	53.13
1320	53.16	53.20	53.23	53.27	53.30	53.34	53.37	53.41	53.44	53.47
1330	53.51	53.54	53.58	53.61	53.65	53.68	53.72	53.75	53.79	53.82
1340	53.85	53.89	53.92	53.96	53.99	54.03	54.06	54.10	54.13	54.16
1350	54.20	54.23	54.27	54.30	54.34	54.37	54.40	54.44	54.47	54.51
1360	54.54	54.57	54.61	54.64	54.68	54.71	54.74	54.78	54.81	54.85
1370	54.88	54.91	....	....	....	....	....	....	....	....

CHROMEL-ALUMEL THERMOCOUPLES

(Electromotive Force in Absolute Millivolts. Temperatures in Degrees F.* Reference Junctions at 32° F.)

°F	0	1	2	3	4	5	6	7	8	9
	Millivolts									
−300	−5.51	−5.52	−5.53	−5.54	−5.54	−5.55	−5.56	−5.57	−5.58	−5.59
−290	−5.41	−5.42	−5.43	−5.44	−5.45	−5.46	−5.47	−5.48	−5.49	−5.50
−280	−5.30	−5.31	−5.32	−5.34	−5.35	−5.36	−5.37	−5.38	−5.39	−5.40
−270	−5.20	−5.21	−5.22	−5.23	−5.24	−5.25	−5.26	−5.27	−5.28	−5.29
−260	−5.08	−5.09	−5.10	−5.12	−5.13	−5.14	−5.15	−5.16	−5.17	−5.18
−250	−4.96	−4.97	−4.99	−5.00	−5.01	−5.02	−5.03	−5.04	−5.06	−5.07
−240	−4.84	−4.85	−4.86	−4.88	−4.89	−4.90	−4.91	−4.92	−4.94	−4.95
−230	−4.71	−4.72	−4.74	−4.75	−4.76	−4.77	−4.79	−4.80	−4.81	−4.82
−220	−4.58	−4.59	−4.60	−4.62	−4.63	−4.64	−4.66	−4.67	−4.68	−4.70
−210	−4.44	−4.45	−4.46	−4.48	−4.49	−4.51	−4.52	−4.53	−4.55	−4.56
−200	−4.29	−4.31	−4.32	−4.34	−4.35	−4.36	−4.38	−4.39	−4.41	−4.42
−190	−4.15	−4.16	−4.18	−4.19	−4.21	−4.22	−4.24	−4.25	−4.26	−4.28
−180	−4.00	−4.01	−4.03	−4.04	−4.06	−4.07	−4.09	−4.10	−4.12	−4.13
−170	−3.84	−3.86	−3.88	−3.89	−3.91	−3.92	−3.94	−3.95	−3.97	−3.98
−160	−3.69	−3.70	−3.72	−3.73	−3.75	−3.76	−3.78	−3.80	−3.81	−3.83
−150	−3.52	−3.54	−3.56	−3.57	−3.59	−3.60	−3.62	−3.64	−3.65	−3.67
−140	−3.36	−3.38	−3.39	−3.41	−3.42	−3.44	−3.46	−3.47	−3.49	−3.51
−130	−3.19	−3.20	−3.22	−3.24	−3.25	−3.27	−3.29	−3.31	−3.32	−3.34
−120	−3.01	−3.03	−3.05	−3.06	−3.08	−3.10	−3.12	−3.13	−3.15	−3.17
−110	−2.83	−2.85	−2.87	−2.89	−2.90	−2.92	−2.94	−2.96	−2.98	−2.99
−100	−2.65	−2.67	−2.69	−2.71	−2.72	−2.74	−2.76	−2.78	−2.80	−2.82
−90	−2.47	−2.49	−2.50	−2.52	−2.54	−2.56	−2.58	−2.60	−2.62	−2.63
−80	−2.28	−2.30	−2.32	−2.34	−2.36	−2.37	−2.39	−2.41	−2.43	−2.45
−70	−2.09	−2.11	−2.13	−2.15	−2.17	−2.18	−2.20	−2.22	−2.24	−2.26
−60	−1.90	−1.92	−1.94	−1.96	−1.97	−1.99	−2.01	−2.03	−2.05	−2.07
−50	−1.70	−1.72	−1.74	−1.76	−1.78	−1.80	−1.82	−1.84	−1.86	−1.88
−40	−1.50	−1.52	−1.54	−1.56	−1.58	−1.60	−1.62	−1.64	−1.66	−1.68
−30	−1.30	−1.32	−1.34	−1.36	−1.38	−1.40	−1.42	−1.44	−1.46	−1.48
−20	−1.10	−1.12	−1.14	−1.16	−1.18	−1.20	−1.22	−1.24	−1.26	−1.28
−10	−0.89	−0.91	−0.93	−0.95	−0.97	−0.99	−1.01	−1.03	−1.06	−1.08
(−)0	−0.68	−0.70	−0.72	−0.75	−0.77	−0.79	−0.81	−0.83	−0.85	−0.87
(+)0	−0.68	−0.66	−0.64	−0.62	−0.60	−0.58	−0.56	−0.54	−0.52	−0.49
10	−0.47	−0.45	−0.43	−0.41	−0.39	−0.37	−0.34	−0.32	−0.30	−0.28
20	−0.26	−0.24	−0.22	−0.19	−0.17	−0.15	−0.13	−0.11	−0.09	−0.07
30	−0.04	−0.02	0.00	0.02	0.04	0.07	0.09	0.11	0.13	0.15
40	0.18	0.20	0.22	0.24	0.26	0.29	0.31	0.33	0.35	0.37
50	0.40	0.42	0.44	0.46	0.48	0.51	0.53	0.55	0.57	0.60
60	0.62	0.64	0.66	0.68	0.71	0.73	0.75	0.77	0.80	0.82
70	0.84	0.86	0.88	0.91	0.93	0.95	0.97	1.00	1.02	1.04
80	1.06	1.09	1.11	1.13	1.15	1.18	1.20	1.22	1.24	1.27
90	1.29	1.31	1.33	1.36	1.33	1.40	1.43	1.45	1.47	1.49
100	1.52	1.54	1.56	1.58	1.61	1.63	1.65	1.68	1.70	1.72
110	1.74	1.77	1.79	1.81	1.84	1.86	1.88	1.90	1.93	1.95
120	1.97	2.00	2.02	2.04	2.06	2.09	2.11	2.13	2.16	2.18
130	2.20	2.23	2.25	2.27	2.29	2.32	2.34	2.36	2.39	2.41
140	2.43	2.46	2.48	2.50	2.52	2.55	2.57	2.59	2.62	2.64
150	2.66	2.69	2.71	2.73	2.75	2.78	2.80	2.82	2.85	2.87
160	2.89	2.92	2.94	2.96	2.98	3.01	3.03	3.05	3.08	3.10
170	3.12	3.15	3.17	3.19	3.22	3.24	3.26	3.29	3.31	3.33
180	3.36	3.38	3.40	3.43	3.45	3.47	3.49	3.52	3.54	3.56
190	3.59	3.61	3.63	3.66	3.68	3.70	3.73	3.75	3.77	3.80
200	3.82	3.84	3.87	3.89	3.91	3.94	3.96	3.98	4.01	4.03
210	4.05	4.08	4.10	4.12	4.15	4.17	4.19	4.21	4.24	4.26
220	4.28	4.31	4.33	4.35	4.38	4.40	4.42	4.44	4.47	4.49
230	4.51	4.54	4.56	4.58	4.61	4.63	4.65	4.67	4.70	4.72
240	4.74	4.77	4.79	4.81	4.83	4.86	4.88	4.90	4.92	4.95
250	4.97	4.99	5.02	5.04	5.06	5.08	5.11	5.13	5.15	5.17
260	5.20	5.22	5.24	5.26	5.29	5.31	5.33	5.35	5.38	5.40
270	5.42	5.44	5.47	5.49	5.51	5.53	5.56	5.58	5.60	5.62
280	5.65	5.67	5.69	5.71	5.73	5.76	5.78	5.80	5.82	5.85
290	5.87	5.89	5.91	5.93	5.96	5.98	6.00	6.02	6.05	6.07
300	6.09	6.11	6.13	6.16	6.18	6.20	6.22	6.25	6.27	6.29
310	6.31	6.33	6.36	6.38	6.40	6.42	6.45	6.47	6.49	6.51
320	6.53	6.56	6.58	6.60	6.62	6.65	6.67	6.69	6.71	6.73
330	6.76	6.78	6.80	6.82	6.84	6.87	6.89	6.91	6.93	6.96
340	6.98	7.00	7.02	7.04	7.07	7.09	7.11	7.13	7.15	7.18
350	7.20	7.22	7.24	7.26	7.29	7.31	7.33	7.35	7.38	7.40

* Based on the International Temperature Scale of 1948.

CALIBRATION TABLES FOR THERMOCOUPLES (Continued)
CHROMEL-ALUMEL THERMOCOUPLES

(Electromotive Force in Absolute Millivolts. Temperatures in Degrees F.* Reference Junctions at 32° F.)

°F	0	1	2	3	4	5	6	7	8	9
	Millivolts									
360	7.42	7.44	7.46	7.49	7.51	7.53	7.55	7.58	7.60	7.62
370	7.64	7.66	7.69	7.71	7.73	7.75	7.78	7.80	7.82	7.84
380	7.87	7.89	7.91	7.93	7.95	7.98	8.00	8.02	8.04	8.07
390	8.09	8.11	8.13	8.16	8.18	8.20	8.22	8.24	8.27	8.29
400	8.31	8.33	8.36	8.38	8.40	8.42	8.45	8.47	8.49	8.51
410	8.54	8.56	8.58	8.60	8.62	8.65	8.67	8.69	8.71	8.74
420	8.76	8.78	8.80	8.82	8.85	8.87	8.89	8.91	8.94	8.96
430	8.98	9.00	9.03	9.05	9.07	9.09	9.12	9.14	9.16	9.18
440	9.21	9.23	9.25	9.27	9.30	9.32	9.34	9.36	9.39	9.41
450	9.43	9.45	9.48	9.50	9.52	9.54	9.57	9.59	9.61	9.63
460	9.66	9.68	9.70	9.73	9.75	9.77	9.79	9.82	9.84	9.86
470	9.88	9.91	9.93	9.95	9.97	10.00	10.02	10.04	10.06	10.09
480	10.11	10.13	10.16	10.18	10.20	10.22	10.25	10.27	10.29	10.31
490	10.34	10.36	10.38	10.40	10.43	10.45	10.47	10.50	10.52	10.54
500	10.57	10.59	10.61	10.63	10.66	10.68	10.70	10.72	10.75	10.77
510	10.79	10.82	10.84	10.86	10.88	10.91	10.93	10.95	10.98	11.00
520	11.02	11.04	11.07	11.09	11.11	11.13	11.16	11.18	11.20	11.23
530	11.25	11.27	11.29	11.32	11.34	11.36	11.39	11.41	11.43	11.45
540	11.48	11.50	11.52	11.55	11.57	11.59	11.61	11.64	11.66	11.68
550	11.71	11.73	11.75	11.78	11.80	11.82	11.84	11.87	11.89	11.91
560	11.94	11.96	11.98	12.01	12.03	12.05	12.07	12.10	12.12	12.14
570	12.17	12.19	12.21	12.24	12.26	12.28	12.30	12.33	12.35	12.37
580	12.40	12.42	12.44	12.47	12.49	12.51	12.53	12.56	12.58	12.60
590	12.63	12.65	12.67	12.70	12.72	12.74	12.76	12.79	12.81	12.83
600	12.86	12.88	12.90	12.93	12.95	12.97	13.00	13.02	13.04	13.06
610	13.09	13.11	13.13	13.16	13.18	13.20	13.23	13.25	13.27	13.30
620	13.32	13.34	13.36	13.39	13.41	13.44	13.46	13.48	13.50	13.53
630	13.55	13.57	13.60	13.62	13.64	13.67	13.69	13.71	13.74	13.76
640	13.78	13.81	13.83	13.85	13.88	13.90	13.92	13.95	13.97	13.99
650	14.02	14.04	14.06	14.09	14.11	14.13	14.15	14.18	14.20	14.22
660	14.25	14.27	14.29	14.32	14.34	14.36	14.39	14.41	14.43	14.46
670	14.48	14.50	14.53	14.55	14.57	14.60	14.62	14.64	14.67	14.69
680	14.71	14.74	14.76	14.78	14.81	14.83	14.85	14.88	14.90	14.92
690	14.95	14.97	14.99	15.02	15.04	15.06	15.09	15.11	15.13	15.16
700	15.18	15.20	15.23	15.25	15.27	15.30	15.32	15.34	15.37	15.39
710	15.41	15.44	15.46	15.48	15.51	15.53	15.55	15.58	15.60	15.62
720	15.65	15.67	15.69	15.72	15.74	15.76	15.79	15.81	15.83	15.86
730	15.88	15.90	15.93	15.95	15.98	16.00	16.02	16.05	16.07	16.09
740	16.12	16.14	16.16	16.19	16.21	16.23	16.26	16.28	16.30	16.33
750	16.35	16.37	16.40	16.42	16.45	16.47	16.49	16.52	16.54	16.56
760	16.59	16.61	16.63	16.66	16.68	16.70	16.73	16.75	16.77	16.80
770	16.82	16.84	16.87	16.89	16.92	16.94	16.96	16.99	17.01	17.03
780	17.06	17.08	17.10	17.13	17.15	17.17	17.20	17.22	17.24	17.27
790	17.29	17.31	17.34	17.36	17.39	17.41	17.43	17.46	17.48	17.50
800	17.53	17.55	17.57	17.60	17.62	17.64	17.67	17.69	17.71	17.74
810	17.76	17.78	17.81	17.83	17.86	17.88	17.90	17.93	17.95	17.97
820	18.00	18.02	18.04	18.07	18.09	18.11	18.14	18.16	18.18	18.21
830	18.23	18.25	18.28	18.30	18.33	18.35	18.37	18.40	18.42	18.44
840	18.47	18.49	18.51	18.54	18.56	18.58	18.61	18.63	18.65	18.68
850	18.70	18.73	18.75	18.77	18.80	18.82	18.84	18.87	18.89	18.91
860	18.94	18.96	18.99	19.01	19.03	19.06	19.08	19.10	19.13	19.15
870	19.18	19.20	19.22	19.25	19.27	19.29	19.32	19.34	19.36	19.39
880	19.41	19.44	19.46	19.48	19.51	19.53	19.55	19.58	19.60	19.63
890	19.65	19.67	19.70	19.72	19.75	19.77	19.79	19.82	19.84	19.86
900	19.89	19.91	19.94	19.96	19.98	20.01	20.03	20.05	20.08	20.10
910	20.13	20.15	20.17	20.20	20.22	20.24	20.27	20.29	20.32	20.34
920	20.36	20.39	20.41	20.43	20.46	20.48	20.50	20.53	20.55	20.58
930	20.60	20.62	20.65	20.67	20.69	20.72	20.74	20.76	20.79	20.81
940	20.84	20.86	20.88	20.91	20.93	20.95	20.98	21.00	21.03	21.05
950	21.07	21.10	21.12	21.14	21.17	21.19	21.21	21.24	21.26	21.28
960	21.31	21.33	21.36	21.38	21.40	21.43	21.45	21.47	21.50	21.52
970	21.54	21.57	21.59	21.62	21.64	21.66	21.69	21.71	21.73	21.76
980	21.78	21.81	21.83	21.85	21.88	21.90	21.92	21.95	21.97	21.99
990	22.02	22.04	22.07	22.09	22.11	22.14	22.16	22.18	22.21	22.23
1000	22.26	22.28	22.30	22.33	22.35	22.37	22.40	22.42	22.44	22.47
1010	22.49	22.52	22.54	22.56	22.59	22.61	22.63	22.66	22.68	22.71
1020	22.73	22.75	22.78	22.80	22.82	22.85	22.87	22.90	22.92	22.94
1030	22.97	22.99	23.01	23.04	23.06	23.08	23.11	23.13	23.16	23.18
1040	23.20	23.23	23.25	23.27	23.30	23.32	23.35	23.37	23.39	23.42
1050	23.44	23.46	23.49	23.51	23.54	23.56	23.58	23.61	23.63	23.65
1060	23.68	23.70	23.72	23.75	23.77	23.80	23.82	23.84	23.87	23.89
1070	23.91	23.94	23.96	23.99	24.01	24.03	24.06	24.08	24.10	24.13
1080	24.15	24.18	24.20	24.22	24.25	24.27	24.29	24.32	24.34	24.36
1090	24.39	24.41	24.44	24.46	24.49	24.51	24.53	24.55	24.58	24.60
1100	24.63	24.65	24.67	24.70	24.72	24.74	24.77	24.79	24.82	24.84

* Based on the International Temperature Scale of 1948.

CALIBRATION TABLES FOR THERMOCOUPLES (Continued)
CHROMEL-ALUMEL THERMOCOUPLES

(Electromotive Force in Absolute Millivolts. Temperatures in Degrees F.* Reference Junctions at 32° F.)

°F	0	1	2	3	4	5	6	7	8	9
	Millivolts									
1100	24.63	24.65	24.67	24.70	24.72	24.74	24.77	24.79	24.82	24.84
1110	24.86	24.89	24.91	24.93	24.96	24.98	25.01	25.03	25.05	25.08
1120	25.10	25.12	25.15	25.17	25.20	25.22	25.24	25.27	25.29	25.31
1130	25.34	25.36	25.38	25.41	25.43	25.46	25.48	25.50	25.53	25.55
1140	25.57	25.60	25.62	25.65	25.67	25.69	25.72	25.74	25.76	25.79
1150	25.81	25.83	25.86	25.88	25.91	25.93	25.95	25.98	26.00	26.02
1160	26.05	26.07	26.09	26.12	26.14	26.16	26.19	26.21	26.24	26.26
1170	26.28	26.31	26.33	26.35	26.38	26.40	26.42	26.45	26.47	26.49
1180	26.52	26.54	26.56	26.59	26.61	26.63	26.66	26.68	26.70	26.73
1190	26.75	26.77	26.80	26.82	26.85	26.87	26.89	26.91	26.94	26.96
1200	26.98	27.01	27.03	27.06	27.08	27.10	27.12	27.15	27.17	27.20
1210	27.22	27.24	27.27	27.29	27.31	27.34	27.36	27.38	27.40	27.43
1220	27.45	27.48	27.50	27.52	27.55	27.57	27.59	27.62	27.64	27.66
1230	27.69	27.71	27.73	27.76	27.78	27.80	27.83	27.85	27.87	27.90
1240	27.92	27.94	27.97	27.99	28.01	28.04	28.06	28.08	28.11	28.13
1250	28.15	28.18	28.20	28.22	28.25	28.27	28.29	28.32	28.34	28.37
1260	28.39	28.41	28.44	28.46	28.48	28.50	28.53	28.55	28.58	28.60
1270	28.62	28.65	28.67	28.69	28.72	28.74	28.76	28.79	28.81	28.83
1280	28.86	28.88	28.90	28.93	28.95	28.97	29.00	29.02	29.04	29.07
1290	29.09	29.11	29.14	29.16	29.18	29.21	29.23	29.25	29.28	29.30
1300	29.32	29.35	29.37	29.39	29.42	29.44	29.46	29.49	29.51	29.53
1310	29.56	29.58	29.60	29.63	29.65	29.67	29.70	29.72	29.74	29.77
1320	29.79	29.81	29.84	29.86	29.88	29.91	29.93	29.95	29.97	30.00
1330	30.02	30.05	30.07	30.09	30.11	30.14	30.16	30.18	30.21	30.23
1340	30.25	30.28	30.30	30.32	30.35	30.37	30.39	30.42	30.44	30.46
1350	30.49	30.51	30.53	30.56	30.58	30.60	30.63	30.65	30.67	30.70
1360	30.72	30.74	30.77	30.79	30.81	30.83	30.86	30.88	30.90	30.93
1370	30.95	30.97	31.00	31.02	31.04	31.07	31.09	31.11	31.14	31.16
1380	31.18	31.21	31.23	31.25	31.28	31.30	31.32	31.34	31.37	31.39
1390	31.42	31.44	31.46	31.48	31.51	31.53	31.55	31.58	31.60	31.62
1400	31.65	31.67	31.69	31.72	31.74	31.76	31.78	31.81	31.83	31.85
1410	31.88	31.90	31.92	31.95	31.97	31.99	32.02	32.04	32.06	32.08
1420	32.11	32.13	32.15	32.18	32.20	32.22	32.25	32.27	32.29	32.31
1430	32.34	32.36	32.38	32.41	32.43	32.45	32.48	32.50	32.52	32.54
1440	32.57	32.59	32.61	32.64	32.66	32.68	32.70	32.73	32.75	32.77
1450	32.80	32.82	32.84	32.86	32.89	32.91	32.93	32.96	32.98	33.00
1460	33.02	33.05	33.07	33.09	33.12	33.14	33.16	33.18	33.21	33.23
1470	33.25	33.28	33.30	33.32	33.34	33.37	33.39	33.41	33.43	33.46
1480	33.48	33.50	33.53	33.55	33.57	33.59	33.62	33.64	33.66	33.69
1490	33.71	33.73	33.75	33.78	33.80	33.82	33.84	33.87	33.89	33.91
1500	33.93	33.96	33.98	34.00	34.03	34.05	34.07	34.09	34.12	34.14
1510	34.16	34.18	34.21	34.23	34.25	34.28	34.30	34.32	34.34	34.37
1520	34.39	34.41	34.43	34.46	34.48	34.50	34.53	34.55	34.57	34.59
1530	34.62	34.64	34.66	34.68	34.71	34.73	34.75	34.77	34.80	34.82
1540	34.84	34.87	34.89	34.91	34.93	34.96	34.98	35.00	35.02	35.05
1550	35.07	35.09	35.11	35.14	35.16	35.18	35.21	35.23	35.25	35.27
1560	35.29	35.32	35.34	35.36	35.39	35.41	35.43	35.45	35.48	35.50
1570	35.52	35.54	35.57	35.59	35.61	35.63	35.66	35.68	35.70	35.72
1580	35.75	35.77	35.79	35.81	35.84	35.86	35.88	35.90	35.93	35.95
1590	35.97	35.99	36.02	36.04	36.06	36.08	36.11	36.13	36.15	36.17
1600	36.19	36.19	36.22	36.24	36.29	36.31	36.33	36.35	36.37	36.40
1610	36.42	36.44	36.46	36.49	36.51	36.53	36.55	36.58	36.60	36.62
1620	36.64	36.67	36.69	36.71	36.73	36.76	36.78	36.80	36.82	36.84
1630	36.87	36.89	36.91	36.93	36.96	36.98	37.00	37.02	37.05	37.07
1640	37.09	37.11	37.14	37.16	37.18	37.20	37.23	37.25	37.27	37.29
1650	37.31	37.34	37.36	37.38	37.40	37.43	37.45	37.47	37.49	37.52
1660	37.54	37.56	37.58	37.60	37.63	37.65	37.67	37.69	37.72	37.74
1670	37.76	37.78	37.81	37.83	37.85	37.87	37.89	37.92	37.94	37.96
1680	37.98	38.01	38.03	38.05	38.07	38.09	38.12	38.14	38.16	38.18
1690	38.20	38.23	38.25	38.27	38.29	38.32	38.34	38.36	38.38	38.40
1700	38.43	38.45	38.47	38.49	38.51	38.54	38.56	38.58	38.60	38.62
1710	38.65	38.67	38.69	38.71	38.73	38.76	38.78	38.80	38.82	38.84
1720	38.87	38.89	38.91	38.93	38.95	38.98	39.00	39.02	39.04	39.06
1730	39.09	39.11	39.13	39.15	39.17	39.20	39.22	39.24	39.26	39.28
1740	39.31	39.33	39.35	39.37	39.39	39.42	39.44	39.46	39.48	39.50
1750	39.53	39.55	39.57	39.59	39.61	39.64	39.66	39.68	39.70	39.72
1760	39.75	39.77	39.79	39.81	39.83	39.86	39.88	39.90	39.92	39.94
1770	39.96	39.99	40.01	40.03	40.05	40.07	40.10	40.12	40.14	40.16
1780	40.18	40.20	40.23	40.25	40.27	40.29	40.31	40.34	40.36	40.38
1790	40.40	40.42	40.44	40.47	40.49	40.51	40.53	40.55	40.58	40.60
1800	40.62	40.64	40.66	40.68	40.71	40.73	40.75	40.77	40.79	40.82
1810	40.84	40.86	40.88	40.90	40.92	40.95	40.97	40.99	41.01	41.03
1820	41.05	41.08	41.10	41.12	41.14	41.16	41.18	41.21	41.23	41.25
1830	41.27	41.29	41.31	41.34	41.36	41.38	41.40	41.42	41.45	41.47
1840	41.49	41.51	41.53	41.55	41.57	41.60	41.62	41.64	41.66	41.68
1850	41.70	41.73	41.75	41.77	41.79	41.81	41.83	41.85	41.88	41.90

* Based on the International Temperature Scale of 1948.

CALIBRATION TABLES FOR THERMOCOUPLES (Continued)
CHROMEL-ALUMEL THERMOCOUPLES

(Electromotive Force in Absolute Millivolts. Temperatures in Degrees F.*
Reference Junctions at 32° F.)

°F	0	1	2	3	4	5	6	7	8	9
	Millivolts									
1860	41.92	41.94	41.96	41.99	42.01	42.03	42.05	42.07	42.09	42.11
1870	42.14	42.16	42.18	42.20	42.22	42.24	42.26	42.29	42.31	42.33
1880	42.35	42.37	42.39	42.42	42.44	42.46	42.48	42.50	42.52	42.55
1890	42.57	42.59	42.61	42.63	42.65	42.67	42.69	42.72	42.74	42.76
1900	42.78	42.80	42.82	42.84	42.87	42.89	42.91	42.93	42.95	42.97
1910	42.99	43.01	43.04	43.06	43.08	43.10	43.12	43.14	43.17	43.19
1920	43.21	43.23	43.25	43.27	43.29	43.31	43.34	43.36	43.38	43.40
1930	43.42	43.44	43.47	43.49	43.51	43.53	43.55	43.57	43.59	43.61
1940	43.63	43.66	43.68	43.70	43.72	43.74	43.76	43.78	43.81	43.83
1950	43.85	43.87	43.89	43.91	43.93	43.95	43.98	44.00	44.02	44.04
1960	44.06	44.08	44.10	44.13	44.15	44.17	44.19	44.21	44.23	44.25
1970	44.27	44.30	44.32	44.34	44.36	44.38	44.40	44.42	44.44	44.47
1980	44.49	44.51	44.53	44.55	44.57	44.59	44.61	44.63	44.66	44.68
1990	44.70	44.72	44.74	44.76	44.78	44.80	44.82	44.85	44.87	44.89
2000	44.91	44.93	44.95	44.97	44.99	45.01	45.03	45.06	45.08	45.10
2010	45.12	45.14	45.16	45.18	45.20	45.22	45.24	45.27	45.29	45.31
2020	45.33	45.35	45.87	45.39	45.41	45.43	45.45	45.48	45.50	45.52
2030	45.54	45.56	45.58	45.60	45.62	45.64	45.66	45.69	45.71	45.73
2040	45.75	45.77	45.79	45.81	45.83	45.85	45.87	45.90	45.92	45.94
2050	45.96	45.98	46.00	46.02	46.04	46.06	46.08	46.11	46.13	46.15
2060	46.17	46.19	46.21	46.23	46.25	46.27	46.29	46.31	46.33	46.36
2070	46.38	46.40	46.42	46.44	46.46	46.48	46.50	46.52	46.54	46.56
2080	46.58	46.60	46.63	46.65	46.67	46.69	46.71	46.73	46.75	46.77
2090	46.79	46.81	46.83	46.85	46.87	46.90	46.92	46.94	46.96	46.98
2100	47.00	47.02	47.04	47.06	47.08	47.10	47.12	47.14	47.17	47.19
2110	47.21	47.23	47.25	47.27	47.29	47.31	47.33	47.35	47.37	47.39
2120	47.41	47.43	47.45	47.47	47.49	47.52	47.54	47.56	47.58	47.60
2130	47.62	47.64	47.66	47.68	47.70	47.72	47.74	47.76	47.78	47.80
2140	47.82	47.84	47.86	47.89	47.91	47.93	47.95	47.97	47.99	48.01
2150	48.03	48.05	48.07	48.09	48.11	48.13	48.15	48.17	48.19	48.21
2160	48.23	48.25	48.27	48.29	48.32	48.34	48.36	48.38	48.40	48.42
2170	48.44	48.46	48.48	48.50	48.52	48.54	48.56	48.58	48.60	48.62
2180	48.64	48.66	48.68	48.70	48.72	48.74	48.76	48.79	48.81	48.83
2190	48.85	48.87	48.89	48.91	48.93	48.95	48.97	48.99	49.01	49.03
2200	49.05	49.07	49.09	49.11	49.13	49.15	49.17	49.19	49.21	49.23
2210	49.25	49.27	49.29	49.31	49.33	49.35	49.37	49.39	49.41	49.43
2220	49.45	49.47	49.49	49.51	49.53	49.55	49.57	49.59	49.61	49.63
2230	49.65	49.67	49.69	49.71	49.73	49.76	49.78	49.80	49.82	49.84
2240	49.86	49.88	49.90	49.92	49.94	49.96	49.98	50.00	50.02	50.04
2250	50.06	50.08	50.10	50.12	50.14	50.16	50.18	50.20	50.22	50.24
2260	50.26	50.28	50.30	50.32	50.34	50.36	50.38	50.40	50.42	50.44
2270	50.46	50.48	50.50	50.52	50.54	50.56	50.57	50.59	50.61	50.63
2280	50.65	50.67	50.69	50.71	50.73	50.75	50.77	50.79	50.81	50.83
2290	50.85	50.87	50.89	50.91	50.93	50.95	50.97	50.99	51.01	51.03
2300	51.05	51.07	51.09	51.11	51.13	51.15	51.17	51.19	51.21	51.23
2310	51.25	51.27	51.29	51.31	51.33	51.35	51.37	51.39	51.41	51.43
2320	51.45	51.47	51.48	51.50	51.52	51.54	51.56	51.58	51.60	51.62
2330	51.64	51.66	51.68	51.70	51.72	51.74	51.76	51.78	51.80	51.82
2340	51.84	51.86	51.88	51.90	51.92	51.94	51.96	51.98	52.00	52.01
2350	52.03	52.05	52.07	52.09	52.11	52.13	52.15	52.17	52.19	52.21
2360	52.23	52.25	52.27	52.29	52.31	52.33	52.35	52.37	52.39	52.41
2370	52.42	52.44	52.46	52.48	52.50	52.52	52.54	52.56	52.58	52.60
2380	52.62	52.64	52.66	52.68	52.70	52.72	52.74	52.76	52.77	52.79
2390	52.81	52.83	52.85	52.87	52.89	52.91	52.93	52.95	52.97	52.99
2400	53.01	53.03	53.05	53.07	53.08	53.10	53.12	53.14	53.16	53.18
2410	53.20	53.22	53.24	53.26	53.28	53.30	53.32	53.34	53.35	53.37
2420	53.39	53.41	53.43	53.45	53.47	53.49	53.51	53.53	53.55	53.57
2430	53.59	53.60	53.62	53.64	53.66	53.68	53.70	53.72	53.74	53.76
2440	53.78	53.80	53.82	53.83	53.85	53.87	53.89	53.91	53.93	53.95
2450	53.97	53.99	54.01	54.03	54.04	54.06	54.08	54.10	54.12	54.14
2460	54.16	54.18	54.20	54.22	54.24	54.25	54.27	54.29	54.31	54.33
2470	54.35	54.37	54.39	54.41	54.43	54.44	54.46	54.48	54.50	54.52
2480	54.54	54.56	54.58	54.60	54.62	54.63	54.65	54.67	54.69	54.71
2490	54.73	54.75	54.77	54.79	54.81	54.82	54.84	54.86	54.88	54.90
2500	54.92	....	....	....	....	....	....	....	....	....

* Based on the International Temperature Scale of 1948.

CALIBRATION TABLES FOR THERMOCOUPLES (Continued)
IRON-CONSTANTAN THERMOCOUPLES (MODIFIED 1913)

(Electromotive Force in Absolute Millivolts. Temperatures in Degrees C (Int. 1948). Reference Junctions at 0° C.)

°C	0	1	2	3	4	5	6	7	8	9
	Millivolts									
−190	−7.66	−7.69	−7.71	−7.73	−7.76	−7.78				
−180	−7.40	−7.43	−7.46	−7.49	−7.51	−7.54	−7.56	−7.59	−7.61	−7.64
−170	−7.12	−7.15	−7.18	−7.21	−7.24	−7.27	−7.30	−7.32	−7.35	−7.38
−160	−6.82	−6.85	−6.88	−6.91	−6.94	−6.97	−7.00	−7.03	−7.06	−7.09
−150	−6.50	−6.53	−6.56	−6.60	−6.63	−6.66	−6.69	−6.72	−6.76	−6.79
−140	−6.16	−6.19	−6.22	−6.26	−6.29	−6.33	−6.36	−6.40	−6.43	−6.46
−130	−5.80	−5.84	−5.87	−5.91	−5.94	−5.98	−6.01	−6.05	−6.08	−6.12
−120	−5.42	−5.46	−5.50	−5.54	−5.58	−5.61	−5.65	−5.69	−5.72	−5.76
−110	−5.03	−5.07	−5.11	−5.15	−5.19	−5.23	−5.27	−5.31	−5.35	−5.38
−100	−4.63	−4.67	−4.71	−4.75	−4.79	−4.83	−4.87	−4.91	−4.95	−4.99
−90	−4.21	−4.25	−4.30	−4.34	−4.38	−4.42	−4.46	−4.50	−4.55	−4.59
−80	−3.78	−3.82	−3.87	−3.91	−3.96	−4.00	−4.04	−4.08	−4.13	−4.17
−70	−3.34	−3.38	−3.43	−3.47	−3.52	−3.56	−3.60	−3.65	−3.69	−3.74
−60	−2.89	−2.94	−2.98	−3.03	−3.07	−3.12	−3.16	−3.21	−3.25	−3.30
−50	−2.43	−2.48	−2.52	−2.57	−2.62	−2.66	−2.71	−2.75	−2.80	−2.84
−40	−1.96	−2.01	−2.06	−2.10	−2.15	−2.20	−2.24	−2.29	−2.34	−2.38
−30	−1.48	−1.53	−1.58	−1.63	−1.67	−1.72	−1.77	−1.82	−1.87	−1.91
−20	−1.00	−1.04	−1.09	−1.14	−1.19	−1.24	−1.29	−1.34	−1.39	−1.43
−10	−0.50	−0.55	−0.60	−0.65	−0.70	−0.75	−0.80	−0.85	−0.90	−0.95
(−)0	0.00	−0.05	−0.10	−0.15	−0.20	−0.25	−0.30	−0.35	−0.40	−0.45
(+)0	0.00	0.05	0.10	0.15	0.20	0.25	0.30	0.35	0.40	0.45
10	0.50	0.56	0.61	0.66	0.71	0.76	0.81	0.86	0.91	0.97
20	1.02	1.07	1.12	1.17	1.22	1.28	1.33	1.38	1.43	1.48
30	1.54	1.59	1.64	1.69	1.74	1.80	1.85	1.90	1.95	2.00
40	2.06	2.11	2.16	2.22	2.27	2.32	2.37	2.42	2.48	2.53
50	2.58	2.64	2.69	2.74	2.80	2.85	2.90	2.96	3.01	3.06
60	3.11	3.17	3.22	3.27	3.33	3.38	3.43	3.49	3.54	3.60
70	3.65	3.70	3.76	3.81	3.86	3.92	3.97	4.02	4.08	4.13
80	4.19	4.24	4.29	4.35	4.40	4.46	4.51	4.56	4.62	4.67
90	4.73	4.78	4.83	4.89	4.94	5.00	5.05	5.10	5.16	5.21
100	5.27	5.32	5.38	5.43	5.48	5.54	5.59	5.65	5.70	5.76
110	5.81	5.86	5.92	5.97	6.03	6.08	6.14	6.19	6.25	6.30
120	6.36	6.41	6.47	6.52	6.58	6.63	6.68	6.74	6.79	6.85
130	6.90	6.96	7.01	7.07	7.12	7.18	7.23	7.29	7.34	7.40
140	7.45	7.51	7.56	7.62	7.67	7.73	7.78	7.84	7.89	7.95
150	8.00	8.06	8.12	8.17	8.23	8.28	8.34	8.39	8.45	8.50
160	8.56	8.61	8.67	8.72	8.78	8.84	8.89	8.95	9.00	9.06
170	9.11	9.17	9.22	9.28	9.33	9.39	9.44	9.50	9.56	9.61
180	9.67	9.72	9.78	9.83	9.89	9.95	10.00	10.06	10.11	10.17
190	10.22	10.28	10.34	10.39	10.45	10.50	10.56	10.61	10.67	10.72
200	10.78	10.84	10.89	10.95	11.00	11.06	11.12	11.17	11.23	11.28
210	11.34	11.39	11.45	11.50	11.56	11.62	11.67	11.73	11.78	11.84
220	11.89	11.95	12.00	12.06	12.12	12.17	12.23	12.28	12.34	12.39
230	12.45	12.50	12.56	12.62	12.67	12.73	12.78	12.84	12.89	12.95
240	13.01	13.06	13.12	13.17	13.23	13.28	13.34	13.40	13.45	13.51
250	13.56	13.62	13.67	13.73	13.78	13.84	13.89	13.95	14.00	14.06
260	14.12	14.17	14.23	14.28	14.34	14.39	14.45	14.50	14.56	14.61
270	14.67	14.72	14.78	14.83	14.89	14.94	15.00	15.06	15.11	15.17
280	15.22	15.28	15.33	15.39	15.44	15.50	15.55	15.61	15.66	15.72
290	15.77	15.83	15.88	15.94	16.00	16.05	16.11	16.16	16.22	16.27
300	16.33	16.38	16.44	16.49	16.55	16.60	16.66	16.71	16.77	16.82
310	16.88	16.93	16.99	17.04	17.10	17.15	17.21	17.26	17.32	17.37
320	17.43	17.48	17.54	17.60	17.65	17.71	17.76	17.82	17.87	17.93
330	17.98	18.04	18.09	18.15	18.20	18.26	18.32	18.37	18.43	18.48
340	18.54	18.59	18.65	18.70	18.76	18.81	18.87	18.92	18.98	19.03
350	19.09	19.14	19.20	19.26	19.31	19.37	19.42	19.48	19.53	19.59
360	19.64	19.70	19.75	19.81	19.86	19.92	19.97	20.03	20.08	20.14
370	20.20	20.25	20.31	20.36	20.42	20.47	20.53	20.58	20.64	20.69
380	20.75	20.80	20.86	20.91	20.97	21.02	21.08	21.13	21.19	21.24
390	21.30	21.35	21.41	21.46	21.52	21.57	21.63	21.68	21.74	21.79
400	21.85	21.90	21.96	22.02	22.07	22.13	22.18	22.24	22.29	22.35
410	22.40	22.46	22.51	22.57	22.62	22.68	22.73	22.79	22.84	22.90
420	22.95	23.01	23.06	23.12	23.17	23.23	23.28	23.34	23.39	23.45
430	23.50	23.56	23.61	23.67	23.72	23.78	23.83	23.89	23.94	24.00
440	24.06	24.11	24.17	24.22	24.28	24.33	24.39	24.44	24.50	24.55
450	24.61	24.66	24.72	24.77	24.83	24.88	24.94	25.00	25.05	25.11
460	25.16	25.22	25.27	25.33	25.38	25.44	25.49	25.55	25.60	25.66
470	25.72	25.77	25.83	25.88	25.94	25.99	26.05	26.10	26.16	26.22
480	26.27	26.33	26.38	26.44	26.49	26.55	26.61	26.66	26.72	26.77
490	26.83	26.89	26.94	27.00	27.05	27.11	27.17	27.22	27.28	27.33
500	27.39	27.45	27.50	27.56	27.61	27.67	27.73	27.78	27.84	27.90
510	27.95	28.01	28.07	28.12	28.18	28.23	28.29	28.35	28.40	28.46
520	28.52	28.57	28.63	28.69	28.74	28.80	28.86	28.91	28.97	29.02
530	29.08	29.14	29.20	29.25	29.31	29.37	29.42	29.48	29.54	29.59
540	29.65	29.71	29.76	29.82	29.88	29.94	29.99	30.05	30.11	30.16
550	30.22	30.28	30.34	30.39	30.45	30.51	30.57	30.62	30.68	30.74

* Based on the International Temperature Scale of 1948.

CALIBRATION TABLES FOR THERMOCOUPLES (Continued)
IRON-CONSTANTAN THERMOCOUPLES (MODIFIED 1913)

(Electromotive Force in Absolute Millivolts. Temperatures in Degrees C (Int. 1948). Reference Junctions at 0° C.)

°C	0	1	2	3	4	5	6	7	8	9
	Millivolts									
560	30.80	30.85	30.91	30.97	31.02	31.08	31.14	31.20	31.26	31.31
570	31.37	31.43	31.49	31.54	31.60	31.66	31.72	31.78	31.83	31.89
580	31.95	32.01	32.06	32.12	32.18	32.24	32.30	32.36	32.41	32.47
590	32.53	32.59	32.65	32.71	32.76	32.82	32.88	32.94	33.00	33.06
600	33.11	33.17	33.23	33.29	33.35	33.41	33.46	33.52	33.58	33.64
610	33.70	33.76	33.82	33.88	33.94	33.99	34.05	34.11	34.17	34.23
620	34.29	34.35	34.41	34.47	34.53	34.58	34.64	34.70	34.76	34.82
630	34.88	34.94	35.00	35.06	35.12	35.18	35.24	35.30	35.36	35.42
640	35.48	35.54	35.60	36.66	35.72	35.78	35.84	35.90	35.96	36.02
650	36.08	36.14	36.20	36.26	36.32	36.38	36.44	36.50	36.56	36.62
660	36.69	36.75	36.81	36.87	36.93	36.99	37.05	37.11	37.18	37.24
670	37.30	37.36	37.42	37.48	37.54	37.60	37.66	37.73	37.79	37.85
680	37.91	37.97	38.04	38.10	38.16	38.22	38.28	38.34	38.41	38.47
690	38.53	38.59	38.66	38.72	38.78	38.84	38.90	38.97	39.03	39.09
700	39.15	39.22	39.28	39.34	39.40	39.47	39.53	39.59	39.65	39.72
710	39.78	39.84	39.91	39.97	40.03	40.10	40.16	40.22	40.28	40.35
720	40.41	40.48	40.54	40.60	40.66	40.73	40.79	40.86	40.92	40.98
730	41.05	41.11	41.17	41.24	41.30	41.36	41.43	41.49	41.56	41.62
740	41.68	41.75	41.81	41.87	41.94	42.00	42.07	42.13	42.19	42.26
750	42.32	42.38	42.45	42.51	42.58	42.64	42.70	42.77	42.83	42.90
760	42.96	....	....	....	....	....	....	....	....	....

(Electromotive Force in Absolute Millivolts. Temperatures in Degrees F.* Reference Junctions at 32° F.)

°F	0	1	2	3	4	5	6	7	8	9
	Millivolts									
−310	−7.66	−7.68	−7.69	−7.70	−7.71	−7.73	−7.74	−7.75	−7.76	−7.78
−300	−7.52	−7.54	−7.55	−7.57	−7.58	−7.59	−7.61	−7.62	−7.64	−7.65
−290	−7.38	−7.39	−7.40	−7.42	−7.44	−7.45	−7.46	−7.48	−7.49	−7.51
−280	−7.22	−7.24	−7.25	−7.27	−7.28	−7.30	−7.31	−7.33	−7.34	−7.36
−270	−7.06	−7.07	−7.09	−7.11	−7.12	−7.14	−7.15	−7.17	−7.19	−7.20
−260	−6.89	−6.90	−6.92	−6.94	−6.96	−6.97	−6.99	−7.01	−7.02	−7.04
−250	−6.71	−6.73	−6.75	−6.77	−6.78	−6.80	−6.82	−6.84	−6.85	−6.87
−240	−6.53	−6.55	−6.57	−6.59	−6.61	−6.62	−6.64	−6.66	−6.68	−6.70
−230	−6.35	−6.37	−6.38	−6.40	−6.42	−6.44	−6.46	−6.48	−6.50	−6.52
−220	−6.16	−6.18	−6.19	−6.21	−6.23	−6.25	−6.27	−6.29	−6.31	−6.33
−210	−5.96	−5.98	−6.00	−6.02	−6.04	−6.06	−6.08	−6.10	−6.12	−6.14
−200	−5.76	−5.78	−5.80	−5.82	−5.84	−5.86	−5.88	−5.90	−5.92	−5.94
−190	−5.55	−5.57	−5.59	−5.61	−5.63	−5.65	−5.67	−5.70	−5.72	−5.74
−180	−5.34	−5.36	−5.38	−5.40	−5.42	−5.44	−5.46	−5.49	−5.51	−5.53
−170	−5.12	−5.14	−5.16	−5.19	−5.21	−5.23	−5.25	−5.27	−5.30	−5.32
−160	−4.90	−4.92	−4.94	−4.97	−4.99	−5.01	−5.03	−5.06	−5.08	−5.10
−150	−4.68	−4.70	−4.72	−4.74	−4.76	−4.79	−4.81	−4.83	−4.86	−4.88
−140	−4.44	−4.47	−4.49	−4.51	−4.54	−4.56	−4.58	−4.61	−4.63	−4.65
−130	−4.21	−4.23	−4.26	−4.28	−4.30	−4.33	−4.35	−4.38	−4.40	−4.42
−120	−3.97	−4.00	−4.02	−4.04	−4.07	−4.09	−4.12	−4.14	−4.16	−4.19
−110	−3.73	−3.76	−3.78	−3.81	−3.83	−3.85	−3.88	−3.90	−3.93	−3.95
−100	−3.49	−3.51	−3.54	−3.56	−3.59	−3.61	−3.64	−3.66	−3.68	−3.71
−90	−3.24	−3.27	−3.29	−3.32	−3.34	−3.36	−3.39	−3.41	−3.44	−3.46
−80	−2.99	−3.02	−3.04	−3.07	−3.09	−3.12	−3.14	−3.17	−3.19	−3.22
−70	−2.74	−2.76	−2.79	−2.81	−2.84	−2.86	−2.89	−2.92	−2.94	−2.97
−60	−2.48	−2.51	−2.53	−2.56	−2.58	−2.61	−2.64	−2.66	−2.69	−2.71
−50	−2.22	−2.25	−2.27	−2.30	−2.33	−2.35	−2.38	−2.40	−2.43	−2.46
−40	−1.96	−1.99	−2.01	−2.04	−2.06	−2.09	−2.12	−2.14	−2.17	−2.20
−30	−1.70	−1.72	−1.75	−1.78	−1.80	−1.83	−1.86	−1.88	−1.91	−1.94
−20	−1.43	−1.46	−1.48	−1.51	−1.54	−1.56	−1.59	−1.62	−1.64	−1.67
−10	−1.16	−1.19	−1.21	−1.24	−1.27	−1.29	−1.32	−1.35	−1.38	−1.40
(−)0	−0.89	−0.91	−0.94	−0.97	−1.00	−1.02	−1.05	−1.08	−1.10	−1.13
+(0)	−0.89	−0.86	−0.83	−0.80	−0.78	−0.75	−0.72	−0.70	−0.67	−0.64
10	−0.61	−0.58	−0.56	−0.53	−0.50	−0.48	−0.45	−0.42	−0.39	−0.36
20	−0.34	−0.31	−0.28	−0.25	−0.22	−0.20	−0.17	−0.14	−0.11	−0.09
30	−0.06	−0.03	0.00	0.03	0.05	0.08	0.11	0.14	0.17	0.19
40	0.22	0.25	0.28	0.31	0.34	0.36	0.39	0.42	0.45	0.48
50	0.50	0.53	0.56	0.59	0.62	0.65	0.67	0.70	0.73	0.76
60	0.79	0.82	0.84	0.87	0.90	0.93	0.96	0.99	1.02	1.04
70	1.07	1.10	1.13	1.16	1.19	1.22	1.25	1.28	1.30	1.33
80	1.36	1.39	1.42	1.45	1.48	1.51	1.54	1.56	1.59	1.62
90	1.65	1.68	1.71	1.74	1.77	1.80	1.83	1.85	1.88	1.91
100	1.94	1.97	2.00	2.03	2.06	2.09	2.12	2.14	2.17	2.20
110	2.23	2.26	2.29	2.32	2.35	2.38	2.41	2.44	2.47	2.50
120	2.52	2.55	2.58	2.61	2.64	2.67	2.70	2.73	2.76	2.79
130	2.82	2.85	2.88	2.91	2.94	2.97	3.00	3.03	3.06	3.08
140	3.11	3.14	3.17	3.20	3.23	3.26	3.29	3.32	3.35	3.38
150	3.41	3.44	3.47	3.50	3.53	3.56	3.59	3.62	3.65	3.68
160	3.71	3.74	3.77	3.80	3.83	3.86	3.89	3.92	3.95	3.98
170	4.01	4.04	4.07	4.10	4.13	4.16	4.19	4.22	4.25	4.28
180	4.31	4.34	4.37	4.40	4.43	4.46	4.49	4.52	4.55	4.58
190	4.61	4.64	4.67	4.70	4.73	4.76	4.79	4.82	4.85	4.88
200	4.91	4.94	4.97	5.00	5.03	5.06	5.09	5.12	5.15	5.18

* Based on the International Temperature Scale of 1948.

CALIBRATION TABLES FOR THERMOCOUPLES (Continued)
IRON-CONSTANTAN THERMOCOUPLES (MODIFIED 1913)

(Electromotive Force in Absolute Millivolts. Temperatures in Degrees F.* Reference Junctions at 32° F.)

°F	0	1	2	3	4	5	6	7	8	9
	Millivolts									
200	4.91	4.94	4.97	5.00	5.03	5.06	5.09	5.12	5.15	5.18
210	5.21	5.24	5.27	5.30	5.33	5.36	5.39	5.42	5.45	5.48
220	5.51	5.54	5.57	5.60	5.63	5.66	5.69	5.72	5.75	5.78
230	5.81	5.84	5.87	5.90	5.93	5.96	5.99	6.02	6.05	6.08
240	6.11	6.14	6.17	6.20	6.24	6.27	6.30	6.33	6.36	6.39
250	6.42	6.45	6.48	6.51	6.54	6.57	6.60	6.63	6.66	6.69
260	6.72	6.75	6.78	6.81	6.84	6.87	6.90	6.93	6.96	7.00
270	7.03	7.06	7.09	7.12	7.15	7.18	7.21	7.24	7.27	7.30
280	7.33	7.36	7.39	7.42	7.45	7.48	7.51	7.54	7.58	7.61
290	7.64	7.67	7.70	7.73	7.76	7.79	7.82	7.85	7.88	7.91
300	7.94	7.97	8.00	8.04	8.07	8.10	8.13	8.16	8.19	8.22
310	8.25	8.28	8.31	8.34	8.37	8.40	8.44	8.47	8.50	8.53
320	8.56	8.59	8.62	8.65	8.68	8.71	8.74	8.77	8.80	8.84
330	8.87	8.90	8.93	8.96	8.99	9.02	9.05	9.08	9.11	9.14
340	9.17	9.20	9.24	9.27	9.30	9.33	9.36	9.39	9.42	9.45
350	9.48	9.51	9.54	9.58	9.61	9.64	9.67	9.70	9.73	9.76
360	9.79	9.82	9.85	9.88	9.92	9.95	9.98	10.01	10.04	10.07
370	10.10	10.13	10.16	10.19	10.22	10.25	10.28	10.32	10.35	10.38
380	10.41	10.44	10.47	10.50	10.53	10.56	10.60	10.63	10.66	10.69
390	10.72	10.75	10.78	10.81	10.84	10.87	10.90	10.94	10.97	11.00
400	11.03	11.06	11.09	11.12	11.15	11.18	11.21	11.24	11.28	11.31
410	11.34	11.37	11.40	11.43	11.46	11.49	11.52	11.55	11.58	11.62
420	11.65	11.68	11.71	11.74	11.77	11.80	11.83	11.86	11.89	11.92
430	11.96	11.99	12.02	12.05	12.08	12.11	12.14	12.17	12.20	12.23
440	12.26	12.30	12.33	12.36	12.39	12.42	12.45	12.48	12.51	12.54
450	12.57	12.60	12.64	12.67	12.70	12.73	12.76	12.79	12.82	12.85
460	12.88	12.91	12.94	12.98	13.01	13.04	13.07	13.10	13.13	13.16
470	13.19	13.22	13.25	13.28	13.31	13.34	13.38	13.41	13.44	13.47
480	13.50	13.53	13.56	13.59	13.62	13.65	13.68	13.72	13.75	13.78
490	13.81	13.84	13.87	13.90	13.93	13.96	13.99	14.02	14.05	14.08
500	14.12	14.15	14.18	14.21	14.24	14.27	14.30	14.33	14.36	14.39
510	14.42	14.45	14.48	14.52	14.55	14.58	14.61	14.64	14.67	14.70
520	14.73	14.76	14.79	14.82	14.85	14.88	14.91	14.94	14.98	15.01
530	15.04	15.07	15.10	15.13	15.16	15.19	15.22	15.25	15.28	15.31
540	15.34	15.37	15.40	15.44	15.47	15.50	15.53	15.56	15.59	15.62
550	15.65	15.68	15.71	15.74	15.77	15.80	15.84	15.87	15.90	15.93
560	15.96	15.99	16.02	16.05	16.08	16.11	16.14	16.17	16.20	16.23
570	16.26	16.30	16.33	16.36	16.39	16.42	16.45	16.48	16.51	16.54
580	16.57	16.60	16.63	16.66	16.69	16.72	16.75	16.78	16.82	16.85
590	16.88	16.91	16.94	16.97	17.00	17.03	17.06	17.09	17.12	17.15
600	17.18	17.21	17.24	17.28	17.31	17.34	17.37	17.40	17.43	17.46
610	17.49	17.52	17.55	17.58	17.61	17.64	17.68	17.71	17.74	17.77
620	17.80	17.83	17.86	17.89	17.92	17.95	17.98	18.01	18.04	18.08
630	18.11	18.14	18.17	18.20	18.23	18.26	18.29	18.32	18.35	18.38
640	18.41	18.44	18.47	18.50	18.54	18.57	18.60	18.63	18.66	18.69
650	18.72	18.75	18.78	18.81	18.84	18.87	18.90	18.94	18.97	19.00
660	19.03	19.06	19.09	19.12	19.15	19.18	19.21	19.24	19.27	19.30
670	19.34	19.37	19.40	19.43	19.46	19.49	19.52	19.55	19.58	19.61
680	19.64	19.67	19.70	19.74	19.77	19.80	19.83	19.86	19.89	19.92
690	19.95	19.98	20.01	20.04	20.07	20.10	20.13	20.16	20.20	20.23
700	20.26	20.29	20.32	20.35	20.38	20.41	20.44	20.47	20.50	20.53
710	20.56	20.59	20.62	20.66	20.69	20.72	20.75	20.78	20.81	20.84
720	20.87	20.90	20.93	20.96	20.99	21.02	21.05	21.08	21.11	21.14
730	21.18	21.21	21.24	21.27	21.30	21.33	21.36	21.39	21.42	21.45
740	21.48	21.51	21.54	21.57	21.60	21.64	21.67	21.70	21.73	21.76
750	21.79	21.82	21.85	21.88	21.91	21.94	21.97	22.00	22.03	22.06
760	22.10	22.13	22.16	22.19	22.22	22.25	22.28	22.31	22.34	22.37
770	22.40	22.43	22.46	22.49	22.52	22.55	22.58	22.62	22.65	22.68
780	22.71	22.74	22.77	22.80	22.83	22.86	22.89	22.92	22.95	22.98
790	23.01	23.04	23.08	23.11	23.14	23.17	23.20	23.23	23.26	23.29
800	23.32	23.35	23.38	23.41	23.44	23.47	23.50	23.53	23.56	23.60
810	23.63	23.66	23.69	23.72	23.75	23.78	23.81	23.84	23.87	23.90
820	23.93	23.96	23.99	24.02	24.06	24.09	24.12	24.15	24.18	24.21
830	24.24	24.27	24.30	24.33	24.36	24.39	24.42	24.45	24.48	24.52
840	24.55	24.58	24.61	24.64	24.67	24.70	24.73	24.76	24.79	24.82
850	24.85	24.88	24.91	24.94	24.98	25.01	25.04	25.07	25.10	25.13
860	25.16	25.19	25.22	25.25	25.28	25.32	25.35	25.38	25.41	25.44
870	25.47	25.50	25.53	25.56	25.59	25.62	25.65	25.68	25.72	25.75
880	25.78	25.81	25.84	25.87	25.90	25.93	25.96	25.99	26.02	26.06
890	26.09	26.12	26.15	26.18	26.21	26.24	26.27	26.30	26.33	26.36
900	26.40	26.43	26.46	26.49	26.52	26.55	26.58	26.61	26.64	26.67
910	26.70	26.74	26.77	26.80	26.83	26.86	26.89	26.92	26.95	26.98
920	27.02	27.05	27.08	27.11	27.14	27.17	27.20	27.23	27.26	27.30
930	27.33	27.36	27.39	27.42	27.45	27.48	27.51	27.54	27.58	27.61
940	27.64	27.67	27.70	27.73	27.76	27.80	27.83	27.86	27.89	27.92
950	27.95	27.98	28.02	28.05	28.08	28.11	28.14	28.17	28.20	28.23

* Based on the International Temperature Scale of 1948.

CALIBRATION TABLES FOR THERMOCOUPLES (Continued)

IRON-CONSTANTAN THERMOCOUPLES (MODIFIED 1913)

(Electromotive Force in Absolute Millivolts. Temperatures in Degrees F.* Reference Junctions at 32° F.)

°F	0	1	2	3	4	5	6	7	8	9
	Millivolts									
960	28.26	28.30	28.33	28.36	28.39	28.42	28.45	28.48	26.52	28.55
970	28.58	28.61	28.64	28.67	28.70	28.74	28.77	28.80	28.83	28.86
980	28.89	28.92	28.96	28.99	29.02	29.05	29.08	29.11	29.14	29.18
990	29.21	29.24	29.27	29.30	29.33	29.37	29.40	29.43	29.46	29.49
1000	29.52	29.56	29.59	29.62	29.65	29.68	29.71	29.75	29.78	29.81
1010	29.84	29.87	29.90	29.94	29.97	30.00	30.03	30.06	30.10	30.13
1020	30.16	30.19	30.22	30.25	30.28	30.32	30.35	30.38	30.41	30.44
1030	30.48	30.51	30.54	30.57	30.60	30.64	30.67	30.70	30.73	30.76
1040	30.80	30.83	30.86	30.89	30.92	30.96	30.99	31.02	31.05	31.08
1050	31.12	31.15	31.18	31.21	31.24	31.28	31.31	31.34	31.37	31.40
1060	31.44	31.47	31.50	31.53	31.56	31.60	31.63	31.66	31.69	31.72
1070	31.76	31.79	31.82	31.85	31.88	31.92	31.95	31.98	32.01	32.05
1080	32.08	32.11	32.14	32.18	32.21	32.24	32.27	32.30	32.34	32.37
1090	32.40	32.43	32.47	32.50	32.53	32.56	32.60	32.63	32.66	32.69
1100	32.72	32.76	32.79	32.82	32.86	32.89	32.92	32.95	32.98	33.02
1110	33.05	33.08	33.11	33.15	33.18	33.21	33.24	33.28	33.31	33.34
1120	33.37	33.41	33.44	33.47	33.50	33.54	33.57	33.60	33.64	33.67
1130	33.70	33.73	33.76	33.80	33.83	33.86	33.90	33.93	33.96	33.99
1140	34.03	34.06	34.09	34.12	34.16	34.19	34.22	34.26	34.29	34.32
1150	34.36	34.39	34.42	34.45	34.49	34.52	34.55	34.58	34.62	34.65
1160	34.68	34.72	34.75	34.78	34.82	34.85	34.88	34.92	34.95	34.98
1170	35.01	35.05	35.08	35.11	35.15	35.18	35.21	35.25	35.28	35.31
1180	35.35	35.38	35.41	35.45	35.48	35.51	35.54	35.58	35.61	35.64
1190	35.68	35.71	35.74	35.78	35.81	35.84	35.88	35.91	35.94	35.98
1200	36.01	36.05	36.08	36.11	36.15	36.18	36.21	36.25	36.28	36.31
1210	36.35	36.38	36.42	36.45	36.48	36.52	36.55	36.58	36.62	36.65
1220	36.69	36.72	36.75	36.79	36.82	36.86	36.89	36.92	36.96	36.99
1230	37.02	37.06	37.09	37.13	37.16	37.20	37.23	37.26	37.30	37.33
1240	37.36	37.40	37.43	37.47	37.50	37.54	37.57	37.60	37.64	37.67
1250	37.71	37.74	37.78	37.81	37.84	37.88	37.91	37.95	37.98	38.02
1260	38.05	38.08	38.12	38.15	38.19	38.22	38.26	38.29	38.32	38.36
1270	38.39	38.43	38.46	38.50	38.53	38.57	38.60	38.64	38.67	38.70
1280	38.74	38.77	38.81	38.84	38.88	38.91	38.95	38.98	39.02	39.05
1290	39.08	39.12	39.15	39.19	39.22	39.26	39.29	39.33	39.36	39.40
1300	39.43	39.47	39.50	39.54	39.57	39.61	39.64	39.68	39.71	39.75
1310	39.78	39.82	39.85	39.89	39.92	39.96	39.99	40.03	40.06	40.10
1320	40.13	40.17	40.20	40.24	40.27	40.31	40.34	40.38	40.41	40.45
1330	40.48	40.52	40.55	40.59	40.62	40.66	40.69	40.73	40.76	40.80
1340	40.83	40.87	40.90	40.94	40.98	41.01	41.05	41.08	41.12	41.15
1350	41.19	41.22	41.26	41.29	41.33	41.36	41.40	41.43	41.47	41.50
1360	41.54	41.58	41.61	41.65	41.68	41.72	41.75	41.79	41.82	41.86
1370	41.90	41.93	41.97	42.00	42.04	42.07	42.11	42.14	42.18	42.22
1380	42.25	42.29	42.32	42.36	42.39	42.43	42.46	42.50	42.53	42.57
1390	42.61	42.64	42.68	42.71	42.75	42.78	42.82	42.85	42.89	42.92
1400	42.96	....	....	....	....	....	....	....	....	....

* Based on the International Temperature Scale of 1948.

CALIBRATION TABLES FOR THERMOCOUPLES (Continued)

TEMPERATURE-E. M. F. VALUES FOR COPPER-CONSTANTAN

E. M. F. values are in millivolts; reference junctions at 0°C.; temperatures are in degrees Centigrade

Roeser and Wensel, National Bureau of Standards

°C	0°	10°	20°	30°	40°	50°	60°	70°	80°	90°
−200°	−5.54									
−100°	−3.35	−3.62	−3.89	−4.14	−4.38	−4.60	−4.82	−5.02	−5.20	−5.38
0°	0	−0.38	−0.75	−1.11	−1.47	−1.81	−2.14	−2.46	−2.77	−3.06
0°	0	0.39	0.79	1.19	1.61	2.03	2.47	2.91	3.36	3.81
100°	4.28	4.75	5.23	5.71	6.20	6.70	7.21	7.72	8.23	8.76
200°	9.29	9.82	10.36	10.91	11.46	12.01	12.57	13.14	13.71	14.28
300°	14.86	15.44	16.03	16.62	17.22	17.82	18.42	19.03	19.64	20.25
400°	20.87									

TEMPERATURE-E. M. F. VALUES FOR COPPER-CONSTANTAN

E. M. F. values are in millivolts; reference junctions at 32°F.; temperatures are in degrees Fahrenheit

Roeser and Wensel, National Bureau of Standards

°F	0°	10°	20°	30°	40°	50°	60°	70°	80°	90°
−300°	−5.28									
−200°	−4.11	−4.25	−4.38	−4.50	−4.63	−4.75	−4.86	−4.97	−5.08	−5.18
−100°	−2.56	−2.73	−2.90	−3.06	−3.22	−3.38	−3.53	−3.68	−3.83	−3.97
0°	−0.67	−0.87	−1.07	−1.27	−1.47	−1.66	−1.84	−2.03	−2.21	−2.39
0°	−0.67	−0.46	−0.26	−0.04	+0.17	0.39	0.61	0.83	1.06	1.29
100°	1.52	1.75	1.99	2.23	2.47	2.71	2.96	3.21	3.46	3.71
200°	3.97	4.22	4.48	4.75	5.01	5.28	5.55	5.82	6.09	6.37
300°	6.64	6.92	7.21	7.49	7.77	8.06	8.35	8.64	8.93	9.23
400°	9.52	9.82	10.12	10.42	10.72	11.03	11.33	11.64	11.95	12.26
500°	12.57	12.89	13.20	13.52	13.83	14.15	14.47	14.79	15.12	15.44
600°	15.77	16.10	16.42	16.75	17.08	17.42	17.75	18.08	18.42	18.75
700°	19.09	19.43	19.77	20.11	20.45	20.80				

REFERENCE TABLE FOR Pt TO Pt—10 PER CENT Rh THERMOCOUPLE

Emfs are expressed in microvolts and temperatures in °C. Cold junctions at 0°C.

ROESER AND WENSEL, NATIONAL BUREAU OF STANDARDS

E(μv)	0	1,000	2,000	3,000	4,000	5,000	6,000	7,000	8,000	E(μv)
0	0	146.9	265.0	373.7	477.7	578.1	675.3	769.5	861.0	0
	17.7	12.5	11.2	10.5	10.2	9.9	9.5	9.3	9.0	
100	17.7	159.4	276.2	384.2	487.9	588.0	684.8	778.8	870.0	100
	16.7	12.3	11.1	10 5	10.2	9.8	9.5	9.2	9.0	
200	34.4	171.7	287.3	394.7	498.1	597.8	694.3	788.0	879.0	200
	15.8	12.1	11.0	10.5	10 1	9 8	9.5	9.2	9.0	
300	50.2	183.8	298.3	405.2	508.2	607.6	703.8	797.2	888.0	300
	15.2	12.0	11.0	10.5	10.1	9.8	9.5	9.2	9.0	
400	65.4	195.8	309.3	415.7	518.3	617.4	713.3	806.4	897.0	400
	14.6	11.8	10.9	10 4	10.1	9.7	9.4	9 2	8.9	
500	80.0	207.6	320.2	426.1	528.4	627.1	722.7	815.6	905.9	500
	14.1	11.7	10.8	10.4	10.0	9.7	9.4	9 1	8.9	
600	94.1	219.3	331.0	436.5	538.4	636.8	732.1	824.7	914.8	600
	13.7	11 6	10 7	10.3	10.0	9.7	9.4	9.1	8.9	
700	107.8	230.9	341.7	446.8	548.4	646.5	741.5	833.8	923.7	700
	13.3	11.5	10 7	10.3	9.9	9.6	9.4	9.1	8.9	
800	121.1	242 4	352.4	457.1	558 3	656.1	750.9	842.9	932.6	800
	13.0	11 3	10.7	10.3	9.9	9.6	9.3	9.1	8.8	
900	134.1	253.7	363.1	467.4	568.2	665.7	760.2	852.0	941.4	900
	12.8	11.3	10.6	10.3	9 9	9.6	9.3	9.0	8.8	
1,000	146.9	265.0	373.7	477.7	578.1	675.3	769.5	861.0	950.2	1,000

REFERENCE TABLE FOR Pt TO Pt—10 PER CENT Rh THERMOCOUPLE (Continued)

Emfs are expressed in microvolts and temperatures in °C. Cold junctions at 0°C.

E(μv)	9,000	10,000	11,000	12,000	13,000	14,000	15,000	16,000	17,000	E(μv)
0	950.2	1,037.2	1,122.3	1,206.4	1,290.0	1,373.8	1,458.0	1,542.6	1,627.8	0
	8.8	8.6	8.5	8.3	8.3	8.4	8.4	8.5	8.6	
100	959.0	1,045.8	1,130.8	1,214.7	1,298.3	1,382.2	1,466.4	1,551.1	1,636.4	100
	8.8	8.6	8.4	8.4	8.4	8.4	8.4	8.5	8.5	
200	967.8	1,054.4	1,139.2	1,223.1	1,306.7	1,390.6	1,474.8	1,559.6	1,644.9	200
	8.7	8.5	8.4	8.3	8.4	8.4	8.5	8.5	8.6	
300	976.5	1,062.9	1,147.6	1,231.4	1,315.1	1,399.0	1,483.3	1,568.1	1,653.5	300
	8.8	8.6	8.4	8.4	8.4	8.4	8.5	8.5	8.6	
400	985.3	1,071.5	1,156.0	1,239.8	1,323.5	1,407.4	1,491.8	1,576.6	1,662.1	400
	8.7	8.5	8.4	8.4	8.3	8.4	8.4	8.5	8.6	
500	994.0	1,080.0	1,164.4	1,248.2	1,331.8	1,415.8	1,500.2	1,585.1	1,670.7	500
	8.7	8.5	8.4	8.3	8.4	8.4	8.5	8.6	8.6	
600	1,002.7	1,088.5	1,172.8	1,256.5	1,340.2	1,424.2	1,508.7	1,593.7	1,679.3	600
	8.6	8.5	8.4	8.4	8.4	8.5	8.5	8.5	8.6	
700	1,011.3	1,097.0	1,181.2	1,264.9	1,348.6	1,432.7	1,517.2	1,602.2	1,687.9	700
	8.7	8.4	8.4	8.3	8.4	8.4	8.4	8.5	8.6	
800	1,020.0	1,105.4	1,189.6	1,273.2	1,357.0	1,441.1	1,525.6	1,610.7	1,696.5	800
	8.6	8.5	8.4	8.4	8.4	8.4	8.5	8.6	8.6	
900	1,028.6	1,113.9	1,198.0	1,281.6	1,365.4	1,449.5	1,534.1	1,619.3	1,705.1	900
	8.6	8.4	8.4	8.4	8.4	8.5	8.5	8.5	8.6	
1,000	1,037.2	1,122.3	1,206.4	1,290.0	1,373.8	1,458.0	1,542.6	1,627.8	1,713.7	1,000

Emfs are expressed in microvolts and temperatures in °F. Cold junctions at 32°F.
ROESER AND WENSEL, NATIONAL BUREAU OF STANDARDS

E(μv)	0	1,000	2,000	3,000	4,000	5,000	6,000	7,000	8,000	E(μv)
0	32.0	296.4	509.0	704.7	891.9	1,072.6	1,247.5	1,417.1	1,581.8	0
	31.9	22.5	20.1	19.0	18.4	17.7	17.1	16.7	16.2	
100	63.9	318.9	529.1	723.7	910.3	1,090.3	1,264.6	1,433.8	1,598.0	100
	30.0	22.1	20.0	18.9	18.3	17.7	17.1	16.6	16.2	
200	93.9	341.0	549.1	742.6	928.6	1,108.0	1,281.7	1,450.4	1,614.2	200
	28.5	21.8	19.9	18.8	18.2	17.7	17.1	16.6	16.2	
300	122.4	362.8	569.0	761.4	946.8	1,125.7	1,298.8	1,467.0	1,630.4	300
	27.3	21.6	19.8	18.8	18.1	17.6	17.1	16.5	16.1	
400	149.7	384.4	588.8	780.2	964.9	1,143.3	1,315.9	1,483.5	1,646.5	400
	26.3	21.3	19.6	18.8	18.1	17.5	17.0	16.5	16.1	
500	176.0	405.7	608.4	799.0	983.0	1,160.8	1,332.9	1,500.0	1,662.6	500
	25.4	21.0	19.4	18.7	18.1	17.5	16.9	16.5	16.0	
600	201.4	426.7	627.8	817.7	1,001.1	1,178.3	1,349.8	1,516.5	1,678.6	600
	24.6	20.9	19.3	18.6	18.0	17.4	16.9	16.4	16.0	
700	226.0	447.6	647.1	836.3	1,019.1	1,195.7	1,366.7	1,532.9	1,694.6	700
	24.0	20.7	19.3	18.5	17.9	17.3	16.9	16.3	16.0	
800	250.0	468.3	666.4	854.8	1,037.0	1,213.0	1,383.6	1,549.2	1,710.6	800
	23.4	20.4	19.2	18.6	17.8	17.3	16.8	16.3	15.9	
900	273.4	488.7	685.6	873.4	1,054.8	1,230.3	1,400.4	1,565.5	1,726.5	900
	23.0	20.3	19.1	18.5	17.8	17.2	16.7	16.3	15.9	
1,000	296.4	509.0	704.7	891.9	1,072.6	1,247.5	1,417.1	1,581.8	1,742.4	1,000

Emfs are expressed in microvolts and temperatures in °F. Cold junctions at 32°F.

E(μv)	9,000	10,000	11,000	12,000	13,000	14,000	15,000	16,000	17,000	E(μv)
0	1,742.4	1,899.0	2,052.1	2,203.5	2,354.0	2,504.8	2,656.4	2,808.7	2,962.0	0
	15.8	15.5	15.2	15.0	15.0	15.2	15.1	15.3	15.4	
100	1,758.2	1,914.5	2,067.3	2,218.5	2,369.0	2,520.0	2,671.5	2,824.0	2,977.4	100
	15.8	15.4	15.2	15.0	15.1	15.1	15.2	15.3	15.4	
200	1,774.0	1,929.9	2,082.5	2,233.5	2,384.1	2,535.1	2,686.7	2,839.3	2,992.8	200
	15.7	15.4	15.2	15.0	15.1	15.1	15.2	15.3	15.5	
300	1,789.7	1,945.3	2,097.7	2,248.5	2,399.2	2,550.2	2,701.9	2,854.6	3,008.3	300
	15.8	15.4	15.1	15.1	15.1	15.1	15.3	15.3	15.5	
400	1,805.5	1,960.7	2,112.8	2,263.6	2,414.3	2,565.3	2,717.2	2,869.9	3,023.8	400
	15.7	15.3	15.1	15.1	15.0	15.1	15.2	15.3	15.5	
500	1,821.2	1,976.0	2,127.9	2,278.7	2,429.3	2,580.4	2,732.4	2,885.2	3,039.3	500
	15.6	15.3	15.1	15.0	15.1	15.2	15.3	15.4	15.4	
600	1,836.8	1,991.3	2,143.0	2,293.7	2,444.4	2,595.6	2,747.7	2,900.6	3,054.7	600
	15.6	15.2	15.2	15.1	15.1	15.2	15.3	15.4	15.5	
700	1,852.4	2,006.5	2,158.2	2,308.8	2,459.5	2,610.8	2,763.0	2,916.0	3,070.2	700
	15.6	15.2	15.1	15.0	15.1	15.2	15.2	15 3	15.5	
800	1,868.0	2,021.7	2,173.3	2,323.8	2,474.6	2,626.0	2,778.2	2,931.3	3,085.7	800
	15.5	15.2	15.1	15.1	15.2	15.2	15.2	15.4	15.5	
900	1,883.5	2,036.9	2,188.4	2,338.9	2,489.7	2,641.2	2,793.4	2,946.7	3,101.2	900
	15.5	15.2	15.1	15.1	15.1	15.2	15.3	15.3	15.5	
1,000	1,899.0	2,052.1	2,203.5	2,354.0	2,504.8	2,656.4	2,808.7	2,962.0	3,116.7	1,000

REFERENCE TABLE FOR Pt TO Pt—13 PER CENT Rh THERMOCOUPLE

Emfs are expressed in microvolts and temperatures in °C. Cold junctions at 0°C.

ROESER AND WENSEL, NATIONAL BUREAU OF STANDARDS

E(μv)	0	1,000	2,000	3,000	4,000	5,000	6,000	7,000	8,000	9,000	E(μv)
0	0	145.3	258.8	361.0	457.4	549.8	638.3	723.5	806.0	886.1	0
	17.9	12.2	10.6	9.9	9.5	9.0	8.7	8.4	8.1	7.9	
100	17.9	157.5	269.4	370.9	466.9	558.8	647.0	731.9	814.1	894.0	100
	16.7	12.0	10.5	9.8	9.4	9.0	8.6	8.4	8.1	7.8	
200	34.6	169.5	279.9	380.7	476.3	567.8	655.6	740.3	822.2	901.8	200
	15.8	11.7	10.4	9.8	9.4	9.0	8.6	8.3	8.1	7.9	
300	50.4	181.2	290.3	390.5	485.7	576.8	644.2	748.6	830.3	909.7	300
	15.1	11.5	10.3	9.7	9.3	8.9	8.6	8.3	8.0	7.8	
400	65.5	192.7	300.6	400.2	495.0	585.7	672.8	756.9	838.3	917.5	400
	14.5	11.4	10.2	9.7	9.3	8.9	8.5	8.2	8.0	7.8	
500	80.0	204.1	310.8	409.9	504.3	594.6	681.3	765.1	846.3	925.3	500
	13.9	11.2	10.2	9.6	9.2	8.8	8.5	8.2	8.0	7.8	
600	93.9	215.3	321.0	419.5	513.5	603.4	689.8	773.3	854.3	933.1	600
	13.4	11.1	10.1	9.5	9.1	8.8	8.5	8.2	8.0	7.8	
700	107.3	226.4	331.1	429.0	522.6	612.2	698.3	781.5	862.3	940.9	700
	13.0	10.9	10.0	9.5	9.1	8.7	8.4	8.2	8.0	7.8	
800	120.3	237.3	341.1	438.5	531.7	620.9	706.7	789.7	870.3	948.7	800
	12.6	10.8	10.0	9.5	9.1	8.7	8.4	8.1	7.9	7.7	
900	132.9	248.1	351.1	448.0	540.8	629.6	715.1	797.8	878.2	956.4	900
	12.4	10.7	9.9	9.4	9.0	8.7	8.4	8.2	7.9	7.7	
1,000	145.3	258.8	361.0	457.4	549.8	638.3	723.5	806.0	886.1	964.1	1,000

Emfs are expressed in microvolts and temperatures in °C. Cold junctions at 0°C.

E(μv)	10,000	11,000	12,000	13,000	14,000	15,000	16,000	17,000	18,000	19,000	E(μv)
0	964.1	1,040.0	1,113.9	1,186.9	1,259.3	1,331.8	1,404.3	1,476.9	1,550.0	1,623.6	0
	7.6	7.4	7.3	7.2	7.3	7.2	7.3	7.3	7.4	7.4	
100	971.7	1,047.4	1,121.2	1,194.1	1,266.6	1,339.0	1,411.6	1,484.2	1,557.4	1,631.0	100
	7.7	7.5	7.4	7.3	7.2	7.2	7.3	7.3	7.4	7.4	
200	979.4	1,054.9	1,128.6	1,201.4	1,273.8	1,346.2	1,418.9	1,491.5	1,564.8	1,638.4	200
	7.6	7.4	7.3	7.3	7.3	7.3	7.2	7.3	7.3	7.4	
300	987.0	1,062.3	1,135.9	1,208.7	1,281.1	1,353.5	1,426.1	1,498.8	1,572.1	1,645.8	300
	7.7	7.4	7.3	7.2	7.2	7.3	7.2	7.3	7.4	7.4	
400	994.7	1,069.7	1,143.2	1,215.9	1,288.3	1,360.8	1,433.3	1,506.1	1,579.5	1,653.2	400
	7.6	7.4	7.3	7.3	7.2	7.2	7.2	7.3	7.3	7.4	
500	1,002.3	1,077.1	1,150.5	1,223.2	1,295.5	1,368.0	1,440.5	1,513.4	1,586.8	1,660.6	500
	7.6	7.3	7.3	7.2	7.2	7.3	7.3	7.3	7.4	7.4	
600	1,009.9	1,084.4	1,157.8	1,230.4	1,302.7	1,375.3	1,447.8	1,520.7	1,594.2	1,668.0	600
	7.6	7.4	7.3	7.2	7.2	7.3	7.2	7.3	7.3	7.4	
700	1,017.5	1,091.8	1,165.1	1,237.6	1,309.9	1,382.6	1,455.0	1,528.0	1,601.5	1,675.4	700
	7.5	7.4	7.2	7.2	7.3	7.2	7.3	7.3	7.4	7.4	
800	1,025.0	1,099.2	1,172.3	1,244.8	1,317.2	1,389.8	1,462.3	1,535.3	1,608.9	1,682.8	800
	7.5	7.3	7.3	7.3	7.3	7.3	7.3	7.4	7.3	7.4	
900	1,032.5	1,106.5	1,179.6	1,252.1	1,324.5	1,397.1	1,469.6	1,542.7	1,616.2	1,690.2	900
	7.5	7.4	7.3	7.2	7.3	7.2	7.3	7.3	7.4	7.4	
1,000	1,040.0	1,113.9	1,186.9	1,259.3	1,331.8	1,404.3	1,476.9	1,550.0	1,623.6	1,697.6	1,000

Emfs are expressed in microvolts and temperatures in °F. Cold junctions at 32°F.

ROESER AND WENSEL, NATIONAL BUREAU OF STANDARDS

E(μv)	0	1,000	2,000	3,000	4,000	5,000	6,000	7,000	8,000	9,000	E(μv)
0	32.0	293.5	497.8	681.8	855.3	1,021.6	1,180.9	1,334.3	1,482.8	1,627.0	0
	32.2	22.0	19.1	17.8	17.0	16.2	15.6	15.1	14.6	14.2	
100	64.2	315.5	516.9	699.6	872.3	1,037.8	1,196.5	1,349.4	1,497.4	1,641.2	100
	30.1	21.6	18.9	17.7	17.0	16.2	15.6	15.1	14.6	14.1	
200	94.3	337.1	535.8	717.3	889.3	1,054.0	1,212.1	1,364.5	1,512.0	1,655.3	200
	28.4	21.1	18.7	17.6	16.9	16.2	15.5	15.0	14.5	14.1	
300	122.7	358.2	554.5	734.9	906.2	1,070.2	1,227.6	1,379.5	1,526.5	1,669.4	300
	27.2	20.7	18.6	17.5	16.8	16.1	15.4	14.9	14.4	14.1	
400	149.9	378.9	573.1	752.4	923.0	1,086.3	1,243.0	1,394.4	1,540.9	1,683.5	400
	26.1	20.4	18.4	17.4	16.7	16.0	15.3	14.8	14.4	14.0	
500	176.0	399.3	591.5	769.8	939.7	1,102.3	1,258.3	1,409.2	1,555.3	1,697.5	500
	25.0	20.2	18.3	17 3	16.6	15.9	15.3	14.8	14.4	14.1	
600	201.0	419.5	609.8	787.1	956.3	1,118.2	1,273.6	1,424.0	1,569.7	1,711.6	600
	24.1	20.0	18.2	17.2	16.4	15.8	15.3	14.8	14.4	14.0	
700	225.1	439.5	628.0	804.3	972.7	1,134.0	1,288.9	1,438.8	1,584.1	1,725.6	700
	23.4	19.7	18.0	17.1	16.4	15.7	15.2	14.7	14.4	14.0	
800	248.5	459.2	646.0	821.4	989.1	1,149.7	1,304.1	1,453.5	1,598.5	1,739.6	800
	22.7	19.4	18.0	17.0	16.3	15.6	15.1	14.6	14.3	13.9	
900	271.2	478.6	664.0	838.4	1,005.4	1,165.3	1,319.2	1,468.1	1,612.8	1,753.5	900
	22.3	19.2	17.8	16.9	16.2	15.6	15.1	14.7	14.2	13.9	
1,000	293.5	497.8	681.8	855.3	1,021.6	1,180.9	1,334.3	1,482.8	1,627.0	1,767.4	1,000

Emfs are expressed in microvolts and temperatures in °F. Cold junctions at 32°F.

E(μv)	10,000	11,000	12,000	13,000	14,000	15,000	16,000	17,000	18,000	19,000	E(μv)
0	1,767.4	1,904.0	2,037.0	2,168.4	2,298.7	2,429.2	2,559.7	2,690.4	2,822.0	2,954.5	0
	13.8	13.4	13.2	13.0	13.1	13.0	13.1	13.2	13.3	13.3	
100	1,781.2	1,917.4	2,050.2	2,181.4	2,311.8	2,442.2	2,572.8	2,703.6	2,835.3	2,967.8	100
	13.7	13.4	13.2	13.1	13.0	13.0	13.1	13.1	13.3	13.3	
200	1,794.9	1,930.8	2,063.4	2,194.5	2,324.8	2,455.2	2,585.9	2,716.7	2,848.6	2,981.1	200
	13.7	13.3	13.2	13.1	13.1	13.1	13.0	13.1	13.2	13.3	
300	1,808.6	1,944.1	2,076.6	2,207.6	2,337.9	2,468.3	2,598.9	2,729.8	2,861.8	2,994.4	300
	13.8	13.3	13.2	13.0	13.0	13.1	13.0	13.2	13.3	13.4	
400	1,822.4	1,957.4	2,089.8	2,220.6	2,350.9	2,481.4	2,611.9	2,743.0	2,875.1	3,007.8	400
	13.7	13.3	13.1	13.1	13.0	13.0	13.0	13.1	13.2	13.3	
500	1,836.1	1,970.7	2,102.9	2,233.7	2,363.9	2,494.4	2,624.9	2,756.1	2,888.3	3,021.1	500
	13.7	13 3	13.1	13.0	13.0	13.1	13.1	13.2	13.3	13.3	
600	1,849.8	1,984.0	2,116.0	2,246.7	2,376.9	2,507.5	2,638.0	2,769.3	2,901.6	3,034.4	600
	13.6	13.3	13.1	13.0	13.0	13.1	13.0	13.1	13.2	13.3	
700	1,863.4	1,997.3	2,129.1	2,259.7	2,389.9	2,520.6	2,651.0	2,782.4	2,914.8	3,047.7	700
	13.6	13.3	13.1	13.0	13.1	13.0	13.1	13.2	13.2	13.3	
800	1,877.0	2,010.6	2,142.2	2,272.7	2,403.0	2,533.6	2,664.1	2,795.6	2,928.0	3,061.0	800
	13.5	13.2	13.1	13.0	13.1	13.1	13.1	13.2	13.2	13.4	
900	1,890.5	2,023.8	2,155.3	2,285.7	2,416.1	2,546.7	2,677.2	2,808.8	2,941.2	3,074.4	900
	13.5	13.2	13.1	13.0	13.1	13.0	13.2	13.2	13.3	13.3	
1,000	1,904.0	2,037.0	2,168.4	2,298.7	2,429.2	2,559.7	2,690.4	2,822.0	2,954.5	3,087.7	1,000

VALUES FOR THE LANGEVIN FUNCTION $\mathcal{L}(u)$

Compiled by Allen L. King

Because of random thermal rotations a dipole or dipole-like element ordinarily has an average dipole moment of zero. If it is placed in an orienting field F however it tends to align itself with the field so that the average component of the dipole moment parallel to the field equals $\bar{p}$. Classically, if the system is in thermal equilibrium,

$$\bar{p} = p_0(\coth u - 1/u) = p_0\mathcal{L}(u)$$

Here p_0 is the moment of the dipole itself and $\mathcal{L}(u)$ is the Langevin function of the argument u which equals Fp_0/kT.

$$\text{For } u \ll 1 \quad \mathcal{L}(u) = \frac{u}{3}$$

The Langevin function applies to permanent electric and magnetic dipoles.

(u)	0	1	2	3	4	5	6	7	8	9	10	Diff.
0.0	0000	0033	0066	0100	0133	0166	0200	0233	0266	0300	0333	33
0.1	0333	0366	0400	0433	0466	0499	0532	0565	0598	0631	0665	33
0.2	0665	0698	0731	0764	0797	0830	0862	0896	0929	0961	0994	33
0.3	0994	1027	1059	1092	1124	1157	1190	1222	1254	1287	1319	32
0.4	1319	1352	1384	1416	1448	1480	1512	1544	1576	1608	1640	32
0.5	1640	1671	1703	1734	1765	1797	1829	1860	1892	1923	1953	31
0.6	1953	1985	2016	2046	2077	2108	2138	2169	2200	2230	2260	31
0.7	2260	2290	2321	2351	2381	2411	2441	2471	2500	2530	2559	30
0.8	2559	2589	2618	2647	2676	2705	2734	2763	2792	2821	2850	29
0.9	2850	2878	2906	2934	2963	2991	3019	3047	3075	3103	3130	28
1.0	3130	3158	3185	3212	3240	3267	3294	3321	3348	3375	3401	27
1.1	3401	3428	3454	3480	3507	3533	3559	3585	3611	3637	3662	26
1.2	3662	3688	3713	3738	3763	3788	3813	3838	3863	3888	3913	25
1.3	3913	3937	3961	3985	4009	4033	4057	4081	4105	4129	4152	24
1.4	4152	4176	4199	4222	4245	4268	4291	4313	4336	4359	4381	23
1.5	4381	4403	4426	4448	4469	4491	4514	4536	4557	4579	4600	22
1.6	4600	4621	4642	4663	4684	4705	4726	4747	4768	4788	4808	21
1.7	4808	4828	4849	4869	4889	4909	4928	4948	4967	4987	5006	20
1.8	5006	5025	5044	5064	5083	5102	5121	5139	5158	5176	5195	19
1.9	5195	5213	5231	5249	5267	5285	5303	5321	5338	5356	5373	18
2.0	5373	5391	5408	5425	5442	5459	5476	5493	5509	5526	5542	17
2.1	5542	5559	5575	5591	5608	5624	5640	5656	5672	5688	5704	16
2.2	5704	5719	5734	5750	5765	5780	5795	5810	5825	5840	5855	15
2.3	5855	5870	5885	5899	5913	5928	5943	5957	5971	5985	6000	14
2.4	6000	6014	6027	6041	6055	6068	6082	6095	6109	6122	6136	13.5
2.5	6136	6149	6162	6175	6188	6201	6214	6227	6239	6252	6265	13
2.6	6265	6278	6290	6302	6314	6326	6339	6351	6363	6375	6387	12
2.7	6387	6399	6411	6422	6434	6446	6457	6469	6480	6492	6503	11.5
2.8	6503	6514	6525	6536	6548	6559	6570	6581	6591	6602	6613	11
2.9	6613	6624	6634	6644	6655	6665	6676	6686	6696	6707	6717	10
3.0	6717	6727	6737	6747	6757	6767	6777	6787	6796	6806	6815	10
3.1	6815	6825	6834	6844	6853	6862	6871	6880	6890	6899	6908	9
3.2	6908	6917	6926	6935	6944	6953	6962	6971	6979	6988	6997	9
3.3	6997	7006	7014	7023	7031	7040	7048	7056	7064	7073	7081	8.5
3.4	7081	7089	7097	7105	7113	7121	7129	7137	7145	7153	7161	8
3.5	7161	7169	7176	7184	7192	7200	7207	7215	7223	7230	7237	8
3.6	7237	7245	7252	7259	7267	7274	7281	7288	7295	7302	7309	7
3.7	7309	7317	7324	7330	7337	7344	7351	7358	7364	7371	7378	7
3.8	7378	7385	7392	7398	7405	7412	7418	7425	7431	7437	7444	6.5
3.9	7444	7450	7457	7463	7470	7476	7482	7488	7494	7501	7507	6

Note: Higher values of the Langevin function are nearly equal to $(u - 1)/u$.

MAGNETIC PROPERTIES AND COMPOSITION OF SOME PERMANENT MAGNETIC ALLOYS

Name	Composition,* Weight percent					Remanence, B_r (Gauss)	Coercive force H_c (Oersteds)	Maximum energy product, $(BH)_{max}$ (Gauss-Oersteds $\times 10^{-6}$)
	Al	Ni	Co	Cu	Other			
U.S.A.								
Alnico I	12	20–22	5	....	......	7,100	440	1.4
Alnico II	10	17	12.5	6	......	7,200	540	1.6
Alnico III	12	24–26	....	3	......	6,900	470	1.35
Alnico IV	12	27–28	5	....	......	5,500	700	1.3
Alnico V†	8	14	24	3	......	12,500	600	5.0
Alnico V DG†	8	14	24	3	......	13,100	640	6.0
Alnico VI†	8	15	24	3	1.25 Ti	10,500	750	3.75
Alnico VII†	8.5	18	24	3	5 Ti	7,200	1,050	2.75
Alnico XII	6	18	35	....	8 Ti	5,800	950	1.6
Carbon steel	....	....	....	....	1 Mn 0.9 C	10,000	50	0.2
Chromium steel	....	....	....	....	3.5 Cr 0.9 C 0.3 Mn	9,700	65	0.3
Cobalt steel	....	....	17	....	2.5 Cr 8 W 0.75 C	9,500	150	0.65
Cunico	....	21	29	50	.....	3,400	660	0.80
Cunife	....	20	....	60	......	5,400	550	1.5
Ferroxdur 1	$BaFe_{12}O_{19}$					2,200	1,800	1.0
Ferroxdur 2	$BaF_{12}O_{19}$ (oriented)					3,840	2,000	3.5
Platinum-Cobalt	....	....	23	....	77 Pt	6,000	4,300	7.5
Remalloy	....	....	12	....	17 Mo	10,500	250	1.1
Silmanol	4.4	....	....	....	86.6 Ag 8.8 Mn	550	6,000	0.075
Tungsten steel	....	....	....	....	5 W 0.3 Mn 0.7 C	10,300	70	0.32
Vicalloy I	....	....	52	....	10 V	8,800	300	1.0
Vicalloy II (wire)	....	....	52	....	14 V	10,000	510	3.5
Germany								
Alni 90	12	21	....	....	......	8,000	350	1.2
Alni 120	13	27	....	....	......	6,000	570	1.2
Alnico 130	12	23	5	....	......	6,300	620	1.4
Alnico 160	11	24	12	4	......	6,200	700	1.6
Alnico 190	12	21	15	4	......	7,000	700	1.8
Alnico 250	8	19	23	4	6 Ti	6,500	1,000	2.2
Alnico 400†	9	15	23	4	......	12,000	650	4.8
Alnico 580† (semicolumnar)	9	15	23	4	......	13,000	700	6.0
Oerstit 800	9	18	19	4	4 Ti	6,600	750	1.95
Great Britain								
Alcomax I	7.5	11	25	3	1.5 Ti	12,000	475	3.5
Alcomax II	8	11.5	24	4.5	......	12,400	575	4.7
Alcomax IISC (semicolumnar)	8	11	22	4.5	......	12,800	600	5.15
Alcomax III	8	13.5	24	3	0.8 Nb	12,500	670	5.10
Alcomax IIISC (semicolumnar)	8	13.5	24	3	0.8 Nb	13,000	700	5.80
Alcomax IV	8	13.5	24	3	2.5 Nb	11,200	750	4.30

* Remainder of unlisted composition is either iron or iron plus trace impurities.
† Cast anisotropic. Unmarked ones are cast isotropic.

MAGNETIC PROPERTIES AND COMPOSITION OF SOME PERMANENT MAGNETIC ALLOYS (Continued)

Name	Composition,* Weight percent					Remanence, B_r (Gauss)	Coercive force H_c (Oersteds)	Maximum energy product, $(BH)_{max}$ (Gauss-Oersteds $\times 10^{-6}$)
	Al	Ni	Co	Cu	Other			
Alcomax IVSC (semicolumnar)	8	13.5	24	3	2.5 Nb	11,700	780	5.10
Alni, high B_r	13	24	....	3.5	......	6,200	480	1.25
Alni, normal	....	....	....	....	......	5,600	580	1.25
Alni, high H_c	12	32	....	....	0–0.5 Ti	5,000	680	1.25
Alnico, high B_r	10	17	12	6	......	8,000	500	1.70
Alnico, normal	....	....	....	....	......	7,250	560	1.70
Alnico, high H_c	10	20	13.5	6	0.25 Ti	6,600	620	1.70
Columax (columnar)	similar to Alcomax III or IV					13,000–14,000	700–800	7.0–8.5
Hycomax	9	21	20	1.6	......	9,500	830	3.3

* Remainder of unlisted composition is either iron or iron plus trace impurities.
† Cast anisotropic. Unmarked ones are cast isotropic.

HIGH PERMEABILITY MAGNETIC ALLOYS

(See separate table for magnetic properties)

Name	Composition,* Weight percent	Sp. gr., gm. per cc	Tensile strength		Remark	Use
			Kg/mm^2**	Form		
Silicon iron AISI M 15	Si 4	7.68–7.64	44.3	....	Annealed 4 hrs 802–1093°C	Low core losses
Silicon iron AISI M 8	Si 3	7.68–7.64	44.2	Grain oriented	Annealed 4 hrs 802–1204°C	
45 Permalloy	Ni 45; Mn 0.3	8.17	....	....	....	Audio transformer, coils, relays
Monimax	Ni 47; Mo 3	8.27	....	....	....	High frequency coils
4-79 Permalloy	Ni 79; Mo 4; Mn 0.3	8.74	55.4	....	H_2 annealed 1121°C	Audio coils, transformers, magnetic shields
Sinimax	Ni 43; Si 3	7.70	....	....	....	High frequency coils
Nu-metal	Ni 75; Cr 2; Cu 5	8.58	44.8	....	H_2 annealed 1221°C	Audio coils, magnetic shields, transformers
Supermalloy	Ni 79; Mo 5; Mn 0.3	8.77	....	....	....	Pulse transformers, magnetic amplifiers, coils
2-V Permendur	Co 40; V 2	8.15	46.3	....	....	D-c electromagnets, pole tips

* Iron is additional alloying metal.
** kg/mm^2 $\times$ 1422.33 = lbs/in.^{2}

CAST PERMANENT MAGNETIC ALLOYS

(See separate table for magnetic properties)

Alloy name, country of manufacture*	Composition,** weight percent	Sp. gr., gm per cc	Thermal expansion		Tensile strength		Remark†	Use
			$\frac{Cm \times 10^{-6}}{cm \times °C}$	Between °C	*** Kg/ mm^2	Form		
Alnico I (USA)*	Al 12; Ni 20–22; Co 5	6.9	12.6	20–300	2.9	Cast	i.	Permanent magnets
Alnico II (USA)*	Al 10; Ni 17; Cu 6; Co 12.5	7.1	12.4	20–300	2.1 45.7	Cast Sintered	i.	Temperature controls, magnetic toys and novelties
Alnico III (USA)*	Al 12; Ni 24–26; Cu 3	6.9	12	20–300	8.5	Cast	i.	Tractor magnetos
Alnico IV (USA)*	Al 12; Ni 27–28; Co 5	7.0	13.1	20–300	6.3 42.1	Cast Sintered	i.	Application requiring high coercive force
Alnico V (USA)*	Al 8; Ni 14; Co 24; Cu 3	7.3	11.3	......	3.8 35	Cast Sintered	a.	Application requiring high energy
Alnico V DG (USA)*	Al 8; Ni 14; Co 24; Cu 3	7.3	11.3	......	....	....	a., c.	
Alnico VI (USA)*	Al 8; Ni 15; Co 24; Cu 3; Ti 1.25	7.3	11.4	......	16.1	Cast	a.	Application requiring high energy
Alnico VII (USA)*	Al 8.5; Ni 18; Cu 3; Co 24; Ti 5	7.17	11.4	......	....	....	a.	
Alnico XII (USA)*	Al 6; Ni 18; Co 35; Ti 8	7.2	11	20–300	....	....	.....	Permanent magnets
Comol (USA)*	Co 12; Mo 17	8.16	9.3	20–300	88.6	....	.....	Permanent magnets
Cunife (USA)*	Cu 60; Ni 20	8.52	....	.....	70.3	....	.....	Permanent magnets
Cunico (USA)*	Cu 50; Ni 21	8.31	....	.....	70.3	....	.....	Permanent magnets
Barium ferrite Feroxdur (USA)*	Ba $Fe_{12}O_{19}$	4.7	10	......	70.3	....	....	Ceramics
Alcomax I (GB)*	Al 7.5; Ni 11; Co 25; Cu 3; Ti 1.5	....	....	......	....	....	a.	Permanent magnets
Alcomax II (GB)*	Al 8; Ni 11.5; Co 24; Cu 4.5	....	....	......	....	....	a.	Permanent magnets
Alcomax II SC(GB)*	Al 8; Ni 11; Co 22; Cu 4.5	7.3	....	......	....	....	a., sc.	
Alcomax III (GB)*	Al 8; Ni 13.5; Co 24; Nb 0.8	7.3	....	......	....	....	a.	Magnets for motors, loudspeakers
Alcomax IV (GB)*	Al 8; Ni 13.5; Cu 3; Co 24; Nb 2.5	....	....	......	....	....	.....	Magnets for cycle-dynamos
Columax (GB)*	Similar to Alcomax III or IV	....	....	......	....	....	a., sc.	Permanent magnets, heat treatable
Hycomax (GB)*	Al 9; Ni 21; Co 20; Cu 1.6	....	....	......	....	....	a.	Permanent magnets
Alnico (high H_c) (GB)*	Al 10; Ni 20; Co 13.5; Cu 6; Ti 0.25	7.3	....	......	....	....	i.	

* USA—United States; GB—Great Britain; Ger—Germany.

** Iron is the additional alloying metal for each of the magnets listed.

*** kg/mm × 1422.33 = lbs/in.²

† i. = isotropic; a. = anisotropic; c. = columnar; sc. = semicolumnar.

CAST PERMANENT MAGNETIC ALLOYS (Continued)

Alloy name, country of manufacture	Composition,** weight percent	Sp. gr., gm per cc	Thermal expansion		Tensile strength		Remark†	Use
			$\frac{\text{Cm} \times 10^{-6}}{\text{cm} \times {}^\circ\text{C}}$	Between °C	*** Kg/mm²	Form		
Alnico (high B_r) (GB)*	Al 10; Ni 17; Co 12; Cu 6	7.3	....	......	....	....	i.	
Alni (high H_c) (GB)*	Al 12; Ni 32; Ti 0–0.5	6.9	....	......	....	....	i.	
Alni (high B_r) (GB)*	Al 13; Ni 24; Cu 3.5	....	....	......	....	....	i.	
Alnico 580 (Ger)*	Al 9; Ni 15; Co 23; Cu 4	....	....	......	....	....	i.	
Alnico 400 (Ger)*	Al 9; Ni 15; Co 23; Cu 4	....	....	......	....	....	a.	
Oerstit 800 (Ger)*	Al 9; Ni 18; Co 19; Cu 4; Ti 4	....	....	......	....	....	i.	Permanent magnets
Alnico 250 (Ger)*	Al 8; Ni 19; Co 23; Cu 4; Ti 6	....	....	......	....	....	i.	
Alnico 190 (Ger)*	Al 12; Ni 21; Cu 4; Co 15	....	....	......	....	....	i.	
Alnico 160 (Austria)	Al 11; Ni 24; Co 12; Cu 4	....	....	......	....	....	i.	Permanent magnets, sintered
Alnico 130 (Ger)*	Al 12; Ni 23; Co 5	....	....	......	....	....	i.	
Alni 120 (Ger)*	Al 13; Ni 27	....	....	......	....	....	i.	
Alni 90 (Ger)*	Al 12; Ni 21	....	....	......	....	....	i.	

* USA—United States; GB—Great Britain; Ger—Germany.

** Iron is the additional alloying metal for each of the magnets listed.

*** kg/mm × 1422.33 = lbs/in.²

† i. = isotropic; a. = anisotropic; c. = columnar; sc. = semicolumnar.

MAGNETIC MATERIALS

Properties of Antiferromagnetic Compounds

Compound	Crystal Symmetry	θ_N(°K)	θ_P(°K)	$(P_A)_{eff}$ μ_B	P_A μ_B
$CoCl_2$	Rhombohedral	25	−38.1	5.18	3.1 ± 0.6
CoF_2	Tetragonal	38	50	5.15	3.0
CoO	Tetragonal	291	330	5.1	3.8
Cr	Cubic	475			
Cr_2O_3	Rhombohedral	307	485	3.73	3.0
CrSb	Hexagonal	723	550	4.92	2.7
$CuBr_2$	Monoclinic	189	246	1.9	
$CuCl_2{\cdot}2H_2O$	Orthorhombic	4.3	4–5	1.9	
$CuCl_2$	Monoclinic	~70	109	2.08	
$FeCl_2$	Hexagonal	24	−48	5.38	4.4 ± 0.7
FeF_2	Tetragonal	79–90	117	5.56	4.64
FeO	Rhombohedral	198	507	7.06	3.32
α-Fe_2O_3	Rhombohedral	953	2940	6.4	5.0
α-Mn	Cubic	95			
$MnBr_2{\cdot}4H_2O$	Monoclinic	2.1	{2.5, 1.3}	5.93	
$MnCl_2{\cdot}4H_2O$	Monoclinic	1.66	1.8	5.94	
MnF_2	Tetragonal	72–75	113.2	5.71	5
MnO	Rhombohedral	122	610	5.95	5.0
β-MnS	Cubic	160	982	5.82	5.0
MnSe	Cubic	~173	361	5.67	
MnTe	Hexagonal	310–323	690	6.07	5.0
$NiCl_2$	Hexagonal	50	−68	3.32	
NiF_2	Tetragonal	78.5–83	115.6	3.5	2.0
NiO	Rhombohedral	533–650	~2000	4.6	2.0
$TiCl_3$		100			
V_2O_3		170			

1. θ_N = Néel temperature, determined from susceptibility maxima or from the disappearance of magnetic scattering.
2. θ_P = a constant in the Curie-Weiss law written in the form $\chi_A = C_A/(T + \theta_P)$, which is valid for antiferromagnetic material for $T > \theta_N$.
3. $(P_A)_{eff}$ = effective moment per atom, derived from the atomic Curie constant $C_A = (P_A)^2_{eff}(N^2/3R)$ and expressed in units of the Bohr magneton, $\mu_B = 0.9273 \times 10^{-20}$ erg gauss^{-1}.
4. P_A = magnetic moment per atom, obtained from neutron diffraction measurements in the ordered state.

SATURATION CONSTANTS AND CURIE POINTS OF FERROMAGNETIC ELEMENTS

Element	σ_s (20°C)	M_s (20°C)	σ_s (0°K)	n_B	Curie Point (°C)
Fe	218.0	1,714	221.9	2.219	770
Co	161	1,422	162.5	1.715	1,131
Ni	54.39	484.1	57.50	0.604	358
Gd	0	0	253.5	7.12	16

σ_s = saturation magnetic moment/gram; M_s = saturation magnetic moment/cm³, in cgs units. n_B = magnetic moment per atom in Bohr magnetons.

From American Institute of Physics Handbook, McGraw-Hill Company (1963) by permission.

MAGNETIC PROPERTIES OF TRANSFORMER STEELS

Ordinary Transformer Steel

B (Gauss)	H (Oersted)	Permeability = B/H
2,000	0.60	3,340
4,000	0.87	4,600
6,000	1.10	5,450
8,000	1.48	5,400
10,000	2.28	4,380
12,000	3.85	3,120
14,000	10.9	1,280
16,000	43.0	372
18,000	149	121

High Silicon Transformer Steel

B	H	Permeability
2,000	0.50	4,000
4,000	0.70	5,720
6,000	0.90	6,670
8,000	1.28	6,250
10,000	1.99	5,020
12,000	3.60	3,340
14,000	9.80	1,430
16,000	47.4	338
18,000	165	109

SATURATION CONSTANTS FOR MAGNETIC SUBSTANCES

Substance	Field intensity (For saturation)	Induced magnetization (For saturation)	Substance	Field intensity (For saturation)	Induced magnetization (For saturation)
Cobalt	9000	1300	Nickel, hard	8000	400
Iron, wrought	2000	1700	annealed	7000	515
cast	4000	1200	Vicker's steel	15000	1600
Manganese steel	7000	200			

INITIAL PERMEABILITY OF HIGH PURITY IRON FOR VARIOUS TEMPERATURES

L. Alberts and B. J. Shepstone

Temperature °C	Permeability (gauss/oersted)
0	920
200	1040
400	1440
600	2550
700	3900
770	12580

MAGNETIC MATERIALS
High-permeability Materials

Material	Form	Approximate percent composition: Fe	Ni	Co	Mo	Other	Typical heat treatment °C	Permeability at $B = 20$ gausses	Maximum permeability	Saturation flux density B gausses	Hysteresis‡ loss, W_h ergs/cm³	Coercive‡ force H_c oersteds	Resistivity microhm-cm	Density, g/cm³
Cold rolled steel	Sheet	98.5	—	—	—	—	950 Anneal	180	2,000	21,000	—	1.8	10	7.88
Iron	Sheet	99.91	—	—	—	—	950 Anneal	200	5,000	21,500	5,000	1.0	10	7.88
Purified iron	Sheet	99.95	—	—	—	—	1480 H_2 + 880	5,000	180,000	21,500	300	.05	10	7.88
4% Silicon-iron	Sheet	96	—	—	—	4 Si	800 Anneal	500	7,000	19,700	3,500	.5	60	7.65
Grain oriented*	Sheet	97	—	—	—	3 Si	800 Anneal	1,500	30,000	20,000	—	.15	47	7.67
45 Permalloy	Sheet	54.7	45	—	—	.3 Mn	1050 Anneal	2,500	25,000	16,000	1,200	.3	45	8.17
45 Permalloy†	Sheet	54.7	45	—	—	.3 Mn	1200 H_2 Anneal	4,000	50,000	16,000	—	.07	45	8.17
Hipernik	Sheet	50	50	—	—	—	1200 H_2 Anneal	4,500	70,000	16,000	220	.05	50	8.25
Monimax	Sheet	—	—	—	—	—	1125 H_2 Anneal	2,000	35,000	15,000	—	.1	80	8.27
Sinimax	Sheet	—	—	—	—	—	1125 H_2 Anneal	3,000	35,000	11,000	—	—	90	—
78 Permalloy	Sheet	21.2	78.5	—	—	.3 Mn	1050 + 600 Q§	8,000	100,000	10,700	200	.05	16	8.60
4–79 Permalloy	Sheet	16.7	79	—	4	.3 Mn	1100 + Q	20,000	100,000	8,700	200	.05	55	8.72
Mu metal	Sheet	18	75	—	—	2 Cr, 5 Cu	1175 H_2	20,000	100,000	6,500	—	.05	62	8.58
Supermalloy	Sheet	15.7	79	—	5	.3 Mn	1300 H_2 + Q	100,000	800,000	8,000	—	.002	60	8.77
Permendur	Sheet	49.7	—	50	—	.3 Mn	800 Anneal	800	5,000	24,500	12,000	2.0	7	8.3
2V Permendur	Sheet	49	—	49	—	2 V	800 Anneal	800	4,500	24,000	6,000	2.0	26	8.2
Hiperco	Sheet	64	—	34	—	Cr	850 Anneal	650	10,000	24,200	—	1.0	25	8.0
2–81 Permalloy	Insulated powder	17	81	—	2	—	650 Anneal	125	130	8,000	—	<1.0	10^6	7.8
Carbonyl iron	Insulated powder	99.9	—	—	—	—	—	55	132	—	—	—	—	7.86
Ferroxcube III	Sintered powder	$MnFe_2O_4 + ZnFe_2O_4$					—	1,000	1,500	2,500	—	.1	10^8	5.0

* Properties in direction of rolling.
† Similar properties for Nicaloi, 4750 alloy, Carpenter 49, Armco 48.
‡ At saturation.
§ Q, quench or controlled cooling.

MAGNETIC MATERIALS (Continued)

Permanent Magnet Alloys

Material	Percent composition (remainder Fe)	Heat treatment* (temperature, °C)	Magnetizing force $H_{max.}$ oersteds	Coercive force H_c oersteds	Residual induction B_r gausses	Energy product $BH_{max.}$ $\times 10^{-6}$	Method of fabrication†	Mechanical properties‡	Weight lb/in.^{3}
Carbon steel	1 Mn, 0.9 C	Q 800	300	50	10,000	.20	HR, M, P	H, S	.280
Tungsten steel	5 W, 0.3 Mn, 0.7 C	Q 850	300	70	10,300	.32	HR, M, P	H, S	.292
Chromium steel	3.5 Cr, 0.9 C, 0.3 Mn	Q 830	300	65	9,700	.30	HR, M, P	H, S	.280
17% Cobalt steel	17 Co, 0.75 C, 2.5 Cr, 8 W	—	1,000	150	9,500	.65	HR, M, P	H, S	—
36% Cobalt steel	36 Co, 0.7 C, 4 Cr, 5 W	Q 950	1,000	240	9,500	.97	HR, M, P	H, S	.296
Remalloy or Comol	17 Mo, 12 Co	Q 1200, B 700	1,000	250	10,500	1.1	HR, M, P	H	.295
Alnico I	12 Al, 20 Ni, 5 Co	A 1200, B 700	2,000	440	7,200	1.4	C, G	H, B	.249
Alnico II	10 Al, 17 Ni, 2.5 Co, 6 Cu	A 1200, B 600	2,000	550	7,200	1.6	C, G	H, B	.256
Alnico II (sintered)	10 Al, 17 Ni, 2.5 Co, 6 Cu	A 1300	2,000	520	6,900	1.4	Sn, G	H	.249
Alnico IV	12 Al, 28 Ni, 5 Co	Q 1200, B 650	3,000	700	5,500	1.3	Sn, C, G	H	.253
Alnico V	8 Al, 14 Ni, 24 Co, 3 Cu	AF 1300, B 600	2,000	550	12,500	4.5	C, G	H, B	.264
Alnico VI	8 Al, 15 Ni, 24 Co, 3 Cu, 1 Ti	—	3,000	750	10,000	3.5	C, G	H, B	.268
Alnico XII	6 Al, 18 Ni, 35 Co, 8 Ti	—	3,000	950	5,800	1.5	C, G	H, B	.26
Vicalloy I	52 Co, 10 V	B 600	1,000	300	8,800	1.0	C, CR, M, P	D	.295
Vicalloy II (wire)	52 Co, 14 V	CW + B 600	2,000	510	10,000	3.5	C, CR, M, P	D	.292
Cunife (wire)	60 Cu, 20 Ni	CW + B 600	2,400	550	5,400	1.5	C, CR, M, P	D, M	.311
Cunico	50 Cu, 21 Ni, 29 Co	—	3,200	660	3,400	.80	C, CR, M, P	D, M	.300
Vectolite	30 Fe_2O_3, 44 Fe_3O_4, 26 C_2O_3	—	3,000	1,000	1,600	.60	Sn, G	W	.113
Silmanal	86.8 Ag, 8.8 Mn, 4.4 Al	—	20,000	6,000[a]	550	.075	C, CR, M, P	D, M	.325
Platinum-cobalt	77 Pt, 23 Co	Q 1200, B 650	15,000	3,600	5,900	6.5	C, CR, M	D	—
Hyflux	Fine powder	—	2,000	390	6,600	.97	—	—	.176

[a] Value given is intrinsic H_c.
* Q—Quenched in oil or water. A—Air cooled. B—Baked. F—Cooled in magnetic field. CW—Cold worked.
† HR—Hot rolled or forged. CR—Cold rolled or drawn. M—Machined. G—Must be ground. P—Punched. C—Cast. Sn—Sintered.
‡ H—Hard. B—Brittle. S—Strong. D—Ductile. M—Malleable. W—Weak.

MAGNETIC SUSCEPTIBILITY OF THE ELEMENTS AND INORGANIC COMPOUNDS

The following table lists the magnetic susceptibilities of one gram formula weight of a number of paramagnetic and diamagnetic inorganic compounds as well as the magnetic susceptibilities of the elements.

In each instance the magnetic moment is expressed in cgs units.

A more extensive listing of the magnetic susceptibilities of inorganic compounds as well as those for organic compounds may be found in Constantes Selectionnees Diamagnetisme et Paramagnetisme Relaxation Paramagnetique, Volume 7. This table is abridged from the above publication by permission of the publishers.

Substance	Formula	Temp. °K.	Susceptibility 10^{-6} cgs
Aluminum (s)	Al	ord.	+16.5
" (*l*)	Al	....	+12.0
Fluoride	AlF_3	302	−13.4
Oxide	Al_2O_3	ord.	−37.0
Sulfate	$Al_2(SO_4)_3$	ord.	−93.0
"	$Al_2(SO_4)_3 \cdot 18H_2O$	ord.	−323.0
Ammonia (g)	NH_3	ord.	−18.0
" (aq)	NH_3	ord.	−17.0
Ammonium			
Acetate	$NH_4C_2H_3O_2$	ord.	−41.1
Bromide	NH_4Br	ord.	−47.0
Carbonate	$(NH_4)_2CO_3$	ord.	−42.50
Chlorate	NH_4ClO_3	ord.	−42.1
Chloride	NH_4Cl	ord.	−36.7
Fluoride	NH_4F	ord.	−23.0
Hydroxide (aq)	NH_4OH	ord.	−31.5
Iodate	NH_4IO_3	ord.	−62.3
Iodide	NH_4I	ord.	−66.0
Nitrate	NH_4NO_3	ord.	−33.6
Sulfate	$(NH_4)_2SO_4$	ord.	−67.0
Thiocyanate	NH_4SCN	ord.	−48.1
Americium (s)	Am	300	+1000.0
Antimony (s)	Sb	293	−99.0
" (*l*)	Sb	....	−2.5
Bromide	$SbBr_3$	ord.	−115.0
Chloride, tri	$SbCl_3$	ord.	−86.7
Chloride, penta	$SbCl_5$	ord.	−120.0
Fluoride	SbF_3	ord.	−46.0
Iodide	SbI_3	ord.	−147.0
Oxide	Sb_2O_3	ord.	−69.4
Sulfide	Sb_2S_3	ord.	−86.0
Argon (g)	A	ord.	−19.6
Arsenic (α)	As	293	−5.5
" (β)	As	293	−23.7
" (γ)	As	293	−23.0
Bromide	$AsBr_3$	ord.	−106.0
Chloride	$AsCl_3$	ord.	−79.9
Iodide	AsI_3	ord.	−142.0
Sulfide	As_2S_3	ord.	−70.0
Arsenious Acid	H_3AsO_3	ord.	−51.2
Barium	Ba	ord.	+20.6
Acetate	$Ba(C_2H_3O_2)_2 \cdot H_2O$	ord.	−100.1
Bromate	$Ba(BrO_3)_2$	ord.	−105.8

Substance	Formula	Temp. °K.	Susceptibility 10^{-6} cgs
Barium (continued)			
Bromide	$BaBr_2$	ord.	−92.0
"	$BaBr_2 \cdot 2H_2O$	ord.	−119.0
Carbonate	$Ba(CO_3)$	ord.	−58.9
Chlorate	$Ba(ClO_3)_2$	ord.	−87.5
Chloride	$BaCl_2$	ord.	−72.6
"	$BaCl_2 \cdot 2H_2O$	ord.	−100.0
Fluoride	BaF_2	ord.	−51.0
Hydroxide	$Ba(OH)_2$	ord.	−53.2
"	$Ba(OH)_2 \cdot 8H_2O$	ord.	−157.0
Iodate	$Ba(IO_3)_2$	ord.	−122.5
Iodide	BaI_2	ord.	−124.0
"	$BaI_2 \cdot 2H_2O$	ord.	−163.0
Nitrate	$Ba(NO_3)_2$	ord.	−66.5
Oxide	BaO	ord.	−29.1
"	BaO_2	ord.	−40.6
Sulfate	$BaSO_4$	ord.	−71.3
Beryllium (s)	Be	ord.	−9.0
Chloride	$BeCl_2$	ord.	−26.5
Hydroxide	$Be(OH)_2$	ord.	−23.1
Nitrate (aq)	$Be(NO_3)_2$	298	−41.0
Oxide	BeO	ord.	−11.9
Sulfate	$BeSO_4$	ord.	−37.0
Bismuth (s)	Bi	ord.	−280.1
" (*l*)	Bi	....	−10.5
Bromide	$BiBr_3$	ord.	−147.0
Chloride	$BiCl_3$	ord.	−26.5
Chromate	$Bi_2(CrO_4)_3$	ord.	+154.0
Fluoride	BiF_3	303	−61.0
Hydroxide	$Bi(OH)_3$	ord.	−65.8
Iodide	BiI_3	ord.	−200.5
Nitrate	$Bi(NO_3)_3$	ord.	−91.0
"	$Bi(NO_3)_3 \cdot 5H_2O$	ord.	−159.0
Oxide	BiO	ord.	−110.0
"	Bi_2O_3	ord.	−83.0
Phosphate	$BiPO_4$	ord.	−77.0
Sulfate	$Bi_2(SO_4)_3$	ord.	−199.0
Sulfide	Bi_2S_3	ord.	−123.0
Boric Acid	H_3BO_3	ord.	−34.1
Boron (s)	B	ord.	−6.7
Chloride	BCl_3	ord.	−59.9
Oxide	B_2O_3	ord.	−39.0
Bromine (*l*)	Br_2	....	−56.4
" (g)	Br_2	....	−73.5

MAGNETIC SUSCEPTIBILITY OF THE ELEMENTS AND INORGANIC COMPOUNDS (Continued)

Substance	Formula	Temp. °K.	Susceptibility 10^{-6} cgs
Bromine (*l*) (continued)			
Fluoride	BrF_3	ord.	−33.9
"	BrF_5	ord.	−45.1
Cadmium (s)	Cd	ord.	−19.8
" (*l*)	Cd	....	−18.0
Acetate	$Cd(C_2H_3O_2)_2$	ord.	−83.7
Bromide	$CdBr_2$	ord.	−87.3
"	$CdBr_2 \cdot 4H_2O$	ord.	−140.0
Carbonate	$CdCO_3$	ord.	−46.7
Chloride	$CdCl_2$	ord.	−68.7
"	$CdCl_2 \cdot 2H_2O$	ord.	−99.0
Chromate	$CdCrO_4$	ord.	−16.8
Cyanide	$Cd(CN)_2$	ord.	−54.0
Fluoride	CdF_2	ord.	−40.6
Hydroxide	$Cd(OH)_2$	ord.	−41.0
Iodate	$Cd(IO_3)_2$	ord.	−108.4
Iodide	CdI_2	ord.	−117.2
Nitrate	$Cd(NO_3)_2$	ord.	−55.1
"	$Cd(NO_3)_2 \cdot 4H_2O$	ord.	−140.0
Oxide	CdO	ord.	−30.0
Phosphate	$Cd_3(PO_4)_2$	ord.	−159.0
Sulfate	$CdSO_4$	ord.	−59.2
Sulfide	CdS	ord.	−50.0
Calcium (s)	Ca	....	+40.0
Acetate	$Ca(C_2H_3O_2)_2$	ord.	−70.5
Bromate	$Ca(BrO_3)_2$	ord.	−84.9
Bromide	$CaBr_2$	ord.	−73.8
"	$CaBr_2 \cdot 3H_2O$	ord.	−115.0
Carbonate	$CaCO_3$	ord.	−38.2
Chloride	$CaCl_2$	ord.	−54.7
Fluoride	CaF_2	ord.	−28.0
Hydroxide	$Ca(OH)_2$	ord.	−22.0
Iodate	$Ca(IO_3)_2$	ord.	−101.4
Iodide	CaI_2	ord.	−109.0
Nitrate (aq)	$Ca(NO_3)_2$	ord.	−45.9
Oxide	CaO	ord.	−15.0
"	CaO_2	ord.	−23.8
Sulfate	$CaSO_4$	ord.	−49.7
"	$CaSO_4 \cdot 2H_2O$	ord.	−74.0
Carbon (dia)	C	ord.	−5.9
" (graph)	C	ord.	−6.0
Dioxide	CO_2	ord.	−21.0
Monoxide	CO	ord.	−9.8
Cerium (α)	Ce	80.5	+5160.0
" (β)	Ce	293	+2450.0
" (β)	Ce	80.5	+6230.0
" (γ)	Ce	287.9	+2420.0
" (γ)	Ce	125.6	+4640.0
" (γ)	Ce	80.5	+5200.0
Chloride	$CeCl_3$	287	+2490.0
Fluoride	CeF_3	293	+2190.0
Nitrate	$Ce(NO_3)_3 \cdot 5H_2O$	292	+2310.0
Oxide	CeO_2	293	+26.0
Sulfate	$CeSO_4$	ord.	+37.0
"	$Ce_2(SO_4)_3 \cdot 5H_2O$	293	+4540.0
"	$Ce(SO_4)_2 \cdot 4H_2O$	293	−97.0
Sulfide	CeS	ord.	+2110.0
"	Ce_2S_3	292	+5080.0
Cesium (s)	Cs	ord.	+29.0
" (*l*)	Cs	....	+26.5
Bromate	$CsBrO_3$	ord.	−75.1
Bromide	CsBr	ord.	−67.2
Carbonate	Cs_2CO_3	ord.	−103.6
Chlorate	$CsClO_3$	ord.	−65.0
Chloride	CsCl	ord.	−86.7
Fluoride	CsF	ord.	−44.5
Iodate	$CsIO_3$	ord.	−83.1
Iodide	CsI	ord.	−82.6
Oxide	CsO_2	293	+1534.0
"	CsO_2	90	+4504.0
Sulfate	Cs_2SO_4	ord.	−116.0
Sulfide	Cs_2S	ord.	−104.0
Chlorine (*l*)	Cl_2	ord.	−40.5
Fluoride, tri	ClF_3	ord.	−26.5
Chromium	Cr	273	+180.0
"	Cr	1713	+224.0
Acetate	$Cr(C_2H_3O_2)_3$	293	+5104.0
Chloride	$CrCl_2$	293	+7230.0
"	$CrCl_3$	293	+6890.0
Fluoride	CrF_3	293	+4370.0
Oxide	Cr_2O_3	300	+1960.0
"	CrO_3	ord.	+40.0
Sulfate	$Cr_2(SO_4)_3$	293	+11,800.0
"	$Cr_2(SO_4)_3 \cdot 8H_2O$	290	+12,700.0
"	$Cr_2(SO_4)_3 \cdot 10H_2O$	290	+12,600.0
"	$Cr_2(SO_4)_3 \cdot 14H_2O$	290	+12,160.0
"	$CrSO_4 \cdot 6H_2O$	293	+9690.0
Sulfide	CrS	ord.	+2390.0
Cobalt	Co	....	ferro
Acetate	$Co(C_2H_3O_2)_2$	293	+11,000.0
Bromide	$CoBr_2$	293	+13,000.0
Chloride	$CoCl_2$	293	+12,660.0
"	$CoCl_2 \cdot 6H_2O$	293	+9710.0
Cyanide	$Co(CN)_2$	303	+3825.0
Cobalt (continued)			
Fluoride	CoF_2	293	+9490.0
"	CoF_3	293	+1900.0
Iodide	CoI_2	293	+10,760.0
Oxide	CoO	260	+4900.0
"	Co_2O_3	ord.	+4560.0
"	Co_3O_4	ord.	+7380.0
Phosphate	$Co_3(PO_4)_2$	291	+28,110.0
Sulfate	$CoSO_4$	293	+10,000.0
"	$Co_2(SO_4)_3$	297	+1000.0
Sulfide	CoS	688	+251.0
"	CoS	293	+225.0
Thiocyanate	$Co(SCN)_2$	303	+11,090.0
Copper (s)	Cu	296	−5.46
" (*l*)	Cu	....	−6.16
Bromide	CuBr	ord.	−49.0
"	$CuBr_2$	341.6	+653.3
"	$CuBr_2$	292.7	+685.5
"	$CuBr_2$	189	+736.9
"	$CuBr_2$	90	+658.7
Chloride	CuCl	ord.	−40.0
"	$CuCl_2$	373.3	+1030.0
"	$CuCl_2$	289	+1080.0
"	$CuCl_2$	170	+1815.0
"	$CuCl_2$	69.25	+2370.0
"	$CuCl_2 \cdot 2H_2O$	293	+1420.0
Cyanide	CuCN	ord.	−24.0
Fluoride	CuF_2	293	+1050.0
"	CuF_2	90	+1420.0
"	$CuF_2 \cdot 2H_2O$	293	+1600.0
Hydroxide	$Cu(OH)_2$	292	+1170.0
Iodide	CuI	ord.	−63.0
Nitrate	$Cu(NO_3)_2 \cdot 3H_2O$	293	+1570.0
"	$Cu(NO_3)_2 \cdot 6H_2O$	293	+1625.0
Oxide	Cu_2O	293	−20.0
"	CuO	780	+259.6
"	CuO	561	+267.3
"	CuO	397	+256.9
"	CuO	289.6	+238.9
"	CuO	120	+156.2
Phosphide	Cu_3P	ord.	−33.0
"	CuP_2	ord.	−35.0
Sulfate	$CuSO_4$	293	+1330.0
"	$CuSO_4 \cdot H_2O$	293	+1520.0
"	$CuSO_4 \cdot 3H_2O$	ord.	+1480.0
"	$CuSO_4 \cdot 5H_2O$	293	+1460.0
Sulfide	CuS	293	−2.0
Thiocyanate	CuSCN	ord.	−48.0
Dysprosium	Dy	293.2	103,500.0
Oxide	Dy_2O_3	287.2	+89,600.0
Sulfate	$Dy_2(SO_4)_3$	293	+91,400.0
"	$Dy_2(SO_4)_3 \cdot 8H_2O$	291.2	+92,760.0
Sulfide	Dy_2S_3	292	+95,200.0
Erbium	Er	291	+44,300.0
Oxide	Er_2O_3	286	+73,920.0
Sulfate	$Er_2(SO_4)_3 \cdot 8H_2O$	293	+74,600.0
Sulfide	Er_2S_3	292	+77,200.0
Europium	Eu	293	+34,000.0
Bromide	$EuBr_2$	292	+26,800.0
Chloride	$EuCl_2$	292	+26,500.0
Fluoride	EuF_2	292	+23,750.0
Iodide	EuI_2	292	+26,000.0
Oxide	Eu_2O_3	298	+10,100.0
Sulfate	$EuSO_4$	293	+25,730.0
"	$Eu_2(SO_4)_3$	293	+10,400.0
"	$Eu_2(SO_4)_3 \cdot 8H_2O$	293	+9,540.0
Sulfide	EuS	293	+23,800.0
"	EuS	195	+35,400.0
Gadolinium	Gd	300.6	+755,000.0
Chloride	$GdCl_3$	293	+27,930.0
Oxide	Gd_2O_3	293	+53,200.0
Sulfate	$Gd_2(SO_4)_3$	285.5	+54,200.0
"	$Gd_2(SO_4)_3 \cdot 8H_2O$	293	+53,280.0
Sulfide	Gd_2S_3	292	+55,500.0
Gallium (s)	Ga	80	−24.4
" (s)	Ga	290	−21.6
" (*l*)	Ga	313	+2.5
Chloride	$GaCl_3$	ord.	−63.0
Iodide	GaI_3	ord.	−149.0
Oxide	Ga_2O	ord.	−34.0
Sulfide	Ga_2S	ord.	−36.0
"	GaS	ord.	−23.0
"	Ga_2S_3	ord.	−80.0
Germanium	Ge	293	−76.84
Chloride	$GeCl_4$	ord.	−72.0
Fluoride	GeF_4	ord.	−50.0
Iodide	GeI_4	ord.	−174.0
Oxide	GeO	ord.	−28.8
"	GeO_2	ord.	−34.3
Sulfide	GeS	ord.	−40.9
"	GeS_2	ord.	−53.3
Gold (s)	Au	296	−28.0
" (*l*)	Au	....	−34.0

MAGNETIC SUSCEPTIBILITY OF THE ELEMENTS AND INORGANIC COMPOUNDS (Continued)

Substance	Formula	Temp. °K.	Susceptibility 10^{-6} cgs
Gold (continued)			
Bromide	$AuBr$	ord.	−61.0
Chloride	$AuCl$	ord.	−67.0
"	$AuCl_3$	ord.	−112.0
Fluoride	AuF_3	ord.	+74.0
Iodide	AuI	ord.	−91.0
Phosphide	AuP_3	ord.	−107.0
Hafnium (s)	Hf	298	+75.0
" (s)	Hf	1673	+104.0
Oxide	HfO_2	ord.	−23.0
Helium (g)	He	ord.	−1.88
Holmium	Ho	....	....
Oxide	Ho_2O_3	293	+88,100.0
Sulfate	$Ho_2(SO_4)_3$	293	+91,700.0
"	$Ho_2(SO_4)_3\cdot 8H_2O$	293	+91,600.0
Hydrogen (g)	H_2	....	−3.98
Bromide (*l*)	HBr	273	....
" (aq)	HBr	ord.	....
Chloride (*l*)	HCl	273	−22.6
" (aq)	HCl	300	−22.0
Fluoride (*l*)	HF	287	−8.6
" (aq)	HF	ord.	−9.3
Iodide (*l*)	HI	281	−47.7
" (*l*)	HI	233	−48.3
" (s)	HI	195	−47.3
" (aq)	HI	ord.	−50.2
Oxide—See Water			
Peroxide	H_2O_2	ord.	−17.7
Sulfide	H_2S	ord.	−25.5
Indium	In	ord.	−64.0
Bromide	$InBr_3$	ord.	−107.0
Chloride	$InCl$	ord.	−30.0
"	$InCl_2$	ord.	−56.0
"	$InCl_3$	ord.	−86.0
Fluoride	InF_3	ord.	−61.0
Oxide	In_2O	ord.	−47.0
"	In_2O_3	ord.	−56.0
Sulfide	In_2S	ord.	−50.0
"	InS	ord.	−28.0
"	In_2S_3	ord.	−98.0
Iodic Acid	HIO_3	ord.	−48.0
metaper	HIO_4	ord.	−56.5
orthoparaper	H_5IO_6	ord.	−71.4
Iodine (s)	I_2	ord.	−88.7
" (atomic)	I	1303	+869.0
" (atomic)	I	1400	+1120.0
Chloride	ICl	ord.	−54.6
"	ICl_3	ord.	−90.2
Fluoride	IF_5	ord.	−58.1
Oxide	I_2O_5	ord.	−79.4
Iridium	Ir	298	+25.6
"	Ir	698	+32.1
Chloride	$IrCl_3$	ord.	−14.4
Oxide	IrO_2	298	+224.0
Iron	Fe	....	ferro
Bromide	$FeBr_2$	ord.	+13,600.0
Carbonate	$FeCO_3$	293	+11,300.0
Chloride	$FeCl_2$	293	+14,750.0
"	$FeCl_2\cdot 4H_2O$	293	+12,900.0
"	$FeCl_3$	293	+13,450.0
"	$FeCl_3$	398	+9,980.0
"	$FeCl_3\cdot 6H_2O$	290	+15,250.0
Fluoride	FeF_2	293	+9,500.0
"	FeF_3	305	+13,760.0
"	$FeF_3\cdot 3H_2O$	293	+7870.0
Iodide	FeI_2	ord.	+13,600.0
Nitrate	$Fe(NO_3)_3\cdot 9H_2O$	293	+15,200.0
Oxide	FeO	293	+7200.0
"	Fe_2O_3	1033	+3586.0
Phosphate	$FePO_4$	ord.	+11,500.0
Sulfate	$FeSO_4$	293	+10,200.0
"	$FeSO_4\cdot H_2O$	290	+10,500.0
"	$FeSO_4\cdot 7H_2O$	293	+11,200.0
Sulfide	FeS	293	+1074.0
Krypton	Kr	....	−28.8
Lanthanum	La	ord.	+118.0
Oxide	La_2O_3	ord.	−78.0
Sulfate	$La_2(SO_4)_3\cdot 9H_2O$	293	−262.0
Sulfide	La_2S_3	292	−37.0
"	La_2S_4	293	−100.0
Lead (s)	Pb	289	−23.0
" (*l*)	Pb	330	−15.5
Acetate	$Pb(C_2H_3O_2)_2$	ord.	−89.1
Bromide	$PbBr_2$	ord.	−90.6
Carbonate	$PbCO_3$	ord.	−61.2
Chloride	$PbCl_2$	ord.	−73.8
Chromate	$PbCrO_4$	ord.	−18.0
Fluoride	PbF_2	ord.	−58.1
Iodate	$Pb(IO_3)_2$	ord.	−131.0
Iodide	PbI_2	ord.	−126.5
Nitrate	$Pb(NO_3)_2$	ord.	−74.0
Oxide	PbO	ord.	−42.0
Phosphate	$Pb_3(PO_4)_2$	ord.	−182.0
Sulfate	$PbSO_4$	ord.	−69.7
Sulfide	PbS	ord.	−84.0
Lead (s) (continued)			
Thiocyanate	$Pb(CNS)_2$	ord.	−82.0
Lithium	Li	ord.	+14.2
Acetate	$LiC_2H_3O_2$	ord.	−34.0
Bromate	$LiBrO_3$	ord.	−39.0
Bromide	$LiBr$	ord.	−34.7
Carbonate	Li_2CO_3	ord.	−27.0
Chlorate (aq)	$LiClO_3$	ord.	−28.8
Chloride	$LiCl$	ord.	−24.3
Fluoride	LiF	ord.	−10.1
Hydride	LiH	ord.	−4.6
Hydroxide	$LiOH$	ord.	−12.3
Iodate	$LiIO_3$	ord.	−47.0
Iodide	LiI	ord.	−50.0
Nitrate	$LiNO_3\cdot 3H_2O$	ord.	−62.0
Sulfate	Li_2SO_4	ord.	−40.0
Lutetium	Lu	ord.	>0.0
Magnesium	Mg	ord.	+13.1
Acetate	$Mg(C_2H_3O_2)_2\cdot 4H_2O$	ord.	−116.0
Bromide	$MgBr_2$	ord.	−72.0
Carbonate	$MgCO_3$	ord.	−32.4
"	$MgCO_3\cdot 3H_2O$	ord.	−72.7
Chloride	$MgCl_2$	ord.	−47.4
Fluoride	MgF_2	ord.	−22.7
Hydroxide	$Mg(OH)_2$	288	−22.1
Iodide	MgI_2	ord.	−111.0
Oxide	MgO	ord.	−10.2
Phosphate	$Mg_3(PO_4)_2\cdot 4H_2O$	ord.	−167.0
Sulfate	$MgSO_4$	294	−50.0
"	$MgSO_4\cdot H_2O$	ord.	−61.0
"	$MgSO_4\cdot 5H_2O$	ord.	−109.0
"	$MgSO_4\cdot 7H_2O$	ord.	−135.7
Manganese (α)	Mn	293	+529.0
" (β)	Mn	293	+483.0
Acetate	$Mn(C_2H_3O_2)_2$	293	+13,650.0
Bromide	$MnBr_2$	294	+13,900.0
Carbonate	$MnCO_3$	293	+11,400.0
Chloride	$MnCl_2$	293	+14,350.0
"	$MnCl_2\cdot 4H_2O$	293	+14,600.0
Fluoride	MnF_2	290	+10,700.0
"	MnF_3	293	+10,500.0
Hydroxide	$Mn(OH)_2$	293	+13,500.0
Iodide	MnI_2	293	+14,400.0
Oxide	MnO	293	+4850.0
"	Mn_2O_3	293	+14,100.0
"	MnO_2	293	+2280.0
"	Mn_3O_4	298	+12,400.0
Sulfate	$MnSO_4$	293	+13,660.0
"	$MnSO_4\cdot H_2O$	293	+14,200.0
"	$MnSO_4\cdot 4H_2O$	293	+14,600.0
"	$MnSO_4\cdot 5H_2O$	293	+14,700.0
Sulfide (α)	MnS	293	+5630.0
" (β)	MnS	293	+3850.0
Mercury (s)	Hg	....	−24.1
" (*l*)	Hg	293	−33.44
" (g)	Hg	....	−78.3
Acetate	$HgC_2H_3O_2$	ord.	−70.5
"	$Hg(C_2H_3O_2)_2$	ord.	−100.0
Bromate	$HgBrO_3$	ord.	−57.7
Bromide	$HgBr$	ord.	−57.2
"	$HgBr_2$	ord.	−94.2
Chloride	$HgCl$	ord.	−52.0
"	$HgCl_2$	ord.	−82.0
Chromate	Hg_2CrO_4	ord.	−63.0
"	$HgCrO_4$	ord.	−12.5
Cyanide	$Hg(CN)_2$	ord.	−67.0
Fluoride	HgF	ord.	−53.0
"	HgF_2	302	−62.0
Hydroxide	$Hg_2(OH)_2$	ord.	−100.0
Iodate	$HgIO_3$	ord.	−92.0
Iodide	HgI	ord.	−83.0
"	HgI_2	ord.	−128.6
Nitrate	$HgNO_3$	ord.	−55.9
"	$Hg(NO_3)_2$	ord.	−74.0
Oxide	Hg_2O	ord.	−76.3
"	HgO	ord.	−44.0
Sulfate	Hg_2SO_4	ord.	−123.0
"	$HgSO_4$	ord.	−78.1
Sulfide	HgS	ord.	−55.4
Thiocyanate	$Hg(SCN)_2$	ord.	−96.5
Molybdenum	Mo	298	+89.0
	Mo	63.8	+108.0
	Mo	20.4	+149.2
Bromide	$MoBr_3$	293	+525.0
"	$MoBr_4$	293	+520.0
	Mo_3Br_6	290.5	−46.0
Chloride	$MoCl_3$	290	+43.0
"	$MoCl_4$	291	+1750.0
"	$MoCl_5$	289	+990.0
Fluoride	MoF_6	ord.	−26.0
Oxide	Mo_2O_3	ord.	−42.0
"	MoO_2	289	+41.0
"	MoO_3	292.5	+3.0
"	Mo_3O_8	ord.	+42.0
Sulfide	MoS_3	289	−63.0

MAGNETIC SUSCEPTIBILITY OF THE ELEMENTS AND INORGANIC COMPOUNDS (Continued)

Substance	Formula	Temp. °K.	Susceptibility 10^{-6} cgs
Neodymium	Nd	287.7	5628.0
Fluoride	NdF_3	293	+4980.0
Nitrate	$Nd(NO_3)_3$	293	+5020.0
Oxide	Nd_2O_3	292.0	+10,200.0
Sulfate	$Nd_2(SO_4)_3$	293	+9990.0
Sulfide	Nd_2S_3	292	+5550.0
Neon	Ne	ord.	−6.74
Nickel	Ni	....	ferro
Acetate	$Ni(C_2H_3O_2)_2$	293	+4690.0
Bromide	$NiBr_2$	293	+5600.0
Chloride	$NiCl_2$	293	+6145.0
"	$NiCl_2 \cdot 6H_2O$	293	+4240.0
Fluoride	NiF_2	293	+2410.0
Hydroxide	$Ni(OH)_2$	ord.	+4500.0
Iodide	NiI_2	293	+3875.0
Nitrate	$Ni(NO_3)_2 \cdot 6H_2O$	293.5	+4300.0
Oxide	NiO	293	+660.0
Sulfate	$NiSO_4$	293	+4005.0
Sulfide	NiS	293	+190.0
"	Ni_3S_2	ord.	+1030.0
Niobium	Nb	298	+195.0
Oxide	Nb_2O_5	ord.	−10.0
Nitric Acid	HNO_3	ord.	−19.9
Nitrogen	N_2	ord.	−12.0
Oxide	N_2O	285	−18.9
" (g)	NO	293	+1461.0
" (g)	NO	203.8	+1895.0
" (g)	NO	146.9	+2324.0
" (*l*)	NO	117.64	+114.2
" (s)	NO	90	+19.8
"	N_2O_3	291	−16.0
"	NO_2	408	+150.0
"	N_2O_4	303.6	−22.1
"	N_2O_4	295.1	−23.0
"	N_2O_4	257	−25.4
" (aq)	N_2O_5	289	−35.6
Osmium	Os	298	+9.9
Chloride	$OsCl_2$	ord.	+41.3
Oxygen (g)	O_2	293	+3449.0
" (*l*)	O_2	90.1	+7699.0
" (*l*)	O_2	70.8	+8685.0
" (s, γ)	O_2	54.3	+10,200.0
" (s, β)	O_2	....	....
" (s, α)	O_2	23.7	+1760.0
Ozone (*l*)	O_3	....	+6.7
Palladium	Pd	288	+567.4
Chloride	$PdCl_2$	291.3	−38.0
Fluoride	PdF_3	293	+1760.0
Hydride	PdH	ord.	+1077.0
"	Pd_4H	ord.	+2353.0
Phosphoric Acid (aq)	H_3PO_4	ord.	−43.8
Phosphorous Acid (aq)	H_3PO_3	ord.	−42.5
Phosphorous (red)	P	....	−20.8
" (black)	P	....	−26.6
Chloride	PCl_3	ord.	−63.4
Platinum	Pt	290.3	+201.9
Chloride	$PtCl_2$	298	−54.0
"	$PtCl_3$	ord.	−66.7
"	$PtCl_4$	ord.	−93.0
Fluoride	PtF_4	293	+455.0
Oxide	Pt_2O_3	ord.	−37.70
Plutonium	Pu	293	+610.0
Fluoride	PuF_4	301	+1760.0
"	PuF_6	295	+173.0
Oxide	PuO_2	300	+730.0
Potassium	K	ord.	+20.8
Acetate (aq)	$KC_2H_3O_2$	28	−45.0
Bromate	$KBrO_3$	ord.	−52.6
Bromide	KBr	ord.	−49.1
Carbonate	K_2CO_3	ord.	−59.0
Chlorate	$KClO_3$	ord.	−42.8
Chloride	KCl	ord.	−39.0
Chromate	K_2CrO_4	ord.	−3.9
"	$K_2Cr_2O_7$	293	+29.4
Cyanide	KCN	ord.	−37.0
Ferricyanide	$K_3Fe(CN)_6$	297	+2290.0
Ferrocyanide	$K_4Fe(CN)_6$	ord.	−130.0
"	$K_4Fe(CN)_6 \cdot 3H_2O$	ord.	−172.3
Fluoride	KF	ord.	−23.6
Hydroxide (aq)	KOH	ord.	−22.0
Iodate	KIO_3	ord.	−63.1
Iodide	KI	ord.	−63.8
Nitrate	KNO_3	ord.	−33.7
Nitrite	KNO_2	ord.	−23.3
Oxide	KO_2	293	+3230.0
"	KO_3	ord.	+1185.0
Permanganate	$KMnO_4$	ord.	+20.0
Sulfate	K_2SO_4	ord.	−67.0
"	$KHSO_4$	ord.	−49.8
Sulfide	K_2S	ord.	−60.0
"	K_2S_2	ord.	−71.0
"	K_2S_3	ord.	−80.0
"	K_2S_4	ord.	−89.0
"	K_2S_5	ord.	−98.0
Sulfite	K_2SO_3	ord.	−64.0
Potassium (continued)			
Thiocyanate	$KSCN$	ord.	−48.0
Praseodymium	Pr	293	+5010.0
Chloride	$PrCl_3$	307.1	+44.5
Oxide	PrO_2	293	+1930.0
"	Pr_2O_3	827	+4000.0
Oxide	Pr_2O_3	294.5	+8994.0
Sulfate	$Pr_2(SO_4)_3$	291	+9660.0
"	$Pr_2(SO_4)_3 \cdot 8H_2O$	289	+9880.0
Sulfide	Pr_2S_3	292	+10,770.0
Rhenium	Re	293	+67.6
Chloride	$ReCl_5$	293	+1225.0
Oxide	ReO_2	ord.	+44.0
"	$ReO_2 \cdot 2H_2O$	295	+74.0
"	ReO_3	ord.	+16.0
"	Re_2O_7	ord.	−16.0
Sulfide	ReS_2	ord.	+38.0
Rhodium	Rh	298	+111.0
"	Rh	723	+123.0
Chloride	$RhCl_3$	298	−7.5
Fluoride	RhF_4	293	+500.0
Oxide	Rh_2O_3	298	+104.0
Sulfate	$Rh_2(SO_4)_3 \cdot 6H_2O$	298	+104.0
"	$Rh_2(SO_4)_3 \cdot 14H_2O$	298	+149.0
Rubidium	Rb	303	+17.0
Bromide	$RbBr$	ord.	−56.4
Carbonate	Rb_2CO_3	ord.	−75.4
Chloride	$RbCl$	ord.	−46.0
Fluoride	RbF	ord.	−31.9
Iodide	RbI	ord.	−72.2
Nitrate	$RbNO_3$	ord.	−41.0
Oxide	RbO_2	293	+1527.0
Sulfate	Rb_2SO_4	ord.	−88.4
Sulfide	Rb_2S	ord.	−80.0
"	Rb_2S_2	ord.	−90.0
Ruthenium	Ru	298	+43.2
"	Ru	723	+50.2
Chloride	$RuCl_3$	290.9	+1998.0
Oxide	RuO_2	298	+162.0
Samarium	Sm	291	+1860.0
"	Sm	195	+2230.0
Bromide	$SmBr_2$	293	+5337.0
"	$SmBr_3$	293	+972.0
Oxide	Sm_2O_3	292	+1988.0
"	Sm_2O_3	170	+1960.0
"	Sm_2O_3	85	+2282.0
Sulfate	$Sm_2(SO_4)_3 \cdot 8H_2O$	293	+1710.0
Sulfide	Sm_2S_3	292	+3300.0
Scandium	Sc	292	+315.0
Selenic Acid	H_2SeO_4	ord.	−51.2
Selenious Acid	H_2SeO_3	ord.	−45.4
Selenium (s)	Se	ord.	−25.0
" (*l*)	Se	900	−24.0
Bromide	Se_2Br_2	ord.	−113.0
Chloride	Se_2Cl_2	ord.	−94.8
Fluoride	SeF_6	ord.	−51.0
Oxide	SeO_2	ord.	−27.2
Silicon	Si	ord.	−3.9
Bromide	$SiBr_4$	ord.	−128.6
Carbide	SiC	ord.	−12.8
Chloride	$SiCl_4$	ord.	−88.3
Hydroxide	$Si(OH)_4$	ord.	−42.6
Oxide	SiO_2	ord.	−29.6
Silver (s)	Ag	296	−19.5
" (*l*)	Ag	....	−24.0
Acetate	$AgC_2H_3O_2$	ord.	−60.4
Bromide	$AgBr$	283	−59.7
Carbonate	Ag_2CO_3	ord.	−80.90
Chloride	$AgCl$	ord.	−49.0
Chromate	Ag_2CrO_4	ord.	−40.0
Cyanide	$AgCN$	ord.	−43.2
Fluoride	AgF	ord.	−36.5
Iodide	AgI	ord.	−80.0
Nitrate	$AgNO_3$	ord.	−45.7
Nitrite	$AgNO_2$	ord.	−42.0
Oxide	Ag_2O	ord.	−134.0
"	AgO	287	−19.6
Permanganate	$AgMnO_4$	300	−63.0
Phosphate	Ag_3PO_4	ord.	−120.0
Sulfate	Ag_2SO_4	ord.	−92.90
Thiocyanate	$AgSCN$	ord.	−61.8
Sodium	Na	ord.	+16.0
Acetate	$NaC_2H_3O_2$	ord.	−37.6
Borate, tetra	$Na_2B_4O_7$	ord.	−85.0
Bromate	$NaBrO_3$	ord.	−44.2
Bromide	$NaBr$	ord.	−41.0
Carbonate	Na_2CO_3	ord.	−41.0
Chlorate	$NaClO_3$	ord.	−34.7
Chloride	$NaCl$	ord.	−30.3
Chromate, di	$Na_2Cr_2O_7$	ord.	+55.0
Fluoride	NaF	ord.	−16.4
Hydroxide (aq)	$NaOH$	300	−16.0
Iodate	$NaIO_3$	ord.	−53.0
Iodide	NaI	ord.	−57.0
Nitrate	$NaNO_3$	ord.	−25.6

MAGNETIC SUSCEPTIBILITY OF THE ELEMENTS AND INORGANIC COMPOUNDS (Continued)

Substance	Formula	Temp. °K.	Susceptibility 10^{-6} cgs
Sodium (continued)			
Nitrite	$NaNO_2$	ord.	−14.5
Oxide	Na_2O	ord.	−19.8
"	Na_2O_2	ord.	−28.10
Phosphate, meta	$NaPO_3$	ord.	−42.5
"	Na_2HPO_4	ord.	−56.6
Sulfate	Na_2SO_4	ord.	−52.0
"	$Na_2SO_4 \cdot 10H_2O$	ord.	−184.0
Sulfide	Na_2S	ord.	−39.0
"	Na_2S_2	ord.	−53.0
"	Na_2S_3	ord.	−68.0
"	Na_2S_4	ord.	−84.0
"	Na_2S_5	ord.	−99.0
Strontium	Sr	ord.	+92.0
Acetate	$Sr(C_2H_3O_2)_2$	ord.	−79.0
Bromate	$Sr(BrO_3)_2$	ord.	−93.5
Bromide	$SrBr_2$	ord.	−86.6
"	$SrBr_2 \cdot 6H_2O$	ord.	−160.0
Carbonate	$SrCO_3$	ord.	−47.0
Chlorate	$Sr(ClO_3)_2$	ord.	−73.0
Chloride	$SrCl_2$	ord.	−63.0
"	$SrCl_2 \cdot 6H_2O$	ord.	−145.0
Chromate	$SrCrO_4$	ord.	−5.1
Fluoride	SrF_2	ord.	−37.2
Hydroxide	$Sr(OH)_2$	ord.	−40.0
"	$Sr(OH)_2 \cdot 8H_2O$	ord.	−136.0
Iodate	$Sr(IO_3)_2$	ord.	−108.0
Iodide	SrI_2	ord.	−112.0
Nitrate	$Sr(NO_3)_2$	ord.	−57.2
"	$Sr(NO_3)_2 \cdot 4H_2O$	ord.	−106.0
Oxide	SrO	ord.	−35.0
"	SrO_2	ord.	−32.3
Sulfate	$SrSO_4$	ord.	−57.9
Sulfur (α)	S	ord.	−15.5
" (β)	S	ord.	−14.9
" (*l*)	S	....	−15.4
" (g)	S	828	+700.0
" (g)	S	1023	+464.0
Chloride	S_2Cl_2	ord.	−62.2
"	SCl_2	ord.	−49.4
"	SCl_3	ord.	−49.4
Fluoride	SF_6	ord.	−44.0
Iodide	SI	ord.	−52.7
Oxide (*l*)	SO_2	ord.	−18.2
Sulfuric Acid	H_2SO_4	ord.	−39.8
Tantalum	Ta	293	+154.0
"	Ta	2143	+124.0
Chloride	$TaCl_5$	304	+140.0
Fluoride	TaF_3	293	+795.0
Oxide	Ta_2O_5	ord.	−32.0
Technetium	Tc	402	250.0
"	Tc	298	270.0
"	Tc	78	290.0
Oxide	$TcO_2 \cdot 2H_2O$	300	+244.0
"	Tc_2O_7	298	−40.0
Tellurium (s)	Te	ord.	−39.5
" (*l*)	Te	....	−6.4
Bromide	$TeBr_2$	ord.	−106.0
Chloride	$TeCl_2$	ord.	−94.0
Fluoride	TeF_6	ord.	−66.0
Terbium	Tb	273	+146,000.0
Oxide	Tb_2O_3	288.1	+78,340.0
Sulfate	$Tb_2(SO_4)_3$	293	+78,200.0
"	$Tb_2(SO_4)_3 \cdot 8H_2O$	293	+76,500.0
Thallium (α)	Tl	ord.	−50.9
" (β)	Tl	>508	−32.3
" (*l*)	Tl	573	−26.8
Acetate	$TlC_2H_3O_2$	ord.	−69.0
Bromate	$TlBrO_3$	ord.	−75.9
Bromide	TlBr	ord.	−63.9
Carbonate	Tl_2CO_3	ord.	−101.6
Chlorate	$TlClO_3$	ord.	−65.5
Chloride	TlCl	ord.	−57.8
Chromate	Tl_2CrO_4	ord.	−39.3
Cyanide	TlCN	ord.	−49.0
Fluoride	TlF	ord.	−44.4
Iodate	$TlIO_3$	ord.	−86.8
Iodide	TlI	ord.	−82.2
Nitrate	$TlNO_3$	ord.	−56.5
Nitrite	$TlNO_2$	ord.	−50.8
Oxide	Tl_2O_3	ord.	+76.0
Phosphate	Tl_3PO_4	ord.	−145.2
Sulfate	Tl_2SO_4	ord.	−112.6
Sulfide	Tl_2S	ord.	−88.8
Thiocyanate	TlCNS	ord.	−66.7
Thorium	Th	293	+132.0
"	Th	90	+153.0
Chloride	$ThCl_4 \cdot 8H_2O$	305.2	−180.0
Nitrate	$Th(NO_3)_4$	ord.	−108.0
Oxide	ThO_2	ord.	−16.0
Thulium	Tm	291	+25,500.0
Oxide	Tm_2O_3	296.5	+51,444.0
Tin (white)	Sn	ord.	+3.1
" (gray)	Sn	280	−37.0
" (gray)	Sn	100	−31.7

Substance	Formula	Temp. °K.	Susceptibility 10^{-6} cgs
Tin (continued)			
" (*l*)	Sn	....	−4.5
Bromide	$SnBr_4$	ord.	−149.0
Chloride	$SnCl_2$	ord.	−69.0
"	$SnCl_2 \cdot 2H_2O$	ord.	−91.4
" (*l*)	$SnCl_4$	ord.	−115.0
Hydroxide	$Sn(OH)_4$	ord.	−60.0
Oxide	SnO	ord.	−19.0
"	SnO_2	ord.	−41.0
Titanium	Ti	293	+153.0
"	Ti	90	+150.0
Bromide	$TiBr_2$	288	+640.0
"	$TiBr_3$	441	+520.0
"	$TiBr_3$	291	+660.0
"	$TiBr_3$	195	+680.0
"	$TiBr_3$	90	+220.0
Carbide	TiC	ord.	+8.0
Chloride	$TiCl_2$	288	+570.0
"	$TiCl_3$	685	+705.0
"	$TiCl_3$	373	+1030.0
"	$TiCl_3$	292	+1110.0
"	$TiCl_3$	212	+690.0
"	$TiCl_3$	90	+220.0
"	$TiCl_4$	ord.	−54.0
Fluoride	TiF_3	293	+1300.0
Iodide	TiI_2	288	+1790.0
"	TiI_3	434	+221.0
"	TiI_3	292	+160.0
"	TiI_3	195	+159.0
"	TiI_3	90	+167.0
Oxide	Ti_2O_3	382	+152.0
"	Ti_2O_3	298	+125.6
"	TiO_3	248	+132.4
"	TiO_2	ord.	+5.9
Sulfide	TiS	ord.	+432.0
Tungsten	W	298	+59.0
Bromide	WBr_5	293	+250.0
Carbide	WC	ord.	+10.0
Chloride	WCl_2	293	−25.0
"	WCl_5	293	+387.0
"	WCl_6	ord.	−71.0
Fluoride	WF_6	ord.	−40.0
Oxide	WO_2	ord.	+57.0
"	WO_3	ord.	−15.8
Sulfide	WS_2	303	+5850.0
Tungstic Acid, ortho	H_2WO_4	ord.	−28.0
Uranium (α)	U	78	+395.0
" (α)	U	298	+409.0
" (α)	U	623	+440.0
" (β)	U	....	....
" (γ)	U	1393	+514.0
Bromide	UBr_3	294	+4740.0
"	UBr_4	293	+3530.0
Chloride	UCl_3	300	+3460.0
"	UCl_4	294	+3680.0
Fluoride	UF_4	300	+3530.0
"	UF_6	ord.	+43.0
Hydrides	UH_3	462	+2821.0
"	UH_3	391	+3568.0
"	UH_3	295	+6244.0
"	UH_3	255	+9306.0
Iodide	UI_3	293	+4460.0
Oxide	UO	293	+1600.0
"	UO_2	293	+2360.0
"	UO_3	ord.	+128.0
Sulfate	$U(SO_4)_2$	ord.	+31.0
Sulfide (α)	US_2	290.5	+3137.0
" (β)	US_2	290.5	+3470.0
"	U_2S_3	ord.	+5206.0
"	U[illegible]S_5	ord.	+11,220.0
Vanadium	V	298	+255.0
Bromide	VBr_2	293	+3230.0
"	VBr_2	195	+3760.0
"	VBr_2	90	+4470.0
"	VBr_3	293	+2890.0
"	VBr_3	195	+4110.0
"	VBr_3	90	+8540.0
Chloride	VCl_2	293	+2410.0
"	VCl_3	293	+3030.0
"	VCl_4	293	+1130.0
"	VCl_4	195	+1700.0
"	VCl_4	90	+4360.0
Fluoride	VF_3	293	+2730.0
Oxide	VO_2	290	+270.0
"	V_2O_3	293	+1976.0
"	V_2O_5	ord.	+128.0
Sulfide	VS	ord.	+600.0
"	V_2S_3	293	+1560.0
Water (g)	H_2O	>373	−13.1
" (*l*)	H_2O	373	−13.09
" (*l*)	H_2O	293	−12.97
" (*l*)	H_2O	273	−12.93
" (s)	H_2O	273	−12.65
" (s)	H_2O	223	−12.31

MAGNETIC SUSCEPTIBILITY OF THE ELEMENTS AND INORGANIC COMPOUNDS (Continued)

Substance	Formula	Temp. °K.	Susceptibility 10^{-6} cgs
Water (continued)			
" (*l*)	DHO	302	−12.97
" (*l*)	D_2O	293	−12.76
" (*l*)	D_2O	276.8	−12.66
" (s)	D_2O	276.8	−12.54
" (s)	D_2O	213	−12.41
Xenon	Xe	ord.	−43.9
Ytterbium	Yb	292	+249.0
"	Yb	90	+639.0
Sulfide	Yb_2S_3	292	+18,300.0
Yttrium	Y	292	+2.15
"	Y	90	+2.43
Oxide	Y_2O_3	293	+44.4
Sulfide	Y_2S_3	ord.	+100.0
Zinc (s)	Zn	ord.	−11.4
" (*l*)	Zn	....	−7.8
Acetate	$Zn(C_2H_3O_2)_2{\cdot}2H_2O$	ord.	−101.0
Carbonate	$ZnCO_3$	ord.	−34.0
Chloride	$ZnCl_2$	296	−65.0
Cyanide	$Zn(CN)_2$	ord.	−46.0
Fluoride	ZnF_2	299.6	−38.2
Hydroxide	$Zn(OH)_2$	ord.	−67.0
Iodide	ZnI_2	ord.	−98.0
Nitrate (aq)	$Zn(NO_3)_2$	ord.	−63.0
Oxide	ZnO	ord.	−46.0
Phosphate	$Zn_3(PO_4)_2$	ord.	−141.0
Sulfate	$ZnSO_4$	ord.	−45.0
"	$ZnSO_4{\cdot}H_2O$	ord.	−63.0
"	$ZnSO_4{\cdot}7H_2O$	ord.	−143.0
Sulfide	ZnS	ord.	−25.0
Zirconium	Zr	293	+122.0
"	Zr	90	+119.0
Carbide	ZrC	ord.	−26.0
Nitrate	$Zr(NO_3)_4{\cdot}5H_2O$	ord.	−77.0
Oxide	ZrO_2	ord.	−13.8

DIAMAGNETIC SUSCEPTIBILITIES OF ORGANIC COMPOUNDS

Compiled by George W. Smith

This table and its supplement contain values for the molar susceptibility of χ_M, specific susceptibility χ, and volumetric susceptibility K. The cgs Gaussian system of units is employed. In the Gaussian units the relation between magnetic induction B and magnetic field strength H is

$$B = H + 4\pi I \qquad (1)$$

where I is the magnetization or magnetic moment per unit volume. Actually, the quantities involved in equation (1) are all vectors, but one may assume that all three are collinear, a reasonable assumption for organic diamagnetic substances in the liquid state. For crystals I may vary with crystal orientation. Equation (1) may be written

$$B = H + 4\pi KH = (I + 4\pi K)H \qquad (2)$$

Here, K is the magnetic susceptibility, often called the volumetric susceptibility and is a unitless quantity.

Other susceptibilities of use to chemists and physicists are the specific or mass susceptibility which is defined

$$\chi = K/\rho \qquad (3)$$

where ρ is the density of the sample in grams per cc., and the molar susceptibility which is defined as

$$\chi_M = M\chi = MK/\rho \qquad (4)$$

where M is the molecular weight of the substance in grams.

Temperatures, when listed, are enclosed in parentheses and are listed in degrees C.

Literature references for values contained in this table may be found in General Motors Research Laboratories Bulletins GMR-317 and GMR-396.

Compound	$-\chi_M \times 10^6$	$-\chi \times 10^6$	$-K \times 10^6$
Acenaphthanthracene	184	.73	
Acenaphthene	109.3	(.709)	(.726) (99°)
Acetal	81.39	.688 (32°)	(.568) (32°)
Acetaldehyde	22.70	$.515_3$	(.403) (18°)
Acetamide	34.1	.577	(.618) (20°)
Acetic acid	31.54	.525 (32°)	(.551) (32°)
Acetic anhydride	(52.8)	.517	(.562) (15°)
Acetoaminofluorene	(141)	.63	
Acetone	33.7_8	$.581_4$	(.460) (20°)
Acetonitrile	28.0	(.682)	(.534) (20°)
Acetonylacetone	62.51	$.547_6$	(.531) (20°)
Acetophenone	72.05	$.599_8$	(.615) (20°)
Acetophenone oxime	79.9_0	$.592_0$	
Acetophenone oxime-O-methyl ether	92.3_1	$.618_8$	
Acetoxime	44.4_2	607_6	(.480) (20°)

DIAMAGNETIC SUSCEPTIBILITIES OF ORGANIC COMPOUNDS

Compound	$-\chi_M \times 10^6$	$-\chi \times 10^6$	$-K \times 10^6$
Acetoxime-O-benzyl ether	104.8_9	$.642_7$	
Acetoxime-O-methyl-ether	54.8_7	$.629_8$	
Acetylacetone	54.88	$.548_1$	(.535) (20°)
Acetyl chloride	38.9	.496	(.548) (20°)
Acetylene	12.5	(.480)	
Acetylphenylacetylene	86.9	.508	
Acetylthiophene	71.7	(.568)	
Acridine	(123.3)	.688	(.757) (20°)
Adonitol	91.30	.600	
Alanine	50.5	(.567)	
Allyl acetate	(56.7)	.566	(.525) (20°)
Allyl alcohol	36.70	.632	.540 (20°)
1-Allylpyrrole	73.80	.685 (20°)	
Aminoazobenzene	(118.3)	.600	
Aminoazotoluene	(142.2)	.631	
o-Aminoazotoluene	(138)	.61 ± .02	
α-Aminobutyric acid	62.1	(.602)	
Aminomethyldiethyldiazine	114.8	(.696)	
4-Aminostilbene	122.5	(.628)	
2-Aminothiazole	56.0	.564	
n-Amyl acetate	89.06	684_5	.5979 (20.7°)
iso-Amyl acetate	89.40	.687	.599 (20°)
n-Amyl alcohol	(67.5)	.766	(.624) (20°)
γ-Amyl alcohol	(71.0)	.8060	(.655) (25°)
Inactive Amyl alcohol	69.06	783_4 (25°)	
iso-Amyl alcohol	68.96	$.782_3$ (25°)	(.64) (15°)
sec-Amyl alcohol	69.1	.785	(.635) (20°)
tert-Amyl alcohol	(70.9)	.804	(.654) (15°)
n-Amylamine	69.4	(.796)	(.606) (20°)
iso-Amylamine	71.6	(.821)	(.616) (20°)
n-Amylbenzene	112.55	(.759)	(.652) (20°)
iso-Amyl bromide	(88.7)	.587	(.706) (20°)
iso-Amyl-n-butyrate	113.52	$.717_4$ (25°)	(.616) (25°)
iso-Amyl chloride	(79.0)	.741	(.662) (20°)
iso-Amyl cyanide	73.4	(.755)	(.609) (20°)
iso-Amylene	53.7	.766	
Amylene bromide	(114.5)	.498	
Amylene chloride	(95.2)	.675	
iso-Amyl ether	(129)	.813	(.635) (15°)
iso-Amyl formate	78.38	$.674_8$	(.591) (25°)
Amylidene chloride	(93.4)	.662	
Amyl iodide	(118.7)	.5996 (18°)	(.910) (20°)
n-Amyl methyl ketone	80.50	$.705_3$	(.580) (15°)
Amyl nitrate	(76.4)	.574	
iso-Amyl propionate	101.73	705_4 (25°)	(.609) (25°)
n-Amyl valerate	124.55	$.723_9$	(.638) (0°)
Anethole	(96.0)	.648	(.644) (15°)
Aniline	62.95	(.676)	(.691) (20°)
Anisidine	(80.5)	.654	(.72) (20°)
Anisole	72.79	(.673)	(.672) (15°)
Anthanthrene	204.2	.739	
Anthanthrone	178.1	.581	
Anthracene	(130)	.731	(.914) (27°)
Anthracenedinitrile	154.6	(.678)	
Anthracenonitrile	142.1	(.700)	
Anthraquinone	(119.6)	.575	(.825) (20°)
Anthrazine	245.7	.646	
Arabinose	85.70	.571	(.905) (20°)
Arbutoside	158.0	(.589)	
Asarone	131.4	(.631)	(.735) (18°)
Asparagine	69.5	(.526)	(.812) (15°)
Aspartic acid	64.2 ± .4	(.482)	(.800) (12°)
Aurin	(161.4)	.556	
p-Azoanisole	(147.7)	.610	
Azobenzene	(106.8)	.586	.611 (70.5°)
p-Azophenetole	(171.7)	.635	
m-Azotoluene	(127.8)	.608	.643 (58°)
Azulene	98.5	.768	
Barbituric acid (Anh.)	53.8	(.420)	
Barbituric acid (.$2H_2O$)	78.6	(4.79)	
Benzalazine	(123.7)	.594	
Benzaldehyde	60.78	(.573)	(.602) (15°)
Benzaldoxime	(69.8)	.576	(.639) (20°)
Benzamide	(72.3)	.597	(.801) (4°)
Benzanthrone	142.9	.620	
Benzene	54.84	.702 (32°)	.611
Benzidine	110.9	.603	(.754) (20°)
Benzil	(118.6)	.564	.616 (100°)
Benzoic acid	70.28	(.575)	(.728) (15°)
Benzoic anhydride	(124.9)	.552	(.662) (15°)
Benzonitrile	65.19	(.632)	(.638) (15°)
Benzophenone	109.60	$.601_3$	(.66) (50°)
3,4-Benzopyrene	135.7	.538	
Benzopyrene	194.0		
Benzoyl acetone	(95.0)	.586	(.639) (60°)
Benzoyl chloride	(75.8)	.539 (20°)	(.657) 15°)
Benzyl acetate	93.18	(.620)	(.655) (16°)
Benzyl alcohol	71.83	(.664)	(.697) (15°)
Benzylamine	75.26	(.702)	(.690) (19°)
Benzyl chloride	81.98	(.647)	(.713) (18°)
Benzyl formate	81.43	(.598)	(.646) (20°)
Benzylideneaniline	(100.4)	.554	
Benzylidene chloride	(97.9)	.608	(.763) (14°)
Benzylidenemethylamine	(73.1)	.613	
Benzyl methyl ketone	83.44	$.621_9$	(.624) (20°)
Bibenzyl	(126.8)	.696	.671 (54.5°)

DIAMAGNETIC SUSCEPTIBILITIES OF ORGANIC COMPOUNDS (Continued)

Compound	$-\chi_M \times 10^6$	$-\chi \times 10^6$	$-K \times 10^6$
3,4-Bis-(p-hydroxyphenyl)-2,4-hexadiene	(157)	.59	
m,m'-Bitolyl	(127.4)	.6993 (27.4°)	(.699) (16°)
m,m'-Bitolyl sulfide	(140.0)	.6530 (27.4°)	
Borneol	126.0	(.817)	(.826) (20°)
Bromobenzene	78.92	.5030 (20°)	(.753) (20°)
Bromobenzenediazocyanide	86.88	(.414)	
Bromochloromethane	55.0 ± .6	(.425)	(.846) (19°)
Bromodichloromethane	66.3 ± .3	(.405)	(.812) (15°)
Bromoform	82.60	.327	.948 (20°)
Bromonaphthalene	(123.8)	.598	
α-Bromonaphthalene	115.90	.560	.840 (20°)
m-Bromotoluene	(93.4)	.546	(.770) (20°)
Bromotrichloromethane	73.1 ± .7	(.369)	(.758) (0°)
Butane	57.4	(.988)	
iso-Butane	51.7	(.890)	
1,4-Butanediol	61.5	(.682)	(.696) (20°)
2-Butene (cis)	42.6	(.759)	
2-Butene (trans)	43.3	(.772)	
1-Butene-3,4-diacetate	95.5	(.555)	
2-Butene-1,4-diacetate (cis)	95.2	(.553)	
2-Butene-1,4-diacetate (trans)	95.1	(.552)	
2-Butene-1,4-diol (cis)	54.3	(.616)	
2-Butene-1,4-diol (trans)	53.5	(.607)	
n-Butyl acetate	77.47	$.666_9$ (25°)	(.583) (25°)
iso-Butyl acetate	78.52	$.676_9$	(.584) (25°)
n-Butyl alcohol	56.536 (20°)	(.7627)	(.6176) (20°)
iso-Butyl alcohol	57.704 (20°)	(.7785)	(.624) (20°)
sec-Butyl alcohol	57.683 (20°)	(.7782)	(.629) (20°)
tert-Butyl alcohol	57.42	$.774_7$ (25°)	(.611) (20°)
n-Butylamine	58.9	(.805)	(.596) (20°)
iso-Butylamine	59.8	(.818)	(.599) (20°)
9-Butyl anthracene	176.0	(.751)	
n-Butylbenzene	100.79	(.751)	(.646) (20°)
iso-Butylbenzene	101.81	(.759)	(.648) (20°)
tert-Butylbenzene	102.5	(.764)	(.662) (20°)
n-Butyl benzoate	116.69	$.654_8$	(.656) (25°)
Butyl bromide	77.14	.563 (20°)	(.730) (20°)
iso-Butyl bromide	79.88	.583 (20°)	(.737) (20°)
1-n-Butyl chloride	67.10	.725	.642 (20°)
2-n-Butyl chloride	67.40	.728	.635 (20°)
n-Butyl cyanide	(62.8)	.7558 (27.4°)	(.606) (20°)
tert-Butyl cyclohexane	115.09	.8205	.6670 (20°)
1,4-Butyl diacetate	103.4	(.594)	
n-Butyl ethyl ketone	80.73	$.707_2$	(.579) (20°)
n-Butyl formate	65.83	$.644_6$	(.571) (25°)
iso-Butyl formate	66.79	$.654_9$	(.574) (25°)
iso-Butylideneazine	(95.8)	.683	
Butyl iodide	(93.6)	.5086 (18°)	(.822) (20°)
iso-Butyl methyl ketone	70.05	$.699_5$	(.561) (20°)
tert-Butyl methyl ketone	69.86	$.697_9$	(.558) (16°)
n-Butyl perfluor-n-butyrate	126.7	(.469)	
p-tert-Butylphenol	108.0	(.719)	(.653) (114°)
Butyl sulfide	(113.7)	.7774 (27.4°)	(.652) (16°)
Butyl thiocyanate	(79.38)	.6891 (27.4°)	(.659) (25°)
2-Butyne-1,4-diacetate	95.9	(.564)	
2-Butyne-1,4-dibenzoate	169.0	(.561)	
2-Butyne-1,4-diol	50.3	(.584)	
n-Butyraldehyde	46.08	$.639_4$	(.522) (20°)
iso-Butyraldehyde	46.38	$.643_6$	(.511) (20°)
iso-Butyraldoxime	56.1_2	$.644_3$	(.576) (20°)
n-Butyric acid	55.10	.625	.598 (20°)
iso-Butyric acid	56.06	$.636_3$ (25°)	(.601) (25°)
Butyronitrile	49.4	(.715)	(.569) (15°)
Butyrylphenylacetylene	(106.4)	.618	
Cacodyl	(99.9)	.476	(.689) (15°)
Cacodylic acid	(79.9)	.579	
Camphor	(103)	.68	(.67) (25°)
Camphoric acid	129.0	(.644)	(.791) (20°)
Camphoric anhydride	(113)	.620	(.740) (20°)
n-Caproic acid	78.55	$.676_2$	(.624) (25°)
Caproylphenylacetylene	(130.4)	.651	
n-Caprylic acid	101.60	.7053	(.642) (20°)
Carbanilide	134.05	(.6316)	(.783) (20°)
Carbazole	117.4	(.702)	
Carbon disulfide	42.2	.554	(.699) (22°)
Carbon tetrabromide	93.73	.2826 (20°)	(.966)
Carbon tetrachloride	66.60	.433	.691 (20°)
Carbon tetraiodide	(136)	.261	(1.13) (20°)
Carvacrol	(109.1)	.726	(.709) (20°)
Carvone	(92.2)	.614	(.590) (20°)
Cetyl alcohol	(183.5)	.757 (17.5°)	(.619) (50°)
Cetyl mercaptan	390.4	(1.510)	
Chloral	(67.7)	.459	(.694) (20°)
Chloranil	(112.6)	.458	
Chloracetic acid	48.1	(.509)	(.804) (20°)
Chloroacetone	(50.9)	.550	(.633) (20°)
Chloroacetylchloride	53.7	(.475)	(.710) (0°)
p-Chloranisole	89.1	(.625)	
Chlorobenzene	69.97	(.6216)	(.688) (20°)
Chlorobenzene diazocyanide	65.02	(.393)	
Chlorodibromomethane	75.1 ± .4	(.361)	(.883) (15°)
Chlorodifluoromethane	38.6	.446	
1-Chloro-2,3-dihydroxypropane	(77.9)	.604	
Chlorodiphenylmethane	(131.9)	(.651)	
Chloroethylene	35.9	.574	(.528) (liq., 15°)
Chloroform	59.30	.497	.740 (20°)
Chlorofumaric acid	67.02	(.445)	
p-Chloroiodo benzene	99.42	(.417)	(.786) (57°)
Chloromaleic acid	67.36	(.448)	
Chloromethylstilbene	144.8	(.633)	
α-Chloronaphthalene	107.60	.661	.789 (20°)
m-Chloronitrobenzene	(74.8)	.475	.638 (48°)
o-Chlorophenol	77.4	(.602)	(.747) (18°)
p-Chlorophenol	77.6	(.604)	(.789) (20°)
1-(o-Chlorophenylazo)-2-naphthol	161.0	(.570)	
1-(p-Chlorophenylazo)-2-naphthol	161.4	(.571)	
Chlorotrifluoroethylene	49.1	.422	
Chlorotrifluoromethane	45.3 ± 1.5	(.434)	
Cholesterol	(284.2)	.735	(.784) (20°)
Chrysene	166.67	.731	
Chrysoidine	(126.3)	.595	
Cinnamic acid	78.36	.529	(.660) (4°)
Cinnamic acid (α-trans)	78.2	(.528)	
Cinnamic acid (β-trans)	79.0	(.533)	
Cinnamic acid (cis-MP 68°)	77.6	(.524)	
Cinnamic acid (cis-MP 58°)	77.9	(.526)	
Cinnamic acid (cis-MP 42°)	83.2	(.562)	
Cinnamic aldehyde	(74.8)	.566	(.629) (15°)
Cinnamyl alcohol	(87.2)	.650	(.679) (20°)
Cinnamylideneaniline	123.2	(.595)	
Citral	(98.9)	.650	(.577) (20°)
Coronene	(243.3)	.810	
Coumarin	82.5	(.565)	(.528) (20°)
o-Cresol	72.90	.675	(.706) (20°)
m-Cresol	72.02	.667 (26°)	(.690) (20°)
p-Cresol	72.1	.667 (26°)	(.690) (20°)
o-Cresylmethyl ether	81.94	.671 (40°)	(.661) (15°)
m-Cresylmethyl ether	77.91	.638 (40°)	(.623) (15°)
p-Cresylmethyl ether	79.13	.648 (40°)	(.629) (19°)
Cumene	89.53	$.744_9$	(.642)
Cyamelide	(56.1)	.435	(.490) (15°)
Cyameluric acid	101.1 (10°)	(.457)	
9-Cyanoanthracene	142.1	(.699)	
Cyanogen	(21.6)	.415	(.359) (liq., 17°)
Cyanuric acid	61.5	.476	(.842) (0°)
Cyclobutanecarboxylic acid	58.16	.5816 (30°)	(.613) (30°)
1,3-Cyclohexadiene	48.6	(.607)	(.510) (20°)
1,4-Cyclohexadiene	48.7	(.608)	(.515) (20°)
Cyclohexane	68.13	.8100 (27.5°)	(.627) (20°)
Cyclohexanecarboxylic acid	83.24	.6499 (30°)	(.668) (30°)
Cyclohexanol	73.40	.732	.694 (20°)
Cyclohexanone	62.0_4	$.632_3$	(.599) (20°)
Cyclohexanone oxime	71.5_2	$.632_1$	
Cyclohexanoneoxime-O-methyl ether	82.9_6	$.652_3$	
Cyclohexene	57.5	(.700)	(.567) (20°)
Cyclohexenol	64.1	(.653)	
Cyclooctane	91.4	(.815)	(.684) (20°)
Cyclooctene	84.6	(.769)	(.654) (20°)
Cyclooctatetraene	(53.9)	.518	
Cyclopentane	59.18	.8439	.6290 (20°)
Cyclopentanecarboxylic acid	73.48	.6446 (30°)	(.677) (30°)
Cyclopentanone	51.63	.6141 (30°)	(.582) (30°)
Cyclopropane	39.9	(.948)	(.683) (−79°)
Cyclopropanecarboxylic acid	45.33	.5271 (30°)	(.569) (30°)
p-Cymene	102.8	.766 (20°)	(.656) (20°)
Decalin	106.70	.7718	.6814 (20°)
cis-Decalin	(107.0)	.774	(.686) (35°)
trans-Decalin	(107.7)	.779	(.670) (35°)
n-Decane	119.74	(.8416)	(.6143) (20°)
1-Deuterio pyrrole	48.75	.716 (20°)	
Deuteroindene	80.88	.690	(.692) (13°)
Diacetal	(153.8)	.668	
Di-iso-amylamine	(133.1)	.846	(.649) (21°)
Diazoacetic ester	57	(.50)	(.54) (24°)
Dibenzocoronene	289.4	.778	
1,2,5,6-Dibenzofluorene	184	.69 ± .03	
3,4,5,6-Dibenzophenanthrene	(203)	.73 ± .03	
Dibenzphenanthrone	200.5	.716	
Dibenzpyrene	213.6	.706	
Dibenzpyrenequinone	183.1	.551	
iso-Dibenzpyrenequinone	194.6	.586	
Dibenzyl ketone	131.70	$.626_2$	
p-Dibromobenzene	(101.4)	.430	.786 (100°)
2,3-Dibromo-2-butene-1,4-diol	94.2	(.383)	
Dibromodichloromethane	81.1 ± .4	(.334)	(.808) (25°)
1,2-Dibromodiiodoethylene	(140.1)	.320	
1,2-Dibromoethylene	(71.7)	.386	(.877) (17.5°)
1,2-Dibromo-2-fluoroethane	(78.0)	.379	(.855) (17°)
Dibromo-4-nitrophenol	(167.5)	.564	
1,2-Dibromotetrachloroethane	(126.0)	.387	(1.049)
Di-n-butylamine	103.7	(.802)	(.767) (20°)
Di-iso-butylamine	105.7	(.817)	(.609) (20°)
Di-sec-butylamine	105.9	(.819)	(.641) (0°)
Di-iso-butyl ketone	104.30	$.733_5$	(.591) (20°)
Di-tert-butyl ketone	104.06	$.732_1$	
2,6-Di-tert-butyl-4-methyl phenol	165.3	(.750)	
2,4-Di-tert-butyl phenol	155.6	(.754)	
Dibutyl phthalate	175.1	(.629)	(.657) (21°)
Di-iso-butyralacetylene	(125.6)	.738	
Dicetyl sulfide	401.7	(.832)	
Dichloroacetic acid	58.2	(.451)	(.705) (20°)
Dichloroacetyl chloride	69.0	(.468)	

Compound	$-\chi_M \times 10^6$	$-\chi \times 10^6$	$-K \times 10^6$
o-Dichlorobenzene	84.26	.5734	(.748) (20°)
m-Dichlorobenzene	83.19	.5661	(.729) (20°)
p-Dichlorobenzene	82.93	.5644	(.823) (20.5°)
1,4-Dichloro-2-butyne	74.2	(.603)	
1,2-Dichloro-1,2-dibromoethane	(108.6)	.423	
1,1-Dichloro-difluoroethylene	60.0	.451	
Dichlorodifluoromethane	52.2	.432	(.642) (−30°)
1,1-Dichloroethylene	49.2	.508	(.635) (15°)
cis-1,2-Dichloroethylene	51.0	.526	(.679) (15°)
trans-1,2-Dichloroethylene	48.9	.504	(.638) (15°)
1,3-Dichloro-2-hydroxypropane	(80.1)	.621	
Dicyandiamide	44.55	(.530)	(.742) (14°)
Dicyclohexanol acetylene	(151.6)	.682	
Dicyclohexyl	129.31	.7776	.6889 (20°)
1,1-Dicyclohexylnonane	231.98	.7930	.7001 (20°)
Diethanolacetylene	(75.3)	.660	
Diethyl acetaldehyde	70.71	$.705_9$	(.576) (20°)
Diethylallylacetophenone	146.2	(.676)	(.663) (16°)
Diethyl allylmalonate	118.8	(.593)	(.602) (14°)
Diethylamine	56.8	(.777)	(.552) (18°)
Diethylcyclohexylamine	(124.5)	.802	(.699) (0°)
Diethyl ethylmalonate	115.2	(.612)	(.614) (20°)
Diethyl ketone	58.1_4	$.675_1$	(.551) (19°)
Diethyl ketoxime	68.3_1	.6754	
Diethyl malonate	(92.6)	.5782	(.611) (20°)
Diethyl-3-(1-methyl butane) ethylmalonate	175	(.677)	
Diethyl oxalate	81.71	.5595	(.603) (15°)
Diethyl phthalate	127.5	(.574)	(.645) (25°)
Diethyl sebacate	(177.0)	.685	(.661) (20°)
Diethylstilbestrol	172.0	(.547)	
Diethylstilbestrol dipropionate	265.2	.720	
Diethyl succinate	105.07	.6035	(.628) (20°)
Diethyl sulfate	(86.8)	.563	(.667) (15°)
Diethyl sulfide	(67.9)	.753	(.630) (20°)
Diethyl tartrate	(113.4)	.550	(.662) (20°)
Difluoroacetamide	(41.2)	.433	
1-Difluoro-2-dibromoethane	(85.5)	.382	(.883) (20°)
1,1-Difluoro-2,2-dichloroethyl amyl ether	129.84	(.587)	(.694) (20°)
1,1-Difluoro-2,2-dichloroethyl butyl ether	119.48	(.577)	(.703) (20°)
1,1-Difluoro-2,2-dichloroethyl ethyl ether	96.13	(.537)	(.723) (20°)
1,1-Difluoro-2,2-dichloroethyl methyl ether	80.68	(.489)	(.696) (20°)
1,1-Difluoro-2,2-dichloroethyl propyl ether	107.19	(.555)	(.701) (20°)
Difluoroethanol	(41.3)	.503	
Di-n-heptylamine	171.5	(.805)	
Di-n-hexylamine	148.9	(.803)	
Dihydronaphthalene	(85.1)	.654	(.652) (12°)
o-Dimethoxybenzene	87.39	.6329	(.686) (25°)
m-Dimethoxybenzene	87.21	.6316	(.682) (0°)
p-Dimethoxybenzene	86.65	.6275	(.661) (55°)
o-(2,5-Dimethoxybenzoyl)-benzoic acid	161.0	(.562)	
Dimethoxymethane	(47.3)	.621	(.532)
Dimethylacetophenone	96.8	(.653)	(.645) (16°)
Dimethylallylacetophenone	122.4	(.650)	(.635) (16°)
Dimethylaniline	89.66	(.740)	
2,2-Dimethylbutane	76.24	.8848	.5744 (20°)
2,3-Dimethylbutane	76.22	.8845	.5853 (20°)
2,3-Dimethyl-2-butene	65.9	(.783)	(.557)
Dimethylcyclohexanone	(84.8)	.672	
1,2 and 1,3 Dimethylcyclopentanes	81.31	.8281	.6224 (20°)
2,5-Dimethyl-2,5-dibromo-3-hexine	(135.6)	.506	
Dimethyl diethylketo tetrahydrofurfurane	(116.2)	.753	
2,5-Dimethyl-1-ethylpyrrole	94.61	.768 (20°)	
2,5-Dimethyl-3-ethylpyrrole	93.87	.762 (20°)	
2,5-Dimethylfuran	66.37	.687 (20°)	(.620) (18°)
Dimethyl furazan	57.27	.584	
2,5-Dimethyl-4-heptene	100.6	(.797)	
2,4-Dimethyl-2,4-hexadiene	(78.7)	.714	
2,3-Dimethylhexane	98.77	.8648	.6164 (20°)
2,5-Dimethylhexane	98.15	.8593	.5969 (20°)
3,4-Dimethylhexane	99.06	.8673	.6240 (20°)
2,6-Dimethyl-4 hexanol	116.9	.812	
2,5-Dimethyl-3-hexine-2,5-diol	(103.0)	.724	
Dimethyl isoxazole	59.7	(.615)	
Dimethylketo tetrahydrofurfurane	(68.5)	.600	
Dimethyl malonate	69.69	.5277	(.609) (20°)
1,6-Dimethylnaphthalene	113.3	(.725)	
2,4-Dimethylnonane	134.68	(.862)	(.636) (20°)
3,4-Dimethylnonane	134.70	(.862)	(.647) (20°)
4,5-Dimethylnonane	134.52	(.861)	(.647) (20°)
2,6-Dimethyl-2,6,8-nonatriene	(108.8)	.724	
Dimethyl-2,4-nonatriene	(148.8)	.990	
2,6-Dimethyloctane	122.54	(.861)	(.627) (20°)
3,4-Dimethyloxadiazole	57.17	.583	
Dimethyl oxalate	(55.7)	.472	(.542) (54°)
Dimethyl oxamide	(63.2)	.544	
2,2-Dimethylpentane	86.97	.8680	.5849 (20°)
2,3-Dimethylpentane	87.51	.8733	.6070 (20°)
2,4-Dimethylpentane	87.48	.8732	.5876 (20°)
2,2-Dimethylpropane	63.1	(.875)	(.536) (0°)
2,5-Dimethyl-3-propyl-pyrrole	106.07	.773 (20°)	
2,4-Dimethylpyrrole	69.64	.732 (20°)	(.679) (14°)

Compound	$-\chi_M \times 10^6$	$-\chi \times 10^6$	$-K \times 10^6$
2,5-Dimethylpyrrole	71.92	.756 (20°)	(.707) (20°)
α,ω-Dimethyl styrene	(90.7)	.686	
Dimethyl succinate	81.50	.5581	(.625) (18°)
Dimethyl sulfate	(62.2)	.493	(.657) (20°)
Dimethyl sulfide	(44.9)	.723	(.612) (21°)
Dimethyltrichloromethylcarbinol	(105)	.59	
N,N-Dimethyl urea	55.1	(.625)	(.784)
N,N′-Dimethyl urea	56.3	(.639)	(.730)
o-Dinitrobenzene	65.98	.3921	(.614) (17°)
m-Dinitrobenzene	70.53	.4197	(.659) (0°)
p-Dinitrobenzene	68.30	.4064	(.660) (30°)
2,4-Dinitrophenol	(73.1)	.397	(.668) (24°)
Dinitroresorcinol	(62.4)	.312	
1,4-Dioxane	52.16	.592 (32°)	(.606) (32°)
Diphenyl	103.25	.6695	(.664) (73°)
1,1-Diphenylallyl-3-chloride	146.1	(.639)	
1,3-Diphenylallyl-3-chloride	140.7	(.615)	
Diphenylamine	(109.7)	.648	.686 (55.5°)
Diphenyl-bis-diazo cyanide	85.03	(.327)	
Diphenylbutadiene	129.6	(.629)	
Diphenylchloroarsine	(145.5)	.550	(.871) (40°)
Diphenyldecapentaene	180.5	(.635)	
Diphenyldiacetylene	(134.6)	.640	
Diphenyldiazomethane	115	(.592)	
Diphenyldihydrotetrazine	129.9	(.545)	
1,1-Diphenylethylene	(118.0)	.655	(.680) (14°)
1,6-Diphenylhexane	171.81	.7208	.6877 (20°)
Diphenylhexatriene	146.9	(.632)	
Diphenylmethane	(115.7)	.688	.684 (35.5°)
Diphenylmethanol	119.1	.647	
1,1-Diphenylnonane	206.32	.7357	.6935 (20°)
Diphenyloctatetraene	164.3	(.636)	
Diphenylphenoxyarsine	(225.2)	.567	
N,N-Diphenyl urea	126.3	(.595)	(.759)
N,N′-Diphenyl urea	127.5	(.600)	(.743) (20°)
Di-n-propyl ketone	80.45	$.705_0$	(.576) (20°)
Di-iso-propyl ketone	81.14	$.711_0$	(.573) (20°)
Dipropyl oxalate	105.27	.6046	(.628) (0°)
Di-iso-propyl oxalate	106.02	.6089	
Dodecyl alcohol	147.70	.7849 (20.7°)	(.652) (24°)
Dulcitol	112.40	.617	(.905) (15°)
Elaidic acid	204.8	(.725)	(.619) (79°)
Erythritol	73.80	.604	(.876) (20°)
Ethane	27.3_7	(.910)	(.511) (−100°)
4-Ethoxy-3-methoxybenzyl acetate	138.5	.619	
4-Ethoxy-3-methoxybenzyl benzoate	177.3	.620	
1-Ethoxynaphthalene	119.9	(.696)	(.738) (20°)
2-Ethoxynaphthalene	119.2	(.692)	(.734) (25°)
Ethyl acetate	54.10	.614	.554 (20°)
Ethyl acetoacetate	71.67	$.550_8$	(.565) (20°)
Ethylacetophenone	95.5	(.644)	(.639) (16°)
Ethyl alcohol	33.60	.728	.575 (20°)
Ethylally acetophenone	122.5	(.651)	(.634) (16°)
Ethyl amylpropiolate	(112.7)	.670	
Ethylaniline	89.30	(.737)	(.709)
9-Ethyl anthracene	153.0	(.741)	(.771) (99°)
Ethylbenzene	77.20	.7272	.6341 (20°)
Ethyl benzoate	93.32	$.621_1$	(.648) (25°)
Ethyl benzoylacetate	(115.3)	.600	(.673) (20°)
Ethyl benzylidenecyanoacetate	(116.3)	.578	
Ethyl benzylmalonate	(154.5)	.6172	(.663) (20°)
Ethyl bromide	54.70	.502	.719 (20°)
Ethyl bromoacetate	(82.8)	.496	(.747) (20°)
Ethyl-1-isobutylacetoacetate	(121.4)	.652	
Ethyl butylmalonate	139.3	$.644_2$	(.629) (20°)
Ethyl-n-butyrate	(77.7)	.6693	(.585) (25°)
Ethyl-iso-butyrate	78.32	$.674_3$	(.583) (25°)
Ethyl chloroacetate	(72.3)	.590	(.684) (20°)
Ethyl cinnamate	(107.5)	.610	(.640) (20°)
Ethyl-iso-cyanate	(45.6)	.642	(.582) (16°)
Ethyl cyanoacetate	(67.3)	.595	(.632) (20°)
Ethylcyclohexane	91.09	.8118	.6324 (20°)
Ethyldiallylacetophenone	147.4	(.646)	(.636) (16°)
Ethyl dibromocinnamate	174.5	.519	
Ethyl dichloroacetate	85.2	(.543)	(.696) (20°)
Ethyl diethylacetoacetate	(117.9)	.6328	(.615) (20°)
Ethyl diethylmalonate	(140.4)	.6492	(.641)(20°)
Ethyl dithiolacetate	(71.0)	.5904 (27.4°)	
Ethylene	12.0	(.428)	(.242) (−102°)
Ethylene	15.30	.546 (32°)	(.309) (−102°)
Ethylene bromide	78.80	.419	.915 (20°)
Ethylene chloride	59.62	.602 (32°)	(.757) (20°)
Ethylenediamine	46.26	.771 (32°)	(.686) (20°)
Ethylene iodide	104.7	.371 (32°)	(.791) (10°)
Ethylene oxide	30.7	(.697)	(.618) (7°)
Ethyl ether	55.10	.743	.534 (20°)
Ethyl ethylacetoacetate	93.9	.5937	(.582) (20°)
Ethyl ethylbutylmalonate	(163.3)	.6683	(.650) (20°)
Ethyl ethylpropylmalonate	(152.4)	.6619	(.648) (20°)
Ethyl formate	43.00	.580	.531 (20°)
Ethyl hexylpropiolate	129.9	.713	
Ethyl hydroxylamine	(43.0)	.704	
Ethylidene chloride	(57.4)	.580	(.681) (20°)
Ethyl iodide	(69.7)	.4470 (17.5°)	(.864) (20°)
Ethyl iodoacetate	(97.6)	.456	(.829) (13°)
Ethyl lactate	(72.6)	.615	(.633) (25°)

Compound	$-\chi_M \times 10^6$	$-\chi \times 10^6$	$-K \times 10^6$
Ethyl methylacetoacetate	(81.9)	.5684	(.569) (20°)
Ethyl methyl ketoxime	57.3_2	$.658_0$	(.530) (20°)
Ethyl-1-methyl-2-oxocyclohexane-carboxylate	112.1	(.608)	
Ethyl methylphenylmalonate	(153.2)	.6121	(.658) (20°)
Ethyl nitrophenylpropiolate	114.6	.523	
Ethyl oxamate	62.0	(.529)	(.427) (19°)
Ethyl perfluor-n-butyrate	103.5	(.427)	
Ethyl phenylacetate	104.27	(.635)	(.656) (20°)
Ethyl phenylmalonate	(142.2)	.6017	(.659) (20°)
Ethyl phenylpropiolate	(104.2)	.598	(.636) (13°)
Ethyl phosphate	(98.2)	.539	(.576) (25°)
Ethyl propionate	(66.5)	.6514	(.584) (15°)
Ethyl propylacetoacetate	(105.7)	.6135	(.593) (20°)
Ethyl-n-propyl ketone	69.03	$.689_1$	(.560) (22°)
Ethyl succinimide	(72.0)	.566	
Ethyl sulfine	(67.0)	.631	
Ethyl sulfite	(75.4)	.546	(.604) (0°)
Ethylsulfone ethyl ether	(81.8)	.592	
Ethyl thiocyanate	(55.7)	.6392 (27.4°)	(.637) (25°)
Ethyl isothiocyanate	(59.0)	.6772 (27.4°)	(.680) (15°)
Ethyl thiolacetate	(62.7)	.6019 (27.4°)	(.586) (25°)
Ethyl thionacetate	(63.5)	.6098 (27.4°)	
Ethyl tribromoacetate	(119.5)	.368	(.821) (20°)
Ethyl trichloroacetate	(99.6)	.520	(.719) (20°)
N-Ethyl urea	55.5	(.630)	(.764) (18°)
Ethyl iso-valerate	(91.1)	.700	(.607) (20°)
Eucalyptol	(116.3)	.754	(.699) (20°)
Eugenol and iso-eugenol	(102.1)	.622	(.663) (20°)
Flavanthrone	241.0	.590	
Fluorene	110.5	.655	
Fluorenone	99.4	.552	(.623) (100°)
Fluorobenzene	(58.4)	.608	(.623) (20°)
Fluorobromoacetic acid	(59.5)	.379	
Fluorodichloromethane	48.8	.474	(.676) (0°)
p-Fluorophenetole	(88.0)	.628	
Fluoro trichloroethylene	72.5	.485	(.742) (25°)
Fluorotrichloromethane	58.7	.427	(.638) (17°)
Formaldehyde	(18.6)	.62	(.51) (−20°)
Formamide	(21.9)	.486	(.551) (20°)
Formic acid	19.90	.432	.527 (20°)
β-Formylpropionic acid	55.3	(.542)	
Fructose	102.60	.570	
Fulvene (Benzene χ_M measured to be 49)	42.9	(.549)	(.452) (20°)
Fulvene $\left(\chi \frac{54.8}{49}\right)$	48.0	(.614)	(.505) (20°)
Fumaric acid	49.11	(.423)	(.692) (20°)
Furan	43.09	.633 (20°)	(.598) (15°)
Furfural	47.1	(.490)	(.568) (20°)
Galactose	103.00	.572	
Gallic acid	90.0	(.529)	(.896) (4°)
Geraniol formate	(119.9)	.658	(.610) (20°)
Glucose	102.60	.570	
D-Glucose	101.5	(.563)	(.869) (25°)
Glutamic acid	78.5	(.533)	
Glycerol	57.06	.619	.779 (20°)
Glycine	40.3	(.537)	(.846) (50°)
Glycol	38.80	.624	.698 (20°)
Guaiacol	(79.2)	.638	(.720) (21°)
Helianthrone	189.9	.497	
1,2-Heptadiene	73.5	(.764)	
2,3-Heptadiene	72.1	(.749)	
Heptaldehyde	81.02	$.709_6$	(.603) (20°)
n-Heptane	85.24	.8507	.5817 (20°)
4-Heptanol	91.5	.789	(.647) (20°)
n-Heptanoic acid	88.60	.680	.626 (20°)
n-Heptyl amine	93.1	(.808)	(.628) (20°)
n-Heptyl benzene	134.41	.7625	.6528 (20°)
Heptyl cyclohexane	147.40	.8084	.6559 (20°)
n-Heptylic acid	89.74	.6900	(.630) (25°)
1-Heptyne	77.0	(.801)	(.584) (25°)
2-Heptyne	79.5	(.826)	(.615) (25°)
Hexabromoethane	(148.0)	.294	(1.124) (20°)
Hexachlorobenzene	(147.5)	.518	(1.059) (24°)
Hexachloroethane	(112.7)	.476	(.995) (20°)
Hexachlorohexatrione	(145.0)	.433	
n-Hexadecane	187.63	.8286	.6421 (20°)
1,5-Hexadiene	(55.1)	.671	(.462) (20°)
2,3-Hexadiene	60.9	(.741)	
n-Hexaldehyde	69.40	$.693_0$	
2,2,4,7,9,9-Hexamethyldecane	191.52	.8458	.6596 (20°)
Hexamethyl disiloxane	118.9	.7324	
Hexamethylene glycol	84.30	.713	
n-Hexane	(74.6)	.8654 (27.4°)	(.565)
Hexene	65.7	(.781)	
Hexestrol	(165)	.61 ± .02	
n-Hexyl alcohol	79.20	.774	.637 (20°)
n-Hexyl benzene	124.23 (20°)	(.767)	(.658) (20°)
n-Hexyl methyl ketone	91.4_2	$.713_1$	(.583)
n-Hexyl methyl ketoxime	102.5_8	$.716_2$	(.634) (20°)
Hexylpropiolamide	(103.7)	.677	
Hydrindene	(78.5)	.664	(.639) (16°)
Hydroquinone	64.63	.587	(.797) (20°)
Hydroxyazobenzene	(99.7)	.503	
p-Hydroxybenzaldehyde	66.8	(.547)	(.618) (130°)
4-Hydroxy-2-butanone	48.5	.55	(.573) (14°)

Compound	$-\chi_M \times 10^6$	$-\chi \times 10^6$	$-K \times 10^6$
Indene (natural)	84.79	(.730)	(.723) 25°)
Indene (synthetic)	80.89	(.696)	(.690) (25°)
Indole	85.0	(.726)	
Iodobenzene	92.00	.451	.826 (20°)
Iodoform (in sol'n)	117.1	.2974 (20°)	(1.192) (17°)
1-Iodo-2-phenylacetylene	(110.1)	.483	
o-Iodotoluene	(112.2)	.5145 (30°)	(.874) (20°)
m-Iodotoluene	(112.3)	.5152 (30°)	(.875) (20°)
p-Iodotoluene	101.31	(.465)	(.780) (40°)
Leucine	84.9	(.647)	
iso-Leucine	84.9	(.647)	
Maleic acid	49.71	(.428)	(.681) (20°)
Maleic anhydride	(35.8)	.365	(.341) (20°)
Malonic acid	(46.3)	.4453	(.726) (15°)
Mannitol	111.20	.610	(.908) (20°)
Mannose	102.90	.571	(.879)
Mesitylene	92.32	.7682 (20°)	(.665) (20°)
Methane	12.2_7	(.765)	
Methione	91.0	(.610)	
p-Methoxyazobenzene	(118.9)	.560	
o-Methoxybenzaldehyde	76.0	(.558)	(.632) (20°)
p-Methoxybenzaldehyde	78.0	(.572)	(.642) (20°)
o-Methoxybenzyl alcohol	87.9	.637	(.664) (25°)
1-Methoxynaphthalene	107.0	(.676)	(.741) (14°)
2-Methoxynaphthalene	107.6	(.680)	
1-(o-Methoxyphenylazo)-2-naphthol	163.6	(.588)	
Methoxysaligenin acetate	110.3	.613	
Methyl acetate	42.60	.575	.537 (20°)
Methyl acetoacetate	59.60	$.513_2$	(.553) (20°)
Methylacetylacetone	(65.0)	.569	
Methyl alcohol	21.40	.668	.530 (20°)
Methylallylaketone	111.9	(1.330)	
Methylamine	(27.0)	.870	(.608) (−11°)
N-Methylaniline	82.74	(.773)	(.762) (20°)
9-Methylanthracene	146.5	.762)	(.812) (99°)
Methyl benzoate	81.59	$.599_3$	(.651) (25°)
Methyl-o-benzoylbenzoate	139.4	(.580)	(.690) (19°)
Methylbenzylaniline	(132.2)	.670	
Methyl bromide	42.8	.451	(1.044) (0°)
2-Methylbutane	64.40	.8925	.5531 (20°)
2-Methyl-2-butene	54.14	(.772)	(.516) (13°)
Methyl butyl ketone	(69.1)	.690	(.563) (15°)
Methyl iso-butyl ketone	(69.3)	.692	(.554) (20°)
Methyl tert-butyl ketone	(70.4)	.703	(.562) (16°)
p-Methyl-o-tert-butylphenol	120.3	(.732)	
Methyl butyrate	(66.4)	.6498	(.588) (16°)
o-Methylcarbanilide	154.0	(.681)	
Methyl chloride	(32.0)	.633	
Methyl chloroacetate	58.1	(.535)	(.661) (20°)
Methylcholanthrene	(182)	.68 ± .04	
3-Methylcholanthrene	194.0	(.723)	
Methylcyclohexane	78.91	.8038	.6181 (20°)
2-Methylcyclohexanone	(74.0)	.660	(.610) (18°)
3-Methylcyclohexanone	(74.8)	.667	(.610) (20°)
4-Methylcyclohexanone	(63.5)	.566	(.516) (24°)
Methylcyclopentane	70.17	.8338	.6245 (20°)
4-Methyl-2,6-di-tert-butylphenol	167.6	(.761)	
Methyl dichloroacetate	73.1	(.511)	
Methyldiphenoxyphosphine oxide	(152.9)	.616	
Methyldiphenyltriazine	(155.1)	.627	
Methylene bromide	65.10	.375	.935 (20°)
Methylene chloride	(46.6)	.549	(.733) (20°)
Methylene iodide	93.10	.348	1.156 (20°)
Methylene succinic acid	57.57	(.443)	(.723)
Methyl ether	26.3	.571	
Methylethylallyacetophenone	133.3	(.659)	(.643) (16°)
Methyl ethyl ketone	45.5_8	$.632_2$	(.509) (20°)
Methyl formate	(32.0)	.5327	(.519) (20°)
Methylfumaric acid	56.98	(.438)	(.642)
3-Methylheptane	97.99	.8580	.6056 (20°)
2-Methyl-4-heptene	88.0	(.784)	
5-Methyl-1,2-hexadiene	73.6	(.765)	(.553) (19°)
2-Methylhexane	86.24	.8607	.5841 (20°)
Methyl hexyl ketone	(93.3)	.728	(.596) (20°)
Methyl-m-hydroxybenzoate	88.4	(.581)	
Methyl-p-hydroxybenzoate	88.7	(.583)	
Methyl iodide	(57.2)	.403	(.918) (20°)
Methylmaleic acid	57.84	(.446)	(.721)
9-Methyl-10-methoxyanthracene	158.1	(.711)	
Methyl-o-methoxybenzoate	95.6	(.575)	(.665) (19°)
Methyl-p-methoxybenzoate	98.6	(.593)	
Methyl-α-methoxy-isobutyrate	(81.9)	.620	
1-Methylnaphthalene	102.8	(.723)	(.741) (14°)
2-Methylnaphthalene	102.6	(.722)	(.743) (20°)
4-Methylnonane	121.39	(.853)	(.625) (20°)
5-Methyl-5-nonene	111.6	(.796)	
4-Methyloctane	109.63	(.855)	(.618) (20°)
Methylol urea	48.3	(.493)	
2-Methylpentane	75.26	.8734	.5705 (20°)
3-Methylpentane	75.52	.8764	.5823 (20°)
4-Methyl-2-pentanol	80.4	.788	(.641) (20°)
Methyl perfluor-n-butyrate	92.5	(.406)	
Methyl phenylacetate	92.73	(.618)	(.645) (16°)
Methyl phenylpropiolate	95.6	.597	
2-Methylpropene	44.4	(.791)	
Methyl propionate	(55.0)	.6240	(.571) (20°)

DIAMAGNETIC SUSCEPTIBILITIES OF ORGANIC COMPOUNDS (Continued)

Compound	$-\chi_M \times 10^6$	$-\chi \times 10^6$	$-K \times 10^6$
Methyl-n-propyl ketone	57.41	$.666_4$	(.541) (15°)
Methyl-iso-propyl ketone	58.45	$.679_0$	(.545) (20°)
1-Methylpyrrole	58.56	.722 (20°)	(.664) (10°)
2-Methylpyrrole	60.10	.741 (20°)	(.700)
Methyl salicylate	86.30	.567	.668 (20°)
Methyl silicone	(172.7)	.730	
α-Methyl styrene	(80.1)	.678	(.620) (20°)
2-Methylthiazole	59.56	.601 (20°)	
2-Methylthiophene	66.35	.676 (20°)	(.689) (20°)
Methyl trichloroacetate	84.2	(.475)	(.707) (19°)
N-Methyl urea	44.6	(.602)	(.725)
Morpholine	55.0	(.631)	(.631)
Myleran	169.7	.69	(.834) (4°)
Myristic acid	176.0	(.771)	(.661) (60°)
Naphthalaldehydic acid	117.6	(.588)	
Naphthalene	(91.9)	.717	(.821) (20°)
Naphthalene picrate	185.9	(.523)	
2-Naphthalenesulfonylamine	127.6	(.616)	
2-Naphthalenesulfonyl chloride	121.91	(.538)	
meso-Naphthodianthrene	214.6	.612	
meso-Naphthodianthrone	221.8	.583	
1-Naphthol	98.2	.681	(.834) (4°)
	97.0	.673	(.819) (4°)
2-Naphthol	98.25	(.682)	(.829) (4°)
α-Naphthonitrile	103.3	(.674)	(.753) (5°)
β-Naphthonitrile	101.0	(.659)	(.721) (60°)
α-Naphthoquinone	73.5	(.465)	(.661)
β-Naphthoquinone	67.9	(.429)	
N-1-Naphthylacetamide	117.8	(.636)	
N-2-Naphthylacetamide	117.8	(.636)	
1-Naphthylamine	98.8	.690	.757 (54°)
2-Naphthylamine	98.00	(.684)	(.726) (98°)
1-Naphthylamine hydrochloride	(127.6)	.710	
Nicotine	113.328	(.699)	(.705) (20°)
o-Nitroaniline	66.47	(.481)	(.694) (15°)
m-Nitroaniline	70.09	(.507)	(.725) (20°)
p-Nitroaniline	66.43	(.481)	(.691) (14°)
o-Nitrobenzaldehyde	68.23	.4517	
m-Nitrobenzaldehyde	68.55	.4538	
p-Nitrobenzaldehyde	66.57	.4407	(.507) (0°)
Nitrobenzene	61.80	.502	.604 (20°)
Nitrobenzene diazo cyanide	59.22	(.336)	
o-Nitrobenzoic acid	76.11	.4556	(.718) (4°)
m-Nitrobenzoic acid	80.22	.4802	(.717) (4°)
p-Nitrobenzoic acid	78.81	.4718	(.731) (32°)
o-Nitrobromobenzene	87.3	(.432)	(.700) (80°)
m-Nitrobromobenzene	89.5	(.443)	(.755) (20°)
p-Nitrobromobenzene	89.6	(.444)	(.859) (22°)
m-Nitro carbanilide	148.1	(.576)	
Nitroethane	(35.4)	.472	(.497) (20°)
Nitromethane	21.1	.3457	(.391) (25°)
1-Nitronaphthalene	98.47	(.569)	(.696) (62°)
o-Nitrophenol	73.3	.527 (24°)	(.873) (20°)
m-Nitrophenol	70.8	.509 (25°)	(.756)
p-Nitrophenol	69.5	.500 (22°)	(.740)
1-(m-Nitrophenylazo)-2-naphthol	142.0	(.484)	
1-(p-Nitrophenylazo)-2-naphthol	141.7	(.483)	
Nitrophenylfluoroform	(84.1).	.440	
2-Nitropropane	45.73	.5135	(.509) (20°)
Nitrosobenzene	59.1	(.552)	
N-Nitrosodiethylamine	59.3	(.580)	(.546) (20°)
p-Nitrosodiethylaniline	92.6	(.520)	(.644) (15°)
p-Nitrosodimethylaniline	73.3	(.488)	
N-Nitrosodiphenylamine	110.7	(.558)	
1-Nitroso-2-naphthol	83.9	(.485)	
2-Nitroso-1-naphthol	82.7	(.478)	
4-Nitroso-1-naphthol	91.8	(.530)	
m-Nitrosonitrobenzene	66.0	(.433)	
p-Nitrosonitrobenzene	65.8	(.433)	
p-Nitrosophenol	50.7	(.412)	
Nitrosopiperidine	(63.4)	.555	(.590) (20°)
p-Nitrosotoluene	70.4	(.581)	
o-Nitrotoluene	72.28	.5272	(.613) (20°)
m-Nitrotoluene	72.71	.5304	(.614) (20°)
p-Nitrotoluene (in sol'n)	72.06	.5257	(.676) (20°)
n-Nonane	108.13	.8431	.6057 (20°)
1,2-Octadiene	83.6	(.759)	
n-Octane	96.63	.8460	.5949 (20°)
Octanonoxime	(102.7)	.717	
Octyl alcohol	102.65	.7766 (20°)	(.640) (20°)
Octyl chloride	(114.9)	.773	(.676) (20°)
Octylcyclohexane	158.09	.8051	.6578 (20°)
Octylene	(89.5)	.798	(.576) (17°)
Octylene bromide	(150.4)	.553	
n-Octyl mercaptan	(115.1)	.7866 (27.4°)	
Oenanthylidene chloride	(116.5)	.689	
Oleic acid	208.5	(.738)	(.661) (18°)
Opianic acid	111.5	(.530)	
Ovalene	353.8	.888	
Oxalic acid (anh.)	33.8	(.375)	
Oxalic acid	60.05	(.4763)	(.787)
Oxamide	(39.0)	.443	(.738)
Palmitic acid	198.6	(.775)	(.661) (62°)
Paraldehyde	(86.2)	.652	(.648) (20°)
Pentabromophenol	(194.0)	.397	
Pentacene	(205.4)	.738	
Pentachloroethane	(99.1)	.490	(.819) (25°)
Pentachlorohexadione	(129.5)	.452	
2,3-Pentadiene	49.1	(.721)	(.501) (20°)
n-Pentane	63.05	.8739	.5472 (20°)
2,4-Pentanediol	70.4	.677	
Perfluoroacetic acid	43.3	(.380)	
Perfluoro-n-butyric acid	81.0	(.378)	
Perfluorobutyric anhydride	149.4	(.387)	
Perfluorocyclooctane oxide	157.6	(.379)	
Perfluoropropionic acid	61.0	(.372)	
Perhydroanthracene	146.01	.7592	.7178 (20°)
Perylene	166.8	.662	
Phenanthrene	(127.9)	.718	(.763) (100°)
Phenanthrenequinone	104.5	.502	(.698)
Phenanthrenonitrile	139.0	(.685)	
o-Phenetidine	(101.7)	.741 (25°)	
p-Phenetidine	(96.8)	.706 (25°)	(.749) (15°)
Phenetole	(84.5)	.692	(.689) (20°)
Phenol	60.21	(.640)	(.675) (45°)
Phenothiazine	114.8	(.576)	
Phenylacetaldehyde	72.01	$.599_4$	(.614) (20°)
Phenyl acetate	82.04	(.603)	(.647) (25°)
Phenylacetic acid	82.72	(.608)	(.657) (80°)
Phenylacetylene	72.01	(.705)	(.655) (20°)
1-Phenylazo-2-naphthol	137.6	(.554)	
2-Phenylbensofuran	130.5	(.672)	
1-Phenyl-4-benzoyl-1,3-butadiene	(140.3)	.599	
Phenylbutadiene	(85.7)	.658	
4-Phenyl-1-butene	93.49	(.7077)	(.6239) (20°)
Phenylbutyl acetate	134.5	.653	
Phenyl n-butyrate	105.46	(.643)	
Phenyl iso-cyanate	(72.7)	.610	(.699) (20°)
o-Phenylenediamine	71.98	.6662	
m-Phenylenediamine	70.53	.6529	(.723) (58°)
p-Phenylenediamine	70.28	.6503	
Phenyl ether	(108.1)	.635	(.681) (20°)
Phenylethyl sulfide	(94.4)	.6826 (27.4°)	
Phenylfluoroform	(77.3)	.529	
Phenylhydrazine	67.82	(.627)	(.688) (23°)
Phenylhydroxylamine	(68.2)	.625	
Phenyl mercaptan	(70.8)	.6425 (27.4°)	(.693) (20°)
1-Phenyl-2-Methylbutane	113.53	(.766)	(.660) (20°)
Phenylmethyl sulfide	(83.2)	.6695 (27.4°)	
Phenylpropiolamide	(83.3)	.574	
Phenyl propionate	93.79	(.625)	(.654) (25°)
Phenylsulfone	(129.0)	.591	(.740) (20°)
Phenyl thiocyanate	(81.5)	.6027 (27.4°)	(.677) (24°)
Phenyl isothiocyanate	(86.0)	.6365 (27.4°)	(.719) (24°)
1-Phenyl-4,6,6-trimethylheptane	173.90	(.796)	(.682) (20°)
N-Phenyl urea	82.1	(.603)	(.785)
Phloroglucinol	(73.4)	.582	
Phthalamide	(91.3)	.556	
Phthalic acid	83.61	.5035	(.802) (20°)
iso-Phthalic acid	84.64	.5097	
tere-Phthalic acid	83.51	.5029	(.759)
Phthalic anhydride	67.31	(.454)	(.694) (4°)
Phthalimide	(78.4)	.533	
Picric acid	84.38	(.368)	(.649)
Piperazine	56.8	(.659)	
Piperidine	64.2	(.754)	(.650) (20°)
Propane	40.5	(.919)	(.538) (−45°)
Propene	31.5	(.749)	(.456) (−47°)
Propionaldehyde	34.32	$.591_0$	(.477) (20°)
Propionic acid	43.50	.586	.582 (20°)
Propionitrile	38.5	(.699)	(.547) (21°)
Propionylphenylacetylene	(95.1)	.601	
Propiophenone	83.73	$.624_0$	(.631) (20°)
n-Propyl acetate	65.91	$.645_3$ (25°)	(.569) (25°)
iso-Propyl acetate	67.04	$.656_4$	(.566) (25°)
n-Propyl alcohol	45.176 (20°)	(.7518)	(.6047) (20°)
iso-Propyl alcohol	45.794 (20°)	(.7621)	(.5985) (20°)
9-Propylanthracene	164.0	(.744)	
n-Propylbenzene	89.24	(.742)	(.640) (20°)
n-Propyl benzoate	105.00	(.640)	(.646) (25°)
n-Propyl bromide	(65.6)	.533	(.721) (20°)
iso-Propyl bromide	(65.1)	.529	(.693) (20°)
Propyl butyrate	(89.4)	.6867	(.604) (15°)
prim-Propyl chloride	56.10	.715	.633 (20°)
iso-Propylcyclohexane	102.65	.8131	.6528 (20°)
Propylenediamine	(58.1)	.784	(.688) (15°)
Propylene oxide	42.5	(.732)	(.629) (0°)
Propyl formate	(55.0)	.6248	(.563) (20°)
Propyl hexylpropiolate	(136.8)	.697	
Propyl iodide	(84.3)	.4958 (30°)	(.864) (20°)
Propyl propionate (extrap.)	(77.95)	.6711	(.593) (20°)
Propyl sulfide	(92.1)	.7787 (27.4°)	(.634) (17°)
N-Propyl urea	67.4	(.660)	
Pseudocumene	(101.6)	.845 (20°)	(.740) (20°)
Pyramidone	149.0	(.645)	
Pyranthrene	266.9	.709	
Pyranthrone	250.3	.616	
Pyrazine	37.6	(.469)	(.484) (61°)
Pyrene	147.9	.731	(.933) (0°)
Pyridine	49.21	(.622)	(.611) (20°)
Pyrocatechol	68.76	.6248	(.857) (15°)
Pyrrole	47.6	(.709)	(.688) (20°)

Compound	$-\chi_M \times 10^6$	$-\chi \times 10^6$	$-K \times 10^6$
Pyrrolidine	54.8	(.771)	(.657) (23°)
Quinoline	86.0	(.666)	(.729) (20°)
Quinone	38.4	(.355)	(.468) (20°)
Quinonoxime	(50.4)	.409	
Resorcinol	67.26	.6112	(.785) (15°)
Rhamnose	99.20	.605	(.890) (20°)
Safrol and iso-Safro	(97.5)	.601	(.66) (20°)
Salicylaldehyde	64.4	(.527)	(.615) (20°)
Salicylic acid	72.23	.523	(.755) (20°)
Saligenin	76.9	.620	(.720) (25°)
Salol	(123.2)	.575	.678 (45°)
Salvarsan dihydrochloride	(246.1)	.518	
Selenophene	66.82	.510	
iso-Selenophene	110–111	.84–.85	
trans-Selenophene	70–77	.53–.59	
Sorbitol	107.80	.592	
Stearic acid	220.8	(.776)	(.657) (69°)
Stilbene	(120.0)	.666	(.646) (125°)
Stilbestrol	(130)	.62, .63	
Styrene	(68.2)	.655	(.594) (20°)
Succinic acid	(57.9)	.4902	(.767) (15°)
Succinic anhydride	(47.5)	.475	(.524)
Succinimide	(47.3)	.477	(.674) (16°)
Sulfamide	44.4	(.462)	(.832)
p-Sulfanilamide	80.15	(.465)	
Terpineol	111.9	(.725)	(.678) (room temp.)
Tetrabenzylmonosilane	266.2	(.678)	
1,1,2,2-Tetrabromoethane	(123.4)	.357	(1.058) (20°)
Tetrabromoethylene	(114.8)	.334	
Tetracene	(168.0)	.736	
1,1,2,2-Tetrachloroethane	(89.8)	.535	(.856) (20°)
Tetrachloroethylene	81.6	.492	(.802) (15°)
Tetrahydroquinoline	(89.0)	.668	(.715) (4°)
Tetraiodoethylene	(164.3)	.309	(.922) (20°)
Tetraiodopyrrole	(188.9)	.331	
Tetramethylketotetrahydrofurfurane	(104.7)	.736	
Tetranitromethane	43.02	.2195	(.360) (20°)
Tetraphenylbutadiene	228.0	(.636)	
Tetraphenyldecapentaene	280.8	(.643)	
Tetraphenylhexatriene	246.4	(.641)	
Tetraphenyloctatetraene	264.1	(.643)	
Tetraphenylrubene	344.0	(.646)	
Tetra-p-tolylmonosilane	276.4	(.704)	
Tetrolic acetal	(97.8)	.688	
Tetronic acid	(52.5)	.525	
Thiacoumerin	93.6	.577	
Thiazole	50.55	.595 (20°)	(.714) (17°)
Thiobarbituric acid	72.9	(.506)	
Thiophene	57.38	.682 (20°)	(.726) (20°)
Tolane	(118.9)	.667	(.644) (100°)
Toluene	66.11	.7176	.6179 (20°)
o-Toluidine	76.0	.710 (24°)	(.709) (20°)
m-Toluidine	74.6	.697 (25°)	(.689) (20°)
p-Toluidine	72.1	.673 (25°)	(.704) (20°)
α-Tolunitrile	76.87	(.656)	(.666) (18°)
1-(o-Tolylazo)-2-naphthol	148.7	(.567)	
1-(p-Tolylazo)-2-naphthol	157.6	(.601)	
Triallylacetophenone	152.5	(.634)	
Tri-iso-amylamine	192	.845	(.647) (25°)
Trianilinophosphine oxide	(201.7)	.624	
1,2,3-Tribromopropane	(117.9)	.420	(1.023) (23°)
Tri-iso-butylamine	(156.8)	.846	(.646) (25°)
Trichloroacetic acid (in sol'n)	73.0	(.44)	(.723) (46°)
Trichlorobenzene	(106.5)	.587	
Trichloro-tert-butyl alcohol (in sol'n)	98.01	.552	
Trichloroethylene	65.8	.501	(.734) (20°)
Trichloronitromethane	(75.3)	.458	(.756) (20°)
Triethylamine	81.4	(.804)	(.586) (20°)
Triethyl citrate	(161.9)	.586	(.666) (20°)
Triethyl phosphate	(125.3)	.688	(.735) (20°)
Triethylphosphine	(90.0)	.762	(.610) (15°)
Triethylphosphine oxide	(91.6)	.683	
Triethyl phosphite	(104.8)	.631	(.611) (20°)
Triethyl triazinetricarbonate	(164.1)	.552	
Trifluorocresol	(83.8)	.517	
Tri-n-heptylamine	251.3	(.806)	
Tri-n-hexylamine	221.7	(.823)	
Trimethylacetophenone	108.2	(.667)	(.648) (16°)
2,2,3-Trimethylbutane	88.36	.8818	.6086 (20°)
2,2,3-Trimethylpentane	99.86	.8743	.6261 (20°)
2,2,4-Trimethylpentane	98.34	.8610	.5958 (20°)
2,3,5-Trimethylpyrrole	82.31	.754 (20°)	
1,3,5-Trinitrobenzene	74.55	(.350)	(.591) (20°)
Triperfluorobutylamine	253.0	(.377)	
Triphenoxyarsine	(195.2)	.551	
Triphenylarsine	(177.0)	.578	
Triphenylarsine dihydroxide	(270.5)	.795	
Triphenylarsine oxide	(199.1)	.618	
Triphenylbismuthine	(196.8)	.447	(.708) (20°)
Triphenylbismuthine dinitrate	(254.5)	.451	
Triphenylcarbinol	(175.7)	.675	(.802) (20°)
Triphenylmethane	(165.6)	.678	.686 (100°)
Triphenylphosphine	(166.8)	.636	(.759)
Triphenyl phosphite	(183.7)	.592	(.701) (18°)
Triphenylstibine	(182.2)	.516	
Triphenylstibine dihydroxide	(238.5)	.616	
N,N',N-Triphenyl urea	176.5	(.613)	
Triquinoyl	(133.0)	.426	
Tropolone	61	.50	
Tryptophan	132.0	(.646)	
Tyrosine	105.3	(.581)	
Undecane	131.84	(.8435)	(.6247) (20°)
Urea	33.4	(.556)	(.742) (20°)
Urethan	(57)	.64	(.63) (21°)
iso-Valeraldehyde	(57.5)	.668	(.536) (17°)
n-Valeric acid	66.85	.6548	(.617) (20°)
iso-Valeric acid	(67.7)	.663	(.621) (15°)
Valerylphenylacetylene	(119.0)	.639	
Valine	74.3	(.634)	
Violanthrene	273.5	.641	
Violanthrone	204.8	.449	
iso-Violanthrone	215.9	.473	
Water	(13.00)	.7218 (20°)	(.7205) (20°)
Water (value usually used as standard)	(12.97)	.720 (20°)	(.719) (20°)
Xanthone	(108.1)	.551	
o-Xylene	77.78	.7327	.6440 (20°)
m-Xylene	76.56	.7212	.6235 (20°)
p-Xylene	76.78	.7232	.6226 (20°)
Xylose	84.80	.565	(.862) (20°)

DIAMAGNETIC SUSCEPTIBILITIES OF ORGANIC COMPOUNDS (Continued)

Supplementary Table

Formula	Compound	$-\chi_M \times 10^6$	$-\chi \times 10^6$	$-K \times 10^6$
$H_6N_3B_3$	Borazine	49.6	.616	
CO	Carbon monoxide	9.8	.350	
COS	Carbon oxysulfide	32.4	.539	
$COCl_2$	Phosgene	48	.485	(.675) (19°)
CO_2	Carbon dioxide	20	.45	
$CNCl$	Cyanogen chloride	32.4	.527	(.642) (4°)
CCl_2S	Thiophosgene	50.6	.440	(.664) (15°)
CH_2N_2	Cyanamide	24.8	.590	(.639)
CH_3SiCl_3	Methyl trichlorsilane	87.45	$.584_8$	
CH_4N_2S	Thiourea	42.4	.557	(.782) (20°)
C_2HF_3	Trifluoroethylene	32.2	.393	
$C_2HF_3Cl_2$	1,2-Dichloro-1,1,2-trifluoroethane	66.2	.433	
$C_2HF_3Br_2$	1,2-Dibromo-1,1,2-trifluoroethane	90.9	.376	
$C_2H_3O_2N_3$	Urazole	46.2	.456	
C_2H_4OS	Thioacetic acid	38.4	.505	(.542) (10°)
$C_2H_4O_2S$	Mercaptoacetic acid	50.0	.543	(.720) (20°)
C_2H_5NS	Thioacetamide	42.45	.565	
$C_2H_5SiCl_3$	Ethyl trichlorosilane	98.84	$.604_6$	
$C_2H_6O_2S_2$	Methyl thiosulfite	62.3	.494	
$C_2H_6O_3S$	Methyl sulfite	54.2	.492	
$C_2H_6N_2S$	N-Methyl thiourea	53.6	.595	
C_2H_6S	Ethyl mercaptan	47.0	.756	(.635)
$C_2H_6SiCl_2$	Dimethyl dichlorosilane	82.45	$.639_2$	
$C_2H_8N_2 \cdot HCl$	1,2-Diamino ethane hydrochloride	76.2	.789	
C_3H_3ON	Pyruvic nitrile	33.2	.481	
$C_3H_5OF_3$	1-Methoxy-1,1,2-trifluoroethane	55.9	.490	
C_3H_5OCl	Propionyl chloride	51	.55	(.59)

Formula	Compound	$-\chi_M \times 10^6$	$-\chi \times 10^6$	$-K \times 10^6$
$C_3H_5O_2Br$	Methylbromoacetate	71.1	.465	
C_3H_5Cl	1-Chloro-2-propene	47.8	.625	
C_3H_5Br	1-Bromo-2-propene	58.6	.484	
C_3H_5I	1-Iodo-2-propene	72.8	.433	
$C_3H_6N_6$	Melamine	61.8	.490	(.771) (250°)
C_3H_7N	1-Amino-2-propene	40.1	.702	
$C_3H_7SiCl_3$	n-Propyl trichlorosilane	110.2	.620$_0$	
$C_3H_8O_2N_2$	N-Hydroxymethyl-N-methylurea	68.0	.653	
$C_3H_8N_2S$	N,N-Dimethyl thiourea	64.2	.616	
$C_3H_8N_2S$	N,N′-Dimethyl thiourea	64.7	.621	
$C_3H_8N_2S$	N-Ethyl thiourea	62.8	.603	
C_3H_8S	Propyl mercaptan	58.5	.768	(.642) (25°)
C_3H_9SiCl	Trimethyl chlorosilane	77.36	.712$_2$	
$C_3H_{12}N_3B_3$	N-Trimethyl borazine	78.6	.641	
$C_4H_5ON_3$	Cytosine	55.8	.502	
$C_4H_6O_2$	Vinylacetate	46.4	.539	(.502)
$C_4H_6O_2NCl_3$	Nitro tri(chloromethyl) methane	107.8	.522	
$C_4H_6O_6$	Tartaric acid	67.5	.402	
C_4H_7OCl	Butyryl chloride	62.1	.582	(.598) (20°)
C_4H_7OCl	iso-Butyryl chloride	63.9	.599	
$C_4H_8N_2S$	N-Allyl thiourea	69.0	.595	(.725) (20°)
$C_4H_9ON_3$	Acetone semicarbazone	66.29	.575$_8$	(.716) (20°)
$C_4H_{10}O_2S_2$	Ethyl thiosulfite	86.2	.559	
$C_4H_{10}S_2$	Ethyl disulfide	83.6	.684	(.679) (20°)
$C_4H_{10}SiCl_2$	Diethyl dichlorosilane	105.80	.673$_5$	
$C_5H_4O_3N_4$	Uric acid	66.2	.394	(.746)
C_5H_5	Cyclopentadiene	44.5	.673	
$C_5H_6O_2N_2$	Thymine	57.1	.453	
$C_5H_8O_2$	Methyl methacrylate	57.3	.572	(.535) (20°)
$C_5H_8N_2$	3,5-Dimethyl pyrazole	56.2	.585	
$C_5H_8Cl_4$	Tetra (chloromethyl) methane	129	.652	
$C_5H_9S_3Na$	Sodium butyl thiocarbonate	104.5	.555	
$C_5H_9Cl_3$	1,1,1-Tri(chloromethyl) ethane	114	.650	
$C_5H_{10}O$	Cyclopentanol	64.0	.743	(.705) (20°)
$C_5H_{10}O_2$	Ethyl carbonate	75.4	.738	
$C_5H_{10}O_5$	D-Ribose	84.6	.564	
$C_5H_{10}NS_2Na$	Sodium diethyldithiocarbomate	99.2	.579	
$C_5H_{10}Cl_2$	2,2-Di-(Chloromethyl) propane	96.9	.687	
$C_5H_{11}ON$	Methyl n-propyl ketoxime	68.82	.679$_6$	(.618) (20°)
$C_5H_{11}ON_3$	Ethyl methyl ketone semicarbazone	77.93	.603$_4$	(.710) (20°)
$C_5H_{12}ON_2$	N,N′-Diethyl urea	74.1	.638	
$C_5H_{12}ON_2$	N,N,N′,N′-Tetramethyl urea	75.7	.652	(.634) (15°)
$C_5H_{14}O_2Si$	Methyl diethoxysilane	92.99	.692$_4$	
$C_6H_3O_4N_2Cl$	1-Chloro-2,4-dinitrobenzene	84.4	.417	(.708) (22°)
$C_6H_3Cl_2I$	3,4-Dichloro-1-iodobenzene	118	.432	
$C_6H_5SiCl_3$	Phenyl trichlorosilane	120.4	.569$_4$	
C_6H_6NBr	o-Bromoaniline	87.32	.508	
C_6H_6NBr	m-Bromoaniline	84.89	.494	(.780) (20.4°)
C_6H_6NBr	p-Bromoaniline	84.06	.489	(.880)
$C_6H_7O_3NS . H_2O$	Sulfanilic acid	90.1	.471	
C_6H_7N	2-Methyl pyridine	60.3	.648	(.616) (15°)
C_6H_7N	3-Methyl pyridine	59.8	.642	(.617) (15°)
C_6H_7N	4-Methyl pyridine	59.8	.642	(.614) (15°)
$C_6H_7N . HI$	Aniline hydroiodide	113.6	.514	
C_6H_{10}	2,3-Dimethyl-1,3-butadiene	57.2	.696	(.539) (0°)
$C_6H_{11}OCl$	2-Ethyl butyryl chloride	87.2	.648	
$C_6H_{11}SiCl_3$	Cyclohexyltrichlorosilane	138.1	.634$_3$	
$C_6H_{12}N_2S_3$	Tetramethyl thiuram monosulfide	118.4	.568	(.795)
$C_6H_{12}N_2S_4$	Tetramethl thiuram disulfide	140.6	.585	(.755) (20°)
$C_6H_{13}ON$	Methyl-n-butyl ketoxime	79.97	.694$_2$	(.623) (20°)
$C_6H_{13}ON_3$	Methyl-n-propyl ketone semicarbazone	89.47	.624$_9$	(.669) (20°)
$C_6H_{13}ON_3$	Diethyl ketone semicarbazone	90.68	.633$_3$	(.730) (20°)
C_6H_{14}	3-Ethyl pentane	86.21	.861$_0$	(.601) (20°)
$C_6H_{14}O$	iso-Propyl ether	79.4	.777	(.564) (20°)
$C_6H_{14}S_2$	Propyl disulfide	106.2	.707	
$C_6H_{16}O_2Si$	Dimethyl diethoxysilane	104.6	.705$_2$	
$C_6H_{18}O_3Si_3$	Hexamethylcyclotrisiloxane	140.7	.632$_4$	
$C_6H_{18}N_3B_3$	Hexamethylborazine	119	.723	
$C_7H_5O_2Cl$	o-Chlorobenzoic acid	83.56	.534	(.824) (20°)
$C_7H_5O_4N$	Quinoleic acid	72.3	.433	
$C_7H_5NS_2$	2-Mercaptobenzothiazole	99.4	.594	(.843) (20°)
$C_7H_5F_2Cl$	Chlorodifluoromethylbenzene	87.2	.536	
$C_7H_7O_2N$	o-Aminobenzoic acid	77.18	.563	
C_7H_7Cl	o-Chlorotoluene	81.98	.648	(.701) (20°)
C_7H_7Cl	m-Chlorotoluene	80.07	.633	(.679) (20°)
C_7H_7Cl	p-Chlorotoluene	80.07	.633	(.677) (20°)
$C_7H_8N_2S$	N-phenylthiourea	87.4	.574	(.75)
C_7H_9ON	o-Anisidine	80.44	.654	(.714) (20°)
C_7H_9ON	m-Anisidine	79.95	.650	(.712) (20°)
C_7H_9ON	p-Anisidine	80.56	.655	(.702) (55°)
C_7H_9N	2,4-Dimethyl pyridine	71.50	.667	(.633) (0°)
C_7H_9N	2,6-Dimethyl pyridine	71.72	.669	(.630) (0°)
$C_7H_{15}ON$	Methyl amyl ketoxime	91.24	.706$_2$	(.630) (20°)
$C_7H_{15}ON_3$	Methyl-n-butyl ketone semicarbazone	100.40	.638$_6$	(.643) (20°)
$C_7H_{16}O$	1-Heptanol	91.7	.790	(.649) (20°)
$C_7H_{18}O_3Si$	Methyl triethoxysilane	120.6	.676$_2$	
C_8H_7OCl	Phenylacetyl chloride	88.5	.572	(.668) (20°)
$C_8H_8O_2$	o-Toluic acid	80.83	.594	(.626) (112°)
$C_8H_{10}ON_2$	N-Methyl, N′Phenyl urea	93.35	.622	
$C_8H_{10}O_4$	Butyne diacetate	95.9	.563	
$C_8H_{11}ON$	m-Phenetidine	90.28	.659	
$C_8H_{11}N$	2,4,6-Trimethyl pyridine	83.22	.687	(.630) (20°)
$C_8H_{16}O_2$	Hexylacetate	100.9	.700	(.623) (0°)
$C_8H_{17}ON_3$	Methyl amyl ketone semicarbazone	112.25	.655$_5$	(.687) (20°)
C_8H_{18}	3-Methyl-3-ethyl pentane	99.9	.875	(.623)
C_8H_{18}	4-Methyl heptane	97.30	.851$_7$	(.614)
C_8H_{18}	3-Ethyl hexane	97.76	.855$_8$	(.614) (20°)
C_8H_{18}	2,3,4-Trimethyl pentane	99.75	.872$_3$	

Formula	Compound	$-\chi_M \times 10^6$	$-\chi \times 10^6$	$-K \times 10^6$
$C_8H_{18}S_2$	Butyl disulfide	129.6	.727	
$C_8H_{18}S_3$	Butyl trisulfide	144.5	.687	
$C_8H_{18}S_4$	Butyl tetrasulfide	158.5	.653	
$C_8H_{20}O_4Si$	Tetraethoxysilane	137.1	.657$_9$	
$C_8H_{24}O_4Si_4$	Octamethyl cyclotetrasiloxane	187.4	.632$_2$	
C_9H_5N	Phenyl propionitrile	78.2	.615	
$C_9H_6O_2$	Phenyl propiolic acid	81.0	.554	
$C_9H_6O_3$	7-Hydroxycoumarin	88.22	.544$_9$	
C_9H_7N	iso-Quinoline	83.9	.650	(.714) (20°)
$C_9H_{10}O_4N_4$	Acetone-2,4-dinitrophenylhydrazone	110.62	.464$_4$	(.653) (20°)
$C_9H_{11}ON$	Benzylideneamino-1-hydroxyethane	91.0	.610	
$C_9H_{11}N$	Benzylideneaminoethane	85.5	.642	
$C_9H_{12}O_3S$	Ethyl-p-toluene sulfonate	115	.574	
$C_9H_{12}O_6N_2$	Unidine	106.9	.438	
$C_9H_{13}O_5N_3$	Cytidine	123.7	.509	
$C_9H_{14}O_3$	1-Acetyl-1 ethoxycarbonyl cyclobutane	103.4	.608	
$C_9H_{14}O_4$	1,1-Di(ethoxycarbonyl) cyclopropane	110.4	.593	
$C_9H_{14}Si$	Trimethylphenylsilane	109.1	.726	
$C_9H_{19}ON_3$	Methyl-n-hexyl ketone semicarbazone	123.60	.666$_2$	(.716) (20°)
$C_9H_{20}O$	Di-isobutyl carbinol	116.9	.810	(.667) (0°)
$C_9H_{20}ON_2$	N,N,N,'N'-Tetraethylurea	122.4	.710	
$C_9H_{20}N_2S$	N,N,N',N'-Tetraethylthiourea	132.6	.704	
$C_9H_{24}N_3B_3$	B-Triethyl-N-Trimethyl borazine	146	.706	
$C_{10}H_6O_7N_2$	5,7-Dinitro-7-hydroxy-4-methylcoumarin	111.9	.420$_6$	
$C_{10}H_7O_3Br$	7-Hydroxy-4-methyl-3-bromocoumarin	127.3	.498$_7$	
$C_{10}H_7O_5N$	5-Nitro-6-hydroxy-4-methylcoumarin	105.6	.477$_9$	
$C_{10}H_7O_5N$	6-Nitro-7-hydroxy-4-methylcoumarin	105.5	.477$_5$	
$C_{10}H_7O_5N$	8-Nitro-7-hydroxy-4-methylcoumarin	106.0	.479$_8$	
$C_{10}H_8O_3$	6-Hydroxy-4-methylcoumarin	98.69	.560$_7$	
$C_{10}H_8O_3$	7-Hydroxy-4-methylcoumarin	99.96	.563$_7$	
$C_{10}H_8O_4$	Benzylidene malonic acid	97.5	.507	
$C_{10}H_8O_4$	5,7-Dihydroxy-4-methylcoumarin	106.9	.556$_7$	
$C_{10}H_8O_4$	7,8-Dihydroxy-4-methylcoumarin	105.1	.550$_4$	
$C_{10}H_8S$	1-Mercaptonaphthalene	109.5	.684	
$C_{10}H_8S_2$	1,8-Dimercaptonaphthalene	118.0	.614	
$C_{10}H_{10}O_4$	Methyl terephthalate	101.6	.523	
$C_{10}H_{10}Os$	Bis cyclopentadienyl osmium	193(20°)	.602	
$C_{10}H_{12}$	1,2,3,4-Tetrahydronaphthalene	93.3	.706	(.685)
$C_{10}H_{12}O_4N_4$	Ethyl methyl ketone-2,4-dinitrophenylhydrazone	124.11	.492$_1$	(.631) (20°)
$C_{10}H_{13}O_4N_5$	Adenosine	137.5	.514	
$C_{10}H_{13}O_5N_5$	Guanosine	149.1	.526	
$C_{10}H_{14}$	Durene	101.2	.754	(.632) (81°)
$C_{10}H_{14}$	sec-Butyl benzene	101.31	.754$_9$	(.651) (20°)
$C_{10}H_{16}O_4$	1,1-Di(ethoxycarbonyl) cyclobutane	118	.589	
$C_{10}H_{19}N$	Camphylamine	122.0	.796	
$C_{11}H_9BrS$	1-Bromo-4-Methylthionaphthalene	147.0	.581	
$C_{11}H_{10}O_2$	4,6-Dimethylcoumarin	106.2	.610$_3$	
$C_{11}H_{10}O_2$	4,7-Dimethylcoumarin	107.6	.619$_0$	
$C_{11}H_{10}O_3$	5-Hydroxy-4,7-dimethylcoumarin	113.5	.597$_3$	
$C_{11}H_{10}O_3$	6-Methoxy-4-methylcoumarin	109.5	.576$_3$	
$C_{11}H_{10}O_3$	7-Methoxy-4-methylcoumarin	110.5	.581$_4$	
$C_{11}H_{10}S$	Methyl-α-naphthyl sulfide	120.2	.690	
$C_{11}H_{14}O_2$	Benzyl butyrate	116.3	.653	(.663) (17.5°)
$C_{11}H_{14}O_4N_4$	Diethyl ketone-2,4-dinitrophenylhydrazone	135.16	.507$_6$	(.670) (20°)
$C_{11}H_{14}O_4N_4$	Methyl-n-propyl ketone-2,4-dinitrophenylhydrazone	135.70	.509$_7$	(.623) (20°)
$C_{11}H_{16}ON_2$	N,N-Diethyl-N'-phenyl urea	126.4	.657$_3$	
$C_{12}H_6O_4N_2Cl_2$	4,4'-Dichloro-2,2'-dinitro-1,1'-biphenyl	150	.479	
$C_{12}H_8$	Acenaphthylene	111.6	.733	(.659) (16°)
$C_{12}H_8O_4N_2$	2,4'-Dinitro-1,1'-biphenyl	116.5	.477	(.703) (20°)
$C_{12}H_8Cl_2$	4,4'-Dichloro-1,1'-biphenyl	133.1	.597	(.859) (20°)
$C_{12}H_8Br_2$	4,4'-Dibromo-1,1'-biphenyl	151.3	.485	(.920) (20°)
$C_{12}H_9O_2N$	2-Nitro-1,1'-biphenyl	109	.547	(.788) (20°)
$C_{12}H_9O_2N$	5-Nitroacenaphthene	116.0	.582	
$C_{12}H_{10}O_4$	Cinnamylidene malonic acid	105.4	.483	
$C_{12}H_{10}O_4$	Quinhydrone	105	.481	(.674) (20°)
$C_{12}H_{10}S$	Phenyl sulfide	119.2	.640	(.716) (20°)
$C_{12}H_{10}S_2$	Phenyl disulfide	122.5	.561	
$C_{12}H_{10}Cl_2Si$	Diphenyl dichlorosilane	153.5	.606$_3$	
$C_{12}H_{12}O_2Si$	Diphenyl silanediol	131.6	.608$_2$	
$C_{12}H_{16}O_2$	Amyl benzoate	128.5	.668	
$C_{12}H_{16}O_4N_4$	Methyl-n-butyl ketone-2,4-dinitrophenylhydrazone	147.65	.526$_8$	(.643) (20°)
$C_{12}H_{18}$	Hexamethyl benzene	122.5	.755	
$C_{12}H_{18}$	Diisopropyl benzene	124.77	.767$_1$	
$C_{12}H_{22}O_{11}$	Sucrose (saccharose)	189.1	.552	(.877) (15°)
$C_{12}H_{25}Cl_3Si$	Dodecyl trichlorosilane	209.2	.688$_9$	
$C_{12}H_{30}N_3B_3$	Hexaethylborazine	189	.759	
$C_{13}H_8OS$	Thioxanthone	130	.612	
$C_{13}H_{10}O_2$	Phenyl benzoate	117.3	.592	
$C_{13}H_{10}S$	Thiobenzophenone	118.1	.596	
$C_{13}H_{11}NS$	N-Thiobenzoyl aniline	123.0	.577	
$C_{13}H_{12}N_2S$	N,N'-Diphenyl thiourea	136.5	.598	
$C_{13}H_{12}N_2S$	N,N-Diphenyl thiourea	134.9	.591	
$C_{13}H_{18}O_4N_4$	Methyl amyl ketone-2,4-dinitrophenylhydrazone	156.84	.532$_9$	(.651) (20°)
$C_{14}H_8N_2S_4$	Di-2-benzothiazolyl disulfide	189.0	.568	
$C_{14}H_{10}O$	Anthrone	118	.608	
$C_{14}H_{12}O_2$	Benzyl benzoate	132.2	.622	(.693) (18°)
$C_{14}H_{12}O_2$	Diphenyl acetic acid	124.5	.587	
$C_{14}H_{14}O_2S_2$	p-Tolyl thiosulfonate	157	.564	
$C_{14}H_{14}N_2$	p-Azotoluene	135.1	.643	
$C_{14}H_{16}Si$	Dimethyl diphenyl silane	146.6	.690	
$C_{14}H_{20}O_4N_4$	Methyl n-hexyl ketone-2,4-dinitrophenylhydrazone	171.73	.557$_0$	(.655) (20°)
$C_{15}H_{10}O_2$	Flavone	120	.540	
$C_{15}H_{12}$	9-Ethylidene fluorene	124.8	.649	
$C_{15}H_{12}O$	Chalcone	125.7	.604	(.646) (62°)
$C_{15}H_{14}O_2$	Benzyl phenylacetate	143.7	.635	
$C_{15}H_{16}ON_2$	N,N'-Dimethyl-N,N'-Diphenyl urea	148.9	.620	
$C_{16}H_{10}$	Fluoranthene	138.0	.682	
$C_{16}H_{12}$	1-Benzylidene indene	130.5	.716	

Formula	Compound	$-\chi_M \times 10^6$	$-\chi \times 10^6$	$-k \times 10^6$
$C_{16}H_{16}$	1,2-Dibenzylamine ethane	164	.787	
$C_{16}H_{33}SiCl_3$	Hexadecyl trichlorosilane	252.4	$.701_3$	
$C_{16}H_{48}O_6Si_7$	Hexadecamethyl heptasiloxane	351.1	$.658_8$	
$C_{17}H_{16}$	9-Butylidene fluorene	147.5	.670	
$C_{18}H_8S_4$	Tetrathiotetracene	202	.573	
$C_{18}H_{10}Cl_2$	5,11-Dichloro tetracene	202	.680	
$C_{18}H_{10}Br_2$	5,11-Dibromo tetracene	216	.559	
$C_{18}H_{12}$	Triphenylene	156.6	.686	
$C_{18}H_{14}$	o-Diphenyl benzene	150.4	.653	
$C_{18}H_{14}$	p-Diphenyl benzene (Terphenyl)	152	.660	
$C_{18}H_{15}SiCl$	Triphenyl chlorosilane	186.7	$.633_3$	(.814) (0°)
$C_{18}H_{16}OSi$	Triphenyl silanol	176.6	$.634_5$	
$C_{18}H_{16}Si$	Triphenylsilane	174	.668	
$C_{18}H_{20}$	5-Cyclohexyl acenaphthene	165.2	.699	
$C_{18}H_{22}$	Dimesitylene	171.8	.721	
$C_{18}H_{24}O_2$	Estradiol	186.6	.717	
$C_{18}H_{37}SiCl_3$	Octadecyl trichlorosilane	273.7	$.705_7$	
$C_{19}H_{14}O$	4-Benzoyl-1,1′-Biphenyl	158	.612	
$C_{19}H_{18}Si$	Methyl triphenyl silane	186	.678	
$C_{20}H_{14}O_2$	1,4-Dibenzoyl benzene	153	.534	
$C_{21}H_{24}N_3B_3$	B-Trimethyl-N-triphenylborazine	234	.667	
$C_{24}H_{18}$	4,4′-Diphenyl-1,1′-Biphenyl	201.3	.657	
$C_{24}H_{20}Si$	Tetraphenyl silane	224	.666	
$C_{24}H_{30}N_3B_3$	B-Triethyl-N-triphenylborazine	264	.693	
$C_{24}H_{40}O_4$	Desoxycholic acid	272.0	.715	
$C_{24}H_{40}O_5$	Cholic acid	282.3	.691	
$C_{25}H_{20}N_2S$	N,N,N′,N′-Tetraphenylthiourea	229.7	.604	
$C_{26}H_{20}$	Tetraphenyl ethylene	217.4	.654	
$C_{28}H_{16}O_2$	Bianthrone (Dianthraquinone)	220	.572	
$C_{28}H_{18}O_2$	Dianthrone (Dianthronol)	229	.593	
$C_{28}H_{20}$	1,1,4,4-Tetraphenyl-1,2,3-butatriene	227	.659	
$C_{28}H_{44}O$	Calciferol	273.3	.689	
$C_{28}H_{44}O$	Ergosterol	279.6	.704	
$C_{36}H_{30}O_3Si_3$	Hexaphenyl cyclotrisiloxane	364.1	$.612_9$	
$C_{48}H_{40}O_4Si_4$	Octaphenyl cyclotetrasiloxane	485.3	$.613_2$	
$C_{60}H_{36}$	Hexabenzocoronene	346.0	.457	

MASS ABSORPTION COEFFICIENTS FOR X AND γ RAYS

Radiation traversing a layer of substance is reduced in intensity by a constant fraction μ per centimeter. After penetrating to a depth x the intensity is $I = I_0 e^{-\mu x}$ where I_0 is the intensity at the surface. μ/ρ is the mass absorption coefficient where ρ is the density of the material.

Values of μ/ρ for λ = .005 Å to λ = 44.6 Å. Where two values of μ/ρ for one value of λ occur they represent the maximum and minimum values at an absorption discontinuity.

Compiled by S. J. M. Allen

λ = 44.6 – 2.74 Å

λ, Å	He	C	N	O	Ne	Al	S	Cl	A
44.6	3600				13100	....	....	....	45700
11.88	....	2170	3850	5765	6850	850			
9.87	....	1063	1796	2540	4310	500	1320	1570	1860
8.32	....	656	1109	1585	2750	330	794	962	1160
7.94	....	....	....	....	....	280	....	....	....
						3700			
6.97	....	390	645	976	1727	2800	500	610	748
5.39	....	185	312	476	865	1450	249	310	360
5.17	....	160	273	413	763	1350	221	277	324
5.01	....	....	....	....	....	....	210	....	....
							2260		
4.38	....	....	....	....	....	....	....	178	....
								1830	
4.36	....	97.8	166	258	478	815	1570	1800	202
4.15	....	84.6	144	222	416	720	1350	1476	174
3.93	....	71.0	121	189	356	635	1175	1256	153
3.87	....	....	....	....	....	....	....	....	148
									1460
3.69	....	....	....	....	....	....	....	....	....
3.59	....	55.2	96	150	279	500	928	966	1215
3.51	....	....	....	....	....	....	....	....	....
3.38	....	46.0	79.5	117	231	425	795	880	1025
3.35	....	43.0	....	....	....	417	780	870	1015
3.24	....	....	....	....	....	....	....	....	....
3.03	....	35.0	....	84.0	175	323	595	670	760
2.74	...	25.0	....	60.0	135	250	454	520	600

λ = 2.50 – .900 Å

λ, Å	Mg	Al	S	Cl	A	Ca	Fe	Ni	Cu
2.50	161	193	355	400	475	620	147	180	197
2.29	....	150	285	315	355	480	115	137	153
1.93	77.2	93.5	173	198	235	306	71.2	89.5	96.2
1.74	....	83.0	....	....	....	...	54	....	....
							465		
1.656	....	60.7	110	126	143	195	410	59.2	63.5
1.539	40.8	49.0	91	103	114	163	325	48.0	50.9
1.484	....	....	....	....	....	...	....	40.5	....
								338	
1.432	....	40.0	75	85	93	130	285	325	42
1.389	31.5	36.8	68.5	76.7	85.7	125	252	275	38.5
1.377	....	....	....	....	....	...	....	....	37.0
									307
1.293	....	29.8	55.3	60	72	102	212	233	260
1.280	....	28.8	....	....	....	...	....	225	252
1.235	21.4	26.3	49.5	55.5	62.5	90	181	208	230
1.104	....	18.6	38.0	44	50	67	135	155	175
1.071	....	....	....	....	....	...	....	....	....
1.038	....	....	....	....	....	...	....	....	....
1.000	11.8	14.12	26.7	29.7	34.5	49	100	121	130
.980	....	....	....	....	....	...	....	....	....
.949	....	12.0	22.0	24.5	....	42	86	99	114
.932	....	....	....	....	....	...	....	....	....
.900	....	10.4	....	....	....	...	74.5	86.5	98.5

λ = 44.6 – 2.74 Å

λ, Å	Fe	Ni	Cu	Zn	Kr	Ag	Sn	Xe	Pt	Au
44.6	...	....	....	...	31800	....	...	6740	....	12500
11.88	...	6900	7550	...	....	....	...	....	....	....
9.87	...	4540	5030	...	....	2700	...	....	2440	....
8.32	...	3140	3450	...	....	1800	...	....	1560	....
7.94	...	....	....	...	....	....	...	....	....	....
6.97	...	2000	2130	...	....	1300	...	....	1190	....
5.39	...	1250	1290	...	....	845	...	....	1645	....
5.17	...	1150	1190	...	....	790	...	....	....	....
5.01	...	....	....	...	....	....	...	....	....	....
4.38	...	....	....	...	....	....	...	....	....	....
4.36	610	715	760	910	....	535	640	....	....	....
4.15	540	630	690	820	....	461	550	....	1290	....
3.93	470	555	610	715	....	408	490	....	....	....
3.87	...	....	....	...	....	....	...	....	....	....
3.69	...	....	....	...	....	354	...	....	....	....
						1410				
3.59	375	450	495	575	....	1360	...	....	1370	....
3.51	...	....	....	...	....	1300	...	....	....	....
						1510				
3.38	320	380	....	495	....	....	...	....	....	....
3.35	312	375	404	480	....	1310	...	....	1120	....
3.24	...	....	....	...	....	1230	...	....	....	....
						1440				
3.03	245	290	315	375	....	1290	...	....	939	....
2.74	185	239	262	283	....	925	...	....	756	....

λ = 2.50 – .900 Å

λ, Å	Zn	Br	Mo	Ag	Sn	I	W	Pt	Au	Pb
2.50	228	...	..	710	850	...	...	596	.....	...
2.29	180	...	..	550	670	...	...	480	.....	...
1.93	110	...	..	405	470	...	300	358	385	428
1.74	.....	...	..	.....	.....	...	...	.....	.....	...
1.656	72.5	...	..	285	.....	...	...	228	.....	...
1.539	58.6	89	..	217	247	290	176	202	213	230
1.484	.....	...	..	.....	.....	...	...	.....	.....	...
1.432	49.3	...	..	192	220	...	130	172	179	202
1.389	45.2	...	..	174	209	...	...	155	166	185
1.377	.....	...	..	.....	.....	...	...	.....	.....	...
1.293	39	...	..	146	176	...	...	132	138	154
1.280	36	...	..	127	146	...	...	.....	.....	...
	287									
1.235	250	...	..	125	140	...	95	115	122	137
1.104	208	...	..	96.5	115	...	...	99	107	120
1.071	.....	...	..	.....	.....	...	...	77.5	.....	...
								198		
1.038	.....	...	..	.....	.....	...	...	.....	76.5	...
									194	
1.000	145	...	52	73.0	86.0	...	...	165	174	75
.980	.....	...	..	.....	.....	...	...	155	168	73
.949	129	...	..	63.0	75.5	...	...	146	156	68
										168
.932	.....	...	..	.....	.....	...	...	136	148	159
								184		
.900	112	150	..	54.2	65.0	...	...	168	134	145
									182	

λ = 2.50 – .900 Å

λ, Å	H	Li	Be	B	C	N	O	Ne	Na
2.50	.52	4.0	6.1	9.1	17.8	....	44.5	100	128
2.29	...	....	....	....	15.0	....	36.4	75.5	....
1.93	.50	2.10	3.05	4.7	8.75	14.0	21.7	49.0	61.3
1.74	...	....	....	....	....	....	....	....	....
1.656	...	....	....	....	....	....	....	....	....
1.539	.48	1.10	1.60	2.45	4.52	7.45	11.1	24.0	32.1
1.484	...	....	....	....	....	....	....	....	....
1.432	...	....	....	....	....	....	....	....	....
1.389	.47	86	1.25	1.87	3.35	5.50	8.1	17.0	23.4
1.377	...	....	....	....	....	....	....	....	....
1.293	...	....	....	....	....	....	....	....	....
1.280	...	....	....	....	....	....	....	....	....
1.235	.46	67	.95	1.35	2.42	3.95	5.7	12.4	17.1
1.104	...	....	....	....	....	....	....	....	....
1.071	...	....	....	....	....	....	....	....	....
1.038	...	....	....	....	....	....	....	....	....
1.000	45	.43	.55	.76	1.36	2.10	3.13	6.5	8.8
.980	...	....	....	....	....	....	....	....	....
.949	...	....	....	....	1.20	....	....	....	....
.932	...	....	....	....	....	....	....	....	....
.900	...	....	....	....	1.05	....	....	....	....

λ = .892 – .184 Å

λ, Å	H	Li	Be	B	C	N	O	Ne	Na	Mg	Al
.892	....	....	....	....	....	....	....	....	....	....	....
.880	.440	.350	.425	.580	.990	1.50	2.20	4.55	6.10	8.34	9.75
.862	....	....	....	....	....	....	....	....	....	....	....
.850	....	....	....	....	.907	....	....	....	....	....	8.85
.814	....	....	....	....	.814	....	....	....	....	....	7.85
.780	....	....	....	....	.750	....	....	....	....	....	6.86
.710	.435	.260	.315	.365	.598	.870	1.22	2.50	3.30	4.30	5.22
.680	....	....	....	....	.550	....	....	....	....	....	4.52
.631	.435	.225	.255	.305	.467	.610	.900	1.80	2.30	3.0	3.73
.618	....	....	....	....	....	....	....	....	....	....	....
.560	....	....	....	....	.370	....	....	....	....	....	2.60
.497	.435	.198	.210	.220	.315	.400	.520	.930	1.18	1.52	1.90
.485	....	....	....	....	.308	....	....	....	....	....	1.77
.476	.430	....	....	.215	.304	....	.485	....	....	....	1.74
.424	....	....	....	....	....	....	....	....	....	....	1.23
.417	.390	.180	.185	.198	.256	.310	.372	.580	.750	.940	1.170
.380	....	....	....	....	.230	....	....	....	....	....	.950
.331	....	....	....	....	....	....	....	....	....	....	....
.260	.385	.156	.166	.175	.185	.200	.210	.270	.305	.343	.402
.220	....	....	....	....	.178	....	....	....	....	....	.300
.200	.375	.151	.160	.165	.175	.180	.183	.210	.225	.250	.270
184	....	....	....	....	.166	....	....	....	....	....	.246

MASS ABSORPTION COEFFICIENTS FOR X AND γ RAYS (Continued)

λ = .892 − .184 A

λ, Å	S	Cl	A	Ca	Fe	Ni	Cu	Zn	Br	Sr	Mo
.892											
.880	18.2	20.7	24.0	34.8	69.5	82	91.2	103			36.0
.862											
.850					63.5	74	84.5	96.5			
.814					57	66	75.7	86			28
.780					50.5	59.5	67.5	77			
.710	9.90	11.6	13.0	18.6	38.5	48.1	51.0	59.0	80	106.	19.9
.680					32.7	41	45.3	52.7			
.631	6.90	8.40	9.80	13.3	27.0	34	36.2	41.0	56.8	72.5	15.0
.618											12.5
											88.0
.560					18.2	24	25.5	30.7			
.497	3.50	4.20	5.0	6.60	13.9	17.9	18.4	21.0	32.0	40.5	50.2
.485					12.4	15.4	16.9	19.5			
.476							16.6				42
.424											
.417	2.10	2.47	2.95	3.97	8.45	10.5	11.45	12.3	19.0	24.0	30.0
.380					6.32	7.70	8.42	9.95			22
.331											
.260	.650	.750	.850	1.10	2.28	2.89	3.16	3.58	5.30	6.50	8.20
.220					1.42	1.80	2.00	2.32			
.200	.400	.445	.500	.630	1.10	1.45	1.55	1.78	2.4	3.32	4.30
.184						1.24					

λ, Å	Ag	Sn	I	Ba	Ta	W	Pt	Au	Pb	Bi	U
.892							165	178	142		
							201				
.880	50	60					195	170	135		
.862							185	163	130		
								193			
.850	46	56					179	186	124		
.814	41	49.5					160	167	111		
									150		
.780	36	44.5					144	150	136		
									166		
.710	27.5	34.0	38.5	42.0	100	104	115	120	136		
.680	23.5	28.4					102	108	120		
.631	19.6	23.0	26.4	31.1	72	75	84.5	87	98		
.618											
.560	13.3	16.2					62	66	75		
.497	10.5	11.8	15.6	17.8	36	38	47	48.5	52.8		
.485	9.8	11.1									
	62.5										
.476	60						42		47.5		
.424	43.5	8.0									
		46.6									
.417	41	45	9.2	10.5	21.5	22.5	27.4	28.4	32.0		
.380	31.2	34				17.3	21.1	22	26.4	27.8	
.331	21.7	24.5		5.4					18.1	19.5	
				28.0							
.260	11.4	12.8	14.2	16.1	6.7	6.85	8.0	8.3	10.0	11.0	
.220	7.05	7.80				4.25	5.25	5.50	5.92	6.4	
.200	5.48	6.20	7.0	8.0	3.4	3.50	4.25	4.40	4.90	5.1	5.40
.184	4.45				2.8		3.45	3.60	4.05	4.2	
					11.8						

λ = .178 − .005 A

λ, Å	H	Li	Be	B	C	N	O	Ne	Na	Mg	Al
.178					.164						.235
.175	.360	.144	.150	.155	.163	.166	.169	.185	.195	.205	.228
.158					.160						.208
.155											
.146	.340				.155		.162		.170	.176	.195
.142	.330				.153						.191
.130	.320	.132		.149	.152		.157		.160	.168	.186
.120					.150		.154			.163	.172
.113	.310				.147		.153		.155	.160	.166
.107											
.098	.280	.125		.138	.142		.144		.150	.152	.156
.080	.255				.137						.146
.072	.250	.118		.132	.136		.137		.139	.140	.143
.064	.245	.110		.126	.130		.130		.130	.130	.130
.050						120					.115
.040	.205				.110						.105
.030	.180				.095						.093
.024	.165				.080						.079
.010	.117				.059						.058
.005	.078				.0385						.0380

λ, Å	S	Cl	A	Ca	Fe	Ni	Cu	Zn	Br	Sr	Mo
.178							1.15				
.175	.335	.341	.400	.460	.800	1.05	1.12	1.26	1.90	2.24	2.95
.158					.640	.815	.862	.990			
.155											
.146	.249	.280		.345	.520		.680				
.142					.515	.630	.670	.780			1.55
.130	.220	.230		.290	.424		.551				
.120	.200				.368	.430	.455	.537			
.113	.189	.195		.230	.337		.422				
.107											
.098	.166	.176		.200	.265		.325				.790
.080		.164			.235	.264	.268	.308			
.072	.150	.158		.180	.202		.232				
.064	.139	.142		.155	.178		.198				.413
.050					.140		.155				
.040					.118		.126				
.030					.095		.100				
.024					.080		.081				
.010					.058		.057				
.005							.0380				

λ, Å	Ag	Sn	I	Ba	Ta	W	Pt	Au	Pb	U
.178						2.7	3.16	3.30	3.55	
						11.3				
.175	3.96	4.50	5.10	5.70	10.0	10.5	2.97	3.13	3.48	3.95
.158	3.00	3.40				8.6	2.45	2.43	2.60	
							9.40			
.155								2.30		
								8.80		
.146	2.48	2.66			6.75		7.60	7.85	2.35	2.70
.142	2.31	2.64				6.75	7.20	7.33	2.10	
									7.75	
.130	1.97	2.12			5.10		6.30	6.40	6.55	2.20
.120	1.61	1.77		2.20		4.60	4.92	4.98	5.20	1.90
.113	1.47	1.60			3.80		4.40	4.50	4.75	
.107										1.62
										4.65
.098	1.05	1.17			2.80		3.15	3.21	3.50	3.90
.080	.73	.79				2.30	2.40	2.42	2.50	2.70
.072	.584	.614			1.75		2.00	2.05	2.10	2.25
.064	.465	.490			1.35		1.52	1.55	1.64	1.80
.050		.320					.86	.88	1.00	
.040		.21							.62	
.030		.13							.38	
.024		.10							.21	
.010		.060							.071	.082
.005		.0385							.0425	.044

MASS ABSORPTION COEFFICIENTS FOR X AND γ RAYS (Continued)

Values of μ/ρ for E = 2.00 Mev to E = 20.0 Mev.

Computed by G. Allen. Reproduced by permission of the National Aeronautics and Space Administration

$$\mu = \gamma + \sigma + \chi$$

where γ is the photoelectric absorption coefficient; σ is the Compton scattering coefficient calculated; and χ is the pair-formation coefficient.

	Atomic number	E 2.00	2.25	2.50	2.75	3.00	3.50	4.00	4.50	5.00	5.50
H	1	.0876	.0819	.0770	.0727	.0691	.0629	.0578	.0537	.0502	.0471
Li	3	.0382	.0358	.0338	.0319	.0304	.0277	.0256	.0238	.0225	.0212
Be	4	.0393	.0368	.0347	.0328	.0312	.0286	.0265	.0247	.0233	.0221
B	5	.0409	.0385	.0363	.0344	.0327	.0300	.0278	.0260	.0246	.0233
C	6	.0443	.0417	.0393	.0373	.0355	.0327	.0304	.0284	.0269	.0256
N	7	.0445	.0417	.0394	.0374	.0356	.0329	.0306	.0288	.0272	.0260
O	8	.0445	.0418	.0395	.0376	.0359	.0331	.0309	.0291	.0275	.0264
F	9	.0422	.0396	.0376	.0357	.0341	.0315	.0295	.0279	.0264	.0254
Na	11	.0427	.0402	.0381	.0363	.0348	.0322	.0302	.0286	.0274	.0263
Mg	12	.0441	.0416	.0394	.0376	.0360	.0335	.0314	.0298	.0285	.0275
Al	13	.0431	.0407	.0386	.0368	.0353	.0329	.0310	.0294	.0282	.0272
Si	14	.0447	.0422	.0400	.0383	.0367	.0342	.0322	.0307	.0295	.0285
P	15	.0435	.0410	.0390	.0372	.0358	.0334	.0316	.0301	.0290	.0281
S	16	.0448	.0423	.0402	.0385	.0370	.0346	.0327	.0313	.0302	.0293
Cl	17	.0431	.0407	.0387	.0371	.0356	.0335	.0317	.0304	.0293	.0284
K	19	.0438	.0414	.0395	.0378	.0365	.0343	.0326	.0313	.0303	.0295
Ca	20	.0451	.0426	.0406	.0390	.0376	.0354	.0337	.0324	.0315	.0308
Sc	21	.0420	.0399	.0380	.0365	.0352	.0332	.0317	.0305	.0297	.0290
Ti	22	.0416	.0394	.0375	.0361	.0348	.0330	.0315	.0304	.0296	.0290
V	23	.0410	.0389	.0371	.0356	.0344	.0326	.0312	.0302	.0294	.0288
Cr	24	.0419	.0398	.0380	.0366	.0354	.0335	.0322	.0311	.0303	.0298
Mn	25	.0414	.0393	.0376	.0363	.0350	.0334	.0319	.0310	.0302	.0297
Fe	26	.0424	.0403	.0385	.0371	.0360	.0342	.0329	.0319	.0312	.0308
Co	27	.0418	.0397	.0381	.0367	.0356	.0339	.0326	.0316	.0311	.0306
Ni	28	.0437	.0414	.0397	.0384	.0372	.0355	.0343	.0333	.0326	.0322
Cu	29	.0418	.0397	.0381	.0368	.0357	.0341	.0330	.0321	.0316	.0311
Zn	30	.0421	.0401	.0384	.0371	.0360	.0345	.0334	.0326	.0321	.0318
Ge	32	.0406	.0386	.0372	.0358	.0348	.0336	.0325	.0318	.0314	.0311
As	33	.0406	.0387	.0372	.0361	.0350	.0336	.0328	.0321	.0317	.0315
Se	34	.0398	.0379	.0365	.0354	.0344	.0331	.0322	.0315	.0313	.0311
Br	35	.0405	.0387	.0372	.0361	.0352	.0340	.0329	.0323	.0321	.0319
Rb	37	.0403	.0384	.0370	.0359	.0350	.0339	.0331	.0325	.0323	.0323
Sr	38	.0403	.0386	.0372	.0361	.0352	.0341	.0334	.0328	.0327	.0326
Zr	40	.0410	.0392	.0378	.0368	.0360	.0349	.0341	.0338	.0337	.0336
Cb	41	.0414	.0397	.0383	.0372	.0363	.0354	.0346	.0342	.0341	.0343
Mo	42	.0411	.0394	.0381	.0370	.0362	.0353	.0346	.0342	.0341	.0343
Pd	46	.0409	.0392	.0380	.0370	.0364	.0353	.0348	.0346	.0346	.0347
Ag	47	.0414	.0398	.0385	.0376	.0368	.0359	.0355	.0352	.0352	.0355
Cd	48	.0407	.0391	.0379	.0369	.0363	.0355	.0350	.0348	.0348	.0351
Sn	50	.0404	.0389	.0376	.0367	.0361	.0353	.0350	.0348	.0350	.0352
Sb	51	.0403	.0387	.0376	.0367	.0360	.0354	.0350	.0348	.0351	.0353
Te	52	.0394	.0379	.0366	.0359	.0352	.0347	.0343	.0342	.0343	.0347
I	53	.0405	.0389	.0377	.0369	.0362	.0357	.0354	.0354	.0355	.0358
Cs	55	.0404	.0388	.0377	.0369	.0363	.0358	.0355	.0355	.0357	.0362
Ba	56	.0399	.0384	.0374	.0366	.0360	.0355	.0353	.0353	.0355	.0360
La	57	.0404	.0388	.0378	.0369	.0363	.0358	.0357	.0358	.0361	.0365
Hf	72	.0424	.0409	.0399	.0393	.0387	.0387	.0388	.0392	.0398	.0407
Ta	73	.0427	.0411	.0401	.0395	.0391	.0390	.0391	.0396	.0402	.0411
W	74	.0428	.0412	.0402	.0395	.0392	.0391	.0394	.0396	.0405	.0412
Os	76	.0429	.0414	.0403	.0397	.0393	.0394	.0395	.0401	.0407	.0416
Pt	78	.0434	.0419	.0409	.0402	.0399	.0399	.0401	.0406	.0413	.0424
Au	79	.0438	.0422	.0412	.0406	.0402	.0402	.0405	.0409	.0418	.0427
Hg	80	.0439	.0424	.0413	.0407	.0402	.0403	.0406	.0411	.0420	.0430
Pb	82	.0441	.0425	.0414	.0408	.0404	.0405	.0409	.0415	.0422	.0432
Bi	83	.0445	.0428	.0418	.0412	.0409	.0409	.0413	.0419	.0427	.0438
U	92	.0459	.0442	.0431	.0425	.0422	.0424	.0428	.0435	.0446	.0457

	Atomic number	E 6.00	7.00	8.00	9.00	10.00	12.0	14.0	16.0	18.0	20.0
H	1	.0445	.0402	.0369	.0341	.0318	.0281	.0253	.0232	.0215	.0201
Li	3	.0202	.0183	.0170	.0158	.0150	.0135	.0124	.0117	.0109	.0105
Be	4	.0210	.0193	.0179	.0168	.0159	.0145	.0135	.0127	.0121	.0116
B	5	.0223	.0205	.0192	.0180	.0172	.0158	.0148	.0141	.0134	.0130
C	6	.0244	.0227	.0212	.0202	.0192	.0178	.0169	.0161	.0155	.0150
N	7	.0248	.0231	.0218	.0208	.0199	.0185	.0177	.0169	.0165	.0160
O	8	.0253	.0236	.0223	.0213	.0206	.0193	.0185	.0178	.0174	.0170
F	9	.0244	.0229	.0217	.0208	.0201	.0190	.0182	.0178	.0174	.0171
Na	11	.0254	.0240	.0230	.0221	.0215	.0205	.0199	.0196	.0193	.0191
Mg	12	.0266	.0252	.0242	.0234	.0228	.0219	.0214	.0211	.0208	.0207
Al	13	.0264	.0251	.0241	.0234	.0229	.0221	.0217	.0214	.0212	.0212
Si	14	.0277	.0265	.0255	.0248	.0243	.0237	.0232	.0231	.0230	.0230
P	15	.0272	.0261	.0253	.0247	.0243	.0237	.0234	.0233	.0232	.0232
S	16	.0285	.0274	.0266	.0260	.0256	.0251	.0248	.0248	.0249	.0249
Cl	17	.0277	.0268	.0261	.0256	.0252	.0249	.0247	.0247	.0248	.0249
K	19	.0289	.0281	.0274	.0271	.0268	.0266	.0266	.0267	.0269	.0272
Ca	20	.0301	.0293	.0287	.0283	.0282	.0281	.0281	.0283	.0286	.0289
Sc	21	.0285	.0278	.0273	.0270	.0269	.0269	.0269	.0273	.0275	.0279
Ti	22	.0285	.0278	.0275	.0272	.0271	.0272	.0274	.0276	.0280	.0284
V	23	.0283	.0278	.0275	.0273	.0273	.0274	.0276	.0280	.0284	.0288
Cr	24	.0293	.0289	.0285	.0284	.0285	.0287	.0289	.0295	.0299	.0304
Mn	25	.0294	.0288	.0286	.0286	.0287	.0289	.0293	0298	.0303	.0309
Fe	26	.0303	.0300	.0298	.0298	.0299	.0303	.0307	.0314	.0319	.0325
Co	27	.0303	.0300	.0298	.0299	.0300	.0305	.0310	.0316	.0322	.0328
Ni	28	.0319	.0316	.0316	.0317	.0319	.0324	.0330	.0338	.0344	.0352
Cu	29	.0309	.0307	.0307	.0308	.0310	.0316	.0323	.0331	.0338	.0345
Zn	30	.0314	.0313	.0314	.0315	.0318	.0325	.0333	.0341	.0349	.0357
Ge	32	.0310	.0309	.0311	.0313	.0317	.0325	.0334	.0343	.0351	.0360
As	33	.0313	.0315	.0315	.0319	.0322	.0332	.0341	.0350	.0360	.0368
Se	34	.0310	.0312	.0313	.0317	.0321	.0331	.0340	.0350	.0360	.0368
Br	35	.0319	.0321	.0324	.0328	.0333	.0343	.0353	.0364	.0373	.0384
Rb	37	.0322	.0326	.0329	.0335	.0340	.0351	.0363	.0375	.0385	.0397
Sr	38	.0326	.0330	.0334	.0341	.0346	.0358	.0371	.0383	.0395	.0406
Zr	40	.0337	.0342	.0348	.0354	.0362	.0376	.0389	.0403	.0416	.0428
Cb	41	.0343	.0348	.0355	.0362	.0370	.0384	.0399	.0414	.0426	.0439
Mo	42	.0345	.0350	.0357	.0364	.0372	.0387	.0403	.0418	.0431	.0445
Pd	46	.0349	.0357	.0363	.0370	.0378	.0393	.0405	.0418	.0429	.0439
Ag	47	.0356	.0364	.0372	.0380	.0387	.0402	.0416	.0429	.0440	.0450
Cd	48	.0353	.0360	.0369	.0378	.0384	.0400	.0414	.0427	.0438	.0449
Sn	50	.0356	.0363	.0372	.0382	.0390	.0406	.0420	.0434	.0446	.0457
Sb	51	.0357	.0366	.0374	.0384	.0393	.0408	.0424	.0438	.0450	.0460
Te	52	.0350	.0360	.0368	.0378	.0387	.0403	.0417	.0432	.0443	.0454
I	53	.0362	.0373	.0383	.0391	.0401	.0418	.0434	.0448	.0461	.0472
Cs	55	.0366	.0376	.0387	.0398	.0408	.0426	.0442	.0458	.0470	.0481
Ba	56	.0364	.0375	.0385	.0397	.0406	.0424	.0441	.0456	.0470	.0481
La	57	.0369	.0382	.0392	.0404	.0415	.0432	.0449	.0466	.0478	.0491
Hf	72	.0415	.0433	.0448	.0464	.0479	.0501	.0523	.0543	.0559	.0573
Ta	73	.0419	.0437	.0454	.0470	.0484	.0506	.0529	.0549	.0566	.0580
W	74	.0420	.0439	.0456	.0472	.0488	.0510	.0533	.0553	.0569	.0584
Os	76	.0425	.0443	.0461	.0478	.0494	.0516	.0540	.0560	.0577	.0592
Pt	78	.0432	.0451	.0470	.0487	.0504	.0526	.0550	.0571	.0589	.0603
Au	79	.0437	.0457	.0475	.0493	.0509	.0533	.0557	.0578	.0595	.0610
Hg	80	.0439	.0459	.0478	.0495	.0511	.0535	.0559	.0581	.0598	.0613
Pb	82	.0442	.0463	.0482	.0500	.0518	.0540	.0566	.0588	.0604	.0619
Bi	83	.0447	.0469	.0488	.0508	.0523	.0547	.0572	.0595	.0611	.0626
U	92	.0468	.0491	.0513	.0533	.0552	.0574	.0600	.0624	.0642	.0657

X-RAY WAVELENGTHS

J. A. Bearden

These tables were originally published as the final report to the U.S. Atomic Energy Commission as Report NYO-10586 in partial fulfillment of Contract AT(30-1)-2543. The tables were later reproduced in *Review of Modern Physics*. The data may also be obtained from the Superintendent of Documents, U.S. Government Printing Office, Washington, D. C. 20402 in the publication NSRDS-NBS 14. Persons seeking discussion of the experimental work, conventions, secondary standards, etc. will find these in *Review of Modern Physics,* Vol. 39, No. 1, 78-124, January 1967.

THE W $K\alpha_1$ WAVELENGTH STANDARD

A wavelength standard should possess characteristics which permit its ready redetermination in other laboratories by different techniques. Considering all of the factors involved in the selection of a wavelength standard, the W $K\alpha_1$ line is superior to any other x-ray or γ-ray wavelength. Its advantages as the x-ray wavelength standard are discussed in *Review of Modern Physics* Vol. 39, page 82 (1967).

$$\lambda \text{W } K\alpha_1 = 0.2090100 \text{ Å} \pm 5 \text{ ppm}.$$

This numerical value of the wavelength of the W $K\alpha_1$ line is used to define the *x-ray wavelength standard* by the relation

$$\lambda(\text{W } K\alpha_1) = 0.2090100 \text{ Å*}.$$

This is a new unit of length which may differ from the angstrom by ± 5 ppm (probable error), *but as a wavelength standard it has no error*. In order to clearly indicate that this unit is not exactly an angstrom, it has been designated Å*.

Wavelengths tabulated normally refer to the pure element in its solid form. However, there are many instances in which such data are not available. For example, rare gases are of necessity almost always used in the gaseous form, while the rare-earth elements were customarily used in the form of salts.

In high precision work there is some ambiguity as to exactly what feature of a line profile should be taken to be the "true wavelength." In double-crystal work the line peak is usually employed. In crystallography the centroid is widely used; in photographic work with visual observation of the plates, there is involved some subjective criterion of the observer which it is difficult to define precisely. In this survey the peak of the line profile has been adopted as the standard criterion.

X-RAY WAVELENGTHS

X-ray wavelengths in Å* units and in keV. The probable error (p.e.) is the error in the last digit of wavelength. Designation indicates both conventional Siegbahn notation (if applicable) and transition, e.g., $\beta_1\ L_{II}M_{IV}$ denotes a transition between the L_{II} and M_{IV} levels, which is the $L\beta_1$ line in Siegbahn notation.

Designation	Å*	p.e.	keV	Å*	p.e.	keV
	3 Lithium			**4 Beryllium**		
$\alpha\ KL$	228.	1	0.0543	114.	1	0.1085
	5 Boron			**6 Carbon**		
$\alpha\ KL$	67.6	3	0.1833	44.7	3	0.277
	7 Nitrogen			**8 Oxygen**		
$\alpha\ KL$	31.6	4	0.3924	23.62	3	0.5249
	9 Fluorine			**10 Neon**		
$\alpha_{1,2}\ KL_{II,III}$	18.32	2	0.6768	14.610	3	0.8486
$\beta\ KM$				14.452	5	0.8579
	11 Sodium			**12 Magnesium**		
$\alpha_{1,2}\ KL_{II,III}$	11.9101	9	1.0410	9.8900	2	1.25360
$\beta\ KM$	11.575	2	1.0711	9.521	2	1.3022
$L_{II,III}M$	407.1	5	0.03045	251.5	5	0.0493
$L_IL_{II,III}$	376	1	0.0330	317	1	0.0392
	13 Aluminum			**14 Silicon**		
$\alpha_2\ KL_{II}$	8.34173	9	1.48627	7.12791	9	1.73938
$\alpha_1\ KL_{III}$	8.33934	9	1.48670	7.12542	9	1.73998
$\beta\ KM$	7.960	2	1.5574	6.753	1	1.8359
$L_{II,III}$	171.4	5	0.0724	135.5	4	0.0915
$L_IL_{II,III}$	290.	1	0.0428			
	15 Phosphorus			**16 Sulfur**		
$\alpha_2\ KL_{II}$	6.160†	1	2.0127	5.37496	8	2.30664
$\alpha_1\ KL_{III}$	6.157†	1	2.0137	5.37216	7	2.30784
$\beta\ KM$	5.796	2	2.1390			
$\beta_1\ KM$				5.0316	2	2.4640
$\beta_x\ KM$				5.0233	3	2.4681
$L_{II,III}M$	103.8	4	0.1194			
$l, \eta\ L_{II,III}M_I$				83.4	3	0.1487
	17 Chlorine			**18 Argon**		
$\alpha_2\ KL_{II}$	4.7307	1	2.62078	4.19474	5	2.95563
$\alpha_1\ KL_{III}$	4.7278	1	2.62239	4.19180	5	2.95770
$\beta\ KM$	4.4034	3	2.8156			
$\beta_{1,3}\ KM_{II,III}$				3.8860	2	3.1905
$\eta\ L_{II}M_I$	67.33	9	0.1841	55.9†	1	0.2217
$l\ L_{III}M_I$	67.90	9	0.1826	56.3†	1	0.2201
	19 Potassium			**20 Calcium**		
$\alpha_2\ KL_{II}$	3.7445	2	3.3111	3.36166	3	3.68809
$\alpha_1\ KL_{III}$	3.7414	2	3.3138	3.35839	3	3.69168
$\beta_{1,3}\ KM_{II,III}$	3.4539	2	3.5896	3.0897	2	4.0127
$\beta_5\ KM_{IV,V}$	3.4413	4	3.6027	3.0746	3	4.0325

Designation	Å*	p.e.	keV	Å*	p.e.	keV
	19 Potassium (*Cont.*)			**20 Calcium** (*Cont.*)		
$\eta\ L_{II}M_I$	47.24	2	0.2625	40.46	2	0.3064
β_1				35.94	2	0.3449
$l\ L_{III}M_I$	47.74	1	0.25971	40.96	2	0.3027
$\alpha_{1,2}\ L_{III}M_{IV,V}$				36.33	2	0.3413
$M_{II,III}N_I$	692	9	0.0179	525.	9	0.0236
	21 Scandium			**22 Titanium**		
$\alpha_2\ KL_{II}$	3.0342	1	4.0861	2.75216	2	4.50486
$\alpha_1\ KL_{III}$	3.0309†	1	4.0906	2.74851	2	4.51084
$\beta_{1,3}\ KM_{II,III}$	2.7796	2	4.4605	2.51391	2	4.93181
$\beta_5\ KM_{IV,V}$	2.7634	3	4.4865	2.4985	2	4.9623
$\eta\ L_{II}M_I$	35.13	2	0.3529	30.89	3	0.4013
$\beta_1\ L_{II}M_{IV}$	31.02	2	0.3996	27.05	2	0.4584
$l\ L_{III}M_I$	35.59	3	0.3483	31.36	2	0.3953
$\alpha_{1,2}\ L_{III}M_{IV,V}$	31.35	3	0.3954	27.42	2	0.4522
	23 Vanadium			**24 Chromium**		
$\alpha_2\ KL_{II}$	2.50738	2	4.94464	2.293606	3	5.40551
$\alpha_1\ KL_{III}$	2.50356	2	4.95220	2.28970	2	5.41472
$\beta_{1,3}\ KM_{II,III}$	2.28440	2	5.42729	2.08487	2	5.94671
$\beta_5\ KM_{IV,V}$	2.26951	6	5.4629	2.07087	6	5.9869
$\beta_{3,4}\ L_IM_{II,III}$	21.19†	9	0.585	18.96	2	0.654
$\eta\ L_{II}M_I$	27.34	3	0.4535	24.30	3	0.5102
$\beta_1\ L_{II}M_{IV}$	23.88	4	0.5192	21.27	1	0.5828
$l\ L_{III}M_I$	27.77	1	0.4465	24.78	1	0.5003
$\alpha_{1,2}\ L_{III}M_{IV,V}$	24.25	3	0.5113	21.64	3	0.5728
$M_{II,III}M_{IV,V}$	337.	9	0.037	309.	9	0.040
	25 Manganese			**26 Iron**		
$\alpha_2\ KL_{II}$	2.10578	2	5.88765	1.939980	9	6.39084
$\alpha_1\ KL_{III}$	2.101820	9	5.89875	1.936042	9	6.40384
$\beta_{1,3}\ KM_{II,III}$	1.91021	2	6.49045	1.75661	2	7.05798
$\beta_5\ KM_{IV,V}$	1.8971	1	6.5352	1.7442	1	7.1081
$\beta_{3,4}\ L_IM_{II,III}$	17.19	2	0.721	15.65	2	0.792
$\eta\ L_{II}M_I$	21.85	2	0.5675	19.75	4	0.628
$\beta_1\ L_{II}M_{IV}$	19.11	2	0.6488	17.26	1	0.7185
$l\ L_{III}M_I$	22.29	1	0.5563	20.15	1	0.6152
$\alpha_{1,2}\ L_{III}M_{IV,V}$	19.45	1	0.6374	17.59	2	0.7050
$M_{II,III}M_{IV,V}$	273.	6	0.045	243.	5	0.051
	27 Cobalt			**28 Nickel**		
$\alpha_2\ KL_{II}$	1.792850	9	6.91530	1.661747	8	7.46089
$\alpha_1\ KL_{III}$	1.788965	9	6.93032	1.657910	8	7.47815
$\beta_{1,3}\ KM_{II,III}$	1.62079	2	7.64943	1.500135	8	8.26466
$\beta_5\ KM_{IV,V}$	1.60891	3	7.7059	1.48862	4	8.3286
$\beta_{3,4}\ L_IM_{II,III}$	14.31	3	0.870	13.18	1	0.941
$\eta\ L_{II}M_I$	17.87	3	0.694	16.27	3	0.762
$\beta_1\ L_{II}M_{IV}$	15.666	8	0.7914	14.271	6	0.8688
$l\ L_{III}M_I$	18.292	8	0.6778	16.693	9	0.7427
$\alpha_{1,2}\ L_{III}M_{IV,V}$	15.972	6	0.7762	14.561	3	0.8515
$M_{II,III}M_{IV,V}$	214.	6	0.058	190.	2	0.0651

X-Ray Wavelengths

TABLE A (*Continued*)

Designation	Å*	p.e.	keV	Å*	p.e.	keV
	29 Copper			**30 Zinc**		
$\alpha_2\ KL_{II}$	1.544390	2	8.02783	1.439000	8	8.61578
$\alpha_1\ KL_{III}$	1.540562	2	8.04778	1.435155	7	8.63886
$\beta_3\ KM_{II}$	1.3926	1	8.9029			
$\beta_{1,3}\ KM_{II,III}$	1.392218	9	8.90529	1.29525	2	9.5720
$\beta_2\ KN_{II,III}$				1.28372	2	9.6580
$\beta_5\ KM_{IV,V}$	1.38109	3	8.9770	1.2848	1	9.6501
$\beta_{3,4}\ L_IM_{II,III}$	12.122	8	1.0228	11.200	7	1.1070
$\eta\ L_{II}M_I$	14.90	2	0.832	13.68	2	0.906
$\beta_1\ L_{II}M_{IV}$	13.053	3	0.9498	11.983	3	1.0347
$l\ L_{III}M_I$	15.286	9	0.8111	14.02	2	0.884
$\alpha_{1,2}\ L_{III}M_{IV,V}$	13.336	3	0.9297	12.254	3	1.0117
$M_{II,III}M_{IV,V}$	173.	3	0.072	157.	3	0.079
	31 Gallium			**32 Germanium**		
$\alpha_2\ KL_{II}$	1.34399	1	9.22482	1.258011	9	9.85532
$\alpha_1\ KL_{III}$	1.340083	9	9.25174	1.254054	9	9.88642
$\beta_3\ KM_{II}$	1.20835	5	10.2603	1.12936	9	10.9780
$\beta_1\ KM_{III}$	1.20789	2	10.2642	1.12894	2	10.9821
$\beta_2\ KN_{II,III}$	1.19600	2	10.3663	1.11686	2	11.1008
$\beta_5\ KM_{IV,V}$	1.1981	2	10.348	1.1195	1	11.0745
$\beta_4\ L_IM_{II}$				9.640	2	1.2861
$\beta_3\ L_IM_{III}$				9.581	2	1.2941
$\beta_{3,4}\ L_IM_{II,III}$	10.359†	8	1.197			
$\eta\ L_{II}M_I$	12.597	2	0.9842	11.609	2	1.0680
$\beta_1\ L_{II}M_{IV}$	11.023	2	1.1248	10.175	1	1.2185
$l\ L_{III}M_I$	12.953	2	0.9572	11.965	4	1.0362
$\alpha_{1,2}\ L_{III}M_{IV,V}$	11.292	1	1.09792	10.4361	8	1.18800
	33 Arsenic			**34 Selenium**		
$\alpha_2\ KL_{II}$	1.17987	1	10.50799	1.10882	2	11.1814
$\alpha_1\ KL_{III}$	1.17588	1	10.54372	1.10477	2	11.2224
$\beta_3\ KM_{II}$	1.05783	5	11.7203	0.99268	5	12.4896
$\beta_1\ KM_{III}$	1.05730	2	11.7262	0.99218	3	12.4959
$\beta_2\ KN_{II,III}$	1.04500	3	11.8642	0.97992	5	12.6522
$\beta_5\ KM_{IV,V}$	1.0488	1	11.822	0.9843	1	12.595
$\beta_{3,4}\ L_IM_{II,III}$	8.929	1	1.3884	8.321†	9	1.490
$\eta\ L_{II}M_I$	10.734	1	1.1550	9.962	1	1.2446
$\beta_1\ L_{II}M_{IV}$	9.4141	8	1.3170	8.7358	5	1.41923
$l\ L_{III}M_I$	11.072	1	1.1198	10.294	1	1.2044
$\alpha_{1,2}\ L_{III}M_{IV,V}$	9.6709	8	1.2820	8.9900	5	1.37910
M_VN_{III}				230.	2	0.0538
	35 Bromine			**36 Krypton**		
$\alpha_2\ KL_{II}$	1.04382	2	11.8776	0.9841	1	12.598
$\alpha_1\ KL_{III}$	1.03974	2	11.9242	0.9801	1	12.649
$\beta_3\ KM_{II}$	0.93327	5	13.2845	0.8790	1	14.104
$\beta_1\ KM_{III}$	0.93279	2	13.2914	0.8785	1	14.112
$\beta_2\ KN_{II,III}$	0.92046	2	13.4695	0.8661	1	14.315
$\beta_5\ KM_{IV,V}$	0.9255	1	13.396	0.8708	2	14.238
$\beta_4\ KN_{IV,V}$				0.8653	2	14.328
$\beta_4\ L_IM_{II}$				7.304	5	1.697
$\beta_3\ L_IM_{III}$				7.264	5	1.707

Designation	Å*	p.e.	keV	Å*	p.e.	keV
	35 Bromine (*Cont.*)			**36 Krypton** (*Cont.*)		
$\beta_{3,4}\ L_IM_{II,III}$	7.767†	9	1.596			
$\eta\ L_{II}M_I$	9.255	1	1.3396			
$\beta_1\ L_{II}M_{IV}$	8.1251	5	1.52590	7.576†	3	1.6366
γ_5				7.279	5	1.703
$l\ L_{III}M_I$	9.585	1	1.2935			
$\alpha_{1,2}\ L_{III}M_{IV,V}$	8.3746	5	1.48043	7.817†	3	1.5860
β_6				7.510	4	1.6510
$L_{II}N_{III}$				7.250	5	1.710
M_IM_{II}	184.6	3	0.0672			
M_IM_{III}	164.7	3	0.0753			
$M_{II}M_{IV}$	109.4	3	0.1133			
$M_{II}N_I$	76.9	2	0.1613			
$M_{III}M_{IV,V}$	113.8	3	0.1089			
	79.8	3	0.1554			
$\zeta_2\ M_{IV}N_{II}$	191.1	2	0.06488			
$M_{IV}N_{III}$	189.5	3	0.0654			
$\zeta_1\ M_VN_{III}$	192.6	2	0.06437			
	37 Rubidium			**38 Strontium**		
$\alpha_2\ KL_{II}$	0.92969	1	13.3358	0.87943	1	14.0979
$\alpha_1\ KL_{III}$	0.925553	9	13.3953	0.87526	1	14.1650
$\beta_3\ KM_{II}$	0.82921	3	14.9517	0.78345	3	15.8249
$\beta_1\ KM_{III}$	0.82868	2	14.9613	0.78292	2	15.8357
$\beta_2\ KN_{II,III}$	0.81645	3	15.1854	0.77081	3	16.0846
$\beta_5\ KM_{IV,V}$	0.8219	1	15.085	0.7764	1	15.969
$\beta_4\ KN_{IV,V}$	0.8154	2	15.205	0.76989	5	16.104
$\beta_4\ L_IM_{II}$	6.8207	3	1.81771	6.4026	3	1.93643
$\beta_3\ L_IM_{III}$	6.7876	3	1.82659	6.3672	3	1.94719
$\gamma_{2,3}\ L_IN_{II,III}$	6.0458	3	2.0507	5.6445	3	2.1965
$\eta\ L_{II}M_I$	8.0415	4	1.54177	7.5171	3	1.64933
$\beta_1\ L_{II}M_{IV}$	7.0759	3	1.75217	6.6239	3	1.87172
$\gamma_5\ L_{II}N_{IV}$	6.7553	3	1.83532	6.2961	3	1.96916
$l\ L_{III}M_I$	8.3636	4	1.48238	7.8362	3	1.58215
$\alpha_2\ L_{III}M_{IV}$	7.3251	3	1.69256	6.8697	3	1.80474
$\alpha_1\ L_{III}M_V$	7.3183	2	1.69413	6.8628	2	1.80656
$\beta_6\ L_{III}N_I$	6.9842	3	1.77517	6.5191	3	1.90181
M_IM_{III}	144.4	3	0.0859			
$M_{II}M_{IV}$	91.5	2	0.1355	85.7	2	0.1447
$M_{II}N_I$	57.0	2	0.2174	51.3	1	0.2416
$M_{III}M_{IV,V}$	96.7	2	0.1282	91.4	2	0.1357
$M_{III}N_I$	59.5	2	0.2083	53.6	1	0.2313
$\zeta_2\ M_{IV}N_{II}$	127.8	2	0.0970			
$M_{IV}N_{III}$	126.8	2	0.0978			
$\zeta_2\ M_{IV}N_{II,III}$				108.0	2	0.1148
$\zeta_1\ M_VN_{III}$	128.7	2	0.0964	108.7	1	0.1140
	39 Yttrium			**40 Zirconium**		
$\alpha_2\ KL_{II}$	0.83305	1	14.8829	0.79015	1	15.6909
$\alpha_1\ KL_{III}$	0.82884	1	14.9584	0.78593	1	15.7751
$\beta_3\ KM_{II}$	0.74126	3	16.7258	0.70228	4	17.654
$\beta_1\ KM_{III}$	0.74072	2	16.7378	0.70173	3	17.6678
$\beta_2\ KN_{II,III}$	0.72864	4	17.0154	0.68993	4	17.970
$\beta_5\ KM_{IV,V}$	0.7345	1	16.879	0.6959	1	17.815

Designation	Å*	p.e.	keV	Å*	p.e.	keV
	39 Yttrium (*Cont.*)			**40 Zirconium** (*Cont.*)		
$\beta_4\ KN_{IV,V}$	0.72776	5	17.036	0.68901	5	17.994
$\beta_4\ L_IM_{II}$	6.0186	3	2.0600	5.6681	3	2.1873
$\beta_3\ L_IM_{III}$	5.9832	3	2.0722	5.6330	3	2.2010
$\gamma_{2,3}\ L_IN_{II,III}$	5.2830	3	2.3468	4.9536	3	2.5029
$\eta\ L_IM_I$	7.0406	3	1.76095	6.6069	3	1.87654
$\beta_1\ L_{III}M_{IV}$	6.2120	3	1.99584	5.8360	3	2.1244
$\gamma_5\ L_{II}N_I$	5.8754	3	2.1102	5.4977	3	2.2551
$\gamma_1\ L_{II}N_{IV}$				5.3843	3	2.3027
$l\ L_{III}M_I$	7.3563	3	1.68536	6.9185	3	1.79201
$\alpha_2\ L_{III}M_{IV}$	6.4558	3	1.92047	6.0778	3	2.0399
$\alpha_1\ L_{III}M_V$	6.4488	2	1.92256	6.0705	2	2.04236
$\beta_6\ L_{III}N_I$	6.0942	3	2.0344	5.7101	3	2.1712
$\beta_{2,15}$				5.5863	3	2.2194
$M_{II}M_{IV}$	81.5	2	0.1522	76.7	2	0.1617
$M_{II}N_I$	46.48	9	0.267			
$M_{III}M_V$				80.9	3	0.1533
$M_{III}N_I$	48.5	2	0.256			
$M_{III}M_{IV,V}$	86.5	2	0.1434			
$\zeta\ M_{IV,V}N_{II,III}$	93.4	2	0.1328	82.1	2	0.1511
$M_{IV,V}O_{II,III}$				70.0	4	0.177
	41 Niobium			**42 Molybdenum**		
$\alpha_2\ KL_{II}$	0.75044	1	16.5210	0.713590	6	17.3743
$\alpha_1\ KL_{III}$	0.74620	1	16.6151	0.709300	1	17.47934
$\beta_3\ KM_{II}$	0.66634	3	18.6063	0.632872	9	19.5903
$\beta_1\ KM_{III}$	0.66576	2	18.6225	0.632288	9	19.6083
β_2^{II}				0.62107	5	19.963
$\beta_2\ KN_{II,III}$	0.65416	4	18.953	0.62099	2	19.9652
$\beta_4\ KN_{IV,V}$	0.65318	5	18.981			
$\beta_5^{II}\ KM_{IV}$				0.62708	5	19.771
$\beta_5^{I}\ KM_V$				0.62692	5	19.776
$\beta_4\ KN_{IV,V}$				0.62001	9	19.996
$\beta_4\ L_IM_{II}$	5.3455	3	2.3194	5.0488	3	2.4557
$\beta_3\ L_IM_{III}$	5.3102	3	2.3348	5.0133	3	2.4730
$\gamma_{2,3}\ L_IN_{II,III}$	4.6542	2	2.6638	4.3800	2	2.8306
$\eta\ L_{II}M_I$	6.2109	3	1.99620	5.8475	3	2.1202
$\beta_1\ L_{II}M_{IV}$	5.4923	3	2.2574	5.17708	8	2.39481
$\gamma_5\ L_{II}N_I$	5.1517	3	2.4066	4.8369	2	2.5632
$\gamma_1\ L_{II}N_{IV}$	5.0361	3	2.4618	4.7258	2	2.6235
$l\ L_{III}M_I$	6.5176	3	1.90225	6.1508	3	2.01568
$\alpha_2\ L_{III}M_{IV}$	5.7319	3	2.1630	5.41437	8	2.28985
$\alpha_1\ L_{III}M_V$	5.7243	2	2.16589	5.40655	8	2.29316
$\beta_6\ L_{III}N_I$	5.3613	3	2.3125	5.0488	5	2.4557
$\beta_{2,15}\ L_{III}N_{IV,V}$	5.2379	3	2.3670	4.9232	2	2.5183
$M_{II}M_{IV}$	72.1	3	0.1718	68.9	2	0.1798
$M_{II}N_I$	38.4	3	0.323	35.3	3	0.351
$M_{II}N_{IV}$	33.1	2	0.375			
$M_{III}M_V$	78.4	2	0.1582	74.9	1	0.1656
$M_{III}N_I$	40.7	2	0.305	37.5	2	0.331
$\gamma\ M_{III}N_{IV,V}$	34.9	2	0.356			
$\zeta\ M_{IV,V}N_{II,III}$	72.19	9	0.1717	64.38	7	0.1926
$M_{IV,V}O_{II,III}$	61.9	2	0.2002	54.8	2	0.2262

Designation	Å*	p.e.	keV	Å*	p.e.	keV
	43 Technetium			**44 Ruthenium**		
$\alpha_2\ KL_{II}$	0.67932†	3	18.2508	0.647408	5	19.1504
$\alpha_1\ KL_{III}$	0.67502†	3	18.3671	0.643083	4	19.2792
$\beta_3\ KM_{II}$	0.60188†	4	20.599	0.573067	4	21.6346
$\beta_1\ KM_{III}$	0.60130†	4	20.619	0.572482	4	21.6568
$\beta_2\ KN_{II,III}$	0.59024†	5	21.005	0.56166	3	22.074
$\beta_5^{II}\ KM_{IV}$				0.5680	2	21.829
$\beta_5^{I}\ KM_V$				0.56785	9	21.834
β_4				0.56089	9	22.104
$\beta_4\ L_{II}M_{IV}$				4.5230	2	2.7411
$\beta_3\ L_IM_{III}$				4.4866	3	2.7634
$\gamma_{2,3}\ L_IN_{II,III}$				3.8977	2	3.1809
$\eta\ L_{II}M_I$				5.2050	2	2.38197
$\beta_1\ L_{II}M_{IV}$	4.8873†	8	2.5368	4.62058	3	2.68323
$\gamma_5\ L_{II}N_I$				4.2873	2	2.8918
$\gamma_1\ L_{II}N_{IV}$				4.1822	2	2.9645
$l\ L_{III}M_I$				5.5035	3	2.2528
$\alpha_2\ L_{III}M_{IV}$				4.85381	7	2.55431
$\alpha_1\ L_{III}M_V$	5.1148†	3	2.4240	4.84575	5	2.55855
$\beta_6\ L_{III}N_I$				4.4866	3	2.7634
$\beta_{2,15}\ L_{III}N_{IV,V}$				4.3718	2	2.8360
$M_{II}M_{IV}$				62.2	1	0.1992
$M_{II}N_I$				32.3	2	0.384
$M_{II}N_{IV}$				25.50	9	0.486
$M_{III}M_V$				68.3	1	0.1814
$\gamma\ M_{III}N_{IV,V}$				26.9	1	0.462
$\zeta\ M_{IV,V}N_{II,III}$				52.34	7	0.2369
$M_{IV,V}O_{II,III}$				44.8	1	0.2768
	45 Rhodium			**46 Palladium**		
$\alpha_2\ KL_{II}$	0.617630	4	20.0737	0.589821	3	21.0201
$\alpha_1\ KL_{III}$	0.613279	4	20.2161	0.585448	3	21.1771
$\beta_3\ KM_{II}$	0.546200	4	22.6989	0.521123	4	23.7911
$\beta_1\ KM_{III}$	0.545605	4	22.7236	0.520520	4	23.8187
$\beta_2^{II}\ KN_{II}$	0.53513	5	23.168			
$\beta_2\ KN_{II,III}$	0.53503	2	23.1728	0.510228	4	24.2991
$\beta_5^{II}\ KM_{IV}$	0.54118	9	22.909			
$\beta_5^{I}\ KM_V$	0.54101	9	22.917			
$\beta_4\ KN_{IV,V}$	0.53401	9	23.217	0.5093	2	24.346
$\beta_5\ KM_{IV,V}$				0.51670	9	23.995
$\beta_4\ L_IM_{II}$	4.2888	2	2.8908	4.0711	2	3.0454
$\beta_3\ L_IM_{III}$	4.2522	2	2.9157	4.0346	2	3.0730
$\gamma_{2,3}\ L_IN_{II,III}$	3.6855	2	3.3640	3.4892	2	3.5533
$\eta\ L_{II}M_I$	4.9217	2	2.5191	4.6605	2	2.6603
$\beta_1\ L_{II}M_{IV}$	4.37414	4	2.83441	4.14622	5	2.99022
$\gamma_5\ L_{II}N_I$	4.0451	2	3.0650	3.8222	2	3.2437
$\gamma_1\ L_{II}N_{IV}$	3.9437	2	3.1438	3.7246	2	3.3287
$l\ L_{III}M_I$	5.2169	3	2.3765	4.9525	3	2.5034
$\alpha_2\ L_{III}M_{IV}$	4.60545	9	2.69205	4.37588	7	2.83329
$\alpha_1\ L_{III}M_V$	4.59743	9	2.69674	4.36767	5	2.83861
$\beta_6\ L_{III}N_I$	4.2417	2	2.9229	4.0162	2	3.0870
$\beta_{2,15}\ L_{III}N_{IV,V}$	4.1310	2	3.0013	3.90887	4	3.17179
$\beta_{10}\ L_IM_{IV}$				3.7988	2	3.2637

Designation	Å*	p.e.	keV	Å*	p.e.	keV
	45 Rhodium (*Cont.*)			**46 Palladium** (*Cont.*)		
$\beta_9\ L_IM_V$				3.7920	2	3.2696
$M_IN_{II,III}$				20.1	2	0.616
$M_{II}M_{IV}$	59.3	1	0.2090	56.5	1	0.2194
$M_{II}N_I$	28.1	2	0.442	26.2	2	0.474
$M_{II}N_{IV}$				22.1	1	0.560
$M_{II}\ M_V$	65.5	1	0.1892	62.9	1	0.1970
$M_{II}^{I}N_I$	29.8	1	0.417	27.9	1	0.445
$\gamma\ M_{III}N_{IV,V}$	25.01	9	0.496	23.3†	1	0.531
$\zeta\ M_{IV,V}N_{II,III}$	47.67	9	0.2601	43.6	1	0.2844
$M_{IV,V}O_{II,III}$	40.9	2	0.303	37.4	2	0.332
	47 Silver			**48 Cadmium**		
$\alpha_2\ KL_{II}$	0.563798	4	21.9903	0.539422	3	22.9841
$\alpha_1\ KL_{III}$	0.5594075	6	22.16292	0.535010	3	23.1736
$\beta_3\ KM_{II}$	0.497685	4	24.9115	0.475730	5	26.0612
$\beta_1\ KM_{III}$	0.497069	4	24.9424	0.475105	6	26.0955
$\beta_2\ KN_{II,III}$	0.487032	4	25.4564	0.465328	7	26.6438
$\beta_5\ KM_{IV,V}$	0.49306	2	25.145			
$\beta_4\ KN_{IV,V}$	0.48598	3	25.512			
$\beta_4\ L_IM_{II}$	3.87023	5	3.20346	3.68203	9	3.36719
$\beta_3\ L_IM_{III}$	3.83313	9	3.23446	3.64495	9	3.40145
$\gamma_2\ L_IN_{II}$	3.31216	9	3.7432	3.1377	2	3.9513
$\gamma_3\ L_IN_{III}$	3.30635	9	3.7498			
$\eta\ L_{II}M_I$	4.4183	2	2.8061	4.19315	9	2.95675
$\beta_1\ L_{II}M_{IV}$	3.93473	3	3.15094	3.73823	4	3.31657
$\gamma_5\ L_{II}N_I$	3.61638	9	3.42832	3.42551	9	3.61935
$\gamma_1\ L_{III}N_{IV}$	3.52260	4	3.51959	3.33564	6	3.71686
$l\ L_{III}M_I$	4.7076	2	2.6337	4.48014	9	2.76735
$\alpha_2\ L_{III}M_{IV}$	4.16294	5	2.97821	3.96496	6	3.12691
$\alpha_1\ L_{III}M_V$	4.15443	3	2.98431	3.95635	4	3.13373
$\beta_6\ L_{III}N_I$	3.80774	9	3.25603	3.61467	9	3.42994
$\beta_{2,15}\ L_{III}N_{IV,V}$	3.70335	3	3.34781	3.51408	4	3.52812
$\beta_{10}\ L_IM_{IV}$	3.61158	9	3.43287	3.4367	2	3.6075
$\beta_9\ L_IM_V$	3.60497	9	3.43917	3.43015	9	3.61445
$M_IN_{II,III}$	18.8	2	0.658			
$M_{II}M_{IV}$	54.0	1	0.2295	52.0	2	0.2384
$M_{II}N_I$				22.9	2	0.540
$M_{II}N_{IV}$	20.66	7	0.600	19.40	7	0.639
$M_{III}M_V$	60.5	1	0.2048	58.7	2	0.2111
$M_{III}N_I$	26.0	1	0.478	24.5	1	0.507
$\gamma\ M_{III}N_{IV,V}$	21.82	7	0.568	20.47	7	0.606
$M_{IV}O_{II,III}$				30.4	1	0.408
$\zeta\ M_{IV,V}N_{II,III}$	39.77	7	0.3117	36.8	1	0.3371
M_VN_I	24.4	2	0.509			
M_VO_{III}				30.8	1	0.403
$M_{IV,V}O_{II,III}$	33.5	3	0.370			
	49 Indium			**50 Tin**		
$\alpha_2\ KL_{II}$	0.516544	3	24.0020	0.495053	3	25.0440
$\alpha_1\ KL_{III}$	0.512113	3	24.2097	0.490599	3	25.2713
$\beta_3\ KM_{II}$	0.455181	4	27.2377	0.435877	5	28.4440

Designation	Å*	p.e.	keV	Å*	p.e.	keV
	49 Indium (*Cont.*)			**50 Tin** (*Cont.*)		
$\beta_1\ KM_{III}$	0.454545	4	27.2759	0.435236	5	28.4860
$\beta_2\ KN_{II,III}$	0.44500	1	27.8608	0.425915	8	29.1093
$KO_{II,III}$	0.44374	3	27.940	0.42467	3	29.195
$\beta_5^{II}\ KM_{IV}$	0.45098	2	27.491	0.43184	3	28.710
$\beta_5^{I}\ KM_V$	0.45086	2	27.499	0.43175	3	28.716
$\beta_4\ KN_{IV,V}$	0.44393	4	27.928	0.42495	3	29.175
$\beta_4\ L_IM_{II}$	3.50697	9	3.5353	3.34335	9	3.7083
$\beta_3\ L_IM_{III}$	3.46984	9	3.5731	3.30585	3	3.7500
$\gamma_{2,3}\ L_IN_{II,III}$	2.9800	2	4.1605	2.8327	2	4.3768
$\gamma_4\ L_IO_{II,III}$	2.9264	2	4.2367	2.7775	2	4.4638
$\eta\ L_{II}M_I$	3.98327	9	3.11254	3.78876	9	3.27234
$\beta_1\ L_{II}M_{IV}$	3.55531	4	3.48721	3.38487	3	3.66280
$\gamma_5\ L_{II}N_I$	3.24907	9	3.8159	3.08475	9	4.0192
$\gamma_1\ L_{II}N_{IV}$	3.16213	4	3.92081	3.00115	3	4.13112
$l\ L_{III}M_I$	4.26873	9	2.90440	4.07165	9	3.04499
$\alpha_2\ L_{III}M_{IV}$	3.78073	6	3.27929	3.60891	4	3.43542
$\alpha_1\ L_{III}M_V$	3.77192	4	3.28694	3.59994	3	3.44398
$\beta_6\ L_{III}N_I$	3.43606	9	3.60823	3.26901	9	3.7926
$\beta_{2,15}\ L_{III}N_{IV,V}$	3.33838	3	3.71381	3.17505	3	3.90486
$\beta_7\ L_{III}O_I$	3.324	4	3.730	3.1564	3	3.9279
$\beta_{10}\ L_IM_{IV}$	3.27404	9	3.7868	3.12170	9	3.9716
$\beta_9\ L_IM_V$	3.26763	9	3.7942	3.11513	9	3.9800
$M_{II}M_{IV}$				47.3	1	0.2621
$M_{II}N_I$				20.0	1	0.619
$M_{II}N_{IV}$				16.93	5	0.733
$M_{III}M_V$				54.2	1	0.2287
$M_{III}N_I$				21.5	1	0.575
$\gamma\ M_{III}N_{IV,V}$				17.94	5	0.691
$M_{IV}O_{II,III}$				25.3	1	0.491
$\zeta\ M_{IV,V}N_{II,III}$				31.24	9	0.397
M_VO_{III}				25.7	1	0.483
	51 Antimony			**52 Tellurium**		
$\alpha_2\ KL_{II}$	0.474827	3	26.1108	0.455784	3	27.2017
$\alpha_1\ KL_{III}$	0.470354	3	26.3591	0.451295	3	27.4723
$\beta_3\ KM_{II}$	0.417737	4	29.6792	0.400659	4	30.9443
$\beta_1\ KM_{III}$	0.417085	3	29.7256	0.399995	5	30.9957
$\beta_2\ KN_{II,III}$	0.407973	5	30.3895	0.391102	6	31.7004
$KO_{II,III}$	0.40666	1	30.4875	0.38974	1	31.8114
$\beta_5^{II}\ KM_{IV}$	0.41388	1	29.9560			
$\beta_5^{I}\ KM_V$	0.41378	1	29.9632			
$\beta_4\ KN_{IV,V}$	0.40702	1	30.4604			
$\beta_4\ L_IM_{II}$	3.19014	9	3.8864	3.04661	9	4.0695
$\beta_3\ L_IM_{III}$	3.15258	9	3.9327	3.00893	9	4.1204
$\gamma_{2,3}\ L_IN_{II,III}$	2.6953	2	4.5999	2.5674	2	4.8290
$\gamma_4\ L_IO_{II,III}$	2.6398	2	4.6967	2.5113	2	4.9369
$\eta\ L_{II}M_I$	3.60765	9	3.43661	3.43832	9	3.60586
$\beta_1\ L_{II}M_{IV}$	3.22567	4	3.84357	3.07677	6	4.02958
$\gamma_5\ L_{II}N_I$	2.93187	9	4.2287	2.79007	9	4.4437
$\gamma_1\ L_{II}N_{IV}$	2.85159	3	4.34779	2.71241	6	4.5709
$l\ L_{III}M_I$	3.88826	9	3.18860	3.71696	9	3.33555
$\alpha_2\ L_{III}M_{IV}$	3.44840	6	3.59532	3.29846	9	3.7588

Designation	Å*	p.e.	keV	Å*	p.e.	keV
	51 Antimony (*Cont.*)			**52 Tellurium** (*Cont.*)		
α_1 $L_{III}M_V$	3.43941	4	3.60472	3.28920	6	3.76933
β_6 $L_{III}N_I$	3.11513	9	3.9800	2.97088	9	4.1732
$\beta_{2,15}$ $L_{III}N_{IV,V}$	3.02335	3	4.10078	2.88217	8	4.3017
β_7 $L_{III}O_I$	3.0052	3	4.1255	2.8634	3	4.3298
β_{10} L_IM_{IV}	2.97917	9	4.1616	2.84679	9	4.3551
β_9 L_IM_V	2.97261	9	4.1708	2.83897	9	4.3671
$M_{II}M_{IV}$	45.2	1	0.2743			
$M_{II}N_I$	18.8	1	0.658	17.6	1	0.703
$M_{II}N_{IV}$	15.98	5	0.776			
$M_{III}M_V$	52.2	1	0.2375	50.3	1	0.2465
$M_{III}N_I$	20.2	1	0.612	19.1	1	0.648
γ $M_{III}N_{IV,V}$	16.92	4	0.733	15.93	4	0.778
$M_{IV}O_{II,III}$				21.34	5	0.581
ζ $M_{IV,V}N_{II,III}$	28.88	8	0.429	26.72	9	0.464
M_VO_{III}				21.78	5	0.569
	53 Iodine			**54 Xenon**		
α_2 KL_{II}	0.437829	7	28.3172	0.42087†	2	29.458
α_1 KL_{III}	0.433318	5	28.6120	0.41634†	2	29.779
β_3 KM_{II}	0.384564	4	32.2394	0.36941†	2	33.562
β_1 KM_{III}	0.383905	4	32.2947	0.36872†	2	33.624
β_2 $KN_{II,III}$	0.37523†	2	33.042	0.36026†	3	34.415
β_4 L_IM_{II}	2.91207	9	4.2575			
β_3 L_IM_{III}	2.87429	9	4.3134			
$\gamma_{2,3}$ $L_IN_{II,III}$	2.4475	2	5.0657			
γ_4 $L_IO_{II,III}$	2.3913	2	5.1848			
η $L_{II}M_I$	3.27979	9	3.7801			
β_1 $L_{II}M_{IV}$	2.93744	6	4.22072			
γ_5 $L_{II}N_I$	2.65710	9	4.6660			
γ_1 $L_{II}N_{IV}$	2.58244	8	4.8009			
l $L_{III}M_I$	3.55754	9	3.48502			
α_2 $L_{III}M_{IV}$	3.15791	6	3.92604			
α_1 $L_{III}M_V$	3.14860	6	3.93765	3.0166†	2	4.1099
β_6 $L_{III}N_I$	2.83672	9	4.3706			
$\beta_{2,15}$ $L_{III}N_{IV,V}$	2.75053	8	4.5075			
β_7 $L_{III}O_I$	2.7288	3	4.5435			
β_{10} L_IM_{IV}	2.72104	9	4.5564			
β_9 L_IM_V	2.71352	9	4.5690			
	55 Cesium			**56 Barium**		
α_2 KL_{II}	0.404835	4	30.6251	0.389668	5	31.8171
α_1 KL_{III}	0.400290	4	30.9728	0.385111	4	32.1936
β_3 KM_{II}	0.355050	4	34.9194	0.341507	4	36.3040
β_1 KM_{III}	0.354364	7	34.9869	0.340811	3	36.3782
β_2 $KN_{II,III}$	0.34611	2	35.822	0.33277	1	37.257
$KO_{II,III}$				0.33127	2	37.426
β_5^{II} KM_{IV}				0.33835	2	36.643
β_5^{I} KM_V				0.33814	2	36.666
β_4 $KN_{IV,V}$				0.33229	2	37.311
β_4 L_IM_{II}	2.6666	2	4.6494	2.5553	2	4.8519
β_3 L_IM_{III}	2.6285	2	4.7167	2.5164	2	4.9269
γ_2 L_IN_{II}	2.2371	2	5.5420	2.1387	2	5.7969
γ_3 L_IN_{III}	2.2328	2	5.5527	2.1342	2	5.8092

Designation	Å*	p.e.	keV	Å*	p.e.	keV
	55 Cesium (*Cont.*)			**56 Tellurium** (*Cont.*)		
γ_4 $L_IO_{II,III}$	2.1741	2	5.7026	2.0756	3	5.9733
η $L_{II}M_I$	2.9932	2	4.1421	2.8627	3	4.3309
β_1 $L_{II}M_{IV}$	2.6837	2	4.6198	2.56821	5	4.82753
γ_5 $L_{II}N_I$	2.4174	2	5.1287	2.3085	3	5.3707
γ_1 $L_{II}N_{IV}$	2.3480	2	5.2804	2.2415	2	5.5311
l $L_{III}M_I$	3.2670	2	3.7950	3.1355	2	3.9541
α_2 $L_{III}M_{IV}$	2.9020	2	4.2722	2.78553	5	4.45090
α_1 $L_{III}M_V$	2.8924	2	4.2865	2.77595	5	4.46626
β_6 $L_{III}N_I$	2.5932	2	4.7811	2.4826	2	4.9939
$\beta_{2,15}$ $L_{III}N_{IV,V}$	2.5118	2	4.9359	2.40435	6	5.1565
β_7 $L_{III}O_I$	2.4849	2	4.9893	2.3806	2	5.2079
β_{10} L_IM_{IV}	2.4920	2	4.9752	2.3869	2	5.1941
β_9 L_IM_V	2.4783	2	5.0026	2.3764	2	5.2171
γ $M_{III}N_{IV,V}$				12.75	3	0.973
$M_{IV}O_{II}$				15.91	5	0.779
$M_{IV}O_{III}$				15.72	9	0.789
ζ M_VN_{III}				20.64	4	0.601
M_VO_{III}				16.20	5	0.765
$N_{IV}O_{II}$	188.6	1	0.06574	163.3	2	0.07590
$N_{IV}O_{III}$	183.8	1	0.06746	159.0	2	0.07796
N_VO_{III}	190.3	1	0.06515	164.6	2	0.07530
	57 Lanthanum			**58 Cerium**		
α_2 KL_{II}	0.375313	2	33.0341	0.361683	2	34.2789
α_1 KL_{III}	0.370737	2	33.4418	0.357092	2	34.7197
β_3 KM_{II}	0.328686	4	37.7202	0.316520	4	39.1701
β_1 KM_{III}	0.327983	3	37.8010	0.315816	2	39.2573
β_2 $KN_{II,III}$	0.320117	7	38.7299	0.30816	1	40.233
$KO_{II,III}$	0.31864	2	38.909	0.30668	2	40.427
β_5^{II} KM_{IV}	0.32563	2	38.074	0.31357	2	39.539
β_5^{I} KM_V	0.32546	2	38.094	0.31342	2	39.558
β_4 $KN_{IV,V}$	0.31931	2	38.828	0.30737	2	40.337
β_4 L_IM_{II}	2.4493	3	5.0620	2.3497	4	5.2765
β_3 L_IM_{III}	2.4105	3	5.1434	2.3109	3	5.3651
γ_2 L_IN_{II}	2.0460	4	6.060	1.9602	3	6.3250
γ_3 L_IN_{III}	2.0410	4	6.074	1.9553	3	6.3409
γ_4 $L_IO_{II,III}$	1.9830	4	6.252	1.8991	4	6.528
η $L_{II}M_I$	2.740	3	4.525	2.6203	4	4.7315
β_1 $L_{II}M_{IV}$	2.45891	5	5.0421	2.3561	3	5.2622
γ_5 $L_{II}N_I$	2.2056	4	5.621	2.1103	3	5.8751
γ_1 $L_{II}N_{IV}$	2.1418	3	5.7885	2.0487	4	6.052
γ_8 $L_{II}O_I$				2.0237	4	6.126
l $L_{III}M_I$	3.006	3	4.124	2.8917	4	4.2875
α_2 $L_{III}M_{IV}$	2.67533	5	4.63423	2.5706	3	4.8230
α_1 $L_{III}M_V$	2.66570	5	4.65097	2.5615	2	4.8402
β_6 $L_{III}N_I$	2.3790	4	5.2114	2.2818	3	5.4334
$\beta_{2,15}$ $L_{III}N_{IV,V}$	2.3030	3	5.3835	2.2087	2	5.6134
β_7 $L_{III}O_I$	2.275	3	5.450	2.1701	2	5.7132
β_{10} L_IM_{IV}	2.290	3	5.415	2.1958	5	5.646
β_9 L_IM_V	2.282	3	5.434	2.1885	3	5.6650
γ $M_{III}N_{IV,V}$	12.08	4	1.027	11.53	1	1.0749
β $M_{IV}N_{VI}$	14.51	5	0.854	13.75	4	0.902
ζ M_VN_{III}	19.44	5	0.638	18.35	4	0.676
α $M_VN_{VI,VII}$	14.88	5	0.833	14.04	2	0.883

Designation	Å*	p.e.	keV	Å*	p.e.	keV
	57 Lanthanum (*Cont.*)			**58 Cerium** (*Cont.*)		
$M_VO_{II,III}$				14.39	5	0.862
$N_{IV,V}O_{II,III}$	152.6	6	0.0812	144.4	6	0.0859
	59 Praseodymium			**60 Neodymium**		
$\alpha_2\ KL_{II}$	0.348749	2	35.5502	0.336472	2	36.8474
$\alpha_1\ KL_{III}$	0.344140	2	36.0263	0.331846	2	37.3610
$\beta_3\ KM_{II}$	0.304975	5	40.6529	0.294027	3	42.1665
$\beta_1\ KM_{III}$	0.304261	4	40.7482	0.293299	2	42.2713
$\beta_2\ KN_{II,III}$	0.29679	2	41.773	0.2861†	1	43.33
$\beta_4\ L_IM_{II}$	2.2550	4	5.4981	2.1669	3	5.7216
$\beta_3\ L_IM_{III}$	2.2172	3	5.5918	2.1268	2	5.8294
$\gamma_2\ L_IN_{II}$	1.8791	4	6.598	1.8013	4	6.883
$\gamma_3\ L_IN_{III}$	1.8740	4	6.616	1.7964	4	6.902
$\gamma_4\ L_IO_{II,III}$	1.8193	4	6.815	1.7445	4	7.107
$\eta\ L_{II}M_I$	2.512	3	4.935	2.4094	4	5.1457
$\beta_1\ L_{II}M_{IV}$	2.2588	3	5.4889	2.1669	2	5.7216
$\gamma_5\ L_{II}N_I$	2.0205	4	6.136	1.9355	4	6.406
$\gamma_1\ L_{II}N_{IV}$	1.9611	3	6.3221	1.8779	2	6.6021
$\gamma_8\ L_{II}O_I$	1.9362	4	6.403	1.8552	5	6.683
$l\ L_{III}M_I$	2.7841	4	4.4532	2.6760	4	4.6330
$\alpha_2\ L_{III}M_{IV}$	2.4729	3	5.0135	2.3807	3	5.2077
$\alpha_1\ L_{III}M_V$	2.4630	2	5.0337	2.3704	2	5.2304
$\beta_6\ L_{III}N_I$	2.1906	4	5.660	2.1039	3	5.8930
$\beta_{2,15}\ L_{III}N_{IV,V}$	2.1194	4	5.850	2.0360	3	6.0894
$\beta_7\ L_{III}O_I$	2.0919	4	5.927	2.0092	3	6.1708
$\beta_{10}\ L_IM_{IV}$	2.1071	4	5.884	2.0237	3	6.1265
$\beta_9\ L_IM_V$	2.1004	4	5.903	2.0165	3	6.1484
$\gamma\ M_{III}N_{IV,V}$	10.998	9	1.1273	10.505	9	1.180
$\beta\ M_{IV}N_{VI}$	13.06	2	0.950	12.44	2	0.997
$\zeta\ M_VN_{III}$	17.38	4	0.714	16.46	4	0.753
$\alpha\ M_VN_{VI,VII}$	13.343	5	0.9292	12.68	2	0.978
$N_{IV,V}N_{VI,VII}$	113.	1	0.1095	107.	1	0.116
$N_{IV,V}O_{II,III}$	136.5	4	0.0908	128.9	7	0.0962
	61 Promethium			**62 Samarium**		
$\alpha_2\ KL_{II}$	0.324803	4	38.1712	0.313698	2	39.5224
$\alpha_1\ KL_{III}$	0.320160	4	38.7247	0.309040	2	40.1181
$\beta_3\ KM_{II}$	0.28363†	4	43.713	0.27376	2	45.289
$\beta_1\ KM_{III}$	0.28290†	3	43.826	0.27301	2	45.413
$\beta_2\ KN_{II,III}$	0.2759†	1	44.94	0.2662	1	46.58
$KO_{II,III}$				0.26491	3	46.801
$\beta_5\ KM_{IV,V}$				0.27111	3	45.731
$\beta_4\ L_IM_{II}$				2.00095	6	6.1963
$\beta_3\ L_IM_{III}$	2.0421	4	6.071	1.96241	3	6.3180
$\gamma_2\ L_IN_{II}$				1.66044	6	7.4668
$\gamma_3\ L_IN_{III}$				1.65601	3	7.4867
$\gamma_4\ L_IO_{II,III}$				1.60728	3	7.7137
$\eta\ L_{II}M_I$				2.21824	3	5.5892
$\beta_1\ L_{II}M_{IV}$	2.0797	4	5.961	1.99806	3	6.2051
$\gamma_5\ L_{II}N_I$				1.77934	3	6.9678
$\gamma_1\ L_{II}N_{IV}$	1.7989	9	6.892	1.72724	3	7.1780
$\gamma_6\ L_{II}O_{IV}$				1.6966	9	7.3076
$l\ L_{III}M_I$				2.4823	4	4.9945

Designation	Å*	p.e.	keV	Å*	p.e.	keV
	61 Promethium (*Cont.*)			**62 Samarium** (*Cont.*)		
$\alpha_2\ L_{III}M_{IV}$	2.2926	4	5.4078	2.21062	3	5.6084
$\alpha_1\ L_{III}M_V$	2.2822	3	5.4325	2.1998	2	5.6361
$\beta_6\ L_{III}N_I$				1.94643	3	6.3697
$\beta_{2,15}\ L_{III}N_{IV,V}$	1.9559	6	6.339	1.88221	3	6.5870
$\beta_7\ L_{III}O_I$				1.85626	3	6.6791
$\beta_5\ L_{III}O_{IV,V}$				1.84700	9	6.7126
$\beta_{10}\ L_IM_{IV}$				1.86990	3	6.6304
$\beta_9\ L_IM_V$				1.86166	3	6.6597
$\gamma\ M_{III}N_{IV,V}$				9.600	9	1.291
$\beta\ M_{IV}N_{VI}$				11.27	1	1.0998
$\zeta\ M_VN_{III}$				14.91	4	0.831
$\alpha\ M_VN_{VI,VII}$				11.47	3	1.081
$N_{IV,V}N_{VI,VII}$				98.	1	0.126
$N_{IV,V}O_{II,III}$				117.4	4	0.1056
	63 Europium			**64 Gadolinium**		
$\alpha_2\ KL_{II}$	0.303118	2	40.9019	0.293038	2	42.3089
$\alpha_1\ KL_{III}$	0.298446	2	41.5422	0.288353	2	42.9962
$\beta_3\ KM_{II}$	0.264332	5	46.9036	0.25534	2	48.555
$\beta_1\ KM_{III}$	0.263577	5	47.0379	0.25460	2	48.697
$\beta_2\ KN_{II,III}$	0.256923	8	48.256	0.24816	3	49.959
$KO_{II,III}$	0.255645	7	48.497	0.24687	3	50.221
$\beta_5\ KM_{IV,V}$				0.25275	3	49.052
$\beta_4\ L_IM_{II}$	1.9255	2	6.4389	1.8540	2	6.6871
$\beta_3\ L_IM_{III}$	1.8867	2	6.5713	1.8150	2	6.8311
$\gamma_2\ L_IN_{II}$	1.5961	2	7.7677	1.5331	2	8.087
$\gamma_3\ L_IN_{III}$	1.5903	2	7.7961	1.5297	2	8.105
$\gamma_4\ L_IO_{II,III}$	1.5439	1	8.0304	1.4839	2	8.355
$\eta\ L_{II}M_I$	2.1315	2	5.8166	2.0494	1	6.0495
$\beta_1\ L_{II}M_{IV}$	1.9203	2	6.4564	1.8468	2	6.7132
$\gamma_5\ L_{II}N_I$	1.7085	2	7.2566	1.6412	2	7.5543
$\gamma_1\ L_{II}N_{IV}$	1.6574	2	7.4803	1.5924	2	7.7858
$\gamma_8\ L_{II}O_I$	1.6346	2	7.5849	1.5707	2	7.894
$\gamma_6\ L_{II}O_{IV}$	1.6282	2	7.6147	1.5644	2	7.925
$l\ L_{III}M_I$	2.3948	2	5.1772	2.3122	2	5.3621
$\alpha_2\ L_{III}M_{IV}$	2.1315	2	5.8166	2.0578	2	6.0250
$\alpha_1\ L_{III}M_V$	2.1209	2	5.8457	2.0468	2	6.0572
$\beta_6\ L_{III}N_I$	1.8737	2	6.6170	1.8054	2	6.8671
$\beta_{2,15}\ L_{III}N_{IV,V}$	1.8118	2	6.8432	1.7455	2	7.1028
$\beta_7\ L_{III}O_I$	1.7851	2	6.9453	1.7203	2	7.2071
$\beta_5\ L_{III}O_{IV,V}$	1.7772	2	6.9763	1.7130	2	7.2374
$\beta_{10}\ L_IM_{IV}$	1.7993	3	6.890	1.7315	3	7.160
$\beta_9\ L_IM_V$	1.7916	3	6.920	1.7240	3	7.192
$L_IO_{IV,V}$				1.4807	3	8.373
$\gamma\ M_{III}N_{IV,V}$	9.211	9	1.346	8.844	9	1.402
$\beta\ M_{IV}N_{VI}$	10.750	7	1.1533	10.254	6	1.2091
$\zeta\ M_VN_{III}$	14.22	2	0.872	13.57	2	0.914
$\alpha\ M_VN_{VI,VII}$	10.96	3	1.131	10.46	3	1.185
$N_{IV,V}O_{II,III}$	112.0	6	0.1107			
	65 Terbium			**66 Dysprosium**		
$\alpha_2\ KL_{II}$	0.283423	2	43.7441	0.274247	2	45.2078
$\alpha_1\ KL_{III}$	0.278724	2	44.4816	0.269533	2	45.9984
$\beta\ KM_{II}$	0.24683	2	50.229	0.23862	2	51.957

Designation	Å*	p.e.	keV	Å*	p.e.	keV
	65 Terbium (*Cont.*)			**66 Dysprosium** (*Cont.*)		
$\beta_1\ KM_{III}$	0.24608	2	50.382	0.23788	2	52.119
$\beta_2\ KN_{II,III}$	0.2397†	2	51.72	0.2317†	2	53.51
$KO_{II,III}$	0.23858	3	51.965	0.23056	3	53.774
$\beta_5\ KM_{IV,V}$				0.23618	3	52.494
$\beta_4\ L_IM_{II}$	1.7864	2	6.9403	1.72103	7	7.2039
$\beta_3\ L_IM_{III}$	1.7472	2	7.0959	1.6822	2	7.3702
$\gamma_2\ L_IN_{II}$	1.4764	2	8.398	1.42278	7	8.7140
$\gamma_3\ L_IN_{III}$	1.4718	2	8.423	1.41640	7	8.7532
$\gamma_4\ L_IO_{II,III}$	1.4276	2	8.685	1.37459	7	9.0195
$\eta\ L_{II}M_I$	1.9730	2	6.2839	1.89743	7	6.5342
$\beta_1\ L_{II}M_{IV}$	1.7768	3	6.978	1.71062	7	7.2477
$\gamma_5\ L_{II}N_I$	1.5787	2	7.8535	1.51824	7	8.1661
$\gamma_1\ L_{II}N_{IV}$	1.5303	2	8.102	1.47266	7	8.4188
$\gamma_8\ L_{II}O_I$	1.5097	2	8.212			
$\gamma_6\ L_{II}O_{IV}$	1.5035	2	8.246	1.44579	7	8.5753
$l\ L_{III}M_I$	2.2352	2	5.5467	2.15877	7	5.7431
$\alpha_2\ L_{III}M_{IV}$	1.9875	2	6.2380	1.91991	3	6.4577
$\alpha_1\ L_{III}M_V$	1.9765	2	6.2728	1.90881	3	6.4952
$\beta_6\ L_{III}N_I$	1.7422	2	7.1163	1.68213	7	7.3705
$\beta_{2,15}\ L_{III}N_{IV,V}$	1.6830	2	7.3667	1.62369	7	7.6357
$\beta_7\ L_{III}O_I$	1.6585	2	7.4753	1.60447	7	7.7272
$\beta_5\ L_{III}O_{IV,V}$	1.6510	2	7.5094	1.58837	7	7.8055
$\beta_{10}\ L_IM_{IV}$	1.6673	3	7.436	1.60743	9	7.7130
$\beta_9\ L_IM_V$				1.59973	9	7.7501
$L_IO_{IV,V}$	1.4228	3	8.714			
$\gamma\ M_{III}N_{IV,V}$	8.486	9	1.461	8.144	9	1.522
$\beta\ M_{IV}N_{VI}$	9.792	6	1.2661	9.357	6	1.3250
$\zeta\ M_VN_{III}$	12.98	2	0.955	12.43	2	0.998
$\alpha\ M_VN_{VI,VII}$	10.00	2	1.240	9.59	2	1.293
$N_{IV,V}N_{VI,VII}$	86.	1	0.144	83.	1	0.149
$N_{IV,V}O_{II,III}$	102.2	4	0.1213	97.2	8	0.128
	67 Holmium			**68 Erbium**		
$\alpha_2\ KL_{II}$	0.265486	2	46.6997	0.257110	2	48.2211
$\alpha_1\ KL_{III}$	0.260756	2	47.5467	0.252365	2	49.1277
$\beta_3\ KM_{II}$	0.23083	2	53.711	0.22341	2	55.494
$\beta_1\ KM_{III}$	0.23012	2	53.877	0.22266	2	55.681
$\beta_2\ KN_{II,III}$	0.2241†	2	55.32	0.2167†	2	57.21
$KO_{II,III}$	0.22305	3	55.584	0.21581	3	57.450
$\beta_5\ KM_{IV,V}$	0.22855	3	54.246	0.22124	3	56.040
$\beta_4\ L_IM_{II}$	1.6595	2	7.4708	1.6007	1	7.7453
$\beta_3\ L_IM_{III}$	1.6203	2	7.6519	1.5616	1	7.9392
$\gamma_2\ L_IN_{II}$	1.3698	2	9.051	1.3210	2	9.385
$\gamma_3\ L_IN_{III}$	1.3643	2	9.087	1.3146	1	9.4309
$\gamma_4\ L_IO_{II,III}$	1.3225	2	9.374	1.2752	2	9.722
$\eta\ L_{II}M_I$	1.8264	2	6.7883	1.7566	1	7.0579
$\beta_1\ L_{II}M_{IV}$	1.6475	2	7.5253	1.5873	1	7.8109
$\gamma_5\ L_{II}N_I$	1.4618	2	8.481	1.4067	3	8.814
$\gamma_1\ L_{II}N_{IV}$	1.4174	2	8.747	1.3641	2	9.089
$\gamma_8\ L_{II}O_I$	1.3983	2	8.867			
$\gamma_6\ L_{II}O_{IV}$	1.3923	2	8.905	1.3397	3	9.255
$l\ L_{III}M_I$	2.0860	2	5.9434	2.015	1	6.152
$\alpha_2\ L_{III}M_{IV}$	1.8561	2	6.6795	1.7955	2	6.9050
$\alpha_1\ L_{III}M_V$	1.8450	2	6.7198	1.78425	9	6.9487
$\beta_6\ L_{III}N_I$	1.6237	2	7.6359	1.5675	2	7.909
$\beta_{2,15}\ L_{III}N_{IV,V}$	1.5671	2	7.911	1.51399	9	8.1890
$\beta_7\ L_{III}O_I$				1.4941	3	8.298
$\beta_5\ L_{III}O_{IV,V}$	1.5378	2	8.062	1.4848	3	8.350

Designation	Å*	p.e.	keV	Å*	p.e.	keV
	67 Holmium (*Cont.*)			**68 Erbium** (*Cont.*)		
$\beta_{10}\ L_IM_{IV}$	1.5486	3	8.006	1.4941	3	8.298
$L_IO_{IV,V}$	1.3208	3	9.387			
$\beta_9\ L_IM_V$				1.4855	5	8.346
$M_{II}N_{IV}$				7.60	1	1.632
$\gamma\ M_{III}N_{IV,V}$	7.865	9	1.576			
$\gamma\ M_{III}\ V$				7.546	8	1.643
$\beta\ M_{IV}N_{VI}$	8.965	4	1.3830	8.592	3	1.4430
$\zeta\ M_VN_{III}$	11.86	1	1.0450	11.37	1	1.0901
$\alpha\ M_VN_{VI,VII}$	9.20	2	1.348	8.82	1	1.406
$N_{IV}N_{VI}$				72.7	9	0.171
$N_VN_{VI,VII}$				76.3	7	0.163
	69 Thulium			**70 Ytterbium**		
$\alpha_2\ KL_{II}$	0.249095	2	49.7726	0.241424	2	51.3540
$\alpha_1\ KL_{III}$	0.244338	2	50.7416	0.236655	2	52.3889
$\beta_3\ KM_{II}$	0.21636	2	57.304	0.2096†	1	59.14
$\beta_1\ KM_{III}$	0.21556	2	57.517	0.20884	8	59.37
$\beta_2\ KN_{II,III}$	0.2098†	2	59.09	0.2033†	2	60.98
$KO_{II,III}$	0.20891	2	59.346	0.20226	2	61.298
$\beta_5\ KM_{IV,V}$	0.21404	2	57.923	0.20739	2	59.782
$\beta_4\ L_IM_{II}$	1.5448	2	8.026	1.49138	3	8.3132
$\beta_3\ L_IM_{III}$	1.5063	2	8.231	1.45233	5	8.5367
$\gamma_2\ L_IN_{II}$	1.2742	2	9.730	1.22879	7	10.0897
$\gamma_3\ L_IN_{III}$	1.2678	2	9.779	1.22232	5	10.1431
$\gamma_4\ L_IO_{II,III}$	1.2294	2	10.084	1.1853	1	10.4603
$\eta\ L_{II}M_I$	1.6963	2	7.3088	1.63560	5	7.5802
$\beta_1\ L_{II}M_{IV}$	1.5304	2	8.101	1.47565	5	8.4018
$\gamma_5\ L_{II}N_I$	1.3558	2	9.144	1.3063	1	9.4910
$\gamma_1\ L_{II}N_{IV}$	1.3153	2	9.426	1.26769	5	9.8701
$\gamma_8\ L_{II}O_I$				1.24923	5	9.9246
$\gamma_6\ L_{II}O_{IV}$	1.2905	2	9.607	1.24271	3	9.9766
$l\ L_{III}M_I$	1.9550	2	6.3419	1.89415	5	6.5455
$\alpha_2\ L_{III}M_{IV}$	1.7381	2	7.1331	1.68285	5	7.3673
$\alpha_1\ L_{III}M_V$	1.7268†	2	7.1799	1.67189	4	7.4156
$\beta_6\ L_{III}N_I$	1.5162	2	8.177	1.4661	1	8.4563
$\beta_{2,15}\ L_{III}N_{IV,V}$	1.4640	2	8.468	1.41550	5	8.7588
$\beta_7\ L_{III}O_I$				1.3948	1	8.8889
$\beta_5\ L_{III}O_{IV,V}$	1.4349	2	8.641	1.38696	7	8.9390
$\beta_{10}\ L_IM_{IV}$	1.4410	3	8.604	1.3915	1	8.9100
$\beta_9\ L_IM_V$	1.4336	3	8.648	1.3838	1	8.9597
L_IO_I				1.1886	1	10.4312
$L_IO_{IV,V}$	1.2263	3	10.110	1.1827	1	10.4833
$L_{II}M_{II}$				1.58844	9	7.8052
$L_{II}O_{II,III}$				1.2453	1	9.9561
$t\ L_{III}M_{II}$				1.83091	9	6.7715
$L_{III}O_{II,III}$				1.3898	1	8.9209
$M_{III}N_I$				8.470	9	1.464
$\gamma\ M_{III}N_V$				7.024	8	1.765
$\beta\ M_{IV}N_{VI}$	8.249	7	1.503	7.909	2	1.5675
$\zeta\ M_VN_{III}$				10.48	1	1.183
$\alpha\ M_VN_{VI,VII}$	8.48	1	1.462	8.149	5	1.5214
$N_{IV}N_{VI}$				65.1	7	0.190
$N_VN_{VI,VII}$				69.3	5	0.179
	71 Lutetium			**72 Hafnium**		
$\alpha_2\ KL_{II}$	0.234081	2	52.9650	0.227024	3	54.6114
$\alpha_1\ KL_{III}$	0.229298	2	54.0698	0.222227	3	55.7902
$\beta_3\ KM_{II}$	0.20309†	4	61.05	0.19686†	4	62.98

Designation	Å*	p.e.	keV	Å*	p.e.	keV
	71 Lutetium (*Cont.*)			**72 Hafnium** (*Cont.*)		
$\beta_1\ KM_{III}$	0.20231†	3	61.283	0.19607†	3	63.234
$\beta_2\ KN_{II,III}$	0.1969†	2	62.97	0.1908†	2	64.98
$KO_{II,III}$	0.19589	2	63.293			
$\beta_5\ KM_{IV,V}$	0.20084	2	61.732			
$\beta_4\ L_IM_{II}$	1.44056	5	8.6064	1.39220	5	8.9054
$\beta_3\ L_IM_{III}$	1.40140	5	8.8469	1.35300	5	9.1634
$\gamma_2\ L_IN_{II}$	1.1853	2	10.460	1.14442	5	10.8335
$\gamma_3\ L_IN_{III}$	1.17953	4	10.5110	1.13841	5	10.8907
$\gamma'_4\ L_IO_{II}$				1.10376	5	11.2326
$\gamma_4\ L_IO_{II,III}$	1.1435	1	10.8425	1.10303	5	11.2401
$\eta\ L_{II}M_I$	1.5779	1	7.8575	1.52325	5	8.1393
$\beta_1\ L_{II}M_{IV}$	1.42359	3	8.7090	1.37410	5	9.0227
$\gamma_5\ L_{II}N_I$	1.2596	1	9.8428	1.21537	5	10.2011
$\gamma_1\ L_{II}N_{IV}$	1.22228	4	10.1434	1.17900	5	10.5158
$\gamma_8\ L_{II}O_I$	1.2047	1	10.2915	1.16138	5	10.6754
$\gamma_6\ L_{II}O_{IV}$	1.1987	1	10.3431	1.15519	5	10.7325
$l\ L_{III}M_I$	1.8360	1	6.7528	1.78145	5	6.9596
$\alpha_2\ L_{III}M_{IV}$	1.63029	5	7.6049	1.58046	5	7.8446
$\alpha_1\ L_{III}M_{IV}$	1.61951	3	7.6555	1.56958	5	7.8990
$\beta_6\ L_{III}N_I$	1.4189	1	8.7376	1.37410	5	9.0227
$\beta_{15}\ L_{III}N_{IV}$	1.3715	1	9.0395	1.32783	5	9.3371
$\beta_2\ L_{III}N_V$	1.37012	3	9.0489	1.32639	5	9.3473
$\beta_7\ L_{III}O_I$	1.34949	5	9.1873	1.30564	5	9.4958
$\beta_5\ L_{III}O_{IV,V}$	1.34183	7	9.2397	1.29761	5	9.5546
L_IM_I				1.43025	9	8.6685
$\beta_{10}\ L_IM_{IV}$	1.3430	2	9.232	1.29819	9	9.5503
$\beta_9 L_IM_V$	1.3358	1	9.2816	1.29025	9	9.6090
L_IN_{IV}	1.16227	9	10.6672	1.12250	9	11.0451
$\gamma_{11}\ L_IN_V$	1.16107	9	10.6782	1.12146	9	11.0553
L_IO_I				1.10664	9	11.2034
L_IO_{IV}				1.10086	9	11.2622
$L_{II}M_{II}$	1.53333	9	8.0858	1.48064	9	8.3735
$\beta_{17}\ L_{II}M_{III}$				1.43643	9	8.6312
$L_{II}N_V$				1.17788	9	10.5258
$v\ L_{II}N_{VI}$				1.15830	9	10.7037
$L_{II}O_{II,III}$	1.2014	1	10.3198			
$t\ L_{III}M_{II}$	1.7760	1	6.9810	1.72305	9	7.1954
$s\ L_{III}M_{III}$				1.66346	9	7.4532
$L_{III}N_{II}$				1.35887	9	9.1239
$L_{III}N_{III}$				1.35053	9	9.1802
$u\ L_{III}N_{VI,VII}$				1.30165	9	9.5249
$L_{III}O_{II,III}$	1.34524	9	9.2163			
$M_{III}N_I$				7.887	9	1.572
$\gamma\ M_{III}N_V$	6.768	6	1.832	6.544	4	1.895
ζ_2				9.686	7	1.2800
$\beta\ M_{IV}N_{VI}$	7.601	2	1.6312	7.303	1	1.6976
ζ_1				9.686	7	1.2800
$\alpha\ M_VN_{VI,VII}$	7.840	2	1.5813	7.539	1	1.6446
$N_{IV}N_{VI}$	63.0	5	0.197			
$N_VN_{VI,VII}$	65.7	2	0.1886			
	73 Tantalum			**74 Tungsten**		
$\alpha_2\ KL_{II}$	0.220305	8	56.277	0.213828	2	57.9817
$\alpha_1\ KL_{III}$	0.215497	4	57.532	0.2090100	Std	59.31824
$\beta_3\ KM_{II}$	0.190890	2	64.9488	0.185181	2	66.9514
$\beta_1\ KM_{III}$	0.190089	4	65.223	0.184374	2	67.2443
$\beta_2^{II}\ KN_{II}$	0.185188	9	66.949	0.17960	1	69.031
$\beta_2^{I}\ KN_{III}$	0.185011	8	67.013	0.179421	7	69.101

Designation	Å*	p.e.	keV	Å*	p.e.	keV
	73 Tantalum (*Cont.*)			**74 Tungsten** (*Cont.*)		
$KO_{II,III}$	0.184031	7	67.370	0.178444	5	69.479
KL_I				0.21592	4	57.42
$\beta_5^{II}\ KM_{IV}$	0.188920	6	65.626	0.183264	5	67.652
$\beta_5^{I}\ KM_V$	0.188757	6	65.683	0.183092	7	67.715
$\beta_4\ KN_{IV,V}$	0.18451	1	67.194	0.17892	2	69.294
$\beta_4\ L_IM_{II}$	1.34581	3	9.2124	1.30162	5	9.5252
$\beta_3\ L_IM_{III}$	1.30678	3	9.4875	1.26269	5	9.8188
$\gamma_2\ L_IN_{II}$	1.1053	1	11.217	1.06806	3	11.6080
$\gamma_3\ L_IN_{III}$	1.09936	4	11.2776	1.06200	6	11.6743
$\gamma'_4\ L_IO_{II}$	1.06544	3	11.6366	1.02863	3	12.0530
$\gamma_4\ L_IO_{III}$	1.06467	3	11.6451	1.02775	3	12.0634
$\eta\ L_{II}M_I$	1.47106	5	8.4280	1.42110	3	8.7243
$\beta_1\ L_{II}M_{IV}$	1.32698	3	9.3431	1.281809	9	9.67235
$\gamma_5\ L_{II}N_I$	1.1729	1	10.5702	1.13235	3	10.9490
$\gamma_1\ L_{II}N_{IV}$	1.13794	3	10.8952	1.09855	3	11.2859
$\gamma_8\ L_{II}O_I$	1.1205	1	11.0646	1.08113	4	11.4677
$\gamma_6\ L_{II}O_{IV}$	1.11388	3	11.1306	1.07448	5	11.5387
$l\ L_{III}M_I$	1.72841	5	7.1731	1.6782	1	7.3878
$\alpha_2\ L_{III}M_{IV}$	1.53293	2	8.0879	1.48743	2	8.3352
$\alpha_1\ L_{III}M_V$	1.52197	2	8.1461	1.47639	2	8.3976
$\beta_6\ L_{III}N_I$	1.33094	8	9.3153	1.28989	7	9.6117
$\beta_{15}\ L_{III}N_{IV}$	1.28619	5	9.6394	1.24631	3	9.9478
$\beta_2\ L_{III}N_V$	1.28454	2	9.6518	1.24460	3	9.9615
$\beta_7\ L_{III}O_I$	1.26385	5	9.8098	1.22400	4	10.1292
$\beta_5\ L_{III}O_{IV,V}$	1.2555	1	9.8750	1.21545	3	10.2004
L_IM_I				1.3365	3	9.277
$\beta_{10}\ L_IM_{IV}$	1.2537	2	9.889	1.21218	3	10.2279
$\beta_9\ L_IM_V$	1.2466	2	9.946	1.20479	7	10.2907
L_IN_I	1.11521	9	11.1173			
L_IN_{IV}	1.08377	7	11.4398	1.0468	2	11.844
$\gamma_{11}\ L_IN_V$	1.08205	7	11.4580	1.0458	1	11.856
$L_IN_{VI,VII}$	1.06357	9	11.6570			
L_IO_I	1.06771	9	11.6118	1.0317	3	12.017
$L_IO_{IV,V}$	1.06192	9	11.6752	1.0250	2	12.095
$L_{II}M_{II}$	1.43048	9	8.6671			
$\beta_{17}\ L_{II}M_{III}$	1.3864	1	8.9428	1.3387	2	9.261
$L_{II}M_V$	1.31897	9	9.3998	1.2728	2	9.741
$L_{II}N_{II}$	1.1600	2	10.688	1.1218	3	11.052
$L_{II}N_{III}$	1.1553	1	10.7316	1.1149	2	11.120
$L_{II}N_V$	1.13687	9	10.9055			
$v\ L_{II}N_{VI}$	1.1158	1	11.1113	1.0771	1	11.510
$L_{II}O_{II}$	1.11789	9	11.0907			
$L_{II}O_{III}$	1.11693	9	11.1001	1.0792	2	11.488
$t\ L_{III}M_{II}$	1.67265	9	7.4123	1.6244	3	7.632
$s\ L_{III}M_{III}$	1.61264	9	7.6881	1.5642	3	7.926
$L_{III}N_{II}$	1.3167	1	9.4158	1.2765	2	9.712
$L_{III}N_{III}$	1.3086	1	9.4742	1.2672	2	9.784
$u\ L_{III}N_{VI,VII}$	1.25778	4	9.8572	1.21868	5	10.1733
$L_{III}O_{II,III}$	1.2601	3	9.839	1.2211	2	10.153
M_IN_{III}	5.40	2	2.295	5.172	9	2.397
$M_IO_{II,III}$				4.44	2	2.79
$M_{II}N_I$				6.28	2	1.973
$M_{II}N_{IV}$	5.570	4	2.226	5.357	4	2.314
$M_{III}N_I$	7.612	9	1.629	7.360	8	1.684
$M_{III}N_{IV}$	6.353	5	1.951	6.134	4	2.021
$\gamma\ M_{III}N_V$	6.312	4	1.964	6.092	3	2.035
$M_{II}O_I$	5.83	2	2.126	5.628	8	2.203
$M_{III}O_{IV,V}$	5.67	3	2.19			

Designation	Å*	p.e.	keV	Å*	p.e.	keV
	73 Tantalum (*Cont.*)			**74 Tungsten** (*Cont.*)		
$\zeta_2\ M_{IV}N_{II}$	9.330	5	1.3288	8.993	5	1.3787
$M_{IV}N_{III}$	8.90	2	1.393	8.573	8	1.446
$\beta\ M_{IV}N_{VI}$	7.023	1	1.7655	6.757	1	1.8349
$M_{IV}O_{II}$	7.09	2	1.748	6.806	9	1.822
$\zeta_1\ M_VN_{III}$	9.316	4	1.3308	8.962	4	1.3835
$\alpha\ M_VN_{VI,VII}$	7.252	1	1.7096			
$\alpha_2\ M_VN_{VI}$				6.992	2	1.7731
$\alpha_1\ M_VN_{VII}$				6.983	1	1.7754
M_VO_{III}	7.30	2	1.700	7.005	9	1.770
$N_{II}N_{IV}$				54.0	2	0.2295
$N_{IV}N_{VI}$	58.2	1	0.2130	55.8	1	0.2221
$N_VN_{VI,VII}$	61.1	2	0.2028			
N_VN_{VI}				59.5	3	0.208
N_VN_{VII}				58.4	1	0.2122
	75 Rhenium			**76 Osmium**		
$\alpha_2\ KL_{II}$	0.207611	1	59.7179	0.201639	2	61.4867
$\alpha_1\ KL_{III}$	0.202781	2	61.1403	0.196794	2	63.0005
$\beta_3\ KM_{II}$	0.179697	3	68.994	0.174431	3	71.077
$\beta_1\ KM_{III}$	0.178880	3	69.310	0.173611	3	71.413
$\beta_2^{II}\ KN_{II}$	0.17425	1	71.151	0.16910	1	73.318
$\beta_2^{I}\ KN_{III}$	0.174054	6	71.232	0.168906	6	73.402
$KO_{II,III}$	0.17308	1	71.633	0.16798	1	73.808
$\beta_5^{II}\ KM_{IV}$	0.17783	1	69.719	0.17262	1	71.824
$\beta_5^{I}\ KM_V$	0.17766	1	69.786	0.17245	1	71.895
$\beta_4\ KN_{IV,V}$	0.17362	2	71.410	0.16842	2	73.615
$\beta_4\ L_IM_{II}$	1.25917	5	9.8463	1.21844	5	10.1754
$\beta_3\ L_IM_{III}$	1.22031	5	10.1598	1.17955	7	10.5108
$\gamma_2\ L_IN_{II}$	1.03233	5	12.0098	0.99805	5	12.4224
$\gamma_3\ L_IN_{III}$	1.02613	7	12.0824	0.99186	5	12.4998
$\gamma'_4\ L_IO_{II}$	0.99334	5	12.4813	0.96033	8	12.910
$\gamma_4\ L_IO_{III}$	0.99249	5	12.4920	0.95938	8	12.923
$\eta\ L_{II}M_I$	1.37342	5	9.0272	1.32785	7	9.3370
$\beta_1\ L_{II}M_{IV}$	1.23858	2	10.0100	1.19727	7	10.3553
$\gamma_5\ L_{II}N_I$	1.09388	5	11.3341	1.05693	5	11.7303
$\gamma_1\ L_{II}N_{IV}$	1.06099	5	11.6854	1.02503	5	12.0953
$\gamma_8\ L_{II}O_I$	1.04398	5	11.8758	1.00788	5	12.3012
$\gamma_6\ L_{II}O_{IV}$	1.03699	9	11.956	1.00107	5	12.3848
$l\ L_{III}M_I$	1.63056	5	7.6036	1.58498	7	7.8222
$\alpha_2\ L_{III}M_{IV}$	1.44396	5	8.5862	1.40234	5	8.8410
$\alpha_1\ L_{III}M_V$	1.43290	4	8.6525	1.39121	5	8.9117
$\beta_6\ L_{III}N_I$	1.25100	5	9.9105	1.21349	5	10.2169
$\beta_{15}\ L_{III}N_{IV}$	1.20819	5	10.2617	1.17167	5	10.5816
$\beta_2\ L_{III}N_V$	1.20660	4	10.2752	1.16979	8	10.5985
$\beta_7\ L_{III}O_I$	1.18610	5	10.4529	1.14933	8	10.7872
$\beta_5\ L_{III}O_{IV,V}$	1.17721	5	10.5318	1.1405	1	10.8711
$\beta_{10}\ L_IM_{IV}$	1.17218	5	10.5770	1.13353	5	10.9376
$\beta_9\ L_IM_V$	1.16487	4	10.6433	1.12637	6	11.0071
L_IN_I	1.0420	1	11.899			
L_IN_{IV}	1.0119	1	12.252	0.9772	3	12.687
$\gamma_{11}\ L_IN_V$	1.0108	1	12.266	0.9765	3	12.696
L_IO_I	0.9965	1	12.442	0.96318	7	12.8721
$L_IO_{IV,V}$	0.9900	1	12.524	0.95603	5	12.9683
$L_{II}M_{II}$	1.3366	1	9.2761	1.2934	2	9.586
$\beta_{17}\ L_{II}M_{III}$	1.2927	1	9.5910	1.2480	2	9.934

Designation	Å*	p.e.	keV	Å*	p.e.	keV
	75 Rhenium (*Cont.*)			**76 Osmium** (*Cont.*)		
$L_{II}M_V$	1.2305	1	10.0753	1.18977	7	10.4205
$L_{II}N_{II}$	1.0839	1	11.438			
$L_{II}N_{III}$	1.0767	1	11.515	1.03973	5	11.9243
$v\ L_{II}N_{VI}$	1.0404	1	11.917	1.0050	2	12.337
$L_{II}O_{III}$	1.0397	1	11.925	1.0047	2	12.340
$t\ L_{III}M_{II}$	1.5789	1	7.8525	1.5347	2	8.079
$s\ L_{III}M_{III}$	1.5178	1	8.1682	1.4735	2	8.414
$L_{III}N_I$				1.20086	7	10.3244
$L_{III}N_{III}$	1.2283	1	10.0933			
$u\ L_{III}N_{VI,VII}$	1.1815	1	10.4931	1.14537	7	10.8245
M_IN_{III}				4.79	2	2.59
$M_{II}N_I$				5.81	2	2.133
$M_{II}N_{IV}$				4.955	4	2.502
$M_{III}N_I$				6.89	2	1.798
$M_{III}N_{IV}$	5.931	5	2.090	5.724	5	2.166
$\gamma\ M_{III}N_V$	5.885	2	2.1067	5.682	4	2.182
$\zeta_2\ M_{IV}N_{II}$	8.664	5	1.4310	8.359	5	1.4831
$M_{IV}N_{III}$	8.239	8	1.505			
$\beta\ M_{IV}N_{VI}$	6.504	1	1.9061	6.267	1	1.9783
$\zeta_1\ M_VN_{III}$	8.629	4	1.4368	8.310	4	1.4919
$\alpha\ M_VN_{VI,VII}$	6.729	1	1.8425	6.490	1	1.9102
$N_{IV}N_{VI}$				51.9	1	0.2388
$N_VN_{VI,VII}$				54.7	2	0.2266
	77 Iridium			**78 Platinum**		
$\alpha_2\ KL_{II}$	0.195904	2	63.2867	0.190381	4	65.122
$\alpha_1\ KL_{III}$	0.191047	2	64.8956	0.185511	4	66.832
$\beta_3\ KM_{II}$	0.169367	2	73.2027	0.164501	3	75.368
$\beta_1\ KM_{III}$	0.168542	2	73.5608	0.163675	3	75.748
$\beta_2^{II}\ KN_{II}$	0.16415	1	75.529	0.15939	1	77.785
$\beta_2^{I}\ KN_{III}$	0.163956	7	75.619	0.15920	1	77.878
$KO_{II,III}$	0.163019	5	76.053	0.15826	1	78.341
$\beta_5^{II}\ KM_{IV}$	0.16759	2	73.980	0.16271	2	76.199
$\beta_5^{I}\ KM_V$	0.167373	9	74.075	0.16255	3	76.27
$\beta_4\ KN_{IV\ V}$	0.16352	2	75.821	0.15881	2	78.069
$\beta_4\ L_IM_{II}$	1.17958	3	10.5106	1.14223	5	10.8543
$\beta_3\ L_IM_{III}$	1.14085	3	10.8674	1.10394	5	11.2308
$\gamma_2\ L_IN_{II}$	0.96545	3	12.8418	0.93427	5	13.2704
$\gamma_3\ L_IN_{III}$	0.95931	5	12.9240	0.92791	5	13.3613
$\gamma'_4\ L_IO_{II}$	0.92831	3	13.3555	0.89747	4	13.8145
$\gamma_4\ L_IO_{III}$	0.92744	3	13.3681	0.89659	4	13.8281
$\eta\ L_{II}M_I$	1.28448	3	9.6522	1.2429	2	9.975
$\beta_1\ L_{II}M_{IV}$	1.15781	3	10.7083	1.11990	2	11.0707
$\gamma_5\ L_{II}N_I$	1.02175	5	12.1342	0.9877	2	12.552
$\gamma_1\ L_{II}N_{IV}$	0.99085	3	12.5126	0.95797	3	12.9420
$\gamma_8\ L_{II}O_I$	0.97409	3	12.7279	0.9411	1	13.173
$\gamma_6\ L_{II}O_{IV}$	0.96708	4	12.8201	0.9342	2	13.271
$l\ L_{III}M_I$	1.54094	3	8.0458	1.4995	2	8.268
$\alpha_2\ L_{III}M_{IV}$	1.36250	5	9.0995	1.32432	2	9.3618
$\alpha_1\ L_{III}M_V$	1.35128	3	9.1751	1.31304	3	9.4423
$\beta_6\ L_{III}N_I$	1.17796	3	10.5251	1.14355	5	10.8418
$\beta_{15}\ L_{III}N_{IV}$	1.13707	3	10.9036			
$\beta_2\ L_{III}N_V$	1.13532	3	10.9203	1.10200	3	11.2505
$\beta_7\ L_{III}O_I$	1.11489	3	11.1205	1.08168	3	11.4619

X-Ray Wavelengths
TABLE A (*Continued*)

Designation	Å*	p.e.	keV	Å*	p.e.	keV
	77 Iridium (*Cont.*)			**78 Platinum** (*Cont.*)		
$\beta_5\ L_{III}O_{IV,V}$	1.10585	3	11.2114	1.0724	2	11.561
L_IM_I	1.2102	2	10.245	1.16962	9	10.6001
$\beta_{10}\ L_IM_{IV}$	1.09702	4	11.3016	1.06183	7	11.6762
$\beta_9\ L_IM_V$	1.08975	5	11.3770	1.05446	5	11.7577
L_IN_I	0.9766	2	12.695	0.9455	2	13.113
L_IN_{IV}	0.9459	2	13.108			
$\gamma_{11}\ L_IN_V$	0.9446	2	13.126	0.9143	2	13.560
$L_IO_{IV,V}$	0.9243	3	13.413			
L_IO_I				0.8995	2	13.784
L_IO_{IV}				0.8943	1	13.864
L_IO_V				0.8934	1	13.878
$L_{II}M_{II}$	1.2502	3	9.917	1.213	1	10.225
$\beta_{17}\ L_{II}M_{III}$	1.2069	2	10.273	1.1667	1	10.6265
$L_{II}M_V$	1.1489	2	10.791	1.1129	2	11.140
$L_{II}N_{II}$	1.0120	2	12.251	0.9792	2	12.661
$L_{II}N_{III}$	1.0054	3	12.332	0.97173	4	12.7588
$v\ L_{II}N_{VI}$	0.97161	6	12.7603	0.93931	5	13.1992
$L_{II}O_{III}$	0.96979	5	12.7843			
$t\ L_{III}M_{II}$	1.4930	3	8.304	1.4530	2	8.533
$s\ L_{III}M_{III}$	1.4318	2	8.659	1.3895	2	8.923
$L_{III}N_{II}$	1.16545	5	10.6380	1.1310	2	10.962
$L_{III}N_{III}$	1.1560	3	10.725	1.1226	2	11.044
$u\ L_{III}N_{VI,VII}$	1.11145	4	11.1549	1.07896	5	11.4908
$L_{III}O_{II,III}$	1.10923	6	11.1772	1.0761	3	11.521
M_IN_{III}	4.631†	9	2.677	4.460	9	2.780
$M_{II}N_{IV}$	4.780	4	2.594	4.601	4	2.695
$M_{III}N_I$	6.669	9	1.859	6.455	9	1.921
$M_{III}N_{IV}$	5.540	5	2.238	5.357	5	2.314
$\gamma\ M_{III}N_V$	5.500	4	2.254	5.319	4	2.331
$M_{III}O_I$				4.876	9	2.543
$M_{III}O_{IV,V}$	4.869	9	2.546	4.694	8	2.641
$\zeta_2\ M_{IV}N_{II}$	8.065	5	1.5373	7.790	5	1.592
$M_{IV}N_{III}$	7.645	8	1.622	7.371	8	1.682
$\beta\ M_{IV}N_{VI}$	6.038	1	2.0535	5.828	1	2.1273
$\zeta_1\ M_VN_{III}$	8.021	4	1.5458	7.738	4	1.6022
$\alpha_2\ M_VN_{VI}$	6.275	3	1.9758	6.058	3	2.047
$\alpha_1\ M_VN_{VII}$	6.262	1	1.9799	6.047	1	2.0505
M_VO_{III}				5.987	9	2.071
$N_{IV}N_{VI}$	50.2	1	0.2470	48.1	2	0.258
$N_VN_{VI,VII}$	52.8	1	0.2348	50.9	1	0.2436
	79 Gold			**80 Mercury**		
$\alpha_2\ KL_{II}$	0.185075	2	66.9895	0.179958	3	68.895
$\alpha_1\ KL_{III}$	0.180195	2	68.8037	0.175068	3	70.819
$\beta_3\ KM_{II}$	0.159810	2	77.580	0.155321	3	79.822
$\beta_1\ KM_{III}$	0.158982	3	77.984	0.154487	3	80.253
$\beta_2^{II}\ KN_{II}$	0.15483	2	80.08	0.15040	2	82.43
$\beta_2^{I}\ KN_{III}$	0.154618	9	80.185	0.15020	2	82.54
$KO_{II,III}$	0.153694	7	80.667	0.14931	2	83.04
KL_I	0.18672	4	66.40			
$\beta_5^{II}\ KM_{IV}$	0.158062	7	78.438			
$\beta_5^{I}\ KM_V$	0.157880	5	78.529			
$\beta_5\ KM_{IV,V}$				0.15353	2	80.75
$\beta_4\ KN_{IV,V}$	0.154224	5	80.391	0.14978	2	82.78
$\beta_4\ L_IM_{II}$	1.10651	3	11.2047	1.07222	7	11.5630
$\beta_3\ L_IM_{III}$	1.06785	9	11.6103	1.03358	7	11.9953

Designation	Å*	p.e.	keV	Å*	p.e.	keV
	79 Gold (*Cont.*)			**80 Mercury** (*Cont.*)		
$\gamma_2\ L_IN_{II}$	0.90434	3	13.7095	0.87544	7	14.162
$\gamma_3\ L_IN_{III}$	0.89783	5	13.8090	0.86915	7	14.265
$\gamma'_4\ L_IO_{II}$	0.86816	4	14.2809	0.84013	7	14.757
$\gamma_4\ L_IO_{III}$	0.86703	4	14.2996	0.83894	7	14.778
$\eta\ L_{II}M_I$	1.20273	3	10.3083	1.1640	1	10.6512
$\beta_1\ L_{II}M_{IV}$	1.08353	3	11.4423	1.04868	5	11.8226
$\gamma_5\ L_{II}N_I$	0.95559	3	12.9743	0.92453	7	13.410
$\gamma_1\ L_{II}N_{IV}$	0.92650	3	13.3817	0.89646	5	13.8301
$\gamma_8\ L_{II}O_I$	0.90989	5	13.6260	0.87995	7	14.090
$\gamma_6\ L_{II}O_{IV}$	0.90297	3	13.7304	0.87319	7	14.199
$l\ L_{III}M_I$	1.45964	9	8.4939	1.4216	1	8.7210
$\alpha_2\ L_{III}M_{IV}$	1.28772	3	9.6280	1.25264	7	9.8976
$\alpha_1\ L_{III}M_V$	1.27640	3	9.7133	1.24120	5	9.9888
$\beta_6\ L_{III}N_I$	1.11092	3	11.1602	1.07975	7	11.4824
$\beta_{15}\ L_{III}N_{IV}$	1.07188	5	11.5667	1.04151	7	11.9040
$\beta_2\ L_{III}N_V$	1.07022	3	11.5847	1.03975	7	11.9241
$\beta_7\ L_{III}O_I$	1.04974	8	11.8106	1.01937	7	12.1625
$\beta_5\ L_{III}O_{IV,V}$	1.04044	3	11.9163	1.00987	7	12.2769
L_IM_I	1.13525	5	10.9210	1.0999	2	11.272
$\beta_{10}\ L_IM_{IV}$	1.02789	7	12.0617	0.9962	2	12.446
$\beta_9\ L_IM_V$	1.02063	7	12.1474	0.9871	2	12.560
L_IN_I	0.9131	1	13.578	0.8827	2	14.045
L_IN_{IV}	0.88563	7	13.999			
$\gamma_{11}\ L_IN_V$	0.88433	7	14.020	0.85657	7	14.474
L_IO_I	0.87074	5	14.2385	0.8452	2	14.670
$L_IO_{IV,V}$	0.86400	5	14.3497	0.8350	2	14.847
$L_{II}M_{II}$	1.1708	1	10.5892	1.1387	5	10.888
$\beta_{17}\ L_{II}M_{III}$	1.12798	5	10.9915	1.0916	5	11.358
$L_{II}M_V$	1.0756	2	11.526			
$L_{II}N_{III}$	0.9402	2	13.186	0.90894	7	13.640
$v\ L_{II}N_{VI}$	0.90837	5	13.6487	0.87885	7	14.107
$L_{II}O_{II}$	0.90746	7	13.662	0.8784	1	14.114
$L_{II}O_{III}$	0.90638	7	13.679	0.8758	1	14.156
$t\ L_{III}M_{II}$	1.41366	7	8.7702	1.3746	2	9.019
$s\ L_{III}M_{III}$	1.35131	7	9.1749	1.3112	2	9.455
$L_{III}N_{II}$	1.09968	7	11.2743	1.0649	2	11.642
$L_{III}N_{III}$	1.09026	7	11.3717	1.0585	1	11.713
$u\ L_{III}N_{VI,VII}$	1.04752	5	11.8357			
$u'\ L_{III}N_{VI}$				1.01769	7	12.1826
$u\ L_{III}N_{VII}$				1.01674	7	12.1940
$L_{III}O_{II,III}$	1.0450	2	11.865			
$L_{III}O_{II}$				1.01558	7	12.2079
$L_{III}O_{III}$				1.01404	7	12.2264
$L_{III}P_{II,III}$	1.03876	7	11.9355			
M_IN_{III}	4.300	9	2.883			
$M_{II}N_{IV}$	4.432	4	2.797			
$M_{III}N_I$	6.259	9	1.981	6.09	2	2.036
$M_{III}N_{IV}$	5.186	5	2.391			
$\gamma\ M_{III}N_V$	5.145	4	2.410	4.984†	2	2.4875
$M_{III}O_I$	4.703	9	2.636			
$M_{III}O_{IV,V}$	4.522	6	2.742			
$\zeta_2\ M_{IV}N_{II}$	7.523	5	1.648			
$M_{IV}N_{III}$	7.101	8	1.746	6.87	2	1.805
$\beta\ M_{IV}N_{VI}$	5.624	1	2.2046	5.4318†	9	2.2825
$\zeta_1\ M_VN_{III}$	7.466	4	1.6605			
$\alpha_2\ M_VN_{VI}$	5.854	3	2.118			

Designation	Å*	p.e.	keV	Å*	p.e.	keV
	79 Gold (*Cont.*)			**80 Mercury** (*Cont.*)		
α_1 M_VN_{VII}	5.840	1	2.1229	5.6476†	9	2.1953
M_VO_{III}	5.767	9	2.150			
$N_{IV}N_{VI}$	46.8	2	0.265	45.2†	3	0.274
$N_VN_{VI,VII}$	49.4	1	0.2510	47.9†	3	0.259
	81 Thallium			**82 Lead**		
α_2 KL_{II}	0.175036	2	70.8319	0.170294	2	72.8042
α_1 KL_{III}	0.170136	2	72.8715	0.165376	2	74.9694
β_3 KM_{II}	0.150980	6	82.118	0.146810	4	84.450
β_1 KM_{III}	0.150142	5	82.576	0.145970	6	84.936
β_2^{II} KN_{II}	0.14614	1	84.836	0.14212	2	87.23
β_2^{I} KN_{III}	0.14595	1	84.946	0.14191	1	87.364
$KO_{II,III}$	0.14509	1	85.451	0.141012	8	87.922
KP				0.1408	1	88.06
β_5 $KM_{IV,V}$	0.14917	1	83.114			
β_5^{II} KM_{IV}				0.14512	2	85.43
β_5^{I} KM_V				0.14495	3	85.53
β_4 $KN_{IV,V}$	0.14553	2	85.19	0.14155	3	87.59
β_4 L_IM_{II}	1.03918	3	11.9306	1.0075	1	12.306
β_3 L_IM_{III}	1.00062	3	12.3904	0.96911	7	12.7933
γ_2 L_IN_{II}	0.84773	5	14.6251	0.8210	2	15.101
γ_3 L_IN_{III}	0.84130	4	14.7368	0.8147	1	15.218
γ'_4 L_IO_{II}	0.81308	5	15.2482	0.78706	7	15.752
γ_4 L_IO_{III}	0.81184	5	15.2716	0.7858	1	15.777
η $L_{II}M_I$	1.12769	3	10.9943	1.09241	7	11.3493
β_1 $L_{II}M_{IV}$	1.01513	4	12.2133	0.98291	3	12.6137
γ_5 $L_{II}N_I$	0.89500	4	13.8526	0.86655	5	14.3075
γ_1 $L_{II}N_{IV}$	0.86752	3	14.2915	0.83973	3	14.7644
γ_8 $L_{II}O_I$	0.8513	2	14.564	0.82365	5	15.0527
γ_6 $L_{II}O_{IV}$	0.8442	2	14.685	0.81683	5	15.1783
$L_{II}P_I$				0.81583	5	15.1969
l $L_{III}M_I$	1.38477	3	8.9532	1.34990	7	9.1845
α_2 $L_{III}M_{IV}$	1.21875	3	10.1728	1.18648	5	10.4495
α_1 $L_{III}M_V$	1.20739	4	10.2685	1.17501	2	10.5515
β_6 $L_{III}N_I$	1.04963	5	11.8118	1.0210	1	12.143
β_{15} $L_{III}N_{IV}$	1.01201	3	12.2510	0.98389	7	12.6011
β_2 $L_{III}N_V$	1.01031	3	12.2715	0.98221	7	12.6226
β_7 $L_{III}O_I$	0.99017	5	12.5212	0.9620	1	12.888
β_5 $L_{III}O_{IV,V}$	0.98058	3	12.6436	0.9526	1	13.015
L_IM_I	1.0644	2	11.648	1.0323	2	12.010
β_{10} L_IM_{IV}	0.96389	7	12.8626	0.9339	2	13.275
β_9 L_IM_V	0.95675	7	12.9585	0.9268	1	13.377
L_IN_I	0.8549	1	14.503	0.82859	7	14.963
L_IN_{IV}	0.83001	7	14.937	0.80364	7	15.427
γ_{11} L_IN_V	0.82879	5	14.9593	0.80233	9	15.453
$L_IN_{VI,VII}$				0.7884	1	15.725
L_IO_I	0.8158	1	15.198	0.7897	1	15.699
$L_IO_{IV,V}$	0.80861	5	15.3327	0.78257	7	15.843
$L_{II}M_{II}$	1.0997	1	11.274	1.0644	2	11.648
β_{17} $L_{II}M_{III}$	1.05609	7	11.7397	1.0223	1	12.127
$L_{II}M_V$	1.00722	5	12.3093	0.9747	1	12.720
$L_{II}N_{II}$	0.882	2	14.057	0.8585	3	14.442

Designation	Å*	p.e.	keV	Å*	p.e.	keV
	81 Thallium (*Cont.*)			**82 Lead** (*Cont.*)		
$L_{II}N_{III}$	0.87996	5	14.0893	0.85192	7	14.553
$L_{II}N_V$				0.8382	2	14.791
v $L_{II}N_{VI}$	0.85048	5	14.5777	0.82327	7	15.060
$L_{II}O_{II}$	0.8490	1	14.604			
$L_{II}O_{III}$				0.8200	1	15.120
t $L_{III}M_{II}$	1.34154	5	9.2417	1.30767	7	9.4811
s $L_{III}M_{III}$	1.27807	5	9.7007	1.24385	7	9.9675
$L_{III}N_{II}$				1.01040	7	12.2705
$L_{III}N_{III}$	1.0286	1	12.053	1.0005	1	12.392
u $L_{III}N_{VI,VII}$	0.9888	1	12.538	0.96133	7	12.8968
$L_{III}O_{II}$	0.98738	5	12.5566	0.9586	1	12.934
$L_{III}O_{III}$	0.98538	5	12.5820	0.9578	1	12.945
$L_{III}P_{II,III}$	0.97926	5	12.6607	0.95118	7	13.0344
M_IN_{III}	4.013	9	3.089	3.872	9	3.202
$M_{II}N_I$				4.655	8	2.664
$M_{II}N_{IV}$	4.116	4	3.013	3.968	5	3.124
$M_{III}N_I$	5.884	8	2.107	5.704	8	2.174
$M_{III}N_{IV}$	4.865	5	2.548	4.715	3	2.630
γ $M_{III}N_V$	4.823	4	2.571	4.674	1	2.6527
$M_{III}O_I$				4.244	9	2.921
$M_{II}O_{IV,V}$	4.216	6	2.941	4.069	6	3.047
ζ_2 $M_{IV}N_{II}$	7.032	5	1.763	6.802	5	1.823
$M_{IV}N_{III}$				6.384	7	1.942
β $M_{IV}N_{VI}$	5.249	1	2.3621	5.076	1	2.4427
$M_{IV}O_{II}$	5.196	9	2.386	5.004	9	2.477
ζ_1 M_VN_{III}	6.974	4	1.778	6.740	3	1.8395
α_2 M_VN_{VI}	5.472	2	2.2656	5.299	2	2.3397
α_1 M_VN_{VII}	5.460	1	2.2706	5.286	1	2.3455
M_VO_{III}				5.168	9	2.399
$N_{IV}N_{VI}$				42.3	2	0.293
$N_VN_{VI,VII}$	46.5	2	0.267	45.0	1	0.2756
$N_{VI}O_{IV}$	115.3	2	0.1075	102.4	1	0.1211
$N_{VI}O_V$	113.0	1	0.10968	100.2	2	0.1237
$N_{VII}O_V$	117.7	1	0.10530	104.3	1	0.1189
	83 Bismuth			**84 Polonium**		
α_2 KL_{II}	0.165717	2	74.8148	0.16130†	1	76.862
α_1 KL_{III}	0.160789	2	77.1079	0.15636†	1	79.290
β_3 KM_{II}	0.142779	7	86.834	0.13892†	2	89.25
β_1 KM_{III}	0.141948	3	87.343	0.13807†	2	89.80
β_2^{II} KN_{II}	0.13817	1	89.733	0.13438†	2	92.26
β_2^{I} KN_{III}	0.13797	1	89.864	0.13418†	2	92.40
$KO_{II,III}$	0.13709	1	90.435			
β_5 $KM_{IV,V}$	0.14111	1	87.860			
β_4 $KN_{IV,V}$	0.13759	2	90.11			
β_4 L_IM_{II}	0.97690	4	12.6912	0.9475	3	13.086
β_3 L_IM_{III}	0.93855	3	13.2098	0.9091	3	13.638
γ_2 L_IN_{II}	0.79565	3	15.5824	0.772	1	16.07
γ_3 L_IN_{III}	0.78917	5	15.7102			
γ'_4 L_IO_{II}	0.76198	3	16.2709			
γ_4 L_IO_{III}	0.76087	3	16.2947			
γ_{13} $L_IP_{II,III}$	0.75690	3	16.3802			
η $L_{II}M_I$	1.05856	3	11.7122			
β_1 $L_{II}M_{IV}$	0.951978	9	13.0235	0.9220	2	13.447
γ_5 $L_{II}N_I$	0.83923	5	14.7732			

Designation	Å*	p.e.	keV	Å*	p.e.	keV
	83 Bismuth (*Cont.*)			**84 Pononium** (*Cont.*)		
$\gamma_1\ L_{II}N_{IV}$	0.81311	2	15.2477	0.78748	9	15.744
$\gamma_8\ L_{II}O_I$	0.7973	1	15.551			
$\gamma_6\ L_{II}O_{IV}$	0.79043	3	15.6853	0.7645	2	16.218
$l\ L_{III}M_I$	1.31610	7	9.4204	1.2829	5	9.664
$\alpha_2\ L_{III}M_{IV}$	1.15536	1	10.73091	1.12548†	5	11.0158
$\alpha_1\ L_{III}M_V$	1.14386	2	10.8388	1.11386	4	11.1308
$\beta_6\ L_{III}N_I$	0.99331	3	12.4816	0.9672	2	12.819
$\beta_{15}\ L_{III}N_{IV}$	0.95702	5	12.9549	0.9312	2	13.314
$\beta_2\ L_{III}N_V$	0.95518	4	12.9799	0.92937	5	13.3404
$\beta_7\ L_{III}O_I$	0.93505	5	13.2593			
$\beta_5\ L_{III}O_{IV,V}$	0.92556	3	13.3953	0.8996	2	13.782
L_IM_I	1.0005	9	12.39			
$\beta_{10}\ L_IM_{IV}$	0.90495	4	13.7002			
$\beta_9\ L_IM_V$	0.89791	3	13.8077			
L_IN_I	0.8022	1	15.456			
L_IN_{IV}	0.7795	5	15.904			
$\gamma_{11}\ L_IN_V$	0.77728	5	15.951			
$L_IN_{VI,VII}$	0.7641	5	16.23			
$L_IO_{IV,V}$	0.75791	5	16.358			
$L_{II}M_{II}$	1.0346	9	11.98			
$\beta_{17}\ L_{II}M_{III}$	0.98913	5	12.5344			
$L_{II}M_V$	0.94419	5	13.1310			
$L_{II}N_{II}$	0.8344	9	14.86			
$L_{II}N_{III}$	0.8248	1	15.031			
$v\ L_{II}N_{VI}$	0.79721	9	15.552			
$L_{II}O_{III}$	0.79384	5	15.6178			
$t\ L_{III}M_{II}$	1.2748	1	9.7252			
$s\ L_{III}M_{III}$	1.2105	1	10.2421			
$L_{III}N_{II}$	0.98280	5	12.6151			
$L_{III}N_{III}$	0.97321	5	12.7394			
$u\ L_{III}N_{VI,VII}$	0.93505	5	13.2593			
$L_{III}O_{II}$	0.9323	2	13.298			
$L_{III}O_{III}$	0.9302	2	13.328			
$L_{III}P_{II,III}$	0.92413	4	13.4159			
M_IN_{II}	3.892	9	3.185			
M_IN_{III}	3.740	9	3.315			
$M_{II}N_{IV}$	3.834	4	3.234			
$M_{III}N_I$	5.537	8	2.239			
$M_{III}N_{IV}$	4.571	5	2.712			
$\gamma\ M_{III}N_V$	4.532	2	2.735			
$M_{III}O_I$	4.105	9	3.021			
$M_{III}O_{IV,V}$	3.932	6	3.153			
$\zeta_2\ M_{IV}N_{II}$	6.585	5	1.883			
$M_{IV}N_{III}$	6.162	8	2.012			
$\beta\ M_{IV}N_{VI}$	4.909	1	2.5255			
$M_{IV}O_{II}$	4.823	3	2.571			
$M_{IV}P_{II,III}$	4.59	2	2.70			
$\zeta_1\ M_VN_{III}$	6.521	4	1.901			
$\alpha_2\ M_VN_{VI}$	5.130	2	2.4170			
$\alpha_1\ M_VN_{VII}$	5.118	1	2.4226			
$N_IP_{II,III}$	13.30	6	0.932			
$N_{VI}O_{IV}$	91.6	1	0.1354			
$N_{VII}O_V$	93.2	1	0.1330			

Designation	Å*	p.e.	keV	Å*	p.e.	keV
	85 Astatine			**86 Radon**		
$\alpha_2\ KL_{II}$	0.15705†	2	78.95	0.15294†	3	81.07
$\alpha_1\ KL_{III}$	0.15210†	2	81.52	0.14798†	3	83.78
$\beta_3\ KM_{II}$	0.13517†	4	91.72	0.13155†	5	94.24
$\beta_1\ KM_{III}$	0.13432†	4	92.30	0.13069†	5	94.87
$\beta_2^{II}\ KN_{II}$	0.13072†	4	94.84	0.12719†	5	97.47
$\beta_2^{I}\ KN_{III}$	0.13052†	4	94.99	0.12698†	5	97.64
$\beta_3\ L_IM_{III}$	0.88135†	9	14.067	0.85436†	9	14.512
$\beta_1\ L_{II}M_{IV}$	0.89349†	9	13.876	0.86605†	9	14.316
$\gamma_1\ L_{II}N_{IV}$	0.76289†	9	16.251	0.73928†	9	16.770
$\alpha_2\ L_{III}M_{IV}$	1.09671†	5	11.3048	1.06899†	5	11.5979
$\alpha_1\ L_{III}M_V$	1.08500†	5	11.4268	1.05723†	5	11.7270
	87 Francium			**88 Radium**		
$\alpha_2\ KL_{II}$	0.14896†	3	83.23	0.14512†	2	85.43
$\alpha_1\ KL_{III}$	0.14399†	3	86.10	0.14014†	2	88.47
$\beta_3\ KM_{II}$	0.12807†	5	96.81	0.12469†	3	99.43
$\beta_1\ KM_{III}$	0.12719†	5	97.47	0.12382†	3	100.13
$\beta_2^{II}\ KN_{II}$	0.12379†	5	100.16	0.12050†	3	102.89
$\beta_2^{I}\ KN_{III}$	0.12358†	5	100.33	0.12029†	3	103.07
$\beta_4\ L_IM_{II}$				0.84071	5	14.7472
$\beta_3\ L_IM_{III}$	0.82789†	9	14.976	0.80273	5	15.4449
$\gamma_2\ L_IN_{II}$				0.68199	5	18.179
$\gamma_3\ L_IN_{III}$				0.67538	5	18.357
$\gamma'_4\ L_IO_{II}$				0.65131	5	19.036
$\gamma_4\ L_IO_{III}$				0.64965	5	19.084
$\gamma_{13}\ L_IP_{II,III}$				0.64513	5	19.218
$\eta\ L_{II}M_I$				0.90742	5	13.6630
$\beta_1\ L_{II}M_{IV}$	0.83940†	9	14.770	0.81375	5	15.2358
$\gamma_5\ L_{II}N_I$				0.71774	5	17.274
$\gamma_1\ L_{II}N_{IV}$	0.71652†	9	17.303	0.69463	5	17.849
γ_8				0.6801	1	18.230
$\gamma_6\ L_{II}O_{IV}$				0.67328	5	18.414
$L_{II}P_I$				0.6724	1	18.439
$l\ L_{III}M_I$				1.16719	5	10.6222
$\alpha_2\ L_{III}M_{IV}$	1.04230	5	11.8950	1.01656	5	12.1962
$\alpha_1\ L_{III}M_V$	1.03049	5	12.0313	1.00473	5	12.3397
$\beta_6\ L_{III}N_I$				0.87088	5	14.2362
$\beta_{15}\ L_{III}N_{IV}$				0.83722	5	14.8086
$\beta_2\ L_{III}N_V$	0.858	2	14.45	0.83537	5	14.8414
$\beta_7\ L_{III}O_I$				0.8162	1	15.190
$\beta_5\ L_{III}O_{IV,V}$				0.80627	5	15.3771
$L_{III}P_I$				0.8050	1	15.402
$\beta_{10}\ L_IM_{IV}$				0.77546	5	15.988
$\beta_9\ L_IM_V$				0.76857	5	16.131
L_IN_I				0.6874	1	18.036
L_IN_{IV}				0.6666	1	18.600
$\gamma_{11}\ L_IN_V$				0.6654	1	18.633
$L_IO_{IV,V}$				0.6468	1	19.167
$\beta_{17}\ L_{II}M_{III}$				0.8438	1	14.692
$L_{II}N_{III}$				0.7043	1	17.604
$L_{II}N_V$				0.6932	1	17.884
$L_{II}O_{II}$				0.6780	1	18.286

Designation	Å*	p.e.	keV	Å*	p.e.	keV
	87 Francium (*Cont.*)			**88 Radium** (*Cont.*)		
$L_{II}O_{III}$				0.6764	1	18.330
$L_{II}P_{II,III}$				0.6714	1	18.466
$L_{III}N_{II}$				0.8618	1	14.387
$L_{III}N_{III}$				0.8512	1	14.566
$u\ L_{III}N_{VI,VII}$				0.8186	1	15.146
$L_{III}P_{II,III}$				0.8038	1	15.425
	89 Actinium			**90 Thorium**		
$\alpha_2\ KL_{II}$	0.14141†	2	87.67	0.137829	2	89.953
$\alpha_1\ KL_{III}$	0.136417†	8	90.884	0.132813	2	93.350
$\beta_3\ KM_{II}$	0.12143†	2	102.10	0.118268	3	104.831
$\beta_1\ KM_{III}$	0.12055†	2	102.85	0.117396	9	105.609
$\beta_2^{II}\ KN_{II}$	0.11732†	2	105.67	0.11426	1	108.511
$\beta_2^{I}\ KN_{III}$	0.11711†	2	105.86	0.114040	9	108.717
$KO_{II,III}$				0.11322	1	109.500
$\beta_5\ KM_{IV,V}$				0.116667	9	106.269
$\beta_4\ KN_{IV,V}$				0.11366	2	109.08
$\beta_4\ L_IM_{II}$				0.79257	4	15.6429
$\beta_3\ L_IM_{III}$	0.77822†	9	15.931	0.75479	3	16.4258
$\gamma_2\ L_IN_{II}$				0.64221	4	19.305
$\gamma_3\ L_IN_{III}$				0.63559	4	19.507
$\gamma'_4\ L_IO_{II}$				0.61251	4	20.242
$\gamma_4\ L_IO_{III}$				0.61098	4	20.292
$\gamma_{13}\ L_IP_{II,III}$				0.60705	8	20.424
$\eta\ L_{II}M_I$				0.85446	4	14.5099
$\beta_1\ L_{II}M_{IV}$	0.78903†	9	15.713	0.765210	9	16.2022
$\gamma_5\ L_{II}N_I$				0.67491	4	18.370
$\gamma_1\ L_{II}N_{IV}$	0.67351†	9	18.408	0.65313	3	18.9825
$\gamma_8\ L_{II}O_I$				0.63898	5	19.403
$\gamma_6\ L_{II}O_{IV}$				0.63258	4	19.599
$L_{II}P_I$				0.6316	1	19.629
$L_{II}P_{IV}$				0.62991	9	19.682
$l\ L_{III}M_I$				1.11508	4	11.1186
$\alpha_2\ L_{III}M_{IV}$	0.99178†	5	12.5008	0.96788	2	12.8096
$\alpha_1\ L_{III}M_V$	0.97993†	5	12.6520	0.95600	3	12.9687
$\beta_6\ L_{III}N_I$				0.82790	8	14.975
$\beta_{15}\ L_{III}N_{IV}$				0.79539	5	15.5875
$\beta_2\ L_{III}N_V$				0.79354	3	15.6237
$\beta_7\ L_{III}O_I$				0.77437	4	16.0105
$\beta_5\ L_{III}O_{IV,V}$				0.76468	5	16.213
$L_{III}P_I$				0.76338	5	16.241
$L_{III}P_{IV,V}$				0.76087	9	16.295
$\beta_{10}\ L_IM_{IV}$				0.7301	1	16.981
$\beta_9\ L_IM_V$				0.7234	1	17.139
L_IN_I				0.64755	5	19.146
L_IN_{IV}				0.6276	1	19.755
$\gamma_{11}\ L_IN_V$				0.62636	9	19.794
$L_IN_{VI,VII}$				0.6160	1	20.128
L_IO_I				0.6146	1	20.174
$L_IO_{IV,V}$				0.6083	1	20.383
$L_{II}M_{II}$				0.8338	1	14.869
$\beta_{17}\ L_{II}M_{III}$				0.79257	4	15.6429
$L_{II}M_V$				0.7579	1	16.359
$L_{II}N_{III}$				0.6620	1	18.729
$L_{II}N_V$				0.6521	1	19.014

Designation	Å*	p.e.	keV	Å*	p.e.	keV
	89 Actinium (*Cont.*)			**90 Thorium** (*Cont.*)		
$v\ L_{II}N_{VI}$				0.64064	9	19.353
$L_{II}O_{II}$				0.6369	1	19.466
$L_{II}O_{III}$				0.6356	1	19.506
$L_{II}P_{II,III}$				0.6312	1	19.642
$t\ L_{III}M_I$				1.08009	9	11.4788
$s\ L_{III}M_{II}$				1.0112	1	12.261
$L_{III}N_{II}$				0.8190	2	15.138
$L_{III}N_{III}$				0.8082	1	15.341
$u\ L_{III}N_{VI,VII}$				0.77661	5	15.964
$L_{III}O_{II}$				0.7713	1	16.074
$L_{III}O_{III}$				0.7690	1	16.123
$L_{III}P_{II,III}$				0.7625	2	16.260
M_IN_{III}				2.934	8	4.23
M_IO_{III}				2.442	9	5.08
$M_{II}N_I$				3.537	9	3.505
$M_{II}N_{IV}$				3.011	2	4.117
$M_{II}O_{IV}$				2.618	5	4.735
$M_{III}N_I$				4.568	5	2.714
$M_{III}N_{IV}$				3.718	3	3.335
$\gamma\ M_{II}N_V$				3.679	2	3.370
$M_{III}O_I$				3.283	9	3.78
$M_{III}O_{IV,V}$				3.131	3	3.959
$\zeta_2\ M_VN_{II}$				5.340	5	2.322
$M_{IV}N_{III}$				4.911	5	2.524
$\beta\ M_{IV}N_{VI}$				3.941	1	3.1458
$M_{IV}O_{II}$				3.808	4	3.256
$\zeta_1\ M_VN_{III}$				5.245	5	2.364
$\alpha_2\ M_VN_{VI}$				4.151	2	2.987
$\alpha_1\ M_VN_{VII}$				4.1381	9	2.9961
M_VP_{III}				3.760	9	3.298
$N\ P_{II}$				9.44	7	1.313
$N\ P_{III}$				9.40	7	1.3119
$N_{II}O_{IV}$				11.56	5	1.072
N_IP_I				11.07	7	1.120
$N_{III}O_V$				13.8	1	0.897
$N_{IV}N_{VI}$				33.57	9	0.3693
$N_VN_{VI,VII}$				36.32	9	0.3414
$N_{VI}O_{IV}$				49.5	1	0.2505
$N_{VI}O_V$				48.2	1	0.2572
$N_{VII}O_V$				50.0	1	0.2479
$O_{III}P_{IV,V}$				68.2	3	0.1817
$O_{IV,V}Q_{II,III}$				181.	5	0.068
	91 Protactinium			**92 Uranium**		
$\alpha_2\ KL_{II}$	0.134343†	9	92.287	0.130968	4	94.665
$\alpha_1\ KL_{III}$	0.129325†	3	95.868	0.125947	3	98.439
$\beta_3\ KM_{II}$	0.11523†	2	107.60	0.112296	4	110.406
$\beta_1\ KM_{III}$	0.114345†	8	108.427	0.111394	5	111.300
$\beta_2^{II}\ KN_{II}$	0.11129†	2	111.40	0.10837	1	114.40
$\beta_2^{I}\ KN_{III}$	0.11107†	2	111.62	0.10818	1	114.60
$KO_{II,III}$				0.10744	1	115.39
$\beta_5\ KM_{IV,V}$				0.11069	1	112.01
$\beta_4\ KN_{IV,V}$				0.10780	2	115.01
$\beta_4\ L_IM_I$	0.7699	1	16.104	0.747985	9	16.5753
$\beta_3\ L_IM_{III}$	0.73230	5	16.930	0.71029	2	17.4550

Designation	Å*	p.e.	keV	Å*	p.e.	keV
	91 Protactinium (*Cont.*)			**92 Uranium** (*Cont.*)		
$\gamma_2\ L_IN_{II}$	0.6239	1	19.872	0.605237	9	20.4847
$\gamma_3\ L_IN_{III}$	0.6169	1	20.098	0.598574	9	20.7127
$\gamma'_4\ L_IO_{II}$				0.576700	9	21.4984
$\gamma_4\ L_IO_{II,III}$	0.5937	1	20.882	0.57499	9	21.562
γ_{13}				0.5706	1	21.729
$\eta\ L_{II}M_I$	0.8295	1	14.946	0.80509	2	15.3997
$\beta_1\ L_{II}M_{IV}$	0.74232	5	16.702	0.719984	8	17.2200
$\gamma_5\ L_{II}N_I$	0.6550	1	18.930	0.63557	2	19.5072
$\gamma_1\ L_{II}N_{IV}$	0.63358†	9	19.568	0.614770	9	20.1671
$\gamma_8\ L_{II}O_I$				0.60125	5	20.621
$\gamma_6\ L_{II}O_{IV}$	0.6133	1	20.216	0.594845	9	20.8426
$L_{II}P_{IV}$				0.59203	5	20.942
$l\ L_{III}M_I$	1.0908	1	11.366	1.06712	2	11.6183
$\alpha_2\ L_{III}M_{IV}$	0.94482†	5	13.1222	0.922558	9	13.4388
$\alpha_1\ L_{III}M_V$	0.93284	5	13.2907	0.910639	9	13.6147
$\beta_6\ L_{III}N_I$	0.8079	1	15.347	0.78838	2	15.7260
$\beta_{15}\ L_{III}N_{IV}$				0.756642	9	16.3857
$\beta_2\ L_{III}N_V$	0.7737	1	16.024	0.754681	9	16.4283
$\beta_7\ L_{III}O_I$	0.7546	2	16.431	0.73602	6	16.845
$\beta_5\ L_{III}O_{IV,V}$	0.7452	2	16.636	0.726305	9	17.0701
$L_{III}P_I$				0.72521	5	17.096
$L_{III}P_{IV,V}$				0.72240	5	17.162
$\beta_{10}\ L_IM\ _V$	0.7088	2	17.492	0.68760	5	18.031
$\beta_9\ L_IM_V$	0.7018	1	17.667	0.681014	8	18.2054
L_IN_{IV}				0.59096	5	20.979
$\gamma_{11}\ L_IN_V$				0.58986	5	21.019
$L_IO_{IV,V}$				0.5725	1	21.657
$\beta_{17}\ L_{II}M_{III}$				0.74503	5	16.641
$L_{II}N_{III}$				0.6228	1	19.907
$v\ L_{II}N_{VI}$				0.6031	1	20.556
$L_{III}O_{III}$				0.59728	5	20.758
$L_{II}P_{II,III}$				0.5930	2	20.906
$t\ L_{III}M_{II}$				1.0347	1	11.982
$s\ L_{III}M_{III}$				0.9636	1	12.866
$L_{III}N_{II}$				0.78017	9	15.892
$L_{III}N_{III}$				0.7691	1	16.120
$u\ L_{III}N_{VI,VII}$				0.738603	9	16.7859
$L_{III}O_{II}$				0.7333	1	16.907
$L_{III}O_{III}$				0.7309	1	16.962
$L_{III}P_{II,III}$				0.72426	5	17.118
M_IN_{II}				2.92	2	4.25
M_IN_{III}				2.753	8	4.50
M_IO_{III}				2.304	7	5.38
M_IP_{III}				2.253	6	5.50
$M_{II}N_I$	3.441	5	3.603	3.329	4	3.724
$M_{II}N_{IV}$	2.910	2	4.260	2.817	2	4.401
$M_{II}O_{IV}$	2.527	4	4.906	2.443	4	5.075
$M_{III}N_I$	4.450	4	2.786	4.330	2	2.863
$M_{III}N_{IV}$	3.614	2	3.430	3.521	2	3.521
$\gamma\ M_{III}N_V$	3.577	1	3.4657	3.479	1	3.563
$M_{III}O_I$	3.245	9	3.82	3.115	7	3.980
$M_{III}O_{IV,V}$	3.038	2	4.081	2.948	2	4.205
$\zeta_2\ M_{IV}N_{II}$	5.193	2	2.3876	5.050	2	2.4548
$M_{IV}N_{III}$				4.625	5	2.681
$\beta\ M_{IV}N_{VI}$	3.827	1	3.2397	3.716	1	3.3367

Designation	Å*	p.e.	keV	Å*	p.e.	keV
	91 Protactinium (*Cont.*)			**92 Uranium** (*Cont.*)		
$M_{IV}O_{II}$	3.691	2	3.359	3.576	1	3.4666
$\zeta_1\ M_VN_{III}$	5.092	2	2.4350	4.946	2	2.507
$\alpha_2\ M_VN_{VI}$	4.035	3	3.072	3.924	1	3.1595
$\alpha_1\ M_VN_{VII}$	4.022	1	3.0823	3.910	1	3.1708
N_IO_{III}				10.09	7	1.229
N_IP_{II}				8.81	7	1.41
N_IP_{III}				8.76	7	1.42
$N_{II}P_I$				10.40	7	1.192
$N_{III}O_V$				12.90	9	0.961
$N_{IV}N_{VI}$				31.8	1	0.390
$N_VN_{VI,VII}$				34.8	1	0.357
$N_{IV}O_{IV}$				43.3	2	0.286
$N_{VI}O_V$				42.1	2	0.295
$N_IP_{IV,V}$				8.60	7	1.44
	93 Neptunium			**94 Plutonium**		
$\beta_4\ L_IM_{II}$	0.72671	2	17.0607	0.70620	2	17.5560
$\beta_3\ L_IM_{III}$	0.68920†	9	17.989	0.66871	2	18.5405
$\gamma_2\ L_IN_{II}$	0.5873	5	21.11	0.57068	2	21.7251
$\gamma_3\ L_IN_{III}$	0.5810	5	21.34	0.564001	9	21.9824
$\gamma'_4\ L_IO_{II}$				0.5432	1	22.823
$\gamma_4\ L_IO_{II,III}$	0.5585	5	22.20	0.5416	1	22.891
$\eta\ L_{II}M_I$	0.7809	2	15.876	0.7591	1	16.333
$\beta_1\ L_{II}M_{IV}$	0.698478	9	17.7502	0.67772	2	18.2937
$\gamma_5\ L_{II}N_I$	0.616	1	20.12	0.5988	1	20.704
$\gamma_1\ L_{II}N_{IV}$	0.596498	9	20.7848	0.578882	9	21.4173
γ_8				0.5658	1	21.914
$\gamma_6\ L_{II}O_{IV}$	0.57699	5	21.488	0.55973	2	22.1502
$l\ L_{III}M_I$	1.0428	6	11.890	1.0226	1	12.124
$\alpha_2\ L_{III}M_{IV}$	0.901045	9	13.7597	0.88028	2	14.0842
$\alpha_1\ L_{III}M_V$	0.889128	9	13.9441	0.86830	2	14.2786
$\beta_6\ L_{III}N_I$	0.769	1	16.13	0.75148	2	16.4983
$\beta_{15}\ L_{III}N_{IV}$				0.7205	1	17.208
$\beta_2\ L_{III}N_V$	0.736230	9	16.8400	0.71851	2	17.2553
$\beta_7\ L_{III}O_I$				0.7003	1	17.705
$\beta_5\ L_{III}O_{IV,V}$	0.70814	2	17.5081	0.69068	2	17.9506
$\beta_{10}\ L_IM_{IV}$				0.6482	1	19.126
$\beta_9\ L_IM_V$				0.6416	1	19.323
$u\ L_{III}N_{VI,VII}$				0.7031	1	17.635
	95 Americium					
$\beta_4\ L_IM_{II}$	0.68639	2	18.0627			
$\beta_3\ L_IM_{III}$	0.64891	2	19.1059			
$\gamma_2\ L_IN_{II}$	0.5544	2	22.361			
$\beta_1\ L_{II}M_{IV}$	0.657655	9	18.8520			
$\gamma_1\ L_{II}N_{IV}$	0.561886	9	22.0652			
$\gamma_6\ L_{II}O_{IV}$	0.54311	2	22.8282			
$l\ L_{III}M_I$	1.0012	6	12.384			
$\alpha_2\ L_{III}M_{IV}$	0.860266	9	14.4119			
$\alpha_1\ L_{III}M_V$	0.848187	9	14.6172			
$\beta_6\ L_{III}N_I$	0.73418	2	16.8870			
$\beta_{15}\ L_{III}N_{IV}$	0.70341	2	17.6258			
$\beta_2\ L_{III}N_V$	0.701390	9	17.6765			
$\beta_5\ L_{III}O_{IV,V}$	0.67383	2	18.3996			

Table B. Wavelengths in numerical order of the emission lines and absorption edges.

Wavelength Å*	p.e.	Element	Designation		keV
0.10723	1	92 U	K	Abs. Edge	115.62
0.10744	1	92 U		$KO_{II,III}$	115.39
0.10780	2	92 U	$K\beta_4$	$KN_{IV,V}$	115.01
0.10818	1	92 U	$K\beta_2^{I}$	KN_{III}	114.60
0.10837	1	92 U	$K\beta_2^{II}$	KN_{II}	114.40
0.11069	1	92 U	$K\beta_5$	$KM_{IV,V}$	112.01
0.11107	2	91 Pa	$K\beta_2^{I}$	KN_{III}	111.62
0.11129	2	91 Pa	$K\beta_2^{II}$	KN_{II}	111.40
0.111394	5	92 U	$K\beta_1$	KM_{III}	111.300
0.112296	4	92 U	$K\beta_3$	KM_{II}	110.406
0.11307	1	90 Th	K	Abs. Edge	109.646
0.11322	1	90 Th		$KO_{II,III}$	109.500
0.11366	2	90 Th	$K\beta_4$	$KN_{IV,V}$	109.08
0.114040	9	90 Th	$K\beta_2^{I}$	KN_{III}	108.717
0.11426	1	90 Th	$K\beta_2^{II}$	KN_{II}	108.511
0.114345	8	91 Pa	$K\beta_1$	KM_{III}	108.427
0.11523	2	91 Pa	$K\beta_3$	KM_{II}	107.60
0.116667	9	90 Th	$K\beta_5$	$KM_{IV,V}$	106.269
0.11711	2	89 Ac	$K\beta_2^{I}$	KN_{III}	105.86
0.11732	2	89 Ac	$K\beta_2^{II}$	KN_{II}	105.67
0.117396	9	90 Th	$K\beta_1$	KM_{III}	105.609
0.118268	3	90 Th	$K\beta_3$	KM_{II}	104.831
0.12029	3	88 Ra	$K\beta_2^{I}$	KN_{III}	103.07
0.12050	3	88 Ra	$K\beta_2^{II}$	KN_{II}	102.89
0.12055	2	89 Ac	$K\beta_1$	KM_{III}	102.85
0.12143	2	89 Ac	$K\beta_3$	KM_{II}	102.10
0.12358	5	87 Fr	$K\beta_2^{I}$	KN_{III}	100.33
0.12379	5	87 Fr	$K\beta_2^{II}$	KN_{II}	100.16
0.12382	3	88 Ra	$K\beta_1$	KM_{III}	100.13
0.12469	3	88 Ra	$K\beta_3$	KM_{II}	99.43
0.125947	3	92 U	$K\alpha_1$	KL_{III}	98.439
0.12698	5	86 Rn	$K\beta_2^{I}$	KN_{III}	97.64
0.12719	5	86 Rn	$K\beta_2^{II}$	KN_{II}	97.47
0.12719	5	87 Fr	$K\beta_1$	KM_{III}	97.47
0.12807	5	87 Fr	$K\beta_3$	KM_{II}	96.81
0.129325	3	91 Pa	$K\alpha_1$	KL_{III}	95.868
0.13052	4	85 At	$K\beta_2^{I}$	KN_{III}	94.99
0.13069	5	86 Rn	$K\beta_1$	KM_{III}	94.87
0.13072	4	85 At	$K\beta_2^{II}$	KN_{II}	94.84
0.130968	4	92 U	$K\alpha_2$	KL_{II}	94.665
0.13155	5	86 Rn	$K\beta_3$	KM_{II}	94.24
0.132813	2	90 Th	$K\alpha_1$	KL_{III}	93.350
0.13418	2	84 Po	$K\beta_2^{I}$	KN_{III}	92.40
0.13432	4	85 At	$K\beta_1$	KM_{III}	92.30
0.134343	9	91 Pa	$K\alpha_2$	KL_{II}	92.287
0.13438	2	84 Po	$K\beta_2^{II}$	KN_{II}	92.26
0.13517	4	85 At	$K\beta_3$	KM_{II}	91.72
0.136417	8	89 Ac	$K\alpha_1$	KL_{III}	90.884
0.13694	1	83 Bi	K	Abs. Edge	90.534
0.13709	1	83 Bi		$KO_{II,III}$	90.435
0.13759	2	83 Bi	$K\beta_4$	$KN_{IV,V}$	90.11
0.137829	2	90 Th	$K\alpha_2$	KL_{II}	89.953
0.13797	1	83 Bi	$K\beta_2^{I}$	KN_{III}	89.864
0.13807	2	84 Po	$K\beta_1$	KM_{III}	89.80
0.13817	1	83 Bi	$K\beta_2^{II}$	KN_{II}	89.733
0.13892	2	84 Po	$K\beta_3$	KM_{II}	89.25
0.14014	2	88 Ra	$K\alpha_1$	KL_{III}	88.47
0.1408	1	82 Pb		KP	88.06
0.140880	5	82 Pb	K	Abs. Edge	88.005
0.141012	8	82 Pb		$KO_{II,III}$	87.922
0.14111	1	83 Bi	$K\beta_5$	$KM_{IV,V}$	87.860
0.14141	2	89 Ac	$K\alpha_2$	KL_{II}	87.67
0.14155	3	82 Pb	$K\beta_4$	$KN_{IV,V}$	87.59
0.14191	1	82 Pb	$K\beta_2^{I}$	KN_{III}	87.364
0.141948	3	83 Bi	$K\beta_1$	KM_{III}	87.343
0.14212	2	82 Pb	$K\beta_2^{II}$	KN_{II}	87.23
0.142779	7	83 Bi	$K\beta_3$	KM_{II}	86.834
0.14399	3	87 Fr	$K\alpha_1$	KL_{III}	86.10
0.14495	1	81 Tl	K	Abs. Edge	85.533
0.14495	3	82 Pb	$K\beta_5^{I}$	KM_{V}	85.53
0.14509	1	81 Tl		$KO_{II,III}$	85.451
0.14512	2	82 Pb	$K\beta_5^{II}$	KM_{IV}	85.43
0.14512	2	88 Ra	$K\alpha_2$	KL_{II}	85.43
0.14553	2	81 Tl	$K\beta_4$	$KN_{IV,V}$	85.19
0.14595	1	81 Tl	$K\beta_2^{I}$	KN_{III}	84.946
0.145970	6	82 Pb	$K\beta_1$	KM_{III}	84.936
0.14614	1	81 Tl	$K\beta_2^{II}$	KN_{II}	84.836
0.146810	4	82 Pb	$K\beta_3$	KM_{II}	84.450
0.14798	3	86 Rn	$K\alpha_1$	KL_{III}	83.78
0.14896	3	87 Fr	$K\alpha_2$	KL_{II}	83.23
0.14917	1	81 Tl	$K\beta_5$	$KM_{IV,V}$	83.114
0.14918	1	80 Hg	K	Abs. Edge	83.109
0.14931	2	80 Hg		$KO_{II,III}$	83.04
0.14978	2	80 Hg	$K\beta_4$	$KN_{IV,V}$	82.78
0.150142	5	81 Tl	$K\beta_1$	KM_{III}	82.576
0.15020	2	80 Hg	$K\beta_2^{I}$	KN_{III}	82.54
0.15040	2	80 Hg	$K\beta_2^{II}$	KN_{II}	82.43
0.150980	6	81 Tl	$K\beta_3$	KM_{II}	82.118
0.15210	2	85 At	$K\alpha_1$	KL_{III}	81.52
0.15294	3	86 Rn	$K\alpha_2$	KL_{II}	81.07
0.15353	2	80 Hg	$K\beta_5$	$KM_{IV,V}$	80.75
0.153593	5	79 Au	K	Abs. Edge	80.720
0.153694	7	79 Au		$KO_{II,III}$	80.667
0.154224	5	79 Au	$K\beta_4$	$KN_{IV,V}$	80.391
0.154487	3	80 Hg	$K\beta_1$	KM_{III}	80.253
0.154618	9	79 Au	$K\beta_2^{I}$	KN_{III}	80.185
0.15483	2	79 Au	$K\beta_2^{II}$	KN_{II}	80.08
0.155321	3	80 Hg	$K\beta_3$	KM_{II}	79.822
0.15636	1	84 Po	$K\alpha_1$	KL_{III}	79.290
0.15705	2	85 At	$K\alpha_2$	KL_{II}	78.95
0.157880	5	79 Au	$K\beta_5^{I}$	KM_{V}	78.529
0.158062	7	79 Au	$K\beta_5^{II}$	KM_{IV}	78.438
0.15818	1	78 Pt	K	Abs. Edge	78.381
0.15826	1	78 Pt		$KO_{II,III}$	78.341
0.15881	2	78 Pt	$K\beta_4$	$KN_{IV,V}$	78.069
0.158982	3	79 Au	$K\beta_1$	KM_{III}	77.984
0.15920	1	78 Pt	$K\beta_2^{I}$	KN_{III}	77.878
0.15939	1	78 Pt	$K\beta_2^{II}$	KN_{II}	77.785
0.159810	2	79 Au	$K\beta_3$	KM_{II}	77.580
0.160789	2	83 Bi	$K\alpha_1$	KL_{III}	77.1079
0.16130	1	84 Po	$K\alpha_2$	KL_{II}	76.862
0.16255	3	78 Pt	$K\beta_5^{I}$	KM_{V}	76.27
0.16271	2	78 Pt	$K\beta_5^{II}$	KM_{IV}	76.199
0.16292	1	77 Ir	K	Abs. Edge	76.101

TABLE B (*Continued*)

Wavelength Å*	p.e.	Element	Designation		keV
0.163019	5	77 Ir		$KO_{II,III}$	76.053
0.16352	2	77 Ir	$K\beta_4$	$KN_{IV,V}$	75.821
0.163675	3	78 Pt	$K\beta_1$	KM_{III}	75.748
0.163956	7	77 Ir	$K\beta_2^{I}$	KN_{III}	75.619
0.16415	1	77 Ir	$K\beta_2^{II}$	KN_{II}	75.529
0.164501	3	78 Pt	$K\beta_3$	KM_{II}	75.368
0.165376	2	82 Pb	$K\alpha_1$	KL_{III}	74.9694
0.165717	2	83 Bi	$K\alpha_2$	KL_{II}	74.8148
0.167373	9	77 Ir	$K\beta_5^{I}$	KM_{V}	74.075
0.16759	2	77 Ir	$K\beta_5^{II}$	KM_{IV}	73.980
0.16787	1	76 Os	K	Abs. Edge	73.856
0.16798	1	76 Os		$KO_{II,III}$	73.808
0.16842	2	76 Os	$K\beta_4$	$KN_{IV,V}$	73.615
0.168542	2	77 Ir	$K\beta_1$	KM_{III}	73.5608
0.168906	6	76 Os	$K\beta_2^{I}$	KN_{III}	73.402
0.16910	1	76 Os	$K\beta_2^{II}$	KN_{II}	73.318
0.169367	2	77 Ir	$K\beta_3$	KM_{II}	73.2027
0.170136	2	81 Tl	$K\alpha_1$	KL_{III}	72.8715
0.170294	2	82 Pb	$K\alpha_2$	KL_{II}	72.8042
0.17245	1	76 Os	$K\beta_5^{I}$	KM_{V}	71.895
0.17262	1	76 Os	$K\beta_5^{II}$	KM_{IV}	71.824
0.17302	1	75 Re	K	Abs. Edge	71.658
0.17308	1	75 Re		$KO_{II,III}$	71.633
0.173611	3	76 Os	$K\beta_1$	KM_{III}	71.413
0.17362	2	75 Re	$K\beta_4$	$KN_{IV,V}$	71.410
0.174054	6	75 Re	$K\beta_2^{I}$	KN_{III}	71.232
0.17425	1	75 Re	$K\beta_2^{II}$	KN_{II}	71.151
0.174431	3	76 Os	$K\beta_3$	KM_{II}	71.077
0.175036	2	81 Tl	$K\alpha_2$	KL_{II}	70.8319
0.175068	3	80 Hg	$K\alpha_1$	KL_{III}	70.819
0.17766	1	75 Re	$K\beta_5^{I}$	KM_{V}	69.786
0.17783	1	75 Re	$K\beta_5^{II}$	KM_{IV}	69.719
0.17837	1	74 W	K	Abs. Edge	69.508
0.178444	5	74 W		$KO_{II,III}$	69.479
0.178880	3	75 Re	$K\beta_1$	KM_{III}	69.310
0.17892	2	74 W	$K\beta_4$	$KN_{IV,V}$	69.294
0.179421	7	74 W	$K\beta_2^{I}$	KN_{III}	69.101
0.17960	1	74 W	$K\beta_2^{II}$	KN_{II}	69.031
0.179697	3	75 Re	$K\beta_3$	KM_{II}	68.994
0.179958	3	80 Hg	$K\alpha_2$	KL_{II}	68.895
0.180195	2	79 Au	$K\alpha_1$	KL_{III}	68.8037
0.183092	7	74 W	$K\beta_5^{I}$	KM_{V}	67.715
0.183264	5	74 W	$K\beta_5^{II}$	KM_{IV}	67.652
0.18394	1	73 Ta	K	Abs. Edge	67.403
0.184031	7	73 Ta		$KO_{II,III}$	67.370
0.184374	2	74 W	$K\beta_1$	KM_{III}	67.2443
0.18451	1	73 Ta	$K\beta_4$	$KN_{IV,V}$	67.194
0.185011	8	73 Ta	$K\beta_2^{I}$	KN_{III}	67.013
0.185075	2	79 Au	$K\alpha_2$	KL_{II}	66.9895
0.185181	2	74 W	$K\beta_3$	KM_{II}	66.9514
0.185188	9	73 Ta	$K\beta_2^{II}$	KN_{II}	66.949
0.185511	4	78 Pt	$K\alpha_1$	KL_{III}	66.832
0.18672	4	79 Au		KL_{I}	66.40
0.188757	6	73 Ta	$K\beta_5^{I}$	KM_{V}	65.683
0.188920	6	73 Ta	$K\beta_5^{II}$	KM_{IV}	65.626
0.18982	5	72 Hf	K	Abs. Edge	65.31
0.190089	4	73 Ta	$K\beta_1$	KM_{III}	65.223
0.190381	4	78 Pt	$K\alpha_2$	KL_{II}	65.122
0.1908	2	72 Hf	$K\beta_2$	$KN_{II,III}$	64.98
0.190890	2	73 Ta	$K\beta_3$	KM_{II}	64.9488
0.191047	2	77 Ir	$K\alpha_1$	KL_{III}	64.8956
0.19585	5	71 Lu	K	Abs. Edge	63.31
0.19589	2	71 Lu		$KO_{II,III}$	63.293
0.195904	2	77 Ir	$K\alpha_2$	KL_{II}	63.2867
0.19607	3	72 Hf	$K\beta_1$	KM_{III}	63.234
0.196794	2	76 Os	$K\alpha_1$	KL_{III}	63.0005
0.19686	4	72 Hf	$K\beta_3$	KM_{II}	62.98
0.1969	2	71 Lu	$K\beta_2$	$KN_{II,III}$	62.97
0.20084	2	71 Lu	$K\beta_5$	$KM_{IV,V}$	61.732
0.201639	2	76 Os	$K\alpha_2$	KL_{II}	61.4867
0.20224	5	70 Yb	K	Abs. Edge	61.30
0.20226	2	70 Yb		$KO_{II,III}$	61.298
0.20231	3	71 Lu	$K\beta_1$	KM_{III}	61.283
0.202781	2	75 Re	$K\alpha_1$	KL_{III}	61.1403
0.20309	4	71 Lu	$K\beta_3$	KM_{II}	61.05
0.2033	2	70 Yb	$K\beta_2$	$KN_{II,III}$	60.89
0.20739	2	70 Yb	$K\beta_5$	$KM_{IV,V}$	59.782
0.207611	1	75 Re	$K\alpha_2$	KL_{II}	59.7179
0.20880	5	69 Tm	K	Abs. Edge	59.38
0.20884	8	70 Yb	$K\beta_1$	KM_{III}	59.37
0.20891	2	69 Tm		$KO_{II,III}$	59.346
0.2090100	Std.	74 W	$K\alpha_1$	KL_{III}	59.31824
0.2096	1	70 Yb	$K\beta_3$	KM_{II}	59.14
0.2098	2	69 Tm	$K\beta_2$	$KN_{II,III}$	59.09
0.213828	2	74 W	$K\alpha_2$	KL_{II}	57.9817
0.21404	2	69 Tm	$K\beta_5$	$KM_{IV,V}$	57.923
0.215497	4	73 Ta	$K\alpha_1$	KL_{III}	57.532
0.21556	2	69 Tm	$K\beta_1$	KM_{III}	57.517
0.21567	1	68 Er	K	Abs. Edge	57.487
0.21581	3	68 Er		$KO_{II,III}$	57.450
0.21592	4	74 W		KL_{I}	57.42
0.21636	2	69 Tm	$K\beta_3$	KM_{II}	57.304
0.2167	2	68 Er	$K\beta_2$	$KN_{II,III}$	57.21
0.220305	8	73 Ta	$K\alpha_2$	KL_{II}	56.277
0.22124	3	68 Er	$K\beta_5$	$KM_{IV,V}$	56.040
0.222227	3	72 Hf	$K\alpha_1$	KL_{III}	55.7902
0.22266	2	68 Er	$K\beta_1$	KM_{III}	55.681
0.22291	1	67 Ho	K	Abs. Edge	55.619
0.22305	3	67 Ho		$KO_{II,III}$	55.584
0.22341	2	68 Er	$K\beta_3$	KM_{II}	55.494
0.2241	2	67 Ho	$K\beta_2$	$KN_{II,III}$	55.32
0.227024	3	72 Hf	$K\alpha_2$	KL_{II}	54.6114
0.22855	3	67 Ho	$K\beta_5$	$KM_{IV,V}$	54.246
0.229298	2	71 Lu	$K\alpha_1$	KL_{III}	54.0698
0.23012	2	67 Ho	$K\beta_1$	KM_{III}	53.877
0.23048	1	66 Dy	K	Abs. Edge	53.793
0.23056	3	66 Dy		$KO_{II,III}$	53.774
0.23083	2	67 Ho	$K\beta_3$	KM_{II}	53.711
0.2317	2	66 Dy	$K\beta_2$	$KN_{II,III}$	53.47
0.234081	2	71 Lu	$K\alpha_2$	KL_{II}	52.9650
0.23618	3	66 Dy	$K\beta_5$	$KM_{IV,V}$	52.494
0.236655	2	70 Yb	$K\alpha_1$	KL_{III}	52.3889
0.23788	2	66 Dy	$K\beta_1$	KM_{III}	52.119
0.23841	1	65 Tb	K	Abs. Edge	52.002

Wavelength Å*	p.e.	Element	Designation		keV
0.23858	3	65 Tb		$KO_{II,III}$	51.965
0.23862	2	66 Dy	$K\beta_3$	KM_{II}	51.957
0.2397	2	65 Tb	$K\beta_2$	$KN_{II,III}$	51.68
0.241424	2	70 Yb	$K\alpha_2$	KL_{II}	51.3540
0.244338	2	69 Tm	$K\alpha_1$	KL_{III}	50.7416
0.24608	2	65 Tb	$K\beta_1$	KM_{III}	50.382
0.24681	1	64 Gd	K	Abs. Edge	50.233
0.24683	2	65 Tb	$K\beta_3$	KM_{II}	50.229
0.24687	3	64 Gd		$KO_{II,III}$	50.221
0.24816	3	64 Gd	$K\beta_2$	$KN_{II,III}$	49.959
0.249095	2	69 Tm	$K\alpha_2$	KL_{II}	49.7726
0.252365	2	68 Er	$K\alpha_1$	KL_{III}	49.1277
0.25275	3	64 Gd	$K\beta_5$	$KM_{IV,V}$	49.052
0.25460	2	64 Gd	$K\beta_1$	KM_{III}	48.697
0.25534	2	64 Gd	$K\beta_3$	KM_{II}	48.555
0.25553	1	63 Eu	K	Abs. Edge	48.519
0.255645	7	63 Eu		$KO_{II,III}$	48.497
0.256923	8	63 Eu	$K\beta_2{}^{I}$	$KN_{II,III}$	48.256
0.257110	2	68 Er	$K\alpha_2$	KL_{II}	48.2211
0.260756	2	67 Ho	$K\alpha_1$	KL_{III}	47.5467
0.263577	5	63 Eu	$K\beta_1$	KM_{III}	47.0379
0.264332	5	63 Eu	$K\beta_3$	KM_{II}	46.9036
0.26464	5	62 Sm	K	Abs. Edge	46.849
0.26491	3	62 Sm		$KO_{II,III}$	46.801
0.265486	2	67 Ho	$K\alpha_2$	KL_{II}	46.6997
0.2662	1	62 Sm	$K\beta_2$	$KN_{II,III}$	46.57
0.269533	2	66 Dy	$K\alpha_1$	KL_{III}	45.9984
0.27111	3	62 Sm	$K\beta_5$	$KM_{IV,V}$	45.731
0.27301	2	62 Sm	$K\beta_1$	KM_{III}	45.413
0.27376	2	62 Sm	$K\beta_3$	KM_{II}	45.289
0.274247	2	66 Dy	$K\alpha_2$	KL_{II}	45.2078
0.27431	5	61 Pm	K	Abs. Edge	45.198
0.2759	1	61 Pm	$K\beta_2$	$KN_{II,III}$	44.93
0.278724	2	65 Tb	$K\alpha_1$	KL_{III}	44.4816
0.28290	3	61 Pm	$K\beta_1$	KM_{III}	43.826
0.283423	2	65 Tb	$K\alpha_2$	KL_{II}	43.7441
0.28363	4	61 Pm	$K\beta_3$	KM_{II}	43.713
0.28453	5	60 Nd	K	Abs. Edge	43.574
0.2861	1	60 Nd	$K\beta_2$	$KN_{II,III}$	43.32
0.288353	2	64 Gd	$K\alpha_1$	KL_{III}	42.9962
0.293038	2	64 Gd	$K\alpha_2$	KL_{II}	42.3089
0.293299	2	60 Nd	$K\beta_1$	KM_{III}	42.2713
0.294027	3	60 Nd	$K\beta_3$	KM_{II}	42.1665
0.29518	5	59 Pr	K	Abs. Edge	42.002
0.29679	2	59 Pr	$K\beta_2$	$KN_{II,III}$	41.773
0.298446	2	63 Eu	$K\alpha_1$	KL_{III}	41.5422
0.303118	2	63 Eu	$K\alpha_2$	KL_{II}	40.9019
0.304261	4	59 Pr	$K\beta_1$	KM_{III}	40.7482
0.304975	5	59 Pr	$K\beta_3$	KM_{II}	40.6529
0.30648	5	58 Ce	K	Abs. Edge	40.453
0.30668	2	58 Ce		$KO_{II,III}$	40.427
0.30737	2	58 Ce	$K\beta_4{}^{I}$	$KN_{IV,V}$	40.337
0.30816	1	58 Ce	$K\beta_2$	$KN_{II,III}$	40.233
0.309040	2	62 Sm	$K\alpha_1$	KL_{III}	40.1181
0.31342	2	58 Ce	$K\beta_5{}^{I}$	KM_{V}	39.558
0.31357	2	58 Ce	$K\beta_5{}^{II}$	KM_{IV}	39.539
0.313698	2	62 Sm	$K\alpha_2$	KL_{II}	39.5224
0.315816	2	58 Ce	$K\beta_1$	KM_{III}	39.2573
0.316520	4	58 Ce	$K\beta_3$	KM_{II}	39.1701
0.31844	5	57 La	K	Abs. Edge	38.934
0.31864	2	57 La		$KO_{II,III}$	38.909
0.31931	2	57 La	$K\beta_4{}^{I}$	$KN_{IV,V}$	38.828
0.320117	7	57 La	$K\beta_2$	$KN_{II,III}$	38.7299
0.320160	4	61 Pm	$K\alpha_1$	KL_{III}	38.7247
0.324803	4	61 Pm	$K\alpha_2$	KL_{II}	38.1712
0.32546	2	57 La	$K\beta_5{}^{I}$	KM_{V}	38.094
0.32563	2	57 La	$K\beta_5{}^{II}$	KM_{IV}	38.074
0.327983	3	57 La	$K\beta_1$	KM_{III}	37.8010
0.328686	4	57 La	$K\beta_3$	KM_{II}	37.7202
0.33104	1	56 Ba	K	Abs. Edge	37.452
0.33127	2	56 Ba		$KO_{II,III}$	37.426
0.331846	2	60 Nd	$K\alpha_1$	KL_{III}	37.3610
0.33229	2	56 Ba	$K\beta_4{}^{II}$	KN_{IV}	37.311
0.33277	1	56 Ba	$K\beta_2$	$KN_{II,III}$	37.257
0.336472	2	60 Nd	$K\alpha_2$	KL_{II}	36.8474
0.33814	2	56 Ba	$K\beta_5{}^{I}$	KM_{V}	36.666
0.33835	2	56 Ba	$K\beta_5{}^{II}$	KM_{IV}	36.643
0.340811	3	56 Ba	$K\beta_1$	KM_{III}	36.3782
0.341507	4	56 Ba	$K\beta_3$	KM_{II}	36.3040
0.344140	2	59 Pr	$K\alpha_1$	KL_{III}	36.0263
0.34451	1	55 Cs	K	Abs. Edge	35.987
0.34611	2	55 Cs	$K\beta_2$	$KN_{II,III}$	35.822
0.348749	2	59 Pr	$K\alpha_2$	KL_{II}	35.5502
0.354364	7	55 Cs	$K\beta_1$	KM_{III}	34.9869
0.355050	4	55 Cs	$K\beta_3$	KM_{II}	34.9194
0.357092	2	58 Ce	$K\alpha_1$	KL_{III}	34.7197
0.3584	5	54 Xe	K	Abs. Edge	34.59
0.36026	3	54 Xe	$K\beta_2$	$KN_{II,III}$	34.415
0.361683	2	58 Ce	$K\alpha_2$	KL_{II}	34.2789
0.36872	2	54 Xe	$K\beta_1$	KM_{III}	33.624
0.36941	2	54 Xe	$K\beta_3$	KM_{II}	33.562
0.370737	2	57 La	$K\alpha_1$	KL_{III}	33.4418
0.37381	1	53 I	K	Abs. Edge	33.1665
0.37523	2	53 I	$K\beta_2$	$KN_{II,III}$	33.042
0.375313	2	57 La	$K\alpha_2$	KL_{II}	33.0341
0.383905	4	53 I	$K\beta_1$	KM_{III}	32.2947
0.384564	4	53 I	$K\beta_3$	KM_{II}	32.2394
0.385111	4	56 Ba	$K\alpha_1$	KL_{III}	32.1936
0.389668	5	56 Ba	$K\alpha_2$	KL_{II}	31.8171
0.38974	1	52 Te		$KO_{II,III}$	31.8114
0.38974	1	52 Te	K	Abs. Edge	31.8114
0.391102	6	52 Te	$K\beta_2$	$KN_{II,III}$	31.7004
0.399995	5	52 Te	$K\beta_1$	KM_{III}	30.9957
0.400290	4	55 Cs	$K\alpha_1$	KL_{III}	30.9728
0.400659	4	52 Te	$K\beta_3$	KM_{II}	30.9443
0.404835	4	55 Cs	$K\alpha_2$	KL_{II}	30.6251
0.40666	1	51 Sb		$KO_{II,III}$	30.4875
0.40668	1	51 Sb	K	Abs. Edge	30.4860
0.40702	1	51 Sb	$K\beta_4{}^{I}$	$KN_{IV,V}$	30.4604
0.407973	5	51 Sb	$K\beta_2$	$KN_{II,III}$	30.3895
0.41378	1	51 Sb	$K\beta_5{}^{I}$	KM_{V}	29.9632
0.41388	1	51 Sb	$K\beta_5{}^{II}$	KM_{IV}	29.9560
0.41634	2	54 Xe	$K\alpha_1$	KL_{III}	29.779
0.417085	3	51 Sb	$K\beta_1$	KM_{III}	29.7256

Wavelength Å*	p.e.	Element	Designation		keV
0.417737	4.	51 Sb	$K\beta_3$	KM_{II}	29.6792
0.42087	2	54 Xe	$K\alpha_2$	KL_{II}	29.458
0.42467	3	50 Sn		$KO_{II,III}$	29.195
0.42467	1	50 Sn	K	Abs. Edge	29.1947
0.42495	3	50 Sn	$K\beta_4^{I}$	$KN_{IV,V}$	29.175
0.425915	8	50 Sn	$K\beta_2$	$KN_{II,III}$	29.1093
0.43175	3	50 Sn	$K\beta_5^{I}$	KM_{V}	28.716
0.43184	3	50 Sn	$K\beta_5^{II}$	KM_{IV}	28.710
0.433318	5	53 I	$K\alpha_1$	KL_{III}	28.6120
0.435236	5	50 Sn	$K\beta_1$	KM_{III}	28.4860
0.435877	5	50 Sn	$K\beta_3$	KM_{II}	28.4440
0.437829	7	53 I	$K\alpha_2$	KL_{II}	28.3172
0.44371	1	49 In	K	Abs. Edge	27.9420
0.44374	3	49 In		$KO_{II,III}$	27.940
0.44393	4	49 In	$K\beta_4^{I}$	$KN_{IV,V}$	27.928
0.44500	1	49 In	$K\beta_2$	$KN_{II,III}$	27.8608
0.45086	2	49 In	$K\beta_5^{I}$	KM_{V}	27.499
0.45098	2	49 In	$K\beta_5^{II}$	KM_{IV}	27.491
0.451295	3	52 Te	$K\alpha_1$	KL_{III}	27.4723
0.454545	4	49 In	$K\beta_1$	KM_{III}	27.2759
0.455181	4	49 In	$K\beta_3$	KM_{II}	27.2377
0.455784	3	52 Te	$K\alpha_2$	KL_{II}	27.2017
0.46407	1	48 Cd	K	Abs. Edge	26.7159
0.465328	7	48 Cd	$K\beta_2$	$KN_{II,III}$	26.6438
0.470354	3	51 Sb	$K\alpha_1$	KL_{III}	26.3591
0.474827	3	51 Sb	$K\alpha_2$	KL_{II}	26.1108
0.475105	6	48 Cd	$K\beta_1$	KM_{III}	26.0955
0.475730	5	48 Cd	$K\beta_3$	KM_{II}	26.0612
0.48589	1	47 Ag	K	Abs. Edge	25.5165
0.4859	9	47 Ag	$K\beta_4$	$KN_{IV,V}$	25.512
0.487032	4	47 Ag	$K\beta_2$	$KN_{II,III}$	25.4564
0.490599	3	50 Sn	$K\alpha_1$	KL_{III}	25.2713
0.49306	2	47 Ag	$K\beta_5$	$KM_{IV,V}$	25.145
0.495053	3	50 Sn	$K\alpha_2$	KL_{II}	25.0440
0.497069	4	47 Ag	$K\beta_1$	KM_{III}	24.9424
0.497685	4	47 Ag	$K\beta_3$	KM_{II}	24.9115
0.5092	1	46 Pd	K	Abs. Edge	24.348
0.5093	2	46 Pd	$K\beta_4$	$KN_{IV,V}$	24.346
0.510228	4	46 Pd	$K\beta_2$	$KN_{II,III}$	24.2991
0.512113	3	49 In	$K\alpha_1$	KL_{III}	24.2097
0.516544	3	49 In	$K\alpha_2$	KL_{II}	24.0020
0.51670	9	46 Pd	$K\beta_5$	$KM_{IV,V}$	23.995
0.520520	4	46 Pd	$K\beta_1$	KM_{III}	23.8187
0.521123	4	46 Pd	$K\beta_3$	KM_{II}	23.7911
0.53395	1	45 Rh	K	Abs. Edge	23.2198
0.53401	9	45 Rh	$K\beta_4^{I}$	$KN_{IV,V}$	23.217
0.535010	3	48 Cd	$K\alpha_1$	KL_{III}	23.1736
0.53503	2	45 Rh	$K\beta_2$	$KN_{II,III}$	23.1728
0.53513	5	45 Rh	$K\beta_2^{II}$	KN_{II}	23.168
0.5365	1	94 Pu	L_I	Abs. Edge	23.109
0.539422	3	48 Cd	$K\alpha_2$	KL_{II}	22.9841
0.54101	9	45 Rh	$K\beta_5^{I}$	KM_{V}	22.917
0.54118	9	45 Rh	$K\beta_5^{II}$	KM_{IV}	22.909
0.5416	1	94 Pu	$L\gamma_4$	L_IO_{III}	22.891
0.54311	2	95 Am	$L\gamma_6$	$L_{II}O_{IV}$	22.8282
0.5432	1	94 Pu	$L\gamma_4'$	L_IO_{II}	22.823
0.545605	4	45 Rh	$K\beta_1$	KM_{III}	22.7236
0.546200	4	45 Rh	$K\beta_3$	KM_{II}	22.6989
0.5544	2	95 Am	$L\gamma_2$	L_IN_{II}	22.361
0.5572	1	94 Pu	L_{II}	Abs. Edge	22.253
0.5585	5	93 Np	$L\gamma_4$	$L_IO_{II,III}$	22.20
0.5594075	6	47 Ag	$K\alpha_1$	KL_{III}	22.16292
0.55973	2	94 Pu	$L\gamma_6$	$L_{II}O_{IV}$	22.1502
0.56051	1	44 Ru	K	Abs. Edge	22.1193
0.56089	9	44 Ru	$K\beta_4$	$KN_{IV,V}$	22.104
0.56166	3	44 Ru	$K\beta_2$	$KN_{II,III}$	22.074
0.561886	9	95 Am	$L\gamma_1$	$L_{II}N_{IV}$	22.0652
0.563798	4	47 Ag	$K\alpha_2$	KL_{II}	21.9903
0.564001	9	94 Pu	$L\gamma_3$	L_IN_{III}	21.9824
0.5658	1	94 Pu	$L\gamma_8$	$L_{II}O_I$	21.914
0.56785	9	44 Ru	$K\beta_5^{I}$	KM_{V}	21.834
0.5680	2	44 Ru	$K\beta_5^{II}$	KM_{IV}	21.829
0.5695	1	92 U	L_I	Abs. Edge	21.771
0.5706	1	92 U	$L\gamma_{13}$	$L_IP_{II,III}$	21.729
0.57068	2	94 Pu	$L\gamma_2$	L_IN_{II}	21.1251
0.572482	4	44 Ru	$K\beta_1$	KM_{III}	21.6568
0.5725	1	92 U		$L_IO_{IV,V}$	21.657
0.573067	4	44 Ru	$K\beta_3$	KM_{II}	21.6346
0.57499	9	92 U	$L\gamma_4$	L_IO_{III}	21.562
0.576700	9	92 U	$L\gamma_4'$	L_IO_{II}	21.4984
0.57699	5	93 Np	$L\gamma_6$	$L_{II}O_{IV}$	21.488
0.578882	9	94 Pu	$L\gamma_1$	$L_{II}N_{IV}$	21.4173
0.5810	5	93 Np	$L\gamma_3$	L_IN_{III}	21.34
0.585448	3	46 Pd	$K\alpha_1$	KL_{III}	21.1771
0.5873	5	93 Np	$L\gamma_2$	L_IN_{II}	21.11
0.58906	1	43 Te	K	Abs. Edge	21.0473
0.589821	3	46 Pd	$K\alpha_2$	KL_{II}	21.0201
0.58986	5	92 U	$L\gamma_{11}$	L_IN_V	21.019
0.59024	5	43 Tc	$K\beta_2$	$KN_{II,III}$	21.005
0.59096	5	92 U		L_IN_{IV}	20.979
0.5919	1	92 U	L_{II}	Abs. Edge	20.945
0.59203	5	92 U		$L_{II}P_{IV}$	20.942
0.5930	2	92 U		$L_{II}P_{II,III}$	20.906
0.5937	1	91 Pa	$L\gamma_4$	$L_IO_{II,III}$	20.882
0.594845	9	92 U	$L\gamma_6$	$L_{II}O_{IV}$	20.8426
0.596498	9	93 Np	$L\gamma_1$	$L_{II}N_{IV}$	20.7848
0.59728	5	92 U		$L_{II}O_{III}$	20.758
0.598574	9	92 U	$L\gamma_3$	L_IN_{III}	20.7127
0.5988	1	94 Pu	$L\gamma_5$	$L_{II}N_I$	20.704
0.60125	5	92 U	$L\gamma_8$	$L_{II}O_I$	20.621
0.60130	4	43 Tc	$K\beta_1$	KM_{III}	20.619
0.60188	4	43 Tc	$K\beta_3$	KM_{II}	20.599
0.6031	1	92 U	Lv	$L_{II}N_{VI}$	20.556
0.605237	9	92 U	$L\gamma_2$	L_IN_{II}	20.4847
0.6059	1	90 Th	L_I	Abs. Edge	20.464
0.60705	8	90 Th	$L\gamma_{13}$	$L_IP_{II,III}$	20.424
0.6083	1	90 Th		$L_IO_{IV,V}$	20.383
0.61098	4	90 Th	$L\gamma_4$	L_IO_{III}	20.292
0.61251	4	90 Th	$L\gamma_4'$	L_IO_{II}	20.242
0.6133	1	91 Pa	$L\gamma_6$	$L_{II}O_{IV}$	20.216
0.613279	4	45 Rh	$K\alpha_1$	KL_{III}	20.2161
0.6146	1	90 Th		L_IO_I	20.174
0.614770	9	92 U	$L\gamma_1$	$L_{II}N_{IV}$	20.1671
0.6160	1	90 Th		$L_IN_{VI,VII}$	20.128

TABLE B (*Continued*)

Wavelength Å*	p.e.	Element	Designation		keV
0.616	1	93 Np	$L\gamma_5$	$L_{II}N_I$	20.12
0.6169	1	91 Pa	$L\gamma_3$	L_IN_{III}	20.098
0.617630	4	45 Rh	$K\alpha_2$	KL_{II}	20.0737
0.61978	1	42 Mo	K	Abs. Edge	20.0039
0.62001	9	42 Mo	$K\beta_4^I$	$KN_{IV,V}$	19.996
0.62099	2	42 Mo	$K\beta_2$	$KN_{II,III}$	19.9652
0.62107	5	42 Mo	$K\beta_2^{II}$	KN_{II}	19.963
0.6228	1	92 U		$L_{II}N_{III}$	19.907
0.6239	1	91 Pa	$L\gamma_2$	L_IN_{II}	19.872
0.62636	9	90 Th	$L\gamma_{11}$	L_IN_V	19.794
0.62692	5	42 No	$K\beta_5^I$	KM_V	19.776
0.62708	5	42 Mo	$K\beta_5^{II}$	KM_{IV}	19.771
0.6276	1	90 Th		L_IN_{IV}	19.755
0.6299	1	90 Th	L_{II}	Abs. Edge	19.683
0.62991	9	90 Th		$L_{II}P_{IV}$	19.682
0.6312	1	90 Th		$L_{II}P_{II,III}$	19.642
0.6316	1	90 Th		$L_{II}P_I$	19.629
0.632288	9	42 Mo	$K\beta_1$	KM_{III}	19.6083
0.63258	4	90 Th	$L\gamma_6$	$L_{II}O_{IV}$	19.599
0.632872	2	42 Mo	$K\beta_3$	KM_{II}	19.5903
0.63358	9	91 Pa	$L\gamma_1$	$L_{II}N_{IV}$	19.568
0.63557	2	92 U	$L\gamma_5$	$L_{II}N_I$	19.5072
0.63559	4	90 Th	$L\gamma_3$	L_IN_{III}	19.507
0.6356	1	90 Th		$L_{II}O_{III}$	19.506
0.6369	1	90 Th		$L_{II}O_{II}$	19.466
0.63898	5	90 Th	$L\gamma_8$	$L_{II}O_I$	19.403
0.64064	9	90 Th	Lv	$L_{II}N_{VI}$	19.353
0.6416	1	94 Pu	$L\beta_9$	L_IM_V	19.323
0.64221	4	90 Th	$L\gamma_2$	L_IN_{II}	19.305
0.643083	4	44 Ru	$K\alpha_1$	KL_{III}	19.2792
0.6445	1	88 Ra	L_I	Abs. Edge	19.236
0.64513	5	88 Ra	$L\gamma_{13}$	$L_IP_{II,III}$	19.218
0.6468	1	88 Ra		$L_IO_{IV,V}$	19.167
0.647408	5	44 Ru	$K\alpha_2$	KL_{II}	19.1504
0.64755	5	90 Th		L_IN_I	19.146
0.6482	1	94 Pu	$L\beta_{10}$	L_IM_{IV}	19.126
0.64891	2	95 Am	$L\beta_3$	L_IM_{III}	19.1059
0.64965	5	88 Ra	$L\gamma_4$	L_IO_{III}	19.084
0.65131	5	88 Ra	$L\gamma_4'$	L_IO_{II}	19.036
0.6521	1	90 Th		$L_{II}N_V$	19.014
0.65298	1	41 Nb	K	Abs. Edge	18.9869
0.65313	3	90 Th	$L\gamma_1$	$L_{II}N_{IV}$	18.9825
0.65318	5	41 Nb	$K\beta_4$	$KN_{IV,V}$	18.981
0.65416	4	41 Nb	$K\beta_2$	$KN_{II,III}$	18.953
0.6550	1	91 Pa	$L\gamma_5$	$L_{II}N_I$	18.930
0.657655	9	95 Am	$L\beta_1$	$L_{II}M_{IV}$	18.8520
0.6620	1	90 Th		$L_{II}N_{III}$	18.729
0.6654	1	88 Ra	$L\gamma_{11}$	L_IN_V	18.633
0.66576	2	41 Nb	$K\beta_1$	KM_{III}	18.6225
0.66634	3	41 Nb	$K\beta_3$	KM_{II}	18.6063
0.6666	1	88 Ra		L_IN_{IV}	18.600
0.66871	2	94 Pu	$L\beta_3$	L_IM_{III}	18.5405
0.6707	1	88 Ra	L_{II}	Abs. Edge	18.486
0.6714	1	88 Ra		$L_{II}P_{II,III}$	18.466
0.6724	1	88 Ra		$L_{II}P_I$	18.439
0.67328	5	88 Ra	$L\gamma_6$	$L_{II}O_{IV}$	18.414
0.67351	9	89 Ac	$L\gamma_1$	$L_{II}N_{IV}$	18.408
0.67383	2	95 Am	$L\beta_5$	$L_{III}O_{IV,V}$	18.3996
0.67491	4	90 Th	$L\gamma_5$	$L_{II}N_I$	18.370
0.67502	3	43 Tc	$K\alpha_1$	KL_{III}	18.3671
0.67538	5	88 Ra	$L\gamma_3$	L_IN_{III}	18.357
0.6764	1	88 Ra		$L_{II}O_{III}$	18.330
0.67772	2	94 Pu	$L\beta_1$	$L_{II}M_{IV}$	18.2937
0.6780	1	88 Ra		$L_{II}O_{II}$	18.286
0.67932	3	43 Tc	$K\alpha_2$	KL_{II}	18.2508
0.6801	1	88 Ra	$L\gamma_8$	$L_{II}O_I$	18.230
0.681014	8	92 U	$L\beta_9$	L_IM_V	18.2054
0.68199	5	88 Ra	$L\gamma_2$	L_IN_{II}	18.179
0.68639	2	95 Am	$L\beta_4$	L_IM_{II}	18.0627
0.6867	1	94 Pu	L_{III}	Abs. Edge	18.054
0.6874	1	88 Ra		L_IN_I	18.036
0.68760	5	92 U	$L\beta_{10}$	L_IM_{IV}	18.031
0.68883	1	40 Zr	K	Abs. Edge	17.9989
0.68901	5	40 Zr	$K\beta_4$	$KN_{IV,V}$	17.994
0.68920	9	93 Np	$L\beta_3$	L_IM_{III}	17.989
0.68993	4	40 Zr	$K\beta_2$	$KN_{II,III}$	17.970
0.69068	2	94 Pu	$L\beta_5$	$L_{III}O_{IV,V}$	17.9506
0.6932	1	88 Ra		$L_{II}N_V$	17.884
0.69463	5	88 Ra	$L\gamma_1$	$L_{II}N_{IV}$	17.849
0.6959	1	40 Zr	$K\beta_5$	$KM_{IV,V}$	17.815
0.698478	9	93 Np	$L\beta_1$	$L_{II}M_{IV}$	17.7502
0.7003	1	94 Pu	$L\beta_7$	$L_{III}O_I$	17.705
0.701390	9	95 Am	$L\beta_2$	$L_{III}N_V$	17.6765
0.70173	3	40 Zr	$K\beta_1$	KM_{III}	17.6678
0.7018	1	91 Pa	$L\beta_9$	L_IM_V	17.667
0.70228	4	40 Zr	$K\beta_3$	KM_{II}	17.654
0.7031	1	94 Pu	Lu	$L_{III}N_{VI,VII}$	17.635
0.70341	2	95 Am	$L\beta_{15}$	$L_{III}N_{IV}$	17.6258
0.7043	1	88 Ra		$L_{II}N_{III}$	17.604
0.70620	2	94 Pu	$L\beta_4$	L_IM_{II}	17.5560
0.70814	2	93 Np	$L\beta_5$	$L_{III}O_{IV,V}$	17.5081
0.7088	2	91 Pa	$L\beta_{10}$	L_IM_{IV}	17.492
0.709300	1	42 Mo	$K\alpha_1$	KL_{III}	17.47934
0.71029	2	92 U	$L\beta_3$	L_IM_{III}	17.4550
0.713590	6	42 Mo	$K\alpha_2$	KL_{II}	17.3743
0.71652	9	87 Fr	$L\gamma_1$	$L_{II}N_{IV}$	17.303
0.71774	5	88 Ra	$L\gamma_5$	$L_{II}N_I$	17.274
0.71851	2	94 Pu	$L\beta_2$	$L_{III}N_V$	17.2553
0.719984	8	92 U	$L\beta_1$	$L_{II}M_{IV}$	17.2200
0.7205	1	94 Pu	$L\beta_{15}$	$L_{III}N_{IV}$	17.208
0.7223	1	92 U	L_{III}	Abs. Edge	17.165
0.72240	5	92 U		$L_{III}P_{IV,V}$	17.162
0.7234	1	90 Th	$L\beta_9$	L_IM_V	17.139
0.72426	5	92 U		$L_{III}P_{II,III}$	17.118
0.72521	5	92 U		$L_{III}P_I$	17.096
0.726305	9	92 U	$L\beta_5$	$L_{III}O_{IV,V}$	17.0701
0.72671	2	93 Np	$L\beta_4$	L_IM_{II}	17.0607
0.72766	5	39 Y	K	Abs. Edge	17.038
0.72776	5	39 Y	$K\beta_4$	$KN_{IV,V}$	17.036
0.72864	4	39 Y	$K\beta_2$	$KN_{II,III}$	17.0154
0.7301	1	90 Th	$L\beta_{10}$	L_IM_{IV}	16.981
0.7309	1	92 U		$L_{III}O_{III}$	16.962
0.73230	5	91 Pa	$L\beta_3$	L_IM_{III}	16.930
0.7333	1	92 U		$L_{III}O_{II}$	16.907

TABLE B (*Continued*)

Wavelength Å*	p.e.	Element	Designation		keV
0.73418	2	95 Am	$L\beta_6$	$L_{III}N_{I}$	16.8870
0.7345	1	39 Y	$K\beta_5$	$KM_{IV,V}$	16.879
0.73602	6	92 U	$L\beta_7$	$L_{III}O_{I}$	16.845
0.736230	9	93 Np	$L\beta_2$	$L_{III}N_{V}$	16.8400
0.738603	9	92 U	Lu	$L_{III}N_{VI,VII}$	16.7859
0.73928	9	86 Rn	$L\gamma_1$	$L_{II}N_{IV}$	16.770
0.74072	2	39 Y	$K\beta_1$	KM_{III}	16.7378
0.74126	3	39 Y	$K\beta_3$	KM_{II}	16.7258
0.74232	5	91 Pa	$L\beta_1$	$L_{II}M_{IV}$	16.702
0.74503	5	92 U	$L\beta_{17}$	$L_{II}M_{III}$	16.641
0.7452	2	91 Pa	$L\beta_5$	$L_{III}O_{IV,V}$	16.636
0.74620	1	41 Nb	$K\alpha_1$	KL_{III}	16.6151
0.747985	9	92 U	$L\beta_4$	$L_{I}M_{II}$	16.5753
0.75044	1	41 Nb	$K\alpha_2$	KL_{II}	16.5210
0.75148	2	94 Pu	$L\beta_6$	$L_{III}N_{I}$	16.4983
0.7546	2	91 Pa	$L\beta_7$	$L_{III}O_{I}$	16.431
0.754681	9	92 U	$L\beta_2$	$L_{III}N_{V}$	16.4283
0.75479	3	90 Th	$L\beta_3$	$L_{I}M_{III}$	16.4258
0.756642	9	92 U	$L\beta_{15}$	$L_{III}N_{IV}$	16.3857
0.75690	3	83 Bi	$L\gamma_{13}$	$L_{I}P_{II,III}$	16.3802
0.7571	1	83 Bi	L_{I}	Abs. Edge	16.376
0.7579	1	90 Th		$L_{II}M_{V}$	16.359
0.75791	5	83 Bi		$L_{I}O_{IV,V}$	16.358
0.7591	1	94 Pu	$L\eta$	$L_{II}M_{I}$	16.333
0.7607	1	90 Th	L_{III}	Abs. Edge	16.299
0.76087	9	90 Th		$L_{III}P_{IV,V}$	16.295
0.76087	3	83 Bi	$L\gamma_4$	$L_{I}O_{III}$	16.2947
0.76198	3	83 Bi	$L\gamma_4'$	$L_{I}O_{II}$	16.2709
0.7625	2	90 Th		$L_{III}P_{II,III}$	16.260
0.76289	9	85 At	$L\gamma_1$	$L_{II}N_{IV}$	16.251
0.76338	5	90 Th		$L_{III}P_{I}$	16.241
0.7641	5	83 Bi		$L_{I}N_{VI,VII}$	16.23
0.7645	2	84 Po	$L\gamma_6$	$L_{II}O_{IV}$	16.218
0.76468	5	90 Th	$L\beta_5$	$L_{III}O_{IV,V}$	16.213
0.765210	9	90 Th	$L\beta_1$	$L_{II}M_{IV}$	16.2022
0.76857	5	88 Ra	$L\beta_9$	$L_{I}M_{V}$	16.131
0.769	1	93 Np	$L\beta_6$	$L_{III}N_{I}$	16.13
0.7690	1	90 Th		$L_{III}O_{III}$	16.123
0.7691	1	92 U		$L_{III}N_{III}$	16.120
0.76973	5	38 Sr	K	Abs. Edge	16.107
0.7699	1	91 Pa	$L\beta_4$	$L_{I}M_{II}$	16.104
0.76989	5	38 Sr	$K\beta_4$	$KN_{IV,V}$	16.104
0.77081	3	38 Sr	$K\beta_2$	$KN_{II,III}$	16.0846
0.7713	1	90 Th		$L_{III}O_{II}$	16.074
0.772	1	84 Po	$L\gamma_2$	$L_{I}N_{II}$	16.07
0.7737	1	91 Pa	$L\beta_2$	$L_{III}N_{V}$	16.024
0.77437	4	90 Th	$L\beta_7$	$L_{III}O_{I}$	16.0105
0.77546	5	88 Ra	$L\beta_{10}$	$L_{I}M_{IV}$	15.988
0.7764	1	38 Sr	$K\beta_5$	$KM_{IV,V}$	15.969
0.77661	5	90 Th	Lu	$L_{III}N_{VI,VII}$	15.964
0.77728	5	83 Bi	$L\gamma_{11}$	$L_{I}N_{V}$	15.951
0.77822	9	89 Ac	$L\beta_3$	$L_{I}M_{III}$	15.931
0.77954	5	83 Bi		$L_{I}N_{IV}$	15.904
0.78017	9	92 U		$L_{III}N_{II}$	15.892
0.7809	2	93 Np	$L\eta$	$L_{II}M_{I}$	15.876
0.78196	5	82 Pb	L_{I}	Abs. Edge	15.855
0.78257	7	82 Pb		$L_{I}O_{IV,V}$	15.843
0.78292	2	38 Sr	$K\beta_1$	KM_{III}	15.8357
0.78345	3	38 Sr	$K\beta_3$	KM_{II}	15.8249
0.7858	1	82 Pb	$L\gamma_4$	$L_{I}O_{III}$	15.777
0.78593	1	40 Zr	$K\alpha_1$	KL_{III}	15.7751
0.78706	7	82 Pb	$L\gamma_4'$	$L_{I}O_{II}$	15.752
0.78748	9	84 Po	$L\gamma_1$	$L_{II}N_{IV}$	15.744
0.78838	2	92 U	$L\beta_6$	$L_{III}N_{I}$	15.7260
0.7884	1	82 Pb		$L_{I}N_{VI,VII}$	15.725
0.7887	1	83 Bi	L_{II}	Abs. Edge	15.719
0.78903	9	89 Ac	$L\beta_1$	$L_{II}M_{IV}$	15.713
0.78917	5	83 Bi	$L\gamma_3$	$L_{I}N_{III}$	15.7102
0.7897	1	82 Pb		$L_{I}O_{I}$	15.699
0.79015	1	40 Zr	$K\alpha_2$	KL_{II}	15.6909
0.79043	3	83 Bi	$L\gamma_6$	$L_{II}O_{IV}$	15.6853
0.79257	4	90 Th	$L\beta_4$	$L_{I}M_{II}$	15.6429
0.79257	4	90 Th	$L\beta_{17}$	$L_{II}M_{III}$	15.6429
0.79354	3	90 Th	$L\beta_2$	$L_{III}M_{V}$	15.6237
0.79384	5	83 Bi		$L_{II}O_{III}$	15.6178
0.79539	5	90 Th	$L\beta_{15}$	$L_{III}N_{IV}$	15.5875
0.79565	3	83 Bi	$L\gamma_2$	$L_{I}N_{II}$	15.5824
0.79721	9	83 Bi	Lv	$L_{II}N_{VI}$	15.552
0.7973	1	83 Bi	$L\gamma_8$	$L_{II}O_{I}$	15.551
0.8022	1	83 Bi		$L_{I}N_{I}$	15.456
0.80233	9	82 Pb	$L\gamma_{11}$	$L_{I}N_{V}$	15.453
0.80273	5	88 Ra	$L\beta_3$	$L_{I}M_{III}$	15.4449
0.8028	1	88 Ra	L_{III}	Abs. Edge	15.444
0.80364	7	82 Pb		$L_{I}N_{IV}$	15.427
0.8038	1	88 Ra		$L_{III}P_{II,III}$	15.425
0.8050	1	88 Ra		$L_{III}P_{I}$	15.402
0.80509	2	92 U	$L\eta$	$L_{II}M_{I}$	15.3997
0.80627	5	88 Ra	$L\beta_5$	$L_{III}O_{IV,V}$	15.3771
0.8079	1	91 Pa	$L\beta_6$	$L_{III}N_{I}$	15.347
0.8081	1	81 Tl	L_{I}	Abs. Edge	15.343
0.8082	1	90 Th		$L_{III}N_{III}$	15.341
0.80861	5	81 Tl		$L_{I}O_{IV,V}$	15.3327
0.81163	9	90 Th		$L_{I}M_{I}$	15.276
0.81184	5	81 Tl	$L\gamma_4$	$L_{I}O_{III}$	15.2716
0.81308	5	81 Tl	$L\gamma_4'$	$L_{I}O_{II}$	15.2482
0.81311	2	83 Bi	$L\gamma_1$	$L_{II}N_{IV}$	15.2477
0.81375	5	88 Ra	$L\beta_1$	$L_{II}M_{IV}$	15.2358
0.8147	1	82 Pb	$L\gamma_3$	$L_{I}N_{III}$	15.218
0.81538	5	82 Pb	L_{II}	Abs. Edge	15.2053
0.8154	2	37 Rb	$K\beta_4$	$KN_{IV,V}$	15.205
0.81554	5	37 Rb	K	Abs. Edge	15.2023
0.8158	1	81 Tl		$L_{I}O_{I}$	15.198
0.81583	5	82 Pb		$L_{II}P_{I}$	15.1969
0.8162	1	88 Ra	$L\beta_7$	$L_{III}O_{I}$	15.190
0.81645	3	37 Rb	$K\beta_2$	$KN_{II,III}$	15.1854
0.81683	5	82 Pb	$L\gamma_6$	$L_{II}O_{IV}$	15.1783
0.8186	1	88 Ra	Lu	$L_{III}N_{VI,VII}$	15.146
0.8190	2	90 Th		$L_{III}N_{II}$	15.138
0.8200	1	82 Pb		$L_{II}O_{III}$	15.120
0.8210	2	82 Pb	$L\gamma_2$	$L_{I}N_{II}$	15.101
0.8219	1	37 Rb	$K\beta_5$	$KM_{IV,V}$	15.085
0.82327	7	82 Pb	Lv	$L_{II}N_{VI}$	15.060
0.82365	5	82 Pb	$L\gamma_8$	$L_{II}O_{I}$	15.0527
0.8248	1	83 Bi		$L_{II}N_{III}$	15.031

TABLE B (*Continued*)

Wavelength Å*	p.e.	Element	Designation		keV
0.82789	9	87 Fr	$L\beta_3$	$L_{I}M_{III}$	14.976
0.82790	8	90 Th	$L\beta_6$	$L_{III}N_{I}$	14.975
0.82859	7	82 Pb		$L_{I}N_{I}$	14.963
0.82868	2	37 Rb	$K\beta_1$	KM_{III}	14.9613
0.82879	5	81 Tl	$L\gamma_{11}$	$L_{I}N_{V}$	14.9593
0.82884	1	39 Y	$K\alpha_1$	KL_{III}	14.9584
0.82921	3	37 Rb	$K\beta_3$	KM_{II}	14.9517
0.8295	1	91 Pa	$L\eta$	$L_{II}M_{I}$	14.946
0.83001	7	81 Tl		$L_{I}N_{IV}$	14.937
0.83305	1	39 Y	$K\alpha_2$	KL_{II}	14.8829
0.8338	1	90 Th		$L_{II}M_{II}$	14.869
0.8344	9	83 Bi		$L_{II}N_{II}$	14.86
0.8350	2	80 Hg		$L_{I}O_{IV,V}$	14.847
0.8353	1	80 Hg	L_{I}	Abs. Edge	14.842
0.83537	5	88 Ra	$L\beta_2$	$L_{III}N_{V}$	14.8414
0.83722	5	88 Ra	$L\beta_{15}$	$L_{III}N_{IV}$	14.8086
0.8382	2	82 Pb		$L_{II}N_{V}$	14.791
0.83894	7	80 Hg	$L\gamma_4$	$L_{I}O_{III}$	14.778
0.83923	5	83 Bi	$L\gamma_5$	$L_{II}N_{I}$	14.7732
0.83940	9	87 Fr	$L\beta_1$	$L_{II}M_{IV}$	14.770
0.83973	3	82 Pb	$L\gamma_1$	$L_{II}N_{IV}$	14.7644
0.84013	7	80 Hg	$L\gamma_4'$	$L_{I}O_{II}$	14.757
0.84071	5	88 Ra	$L\beta_4$	$L_{I}M_{II}$	14.7472
0.84130	4	81 Tl	$L\gamma_3$	$L_{I}N_{III}$	14.7368
0.8434	1	81 Tl	L_{II}	Abs. Edge	14.699
0.8438	1	88 Ra	$L\beta_{17}$	$L_{II}M_{III}$	14.692
0.8442	2	81 Tl	$L\gamma_6$	$L_{II}O_{IV}$	14.685
0.8452	2	80 Hg		$L_{I}O_{I}$	14.670
0.84773	5	81 Tl	$L\gamma_2$	$L_{I}N_{II}$	14.6251
0.848187	9	95 Am	$L\alpha_1$	$L_{III}M_{V}$	14.6172
0.8490	1	81 Tl		$L_{II}O_{II}$	14.604
0.85048	5	81 Tl	Lv	$L_{II}N_{VI}$	14.5777
0.8512	1	88 Ra		$L_{III}N_{III}$	14.566
0.8513	2	81 Tl	$L\gamma_8$	$L_{II}O_{I}$	14.564
0.85192	7	82 Pb		$L_{II}N_{III}$	14.553
0.85436	9	86 Rn	$L\beta_3$	$L_{I}M_{III}$	14.512
0.85446	4	90 Th	$L\eta$	$L_{II}M_{I}$	14.5099
0.8549	1	81 Tl		$L_{I}N_{I}$	14.503
0.85657	7	80 Hg	$L\gamma_{11}$	$L_{I}N_{V}$	14.474
0.858	2	87 Fr	$L\beta_2$	$L_{III}N_{V}$	14.45
0.8585	3	82 Pb		$L_{II}N_{II}$	14.442
0.860266	9	95 Am	$L\alpha_2$	$L_{III}M_{IV}$	14.4119
0.8618	1	88 Ra		$L_{III}N_{II}$	14.387
0.86376	5	79 Au	L_{I}	Abs. Edge	14.3537
0.86400	5	79 Au		$L_{I}O_{IV,V}$	14.3497
0.8653	2	36 Kr	$K\beta_4$	$KN_{IV,V}$	14.328
0.86552	1	36 Kr	K	Abs. Edge	14.3244
0.86605	9	86 Rn	$L\beta_1$	$L_{II}M_{IV}$	14.316
0.8661	1	36 Kr	$K\beta_2$	$KN_{II,III}$	14.315
0.86655	5	82 Pb	$L\gamma_5$	$L_{II}N_{I}$	14.3075
0.86703	4	79 Au	$L\gamma_4$	$L_{I}O_{III}$	14.2996
0.86752	3	81 Tl	$L\gamma_1$	$L_{II}N_{IV}$	14.2915
0.86816	4	79 Au	$L\gamma_4'$	$L_{I}O_{II}$	14.2809
0.86830	2	94 Pu	$L\alpha_1$	$L_{III}M_{V}$	14.2786
0.86915	7	80 Hg	$L\gamma_3$	$L_{I}N_{III}$	14.265
0.87074	5	79 Au		$L_{I}O_{I}$	14.2385
0.8708	2	36 Kr	$K\beta_5$	$KM_{IV,V}$	14.238
0.87088	5	88 Ra	$L\beta_6$	$L_{III}N_{I}$	14.2362
0.8722	1	80 Hg	L_{II}	Abs. Edge	14.215
0.87319	7	80 Hg	$L\gamma_6$	$L_{II}O_{IV}$	14.199
0.87526	1	38 Sr	$K\alpha_1$	KL_{III}	14.1650
0.87544	7	80 Hg	$L\gamma_2$	$L_{I}N_{II}$	14.162
0.8758	1	80 Hg		$L_{II}O_{III}$	14.156
0.8784	1	80 Hg		$L_{II}O_{II}$	14.114
0.8785	1	36 Kr	$K\beta_1$	KM_{III}	14.112
0.87885	7	80 Hg	Lv	$L_{II}N_{VI}$	14.107
0.8790	1	36 Kr	$K\beta_3$	KM_{II}	14.104
0.87943	1	38 Sr	$K\alpha_2$	KL_{II}	14.0979
0.87995	7	80 Hg	$L\gamma_8$	$L_{II}O_{I}$	14.090
0.87996	5	81 Tl		$L_{II}N_{III}$	14.0893
0.88028	2	94 Pu	$L\alpha_2$	$L_{III}M_{IV}$	14.0842
0.88135	9	85 At	$L\beta_3$	$L_{I}M_{III}$	14.067
0.8827	2	80 Hg		$L_{I}N_{I}$	14.045
0.88433	7	79 Au	$L\gamma_{11}$	$L_{I}N_{V}$	14.020
0.88563	7	79 Au		$L_{I}N_{IV}$	13.999
0.8882	2	81 Tl		$L_{II}M_{II}$	13.959
0.889128	9	93 Np	$L\alpha_1$	$L_{III}M_{V}$	13.9441
0.8931	1	78 Pt	L_{I}	Abs. Edge	13.883
0.8934	1	78 Pt		$L_{I}O_{V}$	13.878
0.89349	9	85 At	$L\beta_1$	$L_{II}M_{IV}$	13.876
0.8943	1	78 Pt		$L_{I}O_{IV}$	13.864
0.89500	4	81 Tl	$L\gamma_5$	$L_{II}N_{I}$	13.8526
0.89646	5	80 Hg	$L\gamma_1$	$L_{II}N_{IV}$	13.8301
0.89659	4	78 Pt	$L\gamma_4$	$L_{I}O_{III}$	13.8281
0.89747	4	78 Pt	$L\gamma_4'$	$L_{I}O_{II}$	13.8145
0.89783	5	79 Au	$L\gamma_3$	$L_{I}N_{III}$	13.8090
0.89791	3	83 Bi	$L\beta_9$	$L_{I}M_{V}$	13.8077
0.8995	2	78 Pt		$L_{I}O_{I}$	13.784
0.8996	2	84 Po	$L\beta_5$	$L_{III}O_{IV,V}$	13.782
0.901045	9	93 Np	$L\alpha_2$	$L_{III}M_{IV}$	13.7597
0.90259	5	79 Au	L_{II}	Abs. Edge	13.7361
0.90297	3	79 Au	$L\gamma_6$	$L_{II}O_{IV}$	13.7304
0.90434	3	79 Au	$L\gamma_2$	$L_{I}N_{II}$	13.7095
0.90495	4	83 Bi	$L\beta_{10}$	$L_{I}M_{IV}$	13.7002
0.90638	7	79 Au		$L_{II}O_{III}$	13.679
0.90742	5	88 Ra	$L\eta$	$L_{II}M_{I}$	13.6630
0.90746	7	79 Au		$L_{II}O_{II}$	13.662
0.90837	5	79 Au	Lv	$L_{II}N_{VI}$	13.6487
0.90894	7	80 Hg		$L_{II}N_{III}$	13.640
0.9091	3	84 Po	$L\beta_3$	$L_{I}M_{III}$	13.638
0.90989	5	79 Au	$L\gamma_8$	$L_{II}O_{I}$	13.6260
0.910639	9	92 U	$L\alpha_1$	$L_{III}M_{V}$	13.6147
0.9131	1	79 Au		$L_{I}N_{I}$	13.578
0.9143	2	78 Pt	$L\gamma_{11}$	$L_{I}N_{V}$	13.560
0.9204	1	35 Br	K	Abs. Edge	13.470
0.92046	2	35 Br	$K\beta_2$	$KN_{II,III}$	13.4695
0.9220	2	84 Po	$L\beta_1$	$L_{II}M_{IV}$	13.447
0.922558	9	92 U	$L\alpha_2$	$L_{III}M_{IV}$	13.4388
0.9234	1	83 Bi	L_{III}	Abs. Edge	13.426
0.9236	1	77 Ir	L_{I}	Abs. Edge	13.423
0.92413	4	83 Bi		$L_{III}P_{II,III}$	13.4159
0.9243	3	77 Ir		$L_{I}O_{IV,V}$	13.413
0.92453	7	80 Hg	$L\gamma_5$	$L_{II}N_{I}$	13.410
0.9255	1	35 Br	$K\beta_5$	$KM_{IV,V}$	13.396

Wavelength Å*	p.e.	Element	Designation		keV
0.925553	9	37 Rb	$K\alpha_1$	KL_{III}	13.3953
0.92556	3	83 Bi	$L\beta_5$	$L_{III}O_{IV,V}$	13.3953
0.92650	3	79 Au	$L\gamma_1$	$L_{II}N_{IV}$	13.3817
0.9268	1	82 Pb	$L\beta_9$	L_IM_V	13.377
0.92744	3	77 Ir	$L\gamma_4$	L_IO_{III}	13.3681
0.92791	5	78 Pt	$L\gamma_3$	L_IN_{III}	13.3613
0.92831	3	77 Ir	$L\gamma_4'$	L_IO_{II}	13.3555
0.92937	5	84 Po	$L\beta_2$	$L_{III}N_V$	13.3404
0.92969	1	37 Rb	$K\alpha_2$	KL_{II}	13.3358
0.9302	2	83 Bi		$L_{III}O_{III}$	13.328
0.9312	2	84 Po	$L\beta_{15}$	$L_{III}N_{IV}$	13.314
0.9323	2	83 Bi		$L_{III}O_{II}$	13.298
0.93279	2	35 Br	$K\beta_1$	KM_{III}	13.2914
0.93284	5	91 Pa	$L\alpha_1$	$L_{III}M_V$	13.2907
0.93327	5	35 Br	$K\beta_3$	KM_{II}	13.2845
0.9339	2	82 Pb	$L\beta_{10}$	L_IM_{IV}	13.275
0.93414	5	78 Pt	L_{II}	Abs. Edge	13.2723
0.9342	2	78 Pt	$L\gamma_6$	$L_{II}O_{IV}$	13.271
0.93427	5	78 Pt	$L\gamma_2$	L_IN_{II}	13.2704
0.93505	5	83 Bi	$L\beta_7$	$L_{III}O_I$	13.2593
0.93505	5	83 Bi	Lu	$L_{III}N_{VI,VII}$	13.2593
0.93855	3	83 Bi	$L\beta_3$	L_IM_{III}	13.2098
0.93931	5	78 Pt	Lv	$L_{II}N_{VI}$	13.1992
0.9402	2	79 Au		$L_{II}N_{III}$	13.186
0.9411	1	78 Pt	$L\gamma_8$	$L_{II}O_I$	13.173
0.94419	5	83 Bi		$L_{II}M_V$	13.1310
0.9446	2	77 Ir	$L\gamma_{11}$	L_IN_V	13.126
0.94482	5	91 Pa	$L\alpha_2$	$L_{III}M_{IV}$	13.1222
0.9455	2	78 Pt		L_IN_I	13.113
0.9459	2	77 Ir		$\mathrm{L_IN_{IV}}$	13.108
0.9475	3	84 Po	$L\beta_4$	L_IM_{II}	13.086
0.95073	5	82 Pb	L_{III}	Abs. Edge	13.0406
0.95118	7	82 Pb		$L_{III}P_{II,III}$	13.0344
0.951978	9	83 Bi	$L\beta_1$	$L_{II}M_{IV}$	13.0235
0.9526	1	82 Pb	$L\beta_5$	$L_{III}O_{IV,V}$	13.015
0.95518	4	83 Bi	$L\beta_2$	$L_{III}N_V$	12.9799
0.95559	3	79 Au	$L\gamma_5$	$L_{II}N_I$	12.9743
0.9558	1	76 Os	L_I	Abs. Edge	12.972
0.95600	3	90 Th	$L\alpha_1$	$L_{III}M_V$	12.9687
0.95603	5	76 Os		$L_IO_{IV,V}$	12.9683
0.95675	7	81 Tl	$L\beta_9$	L_IM_V	12.9585
0.95702	5	83 Bi	$L\beta_{15}$	$L_{III}N_{IV}$	12.9549
0.9578	1	82 Pb		$L_{III}O_{III}$	12.945
0.95797	3	78 Pt	$L\gamma_1$	$L_{II}N_{IV}$	12.9420
0.9586	1	82 Pb		$L_{III}O_{II}$	12.934
0.95931	5	77 Ir	$L\gamma_3$	L_IN_{III}	12.9240
0.95938	8	76 Os	$L\gamma_4$	L_IO_{III}	12.923
0.96033	8	76 Os	$L\gamma_4'$	L_IO_{II}	12.910
0.96133	7	82 Pb	Lu	$L_{III}N_{VI,VII}$	12.8968
0.9620	1	82 Pb	$L\beta_7$	$L_{III}O_I$	12.888
0.96318	7	76 Os		L_IO_I	12.8721
0.9636	1	92 U	Ls	$L_{III}M_{III}$	12.866
0.96389	7	81 Tl	$L\beta_{10}$	L_IM_{IV}	12.8626
0.96545	3	77 Ir	$L\gamma_2$	L_IN_{II}	12.8418
0.96708	4	77 Ir	$L\gamma_6$	$L_{II}O_{IV}$	12.8201
0.9671	1	77 Ir	L_{II}	Abs. Edge	12.820
0.9672	2	84 Po	$L\beta_6$	$L_{III}N_I$	12.819
0.96788	2	90 Th	$L\alpha_2$	$L_{III}M_{IV}$	12.8096
0.96911	7	82 Pb	$L\beta_3$	L_IM_{III}	12.7933
0.96979	5	77 Ir		$L_{II}O_{III}$	12.7843
0.97161	6	77 Ir	Lv	$L_{II}N_{VI}$	12.7603
0.97173	4	78 Pt		$L_{II}N_{III}$	12.7588
0.97321	5	83 Bi		$L_{III}N_{III}$	12.7394
0.97409	3	77 Ir	$L\gamma_8$	$L_{II}O_I$	12.7279
0.9747	1	82 Pb		$L_{II}M_V$	12.720
0.9765	3	76 Os	$L\gamma_{11}$	L_IN_V	12.696
0.9766	2	77 Ir		L_IN_I	12.695
0.97690	4	83 Bi	$L\beta_4$	L_IM_{II}	12.6912
0.9772	3	76 Os		L_IN_{IV}	12.687
0.9792	2	78 Pt		$L_{II}N_{II}$	12.661
0.97926	5	81 Tl		$L_{III}P_{II,III}$	12.6607
0.9793	1	81 Tl	L_{III}	Abs. Edge	12.660
0.97974	1	34 Se	K	Abs. Edge	12.6545
0.97992	5	34 Se	$K\beta_2$	$KN_{II,III}$	12.6522
0.97993	5	89 Ac	$L\alpha_1$	$L_{III}M_V$	12.6520
0.9801	1	36 Kr	$K\alpha_1$	KL_{III}	12.649
0.98058	3	81 Tl	$L\beta_5$	$L_{III}O_{IV,V}$	12.6436
0.98221	7	82 Pb	$L\beta_2$	$L_{III}N_V$	12.6226
0.98280	5	83 Bi		$L_{III}N_{II}$	12.6151
0.98291	3	82 Pb	$L\beta_1$	$L_{II}M_{IV}$	12.6137
0.98389	7	82 Pb	$L\beta_{15}$	$L_{III}N_{IV}$	12.6011
0.9841	1	36 Kr	$K\alpha_2$	KL_{II}	12.598
0.9843	1	34 Se	$K\beta_5$	$KM_{IV,V}$	12.595
0.98538	5	81 Tl		$L_{III}O_{III}$	12.5820
0.9871	2	80 Hg	$L\beta_9$	L_IM_V	12.560
0.98738	5	81 Tl		$L_{III}O_{II}$	12.5566
0.9877	2	78 Pt	$L\gamma_5$	$L_{II}N_I$	12.552
0.9888	1	81 Tl	Lu	$L_{III}N_{VI,VII}$	12.538
0.98913	5	83 Bi	$L\beta_{17}$	$L_{II}M_{III}$	12.5344
0.9894	1	75 Re	L_I	Abs. Edge	12.530
0.9900	1	75 Re		$L_IO_{IV,V}$	12.524
0.99017	5	81 Tl	$L\beta_7$	$L_{III}O_I$	12.5212
0.99085	3	77 Ir	$L\gamma_1$	$L_{II}N_{IV}$	12.5126
0.99178	5	89 Ac	$L\alpha_2$	$L_{III}M_{IV}$	12.5008
0.99186	5	76 Os	$L\gamma_3$	L_IN_{III}	12.4998
0.99218	3	34 Se	$K\beta_1$	KM_{III}	12.4959
0.99249	5	75 Re	$L\gamma_4$	L_IO_{III}	12.4920
0.99268	5	34 Se	$K\beta_3$	KM_{II}	12.4896
0.99331	3	83 Bi	$L\beta_6$	$L_{III}N_I$	12.4816
0.99334	5	75 Re	$L\gamma_4'$	L_IO_{II}	12.4813
0.9962	2	80 Hg	$L\beta_{10}$	L_IM_{IV}	12.446
0.9965	1	75 Re		L_IO_I	12.442
0.99805	5	76 Os	$L\gamma_2$	L_IN_{II}	12.4224
1.0005	1	82 Pb		$L_{III}N_{III}$	12.392
1.0005	9	83 Bi		L_IM_I	12.39
1.00062	3	81 Tl	$L\beta_3$	L_IM_{III}	12.3904
1.00107	5	76 Os	$L\gamma_6$	$L_{II}O_{IV}$	12.3848
1.0012	6	95 Am	Ll	$L_{III}M_I$	12.384
1.0014	1	76 Os	L_{II}	Abs. Edge	12.381
1.0047	2	76 Os		$L_{II}O_{III}$	12.340
1.00473	5	88 Ra	$L\alpha_1$	$L_{III}M_V$	12.3397
1.0050	2	76 Os	Lv	$L_{II}N_{VI}$	12.337
1.0054	3	77 Ir		$L_{II}N_{III}$	12.332
1.00722	5	81 Tl		$L_{II}M_V$	12.3093

TABLE B (*Continued*)

Wavelength Å*	p.e.	Element	Designation		keV
1.0075	1	82 Pb	$L\beta_4$	L_IM_{II}	12.306
1.00788	5	76 Os	$L\gamma_8$	$L_{II}O_I$	12.3012
1.0091	1	80 Hg	L_{III}	Abs. Edge	12.286
1.00987	7	80 Hg	$L\beta_5$	$L_{III}O_{IV,V}$	12.2769
1.01031	3	81 Tl	$L\beta_2$	$L_{III}N_V$	12.2715
1.01040	7	82 Pb		$L_{III}N_{II}$	12.2705
1.0108	1	75 Re	$L\gamma_{11}$	L_IN_V	12.266
1.0112	1	90 Th	Ls	$L_{III}M_{III}$	12.261
1.0119	1	75 Re		L_IN_{IV}	12.252
1.0120	2	77 Ir		$L_{II}N_{II}$	12.251
1.01201	3	81 Tl	$L\beta_{15}$	$L_{III}N_{IV}$	12.2510
1.01404	7	80 Hg		$L_{III}O_{III}$	12.2264
1.01513	4	81 Tl	$L\beta_1$	$L_{II}M_{IV}$	12.2133
1.01558	7	80 Hg		$L_{III}O_{II}$	12.2079
1.01656	5	88 Ra	$L\alpha_2$	$L_{III}M_{IV}$	12.1962
1.01674	7	80 Hg	Lu	$L_{III}N_{VII}$	12.1940
1.01769	7	80 Hg	Lu'	$L_{III}N_{VI}$	12.1826
1.01937	7	80 Hg	$L\beta_7$	$L_{III}O_I$	12.1625
1.02063	7	79 Au	$L\beta_9$	L_IM_V	12.1474
1.0210	1	82 Pb	$L\beta_6$	$L_{III}N_I$	12.143
1.02175	5	77 Ir	$L\gamma_5$	$L_{II}N_I$	12.1342
1.0223	1	82 Pb	$L\beta_{17}$	$L_{II}M_{III}$	12.127
1.0226	1	94 Pu	Ll	$L_{III}M_I$	12.124
1.02467	5	74 W	L_I	Abs. Edge	12.0996
1.0250	2	74 W		$L_IO_{IV,V}$	12.095
1.02503	5	76 Os	$L\gamma_1$	$L_{II}N_{IV}$	12.0953
1.02613	7	75 Re	$L\gamma_3$	L_IN_{III}	12.0824
1.02775	3	74 W	$L\gamma_4$	L_IO_{III}	12.0634
1.02789	7	79 Au	$L\beta_{10}$	L_IM_{IV}	12.0617
1.0286	1	81 Tl		$L_{III}N_{III}$	12.053
1.02863	3	74 W	$L\gamma_4'$	L_IO_{II}	12.0530
1.03049	5	87 Fr	$L\alpha_1$	$L_{III}M_V$	12.0313
1.0317	3	74 W		L_IO_I	12.017
1.03233	5	75 Re	$L\gamma_2$	L_IN_{II}	12.0098
1.0323	2	82 Pb		L_IM_I	12.010
1.03358	7	80 Hg	$L\beta_3$	L_IM_{III}	11.9953
1.0346	9	83 Bi		L_IM_{II}	11.98
1.0347	1	92 U	Lt	$L_{III}M_{II}$	11.982
1.03699	9	75 Re	$L\gamma_6$	$L_{II}O_{IV}$	11.956
1.0371	1	75 Re	L_{II}	Abs. Edge	11.954
1.03876	7	79 Au		$L_{III}P_{II,III}$	11.9355
1.03918	3	81 Tl	$L\beta_4$	L_IM_{II}	11.9306
1.0397	1	75 Re		$L_{II}O_{III}$	11.925
1.03973	5	76 Os		L_IN_{III}	11.9243
1.03974	2	35 Br	$K\alpha_1$	KL_{III}	11.9242
1.03975	7	80 Hg	$L\beta_2$	$L_{III}N_V$	11.9241
1.04000	5	79 Au	L_{III}	Abs. Edge	11.9212
1.0404	1	75 Re	Lv	$L_{II}N_{VI}$	11.917
1.04044	3	79 Au	$L\beta_5$	$L_{III}O_{IV,V}$	11.9163
1.04151	7	80 Hg	$L\beta_{15}$	$L_{III}N_{IV}$	11.9040
1.0420	1	75 Re		L_IN_I	11.899
1.04230	5	87 Fr	$L\alpha_2$	$L_{III}M_{IV}$	11.8950
1.0428	6	93 Np	Ll	$L_{III}M_I$	11.890
1.04382	2	35 Br	$K\alpha_2$	KL_{II}	11.8776
1.04398	5	75 Re	$L\gamma_8$	$L_{II}O_I$	11.8758
1.0450	2	79 Au		$L_{III}O_{II,III}$	11.865
1.0450	1	33 As	K	Abs. Edge	11.865
1.04500	3	33 As	$K\beta_2$	$KN_{II,III}$	11.8642
1.0458	1	74 W	$L\gamma_{11}$	L_IN_V	11.856
1.0468	2	74 W		L_IN_{IV}	11.844
1.04752	5	79 Au	Lu	$L_{III}N_{VI,VII}$	11.8357
1.04868	5	80 Hg	$L\beta_1$	$L_{II}M_{IV}$	11.8226
1.0488	1	33 As	$K\beta_5$	$KM_{IV,V}$	11.822
1.04963	5	81 Tl	$L\beta_6$	$L_{III}N_I$	11.8118
1.04974	8	79 Au	$L\beta_7$	$L_{III}O_I$	11.8106
1.05446	5	78 Pt	$L\beta_9$	L_IM_V	11.7577
1.05609	7	81 Tl	$L\beta_{17}$	$L_{II}M_{III}$	11.7397
1.05693	5	76 Os	$L\gamma_5$	$L_{II}N_I$	11.7303
1.05723	5	86 Rn	$L\alpha_1$	$L_{III}M_V$	11.7270
1.05730	2	33 As	$K\beta_1$	KM_{III}	11.7262
1.05783	5	33 As	$K\beta_3$	KM_{II}	11.7203
1.0585	1	80 Hg		$L_{III}N_{III}$	11.713
1.05856	3	83 Bi	$L\eta$	$L_{II}M_I$	11.7122
1.06099	5	75 Re	$L\gamma_1$	$L_{II}N_{IV}$	11.6854
1.0613	1	73 Ta	L_I	Abs. Edge	11.682
1.06183	7	78 Pt	$L\beta_{10}$	L_IM_{IV}	11.6762
1.06192	9	73 Ta		$L_IO_{IV,V}$	11.6752
1.06200	6	74 W	$L\gamma_3$	L_IN_{III}	11.6743
1.06357	9	73 Ta		$L_IN_{VI,VII}$	11.6570
1.0644	2	82 Pb		$L_{II}M_{II}$	11.648
1.0644	2	81 Tl		L_IM_I	11.648
1.06467	3	73 Ta	$L\gamma_4$	L_IO_{III}	11.6451
1.0649	2	80 Hg		$L_{III}N_{II}$	11.642
1.06544	3	73 Ta	$L\gamma_4'$	L_IO_{II}	11.6366
1.06712	2	92 U	Ll	$L_{III}M_I$	11.6183
1.06771	9	73 Ta		L_IO_I	11.6118
1.06785	9	79 Au	$L\beta_3$	L_IM_{III}	11.6103
1.06806	3	74 W	$L\gamma_2$	L_IN_{II}	11.6080
1.06899	5	86 Rn	$L\alpha_2$	$L_{III}M_{IV}$	11.5979
1.07022	3	79 Au	$L\beta_2$	$L_{III}N_V$	11.5847
1.07188	5	79 Au	$L\beta_{15}$	$L_{III}N_{IV}$	11.5667
1.07222	7	80 Hg	$L\beta_4$	L_IM_{II}	11.5630
1.0723	1	78 Pt	L_{III}	Abs. Edge	11.562
1.0724	2	78 Pt	$L\beta_5$	$L_{III}O_{IV,V}$	11.561
1.07448	5	74 W	$L\gamma_6$	$L_{II}O_{IV}$	11.5387
1.0745	1	74 W	L_{II}	Abs. Edge	11.538
1.0756	2	79 Au		$L_{II}M_V$	11.526
1.0761	3	78 Pt		$L_{III}O_{II,III}$	11.521
1.0767	1	75 Re		$L_{II}N_{III}$	11.515
1.0771	1	74 W	Lv	$L_{II}N_{VI}$	11.510
1.07896	5	78 Pt	Lu	$L_{III}N_{VI,VII}$	11.4908
1.0792	2	74 W		$L_{II}O_{III}$	11.488
1.07975	7	80 Hg	$L\beta_6$	$L_{III}N_I$	11.4824
1.08009	9	90 Th	Lt	$L_{III}M_{II}$	11.4788
1.08113	4	74 W	$L\gamma_8$	$L_{II}O_I$	11.4677
1.08168	3	78 Pt	$L\beta_7$	$L_{III}O_I$	11.4619
1.08205	7	73 Ta	$L\gamma_{11}$	L_IN_V	11.4580
1.08353	3	79 Au	$L\beta_1$	$L_{II}M_{IV}$	11.4423
1.08377	7	73 Ta		L_IN_{IV}	11.4398
1.0839	1	75 Re		$L_{II}N_{II}$	11.438
1.08500	5	85 At	$L\alpha_1$	$L_{III}M_V$	11.4268
1.08975	5	77 Ir	$L\beta_9$	L_IM_V	11.3770
1.09026	7	79 Au		$L_{III}N_{III}$	11.3717
1.0908	1	91 Pa	Ll	$L_{III}M_I$	11.366

TABLE B (*Continued*)

Wavelength Å*	p.e.	Element	Designation		keV
1.0916	5	80 Hg	$L\beta_{17}$	$L_{II}M_{III}$	11.358
1.09241	7	82 Pb	$L\eta$	$L_{II}M_{I}$	11.3493
1.09388	5	75 Re	$L\gamma_5$	$L_{II}N_{I}$	11.3341
1.09671	5	85 At	$L\alpha_2$	$L_{III}M_{IV}$	11.3048
1.09702	4	77 Ir	$L\beta_{10}$	$L_{I}M_{IV}$	11.3016
1.09855	3	74 W	$L\gamma_1$	$L_{II}N_{IV}$	11.2859
1.09936	4	73 Ta	$L\gamma_3$	$L_{I}N_{III}$	11.2776
1.0997	1	81 Tl		$L_{II}M_{II}$	11.274
1.0997	1	72 Hf	L_{I}	Abs. Edge	11.274
1.09968	7	79 Au		$L_{III}N_{II}$	11.2743
1.0999	2	80 Hg		$L_{I}M_{I}$	11.272
1.10086	9	72 Hf		$L_{I}O_{IV}$	11.2622
1.10200	3	78 Pt	$L\beta_2$	$L_{III}N_{V}$	11.2505
1.10303	5	72 Hf	$L\gamma_4$	$L_{I}O_{III}$	11.2401
1.10376	5	72 Hf	$L\gamma_4'$	$L_{I}O_{II}$	11.2326
1.10394	5	78 Pt	$L\beta_3$	$L_{I}M_{III}$	11.2308
1.10477	2	34 Se	$K\alpha_1$	KL_{III}	11.2224
1.1053	1	73 Ta	$L\gamma_2$	$L_{I}N_{II}$	11.217
1.1058	1	77 Ir	L_{III}	Abs. Edge	11.212
1.10585	3	77 Ir	$L\beta_5$	$L_{III}O_{IV,V}$	11.2114
1.10651	3	79 Au	$L\beta_4$	$L_{I}M_{II}$	11.2047
1.10664	9	72 Hf		$L_{I}O_{I}$	11.2034
1.10882	2	34 Se	$K\alpha_2$	KL_{II}	11.1814
1.10923	6	77 Ir		$L_{III}O_{II,III}$	11.1772
1.11092	3	79 Au	$L\beta_6$	$L_{III}N_{I}$	11.1602
1.11145	4	77 Ir	Lu	$L_{III}N_{VI,VII}$	11.1549
1.1129	2	78 Pt		$L_{II}M_{V}$	11.140
1.1137	1	73 Ta	L_{II}	Abs. Edge	11.132
1.11386	4	84 Po	$L\alpha_1$	$L_{III}M_{V}$	11.1308
1.11388	3	73 Ta	$L\gamma_6$	$L_{II}O_{IV}$	11.1306
1.11489	3	77 Ir	$L\beta_7$	$L_{III}O_{I}$	11.1205
1.1149	2	74 W		$L_{II}N_{III}$	11.120
1.11508	4	90 Th	Ll	$L_{III}M_{I}$	11.1186
1.11521	9	73 Ta		$L_{I}N_{I}$	11.1173
1.1158	1	73 Ta	Lv	$L_{II}N_{VI}$	11.1113
1.11658	5	32 Ge	K	Abs. Edge	11.1036
1.11686	2	32 Ge	$K\beta_2$	$KN_{II,III}$	11.1008
1.11693	9	73 Ta		$L_{II}O_{III}$	11.1001
1.11789	9	73 Ta		$L_{II}O_{II}$	11.0907
1.1195	1	32 Ge	$K\beta_5$	$KM_{IV,V}$	11.0745
1.11990	2	78 Pt	$L\beta_1$	$L_{II}M_{IV}$	11.0707
1.1205	1	73 Ta	$L\gamma_8$	$L_{II}O_{I}$	11.0646
1.12146	9	72 Hf	$L\gamma_{11}$	$L_{I}N_{V}$	11.0553
1.1218	3	74 W		$L_{II}N_{II}$	11.052
1.12250	9	72 Hf		$L_{I}N_{IV}$	11.0451
1.1226	2	78 Pt		$L_{III}N_{III}$	11.044
1.12548	5	84 Po	$L\alpha_2$	$L_{III}M_{IV}$	11.0158
1.12637	6	76 Os	$L\beta_9$	$L_{I}M_{V}$	11.0071
1.12769	3	81 Tl	$L\eta$	$L_{II}M_{I}$	10.9943
1.12798	5	79 Au	$L\beta_{17}$	$L_{II}M_{III}$	10.9915
1.12894	2	32 Ge	$K\beta_1$	KM_{III}	10.9821
1.12936	9	32 Ge	$K\beta_3$	KM_{II}	10.9780
1.1310	2	78 Pt		$L_{III}N_{II}$	10.962
1.13235	3	74 W	$L\gamma_5$	$L_{II}N_{I}$	10.9490
1.13353	5	76 Os	$L\beta_{10}$	$L_{I}M_{IV}$	10.9376
1.13525	5	79 Au		$L_{I}M_{I}$	10.9210
1.13532	3	77 Ir	$L\beta_2$	$L_{III}N_{V}$	10.9203
1.13687	9	73 Ta		$L_{II}N_{V}$	10.9055
1.13707	3	77 Ir	$L\beta_{15}$	$L_{III}N_{IV}$	10.9036
1.13794	3	73 Ta	$L\gamma_1$	$L_{II}N_{IV}$	10.8952
1.13841	5	72 Hf	$L\gamma_3$	$L_{I}N_{III}$	10.8907
1.1387	5	80 Hg		$L_{II}M_{II}$	10.888
1.1402	1	71 Lu	L_{I}	Abs. Edge	10.8740
1.1405	1	76 Os	$L\beta_5$	$L_{III}O_{IV,V}$	10.8711
1.1408	1	76 Os	L_{III}	Abs. Edge	10.8683
1.14085	3	77 Ir	$L\beta_3$	$L_{I}M_{III}$	10.8674
1.14223	5	78 Pt	$L\beta_4$	$L_{I}M_{II}$	10.8543
1.1435	1	71 Lu	$L\gamma_4$	$L_{I}O_{II,III}$	10.8425
1.14355	5	78 Pt	$L\beta_6$	$L_{III}N_{I}$	10.8418
1.14386	2	83 Bi	$L\alpha_1$	$L_{III}M_{V}$	10.8388
1.14442	5	72 Hf	$L\gamma_2$	$L_{I}N_{II}$	10.8335
1.14537	7	76 Os	Lu	$L_{III}N_{VI,VII}$	10.8245
1.1489	2	77 Ir		$L_{II}M_{V}$	10.791
1.14933	8	76 Os	$L\beta_7$	$L_{III}O_{I}$	10.7872
1.1548	1	72 Hf	L_{II}	Abs. Edge	10.7362
1.15519	5	72 Hf	$L\gamma_6$	$L_{II}O_{IV}$	10.7325
1.1553	1	73 Ta		$L_{II}N_{III}$	10.7316
1.15536	1	83 Bi	$L\alpha_2$	$L_{III}M_{IV}$	10.73091
1.1560	3	77 Ir		$L_{III}N_{III}$	10.725
1.15781	3	77 Ir	$L\beta_1$	$L_{II}M_{IV}$	10.7083
1.15830	9	72 Hf	Lv	$L_{II}N_{VI}$	10.7037
1.1600	2	73 Ta		$L_{II}N_{II}$	10.688
1.16107	9	71 Lu	$L\gamma_{11}$	$L_{I}N_{V}$	10.6782
1.16138	5	72 Hf	$L\gamma_8$	$L_{II}O_{I}$	10.6754
1.16227	9	71 Lu		$L_{I}N_{IV}$	10.6672
1.1640	1	80 Hg	$L\eta$	$L_{II}M_{I}$	10.6512
1.16487	4	75 Re	$L\beta_9$	$L_{I}M_{V}$	10.6433
1.16545	5	77 Ir		$L_{III}N_{II}$	10.6380
1.1667	1	78 Pt	$L\beta_{17}$	$L_{II}M_{III}$	10.6265
1.16719	5	88 Ra	Ll	$L_{III}M_{I}$	10.6222
1.16962	9	78 Pt		$L_{I}M_{I}$	10.6001
1.16979	8	76 Os	$L\beta_2$	$L_{III}N_{V}$	10.5985
1.1708	1	79 Au		$L_{II}M_{II}$	10.5892
1.17167	5	76 Os	$L\beta_{15}$	$L_{III}N_{IV}$	10.5816
1.17218	5	75 Re	$L\beta_{10}$	$L_{I}M_{IV}$	10.5770
1.1729	1	73 Ta	$L\gamma_5$	$L_{II}N_{I}$	10.5702
1.17501	2	82 Pb	$L\alpha_1$	$L_{III}M_{V}$	10.5515
1.17588	1	33 As	$K\alpha_1$	KL_{III}	10.54372
1.17721	5	75 Re	$L\beta_5$	$L_{III}O_{IV,V}$	10.5318
1.1773	1	75 Re	L_{III}	Abs. Edge	10.5306
1.17788	9	72 Hf		$L_{II}N_{V}$	10.5258
1.17796	3	77 Ir	$L\beta_6$	$L_{III}N_{I}$	10.5251
1.17900	5	72 Hf	$L\gamma_1$	$L_{II}N_{IV}$	10.5158
1.17953	4	71 Lu	$L\gamma_3$	$L_{I}N_{III}$	10.5110
1.17955	7	76 Os	$L\beta_3$	$L_{I}M_{III}$	10.5108
1.17958	3	77 Ir	$L\beta_4$	$L_{I}M_{II}$	10.5106
1.17987	1	33 As	$K\alpha_2$	KL_{II}	10.50799
1.1815	1	75 Re	Lu	$L_{III}N_{VI,VII}$	10.4931
1.1818	1	70 Yb	L_{I}	Abs. Edge	10.4904
1.1827	1	70 Yb		$L_{I}O_{IV,V}$	10.4833
1.1853	1	70 Yb	$L\gamma_4$	$L_{I}O_{II,III}$	10.4603
1.1853	2	71 Lu	$L\gamma_2$	$L_{I}N_{II}$	10.460
1.18610	5	75 Re	$L\beta_7$	$L_{III}O_{I}$	10.4529
1.18648	5	82 Pb	$L\alpha_2$	$L_{III}M_{IV}$	10.4495

Wavelength Å*	p.e.	Element		Designation	keV
1.1886	1	70 Yb		L_IO_I	10.4312
1.18977	7	76 Os		$L_{II}M_V$	10.4205
1.1958	1	31 Ga	K	Abs. Edge	10.3682
1.19600	2	31 Ga	$K\beta_2$	$KN_{II,III}$	10.3663
1.19727	7	76 Os	$L\beta_1$	$L_{II}M_{IV}$	10.3553
1.1981	2	31 Ga	$K\beta_5$	$KM_{IV,V}$	10.348
1.1985	1	71 Lu	L_{II}	Abs. Edge	10.3448
1.1987	1	71 Lu	$L\gamma_6$	$L_{II}O_{IV}$	10.3431
1.20086	7	76 Os		$L_{III}N_{II}$	10.3244
1.2014	1	71 Lu		$L_{II}O_{II,III}$	10.3198
1.20273	3	79 Au	$L\eta$	$L_{II}M_I$	10.3083
1.2047	1	71 Lu	$L\gamma_8$	$L_{II}O_I$	10.2915
1.20479	7	74 W	$L\beta_9$	L_IM_V	10.2907
1.20660	4	75 Re	$L\beta_2$	$L_{III}N_V$	10.2752
1.2069	2	77 Ir	$L\beta_{17}$	$L_{II}M_{III}$	10.273
1.20739	4	81 Tl	$L\alpha_1$	$L_{III}M_V$	10.2685
1.20789	2	31 Ga	$K\beta_1$	KM_{III}	10.2642
1.20819	5	75 Re	$L\beta_{15}$	$L_{III}N_{IV}$	10.2617
1.20835	5	31 Ga	$K\beta_3$	KM_{II}	10.2603
1.2102	2	77 Ir		L_IM_I	10.245
1.2105	1	83 Bi	Ls	$L_{III}M_{III}$	10.2421
1.21218	3	74 W	$L\beta_{10}$	L_IM_{IV}	10.2279
1.213	1	78 Pt		$L_{II}M_{II}$	10.225
1.21349	5	76 Os	$L\beta_6$	$L_{III}N_I$	10.2169
1.21537	5	72 Hf	$L\gamma_5$	$L_{II}N_I$	10.2011
1.21545	3	74 W	$L\beta_5$	$L_{III}O_{IV,V}$	10.2004
1.2155	1	74 W	L_{III}	Abs. Edge	10.1999
1.21844	5	76 Os	$L\beta_4$	L_IM_{II}	10.1754
1.21868	5	74 W	Lu	$L_{III}N_{VI,VII}$	10.1733
1.21875	3	81 Tl	$L\alpha_2$	$L_{III}M_{IV}$	10.1728
1.22031	5	75 Re	$L\beta_3$	L_IM_{III}	10.1598
1.2211	2	74 W		$L_{III}O_{II,III}$	10.153
1.22228	4	71 Lu	$L\gamma_1$	$L_{II}N_{IV}$	10.1434
1.22232	5	70 Yb	$L\gamma_3$	L_IN_{III}	10.1431
1.22400	4	74 W	$L\beta_7$	$L_{III}O_I$	10.1292
1.2250	1	69 Tm	L_I	Abs. Edge	10.1206
1.2263	3	69 Tm		$L_IO_{IV,V}$	10.110
1.2283	1	75 Re		$L_{III}N_{III}$	10.0933
1.22879	7	70 Yb	$L\gamma_2$	L_IN_{II}	10.0897
1.2294	2	69 Tm	$L\gamma_4$	$L_IO_{II,III}$	10.084
1.2305	1	75 Re		$L_{II}M_V$	10.0753
1.23858	2	75 Re	$L\beta_1$	$L_{II}M_{IV}$	10.0100
1.24120	5	80 Hg	$L\alpha_1$	$L_{III}M_V$	9.9888
1.24271	3	70 Yb	$L\gamma_6$	$L_{II}O_{IV}$	9.9766
1.2428	1	70 Yb	L_{II}	Abs. Edge	9.9761
1.2429	2	78 Pt	$L\eta$	$L_{II}M_I$	9.975
1.24385	7	82 Pb	Ls	$L_{III}M_{III}$	9.9675
1.24460	3	74 W	$L\beta_2$	$L_{III}N_V$	9.9615
1.2453	1	70 Yb		$L_{II}O_{II,III}$	9.9561
1.24631	3	74 W	$L\beta_{15}$	$L_{III}N_{IV}$	9.9478
1.2466	2	73 Ta	$L\beta_9$	L_IM_V	9.946
1.2480	2	76 Os	$L\beta_{17}$	$L_{II}M_{III}$	9.934
1.24923	5	70 Yb	$L\gamma_8$	$L_{II}O_I$	9.9246
1.2502	3	77 Ir		$L_{II}M_{II}$	9.917
1.25100	5	75 Re	$L\beta_6$	$L_{III}N_I$	9.9105
1.25264	7	80 Hg	$L\alpha_2$	$L_{III}M_{IV}$	9.8976
1.2537	2	73 Ta	$L\beta_{10}$	L_IM_{IV}	9.889
1.254054	9	32 Ge	$K\alpha_1$	KL_{III}	9.88642
1.2553	1	73 Ta	L_{III}	Abs. Edge	9.8766
1.2555	1	73 Ta	$L\beta_5$	$L_{III}O_{IV,V}$	9.8750
1.25778	4	73 Ta	Lu	$L_{III}N_{VI,VII}$	9.8572
1.258011	9	32 Ge	$K\alpha_2$	KL_{II}	9.85532
1.25917	5	75 Re	$L\beta_4$	L_IM_{II}	9.8463
1.2596	1	71 Lu	$L\gamma_5$	$L_{II}N_I$	9.8428
1.2601	3	73 Ta		$L_{III}O_{II,III}$	9.839
1.26269	5	74 W	$L\beta_3$	L_IM_{III}	9.8188
1.26385	5	73 Ta	$L\beta_7$	$L_{III}O_I$	9.8098
1.2672	2	74 W		$L_{III}N_{III}$	9.784
1.26769	5	70 Yb	$L\gamma_1$	$L_{II}N_{IV}$	9.7801
1.2678	2	69 Tm	$L\gamma_3$	L_IN_{III}	9.779
1.2706	1	68 Er	L_I	Abs. Edge	9.7574
1.2728	2	74 W		$L_{II}M_V$	9.741
1.2742	2	69 Tm	$L\gamma_2$	L_IN_{II}	9.730
1.2748	1	83 Bi	Lt	$L_{III}M_{II}$	9.7252
1.2752	2	68 Er	$L\gamma_4$	$L_IO_{II,III}$	9.722
1.27640	3	79 Au	$L\alpha_1$	$L_{III}M_V$	9.7133
1.2765	2	74 W		$L_{III}N_{II}$	9.712
1.27807	5	81 Tl	Ls	$L_{III}M_{III}$	9.7007
1.281809	9	74 W	$L\beta_1$	$L_{II}M_{IV}$	9.67235
1.2829	5	84 Po	Ll	$L_{III}M_I$	9.664
1.2834	1	30 Zn	K	Abs. Edge	9.6607
1.28372	2	30 Zn	$K\beta_2$	$KN_{II,III}$	9.6580
1.28448	3	77 Ir	$L\eta$	$L_{II}M_I$	9.6522
1.28454	2	73 Ta	$L\beta_2$	$L_{III}N_V$	9.6518
1.2848	1	30 Zn	$K\beta_5$	$KM_{IV,V}$	9.6501
1.28619	5	73 Ta	$L\beta_{15}$	$L_{III}N_{IV}$	9.6394
1.28772	3	79 Au	$L\alpha_2$	$L_{III}M_{IV}$	9.6280
1.2892	1	69 Tm	L_{II}	Abs. Edge	9.6171
1.28989	7	74 W	$L\beta_6$	$L_{III}N_I$	9.6117
1.29025	9	72 Hf	$L\beta_9$	L_IM_V	9.6090
1.2905	2	69 Tm	$L\gamma_6$	$L_{II}O_{IV}$	9.607
1.2927	1	75 Re	$L\beta_{17}$	$L_{II}M_{III}$	9.5910
1.2934	2	76 Os		$L_{II}M_{II}$	9.586
1.29525	2	30 Zn	$K\beta_{1,3}$	$KM_{II,III}$	9.5720
1.2972	1	72 Hf	L_{III}	Abs. Edge	9.5577
1.29761	5	72 Hf	$L\beta_5$	$L_{III}O_{IV,V}$	9.5546
1.29819	9	72 Hf	$L\beta_{10}$	L_IM_{IV}	9.5503
1.30162	5	74 W	$L\beta_4$	L_IM_{II}	9.5252
1.30165	9	72 Hf	Lu	$L_{III}N_{VI,VII}$	9.5249
1.30564	5	72 Hf	$L\beta_7$	$L_{III}O_I$	9.4958
1.3063	1	70 Yb	$L\gamma_5$	$L_{II}N_I$	9.4910
1.30678	3	73 Ta	$L\beta_3$	L_IM_{III}	9.4875
1.30767	7	82 Pb	Lt	$L_{III}M_{II}$	9.4811
1.3086	1	73 Ta		$L_{III}N_{III}$	9.4742
1.3112	2	80 Hg	Ls	$L_{III}M_{III}$	9.455
1.31304	3	78 Pt	$L\alpha_1$	$L_{III}M_V$	9.4423
1.3146	1	68 Er	$L\gamma_3$	L_IN_{III}	9.4309
1.3153	2	69 Tm	$L\gamma_1$	$L_{II}N_{IV}$	9.426
1.31610	7	83 Bi	Ll	$L_{III}M_I$	9.4204
1.3167	1	73 Ta		$L_{III}N_{II}$	9.4158
1.31897	9	73 Ta		$L_{II}M_V$	9.3998
1.3190	1	67 Ho	L_I	Abs. Edge	9.3994
1.3208	3	67 Ho		$L_IO_{IV,V}$	9.387
1.3210	2	68 Er	$L\gamma_2$	L_IN_{II}	9.385

TABLE B (*Continued*)

Wavelength Å*	p.e.	Element	Designation		keV
1.3225	2	67 Ho	$L\gamma_4$	$L_{I}O_{II,III}$	9.374
1.32432	2	78 Pt	$L\alpha_2$	$L_{III}M_{IV}$	9.3618
1.32639	5	72 Hf	$L\beta_2$	$L_{III}N_{V}$	9.3473
1.32698	3	73 Ta	$L\beta_1$	$L_{II}M_{IV}$	9.3431
1.32783	5	72 Hf	$L\beta_{15}$	$L_{III}N_{IV}$	9.3371
1.32785	7	76 Os	$L\eta$	$L_{II}M_{I}$	9.3370
1.33094	8	73 Ta	$L\beta_6$	$L_{III}N_{I}$	9.3153
1.3358	1	71 Lu	$L\beta_9$	$L_{I}M_{V}$	9.2816
1.3365	3	74 W		$L_{I}M_{I}$	9.277
1.3366	1	75 Re		$L_{II}M_{II}$	9.2761
1.3386	1	68 Er	L_{II}	Abs. Edge	9.2622
1.3387	2	74 W	$L\beta_{17}$	$L_{II}M_{III}$	9.261
1.3397	3	68 Er	$L\gamma_6$	$L_{II}O_{IV}$	9.255
1.340083	9	31 Ga	$K\alpha_1$	KL_{III}	9.25174
1.3405	1	71 Lu	L_{III}	Abs. Edge	9.2490
1.34154	5	81 Tl	Lt	$L_{III}M_{II}$	9.2417
1.34183	7	71 Lu	$L\beta_5$	$L_{III}O_{IV,V}$	9.2397
1.3430	2	71 Lu	$L\beta_{10}$	$L_{I}M_{IV}$	9.232
1.34399	1	31 Ga	$K\alpha_2$	KL_{II}	9.22482
1.34524	9	71 Lu		$L_{III}O_{II,III}$	9.2163
1.34581	3	73 Ta	$L\beta_4$	$L_{I}M_{II}$	9.2124
1.34949	5	71 Lu	$L\beta_7$	$L_{III}O_{I}$	9.1873
1.34990	7	82 Pb	Ll	$L_{III}M_{I}$	9.1845
1.35053	9	72 Hf		$L_{III}N_{III}$	9.1802
1.35128	3	77 Ir	$L\alpha_1$	$L_{III}M_{V}$	9.1751
1.35131	7	79 Au	Ls	$L_{III}M_{III}$	9.1749
1.35300	5	72 Hf	$L\beta_3$	$L_{I}M_{III}$	9.1634
1.3558	2	69 Tm	$L\gamma_5$	$L_{II}N_{I}$	9.144
1.35887	9	72 Hf		$L_{III}N_{II}$	9.1239
1.36250	5	77 Ir	$L\alpha_2$	$L_{III}M_{IV}$	9.0995
1.3641	2	68 Er	$L\gamma_1$	$L_{II}N_{IV}$	9.089
1.3643	2	67 Ho	$L\gamma_3$	$L_{I}N_{III}$	9.087
1.3692	1	66 Dy	L_{I}	Abs. Edge	9.0548
1.3698	2	67 Ho	$L\gamma_2$	$L_{I}N_{II}$	9.051
1.37012	3	71 Lu	$L\beta_2$	$L_{III}N_{V}$	9.0489
1.3715	1	71 Lu	$L\beta_{15}$	$L_{III}N_{IV}$	9.0395
1.37342	5	75 Re	$L\eta$	$L_{II}M_{I}$	9.0272
1.37410	5	72 Hf	$L\beta_1$	$L_{II}M_{IV}$	9.0227
1.37410	5	72 Hf	$L\beta_6$	$L_{III}N_{I}$	9.0227
1.37459	7	66 Dy	$L\gamma_4$	$L_{I}O_{II,III}$	9.0195
1.3746	2	80 Hg	Lt	$L_{III}M_{II}$	9.019
1.38059	5	29 Cu	K	Abs. Edge	8.9803
1.38109	3	29 Cu	$K\beta_2$	$KM_{IV,V}$	8.9770
1.3838	1	70 Yb	$L\beta_9$	$L_{I}M_{V}$	8.9597
1.38477	3	81 Tl	Ll	$L_{III}M_{I}$	8.9532
1.3862	1	70 Yb	L_{III}	Abs. Edge	8.9441
1.3864	1	73 Ta	$L\beta_{17}$	$L_{II}M_{III}$	8.9428
1.38696	7	70 Yb	$L\beta_5$	$L_{III}O_{IV,V}$	8.9390
1.3895	2	78 Pt	Ls	$L_{III}M_{III}$	8.923
1.3898	1	70 Yb		$L_{III}O_{II,III}$	8.9209
1.3905	1	67 Ho	L_{II}	Abs. Edge	8.9164
1.39121	5	76 Os	$L\alpha_1$	$L_{III}M_{V}$	8.9117
1.3915	1	70 Yb	$L\beta_{10}$	$L_{I}M_{IV}$	8.9100
1.39220	5	72 Hf	$L\beta_4$	$L_{I}M_{II}$	8.9054
1.392218	9	29 Cu	$K\beta_{1,3}$	$KM_{II,III}$	8.90529
1.3923	2	67 Ho	$L\gamma_6$	$L_{II}O_{IV}$	8.905
1.3926	1	29 Cu	$K\beta_3$	KM_{II}	8.9029
1.3948	1	70 Yb	$L\beta_7$	$L_{III}O_{I}$	8.8889
1.3983	2	67 Ho	$L\gamma_8$	$L_{II}O_{I}$	8.867
1.40140	5	71 Lu	$L\beta_3$	$L_{I}M_{III}$	8.8469
1.40234	5	76 Os	$L\alpha_2$	$L_{III}M_{IV}$	8.8410
1.4067	3	68 Er	$L\gamma_5$	$L_{II}N_{I}$	8.814
1.41366	7	79 Au	Lt	$L_{III}M_{II}$	8.7702
1.41550	5	70 Yb	$L\beta_{2,15}$	$L_{III}N_{IV,V}$	8.7588
1.41640	7	66 Dy	$L\gamma_3$	$L_{I}N_{III}$	8.7532
1.4174	2	67 Ho	$L\gamma_1$	$L_{II}N_{IV}$	8.747
1.4189	1	71 Lu	$L\beta_6$	$L_{III}N_{I}$	8.7376
1.42110	3	74 W	$L\eta$	$L_{II}M_{I}$	8.7243
1.4216	1	80 Hg	Ll	$L_{III}M_{I}$	8.7210
1.4223	1	65 Tb	L_{I}	Abs. Edge	8.7167
1.42278	7	66 Dy	$L\gamma_2$	$L_{I}N_{II}$	8.7140
1.4228	3	65 Tb		$L_{I}O_{IV,V}$	8.714
1.42359	3	71 Lu	$L\beta_1$	$L_{II}M_{IV}$	8.7090
1.4276	2	65 Tb	$L\gamma_4$	$L_{I}O_{II,III}$	8.685
1.43025	9	72 Hf		$L_{I}M_{I}$	8.6685
1.43048	9	73 Ta		$L_{II}M_{II}$	8.6671
1.4318	2	77 Ir	Ls	$L_{III}M_{III}$	8.659
1.43290	4	75 Re	$L\alpha_1$	$L_{III}M_{V}$	8.6525
1.4334	1	69 Tm	L_{III}	Abs. Edge	8.6496
1.4336	3	69 Tm	$L\beta_9$	$L_{I}M_{V}$	8.648
1.4349	2	69 Tm	$L\beta_5$	$L_{III}O_{IV,V}$	8.641
1.435155	7	30 Zn	$K\alpha_1$	KL_{III}	8.63886
1.43643	9	72 Hf	$L\beta_{17}$	$L_{II}M_{III}$	8.6312
1.439000	8	30 Zn	$K\alpha_2$	KL_{II}	8.61578
1.44056	5	71 Lu	$L\beta_4$	$L_{I}M_{II}$	8.6064
1.4410	3	69 Tm	$L\beta_{10}$	$L_{I}M_{IV}$	8.604
1.44396	5	75 Re	$L\alpha_2$	$L_{III}M_{IV}$	8.5862
1.4445	1	66 Dy	L_{II}	Abs. Edge	8.5830
1.44579	7	66 Dy	$L\gamma_6$	$L_{II}O_{IV}$	8.5753
1.45233	5	70 Yb	$L\beta_3$	$L_{I}M_{III}$	8.5367
1.4530	2	78 Pt	Lt	$L_{III}M_{II}$	8.533
1.45964	9	79 Au	Ll	$L_{III}M_{I}$	8.4939
1.4618	2	67 Ho	$L\gamma_5$	$L_{II}N_{I}$	8.481
1.4640	2	69 Tm	$L\beta_{2,15}$	$L_{III}N_{IV,V}$	8.468
1.4661	1	70 Yb	$L\beta_6$	$L_{III}N_{I}$	8.4563
1.47106	5	73 Ta	$L\eta$	$L_{II}M_{I}$	8.4280
1.4718	2	65 Tb	$L\gamma_3$	$L_{I}N_{III}$	8.423
1.47266	7	66 Dy	$L\gamma_1$	$L_{II}N_{IV}$	8.4188
1.4735	2	76 Os	Ls	$L_{III}M_{III}$	8.414
1.47565	5	70 Yb	$L\beta_1$	$L_{II}M_{IV}$	8.4018
1.4764	2	65 Tb	$L\gamma_2$	$L_{I}N_{II}$	8.398
1.47639	2	74 W	$L\alpha_1$	$L_{III}M_{V}$	8.3976
1.4784	1	64 Gd	L_{I}	Abs. Edge	8.3864
1.48064	9	72 Hf		$L_{II}M_{II}$	8.3735
1.4807	3	64 Gd		$L_{I}O_{IV,V}$	8.373
1.4835	1	68 Er	L_{III}	Abs. Edge	8.3575
1.4839	2	64 Gd	$L\gamma_4$	$L_{I}O_{II,III}$	8.355
1.4848	3	68 Er	$L\beta_5$	$L_{III}O_{IV,V}$	8.350
1.4855	5	68 Er	$L\beta_9$	$L_{I}M_{V}$	8.346
1.48743	2	74 W	$L\alpha_2$	$L_{III}M_{IV}$	8.3352
1.48807	1	28 Ni	K	Abs. Edge	8.33165
1.48862	4	28 Ni	$K\beta_5$	$KM_{IV,V}$	8.3286
1.49138	3	70 Yb	$L\beta_4$	$L_{I}M_{II}$	8.3132
1.4930	3	77 Ir	Lt	$L_{III}M_{II}$	8.304

Wavelength Å*	p.e.	Element		Designation	keV
1.4941	3	68 Er	$L\beta_7$	$L_{III}O_I$	8.298
1.4941	3	68 Er	$L\beta_{10}$	L_IM_{IV}	8.298
1.4995	2	78 Pt	Ll	$L_{III}M_I$	8.268
1.500135	8	28 Ni	$K\beta_{1,3}$	$KM_{II,III}$	8.26466
1.5023	1	65 Tb	L_{II}	Abs. Edge	8.2527
1.5035	2	65 Tb	$L\gamma_6$	$L_{II}O_{IV}$	8.246
1.5063	2	69 Tm	$L\beta_3$	L_IM_{III}	8.231
1.5097	2	65 Tb	$L\gamma_8$	$L_{II}O_I$	8.212
1.51399	9	68 Er	$L\beta_{2,15}$	$L_{III}N_{IV,V}$	8.1890
1.5162	2	69 Tm	$L\beta_6$	$L_{III}N_I$	8.177
1.5178	1	75 Re	Ls	$L_{III}M_{III}$	8.1682
1.51824	7	66 Dy	$L\gamma_5$	$L_{II}N_I$	8.1661
1.52197	2	73 Ta	$L\alpha_1$	$L_{III}M_V$	8.1461
1.52325	5	72 Hf	$L\eta$	$L_{II}M_I$	8.1393
1.5297	2	64 Gd	$L\gamma_3$	L_IN_{III}	8.105
1.5303	2	65 Tb	$L\gamma_1$	$L_{II}N_{IV}$	8.102
1.5304	2	69 Tm	$L\beta_1$	$L_{II}M_{IV}$	8.101
1.53293	2	73 Ta	$L\alpha_2$	$L_{III}M_{IV}$	8.0879
1.5331	2	64 Gd	$L\gamma_2$	L_IN_{II}	8.087
1.53333	9	71 Lu		$L_{II}M_{II}$	8.0858
1.5347	2	76 Os	Lt	$L_{III}M_{II}$	8.079
1.5368	1	67 Ho	L_{III}	Abs. Edge	8.0676
1.5378	2	67 Ho	$L\beta_5$	$L_{III}O_{IV,V}$	8.062
1.5381	1	63 Eu	L_I	Abs. Edge	8.0607
1.540562	2	29 Cu	$K\alpha_1$	KL_{III}	8.04778
1.54094	3	77 Ir	Ll	$L_{III}M_I$	8.0458
1.5439	1	63 Eu	$L\gamma_4$	$L_IO_{II,III}$	8.0304
1.544390	9	29 Cu	$K\alpha_2$	KL_{II}	8.02783
1.5448	2	69 Tm	$L\beta_4$	L_IM_{II}	8.026
1.5486	3	67 Ho	$L\beta_{10}$	L_IM_{IV}	8.006
1.5616	1	68 Er	$L\beta_3$	L_IM_{III}	7.9392
1.5632	1	64 Gd	L_{II}	Abs. Edge	7.9310
1.5642	3	74 W	Ls	$L_{III}M_{III}$	7.926
1.5644	2	64 Gd	$L\gamma_6$	$L_{II}O_{IV}$	7.925
1.5671	2	67 Ho	$L\beta_{2,15}$	$L_{III}N_{IV,V}$	7.911
1.5675	2	68 Er	$L\beta_6$	$L_{III}N_I$	7.909
1.56958	5	72 Hf	$L\alpha_1$	$L_{III}M_V$	7.8990
1.5707	2	64 Gd	$L\gamma_8$	$L_{II}O_I$	7.894
1.5779	1	71 Lu	$L\eta$	$L_{II}M_I$	7.8575
1.5787	2	65 Tb	$L\gamma_5$	$L_{II}N_I$	7.8535
1.5789	1	75 Re	Lt	$L_{III}M_{II}$	7.8525
1.58046	5	72 Hf	$L\alpha_2$	$L_{III}M_{IV}$	7.8446
1.58498	7	76 Os	Ll	$L_{III}M_I$	7.8222
1.5873	1	68 Er	$L\beta_1$	$L_{II}M_{IV}$	7.8109
1.58837	7	66 Dy	$L\beta_5$	$L_{III}O_{IV,V}$	7.8055
1.58844	9	70 Yb		$L_{II}M_{II}$	7.8052
1.5903	2	63 Eu	$L\gamma_3$	L_IN_{III}	7.7961
1.5916	1	66 Dy	L_{III}	Abs. Edge	7.7897
1.5924	2	64 Gd	$L\gamma_1$	$L_{II}N_{IV}$	7.7858
1.5961	2	63 Eu	$L\gamma_2$	L_IN_{II}	7.7677
1.59973	9	66 Dy	$L\beta_9$	L_IM_V	7.7501
1.6002	1	62 Sm	L_I	Abs. Edge	7.7478
1.6007	1	68 Er	$L\beta_4$	L_IM_{II}	7.7453
1.60447	7	66 Dy	$L\beta_7$	$L_{III}O_I$	7.7272
1.60728	3	62 Sm	$L\gamma_4$	$L_IO_{II,III}$	7.714
1.60743	9	66 Dy	$L\beta_{10}$	L_IM_{IV}	7.7130
1.60815	1	27 Co	K	Abs. Edge	7.70954
1.60891	3	27 Co	$K\beta_5$	$KM_{IV,V}$	7.7059
1.61264	9	73 Ta	Ls	$L_{III}M_{III}$	7.6881
1.61951	3	71 Lu	$L\alpha_1$	$L_{III}M_V$	7.6555
1.6203	2	67 Ho	$L\beta_3$	L_IM_{III}	7.6519
1.62079	2	27 Co	$K\beta_{1,3}$	$KM_{II,III}$	7.64943
1.6237	2	67 Ho	$L\beta_6$	$L_{III}N_I$	7.6359
1.62369	7	66 Dy	$L\beta_{2,15}$	$L_{III}N_{IV,V}$	7.6357
1.6244	3	74 W	Lt	$L_{III}M_{II}$	7.6324
1.6271	1	63 Eu	L_{II}	Abs. Edge	7.6199
1.6282	2	63 Eu	$L\gamma_6$	$L_{II}O_{IV}$	7.6147
1.63029	5	71 Lu	$L\alpha_2$	$L_{III}M_{IV}$	7.6049
1.63056	5	75 Re	Ll	$L_{III}M_I$	7.6036
1.6346	2	63 Eu	$L\gamma_8$	$L_{II}O_I$	7.5849
1.63560	5	70 Yb	$L\eta$	$L_{II}M_I$	7.5802
1.6412	2	64 Gd	$L\gamma_5$	$L_{II}N_I$	7.5543
1.6475	2	67 Ho	$L\beta_1$	$L_{II}M_{IV}$	7.5253
1.6497	1	65 Tb	L_{III}	Abs. Edge	7.5153
1.6510	2	65 Tb	$L\beta_5$	$L_{III}O_{IV,V}$	7.5094
1.65601	3	62 Sm	$L\gamma_3$	L_IN_{III}	7.487
1.6574	2	63 Eu	$L\gamma_1$	$L_{II}N_{IV}$	7.4803
1.657910	8	28 Ni	$K\alpha_1$	KL_{III}	7.47815
1.6585	2	65 Tb	$L\beta_7$	$L_{III}O_I$	7.4753
1.6595	2	67 Ho	$L\beta_4$	L_IM_{II}	7.4708
1.66044	6	62 Sm	$L\gamma_2$	L_IN_{II}	7.467
1.661747	8	28 Ni	$K\alpha_2$	KL_{II}	7.46089
1.66346	9	72 Hf	Ls	$L_{III}M_{III}$	7.4532
1.6673	3	65 Tb	$L\beta_{10}$	L_IM_{IV}	7.436
1.6674	5	61 Pm	L_I	Abs. Edge	7.436
1.67189	4	70 Yb	$L\alpha_1$	$L_{III}M_V$	7.4156
1.67265	9	73 Ta	Lt	$L_{III}M_{II}$	7.4123
1.6782	1	74 W	Ll	$L_{III}M_I$	7.3878
1.68213	7	66 Dy	$L\beta_6$	$L_{III}N_I$	7.3705
1.6822	2	66 Dy	$L\beta_3$	L_IM_{III}	7.3702
1.68285	5	70 Yb	$L\alpha_2$	$L_{III}M_{IV}$	7.3673
1.6830	2	65 Tb	$L\beta_{2,15}$	$L_{III}N_{IV,V}$	7.3667
1.6953	1	62 Sm	L_{II}	Abs. Edge	7.3132
1.6963	2	69 Tm	$L\eta$	$L_{II}M_I$	7.3088
1.6966	9	62 Sm	$L\gamma_6$	$L_{II}O_{IV}$	7.308
1.7085	2	63 Eu	$L\gamma_5$	$L_{II}N_I$	7.2566
1.71062	7	66 Dy	$L\beta_1$	$L_{II}M_{IV}$	7.2477
1.7117	1	64 Gd	L_{III}	Abs. Edge	7.2430
1.7130	2	64 Gd	$L\beta_5$	$L_{III}O_{IV,V}$	7.2374
1.7203	2	64 Gd	$L\beta_7$	$L_{III}O_I$	7.2071
1.72103	7	66 Dy	$L\beta_4$	L_IM_{II}	7.2039
1.72305	9	72 Hf	Lt	$L_{III}M_{II}$	7.1954
1.7240	3	64 Gd	$L\beta_9$	L_IM_V	7.192
1.72724	3	62 Sm	$L\gamma_1$	$L_{II}N_{IV}$	7.178
1.7268	2	69 Tm	$L\alpha_1$	$L_{III}M_V$	7.1799
1.72841	5	73 Ta	Ll	$L_{III}M_I$	7.1731
1.7315	3	64 Gd	$L\beta_{10}$	L_IM_{IV}	7.160
1.7381	2	69 Tm	$L\alpha_2$	$L_{III}M_{IV}$	7.1331
1.7390	1	60 Nd	L_I	Abs. Edge	7.1294
1.7422	2	65 Tb	$L\beta_6$	$L_{III}N_I$	7.1163
1.74346	1	26 Fe	K	Abs. Edge	7.11120
1.7442	1	26 Fe	$K\beta_5$	$KM_{IV,V}$	7.1081
1.7445	4	60 Nd	$L\gamma_4$	$L_IO_{II,III}$	7.107
1.7455	2	64 Gd	$L\beta_{2,15}$	$L_{III}N_{IV,V}$	7.1028

TABLE B (*Continued*)

Wavelength Å*	p.e.	Element	Designation		keV
1.7472	2	65 Tb	$L\beta_3$	L_IM_{III}	7.0959
1.75661	2	26 Fe	$K\beta_{1,3}$	$KM_{II,III}$	7.05798
1.7566	1	68 Er	$L\eta$	$L_{II}M_I$	7.0579
1.7676	5	61 Pm	L_{II}	Abs. Edge	7.014
1.7760	1	71 Lu	Lt	$L_{III}M_{II}$	6.9810
1.7761	1	63 Eu	L_{III}	Abs. Edge	6.9806
1.7768	3	65 Tb	$L\beta_1$	$L_{II}M_{IV}$	6.978
1.7772	2	63 Eu	$L\beta_5$	$L_{III}O_{IV,V}$	6.9763
1.77934	3	62 Sm	$L\gamma_5$	$L_{II}N_I$	6.968
1.78145	5	72 Hf	Ll	$L_{III}M_I$	6.9596
1.78425	9	68 Er	$L\alpha_1$	$L_{III}M_V$	6.9487
1.7851	2	63 Eu	$L\beta_7$	$L_{III}O_I$	6.9453
1.7864	2	65 Tb	$L\beta_4$	L_IM_{II}	6.9403
1.788965	9	27 Co	$K\alpha_1$	KL_{III}	6.93032
1.7916	3	63 Eu	$L\beta_9$	L_IM_V	6.920
1.792850	9	27 Co	$K\alpha_2$	KL_{II}	6.91530
1.7955	2	68 Er	$L\alpha_2$	$L_{III}M_{IV}$	6.9050
1.7964	4	60 Nd	$L\gamma_3$	L_IN_{III}	6.902
1.7989	9	61 Pm	$L\gamma_1$	$L_{II}N_{IV}$	6.892
1.7993	3	63 Eu	$L\beta_{10}$	L_IM_{IV}	6.890
1.8013	4	60 Nd	$L\gamma_2$	L_IN_{II}	6.883
1.8054	2	64 Gd	$L\beta_6$	$L_{III}N_I$	6.8671
1.8118	2	63 Eu	$L\beta_{2,15}$	$L_{III}N_{IV,V}$	6.8432
1.8141	5	59 Pr	L_I	Abs. Edge	6.834
1.8150	2	64 Gd	$L\beta_3$	L_IM_{III}	6.8311
1.8193	4	59 Pr	$L\gamma_4$	$L_IO_{II,III}$	6.815
1.8264	2	67 Ho	$L\eta$	$L_{II}M_I$	6.7883
1.83091	9	70 Yb	Lt	$L_{III}M_{II}$	6.7715
1.8360	1	71 Lu	Ll	$L_{III}M_I$	6.7528
1.8440	1	60 Nd	L_{II}	Abs. Edge	6.7234
1.8450	2	67 Ho	$L\alpha_1$	$L_{III}M_V$	6.7198
1.8457	1	62 Sm	L_{III}	Abs. Edge	6.7172
1.8468	2	64 Gd	$L\beta_1$	$L_{II}M_{IV}$	6.7132
1.84700	9	62 Sm	$L\beta_5$	$L_{III}O_{IV,V}$	6.7126
1.8540	2	64 Gd	$L\beta_4$	L_IM_{II}	6.6871
1.8552	5	60 Nd	$L\gamma_8$	$L_{II}O_I$	6.683
1.8561	2	67 Ho	$L\alpha_2$	$L_{III}M_{IV}$	6.6795
1.85626	3	62 Sm	$L\beta_7$	$L_{III}O_I$	6.679
1.86166	3	62 Sm	$L\beta_9$	L_IM_V	6.660
1.86990	3	62 Sm	$L\beta_{10}$	L_IM_{IV}	6.634
1.8737	2	63 Eu	$L\beta_6$	$L_{III}N_I$	6.6170
1.8740	4	59 Pr	$L\gamma_3$	L_IN_{III}	6.616
1.8779	2	60 Nd	$L\gamma_1$	$L_{II}N_{IV}$	6.6021
1.8791	4	59 Pr	$L\gamma_2$	L_IN_{II}	6.598
1.8821	3	62 Sm	$L\beta_{2,15}$	$L_{III}N_{IV,V}$	6.586
1.8867	2	63 Eu	$L\beta_3$	L_IM_{III}	6.5713
1.8934	5	58 Ce	L_I	Abs. Edge	6.548
1.89415	5	70 Yb	Ll	$L_{III}M_I$	6.5455
1.89643	5	25 Mn	K	Abs. Edge	6.5376
1.8971	1	25 Mn	$K\beta_5$	$KM_{IV,V}$	6.5352
1.89743	7	66 Dy	$L\eta$	$L_{II}M_I$	6.5342
1.8991	4	58 Ce	$L\gamma_4$	$L_IO_{II,III}$	6.528
1.90881	3	66 Dy	$L\alpha_1$	$L_{III}M_V$	6.4952
1.91021	2	25 Mn	$K\beta_{1,3}$	$KM_{II,III}$	6.49045
1.9191	1	61 Pm	L_{III}	Abs. Edge	6.4605
1.91991	3	66 Dy	$L\alpha_2$	$L_{III}M_{IV}$	6.4577
1.9203	2	63 Eu	$L\beta_1$	$L_{II}M_{IV}$	6.4564
1.9255	2	63 Eu	$L\beta_4$	L_IM_{II}	6.4389
1.9255	5	59 Pr	L_{II}	Abs. Edge	6.439
1.9355	4	60 Nd	$L\gamma_5$	$L_{II}N_I$	6.406
1.936042	9	26 Fe	$K\alpha_1$	KL_{III}	6.40384
1.9362	4	59 Pr	$L\gamma_8$	$L_{II}O_I$	6.403
1.939980	9	26 Fe	$K\alpha_2$	KL_{II}	6.39084
1.94643	3	62 Sm	$L\beta_6$	$L_{III}N_I$	6.3693
1.9550	2	69 Tm	Ll	$L_{III}M_I$	6.3419
1.9553	3	58 Ce	$L\gamma_3$	L_IN_{III}	6.3409
1.9559	6	61 Pm	$L\beta_{2,15}$	$L_{III}N_{IV,V}$	6.339
1.9602	3	58 Ce	$L\gamma_2$	L_IN_{II}	6.3250
1.9611	3	59 Pr	$L\gamma_1$	$L_{II}N_{IV}$	6.3221
1.96241	3	62 Sm	$L\beta_3$	L_IM_{III}	6.318
1.9730	2	65 Tb	$L\eta$	$L_{II}M_I$	6.2839
1.9765	2	65 Tb	$L\alpha_1$	$L_{III}M_V$	6.2728
1.9780	5	57 La	L_I	Abs. Edge	6.268
1.9830	4	57 La	$L\gamma_4$	$L_IO_{II,III}$	6.252
1.9875	2	65 Tb	$L\alpha_2$	$L_{III}M_{IV}$	6.2380
1.9967	1	60 Nd	L_{III}	Abs. Edge	6.2092
1.99806	3	62 Sm	$L\beta_1$	$L_{II}M_{IV}$	6.2051
2.00095	6	62 Sm	$L\beta_4$	L_IM_{II}	6.196
2.0092	3	60 Nd	$L\beta_7$	$L_{III}O_I$	6.1708
2.0124	5	58 Ce	L_{II}	Abs. Edge	6.161
2.015	1	68 Er	Ll	$L_{III}M_I$	6.152
2.0165	3	60 Nd	$L\beta_9$	L_IM_V	6.1484
2.0205	4	59 Pr	$L\gamma_5$	$L_{II}N_I$	6.136
2.0237	4	58 Ce	$L\gamma_8$	$L_{II}O_I$	6.126
2.0237	3	60 Nd	$L\beta_{10}$	L_IM_{IV}	6.1265
2.0360	3	60 Nd	$L\beta_{2,15}$	$L_{III}N_{IV,V}$	6.0894
2.0410	4	57 La	$L\gamma_3$	L_IN_{III}	6.074
2.0421	4	61 Pm	$L\beta_3$	L_IM_{III}	6.071
2.0460	4	57 La	$L\gamma_2$	L_IN_{II}	6.060
2.0468	2	64 Gd	$L\alpha_1$	$L_{III}M_V$	6.0572
2.0487	4	58 Ce	$L\gamma_1$	$L_{II}N_{IV}$	6.052
2.0494	1	64 Gd	$L\eta$	$L_{II}M_I$	6.0495
2.0578	2	64 Gd	$L\alpha_2$	$L_{III}M_{IV}$	6.0250
2.0678	5	56 Ba	L_I	Abs. Edge	5.996
2.07020	5	24 Cr	K	Abs. Edge	5.9888
2.07087	6	24 Cr	$K\beta_5$	$KM_{IV,V}$	5.9869
2.0756	3	56 Ba	$L\gamma_4$	$L_IO_{II,III}$	5.9733
2.0791	5	59 Pr	L_{III}	Abs. Edge	5.963
2.0797	4	61 Pm	$L\beta_1$	$L_{II}M_{IV}$	5.961
2.08487	2	24 Cr	$K\beta_{1,3}$	$KM_{II,III}$	5.94671
2.0860	2	67 Ho	Ll	$L_{III}M_I$	5.9434
2.0919	4	59 Pr	$L\beta_7$	$L_{III}O_I$	5.927
2.1004	4	59 Pr	$L\beta_9$	L_IM_V	5.903
2.101820	9	25 Mn	$K\alpha_1$	KL_{III}	5.89875
2.1039	3	60 Nd	$L\beta_6$	$L_{III}N_I$	5.8930
2.1053	5	57 La	L_{II}	Abs. Edge	5.889
2.10578	2	25 Mn	$K\alpha_2$	KL_{II}	5.88765
2.1071	4	59 Pr	$L\beta_{10}$	L_IM_{IV}	5.884
2.1103	3	58 Ce	$L\gamma_5$	$L_{II}N_I$	5.8751
2.1194	4	59 Pr	$L\beta_{2,15}$	$L_{III}N_{IV,V}$	5.850
2.1209	2	63 Eu	$L\alpha_1$	$L_{III}M_V$	5.8457
2.1268	2	60 Nd	$L\beta_3$	L_IM_{III}	5.8294
2.1315	2	63 Eu	$L\eta$	$L_{II}M_I$	5.8166
2.1315	2	63 Eu	$L\alpha_2$	$L_{III}M_{IV}$	5.8166

Wavelength Å*	p.e.	Element	Designation		keV
2.1342	2	56 Ba	$L\gamma_3$	L_IN_{III}	5.8092
2.1387	2	56 Ba	$L\gamma_2$	L_IN_{II}	5.7969
2.1418	3	57 La	$L\gamma_1$	$L_{II}N_{IV}$	5.7885
2.15877	7	66 Dy	Ll	$L_{III}M_I$	5.7431
2.166	1	58 Ce	L_{III}	Abs. Edge	5.723
2.1669	3	60 Nd	$L\beta_4$	L_IM_{II}	5.7216
2.1669	2	60 Nd	$L\beta_1$	$L_{II}M_{IV}$	5.7216
2.1673	5	55 Cs	L_I	Abs. Edge	5.721
2.1701	2	58 Ce	$L\beta_7$	$L_{III}O_I$	5.7132
2.1741	2	55 Cs	$L\gamma_4$	$L_IO_{II,III}$	5.7026
2.1885	3	58 Ce	$L\beta_9$	L_IM_V	5.6650
2.1906	4	59 Pr	$L\beta_6$	$L_{III}N_I$	5.660
2.1958	5	58 Ce	$L\beta_{10}$	L_IM_{IV}	5.646
2.1998	2	62 Sm	$L\alpha_1$	$L_{III}M_V$	5.6361
2.2048	1	56 Ba	L_{II}	Abs. Edge	5.6233
2.2056	4	57 La	$L\gamma_5$	$L_{II}N_I$	5.621
2.2087	2	58 Ce	$L\beta_{2,15}$	$L_{III}N_{IV,V}$	5.6134
2.21062	3	62 Sm	$L\alpha_2$	$L_{III}M_{IV}$	5.6090
2.2172	3	59 Pr	$L\beta_3$	L_IM_{III}	5.5918
2.21824	3	62 Sm	$L\eta$	$L_{II}M_I$	5.589
2.2328	2	55 Cs	$L\gamma_3$	L_IN_{III}	5.5527
2.2352	2	65 Tb	Ll	$L_{III}M_I$	5.5467
2.2371	2	55 Cs	$L\gamma_2$	L_IN_{II}	5.5420
2.2415	2	56 Ba	$L\gamma_1$	$L_{II}N_{IV}$	5.5311
2.253	6	92 U		M_IP_{III}	5.50
2.2550	4	59 Pr	$L\beta_4$	L_IM_{II}	5.4981
2.2588	3	59 Pr	$L\beta_1$	$L_{II}M_{IV}$	5.4889
2.261	1	57 La	L_{III}	Abs. Edge	5.484
2.2691	1	23 V	K	Abs. Edge	5.4639
2.26951	6	23 V	$K\beta_5$	$KM_{IV,V}$	5.4629
2.2737	1	54 Xe	L_I	Abs. Edge	5.4528
2.275	3	57 La	$L\beta_7$	$L_{III}O_I$	5.450
2.282	3	57 La	$L\beta_9$	L_IM_V	5.434
2.2818	3	58 Ce	$L\beta_6$	$L_{III}N_I$	5.4334
2.2822	3	61 Pm	$L\alpha_1$	$L_{III}M_V$	5.4325
2.28440	2	23 V	$K\beta_{1,3}$	$KM_{II,III}$	5.42729
2.28970	2	24 Cr	$K\alpha_1$	KL_{III}	5.41472
2.290	3	57 La	$L\beta_{10}$	L_IM_{IV}	5.415
2.2926	4	61 Pm	$L\alpha_2$	$L_{III}M_{IV}$	5.4078
2.293606	3	24 Cr	$K\alpha_2$	KL_{II}	5.405509
2.3030	3	57 La	$L\beta_{2,15}$	$L_{III}N_{IV,V}$	5.3835
2.304	7	92 U		M_IO_{III}	5.38
2.3085	3	56 Ba	$L\gamma_5$	$L_{II}N_I$	5.3707
2.3109	3	58 Ce	$L\beta_3$	L_IM_{III}	5.3651
2.3122	2	64 Gd	Ll	$L_{III}M_I$	5.3621
2.3139	1	55 Cs	L_{II}	Abs. Edge	5.3581
2.3480	2	55 Cs	$L\gamma_1$	$L_{II}N_{IV}$	5.2804
2.3497	4	58 Ce	$L\beta_4$	L_IM_{II}	5.2765
2.3561	3	58 Ce	$L\beta_1$	$L_{II}M_{IV}$	5.2622
2.3629	1	56 Ba	L_{III}	Abs. Edge	5.2470
2.3704	2	60 Nd	$L\alpha_1$	$L_{III}M_V$	5.2304
2.3764	2	56 Ba	$L\beta_9$	L_IM_V	5.2171
2.3790	4	57 La	$L\beta_6$	$L_{III}N_I$	5.2114
2.3806	2	56 Ba	$L\beta_7$	$L_{III}O_I$	5.2079
2.3807	3	60 Nd	$L\alpha_2$	$L_{III}M_{IV}$	5.2077
2.3869	2	56 Ba	$L\beta_{10}$	L_IM_{IV}	5.1941
2.3880	5	53 I	L_I	Abs. Edge	5.192
2.3913	2	53 I	$L\gamma_4$	$L_IO_{II,III}$	5.1848
2.3948	2	63 Eu	Ll	$L_{III}M_I$	5.1772
2.40435	6	56 Ba	$L\beta_{2,15}$	$L_{III}N_{IV,V}$	5.1565
2.4094	4	60 Nd	$L\eta$	$L_{II}M_I$	5.1457
2.4105	3	57 La	$L\beta_3$	L_IM_{III}	5.1434
2.4174	2	55 Cs	$L\gamma_5$	$L_{II}N_I$	5.1287
2.4292	1	54 Xe	L_{II}	Abs. Edge	5.1037
2.442	9	90 Th		M_IO_{III}	5.08
2.443	4	92 U		$M_{II}O_{IV}$	5.075
2.4475	2	53 I	$L\gamma_{2,3}$	$L_IN_{II,III}$	5.0657
2.4493	3	57 La	$L\beta_4$	L_IM_{II}	5.0620
2.45891	5	57 La	$L\beta_1$	$L_{II}M_{IV}$	5.0421
2.4630	2	59 Pr	$L\alpha_1$	$L_{III}M_V$	5.0337
2.4729	3	59 Pr	$L\alpha_2$	$L_{III}M_{IV}$	5.0135
2.4740	1	55 Cs	L_{III}	Abs. Edge	5.0113
2.4783	2	55 Cs	$L\beta_9$	L_IM_V	5.0026
2.4823	4	62 Sm	Ll	$L_{III}M_I$	4.9945
2.4826	2	56 Ba	$L\beta_6$	$L_{III}N_I$	4.9939
2.4849	2	55 Cs	$L\beta_7$	$L_{III}O_I$	4.9893
2.4920	2	55 Cs	$L\beta_{10}$	L_IM_{IV}	4.9752
2.49734	5	22 Ti	K	Abs. Edge	4.96452
2.4985	2	22 Ti	$K\beta_5$	$KM_{IV,V}$	4.9623
2.50356	2	23 V	$K\alpha_1$	KL_{III}	4.95220
2.50738	2	23 V	$K\alpha_2$	KL_{II}	4.94464
2.5099	1	52 Te	L_I	Abs. Edge	4.9397
2.5113	2	52 Te	$L\gamma_4$	$L_IO_{II,III}$	4.9369
2.5118	2	55 Cs	$L\beta_{2,15}$	$L_{III}N_{IV,V}$	4.9359
2.512	3	59 Pr	$L\eta$	$L_{II}M_I$	4.935
2.51391	2	22 Ti	$K\beta_{1,3}$	$KM_{II,III}$	4.93181
2.5164	2	56 Ba	$L\beta_3$	L_IM_{III}	4.9269
2.527	4	91 Pa		$M_{II}O_{IV}$	4.906
2.5542	5	53 I	L_{II}	Abs. Edge	4.8540
2.5553	2	56 Ba	$L\beta_4$	L_IM_{II}	4.8519
2.5615	2	58 Ce	$L\alpha_1$	$L_{III}M_V$	4.8402
2.5674	2	52 Te	$L\gamma_{2,3}$	$L_IN_{II,III}$	4.8290
2.56821	5	56 Ba	$L\beta_1$	$L_{II}M_{IV}$	4.82753
2.5706	3	58 Ce	$L\alpha_2$	$L_{III}M_{IV}$	4.8230
2.58244	8	53 I	$L\gamma_1$	$L_{II}N_{IV}$	4.8009
2.5926	1	54 Xe	L_{III}	Abs. Edge	4.7822
2.5932	2	55 Cs	$L\beta_6$	$L_{III}N_I$	4.7811
2.618	5	90 Th		$M_{II}O_{IV}$	4.735
2.6203	4	58 Ce	$L\eta$	$L_{II}M_I$	4.7315
2.6285	2	55 Cs	$L\beta_3$	L_IM_{III}	4.7167
2.6388	1	51 Sb	L_I	Abs. Edge	4.6984
2.6398	2	51 Sb	$L\gamma_4$	$L_IO_{II,III}$	4.6967
2.65710	9	53 I	$L\gamma_5$	$L_{II}N_I$	4.6660
2.66570	5	57 La	$L\alpha_1$	$L_{III}M_V$	4.65097
2.6666	2	55 Cs	$L\beta_4$	L_IM_{II}	4.6494
2.67533	5	57 La	$L\alpha_2$	$L_{III}M_{IV}$	4.63423
2.6760	4	60 Nd	Ll	$L_{III}M_I$	4.6330
2.6837	2	55 Cs	$L\beta_1$	$L_{II}M_{IV}$	4.6198
2.6879	1	52 Te	L_{II}	Abs. Edge	4.6126
2.6953	2	51 Sb	$L\gamma_{2,3}$	$L_IN_{II,III}$	4.5999
2.71241	6	52 Te	$L\gamma_1$	$L_{II}N_{IV}$	4.5709
2.71352	9	53 I	$L\beta_9$	L_IM_V	4.5690
2.7196	5	53 I	L_{III}	Abs. Edge	4.5587
2.72104	9	53 I	$L\beta_{10}$	L_IM_{IV}	4.5564

Wavelength Å*	p.e.	Element	Designation		keV
2.7288	3	53 I	$L\beta_7$	$L_{III}O_I$	4.5435
2.740	3	57 La	$L\eta$	$L_{II}M_I$	4.525
2.74851	2	22 Ti	$K\alpha_1$	KL_{III}	4.51084
2.75053	8	53 I	$L\beta_{2,15}$	$L_{III}N_{IV,V}$	4.5075
2.75216	2	22 Ti	$K\alpha_2$	KL_{II}	4.50486
2.753	8	92 U		M_IN_{III}	4.50
2.762	1	21 Sc	K	Abs. Edge	4.489
2.7634	3	21 Sc	$K\beta_5$	$KM_{IV,V}$	4.4865
2.77595	5	56 Ba	$L\alpha_1$	$L_{III}M_V$	4.46626
2.7769	1	50 Sn	L_I	Abs. Edge	4.4648
2.7775	2	50 Sn	$L\gamma_4$	$L_IO_{II,III}$	4.4638
2.7796	2	21 Sc	$K\beta_{1,3}$	$KM_{II,III}$	4.4605
2.7841	4	59 Pr	Ll	$L_{III}M_I$	4.4532
2.78553	5	56 Ba	$L\alpha_2$	$L_{III}M_{IV}$	4.45090
2.79007	9	52 Te	$L\gamma_5$	$L_{II}N_I$	4.4437
2.817	2	92 U		$M_{II}N_{IV}$	4.401
2.8294	5	51 Sb	L_{II}	Abs. Edge	4.3819
2.8327	2	50 Sn	$L\gamma_{2,3}$	$L_IN_{II,III}$	4.3768
2.83672	9	53 I	$L\beta_6$	$L_{III}N_I$	4.3706
2.83897	9	52 Te	$L\beta_9$	L_IM_V	4.3671
2.84679	9	52 Te	$L\beta_{10}$	L_IM_{IV}	4.3551
2.85159	3	51 Sb	$L\gamma_1$	$L_{II}N_{IV}$	4.34779
2.8555	1	52 Te	L_{III}	Abs. Edge	4.3418
2.8627	3	56 Ba	$L\eta$	$L_{II}M_I$	4.3309
2.8634	3	52 Te	$L\beta_7$	$L_{III}O_I$	4.3298
2.87429	9	53 I	$L\beta_3$	L_IM_{III}	4.3134
2.88217	8	52 Te	$L\beta_{2,15}$	$L_{III}N_{IV,V}$	4.3017
2.884	5	92 U	M_{III}	Abs. Edge	4.299
2.8917	4	58 Ce	Ll	$L_{III}M_I$	4.2875
2.8924	2	55 Cs	$L\alpha_1$	$L_{III}M_V$	4.2865
2.9020	2	55 Cs	$L\alpha_2$	$L_{III}M_{IV}$	4.2722
2.910	2	91 Pa		$M_{II}N_{IV}$	4.260
2.91207	9	53 I	$L\beta_4$	L_IM_{II}	4.2575
2.92	2	92 U		M_IN_{II}	4.25
2.9260	1	49 In	L_I	Abs. Edge	4.2373
2.9264	2	49 In	$L\gamma_4$	$L_IO_{II,III}$	4.2367
2.93187	9	51 Sb	$L\gamma_5$	$L_{II}N_I$	4.2287
2.934	8	90 Th		M_IN_{III}	4.23
2.93744	6	53 I	$L\beta_1$	$L_{II}M_{IV}$	4.22072
2.948	2	92 U		$M_{III}O_{IV,V}$	4.205
2.97088	9	52 Te	$L\beta_6$	$L_{III}N_I$	4.1732
2.97261	9	51 Sb	$L\beta_9$	L_IM_V	4.1708
2.97917	9	51 Sb	$L\beta_{10}$	L_IM_{IV}	4.1616
2.9800	2	49 In	$L\gamma_{2,3}$	$L_IN_{II,III}$	4.1605
2.9823	1	50 Sn	L_{II}	Abs. Edge	4.1573
2.9932	2	55 Cs	$L\eta$	$L_{II}M_I$	4.1421
3.0003	1	51 Sb	L_{III}	Abs. Edge	4.1323
3.00115	3	50 Sn	$L\gamma_1$	$L_{II}N_{IV}$	4.13112
3.0052	3	51 Sb	$L\beta_7$	$L_{III}O_I$	4.1255
3.006	3	57 La	Ll	$L_{III}M_I$	4.124
3.00893	9	52 Te	$L\beta_3$	L_IM_{III}	4.1204
3.011	2	90 Th		$M_{II}N_{IV}$	4.117
3.0166	2	54 Xe	$L\alpha_1$	$L_{III}M_V$	4.1099
3.02335	3	51 Sb	$L\beta_{2,15}$	$L_{III}N_{IV,V}$	4.10078
3.0309	1	21 Sc	$K\alpha_1$	KL_{III}	4.0906
3.0342	1	21 Sc	$K\alpha_2$	KL_{II}	4.0861
3.038	2	91 Pa		$M_{III}O_{IV,V}$	4.081
3.04661	9	52 Te	$L\beta_4$	L_IM_{II}	4.0695
3.068	5	90 Th	M_{III}	Abs. Edge	4.041
3.0703	1	20 Ca	K	Abs. Edge	4.0381
3.0746	3	20 Ca	$K\beta_5$	$KM_{IV,V}$	4.0325
3.07677	6	52 Te	$L\beta_1$	$L_{II}M_{IV}$	4.02958
3.08475	9	50 Sn	$L\gamma_5$	$L_{II}N_I$	4.0192
3.0849	1	48 Cd	L_I	Abs. Edge	4.0190
3.0897	2	20 Ca	$K\beta_{1,3}$	$KM_{II,III}$	4.0127
3.094	5	83 Bi	M_I	Abs. Edge	4.007
3.11513	9	50 Sn	$L\beta_9$	L_IM_V	3.9800
3.11513	9	51 Sb	$L\beta_6$	$L_{III}N_I$	3.9800
3.115	7	92 U		$M_{III}O_I$	3.980
3.12170	9	50 Sn	$L\beta_{10}$	L_IM_{IV}	3.9716
3.131	3	90 Th		$M_{III}O_{IV,V}$	3.959
3.1355	2	56 Ba	Ll	$L_{III}M_I$	3.9541
3.1377	2	48 Cd	$L\gamma_2$	L_IN_{II}	3.9513
3.1473	1	49 In	L_{II}	Abs. Edge	3.9393
3.14860	6	53 I	$L\alpha_1$	$L_{III}M_V$	3.93765
3.15258	9	51 Sb	$L\beta_3$	L_IM_{III}	3.9327
3.1557	1	50 Sn	L_{III}	Abs. Edge	3.9288
3.1564	3	50 Sn	$L\beta_7$	$L_{III}O_I$	3.9279
3.15791	6	53 I	$L\alpha_2$	$L_{III}M_{IV}$	3.92604
3.16213	4	49 In	$L\gamma_1$	$L_{II}N_{IV}$	3.92081
3.17505	3	50 Sn	$L\beta_{2,15}$	$L_{III}N_{IV,V}$	3.90486
3.19014	9	51 Sb	$L\beta_4$	L_IM_{II}	3.8364
3.217	5	82 Pb	M_I	Abs. Edge	3.854
3.22567	4	51 Sb	$L\beta_1$	$L_{II}M_{IV}$	3.84357
3.245	9	91 Pa		$M_{III}O_I$	3.82
3.24907	9	49 In	$L\gamma_5$	$L_{II}N_I$	3.8159
3.2564	1	47 Ag	L_I	Abs. Edge	3.8072
3.2670	2	55 Cs	Ll	$L_{III}M_I$	3.7950
3.26763	9	49 In	$L\beta_9$	L_IM_V	3.7942
3.26901	9	50 Sn	$L\beta_6$	$L_{III}N_I$	3.7926
3.27404	9	49 In	$L\beta_{10}$	L_IM_{IV}	3.7868
3.27979	9	53 I	$L\eta$	$L_{II}M_I$	3.7801
3.283	9	90 Th		$M_{III}O_I$	3.78
3.28920	6	52 Te	$L\alpha_1$	$L_{III}M_V$	3.76933
3.29846	9	52 Te	$L\alpha_2$	$L_{III}M_{IV}$	3.7588
3.30585	3	50 Sn	$L\beta_3$	L_IM_{III}	3.7500
3.30635	9	47 Ag	$L\gamma_3$	L_IN_{III}	3.7498
3.31216	9	47 Ag	$L\gamma_2$	L_IN_{II}	3.7432
3.3237	1	49 In	L_{III}	Abs. Edge	3.7302
3.324	4	49 In	$L\beta_7$	$L_{III}O_I$	3.730
3.3257	1	48 Cd	L_{II}	Abs. Edge	3.7280
3.329	4	92 U		$M_{II}N_I$	3.724
3.333	5	92 U	M_{IV}	Abs. Edge	3.720
3.33564	6	48 Cd	$L\gamma_1$	$L_{II}N_{IV}$	3.71686
3.33838	3	49 In	$L\beta_{2,15}$	$L_{III}N_{IV,V}$	3.71381
3.34335	9	50 Sn	$L\beta_4$	L_IM_{II}	3.7083
3.346	5	81 Tl	M_I	Abs. Edge	3.705
3.35839	3	20 Ca	$K\alpha_1$	KL_{III}	3.69168
3.359	5	83 Bi	M_{II}	Abs. Edge	3.691
3.36166	3	20 Ca	$K\alpha_2$	KL_{II}	3.68809
3.38487	3	50 Sn	$L\beta_1$	$L_{II}M_{IV}$	3.66280
3.42551	9	48 Cd	$L\gamma_5$	$L_{II}N_I$	3.61935
3.43015	9	48 Cd	$L\beta_9$	L_IM_V	3.61445
3.43606	9	49 In	$L\beta_6$	$L_{III}N_I$	3.60823

TABLE B (*Continued*)

Wavelength Å*	p.e.	Element	Designation		keV
3.4365	1	19 K	K	Abs. Edge	3.6078
3.4367	2	48 Cd	$L\beta_{10}$	L_IM_{IV}	3.6075
3.437	1	46 Pd	L_I	Abs. Edge	3.607
3.43832	9	52 Te	$L\eta$	$L_{II}M_I$	3.60586
3.43941	4	51 Sb	$L\alpha_1$	$L_{III}M_V$	3.60472
3.441	5	91 Pa		$M_{II}N_I$	3.603
3.4413	4	19 K	$K\beta_5$	$KM_{IV,V}$	3.6027
3.44840	6	51 Sb	$L\alpha_2$	$L_{III}M_{IV}$	3.59532
3.4539	2	19 K	$K\beta_{1,3}$	$KM_{II,III}$	3.5896
3.46984	9	49 In	$L\beta_3$	L_IM_{III}	3.57311
3.478	5	80 Hg	M_I	Abs. Edge	3.565
3.479	1	92 U	$M\gamma$	$M_{III}N_V$	3.563
3.4892	2	46 Pd	$L\gamma_{2,3}$	$L_IN_{II,III}$	3.5533
3.492	5	82 Pb	M_{II}	Abs. Edge	3.550
3.497	5	92 U	M_V	Abs. Edge	3.545
3.5047	1	48 Cd	L_{III}	Abs. Edge	3.5376
3.50697	9	49 In	$L\beta_4$	L_IM_{II}	3.53528
3.51408	4	48 Cd	$L\beta_{2,15}$	$L_{III}N_{IV,V}$	3.52812
3.5164	1	47 Ag	L_{II}	Abs. Edge	3.5258
3.521	2	92 U		$M_{III}N_{IV}$	3.521
3.52260	4	47 Ag	$L\gamma_1$	$L_{II}N_{IV}$	3.51959
3.537	9	90 Th		$M_{II}N_I$	3.505
3.55531	4	49 In	$L\beta_1$	$L_{II}M_{IV}$	3.48721
3.557	5	90 Th	M_{IV}	Abs. Edge	3.485
3.55754	9	53 I	Ll	$L_{III}M_I$	3.48502
3.576	1	92 U		$M_{IV}O_{II}$	3.4666
3.577	1	91 Pa	$M\gamma$	$M_{III}N_V$	3.4657
3.59994	3	50 Sn	$L\alpha_1$	$L_{III}M_V$	3.44398
3.60497	9	47 Ag	$L\beta_9$	L_IM_V	3.43917
3.60765	9	51 Sb	$L\eta$	$L_{II}M_I$	3.43661
3.60891	4	50 Sn	$L\alpha_2$	$L_{III}M_{IV}$	3.43542
3.61158	9	47 Ag	$L\beta_{10}$	L_IM_{IV}	3.43287
3.614	2	91 Pa		$M_{III}N_{IV}$	3.430
3.61467	9	48 Cd	$L\beta_6$	$L_{III}N_I$	3.42994
3.61638	9	47 Ag	$L\gamma_5$	$L_{II}N_I$	3.42832
3.616	5	79 Au	M_I	Abs. Edge	3.428
3.629	5	45 Rh	L_I	Abs. Edge	3.417
3.634	5	81 Tl	M_{II}	Abs. Edge	3.412
3.64495	9	48 Cd	$L\beta_3$	L_IM_{III}	3.40145
3.679	2	90 Th	$M\gamma$	$M_{III}N_V$	3.370
3.68203	9	48 Cd	$L\beta_4$	L_IM_{II}	3.36719
3.6855	2	45 Rh	$L\gamma_{2,3}$	$L_IN_{II,III}$	3.3640
3.691	2	91 Pa		$M_{IV}O_{II}$	3.359
3.6999	1	47 Ag	L_{III}	Abs. Edge	3.35096
3.70335	3	47 Ag	$L\beta_{2,15}$	$L_{III}N_{IV,V}$	3.34781
3.716	1	92 U	$M\beta$	$M_{IV}N_{VI}$	3.3367
3.71696	9	52 Te	Ll	$L_{III}M_I$	3.33555
3.718	3	90 Th		$M_{III}N_{IV}$	3.335
3.7228	1	46 Pd	L_{II}	Abs. Edge	3.33031
3.7246	2	46 Pd	$L\gamma_1$	$L_{II}N_{IV}$	3.3287
3.729	5	90 Th	M_V	Abs. Edge	3.325
3.73823	4	48 Cd	$L\beta_1$	$L_{II}M_{IV}$	3.31657
3.740	9	83 Bi		M_IN_{III}	3.315
3.7414	2	19 K	$K\alpha_1$	KL_{III}	3.3138
3.7445	2	19 K	$K\alpha_2$	KL_{II}	3.3111
3.760	9	90 Th		M_VP_{III}	3.298
3.762	5	78 Pt	M_I	Abs. Edge	3.296
3.77192	4	49 In	$L\alpha_1$	$L_{III}M_V$	3.28694
3.78073	6	49 In	$L\alpha_2$	$L_{III}M_{IV}$	3.27929
3.783	5	80 Hg	M_{II}	Abs. Edge	3.277
3.78876	9	50 Sn	$L\eta$	$L_{II}M_I$	3.27234
3.7920	2	46 Pd	$L\beta_9$	L_IM_V	3.2696
3.7988	2	46 Pd	$L\beta_{10}$	L_IM_{IV}	3.2637
3.80774	9	47 Ag	$L\beta_6$	$L_{III}N_I$	3.25603
3.808	4	90 Th		$M_{IV}O_{II}$	3.256
3.8222	2	46 Pd	$L\gamma_5$	$L_{II}N_I$	3.2437
3.827	1	91 Pa	$M\beta$	$M_{IV}N_{VI}$	3.2397
3.83313	9	47 Ag	$L\beta_3$	L_IM_{III}	3.23446
3.834	4	83 Bi		$M_{II}N_{IV}$	3.234
3.835	5	44 Ru	L_I	Abs. Edge	3.233
3.87023	5	47 Ag	$L\beta_4$	L_IM_{II}	3.20346
3.87090	5	18 A	K	Abs. Edge	3.20290
3.872	9	82 Pb		M_IN_{III}	3.202
3.8860	2	18 A	$K\beta_{1,3}$	$KM_{II,III}$	3.1905
3.88826	9	51 Sb	Ll	$L_{III}M_I$	3.18860
3.892	9	83 Bi		M_IN_{II}	3.185
3.8977	2	44 Ru	$L\gamma_{2,3}$	$L_IN_{II,III}$	3.1809
3.904	5	83 Bi	M_{III}	Abs. Edge	3.176
3.9074	1	46 Pd	L_{III}	Abs. Edge	3.17298
3.90887	4	46 Pd	$L\beta_{2,15}$	$L_{III}N_{IV,V}$	3.17179
3.910	1	92 U	$M\alpha_1$	M_VN_{VII}	3.1708
3.915	5	77 Ir	M_I	Abs. Edge	3.167
3.924	1	92 U	$M\alpha_2$	M_VN_{VI}	3.1595
3.932	6	83 Bi		$M_{III}O_{IV,V}$	3.153
3.93473	3	47 Ag	$L\beta_1$	$L_{II}M_{IV}$	3.15094
3.936	5	79 Au	M_{II}	Abs. Edge	3.150
3.941	1	90 Th	$M\beta$	$M_{IV}N_{VI}$	3.1458
3.9425	5	45 Rh	L_{II}	Abs. Edge	3.1448
3.9437	2	45 Rh	$L\gamma_1$	$L_{II}N_{IV}$	3.1438
3.95635	4	48 Cd	$L\alpha_1$	$L_{III}M_V$	3.13373
3.96496	6	48 Cd	$L\alpha_2$	$L_{III}M_{IV}$	3.12691
3.968	5	82 Pb		$M_{II}N_{IV}$	3.124
3.98327	9	49 In	$L\eta$	$L_{II}M_I$	3.11254
4.013	9	81 Tl		M_IN_{III}	3.089
4.0162	2	46 Pd	$L\beta_6$	$L_{III}N_I$	3.0870
4.022	1	91 Pa	$M\alpha_1$	M_VN_{VII}	3.0823
4.0346	2	46 Pd	$L\beta_3$	L_IM_{III}	3.0730
4.035	3	91 Pa	$M\alpha_2$	M_VN_{VI}	3.072
4.0451	2	45 Rh	$L\gamma_5$	$L_{II}N_I$	3.0650
4.047	1	82 Pb	M_{III}	Abs. Edge	3.0632
4.058	5	43 Te	L_I	Abs. Edge	3.055
4.069	6	82 Pb		$M_{III}O_{IV,V}$	3.047
4.0711	2	46 Pd	$L\beta_4$	L_IM_{II}	3.0454
4.071	5	76 Os	M_I	Abs. Edge	3.045
4.07165	9	50 Sn	Ll	$L_{III}M_I$	3.04499
4.093	5	78 Pt	M_{II}	Abs. Edge	3.029
4.105	9	83 Bi		$M_{III}O_I$	3.021
4.116	4	81 Tl		$M_{II}N_{IV}$	3.013
4.1299	5	45 Rh	L_{III}	Abs. Edge	3.0021
4.1310	2	45 Rh	$L\beta_{2,15}$	$L_{III}N_{IV,V}$	3.0013
4.1381	9	90 Th	$M\alpha_1$	M_VN_{VII}	2.9961
4.14622	5	46 Pd	$L\beta_1$	$L_{II}M_{IV}$	2.99022
4.151	2	90 Th	$M\alpha_2$	M_VN_{VI}	2.987
4.15443	3	47 Ag	$L\alpha_1$	$L_{III}M_V$	2.98431

TABLE B (*Continued*)

Wavelength Å*	p.e.	Element	Designation		keV
4.16294	5	47 Ag	$L\alpha_2$	$L_{III}M_{IV}$	2.97821
4.180	1	44 Ru	L_{II}	Abs. Edge	2.9663
4.1822	2	44 Ru	$L\gamma_1$	$L_{II}N_{IV}$	2.9645
4.19180	5	18 A	$K\alpha_1$	KL_{III}	2.95770
4.19315	9	48 Cd	$L\eta$	$L_{II}M_I$	2.95675
4.19474	5	18 A	$K\alpha_2$	KL_{II}	2.95563
4.198	1	81 Tl	M_{III}	Abs. Edge	2.9535
4.216	6	81 Tl		$M_{III}O_{IV,V}$	2.941
4.236	5	75 Re	M_I	Abs. Edge	2.927
4.2417	2	45 Rh	$L\beta_6$	$L_{III}N_I$	2.9229
4.244	9	82 Pb		$M_{III}O_I$	2.921
4.2522	2	45 Rh	$L\beta_3$	L_IM_{III}	2.9157
4.260	5	77 Ir	M_{II}	Abs. Edge	2.910
4.26873	9	49 In	Ll	$L_{III}M_I$	2.90440
4.2873	2	44 Ru	$L\gamma_5$	$L_{II}N_I$	2.8918
4.2888	2	45 Rh	$L\beta_4$	L_IM_{II}	2.8908
4.300	9	79 Au		M_IN_{III}	2.883
4.304	5	42 Mo	L_I	Abs. Edge	2.881
4.330	2	92 U		$M_{III}N_I$	2.863
4.355	1	80 Hg	M_{III}	Abs. Edge	2.8469
4.36767	5	46 Pd	$L\alpha_1$	$L_{III}M_V$	2.83861
4.369	1	44 Ru	L_{III}	Abs. Edge	2.8377
4.3718	2	44 Ru	$L\beta_{2,15}$	$L_{III}N_{IV,V}$	2.8360
4.37414	4	45 Rh	$L\beta_1$	$L_{II}M_{IV}$	2.83441
4.37588	7	46 Pd	$L\alpha_2$	$L_{III}M_{IV}$	2.83329
4.3800	2	42 Mo	$L\gamma_{2,3}$	$L_IN_{II,III}$	2.8306
4.3971	1	17 Cl	K	Abs. Edge	2.81960
4.4034	3	17 Cl	$K\beta$	KM	2.8156
4.407	5	74 W	M_I	Abs. Edge	2.813
4.4183	2	47 Ag	$L\eta$	$L_{II}M_I$	2.8061
4.432	4	79 Au		$M_{II}N_{IV}$	2.797
4.433	5	76 Os	M_{II}	Abs. Edge	2.797
4.436	1	43 Te	L_{II}	Abs. Edge	2.7948
4.44	2	74 W		$M_IO_{II,III}$	2.79
4.450	4	91 Pa		$M_{III}N_I$	2.786
4.460	9	78 Pt		M_IN_{III}	2.780
4.48014	9	48 Cd	Ll	$L_{III}M_I$	2.76735
4.4866	3	44 Ru	$L\beta_3$	L_IM_{III}	2.7634
4.4866	3	44 Ru	$L\beta_6$	$L_{III}N_I$	2.7634
4.518	1	79 Au	M_{III}	Abs. Edge	2.7439
4.522	6	79 Au		$M_{III}O_{IV,V}$	2.742
4.5230	2	44 Ru	$L\beta_4$	L_IM_{II}	2.7411
4.532	2	83 Bi	$M\gamma$	$M_{III}N_V$	2.735
4.568	5	90 Th		$M_{III}N_I$	2.714
4.571	5	83 Bi		$M_{III}N_{IV}$	2.712
4.572	5	83 Bi	M_{IV}	Abs. Edge	2.711
4.575	5	41 Nb	L_I	Abs. Edge	2.710
4.585	5	73 Ta	M_I	Abs. Edge	2.704
4.59	2	83 Bi		$M_{IV}P_{II,III}$	2.70
4.59743	9	45 Rh	$L\alpha_1$	$L_{III}M_V$	2.69674
4.601	4	78 Pt		$M_{II}N_{IV}$	2.695
4.60545	9	45 Rh	$L\alpha_2$	$L_{III}M_{IV}$	2.69205
4.620	5	75 Re	M_{II}	Abs. Edge	2.684
4.62058	3	44 Ru	$L\beta_1$	$L_{II}M_{IV}$	2.68323
4.625	5	92 U		$M_{IV}N_{III}$	2.681
4.630	1	43 Tc	L_{III}	Abs. Edge	2.6780
4.631	9	77 Ir		M_IN_{III}	2.677
4.6542	2	41 Nb	$L\gamma_{2,3}$	$L_IN_{II,III}$	2.6638
4.655	8	82 Pb		$M_{II}N_I$	2.664
4.6605	2	46 Pd	$L\eta$	$L_{II}M_I$	2.6603
4.674	1	82 Pb	$M\gamma$	$M_{III}N_V$	2.6527
4.686	1	78 Pt	M_{III}	Abs. Edge	2.6459
4.694	8	78 Pt		$M_{III}O_{IV,V}$	2.641
4.703	9	79 Au		$M_{III}O_I$	2.636
4.7076	2	47 Ag	Ll	$L_{III}M_I$	2.6337
4.715	3	82 Pb		$M_{III}N_{IV}$	2.630
4.719	1	42 Mo	L_{II}	Abs. Edge	2.6274
4.7258	2	42 Mo	$L\gamma_1$	$L_{II}N_{IV}$	2.6235
4.7278	1	17 Cl	$K\alpha_1$	KL_{III}	2.62239
4.7307	1	17 Cl	$K\alpha_2$	KL_{II}	2.62078
4.757	5	82 Pb	M_{IV}	Abs. Edge	2.606
4.764	5	83 Bi	M_V	Abs. Edge	2.603
4.780	4	77 Ir		$M_{II}N_{IV}$	2.594
4.79	2	76 Os		M_IN_{III}	2.59
4.815	5	74 W	M_{II}	Abs. Edge	2.575
4.823	3	83 Bi		$M_{IV}O_{II}$	2.571
4.823	4	81 Tl	$M\gamma$	$M_{III}N_V$	2.571
4.8369	2	42 Mo	$L\gamma_5$	$L_{II}N_I$	2.5632
4.84575	5	44 Ru	$L\alpha_1$	$L_{III}M_V$	2.55855
4.85381	7	44 Ru	$L\alpha_2$	$L_{III}M_{IV}$	2.55431
4.861	1	77 Ir	M_{III}	Abs. Edge	2.5505
4.865	5	81 Tl		$M_{III}N_{IV}$	2.548
4.869	9	77 Ir		$M_{III}O_{IV,V}$	2.546
4.876	9	78 Pt		$M_{III}O_I$	2.543
4.879	5	40 Zr	L_I	Abs. Edge	2.541
4.8873	8	43 Tc	$L\beta_1$	$L_{II}M_{IV}$	2.5368
4.909	1	83 Bi	$M\beta$	$M_{IV}N_{VI}$	2.5255
4.911	5	90 Th		$M_{IV}N_{III}$	2.524
4.913	1	42 Mo	L_{III}	Abs. Edge	2.5234
4.9217	2	45 Rh	$L\eta$	$L_{II}M_I$	2.5191
4.9232	2	42 Mo	$L\beta_{2,15}$	$L_{III}N_{IV,V}$	2.5183
4.946	2	92 U	$M\zeta_1$	M_VN_{III}	2.507
4.952	5	81 Tl	M_{IV}	Abs. Edge	2.504
4.9525	3	46 Pd	Ll	$L_{III}M_I$	2.5034
4.9536	3	40 Zr	$L\gamma_{2,3}$	$L_IN_{II,III}$	2.5029
4.955	4	76 Os		$M_{II}N_{IV}$	2.502
4.955	5	82 Pb	M_V	Abs. Edge	2.502
4.984	2	80 Hg	$M\gamma$	$M_{III}N_V$	2.4875
5.004	9	82 Pb		$M_{IV}O_{II}$	2.477
5.0133	3	42 Mo	$L\beta_3$	L_IM_{III}	2.4730
5.0185	1	16 S	K	Abs. Edge	2.47048
5.020	5	73 Ta	M_{II}	Abs. Edge	2.470
5.0233	3	16 S	$K\beta_x$	KM	2.4681
5.031	1	41 Nb	L_{II}	Abs. Edge	2.4641
5.0316	2	16 S	$K\beta_1$	KM	2.46404
5.0361	3	41 Nb	$L\gamma_1$	$L_{II}N_{IV}$	2.4618
5.043	5	76 Os	M_{III}	Abs. Edge	2.458
5.0488	3	42 Mo	$L\beta_4$	L_IM_{II}	2.4557
5.0488	5	42 Mo	$L\beta_6$	$L_{III}N_I$	2.4557
5.050	2	92 U	$M\zeta_2$	$M_{IV}N_{II}$	2.4548
5.076	1	82 Pb	$M\beta$	$M_{IV}N_{VI}$	2.4427
5.092	2	91 Pa	$M\zeta_1$	M_VN_{III}	2.4350
5.1148	3	43 Tc	$L\alpha_1$	$L_{III}M_V$	2.4240
5.118	1	83 Bi	$M\alpha_1$	M_VN_{VII}	2.4226

X-Ray Wavelengths
TABLE B (*Continued*)

Wavelength Å*	p.e.	Element	Designation		keV
5.130	2	83 Bi	$M\alpha_2$	M_VN_{VI}	2.4170
5.145	4	79 Au	$M\gamma$	$M_{III}N_V$	2.410
5.1517	3	41 Nb	$L\gamma_5$	$L_{II}N_I$	2.4066
5.153	5	81 Tl	M_V	Abs. Edge	2.406
5.157	5	80 Hg	M_{IV}	Abs. Edge	2.404
5.168	9	82 Pb		M_VO_{III}	2.399
5.172	9	74 W		M_IN_{III}	2.397
5.17708	8	42 Mo	$L\beta_1$	$L_{II}M_{IV}$	2.39481
5.186	5	79 Au		$M_{III}N_{IV}$	2.391
5.193	2	91 Pa	$M\zeta_2$	$M_{IV}N_{II}$	2.3876
5.196	9	81 Tl		$M_{IV}O_{II}$	2.386
5.2050	2	44 Ru	$L\eta$	$L_{II}M_I$	2.38197
5.217	5	39 Y	L_I	Abs. Edge	2.377
5.2169	3	45 Rh	Ll	$L_{III}M_I$	2.3765
5.230	1	41 Nb	L_{III}	Abs. Edge	2.3706
5.234	5	75 Re	M_{III}	Abs. Edge	2.369
5.2379	3	41 Nb	$L\beta_{2,15}$	$L_{III}N_{IV,V}$	2.3670
5.245	5	90 Th	$M\zeta_1$	M_VN_{III}	2.364
5.249	1	81 Tl	$M\beta$	$M_{IV}N_{VI}$	2.3621
5.2830	3	39 Y	$L\gamma_{2,3}$	$L_IN_{II,III}$	2.3468
5.286	1	82 Pb	$M\alpha_1$	M_VN_{VII}	2.3455
5.299	2	82 Pb	$M\alpha_2$	M_VN_{VI}	2.3397
5.3102	3	41 Nb	$L\beta_3$	L_IM_{III}	2.3348
5.319	4	78 Pt	$M\gamma$	$M_{III}N_V$	2.331
5.340	5	90 Th	$M\zeta_2$	$M_{IV}N_{II}$	2.322
5.3455	3	41 Nb	$L\beta_4$	L_IM_{II}	2.3194
5.357	4	74 W		$M_{II}N_{IV}$	2.314
5.357	5	78 Pt		$M_{III}N_{IV}$	2.314
5.36	1	80 Hg	M_V	Abs. Edge	2.313
5.3613	3	41 Nb	$L\beta_6$	$L_{III}N_I$	2.3125
5.37216	7	16 S	$K\alpha_1$	KL_{III}	2.30784
5.374	5	79 Au	M_{IV}	Abs. Edge	2.307
5.37496	8	16 S	$K\alpha_2$	KL_{II}	2.30664
5.378	1	40 Zr	L_{II}	Abs. Edge	2.3053
5.3843	3	40 Zr	$L\gamma_1$	$L_{II}N_{IV}$	2.3027
5.40	2	73 Ta		M_IN_{III}	2.295
5.40655	8	42 Mo	$L\alpha_1$	$L_{III}M_V$	2.29316
5.41437	8	42 Mo	$L\alpha_2$	$L_{III}M_{IV}$	2.28985
5.4318	9	80 Hg	$M\beta$	$M_{IV}N_{VI}$	2.2825
5.435	1	74 W	M_{III}	Abs. Edge	2.2811
5.460	1	81 Tl	$M\alpha_1$	M_VN_{VII}	2.2706
5.472	2	81 Tl	$M\alpha_2$	M_VN_{VI}	2.2656
5.4923	3	41 Nb	$L\beta_1$	$L_{II}M_{IV}$	2.2574
5.4977	3	40 Zr	$L\gamma_5$	$L_{II}N_I$	2.2551
5.500	4	77 Ir	$M\gamma$	$M_{III}N_V$	2.254
5.5035	3	44 Ru	Ll	$L_{III}M_I$	2.2528
5.537	8	83 Bi		$M_{III}N_I$	2.239
5.540	5	77 Ir		$M_{III}N_{IV}$	2.238
5.570	4	73 Ta		$M_{II}N_{IV}$	2.226
5.579	1	40 Zr	L_{III}	Abs. Edge	2.2225
5.584	5	79 Au	M_V	Abs. Edge	2.220
5.5863	3	40 Zr	$L\beta_{2,15}$	$L_{III}N_{IV,V}$	2.2194
5.59	1	78 Pt	M_{IV}	Abs. Edge	2.217
5.592	5	38 Sr	L_I	Abs. Edge	2.217
5.624	1	79 Au	$M\beta$	$M_{IV}N_{VI}$	2.2046
5.628	8	74 W		$M_{III}O_I$	2.203
5.6330	3	40 Zr	$L\beta_3$	L_IM_{III}	2.2010
5.6445	3	38 Sr	$L\gamma_{2,3}$	$L_IN_{II,III}$	2.1965
5.6476	9	80 Hg	$M\alpha_1$	M_VN_{VII}	2.1953
5.650	5	73 Ta	M_{III}	Abs. Edge	2.194
5.6681	3	40 Zr	$L\beta_4$	L_IM_{II}	2.1873
5.67	3	73 Ta		$M_{III}O_{IV,V}$	2.19
5.682	4	76 Os	$M\gamma$	$M_{III}N_V$	2.182
5.704	8	82 Pb		$M_{III}N_I$	2.174
5.7101	3	40 Zr	$L\beta_6$	$L_{III}N_I$	2.1712
5.724	5	76 Os		$M_{III}N_{IV}$	2.166
5.7243	2	41 Nb	$L\alpha_1$	$L_{III}M_V$	2.16589
5.7319	3	41 Nb	$L\alpha_2$	$L_{III}M_{IV}$	2.1630
5.756	1	39 Y	L_{II}	Abs. Edge	2.1540
5.767	9	79 Au		M_VO_{III}	2.150
5.784	1	15 P	K	Abs. Edge	2.1435
5.796	2	15 P	$K\beta$	KM	2.1391
5.81	2	76 Os		$M_{II}N_I$	2.133
5.81	1	78 Pt	M_V	Abs. Edge	2.133
5.828	1	78 Pt	$M\beta$	$M_{IV}N_{VI}$	2.1273
5.83	2	73 Ta		$M_{III}O_I$	2.126
5.83	1	77 Ir	M_{IV}	Abs. Edge	2.126
5.8360	3	40 Zr	$L\beta_1$	$L_{II}M_{IV}$	2.1244
5.840	1	79 Au	$M\alpha_1$	M_VN_{VII}	2.1229
5.8475	3	42 Mo	$L\eta$	$L_{II}M_I$	2.1202
5.854	3	79 Au	$M\alpha_2$	M_VN_{VI}	2.118
5.8754	3	39 Y	$L\gamma_5$	$L_{II}N_I$	2.1102
5.884	8	81 Tl		$M_{III}N_I$	2.107
5.885	2	75 Re	$M\gamma$	$M_{III}N_V$	2.1067
5.931	5	75 Re		$M_{III}N_{IV}$	2.090
5.962	1	39 Y	L_{III}	Abs. Edge	2.0794
5.9832	3	39 Y	$L\beta_3$	L_IM_{III}	2.0722
5.987	9	78 Pt		M_VO_{III}	2.071
6.008	5	37 Rb	L_I	Abs. Edge	2.063
6.0186	3	39 Y	$L\beta_4$	L_IM_{II}	2.0600
6.038	1	77 Ir	$M\beta$	$M_{IV}N_{VI}$	2.0535
6.0458	3	37 Rb	$L\gamma_{2,3}$	$L_IN_{II,III}$	2.0507
6.047	1	78 Pt	$M\alpha_1$	M_VN_{VII}	2.0505
6.05	1	77 Ir	M_V	Abs. Edge	2.048
6.058	3	78 Pt	$M\alpha_2$	M_VN_{VI}	2.047
6.0705	2	40 Zr	$L\alpha_1$	$L_{III}M_V$	2.04236
6.073	5	76 Os	M_{IV}	Abs. Edge	2.042
6.0778	3	40 Zr	$L\alpha_2$	$L_{III}M_{IV}$	2.0399
6.09	2	80 Hg		$M_{III}N_I$	2.036
6.092	3	74 W	$M\gamma$	$M_{III}N_V$	2.035
6.0942	3	39 Y	$L\beta_6$	$L_{III}N_I$	2.0344
6.134	4	74 W		$M_{III}N_{IV}$	2.021
6.1508	3	42 Mo	Ll	$L_{III}M_I$	2.01568
6.157	1	15 P	$K\alpha_1$	KL_{III}	2.0137
6.160	1	15 P	$K\alpha_2$	KL_{II}	2.0127
6.162	8	83 Bi		$M_{IV}N_{III}$	2.012
6.173	1	38 Sr	L_{II}	Abs. Edge	2.0085
6.2109	3	41 Nb	$L\eta$	$L_{II}M_I$	1.99620
6.2120	3	39 Y	$L\beta_1$	$L_{II}M_{IV}$	1.99584
6.259	9	79 Au		$M_{III}N_I$	1.981
6.262	1	77 Ir	$M\alpha_1$	M_VN_{VII}	1.9799
6.267	1	76 Os	$M\beta$	$M_{IV}N_{VI}$	1.9783
6.275	3	77 Ir	$M\alpha_2$	M_VN_{VI}	1.9758
6.28	2	74 W		$M_{II}N_I$	1.973

TABLE B (*Continued*)

Wavelength Å*	p.e.	Element		Designation	keV
6.2961	3	38 Sr	$L\gamma_5$	$L_{II}N_I$	1.96916
6.30	1	76 Os	M_V	Abs. Edge	1.967
6.312	4	73 Ta	$M\gamma$	$M_{III}N_V$	1.964
6.33	1	75 Re	M_{IV}	Abs. Edge	1.958
6.353	5	73 Ta		$M_{III}N_{IV}$	1.951
6.3672	3	38 Sr	$L\beta_3$	L_IM_{III}	1.94719
6.384	7	82 Pb		$M_{IV}N_{III}$	1.942
6.387	1	38 Sr	L_{III}	Abs. Edge	1.9411
6.4026	3	38 Sr	$L\beta_4$	L_IM_{II}	1.93643
6.4488	2	39 Y	$L\alpha_1$	$L_{III}M_V$	1.92256
6.455	9	78 Pt		$M_{III}N_I$	1.921
6.4558	3	39 Y	$L\alpha_2$	$L_{III}M_{IV}$	1.92047
6.47	1	36 Kr	L_I	Abs. Edge	1.915
6.490	1	76 Os	$M\alpha$	$M_VN_{VI,VII}$	1.9102
6.504	1	75 Re	$M\beta$	$M_{IV}N_{VI}$	1.9061
6.5176	3	41 Nb	Ll	$L_{III}M_I$	1.90225
6.5191	3	38 Sr	$L\beta_6$	$L_{III}N_I$	1.90181
6.521	4	83 Bi	$M\zeta_1$	M_VN_{III}	1.901
6.544	4	72 Hf	$M\gamma$	$M_{III}N_V$	1.895
6.560	5	75 Re	M_V	Abs. Edge	1.890
6.585	5	83 Bi	$M\zeta_2$	$M_{IV}N_{II}$	1.883
6.59	1	74 W	M_{IV}	Abs. Edge	1.880
6.6069	3	40 Zr	$L\eta$	$L_{II}M_I$	1.87654
6.6239	3	38 Sr	$L\beta_1$	$L_{II}M_{IV}$	1.87172
6.644	1	37 Rb	L_{II}	Abs. Edge	1.8661
6.669	9	77 Ir		$M_{III}N_I$	1.859
6.729	1	75 Re	$M\alpha$	$M_VN_{VI,VII}$	1.8425
6.738	1	14 Si	K	Abs. Edge	1.8400
6.740	3	82 Pb	$M\zeta_1$	M_VN_{III}	1.8395
6.7530	1	14 Si	$K\beta$	KM	1.83594
6.755	3	37 Rb	$L\gamma_5$	$L_{II}N_{IV}$	1.83532
6.757	1	74 W	$M\beta$	$M_{IV}N_{VI}$	1.8349
6.768	6	71 Lu	$M\gamma$	$M_{III}N_V$	1.832
6.7876	3	37 Rb	$L\beta_3$	L_IM_{III}	1.82659
6.802	5	82 Pb	$M\zeta_2$	$M_{IV}N_{II}$	1.823
6.806	9	74 W		$M_{IV}O_{II}$	1.822
6.8207	3	37 Rb	$L\beta_4$	L_IM_{II}	1.81771
6.83	1	74 W	M_V	Abs. Edge	1.814
6.862	1	37 Rb	L_{III}	Abs. Edge	1.8067
6.8628	2	38 Sr	$L\alpha_1$	$L_{III}M_V$	1.80656
6.8697	3	38 Sr	$L\alpha_2$	$L_{III}M_{IV}$	1.80474
6.87	1	73 Ta	M_{IV}	Abs. Edge	1.804
6.87	2	80 Hg	δ	$M_{IV}N_{III}$	1.805
6.89	2	76 Os		$M_{III}N_I$	1.798
6.9185	3	40 Zr	Ll	$L_{III}M_I$	1.79201
6.959	5	35 Br	L_I	Abs. Edge	1.781
6.974	4	81 Tl	$M\zeta_1$	M_VN_{III}	1.778
6.983	1	74 W	$M\alpha_1$	M_VN_{VII}	1.7754
6.9842	3	37 Rb	$L\beta_6$	$L_{III}N_I$	1.77517
6.992	2	74 W	$M\alpha_2$	M_VN_{VI}	1.7731
7.005	9	74 W		M_VO_{III}	1.770
7.023	1	73 Ta	$M\beta$	$M_{IV}N_{VI}$	1.7655
7.024	8	70 Yb	$M\gamma$	$M_{III}N_V$	1.765
7.032	5	81 Tl	$M\zeta_2$	$M_{IV}N_{II}$	1.763
7.0406	3	39 Y	$L\eta$	$L_{II}M_I$	1.76095
7.0759	3	37 Rb	$L\beta_1$	$L_{II}M_{IV}$	1.75217
7.09	2	73 Ta		$M_{IV}O_{II,III}$	1.748
7.101	8	79 Au		$M_{IV}N_{III}$	1.746
7.11	1	73 Ta	M_V	Abs. Edge	1.743
7.12542	9	14 Si	$K\alpha_1$	KL_{III}	1.73998
7.12791	9	14 Si	$K\alpha_2$	KL_{II}	1.73938
7.168	1	36 Kr	L_{II}	Abs. Edge	1.7297
7.250	5	36 Kr		$L_{II}N_{III}$	1.710
7.252	1	73 Ta	$M\alpha$	$M_VN_{VI,VII}$	1.7096
7.264	5	36 Kr	$L\beta_3$	L_IM_{III}	1.707
7.279	5	36 Kr	$L\gamma_5$	$L_{II}N_I$	1.703
7.30	2	73 Ta		M_VO_{III}	1.700
7.303	1	72 Hf	$M\beta$	$M_{IV}N_{VI}$	1.6976
7.304	5	36 Kr	$L\beta_4$	L_IM_{II}	1.697
7.3183	2	37 Rb	$L\alpha_1$	$L_{III}M_V$	1.69413
7.3251	3	37 Rb	$L\alpha_2$	$L_{III}M_{IV}$	1.69256
7.3563	3	39 Y	Ll	$L_{III}M_I$	1.68536
7.360	8	74 W		$M_{III}N_I$	1.684
7.371	8	78 Pt		$M_{IV}N_{III}$	1.682
7.392	1	36 Kr	L_{III}	Abs. Edge	1.6772
7.466	4	79 Au	$M\zeta_1$	M_VN_{III}	1.6605
7.503	1	34 Se	L_I	Abs. Edge	1.6525
7.510	4	36 Kr	$L\beta_6$	$L_{III}N_I$	1.6510
7.5171	3	38 Sr	$L\eta$	$L_{II}M_I$	1.64933
7.523	5	79 Au	$M\zeta_2$	$M_{IV}N_{II}$	1.648
7.539	1	72 Hf	$M\alpha$	$M_VN_{VI,VII}$	1.6446
7.546	8	68 Er	$M\gamma$	$M_{III}N_V$	1.643
7.576	3	36 Kr	$L\beta_1$	$L_{II}M_{IV}$	1.6366
7.60	1	68 Er		$M_{III}N_{IV}$	1.632
7.601	2	71 Lu	$M\beta$	$M_{IV}N_{VI}$	1.6312
7.612	9	73 Ta		$M_{III}N_I$	1.629
7.645	8	77 Ir		$M_{IV}N_{III}$	1.622
7.738	4	78 Pt	$M\zeta_1$	M_VN_{III}	1.6022
7.753	5	35 Br	L_{II}	Abs. Edge	1.599
7.767	9	35 Br	$L\beta_{3,4}$	$L_IM_{II,III}$	1.596
7.790	5	78 Pt	$M\zeta_2$	$M_{IV}N_{II}$	1.592
7.817	3	36 Kr	$L\alpha_{1,2}$	$L_{III}M_{IV,V}$	1.5860
7.8362	3	38 Sr	Ll	$L_{III}M_I$	1.58215
7.840	2	71 Lu	$M\alpha$	$M_VN_{VI,VII}$	1.5813
7.865	9	67 Ho	$M\gamma$	$M_{III}N_{IV,V}$	1.576
7.887	9	72 Hf		$M_{III}N_I$	1.572
7.909	2	70 Yb	$M\beta$	$M_{IV}N_{VI}$	1.5675
7.94813	5	13 Al	K	Abs. Edge	1.55988
7.960	2	13 Al	$K\beta$	KM	1.55745
7.984	5	35 Br	L_{III}	Abs. Edge	1.5530
8.021	4	77 Ir	$M\zeta_1$	M_VN_{III}	1.5458
8.0415	4	37 Rb	$L\eta$	$L_{II}M_I$	1.54177
8.065	5	77 Ir	$M\zeta_2$	$M_{IV}N_{II}$	1.5373
8.107	1	33 As	L_I	Abs. Edge	1.5293
8.1251	5	35 Br	$L\beta_1$	$L_{II}M_{IV}$	1.52590
8.144	9	66 Dy	$M\gamma$	$M_{III}N_{IV,V}$	1.522
8.149	5	70 Yb	$M\alpha$	$M_VN_{VI,VII}$	1.5214
8.239	8	75 Re		$M_{IV}N_{III}$	1.505
8.249	7	69 Tm	$M\beta$	$M_{IV}N_{VI}$	1.503
8.310	4	76 Os	$M\zeta_1$	M_VN_{III}	1.4919
8.321	9	34 Se	$L\beta_{3,4}$	$L_IM_{II,III}$	1.490
8.33934	9	13 Al	$K\alpha_1$	KL_{III}	1.48670
8.34173	9	13 Al	$K\alpha_2$	KL_{II}	1.48627
8.359	5	76 Os	$M\zeta_2$	$M_{IV}N_{II}$	1.4831

Wavelength Å*	p.e.	Element	Designation		keV
8.3636	4	37 Rb	Ll	$L_{III}M_{I}$	1.48238
8.3746	5	35 Br	$L\alpha_{1,2}$	$L_{III}M_{IV,V}$	1.48043
8.407	1	34 Se	L_{II}	Abs. Edge	1.4747
8.470	9	70 Yb		$M_{III}N_{I}$	1.464
8.48	1	69 Tm	$M\alpha$	$M_{V}N_{VI,VII}$	1.462
8.486	9	65 Tb	$M\gamma$	$M_{III}N_{IV,V}$	1.461
8.487	5	69 Tm	M_{V}	Abs. Edge	1.4609
8.573	8	74 W		$M_{IV}N_{III}$	1.446
8.592	3	68 Er	$M\beta$	$M_{IV}N_{VI}$	1.4430
8.60	7	92 U		$N_{I}P_{IV,V}$	1.44
8.601	5	68 Er	M_{IV}	Abs. Edge	1.4415
8.629	4	75 Re	$M\zeta_{1}$	$M_{V}N_{III}$	1.4368
8.646	1	34 Se	L_{III}	Abs. Edge	1.4340
8.664	5	75 Re	$M\zeta_{2}$	$M_{IV}N_{II}$	1.4310
8.7358	5	34 Se	$L\beta_{1}$	$L_{II}M_{IV}$	1.41923
8.76	7	92 U		$N_{I}P_{III}$	1.42
8.773	1	32 Ge	L_{I}	Abs. Edge	1.4132
8.81	7	92 U		$N_{I}P_{II}$	1.41
8.82	1	68 Er	$M\alpha$	$M_{V}N_{VI,VII}$	1.406
8.844	9	64 Gd	$M\gamma$	$M_{III}N_{IV,V}$	1.402
8.847	5	68 Er	M_{V}	Abs. Edge	1.4013
8.90	2	73 Ta		$M_{IV}N_{III}$	1.393
8.929	1	33 As	$L\beta_{3,4}$	$L_{I}M_{II,III}$	1.3884
8.962	4	74 W	$M\zeta_{1}$	$M_{V}N_{III}$	1.3835
8.965	4	67 Ho	$M\beta$	$M_{IV}N_{VI}$	1.3830
8.9900	5	34 Se	$L\alpha_{1,2}$	$L_{III}M_{IV,V}$	1.37910
8.993	5	74 W	$M\zeta_{2}$	$M_{IV}N_{II}$	1.3787
9.125	1	33 As	L_{II}	Abs. Edge	1.3587
9.20	2	67 Ho	$M\alpha$	$M_{V}N_{VI,VII}$	1.348
9.211	9	63 Eu	$M\gamma$	$M_{III}N_{IV,V}$	1.346
9.255	1	35 Br	$L\eta$	$L_{II}M_{I}$	1.3396
9.316	4	73 Ta	$M\zeta_{1}$	$M_{V}N_{III}$	1.3308
9.330	5	73 Ta	$M\zeta_{2}$	$M_{IV}N_{II}$	1.3288
9.357	6	66 Dy	$M\beta$	$M_{IV}N_{VI}$	1.3250
9.367	1	33 As	L_{III}	Abs. Edge	1.3235
9.40	7	90 Th		$N_{I}P_{III}$	1.319
9.4141	8	33 As	$L\beta_{1}$	$L_{II}M_{IV}$	1.3170
9.44	7	90 Th		$N_{I}P_{II}$	1.313
9.5122	1	12 Mg	K	Abs. Edge	1.30339
9.517	5	31 Ga	L_{I}	Abs. Edge	1.3028
9.521	2	12 Mg	$K\beta$	KM	1.3022
9.581	2	32 Ge	$L\beta_{3}$	$L_{I}M_{III}$	1.2941
9.585	1	35 Br	Ll	$L_{III}M_{I}$	1.2935
9.59	2	66 Dy	$M\alpha$	$M_{V}N_{VI,VII}$	1.293
9.600	9	62 Sm	$M\gamma$	$M_{III}N_{IV,V}$	1.291
9.640	2	32 Ge	$L\beta_{4}$	$L_{I}M_{II}$	1.2861
9.6709	8	33 As	$L\alpha_{1,2}$	$L_{III}M_{IV,V}$	1.2820
9.686	7	72 Hf	$M\zeta_{2}$	$M_{IV}N_{II}$	1.2800
9.686	7	72 Hf	$M\zeta_{1}$	$M_{V}N_{III}$	1.2800
9.792	6	65 Tb	$M\beta$	$M_{IV}N_{VI}$	1.2661
9.8900	2	12 Mg	$K\alpha_{1,2}$	$KL_{II,III}$	1.25360
9.924	1	32 Ge	L_{II}	Abs. Edge	1.2494
9.962	1	34 Se	$L\eta$	$L_{II}M_{I}$	1.2446
10.00	2	65 Tb	$M\alpha$	$M_{V}N_{VI,VII}$	1.240
10.09	7	92 U		$N_{I}O_{III}$	1.229
10.175	1	32 Ge	$L\beta_{1}$	$L_{II}M_{IV}$	1.2185
10.187	1	32 Ge	L_{III}	Abs. Edge	1.2170
10.254	6	64 Gd	$M\beta$	$M_{IV}N_{VI}$	1.2091
10.294	1	34 Se	Ll	$L_{III}M_{I}$	1.2044
10.359	9	31 Ga	$L\beta_{3,4}$	$L_{I}M_{II,III}$	1.197
10.40	7	92 U		$N_{II}P_{I}$	1.192
10.4361	8	32 Ge	$L\alpha_{1,2}$	$L_{III}M_{IV,V}$	1.18800
10.46	3	64 Gd	$M\alpha$	$M_{V}N_{VI,VII}$	1.185
10.48	1	70 Yb	$M\zeta$	$M_{V}N_{III}$	1.183
10.505	9	60 Nd	$M\gamma$	$M_{III}N_{IV,V}$	1.180
10.711	5	63 Eu	M_{IV}	Abs. Edge	1.1575
10.734	1	33 As	$L\eta$	$L_{II}M_{I}$	1.1550
10.750	7	63 Eu	$M\beta$	$M_{IV}N_{VI}$	1.1533
10.828	5	31 Ga	L_{II}	Abs. Edge	1.1450
10.96	3	63 Eu	$M\alpha$	$M_{V}N_{VI,VII}$	1.131
10.998	9	59 Pr	$M\gamma$	$M_{III}N_{IV,V}$	1.1273
11.013	5	63 Eu	M_{V}	Abs. Edge	1.1258
11.023	2	31 Ga	$L\beta_{1}$	$L_{II}M_{IV}$	1.1248
11.072	1	33 As	Ll	$L_{III}M_{I}$	1.1198
11.07	7	90 Th		$N_{II}P_{I}$	1.120
11.100	1	31 Ga	L_{III}	Abs. Edge	1.1169
11.200	7	30 Zn	$L\beta_{3,4}$	$L_{I}M_{II,III}$	1.1070
11.27	1	62 Sm	$M\beta$	$M_{IV}N_{VI}$	1.0998
11.288	5	62 Sm	M_{IV}	Abs. Edge	1.0983
11.292	1	31 Ga	$L\alpha_{1,2}$	$L_{III}M_{IV,V}$	1.09792
11.37	1	68 Er	$M\zeta$	$M_{V}N_{III}$	1.0901
11.47	3	62 Sm	$M\alpha$	$M_{V}N_{VI,VII}$	1.081
11.53	1	58 Ce	$M\gamma$	$M_{III}N_{IV,V}$	1.0749
11.552	5	62 Sm	M_{V}	Abs. Edge	1.0732
11.56	5	90 Th		$N_{II}O_{IV}$	1.072
11.569	1	11 Na	K	Abs. Edge	1.07167
11.575	2	11 Na	$K\beta$	KM	1.0711
11.609	2	32 Ge	$L\eta$	$L_{II}M_{I}$	1.0680
11.862	1	30 Zn	L_{II}	Abs. Edge	1.04523
11.86	1	67 Ho	$M\zeta$	$M_{V}N_{III}$	1.0450
11.9101	9	11 Na	$K\alpha_{1,2}$	$KL_{II,III}$	1.04098
11.965	2	32 Ge	Ll	$L_{III}M_{I}$	1.0362
11.983	3	30 Zn	$L\beta_{1}$	$L_{II}M_{IV}$	1.0347
12.08	4	57 La	$M\gamma$	$M_{III}N_{IV,V}$	1.027
12.122	3	29 Cu	$L\beta_{3,4}$	$L_{I}M_{II,III}$	1.0228
12.131	1	30 Zn	L_{III}	Abs. Edge	1.02201
12.254	3	30 Zn	$L\alpha_{1,2}$	$L_{III}M_{IV,V}$	1.0117
12.43	2	66 Dy	$M\zeta$	$M_{V}N_{III}$	0.998
12.44	2	60 Nd	$M\beta$	$M_{IV}N_{VI}$	0.997
12.459	5	60 Nd	M_{IV}	Abs. Edge	0.9951
12.597	2	31 Ga	$L\eta$	$L_{II}M_{I}$	0.9842
12.68	2	60 Nd	$M\alpha$	$M_{V}N_{VI,VII}$	0.978
12.737	5	60 Nd	M_{V}	Abs. Edge	0.9734
12.75	3	56 Ba	$M\gamma$	$M_{III}N_{IV,V}$	0.973
12.90	9	92 U		$N_{III}O_{V}$	0.961
12.953	2	31 Ga	Ll	$L_{III}M_{I}$	0.9572
12.98	2	65 Tb	$M\zeta$	$M_{V}N_{III}$	0.955
13.014	1	29 Cu	L_{II}	Abs. Edge	0.95268
13.053	3	29 Cu	$L\beta_{1}$	$L_{II}M_{IV}$	0.9498
13.06	2	59 Pr	$M\beta$	$M_{IV}N_{VI}$	0.950
13.06	1	30 Zn	L_{I}	Abs. Edge	0.9495
13.122	5	59 Pr	M_{IV}	Abs. Edge	0.9448
13.18	2	28 Ni	$L\beta_{3,4}$	$L_{I}M_{II,III}$	0.941
13.288	1	29 Cu	L_{III}	Abs. Edge	0.93306

Table B (*Continued*)

Wavelength Å*	p.e.	Element	Designation		keV
13.30	6	83 Bi		$N_IP_{II,III}$	0.932
13.336	3	29 Cu	$L\alpha_{1,2}$	$L_{III}M_{IV,V}$	0.9297
13.343	5	59 Pr	$M\alpha$	$M_VN_{VI,VII}$	0.9292
13.394	5	59 Pr	M_V	Abs. Edge	0.9257
13.57	2	64 Gd	$M\zeta$	M_VN_{III}	0.914
13.68	2	30 Zn	$L\eta$	$L_{II}M_I$	0.906
13.75	4	58 Ce	$M\beta$	$M_{IV}N_{VI}$	0.902
13.8	1	90 Th		$N_{III}O_V$	0.897
14.02	2	30 Zn	Ll	$L_{III}M_I$	0.884
14.04	2	58 Ce	$M\alpha$	$M_VN_{VI,VII}$	0.883
14.22	2	63 Eu	$M\zeta$	M_VN_{III}	0.872
14.242	5	28 Ni	L_{II}	Abs. Edge	0.8706
14.271	6	28 Ni	$L\beta_1$	$L_{II}M_{IV}$	0.8688
14.3018	1	10 Ne	K	Abs. Edge	0.866889
14.31	3	27 Co	$L\beta_{3,4}$	$L_IM_{II,III}$	0.870
14.39	5	58 Ce		$M_VO_{II,III}$	0.862
14.452	5	10 Ne	$K\beta$	KM	0.8579
14.51	5	57 La	$M\beta$	$M_{IV}N_{VI}$	0.854
14.525	5	28 Ni	L_{III}	Abs. Edge	0.8536
14.561	3	28 Ni	$L\alpha_{1,2}$	$L_{III}M_{IV,V}$	0.8515
14.610	3	10 Ne	$K\alpha_{1,2}$	$KL_{II,III}$	0.8486
14.88	5	57 La	$M\alpha$	$M_VN_{VI,VII}$	0.833
14.90	2	29 Cu	$L\eta$	$L_{II}M_I$	0.832
14.91	4	62 Sm	$M\zeta$	M_VN_{III}	0.831
15.286	9	29 Cu	Ll	$L_{III}M_I$	0.8111
15.56	1	56 Ba	M_{IV}	Abs. Edge	0.7967
15.618	5	27 Co	L_{II}	Abs. Edge	0.7938
15.65	4	26 Fe	$L\beta_{3,4}$	$L_IM_{II,III}$	0.792
15.666	8	27 Co	$L\beta_1$	$L_{II}M_{IV}$	0.7914
15.72	9	56 Ba		$M_{IV}O_{III}$	0.789
15.89	1	56 Ba	M_V	Abs. Edge	0.7801
15.91	5	56 Ba		$M_{IV}O_{II}$	0.779
15.915	5	27 Co	L_{III}	Abs. Edge	0.7790
15.93	4	52 Te	$M\gamma$	$M_{III}N_{IV,V}$	0.778
15.972	6	27 Co	$L\alpha_{1,2}$	$L_{III}M_{IV,V}$	0.7762
15.98	5	51 Sb		$M_{II}N_{IV}$	0.776
16.20	5	56 Ba		M_VO_{III}	0.765
16.27	3	28 Ni	$L\eta$	$L_{II}M_I$	0.762
16.46	4	60 Nd	$M\zeta$	M_VN_{III}	0.753
16.693	9	28 Ni	Ll	$L_{III}M_I$	0.7427
16.7	1	24 Cr	L_I	Abs. Edge	0.741
16.92	4	51 Sb	$M\gamma$	$M_{III}N_{IV,V}$	0.733
16.93	5	50 Sn		$M_{II}N_{IV}$	0.733
17.19	4	25 Mn	$L\beta_{3,4}$	$L_IM_{II,III}$	0.721
17.202	5	26 Fe	L_{II}	Abs. Edge	0.7208
17.26	1	26 Fe	$L\beta_1$	$L_{II}M_{IV}$	0.7185
17.38	4	59 Pr	$M\zeta$	M_VN_{III}	0.714
17.525	5	26 Fe	L_{III}	Abs. Edge	0.7074
17.59	2	26 Fe	$L\alpha_{1,2}$	$L_{III}M_{IV,V}$	0.7050
17.6	1	52 Te		$M_{II}N_I$	0.703
17.87	3	27 Co	$L\eta$	$L_{II}M_I$	0.694
17.94	5	50 Sn	$M\gamma$	$M_{III}N_{IV,V}$	0.691
17.9	1	24 Cr	L_{II}	Abs. Edge	0.691
18.292	8	27 Co	Ll	$L_{III}M_I$	0.6778
18.32	2	9 F	$K\alpha$	KL	0.6768
18.35	4	58 Ce	$M\zeta$	M_VN_{III}	0.676
18.8	1	51 Sb		$M_{II}N_I$	0.658
18.8	2	47 Ag		$M_IN_{II,III}$	0.658
18.96	4	24 Cr	$L\beta_{3,4}$	$L_IM_{II,III}$	0.654
19.11	2	25 Mn	$L\beta_1$	$L_{II}M_{IV}$	0.6488
19.1	1	52 Te		$M_{III}N_I$	0.648
19.40	7	48 Cd		$M_{II}N_{IV}$	0.639
19.44	5	57 La	$M\zeta$	M_VN_{III}	0.638
19.45	1	25 Mn	$L\alpha_{1,2}$	$L_{III}M_{IV,V}$	0.6374
19.66	5	53 I	$M_{IV,V}$	Abs. Edge	0.631
19.75	4	26 Fe	$L\eta$	$L_{II}M_I$	0.628
20.0	1	50 Sn		$M_{II}N_I$	0.619
20.1	2	46 Pd		$M_IN_{II,III}$	0.616
20.15	1	26 Fe	Ll	$L_{III}M_I$	0.6152
20.2	1	51 Sb		$M_{III}N_I$	0.612
20.47	7	48 Cd	$M\gamma$	$M_{III}N_{IV,V}$	0.606
20.64	4	56 Ba	$M\zeta$	M_VN_{III}	0.601
20.66	7	47 Ag		$M_{II}N_{IV}$	0.600
20.7	1	24 Cr	L_{III}	Abs. Edge	0.598
21.19	5	23 Va	$L\beta_{3,4}$	$L_IM_{II,III}$	0.585
21.27	1	24 Cr	$L\beta_1$	$L_{II}M_{IV}$	0.5828
21.34	5	52 Te		$M_{IV}O_{II,III}$	0.581
21.5	1	50 Sn		$M_{III}N_I$	0.575
21.64	3	24 Cr	$L\alpha_{1,2}$	$L_{III}M_{IV,V}$	0.5728
21.78	5	52 Te		M_VO_{III}	0.569
21.82	7	47 Ag	$M\gamma$	$M_{III}N_{IV,V}$	0.568
21.85	2	25 Mn	$L\eta$	$L_{II}M_I$	0.5675
22.1	1	46 Pd		$M_{II}N_{IV}$	0.560
22.29	1	25 Mn	Ll	$L_{III}M_I$	0.5563
22.9	2	48 Cd		$M_{II}N_I$	0.540
23.32	1	8 O	K	Abs. Edge	0.5317
23.3	1	46 Pd	$M\gamma$	$M_{III}N_{IV,V}$	0.531
23.62	3	8 O	$K\alpha$	KL	0.5249
23.88	4	23 Va	$L\beta_1$	$L_{II}M_{IV}$	0.5192
24.25	3	23 Va	$L\alpha_{1,2}$	$L_{III}M_{IV,V}$	0.5113
24.28	5	50 Sn	$M_{IV,V}$	Abs. Edge	0.511
24.30	3	24 Cr	$L\eta$	$L_{II}M_I$	0.5102
24.4	2	47 Ag		M_VN_I	0.509
24.5	1	48 Cd		$M_{III}N_I$	0.507
24.78	1	24 Cr	Ll	$L_{III}M_I$	0.5003
25.01	9	45 Rh	$M\gamma$	$M_{III}N_{IV,V}$	0.496
25.3	1	50 Sn		$M_{IV}O_{II,III}$	0.491
25.50	9	44 Ru		$M_{II}N_{IV}$	0.486
25.7	1	50 Sn		M_VO_{III}	0.483
26.0	1	47 Ag		$M_{III}N_I$	0.478
26.2	2	46 Pd		$M_{II}N_I$	0.474
26.72	9	52 Te	$M\zeta$	$M_{IV,V}N_{II,III}$	0.464
26.9	1	44 Ru	$M\gamma$	$M_{III}N_{IV,V}$	0.462
27.05	2	22 Ti	$L\beta_1$	$L_{II}M_{IV}$	0.4584
27.29	1	22 Ti	$L_{II,III}$	Abs. Edge	0.4544
27.34	3	23 Va	$L\eta$	$L_{II}M_I$	0.4535
27.42	2	22 Ti	$L\alpha_{1,2}$	$L_{III}M_{IV,V}$	0.4522
27.77	1	23 Va	Ll	$L_{III}M_I$	0.4465
27.9	1	46 Pd		$M_{III}N_I$	0.445
28.1	2	45 Rh		$M_{II}N_I$	0.442
28.13	5	48 Cd	$M_{IV,V}$	Abs. Edge	0.4408
28.88	8	51 Sb	$M\zeta$	$M_{IV,V}N_{II,III}$	0.429
29.8	1	45 Rh		$M_{III}N_I$	0.417
30.4	1	48 Cd		$M_{IV}O_{II,III}$	0.408

TABLE B (*Continued*)

Wavelength Å*	p.e.	Element		Designation	keV
30.8	1	48 Cd		$M_{V}O_{III}$	0.403
30.82	5	47 Ag	M_{IV}	Abs. Edge	0.4022
30.89	3	22 Ti	$L\eta$	$L_{II}M_{I}$	0.4013
30.99	1	7 N	K	Abs. Edge	0.4000
31.02	2	21 Sc	$L\beta_1$	$L_{II}M_{IV}$	0.3996
31.14	5	47 Ag	M_V	Abs. Edge	0.3981
31.24	9	50 Sn	$M\zeta$	$M_{IV,V}N_{II,III}$	0.397
31.35	3	21 Sc	$L\alpha_{1,2}$	$L_{III}M_{IV,V}$	0.3954
31.36	2	22 Ti	Ll	$L_{III}M_{I}$	0.3953
31.60	4	7 N	$K\alpha$	KL	0.3924
31.8	1	92 U		$N_{IV}N_{VI}$	0.390
32.3	2	44 Ru		$M_{II}N_{I}$	0.384
33.1	2	41 Nb		$M_{II}N_{IV}$	0.375
33.5	3	47 Ag		$M_{IV,V}O_{II,III}$	0.370
33.57	9	90 Th		$N_{IV}N_{VI}$	0.3693
34.8	1	92 U		$N_{V}N_{VI,VII}$	0.357
34.9	2	41 Nb	$M\gamma$	$M_{III}N_{IV,V}$	0.356
35.13	2	21 Sc	$L\eta$	$L_{II}M_{I}$	0.3529
35.13	1	20 Ca	L_{II}	Abs. Edge	0.3529
35.3	3	42 Mo		$M_{II}N_{I}$	0.351
35.49	1	20 Ca	L_{III}	Abs. Edge	0.34931
35.59	3	21 Sc	Ll	$L_{III}M_{I}$	0.3483
35.63	1	20 Ca	$L_{II,III}$	Abs. Edge	0.34793
35.94	2	20 Ca	$L\beta_1$	$L_{II}M_{IV}$	0.3449
36.32	9	90 Th		$N_{V}N_{VI,VII}$	0.3414
36.33	2	20 Ca	$L\alpha_{1,2}$	$L_{III}M_{IV,V}$	0.3413
36.8	1	48 Cd	$M\zeta$	$M_{IV,V}N_{II,III}$	0.3371
37.4	2	46 Pd		$M_{IV,V}O_{II,III}$	0.332
37.5	2	42 Mo		$M_{III}N_{I}$	0.331
38.4	3	41 Nb		$M_{II}N_{I}$	0.323
39.77	7	47 Ag	$M\zeta$	$M_{IV,V}N_{II,III}$	0.3117
40.46	2	20 Ca	$L\eta$	$L_{II}M_{I}$	0.3064
40.7	2	41 Nb		$M_{III}N_{I}$	0.305
40.9	2	45 Rh		$M_{IV,V}O_{II,III}$	0.303
40.96	2	20 Ca	Ll	$L_{III}M_{I}$	0.3027
42.1	2	92 U		$N_{VI}O_{V}$	0.295
42.1	1	19 K	$L_{II,III}$	Abs. Edge	0.2946
42.3	2	82 Pb		$N_{IV}N_{VI}$	0.293
43.3	2	92 U		$N_{VI}O_{IV}$	0.286
43.6	1	46 Pd	$M\zeta$	$M_{IV,V}N_{II,III}$	0.2844
43.68	1	6 C	K	Abs. Edge	0.28384
44.7	3	6 C	$K\alpha$	KL	0.277
44.8	1	44 Ru		$M_{IV,V}O_{II,III}$	0.2768
45.0	1	82 Pb		$N_{V}N_{VI,VII}$	0.2756
45.2	3	80 Hg		$N_{IV}N_{VI}$	0.274
45.2	1	51 Sb		$M_{II}M_{IV}$	0.2743
46.48	9	39 Y		$M_{II}N_{I}$	0.267
46.5	2	81 Tl		$N_{V}N_{VI,VII}$	0.267
46.8	2	79 Au		$N_{IV}N_{VI}$	0.265
47.24	2	19 K	Ll	$L_{II}M_{I}$	0.2625
47.3	1	50 Sn		$M_{II}M_{IV}$	0.2621
47.67	9	45 Rh	$M\zeta$	$M_{IV,V}N_{II,III}$	0.2601
47.74	1	19 K	Ll	$L_{III}M_{I}$	0.25971
47.9	3	80 Hg		$N_{V}N_{VI,VII}$	0.259
48.1	2	78 Pt		$N_{IV}N_{VI}$	0.258
48.2	1	90 Th		$N_{VI}O_{V}$	0.2572
48.5	2	39 Y		$M_{III}N_{I}$	0.256
49.4	1	79 Au		$N_{V}N_{VI,VII}$	0.2510
49.5	1	90 Th		$N_{VI}O_{IV}$	0.2505
50.0	1	90 Th		$N_{VII}O_{V}$	0.2479
50.2	1	77 Ir		$N_{IV}N_{VI}$	0.2470
50.3	1	52 Te		$M_{III}M_{V}$	0.2465
50.9	1	78 Pt		$N_{V}N_{VI,VII}$	0.2436
51.3	1	38 Sr		$M_{II}N_{I}$	0.2416
51.9	1	76 Os		$N_{IV}N_{VI}$	0.2388
52.0	2	48 Cd		$M_{II}M_{IV}$	0.2384
52.2	1	51 Sb		$M_{III}M_{V}$	0.2375
52.34	7	44 Ru	$M\zeta$	$M_{IV,V}N_{II,III}$	0.2369
52.8	1	77 Ir		$N_{V}N_{VI,VII}$	0.2348
53.6	1	38 Sr		$M_{III}N_{I}$	0.2313
54.0	2	74 W		$N_{II}N_{IV}$	0.2295
54.0	1	47 Ag		$M_{II}M_{IV}$	0.2295
54.2	1	50 Sn		$M_{III}M_{V}$	0.2287
54.7	2	76 Os		$N_{V}N_{VI,VII}$	0.2266
54.8	2	42 Mo		$M_{IV,V}O_{II,III}$	0.2262
55.8	1	74 W		$N_{IV}N_{VI}$	0.2221
55.9	1	18 A	$L\eta$	$L_{II}M_{I}$	0.2217
56.3	1	18 A	Ll	$L_{III}M_{I}$	0.2201
56.5	1	46 Pd		$M_{II}M_{IV}$	0.2194
57.0	2	37 Rb		$M_{II}N_{I}$	0.2174
58.2	1	73 Ta		$N_{IV}N_{VI}$	0.2130
58.4	1	74 W		$N_{V}N_{VII}$	0.2122
58.7	2	48 Cd		$M_{III}M_{V}$	0.2111
59.3	1	45 Rh		$M_{II}M_{IV}$	0.2090
59.5	3	74 W		$N_{V}N_{VI}$	0.208
59.5	2	37 Rb		$M_{III}N_{I}$	0.2083
60.5	1	47 Ag		$M_{III}M_{V}$	0.2048
61.1	2	73 Ta		$N_{V}N_{VI,VII}$	0.2028
61.9	2	41 Nb		$M_{IV,V}O_{II,III}$	0.2002
62.2	1	44 Ru		$M_{II}M_{IV}$	0.1992
62.9	1	46 Pd		$M_{III}M_{V}$	0.1970
63.0	5	71 Lu		$N_{IV}N_{VI}$	0.197
64.38	7	42 Mo	$M\zeta$	$M_{IV,V}N_{II,III}$	0.1926
65.1	7	70 Yb		$N_{IV}N_{VI}$	0.190
65.5	1	45 Rh		$M_{III}M_{V}$	0.1892
65.7	2	71 Lu		$N_{V}N_{VI,VII}$	0.1886
67.33	9	17 Cl	$L\eta$	$L_{II}M_{I}$	0.1841
67.6	3	5 B	$K\alpha$	KL	0.1833
67.90	9	17 Cl	Ll	$L_{III}M_{I}$	0.1826
68.2	3	90 Th		$O_{III}P_{IV,V}$	0.1817
68.3	1	44 Ru		$M_{III}M_{V}$	0.1814
68.9	2	42 Mo		$M_{II}M_{IV}$	0.1798
69.3	5	70 Yb		$N_{V}N_{VI,VII}$	0.179
70.0	4	40 Zr		$M_{IV,V}O_{II,III}$	0.177
72.1	3	41 Nb		$M_{II}M_{IV}$	0.1718
72.19	9	41 Nb	$M\zeta$	$M_{IV,V}N_{II,III}$	0.1717
72.7	9	68 Er		$N_{IV}N_{VI}$	0.171
74.9	1	42 Mo		$M_{III}M_{V}$	0.1656
76.3	7	68 Er		$N_{V}N_{VI,VII}$	0.163
76.7	2	40 Zr		$M_{II}M_{IV}$	0.1617
76.9	2	35 Br		$M_{II}N_{I}$	0.1613
78.4	2	41 Nb		$M_{III}M_{V}$	0.1582
79.8	3	35 Br		$M_{III}N_{I}$	0.1554
80.9	3	40 Zr		$M_{III}M_{V}$	0.1533

X-Ray Wavelengths

TABLE B (*Continued*)

Wavelength Å*	p.e.	Element	Designation		keV
81.5	2	39 Y		$M_{II}M_{IV}$	0.1522
82.1	2	40 Zr	$M\zeta$	$M_{IV,V}N_{II,III}$	0.1511
83.	1	66 Dy		$N_{IV,V}N_{VI,VII}$	0.149
83.4	3	16 S	Ll, η	$L_{II,III}M_{I}$	0.1487
85.7	2	38 Sr		$M_{II}M_{IV}$	0.1447
86.	1	65 Tb		$N_{IV,V}N_{VI,VII}$	0.144
86.5	2	39 Y		$M_{III}M_{IV,V}$	0.1434
91.4	2	38 Sr		$M_{III}M_{IV,V}$	0.1357
91.5	2	37 Rb		$M_{II}M_{IV}$	0.1355
91.6	1	83 Bi		$N_{VI}O_{IV}$	0.1354
93.2	1	83 Bi		$N_{VII}O_{V}$	0.1330
93.4	2	39 Y	$M\zeta$	$M_{IV,V}N_{II,III}$	0.1328
94.	1	15 P	$L_{II,III}$	Abs. Edge	0.132
96.7	2	37 Rb		$M_{III}M_{IV,V}$	0.1282
97.2	8	66 Dy		$N_{IV,V}O_{II,III}$	0.128
98.	1	62 Sm		$N_{IV,V}N_{VI,VII}$	0.126
100.2	2	82 Pb		$N_{VI}O_{V}$	0.1237
102.2	4	65 Tb		$N_{IV,V}O_{II,III}$	0.1213
102.4	1	82 Pb		$N_{VI}O_{IV}$	0.1211
103.8	4	15 P		$L_{II,III}M$	0.1194
104.3	1	82 Pb		$N_{VII}O_{V}$	0.1189
107.	1	60 Nd		$N_{IV,V}N_{VI,VII}$	0.116
108.0	2	38 Sr	$M\zeta_2$	$M_{IV}N_{II,III}$	0.1148
108.7	1	38 Sr	$M\zeta_1$	$M_{V}N_{III}$	0.1140
109.4	3	35 Br		$M_{II}M_{IV}$	0.1133
110.6	5	29 Cu	M_{I}	Abs. Edge	0.1121
111.	1	4 Be	K	Abs. Edge	0.111
112.0	6	63 Eu		$N_{IV,V}O_{II,III}$	0.1107
113.0	1	81 Tl		$N_{VI}O_{V}$	0.10968
113.	1	59 Pr		$N_{IV,V}N_{VI,VII}$	0.1095
113.8	3	35 Br		$M_{III}M_{IV,V}$	0.1089
114.	1	4 Be	$K\alpha$	KL	0.1085
115.3	2	81 Tl		$N_{VI}O_{IV}$	0.1075
117.4	4	62 Sm		$N_{IV,V}O_{II,III}$	0.1056
117.7	1	81 Tl		$N_{VII}O_{V}$	0.10530
123.	1	14 Si	$L_{II,III}$	Abs. Edge	0.1006
126.8	2	37 Rb		$M_{IV}N_{III}$	0.0978
127.8	2	37 Rb	$M\zeta_2$	$M_{IV}N_{II}$	0.0970
128.7	2	37 Rb	$M\zeta_1$	$M_{V}N_{III}$	0.0964
128.9	7	60 Nd		$N_{IV,V}O_{II,III}$	0.0962
135.5	4	14 Si		$L_{II,III}M$	0.0915
136.5	4	59 Pr		$N_{IV,V}O_{II,III}$	0.0908
137.0	5	30 Zn	M_{II}	Abs. Edge	0.0905
142.5	1	13 Al	L_{I}	Abs. Edge	0.08701
143.9	5	30 Zn	M_{III}	Abs. Edge	0.0862
144.4	6	58 Ce		$N_{IV,V}O_{II,III}$	0.0859
144.4	3	37 Rb		$M_{I}M_{III}$	0.0859
152.6	6	57 La		$N_{IV,V}O_{II,III}$	0.0812
157.	3	30 Zn		$M_{II,III}M_{IV,V}$	0.079
159.0	2	56 Ba		$N_{IV}O_{III}$	0.07796
159.5	5	29 Cu	M_{II}	Abs. Edge	0.0777
163.3	2	56 Ba		$N_{IV}O_{II}$	0.07590
164.6	2	56 Ba		$N_{V}O_{III}$	0.07530
164.7	3	35 Br		$M_{I}M_{III}$	0.0753
166.0	5	29 Cu	M_{III}	Abs. Edge	0.0747
170.4	1	13 Al	$L_{II,III}$	Abs. Edge	0.07278
171.4	5	13 Al		$L_{II,III}M$	0.0724
173.	3	29 Cu		$M_{II,III}M_{IV,V}$	0.072
181.	5	90 Th		$O_{IV,V}Q_{II,III}$	0.068
183.8	1	55 Cs		$N_{IV}O_{III}$	0.06746
184.6	3	35 Br		$M_{I}M_{II}$	0.0672
188.4	1	28 Ni	M_{III}	Abs. Edge	0.06581
188.6	1	55 Cs		$N_{IV}O_{II}$	0.06574
189.5	3	35 Br		$M_{IV}N_{III}$	0.0654
190.3	1	55 Cs		$N_{V}O_{III}$	0.06515
190.	2	28 Ni		$M_{II,III}M_{IV,V}$	0.0651
191.1	2	35 Br	$M\zeta_2$	$M_{IV}N_{II}$	0.06488
192.6	2	35 Br	$M\zeta_1$	$M_{V}N_{III}$	0.06437
197.3	1	12 Mg	L_{I}	Abs. Edge	0.06284
202.	5	27 Co	$M_{II,III}$	Abs. Edge	0.061
203.	1	16 S		$L_{I}L_{II,III}$	0.061
214.	6	27 Co		$M_{II,III}M_{IV,V}$	0.058
224.	1	53 I	$N_{IV,V}$	Abs. Edge	0.0552
226.5	1	3 Li	K	Abs. Edge	0.05475
227.8	1	34 Se	M_{V}	Abs. Edge	0.05443
228.	1	3 Li	$K\alpha$	KL	0.0543
230.	2	34 Se		$M_{V}N_{III}$	0.0538
230.	1	26 Fe	$M_{II,III}$	Abs. Edge	0.0538
243.	5	26 Fe		$M_{II,III}M_{IV,V}$	0.051
249.3	1	12 Mg	L_{II}	Abs. Edge	0.04973
250.7	1	12 Mg	L_{III}	Abs. Edge	0.04945
251.5	5	12 Mg		$L_{II,III}M$	0.04929
273.	6	25 Mn		$M_{II,III}M_{IV,V}$	0.045
290.	1	13 Al		$L_{I}L_{II,III}$	0.0428
309.	9	24 Cr		$M_{II,III}M_{IV,V}$	0.040
317.	1	12 Mg		$L_{I}L_{II,III}$	0.0392
337.	9	23 V		$M_{II,III}M_{IV,V}$	0.0368
376.	1	11 Na		$L_{I}L_{II,III}$	0.03299
399.	5	35 Br	N_{I}	Abs. Edge	0.0311
405.	5	11 Na	$L_{II,III}$	Abs. Edge	0.0306
407.1	5	11 Na		$L_{II,III}M$	0.03045
417.	5	17 Cl	M_{I}	Abs. Edge	0.0297
444.	5	53 I	O_{I}	Abs. Edge	0.0279
525.	9	20 Ca		$M_{II,III}N_{I}$	0.0236
692.	9	19 K		$M_{II,III}N_{I}$	0.0179

X-RAY ATOMIC ENERGY LEVELS

J. A. Bearden and A. F. Burr

These tables were originally published as the final report to the U.S. Atomic Energy Commission as Report NYO-2543-1 in partial fulfillment of Contract AT(30-1)-2543. The tables were later reproduced in *Review of Modern Physics*. The data may also be obtained from the Superintendent of Documents, U.S. Government Printing Office, Washington, D. C. 20402 in the publication NSRDS-NBS 14. Persons seeking discussion of the details of calculations, sources of energy level information and the problem of properly interpreting the experimental measurements should refer to the original publication or to *Review of Modern Physics*, Vol. 39, 125–142, January 1967.

All of the x-ray emission wavelengths have recently been reevaluated and placed on a consistent Å* scale. For most elements these data give a highly overdetermined set of equations for energy level differences, which have been solved by least-squares adjustment for each case. This procedure makes "best" use of all x-ray wavelength data, and also permits calculation of the probable error for each energy difference. Photoelectron measurements of absolute energy levels are more precise than x-ray absorption edge data. These have been used to establish the absolute scale for eighty-one elements and, in many cases, to provide additional energy level difference data. The x-ray absorption wavelengths were used for eight elements and ionization measurements for two; the remaining five were interpolated by a Moseley diagram involving the output values of energy levels from adjacent elements. Probable errors are listed on an absolute energy basis. In the original source of the present data, a table of energy levels in Rydberg units is given. Difference tables in volts, Rydbergs, and milli-Å* wavelength units, with the respective probable errors, are also included there.

X-Ray Atomic Energy Levels

Recommended values of the atomic energy levels, and probable errors in eV. Where available, photoelectron direct measurements are listed in brackets [ ] immediately under the recommended values. The measured values of the x-ray absorption energies are shown in parentheses ( ). Interpolated values are enclosed in angle brackets ⟨ ⟩.

Level	1 H	2 He	3 Li	4 Be	5 B	6 C	7 N	8 O
K	13.59811[a]	24.58678[b]	54.75±0.02 (54.75)	111.0±1.0 (111.0)	188.0±0.4 [188.0][c]	283.8±0.4 [283.8][c] (283.8)	401.6±0.4 [401.6][c]	532.0±0.4 [532.0][c]
L_I								23.7±0.4 [23.7][d]
$L_{II,III}$					4.7±0.9	6.4±1.9	9.2±0.6	7.1±0.8
	9 F	**10 Ne**	**11 Na**	**12 Mg**	**13 Al**	**14 Si**	**15 P**	**16 S**
K	685.4±0.4 [685.4][c]	866.9±0.3 (866.9)	1072.1±0.4 [1072.1][c] (1072.)	1305.0±0.4 [1305.0][c] (1303.)	1559.6±0.4 [1559.6][c] (1559.8)	1838.9±0.4 [1838.9][c]	2145.5±0.4 [2145.5][d]	2472.0±0.4 [2472.0][c] (2470.)
L_I	⟨31.⟩	⟨45.⟩	63.3±0.4 [63.3][d]	89.4±0.4 [89.4][d] (63.)	117.7±0.4 [117.7][d] (87.)	148.7±0.4 [148.7][d]	189.3±0.4 [189.3][d]	229.2±0.4 [229.2][d]
$L_{II,III}$	8.6±0.8	18.3±0.4	31.1±0.4 (31.)	51.4±0.5 (50.)	73.1±0.5 (72.8)	99.2±0.5 (100.6)	132.2±0.5 (132.)	164.8±0.7
	17 Cl	**18 Ar**	**19 K**	**20 Ca**	**21 Sc**	**22 Ti**	**23 V**	**24 Cr**
K	2822.4±0.3 [2822.4][c] (2020.)	3202.9±0.3 (3202.9)	3607.4±0.4 [3607.4][c] (3607.8)	4038.1±0.4 [4038.1][c] (4038.1)	4492.8±0.4 [4492.8][c]	4966.4±0.4 [4966.4][d] (4964.5)	5465.1±0.3 [5465.1][c] (5464.)	5989.2±0.3 [5989.2][c] (5989.)
L_I	270.2±0.4 [270.2][d]	320. ⟨320.⟩[d]	377.1±0.4 [377.1][d]	437.8±0.4 [437.8][d]	500.4±0.4 [500.4][d]	563.7±0.4 [563.7][d]	628.2±0.4 [628.2][d]	694.6±0.4 [694.6][d]
L_{II}	201.6±0.3	247.3±0.3	296.3±0.4	350.0±0.4	406.7±0.4	461.5±0.4	520.5±0.3	583.7±0.3
L_{III}	200.0±0.3	245.2±0.3	293.6±0.4	346.4±0.4	402.2±0.4	455.5±0.4	512.9±0.3	574.5±0.3
M_I	17.5±0.4	25.3±0.4	33.9±0.4	43.7±0.4	53.8±0.4	60.3±0.4	66.5±0.4	74.1±0.4
$M_{II,III}$	6.8±0.4	12.4±0.3	17.8±0.4	25.4±0.4	32.3±0.5	34.6±0.4	37.8±0.3	42.5±0.3
$M_{IV,V}$					6.6±0.5	3.7	2.2±0.3	2.3±0.4

X-RAY ATOMIC ENERGY LEVELS (*Continued*)

	25 Mn	26 Fe	27 Co	28 Ni	29 Cu	30 Zn	31 Ga	32 Ge
K	6539.0±0.4 [6539.0][e] (6538.)	7112.0±0.9 [7111.3][e,f] (7111.2)	7708.9±0.3 [7708.9][e] (7709.5)	8332.8±0.4 [8332.8][e] (8331.6)	8978.9±0.4 [8978.9][e,g] (8980.3)	9658.6±0.6 [9658.6][g] (9660.7)	10367.1±0.5 [10367.1][g] (10368.2)	11103.1±0.7 [11103.8][g] (11103.6)
L_I	769.0±0.4 [769.0][d]	846.1±0.4 [846.1][d]	925.6±0.4 [925.6][d]	1008.1±0.4 [1008.1][d]	1096.1±0.4 [1096.0][d]	1193.6±0.9	1297.7±1.1	1414.3±0.7 [1413.6][g]
L_{II}	651.4±0.4	721.1±0.9 (720.8)	793.6±0.3 (793.8)	871.9±0.4 (870.6)	951.0±0.4 [950.0][h] (953.)	1042.8±0.6 (1045.)	1142.3±0.5	1247.8±0.7 (1249.)
L_{III}	640.3±0.4	708.1±0.9 (707.4)	778.6±0.3 (779.0)	854.7±0.4 (853.6)	931.1±0.4 [931.4][h] (933.)	1019.7±0.6 (1022.)	1115.4±0.5 (1117.)	1216.7±0.7 (1217.0)
M_I	83.9±0.5	92.9±0.9	100.7±0.4	111.8±0.6	119.8±0.6	135.9±1.1	158.1±0.5	180.0±0.8
M_{II} M_{III}	48.6±0.4	54.0±0.9 (54.)	59.5±0.3 (61.)	68.1±0.4 (66.)	73.6±0.4 (75.)	86.6±0.6 (86.)	106.8±0.7 102.9±0.5	127.9±0.9 120.8±0.7
$M_{IV,V}$	3.3±0.5	3.6±0.9	2.9±0.3	3.6±0.4	1.6±0.4	8.1±0.6	17.4±0.5	28.7±0.7

	33 As	34 Se	35 Br	36 Kr	37 Rb	38 Sr	39 Y	40 Zr
K	11866.7±0.7 [11866.7][i] (11865.)	12657.8±0.7 [12657.8][g] (12654.5)	13473.7±0.4 (13470.)	14325.6±0.8 (14324.4)	15199.7±0.3 (15202.)	16104.6±0.3 (16107.)	17038.4±0.3 (17038.)	17997.6±0.4 (17999.)
L_I	1526.5±0.8 (1529.)	1653.9±3.5 (1652.5)	1782.0±0.4 [1782.0][j]	1921.0±0.6 [1921.2][k]	2065.1±0.3 [2065.4][j]	2216.3±0.3 [2216.2][l]	2372.5±0.3 [2372.7][l]	2531.6±0.3 [2531.6][l]
L_{II}	1358.6±0.7 (1358.7)	1476.2±0.7 (1474.7)	1596.0±0.4 [1596.2][j]	1727.2±0.5 [1727.2][k] (1730.)	1863.9±0.3 [1863.4][j]	2006.8±0.3 [2006.6][l] (2008.5)	2155.5±0.3 [2155.0][l] (2154.0)	2306.7±0.3 [2306.5][l] (2305.3)
L_{III}	1323.1±0.7 (1323.5)	1435.8±0.7 (1434.0)	1549.9±0.4 [1549.7][j]	1674.9±0.5 [1674.8][k] (1677.)	1804.4±0.3 [1804.6][j]	1939.6±0.3 [1939.9][l] (1941.)	2080.0±0.3 [2080.2][l] (2079.4)	2222.3±0.3 [2222.5][l] (2222.5)
M_I	203.5±0.7	231.5±0.7	256.5±0.4		322.1±0.3	357.5±0.3	393.6±0.3	430.3±0.3
M_{II}	146.4±1.2	168.2±1.3	189.3±0.4	222.7±1.1	247.4±0.3	279.8±0.3	312.4±0.4	344.2±0.4

X-RAY ATOMIC ENERGY LEVELS (*Continued*)

	33 As	34 Se	35 Br	36 Kr	37 Rb	38 Sr	39 Y	40 Zr
M_{III}	140.5±0.8	161.9±1.0	181.5±0.4	213.8±1.1	238.5±0.3	269.1±0.3	300.3±0.4	330.5±0.4
M_{IV}	41.2±0.7	56.7±0.8	70.1±0.4	88.9±0.8	111.8±0.3	135.0±0.3	159.6±0.3	182.4±0.3
M_V			69.0±0.4		110.3±0.3	133.1±0.3	157.4±0.3	180.0±0.3
N_I			27.3±0.5	24.0±0.8	29.3±0.3	37.7±0.3	45.4±0.3	51.3±0.3
N_{II}	2.5±1.0	5.6±1.3	5.2±0.4	10.6±1.9	14.8±0.4	19.9±0.3	25.6±0.4	28.7±0.4
N_{III}			4.6±0.4		14.0±0.3			

	41 Nb	42 Mo	43 Tc	44 Ru	45 Rh	46 Pd	47 Ag	48 Cd
K	18985.6±0.4 (18987.)	19999.5±0.3 (20004.)	21044.0±0.7	22117.2±0.3 (22119.)	23219.9±0.3 (23219.8)	24350.3±0.3 (24348.)	25514.0±0.3 (25516.)	26711.2±0.3 (26716.)
L_I	2697.7±0.3 [2697.7][l]	2865.5±0.3 [2866.0][l]	3042.5±0.4 [3042.5][l]	3224.0±0.3 [3224.3][l]	3411.9±0.3 [3412.0][l] (3417.)	3604.3±0.3 [3604.6][l] (3607.)	3805.8±0.3 [3806.2][m] (3807.)	4018.0±0.3 [4018.1][m] (4019.)
L_{II}	2464.7±0.3 [2464.7][l]	2625.1±0.3 [2624.5][l] (2627.)	2793.2±0.4 [2973.2][l]	2966.9±0.3 [2966.8][l] (2966.3)	3146.1±0.3 [3146.3][l] (3145.)	3330.3±0.3 [3330.3][l] (3330.3)	3523.7±0.3 [3523.6][g,m] (3526.)	3727.0±0.3 [3727.1][m] (3728.)
L_{III}	2370.5±0.3 [2370.6][l]	2520.2±0.3 [2520.2][l] (2523.2)	2676.9±0.4 [2676.9][l]	2837.9±0.3 [2837.7][l] (2837.7)	3003.8±0.3 [3003.5][g,l] (3002.)	3173.3±0.3 [3173.0][g,l] (3173.0)	3351.1±0.3 [3350.8][g] (3351.0)	3537.5±0.3 [3537.3][g] (3537.6)
M_I	468.4±0.3	504.6±0.3		585.0±0.3	627.1±0.3	669.9±0.3	717.5±0.3	770.2±0.3
M_{II}	378.4±0.4	409.7±0.4	444.9±1.5	482.8±0.3	521.0±0.3	559.1±0.3	602.4±0.3	650.7±0.3
M_{III}	363.0±0.4	392.3±0.3	425.0±1.5	460.6±0.3	496.2±0.3	531.5±0.3	571.4±0.3	616.5±0.3
M_{IV}	207.4±0.3	230.3±0.3	256.4±0.5	283.6±0.3	311.7±0.3	340.0±0.3	372.8±0.3	410.5±0.3
M_V	204.6±0.3	227.0±0.3	252.9±0.4	279.4±0.3	307.0±0.3	334.7±0.3	366.7±0.3	403.7±0.3
N_I	58.1±0.3	61.8±0.3		74.9±0.3	81.0±0.3	86.4±0.3	95.2±0.3	107.6±0.3
N_{II}	33.9±0.4	34.8±0.4	38.9±1.9	43.1±0.4	47.9±0.4	51.1±0.4	62.6±0.3	66.9±0.4
N_{III}							55.9±0.3	
$N_{IV,V}$	3.2±0.3	1.8±0.3		2.0±0.3	2.5±0.4	1.5±0.3	3.3±0.3	9.3±0.3

X-Ray Atomic Energy Levels (*Continued*)

	49 In	50 Sn	51 Sb	52 Te	53 I	54 Xe	55 Cs	56 Ba
K	27939.9±0.3	29200.1±0.4 (29195.)	30491.2±0.3 (30486.)	31813.8±0.3 (31811.)	33169.4±0.4 (33167.)	34561.4±1.1 (34590.)	35984.6±0.4 (35987.)	37440.6±0.4 (37452.)
L_I	4237.5±0.3 [4237.7][m] (4237.3)	4464.7±0.3 [4464.5][g] (4464.8)	4698.3±0.3 [4698.3][m] (4698.4)	4939.2±0.3 [4939.3][m] (4939.7)	5188.1±0.3 [5188.1][j]	5452.8±0.4 (5452.8)	5714.3±0.4 [5712.7][j] (5721.)	5988.8±0.4 [5986.8][j] (5996.)
L_{II}	3938.0±0.3 [3937.8][m] (3939.3)	4156.1±0.3 [4156.2][g] (4157.)	4380.4±0.3 [4380.6][m] (4382.)	4612.0±0.3 [4612.0][m] (4612.6)	4852.1±0.3 [4852.0][j]	5103.7±0.4 (5103.7)	5359.4±0.3 [5359.5][j] (5358.)	5623.6±0.3 [5623.6][j] (5623.3)
L_{III}	3730.1±0.3 [3730.0][g] (3730.2)	3928.8±0.3 [3928.8][g] (3928.8)	4132.2±0.3 [4132.2][g] (4132.3)	4341.4±0.3 [4341.2][g] (4341.8)	4557.1±0.3 [4557.1][j]	4782.2±0.4 (4782.2)	5011.9±0.3 [5012.0][j] (5011.3)	5247.0±0.3 [5247.3][j] (5247.0)
M_I	825.6±0.3	883.8±0.3	943.7±0.3	1006.0±0.3	1072.1±0.3		1217.1±0.4	1292.8±0.4
M_{II}	702.2±0.3	756.4±0.4	811.9±0.3	869.7±0.3	930.5±0.3	999.0±2.1	1065.0±0.5	1136.7±0.5
M_{III}	664.3±0.3	714.4±0.3	765.6±0.3	818.7±0.3	874.6±0.3	937.0±2.1	997.6±0.5	1062.2±0.5
M_{IV}	450.8±0.3	493.3±0.3	536.9±0.3	582.5±0.3	631.3±0.3		739.5±0.4	796.1±0.3
M_V	443.1±0.3	484.8±0.3	527.5±0.3	572.1±0.3	619.4±0.3	672.3±0.5	725.5±0.5	780.7±0.3
N_I	121.9±0.3	136.5±0.4	152.0±0.3	168.3±0.3	186.4±0.3		230.8±0.4	253.0±0.5
N_{II} }	77.4±0.4	88.6±0.4	98.4±0.5	110.2±0.5	122.7±0.5	146.7±3.1	172.3±0.6	191.8±0.7
N_{III} }							161.6±0.6	179.7±0.6
N_{IV} }	16.2±0.3	23.9±0.3	31.4±0.3	39.8±0.3	49.6±0.3		78.8±0.5	92.5±0.5
N_V }							76.5±0.5	89.9±0.5
O_I	0.1±4.5	0.9±0.5	6.7±0.5	11.6±0.6	13.6±0.6		22.7±0.5	39.1±0.6
O_{II} }	0.8±0.4	1.1±0.5	2.1±0.4	2.3±0.5	3.3±0.5		13.1±0.5	16.6±0.5
O_{III} }							11.4±0.5	14.6±0.5

	57 La	58 Ce	59 Pr	60 Nd	61 Pm	62 Sm	63 Eu	64 Gd
K	38924.6±0.4 (38934.)	40443.0±0.4 (40453.)	41990.6±0.5 (42002.)	43568.9±0.4 (43574.)	45184.0±0.7 (45198.)	46834.2±0.5 (46849.)	48519.0±0.4 (48519.)	50239.1±0.5 (50233.)

X-Ray Atomic Energy Levels (*Continued*)

	57 La	58 Ce	59 Pr	60 Nd	61 Pm	62 Sm	63 Eu	64 Gd
L_I	6266.3±0.5 [6266.3][n]	6548.8±0.5 [6548.5][n]	6834.8±0.5 [6834.9][n]	7126.0±0.4 [7125.8][n] (7129.)	7427.9±0.8 [7427.9][o]	7736.8±0.5 [7736.2][n] (7748.)	8052.0±0.4 [8051.7][n] (8061.)	8375.6±0.5 [8375.4][n] (8386.)
L_{II}	5890.6±0.4 [5890.7][n]	6164.2±0.4 [6164.3][n]	6440.4±0.5 [6440.2][n]	6721.5±0.4 [6721.8][n] (6723.)	7012.8±0.6 [7012.8][o]	7311.8±0.4 [7312.0][n] (7313.)	7617.1±0.4 [7617.6][n] (7620.)	7930.3±0.4 [7930.5][n] (7931.)
L_{III}	5482.7±0.4 [5482.6][n]	5723.4±0.4 [5723.6][n]	5964.3±0.4 [5964.3][n]	6207.9±0.4 [6208.0][n] (6209.)	6459.3±0.6 [6459.4][o]	6716.2±0.5 [6716.8][n] (6717.)	6976.9±0.4 [6976.7][n] (6981.)	7242.8±0.4 [7242.8][n] (7243.)
M_I	1361.3±0.3	1434.6±0.6	1511.0±0.8	1575.3±0.7		1722.8±0.8	1800.0±0.5	1880.8±0.5
M_{II}	1204.4±0.6	1272.8±0.6	1337.4±0.7	1402.8±0.6	1471.4±6.2	1540.7±1.2	1613.9±0.7	1688.3±0.7
M_{III}	1123.4±0.5	1185.4±0.5	1242.2±0.6	1297.4±0.5	1356.9±1.4	1419.8±1.1	1480.6±0.6	1544.0±0.8
M_{IV}	848.5±0.4	901.3±0.6	951.1±0.6	999.9±0.6	1051.5±0.9	1106.0±0.8	1160.6±0.6	1217.2±0.6
M_V	831.7±0.4	883.3±0.5	931.0±0.6	977.7±0.6	1026.9±1.0	1080.2±0.6	1130.9±0.6	1185.2±0.6
N_I	270.4±0.8	289.6±0.7	304.5±0.9	315.2±0.8		345.7±0.9	360.2±0.7	375.8±0.7
N_{II}	205.8±1.2	223.3±1.1	236.3±1.5	243.3±1.6	} 242.±16.	{ 265.6±1.9	283.9±1.0	288.5±1.2
N_{III}	191.4±0.9	207.2±0.9	217.6±1.1	224.6±1.3	}	{ 247.4±1.5	256.6±0.8	270.9±0.9
$N_{IV,V}$	98.9±0.8	110.0±0.6	113.2±0.7	117.5±0.7	120.4±2.0	129.0±1.2	133.2±0.6	140.5±0.8
$N_{VI,VII}$		0.1±1.2	2.0±0.6	1.5±0.9		5.5±1.1	0.0±3.2	0.1±3.5
O_I	32.3±7.2	37.8±1.3	37.4±1.0	37.5±0.9		37.4±1.5	31.8±0.7	36.1±0.8
$O_{II,III}$	14.4±1.2	19.8±1.2	22.3±0.7	21.1±0.8		21.3±1.5	22.0±0.6	20.3±1.2

	65 Tb	66 Dy	67 Ho	68 Er	69 Tm	70 Yb	71 Lu	72 Hf
K	51995.7±0.5 (52002.)	53788.5±0.5 (53793.)	55617.7±0.5 (55619.)	57485.5±0.5 (57487.)	59389.6±0.5	61332.3±0.5 (61300.)	63313.8±0.5 (63310.)	65350.8±0.6 (65310.)
L_I	8708.0±0.5 [8707.6][n] (8717.)	9045.8±0.5 [9046.5][n]	9394.2±0.4 [9394.3][n] (9399.)	9751.3±0.4 [9751.5][n] (9757.)	10115.7±0.4 [10115.6][n] (10121.)	10486.4±0.4 [10487.3][n] (10490.)	10870.4±0.4 [10870.1][n] (10874.)	11270.7±0.4 [11271.6][o] (11274.)

X-RAY ATOMIC ENERGY LEVELS (*Continued*)

	65 Tb	66 Dy	67 Ho	68 Er	69 Tm	70 Yb	71 Lu	72 Hf
L_{II}	8251.6±0.4 [8251.8][n] (8253.)	8580.6±0.4 [8580.4][n] (8583.)	8917.8±0.4 [8918.2][n] (8916.)	9264.3±0.4 [9264.3][n] (9262.)	9616.9±0.4 [9617.1][n] (9617.1)	9978.2±0.4 [9977.9][n] (9976.)	10348.6±0.4 [10349.0][n] (10345.)	10739.4±0.4 [10738.9][o] (10736.)
L_{III}	7514.0±0.4 [7514.2][n] (7515.)	7790.1±0.4 [7789.6][n] (7789.7)	8071.1±0.4 [8070.6][n] (8068.)	8357.9±0.4 [8357.6][n] (8357.5)	8648.0±0.4 [8647.8][n] (8649.6)	8943.6±0.4 [8942.6][n] (8944.1)	9244.1±0.4 [9243.8][n]	9560.7±0.4 [9560.4][o] (9558.)
M_I	1967.5±0.6	2046.8±0.4	2128.3±0.6	2206.5±0.6	2306.8±0.7	2398.1±0.4	2491.2±0.5	2600.9±0.4
M_{II}	1767.7±0.9	1841.8±0.5	1922.8±1.0	2005.8±0.6	2089.8±1.1	2173.0±0.4	2263.5±0.4	2365.4±0.4
M_{III}	1611.3±0.8	1675.6±0.9	1741.2±0.9	1811.8±0.6	1884.5±1.1	1949.8±0.5	2023.6±0.5	2107.6±0.4
M_{IV}	1275.0±0.6	1332.5±0.4	1391.5±0.7	1453.3±0.5	1514.6±0.7	1576.3±0.4	1639.4±0.4	1716.4±0.4
M_V	1241.2±0.7	1294.9±0.4	1351.4±0.8	1409.3±0.5	1467.7±0.9	1527.8±0.4	1588.5±0.4	1661.7±0.4
N_I	397.9±0.8	416.3±0.5	435.7±0.8	449.1±1.0	471.7±0.9	487.2±0.6	506.2±0.6	538.1±0.4
N_{II}	310.2±1.2	331.8±0.6	343.5±1.4	366.2±1.5	385.9±1.6	396.7±0.7	410.1±1.8	437.0±0.5
N_{III}	385.0±1.0	292.9±0.6	306.6±0.9	320.0±0.7	336.6±1.6	343.5±0.5	359.3±0.5	380.4±0.5
N_{IV}	147.0±0.8 (N_IV, N_V)	154.2±0.5 (N_IV, N_V)	161.0±1.0 (N_IV, N_V)	176.7±1.2	179.6±1.2 (N_IV, N_V)	198.1±0.5	204.8±0.5	223.8±0.4
N_V				167.6±1.5		184.9±1.3	195.0±0.4	213.7±0.5
$N_{VI,VII}$	2.6±1.5	4.2±1.6	3.7±3.0	4.3±1.4	5.3±1.9	6.3±1.0	6.9±0.5	17.1±0.5
O_I	39.0±0.8	62.9±0.5	51.2±1.3	59.8±1.7	53.2±3.0	54.1±0.5	56.8±0.5	64.9±0.4
O_{II}	25.4±0.8 (O_II, O_III)	26.3±0.6 (O_II, O_III)	20.3±1.5 (O_II, O_III)	29.4±1.6 (O_II, O_III)	32.3±1.6 (O_II, O_III)	23.4±0.6 (O_II, O_III)	28.0±0.6 (O_II, O_III)	38.1±0.6
O_{III}								30.6±0.6

	73 Ta	74 W	75 Re	76 Os	77 Ir	78 Pt	79 Au	80 Hg
K	67416.4±0.6 (67403.)	69525.0±0.3 (69508.)	71676.4±0.4 (71658.)	73870.8±0.5	76111.0±0.5	78394.8±0.7 (78381.)	80724.9±0.5 (80720.)	83102.3±0.8
L_I	11681.5±0.3 [11680.2][p] (11682.)	12099.8±0.3 [12098.2][p] (12099.6)	12526.7±0.4 (12530.)	12968.0±0.4 (12972.)	13418.5±0.3 (13423.)	13879.9±0.4 (13883.)	14352.8±0.4 (14353.7)	14839.3±1.0 (14842.)

X-Ray Atomic Energy Levels (*Continued*)

	73 Ta	74 W	75 Re	76 Os	77 Ir	78 Pt	79 Au	80 Hg
L_{II}	11136.1±0.3 [11136.1][p] (11132.)	11544.0±0.3 [11541.4][p] (11538.)	11958.7±0.3 [11956.9][p] (11954.)	12385.0±0.4 (12381.)	12824.1±0.3 [12824.0][e,p] (12820.)	13272.6±0.3 [13272.6][e,p] (13272.3)	13733.6±0.3 [13733.5][e,p] (13736.)	14208.7±0.7 (14215.)
L_{III}	9881.1±0.3 [9880.3][p] (9877.7)	10206.8±0.3 [10204.2][p] (10200.)	10535.3±0.3 [10534.2][p] (10531.)	10870.9±0.3 [10870.7][p] (10868.)	11215.2±0.3 [11215.1][e,p] (11212.)	11563.7±0.3 [11563.7][e,p] (11562.)	11918.7±0.3 [11918:2][e,p] (11921.)	12283.9±0.4 [12284.0][e,p] (12286.)
M_I	2708.0±0.4	2819.6±0.4	2931.7±0.4	3048.5±0.4	3173.7±1.7	3296.0±0.9	3424.9±0.3 [3424.8][p]	3561.6±1.1
M_{II}	2468.7±0.3 [2468.6][p]	2574.9±0.3 [2575.0][p]	2681.6±0.4	2792.2±0.3 [2791.9][p]	2908.7±0.3 [2909.1][p]	3026.5±0.4 [3026.5][p] (3029.)	3147.8±0.4 [3149.5][p]	3278.5±1.3
M_{III}	2194.0±0.3 [2194.1][p]	2281.0±0.3 [2281.0][p]	2367.3±0.3 [2367.3][p]	2457.2±0.4 [2457.4][p]	2550.7±0.3 [2550.5][p] (2550.5)	2645.4±0.4 [2645.5][p] (2645.9)	2743.0±0.3 [2743.1][p] (2744.0)	2847.1±0.4 [2847.1][p]
M_{IV}	1793.2±0.3 [1793.1][p]	1871.6±0.3 [1871.4][p]	1948.9±0.3 [1948.9][p]	2030.8±0.3 [2031.0][p]	2116.1±0.3 [2116.1][p]	2201.9±0.3 [2201.9][p]	2291.1±0.3 [2291.2][p] (2307.)	2384.9±0.3 [2384.9][p]
M_V	1735.1±0.3 [1735.2][p]	1809.2±0.3 [1809.3][p]	1882.9±0.3 [1882.9][p]	1960.1±0.3 [1960.2][p]	2040.4±0.3 [2040.5][p]	2121.6±0.3 [2121.6][p]	2205.7±0.3 [2206.1][p] (2220.)	2294.9±0.3 [2294.9][p]
N_I	565.5±0.5	595.0±0.4	625.0±0.4	654.3±0.5	690.1±0.4	722.0±0.6	758.8±0.4	800.3±1.0
N_{II}	464.8±0.5	491.6±0.4	517.9±0.5	546.5±0.5	577.1±0.4	609.2±0.6	643.7±0.5	676.9±2.4
N_{III}	404.5±0.4	425.3±0.5	444.4±0.5	468.2±0.6	494.3±0.6	519.0±0.6	545.4±0.5	571.0±1.4
N_{IV}	241.3±0.4	258.8±0.4	273.7±0.5	289.4±0.5	311.4±0.4	330.8±0.5	352.0±0.4	378.3±1.0
N_V	229.3±0.3	245.4±0.4	260.2±0.4	272.8±0.6	294.9±0.4	313.3±0.4	333.9±0.4	359.8±1.2
N_{VI}, N_{VII} (combined)	25.0±0.4		40.6±0.4	46.3±0.6				
N_{VI}		36.5±0.4			63.4±0.4	74.3±0.4	86.4±0.4	102.2±0.5
N_{VII}		33.6±0.4			60.5±0.4	71.1±0.5	82.8±0.5	98.5±0.5
O_I	71.1±0.5	77.1±0.4	82.8±0.5	83.7±0.6	95.2±0.4	101.7±0.4	107.8±0.7	120.3±1.3
O_{II}	44.9±0.4	46.8±0.5	45.6±0.7	58.0±1.1	63.0±0.6	65.3±0.7	71.7±0.7	80.5±1.3
O_{III}	36.4±0.4	35.6±0.5	34.6±0.6	45.4±1.0	50.5±0.6	51.7±0.7	53.7±0.7	57.6±1.3
$O_{IV,V}$	5.7±0.4	6.1±0.4	3.5±0.5		3.8±0.4	2.2±1.3	2.5±0.5	6.4±1.4

X-Ray Atomic Energy Levels (*Continued*)

	81 Tl	82 Pb	83 Bi	84 Po	85 At	86 Rn	87 Fr	88 Ra
K	85530.4±0.6	88004.5±0.7 (88005.)	90525.9±0.7 (90534.)	93105.0±3.8	95729.9±7.7	98404.±12.	101137.±13.	103921.9±7.2
L_I	15346.7±0.4 (15343.)	15860.8±0.5 (15855.)	16387.5±0.4 (16376.)	16939.3±9.8	17493.±29.	18049.±38.	18639.±40.	19236.7±1.5 (19236.0)
L_{II}	14697.9±0.3 [14697.3][p] (14699.)	15200.0±0.4 (15205.)	15711.1±0.3 [15708.4][p] (15719.)	16244.3±2.4	16784.7±2.5	17337.1±3.4	17906.5±3.5	18484.3±1.5 (18486.0)
L_{III}	12657.5±0.3 [12656.3][e,p] (12660.)	13035.2±0.3 [13034.9][e,p] (13041.)	13418.6±0.3 [13418.3][e,p] (13426.)	13813.8±1.0 ⟨13813.8⟩	14213.5±2.0 ⟨14213.5⟩	14619.4±3.0 ⟨14619.4⟩	15031.2±3.0 ⟨15031.2⟩	15444.4±1.5 (15444.0)
M_I	3704.1±0.4	3850.7±0.5	3999.1±0.3 [3999.1][p]	4149.4±3.9	⟨4317.⟩	⟨4482.⟩	⟨4652.⟩	4822.0±1.5
M_{II}	3415.7±0.3 [3415.7][p]	3554.2±0.3 [3554.2][p]	3696.3±0.3 [3696.4][p]	3854.1±9.8	4008.±28.	4159.±38.	4327.±40.	4489.5±1.8
M_{III}	2956.6±0.3 [2956.5][p]	3066.4±0.4 [3066.3][p]	3176.9±0.3 [3176.8][p]	3301.9±9.9	3426.±29.	3538.±38.	3663.±40.	3791.8±1.7
M_{IV}	2485.1±0.3 [2485.2][p]	2585.6±0.3 [2585.5][p] (2606.)	2687.6±0.3 [2687.4][p]	2798.0±1.2	2908.7±2.1	3021.5±3.1	3136.2±3.1	3248.4±1.6
M_V	2389.3±0.3 [2389.4][p]	2484.0±0.3 [2484.2][p] (2502.)	2579.6±0.3 [2579.5][p]	2683.0±1.1	2786.7±2.1	2892.4±3.1	2999.9±3.1	3104.9±1.6
N_I	845.5±0.5	893.6±0.7	938.2±0.3 [938.7][p]	995.3±2.9	⟨1042.⟩	⟨1097.⟩	⟨1153.⟩	1208.4±1.6
N_{II}	721.3±0.8	763.9±0.8	805.3±0.3 [805.3][p]	851.±12.	886.±30.	929.±40.	980±42.	1057.6±1.8
N_{III}	609.0±0.5	644.5±0.6	678.9±0.3 [678.9][p]	705.±14.	740.±30.	768.±40.	810±43.	879.1±1.8
N_{IV}	406.6±0.4	435.2±0.5	463.6±0.3 [463.6][p]	500.2±2.4	533.2±3.2	566.6±4.0	603.3±4.1	635.9±1.6
N_V	386.2±0.5	412.9±0.6	440.0±0.3 [440.1][p]	473.4±1.3			577.±34.	602.7±1.7

X-RAY ATOMIC ENERGY LEVELS (*Continued*)

	81 Tl	82 Pb	83 Bi	84 Po	85 At	86 Rn	87 Fr	88 Ra
N_{VI}	122.8±0.4	142.9±0.4	161.9±0.5					
N_{VII}	118.5±0.4	138.1±0.4	157.4±0.6					298.9±2.4
O_{I}	136.3±0.7	147.3±0.8	159.3±0.7					254.4±2.1
O_{II}	99.6±0.6	104.8±1.0	116.8±0.7					200.4±2.0
O_{III}	75.4±0.6	86.0±1.0	92.8±0.6					152.8±2.0
O_{IV}	15.3±0.4	21.8±0.4	26.5±0.5	31.4±3.2				
O_{V}	13.1±0.4	19.2±0.4	24.4±0.6					67.2±1.7
P_{I}		3.1±1.0						43.5±2.2
$P_{II,III}$		0.7±1.0	2.7±0.7					18.8±1.8

	89 Ac	90 Th	91 Pa	92 U	93 Np	94 Pu	95 Am	96 Cm
K	106755.3±5.3	109650.9±0.9	112601.4±2.4	115606.1±1.6	118678.±33.	121818.±44.	125027.±55.	128220
L_{I}	19840.±18.	20472.1±0.5 (20464.)	21104.6±1.8 (21128.)	21757.4±0.3 (21771.)	22426.8±0.9	23097.2±1.6 (23109.)	23772.9±2.0 ⟨23772.9⟩	24460
L_{II}	19083.2±2.8	19693.2±0.4 (19683.)	20313.7±1.5 (20319.)	20947.6±0.3 (20945.)	21600.5±0.4	22266.2±0.7 (22253.)	22944.0±1.0	23779
L_{III}	15871.0±2.0 ⟨15871.0⟩	16300.3±0.3 [16299.6][q] (16299.)	16733.1±1.4 (16733.)	17166.3±0.3 [17168.5][r] (17165.)	17610.0±0.4 (17606.2)	18056.8±0.6 (18053.1)	18504.1±0.9 (18504.1)	18930
M_{I}	⟨5002.⟩	5182.3±0.3 [5182.3][q]	5366.9±1.6	5548.0±0.4	5723.2±3.6	5932.9±1.4	6120.5±7.5	6288
M_{II}	4656.±18.	4830.4±0.4 [4830.6][q]	5000.9±2.3	5182.2±0.4 [5180.9][r]	5366.2±0.7 [5366.4][s]	5541.2±1.7	5710.2±2.1	5895
M_{III}	3909.±18.	4046.1±0.4 [4046.1][q] (4041.)	4173.8±1.8	4303.4±0.3 [4303.6][r] (4299.)	4434.7±0.5 [4434.6][s]	4556.6±1.5	4667.0±2.1	4797
M_{IV}	3370.2±2.1	3490.8±0.3 [3490.7][q] (3485.)	3611.2±1.4 (3608.)	3727.6±0.3 [3728.1][r] (3720.)	3850.3±0.4 [3849.8][s]	3972.6±0.6 [3972.7][t]	4092.1±1.0	4227

X-RAY ATOMIC ENERGY LEVELS (*Continued*)

	89 Ac	90 Th	91 Pa	92 U	93 Np	94 Pu	95 Am	96 Cm
M_{V}	3219.0±2.1	3332.0±0.3 [3332.1][q] (3325.)	3441.8±1.4 (3436.)	3551.7±0.3 [3551.7][r] (3545.)	3665.8±0.4 [3664.2][s]	3778.1±0.6 [3778.0][t]	3886.9±1.0	3971
N_{I}	⟨1269.⟩	1329.5±0.4 [1329.8][q]	1387.1±1.9	1440.8±0.4 [1441.3][r]	1500.7±0.8 [1500.7][s]	1558.6±0.8	1617.1±1.1	1643
N_{II}	1080.±19.	1168.2±0.4 [1168.3][q]	1224.3±1.6	1272.6±0.3 [1272.5][r]	1327.7±0.8 [1327.7][s]	1372.1±1.8	1411.8±8.3	1440
N_{III}	890.±19.	967.3±0.4 [967.6][q]	1006.7±1.7	1044.9±0.3 [1044.9][r]	1086.8±0.7 [1086.8][s]	1114.8±1.6	⟨1135.7⟩	1154
N_{IV}	674.9±3.7	714.1±0.4 [714.4][q]	743.4±2.1	780.4±0.3 [779.7][r]	815.9±0.5 [817.1][s]	848.9±0.6 [848.9][t]	878.7±1.0	
N_{V}		676.4±0.4 [676.4][q]	708.2±1.8	737.7±0.3 [737.6][r]	770.3±0.4 [773.2][s]	801.4±0.6 [801.4][t]	827.6±1.0	
N_{VI}		344.4±0.3 [344.2][q]	371.2±1.6	391.3±0.6	415.0±0.8 [415.0][s]	445.8±1.7		
N_{VII}		335.2±0.4 [335.0][q]	359.5±1.6	380.9±0.9	404.4±0.5 [404.4][s]	432.4±2.1		
O_{I}		290.2±0.8	309.6±4.3	323.7±1.1		351.9±2.4		385
O_{II}		229.4±1.1	222.9±3.9	259.3±0.5	283.4±0.8 [283.4][s]	274.1±4.7		
O_{III}		181.8±0.4 [181.8][q]		195.1±1.3	206.1±0.7 [206.1][s]	206.5±4.7		
O_{IV}		94.3±0.4 [94.4][q]	94.1±2.8	105.0±0.5	109.3±0.7 [108.8][s]	116.0±1.2	115.8±1.3	
O_{V}		87.9±0.3 [88.1][q]		96.3±1.4	101.3±0.5 [101.4][s]	105.4±1.0	103.3±1.1	
P_{I}		59.5±1.1		70.7±1.2				
P_{II}		49.0±2.5		42.3±9.0				
P_{III}		43.0±2.5		32.3±9.0				

X-RAY ATOMIC ENERGY LEVELS (*Continued*)

	97 Bk	98 Cf	99 Es	100 Fm	101 Md	102 No	103 Lw
K	[131590±40][u]	135960	139490	143090	146780	150540	154380
L_I	[25275±17][u]	26110	26900	27700	28530	29380	30240
L_{II}	[24385±17][u]	25250	26020	26810	27610	28440	29280
L_{III}	[19452±20][u]	19930	20410	20900	21390	21880	22360
M_I	[6556±21][u]	6754	6977	7205	7441	7675	7900
M_{II}	[6147±31][u]	6359	6574	6793	7019	7245	7460
M_{III}	[4977±31][u]	5109	5252	5397	5546	5688	5710
M_{IV}	4366	4497	4630	4766	4903	5037	5150
M_V	4132	4253	4374	4498	4622	4741	4860
N_I	[1755±22][u]	1799	1868	1937	2010	2078	2140
N_{II}	1554	1616	1680	1747	1814	1876	1930
N_{III}	1235	1279	1321	1366	1410	1448	1480
O_I	[398±22][u]	419	435	454	472	484	490

[a] J. E. Mack, 1949, as given in C. E. Moore, *Atomic Energy Levels* (U. S. National Bureau of Standards, Washington, D. C., 1949), Vol. 1, p. 1.
[b] G. Herzberg, 1957, as given in C. E. Moore, *Atomic Energy Levels* (U. S. National Bureau of Standards, Washington, D. C., 1958), Vol. 3, p. 238.
[c] See Ref. 18.
[d] A. Fahlman, D. Hamrin, R. Nordberg, C. Nordling, and K. Siegbahn, Phys. Rev. Letters **14**, 127 (1965). See also Ref. 26.
[e] See Ref. 15.
[f] See Ref. 11.
[g] C. Nordling, Arkiv Fysik **15**, 397 (1959).
[h] E. Sokolowski, C. Nordling, and K. Siegbahn, Arkiv Fysik **12**, 301 (1957).
[i] C. Nordling and S. Hagström, Arkiv Fysik **16**, 515 (1960).
[j] I. Andersson and S. Hagström, Arkiv Fysik **27**, 161 (1964).
[k] M. O. Krause, Phys. Rev. **140**, A1845 (1965).
[l] A. Fahlman, O. Hörnfeldt, and C. Nordling, Arkiv Fysik **23**, 75 (1962).
[m] P. Bergvall, O. Hörnfeldt, and C. Nordling, Arkiv Fysik **17**, 113 (1960).
[n] P. Bergvall and S. Hagström, Arkiv Fysik **17**, 61 (1960).
[o] S. Hagström, Z. Physik **178**, 82 (1964).
[p] A. Fahlman and S. Hagström, Arkiv Fysik **27**, 69 (1964).
[q] C. Nordling and S. Hagström, Z. Physik **178**, 418 (1964).
[r] C. Nordling and S. Hagström, Arkiv Fysik **15**, 431 (1959).
[s] S. Hagström, Bull. Am. Phys. Soc. **11**, 389 (1966).
[t] A. Fahlman, K. Hamrin, R. Nordberg, C. Nordling, K. Siegbahn, and L. W. Holm, Phys. Letters **19**, 643 (1966).
[u] J. M. Hollander, M. D. Holtz, T. Novakov, and R. L. Graham ,Arkiv Fysik **28**, 375 (1965).

K AND L X-RAY ABSORPTION EDGES

The wavelengths in Angstroms for the K absorption edge of elements having atomic numbers 3 to 11 and 83 to 100 were calculated from critical X-ray absorption energies listed in the Review of Scientific Instruments, **23,** 523–528. (1952.)

All wavelengths for the L absorption edges were calculated from absorption energies given in the above reference.

Values for the M, N, and O absorption edges are also listed in the above reference. The wavelengths in Angstroms for the K absorption edge of the elements having atomic numbers 12 to 82 are reproduced by permission from Structure of Metals, Barrett, 2nd edition, McGraw-Hill Book Company, Inc. (1952.)

ABSORPTION EDGES—ANGSTROMS

Atomic Number	Element	K	L_I	L_{II}	L_{III}
3	Li	226.62			
4	Be	110.68			
5	B	66.289			
6	C	43.648			
7	N	30.990			
8	O	23.301			
9	F	17.913			
10	Ne	14.183			
11	Na	11.478			
12	Mg	9.5117	197.39	247.92	
13	Al	7.9511	142.48	172.16	
14	Si	6.7446	105.05	126.48	
15	P	5.7866	81.02	96.843	
16	S	5.0182	64.228	76.049	76.519
17	Cl	4.3969	52.084	61.366	61.672
18	A	3.87068	43.192	50.390	50.803
19	K	3.43645	36.352	42.020	42.452
20	Ca	3.07016	31.068	35.417	35.827
21	Sc	2.7572	26.831	30.161	30.457
22	Ti	2.497$_{30}$	23.389	26.831	27.184
23	Va	2.269$_{02}$	20.523	23.702	24.0699
24	Cr	2.070$_{12}$	18.256	21.226	21.5958
25	Mn	1.896$_{36}$	16.268	18.896	19.2484
26	Fe	1.743$_{34}$	14.601	17.169	17.4838
27	Co	1.608$_{11}$	13.343	15.534	15.8314
28	Ni	1.488$_{02}$	12.267	14.135	14.4476
29	Cu	1.380$_{43}$	11.269	12.994	13.2578
30	Zn	1.283$_{3}$	10.330	11.8395	12.1055
31	Ga	1.195$_{67}$	9.535	10.6130	10.8546
32	Ge	1.116$_{52}$	8.729	9.9646	10.2277
33	As	1.044$_{97}$	8.107	9.1281	9.3767
34	Se	0.979$_{78}$	7.467	8.4212	8.6624
35	Br	0.91994	6.925	7.7523	7.9871
36	Kr	0.86546	6.456	7.1653	7.4227
37	Rb	0.81549	6.006	6.6538	6.8752
38	Sr	0.76969	5.604	6.1856	6.3996
39	Y	0.72762	5.19312	5.70981	5.91412
40	Zr	0.68877	4.89380	5.37088	5.57374
41	Nb	0.65291	4.59111	5.02472	5.22596
42	Mo	0.61977	4.32066	4.71330	4.90930
43	Tc	0.58888	4.06426	4.42714	4.62537
44	Ru	0.560$_{47}$	3.84133	4.17654	4.36632
45	Rh	0.533$_{78}$	3.64159	3.94902	4.13889
46	Pd	0.509$_{15}$	3.42999	3.71360	3.89688
47	Ag	0.4858$_{2}$	3.23824	3.49478	3.67288
48	Cd	0.46408	3.08434	3.32243	3.50070
49	In	0.44397	2.93327	3.15500	3.32154
50	Sn	0,42468	2.78875	2.99492	3.16952

Atomic Number	Element	K	L_I	L_{II}	L_{III}
51	Sb	0.40663	2.63296	2.82304	2.99637
52	Te	0.38972	2.50272	2.68253	2.85162
53	I	0.373$_{79}$	2.38982	2.55324	2.71901
54	Xe	0.35849	2.27449	2.43058	2.59330
55	Cs	0.34473	2.17245	2.31268	2.47227
56	Ba	0.33137	2.08336	2.20216	2.36114
57	La	0.31842	1.97892	2.10030	2.25792
58	Ce	0.30647	1.89078	2.00940	2.16410
59	Pr	0.29517	1.81307	1.92305	2.07707
60	Nd	0.28451	1.73759	1.84244	1.99452
61	Pm	0.27425	1.66837	1.76581	1.91888
62	Sm	0.26462	1.60113	1.69436	1.84464
63	Eu	0.25552	1.53815	1.62591	1.77491
64	Gd	0.24680	1.47870	1.56081	1.70955
65	Tb	0.23840	1.42270	1.50108	1.64840
66	Dy	0.23046	1.36926	1.44357	1.59025
67	Ho	0.22290	1.31942	1.38999	1.53529
68	Er	0.21565	1.27086	1.33721	1.48242
69	Tm	0.2089	1.22708	1.28856	1.43140
70	Yb	0.20223	1.18136	1.24146	1.38518
71	Lu	0.19584	1.14007	1.19745	1.34039
72	Hf	0.18981	1.09864	1.15311	1.29570
73	Ta	0.18393	1.06084	1.11264	1.25427
74	W	0.17837	1.02497	1.07436	1.21529
75	Re	0.17311	0.99009	1.03645	1.17720
76	Os	0.16780	0.95574	1.00129	1.14143
77	Ir	0.16286	0.92425	0.96700	1.10599
78	Pt	0.15817	0.89405	0.93484	1.07306
79	Au	0.15344	0.86378	0.90277	1.04028
80	Hg	0.14923	0.83531	0.87790	1.00944
81	Tl	0 14470	0.80787	0.84355	0.97968
82	Pb	0.14077	0.78153	0.81552	0.95112
83	Bi	0.13691	0.75649	0.78910	0.92459
84	Po	0.13306	0.73219	0.76377	0.89761
85	At	0.12949	0.70915	0.73873	0.87234
86	Rn	0.12591	0.68675	0.71529	0.84845
87	Fr	0.12261	0.66537	0.69290	0.82529
88	Ra	0.11931	0.64461	0.67114	0.80284
89	Ac	0.11618	0.62479	0.65002	0.78158
90	Th	0.11290	0.60610	0.63010	0.76151
91	Pa	0.11028	0.58748	0.61064	0.74138
92	U	0.10775	0.56974	0.59186	0.72225
93	Np	0.10487	0.55314	0.57415	0.70391
94	Pu	0.10228	0.53662	0.55712	0.68637
95	Am	0.09972	0.52084	0.54036	0.66933
96	Cm	0.09674	0.50595	0.52458	0.65276
97	Bk	0.09506	0.49131	0.50928	0.63667
98	Cf	0.09278	0.47713	0.49445	0.62135
99		0.09068	0.46357	0.48009	0.60645
100		0.08861	0.45059	0.46654	0.59225

K AND L X-RAY ABSORPTION EDGES

CRITICAL X-RAY ABSORPTION ENERGIES IN KEV

Compiled by J. M. Cork

For values of energies not listed see The Review of Scientific Instruments, Volume 23 No. 10, page 523, 1952)

Z	Element	K	L_1	L_2	L_3	M_1	N_1	O_1
3	Lithium	(0.0547)						
4	Beryllium	*(0.112)						
5	Boron	*(0.187)						
6	Carbon	(0.284)						
7	Nitrogen	(0.400)						
8	Oxygen	(0.532)						
9	Fluorine	*(0.692)						
10	Neon	*(0.874)						
11	Sodium	*(1.08)						
12	Magnesium	1.30	(0.0628)	0.0502				
13	Aluminum	1.559	(0.0870)	0.0720				
14	Silicon	1.838	*(0.118)	0.0977				
15	Phosphorus	2.142	*(0.153)	0.128				
16	Sulfur	2.469	*(0.193)	0.163	0.162			
17	Chlorine	2.822	*(0.238)	0.202	0.201	(0.0297)		
18	Argon	3.200	*(0.287)	*0.246	*0.244	*(0.0350)		
19	Potassium	3.606	*(0.341)	0.295	0.292	*(0.0408)		
20	Calcium	4.038	*(0.399)	0.350	0.346	*(0.0471)		
21	Scandium	4.496	*(0.462)	0.411	0.407	*(0.0538)		
22	Titanium	4.966	*(0.530)	0.462	0.456	*(0.0605)		
23	Vanadium	5.467	*(0.604)	0.523	0.515	*(0.0681)		
24	Chromium	5.988	*(0.679)	0.584	0.574	*(0.0762)		
25	Manganese	6.542	*(0.762)	0.656	0.644	*(0.0847)		
26	Iron	7.113	(0.849)	0.722	0.709	0.0937		
27	Cobalt	7.713	*0.929	0.798	0.783	0.103		
28	Nickel	8.337	1.02	0.877	0.858	0.111		
29	Copper	8.982	1.10	0.954	0.935	0.121		
30	Zinc	9.662	1.20	1.05	1.02	0.137		
31	Gallium	10.39	*1.30	1.17	1.14	0.182		
32	Germanium	11.10	*1.42	1.24	1.21	0.175		
33	Arsenic	11.87	1.529	1.358	1.32	0.202		
34	Selenium	12.65	*1.66	1.472	1.431	0.227		
35	Bromine	13.48	*1.79	1.599	1.552	0.259		
36	Krypton	*14.32	*1.92	*1.729	*1.674	*0.289		
37	Rubidium	15.197	2.064	1.863	1.803	*0.323		
38	Strontium	16.101	2.212	2.004	1.937	0.356	0.0339	
39	Yttrium	17.053	2.387	2.171	2.096	0.411	0.0638	
40	Zirconium	17.998	2.533	2.308	2.224	0.432	0.0516	
41	Niobium	18.986	2.700	2.467	2.372	0.471	0.0584	
42	Molybdenum	20.003	2.869	2.630	2.525	0.509	0.0692	
43	Technetium	*21.050	*3.045	*2.796	*2 680	*0.546	*0.0718	
44	Ruthenium	22.117	3.227	2.968	2.839	0.584	0.0747	
45	Rhodium	23.210	3.404	3.139	2.995	0.619	0.0733	
46	Palladium	24.356	3.614	3.338	3.181	0.677	0.0937	
47	Silver	25.535	3.828	3.547	3.375	0.743	0.118	
48	Cadmium	26.712	4.019	3.731	3.541	0.775	0.111	
49	Indium	27.929	4.226	3.929	3.732	0.823	0.118	
50	Tin	29.182	4.445	4.139	3.911	0.868	0.119	

Z	Element	K	L_1	L_2	L_3	M_1	N_1	O_1
51	Antimony	30.497	4.708	4.391	4.137	0.950	0.159	
52	Tellurium	31.817	4.953	4.621	4.347	1.01	0.175	
53	Iodine	33.164	5.187	4.855	4.559	1.07	0.187	
54	Xenon	*34.551	*5.448	*5.103	*4.783	*1.14	*0.208	
55	Cesium	35.974	5.706	5.360	5.014	1.22	0.231	0.0258
56	Barium	37.432	5.995	5.629	5.250	1.30	0.255	0.0475
57	Lanthanum	38.923	6.264	5.902	5.490	1.371	0.280	0.0407
58	Cerium	40.43	6.556	6 169	5.728	1.439	0.295	0.0434
59	Praseodymium	41.99	6.837	6.446	5.968	1.514	0.310	0.0421
60	Neodymium	43.57	7.134	6.728	6.215	1.583	0.323	0.0421
61	Promethium	*45.19	*7.431	*7.022	*6.462	*1.654	*0.336	*0.0414
62	Samarium	46.85	7.742	7.316	6.720	1.727	0.349	0.0407
63	Europium	48.51	8.059	7.624	6.984	1.808	0.370	0.0394
64	Gadolinium	50.23	8.383	7.942	7.251	1.888	0.388	0.0570
65	Terbium	52.00	8.713	8.258	7.520	1.970	0.399	0.0475
66	Dysprosium	53.77	9.053	8.587	7.795	2.036	0.422	0.0421
67	Holmium	55.61	9.395	8.918	8.074	2.130	0.433	*0.0480
68	Erbium	57.47	9.754	9.270	8.362	2.221	0.451	0.0543
69	Thulium	*59.38	*10.12	*9.622	*8.655	*2.310	*0.471	*0.0640
70	Ytterbium	61.31	10.49	9.985	8.949	2.402	0.491	0.0747
71	Lutetium	63.32	10.87	10.35	9.248	2.493	0.502	0.0584
72	Hafnium	65.37	11.28	10.75	9.567	2.608	0.544	0.0760
73	Tantalum	67.46	11.68	11.14	9.883	2.713	0.572	0.0747
74	Tungsten (Wolfram)	69.51	12.09	11.54	10.20	2.814	0.588	0.0733
75	Rhenium	*71.67	*12.52	*11.96	*10.53	*2.931	*0.619	*0.0784
76	Osmium	*73.87	*12.97	*12.38	*10.86	*3.045	*0.650	*0.0841
77	Iridium	76.11	13.41	12.82	11.21	3.166	0.683	0.0896
78	Platinum	78.35	13.865	13.26	11.55	3.285	0.710	0.0869
79	Gold	80.67	14.351	13.731	11.92	3.424	0.758	0.106
80	Mercury	*83.08	*14.838	*14.205	*12.278	*3.561	0.799	*0.116
81	Thallium	85.52	15.344	14.695	12.65	3.701	0.842	0.126
82	Lead	87.95	15.861	15.200	13.03	3.852	0.893	0.148
83	Bismuth	90.54	16.386	15.709	13.42	4.000	0.938	0.157
84	Polonium	*93.16	*16.925	*16.233	*13.81	*4.153	*0.986	*0.171
85	Astatine	*95.73	*17.481	*16.777	*14.21	4.314	*1.036	*0.185
86	Radon	*98.45	*18.054	*17.331	*14.61	*4.477	*1.090	*0.200
87	Francium	*101.1	*18.628	*17.893	*15.02	*4.644	*1.143	*0.215
88	Radium	*103.9	*19.228	*18.473	*15.44	*4.814	*1.199	*0.232
89	Actinium	*106.7	*19.829	*19.071	*15.86	*4.991	*1.254	*0.249
90	Thorium	109.8	20.452	19.673	16.278	5.163	1.306	0.266
91	Protactinium	*112.4	*21.096	*20.295	*16.720	*5.355	*1.374	*0.294
92	Uranium	115.0	21.757	20.944	17.163	5.546	1.439	0.322
93	Neptunium	†118.2	†22.411	†21.585	†17.606	†5.736	†1.498	
94	Plutonium	†121.2	†23.117	†22.250	†18.062	†5.934	†1.562	
95	Americium	†124.3	†23.795	†22.935	†18.524	†6.136	†1.631	
96	Curium	†127.2	†24.502	†23.629	†18.992	†6.340	†1.698	
97	Berkelium	†131.3	†25.231	†24.344	†19.466	†6.554	†1.769	
98	Californium	†133.6	†26.010	†25.070	†19.954	†6.770	†1.839	

* Computed by interpolation.
† Computed by extrapolation.

X-RAY SPECTRA

Compiled by J. M. Cork

Emission Wave-Lengths in kX Units

1 kX Unit = 1.00202 Å

At. No.	Element	K Series				L Series								
		α_2	α_1	β_1	β_2	l	α_2	α_1	η	β_1	β_4	β_3	β_2	γ_1
11	Sodium	11.885		11.594										
12	Magnesium	9.869		9.539										
13	Aluminum	8.320		7.965										
14	Silicon	7.111		6.7545										
15	Phosphorus	6.142		5.7921										
16	Sulphur	5.3637	5.3613	5.0211										
17	Chlorine	4.7212	4.7182	4.3942										
18	Argon	......	......	......										
19	Potassium	3.7371	3.7337	3.4468										
20	Calcium	3.3549	3.3517	3.0834										
21	Scandium	3.0284	3.0250	2.7739										
22	Titanium	2.7468	2.7432	2.5090										
23	Vanadium	2.5021	2.4984	2.2797	......	......	24.3		......					
24	Chromium	2.2889	2.2850	2.0806	......	......	21.52		......	21.19				
25	Manganese	2.1015	2.0975	1.9062	......	......	19.39		......	19.04				
26	Iron	1.9360	1.9321	1.7530	......	20.12	17.58		19.65	17.22	15.61			
27	Cobalt	1.7892	1.7853	1.6174	......	18.20	15.94		17.77	15.62	.....			
28	Nickel	1.6584	1.6545	1.4970	1.4856	16.55	14.53		16.17	14.24	13.14			
29	Copper	1.5412	1.5374	1.3894	1.3782	15.19	13.306		14.83	13.03	12.10			
30	Zinc	1.4360	1.4322	1.2926	1.2811	13.95	12.23		13.61	11.95	11.16			
31	Gallium	1.3409	1.3372	1.2052	1.1938	12.89	11.27		12.56	11.01	.....			
32	Germanium	1.2552	1.2513	1.1267	1.1146	11.922	10.415		11.587	10.153				

At. No.	Element	K Series				L Series								
		α_2	α_1	β_1	β_2	l	α_2	α_1	η	β_1	β_4	β_3	β_2	γ_1
33	Arsenic	1.1774	1.1734	1.0551	1.0428	11.048	9.652		10.711	......				
34	Selenium	1.1065	1.1025	.99013	.97791	10.272	8.972		9.939	8.718				
35	Bromine	1.0417	1.0376	.93087	.91853	9.564	8.358		9.235	8.109				
37	Rubidium	.9278	.9236	.82696	.81476	......	......		......	......	6.801	6.769		
38	Strontium	.8776	.8734	.78130	.76921	7.822	6.849		7.506	6.610	6.392	6.358		
39	Yttrium	.8313	.8271	.73919	.72713	......	6.436		7.031	6.204	6.008	5.974	.....	.....
40	Zirconium	.7885	.7843	.70028	.68850	6.899	6.057		6.594	5.824	5.652	5.619	5.574	5.374
41	Niobium	.7489	.7446	.66438	.65280	6.510	5.718	5.712	6.196	5.480	5.330	5.297	5.226	5.024
42	Molybdenum	.71210	.70783	.63098	.61970	......	5.401	5.395	5.836	5.166	5.041	5.005	4.910	......
43	Technetium	.6735	.6778	.6014	.5899	......	......	......						
44	Ruthenium	.64606	.64174	.57131	.56051	5.486	4.843	4.836	.....	4.611	4.513	4.476	4.362	4.173
45	Rhodium	.61637	.61202	.54449	.53396	5.2070	4.5956	4.5878	4.9112	4.3640	4.2802	4.2447	4.1221	3.9357
46	Palladium	.58860	.58422	.51961	.50928	4.9396	4.3666	4.3585	4.6502	4.1373	4.0623	4.0257	3.9007	3.7164
47	Silver	.56265	.55824	.49622	.48607	4.6976	4.1538	4.1456	4.4101	3.9266	3.8611	3.8245	3.6938	3.5149
48	Cadmium	.53831	.53388	.47412	.46437	4.4713	3.9564	3.9478	4.1875	3.7301	3.6743	3.6364	3.5064	3.3280
49	Indium	.51547	.51104	.45360	.44407	4.2593	3.7724	3.7637	3.9761	3.5478	3.4990	3.4619	3.3312	3.1513
50	Tin	.49404	.48961	.43434	.42504	4.0633	3.6011	3.5922	3.7818	3.3779	3.3363	3.2989	3.1679	2.9949
51	Antimony	.47383	.46937	.41622	.40713	3.8803	3.4408	3.4318	3.5996	3.2184	3.1843	3.1451	3.0166	2.8451
52	Tellurium	.45483	.45035	.39917	.39029	3.7101	3.2910	3.2820	......	3.0700	3.0400	3.0013	2.8761	2.7065
53	Iodine	.43692	.43242	.38311	.37466	3.5497	3.1509	3.1417	......	2.9309	2.9059	2.8682	2.7461	2.5775
54	Xenon	......	......	......	......	......	......	......	......	......	......	......	......	......
55	Cesium	.40400	.39946	.35363	.34538	3.2596	2.8956	2.8861	2.9833	2.6778	2.6605	2.6299	2.5064	2.3425
56	Barium	.38886	.38431	.34010	.33207	3.1287	2.7790	2.7696	2.8571	2.5622	2.5498	2.5110	2.3993	2.2366
57	Lanthanum	.37452	.36996	.32730	.31944	3.0000	2.6688	2.6597	2.7340	2.4533	2.4438	2.4053	2.2980	2.1372
58	Cerium	.36094	.35636	.31515	.30752	2.8857	2.5651	2.5560	2.6147	2.3510	2.3442	2.3059	2.2041	2.0443
59	Praseodymium	.34803	.34343	.30363	.29617	2.7781	2.4676	2.4577	2.5070	2.2539	2.2501	2.2124	2.1148	1.9568

X-RAY SPECTRA

Compiled by J. M. Cork

Emission Wave-Lengths in kX Units

1 kX Unit = 1.00202 Å

EMISSION WAVE-LENGTHS IN kX UNITS

At. No.	Element	K Series				L Series								
		α_2	α_1	β_1	β_2	1	α_2	α_1	η	β_1	β_4	β_3	β_2	γ_1
60	Neodymium	.33577	.33115	.29268	.28573	2.6703	2.3756	2.3653	2.4042	2.1622	2.1622	2.1222	2.0314	1.8738
61	Promethium	.32368	.31902	.28200	.27503	.....	2.2879	2.2775	.....	2.754	.....	2.0379	1.9518	1.7953
62	Samarium	.31311	.30844	.27250	.26575	2.4770	2.2057	2.1950	2.2140	1.9936	1.9964	1.9580	1.8781	1.7231
63	Europium	.30267	.29795	.26307	.25645	2.3903	2.1273	2.1163	.....	1.9163	1.9221	1.8827	1.8082	1.6543
64	Gadolinium	.29251	.28778	.25394	.24762	2.3071	2.0526	2.0419	.....	1.8425	1.8493	1.8109	1.7419	1.5886
65	Terbium	.28294	.27820	.24551	.23912	2.2290	1.9823	1.9715	.....	1.7727	1.7814	1.7425	1.6790	1.5266
66	Dysprosium	.27369	.26895	.23710	.23128	2.1540	1.9159	1.9048	1.8922	1.7070	1.7175	1.6786	1.6203	1.4696
67	Holmium	.26499	.26030	.....	.....	2.0821	1.8521	1.8410	1.8220	1.6435	1.6553	1.6160	1.5637	1.4142
68	Erbium	.25669	.25198	.22215	.21671	2.0151	1.7914	1.7804	1.7548	1.5834	1.5964	1.5579	1.5106	1.3623
69	Thulium	.24861	.24387	.21487	.....	1.9511	1.7339	1.7228	1.6923	1.5268	1.5412	1.5023	1.4602	1.3127
70	Ytterbium	.24099	.23625	.20834	.20322	1.8900	1.6789	1.6678	1.6310	1.4725	1.4882	1.4494	1.4128	1.2648
71	Lutetium	.23358	.22882	.20171	.19649	1.8318	1.6264	1.6155	1.5738	1.4207	1.4372	1.3982	1.3672	1.2203
72	Hafnium	.22653	.22173	.19515	.19042	1.7774	1.5770	1.5661	1.5197	1.3711	1.3893	1.3497	1.3235	1.1765
73	Tantalum	.21984	.21505	.18969	.18462	1.7249	1.5298	1.5188	1.4679	1.3242	1.3431	1.3041	1.2819	1.1356
74	Tungsten (Wolfram)	.21338	.20857	.18399	.17905	1.6750	1.4844	1.4734	1.4181	1.2792	1.2988	1.2599	1.2420	1.0963
75	Rhenium	.20718	.20236	.17851	.17370	.....	.....	.....	.....	.....	.....	.....	.....	.....
76	Osmium	.20122	.19639	.17326	.16856	.....	1.3987	1.3886	.....	1.1949	.....	.....	1.1688	1.0229
77	Iridium	.19550	.19065	.16819	.16361	.....	1.3598	1.3485	1.2817	1.1554	1.1771	1.1385	1.1329	.98876
78	Platinum	.18999	.18513	.16333	.15886	1.4964	1.3215	1.3103	1.2403	1.1176	1.1398	1.1016	1.0997	.95599
79	Gold	.18469	.17982	.15865	.15429	1.4569	1.2850	1.2737	1.2003	1.0813	1.1042	1.0655	1.0680	.9246
80	Mercury	.....	.....	.....	.....	1.4184	1.2495	1.2386	1.1616	1.0465	1.0692	1.0305	1.0377	.8946
81	Thallium	.17467	.16979	.13983	.14540	1.3819	1.2163	1.2049	1.1254	1.0130	1.0370	.9985	1.0082	.8657
82	Lead	.16994	.16503	.14567	.14162	1.3474	1.1841	1.1726	1.0900	.9808	1.0056	.9672	.9808	.8380
83	Bismuth	.16537	.16045	.14165	.13769	1.3137	1.1530	1.1415	1.0565	.9500	.9750	.9367	.9532	.8114
90	Thorium	.13754	.13254	.11715	.11381	1.1128	.9658	.9540	.8528	.7636	.7919	.7532	.7919	.6517
92	Uranium	.13070	.12569	.11116	.1084	1.0649	.9206	.9087	.8035	.7185	.7464	.7088	.7531	.6136

EMISSION WAVE-LENGTHS IN THE M SERIES, kX UNITS

Atomic Number	Element	α_2	α_1	β	γ
58	Cerium	.....	14.030	13.755	11.511
59	Praseodymium	.....	.....	.....	10.975
60	Neodymium	.....	12.650	12.375	10.483
62	Samarium	11.475	11.406	11.238	9.580
63	Europium	11.003	10.932	10.723	9.192
64	Gadolinium	10.428	10.394	10.233	8.826
65	Terbium	9.946	9.917	9.772	8.468
66	Dysprosium	9.555	9.524	9.345	8.127
67	Holmium	9.165	9.143	8.947	7.849
68	Erbium	8.794	8.783	8.576	7.530
70	Ytterbium	8.138	8.122	7.893	7.009
71	Lutetium	.....	7.824	7.585	6.748
72	Hafnium	.....	7.524	7.289	6.530
73	Tantalum	.....	7.237	7.008	6.299
74	Tungsten (Wolfram)	.....	6.969	6.743	6.076
75	Rhenium	.....	6.715	6.491	5.875
76	Osmium	.....	6.477	6.254	5.670
77	Iridium	6.262	6.249	6.025	5.490
78	Platinum	6.045	6.034	5.816	5.309
79	Gold	5.842	5.828	5.612	5.135
81	Thallium	5.461	5.450	5.239	4.815
82	Lead	5.288	5.274	5.065	4.665
83	Bismuth	5.119	5.108	4.899	4.522
90	Thorium	4.143	4.130	3.934	3.672
92	Uranium	3.916	3.902	3.708	3.473

GRATING SPACE IN CRYSTALS
kX Units*

Crystal	Spacing		Authority
Calcite	3.02904	(1st order)	Siegbahn
	3.02935	(2nd order)	"
Calcium fluoride	5.455	(Cu radiation)	Gerlach
	5.478	(Ni radiation)	"
Mica	9.845	(1st order)	Davis and
	9.958	(7th order)	Terrill
Potassium ferrocyanide	8.408		Siegbahn
Quartz (0001)	1.797		Ingelstam
Rocksalt (plane parallel to face)	2.81400		Siegbahn
Silicon	5.415	(Cu radiation)	Gerlach
	5.410	(Ni radiation)	"
Topaz (100)	2.316		Ingelstam
Zinc blende	5.90	(Cu radiation)	Gerlach

* 1 kX unit = 1.00202 Å.

X-RAY SPECTRA (Continued)

GRATING SPACE IN CRYSTALS
kX Units*

DIFFRACTION DATA FOR CUBIC ISOMORPHS

From Volume 14, pages 689, 690, and 691 of the Analytical Edition of Industrial and Engineering Chemistry. (With permission.)

X Units	Substance
A 4	
3.56	C (diamond)
5.42	Si
5.62	Ge
6.46	α-Sn
A 1	
3.517	Ni
3.554	α-Co
3.60	Taenite (57.7% Fe, 40.8% Ni, 0.5% P)
3.608	Cu
3.63	γ-Fe (1370°K.)
3.797	Rh
3.831	Ir
3.880	Pd
3.88–4.04	Pd-H
3.912	Pt
4.041	Al
4.070	Au
4.077	Ag
4.30	Co-N
4.40–4.46	Ti-H
4.52	Ne (4°K.)
4.66	Zr-H
4.84	β-Tl
4.939	Pb
5.08	Th
5.14	α-Ce
5.296	β-La
5.43	A (4°K.)
5.56	Ca
5.59	Kr (20°K.)
5.70	Kr (92°K.)
6.05	Sr
6.20	X (88°K.)
A 2	
2.861	α-Fe
2.875	α-Cr
2.90	β-Fe (1070°K.)
2.93	δ-Fe (1700°K.)
3.03	V
3.03–3.41	V-C
3.140	Mo
3.157	W
3.295	Cb
3.30	Ta
3.32	β-Ti (1200°K.)
3.46	Li (∼80°K.)
3.50	Li
3.61	β-Zr (1120°K.)
4.24	Na (∼80°K.)
4.29	Na
5.02	Ba
5.20	K (120°K.)
5.33	K
5.62	Rb (∼80°K.)
6.05	Cs (∼80°K.)
B 1	
4.018	LiF
4.065	LiD
4.08	VO
4.09	LiH
4.12	(Li_2TiO_3)
4.12–4.20	(Li_2TiO_3-MgO)
4.13	VN
4.14	CrN
4.14	VC
4.142	($63Li_2Fe_2O_4 \cdot 37Li_2TiO_3$)
4.173	NiO
4.207	MgO
4.282	MgO (1570°K.)
4.225	TiN
4.235	TiO
4.24	80 TiN-20 TiC
4.27	CoO
4.28	V-N
4.283	FeO (160°K.)
4.290	FeO (299°K.)
4.30	VC (ε-phase)
4.315	TiC
4.40	CbC
4.41	CbN
4.426	MnO (117°K.)
4.436	MnO (299°K.)
4.44	ScN
4.446	TaC
4.458	HfC
4.615	NaF
4.62	ZrN
4.69	CdO
4.69	ZrC
4.80	CaO
B 1 (Continued)	
4.82	(Na_2CeO_3)
4.84	(Na_2PrO_3)
4.88	NaH
4.92	AgF
5.006	CaNH
5.13	SrO
5.14	LiCl
5.14	NdN
5.19	MgS
5.192	MnS (130°K.)
5.210	MnS (299°K.)
5.33	KF
5.45	MgSe
5.45	MnSe
5.45	SrNH
5.49	LiBr
5.52	BaO
5.545	AgCl
5.55–5.76	AgCl-AgBr
5.627	NaCl
5.63	RbF
5.68	CaS
5.69	SnAs
5.70	KH
5.755	AgBr
5.76–5.92	AgBr-AgI
5.83	NdP
5.83	NaCN
5.84	BaNH
5.87	SrS
5.91	CaSe
5.94	PbS
5.95	NaBr
5.957	EuS
5.96	NdAs
6.00	PrAs
6.00	LiI
6.01	CsF
6.04	RbH
6.05	β-NaSH (>360°K.)
6.06	CeAs
6.13	LaAs
6.14	PbSe
6.23	SrSe
6.278	KCl
6.285	SnTe
6.31	NdSb
6.345	CaTe
6.35	PrSb
6.36	BaS
6.38	CsH
6.40	CeSb
6.44	PbTe
6.462	NaI
6.45	PrBi
6.48	LaSb
6.49	CeBi
6.53	KCN
6.53	NH_4Cl (>457°K.)
6.56	RbCl
6.57	LaBi
6.58	KBr
6.59	BaSe
6.60	β-KSH (>440°K.)
6.65	SrTe
6.82	RbCN
6.86	RbBr
6.90	NH_4Br (>411°K.)
6.93	β-RbSH (470°K.)
6.99	BaTe
7.052	KI
7.10	β-CsCl (>730°K.)
7.24	NH_4I (>255°K.)
7.325	RbI
H 0_5	
6.96 ± 0.04	$AgClO_4$ (453 ± 20°K.)
7.16 ± 0.10	$NaClO_4$ (618 ± 35°K.)
7.49 ± 0.02	$KClO_4$ (598 ± 15°K.)
7.65 ± 0.05	$TlClO_4$ (553°K.)
7.65 ± 0.02	NH_4ClO_4 (528 ± 15°K.)
7.68 ± 0.03	$RbClO_4$ (583 ± 10°K.)
7.97 ± 0.01	$CsClO_4$ (513 ± 10°K.)
B 3	
4.255	CuF
4.36	CSi IV
4.855	BeS
5.10	BeSe
5.304	$(Cu, Fe, Mo, Sn)_4(S, As, Te)_{3-4}$, cousite
5.41	CuCl
5.425	β-ZnS
5.43	AlP
5.44	GaP
5.58	BeTe
5.60	MnS (red)

DIFFRACTION DATA FOR CUBIC ISOMORPHS
(Continued)

X Units	Substance
B 3 (Continued)	
5.63	AlAs
5.635	GaAs
5.655	ZnSe
5.68	CuBr
5.82	β-CdS
5.84	HgS
5.86	InP
6.04	CdSe
6.04	InAs
6.05	CuI
6.07	HgSe
6.08	ZnTe
6.103	α-Cu_2HgI_4
6.12	AlSb
6.12	GaSb
6.13	SnSb
6.383	α-Ag_2HgI_4
6.40	HgTe
6.43	CdTe
6.45	InSb
6.48	AgI
B 32	
6.195	LiGa
6.209	LiZn
6.36	LiAl
6.687	LiCd
6.786	LiIn
7.297	NaIn
7.373	($CeMg_3$)
7.373	($PrMg_3$)
7.473	NaTl
B 20	
4.437	NiSi
4.438	FeSi
4.438	CoSi
4.548	MnSi
4.620	CrSi
B 2	
2.603	NiBe
2.606	CoBe
2.69	CuBe
2.813	PdBe
2.82	AlNi
2.945	CuZn
2.989	CuPd
3.146	AuZn
3.156	AgZn
3.168	AgLi
3.259	AuMg
3.275	AgMg
3.287	HgLi
3.325	AgCd
3.34	AuCd (670°K.)
3.424	LiTl
3.442	HgMg
3.628	MgTl
3.67	PrZn
3.70	CeZn
3.73	AlNd
3.74	α-RbCl (83°K.)
3.75	LaZn
3.82	TlCN
3.82	PrCd
3.84	TlSb
3.835	TlCl
3.847	CaTl
3.86	NH_4Cl (<457°K.)
3.86	CeCd
3.88	MgPr
3.90	LaCd
3.97	TlBr
3.98	TlBi
4.024	SrTl
4.05	NH_4Br (<411°K.)
4.112	CsCl
4.20	TlI
4.20	CsCl (<720°K.)
4.25	CsCN
4.287	CsBr
4.29	CsSH
4.37	NH_4I (290°K.)
4.56	CsI
D 2_1	
4.07	YB_6
4.07	ErB_6
4.10	NdB_6
4.12	GdB_6
4.12	PrB_6
4.13	CeB_6
4.13	YbB_6
4.14	CaB_6
4.15	LaB_6
4.15	ThB_6
4.19	SrB_6
4.33	BaB_6

X Units	Substance
C 1	
4.33	Be_2C
4.619	Li_2O
5.06	($3ZrO_2 \cdot MgO$)
5.07	ZrO_2
5.08	($95ZrO_2 \cdot 5CeO_2$)
5.13	($95HfO_2 \cdot 5CeO_2$)
5.38	PrO_2
5.40	CeO_2
5.40	CdF_2
5.406	CuF_2
5.45	CaF_2
5.47	UO_2
5.526	($66CaF_2 \cdot 33YF_3$)
5.53	($91CaF_2 \cdot 9ThF_4$)
5.54	HgF_2
5.55	Na_2O
5.58	ThO_2
5.59	Cu_2S
5.704	Li_2S
5.749	Cu_2Se
5.782	SrF_2
5.796	EuF_2
5.838	($66SrF_2 \cdot 33LaF_3$)
5.91	$PtAl_2$
5.91	$PtGa_2$
5.935	β-PbF_2 (520°K.)
5.99	Al_2Au
6.005	Li_2Se
6.06	$AuGa_2$
6.19	BaF_2
6.34	Mg_2Si
6.35	$PtIn_2$
6.368	RaF_2
6.379	Mg_2Ge
6.436	K_2O
6.50	Li_2Te
6.50	$AuIn_2$
6.526	Na_2S
6.763	Mg_2Sn
6.809	Na_2Se
6.81	Mg_2Pb
6.98	$SrCl_2$
7.314	Na_2Te
7.38	K_2S
7.65	RbS_2
7.676	K_2Se
8.152	K_2Te
C 15	
5.94	Be_2Cu
6.287	Be_2Ag
6.435	Be_2Ti
6.96	MgNiZn
7.03	Cu_2Mg
7.61	W_2Zr
7.79	Au_2Na
7.91	Au_2Pb
7.94	Au_2Bi
8.02	Al_2Ca
8.04	Al_2Ce
8.16	Al_2La
9.50	Bi_2K
C 2	
5.41	FeS_2
5.42	(Fe, Ni)S_2 (6.5% Ni)
5.57	RbS_2
5.57	RuS_2
5.57	Bravoite (53.8% NiS_2, 39.1% FeS_2, 7.1% CoS_2)
5.62	OsS_2
5.64	CoS_2
5.65	(Cu, Ni, Co, Fe)(S, Se)$_2$
5.68	PtP_2
5.74	NiS_2
5.85	$CoSe_2$
5.92	$RuSe_2$
5.93	$OsSe_2$
5.94	$PtAs_2$
5.97	$PdAs_2$
6.02	$NiSe_2$
6.096	MnS_2
6.36	$RuTe_2$
6.37	$OsTe_2$
6.43	$PtSb_2$
6.44	$PdSb_2$
6.64	$AuSb_2$
6.94	$MnTe_2$
C 3	
4.25	Cu_2O
4.73	Ag_2O
F 1	
5.55	CoAsS
5.68	NiAsS
5.90	NiSbS
	(Ni, Fe)AsS, plessite
	Ni(As, Sb)S, corynite
	Ni(Sb, Bi)S, kallilite
	(Co, Ni)SbS, willyamite

DIFFRACTION DATA FOR CUBIC ISOMORPHS (Continued)

X Units	Substance
D 5_3	
8.13	Be_3N_2
9.37	$(Mn, Fe)_2O_3$
9.42	Mn_2O_3
9.74	Zn_3N_2
9.79	Sc_2O_3
9.94	Mg_3N_2
10.12	In_2O_3
10.15	Be_3P_2
10.37	Lu_2O_3
10.39	Yb_2O_3
10.52	Tm_2O_3
10.54	Er_2O_3
10.57	Tl_2O_3
10.58	Ho_2O_3
10.60	Y_2O_3
10.63	Dy_2O_3
10.70	Tb_2O_3
10.79	Gd_2O_3
10.79	Cd_3N_2
10.84	Eu_2O_3
10.85	Sm_2O_3
11.05	Nd_2O_3
11.40	α-Ca_3N_2
12.02	Mg_3P_2
12.33	Mg_3As_2
D 6_1	
11.05	As_4O_6
11.14	Sb_4O_6
D 1_1	
10.32	$ZrCl_4$
11.25	$TiBr_4$
(11.34)	(CBr_4) (>320°K.)
(11.62)	(CI_4)
11.89	GeI_4
11.99	SiI_4
12.00	TiI_4
12.23	SnI_4
E 2_1	
3.67	$YAlO_3$
3.75	$CdTiO_3$
3.78	$LaAlO_3$
3.80	$CaTiO_3$
3.83	$NaWO_3$
3.85	$(Na, Ce, Ca)(Ti, Cb)O_3$ Loparite
3.88	$NaTaO_3$
3.89	$LaGaO_3$
3.89	$NaCbO_3$
3.91	$SrTiO_3$
3.92	$CaSnO_3$
3.97	$BaTiO_3$
3.98	$KTaO_3$
3.99	$CaZrO_3$
4.00	$KMgF_3$
4.005	$KNiF_3$
4.01	$KCbO_3$
4.03	$SrSnO_3$
4.05	$KZnF_3$
4.07	$KCoF_3$
4.07	$SrHfO_3$
4.09	$SrZrO_3$
4.18	$BaZrO_3$
4.35	$BaPrO_3$
4.38	$BaCeO_3$
4.46	KIO_3
4.48	$BaThO_3$
4.5	NH_4IO_3
4.52	$RbIO_3$
4.66	$CsIO_3$
5.12	$MgZrO_3$
5.20	$CsCdCl_3$
5.33	$CsCdBr_3$
5.44	$CsHgCl_3$
5.77	$CsHgBr_3$
G 0_3	
6.57	$NaClO_3$
6.71	$NaBrO_3$
G 2_1	
7.60	$Ca(NO_3)_2$
7.81	$Sr(NO_3)_2$
7.84	$Pb(NO_3)_2$
8.11	$Ba(NO_3)_2$
H 1_1	
8.045	$NiAl_2O_4$
8.07	$CuAl_2O_4$
8.07	$CoCo_2O_4$
8.07	$MgAl_2O_4$
8.08	$CoAl_2O_4$
8.08	$ZnAl_2O_4$
8.10	$FeAl_2O_4$
8.11	$(Ni, Co)(Co, Ni)_2O_4$
8.11	$(Zn, Co)Co_2O_4$
8.11	$MgCo_2O_4$
8.27	$MnAl_2O_4$
8.27	$(Mn, Co)(Co, Mn)_2O_4$
H 1_1	
8.28	$MgGa_2O_4$
8.30	$NiCr_2O_4$
8.30	$MgCr_2O_4$
8.31	$ZnCr_2O_4$
8.32	$CoCr_2O_4$
8.32	$ZnGa_2O_4$
8.35	$NiFe_2O_4$
8.35	$Cu_2Cr_2O_4$
8.35	$FeCr_2O_4$
8.36	$MgFe_2O_4$
8.38	$CoFe_2O_4$
8.38	$NiMn_2O_4$
8.40	$ZnFe_2O_4$
8.40	$FeFe_2O_4$
8.42	$(Mn, Mg)Fe_2O_4$
8.42	$TiCo_2O_4$
8.43	$MnCr_2O_4$
8.43	$TiMg_2O_4$
8.43	$TiZn_2O_4$
8.44	$CuFe_2O_4$
8.47	FeV_2O_4
8.49	$MnCr_2O_4$
8.50	$TiFe_2O_4$
8.54	$MnFe_2O_4$
8.58	$CdCr_2O_4$
8.58	$SnMg_2O_4$
8.61	$SnCo_2O_4$
8.63	$SnZn_2O_4$
8.67	$CdFe_2O_4$
8.67	$TiMn_2O_4$
8.81	$MgIn_2O_4$
9.26	Ag_2MoO_4
9.4	$CoCo_2S_4$
9.45	$(Co, Ni)_3S_4$
9.46	$CuCo_2O_4$
9.5	NiN_2S_4
9.92	$ZnCr_2S_4$
10.05	$MnCr_2S_4$
10.19	$CdCr_2S_4$
12.54	$K_2Zn(Cn)_4$
12.76	$K_2Hg(CN)_4$
12.84	$K_2Cd(CN)_4$
H 5_8	
10.08	$2Na_2SO_4 \cdot NaCl \cdot NaF$
H 4_{13}	
12.11	$KCr(SO_4)_2 \cdot 12H_2O$
12.12	$KAl(SO_4)_2 \cdot 12H_2O$
12.15	$NH_4Al(SO_4)_2 \cdot 12H_2O$
12.15	$NH_4Fe(SO_4)_2 \cdot 12H_2O$
12.20	$RbAl(SO_4)_2 \cdot 12H_2O$
12.21	$TlAl(SO_4)_2 \cdot 12H_2O$
12.31	$CsAl(SO_4)_2 \cdot 12H_2O$
12.44	$NH_3 \cdot CH_3Al(SO_4)_2 \cdot 12H_2O$ (β-alum)
Langbeinite	
9.93	$K_2Mg(SO_4)_3$
10.2	$K_2(Ca, Mg)(SO_3)_3$
H 2_1	
6.00	Ag_3PO_4
6.120	Ag_3AsO_4 (90°K.)
6.130	Ag_3AsO_4 (380°K.)
H 2_4	
5.37	Cu_3VS_4
J 1_1	
8.17	K_2SiF_6
8.35	$(NH_4)_2SiF_6$
8.38	$Rb_2CrF_5H_2O$
8.41	$Tl_2CrF_5H_2O$
8.42	$(NH_4)_2VF_5H_2O$
8.42	$Rb_2VF_5H_2O$
8.45	$Tl_2VF_5H_2O$
8.45	Rb_2SiF_6
8.58	Tl_2SiF_6
8.87	Cs_2SiF_6
8.99	Cs_2GeF_6
9.73	K_2PtCl_6
9.73	K_2OsCl_6
9.76	Tl_2PtCl_6
9.84	$(NH_4)_2PtCl_6$
9.86	K_2ReCl_6
9.88	Rb_2PtCl_6
9.92	Rb_2TiCl_6
9.94	$(NH_4)_2SeCl_6$
9.97	K_2SnCl_6
9.97	Tl_2SnCl_6
9.98	Rb_2SeCl_6
10.02	Rb_2PdBr_6
10.04	$(NH_4)_2SnCl_6$
10.08	$Ni(NH_3)_6Cl_2$
10.10	Rb_2SnCl_6
10.10	$Co(NH_3)_6Cl_2$
10.11	Tl_2TeCl_6
10.14	$(NH_4)_2PbCl_6$
10.14	K_2TeCl_6
10.15	$Fe(NH_3)_6Cl_2$
10.16	$Mg(NH_3)_6Cl_2$
10.17	Cs_2PtCl_6

DIFFRACTION DATA FOR CUBIC ISOMORPHS (Continued)

X Units	Substance
J 1₁	
10.18	Rb_2ZrCl_6
10.18	$(NH_4)_2TeCl_6$
10.20	$Mn(NH_3)_6Cl_2$
10.20	Rb_2PbCl_6
10.22	Cs_2TiCl_6
10.23	Rb_2TeCl_6
10.25	$Zn(NH_3)_6(ClO_4)_2$
10.26	Cs_2SeCl_6
10.30	K_2OSBr_6
10.35	Cs_2SnCl_6
10.36	K_2SeBr_6
10.36	K_2PtBr_6
10.39	$Co(NH_3)_6Br_2$
10.4	$Ni(NH_3)_6Br_2$
10.41	Cs_2ZrCl_6
10.42	Cs_2PbCl_6
10.45	Cs_2TeCl_6
10.45	$Co(NH_3)_5H_2OSO_4Br$
10.46	$(NH_4)_2SeBr_6$
10.46	$Zn(NH_3)_6Br_2$
10.47	$Fe(NH_3)_6Br_2$
10.47	$Mg(NH_3)_6Br_2$
10.48	K_2SnBr_6
10.51	$Co(NH_3)_6SO_4Br$
10.52	$Mn(NH_3)_6Br_2$
10.54	$Sr_2Ni(NO_2)_6$
10.55	$Pb_2Ni(NO_2)_6$
10.57	$(NH_4)_2SnBr_6$
10.58	Rb_2SnBr_6
10.62	$Co(NH_3)_5H_2OSO_4I$
10.63	$Co(NH_3)_6SeO_4Br$
10.67	$Ba_2Ni(NO_2)_6$
10.71	$Ca(NH_3)_6Br_2$
10.71	$Co(NH_3)_6SO_4I$
10.77	Cs_2SnBr_6
10.79	$Co(NH_3)_6SeO_4I$
10.9	$Ni(NH_3)_6I_2$
10.91	$Co(NH_3)_6I_2$
10.96	$Zn(NH_3)_6I_2$
10.97	$Fe(NH_3)_6I_2$
10.98	$Mg(NH_3)_6I_2$
11.04	$Mn(NH_3)_6I_2$
11.04	$Cd(NH_3)_6I_2$
11.24	$Ca(NH_3)_6I_2$
11.27	$Ni(NH_3)_6(BF_4)_2$
11.3	$Co(NH_3)_6(BF_4)_2$
(11.3)	$Zn(NH_3)_6(ClO_4)_2$
11.34	$Mg(NH_3)_6(BF_4)_2$
11.34	$Fe(NH_3)_6(BF_4)_2$
11.37	$Mn(NH_3)_6(BF_4)_2$
11.38	$Cd(NH_3)_6(BF_4)_2$
11.41	$Ni(NH_3)_6(ClO_4)_2$
11.43	$Co(NH_3)_6(ClO_4)_2$
11.46	$Ni(NH_3)_6(SO_3F)_2$
11.49	$Co(NH_3)_6(SO_3F)_2$
11.52	$Fe(NH_3)_6(ClO_4)_2$
11.53	$Mg(NH_3)_6(ClO_4)_2$
11.54	$Cd(NH_3)_6Br_2$
11.54	$Fe(NH_3)_6(SO_3F)_2$
11.58	$Mn(NH_3)_6(ClO_4)_2$
11.59	$Cd(NH_3)_6(ClO_4)_2$
11.59	$Mn(NH_3)_6(SO_3F)_2$
11.62	$Cd(NH_3)_6(SO_3F)_2$
11.91	$Ni(NH_3)_6(PF_6)_2$
11.94	$Co(NH_3)_6(PF_6)_2$
12.03	$Ni(NH_2{\cdot}CH_3)_6I_2$
12.05	$Co(NH_2{\cdot}CH_3)_6I_2$
12.19	$[NH(CH_3)_3]_2SnCl_6$
12.41	$[S(CH_3)_3]_2SnCl_6$
12.65	$[N(CH_3)_4]_2PtCl_6$
12.80	$[S(CH_3)_2C_2H_5]_2SnCl_6$
12.87	$[N(CH_3)_4]_2SnCl_6$
13.17	$[N(CH_3)_3C_2H_5]_2SnCl_6$
13.51	$[N(CH_3)(C_2H_5)_3]_2SnCl_6$
13.93	$[P(CH_3)(C_2H_5)_3]_2SnCl_6$
J 2₁ and Related Structures	
8.88	Li_3FeF_6
8.90	$(NH_4)_3AlF_6$
9.01	$(NH_4)_3CrF_6$
9.04	$(NH_4)_3VF_6$
9.10	$(NH_4)_3FeF_6$
9.10	$(NH_4)MoO_3F_3$
9.26	Na_3FeF_6
9.93	K_3FeF_6
9.96	$CuLi_2Fe(CN)_6$
10.0	$CuR_2Fe(CN)_6$ R = Na, K, Rb, NH_4, Tl
10.15	$K_2CdFe(NO_2)_6$
10.17	$K_2CaCo(NO_2)_6$
10.19	$K_2CaFe(NO_2)_6$
10.2	$Fe^{III}RFe^{II}(CN)_6$ R = Na, K, Rb, NH_4
10.22	$K_2HgFe(NO_2)_6$
10.23	$K_2SrCo(NO_2)_6$
10.25	$(NH_4)_2CaFe(NO_2)_6$
10.25	$NaTl_2Co(NO_2)_6$
10.28	$K_2CdNi(NO_2)_6$
10.28	$(NH_4)_2CdFe(NO_2)_6$
10.29	$K_2HgNi(NO_2)_6$
10.30	$K_2SrFe(NO_2)_6$
10.30	$Tl_2CaFe(NO_2)_6$
10.31	$K_2PbFe(NO_2)_6$
10.32	$K_2CaNi(NO_2)_6$
10.34	$(NH_4)_2SrFe(NO_2)_6$
10.37	$(NH_4)_2PbFe(NO_2)_6$
10.37	$Tl_2CdNi(NO_2)_6$
10.39	$Tl_2PbFe(NO_2)_6$
10.39	$NaRb_2Co(NO_2)_6$
10.40	$Tl_2SrFe(NO_2)_6$
10.4	$K_2PbCo(NO_2)_6$
10.41	$(NH_4)_2CdNi(NO_2)_6$
10.42	$Tl_2HgNi(NO_2)_6$
10.43	$K_2BaFe(NO_2)_6$
10.45	$K_2BaCo(NO_2)_6$
10.45	$K_3Co(NO_2)_6$
10.46	$(NH_4)_2HgNi(NO_2)_6$
10.47	$Rb_2HgNi(NO_2)_6$
10.49	$K_2SrNi(NO_2)_6$
10.49	$K_4Ni(NO_2)_6$
10.50	$(NH_4)_2BaFe(NO_2)_6$
10.54	$K_2LiBi(NO_2)_6$
10.55	$K_2PbNi(NO_2)_6$
10.55	$Tl_2BaFe(NO_2)_6$
10.58	$Rb_2CdNi, Cd(NO_2)_6$
10.58	$K_3Ir(NO_2)_6$
10.59	$Rb_2LiBi(NO_2)_6$
10.6	$K_2PbCu(NO_2)_6$
10.63	$K_2Rh(NO_2)_6$
10.63	$(NH_4)_2LiBi(NO_2)_6$
10.64	$Tl_2LiBi(NO_2)_6$
10.67	$K_2BaNi(NO_2)_6$
10.70	$NaCs_2Co(NO_2)_6$
10.70	$Ba_3[Rh(NO_2)_6]_2$
10.72	$Tl_3Co(NO_2)_6$
10.73	$Rb_3Co(NO_2)_6$
10.73	$(NH_4)_3Ir(NO_2)_6$
10.73	$Tl_3Ir(NO_2)_6$
10.77	$Rb_3Ir(NO_2)_6$
10.8	$(NH_4)_2Co(NO_2)_6$
10.81	$Cs_2Cd[Ni, Cd(NO_2)_6]$
10.82	$Co(NH_3)_5H_2OI_3$
10.83	$Rb_3Rh(NO_2)_6$
10.88	$K_2NaBi(NO_2)_6$
10.89	$Co(NH_3)_6I_3$
10.91	$Tl_3Rh(NO_2)_6$
10.91	$(NH_4)_3Rh(NO_2)_6$
10.94	$Cs_2LiBi(NO_2)_6$
10.95	$K_2AgBi(NO_2)_6$
10.98	$Rb_2NaBi(NO_2)_6$
10.99	$(NH_4)_2NaBi(NO_2)_6$
11.01	$Tl_2NaBi(NO_2)_6$
11.05	$Rb_2AgBi(NO_2)_6$
11.06	$Tl_2AgBi(NO_2)_6$
11.10	$(NH_4)_2AgBi(NO_2)_6$
11.15	$Cs_2NaBi(NO_2)_6$
11.15	$Cs_3Co(NO_2)_6$
11.17	$Cs_3Ir(NO_2)_6$
11.19	$Cs_3Bi(NO_2)_6$
11.19	$Cs_2AgBi(NO_2)_6$
11.21	$Co(NH_3)_6(BF_4)_3$
11.30	$Cs_3Rh(NO_2)_6$
11.32	$[Co(NH_3)_5{\cdot}H_2O](ClO_4)_3$
11.39	$Co(NH_3)_6(ClO_4)_3$
11.67	$Co(NH_3)_6(PF_6)$
K 6₁	
7.46	SiP_2O_7
7.80	TiP_2O_7
7.98	SnP_2O_7
8.18	HfP_2O_7
8.20	ZrP_2O_7
8.61	UP_2O_7
S 1₄	
11.51	$Al_2(Mg, Fe)_3(SiO_4)_3$, pyrope
11.51	$Al_2Fe_3(SiO_4)_3$, almandite
11.60	$Al_2Mn_3(SiO_4)_3$, spessartite
11.87	$Al_2Ca_3(SiO_4)_3$, grossularite
11.89	$(Al, Fe)_2Ca_3(SiO_4)_3$, hessonite
11.95	$Cr_2Ca_3(SiO_4)_3$, uvarovite
12.03	$Fe_2Ca_3(SiO_4)_3$, andradite
12.10	$(Na, Li)_3AlF_6$, cryolithionite
12.35–12.46	$(Mg, Mn)_2(Ca, Na)_3AsO_4)_3$, berzelite
S 6₁	
13.68	$NaAlSi_2O_6H_2O$
S 0₈	
13.82	$Al_{13}Si_5O_{20}(OH, F)_{18}Cl$, zunyite
S 6₂	
8.87	$Na_4(AlSiO_4)_3Cl$, sodalite
Tetrahedrite	
10.19	$(Cu, Fe)_{12}As_4S_{13}$, binnite
10.2–10.6	$(Cu, Ag)_{10}(Zn, Fe)_2(Sb, As)_4S_{13}$

LIGHT

PHOTOMETRIC QUANTITIES, UNITS AND STANDARDS

Photometric quantities and units are also given in the section Quantities and Units under the sub-division Light.

Candle (or International Candle). The candle is the unit of luminous intensity. It is a specified fraction of the average horizontal candlepower of a group of 45 carbon-filament lamps preserved at the Bureau of Standards.

Candle (new unit). $\frac{1}{60}$ of the intensity of one square centimeter of a blackbody radiator at the temperature of solidification of platinum (2,046°K).

Foot-candle. Foot-candle = 1 lumen incident per ft^2 = 1.076 milliphots = 10.76 lux.

Lumen. The lumen is the unit of luminous flux. It is equal to the flux through a unit solid angle (steradian) from a uniform point source of one candle, or to the flux on a unit surface all points of which are at unit distance from a uniform point source of one candle.

Illumination. Illumination is the density of the luminous flux on a surface. It is the quotient of the flux by the area of the surface when the latter is uniformly illuminated.

Least Mechanical Equivalent of Light. One lumen at the wavelength of maximum visibility (0.556μ) equals 0.00161 watts (= 0.000385 gram calories per sec.); one watt at the same wavelength equals 621 lumens.

Relative Visibility. The relative visibility factor for a particular wavelength is the ratio of the visibility factor for that wavelength to the maximum visibility factor.

Values of the relative visibility are given as a part of the specification of the standard observer under Colorimetry.

Efficiency of a Source of Light. The efficiency of a source is the ratio of the total luminous flux to the total power consumed. In the case of an electric lamp it is expressed in lumens per watt.

Spherical Candlepower. The spherical candlepower of a lamp is the average candlepower of the lamp in all directions in space. It is equal to the total luminous flux of the lamp in lumens divided by 4π

Lambert is the unit of brightness equal to $1/\pi$ candle per square centimeter.

Foot-lambert is the unit of photometric brightness (luminance) equal to $1/\pi$ candle per square foot.

FLAME STANDARDS

Value of Various Former Standards in International Candles

Standard Pentane Lamp, burning pentane	10.0 candles
Standard Hefner Lamp, burning amyl acetate	0.9 "
Standard Carcel Lamp, burning colza oil	9.6 "

The *Carcel unit* is the horizontal intensity of the carcel lamp, burning 42 grams of colza oil per hour. For a consumption between 38 and 46 grams per hour the intensity may be considered proportional to the consumption.

The *Hefner unit* is the horizontal intensity of the Hefner lamp burning amyl acetate, with a flame 4 cm. high. If the flame is l mm. high, the intensity $I = 1 + 0.027(l - 40)$.

EFFICIENCIES OF ILLUMINANTS

Compiled by J. M. Smith and C. E. Weitz

The rating listed is the commercial rating of the lamp. The absolute efficiency is the equivalent power in light flux (0.556μ) per watt input. Efficiency is given in lumens per watt input.

Lamp	Rating, or Specification	Eff.	Ab. Eff.
Acetylene	1.0 liters per hour	0.67	0.0010
Arc, Electric			
Carbon, Enclosed d.c.	6.6 amps. opal globe & reflector	5.9	0.0087
Carbon, Open d.c.	9.6 amps. clear globe	11.8	0.0173
High Intensity	150 amps. bare arc	18.5	0.0272
Magnetite d.c.	6.6 amps.	21.6	0.0318

EFFICIENCIES OF ILLUMINANTS (Continued)

Lamp	Rating, or Specification	Eff.	Ab. Eff.
Gas burner, open flame	Bray high pressure	0.22	0.00032
Gas mantle, incandescent			
High pressure	0.578 lumens per BTU per hour	2.0	0.0030
Low pressure	0.350 lumens per BTU per hour	1.2	0.0018
Incandescent electric			
Carbon filament			
First commercial	....	1.6	0.0023
Squirted cellulose	....	3.3	0.0048
Metalized	....	4.0	0.0059
Tungsten filaments			
Vacuum	25 watt 120 volt (1,000 hrs. life)	10.6	0.0156
Gas filled	40 watt 120 volt (1,000 hrs. life)	11.6	0.0171
Gas filled	60 watt 120 volt (1,000 hrs. life)	13.9	0.0204
Gas filled	100 watt 120 volt (750 hrs. life)	16.3	0.0239
Gas filled	1000 watt 120 volt (1,000 hrs. life)	21.6	0.0318
Gas filled	5000 watt 120 volt (75 hrs. life)	32.8	0.0482
Fluorescent lamps			
General line	20 watt standard warm white (T12)	50.0	0.0735
General line	40 watt " " " "	64.0	0.0940
General line	90 watt " " " (T17)	58.0	0.0850
Slimline	96T8 (120 ma.) Std. warm white	76.0	0.1115
Slimline	96T12 (425 ma.) " " "	69.0	0.1015
General line	40 W Daylight (T12)	54.0	0.0795
General line	40 W Green (T12)	84.0	0.1235
General line	40 W Blue (T12)	33.0	0.0485
General line	40 W Red (T12)	3.6	0.0053
Mercury lamps			
	400 W (E1)	50.0	0.0735
	1000 W (A6)	65.0	0.0955
Sodium	10,000 lumen	55.0	0.0808

APPROXIMATE BRIGHTNESS OF VARIOUS LIGHT SOURCES

Compiled by J. M. Smith and C. E. Weitz
Brightness of source is given in lamberts

Source		Lamberts*
Natural Sources		
Clear sky	Average brightness	2.5
Sun (as observed from earth's surface)	At meridian	519,000
Sun (" " " " ")	Near horizon	1885
Moon (" " " " ")	Bright spot	0.8
Combustion Sources		
Candle flame (sperm)	Bright spot	3.1
Kerosene flame (flat wick)	Bright spot	3.8
Illuminating—gas flame	Fish-tail burner	1.3
Welsbach mantle	Bright spot	20.0
Acetylene flame	Mees burner	34.0
Incandescent Electric lamps		
Carbon filament		165
Metalized carbon filament (Gem)		300
Tungsten filament	Vacuum lamp 10 lumens per watt	650
Tungsten filament	Gas filled lamp 20 lumens per watt	3800
Tungsten filament	750 watt projector lamp 26 lumens per watt	7500
Fluorescent Lamps		
20 watt T12 Standard warm white		1.67
40 watt T12 " " "		2.10
96T12 " " "		2.052

APPROXIMATE BRIGHTNESS OF VARIOUS LIGHT SOURCES (Continued)

Source		Lamberts*
Electric Arc Lamps		
Plain carbon arc	Positive crater 7 mm. non-rotating	55,000
High intensity carbon arc	Positive crater 8 mm. non-rotating	125,000
High intensity carbon arc	Positive crater 13.6 mm. non-rotating	220,000
High intensity carbon arc	Positive crater	314,000
Mercury Lamps		
Low pressure mercury arc	50″ AC rectified tube	6.6
400 W (H1)		440
1000 W (A6)	Water cooled	94,000
Sodium Lamps	10,000 lumens	18

* To convert lamberts to footlamberts multiply by 929.
To convert lamberts to candles/cm² divide by π.

WAVE LENGTHS OF VARIOUS RADIATIONS

	Ångstroms
Cosmic Rays	0.0005
Gamma Rays	0.005–1.40
X-Rays	0.1–100
Ultra Violet, below	4000
Limit of suns U.V. at earth's surface	2920
Visible Spectrum	4000–7000
Violet, representative, 4100, limits	4000–4240
Blue, representative, 4700, limits	4240–4912
Green, representative, 5200, limits	4912–5750
Maximum visibility	5560
Yellow, representative, 5800, limits	5750–5850
Orange, representative, 6000, limits	5850–6470
Red, representative, 6500, limits	6470–7000
Infra Red, greater than	7000
Hertzian Waves, beyond	2.20×10^6

BRIGHTNESS OF TUNGSTEN

Characteristics of Straight Tungsten Wire in a Vacuum Forsythe and Worthing, 1924).

Temperature °K			Brightness Candles/cm²	$\frac{B}{dB}\frac{dt}{T}$
Absolute	Brightness	Color		
1000	966	1006	0.00012	22.0
1200	1149	1210	0.006	20.0
1400	1330	1414	0.11	17.2
1600	1509	1619	0.92	15.2
1800	1684	1825	5.05	13.7
2000	1857	2033	20.0	12.3
2200	2026	2242	61.3	11.2
2400	2192	2452	157.0	10.3
2600	2356	2663	347.0	9.6
2800	2516	2878	694.0	8.9
3000	2673	3094	1257.0	8.3
3200	2827	3311	2110.0	7.8
3400	2978	3533	3370.0	7.6
3655*	3165	3817	5740.0	7.3

* Melting-point of tungsten.

WAVE LENGTHS OF THE FRAUNHOFER LINES

SUN'S SPECTRUM

At 15° C and 76 cm pressure. Wave length in Ångström units (Fabry and Buisson system).

Line	Due to	Wave length	Line	Due to	Wave length
U	Fe	2947.9	*h*	H	4101.750
t	Fe	2994.4	*g*	Ca	4226.742
T	Fe	3021.067	*G*	Fe	4307.914
s	Fe	3047.623		Ca	4307.749
S_1	Fe	3100.683	*G'*	H	4340.477
S_2	Fe	3100.326	*F*	H	4861.344
	Fe	3099.943	b_4	Fe	5167.510
R	Ca	3181.277		Mg	5167.330
	Ca	3179.343	b_2	Mg	5172.700
Q	Fe	3286.773	b_1	Mg	5183.621
P	Ti	3361.194	E_2	Fe	5269.557
O	Fe	3441.020	D_2	Na	5889.977
N	Fe	3581.210	D_1	Na	5895.944
M	Fe	3727.636	*C*	H	6562.816
L	Fe	3820.438	*B*	O	6869.955
K	Ca	3933.684	*A*	O	7621
H	Ca	3968.494		O	7594
			Z	....	8228.5
			Y	....	8990.0

WAVE LENGTHS FOR SPECTROSCOPE CALIBRATION

Source	Wave Length	Source	Wave Length
Potassium flame	0.7699μ	*E*, solar	0.5270μ
Potassium flame	0.7665	b_1, solar or magnesium flame	0.5184
Mercury I arc	0.6907	b_2, solar or magnesium flame	0.5173
B, solar	0.6869	Mercury I arc	0.4960
Lithium flame	0.6708	Mercury I arc	0.4916
C, solar or hydrogen tube	0.6563	*F*, solar or hydrogen tube	0.4861
Mercury I arc	0.6234	Strontium flame	0.4608
D_1, solar or sodium flame	0.5896	Mercury I arc	0.4358
D_2, solar or sodium flame	0.5890	*G'*, solar or hydrogen tube	0.4340
Mercury I arc	0.5791	Mercury I arc	0.4047
Mercury I arc	0.5770	H_1, solar	0.3969
Mercury I arc	0.5461	*K*, solar	0.3934
Thallium flame	0.5351		

VELOCITY OF LIGHT (IN VACUO)

$(2.99776 \pm .00004) \times 10^{10}$ cm/sec. (Birge, 1941)
$(2.99793 \pm .00002) \times 10^{10}$ cm/sec. (Aslakson, 1949)
$(2.99790 \pm .00001) \times 10^{10}$ cm/sec. (DuMond and Cohen, 1951)
299,776 km/sec. (Birge)
186,272 miles/sec. (Birge)
$(2.997928 \pm .000004) \times 10^{10}$ cm/sec. (Bearden and Thomsen, 1955)

FLAME SPECTRA OF THE ELEMENTS

By Paul T. Gilbert, Jr., Beckman Instruments, Inc.

The following table lists the principal lines and bands emitted by the elements in various flames used in analytical flame spectrophotometry. This table is followed by an index of these emissions according to wavelength, as an aid to identifying an unknown flame spectrum. Since some elements have much stronger or richer spectra than others, it must be kept in mind that of the hundreds of known flame emissions omitted from these tables, many are stronger than certain relatively weak emissions that are included. Among the elements for which no quantitative data are available, osmium, iridium, and thorium are said to emit only continua, while hafnium, tantalum, and tungsten probably do likewise; selenium may emit a useful band spectrum under the proper conditions.

The tabulated intensity is given by $J = 100(R - B)/B$, where B is the blank reading (pure solvent) and R is the reading for a solution containing 10 mg of the element per liter, for the slit width at which, for the blank, shot-effect noise equals flame flicker; the instrument is a Beckman DU spectrophotometer with photomultiplier and flame attachments, optimally adjusted for each element and flame. Intensities are given for air-hydrogen (*AH*), oxyhydrogen (*OH*), and oxyacetylene (*OA*) flames. (In Table II the intensity is for *OH* unless otherwise indicated.) Under *AH*, *OH*, and *OA* are data for aqueous solutions. Under *AHn*, *OHn*, and *OAn* are data for the most favorable known nonaqueous solvent. The solvent is indicated by the symbol following the first of the set of data to which it applies. The symbol in the Character column applies to all data in that line except as indicated ahead of any given intensity value. The slit widths implied above, for aqueous solutions, are about .5 mm for *AH*, .10–.15 mm for *OH*, and .02–.05 mm for *OA*, in the range 330–550 mμ; they are smaller at 280–330 mμ, and wider below 280 and above 550 mμ, being about twice as wide at 620 as at 550 mμ. Nonaqueous solutions permit narrower slits. In working with lines, narrower slits than those implied by the definition of J—usually not over .05 mm—can be used without loss of sensitivity.

Detection limits for various flame spectrophotometers can be estimated from these intensity data. Detection limit D is here defined as the minimal concentration of element in mg/liter (in a solution containing no interfering materials) yielding a net signal equal to total peak-to-peak noise for a time constant of 1 second. For the DU with photomultiplier up to about 700 mμ, $D = 10/J$; for the red-sensitive phototube of the DU, $D = 300/J$ for lines and $10/J$ to $100/J$ for bands (depending on wavelength, flame, and width of the band). For the DU with blue-sensitive phototube, $D = 100/J$ to $200/J$ for lines and $10/J$ to $20/J$ for most bands, up to about 630 mμ, beyond which the red-sensitive tube is used. Intensities marked with *k* apply to a red-sensitive photomultiplier and must be divided by 30 (for lines) before these formulas for D are applied for the red-sensitive phototube. For the Beckman Model B, with either photomultiplier or phototubes, multiply D as found for the DU (with photomultiplier or phototubes, respectively) by 5, down to 350 mμ; at 330 mμ the factor is 10, and at 320 mμ it is 15. For the Beckman DK-2 spectrophotometer, $D = 20/J$ for lines and $10/J$ for bands. For the Beckman DB spectrophotometer, $D = 40/J$ for lines and $20/J$ for bands. The DK-2 and DB are not recommended for flame photometry beyond 700 mμ. The Zeiss PMQ II flame spectrophotometer and the Hitachi EPU-2A spectrophotometer with H-2 flame attachment should yield J and D values similar to those for the DU. The Unicam SP.900 Mark II flame spectrophotometer, the Optica CF-H spectrometer with flame attachment and possibly the Jobin-Yvon research model also give similar values, but they do not resemble the DU as closely as the Zeiss and Hitachi instruments. The Waters Associates flame photometer probably gives somewhat higher values of J, but data are not available. The data below are not applicable to flame photometers with filters or to flame spectrophotometers utilizing other flames or lacking atomizer-burners.

To supplement these tables, the reader is referred to the collections of recorded flame spectrograms in the following publications:

Herrmann and Alkemade, "Flammenphotometrie," 2nd edtion, Springer, Berlin, 1960.
Dean, "Flame Photometry," McGraw-Hill, New York, 1960.
Gilbert, "Symposium on Spectroscopy," ASTM STP 269, p. 73 (1960).
Watanabe and Kendall, *Appl. Spectroscopy 9*, 132 (1955).
Rains, House and Menis, *Anal. Chim. Acta 22*, 315 (1960) (rare earths).

Many of the data in these tables are calculated from unpublished results supplied by V. A. Fassel and R. N. Curry (Iowa State University), J. A. Dean (University of Tennessee), B. E. Buell (Union Oil Co. of California), W. Schmidt (Forschungsinstitut der Feuerfest-Industrie, Bonn), and H. Watanabe and Miss Jean E. Kinnear (Beckman Instruments, Inc., Fullerton, California). Many of the data on rare earths were supplied in advance of publication by O. Menis (Oak Ridge National Laboratory). The kindness of these colleagues in providing this information is gratefully acknowledged.

FLAME SPECTRA OF THE ELEMENTS (Continued)

Abbreviations

a = arc line
AA = acetylacetone
Ac = acetone
b = peak of a band without conspicuous head
Bu = butanol
Bz = benzene

c = continuum or very wide, featureless band; wavelength indicates general region of greatest intensity. If c follows a, b, r, etc., the latter emission is superposed upon an at least equally strong continuum.
C = chloroform
d = double
EA = 1:2:1 ether-alcohol-water
f = intercombination line
G = gasoline
H = hexone (methyl isobutyl ketone)
HAP = 50% hexone, 20% acetone, 20% isopropanol, 10% water
i = line occurring in inner cone of flame
k = intensity for Farnsworth 16PMI photomultiplier
K = kerosene
m = multiple band, resolved with narrow slits
M = methanol
n = flame background band occurring with organic solvents (in the index table).
N = naphtha
Oc = 2-octanone
p = resonance line
P = isopropanol
r = head of band degraded to red
s = spark line
t = triple
u = unclassified line
v = head of band degraded to violet
w = wide or diffuse band
x = with uncoated mirror in burner housing (mirror normally has a silicone coating).
( ) denote doubtful or roughly estimated data
Subscript numerals indicate volume percentage: Ac_{60} = 60% acetone (and 40% water).

* Numbers with asterisks refer to footnotes.

FLAME SPECTRA OF THE ELEMENTS (Continued)

Table I. Flame Spectra Alphabetically by Element

Element	Wave-length	Char-acter	Species	Intensities in Various Flames					
				AH	AHn	OH	OHn	OA	OAn
Aluminum	394.40	p	Al	pc 0.3		0.8	20 H	1	100 H
	396.15	p	Al	pc 0.35		1.0	30	2	200
	464.8	r	AlO			1.7	10	0.13	17
	467.2	r	AlO			2	10	0.15	17
	469.5	r	AlO	rc 0.8		1.7	7	0.14	11
	471.6	r	AlO			1.7	5	0.13	6
	484.2	r	AlO	1.5		3	50	0.3	70
	486.6	r	AlO			3	30	0.25	50
	507.9	r	AlO			2	7	0.15	11
	510.2	r	AlO	rc 1		2	10	0.18	17
	512.3	r	AlO			2	10	0.18	17
	514.3	r	AlO			2	5	0.17	7
Antimony	217.59 x	p	Sb		17 P				
	231.15 x	a	Sb	0.07	25				
	252.85 x	a	Sb	0.01	8			(0.1) i	
	257.4 x	b	(SbO)	1	0.5 P_{40}				
	259.81 x	a	Sb	0.025	7 P			(0.1) i	
Arsenic	228.81 x	a	As		2.5 P				
	234.98 x	a	As		4			(0.1) i	
	249.29 x	a	As		0.7 P_{50}				
	250.4 x	dr	AsO		0.7				
	500	c		2					
Barium	455.40	s	Ba^+	c 8		8	20 Bu_4	5	100 N
	488	b	BaOH	100		30	40	2	5
	493.41	sb	Ba^+	c 80		25	40	5	100
	497	bc	BaOH	80		25	35	1.7	3
	502	bc	BaOH	80		25	35 Bu_4	1.7	3 N
	513	b	BaOH	150		30	50	3	5
	524	bc	BaOH	80		25	35	3	3
	535.0	rc	BaO	80		20	30	2	2.5
	549.3	rc	BaO	80		20	25	1.7	2.5
	553.56	p	Ba	170		40	100	10	50
	564.4	rc	BaO	80		20	22	1.5	2.5
	570.1	rc	BaO	80		20	25	1.7	2.5
	586.5	rc	BaO	70		20	25	1.7	
	604.0	rc	BaO	70		20	25	1.7	
	745	w	BaOH	k 50		10	10 Bz	7	(100) N
	830	b	BaOH	k 200		30	25	15	(300)
	873	b	BaOH	k 80		30	25	15	(300)
Beryllium	470.9	r	BeO			0.7		0.25	
	473.3	r	BeO	rc 0.2		0.5		0.2	
	475.5	r	BeO			0.4		0.11	
	505.4	r	BeO			0.25		0.10	
	507.6	r	BeO	rc 0.1		0.25		0.10	
	509.5	rc	BeO			0.25		0.10	
Bismuth	223.0 x	dp	Bi		7 P				
	227.66 x	p	Bi		0.3				
	306.77	p	Bi	0.2	0.2 P_{85}	0.017		0.01 i[1*]	
	439.4	bc	(BiH)	0.3		c 0.04		c 0.003	
	442.4	bc	(BiH)	0.3		c 0.05		c 0.004	
	472.26	a	Bi	0.5	0.00[2*]	0.25	0.005 P	0.05	
	556.4	rc	BiO	0.5		c 0.10		c 0.01	
Boron	453	b	BO_2	1		7	15 Bu_4	1.2	2.5 M_{50}
	471.5	b	BO_2	2.5		20	30	2.5	5
	494	db	BO_2	5		30	50	5	8
	518.0	b	BO_2	8		50	80	6	15
	547.6	b	BO_2	11		60	90	15	17
	579	b	BO_2	7		30	70	10	10
	603	b	BO_2	2		10	70	7	5
	620	b	BO_2	1.5		7	50	3	3
Cadmium	228.80 x	p	Cd	0.2	20 P	1	17 P_{50}	0.25	
	326.11	f	Cd	20	6	2	5 Ac	0.25	5 Ac
Calcium	393.37	sc	Ca^+			25		30	
	396.85	sc	Ca^+			20		17	
	422.67	p	Ca	250		1000	1700 Bu_4	250	1000 EA
	554	b	CaOH	500		1700	2500	170	700
	572	b	CaOH	25		100	200	20	50
	602	b	CaOH	100		700	1000	250	300
	622	b	CaOH	500		2500	5000	500	1000
	644	b	CaOH	70		300	700	70	120
	824	b	CaO			1.5		5	
	872	b	CaO			2		7	
Cerium	468.4	rc	CeO	c 25		5	2.5 Ac	c 0.5	[3*]
	481	bc	CeO	c 30		5	5	c 0.5	
	494	bc	(CeO)	c 40		7	5	c 0.7	
	550–600	c		70		10	10	0.7	

Table I. Flame Spectra Alphabetically by Element (Continued)

Element	Wave-length	Char-acter	Species	Intensities in Various Flames AH	AHn	OH	OHn	OA	OAn
Cesium	455.54	a	Cs	20		25	30 Bu_4	0.3	
	459.32	a	Cs	5		7	15	0.1	
	852.11 k	p	Cs	1000		1000	2000	1000	
	894.35 k	p	Cs	300		300	700	300	
Chromium	357.87	a	Cr	40	15 P_{25}	80	80 N	20	1000 H
	359.35	a	Cr	35	14	70	70	17	800
	360.53	a	Cr	30	13	55	55	15	500
	425.43	p	Cr	120	15	100	150 Ac_{60}	20	900
	427.48	p	Cr	110	8	80	110	17	550
	428.97	p	Cr	100	4*	70	80	12	300
	520.6	ta	Cr	ac 170		70	70 Bu_4	10	100
	535.6	rc	CrO	200		20	70	10	
	541.7	rc	CrO	200		20	80	10	
	556.4	rc	CrO	200		20	90	10	
	562.3	rc	CrO	200		20	90	10	
	579.4	rc	CrO	250		20	120	15	
	585.2	rc	CrO	250		25	100	17	
	605.2	rc	CrO	250		25	100	20	
	639.4	rc	CrO			10	50	17	
	685	bc	CrO			10	50	15	
Cobalt	340.51	a	Co	40		20	30 Bu_4	5	
	341.25	dpf	Co	40		25	30	3	
	343.2	dap	Co	10		4.5	7	1.5	
	344.3	da	Co	20		10	20	2.5	
	344.93	da	Co	25		12	25	3	
	345.4	dap	Co	45		25	50	7	
	346.5	dap	Co	20		12	25	3	
	347.40	p	Co	17		7	12 Bu_4	2	
	350.23	a	Co	} 30		17	30	3	
	350.63	a	Co			9	20	2	
	351.3	tap	Co	20		15	25	3	
	352.8	tap	Co	35		35	50	7	
	356.94	a	Co	} 20		8	15	1.7	
	357.54	a	Co			15	25	2.5	
	359.49	p	Co	17		10	17	2	
	384.55	a	Co	22		11	17	2	
	387.35	da	Co	30		20	30	4	
	389.45	da	Co	ac 22		10	17	2	
	399.53	f	Co	ac 25		9	15	2	
	412.13	(d)a	Co	ac 25		11	20	2	
	563.5	bc	(CoO)			17			
Copper	324.75	p	Cu	60	40 P_{25}	100	500 M	100	300 K
	327.40	p	Cu	40	25	100	500	100	300
	428.0	r	CuH			3	2.5 Bu_4	0.7	
	505	w	CuOH	50		7	9	2.5	
	510.55	ab	Cu			8	11	a 7	
	524	bc	CuOH	70		12	15	5	
	537	w	CuOH	100		17	20	6	
	605	bc	CuO			5	5	3	
Dysprosium	457.4	b	DyO			8	10 H		3*
	516	mb	DyO			20	30		
	520	bc	DyO			25	30		
	526.3	mr	DyO			70	80		
	540.0	b	DyO			60	70		
	549.3	b	DyO			60	70 H		
	572.9	b	DyO			120	150		
	583.3	b	DyO			120	150		
	605.7	bc	DyO			15	70		
	609.1	bc	DyO			15	55		
Erbium	504	b	ErO			30	55 H		3*
	515	bc	ErO			17	30		
	546	bc	ErO			40	60		
	552	b	ErO			50	80		
	560	bc	ErO			35	70		
	566	bc	ErO			25	50		
Europium	459.40	p	Eu			25	35 H	10	3*
	462.72	p	Eu			20	30	10	
	466.19	p	Eu			17	25	10	
	598	bc	(EuOH)			100	120	(10)	
	601.82	uc	Eu			100	120	(10)	
	623	wc	(EuOH)			70	100	(10)	
	647	wc	(EuOH)			25	150	(10)	
	684	wc	(EuOH)			20	100	(10)	
	702	w	(EuOH)			50	100	(10)	
Gadolinium	461.6	r	GdO			17	30 H		3*
	463.3	r	GdO			15	15		
	489.2	r	GdO			12	15		
	491.0	r	GdO			12	20		
	492.8	r	GdO			7	15		
	540	w	GdO			12	20		

FLAME SPECTRA OF THE ELEMENTS (Continued)

Table I. Flame Spectra Alphabetically by Element (Continued)

Element	Wave-length	Char-acter	Species	Intensities in Various Flames: AH	AHn	OH	OHn	OA	OAn
Gadolinium (continued)	545	bc	GdO			15	25 H		
	569.9	mr	GdO	(3)		35	70		
	580.7	mrc	GdO	(5)		50	120	(10)	
	591.1	mrc	GdO	(10)		80	250	(20)	
	598.7	mrc	GdO	(10)		80	250	(20)	
	612	bc	GdO	(10)		70	200	(20)	
	622	bc	GdO	(10)		70	250	(20)	
Gallium	294.4	da	Ga			(1)		(0.2)	
	403.30	p	Ga	10		100	150 Bu_4	10	
	417.21	p	Ga	20		200	300	20	
Germanium	259.25 x	u	Ge		0.7 P		0.025 P_{50}	0.015 i	0.025 i P_{50}
	265.14 x	dp	Ge	0.00	2	0.01	0.03	0.04 i	0.05 i
	270.96 x	p	Ge		0.4			0.01 i	0.02 i
	275.46 x	u	Ge		0.7			0.008 i	
Gold	242.80	p	Au	0.1		1		1.1	
	267.60	p	Au	0.3		2		1.7	
Holmium	510.5	rc	HoO			35	40 H		3*
	515.7	r	HoO			50	50		
	527	b	HoO			50	50		
	532.0	b	HoO			50	50		
	565.9	b	HoO			120	170		
	585.0	v	HoO			35	50		
Indium	303.94	a	In	5		3	3 Bu_4	3	
	325.61	a	In	10	3 P_{25}	10	10	2.5	
	410.18	p	In	150	50	200	300	50	
	428.3	r	InO			1			
	451.13	p	In	250	80	350	500	70	
Iodine[5*]	469.4	rc	IO			(1)			
	484.5	rc	IO			(1)			
	496.4	rc	IO			(1)			
	513.1	rc	IO			(1)			
	520.9	rc	IO			(1)			
	530.8	rc	IO			(1)			
	553.3	rc	IO			(1)			
	573.0	rc	IO			(1)			
Iron	302.06	a	Fe	4	5 P_{25}	2.5	50 H	1.5	
	344.06	a	Fe	35		6	10 Bu_4	2	10 AA
	358.12	a	Fe	50		6	10	1.2	10
	371.99	p	Fe	80	11	40	50	15	100
	373.71	(d)p	Fe	80	10	30	50	8	70
	374.7	tp	Fe	70	8	25	50	6	50
	382.04	a	Fe	} 50	3.5	11	20	{ 1.7	10
	382.5	da	Fe					2	12
	385.99	(d)p	Fe	70	9	35	50	11	70
	387.86	p	Fe	} 60	{ 4*	7	12	1.2	8
	388.63	p	Fe		4*	12	20	2.5	20
	389.97	p	Fe	pc 50	4*	5	10	0.9	5
	392.2	dp	Fe	} 50	1.7	8	17	{ 1.0	7
	392.9	dp	Fe					1.5	10
	438.35	fc	Fe	50	4*	5	15 Bu_4	0.9	
	526.95	ac	Fe	70		17	60	3.5	1.7 M_{50}
	532.81	ac	Fe	70		15	60	3.5	1.7
	553.1	rc	FeO	c 100		20	80	6	2.5
	561.4	rc	FeO	c 150		35	110	9	3
	564.7	rc	FeO	c 170		40	120	10	3
	579.0	rc	FeO	}		{ 40	110	10	3
	581.9	rc	FeO	c 170		40	120	11	3
	586.8	rc	FeO			40	100	10	
	609.7	bc	FeO	c 100		20	50	12	
	618.1	rc	FeO	c 100		20	50	10	
	621.9	rc	FeO	c 100		20	50	10	
Lanthanum	438.0	mr	LaO	5		25	25 H	17	80 H
	442.3	mr	LaO	5		25	30	17	80
	538.0	r	LaO	}	{	15	25	5	15
	540.6	r	LaO	3		20	30	7	20
	543.1	r	LaO			20	30	7	25
	545.7	r	LaO			17	30	6	20
	560.0	r	LaO	} 17	{	70	100	17	170
	562.6	r	LaO			50	100	17	170
	586.6	r	LaO	}	{	10	25	12	100
	589.3	r	LaO	3		10	25	12	100
	592.1	r	LaO			10	25	12	100
	743 k	mr	LaO	30		20	1000	100	120
	792 k	mr	LaO	25		20	1000	100	150
Lead	217.00 x	a	Pb		1.5 P				
	261.42 x	a	Pb		2.5				

Table 1. Flame Spectra Alphabetically by Element (Continued)

Element	Wave-length	Char-acter	Species	Intensities in Various Flames					
				AH	AHn	OH	OHn	OA	OAn
Lead (continued)	283.31	p	Pb			0.4	2.5 G		4 G
	363.96	a	Pb	1.5	2 G	5	7	0.1	5
	368.35	a	Pb	3	4	10	15	0.2	10
	405.78	a	Pb	3	4	10	15	0.3	10
Lithium	460.29	a	Li			2.5	3 Bu_4	0.5	4
	610.36	a	Li	30		20	30	10	40
	670.78[6]*	p	Li	10,000	1500 P_{25}	50,000	70,000	10,000	50,000
Lutetium	466.2	mr	LuO			30	80 H		[3]*
	517.0	mr	LuO			50	110		
	600	mb	LuO			20	40		
	675	mb	LuO			10	30		
Magnesium	285.21	p	Mg	100	20 P_{25}	100	250 Ac	70	100 EA
	362.4	b	MgOH	100	30	25	25 EA	1	
	370.2	b	MgOH	500	60	100	100	5	
	381–383[7]*	mb	MgOH	500	50	80	80	3.5	
	387.7	b	MgOH	200		50	50	1.7	
	391.2	b	MgOH	200		50	50	1.7	
	500.7	mv	MgO			10	25 Bu_4	1.7	2
	517.0	dac	Mg	}		5	20	0.6	0.5
	518.36	ac	Mg	}					
Manganese	279.48 x	a	Mn			15	17 G	1.8	
	279.83 x	a	Mn			11	15	1.7	
	280.11 x	a	Mn			8	11	1.5	
	363–410	w	MnOH	70		10	8 Bu_4		
	403.2	tp	Mn	500		1000	1500 N	500	2000
	515.8	rc	MnO	}	{	20	30 Bu_4	3	
	519.2	rc	MnO	50	{	25	40	4	
	522.9	rc	MnO	}	{	25	35	3.5	
	535.9	r	MnO	}	{	45	70	8	
	538.9	r	MnO	100	{	50	80	8	
	542.4	r	MnO	}	{	35	70	7	
	558.6	r	MnO	120	{	80	120	17	
	561.0	r	MnO	}	{	60	100	17	
	586.0	r	MnO	70	{	40	50	8	
	588.1	r	MnO	}	{	50	70	8	
	615.5	mrc	MnO	30		25	30	5	
Mercury	253.65 x	f	Hg	1.7	2.5 P	0.3	1.7 Ac	{ 0.2 1.2 i	
Molybdenum	379.83	p	Mo	c 2.5	0.0 P_{80}	c 0.8	10 i N	(0.5) i	20 HAP
	386.41	p	Mo	c 3	0.0	c 0.8	10 i	(0.3) i	20
	390.30	p	Mo	c 3	0.0	c 0.8	5 i	(0.25) i	20
	550–600	c	(MoO_2)	25		10		(3)	
Neodymium	462.0	trc	NdO			3	5 H		[3]*
	531.4	bc	NdO			5	10		
	599	mbc	NdO			10	25	1	
	622	bc	NdO			2.5	20	0.7	
	636	mbc	NdO			2.5	30	2	
	643	tbc	NdO			2.5	40	3	
	660	mb	NdO			10	100	10	
	691	bc	NdO			10	500	5	
	702 k	bc	NdO			10	500	10	
	712 k	tbc	NdO			10	500	10	
Nickel	300.25	a	Ni	4	3.5 P_{25}	3	5 Bu_4	1.1	
	336.96	p	Ni	17	5	9	17	3	
	338.06	a	Ni	12	4	7	10	2.5	
	339.30	dp	Ni	25	7	17	25	5	
	341.48	p	Ni	80	25	50	60	10	
	343.36	p	Ni	15	3	10	12	3	
	344.63	p	Ni	20	5	12	20	5	
	346.0	dp	Ni	45	15	20	30	7	
	347.25	p	Ni	7	3	4	6	2.5	
	349.30	p	Ni	25	10	17	25	7	
	351.03	p	Ni	} 45	15 {	7	10	5	
	351.51	p	Ni	}	{	22	30	9	
	352.45	p	Ni	80	25	50	80	15	(50) C
	356.64	a	Ni	20	5	10	15	3	
	361.05	p	Ni	15	3	7	10	2.5	
	361.94	a	Ni	40	10	22	30	5	
	385.83	f	Ni	10	*	4	10	1.7	
	502.4	rc	NiO			12	30		
	517.4	rc	NiO			15	35		
	520–600	c				17	35		
Niobium	{ 450	c		5		1		0.17	
	{ 550	c		5		1		0.3	
Nitrogen	214.9 x	dv	NO		0.4 P				
	226.3 x	dv	NO		0.4				
	236.3 x	dv	NO		0.5				
	247.1 x	dv	NO		0.2				
	385.1 } 385.5 }	v	CN[8]*				0.17 i N		

FLAME SPECTRA OF THE ELEMENTS (Continued)

Table I. Flame Spectra Alphabetically by Element (Continued)

Element	Wave-length	Char-acter	Species	Intensities in Various Flames					
				AH	AHn	OH	OHn	OA	OAn
Palladium	340.46	a	Pd	45		70	200 H	10	
	342.12	a	Pd	11		10	50	2	
	346.08	a	Pd	6		10	20	2	
	351.69	a	Pd	8		20	40	2	
	360.95	a	Pd	30		45	90	7	
	363.47	a	Pd	55		80	150	10	
	500	c		30					
Phosphorus	237.5 x	vb	PO	0.4	8 P				
	238.3 x	vb	PO	0.4	8				
	239.6 x	v	PO	0.22	4.5				
	246.4 x	v	PO	0.5	10				
	247.8 x	v	PO	0.4	8				
	252.9 x	vb	PO	0.2	4				
	254.0 x	v	PO	0.12	2.5				
	520	c		30					
Platinum	265.95 x	p	Pt	0.11		0.8		0.5	
	306.47	p	Pt	0.25		1		0.7	
Potassium	404.5	da	K	30	1 P_{25}	70	100 Bu_4	10	
	766.49 k	p	K	} 10,000		30,000	50,000	{ 30,000	
	769.90 k	p	K					20,000	
Praseodymium	515.7	rc	PrO			7	3 H		[3]*
	538	mrc	PrO			10	6		
	561	trc	PrO			11	8		
	569.1	trc	PrO			12	10		
	576.3	rc	PrO			15	11		
	603	dr	PrO			10	20		
	629	dbc	PrO			5	11		
	636.3	rc	PrO			5	15		
	648	drc	PrO			5	17		
	695 k	bc	PrO			(10)	50		
	709.5 k	bc	PrO			(10)	50		
	732.1 } k 737.6 }	rc	PrO			(10)	30		
	805 k	bc	PrO			(5)	10		
	851 k	b	PrO			(2)	5		
Promethium	(640)	b	(PmO)			(10)	(50) H	(1)	
	(680)	b	(PmO)			(10)	(50)	(1)	
Radium	381.44	s	Ra^+			(3)			
	468.23	s	Ra^+			(1)			
	482.59	p	Ra			(5)			
	602	w	(RaOH)			(1)			
	627	b	(RaOH)			(5)			
	665	b	(RaOH)			(5)			
Rhenium	488.92	p	Re			(0.1)			
	527.55	p	Re			(0.1)			
Rhodium	332.31	f	Rh	fc 2		2.5	3.5 P	1.7	3 P_{90}
	339.68	p	Rh	pc 4		8	12	5	8
	343.49	p	Rh	2.5		15	25	9	17
	350.25	p	Rh	pc 3.5		8	10	5	8
	352.80	a	Rh	ac 3		7	9	4	7
	358.31	a	Rh	ac 4.5		4	4 P	1.7	3 P_{90}
	359.66	dpf	Rh	pfc 5		5	5	2.2	4.5
	365.80	a	Rh	ac 6		11	15	6	9
	369.24	p	Rh	7		35	40	15	22
	370.09	a	Rh	ac 6		11	11	4.5	8
	385.65	a	Rh	ac 7		4	3.5	1.5	3
	421.11	f	Rh	fc 11		5	3	1.1	1.8
	437.48	f	Rh	fc 15		8	4	1.7	2.2
	542.5	bc	(RhO)	c 40		15		1.	
Rubidium	420.19	a	Rb	20		35	50 Bu_4	2	
	421.56	a	Rb	10		15	20	1	
	780.02 k	p	Rb	3500		3500	5000	2000	
	794.76 k	p	Rb	2500		2500	3000	1700	
Ruthenium	342.83	a	Ru			2	15 M_{50}	1.5	
	343.67	p	Ru			1	10	1	
	349.89	p	Ru			5	30	3	
	359.3	tp	Ru			1	10	1	
	366.14	f	Ru			2	20	2	
	372.80	p	Ru			20	150	30	
	379.9	tp	Ru			10	100	25	
	569.90	a	Ru			0.5	5	1	
Samarium	472.0	bc	(SmO)			10	10 H		[3]*
	582.0	bc	SmO			20	30	0.2	
	587	bc	SmO			30	50	0.3	
	595	tbc	SmO			35	60	0.4	
	603	mwc	SmO			35	70	0.5	

Table I. Flame Spectra Alphabetically by Element (Continued)

Element	Wave-length	Char-acter	Species	Intensities in Various Flames					
				AH	AHn	OH	OHn	OA	OAn
Samarium (continued)	614	mbc	SmO			40	80 H	2	
	624	mbc	SmO			30	80	2.5	
	638	bc	SmO			20	100	2	
	642	tbc	SmO			20	100	2	
	652	mbc	SmO			20	110	3	
	668	wc	SmO			5	60	0.5	
	683	bc	SmO			3	40	0.3	
Scandium	467.3	r	ScO			10	25 H		[3]*
	470.7	r	ScO			12	35		
	474.2	r	ScO			8	25		
	485.8	r	ScO			20	60		
	489.3	r	ScO			15	45		
	509.7	r	ScO			8	30		
	513.4	r	ScO			9	35		
	517.1	r	ScO			8	25		
	573.7	r	ScO			17	60	2	
	577.3	r	ScO			35	120	3	
	581.2	r	ScO			40	150	5	
	584.9	r	ScO			40	150	5	
	588.7	r	ScO			35	120	3	
	592.8	r	ScO			30	100	3	
	601.7	r	ScO	}		170	800	20	
	603.6	r	ScO						
	607.3	tr	ScO			250	1700	30	
	611.0	tr	ScO			200	1200	25	
	615.4	r	ScO			110	700	10	
	619	dr	ScO			50	400	5	
	623	tr	ScO			30	250	3	
	650	mr	ScO			10	70	1	
Silicon	241.4 x	r	SiO		0.00 P_{50}		(0.1)[9]*		
	248.7 x	r	SiO		0.00		(0.1)		
	256.4 x	r	SiO		0.00		(0.1)		
	258.7 x	r	SiO		0.00		(0.1)		
Silver	328.07	p	Ag	100		100	100 EA	50	**30 EA**
	338.29	p	Ag	250		170	170	50	**40**
Sodium	330.3	da	Na	2.5		20	10 Bu_4	10	50
	568.5	dac	Na			30	40	10	
	589.2	dp	Na	30,000		50,000	70,000	25,000	100,000
	819 k	da	Na			5	10	20	
Strontium	407.77	s	Sr^+			30	50 Bu_4	17	
	421.55	s	Sr^+			20	30	12	
	460.73	p	Sr	500	150 P_{25}	1000	2000 Ac_{60}	200	
	595	b	Sr_2O_2	500		90	100 Bu_4	10	
	597	b	Sr_2O_2	500		80	100	10	
	605	b	SrOH	5000	250	1000	1500	100	
	645	b	SrOH	250		100	150	10	
	659	b	SrOH	500		200	300	20	
	666	b	SrOH	500		700	1000	40	
	680	b	SrOH	250		1000	1000	30	
	704	b	SrOH			70	100		
Tellurium	238.32 x	a	Te	} 0.01	1 P			0.000	
	238.58 x	a	Te						
	356.1	rc	TeO	2.5		0.15		c 0.015	
	360.7	rc	TeO	3		0.2		c 0.015	
	366.2	rc	TeO	3.5		0.2		c 0.015	
	371.4	drc	TeO	3.5		0.3		c 0.015	
	377.3	drc	TeO	3.5		0.2		c 0.015	
	382.7	trc	TeO	3.5		0.25		c 0.017	
	388.4	trc	TeO	3.5		0.25		c 0.017	
	394.5	trc	TeO	3.5		0.25		c 0.017	
	400.7	drc	TeO	3.5		0.22		c 0.017	
	407.5	drc	TeO	3.5		0.22		c 0.017	
	413.1	drc	TeO	3.5		0.25		c 0.017	
	420.5	rc	TeO	3.5		0.22		c 0.017	
	426.8	rc	TeO	3		0.22		c 0.017	
	434.3	rc	TeO	3.5		0.22		c 0.017	
	448.7	rc	TeO	3		0.22		c 0.017	
	464.0	rc	TeO	2.5		0.17		c 0.017	
Terbium	461	bc	TbO			8	10 H		[3]*
	534	b	TbO			40	60		
	544	bc	TbO			30	40		
	563.9	bc	TbO			40	70		
	573	bc	TbO			50	100		
	592.1	bc	TbO			80	170	(10)	
	598.0	bc	TbO			70	150	(10)	
	608.0	bc	TbO			40	100	(10)	
	635	bc	TbO			17	30	(5)	
Thallium	276.79 x	a	Tl			(0.5)		0.5	
	351.92	a	Tl	4		3	25 Bu_4	0.7	2 Oc
	377.57	p	Tl	50		100	170	10	30
	535.05	a	Tl	30		70	100	5	20

FLAME SPECTRA OF THE ELEMENTS (Continued)

Table I. Flame Spectra Alphabetically by Element (Continued)

Element	Wave-length	Char-acter	Species	Intensities in Various Flames					
				AH	AHn	OH	OHn	OA	OAn
Thulium	483	b	TmO			22	35 H		[3]*
	491	b	TmO			30	50		
	523	bc	TmO			17	35		
	538	b	TmO			35	60		
	541.5	b	TmO			35	60		
	553	b	TmO			35	60		
Tin	224.61 x	s	Sn^+		8 P		0.6 i P		
	235.48 x	a	Sn		15		1.5 i		
	242.95 x	a	Sn		15		1.5 i		
	270.65 x	a	Sn	0.08	12	0.012		1.7 i	
	284.00	a	Sn	0.08	11	(0.015)		0.2 i	
	286.33	p	Sn	0.11	6	0.017		{ 0.03 0.17 i	
	300.91	p	Sn	0.2	7	0.04		{ 0.017 0.3 i	
	303.41	p	Sn	0.4	12	0.05		{ 0.05 0.4 i	
	317.50	p	Sn	0.12	7			{ 0.017 0.12 i	
	326.23	a, r	Sn, SnO	ar 0.4	a 10	r 0.09	r 0.08 Ac_{60}	{ r 0.17 a 0.4 i	
	332.3	r	SnO	1.5		0.35	0.2	0.2	
	333.06	f	Sn		1.7			0.4 i	
	338.8	r	SnO	2.5		0.7	0.18	0.2	
	341.6	r	SnO	3		0.7	0.18	0.12	
	348.5	r	SnO	1.5	0.5 P_{25}	0.25	0.17	0.2	
	358.5	r	SnO	3.5	0.5	0.9	0.22	0.2	
	369.1	r	SnO	2.5	0.5	0.7	0.15	0.17	
	372.1	r	SnO	2.5	0.35 P_{25}	0.7	0.15 Ac_{60}	0.15	
	380.10	f	Sn	frc 2.5	3 P			0.3 i	
	383.3	r	SnO	rc 2.5		0.6	0.11	0.08	
	386.5	r	SnO	3.5	0.25 P_{25}	0.7	0.11	0.08	
	398.4	r	SnO	3		0.5	0.08	0.07	
	485	b	(SnO)	5	0.25	c 0.22			
Titanium	480.5	rc	TiO			30		c 0.7	0.7 p
	484.8	rc	TiO			30		c 0.8	0.8
	495.5	rc	TiO			35		1.1	1.1
	500.3	rc	TiO			35		c 1.1	1.1
	516.7	rc	TiO			40		1.5	1.5
	544.9	rc	TiO			45		1.5	1.5
	575.9	rc	TiO			45		1.5	1.5
	673	bc	TiO			20		c 10	
	715	bc	TiO			20		10	
Uranium	550	c	(UO_2)	15		5		c	
Vanadium	505.8	rc	VO	70	c 10 P_{50}	25	70 Bu_4	1.2	
	522.9	rc	VO	80	c 18	30	80	3	
	527.6	rc	VO	80	c 20	30	80	3	
	547.0	rc	VO	100	c 17	40	100	4	
	573.7	rc	VO	110	c 30	40	110	6	
	608.7	rc	VO	50	c 25	20	50	3	
	710	b	VO			10		10	
	747	b	VO			7		7	
	800	b	VO			10		10	
Ytterbium	346.44	u	Yb			2		(1)	[3]*
	369.42	s	Yb^+			1.5		(0.5)	
	398.80	a	Yb	70		25	120 H	(10)	
	477.8	b	(YbOH)	100		22	35	(2)	
	485.0	b	(YbOH)	100		25	35	(2)	
	498.1	b	(YbOH)	200		50	50	(5)	
	517.4	bc	(YbOH)	150		40	40	(3)	
	532.5	b	(YbOH)	250		80	80	(5)	
	544.3	bc	(YbOH)	120		35	45	(3)	
	555	b	(YbOH) }	250		70	60	(10)	
	555.65	p	Yb }						
	572.5	b	(YbOH)	300		110	100	(10)	
	587	b	(YbOH)	80		25	35	(2)	
	602	bc	(YbOH)	30		10	c 25	(1)	
Yttrium	465.0	r	YO			8	17 H		[3]*
	467.6	r	YO			10	20		
	470.5	r	YO			9	18		
	481.8	r	YO			30	50		
	484.2	r	YO			22	40		
	505.0	r	YO			7	20		
	507.8	r	YO			6	20		
	574	tr	YO			8	50	1	
	586	mrc	YO			10	50	1	
	599	mr	YO			300	1000	30	70 H
	615	mr	YO			300	800	30	50
Zinc	213.86 x	p	Zn	0.04	0.15 P_{50}	0.06	1 Ac	0.017	
	307.59	f	Zn	0.1	0.00	c 0.02			
	481.05	a	Zn	c 0.3		c 0.06	0.6 i N	i	c 1.7 Ac
	520–600	c		0.5		0.12	12 Ac		4
Zirconium	564	bc	ZrO			1.2		(0.1)	
	574	bc	ZrO			1.2		(0.1)	

*[1] Seen in absorption against an effectively hotter band head of OH. [2] Obscured by C_2 bands. [3] Data to be supplied by Kniseley, Fassel and Curry, *Spectrochimica Acta*. [4] Obscured by CH bands. [5] Bands observed only with iodate ion. [6] With red filter to reduce stray light. [7] The arc lines 382.94, 383.23, 383.83 may appear within this band, especially in hotter flames. [8] The stronger CN bands 386.2, 387.1, 388.3 are interfered with by the CH bands 387.1, 388.9. [9] Emitted by ethyl silicate.

FLAME SPECTRA OF THE ELEMENTS (Continued)

Table II. Index of Flame Spectra by Wavelength

Wavelength	Character	Element	Intensity	Wavelength	Character	Element	Intensity	Wavelength	Character	Element	Intensity	Wavelength	Character	Element	Intensity
213.86	p	Zn	OHn 1	344.06	a	Fe	6	387.35	da	Co	20	471.5	v	C	n
214.9	dv	N	AHn 0.4	344.3	da	Co	10	387.7	b	Mg	50	471.6	r	Al	1.7
217.00	a	Pb	AHn 1.5	344.63	p	Ni	12	387.86	p	Fe	7	472.0	bc	Sm	10
217.59	p	Sb	AHn 17	344.93	da	Co	12	388.3	v	N	OHni 0.2	472.26	a	Bi	0.25
222.83	p	Bi	AHn 7	345.4	dap	Co	25	388.4	trc	Te	0.25	473.3	r	Be	0.5
223.06				345.85	p	Ni	20	388.63	p	Fe	12	473.7	v	C	n
224.61	s	Sn	AHn 8	346.17				388.9	r	C	n	474.2	r	Sc	8
226.3	dv	N	AHn 0.4	346.08	a	Pd	10	389.45	da	Co	10	475.5	r	Be	0.4
227.66	p	Bi	AHn 0.3	346.28	a	Co	12	389.97	p	Fe	5	477.8	b	Yb	22
228.80	p	Cd	AHn 20	346.58	p			390.30	p	Mo	OHni 5	480.5	rc	Ti	30
228.81	a	As	AHn 2.5	346.44	u	Yb	2	391.2	b	Mg	50	481	bc	Ce	5
231.15	a	Sb	AHn 25	347.25	p	Ni	4	392.03	p	Fe	8	481.05	a	Zn	OHni 0.6
234.98	a	As	AHn 4	347.40	p	Co	7	392.29				481.8	r	Y	30
235.48	a	Sn	AHn 15	348.5	r	Sn	0.25	392.79				482.59	p	Ra	5
236.3	dv	N	AHn 0.5	349.30	p	Ni	17	393.03				483	b	Tm	22
237.5	vb	P	AHn 8	349.89	p	Ru	5	393.37	sc	Ca	25	484.2	r	Al	3
238.3	vb	P	AHn 8	350.23	a	Co	17	394.40	p	Al	0.8	484.2	r	Y	22
238.32	a	Te	AHn 1	350.25	p	Rh	8	394.5	trc	Te	0.25	484.5	rc	I	1
238.58				350.63	a	Co	9	396.15	p	Al	1	484.8	rc	Ti	30
239.6	v	P	AHn 4.5	351.03	p	Ni	7	396.85	sc	Ca	20	485	b	Sn	AH 5
241.4	r	Si	OHn 0.1	351.04	a	Co	15	398.4	r	Sn	0.5	485.0	b	Yb	25
242.80	p	Au	1	351.3	dap			398.80	a	Yb	25	485.8	r	Sc	20
242.95	a	Sn	AHn 15	351.51	p	Ni	22	399.53	f	Co	9	486.6	r	Al	3
246.4	v	P	AHn 10	351.69	a	Pd	20	400.7	drc	Te	0.22	488	b	Ba	30
247.1	dv	N	AHn 0.2	351.92	a	Tl	3	403.08	p	Mn	1000	488.92	p	Re	0.1
247.8	v	P	AHn 8	352.45	p	Ni	50	403.4	dp			489.2	r	Gd	12
247.86	a	C	OA	352.68	p	Co	35	403.30	p	Ga	100	489.3	r	Sc	15
248.7	r	Si	OHn 0.1	352.94	dap			404.41	a	K	70	491	b	Tm	30
249.29	a	As	AHn 0.7	352.80	a	Rh	7	404.72				491.0	r	Gd	12
250.4	dr	As	AHn 0.7	356.1	rc	Te	0.15	405.78	a	Pb	10	492.8	r	Gd	7
252.85	a	Sb	AHn 8	356.64	a	Ni	10	407.5	drc	Te	0.22	493.41	sb	Ba	25
252.9	vb	P	AHn 4	356.94	a	Co	8	407.77	s	Sr	30	493	b	B	30
253.65	f	Hg	AHn 2.5	357.54	a	Co	15	410.18	p	In	200	495			
254.0	v	P	AHn 2.5	357.87	a	Cr	80	411.88	a	Co	11	494	bc	Ce	7
256.4	r	Si	OHn 0.1	358.12	a	Fe	6	412.13				495.5	rc	Ti	35
257.4	b	Sb	AH 1	358.31	a	Rh	4	413.1	drc	Te	0.25	496.4	rc	I	1
258.7	r	Si	OHn 0.1	358.5	r	Sn	0.9	417.21	p	Ga	200	497	bc	Ba	25
259.25	u	Ge	AHn 0.7	358.92	p	Ru	1	420.19	a	Rb	35	498.1	b	Yb	50
259.81	a	Sb	AHn 7	359.30				420.5	rc	Te	0.22	499.7	v	Mg	10
261.42	a	Pb	AHn 2.5	359.62				421.11	f	Rh	5	500.7			
265.14	dp	Ge	AHn 2	359.35	a	Cr	70	421.55	s	Sr	20	500.3	rc	Ti	35
265.95	p	Pt	0.8	359.49	p	Co	10	421.56	a	Rb	15	502	bc	Ba	25
267.60	p	Au	2	359.66	dpf	Rh	5	422.67	p	Ca	1000	502.4	rc	Ni	12
270.65	a	Sn	AHn 12	360.53	a	Cr	55	425.43	p	Cr	100	504	b	Er	30
270.96	p	Ge	AHn 0.4	360.7	rc	Te	0.2	426.8	rc	Te	0.22	505	w	Cu	7
275.46	u	Ge	AHn 0.7	360.95	a	Pd	45	427.48	p	Cr	80	505.0	r	Y	7
276.79	a	Tl	0.5	361.05	p	Ni	7	428.0	r	Cu	3	505.4	r	Be	0.25
279.48	a	Mn	15	361.94	a	Ni	22	428.3	r	In	1	505.8	rc	V	25
279.83	a	Mn	11	362.4	b	Mg	25	428.97	p	Cr	70	507.6	r	Be	0.25
280.11	a	Mn	8	363.47	a	Pd	80	431.2	v	C	n	507.8	r	Y	6
283.31	p	Pb	0.4	363.96	a	Pb	5	432.4	b	C	n	507.9	r	Al	2
284.00	a	Sn	AHn 11	365.80	a	Rh	11	434.3	rc	Te	0.22	509.5	rc	Be	0.25
285.21	p	Mg	100	366.14	f	Ru	2	437.48	f	Rh	8	509.7	r	Sc	8
286.33	p	Sn	AHn 6	366.2	rc	Te	0.2	438.0	mr	La	25	509.8	v	C	n
294.4	da	Ga	1	368.35	a	Pb	10	438.35	fc	Fe	5	510.2	r	Al	2
300.25	a	Ni	3	369.1	r	Sn	0.7	438.4	vw	C	n	510.5	rc	Ho	35
300.91	p	Sn	AHn 7	369.24	p	Rh	35	439.4	bc	Bi	AH 0.3	510.55	ab	Cu	8
302.06	a	Fe	2.5	369.42	s	Yb	1.5	442.3	mr	La	25	512.3	r	Al	2
303.41	p	Sn	AHn 12	370.09	a	Rh	11	442.4	bc	Bi	AH 0.3	512.9	v	C	n
303.94	a	In	3	370.2	b	Mg	100	448.7	rc	Te	0.22	513	b	Ba	30
306.47	p	Pt	1	371.4	drc	Te	0.3	451.13	p	In	350	513.1	rc	I	1
306.77	p	Bi	AH 0.2	371.99	p	Fe	40	453	b	B	7	513.4	r	Sc	9
307.59	f	Zn	AH 0.1	372.1	r	Sn	0.7	455.40	s	Ba	8	514.3	r	Al	2
317.50	p	Sn	AHn 7	372.80	p	Ru	20	455.54	a	Cs	25	515	bc	Er	17
324.75	p	Cu	100	373.49	p	Fe	30	457.4	b	Dy	8	515.7	r	Ho	50
325.61	a	In	10	373.71				459.32	a	Cs	7	515.7	rc	Pr	7
326.11	f	Cd	2	374.56	p	Fe	25	459.40	p	Eu	25	515.8	rc	Mn	20
326.2	r	Sn	0.09	374.9	dp			460.29	a	Li	2.5	516	mb	Dy	20
326.23	a	Sn	AHn 10	377.3	drc	Te	0.2	460.73	p	Sr	1000	516.5	v	C	n
327.40	p	Cu	100	377.57	p	Tl	100	461	bc	Tb	8	516.7	rc	Ti	40
328.07	p	Ag	100	379.83	p	Mo	OHni 10	461.6	r	Gd	17	517.0	mr	Lu	50
330.3	da	Na	20	379.9	tp	Ru	10	462.0	trc	Nd	3	517.1	r	Sc	8
332.3	r	Sn	0.35	380.10	f	Sn	AHn 3	462.72	p	Eu	20	517.4	rc	Ni	15
332.31	f	Rh	2.5	380.7	b	Mg	80	463.3	r	Gd	15	517.4	bc	Yb	40
333.06	f	Sn	AHn 1.7	381.4				464.0	rc	Te	0.17	517.0	dac	Mg	5
336.96	p	Ni	9	381.44	s	Ra	3	464.8	r	Al	1.7	518.36	ac		
338.06	a	Ni	7	382.04	a	Fe	11	465.0	r	Y	8	518.0	b	B	50
338.29	p	Ag	170	382.5	da			466.19	p	Eu	17	519.2	rc	Mn	25
338.8	r	Sn	0.7	382.3	b	Mg	80	466.2	r	Lu	30	520	bc	Dy	25
339.11	p	Ni	17	383.4				467.2				520.6	ta	Cr	70
339.30				382.7	trc	Te	0.25	467.2	r	Al	2	520.9	rc	I	1
339.68	p	Rh	8	383.3	r	Sn	0.6	467.3	r	Sc	10	522.9	rc	Mn	25
340.51	a	Co	20	384.55	a	Co	11	467.6	r	Y	10	522.9	rc	V	30
340.46	a	Pd	70	385.1	v	N	OHni 0.17	468.23	s	Ra	1	523	bc	Tm	17
341.25	dpf	Co	25	385.5				468.4	rc	Ce	5	524	bc	Ba	25
341.48	p	Ni	50	385.65	a	Rh	4	468.5	v	C	n	524	bc	Cu	12
341.6	r	Sn	0.7	385.83	f	Ni	4	469.4	rc	I	1	526.3	mr	Dy	70
342.12	a	Pd	10	385.99	(d)p	Fe	35	469.5	r	Al	1.7	526.95	ac	Fe	17
342.83	a	Ru	2	386.2	v	N	OHni 0.2	469.8	v	C	n	527	b	Ho	50
343.2	dap	Co	4	386.41	p	Mo	OHni 10	470.5	r	Y	9	527.55	p	Re	0.1
343.36	p	Ni	10	386.5	r	Sn	0.7	470.7	r	Sc	12	527.6	rc	V	30
343.49	p	Rh	15	387.1	v	N	OHni 0.2	470.9	r	Be	0.7	530.8	rc	I	1
343.67	p	Ru	1	387.2	r	C	n	471.5	b	B	20	531.4	bc	Nd	5

FLAME SPECTRA OF THE ELEMENTS (Continued)

Table II. Index of Flame Spectra by Wavelength (Continued)

Wave-length	Char-acter	Ele-ment	Intensity
532.0	b	Ho	50
532.5	b	Yb	80
532.81	ac	Fe	15
534	b	Tb	40
535.0	rc	Ba	20
535.05	a	Tl	70
535.6	rc	Cr	20
535.9	r	Mn	45
537	w	Cu	17
536.7 538.2	rc	Pr	10
538	b	Tm	35
538.0	r	La	15
538.9	r	Mn	50
540	w	Gd	12
540.0	b	Dy	60
540.6	r	La	20
541.5	b	Tm	35
541.7	rc	Cr	20
542.4	r	Mn	35
542.5	bc	Rh	15
543.1	r	La	20
544	bc	Tb	30
544.3	bc	Yb	35
544.9	rc	Ti	45
545	bc	Gd	15
545.7	r	La	17
546	bc	Er	40
547.0	rc	V	40
547.6	b	B	60
549.3	rc	Ba	20
549.3	b	Dy	60
552	b	Er	50
553	b	Tm	35
553.1	rc	Fe	20
553.3	rc	I	1
553.56	p	Ba	40
554	b	Ca	1700
554.1	v	C	n
555 555.65	b p	Yb	70
556.4	rc	Bi	AH 0.5
556.4	rc	Cr	20
558.5	v	C	n
558.6	r	Mn	80
560	bc	Er	35
560.0	r	La	70
561 0	r	Mn	60
559.7 561.2 562.8	rc	Pr	11
561.4	rc	Fe	35
562.3	rc	Cr	20
562.6	r	La	50
563.5	v	C	n
563.5	bc	Co	17
563.9	bc	Tb	40
564	bc	Zr	1.2
564.4	rc	Ba	20
564.7	rc	Fe	40
565.9	b	Ho	120
566	bc	Er	25
568.27 568.82	ac	Na	30
569.1 570.5	rc	Pr	12
569.9	mr	Gd	35
569.90	a	Ru	0.5
570.1	rc	Ba	20
572	b	Ca	100
572.5	b	Yb	110
572.9	b	Dy	120
573	bc	Tb	50
573.0	rc	I	1
573.7	r	Sc	17
573.7	rc	V	40
573.0 574.7 576.4	r	Y	8
574	bc	Zr	1.2

Wave-length	Char-acter	Ele-ment	Intensity
575.9	rc	Ti	45
576.3	rc	Pr	15
577.3	r	Sc	35
579	b	B	30
579.0	rc	Fe	40
579.4	rc	Cr	20
580.7	mrc	Gd	50
581.2	r	Sc	40
581.9	rc	Fe	40
582.0	bc	Sm	20
583.3	b	Dy	120
584.9	r	Sc	40
585.0	v	Ho	35
585.2	rc	Cr	25
585.9 587.6	rc	Y	10
586.0	r	Mn	40
586.5	rc	Ba	20
586.6	r	La	10
586.8	rc	Fe	40
587	bc	Sm	30
587	b	Yb	25
588.1	r	Mn	50
588.7	r	Sc	35
589.00 589.59	p	Na	50,000
589.3	r	La	10
591.1	mrc	Gd	80
592.1	r	La	10
592.1	bc	Tb	80
592.8	r	Sc	30
595	tbc	Sm	35
595	b	Sr	90
597	b	Sr	80
598	bc	Eu	100
598.0	bc	Tb	70
598.7	mrc	Gd	80
599	mbc	Nd	10
597.2 598.8 600.4 602.0	r	Y	300
600	mb	Lu	20
601.7	r	Sc	40
601.82	uc	Eu	100
602	b	Ca	700
602	w	Ra	1
602	bc	Yb	10
602.0 603.8	r	Pr	10
603	b	B	10
603	mwc	Sm	35
603.6	r	Sc	170
604.0	rc	Ba	20
605	bc	Cu	5
605	b	Sr	1000
605.2	rc	Cr	25
605.7	bc	Dy	15
606.4 607.3 607.9	r	Sc	250
608.0	bc	Tb	40
608.7	rc	V	20
609.1	bc	Dy	15
609.7	bc	Fe	20
610.36	a	Li	20
611.0	tr	Sc	200
612	bc	Gd	70
614	mbc	Sm	40
613.2 614.8 616.5	r	Y	300
615.4	r	Sc	110
615.5	mrc	Mn	25
617.6	rc	Mn	25
618.1	rc	Fe	20
619	dr	Sc	50
620	b	B	7
620.3	rc	Mn	20
621.9	rc	Fe	20

Wave-length	Char-acter	Ele-ment	Intensity
622	b	Ca	2500
622	bc	Gd	70
622	bc	Nd	2.5
623	wc	Eu	70
623	tr	Sc	30
624	mbc	Sm	30
627	b	Ra	5
626.7 629.8	bc	Pr	5
635	bc	Tb	17
636	mbc	Nd	2.5
636.3	rc	Pr	5
638	bc	Sm	20
639.4	rc	Cr	10
642	tbc	Sm	20
643	tbc	Nd	2.5
644	b	Ca	300
645	b	Sr	100
647	wc	Eu	25
647.6 649.5	rc	Pr	5
648.5 649.6 652.6	r	Sc	10
650.5 652.5 655.5 657.5	bc	Sm	20
659	b	Sr	200
658.0 660.0 662.5 665	r tb tr r	Nd	10
665	b	Ra	5
666	b	Sr	700
668	wc	Sm	5
670.78	p	Li	50,000
673	bc	Ti	20
675	mb	Lu	10
680	b	Sr	1000
683	bc	Sm	3
684	wc	Eu	20
685	bc	Cr	10
691	bc	Nd	10
695	bc	Pr	10
702	w	Eu	50
702	bc	Nd	10
704	b	Sr	70
709.5	bc	Pr	10
710	b	V	10
710 712 714	bc	Nd	10
715	bc	Ti	20
732.1 737.6	rc	Pr	10
740.4 743.4 746.5	r	La	20
745	w	Ba	10
747	b	V	7
766.49 769.90	p	K	30,000
780.02	p	Rb	3500
791.1 794.5 798.0	r	La	20
794.76	p	Rb	2500
800	b	V	10
805	bc	Pr	5
818.33 819.48	a	Na	5
824	b	Ca	1.5
830	b	Ba	30
851	b	Pr	2
852.11	p	Cs	1000
872	b	Ca	2
873	b	Ba	30
894.35	p	Cs	300

Primary Standard

Wave length of the red cadmium line in air, 760 mm. pressure 15°C. measures of Benoit, Fabry and Perot 1907,

6438.4696 Ångström units

Secondary Standards

Lines of the Iron Arc

Selected lines from list by Ch. Fabry: *International Critical Tables*, 1929
Wave lengths in international Ångströms, atmospheric pressure, 15°C

Wave-length	Wave-length	Wave-length	Wave-length	Wave-length
3370.789	3935.816	4592.655	5232.948	6065.489
3399.337	3977.744	4602.945	5266.564	6137.697
3485.343	4021.870	4647.437	5371.493	6191.563
3513.821	4076.638	4691.414	5405.779	6230.729
3556.882	4118.549	4707.282	5434.527	6265.141
3606 682	4134.680	4736.782	5455.613	6318.023
3640.392	4147.673	4789.654	5497.520	6335.338
3676.314	4191.436	4878.219	5506.783	6393.606
3677.630	4233.609	4903.318	5569.626	6430.852
3724.381	4282.406	4919.001	5586.763	6494.985
3753 615	4315.087	5001.872	5615.652	6546.245
3805.346	4375.933	5012.072	5658.825	6592.920
3843.261	4427.313	5049 825	5709.396	6677.994
3850.820	4466.556	5083.343	5763.013	6750.157
3865.527	4494.568	5110.414	5857.759 Ni	
3906.482	4531.152	5167.491	5892.882 Ni	
3907.937	4547.851	5192.353	6027.058	

Iron, Wave Lengths in the Ultraviolet Spectrum

The following table presents the results of interferometer measurements made by Meggers and Humphreys and reported in the Jour. Research of B of S. **18**, 543, 1937. The standard iron arc was used as a source and the wave lengths in air at 15°C. and 76 mm are given in international Ångströms.

λ_{air} A	λ_{air} A	λ_{air} A	λ_{air} A
2100.795	2265.053	2404.430	2735.473
2102.349	2270.8601	2406.6593	2739.5467
2108.955	2271.781	2410.5172	2746.4833
2110.233	2272.0670	2411.0663	2746.9823
2112.966	2274.0085	2413.3087	2749.325
2115.168	2276.0247	2431.025	2755.7366
2130.962	2277.098	2438.1811	2763.1078
2132.015	2279.922	2442.5674	2767.5208
2135.957	2283.653	2443.8707	2778.2205
2138.589	2284.087	2447.7086	2781.8347
2139.695	2287.2477	2453.4746	2797.7751
2141.715	2287.632	2457.5956	2804.5200
2145.188	2291.122	2465.1479	2806.9840
2147.787	2292.5227	2468.8782	2813.2861
2150.182	2293.8454	2474.8131	2823 2753
2151.099	2294 4059	2487.0643	2832.4350
2153.004	2296.9247	2496.5324	2838.1193
2154.458	2297.785	2507.8987	2845.5945
2157.792	2299.2180	2519.6279	2851.7970
2161.577	2300.1397	2530.6938	2863.864
2163.368	2301.6818	2542.1007	2869.3075
2163.860	2303.4225	2551.0936	2874.1722
2164.547	2303.579	2562.5348	2877.3005
2165.861	2308.9971	2575.7442	2894.5050
2172.581	2313.1022	2576.1033	2895.0352
2173.212	2320.3561	2584.5349	2899.4156
2176.837	2327.3940	2585.8753	2912.1581
2180.866	2331.3067	2598.3689	2920.6906
2183.979	2332.7972	2611.8725	2929.0081
2186.890	2338.0052	2613.8240	2941.3430
2187.192	2344.2802	2617.6160	2953.9400
2191.202	2354.8888	2621.6690	2957.3654
2196.040	2359.1039	2625.6663	2959.9924
2200.7227	2359.997	2628.2923	2965.2551
2201.117	2360.294	2635.8082	2981.4448
2207.068	2362.019	2643.9972	2987.2919
2210.686	2364.8269	2647.5576	2990.3923
2211.234	2366.592	2651.7059	2999.5123
2228.1704	2368.595	2662.0563	3003.0311
2231.211	2370.497	2673.2127	3009.5698
2240.627	2371.4285	2679.0608	3015.9129
2245.651	2374.517	2689.2117	3024.0330
2248.858	2375.193	2699.1060	3030.1491
2249.177	2379.2756	2706.5812	3037.3891
2253.1251	2380.7591	2711.6548	3040.4281
2255.861	2384.386	2714.413	3047.6059
2259.511	2388.6270	2718.4352	3055.2631
2260.079	2389.9713	2723.5770	3057.4452
2264.3894	2399.2396	2727.540	3059.0874

Iron, Wave Lengths in the Ultraviolet Spectrum (Continued)

λ_{air} A	λ_{air} A	λ_{air} A	λ_{air} A
3067.2433	3196.9288	3284.5892	3383.9808
3075.7204	3200.4741	3286.7538	3396.9772
3083.7419	3205.3992	3298.1328	3399.3343
3091.5777	3215.9398	3305.971	3401.5196
3116.6329	3217.3796	3306.356	3407.4608
3125.653	3222.0682	3314.7421	3413.1335
3134.1113	3225.7883	3323.7374	3427.1207
3143.9896	3236.2226	3328.8669	3443.8774
3157.0388	3239.4362	3337.6655	3445.1506
3160.6582	3244.1887	3340.5659	3465.8622
3175.4465	3254.3628	3347.9262	3476.7035
3178.0137	3257.5937	3355.2285	3485.3415
3184.8948	3271.0014	3370.7845	3490.5746
3191.6583	3280.2613	3380.1111	3497.8418

Iron, Wave Lengths in the Infrared

From interferometer measurements by Meggers, Jour. Research of B. of S. **14**, 33, 1935. The integrated light from the iron arc was used as a source. Wave lengths in air at 15°C and 760 mm are given as well as the intensity.

λ_{air} A	Intensity	λ_{air} A	Intensity	λ_{air} A	Intensity
7164.469	250	8096.874	10	8866.961	60
7187.341	800	8198.951	80	8945.204	10 *n*
7207.406	500	8207.767	40	8975.408	10
7389.425	80	8220.406	1500	8999.561	200
7401.689	4	8232.347	50	9012.098	10
7411.178	100	8239.130	8	9079.599	4
7418.674	5	8248.151	30	9088.326	50
7445.776	200	8293.527	20	9089.413	30
7495.088	400	8327.063	1200	9118.888	25
7511.045	800	8331.941	200	9147.800	2 *n*
7531.171	60	8339.431	80	9210.030	6
7568.925	30	8360.822	8	9258.30	10 *n*
7583.796	50	8365.642	25	9350.46	6
7586.044	150	8387.781	1200	9359.420	3
7620.538	25	8439.603	20	9362.370	4
7661.223	30	8468.413	300	9372.900	6
7664.302	80	8514.075	150	9430.08	3
7710.390	25	8526.685	8	9513.24	8 *n*
7748.281	125	8582.267	15	9569.960	15 *n*
7780.586	300	8611.807	40	9626.562	12 *n*
7832.224	400	8621.612	10	9653.143	15
7912.866	6	8661.908	600	9738.624	100
7937.166	700	8674.751	60	9753.129	10
7945.878	600	8688.633	1500	9763.450	10
7994.473	20	8757.192	25	9763.913	12
7998.972	700	8764.000	20 *n*	9800.335	8 *n*
8028.341	50	8793.376	25 *n*	9861.793	12
8046.073	600	8804.624	......	9889.082	15
8080.668	10 *n*	8824.227	250	10065.080	30
8085.200	500	8838.433	30	10145.601	40
				10216.351	50

Helium

Merrill, Bulletin 14, Bureau of Standards 1917.

λ_{air}	λ_{air}	λ_{air}	λ_{air}
2945.104	3888.646	4387.928	5047.736
3187.743	3964.727	4471.479	5875.620
3613.641	4026.189	4713.143	6678.149
3705.003	4120.812	4921.928	7065.188
3819.606	4143.759	5015.675	7281.349

From Meggers and Humphreys, Jour. Research of B. of S. **13**, 293. 1934.

λ_{air} A	λ_{air} A	λ_{air} A	λ_{air} A
10829.081	10830.250	10830.341	....

Neon

From Meggers and Humphreys, Jour Research of B. of S **13, 293, 1934.** The source was an "end on" Geissler tube Values are given for the wave length in air at 15°C and 760 mm pressure, and referred to the primary standard (Cd 6438.4696 A) and also to the red lines of neon. For complex lines the value for the principal component is given.

λ_{air} A Ne Std.	λ_{air} A Cd Std.	λ_{air} A Ne Std.	λ_{air} A Cd Std	λ_{air} A Ne Std	λ_{air} A Cd Std.
		4865.505	....		
4334.125	....	4866.476	....	5906.429	....
4363.524	....	4884.915	....	5913.633	....
4381.220	....	4892.090	....	....	5944.8340
4395.556	....	4928.235	....	5965.474	....
4422.519	....	4939.041	....	5974.628	....
4424.800	....	4944.987	....	....	5975.5345
4425.400	....	4957.0334	....	5987.9069	....
4433.721	....	4957.122	....	....	6029.9968
4460.175	....	4994.930	....	....	6074.3376
4466.807	....	5005.160	....	....	6096.1630
4475.556	....	5011.003	....	6128.4513	6128.4502
4483.190	....	5022.870	....	....	6143.0627
4488.0928	....	5031.3484	....	....	6163.5937
4500.182	....	5037.7505	....	6182.146	....
4517.736	....	5074.200	....	....	6217.2812
4525.764	....	5080.383	....	....	6266.4952
4537.751	....	5104.705	....	....	6304.7893
4540.376	....	5113.675	....	....	6334.4276
4552.598	....	5116.503	5116.501	....	6382.9914
4565.888	....	5122.257	....	6402.247	6402.248
4573.759	....	5144.9376	....	....	6506.5277
4575.060	....	5151.963	....	....	6532.8824
4582.035	....	5154.422	....	....	6598.9528
4582.450	....	5156.664	....	....	6678.2766
4609.910	....	5188.612	....	....	6717.0430
4614.391	....	5193.130	....	....	6929.4679
4617.837	....	5193.224	....	7024.0508	....
4628.309	....	5203.8950	....	7032.4134	7032.4125
4636.125	....	5208.863	....	7059.109	7059.108
4636.634	....	5210.573	....	7173.9389	7173.9390
4645.416	....	5222.351	....	7245.1668	7245.1668
4649.904	....	5234.028	....	7438.8990	7438.8988
4656.3923	....	5298.190	....	7488.8722	7488.8717
4661.104	....	5304.756	....	7535.7750	7535.774
4670.884	....	5326.396	....	7544.046	....
4678.218	....	5330.7766	5330.778	7943.1802	7943.1802
4679.135	....	5341.091	5341.093	8082.4580	8082.4585
4687.671	....	5343.284	....	8118.5495	8118.549
4704.395	....	5355.422	....	8136.4060	8136.4058
4708.854	....	5360.012	....	8259.380	....
4715.344	....	5374.975	....	8266.076	8266.077
4725.145	....	5400.5619	5400.5620	8300.3258	8300.3257
4749.572	....	5433.649	....	8377.6068	8377.6070
4752.7313	....	5448.508	....	8418.4274	8418.4275
4758.728	....	5562.769	....	8495.3600	8495.3604
4780.338	....	5656.6585	....	8591.2584	8591.2585
4788.9258	....	5662.547	....	8634.6480	8634.649
4790.218	....	5689.8164	5689.817	8654.3835	8654.383
4800.111	....	5719.2254	5719.224	8679.491	....
4810.0625	....	5748.299	5748.298	8681.920	....
4817.636	....	5764.418	5764.4182	8780.6223	8780.6220
4821.924	....	5804.4488	5804.450	8783.755	8783.755
4827.338	....	5820.1548	5820.1553	8853.867	8853.864
4827.587	....	....	5852.4878	8865.759	....
4837.3118	....	5872.828	....	9486.680	....
4852.655	....	....	5881.8950	9535.167	....
4863.074	....	5902.464	5902.4634	9665.424	....

Neon$_I$, Ultraviolet Lines

From Humphreys, Jour. Research of B. of S. **20,** 17, 1938 The source was an "end on" Geissler tube Values are given for the wave length in dry air at 15°C and 760 mm of Hg pressure and referred to the Krypton secondary standards.

λ_{air} A	λ_{air} A	λ_{air} A	λ_{air} A
3369.8086	3450.7653	3501.2165	3609.1793
3369.9081	3454.1952	3510.7214	3633.6646
3375.6498	3460.5245	3515.1908	3682.2428
3417.9036	3464.3389	3520.4717	3685.7359
3418.0066	3466.5786	3593.5259	3701.2250
3423.9127	3472.5711	3593.6398	3754.2160
3447.7029	3498.0644	3600.1693	

Argon

From Meggers and Humphreys, Jour. Research of B. of S. **13,** 293, 1934.

λ_{air} A Ne Std.	λ_{air} A Cd Std.	λ_{air} A Ne Std.	λ_{air} A Cd Std.	λ_{air} A Ne Std.	λ_{air} A Cd Std.
3948.977	....	4887.9465	....	7030.262	....
4044.4173	....	5162.2845	....	7067.2170	7067.2177
4054.5250	....	5187.7458	....	7147.0406	7147.0412
4158.5896	4158.5895	5221.270	....	7272.9357	7272.9356
4164.1788	4164.1789	5252.786	....	7372.117	....
4181.8826	4181.8825	5421.346	....	7383.9800	7383.9800
....	4190.7098	5451.650	....	7503.8676	7503.8667
....	4191.0270	5495.8720	....	7514.6510	7514.653
4198.316	4198.3160	5506.112	....	7635.1053	7635.1055
4200.674	4200.6738	5558.702	....	7723.7597	7723.761
4251.1842	....	5572.548	....	7724.2064	7724.206
4259.3603	4259.3607	5606.732	....	7891.075	....
4266.2853	4266.2855	5650.7034	....	7948.1754	7948.1756
4272.1680	4272.1678	5739.517	....	8006.1556	8006.155
4300.0995	4300.1000	5834.263	....	8014.7856	8014.785
4333.5595	4333.5601	5860.315	....	8053.307	....
4335.3370	4335.3363	5888.592	....	8103.6922	8103.6922
4345.1666	....	5912.084	....	8115.3115	8115.3095
4363.7936	....	5928.805	....	8264.5209	8264.5210
4423.9936	....	6032.124	....	8408.208	8408.207
4510.7322	4510.7324	6043.230	....	8424.647	8424.646
4522.3216	....	6052.721	....	8521.4407	8521.4406
4589.2884	....	6059.373	....	8667.9430	8667.9435
4596.0964	....	6105.645	....	9122.9660	9122.9664
4628.4398	....	6170.183	....	9224.498	9224.498
4702.3151	....	6173.106	....	9354.218	....
4752.9381	....	6416.315	....	9657.7841	....
4768.6716	....	6752.832	....	9784.5010	....
4876.2596	....	6965.4302	6965.4304	10470.051	....

Argon$_I$

From Humphreys, Jour. Research of B. of S. **20.** 17, 1938. The source was an "end on" Geissler tube. Values are given for the wave length in dry air at 15°C and 760 mm of Hg pressure, referred to the Krypton secondary standards.

λ_{air} A	λ_{air} A	λ_{air} A	λ_{air} A
3319.3446	3690.8957	4158.5906	4272.1690
3373.481	3770.3688	4164.1800	4300.1011
3393.7517	3781.3609	4181.8838	4333.5612
3461.0780	3834.6785	4190.7127	4335.3380
3554.3061	3894.6602	4191.0296	4345.1682
3567.6565	3947.5043	4198.3170	4510.7333
3606.5224	3948.9788	4200.6751	4522.3238
3632.6837	4044.4182	4251.1852	4596.0970
3634.4605	4045.9658	4259.3618	4628.4410
3649.8324	4054.5254	4266.2867	4702.3164

Krypton

Meggers, Journal Optical Society of America, 1921.

λ_{air} A	λ_{air} A	λ_{air} A	λ_{air} A
4273.9696	4362.6422	4502.354	6456.290
4282.967	4376.122	4807.065	7587.414
4318.552	4399.969	5562.224	7601.544
4319.580	4453.9174	5570.2872	
4355.478	4463.690	5870.9137	

Krypton$_I$

From Humphreys, Jour. Research of B. of S. **20,** 17, 1938. The source was the "end on" Geissler tube. Values are given for the wave length in air at 15°C and 760 mm of Hg pressure.

λ_{air} A	λ_{air} A	λ_{air} A	λ_{air} A
3424.9433	3615.4755	3796.8839	4184.4726
3431.7217	3628.1570	3800.5437	4263.2881
3434.1423	3632.4896	3812.2155	4302.4455
3495.9900	3665.3259	3837.7028	4410.3685
3502.5537	3668.7363	3837.8162	4416.8838
3503.8981	3679.5609	3845.9778	4418.7626
3511.8963	3679.6111	3982.1699	4425.1908
3522.6747	3698.0452	3991.0797	4550.2985
3539.5416	3773.4241	3991.2581	4812.6367
3540.9538			

STANDARD WAVE LENGTHS

Titanium, Vacuum Arc Spectrum

From the measurements by Kiess. Jour. Research of B. of S. **1**, 75. 1928.
Wave lengths in air at 15°C. 760 mm.

λ_{air} A	Intensity	λ_{air} A	Intensity	λ_{air} A	Intensity
2941.995	60 *r*	3929.875	40	4544.688	30
2948.255	60 *r*	3947.779	40	4548.764	35
2956.133	70 *R*	3948.679	60	4549.622	25
3075.225	40	3956.336	60	4552.453	35
3078.645	45	3958.206	80	4555.486	30
3088.027	60	3962.851	35	4559.920	6
3130.804	15	3964.269	35	4563.761	15
3148.033	12	3981.761	70 *R*	4571.971	15
3161.755	20	3989.758	80 *R*	4599.226	5
3168.519	30	3998.635	100 *R*	4617.269	30
3186.451	60 *r*	4008.926	35	4623.098	25
3190.801	20	4013.587	12 *n*	4629.336	15
3191.994	80 *R*	4015.377	12 *n*	4645.193	12
3199.915	100 *R*	4017.771	15 *n*	4656.468	25
3202.535	12	4024.573	35	4667.585	25
3214.240	12	4026.539	25 *n*	4675.118	10
3217.056	15	4030.512	25 *n*	4681.908	30
3222.843	15	4035.828	10	4691.336	20
3229.397	10	4055.011	20	4698.766	20
3234.517	60	4058.139	7	4710.186	18
3236.573	50	4060.263	20	4715.295	4
3239.037	40	4065.094	15	4731.172	9
3241.984	40	4078.471	30	4742.791	20
3261.596	25	4082.456	20	4758.120	25
3287.657	10	4099.166	8	4759.272	25
3292.078	20	4112.708	20	4778.259	10
3314.422	10	4122.143	10	4792.482	10
3318.024	8	4127.531	15	4799.797	12
3322.936	20	4137.284	10 *n*	4805.416	12
3326.762	5	4150.963	10	4820.410	20
3329.455	20	4159.634	9	4840.874	25
3332.111	8	4163.644	8	4856.012	20
3335.192	20	4171.897	5	4868.264	18
3340.344	15	4186.119	25	4870.129	20
3341.875	50 *r*	4237.889	7	4885.082	20
3349.399	40	4249.114	5 *n*	4899.910	20
3354.634	60 *r*	4256.025	8 *n*	4913.616	20
3358.271	10	4258.523	4 *n*	4919.867	10
3361.213	40	4263.134	15	4921.768	12
3370.436	40 *r*	4274.584	15	4928.342	12
3371.447	80 *R*	4281.371	10	4938.283	8
3377.577	30 *r*	4282.702	12	4975.344	10
3380.278	15	4286.006	25	4978.191	10
3383.761	40	4287.405	22	4981.732	60
3385.944	40 *r*	4289.068	25	4991.067	50
3387.834	15	4294.101	8	4997.099	8
3394.574	15	4295.751	22	4999.504	45
3444.306	15	4298.664	40	5007.209	40
3461.496	20	4305.910	60	5009.652	7
3477.181	15	4307.900	12	5014.185	25
3480.525	12	4312.861	7	5014.277	25
3491.053	8	4314.801	25	5016.162	20
3504.890	8	4318.631	10 *n*	5020.028	25
3510.840	10	4321.655	8 *n*	5022.871	25
3535.408	10	4325.134	9 *n*	5024.842	20
3547.029	15	4337.916	10	5025.570	18
3596.048	10	4346.104	5	5035.908	25
3598.714	15	4360.487	4	5036.468	25
3610.154	12	4369.682	5 *n*	5038.400	25
3624.826	8	4372.383	3	5039.959	22
3635.462	80 *r*	4393.925	8	5052.879	8
3641.330	10	4395.031	25	5062.112	7
3642.675	80 *r*	4399.767	6	5064.654	25
3646.198	12	4417.274	15	5087.055	8
3653.497	100 *r*	4421.754	6	5113.448	10
3654.592	15	4422.823	10	5120.420	12
3658.097	20	4426.054	10	5145.465	12
3660.631	12	4427.098	40	5147.483	10
3668.965	15	4430.366	7	5173.742	30
3671.672	20	4434.003	15	5192.971	35
3685.192	40	4436.586	4	5210.386	40
3689.916	15	4440.345	10	5224.301	15
3694.445	10	4443.802	25	5224.928	8
3717.393	20	4449.143	30	5265.967	10
3722.568	15	4450.896	25	5283.441	8
3724.570	20	4453.312	30	5295.781	4
3725.155	20	4453.708	20	5297.236	6
3729.806	50 *r*	4455.321	30	5298.429	4
3741.059	60 *r*	4457.428	40	5369.635	4
3752.860	80 *r*	4465.807	20	5389.996	3
3759.291	40	4468.493	25	5397.093	4
3761.320	40	4471.238	20	5409.609	6
3771.652	25	4474.852	8	5429.139	6
3786.043	20	4481.261	30	5477.695	8
3798.276	6	4482.688	10	5481.426	6
3866.446	15 *n*	4489.089	20	5488.210	5
3875.262	20 *n*	4496.146	20	5490.151	12
3895.243	30 *n*	4501.270	25	5503.897	8
3900.546	50	4503.762	4 *n*	5512.529	25
3904.785	40 *n*	4512.734	40	5565.476	9
3913.464	40	4518.022	50	5644.137	18
3914.334	35	4522.798	40	5648.570	5
3921.423	30	4527.305	35	5662.154	12
3924.527	50	4533.238	80	5675.413	9

Titanium, Vacuum Arc Spectrum (Continued)

λ_{air} A	Intensity	λ_{air} A	Intensity	λ_{air} A	Intensity
5689.465	10	5922.112	18	6258.103	40
5715.123	9	5937.806	6	6258.706	50
5739.464	9	5941.755	12	6261.101	35
5766.330	4 *n*	5953.162	30	6303.754	10
5774.037	5 *n*	5965.828	30	6312.240	10
5785.979	5 *n*	5978.543	25	6336.104	8
5804.265	5 *n*	5999.668	8	6366.354	8
5866.453	35	6064.631	9	6546.276	20
5899.295	25	6085.228	20	6554.226	30
5903.317	5	6091.175	20	6556.066	25
5918.548	10	6126.217	20	6743.124	10

Xenon

Meggers. Journal Optical Society of America. 1921

λ_{air}	λ_{air}	λ_{air}
4500.978	4624.275	4807.019
4524.680	4671.225	4829.705
4582.746	4697.020	4844.333
4603.028	4734.154	4923.246

Wave Lengths in the Vacuum Ultraviolet

J. C. Boyce, Rev. Med. Phys. 13, 34, (1941)

Copper (Shenstone, 1936)

λA	Intensity	P.E. in .001 A	λA	Intensity	P.E. in .001 A	λA	Intensity	P.E. in .001 A
685.139	8	2	1065.781	20	1	1393.126	10	2
685.396	2	2	1066.133	20	2	1399.355	3	2
724.487	15	2	1069.193	50	2	1402.776	15	2
735.519	20	2	1088.393	20	2	1442.136	15	2
736.031	25	2	1106.446	3	2	1444.131	2	2
810.997	15	2	1109.742	1	2	1473.976	25	2
813.882	20	2	1185.899	2	2	1496.686	35	4
826.995	30	2	1214.553	1	2	1517.630	20	2
866.440	5	2	1219.332	1	2	1519.491	50	2
876.719	20	2	1241.961	2	1	1535.004	25	2
883.837	5	2	1248.790	5	2	1540.391	30	2
884.824	5	2	1250.045	10	2	1541.701	75	2
911.654	1	2	1265.504	15	1	1558.344	30	2
912.022	0	2	1266.308	10	1	1565.925	40	2
992.951	25	1	1274.069	3	2	1566.411	40	2
1001.010	8	1	1274.463	3	2	1569.216	10	2
1004.053	30	1	1275.570	30	2	1590.164	40	3
1008.568	30	1	1281.458	8	4	1593.557	60	3
1011.433	2	1	1297.549	2	2	1598.402	40	3
1012.595	25	1	1298.394	15	2	1602.387	40	3
1018.705	50	1	1299.267	10	2	1604.848	20	3
1019.652	15	1	1308.296	30	2	1606.834	40	3
1022.100	5	1	1309.463	15	2	1608.638	25	3
1027.830	50	1	1314.147	15	2	1610.298	15	3
1028.326	25	1	1314.335	30	2	1617.914	20	3
1031.764	8	1	1325.511	3	2	1621.426	60	3
1035.160	8	1	1326.394	10	2	1649.457	25	3
1036.468	60	1	1350.592	15	2	1656.326	20	3
1039.345	60	2	1351.837	25	2	1660.005	20	3
1044.742	80	1	1355.304	15	2	1663.003	30	3
1049.363	20	2	1359.010	20	2	1944.586	25	4
1049.754	50	1	1359.935	5	2	1970.489	15	4
1054.690	60	1	1362.598	20	2	1979.947	50	4
1055.795	40	2	1363.501	5	2	1989.849	30	4
1058.796	40	2	1370.558	2	2	2000.339	60	4
1059.094	60	1	1371.840	20	2			

Iron (Green, 1939)

	Intensities			Intensities			Intensities	
λA	Schüler tube	Spark	λA	Schüler tube	Spark	λA	Schüler tube	Spark
1550.273	1	..	1570.244	20	1	1584.949	15	
1559.084	20	2	1572.754	1	..	1612.805	20	8
1563.788	25	2	1573.826	5	..	1623.090	8	1
1566.821	20	1	1574.769	0	..	1625.520	20	8
1568.017	8	..	1574.921	20	1	1632.665	1	..
1569.674	12	..	1577.167	1	..	1633.906	15	2

Iron (Green, 1939) (Continued)

λA	Intensities Schüler tube	Intensities Spark	λA	Intensities Schüler tube	Intensities Spark
1637.398	15	2	1702.044	25	25
1640.150	12	2	1709.551	0	..
1643.576	15	2	1712.998	20	25
1652.482	0	..	1718.100	2	..
1654.476	5	1	1720.611	20	20
1658.771	15	2	1724.962	8	1
1659.479	20	10	1726.391	12	8
1663.220	15	2	1815.411	0	1
1670.990	1	..	1818.516	2	1
1674.254	2	1	1826.994	1	1
1676.854	1	..	1833.073	0	..
1685.952	5	1	1842.283	0	..
1686.454	8	1	1848.771	12	2
1691.272	8	1	1851 526	1	..
1693.475	0	..	1859.744	15	10
1693.935	0	..	1898.535	10	2
1696.794	8	..	1903.384	1	..
1699.195	2	..	1904.785	15	5
			2001.025	30	30

Hydrogen (Paschen, 1929)

λA	λA
923.148	949.739
926.222	972.532
930.745	1025.717
937.799	1215.664

Helium (He II)(Paschen, 1929)

λA	λA	λA
234.3452	949.326	1084.940
237.3297	958.696	1215.129
243.0244	972.109	1640.409
256.3145	992.361	
303.7788	1025.270	

Carbon, Nitrogen, Oxygen (More and Rieke, 1936)

El.	λA	Int.	El.	λA	Int.	El.	λA	Int.
O II	832.764	0	N I	1176.502	1	C II	1335.703	18*d*
O II	833.332	1	N I	1199.550	7	N I	1494.670	4
O II	834.467	2	N I	1200.218	6	C I	1560.313	8*d*
O I	990.205	4*d*	N I	1200.707	5	C I	1560.702	15*d*
O I	990.794	3	O I	1217.643	2	C I	1656.271	8*d*
O I	999.493	2	O I	1302.174	8	C I	1656.994	15*d*
N II	1083.996	2	O I	1304.858	8	C I	1657.388	5*d*
N II	1084.582	3	O I	1306.023	6	C I	1657.908	8*d*
N I	1134.168	3	C I	1328.820	3	C I	1658.135	8*d*
N I	1134.417	3	C I	1329.099	5			
N I	1134.979	4	C II	1334.534	15*d*			

PERSISTENT LINES OF THE ELEMENTS

Spectra of the neutral atom, the singly ionized atom, and the doubly ionized atom are indicated by I, II, and III, respectively. The most sensitive lines are indicated by an asterisk (*). The symbol "D" preceding the intensity indicates the discharge-tube spectrum. Wave lengths are given in International Ångstrom units.
Wave lengths and intensities of lines between 2000 and 9999 ångstroms have been given according to the M.I.T. Wavelength Tables.

ARRANGED BY ELEMENTS

Element	Wave length	Intensity Arc	Intensity Spark
A I	*1048.26		
	1066.70		
	6965.430	....	D 400
	7067.217	....	D 400
	7503.867	....	D 700
	8115.311	....	D 5000
Ag I	*3280.683	2000	1000
	3382.891	1000	700
	5209.067	1500	1000
	5465.487	1000	500
Ag II	2246.412	25	300
	*2437.791	60	500
Al I	3082.155	800	800
	3092.713	1000	1000
	3944.032	2000	1000
	*3961.527	3000	2000
Al II	1671.0		
	1856.00		
	1858.13		
	1862.48		
	2631.553	....	40
	*2669.166	3	100
Al II	2816.179	10	100
	6231.76	....	30
	6243.36	....	100
Al III	1854.67		
	1862.90		
As I	1889.9		
	*1890.5		
	1936.9		
	1972.0		
	2288.12	250	5
	2349.84	250	18
	2369.67	40	
	2370.77	50	3
	2456.53	100	8
	2780.197	75	75
	2860.452	50	50
	2898.71	25	40
Au I	*2427.95	400	100
	2675.95	250	100
	2802.19	....	200
B I	2496.778	300	300
	*2497.733	500	400
B II	1362.46		
	3451.41	5	30
	3452.33		
Ba I	3071.591	100	50
	5424.616	100	30
	5519.115	200	60
	*5535.551	1000	200
	5777.665	500	100
Ba II	2304.235	60	80
	2335.269	60	100
	3891.785	18	25
	4130.664	50	60
	*4554.042	1000	200
	4934.086	400	400
Be I	*2348.610	2000	50
	2650.781	25	
	3321.013	50	
	3321.086	100	
	3321.343	1000	30
Be II	*3130.416	200	200
	3131.072	200	150
Bi I	2061.70	300	100
	2276.578	100	40
	2780.521	200	100
	2809.625	200	100
	2897.975	500	500
	2938.298	300	300
	2989.029	250	100
	*3067.716	3000	2000
	4722.552	1000	100
Bi II	1909.41		
Br I	1540.8		
	1633.8		
Br II	*4704.86	....	D 250
	4785.50	....	D 400
	4816.71	....	D 300
C I	*2478.573	400	D 400
C II	1334.54		
	1335.72		
	2836.710	....	200
	2837.602	....	40
	4267.02	....	350
	4267.27	....	500
C III	2296.89	....	200
Ca I	*4226.728	500	50
	4425.441	100	
	4434.960	150	
	4454.781	200	
	4455.880		
	4456.62		
Ca II	3158.869	100	300
	3179.332	100	400
	*3933.666	600	600
	3968.468	500	500
Cb I	*4058.938	1000	400
	4079.729	500	200
	4100.923	300	200
	4123.810	200	125
	4137.095	100	60
Cb II	*3094.183	100	1000
	3130.786	100	100
	3163.402	15	8
	3194.977	30	300
	3225.479	150	800
Cd I	*2288.018	1500	300
	3261.057	300	300
	3403.653	800	500
	3466.201	1000	500
	3610.510	1000	500
	6438.4696	2000	1000
Cd II	*2144.382	50	200
Cd II	2265.017	25	300
	2312.84	1	200
	2573.09	3	150
	2748.58	5	200
Ce II	4012.388	60	20
	4040.762	70	5
	4165.606	40	6
	*4186.599	80	25
Cl I	1379.6		
	1396.5		
Cl II	4794.54	....	D 250
	4810.06	....	D 200
	4819.46	....	D 200
Co I	*3453.505	3000	200
	3465.800	2000	25
	3529.813	1000	30
Co II	*2286.156	40	300
	2307.857	25	50
	2363.787	25	50
	2378.622	25	50
	2388.918	10	35
	2519.822	40	200
	3405.120	2000	150
Cr I	*4254.346	5000	1000
	4274.803	4000	800
	4289.721	3000	800
	5204.518	400	100
	5206.039	500	200
	5208.436	500	100
Cr II	*2835.633	100	400
	2843.252	125	400
	2849.838	80	150
	2855.676	60	200
	2860.934	60	100
Cs I	4555.355	2000	100
	4593.177	1000	50
	*8521.10	5000	
	8943.50	2000	
Cu I	*3247.540	5000	2000
	3273.962	3000	1500
	5105.541	500	
	5153.235	600	
	5218.202	700	
Cu II	*2135.976	25	500
	2192.260	25	500
	2246.995	30	500
Dy I	4000.454	400	300
	4045.983	150	12
Dy I	4077.974	150	100
	4167.966	50	12
	4211.719	200	15
Er I	3499.104	18	15
	3692.652	20	12
	3906.316	25	12
Eu I	*4594.02		
Eu II	4129.737	150	50
	*4205.046	200	50
F I	6856.02	....	D 1000
	6902.46	....	D 500
Fe I	3581.20		
	*3719.935	1000	700
	3737.133	1000	600
	3745.564	500	500
	3745.903	150	100
	3748.264	500	200
Fe II	*2382.039	40	100
	2395.625	50	100
	2404.882	50	100
	2410.517	50	70
	2413.309	60	100
Ga I	2874.244	10	15
	2943.637	10	20
	4032.982	1000	500
	*4172.056	2000	1000
Gd I	3646.196	200	150
	3768.405	20	20
Ge I	*2651.178	40	20
	2651.575	30	20
	2709.626	30	20
	3039.064	1000	1000
	3269.494	300	300
	4226.570	200	50
H I	1215.7		
	4861.327	....	D 500
	6562.79	....	D 3000
He I	584.4		
	*3888.646	....	D 1000
	5875.618	....	D 1000
He II	303.8		
	1640.5		
	4685.75	....	D 300
Hf I	2898.259	50	12
	2904.408	30	6

PERSISTENT LINES OF THE ELEMENTS (Continued)

ARRANGED BY ELEMENTS (Continued)

Element	Wave length	Intensity	
		Arc	Spark
Hf I	2916.481	50	15
	2940.772	60	12
	3072.877	80	18
	4093.161	25	20
Hf II	2513.028	25	70
	2516.881	35	100
	*2641.406	40	125
	2773.357	25	60
	2820.224	40	100
	3134.718	80	125
Hg I	*1849.68		
	2536.519	2000	1000
	3650.146	200	500
	3654.833	....	D 200
Hg I	3663.276	500	400
	4046.561	200	300
	4358.35	3000	500
	5460.740	....	D 2000
Hg II	1649.8		
	1942.3		
Ho I	3748.17	60	40
	3891.02	200	40
Ho II	2936.77	....	1000
I I	1782.9		
	2062.38	...	D 900
I II	5161.188	....	D 300
	5464.61	....	D 900
In I	3039.356	1000	500
	3256.090	1500	600
	3258.564	500	300
	4101.773	2000	1000
	*4511.323	5000	4000
Ir I	2543.97		
	2849.725	40	20
	2924.792	25	15
	*3220.780	100	30
	3437.015	20	15
	3513.645	100	100
K I	4044.140	800	400
	4047.201	400	200
	*7664.907	9000	
	7698.979	5000	
Kr I	5570.2895	....	D 2000
	5870.9158	....	D 3000
La I	5455.146	200	1
	5930.648	250	
	*6249.929	300	
La II	*3949.106	1000	800
	4077.340	600	400
	4123.228	500	500
Li I	3232.61	1000	500
	4603.00	800	
	6103.642	2000	300
	*6707.844	3000	200
Lu I	4518.57	300	40
Lu II	*2615.43		
	2894.84	60	200
	2911.39	100	300
	3397.07	50	20
	3472.48	50	150
	3554.43	50	150
Mg I	*2852.129	300	100
	3829.350	100	150
	3832.306	250	200
	3838.258	300	200
	5167.343	100	50
	5172.699	200	100
	5183.618	500	300
Mg II	*2795.53	150	300
	2802.695	150	300
Mn I	*4030.755	500	20
	4033.073	400	20
	4034.490	250	20
Mn II	*2576.104	300	2000
	2593.729	200	1000
	2605.688	100	500
Mo I	*3798.252	1000	1000
	3864.110	1000	500
	3902.963	1000	500
Mo II	*2816.154	200	300
	2848.232	125	200
	2871.508	100	100
	2890.994	30	50
	2909.116	25	40
N I	1199.5		
	1200.2		
	1200.7		
	*4099.94	....	D 150
	4109.98	....	D 1000
N II	5666.64	....	D 300
	5676.02	....	D 100
	5679.56	....	D 500
N III	989.8		
N III	991.6		
	4097.31	....	D 100
	4103.37	....	D 80
Na I	3302.323	600	300
	3302.988	300	150
	5682.657	80	
	5688.224	300	
	*5889.953	9000	1000
	5895.923	5000	500
Nd I	3951.154	40	30
	4177.321	15	25
	*4303.573	100	40
Ne I	735.95		
	743.73		
	5400.562	....	D 2000
	5832.488		
	5852.488	....	D 2000
	6402.246	....	D 2000
Ni I	*3414.765	1000	50
	3492.956	1000	100
	3515.054	1000	50
	3524.541	1000	100
Ni II	2216.47		
	2253.86	100	300
	2264.457	150	400
	2270.213	100	400
	*2287.084	100	500
O I	1302.27		
	1304.96		
	1306.12		
	*7771.928	....	D 1000
	7774.138	....	D 300
	7775.433	....	D 100
Os I	*2909.061	500	400
	3058.66	500	500
	3262.290	500	50
	3267.945	400	30
	3301.56		
	3752.54		
Os I	3782.20		
	4420.468	400	100
P I	1774.8		
	1782.7		
	1787.5		
	2136.8		
	2149.8		
	2534.01	50	D 20
	2535.65	100	D 30
	2536.38		
	2553.28	80	D 20
	2554.02		
	2554.93	60	D 20
Pb I	2169.994	1000	1000
	2614.178	200	80
	2833.069	500	80
	3639.580	300	50
	3683.471	300	50
	*4057.820	2000	300
Pb II	1682.4		
	*2203.505	50	5000
	5608.8	....	D 40
Pd I	*3404.580	2000	1000
	3421.24	2000	1000
	3516.943	1000	500
	3609.548	1000	700
	3634.695	2000	1000
Pd II	*2296.53		
	2488.921	10	30
	2498.784	4	150
	2505.739	3	30
	2658.722	20	300
	2854.581	4	500
Pr I	4062.817	150	50
Pr II	*4179.422	200	40
	4189.518	100	50
	4225.327	50	40
Pt I	2659.454	2000	500
	2830.295	1000	600
	2929.794	800	200
	2997.967	1000	200
	*3064.712	2000	300
Ra I	*4825.91	....	D 800
Ra II	*3814.42	....	D 2000
	4682.28	....	D 800
Rb I	4201.851	2000	500
	4215.556	1000	300
	*7800.227	9000	
	7947.60	5000	
Re I	*3460.47	1000	
	4889.17	2000	
Rh I	3323.092	1000	200
	3396.85	1000	500
	*3434.893	1000	200
Rh I	3657.987	500	200
	3692.357	500	150
Rn I	7055.42	....	D 400
	7450.00	....	D 600
Ru I	3436.737	300	150
Ru I	*3498.942	500	200
	3596.179	30	100
Ru II	2678.758	100	300
	2692.065	8	200
	2712.410	80	300
	2945.668	60	300
	2965.546	60	200
	2976.586	60	200
S I	1807.4		
	1820.5		
	1826.4		
	4694.13	....	D 500
	4695.45	....	D 30
	4696.25	....	D 15
	9212.91	....	D 200
	9228.11	....	D 200
	9237.49	....	D 200
Sb I	*2068.38	300	3
	2175.890	300	40
	2311.469	150	50
	2528.535	300	200
	2598.062	200	100
	2877.915	250	150
	3232.499	150	250
	3267.502	150	150
Sc I	3907.476	125	25
	*3911.810	150	30
	4020.399	50	20
	4023.688	100	25
	5671.80		
Sc II	*3613.836	40	70
	3630.740	50	70
	3642.785	60	50
Se I	*1960.2		
	2039.851	....	D 1000
	2062.788	....	D 800
	4730.78	....	D 1000
	4739.03	....	D 800
	4742.25	....	D 500
Si I	2506.899	300	200
	2516.123	500	500
	2528.516	400	500
	*2881.578	500	400
	3905.528	20	15
Si II	1526.83		
	*1533.55		
Sm I	*4296.75		
Sm II	3568.27		
	4390.865	150	150
	*4424.342	300	300
	4434.321	200	200
Sn I	*2839.989	300	300
	2863.327	300	300
	3009.147	300	200
	3034.121	200	150
	3175.019	500	400
	3262.328	400	300
	4524.741	500	50
Sn III	*2152.22		
Sr I	*4607.331	1000	50
	4832.075	200	8
	4872.493	25	
	4962.263	40	
Sr II	3380.711	150	200
	3464.57	200	200
	3474.887	80	50
	*4077.714	400	500
	4215.524	300	400
	4305.447	40	
Ta I	*3311.162	300	70
	3318.840	125	35
	3406.664	70	18
Tb I	3509.17	200	200
	3561.74	200	200
	3848.75	100	200
	3874.18	200	200
Te I	*2142.75	60	
	2383.25	500	D 300
	2385.76	600	D 300
	2530.70	....	D 30
	2769.67	....	D 30
Th I	3538.75	....	50
	3601.040	8	10
	4019.137	8	8
Th II	3290.59	....	40
Ti I	3635.463	200	100
	3642.675	300	125
	3653.496	500	200
Ti I	*4981.733	300	125
	4991.066	200	100
	4999.510	200	80
	5007.213	200	40
	5014.25		
Ti II	*3349.035	125	800
	3361.213	100	600
	3372.800	80	400
	3383.761	70	300
Tl I	2767.87	400	300
	2918.32	400	200
	3229.75	2000	800
	3519.24	2000	1000
	3775.72	3000	1000
	*5350.46	5000	2000
Tm I	3462.21	200	100
	3761.333	250	150
	3761.917	200	120
Tm II	*3848.02		
U I	3552.172	8	12
	3672.579	8	15
	4241.669	40	50
V I	3183.406	200	100
	3183.982	500	400
	3185.396	500	400
	*4379.238	200	200
	4384.722	125	125
V I	4389.974	80	60
V II	*3093.108	100	400
	3102.299	70	300
	3110.706	70	300
	3118.383	70	200
	3125.284	80	200
W I	4008.753	45	45
	4294.614	50	50
	*4302.108	60	60
W II	*2204.49		
	2397.091	18	30
	2589.167	15	25
	3215.560	10	9
	3613.790	10	30
Xe I	1295.8		
	1469.9		
	4500.977	....	D 500
	4624.276	....	D 1000
	4671.226	....	D 2000
Y I	4643.695	50	100
	*4674.848	80	100
	5466.47		
Y II	3242.280	60	100
	3600.734	100	300
	3633.123	50	100
	*3710.290	80	150
	3774.332	12	100
	3788.697	30	30
Yb I	3289.37	500	1000
	*3987.994	1000	500
Yb II	*3694.203	500	1000
Zn I	*2138.56	800	500
	3282.333	500	300
	3302.588	800	300
	3345.020	800	300
	4680.138	300	200
	4722.159	400	300
	4810.534	400	300
	6362.347	1000	500
Zn II	*2025.51	200	200
	2061.91	100	100
	2502.001	20	400
	2557.958	10	300
Zr I	3519.605	100	10
	3547.682	200	12
	*3601.193	400	15
	4687.803	125	
	4710.075	60	
	4739.478	100	
	4772.312	100	
	4815.62		
Zr II	*3391.975	300	400
	3438.230	250	200
	3496.210	100	100
	3572.473	60	80

PERSISTENT LINES OF THE ELEMENTS (Continued)

ARRANGED BY WAVE LENGTHS

Wave length	Element	Intensity Arc	Intensity Spark
303.8	He II		
584.4	He I		
735.95	Ne I		
743.73	Ne I		
989.8	N III		
991.6	N III		
*1048.26	A I		
1066.70	A I		
1199.5	N I		
1200.2	N I		
1200.7	N I		
1215.7	H I		
1295.8	Xe I		
1302.27	O I		
1304.96	O I		
1306.12	O I		
1334.54	C II		
1335.72	C II		
1362.46	B II		
1379.6	Cl I		
1396.5	Cl I		
1469.9	Xe I		
1526.83	Si II		
*1533.55	Si II		
1540.8	Br I		
1633.8	Br I		
1640.5	He II		
1649.8	Hg II		
1671.0	Al II		
1682.4	Pb II		
1774.8	P I		
1782.7	P I		
1782.9	I I		
1787.5	P I		
1807.4	S I		
1820.5	S I		
1826.4	S I		
*1849.6	Hg I		
1854.67	Al III		
1856.00	Al II		
1858.13	Al II		
1862.48	Al II		
1862.90	Al III		
1889.9	As I		
*1890.5	As I		
1909.41	Bi II		
1936.9	As I		
1942.3	Hg II		
*1960.2	Se I		
1972.0	As I		
*2025.51	Zn II	200	200
2039.851	Se I	....	D 1000
2061.70	Bi I	300	100
2061.91	Zn II	100	100
2062.38	I I	....	D 900
2062.788	Se I	....	D 800
*2068.38	Sb I	300	3
*2135.976	Cu II	25	500
2136.8	P I		
*2138.56	Zn I	800	500
*2142.75	Te I	60	
*2144.382	Cd II	50	200
2149.8	P I		
*2152.22	Sn II		
2169.994	Pb I	1000	1000
2175.890	Sb I	300	40
2192.260	Cu II	25	500
*2203.505	Pb II	50	5000
*2204.49	W II		
2216.47	Ni II		
2246.412	Ag II	25	300
2246.995	Cu II	30	500
2253.86	Ni II	100	300
2264.457	Ni II	150	400
2265.017	Cd II	25	300
2270.213	Ni II	100	400
2276.578	Bi I	100	40
*2286.156	Co II	40	300
*2287.084	Ni II	100	500
*2288.018	Cd I	1500	300
2288.12	As I	250	5
*2296.53	Pd II		
2296.89	C III	....	200
2304.235	Ba II	60	80
2307.857	Co II	25	50
2311.469	Sb I	150	50
2312.84	Cd II	1	200
2335.269	Ba II	60	100
*2348.610	Be I	2000	50
2349.84	As I	250	18
2363.787	Co II	25	50
2369.67	As I	40	
2370.77	As I	50	3
2378.622	Co II	25	50
*2382.039	Fe II	40	100
2383.25	Te I	500	D 300
2385.76	Te I	600	D 300
2388.918	Co II	10	35
2395.625	Fe II	50	100
2397.091	W II	18	30
2404.882	Fe II	50	100
2410.517	Fe II	50	70
2413.309	Fe II	60	100
*2427.95	Au I	400	100
*2437.791	Ag II	60	500
2456.53	As I	100	8
*2478.573	C I	400	D 400
2488.921	Pd II	10	30
2496.778	B I	300	300
*2497.733	B I	500	400
2498.784	Pd II	4	150
2502.001	Zn II	20	400
2505.739	Pd II	3	30
2506.899	Si I	300	200
2513.028	Hf II	25	70
2516.123	Si I	500	500
2516.881	Hf II	35	100
2519.822	Co II	40	200
2528.516	Si I	400	500
2528.535	Sb I	300	200
2530.70	Te I	....	D 30
2534.01	P I	50	D 20
2535.65	P I	100	D 30
2536.38	P I		
2536.519	Hg I	2000	1000
2543.97	Ir I		
2553.28	P I	80	D 20
2554.02	P I		
2554.93	P I	60	D 20
2557.958	Zn II	10	300
2573.09	Cd II	3	150
*2576.104	Mn II	300	2000
2589.167	W II	15	25
2593.729	Mn II	200	1000
2598.062	Sb I	200	100
2605.688	Mn II	100	500
2614.178	Pb I	200	80
*2615.43	Lu II		
2631.553	Al II	....	40
*2641.406	Hf II	40	125
2650.781	Be I	25	
*2651.178	Ge I	40	20
2651.575	Ge I	30	20
2658.722	Pd II	20	300
2659.454	Pt I	2000	500
*2669.166	Al II	3	100
2675.95	Au I	250	100
2678.758	Ru II	100	300
2692.065	Ru II	8	200
2709.626	Ge I	30	20
2712.410	Ru II	80	300
2748.58	Cd II	5	200
2767.87	Tl I	400	300
2769.67	Te I	....	D 30
2773.357	Hf II	25	60
2780.197	As I	75	75
2780.521	Bi I	200	100
*2795.53	Mg II	150	300
2802.19	Au I		200
2802.695	Mg II	150	300
2809.625	Bi I	200	100
*2816.154	Mo II	200	300
2816.179	Al II	10	100
2820.224	Hf II	40	100
2830.295	Pt I	1000	600
2833.069	Pb I	500	80
*2835.633	Cr II	100	400
2836.710	C II	....	200
2837.602	C II	....	40
*2839.989	Sn I	300	300
2843.252	Cr II	125	400
2848.232	Mo II	125	200
2849.725	Ir I	40	20
2849.838	Cr II	80	150
*2852.129	Mg I	300	100
2854.581	Pd II	4	500
2855.676	Cr II	60	200
2860.452	As I	50	50
2860.934	Cr II	60	100
2863.327	Sn I	300	300
2871.508	Mo II	100	100
2874.244	Ga I	10	15
2877.915	Sb I	250	150
*2881.578	Si I	500	400
2890.994	Mo II	30	50
2894.84	Lu II	60	200
2897.975	Bi I	500	500
2898.259	Hf I	50	12
2898.71	As I	25	40
2904.408	Hf I	30	6
*2909.061	Os I	500	400
2909.116	Mo II	25	40
2911.39	Lu II	100	300
2916.481	Hf I	50	15
2918.32	Tl I	400	200
2924.792	Ir I	25	15
2929.794	Pt I	800	200
2936.77	Ho II	....	1000
2938.298	Bi I	300	300
2940.772	Hf I	60	12
2943.637	Ga I	10	20
2945.668	Ru II	60	300
2965.546	Ru II	60	200
2976.586	Ru II	60	200
2989.029	Bi I	250	100
2997.967	Pt I	1000	200
3009.147	Sn I	300	200
3034.121	Sn I	200	150
3039.064	Ge I	1000	1000
3039.356	In I	1000	500
3058.66	Os I	500	500
*3064.712	Pt I	2000	300
*3067.716	Bi I	3000	2000
3071.591	Ba I	100	50
3072.877	Hf I	80	18
3082.155	Al I	800	800
3092.713	Al I	1000	1000
*3093.108	V II	100	400
*3094.183	Cb II	100	1000
3102.299	V II	70	300
3110.706	V II	70	300
3118.383	V II	70	200
3125.284	V II	80	200
*3130.416	Be II	200	200
3130.786	Cb II	100	100
3131.072	Be II	200	150
3134.718	Hf II	80	125
3158.869	Ca II	100	300
3163.402	Cb II	15	8
3175.019	Sn I	500	400
3179.332	Ca II	100	400
3183.406	V I	200	100
3183.982	V I	500	400
3185.396	V I	500	400
3194.977	Cb II	30	300
3215.560	W I	10	9
*3220.780	Ir I	100	30
3225.479	Cb II	150	800
3229.75	Tl I	2000	800
3232.499	Sb I	150	250
3232.61	Li I	1000	500
3242.280	Y II	60	100
*3247.540	Cu I	5000	2000
3256.090	In I	1500	600
3258.564	In I	500	300
3261.057	Cd I	300	300
3262.290	Os I	500	50
3262.328	Sn I	400	300
3267.502	Sb I	150	150
3267.945	Os I	400	30
3269.494	Ge I	300	300
3273.962	Cu I	3000	1500
*3280.683	Ag I	2000	1000
3282.333	Zn I	500	300
3289.37	Yb I	500	1000
3290.59	Th II	....	40
3301.56	Os I		
3302.323	Na I	600	300
3302.588	Zn I	800	300
3302.988	Na I	300	150
*3311.162	Ta I	300	70
3318.840	Ta I	125	35
3321.013	Be I	50	
3321.086	Be I	100	
3321.343	Be I	1000	30
3323.092	Rh I	1000	200
3345.020	Zn I	800	300
*3349.035	Ti II	125	800
3361.213	Ti II	100	600
3372.800	Ti II	80	400
3380.711	Sr II	150	200
3382.891	Ag I	1000	700
3383.761	Ti II	70	300
*3391.975	Zr II	300	400
3396.85	Rh I	1000	500
3397.07	Lu II	50	20
3403.653	Cd I	800	500
*3404.580	Pd I	2000	1000
3405.120	Co I	2000	150
3406.664	Ta I	70	18
*3414.765	Ni I	1000	50
3421.24	Pd I	2000	1000
*3434.893	Rh I	1000	200
3436.737	Ru I	300	150
3437.015	Ir I	20	15
3438.230	Zr II	250	200
3451.41	B II	5	30
3452.33	B II		
*3453.505	Co I	3000	200
*3460.47	Re I	1000	
3462.21	Tm I	200	100
3464.57	Sr II	200	200
3465.800	Co I	2000	25
3466.201	Cd I	1000	500
3472.48	Lu II	50	150
3474.887	Sr II	80	50
3492.956	Ni I	1000	100
3496.210	Zr II	100	100
*3498.942	Ru I	500	200
3499.104	Er I	18	15
3509.17	Tb I	200	200
3513.645	Ir I	100	100
3515.054	Ni I	1000	50
3516.943	Pd I	1000	500
3519.24	Tl I	2000	1000
3519.605	Zr I	100	10
3524.541	Ni I	1000	100
3529.813	Co I	1000	30
3538.75	Th I	....	50
3547.682	Zr I	200	12
3552.172	U I	8	12
3554.43	Lu II	50	150
3561.74	Tb I	200	200
3568.27	Sm II		
3572.473	Zr II	60	80
3581.20	Fe I		
3596.179	Ru I	30	100
3600.734	Y II	100	300
3601.040	Th I	8	10
*3601.193	Zr I	400	15
3609.548	Pd I	1000	700
3610.510	Cd I	1000	500
3613.790	W II	10	30
*3613.836	Sc II	40	70
3630.740	Sc II	50	70
3633.123	Y II	50	100
3634.695	Pd I	2000	1000
3635.463	Ti I	200	100
3639.580	Pb I	300	50
3642.675	Ti I	300	125
3642.785	Sc II	60	50
3646.196	Gd I	200	150
3650.146	Hg I	200	500
3653.496	Ti I	500	200
3654.833	Hg I	...	D 200
3657.987	Rh I	500	200
3663.276	Hg I	500	400
3672.579	U I	8	15
3683.471	Pb I	300	50
3692.357	Rh I	500	150
3692.652	Er I	20	12
*3694.203	Yb II	500	1000
*3710.290	Y II	80	150
*3719.935	Fe I	1000	700
3737.133	Fe I	1000	600
3745.564	Fe I	500	500
3745.903	Fe I	150	100
3748.17	Ho I	60	40
3748.264	Fe I	500	200
3752.54	Os I		
3761.333	Tm I	250	150
3761.917	Tm I	200	120
3768.405	Gd I	20	20
3774.332	Y II	12	100
3775.72	Tl I	3000	1000
3782.20	Os I		
3788.697	Y II	30	30
*3798.252	Mo I	1000	1000
*3814.42	Ra II	....	D 2000
3829.350	Mg I	100	150
3832.306	Mg I	250	200
3838.258	Mg I	300	200
*3848.02	Tm II		
3848.75	Tb I	100	200
3864.110	Mo I	1000	500
3874.18	Tb I	200	200
*3888.646	He I	....	D 1000
3891.02	Ho I	200	40
3891.785	Ba II	18	25

PERSISTENT LINES OF THE ELEMENTS (Continued)

Arranged by Wave Lengths (Continued)

Wave length	Element	Intensity	
		Arc	Spark
3902.963	Mo I	1000	500
3905.528	Si I	20	15
3906.316	Er I	25	12
3907.476	Sc I	125	25
*3911.810	Sc I	150	30
*3933.666	Ca II	600	600
3944.032	Al I	2000	1000
*3949.106	La II	1000	800
3951.154	Nd I	40	30
*3961.527	Al I	3000	2000
3968.468	Ca II	500	500
*3987.994	Yb I	1000	500
4000.454	Dy I	400	300
4008.753	W I	45	45
4012.388	Ce II	60	20
4019.137	Th I	8	8
4020.399	Sc I	50	20
4023.688	Sc I	100	25
*4030.755	Mn I	500	20
4032.982	Ga I	1000	500
4033.073	Mn I	400	20
4034.490	Mn I	250	20
4040.762	Ce II	70	5
4044.140	K I	800	400
4045.983	Dy I	150	12
4046.561	Hg I	200	300
4047.201	K I	400	200
*4057.820	Pb I	2000	300
*4058.938	Cb I	1000	400
4062.817	Pr I	150	50
4077.340	La II	600	400
*4077.714	Sr II	400	500
4077.974	Dy I	150	100
4079.729	Cb I	500	200
4093.161	Hf II	25	20
4097.31	N III	....	D 100
*4099.94	N I	....	D 150
4100.923	Cb I	300	200
4101.773	In I	2000	1000
4103.37	N III	....	D 80
4109.98	N I	....	D 1000
4123.228	La II	500	500
4123.810	Cb I	200	125
4129.737	Eu II	150	50
4130.664	Ba II	50	60
4137.095	Cb I	100	60
4165.606	Ce II	40	6
4167.966	Dy I	50	12
*4172.056	Ga I	2000	1000
4177.321	Nd I	15	25

Wave length	Element	Intensity	
		Arc	Spark
*4179.422	Pr II	200	40
*4186.599	Ce II	80	25
4189.518	Pr II	100	50
4201.851	Rb I	2000	500
*4205.046	Eu II	200	50
4211.719	Dy I	200	15
4215.524	Sr II	300	400
4215.556	Rb I	1000	300
4225.327	Pr II	50	40
4226.570	Ge I	200	50
*4226.728	Ca I	500	50
4241.669	U I	40	50
*4254.346	Cr I	5000	1000
4267.02	C II	....	350
4267.27	C II	....	500
4274.803	Cr I	4000	800
4289.721	Cr I	3000	800
4294.614	W I	50	50
*4296.75	Sm I		
*4302.108	W I	60	60
*4303.573	Nd I	100	40
4305.447	Sr II	40	
4358.35	Hg I	3000	500
*4379.238	V I	200	200
4384.722	V I	125	125
4389.974	V I	80	60
4390.865	Sm II	150	150
4420.468	Os I	400	100
*4424.342	Sm II	300	300
*4425.441	Ca I	100	
*4434.321	Sm II	200	200
4434.960	Ca I	150	
4454.781	Ca I	200	
4455.880	Ca I		
4456.62	Ca I		
4500.977	Xe I	....	D 500
*4511.323	In I	5000	4000
4518.57	Lu I	300	40
4524.741	Sn I	500	50
*4554.042	Ba II	1000	200
4555.355	Cs I	2000	100
4593.177	Cs I	1000	50
*4594.02	Eu I		
4603.00	Li I	800	
*4607.331	Sr I	1000	50
4624.276	Xe I	....	D 1000
4643.695	Y I	50	100
4671.226	Xe I	....	D 2000
*4674.848	Y I	80	100
4680.138	Zn I	300	200

Wave length	Element	Intensity	
		Arc	Spark
4682.28	Ra II	....	D 800
4685.75	He II	....	D 300
4687.803	Zr I	125	
4694.13	S I	....	D 500
4695.45	S I	....	D 30
4696.25	S I	....	D 15
*4704.86	Br II	....	D 250
4710.075	Zr I	60	
4722.159	Zn I	400	300
4722.552	Bi I	1000	100
4730.78	Se I	....	D 1000
4739.03	Se I	....	D 800
4739.478	Zr I	100	
4742.25	Se I	....	D 500
4772.312	Zr I	100	
4785.50	Br II	....	D 400
4794.54	Cl II	....	D 250
4810.06	Cl II	....	D 200
4810.534	Zn I	400	300
4815.62	Zr I		
4816.71	Br II	....	D 300
4819.46	Cl II	....	D 200
*4825.91	Ra I	....	D 800
4832.075	Sr I	200	8
4861.327	H I	....	D 500
4872.493	Sr I	25	
4889.17	Re I	2000	
4934.086	Ba II	400	400
4962.263	Sr I	40	
*4981.733	Ti I	300	125
4991.066	Ti I	200	100
4999.510	Ti I	200	80
5007.213	Ti I	200	40
5014.25	Ti I		
5105.541	Cu I	500	
5153.235	Cu I	600	
5161.188	I II	....	D 300
5167.343	Mg I	100	50
5172.699	Mg I	200	100
5183.618	Mg I	500	300
5204.518	Cr I	400	100
5206.039	Cr I	500	200
5208.436	Cr I	500	100
5209.067	Ag I	1500	1000
5218.202	Cu I	700	
*5350.46	Tl I	5000	2000
5400.562	Ne I	....	D 2000
5424.616	Ba I	100	30
5455.146	La I	200	1
5460.740	Hg I	....	D 2000
5464.61	I II	....	D 900

Wave length	Element	Intensity	
		Arc	Spark
5465.487	Ag I	1000	500
5466.47	Y I		
5519.115	Ba I	200	60
*5535.551	Ba I	1000	200
5570.2895	Kr I	....	D 2000
5608.8	Pb II	....	D 40
5666.64	N II	....	D 300
5671.80	Sc I		
5676.02	N II	....	D 100
5679.56	N II	....	D 500
5682.657	Na I	80	
5688.224	Na I	300	
5777.665	Ba I	500	100
5832.488	Ne I		
5852.488	Ne I	....	D 2000
5870.9158	Kr I	....	D 3000
5875.618	He I	....	D 1000
*5889.953	Na I	9000	1000
5895.923	Na I	5000	500
5930.648	La I	250	
6103.642	Li I	2000	.300
6231.76	Al II	....	30
6243.36	Al II	....	100
*6249.929	La I	300	
6362.347	Zn I	1000	500
6402.246	Ne I	....	D 2000
6438.4696	Cd I	2000	1000
6562.79	H I	....	D 3000
*6707.844	Li I	3000	200
6856.02	F I	....	D 1000
6902.46	F I	....	D 500
6965.430	A I	....	D 400
7055.42	Rn I	....	D 400
7067.217	A I	....	D 400
7450.00	Rn I	....	D 600
7503.867	A I	....	D 700
*7664.907	K I	9000	
7698.979	K I	5000	
*7771.928	O I	....	D 1000
7774.138	O I	....	D 300
7775.433	O I	....	D 100
*7800.227	Rb I	9000	
7947.60	Rb I	5000	
8115.311	A I	....	D 5000
*8521.10	Cs I	5000	
8943.50	Cs I	2000	
9212.91	S I	....	D 200
9228.11	S I	....	D 200
9237.49	S I	....	D 200

SECONDARY ELECTRON EMISSION

N. R. WHETTEN

General Electric Research Laboratory, Schenectady, New York

By permission from "Methods of Experimental Physics" Vol. IV (1962) Academic Press

The secondary emission yield, or secondary emission ratio, δ, is the average number of secondary electrons emitted from a bombarded material for every incident primary electron. The secondary emission yield is a function of the primary electron energy. δ_{max} is the maximum yield corresponding to a primary electron energy $E_{p\,max}$ (see figure). The two primary electron energies corresponding to a yield of unity are denoted the first and second crossovers (E_I and E_{II}). An insulating target, or a conducting target that is electrically floating, will charge positively or negatively depending on the primary electron energy. For $E_I < E_p < E_{II}$, $\delta > 1$ and the surface charges positively provided there is a collector present that is positive with respect to the target. For $E_p < E_I$, $\delta < 1$, and the surface charges negatively towards the potential of the source of primary electrons. For $E_p > E_{II}$, $\delta < 1$, and the surface charges negatively to the second crossover.

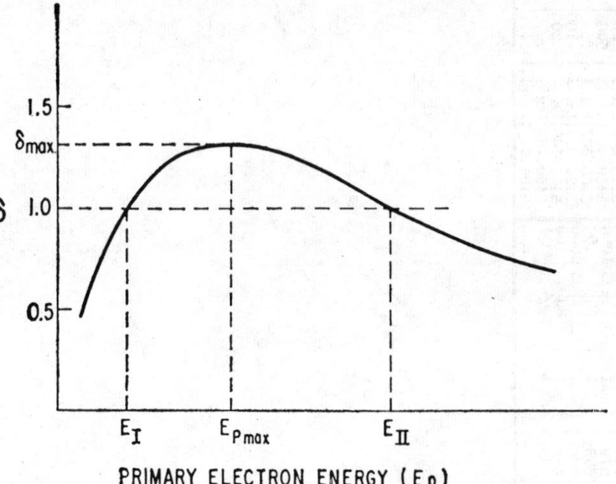

The secondary emission yield is very sensitive to surface contamination, such as oxide films and carbon deposits. Whenever possible, yields believed to be most typical of clean surfaces have been selected. The yields are for measurements at room temperature and normal incidence of the primary electrons.

Table I

Secondary Electron Emission Properties of Elements and Compounds. δ_{max} is the maximum secondary emission yield, $E_{p\,max}$ the primary electron energy for maximum yield, and E_I and E_{II} are the first and second crossovers.

Elements	δ_{max}	$E_{p\,max}$ (ev)	E_I (ev)	E_{II} (ev)	Ref.
Ag	1.5	800	200	>2000	a, b, c
Al	1.0	300	300	300	b
Au	1.4	800	150	>2000	a, c, d
B	1.2	150	50	600	f
Ba	0.8	400	None	None	b
Bi	1.2	550			g, s
Be	0.5	200	None	None	b, h, i, d
C (diamond)	2.8	750		>5000	j
(graphite)	1.0	300	300	300	k
(soot)	0.45	500	None	None	k
Cd	1.1	450	300	700	l, d
Co	1.2	600	200		m, n
Cs	0.7	400	None	None	b, o
Cu	1.3	600	200	1500	a, l, b
Fe	1.3	400	120	1400	n, c, p
Ga	1.55	500	75		q
Ge	1.15	500	150	900	f, r, s
Hg	1.3	600	350	>1200	q
K	0.7	200	None	None	t, u
Li	0.5	85	None	None	b
Mg	0.95	300	None	None	o, b
Mo	1.25	375	150	1200	a, w, c, e, p
Na	0.82	300	None	None	x
Nb	1.2	375	150	1050	a, c
Ni	1.3	550	150	>1500	a, n, m, w, p
Pb	1.1	500	250	1000	g, q
Pd	>1.3	>250	120		v
Pt	1.8	700	350	3000	c
Rb	0.9	350	None	None	t
Sb	1.3	600	250	2000	y
Si	1.1	250	125	500	f
Sn	1.35	500			g, x
Ta	1.3	600	250	>2000	a
Th	1.1	800			b
Ti	0.9	280	None	None	k
Tl	1.7	650	70	>1500	s
W	1.4	650	250	>1500	z, a, a′, c
Zr	1.1	350			k

SECONDARY ELECTRON EMISSION (Continued)

Table I (Continued)

Compounds	δ_{max}	$E_{p\ max}$ (ev)	Ref.
Alkali Halides			
CsCl	6.5		b′
KBr (crystal)	14	1800	c′, e′
KCl (crystal)	12	1600	p′, d′
(layer)	7.5	1200	b′, f′
KI (crystal)	10	1600	c′, d′, e′
(layer)	5.6		b′
LiF (crystal)	8.5		g′
(layer)	5.6	700	b′
NaBr (crystal)	24	1800	g′, h′, d′
(layer)	6.3		b′
NaCl (crystal)	14	1200	i′, e′, c′, d′, g′
(layer)	6.8	600	b′, j′
NaF (crystal)	14	1200	g′
(layer)	5.7		b′
NaI (crystal)	19	1300	g′
(layer)	5.5		b′
RbCl (layer)	5.8		b′
Oxides			
Ag_2O	1.0		l′
Al_2O_3 (layer)	2 to 9		k′, m′, i′, b
BaO (layer)	2.3 to 4.8	400	m′, b
BeO	3.4	2000	m′
CaO	2.2	500	m′
Cu_2O	1.2	400	b′, n′
MgO (crystal)	20 to 25	1500	t′, u′, v′
(layer)	3 to 15	400 to 1500	w, m′, r′, s′, t′
MoO_2	1.2		l′
SiO_2 (quartz)	2.1 to 4	400	k′, g′
SnO_2	3.2	640	o′
Sulfides			
MoS_2	1.1		b′
PbS	1.2	500	p′
WS_2	1.0		b′
ZnS	1.8	350	q′
Others			
BaF_2 (layer)	4.5		b′
CaF_2 (layer)	3.2		b′
$BiCs_3$	6	1000	p′
BiCs	1.9	1000	p′
GeCs	7	700	p′
Rb_3Sb	7.1	450	p′
$SbCs_3$	6	700	p′, x′
Mica	2.4	350	k′
Glasses	2 to 3	300 to 450	k′, y′

References to Table I

a R. Warnecke, J. phys. radium, **7,** 270 (1936).
b H. Bruining and J. H. deBoer, Physica, **5,** 17 (1938).
c R. Kollath, Physik. Z., **38,** 202 (1937).
d R. Suhrmann and W. Kundt, Z. Physik, **121,** 118 (1943).
e P. L. Copeland, J. Franklin Inst., **215,** 593 (1933).
f L. R. Koller and J. S. Burgess, Phys. Rev., **70,** 571 (1946).
g P. M. Morozov, J. Exptl. Theoret. Phys. USSR, **11,** 410 (1941).
h R. Kollath, Ann. Physik, **33,** 285 (1938).
i E. G. Schneider, Phys. Rev., **54,** 185 (1938).
j J. B. Johnson, Phys. Rev., **92,** 843 (1953).
k H. Bruining, Philips Tech. Rev., **3,** 80 (1938).
l R. Suhrmann and W. Kundt, Z. Physik, **120,** 363 (1943).
m D. E. Wooldridge, Phys. Rev., **56,** 1062 (1939).
n L. R. G. Treloar and D. H. Landon, Proc. Soc. (London), **B50,** 625 (1938).
o N. S. Klebnikov, Tech. Phys. USSR, **5,** 593 (1938).
p R. L. Petry, Phys. Rev., **26,** 346 (1925).
q J. J. Brophy, Phys. Rev., **83,** 534 (1951).
r J. B. Johnson and K. G. McKay, Phys. Rev., **93,** 668 (1954).
s H. Gobrecht and F. Spear, Z. Physik, **135,** 602 (1953).
t A. Afanasjewa and P. W. Timofeew, Tech. Phys. USSR, **4,** 953 (1937).
u M. S. Joffe and I. V. Nechlaev, J. Exptl. Theoret. Phys. USSR, **11,** 93 (1941).
v H. E. Farnsworth, Phys. Rev., **25,** 41 (1925).
w G. Blankenfeld, Ann. Physik, **9,** 48 (1951).
x J. Woods, Proc. Phys. Soc. (London), **B67,** 843 (1954).
y R. Kollath, *Handbuch der Physik*, vol. **21** (1956), p. 232.
z R. L. Petry, Phys. Rev., **28,** 362 (1926).
a′ E. A. Coomes, Phys. Rev., **55,** 519 (1939).
b′ H. Bruining and J. H. deBoer, Physica, **6,** 834 (1939).
c′ D. N. Dobretzov and A. S. Titkow, Doklady Acad. Nauk USSR, **100,** 33 (1955).
d′ N. R. Whetten, Bull. Am. Phys. Soc. Ser. II, **5,** 347 (1960).
e′ A. R. Shulman and B. P. Dementyev, J. Tech. Phys. USSR, **25,** 2256 (1955).
f′ M. Knoll, O. Hachenberg, and J. Randmer, Z. Physik, **122,** 137 (1944).
g′ D. N. Dobretzov and T. L. Matskevich, J. Tech. Phys. USSR, **27,** 734 (1957).
h′ T. L. Matskevich, J. Tech. Phys. USSR, **26,** 2399 (1956).
i′ A. R. Shulman, W. L. Makedonsky, and J. D. Yaroshetsky, J. Tech. Phys. USSR, **23,** 1152 (1953).
j′ M. M. Vudinsky, J. Tech. Phys. USSR, **9,** 271 (1939).
k′ H. Salów, Z. Tech. Phys., **21,** 8 (1940).
l′ A. Afanasjewa, P. Timofeew, and A. Ignaton, Phys. Z. Sowjet, **10,** 831 (1936).
m′ K. H. Geyer, Ann Phys., **42,** 241 (1942).
n′ N. B. Gornij, J. Exptl. Theoret. Phys. USSR, **26,** 79 (1954).
o′ H. E. Mendenhall, Phys. Rev., **72,** 532 (1947).
p′ O. Hachenberg and W. Brauer in *Advances in Electronics and Electron Physics*, Vol. XI, Academic Press, New York (1959), p. 438.
q′ N. B. Gornij, J. Exptl. Theoret. Phys. USSR, **26,** 88 (1954).
r′ P. Wargo, B. V. Haxby, and W. G. Shepherd, J. Appl. Phys., **27,** 1311 (1956).
s′ P. Rappaport, J. Appl. Phys., **25,** 288 (1954).
t′ N. R. Whetten and A. B. Laponsky, J Appl. Phys., **30,** 432 (1959).
u′ J. B. Johnson and K. G. McKay, Phys. Rev., **91,** 582 (1953).
v′ R. G. Lye, Phys. Rev., **99,** 1647 (1955).
w′ N. R. Whetten and A. B. Laponsky, J. Appl. Phys., **28,** 515 (1957).
x′ N. D. Morgulis and B. I. Djatlowitskaja, J. Tech. Phys. USSR, **10,** 657 (1940).
y′ C. W. Mueller, J. Appl. Phys., **16,** 453 (1945).

EMISSIVITY OF TUNGSTEN

Wavelengths in μ

Temperature °K	0.25	0.30	0.35	0.40	0.50	0.60	0.70
1600	.448*	.482	.478	.481	.469	.455	.444
1800	.442*	.478*	.476	.477	.465	.452	.44
2000	.436*	.474	.473	.474	.462	.448	.436
2200	.429*	.470	.470	.471	.458	.445	.431
2400	.422	.465	.466	.468	.455	.441	.427
2600	.418	.461	.464	.464	.451	.437	.423
2800	.411	.456	.461	.461	.448	.434	.419

Temperature °K	0.80	0.90	1.0	1.1	1.2	1.3	1.4
1600	.431	.413	.39	.366	.345	.322*	.300*
1800	.425	.407	.385	.364	.344	.323*	.302*
2000	.419	.401	.381	.361	.343	.323	.305
2200	.415	.396	.378	.359	.342	.324	.306
2400	.408	.391	.372	.355	.340	.324	.309
2600	.404	.386	.369	.352	.338	.325	.310
2800	.400	.383	.367	.352	.337	.325	.313

Temperature °K	1.5	1.6	1.8	2.0	2.2	2.4	2.6
1600	.279*	.263*	.234*	.210*	.19*	.175*	.164*
1800	.282	.267*	.241*	.218*	.20*	.182*	.174*
2000	.288	.273	.247	.227	.209	.197	.175
2200	.291	.278	.254	.235	.218	.205	.194
2400	.296	.283	.262	.244	.228	.215	.205
2600	.299	.288	.269	.251	.236	.224	.214
2800	.302	.292	.274	.259	.245	.233	.224

* Values by extrapolation.

LIQUIDS FOR INDEX BY IMMERSION METHOD

Liquid	N_D 24° C
Trimethylene chloride	1.446
Cineole	1.456
Hexahydrophenol	1.466
Decahydronaphthalene	1.477
Isoamylphthalate	1.486
Tetrachloroethane	1.492
Pentachloroethane	1.501
Trimethylene bromide	1.513
Chlorobenzene	1.523
Ethylene bromide + Chlorobenzene	1.533
o-Nitrotoluene	1.544
Xylidine	1.557
o-Toluidine	1.570
Aniline	1.584
Bromoform	1.595
Iodobenzene + Bromobenzene	1.603
Iodobenzene + Bromobenzene	1.613
Quinoline	1.622
α-Chloronaphthalene	1.633
α-Bromonaphthalene + α-Chloronaphthalene	1.640–1.650
α-Bromonaphthalene + α-Iodonaphthalene	1.660–1.690
Methylene iodide + Iodobenzene	1.700–1.730
Methylene iodide	1.738
Methylene iodide saturated with sulfur	1.78
Yellow phosphorus, sulfur and methylene iodide (8:1:1 by weight)	2.06
Can be diluted with methylene iodide to cover range 1.74–2.06. For precautions in use, cf. West, Am. Mineral, **21**, p. 245–9 (1936).	

HEAVY LIQUIDS FOR MINERAL SEPARATION

Liquid	Density
Tetrabromoethane (sym.)	2.964, 20°/4°
Can be diluted with carbon tetrachloride (1.595) or benzene (0.894).	
Methylene iodide	3.325, 20°/4°
Can be diluted with carbon tetrachloride or benzene.	
Thallium formate, aq	3.5
Can be diluted with water.	
Thallium malonate-thallium formate, aq	4.9
Can be diluted with water.	

For preparation and recovery of these liquids, cf. U. S. Bureau Mines, Rept. Inv. #2897 (1928).

INDEX OF REFRACTION

Indices of refraction for elements, inorganic, metal-organic and organic compounds and minerals will be found in the tables of physical constants for the various classes of substances in the section Properties and Physical Constants.

Values for compounds not there listed and data subsequently collected are given below.

Indices not otherwise indicated are for sodium light, $\lambda = 589.3$ mμ. Other wave lengths are indicated by the value in millimicrons or symbol in parentheses which follows the index. Wave lengths are indicated as follows: He, $\lambda = 587.6$ mμ; Li, $\lambda = 670.8$ mμ; Hg, $\lambda = 579.1$ mμ; A, $\lambda = 759.4$ mμ; C, $\lambda = 656.3$ mμ; D, $\lambda = 589.3$ mμ; F, $\lambda = 486.1$ mμ.

Temperatures are understood to be 20°C for liquids, or ordinary room temperatures in the case of solids. Other temperatures appear as superior figures with the index.

Indices for the elements and inorganic compounds will be understood to be for the solid form except as indicated by the abbreviation liq.

See also under Physical Constants of Inorganic Compounds and index of Refraction of Gases.

ELEMENTS

Name	Formula	Index
Bromine (liq.)	Br_2	1.661^{15}
Cadmium (liq.)	Cd	0.82 (579 mμ)
(sol.)		1.13
Chlorine (liq.)	Cl_2	1.385
" (gas)		1.000768
Hydrogen (liq.)	H_2	$1.10974^{-252.83}$(579 mμ)
Iodine (sol.)	I_2	3.34
(gas)		1.001920
Lead	Pb	2.6 (579 mμ)
Mercury (liq.)	Hg	1.6–1.9
Nitrogen (liq.)	N_2	1.2053^{-190}

INDEX OF REFRACTION (Continued)

ELEMENTS

Name	Formula	Index
Oxygen (liq.)	O_2	1.221^{-181}
Phosphorus (yel.) (sol.)		2.1442^{25}
Selenium	Se_8	3.00, 4.04
" (amor.) (sol.)		2.92
Sodium (liq.)	Na	0.0045
(sol.)		4.22
Sulfur (liq.)	S_8	1.929^{110}
" (amor.) (sol.)		1.998
" (rhombic, α)		1.957, 2.0377, 2.2454
Tin (liq.)	Sn	2.1

INORGANIC COMPOUNDS

See also under Physical Constants of Inorganic Compounds.

Name	Formula	Index
Aluminum carbide	Al_4C_3	2.7, 2.75 (700 mμ)
chloride	$AlCl_3.6H_2O$	1.560, 1.507
oxide	Al_2O_3	1.665–1.680, 1.63–1.65
Alums. See under appropriate	element.	
Ammonium antimonyl tartrate	$2(NH_4.SbO.C_4H_4O_6).H_2O$	β1.6229 (C)
*ortho*arsenate, di-H	$NH_4H_2AsO_4$	1.5766, 1.5217
bromide	NH_4Br	1.7108
*per*chlorate	NH_4ClO_4	1.4818, 1.4833, 1.4881
chloroplatinate	$(NH_4)_2PtCl_6$	1.8
fluoride	NH_4F	ω<1.328
" acid	NH_4HF_2	1.385, 1.390, 1.394
hydrogen malate (*d*)	$NH_4C_4H_5O_5$	β1.503
nitrate	NH_4NO_3	1.413, 1.611(He), 1.63
Ammonium sulfate, acid	NH_4HSO_4	1.463, 1.473, 1.510
tartrate (*dl*)	$(NH_4)_2C_4H_4O_6.2H_2O$	β1.564
thiocyanate	NH_4CNS	1.546, 1.685, 1.692
uranyl acetate	$NH_4C_2H_3O_2.UO_2(C_2H_3O_2)_2$	1.4808, 1.4933
Antimony bromide	$SbBr_3$	>1.74+
iodide, tri-	SbI_3	2.78 (Li), 2.36
Barium cadmium bromide	$BaCdBr_4.4H_2O$	β1.702
cadmium chloride	$BaCdCl_4.4H_2O$	β1.651
calcium propionate	$BaCa_2(C_3H_5O_2)_6$	1.4442
fluochloride	$BaCl_2.BaF_2$	1.640, 1.633
fluoride	BaF_2	1.475 also 1.4741
Barium oxide	BaO	1.980
*ortho*phosphate, di-	$BaHPO_4$	1.617, 1.63±, 1.635
propionate	$Ba(C_2H_5CO_2)_2.H_2O$	β1.5175
sulfide, mono-	BaS	2.155
Cadmium ammonium chloride	$CdCl_2.4NH_4Cl$	1.6038, 1.6042
cesium sulfate	$CdSO_4.Cs_2SO_4.6H_2O$	1.498, 1.500, 1.506
fluoride	CdF_2	1.56
magnesium chloride	$(CdCl_2)_2.MgCl_2.12H_2O$	1.49, 1.5331, 1.5769
oxide	CdO	2.49 (Li)
potassium chloride	$CdCl_2.4KCl$	1.5906, 1.5907
" cyanide	$Cd(CN)_2.2KCN$	1.4213
rubidium sulfate	$CdSO_4.Rb_2SO_4.6H_2O$	1.4798, 1.4848, 1.4948
Calcium aluminate	$Ca_3Al_2O_6$	1.710
borate	$CaO.B_2O_3$	1.540, 1.656, 1.682
carbide	CaC_2	>1.75
copper acetate	$CaCu(C_2H_3O_2)_4.6H_2O$	1.436, 1.478
cyanamide	$CaCN_2$	1.60, >1.95
dithionate	$CaS_2O_6.4H_2O$	1.5516, 1.5414
*pyro*phosphate	$Ca_2P_2O_7$	1.585, 1.60±, 1.605
platinocyanide	$CaPt(CN)_4.5H_2O$	1.623, 1.644, 1.767
strontium propionate	$Ca_2Sr(C_3H_5O_2)_6$	1.4871, 1.4956
sulfide (oldhamite)	CaS	2.137
sulfite	$CaSO_3.2H_2O$	1.590, 1.595, 1.628
thiosulfate	$CaS_2O_3.6H_2O$	1.545, 1.560, 1.605
Carbon dioxide (liq.)	CO_2	1.195^{15}
Cerium dithionate	$Ce_2(S_2O_6)_3.15H_2O$	β1.507
Cesium *per*chlorate	$CsClO_4$	1.4752, 1.4788, 1.4804
nitrate	$CsNO_3$	1.55, 1.56
selenate	Cs_2SeO_4	1.5989, 1.5999, 1.6003
thallium chloride	$Cs_3Tl_2Cl_9$	1.784, 1.774
Chromium cesium sulfate	$CrCs(SO_4)_2.12H_2O$	1.4810
oxide (ic)	Cr_2O_3	2.5
potassium cyanide (ic)	$CrK_3(CN)_6$	1.5221, 1.5244, 1.5373
sulfate (ic)	$Cr_2(SO_4)_3.18H_2O$	1.564
thallium sulfate	$CrTl(SO_4)_2.12H_2O$	1.5228
Cobalt acetate	$Co(C_2H_3O_2)_2.4H_2O$	β1.542
aluminate (Thenard's Blue)	$Co(AlO_2)_2$	>1.78 (red), 1.74 (blue)
ammonium selenate	$CoSeO_4.(NH_4)_2SeO_4.6H_2O$	1.5246, 1.5311, 1.5396
cesium sulfate	$CoCs_2(SO_4)_2.6H_2O$	1.5057, 1.5085, 1.5132
chloride (ous)	$CoCl_2.2H_2O$	<1.625, <1.671, >1.67
potassium selenate	$CoSeO_4.K_2SeO_4.6H_2O$	1.5135, 1.5195, 1.5358
rubidium sulfate	$CoSO_4.Rb_2SO_4.6H_2O$	1.4859, 1.4916, 1.5014
selenate	$CoSeO_4.6H_2O$	β1.5225, γ1.5227
Copper ammonium selenate	$CuSeO_4.(NH_4)_2SeO_4.6H_2O$	1.5213, 1.5355, 1.5395
ammonium sulfate	$CuSO_4.(NH_4)_2SO_4.6H_2O$	1.4910, 1.5007, 1.5054
cesium sulfate	$CuSO_4.Cs_2SO_4.6H_2O$	1.5048, 1.5061, 1.5153
chloride (ic)	$CuCl_2.2H_2O$	1.644, 1.684, 1.742
formate	$Cu(CHO_2)_2.4H_2O$	1.4133, 1.5423, 1.5571

Name	Formula	Index
Copper oxide (ous) (cuprite)	Cu_2O	2.705
potassium chloride	$CuCl_2.2KCl.2H_2O$	1.6365, 1.6148
" cyanide (ous)	$CuK_3(CN)_4$	1.5215
" selenate	$CuSeO_4.K_2SeO_4.6H_2O$	1.5096, 1.5235, 1.5387
" sulfate	$CuSO_4.K_2SO_4.6H_2O$	1.4836, 1.4864, 1.5020
strontium formate	$Cu(HCO_2)_2.2[Sr(HCO_2)_2]$ $8H_2O$	1.4995, 1.5199, 1.5801
sulfate (ic)	$CuSO_4$	1.724, 1.733, 1.739
Cyanogen	C_2N_2	1.327^{18} (liq.)
Germanium bromide, tetra-	$GeBr_4$	1.6269
Gold sodium chloride	$AuNaCl_4.2H_2O$	α1.545, γ1.75+
Hafnium oxychloride	$HfOCl_2.8H_2O$	1.557, 1.543
Ice	H_2O	1.3049, 1.3062 (A), 1.3011, 1.3104 (D), 1.3133, 1.3147 (F)
Iron ammonium chloride	$Fe(NH_4)_2Cl_4$	1.6439
ammonium selenate	$FeSeO_4.(NH_4)_2SeO_4.6H_2O$	1.5201, 1.5260, 1.5356
cesium sulfate (ic)	$FeCs(SO_4)_2.12H_2O$	1.4839
" " (ous)	$FeSO_4.Cs_2SO_4.6H_2O$	1.5003, 1.5035, 1.5094
rubidium sulfate	$FeRb(SO_4)_2.12H_2O$	1.48234
sulfate (ic)	$Fe_2(SO_4)_3$	1.802, 1.814, 1.818
thallium sulfate	$FeTl(SO_4)_2.12H_2O$	1.52365
Lanthanum sulfate	$La_2(SO_4)_3.9H_2O$	1.564, 1.569
Lead *ortho*arsenate, di-	$PbHAsO_4$	1.8903, 1.9097, 1.9765
nitrate	$Pb(NO_3)_2$	1.782
Lithium ammonium sulfate	$LiNH_4SO_4$	β1.437 (Li)
ammonium tartrate (*d*)	$LiNH_4(C_4H_4O_6).H_2O$	β1.567, γ1.5673
" " (*dl*)	$LiNH_4(C_4H_4O_6).H_2O$	β1.5287
bromide	$LiBr$	1.784
chloride	$LiCl$	1.662
dithionate	$Li_2S_2O_6.2H_2O$	1.5487, 1.5602, 1.5788
oxide	Li_2O	1.644
potassium sulfate	$LiKSO_4$	1.4723, 1.4717
" tartrate	$LiK(C_4H_4O_6).H_2O$	β1.5226 (red)
rubidium tartrate (*a*)	$LiRb(C_4H_4O_6).H_2O$	β1.552
sodium tartrate (*dl*)	$LiNa(C_4H_4O_6).2H_2O$	β1.4904
Magnesium ammonium selenate	$MgSeO_4.(NH_4)_2SeO_4.6H_2O$	1.5070, 1.5093, 1.5169
ammonium sulfate	$Mg(NH_4)_2.(SO_4)_2.6H_2O$	1.4716, 1.4730, 1.4786
*ortho*borate	$3MgO.B_2O_3$	1.6527, 1.6537, 1.6748
cesium sulfate	$MgCs_2(SO_4)_2.6H_2O$	1.4857, 1.4858, 1.4916
chlorostannate	$MgSnCl_6.6H_2O$	1.5885, 1.5970
fluosilicate	$MgSiF_6.6H_2O$	1.3439, 1.3602
platinocyanide	$MgPt(CN)_4.7H_2O$	1.5608, 1.91
Magnesium potassium selenate	$MgK_2(SeO_4)_2.6H_2O$	1.4969, 1.4991, 1.5139
potassium sulfate	$MgK_2(SO_4)_2.6H_2O$	1.4607, 1.4629, 1.4755
rubidium sulfate	$MgRb_2(SO_4)_2.6H_2O$	1.4672, 1.4689, 1.4779
silicate	$MgSiO_3$	1.651, 1.654 (calc.), 1.660
sulfide	MgS	2.271 also 2.268
Manganese borate	$Mn_3B_4O_9$	1.617, 1.738, 1.776
cesium sulfate	$MnCs_2(SO_4)_2.6H_2O$	1.4946, 1.4966, 1.5025
chloride	$MnCl_2.4H_2O$	1.555, 1.575, 1.607
rubidium sulfate	$MnRb_2(SO_4)_2.6H_2O$	1.4767, 1.4807, 1.4907
sulfate (ous)	$MnSO_4.4H_2O$	1.508, 1.518, 1.522
" "	$MnSO_4.5H_2O$	1.495, 1.508, 1.514
Mercury chloride (ic)	$HgCl_2$	1.725, 1.859, 1.965
cyanide (ic)	$Hg(CN)_2$	1.645, 1.492
iodide (ic) (red)	HgI_2	2.748, 2.455
Nickel ammonium selenate	$Ni(NH_4)_2.(SeO_4)_2.6H_2O$	1.5291, 1.5372, 1.5466
cesium sulfate	$NiCs_2(SO_4)_2.6H_2O$	1.5087, 1.5129, 1.5162
Nickel chloride	$NiCl_2.6H_2O$	α1.535, γ1.61
fluoride, acid	$NiF_2.5HF.6H_2O$	1.392, 1.408
potassium selenate	$NiK_2(SeO_4)_2.6H_2O$	1.5199, 1.5248, 1.5339
rubidium sulfate	$NiRb_2(SO_4)_2.6H_2O$	1.4895, 1.4961, 1.505
selenate	$NiSeO_4.6H_2O$	1.5393, 1.5125
Platinum potassium dibromonitrite	$PtK_2(NO_2)_2Br_2.H_2O$	1.626, 1.6684, 1.757
Potassium carbonate	K_2CO_3	1.426, 1.531, 1.541
carbonate, acid	$KHCO_3$	1.380, 1.482, 1.578
*per*chlorate	$KClO_4$	1.4731, 1.4737, 1.4769
chloroplatinate	K_2PtCl_6	1.827 (577 mμ)
chloroplatinite	K_2PtCl_4	1.64, 1.67
*di*chromate	$K_2Cr_2O_7$	1.7202, 1.7380, 1.8197
cyanide	KCN	1.410
fluoborate	KBF_4	1.3239, 1.3245, 1.3247
fluoride	KF	1.352 (1.361)
"	$KF.2H_2O$	1.345, 1.352, 1.363
fluosilicate	K_2SiF_6	1.3391
*per*iodate	KIO_4	1.6205, 1.6479
lithium ferrocyanide	$K_2Li_2Fe(CN)_6.3H_2O$	1.5883, 1.6007, 1.6316
*hypo*phosphate	$K_2H_2P_2O_6.2H_2O$	1.4893, 1.5314, 1.5363
	$K_2H_2P_2O_6.3H_2O$	1.4768, 1.4843, 1.4870
ruthenium cyanide	$K_4Ru(CN)_6.3H_2O$	β1.5837
silicate	K_2SiO_3	1.520, 1.521, 1.528
thiocyanate	$KCNS$	1.532, 1.660, 1.730
thionate, tetra-	$K_2S_4O_6$	1.5896, 1.6057, 1.6435
" penta-	$2K_2S_5O_6.3H_2O$	1.565, 1.63, 1.655
Rhodium cesium sulfate	$RhCs(SO_4)_2.12H_2O$	1.5077
Rubidium *per*chlorate	$RbClO_4$	1.4692, 1.4701, 1.4731
chromate	Rb_2CrO_4	β1.71, γ1.72
dithionate	$Rb_2S_2O_6$	1.4574, 1.5078
fluoride	RbF	1.396
selenate	Rb_2SeO_4	1.5515, 1.5537, 1.5582
Ruthenium sodium nitrate	$RuNa_2(NO_2)_5.2H_2O$	1.5889, 1.5943, 1.7163
Selenium oxide	SeO_2	>1.76
Silver cyanide	$AgCN$	1.685, 1.94
nitrate	$AgNO_3$	1.729, 1.744, 1.788
phosphate	Ag_2HPO_4	1.8036, 1.7983
potassium cyanide	$AgK(CN)_2$	1.625, 1.63
Sodium ammonium tartrate (*d*)	$NaNH_4(C_4H_4O_6).4H_2O$	1.495, 1.498, 1.499
ammonium tartrate (*dl*)	$NaNH_4(C_4H_4O_6).H_2O$	β1.473 (red)
*ortho*arsenate	$NaH_2AsO_4.H_2O$	1.5382, 1.5535, 1.5607
"	$NaH_2AsO_4.2H_2O$	1.4794, 1.5021, 1.5265
bromide	$NaBr$	1.6412
carbonate	Na_2CO_3	1.415, 1.535, 1.546
Sodium carbonate, acid	$NaHCO_3$	1.376, 1.500, 1.582
cyanide	$NaCN$	1.452
iodide	NaI	1.7745
molybdate	$3Na_2O.7MoO_3.22H_2O$	β1.627
nitrate	$NaNO_3$	1.5874, 1.3361
phosphate	$NaH_2PO_4.2H_2O$	1.4401, 1.4629, 1.4815
"	$Na_2HPO_4.7H_2O$	1.4412, 1.4424, 1.4526
*hypo*phosphate	$Na_3HP_2O_6.9H_2O$	1.4653, 1.4738, 1.4804
silicate	Na_2SiO_3	1.513, 1.520, 1.528
sulfate, acid	$NaHSO_4.H_2O$	1.43, 1.46, 1.47
sulfite	Na_2SO_3	1.565, 1.515
" acid	$NaHSO_3$	1.474, 1.526, 1.685
tartrate, acid (*d*)	$NaH(C_4H_4O_6).H_2O$	β1.533
thiocyanate	$NaCNS$	1.545, 1.625, 1.695
Sodium tungstate	$Na_2WO_4.2H_2O$	1.5526, 1.5533, 1.5695
vanadate	$Na_3VO_4.10H_2O$	1.5305; ω1.5398, ϵ1.5475
"	$Na_3VO_4.12H_2O$	1.5095, 1.5232
Strontium dichromate	$SrCr_2O_7.3H_2O$	1.7146, 1.7174, 1.812
fluoride	SrF_2	1.442 (1.438)
oxide	SrO	1.870
*ortho*phosphate, acid	$SrHPO_4$	1.608, 1.62±, 1.625
sulfide, mono-	SrS	2.107
Sulfur nitride	S_4N_4	α1.908, β2.046
Thallium chloride, mono-	$TlCl$	2.247
iodide, mono-	TlI	2.78
Tin iodide (ic)	SnI_4	2.106
Uranyl potassium sulfate	$UO_2.SO_4.K_2SO_4.2H_2O$	1.5144, 1.5266, 1.57 (580 mμ)
Vanadium ammonium sulfate	$VNH_4(SO_4)_2.12H_2O$	1.475
Zinc ammonium selenate	$Zn(SeO_4).(NH_4)_2SeO_4.6H_2O$	1.5240, 1.5300, 1.5385
bromate	$Zn(BrO_3)_2.6H_2O$	1.5452
cesium sulfate	$ZnCs_2(SO_4)_2.6H_2O$	1.5022, 1.5048, 1.5093
chloride	$ZnCl_2$	1.687, 1.713
fluosilicate	$ZnSiF_6.6H_2O$	1.3824, 1.3956
potassium cyanide	$ZnK_2(CN)_4$	1.4115
" selenate	$ZnK_2(SeO_4)_2.6H_2O$	1.5121, 1.5181, 1.5335
" sulfate	$ZnK_2(SO_4)_2.6H_2O$	1.4775, 1.4833, 1.4969
rubidium sulfate	$ZnRb_2(SO_4)_2.6H_2O$	1.4833, 1.4884, 1.4975
silicate	$ZnSiO_3$	1.616, 1.62±, 1.623
Zirconium ammonium fluoride	$Zr(NH_4)_3F_7$	1.433

ORGANIC COMPOUNDS

See also under Physical Constants of Organic Compounds.

Name	Index
Allontoin, solid	α1.579, γ1.660
Dimethyl thiophene (α, α'), liq	$1.51693^{13.4}$ (He)
" " (β, β'), liq	1.52217^{15} (He)
Ethyl carbylamine, liq	1.3659^{24}
Ethylidene cyanhydrin, liq	$1.40582^{18.4}$
Hexyl acetylene (n), liq	$1.4208^{12.5}$

MISCELLANEOUS

Name	Index	Name	Index
Albite glass	1.4890	Magdala red	1.90
Amber	1.546	Obsidian	1.482–1.496
Anorthite glass	1.5755	Paraffin	$1.43295^{38.3}$ (C)
Asphalt	1.635	Quartz, fused	1.45640 (656 mμ)
Bell metal	1.0052		1.45843 (589 mμ)
Borax, amorphous, fused	1.4630		1.46190 (509 mμ)
Canada balsam	1.530		1.47503 (361 mμ)
Ebonite	1.66 (red)		1.49634 (275 mμ)
Fuchsin	2.70		1.53386 (214 mμ)
Gelatin, Nelson's No. 1	1.530		1.57464 (185 mμ)
Gelatin, various	1.516–1.534	Resin, aloes	1.619 (red)
Gum Arabic	1.480 (1.514) (red)	colophony	1.548 (red)
		copal	1.528 (red)
Hoffman's violet	2.20	mastic	1.535 (red)
Ivory	1.539, 1.541	Peru balsam	1.593

INDEX OF REFRACTION OF ORGANIC COMPOUNDS

(See also that section of this book which contains data of Physical Constants of Organic Compounds)
The following table contains a list of organic compounds arranged in order of increasing refractive index. Measurements were made at 25°C.

Compound	nD	Compound	nD
Trifluoroacetic acid	1.283	2,2,3-Trimethylpentane	1.401
2,2,2-Trifluoroethanol	1.290	1-Chlorobutane	1.401
Octofluoropentanol-1	1.316	β-Methoxypropionitrile	1.401
Dodecafluoroheptanol-1	1.316	3-Methyl butanoic acid	1.402
Methanol	1.326	n-Nonane	1.403
Acetonitrile	1.342	Dipropylamine	1.403
Ethyl ether	1.352	Isoamylacetate	1.403
Acetone	1.357	Cyclopentane	1.404
Ethyl formate	1.358	2-Methyl-2-butanol	1.404
Ethanol	1.359	3-Methyl-1-butanol	1.404
Methyl acetate	1.360	Tetrahydrofuran	1.404
Propionitrile	1.363	Capronitrile	1.405
2,2-Dimethylbutane	1.366	2-Pentanone	1.405
Isopropyl ether	1.367	2-Ethoxyethanol	1.405
2-methylpentane	1.369	2-Heptanone	1.406
Ethyl acetate	1.370	Valeric acid	1.406
Acetic acid	1.370	Diisobutylene	1.407
Propionaldehyde	1.371	Methylcyclopentane	1.407
n-Hexane	1.372	Isoamyl ether	1.407
2,3-Dimethylbutane	1.372	Methylpropyl carbinol	1.407
3-Methylpentane	1.374	Tributyl borate	1.407
2-Propanol	1.375	1-Pentanol	1.408
Isopropyl acetate	1.375	3-Methyl-2-butanol	1.408
Propyl formate	1.375	Diethyl oxalate	1.408
2-Chloropropane	1.376	n-Decane	1.409
2-Butanone	1.377	4-Methyl-2-pentanol	1.409
2-Chloropropane	1.377	3-Isopropyl-2-pentanone	1.409
Methylethyl ketone	1.377	2-Methyl-1-butanol	1.409
Butyraldehyde	1.378	Butyric acid anhydride	1.409
2,4-Dimethylpentane	1.379	Amyl ether	1.410
Propyl ether	1.379	Isoamyl isovalerate	1.410
Acetaldehyde-diethylacetal	1.379	1-Chloropentane	1.410
Butylethyl ether	1.380	2-Propene-1-ol	1.411
Nitromethane	1.380	2,4-Dimethyl dioxane	1.412
Trifluoropropanol	1.381	Ethyl lactate	1.412
2-Methylhexane	1.382	Diethyl malonate	1.412
Butyronitrile	1.382	3-Chloropropene	1.413
Propyl acetate	1.382	Ethyleneglycol diacetate	1.413
Ethyl propionate	1.382	2-Octanone	1.414
2-Methyl-2-propanol	1.383	3-Octanone	1.414
1-Propanol	1.383	3-Methyl-2-heptanone	1.415
Isobutyl formate	1.383	Caproic acid	1.415
Diethyl carbonate	1.383	4-Methyldioxane	1.415
Heptane	1.385	1,2-Propyleneglycol-1-monobutyl ether	1.415
2-Methyl-2-propanol	1.385	Ethylcyanoacetate	1.415
Propionic acid	1.385	Dibutylamine	1.416
3-Methylhexane	1.386	2-Pentanol	1.416
n-Propyl amine	1.386	1,1-Dichloroethane	1.416
1,1-Dimethyl-2-propanone	1.386	Heptachlorodiethyl ether	1.416
1-Chloropropane	1.386	1-Hexanol	1.416
2,2,3-Trimethylbutane	1.387	1-Amino-3-methoxy propane	1.417
Methylpropyl ketone	1.387	Octyl nitrile	1.418
sec-Butyl acetate	1.387	2-Heptanol	1.418
Butyl formate	1.387	2-Propenyl amine	1.419
β-Methylpropyl ethanoate	1.388	1,2-Propyleneglycol carbonate	1.419
2,2,4-Trimethyl pentane	1.389	Methylpentyl carbinol	1.420
2,3-Dimethyl pentane	1.389	2-Ethyl-1-butanol	1.420
Acetic anhydride	1.389	1-Chloro-2-methyl-1-propene	1.420
Diisopropyl amine	1.390	p-Dioxane	1.420
2-Aminobutane	1.390	Methylcyclohexane	1.421
2-Pentanone	1.390	4-Hydroxy-4-methyl-2-pentanone	1.421
3-Pentanone	1.390	1-Heptanol	1.422
Nitroethane	1.390	3-Isopropyl-2-heptanone	1.423
Methyl-b-butyrate	1.391	Cyclohexane	1.424
Butyl acetate	1.392	2-Bromopropane	1.424
2-Nitropropane	1.392	3-Chloro-2-methylprop-1-ene	1.425
4-Methyl-2-pentanone	1.394	Caproic acid	1.426
2-Methyl-1-propanol	1.394	Glycol carbonate	1.426
Octane	1.395	1-Octanol	1.427
1-Amino-2-methylpropane	1.395	1,1-Dimethylhexanol	1.427
Valeronitrile	1.395	N,N-Dimethylformamide	1.427
2-Butanol	1.395	Sulfuric acid	1.427
2-Hexanone	1.395	1-Chlorooctane	1.428
5-Methyl-3-hexanone	1.395	Triisobutylene	1.429
2-Chlorobutane	1.395	N-Methylaniline nitrile	1.429
Butyric acid	1.396	Ethylene glycol	1.429
2,2,2-Trimethylhexane	1.397	1-Chloro-2-ethylhexane	1.430
n-Dibutyl ether	1.397	Ethylcyclohexane	1.431
1-Butanol	1.397	1,2-Propanediol	1.431
Acrolein	1.397	1-Bromopropane	1.431
1-Chloro-2-methylpropane	1.397	2-Methyl-7-ethyl-4-nonanone	1.433
Methacrylonitrile	1.398	Ethyleneglycol-mono-allyl ether	1.434
3-Methyl-2-pentanone	1.398	Butyral lactone	1.434
Triethyl amine	1.399	2-Methyl-7-ethyl-4-undecanone	1.435
n-Butyl amine	1.399	4-n-Propyl-5-ethyldioxane	1.435
1,1,3,3-Tetramethyl-2-propanone	1.399	1,2-Dichloro-2-methylpropane	1.435
Isobutyl-n-butyrate	1.399	1,2-Propyleneglycol sulfite	1.435
1-Nitropropane	1.399	N-Methylmorpholine	1.436
n-Dodecane	1.400	1-Chloro-2-methyl-2-propanol	1.436
Amyl acetate	1.400	Epichlorohydrin	1.436
1-Chlorobutane	1.400	Triethyleneglycol-mono-butyl ether	1.437
2-Methoxy ethanol	1.400	4-Ethyl-7,7,7-trimethyl-1-heptanol	1.438
Propionic acid anhydride	1.400	1-Methyl-3-ethyloctan-1-ol	1.438

INDEX OF REFRACTION OF ORGANIC COMPOUNDS (Continued)

Compound	n_D
1-Ethyl-3-ethylhexan-1-ol	1.438
Diethyl maleate	1.438
1-Butanethiol	1.440
2-Chloroethanal	1.440
Dibutyl sebacate	1.440
1-Ethyl-3-ethyloctan-1-ol	1.441
Dimethylmaleate	1.441
3-Methylpentane-2,4-diol	1.441
Ethyl sulfide	1.442
Mesityl oxide	1.442
Butyl stearate	1.442
Cyclohexane	1.443
1,2-Dichloroethane	1.444
Chloroform	1.444
trans-1,2-Dichloroethylene	1.444
Diethyleneglycol	1.445
cis-1,2-Dichloroethylene	1.445
3-(α-Butyloctyl)-oxypropyl-1-amine	1.446
2-Methylmorpholine	1.446
Dipropyleneglycol-monoethyl ether	1.446
Formamide	1.446
3-Lauryloxypropyl-1-amine	1.447
Cyclohexanone	1.448
1-Aminopropan-1-ol	1.448
Diethyleneglycol-mono-β-oxypropyl ether	1.448
1-Amino-2-methylpentan-1-ol	1.449
Tetrahydrofurfural alcohol	1.450
2-Propylcyclohexa-1-one	1.452
2-Aminoethanol	1.452
2-Butylcyclohexan-1-one	1.453
Ethylenediamine	1.454
2-(β-Methyl)-propylcyclohexan-1-one	1.454
4-Methylcyclohexanol	1.454
3-Methylcyclohexanol	1.455
bis-2-Chloroethyl ether	1.455
Cyclohexylamine	1.456
1,8-Cineol	1.456
2,2'-Dimethyl-2,2'-dipropyldieththanol amine	1.456
1,1',2,2'-Tetramethyldiethanol amine	1.459
1-Aminopropan-3-ol	1.459
Carbon tetrachloride	1.459
3-Methyl-5-ethylheptan-2,4-diol	1.459
2-(β-Ethyl)-butylcyclohexan-1-one	1.461
2-Methylcyclohexanol	1.461
N-(n-Butyl)-diethanol amine	1.461
4,5-Chloro-1,3-dioxolane-2	1.461
2-Butylcyclohexan-1-ol	1.462
N-β-Oxypropyl morpholine	1.462
2-(β-Ethyl)-hexylcyclohexanone	1.463
2-Ethylcyclohexan-1-ol	1.463
Fluorobenzene	1.463
d-α-Pinene	1.464
1-α-Pinene	1.465
Cyclohexanol	1.465
m-Fluorotoluene	1.465
p-Fluorotoluene	1.467
trans-Decahydronaphthalene	1.468
o-Fluorotoluene	1.468
3-Alloxy-2-oxypropylamine-1	1.469
Ethanol-1-methylisopropanol amine	1.470
d-Limonene	1.471
1,2,3-Trichloroisobutane	1.473
Decahydronaphthalene	1.474
1,2,3-Propanetriol	1.474
Trichloroethylene	1.475
N-β-Oxyethylmorpholine	1.476
Dimethylsulfoxide	1.476
cis-Decahydronaphthalene	1.479
N-β-Chlorallylmorpholine	1.481
n-Dodecyl-4-tertiarybutylphenyl ether	1.482
n-Dodecylphenyl ether	1.482
n-Dodecyl-4-methylphenyl ether	1.483
2-Ethylidene cyclohexanone	1.486
n-Butylbenzene	1.487
p-Cymene	1.488
iso-Propylbenzene	1.489
Furfuralcohol	1.489
tert-Butylcumene	1.490
n-Propylbenzene	1.490
sec-Butylbenzene	1.490
tert-Butylbenzene	1.490
Dibutylphthalate	1.490
tert-Butyltoluene	1.491
1-Penyl-1-oxyphenylethane	1.491
n-Hexylcumene	1.492
n-Octyltoluene	1.492
n-Octylcumene	1.492
p-Xylene	1.493
1,31-Diethylbenzene	1.493
Ethylbenzene	1.493
1,3-Dimorpholylpropan-2-ol	1.493
1,12,2-Tetrachloroethane	1.493
Toluene	1.494
Benzylethyl ether	1.494
m-Xylene	1.495
1,4-Diethylbenzene	1.496
2,3-Dichlorodioxane	1.496
Mesitylene	1.497
2-Iodopropane	1.497
Benzene	1.498
Propyl benzoate	1.498
α-Picoline	1.499
1,2-Diethylbenzene	1.501
Pentachloroethane	1.501
1-Iodopropane	1.502
1,2-Dimethylbenzene	1.503
Ethyl benzoate	1.503
β-Picoline	1.504
Tetrachloroethylene	1.504
Phenetole	1.505
Pyridine	1.507
Iodoethane	1.512
Phenylmethallyl ether	1.514
Anisole	1.515
Methyl benzoate	1.515
Diallylphthalate	1.517
Benzylacetate	1.518
2-Methyl-4-tertiarybutylphenetol	1.521
Phenylacetonitrile	1.521
Methyl salicylate	1.522
Chlorobenzene	1.523
Fufural	1.524
Benzonitrile	1.526
Thiophene	1.526
Nonachlorodiethyl ether	1.529
Iodomethane	1.530
4-Phenyldioxane	1.530
3-Phenylpropan-1-ol	1.532
Acetophenone	1.532
Benzyl alcohol	1.538
1,2-Dibromoethane	1.538
1,2,3,4-Tetrahydronaphthalene	1.539
m-Cresol	1.542
1,3-Dichlorobenzene	1.543
Benzaldehyde	1.544
Styrene	1.545
Nitrobenzene	1.550
o-Dichlorobenzene	1.551
Bromobenzene	1.557
o-Nitroanisole	1.560
m-Toluidine	1.566
Benzyl benzoate	1.568
o-Toluidine	1.570
1-Methoxyphenyl-1-phenyl-ethane	1.571
Aniline	1.583
o-Chloroaniline	1.586
Bromoform	1.587
Benzenethiol	1.588
2,4-Bis(β-phenylethyl)-phenylmethyl ether	1.590
Carbondisulfide	1.628
1,12,2-Tetrabromomethane	1.633
Diiodomethane	1.749

MOLAR REFRACTION OF ORGANIC COMPOUNDS

The molar refraction, R, is defined as:

$$R = \left(\frac{n^2 - 1}{n^2 + 2}\right)\left(\frac{M}{d}\right)$$

where n = refractive index; M = molecular weight; d = density in grams per cm³; and (M/d) is the volume occupied by 1 gram molecular weight of the compound. The units of R will then be cm^{-3}.

R is, to a first approximation, independent of temperature or physical state, and it provides an approximate measure of the actual total volume (without free space) of the molecules in one gram mole.

For a very large number of compounds R is approximately additive for the bonds present in the molecule. Using R_D based on n_D (sodium light), the following atomic, group and structural contributions to R_D are based on Vogel's extensive modern measurements published in the Journal of the Chemical Society, 1948.

C	2.591	Cl	5.844	=S	7.921
H	1.028	Br	8.741	C≡N	5.459
=O	2.122	I	13.954	N (primary aliphatic)	2.376
>O	1.643	C_6H_5	25.463	N (secondary aliphatic)	2.582
OH	2.553	$C_{10}H_7$	43.00		
F	0.81	>S	7.729		

Ethylenic bond	1.575	Four membered ring	0.317
Acetylenic bond	1.977	Three membered ring	0.614
N (aromatic)	3.550		

Example:—For C_2H_5COOH: $R_{calc.} = 3(2.591) + 5(1.028) + 2.122 + 2.553 = 17.588$

$R_{obs.}$ for this compound is 17.51

For $C_6H_5NHCH_3$: $R_{calc.} = 25.463 + 3.550 + 2.591 + 4(1.028) = 35.716$

$R_{obs.}$ for this compound is 35.67

INDEX OF REFRACTION OF WATER

Alcohol and Carbon Bisulfide

For sodium light, λ = .5893

Temp. °C	Water, pure relative to air	Ethyl Alcohol 99.8 relative to air	Carbon Bisulfide relative to air
14	1.33348	....	....
15	1.33341		1.62935
16	1.33333	1.36210	1.62858
18	1.33317	1.36129	1.62704
20	1.33299	1.36048	1.62546
22	1.33281	1.35967	1.62387
24	1.33262	1.35885	1.62226
26	1.33241	1.35803	1.62064
28	1.33219	1.35721	1.61902
30	1.33192	1.35639	1.61740
32	1.33164	1.35557	1.61577
34	1.33136	1.35474	1.61413
36	1.33107	1.35390	1.61247
38	1.33079	1.35306	1.61080
40	1.33051	1.35222	1.60914
42	1.33023	1.35138	1.60748
44	1.32992	1.35054	1.60582
46	1.32959	1.34969	....
48	1.32927	1.34885	....
50	1.32894	1.34800	....
52	1.32860	1.34715	....
54	1.32827	1.34629	....
56	1.32792	1.34543	....
58	1.32755	1.34456	....
60	1.32718	1.34368	....
62	1.32678	1.34279	....
64	1.32636	1.34189	....
66	1.32596	1.34096	....
68	1.32555	1.34004	....
70	1.32511	1.33912	....
72	1.32466	1.33820	....
74	1.32421	1.33728	....
76	1.32376	1.33626	....
78	1.32332	....	....
80	1.32287	....	....
82	1.32241	....	....
84	1.32195	....	....
86	1.32148	....	....
88	1.32100	....	....
90	1.32050	....	....
92	1.32000	....	....
94	1.31949	....	....
96	1.31897	....	....
98	1.31842	....	....
100	1.31783		

ABSOLUTE INDEX FOR PURE WATER FOR SODIUM LIGHT

Temperature	Index	Temperature	Index
15° C.	1.33377	60° C.	1.32754
20	1.33335	65	1.32652
25	1.33287	70	1.32547
30	1.33228	75	1.32434
35	1.33157	80	1.32323
40	1.33087	85	1.32208
45	1.33011	90	1.32086
50	1.32930	95	1.31959
55	1.32846	100	1.31819

INDEX OF REFRACTION OF GLASS

Relative to Air

Variety	Wave length in microns.							
	.361	.434	.486	.589 (Na)	.656	.768	1.20	2.00
Zinc crown..........	1.539	1.528	1.523	1.517	1.514	1.511	1.505	1.497
Higher dispersion crown	1.546	1.533	1.527	1.520	1.517	1.514	1.507	1.497
Light flint..........	1.614	1.594	1.585	1.575	1.571	1.567	1.559	1.549
Heavy flint..........	1.705	1.675	1.664	1.650	1.644	1.638	1.628	1.617
Heaviest flint........	...	1.945	1.919	1.890	1.879	1.867	1.848	1.832

INDEX OF REFRACTION OF ROCK SALT, SYLVINE, CALCITE, FLUORITE AND QUARTZ

(Compiled from data of Martens, Paschen, and others.)

Wave length.	Rock salt.	Silvine, KCl.	Fluorite.	Calcspar, ordinary ray.	Calcspar, extraordinary ray.	Quartz, ordinary ray.	Quartz, extraordinary ray.
0.185	1.893	1.827	......	......	......	1.676	1.690
0.198	.....	......	1.496	......	1.578	1.651	1.664
0.340	.....	......	......	1.701	1.506	1.567	1.577
0.589	1.544	1.490	1.434	1.658	1.486	1.544	1.553
0.760	.....	......	1.431	1.650	1.483	1.539	1.548
0.884	1.534	1.481	1.430				
1.179	1.530	1.478	1.428				
1.229	.....	......	......	1.639	1.479		
2.324	.....	......	......	......	1.474	1.516	
2.357	1.526	1.475	1.421				
3.536	1.523	1.473	1.414				
5.893	1.516	1.469	1.387				
8.840	1.502	1.461	1.331				

INDEX OF REFRACTION, AQUEOUS SOLUTIONS

Substance	Density	Temp. °C	Index for λ = .5893 (Na)	Observer
Ammonium chloride	1.067	27.05	1.379	Willigen
Ammonium chloride	1.025	29.75	1.351	Willigen
Calcium chloride	1.398	25.65	1.443	Willigen
Calcium chloride	1.215	22.9	1.397	Willigen
Calcium chloride	1.143	25.8	1.374	Willigen
Hydrochloric acid	1.166	20.75	1.411	Willigen
Nitric acid	1.359	18.75	1.402	Willigen
Potash (caustic)	1.416	11.0	1.403	Frauenhofer
Potassium chloride	Normal	solution	1.343	Bender
Potassium chloride	Double	normal	1.352	Bender
Potassium chloride	Triple	normal	1.360	Bender
Soda (caustic)	1.376	21.6	1.413	Willigen
Sodium chloride	1.189	18.07	1.378	Schutt
Sodium chloride	1.109	18.07	1.360	Schutt
Sodium chloride	1.035	18.07	1.342	Schutt
Sodium nitrate	1.358	22.8	1.385	Willigen
Sulfuric acid	1.811	18.3	1.437	Willigen
Sulfuric acid	1.632	18.3	1.425	Willigen
Sulfuric acid	1.221	18.3	1.370	Willigen
Sulfuric acid	1.028	18.3	1.339	Willigen
Zinc chloride	1.359	26.6	1.402	Willigen
Zinc chloride	1.209	26.4	1.375	Willigen

INDEX OF REFRACTION OF FUSED QUARTZ

λ mμ, 15° C	n, 18° C	λ mμ, 15° C	n, 18° C
185.467	1.57436	434.047	1.46690
193.583	1.55999	435.834	1.46675
202.55	1.54727	467.815	1.46435
214.439	1.53386	479.991	1.46355
219.462	1.52907	486.133	1.46318
226.503	1.52308	508.582	1.46191
231.288	1.51941	533.85	1.46067
250.329	1.50745	546.072	1.46013
257.304	1.50379	589.29	1.45845
274.867	1.49617	643.847	1.45674
303.412	1.48594	656.278	1.45640
340.365	1.47867	706.520	1.45517
396.848	1.47061	794.763	1.45340
404.656	1.46968		

INDEX OF REFRACTION OF AIR (15°C, 76 cm Hg)

Corrections for reducing wavelengths and frequencies in air (15°C, 76 cm Hg) to vacuo

The indices were computed from the Cauchy formula $(n-1)10^7 = 2726.43 + 12.288/(\lambda^2 \times 10^{-8}) + 0.3555/(\lambda^4 \times 10^{-16})$. For 0°C and 76 cm Hg the constants of the equation become 2875.66, 13.412 and 0.3777 respectively, and for 30°C and 76 cm Hg 2589.72, 12.259 and 0.2576. Sellmeier's formula for but one absorption band closely fits the observations: $n^2 = 1 + 0.00057378\lambda^2/(\lambda^2 - 595260)$. If $n - 1$ were strictly proportional to the density, then $(n-1)_0/(n-1)t$ would equal $1 + \alpha t$ where α should be 0.00367. The following values of α were found to hold:

λ	0.85μ	0.75μ	0.65μ	0.55μ	0.45μ	0.35μ	0.25μ
α	0.003672	0.003674	0.003678	0.003685	0.003700	0.003738	0.003872

The indices are for dry air (0.05 ± % CO_2). Corrections to reduce to dry air the indices for moist air may be made for any wavelength by Lorenz's formula, $+0.000041(m/760)$, where m is the vapor pressure in mm. The corresponding frequencies in waves per cm and the corrections to reduce wavelengths and frequencies in air at 15°C and 76 cm Hg pressure to vacuo are given. E.g., a light wave of 5000 angstroms in dry air at 15°C, 76 cm Hg becomes 5001.391 A in vacuo; a frequency of 20.000 waves per cm correspondingly becomes 19994.44.

Wave-length, λ angstroms	Dry air $(n-1) \times 10^7$ 15°C 76 cm Hg	Vacuo correction for λ in air $(n\lambda - \lambda)$ add	Frequency waves per cm $\frac{1}{\lambda}$ in air	Vacuo correction for $\frac{1}{\lambda}$ in air $\left(\frac{1}{n\lambda} - \frac{1}{\lambda}\right)$ subtract	Wave-length, λ angstroms	Dry air $(n-1) \times 10^7$ 15°C 76 cm Hg	Vacuo correction for λ in air $(n\lambda - \lambda)$ add	Frequency waves per cm $\frac{1}{\lambda}$ in air	Vacuo correction for $\frac{1}{\lambda}$ in air $\left(\frac{1}{n\lambda} - \frac{1}{\lambda}\right)$ subtract
2000	3256	.651	50,000	16.27	3500	2850	.998	28,571	8.14
2100	3188	.670	47,619	15.18	3600	2842	1.023	27,777	7.89
2200	3132	.689	45,454	14.23	3700	2835	1.049	27,027	7.66
2300	3086	.710	43,478	13.41	3800	2829	1.075	26,315	7.44
2400	3047	.731	41,666	12.69	3900	2823	1.101	25,641	7.24
2500	3014	.754	40,000	12.05	4000	2817	1.127	25,000	7.04
2600	2986	.776	38,461	11.48	4100	2812	1.153	24,390	6.86
2700	2962	.800	37,037	10.97	4200	2808	1.179	23,809	6.68
2800	2941	.824	35,714	10.50	4300	2803	1.205	23,255	6.52
2900	2923	.848	34,482	10.08	4400	2799	1.232	22,727	6.36
3000	2907	.872	33,333	9.69	4500	2796	1.258	22,222	6.21
3100	2893	.897	32,258	9.33	4600	2792	1.284	21,739	6.07
3200	2880	.922	31,250	9.00	4700	2789	1.311	21,276	5.93
3300	2869	.947	30,303	8.69	4800	2786	1.338	20,833	5.80
3400	2859	.972	29,411	8.41	4900	2784	1.364	20,406	5.68

INDEX OF REFRACTION OF AIR (15°C, 76 cm Hg) (Continued)

Wave-length, λ angstroms	Dry air $(n-1) \times 10^7$ 15°C 76 cm Hg	Vacuo correction for λ in air $(n\lambda - \lambda)$ add	Frequency waves per cm $\frac{1}{\lambda}$ in air	Vacuo correction for $\frac{1}{\lambda}$ in air $\left(\frac{1}{n\lambda} - \frac{1}{\lambda}\right)$ subtract	Wave-length, λ angstroms	Dry air $(n-1) \times 10^7$ 15°C 76 cm Hg	Vacuo correction for λ in air $(n\lambda - \lambda)$ add	Frequency waves per cm $\frac{1}{\lambda}$ in air	Vacuo correction for $\frac{1}{\lambda}$ in air $\left(\frac{1}{n\lambda} - \frac{1}{\lambda}\right)$ subtract
5000	2781	1.391	20,000	5.56	7000	2753	1.927	14,285	3.93
5100	2779	1.417	19,607	5.45	7100	2752	1.954	14,084	3.88
5200	2777	1.444	19,230	5.34	7200	2751	1.981	13,888	3.82
5300	2775	1.471	18,867	5.23	7300	2751	2.008	13,698	3.77
5400	2773	1.497	18,518	5.13	7400	2750	2.035	13,513	3.72
5500	2771	1.524	18,181	5.04	7500	2749	2.062	13,333	3.66
5600	2769	1.551	17,857	4.94	7600	2749	2.089	13,157	3.62
5700	2768	1.578	17,543	4.85	7700	2748	2.116	12,987	3.57
5800	2766	1.604	17,241	4.77	7800	2748	2.143	12,820	3.52
5900	2765	1.631	16,949	4.68	7900	2747	2.170	12,658	3.48
6000	2763	1.658	16,666	4.60	8000	2746	2.197	12,500	3.43
6100	2762	1.685	16,393	4.53	8100	2746	2.224	12,345	3.39
6200	2761	1.712	16,129	4.45	8250	2745	2.265	12,121	3.33
6300	2760	1.739	15,873	4.38	8500	2744	2.332	11,764	3.23
6400	2759	1.766	15,625	4.31	8750	2743	2.400	11,428	3.13
6500	2758	1.792	15,384	4.24	9000	2742	2.468	11,111	3.05
6600	2757	1.819	15,151	4.18	9250	2741	2.536	10,810	2.96
6700	2756	1.846	14,925	4.11	9500	2740	2.604	10,526	2.88
6800	2755	1.873	14,705	4.05	9750	2740	2.671	10,256	2.81
6900	2754	1.900	14,492	3.99	10000	2739	2.739	10,000	2.74

INDEX OF REFRACTION, GASES

Values are relative to a vacuum and for a Temp. of 0° C. and 760 mm. pressure.

(From Smithsonian Tables)

Substance	Kind of light	Indices of refraction	Observer
Acetone	*D*	1.001079–1.001100	
Air	*D*	1.0002926	Perreau
Ammonia	white	1.000381–1.000385	
Ammonia	*D*	1.000373–1.000379	
Argon	*D*	1.000281	Rayleigh
Benzene	*D*	1.001700–1.001823	
Bromine	*D*	1.001132	Mascart
Carbon dioxide	white	1.000449–1.000450	
dioxide	*D*	1.000448–1.000454	
disulfide	white	1.001500	Dulong
disulfide	*D*	1.001478–1.001485	
monoxide	white	1.000340	Dulong
monoxide	white	1.000335	Mascart
Chlorine	white	1.000772	Dulong
Chlorine	*D*	1.000773	Mascart
Chloroform	*D*	1.001436–1.001464	
Cyanogen	white	1.000834	Dulong
Cyanogen	*D*	1.000784–1.000825	
Ethyl alcohol	*D*	1.000871–1.000885	
ether	*D*	1.001521–1.001544	
Helium	*D*	1.000036	Ramsay
Hydrochloric acid	white	1.000449	Mascart
Hydrochloric acid	*D*	1.000447	Mascart
Hydrogen	white	1.000138–1.000143	
Hydrogen	*D*	1.000132	Burton
sulfide	*D*	1.000644	Dulong
sulfide	*D*	1.000623	Mascart
Methane	white	1.000443	Dulong
Methane	*D*	1.000444	Mascart
Methyl alcohol	*D*	1.000549–1.000623	
Methyl ether	*D*	1.000891	Marcast
Nitric oxide	white	1.000303	Dulong
Nitric oxide	*D*	1.000297	Mascart
Nitrogen	white	1.000295–1.000300	
Nitrogen	*D*	1.000296–1.000298	
Nitrous oxide	white	1.000503–1.000507	
Nitrous oxide	*D*	1.000516	Mascart
Oxygen	white	1.000272–1.000280	
Oxygen	*D*	1.000271–1.000272	
Pentane	*D*	1.001711	Mascart
Sulfur dioxide	white	1.000665	Dulong
Sulfur dioxide	*D*	1.000686	Ketteler
Water	white	1.000261	Jamin
Water	*D*	1.000249–1.000259	

COEFFICIENT OF TRANSPARENCY OF UVIOL GLASS FOR THE ULTRA-VIOLET

For a thickness of 1 mm.

Wave length, microns	0.280	0.309	0.325	0.346	0.361	0.383	0.397
Uviol crown	0.56	0.95	0.990	0.996	0.999	1.000	1.000

INDEX OF REFRACTION OF AQUEOUS SOLUTIONS OF SUCROSE (CANE SUGAR)

The table gives the index of refraction for λ = 0.5893 of aqueous **sugar solutions** at 20°C from 0−85% sugar. Corrections for temperatures other than 20° are given at the end of the table.

Per cent sugar		.0	.1	.2	.3	.4	.5	.6	.7	.8	.9
00.	1.3	330	331	333	334	336	337	338	340	341	342
1.		344	345	347	348	350	351	353	355	356	357
2.		359	361	362	363	365	367	368	369	371	373
3.		374	375	377	378	380	381	382	384	385	387
4.		388	389	391	393	394	395	397	399	400	401
5.		403	405	406	407	409	411	412	413	415	417
6.		418	419	421	423	424	425	427	429	430	431
7.		433	435	436	437	439	441	442	443	445	447
8.		448	450	451	453	454	456	458	459	461	462
9.		464	465	467	469	470	471	473	475	476	477
10.		479	481	482	483	485	487	488	489	491	493
11.		494	496	497	499	500	502	504	505	507	508
12.		510	512	513	515	516	518	520	521	523	524
13.		526	527	529	531	532	533	535	537	538	539
14.		541	543	544	546	547	549	551	552	554	555
15.		557	559	560	562	563	565	567	568	570	571
16.		573	575	576	578	580	582	583	585	587	588
17.		590	592	593	595	596	598	600	601	603	604
18.		606	608	609	611	612	614	616	617	619	620
19.		622	624	625	627	629	631	632	634	636	637
20.		639	641	642	644	645	647	649	650	652	653
21.		655	657	658	660	662	663	665	667	669	670
22.		672	674	675	677	679	681	682	684	686	687
23.		689	691	692	694	696	698	699	701	703	704
24.		706	708	709	711	713	715	716	718	720	721
25.		723	725	726	728	730	731	733	735	737	738
26.		740	742	744	745	747	749	751	753	754	756
27.		758	760	761	763	765	767	768	770	772	773
28.		775	777	779	780	782	784	786	788	789	791
29.		793	795	797	798	800	802	804	806	807	809
30.		811	813	815	816	818	820	822	824	825	827
31.		829	831	833	834	836	838	840	842	843	845
32.		847	849	851	852	854	856	858	860	861	863
33.		865	867	869	870	872	874	876	878	879	881
34.		883	885	887	889	891	893	894	896	898	900
35.		902	904	906	907	909	911	913	915	916	918
36.		920	922	924	926	928	929	931	933	935	937
37.		939	941	943	945	947	949	950	952	954	956
38.		958	960	962	964	966	968	970	972	974	976
39.		978	980	982	984	986	987	989	991	993	995
40.		997	999	*001	*003	*005	*007	*008	*010	*012	*014
41.	1.4	016	018	020	022	024	026	028	030	032	034
42.		036	038	040	042	044	046	048	050	052	054
43.		056	058	060	062	064	066	068	070	072	074
44.		076	078	080	082	084	086	088	090	092	094
45.		096	098	100	102	104	107	109	111	113	115
46.		117	119	121	123	125	127	129	131	133	135
47.		137	139	141	143	145	147	150	152	154	156
48.		158	160	162	164	166	169	171	173	175	177
49.		179	181	183	185	187	189	192	194	196	198
50.		200	202	204	206	208	211	213	215	217	219
51.		221	223	225	227	229	231	234	236	238	240
52.		242	244	246	249	251	253	255	257	260	262
53.		264	266	268	270	272	275	277	279	281	283
54.		285	287	289	292	294	296	298	300	303	305
55.		307	309	311	313	316	318	320	322	325	327
56.		329	331	333	336	338	340	342	344	347	349
57.		351	353	355	358	360	362	364	366	369	371
58.		373	375	378	380	382	385	387	389	391	394
59.		396	398	400	403	405	407	409	411	414	416
60.		418	420	423	425	427	429	432	434	436	439
61.		441	443	446	448	450	453	455	457	459	462
62.		464	466	468	471	473	475	477	479	482	484
63.		486	488	491	493	495	497	500	502	504	507
64.		509	511	514	516	518	521	523	525	527	530
65.		532	534	537	539	541	544	546	548	550	553
66.		558	561	563	565	567	570	572	574	577	579
67.		581	584	586	588	591	593	595	598	600	602
68.		605	607	609	612	614	616	619	621	623	625
69.		628	630	632	635	637	639	642	644	646	649
70.		651	653	656	658	661	663	666	668	671	673
71.		676	678	681	683	685	688	690	693	695	698
72.		700	703	705	708	710	713	715	717	720	722
73.		725	727	730	732	735	737	740	742	744	747
74.		749	752	754	757	759	762	764	767	769	772
75.		774	777	779	782	784	787	789	792	794	797
76.		799	802	804	807	810	812	815	817	820	822
77.		825	827	830	832	835	838	840	843	845	848
78.		850	853	855	858	860	863	865	868	871	873
79.		876	878	881	883	886	888	891	893	896	898
80.		901	904	906	909	912	914	917	919	922	925
81.		927	930	933	935	938	941	943	946	949	951
82.		954	956	959	962	964	967	970	972	975	978
83.		980	983	985	988	991	993	996	999	*001	*004
84.	1.5	007	009	012	015	017	020	022	025	028	030
85.		033									

INDEX OF REFRACTION OF AQUEOUS SOLUTIONS OF SUCROSE (CANE SUGAR) (Continued)

Temperature Corrections

Below 20°C the correction should be subtracted from the per cent sugar. Above 20°C the correction is to be added to the per cent sugar.

Temp. °C	Approximate per cent sugar 5	10	15	20	30	40	50	60	70	75
15	0.25	0.27	0.31	0.31	0.34	0.35	0.36	0.37	0.36	0.36
16	0.21	0.23	0.26	0.27	0.29	0.31	0.31	0.32	0.31	0.29
17	0.16	0.18	0.20	0.20	0.22	0.23	0.23	0.23	0.20	0.17
18	0.11	0.12	0.14	0.14	0.15	0.16	0.16	0.15	0.12	0.09
19	0.06	0.07	0.08	0.08	0.08	0.09	0.09	0.08	0.07	0.05
21	0.06	0.07	0.07	0.07	0.07	0.07	0.07	0.07	0.07	0.07
22	0.12	0.14	0.14	0.14	0.14	0.14	0.15	0.14	0.14	0.14
23	0.18	0.20	0.20	0.21	0.21	0.21	0.23	0.21	0.22	0.22
24	0.24	0.26	0.26	0.27	0.28	0.28	0.30	0.28	0.29	0.29
25	0.30	0.32	0.32	0.34	0.36	0.36	0.38	0.36	0.36	0.37
26	0.36	0.39	0.39	0.41	0.43	0.43	0.46	0.44	0.43	0.44
27	0.43	0.46	0.46	0.48	0.50	0.51	0.55	0.52	0.50	0.51
28	0.50	0.53	0.53	0.55	0.58	0.59	0.63	0.60	0.57	0.59
29	0.57	0.60	0.61	0.62	0.66	0.67	0.71	0.68	0.65	0.67
30	0.64	0.67	0.70	0.71	0.74	0.75	0.80	0.76	0.73	0.75

INDEX OF REFRACTION OF MALTOSE HYDRATE SOLUTIONS

McDonald: Journal of Research of the National Bureau of Standards, Vol. **46,** 165, 1951

The following table gives the refractive index of maltose solutions for λ = 5893 A at 20° and 25°C.

Percent	n_D^{20}	n_D^{25}	Δn	$\Delta n/\Delta t$
1	1.33438	1.33389	0.00049	0.00010
2	1.33579	1.33528	.00051	.00010
3	1.33720	1.33668	.00052	.00010
4	1.33862	1.33810	.00052	.00010
5	1.34006	1.33952	.00054	.00011
6	1.34150	1.34095	.00055	.00011
7	1.34295	1.34239	.00056	.00011
8	1.34442	1.34384	.00058	.00012
9	1.34589	1.34530	.00059	.00012
10	1.34738	1.34677	.00061	.00012
11	1.34887	1.34825	.00062	.00012
12	1.35039	1.34974	.00065	.00013
13	1.35190	1.35124	.00066	.00013
14	1.35343	1.35276	.00067	.00013
15	1.35497	1.35428	.00069	.00014
16	1.35652	1.35582	.00070	.00014
17	1.35808	1.35737	.00071	.00014
18	1.35965	1.35893	.00072	.00014
19	1.36124	1.36051	.00073	.00015
20	1.36283	1.36209	.00074	.00015
21	1.36444	1.36369	.00075	.00015
22	1.36606	1.36530	.00076	.00015
23	1.36770	1.36692	.00078	.00016
24	1.36934	1.36856	.00078	.00016
25	1.37100	1.37021	.00079	.00016
26	1.37267	1.37187	.00080	.00016
27	1.37435	1.37355	.00080	.00016
28	1.37604	1.37524	.00080	.00016
29	1.37775	1.37694	.00081	.00016
30	1.37946	1.37865	.00081	.00016
31	1.38120	1.38038	.00082	.00016
32	1.38294	1.38213	.00082	.00016
33	1.38470	1.38388	.00082	.00016
34	1.38647	1.38565	.00081	.00016
35	1.38825	1.38744	.00081	.00016

INDEX OF REFRACTION OF MALTOSE HYDRATE SOLUTIONS (Continued)

Percent	n_D^{20}	n_D^{25}	Δn	$\Delta n/\Delta t$
36	1.39004	1.38924	0.00080	0.00016
37	1.39185	1.39105	.00080	.00016
38	1.39367	1.39287	.00080	.00016
39	1.39551	1.39471	.00080	.00016
40	1.39735	1.39656	.00079	.00016
41	1.39922	1.39843	.00079	.00016
42	1.40109	1.40031	.00078	.00016
43	1.40298	1.40221	.00077	.00016
44	1.40488	1.40411	.00077	.00015
45	1.40680	1.40603	.00077	.00015
46	1.40873	1.40797	.00076	.00015
47	1.41067	1.40992	.00075	.00015
48	1.41263	1.41188	.00075	.00015
49	1.41460	1.41385	.00075	.00015
50	1.41658	1.41584	.00074	.00015
51	1.41858	1.41784	.00074	.00015
52	1.42059	1.41986	.00073	.00015
53	1.42262	1.42189	.00073	.00015
54	1.42466	1.42392	.00074	.00015
55	1.42672	1.42598	.00074	.00015
56	1.42878	1.42804	.00074	.00015
57	1.43087	1.43012	.00075	.00015
58	1.43296	1.43221	.00075	.00015
59	1.43508	1.43431	.00077	.00015
60	1.43720	1.43643	.00077	.00015
61	1.43934	1.43855	.00079	.00016
62	1.44150	1.44069	.00081	.00016
63	1.44367	1.44283	.00084	.00017
64	1.44585	1.44499	.00086	.00017
65	1.44805	1.44716	.00089	.00018

REFLECTION COEFFICIENTS

Coefficients of Reflection of Miscellaneous Surfaces for Monochromatic Radiation in the Visible Spectrum

(J. L. Michaelson)

Material	Wave lengths (μ) 0.400	0.500	0.600	0.700
Carbon Black in Oil	0.003	0.003	0.003	0.003
Clay,				
Kaolin (treated)	0.82	0.81	0.82	0.82
Kaolin (untreated)	0.75	0.79	0.85	0.86
White Georgia	0.94	0.92	0.93	0.94
Magnesium oxide	0.97	0.98	0.98	0.98
Paint,				
Lithopone	0.95	0.98	0.98	0.98
$MgCO_3$-Vynal Acetate Lacquer	0.90	0.88	0.88	0.88
ZnO-Milk	0.74	0.84	0.85	0.86
Paper,				
Blotting	0.64	0.72	0.79	0.79
Calendered	0.64	0.69	0.73	0.76
Crepe, green	0.23	0.49	0.19	0.48
Crepe, red	0.03	0.02	0.21	0.69
Crepe, yellow	0.17	0.44	0.75	0.79
News Print Stock	0.38	0.61	0.63	0.78
Peach,				
Green	0.18	0.17	0.62	0.63
Ripe	0.10	0.10	0.41	0.42
Pear,				
Green	0.04	0.12	0.29	0.41
Ripe	0.08	0.19	0.46	0.53
Pigment,				
Chrome Yellow	0.05	0.13	0.70	0.77
French Ochre	0.06	0.14	0.50	0.56
Porcelain Enamel,				
Blue	0.44	0.10	0.05	0.23
Orange	0.09	0.09	0.59	0.69
Red	0.05	0.03	0.08	0.62
White	0.77	0.73	0.72	0.70
Yellow	0.11	0.46	0.62	0.62
Talcum, Italian	0.94	0.89	0.88	0.88
Wheat Flour	0.75	0.87	0.94	0.97

REFLECTION COEFFICIENTS OF SURFACES FOR "INCANDESCENT" LIGHT

Material	Nature of Surface	Coefficient	Authority
Aluminum, "Alzak"	Diffusing	0.77–0.81	3
"Alzak"	Specular	0.79–0.83	3
on Glass	First Surface	0.82–0.86	4
Polished	Specular	0.69	3
Black Paper	Diffusing	0.05–0.06	4
Chromium	Specular	0.62	4
Copper	Specular	0.63	4
Gold	Specular	0.75	1
Magnesium oxide	Diffusing	0.98	5
Nickel	Specular	0.62–0.64	1, 3
Platinum	Specular	0.62	1
Porcelain Enamel	Glossy	0.76–0.79	3
Porcelain Enamel	Ground	0.81	3
Porcelain Enamel	Matt.	0.72–0.76	3
Silver	Polished	0.93	1
Silvered Glass	Second Surface	0.88–0.93	3
Snow	Diffusing	0.93	2
Steel	Specular	0.55	1
Stellite	Specular	0.58–0.65	4

(1) Hagen and Rubens. (2) Nutting, Jones, and Elliot. (3) J. E. Bock. (4) Frank Benford. (5) J. L. Michaelson.

EMISSIVITY AND ABSORPTION

These data are the result of investigations made by the Bureau of Standards, the British National Physical Laboratory, General Electric Research Laboratories, and several eastern universities, and were collected by W. J. King of the General Electric Company.

Low Temperature Total Emissivities

Material	Emissivity	Material	Emissivity
Silver, highly polished	0.02	Brass, polished	0.60
Platinum " "	0.05	Oxidized copper	0.60
Zinc " "	0.05	Oxidized steel	0.70
Aluminum, " "	0.08	Bronze paint	0.80
*Monel metal, polished	0.09	Black gloss paint	0.90
Nickel "	0.12	White lacquer	0.95
Copper "	0.15	White vitreous enamel	0.95
Stellite "	0.18	Asbestos paper	0.95
Cast iron "	0.25	Green paint	0.95
Monel metal, oxidized	0.43	Gray paint	0.95
Aluminum paint	0.55	Lamp black	0.95

Coefficient of Absorption of Solar Radiation

Material	Coefficient	Material	Coefficient
Silver, highly polished	0.07	Stellite, polished	0.30
Platinum " "	0.10	Light cream paint	0.35
Nickel " "	0.15	Monel metal, polished	0.40
*Aluminum	0.15	Light yellow paint	0.45
Magnesium carbonate	0.15	Light green paint	0.50
Zinc oxide	0.15	Aluminum paint	0.55
*Steel	0.20	Zinc, polished metal	0.55
Copper	0.25	Gray paint	0.75
White lead paint	0.25	Black matte	0.97
Zinc oxide paint	0.30		

* Questionable because of scant or inconsistent data.

TOTAL EMISSIVITY

Roeser and Wensel, National Bureau of Standards

Material	Temp. °C	Emissivity
Alloys		
20Ni-25Cr-55Fe, oxidized	200	0.90
	500	0.97
60Ni-12Cr-28Fe, oxidized	270	0.89
	560	0.82
80Ni-20Cr, oxidized	100	0.87
	600	0.87
	1300	0.89
Aluminum, unoxidized	25	0.022
	100	0.028
	500	0.060
oxidized	200	0.11
	600	0.19
Bismuth, unoxidized	25	0.048
	100	0.061
Brass, oxidized	200	0.61
	600	0.59
unoxidized	25	0.035
	100	0.035
Carbon, unoxidized	25	0.81
	100	0.81
	500	0.81
Chromium, unoxidized	100	0.08
Cobalt, unoxidized	500	0.13
	1000	0.23
Columbium, unoxidized	1500	0.19
	2000	0.24
Copper, unoxidized	100	0.02
	liquid	0.15
oxidized	200	0.6
	1000	0.6
calorized	100	0.26
	500	0.26
calorized, oxidized	200	0.18
	600	0.19
Fire brick	1000	0.75
Gold, unoxidized	100	0.02
	500	0.03
Gold enamel	100	0.37
Iron, unoxidized	100	0.05
oxidized	100	0.74
	500	0.84
	1200	0.89
cast, unoxidized	100	0.21
	liquid	0.29
cast, oxidized	200	0.64
	600	0.78
cast, strongly oxidized	40	0.95
	250	0.95

Material	Temp. °C	Emissivity
Iron, rusted	25	0.65
wrought, dull	25	0.94
oxidized	350	0.94
Lead, unoxidized	100	0.05
oxidized	200	0.63
Mercury, unoxidized	25	0.10
	100	0.12
Molybdenum, unoxidized	1000	0.13
	1500	0.19
	2000	0.24
Monel metal, oxidized	200	0.43
	600	0.43
Nickel, unoxidized	25	0.045
	100	0.06
	500	0.12
	1000	0.19
oxidized	200	0.37
	1200	0.85
Platinum, unoxidized	25	0.037
	100	0.047
	500	0.096
	1000	0.152
	1500	0.191
Silica brick	1000	0.80
	1100	0.85
Silver, unoxidized	100	0.02
	500	0.035
Steel, unoxidized	100	0.08
	liquid	0.28
oxidized	25	0.80
	200	0.79
	600	0.79
Steel plate, rough	40	0.94
	400	0.97
calorized, oxidized	200	0.52
	600	0.57
Tantalum, unoxidized	1500	0.21
	2000	0.26
Tin, unoxidized	25	0.043
	100	0.05
Tungsten, unoxidized	25	0.024
	100	0.032
	500	0.071
	1000	0.15
	1500	0.23
	2000	0.28
Zinc, unoxidized	300	0.05

SPECTRAL EMISSIVITY

Prepared by Roeser and Wensel, National Bureau of Standards

Spectral Emissivity of Materials, Surface Unoxidized for 0.65μ

Element	Solid	Liquid	Element	Solid	Liquid
Beryllium	0.61	0.61	Thorium	0.36	0.40
Carbon	0.80–0.93	....	Titanium	0.63	0.65
Chromium	0.34	0.39	Tungsten	0.43	
Cobalt	0.36	0.37	Uranium	0.54	0.34
Columbium	0.37	0.40	Vanadium	0.35	0.32
Copper	0.10	0.15	Yttrium	0.35	0.35
Erbium	0.55	0.38	Zirconium	0.32	0.30
Gold	0.14	0.22	Steel	0.35	0.37
Iridium	0.30	....	Cast Iron	0.37	0.40
Iron	0.35	0.37	Constantan	0.35	
Manganese	0.59	0.59	Monel	0.37	
Molybdenum	0.37	0.40	Chromel P (90Ni-10Cr)	0.35	
Nickel	0.36	0.37	80Ni-20Cr	0.35	
Palladium	0.33	0.37	60Ni-24Fe-16Cr	0.36	
Platinum	0.30	0.38	Alumel (95Ni; Bal. Al, Mn, Si)	0.37	
Rhodium	0.24	0.30	90Pt-10Rh	0.27	
Silver	0.07	0.07			
Tantalum	0.49	....			

Spectral Emissivity of Oxides

The emissivity of oxides and oxidized metals depends to a large extent upon the roughness of the surface. In general, higher values of emissivity are obtained on the rougher surfaces.

Material	Range of observed values	Probable value for oxide formed on smooth metal	Material	Range of observed values	Probable value for oxide formed on smooth metal
Aluminum oxide	0.22–0.40	0.30	Alumel (oxidized)	....	0.87
Beryllium oxide	0.07–0.37	0.35	Cast Iron (oxidized)	....	0.70
Cerium oxide	0.58–0.80	....	Chromel P (90Ni-10Cr) (oxidized)	....	0.87
Chromium oxide	0.60–0.80	0.70	80Ni-20Cr (oxidized)	....	0.90
Cobalt oxide	....	0.75	60Ni-24Fe-16Cr (oxidized)	....	0.83
Columbium oxide	0.55–0.71	0.70	55Fe-37.5Cr-7.5 Al (oxidized)	....	0.78
Copper oxide	0.60–0.80	0.70	70Fe-23Cr-5Al-2Co (oxidized)	....	0.75
Iron oxide	0.63–0.98	0.70	Constantan (55Cu-45Ni) (oxidized)	....	0.84
Magnesium oxide	0.10–0.43	0.20	Carbon Steel (oxidized)	....	0.80
Nickel oxide	0.85–0.96	0.90	Stainless Steel (18-8) (oxidized)	....	0.85
Thorium oxide	0.20–0.57	0.50	Porcelain	0.25–0.50	
Tin oxide	0.32–0.60	....			
Titanium oxide	....	0.50			
Uranium oxide	....	0.30			
Vanadium oxide	....	0.70			
Yttrium oxide	....	0.60			
Zirconium oxide	0.18–0.43	0.40			

PROPERTIES OF TUNGSTEN

Jones and Langmuir, General Electric Review

Temp. °K	Resistivity microhm cm	Electron emission amp./cm²	Evaporation g/cm² sec	Vapor pressure dynes/cm²	Thermal expansion per cent l_0 at 293°	Atomic heat cal./g. atom./°C.
300	5.65	....	....	....	.003	6.0
400	8.06	....	....	....	.044	6.0
500	10.56	....	....	....	.086	6.1
600	13.23	....	....	....	.130	6.1
700	16.09	....	....	....	.175	6.2
800	19.00	....	....	....	.222	6.2
900	21.94	....	....	....	.270	6.3
1000	24.93	1.07×10^{-15}	5.32×10^{-34}	1.98×10^{-29}	.320	6.4

PROPERTIES OF TUNGSTEN

Jones and Langmuir, General Electric Review

Temp. °K	Resistivity microhm cm	Electron emission amp./cm²	Evaporation g/cm² sec	Vapor pressure dynes/cm²	Thermal expansion per cent l_0 at 293°	Atomic heat cal./g. atom./°C.
1100	27.94	1.52×10^{-13}	2.17×10^{-30}	1.22×10^{-25}	.371	6.4
1200	30.98	9.73×10^{-12}	3.21×10^{-27}	1.87×10^{-22}	.424	6.5
1300	34.08	3.21×10^{-10}	1.35×10^{-24}	$8.18 \times .0^{-20}$	.479	6.7
1400	37.19	6.62×10^{-9}	2.51×10^{-22}	1.62×10^{-17}	.535	6.8
1500	40.36	9.14×10^{-8}	2.37×10^{-20}	1.54×10^{-15}	.593	7.0
1600	43.55	9.27×10^{-7}	1.25×10^{-18}	8.43×10^{-14}	.652	7.1
1700	46.78	7.08×10^{-6}	4.17×10^{-17}	2.82×10^{-12}	.713	7.2
1800	50.05	4.47×10^{-5}	8.81×10^{-16}	6.31×10^{-11}	.775	7.4
1900	53.35	2.28×10^{-4}	1.41×10^{-14}	1.01×10^{-9}	.839	7.6
2000	56.67	1.00×10^{-3}	1.76×10^{-13}	1.33×10^{-8}	.904	7.7
2100	60.06	3.93×10^{-3}	1.66×10^{-12}	1.28×10^{-7}	.971	7.8
2200	63.48	1.33×10^{-2}	1.25×10^{-11}	9.88×10^{-7}	1.039	8.0
2300	66.91	4.07×10^{-2}	8.00×10^{-11}	6.47×10^{-6}	1.109	8.2
2400	70.39	1.16×10^{-1}	4.26×10^{-10}	3.52×10^{-5}	1.180	8.3
2500	73.91	2.98×10^{-1}	2.03×10^{-9}	1.71×10^{-4}	1.253	8.4
2600	77.49	7.16×10^{-1}	8.41×10^{-9}	7.24×10^{-4}	1.328	8.6
2700	81.04	1.63	3.19×10^{-8}	2.86×10^{-3}	1.404	8.7
2800	84.70	3.54	1.10×10^{-7}	9.84×10^{-3}	1.479	8.9
2900	88.33	7.31	3.30×10^{-7}	3.00×10^{-2}	1.561	9.0
3000	92.04	1.42×10	9.95×10^{-7}	9.20×10^{-2}	1.642	9.2
3100	95.76	2.64×10	2.60×10^{-6}	2.50×10^{-1}	1.724	9.4
3200	99.54	4.78×10	6.38×10^{-6}	6.13×10^{-1}	1.808	9.5
3300	103.3	8.44×10	1.56×10^{-5}	1.51	1.893	9.6
3400	107.2	1.42×10^{2}	3.47×10^{-5}	3.41	1.980	9.8
3500	111.1	2.33×10^{2}	7.54×10^{-5}	7.52	2.068	9.9
3600	115.0	3.73×10^{2}	1.51×10^{-4}	1.53×10	2.158	10.1
3655	117.1	4.79×10^{2}	2.28×10^{-4}	2.33×10	2.209	10.2

Roeser and Wensel, National Bureau of Standards

Temp. °K	Normal brightness new candles per cm²	Spectral emissivity 0.65μ	Spectral emissivity 0.467μ	Color emissivity	Total emissivity	Brightness temp. 0.65μ	Color temp.
300	....	0.472	0.505	.....	0.032		
400	....	.....	.....	.....	.042		
500	....	.....	.....	.....	.053		
600	....	.....	.....	.....	.064		
700	....	.....	.....	.....	.076		
800	....	.....	.....	.....	.088		
900	....	.....	.....	.....	.101		
1000	0.0001	.458	.486	.395	.114	966	1007
1100	0.001	.456	.484	.392	.128	1059	1108
1200	0.006	.454	.482	.390	.143	1151	1210
1300	0.029	.452	.480	.387	.158	1242	1312
1400	0.11	.450	.478	.385	.175	1332	1414
1500	0.33	.448	.476	.382	.192	1422	1516
1600	0.92	.446	.475	.380	.207	1511	1619
1700	2.3	.444	.473	.377	.222	1599	1722
1800	5.1	.442	.472	.374	.236	1687	1825
1900	10.4	.440	.470	.371	.249	1774	1928
2000	20.0	.438	.469	.368	.260	1861	2032
2100	36	.436	.467	.365	.270	1946	2136
2200	61	.434	.466	.362	.279	2031	2241
2300	101	.432	.464	.359	.288	2115	2345
2400	157	.430	.463	.356	.296	2198	2451
2500	240	.428	.462	.353	.303	2280	2556
2600	350	.426	.460	.349	.311	2362	2662
2700	500	.424	.459	.346	.318	2443	2769
2800	690	.422	.458	.343	.323	2523	2876
2900	950	.420	.456	.340	.329	2602	2984
3000	1260	.418	.455	.336	.334	2681	3092
3100	1650	.416	.454	.333	.337	2759	3200
3200	2100	.414	.452	.330	.341	2837	3310
3300	2700	.412	.451	.326	.344	2913	3420
3400	3400	.410	.450	.323	.348	2989	3530
3500	4200	.408	.449	.320	.351	3063	3642
3600	5200	.406	.447	.317	.354	3137	3754

TRANSMISSION OF COLORED GLASSES

If I_o is the intensity of radiation entering a layer of some medium and I the intensity reaching the opposite surface, the ratio I/I_o is called the transmittance. In practice the ratio of intensity of radiation passing through a glass sample to that incident on its surface is often measured and plotted as transmission. The transmission is the result of two factors, the transmittance of the glass and the losses by reflection. These losses amount to about 4% for each glass-air surface; the transmission of a sample is about 92% of its transmittance. Since the reflection losses differ slightly with different samples, the correction is often determined and applied when the transmission is measured. Values which are thus corrected are marked * at the head of the column.

In order to obtain the transmittance for thicknesses other than those listed it is convenient to transform the tabular values to terms of βt in the equation $I/I_o = e^{\beta t}$ where t is the thickness (in millimeters) and β a constant for a particular sample. The base 10 is conveniently used in place of e so that βt becomes the common logarithm of the transmittance, or $\beta t = \log I/I_o$. Using the corrected value of the transmittance for a specific thickness, found in the table, the value of βt may be found, changed to the value for the new thickness and the transmittance for the second thickness computed.

For example: The tabular value of transmission for sample **CG** 396 at $\lambda = .46\mu$ is given as 0.80 for a thickness of 2 mm. It is desired to find the transmittance for 5 mm.

The corrected value of the transmittance for 2 mm is 0.80/.92 or about 0.871. Log .871 = 9.94002–10. Writing this as a wholly negative number the equation becomes $\beta t = -.05998$. For $t = 5$ $\beta t = -.05998 \times 5/2 = -.14995$ or changing to the more familiar form gives 9.85005–10 which is the logarithm of the new transmittance which is found to be .708. The transmission will be .92 × .708 or .651.

In order to identify the glasses listed, the manufacturer's number is given preceded by an abbreviation of the maker's name, as follows: **AO**, American Optical Co.; **BL**, Bausch & Lomb Optical Co.; **CE**, Chicago Eye Shield Co.; **CG**, Corning Glass Works. Data for Jena glasses are given separately in section II of the table.

This table has been compiled with the assistance of: Mr. H. P. Gage, Corning Glass Works; Mr. J. Liautaud, Chicago Eye Shield Co.; Mr. W. B. Rayton, Bausch & Lomb Optical Co.; Mr. A. J. Weinstein of the Fish-Shurman Corporation.

Abbreviations Used

abs., absorbing	lant., lantern	sext., sextant
bl., blue	lt., light	sig., signal
col., colorless	med., medium	tr., transmitting
didym., didymium	neut., neutral	u.v., ultra-violet
dk., dark	purp., purple	v., very
fl., fluorescent	pyrom., pyrometer	viol., violet
grn., green	rd red	yel., yellow
ht., heat		

TRANSMISSION OF COLORED GLASSES

Section I.—Glasses of American Manufacture

Wave-length μ	**AO** Crown 1.50 neut. 1.68 mm	**BL** Crookes 1 neut. 2 mm	**BL** Crookes 2 neut. 2 mm	**BL** Crookes 3 neut. 2 mm	**BL** Smoke A neut. 2 mm	**BL** Smoke B neut. 2 mm	**BL** Smoke C neut. 2 mm	**CG** 254 black ht. tr. 1 mm	**CG** 255 sext. red 1 mm	**CG** 241 Se red pyrom. 38%
0.22	...	*	*	*	*	*	*	....	....	....
.24	...	....	...	...	...	...	...	....	....	....
.26	...	....	...	...	...	...	...	....	....	....
.28	0	....	...	...	...	...	...	....	....	....
.30	.10	.00	.00	.00	.00	.00	.00	....	....	....
.32	.56	.00	.00	.00	.00	.00	.00	....	....	....
.34	.83	.00	.00	.00	.44	.14	.00	....	....	....
.36	.89	.06	.22	.20	.83	.60	.32	....	....	....
.38	.91	.72	.70	.65	.89	.74	.61	....	0	....
.40	.92	.86	.80	.74	.90	.76	.64	....	.07	....
.42	.92	.89	.75	.54	.85	.70	.43	....	.05	....
.44	.92	.91	.77	.46	.82	.54	.28	....	.015	....
.46	.92	.93	.80	.47	.83	.53	.28	....	.005	....
.48	.92	.94	.82	.51	.84	.55	.33	....	.005	....
.50	.92	.95	.83	.55	.85	.59	.35	....	.000	....
.52	.92	.96	.85	.57	.85	.59	.34	....	.005	....
.54	.92	.97	.86	.58	.85	.59	.33	....	.005	....
.56	.92	.97	.84	.57	.85	.60	.33	....	.015	....
.58	.92	.80	.69	.49	.85	.60	.33	....	.030	....
.60	.92	.99	.85	.58	.85	.59	.32	....	.040	....
.62	.92	1.00	.89	.61	.85	.58	.32	....	.060	0
.64	.92	1.00	.90	.63	.85	.60	.33	....	.070	.067
.66	.92	1.00	.92	.66	.87	.65	.39	0	.090	.508
.68	.92	1.00	.94	.73	.90	.73	.52	.006	.120	.660
.70	...	.99	.97	.90	.95	.88	.75	.012	.150	.667
.72	...	....	...	...	...	...	...	.018	.200	.660
1.0	...	.98	.92	.85	.93	.75	.74	.724	.860	....
1.5	...	.95	.89	.87	.91	.82	.74	.845	.920	....
2.0	...	.94	.89	.83	.90	.81	.72	.812	.920	....
2.5	...	.89	.82	.80	.88	.80	.74	.818	.915	....
3.0	...	.55	.53	.53	.59	.61	.56	.672	.743	....
3.5	...	.31	.25	.33	.35	.33	.31	.549	.670	....
4.0	...	.26	.34	.32	.31	.28	.28	.325	.390	....
4.5	...	.23	.20	.21	.22	.22	.23	.030	....	....
5.0	...	.11	.09	.09	10	.12	.10	....	....	....

TRANSMISSION OF COLORED GLASSES (Continued)

Section I.—Glasses of American Manufacture (Continued)

Wave-length μ	**CG** 242 Se red dark 100%	**CG** 244 Se red lant. 225%	**CG** 245 Se red traffic 300%	**CG** 246 Se red lt. house 125%	**CG** 346 amber A 2.5 mm	**BL** Kali-chrome A yellow 2 mm	**BL** Kali-chrome C yellow 2 mm	**CG** 348 yel. lant. Y 4	**CG** 351 yel. traffic Y 3	**CG** 338 yel. noviol C
0.22	.....	.....	....	....	....	*	*	.....	.....	....
.24	.....	.....	....	....	....	...	...	.....	.....	....
.26	.....	.....	....	....	....	...	...	.....	.....	....
.28	.....	.....	....	....	....	...	...	.....	.....	....
.30	.....	.....	....	....	....	.00	.00	.....	.....	....
.32	.....	.....	....	....	....	.00	.00	.....	.....	....
.34	.....	.....	....	....	....	.00	.00	.....	.....	....
.36	.....	.....	....	....	....	.00	.00	0	.....	....
.38	.....	.....	....	....	0	.21	.00	.0008	.....	....
.40	.....	.....	....	....	.045	.50	.00	.0010	.....	....
.42	.....	.....	....	....	.105	.65	.00	.0020	.....	....
.44	.....	.....	....	....	.200	.74	.00	.0039	.....	....
.46	.....	.....	....	....	.330	.81	.00	.0075	.....	.000
.48	.....	.....	....	....	.465	.85	.02	.0137	.....	.380
.50	.....	.....	....	....	.585	.90	.46	.0289	0	.650
.52	.....	.....	....	....	.680	.93	.87	.0826	.0003	.765
.54	.....	.....	....	....	.750	.94	.93	.157	.289	.813
.56	0	0	....	....	.795	.94	.97	.544	.784	.820
.58	.0001	.0001	.000	.100	.835	.95	.98	.848	.844	.825
.60	.0005	.024	.148	.730	.855	.96	.98	.882	.864	.825
.62	.019	.680	.722	.828	.875	.97	.97	.890	.874	....
.64	.630	.832	.832	.844	.865	.97	.97	.893	.883	....
.66	.828	.855	.850	.850	.869	.97	.97	.894	.889	....
.68	.847	.863	....	.847	.879	.97	.97	.894	.891	....
.70	.853	.862	.850	.835	.889	.97	.97	.892	.891	....
.72	.854	.859	....	.825	.900	...	...	.888	.890	....
1.0	.605	.....	....	....	....	.96	.87	.....	.....	....
1.5	.703	.....	....	....	....	.95	.89	.....	.....	....
2.0	.805	.....	....	....	....	.95	.90	.....	.....	....
2.5	.730	.....	....	....	....	.89	.85	.....	.....	....
3.0	.330	.....	....	....	....	.59	.60	.....	.....	....
3.5	.275	.....	....	....	....	.30	.34	.....	.....	....
4.0	.230	.....	....	....	....	.22	.27	.....	.....	....
4.5	.023	.....	....	....	....	.16	.17	.....	.....	....
5.0	.....	.....	....	....	....	.08	.10	.....	.....	....

Wave-length μ	**CG** 038 lt. yel. noviol A	**CG** 306 lt. yel. noviol O 2 mm	**CG** 330 sig. yel. 210% 2.5 mm	**CG** 375 U yel. fluor. 5 mm	**BL** Fieuzal A green 2 mm	**BL** Fieuzal B green 2 mm	**BL** Fieuzal C green 2 mm	**BL** Anti-glare green 2 mm	**BL** Green 19 green 2 mm	**BL** Ht. abs. green 2 mm
0.22	....	....	.....	.....	*	*	*	*	*	*
.24	....	....	.....	.....	...	...	...	...	...	...
.26	....	....	.....	.....	...	...	...	...	...	...
.28	....	....	.....	.....	...	...	...	...	...	...
.30	....	....	.....	.....	.00	.00	.00	.00	.00	.00
.32	....	....	.....	.....	.00	.00	.00	.00	.00	.00
.34	....	....	.....	0	.00	.00	.00	.00	.00	.03
.36	....	0	0	.190	.00	.00	.00	.00	.00	.22
.38	....	.120	.008	.323	.11	.00	.00	.02	.00	.43
.40	....	.473	.0075	.0622	.30	.02	.00	.14	.00	.51
.42	0	.635	.0076	.0367	.40	.05	.01	.27	.00	.55
.44	.550	.745	.0113	.184	.49	.14	.05	.29	.00	.58
.46	.702	.795	.0271	.407	.64	.29	.13	.43	.01	.61
.48	.765	.825	.0636	.518	.70	.38	.19	.50	.03	.65
.50	.787	.855	.1324	.581	.77	.48	.25	.55	.05	.66
.52	.802	.875	.241	.833	.82	.57	.33	.57	.14	.67
.54	.813	.880	.367	.858	.86	.65	.40	.58	.19	.67
.56	.820	.885	.492	.863	.86	.67	.43	.57	.17	.65
.58	.825	.885	.589	.868	.86	.46	.43	.56	.09	.62
.60	.825	.880	.654	.878	.85	.64	.41	.53	.04	.58
.62	....	.883	.693	.878	.84	.61	.39	.50	.01	.53
.64	....	....	.713	.880	.83	.59	.31	.45	.01	.48
.66	....	....	.721	.884	.82	.56	.35	.40	.01	.43
.68	....	....	.723	.893	.81	.54	.33	.36	.01	.38
.70	....	....	.720	.900	.80	.53	.31	.31	.01	.33
.72	....	....	.712	.902	...	...	...	...	...	...
1.0	....	....	.....	.910	.58	.29	.12	.11	.19	.09
1.5	....	....	.....	.880	.61	.40	.20	.15	.50	.15
2.0	....	....	.....	.862	.73	.52	.32	.27	.71	.25
2.5	....	....	.....	.705	.73	.51	.37	.34	.85	.32
3.0	....	....	.....	.090	.51	.35	.27	.27	.70	.20
3.5	....	....	.....	.010	.34	.25	.21	.17	.43	.15
4.0	....	....	.....	.....	.24	.24	.21	.15	.41	.09
4.5	....	....	.....	.....	.16	.11	.11	.11	.32	.07
5.0	....	....	.....	.....	.08	.06	.07	.05	.06	.06

TRANSMISSION OF COLORED GLASSES (Continued)

Section I.—Glasses of American Manufacture (Continued)

Wave-length μ	**BL** red free green 2 mm	**CG** 396 green 2 mm	**CG** 428 bl. grn. 2 mm	**CG** 440 sig. grn. 150% 5.78 mm	**CG** 401 sext. green 1 mm	**CG** 502 blue 2 mm	**CG** 503 dk. blue 2 mm	**CG** 556 sig. blue 100% 5 mm	**CG** 554 blue 101% 2.8 mm	**CG** 590 lt. bl. 4.6 mm
0.22	*	....	....	.....	.....	....	....	.....	....	....
.24	...	....	....	.....	.....	....	....	.....	....	....
.26	...	....	....	.....	.....	....	....	.....	....	....
.28	...	....	....	.....	.....	....	....	.....	....	....
.30	.00	0	0	.....	.....	....	....	.....	....	....
.32	.00	.020	.004	.....	.....	....	....	0	....	....
.34	.00	.280	.266	.....	.....	....	....	.060	....	....
.36	.00	.530	.652	0	.....	....	....	.370	....	....
.38	.00	.670	.780	.073	.....	....	....	.650	....	.417
.40	.00	.735	.830	.237	0	.631	.766	.780	.700	.650
.42	.04	.760	.845	.366	.0494	.782	.832	.800	.730	.653
.44	.10	.785	.860	.480	.124	.825	.830	.795	.700	.550
.46	.17	.800	.865	.588	.252	.822	.791	.727	.635	.462
.48	.23	.815	.860	.638	.429	.787	.698	.458	.400	.380
.50	.25	.820	.835	.618	.592	.703	.547	.205	.175	.316
.52	.22	.820	.770	.526	.670	.578	.380	.049	.050	.258
.54	.13	.805	.660	.366	.653	.441	.236	.012	.015	.200
.56	.06	.790	.515	.205	.538	.281	.105	.022	.030	.205
.58	.02	.760	.375	.0952	.397	.171	.050	.005	.005	.170
.60	.01	.720	.255	.0375	.260	.088	.018	.0008	0	.121
.62	.00	.670	.175	.0134	.161	.038	.005	.0006	0	.099
.64	.00	.610	.105	.0051	.107	.035	.005	.0003	0	.080
.66	.00	.555	.075	.0019	.0762	.024	.003	.0003	0	.082
.68	.00	.495	.050	.....	.0632	.082	.005	.002	.015	.095
.70	.00	.430	.040	.....	.0508	.054	.015	.008	.060	.103
.72	...	.370	.030	.....	.049	.269	.173	.009	.080	....
1.0	.07	.068	....	.....	.050	....	....	.....	....	....
1.5	.10	.172	....	.....	.180	....	....	.....	....	....
2.0	.35	.279	....	.....	.525	....	....	.....	....	....
2.5	.54	.400	....	.....	.780	....	....	.....	....	....
3.0	.46	....	....	.....	.250	....	....	.....	....	....
3.5	.29	....	....	.....	.150	....	....	.....	....	....
4.0	.28	....	....	.....	.110	....	....	.....	....	....
4.5	.20	....	....	.....	0	....	....	.....	....	....
5.0	.08	....	....	.....	.....	....	....	.....	....	....

Wave-length μ	**BL** Blue O blue 2 mm	**BL** Blue A blue 2 mm	**BL** Blue B blue 2 mm	**BL** Blue C blue 2 mm	**CG** 511 violet 2.15 mm	**CG** 557 lt. purp. 6.2 mm	**CG** 555 purp. 4.8 mm	**CG** 512 didym. 5 mm	**CG** 507 purp. 3 mm	**CG** 227 gold ruby 3 mm
0.22	*	*	*	*	.....	.....	.....	.....	.....	....
.24	.....	...	...	...	.....	.....	.....	.....	.....	....
.26	.....	...	...	...	.....	.....	.....	.....	.....	....
.28	.....	...	...	...	.....	.....	.....	.....	.....	....
.30	.020	.11	.08	.11	.....	.....	.....	.....	.....	....
.32	.450	.68	.65	.60	0	.....	.....	.....	.....	....
.34	.79	.88	.87	.87	.0018	.....	.....	.....	.212	....
.36	.88	.94	.93	.97	.082	.....	.....	.....	.552	....
.38	.94	.97	.96	.98	.326	.795	.783	.....	.728	....
.40	.98	.98	.97	.98	.502	.861	.845	.731	.726	.090
.42	1.00	.99	.98	.97	.462	.834	.786	.731	.490	.100
.44	1.00	.99	.98	.95	.326	.810	.705	.530	.225	.105
.46	.99	.99	.98	.92	.138	.781	.589	.548	.092	.100
.48	.99	.97	.93	.81	.014	.665	.318	.530	.045	.090
.50	.99	.94	.83	.66	.0003	.497	.1015	.652	.028	.055
.52	.99	.91	.73	.52	.....	.302	.0163	.601	.0299	.015
.54	.98	.89	.69	.45	.....	.185	.0028	.790	.033	.005
.56	.97	.88	.67	.43	.....	.257	.0092	.811	.050	.005
.58	.96	.87	.64	.40	.....	.160	.0017	.0082	.0727	.045
.60	.96	.86	.62	.37	.....	.095	.0002	.184	.100	.185
.62	.96	.84	.61	.35	.....	.109	.0004	.849	.128	.365
.64	.96	.85	.60	.36	.....	.0935	.0002	.851	.156	.515
.66	.96	.87	.64	.39	.....	.132	.0008	.849	.177	.615
.68	.97	.91	.75	.52	.....	.360	.035	.736	.201	.695
.70	.98	.97	.93	.81	.....	.710	.400	.804	.230	.740
.72	.....	...	...	...	.....	.852	.795	.845	.267	.740
1.0	.99	.98	.99	.99	.....	.....	.....	.....	.793	....
1.5	.96	.95	.89	.83	0	.....	.....	.....	.902	....
2.0	.97	.94	.92	.87	.100	.....	.....	.....	.863	....
2.5	.95	.87	.87	.91	.387	.....	.....	.....	.732	....
3.0	.81	.50	.55	.72	.320	.....	.....	.....	.155	....
3.5	.49	.33	.35	.41	.250	.....	.....	.....	.007	....
4.0	.39	.31	.31	.33	.305	.....	.....	.....	.015	....
4.5	.31	.21	.19	.29	.156	.....	.....	.....	0	....
5.0	.11	.09	.12	.11	0	.....	.....	.....	.....	....

TRANSMISSION OF COLORED GLASSES (Continued)

SECTION I.—GLASSES OF AMERICAN MANUFACTURE (Continued)

Wave-length μ	**CG** G 984 B green	**CG** G 985 B blue	**CG** 585 bl. purp. tr. u.v. 1 mm	**CG** 597 rd. purp. tr. u.v. 1 mm	**CG** 587 rd. purp. tr. u.v. 2 mm	**CG** 586 violet tr. u.v. 5 mm	**CG** 584 red tr. u.v. 1 mm	**CG** 774 neut. pyrex 2 mm	**CG** 970 neut. corex D 2 mm	**CG** 980 neut. corex A 2 mm
0.22	...	0	....	....	....	.....	....	....	0	.0115
.24	0	.02	....	....	....	.....	....	....	.0027	.237
.26	.38	.18	....	....	....	.....	....	....	.039	.531
.28	.34	.46	0	0	0	.....	....	0	.305	.753
.30	.12	.68	.153	.116	.039	0	.129	.120	.700	.840
.32	.36	.78	.650	.570	.342	.0004	.555	.610	.860	.879
.34	.73	.84	.860	.850	.640	.097	.791	.830	.902	.884
.36	.79	.86	.910	.913	.795	.290	.830	.900	.....	.885
.38	.75	.86	.925	.890	.700	.080	.714	.913	.....	.....
.40	.45	.81	.900	.714	.265	0	.115	.913	.....	.887
.42	.11	.67	.840	.260	.045	.....	0	....	.....	.....
.44	.02	.41	.765	.065	.010	.....	....	.913	.....	.893
.46	.01	.16	.620	.025	.010	.....	....	....	.....	.....
.48	.03	.04	.390	.025	.005	.....	....	.917	.....	.897
.50	...	...	.199	.026	.000	.....	....	....	.....	.....
.52	...	...	.090	.026	.005	.....	....	.918	.....	.900
.54	...	...	.053	.028	.005	.....	....	....	.....	.....
.56	...	...	.104	.027	.010	.....	....	.918	.....	.900
.58	...	...	.029	.030	.005	.....	....	....	.....	.....
.60	...	...	.014	.065	.000	.....	....	.919	.....	.900
.62	...	...	.020	.034	.000	.....	....	....	.....	.....
.64	...	...	.018	.045	.000	.....	0	.919	.....	.900
.66	...	...	.033	.160	.005	.....	.010	....	.....	.....
.68	...	...	.146	.350	.035	.....	.080	.919	.....	.900
.70	...	...	.505	.505	.255	.....	.290	.919	.....	.....
.72	...	...	.760	.595	.520	.....	.405	....	.....	.900
1.0	...	...	.850	.450	....	.....	....	.915	.....	.900
1.5	...	...	.425	.610	....	.....	....	.914	.....	.898
2.0	...	...	.645	.415	....	.....	....	.904	.....	.857
2.5	...	...	.805	.450	....	.....	....	....	.....	.744
3.0	...	...	.595	.485	....	.....	....	....	.....	.018
3.5	...	...	.520	.393	....	.....	....	....	.....	.017
4.0	...	...	.580	.422	....	.....	....	....	.....	.....
4.5	...	...	.310	....	....	.....	....	....	.....	.....

Wave-length μ	**CG** 986 rd. purp. corex A 3 mm	**BL** Weld. 3 yel. grn. 1 mm	**BL** Weld. 4 yel. grn. 1 mm	**BL** Weld. 5 yel. grn. 0.5 mm	**BL** Weld. 8 yel. grn. 0.5 mm	**BL** Weld. 12 yel. grn. 0.5 mm	**CE** Cesco-weld no. 3 ht. abs.	**CE** Cesco-weld no. 4 ht. abs.	**CE** Cesco-weld no. 5 ht. abs.	**CE** Cesco-weld no. 6 ht. abs.
0.22	....	*	*	*	*	*	....	....	....	....
.24	0	...	...	...	...	...	....	....	....	....
.26	.135	...	...	...	...	...	....	....	....	....
.28	.412	...	...	...	...	...	....	....	....	....
.30	.625	.00	.00	.00	.00	.00	....	....	....	....
.32	.793	.00	.00	.00	.00	.00	....	....	....	....
.34	.814	.00	.00	.00	.00	.00	....	....	....	....
.36	.797	.03	.03	.06	.11	.00	....	....	....	....
.38	.430	.06	.09	.13	.18	.01	....	.000	.000	....
.40	.075	.09	.12	.14	.21	.03	....	.001	.001	....
.42	.023	.10	.13	.14	.16	.02	0	.002	.001	....
.44	.000	.09	.14	.15	.15	.02	.001	.004	.000	....
.46	.000	.13	.19	.21	.17	.03	.013	.009	.001	.000
.48	.000	.22	.27	.27	.20	.05	.032	.025	.005	.001
.50	.000	.31	.25	.25	.25	.07	.070	.047	.010	.003
.52	.000	.40	.42	.41	.29	.08	.160	.082	.031	.008
.54	.000	.46	.44	.45	.31	.09	.180	.094	.032	.013
.56	.000	.46	.43	.45	.32	.10	.174	.087	.030	.012
.58	.000	.43	.39	.43	.31	.09	.157	.063	.026	.010
.60	.001	.36	.31	.38	.27	.08	.120	.037	.014	.006
.62	.003	.29	.25	.31	.23	.07	.068	.020	.007	.002
.64	.010	.24	.20	.26	.20	.06	.042	.010	.004	.001
.66	.025	.19	.15	.21	.19	.05	.025	.004	.001	.000
.68	.128	.15	.12	.18	.17	.05	.016	.001	.000	.000
.70	.231	.12	.09	.16	.15	.04	.009	.001	.000	.000
.72	.240	...	...	...	...	...	.008	....	.001	.000
1.0	.095	.07	.04	.09	.04	.03	.002	.000	.000	.000
1.5	.009	.03	.02	.04	.07	.02	.000	.000	.000	.000
2.0	.026	.07	.04	.10	.12	.04	.002	.000	.000	.000
2.5	.081	.11	.10	.15	.19	.07	.010	.002	.000	.000
3.0	0	.14	.11	.20	.22	.11	.012	.001	.001	.000
3.5	....	.16	.13	.14	.28	.14	.014	.003	.002	.000
4.0	....	.20	.17	.30	.33	.21	.024	.008	.007	.001
4.5	....	.21	.17	.37	.29	.26	.022	.005	.012	.004
5.0	....	.06	.05	.20	.15	.30	.000	.000	.001	.000

TRANSMISSION OF COLORED GLASSES (Continued)

Section II.—Jena Glasses*

Wave-length μ	UG 1 dk. viol. tr. u.v. 1 mm	UG 2 dk. viol. tr. u.v. 1 mm	UG 3 violet 1 mm	UG 4 dark violet 1 mm	BG 1 blue tr. u.v. 1 mm	BG 2 blue tr. u.v. 1 mm	BG 3 blue tr. u.v. 1 mm	BG 5 dk. blue abs. red 1 mm	BG 6† dk. blue 1 mm
.281	....	0.01	....	....	0.04	....	0.13	....	....
.302	0.17	.27	....	0.09	.40	0.04	.69	....	....
.312	.37	.50	0.04	.22	.64	.16	.77	....	....
.334	.69	.80	.55	.59	.93	.63	.92	....	....
.366	.85	.84	.91	.77	.97	.84	.96	0.09	0.06
.405	.08	.02	.85	.07	.97	.70	.86	.33	.34
.436	....	....	.50	.01	.86	.42	.63	.47	.48
.480	....	....	.17	.02	.44	.02	.10	.39	.33
.509	....	....	.12	.01	.14	....	.01	.19	.12
.546	....	....	.15	.02	.04	....	....	.06	.02
.578	....	....	.21	....	.05	....	....	.04	.02
.644	....	....	.35	....	.01	....	....	.01	.01
.700	.01	.12	.46	.21	.51	.32	.06	.13	.08
.775	.34	.30	.64	.52	.94	.84	.90	.13	.14
.85	.22	.19	.79	.39	.97	.83	.98	.10	.10
.95	.11	.12	.93	.39	.93	.74	.94	.12	.12
1.05	.07	.09	.97	.37	.86	.61	.81	.15	.15
1.15	.05	.07	.98	.32	.76	.50	.64	.19	.18
1.30	.04	.06	.99	.28	.58	.41	.39	.24	.23
1.45	.04	.06	.99	.27	.45	.38	.27	.29	.27
1.60	.03	.06	.99	.26	.40	.37	.19	.34	.31
1.80	.04	.06	.99	.25	.44	.38	.20	.42	.38
2.00	.04	.06	.99	.32	.50	.41	.28	.49	.46
2.40	.11	.11	.98	.51	.69	.51	.47	.68	.66
3.00	.17	.19	.86	.35	.55	.35	.49	.37	.36

Wave-length μ	BG 7 dk. blue abs. red 1 mm	BG 9 blue-green 1 mm	BG 10 blue-green 1 mm	BG 11 lt. blue (Nd) 1 mm	BG 12 blue abs. red 1 mm	BG 13 med. blue 1 mm	BG 14 bright blue 1 mm	BG 15 blue 1 mm	BG 16 blue 1 mm
.281	....	....	....	0.02	....	....	....	....	....
.302	....	....	....	.21	....	....	....	....	....
.312	....	0.01	....	.38	0.02	....	0.13	0.03	....
.334	0.02	.21	0.12	.80	.39	....	.71	.33	0.04
.366	.26	.67	.60	.95	.75	0.35	.92	.73	.57
.405	.51	.90	.78	.97	.86	.77	.98	.86	.80
.436	.70	.94	.84	.96	.85	.87	.98	.89	.84
.480	.79	.96	.88	.93	.48	.87	.98	.90	.86
.509	.73	.96	.88	.95	.12	.85	.96	.90	.86
.546	.50	.95	.87	.99	.02	.77	.90	.90	.84
.578	.26	.93	.86	.70	.01	.64	.75	.89	.81
.644	.03	.86	.74	1.00	....	.35	.46	.80	.68
.700	....	.74	.63	0.99	.04	.21	.30	.65	.54
.775	....	.56	.42	.90	.08	.14	.24	.47	.36
.85	.01	.46	.31	.92	.08	.14	.27	.36	.25
.95	.02	.38	.25	.98	.12	.19	.35	.28	.20
1.05	.04	.34	.24	.99	.18	.24	.45	.24	.18
1.15	.08	.33	.26	.99	.18	.30	.55	.24	.17
1.30	.15	.36	.31	.98	.16	.43	.69	.30	.21
1.45	.24	.42	.39	.97	.19	.52	.78	.41	.28
1.60	.35	.49	.47	.96	.22	.61	.84	.45	.38
1.80	.48	.56	.53	.97	.27	.73	.90	.49	.44
2.00	.58	.59	.55	.97	.53	.81	.93	.54	.44
2.40	.74	.64	.58	.93	.79	.88	.94	.61	.55
3.00	.56	.50	.46	.54	.55	.59	.63	.51	.44

* Data furnished by courtesy of the Fish-Schurman Corp., New York, Importers of Jena Glasses. All values are corrected for reflection losses.
† Discontinued.

Section II.—Jena Glasses (Continued)

Wave-length μ	BG 17 col. abs. heat 1 mm	VG 2* yellow-green 1 mm	VG 3 green (Nd) 1 mm	GG 1 col. abs. u.v. 1 mm	GG 2 col. abs. u.v. 1 mm	GG 3 light yellow 1 mm	GG 4 light green 1 mm	GG 5 bright yellow 1 mm	GG 6 medium yellow 1 mm
.281	....	....	....	....	....	....	....	....	....
.302	0.03	....	....	0.02	....	0.03	....	0.01	0.02
.312	.11	....	....	.13	....	.05	....	.02	.05
.334	.46	....	0.01	.47	....	.09	0.03	.03	.11
.366	.89	....	.49	.86	0.72	.16	.01	.09	.14
.405	.97	....	.27	.98	.98	.21	.40	.14	.21
.436	.98	0.05	.06	.99	.99	.85	.67	.40	.36
.480	.98	.37	.02	1.00	1.00	.99	.92	.99	.73
.509	.98	.59	.38	1.00	1.00	1.00	.95	1.00	.98
.546	.98	.61	.06	1.00	1.00	1.00	.97	1.00	1.00
.578	.98	.44	.69	1.00	1.00	1.00	.96	1.00	1.00
.644	.97	.10	....	1.00	1.00	1.00	.94	1.00	1.00
.700	.93	.08	.85	1.00	1.00	1.00	.96	1.00	0.99
.775	.78	.07	.59	1.00	1.00	0.99	.99	0.99	.98
.85	.61	.06	.32	1.00	1.00	.98	.99	.99	.97
.95	.42	.09	.36	1.00	1.00	.98	.99	.98	.96
1.05	.32	.13	.22	1.00	1.00	.97	.99	.98	.95
1.15	.30	.18	.20	1.00	1.00	.97	.99	.98	.95
1.30	.33	.27	.39	1.00	1.00	.97	.99	.97	.95
1.45	.36	.37	.34	1.00	1.00	.98	.99	.97	.96
1.60	.43	.47	.22	1.00	1.00	.98	.99	.98	.96
1.80	.46	.57	.38	1.00	1.00	.98	.99	.98	.97
2.00	.42	.65	.47	1.00	0.99	.98	.99	.98	.97
2.40	.39	.75	.30	0.96	.98	.96	.98	.97	.94
3.00	.27	.54	.12	.50	.70	.62	.64	.62	.62

Wave-length μ	GG 7 medium yellow 1 mm	GG 8 greenish-yellow 1 mm	GG 9 greenish-yellow 1 mm	GG 10 dk. grn.-yellow 1 mm	GG 11 deep yellow 1 mm	GG 12 yellow, grn. fl. 1 mm	OG 1 yellow-orange 1 mm	OG 2 orange 1 mm	OG 3 red-orange 1 mm
.281	....	....	....	....	....	....	....	....	....
.302	....	....	0.01	....	....	....	....	....	....
.312	0.01	....	.01	....	....	....	....	....	....
.334	.04	....	....	....	....	0.03	....	....	....
.366	.05	....	....	....	....	.74	....	....	....
.405	.12	0.05	.01	....	....	.53	....	....	....
.436	.16	.26	.14	0.09	0.01	.71	....	....	....
.480	.65	.75	.78	.59	.24	.90	....	....	....
.509	.99	.88	.88	.85	.97	.96	0.01	....	....
.546	.99	.99	.93	.97	.99	.99	.91	0.14	....
.578	1.00	.99	.92	.95	.99	1.00	.99	.94	0.48
.644	1.00	.92	.89	.83	.99	1.00	.99	.99	1.00
.700	1.00	.92	.91	.86	.99	1.00	.99	.99	1.00
.775	0.99	.98	.95	.96	.98	1.00	.99	.99	0.99
.85	.99	.98	.98	.99	.97	1.00	.99	.99	.99
.95	.98	.99	.99	.99	.96	1.00	.99	.99	.98
1.05	.98	1.00	.99	.99	.96	1.00	.99	.99	.98
1.15	.97	1.00	.99	1.00	.96	1.00	.99	.99	.98
1.30	.96	1.00	.99	1.00	.96	1.00	.99	.99	.98
1.45	.96	1.00	.99	0.99	.96	1.00	.99	.99	.98
1.60	.97	1.00	.99	.99	.96	0.99	.99	.99	.98
1.80	.98	.99	.99	.99	.97	.99	.98	.99	.98
2.00	.98	.99	.99	.99	.97	.99	.98	.99	.99
2.40	.95	.96	.97	.94	.95	.98	.97	.97	.97
3.00	.65	.68	.69	.44	.66	.67	.62	.60	.63

* Discontinued.

SECTION II.—JENA GLASSES (Continued)

Wave-length μ	RG 1 bright red 1 mm	RG 2 pure red 1 mm	RG 3 dark red 1 mm	RG 4 dark red 1 mm	RG 5 dark red 1 mm	RG 6 blood color 1 mm	RG 7 infra-red 1 mm	RG 8 v. dk red. 1 mm	RG 9 infra-red 1 mm
.281	....	....	....	....	....	....	....	....	....
.302	....	....	....	....	....	0.12	....	....	....
.312	....	....	....	....	....	.20	....	....	....
.334	....	....	....	....	....	.34	....	....	....
.366	....	....	....	....	....	.44	....	....	....
.405	....	....	....	....	....	.49	....	....	....
.436	....	....	....	....	....	.50	....	....	....
.480	....	....	....	....	....	.44	....	....	....
.509	....	....	....	0.02	....	.31	....	....	....
.546	....	....	....	.05	....	.18	....	....	....
.578	....	....	....	.10	....	.40	....	....	....
.644	0.96	0.92	0.54	.55	0.02	.93	....	0.01	....
.700	.98	.98	.70	.73	.96	.98	0.02	.71	0.20
.775	.98	.98	.81	.82	.98	.99	.18	.99	0.98
.85	.98	.98	.88	.87	.98	.99	.41	.99	.97
.95	.98	.98	.92	.91	.98	.99	.74	.99	.92
1.05	.98	.98	.94	.93	.98	.99	.91	.99	.80
1.15	.98	.98	.95	.94	.99	.99	.96	.99	.62
1.30	.98	.98	.97	.96	.99	.99	.97	.99	.38
1.45	.98	.98	.97	.97	.99	.99	.98	.99	.25
1.60	.98	.98	.98	.98	.99	.99	.99	.99	.19
1.80	.98	.98	.98	.98	.99	.99	.99	.99	.20
2.00	.98	.98	.98	.99	.99	.99	.99	.99	.27
2.40	.96	.95	.94	.97	.97	.96	.99	.97	.45
3.00	.65	.65	.56	.67	.58	.63	.85	.58	....

Wave-length μ	NG 1 v. dk. neut. 0.1 mm	NG 2 med. dk. neut. 0.1 mm	NG 3 dk. neut. 1 mm	NG 4 med. neut. 1 mm	NG 5 neut. 1 mm	NG 6 light neut. 1 mm	NG 7 blue-gray 1 mm
.281	....	0.05	....	....	....	....	....
.302	....	.30	....	....	....	....	....
.312	....	.36	....	....	....	....	....
.334	....	.46	....	....	0.02	0.07	0.01
.366	0.21	.47	0.02	0.07	.22	.53	.17
.405	.29	.45	.08	.18	.37	.89	.37
.436	.32	.46	.12	.23	.43	.92	.44
.480	.36	.50	.15	.29	.46	.89	.34
.509	.36	.52	.17	.30	.46	.89	.29
.546	.36	.56	.17	.31	.46	.89	.33
.578	.36	.58	.17	.31	.46	.89	.26
.644	.37	.62	.17	.30	.46	.87	.19
.700	.45	.77	.16	.28	.43	.86	.34
.775	.50	.80	.12	.21	.34	.85	.34
.85	.49	.80	.08	.17	.27	.86	.29
.95	.48	.81	.07	.14	.21	.87	.24
1.05	.50	.85	.07	.14	.19	.89	.21
1 15	.54	.85	.06	.16	.19	.91	.21
1.30	.59	.86	.09	.21	.22	.93	.21
1.45	.63	.88	.15	.26	.26	.95	.22
1.60	.67	.89	.20	.30	.31	.96	.26
1.80	.71	.90	.22	.34	.38	.96	.34
2.00	.73	.91	.26	.37	.43	.97	.40
2.40	.78	.93	.34	.39	.48	.96	.44
3.00	.72	.87	.20	.19	.21	.72	.25

SECTION II.—JENA GLASSES (Continued)

Wave-length μ	UG 5 dk. viol. tr. u. v. 1 mm	UG 6 tr. infra-red 1 mm	UG 7 tr. infra-red 1 mm	BG 18 blue-green 1 mm	BG 19 nearly neutral 1 mm	BG 20 didymium 1 mm	BG 21 neutral 1 mm	BG 22 light blue-green 1 mm	BG 23 blue 1 mm
.281	0.93	....	....	....	0.04	....	0.01	....	....
.302	.98	....	....	....	.38	0.04	.26	....	....
.312	.99	....	....	....	.60	.09	.46	0.03	0.01
.334	.98	....	....	0.03	.87	.41	.75	.42	.33
.366	.91	....	....	.21	.95	.91	.93	.83	.81
.405	.39	....	....	.50	.96	.95	.97	.92	.93
.436	.11	....	....	.70	.97	.89	.98	.94	.94
.480	.03	....	....	.87	.96	.82	.99	.94	.94
.509	.01	....	....	.90	.95	.81	.99	.94	.91
.546	....	....	....	.87	.94	.99	.99	.94	.76
.578	....	....	....	.68	.92	.33	.98	.93	.50
.644	.01	....	0.01	.25	.83	.99	.93	.83	.19
.700	.72	....	.05	.12	.67	.99	.85	.73	.10
.775	.57	....	.25	.08	.43	.98	.65	.58	.05
.85	.50	0.10	.54	.07	.21	.98	.48	.49	.06
.95	.51	.50	.80	.09	.08	.99	.29	.44	.10
1.05	.53	.77	.91	.13	.04	.99	.19	.44	.17
1.15	.46	.89	.95	.20	.02	.99	.16	.45	.26
1.30	.30	.96	.98	.32	.02	.98	.15	.51	.40
1.45	.25	.98	.99	.45	.04	.89	.18	.59	.53
1.60	.27	.99	.99	.59	.06	.90	.23	.66	.65
1.80	.31	.99	.99	.72	.09	.94	.29	.71	.77
2.00	.36	.99	.99	.80	.10	.91	.30	.71	.84
2.40	.34	.99	.99	.80	.09	.90	.29	.75	.92
3.00	.08	.70	.65	.32	.05	.82	.12	.43	.72

Wave-length μ	BG 24 magenta 1 mm	VG 4 greenish yellow 1 mm	VG 5 bright greenish yellow 1 mm	VG 6 bright blue-green 1 mm	VG 7 greenish yellow 1 mm	VG 8 darker blue-green 1 mm	VG 9 yellow green 1 mm	GG 13 neutral abs. u. v. 1 mm
.281	0.92	....	....	....	....	....	....	....
.302	.96	....	....	....	....	....	....	....
.312	.97	....	....	....	....	....	....	....
.334	.99	....	....	....	....	....	....	....
.366	.99	....	....	....	....	....	....	0.20
.405	.97	....	....	0.07	0.02	0.03	....	.92
.436	.89	0.06	0.03	.32	.15	.16	0.03	.98
.480	.35	.39	.55	.69	.50	.55	.47	.99
.509	.19	.69	.83	.80	.75	.73	.69	.99
.546	.11	.85	.87	.77	.84	.71	.64	.99
.578	.04	.77	.79	.60	.78	.48	.44	.99
.644	.24	.47	.50	.23	.58	.13	.10	.99
.700	.91	.48	.46	.17	.61	.08	.07	.99
.775	.95	.55	.48	.15	.67	.08	.06	.99
.85	.92	.59	.52	.15	.70	.09	.06	.99
.95	.88	.63	.59	.21	.75	.14	.09	.99
1.05	.84	.67	.66	.30	.79	.21	.16	.99
1.15	.81	.71	.72	.39	.83	.31	.25	.99
1.30	.75	.78	.80	.54	.87	.45	.39	.99
1.45	.73	.83	.87	.65	.90	.57	.54	.99
1.60	.72	.87	.91	.74	.92	.65	.65	.99
1.80	.71	.91	.94	.83	.94	.75	.76	.99
2.00	.70	.94	.95	.87	.95	.82	.82	.99
2.40	.58	.96	.96	.91	.95	.92	.90	.98
3.00	.21	.85	.79	.66	.63	.75	.65	.68

SECTION II.—JENA GLASSES (Continued)

Wave-length μ	GG 14 dark yellow 1 mm	GG 15 bright yellow 1 mm	OG 4 orange yellow 1 mm	OG 5 orange red 1 mm	RG 9 infra-red 1 mm	RG 10 infra-red 1 mm	NG 8 neutral 1 mm	NG 9 neutral 1 mm
.281	....	0.01	....	....	....	....	....	....
.302	....	.05	....	....	....	....	....	....
.312	....	.08	....	....	....	....	....	....
.334	....	.15	....	....	....	....	0.17	....
.366	....	.26	....	....	....	....	.65	....
.405	....	.46	....	....	....	....	.78	0.02
.436	....	.94	....	....	....	....	.80	.04
.480	....	.99	0.02	....	....	....	.83	.04
.509	0.95	.99	.64	....	....	....	.84	.04
.546	.99	.99	.97	0.92	....	....	.87	.04
.578	.99	.99	.99	.99	....	....	.86	.04
.644	.99	.99	.99	.99	....	0.09	.83	.04
.700	.99	.99	.99	.99	0.23	.20	.82	.08
.775	.99	.99	.99	.99	.94	.75	.79	.09
.85	.99	.99	.99	.99	.95	.92	.73	.06
.95	.99	.99	.99	.99	.83	.94	.67	.04
1.05	.99	.99	.99	.99	.65	.96	.65	.03
1.15	.99	.99	.99	.99	.42	.98	.65	.04
1.30	.99	.99	.99	.99	.25	.99	.69	.07
1.45	.99	.99	.99	.99	.22	.99	.76	.12
1.60	.99	.99	.99	.99	.23	.99	.79	.16
1.80	.99	.99	.99	.99	.28	.99	.80	.18
2.00	.99	.99	.99	.99	.37	.99	.81	.20
2.40	.98	.97	.99	.99	.61	.99	.84	.30
3.00	.68	.74	.72	.65	.45	.69	.63	.13

Wave-length μ	WG 1 neutral 1 mm	WG 2 neutral 1 mm	WG 3 neutral 1 mm	WG 4 neutral 1 mm	WG 5 neutral 1 mm	WG 6 neutral 1 mm	WG 7 neutral 1 mm	WG 8 neutral 1 mm
.281	....	....	....	....	....	0.25	0.60	0.96
.302	....	....	....	....	0.23	.73	.89	.98
.312	....	....	....	0.24	.68	.87	.94	.99
.334	....	0.03	0.60	.88	.96	.97	.99	1.00
.366	0.46	.86	.97	.99	1.00	1.00	1.00	1.00
.405	.98	.99	1.00	1.00	1.00	1.00	1.00	1.00
.436	1.00	1.00	1.00	1.00	1.00	1.00	1.00	1.00
.480	1.00	1.00	1.00	1.00	1.00	1.00	1.00	1.00
.509	1.00	1.00	1.00	1.00	1.00	1.00	1.00	1.00
.546	1.00	1.00	1.00	1.00	1.00	1.00	1.00	1.00
.578	1.00	1.00	1.00	1.00	1.00	1.00	1.00	1.00
.644	1.00	1.00	1.00	1.00	1.00	1.00	1.00	1.00
.700	1.00	1.00	1.00	1.00	1.00	1.00	1.00	1.00
.775	1.00	1.00	1.00	1.00	1.00	1.00	1.00	1.00
.85	1.00	1.00	1.00	1.00	1.00	1.00	1.00	1.00
.95	1.00	1.00	1.00	1.00	1.00	1.00	1.00	1.00
1.05	1.00	1.00	1.00	1.00	1.00	1.00	1.00	1.00
1.15	1.00	1.00	1.00	1.00	1.00	1.00	1.00	1.00
1.30	1.00	1.00	1.00	1.00	1.00	1.00	1.00	1.00
1.45	1.00	1.00	1.00	1.00	1.00	1.00	1.00	.99
1.60	1.00	1.00	1.00	1.00	1.00	1.00	1.00	.98
1.80	1.00	1.00	1.00	1.00	1.00	1.00	.99	.96
2.00	1.00	1.00	1.00	.99	.99	.99	.99	.92
2.40	1.00	.99	1.00	.98	.98	.97	.97	.79
3.00	.84	.81	.82	.79	.77	.57	.66	.51

TRANSMISSION OF WRATTEN FILTERS

Compiled by Allie C. Peed, Jr. for The Eastman Kodak Company

Data condensed from Kodak Wratten Filters for Scientific and Technical Use published by the Eastman Kodak Company, manufacturers of the filters.

The following pages give (1) percentage luminous transmittance at wave lengths from 400 to 700μ at intervals of 10μ for the standard illuminant "C" adopted by the International Commission of Illumination, (2) dominant wave length in millimicrons, and (3) percentage of excitation purity. Values of wave length followed by "c" indicate the complementary wave lengths of purple filters which do not have a dominant wave length.

All colorimetric specifications are based on the 1931 standard ICI colorimetric and luminosity data.

The transmittance data are given as representing standard samples of the filters. They are intended only for the information of users in choosing filters which will meet their requirements. Values taken from the tables of data should not be used by research workers as representing precisely the absorption characteristics of a particular filter. If such precise data are needed, they should be determined for the particular filter being used.

Where the spectra extend into the ultraviolet this fact is indicated by an asterisk (*) in the transmission tables immediately beneath the filter number, and quantitative data are not given. The manufacturer should be consulted for this information. Transmission in the ultraviolet of wave lengths less than 330μ will be eliminated in the case of cemented filters, as glass absorbs ultraviolet radiation of wave lengths shorter than about 330μ.

Stability ratings are given as three letter combinations following the filter description in the table below. In establishing the stability classifications each filter is exposed to a selected light source for a specific time interval. The following grading system is used to describe the result:

Class A—stable
Class B—relatively stable
Class C—somewhat unstable
Class D—unstable

The classification letters, for example, AAA, describe the stability to the following three exposure tests in this order:

1. Two weeks' exposure to daylight in a south window
2. Twenty-four hours' exposure to a "Fade-Ometer"
3. Two weeks' exposure at two feet from a 1000-watt tungsten lamp.

Filters are supplied in two forms; as lacquered gelatin film, or as a gelatin film cemented between pieces of optical glass. Filters in glass are cemented between sheets of plane-parallel glass, which is surfaced in quantities and is of sufficient accuracy for general photographic work, and for most scientific purposes.

Most Wratten Gelatin Filters are stocked in 2- or 3-inch squares. Stocks of 2- or 3-inch square filters cemented in glass are maintained only in filters usually used for general photographic work.

The booklet "Kodak Filters and Lens Attachments" gives more valuable information on this subject.

FILTER DATA

No.	Description, use, and stability
	Colorless
0	For compensating thickness of other gelatin filters in optical systems, AAA.
1	Absorbs ultraviolet below 360 mμ, DDD.
1A	Kodak Skylight Filter—Reduces excess bluishness in outdoor color photographs in open shade under a clear, blue sky, ACA.
	Yellows
2B	Absorbs ultraviolet below 410 mμ, ACA.
3	Light yellow, CCD.
3N5	No. 3 plus 0.5 neutral density, AAA.
4	Light yellow—Approximate correction on panchromatic materials for outdoor scenes, including sky, CCC.
6	K1—Light yellow—Partial correction outdoors, BBA.
8	K2—Yellow—Full correction outdoors on Type B panchromatic materials. Widely used for proper sky, cloud, and foliage rendering. Green separation for Fluorescence Process, AAA.
8N5	No. 8 plus 0.5 neutral density, AAA.
9	K3—Deep yellow. Moderate contrast in outdoor photography (with black-and-white films), AAA.
11	X1—Greenish yellow. Correction for tungsten light on Type B panchromatic materials; also for daylight correction with Type C panchromatic materials in making outdoor portraits, darkening skies, or lightening foliage, AAA.
12	Minus blue. Haze cutting in aerial photography, AAA.
13	X2—Yellow green. Correction for Type C panchromatic materials in tungsten light, ABA.
15	G—Deep yellow. Overcorrection in landscape photography. Contrast control in copying and in aerial infrared photography, AAA.
16	Blue absorption, AAB.
18A	Transmits ultraviolet and infrared only (glass), AAA.
	Oranges and Reds
21	Blue and blue-green absorption, CBB.
22	Yellow-orange. For increasing contrast in blue preparations in microscopy. Mercury yellow, BAC.
23A	Light red. Two-color projection—contrast effects, BAB.
24	Red for two-color photography (daylight or tungsten). White-flame-arc tricolor projection, AAB.
25	A—Tricolor red for direct color separation. Contrast effects in commercial photography and in outdoor scenes. Two-color general viewing. Aerial infrared photography and haze cutting, AAA.
26	Stereo red, AAA.
29	Red color separation from transparencies and for the Kodak Fluorescence Process. Strong contrast effects. Copying blueprints. Tungsten tricolor projection, AAA.
	Magentas and Violets
30	Green absorption, BBC.
31	Green absorption, CCA.
32	Minus green, CCD.
33	Strong green absorption, CCB.
34	Violet, CDD.
34A	Blue separation—Kodak Fluorescence Process, DCC.
35	Contrast in microscopy, CDD.
36	Dark violet, CCC.
	Blues and Blue-greens
38	Red absorption, BCA.
38A	Red absorption. Increasing contrast in visual microscopy, BBB.
39	Contrast control in printing motion-picture duplicates (glass), AAA.
40	Green for two-color photography (tungsten), CBC.
44	Minus red—Two-color general viewing, DDD.
44A	Minus red, DDD.
45	Contrast in microscopy, DDD.
45A	Highest resolving power in visual microscopy, CDC.
46	Blue projection (experimental), DDD.
47	Tricolor blue for direct color separation and from Kodak Ektacolor Film for Dye Transfer. Contrast effects in commercial photography. Tungsten and white-flame-arc tricolor projection, BBC.
47B	Tricolor blue for color separation from transparencies and from Kodak Ektacolor Film for Graphic Arts, BBB.
48	Green and red absorption, CBC.
48A	Green and red absorption, AAB.
49	Dark blue, BCB.
49B	Very dark blue, BBB.
50	Very dark blue. Mercury violet, CCC.
	Greens
52	Light green, AAB.
53	Medium green, CCB.
54	Very dark green, AAA.
55	Stereo green, BBC.
56	Very light green, CBC.

FILTER DATA

No.	Description, use, and stability
	Greens (Continued)
57	Green for two-color photography (daylight), CBC.
57A	Light green, BBC.
58	Tricolor green for direct color separation. Contrast effects in commercial photography and microscopy, BBC.
59	Green for tricolor projection (white-flame-arc), BBB.
59A	Very light green, BBB.
60	Green for two-color photography (tungsten), BDC.
61	Green color separation from transparencies and Kodak Ekta-color Film. Tricolor projection (tungsten), ABC.
64	Red absorption (light), CDB.
65	Red absorption, ADB.
65A	Red absorption, CCD.
66	Contrast effects in microscopy and medical photography, DDC.
67A	Red absorption (light). Two-color projection, CDC.
	Narrow-band
70	Dark red. Infrared photography. Color separation for Kodak Ektacolor Film (with tungsten), ABC.
72B	Dark orange-yellow, CCC.
73	Dark yellow-green, ABB.
74	Dark green. Mercury green, BBC.
75	Dark blue-green, ACC.
76	Dark violet (compound filter), DDD.
77	Transmits 546 $m\mu$ mercury line (glass plus gelatin), AAA.
77A	Transmits 546 $m\mu$ mercury line (glass plus gelatin), AAA.
	Photometrics
78	Bluish. Photometric filter (visual), BAB.
78AA	Bluish. Photometric filter (visual), BAA.
78A	Bluish. Photometric filter (visual), AAA.
78B	Bluish. Photometric filter (visual), AAA.
78C	Bluish. Photometric filter (visual), BAA.
86	Yellowish. Photometric filter (visual), BBA.
86A	Yellowish. Photometric filter (visual), AAA.
†86B	Yellowish. Photometric filter (visual), BCA.
†86C	Yellowish. Photometric filter (visual), AAA.

No.	Description, use, and stability
	Light Balancing
80A	For Kodachrome Film, Daylight Type, and photographic flood lamps, ABA.
81	Yellowish. For warmer color rendering.
81A	Yellowish. For Kodak Ektachrome Film, Type B, with photographic flood lamps.
81B	Yellowish. For warmer color rendering.
81C	Yellowish. For Kodachrome Film, Type A, with flash lamps.
81D	Yellowish. For Kodachrome Film, Type A, with flash lamps.
81EF	Yellowish. For Kodak Ektachrome Film, Type B, with flash lamps.
82	Bluish. For cooler color rendering.
82A	Bluish. For Kodachrome Film, Type A, with 3200 K lamps.
82B	Bluish. For cooler color rendering.
82C	Bluish. For cooler color rendering.
83	Yellowish. For 16 mm Commercial Kodachrome Film and daylight exposure, BBB.
85	Orange. For Type A Kodak color films and daylight exposure, BAA.
85B	Orange. For Kodak Ektachrome Film, Type B, and daylight exposure, BAB.
	Miscellaneous
79	Photographic sensitometry. Corrects 2360 K to 5500 K, AAA.
87	For infrared photography. Absorbs visual.
87C	Absorbs visual, transmits infrared.
88A	For infrared photography. Absorbs visual.
89B	For infrared photography, AAA.
90	Narrow-band viewing filter for judging brightness scale of scenes, CCD.
96	Neutral filters for controlling luminance, AAB.
97	Dichroic absorption, AAA.
102	Correction filter for Barrier-layer photocell, ABA.
106	Correction filter for S-4 type photocell, AAA.

Wave length	Percent transmittance												
	No. 0 *	No. 1 *	No. 1A *	No. 2B *	No. 3	No. 3N5 *	No. 4	No. 6 *	No. 8	No. 8N5	No. 9	No. 11 *	No. 12 *
400	88.0	85.0	59.0	19.0	.....	......	.....	7.40	......	......	......	......	......
10	88.5	85.5	76.0	48.0	.....	......	.....	8.32	......	......	......	......	......
20	88.9	86.0	82.0	67.0	.....	......	.....	10.4	......	......	......	0.16	......
30	89.3	86.5	84.6	75.3	0.36	......	.....	13.5	......	......	......	0.29	......
40	89.6	87.0	86.0	80.0	1.78	......	.....	18.9	......	......	......	0.56	......
50	89.8	87.4	86.8	83.0	11.5	1.59	.....	27.6	......	......	......	1.32	......
60	89.9	87.8	87.2	85.2	38.0	9.40	6.9	39.0	0.25	0.16	......	4.00	......
70	90.1	88.2	87.5	86.7	68.0	18.5	42.0	52.3	5.50	2.0	1.78	12.0	......
80	90.3	88.5	87.3	88.1	80.8	23.5	74.0	65.8	19.0	6.3	8.31	26.0	......
90	90.4	88.7	86.8	88.8	85.2	25.5	84.7	76.8	41.0	13.2	20.7	43.7	......
500	90.5	88.9	86.3	89.5	86.9	26.3	87.5	83.5	63.5	20.0	34.5	55.0	1.50
10	90.6	89.1	85.5	89.9	87.8	26.7	88.5	87.0	78.0	24.3	48.8	60.0	17.3
20	90.7	89.3	84.8	90.3	88.4	27.0	89.1	88.4	84.1	26.7	62.0	60.2	55.0
30	90.7	89.5	84.3	90.5	89.0	27.2	89.4	89.0	86.5	28.0	76.0	57.8	77.8
40	90.8	89.7	84.0	90.6	89.5	27.5	89.6	89.4	87.7	28.6	83.8	54.2	86.0
50	90.8	89.9	83.9	90.7	89.8	27.8	89.8	89.7	88.4	29.0	87.0	50.0	88.4
60	90.9	90.1	84.1	90.8	90.1	27.9	90.0	89.9	88.8	29.3	88.3	44.8	89.4
70	90.9	90.2	84.8	90.9	90.4	28.0	90.2	90.1	89.2	29.5	88.8	38.9	89.7
80	90.9	90.3	86.0	90.9	90.6	28.4	90.4	90.3	89.5	29.6	89.1	33.1	90.1
90	91.0	90.4	87.4	91.0	90.7	29.0	90.6	90.5	89.8	29.8	89.3	27.6	90.3
600	91.0	90.5	88.5	91.1	90.8	29.5	90.8	90.6	90.1	29.9	89.5	22.7	90.4
10	91.0	90.5	89.5	91.2	90.9	29.5	90.9	90.7	90.3	29.6	89.7	19.0	90.5
20	91.0	90.6	90.2	91.3	91.0	29.3	91.0	90.8	90.5	29.4	89.8	14.9	90.7
30	91.0	90.6	90.6	91.3	91.0	29.1	91.1	90.9	90.7	29.1	89.9	11.4	90.8
40	91.1	90.7	90.8	91.4	91.1	29.0	91.2	91.0	90.9	28.8	90.0	9.10	90.9
50	91.1	90.7	91.0	91.4	91.2	29.4	91.3	91.1	91.0	28.9	90.1	8.05	91.0
60	91.1	90.8	91.1	91.5	91.3	29.6	91.4	91.2	91.1	29.2	90.1	7.50	91.1
70	91.1	90.8	91.1	91.5	91.4	29.8	91.5	91.2	91.2	29.4	90.2	7.05	91.2
80	91.1	90.9	91.1	91.6	91.5	30.0	91.5	91.3	91.3	29.5	90.2	6.50	91.2
90	91.1	90.9	91.1	91.7	91.6	30.2	91.6	91.4	91.4	29.7	90.3	6.10	91.2
700	91.1	91.0	91.1	91.8	91.7	31.0	91.6	91.5	91.5	30.2	90.3	6.20	91.3
Luminous transmit.	90.8	89.9	85.9	90.5	88.3	27.4	87.8	87.5	82.7	27.0	76.6	40.2	73.8
Dominant wave lgth.	571.0	575.0	498.0	570.0	569.5	570.5	569.5	570.3	571.8	572.0	574.4	550.3	576.1
Excitation purity.	0.8	1.5	1.2	5.7	50.0	56.3	64.0	44.7	85.2	84.0	91.4	60.7	97.8

* Some transmission below 400 $m\mu$. Consult the manufacturer.

TRANSMISSION OF WRATTEN FILTERS (Continued)

Wave length	Percent transmittance												
	No. 13 *	No. 15 *	No. 16	No. 18A *	No. 21	No. 22	No. 23A	No. 24	No. 25	No. 26	No. 29	No. 30 *	No. 31 *
400	.....	.....	.....	.....	.....	.....	.....	.....	.....	.....	.....	48.6	13.8
10	.....	.....	.....	.....	.....	.....	.....	.....	.....	.....	.....	47.4	14.5
20	.....	.....	.....	.....	.....	.....	.....	.....	.....	.....	.....	48.5	16.4
30	0.18	.....	.....	.....	.....	.....	.....	.....	.....	.....	.....	50.1	25.5
40	0.50	.....	.....	.....	.....	.....	.....	.....	.....	.....	.....	49.4	42.7
50	1.35	.....	.....	.....	.....	.....	.....	.....	.....	.....	.....	43.0	50.2
60	4.08	.....	.....	.....	.....	.....	.....	.....	.....	.....	.....	26.5	40.4
70	11.0	.....	.....	.....	.....	.....	.....	.....	.....	.....	.....	13.8	22.6
80	23.5	.....	.....	.....	.....	.....	.....	.....	.....	.....	.....	5.00	8.20
90	39.0	.....	.....	.....	.....	.....	.....	.....	.....	.....	.....	0.63	1.85
500	50.8	.....	.....	.....	.....	.....	.....	.....	.....	.....	.....	.....	0.12
10	55.2	1.00	.....	.....	.....	.....	.....	.....	.....	.....	.....	.....	.....
20	56.5	16.0	3.00	.....	.....	.....	.....	.....	.....	.....	.....	.....	.....
30	55.0	52.1	22.0	.....	.....	.....	.....	.....	.....	.....	.....	.....	.....
40	51.0	70.7	48.0	.....	2.50	.....	.....	.....	.....	.....	.....	.....	.....
50	46.0	84.3	69.5	.....	29.0	0.25	.....	.....	.....	.....	.....	.....	.....
60	39.2	87.5	79.5	.....	65.0	19.0	.....	.....	.....	.....	.....	0.10	.....
70	32.0	88.7	84.0	.....	80.6	60.0	11.0	.....	.....	.....	.....	10.0	.....
80	25.1	89.3	86.3	.....	85.4	81.0	47.0	4.55	.....	.....	.....	45.0	.....
90	18.2	89.7	87.8	.....	87.3	87.0	69.6	37.3	12.6	2.90	.....	76.0	0.63
600	13.5	90.0	89.0	.....	88.1	88.5	82.7	72.3	50.0	30.0	.....	87.4	26.0
10	9.60	90.1	89.6	.....	88.7	89.0	85.8	82.9	75.0	63.2	10.0	89.5	67.2
20	6.40	90.2	90.0	.....	89.0	89.5	87.2	86.4	82.6	78.9	45.3	90.2	84.0
30	3.66	90.3	90.2	.....	89.5	89.8	87.9	87.8	85.5	84.0	71.4	90.5	88.1
40	2.20	90.4	90.3	.....	89.9	90.0	88.5	88.5	86.7	86.1	82.7	90.7	89.8
50	1.58	90.5	90.4	.....	90.2	90.1	89.0	89.0	87.6	87.2	86.6	90.8	90.2
60	1.74	90.6	90.5	.....	90.4	90.2	89.4	89.3	88.2	88.1	88.4	90.9	90.4
70	2.62	90.6	90.6	.....	90.5	90.3	89.6	89.7	88.5	88.5	89.4	91.0	90.5
80	3.55	90.7	90.7	.....	90.5	90.4	89.8	89.9	89.0	88.9	90.0	91.1	90.7
90	4.48	90.7	90.8	0.25	90.6	90.5	90.0	90.2	89.3	89.2	90.3	91.1	90.8
700	5.25	90.8	90.8	1.20	90.6	90.6	90.2	90.3	89.5	89.5	90.4	91.1	91.0
Luminous transmit.	34.5	66.2	57.7	0.0014	45.6	35.8	25.0	17.8	14.0	11.7	6.3	26.6	12.9
Dominant wave lgth.	542.0	579.3	582.7	700.0	588.9	595.1	602.7	610.6	615.1	619.0	631.6	498.6$_c$	513.1$_c$
Excitation purity.	57.5	99.0	99.3	100.0	99.9	99.9	100.0	100.0	100.0	100.0	100.0	62.4	81.9

Wave length	Percent transmittance												
	No. 32 *	No. 33 *	No. 34 *	No. 34A *	No. 35 *	No. 36 *	No. 38 *	No. 38A *	No. 39 *	No. 40	No. 44 *	No. 44A *	No. 45
400	38.0	0.85	64.0	.....	48.0	36.5	60.5	33.4	85.2	.....	0.44	2.52	.....
10	37.9	0.71	70.1	0.1	57.0	45.5	66.5	41.2	78.2	.....	0.36	3.39	.....
20	40.0	1.17	72.0	40.0	57.6	45.5	72.5	53.0	70.5	.....	0.63	6.30	.....
30	43.0	1.69	68.4	69.7	47.5	32.7	75.3	58.0	63.3	.....	3.63	17.4	.....
40	55.5	5.36	58.2	68.7	29.5	15.2	76.2	58.8	53.6	.....	13.1	32.7	5.00
50	66.0	14.3	42.3	56.2	12.3	3.7	75.9	57.6	42.5	.....	25.4	41.8	19.0
60	66.0	12.4	25.2	40.5	3.5	0.35	74.8	55.2	28.5	3.16	36.5	48.1	29.5
70	57.0	5.00	12.1	23.8	0.25	.....	73.4	51.9	17.3	21.6	46.5	51.7	34.4
80	40.0	0.50	2.7	9.2	.....	.....	71.6	48.5	10.2	44.7	53.6	52.9	35.7
90	21.0	.....	0.2	2.3	.....	.....	69.5	44.6	4.00	61.4	56.8	52.2	34.5
500	9.56	.....	.....	0.33	.....	.....	66.7	40.2	1.33	70.2	55.8	49.8	29.7
10	2.51	.....	.....	.....	.....	.....	63.9	35.8	0.35	72.4	50.9	44.8	21.5
20	0.13	.....	.....	.....	.....	.....	60.8	31.7	.....	70.5	42.1	36.8	11.5
30	.....	.....	.....	.....	.....	.....	57.0	27.2	.....	64.8	30.5	26.8	3.80
40	.....	.....	.....	.....	.....	.....	52.6	22.3	.....	55.5	18.6	16.8	0.85
50	.....	.....	.....	.....	.....	.....	48.0	17.6	.....	44.2	8.99	8.20	.....
60	.....	.....	.....	.....	.....	.....	42.8	12.9	.....	32.5	3.59	2.95	.....
70	.....	.....	.....	.....	.....	.....	37.0	8.78	.....	20.3	0.80	0.91	.....
80	.....	.....	.....	.....	.....	.....	30.6	5.65	.....	9.56	.....	0.10	.....
90	.....	.....	.....	.....	.....	.....	25.5	3.48	.....	3.20	.....	.....	.....
600	6.04	.....	.....	.....	.....	.....	20.9	2.09	.....	1.10	.....	.....	.....
10	41.0	0.80	.....	0.13	.....	.....	16.8	1.15	.....	0.32	.....	.....	.....
20	75.0	24.9	.....	1.0	.....	.....	12.9	0.59	.....	.....	.....	.....	.....
30	86.1	60.8	.....	6.3	.....	.....	10.0	0.28	.....	.....	.....	.....	.....
40	89.0	78.0	0.4	22.0	.....	.....	7.79	0.13	.....	.....	.....	.....	.....
50	90.0	85.0	4.0	45.0	0.1	.....	6.68	.....	.....	.....	.....	.....	.....
60	90.6	87.5	20.7	65.0	3.0	0.21	6.20	.....	.....	.....	.....	.....	.....
70	90.7	88.7	45.2	77.3	19.0	7.5	5.91	.....	.....	.....	.....	.....	.....
80	90.8	89.4	66.5	85.0	43.5	29.0	5.41	.....	0.50	0.80	.....	.....	.....
90	90.9	89.8	78.8	88.2	66.0	55.0	4.90	.....	4.06	6.99	0.18	.....	.....
700	91.0	90.0	85.0	89.8	77.7	71.3	5.00	.....	17.8	23.5	1.60	.....	1.00
Luminous transmit.	12.5	5.2	1.3	2.9	0.45	0.25	42.5	17.3	1.2	33.6	15.6	14.4	5.2
Dominant wave lgth.	551.7$_c$	498.0$_c$	424.0	564.8$_c$	566.8$_c$	566.4$_c$	483.5	478.9	450.6	516.2	589.1	483.4	481.5
Excitation purity.	79.6	88.3	94.4	91.4	96.3	97.8	41.8	69.8	98.9	48.5	72.9	77.2	88.4

* Some transmission below 400 mμ. Consult the manufacturer.

TRANSMISSION OF WRATTEN FILTERS (Continued)

Wave length	Percent transmittance												
	No. 45A	No. 46 *	No. 47 *	No. 47B *	No. 48 *	No. 48A *	No. 49 *	No. 49B *	No. 50 *	No. 52 *	No. 53 *	No. 54	No. 55
400	....	1.20	7.80	16.0	0.96	5.65	3.30	1.70	0.45	2.18	....	....	....
10	....	0.60	17.4	29.5	3.16	10.0	4.28	2.00	0.39	1.51	....	....	....
20	....	0.80	34.0	43.6	8.25	16.0	6.93	3.55	0.59	0.80	....	....	....
30	1.00	5.98	47.0	50.0	15.0	21.0	11.2	7.00	2.63	0.44	....	....	....
40	8.81	19.0	50.3	47.2	22.6	25.0	18.9	13.0	8.90	0.41	....	....	....
50	17.4	30.1	48.3	36.0	30.3	26.2	25.6	17.4	14.0	0.69	....	....	....
60	20.9	33.8	43.4	25.0	33.2	22.9	24.0	14.8	12.3	1.45	....	....	0.20
70	21.6	32.1	36.2	13.2	29.6	16.5	15.7	7.60	5.36	2.70	0.10	....	2.90
80	20.5	27.0	28.5	4.5	22.4	9.55	6.93	2.76	1.55	4.90	0.71	....	13.1
90	18.0	20.2	19.6	1.3	14.1	4.27	2.14	0.40	0.10	8.50	2.14	....	34.2
500	14.4	11.1	11.3	0.17	7.30	1.58	0.46	....	....	13.3	4.47	....	53.4
10	10.1	4.39	5.64	....	2.64	0.48	....	....	....	18.2	7.24	0.10	67.0
20	5.60	1.66	1.91	....	0.50	....	....	....	....	23.7	10.7	0.31	69.3
30	2.52	0.35	0.36	....	....	....	....	....	....	28.5	14.0	0.64	65.1
40	0.64	....	....	....	....	....	....	....	....	32.1	16.6	0.89	56.7
50	0.10	....	....	....	....	....	....	....	....	33.1	17.3	0.93	45.0
60	....	....	....	....	....	....	....	....	....	31.0	15.4	0.62	33.1
70	....	....	....	....	....	....	....	....	....	25.6	11.4	0.21	20.7
80	....	....	....	....	....	....	....	....	....	19.1	6.90	....	9.00
90	....	....	....	....	....	....	....	....	....	12.6	3.60	....	2.70
600	....	....	....	....	....	....	....	....	....	7.78	1.41	....	0.40
10	....	....	....	....	....	....	....	....	....	4.17	0.40	....	....
20	....	....	....	....	....	....	....	....	....	2.34	0.15	....	....
30	....	....	....	....	....	....	....	....	....	1.38	....	....	....
40	....	....	....	....	....	....	....	....	....	0.80	....	....	....
50	....	....	....	....	....	....	....	....	....	0.54	....	....	....
60	....	....	....	....	....	....	....	....	....	0.36	....	....	....
70	....	....	....	....	....	....	....	....	....	0.27	....	....	....
80	....	....	....	....	....	....	....	....	....	0.23	....	....	0.66
90	0.20	0.25	....	....	....	....	....	....	....	0.19	....	....	6.90
700	2.24	0.85	....	....	....	....	....	....	....	0.17	....	....	27.8
Luminous transmit.	2.8	2.4	2.8	0.78	1.86	0.88	0.69	0.36	0.26	20.1	9.0	0.032	31.4
Dominant wave lgth.	477.6	470.4	463.7	479.8	466.5	458.0	457.9	455.5	455.9	553.3	551.1	546.1	530.2
Excitation purity.	89.7	94.9	95.8	69.1	96.1	98.3	98.9	99.3	99.4	77.3	89.7	97.0	68.4

Wave length	Percent transmittance												
	No. 56	No. 57	No. 57A	No. 58	No. 59 *	No. 59A *	No. 60	No. 61	No. 64 *	No. 65	No. 65A *	No. 66 *	No. 67A *
400	....	....	....	....	....	....	....	....	9.00	....	....	12.3	1.10
10	....	....	....	....	....	....	....	....	9.20	....	....	13.0	0.93
20	....	....	....	....	....	....	....	....	8.75	0.23	....	15.0	1.28
30	....	....	....	....	....	0.16	....	....	9.20	0.61	0.16	18.4	3.16
40	....	....	0.19	....	....	0.37	....	....	11.3	1.58	1.32	23.2	6.40
50	....	....	0.87	....	0.40	1.26	0.19	....	15.5	4.10	5.50	31.2	10.5
60	0.16	0.44	2.56	....	1.90	4.57	1.38	....	23.3	9.00	13.0	42.2	17.7
70	3.12	3.10	7.80	0.23	7.70	13.2	5.38	....	34.4	16.8	24.9	55.5	28.5
80	13.0	13.1	21.6	1.38	21.0	30.0	15.0	0.33	46.8	24.9	36.6	68.4	41.4
90	34.5	31.9	41.7	4.90	41.5	50.8	32.0	4.00	56.6	31.3	45.1	77.6	52.1
500	59.0	50.5	58.8	17.7	59.0	66.0	48.4	16.6	62.1	33.7	45.8	82.7	57.9
10	73.0	60.6	67.9	38.8	67.7	73.0	57.2	32.3	62.9	32.4	39.7	84.6	58.8
20	79.0	63.3	70.1	52.2	69.8	75.1	59.2	40.0	59.1	27.5	29.7	84.0	55.4
30	79.9	61.0	67.6	53.6	67.2	73.2	55.5	39.6	51.6	20.7	17.8	82.6	47.5
40	77.5	55.0	61.8	47.6	61.5	68.5	47.5	34.5	41.3	13.7	7.90	79.1	36.6
50	72.6	47.1	53.5	38.4	54.0	62.0	36.8	26.3	28.0	6.50	2.40	73.7	25.0
60	66.1	37.3	43.3	27.8	45.0	54.4	25.2	17.3	16.2	1.66	0.32	67.1	14.2
70	58.0	26.5	31.6	17.4	35.0	44.5	14.4	9.70	7.95	0.40	....	58.8	5.50
80	46.1	16.6	19.4	9.0	24.0	33.0	6.3	4.40	3.10	....	....	47.2	1.40
90	33.8	8.69	9.70	3.50	14.0	22.0	1.82	1.66	0.80	....	....	34.5	0.28
600	24.0	3.70	4.50	1.50	7.95	14.6	0.48	0.38	....	....	....	24.4	....
10	18.7	1.60	2.00	0.41	4.90	10.5	0.10	....	....	....	....	18.5	....
20	13.2	0.49	0.87	....	2.70	6.92	....	....	....	....	....	13.7	....
30	7.22	....	0.22	....	1.00	3.16	....	....	....	....	....	7.70	....
40	3.02	....	....	....	0.17	1.07	....	....	....	....	....	3.00	....
50	1.48	....	....	....	....	0.50	....	....	....	....	....	1.46	....
60	1.91	....	....	....	....	0.91	....	....	....	....	....	1.91	....
70	7.95	....	....	....	0.63	3.00	....	....	....	....	....	6.17	....
80	23.0	....	0.16	....	4.00	10.0	....	....	....	....	....	19.9	....
90	44.1	....	1.15	....	12.0	20.0	2.10	....	0.10	....	0.20	42.6	....
700	64.8	....	3.17	0.53	22.6	30.0	8.70	....	4.50	....	2.18	63.1	0.40
Luminous transmit.	52.8	32.5	37.2	23.7	38.7	45.8	26.1	16.8	25.0	9.6	9.8	58.3	22.4
Dominant wave lgth.	552.3	536.4	534.0	540.2	538.3	541.4	525.7	536.8	497.3	496.6	492.7	512.3	499.8
Excitation purity.	78.2	69.2	62.1	88.1	66.0	59.3	62.2	85.4	55.0	67.8	77.4	21.5	55.8

* Some transmission below 400 mμ. Consult the manufacturer.

TRANSMISSION OF WRATTEN FILTERS (Continued)

Wave length	Percent transmittance											
	No. 70	No. 72B	No. 73	No. 74	No. 75	No. 76 *	No. 77	No. 77A	No. 78 *	No. 78AA *	No. 78A *	No. 78B *
400	......	......	......	......	......	0.22	......	......	37.2	43.0	56.0	64.1
10	......	......	......	......	......	0.18	......	......	41.7	46.0	58.6	66.5
20	......	......	......	......	......	0.29	......	......	44.2	48.7	61.0	68.4
30	......	......	......	......	......	1.38	......	......	44.6	49.8	61.8	69.5
40	......	......	......	......	......	3.50	......	......	44.2	49.7	61.8	70.0
50	......	......	......	......	......	3.50	......	......	41.7	48.0	61.0	69.4
60	......	......	......	......	1.97	1.92	......	......	38.0	44.9	58.7	67.5
70	......	......	......	......	10.0	0.51	......	......	33.8	40.3	55.0	65.4
80	......	......	......	......	17.4	......	......	......	27.5	35.6	51.0	62.9
90	......	......	......	......	18.0	......	......	......	23.5	30.9	47.1	59.8
500	......	......	......	......	13.0	......	......	......	19.5	26.5	43.5	57.0
10	......	......	......	0.96	7.35	......	0.30	0.10	15.8	23.4	40.0	54.2
20	......	......	......	7.95	3.20	......	9.10	5.35	13.8	20.3	36.9	51.4
30	......	......	......	14.6	0.83	......	13.5	1.90	11.8	17.8	34.4	49.3
40	......	......	......	12.9	0.14	......	46.1	35.0	10.5	16.6	32.7	48.1
50	......	......	......	7.60	......	......	78.0	71.8	9.56	14.9	31.2	46.7
60	......	......	2.24	3.06	......	......	75.8	63.1	8.53	13.2	29.4	45.0
70	......	......	5.97	0.83	......	......	8.00	......	7.77	12.1	28.0	43.6
80	......	......	4.56	0.12	......	......	1.00	......	7.41	11.6	27.5	43.1
90	......	1.26	2.00	......	......	......	0.32	......	6.93	11.1	27.0	42.9
600	......	5.89	0.56	......	......	......	16.2	1.60	6.45	10.40	26.0	41.8
10	......	5.25	0.10	......	......	......	52.1	32.1	5.50	9.20	24.1	40.0
20	......	2.88	......	......	......	......	83.0	78.0	4.80	7.70	21.8	37.6
30	......	1.26	......	......	......	......	84.9	79.5	3.94	6.50	19.7	35.5
40	......	0.48	......	......	......	......	88.1	86.5	3.46	5.60	18.6	34.2
50	0.63	0.14	......	......	......	......	89.8	89.2	3.24	5.50	18.4	33.6
60	10.5	......	......	......	......	......	89.8	89.0	3.16	5.60	18.5	34.0
70	35.0	......	......	......	......	......	85.5	79.5	3.39	5.80	18.7	34.1
80	55.2	......	......	......	......	......	76.1	62.5	3.45	6.10	19.0	34.5
90	70.0	......	......	......	......	0.13	75.0	62.4	3.51	6.10	19.3	34.8
700	79.0	......	......	......	0.14	1.24	86.5	83.0	3.90	6.50	20.2	36.0
Luminous transmit.	0.31	0.74	1.3	4.0	1.9	0.046	32.3	25.5	10.7	15.8	31.6	46.7
Dominant wave lgth.	675.6	604.9	574.9	538.6	487.7	449.2	579.9	581.5	471.1	473.4	475.7	477.2
Excitation purity.	100.0	100.0	100.0	96.7	90.4	99.7	99.0	99.1	63.0	54.5	33.7	20.7

Wave length	Percent transmittance											
	No. 78C *	No. 79 *	No. 80A *	No. 81 *	No. 81A *	No. 81B *	No. 81C *	No. 81D *	No. 81EF *	No. 82 *	No. 82A *	No. 82B *
400	74.9	24.0	67.6	77.7	65.1	55.1	46.1	38.2	30.7	83.0	80.1	76.7
10	76.6	26.0	73.1	78.1	65.9	55.8	46.6	38.4	31.5	83.7	80.8	78.0
20	77.9	29.0	76.8	79.0	67.6	57.7	49.0	41.0	34.3	84.6	81.6	79.2
30	78.9	31.0	7.77	80.5	70.2	61.0	52.5	45.5	38.6	85.1	82.2	79.7
40	79.4	32.2	76.5	81.9	72.8	64.5	57.2	50.0	43.2	85.4	82.4	79.7
50	79.5	32.7	73.0	83.0	74.8	67.2	60.5	53.9	47.4	85.4	82.4	79.2
60	79.3	31.4	69.0	83.7	76.0	69.1	63.0	56.5	50.2	85.0	81.7	78.0
70	78.6	28.8	63.6	84.3	77.1	70.6	64.2	58.1	52.0	84.6	80.7	76.3
80	77.8	25.6	57.6	84.6	77.8	71.3	65.0	59.0	53.0	84.0	79.3	74.4
90	76.7	22.2	51.3	84.9	78.3	71.8	65.7	60.0	54.0	83.3	78.0	72.1
500	75.5	19.3	45.2	85.3	78.6	72.6	66.4	60.8	55.4	82.6	76.6	70.2
10	74.2	16.8	39.4	85.4	79.0	72.9	66.5	61.1	56.2	82.0	75.3	68.3
20	73.0	14.2	34.2	85.5	79.5	73.2	67.0	61.6	57.0	81.4	74.0	66.5
30	72.1	12.7	30.0	86.0	80.4	74.5	68.8	62.5	59.5	81.0	73.1	65.5
40	71.5	11.8	27.1	86.5	81.5	76.0	71.0	66.1	62.7	80.8	72.7	65.0
50	70.7	11.0	24.8	86.8	82.3	77.0	72.0	67.3	64.5	80.6	72.4	64.5
60	69.8	9.76	23.5	87.0	82.6	77.6	72.5	68.0	65.3	80.4	71.8	63.8
70	69.0	8.81	22.6	87.1	82.7	77.8	72.7	68.3	65.8	80.2	71.5	63.2
80	68.8	8.50	22.6	87.1	82.8	78.0	73.0	68.5	66.0	80.2	71.5	63.2
90	68.6	8.29	23.2	87.4	83.1	78.2	74.0	69.5	66.5	80.3	71.7	63.4
600	68.0	7.56	23.7	87.6	84.0	79.1	75.6	72.0	68.1	80.2	71.5	63.0
10	66.7	6.45	23.2	88.1	85.0	81.0	78.5	75.0	71.6	79.3	70.3	61.5
20	65.0	5.13	21.0	88.8	86.1	83.1	80.8	78.0	74.7	78.4	68.5	59.0
30	63.8	4.17	18.2	89.2	87.0	84.2	82.1	79.8	77.0	77.5	66.9	56.9
40	63.0	3.47	15.8	89.4	87.4	85.1	83.0	80.8	78.4	76.8	65.5	55.0
50	62.7	3.16	14.5	89.5	87.7	85.6	83.5	81.5	79.2	76.5	64.8	54.1
60	63.0	3.09	13.8	89.8	88.0	86.0	84.1	82.1	80.1	76.2	64.6	53.7
70	63.3	3.16	13.4	90.0	88.2	86.5	84.8	83.0	80.9	76.1	64.5	53.7
80	63.4	3.16	12.7	90.1	88.5	87.0	85.5	83.7	81.8	76.1	64.4	53.5
90	63.6	3.16	11.7	90.3	89.0	87.5	86.1	84.6	82.9	76.2	64.2	53.4
700	65.0	3.31	11.5	90.5	89.2	88.0	86.8	85.5	84.0	77.1	64.6	54.1
Luminous transmit.	70.4	11.3	28.4	86.8	82.0	76.9	72.0	67.4	64.0	80.7	72.5	64.6
Dominant wave lgth.	479.8	474.8	471.7	576.7	577.5	577.8	577.4	579.5	579.0	477.5	476.6	475.6
Excitation purity.	6.8	52.8	45.9	2.9	6.0	8.7	10.7	14.7	19.0	3.0	6.3	10.2

* Some transmission below 400 mμ. Consult the manufacturer.

TRANSMISSION OF WRATTEN FILTERS (Continued)

Wave length	Percent transmittance												
	No. 82C *	No. 83 *	No. 85 *	No. 85B *	No. 86 *	No. 86A *	No. 86B *	No. 86C *	No. 89B	No. 90	No. 96 *	No. 97	No. 102 *
400	73.4	13.5	6.0	1.59	0.50	8.00	20.0	44.0	.....		4.28	....	1.12
10	75.0	13.1	18.0	9.32	0.81	12.2	26.1	55.0	.....		4.91	....	0.96
20	76.4	13.5	28.4	15.5	1.55	16.7	31.6	62.0	.....		5.50	....	0.89
30	77.2	14.1	33.4	19.0	2.88	21.5	37.5	66.6	.....		6.17	....	0.96
40	77.2	15.6	36.2	20.8	5.50	27.8	44.0	70.8	.....		6.92	....	1.23
50	76.6	17.8	38.1	22.1	9.10	34.2	50.1	74.3	.....		7.50	....	1.86
60	75.2	21.0	40.4	24.3	13.5	40.4	55.4	76.8	.....		7.81	....	3.23
70	73.2	25.5	43.0	27.5	17.8	45.0	59.5	78.7	.....		8.15	0.22	6.45
80	70.7	30.2	45.3	30.9	21.3	48.7	62.5	80.2	.....		8.47	0.43	14.0
90	68.1	35.8	47.2	34.3	24.5	51.2	64.6	81.2	.....		8.60	0.39	21.6
500	65.7	43.5	48.9	38.3	26.8	52.8	66.0	81.7	.....		8.73	0.15	30.7
10	63.5	46.3	49.2	40.7	27.9	53.4	66.4	81.9	.....		8.85	....	41.4
20	61.5	47.2	48.2	40.6	28.6	53.7	66.6	82.0	.....		8.90	....	51.3
30	59.9	48.3	48.3	40.7	30.4	55.0	67.6	82.4	.....		9.01	....	59.4
40	59.1	49.6	49.2	41.6	32.5	56.5	69.0	83.0	.....		9.07	....	64.2
50	58.3	51.8	51.0	43.2	35.0	58.5	70.2	83.5	.....		9.20	....	66.7
60	57.2	56.5	55.8	47.1	41.2	63.0	73.0	84.6	.....	9.00	9.30	....	66.3
70	56.2	65.0	64.5	56.0	53.0	70.9	78.1	86.8	.....	30.5	9.20	....	63.0
80	56.1	75.5	75.0	68.1	67.5	79.0	84.0	88.9	.....	34.3	9.19	....	58.0
90	56.0	83.0	83.0	78.1	76.5	85.2	87.5	89.9	.....	25.2	9.54	....	51.9
600	55.0	87.3	87.2	85.0	85.0	88.1	89.3	90.6	.....	16.1	9.64	....	45.2
10	53.0	89.3	88.9	88.0	88.1	89.8	90.3	91.0	.....	11.3	9.73	....	37.8
20	50.2	90.4	90.0	89.6	89.6	90.5	90.7	91.1	.....	7.40	9.56	....	30.5
30	47.4	90.8	90.5	90.3	90.4	90.8	90.9	91.2	.....	2.94	9.27	....	25.0
40	45.2	91.0	90.7	90.7	90.7	91.1	91.1	91.3	.....	0.76	9.10	....	20.6
50	44.1	91.1	90.9	90.9	91.0	91.2	91.2	91.4	.....	0.29	9.07	....	17.5
60	43.6	91.3	91.0	91.0	91.1	91.3	91.3	91.5	.....	0.41	9.00	....	15.2
70	43.5	91.5	91.0	91.2	91.2	91.4	91.4	91.6	.....	2.30	9.13	....	13.7
80	43.1	91.5	91.0	91.3	91.3	91.4	91.5	91.6	0.10	9.52	9.08	0.44	12.8
90	42.8	91.5	91.0	91.3	91.3	91.5	91.6	91.6	1.58	28.5	9.21	5.02	12.1
700	43.5	91.5	91.0	91.3	91.3	91.5	91.6	91.6	11.2	51.9	9.52	18.7	12.0
Luminous transmit.	58.1	61.4	62.5	55.5	49.7	67.1	75.5	85.4	0.017	9.8	9.1	0.041	50.8
Dominant wave lgth.	477.2	581.5	587.7	585.7	585.7	581.7	579.6	577.6	700	583.1	572.4	555.0c	564.9
Excitation purity.	14.5	55.4	30.3	48.0	69.7	37.1	24.1	9.0	100	100.0	12.1	48.0	80.0

Wave length	Percent transmittance												
	No. 106	CC-05R *	CC-10R *	CC-20R *	CC-30R *	CC-40R *	CC-50R *	CC-05B *	CC-10B *	CC-20B *	CC-30B *	CC-40B *	CC-50B *
400	......	81.0	73.0	61.5	51.6	42.5	36.4	87.0	85.5	82.2	80.2	77.0	74.1
10	......	81.0	72.4	60.0	50.0	40.0	33.9	87.5	86.4	84.0	82.5	80.3	78.4
20	0.10	81.0	72.0	58.6	48.2	38.2	31.9	87.7	87.2	85.0	84.0	82.2	80.7
30	0.20	81.1	71.6	57.7	47.0	36.8	30.5	88.0	87.5	85.3	84.3	82.5	81.1
40	0.35	81.2	71.5	57.2	46.4	36.0	29.7	88.1	87.5	85.0	83.5	81.3	79.8
50	0.58	81.4	71.6	57.2	46.4	36.1	29.6	88.1	87.2	83.9	81.9	78.7	76.6
60	0.98	81.7	72.4	58.5	47.5	37.5	31.0	87.9	86.4	82.5	79.5	75.9	72.9
70	1.5	82.3	73.7	60.6	49.9	40.0	33.6	87.5	85.3	80.3	76.2	72.0	67.9
80	2.3	82.8	74.9	62.0	52.0	42.5	35.9	87.0	84.0	77.8	72.5	67.5	62.7
90	3.5	83.2	75.8	63.5	53.9	44.8	37.9	86.2	82.4	74.2	68.3	62.3	56.6
500	5.2	83.3	76.6	64.6	55.2	46.1	39.4	85.2	80.5	71.2	63.8	56.7	50.1
10	7.7	83.0	76.1	64.0	54.5	46.0	38.5	84.4	78.6	67.7	58.7	51.0	44.5
20	10.7	82.4	74.9	61.5	51.6	42.5	35.0	83.5	77.0	64.0	54.4	46.0	38.6
30	15.1	81.6	73.5	59.4	48.5	38.5	31.5	82.6	75.2	61.5	50.7	41.6	34.1
40	20.2	81.2	72.5	57.8	46.5	36.6	29.4	82.1	73.9	59.5	48.3	39.0	31.3
50	25.7	81.1	72.4	57.1	45.6	35.6	28.7	81.5	73.0	58.0	46.6	36.9	29.5
60	31.0	81.4	72.8	58.0	46.9	36.4	29.7	81.4	72.7	57.5	45.9	35.9	28.6
70	35.6	82.5	74.5	60.6	49.2	39.2	32.7	81.4	73.0	57.9	46.3	36.1	28.7
80	43.2	83.9	77.3	65.0	54.6	45.0	38.6	81.9	73.9	59.3	47.9	37.8	30.5
90	53.8	85.7	80.7	71.0	61.8	53.5	47.6	82.7	75.1	61.6	50.3	40.8	33.4
600	65.6	87.6	84.0	77.0	70.5	64.0	58.7	83.4	76.3	63.5	53.0	43.5	36.0
10	77.0	89.0	86.5	82.0	77.8	73.0	69.8	83.6	76.7	64.5	54.6	44.7	37.8
20	82.8	90.0	88.9	86.2	83.5	80.8	78.3	83.5	76.5	64.3	54.4	44.3	37.5
30	86.0	90.6	89.9	88.1	87.2	85.1	84.2	83.2	76.6	63.1	53.2	42.5	35.6
40	87.6	91.1	90.5	89.8	89.2	88.0	87.5	82.8	74.5	61.6	51.5	40.2	33.7
50	88.7	91.2	90.8	90.5	90.3	89.5	89.2	82.5	74.0	60.6	50.3	39.0	32.4
60	89.5	91.3	91.1	90.8	90.6	90.4	90.1	82.5	73.8	60.1	49.5	38.4	31.8
70	90.0	91.4	91.3	91.0	90.9	90.8	90.7	82.3	73.3	59.6	49.0	37.7	31.0
80	90.5	91.6	91.5	91.2	91.1	91.0	91.1	82.0	72.8	58.6	48.2	36.5	30.0
90	90.8	91.7	91.7	91.4	91.4	91.3	91.1	81.9	72.5	58.1	47.2	35.4	29.0
700	91.0	91.9	91.9	91.5	91.5	91.4	91.2	82.2	73.0	58.5	47.5	35.6	29.0
Luminous transmit.	34.6	83.7	77.0	65.3	55.9	47.3	41.3	82.8	75.5	62.3	52.0	42.8	35.7
Dominant wave lgth.	589.4	605.0	597.8	604.2	605.8	605.5	608.5	459.0	462.0	460.0	461.0	463.2	462.5
Excitation purity.	95.2	2.0	4.7	8.5	12.3	17.3	21.4	2.8	6.3	13.2	20.2	27.7	34.2

* Some transmission below 400 mμ. Consult the manufacturer.

TRANSMISSION OF WRATTEN FILTERS (Continued)

Wave length	Percent transmittance											
	CC-05G *	CC-10G *	CC-20G *	CC-30G *	CC-40G *	CC-50G *	CC-05Y *	CC-10Y *	CC-20Y *	CC-30Y *	CC-40Y *	CC-50Y *
400	80.0	73.1	58.8	48.0	39.7	32.0	81.0	74.5	61.3	50.5	43.0	34.5
10	80.7	72.9	57.8	46.5	38.1	30.3	80.6	73.2	59.0	47.4	39.5	30.5
20	81.0	72.8	57.3	45.8	37.3	29.5	80.4	72.6	57.8	46.0	37.5	29.0
30	81.4	72.7	57.0	45.5	36.5	29.0	80.4	72.5	57.5	45.6	36.5	28.7
40	81.6	73.0	57.3	45.8	36.6	29.1	80.6	72.8	57.8	46.5	36.8	29.5
50	82.1	73.9	58.4	46.9	38.1	30.6	81.2	74.0	59.5	48.5	38.5	31.5
60	83.0	75.5	61.4	50.3	41.5	34.3	82.5	76.0	63.0	52.5	42.5	36.2
70	84.4	78.0	65.4	55.8	47.0	40.5	83.9	78.5	67.5	48.2	48.8	43.5
80	85.6	80.4	70.0	61.8	53.5	47.8	85.3	81.2	72.3	64.9	56.2	54.0
90	86.8	83.0	75.2	68.9	61.3	57.0	87.0	84.4	78.0	72.4	66.0	64.0
500	87.9	85.9	80.3	76.4	70.7	68.0	88.4	87.2	84.0	81.0	77.0	75.5
10	88.7	87.5	83.8	80.9	77.8	75.3	89.5	89.0	88.0	86.6	85.5	84.2
20	89.0	88.1	84.9	82.3	79.5	77.5	90.0	90.0	89.6	89.1	89.0	88.5
30	89.0	88.0	84.6	81.7	79.4	77.0	90.4	90.4	90.0	89.7	89.9	89.6
40	89.0	87.6	83.7	80.5	77.8	74.8	90.7	90.7	90.6	90.4	90.2	90.0
50	88.6	87.1	82.4	78.6	75.8	72.2	90.9	90.9	90.8	90.6	90.4	90.3
60	88.1	86.3	80.9	76.2	72.9	68.8	91.0	91.0	90.9	90.8	90.7	90.6
70	87.5	85.3	79.0	73.5	69.3	64.8	91.3	91.3	91.0	90.9	90.8	90.7
80	87.0	84.1	77.0	70.4	65.3	60.3	91.4	91.4	91.1	91.0	90.8	90.7
90	86.4	82.8	74.5	67.2	61.9	55.9	91.4	91.4	91.2	91.1	90.9	90.8
600	85.7	81.5	72.0	64.1	57.7	51.7	91.4	91.4	91.3	91.2	90.9	90.8
10	85.0	80.0	69.5	60.7	53.7	47.3	91.4	91.4	91.3	91.2	90.9	90.9
20	84.0	78.5	66.5	57.2	49.8	42.5	91.4	91.4	91.3	91.2	91.0	90.9
30	83.0	76.9	63.8	53.7	45.5	38.0	91.5	91.5	91.4	91.3	91.0	91.0
40	82.2	75.6	61.5	50.8	42.0	34.6	91.5	91.5	91.4	91.3	91.0	91.0
50	81.9	74.9	60.1	49.1	40.2	32.5	91.5	91.5	91.4	91.3	91.1	91.1
60	81.5	74.4	59.4	48.1	39.5	31.5	91.5	91.5	91.4	91.3	91.1	91.1
70	81.4	74.0	58.8	47.5	38.5	31.0	91.5	91.5	91.4	91.4	91.2	91.2
80	81.1	73.5	58.1	46.6	37.6	29.9	91.5	91.5	91.4	91.4	91.2	91.2
90	81.1	73.2	57.6	46.0	36.6	28.9	91.5	91.5	91.4	91.4	91.3	91.3
700	81.5	73.5	58.0	46.4	36.5	28.7	91.5	91.5	91.4	91.4	91.3	91.3
Luminous transmit.	87.2	84.5	77.8	72.2	67.7	63.3	90.4	90.1	89.1	88.2	87.4	86.9
Dominant wave lgth.	553.0	555.5	555.0	554.0	554.3	553.4	572.0	571.3	571.4	571.3	571.3	571.2
Excitation purity.	2.3	5.2	10.9	15.8	21.1	25.9	5.3	9.6	18.8	28.3	35.7	42.0

Wave length	Percent transmittance											
	CC-05M *	CC-10M *	CC-20M *	CC-30M *	CC-40M *	CC-50M *	CC-05C *	CC-10C *	CC-20C *	CC-30C *	CC-40C *	CC-50C *
400	87.6	86.6	85.6	84.2	82.3	80.9	87.3	86.0	83.9	82.3	80.4	78.8
10	88.2	87.7	86.6	85.7	84.6	83.6	88.2	87.5	85.2	84.5	83.4	82.7
20	88.6	88.0	87.0	85.9	85.2	84.4	88.7	88.1	86.5	86.0	85.3	84.8
30	88.7	88.0	86.9	85.6	84.4	83.6	89.0	88.6	87.5	87.0	86.3	85.9
40	88.7	87.9	86.0	84.7	82.5	81.4	89.3	89.0	87.7	87.3	86.6	86.1
50	88.6	87.5	84.9	82.8	80.0	78.1	89.5	89.1	87.8	87.5	86.6	86.0
60	88.4	86.5	83.1	80.0	76.1	73.7	89.6	89.1	87.7	87.3	86.4	85.7
70	87.8	85.2	80.8	76.4	71.3	68.0	89.7	89.0	87.5	87.0	85.8	85.2
80	87.0	83.6	77.9	72.1	65.8	61.7	89.7	89.0	87.2	86.5	85.3	84.3
90	86.0	81.8	74.4	67.0	60.0	55.0	89.7	89.0	87.0	86.0	84.4	83.4
500	85.0	79.7	70.5	61.7	53.7	48.1	89.6	89.0	86.5	85.2	83.5	82.3
10	83.8	77.5	66.7	56.5	47.7	41.6	89.6	88.7	86.0	84.4	82.4	80.8
20	82.7	75.3	63.4	52.0	42.8	36.3	89.5	88.5	85.2	83.5	81.1	79.2
30	81.8	73.7	60.5	48.6	39.0	31.9	89.4	88.0	84.3	82.4	79.6	77.3
40	81.3	72.5	58.6	46.6	36.7	29.8	89.2	87.5	83.4	81.0	77.7	75.0
50	81.2	72.2	58.0	46.0	36.0	29.1	88.9	87.0	82.3	79.0	75.3	72.2
60	81.5	72.8	58.3	46.5	36.7	29.7	88.5	86.1	80.5	76.7	72.7	69.0
70	82.5	74.6	60.5	49.8	40.2	32.3	88.0	85.0	78.5	74.0	69.3	65.0
80	84.0	77.3	64.9	55.6	46.2	39.0	87.5	83.8	76.1	70.9	65.4	60.5
90	85.8	80.8	70.6	63.3	54.9	48.7	87.0	82.5	73.9	67.5	61.6	55.8
600	88.0	84.5	77.1	71.6	64.9	59.9	86.4	81.0	71.2	64.1	57.6	51.3
10	89.3	87.0	82.2	79.2	74.9	70.7	85.5	79.5	68.5	60.4	53.4	36.2
20	90.2	88.9	86.1	84.1	81.4	79.2	84.5	77.9	65.5	56.7	49.2	42.3
30	90.6	90.0	88.7	87.4	86.0	84.5	83.8	76.3	62.7	53.1	45.0	38.0
40	90.8	90.5	90.0	89.3	88.7	87.6	83.3	75.1	60.8	50.4	42.0	34.9
50	91.0	90.7	90.5	90.2	90.0	89.7	82.8	74.4	59.5	48.8	40.2	32.9
60	91.1	91.0	90.8	90.8	90.4	90.4	82.5	74.0	58.5	48.0	39.4	32.0
70	91.2	91.2	91.0	91.0	90.7	90.7	82.4	73.6	57.9	47.2	38.6	31.0
80	91.3	91.3	91.2	91.1	91.0	91.0	82.0	73.0	57.5	46.0	37.5	29.9
90	91.4	91.4	91.4	91.3	91.3	91.3	82.0	72.8	57.4	45.5	36.7	29.1
700	91.5	91.5	91.5	91.5	91.5	91.5	82.5	74.0	58.5	46.4	37.3	29.8
Luminous transmit.	84.2	77.9	67.1	58.1	50.0	44.0	88.0	85.1	78.9	74.8	70.5	66.7
Dominant wave lgth.	541.0_c	547.5_c	551.2_c	550.0_c	550.3_c	551.2_c	489.2	487.5	486.5	486.2	486.1	485.5
Excitation purity.	3.5	7.4	14.4	21.5	28.3	34.0	1.6	4.1	8.9	12.8	17.5	20.2

* Some transmission below 400 mμ. Consult the manufacturer.

TRANSMISSION OF WRATTEN FILTERS (Continued)

Wave length	Percent transmittance			
	No. 87	No. 87C	No. 88A	No. 89B
700	.....	.....	....	11.2
10	.....	.....	....	32.4
20	.....	.....	....	57.6
30	.....	.....	7.4	69.1
40	0.10	.....	32.8	77.6
50	2.19	.....	56.3	83.1
60	7.95	.....	69.2	85.0
70	17.4	.....	74.2	86.1
80	31.6	.....	77.6	87.0
90	43.7	.....	79.7	87.7
800	53.8	0.32	81.4	88.1
10	61.7	3.20	82.6	88.4
20	69.2	8.90	83.7	88.6
30	74.1	17.8	84.7	88.8
40	77.7	28.2	85.5	89.0
50	81.4	41.0	86.1	89.2
60	84.0	53.8	86.6	89.4
70	85.4	61.6	87.2	89.6
80	86.8	69.2	87.5	89.8
90	87.8	74.1	87.8	89.9
900	88.4	78.5	88.0	90.0
10	88.8	81.5	88.2	90.1
20	89.1	83.6	88.4	90.2
30	89.1	85.1	88.6	90.3
40	89.1	86.0	88.8	90.4
50	89.1	87.0	89.0	90.5

* Some transmission below 400 mμ. Consult the manufacturer.

TRANSMISSIBILITY FOR RADIATIONS

Ratio of the transmitted light to the incident light for a definite thickness of the substance, usually 1 cm.

GLASS.

Glass in general is opaque to the ultra-violet and infra-red. Uviol glass is transparent to the longer radiations of the ultra-violet.
Coefficient of transparency of glass for visible and ultra-violet radiations.

	Normal incidence, thickness 1 cm.								
Wave length microns.......	0.309	0.330	0.347	0.357	0.361	0.375	0.384	0.388	0.396
Crown, ordinary..	...	...	...	...	...	.947			
Crown, borosilicate..........	0.08	0.65	0.88	...	0.95	...	0.972	0.975	0.986
Flint, ordinary...	...	...	...	0.72	...	...	...	0.904	
Flint, heavy.....	...	...	0.01	...	0.16	...	0.58		
	Normal incidence, thickness 1 cm.								
Wave length, microns.......	0.400	0.415	0.419	0.425	0.434	0.455	0.500	0.580	0.677
Crown, ordinary..	0.964	...	0.952	...	0.960	0.981	...	0.986	0.990
Crown, borosilicate..........	...	0.985	...	0.993	...	...	0.993		
Flint, ordinary...	...	0.959	...	...	...	...	1.00		
Flint, heavy.....	...	...	...	0.905					

QUARTZ

Quartz is very transparent to the ultra-violet and to the visible spectrum, but opaque for the infra-red beyond 7.0μ.

(Pflüger.)

Wave length, microns.........	0.19	0.20	0.21	0.22
Transmission for 1 mm..........	.67	.84	.92	.94

FLUORITE

Fluorite is very transparent to the ultra-violet, nearly to 0.10μ. Coefficient of transparency at $\lambda=186$ is found by Pflüger to be 0.80.
For the infra-red the values are given in a table below.

ROCK SALT AND SYLVINE AND FLUORITE
TRANSPARENCY FOR THE INFRA-RED.

Thickness 1 cm.

Wave length, microns.	Rock salt	Sylvine KCl	Fluorite
8.	....	....	.844
9.	0.995	1.000	.543
10.	.995	.988	.164
12.	.993	.995	.010
14.	.931	.975	.000
16.	.661	.936	
18.	.275	.862	
19.	.096	.758	
20.7	.006	.585	
23.7	.000	.155	

COLORIMETRY

Selected from Judd, Jour. Opt. Soc. Amer. 23, 359 (1933)

Recommendations of the International Commission on Illumination

Standard Illuminants

A. Gas-filled tungsten incandescent lamp of color temperature 2848° K.

B. Noon Sunlight. Lamp as above in combination with the Davis-Gibson filter for converting color temperature 2848° to 4800° K.

The filter is to be composed of a layer one centimeter thick of each of two separate solutions B_1 and B_2, contained in a double cell of colorless optical glass.

Solution B_1

Copper sulfate ($CuSO_4{\cdot}5H_2O$)	2.452	g
Mannite ($C_6H_8(OH)_6$)	2.452	g
Pyridine (C_5H_5N)	30.0	cc
Distilled water to make	1000	cc

Solution B_2

Cobalt ammonium sulfate ($CoSO_4{\cdot}(NH_4)_2SO_4{\cdot}6H_2O$)	21.71	g
Copper sulfate ($CuSO_4{\cdot}5H_2O$)	16.11	g
Sulfuric acid (density 1.835)	10.0	cc
Distilled water to make	1000	cc

C. Average Daylight. Lamp as in **A** in combination with Davis-Gibson filter for converting color temperature 2848° to 6500° K.

The filter is composed of a layer one centimeter thick of each of two separate solutions C_1 and C_2, contained in a double cell made of colorless optical glass.

Solution C_1

Copper sulfate ($CuSO_4{\cdot}5H_2O$)	3.412	g
Mannite ($C_6H_8(OH)_6$)	3.412	g
Pyridine (C_5H_5N)	30.0	cc
Distilled water to make	1000	cc

Solution C_2

Cobalt ammonium sulfate ($CoSO_4{\cdot}(NH_4)_2SO_4{\cdot}6H_2O$)	30.580	g
Copper sulfate ($CuSO_4{\cdot}5H_2O$)	22.520	g
Sulfuric acid (density 1.835)	10.0	cc
Distilled water to make	1000	cc

See R. Davis and K. S. Gibson Bur. Stds. Misc. Pub. No. 114, Jan. 1931 or Bur. Stds. Jour. Research **7,** 796 (1931).

Standard Coordinate System

The tristimulus system of color specification is based on four chosen stimuli consisting of homogeneous radiant energy of wave lengths

700.0 **546.1** **435.8**

mμ and of standard illuminant B (see above).

To establish the system of specification coordinates are assigned as follows:

Stimulus	x	y	z
700.0 mμ	0.73467	0.26533	0.00000
546.1 mμ	0.27376	0.71741	0.00883
435.8 mμ	0.16658	0.00886	0.82456
Standard illuminant B:	0.34842	0.35161	0.29997

The Standard Observer

The "standard observer" is determined below by the specification for the equal energy spectrum both in fractions, x, y, z of the total amount for each wave length interval of 5 mμ and directly $\bar{x}$, $\bar{y}$, $\bar{z}$. The fractional values are known as the **trilinear coordinates** or **trichromatic coefficients** of the spectrum; the direct values as the **distribution functions** or coefficients.

The sum of the trichromatic coefficients is unity, that is $x + y + z = 1$. Therefore the value of z may be and often is omitted from a specification.

Relative Visibility

The value of $\bar{y}$ given in the table is the standard visibility function or relative visibility.

Wave length mμ	Trichromatic coefficients			Distribution coefficients for equal energy			Wave length mμ
	x	y	z	$\bar{x}$	$\bar{y}$ (Rel. Vis.)	$\bar{z}$	
380	0.1741	0.0050	0.8209	0.0014	0.0000	0.0065	380
385	0.1740	0.0050	0.8210	0.0022	0.0001	0.0105	385
390	0.1738	0.0049	0.8213	0.0042	0.0001	0.0201	390
395	0.1736	0.0049	0.8215	0.0076	0.0002	0.0362	395
400	0.1733	0.0048	0.8219	0.0143	0.0004	0.0679	400
405	0.1730	0.0048	0.8222	0.0232	0.0006	0.1102	405
410	0.1726	0.0048	0.8226	0.0435	0.0012	0.2074	410
415	0.1721	0.0048	0.8231	0.0776	0.0022	0.3713	415
420	0.1714	0.0051	0.8235	0.1344	0.0040	0.6456	420

The Standard Observer (Continued)

Wave length mμ	Trichromatic coefficients			Distribution coefficients for equal energy			Wave length mμ
	x	y	z	$\bar{x}$	$\bar{y}$ (Rel. Vis.)	$\bar{z}$	
425	0.1703	0.0058	0.8239	0.2148	0.0073	1.0391	425
430	0.1689	0.0069	0.8242	0.2839	0.0116	1.3856	430
435	0.1669	0.0086	0.8245	0.3285	0.0168	1.6230	435
440	0.1644	0.0109	0.8247	0.3483	0.0230	1.7471	440
445	0.1611	0.0138	0.8251	0.3481	0.0298	1.7826	445
450	0.1566	0.0177	0.8257	0.3362	0.0380	1.7721	450
455	0.1510	0.0227	0.8263	0.3187	0.0480	1.7441	455
460	0.1440	0.0297	0.8263	0.2908	0.0600	1.6692	460
465	0.1355	0.0399	0.8246	0.2511	0.0739	1.5281	465
470	0.1241	0.0578	0.8181	0.1954	0.0910	1.2876	470
475	0.1096	0.0868	0.8036	0.1421	0.1126	1.0419	475
480	0.0913	0.1327	0.7760	0.0956	0.1390	0.8130	480
485	0.0687	0.2007	0.7306	0.0580	0.1693	0.6162	485
490	0.0454	0.2950	0.6596	0.0320	0.2080	0.4652	490
495	0.0235	0.4127	0.5638	0.0147	0.2586	0.3533	495
500	0.0082	0.5384	0.4534	0.0049	0.3230	0.2720	500
505	0.0039	0.6548	0.3413	0.0024	0.4073	0.2123	505
510	0.0139	0.7502	0.2359	0.0093	0.5030	0.1582	510
515	0.0389	0.8120	0.1491	0.0291	0.6082	0.1117	515
520	0.0743	0.8338	0.0919	0.0633	0.7100	0.0782	520
525	0.1142	0.8262	0.0596	0.1096	0.7932	0.0573	525
530	0.1547	0.8059	0.0394	0.1655	0.8620	0.0422	530
535	0.1929	0.7816	0.0255	0.2257	0.9149	0.0298	535
540	0.2296	0.7543	0.0161	0.2904	0.9540	0.0203	540
545	0.2658	0.7243	0.0099	0.3597	0.9803	0.0134	545
550	0.3016	0.6923	0.0061	0.4334	0.9950	0.0087	550
555	0.3373	0.6589	0.0038	0.5121	1.0002	0.0057	555
560	0.3731	0.6245	0.0024	0.5945	0.9950	0.0039	560
565	0.4087	0.5896	0.0017	0.6784	0.9786	0.0027	565
570	0.4441	0.5547	0.0012	0.7621	0.9520	0.0021	570
575	0.4788	0.5202	0.0010	0.8425	0.9154	0.0018	575
580	0.5125	0.4866	0.0009	0.9163	0.8700	0.0017	580
585	0.5448	0.4544	0.0008	0.9786	0.8163	0.0014	585
590	0.5752	0.4242	0.0006	1.0263	0.7570	0.0011	590
595	0.6029	0.3965	0.0006	1.0567	0.6949	0.0010	595
600	0.6270	0.3725	0.0005	1.0622	0.6310	0.0008	600
605	0.6482	0.3514	0.0004	1.0456	0.5668	0.0006	605
610	0.6658	0.3340	0.0002	1.0026	0.5030	0.0003	610
615	0.6801	0.3197	0.0002	0.9384	0.4412	0.0002	615
620	0.6915	0.3083	0.0002	0.8544	0.3810	0.0002	620
625	0.7006	0.2993	0.0001	0.7514	0.3210	0.0001	625
630	0.7079	0.2920	0.0001	0.6424	0.2650	0.0000	630
635	0.7140	0.2859	0.0001	0.5419	0.2170	0.0000	635
640	0.7190	0.2809	0.0001	0.4479	0.1750	0.0000	640
645	0.7230	0.2770	0.0000	0.3608	0.1382	0.0000	645
650	0.7260	0.2740	0.0000	0.2835	0.1070	0.0000	650
655	0.7283	0.2717	0.0000	0.2187	0.0816	0.0000	655
660	0.7300	0.2700	0.0000	0.1649	0.0610	0.0000	660
665	0.7311	0.2689	0.0000	0.1212	0.0446	0.0000	665
670	0.7320	0.2680	0.0000	0.0874	0.0320	0.0000	670
675	0.7327	0.2673	0.0000	0.0636	0.0232	0.0000	675
680	0.7334	0.2666	0 0000	0.0468	0.0170	0.0000	680
685	0.7340	0.2660	0.0000	0.0329	0.0119	0.0000	685
690	0.7344	0.2656	0.0000	0.0227	0.0082	0.0000	690
695	0.7346	0.2654	0.0000	0.0158	0.0057	0.0000	695
700	0.7347	0.2653	0.0000	0.0114	0.0041	0.0000	700
705	0.7347	0.2653	0.0000	0.0081	0.0029	0.0000	705
710	0.7347	0.2653	0.0000	0.0058	0.0021	0.0000	710
715	0.7347	0.2653	0.0000	0.0041	0.0015	0.0000	715
720	0.7347	0.2653	0.0000	0.0029	0.0010	0.0000	720
725	0.7347	0.2653	0.0000	0.0020	0.0007	0.0000	725
730	0.7347	0.2653	0.0000	0.0014	0.0005	0.0000	730
735	0.7347	0.2653	0.0000	0.0010	0.0004	0.0000	735
740	0.7347	0.2653	0.0000	0.0007	0.0003	0.0000	740
745	0.7347	0.2653	0.0000	0.0005	0.0002	0.0000	745
750	0.7347	0.2653	0.0000	0.0003	0.0001	0.0000	750
755	0.7347	0.2653	0.0000	0.0002	0.0001	0.0000	755
760	0.7347	0.2653	0.0000	0.0002	0.0001	0.0000	760
765	0.7347	0.2653	0.0000	0.0001	0.0000	0.0000	765
770	0.7347	0.2653	0.0000	0.0001	0.0000	0.0000	770
775	0.7347	0.2653	0.0000	0.0000	0.0000	0.0000	775
780	0.7347	0.2653	0.0000	0.0000	0.0000	0.0000	780
			Totals	21.3713	21.3714	21.3715	

SPECIFIC ROTATION

Specific rotation or rotatory power is given in degrees per decimeter for liquids and solutions and in degrees per millimeter for solids; + signifies right handed rotation, − left. Specific rotation varies with the wave length of light used, with temperature and, in the case of solutions, with the concentration. When sodium light is used, indicated by D in the wave length column, a value of $\lambda = 0.5893$ may be assumed.

Optical rotatory power for a large number of organic compounds will be found in the International Critical Tables, Vol. VII; for sugars, Vol. II.

Solids

Substance	Wave length μ	Rotation deg./mm	Substance	Wave length μ	Rotation deg./mm
Cinnabar (HgS)	D	+32.5	Quartz (continued)	0.3609	+63.628
Lead hyposulfate	D	5.5		0.3582	64.459
Potassium hyposulphate	D	8.4		0.3466	69.454
Quartz	0.7604	12.668		0.3441	70.587
	0.7184	14.304		0.3402	72.448
	0.6867	15.746		0.3360	74.571
	0.6562	17.318		0.3286	78.579
	0.5895	21.684		0.3247	80.459
	0.5889	21.727		0.3180	84.972
	0.5269	27.543		0.2747	121.052
	0.4861	32.773		0.2571	143.266
	0.4307	42.604		0.2313	190.426
	0.4101	47.481		0.2265	201.824
	0.3968	51.193		0.2194	220.731
	0.3933	52.155		0.2143	235.972
	0.3820	55.625	Sodium bromate	D	2.8
	0.3726	58.894	Sodium chlorate	D	3.13

Liquid

Liquid	Temp. °C	Wave length μ	Specific rotation deg./dm
Amyl alcohol	....	D	− 5.7
Camphor	204	D	+ 70.33
Cedar oil	15	D	− 30 to −40
Citron oil	15	D	+ 62
Ethyl malate $(C_2H_5)_2C_4H_4O_5$	11	D	− 10.3 to −12.4
Menthol	35.2	D	− 49.7
Nicotine $C_{10}H_{14}N_2$	10–30	D	−162
	20	0.6563	−126
	20	0.5351	−207.5
	20	0.4861	−253.5
Turpentine $C_{10}H_{16}$	20	D	− 37
	20	0.6563	− 29.5
	20	0.5351	− 45
	20	0.4861	− 54.5

Solutions

Corrections for values of the specific rotation for concentration are given in the last column. c indicates concentration in grams per 100 milliliters of solution; d indicates the concentration in grams per 100 grams of solution.

Substance	Solvent	Temp. °C	Wave length μ	Specific rotation deg./dm	Correction for concentration or temperature
Albumen	water	..	D	− 25 to −38	
Arabinose	water	20	D	− 105.0	
Camphor	alcohol	20	D	+ 54.4	− .135d for d = 45–91
	benzene	20	D	+ 56	− .166d for d = 47–90
	ether	..	D	+ 57	
Dextrose d-glucose $C_6H_{12}O_6$	water	20	D	+ 52.5	+ .025d for d = 1–18
			.5461	+ 62.03	+ .04257c for c = 6–32
Galactose	water	..	D	+ 83.9	+ .078d − .21t for d = 4–36 and t = 10–30°C
l-Glucose (β)	water	20	D	− 51.4	
Invert sugar $C_6H_{12}O_6$	water	20	D	− 19.7	− .036c for c = 9–35; $\alpha_t = \alpha_{20} + .304(t - 20) + .00165(t - 20)^2$ for t = 3–30°C
		25	5461	− 21.5	
Lactose	water	20	D	+ 52.4	+ .072 (20° − t) for c = 5
			.5461	+ 61.9	+ .085(20° − t) for c = 5
Levulose fruit sugar	water	25	D	− 88.5	− .145d for d = 2.6–18.6
		25	.5461	− 105.30	

SPECIFIC ROTATION (Continued)

Solutions (Continued)

Substance	Solvent	Temp °C	Wave length μ	Specific rotation deg./dm	Correction for concentration or temperature
Maltose	water	20	D	+ 138.48	− .01837d for d = 5–35
		25	5461	+ 153.75	
Mannose	water	20	D	+ 14.1	c = 10.2
Nicotine	water	20	D	− 77	for d = 1–16
	benzene	20	D	− 164	for d = 8–100
Potassium tartrate	water	20	D	+ 27.14	+ .0992c − .00094c^2 for c = 8–50
Quinine sulfate	water	17	D	− 214	
Santonin	alcohol	20	D	− 161.0	c = 1.78
		20	D	+ 693	c = 4.05
	chloroform	20	D	− 202.7	+ .309d for d = 75–96.5
	alcohol	20	.6867	+ 442	c = 4.05
			.5269	+ 991	c = 4.05
			.4861	+1323	c = 4.05
Sodium potassium tartrate (Rochelle salt)	water	20	D	+ 29.75	− .0078c
Sucrose (cane sugar) $C_{12}H_{22}O_{11}$	water	20	D	+ 66.412	+ .01267d − .000376d^2 for d = 0–50; $\alpha_t = \alpha_{20}[1 - .00037(t - 20)]$ for t = 14–30°C

Sucrose dissolved in water, 20°C.

μ	Spec. rot.	μ	Spec. rot.	μ	Spec. rot
670.8 (Li)	+50.51	510.6 (Cu)	+90.46	435.3 (Fe)	+128.5
643.8 (Cd)	55.04	508.6 (Cd)	91.16	433.7 (Fe)	129.8
636.2 (Zn)	56.51	481.1 (Zn)	103.07	431.5 (Fe)	130.7
589.3 (Na)	66.45	480.0 (Cd)	103.62	428.2 (Fe)	133.6
578.2 (Cu)	69.10	472.2 (Zn)	107.38	427.2 (Fe)	134.2
578.0 (Hg)	69.22	468.0 (Zn)	109.49	426.1 (Fe)	134.9
570.0 (Cu)	71.24	467.8 (Cd)	109.69	419.1 (Fe)	140.0
546.1 (Hg)	78.16	438.4 (Fe)	126.5	414.4 (Fe)	144.2
521.8 (Cu)	86.21	437.6 (Fe)	127.2	388.9 (Fe)	166.7
515.3 (Cu)	88.68	435.8 (Hg)	128.49	383.3 (Fe)	171.8
				382.6 (Fe)	173.1

Substance	Solvent	°C	μ	Spec. rot.	Correct.
Tartaric acid (ord.)	water	20	D	+15.06	− .131c
		20	.6563	7.75	for d = 41
		20	D	8.86	for d = 41
		20	.5351	9.65	for d = 41
		20	.4861	9.37	for d = 41
Turpentine	alcohol	20	D	−37	− .00482d − .00013d^2 for d = 0–90
	benzene	20	D	−37	− .0265d for d = 0–91
Xylose	water	20	D	+19.13	d = 2.7

OPTICAL ROTATION OF ACIDS AND BASES

Optical rotation of acids and bases commonly used in the resoluti of racemic substances. Compiled by F. E. Ray.

Name	Formula	Solvent	Conc. %	a_D
Bromocamphor-sulfonic acid. K salt	$C_{10}H_{15}O_4BrS$	H_2O	....	72.
Camphorsulfonic acid	$C_{10}H_{16}O_4S$	H_2O	....	23.
Chlorocamphor-sulfonic acid	$C_{10}H_{15}ClO_4S$	H_2O	....	49.
Codeinesulfonic acid	$C_{18}H_{21}NO_6S$	H_2O	3	−190.
Hydroxybutyric acid	$C_4H_8O_3$	H_2O	3.3	− 24.
Lactic acid	$C_3H_6O_3$	H_2O	10.5	3.
Malic acid	$C_4H_6O_5$	H_2O	....	2.
Mandelic acid	$C_8H_8O_3$	H_2O	2.01	155.
Methylene-camphor	$C_{11}H_{16}O$	C_2H_5OH	....	127
Phenylsuccinic acid	$C_{10}H_{10}O_4$	C_2H_5OH	1.5	148
Tartaric acid	$C_4H_6O_6$	C_2H_5OH and H_2O	....	3 to 25*
Brucine	$C_{23}H_{26}N_2O_4$	C_2H_5OH	5.4	− 85
Cinchonidine	$C_{19}H_{22}N_2O$	C_2H_5OH	1.0	−111.
Cinchonine	$C_{19}H_{22}N_2O$	$CHCl_3$	0.6	+209.
Cocaine	$C_{17}H_{21}NO_4$	50 % C_2H_5OH	1.1	− 35.
Coniine	$C_8H_{17}N$	$CHCl_3$	4	8.
Codeine	$C_{18}H_{21}NO_3$	C_2H_5OH	5	−135.
Hydrastine	$C_{21}H_{21}NO_6$	50 % C_2H_5OH	0.2	115
Menthol	$C_{10}H_{20}O$	C_2H_5OH	9.6	− 50.
Menthylamine	$C_{10}H_{21}N$	C_2H_5OH	11.3	− 31.
Narcotine	$C_{22}H_{23}NO_7$	$CHCl_3$	2.6	±200.
Quinidine	$C_{20}H_{24}N_2O_4$	C_2H_5OH	1.0	+233.
Quinine	$C_{20}H_{24}N_2O_2$	C_6H_6	0.6	−136
Thebaine	$C_{19}H_{21}NO_3$	$CHCl_3$	5	−229.
Strychnine	$C_{21}H_{22}N_2O_2$	C_2H_5OH	0.9	−128

* Varies greatly with temperature, solvent, and conc.

MAGNETIC ROTATORY POWER

The rotation of the plane of polarization of light by transparent substances subjected to a magnetic field may be applied to problems of molecular structure. This rotatory effect is known as the Faraday effect. Investigations of this effect by E. Verdet showed the angle of rotation (α) to depend on the nature of the substance and to be proportional to the length (l) of the column of the substance which the light traverses and to the strength (H) of the magnetic field. Thus

$$\alpha = \Lambda l H$$

where Λ is the Verdet constant for the experimental material.

Other symbols in the table have the following significance:

λ = Wavelength in μ

t = Degrees centigrade

ρ_M = Molecular magnetic rotation of the substance under consideration compared to that of water determined in the same apparatus in the same magnetic field.

Thus

$$\rho_M = M'\alpha'\rho'/M'\alpha'\rho$$

where M is the molecular weight, α the angle of rotation and ρ the density of the given substance, and M′, α', and ρ' are the same quantities of water.

$[\Lambda]_M^{\lambda,t}$ = Molecular rotatory value of the substance at the given temperature and at the wavelength λ. Values are in (radians)(gauss^{-1})(cm^{-1}).

$\Lambda^{\lambda,t}$ = Verdet constant at the temperature t and at the wavelength λ. Values are in (minutes)(gauss^{-1})(cm^{-1}).

Values in this table are reproduced by permission from Vol ume 3, Tables de Constantes et Donnees Nuimeriques, "Pouvoir Rotatorie Magnetique, Effet Magneto-Optique de Kerr."

A much greater listing of values for both organic and inorganic compounds is in the above publication.

Formula	Name	λ	t	ρ_M	$10^5[\Lambda]_M^{\lambda,t}$	$10^5\Lambda^{\lambda,t}$ Verdet
CH_4	Methane	578	0	......	....	*17.4
CCl_4	Tetrachloromethane	589	25.1	6.58	45.3	1.60
$CHCl_3$	Trichloromethane	589	20.0	5.535	37.98	1.60
$CHBr_3$	Tribromomethane	589	17.9	11.63	80.00	3.13
CH_2O_2	Formic acid	589	20.8	1.671	11.50	1.046
CH_2Cl_2	Dichloromethane	589	11.9	4.31	29.7	1.60
CH_2Br_2	Dibromomethane	589	15.9	8.11	55.8	2.74
CH_2I_2	Diiodomethane	589	15.0	10.83	129.5	1.51
CH_3Cl	Monochloromethane	589	23.0	2.99	20.5	1.37
CH_3Br	Monobromomethane	589	1.5	4.64	31.9	2.04
CH_3I	Monoiodomethane	589	19.5	9.01	63.2	3.35
CH_4O	Methylalcohol	589	18.7	1.640	11.28	0.958
CH_5N	Methylamine	578	.....	......	....	*22.7
CF_2Cl_2	Difluorodichloromethane (Freon)	...	.....	......	....	*32.7
CH_3O_2N	Mononitromethane	589	9.9	1.86	12.8	0.826
CH_4ON_2	Urea (carbamide) 40 % aqueous solution	578	20.0	2.38	22.7	....
C_2H_2	Acetylene (ethyne)	578	.....	......	....	*33.0
C_2H_4	Ethylene (ethene)	578	.....	......	....	*34.5
C_2H_6	Ethane	578	.....	......	....	*23.5
C_2N_2	Cyanogen	...	.....	......	....	....
$C_2H_2O_4$	Oxalic acid (aqueous sol. 8.3 %)	578	20	2.88	20.6	....
$C_2H_2O_4 \cdot 2H_2O$	Oxalic acid (alcohol sol. 16.5 %)	578	24	2.82	20.2	....
C_2H_3Br	Vinylbromide	589	7.8	6.22	42.8	2.10
C_2H_3N	Acetonitrile (ethanonitrile)	589	25.0	2.32	16.0	*21.0
C_2H_4O	Ethyleneoxide (1,2-epoxyethane)	589	8.0	1.935	13.3	0.92
C_2H_4O	Acetaldehyde (ethanal)	589	16.3	2.38	16.4	1.00
$C_2H_4O_2$	Acetic acid	589	21.0	2.525	17.37	1.044
$C_2H_4O_2$	Formic acid methylester (methylformate)	589	16.5	2.49	17.1	0.96
$C_2H_4Cl_2$	Ethylidenechloride (1,1-dichloroethane)	589	14.4	5.33	36.7	1.51
$C_2H_4Cl_2$	Ethylenechloride (1,2-dichloroethane)	589	14.4	5.49	37.7	1.65
$C_2H_4Br_2$	Ethylenebromide (1,2-dibromoethane)	589	15.2	9.70	66.7	2.66
C_2H_5Cl	Monochloroethane	589	5.0	4.04	27.8	1.36
C_2H_5Br	Monobromoethane	589	19.7	5.85	40.2	1.82
C_2H_5I	Monoiodoethane	589	18.1	10.07	69.3	2.95
C_2H_6O	Ethylalcohol (ethanol)	589	16.8	2.780	19.13	1.131
$C_2H_6O_2$	Glycol (1,2-ethanediol)	589	15.1	2.94	20.2	1.25
C_2H_6S	Ethylmercaptan	578	16.0	5.52	39.5	1.85
C_2H_7N	Ethylamine (aminoethane)	589	5.8	3.61	24.8	*34.5
$C_2H_2O_2Cl_2$	Dichloroacetic acid (dichloroethanoic acid)	589	13.5	5.30	3.65	1.52
$C_2H_3O_2Cl$	Chloroacetic acid (chloroethanoic acid)	589	64.5	3.89	26.7	1.33
$C_2H_3O_2Cl_3$	Chloralhydrate (2-2-2-trichloro-1,1-ethanediol)	589	54.6	7.10	48.8	1.65
$C_2H_5O_2N$	Mononitroethane	589	10.2	2.84	19.5	0.946
C_3H_3	Propane	578	.....	......	....	*34.0
C_3H_4O	Acrolein (propenal)	578	20.0	4.74	34.0	1.76
$C_3H_4O_3$	Pyruvic acid (2-oxopropanoic acid)	589	14.5	3.56	24.2	1.21
$C_3H_4O_4$	Malonic acid (propandioic acid) 2*n*-aqueous sol.	589	23.0	3.47	....	....
C_3H_6O	Allyl alcohol	589	18.3	4.68	32.2	1.60
C_3H_6O	Propyl alcohol (1-propanol)	589	13.6	3.33	22.9	1.09
C_3H_6O	Acetone (2-propanone)	589	20.0	3.472	23.89	1.1136
$C_3H_6O_2$	Propionic acid (propanoic acid)	589	20.3	3.462	23.82	1.10
$C_3H_6O_2$	Formic acid ethylester (ethylmethanoate)	589	18.8	3.56	24.5	1.05
$C_3H_6O_2$	Acetic acid methylester (methylacetate)	589	20.0	3.42	23.5	1.03
C_3H_7Cl	Propylchloride (1-chloropropane)	589	16.1	5.04	34.7	1.34
C_3H_7Cl	Isopropylchloride (2-chloropropane)	589	17.2	5.16	35.5	1.34
C_3H_7Br	Propylbromide (1-bromopropane)	589	19.2	6.88	47.8	1.79
C_3H_7Br	Isopropylbromide (2-bromopropane)	589	17.1	7.00	48.2	1.77
C_3H_7I	Propyliodide (1-iodopropane)	589	18.1	11.08	76.2	2.69
C_3H_7I	Isopropyliodide (2-iodopropane)	589	26.3	11.18	76.9	2.63
C_3H_8O	*n*-Propyl alcohol (1-propanol)	589	15.6	3.77	25.9	1.20
C_3H_8O	Isopropyl alcohol (2-propanol)	589	20.0	3.90	26.8	1.23
$C_3H_8O_3$	Glycerine (1,2,3-propanetriol)	589	16.0	4.11	28.3	1.33
C_3H_9N	*n*-Propylamine	589	9.6	4.56	31.4	1.33
$C_3H_5O_9N_3$	Nitroglycerine	589	13.5	5.405	37.2	0.900

* Verdet constant factor = 10^6.

MAGNETIC ROTATORY POWER (Continued)

Formula	Name	λ	t	ρ_M	$10^5[\Lambda]_M^{\lambda,t}$	$10^5\Lambda^{\lambda,t}$ Verdet
$C_3H_7O_2N$	1-Nitropropane	589	18.9	3.82	26.3	1.018
C_4H_6	1,3-Butadiene (erythrene)	589	15.0	7.94	54.6	2.16
C_4H_8	1-Butene (α-butylene)	589	15.0	5.53	38.0	1.39
C_4H_8	cis-2-Butene (β-butylene)	589	15.0	5.27	36.3	1.38
C_4H_8	trans-2-Butene	589	15.0	5.07	34.9	1.29
C_4H_{10}	Butane	589	15.0	4.59	31.6	1.09
C_4H_{10}	Isobutane (2-methylpropane)	589	15.0	4.87	33.5	1.11
$C_4H_2O_3$	Maleic anhydride (cis-butenedioic anhydride)	589	25.0	4.5	31.0	....
C_4H_4O	Furan (furfuran)	589	20.0	5.48	37.7	1.78
$C_4H_4O_4$	Maleic acid (cis-butenedioic acid) 2*n* aqueous solution	589	25.0	5.63	38.7	....
C_4H_4S	Thiophene (thiofuran)	589	20.0	9.40	64.7	2.83
$C_4H_6O_3$	Acetic anhydride (ethanoic anhydride)	589	20.0	4.28	29.5	1.115
$C_4H_6O_4$	Succinic acid (butanedioic acid) 5.9 % aqueous sol.	578	20.0	4.68	33.5	....
$C_4H_6O_6$	Tartaric acid (47.8 % aqueous sol.)	578	20.0	4.79	34.3	....
$C_4H_8O_2$	*n*-Butyric acid (butanoic acid)	589	18.8	4.47	30.8	1.15
$C_4H_8O_2$	Ethylacetate (ethyl ethanoate)	589	20.0	4.47	30.8	1.08
$C_4H_8O_2$	Propionic acid methylester (methylpropanoate)	589	20.0	4.37	30.1	1.07
$C_4H_8O_3$	Lactic acid methylester (methyl 2-hydroxypropanoate)	589	.....	4.66	32.1	....
$C_4H_{10}O$	Ethyl ether (ethoxyethane)	589	20.0	4.78	32.9	1.09
$C_4H_{10}O$	*n*-Butyl alcohol (1-butanol)	589	20.0	4.60	31.6	1.23
$C_4H_{10}O$	Isobutyl alcohol (2-methyl-1-propanol)	589	17.7	4.94	34.0	1.27
$C_4H_{10}O$	sec-Butyl alcohol (methylethylcarbinol)	589	20.0	4.91	33.8	1.27
C_5H_6	Cyclopentadiene	589	15.0	7.03	48.8	2.02
C_5H_8	1,3-Pentadiene	589	15.0	8.80	60.5	2.08
C_5H_8	Isoprene (2-methyl-1,3-butadiene)	589	15.0	8.80	60.5	2.08
C_5H_8	Cyclopentene	589	15.0	5.69	39.1	1.52
C_5H_{10}	1-Pentene	589	15.0	6.45	44.4	1.39
C_5H_{10}	Isopentane (2-methyl-1-butane)	589	15.0	6.36	43.8	1.39
C_5H_{10}	Cyclopentane	589	20.0	4.89	33.6	1.23
C_5H_{12}	Pentane	589	15.0	5.60	38.5	1.15
C_5H_{12}	Isopentane (2-methylbutane)	589	15.0	5.75	39.6	1.17
$C_5H_4O_2$	Furfural (2-furancarbonal)	578	20.0	7.01	50.2	2.06
C_5H_5N	Pyridine	589	11.9	8.76	60.3	2.58
$C_5H_8O_4$	Glutaric acid (pentanedioic acid) 2 N aqueous sol.	589	16.0	5.48	37.7	....
$C_5H_{10}O_2$	Propionic acid ethylester	589	20.0	5.46	37.6	1.13
$C_5H_{10}O_2$	Acetic acid propylester	589	15.7	5.45	37.5	1.13
$C_5H_{14}N_2$	Cadaverine (1,5-pentanediamine)	589	14.7	7.49	51.6	1.53
C_6H_6	Benzene	589	15.0	11.27	77.5	3.00
C_6H_{12}	Cyclohexane	589	20.0	5.66	39.0	1.24
C_6H_{14}	Hexane	589	15.0	6.62	45.5	1.20
$C_6H_4Cl_2$	1,4-Dichlorobenzene (*p*-dichlorobenzene)	589	64.5	13.55	93.2	2.69
C_6H_5F	Fluorobenzene (phenylfluoride)	589	19.0	9.96	68.5	2.51
C_6H_5Cl	Chlorobenzene (phenylchloride)	589	15.0	12.51	86.1	2.92
C_6H_5Br	Bromobenzene (phenylbromide)	589	15.0	14.51	99.8	3.26
C_6H_5I	Iodobenzene (phenyliodide)	589	15.0	19.11	131.4	4.06
C_6H_6O	Phenol (hydroxybenzene)	589	39.0	12.07	83.5	3.21
C_6H_7N	Aniline (aminobenzene)	589	15.0	16.08	110.6	4.18
$C_6H_{11}Cl$	Chlorocyclohexane (cyclohexylchloride)	589	13.0	7.50	51.6	1.46
$C_6H_{12}O_3$	Paraldehyde (paraacetaldehyde)	589	17.3	6.66	45.8	1.19
$C_6H_{12}O_6$	Glucose $11H_2O$ (dextrose) (1 M aqueous sol.)	589	15.0	6.72	46.2	....
$C_6H_{12}O_6$	Galactose $10H_2O$ (1 M aqueous sol.)	589	15.0	6.89	47.4	....
$C_6H_{12}O_6$	Fructose $10H_2O$ (levulose) (1 M aqueous sol.)	589	15.0	6.73	46.3	....
$C_6H_{14}O$	2-Hexanol (butylmethylcarbinol)	589	20.0	6.89	47.4	1.31
$C_6H_{14}O$	3-Hexanol (ethylpropylcarbinol)	589	20.0	6.85	47.1	1.30
$C_6H_{14}O$	2-Methyl-3-Pentanol (ethylisopropylcarbinol)	589	20.0	6.90	47.5	1.32
$C_6H_4O_4N_2$	1,3-Dinitrobenzene (*m*-Dinitrobenzene)	589	17.1	9.65	66.4	2.17
$C_6H_5O_2N$	Nitrobenzene	589	15.0	9.36	64.4	2.17
C_7H_8	Toluene (methylbenzene)	589	15.0	12.16	83.7	2.71
C_7H_{14}	1-Heptene (α-heptylene)	589	18.0	8.48	58.3	1.43
C_7H_{16}	Heptane	589	15.0	7.61	52.7	1.23
C_7H_5N	Benzonitrile (benzenecarbonitrile)	589	15.7	11.85	81.5	2.74
$C_7H_6O_2$	Benzoic acid (20 % alcohol sol.)	578	20.0	11.8	84.7	....
C_7H_7Cl	*o*-Chlorotoluene (2-chloro-1-methylbenzene)	589	15.4	13.72	94.3	2.95
C_7H_7Cl	*p*-Chlorotoluene (4-chloro-1-methylbenzene)	589	15.2	13.25	90.8	2.65
C_7H_7Br	*o*-Bromotoluene (2-bromo-1-methylbenzene)	589	16.7	15.67	107.3	3.08
C_7H_7Br	*p*-Bromotoluene (4-bromo-1-methylbenzene)	589	39.0	15.09	103.6	2.88
C_7H_8O	*o*-Cresol (*o*-methylphenol)	589	16.0	13.38	92.1	3.07
C_7H_8O	*m*-Cresol (*m*-methylphenol)	589	17.9	12.77	87.6	2.89
C_7H_8O	*p*-Cresol (*p*-methylphenol)	589	17.0	12.86	88.5	2.91
C_7H_9N	*o*-Toluidine (*o*-methylaniline)	589	17.3	17.18	118.2	3.79
C_7H_9N	*m*-Toluidine (*m*-methylaniline)	589	15.0	16.21	111.5	3.56
C_7H_9N	*p*-Toluidine (*p*-methylaniline)	589	50.0	15.92	109.5	3.37
$C_7H_{14}O$	Enanthaldehyde (heptanal)	589	16.2	7.42	51.1	1.26
$C_7H_{16}O$	1-Heptanol (*n*-heptylalcohol)	589	12.6	7.85	54.0	1.33
$C_7H_{16}O$	2-Heptanol (amylmethylcarbinol)	589	20.0	7.94	54.6	1.32
$C_7H_{16}O$	3-Heptanol (butylethylcarbinol)	589	20.0	7.86	54.1	1.37
$C_7H_7O_2N$	*o*-Nitrotoluene	589	18.0	10.80	74.3	2.16
$C_7H_7O_2N$	*p*-Nitrotoluene	589	54.3	10.20	70.2	1.97
C_8H_{10}	Ethylbenzene (phenylethane)	589	15.0	13.41	92.3	2.80
C_8H_{10}	*o*-Xylene (1,2-dimethylbenzene)	589	15.0	13.36	91.9	2.62
C_8H_{10}	*m*-Xylene (1,3-dimethylbenzene)	589	15.0	12.82	88.2	2.47
C_8H_{10}	*p*-Xylene (1,4-dimethylbenzene)	589	15.0	12.80	88.1	2.46
C_8H_{16}	1-Octene (α-octylene)	589	15.0	9.00	65.5	1.44
C_8H_{16}	2-Octene (β-octylene)	589	15.0	9.33	64.2	1.43
C_8H_{18}	Octane	589	15.0	8.65	59.5	1.26
$C_8H_{18}O$	1-Octanol (*n*-octylalcohol)	589	20.0	8.88	61.1	1.33
$C_8H_{18}O$	2-Octanol (methylhexylcarbinol)	589	20.0	9.00	61.8	1.34
$C_8H_{18}O$	3-Octanol (ethylamylcarbinol)	589	20.0	8.90	61.2	1.33
C_9H_{12}	*o*-Ethyltoluene (1-ethyl-2-ethylbenzene)	589	15.0	14.56	100.2	2.32
C_9H_{12}	*m*-Ethyltoluene (1-ethyl-3-ethylbenzene)	589	15.0	14.18	97.6	2.91
C_9H_{12}	*p*-Ethylbenzene (1-ethyl-4-ethylbenzene)	589	15.0	13.98	96.2	2.37
C_9H_{12}	Mesitylene (1-3-5-trimethylbenzene)	589	15.0	13.36	91.9	2.28
C_9H_{20}	Nonane	589	15.0	9.70	66.7	1.28
$C_{10}H_8$	Naphthalene	589	89.5	24.98	171.8	4.47
$C_{10}H_{20}$	1-Decene (*n*-decylene)	589	21.0	11.65	80.1	1.45
$C_{10}H_{22}$	Decane	589	15.0	10.70	73.6	1.30

MAGNETIC ROTATORY POWER (Continued)

Formula	Name	λ	t	ρ_M	$10^5[\Lambda]_M^{\lambda,t}$	$10^5\Lambda^{\lambda,t}$ Verdet
$C_{10}H_7Cl$	1-Chloronaphthalene (α-chloronaphthalene)	578	18.0	28.15	201.5	4.91
$C_{10}H_7Br$	1-Bromonaphthalene (α-bromonaphthalene)	578	20.0	31.05	222.0	5.19
$C_{10}H_8O$	β-Naphthol (2-hydroxynaphthalene)	578	136.0	27.1	194.0	4.80
$C_{10}H_9N$	1-Naphthylamine (α-naphthylamine)	589	32.6	37.23	256.1	6.84
$C_{10}H_{12}O_2$	Isoeugenol (4-propenylguaiacol)	589	19.3	21.44	147.5	3.55
$C_{10}H_{12}O_2$	Eugenol (4-allylguaiacol)	589	15.4	18.72	128.8	2.88
$C_{10}H_{12}O_2$	Benzoic acid propylester (*n*-propylbenzoate)	589	15.4	14.87	102.3	2.20
$C_{10}H_{12}O_2$	*o*-Toluic acid ethylester	589	15.2	15.06	103.6	2.25
$C_{10}H_{12}O_2$	*p*-Toluic acid ethylester	589	15.0	14.74	101.4	2.18
$C_{10}H_{12}O_2$	α-Toluic acid ethylester (ethylphenylacetate)	589	14.0	14.99	103.1	2.25
$C_{10}H_{12}O_3$	Methylsalicylic acid ethylester	589	18.6	17.14	117.9	2.50
$C_{10}H_{15}N$	N,N-Diethylaniline (N-phenyldiethylamine)	589	15.3	25.16	173.1	3.74
$C_{10}H_{16}O$	Camphor	589	14.0	9.26	63.7	....
$C_{10}H_{18}O$	α-Terpineol	589	16.0	10.84	74.6	1.56
$C_{10}H_{18}O$	Citronellal	589	14.5	11.48	79.0	1.51
$C_{10}H_{18}O_4$	Dipropylsuccinate	589	11.4	10.36	71.3	1.22
$C_{10}H_{18}O_6$	Tartaric acid dipropylester (propyltartrate)	589	15.4	10.83	74.5	1.24
$C_{10}H_{20}O$	Menthol	589	45.2	10.51	72.3	1.40
$C_{11}H_{24}$	Undecane	589	20.5	11.65	80.1	1.31
$C_{12}H_{26}$	Dodecane	589	21.5	12.71	87.4	1.32
$C_{12}H_{22}O_{11}$	Saccharose $19H_2O$ (1 M aqueous sol.)	589	15.0	12.59	86.6	....
$C_{12}H_{22}O_{11}$	Maltose $20H_2O$ (1 M aqueous sol.)	589	15.0	12.69	87.3	....
$C_{12}H_{22}O_{11}$	Lactose $41H_2O$ (1 M aqueous sol.)	589	18.4	12.71	87.5	....
$C_{14}H_{10}$	Phenanthrene	578	100.0	39.7	284.0	5.84
$C_{16}H_{34}$	Hexadecane	589	15.0	16.8	115.6	1.35
$C_{18}H_{14}$	1,2-Diphenylbenzene	589	15.0	40.2	276.3	4.70
$C_{18}H_{14}$	1,3-Diphenylbenzene (*m*-phenyldiphenyl)	589	15.0	41.0	282.4	....
$C_{18}H_{22}$	1,6-Diphenylhexane	589	20.0	......	....	2.75

TRANSPARENCY TO OPTICAL DENSITY CONVERSION TABLE

Transparency of a layer of material is defined as the ratio of the intensity of the transmitted light to that of the incident light. Opacity is the reciprocal of the transparency. Optical density is the common logarithm of the opacity.

Thus,

$$\text{Transparency} = \frac{I_t}{I_i}$$

$$\text{Opacity} = \frac{1}{\text{Transparency}} = \frac{I_i}{I_t}$$

$$\text{Optical density} = \log_{10}\left(\frac{I_i}{I_t}\right)$$

where I_i = Intensity of incident light
I_t = Intensity of transmitted light.

Trans.	Density	Trans.	Density	Trans.	Density	Trans.	Density	Trans.	Density	Trans.	Density	Trans.	Density	Trans.	Density
0.000	—	**.030**	1.523	**.060**	1.222	**.090**	1.046	**.120**	.9208	**.150**	.8239	**.180**	.7447	**.210**	.6778
.001	3.000	.031	1.509	.061	1.215	.091	1.041	.121	.9172	.151	.8210	.181	.7423	.211	.6757
.002	2.699	.032	1.495	.062	1.208	.092	1.036	.122	.9137	.152	.8182	.182	.7399	.212	.6737
.003	2.523	.033	1.482	.063	1.201	.093	1.032	.123	.9101	.153	.8153	.183	.7375	.213	.6716
.004	2.398	.034	1.469	.064	1.194	.094	1.027	.124	.9066	.154	.8125	.184	.7352	.214	.6696
.005	2.301	.035	1.456	.065	1.187	.095	1.022	.125	.9031	.155	.8097	.185	.7328	.215	.6676
.006	2.222	.036	1.444	.066	1.180	.096	1.018	.126	.8996	.156	.8069	.186	.7305	.216	.6655
.007	2.155	.037	1.432	.067	1.174	.097	1.013	.127	.8962	.157	.8041	.187	.7282	.217	.6635
.008	2.097	.038	1.420	.068	1.168	.098	1.009	.128	.8928	.158	.8013	.188	.7258	.218	.6615
.009	2.046	.039	1.409	.069	1.161	.099	1.004	.129	.8894	.159	.7986	.189	.7235	.219	.6596
.010	2.000	**.040**	1.398	**.070**	1.155	**.100**	1.000	**.130**	.8861	**.160**	.7959	**.190**	.7212	**.220**	.6576
.011	1.959	.041	1.387	.071	1.149	.101	.9957	.131	.8827	.161	.7932	.191	.7190	.221	.6556
.012	1.921	.042	1.377	.072	1.143	.102	.9914	.132	.8794	.162	.7905	.192	.7167	.222	.6536
.013	1.886	.043	1.367	.073	1.137	.103	.9872	.133	.8761	.163	.7878	.193	.7144	.223	.6517
.014	1.854	.044	1.357	.074	1.131	.104	.9830	.134	.8729	.164	.7852	.194	.7122	.224	.6498
.015	1.824	.045	1.347	.075	1.125	.105	.9788	.135	.8697	.165	.7825	.195	.7100	.225	.6478
.016	1.796	.046	1.337	.076	1.119	.106	.9747	.136	.8665	.166	.7799	.196	.7077	.226	.6459
.017	1.770	.047	1.328	.077	1.114	.107	.9706	.137	.8633	.167	.7773	.197	.7055	.227	.6440
.018	1.745	.048	1.319	.078	1.108	.108	.9666	.138	.8601	.168	.7747	.198	.7033	.228	.6421
.019	1.721	.049	1.310	.079	1.102	.109	.9626	.139	.8570	.169	.7721	.199	.7011	.229	.6402
.020	1.699	**.050**	1.301	**.080**	1.097	**.110**	.9586	**.140**	.8539	**.170**	.7696	**.200**	.6990	**.230**	.6383
.021	1.678	.051	1.292	.081	1.092	.111	.9547	.141	.8508	.171	.7670	.201	.6968	.231	.6364
.022	1.658	.052	1.284	.082	1.086	.112	.9508	.142	.8477	.172	.7645	.202	.6946	.232	.6345
.023	1.638	.053	1.276	.083	1.081	.113	.9469	.143	.8447	.173	.7620	.203	.6925	.233	.6326
.024	1.620	.054	1.268	.084	1.076	.114	.9431	.144	.8416	.174	.7594	.204	.6904	.234	.6308
.025	1.602	.055	1.260	.085	1.071	.115	.9393	.145	.8386	.175	.7570	.205	.6882	.235	.6289
.026	1.585	.056	1.252	.086	1.066	.116	.9356	.146	.8356	.176	.7545	.206	.6861	.236	.6271
.027	1.569	.057	1.244	.087	1.060	.117	.9318	.147	.8327	.177	.7520	.207	.6840	.237	.6253
.028	1.553	.058	1.237	.088	1.055	.118	.9281	.148	.8297	.178	.7496	.208	.6819	.238	.6234
.029	1,538	.059	1.229	.089	1.051	.119	.9244	.149	.8268	.179	.7471	.209	.6799	.239	.6216

TRANSPARENCY TO OPTICAL DENSITY CONVERSION TABLE (Continued)

Trans.	Density	Trans.	Density	Trans.	Density	Trans.	Density	Trans.	Density
.240	.6198	.315	.5017	**.390**	.4089	.465	.3325	**.540**	.2676
.241	.6180	.316	.5003	.391	.4078	.466	.3316	.541	.2668
.242	.6162	.317	.4989	.392	.4067	.467	.3307	.542	2660
.243	.6144	.318	.4976	.393	.4056	.468	.3298	.543	.2652
.244	.6126	.319	.4962	.394	.4045	.469	.3288	.544	.2644
.245	.6108	**.320**	.4949	.395	.4034	**.470**	.3279	.545	.2636
.246	.6091	.321	.4935	.396	.4023	.471	.3270	.546	.2628
.247	.6073	.322	.4921	.397	.4012	.472	.3260	.547	.2620
.248	.6056	.323	.4908	.398	.4001	.473	.3251	.548	.2612
.249	.6038	.324	.4895	.399	.3990	.474	.3242	.549	.2604
.250	.6021	.325	.4881	**.400**	.3979	.475	.3233	**.550**	.2596
.251	.6003	.326	.4868	.401	.3969	.476	.3224	.551	.2589
.252	.5986	.327	.4855	.402	.3958	.477	.3215	.552	.2581
.253	.5969	.328	.4841	.403	.3947	.478	.3206	.553	.2573
.254	.5952	.329	.4828	.404	.3936	.479	.3197	.554	.2565
.255	.5935	**.330**	.4815	.405	.3925	**.480**	.3188	.555	.2557
.256	.5918	.331	.4802	.406	.3915	.481	.3179	.556	.2549
.257	.5901	.332	.4789	.407	.3904	.482	.3170	.557	.2541
.258	.5884	.333	.4776	.408	.3893	.483	.3161	.558	.2534
.259	.5867	.334	.4763	.409	.3883	.484	.3152	.559	.2526
.260	.5850	.335	.4750	**.410**	.3872	.485	.3143	**.560**	.2518
.261	.5834	.336	.4737	.411	.3862	.486	.3134	.561	.2510
.262	.5817	.337	.4724	.412	.3851	.487	.3125	.562	.2503
.263	.5800	.338	.4711	.413	.3840	.488	.3116	.563	.2495
.264	.5784	.339	.4698	.414	.3830	.489	.3107	.564	.2487
.265	.5768	**.340**	.4685	.415	.3819	**.490**	.3098	.565	.2479
.266	.5751	.341	.4673	.416	.3809	.491	.3089	.566	.2472
.267	.5735	.342	.4660	.417	.3799	.492	.3080	.567	.2464
.268	.5719	.343	.4647	.418	.3788	.493	.3072	.568	.2457
.269	.5702	.344	.4634	.419	.3778	.494	.3063	.569	.2449
.270	.5686	.345	.4622	**.420**	.3768	.495	.3054	**.570**	.2441
.271	.5670	.346	.4609	.421	.3757	.496	.3045	.571	.2434
.272	.5654	.347	.4597	.422	.3747	.497	.3036	.572	.2426
.273	.5638	.348	.4584	.423	.3737	.498	.3028	.573	.2418
.274	.5622	.349	.4572	.424	.3726	.499	.3019	.574	.2411
275	.5607	**.350**	.4559	.425	.3716	**.500**	.3010	.575	.2403
.276	.5591	.351	.4547	.426	.3706	.501	.3002	.576	.2396
.277	.5575	.352	.4535	.427	.3696	.502	.2993	.577	.2388
.278	.5560	.353	.4522	.428	.3685	.503	.2984	.578	.2381
.279	.5544	.354	.4510	.429	.3675	.504	.2975	.579	.2373
.280	.5528	.355	.4498	**.430**	.3665	.505	.2967	**.580**	.2366
.281	.5513	.356	.4486	.431	.3655	.506	.2959	.581	.2358
282	.5498	.357	.4473	.432	.3645	.507	.2950	.582	.2351
283	.5482	.358	.4461	.433	.3635	.508	.2941	.583	.2343
284	.5467	.359	.4449	.434	.3625	.509	.2933	.584	.2336
.285	.5452	**.360**	.4437	.435	.3615	**.510**	.2924	.585	.2328
.286	.5436	.361	.4425	.436	.3605	.511	.2916	.586	.2321
.287	.5421	.362	.4413	.437	.3595	.512	.2907	.587	.2314
.288	.5406	.363	.4401	.438	.3585	.513	.2899	.588	.2306
.289	.5391	.364	.4389	.439	.3575	.514	.2890	.589	.2299
.290	.5376	.365	.4377	**.440**	.3565	.515	.2882	**.590**	.2291
.291	.5361	.366	.4365	.441	.3556	.516	.2873	.591	.2284
.292	.5346	.367	.4353	.442	.3546	.517	.2865	.592	.2277
.293	.5331	.368	.4342	.443	.3536	.518	.2857	.593	.2269
.294	.5317	.369	.4330	.444	.3526	.519	.2848	.594	.2262
295	.5302	**.370**	.4318	.445	.3516	**.520**	.2840	.595	.2255
.296	.5287	.371	.4306	.446	.3507	.521	.2831	.596	.2248
.297	.5272	.372	.4295	.447	.3497	.522	.2823	.597	.2240
.298	.5258	.373	.4283	.448	.3487	.523	.2815	.598	.2233
.299	.5243	.374	.4271	.449	.3478	.524	.2807	.599	.2226
.300	.5229	.375	.4260	**.450**	.3468	.525	.2798	**.600**	.2219
.301	.5215	.376	.4248	.451	.3458	.526	.2790	.601	.2211
.302	.5200	.377	.4237	.452	.3449	.527	.2782	.602	.2204
.303	.5186	.378	.4225	.453	.3439	.528	.2774	.603	.2197
.304	.5171	.379	.4214	.454	.3429	.529	.2766	.604	.2190
.305	.5157	**.380**	.4202	.455	.3420	**.530**	.2757	.605	.2182
.306	.5143	.381	.4191	.456	.3410	.531	.2749	.606	.2175
.307	.5128	.382	.4179	.457	.3401	.532	.2741	.607	.2168
.308	.5114	.383	.4168	.458	.3391	.533	.2733	.608	.2161
.309	.5100	.384	.4157	.459	.3382	.534	.2725	.609	.2154
.310	.5086	.385	.4145	**.460**	.3372	.535	.2717	**.610**	.2147
.311	.5072	.386	.4134	.461	.3363	.536	.2708	.611	.2140
.312	.5058	.387	.4123	.462	.3354	.537	.2700	.612	.2132
.313	.5045	.388	.4112	.463	.3344	.538	.2692	.613	.2125
.314	.5031	.389	.4101	.464	.3335	.539	.2684	.614	.2118

TRANSPARENCY TO OPTICAL DENSITY CONVERSION TABLE (Continued)

Trans.	Density	Trans.	Density	Trans.	Density	Trans.	Density	Trans.	Density
.615	.2111	.695	.1580	.775	.1107	.855	.0680	.935	.0292
.616	.2104	.696	.1574	.776	.1102	.856	.0675	.936	.0287
.617	.2097	.697	.1568	.777	.1096	.857	.0670	.937	.0282
.618	.2090	.698	.1562	.778	.1090	.858	.0665	.938	.0278
.619	.2083	.699	.1555	.779	.1085	.859	.0660	.939	.0263
.620	2076	**.700**	.1549	**.780**	.1079	**.860**	.0655	**.940**	.0269
.621	2069	.701	.1543	.781	.1073	.861	.0650	.941	.0264
.622	.2062	.702	.1537	.782	.1068	.862	.0645	.942	.0260
.623	.2055	.703	.1531	.783	.1062	.863	.0640	.943	.0255
.624	.2048	.704	.1524	.784	.1057	.864	.0635	.944	.0250
.625	.2041	.705	.1518	.785	.1051	.865	.0630	.945	.0246
.626	.2034	.706	.1512	.786	.1046	.866	.0625	.946	.0241
.627	.2027	.707	.1506	.787	.1040	.867	.0620	.947	.0237
.628	.2020	.708	.1500	.788	.1035	.868	.0615	.948	.0232
.629	.2013	.709	.1493	.789	.1029	.869	.0610	.949	.0227
.630	.2007	**.710**	.1487	**.790**	.1024	**.870**	.0605	**.950**	.0223
.631	.2000	.711	.1481	.791	.1018	.871	.0600	.951	.0218
.632	.1993	.712	.1475	.792	.1013	.872	.0595	.952	.0214
.633	.1986	.713	.1469	.793	.1007	.873	.0590	.953	.0209
.634	.1979	.714	.1463	.794	.1002	.874	.0585	.954	.0204
.635	.1972	.715	.1457	.795	.0996	.875	.0580	.955	.0200
.636	.1965	.716	.1451	.796	.0991	.876	.0575	.956	.0195
.637	.1959	.717	.1445	.797	.0985	.877	.0570	.957	.0191
.638	.1952	.718	.1439	.798	.0980	.878	.0565	.958	.0186
.639	.1945	.719	.1433	.799	0975	.879	.0560	.959	.0182
.640	.1938	**.720**	.1427	**.800**	.0969	**.880**	.0555	**.960**	.0177
.641	.1932	.721	.1421	.801	.0964	.881	.0550	.961	.0173
.642	.1925	.722	.1415	.802	.0958	.882	.0545	.962	.0168
.643	.1918	.723	.1409	.803	.0953	.883	.0540	.963	.0164
.644	.1911	.724	.1403	.804	.0948	.884	.0535	.964	.0159
.645	.1904	.725	.1397	.805	.0942	.885	.0530	.965	.0155
.646	.1898	.726	.1391	.806	.0937	.886	.0526	.966	.0150
.647	.1891	.727	.1385	.807	.0931	.887	.0521	.967	.0146
.648	.1884	.728	.1379	.808	.0926	.888	.0516	.968	.0141
.649	.1877	.729	.1373	.809	.0921	.889	.0511	.969	.0137
.650	.1871	**.730**	.1367	**.810**	.0915	**.890**	.0506	**.970**	.0132
.651	.1864	.731	.1361	.811	.0910	.891	.0501	.971	.0128
.652	.1857	.732	.1355	.812	.0904	.892	.0496	.972	.0123
653	.1851	.733	.1349	.813	.0899	.893	.0491	.973	.0119
.654	.1844	.734	.1343	.814	.0894	.894	.0487	.974	.0114
.655	.1838	.735	.1337	.815	.0888	.895	.0482	.975	.0110
.656	.1831	.736	.1331	.816	.0883	.896	.0477	.976	.0106
.657	.1824	.737	.1325	.817	.0878	.897	.0472	.977	.0101
.658	.1818	.738	.1319	.818	.0872	.898	.0467	.978	.0097
.659	.1811	.739	.1314	.819	.0867	.899	.0462	.979	.0092
.660	.1805	**.740**	.1308	**.820**	.0862	**.900**	.0458	**.980**	.0088
.661	.1798	.741	.1302	.821	.0856	.901	.0453	.981	.0083
.662	.1791	.742	.1296	.822	.0851	.902	.0448	.982	.0079
.663	.1785	.743	.1290	.823	.0846	.903	.0443	.983	.0074
.664	.1778	.744	.1284	.824	.0841	.904	.0438	.984	.0070
.665	.1772	.745	.1278	.825	.0835	.905	.0434	.985	.0066
.666	.1765	.746	.1273	.826	.0830	.906	.0429	.986	.0061
.667	.1759	.747	.1267	.827	.0825	.907	.0424	.987	.0057
.668	.1752	.748	.1261	.828	.0820	.908	.0419	.988	.0052
.669	.1746	.749	.1255	.829	.0815	.909	.0414	.989	.0048
.670	.1739	**.750**	.1249	**.830**	.0809	**.910**	.0410	**.990**	.0044
.671	.1733	.751	.1244	.831	.0804	.911	.0405	.991	.0039
.672	.1726	.752	.1238	.832	.0799	.912	.0400	.992	.0035
.673	.1720	.753	.1232	.833	.0794	.913	.0395	.993	.0030
.674	.1713	.754	.1226	.834	.0788	.914	.0391	.994	.0026
.675	.1707	.755	.1221	.835	.0783	.915	.0386	.995	.0022
.676	.1701	.756	.1215	.836	.0778	.916	.0381	.996	.0017
.677	.1694	.757	.1209	.837	.0773	.917	.0376	.997	.0013
.678	.1688	.758	.1203	.838	.0767	.918	.0371	.998	.0009
.679	.1681	.759	.1198	.839	.0762	.919	.0367	.999	.0004
.680	.1675	**.760**	.1192	**.840**	.0757	**.920**	.0362	**1.000**	.0000
.681	.1668	.761	.1186	.841	.0752	.921	.0357	.....	.....
.682	.1662	.762	.1180	.842	.0747	.922	.0353	.....	.....
.683	.1655	.763	.1175	.843	.0742	.923	.0348	.....	.....
.684	.1649	.764	.1169	.844	.0736	.924	.0343	.....	.....
.685	.1643	.765	.1163	.845	.0731	.925	.0339	.....	.....
.686	.1637	.766	.1158	.846	.0726	.926	.0334	.....	.....
.687	.1630	.767	.1152	.847	.0721	.927	.0329	.....	.....
.688	.1624	.768	.1146	.848	.0716	.928	.0325	.....	.....
.689	.1618	.769	.1141	.849	.0711	.929	.0320	.....	.....
.690	.1612	**.770**	.1135	**.850**	.0706	**.930**	.0315	.....	.....
.691	.1605	.771	.1129	.851	.0701	.931	.0310	.....	.....
.692	.1599	.772	.1124	.852	.0696	.932	.0306	.....	.....
.693	.1593	.773	.1118	.853	.0690	.933	.0301	.....	.....
.694	.1586	.774	.1113	.854	.0685	.934	.0296	.....	.....

DENSITY OF VARIOUS SOLIDS

The approximate density of various solids at ordinary atmospheric temperature.

In the case of substances with voids such as paper or leather the bulk density is indicated rather than the density of the solid portion.

(Selected principally from the Smithsonian Tables.)

Substance	Grams per cu. cm	Pounds per cu. ft.
Agate	2.5–2.7	156–168
Alabaster, carbonate	2.69–2.78	168–173
sulfate	2.26–2.32	141–145
Albite	2.62–2.65	163–165
Amber	1.06–1.11	66–69
Amphiboles	2.9–3.2	180–200
Anorthite	2.74–2.76	171–172
Asbestos	2.0–2.8	125–175
Asbestos slate	1.8	112
Asphalt	1.1–1.5	69–94
Basalt	2.4–3.1	150–190
Beeswax	0.96–0.97	60–61
Beryl	2.69–2.7	168–169
Biotite	2.7–3.1	170–190
Bone	1.7–2.0	106–125
Brick	1.4–2.2	87–137
Butter	0.86–0.87	53–54
Calamine	4.1–4.5	255–280
Calcspar	2.6–2.8	162–175
Camphor	0.99	62
Caoutchouc	0.92–0.99	57–62
Cardboard	0.69	43
Celluloid	1.4	87
Cement, set	2.7–3.0	170–190
Chalk	1.9–2.8	118–175
Charcoal, oak	0.57	35
pine	0.28–0.44	18–28
Cinnabar	8.12	507
Clay	1.8–2.6	112–162
Coal, anthracite	1.4–1.8	87–112
bituminous	1.2–1.5	75–94
Cocoa butter	0.89–0.91	56–57
Coke	1.0–1.7	62–105
Copal	1.04–1.14	65–71
Cork	0.22–0.26	14–16
Cork linoleum	0.54	34
Corundum	3.9–4.0	245–250

Substance	Grams per cu. cm	Pounds per cu. ft.
Diamond	3.01–3.52	188–220
Dolomite	2.84	177
Ebonite	1.15	72
Emery	4.0	250
Epidote	3.25–3.50	203–218
Feldspar	2.55–2.75	159–172
Flint	2.63	164
Fluorite	3.18	198
Galena	7.3–7.6	460–470
Gamboge	1.2	75
Garnet	3.15–4.3	197–268
Gas carbon	1.88	117
Gelatin	1.27	79
Glass, common	2.4–2.8	150–175
flint	2.9–5.9	180–370
Glue	1.27	79
Granite	2.64–2.76	165–172
Graphite*	2.30–2.72	144–170
Gum arabic	1.3–1.4	81–87
Gypsum	2.31–2.33	144–145
Hematite	4.9–5.3	306–330
Hornblende	3.0	187
Ice	0.917	57.2
Ivory	1.83–1.92	114–120
Leather, dry	0.86	54
Lime, slaked	1.3–1.4	81–87
Limestone	2.68–2.76	167–171
Linoleum	1.18	74
Magnetite	4.9–5.2	306–324
Malachite	3.7–4.1	231–256
Marble	2.6–2.84	160–177
Meerschaum	0.99–1.28	62–80
Mica	2.6–3.2	165–200
Muscovite	2.76–3.00	172–187
Ochre	3.5	218
Opal	2.2	137
Paper	0.7 –1.15	44–72
Paraffin	0.87–0.91	54–57

Substance	Grams per cu. cm	Pounds per cu. ft.
Peat blocks	0.84	52
Pitch	1.07	67
Porcelain	2.3–2.5	143–156
Porphyry	2.6–2.9	162–181
Pressed wood pulp board	0.19	12
Pyrite	4.95–5.1	309–318
Quartz	2.65	165
Resin	1.07	67
Rock salt	2.18	136
Rubber, hard	1.19	74
Rubber, soft commercial	1.1	69
pure gum	0.91–0.93	57–58
Sandstone	2.14–2.36	134–147
Serpentine	2.50–2.65	156–165
Silica, fused transparent	2.21	138
translucent	2.07	129
Slag	2.0–3.9	125–240
Slate	2.6–3.3	162–205
Soapstone	2.6–2.8	162–175
Spermaceti	0.95	59
Starch	1.53	95
Sugar	1.59	99
Talc	2.7–2.8	168–174
Tallow, beef	0.94	59
mutton	0.94	59
Tar	1.02	66
Topaz	3.5–3.6	219–223
Tourmaline	3.0–3.2	190–200
Wax, sealing	1.8	112
Wood (seasoned)		
alder	0.42–0.68	26–42
apple	0.66–0.84	41–52
ash	0.65–0.85	40–53
balsa	0.11–0.14	7–9
bamboo	0.31–0.40	19–25

Substance	Grams per cu. cm	Pounds per cu. ft.
basswood	0.32–0.59	20–37
beech	0.70–0.90	43–56
birch	0.51–0.77	32–48
blue gum	1.00	62
box	0.95–1.16	59–72
butternut	0.38	24
cedar	0.49–0.57	30–35
cherry	0.70–0.90	43–56
dogwood	0.76	47
ebony	1.11–1.33	69–83
elm	0.54–0.60	34–37
hickory	0.60–0.93	37–58
holly	0.76	47
juniper	0.56	35
larch	0.50–0.56	31–35
lignum vitae	1.17–1.33	73–83
locust	0.67–0.71	42–44
logwood	0.91	57
mahogany		
Honduras	0.66	41
Spanish	0.85	53
maple	0.62–0.75	39–47
oak	0.60–0.90	37–56
pear	0.61–0.73	38–45
pine, pitch	0.83–0.85	52–53
white	0.35–0.50	22–31
yellow	0.37–0.60	23–37
plum	0.66–0.78	41–49
poplar	0.35–0.5	22–31
satinwood	0.95	59
spruce	0.48–0.70	30–44
sycamore	0.40–0.60	24–37
teak, Indian	0.66–0.88	41–55
African	0.98	61
walnut	0.64–0.70	40–43
water gum	1.00	62
willow	0.40–0.60	24–37

* Some values reported as low as 1.6

DENSITY OF WATER

The temperature of maximum density for pure water, free from air = **3.98**°C.

The density at this temperature = **0.999973** g/cm³.

The density of water at 3.98° C is 1.000000 g/ml.

(International Bureau of Weights and Measures, 1910.)

WEIGHT OF 1 GALLON OF WATER

The weights in air are for dry air at the same temperature as the water up to 40°C. and at a (corrected) barometric pressure of 760 mm and against brass weights of 8.4 density at 0°C. Above 40°C. the temperature of the air is assumed to be 20°C., i. e., the water is allowed to cool to 20°C. before the weighings are made. The volumetric computations are based on the relation that 1 liter = 1.000027 cubic decimeters and 1 cubic decimeter = 61.023378 cubic inches.

Temperature °C.	Weight in vacuo		Weight in air	
	Grams	Pounds	Grams	Pounds
0	3784.833	8.34412	3780.520	8.33461
1	5.055	.34461	0.758	.33513
2	5.211	.34495	0.930	.33551
3	5.303	.34515	1.037	.33575
4	5.332	.34522	1.082	.33585
5	5.302	.34515	1.067	.33582
6	5.212	.34495	0.992	.33565
7	5.066	.34463	0.861	.33536
8	4.864	.34419	0.675	.33495
9	4.610	.34363	0.435	.33442
10	4.303	.34295	0.144	.33378
11	3.943	.34216	3779.798	.33302
12	3.534	.34125	9.403	.33215
13	3.076	.34025	8.960	.33117
14	2.574	.33914	8.472	.33009
15	2.026	.33793	7.939	.32892
($15\frac{5}{9}$) 60° F.	(3781.703)	(8.33722)	(3777.623)	(8.32823)
16	1.435	.33663	7.362	.32765
($16\frac{2}{3}$) 62° F.	(3781.017)	(8.33571)	(3776.953)	(8.32675)
17	0.801	.33523	6.741	.32628
18	0.125	.33374	6.080	.32482
19	3779.407	.33216	5.375	.32327
20	8.649	.33049	4.630	.32162
21	7.850	.32872	3.845	.31989
22	7.012	.32688	3.021	.31808
23	6.135	.32494	2.157	.31617

Temperature °C.	Weight in vacuo		Weight in air	
	Grams	Pounds	Grams	Pounds
24	5.220	.32293	1.256	.31419
25	4.268	.32083	0.317	.31212
26	3.279	.31865	3769.341	.30996
27	2.254	.31639	8.329	.30773
28	1.195	.31405	7.283	.30543
29	0.100	.31164	6.201	.30304
30	3768.972	.30915	5.086	.30058
31	7.812	.30659	3.938	.29805
32	6.618	.30396	2.757	.29545
33	5.393	.30126	1.545	.29278
34	4.137	.29849	0.301	.29003
35	2.851	.29566	3759.027	.28723
40	3755.995	.28054	2.232	.27225
45	3748.39	.2638	3744.40	.2550
50	3740.17	8.2457	3736.20	8.2369
55	3731.32	.2261	3727.35	.2174
60	3721.89	.2054	3717.93	.1966
65	3711.86	.1832	3707.91	.1745
70	3701.33	.1600	3697.40	.1514
75	3690.28	.1357	3686.36	.1270
80	3678.70	.1101	3674.79	.1015
85	3666.66	.0836	3662.76	.0750
90	3654.13	.0560	3650.25	.0474
95	3641.19	.0274	3637.32	.0189
100	3627.79	7.9979	3623.93	7.9894

TEMPERATURE CORRECTION FOR VOLUMETRIC SOLUTIONS

This table gives the correction to various observed volumes of water, measured at the designated temperatures to give the volume at the standard temperature, 20°C. Conversely, by subtracting the corrections from the volume desired at 20°C., the volume that must be measured out at the designated temperatures in order to give the desired volume at 20°C., will be obtained. It is assumed that the volumes are measured in glass apparatus having a coefficient of cubical expansion of 0.000025 per degree centigrade. The table is applicable to dilute aqueous solutions having the same coefficient of expansion as water.

Temperature of measurement, °C.	Capacity of apparatus in milliliters at 20°C.						
	2,000	1,000	500	400	300	250	150
	Correction in milliliters to give volume of water at 20°C.						
15	+1.54	+0.77	+0.38	+0.31	+0.23	+0.19	+0.12
16	+1.28	+ .64	+ .32	+ .26	+ .19	+ .16	+ .10
17	+ .99	+ .50	+ .25	+ .20	+ .15	+ .12	+ .07
18	+ .68	+ .34	+ .17	+ .14	+ .10	+ .08	+ .05
19	+ .35	+ .18	+ .09	+ .07	+ .05	+ .04	+ .03
21	− .37	− .18	− .09	− .07	− .06	− .05	− .03
22	− .77	− .38	− .19	− .15	− .12	− .10	− .06

Temperature of measurement, °C.	Capacity of apparatus in milliliters at 20°C.						
	2,000	1,000	500	400	300	250	150
	Correction in milliliters to give volume of water at 20°C.						
23	−1.18	− .59	− .30	− .24	− .18	− .15	−.09
24	−1.61	− .81	− .40	− .32	− .24	− .20	−.12
25	−2.07	−1.03	− .52	− .41	− .31	− .26	−.15
26	−2.54	−1.27	− .64	− .51	− .38	− .32	−.19
27	−3.03	−1.52	− .76	− .61	− .46	− .38	−.23
28	−3.55	−1.77	− .89	− .71	− .53	− .44	−.27
29	−4.08	−2.04	−1.02	− .82	− .61	− .51	−.31
30	−4.62	−2.31	−1.16	− .92	− .69	− .58	−.35

In using the above table to correct the volume of certain standard solutions to 20°C. more accurate results will be obtained if the numerical values of the corrections are increased by the percentages given below:

Solution	Normality		
	N	N/2	N/10
HNO_3	50	25	6
H_2SO_4	45	25	5
NaOH	40	25	5
KOH	40	20	4

TEMPERATURE CORRECTION FOR GLASS VOLUMETRIC APPARATUS

This table gives the correction to be added to actual capacity (determined at certain temperatures) to give the capacity at the standard temperature, 20°C. Conversely, by subtracting the corrections from the indicated capacity of an instrument standard at 20°C. the corresponding capacity at other temperatures is obtained. The table assumes for the cubical coefficient of expansion of glass 0.000025 per degree centigrade. The coefficients of expansion of glasses used for volumetric instruments vary from 0.000023 to 0.000028.

Temperature in degrees C.	2,000 ml	1,000 ml	500 ml	400 ml	300 ml	250 ml
15	+0.25	+0.12	+0.06	+0.05	+0.04	+0.031
16	+ .20	+ .10	+ .05	+0.04	+ .03	+ .025
17	+ .15	+ .08	+ .04	+ .03	+ .02	+ .019
18	+ .10	+ .05	+ .02	+ .02	+ .02	+ .012
19	+ .05	+ .02	+ .01	+ .01	+ .01	+ .006
21	− .05	− .02	− .01	− .01	− .01	− .006
22	− .10	− .05	− .02	− .02	− .02	− .012

Temperature in degrees C.	2,000 ml	1,000 ml	500 ml	400 ml	300 ml	250 ml
23	− .15	− .08	− .04	− .03	− .02	− .019
24	− .20	− .10	− .05	− .04	− .03	− .025
25	− .25	− .12	− .06	− .05	− .04	− .031
26	− .30	− .15	− .08	− .06	− .04	− .038
27	− .35	− .18	− .09	− .07	− .05	− .044
28	− .40	− .20	− .10	− .08	− .06	− .050
29	− .45	− .22	− .11	− .09	− .07	− .056
30	− .50	− .25	− .12	− .10	− .08	− .062

DENSITY OF VARIOUS LIQUIDS

(Selected from Smithsonian Tables.)

Liquid	Grams per cu. cm	Pounds per cu. ft.	Temp °C
Acetone	0.792	49.4	20°
Alcohol, ethyl	0.791	49.4	20
methyl	0.810	50.5	0
Benzene	0.899	56.1	0
Carbolic acid	0.950–0.965	59.2–60.2	15
Carbon disulfide	1.293	80.7	0
tetrachloride	1.595	99.6	20
Chloroform	1.489	93.0	20
Ether	0.736	45.9	0
Gasoline	0.66–0.69	41.0–43.0	
Glycerin	1.260	78.6	0
Kerosene	0.82	51.2	
Mercury	13.6	849.0	

Liquid	Grams per cu. cm	Pounds per cu. ft.	Temp. °C
Milk	1.028–1.035	64.2–64.6	
Naphtha, petroleum ether	0.665	41.5	15
wood	0.848–0.810	52.9–50.5	0
Oils:			
castor	0.969	60.5	15
cocoanut	0.925	57.7	15
cotton seed	0.926	57.8	16
creosote	1.040–1.100	64.9–68.6	15
linseed, boiled	0.942	58.8	15
olive	0.918	57.3	15
Sea water	1.025	63.99	15
Turpentine (spirits)	0.87	54.3	
Water	1.00	62.43	4

DENSITY OF ALCOHOL

DENSITY OF ETHYL ALCOHOL IN GRAMS PER CUBIC CENTIMETER, COMPUTED FROM MENDELEEFF'S FORMULA

(Selected from Smithsonian Tables.)

Temp. °C	0	1	2	3	4	5	6	7	8	9
0	.80625	.80541	.80457	.80374	.80290	.80207	.80123	.80039	.79956	.79872
10	.79788	.79704	.79620	.79535	.79451	.79367	.79283	.79198	.79114	.79029
20	.78945	.78860	.78775	.78691	.78606	.78522	.78437	.78352	.78267	.78182
30	.78097	.78012	.77927	.77841	.77756	.77671	.77585	.77500	.77414	.77329

HYDROMETERS AND DENSITY UNITS

Alcoholometer. — For testing alcoholic solutions; the scale shows the per cent of alcohol by volume; 0°–100° is the per cent.

Ammoniameter. — For testing ammonia solutions; scale 0°–40°; to convert to sp. gr. multiply by 3 and deduct from 1000.

Barktrometer or *Barkometer.* — For testing tanning liquor; scale 0°–80° Bk; the number to the right of the decimal point of the sp. gr. is the degree Bk; thus, 1.025 sp. gr. is 25° Bk.

Baumé. — There are two kinds in use; heavy Bé, for liquids heavier than water and light Bé for liquids lighter than water. In the former, 0° corresponds to a sp. gr. 1.000 (water at 4°C.) and 66° corresponds to a sp. gr. 1.842; in the lighter than water scale, 0° Bé is equivalent to the gravity of a 10% solution of sodium chloride and 60° Bé corresponds to a sp. gr. of 0.745. For Baumé degrees on the scale of densities greater than unity, the following equation gives the means of conversion:

$$\text{Sp. gr.} = \frac{m}{m - d}$$ where m = 145 (in the United States)
m = 144 (old scale used in Holland)
m = 146.78 (New scale or Gerlach scale)
d = Baumé reading

Beck's Hydrometer has 0° corresponding to sp. gr. 1.000 and 30° to sp. gr. 0.850; equal divisions on the scale are continued as far as required in both directions.

Brix Saccharometer or *Balling Saccharometer* shows directly the per cent of sugar (sucrose) by weight at the temperature indicated on the instrument, usually 17.5°C.; i.e., degrees Brix is the per cent sugar.

Cartier's Hydrometer floats in water at the 10° scale division and at 30° corresponds to 32° Bé.

Oleometer.—For vegetable and sperm oils; scale 50°–0° corresponds to sp. gr. 0.870–0.970.

Soxhlet's Lactometer, for determining the density of milk, has a scale from 25° (sp. gr. 1.025) to 35° (sp. gr. 1.035) divided into suitable scale divisions.

Twaddell Hydrometers have the scale so arranged that the reading multiplied by 5 and added to 1000 gives the sp. gr. with reference to water as 1000; it is always used for densities greater than water.

HYDROMETER CONVERSION TABLES

SHOWING THE RELATION BETWEEN DENSITY (C. G. S.) AND DEGREES BAUMÉ FOR DENSITIES LESS THAN UNITY.

Density.	Degrees Baumé. .00	.01	.02	.03	.04
0.60	103.33	99.51	95.81	92.22	88.75
.70	70.00	67.18	64.44	61.78	59.19
.80	45.00	42.84	40.73	38.68	36.67
.90	25.56	23.85	22.17	20.54	18.94
1.00	10.00	.....	.....	.....	.....

Density.	Degrees Baumé. .05	.06	.07	.08	.09
0.60	85.38	82.12	78.95	75.88	72.90
.70	56.67	54.21	51.82	49.49	47.22
.80	34.71	32.79	30.92	29.09	27.30
.90	17.37	15.83	14.33	12.86	11.41
1.00	.....	.....	.....	.....	.....

HYDROMETER CONVERSION TABLES
(Continued)

SHOWING THE RELATION BETWEEN DENSITY (C. G. S.) AND BAUMÉ AND TWADDELL SCALES FOR DENSITIES ABOVE UNITY.

Density	Degrees Baumé.	Degrees Twaddell.	Density.	Degrees Baume.	Degrees Twaddell.	Density	Degrees Baumé.	Degrees Twaddell.	Density.	Degrees Baumé.	Degrees Twaddell.
1.00	0.00	0	1.20	24.17	40	1.41	42.16	82	1.61	54.94	122
1.01	1.44	2	1.21	25.16	42	1.42	42.89	84	1.62	55.49	124
1.02	2.84	4	1.22	26.15	44	1.43	43.60	86	1.63	56.04	126
1.03	4.22	6	1.23	27.11	46	1.44	44.31	88	1.64	56.58	128
1.04	5.58	8	1.24	28.06	48	1.45	45.00	90	1.65	57.12	130
1.05	6.91	10	1.25	29.00	50	1.46	45.68	92	1.66	57.65	132
1.06	8.21	12	1.26	29.92	52	1.47	46.36	94	1.67	58.17	134
1.07	9.49	14	1.27	30.83	54	1.48	47.03	96	1.68	58.69	136
1.08	10.74	16	1.28	31.72	56	1.49	47.68	98	1.69	59.20	138
1.09	11.97	18	1.29	32.60	58	1.50	48.33	100	1.70	59.71	140
1.10	13.18	20	1.30	33.46	60	1.51	48.97	102	1.71	60.20	142
1.11	14.37	22	1.31	34.31	62	1.52	49.60	104	1.72	60.70	144
1.12	15.54	24	1.32	35.15	64	1.53	50.23	106	1.73	61.18	146
1.13	16.68	26	1.33	35.98	66	1.54	50.84	108	1.74	61.67	148
1.14	17.81	28	1.34	36.79	68	1.55	51.45	110	1.75	62.14	150
1.15	18.91	30	1.35	37.59	70	1.56	52.05	112	1.76	62.61	152
1.16	20.00	32	1.36	38.38	72	1.57	52.64	114	1.77	63.08	154
1.17	21.07	34	1.37	39.16	74	1.58	53.23	116	1.78	63.54	156
1.18	22.12	36	1.38	39.93	76	1.59	53.80	118	1.79	63.99	158
1.19	23.15	38	1.39	40.68	78	1.60	54.38	120	1.80	64.44	160
			1.40	41.43	80				....	.....	...

ABSOLUTE DENSITY OF WATER

DENSITY IN GRAMS PER CUBIC CENTIMETER, COMPUTED FROM THE RELATIVE VALUES BY THIESEN, SCHEEL AND DISSELHORST (1900), AND THE ABSOLUTE VALUE AT 3.98° C. BY THE INTERNATIONAL BUREAU OF WEIGHTS AND MEASURES (1910).

Degrees	0	1	2	3	4	5	6	7	8	9
0	0.999841	847	854	860	866	872	878	884	889	895
1	900	905	909	914	918	923	927	930	934	938
2	941	944	947	950	953	955	958	960	962	964
3	965	967	968	969	970	971	972	972	973	973
4	973	973	973	972	972	972	970	969	968	966
5	965	963	961	959	957	955	952	950	947	944
6	941	938	935	931	927	924	920	916	911	907
7	902	898	893	888	883	877	872	866	861	855
8	849	843	837	830	824	817	810	803	796	789
9	781	774	766	758	751	742	734	726	717	709
10	700	691	682	673	664	654	645	635	625	615
11	605	595	585	574	564	553	542	531	520	509
12	498	486	475	463	451	439	427	415	402	390
13	377	364	352	339	326	312	299	285	272	258
14	244	230	216	202	188	173	159	144	129	114

Degrees	0	1	2	3	4	5	6	7	8	9
15	099	084	069	054	038	023	007	*991	*975	*959
16	0.998943	926	910	893	877	860	843	826	809	792
17	774	757	739	722	704	686	668	650	632	613
18	595	576	558	539	520	501	482	463	444	424
19	405	385	365	345	325	305	285	265	244	224
20	203	183	162	141	120	099	078	056	035	013
21	0.997992	970	948	926	904	882	860	837	815	792
22	770	747	724	701	678	655	632	608	585	561
23	538	514	490	466	442	418	394	369	345	320
24	296	271	246	221	196	171	146	120	095	069
25	044	018	*992	*967	*941	*914	*888	*862	*836	*809
26	0.996783	756	729	703	676	649	621	594	567	546
27	512	485	457	429	401	373	345	317	289	261
28	232	204	175	147	118	089	060	031	002	*973
29	0.995944	914	885	855	826	796	766	736	706	676
30	646	616	586	555	525	494	464	433	402	371

DENSITY OF D_2O

G. S. Kell

t, ° C.	ρ, G./Cc.	t, ° C.	ρ, G./Cc.	t, ° C.	ρ, G./Cc.	t, ° C.	ρ, G./Cc.
0	1.10469	20	1.10534	50	1.09570	80	1.07824
3.813	1.10546	25	1.10445	55	1.09325	85	1.07475
5	1.10562	30	1.10323	60	1.09060	90	1.07112
10	1.10599	35	1.10173	65	1.08777	95	1.06736
11.185	1.10600	40	1.09996	70	1.08475	100	1.06346
15	1.10587	45	1.09794	75	1.08158	101.431	1.06232

VOLUME PROPERTIES OF ORDINARY WATER

G. S. Kell

(Specific volume v, density ρ, thermal expansivity $\alpha = \mathrm{d}\ln v/\mathrm{d}t = -\mathrm{d}\ln \rho/\mathrm{d}t$, compressibility $\kappa = -\mathrm{d}\ln v/\mathrm{d}p = \mathrm{d}\ln \rho/\mathrm{d}p$)

t, °C.	v, Cc./G.	ρ, G./Ml.	$10^6\alpha$, Deg.$^{-1}$	$10^6\kappa$, Bar^{-1}	t, °C.	v, Cc./G.	ρ, G./Ml.	$10^6\alpha$, Deg.$^{-1}$	$10^6\kappa$, Bar^{-1}
−20	1.00658	0.99349	−678.48	61.94	−6	1.000926	0.999102	−192.06	53.50
−18	1.00532	0.99474	−580.83	60.48	−5	1.000746	0.999283	−169.22	53.04
−16	1.00424	0.99581	−495.74	59.11	−4	1.000587	0.999441	−147.34	52.60
−14	1.00332	0.99672	−420.85	57.83	−3	1.000451	0.999578	−126.36	52.17
−12	1.00254	0.99749	−354.33	56.64	−2	1.000334	0.999694	−106.20	51.76
−10	1.001895	0.998137	−294.73	55.52	−1	1.000238	0.999790	−86.81	51.36
−9	1.001614	0.998417	−267.18	54.99	0	1.000160	0.999868	−68.14	50.98
−8	1.001359	0.998671	−240.95	54.48	1	1.000101	0.999927	−50.14	50.61
−7	1.001131	0.998899	−215.94	53.98	2	1.000060	0.999968	−32.77	50.26
3	1.000036	0.999992	−15.98	49.92	57	1.015522	0.984743	504.30	44.37
4	1.000028	1.000000	0.26	49.59	58	1.016038	0.984243	510.71	44.41
5	1.000036	0.999992	15.98	49.28	59	1.016560	0.983737	517.07	44.45
6	1.000060	0.999968	31.23	48.98	60	1.017089	0.983226	523.38	44.50
7	1.000098	0.999930	46.01	48.69	61	1.017625	0.982708	529.64	44.55
8	1.000151	0.999877	60.37	48.41	62	1.018167	0.982185	535.85	44.61
9	1.000219	0.999809	74.33	48.15	63	1.018716	0.981655	542.02	44.66
10	1.000300	0.999728	87.90	47.89	64	1.019271	0.981120	548.14	44.72
11	1.000395	0.999634	101.12	47.65	65	1.019833	0.980580	554.22	44.79
12	1.000502	0.999526	113.99	47.42	66	1.020402	0.980034	560.26	44.85
13	1.000623	0.999406	126.54	47.19	67	1.020977	0.979482	566.26	44.92
14	1.000755	0.999273	138.78	46.98	68	1.021558	0.978924	572.22	45.00
15	1.000900	0.999129	150.73	46.78	69	1.022146	0.978361	578.15	45.07
16	1.001057	0.998972	162.41	46.59	70	1.022740	0.977793	584.04	45.15
17	1.001225	0.998804	173.82	46.40	71	1.023340	0.977219	589.89	45.23
18	1.001405	0.998625	184.99	46.23	72	1.023947	0.976640	595.72	45.32
19	1.001596	0.998435	195.91	46.06	73	1.024560	0.976056	601.51	45.40
20	1.001797	0.998234	206.61	45.91	74	1.025180	0.975466	607.27	45.49
21	1.002010	0.998022	217.10	45.76	75	1.025805	0.974871	613.00	45.59
22	1.002232	0.997801	227.37	45.62	76	1.026437	0.974271	618.71	45.68
23	1.002465	0.997569	237.45	45.48	77	1.027076	0.973665	624.39	45.78
24	1.002708	0.997327	247.34	45.36	78	1.027720	0.973055	630.04	45.88
25	1.002961	0.997075	257.05	45.24	79	1.028371	0.972439	635.67	45.99
26	1.003224	0.996814	266.59	45.13	80	1.029027	0.971819	641.27	46.10
27	1.003496	0.996544	275.96	45.02	81	1.029690	0.971193	646.86	46.21
28	1.003778	0.996264	285.17	44.93	82	1.030360	0.970562	652.42	46.32
29	1.004069	0.995976	294.23	44.84	83	1.031035	0.969926	657.96	46.44
30	1.004369	0.995678	303.14	44.75	84	1.031716	0.969286	663.48	46.56
31	1.004678	0.995372	311.92	44.67	85	1.032404	0.968640	668.98	46.68
32	1.004995	0.995057	320.55	44.60	86	1.033098	0.967990	674.47	46.81
33	1.005322	0.994734	329.06	44.54	87	1.033797	0.967335	679.94	46.94
34	1.005657	0.994403	337.44	44.48	88	1.034503	0.966674	685.40	47.07
35	1.006000	0.994063	345.71	44.42	89	1.035216	0.966009	690.84	47.20
36	1.006352	0.993716	353.85	44.37	90	1.035934	0.965340	696.26	47.34
37	1.006713	0.993360	361.89	44.33	91	1.036658	0.964665	701.68	47.48
38	1.007081	0.992997	369.81	44.29	92	1.037389	0.963986	707.08	47.63
39	1.007457	0.992626	377.64	44.25	93	1.038125	0.963302	712.48	47.77
40	1.007842	0.992247	385.36	44.22	94	1.038868	0.962613	717.86	47.93
41	1.008234	0.991861	392.99	44.20	95	1.039617	0.961920	723.24	48.08
42	1.008634	0.991467	400.52	44.18	96	1.040372	0.961222	728.60	48.24
43	1.009042	0.991067	407.97	44.16	97	1.041133	0.960519	733.96	48.40
44	1.009458	0.990659	415.33	44.15	98	1.041900	0.959812	739.32	48.56
45	1.009881	0.990244	422.60	44.15	99	1.042673	0.959100	744.67	48.73
46	1.010311	0.989822	429.80	44.14	100	1.043453	0.958384	750.01	48.90
47	1.010749	0.989393	436.91	44.15	101	1.044239	0.957662	755.36	49.07
48	1.011194	0.988957	443.95	44.15	102	1.045030	0.956937	760.70	49.25
49	1.011647	0.988515	450.92	44.16	103	1.045828	0.956207	766.03	49.43
50	1.012107	0.988066	457.81	44.17	104	1.046633	0.955472	771.37	49.62
51	1.012574	0.987610	464.64	44.19	105	1.047443	0.954733	776.71	49.80
52	1.013048	0.987148	471.40	44.21	106	1.048260	0.953989	782.05	50.00
53	1.013529	0.986680	478.10	44.24	107	1.049083	0.953240	787.39	50.19
54	1.014017	0.986205	484.74	44.26	108	1.049912	0.952488	792.73	50.39
55	1.014512	0.985723	491.32	44.29	109	1.050747	0.951730	798.07	50.59
56	1.015014	0.985236	497.84	44.33	110	1.051589	0.950968	803.42	50.80

DENSITY AND VOLUME OF MERCURY

BASED ON THE DENSITY OF MERCURY AT 0° C. BY THIESEN AND SCHEEL

(Selected from Smithsonian Tables.)

Temp. ° C.	Mass in gr. per ml.	Vol. of 1 gr. in ml.	Temp. ° C.	Mass in gr. per ml.	Vol. of 1 gr. in ml.	Temp. ° C.	Mass in gr. per ml.	Vol. of 1 gr in ml.	Temp. ° C.	Mass in gr. per ml.	Vol. of 1 gr. in ml.
−10	13.6202	0.0734205	30°	13.5217	0.0739552	11	5684	7011	150	2330	5688
−9	6177	4338	31	5193	9686	12	5659	7145	160	2093	7044
−8	6152	4472	32	5168	9820	13	5634	7278	170	1856	8402
−7	6128	4606	33	5144	9953	14	5610	7412	180	1620	9764
−6	6103	4739	34	5119	40087	15	13.5585	0.0737546	190	13.1384	0.0761128
−5	13.6078	0.0734873	35	13.5095	0.0740221	16	5561	7680	200	1148	2495
−4	6053	5006	36	5070	0354	17	5536	7813	210	0913	3865
−3	6029	5140	37	5046	0488	18	5512	7947	220	0678	5239
−2	6004	5273	38	5021	0622	19	5487	8081	230	0443	6616
−1	5979	5407	39	4997	0756	20	13.5462	0.0738215	240	13.0209	0.0767996
0	13.5955	0.0735540	40	13.4973	0.0740891	21	5438	8348	250	12.9975	9381
1	5930	5674	50	4729	2229	22	5413	8482	260	9741	70769
2	5906	5808	60	4486	3569	23	5389	8616	270	9507	2161
3	5881	5941	70	4244	4910	24	5364	8750	280	9273	3558
4	5856	6075	80	4003	6252	25	13.5340	0.0738883	290	12.9039	0.0774958
5	13.5832	0.0736209	90	13.3762	0.0747594	26	5315	9017	300	8806	6364
6	5807	6342	100	3522	8939	27	5291	9151	310	8572	7774
7	5782	6476	110	3283	50285	28	5266	9285	320	8339	9189
8	5758	6610	120	3044	1633	29	5242	9419	330	8105	80609
9	5733	6744	130	2805	2982	30	13.5217	0.0739552	340	12.7872	0.0782033
10	13.5708	0.0736877	140	13.2567	0.0754334				350	7638	3464
									360	7405	4900

SULFURIC ACID

Specific Gravity of Aqueous Sulfuric Acid Solutions

at $\frac{20°}{4°}$ C.

Be.	Sp. gr.	Per cent H_2SO_4	G. per liter	Lbs. per cu. ft.	Lbs. per gal.	Be.	Sp. gr.	Per cent H_2SO_4	G. per liter	Lbs. per cu. ft.	Lbs. per gal.
0.7	1.0051	1	10.05	0.6275	0.0839	41.8	1.4049	51	716.5	44.73	5.979
1.7	1.0118	2	20.24	1.263	0.1689	42.5	1.4148	52	735.7	45.93	6.140
2.6	1.0184	3	30.55	1.907	0.2550	43.2	1.4248	53	755.1	47.14	6.302
3.5	1.0250	4	41.00	2.560	0.3422	44.0	1.4350	54	774.9	48.37	6.467
4.5	1.0317	5	51.59	3.220	0.4305	44.7	1.4453	55	794.9	49.62	6.634
5.4	1.0385	6	62.31	3.890	0.5200	45.4	1.4557	56	815.2	50.89	6.803
6.3	1.0453	7	73.17	4.568	0.6106	46.1	1.4662	57	835.7	52.17	6.974
7.2	1.0522	8	84.18	5.255	0.7025	46.8	1.4768	58	856.5	53.47	7.148
8.1	1.0591	9	95.32	5.950	0.7955	47.5	1.4875	59	877.6	54.79	7.324
9.0	1.0661	10	106.6	6.655	0.8897	48.2	1.4983	60	899.0	56.12	7.502
9.9	1.0731	11	118.0	7.369	0.9851	48.9	1.5091	61	920.6	57.47	7.682
10.8	1.0802	12	129.6	8.092	1.082	49.6	1.5200	62	942.4	58.83	7.865
11.7	1.0874	13	141.4	8.825	1.180	50.3	1.5310	63	964.5	60.21	8.049
12.5	1.0947	14	153.3	9.567	1.279	51.0	1.5421	64	986.9	61.61	8.236
13.4	1.1020	15	165.3	10.32	1.379	51.7	1.5533	65	1010	63.03	8.426
14.3	1.1094	16	177.5	11.08	1.481	52.3	1.5646	66	1033	64.46	8.618
15.2	1.1168	17	189.9	11.85	1.584	53.0	1.5760	67	1056	65.92	8.812
16.0	1.1243	18	202.4	12.63	1.689	53.7	1.5874	68	1079	67.39	9.008
16.9	1.1318	19	215.0	13.42	1.795	54.3	1.5989	69	1103	68.87	9.207
17.7	1.1394	20	227.9	14.23	1.902	55.0	1.6105	70	1127	70.38	9.408
18.6	1.1471	21	240.9	15.04	2.010	55.6	1.6221	71	1152	71.90	9.611
19.4	1.1548	22	254.1	15.86	2.120	56.3	1.6338	72	1176	73.44	9.817
20.3	1.1626	23	267.4	16.69	2.231	56.9	1.6456	73	1201	74.99	10.02
21.1	1.1704	24	280.9	17.54	2.344	57.5	1.6574	74	1226	76.57	10.24
21.9	1.1783	25	294.6	18.39	2.458	58.1	1.6692	75	1252	78.15	10.45
22.8	1.1862	26	308.4	19.25	2.574	58.7	1.6810	76	1278	79.75	10.66
23.6	1.1942	27	322.4	20.13	2.691	59.3	1.6927	77	1303	81.37	10.88
24.4	1.2023	28	336.6	21.02	2.809	59.9	1.7043	78	1329	82.99	11.09
25.2	1.2104	29	351.0	21.91	2.929	60.5	1.7158	79	1355	84.62	11.31
26.0	1.2185	30	365.6	22.82	3.051	61.1	1.7272	80	1382	86.26	11.53
26.8	1.2267	31	380.3	23.74	3.173	61.6	1.7383	81	1408	87.90	11.75
27.6	1.2349	32	395.2	24.67	3.298	62.1	1.7491	82	1434	89.54	11.97
28.4	1.2432	33	410.3	25.61	3.424	62.6	1.7594	83	1460	91.16	12.19
29.1	1.2515	34	425.5	26.56	3.551	63.0	1.7693	84	1486	92.78	12.40
29.9	1.2599	35	441.0	27.53	3.680	63.5	1.7786	85	1512	94.38	12.62
30.7	1.2684	36	456.6	28.51	3.811	63.9	1.7872	86	1537	95.95	12.83
31.4	1.2769	37	472.5	29.49	3.943	64.2	1.7951	87	1562	97.49	13.03
32.2	1.2855	38	488.5	30.49	4.077	64.5	1.8022	88	1586	99.01	13.23
33.0	1.2941	39	504.7	31.51	4.212	64.8	1.8087	89	1610	100.5	13.42
33.7	1.3028	40	521.1	32.53	4.349	65.1	1.8144	90	1633	101.9	13.63
34.5	1.3116	41	537.8	33.57	4.488	65.3	1.8195	91	1656	103.4	13.82
35.2	1.3205	42	554.6	34.62	4.628	65.5	1.8240	92	1678	104.8	14.00
35.9	1.3294	43	571.6	35.69	4.770	65.7	1.8279	93	1700	106.1	14.19
36.7	1.3384	44	588.9	36.76	4.914	65.8	1.8312	94	1721	107.5	14.36
37.4	1.3476	45	606.4	37.86	5.061	65.9	1.8337	95	1742	108.7	14.54
38.1	1.3569	46	624.2	38.97	5.209	66.0	1.8355	96	1762	110.0	14.70
38.9	1.3663	47	642.2	40.09	5.359	66.0	1.8364	97	1781	111.2	14.87
39.6	1.3758	48	660.4	41.23	5.511	66.0	1.8361	98	1799	112.3	15.02
40.3	1.3854	49	678.8	42.38	5.665	65.9	1.8342	99	1816	113.4	15.15
41.1	1.3951	50	697.6	43.55	5.821	65.8	1.8305	100	1831	114.3	15.28

DENSITY AND COMPOSITION OF FUMING SULFURIC ACID

Actual H_2SO_4, %	Specific gravity	Equiv. H_2SO_4, %	Weight, lb./cu. ft.	Weight, lb. per U.S. gal.	Comb. H_2O, %	Free SO_3, %	Total SO_3, %	SO_3, lb./cu. ft.
100	1.839	100.00	114.70	15.33	18.37	0	81.63	93.63
99	1.845	100.22	115.07	15.38	18.19	1	81.81	94.14
98	1.851	100.45	115.33	15.41	18.00	2	82.00	94.57
97	1.855	100.67	115.70	15.46	17.82	3	82.18	95.08
96	1.858	100.89	115.88	15.49	17.64	4	82.36	95.44
95	1.862	101.13	116.13	15.52	17.45	5	82.55	95.87
94	1.865	101.35	116.32	15.55	17.27	6	82.73	96.23
93	1.869	101.58	116.57	15.58	17.08	7	82.92	96.66
92	1.873	101.80	116.82	15.61	16.90	8	83.10	97.12
91	1.877	102.02	117.07	15.64	16.72	9	83.28	97.50
90	1.880	102.25	117.26	15.67	16.57	10	83.47	97.88
89	1.884	102.47	117.51	15.70	16.35	11	83.65	98.30
88	1.887	102.71	117.69	15.73	16.17	12	83.83	98.66
87	1.891	102.92	117.94	15.76	15.98	13	84.02	99.09
86	1.895	103.15	118.19	15.79	15.80	14	84.20	99.52
85	1.899	103.38	118.44	15.82	15.61	15	84.39	99.95
84	1.902	103.60	118.63	15.86	15.43	16	84.57	100.33
83	1.905	103.82	118.81	15.89	15.25	17	84.75	100.69
82	1.909	104.05	119.06	15.92	15.06	18	84.94	101.13
81	1.911	104.28	119.28	15.95	14.88	19	85.12	101.45
80	1.915	104.50	119.50	15.98	14.70	20	85.30	101.93
79	1.920	104.73	119.75	16.01	14.51	21	85.49	102.37
78	1.923	104.95	119.94	16.04	14.33	22	85.67	102.75
77	1.927	105.18	120.19	16.07	14.14	23	85.86	103.20
76	1.931	105.40	120.44	16.10	13.96	24	86.04	103.63
75	1.934	105.62	120.62	16.12	13.78	25	86.22	104.00
74	1.939	105.85	120.94	16.16	13.59	26	86.41	104.50
73	1.943	106.08	121.18	16.19	13.41	27	86.59	104.93
72	1.946	106.29	121.37	16.22	13.28	28	86.72	105.31
71	1.949	106.53	121.56	16.25	13.04	29	86.96	105.71
70	1.952	106.75	121.75	16.28	12.86	30	87.14	106.09
69	1.955	106.97	121.93	16.30	12.68	31	87.32	106.47
68	1.958	107.20	122.12	16.33	12.49	32	87.51	106.87
67	1.961	107.42	122.31	16.35	12.31	33	87.69	107.25
66	1.965	107.65	122.56	16.38	12.12	34	87.88	107.71
65	1.968	107.87	122.74	16.40	11.94	35	88.06	108.08
64	1.972	108.10	122.99	16.43	11.76	36	88.24	108.53
63	1.976	108.33	123.24	16.46	11.57	37	88.43	108.98
62	1.979	108.55	123.43	16.50	11.39	38	88.61	109.37
61	1.981	108.77	123.55	16.52	11.21	39	88.79	109.70
60	1.983	109.00	123.74	16.54	11.02	40	88.98	110.10
59	1.985	109.22	123.80	16.55	10.84	41	89.16	110.38
58	1.987	109.45	123.93	16.56	10.65	42	89.35	110.83
57	1.989	109.68	124.05	16.58	10.47	43	89.53	111.06
56	1.991	109.90	124.18	16.60	10.29	44	89.71	111.40
55	1.993	110.13	124.30	16.62	10.10	45	89.90	111.75
50	2.001	111.25	124.80	16.68	9.18	50	90.72	113.34
40	2.102	113.50	131.10	17.53	7.35	60	92.65	121.46
30	1.982	115.75	123.62	16.50	5.51	70	94.49	116.81
20	1.949	118.00	121.56	16.25	3.67	80	96.33	117.10
10	1.911	120.25	119.19	15.92	1.84	90	98.16	117.00
0	1.857	122.50	115.83	15.50	0.00	100	100.00	115.83

* By permission from the 7th edition of Chemical Plant Control Data, Chemical Construction Corporation (1957).

DENSITY OF MOIST AIR

The density of dry air may be determined by computation from the general relation $D = D_0(T_0/T)(P/P_0)$ where D_0 represents a known density at absolute temperature T_0 and pressure P_0 and D, the density at absolute temperature T and pressure P.

The density of **moist** air may be determined by a similar relation: $D = 1.2929\ (273.13/T)\ [(B - 0.3783e)/760]$ where T is the absolute temperature; B, the barometric pressure in mm, and e the vapor pressure of the moisture in the air in mm. The density will then be the product of two terms, each of which may be found by use of the tables which follow.

The **first factor,** 1.2929 (273.13/T), may be found directly in Table I for various temperatures. For convenience, temperatures are given in the table in °C although the values of the factor have been computed with absolute temperatures. The tabular values actually represent the density of dry air at various temperatures and 760 mm pressure.

The **second factor,** [(B − 0.3783e)/760], must be obtained in two steps: **First**—the numerator of the expression is obtained by subtracting 0.3783e from the barometric pressure. The quantity 0.3783e may be found directly from the dew point in Table II. If the wet and dry bulb thermometer readings are known e may be found in the table Reduction of Psychrometric Observations given in the section Hygrometric and Barometric Tables. 0.3783e may then be found by calculation or read from the table. **Second**—the value of the whole factor for any value of B − 0.3783e may be obtained from Table III.

The product of the above two factors will give the required density in g/l.

To facilitate obtaining **approximate values** of the density for ordinary pressures and temperatures, a table of products is given which may be entered with the temperature in °C and the corrected (for moisture) value of the barometric pressure in mm to obtain density.

As an illustration of the use of the tables, let it be desired to find the density of air for a barometric pressure of 750 mm, a dew point of 10° C, and air temperature of 20° C.

From the dew point, the value of 0.3783e is found in Table II to be 3.48 mm. 750 − 3.48 = 746.52, the corrected pressure. The pressure factor for this value found in Table III by interpolation is 0.98226.

The temperature factor from Table I is 1.2047.

$$1.2047 \times 0.98224 = 1.1833 \text{ g/l.}$$

To obtain the value directly from Table IV, enter it for **20° C and 746.5 mm** which gives by interpolation **1.183 g/l.**

TABLE I

(1.2929 × 273.13/T)

(Besides being a necessary part of the determination of the density of moist air, the values in this table are actually the density of dry air in g/l at 760 mm pressure for various temperatures.)

Temp. °C	0	1	2	3	4	5	6	7	8	9
−50	1.5 826	897	969	*042	*115	*189	*264	*339	*415	*491
−40	1.5 147	213	278	345	412	479	547	616	686	756
−30	1.4 524	584	645	706	767	829	892	955	*019	*083
−20	1.3 951	*006	*062	*118	*175	*232	*289	*347	*406	*465
−10	1.3 420	472	523	575	628	680	734	787	841	896
− 0	1.2 929	977	*024	*073	*121	*170	*219	*269	*319	*370
+ 0	1.2 929	882	835	789	742	697	651	606	561	517
10	1.2 472	428	385	342	299	256	214	171	130	088
20	1.2 047	006	*965	*925	*885	*845	*805	*766	*727	*688
30	1.1 649	611	573	535	498	460	423	387	350	314
40	1.1 277	242	206	170	135	100	065	031	*996	*962
50	1.0 928	895	861	828	795	762	729	697	664	632
60	1.0 600	569	537	506	475	444	413	382	352	322

TABLE II

Vapor Pressure—Value of 0.3783e

Dew point °C	Vap. press. e mm (ice)	0.3783e	Dew point °C	Vap. press. e mm (water)	0.3783e	Dew point °C	Vap. press. e mm (water)	0.3783e	Dew point °C	Vap. press. e mm (ice)	0.3783e	Dew point °C	Vap. press. e mm (water)	0.3783e	Dew point °C	Vap. press. e mm (water)	0.3783e
−50	0.029	0.01	**−15**	1.252	0.47	**0**	4.58	1.73	**15**	12.79	4.84	**30**	31.86	12.05	**45**	71.97	27.23
−45	.054	.02	−14	1.373	.52	1	4.92	1.86	16	13.64	5.16	31	33.74	12.76	46	75.75	28.66
−40	.096	.04	−13	1.503	.57	2	5.29	2.00	17	14.54	5.50	32	35.70	13.51	47	79.70	30.15
−35	.169	.06	−12	1.644	.62	3	5.68	2.15	18	15.49	5.86	33	37.78	14.29	48	83.83	31.71
−30	.288	.11	−11	1.798	.68	4	6.10	2.31	19	16.49	6.24	34	39.95	15.11	49	88.14	33.34
−25	0.480	0.18	**−10**	1.964	0.74	**5**	6.54	2.47	**20**	17.55	6.64	**35**	42.23	15.98	**50**	92.6	35.03
−24	.530	.20	− 9	2.144	.81	6	7.01	2.65	21	18.66	7.06	36	44.62	16.88	51	97.3	36.81
−23	.585	.22	− 8	2.340	.89	7	7.51	2.84	22	19.84	7.51	37	47.13	17.83	52	102.2	38.66
−22	.646	.24	− 7	2.550	.96	8	8.04	3.04	23	21.09	7.98	38	49.76	18.82	53	107.3	40.59
−21	.712	.27	− 6	2.778	1.05	9	8.61	3.26	24	22.40	8.47	39	52.51	19.86	54	112.7	42.63
−20	0.783	0.30	**− 5**	3.025	1.14	**10**	9.21	3.48	**25**	23.78	9.00	**40**	55.40	20.96	**55**	118.2	44.72
−19	.862	.33	− 4	3.291	1.24	11	9.85	3.73	26	25.24	9.55	41	58.42	22.10	56	124.0	46.91
−18	.947	.36	− 3	3.578	1.35	12	10.52	3.98	27	26.77	10.13	42	61.58	23.30	57	130.0	49.18
−17	1.041	.39	− 2	3.887	1.47	13	11.24	4.25	28	28.38	10.74	43	64.89	24.55	58	136.3	51.56
−16	1.142	.43	− 1	4.220	1.60	14	11.99	4.54	29	30.08	11.38	44	68.35	25.86	59	142.8	54.02
															60	**149.6**	**56.59**

TABLE III

Pressure Factor.—[(B − 0.3783e)/760]

The figures in the body of the table give values of the whole term (B − 0.3783e)/760] for various values of the numerator (B − 0.3783e) expressed at the left and top.

Press. mm corr.	0	1	2	3	4	5	6	7	8	9
80	.10526	.10658	.10789	.10921	.11053	.11184	.11316	.11447	.11579	.11711
90	.11842	.11974	.12105	.12237	.12368	.12500	.12632	.12763	.12895	.13026
100	.13158	.13289	.13421	.13553	.13684	.13816	.13947	.14079	.14211	.14342
110	.14474	.14605	.14737	.14868	.15000	.15132	.15263	.15395	.15526	.15658
120	.15789	.15921	.16053	.16184	.16316	.16447	.16579	.16711	.16842	.16974
130	.17105	.17237	.17368	.17500	.17632	.17763	.17895	.18026	.18158	.18289
140	.18421	.18553	.18684	.18816	.18947	.19079	.19211	.19342	.19474	.19605
150	.19737	.19868	.20000	.20132	.20263	.20395	.20526	.20658	.20789	.20921
160	.21053	.21184	.21316	.21447	.21579	.21711	.21842	.21974	.22105	.22237
170	.22368	.22500	.22632	.22763	.22895	.23026	.23158	.23289	.23421	.23553
180	.23684	.23816	.23947	.24079	.24211	.24342	.24474	.24605	.24737	.24868
190	.25000	.25132	.25263	.25395	.25526	.25658	.25789	.25921	.26053	.26184

Press. mm corr.	0	1	2	3	4	5	6	7	8	9
...	.....	.....	.....	....	.....	.....	.....	.....	.....	.....
...	.....	.....	.....	....	.....	.....	.....	.....	.....	.....
200	.26316	.26447	.26579	.26711	.26842	.26974	.27105	.27237	.27368	.27500
210	.27632	.27763	.27895	.28026	.28158	.28289	.28421	.28553	.28684	.28816
220	.28947	.29079	.29211	.29342	.29474	.29605	.29737	.29868	.30000	.30132
230	.30263	.30395	.30526	.30658	.30789	.30921	.31053	.31184	.31316	.31447
240	.31579	.31711	.31842	.31974	.32105	.32237	.32368	.32500	.32632	.32763
250	.32895	.33026	.33158	.33289	.33421	.33553	.33684	.33816	.33947	.34079
260	.34211	.34342	.34474	.34605	.34737	.34868	.35000	.35132	.35263	.35395
270	.35526	.35658	.35789	.35921	.36053	.36184	.36316	.36447	.36579	.36711
280	.36842	.36974	.37105	.37237	.37368	.37500	.37632	.37763	.37895	.38026
290	.38158	.38289	.38421	.38553	.38684	.38816	.38947	.39079	.39211	.39342

TABLE III (Continued)

Press. mm corr.	0	1	2	3	4	5	6	7	8	9	Press. mm corr.	0	1	2	3	4	5	6	7	8	9
300	.39474	.39605	.39737	.39868	.40000	.40132	.40263	.40395	.40526	.40658	**550**	.72368	.72500	.72632	.72763	.72895	.73026	.73158	.73290	.73421	.73553
310	.40789	.40921	.41053	.41184	.41316	.41447	.41579	.41711	.41842	.41974	560	.73684	.73816	.73947	.74079	.74211	.74342	.74474	.74605	.74737	.74868
320	.42105	.42237	.42368	.42500	.42632	.42763	.42895	.43026	.43158	.43289	570	.75000	.75132	.75263	.75395	.75526	.75658	.75790	.75921	.76053	.76184
330	.43421	.43553	.43684	.43816	.43947	.44079	.44211	.44342	.44474	.44605	580	.76316	.76447	.76579	.76711	.76842	.76974	.77105	.77237	.77368	.77500
340	.44737	.44868	.45000	.45132	.45263	.45395	.45526	.45658	.45789	.45921	590	.77632	.77763	.77895	.78026	.78158	.78290	.78421	.78553	.78684	.78816
350	.46053	.46184	.46316	.46447	.46579	.46711	.46842	.46974	.47105	.47237	**600**	.78947	.79079	.79211	.79342	.79474	.79605	.79737	.79868	.80000	.80132
360	.47368	.47500	.47632	.47763	.47895	.48026	.48158	.48289	.48421	.48553	610	.80263	.80395	.80526	.80658	.80790	.80921	.81053	.81184	.81316	.81447
370	.48684	.48816	.48947	.49079	.49211	.49342	.49474	.49605	.49737	.49868	620	.81579	.81711	.81842	.81974	.82105	.82237	.82368	.82500	.82632	.82763
380	.50000	.50132	.50263	.50395	.50526	.50658	.50789	.50921	.51053	.51184	630	.82895	.83026	.83158	.83290	.83421	.83553	.83684	.83816	.83947	.84079
390	.51316	.51447	.51579	.51711	.51842	.51974	.52105	.52237	.52368	.52500	640	.84211	.84342	.84474	.84605	.84737	.84868	.85000	.85132	.85263	.85395
400	.52632	.52763	.52895	.53026	.53158	.53289	.53421	.53553	.53684	.53816	**650**	.85526	.85658	.85790	.85921	.86053	.86184	.86316	.86447	.86579	.86711
410	.53947	.54079	.54211	.54342	.54474	.54605	.54737	.54868	.55000	.55132	660	.86842	.86974	.87105	.87237	.87368	.87500	.87632	.87763	.87895	.88026
420	.55263	.55395	.55526	.55658	.55789	.55921	.56053	.56184	.56316	.56447	670	.88158	.88290	.88421	.88553	.88684	.88816	.88947	.89079	.89211	.89342
430	.56579	.56711	.56842	.56974	.57105	.57237	.57368	.57500	.57632	.57763	680	.89474	.89605	.89737	.89868	.90000	.90132	.90263	.90395	.90526	.90658
440	.57895	.58026	.58158	.58289	.58421	.58553	.58684	.58816	.58947	.59079	690	.90790	.90921	.91053	.91184	.91316	.91447	.91579	.91711	.91842	.91974
450	.59211	.59342	.59474	.59605	.59737	.59868	.60000	.60132	.60263	.60395	**700**	.92105	.92237	.92368	.92500	.92632	.92763	.92895	.93026	.93158	.93290
460	.60526	.60658	.60789	.60921	.61053	.61184	.61316	.61447	.61579	.61711	710	.93421	.93553	.93684	.93816	.93947	.94079	.94211	.94342	.94474	.94605
470	.61842	.61974	.62105	.62237	.62368	.62500	.62632	.62763	.62895	.63026	720	.94737	.94868	.95000	.95132	.95263	.95395	.95526	.95658	.95790	.95921
480	.63158	.63289	.63421	.63553	.63684	.63816	.63947	.64079	.64211	.64342	730	.96053	.96184	.96316	.96447	.96579	.96711	.96842	.96974	.97105	.97237
490	.64474	.64605	.64737	.64868	.65000	.65132	.65263	.65395	.65526	.65658	740	.97368	.97500	.97632	.97763	.97895	.98026	.98158	.98290	.98421	.98553
500	.65790	.65921	.66053	.66184	.66316	.66447	.66579	.66711	.66842	.66974	**750**	.98684	.98816	.98947	.99079	.99211	.99342	.99474	.99605	.99737	.99868
510	.67105	.67237	.67368	.67500	.67632	.67763	.67895	.68026	.68158	.68290	760	1.0000	1.0013	1.0026	1.0039	1.0053	1.0066	1.0079	1.0092	1.0105	1.0118
520	.68421	.68553	.68684	.68816	.68947	.69079	.69211	.69342	.69474	.69605	770	1.0132	1.0145	1.0158	1.0171	1.0184	1.0197	1.0211	1.0224	1.0237	1.0250
530	.69737	.69868	.70000	.70132	.70263	.70395	.70526	.70658	.70790	.70921	780	1.0263	1.0276	1.0289	1.0303	1.0316	1.0329	1.0342	1.0355	1.0368	1.0382
540	.71053	.71184	.71316	.71447	.71579	.71711	.71842	.71974	.72105	.72237	790	1.0395	1.0408	1.0421	1.0434	1.0447	1.0461	1.0474	1.0487	1.0500	1.0513

TABLE IV

Density of Moist Air

Values in the body of the table give the density of moist air in g/l for a limited range of temperatures and corrected pressure values ($B - 0.3783e$). The latter may be obtained by use of Table II.

°C	600	610	620	630	640	650	660	670	680	690	700	710	720	730	740	750	760	770	780	790
5	1.0024	1.0191	1.0358	1.0525	1.0692	1.0859	1.1026	1.1193	1.1361	1.1528	1.1695	1.1862	1.2029	1.2196	1.2363	1.2530	1.2697	1.2864	1.3031	1.3198
6	.99876	1.0154	1.0321	1.0487	1.0654	1.0820	1.0986	1.1153	1.1319	1.1486	1.1652	1.1819	1.1985	1.2152	1.2318	1.2485	1.2651	1.2817	1.2984	1.3150
7	.99521	1.0118	1.0284	1.0450	1.0616	1.0781	1.0947	1.1113	1.1279	1.1445	1.1611	1.1777	1.1943	1.2108	1.2274	1.2440	1.2606	1.2772	1.2938	1.3104
8	.99165	1.0082	1.0247	1.0412	1.0578	1.0743	1.0908	1.1074	1.1239	1.1404	1.1569	1.1735	1.1900	1.2065	1.2230	1.2396	1.2561	1.2726	1.2892	1.3057
9	.98818	1.0047	1.0211	1.0376	1.0541	1.0705	1.0870	1 1035	1.1199	1.1364	1.1529	1.1694	1.1858	1.2023	1.2188	1.2352	1.2517	1.2682	1.2846	1.3011
10	.98463	1.0010	1.0175	1.0339	1.0503	1.0667	1.0831	1.0995	1.1159	1.1323	1.1487	1.1651	1.1816	1.1980	1.2144	1.2308	1.2472	1.2636	1.2800	1.2964
11	.98115	.99751	1.0139	1.0302	1.0466	1.0629	1.0793	1.0956	1.1120	1.1283	1.1447	1.1610	1.1774	1.1937	1.2101	1.2264	1.2428	1.2592	1.2755	1.2919
12	.97776	.99406	1.0104	1.0267	1.0430	1.0592	1.0755	1.0918	1.1081	1.1244	1.1407	1.1570	1.1733	1.1896	1.2059	1.2222	1.2385	1.2548	1.2711	1.2874
13	.97436	.99061	1.0068	1.0231	1.0393	1.0556	1.0718	1.0880	1.1043	1.1205	1.1368	1.1530	1.1692	1.1855	1.2017	1.2180	1.2342	1.2504	1.2667	1.2829
14	.97097	.98715	1.0033	1.0195	1.0357	1.0519	1.0681	1.0843	1.1004	1.1166	1.1328	1.1490	1.1652	1.1814	1.1975	1.2137	1.2299	1.2461	1.2623	1.2784
15	.96757	.98370	.99983	1.0160	1.0321	1.0482	1.0643	1.0805	1.0966	1.1127	1.1288	1.1450	1.1611	1.1772	1.1933	1.2095	1.2256	1.2417	1.2579	1.2740
16	.96426	.98033	.99641	1.0125	1.0286	1.0446	1.0607	1.0768	1.0928	1.1089	1.1250	1.1410	1.1571	1.1732	1.1893	1.2053	1.2214	1.2375	1.2535	1.2696
17	.96086	.97688	.99290	1.0089	1.0249	1.0409	1.0570	1.0730	1.0890	1.1050	1.1210	1.1370	1.1530	1.1691	1.1851	1.2011	1.2171	1.2331	1.2491	1.2651
18	.95763	.97359	.98955	1.0055	1.0215	1.0374	1.0534	1.0694	1.0853	1.1013	1.1172	1.1332	1.1492	1.1651	1.1811	1.1970	1.2130	1.2290	1.2449	1.2609
19	.95431	.97022	.98613	1.0020	1.0179	1.0338	1.0497	1.0656	1.0816	1.0975	1.1134	1.1293	1.1452	1.1611	1.1770	1.1929	1.2088	1.2247	1.2406	1.2565
20	.95107	.96693	.98278	.99864	1.0145	1.0303	1.0462	1.0620	1.0779	1.0937	1.1096	1.1254	1.1413	1.1572	1.1730	1.1888	1.2047	1.2206	1.2364	1.2522
21	.94784	.96364	.97944	.99524	1.0110	1.0294	1.0426	1.0584	1.0742	1.0900	1.1058	1.1216	1.1374	1.1532	1.1690	1.1848	1.2006	1.2164	1.2322	1.2480
22	.94460	.96035	.97609	.99184	1.0076	1.0233	1.0391	1.0548	1.0706	1.0863	1.1020	1.1178	1.1335	1.1493	1.1650	1.1808	1.1965	1.2122	1.2280	1.2437
23	.94144	.95714	.97283	.98852	1.0042	1.0199	1.0356	1.0513	1.0670	1.0827	1.0984	1.1140	1.1297	1.1454	1.1611	1.1768	1.1925	1.2082	1.2239	1.2396
24	.93829	.95393	.96957	.98521	1.0008	1.0165	1.0321	1.0478	1.0634	1.0790	1.0947	1.1103	1.1259	1.1416	1.1572	1.1729	1.1885	1.2041	1.2198	1.2354
25	.93513	.95072	.96630	.98189	.99748	1.0131	1.0286	1.0442	1.0598	1.0754	1.0910	1.1066	1.1222	1.1377	1.1533	1.1689	1.1845	1.2001	1.2157	1.2313
26	.93197	.94750	.96304	.97858	.99411	1.0096	1.0252	1.0407	1.0562	1.0718	1.0873	1.1028	1.1184	1.1339	1.1494	1.1650	1.1805	1.1960	1.2116	1.2271
27	.92889	.94437	.95986	.97534	.99083	1.0063	1.0218	1.0373	1.0528	1.0682	1.0837	1.0992	1.1147	1.1302	1.1456	1.1611	1.1766	1.1921	1.2076	1.2230
28	.92581	.94124	.95668	.97211	.98754	1.0030	1.0184	1.0338	1.0493	1.0647	1.0801	1.0955	1.1110	1.1264	1.1418	1.1573	1.1727	1.1881	1.2036	1.2190
29	.92273	.93811	.95350	.96888	.98426	.99963	1.0150	1.0304	1.0458	1.0612	1.0765	1.0919	1.1073	1.1227	1.1380	1.1534	1.1688	1.1842	1.1996	1.2149
30	.91965	.93498	.95031	.96564	.98097	.99629	1.0116	1.0270	1.0423	1.0576	1.0729	1.0883	1.1036	1.1189	1.1342	1.1496	1.1649	1.1802	1.1956	1.2109
31	.91665	.93193	.94721	.96249	.97777	.99304	1.0083	1.0236	1.0389	1.0542	1.0694	1.0847	1.1000	1.1153	1.1305	1.1458	1.1611	1.1764	1.1917	1.2069
32	.91365	.92888	.94411	.95934	.97457	.98979	1.0050	1.0203	1.0355	1.0507	1.0659	1.0812	1.0964	1.1116	1.1268	1.1421	1.1573	1.1725	1.1878	1.2030
33	.91065	.92583	.94101	.95619	.97137	.98654	1.0017	1.0169	1.0321	1.0473	1.0624	1.0776	1.0928	1.1080	1.1231	1.1383	1.1535	1.1687	1.1839	1.1990
34	.90773	.92286	.93800	.95313	.96826	.98338	.99851	1.0136	1.0288	1.0439	1.0590	1.0742	1.0893	1.1044	1.1195	1.1347	1.1498	1.1649	1.1801	1.1952
35	.90473	.91981	.93490	.94998	.96506	.98013	.99521	1.0103	1.0254	1.0405	1.0555	1.0706	1.0857	1.1008	1.1158	1.1309	1.1460	1.1611	1.1762	1.1912

DENSITY OF DRY AIR

AT THE TEMPERATURE t, AND UNDER THE PRESSURE H CM OF MERCURY THE DENSITY OF AIR

$$=\frac{0.001293}{1+0.00367\,t}\,\frac{H}{76}.$$

Units of this table are grams per milliliter
(From Miller's Laboratory Physics, Ginn & Co., publishers, by permission.)

t	Pressure H in Centimeters						Proportional Parts	
	72.0	73.0	74.0	75.0	76.0	77.0		
°							17	
10	0.001182	0.001198	0.001215	0.001231	0.001247	0.001264	cm	
11	178	193	210	227	243	259	0.1	2
12	173	190	206	222	239	255	0.2	3
13	169	186	202	218	234	251	0.3	5
14	165	181	198	214	230	246	0.4	7
							0.5	8
							0.6	10
							0.7	12
15	0.001161	0.001177	0.001193	0.001210	0.001226	0.001242	0.8	14
16	157	173	189	205	221	238	0.9	15
17	153	169	185	201	217	233	16	
18	149	165	181	197	213	229	cm	
19	145	161	177	193	209	225	0.1	2
							0.2	3
							0.3	5
							0.4	6
20	0.001141	0.001157	0.001173	0.001189	0.001205	0.001221	0.5	8
21	137	153	169	185	201	216	0.6	10
22	134	149	165	181	197	212	0.7	11
23	130	145	161	177	193	208	0.8	13
24	126	142	157	173	189	204	0.9	14
							15	
							cm	
25	0.001122	0.001138	0.001153	0.001169	0.001185	0.001200	0.1	1
26	118	134	149	165	181	196	0.2	3
27	115	130	146	161	177	192	0.3	4
28	111	126	142	157	173	188	0.4	6
29	107	123	138	153	169	184	0.5	7
							0.6	9
							0.7	10
							0.8	12
30	0.001104	0.001119	0.001134	0.001150	0.001165	0.001180	0.9	13

ISOTHERMAL COMPRESSIBILITY OF LIQUIDS

Compiled by John C. McGowan

The figures in this table are for isothermal compressibilities in c.g.s. units. The compiler suggests, that provided the pressure is not too high, the reciprocal of the isothermal compressibility varies linearly with the pressure. This suggestion also appears in papers by J. R. Macdonald, Reviews of Modern Physics **38**, 669, (1966) and O. L. Anderson, J. Phys. Chem. Solids **27**, 547, (1966). The papers vary somewhat as to how far this linearity persists.

Liquid	Temp. °C	Isothermal Compressibility sq. cms. per dyne × 10^{11}		Ref.
		At 1 atm.	At 1000 atms.	
Acetic Acid	15	8.75	....	1
	20	9.08	....	1
	30	9.72	....	1
	40	10.37	....	1
	50	11.11	....	1
	60	11.91	....	1
	70	12.77	....	1
	80	13.68	....	1
Acetic acid, ethyl ester	0	9.78	....	1
	10	10.36	....	1
	20	11.32	....	1
	30	12.37	....	1
	40	13.52	....	1
	50	14.78	....	1
	60	16.21	....	1
	70	17.90	....	1
Acetone	20	12.75	....	3
	20	12.62	....	10
	25	12.39	6.02	4
	30	13.80	....	10
	40	15.61	....	3
	40	15.03	....	10
	50	16.45	....	10
Aniline	0	4.08	....	1
	10	4.30	....	1
	20	4.53	....	1
	25	4.67	3.23	5
	40	5.04	....	1
	45	5.22	3.48	5
	50	5.33	....	1
	60	5.64	....	1
	65	5.84	3.76	5
	70	5.97	....	1
	80	6.32	....	1
	85	6.56	4.04	5
	90	6.70	....	1
Anisole	21	6.67	....	2
	30	7.04	....	2
	45	7.72	....	2
	60	8.50	....	2
	81	9.79	....	2
	100	11.25	....	2
	120	13.07	....	2
	140	15.45	....	2
Benzene	0	8.09	....	1
	10	8.73	....	1
	10	8.64	....	3
	20	9.44	....	1
	20	9.37	....	3
	20	9.81	....	10
	25	9.67	5.07	6
	25	9.7	....	15
	30	10.55	....	10
	30	10.18	....	1
	30	10.12	....	3
	35	10.43	5.28	6
	39.5	10.91	....	15
	40	11.35	....	10
	40	11.00	....	1
	40	10.96	....	3
	45	11.32	5.50	6
	50	12.22	....	10
	50	11.89	....	1
	50	11.83	....	3
	50.1	11.91	....	15
	55	12.29	5.73	6
	60	13.12	....	10
	60	12.83	....	1
	60	12.96	....	15
	65	13.39	5.98	6
	70	14.12	....	10
	70	14.13	....	1
	75.9	14.95	....	15
	80	15.44	....	1
Benzene, bromo-	25	6.68	4.09	5
	45	7.52	4.39	5
	65	8.50	4.72	5
	85	9.65	5.06	5
Benzene, chloro-	0	6.61	....	1
	10	7.02	....	1
	20	7.45	....	1
	20	7.38	....	3
	25	7.51	4.39	5
	30	7.89	....	1
	30	7.84	....	3
	40	8.39	....	1
	40	8.32	....	3
	45	8.55	4.73	5
	50	8.92	....	1
	50	8.83	....	3
	60	9.50	....	1
	65	9.76	5.10	5
	70	10.13	....	1
	80	10.79	....	1
	85	11.23	5.49	5
Benzene, nitro-	0	4.41	....	1
	10	4.67	....	1
	20	4.93	....	1
	25	5.03	3.39	5
	30	5.23	....	1
	40	5.49	....	1
	45	5.59	3.64	5
	65	6.24	3.91	5
	85	6.99	4.20	5
n-Butyl alcohol	0	8.10	....	11
Carbon disulphide	0	8.04	....	1
	0	7.95	....	3
	10	8.64	....	1
	10	8.54	....	3
	20	9.38	....	10
	20	9.26	....	1
	20	9.19	....	3
	30	10.07	....	10
	30	9.92	....	1
	30	9.96	....	3
	40	10.65	....	1
	40	10.84	....	10
	40	10.89	....	3
	50	11.48	....	1
	50	11.95	....	3
Carbon tetrachloride	−22.9	7.84	....	12
	−13.1	8.31	....	12
	− 3.1	8.87	....	12
	0	8.98	....	1
	0	8.85	....	3
	6.9	9.59	....	12
	10	9.70	....	1
	10	9.57	....	3
	10	9.45	....	16
	16.9	10.35	....	12
	20	10.46	....	1
	20	10.34	....	3
	20	10.68	....	10
	25	10.67	5.30	7
	25	10.77	....	15
	25	10.58	....	16
	26.9	11.15	....	12
	30	11.58	....	10
	30	11.29	....	1
	30	11.18	....	3
	35	11.95	5.52	7
	37.5	11.99	....	15
	40	12.51	....	10
	40	12.23	....	1
	40	12.16	....	3
	40	11.96	....	16
	45	12.54	5.75	7
	50	13.55	....	10
	50	13.32	....	1
	50	13.26	....	3
	50.3	13.28	....	15
	55	13.63	5.97	7
	55	13.51	....	15
	60	14.64	....	10
	60	14.52	....	1
	62.6	14.84	....	15
	65	14.87	6.22	7
	70	15.84	....	10
	70	15.77	....	1
	75	16.70	....	15

ISOTHERMAL COMPRESSIBILITY OF LIQUIDS
(*Continued*)

Liquid	Temp. °C	Isothermal Compressibility sq. cms. per dyne × 10^{11}		Ref.
		At 1 atm.	At 1000 atms.	
Chloroform	−33.1	7.13	....	12
	−23.1	7.61	....	12
	−13.1	8.12	....	12
	− 3.1	8.66	....	12
	0	8.48	....	1
	0	8.55	....	3
	6.9	9.30	....	12
	10	9.17	....	1
	10	9.19	....	3
	16.9	10.35	....	12
	20	9.98	....	1
	20	9.94	....	3
	20	10.42	....	10
	25	9.74	5.34	4
	26.9	11.15	....	12
	30	10.86	....	1
	30	10.81	....	3
	30	11.28	....	10
	40	11.84	....	1
	40	11.79	....	3
	40	12.22	....	10
	50	12.90	....	1
	50	13.29	....	10
	60	14.06	....	1
Cyclohexane	10	9.88	....	16
	25	11.10	....	16
	25	11.40	....	15
	35	12.02	....	16
	37.6	12.67	....	15
	40	12.56	....	16
	45	13.14	....	16
	50.1	14.15	....	15
	55	14.35	....	16
	60	14.88	....	16
	62.4	15.76	....	15
	75	17.84	....	15
Dodecane	37.8	9.9	5.3	13
	60.0	11.3	5.7	13
	79.4	12.8	6.1	13
	98.9	14.4	6.4	13
	115.0	16.1	6.8	13
	135.0	18.3	7.3	13
Ethane 1,1,2,2-tetrachloro-	25	6.17	3.88	4
Ethyl alcohol	0	9.87	....	1
	0	9.87	....	11
	0	9.63	....	3
	10	10.49	....	1
	10	10.30	....	3
	20	11.19	....	1
	20	10.98	....	3
	30	11.91	....	1
	30	11.80	....	3
	40	12.74	....	1
	40	12.61	....	3
	50	13.70	....	1
	50	13.60	....	3
	60	14.74	....	1
	70	15.93	....	1
	75	16.67	....	1
Ethyl bromide	0	10.76	....	1
	10	11.78	....	1
	20	12.94	....	1
	30	14.23	....	1
	40	15.52	....	1
Ethylene, 1,2-dichloro-(trans)	25	11.19	5.62	4
Ethylene, tetrachloro-	25	7.56	4.45	4
Ethylene, trichloro-	25	8.57	4.99	4
Ethylene chloride	0	6.91	....	1
	10	7.42	....	1
	20	7.97	....	1
	20	8.03	....	10
	25	7.78	4.54	4
	30	8.64	....	10
	30	8.58	....	1
	40	9.34	....	10
	40	9.25	....	1
	50	10.13	....	10
	50	9.99	....	1
	60	10.96	....	10
	60	10.83	....	1
	70	11.86	....	10
	70	11.76	....	1
	80	12.79	....	1

Liquid	Temp. °C	Isothermal Compressibility sq. cms. per dyne × 10^{11}		Ref.
		At 1 atm.	At 1000 atms.	
Ethyl ether	0	15.10	....	1
	0	15.07	....	3
	10	16.81	....	1
	10	16.52	....	3
	20	18.65	....	1
	20	18.44	....	3
	30	20.90	....	1
	30	20.80	....	3
	35	24.15	....	1
Ethyl iodide	0	8.45	....	1
	10	9.12	....	1
	20	9.82	....	1
	30	10.59	....	1
	40	11.44	....	1
	50	12.38	....	1
	60	13.40	....	1
	70	14.49	....	1
Glycol	25	3.72	2.73	7
	45	4.00	2.89	7
	65	4.32	3.05	7
	85	4.70	3.24	7
	105	5.14	3.44	7
n-Heptane	0	11.80	5.64	8
	25	14.24	6.18	8
	40	15.96	6.50	8
	60	18.67	6.94	8
1-Heptanol	0	7.05	....	11
n-Hexane	0	13.04	5.92	8
	25	16.06	6.51	8
	40	18.31	6.89	8
	60	21.93	8.87	8
1-Hexanol	0	7.47	....	11
Mercury	0	0.40	....	14
	20	0.40	0.39	14
	40	0.41	....	14
	80	0.42	....	14
	120	0.44	....	14
	160	0.46	....	14
Methanol	0	10.62	....	1
	0	10.68	....	11
	0	10.78	....	3
	10	11.34	....	1
	10	11.45	....	3
	20	12.11	....	1
	20	12.18	....	3
	30	12.93	....	1
	30	12.98	....	3
	40	13.85	....	1
	40	13.82	....	3
	50	14.76	....	3
Methylene bromide	−33.1	4.95	....	12
	−23.1	5.11	....	12
	−13.1	5.50	....	12
	− 3.1	5.81	....	12
	6.9	6.13	....	12
	16.9	6.47	....	12
	26.9	6.85	....	12
Methylene chloride	25	9.74	5.31	4
Methyl iodide	−33.1	6.86	....	12
	−23.1	7.41	....	12
	−13.1	7.97	....	12
	− 3.1	8.55	....	12
	6.9	9.13	....	12
	16.9	9.71	....	12
	26.9	10.33	....	12
n-Octadecane	60	9.4	5.1	13
	79.4	10.4	5.5	13
	98.9	11.6	5.8	13
	115.0	12.8	6.1	13
	135	14.4	6.4	13
n-Octane	0	9.99	5.11	8
	25	11.98	5.69	8
	40	13.36	6.00	8
	60	15.53	6.40	8
1-Octanol	0	6.82	....	11
n-Pentadecane	37.8	9.1	....	13
	60	10.2	5.2	13
	79.4	11.7	5.5	13

ISOTHERMAL COMPRESSIBILITY OF LIQUIDS
(*Continued*)

Liquid	Temp. °C	Isothermal Compressibility sq. cms. per dyne × 10^{11}		Ref.
		At 1 atm.	At 1000 atms.	
n-Pentadecane.......	98.9	13.2	5.8	13
(cont.)...........	115	14.7	6.1	13
	135	16.8	6.4	13
Pentanol...........	0	7.71	....	11
Phenol.............	46	5.61	....	2
	60	6.05	....	2
	80	6.78	....	2
	110	8.12	....	2
	125	8.88	....	2
	150	10.30	....	2
	175	12.35	....	2
n-Propyl alcohol.....	0	8.43	....	11
Toluene.............	−59.3	5.27	....	9
	−41.1	5.93	....	9
	−19.2	6.87	....	9
	0	7.83	....	1
	0	7.97	....	3
	0	7.84	....	9
	10	8.38	....	1
	10	8.44	....	3
	20	8.96	....	1
	20	8.94	....	3
	30	9.60	....	1
	30	9.49	....	3
	40	10.33	....	1
	40	10.14	....	3
	50	11.13	....	1
	50	10.90	....	3
	60	11.99	....	1
	70	12.95	....	1
Water..............	0	5.01	....	1
	0	5.04	....	9
	10	4.78	....	1
	20	4.58	....	1
	25	4.57	3.48	7
	30	4.46	....	1
	34.8	4.41	....	9
	35	4.48	3.42	7
	40	4.41	....	1
	45	4.44	3.40	7
	50	4.40	....	1
	55	4.44	3.40	7
	60	4.43	....	1
	65	4.48	3.42	7
	70	4.49	....	1
	75	4.55	3.47	7
	80	4.57	....	1
	85	4.65	3.53	7
	90	4.68	....	1
	100	4.80	....	1
m-Xylene...........	0	7.44	....	1
	10	7.94	....	1
	20	8.46	....	1
	30	9.03	....	1
	40	9.63	....	1
	50	10.25	....	1
	60	11.01	....	1
	70	11.77	....	1
	80	12.56	....	1

References

(1) D. Tyrer, J. Chem. Soc. **105**, 2534, 1914.
(2) A. E. Lutskii and V. N. Solon ko, Russian Journal of Physical Chemistry **38**, 602, 1964.
(3) E. B. Fryer, J. C. Hubbard and D. H. Andrews, J. Amer. Chem. Soc. **51**, 759, 1929.
(4) D. M. Newitt and K. E. Weale, J. Chem. Soc., 3092, 1951.
(5) R. E. Gibson and O. H. Loeffler, J. Amer. Chem. Soc. **61**, 2515, 1939.
(6) R. E. Gibson and J. F. Kincaid, J. Amer. Chem. Soc. **60**, 511, 1938.
(7) R. E. Gibson and O. H. Loeffler, J. Amer. Chem. Soc. **63**, 898, 1941.
(8) H. E. Eduljee, D. M. Newitt and K. E. Weale, J. Chem. Soc. 3086, 1951.
(9) J. G. Marshall, L. A. K. Staveley and K. R. Hart, Trans. Farad. Soc. **52**, 23, 1956.
(10) L. A. K. Staveley, W. I. Tupman and K. R. Hart, Trans. Farad. Soc. **51**, 323, 1955.
(11) W. P. McKinney, G. F. Skinner and L. A. K. Staveley, J. Chem. Soc. 2415, 1959.
(12) D. Harrison and E. A. Moelwyn-Hughes, Proceedings of the Royal Society, **239A**, 230, 1957.
(13) W. G. Cutler, R. H. McMickle, W. Webb and R. W. Schiessler, Journal Chemical Phys. **29**, 727, 1958.
(14) E. A. Moelwyn-Hughes, Journal Physical and Colloid Chemistry, **55**, 1246, 1951.
(15) G. A. Holder and E. Whalley, Trans. Faraday Soc. 58, 2095, 1962.
(16) M. Diaz Pena and M. L. McGlashan. Trans. Faraday Soc. 57, 1511, 1961.

COEFFICIENT OF FRICTION

Compiled by Harold Minshall

The coefficient of friction between two surfaces is the ratio of the force required to move one over the other to the total force pressing the two together.

If F is the force required to move one surface over another and W, the force pressing the surfaces together, the coefficient of friction,

$$\mu = \frac{F}{W}$$

Materials	Condition	Temperature °C	μ (Static)
A. STATIC FRICTION			
Non Metals			
Glass on glass	clean	—	0.9–1.0
" " "	lubricated with paraffin oil	—	0.5–0.6
" " "	" " liquid fatty acids	—	0.3–0.6
" " "	" " solid hydrocarbons, alcohols or fatty acids	—	0.1
" " metal	clean	—	0.5–0.7
" " "	lubricated	—	0.2–0.3
Diamond on diamond	clean	—	0.1
" " "	lubricated	—	0.05–0.1
" " metal	clean	—	0.1–0.15
" " "	lubricated	—	0.1
Sapphire on sapphire	clean or lubricated	—	0.2
" " steel	" " "	—	0.15
Hard carbon on carbon	clean	—	0.16
" " " "	lubricated	—	0.12–0.14
Graphite on graphite	clean or lubricated	—	0.1
" " "	outgassed	—	0.5–0.8
" " steel	clean or lubricated	—	0.1
Mica on mica	freshly cleaved	—	1.0
" " "	contaminated	—	0.2–0.4
Crystals of $NaNO_3$, KNO_3, NH_4Cl on self	clean	—	0.5
" " " " " " "	lubricated with long chain polar compounds	—	0.12
Tungsten carbide on tungsten carbide	clean	room	0.17
Tungsten carbide on tungsten carbide	outgassed	room	0.58
" " " " "	clean	820	0.35
" " " " "	"	970	0.40
" " " " "	"	1010	0.45
" " " " "	"	1160	0.5
" " " " "	"	1220	0.7
" " " " "	"	1440	1.2
" " " " "	"	1600	1.8
" " " graphite	outgassed	room	0.62
" " " "	clean	"	0.15
" " " "	"	800	0.32
" " " "	"	910	0.30
" " " "	"	1000	0.25
" " " "	"	1120	0.29
" " " "	"	1220	0.26
" " " "	"	1300	0.25
" " " "	"	1410	0.25
" " " "	"	1800	0.24
" " " "	"	2030	0.25
" " " steel	"	—	0.4–0.6
" " " "	lubricated	—	0.1–0.2
Polymethyl methacrylate on self	clean	—	0.8
" " " steel	"	—	0.4–0.5
Polystyrene on self	"	—	0.5
" " steel	"	—	0.3–0.35
Polyethylene on self	"	—	0.2
" " steel	"	—	0.2
Polytetrafluoroethylene on self	"	—	0.04
" " steel	"	—	0.04
Nylon on nylon[1]	"	—	0.15–0.25
Silk on silk	commercially clean	—	0.2–0.3
Cotton on cotton (thread)	" "	—	0.3
" " " (from cotton wool)	" "	—	0.6
Rubber on solids	" "	—	1–4
Wood on wood	" " and dry	—	0.25–0.5
" " "	" " " wet	—	0.2
" " metals	" " " dry	—	0.2–0.6
" " "	" " " wet	—	0.2
" " brick	" "	—	0.6
" " leather	" "	—	0.3–0.4
Leather on metal	" "	—	0.6
" " "	" " and wet	—	0.4
" " "	greasy	—	0.2
Brake material on cast iron	commercially clean	—	0.4
" " " " "	" " and wet	—	0.2
" " " " "	lubricated with mineral oil	—	0.1
Wool fiber on horn	clean (against scales)	—	0.8–1.0
" " " "	" (with scales)	—	0.4–0.6
" " " "	greasy (against scales)	—	0.5–0.8
" " " "	" (with scales)	—	0.3–0.4
Metals			
Steel on steel	clean	20	0.58
" " "	vegetable oil lubricant (a) castor oil	20	0.095
		100	0.105
" " "	(b) rape	20	0.105
		100	0.105
" " "	(c) olive	20	0.105
		100	0.105
" " "	(d) coconut	20	0.08
		100	0.08

[1] Registered trade name.

COEFFICIENT OF FRICTION (Continued)

Materials	Condition	Temperature °C	μ (Static)
A. STATIC FRICTION (Cont.)			
Metals (Cont.)			
Steel on steel	Animal oil lubricant		
" " "	(a) sperm	20	0.10
		100	0.10
" " "	(b) pale whale	20	0.095
		100	0.095
" " "	(c) neatsfoot	20	0.095
		100	0.095
" " "	(d) lard	20	0.085
		100	0.085
" " "	Mineral oil lubricant		
	(a) light machine	20	0.16
		100	0.19
" " "	(b) thick gear	20	0.125
		100	0.15
" " "	(c) solvent refined	20	0.15
		100	0.20
" " "	(d) heavy motor	20	0.195
		100	0.205
" " "	(e) extreme pressure	20	0.09–0.1
		100	0.09–0.1
" " "	(f) graphited oil	20	0.13
		100	0.15
" " "	(g) B.P. Paraffin	20	0.18
		100	0.22
" " "	lubricated with trichloroethylene	20	0.33
" " "	" " benzene	20	0.48
" " "	" " glycerol	20	0.2
" " "	" " ethyl alcohol	20	0.43
" " "	" " butyl alcohol	room	0.3
" " "	" " octyl	"	0.23
" " "	" " decyl	"	0.16
" " "	" " cetyl	"	0.10
" " "	lubricated with nonane	room	0.26
" " "	" " decane	"	0.23
" " "	" " acetic acid	"	0.5
" " "	" " proprionic acid	"	0.4
" " "	" " valeric acid	"	0.17
" " "	" " caproic acid	"	0.12
" " "	" " pelargonic acid	"	0.11
" " "	" " capric acid	"	0.11
" " "	" " lauric acid	"	0.11
" " "	" " myristic acid	"	0.11
" " "	" " oleic acid	20–100	0.08
" " "	" " palmitic acid	room	0.11
" " "	" " stearic acid	"	0.10
" " hard steel	" " rape oil	—	0.14
" " " "	" " castor oil	—	0.12
" " " "	" " mineral oil	—	0.16
" " " "	" " long chain fatty acid	—	0.09
" " cast iron	" " rape oil	—	0.11
" " " "	" " castor oil	—	0.15
" " " "	" " mineral oil	—	0.21
" " " "	clean	—	0.4
" " gun metal	lubricated with rape oil	—	0.15
" " " "	" " castor oil	—	0.16
" " " "	" " mineral oil	—	0.21
" " bronze	" " rape oil	—	0.12
" " "	" " castor oil	—	0.12
" " "	" " mineral oil	—	0.16
" " lead	" " " "	—	0.5
" " "	" " long chain fatty acid	—	0.22
" " " base white metal	" " mineral oil	—	0.1
" " " " " "	" " long chain fatty acid	—	0.08
" " "	clean	—	0.55
" " tin	lubricated with mineral oil	—	0.6
" " "	" " long chain fatty acid	—	0.21
" " white metal, tin base	" " mineral oil	—	0.1
" " " " " "	" " long chain fatty acid	—	0.07
" " " " " "	clean	—	0.8
" " sintered bronze	lubricated with mineral oil	—	0.13
" " brass	" " " "	—	0.19
" " "	" " castor oil	—	0.11
" " "	" " long chain fatty acid	—	0.13
" " "	clean	—	0.35
" " copper-lead alloy	"	—	0.22
" " Wood's alloy	"	—	0.7
" " phosphor bronze	"	—	0.35
" " aluminum bronze	"	—	0.45
" " constantan	"	—	0.4
[2] " " indium film deposited on steel	4 kg load, clean	—	0.08
[2] " " " " " " "	8 kg " , "	—	0.04
[2] " " " " " " silver	4 kg " , "	—	0.1
[2] " " " " " " "	8 kg " , "	—	0.07
[2] " " lead film deposited on copper	4 kg " , "	—	0.18
[2] " " " " " " "	8 kg " , "	—	0.12
[2] " " copper film deposited on steel	4 kg " , "	—	0.3
[2] " " " " " " "	8 kg " , "	—	0.2
[3] Al on Al	in air or O_2	—	1.9
" " "	" H_2O vapor	—	1.1
[3] Cu on Cu	" H_2 or N_2	—	4.0
	" air or O_2	—	1.6

[2] Hemispherical steel slider having 0.6 cm. diameter. The thin, 10^{-3} to 10^{-4} cm., thin metallic films were deposited on various substrates as indicated. Amonton's Law is not obeyed in this case.

[3] The metals which were spectrosopically pure were outgassed in a vacuum prior to other gases being admitted. When clean and in vacuum there is gross seizure.

COEFFICIENT OF FRICTION (Continued)

Materials	Condition	Temperature °C	μ (Static)
A. STATIC FRICTION (Cont.)			
Metals (Cont.)			
[3]Au on Au	in H_2 or N_2	—	4.0
" " "	" air or O_2	—	2.8
" " "	" H_2O vapor	—	2.5
[3]Fe on Fe	" air or O_2	—	1.2
" " "	" H_2O vapor	—	1.2
[3]Mo on Mo	" air or O_2	—	0.8
" " "	" H_2O vapor	—	0.8
[3]Ni on Ni	" H_2 or N_2	—	5.0
" " "	" air or O_2	—	3.0
" " "	" H_2O vapor	—	1.6
[3]Pt on Pt	" air or O_2	—	3.0
" " "	" H_2O vapor	—	3.0
[3]Ag on Ag	" air or O_2	—	1.5
" " "	" H_2O vapor	—	1.5
Various Materials on Snow and Ice			
Ice on ice	clean	0	0.05–0.15
" " "	"	−12	0.3
" " "	"	−71	0.5
" " "	"	−82	0.5
		−110	0.5
Polymethylmethylacrylate	on wet snow	0	0.5
"	" dry "	0	0.3
"	" " "	−10	0.34
		−32	0.4
Polyester of teraphthalic acid and ethylene glycol	" wet "	0	0.5
Polyester of teraphthalic acid and ethylene glycol	" dry "	0	0.35
Polyester of teraphthalic acid and ethylene glycol	" " "	−10	0.38
[1]Nylon	on wet snow	0	0.4
"	" dry "	0	0.3
		−10	0.3
Polytetrafluorethylene	" wet "	0	0.05
"	" dry "	0	0.02
"	" " "	−10	0.08
		−32	0.1
Paraffin wax	" wet "	0	0.06
" "	" dry "	0	0.06
" "	" " "	−10	0.35
		−32	0.4
Swiss wax	" wet "	0	0.05
" "	" dry "	0	0.03
" "	" " "	−10	0.2
		−32	0.2
Ski wax	" wet "	0	0.1
" "	" dry "	0	0.04
" "	" " "	−10	0.2
		−32	0.2
" laquer	" wet "	0	0.2
" "	" dry "	0	0.1
" "	" " "	−10	0 4
		−32	0.4
Aluminum	" wet "	0	0.4
"	" dry "	0	0.35
"	" " "	−10	0.38

Materials	Condition	Temperature °C	μ (Kinetic)
B. KINETIC FRICTION Various Materials			
Unwaxed hickory	4m/sec on dry snow	−3	0.08
Waxed "	0.1m/sec " wet "	0	0.14
" "	0.1m/sec " dry "	0	0.04
" "	0.1m/sec " " "	−3	0.09
	4m/sec " " "	−3	0.03
Waxed hickory	0.1m/sec on dry snow	−10	0.18
" "	0.1m/sec " " "	−40	0.4
Ice on ice	4m/sec, clean	0	0.02
" " "	" , "	−10	0.035
" " "	" , "	−20	0.050
" " "	" , "	−40	0.075
" " "	" , "	−60	0.085
		−80	0.09
Ebonite	4m/sec on ice	0	0.02
"	" " "	−10	0.05
"	" " "	−20	0.065
"	" " "	−40	0.085
"	" " "	−60	0.10
		−80	0.11
Brass	4m/sec on ice	0	0.02
"	" " "	−10	0.075
"	" " "	−20	0.085
"	" " "	−40	0.115
"	" " "	−60	0.14
		−80	0.15
Natural rubber, vulcanized	100m/min on ground glass, clean	—	1.07
" " , "	100m/min " " ", wetted with water	—	0.94
" " , "	100m/min on concrete, clean	—	1 02
" " , "	100m/min " ", wetted with water	—	0.97
" " , "	100m/min on bitumen, clean	—	1.07
" " , "	100m/min " , wetted with water	—	0.95
" " , "	100m/min on rubber flooring or rubber tread vulcanisate, clean	—	1.16
" " , "	100m/min on bitumen containing rubber powder, clean	—	1.15 (varies with quantity of powder)
" " , "	100m/min on bitumen containing rubber powder, wetted with water	—	1.03

[1] Registered trade name. [3] The metals which were spectroscopically pure were outgassed in a vacuum prior to other gases being admitted. When clean and in vacuum there is gross seizure.

LOW MELTING POINT ALLOYS

Melting point °C	Name	Composition, wt %				
−48	Binary Eutectic	Cs	K			
		77.0	23.0			
−40	Binary Eutectic	Cs	Rb			
		87.0	13.0			
−30	Binary Eutectic	Cs	Na			
		95.0	5.0			
−11	Binary Eutectic	K	Na			
		78.0	22.0			
−8	Binary Eutectic	Rb	Na			
		92.0	8.0			
10.7	Ternary Eutectic	Ga	In	Sn		
		62.5	21.5	16.0		
10.8	Ternary Eutectic	Ga	In	Sn		
		69.8	17.6	12.5		
17	Ternary Eutectic	Ga	Sn	Zn		
		82.0	12.0	6.0		
33	Binary Eutectic	Rb	K			
		68.0	32.0			
46.5	Quinternary Eutectic	Sn	Bi	Pb	In	Cd
		10.65	40.63	22.11	18.1	8.2
47	Quinternary Eutectic	Bi	Pb	Sn	Cd	In
		44.7	22.6	8.3	5.3	19.1
58.2	Quaternary Eutectic	Bi	Pb	Sn	In	
		49.5	17.6	11.6	21.3	
60.5		In	Bi	Sn		
		51.0	32.5	16.5		
70	Wood's Metal	Bi	Pb	Sn	Cd	
		50.0	25.0	12.5	12.5	
70	Lipowitz's Metal	Bi	Pb	Sn	Cd	
		50.0	26.7	13.3	10.0	
70	Binary Eutectic	In	Bi			
		67.0	33.0			
91.5	Ternary Eutectic	Bi	Pb	Cd		
		51.6	40.2	8.2		
95	Ternary Eutectic	Bi	Pb	Sn		
		52.5	32.0	15.5		
97	Newton's Metal	Bi	Sn	Pb		
		50.0	18.8	31.2		
98	D'Arcet's Metal	Bi	Sn	Pb		
		50.0	25.0	25.0		
100	Onion's or Lichtenberg's Metal	Bi	Sn	Pb		
		50.0	20.0	30.0		
102.5	Ternary Eutectic	Bi	Sn	Cd		
		54.0	26.0	20.0		
109	Rose's Metal	Bi	Pb	Sn		
		50.0	28.0	22.0		
117	Binary Eutectic	In	Sn			
		52.0	48.0			
120	Binary Eutectic	In	Cd			
		75.0	25.0			
123	Malotte's Metal	Bi	Sn	Pb		
		46.1	34.2	19.7		
124	Binary Eutectic	Bi	Pb			
		55.5	44.5			
130	Ternary Eutectic	Bi	Sn	Zn		
		56.0	40.0	4.0		
140	Binary Eutectic	Bi	Sn			
		58.0	42.0			
140	Binary Eutectic	Bi	Cd			
		60.0	40.0			
183	Eutectic solder	Sn	Pb			
		63.0	37.0			
185	Binary Eutectic	Tl	Bi			
		52.0	48.0			
192	Soft solder	Sn	Pb			
		70.0	30.0			
198	Binary Eutectic	Sn	Zn			
		91.0	9.0			
199	Tin foil	Sn	Zn			
		92.0	8.0			
199	White Metal	Sn	Sb			
		92.0	8.0			
221	Binary Eutectic	Sn	Ag			
		96.5	3.5			
226	Matrix	Bi	Pb	Sn	Sb	
		48.0	28.5	14.5	9.0	
227	Binary Eutectic	Sn	Cu			
		99.25	0.75			
240	Antimonial Tin solder	Sn	Sb			
		95.0	5.0			
245	Tin-silver solder	Sn	Ag			
		95.0	5.0			

MOHS HARDNESS SCALE

Hardness number	Original scale	Modified scale
1	Talc	Talc
2	Gypsum	Gypsum
3	Calcite	Calcite
4	Fluorite	Fluorite
5	Apatite	Apatite
6	Orthoclase	Orthoclase
7	Quartz	Vitreous silica
8	Topaz	Quartz or Stellite
9	Corundum	Topaz
10	Diamond	Garnet
11	....	Fused Zirconia
12	....	Fused Alumina
13	....	Silicon Carbide
14	....	Boron Carbide
15	....	Diamond

HARDNESS OF MATERIALS

Material	Hardness	Material	Hardness
Agate	6–7	Indium	1.2
Alabaster	1.7	Iridium	6–6.5
Alum	2–2.5	Iridosmium	7
Aluminum	2–2.9	Iron	4–5
Alundum	9+	Kaolinite	2.0–2.5
Amber	2–2.5	Lead	1.5
Andalusite	7.5	Lithium	0.6
Anthracite	2.2	Loess (0°)	0.3
Antimony	3.0–3.3	Magnesium	2.0
Apatite	5	Magnetite	6
Aragonite	3.5	Manganese	5.0
Arsenic	3.5	Marble	3–4
Asbestos	5	Meerschaum	2–3
Asphalt	1–2	Mica	2.8
Augite	6	Opal	4–6
Barite	3.3	Orthoclase	6
Bell-metal	4	Osmium	7.0
Beryl	7.8	Palladium	4.8
Bismuth	2.5	Phosphorus	0.5
Boric acid	3	Phosphorbronze	4
Boron	9.5	Platinum	4.3
Brass	3–4	Plat-iridium	6.5
Cadmium	2.0	Potassium	0.5
Calamine	5	Pumice	6
Calcite	3	Pyrite	6.3
Calcium	1.5	Quartz	7
Carbon	10.0	Rock salt (halite)	2
Carborundum	9–10	Ross' metal	2.5–3.0
Cesium	0.2	Rubidium	0.3
Chromium	9.0	Ruthenium	6.5
Copper	2.5–3	Selenium	2.0
Corundum	9	Serpentine	3–4
Diamond	10	Silicon	7.0
Diatomaceous earth	1–1.5	Silver	2.5–7
Dolomite	3.5–4	Silver chloride	1.3
Emery	7–9	Sodium	0.4
Feldspar	6	Steel	5–8.5
Flint	7	Stibnite	2
Fluorite	4	Strontium	1.8
Galena	2.5	Sulfur	1.5–2.5
Gallium	1.5	Talc	1
Garnet	6.5–7	Tellurium	2.3
Glass	4.5–6.5	Tin	1.5–1.8
Gold	2.5–3	Topaz	8
Graphite	0.5–1	Tourmaline	7.3
Gypsum	1.6–2	Wax (0°)	0.2
Hematite	6	Wood's metal	3
Hornblende	5.5	Zinc	2.5

COMPARISON OF HARDNESS VALUES OF VARIOUS MATERIALS ON MOHS AND KNOOP SCALES*

Compiled by Laurence S. Foster

Substance	Formula	Mohs value	Knoop value
Talc	$3MgO \cdot 4SiO_2 \cdot H_2O$	1	
Gypsum	$CaSO_4 \cdot 2H_2O$	2	32
Cadmium	Cd	..	37
Silver	Ag	..	60
Zinc	Zn	..	119
Calcite	$CaCO_3$	3	135
Fluorite	CaF_2	4	163
Copper	Cu	..	163
Magnesia	MgO	..	370
Apatite	$CaF_2 \cdot 3Ca_3(PO_4)_2$	5	430
Nickel	Ni	..	557
Glass (soda lime)	....	..	530
Feldspar (orthoclase)	$K_2O \cdot Al_2O_3 \cdot 6SiO_2$	6	560
Quartz	SiO_2	7	820
Chromium	Cr	..	935
Zirconia	ZrO_2	..	1160
Beryllia	BeO	..	1250
Topaz	$(AlF)_2SiO_4$	8	1340
Garnet	$Al_2O_3 \cdot 3FeO \cdot 3SiO_2$	..	1360
Tungsten carbide alloy	WC, Co	..	1400–1800
Zirconium boride	ZrB_2	..	1550
Titanium nitride	TiN	9	1800
Tungsten carbide	WC	..	1880
Tantalum carbide	TaC	..	2000
Zirconium carbide	ZrC	..	2100
Alumina	Al_2O_3	..	2100
Beryllium carbide	Be_2C	..	2410
Titanium carbide	TiC	..	2470
Silicon carbide	SiC	..	2480
Aluminum boride	AlB	..	2500
Boron carbide	B_4C	..	2750
Diamond	C	10	7000

* Acknowledgment is made to N. W. Thibault, Norton Company, Worcester, Massachusetts, for many of Knoop hardness values. *Cf.* R. F. Geller, "A Study of Ceramics for Nuclear Reactors," Nucleonics, Vol. 7, No. 4, Table 1, pp. 8–9 (Oct. 1950). V. E. Lysaght, *Indentation Hardness Testing,* Reinhold 1949.

SURFACE TENSION OF LIQUID ELEMENTS

Element	Gas	Temperature °C	Dynes/cm	Method	References
Aluminum	Ar	660 (m.p.)	860	Bubble pressure	5
	Ar	660	914	"	Zhivov in VKS (1937)
	vac.	660	865	Sessile drop	8
	Ar	660	825	Bubble pressure	59
	Ar	700	850	"	4
		700	900		Pelzel in VKS (1947)
	Ar	706	494	Bubble pressure	Zhivov in VKS (1937)
		712	502		Klyachko in VKS (1937)
	Ar	759	489	Bubble pressure	Zhivov in VKS (1937)
	Ar	794	483	"	"
	vac.	800	845	Sessile drop	6
		800	865		Pelzel in VKS (1947)
	Ar	816	480	Bubble pressure	Zhivov in VKS (1937)
	Ar	906	466	"	"
	Ar	935	463	"	"
	vac.	950	840	"	7
Antimony	H_2, N_2	635	383	"	16
		635	383		Greenaway in VKS
	vac.	640	367.9	Drop weight	15
		640	349		Bircumshaw in VKS
		640	367.9		Matuyama in VKS
		640	384		Greenaway in VKS
		650	384		"
		675	350		Bircumshaw in VKS
		675	384		Greenaway in VKS
		700	349		Bircumshaw in VKS
		700	382		Greenaway in VKS
		725	383		"
		750	368		Draht, Sauerwald in VKS
		750	383		Greenaway in VKS
	vac.	762	364.9	Drop weight	15
		762	364.9	"	Matuyama in VKS
	H_2, N_2	800	380	Bubble pressure	16
		800	367		Draht, Sauerwald in VKS
		800	347		Bircumshaw in VKS
		800	380		Greenaway in VKS
	N_2	800	359	Bubble pressure	58
		900	361		Draht, Sauerwald in VKS
		974	340		Bircumshaw in VKS
		1000	355		Draht, Sauerwald in VKS
	N_2	1000	351	Bubble pressure	58
	N_2	1100	345	"	58
		1100	348		Draht, Sauerwald in VKS
Barium	Ar	720	224	Bubble pressure	10
Beryllium	vac.	1500	1100	Sessile drop	11
Bismuth	vac.	271	376	Drop pressure	12
	"	271	378	"	13
	"	271	370	"	14
	"	271	393.4		Matuyama in VKS
		280	392	Ring removal	Klyachko, Kunin in VKS
	vac.	293	389.5		Matuyama in VKS
		300	376		MH
		300	388	Plate removal	Bircumshaw in VKS
		320	375		Hogness in VKS (1921)
		340	384	Ring removal	Klyatchko, Kunin in VKS
		350	381.6		Matuyama in VKS
		365	371		Hogness in VKS (1921)
		400	380	Plate removal	Bircumshaw in VKS

Element	Gas	Temperature °C	Dynes/cm	Method	References
Bismuth (cont.)	vac.	400	373.8		Matuyama in VKS
		426	367		Hogness in VKS (1921)
	vac.	450	358	Drop pressure	14
		450	365.9		Matuyama in VKS
		472	365		Hogness in VJS (1921)
		500	363		MH
	vac.	500	353.2		Matuyama in VKS
		550	351.2		"
		600	367	Plate removal	Bircumshaw in VKS
	vac.	600	343.4		Matuyama in VKS
	H_2	800	343	Bubble pressure	2
		802	353	Plate removal	Bircumshaw in VKS
		902	343	"	"
	H_2	1000	328	Bubble pressure	2
Calcium	Ar	850	337	"	10
Cadmium	vac.	320	666.1	Drop weight	15
		320	693		Qunincke in VKS
		320	666.1		Matuyama in VKS
	H_2, N_2	330	564	Bubble pressure	16
	H_2	330	570	Bubble pressure	17
	H_2	330	570		Bircumshaw in VKS
	H_2	339	606		Hogness in VKS (1921)
		340	577		Greenaway in VKS
	H_2	350	583		Bircumshaw in VKS
		350	584		Greenaway in VKS
		355	637.7		Matuyama in VKS
		360	596		Greenaway in VKS
		366	616		Hogness in VKS (1921)
		370	608		Greenaway in VKS
	H_2	374	592		Bircumshaw in VKS
		378	617		Hogness in VKS (1921)
		380	606		Greenaway in VKS
		390	604		"
	H_2	400	597	Bubble pressure (non linear)	17
	H_2	400	597		Bircumshaw in VKS
		400	609		Greenaway in VKS
		410	600		Matuyama in VKS
		413	640.6		"
		420	773		"
		420	622		Klyatchko in VKS
		420	598		Greenaway in VKS
		421	622		Hogness in VKS (1921)
		430	608		Greenaway in VKS
		440	606		"
		440	816		Pelzel in VKS
		444	628.8		Matuyama in VKS
		448	621		Hogness in VKS (1921)
	H_2, N_2	450	611	Bubble pressure (non linear)	16
		450	611		Greenaway in VKS
		460	608		"
		466	612.1		Matuyama in VKS
		470	612		Greenaway in VKS
		480	607		"
		490	599		"
	H_2	500	592		Bircumshaw in VKS
		500	600		Greenaway in VKS

SURFACE TENSION OF LIQUID ELEMENTS (Continued)

Element	Gas	Temperature °C	Dynes/cm	Method	References
Cadmium (cont.)	H_2, N_2	500	600	Bubble pressure	16
		508	619		Hogness in VKS
		517	617.1		Matuyama in VKS
		544	619		Hogness in VKS (1921)
	vac.	589	590.6	Drop weight	15
		589	590.6		Matuyama in VKS
		600	585	Bubble pressure (non linear)	17
	H_2	600	585		Bircumshaw in VKS
Cesium		28.6	60		23
	Ar	62	68.4	Pendant drop	60
Chromium	vac.	1950	1590	Sessile drop	22
Cobalt	H_2, He	1495	1520	Bubble pressure	20
	vac.	1495	1880	Pendant drop	21
	vac.	1500	1870	Sessile drop	19
	He	1500	1560	" on MgO	7
	vac.	1520	1800	" " Al_2O_3	7
	He	1520	1630	Sessile drop	7
	He	1520	1640	" on BeO	7
	H_2	1520	1780	" " Al_2O_3	7
	He	1520	1620	Bubble pressure BeO	7
	H_2	1520	1590	" "	7
	Ar	1550	1836	Sessile drop	18
Copper	vac.	1083	1270	"	19
	vac.	1083	1300	Pendant drop	21
	H_2	1100	1301	Bubble pressure	2
	H_2, He	1100	1350	"	Er
	vac.	1100	1280	Sessile drop	61
	N_2	1100	1341	Bubble pressure	58
	vac.	1120	1285	Sessile drop	24
		1131	1103		Sauerwald in VKS (1927)
		1150	1145		"
	N_2	1200	1335	Bubble pressure	58
		1200	1154		Sauerwald in VKS (1927)
		1215	1166		"
	H_2	1255	1287	Bubble pressure	2
Gallium	H_2, CO_2	30.5	735	Sessile drop	26
		30.1	704		Er
Germanium	He	959	600	Pendant drop	27
		961	621.4	Drop weight	28
Gold	vac.	1063	754	Pendant drop	3
	H_2, Ar	1120	1128	Bubble pressure	9
		1120	1128		Sauerwald in VKS (1927)
		1150	1125		"
		1200	1120		Sauerwald in VKS (1927)
		1250	1115		"
		1300	1110		"
		1310	1109		"
	H_2, Ar	1310	1109		9
Hafnium	vac.	2230 (m.p.)	1630	Pendant drop	21
	vac.	2230	1460	"	29
Indium	H_2	156	559	Capillary method	34
	Ar, He	156	569.3	"	35
		170	554.7	Bubble pressure	Er
		200	552.1	"	Er
		250	548.4	"	Er
		300	544.2	"	Er
		350	540.3	"	Er

Element	Gas	Temperature °C	Dynes/cm	Method	References
Indium (cont.)		400	535.3	Bubble pressure	Er
		450	530.4	"	Er
		500	527.2	"	Er
	H_2	600	515	Capillary method	34
Iridium	vac.	2454 (m.p.)	2250	Pendant drop	21
Iron	H_2, He	1535 (m.p.)	1650	Bubble pressure	20
	vac.	1535	1700	Sessile drop	19
	vac.	1535	1880	Pendant drop	21
	vac.	1535	1510	Drop weight	63
	He	1550	1860	Sessile drop	62
	Ar	1550	1788	"	18
	He, vac.	1550	1865	"	25
	He	1650	1610	" on BeO	7
	He	1650	1430	Bubble pressure BeO	7
Lead		327 (m.p.)	469.9		Matuyama in VKS
	H_2	327	480	Capillary method	34
	vac.	330	451	Drop pressure	45
		330	448		Quincke in VKS
		330	451	Vapor pressure	7
		340	448		Greenaway in VKS
		350	441		Bircumshaw in VKS
		350	453		"
		350	449		Greenaway in VKS
		360	452	Plate removal	Klyachko in VKS (1937)
		360	446		Greenaway in VKS
		366	442	Plate removal	Hogness in VKS (1921)
		370	465.9		Matuyama in VKS
		370	444		Greenaway in VKS
		380	443		"
		390	442		"
		400	438		Bircumshaw in VKS
		400	446		"
		400	443		Greenaway in VKS
		410	442		"
		420	442		"
		425	440	Vapor pressure	Konstantinov in VKS
		425	440	Drop pressure	45
		425	447.3		Matuyama in VKS
		430	441		Greenaway in VKS
		440	439		Greenaway in VKS
		444	433	Plate removal	Hogness in VKS (1921)
		472	443.4		Matuyama in VKS
		480	431		MH
		507	440.5		Matuyama in VKS
		522	429	Plate removal	Hogness in VJS (1921)
		525	424		MH
		550	423		MH
		557	430.7		Matuyama in VKS
		600	436		Bircumshaw in VKS
		600	426		"
		610	421.8		Matuyama in VKS
		657	415.9		"
		750	423	Bubble pressure	Draht in VKS (1926)
		800	409		Bircumshaw in VKS
		800	425		"
		800	418	Bubble pressure	Draht in VKS (1926)
		900	407	"	"
		982	415		Bircumshaw in VKS

SURFACE TENSION OF LIQUID ELEMENTS (Continued)

Element	Gas	Temperature °C	Dynes/cm	Method	References
Lead (cont.)	H_2	1000	388	Bubble pressure	2
		1000	401	"	Draht in VKS (1926)
		1036	397	"	"
Lithium	Ar	180	397.5	"	36
	Ar	500	351.5	"	36
Magnesium		650	515	"	38
	Ar	681	563	"	39
	Ar	681	563	"	Zhivov in VKS (1937)
		700	550	Bubble pressure	37
	Ar	714	552	"	Zhivov in VKS (1937)
	Ar	723	549	"	"
	Ar	766	538	"	"
	Ar	789	532	"	"
	Ar	824	520	"	"
	Ar	838	517	"	"
	Ar	894	502	"	"
Mercury	Air	20	435.5	Oscillating jet	32
	vac.	20	484	Drop pressure	33
		20	465		MH
	vac.	25	484	Sessile drop	31
	vac.	25	483.5	"	30
		112	454		MH
		200	436		"
		300	405		"
		354	394		"
Molybdenum	vac.	2620 (m.p.)	2250	Pendant drop	21
	vac.	2620	2080	"	40
	vac.	2622	2080	Drop weight	63
Nickel	vac.	1455 (m.p.)	1756	Pendant drop	3
	vac.	1455	1780	"	21
	vac.	1455	1670	Drop weight	63
	vac.	1470	1735	Sessile drop on ZrO_2	43
	vac.	1470	1615	" " MgO	43
	vac.	1470	1760	" " Al_2O_3	43
	vac.	1470	1680	Sessile drop on BeO	43
	vac.	1470	1165	" " TiO_2	43
	He	1470	1530	" " MgO	7
	He	1470	1500	" " BeO	7
	He	1470	1490	" " BeO	7
	He	1470	1530	" " MgO	7
	vac.	1520	1740	" " Al_2O_3	7
	He	1520	1770	" "	7
	H_2	1520	1600	" "	7
	H_2	1530	1650	Bubble pressure BeO	7
	Ar	1550	1934	Sessile drop	18
Niobium	vac.	2415 (m.p.)	1900	Pendant drop	21
Osmium	vac.	2700 (m.p.)	2500	"	21
Palladium	vac.	1549 (m.p.)	1500	"	21
	vac.	1549	1470	Sessile drop	64
Phosphorus, white		68.7	64.95	Bubble pressure	44
Platinum	vac.	1773 (m.p.)	1740	Drop volume	46
	vac.	1773	1865	Pendant drop	21
	Ar	1800	1699	Sessile drop	47
Potassium		64	101	Bubble pressure	36
	Ar	79	400.5	"	Zhivov in VKS (1937)
	Ar	98	399.3	"	"
	Ar	112	398.9	"	"
	Ar	120	397.8	"	"

SURFACE TENSION OF LIQUID ELEMENTS (Continued)

Element	Gas	Temperature °C	Dynes/cm	Method	References
Potassium (cont.)	Ar	133	397.3	Bubble pressure	Zhivov in VKS (1937)
	Ar	163	395.5	"	"
	Ar	195	393.4	"	"
	Ar	228	391.5	Bubble pressure	"
Rhenium	vac.	3167 ± 60 (m.p.)	2700	Pendant drop	21
Rhodium	vac.	1966 (m.p.)	2000	"	21
Rubidium		39	76		23
Ruthenium	vac.	2450 (m.p.)	2250	Pendant drop	21
Selenium		220	105.5		49
Silicon	vac.	1410 (m.p.)	730	Sessile drop	43
	He	1450	725	Pendant drop	27
Silver	vac.	961 (m.p.)	785	"	3
		995	923		Sauerwald in VKS (1927)
	H_2	1000	893	Bubble pressure	1
	H_2	1000	916	"	2
		1050	916		Sauerwald in VKS (1927)
		1100	909		"
	H_2	1100	907	Bubble pressure	58
	H_2	1100	894	"	58
		1163	902		Sauerwald in VKS (1927)
	H_2	1200	876	Bubble pressure	58
	H_2	1250	849	Bubble pressure	1
Sodium	Ar	98	191	"	42
	Ar	110	205.7	"	Zhivov in VKS (1937)
	vac.	123	198	Drop volume	41
	Ar	131	204.5	Bubble pressure	Zhivov in VKS (1927)
	Ar	145	203.8	"	"
	Ar	176	202.6	"	"
	Ar	206	201.1	"	"
	Ar	229	199.7	"	"
	Ar	249	199.0	Bubble pressure	Zhivov in VKS (1927)
	Ar	263	198.2	"	"
Sulfur	vac.	120	60.9	Pendant drop	48
Tantalum	vac.	2996 ± 50 (m.p.)	2150	"	21
	vac.	2996 ± 50	2360	"	21
Tellurium	vac.	460	178.4	Capillary method	52
Thallium	H_2	318	436–444	Bubble pressure	9
Tin		232	612		Siedentopf in VKS
		232	480		Smith in VKS
		232	524		Coffman in VKS
	H_2	232	566	Capillary method	34
	Air	232	662.2		Matayuma in VKS
	vac.	232	578.8		"
		232	543.7	Bubble pressure	65
		240	527		Coffman in VKS
	vac.	250	575.8		Matuyama in VKS
	vac.	250	549	Drop pressure	12
	H_2	253	526	"	Bircumshaw in VKS (1926)
	Air	259	652.4		Matuyama in VKS
		280	523	Ring removal	Klyachko in VKS (1949)
	Air	280	523	"	50
	H_2	299	527	Drop pressure	Bircumshaw in VKS (1926)
		299	640.6		Matuyama in VKS
		300	566		"
		319	519	Drop pressure	Hogness in VKS (1921)
		325	640.6		Matuyama in VKS
		340	520	Ring removal	Klyachko in VKS
		340	520	"	50
		347	627.8		Matuyama in VKS

SURFACE TENSION OF LIQUID ELEMENTS (Continued)

Element	Gas	Temperature °C	Dynes/cm	Method	References
Tin (cont.)		350	556.2		Matuyama in VKS
		352	632.7		"
		362	580	Drop pressure	Hogness in VKS (1921)
		364	632.7		Matuyama in VKS
		396	520	Drop pressure	Hogness in VKS (1921)
		399	617.1		Matuyama in VKS
		400	545.4		"
	H_2	401	526	Drop pressure	Bircumshaw in VJS (1926)
		418	617.1		Matuyama in VKS
		450	535.6		"
		500	528.8		"
		550	516		"
		600	506.2		"
	vac.	600	526	Drop pressure	12
	H_2	600	525	"	Bircumshaw in VKS (1926)
	H_2	740	508	Bubble pressure	1
	H_2	800	520	Drop pressure	Bircumshaw in VKS (1926)
	H_2 + Ar	878	508	Bubble pressure	Draht in VKS (1926)
	H_2 + Ar	900	506	"	"
	H_2	964	514	Drop pressure	Bircumshaw in VKS (1926)
	H_2 + Ar	1000	497	Bubble pressure	Draht in VKS (1926)
	H_2 + Ar	1050	492	"	"
	H_2	1115	479.5	"	1
Titanium	vac.	1725 ± 10 (m.p.)	1650	Pendant drop	21
	vac.	1725 ± 10	1460		40
	vac.	1725 ± 10	1510	Capillary method	53
Tungsten	vac.	3410 (m.p.)	2310	Pendant drop	54
	vac.	3410	2500	"	21
Vanadium	vac.	1710 (m.p.)	1950	"	21
Zinc		420	773		Matuyama in VKS (1926)
		440	816		Pelzel in VKS
		460	808		"
		462	782	Ring removal	Klyachko in VKS (1937)
	Air	462	782		56
		477	753		Hogness in VKS (1921)
		490	802		Pelzel in VKS
		500	798		"
		505	762.2		Matuyama in VKS (1926)
	H_2, Ar	510	785	Bubble pressure	9
		510	785		Krause in VKS
		543	747		Hogness in VKS (1921)
		545	749.5		Matuyama in VKS (1926)
		550	778		Krause in VKS
		560	784		Pelzel in VKS
	H_2	600	787	Bubble pressure	55
		600	787		Bircumshaw in VKS (1926)
		600	768		Krause in VKS
		600	774		Pelzel in VKS
		616	379.7		Matuyama in VKS (1926)
		635	728.9		Matuyama in VKS (1926)
	H_2, Ar	640	761	Bubble pressure	9
		640	761		Krause in VKS
		670	756		Pelzel in VKS
	H_2	700	763		55
		700	763		Bircumshaw in VKS
Zirconium	vac.	1857 (m.p.)	1411	Drop weight	?
	vac.	1857	1400	"	29
		1857	1480	Pendant drop	21
		1857	1390	"	40

Reference Key

1. Lauermann, Metzger, Sauerwald	Z. Phys. Chemie, 216 (1961), p. 42
2. Metzger	Z. Phys. Chemie, 211 (1959), p. 1
3. Smirnova, Ormont	Sh. Fis. Chimii, 33 (1959), p. 771
4. Kubitschek	Izv. A. N., OTN, 2 (1959), p. 96
5. Korolkow	Izv. A. N., OTN, 2 (1956), p. 35
6. Naiditsch, Eremenko	Fis. Met. i. Metall., 11 (1961), p. 883
7. Eremenko, Iwaschenko, Nischenko, Fessenko	Izv. A.N., OTN, 7 (1958), p. 144
8. Eremenko, Nischenko, Naiditsch	Izv. A.N., OTN, 3 (1961), p. 150
9. Krause, Sauerwald, Michalke	Z. an. allg. Chem., 181 (1929), p. 353
10. Addison, Coldrey, Pulham	J. Chem. Soc., 1963, p. 1227
11. Eremenko, Nischenko, Tay-Schou-Wej	Izv. A.N., OTN, 3 (1960), p. 116
12. Pokrowski, Saidow	Fis. Met. i. Metall., 2 (1956), p. 546
13. Pokrowski, Kristian	Sh. Fis. Chimii, 28 (1954)
14. Pugatschewitsch, Altynow	Doklady A.N., 86 (1952), p. 117
15. Matuyama	Sci. Rep. R.I.T.U., 16 (1927), p. 555
16. Greenaway	J. Inst. Metals, 74 (1947), p. 133
17. Bircumshaw	Philos, Mag., 2 (1926), p. 341
18. Kozakevitch, Urbain	J. Iron and Steel Inst., 186 (1957), p. 186
19. Allen, Kingery	Trans. A.I.M.E., 215 (1959), p. 30
20. Fessenko	Sh. Fis. Chimii, 35 (1961), p. 707
21. Allen	Trans. A.I.M.E., 227 (1963), p. 1175
22. Eremenko, Naiditsch	Izv. A.N., OTN, 2 (1959), p. 111
23. Korolkow, Igumnowa	Izv. A.N., OTN, 6 (1961), p. 95
24. Urbain, Lucas	Proc. N.F.L. Teddington, 9 (1959) Paper 4E
25. Zein-Tan Wan, Karasew, Samarin	Izv. A.N., OTN, 1 (1960), p. 30, 49
26. Mack, Davis, Bartell	J. Phys. Chem., 45 (1941), p. 846
27. Keck, van Horn	Phys. Review, 91 (1953), p. 512
28. Lasarew, Pugatschewitsch	Doklady A.N., 134 (1960), p. 132
29. Peterson, Kedesday, Keck, Schwarz	J. Appl. Phys., 29 (1958), p. 213
30. Nicholas, Joyner, Tessem, Olson	J. Phys. Chem., 65 (1961), p. 1375
31. Kemball	Trans. Farad. Soc., 42 (1946), p. 526
32. Schmidt	Ann. d. Physik, 39 (1912), p. 1108
33. Bering, Ioilewa	Doklady A.N., 93 (1953), p. 85
34. Melford, Hoar	J. Inst. Metals, 85 (1957), p. 197
35. Timofejewitschewa, Pugatschewitsch	Doklady A.N., 124 (1959), p. 1093
36. Taylor	Philos. Mag., 46 (1955), p. 867
37. Kubitschek, Malzew	Izv. A.N., OTN, 3 (1959), p. 144
38. Korolkow, Bytschkowa	Issled. Spl. Tsvetn. Met., 2 (1960), p. 122
39. Schiwow	Trudy W.A.M.I., 14 (1937), p. 99
40. Samba, Isobe	Proc. 4 Symp. Electron-Beam Techn., 1962, p. 96
41. Addison, Addison, Kerridge, Lewis	J. Chem. Soc., 1955, p. 2262
42. Taylor	J. Inst. Metals, 83 (1954), p. 143
43. Kingery, Humenik	J. Phys. Chem., 57 (1953), p. 359
44. Hutchinson	Trans. Far. Soc., 39 (1943), p. 229
45. Konstantinow	Diss. Staatl. Univ. Moskau, 1950
46. Eremenko, Naiditsch	Izv. A.N., OTN, 6 (1959), p. 129
47. Kozakevitsch, Urbain	C.R. Paris, 253 (1961), p. 2229
48. Ono, Matsishima	Sci. Rep. R.I.T.U., 9 (1957), p. 309
49. Astachow, Panin, Dobkina	Sh. Fis. Chimii, 20 (1946), p. 403
50. Klyatchko, Kunin	Zavodsk. Labor., 14 (1948), p. 66
51. Kelly, Calverley	SERL-Report, 80 (1959), p. 53
52. Smith, Spitzer	J. Phys. Chem., 66 (1962), p. 946
53. Eljutin, Maurach	Izw. A.N., OTN, 4 (1956), p. 129
54. Calverley	Proc. Phys. Soc., 70 (1957), p. 1040
55. Bircumshaw	Philos. Mag., 3 (1927), p. 1286
56. Klyatchko	Zavodsk. Labor., 6 (1937), p. 1376
57. Shunk, Burr	Trans. Amer. Soc. Met., 55 (1962), p. 786
58. Lauermann, Sauerwald	Zeitschrift f. Metallkunde, 55 (1964), p. 605
59. de l. Davies, West	J. Inst. Metals, 92 (1964), p. 208

60. Popel, Derjabin — Izv. Vys. Ut. Sav. T. M, 6 (1963), p. 16
61. Gans, Pawlek, Roepenack — Zeitschrift f. Metallkunde, 54 (1963), p. 147
62. Kurotschkin, Baum, Borodulin — Fis. Met. i. Metall., 15 (1963), p. 461
63. Namba, Isobe — Sci. Pap. Inst. Phys. Chem. Res. Tokyo, 57 (1963), p. 51
64. Eremenko, Naiditsch — Izv. A.N., OTN, 6 (1961), p. 100
65. Cahill, Kirshenbaum — Journ. of Inorg. Nuclear Chemistry, 26 (1964), p. 206

Abbreviations

MH	Metals Handbook, Vol. I., 1960
VKS	V.K. Sementchenko: "Surface Phenomena in Metals and Alloys" 1962
Er	Eremenko: "The Role of Surface Phenomena in Metallurgy"
Z. Phys. Chem.	Zeitschrift fuer physikalische Chemie (Germany)
Sh. Fis. Chimii	Shurnal Fisiceskoi Chimii, (USSR)
Izv. A.N., OTN	Izvestija akademii nauk, otdelenie technitscheskich nauk, (USSR)
Fis. Met. i. Metall.	Fisika metallov i metallovedenia. (USSR)
Z. an. allg. Chem.	Zeitschrift fuer anorganische und allgemeine Chemie (German democratic republic)
J. Chem. Soc.	Journal of the Chemical Society, (London)
Doklady A.N.	Doklady akademii nauk. (USSR)
Sci. Rep. R.I.T.U.	Science report of the research institute of the Tohoku university, (Japan)
J. Inst. Met.	Journal of the Institute of metals, (London)
Philos. Mag.	Philosophical Magazine (Edinburgh)
J. Iron and Steel Inst.	Journal of the Iron and Steel Institute (Great Britain)
Trans. A.I.M.E.	Transactions of the A.I.M.E. (New York)
J. Phys. Chem.	Journal of Physical Chemistry (Washington, D.C.)
Phys. Review	Physical Review (New York)
J. Appl. Phys.	Journal of Applied Physics (New York)
Trans. Farad. Soc.	Transactions of the Faraday Society (Great Britain)
Ann. d. Physik	Annalen der Physik, (Germany)
Issled. Spl. Tsvetn. Met.	Issledovanie splavov tsvetnych metallov (USSR)
Trudy W.A.M.I.	Trudy wsesoyuznoi assotsiatsii metallurgitscheskich Inzhenerov (USSR)
C.R. Paris	Compted rendues hebdomadaires des seances de l'academie des sciences, (Paris)
Zavodsk. Labor.	Zavodskaja laboratorija, (USSR)
Proc. Phys. Soc.	Proceedings of the Physical Society, (London)
Trans. Amer. Soc. Met.	Transactions of the American Society of Metals
Izv. Vys. Ut. Sav T.M.	Izvestija vysschich utschebnich savadenie tscherna metallurgija (USSR)

SURFACE TENSION

SURFACE TENSION OF INORGANIC SOLUTES IN WATER

% = Weight % of solute
γ = Surface tension in dynes/cm.

Solute	T°C	% γ							
HCl	20	%	1.78	3.52	6.78	12.81	16.97	23.74	35.29
		γ	72.55	72.45	72.25	71.85	71.75	70.55	65.75
HNO_3	20	%	4.21	8.64	14.99	34.87			
		γ	72.15	71.65	70.95	68.75			
H_2SO_4	25	%	4.11	8.26	12.18	17.66	21.88	29.07	33.63
		γ	72.21	72.55	72.80	73.36	73.91	74.80	75.29
$HClO_4$	25	%	4.86	10.01	20.38	30.36	53.74	63.47	72.25
		γ	71.18	70.34	69.21	68.57	69.02	69.73	69.01
KOH	18	%	2.73	5.31	10.08	17.57			
		γ	73.95	74.85	76.55	79.75			
NaOH	18	%	2.72	5.66	16.66	30.56	35.90		
		γ	74.35	75.85	83.05	96.05	101.05		
NH_4OH	18	%	1.72	3.39	4.99	9.51	17.37	34.47	54.37
		γ	71.65	70.65	69.95	67.85	65.25	61.05	57.05
KCl	20	%	0.74	3.60	6.93	13.88	18.77	22.97	24.70
		γ	72.99	73.45	74.15	75.55	76.95	78.25	78.75
LiCl	25	%	5.46	7.37	10.17	13.95			
		γ	74.23	75.10	76.30	78.10			

Solute	T°C	% γ							
NaCl	20	%	0.58	2.84	5.43	10.46	14.92	22.62	25.92
		γ	72.92	73.75	74.39	76.05	77.65	80.95	82.55
PbCl	25	%	21.57	28.52	37.74				
		γ	75.20	76.80	79.20				
$BaCl_2$	25	%	9.26	16.73	25.58				
		γ	73.50	74.93	76.38				
$MgCl_2$	20	%	0.94	4.55	8.69	16.00	22.30	25.44	
		γ	73.07	74.00	75.75	79.15	82.95	85.75	
NaBr	20	%	4.89	9.33	13.37	23.00			
		γ	73.45	74.05	74.75	76.55			
$Al_2(SO_4)_3$	25	%	2.54	4.06	9.40	14.60	19.32	23.54	25.50
		γ	72.32	72.92	73.51	74.71	76.06	78.30	79.73
$MgSO_4$	20	%	1.19	5.68	10.75	19.41	24.53		
		γ	73.01	73.78	74.85	77.35	79.25		
Na_2SO_4	20	%	2.76	6.63	12.44				
		γ	73.25	74.15	75.45				
Na_2CO_3	20	%	2.58	5.03	9.59	13.72			
		γ	73.45	74.05	75.45	76.75			
$NaNO_3$	20	%	0.85	4.08	7.84	14.53	29.82	37.30	47.06
		γ	72.87	73.75	73.95	75.15	78.35	80.25	87.05

SURFACE TENSION OF INORGANIC SOLUTES IN ORGANIC SOLVENTS

% = Mol %
γ = Surface tension in dynes/cm.

Solute	Solvent	T°C	% γ						
LiCl	Ethyl alcohol	14	%	0.72	2.30	4.62			
			γ	22.90	23.17	23.26			
LiBr	Ethyl alcohol	14	%	0.95	2.60				
			γ	23.08	23.35				
LiI	Ethyl alcohol	14	%	1.43	2.87	5.08	10.21	19.47	26.92
			γ	23.11	23.56	24.39	26.03	28.87	31.95
KI	Methyl alcohol	14	%	0.81	1.52	2.68			
			γ	23.76	24.11	24.71			
$CaCl_2$	Ethyl alcohol	24	%	0.94	1.98	3.79	7.47		
			γ	22.62	22.58	23.23	23.97		
NaI	Methyl alcohol	14	%	0.76	1.48	4.33	8.55	12.53	
			γ	22.83	23.29	24.85	27.41	29.75	

Solute	Solvent	T°C	% γ						
NaI	Ethyl alcohol	24	%	0.45	1.80	3.63	4.54	6.02	10.
			γ	22.47	22.82	23.41	23.52	24.00	25.
NaI	Acetone	14	%	0.93	2.08	5.07	6.53		
			γ	24.22	24.40	25.04	25.12		
ZnI_2	Methyl alcohol	22	%	0.90	2.79	5.07			
			γ	22.97	24.23	25.84			
ZnI_2	Ethyl alcohol	24	%	0.41	1.72	3.42	6.90		
			γ	22.70	22.90	23.71	25.49		
H_2SO_4	Ethyl ether	17	%	3	10	40	75	90	
			γ	17.30	19.55	32.50	46.30	46.83	
H_2SO_4	Nitrobenzene	17	%	3	10	40	75	90	
			γ	42.73	43.96	46.05	47.52	48.25	

SURFACE TENSION OF ORGANIC COMPOUNDS IN WATER

% = Weight % of solute
γ = Surface tension in dynes/cm.

Solute	T° C	% γ							
Acetic acid...	30	%	1.00	2.475	5.001	10.01	30.09	49.96	69.91
		γ	68.00	64.40	60.10	54.60	43.60	38.40	34.30
Acetone......	25	%	5.00	10.00	20.00	50.00	75.00	95.00	100.00
		γ	55.50	48.90	41.10	30.40	26.80	24.20	23.00
Acetonitrile..	20	%	1.13	3.35	11.77	20.20	37.58	61.33	81.22
		γ	69.02	63.03	47.61	39.06	31.84	30.02	29.02
o-Aminobenzoic acid	25	%	12.35	22.36	30.45	37.44			
		γ	71.96	73.23	74.54	75.79			
m-Aminobenzoic acid	25	%	12.35	22.36	30.45	37.44			
		γ	73.30	74.59	76.16	77.89			

Solute	T° C	% γ							
p-Aminobenzoic acid	25	%	12.35	22.36	30.45	37.44			
		γ	73.38	74.79	76.32	78.20			
Aminobutyric acid	25	%	4.96	9.34	13.43				
		γ	71.91	71.67	71.40				
Ammonium lactate	29	%	30.00	50.00	60.00	70.00	80.00	90.00	
		γ	35.40	34.40	35.40	35.60	38.20	44.50	
n-Butanol....	30	%	0.04	0.41	9.53	80.44	86.05	94.20	97.40
		γ	69.33	60.38	26.97	23.69	23.47	23.29	22.25
n-Butyric acid	25	%	0.14	0.31	1.05	8.60	25.00	79.00	100.00
		γ	69.00	65.00	56.00	33.00	28.00	27.00	26.00

SURFACE TENSION OF ORGANIC COMPOUNDS IN WATER (Continued)

Solute	T°C	% / γ							
Dioxan.......	26	%	0.44	2.20	4.70	11.14	20.17	35.20	55.00
		γ	69.83	65.64	62.45	56.90	51.57	45.30	39.27
Dioxan......	26	%	67.68	76.45	83.02	91.90	95.60	97.77	
		γ	36.95	35.80	35.00	33.95	33.60	33.10	
Formic acid..	30	%	1.00	5.00	10.00	25.00	50.00	75.00	100.00
		γ	70.07	66.20	62.78	56.29	49.50	43.40	36.51
Glycerol.....	18	%	5.00	10.00	20.00	30.00	50.00	85.00	100.00
		γ	72.90	72.90	72.40	72.00	70.00	66.00	63.00
Glycine......	25	%	3.62	6.98	10.12	13.10			
		γ	72.54	73.11	73.74	74.18			
Hydrocinnamic acid	21.5	%	7.02	12.62	18.39	26.09	31.06	38.25	47.93
		γ	69.08	66.49	63.63	59.25	56.14	52.96	47.24
Methyl acetate	25	%	0.66	1.29	2.29	3.56			
		γ	66.33	62.92	58.22	55.08			

Solute	T°C	% / γ							
Morpholine...	20	%	8.56	19.39	30.41	50.45	69.93	80.14	92.00
		γ	67.80	62.62	59.15	52.85	47.05	43.62	41.60
Potassium lactate	29	%	40.00	50.00	60.00	70.00			
		γ	66.40	66.40	65.40	63.40			
Phenol.......	20	%	0.024	0.047	0.118	0.417	0.941	3.76	5.62
		γ	72.60	72.20	71.30	66.50	61.10	46.00	42.30
n-Propanol...	25	%	0.1	0.5	1.0	50.0	60.0	80.0	90.0
		γ	67.10	56.18	49.30	24.34	24.15	23.66	23.41
Propionic acid	25	%	1.91	5.84	9.80	21.70	49.80	73.90	100.00
		γ	60.00	49.00	44.00	36.00	32.00	30.00	26.00
Sodium lactate	29	%	1.00	10.00	30.00	40.00	50.00	60.00	70.00
		γ	70.40	69.60	68.50	64.80	45.40	56.70	60.70
Sucrose......	25	%	10.00	20.00	30.00	40.00	55.00		
		γ	72.50	73.00	73.40	74.10	75.70		

SURFACE TENSION OF ORGANIC COMPOUNDS IN ORGANIC SOLVENTS

% = Weight % of solvent
γ = Surface tension dynes/cm.

Solute	Solvent	T°C	% / γ					
Acetic acid...	Benzene	35	%	10.45	25.53	34.28	43.93	68.77
			γ	25.40	25.21	25.32	25.43	25.99
Acetic acid...	Acetone	25	%	25.63	50.83	75.62		
			γ	27.50	26.61	24.90		
Acetone......	Ethyl ether	30	%	21.83	40.98	61.29	75.69	89.03
			γ	16.75	17.49	19.15	19.80	21.00
Acetonitrile...	Ethyl alcohol	20	%	10.83	32.42	49.90	69.78	80.19
			γ	22.92	23.92	24.36	25.08	26.51
Aniline.......	Cyclohexane	32	%	14.35	37.65	50.67	72.28	96.46
			γ	24.21	24.51	24.50	25.61	37.45
Carbontetrachloride	Benzene	50	%	30.40	51.69	62.38	78.29	88.62
			γ	24.39	24.09	23.78	23.47	23.21
Cyclohexane..	Nitrobenzene	15	%	5.00	10.00			
			γ	37.59	33.03			
Naphthalene..	Benzene	79.5	%	20.00	40.00	50.00	60.00	80.00
			γ	23.42	26.70	27.80	29.20	31.70
Naphthalene..	*p*-Nitrophenol	121	%	10.21	29.74	45.94	58.01	67.57
			γ	29.90	31.80	33.70	34.80	41.80

SURFACE TENSION OF METHYL ALCOHOL IN WATER

% = Volume % of alcohol
γ = Surface tension dynes/cm.

T°C	%	7.5	10.00	25.0	50.0	60.0	80.0	90.0	100.0
20	γ	60.90	59.04	46.38	35.31	32.95	27.26	25.36	22.65
30	γ	59.33	57.27	45.30	34.52	32.26	26.48	24.42	21.58
50	γ	56.19	55.01	43.24	32.95	30.79	25.01	22.55	19.52

SURFACE TENSION (Continued)

SURFACE TENSION OF ETHYL ALCOHOL IN WATER

% = Volume % of alcohol
γ = Surface tension dynes/cm.

T°C	%	5.00	10.00	24.00	34.00	48.00	60.00	72.00	80.00	96.00
20	γ	...	...	...	33.24	30.10	27.56	26.28	24.91	23.04
40	γ	54.92	48.25	35.50	31.58	28.93	26.18	24.91	23.43	21.38
50	γ	53.35	46.77	34.32	30.70	28.24	25.50	24.12	22.56	20.40

WATER AGAINST AIR

Temperature °C	Surface tension dynes/cm.	Temperature °C	Surface tension dynes/cm.	Temperature °C	Surface tension dynes/cm.
−8	77.0	15	73.49	40	69.56
−5	76.4	18	73.05	50	67.91
0	75.6	20	72.75	60	66.18
5	74.9	25	71.97	70	64.4
10	74.22	30	71.18	80	62.6
				100	58.9

INTERFACIAL TENSION

Surface Tension at the Interface Between Two Liquids
(Each liquid saturated with the other)

Liquids	Temperature °C	γ	Liquids	Temperature °C	γ
Benzene- -Mercury...	20	357	Water- -Heptylic acid	20	7.0
Ethyl ether- -Mercury	20	379	Water- -n-Hexane....	20	51.1
Water- -Benzene.....	20	35.00	Water- -Mercury.....	20	375.
Water- -Carbon tetrachloride	20	45.	Water-n-Octane......	20	50.8
Water- -Ethyl ether..	20	10.7	Water-n-Octyl alcohol	20	8.5

SURFACE TENSION OF VARIOUS LIQUIDS

Substance		In contact with	Temperature °C	Surface tension dynes/cm.	References
Name	Formula				
Acetaldehyde	C_2H_4O	vapor	20	21.2	AC(**12**)
Acetaldoxime	C_2H_5NO	vapor	35	30.1	JP(**1**); JSG(**1**)
Acetamide	C_2H_5NO	vapor	85	39.3	JS(**3**)
Acetanilide	C_2H_5NO	vapor	120	35.6	JS(**3**)
Acetic acid	$C_2H_4O_2$	vapor	10	28.8	AC(**22, 23, 25**); GC(**1**); JS(**14**); tPRS(**1**); ZC(**1, 6**)
" "	C_2H_4O	vapor	20	27.8	
" "	C_2H_4O	vapor	50	24.8	
Acetic anhydride	$C_4H_6O_3$	vapor	20	32.7	GC(**1**); JS(**4**)
Acetone	C_3H_6O	air or vapor	0	26.21	AC(**20, 24, 25**); AdC(**1**); BF(**1**); JP(**5**); JS(**4, 14**); ZC(**6**)
"	C_3H_6O	air or vapor	20	23.70	
"	C_3H_6O	air or vapor	40	21.16	
Acetonitrile	C_2H_3N	vapor	20	29.30	AC(**16**); BF(**1**); GC(**1**); JP(**5**)
Acetophenone	C_8H_8O	vapor	20	39.8	AC(**27**); AS(**1**); JP(**2**)
Acetyl chloride	C_2H_3ClO	vapor	14.8	26.7	JS(**4**)
Acetylene	C_2H_2	vapor	−70.5	16.4	AC(**14**)
Acetylsalicylic acid (in aq. sol.)	$C_9H_8O_4$	vapor	25.9	60.06	GC(**2**)
Allyl alcohol	C_3H_6O	air or vapor	20	25.8	AC(**27**); AdC(**1**); JS(**4**)
Allyl isothiocyanate	C_4H_5NS	air or vapor	20	34.5	AC(**16**); GC(**1**)
Ammonia	NH_3	vapor	11.1	23.4	JP(**7**)
"	NH_3	vapor	34.1	18.1	JP(**7**)
Aniline	C_6H_7N	air	10	44.10	AC(**6, 17, 28**); GC(**1**); JP(**5**); JS(**12**)
"	C_6H_7N	vapor	20	42.9	
"	C_6H_7N	air	50	39.4	
Argon	A	vapor	−188	13.2	JS(**15**)
Azoxybenzene	$C_{12}H_{10}N_2O$	vapor	51	43.34	JS(**5**)
Benzaldehyde	C_7H_6O	air	20	40.04	AC(**6, 24**)
Benzene	C_6H_6	air	10	30.22	AC(**3, 5, 31, 32, 34**); BF(**2**); JP(**5**); JS(**4, 9, 10, 11, 14**); PRS(**2**); tRIA(**1**); tPRS(**1**); ZC(**4, 5**)
"	C_6H_6	air	20	28.85	
"	C_6H_6	saturated with vapor	20	28.89	
"	C_6H_6	air	30	27.56	
Benzonitrile	C_7H_5N	air	20	39.05	AC(**28**); AS(**2**); GC(**1**); JP(**5**); JS(**3**)
Benzophenone	$C_{13}H_{10}O$	air or vapor	20	45.1	AC(**27**); AS(**1**); ZC(**5**)
Benzylamine	C_7H_9N	vapor	20	39.5	JS(**3**)
Benzyl alcohol	C_7H_8O	air or vapor	20	39.0	AC(**33**); JS(**6**); PRS(**1**)
Bromine	Br_2	air or vapor	20	41.5	AC(**17**); AdP(**3**); GC(**1**)
Bromobenzene	C_6H_5Br	air	20	36.5	AC(**8, 17**); GC(**1**); JS(**12**)
Bromoform	$CHBr_3$	vapor	20	41.53	AC(**8**); JP(**2**)
p-Bromophenol	C_6H_5BrO	vapor	74.4	42.36	JS(**6**)
d-sec-Butyl alcohol	$C_4H_{10}O$	vapor	10	23.5	JS(**7**)
n-Butyl alcohol	$C_4H_{10}O$	air or vapor	0	26.2	AC(**27, 33**); JS(**4**)
" " "	$C_4H_{10}O$	air or vapor	20	24.6	AC(**27, 33**); JS(**4**)
" " "	$C_4H_{10}O$	air or vapor	50	22.1	AC(**27, 33**); JS(**4**)
tert-Butyl alcohol	$C_4H_{10}O$	air or vapor	20	20.7	AC(**33**); JS(**8**)
n-Butylamine	$C_4H_{11}N$	nitrogen	41	19.7	ZA(**1**)
n-Butyric acid	$C_4H_8O_2$	air	20	26.8	AC(**27**); GC(**1**); JS(**4**)
Carbon bisulfide	CS_2	vapor	20	32.33	AC(**17, 28**); GC(**1**); BF(**2**); JS(**14**); PRS(**2**); ZC(**6**)
Carbon dioxide	CO_2	vapor	20	1.16	VK(**1, 2**)
" "	CO_2	vapor	−25	9.13	VK(**1, 2**)
Carbon tetrachloride	CCl_4	vapor	20	26.95	AC(**3, 5, 6, 28, 31**); PRS(**1, 2**); ZC(**5**)
" "	CCl_4	vapor	100	17.26	
" "	CCl_4	vapor	200	6.53	
Carbon monoxide	CO	vapor	−193	9.8	JS(**15**)
" "	CO	vapor	−203	12.1	JS(**15**)
Chloral	C_2HCl_3O	vapor	19.4	25.34	JS(**4**)
Chlorine	Cl_2	vapor	20	18.4	AC(**11**); JP(**3**)
"	Cl_2	vapor	−30	25.4	AC(**11**); JP(**3**)
"	Cl_2	vapor	−40	27.3	AC(**11**); JP(**3**)
"	Cl_2	vapor	−50	29.2	AC(**11**); JP(**3**)
"	Cl_2	vapor	−60	31.2	AC(**11**); JP(**3**)
Chloroacetic acid	$C_2H_2Cl_2O_2$	nitrogen	25.7	35.4	ZA(**1**)
Chlorobenzene	C_4H_5Cl	vapor	20	33.56	AC(**6, 20, 28**); JP(**5**); JS(**11**); PRS(**2**); tRIA(**1**); tPRS(**1**); ZC(**5**)
Chloroform	$CHCl_3$	air	20	27.14	AC(**6, 28, 31**); AdC(**1**); PRS(**2**); tRIA(**1**); ZC(**6**)
o-Chlorophenol	C_6H_5ClO	vapor	12.7	42.25	JS(**6**)
Cyclohexane	C_6H_{12}	air	20	25.5	PRS(**1**); ZA(**1**)
Dichloroacetic acid	$C_2H_2Cl_2O_2$	nitrogen	25.7	35.4	ZA(**1**)
Dichloroethane	$C_2H_4Cl_2$	air	35.0	23.4	AC(**17**); AdC(**1**)
Diethylamine	$C_4H_{11}N$	air	56	16.4	GC(**1**)
Diethylaniline	$C_{10}H_{15}N$	vapor	20	34.2	AC(**31**); AS(**1**); MfC(**1**)
Diethyl carbonate	$C_5H_{10}O$	air	20	26.31	AC(**4**); JS(**5**)
Diethyl oxalate	$C_6H_{10}O_4$	vapor	20	32.0	AC(**16**); GC(**1**)
Diethyl phthalate	$C_{12}H_{14}O_4$	vapor	20	37.5	AC(**21**); PRS(**1**); ZA(**5**)
Diethyl sulfate	$C_4H_{12}O_4S$	air	13	34.61	JS(**5**)
Dimethylamine	C_2H_7N	nitrogen	0	18.1	ZA(**1**)
"	C_2H_7N	nitrogen	5	17.7	ZA(**1**)
Dimethylaniline	C_8H_{11}	air or vapor	20	36.6	AC(**4, 17, 31**); JP(**5**)
1,5-Dimethyl-2-phenyl-3-pyrazolone	$C_{11}H_{12}N_2O$	vapor	25.9	63.63	GC(**2**)
Dimethyl sulfate	$C_2H_6O_4S$	air	18	40.12	JS(**5**)
Diphenylamine	$C_{12}H_{11}N$	air or vapor	80	37.7	JP(**2, 9**); JS(**3**); ZA(**1**)
Ethyl acetate	$C_4H_8O_2$	air	0	26.5	AC(**26, 33**); AdC(**1**); AS(**2**); JP(**5**); tPRS(**1**); ZC(**6**)
" "	$C_4H_8O_2$	air	20	23.9	
" "	$C_4H_8O_2$	air	50	20.2	
Ethyl acetoacetate	$C_6H_{10}O_3$	air or vapor	20	32.51	AC(**16**); BD(**2**); GC(**1**); JS(**4**)
Ethyl alcohol	C_2H_6O	air	0	24.05	AC(**22, 23, 25, 32**); BF(**2**); JP(**5**); tRIA(**1**); tPRS(**1**)
" "	C_2H_6O	vapor	10	23.61	
" "	C_2H_6O	vapor	20	22.75	
" "	C_2H_6O	vapor	30	21.89	
Ethylamine	C_2H_7N	nitrogen	0	21.3	PRA(**4**); ZA(**1**)
"	C_2H_7N	nitrogen	9.9	20.4	PRA(**4**); ZA(**1**)
Ethylaniline	$C_8H_{11}N$	air or vapor	20	36.6	AC(**17, 33**); AS(**1**); MfC(**1**)

SURFACE TENSION OF VARIOUS LIQUIDS

Substance		In contact with	Temperature °C	Surface tension dynes/cm.	References
Name	Formula				
Ethylbenzene	C_8H_{10}	--vapor	20	29.20	AC(**17, 33, 34**); AdC(**1**); PRS(**1**)
Ethylbenzoate	$C_9H_{10}O_2$	--vapor	20	35.5	GC(**1**); ZA(**1**); MfC(**1**)
Ethyl bromide	C_2H_5Br	--vapor	20	24.15	AC(**6, 33**); GC(**1**)
Ethyl chloroformate	$C_3H_5ClO_2$	--vapor	15.1	27.5	JS(**4**)
Ethyl Cinnamate	$C_{11}H_{12}O_2$	--air	20	38.37	AC(**16**); PRS(**1**); ZC(**5**)
Ethylene bromide	$C_2H_4Br_2$	--vapor	20	38.37	AC(**5, 28, 33**); GC(**1**); JP(**2**); PRS(**2**); tRIA(**1**)
Ethylene chloride	$C_2H_4Cl_2$	--air	20	24.15	AC(**17, 33**); AdC(**1**); JS(**14**); ZC(**6**)
Ethylene oxide	C_2H_4O	--vapor	−20	30.8	
" "	C_2H_4O	--vapor	0.0	27.6	
" "	C_2H_4O	--vapor	20	24.3	
Ethyl ether	$C_4H_{10}O$	--vapor	20	17.01	
" "	$C_4H_{10}O$	--vapor	50	13.47	} AC(**4, 15, 28, 31**); AdC(**1**); tPRS(**1**)
Ethyl formate	$C_3H_6O_2$	--air or vapor	20	23.6	AC(**20, 26**); AdC(**1**); PRS(**2**)
Ethyl iodide	C_2H_5I	--vapor	20	29.4	AC(**33**); GC(**1**); JS(**4**); PRS(**1**); ZC(**6**)
Ethyl nitrate	$C_2H_5NO_3$	--air or vapor	20	28.7	AC(**27**); GC(**1**)
dl-Ethyl lactate	$C_5H_{10}O_3$	--air	20	29.9	AC(**20, 21**); JP(**8**)
Ethyl mercaptan	C_2H_6S	--air or vapor	20	22.5	AC(**4, 16**); JS(**4**)
Ethyl salicylate	$C_9H_{10}O_3$	--vapor	20.5	38.33	JS(**6**)
Formamide	CH_3NO	--vapor	20	58.2	AC(**27**); JS(**3**); ZC(**4**)
Formic acid	CH_2O_2	--air	20	37.6	AC(**23, 27**); GC(**1**); JS(**4**); MfC(**1**); ZA(**1**)
Furfural	$C_5H_4O_2$	--air or vapor	20	43.5	AC(**27**); GC(**1**)
Gelatin solution (1%)		--water	2.85	8.3	ZK(**1**)
Glycerol	$C_3H_8O_3$	--air	20	63.4	
"	$C_3H_8O_3$	--air	90	58.6	} JR(**1**); MB(**1**); ZA(**1**); ZC(**3**)
"	$C_3H_8O_3$	--air	150	51.9	
Glycol	$C_2H_6O_2$	--air or vapor	20	47.7	AC(**29**); JS(**4**)
Helium	He	--vapor	−269	.12	cUL(**2**); PRA(**2**)
"	He	--vapor	−270	.239	cUL(**2**); PRA(**2**)
"	He	--vapor	−271.5	.353	cUL(**2**); PRA(**2**)
n-Hexane	C_6H_{14}	--air	20	18.43	AC(**5, 6, 16**); AdC(**1**); AS(**1**)
Hydrazine	N_2H_4	--vapor	25	91.5	AC(**1**)
Hydrogen	H_2	--vapor	−255	2.31	cUL(**1**); PRA(**1**)
Hydrogen cyanide	HCN	--vapor	17	18.2	ZE(**1**)
Hydrogen peroxide	H_2O_2	--vapor	18.2	76.1	AC(**13**)
Isobutyl alcohol	$C_4H_{10}O$	--vapor	20	23.0	AC(**4, 27, 32**); AdC(**1**); JP(**5**); JS(**4**)
Isobutylamine	$C_4H_{11}N$	--air	68	17.6	GC(**1**)
Isobutyl chloride	C_4H_9Cl	--air	20	21.94	AC(**6**); GC(**1**)
Isobutyric acid	$C_4H_8O_2$	--air or vapor	20	25.2	AC(**27**); GC(**1**); JS(**4**)
Isopentane	C_5H_{12}	--air	20	13.72	AC(**6**)
Isopropyl alcohol	C_3H_8O	--air or vapor	20	21.7	JS(**4**); AdC(**1**)
Methyl acetate	$C_3H_6O_2$	--air or vapor	20	24.6	AC(**26, 33**); AdC(**1**); PRS(**2**)
Methyl alcohol	CH_4O	--air	0	24.49	
" "	CH_4O	--air	20	22.61	} AC(**22, 23, 25, 32**); tPRS(**1**)
" "	CH_4O	--vapor	50	20.14	
Methylamine	CH_3NH_2	--nitrogen	−12	22.2	PRA(**1**); ZA(**1**)
"	CH_3NH_2	--vapor	−20	23.0	PRA(**1**); ZA(**1**)
"	CH_3NH_2	--nitrogen	−70	29.2	PRA(**1**); ZA(**1**)
N-Methylaniline	C_7H_9N	--air or vapor	20	39.6	AC(**17, 32**); AS(**1**)
Methyl benzoate	$C_8H_8O_2$	--air or vapor	20	37.6	GC(**1**); JP(**5**)
Methyl chloride	CH_3Cl	--air	20	16.2	WN(**1**)
Methyl ether	C_2H_6O	--vapor	−10	16.4	AC(**12**)
" "	C_2H_6O	--vapor	−40	21	AC(**12**)
Methylene chloride	CH_2Cl_2	--air	20	26.52	AC(**6**)
Methylene iodide	CH_2I_2	--air	20	50.76	AC(**8**)
Methyl ethyl ketone	C_4H_8O	--air or vapor	20	24.6	AC(**24**); AS(**1**)
Methyl formate	$C_2H_4O_2$	--vapor	20	25.08	AC(**26**); tPRS(**1**); GC(**1**)
Methyl iodide	CH_3I	--air	43.5	25.8	GC(**1**)
Methyl propionate	$C_4H_8O_2$	--air or vapor	20	24.9	AC(**20, 26**); AdC(**1**); PRS(**2**)
Methyl salicylate	$C_8H_8O_3$	--nitrogen	94	31.9	ZA(**1**)
Methyl sulfide	C_2H_6S	--vapor	11.1	26.50	JP(**6**)
Naphthalene	$C_{10}H_8$	--air or vapor	127	28.8	AS(**1**); JP(**2**)
Neon	Ne	--vapor	−248	5.50	cUL(**3**); PRA(**3**)
Nitric acid (98.8%)	HNO_3	--air	11.6	42.7	JS(**2**)
Nitrobenzene	$C_6H_5NO_2$	--air or vapor	20	43.9	AC(**6, 23**); AS(**2**); BF(**2**); GC(**1**); JS(**4, 12**)
Nitroethane	$C_2H_5NO_2$	--air or vapor	20	32.2	GC(**1**); JS(**4**)
Nitrogen	N_2	--vapor	−183	6.6	JS(**15**)
"	N_2	--vapor	−193	8.27	JS(**15**)
"	N_2	--vapor	−203	10.53	JS(**15**)
Nitrogen tetra oxide	N_2O_4	--vapor	19.8	27.5	JS(**4**)
Nitromethane	CH_3NO_2	--vapor	20	36.82	AC(**6, 27**); GC(**1**)
Nitrous oxide	N_2O	--vapor	20	1.75	VK(**1**)
n-Octane	C_8H_{18}	--vapor	20	21.80	AC(**4, 5, 34**); JS(**4**)
n-Octyl alcohol	$C_8H_{18}O$	--air	20	27.53	AC(**4, 5**)
Oleic acid	$C_{18}H_{34}O_2$	--air	20	32.50	PRS(**1**); AC(**4**)
Oxygen	O_2	--vapor	−183	13.2	JS(**15**)
Oxygen (65%)	O_2	--air	−190.5	12.2	AD(**1**)
"	O_2	--vapor	−193	15.7	JS(**15**)
"	O_2	--vapor	−203	18.3	JS(**15**)
Paraldehyde	$C_6H_{12}O_3$	--air	20	25.9	AC(**27**); AdC(**1**)
Phenetole	$C_8H_{10}O$	--vapor	20	32.74	AC(**6, 28**); GC(**1**); JP(**5**)
Phenol	C_6H_6O	--air or vapor	20	40.9	
"	C_6H_6O	--air or vapor	30	39.88	} AC(**18, 19, 25**); JS(**2, 6, 13**); JP(**4**)
Phenylhydrazine	$C_6H_8N_2$	--vapor	20	46.1	JS(**3**)
Phosphorus tribromide	PBr_3	--air	24	45.8	JS(**5**)
Phosphorus trichloride	PCl_3	--vapor	20	29.1	AC(**17**); GC(**1**); JP(**2**); JS(**4**)
Phosphorus triiodide	PI_3	--vapor	75.3	56.5	ZA(**1**)
Propionic acid	$C_3H_6O_2$	--vapor	20	26.7	AC(**27**); GC(**1**); JS(**4**)
n-Propyl acetate	$C_5H_{10}O_2$	--air or vapor	20	24.3	AC(**20, 26**); AdC(**1**); PRS(**2**)
n-Propyl alcohol	C_3H_8O	--vapor	20	23.78	AC(**24, 33**); AdC(**1**); JP(**5**); JS(**4**)
n-Propylamine	C_3H_9N	--air	20	22.4	GC(**1**); JS(**3**)

SURFACE TENSION OF VARIOUS LIQUIDS (Continued)

Substance Name	Formula	In contact with	Temperature °C	Surface tension dynes/cm.	References
n-Propyl bromide	C_3H_7Br	vapor	71	19.65	GC(**1**)
n-Propyl chloride	C_3H_7Cl	air	47	18.2	AdC(**1**)
n-Propyl formate	$C_4H_8O_2$	vapor	20	24.5	AC(**26, 33**); AdC(**1**); PRS(**2**)
Pyridine	C_5H_5N	air	20	38.0	AC(**17, 26, 28**); GC(**1**); JP(**5**); JS(**4, 14**)
Quinoline	C_9H_7N	air	20	45.0	AC(**17, 22**); GC(**1**); JP(**4, 5**)
Ricinoleic acid	$C_{18}H_{34}O_3$	air	16	35.81	PRS(**1**)
Selenium	Se	air	217	92.4	AdP(**3**)
Styrene	C_8H_8	air	19	32.14	PRS(**1**)
Sulfuric acid (98.5%)	H_2SO_4	air or vapor	20	55.1	AC(**17a**); AdP(**7**); JS(**2**)
Tetrabromoethane 1,1,2,2-	$C_2H_2Br_4$	air	20	49.67	AC(**6**); ZC(**2**)
Tetrachloroethane 1,1,2,2-	$C_2H_2Cl_4$	air	22.5	36.03	ZC(**2**)
Tetrachloroethylene	C_2Cl_4	vapor	20	31.74	AC(**8**); GC(**1**)
Toluene	C_7H_8	vapor	10	27.7	AC(**4, 17, 20, 31**
"	C_7H_8	vapor	20	28.5	AdC(**1**); BF(**2**)
"	C_7H_8	vapor	30	27.4	JP(**5**); PRS(**2**); ZC(**5, 6**)
m-Toluidine	C_7H_9N	vapor	20	36.9	AC(**33**)
o-Toluidine	C_7H_9N	air or vapor	20	40.0	AC(**27, 33**); AS(**1, 3**); ZA(**1**)
p-Toluidine	C_7H_9N	air	50	34.6	AC(**27**); AS(**1**); JS(**12**)
Trichloroacetic acid	$C_2HCl_3O_2$	nitrogen	80.2	27.8	ZA(**1**)
Trichloroethane 1,1,2-	$C_2H_3Cl_3$	air	114	22.0	GC(**1**)
Triethyl phosphate	$C_6H_{15}O_4P$	air	15.5	30.61	JS(**5**)
Trimethylamine	C_3H_9N	nitrogen	−4	17.3	ZA(**1**)
Triphenylcarbinol	$C_{19}H_{16}O$	vapor	165.8	30.38	JS(**6**)
Vinyl acetate	$C_4H_6O_2$	vapor	20	23.95	
" "	$C_4H_6O_2$	vapor	25	23.16	
" "	$C_4H_6O_2$	vapor	30	22.54	
Water	H_2O	air	18	73.05	
m-Xylene	C_8H_{10}	vapor	20	28.9	AC(**4, 33, 34**); AdC(**1**); JP(**5**); ZC(**5**)
o-Xylene	C_8H_{10}	air	20	30.10	AC(**6, 17, 34**)
p-Xylene	C_8H_{10}	vapor	20	28.37	AC(**6, 17, 33, 34**); AdC(**1**)

REFERENCE KEY

AC—

Journal of the American Chemical Society. (**1**) Baker and Gilbert, 62: 2479–2480; 40. (**2**) H. Brown, 56: 2564–2568; 38. (**3**) Harkins and Brown, 41: 449; 19. (**4**) Harkins, Brown and Davies, 39: 354; 17. (**5**) Harkins and Cheng, 43: 35; 21. (**6**) Harkins, Clark and Roberts, 42: 700; 20. (**7**) Harkins and Ewing, 42: 2539; 20. (**8**) Harkins and Feldman, 44: 2665; 22. (**9**) Harkins and Grafton, 42: 2534; 20. (**10**) Hogness, 43: 1621; 21. (**11**) Johnson and McIntosh, 31: 1139; 09. (**12**) Maass and Boomer, 44: 1709; 22. (**13**) Maass and Hatcher, 42: 2548; 20. (**14**) Maass and McIntosh, *36*, 741 (1914). (**15**) Maass and Wright, 43: 1098; 21. (**16**) Morgan and Chazel, 35: 1821, 13. (**17**) Morgan and Daghlian, 33: 672; 11. (**17a**) Morgan and Davis, 38: 555; 16. (**18**) Morgan and Egloff, 38; 844: 16. (**19**) Morgan and Evans, 39: 2151; 17. (**20**) Morgan and Griggs, 39: 2261; 17. (**21**) Morgan and Kramer, 35: 1834; 13. (**22**) Morgan and McAfee, 33: 1275; 11. (**23**) Morgan and Neidle, 35: 1856; 13. (**24**) Morgan and Owen, 33: 1713; 11. (**25**) Morgan and Scarlett, 39, 2275; 17. (**26**) Morgan and Schartz, 33: 1041; 11. (**27**) Morgan and Stone, 35: 1505; 13. (**28**) Morgan and Thomssen, 33: 657; 11. (**29**) Morgan and Woodward, 35: 1249; 13. (**30**) Richards and Boyer, 43: 274; 21. (**31**) Richards and Carver 43: 827; 21. (**32**) Richards and Coombs, 37: 1656; 15. (**33**) Richards and Matthews, 30: 8; 08. (**34**) Richards, Speyers and Carver, 46: 1196: 24.

AD—

Atti della reale accademia nazionale dei Lincei. (**1**) Magini, 19 II: 184; 10.

AdC—Annalen der Chemie, Justus Liebig's. (**1**) Schiff. 223: 47; 84

AdP—

Annalen der Physik. (**1**) Heydweiller, 62: 694; 97. (**2**) Quincke, 134: 356; 68. (**3**) Quincke, 135: 621; 68. (**4**) Gradenwitz, 67: 467; 99. (**5**) Meyer, 66: 523; 98. (**6**) Stöckle 66: 499; 98. (**7**) Rontgen and Schneider, 29: 165; 86.

AS—

Archives des sciences physicques at naturelles. (**1**) Dutoit and Friederich, 9: 105; 00. (**2**) Guye and Baud, 11: 449; 01. (**3**) Herzen, 14: 232; 02.

BD—

Berichte der deutschen Chemischen Gesellschaft. (**1**) Lorenz and Kauffler, 41: 3727; 08. (**2**) Schenck and Ellenberger, 37: 3443; 04.

BF—

Bulletin de la société Chimique de France. (**1**) Dutoit and Frederich, 19: 321; 98. (**2**) Sentis Ann. Univ. Grenoble, 27: 593; 04.

CR—

Comptes Rendus. (**1**) A Portevin and P. Bastien, 202; 1072–1074; 36. (**2**) Popesco, 172: 1474; 21.

cUL—

Communications from the Physical Laboratory at the University of Leiden. (**1**) No. 142. (**2**) No. 179a. (**3**) No. 182b.

GC—

Gazzetta chimica italiana. (**1**) Schiff, 14: 368; 84. (**2**) A. Giacolone and D. DiMaggio, 3: 198–206: 39

JI—

Journal of the Institute of Metals, London. (**1**) Smith, 12: 168; 14.

JP—

Journal de Chimie physique. (**1**) Dutoit and Fath, 1: 358; 03. (**2**) Dutoit and Mojoiu, 7: 169; 09. (**3**) Marchand, 11: 573; 13. (**4**) Bolle and Guye, 3: 38; 05. (**5**) Renard and Guye, 5: 81; 07. (**6**) Berthoud, 21: 143; 24. (**7**) Berthoud, 16: 429; 18. (**8**) Homfray and Guye, 1: 505; 03. (**9**) Przyluska. 7: 511: 09.

JR—

Journal of the Russian Physico Chem. Soc. (Chem. part). (**1**) Elisseev and Kurbatov, 41: 1426; 09.

JS—

Journal of the Chemical Society of London. (**1**) Kellas, 113: 903; 18. (**2**) Aston and Ramsay, 65: 167; 94. (**3**) Turner and Merry, 97: 2069; 10. (**4**) Ramsay and Shields, 63: 1089; 93. (**5**) Sugden, Reed and Wilkins, 127: 1525; 25. (**6**) Hewitt and Winmill, 91: 441; 07. (**7**) Smith, 105: 1703; 14. (**8**) Atkins, 99: 10; 11. (**9**) Sugden, 119: 1483; 21. (**10**) Sugden, 121: 858, 22. (**11**) Sugden, 125: 32; 24. (**12**) Sugden, 125: 1167; 24. (**13**) Worley, 105; 260; 14. (**14**) Worley, 105: 273; 14. (**15**) Baly and Donnan, 81: 907; 02.

JSG—Jahresb. Schles. Ges. Vaterl. Kultur. (**1**) Wilborn, 1912: 56.

MB—Metron. Beit. (**1**) Weinstein, No. 6: 89.

MfC—

Monatshefte für Chemie und verwandte Teile anderer Wissenschaften. (**1**) Kremann and Meingast, 35: 1323: 14.

PM—

Philosophical Magazine and Journal of Science, Lond, Edinburg and Dublin. (**1**) Bircumshaw, 2: 341; 26. 3: 1286; 27. (**2**) R. C. Brown. 13: 578–584; 32. (**3**) A. E. Bate, 28: 252–255; 39.

PR—The Physical Review. (**1**) Poindexter, 27: 820; 26.

PRA—

Proceedings of the Royal Academy of Sciences of Amsterdam. (**1**) Kamerlingh Onnes and Kuypers, 17: 528; 14. (**2**) Van Urk, Keesom and Onnes, 28: 958; 25. (**3**) Van Urk, Keesom and Nijhoff 29: 914; 26. (**4**) Jaeger and Kahn, 18: 75; 15.

PRS—

Proceedings of the Royal Society, London. (**1**) Hardy 88: 303; 13. (**2**) Ramsay and Aston, 56: 162; 182; 94.

tPRS—

Philosophical Transactions Royal Society of London, Series A. (**1**) Ramsay and Shields, 184: 647; 93.

tRIA—

Royal Irish Academy Transactions. (**1**) Ramsay and Aston, 32A: 93; 02.

VK—

Verslag koninklijke Academie van Wetenschappen te Amsterdam. (**1**) Verschaffelt, No. 18: 1895; 74. (**2**) Verschaffelt, No. 28, 1896: 94.

WN—

Wissenschaftliche Natuurk. Tydschr. (**1**) Verschaffelt, 2: 231, 25.

ZA—

Zeitschrift fur anorganische ana allgemeine Chemie. (**1**) Jaeger, 101: 1; 17. (**2**) Motylewski, 38: 410, 04. (**3**) Lorenz, Liebmann and Hochberg, 94: 301; 16. (**4**) Sauerwald and Drath 154: 79; 26.

ZC—

Zeitschrift für physikalische Chemie. (**1**) Bennett and Mitchell, 84: 475; 13. (**2**) Walden and Swinne, 82: 271; 13. (**3**) Drucker, 52: 641; 05. (**4**) Walden, 75: 555; 10. (**5**) Walden and Swinne, 79: 700; 12. (**6**) Whatmough, 39: 129: 02.

ZE—

Zeitschrift fur Elektrochemie und angewandte physikalische Chemie. (**1**) Bredig and Tiechmann. 31: 449: 25.

ZK—

Zeitschrift Kolloid. (**1**) N. Jermolanko, 48: 14–146; 29.

VISCOSITY

Viscosity.—All fluids possess a definite resistance to change of form and many solids show a gradual yielding to forces tending to change their form. This property, a sort of internal friction, is called viscosity; it is expressed in dyne-seconds per cm^2 or poises. Dimensions,—$[m\ l^{-1}\ t^{-1}]$. If the tangential force per unit area, exerted by a layer of fluid upon one adjacent is one dyne for a space rate of variation of the tangential velocity of unity, the viscosity is one poise.

Kinematic viscosity is the ratio of viscosity to density. The c. g. s. unit of kinematic viscosity is the **stoke.**

Flow of liquids through a tube; where l is the length of the tube, r its radius, p the difference of pressure at the ends, η the coefficient of viscosity, the volume escaping per second,

$$v = \frac{\pi p r^4}{8 l \eta} \text{ (Poiseuille).}$$

The volume will be given in cm^3 per second if l and r are in cm, p in dynes per cm^2 and η in poises or dyne-seconds per cm^2.

VISCOSITY OF WATER BELOW 0°C

White-Twining 1914

Temperature	Viscosity centipoises	Temperature	Viscosity centipoises
0°C	1.798	−7.23	2.341
−2.10	1.930	−8.48	2.458
−4.70	2.121	−9.30	2.549
−6.20	2.250		

ABSOLUTE VISCOSITY OF WATER AT 20°C

Swindells, J. R. Coe, Jr., and T. B. Godfrey, Journal of Research, National Bureau of Standards **48**, 1, 1952.

The value found for the viscosity of water at 20°C was 0.010019 ± 0.000003 poise.

The value **0.01002** poise is to be used as the absolute value of the viscosity of water for calibration purposes.

VISCOSITY CONVERSION TABLE

Poise = c.g.s. unit of absolute viscosity = $\frac{gm}{sec \times cm}$

Stoke = c.g.s. unit of kinematic viscosity = $\frac{gm}{sec \times cm \times density_{(t°F)}}$

Centipoise = 0.01 poise

Centistoke = 0.01 stoke

Centipoises = Centistokes × density (at given temperature)

To convert poises to $\frac{lb}{sec \times ft}$ **or** $\frac{lb}{hr \times ft}$ **multiply by 0.0672 or 242 respectively.**

Centistokes	Saybolt Seconds at 100°F	Saybolt Seconds at 130°F	Saybolt Seconds at 210°F	Redwood Seconds at 70°F	Redwood Seconds at 140°F	Redwood Seconds at 200°F	Engler Degrees at all temps.
2.0	32.6	32.7	32.8	30.2	31.0	31.2	1.14
3.0	36.0	36.1	36.3	32.7	33.5	33.7	1.22
4.0	39.1	39.2	39.4	35.3	36.0	36.3	1.31
5.0	42.3	42.4	42.6	37.9	38.5	38.9	1.40
6.0	45.5	45.6	45.8	40.5	41.0	41.5	1.48
7.0	48.7	48.8	49.0	43.2	43.7	44.2	1.56
8.0	52.0	52.1	52.4	46.0	46.4	46.9	1.65
9.0	55.4	55.5	55.8	48.9	49.1	49.7	1.75
10.0	58.8	58.9	59.2	51.7	52.0	52.6	1.84
11.0	62.3	62.4	62.7	54.8	55.0	55.6	1.93
12.0	65.9	66.0	66.4	57.9	58.1	58.8	2.02
14.0	73.4	73.5	73.9	64.4	64.6	65.3	2.22
16.0	81.1	81.3	81.7	71.0	71.4	72.2	2.43
18.0	89.2	89.4	89.8	77.9	78.5	79.4	2.64
20.0	97.5	97.7	98.2	85.0	85.8	86.9	2.87
22.0	106.0	106.2	106.7	92.4	93.3	94.5	3.10
24.0	114.6	114.8	115.4	99.9	100.9	102.2	3.34
26.0	123.3	123.5	124.2	107.5	108.6	110.0	3.58
28.0	132.1	132.4	133.0	115.3	116.5	118.0	3.82
30.0	140.9	141.2	141.9	123.1	124.4	126.0	4.07
32.0	149.7	150.0	150.8	131.0	132.3	134.1	4.32
34.0	158.7	159.0	159.8	138.9	140.2	142.2	4.57
36.0	167.7	168.0	168.9	146.9	148.2	150.3	4.83
38.0	176.7	177.0	177.9	155.0	156.2	158.3	5.08
40.0	185.7	186.0	187.0	163.0	164.3	166.7	5.34
42.0	194.7	195.1	196.1	171.0	172.3	175.0	5.59
44.0	203.8	204.2	205.2	179.1	180.4	183.3	5.85
46.0	213.0	213.4	214.5	187.1	188.5	191.7	6.11
48.0	222.2	222.6	223.8	195.2	196.6	200.0	6.37
50.0	231.4	231.8	233.0	203.3	204.7	208.3	6.63
60.0	277.4	277.9	279.3	243.5	245.3	250.0	7.90
70.0	323.4	324.0	325.7	283.9	286.0	291.7	9.21
80.0	369.6	370.3	372.2	323.9	326.6	333.4	10.53
90.0	415.8	416.6	418.7	364.4	367.4	375.0	11.84
*100.0	462.0	462.9	465.2	404.9	408.2	416.7	13.16

*** At higher values use the same ratio as above for 100 centistokes; e.g., 110 centistokes = 110 × 4.620 Saybolt seconds at 100°F.**

To obtain the Saybolt Universal viscosity equivalent to a kinematic viscosity determined at t°F, multiply the equivalent Saybolt Universal viscosity at 100°F by $1 + (t - 100)0.000064$; e.g., 10 centistokes at 210°F are equivalent to 58.8 × 1.0070, or 59.2 Saybolt Universal seconds at 210°F.

VISCOSITY CONVERSION

Kinematic

To convert from	To	Multiply by
cm^2/sec (Stokes)	Centistokes	10^2
	ft^2/hr.	3.875
	ft^2/sec.	1.076×10^{-3}
	in.^{2}/sec.	1.550×10^{-1}
	m^2/hr.	3.600×10^{-1}
cm^2/sec $\times 10^2$ (Centistokes)	cm^2/sec. (Stokes)	1×10^{-2}
	ft^2/hr.	3.875×10^{-2}
	ft^2/sec.	1.076×10^{-5}
	in.^{2}/sec.	1.550×10^{-3}
	m^2/hr.	3.600×10^{-3}
ft^2/hr	cm^2/sec (Stokes)	2.581×10^{-1}
	cm/sec $\times 10^2$ (centistokes)	2.581×10
	ft^2/sec.	2.778×10^{-4}
	in.^{2}/sec.	4.00×10^{-2}
	m^2/hr.	9.290×10^{-2}
ft^2/sec	cm^2/sec. (Stokes)	9.29×10^2
	cm^2/sec.$\times 10^2$ (centistokes)	9.29×10^4
	ft^2/hr	3.60×10^3
	in.^{2}/sec.	1.44×10^2
	m^2/hr.	3.345×10^2
in.^{2}/sec	cm^2/sec. (Stokes)	6.452
	cm^2/sec.$\times 10^2$ (centistokes)	6.452×10^2
	ft^2/hr.	2.50×10
	ft^2/sec.	6.944×10^{-3}
	m^2/hr.	2.323
m^2/hr	cm^2/sec. (Stokes)	2.778
	cm^2/sec.$\times 10^2$ (centistokes)	2.778×10^2
	ft^2/hr.	1.076×10
	ft^2/sec.	2.990×10^{-3}
	in.^{2}/sec.	4.306×10^{-1}

VISCOSITY CONVERSION

Absolute

Absolute viscosity = kinematic viscosity × density; lb = mass pounds; lb$_F$ = force pounds

To convert from	To	Multiply by
gm/(cm)(sec) [Poise]	gm/(cm)(sec)(10^2) [Centipoise]	10^2
	kg/(m)(hr)	3.6×10^2
	lb/(ft)(sec)	6.72×10^{-2}
	lb/(ft)(hr)	2.419×10^2
	lb/(in.)(sec)	5.6×10^{-3}
	(gm$_F$)(sec)/cm^2	1.02×10^{-3}
	(lb$_F$)(sec)/in.^{2} [Reyn]	1.45×10^{-5}
	(lb$_F$)(sec)/ft^2	2.089×10^{-3}
gm/(cm)(sec)(10^2) [Centipoise]	gm/(cm)(sec) [Poise]	10^{-2}
	kg/(m)(hr)	3.6
	lb/(ft)(sec)	6.72×10^{-4}
	lb/(ft)(hr)	2.419
	lb/(in.)(sec)	5.60×10^{-5}
	(gm$_F$)(sec)/cm^2	1.02×10^{-5}
	(lb$_F$)(sec)/in.^{2} [Reyn]	1.45×10^{-7}
	(lb$_F$)(sec)/ft^2	2.089×10^{-5}
kg/(m)(hr)	gm/(cm)(sec)	2.778×10^{-3}
	gm/(cm)(sec)(10^2) [Centipoise]	2.778×10^{-1}
	lb/(ft)(sec)	1.867×10^{-4}
	lb/(ft)(hr)	6.720×10^{-1}
	lb/(in.)(sec)	1.555×10^{-5}
	(gm$_F$)(sec)/cm^2	2.833×10^{-4}

VISCOSITY CONVERSION

Absolute (Continued)

To convert from	To	Multiply by
kg/(m)(hr) (Cont.)	(lb$_F$)(sec)/in.^{2} [Reyn]	4.029×10^{-8}
	(lb$_F$)(sec)/ft^2	5.801×10^{-6}
lb/(ft)(sec)	gm/(cm)(sec) [Poise]	1.488×10^1
	gm/(cm)(sec)(10^2) [Centipoise]	1.488×10^3
	kg/(m)(hr)	5.357×10^3
	lb/(ft)(hr)	3.60×10^3
	lb/(in.)(sec)	8.333×10^{-2}
	(gm$_F$)(sec)/cm^2	1.518×10^{-2}
	(lb$_F$)(sec)/in.^{2} [Reyn]	2.158×10^{-4}
	(lb$_F$)(sec)/ft^2	3.108×10^{-2}
lb/(ft)(hr)	gm/(cm)(sec) [Poise]	4.134×10^{-3}
	gm/(cm)(sec)(10^2) [Centipoise]	4.134×10^{-1}
	kg/(m)(hr)	1.488
	lb/(ft)(sec)	2.778×10^{-4}
	lb/(in.)(sec)	2.315×10^{-5}
	(gm$_F$)(sec)/cm^2	4.215×10^{-6}
	(lb$_F$)(sec)/in.^{2} [Reyn]	5.996×10^{-8}
	(lb$_F$)(sec)/ft^2	8.634×10^{-6}
lb/(in.)(sec)	gm/(cm)(sec) [Poise]	1.786×10^2
	gm/(cm)(sec)(10^2) [Centipoise]	1.786×10^4
	kg/(m)(hr)	6.429×10^4
	lb/(ft)(sec)	1.20×10
	lb/(ft)(hr)	4.32×10^4
	(gm$_F$)(sec)/cm^2	1.821×10^{-1}
	(lb$_F$)(sec)/in.^{2} [Reyn]	2.590×10^{-3}
	(lb$_F$)(sec)/ft^2	3.73×10^{-1}
(gm$_F$)(sec)/cm^2	gm/(cm)(sec)	9.807×10^2
	gm/(cm)(sec)(10^2) [Centipoise]	9.807×10^4
	kg/(m)(hr)	3.530×10^5
	lb/(ft)(sec)	6.590×10
	lb/(ft)(hr)	2.372×10^5
	lb/(in.)(sec)	5.492
	(lb$_F$)(sec)/in.^{2} [Reyn]	1.422×10^{-2}
	(lb$_F$)(sec)/ft^2	2.048
(lb$_F$)(sec)/in.^{2} [Reyn]	gm/(cm)(sec) [Poise]	6.895×10^4
	gm/(cm)(sec)(10^2) [Centipoise]	6.895×10^6
	kg/(m)(hr)	2.482×10^7
	lb/(ft)(sec)	4.633×10^3
	lb/(ft)(hr)	1.668×10^7
	lb/(in.)(sec)	3.861×10^2
	(gm$_F$)(sec)/cm^2	7.031×10
	(lb$_F$)(sec)/ft^2	1.440×10^2
(lb$_F$)(sec)/ft^2	gm/(cm)(sec) [Poise]	4.788×10^2
	gm/(cm)(sec)(10^2) [Centipoise]	4.788×10^4
	kg/(m)(hr)	1.724×10^5
	lb/(ft)(sec)	3.217×10
	lb/(ft)(hr)	1.158×10^5
	lb/(in.)(sec)	2.681
	(gm$_F$)(sec)/cm^2	4.882×10^{-1}
	(lb$_F$)(sec)/in.^{2} [Reyn]	6.944×10^{-3}

THE VISCOSITY OF WATER 0°C TO 100°C

°C	η(cp)	°C	η(cp)	°C	η(cp)	°C	η(cp)
0	1.787	26	0.8705	52	0.5290	78	0.3638
1	1.728	27	.8513	53	.5204	79	.3592
2	1.671	28	.8327	54	.5121	80	.3547
3	1.618	29	.8148	55	.5040	81	.3503
4	1.567	30	.7975	56	.4961	82	.3460
5	1,519	31	.7808	57	.4884	83	.3418
6	1.472	32	.7647	58	.4809	84	.3377
7	1.428	33	.7491	59	.4736	85	.3337
8	1.386	34	.7340	60	.4665	86	.3297
9	1.346	35	.7194	61	.4596	87	.3259
10	1.307	36	.7052	62	.4528	88	.3221
11	1.271	37	.6915	63	.4462	89	.3184
12	1.235	38	.6783	64	.4398	90	.3147
13	1.202	39	.6654	65	.4335	91	.3111
14	1.169	40	.6529	66	.4273	92	.3076
15	1.139	41	.6408	67	.4213	93	.3042
16	1.109	42	.6291	68	.4155	94	.3008
17	1.081	43	.6178	69	.4098	95	.2975
18	1.053	44	.6067	70	.4042	96	.2942
19	1.027	45	.5960	71	.3987	97	.2911
20	1.002	46	.5856	72	.3934	98	.2879
21	0.9779	47	.5755	73	.3882	99	.2848
22	.9548	48	.5656	74	.3831	100	.2818
23	.9325	49	.5561	75	.3781		
24	.9111	50	.5468	76	.3732		
25	.8904	51	.5378	77	.3684		

The above table was calculated from the following empirical relationships derived from measurements in viscometers calibrated with water at 20°C (and one atmosphere), modified to agree with the currently accepted value for the viscosity at 20° of 1.002 cp:

$$0° \text{ to } 20°\text{C}: \log_{10} \eta_T = \frac{1301}{998.333 + 8.1855(T-20) + 0.00585(T-20)^2} - 3.30233$$

(R. C. Hardy and R. L. Cottingham, J.Res.NBS *42*, 573 (1949).)

$$20° \text{ to } 100°\text{C}: \log_{10} \frac{\eta_T}{\eta_{20}} = \frac{1.3272(20-T) - 0.001053(T-20)^2}{T + 105}$$

(J. F. Swindells, NBS, unpublished results.)

VISCOSITY OF LIQUIDS

Viscosity of liquids in centipoises (cp) including elements, inorganic and organic compounds and mixtures.

Liquid	Temp. °C	Viscosity cp
Acetaldehyde	0	.2797
	10	.2557
	20	.22
Acetanilide	120	2.22
	130	1.90
Acetic acid	15	1.31
	18	1.30
	25.2	1.155
	30	1.04
	41	1.00
	59	.70
	70	.60
	100	.43
anhydride	0	1.24
	15	.971
	18	.90
	30	.783
	100	.49
Acetone	−92.5	2.148
	−80.0	1.487
	−59.6	.932
	−42.5	.695
	−30.0	.575
	−20.9	.510
	−13.0	.470
	−10.0	.450
	0	.399
	15	.337
	25	.316
	30	.295
	41	.280
Acetonitrile	0	.442
	15	.375
	25	.345
Acetophenone	11.9	2.28
	23.5	1.59
	25.0	1.617
	50.0	1.246
	80.0	.734
Air, liq	−192.3	.172
Alcohol. See *Ethyl*, *Methyl*, etc.		
Allyl alcohol	0	2.145
	15	1.49
	20	1.363
	30	1.07
	40	.914
	70	.553
Allylamine	130	.506
Allyl chloride	15	.347
	30	.300
Ammonia	−69	.475
	−50	.317
	−40	.276
	−33.5	.255

Liquid	Temp. °C	Viscosity cp
n-Amyl acetate	11	1.58
	45	.805
alcohol	15	4.65
	30	2.99
ether	15	1.188
Aniline	−6	13.8
	0	10.2
	5	8.06
	10	6.50
	15	5.31
	20	4.40
	25	3.71
	30	3.16
	35	2.71
	40	2.37
	50	1.85
	60	1.51
	70	1.27
	80	1.09
	90	.935
	100	.825
Anisol	0	1.78
	20	1.32
	40	1.12
Antimony, liq	645	1.55
	700	1.26
	800	1.08
	850	1.05
Benzaldehyde	25	1.39
Benzene	0	.912
	10	.758
	20	.652
	30	.564
	40	.503
	50	.442
	60	.392
	70	.358
	80	.329
Benzonitrile	25	1.24
Benzophenone	55	4.79
	120	1.38
Benzyl alcohol	20	5.8
Benzylamine	25	1.59
Benzylaniline	33	2.18
	130	1.20
Benzyl ether	0	10.5
	20	5.33
	40	3.21
Bismuth	285	1.61
	304	1.662
	365	1.46
	451	1.280
	600	.998
Bromine, liq	−4.3	1.31
	0	1.241

VICOSITY OF LIQUIDS (Continued)

Liquid	Temp. °C	Viscosity cp
Bromine, liq	12.6	1.07
	16	1.0
	19.5	.995
	28.9	.911
o-Bromoaniline	40	3.19
m-Bromoaniline	20	6.81
	40	3.70
	80	1.70
p-Bromoaniline	80	1.81
Bromobenzene	15	1.196
	30	.985
Bromoform	15	2.152
	25	1.89
	30	1.741
Butyl acetate	0	1.004
	20	.732
	40	.563
n-Butyl alcohol	−50.9	36.1
	−30.1	14.7
	−22.4	11.1
	−14.1	8.38
	0	5.186
	15	3.379
	20	2.948
	30	2.30
	40	1.782
	50	1.411
	70	.930
	100	.540
sec-Butyl alcohol	15	4.21
n-Butyl bromide	15	.626
n-Butyl chloride	15	.469
Butyl chloride, tertiary	15	.543
n-Butyl formate	0	.940
	20	.689
Butyric acid	0	2.286
	15	1.81
	20	1.540
	40	1.120
	50	.975
	70	.760
	100	.551
Cadmium, liq	349	1.44
	506	1.18
	603	1.10
Carbolic acid. See *Phenol.*		
Carbon dioxide, liq., pressure that of saturated vapor	0	.099
	10	.085
	20	.071
	30	.053
disulfide	−13	.514
	−10	.495
	0	.436
	5	.380
	20	.363
	40	.330
Carbon tetrachloride	0	1.329
	15	1.038

Liquid	Temp. °C	Viscosity cp
Carbon tetrachloride	20	.969
	30	.843
	40	.739
	50	.651
	60	.585
	70	.524
	80	.468
	90	.426
	100	.384
Cetyl alcohol	50	13.4
Chlorine, liq	−76.5	.729
	−70.5	.680
	−60.2	.616
	−52.4	.566
	−35.4	.494
	0	.385
Chlorobenzene	15	.900
	20	.799
	40	.631
	80	.431
	100	.367
Chloroform	−13	.855
	0	.700
	8.1	.643
	15	.596
	20	.58
	25	.542
	30	.514
	39	.500
o-Chlorophenol	25	4.11
	50	2.015
m-Chlorophenol	25	11.55
p-Chlorophenol	50	4.99
Copper, liq	1,085	3.36
	1,100	3.33
	1,150	3.22
	1,200	3.12
o-Cresol	40	4.49
m-Cresol	10	43.9
	20	20.8
	40	6.18
p-Cresol	40	7.00
Creosote	20	12.0
Cycloheptane	13.5	1.64
Cyclohexane	17	1.02
Cyclohexanol	20	68
Cyclohexene	13.5	.696
	20	.66
Cyclooctane	13.5	2.35
Cyclopentane	13.5	.493
*n*Decane	20	.92
Diethylamine	25	.346
	25	.367
Diethylaniline	.5	3.84
	20.0	2.18
	25.0	1.95
Diethylcarbinol	15.0	7.34
Diethylketone	15	.493

VISCOSITY OF LIQUIDS (Continued)

Liquid	Temp. °C	Viscosity cp
Dimethylaniline	10	1.69
	20	1.41
	25	1.285
	30	1.17
	40	1.04
	50	.91
Dimethyl-α-naphthylamine	130	.868
Dimethyl-β-naphthylamine	130	.952
Diphenyl	70	1.49
	100	.97
Diphenylamine	130	1.04
Dodecane	25	1.35
Ether (diethyl-)	−100	1.69
	−80	.958
	−60	.637
	−40	.461
	−20	.362
	0	.2842
	17	.240
	20	.2332
	25	.222
	40	.197
	60	.166
	80	.140
	100	.118
Ethyl acetate	0	.582
	8.96	.516
	10	.512
	15	.473
	20	.455
	25	.441
	30	.400
	50	.345
	75	.283
alcohol	−98.11	44.0
	−89.8	28.4
	−71.5	13.2
	−59.42	8.41
	−52.58	6.87
	−32.01	3.84
	−17.59	2.68
	−.30	1.80
	0	1.773
	10	1.466
	20	1.200
	30	1.003
	40	.834
	50	.702
	60	.592
	70	.504
alcohol, anh	−148	8,470
	−146	5,990
	−130	467
aniline	25	2.04
Ethylbenzene	17	.691
benzoate	20	2.24
bromide	−120	5.6
	−100	2.89

Liquid	Temp. °C	Viscosity cp
Ethylbenzene, bromide	−80	1.81
	0	.487
	10	.441
	15	.418
	20	.402
	30	.348
n-Ethyl butyrate	15	.711
Ethyl carbonate	15	.868
Ethylene bromide	0	2.438
	17	1.95
	20	1.721
	40	1.286
	67.3	.922
	70	.903
	82.2	.750
	99.0	.648
chloride	0	1.077
	15	.887
	19.4	.800
	40	.652
	50	.565
	70	.479
glycol	20	19.9
	40	9.13
	60	4.95
	80	3.02
	100	1.99
oxide	−49.8	.577
	−38.2	.488
	−21.0	.394
	0	.320
Ethyl formate	20	.402
iodide	0	.727
	15	.617
	20	.592
	40	.495
	70	.391
malate	24.7	3.016
oxalate	15	2.31
propionate	15	.564
Eugenol	0	29.9
	20	9.22
	40	4.22
Fluorobenzene	20	.598
	40	.478
	60	.389
	80	.329
	100	.275
Formamide	0	7.55
	25	3.30
Formic acid	7.59	2.3868
	10	2.262
	20	1.804
	30	1.465
	40	1.219
	70	.780
	100	.549

VISCOSITY OF LIQUIDS (Continued)

Liquid	Temp. °C	Viscosity cp
Furfurol	0	2.48
	25	1.49
Glucose	22	9.1×10^{15}
	30	6.6×10^{13}
	40	2.8×10^{11}
	60	9.3×10^{7}
	80	6.6×10^{5}
	100	2.5×10^{4}
Glycerin	−42	6.71×10^{6}
	−36	2.05×10^{6}
	−25	2.62×10^{5}
	−20	1.34×10^{5}
	−15.4	6.65×10^{4}
	−10.8	3.55×10^{4}
	−4.2	1.49×10^{4}
	0	12,110
	6	6,260
	15	2,330
	20	1,490
	25	954
	30	629
Glycerin trinitrate	10	69.2
	20	36.0
	30	21.0
	40	13.6
	60	6.8
Heptane	0	.524
	17	.461
	20	.409
	25	.386
	40	.341
	70	.262
n-Heptyl alcohol	15	8.53
Hexadecane	20	3.34
Hexane	0	.401
	17	.374
	20	.326
	25	.294
	40	.271
	50	.248
Hydrazine	1	1.29
	10	1.12
	20	.97
Hydrogen, liq		.011
Iodine, liq	116	2.27
Iodobenzene	15	1.74
Iron, 2.5% carbon, liq	1,400	2.25
Isoamyl acetate	8.97	1.030
	19.91	.872
alcohol	10	6.20
amine	25	.724
Isobutyl alcohol	15	4.703
amine	25	.553
Isobutyric acid	15	1.44
	30	1.13
Isoeugenol	25	26.72
Isoheptane	0	.481
	20	.384
Isoheptane	40	.315
Isohexane	0	.376
	20	.306
	40	.254
Isopentane	0	.273
	20	.223
Isopropyl alcohol	15	2.86
	30	1.77
Isoquinoline	25	3.57
Isosafrol	25	3.981
Lead, liq	350	2.58
	400	2.33
	441	2.116
	500	1.84
	551	1.70
	600	1.38
	703	1.349
	844	1.185
Menthol, liq	55.6	6.29
	74.6	2.47
	99.0	1.04
Mercury	−20	1.855
	−10	1.764
	0	1.685
	10	1.615
	19.02	1.56
	20	1.554
	20.2	1.55
	30	1.499
	40	1.450
	40.8	1.45
	41.86	1.44
	50	1.407
	60	1.367
	70	1.331
	80	1.298
	90	1.268
	100	1.240
	150	1.130
	200	1.052
	250	.995
	300	.950
	340	.921
Methyl acetate	0	.484
	20	.381
	40	.320
Methyl alcohol (Methanol)	−98.30	13.9
	−84.23	6.8
	−72.55	4.36
	−44.53	1.98
	−22.29	1.22
	0	.82
	15	.623
	20	.597
	25	.547
	30	.510
	40	.456
	50	.403

VISCOSITY OF LIQUIDS (Continued)

Liquid	Temp. °C	Viscosity cp	Liquid	Temp. °C	Viscosity cp
Methyl amine	0	.236	Oil, olive	10	138.0
aniline	25	2.02		20	84.0
	30	1.55		40	36.3
chloride	20	.1834		70	12.4
Methylene bromide	15	1.09	rape	0	2,530
	30	0.92		10	385
chloride	15	.449		20	163
	30	.393		30	96
Methyl iodide	0	.606	soya bean	20	69.3
	15	.518		30	40.6
	20	.500		50	20.6
	30	.460		90	7.8
	40	.424	sperm	15.6	42.0
Naphthalene	80	.967		37.8	18.5
	100	.776		100.0	4.6
Nitric acid	0	2.275	Oleic acid	30	25.6
	10	1.770	Pentadecane	22	2.81
Nitrobenzene	2.95	2.91	Pentane	0	.289
	5.69	2.71		20	.240
	5.94	2.71	*o*-Phenetidine	0	16.5
	9.92	2.48		20	6.08
	14.94	2.24		30	4.22
	20.00	2.03	*m*-Phenetidine	30	12.9
Nitromethane	0	.853	*p*-Phenetidine	20	12.9
	25	.620		30	8.3
o-Nitrotoluene	0	3.83	Phenol	18.3	12.7
	20	2.37		50	3.49
	40	1.63		60	2.61
	60	1.21		70	2.03
m-Nitrotoluene	20	2.33		90	1.26
	40	1.60	Phenylcyanide	.28	1.96
	60	1.18		20.0	1.33
p-Nitrotoluene	60	1.20	Phosphorus, liq	21.5	2.34
n-Nonane	20	.711		31.2	2.01
n-Octane	0	.706		43.2	1.73
	16	.574		50.5	1.60
	20	.542		60.2	1.45
	40	.433		69.7	1.32
Octodecane	40	2.86		79.9	1.21
n-Octylalcohol	15	10.6	Potassium bromide, liq	745	1.48
Oil, castor	10	2,420		775	1.34
	20	986		805	1.19
	30	451	nitrate, liq	334	2.1
	40	231		358	1.7
	100	16.9		333	2.97
cottonseed	20	70.4		418	2.00
cylinder, filtered	37.8	240.6	Propionic acid	10	1.289
	100	18.7		15	1.18
cylinder, dark	37.8	422.4		20	1.102
	100	24.0		40	.845
linseed	30	33.1	Propyl acetate	10	.66
	50	17.6		20	.59
	90	7.1		40	.44
machine, light	15.6	113.8	*n*-Propyl alcohol	0	3.883
	37.8	34.2		15	2.52
	100	4.9		20	2.256
machine, heavy	15.6	660.6		30	1.72
	37.8	127.4		40	1.405

VISCOSITY OF LIQUIDS (Continued)

Liquid	Temp. °C	Viscosity cp
n-Propyl alcohol	50	1.130
	70	.760
Propyl aldehyde	10	.47
	20	.41
	40	.33
bromide	0	.651
	20	.524
	40	.433
chloride	0	.436
	20	.352
	40	.291
n-Propyl ether	15	.448
Pyridine	20	.974
Salicylic acid	10	3.20
	20	2.71
	40	1.81
Salol	45	.746
Sodium bromide	762	1.42
	780	1.28
chloride, liq	841	1.30
	896	1.01
	924	.97
nitrate, liq	308	2.919
	348	2.439
	398	1.977
	418	1.828
Stearic acid	70	11.6
Sucrose (cane sugar)	109	2.8×10^6
	124.6	1.9×10^5
Sulfur (gas free)	123.0	10.94
	135.5	8.66
	149.5	7.09
	156.3	7.19
	158.2	7.59
	159.2	9.48
	159.5	14.45
	160.0	22.83
	160.3	77.32
	165.0	500.0
	171.0	4,500.0
	184.0	16,000.00
	190.5	19,700.0
	197.5	21,300.0
	200.0	21,500.0
	210.0	20,500.0
	217.0	19,100.0
	220.0	18,600.0
Sulfur dioxide, liq	−33.5	.5508
	−10.5	.4285
	0.1	.3936
Sulfuric acid	0	48.4
	15	32.8
	20	25.4
	30	15.7
	40	11.5
Sulfuric acid	50	8.82
	60	7.22
	70	6.09
	80	5.19
Tetrachloroethane	15	1.844
Tetradecane	20	2.18
Tin, liq	240	2.12
	280	1.678
	300	1.73
	301	1.680
	400	1.43
	450	1.270
	500	1.20
	600	1.08
	604	1.045
	750	.905
Toluene	0	.772
	17	.61
	20	.590
	30	.526
	40	.471
	70	.354
o-Toluidine	20	4.39
m-Toluidine	20	3.81
p-Toluidine	50	1.80
Triacetin	17	28.0
Tributyrin	20	11.6
Trichlorethane	20	1.2
Tridecane	23.3	1.55
Triethylcarbinol	20	6.75
Tripalmitin	70	16.8
Tristearin	75	18.5
Turpentine	0	2.248
	10	1.783
	20	1.487
	30	1.272
	40	1.071
	70	.728
Turpentine, Venice	17.3	1.3×10^5
n-Undecane	20	1.17
o-Xylene (xylol)	0	1.105
	16	.876
	20	.810
	40	.627
m-Xylene (xylol)	0	.806
	15	.650
	20	.620
	40	.497
p-Xylene (xylol)	16	.696
	20	.648
	40	.513
Zinc, liq	280	1.68
	357	1.42
	389	1.31

Gas or vapor	Temp. °C	Viscosity micro-poises
Acetic acid, vap	119.1	107.0
Acetone, vap	100	93.1
	119.0	99.1
	190.4	118.6
	247.7	133.4
	306.4	148.1
Acetylene	0	93.5
Air	−194.2	55.1
	−183.1	62.7
	−104.0	113.0
	−69.4	133.3
	−31.6	153.9
	0	170.8
	18	182.7
	40	190.4
	54	195.8
	74	210.2
	229	263.8
	334	312.3
	357	317.5
	409	341.3
	466	350.1
	481	358.3
	537	368.6
	565	375.0
	620	391.6
	638	401.4
	750	426.3
	810	441.9
	923	464.3
	1034	490.6
	1134	520.6
Alcohol. See *Ethyl*, *Methyl*, etc.		
Ammonia	−78.5	67.2
	0	91.8
	20	98.2
	50	109.2
	100	127.9
	132.9	139.9
	150	146.3
	200	164.6
	250	181.4
	300	198.7
Argon	0	209.6
	20	221.7
	100	269.5
	200	322.3
	302	368.5
	401	411.5
	493	448.4
	584	481.5
	714	525.7
	827	563.2
Arsenic hydride (Arsine)	0	145.8
	15	114.0
	100	198.1

Gas or vapor	Temp. °C	Viscosity micro-poises
Benzene, vap	14.2	73.8
	131.2	103.1
	194.6	119.8
	252.5	134.3
	312.8	148.4
Bromine, vap	12.8	151
	65.7	170
	99.7	188
	139.7	208
	179.7	227
	220.3	248
Bromoform, vap	151.2	253.0
Butyl alcohol, *n*, vap	116.9	143
tert, vap	82.9	160
chloride, *n*, vap	78	149.5
iodide, vap	130	202
β-Butylene	18.8	74.4
	100.4	94.5
	200	119.2
Butyric acid, vap	161.7	130.0
Carbon dioxide	−97.8	89.6
	−78.2	97.2
	−60.0	106.1
	−40.2	115.5
	−21	129.4
	−19.4	126.0
	0	139.0
	15	145.7
	19	149.9
	20	148.0
	30	153
	32	155
	35	156
	40	157
	99.1	186.1
	104	188.9
	182.4	222.1
	235	241.5
	302.0	268.2
	490	330.0
	685	380.0
	850	435.8
	1052	478.6
disulfide, vap	0	91.1
	14.2	96.4
	114.3	130.3
	190.2	156.1
	309.8	196.6
monoxide	−191.5	56.1
	−78.5	127
	0	166
	15	172
	21.7	175.3
	126.7	218.3
	227.0	254.8
	276.9	271.4
tetrachloride, vap	76.7	195.0

Gas or vapor	Temp. °C	Viscosity micro-poises
Carbon tetrachloride, vap	127.9	133.4
	200.2	156.2
	314.9	190.2
Chlorine	12.7	129.7
	20	132.7
	50	146.9
	100	167.9
	150	187.5
	200	208.5
	250	227.6
Chloroform, vap	0	93.6
	14.2	98.9
	100	129
	121.3	135.7
	189.1	157.9
	250.0	177.6
	307.5	194.7
Cyanogen	0	92.8
	17	98.7
	100	127.1
Ethane	−78.5	63.4
	0	84.8
	17.2	90.1
	50.8	100.1
	100.4	114.3
	200.3	140.9
Ether (diethyl), vap	0	67.8
	14.2	71.6
	100	95.5
	121.8	98.3
	159.4	107.9
	189.9	115.2
	251.0	130.0
	277.8	135.8
Ethyl acetate, vap	0	68.4
	100	94.3
	128.1	101.8
	158.6	109.8
	192.9	119.5
	212.5	126
alcohol, vap	100	108
	130.2	117.3
	170.7	129.3
	191.8	135.5
	212.5	140
	251.7	151.9
	308.7	167.0
bromide, vap	38.4	186.5
butyrate, vap	119.8	160.0
chloride, vap	0	93.7
Ethylene	−75.7	69.9
	−44.1	76.9
	−38.6	78.5
	0	90.7
	13.8	95.4
	20	100.8
Ethylene	50	110.3
	100	125.7
	150	140.3
	200	154.1
	250	166.6
bromide, vap	131.6	221.0
chloride, vap	83.5	168.0
Ethyl formate, vap	99.8	92
iodide, vap	72.3	216.0
Helium	−257.4	27.0
	−252.6	35.0
	−191.6	87.1
	0	186.0
	20	194.1
	100	228.1
	200	267.2
	250	285.3
	282	299.2
	407	343.6
	486	370.6
	606	408.7
	676	430.3
	817	471.3
Hydrogen	−257.7	5.7
	−252.5	8.5
	−198.4	33.6
	−183.4	38.8
	−113.5	57.2
	−97.5	61.5
	−31.6	76.7
	0	83.5
	20.7	87.6
	28.1	89.2
	129.4	108.6
	229.1	126.0
	299	138.1
	412	155.4
	490	167.2
	601	182.9
	713	198.2
	825	213.7
bromide	18.7	181.9
	100.2	234.4
chloride	12.5	138.5
	16.5	140.7
	18	142.6
	100.3	182.2
iodide	20	165.5
	50	201.8
	100	231.6
	150	262.7
	200	292.4
	250	318.9
phosphide	0	106.1
	15	112.0
	100	143.8
sulfide	0	116.6

Gas or vapor	Temp. °C	Viscosity micro-poises
Hydrogen sulfide	17	124.1
	100	158.7
Iodine, vap	124.0	184
	170.0	204
	205.4	220
	247.1	240
Isobutyl acetate, vap	16.1	76.4
	116.4	155.0
alcohol, vap	108.4	144.5
bromide, vap	92.3	179.5
butyrate, vap	156.9	167.0
chloride, vap	68.5	150.0
iodide, vap	120	204.7
Isopentane, vap	25	69.5
	100	86.0
Isopropyl alcohol, vap	99.8	109
	120.3	103.1
	198.4	124.8
	293.1	148.8
bromide, vap	60	176.0
chloride, vap	37.0	148.5
iodide, vap	89.3	201.5
Krypton	0	232.7
	15	246
Mercury, vap	273	494
	313	551
	369	641
	380	654
Methane	−181.6	34.8
	−78.5	76.0
	0	102.6
	20	108.7
	100.0	133.1
	200.5	160.5
	284	181.3
	380	202.6
	499	226.4
Methyl acetate, vap	99.8	98
	100	100
	143.3	113.9
	218.5	134.8
alcohol, vap	66.8	135.0
	111.3	125.9
	217.5	162.0
	311.5	192.1
chloride	−15.3	92
	0	96.9
	15.0	104
	99.1	137
	182.4	168
	302.0	211
iodide, vap	44	232
Neon	0	297.3
	20	311.1
	100	364.6
	200	424.8
	250	453.2
Neon	285	470.8
	429	545.4
	502	580.2
	594	623.0
	686	662.6
	827	721.0
Nitric oxide (NO)	0	178
	20	187.6
	100	227.2
	200	268.2
Nitrogen	−21.5	156.3
	10.9	170.7
	27.4	178.1
	127.2	219.1
	226.7	255.9
	299	279.7
	490	337.4
	825	419.2
Nitrosyl chloride	15	113.9
	100	150.4
	200	192.0
Nitrous oxide (N_2O)	0	135
	26.9	148.8
	126.9	194.3
n-Nonane, vap	100.3	63.3
	202.1	78.1
n-Octane, vap	100.4	67.5
	202.2	84.8
Oxygen	0	189
	19.1	201.8
	127.7	256.8
	227.0	301.7
	283	323.3
	402	369.3
	496	401.3
	608	437.0
	690	461.2
	829	501.2
n-Pentane, vap	25	67.6
	100	84.1
Propane	17.9	79.5
	100.4	100.9
	199.3	125.1
n-Propyl alcohol, vap	99.9	93
	121.7	102.5
	209.7	126.7
	273.0	143.4
bromide, vap	99.8	119
Propylene	16.7	83.4
	49.9	93.5
	100.1	107.6
	199.4	133.8
Propyl iodide, vap	102	210.0
Sulfur dioxide	−75.0	85.8
	−20.0	107.8
	0	115.8
	0	117

VISCOSITY OF GASES (Continued)

Gas or vapor	Temp. °C	Viscosity micro-poises	Gas or vapor	Temp. °C	Viscosity micro-poises
Sulfur dioxide	18	124.2	Water, vap	150	144.5
	20.5	125.4		200	163.5
	100.4	161.2		250	182.7
	199.4	203.8		300	202.4
	293	244.7		350	221.8
	490	311.5		400	241.2
Trimethylbutane. (2,2,3-),			Xenon	0	210.1
vap	70.3	73.4		16.5	223.5
	132.2	82.7		20	226.0
	262.1	104.8		127	300.9
Trimethylethylene, vap	25	70.1		177	335.1
	100	86.9		227	365.2
Water, vap	100	125.5		277	395.4

DIFFUSION COEFFICIENTS IN AQUEOUS SOLUTIONS AT 25°

The diffusion coefficient D may be defined by either of the equations

$$J = -D\frac{\partial c}{\partial x}$$

or

$$\frac{\partial c}{\partial t} = D\frac{\partial^2 c}{\partial x^2}$$

when diffusion occurs in the x-direction only. Here J is the diffusion-flux across unit area normal to the x-direction, $\frac{\partial c}{\partial x}$ is the concentration-gradient at a fixed time, $\frac{\partial c}{\partial t}$ is the rate of change of concentration with time at a fixed distance. If J is expressed in mole cm^{-2} sec^{-1} and c in mole cm^{-3}, x in cm, and t in sec, D will be given in units of cm^2 sec^{-1}. In general D varies somewhat with concentration. The values below are a selection from measurements by modern high-precision methods, mainly by H. S. Harned and collaborators, R. H. Stokes and collaborators, L. J. Gosting and collaborators, and L. G. Longsworth.

For strong electrolytes at infinite dilution, limiting diffusion coefficients may be calculated by the Nernst relation:

$$D = \frac{RT}{F^2}\left[\frac{(\nu_1 + \nu_2)(\lambda_1^0 + \lambda_2^0)}{\nu_1|Z_1|(\lambda_1^0 + \lambda_2^0)}\right]$$

where R = gas constant, F = Faraday, T = absolute temperature, λ_1^0 and λ_2^0 are cation and anion limiting equivalent conductances, ν_1 and ν_2 are the numbers of cations and anions formed from one "molecule" of electrolyte, and Z_1 is the cation valency. Concentrations, unless expressed otherwise are as molarities and the diffusion coefficients are expressed as $10^5 D/cm^2\ sec^{-1}$ at 25°C.

DIFFUSION COEFFICIENTS OF STRONG ELECTROLYTES

Molarity

Solute	0.01	0.1	1.0
HCL	....	3.050	3.436
HBr	....	3.156	3.87
LiCl	1.312	1.269	1.302
LiBr	....	1.279	1.404
$LiNO_3$	1.276	1.240	1.293
NaCl	1.545	1.483	1.484
NaBr	....	1.517	1.596
NaI	....	1.520	1.662
KCl	1.917	1.844	1.892
KBr	....	1.874	1.975
KI	....	1.865	2.065
KNO_3	1.846	....	....
$KClO_4$	1.790	....	....
$CaCl_2$	1.188	1.110	1.203

C = 0.005M

Solute	0.01	0.1	1.0
$BaCl_2$	1.265	1.159	1.179
Na_2SO_4	1.123	....	....
$MgSO_4$	0.710	....	....
$LaCl_3$	1.105	....	....
$K_4Fe(CN)_6$	1.183	....	....

DIFFUSION COEFFICIENTS OF WEAK AND NON-ELECTROLYTES

Concentration

Solute	Concentration	Coefficient
Glucose	0.39%	0.673
Sucrose	0.38%	0.521
Raffinose	0.38%	0.434
Sucrose	Zero	0.5226
Mannitol	Zero	0.682
Penta-erythritol	Zero	0.761
Glycolamide	Zero	1.142
Glycine	Zero	1.064
α-alanine	0.32%	0.910
β-alanine	0.31%	0.933
Amino-benzoic acid ortho	0.24%	0.840
Amino-benzoic acid meta	0.24%	0.774
Amino-benzoic acid para	0.23%	0.843
Citric acid	0.1 M	0.661

DIFFUSION OF GASES INTO AIR

Gas or vapor	Temp. °C	Coefficient of diffusion, sq. cm/sec	Observer
Alcohol, vapor..........	40.4	0.137	Winkelmann
Carbon dioxide..........	0.0	0.139	Mean of various
Carbon disulfide..........	19.9	0.102	Winkelmann
Ether, vapor..........	19.9	0.089	Winkelmann
Hydrogen..........	0.0	0.634	Obermayer
Oxygen..........	0.0	0.178	Obermayer
Water, vapor..........	8.0	0.239	Guglielmo

TRACER DIFFUSION IN METALS
Self Diffusion

John Askill

These tables include data on radio tracer self diffusion in metals reported between 1940 and January 1964.

No distinction has been made between the data as far as reliability is concerned although, in general, the most reliable measurements are the more recent ones using high purity materials, the serial sectioning analysis and giving D_0 values in the range 0.1 to 10. The diffusion coefficient D_T at a temperature T (°K) is given by the relation $D_T = D_0 e^{-Q/RT}$.

The following abbreviations are used:—

A.R.G. = Autoradiography.
R.A. = Residual Activity.
S.D. = Surface Decrease.
S.S. = Serial Sectioning.
P. = Polycrystalline.
S. = Single crystals.
⊥c. = Perpendicular to c direction.
99.95 = 99.95% purity.

Material		Tracer	Temperature range (°C)	Form of analysis	Activation energy Q (Kcal/gm atom)	Frequency factor, D_0 (cm²/sec)	Reference number and year
Aluminum							
P	99.99	Al^{26}	450–650	S.S.	34.0	1.71	1 (1962)
Cadmium							
S, ⊥c	99.5	Cd^{115}	130–280	S.S.	19.1	0.10	2 (1955)
S, ‖c	99.5	Cd^{115}	130–280	S.S.	18.2	0.05	2 (1955)
S, P		Cd^{115}	200–285	S.D.	19.7	0.14	3 (1958)
Carbon (Graphite)							
Natural crystals		C^{14}	2000–2200	....	163	0.4–14.4	4 (1957)
Chromium							
P	99.94	Cr^{51}	1000–1350	A.R.G.	76	....	5 (1957)
P		Cr^{51}	1000–1350	....	85.0	45.0	6 (1959)
P	99.8	Cr^{51}	950–1260	S.D., R.A.	52.7	1.5×10^{-4}	7 (1959)
P	99.87	Cr^{51}	1080–1340	A.R.G.	62.4	1.65×10^{-3}	8 (1960)
P	99.96	Cr^{51}	1060–1400	R.A.	59.2	6.47×10^{-2}	9 (1962)
P	99.99	Cr^{51}	1200–1600	R.A.	73.2	0.28	10 (1962)
β-Cobalt							
P	98.7	Co^{60}	1050–1250	S.D.	67.0	0.367	11 (1951)
P	99.9	Co^{60}	1000–1250	S.D.	61.9	0.032	12 (1951)
P	98.4	Co^{60}	1000–1300	R.A.	62.0	0.2	13 (1952)
P	99.9	Co^{60}	1100-1405	S.S.	67.7	0.83	14 (1955)
Copper							
P		Cu^{64}	750–850	....	57.2	11	15 (1939)
P		Cu^{64}	860–970	S.S.	45.1	0.10	16 (1942)
S		Cu^{64}	860–970	S.S.	49.0	0.6	16 (1942)
P		Cu^{64}	650–850	S.D.	46.8	0.3	17 (1942)
P		Cu^{64}	685–1060	R.A.	47.12	0.2	18 (1954)
P		Cu^{64}	850–1050	S.S.	49.56	0.621	19 (1955)
Germanium							
S		Ge^{71}	780–930	S.S.	73.5	87	20 (1954)
S		Ge^{71}	766–928	R.A.	68.5	7.8	21 (1956)
S		Ge^{71}	730–916	R.A.	69.4	10.8	22 (1961)
S		Ge^{71}	730–916	R.A.	77.5	44	22 (1961)
Gold							
P		Au^{198}	720–970	R.A.	51.0	2	23 (1938)
P		Au^{198}	800–1000	A.R.G.	45.0	....	24 (1952)

Material		Tracer	Temperature range (°C)	Form of analysis	Activation Energy Q (Kcal/gm atom)	Frequency factor, D_0 (cm^2/sec)	Reference number and year
P		Au^{198}	775–1060	A.R.G.	45.3	0.265	25 (1954)
P	99.96	Au^{198}	730–1030	A.R.G.	45.3	0.26	26 (1955)
S	99.999	Au^{198}	600–954	S.S.	39.36	0.031	27 (1956)
P	99.99	Au^{198}	720–1000	S.S.	42.9	0.14	28 (1957)
P	99.95	Au^{198}	700–1050	R.A.	41.7	0.091	29 (1957)
P	99.93	Au^{198}	700–900	R.A.	42.1	0.117	30 (1963)
Indium							
P, S	99.998	In^{114}	50–160	S.S.	17.9	1.02	31 (1952)
S, ⊥c	99.99	In^{114}	44–144	S.S.	18.7	3.7	32 (1959)
S, ∥c	99.99	In^{114}	44–144	S.S.	18.7	2.7	32 (1959)
α-Iron							
P		Fe^{55}, Fe^{59}	715–890	R.A.	77.2	34000	33 (1948)
P		Fe^{55}, Fe^{59}	700–900	R.A.	73.2	2300	34 (1950)
P	99.97	Fe^{55}, Fe^{59}	800–900	S.D.	59.7	5.8	35 (1951)
P		Fe^{55}, Fe^{59}	800–900	R.A.	48.0	0.1	36 (1952)
P		Fe^{55}, Fe^{59}	700–900	....	47.4	0.019	37 (1955)
P	99.96	Fe^{55}, Fe^{59}	650–850	S.D.	67.1	530	38 (1955)
P, S		Fe^{55}, Fe^{59}	775–900	....	64.1	18	39 (1958)
P		Fe^{55}, Fe^{59}	800–900	R.A.	67.24	118	40 (1958)
P, S	99.97	Fe^{55}, Fe^{59}	700–790	S.S., R.A.	60.0	2.0	41 (1961)
P, S	99.97	Fe^{55}, Fe^{59}	790–900	S.S., R.A.	57.2	1.9	41 (1961)
P	99.998	Fe^{55}, Fe^{59}	860–900	A.R.G.	57.3	2.0	42 (1963)
γ-Iron							
P		Fe^{55}, Fe^{59}	935–1110	R.A.	48.0	1.0×10^{-3}	33 (1948)
P		Fe^{55}, Fe^{59}	950–1400	S.D.	67.9	0.58	35 (1951)
P		Fe^{55}, Fe^{59}	1000–1360	R.A.	74.2	5.8	34 (1950)
P		Fe^{55}, Fe^{59}	1000–1300	R.A.	68.0	0.76	36 (1952)
P	99.1	Fe^{55}, Fe^{59}	950–1250	R.A.	68.0	0.7	43 (1953)
P		Fe^{55}, Fe^{59}	900–1200	....	67.9	1.3	37 (1955)
P		Fe^{55}, Fe^{59}	1000–1300	S.S.	67.0	0.44	44 (1955)
P		Fe^{55}, Fe^{59}	900–1200	....	64.0	0.16	45 (1957)
P		Fe^{55}, Fe^{59}	1000–1250	R.A.	63.5	0.11	46 (1958)
P, S	99.97	Fe^{55}, Fe^{59}	1044–1195	S.D., R.A.	64.5	0.18	41 (1961)
P		Fe^{55}, Fe^{59}	950–1200	S.S.	67.7	2.5	47 (1962)
P	99.998	Fe^{55}, Fe^{59}	1156–1350	A.R.G.	64.0	0.22	42 (1963)
δ-Iron							
P		Fe^{55}, Fe^{59}	1405–1520	R.A.	42.4	0.019	48 (1961)
P	99.96	Fe^{55}, Fe^{59}	1415–1510	S.S., R.A.	57.0	1.9	49 (1963)
P	99.998	Fe^{55}, Fe^{59}	1405–1515	A.R.G.	61.7	6.8	42 (1963)
Lead							
P		Pb^{212}	260–320	S.D.	27.9	6.56	50 (1932)
S, P		Pb^{212}	190–270	S.S.	25.7	1.17	51 (1954)
S	99.999	Pb^{212}	175–322	S.S.	24.21	0.281	52 (1955)
S	99.99	Pb^{212}	207–323	S.S.	26.06	1.372	53 (1961)
Magnesium							
P	99.92	Mg^{28}	470–630	S.S.	32.0	1.0	54 (1954)
S, ⊥c		Mg^{28}	467–635	S.S.	32.5	1.5	55 (1956)
S, ∥c		Mg^{28}	467–635	S.S.	32.2	1.0	55 (1956)
Molybdenum							
P		Mo^{99}	1800–2200	R.A.	115	4	56 (1959)
P		Mo^{99}	1800–2200	R.A.	114	4	6 (1959)
P		Mo^{99}	1700–1900	....	111	2.77	57 (1960)
P		Mo^{99}	1600–2200	....	100.8	0.38	58 (1961)
S	99.99	Mo^{99}	1850–2350	S.S.	92.2	0.1	59 (1963)
P	99.98	Mo^{99}	1850–2350	S.S.	96.9	0.5	59 (1963)

Material		Tracer	Temperature range (°C)	Form of analysis	Activation energy Q (Kcal/gm atom)	Frequency factor, D_0 (cm^2/sec)	Reference number and year
Nickel							
P	99.93	Ni^{63}	900–1250	S.D., R.A.	66.8	1.27	60 (1956)
P		Ni^{63}	1100–1175	S.S.	63.8	0.4	61 (1957)
S	99.95	Ni^{63}	700–1100	A.R.G.	65.9	0.48	62 (1958)
P		Ni^{63}	1150–1400	S.S.	69.8	3.36	63 (1959)
P	99.92	Ni^{63}	1150–1400	R.A.	71	5.12	64 (1959)
S		Ni^{63}	680–830	R.A.	69.7	5.8	65 (1960)
S	99.999	Ni^{63}	950–1020	S.D.	66.8	1.9	66 (1961)
P	99.98	Ni^{63}	1085–1300	S.S.	69.5	2.59	67 (1963)
Niobium							
P	99.4	Nb^{95}	1535–2120	S.S.	105	12.4	68 (1960)
P	99.8	Nb^{95}	1700–2100	A.R.G.	95	1.3	69 (1962)
Phosphorus							
P		P^{32}	0–44	S.S.	9.4	1.07×10^{-3}	70 (1955)
Platinum							
P	99.99	Pt^{195}, Pt^{199}	1325–1600	S.S.	68.2	0.33	71 (1958)
P		Pt^{195}, Pt^{199}	1250–1725	S.D.	66.5	0.22	72 (1962)
Selenium							
P		Se^{75}	35–140	....	11.7	1.4×10^{-4}	73 (1957)
Silver							
P		Ag^{110}	500–900	S.S.	45.95	0.895	74 (1941)
S, P		Ag^{110}	500–950	....	46.0	0.9	75 (1949)
P, S	99.99	Ag^{110}	500–875	S.S., S.D.	45.95	0.895	76 (1951)
P	99.99	Ag^{110}	670–940	S.S.	45.50	0.724	77 (1952)
P, S		Ag^{110}	640–905	S.S.	40.8	0.11	78 (1952)
P		Ag^{110}	725–925	....	47.4	1.8	79 (1953)
P		Ag^{110}	700–860	A.R.G.	45.0	0.905	80 (1954)
P		Ag^{110}	750–925	....	45.0	0.53	81 (1955)
P		Ag^{110}	700–900	R.A.	45.4	0.65	82 (1955)
S		Ag^{110}	550–900	S.S.	44.05	0.40	83 (1955)
P		Ag^{110}	707–880	A.R.G.	44.9	0.834	84 (1955)
S	99.99	Ag^{110}	630–940	S.S.	44.09	0.395	85 (1956)
P		Ag^{110}	660–740	....	41.8	2.78	86 (1956)
S		Ag^{110}	715–940	S.S.	43.7	0.27	87 (1957)
P		Ag^{110}	500–900	....	44.6	0.86	88 (1957)
P		Ag^{110}	750–950	S.S.	41.53	0.094	89 (1958)
P		Ag^{110}	650–900	....	44.8	1.08	90 (1958)
Sodium							
P		Na^{22}	10–95	S.S.	10.45	0.242	91 (1952)
Sulphur							
P, ⊥c		S^{35}	310–370	....	3.08	8.3×10^{-12}	92 (1951)
P, ‖c		S^{35}	310–370	....	78.0	1.7×10^{36}	92 (1951)
Tantalum							
P		Ta^{182}	1830–2530	....	110 (89.5)	2 (0.03)	93 (1953)
P		Ta^{182}	1200–1350	R.A.	110	1300	94 (1955)
α-Thallium							
S, ⊥c	99.9	Tl^{204}	135–230	S.S.	22.6	0.4	95 (1955)
S, ‖c	99.9	Tl^{204}	135–230	S.S.	22.9	0.4	95 (1955)
β-Thallium							
S	99.9	Tl^{204}	230–280	S.S.	20.7	0.7	95 (1955)
β-Tin							
P, ⊥c		Sn^{113}, Sn^{123}	180–225	S.S.	10.5	8.4×10^{-4}	96 (1949)
P, ‖c		Sn^{113}, Sn^{123}	180–225	S.S.	5.9	3.7×10^{-8}	96 (1949)
S, ⊥c		Sn^{113}, Sn^{123}	180–223	S.S.	6.7	9.2×10^{-8}	97 (1950)
S, ‖c		Sn^{113}, Sn^{123}	180–223	S.S.	9.4	3.6×10^{-6}	97 (1950)

Material		Tracer	Temperature range (°C)	Form of analysis	Activation Energy Q (Kcal/gm atom)	Frequency factor, D_0 (cm²/sec)	Reference number and year
S, ⊥c		Sn^{113}, Sn^{123}	178–222	S.S.	23.3	1.4	98 (1960)
S, ‖c		Sn^{113}, Sn^{123}	178–222	S.S.	25.6	8.2	98 (1960)
S, P		Sn^{113}, Sn^{123}	140–217	R.A.	10.8	9.9×10^{-4}	99 (1960)
P	99.99	Sn^{113}, Sn^{123}	130–255	....	22.4	0.78	100 (1961)
α-Titanium							
P		Ti^{44}	690–850	R.A.	29.3	6.4×10^{-8}	101 (1963)
Tungsten							
P		W^{185}	1290–1450	S.D.	135.8	6.3×10^{7}	102 (1950)
P		W^{185}	2000–2700	....	120.5	0.54	103 (1961)
S		W^{185}	2670–3225	S.S.	153.1	42.8	104 (1963)
α-Uranium							
P		U^{234}	580–650	....	40.0	2×10^{-3}	105 (1961)
β-Uranium							
P		U^{234}	700–760	R.A.	42.0	0.0135	106 (1959)
γ-Uranium							
P		U^{233}, U^{234}, U^{235}	800–1050	A.R.G.	21.0	1.4×10^{-4}	107 (1958)
P		U^{233}, U^{234}, U^{235}	800–1050	S.S.	26.6	1.17×10^{-3}	107 (1958)
P		U^{233}, U^{234}, U^{235}	800–1040	S.S.	27.5	1.8×10^{-3}	108 (1958)
P	99.99	U^{233}, U^{234}, U^{235}	800–1070	S.S.	28.5	2.33×10^{-3}	109 (1960)
Zinc							
S, ⊥c		Zn^{65}	340–410	S.S.	19.6	0.02	110 (1941)
S, ⊥c		Zn^{65}	340–410	S.S.	31.0	93	111 (1942)
S, ‖c		Zn^{65}	340–410	S.S.	20.4	0.05	111 (1942)
S, ⊥c	99.999	Zn^{65}	300–400	S.S.	25.4	1.3	112 (1952)
S, ‖c	99.999	Zn^{65}	300–400	S.S.	21.7	0.1	112 (1952)
S, ⊥c	99.999	Zn^{65}	240–400	S.S.	24.3	0.58	113 (1953)
S, ‖c	99.999	Zn^{65}	240–400	S.S.	21.8	0.13	113 (1953)
S, ⊥c		Zn^{65}	260–400	S.S.	25.9	1.6	114 (1954)
S, ‖c		Zn^{65}	260–400	S.S.	19.6	0.02	114 (1954)
P		Zn^{65}	260–400	S.S.	23.8	0.41	114 (1954)
S, ⊥c		Zn^{65}	200–415	S.S., S.D.	24.9	0.39	115 (1956)
S, ‖c		Zn^{65}	200–415	S.S., S.D.	22.0	0.08	115 (1956)
P, ⊥c		Zn^{65}	200–415	S.S., S.D.	22.7	0.19	115 (1956)
P, ‖c		Zn^{65}	200–415	S.S., S.D.	23.5	0.11	115 (1956)
S, ⊥c		Zn^{65}	270–370	....	26.0	2.78	116 (1957)
S, ‖c		Zn^{65}	270–370	....	19.0	0.013	116 (1957)
α-Zirconium							
P	99.96	Zr^{95}	700–800	R.A.	22.0	5×10^{-8}	117 (1958)
P	99.6	Zr^{95}	650–825	R.A.	52.0	5.9×10^{-2}	119 (1959)
β-Zirconium							
P	99.9	Zr^{95}	1000–1250	R.A.	26.0	4.0×10^{-5}	118 (1959)
P	99.6	Zr^{95}	1115–1500	R.A.	38.0	2.4×10^{-3}	119 (1959)
P	99.6	Zr^{95}	900–1200	R.A.	27.0	1×10^{-4}	120 (1960)
P		Zr^{95}	900–1240	S.S.	24.0	4.2×10^{-5}	121 (1960)
P	99.89	Zr^{95}	1100–1500	S.S.	30.1	2.4×10^{-4}	122 (1961)
P	99.94	Zr^{95}	900–1750	S.S.	20.7–46.9	4.8×10^{-6} –2.5×10^{-2}	123*(1963)

* Values of D_0 and Q given for a nonlinear log D_T versus 1/T plot.

References

(1) Lundy, T. S., Murdock, J. F., J. Appl. Phys. **33** (5), 1671 (1962).
(2) Wajda, E. S., Shirn, G. A., Huntington, H. B., Acta. Met. **3,** 39 (1955).
(3) Mahmoud, K. A., Kamel, R., Radioisotopes in Scientific Research Vol. 1, page 271, Pergamon Press, New York (1958).
(4) Kanter, M. A., Phys. Rev. **107,** 655 (1957).
(5) Bokstein, S. Z., Kishkin, S. T., Moroz, L. M., Zav. Lab. **23** (3), 316 (1957).
(6) Gruzin, P. L., Pavlinov, L. V., Tyutyunnik, A. D., Iz. Akad. Nauk. S.S.S.R. **5,** 155 (1959).
(7) Paxton, H. W., Gondolf, E. G., Archiv. Eisen hutten wessen. **30,** 55 (1959).
(8) Bogdanov, N. A., Russian Metallurgy and Fuels **3,** 95 (1960).
(9) Ivanov, L. I., Matveeva, M. P., Morozov, V. A., Prokoshkin, D. A., Russian Metallurgy and Fuels **2,** 63 (1962).
(10) Hagel, W. C., Trans. A.I.M.E. **224** (3), 430 (1962).
(11) Nix, F. C., Jaumot, F. E., Phys. Rev. **82,** 72 (1951).
(12) Ruder, R. C., Birchenall, C. E., Trans. A.I.M.E. **191,** 142 (1951).
(13) Gruzin, P. L., Dokl. Akad. Nauk. S.S.S.R. **86,** 289 (1952).
(14) Mead, H. W., Birchenall, C. E., Trans. A.I.M.E. **203** (9), 994 (1955).
(15) Shockley, W., Nix, F. C., Phys. Rev. **56,** 13 (1939).
(16) Maier, M. S., Nelson, H. R., Trans. A.I.M.E. **147,** 39 (1942).
(17) Raynor, C. L., Thomassen, L., Rousse, J., Trans. A.S.M. **30,** 313 (1942).
(18) Kuper, A., Letaw, H., Slifkin, L., Sonder, E., Tomizuka, C. T., Phys. Rev. **96,** 1224 (1954), errata **98,** 1870 (1956).
(19) Mercer, W. L., Ph.D. Thesis. Leeds Univ. England (1955).
(20) Letaw, H., Slifkin, L. M., Portnoy, W. M., Phys. Rev. **93** (4), 892 (1954).
(21) Letaw, H., Portnoy, W. M., Slifkin, L., Phys. Rev. **102,** 636 (1956).
(22) Widner, H., Gunther-Mohr, G. R., Helv. Phys. Acta. **34** (6), 635 (1961).
(23) Mckay, H. A. C., Trans. Far. Soc. **34,** 845 (1938).
(24) Gatos, H. C., Azzam, A., Trans. A.I.M.E. **194,** 407 (1952).
(25) Gatos, H. C., Kurtz, A. D., Trans. A.I.M.E. **200,** 616 (1954).
(26) Kurtz, A. D., Averbach, B. L., Cohen, M., Acta. Met. **3,** 442 (1955).
(27) Okkerse, B., Phys. Rev. **103** (5), 1246 (1956).
(28) Mead, H. W., Birchenall, C. E., Trans. A.I.M.E. **209** (7), 874 (1957).
(29) Makin, S. M., Rowe, A. H., LeClaire, A. D., Proc. Phys. Soc. **B70** (6), 545 (1957).
(30) Duhl, D., Hirano, K., Cohen, M., Acta Met. **11** (1) 1, (1963).
(31) Eckert, R. E., Drickamer, H. G., J. Chem. Phys. **20,** 13 (1952).
(32) Dickey, J. E., Acta Met. **7,** 350 (1959).
(33) Birchenall, C. E., Mehl, R. F., J. Appl. Phys. **19,** 217 (1948).
(34) Birchenall, C. E., Mehl, R. F., Trans. A.I.M.E. **188,** 144 (1950).
(35) Buffington, F. S., Bakalar, I. D., Cohen, M., Physics of Powder Metallurgy p. 92, McGraw-Hill, New York (1951).
(36) Gruzin, P. L., Prob. Metalloved i Fiz. Mettallov. **3,** 201 (1952).
(37) Zhukhovitsky, A. A., Geodakyan, V. A., Primenenie Radioaktiv. Izotopov v Metall. **34,** 267 (1955); AEC-tr-3100 Pt2, p. 52.
(38) Golikov, V. M., Borisov, V. T., Problems of met. and phys. of metals. 4th Symposium (1955) Cons. Bureau. Inc. New York. (1957); AEC-tr-2924 (1958).
(39) Leymonie, C., Lacombe, P., Rev. Met. **55,** 524 (1958); Comptes Rendus **245,** 1922 (1957); Met. Corr. Ind. **45** (1960).
(40) Borg, R. J., Birchenall, C. E., Trans. A.I.M.E. **218,** 980 (1960).
(41) Buffington, F. S., Hirano, K., Cohen, M., Acta. Met. **9** (5), 434 (1961).
(42) Graham, D., Tomlin, D. H., Phil. Mag. **8,** 1581 (1963).
(43) Gruzin, P. L., Izvest. Akad. Nauk. S.S.S.R. Otd. Tekn. N. **3,** 383 (1953).
(44) Mead, H. W., Birchenall, C. E., Trans. A.I.M.E. **206,** 1336 (1956).
(45) Bokshtein, S. Z., Kishkin, S. T., Moroz, L. M., Metallovedemie i Obrab. Metallov. **2,** 2 (1957).
(46) Gerstriken, S. D., Pryanishnikov, M. P., Voprosa Fiziki met. i metallov. **9,** 147 (1958).
(47) Bogdanov, N. A., Russian Metallurgy and Fuel **2,** 61 (1962).
(48) Staffansson, L. I., Birchenall, C. E., U.S. Report. AFOSR-733 (1961).
(49) Borg, R. J., Lai, D. Y. F., Krikorian, O., Acta. Met. **11** (8), 867 (1963).
(50) Hevesy, G. Von., Seith, W., Keil, A., Z. Physik **79,** 197 (1932), Z. Metallkunde **25,** 104 (1935).
(51) Okkerse, B., Acta. Met. **2,** 551 (1954).
(52) Nachtrieb, N. H., Handler, G. S., J. Chem. Phys. **23** (9), 1569 (1955).
(53) Resing, H. A., Nachtrieb, N. H., J. Phys. Chem. Solids **21,** 40 (1961).
(54) Shewmon, P. G., Rhines, F. N., Trans. A.I.M.E. **200,** 1021 (1954).
(55) Shewmon, P. G., Trans. A.I.M.E. **206,** 918 (1956).

References (Continued)

(56) Borisov, Y. V., Gruzin, P. L., Pavlinov, L. V., Metall i Metalloved. **1,** 213 (1959); U.S. Report JPRS-5195.
(57) Bronfin, M. B., Bokstein, S. Z., Zhukhovitsky, A. A., Zav. Lab. **26** (7), 828 (1960).
(58) Danneberg, W., Krautz, E. Z., Naturforch. **16a** (a), 854 (1961).
(59) Askill, J., Tomlin, D. H., Phil. Mag. **8** (90), 997 (1963).
(60) Hoffman, R. E., Pikus, F. W., Ward, R. A., Trans. A.I.M.E. **206,** 483 (1956).
(61) Reynolds, J. E., Averbach, B. L., Cohen, M., Acta. Met. **5,** 29 (1957).
(62) Upthegrove, W. R., Sinnott, M. J., Trans. A.S.M. **50,** 1031 (1958).
(63) MacEwan, J. R., MacEwan, J. U., Yaffe, L., Canad. J. Chem. **37** (10), 1623 (1959).
(64) MacEwan, J. R., MacEwan, J. U., Yaffe, L., Canad. J. Chem. **37** (10), 1629 (1959).
(65) Messner, A., Benson, R., Dorn, J. E., ASM No. 193 (1960).
(66) Wazzam, A. R., Mote, J., Dorn, J. E., Report AD-257188 (1961).
(67) Shinyaev, A. Y., Physics of metals and metallography **15** (1), 100 (1963).
(68) Resnick, R., Castleman, L. S., Trans. A.I.M.E. **218,** 307 (1960).
(69) Peart, R. F., Graham, D., Tomlin, D. H., Acta. Met. **10,** 519 (1962).
(70) Nachtrieb, N. H., Handler, G. S., J. Chem. Phys. **23,** 1187 (1955).
(71) Kidson, G. V., Ross, R., Proc. 1st. UNESCO. Int. Conf. Radioisotopes in Scientific Res. p. 185 (1958).
(72) Cattaneo, F., Germagnoli, E., Grasso, F., Phil. Mag. **7,** 1373 (1962).
(73) Boltaks, B. I., Plachenov, B. T., Soviet Physics—Tech. Phys. **27** (10), 2071 (1957).
(74) Johnson, W. A., Trans. A.I.M.E. **143,** 107 (1941).
(75) Turnbull, D., Phys. Rev. **76,** 471 (1949).
(76) Hoffman, R. E., Turnbull, D., J. Appl. Phys. **22,** 634 (1951), errata p. 984.
(77) Slifkin, L., Lazarus, D., Tomizuka, T., J. Appl. Phys. **23,** 1032 (1952).
(78) Johnson, R. D., Martin, A. B., Phys. Rev. **86,** 642 (1952); U.S. report NAA-SR-170 (1952).
(79) Kriukov, S. N., Zhukhovitsky, A. A., Dokl. Akad. Nauk. S.S.S.R. **90,** 379 (1953).
(80) Finkelshtein, B. N., Yamashchikova, A. I., Dokl. Ahad. Nauk. S.S.S.R. **98,** 781 (1954).
(81) Zhuklovitsky, A. A., Geodakyan, V. A., Dokl. Akad. Nauk. S.S.S.R. **102,** 301 (1955); AEC-tr-2265 (1955).
(82) Gertsriken, S. D., Pryanishnikov, M. P., Voprosa. Fiz. i Fiziki metallov **4,** 528 (1955).
(83) Sonder, E., Phys. Rev. **100,** 1662 (1955).
(84) Kruegar, H., Hersh, H. N., Trans. A.I.M.E. **203,** 125 (1955).
(85) Tomizuka, C. T., Sonder, E., Phys. Rev. **103** (5), 1182 (1956).
(86) Vasilev, V. P., Dokl. Akad. Nauk. S.S.S.R. **110,** 61 (1956); U.S. report A.E.C.-tr-2972.
(87) Nachtrieb, N. H., Petit, J., Wehrenberg, J. J., Chem. Phys. **26** (1), 106 (1957).
(88) Gertsriken, S. D., Yatsenko, T. K., Voprosa Fiziki metall. i metallov **8,** 101 (1957).
(89) Airoldi, G., Germagnoli, E., Energia Nucleare (Italy) **5,** 445 (1958).
(90) Gertsriken, S. D., Tsitsiliano, D. D., Phys. of metals and metallography **6,** 80 (1958).
(91) Nachtrieb, N. H., Catalano, E., Weil, J. A., J. Chem. Phys. **20** (8), 1185 (1952).
(92) Cuddeback, R. B., Drickamer, H. G., J. Chem. Phys. **19,** 790 (1951).
(93) Eager, R. L., Langmuir, D. B., Phys. Rev. **89,** 911 (1953); errata p. 890. Langmuir, D. B., Phys. Rev. **86,** 642 (1952).
(94) Gruzin, P. L., Meshkov, V. I., Problemi metalloved i fiz metallov. 570 (1955). U.S. report AEC-tr-2926.
(95) Shirn, G. A., Acta. Met. **3,** 87 (1955).
(96) Boas, W., Fensham, P. J., Nature **164,** 1127 (1949).
(97) Fensham, P. J., Aust. J. Sci. Res. **A3,** 91, 105 (1950); errata **A4** 229 (1951).
(98) Meakin, J. D., Klokholm, E., Trans. A.I.M.E. **218** (3), 463 (1960).
(99) Chomba, W., Andrusz Kiewicz, J., Nukleonika (Poland) **5** (10), 611 (1960).
(100) Lange, W., Hassner, A., Phys. Status. Solidi **1,** 50 (1961).
(101) Libanati, C. M., Dyment, S. F., Acta. Met. **11,** 1263 (1963).
(102) Vasilev, V. P., Chernomorchenko, S. G., Zav. Lab. **22,** 688 (1956); U.S. AEC-tr-4276.
(103) Danneberg, W., Metall. **15,** 977 (1961).
(104) Andelin, R. L., Los Alamos Scientific Lab. report LA 2880 (1963).
(105) Adda, Y., Kirianenko, A., Mairy, C., C. R. Acad-Sci. (France) **253,** 445 (1961); J. Nuclear Materials **6** (1), 130 (1962).
(106) Adda, Y., Kirianenko, A., Mairy, C., J. Nuclear Materials. **1** (3), 300 (1959).
(107) Bochvar, A. A., Kuznetsova, V. G., Sergeev, V. S., 2nd. Int. Conf. on Peaceful use of atomic energy **6,** 8 (1958).
(108) Adda, Y., Kirianenko, A., Compte Rendus. **247** (9), 744 (1958).
(109) Rothman, S. J., Lloyd, L. T., Harkness, A. L., Trans. A.I.M.E. **218** (4), 605 (1960).
(110) Banks, F. R., Phys. Rev. **59,** 376 (1941).

References (Continued)

(111) Miller, P. H., Banks, F. R., Phys. Rev. **61,** 648 (1942).

(112) Huntington, H. B., Shirn, G. A., Wajda, E. S., Phys. Rev. **87,** 211 (1952).

(113) Shirn, G. A., Wajda, E. S., Huntington, H. B., Acta. Met. **1,** 513 (1953).

(114) Liu, T., Drickamer, H. G., J. Chem. Phys. **22,** 314 (1954).

(115) Jaumot, F. E., Smith, R. L., Trans. A.I.M.E. **206,** 137 (1956).

(116) Naskidashvili, I. A., Dolidze, V. M., Akad. Nauk. Gruz. S.S.R. **18,** 671 (1957).

(117) Gruzin, P. L., Emelyanova, V. S., Ryabova, G. G., 2nd Int. Conf. on the peaceful use of atomic energy **19,** 187 (1958).

(118) Federov, G. B., Gulyakin, V. D., Met. i metalloved chistykh metal Sbornik Nauch. Rabot No. 1. 170 (1959).

(119) Lyashenko, V. S., Bikov, B. N., Pavlinov, L. V., Physics of metals and metallography. **8** (3), 40 (1959).

(120) Borisov, Y. V., Godin, Yu. G., Gruzin, P. L., Metallurgy and metallography NP-TR-448, p. 196 (1960).

(121) Volokoff, D., May, S., Adda, Y., Compte. Rend. **251** (3), 2341 (1960).

(122) Kidson, G. V., McGurn, J., Canad. J., Phys. **39** (8), 1146 (1961).

(123) Federer, J. I., Lundy, T. S., Trans. A.I.M.E. **227,** 592 (1963).

DIFFUSION OF METALS INTO METALS

Compiled by Rudolf Loebel

This table is prepared in an alphabetical listing with respect to the symbol for the diffusing metal.

Diffusing metal	Metal diffused into	Temperature °C	D cm²/hr.
Ag	Al	466	$6.84 - 8.1 \times 10^{-7}$
"	"	500	$7.2 - 3.96 \times 10^{-6}$
"	"	573	1.26×10^{-5}
"	Pb	220	5.40×10^{-5}
"	"	250	1.08×10^{-4}
"	"	285	3.29×10^{-4}
"	Sn	500	1.73×10^{-1}
Al	Cu	500	6.12×10^{-9}
"	"	850	7.92×10^{-6}
As	Si		$0.32^{-82,000/RT}$
Au	Ag	456	1.76×10^{-9}
"	"	491	$0.92 - 2.38 \times 10^{-13}$
"	"	585	3.6×10^{-8}
"	"	601	3.96×10^{-8}
"	"	624	$2.5 - 5 \times 10^{-11}$
"	"	717	$1.04 - 2.25 \times 10^{-9}$
"	"	729	1.76×10^{-9}
"	"	767	1.15×10^{-6}
"	"	847	2.30×10^{-6}
"	"	858	3.63×10^{-8}
"	"	861	3.92×10^{-8}
"	"	874	3.92×10^{-8}
"	"	916	5.40×10^{-6}
"	"	1040	1.17×10^{-6}
"	"	1120	2.29×10^{-5}
"	"	1189	5.42×10^{-6}
"	Au	800	1.17×10^{-8}
"	"	900	9×10^{-8}
"	"	1020	5.4×10^{-7}
"	Bi	500	1.88×10^{-1}
"	Cu	970	5.04×10^{-6}
"	Hg	11	3×10^{-2}
"	Pb	100	8.28×10^{-6}
"	"	150	1.80×10^{-4}
"	"	200	3.10×10^{-4}
"	"	240	1.58×10^{-3}
"	"	300	5.40×10^{-3}
"	"	500	1.33×10^{-1}
"	Si		$0.0011^{-25,800/RT}$
"	Sn	500	1.94×10^{-1}
B	Si		$10.5^{-85,000/RT}$
Ba	Hg	7.8	2.17×10^{-2}
Bi	Si		$1030^{-107,000/RT}$
"	Pb	220	1.73×10^{-7}
"	"	250	1.33×10^{-6}
"	"	285	1.58×10^{-6}
C	W	1700	1.87×10^{-9}
"	Fe	930	$7.51 - 9.18 \times 10^{-9}$
Ca	Hg	10.2	2.25×10^{-2}
Cd	Ag	650	9.36×10^{-7}
"	"	800	4.68×10^{-6}
"	"	900	2.23×10^{-5}
"	Hg	8.7	6.05×10^{-2}
"	"	15	6.51×10^{-2}
"	"	20	5.47×10^{-2}
"	"	99.1	1.23×10^{-1}
" 1 atom %	Pb	167	1.66×10^{-7}
"	"	200	4.59×10^{-7}
"	"	252	3.10×10^{-6}
Ce	W	1727	3.42×10^{-6}
Cs	Hg	7.3	1.88×10^{-2}
"	W	27	4.32×10^{-8}
"	"	227	5.40×10^{-4}
"	"	427	2.88×10^{-2}
"	"	540	1.44×10^{-1}
Cu	Al	440	1.8×10^{-7}
"	"	457	2.88×10^{-7}
"	"	540	5.04×10^{-6}
"	"	565	$4.68 - 5.00 \times 10^{-4}$
"	Ag	650	1.04×10^{-6}
"	"	760	1.30×10^{-6}
"	"	895	3.38×10^{-6}
"	Au	301	5.40×10^{-10}
"	"	443	8.64×10^{-9}
"	"	560	3.38×10^{-7}
"	"	604	5.10×10^{-7}
"	"	616	7.92×10^{-7}
"	"	740	3.35×10^{-6}
Cu	Cu	650	1.15×10^{-8}
"	"	750	2.34×10^{-8}
"	"	830	1.44×10^{-7}
"	"	850	9.36×10^{-7}
"	"	950	2.30×10^{-6}
"	"	1030	1.01×10^{-5}
"	Ge	700–900	$1.01 \pm 0.1 \times 10^{-1}$
"	Pt	1041	$7.83 - 9 \times 10^{-8}$
"	"	1213	5.04×10^{-7}
"	"	1401	6.12×10^{-6}
Fe	Au	753	1.94×10^{-6}
"	"	1003	2.70×10^{-5}
"	Si		$0.0062^{-20,000/RT}$
Ga	Si		$3.6^{-81,000/RT}$
Ge	Al	630	3.31×10^{-1}
"	Au	529	1.84×10^{-1}
"	"	563	2.80×10^{-1}
"	Ge	766–928	$7.8^{-68,509/RT}$
"	"	1060–1200°K	$87^{-73,000/RT}$
Hg	Cd	156	9.36×10^{-7}
"	"	176	2.55×10^{-6}
"	"	202	9×10^{-6}
"	Pb	177	8.34×10^{-8}
"	"	197	2.09×10^{-5}
In	Ag	650	1.04×10^{-6}
"	"	800	6.84×10^{-6}
"	"	895	4.68×10^{-5}
"	Si		$16.5^{-90/000/RT}$
K	Hg	10.5	2.21×10^{-2}
"	W	207	2.05×10^{-2}
"	"	317	3.6×10^{-1}
"	"	507	$1.1 \times 10^{+1}$
Li	Hg	8.2	2.75×10^{-2}
Mg	Al	365	3.96×10^{-8}
"	"	395	$1.98 - 2.41 \times 10^{-7}$
"	"	420	$2.38 - 2.74 \times 10^{-7}$
"	"	440	1.19×10^{-7}
"	"	447	9.36×10^{-7}
"	"	450	6.84×10^{-6}
"	"	500	$3.96 - 7.56 \times 10^{-6}$
"	"	577	1.58×10^{-5}
"	Pb	220	4.32×10^{-7}
Mn	Cu	400	7.2×10^{-10}
"	"	850	4.68×10^{-7}
Mo	W	1533	9.36×10^{-10}
"	"	1770	4.32×10^{-9}
"	"	2010	7.92×10^{-8}
"	"	2260	2.81×10^{-7}
Na	Hg	9.6	2.67×10^{-2}
"	W	20	2.88×10^{-2}
"	"	227	1.80
"	"	417	9.72
"	"	527	1.19×10^{-1}
Ni	Au	800	2.77×10^{-6}
"	"	1003	2.48×10^{-5}
"	Cu	550	2.56×10^{-9}
"	"	950	7.56×10^{-7}
" 3 atom %	Pb	252	1.25×10^{-7}
" 1 atom %	Pb	285	8.34×10^{-7}
"	"	320	1.26×10^{-6}
"	Pt	1043	1.81×10^{-8}
"	"	1241	1.73×10^{-6}
"	"	1401	5.40×10^{-6}
Pb	Cd	252	2.88×10^{-8}
" 2 atom %	Hg	9.4	6.46×10^{-2}
" 2 atom %	"	15.6	5.71×10^{-2}
" 2 atom %	"	99.2	8×10^{-2}
"	Pb	250	5.42×10^{-8}
"	"	285	2.92×10^{-7}
"	Sn	500	1.33×10^{-1}
Pd	Ag	444	4.68×10^{-9}
"	"	571	1.33×10^{-7}
"	"	642	4.32×10^{-7}
"	"	917	4.32×10^{-6}
"	Au	727	2.09×10^{-8}
"	"	970	1.15×10^{-6}
"	Cu	490	3.24×10^{-9}
"	"	950	$9.0 - 10.44 \times 10^{-7}$

DIFFUSION OF METALS INTO METALS (Continued)

* Saturated FeC alloy.

Diffusing metal	Metal diffused into	Temperature °C	D cm^2/hr.
Po	Au	470	4.59×10^{-11}
Po	Al	20	1.08×10^{-9}
"	"	500	1.80×10^{-7}
"	Bi	150	1.80×10^{-7}
"	"	200	1.80×10^{-6}
"	Pb	150	4.59×10^{-11}
"	"	200	4.59×10^{-9}
"	"	310	5.41×10^{-7}
Pt	Au	740	1.69×10^{-8}
"	"	986	$6.12 - 10.08 \times 10^{-7}$
"	Cu	490	2.01×10^{-9}
"	"	960	$3.96 - 8.28 \times 10^{-7}$
"	Pb	490	7.04×10^{-2}
Ra (B + C)	Ag	470	1.57×10^{-8}
"	Au	470	1.42×10^{-8}
"	Pt	470	3.42×10^{-8}
Rb	Hg	7.3	1.92×10^{-2}
Rh	Pb	500	1.27×10^{-1}
Sb	Ag	650	1.37×10^{-6}
"	"	760	5.40×10^{-6}
"	"	895	1.55×10^{-5}
"	Si		$5.6^{-91,000/RT}$
Si	Al	465	1.22×10^{-6}
"	"	510	7.2×10^{-6}
"	"	600	3.35×10^{-5}
"	"	667	1.44×10^{-1}
"	"	697	3.13×10^{-1}
"	Fe + C*	1400–1600	$3.24 - 5.4 \times 10^{-2}$
Sn	Ag	650	2.23×10^{-6}
"	"	895	2.63×10^{-5}
"	Cu	400	1.69×10^{-9}
"	"	650	2.48×10^{-7}
"	"	850	1.40×10^{-5}
"	Hg	10.7	6.38×10^{-2}
"	Pb	245	1.12×10^{-7}
Sn	"	250	1.83×10^{-7}
"	"	285	5.76×10^{-7}
Sr	Hg	9.4	1.96×10^{-2}
Th	Mo	1615	1.30×10^{-6}
"	"	2000	3.60×10^{-3}
"	Tl	285	8.76×10^{-7}
"	W	1782	3.96×10^{-7}
"	"	2027	4.03×10^{-6}
"	"	2127	1.29×10^{-5}
"	"	2227	2.45×10^{-5}
" *B*	Pb	165	2.54×10^{-12}
" *B*	"	260	2.54×10^{-8}
" *B*	"	324	5.84×10^{-6}
Tl	Hg	11.5	3.63×10^{-2}
"	Pb	220	1.01×10^{-7}
"	"	250	7.92×10^{-7}
"	"	270	3.96×10^{-7}
"	"	285	1.12×10^{-6}
"	"	315	2.09×10^{-6}
"	Si		$16.5^{-90,000/RT}$
U	W	1727	4.68×10^{-8}
Y	W	1727	6.55×10^{-5}
Zn	Ag	750	1.66×10^{-5}
"	"	850	4.37×10^{-5}
"	Al	415	9×10^{-7}
"	"	473	1.91×10^{-6}
"	"	500	$7.2 - 13.68 \times 10^{-6}$
"	"	555	1.8×10^{-5}
"	Hg	11.5	9.09×10^{-2}
"	"	15	8.72×10^{-2}
"	"	99.2	1.20×10^{-1}
"	Pb	285	$5.84 \times$
Zr	W	1727	1.17×10^{-5}

DIFFUSION OF SOLUTE ELEMENTS IN BODY CENTERED CUBIC TITANIUM AND TITANIUM ALLOYS

Compiled by D. H. Tomlin and R. F. Peart

Diffusion zone analysis by autoradiographic techniques.[1]

Solvent	Solute	Temperature range (°C)	Activation energy Q (k. cal/g. mol.)	Frequency factor D_0 (cm^2/sec)
Iodide Ti[6]	Cr^{51}	926–1193	37.7 ± 2.2[1]	$\left(10\,{}^{+13}_{-6}\right)\cdot 10^{-3}$
Iodide Ti[6]	Fe^{55}	910–1195	31.6 ± 0.4[2]	$\left(7.8\,{}^{+1.3}_{-1.1}\right)\cdot 10^{-3}$
Iodide Ti[6]	Fe^{59}	904–1284	32.0 ± 0.7[2]	$\left(8.5\,{}^{+1.9}_{-1.6}\right)\cdot 10^{-3}$
Commercial Ti[3]	Cr^{51}	980–1178	35.3 ± 2.7[1]	$\left(5\,{}^{+9}_{-3}\right)\cdot 10^{-3}$
Commercial Ti[4]	Fe^{59}	965–1284	32.9 ± 0.9[2]	$\left(12.3\,{}^{+4.0}_{-3.0}\right)\cdot 10^{-3}$
Commercial Ti[5]	Ni^{63}	934–1235	29.6 ± 0.5[2]	$\left(9.2\,{}^{+2.1}_{-1.8}\right)\cdot 10^{-3}$
Commercial Ti[5]	Nb^{95}	996–1264	39.3 ± 0.9[2]	$\left(5.0\,{}^{+2.0}_{-1.5}\right)\cdot 10^{-3}$
Ti + 10 % Cr (by weight)	Cr^{51}	926–1177	40.2 ± 3.9[1]	$\left(20\,{}^{+6}_{-1}\right)\cdot 10^{-3}$
Ti + 18 % Cr	Cr^{51}	925–1178	44.5 ± 1.6[1]	$\left(9\,{}^{+8}_{-4}\right)\cdot 10^{-2}$
Ti + 5.8 % Fe	Fe^{55}	846–1264	39.6 ± 0.4[2]	$\left(9.2\,{}^{+1.5}_{-1.2}\right)\cdot 10^{-2}$
Ti + 11.7 Fe	Fe^{55}	849–1196	48.6 ± 0.3[2]	$\left(2.14\,{}^{+0.35}_{-0.30}\right)$
Ti + 17.5 Fe	Fe^{55}	846–1099	58.1 ± 0.3[2]	$\left(52.5\,{}^{+7.6}_{-5.4}\right)$

[1] A. J. Mortlock and D. H. Tomlin, Phil. Mag., 4, 41, 628, May 1959.
[2] R. F. Peart and D. H. Tomlin, Department of Physics, Reading University. To be published.
[3] Metal purity 99.2 %, major impurities 0.3 % C, 0.15 % Fe, 0.29 % O
[4] Metal purity 99.6 %, major impurities 0.3 % Fe, 0.1 % C.
[5] Metal purity 99.7 %, major impurities 0.2 % Fe, 0.1 % C.
[6] Metal purity > 99.9 %.

PHYSICAL CONSTANTS OF OZONE AND OXYGEN

Armour Research Foundation, 1958

Physical Constant	Ozone (O_3)	Oxygen (O_2)
Molecular Weight	48.0 gg/mol	32.0 g/g mol
Boiling Point (760 mm)	−111.9 ± 0.3°C	−182.96°C
Melting Point	−192.7 ± 0.2°C	−218.4°C
Critical Temperature	−12.1 ± 0.1°C	−118.8°C
Critical Pressure	54.6 atm	49.7 atm
Critical Density	0.437 g/cc	0.430 g/cc
Critical Volume	147.1 cc/mol	74.5 cc/mol
Gas Density (0°C) (760 mm pressure)	2.144 g/liter	1.429 g/liter
Liquid Density		
−112°C	1.358 g/cc	
−183°C	1.571 g/cc	1.14 g/cc
−195.4°C	1.614 g/cc	1.201 g/cc
Surface Tension		
−195°C	43.8 ± 0.1 dyne/cm	
−182.7°C	38.1 ± 0.2 dyne/cm	
−183.0°C	38.4 ± 0.7 dyne/cm	13.2 dyne/cm
Heat Capacity of Liquid		
−183 to −145°C	0.45 cal/g°C	
Heat Capacity of Gas		
−173°C	7.95 cal/g mol°C	
0°C	9.10 cal/g mol°C	
25°C	9.37 cal/g mol°C	
100°C		6.979 cal/g mol°C
127°C	10.44 cal/g mol°C	
Viscosity of Liquid		
−195.6°C	4.14 ± 0.05 cp	
−183.0°C	1.57 ± 0.02 cp	0.189 cp
Heat of Vaporization		
−112°C	75.6 cal/g	
−182.9°C		50.9 cal/g
Heat of Formation		
25°C	−34.4 Kg cal/mol	
Free Energy		
25°C	32.4 Kg cal/mol	
Van der Waals Constant (a)	3.545 atm liter²/mol²	1.36 atm liter²/mol²
Van der Waals Constant (b)	0.04903 liter/mol	0.03803 liter/mol
Magnetic Susceptibility		
gas (× 10^{-6})	0.002 cgs units	10.6.2 cgs units
liq (× 10^{-6})	0.150 cgs units	260.0 cgs units
Thermal Conductivity of Liquid		
−195.8°C	5.21 cal/sec cm°C × 10^4	
−183.0°C	5.31 cal/sec cm°C × 10^4	
−165.0°C	5.42 cal/sec cm°C × 10^4	
−128.0°C	5.52 cal/sec cm°C × 10^4	

Phase Boundaries—Ozone-Oxygen System
- −183°C 29.8 and 72.4 wt % O_3
- −195.4°C 9 and 90.8 wt % O_3

Consolute Temperature—Ozone-Oxygen System
- −180 ± 0.5°C

Coefficient of Thermal Expansion for Liquid Ozone

Temp. °C	α
−195.6	1.62
−183.0	1.58
−148.0	1.47
−123.0	1.41
−112.0	1.35
− 98.8	1.31

PHYSICAL CONSTANTS OF CLEAR FUSED QUARTZ

Based on information contained in Fused Quartz Catalogue Q-7A General Electric Company.

Property	Clear fused quartz	Property	Clear fused quartz
Density	2.2 g./c.c.	Annealing Point	(approx.) 1140°C
Hardness	4.9 (Mohs')	Strain Point	1070°C
Tensile Strength	7,000 p.s.i.	Electrical Resistance	9.5 $\log_{10}$ R for cm.³ at 350°C
Compressive Strength	>160,000 p.s.i.	Dielectric Constant	3.75 at 20°C. 1 Mc.
Bulk Modulus	(approx.) 5.3 × 10^6 p.s.i.	Dielectric Loss Factor	less than .0004 at 20°C. 1 Mc.
Rigidity Modulus	4.5 × 10^6 p.s.i.	Dissipation Factor	less than .0001 at 20°C. 1 Mc.
Young's Modulus	10.4 × 10^6 p.s.i.	Index of Refraction	1.4585
Poisson's Ratio	.16	Velocity of Sound—Shear Wave	3.75 × 10^5 cm./sec.
Coefficient of Thermal Expansion	(av.) 5.5 × 10^{-7} cm./cm./°C {20°C–320°C}	Velocity of Sound—Compressional Wave	5.90 × 10^5 cm./sec.
Thermal Conductivity	.0033 g. cal./cm.²/sec./°C/cm.	Sonic Attenuation	less than .033 db/ft./mc.
Specific Heat	.18 g. cal./gm.		
Softening Point	(approx.) 1665°C		

PHYSICAL PROPERTIES OF SODIUM, POTASSIUM AND Na-K ALLOYS

Temperature °C	Density (g/cm³)								Viscosity (centipoise)		
	Na			K	Alloys (wt % K)				Na	Alloys (wt % K)	
					43.4		78.6			43.3	66.0
	(a)	(b)	(c)		Experimental	(d) Calculated	Experimental	(d) Calculated			
100	.927	.927	.9265	.819	.887	.890	.847	.850	.705	.540	.529
200	.904	.904	.9037	.795	.862	.867	.823	.827	.450	.379	.354
300	.882	.880	.8805	.771	.838	.843	.799	.802	.345	.299	.276
400	.859	.856	.8570	.747	.814	.818	.775	.778	.284	.245	.229
500	.834	.831	.8331	.723	.789	.794	.751	.754	.234	.207	.195[e]
600	.809	.808	.8089	.701	.765	.771	.727	.732	.210	.178[e]	.168[e]
700	.783	.784	.....	.676	.740	.745	703	705	.186[e]	.257[e]	.146[e]
800	.757	.760	.....	....	....	.....	....	....	.165[e]	......	......
900	....	....	.....	....	....	....	....	....	.150[e]	......	......

Temperature, °C	Thermal conductivity (watts/cm²-°C/cm)				Electrical resistivity (microohms)			Heat capacity[g] (cal/°C-g)		
	Na		Alloy (wt % K)		Na[f]	Alloy (wt % K)		K	Alloy (wt % K)	
	Experimental	Calculated	56.5	77.0		56.5	78.0		44.8	78.26
100	......	....	....	.238[h]	8.99	41.61	45.63	.1940	.2690	.2248
200	.815	.808	.249	.247	13.52	47.23	51.33	.1887	.2612	.2169
300	.757	.755	.262	.259	17.52	54.33	58.58	.1894	.2553	.2122
400	.712	.710	.269	.262	21.93	62.21	65.65	.126	.2512	.2097
500	.668	.672	.271	.259	26.96	69.37	73.48	.1818	.2498	.2088
600	.627[e]	.639	....	.255	32.65	78.29	82.61	.1825	.2484	.2092
700	.590[e]	.610	....	......	39.05	88.23	91.76	.1846	.2497	.2108
800	.547[e]	.583	....	......	46.15	99.68	104.51	.1883	.2529	.2133
900	......	....	....	......	.....	.....	......	.....	.....	.....

Vapor Pressure, mm Hg

Temperature °C	Na	K	Alloys (wt % K)	
			56	78
127	2.23×10^{-6}	....	....	....
227	1.15×10^{-3}	2.88×10^{-2}	1.57×10^{-3}	1.81×10^{-3}
327	5.03×10^{-2}	9.27×10^{-1}	5.73×10^{-2}	6.14×10^{-2}
427	.881	9.26	3.53	5.06
527	7.53	52.22	23.0	31.87
627	39.98	201.25	101.35	136.50
727	148.5	588.62	328.7	431.85
827	453.7	1421.0	864.2	1099.52
927	1127.8	....	....	....
1027	2522.4	....	....	....
1127	4696.8	....	....	....

NOTES: (a) From plotted data.

(b) Epstein equation: $d_t = 0.9514 - 2.392 \times 10^{-4}\, t°C$.

(c) Thomson and Garelis: $d_t = 0.9490 - 22.3 \times 10^{-5} t°C - 1.75 \times 10^{-8} t^2 °C$.

(d) Formula to calculate density: $V = M_K \cdot V_K + M_{Na} \cdot V_{Na}$ (where V, V_K and V_{Na} are the specific volumes (reciprocal of density) of the alloy, K and Na respectively, M_K and M_{Na} the mole fraction of the elements).

(e) Extrapolated by calculation, Epstein equation:

$$K = \frac{2.433 \times 10^{-2}(t + 273.16)}{6.8393 + 3.3873 \times 10^{-2} t + 1.7235 \times 10^{-5} t^2}$$

(f) Epstein equation: $r_t = 10.892 + 0.015272t + 3.6746 \times 10^{-5} t^2 - \dfrac{379.26}{t}$.

(g) Formula to calculate heat capacity: $C = W_{Na} \cdot C_{Na} + W_K \cdot C_K$ (where C, C_{Na} and C_K are the heat capacity of the alloy, Na and K respectively, W_{Na} and W_{Ka} the weight fractions of Na and K respectively in the alloy).

(h) 150°C.

MECHANICAL AND PHYSICAL PROPERTIES OF WHISKERS

From NASA SP-5055

This table lists some nominal values for physical and mechanical properties of some whiskers. The strength of whiskers is influenced by temperature, time, surface conditions, surface films or corrosion, crystallographic orientation, impurities and testing techniques.

Material	Tensile Strength (σ_f) lb/in.^{2} × 10^{-6}	Young's Modulus (E) lb/in.^{2} × 10^{-6}	Specific Gravity (S) lb/in.^{2} × 10^{-6}	$\frac{(\sigma_f)}{(S)}$ inch	$\frac{(E)}{(S)}$ inch
Graphite	3.0	..	137	21,700	....
Al_2O_3	3.0	76	250	12,000	300,000
Iron	1.8	28	485	3,700	58,000
Si_3N_4	2.0	55	193	10,000	285,000
SiC	3.0	70	187	16,000	380,000
Si	1.1	26	143	7,000	182,000

ABSOLUTE VISCOSITY OF LIQUID SODIUM AND POTASSIUM

Sodium

T (°K)	η (10^{-2} poise)	ν (cm^3/g)	$\eta\nu^{1/3}$ (10^3 poise cm/g$^{-1/3}$)	$1/T\nu$ [10^3g/(cm^3 · °K)]
Experimental range				
371.00	0.690	1.078,75	7.0766	2.4987
473	.450	1.106,56	4.6544	1.9106
573	.340	1.135,72	3.5482	1.5366
673	.278	1.166,86	2.9268	1.2734
773	.239	1.200,34	2.54057	1.0776
873	.212	1.236,25	2.2754	0.9266
973	.193	1.274,37	2.0925	.8065
1073	.179	1.315,79	1.9615	.7083
1173	.167	1.360,54	1.8505	.6266
1203	.164	1.373,62	1.8230	.6052

Potassium

T (°K)	η (10^{-2} poise)	ν (cm^3/g)	$\eta\nu^{1/3}$ (10^3 poise cm/g$^{-1/3}$)	$1/T\nu$ [10^3g/(cm^3 · °K)]
Experimental range				
336.9	0.560	1.2062_7	5.961_2	2.4606
400	.384	1.22911	4.113_4	2.0340
500	.276	1.26711	2.986_6	1.5784
600	.221	1.3075_3	2.416_6	1.2747
700	.185	1.3506_2	2.045_0	1.0577
800	.162	1.3966_5	1.810_8	0.8949_7
900	.147	1.4459_2	1.6623	$.7684_6$
1000	.132	1.4988_0	1.5106	$.6672_0$
1100	.121	1.5556_9	1.4020	$.5843_5$
1200	.113	1.6170_7	1.3264	$.5153_3$
1300	.106	1.6835_0	1.2610	$.4569_1$
1400	.100	1.7556_1	1.2064	$.4068_7$

PHYSICAL PROPERTIES OF PIGMENTS

From Rutherford J. Gettens and George L. Stout. *Painting Materials*, Dover Publications, Inc., New York, 1966. Reprinted by permission of the Publisher.

OPAQUE WHITE PIGMENTS

Pigment Name and Chemical Composition[1]	Specific Gravity[2]	Particle Characteristics[3]	Refractive Index[4]
Titanium calcium white, TiO_2 (25%) + $CaSO_4$ (75%)	3.10	prism. or ragged gr.	mostly 1.8–2.0 (irr.) (bi.) [M*]
Titanium dioxide (rutile) + $CaSO_4$	3.25	. . .	Av. 1.98 [TPC]
Titanium barium white, TiO_2 (25%) + $BaSO_4$ (75%)	4.30	min. round. gr.	$n_{\Sigma c}$ 1.7–2.5 [M*]
White lead (basic sulfate) $PbSO_4 \cdot PbO$	6.46	. . .	Av. 1.93 [TPC]
Lithopone	4.3	. . .	Av. 1.84 [TPC]
Lithopone (regular), ZnS (28–30%), $BaSO_4$ (72–70%)	4.30	fine comp. gr.	2.3 (ZnS)–1.64 ($BaSO_4$) [M]
Zinc white (ordinary), ZnO	5.65	v. fine cryst. gr.	ϵ2.02, ω2.00 [M]
(acicular), ZnO	. . .	spicules, fourlets	ϵ2.02, ω2.00 [M*]
White lead (basic carbonate), $2PbCO_3 \cdot Pb(OH)_2$	6.70	v. fine cryst.	ϵ1.94, ω2.09 [M]
Antimony oxide, Sb_2O_3	5.75	v. fine cryst.	valentinite, α2.18, γ and β2.35 [LB, M*] senarmonite, 2.09 (isot.)
Zirconium oxide (baddeleyite), ZrO_2	5.69	. . .	χ2.13, γ2.20, β2.19 [LB], Av. 2.40 [TPC]
Titanium dioxide (anatase), TiO_2	3.9	min. round. gr.	ϵ and ω 2.5 (w. bi.) [M*]
(rutile), TiO_2	4.2	round. or prism. gr.	ϵ2.9, ω2.6 [M*]

TRANSPARENT WHITE PIGMENTS†

Pigment Name and Chemical Composition	Specific Gravity	Particle Characteristics	Refractive Index
Diatomaceous earth, SiO_2	2.31	min. fossil forms	n mostly 1.435, some 1.40 [M*]
Aluminum stearate, $Al(C_{18}H_{35}O_2)_3$	0.99	agg. of spher. gr.	1.49 (w. bi.) [W]
Pumice (volcanic glass), Na, K, Al, silicate	. . .	vesicular vitr. frag.	c 1.50 (isot.) [M*]
Aluminum hydrate, $Al(OH)_3$	2.45	v. fine amorph. part.	$n_{\Sigma c}$ 1.50–1.56 [M*]
Gypsum, $CaSO_4 \cdot 2H_2O$	2.36	fine cryst. gr.	α1.520, γ1.530, β1.523 [LB]
Silica quartz), SiO_2	2.66	cryst. frag.	ϵ1.553, ω1.544 [LB]
(chalcedony), SiO_2	2.6	crypt. agg.	ϵ, ω1.54 [LB, M*]
China clay (kaolinite), $Al_2O_3 \cdot 2SiO_2 \cdot 2H_2O$	2.60	fine, vermicular cryst.	α1.558, γ1.565, β1.564 (all ± .005) [LB, M*]
Talc, $3MgO \cdot 4SiO_2 \cdot H_2O$	2.77	platy frag.	α1.539, γ1.589, β1.589 [LB]
Mica (muscovite), $H_2KAl_3(SiO_4)_3$	2.89	platy frag.	α1.563, γ1.604, β1.599 [LB]
Anhydrite, $CaSO_4$	2.93	cryst. frag.	α1.570, γ1.614, β1.575 [LB]
Chalk (whiting), $CaCO_3$	2.70	hollow spherulites	$\epsilon_{\Sigma c}$ 1.510, $\omega_{\Sigma c}$ 1.645 [M*]
Barytes (barite, nat.), $BaSO_4$	4.45	cryst. frag.	α1.636, γ1.648, β1.637 [LB]
(blanc fixe, art.), $BaSO_4$	4.36	v. fine cryst. agg.	1.62–1.64 [M*]
Barium carbonate, $BaCO_3$ (witherite)	4.3	. . .	α1.529, γ1.677, β1.676 [LB]

†Also called "Extender" or "Inert" White Pigments.

IRON OXIDE PIGMENTS

Pigment Name and Chemical Composition	Specific Gravity	Particle Characteristics	Refractive Index
Ochre, yellow (goethite), $Fe_2O_3 \cdot H_2O$, clay, etc.	2.9–4.0	irr. spherulites	n_Σ2.0 (isot. part); $(\alpha, \beta)_\Sigma$ 2.05–2.31; γ_Σ 2.08–2.40 (bi. part) [M*]
Sienna, raw (goethite), $Fe_2O_3 \cdot H_2O$, clay, etc.	3.14	uneven spherulites	1.87–2.17 (mostly 2.06) (isot.) [M*]
Sienna, burnt, Fe_2O_3, clay, etc.	3.56	uneven, round. part.	c1.85 (var.) (isot.) [M]
Umber, raw, $Fe_2O_3 + MnO_2 + H_2O$, clay, etc.	3.20	uneven, round. gr.	mostly 1.87–2.17 [M*]
Umber, burnt, $Fe_2O_3 + MnO_2$, clay, etc.	3.64	uneven, round. gr.	mostly 2.2–2.3 [M*]
Iron oxide red (haematite), Fe_2O_3	5.2	min. cryst.	ϵ_{Li} 2.78, ω_{Li} 3.01 [M]

PHYSICAL PROPERTIES OF PIGMENTS (Continued)

RED AND ORANGE PIGMENTS

Pigment Name and Chemical Composition[1]	Specific Gravity[2]	Particle Characteristics[3]	Refractive Index[4]
Red lead, Pb_3O_4 (*c*95%)	8.73	crypt. agg.	2.42_{Li} (w. bi.; pleo.) [M]
**Realgar, As_2S_2	3.56	cryst. frag.	α_{Li} 2.46, γ_{Li} 2.61, β_{Li} 2.59 [LB]
Molybdate orange, Pb(Mo,S,Cr,P)O_4	...	min. round. gr.	β_{Li} 2.55 (s. bi.) [M*]
Chrome orange, $PbCrO_4 \cdot Pb(OH)_2$	6.7	tabular cryst.	α2.42, γ2.7 +, β2.7 [M*]
Cadmium red lithopone, CdS(Se) + $BaSO_4$	4.30	min. round. gr.	2.50–2.76 (for CdS(Se)part) (isot.) [M*]
Cadmium red, CdS(Se)	4.5	min. round. gr.	2.64 (bright red)–2.77 (deep red) (isot.) [M*]
Antimony vermilion, Sb_2S_3	...	v. fine red glob.	n_{Li} 2.65 (isot.) [M*]
Vermilion (art.), HgS	8.09	hexagonal gr. and prisms	ϵ_{Li} 3.14, ω_{Li} 2.81 [M]
**(nat., cinnabar), HgS	8.1	cryst. frag.	ϵ_{Li} 3.146, ω_{Li} 2.819 [LB]
Quinacridone red, $C_{20}H_{12}O_2N_2$ (gamma)	1.5	thin plates	n_{Na} Av. 2.04 [Du P]

YELLOW PIGMENTS

Pigment Name and Chemical Composition[1]	Specific Gravity[2]	Particle Characteristics[3]	Refractive Index[4]
**Gamboge, organic resin	...	irr. amorph. part.	1.582–1.586 [W]
**Indian yellow, $C_{19}H_{18}O_{11}Mg \cdot 5H_2O$	...	prisms, plates	1.67 (w. bi.) [M*]
Cobalt yellow, $CoK_3(NO_2)_6 \cdot H_2O$	...	fine dendritic cryst.	1.72–1.76 (isot.) [W]
Zinc yellow, $4ZnO \cdot 4CrO_3 \cdot K_2O \cdot 3H_2O$	3.46	min. spher. gr.	1.84–1.9 (irr.; bi.) [M*]
Strontium yellow, $SrCrO_4$	...	small needles	α, β (or ω) 1.92, γ (or ϵ) 2.01 (‖ext.) [M*]
Barium yellow, $BaCrO_4$	4.49	v. fine cryst. gr.	1.94–1.98 (bi.) [M]
**Naples yellow, $Pb_3(SbO_4)_2$	...	round. gr.	2.01–2.28 (isot.) [M*]
Chrome yellow (med.), $PbCrO_4$	5.96	fine prism. gr.	$\alpha_{620m\mu} < 2.31$, $\gamma_{650m\mu}$ 2.49 [M]
Cadmium yellow lithopone, CdS + $BaSO_4$	4.25	fine comp. gr.	2.39–2.40 (for CdS part) [M*]
Cadmium yellow, CdS	4.35	min. round. gr.	2.35–2.48 (isot.) [M*]
Massicot (litharge), PbO	9.40	min. flakes	α_{Li} 2.51, γ_{Li} 2.71, β_{Li} 2.61 [M]
**Orpiment, As_2S_3	3.4	min. flakes	α_{Li} 2.4 ±, γ_{Li} 3.02, β_{Li} 2.81 [LB]

GREEN PIGMENTS

Pigment Name and Chemical Composition[1]	Specific Gravity[2]	Particle Characteristics[3]	Refractive Index[4]
Phthalocyanine green, chloro-copper phthalocyanine	2.1	laths	$n_{580m\mu}$ 1.40 [ACC]
**Verdigris (copper basic acetate), $Cu(C_2H_3O_2)_2 \cdot 2Cu(OH)_2$	...	cryst. frag.	α1.53, γ1.56 [M]
**Chrysocolla, $CuSiO_3 \cdot \eta H_2O$	2.4	crypt. agg.	α1.575, γ1.598, β1.597 [LB]
Green earth (celadonite and glauconite), Fe, Mg, Al, K, hydrosilicate	2.5–2.7	round. irr. gr.	*n* var. *c* 1.62, (porous) [M*]
Emerald green (Paris green), $Cu(C_2H_3O_2)_2 \cdot 3Cu(AsO_2)_2$	3.27	spherulites and disks	α_Σ1.71, γ_Σ1.78 (w. pleo.) [M*]
**Malachite, $CuCO_3 \cdot Cu(OH)_2$	4.0	cryst. frag.	α1.655, γ1.909, β1.875 [LB]
Cobalt green, $CoO \cdot \eta ZnO$	...	spher. gr.	1.94–2.0 (w. bi.) [M*]
Viridian (chromium oxide, transparent), $Cr_2O_3 \cdot 2H_2O$	3.32	spherul. gr.	α, β_Σ 1.82, γ_Σ 2.12 [M*]
Chrome green (med.), $Fe_4[Fe(CN)_6]_3 + PbCrO_4$	4.06	fine green agg.	*c* 2.4 (cf. Prussian blue and chrome yellow)
Chromium oxide green, opaque, Cr_2O_3	5.10	fine cryst. agg.	n_{Li} 2.5 [M]

BLUE PIGMENTS

Pigment Name and Chemical Composition[1]	Specific Gravity[2]	Particle Characteristics[3]	Refractive Index[4]
Phthalocyanine blue, copper phthalocyanine	1.6	laths	Av. 1.38 [DuP]
Ultramarine blue (art.), $Na_{8-10}Al_6Si_6O_{24}S_{2-4}$	2.34	uniform small round. gr.	n 1.51 $_{green}$, 1.63 $_{red}$ (isot.) [M]
**(nat., lazurite), $3Na_2O \cdot 3Al_2O_3 \cdot 6SiO_2 \cdot 2Na_2S$	2.4	angular, broken frag.	1.50± (isot.) [LB]
**Maya blue, Fe, Mg, Ca, Al, silicate (?)	...	porous irr. agg.	β_Σ1.54 (irr.; bi. and pleo.) [M*]
Smalt, K, Co(Al), silicate (glass)	...	splintery, vitr. frag.	1.49–1.52 [M*]
Prussian blue, $Fe_4[Fe(CN)_6]_3$	1.83	colloidal agg.	$1.56_{460m\mu}$ [M*]
**Egyptian blue, $CaO \cdot CuO \cdot 4SiO_2$	...	cryst. frag.	ϵ1.605, ω1.635 [APL]
Manganese blue, $BaMnO_4 + BaSO_4$	...	gr. and stubby prisms	c 1.65 [W]
**Blue verditer, $2CuCO_3 \cdot Cu(OH)_2$	...	fibrous agg.	α_Σ1.72, $\gamma_\Sigma$slightly > 1.74 [M*]
Cobalt blue, $CoO \cdot Al_2O_3$	3.83	round. gr.	n var.; max. c 1.74_{blue} (isot.) [M]
**Azurite, $2CuCO_3 \cdot Cu(OH)_2$	3.80	cryst. frag.	α1.730, γ1.838, β1.758 [LB]
Cerulean blue, $CoO \cdot \eta SnO_2$	...	round gr.	1.84 (isot.) [M*]

VIOLET PIGMENTS

Pigment Name and Chemical Composition	Specific Gravity	Particle Characteristics	Refractive Index
Ultramarine violet	...	round. gr. (blue, rose and violet)	c 1.56 (isot.) [M*]
Cobalt violet, $Co_3(PO_4)_2$	...	round. gr.	ϵ1.65–1.79 (dull violet), ω1.68–1.81 (salmon) (s. bi.) [M*]
Manganese violet, $(NH_4)_2Mn_2(P_2O_7)_2$	...	fine cryst. gr.	α1.67, γ1.75, β1.72 (for violet) [M]
Quinacridone violet, $C_{20}H_{12}O_2N_2$ (beta)	1.5	thin plates	Av. 2.02 [DuP]

BROWN PIGMENTS

Pigment Name and Chemical Composition	Specific Gravity	Particle Characteristics	Refractive Index
Sepia (organic)	...	angular frag.	(opaque) [M*]
Asphaltum (bitumen)	...	irr. amorph. part.	1.64–1.66 [M*]
Van Dyke brown (bituminous earth)	1.66	irr. amorph. part.	1.62–1.69 [M*]

BLACK PIGMENTS

Pigment Name and Chemical Composition	Specific Gravity	Particle Characteristics	Refractive Index
Bone black, $C + Ca_3(PO_4)_2$	2.29	irr. coarse grains	1.65–1.70 (for larger translucent gr.) [M]
Lamp black, C	1.77	min. round. part.	(opaque)
Charcoal black, C	...	irr. splintery part.	(opaque)
Graphite, C	2.36	irr. plates	(opaque) [M]

[1]Abbreviations: art. = artificial; med. = medium; nat. = natural. The chemical formulas are those commonly accepted in chemical and mineralogical literature, but they may not compare exactly with structural formulas based on x-ray diffraction data or even on critical chemical analysis.

[2]The figures for specific gravity of the artificial pigments are mainly from H. A. Gardner, pp. 710–712, and those on the mineral pigments are chiefly from E. S. Larsen and H. Berman.

[3]Symmetry terms (monoclinic, orthorhombic, etc.) are omitted because pigments are so finely divided that it is rare when observations on crystal symmetry can be made. The term "spherulitic," as used here means aggregates that tend toward radial structure and spherical shape. "Amorphous" describes materials that are microscopically formless but may be truly crystalline on the basis of x-ray diffraction data. Abbreviations: agg. = aggregate(s); amorph. = amorphous; comp. = composite; crypt. = cryptocrystalline; cryst. = crystal(s); frag. = fragment(s); glob. = globule(s); gr. = grain(s); irr. = irregular; min. = minute; part. = particle(s); prism. = prismatic; round. = rounded; spher. = spheroidal; spherul. = spherulitic; var. = variable; v. = very; vitr. = vitreous.

[4]Unless otherwise indicated, all refractive index measurements are by sodium light. Σ is the symbol used by H. E. Merwin to indicate greater or less indefiniteness or irregularity in the case of aggregates, especially in respect to refractive index. Abbreviations: bi. = birefringent; c = *circa*; ext. = extinction; isot. = isotropic; ‖ = parallel; pleo. = pleochroic; s. = strongly; w. = weakly. The letters in brackets refer to the authorities for the refractive index data: M = H. E. Merwin; M* = H. E. Merwin, data by private communication, hitherto unpublished; W = C. D. West data by private communication, hitherto unpublished; LB = E. S. Larsen and H. Berman; APL = A. P. Laurie and co-authors; ACC = A. C. Cooper, "The refractive index of organic pigments. Its determination and significance," *Journal Oil & Colour Chemists Association*, Vol. 31 (1948), pp. 343–357; TPC = Titanium Pigment Corporation; Du P = E. I. Du Pont de Nemours & Co.

**Chiefly of historical interest.

CRITICAL TEMPERATURES AND PRESSURES

Compiled by Rudolfh Loebel

Table I —Organic compounds
Table II—Inorganic compounds

Table I

Formula	Name	Critical temp. T_c °C	Critical press. P_c atm.
$CHClF_2$	Methane, monochlorodifluoro-	96	48.5
$CHCl_2F$	Methane, dichloromonofluoro-	178.5	51
$CHCl_3$	Methane, trichloro- (Chloroform)	263	54
CHF_3	Methane, trifluoro- (Fluoroform)	25.9	46.9
CH_2Cl_2	Methylene chloride	237	60
CH_3NO_2	Methane, nitro-	314.8	62.3
CH_3Br	Methane, monobromide-	194	83.4
CH_3Cl	Methane, monochloro-	143.8	65.9
CH_3F	Methane, monofluoro-	44.6	58
CH_3I	Methane, monoiodo-	254.8	72.7
CH_4	Methane	−82.1	45.8
CH_4O	Methanol (Methyl alcohol)	240	78.5
CH_4S	Methylmercaptan	196.8	71.4
CH_5N	Methylamine	156.9	40.2
$CBrF_3$	Methane, monobromotrifluoro-	67	50.3
$CClF_3$	Methane, monochlorotrifluoro-	28.85	38.2
CCl_2F_2	Methane, dichlorodifluoro-	111.5	39.6
CCl_3F	Methane, trichloromonofluoro-	198	43.2
CCl_4	Methane, tetrachloro- (Carbon tetrachloride)	283.1	45
CF_4	Methane, tetrafluoro-	−45.7	41.4
C_2H_2	Acetylene	35.5	61.6
C_2H_2	Ethyne (see Acetylene)	35.5	61.6
$C_2H_2F_2$	Ethylene, 1,1-difluoro-	30.1	—
C_2H_3N	Acetonitrile	274.7	47.7
$C_2H_3F_3$	Ethane, 1,1,1-trifluoro-	73.1	—
C_2H_4	Ethene	9.9	50.5
C_2H_4	Ethylene (see Ethene)	9.9	50.5
C_2H_4O	Acetaldehyde	187.8	54.7
C_2H_4O	Ethylene oxide	195.8	71
$C_2H_4O_2$	Acetic acid	321.6	57.1
$C_2H_4O_2$	Formic acid, methyl- (Methyl formate)	214	59.2
$C_2H_4Cl_2$	Ethane, 1,1-dichloro-	249.8	50
$C_2H_4F_2$	Ethane, 1,1-difluoro-	386.7	—
C_2H_5Br	Ethane, monobromo-	230.8	61.5
C_2H_5Cl	Ethane, monochloro-	187.2	52
C_2H_5F	Ethane, monofluoro-	102.16	49.6
C_2H_6	Ethane	32.2	48.2
C_2H_6O	Ether, dimethyl-	127	52.6
C_2H_6O	Ethanol (Ethyl alcohol)	243	63
C_2H_6O	Glycol, ethylene-	(374)*	—
C_2H_6S	Dimethylsulfide	229.9	54.5
C_2H_6S	Ethylmercaptan	225.5	54.2
C_2H_7N	Dimethylamine	164.6	52.4
C_2H_7N	Ethylamine	183.2	55.5
$C_2Br_2F_4$	Ethane, dibromotetrafluoro-	214.5	—
C_2ClF_5	Ethane, chloropentafluoro-	80	—
$C_2Cl_2F_4$	Ethane, 1,2-dichlorotetrafluoro-	145.7	—
$C_2Cl_3F_3$	Ethane, trichlorotrifluoro-	214.2	33.7
$C_2Cl_4F_2$	Ethane, tetrachlorodifluoro-	278	32.9
C_2F_6	Ethane, hexafluoro-	24.3	—
C_3H_4	Propadiene	120	43.6
C_3H_4	Allene (see Propadiene)	120	43.6
C_3H_4	Acetylene, methyl-	127.8	52.8
C_3H_4	Propyne (see Acetylene, methyl-)	127.8	52.8
C_3H_5N	Ethyl cyanide	290.8	41.3
C_3H_5N	Propionitrile (see Ethyl cyanide)	290.8	41.3
C_3H_5OCl	Epichlorohydrin	(323)	—
C_3H_5Cl	Propene, 3-chloro- (allyl-chloride)	240.3	46.5
C_3H_6	Propylene	91.9	45.4
C_3H_6	Cyclopropane	124.7	—
C_3H_6O	Acetone	235.5	47
C_3H_6O	Allylalcohol	272	55.5
C_3H_6O	Propylene oxide	209	48.6
$C_3H_6O_2$	Formic acid, ethyl- (Ethyl formate)	235.3	46.3
$C_3H_6O_2$	Acetic acid, methyl- (Methyl acetate)	233.7	46.3
$C_3H_6O_2$	Propanoic acid	337.6	53
C_3H_7Cl	n-Propane, monochloro-	230	45.2
C_3H_8	Propane	96.8	42
C_3H_8O	Ether, ethyl methyl- (Methoxyethane)	164.7	43.4
C_3H_8O	Glycol, 1,2-propylene-	351	—
C_3H_8O	Isopropyl alcohol	235	47
C_3H_8O	n-Propyl alcohol	263.6	51
$C_3H_8O_3$	Glycerol	(452)	—
C_3H_9N	Isopropylamine	209.7	—
C_3H_9N	n-Propylamine	223.8	46.8
	*( ) uncertain		
C_4H_4O	Furan	213.8	52.5
C_4H_4S	Thiophene	307	56.2
C_4H_6	1,2-Butadiene	171	44.4
C_4H_6	1,3-Butadiene	152	42.7
C_4H_8	n-Butene	146	39.7
C_4H_8	2-Butene, cis-	160	40.5
C_4H_8	2-Butene, trans-	155	41.5
C_4H_8O	1,2-Butylene oxide	243	—
C_4H_8O	Ketone, ethyl methyl- (2-Butanone)	262	41
$C_4H_8O_2$	Butanoic acid	355	52
$C_4H_8O_2$	p-Dioxane	314.8	51.4
$C_4H_8O_2$	Acetic acid, ethyl- (Ethylacetate)	250.4	37.8
$C_4H_8O_2$	Propanoic acid, methyl- (Methylpropionate)	257.4	39.3
$C_4H_8O_2$	Formic acid, propyl-(Propyl formate)	264.9	40.1
C_4H_9Cl	n-Butane, monochloride-	269	—
C_4H_{10}	n-Butane	152	37.5
C_4H_{10}	Isobutane	135	36
$C_4H_{10}O$	Butanol (n-Butyl alcohol)	289.8	43.6
$C_4H_{10}O$	sec-Butyl alcohol	263	41.4
$C_4H_{10}O$	tert-Butyl alcohol	235	39.2
$C_4H_{10}O$	Ether, diethyl-	192.6	35.6
$C_4H_{10}O$	Isobutyl alcohol	277	42.4
$C_4H_{10}O_3$	Glycol, diethylene-	407	—
$C_4H_{10}S$	Diethyl sulfide	283.8	39.1
$C_4H_{11}N$	Butyl amine	287.9	—
$C_4H_{11}N$	Diethyl amine	223.3	36.6
$C_4H_{11}N$	Isobutyl amine	266.7	—
C_4F_{10}	Butane, perfluoro-	113.2	23
C_5H_5N	Pyridine	346.8	—
C_5H_8	Cyclopentene	232.94	47.2
C_5H_{10}	Cyclopentane	238.6	44.6
C_5H_{10}	1-Pentene	191	39.9
$C_5H_{10}O$	Ketone, diethyl-	287.8	36.9
$C_5H_{10}O$	Propanoic acid, ethyl- (Ethyl propionate)	272.9	33
$C_5H_{10}O_2$	n-Butanoic acid, methyl- (n-Methyl butyrate)	281.3	34.3
$C_5H_{10}O_2$	Acetic acid, n-propyl (n-Propyl acetate)	276	32.9
$C_5H_{10}O_2$	n-Valeric acid	378	37.6
$C_5H_{11}N$	Piperidine	320.8	44.1
C_5H_{12}	Butane, 2-methyl- (See Isopentane)	187.8	32.9
C_5H_{12}	Isopentane	187.8	32.9
C_5H_{12}	Neopentane	160.6	31.6
C_5H_{12}	n-Pentane	196.6	33.3
C_5H_{12}	Propane, 2,2-dimethyl (see Neopentane)	160.6	31.6
$C_5H_{12}O$	Isoamyl alcohol	309.77	—
C_6H_5Br	Benzene, bromo-	397	44.6
C_6H_5Cl	Benzene, chloro-	359.2	44.6
C_6H_5F	Benzene, fluoro-	286.95	44.6
C_6H_6	Benzene	288.9	48.6
C_6H_6O	Phenol	421.1	60.5
C_6H_7N	Aniline	425.6	52.3
C_6H_7N	α-Picoline	348	—
C_6H_7N	β-Picoline	371.7	—
C_6H_7N	γ-Picoline	372.5	—
C_6H_{10}	Cyclohexene	287.3	—
C_6H_{10}	1,5-Hexadiene	234.4	32.6
C_6H_{12}	Cyclohexane	280.4	40
C_6H_{12}	Cyclopentane, methyl-	259.5	37.4
C_6H_{12}	1-Hexene	231	31.1
$C_6H_{12}O_2$	Formic acid, n-amyl (n-Amyl formate)	302.6	34.1
$C_6H_{12}O_2$	Acetic acid, n-butyl- (n-Butyl acetate)	305.9	30.7
$C_6H_{12}O_2$	Butanoic acid, ethyl- (Ethyl butyrate)	293	30.2
$C_6H_{12}O_2$	Propanoic acid, propyl- (Propyl propionate)	304.8	30.7
$C_6H_{12}O_3$	Paraldehyde	290	—
C_6H_{14}	Butane, 2,3-dimethyl-	226.8	30.9
C_6H_{14}	n-Hexane	234.2	29.9
C_6H_{14}	Pentane, 2-methyl-	224.3	30
$C_6H_{14}O$	Ether, isopropyl-	226.9	28.4
$C_6H_{14}O$	1-Hexyl alcohol	313.5	—
$C_6H_{14}O_4$	Glycol, triethylene-	(437)	—
$C_6H_{15}N$	Dipropyl amine	277	31
$C_6H_{15}N$	Hexyl amine	318.8	—
$C_6H_{15}N$	Triethyl amine	258.9	30
C_7H_5N	Benzonitrile	426.2	41.6
C_7H_8	Benzene, methyl- (see Toluene)	320.8	41.6

CRITICAL TEMPERATURES AND PRESSURES (Continued)

Table I (Continued)

Formula	Name	Critical temp. T_c °C	Critical press. P_c atm.
C_7H_8	Toluene	320.8	41.6
C_7H_8O	Anisole, (see Benzene, methoxy-)	368.5	41.3
C_7H_8O	Benzene, methoxy-	368.5	41.3
C_7H_8O	o-Cresol	424.4	49.4
C_7H_8O	m-Cresol	432	45
C_7H_8O	p-Cresol	431.4	50.8
C_7H_9N	Aniline, methyl-	428.4	51.3
C_7H_9N	2,3-Lutidine	382.3	—
C_7H_9N	2,4-Lutidine	374	—
C_7H_9N	2,6-Lutidine	350.6	—
C_7H_9N	3,4-Lutidine	410.6	—
C_7H_9N	3,5-Lutidine	394.1	—
C_7H_{14}	Cyclohexane, methyl-	299.1	34.3
C_7H_{14}	Cyclopentane, ethyl-	296.3	33.5
C_7H_{14}	1-Heptene	264.1	—
$C_7H_{14}O_2$	Acetic acid, isoamyl- (Isoamyl acetate)	326.1	28
C_7H_{16}	Butane, 2,2,3-trimethyl-	258.3	29.8
C_7H_{16}	n-Heptane	267.1	27
C_7H_{16}	Hexane, 2-methyl-	257.9	27.2
C_7H_{16}	Pentane, 3-ethyl-	267.6	28.6
C_7H_{16}	Pentane, 2,4-dimethyl-	247.1	27.4
$C_7H_{16}O$	1-Heptyl alcohol	365.3	29.4
C_7F_{16}	n-Heptane, perfluoro-	201.7	16
C_8H_8	Styrene	374.4	39.4
C_8H_{10}	Benzene, ethyl-	343.9	36.9
C_8H_{10}	o-Xylene	359	35.7
C_8H_{10}	m-Xylene	346	34.7
C_8H_{10}	p-Xylene	345	33.9
$C_8H_{10}O$	Xylenol	449.7	56.4
$C_8H_{10}O$	Phenetol (see Benzene, ethoxy-)	374	33.8
$C_8H_{10}O$	Benzene, ethoxy-	374	33.8
$C_8H_{11}N$	Aniline, N,N-dimethyl-	414.4	35.8
$C_8H_{11}N$	Aniline, N-ethyl-	425.4	—
C_8H_{16}	n-Octene	305	25.5
C_8H_{18}	Butane, 2,2,3,3-tetramethyl-	270.8	24.5
C_8H_{18}	Heptane, 4-methyl-	290	25.6
C_8H_{18}	Hexane, 2,4-dimethyl-	282	25.8
C_8H_{18}	n-Octane	296	24.8
C_8H_{18}	Pentane, 2,2,3-trimethyl-	294	28.2
$C_8H_{18}O$	1-Octyl alcohol	385.5	26.5
$C_8H_{19}N$	Dibutyl amine	322.6	—
C_9H_7N	Quinoline	508.8	—
C_9H_{12}	Benzene, n-propyl-	365	31.2
C_9H_{12}	Benzene, 1,2,3-trimethyl-	391.3	31
C_9H_{12}	Cumene (see Benzene, isopropyl-)	362.7	31.2
C_9H_{12}	Benzene, isopropyl-	362.7	31.2
C_9H_{20}	n-Nonane	321	22.5
C_9H_{21}	Tripropyl amine	304.3	
$C_{10}H_8$	Naphthalene	474.8	40.6
$C_{10}H_{12}O$	Ether, diphenyl-	494	30.9
$C_{10}H_{14}$	Benzene, 1-isopropyl-4-methyl-	385.5	27.7
$C_{10}H_{14}$	Benzene, 1,2,3,5-tetramethyl-	402.8	28.6
$C_{10}H_{14}$	p-Cymene, (see Benzene, 1-isopropyl-4-methyl-)	—	—
$C_{10}H_{14}$	Isodurene, (see Benzene, 1,2,3,5-tetramethyl-)	—	—
$C_{10}H_{14}O$	3-p-Cymenol	425.1	33
$C_{10}H_{14}O$	Thymol, (see 3-p-Cymenol)	425.1	33
$C_{10}H_{18}$	Decalin, cis-	418	28.7
$C_{10}H_{18}$	Decalin, trans-	408	28.7
$C_{10}H_{22}$	n-Decane	344.4	20.8
$C_{11}H_{10}$	Naphthalene, 1-methyl-	498.8	32.1
$C_{12}H_{10}$	Biphenyl	495	31.8
$C_{12}H_{11}N$	Diphenylamine	615.5	—
$C_{12}H_{18}$	Benzene, hexamethyl-	494	23.5
$C_{12}H_{18}$	Mellitine, (see Benzene, hexamethyl-)	—	—
$C_{12}H_{26}$	n-Dodecane	386	17.9
$C_{12}H_{27}N$	Tributyl amine	365.2	—

Table II

Name	Formula	Critical temp. T_c °C	Critical press. P_c atm.
Ammonia	NH_3	132.5	112.5
Argon	A	−122.3	48
Boron tribromide	BBr_3	300	—
Boron trichloride	BCl_3	178.8	38.2
Boron trifluoride	BF_3	−12.26	49.2
Carbon dioxide	CO_2	31	72.9
Carbon disulfide	CS_2	279	78
Carbon monoxide	CO	−140	34.5
Carbonyl sulfide	COS	104.8	65
Chlorine	Cl_2	144	76.1
Cyanogen	C_2N_2	126.6	—
Deuterium	D_2	−234.8	16.4
Fluorine	F_2	−129	55
Germanium tetrachloride	$GeCl_4$	276.9	38
Helium	He	−267.9	2.26
Hydrogen	H_2	−239.9	12.8
Hydrogen bromide	HBr	90	84.5
Hydrogen chloride	HCl	51.4	82.1
Hydrogen deuteride	HD	237.3	14.6
Hydrogen cyanide	HCN	183.5	48.9
Hydrogen fluoride	HF	188	64
Hydrogen iodide	HI	150	81.9
Hydrogen sulfide	H_2S	100.4	88.9
Hydrazine	N_2H_2	380	—
Iodine	I_2	512	116
Krypton	Kr	−63.8	54.3
Neon	Ne	−228.7	26.9
Nitric oxide	NO	−93	64
Nitrogen dioxide	NO_2	157.8	100
Nitrogen	N_2	−147	33.5
Nitrous oxide	N_2O	36.5	71.7
Oxygen	O_2	−118.4	50.1
Ozone	O_3	−5.16	67
Phosphine	PH_3	51.3	64.5
Radon	Rn	104.04	62
Silane, chlorotrifluoro-	$SiClF_3$	34.5	34.2
Silane	SiH_4	−3.46	47.8
Silicon tetrachloride	$SiCl_4$	232.8	—
Silicon tetrafluoride	SiF_4	−14.06	36.7
Sulfur dioxide	SO_2	157.8	77.7
Stannic chloride	$SnCl_4$	318.7	37
Water	H_2O	374.1	218.3
Xenon	Xe	16.6	58

DISSOCIATION PRESSURE OF CALCIUM CARBONATE

Temp. °C	mm/Hg	Temp. °C	mm/Hg	Temp. °C	mm/Hg	Temp. °C	mm/Hg
550	0.41	727	44	819	235	894	716
587	1.0	736	54	830	255	898	760 atm.
605	2.3	743	60	840	311	906.5	1.151
671	13.5	748	70	852	381	937	1.770
680	15.8	749	72	857	420	1082.5	8.892
691	19.0	777	105	871	537	1157.7	18.687
701	23.0	786	134	881	603	1226.3	34.333
703	25.5	795	150	891	684	1241	39.094
711	32.7	800	183				

DEFINITIONS AND FORMULAS

The chemical terms have been compiled with the collaboration of
B. Clifford Hendricks

AB-.—A prefix attached to the names of the practical electric units to indicate the corresponding unit in the cgs electromagnetic system (emu), e.g. abampere, abvolt.

Abcoulomb.—The abcoulomb, the emu of charge, is defined as the charge which passes a given surface in one second if a steady current of one abampere flows across the surface. Its dimensions are, therefore, $cm^{\frac{1}{2}}\ gm^{\frac{1}{2}}$, which differ from the dimensions of the statcoulomb by a factor which has the dimensions of a speed. This relationship is connected with the fact that the ratio $2K_e/K_m$ must have the value of the square of the speed of light in any consistent system of units. It follows further that

$$1 \text{ abcoulomb} = 2.99793 \times 10^{10} \text{ statcoulomb},$$

the speed of light in vacuo being $(2.99793 \pm 0.000003) \times 10^{10}$ cm/sec.

Abegg's rule.—For use in regard to a helical periodic system. If the maximum positive valence exhibited by an element be numerically added to its maximum negative valence, there is evidently a tendency for the sum to equal 8. This tendency is exhibited especially by the elements of the 4th, 5th, 6th and 7th groups and is known as Abegg's rule.

Absolute humidity.—See *Humidity*

Absolute pressure.—See *Pressure*

Absolute temperature.—Temperature reckoned from the absolute zero. See *Temperature*

Absolute units.—A system of units based on the smallest possible number of independent units. Specifically, units of force, work, energy and power not derived from or dependent on gravitation.

Absolute zero.—The temperature at which a gas would show no pressure if the general law for gases would hold for all temperatures. It is equal to −273.15°C or −459.67°F.

Absorption.—1. Penetration of a substance into the body of another. 2. Transformation into other forms suffered by radiant energy passing through a material substance.

Absorption coefficient.—See *Absorption factor*

Absorption factor.—The ratio of the intensity loss by absorption to the total original intensity of radiation. If I_o represents the original intensity, I_r, the intensity of reflected radiation, I_t, the intensity of the transmitted radiation, the absorption factor is given by the expression

$$\frac{I_o - (I_r + I_t)}{I_o}$$

Also called coefficient of absorption.

Absorption, Lambert's law.—If I_o is the original intensity, I the intensity after passing through a thickness x of a material whose absorption coefficient is k,

$$I = I_o e^{-kx}$$

The **index of absorption** k' is given by the relation $k = (4\pi k'n)/\lambda$ where n is the index of refraction and λ the wave length in vacuo. The **mass absorption** is given by k/d when d is the density. The transmission factor is given by I/I_o.

Absorption spectrum.—The spectrum obtained by the examination of light from a source, itself giving a continuous spectrum, after this light has passed through an absorbing medium in the gaseous state. The absorption spectrum will consist of dark lines or bands, being the reverse of the emission spectrum of the absorbing substance.

When the absorbing medium is in the solid or liquid state the spectrum of the transmitted light shows broad dark regions which are not resolvable into lines and have no sharp or distinct edges.

Absorptive power or absorptivity for any body is measured by the fraction of the radiant energy falling upon the body which is absorbed or transformed into heat. This ratio varies with the character of the surface and the wave length of the incident energy. It is the ratio of the radiation absorbed by any substance to that absorbed under the same conditions by a black body.

Abvolt.—The cgs electromagnetic unit of potential difference and electromotive force. It is the potential difference that must exist between two points in order that one erg of work be done when one abcoulomb of charge is moved from one point to the other. One abvolt is 10^{-8} volt.

Acceleration.—The time rate of change of velocity in either speed or direction. Cgs unit,—one centimeter per second per second. Dimensions,—$[l\ t^{-2}]$.—See *Angular acceleration*

Acceleration due to gravity.—The acceleration of a body freely falling in a vacuum. The International Committee on Weights and Measures has adopted as a standard or accepted value, 980.665 cm/sec² or 32.174 ft/sec².

Acceleration due to gravity at any latitude and elevation.—If ϕ is the latitude and H the elevation in centimeters the acceleration in cgs units is, $g = 980.616 - 2.5928 \cos 2\phi + 0.0069 \cos^2 2\phi - 3.086 \times 10^{-6} H$. (Helmert's equation)

Accelerators.—Machines for speeding up subatomic particles to energies running into millions of electron volts.—See *Betatron, cyclotron*, etc.

Achromatic.—A term applied to lenses signifying their more or less complete correction for chromatic aberration.

Acid.—For many purposes it is sufficient to say that an acid is a hydrogen-containing substance which dissociates on solu-

Permission was granted by D. Van Nostrand Company, Inc. publishers of The Dictionary of Physics and Electronics for the inclusion of 18 definitions in this section.

DEFINITIONS AND FORMULAS
The chemical terms have been compiled with the collaboration of
B. Clifford Hendricks

tion in water to produce one or more hydrogen ions. More generally, however, acids are defined according to other concepts. The Brönsted concept states that an acid is any compound which can furnish a proton. Thus NH_4^+ is an acid since it can give up a proton:

$$NH_4^+ \rightleftarrows NH_3 + H^+$$

and NH_3 is a base since it accepts a proton.

A still more general concept is that of G. N. Lewis which defines an acid as anything which can attach itself to something with an unshared pair of electrons. Thus in the reaction

$$H^+ + :N(H)_2{-}H \rightleftarrows NH_4^+$$

the NH_3 is a base because it possesses an unshared pair of electrons. This latter concept explains many phenomena, such as the effect of certain substances other than hydrogen ions in the changing of the color of indicators. It also explains acids and bases in non-aqueous sytems as liquid NH_3 and SO_2.

Actinide Series.—Elements of atomic numbers 89 to 103 analogous to the lanthanide series of the so-called rare earths.

Action is measured by the product of work by time. Cgs units of action are the erg-second and the joule-second. Dimensions,—$[m\ l^2\ t^{-1}]$. Planck's quantum or constant of action is $(6.62517 - 0.00023) \times 10^{-27}$ erg-sec.

Active mass of a substance is the number of gram molecular weights per liter in solution, or in gaseous form.

Activity coefficient.—A factor which, when multiplied by the molecular concentration yields the active mass. The activity coefficient is evaluated by thermodynamic calculations, usually from data on the emf of certain cells, or the lowering of the freezing point of certain solutions. It is a correction factor which makes the thermodynamic calculations correct.

Adiabatic.—A body is said to undergo an adiabatic change when its condition is altered without gain or loss of heat. The line on the pressure volume diagram representing the above change is called an adiabatic line.

Adsorption.—The condensation of gases, liquids, or dissolved substances on the surfaces of solids is called adsorption.

Air columns, frequency of vibration in—See *Organ pipes*

Allobar.—A form of an element differing in isotopic composition from the naturally occurring form.

Allotropy.—The property shown by certain elements or being capable of existence in more than one form, due to differences in the arrangement of atoms or molecules.—See *Monotropic* and *Enantiotropic*

Alpha (α)-particle, or alpha-ray.—One of the particles emitted in radioactive decay. It is identical with the nucleus of the helium atom and consists, therefore, of two protons plus two neutrons bound together. A moving alpha particle is strongly ionizing and so loses energy rapidly in traversing through matter. Natural alpha particles will traverse only a few centimeters of air before coming to rest.

Alternating current, (A-C).—Current in which the charge-flow periodically reverses, as opposed to direct current, and whose average value is zero. Alternating current usually implies a sinusoidal variation of current and voltage. This behavior is represented mathematically in various ways:

$$I = I_0 \cos (2\pi ft + \phi)$$
$$I = I_0 \angle\phi$$
$$I = I_1 e^{j\omega t}$$

where f is the frequency; $\omega \equiv 2\pi f$, the pulsatance, or radian frequency; ϕ the phase angle; I_0 the amplitude; and I_1 the complex amplitude. In the complex rotation, it is understood that the actual current is the real part of I. For circuits involving also a capacitance C in farads and L in henrys, the impedance becomes,

$$\sqrt{R^2 + \left(2\pi fL - \frac{1}{2\pi fC}\right)^2}$$

Altitudes with the barometer.—If b_1 and b_2 denote the corrected barometer readings at two stations, t the mean of the temperatures, t_1 and t_2 of the air at the two stations, e_1 and e_2 the tension of water vapor at the two stations, h the mean height above sea level, ϕ the latitude; then the difference in elevation in centimeters is $H = 1{,}843{,}000 (\log b_1 - \log b_2) (1 + 0.00367t) (1 + 0.0026 \cos 2\phi + 0.00002h + \frac{3}{8}k)$, where

$$k = \tfrac{1}{2}\left(\frac{e_1}{b_1} + \frac{e_2}{b_2}\right)$$

An approximate formula, sufficient for differences not over 1000 meters is

$$H = 1{,}600{,}000 \frac{b_1 - b_2}{b_1 + b_2} (1 + 0.004t).$$

Amorphous.—Without definite form, not crystallized.

Ampere's rule.—A positive charge moving horizontally is deflected by a force to the right if it is moving in a region where the magnetic field is vertically upward. This may be generalized to currents in wires by recalling that a current in a certain direction is equivalent to the motion of positive charges in that direction. The force felt by a negative charge is opposite to that felt by a positive charge.

Amplitude.—The maximum value of the displacement in an oscillatory motion.

DEFINITIONS AND FORMULAS (Continued)

AMU.—The atomic mass unit, a unit of mass equal to one-sixteenth the mass of the atom of oxygen of mass number 16.

1 amu = $(1.65979 \pm 0.00004) \times 10^{-24}$ grams (Physical)
9.31141×10^8 e.v. "
$(1.66024 \pm 0.00004) \times 10^{-24}$ (Chemical)
9.31395×10^8 e.v. "

In terms of energy, 1 amu = 931 mev = 1.49×10^{-3} ergs.

Angle.—The ratio between the arc and the radius of the arc. Units of angle,—the radian, the angle subtended by an arc equal to the radius; the degree, $\frac{1}{360}$ part of the total angle about a point. Dimensions,—a numeric.

Angstrom.—A unit of length, used especially in expressing the length of light waves, equal to one ten-thousandth of a micron, or one hundred-millionth of a centimeter (1×10^{-8} cm).

Angular acceleration.—The time rate of change of angular velocity either in angular speed or in direction of the axis of rotation (precession). Cgs unit,—one radian per second per second. Dimensions,—$[t^{-2}]$.

If the initial angular velocity is ω_o, and the velocity after time t is ω_t, the angular acceleration,

$$\alpha = \frac{\omega_t - \omega_o}{t}$$

The angular velocity after time t,

$$\omega_t = \omega_o + \alpha t$$

The angle swept out in time t,

$$\theta = \omega_o t + \tfrac{1}{2}\alpha t^2$$

The angular velocity after movement through the arc θ,

$$\omega = \sqrt{\omega_o^2 + 2\alpha\theta}$$

In the above equations, for angular displacement in radians, angular velocity will be in radians per second and angular acceleration in radians per second per second.

Angular aperture of an objective is the largest angular extent of wave surface which it can transmit.

Angular harmonic motion or harmonic motion of rotation.—Periodic, oscillatory angular motion in which the restoring torque is proportional to the angular displacement. Torsional vibration.

Angular momentum or moment of momentum.—Quantity of angular motion measured by the product of the angular velocity and the moment of inertia. Cgs unit,—unnamed, its nature is expressed by g-cm²/sec. Dimensions, —$[m\, l^2\, t^{-1}]$.

The angular momentum of a mass whose moment of inertia is I, rotating with angular velocity ω, is $I\omega$.

Angular velocity.—Time rate of angular motion about an axis. Cgs unit,—one radian per second. Dimensions,—$[t^{-1}]$.

If the angle described in time t is θ, the angular velocity,

$$\omega = \frac{\theta}{t}$$

θ in radians and t in seconds gives ω in radians per second.

Anhydride (of acid or base).—An oxide which when combined with water gives an acid or base.

Anion.—A negatively charged ion.

Anode.—The electrode at which oxidation occurs in a cell. It is also the electrode toward which anions travel due to the electrical potential. In spontaneous cells the anode is considered negative. In non-spontaneous or electrolytic cells the anode is considered positive.

Apochromat.—A term applied to photographic and microscope objectives indicating the highest degree of color correction.

Archimedes principle.—A body wholly or partly immersed in a fluid is buoyed up by a force equal to the weight of the fluid displaced. A body of volume V cm³ immersed in a fluid of density ρ grams per cm³ is buoyed up by a force in dynes,

$$F = \rho g V$$

where g is the accleration due to gravity.

A floating body displaces its own weight of liquid.

Area, unit of.—The square centimeter. The area of a square whose sides are one centimeter in length. Other units of area are similarly derived. Dimensions,—$[l^2]$.

Arrhenius theory of electrolytic dissociation states that the molecule of an electrolyte can give rise to two or more electrically charged atoms or ions.

Astigmatism is an error of spherical lenses peculiar to the formation of images by oblique pencils. The image of a point when astigmatism is present will consist of two focal lines at right angles to each other and separated by a measurable distance along the axis of the pencil. The error is not eliminated by reduction of aperture as is spherical aberration.

Atom.—The smallest particle of an element which can enter into a chemical *combination.* All chemical compounds are formed of atoms, the difference between compounds being attributable to the nature, number, and arrangement of their constituent atoms.—See *isotopes, nuclear atom*

Atomic bomb.—An explosive that derives its energy from the fission or fusion of atomic nuclei.

Atomic energy.—1. The constitutive internal energy of the atom which was absorbed when it was formed. 2. Energy derived from the mass converted into energy in nuclear transformations.—See *Einstein's formula*

Atomic mass (**atomic weight**).—The mass of a neutral atom of a nuclide. It is usually expressed in terms of the physical scale of atomic masses, that is, in atomic mass units (amu). See AMU

Atomic number.—The number (Z) of protons within the atomic nucleus. The electrical charge of these protons determines the number and arrangement of the outer electrons of the

DEFINITIONS AND FORMULAS (Continued)

atom, and thereby the chemical and physical properties of the element.

Atomic structure.—According to the currently accepted view, the atom consists of a central part, called nucleus, and a number of *electrons* (called orbital or planetary electrons) circling about the latter, like planets about the sun. The nucleus is of a high specific weight; it contains most of the mass of the entire atom (its mass is considered equal to the atomic mass) and is composed of positively charged particles, called *protons* (the number of which always equals the atomic number, (Z), and particles of 0 charge, called neutrons (the number of which equals the difference between the atomic weight and the atomic number, $A - Z$). The diameter of the nucleus is between 10^{-13} and 10^{-12} cm, and the relatively vast distance in which the orbital electrons circle about it is illustrated by the fact that this nuclear diameter is only 10^{-4} to 10^{-5} of the entire atomic diameter. While the nucleus carries an integral number of positive charges (an integral number of protons) each of $1{\cdot}6 \times 10^{-19}$ coulomb, each electron carries one negative charge of $1{\cdot}6 \times 10^{-19}$ coulomb, and the number of orbital electrons is equal to the number of protons in the nucleus (i.e. to the atomic number, Z), so that the atom as a whole has a net charge of 0. The electrons are arranged in successive shells (q.v.) around the nucleus; the maximum number of electrons in each shell is determined by natural laws, and the extranuclear electronic structure of the atom is characteristic of the element. The electrons in the inner shells are tightly bound to the nucleus; this inner structure can be altered by high-energy particles, γ-rays of radium, or x-rays. The electrons in the outer shells are responsible for the chemical properties of the element.—See *Bohr's atomic theory, Heisenberg's theory, shell and subshell*

Atomic theory.—All elementary forms of matter are composed of very small unit quantities called atoms. The atoms of a given element all have the same size and weight. The atoms of different elements have different sizes and weights. Atoms of the same or different elements unite with each other to form very small unit quantities of compound substances called molecules.

Atomic weight.—Atomic weight is the relative weight of the atom on the basis of oxygen as 16. For a pure isotope, the atomic weight rounded off to the nearest integer gives the total number of nucleons (neutrons and protons) making up the atomic nucleus. If these weights are expressed in grams they are called grám atomic weights.—See *Isotopes and Atomic Mass*

Avogadro's law.—Equal volumes of different gases at the same pressure and temperature contain the same number of molecules.

Avogadro's number.—The number of molecules in one mole or gram-molecular weight of a substance. A number of values of the Avogadro number, which is usually denoted by N, have been found by various methods, generally lying within a range of 1% about the value 6.02252×10^{23} per gram mole.

Avogadro's principle (or theory).—The numbers of molecules present in equal volumes of gases at the same temperature and pressure are equal.

Babo's law.—The addition of a non-volatile solid to a liquid in which it is soluble lowers the vapor pressure of the solvent in proportion to the amount of substance dissolved.

Balmer series of spectral lines. The wave lengths of a series of lines in the spectrum of hydrogen are given in angstroms by the equation

$$\lambda = 3646 \frac{N^2}{N^2 - 4}$$

where N is an integer having values greater than 2.

Bar.—International unit of pressure 10^6 dyne/cm². Unfortunately some writers have used this term for 1 dyne/cm². 1 bar = 0.987 atmosphere.

Barn.—Unit for measuring capture cross sections (q.v.) of elements. One barn = 10^{-24} cm² per nucleus.

Barye.—Cgs pressure unit = one dyne/cm².

Bases.—For many purposes it is sufficient to say that a base is a substance which dissociates on solution in water to produce one or more hydroxyl ions. More generally, however, bases are defined according to other concepts. The Brönsted concept states that a base is any compound which can accept a proton. Thus NH_3 is a base since it can accept a proton to form ammonium ions.

$$NH_3 + H^+ \rightleftarrows NH_4^+$$

A still more general concept is that of G. N. Lewis which defines a base as anything which has an unshared pair of electrons. Thus in the reaction

$$H^+ + :N(H)(H)—H \rightleftarrows NH_4^+$$

the NH_3 is a base because it possesses an unshared pair of electrons. This latter concept explains many phenomena, such as the effect of certain substances other than hydrogen ions in the changing of the color of indicators. It also explains acids and bases in non-aqueous systems as liquid NH_3 and SO_2.

Beat(s).—Two vibrations of slightly different frequencies f_1 and f_2 when added together, produce in a detector sensitive to both these frequencies, a regularly varying response which rises and falls at the "beat" frequency $f_b = |f_1 - f_2|$. It is important to note that a resonator which is sharply tuned to f_b alone will not resound at all in the presence of these two beating frequencies.—See *Combination Frequencies*

Beat frequencies.—The beat of two different frequencies of signals on a non-linear circuit when they combine or beat together. It has a frequency equal to the difference of the two applied frequencies.

Beer's law.—If two solutions of the same colored compound be made in the same solvent, one of which is, say, twice the concentration of the other, the absorption due to a given thickness of the first solution should be equal to that of twice the thickness of the second.

Mathematically this may be expressed $l_1c_1 = l_2c_2$ when the intensity of light passing through the two solutions is a constant and if the intensity and wave length of light incident upon each solution are the same.

Bernoulli's theorem.—At any point in a tube through which a liquid is flowing the sum of the pressure energy, potential energy, and kinetic energy is constant. If p is pressure; h, height above a reference plane; d, density of the liquid, and v, velocity of flow,

$$p + hdg + \frac{1}{2}dv^2 = \text{a constant.}$$

Berthelot principle of maximum work.—Of all possible chemical processes which can proceed without the aid of external energy, that process always takes place which is accompanied by the greatest evolution of heat. This law holds good for low temperatures only and does not account for endothermic reactions.

Beta (β)**-particle, (Beta ray).**—One of the particles which can be emitted by a radioactive atomic nucleus. It has a mass about $\frac{1}{1837}$ that of the proton. The negatively charged beta particle is identical with the ordinary electron, while the positively charged type (positron) differs from the electron in having equal but opposite electrical properties. The emission of an electron entails the change of a neutron into a proton inside the nucleus. The emission of a positron is similarly associated with the change of a proton into a neutron. Beta particles have no independent existance inside the nucleus, but are created at the instant of emission.—See *Neutrino*

Betatron.—An accelerator used to impart high velocities to electrons (beta particles). Propellant is an electromagnetic field. A five to six Mev betatron can produce X-rays equivalent to the gamma readiation of 10 to 20 grams of radium.

Bevatron.—A six or more billion electron volt accelerator of protons and other atomic particles. Makes use of a Cockcroft-Walton transformer cascade accelerator and a linear (q.v.) as well as an electromagnetic field in the build-up.

Black body.—If, for all values of the wave length of the incident radiant energy, all of the energy is absorbed the body is called a black body.

Bohr's atomic theory.—The theory that atoms can exist for a duration solely in certain states, characterized by definite electronic orbits, i.e., by definite energy levels of their extranuclear electrons, and in these stationary states they do not emit radiation; the jump of an electron from an orbit to another of a smaller radius is accompanied by monochromatic radiation.

Boyle's law for gases.—At a constant temperature the volume of a given quantity of any gas varies inversely as the pressure to which the gas is subjected. For a perfect gas, changing from pressure p and volume v to pressure p' and volume v' without change of temperature,

$$pv = p'v'$$

Breeder, Reactor (Breeder pile).—A nuclear chain reactor in which transmutation produces a greater number of fissionable atoms than the number of parent atoms consumed.

Brewster's law.—The tangent of the polarizing angle for a substance is equal to the index of refraction. The polarizing angle is that angle of incidence for which the reflected polarized ray is at right angles to the refracted ray. If n is the index of refraction and θ the polarizing angle, $n = \tan\theta$.

Brightness is measured by the flux emitted per unit emissive area as projected on a plane normal to the line of sight. The unit of brightness is that of a perfectly diffusing surface giving out one lumen per square centimeter of projected surface and is called the lambert. The millilambert (0.001 lambert) is a more convenient unit. **Candle per square centimeter** is the brightness of a surface which has, in the direction considered, a luminous intensity of one candle per cm².

British thermal unit.—The quantity of heat required to raise the temperature of one pound of water one degree Fahrenheit at, or near, its point of maximum density (39.1°F). The Btu is equivalent to 0.252 kilogram-calorie.

Brownian movement.—A continuous agitation of particles in a colloidal solution caused by unbalanced impacts with molecules of the surrounding medium. The motion may be observed with a microscope when a strong beam of light is caused to traverse the solution across the line of sight.

Bulk modulus.—The modulus of volume elasticity,

$$M_B = \frac{p_2 - p_1}{\frac{v_1 - v_2}{v_1}}$$

where p_1, p_2; v_1, v_2 are the initial and final pressure and volume respectively.

Calorie.—The amount of heat necessary to raise 1 gram of water at 15°C, 1°C. There are various calories depending upon the interval chosen. Sometimes the unit is written as the gram-calorie or the kilogram calorie, the meaning of which is evident. The calorie may be defined in terms of its mechanical equivalent. The National Bureau of Standards defines the calorie as

4.18400 joules. At the International Steam Table Conference held in London in 1929 the international calorie was defined at $\frac{1}{860}$ of the international watt hour, which makes it equal to 4.1860 international joules.

With the adoption of the absolute sytem of electrical units, this becomes 1/859.858 watt hours or 4.18674 joules. The Btu was defined at the same time as 251.996 international calories.

Calutron.—An apparatus operating on the principle of the mass spectrograph and used for separating U^{235} from U^{238}.

Candle (or International Candle).—The candle is the unit of luminous intensity. It is a specified fraction of the average horizontal candlepower of a group of 45 carbon-filament lamps preserved at the Bureau of Standards.

Candle (new unit).—$\frac{1}{60}$ of the intensity of one square centimeter of a blackbody radiator at the temperature of solidification of platinum (2,046°K).

Capacitance is measured by the charge which must be communicated to a body to raise its potential one unit. Electrostatic unit capacitance is that which requires one electrostatic unit of charge to raise the potential one electrostatic unit. The farad = 9×10^{11} electrostatic units. A capacitance of one farad requires one coulomb of electricity to raise its potential one volt. Dimensions,—$[\epsilon\, l]$; $[\mu^{-1}\, l^{-1}\, t^2]$.

A conductor charged with a quantity Q to a potential V has a capacitance,

$$C = \frac{Q}{V}$$

Capacitance of a spherical conductor of radius r,

$$C = Kr$$

Capacitance of two concentric spheres of radii r and r'

$$C = K\frac{rr'}{r - r'}$$

Capacitance of a parallel plate condenser, the area of whose plates is A and the distance between them d,

$$C = \frac{KA}{4\pi d}$$

Capacitances will be given in electrostatic units if the dimensions of condensers are substituted in cm. K is the dielectric constant of the medium.

Capillary constant or specific cohesion,

$$a^2 = \frac{2T}{(d_1 - d_2)g} = hr$$

where T is surface tension, d_1 and d_2, the densities of the two fluids, g the acceleration due to gravity, h the height of rise in a capillary tube of radius r.—See *Surface tension*

Carnot cycle.—A sequence of operations forming the working cycle of an ideal heat engine of maximum thermal efficiency. It consists of isothermal expansion, adiabatic expansion, isothermal compression, and adiabatic compression to the initial state.

Catalytic agent.—A substance which by its mere presence alters the velocity of a reaction, and may be recovered unaltered in nature or amount at the end of the reaction.

Cathode.—The electrode at which reduction occurs. It is the negative electrode in a cell through which current is being forced, but it is the positive pole of a battery. In a vacuum tube, the cathode is the electrode from which electrons are liberated.—See *Anode*

Cation.—A positively charged ion.

Cauchy's dispersion formula.

$$n = A + \frac{B}{\lambda^2} + \frac{C}{\lambda^4} + \cdots$$

An empirical expression giving an approximate relation between the refractive index n of a medium and the wavelength λ of the light; A, B, and C being constants for a given medium.

Celsius.—See temperature, Celsius, in this section.

Centipoise.—A standard unit of *viscosity*, equal to 0.01 poise, the c.g.s. unit of viscosity. Water at 20°C has a viscosity of 1.002.

Centripetal force.—The force required to keep a moving mass in a circular path. Centrifugal force is the name given to the reaction against centripetal force.

Chain reaction.—In general, any self-sustaining process, whether molecular or nuclear, the products of which are instrumental in, and directly contribute to the propagation of the process. Specifically, a *fission chain reaction*, where the energy liberated or particles produced (fission products) by the fission of an atom cause the fission of other atomic nuclei, which in turn propagate the fission recation in the same manner.

Charles' law or Gay-Lussac's law.—The volumes assumed by a given mass of a gas at different temperatures, the pressure remaining constant, are, within moderate ranges of temperature, directly proportional to the corresponding absolute temperatures.

Chemiluminescence.—Emission of light during a chemical reaction.

Christiansen effect.—When finely powdered substances, such as glass or quartz, are immersed in a liquid of the same index of refraction complete transparency can only be obtained for monochromatic light. If white light is employed the transmitted color corresponds to the particular wave-length for which the two substances, solid and liquid have exactly the same index of refraction. Due to differences in dispersion the indices of refraction will match for only a narrow band of the spectrum.

DEFINITIONS AND FORMULAS (Continued)

Chromatic aberration.—Due to the difference in the index of refraction for different wave lengths, light of various wave lengths from the same source cannot be focused at a point by a simple lens. This is called chromatic aberration.

Cloud chamber.—An apparatus containing moist air or other gas which on sudden expansion condenses moisture to droplets on dust particles or other nuclei. Thus charged particles or ions in the space become nuclei and their numbers and behavior, when properly illuminated, may be studied.

Colligative property.—A property numerically the same for a group of substances, independent of their chemical nature.

Colloid.—A phase dispersed to such a degree that the surface forces become an important factor in determining its properties.

In general particles of colloidal dimensions are approximately 10 angstroms to 1 micron in size. Colloidal particles are often best distinguished from ordinary molecules due to the fact that colloidal particles cannot diffuse through membranes which do allow ordinary molecules and ions to pass freely.

Coma.—An aberration of spherical lenses, occurring in the case of oblique incidence, when the bundle of rays forming the image is unsymmetrical. The image of a point is comet shaped, hence the name.

Combination frequencies.—Two vibrations of arbitrary frequencies f_1 and f_2 when applied simultaneously to a nonlinear (distorting) device will excite it to a motion containing not only the original frequencies, but also members of a set of "combination" frequencies given by $f_c = mf_1 + nf_2$ where m and n are integers. A resonator sharply tuned to any one of these frequencies which may be produced in the nonlinear device will resound to it with an amplitude depending on the type of nonlinearity. The superheterodyne radio receiver depends on this phenomenon.

Combining volumes.—Under comparable conditions of pressure and temperature the volume ratios of gases involved in chemical reactions are simple whole numbers.

Combining weight of an element or radical is its atomic weight divided by its valence.

Combining weights, law of.—If the weights of elements which combine with each other be called their "combining weights," then elements always combine either in the ratio of their combining weights or of simple multiples of these weights.

Component substances, law of.—Every material consists of one substance, or is a mixture of two or more substances, each of which exhibits a specific set of properties, independent of the other substances.

Compounds are substances containing more than one constituent element and having properties, on the whole, different from those which their constituents had as elementary substances. The composition of a given pure compound is perfectly definite, and is always the same no matter how that compound may have been formed.

Compressibility.—Reciprocal of the bulk modulus.

Compton effect, (Compton recoil effect).—Elastic scattering of photons by electrons results in decrease in frequency and increase of wave length of x-rays and gamma-rays when scattered by free electrons.

Concentration.—The amount of a substance in weight, moles, or equivalents contained in unit volume.

Condensers in parallel and series.—If c_1, c_2, c_3, etc. represent the capacitances of a series of condensers and C their combined capacitance,—

when in parallel, $C = c_1 + c_2 + c_3 \cdots$

when in series, $\frac{1}{C} = \frac{1}{c_1} + \frac{1}{c_2} + \frac{1}{c_3} \cdots$

Conductance, the reciprocal of resistance, is measured by the ratio of the current flowing through a conductor to the difference of potential between its ends. The practical unit of conductance, the mho, the conductance of a body through which one ampere of current flows when the potential difference is one volt. The conductance of a body in mho is the reciprocal of the value of its resistance in ohms. Dimensions,—$[\epsilon\, l\, t^{-1}]$. $[\mu^{-1}\, l^{-1}\, t]$.

Conductivity, electrical, is measured by the quantity of electricity transferred across unit area, per unit potential gradient per unit time. Reciprocal of resistivity. **Volume conductivity** or specific conductance, $k = 1/\rho$ where ρ is the volume resistivity. **Mass conductivity** $= k/d$ where d is density. **Equivalent conductivity** $\Lambda = k/c$ where c is the number of equivalents per unit volume of solution. **Molecular conductivity** $\mu = k/m$ where m is the number of moles per unit volume of solution. Dimensions: volume conductivity,—$[\epsilon\, t^{-1}]$; $[\mu^{-1}\, l^{-2}\, t]$,—mass conductivity,—$[\epsilon\, m^{-1}\, l^3\, t^{-1}]$; $[\mu^{-1}\, m^{-1}\, lt]$.

Conductivity, thermal.—Time rate of transfer of heat by conduction, through unit thickness, across unit area for unit difference of temperature. It is measured as calories per second per square centimeter for a thickness of one centimeter and a difference of temperature of 1°C. Dimensions,—$[m\, l\, t^{-3}\, \theta^{-1}]$.

If the two opposite faces of a rectangular solid are maintained at temperatures t_1 and t_2 the heat conducted across the solid of section a and thickness d in a time T will be,

$$Q = \frac{K(t_2 - t_1)aT}{d}$$

K is a constant depending on the nature of the substance, designated as the specific heat conductivity. K is usually given for Q in calories, t_1 and t_2 in °C, a in cm², T in sec, and d in cm.—See *Heat conductivity*

DEFINITIONS AND FORMULAS (Continued)

Conductors.—A class of bodies which are incapable of supporting electric strain. A charge given to a conductor spread to all parts of the body.

Conjugate foci.—Under proper conditions light divergent from a point on or near the axis of a lens or spherical mirror is focused at another point. The point of convergence and the position of the source are interchangeable and are called conjugate foci.

Conservation of energy, (Chem).—In a chemical change there is no loss or gain but merely a transformation of energy from one form to another.

Conservation of energy, law of.—Energy can neither be created nor destroyed and therefore the total amount of energy in the universe remains constant.

Conservation of mass.—In all ordinary chemical changes, the total of the reactants is always equal to the total mass of the products.

Conservation of momentum, law of.—For any collision, the vector sum of the momenta of the colliding bodies after collison equals the vector sum of their momenta before collision. If two bodies of masses m_1 and m_2 have, before impact velocities v_1 and v_2 and after impact velocities u_1 and u_2

$$m_1u_1 + m_2u_2 = m_1v_1 + m_2v_2$$

Constitutive property.—A property which depends on the constitution or structure of the molecule.

Cooling.—Processing highly radioactive materials to attain lesser radioactivity for subsequent use or handling.

Cosmic rays.—Highly penetrating radiations which strike the earth, assumed to originate in interstellar space. They are classed as: primary, coming from the assumed source, and secondary, those induced in upper atmospheric nuclei by collisions with primary cosmic rays.

Cosmotron.—A particle accelerator capable of giving them energies to billions of electron volts.

Coulomb.—A unit quantity of electricity. It is the quantity of electricity which must pass through a circuit to deposit 0.0011180 grams of silver from a solution of silver nitrate. An ampere is one coulomb per second. A coulomb is also the quantity of electricity on the positive plate of a condenser of one-farad capacity when the electromotive force is one volt.

Couple.—Two equal and oppositely directed parallel but not colinear forces acting upon a body form a couple. The moment of the couple or torque is given by the product of one of the forces by the perpendicular distance between them. Dimension, —$[m\ l^2\ t^{-2}]$.

Couple acting on a magnet of magnetic moment ml in a field of strength H. If the magnet is perpendicular to the direction of the field

$$C = Hml = HM$$

If the angle between the magnet and the field is θ

$$C = Hml \sin \theta$$

The couple will be in dyne-cm for cgs electromagnetic units of H, m and l.

Critical mass.—The minimum mass the fissile material must have in order to maintain a spontaneous fission chain reaction. For pure U^{235} it is computed to be about 20 pounds.

Critical temperature is that temperature above which a gas cannot be liquefied by pressure alone. The pressure under which a substance may exist as a gas in equilibrium with the liquid at the critical temperature is the **critical pressure.**

Cross section, (Nuclear cross section).—A measure of the probability of a particular process. The nuclear cross section is expressed by a/bc, where a is the number of processes occurring, b the number of incident particles, and c the number of target nuclei per cm^2. There are nuclear cross sections for fission, for slow neutron capture, for Compton collision, and for ionization by electron impact.

Cryohydrate.—The solid which separates when a saturated solution freezes. It contains the solvent and the solute in the same proportions as they were in the saturated solution.

Crystal.—The "ideal crystal" is a homogeneous portion of crystalline matter, (q.v.) whether bounded by faces or not.

Crystalline matter is matter that possesses a triperiodic structure on the atomic scale. It is characterized by discontinuous vectorial properties that give rise to "crystal planes" [(1) crystal growth (faces); (2) cohesion (cleavage planes); (3) twinning (twin planes); (4) gliding (gliding planes); (5) x-ray, electron, or neutron diffraction ("reflecting" planes); all of which are parallel to lattice planes.]

Curie.—Unit for measuring radioactivity. One Curie = that quantity of any radioactive isotope undergoing 3.7×10^{10} disintegrations per second.

Curie's law.—The intensity of magnetization,

$$I = \frac{AH}{T}$$

where H, is the magnetic field strength, T the absolute temperature and A Curie's constant. Used for paramagnetic substances.

Curie point.—All ferro-magnetic substances have a definite temperature of transition at which the phenomena of ferro-magnetism disappear and the substances become merely paramagnetic. This temperature is called the "Curie Point" and is usually lower than the melting point.

Curie-Weiss law.—The Curie law was modified by Weiss to state that the susceptibility of a paramagnetic substance above the Curie point varies inversely as the excess of the temperature above that point.

This law is not valid at or below the Curie point.

DEFINITIONS AND FORMULAS (Continued)

Current (**electric**).—The rate of transfer of electricity. The transfer at the rate of one electrostatic unit of electricity in one second is the electrostatic unit of current. The electromagnetic unit of current is a current of such strength that one centimeter of the wire in which it flows is pushed sideways with a force of one dyne when the wire is at right angles to a magnetic field of unit intensity. The practical unit of current is the **ampere,** a transfer of one coulomb per second, which is one tenth the electromagnetic unit. The **international ampere** is the unvarying electric current which, when passed through a solution of silver nitrate in accordance with certain specifications, deposits silver at the rate of 0.00111800 gram per second. The international ampere is equivalent to 0.999835 absolute ampere. The **ampere-turn** is the magnetic potential produced between the two faces of a coil of one turn carrying one ampere. Dimensions,—$[\epsilon^{\frac{1}{2}}\, m^{\frac{1}{2}}\, l^{\frac{3}{2}}\, t^{-2}]$; $[\mu^{-\frac{1}{2}}\, m^{\frac{1}{2}}\, l^{\frac{1}{2}}\, t^{-1}]$.

Current in a simple circuit.—The current in a circuit including an external resistance R and a cell of electromotive force E and internal resistance r,

$$I = \frac{E}{R + r}$$

If E is in volts and r and R in ohms the current will be in amperes.

For two cells in parallel,

$$I = \frac{E}{R + \frac{r}{2}}$$

For two cells in series,

$$I = \frac{2E}{R + 2r}$$

Cyclotron.—The magnetic resonance accelerator for imparting very great velocities to heavier nuclear particles without the use of excessive voltages.

Dalton's law of partial pressures.—The pressure exerted by a mixture of gases is equal to the sum of the separate pressures which each gas would exert if it alone occupied the whole volume. This fact is expressed in the following formula:

$$PV = V(p_1 + p_2 + p_3, \text{etc.})$$

Decay.—Diminution of a radioactive substance due to nuclear emission of alpha or beta particles, gamma rays or positrons.

Declination.—The angle between the vertical plane containing the direction of the earth's field at any point and a plane containing the geographic north and south meridian.

Decomposition is the chemical separation of a substance into two or more substances, which may differ from each other and from the original substances.

Definite proportions, law of.—In every sample of each compound substance the proportions by weight of the constituent elements are always the same.

Degree of freedom.—The number of the variables determining the state of a system (usually pressure, temperature, and concentrations of the components) to which arbitrary values can be assigned.

Density.—Concentration of matter, measured by the mass per unit volume. Dimensions,—$[m\, l^{-3}]$.

Deuteron.—Nucleus of the deuterium atom or the ion of deuterium. Its structure—one neutron and one proton.

Dew point.—The temperature at which condensation of water vapor in the air takes place.

Diamagnetic materials.—Are those within which an externally applied magnetic field is slightly reduced because of an alteration of the atomic electron orbits produced by the field. Diamagnetism is an atomic-scale consequence of the Lenz law of induction. The permeability of diagmagnetic materials is slightly less than that of empty space.

Dielectric constant of a medium is defined by ϵ in the equation

$$F = \frac{QQ'}{\epsilon r^2}$$

where F is the force of attraction between two charges Q and Q' separated by a distance r in a uniform medium.

Dielectrics or insulators or non-conductors.—A class of bodies supporting an electric strain. A charge on one part of a non-conductor is not communicated to any other part.

Diffraction.—That phenomena produced by the spreading of waves around and past obstacles which are comparable in size to their wavelength.

Diffraction grating.—If s is the distance between the rulings, d the angle of diffraction, then the wave length where the angle of incidence is 90° is (for the nth order spectrum),

$$\lambda = \frac{s \sin d}{n}$$

If i is the angle of incidence, d the angle of diffraction, s the distance between the rulings, n the order of the spectrum, the wave length is, $\lambda = \frac{s}{n}(\sin i + \sin d)$.

Diffusion.—If the concentration (mass of solid per unit volume of solution) at one surface of a layer of liquid is d_1 and at the other surface d_2, the thickness of the layer h and the area under consideration A, then the mass of the substance which diffuses through the cross-section A in time t is,

$$m = \Delta A \frac{(d_2 - d_1)t}{h}$$

where Δ is the coefficient of diffusion.

DEFINITIONS AND FORMULAS (Continued)

Diffusivity or coefficient of diffusion is also given by Δ in the equation

$$\frac{dQ}{dt} = -\Delta\left(\frac{dc}{dx}\right) dy\, dz$$

where dQ is the amount passing through an area $dy\, dz$ in the direction of x in a time dt where dc/dx is the rate of increase of volume concentration in the direction of x. Dimensions,—$[l^2\, t^{-1}]$.

Diffusivity of heat is given by Δ in the equation

$$\frac{dH}{dt} = -\Delta sd \frac{dT}{dx} dy\, dz$$

where dH is the quantity of heat passing through the area $dy\, dz$ in the direction of x in a time dt. The rate of variation of temperature along x is given by dT/dx, s is specific heat and d, density. Dimensions,—$[l^2\, t^{-1}]$.

Dimensional formulae.—If mass, length, and time are considered fundamental quantities, the relation of other physical quantities and their units to these three may be expressed by a formula involving the symbols l, m and t respectively, with appropriate exponents. For example; the dimensional formula for volume would be expressed,—$[l^3]$; velocity,—$[lt^{-1}]$; force—$[m\, lt^{-2}]$. Other fundamental quantities used in dimensional formulae may be indicated as follows: θ, temperature, ϵ the dielectric constant of a vacuum; μ, the magnetic permeability of a vacuum.

Diminution of pressure at the side of a moving stream. If a fluid of density d moves with a velocity v, the diminution of pressure due to the motion is (neglecting viscosity),

$$p = \tfrac{1}{2}\, dv^2$$

Dip.—The angle measured in a vertical plane between the direction of the earth's magnetic field and the horizontal.

Dipole.—(1) A combination of two electrically or magnetically charged particles of opposite sign which are separated by a very small distance. (2) Any system of charges, such as a circulating current, which has the properties that: (a) no forces act on it in a uniform field; (b) a torque proportional to sin θ, where θ is the angle between the dipole axis and a uniform field, does act on it; (c) it produces a potential which is proportional to the inverse square of the distance from it.

Dipole moment.—A mathematical entity; the product of one of the charges of a dipole unit by the distance separating the two dipolar charges. In terms of the definition of a dipole (2), the dipole moment **p** is related to the torque **T**, and the field strength **E** (or **B**) through the equation:

$$\mathbf{T} = \mathbf{p} \times \mathbf{E}.$$

Dipole moment, molecular.—It is found from measurements of dielectric constant (i.e. by its temperature dependence, as in the **Debye equation for total polarization**) that certain molecules have permanent dipole moments. These moments are associated with transfer of charge within the molecule, and provide valuable information as to the molecular structure.

Dispersion.—The difference between the index of refraction of any substance for any two wave lengths is a measure of the dispersion for these wave lengths, called the coefficient of dispersion.

Dispersion forces.—The force of attraction between molecules possessing no permanent **dipole.** The interaction energy is given by

$$U_D = -\frac{3}{4} h \frac{V_0 \alpha^2}{r^6}$$

where h is Planck's constant, V_0 a characteristic frequency of the molecule, r the distance between the molecules, and α the polarizability.

Dispersive power.—If n_1 and n_2 are the indices of refraction for wave lengths λ_1 and λ_2 and n the mean index or that for sodium light, the dispersive power for the specified wave length is,

$$\omega = \frac{n_2 - n_1}{n - 1}$$

Displacement is a reaction in which an elementary substance displaces and sets free a constituent element from a compound.

Displacement or elongation at any instant. The distance of a vibrating or oscillating particle from its position of equilibrium.

Distribution law.—A substance distributes itself between two immiscible solvents so that the ratio of its concentrations in the two solvents is approximately a constant (and equal to the ratio of the solubilities of the substance in each solvent). The above statement requires modification if more than one molecular species is formed.

Doppler effect, (Light).—The apparent change in the wave-length of light produced by the motion in the line of sight of either the observer or the source of light.

Doppler effects.—Effects on the apparent frequency of a wave train produced (1) by motion of the source toward or away from the stationary observer, and (2) by motion of the observer toward or from the stationary source; the motion in each case being with reference to the (supposedly stationary) medium.

For sound waves, the observed frequency f_0, in cycles/sec., is given by

$$f_0 = \frac{v + w - v_0}{v + w - v_s} f_s$$

where v is the velocity of sound in the medium, v_0 is the velocity

DEFINITIONS AND FORMULAS (Continued)

of the observer, v_s is the velocity of the source, w is the velocity of the wind in the direction of sound propagation, and f_s is the frequency of source.

For optical waves

$$f_0 = f_s \sqrt{\frac{c + v_r}{c - v_r}}$$

where v_r is the velocity of the source relative to the observer and c is the speed of light.

Dosimeter.—A gadget worn by those who work about radioactive sources to record the amount of exposure the wearer has had to radioactivity.

Double decomposition consists of a simple exchange of the parts of two substances to form two new substances.

Dulong and Petit, law of.—The specific heats of the several elements are inversely proportional to their atomic weights. The atomic heats of solid elements are constant and approximately equal to 6.3. Certain elements of low atomic weight and high melting point have, however, much lower atomic heats at ordinary temperatures.

Dyne.—The force necessary to give acceleration of one centimeter per second per second to one gram of mass.

Eddy current.—A current induced in a mass of conducting material by a varying magnetic field. Also called *Foucault current.*

Einstein theory for mass-energy equivalence.—The equivalence of a quantity of mass m and a quantity of energy E by the formula $E = mc^2$. The conversion factor c^2 is the square of the velocity of light.

Elasticity.—The property by virtue of which a body resists and recovers from deformation produced by force.

Elastic limit.—The smallest value of the stress producing permanent alteration.

Elastic moduli.

Young's modulus by stretching.—If an elongation s is produced by the weight of the mass m, in a wire of length l, and radius r, the modulus,

$$M = \frac{mgl}{\pi r^2 s}$$

Young's modulus by bending, bar supported at both ends. If a flexure s is produced by the weight of mass m, added midway between the supports separated by a distance l, for a rectangular bar with vertical dimensions of cross-section a and horizontal dimension b, the modulus is,

$$M = \frac{mgl^3}{4sa^3b}$$

For a cylindrical bar of radius r,

$$M = \frac{mgl^3}{12\pi r^4 s}$$

For a bar supported at one end. In the case of a rectangular bar as described above,

$$M = \frac{4mgl^3}{sa^3b}$$

For a round bar supported at one end,

$$M = \frac{4mgl^3}{3\pi r^4 s}$$

Modulus of rigidity.—If a couple C $(= mgx)$ produces a twist of θ radians in a bar of length l and radius r, the modulus is

$$M = \frac{2Cl}{\pi r^4 \theta}$$

The substitution in the above formulae for the elastic coefficients of m in grams, g in cm per sec², l, a, b, and r in cm, s in cm, and C in dyne-cm will give moduli in dynes per cm². The dimensions of elastic moduli are the same as of stress,—$[m\, l^{-1}\, t^{-2}]$.

Coefficient of restitution.—Two bodies moving in the same straight line, with velocities v_1 and v_2 respectively, collide and after impact move with velocities v_3 and v_4. The coefficient of restitution is

$$C = \frac{v_4 - v_3}{v_2 - v_1}$$

Electric field intensity is measured by the force exerted on unit charge. Unit field intensity is the field which exerts the force of one dyne on unit positive charge. Dimensions,—$[\epsilon^{-\frac{1}{2}}\, m^{\frac{1}{2}}\, l^{-\frac{1}{2}}\, t^{-1}]$; $[\mu^{\frac{1}{2}}\, m^{\frac{1}{2}}\, l^{\frac{1}{2}}\, t^{-2}]$.

The field intensity or force exerted on unit charge at a point distant r from a charge q in a vacuum

$$H = \frac{q}{r^2}$$

If the dielectric in the above cases is not a vacuum the dielectric constant ϵ must be introduced. The formula becomes

$$H = \frac{q}{\epsilon r^2}$$

The value of ϵ is frequently considered unity for air. If the dielectric constant of a vacuum is considered unity the value for air at 0°C and 760 mm pressure is 1.000576.

Electrochemical equivalent of an ion is the mass liberated by the passage of unit quantity of electricity.

DEFINITIONS AND FORMULAS (Continued)

Electrolysis.—If a current i flows for a time t and deposits a metal whose electrochemical equivalent is e, the mass deposited is

$$m = eit$$

The value of e is usually given for mass in grams, i in amperes and t in seconds.

Electrolytic dissociation or ionization theory.—When an acid, base or salt is dissolved in water or any other dissociating solvent, a part or all of the molecules of the dissolved substance are broken up into parts called ions, some of which are charged with positive electricity and are called cations, and an equivalent number of which are charged with negative electricity and are called anions.

Electrolytic solution tension theory (or the Helmholtz double layer theory).—When a metal, or any other substance capable of existing in solution as ions, is placed in water or any other dissociating solvent, a part of the metal or other substances passes into solution in the form of ions, thus leaving the remainder of the metal or substances charged with an equivalent amount of electricity of opposite sign from that carried by the ions. This establishes a difference in potential between the metal and the solvent in which it is immersed.

Electromotive force is defined as that which causes a flow of current. The electromotive force of a cell is measured by the maximum difference of potential between its plates. The electromagnetic unit of potential difference is that against which one erg of work is done in the transfer of electromagnetic unit quantity. The **volt** is that potential difference against which one joule of work is done in the transfer of one coulomb. One volt is equivalent to 10^8 electromagnetic units of potential. The **international** volt is the electrical potential which when steadily applied to a conductor whose resistance is one international ohm will cause a current of one international ampere to flow. The international volt = 1.00033 absolute volts. The electromotive force of a Weston standard cell is 1.0183 int. volts at 20°C. Dimensions,—$[\epsilon^{-\frac{1}{2}} m^{\frac{1}{2}} l^{\frac{1}{2}} t^{-1}]$, $[\mu^{\frac{1}{2}} m^{\frac{1}{2}} l^{\frac{3}{2}} t^{-2}]$.

Electromotive series is a list of the metals arranged in the decreasing order of their tendencies to pass into ionic form by losing electrons.

Electron.—The electron is a small particle having a unit negative electrical charge, a small mass, and a small diameter. Its charge is $(4.80294 \pm .00008) \times 10^{-10}$ absolute electrostatic units, its mass $\frac{1}{1837}$ that of the hydrogen nucleus, and its diameter about 10^{-12} cm. Every atom consists of one nucleus and one or more electrons. Cathode rays and Beta rays are electrons.

Electron-volt (ev).—Energy acquired by any charged particle carrying unit electronic charge when it falls through a potential difference of one volt. 1 electron-volt = (1.60207 ± .00007 × 10^{-12} erg. Multiples of this unit are also in common use: the kilo-, million-, and billion electron-volt. 1 kev = 10^3 ev; 1 mev = 10^6 ev; and 1 bev = 10^9 ev.

An ev is associated through the Planck constant with a photon of wave length 1.2395 microns.

Elements are substances which cannot be decomposed by the ordinary types of chemical change, or made by chemical union.

Emissive power or emissivity is measured by the energy radiated from unit area of a surface in unit time for unit difference of temperature between the surface in question and surrounding bodies. For the cgs system the emissive power is given in ergs per second per square centimeter with the radiating surface at 1° absolute and the surroundings at absolute zero.—See *Radiation formula*

Enantiotropic.—Crystal forms capable of existing in reversible equilibrium with each other.

Energy.—The capability of doing work. **Potential energy** is energy due to position of one body with respect to another or to the relative parts of the same body. **Kinetic energy** is energy due to motion. Cgs units,—the erg, the energy expended when a force of one dyne acts through a distance of one centimeter; the joule is 1×10^7 ergs. Dimensions,—$[m\, l^2\, t^{-2}]$.

The potential energy of a mass m, raised through a distance h, where g is the acceleration due to gravity is

$$E = mgh.$$

The kinetic energy of mass m, moving with a velocity v, is

$$E = \tfrac{1}{2}mv^2$$

Energy will be given in ergs if m is in grams, g in cm per sec^2, h in cm and v in cm per sec.

Energy of a charge in ergs where Q is the charge and V the potential in electrostatic units.

$$E = \tfrac{1}{2}QV.$$

Energy of the electric field.—If H is the electric field intensity in electrostatic units and K the specific inductive capacity, the energy of the field in ergs per cm^3 is

$$E = \frac{KH^2}{8\pi}$$

Energy of rotation.—If a mass whose moment of inertia about an axis is I, rotates with angular velocity ω about this axis, the kinetic energy of rotation will be,

$$E = \tfrac{1}{2}I\omega^2$$

Energy will be given in ergs if I is in g-cm^2 and ω in radians per sec.

Enthalpy, or heat content, is a thermodynamic quantity. It is equal to the sum of the internal energy of a system plus

DEFINITIONS AND FORMULAS (Continued)

the product of the pressure-volume work done on the system. Thus

$$H = E + pv$$

where H = enthalpy or heat content
E = internal energy of the system
p = pressure
v = volume.

Entropy.—Entropy is the capacity factor for isothermally unavailable energy. The increase in the entropy of a body during an infinitesimal stage of a reversible process is equal to the infinitesimal amount of heat absorbed divided by the absolute temperature of the body. Thus for a reversible process

$$dS = \frac{Q}{T}$$

Every spontaneous process in nature is characterized by an increase in the total entropy of the bodies concerned in the process.

Equilibrium, chemical.—A state of affairs in which a chemical reaction and its reverse reaction are taking place at equal velocities, so that the concentrations of reacting substances remain constant.

Equilibrium constant.—The product of the concentrations (or activities) of the substances produced at equilibrium in a chemical reaction divided by the product of concentrations of the reacting substances, each concentration raised to that power which is the coefficient of the substance in the chemical equation.

Equivalent conductance of an electrolyte is defined as the conductance of a volume of solution containing one equivalent weight of dissolved substance when placed between two parallel electrodes 1 cm apart, and large enough to contain between them all of the solution. Λ is never determined directly, but is calculated from the specific conductance. If C is the concentration of a solution in gram equivalents per liter, then the concentration per cubic centimeter is C/1000, and the volume containing one equivalent of the solute is, therefore, 1000/C. Since L_s is the conductance of a centimeter cube of the solution, the conductance of 1000/C cc, and hence Λ will be

$$\Lambda = \frac{1000L_s}{C}$$

Equivalent weight or combining weight of an element or ion is its atomic or formula weight divided by its valence. Elements entering into combination always do so in quantities proportional to their equivalent weights.

In oxidation-reduction reactions the equivalent weight of the reacting substances is dependent upon the change in oxidation number of the particular substance.

Ettinghausen's effect (Von Ettinghausen's).—When an electric current flows across the lines of force of a magnetic field an electromotive force is observed which is at right angles to both the primary current and the magnetic field: a temperature gradient is observed which has the opposite direction to the Hall electromotive force.

Eutectic.—A term applied to the mixture of two or more substances which has the lowest melting point.

Expansion of gases.

Charles' law or Gay-Lussac's law.—The volume of a gas at constant pressure increases proportionately to the absolute temperature. If V_1 and V_2 are volumes of the same mass of gas at absolute temperatures, T_1 and T_2,

$$\frac{V_1}{V_2} = \frac{T_1}{T_2}$$

For an original volume V_o at 0°C the volume at t°C (at constant pressure) is

$$V_t = V_o(1 + 0.00367t).$$

General law for gases.

$$p_t v_t = p_o v_o \left(1 + \frac{t}{273}\right)$$

where p_o, v_o, p_t, v_t represent the pressure and value at 0° and t°C or

$$\frac{p_1 v_1}{T_1} = \frac{p_2 v_2}{T_2}$$

where p_1, v_1 and T_1 represent pressure, volume and absolute temperature in one case and p_2, v_2 and T_2 the same quantities for the same mass of gas in another.

The law may also be expressed:

$$pv = RmT$$

where m is the mass of gas at absolute temperature T. R is the **gas constant** which depends on the units used. **Boltzmann's molecular gas constant** is obtained by expressing m in terms of the number of molecules.

For volume in cm³, pressure in dynes per cm² and temperature in Centigrade degrees on the absolute scale $R = 8.3136 \times 10^7$.

Reduction of a gas volume to 0°C, 760 mm pressure.—If V is the original volume of a gas at temperature t and pressure H, the volume at 0°C and 760 mm pressure will be,

$$V_o = \frac{V}{(1 + \alpha t)}\frac{H}{760}$$

If d is the original density the density at 0°C and 760 mm pressure will be

$$d_o = d(1 + \alpha t)\frac{760}{H}$$

$$\alpha = 0.00367 \text{ approximately.}$$

DEFINITIONS AND FORMULAS (Continued)

Falling bodies.—For bodies falling from rest conditions are as for uniformly accelerated motion except that $v_o = O$ and g is the acceleration due to gravity. The formulae become,—air resistance neglected,

$$v_t = gt,\ s = \tfrac{1}{2}gt^2,\ v_s = \sqrt{2gs}.$$

For bodies projected vertically upward,—if v is the velocity of projection, the time to reach greatest height, neglecting the resistance of the air,

$$t = \frac{v}{g}$$

Greatest height,

$$h = \frac{v^2}{2g}$$

—See *Projectiles*

Faraday's laws.—In the process of electrolytic changes equal quantities of electricity charge or discharge equivalent quantities of ions at each electrode.

One gram equivalent weight of matter is chemically altered at each electrode for 96,501 int. coulombs, or one faraday, of electricity passed through the electrolyte.

Faraday effect.—The rotation of the plane of polarization produced when plane-polarized light is passed through a substance in a magnetic field, the light traveling in a direction parallel to the lines of force. For a given substance, the rotation is proportional to the thickness traversed by the light and to the magnetic field strength.

Faraday, new determination of (1960).—By the National Bureau of Standards which uses an electrochemical method that dissolves, rather than deposits, silver from a solution. The new value, 96,516 ± 2 coulombs (physical scale) or 96,489 ± 2 coulombs (chemical scale). NBS used its mass, time, and electrical standards in measuring the faraday, and have found that its value agreed within 22 p.p.m. with the one obtained by an independent physical method using the omegatron.

Fermat's principle.—The path followed by light (or other waves) passing through any collection of media from one specified point to another, is that path for which the time of travel is least.

Fission.—A nuclear reaction from which the atoms produced are each approximately half the mass of the parent nucleus. In other words, the atom is split into two approximately equal masses. There is also the emission of extremely great quantities of energy since the sum of the masses of the two new atoms is less than the mass of the parent heavy atom. The energy released is expressed by Einstein's equation.

Fleming's rule.—A simple rule for relating the directions of the flux, motion, and e.m.f. in an electric machine. The forefinger, second finger and thumb, placed at right-angles to each other, represent respectively the directions of flux, e.m.f., and motion or torque. If the right hand is used the conditions are those obtaining in a generator and if the left hand is used the conditions are those obtaining in a motor.

Fluidity.—The reciprocal of viscosity. The cgs unit is the rhe, the reciprocal of the poise. Dimensions,—$[m^{-1}\,lt]$.

Fluorescence.—The property of emitting radiation as the result of absorption of radiation from some other source. The emitted radiation persists only as long as the exposure is subjected to radiation which may be either electrified particles or waves. The fluorescent radiation generally has a longer wave length than that of the absorbed radiation. If the fluorescent radiation includes waves of the same length as that of the absorbed radiation it is termed resonance radiation.

Foot-candle.—Foot-candle = 1 lumen incident per foot^2 = 1.076 milliphots = 10.76 lux.

Foot-lambert is the unit of photometric brightness (luminance) equal to $1/\pi$ candle per square foot.

Force.—That which changes the state of rest or motion in matter, measured by the rate of change of momentum. Absolute unit,—the **dyne,** the force which will produce an acceleration of one centimeter per second per second in a gram mass. The gram weight or weight of a gram mass is the cgs gravitational unit. The poundal is that force which will give an acceleration of one foot per second per second to a pound mass. Dimensions,—$[m\,lt^{-2}]$.

The force F required to produce an acceleration a in a mass m is given by

$$F = ma.$$

If m is substituted in grams and a in cm per sec^2, F will be given in dynes.

Force between two charges, Coulomb's law.—If two charges q and q' are at a distance r in a vacuum, the force between them is,

$$F = \frac{qq'}{r^2}$$

The force will be given in dynes if q and q' are in electrostatic units and r in cm.

Force between two magnetic poles.—If two poles of strength m and m' are separated by a distance r in a medium whose permeability is μ (unity for a vacuum), the force between them is,

$$F = \frac{mm'}{\mu r^2}$$

Force will be given in dynes if r is in cm and m and m' are in cgs units of pole strength.

DEFINITIONS AND FORMULAS (Continued)

The strength of a magnetic field at a point distance r from an isolated pole of strength m is

$$H = \frac{m}{\mu r^2}$$

The field will be given in gauss if m and r are in cgs units.

Formula, chemical.—A combination of symbols with their subscripts representing the constituents of a substance and their proportions by weight.

Fraunhofer's lines.—When sunlight is examined through a spectroscope it is found that the spectrum is traversed by an enormous number of dark lines parallel to the length of the slit. These dark lines are known as Fraunhofer's lines. Kirchoff conceived the idea that the sun is surrounded by layers of vapors which act as filters of the white light arising from incandescent solids within and which abstract those rays which correspond in their periods of vibration to those of the components of the vapors. Thus reversed or dark lines are obtained due to the absorption by the vapor envelop, in place of the bright lines found in the emission spectrum.

Frequency in uniform circular motion or in any periodic motion is the number of revolutions or cycles completed in unit time. Cgs units,—cycles per second. Dimension,—$[t^{-1}]$.

Frequency of vibrating strings.—The fundamental frequency of a stretched string is given by

$$n = \frac{1}{2l}\sqrt{\frac{T}{m}}$$

where l is the length, T, the tension and m the mass per unit length.

For a string or wire of circular section of length l, tension T, density d, and radius r, the frequency of the fundamental is

$$n = \frac{1}{2rl}\sqrt{\frac{T}{\pi d}}$$

The frequency in vibrations per second will be given if T is in dynes, r and l in cm and d in g per cm³.

Friction, coefficient of.—The coefficient of friction between two surfaces is the ratio of the force required to move one over the other to the total force pressing the two together.

If F is the force required to move one surface over another and W, the force pressing the surfaces together, the coefficient of friction,

$$k = \frac{F}{W}$$

Fundamental units.—See *Mass, Length and Time*

Fusion (atomic).—A nuclear reaction involving the combination of smaller atomic nuclei or particles into larger ones with the release of energy from mass transformation. This is also called a thermo-nuclear reaction by reason of the extremely high temperature required to initiate it.

Gal.—1 gal = cm/sec./sec. Therefore, where the value of gravity is 980 this is the same as 980 gals. The milligal is now quite commonly used since it is approximately one part in a million of the normal gravity of the earth.

Gamma (γ) **rays (nuclear x-rays).**—May be emitted from radioactive substances. They are quanta of electromagnetic wave energy similar to but of much higher energy than ordinary x-rays. The energy of a quantum is equal to $h\nu$ ergs, where h is Planck's constant (6.6254×10^{-27} erg sec) and ν is the frequency of the radiation. Gamma rays are highly penetrating, an appreciable fraction being able to traverse several centimeters of lead.

Gas.—A state of matter in which the molecules are practically unrestricted by cohesive forces. A gas has neither definite shape nor volume.

Gas thermometer.—Where P_o, P_s and P_x represent the total pressure with the bulb at 0°C, at the boiling-point of water and at the unknown temperature respectively, t_s the temperature of steam and t_x the unknown temperature,

$$t_x = t_s\frac{P_x - P_o}{P_s - P_o}$$

(approximately). The total pressure on the gas in the bulb is the algebraic sum of barometric pressure at the time and that measured by the manometer.

Gauss.—The cgs emu of magnetic induction (flux density). It is equal to 1 maxwell per cm². It has such a value that if a conductor 1 cm long moves through a magnetic field at a velocity of 1 cm, in an induction mutually perpendicular, the induced emf is one abvolt.

Gay-Lussac's law.—See *Charles' law*

Gay-Lussac's law of combining volumes.—If gases interact and form a gaseous product, the volumes of the reacting gases and the volumes of the gaseous products are to each other in simple proportions, which can be expressed by small whole numbers.

Geiger counter.—Detector for radioactivity depending upon ionized particles that affect its mechanism. As its name indicates, it both detects and makes a count of them possible.

Gibbs' phase rule.—$F = C + 2 - P$, F, the number of degrees of freedom of a system, is the number of variable factors (temperature, pressure and concentration) of the components, which must be arbitrarily fixed in order that the condition of the system may be perfectly defined. C, the number of the components of the system, is chosen equal to the smallest number of independently variable constituents by means of which the composition of each phase participating

DEFINITIONS AND FORMULAS (Continued)

in the state of equilibrium can be expressed in the form of a chemical equation; the components must be chosen from among the constituents which are present when the system is in a state of true equilibrium and which take part in that equilibrium; as components are chosen the smallest number of such constituents necessary to express the composition of each phase participating in the equilibrium, zero and negative quantities of components being permissible; in any system the number of components is definite, but may alter with changes in conditions of experiment; a qualitative but not quantitative freedom of selection of components is allowed, the choice being influenced by suitability and simplicity of application. *P*, the number of phases of the system, are the homogeneous, mechanically separable and physically distinct portions of a heterogeneous system; the number of phases capable of existence varies greatly in different systems; there can never be more than one gas or vapor phase since all gases are miscible in all proportions, a heterogeneous mixture of solid substances forms as many phases as there are substances present.

Gilbert.—The cgs emu of magnetomotive force. 1 gilbert $= 10/4\pi$ ampere-turns.

Graham's law.—The relative rates of diffusion of gases under the same conditions are inversely proportional to the square roots of the densities of those gases.

Gram atom or gram atomic weight.—The mass in grams numerically equal to the atomic weight.

Gram equivalent of a substance is the weight of a substance displacing or otherwise reacting with 1.008 grams of hydrogen or combining with one-half of a gram atomic weight (8.00 grams) of oxygen.

Gram mole, gram formula weight, gram equivalent.—Mass in grams numerically equal to the molecular weight, formula weight or chemical equivalent, respectively.

Gram molecular weight or gram molecule.—A mass in grams of a substance numerically equal to its molecular weight. Gram mole.

Gravitation.—The universal attraction existing between all material bodies. The force of attraction between two masses m and m', separated by a distance r, k being the constant of gravitation,

$$F = k\frac{mm'}{r^2}$$

(If m and m' are given in grams, and r in centimeters, F will be in dynes if $k = 6.670 \times 10^{-8}$.)

Half-life.—Used to measure the rate of radioactive decay of disintegration. The time lapse during which a radioactive mass loses one half of its radioactivity.

Hall effect.—When a steady current is flowing in a steady magnetic field, electromotive forces are developed which are at right angles both to the magnetic force and to the current and are proportional to the product of the intensity of the current, the magnetic force and the sine of the angle between the directions of these quantities.

Hardness.—Property of substances determined by their ability to abrade or indent one another. An arbitrary scale of hardness is based upon ten selected minerals. For metals the diameter of the indentation made by a hardened steel sphere (Brinnell) or the height of rebound of a small drop hammer (Shore Scleroscope) serve to measure hardness.

Harmonic motion.—See *Simple harmonic motion and Angular harmonic motion*

Heat capacity.—That quantity of heat required to increase the temperature of a system or substance one degree of temperature. It is usually expressed in calories per degree centigrade.

Molar heat capacity is the quantity of heat necessary to raise the temperature of one molecular weight of the substance one degree.

Heat effect.—The heat in calories developed in a circuit by an electric current of I amperes flowing through a resistance of R ohms, with a difference of potential E volts for a time t seconds.

$$H = \frac{RI^2t}{4.18} = \frac{Eit}{4.18}$$

Heat equivalent, or latent heat, of fusion.—The quantity of heat necessary to change one gram of solid to a liquid with no temperature change. Dimensions,—$[l^2 t^{-2}]$.

Heat of combustion of a substance is the amount of heat evolved by the combustion of 1 gram molecular weight of the substance.

Heat quantity.—The cgs unit of heat is the **calorie,** the quantity of heat necessary to change the temperature of one gram of water from 3.5°C to 4.5°C (called a small calorie). If the temperature change involved is from 14.5 to 15.5°C, the unit is the normal calorie. The mean calorie is $\frac{1}{100}$ the quantity of heat necessary to raise one gram of water from 0°C to 100°C. The large calorie is equal to 1000 small calories. The British thermal unit is the heat required to raise the temperature of one pound of water at its maximum density, 1°F. It is equal to about 252 calories. Dimensions of energy,—$[m\, l^2 t^{-2}]$.

Hehner number (value).—A number expressing the percentage (i.e. grams per hundred grams) of water-insoluble fatty acids in an oil or fat.

Heisenberg's theory of atomic structure.—The currently accepted view of the structure of atom, formulated by Heisenberg in 1934, according to which the atomic nuclei are built of nucleons, which may be protons or neutrons, while the extranuclear shells consist of electrons only. The nucleons

DEFINITIONS AND FORMULAS (Continued)

are held together by nuclear forces of attraction, with exchange forces operating between them. The number of protons is equal to the atomic number (Z) of the element, the number of neutrons is equal to the difference between the mass number and the atomic number ($A - Z$). The number of excess neutrons, i.e. the excess of neutrons over protons, is of paramount importance for the radioactive properties or stability of the element.

Henry's law.—The mass of a slightly soluble gas that dissolves in a definite mass of a liquid at a given temperature is very nearly directly proportional to the partial pressure of that gas. This holds for gases which do not unite chemically with the solvent.

Hess' law of constant heat summation.—The amount of heat generated by a chemical reaction is the same whether reaction takes place in one step or in several steps, or all chemical reactions which start with the same original substances and end with the same final substances liberate the same amounts of heat, irrespective of the process by which the final state is reached.

Hooke's law.—Within the elastic limit of any body the ratio of the stress to the strain produced is constant.

Humidity, absolute.—Mass of water vapor present in unit volume of the atmosphere, usually measured as grams per cubic meter. It may also be expressed in terms of the actual pressure of the water vapor present.

Huygens' theory of light.—This theory states that light is a disturbance traveling through some medium, such as the ether. Thus light is due to wave motion in ether.

Every vibrating point on the wave-front is regarded as the center of a new disturbance. These secondary disturbances traveling with equal velocity, are enveloped by a surface identical in its properties with the surface from which the secondary disturbances start and this surface forms the new wave-front.

Hydrogen equivalent of a substance is the number of replaceable hydrogen atoms in 1 molecule or the number of atoms of hydrogen with which 1 molecule could react.

Hydrogen ion concentration.—The concentration of hydrogen ions in solution when the concentration is expressed as gram-ionic weights per liter. A convenient form of expressing hydrogen ion concentration is in terms of the negative logarithm of this concentration. The negative logarithm of the hydrogen ion concentration is called pH. The significance of pH is still in dispute (ref. J. Am. Chem. Soc. 60, 1094 (1938)).

Water at 25°C has a concentration of H ion of 10^{-7} and of OH ion of 10^{-7} moles per liter. Thus the pH of water is 7 at 24°C. A greater accuracy is obtained if one substitutes the thermodynamic activity of the ion for its concentration.

Hydrolysis is a double decomposition reaction involving the splitting of water into its ions and the formation of a weak acid or base or both.

Hydrostatic pressure at a distance h from the surface of a liquid of density d,

$$P = hdg.$$

The total force on an area A due to hydrostatic pressure,

$$F = PA = Ahdg.$$

Force in dynes and pressure in dynes per cm² will be given if h is in cm, d in g per cm³ and g in cm per sec².

Hyperon.—Any article with mass intermediate between that of the neutron and the deuteron.—See *Meson*

Hysteresis.—The magnetization of a sample of iron or steel due to a magnetic field which is made to vary through a cycle of values, lags behind the field. This phenomenon is called hysteresis.

Steinmetz' equation for hysteresis gives the loss of energy in ergs per cycle per cm³,

$$W = \eta B^{1.6}$$

where B is the maximum induction in maxwells per cm² and η the coefficient of hysteresis.

Illumination on any surface is measured by the luminous flux incident on unit area. The units in use are: the **lux,** one lumen per square meter; the **phot,** one lumen per square centimeter and the lumen per square foot. Since at unit distance from a point source of unit intensity the illumination is unity, unit illumination may be defined as that produced by a unit source at unit distance, hence the **meter-candle** or **candle-meter** which is equal to the lux and the **foot-candle** equivalent to one lumen per square foot.

Indeterminancy principle (Uncertainty principle).—The postulate that it is impossible to determine simultaneously both the exact position and the exact momentum of an electron. So this aspect of electronics can only be expressed as a probability.

Index of refraction for any substance is the ratio of the velocity of light in a vacuum to its velocity in the substance. It is also the ratio of the sine of the angle of incidence to the sine of the angle of refraction. In general, the index of refraction for any substance varies with the wave length of the refracted light.

Indicators are substances which change from one color to another when the hydrogen ion concentration reaches a certain value, different for each indicator.

Induced electromotive force in a circuit is proportional to the rate of change of magnetic flux through the circuit.

$$E = -\frac{d\phi}{dt}$$

DEFINITIONS AND FORMULAS (Continued)

where $d\phi$ is the change of magnetic flux in a time dt. The induced current will be given by

$$I = \frac{d\phi}{Rdt}$$

where R is the resistance of the circuit.

Inductance.—The change in magnetic field due to the variation of a current in a conducting circuit causes an induced counter electromotive force in the circuit itself. This phenomenon is known as **self-induction.** If an electromotive force is induced in a neighboring circuit the term mutual induction is used. Inductance may thus be distinguished as self- or mutual and is measured by the electromotive force produced in a conductor by unit rate of variation of the current. Units of inductance are the centimeter (absolute electromagnetic) and the henry, which is equal to 10^9 centimeters of inductance. The **henry** is that inductance in which an induced electromotive force of one volt is produced when the inducing current is changed at the rate of one ampere per second. Dimensions,—$[\epsilon^{-1} l^{-1} t^2]$; $[\mu l]$.

Induction.—Any change in the intensity or direction of a magnetic field causes an electromotive force in any conductor in the field. The induced electromotive force generates an induced current if the conductor forms a closed circuit.

Inertia.—The resistance offered by a body to a change of its state of rest or motion, a fundamental property of matter. Dimension,—$[m]$.

Intensity of illumination in candle meters of a screen illuminated by a source of illuminating power P candles at a distance r meters, for normal incidence,

$$I = \frac{P}{r^2}$$

If two sources of illuminating power P_1 and P_2 produce equal illumination on a screen when at distances r_1 and r_2 respectively,

$$\frac{P_1}{r_1^2} = \frac{P_2}{r_2^2} \quad \text{or} \quad \frac{P_1}{P_2} = \frac{r_1^2}{r_2^2}$$

If I_o is the intensity of illumination when the screen is normal to the incident light, I the intensity when an angle θ

$$I = I_o \cos \theta$$

Intensity of magnetization is given by the quotient of the magnetic moment of a magnet by its volume. Unit intensity of magnetization is the intensity of a magnet which has unit magnetic moment per cubic centimeter. Dimensions,—$[\epsilon^{-\frac{1}{2}} m^{\frac{1}{2}} l^{-\frac{3}{2}}]$; $[\mu^{\frac{1}{2}} m^{\frac{1}{2}} l^{-\frac{3}{2}} t^{-1}]$.

Intensity of radiation is the rate of transfer of energy across unit areas by the radiation. In all forms of energy transfer by waves (radiation) the intensity I is given by $I = Uv$

where U is the energy density of the wave in the medium, and v is the velocity of propagation of the wave. The energy density U is always proportional to the square of the wave amplitude.

Intensity of sound depends upon the energy of the wave motion. The intensity is measured by the energy in ergs transmitted per second through one square centimeter of surface. The energy in ergs per cm^3 in a sound wave is given by

$$E = 2\pi^2 dn^2 a^2$$

where d is density in g per cm^3, n is frequency in vib. per sec and a is amplitude in cm. The energy reaching the ear in unit time will also be proportional to the velocity of propagation.

Iodine number (value).—A number expressing the percentage (i.e. grams per 100 grams) of iodine absorbed by a substance. It is a measure of the proportion of unsaturated linkages present and is usually determined in the analysis of oils and fats.

Ion.—An ion is an atom or group of atoms that is not electrically neutral but instead carries a positive or negative electric charge. Positive ions are formed when neutral atoms or molecules lose valence electrons; negative ions are those which have gained electrons.

Ionization potential.—The work (expressed in electron volts) required to remove a given electron from its atomic orbit and place it at rest at an infinite distance. It is customary to list values in electron volts (ev.) 1 ev. = 23,053 calories per mole.

Isobars.—For chemistry, elements of the same atomic mass but of different atomic numbers. The sum of their nucleons is the same but there are more protons in one than in the other.

Isomerism.—Existence of molecules having the same number and kinds of atoms but in different configurations.

Isothermal.—When a gas passes through a series of pressure and volume variations without change of temperature the changes are called isothermal. A line on a pressure-volume diagram representing these changes is called an isothermal line.

Isotopes.—Two or more nuclides having the same atomic number, hence constituting the same element, but differing in mass number. Isotopes of a given element have the same number of nuclear protons but differing numbers of neutrons. Naturally occurring chemical elements are usually mixtures of isotopes so that observed (non-integer) atomic weights are average values for the mixture.—See *Atomic number*

Joule-Thomson effect.—The cooling which occurs when a highly compressed gas is allowed to expand in such a way that no external work is done is known as the Joule-Thomson effect. This cooling is inversely proportional to the square of the absolute temperature.

Kepler's laws.

I. The planets move about the sun in ellipses, at one focus of which the sun is situated.

II. The radius vector joining each planet with the sun describes equal areas in equal times.

III. The cubes of the mean distances of the planets from the sun are proportional to the squares of their times of revolution about the sun.

Kerr effect.—When plane polarized light is incident on the pole of an electromagnet, polished so as to act like a mirror, the plane of polarization of the reflected light is not the same when the magnet is "on" as when it is "off." It was found that the direction of rotation was opposite to that of the currents exciting the pole from which the light was reflected.

Kinetic theory, expression for pressure.

$$P = \frac{1}{3}Nmv^2$$

where N is the number of molecules in unit volume, m the mass of each molecule and v^2 the mean square of the velocity of the molecules.

Kinetic theory of gases.—Gases are considered to be made up of minute, perfectly elastic particles which are ceaselessly moving about with high velocities, colliding with each other and with the walls of the containing vessel. The pressure exerted by a gas is due to the combined effect of the impacts of the moving molecules upon the walls of the containing vessel, the magnitude of the pressure being dependent upon the kinetic energy of the molecules and their number.

Kirchhoff's laws.

I. The algebraic sum of the currents which meet at any point is zero.

II. In any closed circuit the algebraic sum of the products of the current and the resistance in each conductor in the circuit is equal to the electromotive force in the circuit.

Kirchhoff's laws of radiation.—The relation between the powers of emission and the powers of absorption for rays of the same wave-length is constant for all bodies at the same temperature. First, a substance when excited by some means or other possess a certain power of emission; it tends to emit definite rays, whose wave-lengths depend upon the nature of the substance and upon the temperature. Second, the substance exerts a definite absorptive power, which is a maximum for the rays it tends to emit. Third, at a given temperature the ratio between the emissive and the absorptive power for a given wave-length is the same for all bodies, and is equal to the emissive power of a perfectly black body.

Kohlrausch's law.—When ionization is complete, the conductivity of an electrolyte is equal to the sum of the conductivities of the ions into which the substance dissociates.

Kundt's law.—On approaching an absorption band from the red side of the spectrum the refractive index is abnormally increased by the presence of the band, while the approach is from the blue side and the index is abnormally decreased.

Lambert is the unit of brightness equal to $1/\pi$ candle per square centimeter.

Lambert's law of absorption.—Each layer of equal thickness absorbs an equal fraction of the light which traverses it.

Lambert's law of illumination.—The illumination of a surface on which the light falls normally from a point source is inversely proportional to the square of the distance of the surface from the source. If the normal to the surface makes an angle with the direction of the rays, the illumination is proportional to the cosine of that angle.

Lanthanide series (Lanthanides).—Rare earth elements of atomic numbers 57 through 71, which have chemical properties similar to lanthanum (#57).

Latent heat of vaporization.—The quantity of heat necessary to change one gram of liquid to vapor without change of temperature, measured as calories per gram. Dimensions,—$[l^2 t^{-2}]$.

Lattice energy.—The energy required to separate the ions of a crystal to an infinite distance from each other.

LeChatelier's principle.—If some stress is brought to bear upon a system in equilibrium, a change occurs, such that the equilibrium is displaced in a direction which tends to undo the effect of the stress.

Length, units of.—The centimeter, one of the three fundamental units of the cgs system, is one one-hundredth the length of the International Prototype Meter, at Paris, at zero degrees centigrade. The meter is 1,553,164.13 times the wave length of the red cadmium line in air, 760 mm pressure, 15°C. The standard in the British system is the yard, the prototype of which is kept by the British government. The United States standard yard is defined as $\frac{3600}{3937}$ meter.

Lenses.—For a single thin lens whose surfaces have radii of curvature r_1 and r_2 whose principal focus is F, the index of refraction n, and conjugate focal distances f_1 and f_2,

$$\frac{1}{F} = \frac{1}{f_1} + \frac{1}{f_2} = (n-1)\left(\frac{1}{r_1} + \frac{1}{r_2}\right)$$

For a thick lens, of thickness t,

$$F = \frac{nr_1r_2}{(n-1)[n(r_1+r_2) - t(n-1)]}$$

Combinations of lenses.—If f_1 and f_2 are the focal lengths of two thin lenses separated by a distance d the focal length of the system,

$$F = \frac{f_1f_2}{f_1 + f - d}$$

Lenz's law.—When an electromotive force is induced in a conductor by any change in the relation between the conductor and the magnetic field, the direction of the electromotive force is such as to produce a current whose magnetic field will oppose the change.

Line of force.—A term employed in the description of an electric or magnetic field. A line such that its direction at every point is the same as the direction of the force which would act on a small positive charge (or pole) placed at that point. A line of force is defined as starting from a positive charge (or pole) and ending on a negative charge (or pole).

The line (of force) is also used as a unit of magnetic flux, equivalent to the maxwell.

Lissajous figures.—The path described by a particle which is simultaneously displaced by two simple harmonic motions at right angles, when the periods of the two motions are in the ratio of two small whole numbers, shows a variety of characteristic curves called Lissajous figures.

Liquid.—A state of matter in which the molecules are relatively free to change their positions with respect to each other but restricted by cohesive forces so as to maintain a relatively fixed volume.

Loschmidt's number.—The number of molecules per unit volume of an ideal gas at 0°C and normal atmospheric pressure.

$$n_0 = (2.68719 \pm 0.0001) \times 10^{19}\ \text{cm}^{-3}$$

Loudness.—The psychological response of the ear which is related to the physical quantity *intensity*. The loudness of a sound depends on frequency also since the ear responds more strongly to some frequency bands than to others. The loudness is roughly related to the cube root of the intensity, and for many purposes it is convenient to represent loudness as proportional to the logarithm of the intensity.

Lumen.—The lumen is the unit of luminous flux. It is equal to the luminous flux through a unit-solid angle (steradian) from a uniform point source of one candle, or to the flux on a unit surface all points of which are at unit distance from a uniform point source of one candle.

Luminous flux.—The total visible energy emitted by a source per unit time is called the total luminous flux from the source. The unit of flux, the **lumen,** is the flux emitted in unit solid angle (steradian) by a point source of one candle luminous intensity. A uniform point source of one candle intensity thus emits 4π lumens.

Luminous intensity or candle-power is the property of a source of emitting luminous flux and may be measured by the luminous flux emitted per unit solid angle. The accepted unit of luminous intensity is the **international candle.** The **Hefner unit,** which is equivalent to 0.9 international candles, is the intensity of a lamp of specified design burning amyl acetate, called the Hefner lamp.

The mean horizontal candle-power is the average intensity measured in a horizontal plane passing through the source. The mean spherical candle-power is the average candle-power measured in all directions and is equal to the total luminous flux in lumens divided by 4π.

Magnetic field due to a current.—The intensity of the magnetic field in oersted at the center of a circular conductor of radius r in which a current I in absolute electromagnetic units is flowing,

$$H = \frac{2\pi I}{r}$$

If the circular coil has n turns the magnetic intensity at the center is,

$$H = \frac{2\pi n I}{r}$$

The magnetic field in a long solenoid of n turns per centimeter carrying a current I in absolute electromagnetic units

$$H = 4\pi n I$$

If I is given in amperes the above formulae become,—

$$H = \frac{2\pi I}{10r}, \qquad H = \frac{2\pi n I}{10r}, \qquad H = \frac{4\pi n I}{10}$$

Magnetic field due to a magnet.—At a point on the magnetic axis prolonged, at a distance r cm from the center of the magnet of length $2l$ whose poles are $+m$ and $-m$ and magnetic moment M, the field strength in oersted is,

$$H = \frac{4mlr}{(r^2 - l^2)^2}$$

If r is large compared with l,

$$H = \frac{2M}{r^3}$$

At a point on a line bisecting the magnet at right angles, with corresponding symbols,

$$H = \frac{2ml}{(r^2 + l^2)^{\frac{3}{2}}}$$

For large value of r,

$$H = \frac{M}{r^3}$$

Magnetic field intensity or magnetizing force.—Is measured by the force acting on unit pole. Unit field intensity, the oersted, is that field which exerts a force of one dyne on unit magnetic pole. The field intensity is also specified by the number of lines of force intersecting unit area normal to the field, equal numerically to the field strength in oersted. Magnetizing force is measured by the space rate of variation of magnetic potential and as such its unit may be the **gilbert per centimeter.** The gamma (γ) is equivalent to 0.00001 oersted. Dimensions,—$[\epsilon^{\frac{1}{2}}\, m^{\frac{1}{2}}\, l^{\frac{1}{2}}\, t^{-2}]$; $[\mu^{-\frac{1}{2}}\, m^{\frac{1}{2}}\, l^{-\frac{1}{2}}\, t^{-1}]$.

Magnetic flux through any area perpendicular to a magnetic field is measured as the product of the area by the field strength. The units of magnetic flux, the **maxwell** is the flux through a square centimeter normal to a field of one gauss. The line is also a unit of flux. It is equivalent to the maxwell. Dimensions,—$[\epsilon^{-\frac{1}{2}}\, m^{\frac{1}{2}}\, l^{\frac{1}{2}}]$; $[\mu^{\frac{1}{2}}\, m^{\frac{1}{2}}\, l^{\frac{3}{2}}\, t^{-1}]$.

Magnetic induction resulting when any substance is subjected to a magnetic field is measured as the magnetic flux per unit area taken perpendicular to the direction of the flux. The unit is the maxwell per square centimeter or its equivalent, the gauss. Dimensions,—$[\epsilon^{-\frac{1}{2}}\, m^{\frac{1}{2}}\, l^{-\frac{3}{2}}]$; $[\mu^{\frac{1}{2}}\, m^{\frac{1}{2}}\, l^{-\frac{1}{2}}\, t^{-1}]$.

If a substance of permeability μ is placed in a magnetic field H the magnetic induction in the substance,

$$\mathbf{M} = \mu H.$$

If I is the magnetic moment for unit volume, or intensity of magnetization,

$$\mathbf{M} = H + 4\pi I.$$

The susceptibility,

$$\kappa = \frac{I}{H}, \qquad \mu = 1 + 4\pi\kappa.$$

Magnetic moment of a magnet is measured by the torque experienced when it is at right angles to a uniform field of unit intensity. The value of the magnetic moment is given by the product of the magnetic pole strength by the distance between the poles. Unit magnetic moment is that possessed by a magnet formed by two poles of opposite sign and of unit strength, one centimeter apart. Dimensions,—$[\mu^{\frac{1}{2}}\, m^{\frac{1}{2}}\, l^{\frac{5}{2}}\, t^{-1}]$; $[\epsilon^{-\frac{1}{2}}\, m^{\frac{1}{2}}\, l^{\frac{3}{2}}]$.

If the poles are separated by a distance which is great compared with the dimensions of the magnet, the magnetic moment of a magnet of length 1 whose poles have values of $+m$ and $-m$ is,

$$\mathrm{m} = ml.$$

Magnetic permeability is a property of materials modifying the action of magnetic poles placed therein and modifying the magnetic induction resulting when the material is subjected to a magnetic field or magnetizing force. The permeability of a substance may be defined as the ratio of the magnetic induction in the substance to the magnetizing field to which it is subjected. The permeability of a vacuum is unity. Dimensions,—$[\epsilon^{-1}\, l^{-2}\, t^{2}]$; $[\mu]$.

Magnetic pole or quantity of magnetism.—Two unit quantities of magnetism concentrated at points unit distance apart in a vacuum repel each other with unit force. If the distance involved is one centimeter and the force one dyne, the quantity of magnetism at each point is one cgs unit of magnetism. Dimensions,—$[\epsilon^{-\frac{1}{2}}\, m^{\frac{1}{2}}\, l^{\frac{1}{2}}]$; $[\mu^{\frac{1}{2}}\, m^{\frac{1}{2}}\, l^{\frac{3}{2}}\, t^{-1}]$.

Magnetic potential or magnetomotive force at a point is measured by the work required to bring unit positive pole from an infinite distance (zero potential) to the point. The unit is the *gilbert,* that magnetic potential against which an erg of work is done when unit magnetic pole is transferred. Dimensions,—$[\epsilon^{\frac{1}{2}}\, m^{\frac{1}{2}}\, l^{\frac{3}{2}}\, t^{-2}]$; $[\mu^{-\frac{1}{2}}\, m^{\frac{1}{2}}\, l^{\frac{1}{2}}\, t^{-1}]$.

Magnifying power of an optical instrument is the ratio of the angle subtended by the image of the object seen through the instrument to the angle subtended by the object when seen by the unaided eye. In the case of the microscope or simple magnifier the object as viewed by the unaided eye is supposed to be a distance of 25 cm (10 in.).

Mass.—Quantity of matter. **Units of mass**—the gram is $\frac{1}{1000}$ the quantity of matter in the International Prototype Kilogram; one of the three fundamental units of the cgs system. The British standard of mass is the pound, of which a standard is preserved by the government. The United States standard mass is the avoirdupois pound defined as 1/2.20462 kilogram.

Mass action, law of.—At a constant temperature the product of the active masses on one side of a chemical equation when divided by the product of the active masses on the other side of the chemical equation is a constant, regardless of the amounts of each substance present at the beginning of the action.

At constant temperature the rate of the reaction is proportional to the concentration of each kind of substance taking part in the reaction.

Mass by weighing on a balance with unequal arms.—If W_1 is the value for one side, W_2 the value for the other, the true mass,

$$W = \sqrt{W_1 W_2}.$$

Mass defect.—Difference between atomic mass and mass number of a nuclide.—See *Packing fraction*

Mass-energy equivalence.—The equivalence of a quantity of mass and a quantity of energy when the two quantities are related by the equation $E = mc^2$. The conversion factor c^2 is the square of the velocity of light. The relationship was developed from **relativity theory,** but has been experimentally confirmed.

Maxwell.—The cgs emu magnetic flux is the flux through a cm^2 normal to a field at 1 cm from a unit magnetic pole.

Maxwell's rule.—A law stating that every part of an electric circuit is acted upon by a force tending to move it in such a direction as to enclose the maximum amount of magnetic flux.

DEFINITIONS AND FORMULAS (Continued)

Mechanical equivalent of heat is the quantity of energy which, when transformed into heat, is equivalent to unit quantity of heat; 4.18×10^7 ergs = 1 calorie (20°C).

Meson.—Two types of particles of mass intermediate between that of the electron and proton have been discovered in cosmic radiation and in the laboratory. The one particle with mass about $215m_e$ is called μ-meson, the other with about $280m_e$ π-meson. Mesons of both positive and negative charge have been found and there is now reasonably good evidence for neutral mesons. Both types of mesons decay spontaneously. Some evidence exists for a meson of mass about $1000m_e$.

Metallic elements in general are distinguished from the non-metallic elements by their lustre, malleability, conductivity and usual ability to form positive ions. **Non-metallic** elements are not malleable, have low conductivity and never form positive ions.

Mev.—Mev = million electron volts.

Minimum deviation.—The deviation or change of direction of light passing through a prism is a minimum when the angle of incidence is equal to the angle of emergence. If D is the angle of minimum deviation and A the angle of the prism, the index of refraction of the prism for the wave length used is,

$$n = \frac{\sin \frac{1}{2}(A + D)}{\sin \frac{1}{2}A}$$

Mixtures consist of two or more substances intermingled with no constant percentage composition, and with each component retaining its essential original properties.

Moderator.—A material used for slowing down neutrons in an atomic pile or reactor. Usually graphite or "heavy water" (deuterium oxide).

Modulus of elasticity.—The stress required to produce unit strain, which may be a change of length (Young's modulus): a twist or shear (modulus of rigidity or modulus of torsion) or a change of volume (bulk modulus), expressed in dynes per square centimeter. Dimensions,—the same as of stress, $[m\, l^{-1}\, t^{-2}]$.

Mol volume.—The volume occupied by a mol or a gram molecular weight of any gas measured at standard conditions is 22.414 liters.

A **molal solution** contains one mole per 1000 grams of solvent.

A **molar solution** contains one mole or gram molecular weight of the solute in one liter of solution.

Mole.—Mass numerically equal to the molecular weight. It is most frequently expressed as the gram molecular weight, i. e. as the weight of one mole expressed in grams.

Molecular volume.—Volume occupied by one mole. Numerically equal to the molecular weight divided by the density.

Molecular weight.—The sum of the atomic weights of all the atoms in a molecule.

Molecule.—The smallest unit quantity of matter which can exist by itself and retain all the properties of the original substance.

Moment of force or torque.—The effectiveness of a force to produce rotation about an axis, measured by the product of the force and the perpendicular distance from the line of action of the force to the axis. Cgs unit—the dyne-centimeter. Dimensions,—$[m\, l^2\, t^{-2}]$. If a force F acts to produce rotation about a center at a distance d from the line in which the force acts, the force has a torque,

$$L = Fd.$$

Moment of inertia.—A measure of the effectiveness of mass in rotation. In the rotation of a rigid body not only the body's mass, but the distribution of the mass about the axis of rotation determines the change in the angular velocity resulting from the action of a given torque for a given time. Moment of inertia in rotation is analogous to mass (inertia) in simple translation. The cgs unit is g-cm². Dimensions,—$[m\, l^2]$.

If m_1, m_2, m_3, etc. represent the masses of infinitely small particles of a body; r_1, r_2, r_3, etc. their respective distances from an axis of rotation, the moment of inertia about this axis will be

$$I = (m_1r_1^2 + m_2r_2^2 + m_3r_3^2 + \cdots)$$

or

$$I = \Sigma(mr^2)$$

Momentum.—Quantity of motion measured by the product of mass and velocity. Cgs unit—, one gram-centimeter per second. Dimensions,—$[m\, l\, t^{-1}]$.

A mass m moving with velocity v has a momentum,

$$M = mv.$$

If a mass m has its velocity changed from v_1 to v_2 by the action of a force F for a time t,

$$mv_2 - mv_1 = Ft.$$

Monochromatic emissive power is the ratio of the energy of certain defined wave lengths radiated at definite temperatures to the energy of the same wave lengths radiated by a black body at the same temperature and under the same conditions.

Monotropic.—Crystal forms one of which is always metastable with respect to the other.

Mosley's law.—The frequencies of the characteristic X-rays of the elements show a strict linear relationship with the square of the atomic number.

DEFINITIONS AND FORMULAS (Continued)

Motion, laws of.—See *Newton's law of motion*

Multiple proportions, law of.—If two elements form more than one compound, the weights of the first element which combine with a fixed weight of the second element are in the ratio of integers to each other.

Negatron.—1. A term used for electron when it is necessary to distinguish between (negative) electrons and positrons. 2. A four element vacuum tube which displays a negative resistance characteristic.

Nernst effect.—When heat flows across the lines of magnetic force, there is observed an electromotive force in the mutually perpendicular direction.

Neutralization is a reaction in which the hydrogen ion of an acid and the hydroxyl ion of a base unite to form water, the other product being a salt.

Neutrino.—An electrically neutral particle of very small (probably zero) rest mass and of spin quantum number $\frac{1}{2}$. When the spin is oriented parallel to the linear momentum the particle is the antineutrino. When the spin is oriented antiparallel to the linear momentum the particle is the neutrino. Postulated by Pauli in explaining the beta decay process.

Whenever a beta (positron) particle is created in a radioactive decay so is an antineutrino (neutrino). The two particles and the parent nucleus share between them the available energy and momentum. Neutrinos and antineutrinos can penetrate amounts of matter measured in light years without appreciable attenuation. Detected by Reines and Cowan using antineutrinos from fission reactors and large scintillation detectors.

Neutron.—A neutral elementary particle of mass number 1. It is believed to be a constituent particle of all nuclei of mass number greater than 1. It is unstable with respect to beta-decay, with a half life of about 12 minutes. It produces no detectable primary ionization in its passage through matter, but interacts with matter predominantly by collisions and, to a lesser extent, magnetically. Some properties of the neutron are: rest mass, 1.00894 atomic mass unit; charge, 0; spin quantum number, $\frac{1}{2}$; magnetic moment, −1.9125 nuclear Bohr *magnetrons*.

Neutron cross section.—See *cross section*

Newton.—The force necessary to give acceleration of one meter per second per second to one kilogram of mass.

Newton's law of cooling.—The rate of cooling of a body under given conditions is proportional to the temperature difference between the body and its surroundings.

Newton's law of motion.

I. Every body continues in its state of rest or of uniform motion in a straight line except in so far as it may be compelled to change that state by the action of some outside force.

II. Change of motion is proportional to force applied and takes place in the direction of the line of action of the force.

III. To every action there is always an equal and opposite reaction.

Nodal points.—Two points on the axis of a lens such that a ray entering the lens in the direction of one, leaves as if from the other and parallel to the original direction.

A **normal salt** is an ionic compound containing neither replaceable hydrogen nor hydroxyl ions.

A **normal solution** contains one gram molecular weight of the dissolved substance divided by the hydrogen equivalent of the substance (that is, one gram equivalent) per liter of solution.

Nuclear atom.—The atom of each element consists of a small dense nucleus which includes most of the mass of the atom. The nucleus is made up of roughly equal numbers of neutrons and protons. The positive charges of the protons enables the nucleus to surround itself with a set of negatively charged electrons which move around the nucleus in complicated orbits with well defined energies. The outermost electrons which are least tightly bound to the nucleus play the dominant part in determining the physical and chemical properties of the atom. There are as many electrons in orbits as there are protons in the nucleus.

Nuclear fusion.—See *fusion*

Nuclear isomers.—Isotopes of elements having the same mass number and atomic number but differing in radioactive properties such as half-life period.

Nucleon.—Any particle found in the structure of an atom's nucleus. The most plentiful ones are neutrons and protons.

Nucleus.—The dense central core of the atom, in which most of the mass and all of the positive charge is concentrated. The charge on the nucleus, an integral multiple of Z of the electronic charge, is the essential factor which distinguishes one element from another. Z is called the atomic number and gives the number of protons in the nucleus, which includes a roughly equal number of neutrons. The mass number A gives the total number of neutrons plus protons.—See *Isotopes and Nuclear Atom*

Nuclide.—A species of atom distinguished by the constitution of its nucleus. The nuclear constitution is specified by the number of protons, Z; number of neutrons, N; and energy content. (Or, by the atomic number, Z; mass number A $(= N + Z)$ and atomic mass.)

Numerical aperture is the sine of half the angular aperture, used as a measure of the optical power of an objective.

Oersted.—The cgs emu of magnetic intensity exists at a point where a force of 1 dyne acts upon a unit magnetic pole at that point, i.e., the intensity 1 cm from a unit magnetic pole.

DEFINITIONS AND FORMULAS (Continued)

Ohm's law.—Current in terms of electromotive force E and resistance R.

$$I = \frac{E}{R}.$$

The current is given in amperes when E is in volts and R in ohms.

Organ pipes.—The frequency of vibration of a closed pipe or other air column of length l, where V is the velocity of sound in air, for the fundamental and first three overtones respectively is,

$$n_o = \frac{V}{4l}, \quad n_1 = \frac{3V}{4l}, \quad n_2 = \frac{5V}{4l}, \quad n_3 = \frac{7V}{4l}$$

For an open pipe,

$$n_o = \frac{V}{2l}, \quad n_1 = \frac{2V}{2l}, \quad n_2 = \frac{3V}{2l}, \quad n_3 = \frac{4V}{2l}$$

Oxidation is any process which increases the proportion of oxygen or acid-forming element or radical in a compound.

Packing fraction.—Packing fraction is the difference between the actual mass of the isotope and the nearest whole number divided by the mass number. Thus the packing fraction is equal to $(M - A)/A$ where M is the actual mass and A is the mass number. For example, one of the chlorine isotopes has a mass of 32.9860. The packing fraction for this isotope is

$$\frac{32.9860 - 33.0000}{33.0000} = -0.00042$$

Packing fractions are usually expressed as parts per 10,000 and so the packing fraction for this isotope of chlorine is written as −4.2. Since oxygen is taken as the standard, elements with positive packing fractions are less stable than oxygen, those with negative packing fractions are more stable.

It is positive for nuclides with mass number less than 16 or greater than 180, and negative for most others.

Paramagnetic materials.—Those within which an applied magnetic field is slightly increased by the alignment of electron orbits. The slight diamagnetic effect in materials having magnetic dipole moments is overshadowed by this paramagnetic alignment. As the temperature increases this paramagnetism disappears leaving only diamagnetism. The permeability of paramagnetic materials is slightly greater than that of empty space.

Pascal's law.—Pressure exerted at any point upon a confined liquid is transmitted undiminished in all directions.

Peltier effect.—When a current flows across the junction of two unlike metals it gives rise to an absorption or liberation of heat. If the current flows in the same direction as the current at the hot junction in a thermoelectric circuit of the two metals, heat is absorbed; if it flows in the same direction as the current at the cold junction of the thermoelectric circuit heat is liberated.

Pendulum.—For a simple pendulum of length l, for a small amplitude, the complete period,

$$T = 2\pi \sqrt{\frac{l}{g}} \quad \text{or} \quad g = 4\pi^2 \frac{l}{T^2}$$

T will be given in seconds if l is in cm and g in cm per sec². For a sphere suspended by a wire of negligible mass where d is the distance from the knife edge to the center of the sphere whose radius is r, the length of the equivalent simple pendulum,

$$l = d + \frac{2r^2}{5d}$$

If the period is P for an arc θ, the time of vibration in an infinitely small arc is approximately

$$T = \frac{P}{1 + \frac{1}{4}\sin^2 \frac{\theta}{4}}$$

For a compound pendulum, if a body of mass m be suspended from a point about which its moment of inertia is I with its center of gravity a distance h below the point of suspension, the period

$$T = 2\pi \sqrt{\frac{I}{mgh}}$$

Period in uniform circular motion is the time of one complete revolution. In any oscillatory motion it is the time of a complete oscillation. Dimension,—$[t]$.

Periodic law.—Elements when arranged in the order of their atomic weights or atomic numbers show regular variations in most of their physical and chemical properties.

Permeance, the reciprocal of reluctance. Unit permeance is the permeance of a cylinder one square centimeter cross-section and one centimeter length taken in a vacuum. Dimensions—$[\epsilon^{-1} l^{-1} t^2]$; $[\mu l]$.

Phase of oscillatory motion.—The fraction of a whole period which has elapsed since the moving particle last passed through its middle position in a positive direction.

Photographic density.—The density D of silver deposit on a photographic plate or film is defined by the relation

$$D = \log O$$

where O is the opacity. If I_o and I are the incident and transmitted intensities respectively the opacity is given by I_o/I. The transparency is the reciprocal of the opacity or I/I_o.

Photon.—A photon (or γ-ray) is a quantum of electromagnetic radiation which has zero rest mass and an energy of h

DEFINITIONS AND FORMULAS (Continued)

(Planck's constant) times the frequency of the radiation. Photons are generated in collisions between nuclei or electrons and in any other process in which an electrically charged particle changes its momentum. Conversely photons can be absorbed (i.e., annihilated) by any charged particle.

Piezo-electric effect.—The phenomenon exhibited by certain crystals of expansion along one axis and contraction along another when subjected to an electric field. Conversely compression of certain crystals generate an electrostatic voltage across the crystal. Piezoelectricity is only possible in crystal classes which do not possess a center of symmetry.

Pinch effect.—When an electric current, either direct or alternating, passes through a liquid conductor, that conductor tends to contract in cross-section, due to *electromagnetic* forces.

Pitch.—Psychological response of the ear, primarily dependent upon the frequency of vibration of the air. The intensity of the sound also has a certain effect on the pitch. Pitch of a screw is the axial distance between adjacent turns of a single thread on the screw.

Planck's constant (h).—A universal constant of nature which relates the energy of a quantum of radiation to the frequency of the oscillator which emitted it. It has the dimensions of action (energy × time). Expressed by $E = h\nu$ where E is the energy of the quantum and ν is its frequency. Its numerical value is $(6.62517 \pm 0.00023) \times 10^{-27}$ erg sec.

Plutonium.—A fissile element, artificially produced in the pile by neutron bombardment of U^{238}.

Poise.—A unit of coefficient of viscosity, defined as the tangential force per unit area (dynes/cm²) required to maintain unit difference in velocity (1 cm/sec) between two parallel planes separated by 1 cm of fluid;

$$1 \text{ poise} = 1 \text{ dyne sec/cm}^2 = 1 \text{ gm/cm sec.}$$

Poisson's ratio is the ratio of the transverse contraction per unit dimension of a bar of uniform cross-section to its elongation per unit length, when subjected to a tensile stress.

Polarized light.—Light which exhibits different properties in different directions at right angles to the line of propagation is said to be polarized. Specific rotation is the power of liquids to rotate the plane of polarization. It is stated in terms of specific rotation or the rotation in degrees per decimeter per unit density.

Polymorphism.—The ability to exist in two or more crystalline forms.

Positron.—A particle of the same mass M_e, as an ordinary electron. It has a positive electrical charge of exactly the same amount as that of an ordinary electron (which is sometimes called negatron). Positrons are created either by the radioactive decay of certain unstable nuclei or, together with a negatron, in a collision between an energetic (more than one Mev) photon and an electrically charged particle (or another photon). A positron does not decay spontaneously but on passing through matter it sooner or later collides with an ordinary electron and in this collision the positron-negatron pair is annihilated. The rest energy of the two particles, which is given by Einstein's relation $E = mc^2$ and amounts to 1.0216 mev altogether, is converted into electromagnetic radiation in the form of one or more photons.

Potential (electric) at any point is measured by the work necessary to bring unit positive charge from an infinite distance. Difference of potential between two points is measured by the work necessary to carry unit positive charge from one to the other. If the work involved is one erg we have the electrostatic unit of potential. Dimensions,—$[\epsilon^{-\frac{1}{2}} m^{\frac{1}{2}} l^{\frac{1}{2}} t^{-1}]$, $[\mu^{\frac{1}{2}} m^{\frac{1}{2}} l^{\frac{3}{2}} t^{-2}]$.

The potential at a point due to a charge q at a distance r in a medium whose dielectric constant is ϵ is,

$$V = \frac{q}{\epsilon r}$$

Power.—The time rate at which work is done. Units of power,—the watt, one joule (ten million ergs) per second; the kilowatt is equal to 1000 watts; the horse-power, 33,000 foot-pounds per minute, is equal to 746 watts. Dimensions,—$[m\, l^2\, t^{-3}]$.

If an amount of work W is done in time t the power or rate of doing work is

$$P = \frac{W}{t}$$

Power will be obtained in watts if W is expressed in joules (10^7 ergs) and t in sec.

Power in watts for alternating current,

$$P = EI \cos \phi$$

where E and I are the effective values of the electromotive force and current in volts and amperes respectively and ϕ the phase angle between the current and the impressed electromotive force.

The ratio,

$$\frac{P}{EI} = \cos \phi$$

is called the power factor.

Power developed by a direct current.—The power in watts developed by an electric current flowing in a conductor, where E is the difference of potential at its terminals in volts, R its resistance in ohms, and I the current in amperes,

$$P = EI = RI^2$$

The work done in joules in a time t sec is,

$$W = EIt{:}RI^2t.$$

Power ratios in telephone engineering are measured in **decibels.** The gain or loss of power expressed in decibels is ten times the logarithm of the power ratio. By reference to an arbitrarily chosen "power level" the actual power may be expressed in decibels. The numerical values thus used will not be proportional to the actual power level but roughly to the sensation on the ear produced when the electrical power is converted into sound. A difference of 1 decibel in the power supply to a telephone receiver produces approximately the smallest change in volume of sound which a normal ear can detect.

Pressure.—Force applied to, or distributed, over a surface; measured as force per unit area. Cgs unit,—the barye, one dyne per square centimeter. The megabarye is equal to 10^6 dynes per square centimeter. Pressure is also measured by the height of the column of mercury or water which it supports. Dimensions,—$[m\, l^{-1}\, t^{-2}]$.

The pressure due to a force F distributed over an area A,

$$P = \frac{F}{A}$$

Absolute pressure.—Pressure measured with respect to zero pressure. *Gauge pressure*—pressure measured with respect to that of the atmosphere.

Principal focus of a lens or spherical mirror is the point of convergence of light coming from a source at an infinite distance.

Projectiles.—For bodies projected with velocity v at an angle a above the horizontal, the time to highest point of flight.

$$t = \frac{v \sin a}{g}$$

Total time of flight to reach the original horizontal plane,

$$T = \frac{2v \sin a}{g}$$

Maximum height,

$$h = \frac{v^2 \sin^2 a}{2g}$$

Horizontal range,

$$R = \frac{v^2 \sin 2a}{g}$$

In the above equations the resistance of the air is neglected. g is the acceleration due to gravity.

Proton.—An elementary particle having a positive charge equivalent to the negative charge of the electron but possessing a mass approximately 1837 times as great. The proton is in effect the positive nucleus of the hydrogen atom.

Purkinje effect.—A phenomenon associated with the human eye, making it more sensitive to blue light when the illumination is poor (less than about 0.1 lumen per sq. ft.) and to yellow light when the illumination is good.

Quality or timbre of a sound depends on the coexistence with the fundamental of other vibrations of various frequencies and amplitudes.

Quantity of electricity or charge.—The electrostatic unit of charge, the quantity which when concentrated at a point and placed at unit distance from an equal and similarly concentrated quantity, is repelled with unit force. If the distance is one centimeter and force of repulsion one dyne and the surrounding medium a vacuum, we have the electrostatic unity of quantity. The electrostatic unit of quantity may be defined as that transferred by electrostatic unit current in unit time. The quantity transferred by one ampere in one second is the coulomb, the practical unit. The faraday is the electrical charge carried by one gram equivalent. The coulomb = 3×10^9 electrostatic units. Dimensions,—

$$[\epsilon^{\frac{1}{2}}\, m^{\frac{1}{2}}\, l^{\frac{3}{2}}\, t^{-1}];\ [\mu^{-\frac{1}{2}}\, m^{\frac{1}{2}}\, l^{\frac{1}{2}}].$$

Quantum.—Unit quantity of energy postulated in the quantum theory. The *photon* is a quantum of the electromagnetic field, and in nuclear field theories, the *meson* is considered to be the quantum of the nuclear field.

Radiation.—The emission and propagation of energy through space or through a material medium in the form of waves.

The term may be extended to include streams of sub-atomic particles as alpha-rays, or beta-rays, and cosmic rays as well as electromagnetic radiation. Often used to designate the energy alone without reference to its character. In the case of light this energy is transmitted in bundles (photons).

Radiation formula, Planck's.—The emissive power of a black body at wave length λ may be written

$$E_\lambda = \frac{c_1 \lambda^{-5}}{e^{c_2/\lambda T} - 1}$$

where c_1 and c_2 are constants with c_1 being 3.7403×10^{10} microwatts microns^4 per cm^2 or 3.7403×10^{-12} watt cm^2, c_2 being 14384 micron degrees and T the absolute temperature.

Radioactive nuclides.—Atoms that disintegrate by emission of corpuscular or electromagnetic radiations. The rays most commonly emitted are alpha or beta or gamma rays. The three classes are:

1. Primary, which have half-life times exceeding 10^8 years. These may be alpha-emitters or beta-emitters.
2. Secondary, which are formed in radioactive transformations starting with U^{238}, U^{235}, or Th^{232}.
3. Induced, having geologically short lifetimes and formed by induced nuclear reactions occurring in nature. All these reactions result in transmutation.

Radius of gyration may be defined as the distance from the axis of rotation at which the total mass of a body might be concentrated without changing its moment of inertia. The product of total mass and the square of the radius of gyration will give (the) moment of inertia.

Rankine scale of temperature.—The absolute Fahrenheit scale.

$$°F + 459.69 = °R; \text{ thus } 0°F \text{ Rankine} = -459.69°F.$$

Raoult's law.—Molar weights of non-volatile non-electrolytes when dissolved in a definite weight of a given solvent under the same conditions lower the solvent's freezing point, elevate its boiling point and reduce its vapor pressure equally for all such solutes.

Reduction is any process which increases the proportion of hydrogen or base-forming elements or radicals in a compound. Reduction is also the gaining of electrons by an atom, an ion, or an element thereby reducing the positive valence of that which gained the electron.

Reflection coefficient or reflectivity is the ratio of the light reflected from a surface to the total incident light. The coefficient may refer to diffuse or to specular reflection. In general it varies with the angle of incidence and with the wavelength of the light.

Reflection of light by a transparent medium in air (Fresnel's formulae).—If i is the angle of incidence, r the angle of refraction, n_1 the index of refraction for air (nearly equal to unity), n_2 index of refraction for a medium, then the ratio of the reflected light to the incident light is,

$$R = \tfrac{1}{2}\left(\frac{\sin^2 (i - r)}{\sin^2 (i + r)} + \frac{\tan^2 (i - r)}{\tan^2 (i + r)}\right)$$

If $i = 0$ (normal incidence), and $n_1 = 1$ (approximate for air),

$$R = \left(\frac{n_2 - 1}{n_2 + 1}\right)^2$$

Refraction.—See *Index of refraction; Snell's law.*

Refraction at a spherical surface.—If u be the distance of a point source, v the distance of the point image or the intersection of the refracted ray with the axis, n_1 and n_2 the indices of refraction of the first and second medium, and r the radius of curvature of the separating surface,

$$\frac{n_2}{v} + \frac{n_1}{u} = \frac{n_2 - n_1}{r}$$

If the first medium is air the equation becomes,

$$\frac{n}{v} + \frac{1}{u} = \frac{n - 1}{r}$$

Refractivity is given by $(n - 1)$ when n is the index of refraction; the **specific refractivity** is given by $\frac{n-1}{d}$ where d is the density. **Molecular refractivity** is the product of specific refractivity by the molecular weight.

Relative humidity.—The ratio of the quantity of water vapor present in the atmosphere to the quantity which would saturate at the existing temperature. It is also the ratio of the pressure of water vapor present to the pressure of saturated water vapor at the same temperature.

Reluctance is that property of a magnetic circuit which determines the total magnetic flux in the circuit when a given magnetomotive force is applied. Unit, the reluctance of one centimeter length and one square centimeter cross-section of space taken in a vacuum. Dimensions,—$[\epsilon l\, t^{-2}]$; $[\mu^{-1} l^{-1}]$.

Reluctivity or specific reluctance is the reciprocal of magnetic permeability. The reluctivity of empty space is taken as unity. Dimensions,—$[\epsilon l^2 t^{-2}]$; $[\mu^{-1}]$.

Resistance is a property of conductors depending on their dimensions, material and temperature which determines the current produced by a given difference of potential. The practical unit of resistance, the **ohm** is that resistance through which a difference of potential of one volt will produce a current of one ampere. The **international ohm** is the resistance offered to an unvarying current by a column of mercury at 0°C, 14.4521 grams in mass, of constant cross-sectional area and 106.300 centimeters in length, sometimes called the legal ohm. Dimensions,—$[\epsilon^{-1} l^{-1} t]$; $[\mu l\, t^{-1}]$.

Resistance of a conductor at 0°C, of length l, cross-section s and specific resistance ρ

$$R_o = \rho \frac{l}{s}$$

The resistivity may be expressed as ohm-cm when R is in ohms, l in cm and s in cm^2.

Resistance of a conductor at a temperature t whose resistance at 0°C is R_o and whose temperature resistance coefficient is α

$$R_t = R_o(1 + \alpha t)$$

Resistance of conductors in series and parallel.—The total resistance of any number of resistances joined in series is the sum of the separate resistances. The total resistance of conductors in parallel whose separate resistances are r_1, r_2, r_3 . . . r_n is given by the formula

$$\frac{1}{R} = \frac{1}{r_1} + \frac{1}{r_2} + \frac{1}{r_3} \cdots + \frac{1}{r_n}$$

Where R is the total resistance. For two terms this becomes,

$$R = \frac{r_1 r_2}{r_1 + r_2}$$

DEFINITIONS AND FORMULAS (Continued)

Resistance, specific (Resistivity).—A proportionality factor characteristic of different substances equal to the **resistance** that a centimeter cube of the substance offers to the passage of electricity, the current being perpendicular to two parallel faces. It is defined by the expression:

$$R = \rho \frac{l}{A}$$

where R is the resistance of a uniform conductor, l is its length, A is its cross sectional area, and ρ is its resistivity. Resistivity is usually expressed in ohm-centimeters.

Resolving power of a telescope or microscope is indicated by the minimum separation of two objects for which they appear distinct and separate when viewed through the instrument.

Resonance (chemical).—The moving of electrons from one atom of a molecule or ion to another atom of that molecule or ion. It is simply the oriented movement of the bonds between atoms.

Restitution, coefficient of, for two bodies on impact.—The ratio of the difference in velocity, after impact to the difference before impact.

Reversible reaction.—One which can be caused to proceed in either direction by suitable variation in the conditions of temperature, volume, pressure or of the quantities of reacting substances.

Roentgen.—A radiation unit, that amount of radiation that will produce one electrostatic unit of ions per cubic centimeter volume.

Rotatory power is the power of rotating the plane of polarized light, given in general by θ/l where θ is the total rotation which occurs in a distance l.

The **molecular** or **atomic rotatory power** is the product of the specific rotatory power by the molecular or atomic weight. Magnetic rotatory power is given by

$$\theta/e\ H \cos \alpha$$

where H the intensity of the magnetic field, and α the angle between the field and the direction of the light.

Rydberg formula.—A formula, similar to that of Balmer, for expressing the wave-numbers (ν) of the lines in a spectral series:

$$\nu = R\left[\frac{1}{(n+a)^2} - \frac{1}{(m+b)^2}\right]$$

where n and m are integers and $m > n$, a and b are constants for a particular series, and R is the *Rydberg constant,* 109677.8 cm^{-1} for hydrogen.

Salt.—Any substance which yields ions, other than hydrogen or hydroxyl ions. A salt is obtained by displacing the hydrogen of an acid by a metal.

Scintillometer.—An instrument which detects radiation by emitting flashes of light.

Seebeck effect.—If a circuit consists of two metals, one junction hotter than the other, a current flows in the circuit. The direction of the flow depends on the metals and the temperature of the junctions.

Sensitiveness of a balance.—Assuming the three knife edges of a balance to lie on a straight line,—if M is the weight of the beam, h the distance of the center of gravity below the knife edge, L the length of the balance arms and m a small mass added to one pan, the deflection θ produced is given by

$$\tan \theta = \frac{mL}{Mh}.$$

Shell.—According to *Pauli's exclusion principle* (q.v.), the extranuclear electrons do not circle around the nucleus all in orbits of the same radius, but are arranged in orbits at various distances from the nucleus. The extranuclear orbital electrons are thus assumed to be arranged in a series of concentric spheres, called *shells,* which are designated, in the order of increasing distance from the nucleus, as K, L, M, N, O, P, and Q shells. The number of the electrons which each of these shells can contain is limited. All electrons arranged in the same shell have the same principal quantum number. The electrons in the same shell are grouped into various subshells (q.v.), and all the electrons in the same subshell have the same orbital angular momentum.—See *Subshell*

Simple harmonic motion.—Periodic oscillatory motion in a straight line in which the restoring force is proportional to the displacement. If a point moves uniformly in a circle, the motion of its projection on the diameter (or any straight line in the same plane) is simple harmonic motion.

If r is the radius of the reference circle, ω the angular velocity of the point in the circle, θ the angular displacement at the time t after the particle passes the mid-point of its path, the linear displacement,

$$x = r \sin \theta = r \sin \omega t.$$

The velocity at the same instant,

$$v = r\omega \cos \theta = \omega \sqrt{r^2 - x^2}$$

The acceleration,

$$a = -\omega^2 x.$$

The force for a mass m,

$$F = m\omega^2 x = -\frac{4\pi^2 mx}{T^2}$$

The period,

$$T = 2\pi \sqrt{\frac{x}{a}}$$

DEFINITIONS AND FORMULAS (Continued)

In the above equations the cgs system calls for x and r in cm, v in cm per sec, a in cm per sec^2, T in sec, m in grams, F in dynes, θ in radians and ω in radians per sec.

Simple machine.—A contrivance for the transfer of energy and for increased convenience in the performance of work.

Mechanical advantage is the ratio of the resistance overcome to the force applied. Velocity ratio is the ratio of the distance through which force is applied to the distance through which resistance is overcome.

Efficiency is the ratio of the work done by a machine to the work done upon it.

If a force f applied to a machine through a distance S results in a force F exerted by the machine through a distance s, neglecting friction,

$$fS = Fs.$$

The theoretical mechanical advantage or velocity ratio in the above case is, $\frac{S}{s}$.

Actually the force obtained from the machine will have a smaller value than will satisfy the equation above. If F' be the actual force obtained, the practical mechanical advantage will be, $\frac{F'}{f}$.

The efficiency of the machine,

$$E = \frac{F's}{fS}.$$

Snell's law of refraction.—If i is the angle of incidence, r the angle of refraction, v the velocity of light in the first medium, v' the velocity in the second medium, the index of refraction n,

$$n = \frac{\sin i}{\sin r} = \frac{v}{v'}$$

Solid.—A state of matter in which the relative motion of the molecules is restricted and they tend to retain a definite fixed position relative to each other, giving rise to crystal structure. A solid may be said to have a definite shape and volume.

Solid angle.—Measured by the ratio of the surface of the portion of a sphere enclosed by the conical surface forming the angle, to the square of the radius of the sphere. Unit of solid angle,—the steradian, the solid angle which encloses a surface on the sphere equivalent to the square of the radius. Dimensions,—unity.

Solubility of one liquid or solid in another is the mass of a substance contained in a solution which is in equilibrium with an excess of the substance. Under these conditions the solution is said to be saturated. Solubility of a gas is the ratio of concentration of gas in the solution to the concentration of gas above the solution.

Solubility product or precipitation value is the product of the concentrations of the ions of a substance in a saturated solution of the substance. These concentrations are frequently expressed as moles of solute per liter of solution.

Solute.—That constituent of a solution which is considered to be dissolved in the other, the solvent. The solvent is usually present in larger amount than the solute.

A **solution** is **saturated** if it contains at given temperature as much of a solute as it can retain in the presence of an excess of that solute.

A **true solution** is a mixture, liquid, solid or gaseous, in which the components are uniformly distributed throughout the mixture. The proportion of the constituents may be varied within certain limits.

Solvent.—That constituent of a solution which is present in larger amount; or, the constituent which is liquid in the pure state, in the case of solutions of solids or gases in liquids.

Specific gravity.—The ratio of the mass of a body to the mass of an equal volume of water at 4°C or other specified temperature. Dimensions,—unity.

Specific heat of a substance is the ratio of its thermal capacity to that of water at 15°C. Dimensions,—unity.

If a quantity of heat H calories is necessary to raise the temperature of m grams of a substance from t_1 to t_2°C, the specific heat, or more properly, thermal capacity of the substance,

$$s = \frac{H}{m(t_2 - t_1)}.$$

Specific heat by the method of mixtures.—Where a mass m_1 of the substance is heated to a temperature t_1, then placed in a mass of water m_2 at a temperature t_2 contained in a calorimeter with stirrer (of same material) of mass m_3, specific heat of the calorimeter c, t_3 the final temperature

$$m_1s(t_1 - t_3) = m_3c + m_2(t_3 - t_2).$$

Black's ice calorimeter.—If a body of mass m and temperature t melts a mass m' of ice, its temperature being reduced to 0°C, the specific heat of the substance is,

$$s = \frac{80\ 1m'}{mt}$$

Bunsen's ice calorimeter.—A body of mass m at temperature t causes a motion of the mercury column of l centimeters in a tube whose volume per unit length is v. The specific heat is

$$s = \frac{884lv}{mt}$$

Specific inductive capacity.—The ratio of the capacitance of a condenser with a given substance as dielectric to the

DEFINITIONS AND FORMULAS (Continued)

capacitance of the same condenser with air or a vacuum as dielectric is called the specific inductive capacity. The ratio of the dielectric constant of a substance to that of a vacuum.

Specific rotation.—If there are n grams of active substance in v cubic centimeters of solution and the light passes through l centimeters, r being the observed rotation in degrees, the specific rotation (for 1 centimeter),

$$[\alpha] = \frac{rv}{nl}$$

Specific volume is the reciprocal of density. Dimensions,—$[m^{-1} l^3]$.

Spectral series are spectral lines or groups of lines which occur in an orderly sequence.

Speed.—Time rate of motion measured by the distance moved over in unit time. Cgs unit,—one centimeter per second. Dimension,—$[l\, t^{-1}]$.

Spherical aberration.—When large surfaces of spherical mirrors or lenses are used the light divergent from a point source is not exactly focused at a point. The phenomenon is known as spherical aberration. For axial pencils the error is known as axial spherical aberration; for oblique pencils, coma.

Spherical mirrors.—If R is the radius of curvature, F principal focus, and f_1 and f_2 any two conjugate focal distances,

$$\frac{1}{f_1} + \frac{1}{f_2} = \frac{1}{F} = \frac{2}{R}$$

If the linear dimensions of the object and image be O and I respectively and u and v their distances from the mirror,

$$\frac{O}{I} = \frac{u}{v}$$

Spin.—In nuclear physics, used to describe the angular momentum of elementary particles or of nuclei.

Standard conditions for gases.—Measured volumes of gases are quite generally recalculated to 0 degrees C temperature and 760 mm pressure, which have been arbitrarily chosen as standard conditions.

Stark effect.—The splitting of a single spectrum line into multiple lines which occurs when the emitting material is placed in a strong electric field. The observed effect depends on the angle between the direction of the field and the direction of observation. The effect is due to the shifting of the energy states of certain orbits which all have the same energy in zero field.

Statcoulomb.—The unit of electric charge in the metric system. 3×10^9 statcoulombs = 1 coulomb.

Stationary or standing waves are produced in a medium by the simultaneous transmission, in opposite directions of two similar wave motions. Fixed points of minimum amplitude are called **nodes.** A **segment** extends from one node to the next. An **antinode** or **loop** is the point of maximum amplitude between two nodes.

Stefan-Boltzmann law of radiation.—The energy radiated in unit time by a black body is given by, $E = K(T^4 - T_o^4)$, where T is the absolute temperature of the body, T_o the absolute temperature of the surroundings, and K a constant.

Stoichiometric.—Pertaining to weight relations in chemical reactions.

Stoke.—See under viscosity.

Stokes' law.—1. Gives the rate of fall of a small sphere in a viscous fluid. When a small sphere falls under the action of gravity through a viscous medium it ultimately acquires a constant velocity,

$$V = \frac{2ga^2(d_1 - d_2)}{9\eta}$$

where a is the radius of the sphere, d_1 and d_2 the densities of the sphere and the medium respectively, and η the coefficient of viscosity. V will be in cm per sec if g is in cm per sec², a in cm, d_1 and d_2 in g per cm³ and η in dyne-sec per cm² or poises.

2. The empirical law stating that the wavelength of light emitted by a fluorescent material is longer than that of the radiation used to excite the fluorescence. In modern language the emitted photons carry off less energy than is brought in by the exciting photons; the details accord with the energy conservation principle.

Strain.—The deformation resulting from a stress measured by the ratio of the change to the total value of the dimension in which the change occurred. Dimensions,—unity.

Stress.—The force producing or tending to produce deformation in a body measured by the force applied per unit area Cgs units,—one dyne per square centimeter. Dimensions,—$[m\, l^{-1} t^{-2}]$.

Subshell.—The electrons within the same shell (energy level) of the atom are characterized by the same principal quantum number (n), and are further divided into groups according to the value of their azimuthal quantum numbers (l); the electrons which possess the same azimuthal quantum number for the same principal quantum number are considered to occupy the same subshell (or sublevel). The individual subshells are designated with the letters s, p, d, f, g, and h, as follows:

l value	designation of subshell
0	s
1	p
2	d
3	f
4	g
5	h

An electron assigned to the *s*-subshell is called an *s*-electron, one assigned to the *p*-subshell is referred to as a *p*-electron, etc. In formulae of electron structure, the value of the principal quantum number (n) is prefixed to the letter indicating the azimuthal quantum number (l) of the electron; thus, e.g., a $4f$-electron is an electron which has the principal quantum number 4 (i.e. assigned to the *N*-shell) and the orbital angular momentum 3 (*f*-subshell).

Surface density of electricity.—Quantity of electricity per unit area. Dimensions,—$[\epsilon^{\frac{1}{2}} m^{\frac{1}{2}} l^{-\frac{1}{2}} t^{-1}]$; $[\mu^{-\frac{1}{2}} m^{\frac{1}{2}} l^{-\frac{3}{2}}]$.

Surface density of magnetism.—Quantity of magnetism per unit area. Dimensions,—$[\epsilon^{-1} m^{\frac{1}{2}} l^{-\frac{3}{2}}]$, $[\mu^{\frac{1}{2}} m^{\frac{1}{2}} l^{-\frac{1}{2}} t^{-1}]$.

Surface tension.—Two fluids in contact exhibit phenomena, due to molecular attractions which appear to arise from a tension in the surface of separation. It may be expressed as dynes per cm or as ergs per square centimeter. Dimensions, —$[m\, t^{-2}]$.

The total force along a line of length l on the surface of a liquid whose surface tension is T,

$$F = lT.$$

Capillary tubes.—If a liquid of density d rises a height h in a tube of internal radius r the surface tension is,

$$T = \frac{rhdg}{2}$$

The tension will be in dynes per cm if r and h are in cm, d in g per cm³ and g in cm per sec².

Drops and bubbles.—Pressure in dynes per cm² due to surface tension on a drop of radius r cm for a liquid whose surface tension is T dynes per cm,

$$P = \frac{2T}{r}$$

For a bubble of mean radius r cm,

$$P = \frac{4T}{r}.$$

Susceptibility (magnetic) is measured by the ratio of the intensity of magnetization produced in a substance to the magnetizing force or intensity of field to which it is subjected. The susceptibility of a substance will be unity when unit intensity of magnetization is produced by a field of one gauss. Dimensions,—$[\epsilon^{-1} l^{-2} t^{2}]$, $[\mu]$.

Tangent galvanometer.—A tangent galvanometer with n turns, of radius r, in the earth's field H, has a deflection θ. The current flowing is,

$$i = \frac{Hr}{2\pi n} \tan \theta$$

If $\frac{2\pi n}{r} = G$ (the galvanometer constant).

$$i = \frac{H}{G} \tan \theta$$

Temperature may be defined as the condition of a body which determines the transfer of heat to or from other bodies. Particularly it is a manifestation of the average translational kinetic energy of the molecules of a substance due to heat agitation. The customary unit of temperature is the **Centigrade** degree, $\frac{1}{100}$ the difference between the temperature of melting ice and that of water boiling under standard atmospheric pressure. **Celsius scale.**—The Celsius temperature scale is a designation of the scale also known as the centigrade scale. The degree **Fahrenheit** is $\frac{1}{180}$, and the degree **Reaumur** $\frac{1}{80}$ the same difference of temperature.

The fundamental temperature scale is the absolute, thermodynamic or **Kelvin scale** in which the temperature measure is based on the average kinetic energy per molecule of a perfect gas. The zero of the Kelvin scale is −273.15°C. The temperature scale adopted by the International Bureau of Weights and Measures is that of the constant volume hydrogen gas thermometer. The magnitude of the degree in both these scales is defined as $\frac{1}{100}$ the difference between the temperature of melting ice and that of boiling water at 760 mm pressure. Frequently the Kelvin scale is defined as degrees C + 273.15 and the Rankine scale as degrees F + 459.67°F.

The fundamental temperature scale is now defined by means of the equation

$$\theta(\mathrm{X}) = 273.15^\circ\mathrm{K}\, \frac{\mathrm{X}}{\mathrm{X}_3}$$

where θ denotes the temperature; X the thermometric property (P, V, . . .); the subscript 3 refers to the triple point of water; and 273.16°K is the arbitrary fixed point for the temperature associated with the triple point of water.

The ideal gas temperature θ, (numerically equal to the Kelvin temperature), in particular, is defined by either of the two equations:

$$\theta = \begin{cases} 273.15^\circ \lim\limits_{\mathrm{P}_3 \to 0} \frac{\mathrm{P}}{\mathrm{P}_3}, \text{ const. V} \\ 273.15^\circ \lim\limits_{\mathrm{P} \to 0} \frac{\mathrm{V}}{\mathrm{V}_3}, \text{ const. P} \end{cases}$$

Temperature resistance coefficient.—The ratio of the change of resistance in a wire due to a change of temperature of 1°C to its resistance at 0°C. Dimension,—$[\theta^{-1}]$.

Thermal capacity of a substance is the quantity of heat necessary to produce unit change of temperature in unit mass.

DEFINITIONS AND FORMULAS (Continued)

It is ordinarily expressed as calories per gram per degree Centigrade. Numerically equivalent to specific heat.

Thermal capacity or water equivalent.—The total quantity of heat necessary to raise any body or system unit temperature, measured as calories per degree centigrade in the cgs system. Dimension,—$[m]$.

Thermal expansion.—The coefficient of linear expansion or expansivity is the ratio of the change in length per degree C to the length at 0°C. The coefficient of volume expansion (for solids) is approximately three times the linear coefficient. The coefficient of volume expansion for liquids is the ratio of the change in volume per degree to the volume at 0°C. The value of the coefficient varies with temperature. The coefficient of volume expansion for a gas under constant pressure is nearly the same for all gases and temperatures and is equal to 0.00367 for 1°C. Dimension,—$[\theta^{-1}]$.

If l_o is the length at 0°C, α the coefficient of linear expansion, the length at t°C is,

$$l_t = l_o(1 + \alpha t).$$

General formula for thermal expansion.—The rate of thermal expansion varies with the temperature. The general equation giving the magnitude m_t (length or volume) at a temperature t, where m_o is the magnitude at 0°C, is

$$m_t = m_o(1 + \alpha t + \beta t^2 + \gamma t^3 \cdot \cdot \cdot)$$

where α, β, γ, etc. are empirically determined coefficients.

Volume expansion.—If V represents volume and β the coefficient of expansion,

$$V_t = V_o(1 + \beta t).$$

For solids,

$$\beta = 3\alpha \text{ (approximately)}.$$

Thermal neutrons.—Neutrons slowed down by a moderator to an energy of a fraction of an electron volt—about 0.025 ev. at 15°.

Thermionic emission.—Electron or ion emission due to the temperature of the emitter. The rate of emission increases rapidly with the increase of temperature. It is also very sensitive to the state of the surface.

Thermodynamics, law of.

I. When mechanical work is transformed into heat or heat into work, the amount of work is always equivalent to the quantity of heat.

II. It is impossible by any continuous self-sustaining process for heat to be transferred from a colder to a hotter body.

Thermoelectric power is measured by the electromotive force produced by a thermocouple for unit difference of temperature between the two junctions. It varies with the average temperature and is usually expressed in microvolts per degree C. It is customary to list the thermoelectric power of the various metals with respect to lead. Dimensions,—$[\epsilon^{-\frac{1}{2}} m^{\frac{1}{2}} l^{\frac{1}{2}} t^{-1} \theta^{-1}]$; $[\mu^{\frac{1}{2}} m^{\frac{1}{2}} t^{-1} \theta^{-1}]$.

Thomson thermoelectric effect is the designation of the potential gradient along a conductor which accompanies a temperature gradient. The magnitude and direction of the potential varies with the substance.

The coefficient of the Thomson effect or specific heat of electricity is expressed in joules per coulomb per degree Centigrade. Dimensions,—$[\epsilon^{-\frac{1}{2}} m^{\frac{1}{2}} l^{\frac{1}{2}} t^{-1} \theta^{-1}]$, $[\mu^{\frac{1}{2}} m^{\frac{1}{2}} l^{\frac{3}{2}} t^{-2} \theta^{-1}]$.

Time, unit of.—The fundamental invariable unit of time is the ephemeris second, which is defined as 1/31,556,925.9747 of the tropical year for 1900 January 0^d12^h ephemeris time. The ephemeris day is 86,400 ephemeris seconds.

The former unit of time was the mean solar second, defined as 1/86,400 of the mean solar day. (See supplementary list).

Torque produced by the action of one magnet on another.—The turning moment experienced by a magnet of pole strength m' and length $2l'$ placed at a distance r from another magnet of length $2l$ and pole strength m, where the center of the first magnet is on the axis (extended) of the second and the axis of the first is perpendicular to the axis of the second,

$$C = 8\frac{mm'll'}{r^3} = \frac{2MM'}{r^3}$$

If the first magnet is deflected through an angle θ, the expression becomes,

$$C = \frac{2MM'}{r^3}\cos\theta$$

Torsional vibration.—See *Angular harmonic motion*

Total reflection.—When light passes from any medium to one in which the velocity is greater, refraction ceases and total reflection begins at a certain critical angle of incidence θ such that

$$\sin\theta = \frac{1}{n}$$

where n is the index of the first medium with respect to the second. If the second medium is air n has the ordinary value for the first medium. For any other second medium,

$$n = \frac{n_1}{n_2}$$

where n_1 and n_2 are the ordinary indices of refraction for the first and second medium respectively.

Tractive force of a magnet.—If a magnet with induction B has a pole face of area A the force, is,

$$F = \frac{B^2A}{8\pi}$$

If B and A are in cgs units, F will be in dynes.

DEFINITIONS AND FORMULAS (Continued)

Transmutation.—A nuclear change producing a new element from an old one.

Transuranic elements.—Elements of atomic numbers above 92. All of them are radioactive and are products of artificial nuclear changes. All are members of the actinide group.

Triangle or polygon of forces.—If three or more forces acting on the same point are in equilibrium, the vectors representing them form, when added, a closed figure.

Tritium.—An isotope of hydrogen with a mass of three, structure, two neutrons, and one proton in its nucleus.

Uncertainty principle.—See *Indeterminancy*

Uniform circular motion.—If r is the radius of a circle, v the linear speed in the arc, ω the angular velocity and T the period or time of one revolution,

$$\omega = \frac{v}{r} = \frac{2\pi}{T}$$

The acceleration toward the center is

$$a = \frac{v^2}{r} = \omega^2 r = \frac{4\pi^2 r}{T^2}$$

The centrifugal force for a mass m,

$$F = \frac{mv^2}{r} = m\omega^2 r = \frac{4\pi^2 mr}{T^2}$$

In the above equations ω will be in radians per second and a in cm per sec² if r is in cm, v in cm per sec and T in sec. F will be in dynes if mass is in grams and other units as above.

Application to the solar system.—If M is the mass of the sun, G the constant of gravitation, P the period of the planet and r the distance of the planet from the sun, then the mass of the sun

$$M = \frac{4\pi^2 r^3}{GP^2}, \; (G = 6.670 \times 10^{-8} \text{ for cgs units}).$$

If P is the period and r the distance of a satellite revolving around the planet, the above expression for M gives the mass of the planet. The formula is written on the assumption that the orbit of the planet or satellite is circular, which is only approximately true.

Uniformly accelerated rectilinear motion.—If v_o is the initial velocity, v_t the velocity after time t, the acceleration,

$$a = \frac{v_t - v_o}{t}$$

The velocity after time t,

$$v_t = v_o + at$$

Space passed over in time t,

$$s = v_o t + \tfrac{1}{2}at^2$$

Velocity after passing over space s,

$$v = \sqrt{v_o^2 + 2as}$$

Space passed over in the nth second

$$s = v_o + \tfrac{1}{2}a(2n - 1).$$

In the above and following similar equations the values of the space, velocity, and acceleration must be substituted in the same system. For space in *cm*, velocity will be in *cm* per sec and acceleration in *cm* per sec per sec.

Unit.—Specific magnitude of a quantity, set apart by appropriate definition, which is to serve as a basis of comparison or measurement for other quantities of the same nature.

Valence of an atom of an element is that property which is measured by the number of atoms of hydrogen (or its equivalent) one atom of that element can hold in combination if negative, or can displace in a reaction if it is positive.

Valence electrons of the atom are electrons which are gained, lost or shared in chemical reactions.

Van der Waals' equation of state.—This equation is expressed by:

$$\left(p + \frac{a}{v^2}\right)(v - b) = RT$$

It makes allowance both for the volume occupied by the molecules and for the attractive force between the molecules. b is the effective volume of molecules in one mole of gas. a is a measure of the attractive force between the molecules. For values of R, a, and b see index for table of Van der Waal's constants for gasses.

Van't Hoff's principle.—If the temperature of interacting substances in equilibrium is raised, the equilibrium concentrations of the reaction are changed so that the products of that reaction which absorb heat are increased in quantity, or if the temperature for such an equilibrium is lowered, the products which evolve heat in their formation are increased in amounts.

Vapor.—The words **vapor** and **gas** are often used interchangeably. **Vapor** is more frequently used for a substance which, though present in the gaseous phase, generally exists as a liquid or solid at room temperature. **Gas** is more frequently used for a substance that generally exists in the gaseous phase at room temperature. Thus one would speak of iodine or carbon tetrachloride vapors and of oxygen gas.

Vapor pressure.—The pressure exerted when a solid or liquid is in equilibrium with its own vapor. The vapor pressure is a function of the substance and of the temperature.

DEFINITIONS AND FORMULAS (Continued)

Vectors, composition of.—If the angle between two vectors is A, and their magnitude a and b, their resultant,

$$C = \sqrt{a^2 + b^2 - 2ab \cos A}.$$

Velocity.—Time rate of motion in a fixed direction. Cgs units,—one centimeter per second. Dimensions,—$[l\,t^{-1}]$. If s is space passed over in time t, the velocity,

$$v = \frac{s}{t}$$

Velocity of a compressional wave.—The velocity of a compressional wave in an elastic medium, in terms of elasticity E (bulk modulus) and density d,

$$V = \sqrt{\frac{E}{d}}$$

For the velocity of sound in air, where p is the pressure and d the density,

$$V = \sqrt{\frac{1.4p}{d}}$$

Velocity of efflux of a liquid.—If h is the distance from the opening to the free surface of the liquid, the velocity of efflux is

$$V = \sqrt{2gh}$$

The above is the theoretical discharge velocity disregarding friction and the shape of orifice. For water issuing through a circular opening with sharp edges of area, A, the volume discharged per second is given approximately by,

$$Q = 0.62A\,\sqrt{2gh}$$

Velocity of sound, variation with temperature.—The velocity in meters per sec at any temperature t in °C is given approximately by

$$V = V_o\sqrt{1 + \frac{t}{273}}$$

$$V = 331.5 + .607t$$

The **variation with humidity** is given by the equation

$$V_d = V_h\sqrt{1 - \frac{e}{p}\left(\frac{\gamma_\omega}{\gamma_a} - \frac{5}{8}\right)}$$

where V_d is the velocity in dry air, V_h that in air at barometric pressure p in which the pressure of water vapor is e. γ_ω and γ_a are the specific heat ratios for water vapor and for air respectively.

Velocity of a transverse wave in a stretched cord. If T is the tension of the cord and m the mass per unit length,

$$V = \sqrt{\frac{T}{m}}$$

Velocity of water waves.—If the depth h is small compared with the wave length, the velocity,

$$V = \sqrt{gh}$$

In deep water for a wave length λ,

$$V = \sqrt{\frac{g\lambda}{2\pi}}$$

If the wavelength is very small, less than about 1.6 cm, the velocity increases as the wave length decreases and is expressed by the following,

$$V = \sqrt{\frac{2\pi T}{\lambda d} + \frac{g\lambda}{2\pi}}$$

where T is the surface tension and d the density of the liquid V will be given in cm per sec if h and λ are in cm, g in cm per sec^2, T in dynes per cm and d in g per cm^3.

Velocity of a wave.—The velocity of propagation in terms of wavelength λ and the period T or frequency n is,

$$V = \frac{\lambda}{T} = n\lambda$$

Viscosity.—All fluids possess a definite resistance to change of form and many solids show a gradual yielding to forces tending to change their form. This property, a sort of internal friction, is called viscosity; it is expressed in dyne-seconds per cm^2 or poises. Dimensions,—$[m\,l^{-1}\,t^{-1}]$. If the tangential force per unit area, exerted by a layer of fluid upon one adjacent is one dyne for a space rate of variation of the tangential velocity of unity, the viscosity is one poise.

Kinematic viscosity is the ratio of viscosity to density. The c. g. s. unit of kinematic viscosity is the **stoke.**

Flow of liquids through a tube; where l is the length of the tube, r its radius, p the difference of pressure at the ends, η the coefficient of viscosity, the volume escaping per second,

$$v = \frac{\pi p r^4}{8l\eta} \text{ (Poiseuille).}$$

The volume will be given in cm^3 per second if l and r are in cm, p in dynes per cm^2 and η in poises or dyne-seconds per cm^2.

Visibility is measured by the ratio of the luminous flux in lumens to the total radiant energy in ergs per second or in watts.

DEFINITIONS AND FORMULAS (Continued)

Volt.—The unit of electromotive force. It is the difference in potential required to make a current of one ampere flow through a resistance of one ohm.

Volume, unit of.—The cubic centimeter, the volume of a cube whose edges are one centimeter in length. Other units of volume are derived in a similar manner. Dimension,—$[l^3]$.

Wave motion.—A progressive disturbance propagated in a medium by the periodic vibration of the particles of the medium. Transverse wave motion is that in which the vibration of the particles is perpendicular to the direction of propagation. Longitudinal wave motion is that in which the vibration of the particles is parallel to the direction of propagation.

Weight.—The force with which a body is attracted toward the earth. Cgs unit,—the dyne. Dimensions,—$[m\,l\,t^{-2}]$.

Although the weight of a body varies with its location, the weights of various standards of mass are often used as units of force as,—pound weight, or pound force, gram weight, etc. The weight of mass m, where g is the acceleration due to gravity,

$$W = mg.$$

The weight will be given in dynes when m is in grams and g in cm per sec^2.

Wheatstone's bridge.—If the resistances r_1, r_2, r_3, and r_4 form the arms of a Wheatstone's bridge in order as the circuit (omitting cell and galvanometer connections) is traced, when the bridge is balanced,

$$\frac{r_1}{r_2} = \frac{r_4}{r_3} \quad \text{or} \quad \frac{r_1}{r_4} = \frac{r_2}{r_3}$$

Wien's displacement law.—When the temperature of a radiating black body increases, the wave length corresponding to maximum energy decreases in such a way that the product of the absolute temperature and wave length is constant.

$$\lambda_{max}T = w$$

w is known as **Wien's displacement constant.**

Work.—When a force acts against resistance to produce motion in a body the force is said to do work. Work is measured by the product of the force acting and the distance moved through against the resistance. Cgs units of work,—the erg, a force of one dyne acting through a distance of one centimeter. The joule is 1×10^7 ergs. Dimensions,—$[m\,l^2\,t^{-2}]$. The foot-pound is the work required to raise a mass of one pound a vertical distance of one foot where $g = 32.174$ ft./sec^2. The foot-poundal is the work done by a force of one poundal acting through a distance of one foot. The International joule, a unit of electrical energy, is the work expended per second by a current of one International ampere flowing through one International ohm. The kilowatt-hour is the total amount of energy developed in one hour by a power of one kilowatt.

If a force F act through a space s, the work done is

$$W = Fs$$

Work will be given in ergs if F is in dynes and s in cm.

Work done in rotation. If a torque L dyne-cm acts through an angle θ radians, the work done in ergs is

$$W = L\theta$$

X-rays.—A type of radiation of higher frequency than visible light but lower than gamma rays. Usually produced by high energy electrons impinging upon a metal target.

X units.—X-ray wavelengths have been measured in two kinds of units. The older measurements are given in X units (XU) which are based on the effective lattice constant of rock salt being 2,814.00 XU. More recently X-ray wavelengths have been directly connected, through measurements with ruled gratings, to the wavelengths in the optical region and through them to the standard meter. It turned out that the XU which was originally intended as 10^{-11} cm was 0.202 per cent larger than this value. It has become customary to give X-ray wavelengths in Angstrom units (Å) when the absolute scale is used (1 Å = 10^{-8} cm). The two are related by

$$1{,}000 \text{ XU} = 1.00202 \pm 0.00003) \text{ Å}$$

and wavelengths given in XU must be multiplied by 1.00202 and then divided by 1,000 in order to convert them into Angstrom units.

Zeeman effect.—The splitting of a spectrum line into several symmetrically disposed components, which occurs when the source of light is placed in a strong magnetic field. The components are polarized, the directions of polarization and the appearance of the effect depending on the direction from which the source is viewed relative to the lines of force.

Supplementary Tables

Units for a System of Measures for International Relations

Length	meter	m
Mass	kilogram	kg
Time	second	s
Electric current	ampere	A
Temperature	Kelvin	K
Luminous intensity	candela	cd

Prefix Names of Multiples and Submultiples of Units

Factor by which unit is multiplied	Prefix	Symbol	Factor by which unit is multiplied	Prefix	Symbol
10^{12}	tera	T	10^{-2}	centi	c
10^{9}	giga	G	10^{-3}	milli	m
10^{6}	mega	M	10^{-6}	micro	μ
10^{3}	kilo	k	10^{-9}	nano	n
10^{2}	hecto	h	10^{-12}	pico	p
10	deka	da	10^{-15}	femto	f
10^{-1}	deci	d	10^{-18}	atto	a

Derived Units

Acceleration	meter per second squared	m/s^2	
Activity (of radioactive source)	1 per second	s^{-1}	
Angular acceleration	radian per second squared	rad/s^{-1}	
Angular velocity	radian per second	rad/s	
Area	square meter	m^2	
Density	kilogram per cubic meter	kg/m^3	
Dynamic viscosity	newton-second per sq meter	$N{\cdot}s/m^2$	
Electric capacitance	farad	F	(A·s/V)
Electric charge	coulomb	C	(A·s)
Electric field strength	volt per meter	V/m	
Electric resistance	ohm		(V/A)
Entropy	joule per kelvin	J/K	
Force	newton	N	$(kg{\cdot}m/s^2)$
Frequency	hertz	Hz	(s^{-1})
Illumination	lux	lx	(lm/m^2)
Inductance	henry	H	(V·s/A)
Kinematic viscosity	sq meter per second	m^2/s	
Luminance	candela per sq meter	cd/m^2	
Luminous flux	lumen	lm	(cd·sr)
Magnetomotive force	ampere	A	
Magnetic field strength	ampere per meter	A/m	
Magnetic flux	weber	Wb	(V·s)
Magnetic flux density	tesla	T	(Wb/m^2)
Power	watt	W	(J/s)
Pressure	newton per square meter	N/m^2	
Radiant intensity	watt per steradian	W/sr	
Specific heat	joule per kilogram kelvin	J/kg K	
Thermal conductivity	watt per meter kelvin	W/m K	
Velocity	meter per second	m/s	
Volume	cubic meter	m^3	
Voltage, Potential difference, Electromotive force	volt	V	(W/A)
Wave number	1 per meter	m^{-1}	
Work, energy, quantity of heat	joule	J	(N·m)

Supplementary Tables

Definition of Most Important International System (SI) Units

The *ampere* (unit of electric current) is the constant current which, if maintained in two straight parallel conductors of infinite length, of negligible circular sections, and placed 1 meter apart in a vacuum, will produce between these conductors a force equal to 2×10^{-7} newton per meter of length.

Antiferromagnetic Materials: Those in which the magnetic moments of atoms or ions tend to assume an ordered arrangement in zero applied field, such that the vector sum of the moments is zero, below a characteristic temperature called the Néel Point. The permeability of antiferromagnetic materials is comparable to that of paramagnetic materials. Above the Néel Point, these materials become paramagnetic.

Candela. The candela is the luminous intensity, in the direction of the normal, of a black body surface 1/600,000 square meter in area, at the temperature of solidification of platinum under a pressure of 101,325 newtons per square meter.

Compensation Point: The temperature (below the Néel Point) at which, in some ferrimagnetic compounds, the saturation magnetization becomes zero.

The *coulomb* (unit of quantity of electricity) is the quantity of electricity transported in 1 second by a current of 1 ampere.

Diamagnetic Materials: Those within which the magnetic induction is slightly less than the applied magnetic field. Diamagnetism is an atomic scale consequence of the Lenz law of induction. For diamagnetic materials, the permeability is slightly less than that of empty space, and the magnetic susceptibility is negative and small.

The ephemeris *second* (unit of time) is exactly 1/31 556 925.974 7 of the tropical year of 1900, January, 0 days, and 12 hours ephemeris time.

The *farad* (unit of electric capacitance) is the capacitance of a capacitor between the plates of which there appears a difference of potential of 1 volt when it is charged by a quantity of electricity equal to 1 coulomb.

Ferrimagnetic Materials: Those in which the magnetic moments of atoms or ions tend to assume an ordered but nonparallel arrangement in zero applied field, below a characteristic temperature called the Néel Point. In the usual case, within a magnetic domain, a substantial net magnetization results from the antiparallel alignment of neighboring *nonequivalent* sublattices. The macroscopic behavior is similar to that in ferromagnetism. Above the Néel Point, these materials become paramagnetic.

Ferromagnetic Materials: Those in which the magnetic moments of atoms or ions in a magnetic domain tend to be aligned parallel to one another in zero applied field, below a characteristic temperature called the Curie Point. Complete ordering is achieved only at the absolute zero of temperature. Within a magnetic domain, at absolute zero, the magnetization is equal to the sum of the magnetic moments of the atoms or ions per unit volume. Bulk matter, consisting of many small magnetic domains, has a net magnetization which depends upon the magnetic history of the specimen (hysteresis effect). The permeability depends on the magnetic field, and can reach values of the order of 10^6 times that of free space. Above the Curie Point, these materials become paramagnetic.

The *henry* (unit of electric inductance) is the inductance of a closed circuit in which an electromotive force of 1 volt is produced when the electric current in the circuit varies uniformly at a rate of 1 ampere per second.

International Practical Kelvin Temperature Scale of 1960 and the *International Practical Celsius Temperature Scale* of 1960 are defined by a set of interpolation equations based on the following reference temperatures:

	°*K*	°*C*
Oxygen, liquid-gas equilibrium	90.18	−182.97
Water, solid-liquid equilibrium	273.15	0.00
Water, solid-liquid-gas equilibrium	273.16	0.01
Water, liquid-gas equilibrium	373.15	100.00
Zinc, solid-liquid equilibrium	692.655	419.505
Sulphur, liquid-gas equilibrium	717.75	444.6
Silver, solid-liquid equilibrium	1233.95	960.8
Gold, solid-liquid equilibrium	1336.15	1063.0

The *joule* (unit of energy) is the work done when the point of application of 1 newton is displaced a distance of 1 meter in the direction of the force.

Kelvin: The kelvin, the unit of thermodynamic temperature, is the fraction 1/273.16 of the thermodynamic temperature of the triple point of water. The decision was made at the 13th General Conference on Weights and Measures on October 13, 1967 that the name of the unit of thermodynamic temperature would be changed from *degree Kelvin* (symbol: °*K*) to *kelvin* (symbol: *K*). The name (*kelvin*) and symbol (*K*) are to be used for expressing temperature intervals. The former convention which expressed a temperature interval in *degrees Kelvin* or, abbreviated, *deg. K* is dropped. However, the old designations are acceptable temporarily as alternatives to the new ones. One may also express temperature intervals in *degrees Celsius.*

The *kilogram* (unit of mass) is the mass of a particular cylinder of platinum- iridium alloy, called the International Prototype Kilogram, which is preserved in a vault at Sèvres, France, by the International Bureau of Weights and Measures.

Length: The name "micron", for a unit of length equal to 10^{-6} meter, and the symbol "μ" which has been used for it were dropped by action of the 13th General Conference on Weights and Measures on October 13, 1967. The symbol "μ" is to be used solely as an abbreviation for the prefix "micro-", standing for the multiplication by 10^{-6}. Thus the length previously designated as 1 micron, should be designated 1 μm.

The *lumin* (unit of luminous flux) is the luminous flux emitted in a solid angle of 1 steradian by a uniform point source having an intensity of 1 candela.

Magnetic Anisotropy: In ferro- or ferrimagnetic crystals, it is found that the magnetization prefers to lie along certain crystal directions. These are termed *easy directions* of magnetization. Work must be expended to turn the magnetization away from these easy directions. That work as a function of crystal direction defines the *anisotropy energy surface.* Directions associated with a maximum of the anisotropy energy are termed *hard directions* of magnetization. In general, the energy difference between easy and hard directions decreases as the temperature is increased, and vanishes at the Curie or Néel point.

Magnetic Domains: The magnetization of a ferromagnetic or a ferrimagnetic material tends to break up into regions called *domains* separated by thin transition regions called *domain walls.* Within the volume of a domain, the magnetization has its saturation value, and is directed along a single direction. The magnetizations of other domains are directed along different directions in such a way that the net magnetization of the whole sample may be zero. The application of an external magnetic field first causes some domains to grow by the motion of their walls. At higher fields, the magnetizations of the resulting domains rotate toward parallelism with the field.

Magnetostriction: Change in sample dimensions as the magnitude or the direction of the magnetization in a crystal is changed.

Metamagnetic Materials: Those which are antiferromagnetic in weak fields, but which become ferromagnetically ordered in strong applied fields.

The *meter* (unit of length) is the length of exactly 1 650 763.73 wavelengths of the radiation in vacuum corresponding to the unperturbed transition between the levels $2p_{10}$ and $5d_5$ of the atom of Krypton 86, the orange-red line.

Néel Point: The temperature at which ferrimagnetic and antiferromagnetic materials become paramagnetic.

The *newton* (unit of force) is that force which gives to a mass of 1 kilogram an acceleration of 1 meter per second.

The *ohm* (unit of electric resistance) is the electric resistance between two points of a conductor when a constant difference of potential of 1 volt, applied between these two points, produces in this conductor a current of 1 ampere, this conductor not being the source of any electromotive force.

Paramagnetic Materials: Those within which the magnetic induction is slightly greater than the applied magnetic field. Paramagnetism arises from the partial alignment of the permanent magnetic dipole moments of atoms or ions. The permeability is slightly greater than that of empty space, and the magnetic susceptibility is positive and small.

Second: The second is the unit of time of the International System of Units. The definition adopted at the October 13, 1967 meeting of the 13th General Conference on Weights and Measures is: "The second is the duration of 9,192,631,770 periods of the radiation corresponding to the transition between the two hyperfine levels of the fundamental state of the atom of cesium 133." The frequency (9,192,631,770 Hz) which the definition assigns to the cesium radiation was carefully chosen to make it impossible, by any existing experimental evidence, to distinguish the new second from the "ephemeris second" based on the earth's motion. Therefor no changes need to be made in data stated in terms of the old standard in order to convert them to the new one. The atomic definition has two important advantages over the previous definition: (1) it can be realized (i.e., generated by a suitable clock) with sufficient precision, ± 1 part per hundred billion (10^{11}) or better, to meet the most exacting demands of modern metrology; and (2) it is available to anyone who has access to or who can build an atomic clock controlled by the specified cesium radiation. (A description of such clocks is given in "Atomic Frequency Standards," *NBS Tech. News Bull.* 45, 8–11 (Jan., 1961). For more recent developments and technical details, see R. E. Beehler, R. C. Mockler, and J. M. Richardson, "Cesium Beam Atomic Time and Frequency Standards," *Metrologia* 1, 114–131 (July, 1965)). In addition one can compare other high-precision clocks directly with such a standard in a relatively short time—an hour or so compared against years with the astronomical standard. Laboratory-type atomic clocks are complex and expensive, so that most clocks and frequency generators will continue to be calibrated against a standard such as the NBS Frequency Standard, controlled by a cesium atomic beam, at the Radio Standards Laboratory in Boulder, Colorado. In most cases the comparison will be by way of the standard-frequency and time-interval signals broadcast by NBS radio stations WWV, WWVH, WWVB, and WWVL.

The *volt* (unit of electric potential difference and electromotive force) is the difference of electric potential between two points of a conducting wire carrying a constant current of 1 ampere, when the power dissipated between these points is equal to 1 watt.

The *watt* (unit of power) is the power which gives rise to the production of energy at the rate of 1 joule per second.

The *weber* (unit of magnetic flux) is the magnetic flux which, linking a circuit of one turn, produces in it an electromotive force of 1 volt as it is reduced to zero at a uniform rate in 1 second.

Definitions of the most important SI Units are given in the above Supplementary Tables. These definitions have been extracted from the records of the International Committee and the General Conferences.

TEMPERATURE CONVERSION TABLE

This table permits one to convert from degrees Celcius to degrees Fahrenheit or from degrees Fahrenheit to degrees Celcius. The conversion is accomplished by first locating in a column printed in bold face type the number that is to be converted. If the number to be converted is in degrees Fahrenheit, one may find its equivalent in degrees Celcius by reading to the left. If the number to be converted is in degrees Celcius, one may find its equivalent in degrees Fahrenheit by reading to the right. Degrees Celcius are identical to degrees Centigrade. However, the word Celcius is preferred for international use.

The approved international symbolic abbreviation for degrees Celcius is °C, whereas for degrees Fahrenheit it is °F. Absolute zero on the Celcius scale is −273.15°C; on the Fahrenheit scale it is −459.67°F. The relation between degrees Fahrenheit and degrees Celcius may be expressed by

$$°C = 5/9(°F - 32) \text{ or}$$
$$°F = 9/5(°C) + 32.$$

To Convert			To Convert			To Convert		
To °C	←°F or °C→	To °F	To °C	←°F or °C→	To °F	To °C	←°F or °C→	To °F
−273.15	**−459.67**	—	−245.56	**−410**	—	−217.78	**−360**	—
−272.78	**−459**	—	−245	**−409**	—	−217.22	**−359**	—
−272.22	**−458**	—	−244.44	**−408**	—	−216.67	**−358**	—
−271.67	**−457**	—	−243.89	**−407**	—	−216.11	**−357**	—
−271.11	**−456**	—	−243.33	**−406**	—	−215.56	**−356**	—
−270.56	**−455**	—	−242.78	**−405**	—	−215	**−355**	—
−270	**−454**	—	−242.22	**−404**	—	−214.44	**−354**	—
−269.44	**−453**	—	−241.67	**−403**	—	−213.89	**−353**	—
−268.89	**−452**	—	−241.11	**−402**	—	−213.33	**−352**	—
−268.33	**−451**	—	−240.56	**−401**	—	−212.78	**−351**	—
−267.78	**−450**	—	−240	**−400**	—	−212.22	**−350**	—
−267.22	**−449**	—	−239.44	**−399**	—	−211.67	**−349**	—
−266.67	**−448**	—	−238.89	**−398**	—	−211.11	**−348**	—
−266.11	**−447**	—	−238.33	**−397**	—	−210.56	**−347**	—
−265.56	**−446**	—	−237.78	**−396**	—	−210	**−346**	—
−265	**−445**	—	−237.22	**−395**	—	−209.44	**−345**	—
−264.44	**−444**	—	−236.67	**−394**	—	−208.89	**−344**	—
−263.89	**−443**	—	−236.11	**−393**	—	−208.33	**−343**	—
−263.33	**−442**	—	−235.56	**−392**	—	−207.78	**−342**	—
−262.78	**−441**	—	−235	**−391**	—	−207.22	**−341**	—
−262.22	**−440**	—	−234.44	**−390**	—	−206.67	**−340**	—
−261.67	**−439**	—	−233.89	**−389**	—	−206.11	**−339**	—
−261.11	**−438**	—	−233.33	**−388**	—	−205.56	**−338**	—
−260.56	**−437**	—	−232.78	**−387**	—	−205	**−337**	—
−260	**−436**	—	−232.22	**−386**	—	−204.44	**−336**	—
−259.44	**−435**	—	−231.67	**−385**	—	−203.89	**−335**	—
−258.89	**−434**	—	−231.11	**−384**	—	−203.33	**−334**	—
−258.33	**−433**	—	−230.56	**−383**	—	−202.78	**−333**	—
−257.78	**−432**	—	−230	**−382**	—	−202.22	**−332**	—
−257.22	**−431**	—	−229.44	**−381**	—	−201.67	**−331**	—
−256.67	**−430**	—	−228.89	**−380**	—	−201.11	**−330**	—
−256.11	**−429**	—	−228.33	**−379**	—	−200.56	**−329**	—
−255.56	**−428**	—	−227.78	**−378**	—	−200	**−328**	—
−255	**−427**	—	−227.22	**−377**	—	−199.44	**−327**	—
−254.44	**−426**	—	−226.67	**−376**	—	−198.89	**−326**	—
−253.89	**−425**	—	−226.11	**−375**	—	−198.33	**−325**	—
−253.33	**−424**	—	−225.56	**−374**	—	−197.78	**−324**	—
−252.78	**−423**	—	−225	**−373**	—	−197.22	**−323**	—
−252.22	**−422**	—	−224.44	**−372**	—	−196.67	**−322**	—
−251.67	**−421**	—	−223.89	**−371**	—	−196.11	**−321**	—
−251.11	**−420**	—	−223.33	**−370**	—	−195.56	**−320**	—
−250.56	**−419**	—	−222.78	**−369**	—	−195	**−319**	—
−250	**−418**	—	−222.22	**−368**	—	−194.44	**−318**	—
−249.44	**−417**	—	−221.67	**−367**	—	−193.89	**−317**	—
−248.89	**−416**	—	−221.11	**−366**	—	−193.33	**−316**	—
−248.33	**−415**	—	−220.56	**−365**	—	−192.78	**−315**	—
−247.78	**−414**	—	−220	**−364**	—	−192.22	**−314**	—
−247.22	**−413**	—	−219.44	**−363**	—	−191.67	**−313**	—
−246.67	**−412**	—	−218.89	**−362**	—	−191.11	**−312**	—
−246.11	**−411**	—	−218.33	**−361**	—	−190.56	**−311**	—

TEMPERATURE CONVERSION TABLE (Continued)

To °C	←°F or °C→	To °F
−190	**−310**	—
−189.44	**−309**	—
−188.89	**−308**	—
−188.33	**−307**	—
−187.78	**−306**	—
−187.22	**−305**	—
−186.67	**−304**	—
−186.11	**−303**	—
−185.56	**−302**	—
−185	**−301**	—
−184.44	**−300**	—
−183.89	**−299**	—
−183.33	**−298**	—
−182.78	**−297**	—
−182.22	**−296**	—
−181.67	**−295**	—
−181.11	**−294**	—
−180.56	**−293**	—
−180	**−292**	—
−179.44	**−291**	—
−178.89	**−290**	—
−178.33	**−289**	—
−177.78	**−288**	—
−177.22	**−287**	—
−176.67	**−286**	—
−176.11	**−285**	—
−175.56	**−284**	—
−175	**−283**	—
−174.44	**−282**	—
−173.89	**−281**	—
−173.33	**−280**	—
−172.78	**−279**	—
−172.22	**−278**	—
−171.67	**−277**	—
−171.11	**−276**	—
−170.56	**−275**	—
−170	**−274**	—
—	**−273.15**	−459.67
−169.44	**−273**	−459.4
−168.89	**−272**	−457.6
−168.33	**−271**	−455.8
−167.78	**−270**	−454
−167.22	**−269**	−452.2
−166.67	**−268**	−450.4
−166.11	**−267**	−448.6
−165.56	**−266**	−446.8
−165	**−265**	−445
−164.44	**−264**	−443.2
−163.89	**−263**	−441.4
−163.33	**−262**	−439.6
−162.78	**−261**	−437.8
−162.22	**−260**	−436
−161.67	**−259**	−434.2
−161.11	**−258**	−432.4
−160.56	**−257**	−430.6
−160	**−256**	−428.8
−159.44	**−255**	−427
−158.89	**−254**	−425.2
−158.33	**−253**	−423.4
−157.78	**−252**	−421.6
−157.22	**−251**	−419.8

To °C	←°F or °C→	To °F
−156.67	**−250**	−418
−156.11	**−249**	−416.2
−155.56	**−248**	−414.4
−155	**−247**	−412.6
−154.44	**−246**	−410.8
−153.89	**−245**	−409
−153.33	**−244**	−407.2
−152.78	**−243**	−405.4
−152.22	**−242**	−403.6
−151.67	**−241**	−401.8
−151.11	**−240**	−400
−150.56	**−239**	−398.2
−150	**−238**	−396.4
−149.44	**−237**	−394.6
−148.89	**−236**	−392.8
−148.33	**−235**	−391
−147.78	**−234**	−389.2
−147.22	**−233**	−387.4
−146.67	**−232**	−385.6
−146.11	**−231**	−383.8
−145.56	**−230**	−382
−145	**−229**	−380.2
−144.44	**−228**	−378.4
−143.89	**−227**	−376.6
−143.33	**−226**	−374.8
−142.78	**−225**	−373
−142.22	**−224**	−371.2
−141.67	**−223**	−369.4
−141.11	**−222**	−367.6
−140.56	**−221**	−365.8
−140	**−220**	−364
−139.44	**−219**	−362.2
−138.89	**−218**	−360.4
−138.33	**−217**	−358.6
−137.78	**−216**	−356.8
−137.22	**−215**	−355
−136.67	**−214**	−353.2
−136.11	**−213**	−351.4
−135.56	**−212**	−349.6
−135	**−211**	−347.8
−134.44	**−210**	−346
−133.89	**−209**	−344.2
−133.33	**−208**	−342.4
−132.78	**−207**	−340.6
−132.22	**−206**	−338.8
−131.67	**−205**	−337
−131.11	**−204**	−335.2
−130.56	**−203**	−333.4
−130	**−202**	−331.6
−129.44	**−201**	−329.8
−128.89	**−200**	−328
−128.33	**−199**	−326.2
−127.78	**−198**	−324.4
−127.22	**−197**	−322.6
−126.67	**−196**	−320.8
−126.11	**−195**	−319
−125.56	**−194**	−317.2
−125	**−193**	−315.4
−124.44	**−192**	−313.6
−123.89	**−191**	−311.8

To °C	←°F or °C→	To °F
−123.33	**−190**	−310
−122.78	**−189**	−308.2
−122.22	**−188**	−306.4
−121.67	**−187**	−304.6
−121.11	**−186**	−302.8
−120.56	**−185**	−301
−120	**−184**	−299.2
−119.44	**−183**	−297.4
−118.89	**−182**	−295.6
−118.33	**−181**	−293.8
−117.78	**−180**	−292
−117.22	**−179**	−290.2
−116.67	**−178**	−288.4
−116.11	**−177**	−286.6
−115.56	**−176**	−284.8
−115	**−175**	−283
−114.44	**−174**	−281.2
−113.89	**−173**	−279.4
−113.33	**−172**	−277.6
−112.78	**−171**	−275.8
−112.22	**−170**	−274
−111.67	**−169**	−272.2
−111.11	**−168**	−270.4
−110.56	**−167**	−268.6
−110	**−166**	−266.8
−109.44	**−165**	−265
−108.89	**−164**	−263.2
−108.33	**−163**	−261.4
−107.78	**−162**	−259.6
−107.22	**−161**	−257.8
−106.67	**−160**	−256
−106.11	**−159**	−254.2
−105.56	**−158**	−252.4
−105	**−157**	−250.6
−104.44	**−156**	−248.8
−103.89	**−155**	−247
−103.33	**−154**	−245.2
−102.78	**−153**	−243.4
−102.22	**−152**	−241.6
−101.67	**−151**	−239.8
−101.11	**−150**	−238
−100.56	**−149**	−236.2
−100	**−148**	−234.4
−99.44	**−147**	−232.6
−98.89	**−146**	−230.8
−98.33	**−145**	−229
−97.78	**−144**	−227.2
−97.22	**−143**	−225.4
−96.67	**−142**	−223.6
−96.11	**−141**	−221.8
−95.56	**−140**	−220
−95	**−139**	−218.2
−94.44	**−138**	−216.4
−93.89	**−137**	−214.6
−93.33	**−136**	−212.8
−92.78	**−135**	−211
−92.22	**−134**	−209.2
−91.67	**−133**	−207.4
−91.11	**−132**	−205.6
−90.56	**−131**	−203.8

TEMPERATURE CONVERSION TABLE (Continued)

To Convert			To Convert			To Convert		
To °C	←°F or °C→	To °F	To °C	←°F or °C→	To °F	To °C	←°F or °C→	To °F
−90	**−130**	−202	−56.67	**−70**	−94	−23.33	**−10**	14
−89.44	**−129**	−200.2	−56.11	**−69**	−92.2	−22.78	**−9**	15.8
−88.89	**−128**	−198.4	−55.56	**−68**	−90.4	−22.22	**−8**	17.6
−88.33	**−127**	−196.6	−55	**−67**	−88.6	−21.67	**−7**	19.4
−87.78	**−126**	−194.8	−54.44	**−66**	−86.8	−21.11	**−6**	21.2
−87.22	**−125**	−193	−53.89	**−65**	−85	−20.56	**−5**	23
−86.67	**−124**	−191.2	−53.33	**−64**	−83.2	−20	**−4**	24.8
−86.11	**−123**	−189.4	−52.78	**−63**	−81.4	−19.44	**−3**	26.6
−85.56	**−122**	−187.6	−52.22	**−62**	−79.6	−18.89	**−2**	28.4
−85	**−121**	−185.8	−51.67	**−61**	−77.8	−18.33	**−1**	30.2
−84.44	**−120**	−184	−51.11	**−60**	−76	−17.78	**0**	32
−83.89	**−119**	−182.2	−50.56	**−59**	−74.2	−17.22	**1**	33.8
−83.33	**−118**	−180.4	−50	**−58**	−72.4	−16.67	**2**	35.6
−82.78	**−117**	−178.6	−49.44	**−57**	−70.6	−16.11	**3**	37.4
−82.22	**−116**	−176.8	−48.89	**−56**	−68.8	−15.56	**4**	39.2
−81.67	**−115**	−175	−48.33	**−55**	−67	−15	**5**	41
−81.11	**−114**	−173.2	−47.78	**−54**	−65.2	−14.44	**6**	42.8
−80.56	**−113**	−171.4	−47.22	**−53**	−63.4	−13.89	**7**	44.6
−80	**−112**	−169.6	−46.67	**−52**	−61.6	−13.33	**8**	46.4
−79.44	**−111**	−167.8	−46.11	**−51**	−59.8	−12.78	**9**	48.2
−78.89	**−110**	−166	−45.56	**−50**	−58	−12.22	**10**	50
−78.33	**−109**	−164.2	−45	**−49**	−56.2	−11.67	**11**	51.8
−77.78	**−108**	−162.4	−44.44	**−48**	−54.4	−11.11	**12**	53.6
−77.22	**−107**	−160.6	−43.89	**−47**	−52.6	−10.56	**13**	55.4
−76.67	**−106**	−158.8	−43.33	**−46**	−50.8	−10	**14**	57.2
−76.11	**−105**	−157	−42.78	**−45**	−49	−9.44	**15**	59
−75.56	**−104**	−155.2	−42.22	**−44**	−47.2	−8.89	**16**	60.8
−75	**−103**	−153.4	−41.67	**−43**	−45.4	−8.33	**17**	62.6
−74.44	**−102**	−151.6	−41.11	**−42**	−43.6	−7.78	**18**	64.4
−73.89	**−101**	−149.8	−40.56	**−41**	−41.8	−7.22	**19**	66.2
−73.33	**−100**	−148	−40	**−40**	−40	−6.67	**20**	68
−72.78	**−99**	−146.2	−39.44	**−39**	−38.2	−6.11	**21**	69.8
−72.22	**−98**	−144.4	−38.89	**−38**	−36.4	−5.56	**22**	71.6
−71.67	**−97**	−142.6	−38.33	**−37**	−34.6	−5	**23**	73.4
−71.11	**−96**	−140.8	−37.78	**−36**	−32.8	−4.44	**24**	75.2
−70.56	**−95**	−139	−37.22	**−35**	−31	−3.89	**25**	77
−70	**−94**	−137.2	−36.67	**−34**	−29.2	−3.33	**26**	78.8
−69.44	**−93**	−135.4	−36.11	**−33**	−27.4	−2.78	**27**	80.6
−68.89	**−92**	−133.6	−35.56	**−32**	−25.6	−2.22	**28**	82.4
−68.33	**−91**	−131.8	−35	**−31**	−23.8	−1.67	**29**	84.2
−67.78	**−90**	−130	−34.44	**−30**	−22	−1.11	**30**	86
−67.22	**−89**	−128.2	−33.89	**−29**	−20.2	−0.56	**31**	87.8
−66.67	**−88**	−126.4	−33.33	**−28**	−18.4	0	**32**	89.6
−66.11	**−87**	−124.6	−32.78	**−27**	−16.6	.56	**33**	91.4
−65.56	**−86**	−122.8	−32.22	**−26**	−14.8	1.11	**34**	93.2
−65	**−85**	−121	−31.67	**−25**	−13	1.67	**35**	95
−64.44	**−84**	−119.2	−31.11	**−24**	−11.2	2.22	**36**	96.8
−63.89	**−83**	−117.4	−30.56	**−23**	−9.4	2.78	**37**	98.6
−63.33	**−82**	−115.6	−30	**−22**	−7.6	3.33	**38**	100.4
−62.78	**−81**	−113.8	−29.44	**−21**	−5.8	3.89	**39**	102.2
−62.22	**−80**	−112	−28.89	**−20**	−4	4.44	**40**	104
−61.67	**−79**	−110.2	−28.33	**−19**	−2.2	5	**41**	105.8
−61.11	**−78**	−108.4	−27.78	**−18**	−0.4	5.56	**42**	107.6
−60.56	**−77**	−106.6	−27.22	**−17**	1.4	6.11	**43**	109.4
−60	**−76**	−104.8	−26.67	**−16**	3.2	6.67	**44**	111.2
−59.44	**−75**	−103	−26.11	**−15**	5	7.22	**45**	113
−58.89	**−74**	−101.2	−25.56	**−14**	6.8	7.78	**46**	114.8
−58.33	**−73**	−99.4	−25	**−13**	8.6	8.33	**47**	116.6
−57.78	**−72**	−97.6	−24.44	**−12**	10.4	8.89	**48**	118.4
−57.22	**−71**	−95.8	−23.89	**−11**	12.2	9.44	**49**	120.2

TEMPERATURE CONVERSION TABLE (Continued)

To Convert			To Convert			To Convert		
To °C	←°F or °C→	To °F	To °C	←°F or °C→	To °F	To °C	←°F or °C→	To °F
10	**50**	122	43.33	**110**	230	76.67	**170**	338
10.56	**51**	123.8	43.89	**111**	231.8	77.22	**171**	339.8
11.11	**52**	125.6	44.44	**112**	233.6	77.78	**172**	341.6
11.67	**53**	127.4	45	**113**	235.4	78.33	**173**	343.4
12.22	**54**	129.2	45.56	**114**	237.2	78.89	**174**	345.2
12.78	**55**	131	46.11	**115**	239	79.44	**175**	347
13.33	**56**	132.8	46.67	**116**	240.8	80	**176**	348.8
13.89	**57**	134.6	47.22	**117**	242.6	80.56	**177**	350.6
14.44	**58**	136.4	47.78	**118**	244.4	81.11	**178**	352.4
15	**59**	138.2	48.33	**119**	246.2	81.67	**179**	354.2
15.56	**60**	140	48.89	**120**	248	82.22	**180**	356
16.11	**61**	141.8	49.44	**121**	249.8	82.78	**181**	357.8
16.67	**62**	143.6	50	**122**	251.6	83.33	**182**	359.6
17.22	**63**	145.4	50.56	**123**	253.4	83.89	**183**	361.4
17.78	**64**	147.2	51.11	**124**	255.2	84.44	**184**	363.2
18.33	**65**	149	51.67	**125**	257	85	**185**	365
18.89	**66**	150.8	52.22	**126**	258.8	85.56	**186**	366.8
19.44	**67**	152.6	52.78	**127**	260.6	86.11	**187**	368.6
20	**68**	154.4	53.33	**128**	262.4	86.67	**188**	370.4
20.56	**69**	156.2	53.89	**129**	264.2	87.22	**189**	372.2
21.11	**70**	158	54.44	**130**	266	87.78	**190**	374
21.67	**71**	159.8	55	**131**	267.8	88.33	**191**	375.8
22.22	**72**	161.6	55.56	**132**	269.6	88.89	**192**	377.6
22.78	**73**	163.4	56.11	**133**	271.4	89.44	**193**	379.4
23.33	**74**	165.2	56.67	**134**	273.2	90	**194**	381.2
23.89	**75**	167	57.22	**135**	275	90.56	**195**	383
24.44	**76**	168.8	57.78	**136**	276.8	91.11	**196**	384.8
25	**77**	170.6	58.33	**137**	278.6	91.67	**197**	386.6
25.56	**78**	172.4	58.89	**138**	280.4	92.22	**198**	388.4
26.11	**79**	174.2	59.44	**139**	282.2	92.78	**199**	390.2
26.67	**80**	176	60	**140**	284	93.33	**200**	392
27.22	**81**	177.8	60.56	**141**	285.8	93.89	**201**	393.8
27.78	**82**	179.6	61.11	**142**	287.6	94.44	**202**	395.6
28.33	**83**	181.4	61.67	**143**	289.4	95	**203**	397.4
28.89	**84**	183.2	62.22	**144**	291.2	95.56	**204**	399.2
29.44	**85**	185	62.78	**145**	293	96.11	**205**	401
30	**86**	186.8	63.33	**146**	294.8	96.67	**206**	402.8
30.56	**87**	188.6	63.89	**147**	296.6	97.22	**207**	404.6
31.11	**88**	190.4	64.44	**148**	298.4	97.78	**208**	406.4
31.67	**89**	192.2	65	**149**	300.2	98.33	**209**	408.2
32.22	**90**	194	65.56	**150**	302	98.89	**210**	410
32.78	**91**	195.8	66.11	**151**	303.8	99.44	**211**	411.8
33.33	**92**	197.6	66.67	**152**	305.6	100	**212**	413.6
33.89	**93**	199.4	67.22	**153**	307.4	100.56	**213**	415.4
34.44	**94**	201.2	67.78	**154**	309.2	101.11	**214**	417.2
35	**95**	203	68.33	**155**	311	101.67	**215**	419
35.56	**96**	204.8	68.89	**156**	312.8	102.22	**216**	420.8
36.11	**97**	206.6	69.44	**157**	314.6	102.78	**217**	422.6
36.67	**98**	208.4	70	**158**	316.4	103.33	**218**	424.4
37.22	**99**	210.2	70.56	**159**	318.2	103.89	**219**	426.2
37.78	**100**	212	71.11	**160**	320	104.44	**220**	428
38.33	**101**	213.8	71.67	**161**	321.8	105	**221**	429.8
38.89	**102**	215.6	72.22	**162**	323.6	105.56	**222**	431.6
39.44	**103**	217.4	72.78	**163**	325.4	106.11	**223**	433.4
40	**104**	219.2	73.33	**164**	327.2	106.67	**224**	435.2
40.56	**105**	221	73.89	**165**	329	107.22	**225**	437
41.11	**106**	222.8	74.44	**166**	330.8	107.78	**226**	438.8
41.67	**107**	224.6	75	**167**	332.6	108.33	**227**	440.6
42.22	**108**	226.4	75.56	**168**	334.4	108.89	**228**	442.4
42.78	**109**	228.2	76.11	**169**	336.2	109.44	**229**	444.2

TEMPERATURE CONVERSION TABLE (Continued)

To Convert			To Convert			To Convert		
To °C	←°F or °C→	To °F	To °C	←°F or °C→	To °F	To °C	←°F or °C→	To °F
110	**230**	446	143.33	**290**	554	176.67	**350**	662
110.56	**231**	447.8	143.89	**291**	555.8	177.22	**351**	663.8
111.11	**232**	449.6	144.44	**292**	557.6	177.78	**352**	665.6
111.67	**233**	451.4	145	**293**	559.4	178.33	**353**	667.4
112.22	**234**	453.2	145.56	**294**	561.2	178.89	**354**	669.2
112.78	**235**	455	146.11	**295**	563	179.44	**355**	671
113.33	**236**	456.8	146.67	**296**	564.8	180	**356**	672.8
113.89	**237**	458.6	147.22	**297**	566.6	180.56	**357**	674.6
114.44	**238**	460.4	147.78	**298**	568.4	181.11	**358**	676.4
115	**239**	462.2	148.33	**299**	570.2	181.67	**359**	678.2
115.56	**240**	464	148.89	**300**	572	182.22	**360**	680
116.11	**241**	465.8	149.44	**301**	573.8	182.78	**361**	681.8
116.67	**242**	467.6	150	**302**	575.6	183.33	**362**	683.6
117.22	**243**	469.4	150.56	**303**	577.4	183.89	**363**	685.4
117.78	**244**	471.2	151.11	**304**	579.2	184.44	**364**	687.2
118.33	**245**	473	151.67	**305**	581	185	**365**	689
118.89	**246**	474.8	152.22	**306**	582.8	185.56	**366**	690.8
119.44	**247**	476.6	152.78	**307**	584.6	186.11	**367**	692.6
120	**248**	478.4	153.33	**308**	586.4	186.67	**368**	694.4
120.56	**249**	480.2	153.89	**309**	588.2	187.22	**369**	696.2
121.11	**250**	482	154.44	**310**	590	187.78	**370**	698
121.67	**251**	483.8	155	**311**	591.8	188.33	**371**	699.8
122.22	**252**	485.6	155.56	**312**	593.6	188.89	**372**	701.6
122.78	**253**	487.4	156.11	**313**	595.4	189.44	**373**	703.4
123.33	**254**	489.2	156.67	**314**	597.2	190	**374**	705.2
123.89	**255**	491	157.22	**315**	599	190.56	**375**	707
124.44	**256**	492.8	157.78	**316**	600.8	191.11	**376**	708.8
125	**257**	494.6	158.33	**317**	602.6	191.67	**377**	710.6
125.56	**258**	496.4	158.89	**318**	604.4	192.22	**378**	712.4
126.11	**259**	498.2	159.44	**319**	606.2	192.78	**379**	714.2
126.67	**260**	500	160	**320**	608	193.33	**380**	716
127.22	**261**	501.8	160.56	**321**	609.8	193.89	**381**	717.8
127.78	**262**	503.6	161.11	**322**	611.6	194.44	**382**	719.6
128.33	**263**	505.4	161.67	**323**	613.4	195	**383**	721.4
128.89	**264**	507.2	162.22	**324**	615.2	195.56	**384**	723.2
129.44	**265**	509	162.78	**325**	617	196.11	**385**	725
130	**266**	510.8	163.33	**326**	618.8	196.67	**386**	726.8
130.56	**267**	512.6	163.89	**327**	620.6	197.22	**387**	728.6
131.11	**268**	514.4	164.44	**328**	622.4	197.78	**388**	730.4
131.67	**269**	516.2	165	**329**	624.2	198.33	**389**	732.2
132.22	**270**	518	165.56	**330**	626	198.89	**390**	734
132.78	**271**	519.8	166.11	**331**	627.8	199.44	**391**	735.8
133.33	**272**	521.6	166.67	**332**	629.6	200	**392**	737.6
133.89	**273**	523.4	167.22	**333**	631.4	200.56	**393**	739.4
134.44	**274**	525.2	167.78	**334**	633.2	201.11	**394**	741.2
135	**275**	527	168.33	**335**	635	201.67	**395**	743
135.56	**276**	528.8	168.89	**336**	636.8	202.22	**396**	744.8
136.11	**277**	530.6	169.44	**337**	638.6	202.78	**397**	746.6
136.67	**278**	532.4	170	**338**	640.4	203.33	**398**	748.4
137.22	**279**	534.2	170.56	**339**	642.2	203.89	**399**	750.2
137.78	**280**	536	171.11	**340**	644	204.44	**400**	752
138.33	**281**	537.8	171.67	**341**	645.8	205	**401**	753.8
138.89	**282**	539.6	172.22	**342**	647.6	205.56	**402**	755.6
139.44	**283**	541.4	172.78	**343**	649.4	206.11	**403**	757.4
140	**284**	543.2	173.33	**344**	651.2	206.67	**404**	759.2
140.56	**285**	545	173.89	**345**	653	207.22	**405**	761
141.11	**286**	546.8	174.44	**346**	654.8	207.78	**406**	762.8
141.67	**287**	548.6	175	**347**	656.6	208.33	**407**	764.6
142.22	**288**	550.4	175.56	**348**	658.4	208.89	**408**	766.4
142.78	**289**	552.2	176.11	**349**	660.2	209.44	**409**	768.2

TEMPERATURE CONVERSION TABLE (Continued)

To Convert			To Convert			To Convert		
To °C	←°F or °C→	To °F	To °C	←°F or °C→	To °F	To °C	←°F or °C→	To °F
210	**410**	770	243.33	**470**	878	276.67	**530**	986
210.56	**411**	771.8	243.89	**471**	879.8	277.22	**531**	987.8
211.11	**412**	773.6	244.44	**472**	881.6	277.78	**532**	989.6
211.67	**413**	775.4	245	**473**	883.4	278.33	**533**	991.4
212.22	**414**	777.2	245.56	**474**	885.2	278.89	**534**	993.2
212.78	**415**	779	246.11	**475**	887	279.44	**535**	995
213.33	**416**	780.8	246.67	**476**	888.8	280	**536**	996.8
213.89	**417**	782.6	247.22	**477**	890.6	280.56	**537**	998.6
214.44	**418**	784.4	247.78	**478**	892.4	281.11	**538**	1000.4
215	**419**	786.2	248.33	**479**	894.2	281.67	**539**	1002.2
215.56	**420**	788	248.89	**480**	896	282.22	**540**	1004
216.11	**421**	789.8	249.44	**481**	897.8	282.78	**541**	1005.8
216.67	**422**	791.6	250	**482**	899.6	283.33	**542**	1007.6
217.22	**423**	793.4	250.56	**483**	901.4	283.89	**543**	1009.4
217.78	**424**	795.2	251.11	**484**	903.2	284.44	**544**	1011.2
218.33	**425**	797	251.67	**485**	905	285	**545**	1013
218.89	**426**	798.8	252.22	**486**	906.8	285.56	**546**	1014.8
219.44	**427**	800.6	252.78	**487**	908.6	286.11	**547**	1016.6
220	**428**	802.4	253.33	**488**	910.4	286.67	**548**	1018.4
220.56	**429**	804.2	253.89	**489**	912.2	287.22	**549**	1020.2
221.11	**430**	806	254.44	**490**	914	287.78	**550**	1022
221.67	**431**	807.8	255	**491**	915.8	288.33	**551**	1023.8
222.22	**432**	809.6	255.56	**492**	917.6	288.89	**552**	1025.6
222.78	**433**	811.4	256.11	**493**	919.4	289.44	**553**	1027.4
223.33	**434**	813.2	256.67	**494**	921.2	290	**554**	1029.2
223.89	**435**	815	257.22	**495**	923	290.56	**555**	1031
224.44	**436**	816.8	257.78	**496**	924.8	291.11	**556**	1032.8
225	**437**	818.6	258.33	**497**	926.6	291.67	**557**	1034.6
225.56	**438**	820.4	258.89	**498**	928.4	292.22	**558**	1036.4
226.11	**439**	822.2	259.44	**499**	930.2	292.78	**559**	1038.2
226.67	**440**	824	260	**500**	932	293.33	**560**	1040
227.22	**441**	825.8	260.56	**501**	933.8	293.89	**561**	1041.8
227.78	**442**	827.6	261.11	**502**	935.6	294.44	**562**	1043.6
228.33	**443**	829.4	261.67	**503**	937.4	295	**563**	1045.4
228.89	**444**	831.2	262.22	**504**	939.2	295.56	**564**	1047.2
229.44	**445**	833	262.78	**505**	941	296.11	**565**	1049
230	**446**	834.8	263.33	**506**	942.8	296.67	**566**	1050.8
230.56	**447**	836.6	263.89	**507**	944.6	297.22	**567**	1052.6
231.11	**448**	838.4	264.44	**508**	946.4	297.78	**568**	1054.4
231.67	**449**	840.2	265	**509**	948.2	298.33	**569**	1056.2
232.22	**450**	842	265.56	**510**	950	298.89	**570**	1058
232.78	**451**	843.8	266.11	**511**	951.8	299.44	**571**	1059.8
233.33	**452**	845.6	266.67	**512**	953.6	300	**572**	1061.6
233.89	**453**	847.4	267.22	**513**	955.4	300.56	**573**	1063.4
234.44	**454**	849.2	267.78	**514**	957.2	301.11	**574**	1065.2
235	**455**	851	268.33	**515**	959	301.67	**575**	1067
235.56	**456**	852.8	268.89	**516**	960.8	302.22	**576**	1068.8
236.11	**457**	854.6	269.44	**517**	962.6	302.78	**577**	1070.6
236.67	**458**	856.4	270	**518**	964.4	303.33	**578**	1072.4
237.22	**459**	858.2	270.56	**519**	966.2	303.89	**579**	1074.2
237.78	**460**	860	271.11	**520**	968	304.44	**580**	1076
238.33	**461**	861.8	271.67	**521**	969.8	305	**581**	1077.8
238.89	**462**	863.6	272.22	**522**	971.6	305.56	**582**	1079.6
239.44	**463**	865.4	272.78	**523**	973.4	306.11	**583**	1081.4
240	**464**	867.2	273.33	**524**	975.2	306.67	**584**	1083.2
240.56	**465**	869	273.89	**525**	977	307.22	**585**	1085
241.11	**466**	870.8	274.44	**526**	978.8	307.78	**586**	1086.8
241.67	**467**	872.6	275	**527**	980.6	308.33	**587**	1088.6
242.22	**468**	874.4	275.56	**528**	982.4	308.89	**588**	1090.4
242.78	**469**	876.2	276.11	**529**	984.2	309.44	**589**	1092.2

TEMPERATURE CONVERSION TABLE (Continued)

To Convert			To Convert			To Convert		
To °C	←°F or °C→	To °F	To °C	←°F or °C→	To °F	To °C	←°F or °C→	To °F
310	**590**	1094	343.33	**650**	1202	376.67	**710**	1310
310.56	**591**	1095.8	343.89	**651**	1203.8	377.22	**711**	1311.8
311.11	**592**	1097.6	344.44	**652**	1205.6	377.78	**712**	1313.6
311.67	**593**	1099.4	345	**653**	1207.4	378.33	**713**	1315.4
312.22	**594**	1101.2	345.56	**654**	1209.2	378.89	**714**	1317.2
312.78	**595**	1103	346.11	**655**	1211	379.44	**715**	1319
313.33	**596**	1104.8	346.67	**656**	1212.8	380	**716**	1320.8
313.89	**597**	1106.6	347.22	**657**	1214.6	380.56	**717**	1322.6
314.44	**598**	1108.4	347.78	**658**	1216.4	381.11	**718**	1324.4
315	**599**	1110.2	348.33	**659**	1218.2	381.67	**719**	1326.2
315.56	**600**	1112	348.89	**660**	1220	382.22	**720**	1328
316.11	**601**	1113.8	349.44	**661**	1221.8	382.78	**721**	1329.8
316.67	**602**	1115.6	350	**662**	1223.6	383.33	**722**	1331.6
317.22	**603**	1117.4	350.56	**663**	1225.4	383.89	**723**	1333.4
317.78	**604**	1119.2	351.11	**664**	1227.2	384.44	**724**	1335.2
318.33	**605**	1121	351.67	**665**	1229	385	**725**	1337
318.89	**606**	1122.8	352.22	**666**	1230.8	385.56	**726**	1338.8
319.44	**607**	1124.6	352.78	**667**	1232.6	386.11	**727**	1340.6
320	**608**	1126.4	353.33	**668**	1234.4	386.67	**728**	1342.4
320.56	**609**	1128.2	353.89	**669**	1236.2	387.22	**729**	1344.2
321.11	**610**	1130	354.44	**670**	1238	387.78	**730**	1346
321.67	**611**	1131.8	355	**671**	1239.8	388.33	**731**	1347.8
322.22	**612**	1133.6	355.56	**672**	1241.6	388.89	**732**	1349.6
322.78	**613**	1135.4	356.11	**673**	1243.4	389.44	**733**	1351.4
323.33	**614**	1137.2	356.67	**674**	1245.2	390	**734**	1353.2
323.89	**615**	1139	357.22	**675**	1247	390.56	**735**	1355
324.44	**616**	1140.8	357.78	**676**	1248.8	391.11	**736**	1356.8
325	**617**	1142.6	358.33	**677**	1250.6	391.67	**737**	1358.6
325.56	**618**	1144.4	358.89	**678**	1252.4	392.22	**738**	1360.4
326.11	**619**	1146.2	359.44	**679**	1254.2	392.78	**739**	1362.2
326.67	**620**	1148	360	**680**	1256	393.33	**740**	1364
327.22	**621**	1149.8	360.56	**681**	1257.8	393.89	**741**	1365.8
327.78	**622**	1151.6	361.11	**682**	1259.6	394.44	**742**	1367.6
328.33	**623**	1153.4	361.67	**683**	1261.4	395	**743**	1369.4
328.89	**624**	1155.2	362.22	**684**	1263.2	395.56	**744**	1371.2
329.44	**625**	1157	362.78	**685**	1265	396.11	**745**	1373
330	**626**	1158.8	363.33	**686**	1266.8	396.67	**746**	1374.8
330.56	**627**	1160.6	363.89	**687**	1268.6	397.22	**747**	1376.6
331.11	**628**	1162.4	364.44	**688**	1270.4	397.78	**748**	1378.4
331.67	**629**	1164.2	365	**689**	1272.2	398.33	**749**	1380.2
332.22	**630**	1166	365.56	**690**	1274	398.89	**750**	1382
332.78	**631**	1167.8	366.11	**691**	1275.8	399.44	**751**	1383.8
333.33	**632**	1169.6	366.67	**692**	1277.6	400	**752**	1385.6
333.89	**633**	1171.4	367.22	**693**	1279.4	400.56	**753**	1387.4
334.44	**634**	1173.2	367.78	**694**	1281.2	401.11	**754**	1389.2
335	**635**	1175	368.33	**695**	1283	401.67	**755**	1391
335.56	**636**	1176.8	368.89	**696**	1284.8	402.22	**756**	1392.8
336.11	**637**	1178.6	369.44	**697**	1286.6	402.78	**757**	1394.6
336.67	**638**	1180.4	370	**698**	1288.4	403.33	**758**	1396.4
337.22	**639**	1182.2	370.56	**699**	1290.2	403.89	**759**	1398.2
337.78	**640**	1184	371.11	**700**	1292	404.44	**760**	1400
338.33	**641**	1185.8	371.67	**701**	1293.8	405	**761**	1401.8
338.89	**642**	1187.6	372.22	**702**	1295.6	405.56	**762**	1403.6
339.44	**643**	1189.4	372.78	**703**	1297.4	406.11	**763**	1405.4
340	**644**	1191.2	373.33	**704**	1299.2	406.67	**764**	1407.2
340.56	**645**	1193	373.89	**705**	1301	407.22	**765**	1409
341.11	**646**	1194.8	374.44	**706**	1302.8	407.78	**766**	1410.8
341.67	**647**	1196.6	375	**707**	1304.6	408.33	**767**	1412.6
342.22	**648**	1198.4	375.56	**708**	1306.4	408.89	**768**	1414.4
342.78	**649**	1200.2	376.11	**709**	1308.2	409.44	**769**	1416.2

TEMPERATURE CONVERSION TABLE (Continued)

To Convert			To Convert			To Convert		
To °C	←°F or °C→	To °F	To °C	←°F or °C→	To °F	To °C	←°F or °C→	To °F
410	**770**	1418	443.33	**830**	1526	476.67	**890**	1634
410.56	**771**	1419.8	443.89	**831**	1527.8	477.22	**891**	1635.8
411.11	**772**	1421.6	444.44	**832**	1529.6	477.78	**892**	1637.6
411.67	**773**	1423.4	445	**833**	1531.4	478.33	**893**	1639.4
412.22	**774**	1425.2	445.56	**834**	1533.2	478.89	**894**	1641.2
412.78	**775**	1427	446.11	**835**	1535	479.44	**895**	1643
413.33	**776**	1428.8	446.67	**836**	1536.8	480	**896**	1644.8
413.89	**777**	1430.6	447.22	**837**	1538.6	480.56	**897**	1646.6
414.44	**778**	1432.4	447.78	**838**	1540.4	481.11	**898**	1648.4
415	**779**	1434.2	448.33	**839**	1542.2	481.67	**899**	1650.2
415.56	**780**	1436	448.89	**840**	1544	482.22	**900**	1652
416.11	**781**	1437.8	449.44	**841**	1545.8	482.78	**901**	1653.8
416.67	**782**	1439.6	450	**842**	1547.6	483.33	**902**	1655.6
417.22	**783**	1441.4	450.56	**843**	1549.4	483.89	**903**	1657.4
417.78	**784**	1443.2	451.11	**844**	1551.2	484.44	**904**	1659.2
418.33	**785**	1445	451.67	**845**	1553	485	**905**	1661
418.89	**786**	1446.8	452.22	**846**	1554.8	485.56	**906**	1662.8
419.44	**787**	1448.6	452.78	**847**	1556.6	486.11	**907**	1664.6
420	**788**	1450.4	453.33	**848**	1558.4	486.67	**908**	1666.4
420.56	**789**	1452.2	453.89	**849**	1560.2	487.22	**909**	1668.2
421.11	**790**	1454	454.44	**850**	1562	487.78	**910**	1670
421.67	**791**	1455.8	455	**851**	1563.8	488.33	**911**	1671.8
422.22	**792**	1457.6	455.56	**852**	1565.6	488.89	**912**	1673.6
422.78	**793**	1459.4	456.11	**853**	1567.4	489.44	**913**	1675.4
423.33	**794**	1461.2	456.67	**854**	1569.2	490	**914**	1677.2
423.89	**795**	1463	457.22	**855**	1571	490.56	**915**	1679
424.44	**796**	1464.8	457.78	**856**	1572.8	491.11	**916**	1680.8
425	**797**	1466.6	458.33	**857**	1574.6	491.67	**917**	1682.6
425.56	**798**	1468.4	458.89	**858**	1576.4	492.22	**918**	1684.4
426.11	**799**	1470.2	459.44	**859**	1578.2	492.78	**919**	1686.2
426.67	**800**	1472	460	**860**	1580	493.33	**920**	1688
427.22	**801**	1473.8	460.56	**861**	1581.8	493.89	**921**	1689.8
427.78	**802**	1475.6	461.11	**862**	1583.6	494.44	**922**	1691.6
428.33	**803**	1477.4	461.67	**863**	1585.4	495	**923**	1693.4
428.89	**804**	1479.2	462.22	**864**	1587.2	495.56	**924**	1695.2
429.44	**805**	1481	462.78	**865**	1589	496.11	**925**	1697
430	**806**	1482.8	463.33	**866**	1590.8	496.67	**926**	1698.8
430.56	**807**	1484.6	463.89	**867**	1592.6	497.22	**927**	1700.6
431.11	**808**	1486.4	464.44	**868**	1594.4	497.78	**928**	1702.4
431.67	**809**	1488.2	465	**869**	1596.2	498.33	**929**	1704.2
432.22	**810**	1490	465.56	**870**	1598	498.89	**930**	1706
432.78	**811**	1491.8	466.11	**871**	1599.8	499.44	**931**	1707.8
433.33	**812**	1493.6	466.67	**872**	1601.6	500	**932**	1709.6
433.89	**813**	1495.4	467.22	**873**	1603.4	500.56	**933**	1711.4
434.44	**814**	1497.2	467.78	**874**	1605.2	501.11	**934**	1713.2
435	**815**	1499	468.33	**875**	1607	501.67	**935**	1715
435.56	**816**	1500.8	468.89	**876**	1608.8	502.22	**936**	1716.8
436.11	**817**	1502.6	469.44	**877**	1610.6	502.78	**937**	1718.6
436.67	**818**	1504.4	470	**878**	1612.4	503.33	**938**	1720.4
437.22	**819**	1506.2	470.56	**879**	1614.2	503.89	**939**	1722.2
437.78	**820**	1508	471.11	**880**	1616	504.44	**940**	1724
438.33	**821**	1509.8	471.67	**881**	1617.8	505	**941**	1725.8
438.89	**822**	1511.6	472.22	**882**	1619.6	505.56	**942**	1727.6
439.44	**823**	1513.4	472.78	**883**	1621.4	506.11	**943**	1729.4
440	**824**	1515.2	473.33	**884**	1623.2	506.67	**944**	1731.2
440.56	**825**	1517	473.89	**885**	1625	507.22	**945**	1733
441.11	**826**	1518.8	474.44	**886**	1626.8	507.78	**946**	1734.8
441.67	**827**	1520.6	475	**887**	1628.6	508.33	**947**	1736.6
442.22	**828**	1522.4	475.56	**888**	1630.4	508.89	**948**	1738.4
442.78	**829**	1524.2	476.11	**889**	1632.2	509.44	**949**	1740.2

TEMPERATURE CONVERSION TABLE (Continued)

To Convert			To Convert			To Convert		
To °C	←°F or °C→	To °F	To °C	←°F or °C→	To °F	To °C	←°F or °C→	To °F
510	**950**	1742	543.33	**1010**	1850	576.67	**1070**	1958
510.56	**951**	1743.8	543.89	**1011**	1851.8	577.22	**1071**	1959.8
511.11	**952**	1745.6	544.44	**1012**	1853.6	577.78	**1072**	1961.6
511.67	**953**	1747.4	545	**1013**	1855.4	578.33	**1073**	1963.4
512.22	**954**	1749.2	545.56	**1014**	1857.2	578.89	**1074**	1965.2
512.78	**955**	1751	546.11	**1015**	1859	579.44	**1075**	1967
513.33	**956**	1752.8	546.67	**1016**	1860.8	580	**1076**	1968.8
513.89	**957**	1754.6	547.22	**1017**	1862.6	580.56	**1077**	1970.6
514.44	**958**	1756.4	547.78	**1018**	1864.4	581.11	**1078**	1972.4
515	**959**	1758.2	548.33	**1019**	1866.2	581.67	**1079**	1974.2
515.56	**960**	1760	548.89	**1020**	1868	582.22	**1080**	1976
516.11	**961**	1761.8	549.44	**1021**	1869.8	582.78	**1081**	1977.8
516.67	**962**	1763.6	550	**1022**	1871.6	583.33	**1082**	1979.6
517.22	**963**	1765.4	550.56	**1023**	1873.4	583.89	**1083**	1981.4
517.78	**964**	1767.2	551.11	**1024**	1875.2	584.44	**1084**	1983.2
518.33	**965**	1769	551.67	**1025**	1877	585	**1085**	1985
518.89	**966**	1770.8	552.22	**1026**	1878.8	585.56	**1086**	1986.8
519.44	**967**	1772.6	552.78	**1027**	1880.6	586.11	**1087**	1988.6
520	**968**	1774.4	553.33	**1028**	1882.4	586.67	**1088**	1990.4
520.56	**969**	1776.2	553.89	**1029**	1884.2	587.22	**1089**	1992.2
521.11	**970**	1778	554.44	**1030**	1886	587.78	**1090**	1994
521.67	**971**	1779.8	555	**1031**	1887.8	588.33	**1091**	1995.8
522.22	**972**	1781.6	555.56	**1032**	1889.6	588.89	**1092**	1997.6
522.78	**973**	1783.4	556.11	**1033**	1891.4	589.44	**1093**	1999.4
523.33	**974**	1785.2	556.67	**1034**	1893.2	590	**1094**	2001.2
523.89	**975**	1787	557.22	**1035**	1895	590.56	**1095**	2003
524.44	**976**	1788.8	557.78	**1036**	1896.8	591.11	**1096**	2004.8
525	**977**	1790.6	558.33	**1037**	1898.6	591.67	**1097**	2006.6
525.56	**978**	1792.4	558.89	**1038**	1900.4	592.22	**1098**	2008.4
526.11	**979**	1794.2	559.44	**1039**	1902.2	592.78	**1099**	2010.2
526.67	**980**	1796	560	**1040**	1904	593.33	**1100**	2012
527.22	**981**	1797.8	560.56	**1041**	1905.8	593.89	**1101**	2013.8
527.78	**982**	1799.6	561.11	**1042**	1907.6	594.44	**1102**	2015.6
528.33	**983**	1801.4	561.67	**1043**	1909.4	595	**1103**	2017.4
528.89	**984**	1803.2	562.22	**1044**	1911.2	595.56	**1104**	2019.2
529.44	**985**	1805	562.78	**1045**	1913	596.11	**1105**	2021
530	**986**	1806.8	563.33	**1046**	1914.8	596.67	**1106**	2022.8
530.56	**987**	1808.6	563.89	**1047**	1916.6	597.22	**1107**	2024.6
531.11	**988**	1810.4	564.44	**1048**	1918.4	597.78	**1108**	2026.4
531.67	**989**	1812.2	565	**1049**	1920.2	598.33	**1109**	2028.2
532.22	**990**	1814	565.56	**1050**	1922	598.89	**1110**	2030
532.78	**991**	1815.8	566.11	**1051**	1923.8	599.44	**1111**	2031.8
533.33	**992**	1817.6	566.67	**1052**	1925.6	600	**1112**	2033.6
533.89	**993**	1819.4	567.22	**1053**	1927.4	600.56	**1113**	2035.4
534.44	**994**	1821.2	567.78	**1054**	1929.2	601.11	**1114**	2037.2
535	**995**	1823	568.33	**1055**	1931	601.67	**1115**	2039
535.56	**996**	1824.8	568.89	**1056**	1932.8	602.22	**1116**	2040.8
536.11	**997**	1826.6	569.44	**1057**	1934.6	602.78	**1117**	2042.6
536.67	**998**	1828.4	570	**1058**	1936.4	603.33	**1118**	2044.4
537.22	**999**	1830.2	570.56	**1059**	1938.2	603.89	**1119**	2046.2
537.78	**1000**	1832	571.11	**1060**	1940	604.44	**1120**	2048
538.33	**1001**	1833.8	571.67	**1061**	1941.8	605	**1121**	2049.8
538.89	**1002**	1835.6	572.22	**1062**	1943.6	605.56	**1122**	2051.6
539.44	**1003**	1837.4	572.78	**1063**	1945.4	606.11	**1123**	2053.4
540	**1004**	1839.2	573.33	**1064**	1947.2	606.67	**1124**	2055.2
540.56	**1005**	1841	573.89	**1065**	1949	607.22	**1125**	2057
541.11	**1006**	1842.8	574.44	**1066**	1950.8	607.78	**1126**	2058.8
541.67	**1007**	1844.6	575	**1067**	1952.6	608.33	**1127**	2060.6
542.22	**1008**	1846.4	575.56	**1068**	1954.4	608.89	**1128**	2062.4
542.78	**1009**	1848.2	576.11	**1069**	1956.2	609.44	**1129**	2064.2

TEMPERATURE CONVERSION TABLE (Continued)

To Convert			To Convert			To Convert		
To °C	←°F or °C→	To °F	To °C	←°F or °C→	To °F	To °C	←°F or °C→	To °F
610	**1130**	2066	643.33	**1190**	2174	676.67	**1250**	2282
610.56	**1131**	2067.8	643.89	**1191**	2175.8	677.22	**1251**	2283.8
611.11	**1132**	2069.6	644.44	**1192**	2177.6	677.78	**1252**	2285.6
611.67	**1133**	2071.4	645	**1193**	2179.4	678.33	**1253**	2287.4
612.22	**1134**	2073.2	645.56	**1194**	2181.2	678.89	**1254**	2289.2
612.78	**1135**	2075	646.11	**1195**	2183	679.44	**1255**	2291
613.33	**1136**	2076.8	646.67	**1196**	2184.8	680	**1256**	2292.8
613.89	**1137**	2078.6	647.22	**1197**	2186.6	680.56	**1257**	2294.6
614.44	**1138**	2080.4	647.78	**1198**	2188.4	681.11	**1258**	2296.4
615	**1139**	2082.2	648.33	**1199**	2190.2	681.67	**1259**	2298.2
615.56	**1140**	2084	648.89	**1200**	2192	682.22	**1260**	2300
616.11	**1141**	2085.8	649.44	**1201**	2193.8	682.78	**1261**	2301.8
616.67	**1142**	2087.6	650	**1202**	2195.6	683.33	**1262**	2303.6
617.22	**1143**	2089.4	650.56	**1203**	2197.4	683.89	**1263**	2305.4
617.78	**1144**	2091.2	651.11	**1204**	2199.2	684.44	**1264**	2307.2
618.33	**1145**	2093	651.67	**1205**	2201	685	**1265**	2309
618.89	**1146**	2094.8	652.22	**1206**	2202.8	685.56	**1266**	2310.8
619.44	**1147**	2096.6	652.78	**1207**	2204.6	686.11	**1267**	2312.6
620	**1148**	2098.4	653.33	**1208**	2206.4	686.67	**1268**	2314.4
620.56	**1149**	2100.2	653.89	**1209**	2208.2	687.22	**1269**	2316.2
621.11	**1150**	2102	654.44	**1210**	2210	687.78	**1270**	2318
621.67	**1151**	2103.8	655	**1211**	2211.8	688.33	**1271**	2319.8
622.22	**1152**	2105.6	655.56	**1212**	2213.6	688.89	**1272**	2321.6
622.78	**1153**	2107.4	656.11	**1213**	2215.4	689.44	**1273**	2323.4
623.33	**1154**	2109.2	656.67	**1214**	2217.2	690	**1274**	2325.2
623.89	**1155**	2111	657.22	**1215**	2219	690.56	**1275**	2327
624.44	**1156**	2112.8	657.78	**1216**	2220.8	691.11	**1276**	2328.8
625	**1157**	2114.6	658.33	**1217**	2222.6	691.67	**1277**	2330.6
625.56	**1158**	2116.4	658.89	**1218**	2224.4	692.22	**1278**	2332.4
626.11	**1159**	2118.2	659.44	**1219**	2226.2	692.78	**1279**	2334.2
626.67	**1160**	2120	660	**1220**	2228	693.33	**1280**	2336
627.22	**1161**	2121.8	660.56	**1221**	2229.8	693.89	**1281**	2337.8
627.78	**1162**	2123.6	661.11	**1222**	2231.6	694.44	**1282**	2339.6
628.33	**1163**	2125.4	661.67	**1223**	2233.4	695	**1283**	2341.4
628.89	**1164**	2127.2	662.22	**1224**	2235.2	695.56	**1284**	2343.2
629.44	**1165**	2129	662.78	**1225**	2237	696.11	**1285**	2345
630	**1166**	2130.8	663.33	**1226**	2238.8	696.67	**1286**	2346.8
630.56	**1167**	2132.6	663.89	**1227**	2240.6	697.22	**1287**	2348.6
631.11	**1168**	2134.4	664.44	**1228**	2242.4	697.78	**1288**	2350.4
631.67	**1169**	2136.2	665	**1229**	2244.2	698.33	**1289**	2352.2
632.22	**1170**	2138	665.56	**1230**	2246	698.89	**1290**	2354
632.78	**1171**	2139.8	666.11	**1231**	2247.8	699.44	**1291**	2355.8
633.33	**1172**	2141.6	666.67	**1232**	2249.6	700	**1292**	2357.6
633.89	**1173**	2143.4	667.22	**1233**	2251.4	700.56	**1293**	2359.4
634.44	**1174**	2145.2	667.78	**1234**	2253.2	701.11	**1294**	2361.2
635	**1175**	2147	668.33	**1235**	2255	701.67	**1295**	2363
635.56	**1176**	2148.8	668.89	**1236**	2256.8	702.22	**1296**	2364.8
636.11	**1177**	2150.6	669.44	**1237**	2258.6	702.78	**1297**	2366.6
636.67	**1178**	2152.4	670	**1238**	2260.4	703.33	**1298**	2368.4
637.22	**1179**	2154.2	670.56	**1239**	2262.2	703.89	**1299**	2370.2
637.78	**1180**	2156	671.11	**1240**	2264	704.44	**1300**	2372
638.33	**1181**	2157.8	671.67	**1241**	2265.8	705	**1301**	2373.8
638.89	**1182**	2159.6	672.22	**1242**	2267.6	705.56	**1302**	2375.6
639.44	**1183**	2161.4	672.78	**1243**	2269.4	706.11	**1303**	2377.4
640	**1184**	2163.2	673.33	**1244**	2271.2	706.67	**1304**	2379.2
640.56	**1185**	2165	673.89	**1245**	2273	707.22	**1305**	2381
641.11	**1186**	2166.8	674.44	**1246**	2274.8	707.78	**1306**	2382.8
641.67	**1187**	2168.6	675	**1247**	2276.6	708.33	**1307**	2384.6
642.22	**1188**	2170.4	675.56	**1248**	2278.4	708.89	**1308**	2386.4
642.78	**1189**	2172.2	676.11	**1249**	2280.2	709.44	**1309**	2388.2

TEMPERATURE CONVERSION TABLE (Continued)

To Convert			To Convert			To Convert		
To °C	←°F or °C→	To °F	To °C	←°F or °C→	To °F	To °C	←°F or °C→	To °F
710	**1310**	2390	743.33	**1370**	2498	776.67	**1430**	2606
710.56	**1311**	2391.8	743.89	**1371**	2499.8	777.22	**1431**	2607.8
711.11	**1312**	2393.6	744.44	**1372**	2501.6	777.78	**1432**	2609.6
711.67	**1313**	2395.4	745	**1373**	2503.4	778.33	**1433**	2611.4
712.22	**1314**	2397.2	745.56	**1374**	2505.2	778.89	**1434**	2613.2
712.78	**1315**	2399	746.11	**1375**	2507	779.44	**1435**	2615
713.33	**1316**	2400.8	746.67	**1376**	2508.8	780	**1436**	2616.8
713.89	**1317**	2402.6	747.22	**1377**	2510.6	780.56	**1437**	2618.6
714.44	**1318**	2404.4	747.78	**1378**	2512.4	781.11	**1438**	2620.4
715	**1319**	2406.2	748.33	**1379**	2514.2	781.67	**1439**	2622.2
715.56	**1320**	2408	748.89	**1380**	2516	782.22	**1440**	2624
716.11	**1321**	2409.8	749.44	**1381**	2517.8	782.78	**1441**	2625.8
716.67	**1322**	2411.6	750	**1382**	2519.6	783.33	**1442**	2627.6
717.22	**1323**	2413.4	750.56	**1383**	2521.4	783.89	**1443**	2629.4
717.78	**1324**	2415.2	751.11	**1384**	2523.2	784.44	**1444**	2631.2
718.33	**1325**	2417	751.67	**1385**	2525	785	**1445**	2633
718.89	**1326**	2418.8	752.22	**1386**	2526.8	785.56	**1446**	2634.8
719.44	**1327**	2420.6	752.78	**1387**	2528.6	786.11	**1447**	2636.6
720	**1328**	2422.4	753.33	**1388**	2530.4	786.67	**1448**	2638.4
720.56	**1329**	2424.2	753.89	**1389**	2532.2	787.22	**1449**	2640.2
721.11	**1330**	2426	754.44	**1390**	2534	787.78	**1450**	2642
721.67	**1331**	2427.8	755	**1391**	2535.8	788.33	**1451**	2443.8
722.22	**1332**	2429.6	755.56	**1392**	2537.6	788.89	**1452**	2645.6
722.78	**1333**	2431.4	756.11	**1393**	2539.4	789.44	**1453**	2647.4
723.33	**1334**	2433.2	756.67	**1394**	2541.2	790	**1454**	2649.2
723.89	**1335**	2435	757.22	**1395**	2543	790.56	**1455**	2651
724.44	**1336**	2436.8	757.78	**1396**	2544.8	791.11	**1456**	2652.8
725	**1337**	2438.6	758.33	**1397**	2546.6	791.67	**1457**	2654.6
725.56	**1338**	2440.4	758.89	**1398**	2548.4	792.22	**1458**	2656.4
726.11	**1339**	2442.2	759.44	**1399**	2550.2	792.78	**1459**	2658.2
726.67	**1340**	2444	760	**1400**	2552	793.33	**1460**	2660
727.22	**1341**	2445.8	760.56	**1401**	2553.8	793.89	**1461**	2661.8
727.78	**1342**	2447.6	761.11	**1402**	2555.6	794.44	**1462**	2663.6
728.33	**1343**	2449.4	761.67	**1403**	2557.4	795	**1463**	2665.4
728.89	**1344**	2451.2	762.22	**1404**	2559.2	795.56	**1464**	2667.2
729.44	**1345**	2453	762.78	**1405**	2561	796.11	**1465**	2669
730	**1346**	2454.8	763.33	**1406**	2562.8	796.67	**1466**	2670.8
730.56	**1347**	2456.6	763.89	**1407**	2564.6	797.22	**1467**	2672.6
731.11	**1348**	2458.4	764.44	**1408**	2566.4	797.78	**1468**	2674.4
731.67	**1349**	2460.2	765	**1409**	2568.2	798.33	**1469**	2676.2
732.22	**1350**	2462	765.56	**1410**	2570	798.89	**1470**	2678
732.78	**1351**	2463.8	766.11	**1411**	2571.8	799.44	**1471**	2679.8
733.33	**1352**	2465.6	766.67	**1412**	2573.6	800	**1472**	2681.6
733.89	**1353**	2467.4	767.22	**1413**	2575.4	800.56	**1473**	2683.4
734.44	**1354**	2469.2	767.78	**1414**	2577.2	801.11	**1474**	2685.2
735	**1355**	2471	768.33	**1415**	2579	801.67	**1475**	2687
735.56	**1356**	2472.8	768.89	**1416**	2580.8	802.22	**1476**	2688.8
736.11	**1357**	2474.6	769.44	**1417**	2582.6	802.78	**1477**	2690.6
736.67	**1358**	2476.4	770	**1418**	2584.4	803.33	**1478**	2692.4
737.22	**1359**	2478.2	770.56	**1419**	2586.2	803.89	**1479**	2694.2
737.78	**1360**	2480	771.11	**1420**	2588	804.44	**1480**	2696
738.33	**1361**	2481.8	771.67	**1421**	2589.8	805	**1481**	2697.8
738.89	**1362**	2483.6	772.22	**1422**	2591.6	805.56	**1482**	2699.6
739.44	**1363**	2485.4	772.78	**1423**	2593.4	806.11	**1483**	2701.4
740	**1364**	2487.2	773.33	**1424**	2595.2	806.67	**1484**	2703.2
740.56	**1365**	2489	773.89	**1425**	2597	807.22	**1485**	2705
741.11	**1366**	2490.8	774.44	**1426**	2598.8	807.78	**1486**	2706.8
741.67	**1367**	2492.6	775	**1427**	2600.6	808.33	**1487**	2708.6
742.22	**1368**	2494.4	775.56	**1428**	2602.4	808.89	**1488**	2710.4
742.78	**1369**	2496.2	776.11	**1429**	2604.2	809.44	**1489**	2712.2

TEMPERATURE CONVERSION TABLE (Continued)

To Convert			To Convert			To Convert		
To °C	←°F or °C→	To °F	To °C	←°F or °C→	To °F	To °C	←°F or °C→	To °F
810	**1490**	2714	843.33	**1550**	2822	876.67	**1610**	2930
810.56	**1491**	2715.8	843.89	**1551**	2823.8	877.22	**1611**	2931.8
811.11	**1492**	2717.6	844.44	**1552**	2825.6	877.78	**1612**	2933.6
811.67	**1493**	2719.4	845	**1553**	2827.4	878.33	**1613**	2935.4
812.22	**1494**	2721.2	845.56	**1554**	2829.2	878.89	**1614**	2937.2
812.78	**1495**	2723	846.11	**1555**	2831	879.44	**1615**	2939
813.33	**1496**	2724.8	846.67	**1556**	2832.8	880	**1616**	2940.8
813.89	**1497**	2726.6	847.22	**1557**	2834.6	880.56	**1617**	2942.6
814.44	**1498**	2728.4	847.78	**1558**	2836.4	881.11	**1618**	2944.4
815	**1499**	2730.2	848.33	**1559**	2838.2	881.67	**1619**	2946.2
815.56	**1500**	2732	848.89	**1560**	2840	882.22	**1620**	2948
816.11	**1501**	2733.8	849.44	**1561**	2841.8	882.78	**1621**	2949.8
816.67	**1502**	2735.6	850	**1562**	2843.6	883.33	**1622**	2951.6
817.22	**1503**	2737.4	850.56	**1563**	2845.4	883.89	**1623**	2953.4
817.78	**1504**	2739.2	851.11	**1564**	2847.2	884.44	**1624**	2955.2
818.33	**1505**	2741	851.67	**1565**	2849	885	**1625**	2957
818.89	**1506**	2742.8	852.22	**1566**	2850.8	885.56	**1626**	2958.8
819.44	**1507**	2744.6	852.78	**1567**	2852.6	886.11	**1627**	2960.6
820	**1508**	2746.4	853.33	**1568**	2854.4	886.67	**1628**	2962.4
820.56	**1509**	2748.2	853.89	**1569**	2856.2	887.22	**1629**	2964.2
821.11	**1510**	2750	854.44	**1570**	2858	887.78	**1630**	2966
821.67	**1511**	2751.8	855	**1571**	2859.8	888.33	**1631**	2967.8
822.22	**1512**	2753.6	855.56	**1572**	2861.6	888.89	**1632**	2969.6
822.78	**1513**	2755.4	856.11	**1573**	2863.4	889.44	**1633**	2971.4
823.33	**1514**	2757.2	856.67	**1574**	2865.2	890	**1634**	2973.2
823.89	**1515**	2759	857.22	**1575**	2867	890.56	**1635**	2975
824.44	**1516**	2760.8	857.78	**1576**	2868.8	891.11	**1636**	2976.8
825	**1517**	2762.6	858.33	**1577**	2870.6	891.67	**1637**	2978.6
825.56	**1518**	2764.4	858.89	**1578**	2872.4	892.22	**1638**	2980.4
826.11	**1519**	2766.2	859.44	**1579**	2874.2	892.78	**1639**	2982.2
826.67	**1520**	2768	860	**1580**	2876	893.33	**1640**	2984
827.22	**1521**	2769.8	860.56	**1581**	2877.8	893.89	**1641**	2985.8
827.78	**1522**	2771.6	861.11	**1582**	2879.6	894.44	**1642**	2987.6
828.33	**1523**	2773.4	861.67	**1583**	2881.4	895	**1643**	2989.4
828.89	**1524**	2775.2	862.22	**1584**	2883.2	895.56	**1644**	2991.2
829.44	**1525**	2777	862.78	**1585**	2885	896.11	**1645**	2993
830	**1526**	2778.8	863.33	**1586**	2886.8	896.67	**1646**	2994.8
830.56	**1527**	2780.6	863.89	**1587**	2888.6	897.22	**1647**	2996.6
831.11	**1528**	2782.4	864.44	**1588**	2890.4	897.78	**1648**	2998.4
831.67	**1529**	2784.2	865	**1589**	2892.2	898.33	**1649**	3000.2
832.22	**1530**	2786	865.56	**1590**	2894	898.89	**1650**	3002
832.78	**1531**	2787.8	866.11	**1591**	2895.8	899.44	**1651**	3003.8
833.33	**1532**	2789.6	866.67	**1592**	2897.6	900	**1652**	3005.6
833.89	**1533**	2791.4	867.22	**1593**	2899.4	900.56	**1653**	3007.4
834.44	**1534**	2793.2	867.78	**1594**	2901.2	901.11	**1654**	3009.2
835	**1535**	2795	868.33	**1595**	2903	901.67	**1655**	3011
835.56	**1536**	2796.8	868.89	**1596**	2904.8	902.22	**1656**	3012.8
836.11	**1537**	2798.6	869.44	**1597**	2906.6	902.78	**1657**	3014.6
836.67	**1538**	2800.4	870	**1598**	2908.4	903.33	**1658**	3016.4
837.22	**1539**	2802.2	870.56	**1599**	2910.2	903.89	**1659**	3018.2
837.78	**1540**	2804	871.11	**1600**	2912	904.44	**1660**	3020
838.33	**1541**	2805.8	871.67	**1601**	2913.8	905	**1661**	3021.8
838.89	**1542**	2807.6	872.22	**1602**	2915.6	905.56	**1662**	3023.6
839.44	**1543**	2809.4	872.78	**1603**	2917.4	906.11	**1663**	3025.4
840	**1544**	2811.2	873.33	**1604**	2919.2	906.67	**1664**	3027.2
840.56	**1545**	2813	873.89	**1605**	2921	907.22	**1665**	3029
841.11	**1546**	2814.8	874.44	**1606**	2922.8	907.78	**1666**	3030.8
841.67	**1547**	2816.6	875	**1607**	2924.6	908.33	**1667**	3032.6
842.22	**1548**	2818.4	875.56	**1608**	2926.4	908.89	**1668**	3034.4
842.78	**1549**	2820.2	876.11	**1609**	2928.2	909.44	**1669**	3036.2

TEMPERATURE CONVERSION TABLE (Continued)

To Convert			To Convert			To Convert		
To °C	←°F or °C→	To °F	To °C	←°F or °C→	To °F	To °C	←°F or °C→	To °F
910	**1670**	3038	943.33	**1730**	3146	976.67	**1790**	3254
910.56	**1671**	3039.8	943.89	**1731**	3147.8	977.22	**1791**	3255.8
911.11	**1672**	3041.6	944.44	**1732**	3149.6	977.78	**1792**	3257.6
911.67	**1673**	3043.4	945	**1733**	3151.4	978.33	**1793**	3259.4
912.22	**1674**	3045.2	945.56	**1734**	3153.2	978.89	**1794**	3261.2
912.78	**1675**	3047	946.11	**1735**	3155	979.44	**1795**	3263
913.33	**1676**	3048.8	946.67	**1736**	3156.8	980	**1796**	3264.8
913.89	**1677**	3050.6	947.22	**1737**	3158.6	980.56	**1797**	3266.6
914.44	**1678**	3052.4	947.78	**1738**	3160.4	981.11	**1798**	3268.4
915	**1679**	3054.2	948.33	**1739**	3162.2	981.67	**1799**	3270.2
915.56	**1680**	3056	948.89	**1740**	3164	982.22	**1800**	3272
916.11	**1681**	3057.8	949.44	**1741**	3165.8	982.78	**1801**	3273.8
916.67	**1682**	3059.6	950	**1742**	3167.6	983.33	**1802**	3275.6
917.22	**1683**	3061.4	950.56	**1743**	3169.4	983.89	**1803**	3277.4
917.78	**1684**	3063.2	951.11	**1744**	3171.2	984.44	**1804**	3279.2
918.33	**1685**	3065	951.67	**1745**	3173	985	**1805**	3281
918.89	**1686**	3066.8	952.22	**1746**	3174.8	985.56	**1806**	3282.8
919.44	**1687**	3068.6	952.78	**1747**	3176.6	986.11	**1807**	3284.6
920	**1688**	3070.4	953.33	**1748**	3178.4	986.67	**1808**	3286.4
920.56	**1689**	3072.2	953.89	**1749**	3180.2	987.22	**1809**	3288.2
921.11	**1690**	3074	954.44	**1750**	3182	987.78	**1810**	3290
921.67	**1691**	3075.8	955	**1751**	3183.8	988.33	**1811**	3291.8
922.22	**1692**	3077.6	955.56	**1752**	3185.6	988.89	**1812**	3293.6
922.78	**1693**	3079.4	956.11	**1753**	3187.4	989.44	**1813**	3295.4
923.33	**1694**	3081.2	956.67	**1754**	3189.2	990	**1814**	3297.2
923.89	**1695**	3083	957.22	**1755**	3191	990.56	**1815**	3299
924.44	**1696**	3084.8	957.78	**1756**	3192.8	991.11	**1816**	3300.8
925	**1697**	3086.6	958.33	**1757**	3194.6	991.67	**1817**	3302.6
925.56	**1698**	3088.4	958.89	**1758**	3196.4	992.22	**1818**	3304.4
926.11	**1699**	3090.2	959.44	**1759**	3198.2	992.78	**1819**	3306.2
926.67	**1700**	3092	960	**1760**	3200	993.33	**1820**	3308
927.22	**1701**	3093.8	960.56	**1761**	3201.8	993.89	**1821**	3309.8
927.78	**1702**	3095.6	961.11	**1762**	3203.6	994.44	**1822**	3311.6
928.33	**1703**	3097.4	961.67	**1763**	3205.4	995	**1823**	3313.4
928.89	**1704**	3099.2	962.22	**1764**	3207.2	995.56	**1824**	3315.2
929.44	**1705**	3101	962.78	**1765**	3209	996.11	**1825**	3317
930	**1706**	3102.8	963.33	**1766**	3210.8	996.67	**1826**	3318.8
930.56	**1707**	3104.6	963.89	**1767**	3212.6	997.22	**1827**	3320.6
931.11	**1708**	3106.4	964.44	**1768**	3214.4	997.78	**1828**	3322.4
931.67	**1709**	3108.2	965	**1769**	3216.2	998.33	**1829**	3324.2
932.22	**1710**	3110	965.56	**1770**	3218	998.89	**1830**	3326
932.78	**1711**	3111.8	966.11	**1771**	3219.8	999.44	**1831**	3327.8
933.33	**1712**	3113.6	966.67	**1772**	3221.6	1000	**1832**	3329.6
933.89	**1713**	3115.4	967.22	**1773**	3223.4	1000.56	**1833**	3331.4
934.44	**1714**	3117.2	967.78	**1774**	3225.2	1001.11	**1834**	3333.2
935	**1715**	3119	968.33	**1775**	3227	1001.67	**1835**	3335
935.56	**1716**	3120.8	968.89	**1776**	3228.8	1002.22	**1836**	3336.8
936.11	**1717**	3122.6	969.44	**1777**	3230.6	1002.78	**1837**	3338.6
936.67	**1718**	3124.4	970	**1778**	3232.4	1003.33	**1838**	3340.4
937.22	**1719**	3126.2	970.56	**1779**	3234.2	1003.89	**1839**	3342.2
937.78	**1720**	3128	971.11	**1780**	3236	1004.44	**1840**	3344
938.33	**1721**	3129.8	971.67	**1781**	3237.8	1005	**1841**	3345.8
938.89	**1722**	3131.6	972.22	**1782**	3239.6	1005.56	**1842**	3347.6
939.44	**1723**	3133.4	972.78	**1783**	3241.4	1006.11	**1843**	3349.4
940	**1724**	3135.2	973.33	**1784**	3243.2	1006.67	**1844**	3351.2
940.56	**1725**	3137	973.89	**1785**	3245	1007.22	**1845**	3353
941.11	**1726**	3138.8	974.44	**1786**	3246.8	1007.78	**1846**	3354.8
941.67	**1727**	3140.6	975	**1787**	3248.6	1008.33	**1847**	3356.6
942.22	**1728**	3142.4	975.56	**1788**	3250.4	1008.89	**1848**	3358.4
942.78	**1729**	3144.2	976.11	**1789**	3252.2	1009.44	**1849**	3360.2

TEMPERATURE CONVERSION TABLE (Continued)

To Convert			To Convert			To Convert		
To °C	←°F or °C→	To °F	To °C	←°F or °C→	To °F	To °C	←°F or °C→	To °F
1010	**1850**	3362	1043.33	**1910**	3470	1076.67	**1970**	3578
1010.56	**1851**	3363.8	1043.89	**1911**	3471.8	1077.22	**1971**	3579.8
1011.11	**1852**	3365.6	1044.44	**1912**	3473.6	1077.78	**1972**	3581.6
1011.67	**1853**	3367.4	1045	**1913**	3475.4	1078.33	**1973**	3583.4
1012.22	**1854**	3369.2	1045.56	**1914**	3477.2	1078.89	**1974**	3585.2
1012.78	**1855**	3371	1046.11	**1915**	3479	1079.44	**1975**	3587
1013.33	**1856**	3372.8	1046.67	**1916**	3480.8	1080	**1976**	3588.8
1013.89	**1857**	3374.6	1047.22	**1917**	3482.6	1080.56	**1977**	3590.6
1014.44	**1858**	3376.4	1047.78	**1918**	3484.4	1081.11	**1978**	3592.4
1015	**1859**	3378.2	1048.33	**1919**	3486.2	1081.67	**1979**	3594.2
1015.56	**1860**	3380	1048.89	**1920**	3488	1082.22	**1980**	3596
1016.11	**1861**	3381.8	1049.44	**1921**	3489.8	1082.78	**1981**	3597.8
1016.67	**1862**	3383.6	1050	**1922**	3491.6	1083.33	**1982**	3599.6
1017.22	**1863**	3385.4	1050.56	**1923**	3493.4	1083.89	**1983**	3601.4
1017.78	**1864**	3387.2	1051.11	**1924**	3495.2	1084.44	**1984**	3603.2
1018.33	**1865**	3389	1051.67	**1925**	3497	1085	**1985**	3605
1018.89	**1866**	3390.8	1052.22	**1926**	3498.8	1085.56	**1986**	3606.8
1019.44	**1867**	3392.6	1052.78	**1927**	3500.6	1086.11	**1987**	3608.6
1020	**1868**	3394.4	1053.33	**1928**	3502.4	1086.67	**1988**	3610.4
1020.56	**1869**	3396.2	1053.89	**1929**	3504.2	1087.22	**1989**	3612.2
1021.11	**1870**	3398	1054.44	**1930**	3506	1087.78	**1990**	3614
1021.67	**1871**	3399.8	1055	**1931**	3507.8	1088.33	**1991**	3615.8
1022.22	**1872**	3401.6	1055.56	**1932**	3509.6	1088.89	**1992**	3617.6
1022.78	**1873**	3403.4	1056.11	**1933**	3511.4	1089.44	**1993**	3619.4
1023.33	**1874**	3405.2	1056.67	**1934**	3513.2	1090	**1994**	3621.2
1023.89	**1875**	3407	1057.22	**1935**	3515	1090.56	**1995**	3623
1024.44	**1876**	3408.8	1057.78	**1936**	3516.8	1091.11	**1996**	3624.8
1025	**1877**	3410.6	1058.33	**1937**	3518.6	1091.67	**1997**	3626.6
1025.56	**1878**	3412.4	1058.89	**1938**	3520.4	1092.22	**1998**	3628.4
1026.11	**1879**	3414.2	1059.44	**1939**	3522.2	1092.78	**1999**	3630.2
1026.67	**1880**	3416	1060	**1940**	3524	1093.33	**2000**	3632
1027.22	**1881**	3417.8	1060.56	**1941**	3525.8	1093.89	**2001**	3633.8
1027.78	**1882**	3419.6	1061.11	**1942**	3527.6	1094.44	**2002**	3635.6
1028.33	**1883**	3421.4	1061.67	**1943**	3529.4	1095	**2003**	3637.4
1028.89	**1884**	3423.2	1062.22	**1944**	3531.2	1095.56	**2004**	3639.2
1029.44	**1885**	3425	1062.78	**1945**	3533	1096.11	**2005**	3641
1030	**1886**	3426.8	1063.33	**1946**	3534.8	1096.67	**2006**	3642.8
1030.56	**1887**	3428.6	1063.89	**1947**	3536.6	1097.22	**2007**	3644.6
1031.11	**1888**	3430.4	1064.44	**1948**	3538.4	1097.78	**2008**	3646.4
1031.67	**1889**	3432.2	1065	**1949**	3540.2	1098.33	**2009**	3648.2
1032.22	**1890**	3434	1065.56	**1950**	3542	1098.89	**2010**	3650
1032.78	**1891**	3435.8	1066.11	**1951**	3543.8	1099.44	**2011**	3651.8
1033.33	**1892**	3437.6	1066.67	**1952**	3545.6	1100	**2012**	3653.6
1033.89	**1893**	3439.4	1067.22	**1953**	3547.4	1100.56	**2013**	3655.4
1034.44	**1894**	3441.2	1067.78	**1954**	3549.2	1101.11	**2014**	3657.2
1035	**1895**	3443	1068.33	**1955**	3551	1101.67	**2015**	3659
1035.56	**1896**	3444.8	1068.89	**1956**	3552.8	1102.22	**2016**	3660.8
1036.11	**1897**	3446.6	1069.44	**1957**	3554.6	1102.78	**2017**	3662.6
1036.67	**1898**	3448.4	1070	**1958**	3556.4	1103.33	**2018**	3664.4
1037.22	**1899**	3450.2	1070.56	**1959**	3558.2	1103.89	**2019**	3666.2
1037.78	**1900**	3452	1071.11	**1960**	3560	1104.44	**2020**	3668
1038.33	**1901**	3453.8	1071.67	**1961**	3561.8	1105	**2021**	3669.8
1038.89	**1902**	3455.6	1072.22	**1962**	3563.6	1105.56	**2022**	3671.6
1039.44	**1903**	3457.4	1072.78	**1963**	3565.4	1106.11	**2023**	3673.4
1040	**1904**	3459.2	1073.33	**1964**	3567.2	1106.67	**2024**	3675.2
1040.56	**1905**	3461	1073.89	**1965**	3569	1107.22	**2025**	3677
1041.11	**1906**	3462.8	1074.44	**1966**	3570.8	1107.78	**2026**	3678.8
1041.67	**1907**	3464.6	1075	**1967**	3572.6	1108.33	**2027**	3680.6
1042.22	**1908**	3466.4	1075.56	**1968**	3574.4	1108.89	**2028**	3682.4
1042.78	**1909**	3468.2	1076.11	**1969**	3576.2	1109.44	**2029**	3684.2

TEMPERATURE CONVERSION TABLE (Continued)

To Convert			To Convert			To Convert		
To °C	←°F or °C→	To °F	To °C	←°F or °C→	To °F	To °C	←°F or °C→	To °F
1110	**2030**	3686	1143.33	**2090**	3794	1176.67	**2150**	3902
1110.56	**2031**	3687.8	1143.89	**2091**	3795.8	1177.22	**2151**	3903.8
1111.11	**2032**	3689.6	1144.44	**2092**	3797.6	1177.78	**2152**	3905.6
1111.67	**2033**	3691.4	1145	**2093**	3799.4	1178.33	**2153**	3907.4
1112.22	**2034**	3693.2	1145.56	**2094**	3801.2	1178.89	**2154**	3909.2
1112.78	**2035**	3695	1146.11	**2095**	3803	1179.44	**2155**	3911
1113.33	**2036**	3696.8	1146.67	**2096**	3804.8	1180	**2156**	3912.8
1113.89	**2037**	3698.6	1147.22	**2097**	3806.6	1180.56	**2157**	3914.6
1114.44	**2038**	3700.4	1147.78	**2098**	3808.4	1181.11	**2158**	3916.4
1115	**2039**	3702.2	1148.33	**2099**	3810.2	1181.67	**2159**	3918.2
1115.56	**2040**	3704	1148.89	**2100**	3812	1182.22	**2160**	3920
1116.11	**2041**	3705.8	1149.44	**2101**	3813.8	1182.78	**2161**	3921.8
1116.67	**2042**	3707.6	1150	**2102**	3815.6	1183.33	**2162**	3923.6
1117.22	**2043**	3709.4	1150.56	**2103**	3817.4	1183.89	**2163**	3925.4
1117.78	**2044**	3711.2	1151.11	**2104**	3819.2	1184.44	**2164**	3927.2
1118.33	**2045**	3713	1151.67	**2105**	3821	1185	**2165**	3929
1118.89	**2046**	3714.8	1152.22	**2106**	3822.8	1185.56	**2166**	3930.8
1119.44	**2047**	3716.6	1152.78	**2107**	3824.6	1186.11	**2167**	3932.6
1120	**2048**	3718.4	1153.33	**2108**	3826.4	1186.67	**2168**	3934.4
1120.56	**2049**	3720.2	1153.89	**2109**	3828.2	1187.22	**2169**	3936.2
1121.11	**2050**	3722	1154.44	**2110**	3830	1187.78	**2170**	3938
1121.67	**2051**	3723.8	1155	**2111**	3831.8	1188.33	**2171**	3939.8
1122.22	**2052**	3725.6	1155.56	**2112**	3833.6	1188.89	**2172**	3941.6
1122.78	**2053**	3727.4	1156.11	**2113**	3835.4	1189.44	**2173**	3943.4
1123.33	**2054**	3729.2	1156.67	**2114**	3837.2	1190	**2174**	3945.2
1123.89	**2055**	3731	1157.22	**2115**	3839	1190.56	**2175**	3947
1124.44	**2056**	3732.8	1157.78	**2116**	3840.8	1191.11	**2176**	3948.8
1125	**2057**	3734.6	1158.33	**2117**	3842.6	1191.67	**2177**	3950.6
1125.56	**2058**	3736.4	1158.89	**2118**	3844.4	1192.22	**2178**	3952.4
1126.11	**2059**	3738.2	1159.44	**2119**	3846.2	1192.78	**2179**	3954.2
1126.67	**2060**	3740	1160	**2120**	3848	1193.33	**2180**	3956
1127.22	**2061**	3741.8	1160.56	**2121**	3849.8	1193.89	**2181**	3957.8
1127.78	**2062**	3743.6	1161.11	**2122**	3851.6	1194.44	**2182**	3959.6
1128.33	**2063**	3745.4	1161.67	**2123**	3853.4	1195	**2183**	3961.4
1128.89	**2064**	3747.2	1162.22	**2124**	3855.2	1195.56	**2184**	3963.2
1129.44	**2065**	3749	1162.78	**2125**	3857	1196.11	**2185**	3965
1130	**2066**	3750.8	1163.33	**2126**	3858.8	1196.67	**2186**	3966.8
1130.56	**2067**	3752.6	1163.89	**2127**	3860.6	1197.22	**2187**	3968.6
1131.11	**2068**	3754.4	1164.44	**2128**	3862.4	1197.78	**2188**	3970.4
1131.67	**2069**	3756.2	1165	**2129**	3864.2	1198.33	**2189**	3972.2
1132.22	**2070**	3758	1165.56	**2130**	3866	1198.89	**2190**	3974
1132.78	**2071**	3759.8	1166.11	**2131**	3867.8	1199.44	**2191**	3975.8
1133.33	**2072**	3761.6	1166.67	**2132**	3869.6	1200	**2192**	3977.6
1133.89	**2073**	3763.4	1167.22	**2133**	3871.4	1200.56	**2193**	3979.4
1134.44	**2074**	3765.2	1167.78	**2134**	3873.2	1201.11	**2194**	3981.2
1135	**2075**	3767	1168.33	**2135**	3875	1201.67	**2195**	3983
1135.56	**2076**	3768.8	1168.89	**2136**	3876.8	1202.22	**2196**	3984.8
1136.11	**2077**	3770.6	1169.44	**2137**	3878.6	1202.78	**2197**	3986.6
1136.67	**2078**	3772.4	1170	**2138**	3880.4	1203.33	**2198**	3988.4
1137.22	**2079**	3774.2	1170.56	**2139**	3882.2	1203.89	**2199**	3990.2
1137.78	**2080**	3776	1171.11	**2140**	3884	1204.44	**2200**	3992
1138.33	**2081**	3777.8	1171.67	**2141**	3885.8	1205	**2201**	3993.8
1138.89	**2082**	3779.6	1172.22	**2142**	3887.6	1205.56	**2202**	3995.6
1139.44	**2083**	3781.4	1172.78	**2143**	3889.4	1206.11	**2203**	3997.4
1140	**2084**	3783.2	1173.33	**2144**	3891.2	1206.67	**2204**	3999.2
1140.56	**2085**	3785	1173.89	**2145**	3893	1207.22	**2205**	4001
1141.11	**2086**	3786.8	1174.44	**2146**	3894.8	1207.78	**2206**	4002.8
1141.67	**2087**	3788.6	1175	**2147**	3896.6	1208.33	**2207**	4004.6
1142.22	**2088**	3790.4	1175.56	**2148**	3898.4	1208.89	**2208**	4006.4
1142.78	**2089**	3792.2	1176.11	**2149**	3900.2	1209.44	**2209**	4008.2

TEMPERATURE CONVERSION TABLE (Continued)

To Convert			To Convert			To Convert		
To °C	←°F or °C→	To °F	To °C	←°F or °C→	To °F	To °C	←°F or °C→	To °F
1210	**2210**	4010	1243.33	**2270**	4118	1276.67	**2330**	4226
1210.56	**2211**	4011.8	1243.89	**2271**	4119.8	1277.22	**2331**	4227.8
1211.11	**2212**	4013.6	1244.44	**2272**	4121.6	1277.78	**2332**	4229.6
1211.67	**2213**	4015.4	1245	**2273**	4123.4	1278.33	**2333**	4231.4
1212.22	**2214**	4017.2	1245.56	**2274**	4125.2	1278.89	**2334**	4233.2
1212.78	**2215**	4019	1246.11	**2275**	4127	1279.44	**2335**	4235
1213.33	**2216**	4020.8	1246.67	**2276**	4128.8	1280	**2336**	4236.8
1213.89	**2217**	4022.6	1247.22	**2277**	4130.6	1280.56	**2337**	4238.6
1214.44	**2218**	4024.4	1247.78	**2278**	4132.4	1281.11	**2338**	4240.4
1215	**2219**	4026.2	1248.33	**2279**	4134.2	1281.67	**2339**	4242.2
1215.56	**2220**	4028	1248.89	**2280**	4136	1282.22	**2340**	4244
1216.11	**2221**	4029.8	1249.44	**2281**	4137.8	1282.78	**2341**	4245.8
1216.67	**2222**	4031.6	1250	**2282**	4139.6	1283.33	**2342**	4247.6
1217.22	**2223**	4033.4	1250.56	**2283**	4141.4	1283.89	**2343**	4249.4
1217.78	**2224**	4035.2	1251.11	**2284**	4143.2	1284.44	**2344**	4251.2
1218.33	**2225**	4037	1251.67	**2285**	4145	1285	**2345**	4253
1218.89	**2226**	4038.8	1252.22	**2286**	4146.8	1285.56	**2346**	4254.8
1219.44	**2227**	4040.6	1252.78	**2287**	4148.6	1286.11	**2347**	4256.6
1220	**2228**	4042.4	1253.33	**2288**	4150.4	1286.67	**2348**	4258.4
1220.56	**2229**	4044.2	1253.89	**2289**	4152.2	1287.22	**2349**	4260.2
1221.11	**2230**	4046	1254.44	**2290**	4154	1287.78	**2350**	4262
1221.67	**2231**	4047.8	1255	**2291**	4155.8	1288.33	**2351**	4263.8
1222.22	**2232**	4049.6	1255.56	**2292**	4157.6	1288.89	**2352**	4265.6
1222.78	**2233**	4051.4	1256.11	**2293**	4159.4	1289.44	**2353**	4267.4
1223.33	**2234**	4053.2	1256.67	**2294**	4161.2	1290	**2354**	4269.2
1223.89	**2235**	4055	1257.22	**2295**	4163	1290.56	**2355**	4271
1224.44	**2236**	4056.8	1257.78	**2296**	4164.8	1291.11	**2356**	4272.8
1225	**2237**	4058.6	1258.33	**2297**	4166.6	1291.67	**2357**	4274.6
1225.56	**2238**	4060.4	1258.89	**2298**	4168.4	1292.22	**2358**	4276.4
1226.11	**2239**	4062.2	1259.44	**2299**	4170.2	1292.78	**2359**	4278.2
1226.67	**2240**	4064	1260	**2300**	4172	1293.33	**2360**	4280
1227.22	**2241**	4065.8	1260.56	**2301**	4173.8	1293.89	**2361**	4281.8
1227.78	**2242**	4067.6	1261.11	**2302**	4175.6	1294.44	**2362**	4283.6
1228.33	**2243**	4069.4	1261.67	**2303**	4177.4	1295	**2363**	4285.4
1228.89	**2244**	4071.2	1262.22	**2304**	4179.2	1295.56	**2364**	4287.2
1229.44	**2245**	4073	1262.78	**2305**	4181	1296.11	**2365**	4289
1230	**2246**	4074.8	1263.33	**2306**	4182.8	1296.67	**2366**	4290.8
1230.56	**2247**	4076.6	1263.89	**2307**	4184.6	1297.22	**2367**	4292.6
1231.11	**2248**	4078.4	1264.44	**2308**	4186.4	1297.78	**2368**	4294.4
1231.67	**2249**	4080.2	1265	**2309**	4188.2	1298.33	**2369**	4296.2
1232.22	**2250**	4082	1265.56	**2310**	4190	1298.89	**2370**	4298
1232.78	**2251**	4083.8	1266.11	**2311**	4191.8	1299.44	**2371**	4299.8
1233.33	**2252**	4085.6	1266.67	**2312**	4193.6	1300	**2372**	4301.6
1233.89	**2253**	4087.4	1267.22	**2313**	4195.4	1300.56	**2373**	4303.4
1234.44	**2254**	4089.2	1267.78	**2314**	4197.2	1301.11	**2374**	4305.2
1235	**2255**	4091	1268.33	**2315**	4199	1301.67	**2375**	4307
1235.56	**2256**	4092.8	1268.89	**2316**	4200.8	1302.22	**2376**	4308.8
1236.11	**2257**	4094.6	1269.44	**2317**	4202.6	1302.78	**2377**	4310.6
1236.67	**2258**	4096.4	1270	**2318**	4204.4	1303.33	**2378**	4312.4
1237.22	**2259**	4098.2	1270.56	**2319**	4206.2	1303.89	**2379**	4314.2
1237.78	**2260**	4100	1271.11	**2320**	4208	1304.44	**2380**	4316
1238.33	**2261**	4101.8	1271.67	**2321**	4209.8	1305	**2381**	4317.8
1238.89	**2262**	4103.6	1272.22	**2322**	4211.6	1305.56	**2382**	4319.6
1239.44	**2263**	4105.4	1272.78	**2323**	4213.4	1306.11	**2383**	4321.4
1240	**2264**	4107.2	1273.33	**2324**	4215.2	1306.67	**2384**	4323.2
1240.56	**2265**	4109	1273.89	**2325**	4217	1307.22	**2385**	4325
1241.11	**2267**	4110.8	1274.44	**2326**	4218.8	1307.78	**2386**	4326.8
1241.67	**2267**	4112.6	1275	**2327**	4220.6	1308.33	**2387**	4328.6
1242.22	**2268**	4114.4	1275.56	**2328**	4222.4	1308.89	**2388**	4330.4
1242.78	**2269**	4116.2	1276.11	**2329**	4224.2	1309.44	**2389**	4332.2

TEMPERATURE CONVERSION TABLE (Continued)

To Convert			To Convert			To Convert		
To °C	←°F or °C→	To °F	To °C	←°F or °C→	To °F	To °C	←°F or °C→	To °F
1310	**2390**	4334	1343.33	**2450**	4442	1376.67	**2510**	4550
1310.56	**2391**	4335.8	1343.89	**2451**	4443.8	1377.22	**2511**	4551.8
1311.11	**2392**	4337.6	1344.44	**2452**	4445.6	1377.78	**2512**	4553.6
1311.67	**2393**	4339.4	1345	**2453**	4447.4	1378.33	**2513**	4555.4
1312.22	**2394**	4341.2	1345.56	**2454**	4449.2	1378.89	**2514**	4557.2
1312.78	**2395**	4343	1346.11	**2455**	4451	1379.44	**2515**	4559
1313.33	**2396**	4344.8	1346.67	**2456**	4452.8	1380	**2516**	4560.8
1313.89	**2397**	4346.6	1347.22	**2457**	4454.6	1380.56	**2517**	4562.6
1314.44	**2398**	4348.4	1347.78	**2458**	4456.4	1381.11	**2518**	4564.4
1315	**2399**	4350.2	1348.33	**2459**	4458.2	1381.67	**2519**	4566.2
1315.56	**2400**	4352	1348.89	**2460**	4460	1382.22	**2520**	4568
1316.11	**2401**	4353.8	1349.44	**2461**	4461.8	1382.78	**2521**	4569.8
1316.67	**2402**	4355.6	1350	**2462**	4463.6	1383.33	**2522**	4571.6
1317.22	**2403**	4357.4	1350.56	**2463**	4465.4	1383.89	**2523**	4573.4
1317.78	**2404**	4359.2	1351.11	**2464**	4467.2	1384.44	**2524**	4575.2
1318.33	**2405**	4361	1351.67	**2465**	4469	1385	**2525**	4577
1318.89	**2406**	4362.8	1352.22	**2466**	4470.8	1385.56	**2526**	4578.8
1319.44	**2407**	4364.6	1352.78	**2467**	4472.6	1386.11	**2527**	4580.6
1320	**2408**	4366.4	1353.33	**2468**	4474.4	1386.67	**2528**	4582.4
1320.56	**2409**	4368.2	1353.89	**2469**	4476.2	1387.22	**2529**	4584.2
1321.11	**2410**	4370	1354.44	**2470**	4478	1387.78	**2530**	4586
1321.67	**2411**	4371.8	1355	**2471**	4479.8	1388.33	**2531**	4587.8
1322.22	**2412**	4373.6	1355.56	**2472**	4481.6	1388.89	**2532**	4589.6
1322.78	**2413**	4375.4	1356.11	**2473**	4483.4	1389.44	**2533**	4591.4
1323.33	**2414**	4377.2	1356.67	**2474**	4485.2	1390	**2534**	4593.2
1323.89	**2415**	4379	1357.22	**2475**	4487	1390.56	**2535**	4595
1324.44	**2416**	4380.8	1357.78	**2476**	4488.8	1391.11	**2536**	4596.8
1325	**2417**	4382.6	1358.33	**2477**	4490.6	1391.67	**2537**	4598.6
1325.56	**2418**	4384.4	1358.89	**2478**	4492.4	1392.22	**2538**	4600.4
1326.11	**2419**	4386.2	1359.44	**2479**	4494.2	1392.78	**2539**	4602.2
1326.67	**2420**	4388	1360	**2480**	4496	1393.33	**2540**	4604
1327.22	**2421**	4389.8	1360.56	**2481**	4497.8	1393.89	**2541**	4605.8
1327.78	**2422**	4391.6	1361.11	**2482**	4499.6	1394.44	**2542**	4607.6
1328.33	**2423**	4393.4	1361.67	**2483**	4501.4	1395	**2543**	4609.4
1328.89	**2424**	4395.2	1362.22	**2484**	4503.2	1395.56	**2544**	4611.2
1329.44	**2425**	4397	1362.78	**2485**	4505	1396.11	**2545**	4613
1330	**2426**	4398.8	1363.33	**2486**	4506.8	1396.67	**2546**	4614.8
1330.56	**2427**	4400.6	1363.89	**2487**	4508.6	1397.22	**2547**	4616.6
1331.11	**2428**	4402.4	1364.44	**2488**	4510.4	1397.78	**2548**	4618.4
1331.67	**2429**	4404.2	1365	**2489**	4512.2	1398.33	**2549**	4620.2
1332.22	**2430**	4406	1365.56	**2490**	4514	1398.89	**2550**	4622
1332.78	**2431**	4407.8	1366.11	**2491**	4515.8	1399.44	**2551**	4623.8
1333.33	**2432**	4409.6	1366.67	**2492**	4517.6	1400	**2552**	4625.6
1333.89	**2433**	4411.4	1367.22	**2493**	4519.4	1400.56	**2553**	4627.4
1334.44	**2434**	4413.2	1367.78	**2494**	4521.2	1401.11	**2554**	4629.2
1335	**2435**	4415	1368.33	**2495**	4523	1401.67	**2555**	4631
1335.56	**2436**	4416.8	1368.89	**2496**	4524.8	1402.22	**2556**	4632.8
1336.11	**2437**	4418.6	1369.44	**2497**	4526.6	1402.78	**2557**	4634.6
1336.67	**2438**	4420.4	1370	**2498**	4528.4	1403.33	**2558**	4636.4
1337.22	**2439**	4422.2	1370.56	**2499**	4530.2	1403.89	**2559**	4638.2
1337.78	**2440**	4424	1371.11	**2500**	4532	1404.44	**2560**	4640
1338.33	**2441**	4425.8	1371.67	**2501**	4533.8	1405	**2561**	4641.8
1338.89	**2442**	4427.6	1372.22	**2502**	4535.6	1405.56	**2562**	4643.6
1339.44	**2443**	4429.4	1372.78	**2503**	4537.4	1406.11	**2563**	4645.4
1340	**2444**	4431.2	1373.33	**2504**	4539.2	1406.67	**2564**	4647.2
1340.56	**2445**	4433	1373.89	**2505**	4541	1407.22	**2565**	4649
1341.11	**2446**	4434.8	1374.44	**2506**	4542.8	1407.78	**2566**	4650.8
1341.67	**2447**	4436.6	1375	**2507**	4544.6	1408.33	**2567**	4652.6
1342.22	**2448**	4438.4	1375.56	**2508**	4546.4	1408.89	**2568**	4654.4
1342.78	**2449**	4440.2	1376.11	**2509**	4548.2	1409.44	**2569**	4656.2

TEMPERATURE CONVERSION TABLE (Continued)

To Convert			To Convert			To Convert		
To °C	←°F or °C→	To °F	To °C	←°F or °C→	To °F	To °C	←°F or °C→	To °F
1410	**2570**	4658	1443.33	**2630**	4766	1476.67	**2690**	4874
1410.56	**2571**	4659.8	1443.89	**2631**	4767.8	1477.22	**2691**	4875.8
1411.11	**2572**	4661.6	1444.44	**2632**	4769.6	1477.78	**2692**	4877.6
1411.67	**2573**	4663.4	1445	**2633**	4771.4	1478.33	**2693**	4879.4
1412.22	**2574**	4665.2	1445.56	**2634**	4773.2	1478.89	**2694**	4881.2
1412.78	**2575**	4667	1446.11	**2635**	4775	1479.44	**2695**	4883
1413.33	**2576**	4668.8	1446.67	**2636**	4776.8	1480	**2696**	4884.8
1413.89	**2577**	4670.6	1447.22	**2637**	4778.6	1480.56	**2697**	4886.6
1414.44	**2578**	4672.4	1447.78	**2638**	4780.4	1481.11	**2698**	4888.4
1415	**2579**	4674.2	1448.33	**2639**	4782.2	1481.67	**2699**	4890.2
1415.56	**2580**	4676	1448.89	**2640**	4784	1482.22	**2700**	4892
1416.11	**2581**	4677.8	1449.44	**2641**	4785.8	1482.78	**2701**	4893.8
1416.67	**2582**	4679.6	1450	**2642**	4787.6	1483.33	**2702**	4895.6
1417.22	**2583**	4681.4	1450.56	**2643**	4789.4	1483.89	**2703**	4897.4
1417.78	**2584**	4683.2	1451.11	**2644**	4791.2	1484.44	**2704**	4899.2
1418.33	**2585**	4685	1451.67	**2645**	4793	1485	**2705**	4901
1418.89	**2586**	4686.8	1452.22	**2646**	4794.8	1485.56	**2706**	4902.8
1419.44	**2587**	4688.6	1452.78	**2647**	4796.6	1486.11	**2707**	4904.6
1420	**2588**	4690.4	1453.33	**2648**	4798.4	1486.67	**2708**	4906.4
1420.56	**2589**	4692.2	1453.89	**2649**	4800.2	1487.22	**2709**	4908.2
1421.11	**2590**	4694	1454.44	**2650**	4802	1487.78	**2710**	4910
1421.67	**2591**	4695.8	1455	**2651**	4803.8	1488.33	**2711**	4911.8
1422.22	**2592**	4697.6	1455.56	**2652**	4805.6	1488.89	**2712**	4913.6
1422.78	**2593**	4699.4	1456.11	**2653**	4807.4	1489.44	**2713**	4915.4
1423.33	**2594**	4701.2	1456.67	**2654**	4809.2	1490	**2714**	4917.2
1423.89	**2595**	4703	1457.22	**2655**	4811	1490.56	**2715**	4919
1424.44	**2596**	4704.8	1457.78	**2656**	4812.8	1491.11	**2716**	4920.8
1425	**2597**	4706.6	1458.33	**2657**	4814.6	1491.67	**2717**	4922.6
1425.56	**2598**	4708.4	1458.89	**2658**	4816.4	1492.22	**2718**	4924.4
1426.11	**2599**	4710.2	1459.44	**2659**	4818.2	1492.78	**2719**	4926.2
1426.67	**2600**	4712	1460	**2660**	4820	1493.33	**2720**	4928
1427.22	**2601**	4713.8	1460.56	**2661**	4821.8	1493.89	**2721**	4929.8
1427.78	**2602**	4715.6	1461.11	**2662**	4823.6	1494.44	**2722**	4931.6
1428.33	**2603**	4717.4	1461.67	**2663**	4825.4	1495	**2723**	4933.4
1428.89	**2604**	4719.2	1462.22	**2664**	4827.2	1495.56	**2724**	4935.2
1429.44	**2605**	4721	1462.78	**2665**	4829	1496.11	**2725**	4937
1430	**2606**	4722.8	1463.33	**2666**	4830.8	1496.67	**2726**	4938.8
1430.56	**2607**	4724.6	1463.89	**2667**	4832.6	1497.22	**2727**	4040.6
1431.11	**2608**	4726.4	1464.44	**2668**	4834.4	1497.78	**2728**	4942.4
1431.67	**2609**	4728.2	1465	**2669**	4836.2	1498.33	**2729**	4944.2
1432.22	**2610**	4730	1465.56	**2670**	4838	1498.89	**2730**	4946
1432.78	**2611**	4731.8	1466.11	**2671**	4839.8	1499.44	**2731**	4947.8
1433.33	**2612**	4733.6	1466.67	**2672**	4841.6	1500	**2732**	4949.6
1433.89	**2613**	4735.4	1467.22	**2673**	4843.4	1500.56	**2733**	4951.4
1434.44	**2614**	4737.2	1467.78	**2674**	4845.2	1501.11	**2734**	4953.2
1435	**2615**	4739	1468.33	**2675**	4847	1501.67	**2735**	4955
1435.56	**2616**	4740.8	1468.89	**2676**	4848.8	1502.22	**2736**	4956.8
1436.11	**2617**	4742.6	1469.44	**2677**	4850.6	1502.78	**2737**	4958.6
1436.67	**2618**	4744.4	1470	**2678**	4852.4	1503.33	**2738**	4960.4
1437.22	**2619**	4746.2	1470.56	**2679**	4854.2	1503.89	**2739**	4962.2
1437.78	**2620**	4748	1471.11	**2680**	4856	1504.44	**2740**	4964
1438.33	**2621**	4749.8	1471.67	**2681**	4857.8	1505	**2741**	4965.8
1438.89	**2622**	4751.6	1472.22	**2682**	4859.6	1505.56	**2742**	4967.6
1439.44	**2623**	4753.4	1472.78	**2683**	4861.4	1506.11	**2743**	4969.4
1440	**2624**	4755.2	1473.33	**2684**	4863.2	1506.67	**2744**	4971.2
1440.56	**2625**	4757	1473.89	**2685**	4865	1507.22	**2745**	4973
1441.11	**2626**	4758.8	1474.44	**2686**	4866.8	1507.78	**2746**	4974.8
1441.67	**2627**	4760.6	1475	**2687**	4868.6	1508.33	**2747**	4976.6
1442.22	**2628**	4762.4	1475.56	**2688**	4870.4	1508.89	**2748**	4978.4
1442.78	**2629**	4764.2	1476.11	**2689**	4872.2	1509.44	**2749**	4980.2

To Convert			To Convert			To Convert		
To °C	←°F or °C→	To °F	To °C	←°F or °C→	To °F	To °C	←°F or °C→	To °F
1510	**2750**	4982	1543.33	**2810**	5090	1576.67	**2870**	5198
1510.56	**2751**	4983.8	1543.89	**2811**	5091.8	1577.22	**2871**	5199.8
1511.11	**2752**	4985.6	1544.44	**2812**	5093.6	1577.78	**2872**	5201.6
1511.67	**2753**	4987.4	1545	**2813**	5095.4	1578.33	**2873**	5203.4
1512.22	**2754**	4989.2	1545.56	**2814**	5097.2	1578.89	**2874**	5205.2
1512.78	**2755**	4991	1546.11	**2815**	5099	1579.44	**2875**	5207
1513.33	**2756**	4992.8	1546.67	**2816**	5100.8	1580	**2876**	5208.8
1513.89	**2757**	4994.6	1547.22	**2817**	5102.6	1580.56	**2877**	5210.6
1514.44	**2758**	4996.4	1547.78	**2818**	5104.4	1581.11	**2878**	5212.4
1515	**2759**	4998.2	1548.33	**2819**	5106.2	1581.67	**2879**	5214.2
1515.56	**2760**	5000	1548.89	**2820**	5108	1582.22	**2880**	5216
1516.11	**2761**	5001.8	1549.44	**2821**	5109.8	1582.78	**2881**	5217.8
1516.67	**2762**	5003.6	1550	**2822**	5111.6	1583.33	**2882**	5219.6
1517.22	**2763**	5005.4	1550.56	**2823**	5113.4	1583.89	**2883**	5221.4
1517.78	**2764**	5007.2	1551.11	**2824**	5115.2	1584.44	**2884**	5223.2
1518.33	**2765**	5009	1551.67	**2825**	5117	1585	**2885**	5225
1518.89	**2766**	5010.8	1552.22	**2826**	5118.8	1585.56	**2886**	5226.8
1519.44	**2767**	5012.6	1552.78	**2827**	5120.6	1586.11	**2887**	5228.6
1520	**2768**	5014.4	1553.33	**2828**	5122.4	1586.67	**2888**	5230.4
1520.56	**2769**	5016.2	1553.89	**2829**	5124.2	1587.22	**2889**	5232.2
1521.11	**2770**	5018	1554.44	**2830**	5126	1587.78	**2890**	5234
1521.67	**2771**	5019.8	1555	**2831**	5127.8	1588.33	**2891**	5235.8
1522.22	**2772**	5021.6	1555.56	**2832**	5129.6	1588.89	**2892**	5237.6
1522.78	**2773**	5023.4	1556.11	**2833**	5131.4	1589.44	**2893**	5239.4
1523.33	**2774**	5025.2	1556.67	**2834**	5133.2	1590	**2894**	5241.2
1523.89	**2775**	5027	1557.22	**2835**	5135	1590.56	**2895**	5243
1524.44	**2776**	5028.8	1557.78	**2836**	5136.8	1591.11	**2896**	5244.8
1525	**2777**	5030.6	1558.33	**2837**	5138.6	1591.67	**2897**	5246.6
1525.56	**2778**	5032.4	1558.89	**2838**	5140.4	1592.22	**2898**	5248.4
1526.11	**2779**	5034.2	1559.44	**2839**	5142.2	1592.78	**2899**	5250.2
1526.67	**2780**	5036	1560	**2840**	5144	1593.33	**2900**	5252
1527.22	**2781**	5037.8	1560.56	**2841**	5145.8	1593.89	**2901**	5253.8
1527.78	**2782**	5039.6	1561.11	**2842**	5147.6	1594.44	**2902**	5255.6
1528.33	**2783**	5041.4	1561.67	**2843**	5149.4	1595	**2903**	5257.4
1528.89	**2784**	5043.2	1562.22	**2844**	5151.2	1595.56	**2904**	5259.2
1529.44	**2785**	5045	1562.78	**2845**	5153	1596.11	**2905**	5261
1530	**2786**	5046.8	1563.33	**2846**	5154.8	1596.67	**2906**	5262.8
1530.56	**2787**	5048.6	1563.89	**2847**	5156.6	1597.22	**2907**	5264.6
1531.11	**2788**	5050.4	1564.44	**2848**	5158.4	1597.78	**2908**	5266.4
1531.67	**2789**	5052.2	1565	**2849**	5160.2	1598.33	**2909**	5268.2
1532.22	**2790**	5054	1565.56	**2850**	5162	1598.89	**2910**	5270
1532.78	**2791**	5055.8	1566.11	**2851**	5163.8	1599.44	**2911**	5271.8
1533.33	**2792**	5057.6	1566.67	**2852**	5165.6	1600	**2912**	5273.6
1533.89	**2793**	5059.4	1567.22	**2853**	5167.4	1600.56	**2913**	5275.4
1534.44	**2794**	5061.2	1567.78	**2854**	5169.2	1601.11	**2914**	5277.2
1535	**2795**	5063	1568.33	**2855**	5171	1601.67	**2915**	5279
1535.56	**2796**	5064.8	1568.89	**2856**	5172.8	1602.22	**2916**	5280.8
1536.11	**2797**	5066.6	1569.44	**2857**	5174.6	1602.78	**2917**	5282.6
1536.67	**2798**	5068.4	1570	**2858**	5776.4	1603.33	**2918**	5284.4
1537.22	**2799**	5070.2	1570.56	**2859**	5178.2	1603.89	**2919**	5286.2
1537.78	**2800**	5072	1571.11	**2860**	5180	1604.44	**2920**	5288
1538.33	**2801**	5073.8	1571.67	**2861**	5181.8	1605	**2921**	5289.8
1538.89	**2802**	5075.6	1572.22	**2862**	5183.6	1605.56	**2922**	5291.6
1539.44	**2803**	5077.4	1572.78	**2863**	5185.4	1606.11	**2923**	5293.4
1540	**2804**	5079.2	1573.33	**2864**	5187.2	1606.67	**2924**	5295.2
1540.56	**2805**	5081	1573.89	**2865**	5189	1607.22	**2925**	5297
1541.11	**2806**	5082.8	1574.44	**2866**	5190.8	1607.78	**2926**	5298.8
1541.67	**2807**	5084.6	1575	**2867**	5192.6	1608.33	**2927**	5300.6
1542.22	**2808**	5086.4	1575.56	**2868**	5194.4	1608.89	**2928**	5302.4
1542.78	**2809**	5088.2	1576.11	**2869**	5196.2	1609.44	**2929**	5304.2

TEMPERATURE CONVERSION TABLE (Continued)

To Convert			To Convert			To Convert		
To °C	←°F or °C→	To °F	To °C	←°F or °C→	To °F	To °C	←°F or °C→	To °F
1610	**2930**	5306	1643.33	**2990**	5414	1676.67	**3050**	5522
1610.56	**2931**	5307.8	1643.89	**2991**	5415.8	1677.22	**3051**	5523.8
1611.11	**2932**	5309.6	1644.44	**2992**	5417.6	1677.78	**3052**	5525.6
1611.67	**2933**	5311.4	1645	**2993**	5419.4	1678.33	**3053**	5527.4
1612.22	**2934**	5313.2	1645.56	**2994**	5421.2	1678.89	**3054**	5529.2
1612.78	**2935**	5315	1646.11	**2995**	5423	1679.44	**3055**	5531
1613.33	**2936**	5316.8	1646.67	**2996**	5424.8	1680	**3056**	5532.8
1613.89	**2937**	5318.6	1647.22	**2997**	5426.6	1680.56	**3057**	5534.6
1614.44	**2938**	5320.4	1647.78	**2998**	5428.4	1681.11	**3058**	5536.4
1615	**2939**	5322.2	1648.33	**2999**	5430.2	1681.67	**3059**	5538.2
1615.56	**2940**	5324	1648.89	**3000**	5432	1682.22	**3060**	5540
1616.11	**2941**	5325.8	1649.44	**3001**	5433.8	1682.78	**3061**	5541.8
1616.67	**2942**	5327.6	1650	**3002**	5435.6	1683.33	**3062**	5543.6
1617.22	**2943**	5329.4	1650.56	**3003**	5437.4	1683.89	**3063**	5545.4
1617.78	**2944**	5331.2	1651.11	**3004**	5439.2	1684.44	**3064**	5547.2
1618.33	**2945**	5333	1651.67	**3005**	5441	1685	**3065**	5549
1618.89	**2946**	5334.8	1652.22	**3006**	5442.8	1685.56	**3066**	5550.8
1619.44	**2947**	5336.6	1652.78	**3007**	5444.6	1686.11	**3067**	5552.6
1620	**2948**	5338.4	1653.33	**3008**	5446.4	1686.67	**3068**	5554.4
1620.56	**2949**	5340.2	1653.89	**3009**	5448.2	1687.22	**3069**	5556.2
1621.11	**2950**	5342	1654.44	**3010**	5450	1687.78	**3070**	5558
1621.67	**2951**	5343.8	1655	**3011**	5451.8	1688.33	**3071**	5559.8
1622.22	**2952**	5345.6	1655.56	**3012**	5453.6	1688.89	**3072**	5561.6
1622.78	**2953**	5347.4	1656.11	**3013**	5455.4	1689.44	**3073**	5563.4
1623.33	**2954**	5349.2	1656.67	**3014**	5457.2	1690	**3074**	5565.2
1623.89	**2955**	5351	1657.22	**3015**	5459	1690.56	**3075**	5567
1624.44	**2956**	5352.8	1657.78	**3016**	5460.8	1691.11	**3076**	5568.8
1625	**2957**	5354.6	1658.33	**3017**	5462.6	1691.67	**3077**	5570.6
1625.56	**2958**	5356.4	1658.89	**3018**	5464.4	1692.22	**3078**	5572.4
1626.11	**2959**	5358.2	1659.44	**3019**	5466.2	1692.78	**3079**	5574.2
1626.67	**2960**	5360	1660	**3020**	5468	1693.33	**3080**	5576
1627.22	**2961**	5361.8	1660.56	**3021**	5469.8	1693.89	**3081**	5577.8
1627.78	**2962**	5363.6	1661.11	**3022**	5471.6	1694.44	**3082**	5579.6
1628.33	**2963**	5365.4	1661.67	**3023**	5473.4	1695	**3083**	5581.4
1628.89	**2964**	5367.2	1662.22	**3024**	5475.2	1695.56	**3084**	5583.2
1629.44	**2965**	5369	1662.78	**3025**	5477	1696.11	**3085**	5585
1630	**2966**	5370.8	1663.33	**3026**	5478.8	1696.67	**3086**	5586.8
1630.56	**2967**	5372.6	1663.89	**3027**	5480.6	1697.22	**3087**	5588.6
1631.11	**2968**	5374.4	1664.44	**3028**	5482.4	1697.78	**3088**	5590.4
1631.67	**2969**	5376.2	1665	**3029**	5484.2	1698.33	**3089**	5592.2
1632.22	**2970**	5378	1665.56	**3030**	5486	1698.89	**3090**	5594
1632.78	**2971**	5379.8	1666.11	**3031**	5487.8	1699.44	**3091**	5595.8
1633.33	**2972**	5381.6	1666.67	**3032**	5489.6	1700	**3092**	5597.6
1633.89	**2973**	5383.4	1667.22	**3033**	5491.4	1700.56	**3093**	5599.4
1634.44	**2974**	5385.2	1667.78	**3034**	5493.2	1701.11	**3094**	5601.2
1635	**2975**	5387	1668.33	**3035**	5495	1701.67	**3095**	5603
1635.56	**2976**	5388.8	1668.89	**3036**	5496.8	1702.22	**3096**	5604.8
1636.11	**2977**	5390.6	1669.44	**3037**	5498.6	1702.78	**3097**	5606.6
1636.67	**2978**	5392.4	1670	**3038**	5500.4	1703.33	**3098**	5608.4
1637.22	**2979**	5394.2	1670.56	**3039**	5502.2	1703.89	**3099**	5610.2
1637.78	**2980**	5396	1671.11	**3040**	5504	1704.44	**3100**	5612
1638.33	**2981**	5397.8	1671.67	**3041**	5505.8	1705	**3101**	5613.8
1638.89	**2982**	5399.6	1672.22	**3042**	5507.6	1705.56	**3102**	5615.6
1639.44	**2983**	5401.4	1672.78	**3043**	5509.4	1706.11	**3103**	5617.4
1640	**2984**	5403.2	1673.33	**3044**	5511.2	1706.67	**3104**	5619.2
1640.56	**2985**	5405	1673.89	**3045**	5513	1707.22	**3105**	5621
1641.11	**2986**	5406.8	1674.44	**3046**	5514.8	1707.78	**3106**	5622.8
1641.67	**2987**	5408.6	1675	**3047**	5516.6	1708.33	**3107**	5624.6
1642.22	**2988**	5410.4	1675.56	**3048**	5518.4	1708.89	**3108**	5626.4
1642.78	**2989**	5412.2	1676.11	**3049**	5520.2	1709.44	**3109**	5628.2

To Convert			To Convert			To Convert		
To °C	←°F or °C→	To °F	To °C	←°F or °C→	To °F	To °C	←°F or °C→	To °F
1710	**3110**	5630	1743.33	**3170**	5738	1776.67	**3230**	5846
1710.56	**3111**	5631.8	1743.89	**3171**	5739.8	1777.22	**3231**	5847.8
1711.11	**3112**	5633.6	1744.44	**3172**	5741.6	1777.78	**3232**	5849.6
1711.67	**3113**	5635.4	1745	**3173**	5743.4	1778.33	**3233**	5851.4
1712.22	**3114**	5637.2	1745.56	**3174**	5745.2	1778.89	**3234**	5853.2
1712.78	**3115**	5639	1746.11	**3175**	5747	1779.44	**3235**	5855
1713.33	**3116**	5640.8	1746.67	**3176**	5748.8	1780	**3236**	5856.8
1713.89	**3117**	5642.6	1747.22	**3177**	5750.6	1780.56	**3237**	5858.6
1714.44	**3118**	5644.4	1747.78	**3178**	5752.4	1781.11	**3238**	5860.4
1715	**3119**	5646.2	1748.33	**3179**	5754.2	1781.67	**3239**	5862.2
1715.56	**3120**	5648	1748.89	**3180**	5756	1782.22	**3240**	5864
1716.11	**3121**	5649.8	1749.44	**3181**	5757.8	1782.78	**3241**	5865.8
1716.67	**3122**	5651.6	1750	**3182**	5759.6	1783.33	**3242**	5867.6
1717.22	**3123**	5653.4	1750.56	**3183**	5761.4	1783.89	**3243**	5869.4
1717.78	**3124**	5655.2	1751.11	**3184**	5763.2	1784.44	**3244**	5871.2
1718.33	**3125**	5657	1751.67	**3185**	5765	1785	**3245**	5873
1718.89	**3126**	5658.8	1752.22	**3186**	5766.8	1785.56	**3246**	5874.8
1719.44	**3127**	5660.6	1752.78	**3187**	5768.6	1786.11	**3247**	5876.6
1720	**3128**	5662.4	1753.33	**3188**	5770.4	1786.67	**3248**	5878.4
1720.56	**3129**	5664.2	1753.89	**3189**	5772.2	1787.22	**3249**	5880.2
1721.11	**3130**	5666	1754.44	**3190**	5774	1787.78	**3250**	5882
1721.67	**3131**	5667.8	1755	**3191**	5775.8	1788.33	**3251**	5883.8
1722.22	**3132**	5669.6	1755.56	**3192**	5777.6	1788.89	**3252**	5885.6
1722.78	**3133**	5671.4	1756.11	**3193**	5779.4	1789.44	**3253**	5887.4
1723.33	**3134**	5673.2	1756.67	**3194**	5781.2	1790	**3254**	5889.2
1723.89	**3135**	5675	1757.22	**3195**	5783	1790.56	**3255**	5891
1724.44	**3136**	5676.8	1757.78	**3196**	5784.8	1791.11	**3256**	5892.8
1725	**3137**	5678.6	1758.33	**3197**	5786.6	1791.67	**3257**	5894.6
1725.56	**3138**	5680.4	1758.89	**3198**	5788.4	1792.22	**3258**	5896.4
1726.11	**3139**	5682.2	1759.44	**3199**	5790.2	1792.78	**3259**	5898.2
1726.67	**3140**	5684	1760	**3200**	5792	1793.33	**3260**	5900
1727.22	**3141**	5685.8	1760.56	**3201**	5793.8	1793.89	**3261**	5901.8
1727.78	**3142**	5687.6	1761.11	**3202**	5795.6	1794.44	**3262**	5903.6
1728.33	**3143**	5689.4	1761.67	**3203**	5797.4	1795	**3263**	5905.4
1728.89	**3144**	5691.2	1762.22	**3204**	5799.2	1795.56	**3264**	5907.2
1729.44	**3145**	5693	1762.78	**3205**	5801	1796.11	**3265**	5909
1730	**3146**	5694.8	1763.33	**3206**	5802.8	1796.67	**3266**	5910.8
1730.56	**3147**	5696.6	1763.89	**3207**	5804.6	1797.22	**3267**	5912.6
1731.11	**3148**	5698.4	1764.44	**3208**	5806.4	1797.78	**3268**	5914.4
1731.67	**3149**	5700.2	1765	**3209**	5808.2	1798.33	**3269**	5916.2
1732.22	**3150**	5702	1765.56	**3210**	5810	1798.89	**3270**	5918
1732.78	**3151**	5703.8	1766.11	**3211**	5811.8	1799.44	**3271**	5919.8
1733.33	**3152**	5705.6	1766.67	**3212**	5813.6	1800	**3272**	5921.6
1733.89	**3153**	5707.4	1767.22	**3213**	5815.4	1800.56	**3273**	5923.4
1734.44	**3154**	5709.2	1767.78	**3214**	5817.2	1801.11	**3274**	5925.2
1735	**3155**	5711	1768.33	**3215**	5819	1801.67	**3275**	5927
1735.56	**3156**	5712.8	1768.89	**3216**	5820.8	1802.22	**3276**	5928.8
1736.11	**3157**	5714.6	1769.44	**3217**	5822.6	1802.78	**3277**	5930.6
1736.67	**3158**	5716.4	1770	**3218**	5824.4	1803.33	**3278**	5932.4
1737.22	**3159**	5718.2	1770.56	**3219**	5826.2	1803.89	**3279**	5934.2
1737.78	**3160**	5720	1771.11	**3220**	5828	1804.44	**3280**	5936
1738.33	**3161**	5721.8	1771.67	**3221**	5829.8	1805	**3281**	5937.8
1738.89	**3162**	5723.6	1772.22	**3222**	5831.6	1805.56	**3282**	5939.6
1739.44	**3163**	5725.4	1772.78	**3223**	5833.4	1806.11	**3283**	5941.4
1740	**3164**	5727.2	1773.33	**3224**	5835.2	1806.67	**3284**	5943.2
1740.56	**3165**	5729	1773.89	**3225**	5837	1807.22	**3285**	5945
1741.11	**3166**	5730.8	1774.44	**3226**	5838.8	1807.78	**3286**	5946.8
1741.67	**3167**	5732.6	1775	**3227**	5840.6	1808.33	**3287**	5948.6
1742.22	**3168**	5734.4	1775.56	**3228**	5842.4	1808.89	**3288**	5950.4
1742.78	**3169**	5736.2	1776.11	**3229**	5844.2	1809.44	**3289**	5952.2

TEMPERATURE CONVERSION TABLE (Continued)

To Convert			To Convert			To Convert		
To °C	←°F or °C→	To °F	To °C	←°F or °C→	To °F	To °C	←°F or °C→	To °F
1810	**3290**	5954	1954.44	**3550**	6422	2121.11	**3850**	6962
1810.56	**3291**	5955.8	1957.22	**3555**	6431	2123.89	**3855**	6971
1811.11	**3292**	5957.6	1960.00	**3560**	6440	2126.67	**3860**	6980
1811.67	**3293**	5959.4	1962.78	**3565**	6449	2129.44	**3865**	6989
1812.22	**3294**	5961.2	1965.56	**3570**	6458	2132.22	**3870**	6998
1812.78	**3295**	5963	1968.33	**3575**	6467	2135.00	**3875**	7007
1813.33	**3296**	5964.8	1971.11	**3580**	6476	2137.78	**3880**	7016
1813.89	**3297**	5966.6	1973.89	**3585**	6485	2140.56	**3885**	7025
1814.44	**3298**	5968.4	1976.67	**3590**	6494	2143.33	**3890**	7034
1815	**3299**	5970.2	1979.44	**3595**	6503	2145.11	**3895**	7043
1815.56	**3300**	5972	1982.22	**3600**	6512	2148.89	**3900**	7052
1818.33	**3305**	5981	1985.00	**3605**	6521	2151.67	**3905**	7061
1821.11	**3310**	5990	1987.78	**3610**	6530	2154.44	**3910**	7070
1823.89	**3315**	5999	1990.56	**3615**	6539	2157.22	**3915**	7079
1826.67	**3320**	6008	1993.33	**3620**	6548	2160.00	**3920**	7088
1829.44	**3325**	6017	1996.11	**3625**	6557	2162.78	**3925**	7097
1832.22	**3330**	6026	1998.89	**3630**	6566	2165.56	**3930**	7106
1835.00	**3335**	6035	2001.67	**3635**	6575	2168.33	**3935**	7115
1837.78	**3340**	6044	2004.44	**3640**	6584	2171.11	**3940**	7124
1840.56	**3345**	6053	2007.22	**3645**	6593	2173.89	**3945**	7133
1843.33	**3350**	6062	2010.00	**3650**	6602	2176.67	**3950**	7142
1846.11	**3355**	6071	2012.78	**3655**	6611	2179.44	**3955**	7151
1848.89	**3360**	6080	2015.56	**3660**	6620	2182.22	**3960**	7160
1851.67	**3365**	6089	2018.33	**3665**	6629	2185.00	**3965**	7169
1854.44	**3370**	6098	2021.11	**3670**	6638	2187.78	**3970**	7178
1857.22	**3375**	6107	2023.89	**3675**	6647	2190.56	**3975**	7187
1860.00	**3380**	6116	2026.67	**3680**	6656	2193.33	**3980**	7196
1862.78	**3385**	6125	2029.44	**3685**	6665	2196.11	**3985**	7205
1865.56	**3390**	6134	2032.22	**3690**	6674	2198.89	**3990**	7214
1868.33	**3395**	6143	2035.00	**3695**	6683	2201.67	**3995**	7223
1871.11	**3400**	6152	2037.78	**3700**	6692	2204.44	**4000**	7232
1873.89	**3405**	6161	2040.56	**3705**	6701	2207.22	**4005**	7241
1876.67	**3410**	6170	2043.33	**3710**	6710	2210.00	**4010**	7250
1879.44	**3415**	6179	2046.11	**3715**	6719	2212.78	**4015**	7259
1882.22	**3420**	6188	2048.89	**3720**	6728	2215.56	**4020**	7268
1885.00	**3425**	6197	2051.67	**3725**	6737	2218.33	**4025**	7277
1887.78	**3430**	6206	2054.44	**3730**	6746	2221.11	**4030**	7286
1890.56	**3435**	6215	2057.22	**3735**	6755	2223.89	**4035**	7295
1893.33	**3440**	6224	2060.00	**3740**	6764	2226.67	**4040**	7304
1896.11	**3445**	6233	2062.78	**3745**	6773	2229.44	**4045**	7313
1898.89	**3450**	6242	2065.56	**3750**	6782	2232.22	**4050**	7322
1901.67	**3455**	6251	2068.33	**3755**	6791	2235.00	**4055**	7331
1904.44	**3460**	6260	2071.11	**3760**	6800	2237.78	**4060**	7340
1907.22	**3465**	6269	2073.89	**3765**	6809	2240.56	**4065**	7349
1910.00	**3470**	6278	2076.67	**3770**	6818	2243.33	**4070**	7358
1912.78	**3475**	6287	2079.44	**3775**	6827	2246.11	**4075**	7367
1915.56	**3480**	6296	2082.22	**3780**	6836	2248.89	**4080**	7376
1918.33	**3485**	6305	2085.00	**3785**	6845	2251.67	**4085**	7385
1921.11	**3490**	6314	2087.78	**3790**	6854	2254.44	**4090**	7394
1923.89	**3495**	6323	2090.56	**3795**	6863	2257.22	**4095**	7403
1926.67	**3500**	6332	2093.33	**3800**	6872	2260.00	**4100**	7412
1929.44	**3505**	6341	2096.11	**3805**	6881	2262.78	**4105**	7421
1932.22	**3510**	6350	2098.89	**3810**	6890	2265.56	**4110**	7430
1935.00	**3515**	6359	2101.67	**3815**	6899	2268.33	**4115**	7439
1937.78	**3520**	6368	2104.44	**3820**	6908	2271.11	**4120**	7448
1940.56	**3525**	6377	2107.22	**3825**	6917	2273.89	**4125**	7457
1943.33	**3530**	6386	2110.00	**3830**	6926	2276.67	**4130**	7466
1946.11	**3535**	6395	2112.78	**3835**	6935	2279.44	**4135**	7475
1948.89	**3540**	6404	2115.56	**3840**	6944	2282.22	**4140**	7484
1951.67	**3545**	6413	2118.33	**3845**	6953	2285.00	**4145**	7493

TEMPERATURE CONVERSION TABLE (Continued)

To Convert			To Convert			To Convert		
To °C	←°F or °C→	To °F	To °C	←°F or °C→	To °F	To °C	←°F or °C→	To °F
2287.78	**4150**	7502	2354.44	**4270**	7718	2421.11	**4390**	7934
2290.56	**4155**	7511	2357.22	**4275**	7727	2423.89	**4395**	7943
2293.33	**4160**	7520	2360.00	**4280**	7736	2426.67	**4400**	7952
2296.11	**4165**	7529	2362.78	**4285**	7745	2429.44	**4405**	7961
2298.89	**4170**	7538	2365.56	**4290**	7754	2432.22	**4410**	7970
2301.67	**4175**	7547	2368.33	**4295**	7763	2435.00	**4415**	7979
2304.44	**4180**	7556	2371.11	**4300**	7772	2437.78	**4420**	7988
2307.22	**4185**	7565	2373.89	**4305**	7781	2440.56	**4425**	7997
2310.00	**4190**	7574	2376.67	**4310**	7790	2443.33	**4430**	8006
2312.78	**4195**	7583	2379.44	**4315**	7799	2446.11	**4435**	8015
2315.56	**4200**	7592	2382.22	**4320**	7808	2448.89	**4440**	8024
2318.33	**4205**	7601	2385.00	**4325**	7817	2451.67	**4445**	8033
2321.11	**4210**	7610	2387.78	**4330**	7826	2454.44	**4450**	8042
2323.89	**4215**	7619	2390.56	**4335**	7835	2457.22	**4455**	8051
2326.67	**4220**	7628	2393.33	**4340**	7844	2460.00	**4460**	8060
2329.44	**4225**	7637	2396.11	**4345**	7853	2462.78	**4465**	8069
2332.22	**4230**	7646	2398.89	**4350**	7862	2465.56	**4470**	8078
2335.00	**4235**	7655	2401.67	**4355**	7871	2468.33	**4475**	8087
2337.78	**4240**	7664	2404.44	**4360**	7880	2471.11	**4480**	8096
2340.56	**4245**	7673	2407.22	**4365**	7889	2473.89	**4485**	8105
2343.33	**4250**	7682	2410.00	**4370**	7898	2476.67	**4490**	8114
2346.11	**4255**	7691	2412.78	**4375**	7907	2479.44	**4495**	8123
2348.89	**4260**	7700	2415.56	**4380**	7916	2482.22	**4500**	8132
2351.67	**4265**	7709	2418.33	**4385**	7925			

STANDARD TYPES OF STAINLESS AND HEAT RESISTING STEELS

Chemical Ranges and Limits

Subject to Tolerances for Check Analyses

By permission of American Iron and Steel Institute

Type Number	Chemical Composition, per cent													
	C	Mn Max.	P Max.	S Max.	Si Max.	Cr	Ni	Mo	Zr	Se	Cb-Ta	Ta	Al	N
-201	0.15 Max.	5.50/ 7.50	0.060	0.030	1.00	16.00/ 18.00	3.50/ 5.50							0.25 Max.
-202	0.15 Max.	7.50/ 10.00	0.060	0.030	1.00	17.00/ 19.00	4.00/ 6.00							0.25 Max.
-301	0.15 Max.	2.00	0.045	0.030	1.00	16.00/ 18.00	6.00/ 8.00							
-302	0.15 Max.	2.00	0.045	0.030	1.00	17.00/ 19.00	8.00/ 10.00							
-302B	0.15 Max.	2.00	0.045	0.030	2.00/ 3.00	17.00/ 19.00	8.00/ 10.00							
-303	0.15 Max.	2.00	0.20	0.15 Min.	1.00	17.00/ 19.00	8.00/ 10.00	0.60* Max.	0.60* Max.					
-303 Se	0.15 Max.	2.00	0.20	0.06	1.00	17.00/ 19.00	8.00/ 10.00			0.15 Min.				
-304	0.08 Max	2.00	0.045	0.030	1.00	18.00/ 20.00	8.00/ 12.00							
-304L	0.03 Max.	2.00	0.045	0.030	1.00	18.00/ 20.00	8.00/ 12.00							
-305	0.12 Max.	2.00	0.045	0.030	1.00	17.00/ 19.00	10.00/ 13.00							
-308	0.08 Max.	2.00	0.045	0.030	1.00	19.00/ 21.00	10.00/ 12.00							
-309	0.20 Max.	2.00	0.045	0.030	1.00	22.00/ 24.00	12.00/ 15.00							
-309S	0.08 Max.	2.00	0.045	0.030	1.00	22.00/ 24.00	12.00/ 15.00							
-310	0.25 Max.	2.00	0.045	0.030	1.50	24.00/ 26.00	19.00/ 22.00							
-310S	0.08 Max.	2.00	0.045	0.030	1.50	24.00/ 26.00	19.00/ 22.00							
-314	0.25 Max.	2.00	0.045	0.030	1.50/ 3.00	23.00/ 26.00	19.00/ 22.00							
-316	0.08 Max.	2.00	0.045	0.030	1.00	16.00/ 18.00	10.00/ 14.00	2.00/ 3.00						
-316L	0.03 Max.	2.00	0.045	0.030	1.00	16.00/ 18.00	10.00/ 14.00	2.00 3.00						
-317	0.08 Max.	2.00	0.045	0.030	1.00	18.00/ 20.00	11.00/ 15.00	3.00/ 4.00						
-321	0.08 Max.	2.00	0.045	0.030	1.00	17.00/ 19.00	9.00/ 12.00				5 × C Min.			
-347	0.08 Max.	2.00	0.045	0.030	1.00	17.00/ 19.00	9.00/ 13.00					10 × C Min.		
-348	0.08 Max.	2.00	0.045	0.030	1.00	17.00/ 19.00	9.00/ 13.00				10 × C Min.	0.10 Max.		
**403	0.15 Max.	1.00	0.040	0.030	0.50	11.50/ 13.00								
°405	0.08 Max.	1.00	0.040	0.030	1.00	11.50/ 14.50							0.10/ 0.30	
**410	0.15 Max.	1.00	0.040	0.030	1.00	11.50/ 13.50								
**414	0.15 Max.	1.00	0.040	0.030	1.00	11.50/ 13.50	1.25/ 2.50							
**416	0.15 Max.	1.25	0.06	0.15 Min.	1.00	12.00/ 14.00		0.60* Max.	0.60* Max.					
**416 Se	0.15 Max.	1.25	0.06	0.06	1.00	12.00/ 14.00				0.15 Min.				
**420	Over 0.15	1.00	0.040	0.030	1.00	12.00/ 14.00								

* At producer's option; reported only when intentionally added.
** Heat treatable.
- Not heat treatable.
° Essentially not heat treatable.

Chemical Composition, per cent

Type Number	C	Mn Max.	P Max.	S Max.	Si Max.	Cr	Ni	Mo	Zr	Se	Cb-Ta	Ta	Al	N
°430	0.12 Max.	1.00	0.040	0.030	1.00	14.00/ 18.00								
°430F	0.12 Max.	1.25	0.06	0.15 Min.	1.00	14.00/ 18.00		0.60* Max.	0.60* Max.					
°430F Se	0.12 Max.	1.25	0.06	0.06	1.00	14.00/ 18.00				0.15 Min.				
**431	0.20 Max.	1.00	0.040	0.030	1.00	15.00/ 17.00	1.25/ 2.50							
**440A	0.60/ 0.75	1.00	0.040	0.030	1.00	16.00/ 18.00		0.75 Max.						
**400B	0.75/ 0.95	1.00	0.040	0.030	1.00	16.00/ 18.00		0.75 Max.						
**440C	0.95/ 1.20	1.00	0.040	0.030	1.00	16.00/ 18.00		0.75 Max.						
°446	0.20 Max.	1.50	0.040	0.030	1.00	23.00/ 27.00								0.25 Max.
**501	Over 0.10	1.00	0.040	0.030	1.00	4.00/ 6.00		0.40/ 0.65						
**502	0.10 Max.	1.00	0.040	0.030	1.00	4.00/ 6.00		0.40/ 0.65						

* At producer's option; reported only when intentionally added.
** Heat treatable.
- Not heat treatable.
° Essentially not heat treatable.

WIRE TABLES

COMPARISON OF WIRE GAUGES

Diameter of Wire in Inches

Gauge No.	Brown & Sharpe	Bir-ming-ham or Stubs'.	Wash-burn & Moen	Imperial or Brit. Std.	Stubs' Steel	U. S. Std. plate
00000000	......	......	......	......	......	......
0000000	......	......	......	.500	......	......
000000	......	......	......	.464	......	.46875
00000	......	......	......	.432	......	.4375
0000	.4600	.454	.3938	.400	......	.40625
000	.4096	.425	.3625	.372	......	.375
00	.3648	.380	.3310	.348	......	.34375
0	.3249	.340	.3065	.324	......	.3125
1	.2893	.300	.2830	.300	.227	.28125
2	.2576	.284	.2625	.276	.219	.265625
3	.2294	.259	.2437	.252	.212	.25
4	.2043	.238	.2253	.232	.207	.234375
5	.1819	.220	.2070	.212	.204	.21875
6	.1620	.203	.1920	.192	.201	.203125
7	.1443	.180	.1770	.176	.199	.1875
8	.1285	.165	.1620	.160	.197	.171875
9	.1144	.148	.1483	.144	.194	.15625
10	.1019	.134	.1350	.128	.191	.140625
11	.09074	.120	.1205	.116	.188	.125
12	.08081	.109	.1055	.104	.185	.109375
13	.07196	.095	.0915	.092	.182	.09375
14	.06408	.083	.0800	.080	.180	.078125
15	.05707	.072	.0720	.072	.178	.0703125
16	.05082	.065	.0625	.064	.175	.0625
17	.04526	.058	.0540	.056	.172	.05625
18	.04030	.049	.0475	.048	.168	.05
19	.03589	.042	.0410	.040	.164	.04375
20	.03196	.035	.0348	.036	.161	.0375
21	.02846	.032	.0318	.032	.157	.034375
22	.02535	.028	.0286	.028	.155	.03125
23	.02257	.025	.0258	.024	.153	.028125
24	.02010	.022	.0230	.022	.151	.025

Gauge No.	Brown & Sharpe	Bir-ming-ham or Stubs'.	Wash-burn & Moen	Impe-rial or Brit. Std.	Stubs' steel	U. S. Std. plate
25	0.01790	0.020	0.0204	0.020	0.148	0.021875
26	0.01594	0.018	0.0181	0.018	0.146	0.01875
27	0.01419	0.016	0.0173	0.0164	0.143	0.0171875
28	0.01264	0.014	0.0162	0.0149	0.139	0.015625
29	0.01126	0.013	0.0150	0.0136	0.134	0.0140625
30	0.01003	0.012	0.0140	0.0124	0.127	0.0125
31	0.008928	0.010	0.0132	0.0116	0.120	0.0109375
32	0.007950	0.009	0.0128	0.0108	0.115	0.01015625
33	0.007080	0.008	0.0118	0.0100	0.112	0.009375
34	0.006304	0.007	0.0104	0.0092	0.110	0.00859375
35	0.005614	0.005	0.0095	0.0084	0.108	0.0078125
36	0.005000	0.004	0.0090	0.0076	0.106	0.00703125
37	0.004453	......	0.0085	0.0068	0.103	0.006640625
38	0.003965	......	0.0080	0.0060	0.101	0.00625
39	0.003531	......	0.0075	0.0052	0.099	......
40	0.003145	......	0.0070	0.0048	0.097	......
41	......	......	0.0066	0.0044	0.095	......
42	......	......	0.0062	0.0040	0.092	......
43	......	......	0.0060	0.0036	0.088	......
44	......	......	0.0058	0.0032	0.085	......
45	......	......	0.0055	0.0028	0.081	......
46	......	......	0.0052	0.0024	0.079	......
47	......	......	0.0050	0.0020	0.077	......
48	......	......	0.0048	0.0016	0.075	......
49	......	......	0.0046	0.0012	0.072	......
50	......	......	0.0044	0.0010	0.069	......

Diameter of Wire in Centimeters

Gauge No.	Brown & Sharpe	Bir-ming-ham or Stubs.	Wash-burn & Moen	Impe-rial or Brit. Std.	Stubs' steel	U. S. Std. plate
00000000	......	......	......	......	......	......
0000000	......	......	1.245	1.27	......	1.27
000000	......	......	1.172	1.18	......	1.191
00000	......	......	1.093	1.10	......	1.111
0000	1.168	1.15	1.000	1.02	......	1.032
000	1.040	1.08	0.9208	0.945	......	0.9525
00	0.9266	0.965	0.8407	0.884	......	0.8731
0	0.8252	0.864	0.7785	0.823	......	0.7938
1	0.7348	0.762	0.7188	0.762	0.577	0.7144
2	0.6543	0.721	0.6668	0.701	0.556	0.6747
3	0.5827	0.658	0.6190	0.640	0.538	0.6350
4	0.5189	0.605	0.5723	0.589	0.526	0.5953
5	0.4620	0.559	0.5258	0.538	0.518	0.5556
6	0.4115	0.516	0.4877	0.488	0.511	0.5159
7	0.3665	0.457	0.4496	0.447	0.505	0.4763
8	0.3264	0.419	0.4115	0.406	0.500	0.4366
9	0.2906	0.376	0.3767	0.366	0.493	0.3969
10	0.2588	0.340	0.3429	0.325	0.485	0.3572
11	0.2305	0.305	0.3061	0.295	0.478	0.3175
12	0.2053	0.277	0.2680	0.264	0.470	0.2778
13	0.1828	0.241	0.232	0.234	0.462	0.2381
14	0.1628	0.211	0.203	0.203	0.457	0.1984
15	0.1450	0.183	0.183	0.183	0.452	0.1786
16	0.1291	0.165	0.159	0.163	0.445	0.1588
17	0.1150	0.147	0.137	0.142	0.437	0.1429
18	0.1024	0.124	0.121	0.122	0.427	0.1270
19	0.09116	0.107	0.104	0.102	0.417	0.1111
20	0.08118	0.089	0.0884	0.0914	0.409	0.09525
21	0.07229	0.081	0.0808	0.0813	0.399	0.08731
22	0.06439	0.071	0.0726	0.0711	0.394	0.07938
23	0.05733	0.064	0.0655	0.0610	0.389	0.07144
24	0.05105	0.056	0.0584	0.0559	0.384	0.06350

Gauge No.	Brown & Sharpe	Bir-ming-ham or Stubs'.	Wash-burn & Moen	Impe-rial or Brit. Std.	Stubs' steel	U. S. Std. plate
25	0.04547	0.051	0.0518	0.0508	0.376	0.05556
26	0.04049	0.046	0.0460	0.0457	0.371	0.04763
27	0.03604	0.041	0.0439	0.0417	0.363	0.04366
28	0.03211	0.036	0.0411	0.0378	0.353	0.03969
29	0.02860	0.033	0.0381	0.0345	0.340	0.03572
30	0.02548	0.030	0.0356	0.0315	0.323	0.03175
31	0.02268	0.025	0.0335	0.0295	0.305	0.02778
32	0.02019	0.023	0.0325	0.0274	0.292	0.02580
33	0.01798	0.020	0.0300	0.0254	0.284	0.02381
34	0.01601	0.018	0.0264	0.0234	0.279	0.02183
35	0.01426	0.013	0.024	0.0213	0.274	0.01984
36	0.01270	0.010	0.023	0.0193	0.269	0.01786
37	0.01131	......	0.022	0.0173	0.262	0.01687
38	0.01007	......	0.020	0.0152	0.257	0.01588
39	0.008969	......	0.019	0.0132	0.251	......
40	0.007988	......	0.018	0.0122	0.246	......
41	......	......	0.017	0.0112	0.241	......
42	......	......	0.016	0.0102	0.234	......
43	......	......	0.015	0.0091	0.224	......
44	......	......	0.015	0.0081	0.216	......
45	......	......	0.014	0.0071	0.206	......
46	......	......	0.013	0.0061	0.201	......
47	......	......	0.013	0.0051	0.196	......
48	......	......	0.012	0.0041	0.191	......
49	......	......	0.012	0.0030	0.183	......
50	......	......	0.011	0.0025	0.175	......

TWIST DRILL AND STEEL WIRE GAUGE

Inches

No.	Size	No.	Size	No.	Size	No.	Size	No.	Size
1	0.2280	17	0.1730	33	0.1130	49	0.0730	65	0.0350
2	0.2210	18	0.1695	34	0.1110	50	0.0700	66	0.0330
3	0.2130	19	0.1660	35	0.1100	51	0.0670	67	0.0320
4	0.2090	20	0.1610	36	0.1065	52	0.0635	68	0.0310
5	0.2055	21	0.1590	37	0.1040	53	0.0595	69	0.02925
6	0.2040	22	0.1570	38	0.1015	54	0.0550	70	0.0280
7	0.2010	23	0.1540	39	0.0995	55	0.0520	71	0.0260
8	0.1990	24	0.1520	40	0.0980	56	0.0465	72	0.0250
9	0.1960	25	0.1495	41	0.0960	57	0.0430	73	0.0240
10	0.1935	26	0.1470	42	0.0935	58	0.0420	74	0.0225
11	0.1910	27	0.1440	43	0.0890	59	0.0410	75	0.0210
12	0.1890	28	0.1405	44	0.0860	60	0.0400	76	0.0200
13	0.1850	29	0.1360	45	0.0820	61	0.0390	77	0.0180
14	0.1820	30	0.1285	46	0.0810	62	0.0380	78	0.0160
15	0.1800	31	0.1200	47	0.0785	63	0.0370	79	0.0145
16	0.1770	32	0.1160	48	0.0760	64	0.0360	80	0.0135

Centimeters

No.	Size	No.	Size	No.	Size	No.	Size	No.	Size
1	0.5791	17	0.4394	33	0.2870	49	0.1854	65	0.0889
2	0.5613	18	0.4305	34	0.2819	50	0.1778	66	0.0838
3	0.5410	19	0.4216	35	0.2794	51	0.1702	67	0.0813
4	0.5309	20	0.4089	36	0.2705	52	0.1613	68	0.0787
5	0.5220	21	0.4039	37	0.2642	53	0.1511	69	0.0743
6	0.5182	22	0.3988	38	0.2578	54	0.1397	70	0.0711
7	0.5105	23	0.3912	39	0.2527	55	0.1321	71	0.0660
8	0.5055	24	0.3861	40	0.2489	56	0.1181	72	0.0635
9	0.4978	25	0.3797	41	0.2438	57	0.1092	73	0.0610
10	0.4915	26	0.3734	42	0.2375	58	0.1067	74	0.0572
11	0.4851	27	0.3658	43	0.2261	59	0.1041	75	0.0533
12	0.4801	28	0.3569	44	0.2184	60	0.1016	76	0.0508
13	0.4699	29	0.3454	45	0.2083	61	0.0991	77	0.0457
14	0.4623	30	0.3264	46	0.2057	62	0.0965	78	0.0406
15	0.4572	31	0.3048	47	0.1994	63	0.0940	79	0.0368
16	0.4496	32	0.2946	48	0.1930	64	0.0914	80	0.0343

DIMENSIONS OF WIRE

Stubs' Gauge

Giving the diameter and cross-section in English and metric system for the Birmingham or Stubs' gauge.

Gauge No.	Diameter in in.	Section in sq. in.	Diameter in cm	Section in sq. cm
0000	0.454	0.16188	1.1532	1.0444
000	.425	.14186	1.0795	0.9152
00	.380	.11341	0.9652	.7317
0	0.340	0.09079	0.8636	0.5858
1	.300	.07069	.7620	.4560
2	.284	.06335	.7214	.4087
3	.259	.05269	.6579	.3399
4	.238	.04449	.6045	.2870
5	0.220	0.03801	0.5588	0.2452
6	.203	.03237	.5156	.20881
7	.180	.02545	.4572	.16147
8	.165	.02138	.4191	.13795
9	.148	.01720	.3759	.11099
10	0.134	0.01410	0.3404	0.09098
11	.120	.011310	.3048	.07297
12	.109	.009331	.2769	.06160
13	.095	.007088	.2413	.04573
14	.083	.005411	.2108	.03491
15	0.072	0.004072	0.1829	0.02627
16	.065	.0033183	.16510	.021409
17	.058	.0026421	.14732	.017046
18	.049	.0018857	.12446	.012166
19	.042	.0013854	.10668	.008938
20	0.035	0.0009621	0.08890	0.006207
21	.032	.0008042	.08128	.005189
22	.028	.0006158	.07112	.003973
23	.025	.0004909	.06350	.003167
24	.022	.0003801	.05588	.002452
25	0.020	0.0003142	0.05080	0.002027
26	.018	.0002545	.04572	.0016417
27	.016	.0002011	.04064	.0012972
28	.014	.0001539	.03556	.0009932
29	.013	.0001327	.03302	.0008563
30	0.012	0.0001181	0.03048	0.0007297
31	.010	.00007854	.02540	.0005067
32	.009	.00006362	.02286	.0004104
33	.008	.00005027	.02032	.0003243
34	.007	.00003848	.01778	.0002483
35	0.005	0.00001963	0.01270	0.0001267
36	.004	.00001257	.01016	.0000811

British Standard Gauge

Giving the diameter and cross-section in English and metric system for the British Standard Gauge.

Gauge No.	Diameter in in.	Section in sq. in.	Diameter in cm	Section in sq. cm
0000000	0.500	0.1963	1.2700	1.267
000000	.464	.1691	1.1786	1.091
00000	0.432	0.1466	1.0973	0.9456
0000	.400	.1257	1.0160	.8107
000	.372	.1087	0.9449	.7012
00	.348	.0951	.8839	.6136
0	0.324	0.0825	0.8230	0.5319
1	.300	.07069	.7620	.4560
2	.276	.05983	.7010	.3858
3	.252	.04988	.6401	.3218
4	.232	.04227	.5893	.2727
5	0.212	0.03530	0.5385	0.2277
6	.192	.02895	.4877	.18679
7	.176	.02433	.4470	.15696
8	.160	.02010	.4064	.12973
9	.144	.01629	.3658	.10507
10	0.128	0.01287	0.3251	0.08302
11	.116	.010568	.2946	.06818
12	.104	.008495	.2642	.05480
13	.092	.006648	.2337	.04289
14	.080	.005027	.2032	.03243
15	0.072	0.004071	0.1829	0.02627
16	.064	.003217	.16256	.020755
17	.056	.002463	.14224	.015890
18	.048	.001810	.12192	.011675
19	.040	.001257	.10160	.008107
20	0.036	0.001018	0.09144	0.006567
21	.032	.0008042	.08128	.005189
22	.028	.0006158	.07112	.003973
23	.024	.0004524	.06096	.002922
24	.022	.0003801	.05588	.002452
25	0.020	0.0003142	0.05080	0.002027
26	.0180	.0002545	.04572	.0016417
27	.0164	.0002112	.04166	.0013628
28	.0148	.0001728	.03759	.0011099
29	.0136	.0001453	.03454	.0009363
30	0.0124	0.0001208	0.03150	0.0007791
31	.0116	.00010568	.02946	.0006818
32	.0108	.00009161	.02743	.0005910
33	.0100	.00007854	.02540	.0005067
34	.0092	.00006648	.02337	.0004289
35	0.0084	0.00005542	0.02134	0.0003575
36	.0076	.00004536	.01930	.0002927
37	.0068	.00003632	.01727	.0002343
38	.0060	.00002827	.01524	.0001824
39	.0052	.00002124	.01321	.0001370
40	0.0048	0.00001810	0.01219	0.0001167
41	.0044	.00001521	.01118	.0000982
42	.0040	.00001257	.01016	.0000811
43	.0036	.00001018	.00914	.0000656
44	.0032	.00000804	.00813	.0000519
45	0.0028	0.00000616	0.00711	0.0000397
46	.0024	.00000452	.00610	.0000212
47	.0020	.00000314	.00508	.0000203
48	.0016	.00000201	.00406	.0000129
49	.0012	.00000113	.00305	.0000073
50	0.0010	0.00000079	0.00254	0.0000051

PLATINUM WIRE

Mass in Grams per Foot

B. & S. Gauge	Diameter, inches	Mass, g per ft.	B. & S. Gauge	Diameter, inches	Mass, g per ft.
10	.1019	37.5	23	.02257	1.8
11	.09074	28.0	24	.02010	1.4
12	.08081	22.0	25	.01790	1.1
13	.07196	17.5	26	.01594	0.9
14	.06408	14.0	27	.01420	0.7
15	.05707	11.0	28	.01264	0.6
16	.05082	9.0	29	.01126	0.45
17	.04526	7.0	30	.01003	0.35
18	.04030	5.7	31	.008928	0.28
19	.03589	4.4	32	.007950	0.22
20	.03196	3.4	33	.007080	0.17
21	.02846	2.9	34	.006305	0.15
22	.02535	2.3	35	.005615	0.11

ALLOWABLE CARRYING CAPACITIES OF CONDUCTORS

(National Electrical Code)

The ratings in the following tabulation are those permitted by the National Electrical Code for flexible cords and for interior wiring of houses, hotels, office buildings, industrial plants, and other buildings.

The values are for copper wire. For aluminum wire the allowable carrying capacities shall be taken as 84% of those given in the table for the respective sizes of copper wire with the same kind of covering.

Size A.W.G.	Area Circular Mils	Diameter of Solid Wires Mils	Rubber Insulation Amperes	Varnished Cambric Insulation Amperes	Other Insulations and Bare Conductors Amperes
18	1,624.	40.3	3*	...	6†
16	2,583.	50.8	6*	...	10†
14	4,107.	64.1	15	18	20
12	6,530.	80.8	20	25	30
10	10,380.	101.9	25	30	35
8	16,510.	128.5	35	40	50
6	26,250.	162.0	50	60	70
5	33,100.	181.9	55	65	80
4	41,740.	204.3	70	85	90
3	52,630.	229.4	80	95	100
2	66,370.	257.6	90	110	125
1	83,690.	289.3	100	120	150
0	105,500.	325.0	125	150	200
00	135,100.	364.8	150	180	225
000	167,800.	409.6	175	210	275
0000	211,600.	460	225	270	325

1 Mil = 0.001 inch.

* The allowable carrying capacities of No. 18 and 16 are 5 and 7 amperes respectively, when in flexible cords.

† The allowable carrying capacities of No. 18 and 16 are 10 and 15 amperes respectively, when in cords for portable heaters. Types AFS, AFSJ, HC, HPD, and HSJ.

WIRE TABLE, STANDARD ANNEALED COPPER

American Wire Gauge (B. & S.) English Units

Gauge No.	Diameter in mils at 20°C	Cross section at 20°C		Ohms per 1000 feet*			
		Circular mils	Sq. inches	0°C (32°F)	20°C (68°F)	50°C (122°F)	75°C (167°F)
0000	460.0	211600	0.1662	0.04516	0.04901	0.05479	0.05961
000	409.6	167800	.1318	.05695	.06180	.06909	.07516
00	364.8	133100	.1045	.07181	.07793	.08712	.09478
0	324.9	105500	.08289	.09055	.09827	.1099	.1195
1	289.3	83690	.06573	.1142	.1239	.1385	.1507
2	257.6	66370	.05213	.1440	.1563	.1747	.1900
3	229.4	52640	.04134	.1816	.1970	.2203	.2396
4	204.3	41740	.03278	.2289	.2485	.2778	.3022
5	181.9	33100	.02600	.2887	.3133	.3502	.3810
6	162.0	26250	.02062	.3640	.3951	.4416	.4805
7	144.3	20820	.01635	.4590	.4982	.5569	.6059
8	128.5	16510	.01297	.5788	.6282	.7023	.7640
9	114.4	13090	.01028	.7299	.7921	.8855	.9633
10	101.9	10380	.008155	.9203	.9989	1.117	1.215
11	90.74	8234	.006467	1.161	1.260	1.408	1.532
12	80.81	6530	.005129	1.463	1.588	1.775	1.931
13	71.96	5178	.004067	1.845	2.003	2.239	2.436
14	64.08	4107	.003225	2.327	2.525	2.823	3.071
15	57.07	3257	.002558	2.934	3.184	3.560	3.873
16	50.82	2583	.002028	3.700	4.016	4.489	4.884
17	45.26	2048	.001609	4.666	5.064	5.660	6.158
18	40.30	1624	.001276	5.883	6.385	7.138	7.765
19	35.89	1288	.001012	7.418	8.051	9.001	9.792
20	31.96	1022	.0008023	9.355	10.15	11.35	12.35
21	28.45	810.1	.0006363	11.80	12.80	14.31	15.57
22	25.35	642.4	.0005046	14.87	16.14	18.05	19.63
23	22.57	509.5	.0004002	18.76	20.36	22.76	24.76
24	20.10	404.0	.0003173	23.65	25.67	28.70	31.22
25	17.90	320.4	.0002517	29.82	32.37	36.18	39.36
26	15.94	254.1	.0001996	37.61	40.81	45.63	49.64
27	14.20	201.5	.0001583	47.42	51.47	57.53	62.59
28	12.64	159.8	.0001255	59.80	64.90	72.55	78.93
29	11.26	126.7	.00009953	75.40	81.83	91.48	99.52
30	10.03	100.5	.00007894	95.08	103.2	115.4	125.5
31	8.928	79.70	.00006260	119.9	130.1	145.5	158.2
32	7.950	63.21	.00004964	151.2	164.1	183.4	199.5
33	7.080	50.13	.00003937	190.6	206.9	231.3	251.6
34	6.305	39.75	.00003122	240.4	260.9	291.7	317.3
35	5.615	31.52	.00002476	303.1	329.0	367.8	400.1
36	5.000	25.00	.00001964	382.2	414.8	463.7	504.5
37	4.453	19.83	.00001557	482.0	523.1	584.8	636.2
38	3.965	15.72	.00001235	607.8	659.6	737.4	802.2
39	3.531	12.47	.000009793	766.4	831.8	929.8	1012
40	3.145	9.888	.000007766	966.5	1049	1173	1276

* Resistance at the stated temperatures of a wire whose length is 1000 feet at 20°C.

WIRE TABLE, STANDARD ANNEALED COPPER

American Wire Gauge (B. & S.) English Units (Continued)

Gauge No.	Pounds per 1000 feet	Feet per pound	Feet per ohm*			
			0°C (32°F)	20°C (68°F)	50°C (122°F)	75°C (167°F)
0000	640.5	1.561	22140	20400	18250	16780
000	507.9	1.968	17560	16180	14470	13300
00	402.8	2.482	13930	12830	11480	10550
0	319.5	3.130	11040	10180	9103	8367
1	253.3	3.947	8758	8070	7219	6636
2	200.9	4.977	6946	6400	5725	5262
3	159.3	6.276	5508	5075	4540	4173
4	126.4	7.914	4368	4025	3600	3309
5	100.2	9.980	3464	3192	2855	2625
6	79.46	12.58	2747	2531	2264	2081
7	63.02	15.87	2179	2007	1796	1651
8	49.98	20.01	1728	1592	1424	1309
9	39.63	25.23	1370	1262	1129	1038
10	31.43	31.82	1087	1001	895.6	823.2
11	24.92	40.12	861.7	794.0	710.2	652.8
12	19.77	50.59	683.3	629.6	563.2	517.7
13	15.68	63.80	541.9	499.3	446.7	410.6
14	12.43	80.44	429.8	396.0	354.2	325.6
15	9.858	101.4	340.8	314.0	280.9	258.2
16	7.818	127.9	270.3	249.0	222.8	204.8
17	6.200	161.3	214.3	197.5	176.7	162.4
18	4.917	203.4	170.0	156.6	140.1	128.8
19	3.899	256.5	134.8	124.2	111.1	102.1
20	3.092	323.4	106.9	98.50	88.11	80.99
21	2.452	407.8	84.78	78.11	69.87	64.23
22	1.945	514.2	67.23	61.95	55.41	50.94
23	1.542	648.4	53.32	49.13	43.94	40.39
24	1.223	817.7	42.28	38.96	34.85	32.03
25	0.9699	1031	33.53	30.90	27.64	25.40
26	.7692	1300	26.59	24.50	21.92	20.15
27	.6100	1639	21.09	19.43	17.38	15.98
28	.4837	2067	16.72	15.41	13.78	12.67
29	.3836	2607	13.26	12.22	10.93	10.05
30	.3042	3287	10.52	9.691	8.669	7.968
31	.2413	4145	8.341	7.685	6.875	6.319
32	.1913	5227	6.614	6.095	5.452	5.011
33	.1517	6591	5.245	4.833	4.323	3.974
34	.1203	8310	4.160	3.833	3.429	3.152
35	.09542	10480	3.299	3.040	2.719	2.499
36	.07568	13210	2.616	2.411	2.156	1.982
37	.06001	16660	2.075	1.912	1.710	1.572
38	.04759	21010	1.645	1.516	1.356	1.247
39	.03774	26500	1.305	1.202	1.075	0.9886
40	.02993	33410	1.035	0.9534	0.8529	.7840

* Length at 20°C of a wire whose resistance is 1 ohm at the stated temperatures.

WIRE TABLE, STANDARD ANNEALED COPPER (Continued)

American Wire Gauge (B. & S.) English Units

Gauge No.	Diameter in mils at 20°C	Ohms per pound 0°C (32°F)	Ohms per pound 20°C (68°F)	Ohms per pound 50°C (122°F)	Lbs. per ohm 20°C (68°F)
0000	460.0	0.00007051	0.00007652	0.00008554	13070
000	409.6	.0001121	.0001217	.0001360	8219
00	364.8	.0001783	.0001935	.0002163	5169
0	324.9	.0002835	.0003076	.0003439	3251
1	289.3	.0004507	.0004891	.0005468	2044
2	257.6	.0007166	.0007778	.0008695	1286
3	229.4	.001140	.001237	.001383	808.6
4	204.3	.001812	.001966	.002198	508.5
5	181.9	.002881	.003127	.003495	319.8
6	162.0	.004581	.004972	.005558	201.1
7	144.3	.007284	.007905	.008838	126.5
8	128.5	.01158	.01257	.01405	79.55
9	114.4	.01842	.01999	.02234	50.03
10	101.9	.02928	.03178	.03553	31.47
11	90.74	.04656	05053	05649	19.79
12	80.81	.07404	.08035	.08983	12.45
13	71.96	.1177	.1278	.1428	7.827
14	64.08	.1872	.2032	.2271	4.922
15	57.07	.2976	.3230	.3611	3.096
16	50.82	.4733	.5136	.5742	1.947
17	45.26	.7525	.8167	.9130	1.224
18	40.30	1.197	1.299	1.452	0.7700
19	35.89	1.903	2.065	2.308	.4843
20	31.96	3.025	3.283	3.670	.3046
21	28.46	4.810	5.221	5.836	.1915
22	25.35	7.649	8.301	9.280	.1205
23	22.57	12.16	13.20	14.76	.07576
24	20.10	19.34	20.99	23.46	.04765
25	17.90	30.75	33.37	37.31	.02997
26	15.94	48.89	53.06	59.32	.01885
27	14.20	77.74	84.37	94.32	.01185
28	12.64	123.6	134.2	150.0	.007454
29	11.26	196.6	213.3	238.5	.004688
30	10.03	312.5	339.2	379.2	.002948
31	8.928	497.0	539.3	602.9	.001854
32	7.950	790.2	857.6	958.7	.001166
33	7.080	1256	1364	1524	.0007333
34	6.305	1998	2168	2424	.0004612
35	5.615	3177	3448	3854	.0002901
36	5.000	5051	5482	6128	.0001824
37	4.453	8032	8717	9744	.0001147
38	3.965	12770	13860	15490	.00007215
39	3.531	20310	22040	24640	.00004538
40	3.145	32290	35040	39170	.00002854

Gauge No.	Diameter in mm at 20°C	Cross section in mm² at 20°C	Ohms per kilometer* 0°C	Ohms per kilometer* 20°C	Ohms per kilometer* 50°C	Ohms per kilometer* 75°C
0000	11.68	107.2	0.1482	0.1608	0.1798	0.1956
000	10.40	85.03	.1868	.2028	.2267	.2466
00	9.266	67.43	.2356	.2557	.2858	.3110
0	8.252	53.48	.2971	.3224	.3604	.3921
1	7.348	42.41	.3746	.4066	.4545	.4944
2	6.544	33.63	.4724	.5127	.5731	.6235
3	5.827	26.67	.5956	.6465	.7227	.7862
4	5.189	21.15	.7511	.8152	.9113	.9914
5	4.621	16.77	.9471	1.028	1.149	1.250
6	4.115	13.30	1.194	1.296	1.449	1.576
7	3.665	10.55	1.506	1.634	1.827	1.988
8	3.264	8.366	1.899	2.061	2.304	2.506
9	2.906	6.634	2.395	2.599	2.905	3.161
10	2.588	5.261	3.020	3.277	3.663	3.985
11	2.305	4.172	3.807	4.132	4.619	5.025
12	2.053	3.309	4.801	5.211	5.825	6.337
13	1.828	2.624	6.054	6.571	7.345	7.991
14	1.628	2.081	7.634	8.285	9.262	10.08
15	1.450	1.650	9.627	10.45	11.68	12.71
16	1.291	1.309	12.14	13.17	14.73	16.02
17	1.150	1.038	15.31	16.61	18.57	20.20
18	1.024	0.8231	19.30	20.95	23.42	25.48
19	0.9116	.6527	24.34	26.42	29.53	32.12
20	.8118	.5176	30.69	33.31	37.24	40.51
21	.7230	.4105	38.70	42.00	46.95	51.08
22	.6438	.3255	48.80	52.96	59.21	64.41
23	.5733	.2582	61.54	66.79	74.66	81.22
24	.5106	.2047	77.60	84.21	94.14	102.4
25	.4547	.1624	97.85	106.2	118.7	129.1
26	.4049	.1288	123.4	133.9	149.7	162.9
27	.3606	.1021	155.6	168.9	188.8	205.4
28	.3211	.08098	196.2	212.9	238.0	258.9
29	.2859	.06422	247.4	268.5	300.1	326.5
30	.2546	.05093	311.9	338.6	378.5	411.7
31	.2268	.04039	393.4	426.9	477.2	519.2
32	.2019	.03203	496.0	538.3	601.8	654.7
33	.1798	.02540	625.5	678.8	758.8	825.5
34	.1601	.02014	788.7	856.0	956.9	1041
35	.1426	.01597	994.5	1079	1207	1313
36	.1270	.01267	1254	1361	1522	1655
37	.1131	.01005	1581	1716	1919	2087
38	.1007	.007967	1994	2164	2419	2632
39	.08969	.006318	2514	2729	3051	3319
40	.07987	.005010	3171	3441	3847	4185

* Resistance at the stated temperatures of a wire whose length is 1 kilometer at 20°C.

Gauge No.	Diameter in mm at 20°C	Kilograms per kilometer	Meters per gram	Meters per ohm* 0°C	Meters per ohm* 20°C	Meters per ohm* 50°C	Meters per ohm* 75°C
0000	11.68	953.2	0.001049	6749	6219	5563	5113
000	10.40	755.9	.001323	5352	4932	4412	4055
00	9.266	599.5	.001668	4245	3911	3499	3216
0	8.252	475.4	.002103	3366	3102	2774	2550
1	7.348	377.0	.002652	2669	2460	2200	2022
2	6.544	299.0	.003345	2117	1951	1745	1604
3	5.827	237.1	.004217	1679	1547	1384	1272
4	5.189	188.0	.005318	1331	1227	1097	1009
5	4.621	149.1	.006706	1056	972.9	870.2	799.9
6	4.115	118.2	.008457	837.3	771.5	690.1	634.4
7	3.665	93.78	.01066	664.0	611.8	547.3	503.1
8	3.264	74.37	.01345	526.6	485.2	434.0	399.0
9	2.906	58.98	.01696	417.6	384.8	344.2	316.4
10	2.588	46.77	.02138	331.2	305.1	273.0	250.9
11	2.305	37.09	.02696	262.6	242.0	216.5	199.0
12	2.053	29.42	.03400	208.3	191.9	171.7	157.8
13	1.828	23.33	.04287	165.2	152.2	136.1	125.1
14	1.628	18.50	.05406	131.0	120.7	108.0	99.24
15	1.450	14.67	.06816	103.9	95.71	85.62	78.70
16	1.291	11.63	.08595	82.38	75.90	67.90	62.41
17	1.150	9.226	.1084	65.33	60.20	53.85	49.50
18	1.024	7.317	.1367	51.81	47.74	42.70	39.25
19	0.9116	5.803	.1723	41.09	37.86	33.86	31.13
20	.8118	4.602	.2173	32.58	30.02	26.86	24.69
21	.7230	3.649	.2740	25.84	23.81	21.30	19.58
22	.6438	2.894	.3455	20.49	18.88	16.89	15.53
23	.5733	2.295	.4357	16.25	14.97	13.39	12.31
24	.5106	1.820	.5494	12.89	11.87	10.62	9.764
25	.4547	1.443	.6928	10.22	9.417	8.424	7.743
26	.4049	1.145	.8736	8.105	7.468	6.680	6.141
27	.3606	0.9078	1.102	6.428	5.922	5.298	4.870
28	.3211	.7199	1.389	5.097	4.697	4.201	3.862
29	.2859	.5709	1.752	4.042	3.725	3.332	3.063
30	.2546	.4527	2.209	3.206	2.954	2.642	2.429
31	.2268	.3590	2.785	2.542	2.342	2.095	1.926
32	.2019	.2847	3.512	2.016	1.858	1.662	1.527
33	.1798	.2258	4.429	1.599	1.473	1.318	1.211
34	.1601	.1791	5.584	1.268	1.168	1.045	0.9606
35	.1426	.1420	7.042	1.006	0.9265	0.8288	.7618
36	.1270	.1126	8.879	0.7974	.7347	.6572	.6041
37	.1131	.08931	11.20	.6324	.5827	.5212	.4791
38	.1007	.07083	14.12	.5015	.4621	.4133	.3799
39	.08969	.05617	17.80	.3977	.3664	.3278	.3013
40	.07987	.04454	22.45	.3154	.2906	.2600	.2390

* Length at 20°C of a wire whose resistance is 1 ohm at the stated temperatures.

WIRE TABLE, STANDARD ANNEALED COPPER (Continued)

American Wire Gauge (B. & S.) Metric Units

Gauge No.	Ohms per kilogram 0°C	Ohms per kilogram 20°C	Ohms per kilogram 50°C	Grams per ohm 20°C
0000	0.0001554	0.0001687	0.0001886	5928000
000	.0002472	.0002682	.0002999	3728000
00	.0003930	.0004265	.0004768	2344000
0	.0006249	.0006782	.0007582	1474000
1	.0009936	.001078	.001206	927300
2	.001580	.001715	.001917	583200
3	.002512	.002726	.003048	366800
4	.003995	.004335	.004846	230700
5	.006352	.006893	.007706	145100
6	.01010	.01096	.01225	91230
7	.01606	.01743	.01948	57380
8	.02553	.02771	.03098	36080
9	.04060	.04406	.04926	22690
10	.06456	.07007	.07833	14270
11	.1026	.1114	.1245	8976
12	.1632	.1771	.1980	5645
13	.2595	.2817	.3149	3550
14	.4127	.4479	.5007	2233
15	.6562	.7122	.7961	1404
16	1.043	1.132	1.266	883.1
17	1.659	1.801	2.013	555.4
18	2.638	2.863	3.201	349.3
19	4.194	4.552	5.089	219.7
20	6.670	7.238	8.092	138.2
21	10.60	11.51	12.87	86.88
22	16.86	18.30	20.46	54.64
23	26.81	29.10	32.53	34.36
24	42.63	46.27	51.73	21.61
25	67.79	73.57	82.25	13.59
26	107.8	117.0	130.8	8.548
27	171.4	186.0	207.9	5.376
28	272.5	295.8	330.6	3.381
29	433.3	470.3	525.7	2.126
30	689.0	747.8	836.0	1.337
31	1096	1189	1329	0.8410
32	1742	1891	2114	.5289
33	2770	3006	3361	.3326
34	4404	4780	5344	.2092
35	7003	7601	8497	.1316
36	11140	12090	13510	.08274
37	17710	19220	21480	.05204
38	28150	30560	34160	.03273
39	44770	48590	54310	.02058
40	71180	77260	86360	.01294

ALUMINUM WIRE TABLE

Hard-Drawn Aluminum Wire at 20°C (or, 68°F)

American Wire Gauge (B. & S.) English Units

Gauge No.	Diameter in mils	Cross section Circular mils	Cross section Square inches	Ohms per 1000 ft.	Pounds per 1000 ft.	Pounds per ohm	Feet per ohm
0000	460	212000	0.166	0.0804	195	2420	12400
000	410	168000	.132	.101	154	1520	9860
00	365	133000	.105	.128	122	957	7820
0	325	106000	.0829	.161	97.0	602	6200
1	289	83700	.0657	.203	76.9	379	4920
2	258	66400	.0521	.256	61.0	238	3900
3	229	52600	.0413	.323	48.4	150	3090
4	204	41700	.0328	.408	38.4	94.2	2450
5	182	33100	.0260	.514	30.4	59.2	1950
6	162	26300	.0206	.648	24.1	37.2	1540
7	144	20800	.0164	.817	19.1	23.4	1220
8	128	16500	.0130	1.03	15.2	14.7	970
9	114	13100	.0103	1.30	12.0	9.26	770

ALUMINUM WIRE TABLE (Continued)

Gauge No.	Diameter in mils	Cross section Circular mils	Cross section Square inches	Ohms per 1000 ft.	Pounds per 1000 ft.	Pounds per ohm	Feet per ohm
10	102	10400	.00815	1.64	9.55	5.83	610
11	91	8230	.00647	2.07	7.57	3.66	484
12	81	6530	.00513	2.61	6.00	2.30	384
13	72	5180	.00407	3.29	4.76	1.45	304
14	64	4110	.00323	4.14	3.78	0.911	241
15	57	3260	.00256	5.22	2.99	.573	191
16	51	2580	.00203	6.59	2.37	.360	152
17	45	2050	.00161	8.31	1.88	.227	120
18	40	1620	.00128	10.5	1.49	.143	95.5
19	36	1290	.00101	13.2	1.18	.0897	75.7
20	32	1020	.000802	16.7	0.939	.0564	60.0
21	28.5	810	.000636	21.0	.745	.0355	47.6
22	25.3	642	.000505	26.5	.591	.0223	37.8
23	22.6	509	.000400	33.4	.468	.0140	29.9
24	20.1	404	.000317	42.1	.371	.00882	23.7
25	17.9	320	.000252	53.1	.295	.00555	18.8
26	15.9	254	.000200	67.0	.234	.00349	14.9
27	14.2	202	.000158	84.4	.185	.00219	11.8
28	12.6	160	.000126	106.	.147	.00138	9.39
29	11.3	127	.0000995	134.	.117	.000868	7.45
30	10.0	101	.0000789	169.	.0924	.000546	5.91
31	8.9	79.7	.0000626	213.	.0733	.000343	4.68
32	8.0	63.2	.0000496	269.	.0581	.000216	3.72
33	7.1	50.1	.0000394	339.	.0461	.000136	2.95
34	6.3	39.8	.0000312	428.	.0365	.0000854	2.34
35	5.6	31.5	.0000248	540.	.0290	.0000537	1.85
36	5.0	25.0	.0000196	681.	.0230	.0000338	1.47
37	4.5	19.8	.0000156	858.	.0182	.0000212	1.17
38	4.0	15.7	.0000123	1080.	.0145	.0000134	0.924
39	3.5	12.5	.00000979	1360.	.0115	.00000840	.733
40	3.1	9.9	.00000777	1720.	.0091	.00000528	.581

Gauge No.	Diameter in mm	Cross section in mm²	Ohms per kilometer	Kilograms per kilometer	Grams per ohm	Meters per ohm
0000	11.7	107	0.264	289	1100000	3790
000	10.4	85.0	.333	230	690000	3010
00	9.3	67.4	.419	182	434000	2380
0	8.3	53.5	.529	144	273000	1890
1	7.3	42.4	.667	114.	172000	1500
2	6.5	33.6	.841	90.8	108000	1190
3	5.8	26.7	1.06	72.0	67900	943
4	5.2	21.2	1.34	57.1	42700	748
5	4.6	16.8	1.69	45.3	26900	593
6	4.1	13.3	2.13	35.9	16900	470
7	3.7	10.5	2.68	28.5	10600	373
8	3.3	8.37	3.38	22.6	6680	296
9	2.91	6.63	4.26	17.9	4200	235
10	2.59	5.26	5.38	14.2	2640	186
11	2.30	4.17	6.78	11.3	1660	148
12	2.05	3.31	8.55	8.93	1050	117
13	1.83	2.62	10.8	7.08	657	92.8
14	1.63	2.08	13.6	5.62	413	73.6
15	1.45	1.65	17.1	4.46	260	58.4
16	1.29	1.31	21.6	3.53	164	46.3
17	1.15	1.04	27.3	2.80	103	36.7
18	1.02	0.823	34.4	2.22	64.7	29.1
19	0.91	.653	43.3	1.76	40.7	23.1
20	.81	.518	54.6	1.40	25.6	18.3
21	.72	.411	68.9	1.11	16.1	14.5
22	.64	.326	86.9	0.879	10.1	11.5
23	.57	.258	110	.697	6.36	9.13
24	.51	.205	138	.553	4.00	7.24
25	.45	.162	174	.438	2.52	5.74
26	.40	.129	220	.348	1.58	4.55
27	.36	.102	277	.276	0.995	3.61
28	.32	.0810	349	.219	.626	2.86
29	.29	.0642	440	.173	.394	2.27
30	.25	.0509	555	.138	.248	1.80
31	.227	.0404	700	.109	.156	1.43
32	.202	.0320	883	.0865	.0979	1.13
33	.180	.0254	1110	.0686	.0616	0.899
34	.160	.0201	1400	.0544	.0387	.712
35	.143	.0160	1770	.0431	.0244	.565
36	.127	.0127	2230	.0342	.0153	.448
37	.113	.0100	2820	.0271	.00963	.355
38	.101	.0080	3550	.0215	.00606	.282
39	.090	.0063	4480	.0171	.00381	.223
40	.080	.0050	5640	.0135	.00240	.177

CROSS-SECTION AND MASS OF WIRES

U. S. Measure

Diameters are given in mils (1 mil = .001 in.), and area in square mils (1 sq. mil = .000001 sq. in.). For sections and masses for one-tenth the diameters given, divide by 100 and for sections and masses for ten times the diameter multiply by 100.

Diam. in mils	Cross-sec. in sq. mils	Pounds per foot			
		Copper, density 8.90	Iron, density 7.80	Brass, density 8.56	Aluminum, density 2.67
10	78.54	0.000303	0.0002656	0.0002915	0.0000909
11	95.03	0367	03214	03527	01100
12	113.10	0436	03825	04197	01309
13	132.73	0512	04488	04926	01536
14	153.94	0594	05206	05713	01782
15	176.71	0.000682	0.0005976	0.0006558	0.0002045
16	201.06	0776	06799	07461	02327
17	226.98	0876	07675	08423	02627
18	254.47	0982	08605	09443	02946
19	283.53	1094	09588	10522	03282
20	314.16	0.001212	0.001062	0.001166	0.0003636
21	346.36	1336	1171	1285	04009
22	380.13	1467	1286	1411	04400
23	415.48	1603	1405	1542	04809
24	452.39	1746	1530	1679	05237
25	490.87	0.001894	0.001660	0.001822	0.0005682
26	530.93	2046	1795	1970	06147
27	572.56	2209	1936	2125	06628
28	615.75	2376	2082	2285	07127
29	660.52	2549	2234	2451	07646
30	706.86	0.002727	0.002390	0.002623	0.0008182
31	754.77	2912	2552	2801	08737
32	804.25	3103	2720	2985	09309
33	855.30	3300	2892	3174	09900
34	907.92	3503	3070	3369	10509
35	962.11	0.003712	0.003253	0.003570	0.001114
36	1017.88	3927	3442	3777	1178
37	1075.21	4149	3636	3990	1245
38	1134.11	4376	3844	4218	1316
39	1194.59	4609	4040	4433	1383
40	1256.64	0.004849	0.004249	0.004664	0.001455
41	1320.25	5094	4465	4900	1528
42	1385.44	5346	4685	5141	1604
43	1452.20	5603	4911	5389	1681
44	1520.53	5867	5142	5643	1760
45	1590.43	0.006137	0.005378	0.005902	0.001841
46	1661.90	6412	5620	6167	1924
47	1734.94	6694	5867	6438	2008
48	1809.56	6982	6119	6715	2095
49	1885.74	7276	6377	6998	2183
50	1963.50	0.007576	0.006640	0.007287	0.002273
51	2042.82	7882	6908	7581	2365
52	2123.72	8194	7181	7881	2458
53	2206.18	8512	7460	8187	2554
54	2290.22	8837	7744	8499	2651
55	2375.83	0.009167	0.008034	0.008817	0.002750
56	2463.01	09504	08329	09140	2851
57	2551.76	09846	08629	09470	2954
58	2642.08	10195	08934	09805	3058
59	2733.97	10549	09245	10146	3165
60	2827.43	0.01091	0.00956	0.01049	0.003273
61	2922.47	1128	0988	1085	3383
62	3019.07	1165	1021	1120	3495
63	3117.25	1203	1054	1157	3608
64	3216.99	1241	1088	1194	3724
65	3318.31	0.01280	0.01122	0.01231	0.003841
66	3421.19	1320	1157	1270	3960
67	3525.65	1360	1192	1308	4081
68	3631.68	1401	1228	1348	4204
69	3739.28	1443	1264	1388	4328
70	3848.45	0.01485	0.01302	0.01429	0.004456
71	3959.19	1528	1339	1469	4583
72	4071.50	1571	1377	1511	4713
73	4185.39	1615	1415	1553	4845
74	4300.84	1660	1454	1596	4978
75	4417.86	0.01705	0.01494	0.01639	0.005114
76	4536.46	1751	1534	1684	5251
77	4656.63	1797	1575	1728	5390
78	4778.36	1844	1616	1773	5531
79	4901.67	1892	1658	1819	5674
80	5026.55	0.01939	0.01700	0.01865	0.005818
81	5153.00	1988	1743	1912	5965
82	5281.02	2038	1786	1960	6113
83	5410.61	2088	1830	2008	6263
84	5541.77	2138	1874	2057	6415
85	5674.50	0.02189	0.01919	0.02106	0.006568
86	5808.80	2241	1964	2156	6724
87	5944.68	2294	2010	2206	6881
88	6082.12	2347	2057	2257	7040
89	6221.14	2400	2104	2309	7201
90	6361.73	0.02455	0.02151	0.02360	0.007364
91	6503.88	2509	2199	2414	7528
92	6647.61	2565	2248	2467	7695
93	6792.91	2621	2297	2521	7863
94	6939.78	2678	2347	2575	8033
95	7088.22	0.02735	0.02397	0.02630	0.008205
96	7238.23	2793	2448	2686	8378
97	7389.81	2851	2499	2742	8554
98	7542.96	2910	2551	2799	8731
99	7697.69	2970	2603	2857	8910
100	7853.98	0.03030	0.02656	0.02915	0.009091

CROSS-SECTION AND MASS OF WIRES (Continued)

Metric Measure

Diameters are given in thousandths of a centimeter and area of section in square thousandths of a centimeter. 1 $(cm/1000)^2$ = .000001 sq. cm For sections and masses for diameters 1/10 or 10 times those of the table, divide or multiply by 100.

Diam. in thousandths of a cm	Cross-section in square thousandths of a cm	Grams per meter			
		Copper, density 8.90	Iron, density 7.80	Brass, density 8.56	Aluminum, density 2.67
10	78.54	0.06990	0.06126	0.06723	0.02097
11	95.03	.08458	.07412	.08135	.02537
12	113.10	.10065	.08822	.09681	.03020
13	132.73	.11813	.10353	.11362	.03544
14	153.94	.13701	.12008	.13177	.04110
15	176.71	0.1573	0.1378	0.1513	0.04718
16	201.06	.1789	.1568	.1721	.05368
17	226.98	.2020	.1770	.1943	.06060
18	254.47	.2265	.1985	.2178	.06794
19	283.53	.2523	.2212	.2427	.07570
20	314.16	0.2796	0.2450	0.2689	0.08388
21	346.36	.3083	.2702	.2965	.09248
22	380.13	.3383	.2965	.3254	.10149
23	415.48	.3698	.3241	.3557	.11093
24	452.39	.4026	.3529	.3872	.12079
25	490.87	0.4369	0.3829	0.4202	0.1311
26	530.93	.4725	.4141	.4545	.1418
27	572.56	.5096	.4466	.4901	.1529
28	615.75	.5480	.4803	.5271	.1644
29	660.52	.5879	.5152	.5654	.1764
30	706.86	0.6291	0.5514	0.6051	0.1887
31	754.77	.6717	.5887	.6461	.2015
32	804.25	.7158	.6273	.6884	.2147
33	855.30	.7612	.6671	.7321	.2284
34	907.92	.8081	.7082	.7772	.2424
35	962.11	0.856	0.7504	0.8236	0.2569
36	1017.88	.906	.7939	.8713	.2718
37	1075.21	.957	.8387	.9204	.2871
38	1134.11	1.012	.8866	.9730	.3035
39	1194.59	.063	.9318	1.0230	.3190
40	1256.64	1.118	0.980	1.076	0.3355
41	1320.25	.175	1.030	.130	.3525
42	1385.44	.233	.081	.186	.3699
43	1452.20	.292	.133	.243	.3877
44	1520.53	.353	.186	.302	.4060
45	1590.43	1.415	1.241	1.361	0.4246
46	1661.90	.479	.296	.423	.4437
47	1734.94	.544	.353	.485	.4632
48	1809.56	.611	.411	.549	.4832
49	1885.74	.678	.471	.614	.5035
50	1963.50	1.748	1.532	1.681	.5243
51	2042.82	.818	.593	.753	.5454
52	2123.72	.890	.657	.818	.5670
53	2206.18	.964	.721	.888	.5891
54	2290.22	2.038	.786	.960	.6115

CROSS-SECTION AND MASS OF WIRES (Continued)

Metric Measure (Continued)

Diameters are given in thousandths of a centimeter and area of section in square thousandths of a centimeter. 1 $(cm/1000)^2$ = .000001 sq. cm. For sections and masses for diameters 1/10 or 10 times those of the table, divide or multiply by 100.

Diam. in thousandths of a cm	Cross-section in square thousandths of a cm	Grams per meter: Copper, density 8.90	Iron, density 7.80	Brass, density 8.56	Aluminum, density 2.67
55	2375.83	2.114	1.853	2.034	0.6343
56	2463.01	.192	.921	.108	.6576
57	2551.76	.271	.990	.184	.6813
58	2642.08	.351	2.061	.262	.7054
59	2733.97	.433	.132	.340	.7300
60	2827.43	2.516	2.205	2.420	0.7549
61	2922.47	.601	.280	.502	.7803
62	3019.07	.687	.355	.584	.8061
63	3117.25	.774	.431	.668	.8323
64	3216.99	.863	.509	.760	.8589
65	3318.31	2.953	2.588	2.840	0.8860
66	3421.19	3.045	.669	.929	.9135
67	3525.65	.138	.750	3.018	.9413
68	3631.68	.232	.833	.109	.9697
69	3739.28	.328	.917	.201	.9984
70	3848.45	3.426	3.003	3.295	1.028
71	3959.19	.524	.088	.389	.057
72	4071.50	.624	.176	.485	.087
73	4185.39	.725	.265	.583	.117
74	4300.84	.828	.355	.682	148
75	4417.86	3.932	3.446	3.782	1.180
76	4536.46	4.037	.538	.883	.211
77	4656.63	.144	.632	.986	.243
78	4778.36	.253	.727	4.090	.276
79	4901.67	.362	.823	.177	.309
80	5026.55	4.474	3.921	4.303	1.342
81	5153.00	.586	4.019	.411	.376
82	5281.02	.700	.119	.521	.410
83	5410.61	.815	.220	.631	.445
84	5541.77	.932	.323	.744	.480
85	5674.50	5.050	4.426	4.857	1.515
86	5808.80	.170	.531	.972	.551
87	5944.68	.291	.637	5.089	.587
88	6082.12	.413	.744	.206	.624
89	6221.14	.537	.852	.325	.661
90	6361.73	5.662	4.962	5.446	1.699
91	6503.88	.788	5.073	.567	.737
92	6647.61	.916	.185	.690	.775
93	6792.91	6.046	.298	.815	.814
94	6939.78	.176	.413	.940	.853
95	7088.22	6.309	5.529	6.068	1.893
96	7238.23	.442	.646	.196	.933
97	7389.81	.577	.764	.326	.973
98	7542.96	.713	.884	.457	2.014
99	7697.69	.851	6.004	.589	.055
100	7853.98	6.990	6.126	6.723	2.097

RESISTANCE OF WIRES

The following table gives the approximate resistance of various metallic conductors. The values have been computed from the resistivities at 20°C, except as otherwise stated, and for the dimensions of wire indicated. Owing to differences in purity in the case of elements and of composition in alloys, the values can be considered only as approximations.

The following dimensions have been adopted in the computations.

B. & S. gauge	Diameter mm	Diameter mils 1 mil = .001 in.
10	2.588	101.9
12	2.053	80.81
14	1.628	64.08
16	1.291	50.82
18	1.024	40.30
20	0.8118	31.96
22	0.6438	25.35
24	0.5106	20.10
26	0.4049	15.94
27	0.3606	14.20
28	0.3211	12.64
30	0.2546	10.03
32	0.2019	7.950
34	0.1601	6.305
36	0.1270	5.000
40	0.07987	3.145

RESISTANCE OF WIRES

*Advance (0°C) $\rho = 48. \times 10^{-6}$ ohm cm

B. & S. No.	Ohms per cm	Ohms per ft.
10	.000912	.0278
12	.00145	.0442
14	.00231	.0703
16	.00367	.112
18	.00583	.178
20	.00927	.283
22	.0147	.449
24	.0234	.715
26	.0373	1.14
27	.0470	1.43
28	.0593	1.81
30	.0942	2.87
32	.150	4.57
34	.238	7.26
36	.379	11.5
40	.958	29.2

Aluminum $\rho = 2.828 \times 10^{-6}$ ohm cm

B. & S. No.	Ohms per cm	Ohms per ft.
10	.0000538	.00164
12	.0000855	.00260
14	.000136	.00414
16	.000216	.00658
18	.000344	.0105
20	.000546	.0167
22	.000869	.0265
24	.00138	.0421
26	.00220	.0669
27	.00277	.0844
28	.00349	.106
30	.00555	.169
32	.00883	.269
34	.0140	.428
36	.0223	.680
40	.0564	1.72

Brass $\rho = 7.00 \times 10^{-6}$ ohm cm

B. & S. No.	Ohms per cm	Ohms per ft.
10	.000133	.00406
12	.000212	.00645
14	.000336	.0103
16	.000535	.0163
18	.000850	.0259
20	.00135	.0412
22	.00215	.0655
24	.00342	.104
26	.00543	.166
27	.00686	.209
28	.00864	.263
30	.0137	.419
32	.0219	.666
34	.0348	1.06
36	.0552	1.68
40	.140	4.26

Climax $\rho = 87. \times 10^{-6}$ ohm cm

B. & S. No.	Ohms per cm	Ohms per ft.
10	.00165	.0504
12	.00263	.0801
14	.00418	.127
16	.00665	.203
18	.0106	.322
20	.0168	.512
22	.0267	.815
24	.0425	1.30
26	.0675	2.06
27	.0852	2.60
28	.107	3.27
30	.171	5.21
32	.272	8.28
34	.432	13.2
36	.687	20.9
40	1.74	52.9

Constantan (0°C) $\rho = 44.1 \times 10^{-6}$ ohm cm

B. & S. No.	Ohms per cm	Ohms per ft.
10	.000838	.0255
12	.00133	.0406
14	.00212	.0646
16	.00337	.103
18	.00536	.163
20	.00852	.260
22	.0135	.413
24	.0215	.657
26	.0342	1.04
27	.0432	1.32
28	.0545	1.66
30	.0866	2.64
32	.138	4.20
34	.219	6.67
36	.348	10.6
40	.880	26.8

Copper, annealed $\rho = 1.724 \times 10^{-6}$ ohm cm

B. & S. No.	Ohms per cm	Ohms per ft.
10	.0000328	.000999
12	.0000521	.00159
14	.0000828	.00253
16	.000132	.00401
18	.000209	.00638
20	.000333	.0102
22	.000530	.0161
24	.000842	.0257
26	.00134	.0408
27	.00169	.0515
28	.00213	.0649
30	.00339	.103
32	.00538	.164
34	.00856	.261
36	.0136	.415
40	.0344	1.05

* Trade mark.

RESISTANCE OF WIRES (Continued)

Eureka (0°C) $\rho = 47. \times 10^{-6}$ ohm cm

B. & S. No.	Ohms per cm	Ohms per ft.
10	.000893	.0272
12	.00142	.0433
14	.00226	.0688
16	.00359	.109
18	.00571	.174
20	.00908	.277
22	.0144	.440
24	.0230	.700
26	.0365	1.11
27	.0460	1.40
28	.0580	1.77
30	.0923	2.81
32	.147	4.47
34	.233	7.11
36	.371	11.3
40	.938	28.6

German silver $\rho = 33. \times 10^{-6}$ ohm cm

B. & S. No.	Ohms per cm	Ohms per ft.
10	.000627	.0191
12	.000997	.0304
14	.00159	.0483
16	.00252	.0768
18	.00401	.122
20	.00638	.194
22	.0101	.309
24	.0161	.491
26	.0256	.781
27	.0323	.985
28	.0408	1.24
30	.0648	1.97
32	.103	3.14
34	.164	4.99
36	.260	7.94
40	.659	20.1

Iron $\rho = 10. \times 10^{-6}$ ohm cm

B. & S. No.	Ohms per cm	Ohms per ft.
10	.000190	.00579
12	.000302	.00921
14	.000481	.0146
16	.000764	.0233
18	.00121	.0370
20	.00193	.0589
22	.00307	.0936
24	.00489	.149
26	.00776	.237
27	.00979	.299
28	.0123	.376
30	.0196	.598
32	.0312	.952
34	.0497	1.51
36	.0789	2.41
40	.200	6.08

Magnesium $\rho = 4.6 \times 10^{-6}$ ohm cm

B. & S. No.	Ohms per cm	Ohms per ft.
10	.0000874	.00267
12	.000139	.00424
14	.000221	.00674
16	.000351	.0107
18	.000559	.0170
20	.000889	.0271
22	.00141	.0431
24	.00225	.0685
26	.00357	.109
27	.00451	.137
28	.00568	.173
30	.00903	.275
32	.0144	.438
34	.0228	.696
36	.0363	1.11
40	.0918	2.80

Molybdenum $\rho = 5.7 \times 10^{-6}$ ohm cm

B. & S. No.	Ohms per cm	Ohms per ft.
10	.000108	.00330
12	.000172	.00525
14	.000274	.00835
16	.000435	.0133
18	.000693	.0211
20	.00110	.0336
22	.00175	.0534

Excello $\rho = 92. \times 10^{-6}$ ohm cm

B. & S. No.	Ohms per cm	Ohms per ft.
10	.00175	.0533
12	.00278	.0847
14	.00442	.135
16	.00703	.214
18	.0112	.341
20	.0178	.542
22	.0283	.861
24	.0449	1.37
26	.0714	2.18
27	.0901	2.75
28	.114	3.46
30	.181	5.51
32	.287	8.75
34	.457	13.9
36	.726	22.1
40	1.84	56.0

Gold $\rho = 2.44 \times 10^{-6}$ ohm cm

B. & S. No.	Ohms per cm	Ohms per ft.
10	.0000464	.00141
12	.0000737	.00225
14	.000117	.00357
16	.000186	.00568
18	.000296	.00904
20	.000471	.0144
22	.000750	.0228
24	.00119	.0363
26	.00189	.0577
27	.00239	.0728
28	.00301	.0918
30	.00479	.146
32	.00762	.232
34	.0121	.369
36	.0193	.587
40	.0487	1.48

Lead $\rho = 22. \times 10^{-6}$ ohm cm

B. & S. No.	Ohms per cm	Ohms per ft.
10	.000418	.0127
12	.000665	.0203
14	.00106	.0322
16	.00168	.0512
18	.00267	.0815
20	.00425	.130
22	.00676	.206
24	.0107	.328
26	.0171	.521
27	.0215	.657
28	.0272	.828
30	.0432	1.32
32	.0687	2.09
34	.109	3.33
36	.174	5.29
40	.439	13.4

Manganin $\rho = 44. \times 10^{-6}$ ohm cm

B. & S. No.	Ohms per cm	Ohms per ft.
10	.000836	.0255
12	.00133	.0405
14	.00211	.0644
16	.00336	.102
18	.00535	.163
20	.00850	.259
22	.0135	.412
24	.0215	.655
26	.0342	1.04
27	.0431	1.31
28	.0543	1.66
30	.0864	2.63
32	.137	4.19
34	.218	6.66
36	.347	10.6
40	.878	26.8

Monel Metal $\rho = 42. \times 10^{-6}$ ohm cm

B. & S. No.	Ohms per cm	Ohms per ft.
10	.000798	.0243
12	.00127	.0387
14	.00202	.0615
16	.00321	.0978
18	.00510	.156
20	.00811	.247
22	.0129	.393

Molybdenum $\rho = 5.7 \times 10^{-6}$ ohm cm

B. & S. No.	Ohms per cm	Ohms per ft.
24	.00278	.0849
26	.00443	.135
27	.00558	.170
28	.00704	.215
30	.0112	.341
32	.0178	.542
34	.0283	.863
36	.0450	1.37
40	.114	3.47

***Nichrome $\rho = 150. \times 10^{-6}$ ohm cm**

B. & S. No.	Ohms per cm	Ohms per ft.
10	.0021281	.06488
12	.0033751	.1029
14	.0054054	.1648
16	.0085116	.2595
18	.0138383	.4219
20	.0216218	.6592
22	.0346040	1.055
24	.0548088	1.671
26	.0875760	2.670
28	.1394328	4.251
30	.2214000	6.750
32	.346040	10.55
34	.557600	17.00
36	.885600	27.00
38	1.383832	42.19
40	2.303872	70.24

Platinum $\rho = 10. \times 10^{-6}$ ohm cm

B. & S. No.	Ohms per cm	Ohms per ft.
10	.000190	.00579
12	.000302	.00921
14	.000481	.0146
16	.000764	.0233
18	.00121	.0370
20	.00193	.0589
22	.00307	.0936
24	.00489	.149
26	.00776	.237
27	.00979	.299
28	.0123	.376
30	.0196	.598
32	.0312	.952
34	.0497	1.51
36	.0789	2.41
40	.200	6.08

Steel, piano wire (0°C) $\rho = 11.8 \times 10^{-6}$ ohm cm

B. & S. No.	Ohms per cm	Ohms per ft.
10	.000224	.00684
12	.000357	.0109
14	.000567	.0173
16	.000901	.0275
18	.00143	.0437
20	.00228	.0695
22	.00363	.110
24	.00576	.176
26	.00916	.279
27	.0116	.352
28	.0146	.444
30	.0232	.706
32	.0368	1.12
34	.0586	1.79
36	.0931	2.84
40	.236	7.18

Tantalum $\rho = 15.5 \times 10^{-6}$ ohm cm

B. & S. No.	Ohms per cm	Ohms per ft.
10	.000295	.00898
12	.000468	.0143
14	.000745	.0227
16	.00118	.0361
18	.00188	.0574
20	.00299	.0913
22	.00476	.145
24	.00757	.231
26	.0120	.367
27	.0152	.463
28	.0191	.583
30	.0304	.928
32	.0484	1.47
34	.0770	2.35
36	.122	3.73
40	.309	9.43

Monel Metal $\rho = 42. \times 10^{-6}$ ohm cm

B. & S. No.	Ohms per cm	Ohms per ft.
24	.0205	.625
26	.0326	.994
27	.0411	1.25
28	.0519	1.58
30	.0825	2.51
32	.131	4.00
34	.209	6.36
36	.331	10.1
40	.838	25.6

Nickel $\rho = 7.8 \times 10^{-6}$ ohm cm

B. & S. No.	Ohms per cm	Ohms per ft.
10	.000148	.00452
12	.000236	.00718
14	.000375	.0114
16	.000596	.0182
18	.000948	.0289
20	.00151	.0459
22	.00240	.0730
24	.00381	.116
26	.00606	.185
27	.00764	.233
28	.00963	.294
30	.0153	.467
32	.0244	.742
34	.0387	1.18
36	.0616	1.88
40	.156	4.75

Silver (18°C) $\rho = 1.629 \times 10^{-6}$ ohm cm

B. & S. No.	Ohms per cm	Ohms per ft.
10	.0000310	.000944
12	.0000492	.00150
14	.0000783	.00239
16	.000124	.00379
18	.000198	.00603
20	.000315	.00959
22	.000500	.0153
24	.000796	.0243
26	.00126	.0386
27	.00160	.0486
28	.00201	.0613
30	.00320	.0975
32	.00509	.155
34	.00809	.247
36	.0129	.392
40	.0325	.991

Steel, invar (35 % Ni) $\rho = 81. \times 10^{-6}$ ohm cm

B. & S. No.	Ohms per cm	Ohms per ft.
10	.00154	.0469
12	.00245	.0746
14	.00389	.119
16	.00619	.189
18	.00984	.300
20	.0156	.477
22	.0249	.758
24	.0396	1.21
26	.0629	1.92
27	.0793	2.42
28	.100	3.05
30	.159	4.85
32	.253	7.71
34	.402	12.3
36	.639	19.5
40	1.62	49.3

Tin $\rho = 11.5 \times 10^{-6}$ ohm cm

B. & S. No.	Ohms per cm	Ohms per ft.
10	.000219	.00666
12	.000348	.0106
14	.000553	.0168
16	.000879	.0268
18	.00140	.0426
20	.00222	.0677
22	.00353	.108
24	.00562	.171
26	.00893	.272
27	.0113	.343
28	.0142	.433
30	.0226	.688
32	.0359	1.09
34	.0571	1.74
36	.0908	2.77
40	.230	7.00

RESISTANCE OF WIRES (Continued)

B. & S. No.	Ohms per cm	Ohms per ft.	B. & S. No.	Ohms per cm	Ohms per ft.
Tungsten $\rho = 5.51 \times 10^{-6}$ ohm cm			Zinc (0°C) $\rho = 5.75 \times 10^{-6}$ ohm cm		
10	.000105	.00319	10	.000109	.00333
12	.000167	.00508	12	.000174	.00530
14	.000265	.00807	14	.000276	.00842
16	.000421	.0128	16	.000439	.0134
18	.000669	.0204	18	.000699	.0213
20	.00106	.0324	20	.00111	.0339
22	.00169	.0516	22	.00177	.0538
24	.00269	.0820	24	.00281	.0856
26	.00428	.130	26	.00446	.136
27	.00540	.164	27	.00563	.172
28	.00680	.207	28	.00710	.216
30	.0108	.330	30	.0113	.344
32	.0172	.524	32	.0180	.547
34	.0274	.834	34	.0286	.870
36	.0435	1.33	36	.0454	1.38
40	.110	3.35	40	.115	3.50

COLOR CODE FOR RESISTORS

COLOR CODE FOR RESISTORS

GENERAL REQUIREMENTS

Code Colors. Colors used for color coding shall conform to Standard MIL–STD–174, and shall be permanent and nonfading.

The color-code marking shall remain legible after the resistor has been subjected to all the tests specified in the individual resistor specification.

Body Colors (Background). The exterior body color of resistors shall be any color other than black; for composition-type resistors, tan is preferred, for wire-wound-type resistors, brown is preferred.

Conflict of Colors. When the body color is the same as any of the band colors, then either the body color or the band color shall be differentiated by shade or gloss.

COLOR CODE FOR RESISTORS (Not wire wound)

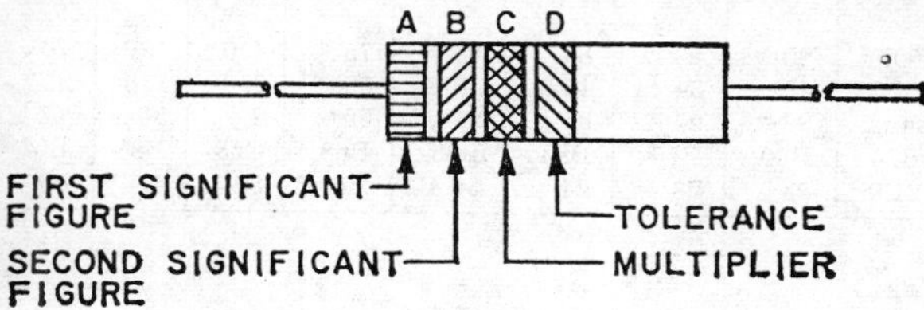

COLOR CODE FOR RESISTORS (Wire wound)

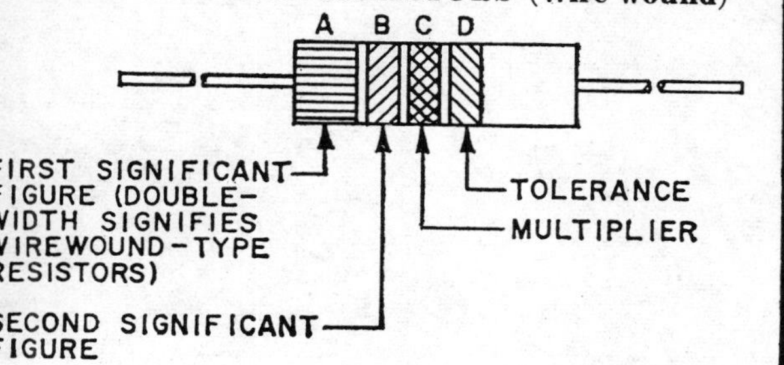

Band A		Band B		Band C		Band D	
Color	First significant figure	Color	Second significant figure	Color	Multiplier	Color	Resistance tolerance (per cent)
Black..	0	Black.....	0	Black...	1	Silver..	±10
Brown..	1	Brown....	1	Brown...	10	Gold...	±5
Red....	2	Red......	2	Red.....	100		
Orange..	3	Orange...	3	Orange..	1,000		
Yellow..	4	Yellow...	4	Yellow..	10,000		
Green..	5	Green....	5	Green...	100,000		
Blue...	6	Blue.....	6	Blue....	1,000,000		
Purple (viole.	7	Purple (violet)..	7	...	...		
Gray...	8	Gray.....	8	Silver...	0.01		
White..	9	White....	9	Gold....	0.1		

Band A—The first significant figure of the resistance value. (For composition-type resistors, all bands shall be of equal width. For wire-wound-type resistors, band A shall be double width.)

Band B—The second significant figure of the resistance value.

Band C—The multiplier. (The multiplier is the factor by which the two significant figures are multiplied to yield the nominal resistance value.)

Band D—The resistance tolerance.

Examples of color coding

3,900 ohms ±10 percent—Band A. orange; band B. white; band C, red; band D, silver.

43,000 ohms ±5 percent—Band A, yellow; band B, orange; band C orange; band D, gold.

PHYSICAL PROPERTIES OF GLASS SEALING AND LEAD WIRE MATERIALS

From General Electric Company

	Glass Sealing Materials												
	Copper		Dumet	42 Ni	Gas Free 42% Ni	42-6 Ni Cr.	46 Ni	52 Ni.	27 Cr.	® Kovar	W	Mo	Nickel 200
	OFC	0.02P											
Analysis: Carbon				0.10	0.05	0.10	0.10	0.10	0.15	0.02			0.06
Manganese				0.50	0.50	0.50	0.50	0.50	0.60	0.30			0.25
Silicon			*Copper Clad*	0.25	0.25	0.25	0.25	0.25	0.40	0.20			0.05
Chromium			*42% Nickel*	—	—	5.75	—	—	28.00	—			
Nickel			*Iron*	42	42	42.5	46	51	0.50	29			99.5
Copper	99.95	99.90	*See GE Spec.*	—	—	—	—	—	—	—			0.05
		P	*DS 8311-01*		Ti0.4		Ti0.4			Co 17			
Other		0.02		Bal.Fe	Bal.Fe	Bal.Fe	Bal.Fe	Bal.Fe	Bal.Fe	Bal.Fe			
Density: grams/cc	8.94	8.94	8.26 to 8.32	8.12	8.12	8.12	8.17	8.30	7.60	8.36	19.3	10.2	8.89
lbs. per cu. in.	0.323	0.323	0.298—0.301	0.293	0.293	0.293	0.295	0.300	0.274	0.302	0.697	0.369	0.321
Thermal Conductivity 20-100°/C													
Cal/cm/sec/cm²/°C	0.948	0.8	0.2—0.3	0.025	0.025	0.029	0.028	0.032	0.054	0.04	0.31	0.34	0.15
Btu/in/hr/sq. ft/°F	2750	2320	580—870	74	74	84	81	93	158	116	900	1000	435
Electrical Resistivity (20°C) Microhm—cm	1.71	2.03	7.3 to 12.0	72	72	95	46	43	63	49	5.5	5.2	9.5
Ohm per cir. mil ft.	10.3	12.2	44 to 72	430	430	570	275	258	380	294	33	31	57
Elec. Cond. % IACS	101	85	23 to 14	2.3	2.3	1.8	3.6	3.9	2.8	3.4	31	33	18
Curie Temperature °C	—	—	380	380	380	295	460	530	610	435	—	—	360
Melting Temperature °C	1083	1083	— —	1425	1425	1425	1425	1425	1425	1450	3410	2610	1455
°F	1981	1981	— —	2597	2597	2597	2597	2597	2597	2642	6170	4730	2651
Specific Heat cal/gr.	0.092	0.092	0.11	0.12	0.12	0.12	0.12	0.12	0.14	0.11	0.033	0.066	0.13
Thermal Expansion in/in/°Cx10⁷			Radial										
25-100°C	168	168	60 to 80	50.1	43.4	65.5	71.0	99.5	94.6	58.6	45	51	133
25-200°C	172	172		47.1	44.1	70.8	73.7	101.0	100.5	52.0	46		139
25-300°C	177	177	60 to 80	47.6	46.1	82.6	75.0	101.0	105.3	51.3	46		144
25-350°C	178	178	65 to 85	50.5		90.4	74.4	100.0	107.0	48.9	46		146
25-400°C	181	181	80 to 100	62.5	64.1	100.0	74.3	100.0	107.8	50.6	46		148
25-500°C	183	183	100 to 140	83.2	85.6	115.0	86.8	102.1	111.2	61.5	46	57	172
25-600°C	188	188		99.0	100.1	125.8	100.2	110.0	112.6	78.0			
Mechanical Properties (Annealed)													
Ultimate Str. (1000 psi)	35	35	74	82	80	80	82	80	85	75	490	120	70
Yield Str. (1000 psi)			50	34	34	40	34	40	55	50	360	110	25
% Elong. (2″)			30	30	30	30	27	35	25	30	8	30	45
Rockwell Hardness				B76	B76	B80	B76	B83	B85	B68	C25	B88	B62
Elastic Modulus (10^6 psi)	16	16		21.5	21.0	23	23	24	30	20	50	47	30

PHYSICAL PROPERTIES OF GLASS SEALING AND LEAD WIRE MATERIALS (Continued)

Metal Alloy Lead Wires									Clad Materials (Also see Dumet)			Pure Metals (Also see W and Mo glass sealing materials)						
Nickel 211	AISI Type 302	AISI Type 316	AISI Type 430	400 Monel	600 Inco-nel	Ad-vance ®	CDA 752	70-30 Brass	Kul-grid ®	40% CCFe	30% CCFe	Au	Ag	Al	Pt	Ta	Fe	Ti
.10	0.15Max	0.08Max	0.12Max	0.12	0.04													
.75	2.00Max	2.00Max	1.00Max	0.90	0.20		0.50Max											
.05	1.00Max	1.00Max	1.00Max	0.15														
	17–19	16–18	14–18		16													
5.0	8–10	10–14	0.50Max	66	76	43	18		27									
		—		31.5		Bal.	65	70		37.5	26							
		Mo 2–3			Fe 7.20		Bal. Zn	Bal. Zn	Cu Core	Fe Core	Fe Core							
.72	7.9	7.9	7.7	8.84	8.41	8.9	8.73	8.53	8.89	8.15	8.15	19.3	10.5	2.69	21.45	16.6	7.87	4.51
.315	0.285	0.285	0.278	0.319	0.304	0.322	0.316	0.308	0.321	0.294	0.294	0.698	0.379	0.097	0.775	0.600	0.284	0.163
12	0.04	0.04	0.05	0.062	0.036	0.051	0.08	0.29		0.46	0.38	0.71	1.00	0.57	0.165	0.13	0.18	0.43
50	116	116	145	180	104	148	232	845		1330	1100	2050	2900	1650	480	380	520	1250
3.3	72	74	60	48.2	98.1	49	28.7	6.16	2.3	4.4	5.9	2.19	1.629	2.65	9.83	12.45	9.71	7.0
10	433	445	361	290	590	294	173	37	14	26.4	35.3	13	9.8	16	59	75	58	42
	2.3	2.3	2.9	3.6	1.7	3.5	6	28	70	40	30	78	105	65	16	14	18	25
52			610	43/60	-125	—	—	—		770	770	—	—	—	—	—	770	—
427	1421	1399	1510	1349	1427	1210	1110	954				1063	960	660	1769	3000	1536	1668
500	2590	2550	2750	2460	2600	2210	2030	1750				1945	1760	1220	3217	5425	2797	3035
13	0.12	0.12	0.11	0.102	0.109	0.094	0.09	0.09		0.10	0.10	0.031	0.056	0.225	0.031	0.034	0.11	0.124
3	166	166	101	140	115	149						142	196	239	91	65	122	88
9				145										243			129	91
4	171	175	110	150			160	199						253				
6																		
8				155													138	94
3	180	180	120	160		163						152	206		96	66		
				165										287			145	97
	90	85	75	85	100	60	60	50				20	22	13	40	55	40	
	37	35	45	40	45		30	—					8	5	12		20	
	55	55	30	40	40		40	60				45	48	40	30	30	45	
	B82	B80	B82	B68	B74		—	—										
	29	29	29	26	31	18	18	16		24	24	11.6	11	9	21.3	27	30	16.8

ELECTRICAL RESISTIVITY AND TEMPERATURE COEFFICIENTS OF ELEMENTS

Element	Temperature °C	Microhm-Cm	Temperature Coefficient per °C
Aluminum, 99.996%	20	2.6548	$0.00429^{20\,z}$
Antimony	0	39.0	
Arsenic	20	33.3	
Beryllium[a]	20	4.0	$0.025^{20\,z}$
Bismuth	0	106.8	
Boron	0	1.8×10^{12}	
Cadmium	0	6.83	$0.0042^{0\,z}$
Calcium	0	3.91	$0.00416^{0\,z}$
Carbon[b]	0	1375.0	
Cerium	25	75.0	0.00087^{0-25}
Cesium	20	20	
Chromium	0	12.9	$0.003^{0\,z}$
Cobalt	20	6.24	0.00604^{0-100}
Copper	20	1.6730	$0.0068^{20\,z}$
Dysprosium[c]	25	57.0	0.00119^{0-25}
Erbium	25	107.0	0.00201^{0-25}
Europium	25	90.0	
Gadolinium	25	140.5	0.00176^{0-25}
Gallium[d]	20	17.4	
Germanium[e]	22	46×10^{6}	
Gold	20	2.35	0.004^{0-100}
Hafnium	25	35.1	$0.0038^{25\,z}$
Holmium	25	87.0	0.00171^{0-25}
Indium	20	8.37	
Iodine	20	1.3×10^{15}	
Iridium	20	5.3	0.003925^{0-100}
Iron, 99.99%	20	9.71	$0.00651^{20\,z}$
Lanthanum	25	5.70	0.00218^{0-25}
Lead	20	20.648	0.00336^{20-40}
Lithium	0	8.55	
Lutetium	25	79.0	0.00240^{0-25}
Magnesium[f]	20	4.45	$0.0165^{20\,z}$
Manganese α	23–100	185.0	
Mercury	50	98.4	
Molybdenum	0	5.2	
Neodymium	25	64.0	0.00164^{0-25}
Nickel	20	6.84	0.0069^{0-100}
Niobium (Columbium)[g]	0	12.5	
Osmium	20	9.5	0.0042^{0-100}
Palladium	20	10.8	0.00377^{0-100}
Phosphorus, white	11	1×10^{17}	
Platinum, 99.85%	20	10.6	0.003927^{0-100}
Plutonium	107	141.4	
Potassium	0	6.15	
Praseodymium	25	68	0.00171^{0-25}
Rhenium	20	19.3	0.00395^{0-100}
Rhodium	20	4.51	0.0042^{0-100}
Rubidium	20	12.5	
Ruthenium	0	7.6	
Samarium	25	88.0	0.00184^{0-25}
Scandium[h]	22	61.0	0.00282^{0-25}
Selenium	0	12.0	
Silicon	0	10.0	
Silver	20	1.59	0.0041^{0-100}
Sodium	0	4.2	
Sulfur, yellow	20	2×10^{23}	
Strontium	20	23.0	
Tantalum	25	12.45	0.00383^{0-100}

ELECTRICAL RESISTIVITY AND TEMPERATURE COEFFICIENTS OF ELEMENTS (Continued)

Element	Temperature °C	Microhm-Cm	Temperature Coefficient per °C
Tellerium	25	4.36×10^5	
Thallium	0	18.0	
Thorium	0	13.0	0.0038^{0-100}
Thulium	25	79.0	0.00195^{0-25}
Tin	0	11.0	0.0047^{0-100}
Titanium	20	42.0	
Tungsten	27	5.65	
Uranium		30.0	
Vanadium	20	24.8–26.0	
Ytterbium	25	29.0	0.0013^{0-25}
Yttrium	25	57.0	0.0027^{0-25}
Zinc	20	5.916	0.00419^{0-100}
Zironium	20	40.0	$0.0044^{20\ z}$

[a] Annealed, comm. pure.
[b] Graphite.
[c] Polycrystalline.
[d] Hard Wire.
[e] Intrinsic Ge.
[f] Polycrystalline.
[g] High Purity.
[h] Zone refined bar.
[z] Data not available to indicate range over which coefficient is valid.

ELECTRICAL RESISTIVITY OF THE ALKALI METALS

Metal	At. No.	At. Wt.	Sp. Gr.	M.P., °C	Resistivity μ-ohm cm	at °C
Li	3	6.939	0.534^{20}	179	12.17	86.6
					13.36	120.5
					14.73	153.3
					15.54	178.2
Na	11	22.9898	0.971^{20}	97.81	5.23	29.4
					5.72	50.6
					6.53	83.3
					6.71	91.6
					6.83	97.1
K	19	39.102	0.86^{20}	63.65	7.01	22.8
					7.32	35.5
					7.54	41.1
					8.05	56.1
Rb	37	85.47	$1.475^{38.89}$	38.89	11.28	0
					12.52	53
Cs	55	132.905	1.8785^{15}	28.5	20.29	18.2
					37.39	28.1

LOW MELTING POINT SOLDERS

(From N.B.S. Circular 492)

Nominal Composition, Weight Percent			Liquidus Temperature, °F
Pb	Sn	Bi	
25	25	50	266
50	37.5	12.5	374
25	50	25	336

BRAZING FILLER METALS (SOLDERS)

AWS-ASTM Classification	Solidus, °F	Liquidus, °F	Brazing Temperature Range, °F	Ag	Al	As	Au	B	Be	Bi	C	Cd	Cr	Cu	Fe	Li	Mg	Mn	Ni	P	Pb	Sb	Si	Sn	Ti	Zn	Other Each	Other Total	AWS-ASTM Classification
ALUMINUM-SILICON																													
BAlSi-2	1070	1135	1110-1150		Bal.									0.25	0.8			0.10					6.8-8.2			0.20	0.05	0.15	BAlSi-2
BAlSi-3	970	1085	1060-1120		Bal.								0.15	3.3-4.7	0.8		0.15	0.15					9.3-10.7			0.20	0.05	0.15	BAlSi-3
BAlSi-4	1070	1080	1080-1120		Bal.									0.30	0.8		0.10	0.15					11.0-13.0			0.20	0.05	0.15	BAlSi-4
BAlSi-5	1070	1095	1090-1120		Bal.									0.30	0.8		0.05	0.05					9.0-11.0		0.20	0.10	0.05	0.15	BAlSi-5
COPPER-PHOSPHORUS																													
BCuP-1	1310	1650	1450-1700											Bal.						4.75-5.25								0.15	BCuP-1
BCuP-2	1310	1460	1350-1550											Bal.						7.00-7.25								0.15	BCuP-2
BCuP-3	1190	1485	1300-1500	4.75-5.25										Bal.						5.75-6.25								0.15	BCuP-3
BCuP-4	1190	1335	1300-1450	5.75-6.25										Bal.						7.00-7.50								0.15	BCuP-4
BCuP-5	1190	1475	1300-1500	14.50-15.50										Bal.						4.75-5.25								0.15	BCuP-5
SILVER																													
BAg-1	1125	1145	1145-1400	44-46								23-25		14-16												14-18		0.15	BAg-1
BAg-1a	1160	1175	1175-1400	49-51								17-19		14.5-16.5												14.5-18.5		0.15	BAg-1a
BAg-2	1125	1295	1295-1550	34-36								17-19		25-27												19-23		0.15	BAg-2
BAg-3	1170	1270	1270-1500	49-51								15-17		14.5-16.5					2.5-3.5							13.5-17.5		0.15	BAg-3
BAg-4	1240	1435	1435-1650	39-41										29-31					1.5-2.5							26-30		0.15	BAg-4
BAg-5	1250	1370	1370-1550	44-46										29-31												23-27		0.15	BAg-5
BAg-6	1270	1425	1425-1600	49-51										33-35												14-18		0.15	BAg-6
BAg-7	1145	1205	1205-1400	55-57										21-23										4.5-5.5		15-19		0.15	BAg-7
BAg-8	1435	1435	1435-1650	71-73										Bal.														0.15	BAg-8
BAg-8a	1410	1410	1410-1600	71-73										Bal.		0.15-0.3												0.15	BAg-8a
BAg-13	1325	1575	1575-1775	53-55										Bal.												4.0-6.0		0.15	BAg-13
BAg-18	1115	1125	1325-1550	59-61										Bal.						0.025				9.5-10.5				0.15	BAg-18
BAg-19	1435	1635	1610-1800	92-93										Bal.		0.15-0.3												0.15	BAg-19
PRECIOUS METALS																													
BAu-1	1815	1860	1860-2000				37.0+1 −0							Bal.														0.15	BAu-1
BAu-2	1635	1635	1635-1850				79.5+1 −0							Bal.														0.15	BAu-2
BAu-3	1785	1885	1885-1995				34.5+1 −0							Bal.					2.5-3.5									0.15	BAu-3
BAu-4	1740	1740	1740-1840				81.5+1 −0												Bal.									0.15	BAu-4
COPPER AND COPPER-ZINC																													
BCu-1	1980	1980	2000-2100		0.01									99.90 min.						0.075	0.02							0.10	BCu-1
BCu-1a	1980	1980	2000-2100											99.0 min.														0.30[a]	BCu-1a
BCu-2	1980	1980	2000-2100											86.5 min.														0.50[c]	BCu-2[b]
RBCuZn-A	1630	1650	1670-1750		0.01*									57-61	*			*			0.05*		*	0.25-1.00		Bal.		0.50[d]	RBCuZn-A[e]
RBCuZn-D	1690	1715	1720-1800		0.01*									46-50					9.0-11.0	0.25	0.05*		0.04-0.25			Bal.		0.50[d]	RBCuZn-D[e]
MAGNESIUM																													
BMg-1	830	1110	1120-1160		8.3-9.7				0.0002-0.0008					0.05	0.005		Bal.	0.15 min.	0.005				0.05			1.7-2.3		0.30	BMg-1
BMg-2	770	1050	1080-1130		11.0-13.0												Bal.									4.5-5.5		0.30	BMg-2
BMg-2a	770	1050	1080-1130		11.0-13.0				0.0002-0.0008								Bal.									4.5-5.5		0.30	BMg-2a

[a] Total other elements requirement pertains only to the metallic elements for this filler metal.

[b] These chemical requirements pertain only to the copper oxide and do not include requirements for the organic vehicle in which the copper oxide is suspended.

[c] Total other elements requirement pertains only to metallic elements for this filler metal. The following limitations are placed on the nonmetallic elements:

Constituent	per cent (max)
Chlorides	0.4
Sulfates	0.1
Oxygen	remainder
Nitric acid insoluble	0.3
Acetone soluble matter	0.5

[d] Total other elements, including the elements marked with an asterisk (*), shall not exceed the value specified.

[e] This AWS-ASTM classification is intended to be identical with the same classification that appears in the Specification for Copper and Copper-Alloy Welding Rods (AWS Designation A5.7; ASTM Designation B 259).[3]

BRAZING FILLER METALS (SOLDERS) (Continued)

AWS-ASTM Classification	Solidus, °F	Liquidus, °F	Brazing Temperature Range, °F	Ag	Al	As	Au	B	Be	Bi	C	Cd	Cr	Cu	Fe	Li	Mg	Mn	Ni	P	Pb	Sb	Si	Sn	Ti	Zn	Other Each	Other Total	AWS-ASTM Classification
NICKEL																													
BNi-1	1790	1900	1950-2200					2.75-4.00			0.6-0.9		13.0-15.0		4.0-5.0				Bal.				3.0-5.0					0.50	BNi-1
BNi-2	1780	1830	1850-2150					2.75-3.5			0.15		6.0-8.0		2.0-4.0				Bal.				4.0-5.0					0.50	BNi-2
BNi-3	1800	1900	1850-2150					2.75-3.5			0.06				1.5				Bal.				4.0-5.0					0.50	BNi-3
BNi-4	1800	1950	1850-2150					1.0-2.2			0.06				1.5				Bal.				3.0-4.0					0.50	BNi-4
BNi-5	1975	2075	2100-2200								0.15		18.0-20.0						Bal.				9.75-10.5					0.50	BNi-5
BNi-6	1610	1610	1700-1875								0.15								Bal.	10.0-12.0								0.50	BNi-6
BNi-7	1630	1630	1700-1900										11.0-15.0						Bal.	9.0-11.0								0.50	BNi-7
SOLDERS—ASTM DESIGNATION B-32-60T, REVISED 1966																													
70A	361	378			0.005 max	0.03 max				0.25 max				0.08 max	0.02 max						30			70		0.005 max			70A
70B					0.005 max	0.03 max				0.25 max				0.08 max	0.02 max						30	0.2-0.5		70		0.005 max			70B
63A	361	361			0.005 max	0.03 max				0.25 max				0.08 max	0.02 max						37	0.12 max		63		0.005 max			63A
63B					0.005 max	0.03 max				0.25 max				0.08 max	0.02 max						37	0.2-0.5		63		0.005 max			63B
60A	361	374			0.005 max	0.03 max				0.25 max				0.08 max	0.02 max						40	0.12 max		60		0.005 max			60A
60B					0.005 max	0.03 max				0.25 max				0.08 max	0.02 max						40	0.2-0.5		60		0.005 max			60B
50A	361	421			0.005 max	0.03 max				0.25 max				0.08 max	0.02 max						50	0.12 max		50		0.005 max			50A
50B	360	420			0.005 max	0.03 max				0.25 max				0.08 max	0.02 max						50	0.2-0.5		50		0.005 max			50B
45A	361	441			0.005 max	0.03 max				0.25 max				0.08 max	0.02 max						55	0.12 max		45		0.005 max			45A
45B					0.005 max	0.03 max				0.25 max				0.08 max	0.02 max						55	0.2-0.5		45		0.005 max			45B
40A	361	460			0.005 max	0.03 max				0.25 max				0.08 max	0.02 max						60	0.12 max		40		0.005 max			40A
40B	360	460			0.005 max	0.03 max				0.25 max				0.08 max	0.02 max						60	0.2-0.5		40		0.005 max			40B
40C	365	448			0.005 max	0.03 max				0.25 max				0.08 max	0.02 max						58	1.8-2.4		40		0.005 max			40C
35A	361	477			0.005 max	0.03 max				0.25 max				0.08 max	0.02 max						65	0.25 max		35		0.005 max			35A
35B					0.005 max	0.03 max				0.25 max				0.08 max	0.02 max						65	0.2-0.5		35		0.005 max			35B
35C	365	470			0.005 max	0.03 max				0.25 max				0.08 max	0.02 max						63.2	1.6-2.0		35		0.005 max			35C
30A	361	491			0.005 max	0.03 max				0.25 max				0.08 max	0.02 max						70	0.25 max		30		0.005 max			30A
30B					0.005 max	0.03 max				0.25 max				0.08 max	0.02 max						70	0.2-0.5		30		0.005 max			30B
30C	364	482			0.005 max	0.03 max				0.25 max				0.08 max	0.02 max						68.4	1.4-1.8		30		0.005 max			30C
25A	361	511			0.005 max	0.03 max				0.25 max				0.08 max	0.02 max						75	0.25 max		25		0.005 max			25A
25B	360	510			0.005 max	0.03 max				0.25 max				0.08 max	0.02 max						75	0.2-0.5		25		0.005 max			25B
25C					0.005 max	0.03 max				0.25 max				0.08 max	0.02 max						73.7	1.1-1.5		25		0.005 max			25C
20B	361	531			0.005 max	0.03 max				0.25 max				0.08 max	0.02 max						80	0.2-0.5		20		0.005 max			20B
20C	363	517			0.005 max	0.03 max				0.25 max				0.08 max	0.02 max						79	0.8-1.2		20		0.005 max			20C
15B	440	550			0.005 max	0.03 max				0.25 max				0.08 max	0.02 max						85	0.2-0.5		15		0.005 max			15B
10B					0.005 max	0.03 max				0.25 max				0.08 max	0.02 max						90	0.2-0.5		10		0.005 max			10B
5A	518	594			0.005 max	0.03 max				0.25 max				0.08 max	0.02 max						95	0.12 max		5*		0.005 max			5A
5B					0.005 max	0.03 max				0.25 max				0.08 max	0.02 max						95	0.2-0.5		5*		0.005 max			5B
2A					0.005 max	0.03 max				0.25 max				0.08 max	0.02 max						98	0.12 max		2**		0.005 max			2A
2B					0.005 max	0.03 max				0.25 max				0.08 max	0.02 max						98	0.2-0.5		2**		0.005 max			2B
2.5S	579	579			0.005 max	0.03 max				0.25 max				0.08 max	0.02 max						97.5	0.4 max		0***		0.005 max			2.5S
1.5S	588	588		2.3-2.7	0.005 max	0.03 max				0.25 max				0.08 max	0.02 max						97.5	0.4 max		1****		0.005 max			1.5S
95TA	452	464		1.3-1.7	0.005 max	0.03 max				0.25 max				0.08 max	0.02 max						0.2	4.5-5.5		95		0.005 max			95TA
96.5TS				3.3-3.7	0.005 max	0.03 max				0.25 max				0.08 mas	0.02 max						0.2	0.2-0.5		96.5		0.005 max			96.5TS

* Permissible tin range, 4.5–5.5%.
** Permissible tin range, 1.5–2.5%.
*** Tin maximum, 0.25%.
**** Permissible tin range, 0.75–1.25%.

ACCELERATION DUE TO GRAVITY AND LENGTH OF THE SECONDS PENDULUM

For Sea Level at Various Latitudes

Based on the formula of the U. S. Coast and Geodetic Survey. The length of the simple pendulum whose period is two seconds, that is which beats seconds, is computed in each case from the corresponding value of the acceleration.

Latitude	Acceleration due to gravity		Length of seconds pendulum	
°	cm/sec.²	ft./sec.²	cm	in.
0	978.039	32.0878	99.0961	39.0141
5	978.078	32.0891	99.1000	39.0157
10	978.195	32.0929	99.1119	39.0204
15	978.384	32.0991	99.1310	39.0279
20	978.641	32.1076	99.1571	39.0382
25	978.960	32.1180	99.1894	39.0509
30	979.329	32.1302	99.2268	39.0656
31	979.407	32.1327		
32	979.487	32.1353		
33	979.569	32.1380		
34	979.652	32.1407		
35	979.737	32.1435	99.2681	39.0819
36	979.822	32.1463		
37	979.908	32.1491		
38	979.995	32.1520		
39	980.083	32.1549		
40	980.171	32.1578	99.3121	39.0992
41	980.261	32.1607		
42	980.350	32.1636		
43	980.440	32.1666		
44	980.531	32.1696		
45	980.621	32.1725	99.3577	39.1171
46	980.711	32.1755		
47	980.802	32.1785		
48	980.892	32.1814		
49	980.981	32.1844		
50	981.071	32.1873	99.4033	39.1351
51	981.159	32.1902		
52	981.247	32.1931		
53	981.336	32.1960		
54	981.422	32.1988		
55	981.507	32.2016	99.4475	39.1525
56	981.592	32.2044		
57	981.675	32.2071		
58	981.757	32.2098		
59	981.839	32.2125		
60	981.918	32.2151	99.4891	39.1689
65	982.288	32.2272	99.5266	39.1836
70	982.608	32.2377	99.5590	39.1964
75	982.868	32.2463	99.5854	39.2068
80	983.059	32.2525	99.6047	39.2144
85	983.178	32.2564	99.6168	39.2191
90	983.217	32.2577	99.6207	39.2207

Free Air Correction for Altitude

−0.0003086 cm/sec.²/m for altitude in meters.
−0.000003086 ft./sec.²/ft. for altitude in feet.

Altitude meters	Correction cm/sec.²	Altitude feet	Correction ft./sec.²
200	−0.0617	200	−0.000617
300	.0926	300	.000926
400	.1234	400	.001234
500	.1543	500	.001543
600	.1852	600	.001852
700	.2160	700	.002160
800	.2469	800	.002469
900	.2777	900	.002777

DATA IN REGARD TO THE EARTH

Equatorial radius, 6,378,390 meters, 3,963.34 miles.
Polar radius, 6,356,910 meters, 3,949.99 miles.
Radius of sphere having same volume, 6,371,220 meters, 3,958.89 miles.
Quadrant of the equator, 10,019,150 meters, 6,225.60 miles.
Quadrant of the meridian, 10,002,290 meters, 6,215.12 miles.
1° latitude at the equator = 69.41 miles
1° latitude at the pole = 68.70 miles.
Mean density of the earth, 5.522 g/cm³, 344.7 lb./ft.³
Mass of the earth, 5.983×10^{24} kg, 6.595×10^{21} tons.
Mean surface density of the continents, 2.67 g/cm³, 166.7 lb./ft.³
Mean linear velocity of the earth in its orbit, 29.77 km/sec., 18.50 mi./sec.
Mean linear velocity of rotation of the surface at the equator, 0.465 km/sec., 0.289 mi./sec.

DATA IN REGARD TO THE EARTH (Continued)

Land area, 148.847×10^6 km², 57.470×10^6 sq. mi.
Ocean area, 361.254×10^6 km², 139.480×10^6 sq. mi.
Highest mountain, Everest, 8840 meters, 29,003 ft.
Greatest sea depth, 10,430 meters, 34,219 ft.
Thermal gradient of the earth, higher at increasing depths, 30° C per km, 48° C per mi. (uncertain).
Mean distance to the sun, 149,500,000 km or 92,900,000 mi.
Mean distance to the moon, 384,393 km or 238,854 mi.

THE COMMONER CHEMICAL ELEMENTS IN THE EARTH'S CRUST

Reprinted from "Principles of Geochemistry" (1952) with the permission of Brian Mason, author, and John Wiley and Sons, publishers.
The "Earth's Crust" refers to the rocks only and does not include atmosphere or the oceans. The atom percent column is obtained by dividing the weight percent by the atomic weights and reducing to 100%. The radius is the ionic radius. The volume percent is the atomic percent multiplied by $\frac{4}{3}\pi r^3$ and reducing to 100%.

Element	Weight %	Atom %	Ion Radius (Å)	Volume %
O	46.60	62.55	1.32	91.97
Si	27.72	21.22	0.39	0.80
Al	8.13	6.47	0.57	0.77
Fe	5.00	1.92	0.82	0.68
Mg	2.09	1.84	0.78	0.56
Ca	3.63	1.94	1.06	1.48
Na	2.83	2.64	0.98	1.60
K	2.59	1.42	1.33	2.14

THE AVERAGE AMOUNTS OF THE ELEMENTS IN EARTH'S CRUST IN GRAMS PER METRIC TON OR PARTS PER MILLION

Reprinted from "Principles of Geochemistry" (1952) with the permission of Brian Mason, author, and John Wiley and Sons, publishers.

O	466,000	N	46	Br	1.6
Si	277,200	Ce	46	Ho	1.2
Al	81,300	Sn	40	Eu	1.1
Fe	50,000	Y	28	Sb	1?
Ca	36,300	Nd	24	Tb	0.9
Na	28,300	Nb	24	Lu	0.8
K	25,900	Co	23	Tl	0.6
Mg	20,900	La	18	Hg	0.5
Ti	4,400	Pb	16	I	0.3
H	1,400	Ga	15	Bi	0.2
P	1,180	Mo	15	Tm	0.2
Mn	1,000	Th	12	Cd	0.15
S	520	Cs	7	Ag	0.1
C	320	Ge	7	In	0.1
Cl	314	Sm	6.5	Se	0.09
Rb	310	Gd	6.4	A	0.04
F	300	Be	6	Pd	0.01
Sr	300	Pr	5.5	Pt	0.005
Ba	250	Sc	5	Au	0.005
Zr	220	As	5	He	0.003
Cr	200	Hf	4.5	Te	0.002?
V	150	Dy	4.5	Rh	0.001
Zn	132	U	4	Re	0.001
Ni	80	B	3	Ir	0.001
Cu	70	Yb	2.7	Os	0.001?
W	69	Er	2.5	Ru	0.001?
Li	65	Ta	2.1		

CHEMICAL COMPOSITION OF ROCKS

Reprinted from "Sedimentary Rocks" (1948) with the permission of F. Pettijohn, author, and Harper Brothers, publishers.

Element	Average igneous rock	Average shale	Average sandstone	Average limestone	Averag sedimen
SiO_2	59.14	58.10	78.33	5.19	57.95
TiO_2	1.05	0.65	0.25	0.06	0.57
Al_2O_3	15.34	15.40	4.77	0.81	13.39
Fe_2O_3	3.08	4.02	1.07	0.54	3.47
FeO	3.80	2.45	0.30		2.08
MgO	3.49	2.44	1.16	7.89	2.65
CaO	5.08	3.11	5.50	42.57	5.89
Na_2O	3.84	1.30	0.45	0.05	1.13
K_2O	3.13	3.24	1.31	0.33	2.86
H_2O	1.15	5.00	1.63	0.77	3.23
P_2O_5	0.30	0.17	0.08	0.04	0.13
CO_2	0.10	2.63	5.03	41.54	5.38
SO_3		0.64	0.07	0.05	0.54
BaO	0.06	0.05	0.05		
C		0.80			0.66
	99 56	100.00	100.00	99.84	99.93

ELEMENTS PRESENT IN SOLUTION IN SEA WATER EXCLUDING DISSOLVED GASES

Reprinted by permission of the publishers from "The Oceans" by Sverdrup, Johnson, and Fleming. Copyright, 1942, by Prentice-Hall, Inc.

Element	Concentration (grams/metric ton) or parts per million
Cl	18,980
Na	10,561
Mg	1,272
S	884
Ca	400
K	380
Br	65
C (inorganic)	28
Sr	13
(SiO_2)	0.01–7.0
B	4.6
Si	0.02–4.0
C (organic)	1.2–3.0
Al	0.16–1.9
F	1.4
N (as nitrate)	0.001–0.7
N (as organic nitrogen)	0.03–0.2
Rb	0.2
Li	0.1
P (as phosphate)	$>$0.001–0.10
Ba	0.05
I	0.05
N (as nitrite)	0.0001–0.05
N (as ammonia)	$>$0.005–0.05
As (as arsenite)	0.003–0.024
Fe	0.002–0.02
P (as organic phosphorus)	0–0.016
Zn	0.005–0.014
Cu	0.001–0.09
Mn	0.001–0.01
Pb	0.004–0.005
Se	0.004
Sn	0.003
Cs	0.002 (approximate)
U	0.00015–0.0016
Mo	0.0003–0.002
Ga	0.0005
Ni	0.0001–0.0005
Th	$<$0.0005
Ce	0.0004
V	0.0003
La	0.0003
Y	0.0003
Hg	0.0003
Ag	0.00015–0.0003
Bi	0.0002
Co	0.0001
Sc	0.00004
Au	0.000004–0.000008
Fe (in true solution)	$<10^{-9}$
Ra	2.10^{-11}–3.10^{-10}
Ge	Present
Ti	Present
W	Present
Cd	Present in marine organisms
Cr	Present in marine organisms
Tl	Present in marine organisms
Sb	Present in marine organisms
Zr	Present in marine organisms
Pt	Present in marine organisms

THE pH OF NATURAL MEDIA AND ITS RELATION TO THE PRECIPITATION OF HYDROXIDES

Reprinted from "Principles of Geochemistry" (1952) with the permission of Brian Mason, author, and John Wiley and Sons, publishers.

pH	Precipitation of hydroxides	Natural media	pH
11	Magnesium		11
10		Alkali soils	10
9	Bivalent manganese		9
8		Seawater	8
7	Bivalent iron	River water	7
6	Zinc copper	Rain water	6
5	Aluminum		5
4		Peat water	4
3	Trivalent iron	Mine waters	3
2		Acid thermal springs	2
1			1

PROPERTIES OF THE EARTH'S ATMOSPHERE AT ELEVATIONS UP TO 160 KILOMETERS

The average atmosphere up to 160 km based on pressure and density data obtained on rocket flights above White Sands, New Mexico.
Havens, Koll and LaGow, Journal of Geophysical Research, March, 1952

Altitude km above sea level	Pressure mm Hg	Density gm/meters³	Temperatures °K (N_2, O_2) M = 29	Temperatures °K (N_2, O) M = 24	Velocity of Sound m. sec	Mean Free Path cm (N_2)
0	760	1220	290		345	6.5×10^{-6}
10	210	425	230		310	1.9×10^{-5}
20	42	92	210		295	8.6×10^{-5}
30	9.5	19	235		315	4.2×10^{-4}
40	2.4	4.3	260		325	1.8×10^{-3}
50	7.5×10^{-1}	1.3	270		330	6.1×10^{-3}
60	2.1×10^{-1}	3.8×10^{-1}	260		325	2.1×10^{-2}
70	5.4×10^{-2}	1.2×10^{-1}	210		295	6.6×10^{-1}
80	1.0×10^{-2}	2.5×10^{-2}	190		280	3.2×10^{-1}
90	1.9×10^{-3}	4.0×10^{-3}	210		295	2.0
100	4.2×10^{-4}	$8. \times 10^{-4}$	240		315	10.0
110	1.2×10^{-4}	2.0×10^{-4}	270	220	330	40
120	3.5×10^{-5}	5.0×10^{-5}	330	270	370	1.5×10^{2}
130	1.5×10^{-5}	2.0×10^{-5}	390	320	400	4×10^{2}
140	7×10^{-6}	7×10^{-6}	450	370	430	1×10^{3}
150	3×10^{-6}	3.0×10^{-6}	510	420	460	2.5×10^{3}
160	2×10^{-6}	1.5×10^{-6}	570	470	480	5×10^{3}

Velocity of Seismic Waves

Depth km	Longitudinal or condensational km/sec.	Transverse or distortional km/sec.
0–20	5.4 –5.6	3.2
20–45	6.25–6.75	3.5
1300	12.5	6.9
2400	13.5	7.5

DATA CONCERNING THE SOLAR SYSTEM

Body	Mass Earth = 1	Distance from Sun		Sidereal period, days	Mean specific gravity
		km	miles		
Sun	329390.	....	....	....	1.42
Mercury	0.0549	58×10^{6}	36.0×10^{6}	87.97	5.61
Venus	0.8073	108×10^{6}	67.1×10^{6}	224.70	5.16
Earth	1.000	149×10^{6}	92.9×10^{6}	365.26	5.52
Mars	0.1065	228×10^{6}	141.7×10^{6}	686.98	3.95
Jupiter	314.5	778×10^{6}	483.4×10^{6}	4332.59	1.34
Saturn	94.07	1426×10^{6}	886.1×10^{6}	10759.20	0.69
Uranus	14.40	2869×10^{6}	1782.7×10^{6}	30685.93	1.36
Neptune	16.72	4495×10^{6}	2793.1×10^{6}	60187.64	1.30
Pluto	....	5900×10^{6}	3666.1×10^{6}	90885.	....
Moon	0.01228	$^{*}38 \times 10^{4}$	$^{*}23.9 \times 10^{4}$	27.32	3.36

Body	Diameter		Acceleration due to gravity at surface		Albedo visual
	km	miles	cm/sec.²	ft./sec.²	
Sun	1390600	864100	27440	900.3	.....
Mercury	5140	3194	392	12.9	0.069
Venus	12620	7842	882	28.9	.59
Earth	12756	7926	980	32.2	.....
Mars	6860	4263	392	12.9	.154
Jupiter	143600	89229	2646	86.8	.56
Saturn	120600	74937	1176	38.6	.63
Uranus	53400	33181	980	32.2	.63
Neptune	49700	30882	980	32.2	.73
Moon	3476	2159.9	167	5.47	.073

ATMOSPHERIC AND METEOROLOGICAL DATA

Total mass of the atmosphere, estimated by Ekholm, 5.2×10^{21} g, 11.4×10^{18} pounds, 5.70×10^{16} tons.

Evidence of extent: twilight, 63 km, 39 mi.; meteors, 200 km, 124 mi.; aurora 44-360 km, 27-224 mi.

* Distance to Earth.

INTERPLANETARY ORBITS

Compiled by D. F. Lawden

Cm., gm., sec., units are to be assumed everywhere unless otherwise stated.

Table A

If M is the mass of an astronomical body and γ is the universal gravitational constant, the attraction on unit mass, distant r from the centre of the body, is

$$\frac{\gamma M}{r^2} = \frac{\mu}{r^2}$$

For the Sun $\mu = 1.33 \times 10^{26}$, and for the Moon $\mu = 4.90 \times 10^{18}$. The values taken by μ for the planets will be found in the table.

R is the radius of an astronomical body. For the Sun $R = 6.95 \times 10^{10}$ and for the Moon $R = 1.74 \times 10^{8}$.

V is the velocity a planet would have if it were placed in a circular orbit about the Sun, the radius of this orbit being equal to the semi-major axis of its actual orbit.

V_e is the escape velocity from the visible surface of the planet.

a is the semi-major axis of a planet's orbit. For the Earth, $a = a_0 = 1.4967 \times 10^{13}$. For the Moon's orbit about the Earth $a = 3.82 \times 10^{10}$. a/a_0 for the planets is given in the table.

T is the sidereal period of the planet's motion in *days*.

e is the eccentricity of a planet's orbit.

i is the inclination of the plane of a planetary orbit to the ecliptic.

Ω is the longitude of the ascending node of a planet's orbit.

ω is the longitude of perihelion.

Γ g is the gravitational acceleration at the Earth's surface ng is the same quantity for a planet measured at its visible surface. g = 981.

Planets	R	V	μ	V_e	a/a_0
Mercury	2.4×10^8	4.790×10^6	1.47×10^{19}	3.5×10^5	0.387
Venus	6.1×10^8	3.504×10^6	3.29×10^{20}	1.04×10^6	0.723
Earth	6.379×10^8	2.980×10^6	3.991×10^{20}	1.119×10^6	1.000
Mars	3.4×10^8	2.414×10^6	4.30×10^{19}	5.03×10^5	1.524
Jupiter	7.14×10^9	1.306×10^6	1.27×10^{23}	5.97×10^6	5.203
Saturn	6.04×10^9	0.965×10^6	3.79×10^{22}	3.54×10^6	9.539
Uranus	2.49×10^9	0.680×10^6	5.81×10^{21}	2.16×10^6	19.191
Neptune	2.66×10^9	0.543×10^6	6.89×10^{21}	2.28×10^6	30.071
Pluto	?	0.474×10^6	?	?	39.457

T	e	i	Ω	ω	n	Planets
88	0.2056	7°0′	47°40′	76°35′	0.26	Mercury
225	0.0068	3°24′	76°11′	130°47′	0.90	Venus
365	0.0167	—	—	101°59′	1.00	Earth
687	0.0934	1°51′	49°8′	335°2′	0.38	Mars
4333	0.0484	1°18′	99°53′	13°25′	2.54	Jupiter
10,759	0.0557	2°29′	113°10′	91°57′	1.06	Saturn
30,685	0.0472	0°46′	73°43′	169°45′	0.96	Uranus
60,181	0.0086	1°47′	131°10′	44°7′	0.99	Neptune
90,470	0.2485	17°9′	109°33′	223°26′	?	Pluto

Table B

In each cell of this table, the upper figure represents the net velocity increment in *Km/sec* which must be given to a space vehicle by its motor to effect transfer between two planets and the lower figure is the time of transit between the planets in days. In computing these results, the actual planetary orbits were approximated by circles in the plane of the ecliptic or radii equal to the semi-major axes of the actual orbits and the transfer orbit was assumed to be an ellipse tangential at its apsis to the two terminal orbits. The transfer was assumed to take place between a launching point on the visible surface of the planet of departure and an arrival point on the visible surface of the target planet. One impulsive thrust from the motor was supposed applied at the planet of departure and one at the planet of arrival, the motor being inoperative at all other times. Atmospheric resistances were neglected.

Mercury	Mercury						
Venus	19.7 **76**	Venus					
Earth	23.9 **106**	22.2 **146**	Earth				
Mars	23.6 **171**	17.9 **218**	17.3 **259**	Mars			
Jupiter	78 **853**	75 **932**	74 **998**	67 **1,130**	Jupiter		
Saturn	55 **2,020**	52 **2,120**	51 **2,210**	45 **2,380**	94 **3,650**	Saturn	
Uranus	42 **5,590**	39 **5,740**	38 **5,860**	32 **6,090**	82 **7,770**	57 **9,940**	Uranus
Neptune	43 **10,900**	41 **11,000**	39 **11,200**	34 **11,500**	83 **13,500**	58 **16,100**	44 **22,300**

STANDARD ATMOSPHERE

U. S. Extension to International Civil Aviation Organization Standard Atmosphere, 1958

The atmosphere was classified in 1958 into three altitude regions designated as

a. Standard 0 to 32 standard geopotential kilometers
b. Tentative 32 to 75 " " "
c. Speculative 75 to 300 " " "

Properties of the atmosphere were calculated as functions of geometric altitude as well as geopotential, the potential being established for the latitude where the acceleration of gravity has a sea-level value of 9.80665 meters per second per second. Symbols and abbreviations used in these tables are as follows:

H = Altitude in geopotential measure
L_M = Molecular-scale-temperature gradient
M = Mean molecular weight of air
m = Meter
m′ = Standard geopotential meter

P = Pressure
ρ = Mass density
T = Temperature in absolute thermodynamic scales
T_M = Molecular-scale temperature in absolute thermodynamic scales
Z = Altitude in geometric measure

Relationship between Geopotential and Geometric Altitude

The concept originally introduced by Bjerknes expresses vertical displacement in units of geopotential. Geopotential at an altitude Z is the potential energy of a unit mass at that altitude relative to the potential energy of that same unit mass at sea level. *Geopotential* H of a point at altitude Z may be rigorously defined as the increase in potential energy of a unit mass lifted from mean sea level to Z against the local force of gravity. Mathematically this definition becomes,

$$GH = \frac{\Delta E}{m} = \int_o^Z g(Z)\mathrm{d}Z$$

where

ΔE = increase in potential energy in joules,
m = mass of the body in kilograms,
$g(Z)$ = acceleration of gravity in m sec^{-2} expressed as a function of Z,
H = geopotential of a point at altitude Z,
G = proportionality factor depending on the units of H.

Solving this equation for H in terms of Z yields:

$$H = \frac{1}{G}\int_o^Z g(Z)\mathrm{d}Z.$$

The differential form of this relationship to be used later is

$$G\mathrm{d}H = g(Z)\mathrm{d}Z.$$

The basic unit of geopotential in these tables is one *standard geopotential* meter, m′, which is defined as 9.80665 m^2 sec^{-2} and which is equal to the vertical distance through which a one-kilogram mass must be lifted against the local force of gravity to increase its potential energy by 9.80665 joules. If the acceleration of gravity were constant at 9.80665 m sec^{-2} over an altitude interval of one geometric meter, one *standard geopotential* meter would be exactly equal to one geometric meter; this condition is true within two parts in a million at 45° 32′ 33″ at sea level, where g is equal to 9.80665 m sec^{-2}.

Relationship between Temperature and Molecular-scale Temperature

The molecular-scale temperature introduced by Minzner and Ripley is the defining atmospheric property of this Extension. This property is a composite of temperature and molecular weight, and is defined by the equation:

$$T_M = \left(\frac{T}{M}\right) M_o$$

where

T = temperature in the absolute thermodynamic scales,
T_M = molecular-scale temperature in the absolute thermodynamic scales,
M = molecular weight (nondimensional),
M_o = sea-level value of molecular weight.

No direct measurements of temperature have been made at altitudes above those which are reached by balloons; instead, the temperature is derived from values of the velocity of sound, or by substitution of measured pressures or densities into the barometric equation. The molecular-scale temperature can be derived in this way without specifying the molecular weight, whereas the temperature can be derived only if the molecular weight is known. Since the molecular weight is not well known at altitudes above 90 km, the molecular-scale temperature is more precisely known than temperature. Thus, the introduction of molecular-scale temperature increases the validity of some of the tabulated properties while simultaneously decreasing the complexity of the mathematics relating the basic atmospheric properties. The use of molecular-scale temperature also avoids the necessity for changing the defined atmosphere each time new values for the inadequately known molecular-weight distribution may be adopted.

Sea-level Atmospheric Composition for a Dry Atmosphere†

Constituent Gas	Mol. Fraction Per Cent	Molecular Weight (O = 16,000)
Nitrogen (N_2)	78.09	28.016
Oxygen (O_2)	20.95	32.0000
Argon (A)	0.93	39.944
Carbon dioxide (CO_2)	0.03	44.010
Neon (Ne)	1.8×10^{-3}	20.183
Helium (He)	5.24×10^{-4}	4.003
Krypton (Kr)	1.0×10^{-4}	83.7
Hydrogen (H_2)	5.0×10^{-5}	2.0160
Xenon (Xe)	8.0×10^{-6}	131.3
Ozone (O_3)	1.0×10^{-6}	48.0000
Radon (Rn)	6.0×10^{-18}	222.

† These values are taken as standard and do not necessarily indicate the exact condition of the atmosphere. Ozone and Radon particularly are known to vary at sea level and above, but these variations would not appreciably affect the value of M_o.

Sea Level Values of ICAO Atmosphere

Property	Metric Units
Collision frequency	6.9204049×10^{9} sec^{-1}
Conductivity, thermal	2.5339053×10^{-2} joule m^{-1} sec^{-1} (°K)$^{-1}$
Conductivity, thermal	6.0532182×10^{-6} k-cal m^{-1} sec^{-1} (°K)$^{-1}$
Conductivity, thermal	2.5838643×10^{-3} kg$f$sec^{-1} (°K)$^{-1}$*
Density, mass	1.2250140 kg m^{-3}
Density, mass	0.12491666 kg$f$sec^2 m^{-4}*
Gravitational acceleration	9.80665 m sec^{-2}
Kinematic viscosity	1.4607413×10^{-5} m^2 sec^{-1}
Mean free path	6.6317223×10^{-8} m
Molar volume	23.645444 m^3 (kg-mol)$^{-1}$
Molar volume	231.88259 m^3 [(kg$f$sec^2 m^{-1})-mol]*
Molecular weight	28.966 (dimensionless)
Number density	2.5475521×10^{25} m^{-3}
Particle speed	458.94204 m sec^{-1}
Pressure	0.760 m Hg
Pressure	1013.2500 mb
Pressure	101,325.00 nt m^{-2}
Pressure	10,332.275 kg$f$/m^{-2}*
Scale height	8434.4134 m
Sound speed	340.29205 m sec^{-1}
Specific weight	12.013284 kg m^{-2} sec^{-2}
Specific weight	1.2250140 kg$f$m^{-3}*
Temperature	15.0°C
Temperature, absolute	288.16°K
Temperature, molecular scale	288.16°K
Viscosity, coefficient of	1.7894285×10^{-5} kg m^{-1} sec^{-1}
Viscosity, coefficient of	1.8247093×10^{-6} kg$f$sec m^{-2}*

* kg$f$ = kilogram (force)

Abbreviated Metric Table of the U. S. Extension to the ICAO Standard Atmosphere

H m′	Z m	L_M °K m′$^{-1}$	T_M °K	T °K	M	P mb	ρ kg m^{-3}
−5,000	−4,996.070		320.66	320.66	28.966	1.7776×10^3	1.9312
		−0.0065					
0	0		288.16	288.16	28.966	1.01325×10^3	1.2250
		−0.0065					
11,000	11,019.067		216.66	216.66	28.966	2.2632×10^2	3.6391×10^{-1}
		zero					
20,000	20,063.124		216.66	216.66	28.966	5.4748×10^1	8.8034×10^{-2}
		zero					
25,000	25,098.710		216.66	216.66	28.966	2.4886×10^1	4.0016×10^{-2}
		+0.0030					
32,000	32,161.906		237.66	237.66	28.966	8.6776×10^0	1.2721×10^{-2}
		+0.0030					
47,000	47,350.101		282.66	282.66	28.966	1.2044×10^0	1.4845×10^{-3}
		zero					
53,000	53,445.620		282.66	282.66	28.966	5.8320×10^{-1}	7.1881×10^{-4}
		−0.0039					
75,000	75,895.488		196.86	196.86	28.966	2.452×10^{-2}	4.339×10^{-5}
		zero					
90,000	91,292.601		196.86	196.86	28.966	1.815×10^{-3}	3.213×10^{-6}
		+0.0035					
126,000	128,548.193		322.86	273.6	24.54	1.451×10^{-5}	1.566×10^{-8}
		+0.0100					
175,000	179,954.614		812.86	669.0	23.84	6.190×10^{-7}	2.655×10^{-10}
		+0.0058					
300,000	314,862.257		1,537.86	973.5	18.34	1.447×10^{-8}	3.279×10^{-12}

Abbreviated English Table of the U. S. Extension to the ICAO Standard Atmosphere

H ft′	Z ft	L_M °R ft′$^{-1}$	T_M °R	T °R	M	P lb/ft^{-2}	ρ slugs ft^{-3}
−16,404.199	−16,391.307		577.188	577.188	28.966	3.7110×10^3	3.7457×10^{-3}
		−0.003566160					
0	0		518.688	518.688	28.966	2.1162×10^3	2.3769×10^{-3}
		−0.003566160					
36,089.239	36,151.798		389.988	389.988	28.966	4.7268×10^2	7.0611×10^{-4}
		zero					
65,616.798	65,823.897		389.988	389.988	28.966	1.1548×10^2	1.7251×10^{-4}
		zero					
82,020.997	82,344.849		389.988	389.988	28.966	3.1975×10^1	7.7644×10^{-5}
		+0.001645920					
104,986.877	105,518.055		427.788	427.788	28.966	1.8124×10^1	2.4682×10^{-5}
		+0.001645920					
154,199.475	155,348.103		508.788	508.788	28.966	2.5155×10^0	2.8803×10^{-6}
		zero					
173,884.514	175,346.523		508.788	508.788	28.966	1.2180×10^0	1.3947×10^{-6}
		−0.002139696					
246,062.992	249,000.945		354.348	354.348	28.966	5.121×10^{-2}	8.420×10^{-8}
		zero					
295,275.591	299,516.408		354.348	354.348	28.966	3.792×10^{-3}	6.234×10^{-9}
		+0.001920240					
413,385.827	421,746.041		581.148	492.4	24.54	3.031×10^{-5}	3.038×10^{-11}
		+0.005486400					
574,146.982	590,402.278		1,463.148	1,204.000	23.84	1.293×10^{-6}	5.147×10^{-13}
		+0.003182112					
984,251.969	1,033,012.654		2,768.148	1,752.000	18.3	3.023×10^{-8}	6.362×10^{-15}

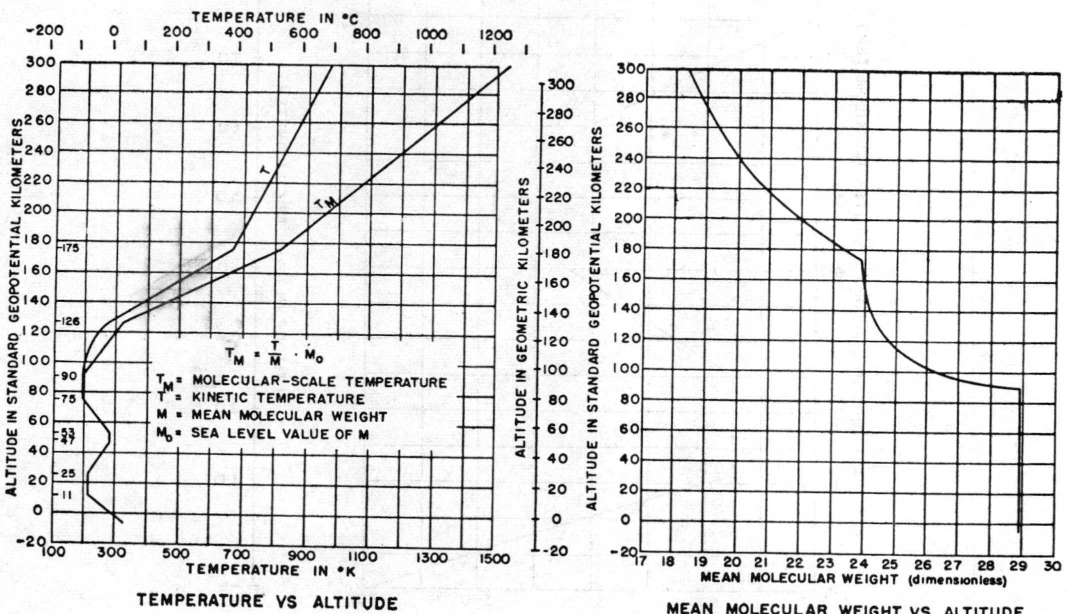

TEMPERATURE VS ALTITUDE

MEAN MOLECULAR WEIGHT VS ALTITUDE

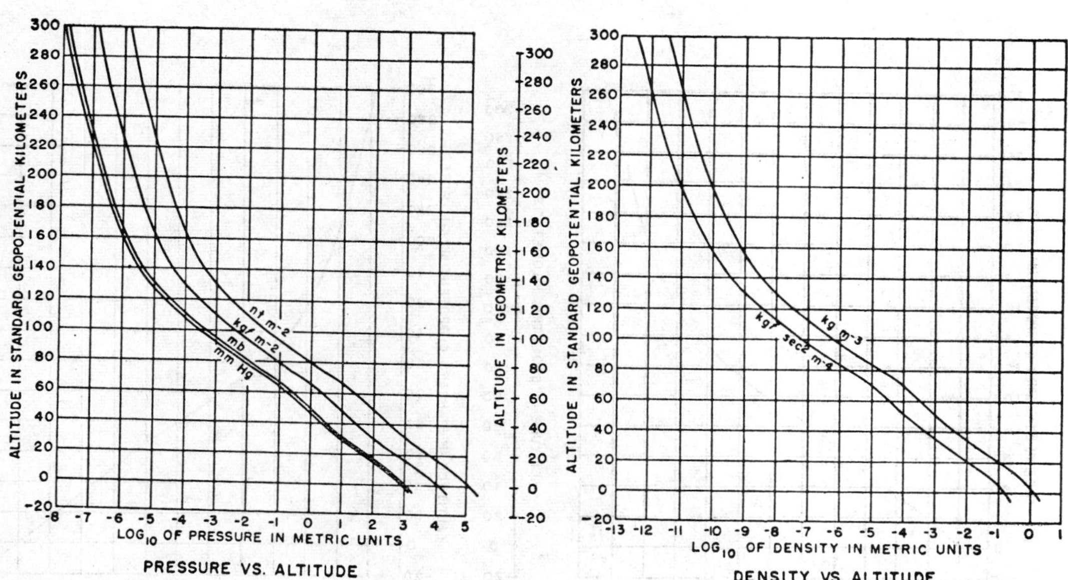

PRESSURE VS. ALTITUDE

DENSITY VS. ALTITUDE

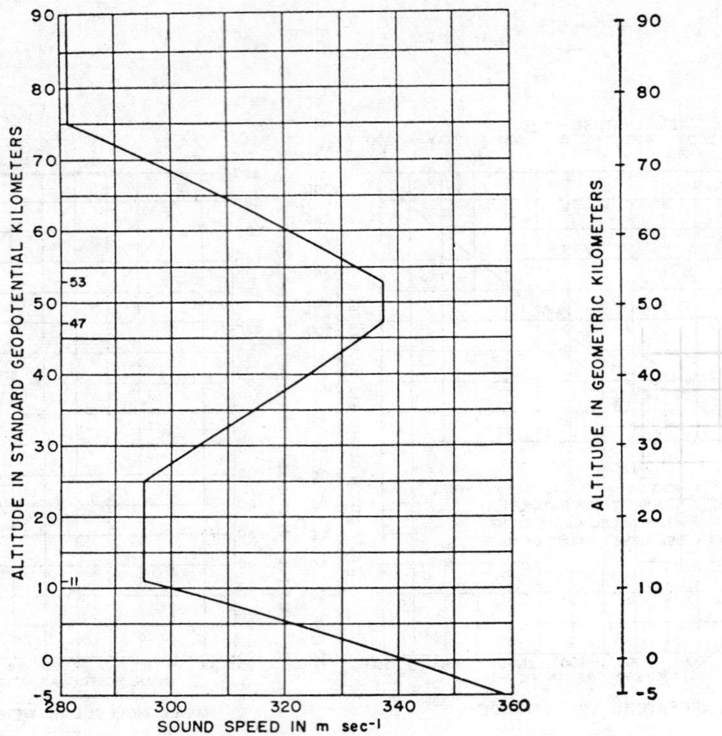

SPEED OF SOUND VS. ALTITUDE

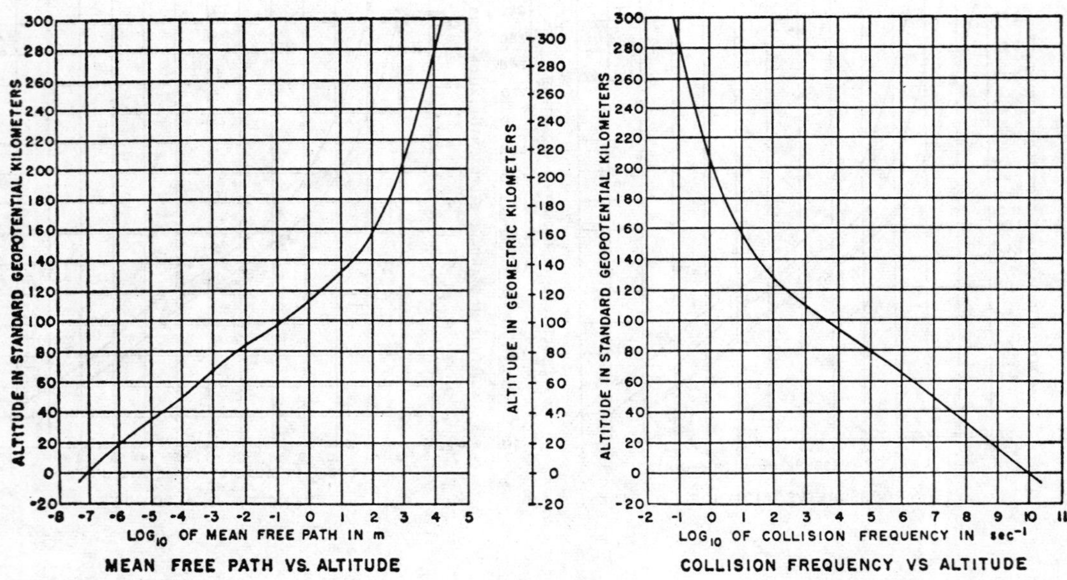

MEAN FREE PATH VS. ALTITUDE

COLLISION FREQUENCY VS ALTITUDE

STANDARD ATMOSPHERE (Continued)

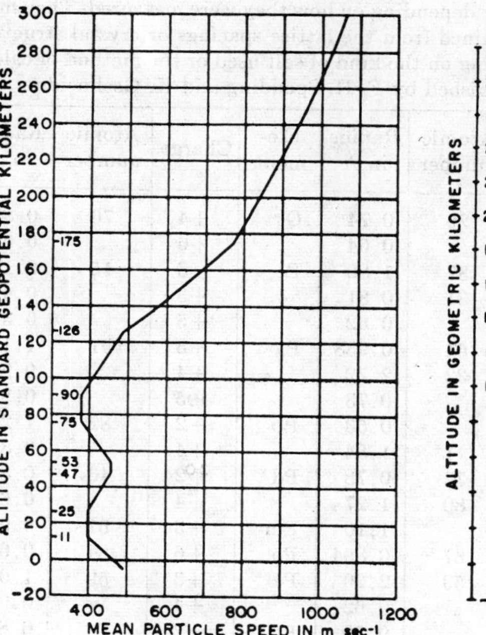

MEAN PARTICLE SPEED VS. ALTITUDE

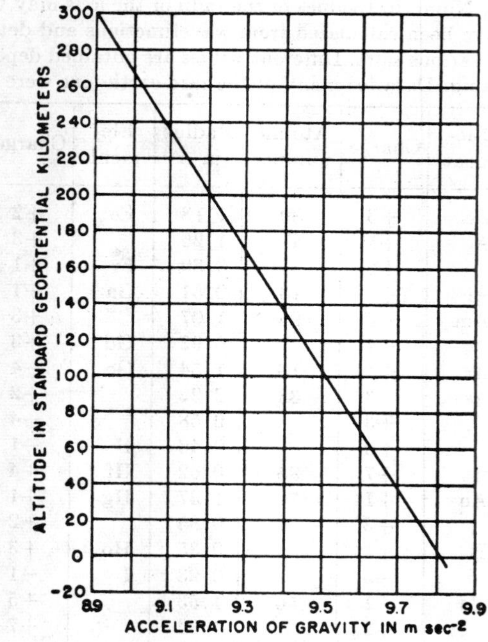

ACCELERATION OF GRAVITY VS. ALTITUDE

Atmospheric Electricity

Normal potential gradient:
Over land, 67 to 317 volts/m.
Over sea, 128 volts/m.
Potential gradient at Earth's surface beneath thunder cloud, 1×10^4 volts/m. Quantity discharged by lightning flash, 10–50 coulombs.
Energy of flash, 1×10^{17} ergs.
Potential difference between discharge points, 1×10^9 volts.

Angular Radius of Halos and Rainbows

Coronae due to small water drops	1° to 10°
Small halo, due to 60° angles of ice crystals	22°
Large halo, due to 90° angles of ice crystals	46°
Rainbow, primary	41° 20′
Rainbow, secondary	52° 15′

Solar Constant

The energy falling on one sq. cm. area at normal incidence, outside the earth's atmosphere, at the mean distance of the earth from the sun equals 2.00 small calories per minute. This value varies ±2%.

COMPONENTS OF ATMOSPHERIC AIR

(Exclusive of water vapor)

Constituent	Content (per cent) by volume	Content (ppm) by volume
N_2	78.084 ± 0.004	
O_2	20.946 ± 0.002	
CO_2	0.033 ± 0.001	
A	0.934 ± 0.001	
Ne		18.18 ± 0.04
He		5.24 ± 0.004
Kr		1.14 ± 0.01
Xe		0.087 ± 0.001
H_2		0.5
CH_4		2
N_2O		0.5 ± 0.1

MOLECULAR CONSTANTS

The following gives the arithmetical average velocity, the mean free path, molecular diameter and collision frequency for the temperatures indicated and at standard pressure, 760 mm Hg. except as otherwise stated.

Gas	Average velocity in cm/sec.		Mean free path in cm at 75 cm Hg.		
			Boltzmann		Meyer
	0° C	20° C	0° C	20° C	20° C
Air	447×10^2	463×10^2	....	....	....
Ammonia	583	604	5.92×10^{-6}	6.60×10^{-6}	5.83×10^{-6}
Argon	381	395	8.98	9.88	8.73
Carbon monoxide	454	471	8.46	9.23	8.16
Carbon dioxide	362	376	5.56	6.15	5.44
Helium	1208	1252	25.25	27.45	33.10
Hydrogen	1696	1755	16.00	17.44	15.40
Krypton	263	272	9.5		
Mercury	170	176	....	(14.70)	(13.0)
Neon	538	557			....
Nitrogen	454	471	8.50	9.29	8.21
Oxygen	425	440	9.05	9.93	8.78
Water vapor	566	587	....	....	....
Xenon	210	218	5.6	....	....

Gas	Collision frequency 20° C	Molecular diameter, cm		
		From viscosity	From van der Waal's equation	From heat conductivity
Ammonia	9150×10^6	2.97×10^{-8}	3.08×10^{-8}	....
Argon	4000	2.88	2.94	2.86×10^{-8}
Carbon monoxide	5100	3.19	3.12	....
Carbon dioxide	6120	3.34	3.23	3.40
Helium	4540	1.90	2.65	2.30
Hydrogen	10060	2.40	2.34	2.32
Krypton	....	....	(3.69)	3.14
Mercury	....	....	3.01	....
Nitrogen	5070	3.15	3.15	3.53
Oxygen	4430	2.98	2.92	....
Xenon	....	....	4.02	3.42

CRYSTAL IONIC RADII OF THE ELEMENTS

Numerical values of the radii of the ions may vary depending on how they were measured. They may have been calculated from wavefunctions and determined from the lattice spacings or crystal structure of various salts. Different values are obtained depending on the kind of salt used or the method of calculating. Data for many of the rare-earth ions were furnished by F. H. Spedding and K. Gschneidner.

Element	Charge	Atomic number	Radius in A	Element	Charge	Atomic number	Radius in A	Element	Charge	Atomic number	Radius in A
Ac	+3	89	1.18	Fe	+2	26	0.74	Os	+4	76	0.88
Ag	+1	47	1.26		+3		0.64		+6		0.69
	+2		0.89	Fr	+1	87	1.80	P	−3	15	2.12
Al	+3	13	0.51	Ga	+1	31	0.81		+3		0.44
Am	+3	95	1.07		+3		0.62		+5		0.35
	+4		0.92	Gd	+3	64	0.938	Pa	+3	91	1.13
Ar	+1	18	1.54	Ge	−4	32	2.72		+4		0.98
As	−3	33	2.22		+2		0.73		+5		0.89
	+3		0.58		+4		0.53	Pb	+2	82	1.20
	+5		0.46	H	−1	1	1.54		+4		0.84
At	+7	85	0.62	Hf	+4	72	0.78	Pd	+2	46	0.80
Au	+1	79	1.37	Hg	+1	80	1.27		+4		0.65
	+3		0.85		+2		1.10	Pm	+3	61	0.979
B	+1	5	0.35	Ho	+3	67	0.894	Po	+6	84	0.67
	+3		0.23	I	−1	53	2.20	Pr	+3	59	1.013
Ba	+1	56	1.53		+5		0.62		+4		0.90
	+2		1.34		+7		0.50	Pt	+2	78	0.80
Be	+1	4	0.44	In	+3	49	0.81		+4		0.65
	+2		0.35	Ir	+4	77	0.68	Pu	+3	94	1.08
Bi	+1	83	0.98	K	+1	19	1.33		+4		0.93
	+3		0.96	La	+1	57	1.39	Ra	+2	88	1.43
	+5		0.74		+3		1.016	Rb	+1	37	1.47
Br	−1	35	1.96	Li	+1	3	0.68	Re	+4	75	0.72
	+5		0.47	Lu	+3	71	0.85		+7		0.56
	+7		0.39	Mg	+1	12	0.82	Rh	+3	45	0.68
C	−4	6	2.60		+2		0.66	Ru	+4	44	0.67
	+4		0.16	Mn	+2	25	0.80	S	−2	16	1.84
Ca	+1	20	1.18		+3		0.66		+2		2.19
	+2		0.99		+4		0.60		+4		0.37
Cd	+1	48	1.14		+7		0.46		+6		0.30
	+2		0.97	Mo	+1	42	0.93	Sb	−3	51	2.45
Ce	+1	58	1.27		+4		0.70		+3		0.76
	+3		1.034		+6		0.62		+5		0.62
	+4		0.92	N	−3	7	1.71	Sc	+3	21	0.732
Cl	−1	17	1.81		+1		0.25	Se	−2	34	1.91
	+5		0.34		+3		0.16		−1		2.32
	+7		0.27		+5		0.13		+1		0.66
Co	+2	27	0.72	NH_4	+1		1.43		+4		0.50
	+3		0.63	Na	+1	11	0.97		+6		0.42
Cr	+1	24	0.81	Nb	+1	41	1.00	Si	−4	14	2.71
	+2		0.89		+4		0.74		−1		3.84
	+3		0.63		+5		0.69		+1		0.65
	+6		0.52	Nd	+3	60	0.995		+4		0.42
Cs	+1	55	1.67	Ne	+1	10	1.12	Sm	+3	62	0.964
Cu	+1	29	0.96	Ni	+2	28	0.69	Sn	−4	50	2.94
	+2		0.72	Np	+3	93	1.10		−1		3.70
Dy	+3	66	0.908		+4		0.95		+2		0.93
Er	+3	68	0.881		+7		0.71		+4		0.71
Eu	+3	63	0.950	O	−2	8	1.32	Sr	+2	38	1.12
	+2		1.09		−1		1.76	Ta	+5	73	0.68
F	−1	9	1.33		+1		0.22	Tb	+3	65	0.923
	+7		0.08		+6		0.09		+4		0.84

CRYSTAL IONIC RADII OF THE ELEMENTS (Continued)

Element	Charge	Atomic number	Radius in A	Element	Charge	Atomic number	Radius in A	Element	Charge	Atomic number	Radius in A
Tc	+7	43	0.979	Ti	+4		0.68	W	+4	74	0.70
Te	−2	52	2.11	Tl	+1	81	1.47		+6		0.62
	−1		2.50		+3		0.95	Y	+3	39	0.893
	+1		0.82	Tm	+3	69	0.87	Yb	+2	70	0.93
	+4		0.70	U	+4	92	0.97		+3		0.858
	+6		0.56		+6		0.80	Zn	+1	30	0.88
Th	+4	90	1.02	V	+2	23	0.88		+2		0.74
Ti	+1	22	0.96		+3		0.74	Zr	+1	40	1.09
	+2		0.94		+4		0.63		+4		0.79
	+3		0.76		+5		0.59				

BOND LENGTHS BETWEEN CARBON AND OTHER ELEMENTS

Prepared by Olga Kennard.

The tables are based on bond distance determinations, by experimental methods, mainly X-ray and electron diffraction, and include values published up to January 1, 1956. In the present tables, for the sake of completeness individual values of bond distances of lower accuracy are quoted with limits of error indicated where possible. Values for tungsten and bismuth should be treated with particular caution.

According to the statistical theory of errors if an average quantity $\bar{\mu}$ and a standard deviation σ can be evaluated there is a 95% probability that the true value lies within the interval $\bar{\mu} \pm 2\sigma$. Too much reliance should, however, not be placed on σ values in bond distance determinations since the derivation of these certain sources of error may have been neglected.

Values are given in Ångstrom units and limits of error quoted in units in the last place thus: 2.07 ± ≡2.07 ± 0.01.

Reproduced by permission from International Tables for X-ray Crystallography.

BOND LENGTHS BETWEEN CARBON AND OTHER ELEMENTS

Reference: HCP and "Tables of interatomic distances" Chem. Soc. of London, 1958

Group	Bond type	Element						
I		H**						
	All types	1.056 − 1.115						
II		Be	Hg					
		1.93	2.07 ± 0.01					
III		B	Al	In				
		1.56 ± 0.01	2.24 ± 0.04	2.16 ± 0.04				
IV		C**	Ge	Si	Sn	Pb		
	All types	1.54 − 1.20						
	Alkyls (CH_3XH_3)		1.98 ± 0.03	1.865 ± 0.008	2.143 ± 0.008	2.29 ± 0.05		
	Aryl ($C_6H_5XH_3$)			1.84 ± 0.01				
	Neg. Subst. (CH_3XCl_3)			1.88 ± 0.01	2.18 ± 0.02			
V		N**	P	As	Sb	Bi		
	All types	1.47 − 1.1						
	Paraffinic $(CH_3)_3X$		1.87 ± 0.02	1.98 ± 0.02	2.202 ± 0.016	2.30*		
VI		O**	S**	Cr	Se	Te	Mo	W
		1.43 − 1.15	1.81 − 1.55	1.92 ± 0.04	1.98 − 1.71	2.05 ± 0.14	2.08 ± 0.04	2.06 ± 0.01*
VII		F	Cl	Br	I			
	Paraffinic (monosubstituted) (CH_3X)	1.381 ± 0.005	1.767 ± 0.002	1.937 ± 0.003	2.13_5 ± 0.01			
	Paraffinic (disubstituted) (CH_2X_2)	1.334 ± 0.004	1.767 ± 0.002	1.937 ± 0.003	2.13_5 ± 0.1			
	Olefinic (CH_2:CHX)	1.32_5 ± 0.1	1.72 ± 0.01	1.89 ± 0.01	2.092 ± 0.005			
	Aromatic (C_6H_5X)	1.30 ± 0.01	1.70 ± 0.01	1.85 ± 0.01	2.05 ± 0.01			
	Acetylenic (HC:CX)		1.635 ± 0.004	1.79_5 ± 0.01	1.99 ± 0.02			
VIII		Fe	Co	Ni	Pd			
		1.84 ± 0.02	1.83 ± 0.02	1.82 ± 0.03	2.27 ± 0.04			

* Error uncertain.
** See following individual tables.

CARBON-CARBON

Single Bond

Paraffinic	1.541 ± 0.003
In diamond (18°C)	1.54452 ± 0.00014

Partial Double Bond

(1) Shortening of single bond in presence of carbon carbon double bond, e.g. $(CH_3)_2$-C:CH_2; or of aromatic ring e.g. C_6H_5.-CH_3	1.53 ± 0.01
(2) Shortening in presence of a carbon oxygen double bond e.g. CH_3CHO	1.516 ± 0.005
(3) Shortening in presence of two carbon-oxygen double bonds, e.g. $(CO_2H)_2$	1.49 ± 0.01
(4) Shortening in presence of one carbon-carbon triple bond, e.g. CH_3.C:CH	1.460 ± 0.003
(5) In compounds with tendency to dipole formation, e.g. C:C.C:N	1.44 ± 0.01
(6) In graphite (at 15°C)	1.4210 ± 0.0001
(7) In aromatic compounds	1.395 ± 0.003
(8) In presence of two carbon carbon triple bonds, e.g. HC:C.C:CH	1.373 ± 0.004

Double Bond

(1) Simple	1.337 ± 0.006
(2) Partial triple bond, e.g. CH_2:C:CH_2	1.309 ± 0.005

Triple Bond

(1) Simple, e.g. C_2H_2	1.204 ± 0.002
(2) Conjugated, e.g. CH_3.(C:C)$_2$.H	1.206 ± 0.004

CARBON-HYDROGEN

(1) Paraffinic (a) in methane	1.091
(b) in monosubstituted carbon	1.101 ± 0.003
(c) in disubstituted carbon	1.073 ± 0.004
(d) in trisubstituted carbon	1.070 ± 0.007
(2) Olefinic, e.g. CH_2:CH_2	1.07 ± 0.01
(3) Aromatic in C_6H_6	1.084 ± 0.006
(4) Acetylenic, e.g. CH:C.X	1.056 ± 0.003
(5) Shortening in presence of a carbon triple bond, e.g. CH_3CN	1.115 ± 0.004
(6) In small rings, e.g. $(CH_2)_2S$	1.081 ± 0.007

BOND LENGTHS BETWEEN CARBON AND OTHER ELEMENTS (Continued)

CARBON-NITROGEN

Single Bond

(1) Paraffinic (a) 4 co-valent nitrogen	1.479	± 0.005
(b) 3 co-valent nitrogen	1.472	± 0.005
(2) In C—N= e.g. CH_3NO_2	1.475	± 0.010
(3) Aromatic in $C_6H_5NHCOCH_3$	1.426	± 0.012
(4) Snortened (partial double bond) in heterocyclic systems, e.g. C_5H_5N	1.352	± 0.005
(5) Shortened (partial double bond) in N—C=O e.g. $HCONH_2$	1.322	± 0.003

Triple Bond

(1) In R.C⋮N	1.158	± 0.002

CARBON-OXYGEN

Single Bond

(1) Paraffinic	1.43	± 0.01
(2) Strained e.g. epoxides	1.47	± 0.01
(3) Shortened (partial double bond) as in carboxylic acids or through influence of aromatic ring, e.g. salicylic acid	1.36	± 0.01

Double Bond

(1) In aldehydes, ketones, carboxylic acids, esters	1.23	± 0.01
(2) In zwitterion forms, e.g. DL serine	1.26	± 0.01
(3) Shortened (partial triple bond) as in conjugated systems	1.207	± 0.006
(4) Partial triple bond as in acyl halides or isocyanates	1.17	± 0.01

CARBON-SULPHUR

Single Bond

(1) Paraffinic, e.g. CH_3SH	1.81(5)	± 0.01
(2) Lengthened in presence of fluorine, e.g. $(CF_3)_2S$	1.83(5)	± 0.01
(3) Shortened (partial double bond) as in heterocyclic systems, e.g. C_4H_4S	1.73	± 0.01

Double Bond

(1) In ethylene thiourea	1.71	± 0.02
(2) Shortened (partial triple bond) in presence of second carbon double bond, e.g. COS	1.558	± 0.003

BOND LENGTHS OF ELEMENTS

Element	Bond	Å
Ac	Ac—Ac	3.756
Ag (25°C)	Ag—Ag	2.8894
Al (25°C)	Al—Al	2.863
As	As—As	2.49
As_4	As—As	2.44 ± 0.03
Au (25°C)	Au—Au	2.8841
B_2	B—B	1.589
Ba (room temp.)	Ba—Ba	4.347
Be (α-form, 20°C)	Be—Be	2.2260
Bi (25°C)	Bi—Bi	3.09
Br_2	Br—Br	2.290
Ca (α-form, 18°C)	Ca—Ca	3.947 (f.c.c.)
(β-form, 500°C		3.877 (b.c.c.)
Cd (21°C)	Cd—Cd	2.9788
Cl_2	Cl—Cl	1.988
Ce	Ce—Ce	3.650
Co (18°C)	Co—Co	2.5061
Cr (α-form, 20°C)	Cr—Cr	2.4980
(β-form, >1850°C		2.61
Cs (−10°C)	Cs—Cs	5.309
Cu (20°C)	Cu—Cu	2.5560
Dy	Dy—Dy	3.503

BOND LENGTHS OF ELEMENTS (Continued)

Element	Bond	Å
Er	Er—Er	3.468
Eu	Eu—Eu	3.989
F_2	F—F	1.417 ± 0.001
Fe (α-form, 20°C)	Fe—Fe	2.4823 (b.c.c.)
(γ-form, 916°C)		2.578 (f.c.c.)
(δ-form, 1394°C)		2.539 (b.c.c.)
Ga (20°C)	Ga—Ga	2.442
Gd (20°C)	Gd—Gd	3.573
Ge (20°C)	Ge—Ge	2.4498
H_2	H—H in H_2	0.74611
	H—D in HD	0.74136
	D—D in D_2	0.74164
He	He—He in $[He_2]^+$	1.08_0
Hf (α-form, 24°C)	Hf—Hf	3.1273 (h.c.p.)
Hg (−46°C)	Hg—Hg	3.005
Ho	Ho—Ho	3.486
I_2	I—I	2.662
In (20°C)	In—In	3.2511
Ir (room temp.)	Ir—Ir	2.714
K (78°K)	K—K	4.544
La (α-form)	La—La	3.739 (h.c.p.)
(β-form)		3.745 (f.c.c.)
Li (20°C)	Li—Li	3.0390
Lu	Lu—Lu	3.435
Mg (25°C)	Mg—Mg	3.1971
Mn (γ-form, 1095°C)	Mn—Mn	2.7311 (f.c.c.)
(δ-form, 1134°C)		2.6679 (b.c.c.)
Mo (20°C)	Mo—Mo	2.7251
N_2	N—N	1.0975_8 ± 0.0001
Na (20°C)	Na—Na	3.7157
Nb (20°C)	Nb—Nb	2.8584
Nd	Nd—Nd	3.628
Ni (18°C)	Ni—Ni	2.4916
Np (α-form, 20°C)	Np—Np	2.60 (orthorhombic)
(β-form, 313°C)		2.76 (tetragon.)
(γ-form, 600°C)		3.05 (b.c.c.)
O_2	O—O	1.208
O_3 angle 116.8 ± 0.5°		1.278 ± 0.003
Os (20°C)	Os—Os	2.6754
P black	P—P	2.18
P_4	P—P	2.21 ± 0.02
Pa	Pa—Pa	3.212
Pb (25°C)	Pb—Pb	3.5003
Pd (25°C)	Pd—Pd	2.7511
Po (α-form, 10°C)	Po—Po	3.345 (cubic)
(β-form, 75°C)		3.359 (rh. hedr.)
Pr (α-form)	Pr—Pr	3.640 (tetrag.)
(β-form)		3.649 (f.c.c.)
Pt (20°C)	Pt—Pt	2.746
Pu (γ-form, 235°C)	Pu—Pu	3.026 (f.c.c.)
(δ-form, 313°C)		3.279 (f.c.c.)
(ε-form, 500°C)		3.150 (b.c.c.)
Rb (20°C)	Rb—Rb	4.95
Re (room temp.)	Re—Re	2.741
Rh (20°C)	Rh—Rh	2.6901
Ru (25°C)	Ru—Ru	2.6502
S_2	S—S	1.887
S_8	S—S	2.07 ± 0.02
Sb (25°C)	Sb—Sb	2.90
Sc (room temp.)	Sc—Sc	3.212
Se (20°C)	Se—Se	2.321
Se_2	Se—Se	2.152 ± 0.003
Se_8	Se—Se	2.32 ± 0.003
Si (20°C)	Si—Si	2.3517
Sn (α-form, 20°C)	Sn—Sn diamond type lattice	2.8099
(β-form, 25°C)		3.022 (tetrag.)
Sr (α-form, 25°C)	Sr—Sr	4.302 (f.c.c.)
(β-form, 248°C)		4.32 (h.c.p.)
(γ-form, 614°C)		4.20 (b.c.c.)
Ta (20°C)	Ta—Ta	2.86
Tb	Tb—Tb	3.525
Tc (room temp.)	Tc—Tc	2.703
Te (25°C)	Te—Te	2.864
Th (α-form, 25°C)	Th—Th	3.595 (f.c.c.)
(β-form, 1450°C)		3.56 (b.c.c.)
Ti (α-form, 25°C)	Ti—Ti	2.8956 (h.c.p.)
(β-form, 900°C)		2.8636 (b.c.c.)
Tl (α-form, 18°C)	Tl—Tl	3.4076 (h.c.p.)
(β-form, 262°C)		3.362 (b.c.c.)
Tm	Tm—Tm	3.447
U (α-form)	U—U	2.77
(β-form, 805°C)		3.058 (b.c.c.)
V (30°C)	V—V	2.6224
W (25°C)	W—W	2.7409
Y	Y—Y	3.551
Yb	Yb—Yb	3.880
Zn (25°C)	Zn—Zn	2.6694
Zr	Zr—Zr	3.179

BOND LENGTHS AND ANGLES OF CHEMICAL COMPOUNDS

BOND LENGTH AND ANGLE VALUES BETWEEN ELEMENTS

Elements	In	Bond length (Å)	Bond angle (°)
Boron			
B—B	B_2H_6	1.770 ± 0.013	H—B—H 121.5 ± 7.5
B—Br	BBr	1.88_7	—
	BBr_3	1.87 ± 0.02	Br—B—Br 120 ± 6
B—Cl	BCl	1.715_7	—
	BCl_3	1.72 ± 0.01	Cl—B—Cl 120 ± 3
B—F	BF	1.262	—
	BF_3	1.29_5 ± 0.01	F—B—F 120
B—H	Hydrides	1.21 ± 0.02	—
B—H Bridge	Hydrides	1.39 ± 0.02	—
B—N	$(BClNH)_3$	1.42 ± 0.01	B—N—B 121
B—O	BO	1.2049	—
	$B(OH)_3$	1.362 ± 0.005 (av.)	O—B—O 119.7
Nitrogen			
N—Cl	NO_2Cl	1.79 ± 0.02	—
N—F	NF_3	1.36 ± 0.02	F—N—F 102.5 ± 1.5
N—H	$[NH_4]^+$	1.034 ± 0.003	—
	NH	1.038	—
	ND	1.041	—
	HNCS	1.013 ± 0.005	H—N—C 130.25 ± 0.25
N—N	N_3H	1.02 ± 0.01	H—N—N′ 112.65 ± 0.5
	N_2O	1.126 ± 0.002	—
	$[N_2]^+$	1.116_2	—
N—O	NO_2Cl	1.24 ± 0.01	O—N—O 126 ± 2
	NO_2	1.188 ± 0.005	O—N—O 134.1 ± 0.25
N=O	N_2O	1.186 ± 0.002	—
	$[NO]^+$	1.0619	—
N—Si	SiN	1.572	—
Oxygen			
O—H	$[OH]^+$	1.0289	—
	OD	0.9699	—
	H_2O_2	0.960 ± 0.005	O—O—H 100 ± 2
O—O	H_2O_2	1.48 ± 0.01	—
	$[O_2]^+$	1.227	—
	$[O_2]^-$	1.26 ± 0.02	—
	$[O_2]^{--}$	1.49 ± 0.02	—
Phosphorus			
P—D	PD	1.429	—
P—H	$[PH_4]^+$	1.42 ± 0.02	—
P—N	PN	1.4910	—
P—S	$PSBr_3(Cl_2, F_3)$	1.86 ± 0.02	—
Sulfur			
S—Br	$SOBr_2$	2.27 ± 0.02	Br—S—Br 96 ± 2
S—F	SOF_2	1.585 ± 0.005	F—S—F 92.8 ± 1
S—D	SD	1.3473	—
	SD_2	1.345	—
S—O	SO_2	1.4321	O—S—O 119.54
	$SOCl_2$	1.45 ± 0.02	—
S—S	S_2Cl_2	2.04 ± 0.01	—
Silicon			
Si—Br	$SiBr_4$	2.17 ± 1.01	—
Si—Cl	$SiCl_4$	2.03 ± 1.01 (av.)	—
Si—F	SiF_4	1.561 ± 0.003 (av.)	—
SiH	SiH_4	1.480 ± 0.005	—
Si—O	$[SiO]^+$	1.504	—
Si—Si	Si_2Cl_2	2.30 ± 0.02	—

BOND LENGTHS AND ANGLES OF CHEMICAL COMPOUNDS

A. Inorganic Compounds

Compound	Formula	Bond lengths in Å	Bond angles (°)
Ammonia	NH_3	N—H 1.008 ± 0.004	H—N—H 107.3 ± 0.2
Antimony tribromide	$SbBr_3$	Sb—Br 2.51 ± 0.02	Br—Sb—Br 97 ± 2
Antimony trichloride	$SbCl_3$	Sb—Cl 2.352 ± 0.005	Cl—Sb—Cl 99.5 ± 1.5
Antimony triiodide	SbI_3	Sb—I 2.67 ± 0.03	I—Sb—I 99.0 ± 1
Arsenic tribromide	$AsBr_3$	As—Br 2.33 ± 0.02	Br—As—Br 100.5 ± 1.5
Arsenic trichloride	$AsCl_3$	As—Cl 2.161 ± 0.004	Cl—As—Cl 98.4 ± 0.5
Arsenic trifluoride	AsF_3	As—F 1.712 ± 0.005	F—As—F 102.0 ± 2
Arsenic triiodide	AsI_3	As—I 2.55 ± 0.03	I—As—I 101.0 ± 1.5
Arsenic trioxide	As_4O_6	As—O 1.78 ± 0.02	O—As—O 99.0 ± 2
			As—O—As 128.0 ± 2
Arsine	AsH_3	As—H 1.5192 ± 0.002	H—As—H 91.83 ± 0.33
Bismuthum tribromide	$BiBr_3$	Bi—Br 2.63 ± 0.02	Br—Bi—Br 100.0 ± 4
Bismuthum trichloride	$BiCl_3$	Bi—Cl 2.48 ± 0.02	Cl—Bi—Cl 100.0 ± 6
Bromosilane	SiH_3Br	Si—H 1.57 ± 0.03	H—Si—H 111.3 ± 1
Chlorine dioxide	ClO_2	Cl—O 1.49	O—Cl—O 118.5
Chlorogermane	GeH_3Cl	Ge—H 1.52 ± 0.03	H—Ge—H 110.9 ± 1.5
Chlorosilane	SiH_3Cl	Si—H 1.483 ± 0.001	H—Si—H 110 ± 0.03
		Si—Cl 2.0479 ± 0.0007	
Chromium oxychloride	$Cr(OCl)_2$	Cr—O 1.57 ± 0.03	O—Cr—O 105 ± 4
		Cr—Cl 2.12 ± 0.02	Cl—Cr—Cl 113 ± 3
			Cl—Cr—O 109 ± 3
Cyanuric triazide	C_3N_{12}	C—N 1.38	C—N=C 113.0
Dichlorosilane	SiH_2Cl_2	Si—H 1.46	
		SiCl 2.02 ± 0.03	Cl—Si—Cl 110 ± 1
Difluorodiazine	N_2F_2	N—F 1.44 ± 0.04	
		N—N 1.25 ± 0.02	F—N=N 115 ± 5
Difluoromethylsilane	Ch_3SiHf_2		F—Si—F 106 ± 0.5
			H—Si—C 116.2 ± 1
			C—Si—F 109.8 ± 0.5
Disilicon hexachloride	Si_2Cl_6	Si—Cl 2.02 ± 0.02	Cl—Si—Cl 109.5 ± 1
(hexachlorosilane)		Si—Si 2.34 ± 0.06	
Fluorosilane	SiH_3F	Si—H 1.460 ± 0.01	H—Si—H 109.3 ± 0.3
		Si—F 1.595 ± 0.002	
Hydrogen phosphide	PH_3	P—H 1.415 ± 0.003	H—P—H 93.3 ± 0.2
Hydrogen selenide	SeH_2	Se—H 1.47	H—Se—H 91.0
Hydrogen sulfide	SH_2	S—H 1.3455	H—S—H 93.3
Hydrogen telluride	$Te—H_2$		H—Te—H 89.5 ± 1
Iodo silane	$Si—H_3I$	Si—H 1.48 ± 0.01	H—Si—H 109.9 ± 0.4
		Si—I 2.45 ± 0.09	
Methylgermane	Ch_3GeH_3		H—C—H 108.2 ± 0.5
			H—Ge—H 108.6 ± 0.5
Nitrosyl bromide	NOBr	O—N 1.15 ± 0.04	Br—N=O 117 ± 3
		N—Br 2.14 ± 0.02	
Nitrosyl chloride	NOCl	N—O 1.14 ± 0.02	Cl—N=O 113.0 ± 2
		N—Cl 1.97 ± 0.01	
Nitrosyl fluoride	NOF	N—O 1.13	F—N=O 110.0
		N—F 1.52	
cis-Nitrous acid	NO(OH)	H—O 0.98	O—N=O 114 ± 2
		O—N 1.46	
		N—O′ 1.20	
trans-Nitrous acid	NO(OH)	H—O 0.98	O—N=O 118 ± 2
		O—N 1.46	
		NO′ 1.20	
Oxygen chloride	OCl_2	O—Cl 1.70_1 ± 0.02	Cl—O—Cl 110.8 ± 1
Oxygen fluoride	OF_2	O—F 1.418	F—O—F 103.2
Phosphorus oxychloride	$POCl_3$	P—Cl 1.99_5 ± 0.02	Cl—P—Cl 103.5 ± 1
Phosphorus oxysulfide	$P_4O_6S_4$	P—O 1.61 ± 0.02	P—O—P 128.5 ± 1.5
			O—P—O 101.5 ± 1
		P—S 1.85 ± 0.02	O—P—S 116.5 ± 1
Phosphorus pentoxide	P_4O_{10}	P—O 1.62 ± 0.02	O—P—O 101.5 ± 1
		P—O′ 1.38 ± 0.02	O—P—O′ 116.5 ± 1
			P—O—P 123.5 ± 1

A. Inorganic Compounds (Continued)

Compound	Formula	Bond lengths in Å	Bond angles (°)
Phosphorus tribromide	PBr_3	P—Br 2.18 ± 0.03	Br—P—Br 101.5 ± 1.5
Phosphorus trichloride	PCl_3	P—Cl 2.043 ± 0.003	Cl—P—Cl 100.1 ± 0.3
Phosphorus trifluoride	PF_3	P—F 1.535	F—P—F 100.0
Phosphorus trioxide	P_4O_6	P—O 1.65 ± 0.02	O—P—O 99.0 ± 1 P—O—P 127.5 ± 1
Stibine	SbH_3	Sb—H 1.7073 ± 0.0025	H—Sb—H 91.3 ± 0.33
Sulfur dichloride	SCl_2	S—Cl 1.99 ± 0.03	Cl—S—Cl 101.0 ± 4
Sulfur dioxide	SO_2	S—O 1.4321	O—S—O 119.536
Sulfur monochloride	S_2Cl_2	S—Cl 2.01 ± 0.07	Cl—S—S 104.5 ± 0.25
Sulfurylchloride	SO_2Cl_2	S—O 1.43 ± 0.02 S—Cl 1.99 ± 0.02	O—S—O 119.75 ± 5 Cl—S—Cl 111.20 ± 2 Cl—O—O 106.5 ± 2
Tellurium bromide	$TeBr_2$	Te—Br 2.51 ± 0.02	Br—Te—Br 98.0 ± 3
Tribromo silane	$SiHBr_3$	Si—Br 2.16 ± 0.03	Br—Si—Br 110.5 ± 1.5
Trichloro germane	$GeHCl_3$	Ge—H 1.55 ± 0.04	Cl—Ge—Cl 108.3 ± 0.2
Trichloro silane	Si—HCl_3	Si—H 1.47	Cl—Si—Cl 109.4 ± 0.3
Trifluorochlorosilane	$SiClF_3$	Si—F 1.560 ± 0.005 Si—Cl 1.989 ± 0.018	F—Si—F 108.5 ± 1
Trifluorochlorogermane	$GeClF_3$	Ge—F 1.688 ± 0.0017 Gr—Cl 2.067 ± 0.005	F—Ge—F 107.7 ± 1.5
Trifluorosilane	$SiHF_3$	Si—H 1.455 ± 0.01 Si—F 1.565 ± 0.005	F—Si—F 108.3 ± 0.5
Vanadium oxytrichloride	$VOCl_3$	V—O 1.56 ± 0.04 V—Cl 2.12 ± 0.03	Cl—V—Cl 111.2 ± 2 Cl—V—O 108.2 ± 2
Water	H_2O	O—H 0.958_4	H—O—H 104.45

B. Organic Compounds

Compound	Formula	Bond lengths in Å	Bond angles (°)
Acetaldehyde	CH_3COH	C—H 1.09 C—C 1.50 ± 0.02 C—O 1.22 ± 0.02	C—C=O 121 ± 2
Bromomethane	Ch_3Br	C—H 1.11 ± 0.01 C—Br 1.929	H—C—H 111.2 ± 0.5
Carbon tetrachloride	CCl_4	C—Cl 1.766 ± 0.003 Cl—Cl 2.887 ± 0.004	Cl—C—Cl 109.5
Chloromethane	CH_3Cl	C—H 1.11 ± 0.01 C—Cl 1.784 ± 0.003	H—C—H 110 ± 2
Dichloromethane	CH_2Cl_2	C—H 1.068 ± 0.005 C—Cl 1.7724 ± 0.0005	H—C—H 112 ± 0.3 Cl—C—Cl 111.8
Difluorochloromethane	$CHCLF_2$	C—H 1.06 C—Cl 1.73 ± 0.03 C—F 1.36 ± 0.03	F—C—F 110.5 ± 1 Cl—C—Cl 110.5 ± 1
1,1-Difluoroethylene	$C_2H_2F_2$	C—H 1.07 ± 0.02 C—F 1.321 ± 0.015 C—C 1.311 ± 0.035	F—C=C 125.2 ± 0.2 H—C—H 117 ± 7
Difluoromethane	CH_2F_2	C—F 1.360 ± 0.005 C—H 1.09 ± 0.03	F—C—F 108.2 ± 0.8 H—C—H 112.5 ± 6
p-Dinitrobenzene	$C_6H_4(NO_2)_2$	C—C 1.38 C—N 1.48 N—O 1.21	O—N=O 124.0
Dithio oxamide	$NH_2CSCSNH_2$	C—C 1.54 C—N 1.30 C—S 1.66	N—C=S 124.8 S—C—C 124.87 N—C—C 115.25
Ethane	C_2H_6	C—H 1.107 C—C 1.536	H—C—H 109.3
Ethylidene fluoride	CH_3CHF_2	C—F 1.345 ± 0.001 C—C 1.540 C—H 1.100	F—C—F 109.15 ± 0.001 C—C—F 109.4 H—C—C 110.2

B. Organic Compounds (Continued)

Compound	Formula	Bond lengths in Å	Bond angles (°)
Fluorochloromethane	CH_2ClF	C—H 1.078 ± 0.005 C—Cl 1.759 ± 0.003 C—F 1.378 ± 0.006	Cl—C—F 100.0 ± 0.1
Fluorotrichloromethane	$CFCl_3$	C—Cl 1.76 ± 0.02 C—F 1.44 ± 0.04	Cl—C—Cl 113 ± 3
Formaldehyde	CH_2O	C—H 1.060 ± 0.038 C—O 1.230 ± 0.017	H—C—H 125.8 ± 7
Formamide	$HCONH_2$	C—O 1.25_6 C—N 1.300	N—C=O 121.5
Formic acid	$HCOOH$	C—O′ 1.245 C—O 1.312 C—H 1.085 O—H 0.95	O—C=O′ 124.3 H—C—O′ 117.8 C—O—H 107.8
Glycine	NH_2CH_2COOH	C—C 1.52 C—O 1.27 C—N 1.39	C—C—O 119.0 O—C—O 122.0 C—C—N 112.0
Hexachloroethane	C_2Cl_6	C—Cl 1.74 ± 0.01 C—C 1.57 ± 0.06	Cl—C—Cl 109.3 ± 0.01
Iodomethane	CH_3I	C—H 1.11 ± 0.01 C—I 2.139	H—C—H 111.4 ± 0.1
Methane	CH_4	C—H 1.091	
Methanethiol	CH_3SH	C—H 1.1039 ± 0.002 C—S 1.8177 ± 0.0002 S—H 1.329 ± 0.004	110.3 ± 0.2 H—S—C 100.3 ± 0.2
Methanol	CH_3OH	C—H 1.096 ± 0.01 C—O 1.427 ± 0.007 O—H 0.956 ± 0.015	H—C—H 109.3 ± 0.75 C—O—H 108.9 ± 2
Methylamine	CH_3NH_2	C—H 1.093 C—N 1.474 ± 0.005 N—H 1.014	H—C—H 109.5 ± 1 H—N—H 105.8 ± 1
Methylether	$(CH_3)_2O$	C—O 1.43 ± 0.03 C—H 1.00	C—O—C 110.0 ± 3
Methylnitrite	CH_3NO_2	C—N 1.49 ± 0.02 N—O 1.22 ± 0.1	O—N=O 127 ± 4
Methylsulfide	$(CH_3)_2S$	C—S 1.82 ± 0.01 C—H 1.06	C—S—C 105 ± 3 H—C—H 109.5
Oxamide	$NH_2COCONH_2$	C—C 1.54 C—O 1.24 C—N 1.32	N—C=O 125.7 ± 0.3
Phosgene	CCl_2O	C—Cl 1.746 ± 0.004 C—O 1.166 ± 0.002	Cl—C=O 124.3 ± 0.1 Cl—C—Cl 111.3 ± 0.1
Propylene	C_3H_6		C—C=C 124.75 ± 0.3
Propynal	CHC.COH	C_1—C_2 1.204 C_2—C_3 1.46 C—O 1.21 C_1—H 1.06 C_3—H 1.08	C—C=O 123.0 C—C—H 120.0
Tribromomethane	$CHBr_3$	C—H 1.068 ± 0.01 C—Br 1.930 ± 0.003	Br—C—Br 110.8 ± 0.3
Trichlorobromomethane	$CBrCl_3$	C—Br 1.936 C—Cl 1.764	Cl—C—Cl 111.2 ± 1
Trifluorochloromethane	$CClF_3$	C—Cl 1.751 ± 0.004 C—F 1.328 ± 0.02	F—C—F 108.6 ± 0.4
Trifluoromethane	CHF_3	C—H 1.098 C—F 1.332 ± 0.008	F—C—F 108 ± 0.75
Triiodomethane	CHI_3	C—I 2.12 ± 0.04	I—C—I 113.0 ± 1
Trimethylamine	$(CH_3)_3N$	C—N 1.47 ± 0.01 C—H 1.06	C—N—C 108.0 ± 4 H—C—H 109.5
Trimethylarsine	$(CH_3)_3As$	C—H 1.09 C—As 1.98 ± 0.02	C—As—C 96 ± 5
Trimethylphosphine	$(CH_3)_3P$	C—H 1.09 C—P 1.87 ± 0.02	C—P—C 100.0 ± 4

BOND LENGTHS AND ANGLES OF CHEMICAL COMPOUNDS (Continued)

STRENGTHS OF CHEMICAL BONDS

J. A. Kerr and A. F. Trotman-Dickenson
University of Wales, Aberystwyth
(Revised to 1 January 1967)

The strength of a chemical bond, D(R—X), often known as the bond dissociation energy, is defined as the heat of the reaction: RX → R + X. It is given by: D(R—X) = ΔHf°(R) + ΔHf°(X) − ΔHf°(RX). Some authors list bond strengths for 0°K but here the values for 25°C are given because more thermodynamic data are available for this temperature. Bond strengths, or bond dissociation energies, are not equal to, and may differ considerably from, mean bond energies derived solely from thermochemical data on molecules and atoms.

BOND STRENGTHS IN DIATOMIC MOLECULES

These have usually been measured spectroscopically and many are known to ±0.1 kcal. mole^{-1}. All bond strengths are in kcal. mole^{-1} at 25°C. The element of lower atomic number is listed first.

Many references are to: Herzberg, Molecular Spectra and Molecular Structure, I. Spectra of Diatomic Molecules, 2nd Ed., New York, 1950 (Herzberg); Gaydon, Dissociation Energies and Spectra of Diatomic Molecules, 2nd Ed. London, 1953 (Gaydon). The references have been chosen primarily as a key to the literature. It should not be assumed that the author referred to was responsible for the determination quoted. The reference may be only to a review article.

Molecule	Kcal mole^{-1}	Ref.	Molecule	Kcal mole^{-1}	Ref.	Molecule	Kcal mole^{-1}	Ref.
H—H	104.18	54	Li—I	84.6	20	O—Cl	64.5	15
H—Li	56.91	76	Be—Be	17	42	O—Ca	91.5	75
H—Be	54	54	Be—O	107	15	O—Sc	162	29
H—B	79	55	Be—F	185	73	O—Ti	157	18
H—C	80.9	54	Be—Cl	137	66	O—V	149	17
H—N	86	47	Be—Au	∼67	28	O—Cr	102.0	51
H—O	102.4	49	B—B	∼66	92	O—Mn	97	67
H—F	153	47	B—N	152	44	O—Fe	98	21
H—Na	48	47	B—O	173	62	O—Ni	87.4	51
H—Mg	47	47	B—F	182	90	O—Cu	114	21
H—Al	68	47	B—S	119	47	O—Zn	93	21
H—Si	74.6	10	B—Cl	128	7	O—Ga	75	93
H—S	82.3	58	B—Br	104	7	O—Ge	159	12
H—Cl	103.1	47	C—C	144	42	O—As	114	47
H—K	44	47	C—N	174	16	O—Se	101	11
H—Ni	61	47	C—O	256.7	40	O—Br	56.1	43
H—Cu	67	26	C—F	107	47	O—Sr	95.0	75
H—Zn	20.5	47	C—Si	104	41	O—Y	170	29
H—Ge	76.5	10	C—P	140	47	O—Zr	151	47
H—Se	73	52	C—S	175	60	O—Nb	93	47
H—Br	87.4	47	C—V	133	46	O—Mo	117	36
H—Rb	40	47	C—Se	116	47	O—Ag	33	47
H—Sr	39	47	C—Br	95.6	39	O—Cd	89	21
H—Ag	59	47	N—N	226.8	45	O—In	26	47
H—Cd	16.5	47	N—O	149.7	27	O—Sn	133	22
H—In	59	47	N—F	62.6	6	O—Sb	75	47
H—Te	64	52	N—Si	105	47	O—Te	91	11
H—I	71.4	47	N—P	139	47	O—I	43	43
H—Cs	42	47	N—S	116	47	O—Ba	116	75
H—Ba	42	47	N—As	116	47	O—La	192	29
H—Au	75	94	N—Br	67	65	O—Ce	186	77
H—Hg	9.5	47	N—Sb	73	47	O—Pr	172	77
H—Tl	47	47	N—Xe	5.5	53	O—Nd	167	49
H—Pb	43	47	O—O	118.86	23	O—Lu	100	47
H—Bi	59	47	O—F	26	38	O—Ta	195	56
Li—Li	26.7	92	O—Mg	94	5	O—W	155	36
Li—O	84	15	O—Al	120	37	O—Os	123	61
Li—F	137.5	20	O—Si	192	15	O—Pb	99	21
Li—Cl	111.9	20	O—P	144	15	O—Bi	86	70
Li—Br	100.2	20	O—S	124.4	11	O—Th	197	80

BOND STRENGTHS IN DIATOMIC MOLECULES (Continued)

Molecule	Kcal. mole⁻¹	Ref.	Molecule	Kcal. mole⁻¹	Ref.	Molecule	Kcal. mole⁻¹	Ref.
O—U	180	36	S—Cd	80.5	63	Ge—Te	96	30
F—F	37.72	72	S—Sn	111.0	29	As—As	92	47
F—Na	114.0	20	S—Te	81	11	Se—Se	73.84	11
F—Mg	107	81	S—Ba	95.6	34	Se—Sn	96.0	31
F—Al	159	82	S—La	138	87	Se—Te	64	11
F—Si	130	57	S—Hg	51	63	Se—Pb	62.4	68
F—Cl	61.4	47	S—Pb	82.7	11	Se—Bi	56	69
F—K	118.9	9	S—Bi	69	35	Br—Br	46.08	47
F—Ca	125	19	Cl—Cl	57.87	47	Br—Rb	90.4	20
F—Cr	106.4	88	Cl—K	101.3	20	Br—Ag	70	47
F—Mn	101.2	95	Cl—Ca	81	14	Br—In	93	7
F—Cu	72	47	Cl—Ti	26	47	Br—Sn	47	47
F—Ga	145	7	Cl—Mn	70	47	Br—I	42.80	47
F—Ge	116	83	Cl—Cu	89	47	Br—Cs	96.5	20
F—Br	57	24	Cl—Ga	114.5	7	Br—Hg	17.3	78
F—Rb	116.1	20	Cl—Br	53.0	47	Br—Tl	77	7
F—Sr	127	81	Cl—Rb	100.7	20	Br—Pb	59	47
F—Cd	73	89	Cl—Ag	73	54	Br—Bi	64.1	47
F—In	127	7	Cl—Cd	49.9	25	Rb—Rb	12.2	92
F—Sn	77	47	Cl—In	103	7	Rb—I	76.7	20
F—Sb	93	47	Cl—I	50.5	47	Sr—I	47	47
F—Xe	11	84	Cl—Cs	106.2	20	Y—Y	37.3	32
F—Cs	119.6	20	Cl—Ba	118	86	Y—La	47.3	32
F—Ba	135	81	Cl—Au	66	47	Pd—Au	34.2	4
F—Tl	110	47	Cl—Hg	24	47	Ag—Ag	38.5	3
F—Pb	75	47	Cl—Tl	90.5	7	Ag—Sn	32.5	1
F—Bi	75	47	Cl—Pb	73	47	Ag—I	61	47
Na—Na	18.4	92	Cl—Bi	68	47	Ag—Au	48.5	3
Na—Cl	97.5	20	K—K	12.8	92	Cd—Cd	3	47
Na—K	15.2	47	K—Br	90.9	20	Cd—I	33	47
Na—Br	86.7	20	K—I	76.8	20	In—In	23.3	37
Na—Rb	15	47	Ca—I	59	47	In—Sb	36.3	37
Na—I	72.7	20	Ca—Au	18	71	In—I	80	7
Mg—Cl	63	47	Sc—Sc	25.9	32	Sn—Sn	46.7	1
Mg—Br	59	47	Ti—Ti	~58	96	Sn—Te	80.2	30
Al—Al	~40	92	Cr—Cr	~44	96	Sn—Au	58.4	1
Al—Cl	118	7	Cr—Au	51.3	4	Sb—Sb	71.5	37
Al—Br	106.6	8	Mn—Mn	~21	96	Sb—Te	61	69
Al—I	88	7	Mn—Br	59	47	Sb—Bi	70	47
Al—Au	65	13	Mn—I	29	47	Te—Te	54.9	11
Si—Si	76	41	Co—Co	39	96	Te—Pb	53.6	30
Si—S	151	59	Ni—Ni	54.5	83	Te—Bi	49	69
Si—Cl	77	47	Cu—Cu	47	42	I—I	36.06	47
Si—Se	135	47	Cu—Br	79	47	I—Cs	82.4	20
Si—Br	70	47	Cu—Ag	41.6	3	I—Hg	9	78
Si—Te	123	47	Cu—Sn	42.3	1	I—Tl	66	7
P—P	116.7	47	Cu—Au	55.4	3	I—Pb	47	47
P—S	70	85	Zn—Zn	6	92	I—Bi	59	47
P—Ga	84	91	Zn—Te	49	47	Cs—Cs	11.3	92
P—W	73	48	Zn—I	33	47	La—La	57.6	32
S—S	101.9	11	Ga—Ga	~33	92	Au—Au	52.5	3
S—Ca	73.7	34	Ga—As	110	64	Hg—Hg	4	79
S—Mn	72	2	Ga—Br	100	7	Hg—Tl	1	54
S—Fe	78	63	Ga—I	81	7	Tl—Tl	15	42
S—Zn	51	63	Ge—Ge	65	42	Pb—Pb	24	42
S—Ge	131.7	11	Ge—Se	116	47	Bi—Bi	48	92
S—Se	91	11	Ge—Br	61	47	At—At	19	42
S—Sr	75.0	34						

STRENGTHS OF CHEMICAL BONDS (Continued)

References

1. Ackerman, Drowart, Stafford, and Verhaegen, J. Chem. Phys., **36,** 1557 (1962).
2. Wiedemeier and Giles, J. Chem. Phys., **42,** 2765 (1965).
3. Ackerman, Stafford, and Drowart, J. Chem. Phys., **33,** 1784 (1960).
4. Ackerman, Stafford and Verhaegen, J. Chem. Phys., **36,** 1560 (1962).
5. Alexander, Ogden, and Levy, J. Chem. Phys., **39,** 3057 (1963).
6. Armstrong, Marantz, and Coyle, J. Am. Chem. Soc., **81,** 3798 (1959).
7. Barrow, Trans. Faraday Soc., **56,** 952 (1960).
8. Barrow, Nature, **189,** 480 (1961).
9. Barrow and Caunt, Proc. Roy. Soc., A **219,** 120 (1953).
10. Barrow and Deutsch, Proc. Chem. Soc., 122 (1960).
11. Drowart and Goldfinger, Quart. Rev. (London) **20,** 545 (1966).
12. Barrow and Rowlinson, Proc. Roy. Soc., A **224,** 374 (1954).
13. Barrow and Travis, Proc. Roy. Soc., A **273,** 133 (1963).
14. Bautista, J. Phys. Chem., **67,** 2411 (1963).
15. Berkowitz, J. Chem. Phys., **30,** 858 (1959).
16. Berkowitz, J. Chem. Phys., **36,** 2533 (1962).
17. Berkowitz, Chupka and Inghram, J. Chem. Phys., **27,** 87 (1957).
18. Berkowitz, Chupka and Inghram, J. Phys. Chem., **61,** 1569 (1957).
19. Blue, Green, Bautista and Margrave, J. Phys. Chem., **67,** 877 (1963).
20. Brewer and Brackett, Chem. Rev., **61,** 425 (1961).
21. Brewer and Mastick, J. Chem. Phys., **19,** 834 (1951).
22. Brewer and Porter, J. Chem. Phys., **21,** 2012 (1953).
23. Brix and Herzberg, J. Chem. Phys., **21,** 2240 (1953).
24. Broderson and Sicre, Z. Physik, **141,** 515 (1955).
25. Bruner and Corbett, U. S. Atomic Energy Comm. I.S. 739 (1963).
26. Bulewicz and Sugden, Trans. Faraday Soc., **52,** 1475 (1956).
27. Callear and Smith, Disc. Faraday Soc., **37,** 96 (1964).
28. Barrow, Gissane and Travis, Proc. Roy. Soc., A **287,** 240 (1965).
29. Smoes, Drowart and Verhaegen, J. Chem. Phys., **43,** 732 (1965).
30. Brebrick, J. Chem. Phys., **41,** 1140 (1964).
31. Colin and Drowart, Trans. Faraday Soc., **60,** 673 (1964).
32. Verhaegen, Smoes, and Drowart, U.S. At. Energy Comm. WADD TR60–782 (Pt. XIX), 1963.
33. Colin, Goldfinger and Juenehomme, Nature, **194,** 282 (1962).
34. Colin, Goldfinger and Juenehomme, Trans. Faraday Soc., **60,** 306 (1964).
35. Cubicciotti, J. Phys. Chem., **67,** 118 (1963).
36. de Maria, Burns, Drowart and Inghram, J. Chem. Phys., **32,** 1373 (1960).
37. de Maria, Drowart and Inghram, J. Chem. Phys., **30,** 318 (1959).
38. Dibeler, Reese and Franklin, J. Chem. Phys., **27,** 1296 (1957).
39. Dixon and Kroto, Trans. Faraday Soc., **59,** 1484 (1963).
40. Douglas, J. Phys. Chem., **59,** 109 (1955).
41. Drowart, de Maria and Inghram, J. Chem. Phys., **29,** 1015 (1958).
42. Drowart and Honig, J. Phys. Chem., **61,** 980 (1957).
43. Durie and Ramsay, Can. J. Phys., **36,** 35 (1958).
44. Fesenko, Poroshkovaya Met. Akad. Nauk. Ukr. SSR, **1,** 80 (1961).
45. Frost and McDowell, Proc. Roy. Soc., A **236,** 278 (1956).
46. Fijishiro, Trans. Japan Inst. Metals, **1,** 125 (1960).
47. Gaydon.
48. Gingerich, J. Phys. Chem., **68,** 768 (1964).
49. Goldstein, Walsh and White, J. Phys. Chem., **65,** 1400 (1961).
50. Gray, Trans. Faraday Soc., **55,** 408 (1959).
51. Grimley, Burns and Inghram, J. Chem. Phys., **34,** 664 (1961).
52. Gunn, J. Phys. Chem., **68,** 949 (1964).
53. Herman and Herman, J. Phys. Radium, **24,** 73 (1963).
54. Herzberg.
55. Hurley, Proc. Phys. Soc., A **261,** 237 (1961).
56. Inghram, Chupka and Berkowitz, J. Chem. Phys., **27,** 569 (1957).
57. Ehlert and Margrave, J. Chem. Phys., **41,** 1066 (1964).
58. Johns and Ramsay, Can. J. Phys., **39,** 210 (1961).
59. Lagerqvist, Nilheden and Barrow, Proc. Phys. Soc. Lond., A **65,** 419 (1952).
60. Lagerqvist, Westerhind, Wright and Barrow, Arkiv. Fysik., **14,** 387 (1958).

References (Continued)

61. Long, Quart. Rev. Chem. Soc., **7,** 134 (1953).
62. Mal'tsev, Kataev and Tatevskii, Friz. Probl. Spektroskopii Akad. Nauk. SSR, **1,** 194 (1960).
63. Marquart and Berkowitz, J. Chem. Phys., **39,** 283 (1963).
64. Miller, Harper and Petty, J. Electrochem. Soc., **108,** 1123 (1961).
65. Milton, Dunford and Douglas, J. Phys. Chem., **35,** 1202 (1961).
66. Novikov and Turnitski, Optika i Spektrosckopiya, **8,** 752 (1960).
67. Padley and Sudgen, Trans. Faraday Soc., **55,** 2054 (1959).
68. Porter, J. Chem. Phys., **34,** 583 (1961).
69. Porter and Spencer, J. Chem. Phys., **32,** 943 (1960).
70. Scari, Magyar Tudomanyos Akad. Kozponti Fiz. Kutato Interzeteek Kozlemenyei, **3,** 341 (1955).
71. Schultz, Compt. rend., **252,** 1750 (1961).
72. Stamper and Barrow, Trans. Faraday Soc., **54,** 1592 (1958).
73. Tatevski, Tunitski and Novikov, Optika i Spektrosckopiya, **5,** 520 (1958).
74. Thorn and Winslow, J. Phys. Chem., **65,** 1297 (1961).
75. Kalff, Hollander, and Alkemande, J. Chem. Phys., **43,** 2299 (1965).
76. Velasco, Can. J. Phys., **35,** 1204 (1957).
77. Walsh, Dever and White, J. Phys. Chem., **65,** 1410 (1961).
78. Wieland, Z. Elektrochem., **64,** 761 (1960).
79. Winans and Heitz, Z. Phys., **135,** 406 (1953).
80. Darnell, McCollum and Milne, J. Phys. Chem., **64,** 341 (1960).
81. Ehlert, Blue, Green, and Margrave, J. Chem. Phys., **41,** 2250 (1964).
82. Ehlert, and Margrave, J. Amer. Chem. Soc., **86,** 3901 (1964).
83. Kant, J. Chem. Phys., **41,** 1876 (1964).
84. Johnson and Woolfolk, J. Chem. Phys., **41,** 269 (1964).
85. Mole and McCoubrey, Nature, **202,** 450 (1964).
86. Gurvich and Ryabova, Teplofiz. Vysokikh Temperatur, Akad Nauk SSSR, **2,** 215 (1964).
87. Cater, Lee, Johnson, Rauh, and Eick, J. Phys. Chem., **69,** 2684 (1965).
88. Kent and Margrave, J. Am. Chem. Soc., **87,** 3582 (1965).
89. Besenbruch, Kana'an, and Margrave, J. Phys. Chem., **69,** 3174 (1965).
90. Hildenbrand and Murad, J. Chem. Phys., **43,** 1400 (1965).
91. Lee and Schoonmaker, Condensation Evaporation Solids, Proc. Intern. Symp. Dayton, Ohio, 1962, 379.
92. Siegel, Quart. Rev. (London) **19,** 77 (1965).
93. Gurvich, Novikov, and Ryabova, Optika i Spektrosckopiya, **18,** 132 (1965).
94. Ringstrom, Arkiv. Fysik., **27,** 227 (1964).
95. Kent, Ehlert and Margrave, J. Am. Chem. Soc., **86,** 5090 (1964).
96. Kant and Strauss, J. Chem. Phys., **41,** 3806 (1964).

HEATS OF FORMATION OF GASEOUS ATOMS FROM ELEMENTS IN THEIR STANDARD STATES

J. A. Kerr and A. F. Trotman-Dickenson

For elements which are diatomic gases in their standard states these are readily obtained from the bond strength. For elements which are crystalline in their standard states they are derived from vapor pressure data. All values are given in kcal. mole^{-1} at 25°C.

Element	Kcal mole^{-1}	Ref.	Element	Kcal mole^{-1}	Ref.	Element	Kcal mole^{-1}	Ref.
H_2(g)	52.10	4	Fe(c)	99.5	4	Sn(c)	72.0	4
Li(c)	38.4	4	Co(c)	101.6	4	Sb(c)	63	4
Be(c)	78.25	4	Ni(c)	102.8	4	Fe(c)	48	4
B(c)	132.6	4	Cu(c)	81.1	4	I_2(c)	25.54	4
C(c)	170.9	1	Zn(c)	31.2	4	Cs(c)	18.7	4
N_2(g)	113.0	4	Ga(c)	69.0	4	Ba(c)	42.5	4
O_2(g)	59.56	4	Ge(c)	90	4	Er(c)	65	3
F_2(g)	18.86	5	As(c)	69	4	Hf(c)	160	4
Na(c)	25.8	4	Se(c)	49.4	4	Ta(c)	186.8	4
Mg(c)	35.1	4	Br_2(l)	26.74	4	W(c)	201.8	4
Al(c)	78.0	4	Rb(c)	19.5	4	Os(c)	187	2
Si(c)	108.4	4	Sr(c)	39.1	4	Ir(c)	159	4
P(c) yellow	75.5	4	Y(c)	98	4	Pt(c)	135.2	4
S(c)	65.65	4	Zr(c)	145.4	4	Au(c)	88.3	4
Cl_2(g)	28.92	4	Nb(c)	173	4	Hg(c)	14.65	4
K(c)	21.3	4	Mo(c)	157.5	4	Tl(c)	43.0	4
Ca(c)	42.81	4	Ru(c)	153	2	Pb(c)	46.75	4
Sc(c)	88	4	Rh(c)	133	4	Bi(c)	49.5	4
Ti(c)	112.5	4	Pd(c)	91	4	U(c)	115	4
V(c)	123	4	Ag(c)	68.4	4	Th(c)	136.6	4
Cr(c)	95	4	Cd(c)	26.75	4	Pu	82.3	7
Mn(c)	66.7	4	In(c)	58.24	6			

References

1. Brewer and Searcy, Ann. Rev. Phys. Chem., **7,** 259 (1956).
2. Panish and Reif, J. Chem. Phys., **37,** 128 (1962).
3. Savitski, Tsvetnye Metally, **33,** 43 (1960).
4. Skinner and Pilcher, Quat. Rev. (London), **17,** 264 (1963).
5. Stamper and Barrow, Trans. Faraday Soc., **54,** 1592 (1958).
6. Herrick, Trans. A. I. M. E., **230,** 1439 (1964).
7. Milford, U.S. At. Energy Comm., LA DC 6926 (1965).

STRENGTHS OF CHEMICAL BONDS (Continued)

BOND STRENGTHS IN POLYATOMIC MOLECULES

J. A. Kerr and A. F. Trotman-Dickenson

These are of very variable reliability. Few are known to better than ±1 kcal. mole^{-1}. An attempt has been made to list all important values obtained by methods that are considered to be valid.

There are some bond strengths which are frequently required in calculations but which are not well established. These are given with a query.

Bond	Kcal. mole^{-1}	Ref.	Bond	Kcal. mole^{-1}	Ref.	Bond	Kcal. mole^{-1}	Ref.
H—CH	108	3	H—nC_3F_7	104	33	$PhCH_2$—COOH	55	28
H—CH_2	104	3	H—CCl_3	95.7	13	CH_3—F	108	8
H—CH_3	104.0	48	H—NH_2	103	26	CH_3—HgMe	51	31
H—C_2H_3	104	46	H—$NHCH_3$	92	26	CH_3—I	56.3	15
H—C_2H_5	98.0	44	H—$N(CH_3)_2$	86	26	CF_3—I	53.5	36
H—C_3H_5	85	10	H—NHC_6H_5	80	12	ON—NO_2	9.6	2
H—nC_3H_7	98	9	H—$N(CH_3)C_6H_5$	74	49	O_2N—NO_2	13	19
H—iC_3H_7	94.5	14	H—N_3	85	20	NF_2—NF_2	19.9	23
H—sC_4H_9	94.6	14	H—NO	49	7	O—NN	40	5
H—tC_4H_9	90.9	38	H—OH	119	43	O—NO	73	39
H—$neoC_5H_{11}$	99.3	14	H—OCH_3	102	35	HO—OH	51	43
H—C_6H_5	104	9	H—$OCMe_3$	106	1	MeO—OMe	36	21
H—$CH_2C_6H_5$	85	11	H—O_2CMe	112	22	EtO—OEt	34	21
H—CN	129	27	H—O_2CEt	110	22	Bu^tO—OBu^t	37	1
H—CH_2CN	86	30	H—O_2CPr	103	22	$MeCO_2$—O_2CMe	30	34
H—CHO	87	45	H—O_2H	90	16	O—ClO	58	17
H—CH_2OH	92	47	H—SH	90?	9	Cl—O_2	8	4
H—$COCH_3$	88	25	H—SMe	88?	9	O=PF_3	130	6
H—$CH(CH_3)OH$	90	47	HC≡CH	230	40	O=PCl_3	122	6
H—$C(CH_3)_2OH$	90	29	H_2C=CH_2	167	41	O=PBr_3	119	6
H—COC_6H_5	74	37	H_3C—CH_3	88	42	HO—Br	57	18
H—CF_3	106.2	32	NC—CN	145	27	F—NO	55	24
H—C_2H_5	103	33	O=CO	128	39			

References

1. Batt and Benson, J. Chem. Phys., **36,** 895 (1962).
2. Beattie and Bell, J. Chem. Soc., 1681 (1957).
3. Bell and Kistiakowsky, J. Am. Chem. Soc., **84,** 3417 (1951).
4. Benson and Buss, J. Chem. Phys., **27,** 1382 (1957).
5. Carlton-Sutton, Amber and Williams, Proc. Phys. Soc., **48,** 189 (1936).
6. Charnley, Skinner, J. Chem. Soc., 450 (1953).
7. Clement and Ramsay, Can. J. Phys., **39,** 205 (1961).
8. Dibeler and Reese, J. Res. Nat. Stand., **54,** 127 (1955).
9. Kerr, Chem. Rev., **66,** 465 (1966).
10. Egger, Golden and Benson, J. Amer. Chem. Soc., **86,** 5420 (1964).
11. Esteban, Kerr and Trotman-Dickenson, J. Chem. Soc., 3873 (1963).
12. Esteban, Kerr and Trotman-Dickenson, J. Chem. Soc., 3879 (1963).
13. Benson, J. Chem. Phys., **43,** 2044 (1965).
14. Fettis and Trotman-Dickenson, J. Chem. Soc., 3037 (1961).
15. Flowers and Benson, J. Chem. Phys., **38,** 882 (1963).
16. Foner and Hudson, J. Chem. Phys., **36,** 2681 (1962).
17. Gaydon.
18. Gelles, Trans. Faraday Soc., **47,** 1158 (1951).
19. Giauque and Kemp, J. Chem. Phys., **6,** 40 (1938).
20. Gray, Quart. Rev., **17,** 441 (1963).
21. Hanst and Calvert, J. Phys. Chem., **63,** 104 (1959).
22. Jaffé, Prosen and Szwarc, J. Chem. Phys., **27,** 416 (1957).
23. Johnson and Colburn, J. Amer. Chem. Soc., **83,** 3043 (1961).
24. Johnston and Bertin, J. Amer. Chem. Soc., **81,** 640 (1959).
25. Kerr and Calvert, J. Phys. Chem., in press.

BOND STRENGTHS IN POLYATOMIC MOLECULES (Continued)

References (Continued)

26. Kerr, Sekhar and Trotman-Dickenson, J. Chem. Soc., 3217 (1963).
27. Knight and Rink, J. Chem. Phys., **35,** 199 (1961).
28. Back and Sehon, Can. J. Chem., **38,** 1261 (1960).
29. Walsh and Benson, J. Am. Chem. Soc., **88,** 3480 (1966).
30. Hunt, Kerr and Trotman-Dickenson, J. Chem. Soc., 5074 (1965).
31. Price and Trotman-Dickenson, Trans. Faraday Soc., **53,** 939 (1957).
32. Amphlett, Coomber and Whittle, J. Phys. Chem., (1966).
33. Pritchard and Thommarson, J. Phys. Chem., **68,** 568 (1964).
34. Rembaum and Szwarc, J. Chem. Phys., **23,** 909 (1955).
35. Shaw and Trotman-Dickenson, J. Chem. Soc., 3210 (1960).
36. Boyd et al., J. Phys. Chem., **67,** 719 (1963).
37. Szwarc and Taylor, J. Chem. Phys., **22,** 270 (1954).
38. Teranishi and Benson, J. Amer. Chem. Soc., **85,** 2887 (1963).
39. Thermochemistry.
40. Thermochemistry D(C—H).
41. Thermochemistry D(CH_2—H).
42. Thermochemistry D(CH_3—H).
43. Thermochemistry D(O—H).
44. Trotman-Dickenson, Chem. and Ind., 379 (1965).
45. Walsh and Benson, J. Am. Chem. Soc., **88,** 4570 (1966).
46. Trotman-Dickenson and Verbeke, J. Chem. Soc., 2580 (1961).
47. Tarr and Whittle, Trans. Faraday Soc., **60,** 2039 (1964).
48. Golden, Walsh and Benson, J. Amer. Chem. Soc., **87,** 4053 (1965).
49. Kerr, Trotman-Dickenson and Wolter, J. Chem. Soc., 3584 (1964).

HEATS OF FORMATION OF FREE RADICALS

J. A. Kerr and A. F. Trotman-Dickenson

These are given in kcal. mole^{-1} at 25°C. They are self-consistent with the tables of bond strengths to which reference should be made for the source of the data.

Radical	Kcal. mole^{-1}	Radical	Kcal. mole^{-1}
CH_2	90	CH_3O	2
CH_3	34.0	CH_3CO	−4
C_2H_3	64	CH_3CO_2	−45
C_2H_5	25.7	$C_2H_5CO_2$	−54
C_3H_5	38	$(CH_3)_3CO$	−24
nC_3H_7	21	C_6H_5CO	16
iC_3H_7	17.6	$C_6H_5CO_2$	−64
tC_4H_9	6.7	CF_3	−111
$neoC_5H_{11}$	7.5	CCl_3	19
C_6H_5	72	NH_2	40
$C_6H_5CH_2$	44.6	NH_2	9
CN	109	OH	9.3
CH_3NH	34	HO_2	5
$(CH_3)_2N$	29	HS	33?
C_6H_5NH	48	CH_3S	30?
CHO	7		

BOND STRENGTH OF SOME ORGANIC MOLECULES

Bond strengths, expressed as Kcal mole^{-1}, for some organic molecules of the general formula R—X are presented below. Due to the arrangement of the table and to the species of radicals and atoms involved, one may also find the bond strengths for ammonia, water and the hydrogen halides.

X / R	H	F	Cl	Br	I	OH	NH_2	OCH_3	CH_3
H	104.18	153	103.1	87.4	71.4	119	103	102	104
CH_3	104	108	84	70	56	91	79	80	88
C_2H_5	98	106	81	69	53	91	78	80	85
i–C_3H_7	94.5	105	81	68	53	92	77	81	83
t–C_4H_9	91	—	79	63	50	91	77	78	80
C_6H_5	104	117	—	71	—	103	92	92	93
$C_6H_5CH_2$	85	—	68	51	40	77	—	66	72
CCl_3	96	106	73	54	—	—	—	—	—

ENERGY, MASS, AND VELOCITY RELATIONS FOR THE ELECTRON

Energy mev	Electron mass g	Electron mass m_0	β	Velocity cm/sec
very small	9.1066×10^{-28}			
.018	9.42×10^{-28}	1.035	.25	$.75 \times 10^{10}$
.05	10.00×10^{-28}	1.10	.42	1.26×10^{10}
.1	10.90×10^{-28}	1.20	.548	1.65×10^{10}
.5	18.02×10^{-28}	1.98	.863	2.585×10^{10}
1	26.93×10^{-28}	2.96	.94	2.818×10^{10}
5	98.24×10^{-28}	10.8	.996	2.985×10^{10}
7	133.89×10^{-28}	14.7	.9976	2.990×10^{10}
10	187.38×10^{-28}	20.6	.9988	2.994×10^{10}
20	365.64×10^{-28}	40.1	.9992	near the velocity of light
100	1791.8×10^{-28}	196.6	.9998	near the velocity of light
1000	17839×10^{-28}	1960	.999999	near the velocity of light
10000	178160×10^{-28}	19580	.9999999	near the velocity of light

LATTICE ENERGIES OF THE ALKALI HALIDES

Values are in kilocalories per gram formula weight.

Elements	Fluoride	Chloride	Bromide	Iodide
Lithium	240.1	199.2	188.5	174.1
Sodium	213.4	183.1	174.5	163.9
Potassium	189.7	165.4	159.3	150.8
Rubidium	181.6	160.7	153.5	145.3
Cesium	173.7	152.2	146.3	139.1

CONSTANTS FOR SATELLITE GEODESY

Defining Constants

1. Number of ephemeris seconds in 1 tropical year (1900) .... s = 31 556 925·9747
2. Gaussian gravitational constant, defining the a. u. .... k = 0·017 202 09895

Primary Constants

3. Measure of 1 a. u. in meters .... A = 149 600 × 10^6
4. Velocity of light in meters per second .... c = 299 792·5 × 10^3
5. Equatorial radius for Earth in meters .... a_e = 6 378 160
6. Dynamical form-factor for Earth .... J_2 = 0·001 082 7
7. Geocentric gravitational constant (units: m^3s^{-2}) .... GE = 398 603 × 10^9
8. Ratio of the masses of the Moon and Earth .... μ = 1/81·30
9. Sidereal mean motion of Moon in radians per second (1900) .... $n_{☾}^*$ = 2·661 699 489 × 10^{-6}
10. General precession in longitude per tropical century (1900) .... p = 5025″.64
11. Obliquity of the ecliptic (1900) .... ϵ = 23°27′08″.26
12. Constant of nutation (1900) .... N = 9″.210

Auxiliary Constants and Factors

$k/86400$, for use when the unit of time is 1 second .... k' = 1·990 983 675 × 10^{-7}
Number of seconds of arc in 1 radian .... = 206 264·806
Factor for constant of aberration (note 15) .... F_1 = 1·000 142
Factor for mean distance of Moon (note 20) .... F_2 = 0·999 093 142
Factor for parallactic inequality (note 23) .... F_3 = 49853″.2

Derived Constants

13. Solar parallax .... $\arcsin(a_e/A) = \pi_\odot$ = 8″.79405 (8″.794)
14. Light-time for unit distance .... $A/c' = \tau_A$ = 499^s.012
= 1^s/0·002 003 96
15. Constant of aberration .... $F_1 k' \tau_A = \kappa$ = 20″.4958 (20″.496)
16. Flattening factor for Earth .... f = 0·003 352 9
= 1/298·25
17. Heliocentric gravitational constant (units: m^3s^{-2}) .... $A^3k'^2 = GS$ = 132 718 × 10^{15}
18. Ratio of masses of Sun and Earth .... $(GS)/(GE)$ = S/E = 332 958
19. Ratio of masses of Sun and Earth + Moon .... $S/E(1+\mu)$ = 328 912
20. Perturbed mean distance of Moon, in meters .... $F_2(GE(1+\mu)/n_{☾}^{*2})^{\frac{1}{3}} = a_{☾}$ = 384 400 × 10^3
21. Constant of sine parallax for Moon .... $a_e/a_{☾} = \sin \pi_{☾}$ = 3422″.451
22. Constant of lunar inequality .... $\dfrac{\mu}{1+\mu}\dfrac{a_{☾}}{A} = L$ = 6″.43987 (6″.440)
23. Constant of parallactic inequality .... $F_3 \dfrac{1-\mu}{1+\mu}\dfrac{a_{☾}}{A} = P_{☾}$ = 124″.986

System of Planetary Masses

	Reciprocal mass		Reciprocal mass
24. Mercury	6 000 000	Jupiter	1 047·355
Venus	408 000	Saturn	3 501·6
Earth + Moon	329 390	Uranus	22 869
Mars	3 093 500	Neptune	19 314
		Pluto	360 000

PHYSICAL CONSTANTS

The following lists of physical constants are recommended by the National Academy of Sciences and have been adopted by the National Bureau of Standards. The lists are taken from the National Bureau of Standards Technical News Bulletin, October 1963.

Adjusted Values of Constants

Constant	Symbol	Value	Est.‡ error limit	Unit			
				Système International (MKSA)		Centimeter-gram-second (CGS)	
Speed of light in vacuum	c	2.997925	3	$\times 10^{8}$	$m\ s^{-1}$	$\times 10^{10}$	$cm\ s^{-1}$
Elementary charge	e	1.60210	7	10^{-19}	C	10^{-20}	$cm^{\frac{1}{2}}\ g^{\frac{1}{2}}$ *
		4.80298	20	......	....	10^{-10}	$cm^{\frac{3}{2}}\ g^{\frac{1}{2}}\ s^{-1}$ †
Avogadro constant	N_A	6.02252	28	10^{23}	mol^{-1}	10^{23}	mol^{-1}
Electron rest mass	m_e	9.1091	4	10^{-31}	kg	10^{-28}	g
		5.48597	9	10^{-4}	u	10^{-4}	u
Proton rest mass	m_p	1.67252	8	10^{-27}	kg	10^{-24}	g
		1.00727663	24	10^{0}	u	10^{0}	u
Neutron rest mass	m_n	1.67482	8	10^{-27}	kg	10^{-24}	g
		1.0086654	13	10^{0}	u	10^{0}	u
Faraday constant	F	9.64870	16	10^{4}	$C\ mol^{-1}$	10^{3}	$cm^{\frac{1}{2}}\ g^{\frac{1}{2}}\ mol^{-1}$ *
		2.89261	5	......	....	10^{14}	$cm^{\frac{3}{2}}\ g^{\frac{1}{2}}\ s^{-1}\ mol^{-1}$ †
Planck constant	h	6.6256	5	10^{-34}	J s	10^{-27}	erg s
	$\hbar$	1.05450	7	10^{-34}	J s	10^{-27}	erg s
Fine structure constant	α	7.29720	10	10^{-3}	....	10^{-3}	
	$1/\alpha$	1.370388	19	10^{2}	....	10^{2}	
	$\alpha/2\pi$	1.161385	16	10^{-3}	....	10^{-3}	
	α^2	5.32492	14	10^{-5}	....	10^{-5}	
Charge to mass ratio for electron	e/m_e	1.758796	19	10^{11}	$C\ kg^{-1}$	10^{7}	$cm^{\frac{1}{2}}\ g^{-\frac{1}{2}}$ *
		5.27274	6	......	....	10^{17}	$cm^{\frac{3}{2}}\ g^{-\frac{1}{2}}\ s^{-1}$ †
Quantum-charge ratio	h/e	4.13556	12	10^{-15}	$J\ s\ C^{-1}$	10^{-7}	$cm^{\frac{3}{2}}\ g^{\frac{1}{2}}\ s^{-1}$ *
		1.37947	4	......	....	10^{-17}	$cm^{\frac{1}{2}}\ g^{\frac{1}{2}}$ †
Compton wavelength of electron	λ_C	2.42621	6	10^{-12}	m	10^{-10}	cm
	$\lambda_C/2\pi$	3.86144	9	10^{-13}	m	10^{-11}	cm
Compton wavelength of proton	$\lambda_{C,p}$	1.32140	4	10^{-15}	m	10^{-13}	cm
	$\lambda_{C,p}/2\pi$	2.10307	6	10^{-16}	m	10^{-14}	cm
Rydberg constant	R_∞	1.0973731	3	10^{7}	m^{-1}	10^{5}	cm^{-1}
Bohr radius	a_0	5.29167	7	10^{-11}	m	10^{-9}	cm
Electron radius	r_e	2.81777	11	10^{-15}	m	10^{-13}	cm
	r_e^2	7.9398	6	10^{-30}	m^2	10^{-26}	cm^2
Thomson cross section	$8\pi r_e^2/3$	6.6516	5	10^{-29}	m^2	10^{-25}	cm^2
Gyromagnetic ratio of proton	γ	2.67519	2	10^{8}	$rad\ s^{-1}\ T^{-1}$	10^{4}	$rad\ s^{-1}\ G^{-1}$ *
	$\gamma/2\pi$	4.25770	3	10^{7}	$Hz\ T^{-1}$	10^{3}	$s^{-1}\ G^{-1}$ *
(uncorrected for diamagnetism, H_2O)	γ'	2.67512	2	10^{8}	$rad\ s^{-1}\ T^{-1}$	10^{4}	$rad\ s^{-1}\ G^{-1}$ *
	$\gamma'/2\pi$	4.25759	3	10^{7}	$Hz\ T^{-1}$	10^{3}	$s^{-1}\ G^{-1}$ *
Bohr magneton	μ_B	9.2732	6	10^{-24}	$J\ T^{-1}$	10^{-21}	$erg\ G^{-1}$ *
Nuclear magneton	μ_N	5.0505	4	10^{-27}	$J\ T^{-1}$	10^{-24}	$erg\ G^{-1}$ *
Proton moment	μ_p	1.41049	13	10^{-26}	$J\ T^{-1}$	10^{-23}	$erg\ G^{-1}$ *
	μ_p/μ_N	2.79276	7	10^{0}	....	10^{0}	
(uncorrected for diamagnetism, H_2O)	μ'_p/μ_N	2.79268	7	10^{0}	....	10^{0}	
Anomalous electron moment correction	$(\mu_e/\mu_0) - 1$	1.159615	15	10^{-3}	....	10^{-3}	
Zeeman splitting constant	μ_B/hc	4.66858	4	10^{1}	$m^{-1}\ T^{-1}$	10^{-5}	$cm^{-1}\ G^{-1}$ *
Gas constant	R	8.3143	12	10^{0}	$J\ {}^\circ K^{-1}\ mol^{-1}$	10^{7}	$erg\ {}^\circ K^{-1}\ mol^{-1}$
Normal volume perfect gas	V_0	2.24136	30	10^{-2}	$m^3\ mol^{-1}$	10^{4}	$cm^3\ mol^{-1}$
Boltzmann constant	k	1.38054	18	10^{-23}	$J\ {}^\circ K^{-1}$	10^{-16}	$erg\ {}^\circ K^{-1}$
First radiation constant ($2\pi hc^2$)	c_1	3.7405	3	10^{-16}	$W\ m^2$	10^{-5}	$erg\ cm^2\ s^{-1}$
Second radiation constant	c_2	1.43879	19	10^{-2}	$m\ {}^\circ K$	10^{0}	$cm\ {}^\circ K$
Wien displacement constant	b	2.8978	4	10^{-3}	$m\ {}^\circ K$	10^{-1}	$cm\ {}^\circ K$
Stefan-Boltzmann constant	σ	5.6697	29	10^{-8}	$W\ m^{-2}\ {}^\circ K^{-4}$	10^{-5}	$erg\ cm^{-2}\ s^{-1}\ {}^\circ K^{-4}$
Gravitational constant	G	6.670	15	10^{-11}	$N\ m^2\ kg^{-2}$	10^{-8}	$dyn\ cm^2\ g^{-2}$
Electron-volt	eV	1.60210	7	$\times 10^{-19}$	$J(eV)^{-1}$	$\times 10^{-12}$	$erg\ (eV)^{-1}$
Energy associated with:							
Unified atomic mass unit	c^2/Ne	9.31478	15	10^{8}	$eV\ u^{-1}$	10^{8}	$eV\ u^{-1}$
Proton mass	m_pc^2/e	9.38256	15	10^{8}	$eV\ m_p^{-1}$	10^{8}	$eV\ m_p^{-1}$
Neutron mass	m_nc^2/e	9.39550	15	10^{8}	$eV\ m_n^{-1}$	10^{8}	$eV\ m_n^{-1}$
Electron mass	m_ec^2/e	5.11006	5	10^{5}	$eV\ m_e^{-1}$	10^{5}	$eV\ m_e^{-1}$
Cycle	e/h	2.41804	7	10^{14}	$Hz(eV)^{-1}$	10^{14}	$s^{-1}(eV)^{-1}$
Wavelength	ch/e	1.23981	4	10^{-6}	eV m	10^{-4}	eV cm
Wave number	e/ch	8.06573	23	10^{5}	$m^{-1}(eV)^{-1}$	10^{3}	$cm^{-1}(eV)^{-1}$
°K	e/k	1.16049	16	10^{4}	$^\circ K(eV)^{-1}$	10^{4}	$^\circ K(eV)^{-1}$

‡ Based on 3 standard deviations; applied to last digits in preceding column.
* Electromagnetic system.
† Electrostatic system.
Abbreviations: C—coulomb; J—joule; Hz—hertz; W—watt; N—newton; T—tesla; G—gauss.

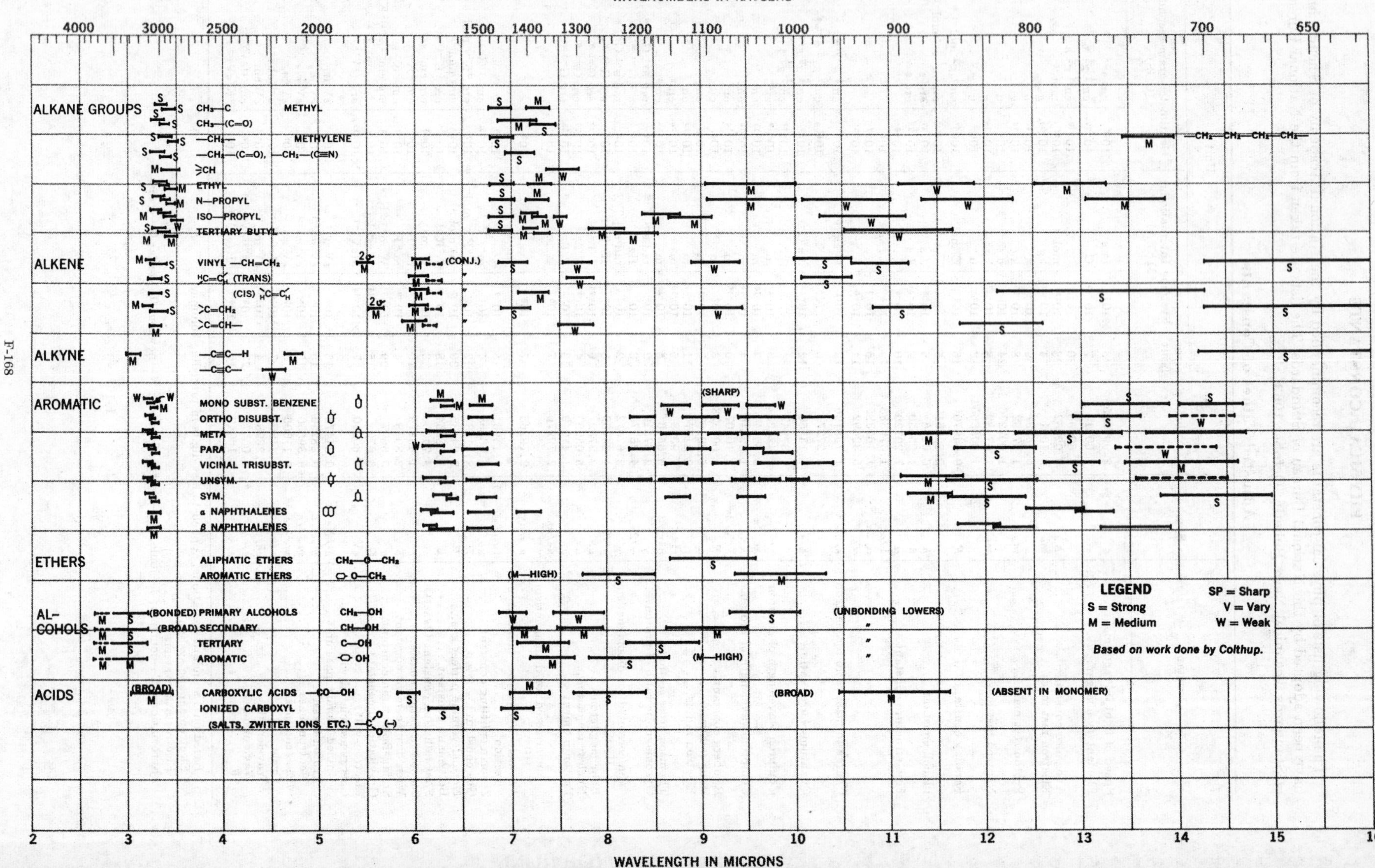

INFRARED CORRELATION CHART No. 1
Prepared from information supplied by Beckman Instruments
WAVENUMBERS IN KAYSERS
4000
3000
2500
2000
1500
1400
1300
1200
1100
1000
900
800
700
650
ALKANE GROUPS
CH₃—C
CH₃—(C=O)
METHYL
—CH₂—
METHYLENE
—CH₂—(C=O), —CH₂—(C≡N)
⋟CH
ETHYL
N—PROPYL
ISO—PROPYL
TERTIARY BUTYL
—CH₂—CH₂—CH₂—CH₂—
ALKENE
VINYL —CH=CH₂
(TRANS)
(CIS)
⋟C=CH₂
⋟C=CH—
(CONJ.)
ALKYNE
—C≡C—H
—C≡C—
AROMATIC
MONO SUBST. BENZENE
ORTHO DISUBST.
META
PARA
VICINAL TRISUBST.
UNSYM.
SYM.
α NAPHTHALENES
β NAPHTHALENES
(SHARP)
ETHERS
ALIPHATIC ETHERS
CH₂—O—CH₂
AROMATIC ETHERS
(M—HIGH)
AL-COHOLS
(BONDED) PRIMARY ALCOHOLS
CH₂—OH
(BROAD) SECONDARY
CH—OH
TERTIARY
C—OH
AROMATIC
(M—HIGH)
(UNBONDING LOWERS)
ACIDS
(BROAD)
CARBOXYLIC ACIDS —CO—OH
IONIZED CARBOXYL
(SALTS, ZWITTER IONS, ETC.)
(BROAD)
(ABSENT IN MONOMER)
LEGEND
S = Strong
M = Medium
SP = Sharp
V = Vary
W = Weak
Based on work done by Colthup.
2
3
4
5
6
7
8
9
10
11
12
13
14
15
16
WAVELENGTH IN MICRONS

INFRARED CORRELATION CHART No. 1 (Con't.)

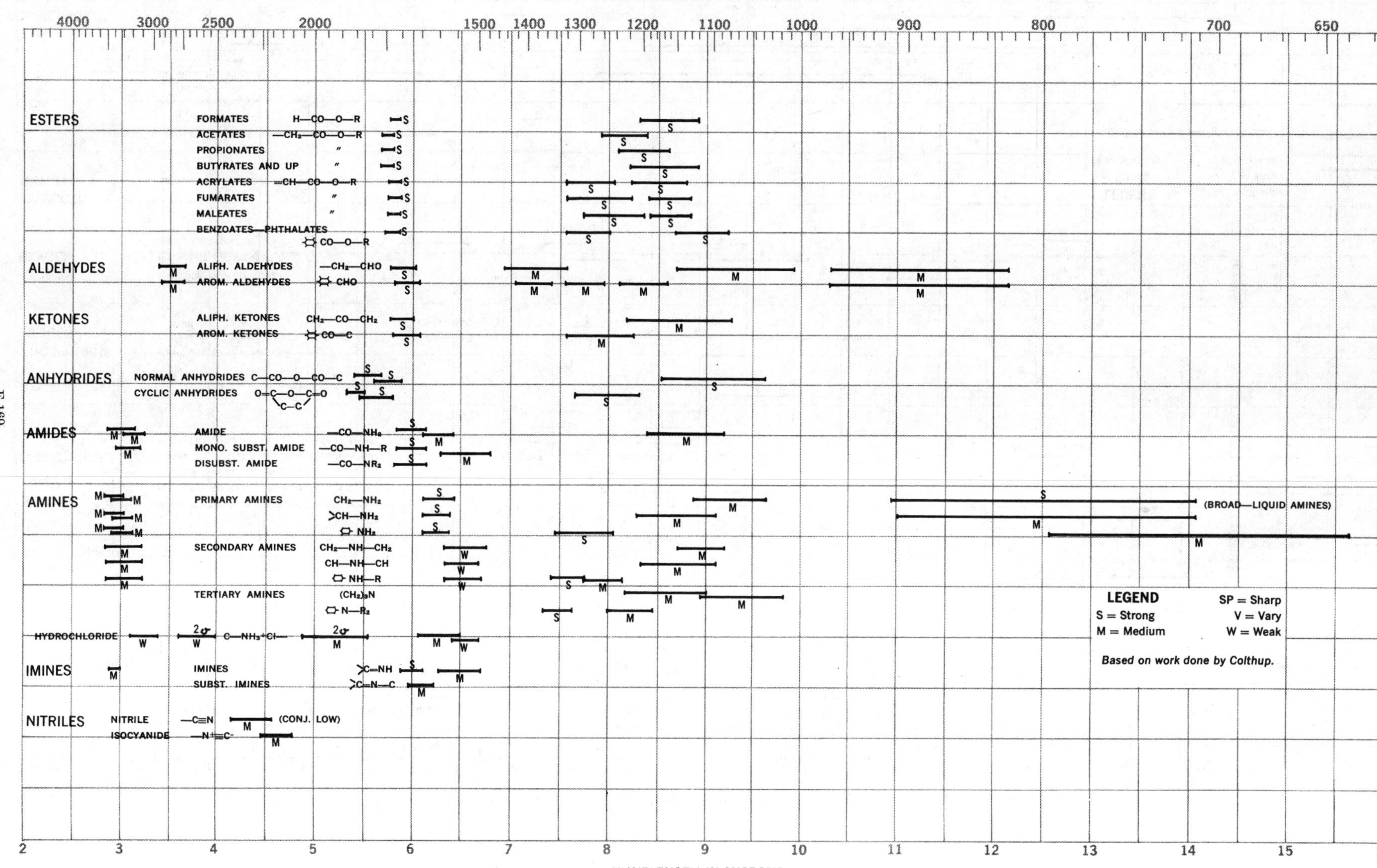

INFRARED CORRELATION CHART No. 1 (Con't.)

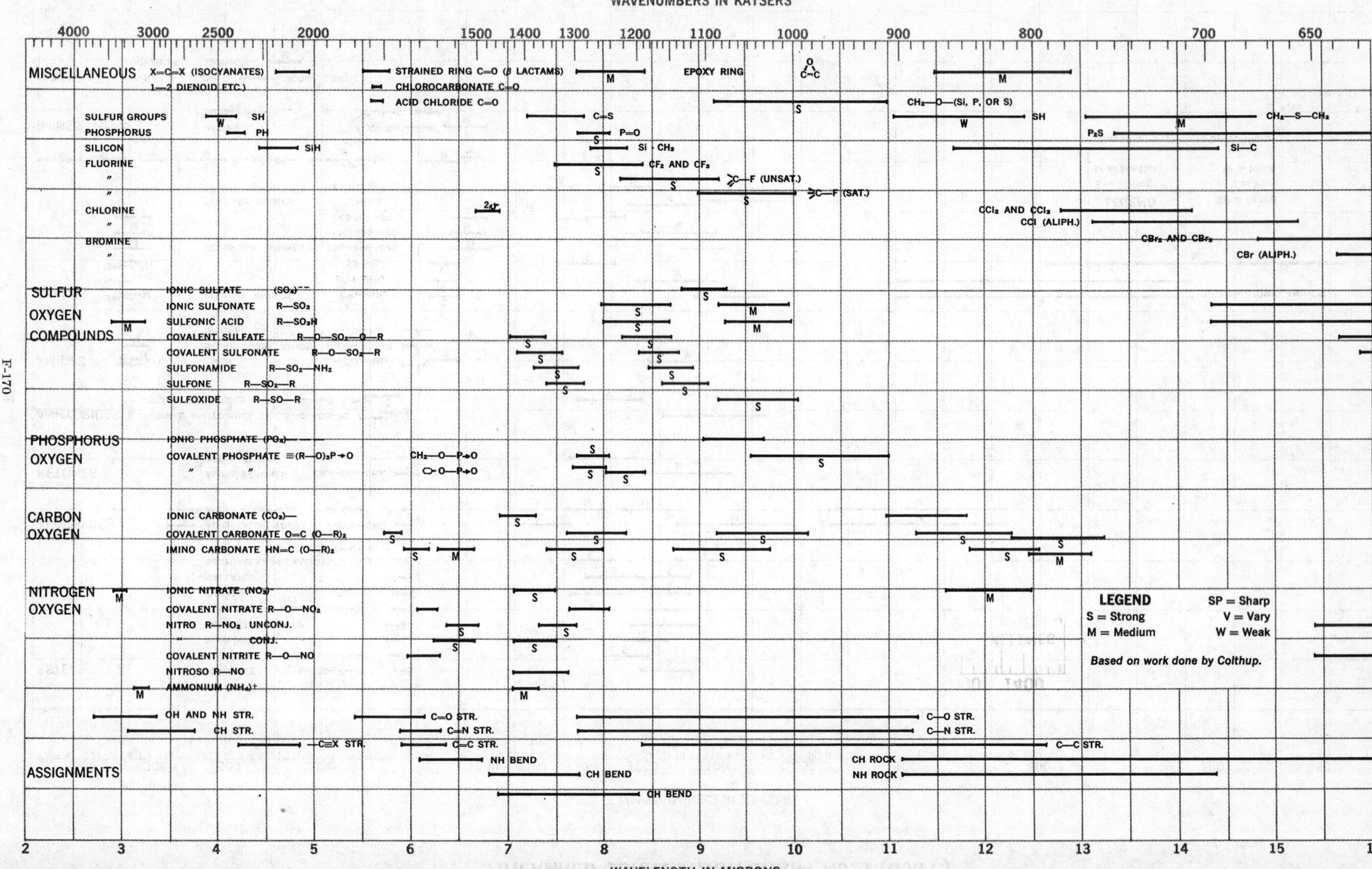

INFRARED CORRELATION CHART No. 2

Prepared from information supplied by Beckman Instruments

This chart presents some information regarding structure, double-bond vibrations, hydrogen stretching and triple-bond vibrations.

HYDROGEN STRETCHING AND TRIPLE-BOND VIBRATIONS, 3750-2000 CM.$^{-1}$

DOUBLE-BOND VIBRATIONS, ETC. 2000-1500 CM.$^{-1}$

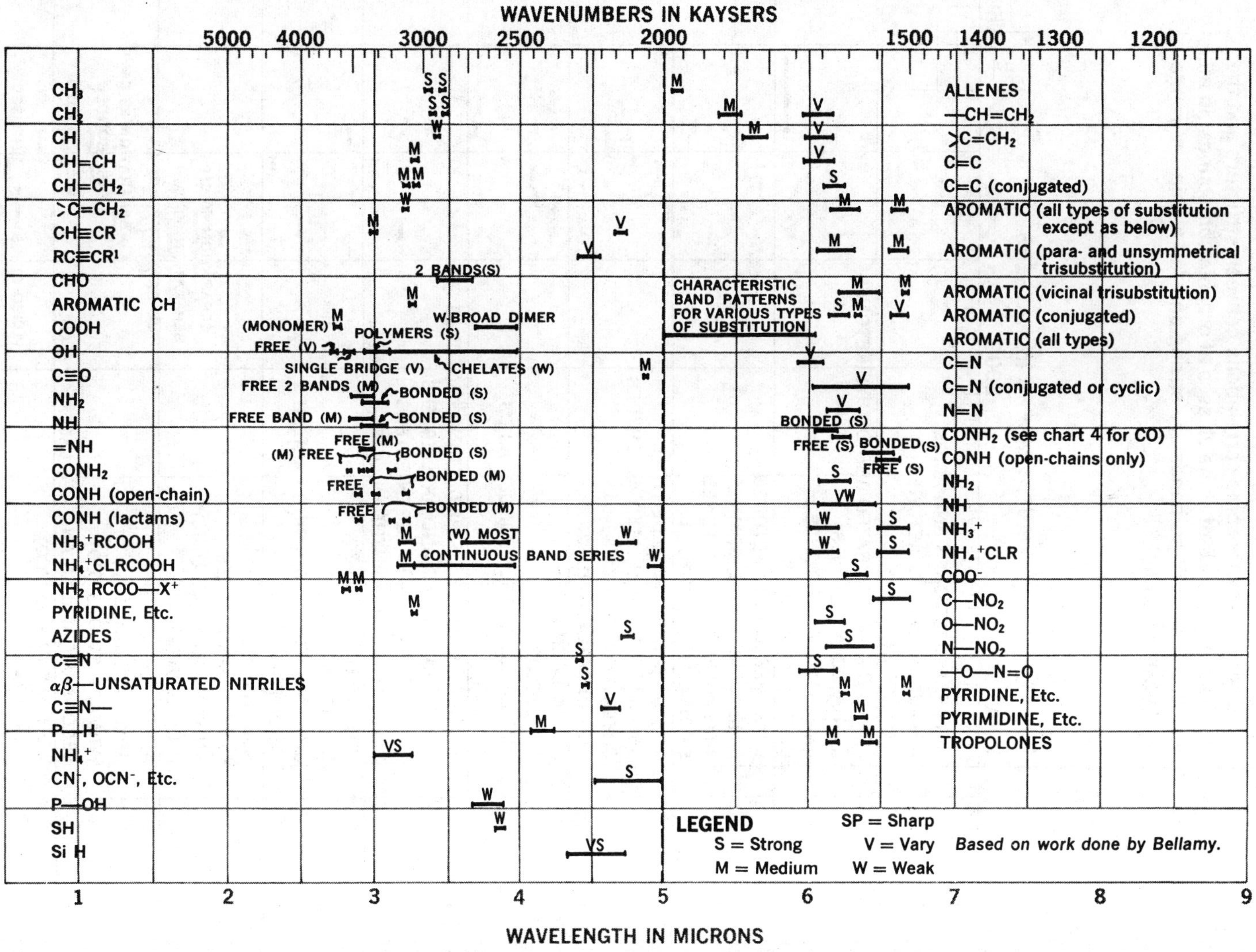

WAVENUMBERS IN KAYSERS

5000 4000 3000 2500 2000 1500 1400 1300 1200

CH_3
CH_2
CH
CH=CH
$CH=CH_2$
$>C=CH_2$
CH≡CR
$RC≡CR^1$
CHO
AROMATIC CH
COOH
OH
C≡O
NH_2
NH
=NH
$CONH_2$
CONH (open-chain)
CONH (lactams)
NH_3^+RCOOH
$NH_4^+CLRCOOH$
NH_2 RCOO—X^+
PYRIDINE, Etc.
AZIDES
C≡N
$\alpha\beta$—UNSATURATED NITRILES
C≡N—
P—H
NH_4^+
CN^-, OCN^-, Etc.
P—OH
SH
Si H

2 BANDS(S)
(MONOMER)
W-BROAD DIMER
POLYMERS (S)
FREE (V)
SINGLE BRIDGE (V)
CHELATES (W)
FREE 2 BANDS (M)
BONDED (S)
FREE BAND (M)
BONDED (S)
FREE (M)
(M) FREE
BONDED (S)
BONDED (M)
FREE
FREE
BONDED (M)
(W) MOST
CONTINUOUS BAND SERIES

CHARACTERISTIC BAND PATTERNS FOR VARIOUS TYPES OF SUBSTITUTION

BONDED (S)
FREE (S)
BONDED(S)
FREE (S)

ALLENES
—$CH=CH_2$
$>C=CH_2$
C=C
C=C (conjugated)
AROMATIC (all types of substitution except as below)
AROMATIC (para- and unsymmetrical trisubstitution)
AROMATIC (vicinal trisubstitution)
AROMATIC (conjugated)
AROMATIC (all types)
C=N
C=N (conjugated or cyclic)
N=N
$CONH_2$ (see chart 4 for CO)
CONH (open-chains only)
NH_2
NH
NH_3^+
NH_4^+CLR
COO^-
C—NO_2
O—NO_2
N—NO_2
—O—N=O
PYRIDINE, Etc.
PYRIMIDINE, Etc.
TROPOLONES

LEGEND

S = Strong
M = Medium
SP = Sharp
V = Vary
W = Weak

Based on work done by Bellamy.

1 2 3 4 5 6 7 8 9

WAVELENGTH IN MICRONS

INFRARED CORRELATION CHART No. 3

Prepared from information supplied by Beckman Instruments

This chart presents some correlations between structure and the carbonyl vibrations of some classes of organic compounds. In all cases the absorption bands are strong and fall within the range of 1900-1500 cm^{-1}.

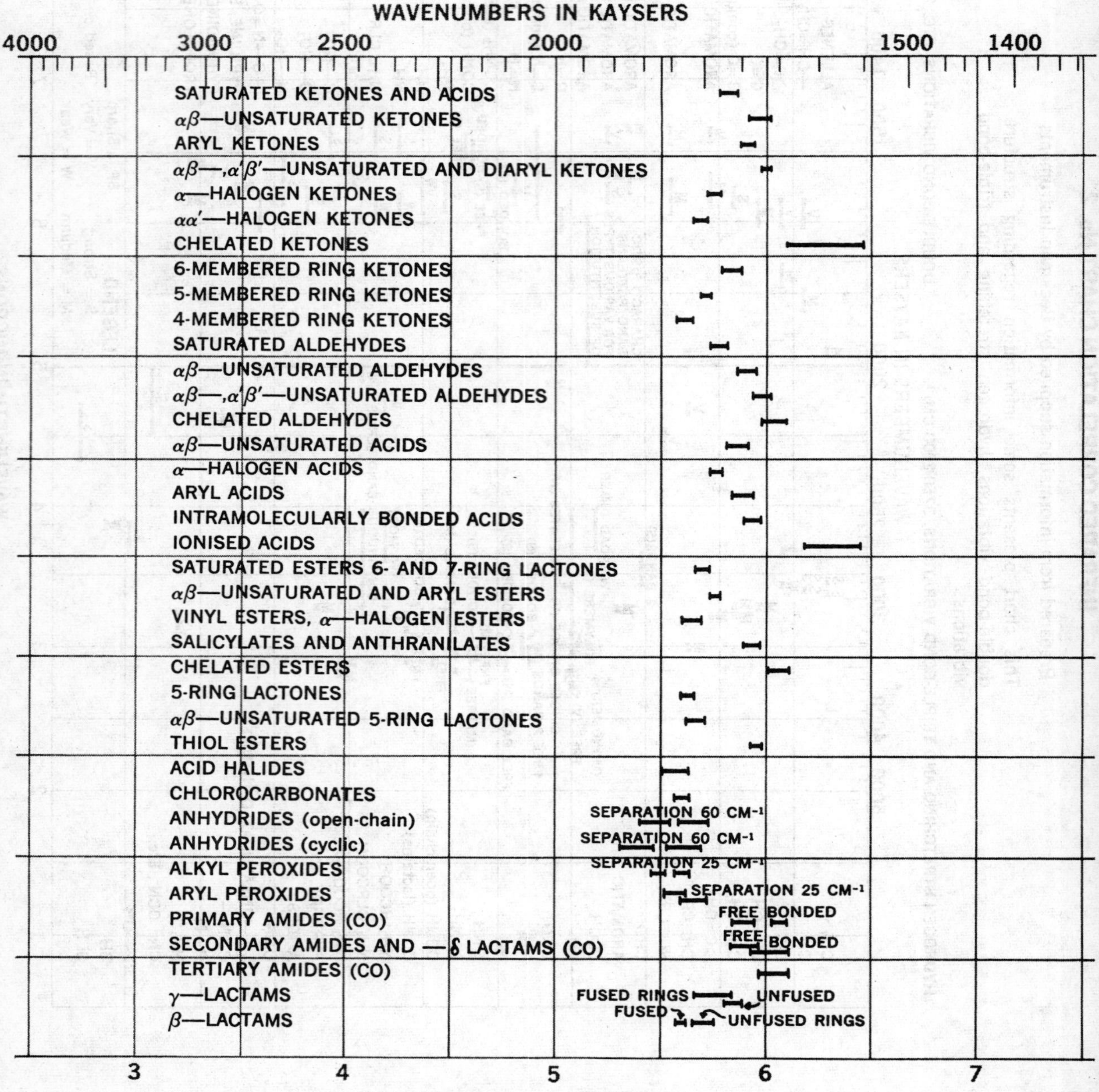

WAVENUMBERS IN KAYSERS

4000 3000 2500 2000 1500 1400

SATURATED KETONES AND ACIDS
$\alpha\beta$—UNSATURATED KETONES
ARYL KETONES
$\alpha\beta$—,$\alpha'\beta'$—UNSATURATED AND DIARYL KETONES
α—HALOGEN KETONES
$\alpha\alpha'$—HALOGEN KETONES
CHELATED KETONES
6-MEMBERED RING KETONES
5-MEMBERED RING KETONES
4-MEMBERED RING KETONES
SATURATED ALDEHYDES
$\alpha\beta$—UNSATURATED ALDEHYDES
$\alpha\beta$—,$\alpha'\beta'$—UNSATURATED ALDEHYDES
CHELATED ALDEHYDES
$\alpha\beta$—UNSATURATED ACIDS
α—HALOGEN ACIDS
ARYL ACIDS
INTRAMOLECULARLY BONDED ACIDS
IONISED ACIDS
SATURATED ESTERS 6- AND 7-RING LACTONES
$\alpha\beta$—UNSATURATED AND ARYL ESTERS
VINYL ESTERS, α—HALOGEN ESTERS
SALICYLATES AND ANTHRANILATES
CHELATED ESTERS
5-RING LACTONES
$\alpha\beta$—UNSATURATED 5-RING LACTONES
THIOL ESTERS
ACID HALIDES
CHLOROCARBONATES
ANHYDRIDES (open-chain) — SEPARATION 60 CM^{-1}
ANHYDRIDES (cyclic) — SEPARATION 60 CM^{-1}
ALKYL PEROXIDES — SEPARATION 25 CM^{-1}
ARYL PEROXIDES — SEPARATION 25 CM^{-1}
PRIMARY AMIDES (CO) — FREE, BONDED
SECONDARY AMIDES AND —δ LACTAMS (CO) — FREE, BONDED
TERTIARY AMIDES (CO)
γ—LACTAMS — FUSED RINGS, UNFUSED
β—LACTAMS — FUSED, UNFUSED RINGS

3 4 5 6 7

WAVELENGTH IN MICRONS

INFRARED CORRELATION CHART No. 4

Prepared from information supplied by Beckman Instruments

This chart presents some correlations between structure and single-bond vibrations for a number of classes of compounds having absorption between 1500-650 cm^{-1}.

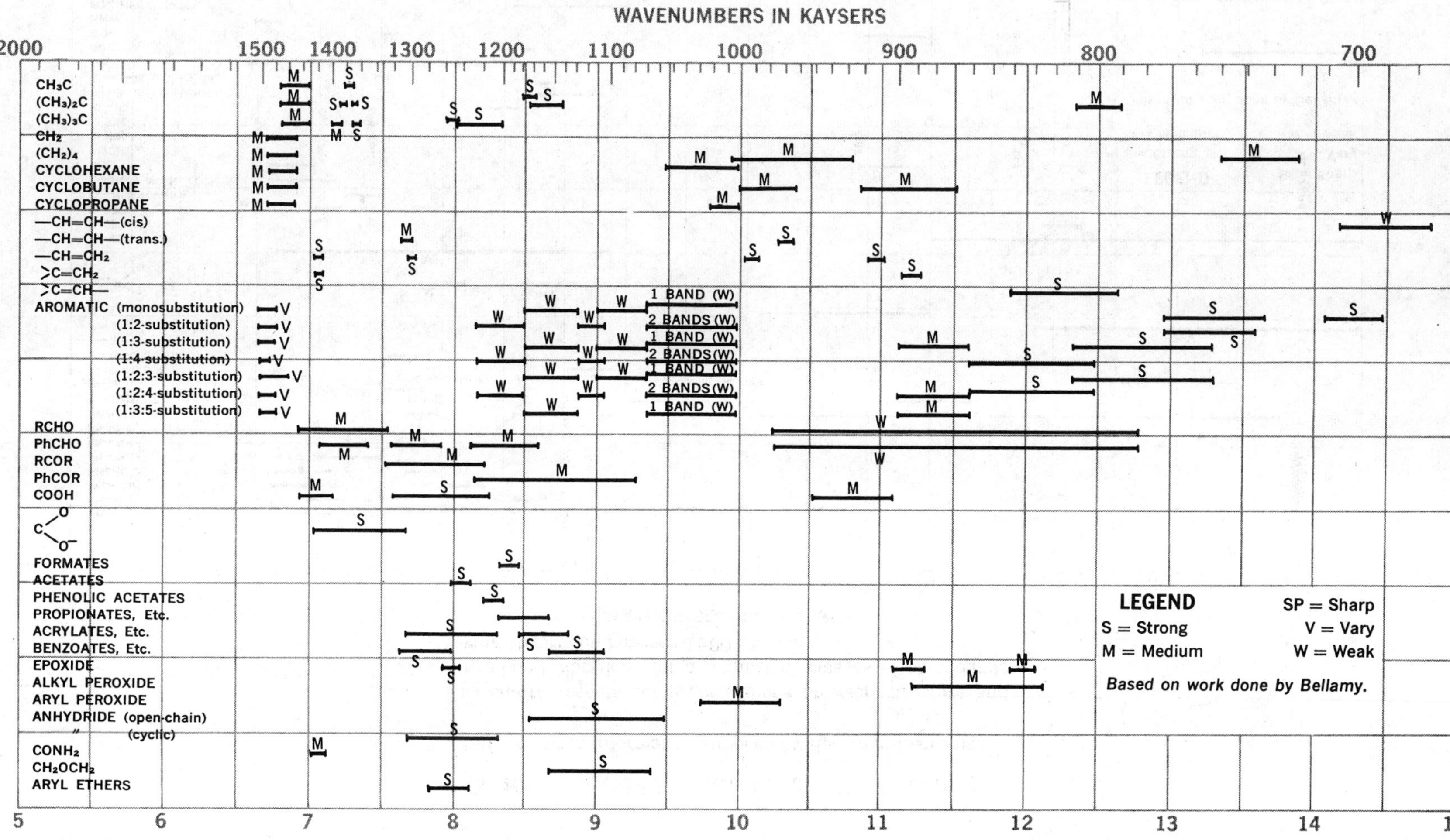

INFRARED CORRELATION CHART No. 4 (Con't.)

Prepared from information supplied by Beckman Instruments

This chart presents some correlations between structure and single-bond vibrations for a number of classes of compounds having absorption between 1500-650 cm^{-1}.

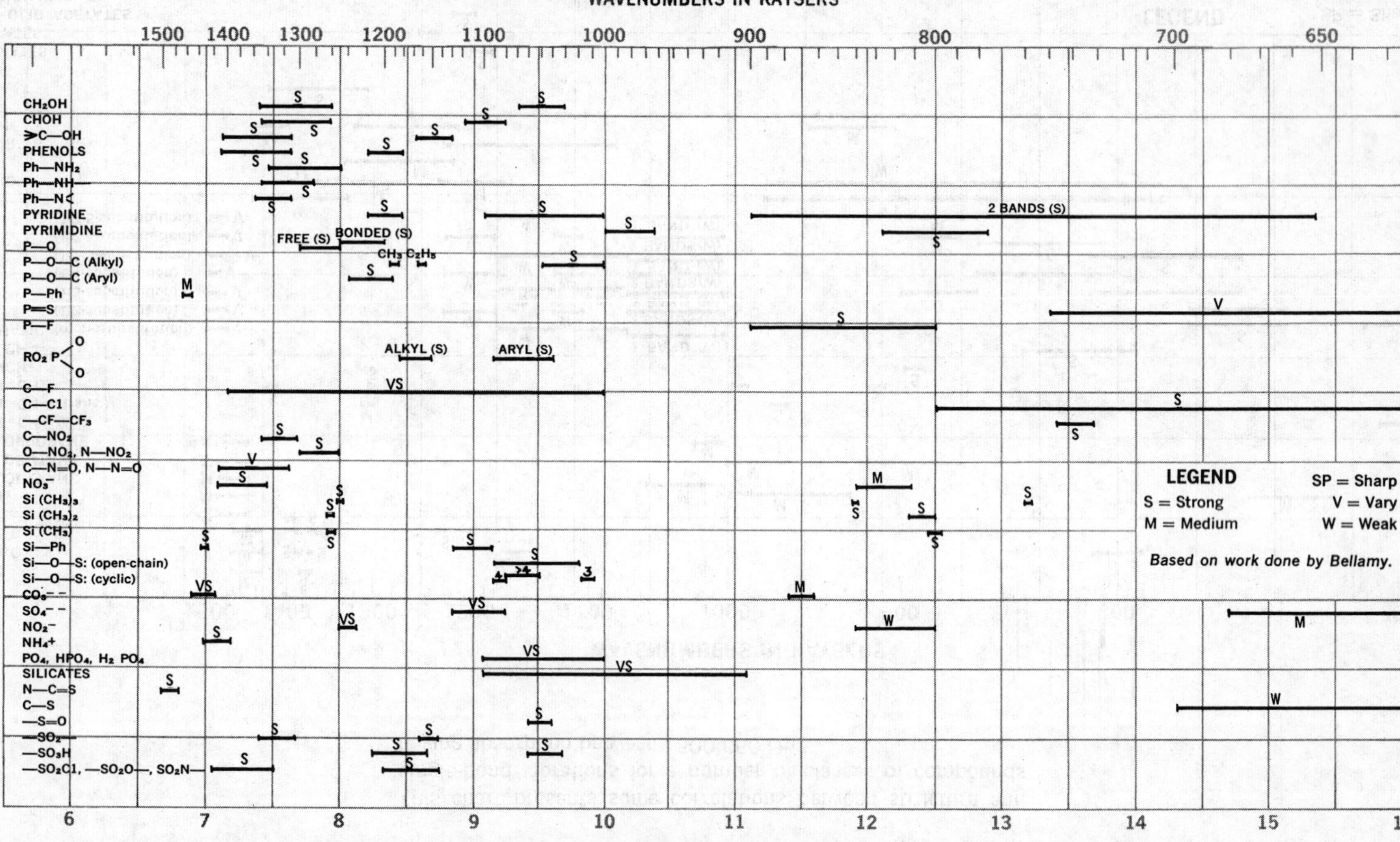

FAR INFRARED VIBRATIONAL FREQUENCY CORRELATION CHART

Based on evidence compiled by James E. Stewart of Beckman Instruments. This chart shows the vibrational frequency correlation in the far infrared region. Because research is continuing in the far infrared region, this chart is not all-inclusive.

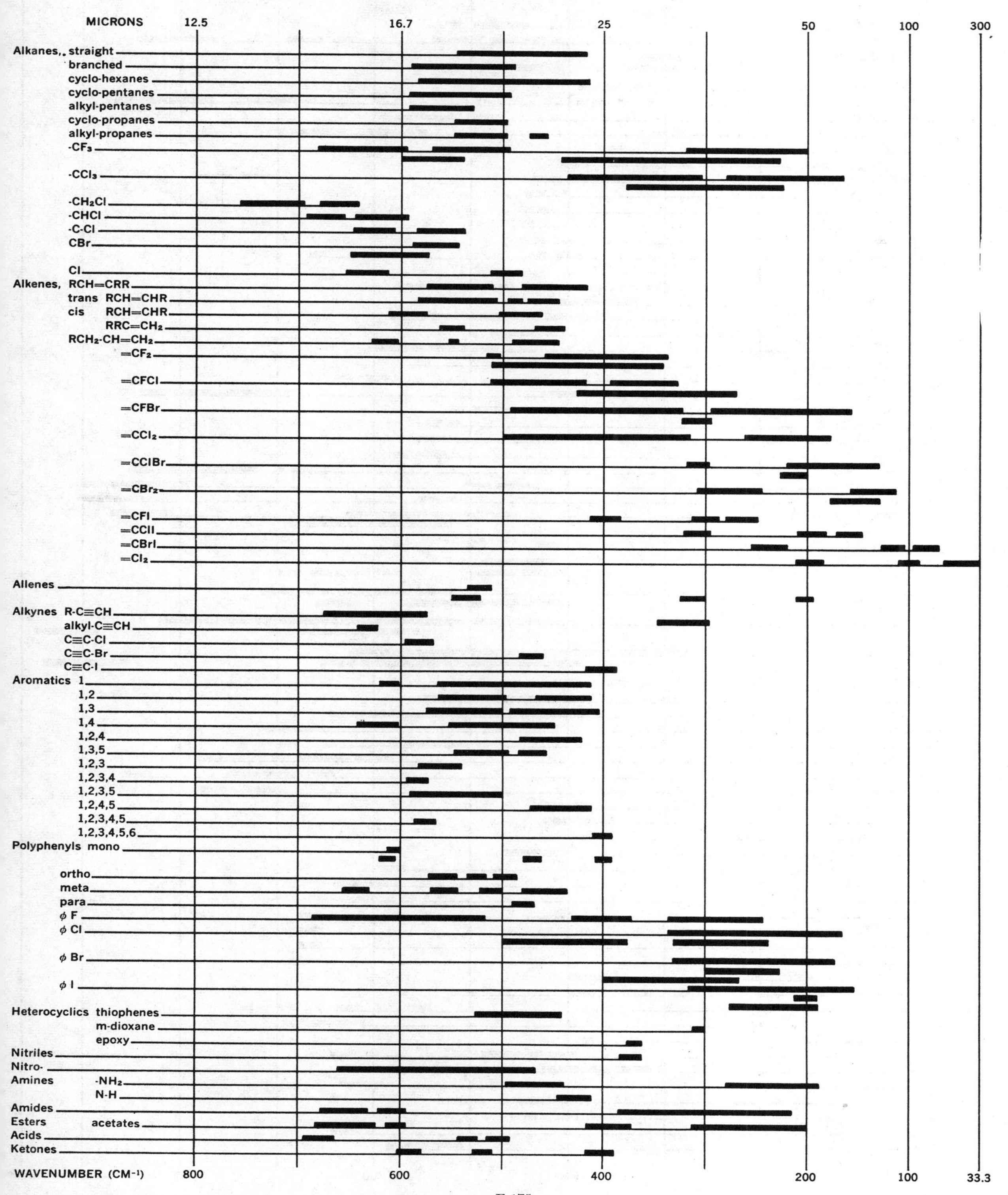

FAR INFRARED VIBRATIONAL FREQUENCY CORRELATION CHART (Con't.)

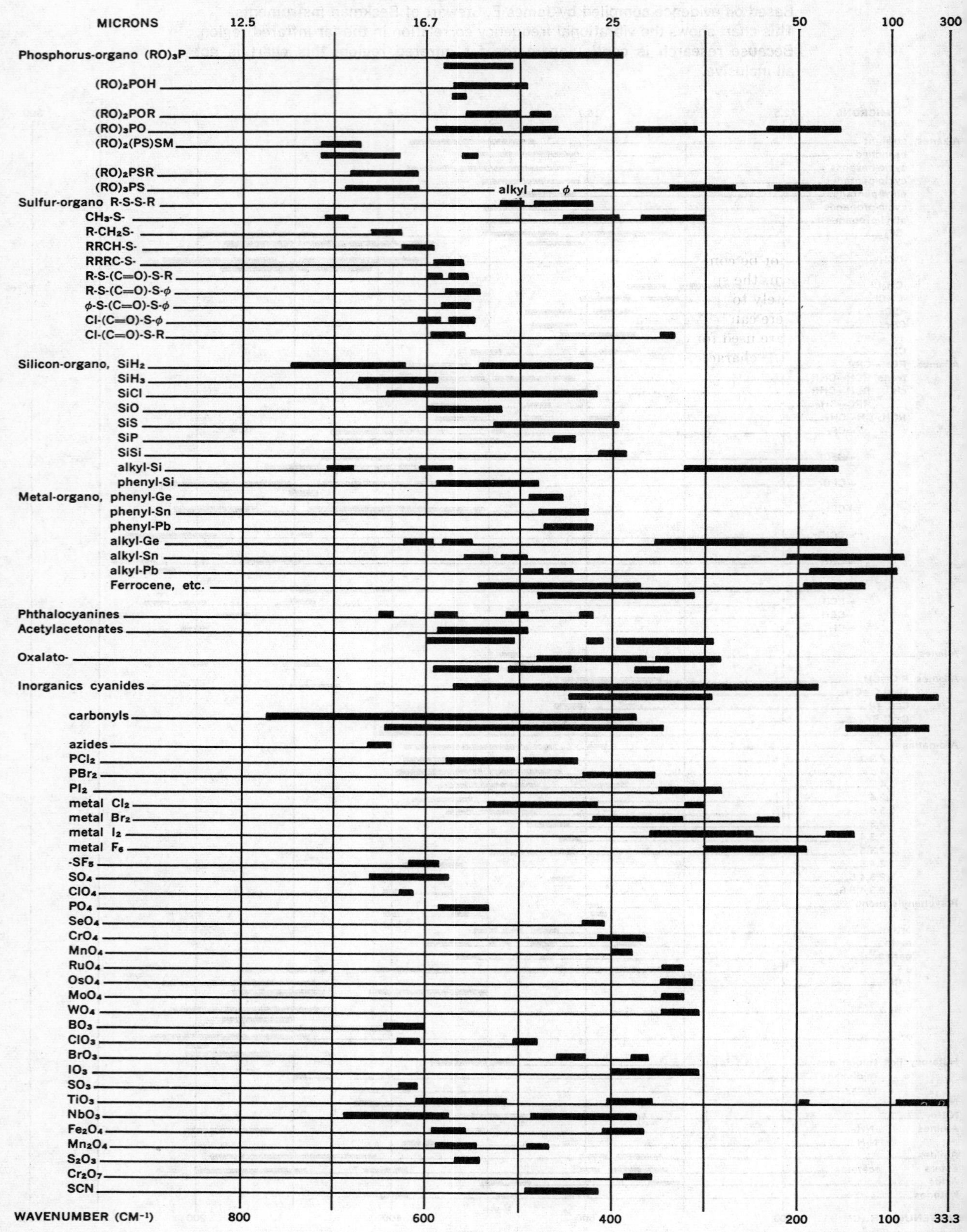

CHART OF CHARACTERISTIC FREQUENCIES BETWEEN ~ 700–300 cm^{-1}

Freeman F. Bentley, Lee D. Smithson, and Adele L. Rozek

This chart summarizes the characteristic frequencies known to occur between approximately 700–300^{-1}. Those who anticipate using this region of the spectrum should consult "Infrared Spectra and Characteristic Frequencies between ~700–300 cm^{-1}" by Interscience Publishers, a division of John Wiley and Sons, Inc. for a complete discussion of the characteristic frequencies summarized in this chart, a large collection of infrared spectra (700–300 cm^{-1}) of most of the common organic and inorganic compounds, and an extensive bibliography of references to infrared data below ~700 cm^{-1}.

In this chart the black horizontal bars indicate the range of the spectrum in which the characteristic frequencies have been observed to occur in the compounds investigated. The number of compounds investigated is given immediately to the right of the names or structures of the compounds. Obviously those characteristic frequency ranges based upon a limited number of compounds should be used with caution.

The letters above the bars indicate the relative intensities of the absorption bands. These intensities are based upon the strongest band in the spectra (700–300 cm^{-1}) of specific classes of compounds investigated, and they cannot be compared accurately with the intensities given for other classes.

When known, the specific vibration giving rise to the characteristic frequency is printed in abbreviated form immediately to the right of the bar indicating the frequency range except when lack of space prevents this. When there can be no ambiguity, this information may be printed other than to the right. In doubtful cases, arrows are used for clarification.

Naturally, the characteristic frequencies vary in their specificity and analytical value. The user is, therefore, cautioned to use this chart with some reserve. After reviewing this chart, the reader should be aware that there are many characteristic frequencies in the 700–300 cm^{-1} region. Used cautiously, this chart can be of considerable value in the elucidation of structures of unknown compounds.

It is important to emphasize that the region of the infrared between ~700–300 cm^{-1} should be used in conjunction with the more conventional 5000–700 cm^{-1} region. Much of the value of the 700–300 cm^{-1} region can only be realized after interpreting the spectrum between 5000–700 cm^{-1}.

The following symbols and abbreviations are used:

Symbol or Abbreviation	Definition
αCCC	In-plane bending of benzene ring
Antisym.	Antisymmetrical (Asymmetrical)
~	Approximately
β	In-plane bending of ring substituent bond
δ	In-plane bending
γ	Out-of-plane bending
i.p.	In-plane
m	Medium
ν	Stretching
ν_s	Symmetrical stretching
ν_{as}	Antisymmetrical stretching
o.p.	Out-of-plane
$\parallel$	Parallel
$\perp$	Perpendicular
ϕ	Phenyl
ϕCC	Out-of-plane bending of aromatic ring
r	Rocking
s	Strong
sh	Shoulder
Sym.	Symmetrical
v	Variable
w	Weak
"X" Sensitive	An aromatic vibrational mode whose frequency position is greatly dependent on the nature of the substituent.

CHART OF CHARACTERISTIC FREQUENCIES BETWEEN ~ 700–300 cm^{-1}

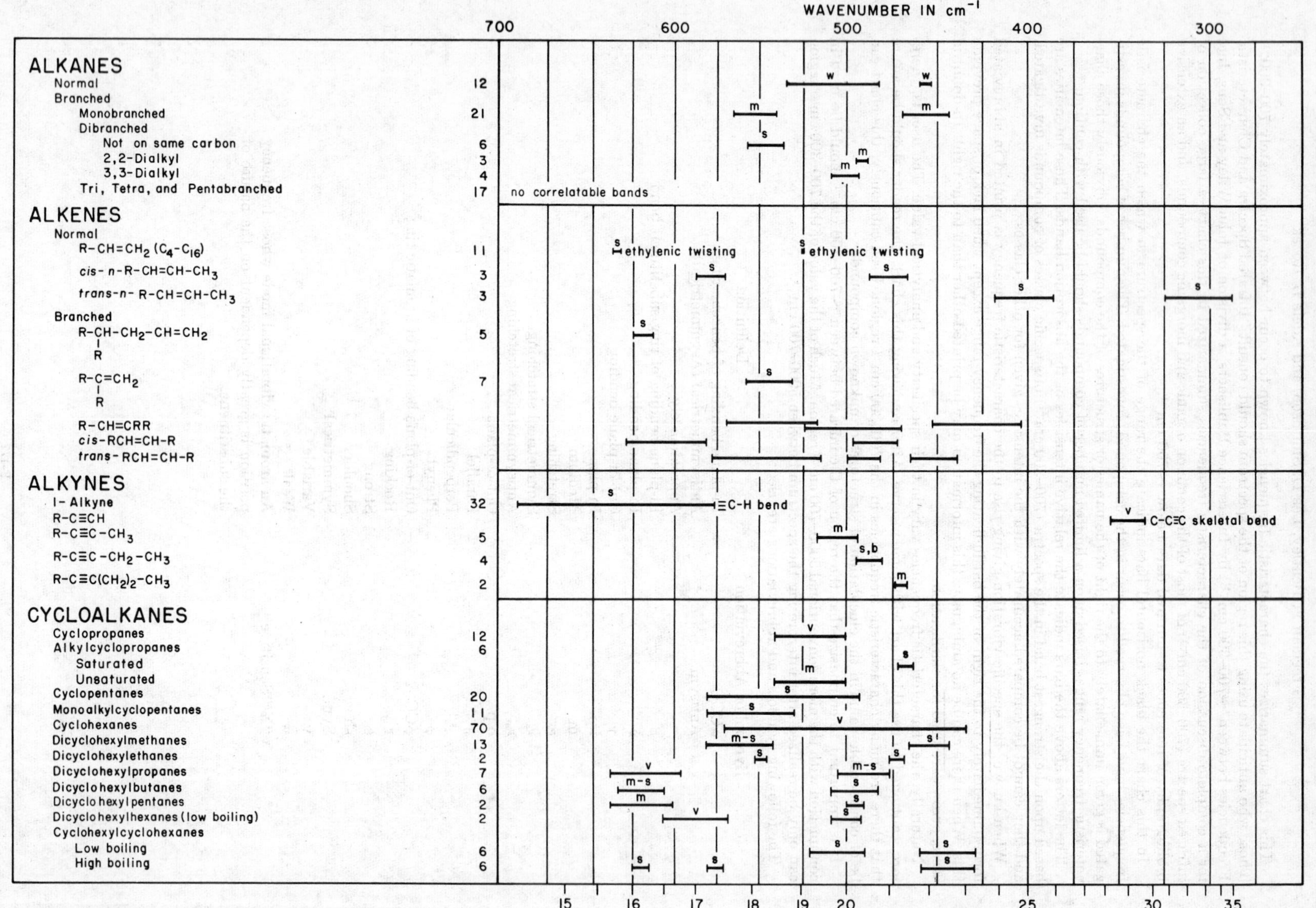

CHART OF CHARACTERISTIC FREQUENCIES BETWEEN ∽ 700–300 cm⁻¹

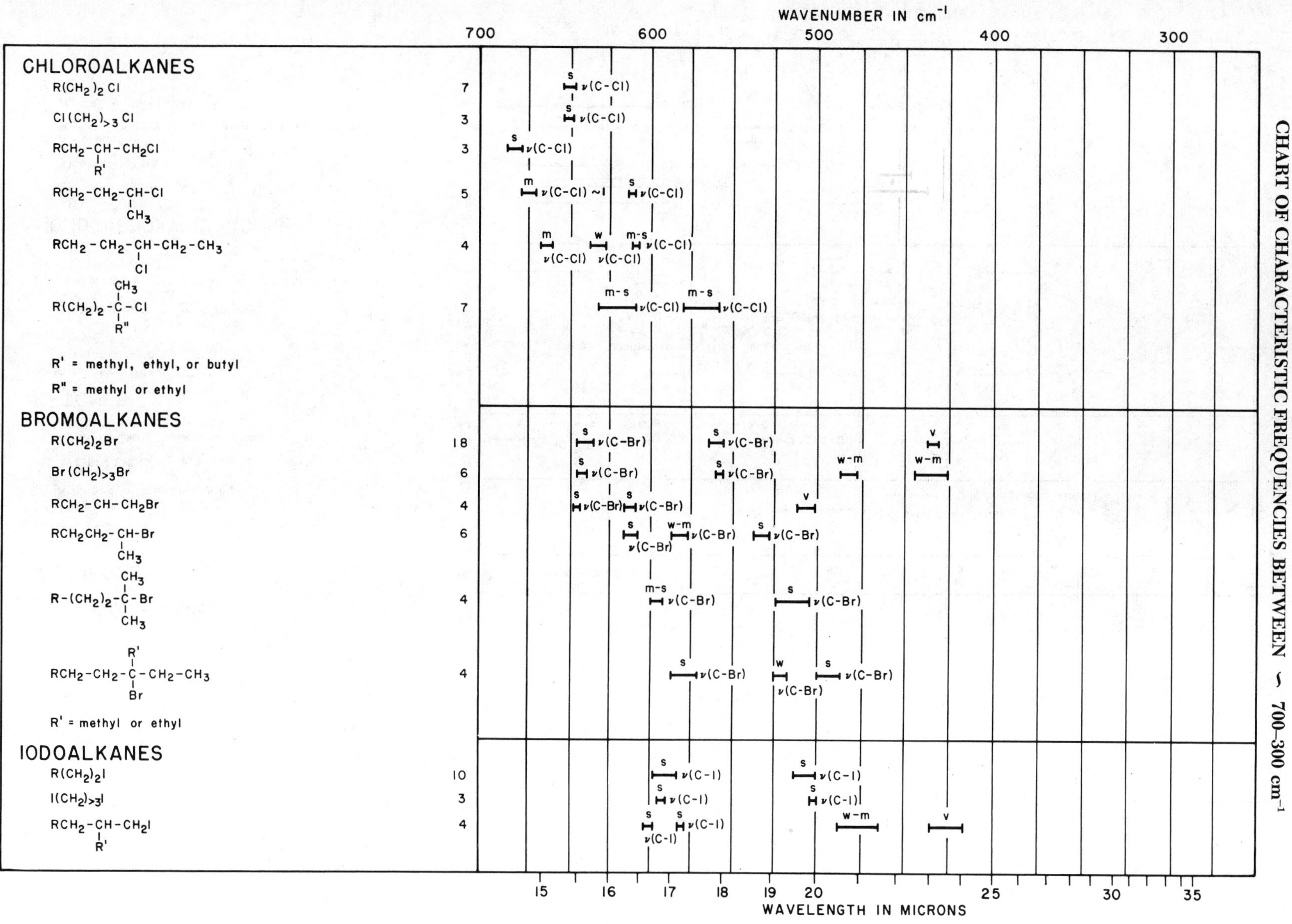

CHART OF CHARACTERISTIC FREQUENCIES BETWEEN ∽ 700–300 cm^{-1}

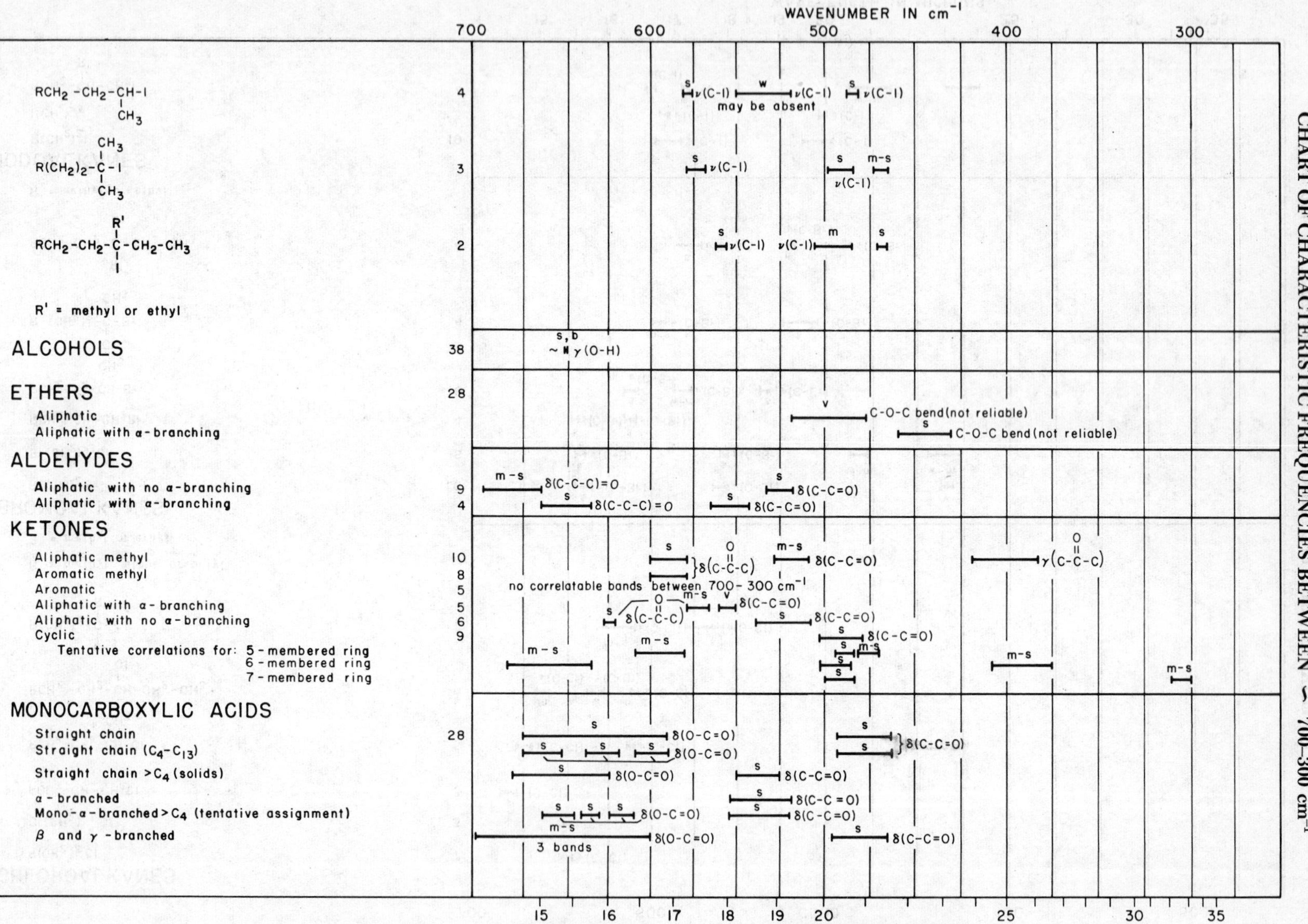

CHART OF CHARACTERISTIC FREQUENCIES BETWEEN ~ 700–300 cm^{-1}

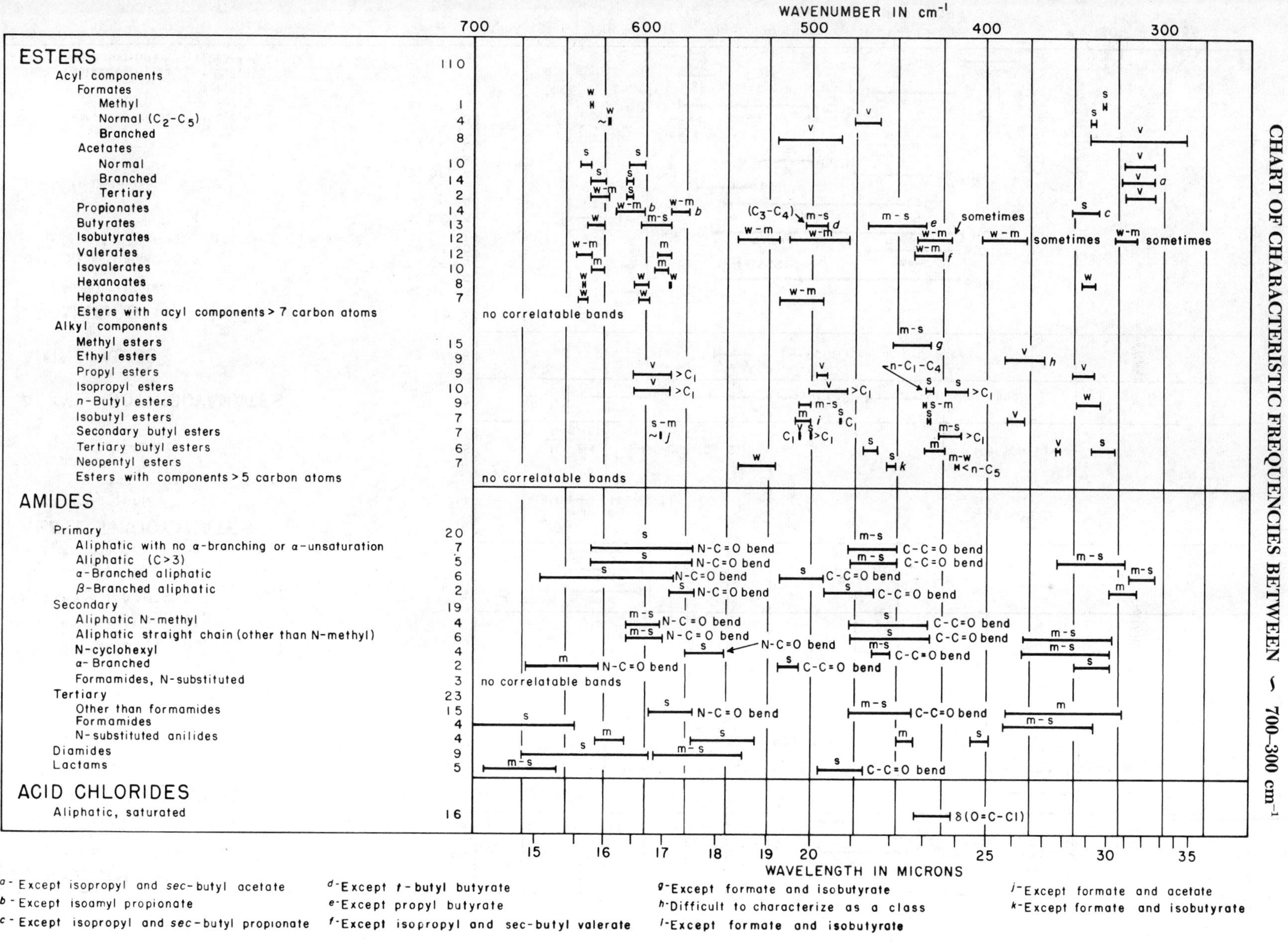

CHART OF CHARACTERISTIC FREQUENCIES BETWEEN ~ 700–300 cm⁻¹

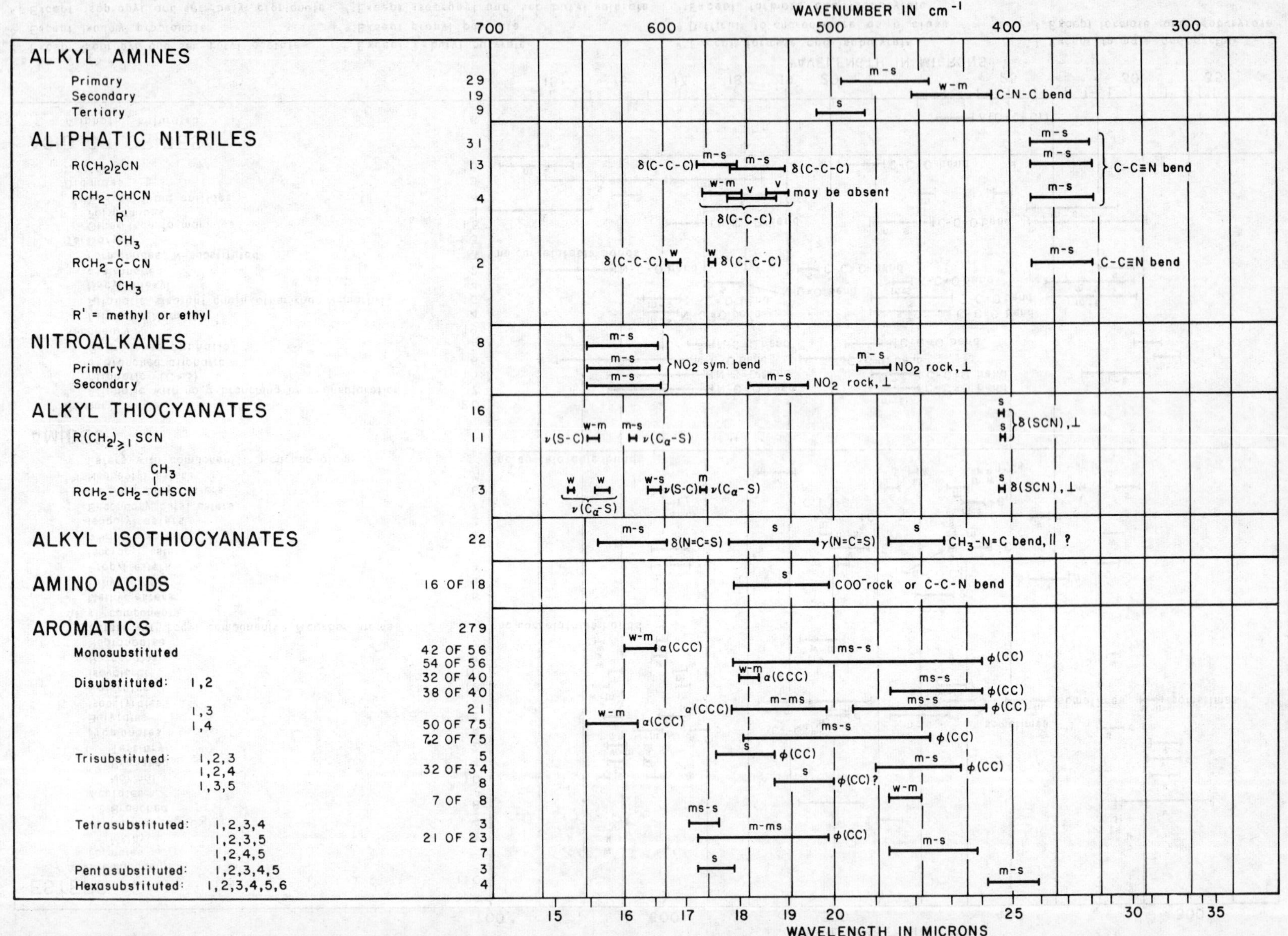

CHART OF CHARACTERISTIC FREQUENCIES BETWEEN ~ 700–300 cm^{-1}

WAVENUMBER IN cm^{-1}: 700, 600, 500, 400, 300

Group	No.	Bands (intensity, assignment)
Mono- and disubstituted benzenes with the following substituents:		
-COOH	7	s r(CO_2) o.p.
-COOR (R≠H)	10	v r(CO_2) or δ(OCO) ?
$-\overset{O}{\overset{\Vert}{C}}-CH_3$	5	s $\delta\left(\overset{O}{\overset{\Vert}{CCC}}\right)$ i.p.
-CHO	8 OF 9	s δ(CCO) i.p.
$-\overset{O}{\overset{\Vert}{C}}-NH_2$	2	s N-C=O bend
$-\overset{O}{\overset{\Vert}{C}}-O$	17	m-s,b; $\delta\left(-\overset{O}{\overset{\Vert}{C}}-O-\right)$?
$-NO_2$	23	w-m δ(NO_2)
$-OCH_3$	11	m-s δ(COC)
-CN	10	s φ(CC) + δ(CN); m β(CCN)
-OH	15 OF 22	←720 s, b γ(OH)
	22	w β(C-OH)
$-NH_2$	9	w β(C-NH_2)
-F	8	mw-s β(C-F)
-Cl	14 OF 24	m-s β(C-Cl)
-Br	12	s β(C-Br) and α(CCC)
$-CF_3$	4	s r(CF_3)
$-CH_3$	33	m β(CX)
$-C_2H_5$	7	m-s δ(CCC
$-C_3H_7$, $-C_4H_9$, $-C_{10}H_{21}$	4	m-s; m-s
$-iC_3H_7$	37 OF 43	m δ(CCC)
BIPHENYLS	9	m-s CC i.p. ring; m-s CC o.p. ring; m-s CC o.p. ring
NAPHTHALENES	18	m-s skeletal distortion i.p.; v o.p. skeletal bend
HETEROCYCLICS		
Pyridines: 2-substituted	14	m-s α(CCC); s φ(CC)
Pyridines: 3-substituted	13	m-s α(CCC); s φ(CC)
Pyrimidines: 2-substituted	11	m-s; m-s; m-s
Pyrimidines: 4-substituted	8	v; m-s; s
Imidazoles: 4 (or 5)-substituted	6	s o.p. ring bend; s γ(N-H); m; m
Imidazoles: 4,5-disubstituted	9	m-s; m-s
Imidazoles: 1,4,5-trisubstituted	11	m-s; w-m
Purines: 2-substituted	7	m-s; m-s γ(CH); v A" o.p. ring bend
Purines: 6-substituted	32	v; s
Purines: 8-substituted	8	v; s
Purines: 2,6-disubstituted	21	s; v

WAVELENGTH IN MICRONS: 15, 16, 17, 18, 19, 20, 25, 30, 35

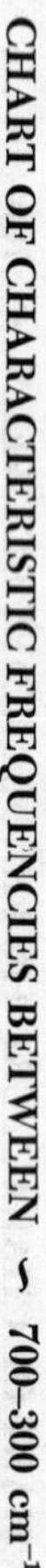

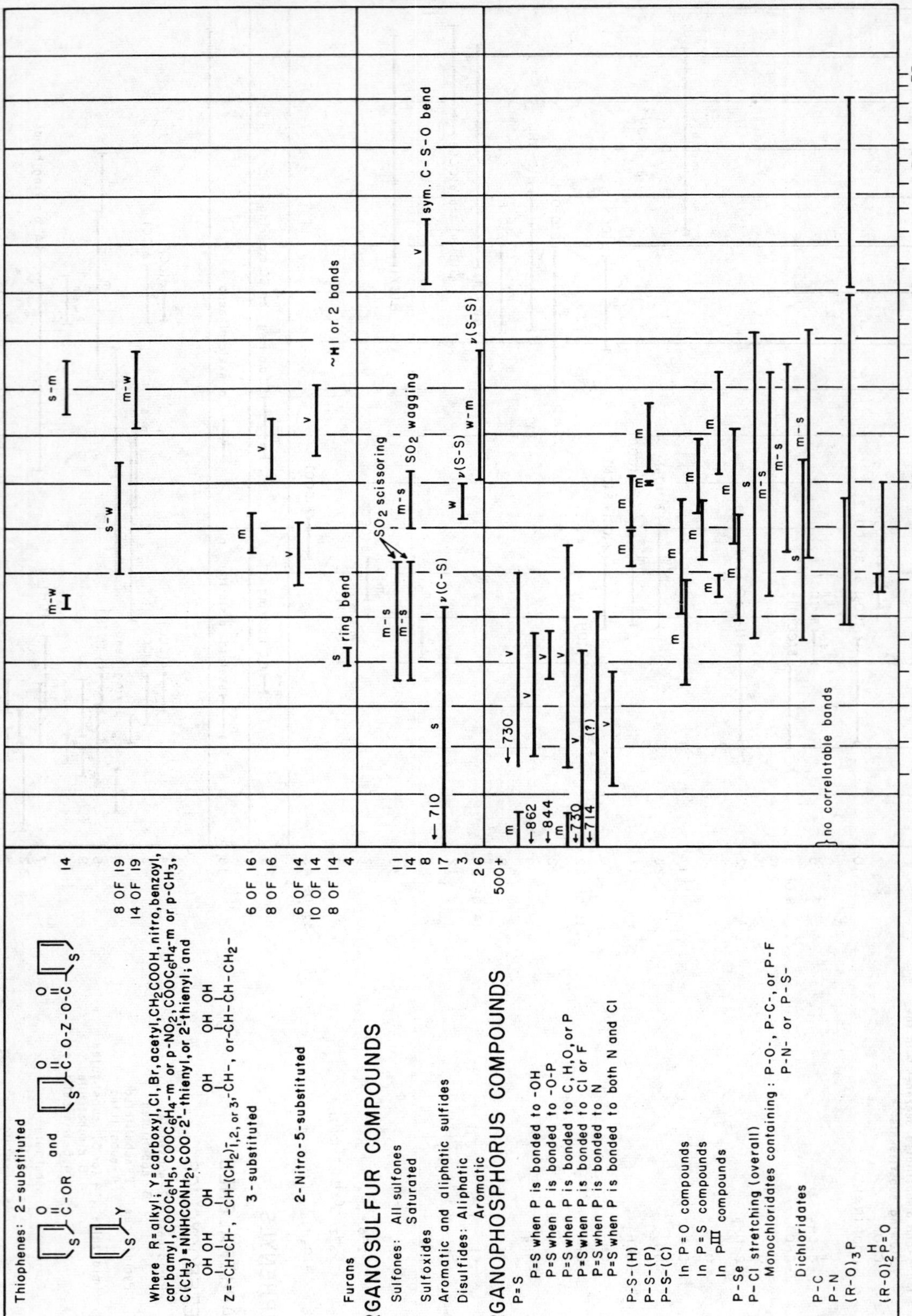

CHART OF CHARACTERISTIC FREQUENCIES BETWEEN ∽ 700–300 cm^{-1}

WAVENUMBER IN cm^{-1}: 700, 600, 500, 400, 300

WAVELENGTH IN MICRONS: 15, 16, 17, 18, 19, 20, 25, 30, 35

Thiophenes: 2-substituted

(2-thienyl)–C(=O)–OR and (2-thienyl)–C(=O)–O–Z–O–C(=O)–(2-thienyl) — 14 — m-w; s-m

(2-thienyl)–Y — 8 OF 19 — s-w; 14 OF 19 — m-w

Where R=alkyl; Y=carboxyl, Cl, Br, acetyl, CH_2COOH, nitro, benzoyl, carbamyl, $COOC_6H_5$, $COOC_6H_4$-m or p-NO_2, $COOC_6H_4$-m or p-CH_3, $C(CH_3)=NNHCONH_2$, COO-2′-thienyl, or 2′-thienyl; and

Z = –CH(OH)–CH(OH)–, –CH(OH)–$(CH_2)_{1,2,\text{ or }3}$–CH(OH)–, or –CH(OH)–CH(OH)–CH–CH_2–

Compounds	No.	Bands
Thiophenes: 3-substituted	6 OF 16	m
	8 OF 16	v
2-Nitro-5-substituted	6 OF 14	v
	10 OF 14	v
	8 OF 14	~1 or 2 bands
Furans	4	s ring bend

ORGANOSULFUR COMPOUNDS

Compounds	No.	Bands
Sulfones: All sulfones	11	m-s (SO_2 scissoring)
Saturated	14	m-s (SO_2 scissoring); m-s SO_2 wagging
Sulfoxides	8	v sym. C-S-O bend
Aromatic and aliphatic sulfides	17	← 710 s ν(C-S)
Disulfides: Aliphatic	3	w ν(S-S)
Aromatic	26	w-m ν(S-S)

ORGANOPHOSPHORUS COMPOUNDS

Compounds	No.	Bands
P=S	500+	m; ← 730 v
P=S when P is bonded to -OH		← 862 v
P=S when P is bonded to -O-P		← 844 v
P=S when P is bonded to C, H, O, or P		m; v
P=S when P is bonded to Cl or F		← 730 v
P=S when P is bonded to N		← 714 (?)
P=S when P is bonded to both N and Cl		v
P-S-(H)		m; m
P-S-(P)		m; m
P-S-(C)		
In P=O compounds		m; m
In P=S compounds		m; m
In P^{III} compounds		m; m
P-Se		m; m
P-Cl stretching (overall)		s
Monochloridates containing: P-O-, P-C-, or P-F		m-s
P-N- or P-S-		m-s
Dichloridates		s; m-s
P-C		no correlatable bands
P-N		no correlatable bands
$(R-O)_3P$		
$(R-O)_2P(H)=O$		

CHART OF CHARACTERISTIC FREQUENCIES BETWEEN ~ 700–300 cm⁻¹

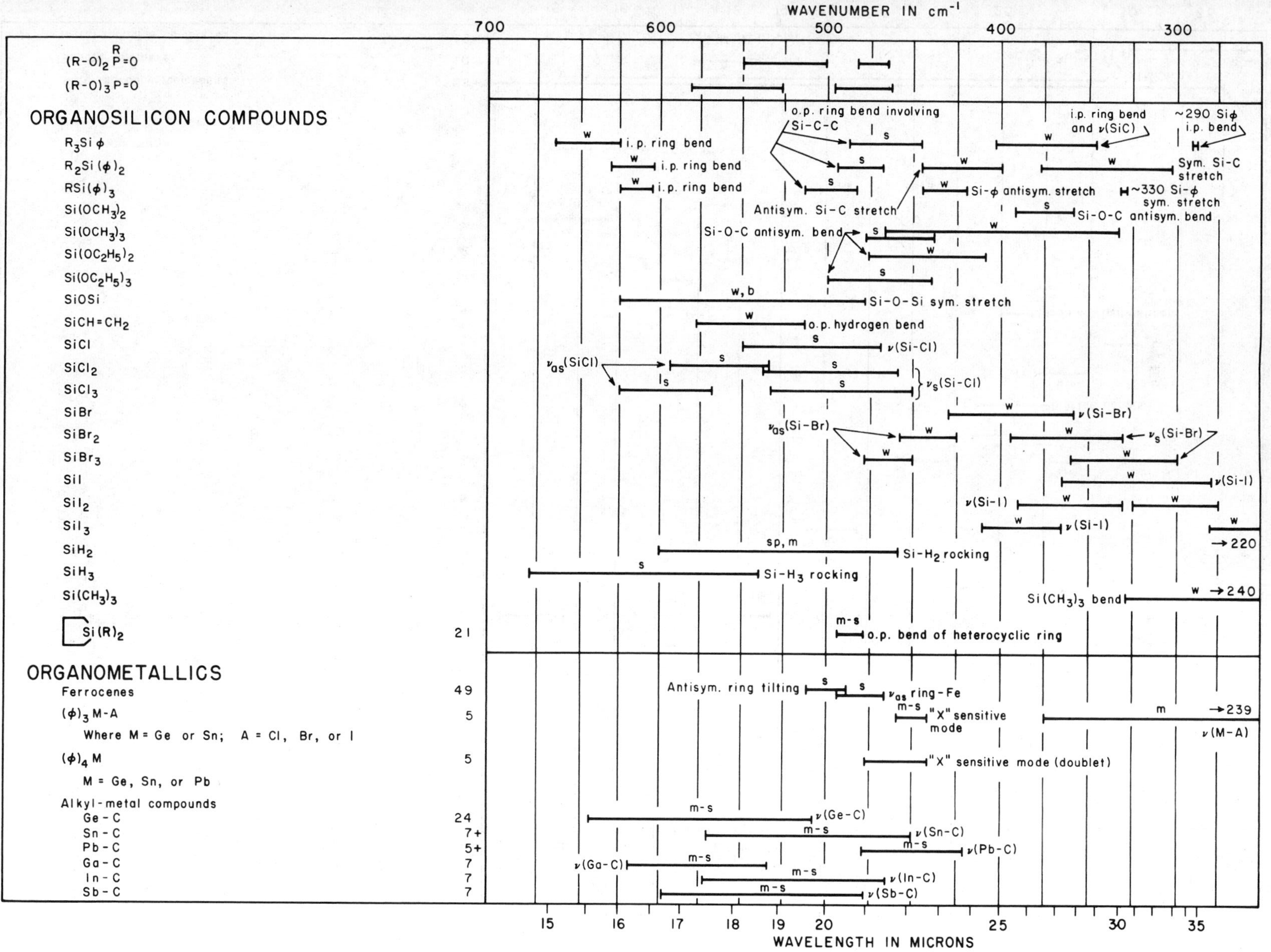

CHART OF CHARACTERISTIC FREQUENCIES BETWEEN ∽ 700–300 cm⁻¹

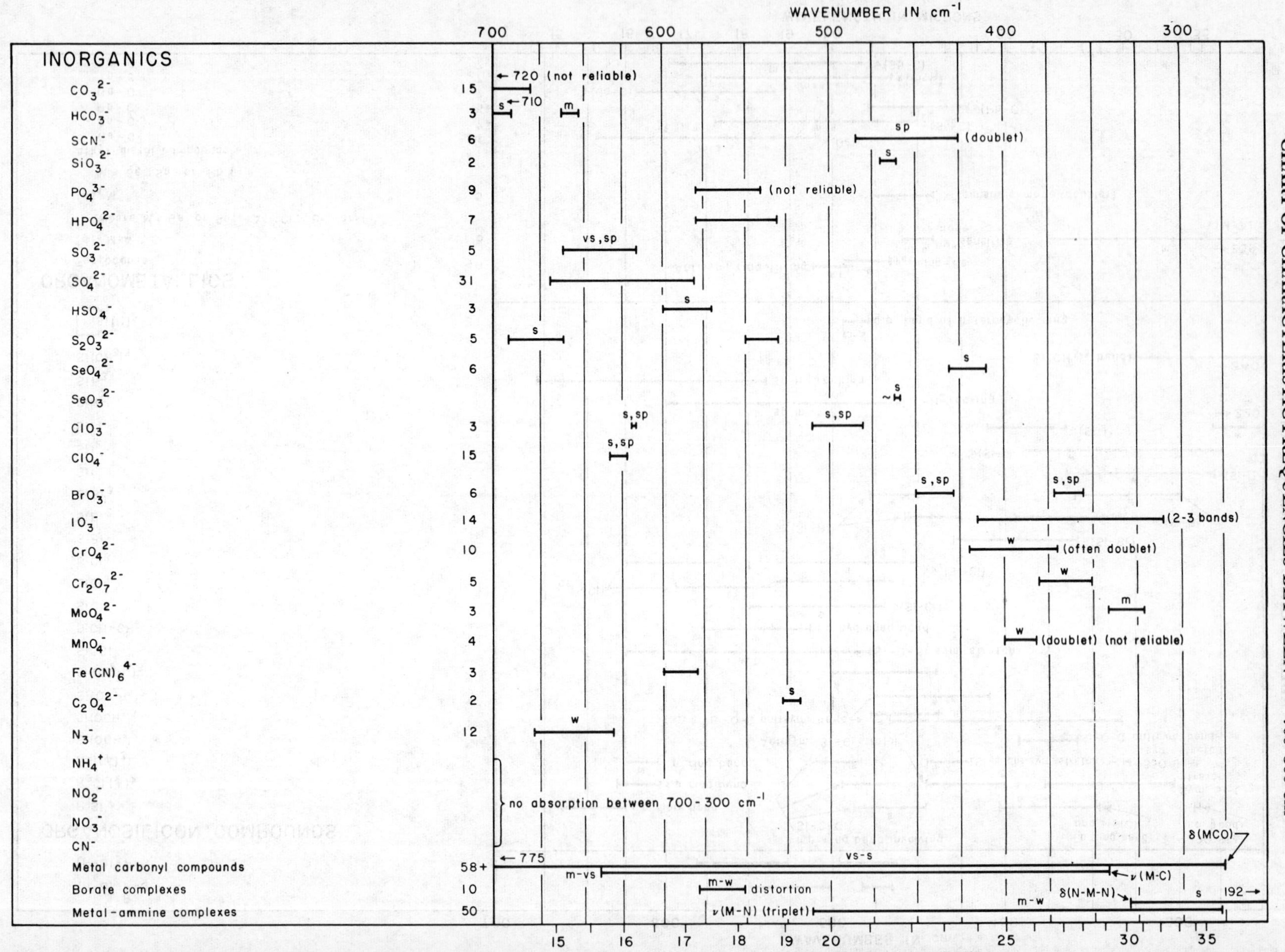

CHARACTERISTIC NMR SPECTRAL POSITIONS FOR HYDROGEN IN ORGANIC STRUCTURES

By permission from Erno Mohacsi, J. of Chemical Education, 41, 38 (1964)

This table is useful for quick qualitative determination of proton spectrum lines by providing a tabulation of line positions obtained using tetramethylsilane as an internal reference. The listing has been kept as simple as possible for this purpose. The proton spectrum lines are arranged according to the chemical shift relative to tetramethylsilane and are given in values of τ and σ. The purpose of this table is to supplement tables available in standard references and to summarize information available in the literature.

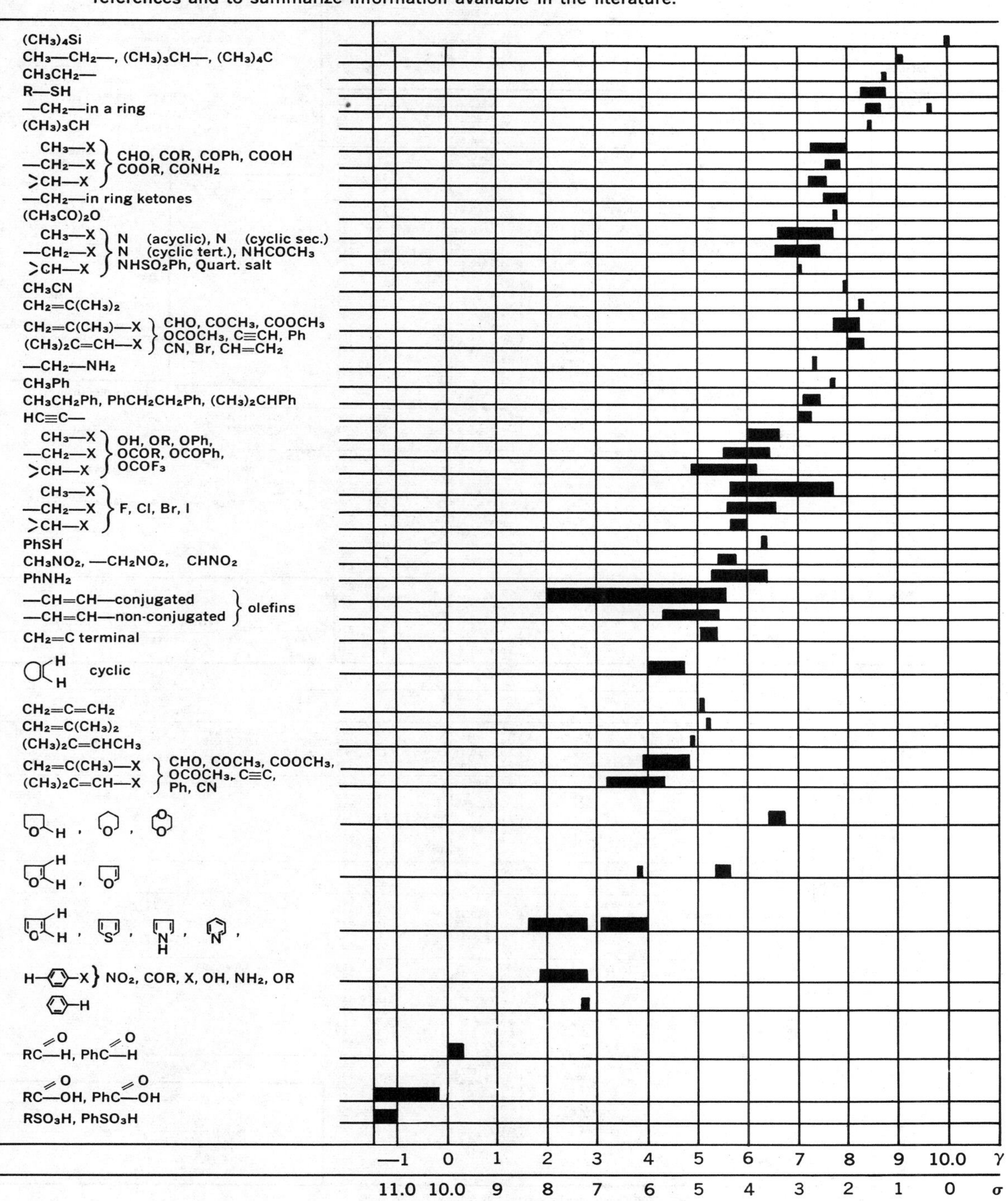

ELEMENTARY PARTICLES

Table S: STABLE PARTICLES. January, 1968.

From *Data on Particles and Resonant States* · REVIEW OF MODERN PHYSICS · JANUARY 1968

A. H. Rosenfeld, N. Barash-Schmidt, A. Barbaro-Galtieri, L. R. Price, Matts Roos, Paul Söding, W. J. Willis, C. G. Wohl

Quantities in italics have changed by more than one standard deviation since January, 1967.

Particle	$I^G(J^P)C$	Mass (MeV)	Mass difference (MeV)	Mean life (sec) $c\tau$(cm)	Mass^2 (GeV)^2	Decays: Partial mode	Decays: Fraction	Q(MeV)	p or p_{max} (MeV/c) ‡	Partial Rates (sec^{-1})	Magnetic Moment ($e\hbar/2m_pc$)	Decay Parameters† Measured: α	Measured: Φ(degree)	Derived: γ	Derived: Δ(degree)
γ	$0,1(1^-)^-$	0		stable	0	stable									
ν (ν_e, ν_μ)	$J=\frac{1}{2}$	0(<0.2 keV) 0(<2.1 MeV)		stable	0	stable									
e	$J=\frac{1}{2}$	0.511006 ±.000002		stable ($>2\times10^{21}$y)	0.000	stable	μ_e = 1.001159596 ±.000000023 $\frac{e\hbar}{2m_ec}$								
μ	$J=\frac{1}{2}$	105.659 ±.002		2.1983×10^{-6} ±.0008	0.011	$e\nu\bar{\nu}$	100	105	53						
						$e\gamma\gamma$	(<1.6)10^{-5}	105	53						
				$c\tau=6.592\times10^4$		3e	(<1.3)10^{-7}	104	53						
						$e\gamma$	(<6)10^{-9}	105	53						
	μ_μ = 1.0011666 ±.0000005	$\frac{e\hbar}{2m_\mu c}$	-33.920 ±.014												
$\pi^\pm$	$1(0^-)$	139.579 ±.014		2.604×10^{-8} ±.007, S=2.3*	0.019	$\mu\nu$	100 %	34	30						
				$c\tau=781$		$e\nu$	(1.24±0.03)10^{-4}	139	70						
				$(\tau^+-\tau^-)/\bar{\tau}$=(.4±.2)%		$\mu\nu\gamma$	(1.24±0.25)10^{-4}	34	30						
			4.6041 ±.0037	(test of CPT)		$\pi^0e\nu$	(1.03±0.07)10^{-8}	4	5						
						$e\nu\gamma$	(3.0 ±0.5)10^{-8}	139	70						
π^0	$1^-(0^-)^+$	134.975		d 0.89×10^{-16} ±.18, S=1.6*	0.018	$\gamma\gamma$	(98.83±0.04)%	135	67						
				$c\tau=2.67\times10^{-6}$		γe^+e^-	(1.17±0.04)%	134	67						
						$\gamma\gamma\gamma$	(< 5)10^{-6}	135	67						
						$e^+e^-e^+e^-$	b(3.47)10^{-5}	133	67						
$K^\pm$	$\frac{1}{2}(0^-)$	493.83 ±0.11		1.235×10^{-8} ±.005 S=2.1*	0.244	$\mu\nu$	c(63.58±0.29)%	388	236	c(51.49 ±0.27)10^6 S=1.1*					
				$c\tau$ = 370		$\pi\pi^0$	(20.94±0.28)% S=1.1*	219	205	(6.96 ±0.23)10^6					
				$(\tau^+-\tau^-)/\bar{\tau}$=(.09±.12)%		$\pi\pi^-\pi^+$	(5.57±0.04)%	75	126	(4.51 ±0.03)10^6					
				(test of CPT)		$\pi\pi^0\pi^0$	(1.70±0.05)%	84	133	(1.38 ±0.04)10^6					
				S=1.3*		$\mu\pi^0\nu$	(3.38±0.17)% S=1.6*	253	215	(2.74 ±0.14)10^6 S=1.6*					
						$e\pi^0\nu$	(4.83±0.12)% S=1.4*	358	229	(3.91 ±0.10)10^6 S=1.4*					
						$\pi\pi^\mp e^\pm\nu$	(3.8 ±0.8)10^{-5}	214	204						
			-3.93 ±0.17			$\pi\pi^\pm e^\mp\nu$	(<2)10^{-6}	214	204						
						$\pi\pi^\mp\mu^\pm\nu$	(≤1.4)10^{-5}	109	151						
						$\pi\pi^\pm\mu^\mp\nu$	(<3)10^{-6}	109	151						
						$e\nu$	(1.24±0.40)10^{-5}	493	247						
						$\pi\pi^0\gamma$	(2.2 ±0.7)10^{-4}	219	205						
						$\pi\pi^+\pi^-\gamma$	(10 ±4)10^{-5}	75	126						
						$\pi e\nu\gamma$	(6 ±4)10^{-4}	354	227						
						πe^+e^-	(<1.1)10^{-6}	353	227						
						$\pi\mu^+\mu^-$	(<2.4)10^{-6}	143	172						
						$\pi\gamma\gamma$	(<1.1)10^{-4}	354	227						
K^0	$\frac{1}{2}(0^-)$	497.75 ±0.18		50%K_{short}, 50%K_{long}											
K^0_S	$\frac{1}{2}(0^-)$			0.874×10^{-10} ±.011 S=1.3*	0.248	$\pi^+\pi^-$	(68.4 ±1.0)%	219	206	(0.783±0.015)10^{10}					
			$-0.48\times\frac{1}{\tau_S}$ ±0.02	$c\tau$ = 2.61		$\pi^0\pi^0$	(31.6 ±1.0)%	228	209	(0.362±0.012)10^{10}					
K^0_L	$\frac{1}{2}(0^-)$			5.30×10^{-8} ±.13	0.248	$\pi^0\pi^0\pi^0$	c(25.5 ±1.9)%	93	139	c(4.81 ±0.41)10^6					
				$c\tau$ = 1593		$\pi^+\pi^-\pi^0$	(*12.1 ± 0.4*)%	84	133	(2.29 ±0.09)10^6					
						$\pi\mu\nu$	(27.3 ±1.3)%	253	216	(5.15 ±0.26)10^6					
						$\pi e\nu$	(*35.1 ±1.4*)%	358	229	(6.62 ±0.29)10^6					
						$\pi^+\pi^-$	(0.149±0.006)%	219	206	(0.028±0.001)10^6					
						d) $\pi^0\pi^0$	still uncertain	228	209						
						$\pi^+\pi^-\gamma$	(< 0.3)%	219	206						
						d) $\gamma\gamma$	still uncertain	498	249						
						$e\mu$	(< 0.8)10^{-5}	392	238						
						$\mu^+\mu^-$	(< 1.5)10^{-6}	287	225						
						e^+e^-	(< 1.7)10^{-5}	497	249						
η	$0^+(0^-)^+$	548.8 ±0.6		Γ=(2.3±0.5)keV	Neutral decays 71.0%	$\gamma\gamma$	(*42 → 34*)% (d)	549	274						
						$\pi^0\gamma\gamma$	(*1 → 19*)% (d)	414	258						
						$3\pi^0$	(*28 → 18*)% (d)	144	179						
					Charged decays 29.0%	$\pi^+\pi^-\pi^0$	(*23.4 ± 1.1*)%	135	174						
						$\pi^+\pi^-\gamma$	(5.5 ±0.5)%	269	236						
						$\pi^0e^+e^-$	(< 0.04)%	413	258						
						$\pi^+\pi^-e^+e^-$	(0.1 ±0.1)%	268	236						
p	$\frac{1}{2}(\frac{1}{2}^+)$	938.256 ±0.005	-1.2933 ±.0001	stable ($>6\times10^{27}$y)	0.880						2.792763 ±.000030				
n	$\frac{1}{2}(\frac{1}{2}^+)$	939.550 ±0.005		(1.01±0.03)10^3 $c\tau=3.03\times10^{13}$	0.882	$pe^-\nu$	100 %	1	1		-1.913148 ±.000066	$\frac{g_A}{g_V}$ = -1.198 ± .022			
Λ	$0(\frac{1}{2}^+)$	1115.50 ±0.08		2.52×10^{-10} ±.04 S=1.4*	1.245	$p\pi^-$	(65.3 ±1.2)%	38	100		-0.73	0.647±0.016	(-6±7)°	0.75	(7±8)°
				$c\tau$ = 7.61		$n\pi^0$	(34.7 ±1.2)%	41	104		±.16	0.71 ±0.18	S=1.2*		
						$pe\nu$	(0.88±0.15)10^{-3}	177	163		S=1.2*	0.06 ±0.19	For $\Lambda\to pe\nu$, $\frac{g_A}{g_V}=-1.14^{+0.23}_{-0.33}$†		
						$p\mu\nu$	(1.35±0.60)10^{-4}	72	131						
Σ^+	$1(\frac{1}{2}^+)$	1189.47 ±0.08		0.810×10^{-10} ±.013	1.412	$p\pi^0$	(52.8 ±1.5)%	116	189		2.5	-.955±0.070 ← S=1.1*			
				$c\tau$ = 2.43		$n\pi^+$	(47.2 ±1.5)%	110	185		±.7	+.017±0.037	(180±30)°	-0.99	(0±85)°
						$p\gamma$	(1.9 ±0.4)10^{-3}	251	225						
						$n\pi^+\gamma$	(≈1.)10^{-3}	110	185						
						$\Lambda e^+\nu$	(2.2 ±0.7)10^{-5}	73	72						
				$\frac{\Gamma(\Sigma^+\to\ell^+n\nu)}{\Gamma(\Sigma^-\to\ell^-n\nu)}=.04^{+.04}_{-.02}$ ←		$n\mu^+\nu$	(< 0.7)10^{-4}	144	202						
			-7.95 ±.11			$ne^+\nu$	(< 0.3)10^{-4}	249	224						
Σ^0	$1(\frac{1}{2}^+)$	1192.54 =0.10		$<1.0\times10^{-14}$	1.422	$\Lambda\gamma$	100 %	77	75						
			-4.88 ±.06	$c\tau<3\times10^{-4}$		Λe^+e^-	b(5.45)10^{-3}								
Σ^-	$1(\frac{1}{2}^+)$	1197.41 ±0.09		1.66×10^{-10} ±.03 S=1.3*	1.434	$n\pi^-$	100 %	118	193			-.06 ±0.05	(22±30)°	0.90	
				$c\tau$ = 4.95		$ne^-\nu$	(1.25±0.17)10^{-3}	257	230						
						$n\mu^-\nu$	(0.62±0.12)10^{-3}	152	210			S=1.6*			
						$\Lambda e^-\nu$	(0.66±0.11)10^{-4}	81	79			For $\Sigma^-\to\Lambda e\nu$, $\frac{g_V}{g_A}$ = 0.3 ± 0.3†			
						$n\pi^-\gamma$	(≈1)10^{-3}	118	193						
Ξ^0	$\frac{1}{2}(\frac{1}{2}^+)$	1314.9 ±0.8		2.9×10^{-10} ±.4 S=1.2*	1.728	$\Lambda\pi^0$	100 %	64	135			-.33 ±0.10			
				$c\tau$=8.85		$p\pi^-$	(< 0.5)%	237	299						
						$pe^-\nu$	(< 0.6)%	376	323						
						$\Sigma^+e^-\nu$	(< 0.7)%	125	119						
			-6.3 ±.7			$\Sigma^-e^+\nu$	(< 0.6)%	117	112						
						$\Sigma^+\mu^-\nu$	(< 0.7)%	20	64						
						$\Sigma^-\mu^+\nu$	(< 0.6)%	12	49						
						$p\mu^-\nu$	(< 0.6)%	271	309						
Ξ^-	$\frac{1}{2}(\frac{1}{2}^+)$	1321.3 ±0.2		1.73×10^{-10} ±.05	1.746	$\Lambda\pi^-$	100 %	66	139			-.402±.051	(6±8)°		
				$c\tau$=5.20		$\Lambda e^-\nu$	($0.90^{+0.71}_{-0.43}$)10^{-3}	205	190			S=1.4*	S=1.3*		
						$n\pi$	(< 5)10^{-3}	242	303						
						$\Lambda\mu^-\nu$	(< 1.2)%	100	163						
						$\Sigma^0e^-\nu$	(< 0.3)%	128	122						
						$\Sigma^0\mu^-\nu$	(< 0.5)%	23	70						
						$ne^-\nu$	(< 1.0)%	381	327						
Ω^-	$0(\frac{3}{2}^+)$? ?	1672±1		$1.1^{+0.6}_{-0.5}\times10^{-10}$	2.795	$\Xi^0\pi^-$		217	293						
						$\Xi^-\pi^0$		216	289						
				$c\tau$=3.3		ΛK^-		63	210						

General Atomic and Nuclear Constants[a]

- N $= 6.02252\times10^{23}$ mole^{-1} (based on $A_C{}^{12}=12$)
- c $= 2.997925\times10^{10}$ cm sec^{-1}
- e $= 4.80298\times10^{-10}$ esu $= 1.60210\times10^{-19}$ coulomb
- 1 MeV $= 1.60210\times10^{-6}$ erg
- $\hbar$ $= 6.5819\times10^{-22}$ MeV sec $= 1.05449\times10^{-27}$ erg sec
- $\hbar c$ $= 1.9732\times10^{-11}$ MeV cm = 197.32 MeV fermi
- k $= 8.6171\times10^{-11}$ MeV deg^{-1} (Boltzmann const.)
- α $= e^2/\hbar c = 1/137.0388$
- m_e $= 0.511006$ MeV/c^2 $= 1/1836.10\ m_p$
- m_p $= 938.256$ MeV/c^2 $= 1836.10\ m_e = 6.721\ m_{\pi^\pm}$ $= 1.00727663\ m_1$ (where $m_1 = 1$ amu $= \frac{1}{12}m_{C^{12}}$ = 931.478 MeV/c^2)
- r_e $= e^2/m_ec^2 = 2.81777$ fermi (1 fermi $= 10^{-13}$ cm)
- $\lambda\!\!\!^{-}_e$ $= \hbar/m_ec = r_e\alpha^{-1} = 3.86144\times10^{-11}$ cm
- a_∞ Bohr $= \hbar^2/m_ee^2 = r_e\alpha^{-2} = 0.529167$ A (1 A $= 10^{-8}$ cm)
- $\sigma_{Thomson}$ $= \frac{8}{3}\pi r_e^2 = 0.66516\times10^{-24}$ cm^2 = 0.66516 barn
- μ_{Bohr} $= e\hbar/2m_ec = 0.578817\times10^{-14}$ Mev gauss^{-1}
- μ_{nucl} $= e\hbar/2m_pc = 3.1524\times10^{-18}$ Mev gauss^{-1}
- $\frac{1}{2}\omega_{cyclotron}$ (rad sec^{-1} gauss^{-1}) $= e/2m_ec = 8.79404\times10^6$; $= e/2m_pc = 4.7895\times10^3$

[a] Based mainly on E. R. Cohen and J. W. M. DuMond, Rev. Mod. Phys. 37, 537 (1965).

Note that there are some disagreements with the results of recent experiments.

CP violation parameters

$$\eta_{+-\,(00)} = \frac{A(K_L\to\pi^{+(0)}\pi^{-(0)})}{A(K_S\to\pi^{+(0)}\pi^{-(0)})} = |\eta_{+-\,(00)}|\,e^{i\Phi_{+-\,(00)}}$$

$|\eta_{+-}| = (1.89\pm0.09)\times10^{-3}$

$\Phi_{+-} = (65\pm11)^\circ$ S=1.1*

$|\eta_{00}|$ still uncertain, see data listings

†The definition of these quantities is as follows:

$$\alpha = \frac{2\,\mathrm{Re}(S^*P)}{|S|^2+|P|^2};\quad \beta = \frac{2\,\mathrm{Im}(S^*P)}{|S|^2+|P|^2};\quad \gamma = \frac{|S|^2-|P|^2}{|S|^2+|P|^2}$$

$\tan\Phi = \beta/\gamma \qquad \tan\Delta = -\beta/\alpha$

g_A/g_V defined by $M \propto \langle B_f|\gamma_\lambda(g_V - g_A\gamma_5)|B_i\rangle$

* S = Scale factor = $\sqrt{\chi^2/(N-1)}$ where N ≈ number of experiments. S should be ≈ 1. If S > 1, we have enlarged the error of the mean, δx, i.e., δx→S δx. This new convention is still inadequate, since if S > 1, the real uncertainty is probably even greater than Sδx. See text of January 1967 edition.

‡ In decays with more than two bodies, P_{max} is the maximum momentum that <u>any</u> particle can have.

b. Theoretical value, see also data card listings.

d. See note in data card listings.

Quantities in italics have changed by more than one standard deviation since January, 1967.

Particle	Symbol (J^P)	$I^G(J^P)C_n$ (estab.; ? = guess)	Mass M (MeV)	Width Γ (MeV)	$M^2 \pm \Gamma M$[(a)] $(GeV)^2$	Partial decay modes: Mode	Fraction %	Q (MeV)	p or p_{max}[(b)] (MeV/c)
$\pi^{\pm}(140)$ $\pi^0(135)$	$\pi(0^-)$	$1^-(0^-)+$	139.58 134.98		0.019 0.018	See Table S			
η(549)	$\eta(0^-)$	$0^+(0^-)+$	548.8 ±0.6	2.3 kev ±.5 keV	0.301 ±.000	all neutral	71 } See Table S		
						$\pi^+\pi^-\pi^0+\pi^+\pi^-\gamma$	29		
$\rho^{\pm}(765)$	$\rho(1^-)$	$1^+(1^-)-$	755-775	110-140	0.585 ±.095	ππ	≈ 100	491	359
						$\pi^{\pm}\pi^+\pi^-\pi^0$	< 0.2	212	247
						$\pi^+\pi^-\pi^+\pi^-$	< 0.15	212	247
						$\pi^0\gamma$	< 0.4	630	372
$\rho^0(765)$			760-780	90-150	0.590 ±.090	$\eta\pi^{\pm}$	< 0.8	82	146
						e^+e^-	.0055±.0007	769	385
						$\mu^+\mu^-$	.0058±.0010	559	370
ω(783)	$\phi'(1^-)$	$0^-(1^-)-$	783.3 ±0.7 S=1.9*	12.2 ±1.3	0.614 ±.009	$\pi^+\pi^-\pi^0$	≈ 90	369	328
						$\pi^+\pi^-$	seen	504	366
						$\pi^0\gamma$	9.5±0.9	648	380
						η+ neutral	< 1.5	234	199
						$\pi^+\pi^-\gamma$	< 5	504	366
						$\pi^0\pi^0\gamma$	< 1	513	368
						e^+e^-	$.0053^{+.0015}_{-.0020}$	782	392
						$\mu^+\mu^-$	< 0.10	572	377
η'(958) or X^0	$\eta'(0^-)$	$0^+(0^-)+$	958.3 ±0.8	<4	0.918 <.004	ηππ	67 ± 10	131	232
						$\pi^+\pi^-\gamma$(incl. $\rho^0\gamma$)	22 ± 3	679	458
						neutrals (excl. ηππ)	11 ± 9		
δ(962) Seen in only one experiment	?( )	≥1 ( )	962 ±5	<5	0.927 <.005	$\delta^{\pm}$ → 1 charged+neutral(s)	≈ 60		
						$\delta^{\pm}$ → ≥ 3 charged+neutral(s)	≈40		
H(990) existence in doubt	φ(A) ?	$0^-(1^+)-$??	≈990	≲80	0.98	3π	only mode seen	575	440
$\pi_V(1016)$ →KK̄	$\pi(0^+)$	$1^-(0^+)+$	1016 ±10 (if resonance)	*25* *±5*	1.032 ±.025	$K^{\pm}K^0$	only mode seen	24	110
						ηπ	< 70	328	342
Resonance, virtual bound state, or antibound state, still not distinguished.									
φ(1019)	$\phi(1^-)$	$0^-(1^-)-$	*1019.3* *±0.6* S = 1.5*	3.4 ±0.8	1.039 ±.004	K^+K^-	47.3±3.2	32	126
						K_LK_S	38.9±3.1	24	109
						$\pi^+\pi^-\pi^0$ (incl. ρπ)	13.8±4.3	605	462
						e^+e^- indic. seen,	< 0.2	1018	510
						$\mu^+\mu^-$ seen,	< 0.5	808	499
$\eta_V(1070)$ $\to K_SK_S$	$\eta(0^+)$	$0^+(0^+)+$	1069 ±10 S=2.3*	~ 80 see note (l)	1.14 ±.09	ππ	< 70	793	516
(Some data still favor large scattering length)						KK̄	>30	76	198
A1(1070) Existence in doubt	$\pi(1^+)$	$1^-(1^+)+$ ($J^P = 2^-$ not yet excluded)	1070 ±20	80 ±35	1.12	3π (see note (n))	≈ 100	651	488
						KK̄	< 0.25, $G=(-1)^{\ell+I}$ forbids this		
B(1220)	ρ(A) ?	$1^+(1^+)-$?? ($J^P = 2^-$ not excluded)	1220 ±11	129 ±14 S=1.3*	1.46 ±.14	ωπ	≈ 100	297	339
f(1260)	$\eta'(2^+)$	$0^+(2^+)+$	1263 ± 10	*141* *±13* *S = 2.3**	1.57 ±.15	ππ	large	984	616
						$2\pi^+2\pi^-$	< 4	705	552
						KK̄ indic. seen,	< *2.5*	267	388
D(1285)	η'(A)	$0^+(1^+)+$	1285 ±4	32 ±8	1.65 ±.04	KK̄π (mainly $\pi_V(1016)\pi$)	only mode seen	154	304
($J^P = 0^-, 2^-$ not excluded)						$K^*\bar{K}+\bar{K}^*K$		-100	
						πρ	not seen	256	356

Quantities in italics have changed by more than one standard deviation since January, 1967.

Particle	Symbol (J^P)	$I^G(J^P)C_n$ (estab.; ? = guess)	Mass M (MeV)	Width Γ (MeV)	$M^2 \pm \Gamma M$[(a)] $(GeV)^2$	Partial decay modes: Mode	Fraction %	Q (MeV)	p or p_{max}[(b)] (MeV/c)
A2(1300) Some evidence for > 1 meson	$\pi(2^+)$	$1^-(2^+)+$	1305 ± 10 S=2.2*	90 ± 10 S=1.2*	1.70 ±.12	ρπ	*85.1 ± 2.2*	395	410
						KK̄	4.0±1.1 S=1.1*	313	424
						ηπ	*10.5 ± 2.1*	617	526
						η'π	< *10*	207	275
						$\pi^+\pi^-\pi^0$(excl. ρπ)	< 17	891	615
E(1420)	η(A)	$0^+(0^-)+$	1424 ±6	71 ±10 S=1.2*	2.03 ±.10	$K^*\bar{K}+\bar{K}^*K$	50 ±10	38	157
						$\pi_V(1016)\pi$	50 ±10	268	328
						ππη	< 88	596	569
($J^P = 1^+$ not yet excluded)						ππρ	not seen	384	455
f'(1514)	$\eta(2^+)$	$0^+(2^+)+$	1514 ±5	73 ±23 S=1.8*	2.29 ±.11	KK̄	72 ± 12	518	570
						$K^*\bar{K}+\bar{K}^*K$	10 ± 10	128	294
						ππ	< 14	1235	744
						ηππ	18 ± 10	686	624
						ηη	< 40	417	522
$\pi_A(1640)$ → 3π	π(A) ?	$1^-(A)+$?	1654 ±12 S=1.6*	109 ±30	2.73 ±.18	3π	appears dominant	1245	797
						[ρπ †	< 40]	745	637
						[fπ †	appears dominant]	252	320
not yet a completely established resonance						KK̄	< 40	654	658
$\rho_V(1650)$ g→ 2π	ρ(V)	$1^+(V)-$	1660 ±17 S=1.4*	169 ±30	2.76 ±.28	2π	appears dominant	1385	820
						4π }	probably observed	1106	773
($J^P = 1^-, 3^-, \ldots$ with 3^- favored)						ρππ }		615	617
						KK̄ indication seen,	< 10	668	667
						ωπ indication seen		741	638

→The remaining data, --e. g., R, S, T, U mesons--in this mass region are too confused to tabulate.

Particle	Symbol (J^P)	$I(J^P)$	Mass M (MeV)	Width Γ (MeV)	$M^2 \pm \Gamma M$[(a)] $(GeV)^2$	Mode	Fraction %	Q (MeV)	p or p_{max}[(b)] (MeV/c)
$K^+(494)$ $K^0(498)$	$K(0^-)$	$1/2(0^-)$	493.83 497.75		0.244 0.248	See Table S			
K*(890)	$K(1^-)$	$1/2(1^-)$	892.7 ±0.5 S=1.2*	49.2 ±1.0	0.797 ±.044	Kπ	≈ 100	259	289
$m_0 - m_{\pm} = 6.3 \pm 4.1$						Kππ	< 0.2	120	216
$K_A(1230)$ or C	K(A)	$1/2(1^+)$	1230	60	1.51 ±.07				
Not yet completely established. See footnote (p). $J^P = 2^-$ not yet ruled out.									
$K_A(1320)$	K(A)	$1/2(1^+)$	1320	60	1.74 ±.08				
$J^P = 2^-$ not yet completely ruled out.									
$K_V(1420)$	$K(2^+)$	$1/2(2^+)$	1418.6 ±3.2 S=1.3*	89.1 ±5.1 S=1.1*	2.012 ±.126	Kπ	51 ± 3 } S = 1.5*	787	615
						K*π	33 ± 3	388	413
						Kρ	12 ± 3	156	318
						Kω	2.4± 1	143	304
						Kη	2.1± 1	378	482
$K_A(1800)$ or L	K(A) ?	1/2(A) ?	1781 ±14 S=1.7*	72 ±24 S=1.3*	3.17 ±.13	Kπ	< 2.3	1167	825
						K*π	24 ± 8	772	670
						$K_V(1420)\pi$	16 ± 8	245	317
						Kρ	10 ± 6	532	632
						Remaining Kππ	45 ± 15	1032	807
						Kω	4.8± 2	519	624
						Kη	< 1	754	732

→ The following bumps, excluded above, are listed among the data cards: σ(410); ε(730); $A2_{I=2}(1320) \to \rho^-\pi^-$; ρρ(1410); $K_SK_S(1440)$; η(1600) → 4π; ρ(1700) → 4π, $K^*\bar{K}$; $R_{1-4}(1630\text{-}1830)$; η(1830) → 4π; φ(1830) → ωππ; S(1930); T(2200); $N\bar{N}_{I=0}(2380)$, U(2380); κ(725); $K_V(1080)$; $K_A(I{=}3/2)(1175)$; $K_A(I{=}3/2)(1265)$; $K_A(I{=}1/2)(1280)$; $K_V(I{=}1/2)(1660)$.

* Quoted error includes scale factor $S = \sqrt{\chi^2/(N-1)}$. See footnote to Table S.

† Square brackets indicate a subreaction of the previous (unbracketed) decay mode.

(a) ΓM is the half-width of the resonance when plotted against M^2.

(b) For decay modes into ≥ 3 particles p_{max} is the maximum momentum that any of the particles in the final state can have. The momenta have been calculated using the averaged central mass values, without taking into account the widths of the resonances.

BARYONS - January, 1968.

	Particle or resonance	I(J^P) —— = estab.	Beam π, K (BeV) (BeV/c)	Mass (MeV)	Γ (MeV)	$M^2 \pm \Gamma M$ (BeV^2)	Partial decay modes Mode	 Fraction (%)	 p or $p^{‡}_{max}$ (MeV/c)	$4\pi ƛ^2$ (mb)
N	p n	1/2(1/2⁺)		938.3 939.6		0.880 0.883		See Table S		
N	N' (1470)	1/2(1/2⁺) P_{11}	T=0.53πp p=0.66	1470	210	2.16 ±0.31	Nπ Nππ [Nσ][a]	65 35 [domin]	420	27.8
N (See footnote c.)	N(1518)	1/2(3/2⁻) D_{13}	T=0.62 p=0.75	1525	115	2.33 ±0.18	Nπ Nππ [Δ(1236)π][a] Nη	55 45 [domin] ~0.5	460 414 229 161	23.2
N (See footnote c.)	N(1550)	1/2(1/2⁻) S_{11}	T=0.66 p=0.79	1550	130	2.40 ±0.20	Nπ Nη Nππ	30 70 small	477 210 434	21.5
N	N(1680)	1/2(5/2⁻) D_{15}	T=0.88 p=1.02	1680	170	2.82 ±0.29	Nπ Nππ [Δ(1236)π][a] ΛK Nη	40 dom. inel. [?] <1.6 <2.5	567 533 365 218 379	15.2
N	N(1688)	1/2(5/2⁺) F_{15}	T=0.90 p=1.03	1690	130	2.86 ±0.22	Nπ Nππ [Δ(1236)π][a] ΛK Nη	65 dom. inel. [?] <.13 <1.5	574 540 374 234 390	14.9
N	N' (1710)	1/2(1/2⁻) S_{11}	T=0.94 p=1.07	1710	300	2.92 ±0.51	Nπ	80	587	14.2
N	N(2190)	1/2(7/2⁻) G_{17}	T=1.96 p=2.10	2200	250	4.84 ±0.55	Nπ	30	894	6.13
N	N(2650)	1/2(?⁻)	T=3.12 p=3.26	2650	360	7.02 ±0.95	Nπ	(J+1/2)x=0.45[b]	1154	3.67
N	N(3030) ‡‡	1/2(?)	T=4.26 p=4.40	3030	400	9.18 ±1.21	Nπ	(J+1/2)x=0.05[b]	1377	2.62
Δ (See footnote c.)	Δ(1236)	3/2(3/2⁺) P_{33}	T=0.195 p=0.304	(++) 1236.0 ±0.6	120 ±2	1.53 ±0.15	Nπ Nπ⁺π⁻	100 0	231 89	91.9
	$m_0 - m_{++}$ = 0.45±0.85	$m_- - m_{++}$ = 7.9±6.8								
Δ (See footnote c.)	Δ(1640)	3/2(1/2⁻) S_{31}	T=0.81 p=0.94	1640	180	2.69 ±0.30	Nπ Nππ	30 dom. inel.	540	16.8
Δ	Δ(1920)	3/2(7/2⁺) F_{37}	T=1.41 p=1.54	1950	220	3.80 ±0.43	Nπ ΣK	40 seen	741 453	8.91
Δ	Δ(2420)	3/2(11/2⁺)	T=2.50 p=2.64	2420	310	5.86 ±0.75	Nπ	11	1024	4.67
Δ	Δ(2850)	3/2(?⁺)	T=3.71 p=3.85	2850	400	8.12 ±1.14	Nπ	(J+1/2)x=0.25[b]	1266	3.05
Δ	Δ(3230)	3/2(?)	T=4.94 p=5.08	3230	440	10.4 ±1.4	Nπ	(J+1/2)x=0.05[b]	1475	2.24
Z →	Z_0(1865)	0(?)	p=1.15 K⁺p	1865	180	3.47 ±0.34	NK	(J+1/2)x=0.35[b]	579	14.6
	Resonance interpretation not established.									
Λ	Λ	0(1/2⁺)		1115.5		1.24		See Table S		
Λ	Λ(1405)	0(1/2⁻) S_{01}	p<0 K⁻p	1405	50	1.97 ±0.07	Σπ	100	140	
Λ	Λ(1520)	0(3/2⁻) D_{03}	p=0.392	1518.8 ±1.5	16 ±2	2.31 ±0.02	NK̄ Σπ Λππ	45±4 }S=1.8* 45±4 10±1	235 258 251	83.6
Λ	Λ' (1670)	0(1/2⁻) S_{01}	p=0.74	1670	18	2.79 ±0.03	NK̄ Λη	K⁻p→Λη seen	410 66	28.5
Λ	Λ' (1690)	0(3/2⁻) D_{03}	p=0.78	1690	45	2.86 ±0.08	NK̄ Σπ	20 58	429 403	26.1
Λ	Λ(1815)	0(5/2⁺) F_{05}	p=1.05	1816 ±2 S=1.3*	74 ±5	3.30 ±0.14	NK̄ Σπ Σ(1385)π Λη	63 11 11 ~ 1	538 500 359 346	16.7
Λ →	Λ(1830)	0(5/2⁻) D_{05}	p=1.08	1827	76	3.34 ±0.14	NK̄ Σπ	8 42	547 508	16.0
Λ	Λ(2100)	0(7/2⁻) G_{07}	p=1.68	2100	140	4.41 ±0.29	NK̄ Σπ Λη ΞK Λω	33 4 < 3 ~ 1 <10	748 699 617 483 443	8.68
Λ	Λ(2350)	0(?)	p=2.29	2350	210	5.52 ±0.49	NK̄	(J+1/2)x=0.7[b]	913	5.85
	Seen in total c. s.									
Σ	Σ	1(1/2⁺)		(+)1189.5 (0)1192.5 (-)1197.4		1.41 1.42 1.43		See Table S		
Σ	Σ(1385)	1(3/2⁺) P_{13}	p<0 K⁻p	(+)1382.2±0.8 S=1.6*	(+)37±3 S=2.1*	1.92 ±0.05	Λπ Σπ	91±3 9±3 S=1.4*	208 117	
	S=4.8* ! ←→(-) 1388.0±3.0 (-)38±8, S=3.7* !									
Σ	Σ(1660)	1(3/2⁻) D_{13}	p=0.72	1660	50	2.76 ±0.08	Λ(1405)π NK̄	large small for both	197 400	29.9
	Decay modes of these two states not separated yet.									
Σ	Σ(1690)	1 (? ?)	p=0.80	1690	120	2.89 ±0.19	Λπ Σπ	large not disentangled	455 395	25.1
Σ	Σ(1770)	1(5/2⁻) D_{15}	p=0.95	1767 ±4 S=1.5*	95 ±12 S=2.3*	3.13 ±0.16	NK̄ Λπ Λ(1520)π Σ(1385)π Ση Σπ	46 15 15 14 0.5 ~1	497 519 190 317 140 463	19.4
Σ →	Σ(1910)	1(5/2⁺) F_{15}	p=1.25	1910	60	3.65 ±0.11	NK̄ Λπ Σπ	8 10 3	612 619 568	12.9
	A small effect. Not established beyond all question.									
Σ	Σ(2030)	1(7/2⁺) F_{17}	p=1.52	2030	120	4.12 ±0.24	NK̄ Λπ Σπ ΞK	11 36 9 < 2	700 700 652 412	9.92
Σ	Σ(2250)	1(?)	p=2.04	2250	200	5.06 ±0.45	NK̄	(J+1/2)x=0.3[b]	849	6.76
	Seen in total c. s.									
Σ	Σ(2455)	1(?)	p=2.57	2455	~140	6.03 ±0.34	NK̄	(J+1/2)x=0.26[b]	979	5.08
	Seen in total c.s.									
Σ →	Σ(2595)	1(?)	p=2.95	2595	~ 140	6.73 ±0.36	NK̄	(J+1/2)x=0.26[b]	1064	4.30
	Seen in total c.s.									
Ξ	Ξ	1/2(1/2⁺)		(0)1314.7 (-)1321.2		1.73 1.75		See Table S		
Ξ	Ξ(1530)	1/2(3/2⁺) p-wave		(0)1528.9±1.1 (-)1533.8±1.9	7.3 ±1.7	2.34 ±0.01	Ξπ	100	145	
Ξ →	Ξ(1815)	1/2(?)		1815 ±3	16 ±8 S=2.2*	3.29 ±0.03	ΛK̄ Ξπ Ξππ [Ξ(1530)π][a] ΣK̄	~65 ~ 10 ~25 [~20] <2	391 409 351 220 299	
	Branching ratios poorly known.									
Ξ	Ξ(1930)	1/2(?)		1930	140	3.72 ±0.27	Ξπ ΛK̄	seen	501 504	
	Needs confirmation.									
Ω^-	Ω^-	0(3/2⁺)		1672		2.80		See Table S		

→ at left of Table indicates a candidate that has been omitted because the evidence for the existence of the effect and/or for its interpretation as a resonance is open to considerable question. See listings for information on the following: $N_?$(3245), N(3690), Z_1^*(1900), Λ(1860) F_{07}, Σ(1780), Σ(3000), and Ξ(1705).

* Quoted error includes an S(scale) factor. See footnote to Table S.

‡ For decay modes into ≥ 3 particles p_{max} is the maximum momentum that any of the particles in the final state can have. The momenta have been calculated using the averaged central mass values, without taking into account the widths of the resonances.

a. Square brackets indicate a sub-reaction of the previous unbracketed decay mode.

b. J is not known; x is Γ_{el}/Γ.

ABBREVIATIONS AND SYMBOLS

Abbreviations

The following list of abbreviations is intended to cover those in common use in chemistry and physics. Symbols are presented in a separate list following the abbreviations.

A.	Acre
Å	Ångström unit
a	Are
a.	Acid
abs.	Absolute
abt.	About
a.c.	Alternating current
acet.	Acetone
acet. a.	Acetic acid
al.	Alcohol
alk.	Alkali
alt.	Altitude
amal.	Amalgam; amalgamated
amor. *or* amorph.	Amorphous
amp.	Ampere
anh.	Anhydrous
antilog	Antilogarithm
ap.	Apothecaries'
appr.	Approximately
aq.	Aqua; aqueous; water
aq. reg.	Aqua regia
asym.	Asymmetrical
atm. *or* atmos.	Atmosphere (atmospheric)
At. No.	Atomic number
At. Wt.	Atomic weight
aux.	Auxiliary
Av.	Average
av. *or* avoir.	Avoirdupois
bar.	Barometer
bbl.	Barrel
bd.	Board
Bé	Beaumé (degrees)
B.G.	Birmingham gauge (hoop and sheet)
b.h.p.	Brake horse power
bl.	Blue
blk.	Black
B.M.	Board measure
b.p.	Boiling point
br.	Brown
BTU	British thermal unit
bu.	Bushel
B.W.G.	Birmingham wire gauge
bz.	Benzene
C	Centigrade
c	Carat; centi-
c.	Cold
ca	Candle
ca.	*Circa*, about; approximately
cal.	Calorie (gram)
cc. *or* c.c.	Cubic centimeter
cd.	Cord
c. cm	Cubic centimeter
Cent.	Centigrade
centi-	Prefix meaning 1/100
cf.	*Confer*, compare
c.f.m.	Cubic foot per minute
cgs	Centimeter-gram-second system of units
cgse	Cgs electrostatic system
cgsm	Cgs electromagnetic system
ch.	Chain
chl.	Chloroform
cir.	Circular
circum.	Cirumference
cl	Centiliter
cm	Centimeter
cm^2	Square centimeter
cm^3	Cubic centimeter
c.m.	Circular mil
coef.	Coefficient
colog	Cologarithm
colorl.	Colorless
comm'l	Commercial
conc.	Concentrated
cond.	Condensing
const.	Constant
cos	Cosine
cos^{-1}	Arc or angle whose cosine is...; anticosine of; inverse cosine of
cosec	Cosecant
cosh	Hyperbolic cosine
cosh^{-1}	Inverse hyperbolic cosine
cot	Cotangent
cot^{-1}	Arc or angle whose cotangent is . . .
coth	Hyperbolic cotangent
coth^{-1}	Inverse hyperbolic cotangent
covers	Coversed sine
c.p.	Candle power; circular pitch; center of pressure
cry. *or* cryst.	Crystalline; crystals
csc	Cosecant
csc^{-1}	Arc or angle whose cosecant is...
csch	Hyperbolic cosecant
csch^{-1}	Inverse hyperbolic cosecant
CTU	Centigrade thermal unit
cu.	Cubic
cu. cm	Cubic centimeter
cu. ft.	Cubic foot
cu. in.	Cubic inch
cu. m	Cubic meter
cu. yd.	Cubic yard
cwt.	Hundredweight
cyl.	Cylinder
d	Derivative; deci-
d.	Decomposes; day
d.	Dextrorotary
d.c.	Direct current
dec.	Decomposes
deci-	Prefix meaning 1/10
def.	Definition (s)
deg	Thermometric degree; absolute C unless contrary is indicated
deka-	Prefix meaning 10
deliq.	Deliquescent
den. *or* dens.	Density
dg	Decigram
diam.	Diameter
dil.	Dilute
dissd.	Dissolved
dk	Deka-
dk.	Dark
dkg.	Dekagram
dkl	Dekaliter
dkm	Dekameter
dkm^2	Square dekameter
dkm^3	Cubic dekameter
dks	Dekastere
dl	Deciliter
dm	Decimeter
dm^2	Square decimeter
dm^3	Cubic decimeter
d.p.	Diametral pitch; double pole
dr.	Dram
dr. ap. *or* ʒ ap.	Dram, apothecaries'
dr. av. *or* ʒ av.	Dram, avoirdupois
dr. fl. *or* ʒ fl.	Dram, fluid
dr. t. *or* ʒt.	Dram, troy
ds	Decistere
dwt.	Pennyweight
efflor.	Efflorescent
e.g.	*Exempli gratia*, for example
e.h.p.	Effective horse power
E.L.	Elastic limit
em	Cgsm unit of quantity of electricity
emf *or* e.m.f.	Electromotive force
es	Electrostatic or cgse unit of quantity of electricity
etc.	*Et. cetera*, and so forth
eth.	Ether
eth. acet.	Ethyl acetate
et. seq.	*Et sequentes*, and the following
evap.	Evaporation
ex.	Excess
exp	Exponential function
exp.	Explodes
exsec	Exterior secant
F	Fahrenheit
f.	From
fahr.	Fahrenheit
fath.	Fathom
feath.	Feathery
f.h.p.	Friction horse power
fir.	Firkin
fl.	Fluid
fl. dr.	Dram, fluid
fl. oz.	Ounce, fluid
fluores.	Fluorescent
fps	Foot-pound-second system of units
fpse	Foot-pound-second electrostatic system
fpsm	Foot-pound-second electromagnetic system
F.S.	Factor of safety
ft.	Foot
ft.^{2}	Square foot
ft.^{3}	Cubic foot
ft.-lb.	Foot-pound
fur.	Furlong
G	Gravitation constant
g	Gram
g-cal. *or* g.-cal.	Gram calorie
gal.	Gallon
gel.	Gelatinous
gi.	Gill
glac.	Glacial
glit.	Glittering
glyc.	Glycerine
gm.	Gram
gr.	Gray; grain
grn.	Green
gyr.	Gyration
h	Hecto-
h.	Hot; hour
ha	Hectare
hecto-	Prefix meaning 100
hex.	Hexagonal
hg	Hectogram
hhd.	Hogshead
hl	Hectoliter
hm	Hectometer
hm^2	Square hectometer
hm^3	Cubic hectometer
hor. *or* horiz.	Horizontal
h.-p.	High-Pressure
HP *or* h.p.	Horse power
h.p.-hr.	Horse power-hour
hr.	Hour
hyg.	Hygroscopic
i.	Insoluble
ibid.	*Ibidem*, in the same place
i.e.	*Id est*, that is
ign.	Ignites
i.h.p.	Indicated horse power
in.	Indigo; inch
in.^{2}	Square inch
in.^{3}	Cubic inch
inc.	Inclusive
in.-lb.	Inch-pound
insol.	Insoluble
Int.	International
iso.	Isotropic
isom.	Isometric
isoth.	Isothermal
k	Kilo-
kg	Kilogram
kg-cal.	Kilogram-calorie
kg-m	Kilogram-meter
kilo-	Prefix meaning 1,000
kl	Kiloliter
km	Kilometer
km^2	Square kilometer
km^3	Cubic kilometer
kva.	Kilovolt-ampere
kw.	Kilowatt
kw.-hr.	Kilowatt-hour
l	Liter
l.	Long
l	Laevorotary
lat.	Latitude
lb.	Pound
lb. ap.	Pound, apothecaries'
lb. av.	Pound, avoirdupois
lb. t.	Pound, troy
leaf.	Leaflets
lgr.	Ligroin
li.	Link
lin.	Linear
liq.	Liquid
lim.	Limit
ln	Natural hyperbolic or Napierian logarithm
log *or* log.	Logarithm
$\log_e$	Logarithm to the base *e*; natural, hyperbolic or Napierian logarithm
$\log_{10}$	Common logarithm; logarithm to the base 10
long.	Longitude
lng.	Long
l.-p.	Low-pressure
lt.	Light
lust.	Lustrous
♏	Minim or drop
m	Meter; milli-
m^2	Square meter
m^3	Cubic meter
m.	Minute
m.	Meta-
max.	Maximum
med.	Medium
meth.	Methyl
meth. al.	Methyl alcohol
m.e.p.	Mean effective pressure
met.	Metallic
mg	Milligram
m.h.c.p.	Mean horizontal candle power
mi.	Mile
mic.	Microscopic
micro-	Prefix meaning 1/1,000,000 or 10^{-6}
micro-micro	Prefix meaning 10^{-12}
milli-	Prefix meaning 1/1,000
milli-micro-	Prefix meaning 10^{-9}
min *or* min.	Minute
min.	Minim; minimum; mineral
ml	Milliliter
m.l.h.c.p.	Mean lower hemispherical candle power
mm	Millimeter
mm^2	Square millimeter
mm^3	Cubic millimeter
mmf *or* m.m.f.	Magnetomotive force
mol.	Molecule
Mol. Wt.	Molecular weight
monocl.	Monoclinic
m.p.	Melting point
m.s.c.p.	Mean spherical candle power
myria-	Prefix meaning 10,000 or 10^4
mμ	Millimicron; millimicro-
N	Numeric; number (in mathematical tables)
n.	Normal
n	Refractive index
need.	Needles
o	Ortho-
Obs.	Observer
octahdr.	Octahedral
oil turp.	Oil of turpentine
or.	Orange
oz.	Ounce
oz. ap. *or* ℥ ap.	Ounce, apothecaries'
oz. av. *or* ℥ av.	Ounce, avoirdupois
oz. fl. *or* ℥ fl.	Ounce, fluid
oz. t. *or* ℥ t.	Ounce, troy
p	Para-
pa.	Pale
p. ct.	Per cent
perp.	Perpendicular
p.f.	Power factor
pk.	Peck
pl.	Plates
powd.	Powder
pr.	Prisms
precip. *or* p'p't'd	Precipitated
p. sol.	Partly soluble
pt.	Point; pint
purp.	Purple
pyr.	Pyridine
Q	Quantity
q	Quintal
qt.	Quart
q.v.	*Quod vide*, which see
R	Réaumur; radioactive mineral
rac	Racemic
rad	Radian, measure of angle
rad.	Radius
rd.	Rod
reg.	Regular
rev.	Revolution
rhbdr.	Rhombohedral
rhomb.	Rhombic or orthorhombic
R.M.S.	Square root of mean square
r.p.m.	Revolutions per minute
s	Stere
s.	Scruple; soluble; second
s. ap. *or* ℈	Scruple, apothecaries'
sat. *or* sat'd	Saturated
sc.	Scales
S.E.	Siemens unit
sec *or* sec.	Second (mean solar unless contrary is stated)
sec	Secant
sec^{-1}	Arc or angle whose secant is . . .
sech	Hyperbolic secant
sech^{-1}	Inverse hyperbolic secant
segm.	Segment
sh.	Short
sin	Sine
sin^{-1}	Arc or angle whose sine is . . .
sinh	Hyperbolic sine
sinh^{-1}	Inverse hyperbolic sine
sl.	Slightly
sm.	Small
sol.	Solution; soluble
soln.	Solution
sp.	Specific
specif.	Specification
sp. gr.	Specific gravity
sq.	Square
sq. ch.	Square chain
sq. ft.	Square foot
sq. in.	Square inch
sq. mi.	Square mile
sq. rd.	Square rod
sq. yd.	Square yard
std.	Standard
subl.	Sublimes
sym.	Symmetrical
t	Metric ton
t.	Troy
tab. *or* tabl.	Tablets
tan	Tangent
tan^{-1}	Arc or angle whose tangent is . . .
tanh	Hyperbolic tangent
tanh^{-1}	Inverse hyperbolic tangent
temp.	Temperature
tetr. *or* tetrag.	Tetragonal
tn.	Ton
tr.	Transition
tricl.	Triclinic
trig.	Trigonal
trim.	Trimetric
T.S.	Tensile strength
turp.	Turpentine
Tw.°	Degrees Twaddell, hydrometer scale
ult.	Ultimate
uns.	Unsymmetrical
U.S.	United States of America; universal system of lens apertures
v.	Very
v.	*Vide*, see
vel. *or* veloc.	Velocity
vers	Versed sine
vert	Vertical
visc.	Viscous
vol.	Volume
volt.	Volatilizes
w.	Water
wh.	White
wt.	Weight
yd.	Yard
yel.	Yellow
yr.	Year
μ	Micron; micro-
$\mu\mu$	Micromicron; micromicro-

SPELLING AND SYMBOLS FOR UNITS

From "Units of Weight and Measure"
L. B. Chisholm, National Bureau of Standards
Miscellaneous Publication 286 (May, 1967)

The spelling of the names of units as adopted by the National Bureau of Standards is that given in the list below. The spelling of the metric units is in accordance with that given in the law of July 28, 1866, legalizing the Metric System in the United States.

Following the name of each unit in the list below is given the symbol that the Bureau has adopted. Attention is particularly called to the following principles:

1. No period is used with symbols for units. Whenever "in" for inch might be confused with the preposition "in", "inch" should be spelled out.

2. The exponents "²" and "³" are used to signify "square" and "cubic," respectively, instead of the symbols "sq" or "cu," which are, however, frequently used in technical literature for the U. S. Customary units.

3. The same symbol is used for both singular and plural.

Some Units and Their Symbols

Unit	Symbol	Unit	Symbol	Unit	Symbol
acre	acre	fathom	fath	millimeter	mm
are	a	foot	ft	minim	minim
barrel	bbl	furlong	furlong	ounce	oz
board foot	fbm	gallon	gal	ounce, avoirdupois	oz avdp
bushel	bu	grain	grain	ounce, liquid	liq oz
carat	c	gram	g	ounce, troy	oz tr
Celsius, degree	°C	hectare	ha	peck	peck
centare	ca	hectogram	hg	pennyweight	dwt
centigram	cg	hectoliter	hl	pint, liquid	liq pt
centiliter	cl	hectometer	hm	pound	lb
centimeter	cm	hogshead	hhd	pound, avoirdupois	lb avdp
chain	ch	hundredweight	cwt	pound, troy	lb tr
cubic centimeter	cm^3	inch	in	quart, liquid	liq qt
cubic decimeter	dm^3	International		rod	rod
cubic dekameter	dam^3	Nautical Mile	INM	second	s
cubic foot	ft^3	Kelvin, degree	°K	square centimeter	cm^2
cubic hectometer	hm^3	kilogram	kg	square decimeter	dm^2
cubic inch	in^3	kiloliter	kl	square dekameter	dam^2
cubic kilometer	km^3	kilometer	km	square foot	ft^2
cubic meter	m^3	link	link	square hectometer	hm^2
cubic mile	mi^3	liquid	liq	square inch	in^2
cubic millimeter	mm^3	liter	liter	square kilometer	km^2
cubic yard	yd^3	meter	m	square meter	m^2
decigram	dg	microgram	μg	square mile	mi^2
deciliter	dl	microinch	μin	square millimeter	mm^2
decimeter	dm	microliter	μl	square yard	yd^2
dekagram	dag	micron	μm	stere	stere
dekaliter	dal	mile	mi	ton, long	long ton
dekameter	dam	milligram	mg	ton, metric	t
dram, avoirdupois	dr avdp	milliliter	ml	ton, short	short ton
				yard	yd

SYMBOLS, UNITS AND NOMENCLATURE IN PHYSICS

From Document U.I.P. 11 (S.U.N. 65–3)
International Union of Pure and Applied Physics
Used by permission of the Secretary

1. PHYSICAL QUANTITIES—GENERAL RECOMMENDATIONS.

1.1 The symbol for a ***physical quantity*** (french: 'grandeur physique', german: 'physikalische Grösse', american sometimes: 'physical magnitude') is equivalent to the product of the *numerical value* (or the measure), a pure number, and a *unit,* i.e.

physical quantity = numerical value × unit.

For dimensionless physical quantities the unit often has no name or symbol and is not explicitly indicated.

Examples: $E = 200$ erg, $F = 27$ N, $n = 1.55$ (for quartz), $\nu = 3 \times 10^8\ \text{s}^{-1}$

1.2 Symbols for physical quantities—General rules.

1. Symbols for physical quantities should be *single letters* of the latin or greek alphabet with or without modifying signs: subscripts, superscripts, dashes, etc.

Remark:

a. An exception to this rule consists of the two letter symbols, which are sometimes used to represent dimensionless combinations of physical quantities. If such a symbol, composed of two letters, appears as a factor in a product, it is recommended to separate this symbol from the other symbols by a dot or by brackets or by a space.

b. Abbreviations, i.e. shortened forms of names or expressions, such as p.f. for partition function, should not be used in physical equations. These abbreviations in the text should be written in ordinary roman type.

2. Symbols for physical quantities should be printed in *italic* (or *sloping*) *type.*

Remark:

It is recommended to consider as a guiding principle for the printing of indices the criterion: only indices which are symbols for physical quantities should be printed in italic (sloping) type. *Examples:*

Upright indices:	*Sloping indices:*
C_g (g = gas)	p in C_p
g_n (n = normal)	n in $\Sigma_n a_n \varphi_n$
μ_r (r = relative)	x in $\Sigma_x a_x b_x$
E_k (k = kinetic)	i, k in g_{ik}
χ_e (e = electric)	x in p_x

3. Symbols for vectors and tensors: To avoid the usage of subscripts it is often convenient to indicate vectors and tensors of the second rank by letters of a special type. The following choice is recommended:

a. Vectors should be printed in bold type, by preference bold italic (sloping) type, e.g. ***A, a.***

b. Tensors of the second rank should be printed in bold face sans serif type, e.g. $\mathsf{\mathbf{S}}$, $\mathsf{\mathbf{T}}$.

Remark:

When this is not available, vectors may be indicated by an arrow and tensors by a double arrow on top of the symbol.

1.3 Simple mathematical operations.

1. Addition and subtraction of two physical quantities are indicated by:

$a + b$ and $a - b$

2. Multiplication of two physical quantities may be indicated in one of the following ways:

ab $\quad a\ b$ $\quad a\,.\,b$ $\quad a \cdot b$ $\quad a \times b$

Remark: The various products of vectors and tensors may be written in the following ways:

scalar product of vectors $\boldsymbol{A}$ and $\boldsymbol{B}$:	$\boldsymbol{A}\,.\,\boldsymbol{B}$	$\boldsymbol{A} \cdot \boldsymbol{B}$
vector product of vectors $\boldsymbol{A}$ and $\boldsymbol{B}$:	$\boldsymbol{A} \wedge \boldsymbol{B}$	$\boldsymbol{A} \times \boldsymbol{B}$
dyadic product of vectors $\boldsymbol{A}$ and $\boldsymbol{B}$:	$\boldsymbol{AB}$	
scalar product of tensors $\mathsf{\mathbf{S}}$ and $\mathsf{\mathbf{T}}$ $(\Sigma_{i,k} S_{ik} T_{ki})$	$\mathsf{\mathbf{S}} : \mathsf{\mathbf{T}}$	
tensor product of tensors $\mathsf{\mathbf{S}}$ and $\mathsf{\mathbf{T}}$ $(\Sigma_k S_{ik} T_{kl})$	$\mathsf{\mathbf{S}} \cdot \mathsf{\mathbf{T}}$	$\mathsf{\mathbf{S}}\,.\,\mathsf{\mathbf{T}}$
product of tensor $\mathsf{\mathbf{S}}$ and vector $\boldsymbol{A}$ $(\Sigma_k S_{ik} A_k)$	$\mathsf{\mathbf{S}} \cdot \boldsymbol{A}$	$\mathsf{\mathbf{S}}\,.\,\boldsymbol{A}$

3. Division of one quantity by another quantity may be indicated in one of the following ways:

$\frac{a}{b}$ $\quad a/b$ $\quad a\ b^{-1}$

or in any other way of writing the product of a and b^{-1}.

These procedures can be extended to cases where one of the quantities or both are themselves products, quotients, sums or differences of other quantities.
If necessary brackets have to be used in accordance with the rules of mathematics. If the solidus is used to separate the numerator from the denominator and if there is any doubt where the numerator starts or where the denominator ends, brackets should be used.

Examples:

Expressions with a horizontal bar:	*Same expressions with a solidus:*
$\frac{a}{bcd}$	a/bcd
$\frac{2}{9}\sin kx,\ \frac{1}{2}RT$	$(2/9)\sin kx,\ (1/2)RT$ or $RT/2$
$\frac{a}{b} - c$	$a/b - c$
$\frac{a}{b-c}$	$a/(b-c)$
$\frac{a-b}{c-d}$	$(a-b)/(c-d)$
$\frac{a}{c} - \frac{b}{d}$	$a/c - b/d$

Remark: It is recommended that in expressions like:

$$\sin\{2\pi(x - x_0)/\lambda\} \qquad \exp\{(r - r_0)/\sigma\}$$
$$\exp\{-V(r)/kT\} \qquad \sqrt{(\epsilon/c^2)}$$

the argument should always be placed between brackets, except when the argument is a simple product of two quantities: e.g. sin kx. When the horizontal bar above the square root is used no brackets are needed.

2. UNITS—GENERAL RECOMMENDATIONS.

2.1 Symbols for units—General rules.

1. *Symbols for units* of physical quantities should be printed in *roman* (*upright*) *type.*

2. *Symbols for units* should not contain a final full stop and should remain unaltered in the plural, e.g.: 7 cm and *not* 7 cms.

3. *Symbols for units* should be printed in *lower case* roman (upright) type. However, the symbol for a unit, derived from a proper name, should start with a capital roman letter, e.g.: m (metre); A (ampere); Wb (weber); Hz (hertz).

2.2 Prefixes—General rules.

1. The following *prefixes* should be used to indicate decimal fractions or multiples of a unit

deci; *déci*	$(= 10^{-1})$	d			
centi; *centi*	$(= 10^{-2})$	c			
milli; *milli*	$(= 10^{-3})$	m	kilo; *kilo*	$(= 10^{3})$	k
micro; *micro*	$(= 10^{-6})$	μ	mega; *méga*	$(= 10^{6})$	M
nano; *nano*	$(= 10^{-9})$	n	giga; *giga*	$(= 10^{9})$	G
pico; *pico*	$(= 10^{-12})$	p	tera; *téra*	$(= 10^{12})$	T
femto; *femto*	$(= 10^{-15})$	f			
atto; *atto*	$(= 10^{-18})$	a			

2. The use of *double prefixes* should be avoided when single prefixes are available.

Not: mμs,	*but:* ns	(nanosecond)
Not: kMW	*but:* GW	(gigawatt)
Not: μμF,	*but:* pF	(picofarad)

3. When a prefix is placed before the symbol of a unit, the *combination of prefix and symbol* should be considered as *one new symbol*, which can be raised to a positive or negative power without using brackets.
Examples:

cm³ , mA² , μs⁻¹

Remark:

cm^3 *means always* $(0.01\ \mathrm{m})^3$ *but never* $0.01\ \mathrm{m}^3$
$\mu\mathrm{s}^{-1}$ *means always* $(10^{-6}\mathrm{s})^{-1}$ *but never* $10^{-6}\ \mathrm{s}^{-1}$

2.3 Mathematical operations.

1. *Multiplication* of two units may be indicated in one of the following ways:

   Nm N m N · m N . m

2. *Division* of one unit by another unit may be indicated in one of the following ways:

   $\frac{\mathrm{m}}{\mathrm{s}}$ m/s m s^{-1}

   or by any other way of writing the product of m and s^{-1}.
   Not more than one solidus should be used. *Examples:*

Not: cm/s/s,	*but:* cm/s² = cm s^{-2}
Not: 1 poise = 1 g/s/cm,	*but:* 1 poise = 1 g/s cm = 1 $\mathrm{g/s^{-1}\ cm^{-1}}$
Not: J/°K/mol,	*but:* J/°K mol = $\mathrm{J°K^{-1}\ mol^{-1}}$

3. NUMBERS.

1. *Numbers* should be printed in *upright type.*
2. The *decimal sign* between digits in a number should be a comma (,) or (but *only* in *English* texts) a point (.).
3. The *multiplication* sign between numbers should be a cross (×) or (but *only in non-English* texts) a centered dot, e.g. 2.3 × 3.4 or 2,3 · 3,4.
4. *Division* of one number by another number may be indicated in the following ways:

   $\frac{136}{273.15}$ 136/273.15

   or by writing it as the product of numerator and the inverse first power of the denominator. In such cases the number under the inverse power should always be placed between brackets.

   Remark: When the solidus is used and when there is any doubt where the numerator starts or the denominator ends, brackets should be used, as in the case of quantities (see 1.3.3).
5. To facilitate the reading of *long numbers,* the digits may be grouped in *groups of three,* but *no* comma or point should be used except for the decimal sign.
   Example: 2 573. 421 736.

4. SYMBOLS FOR CHEMICAL ELEMENTS, NUCLIDES AND PARTICLES.

1. *Symbols for chemical elements* should be written in *roman* (upright) *type.* The symbol is not followed by full stop.

   Examples: Ca C H He
2. The attached numerals specifying a *nuclide* are:

   mass number $^{14}\mathrm{N}_{2\ \mathrm{atoms/molecule}}$

   Remark: The atomic number may be placed as a left subscript, if desired. The right superscript position should be used, if required, for indicating a state of ionisation (e.g. $\mathrm{Ca^{2+}}$, $\mathrm{PO_4^{3-}}$) or an excited state (e.g. $^{110}\mathrm{Ag^m}$, He*).
3. *Symbols for particles and quanta.*
   It is recommended that the following notation be used:

proton:	p	pion	π
neutron:	n	K-meson	K
Λ-particle	Λ	electron	e
Σ-particle	Σ	muon	μ
Ξ-particle	Ξ	neutrino	ν
deuteron	d		
triton	t	photon	γ
α-particle	α		

   A nucleon (proton or neutron) is indicated by N
   It is recommended that the charge of particles may be indicated by adding the superscript +, − or 0.
   Examples:

   $\pi^+\ \pi^-\ \pi^0$, $\mathrm{p^+\ p^-}$, $\mathrm{e^+\ e^-}$.

If in connection with the symbols p and e no charge is indicated, these symbols should refer to the positive proton and the negative electron respectively.
The tilde ~ above the symbol of a particle is used to indicate the corresponding anti-particle.

Examples:

$$\tilde{n}, \tilde{\nu}, \sim p, \tilde{N}$$

5. QUANTUM STATES.

5.1 General rules.

A letter-symbol indicating the quantum state of a *system* should be printed in capital upright type.
A letter-symbol indicating the quantum state of a *single particle* should be printed in lower case upright type.

5.2 Atomic spectroscopy.

The letter-symbols indicating atomic quantum states are:

$L, l = 0$: S, s	$L, l = 4$: G, g	$L, l = 8$: L, l
$= 1$: P, p	$= 5$: H, h	$= 9$: M, m
$= 2$: D, d	$= 6$: I, i	$= 10$: N, n
$= 3$: F, f	$= 7$: K, k	$= 11$: O, o

A right hand subscript indicates the total angular momentum quantum number J or j.
A left hand superscript indicates the spin multiplicity $2S + 1$.

Example: $^2P_{3/2}$—state ($J = 3/2$, multiplicity 2)
$p_{3/2}$—electron ($j = 3/2$)

An atomic electron configuration is indicated symbolically by:

$$(nl)^{\kappa} \quad (n'l')^{\kappa'} \ . \ . \ .$$

Instead of $l = 0, 1, 2, 3$ . . . one uses the quantum state symbol s, p, d, f, . . .

Example: the atomic configuration: $(1s)^2\,(2s)^2\,(2p)^3$.

5.3 Molecular spectroscopy.

The letter symbols, indicating molecular electronic quantum states are in the case of *linear molecules:*

$$\Lambda, \lambda = 0: \Sigma, \sigma$$
$$= 1: \Pi, \pi$$
$$= 2: \Delta, \delta$$

and for *non-linear molecules*

A, a ; B, b ; E, e ; etc.

Remarks: A left hand superscript indicates the spin multiplicity. For molecules having a symmetry center the parity symbol g or u, indicating respectively symmetric or antisymmetric behaviour on inversion, is attached as a right hand subscript. A + or − sign attached as a right hand superscript indicates the symmetry as regards reflection in any plane through the symmetry axis of the molecules.

Examples: Σ_g^+, Π_u, $^2\Sigma$, $^3\Pi$, etc.

The letter symbols indicating the vibrational angular momentum states in the case of *linear molecules* are

$$l = 0: \Sigma$$
$$= 1: \Pi$$
$$= 2: \Delta$$

5.4 Nuclear spectroscopy.

The spin and parity assignment of a nuclear state is

$$J^{\pi}$$

where the parity symbol π is + for even and − for odd parity.

Examples: 3^+, 2^-, etc.

A shell model configuration is indicated symbolically by:

$$(nlj)^{\kappa} \quad (n'l'j')^{\kappa'}$$

where the first bracket refers to the proton shell and the second to the neutron shell. Negative values of κ or κ' indicate holes in a completed shell. Instead of $l = 0, 1, 2, 3,$ . . . one uses the quantum state symbols s, p, d, f, . . .

Example: the nuclear configuration: $(1\,d\,3/2)^3\,(1\,f\,7/2)^2$

5.5 Spectroscopic transitions.

1. The upper level and the lower level are indicated by ′ and ″ respectively.

Examples: $h\nu = E' - E''$ $\quad \sigma = T' - T''$

2. A spectroscopic transition should be indicated by writing the upper state first and the lower state second, connected by a dash inbetween.

Examples:

$^2P_{1/2} - {}^2S_{1/2}$	for an electronic transition
$(J', K') - (J'', K'')$	for a rotational transition
$v' - v''$	for a vibrational transition.

3. Absorption transition and emission transition may be indicated by arrows ← and → respectively.

Examples:

$^2P_{1/2} \rightarrow {}^2S_{1/2}$ emission from $^2P_{1/2}$ to $^2S_{1/2}$
$(J', K') \leftarrow (J'', K'')$ absorption from (J'', K'') to (J', K')

4. The difference Δ between two quantum numbers should be that of the upper state minus that of the lower state.

Example:

$$\Delta J = J' - J''$$

5. The indications of the branches of the rotation band should be as follows:

$\Delta J = J' - J'' = -2$:	O – branch
$= -1$:	P – branch
$= 0$:	Q – branch
$= +1$:	R – branch
$= +2$:	S – branch

6. NOMENCLATURE.

1. *Use of the word specific.*

The word 'specific' in English names for physical quantities should be restricted to the meaning 'divided by mass'. *Examples:*

specific volume	volume/mass
specific energy	energy/mass
specific heat capacity	heat capacity/mass

2. *Notation for covariant character of coupling.*

S	Scalar coupling	A	Axial vector coupling
V	Vector coupling	P	Psuedoscalar coupling
T	Tensor coupling		

3. *Abbreviated notation for a nuclear reaction.*

The meaning of the symbolic expression indicating a nuclear reaction should be the following:

initial nuclide (incoming particle(s) or quanta, outgoing particle(s) or quanta) final nuclide

Examples:

$^{14}N(\alpha,p)^{17}O$	$^{59}Co(n,\gamma)^{60}Co$
$^{23}Na(\gamma,3n)^{20}Na$	$^{31}P(\gamma,pn)^{29}Si$

4. *Character of transitions.*

Multipolarity of transition:

electric or magnetic monopole	E0 or M0
" " " dipole	E1 or M1
" " " quadrupole	E2 or M2
" " " octupole	E3 or M3
" " " 2^n-pole	E*n* or M*n*

parity change in transition:

transition *with* parity change:	yes
transition *without* parity change:	no

5. *Nuclide:* A species of *atoms,* identical as regards atomic number (proton number) and mass number (nucleon number) should be indicated by the word *nuclide,* not by the word isotope.
Different nuclides having the same atomic number should be indicated as *isotopes* or *isotopic nuclides.*
Different nuclides having the same mass number should be indicated as *isobars* or *isobaric nuclides.*

6. *Sign of polarization vector (Basel Convention).*

In nuclear interactions the positive polarization of particles with spin ½ is taken in the direction of the vector product

$$\boldsymbol{k}_i \times \boldsymbol{k}_0,$$

where $\boldsymbol{k}_i$ and $\boldsymbol{k}_0$ are the circular wave vectors of the incoming and outgoing particles respectively.

7. RECOMMENDED SYMBOLS FOR PHYSICAL QUANTITIES.

Remark:

(1) Where several symbols are given for one quantity, and no special indication is made, they are on equal footing.
(2) In general, no special attention is paid to the name of the quantity.

7.1 *Space and time.*

space coordinates; *coordonnées d'espace*	(x, y, z)	
position vector; *vecteur de position*	$\boldsymbol{r}$	
length; *longueur*	l	
breadth; *largeur*	b	
height; *hauteur*	h	
radius; *rayon*	r	
thickness; *épaisseur*	d, δ	
diameter; *diamètre:* $d = 2r$	d	
path; *parcours:* $L = \int \mathrm{d}s$	L, s	
area; *aire; superficie*	A, S	
volume; *volume*	$V, (v)$	
plane angle, *angle plan*	$\alpha, \beta, \gamma, \theta, \vartheta, \varphi$	
solid angle; *angle solide*	ω, Ω	
wave length; *longueur d'onde*	λ	
wave number; *nombre d'onde* $\sigma = 1/\lambda$	σ	*)
wave vector; *vecteur d'onde*	$\boldsymbol{\sigma}$	
circular wave number; *nombre d'onde circulaire:* $k = 2\pi/\lambda$	k	
circular wave vector; *vecteur d'onde circulaire*	$\boldsymbol{k}$	
attenuation coefficient; *constante d'affaiblissement:* $F(x) = \exp(-\alpha x) \cos \beta x$	α	
phase coefficient; *constante de phase*	β	
propagation coefficient; *constante de propagation:* $\gamma = \alpha + \mathrm{i}\beta$	γ	
time; *temps*	t	
period; *période*	T	
frequency; *fréquence:* $\nu = 1/T$	ν, f	
pulsatance; *pulsation:* $\omega = 2\pi\nu$	ω	**)
relaxation time; *temps de relaxation:* $F(t) = \exp(-t/\tau)$	τ	
damping coefficient; *coefficient d'amortissement:* $F(t) = \exp(-\delta t) \sin \omega t$	δ	
logarithmic decrement; *décrément logarithmique:* $\Lambda = T\delta = T/\tau$	Λ	
velocity; *vitesse:* $v = \mathrm{d}s/\mathrm{d}t$	u, v	
angular velocity; *vitesse angulaire:* $\omega = \mathrm{d}\varphi/\mathrm{d}t$	ω	
acceleration; *accélération:* $a = \mathrm{d}v/\mathrm{d}t$	a	
angular acceleration; *accélération angulaire:* $\alpha = \mathrm{d}\omega/\mathrm{d}t$	α	
acceleration of free fall; *accélération de la pesanteur*	g	
standard — —; — — *normale*	g_n	
speed of light in empty space; *vitesse de la lumière dans le vide*	c	
v/c	β	
relativistic coordinates; *coordonnées relativistes:*	$(x_0 x_1 x_2 x_3)$	
$x_0 = ct,\ x_1 = x,\ x_2 = y,\ x_3 = z,\ x_4 = \mathrm{i}ct$	$(x_1 x_2 x_3 x_4)$	

7.2 *Mechanics.*

mass; *masse*	m
(mass) density; *masse volumique:* $\rho = m/V$	ρ
relative density; *densité relative:* $d = \rho/\rho(H_2O)$	d
specific volume; *volume massique:* $v = V/m = 1/\rho$	v
reduced mass; *masse réduite:* $\mu = m_1 m_2/(m_1 + m_2)$	μ
momentum; *quantité de mouvement:* $\boldsymbol{p} = m\boldsymbol{v}$	$\boldsymbol{p}$
angular momentum; *moment cinetiqué:* $\boldsymbol{L} = \boldsymbol{r} \times \boldsymbol{p}$	$\boldsymbol{L}$
second moment of plane area; *moment quadratique d'une aire plane:* $I_{a,y} = \int x^2 \mathrm{d}x\mathrm{d}y$	I_a
second polar moment of plane area; *moment quadratique polaire d'une aire plane:* $I_p = \int (x^2 + y^2)\, \mathrm{d}x\mathrm{d}y$	I_p
moment of inertia; *moment d'inertie:* $I_z = \int (x^2 + y^2)\, \mathrm{d}m$	I, J
force; *force*	$\boldsymbol{F}$
torque, moment of a couple; *torque, moment d'un couple*	$\boldsymbol{T}$
weight; *poids*	$G, (W, P)$
moment of force; *moment d'une force*	$\boldsymbol{M}$
pressure; *pression*	p
normal stress; *tension normale*	σ
shear stress; *tension de cisaillement*	τ
gravitational constant; *constante de gravitation:* $F(r) = G\, m_1 m_2/r^2$	G
linear strain, relative elongation; *dilation linéique relative:* $\epsilon = \Delta l/l_0$	ϵ
modulus of elasticity, Young's modulus; *module d' élasticité:* $\sigma = E\, \epsilon$	E
shear strain; shear angle; *glissement unitaire*	γ
shear modulus; *module de torsion:* $\tau = G \operatorname{tg} \gamma$	G
volume strain, bulk strain; *dilation volumique relative:* $\theta = \Delta V/V_0$	θ
bulk modulus; *module de compression:* $p = -K\theta$	K
Poisson ratio, *rapport de Poisson*	μ, ν
viscosity; *viscosité*	$\eta, (\mu)$
kinematic viscosity; *viscosité cinématique:* $\nu = \eta/\rho$	ν
friction coefficient; *coefficient de frottement*	$\mu, (f)$
surface tension; *tension superficielle*	γ, σ
energy; *énergie*	E, W
potential energy; *énergie potentielle*	E_p, V, Φ
kinetic energy; *énergie cinétique*	E_k, T, K
work; *travail*	W, A

*) In molecular spectroscopy: $\bar{\nu}$. **) Also called angular frequency.

power; *puissance*	P
efficiency; *rendement*	η
Hamiltonian function; *fonction de Hamilton*	H
Lagrangian function; *fonction de Lagrange*	L
principal function of Hamilton; *fonction principale de Hamilton:* $W = \int L\mathrm{d}t$	W, S_p
characteristic function of Hamilton; *fonction caractéristique de Hamilton:* $S = 2\int T\mathrm{d}t$	S
generalized coordinate; *coordonnée généralisée*	q, q_i
generalized momentum; *moment généralisé*	p, p_i
action integral; *intégrale d'action:* $J = \oint p\mathrm{d}q$	J

7.3 *Molecular physics.*

number of molecules; *nombre de molécules*	N
number density of molecules; *nombre volumique de molécules* $n = N/V$	n
Avogadro constant; *constante d'Avogadro*	L, N_A
molecular mass; *masse moléculaire*	m
molecular velocity vector with components; *vecteur vitesse moléculaire et ses composantes*	$\boldsymbol{c}$, (c_x, c_y, c_z) $\boldsymbol{u}$, (u_x, u_y, u_z)
molecular position vector with components; *vecteur position moléculaire et ses composantes*	$\mathbf{r}$, (x, y, z)
molecular momentum vector with components; *vecteur quantité de mouvement moléculaire et ses composantes*	$\boldsymbol{p}$, (p_x, p_y, p_z)
average velocity; *vitesse moyenne*	$\mathbf{c}_0$, $\boldsymbol{u}_0$, $\langle \boldsymbol{c} \rangle$, $\langle \boldsymbol{u} \rangle$
average speed; *vitesse moyenne*	$\bar{c}$, $\bar{u}$, $\langle c \rangle$, $\langle u \rangle$
most probable speed; *vitesse la plus probable*	$\hat{c}$, $\hat{u}$
mean free path; *libre parcours moyen*	l
molecular attraction energy; *énergie d'attraction moléculaire*	ϵ
interaction energy between molecules i and j; *énergie d'interaction entre les molécules i et j*	φ_{ij}, V_{ij}
velocity distribution function; *fonction de distribution des vitesses:* $n = \int f\, \mathrm{d}c_x\mathrm{d}c_y\mathrm{d}c_z$	$f(c)$
Boltzmann function; *fonction de Boltzman*	H
generalized coordinate; *coordonnée généralisée*	q
generalized momentum; *moment généralisé*	p
volume in γ phase space; *volume dans l'espace* γ	Ω
thermodynamic temperature; *température thermodynamique*	T
Boltzmann constant; *constante de Boltzmann*	k
$1/kT$ (in exponential functions; *dans les fonctions exponentielles*)	β
molar gas constant; *constante molaire des gaz*	R
partition function; *fonction de partitions*	Q, Z
symmetry number; *facteur de symétrie*	s
diffusion coefficient; *coefficient de diffusion*	D
thermal diffusion coefficient; *coefficient de thermodiffusion*	D_T
thermal diffusion ratio; *rapport de thermodiffusion*	K_T
thermal diffusion factor; *facteur de thermodiffusion*	α_T
characteristic temperature; *température caractéristique*	Θ
Debye temperature; *température de Debye:* $\Theta_\mathrm{D} = h\nu_\mathrm{D}/k$	Θ_D
Einstein temperature; *température d'Einstein:* $\Theta_\mathrm{E} = h\nu_\mathrm{E}/k$	Θ_E
rotational temperature; *température de rotation:* $\Theta_\mathrm{r} = h^2/8\pi^2 Ik$	Θ_r
vibrational temperature; *température de vibration:* $\Theta_\mathrm{v} = h\nu/k$	Θ_v

7.4 *Thermodynamics.*†)

quantity of heat; *quantité de chaleur*	Q	
work; *travail*	W, A	
temperature; *température*	t, (ϑ)	*)
thermodynamic temperature; *température thermodynamique*	T, (Θ)	**)
entropy; *entropie*	S	
internal energy; *énergie interne*	U	
Helmholtz function; *fonction de Helmholtz, énergie libre:* $F = U - TS$	F	
enthalpy; *enthalpie:* $H = U + pV$	H	
Gibbs function; *fonction de Gibbs, enthalpie libre:* $G = H - TS$	G	
pressure coefficient; *coefficient d'augmentation de pression:* $\beta = (1/p)(\partial p/\partial T)_V$	β	
compressibility; *compressibilité:* $\kappa = -(1/V)(\partial V/\partial p)_T$	κ	
linear expansion coefficient; *coefficient de dilatation linéique*	α	
cubic expansion coefficient; *coefficient de dilatation volumique*	γ	
thermal conductivity; *conductivité thermique*	λ	
specific heat capacity; *chaleur massique:* $c = C/m$	c_p, c_v	
heat capacity; *capacité thermique*	C_p, C_v	
Joule-Thomson coefficient; *coefficient de Joule-Thomson*	μ	
ratio of specific heat capacities; *rapport des chaleurs massiques*	κ, γ	
heat flow rate; *flux thermique*	Φ, (q)	
heat current density; *densité de flux thermique*	$\boldsymbol{q}$, (φ)	
thermal diffusivity; *diffusivité thermique:* $a = \lambda/\rho c_p$	a	

†) The index m is added in the case of molar quantities, if needed, to distinguish them from quantities referring to the whole system. For specific quantities (see 6.1) lower case letters are used.
*) Preferred symbol: t. **) Preferred symbol: T.

7.5 Electricity and magnetism.††)

quantity of electricity; *quantité d' électricité*	Q
charge density; *charge volumique*	ρ
surface charge density; *charge surfacique*	σ
electric potential; *potential électrique*	V, φ
potential difference, tension; *différence de potentiel, tension*	U, V
electromotive force; *force électromotrice*	E
electric field strength; *champ électrique*	$\boldsymbol{E}$
electric flux; *flux électrique*	Ψ
electric displacement; *déplacement électrique*	$\boldsymbol{D}$
capacitance; *capacité*	C
permittivity; *permittivité:* $\boldsymbol{D} = \epsilon\boldsymbol{E}$	ϵ
permittivity of vacuum; *permittivité du vide*	ϵ_0
relative permittivity; *permittivité relative* $\epsilon_r = \epsilon/\epsilon_0$	ϵ_r
dielectric polarization: *polarisation diélectrique:* $\boldsymbol{D} = \epsilon_0\boldsymbol{E} + \boldsymbol{P}$	$\boldsymbol{P}$
electric susceptibility; *susceptibilité électrique*	χ_e
polarizability; *polarisabilité*	α, γ
electric dipole moment; *moment dipolaire électrique*	$\boldsymbol{p}$
electric current; *courant électrique*	I
electric current density; *densité de courant électrique*	$\boldsymbol{j}$
magnetic field strength; *champ magnétique*	$\boldsymbol{H}$
magnetic potential difference; *différence de potential magnétique*	U_m
magnetomotive force; *force magnétomotrice:* $F_m = \oint H_s ds$	F_m
magnetic induction, magnetic flux density; *induction magnétique, densité de flux magnétique*	$\boldsymbol{B}$
magnetic flux; *flux magnétique*	Φ
permeability; *perméabilité:* $\boldsymbol{B} = \mu\mathbf{H}$	μ
permeability of vacuum; *perméabilité du vide*	μ_0
relative permeability; *perméabilité relative:* $\mu_r = \mu/\mu_0$	μ_r
magnetization; *aimantation:* $\boldsymbol{B} = \mu_0(\mathbf{H} + \mathbf{M})$	$\mathbf{M}$
magnetic susceptibility; *susceptibilité magnétique*	χ_m
electromagnetic moment; *moment électromagnétique:* $E_p = -\boldsymbol{m}.\ \boldsymbol{B}$	$\mu, \boldsymbol{m}$
magnetic polarization; *polarisation magnétique:* $\boldsymbol{B} = \mu_0\mathbf{H} + \boldsymbol{J}$	$\boldsymbol{J}$
resistance; *résistance*	R
reactance; *réactance*	X
quality factor: *facteur de qualité:* $Q = \lvert X\rvert/R$	Q
impedance; *impédance:* $Z = R + \mathrm{i}\,X$	Z
admittance; *admittance:* $Y = 1/Z = G + \mathrm{i}B$	Y
conductance; *conductance*	G
susceptance; *susceptance*	B
resistivity; *résistivité*	ρ
conductivity; *conductivité* $\gamma = 1/\rho$	γ, σ
self inductance; *inductance propre*	L
mutual inductance; *inductance mutuelle*	M, L_{12}
coupling coefficient; *coefficient de couplage:* $k = L_{12}/(L_1L_2)^{1/2}$	k
phase number; *nombre de phases*	m
loss angle; *angle de pertes*	δ
number of turns; *nombre de tours*	N
power: *puissance*	P
electromagnetic energy density; *énergie électromagnetique volumique*	w
Poynting vector; *vecteur de Poynting*	$\boldsymbol{S}$
magnetic vector potential; *potentiel vecteur magnétique*	$\boldsymbol{A}$

7.6 Radiation, light.†)

radiant energy; *énergie rayonnante*	$Q, (Q_e), W$
radiant flux, radiant power; *flux énergétique, puissance rayonnante*	$\Phi, (\Phi_e), P$
radiant intensity; *intensité énergétique*	$I, (I_e)$
irradiance; *irradiance, éclairement énergétique*	$E, (E_e)$
radiance; *luminance énergétique, radiance*	$L, (L_e)$
radiant emittance; *émittance énergétique*	$M, (M_e)$
quantity of light; *quantité de lumière*	$Q, (Q_V)$
luminous flux; *flux lumineux*	$\Phi, (\Phi_V)$
luminous intensity; *intensité lumineuse:* $d\Phi/d\omega$	$I, (I_V)$
illuminance, illumination; *éclairement lumineux:* $d\Phi/dS$	$E, (E_V)$
luminance; *luminance:* $dI/dS \cos\vartheta$	$L, (L_V)$
luminous emittance; *émittance lumineuse:* $d\Phi/dS$	$M, (M_V)$
absorption factor; *facteur d'absorption:* Φ_a/Φ_0	α
reflection factor; *facteur de réflexion* Φ_r/Φ_0	ρ
transmission factor; *facteur de transmission* Φ_{tr}/Φ_0	τ
(linear) absorption coefficient; *coefficient d'absorption (linéique)*	a
(linear) extinction coefficient; *coefficient d'atténuation (linéique)*	μ
speed of light in empty space; *vitesse de la lumière dans le vide*	c
refractive index; *indice de réfraction:* $n = c/c_n$	n

††) Written in rationalized, 4-dimensional form. See Appendix, section 2.

†) The symbols between brackets are reserve symbols and are to be used whenever it is necessary to distinguish between radiation and light quantities.

7.7 *Acoustics.*

velocity of sound; *vitesse du son*	c	
velocity of longitudinal waves; *vitesse longitudinale*	c_l	
velocity of transversal waves; *vitesse transversale*	c_t	
group velocity; *vitesse de groupe*	c_g	
sound energy flux; *flux énergétique du son*	P	
reflexion factor; *facteur de réflexion:* P_r/P_0	ρ	
acoustic absorption factor; *facteur d'absorption acoustique* $1 - \rho$	α_a (α)	
transmission factor; *facteur de transmission:* P_{tr}/P_c	τ	
dissipation factor; *facteur de dissipation:* $\alpha_a - \tau$	δ	
loudness level; *niveau d'isosonie*	L_N, (Λ)	

7.8 *Quantum mechanics.*

complexe conjugate of Ψ; *complexe conjugué de* Ψ	Ψ^*	
probability density; *densité de probabilité:* $P = \Psi^*\Psi$	P	
probability current density: *densité de courant de probabilité:* $\boldsymbol{S} = (\hbar/2\mathrm{i}m)(\Psi^*\Delta\Psi - \Psi\Delta\Psi^*)$	$\boldsymbol{S}$	
charge density of electrons; *charge volumique d'électrons:* $\rho = -eP$	ρ	
electric current density of electrons; *densité de courant électrique d'électrons:* $\boldsymbol{j} = -e\boldsymbol{S}$	$\boldsymbol{j}$	
expectation value of A; *valeur moyenne de* A	$\langle A\rangle$, $\bar{A}$	
commutator of A and B, *commutateur de* A *et* B: $[A,B] = AB - BA$	$[A,B]$, $[A,B]$	
anticommutator of A and B; *anticommutateur de* A *et* B: $[A,B]_+ = AB + BA$	$[A,B]_+$	
matrix element; *élément de matrice:* $A_{ij} = \int \psi_i^*(A\psi_j)\,\mathrm{d}\tau$	A_{ij}	
Hermitian conjugate of operator A; *conjugué Hermitien de l'opérateur* A	$A\dagger$	
momentum operator in coordinate representation; *opérateur de quantité de mouvement*	$+(\hbar/\mathrm{i})\Delta$	
annihilation operators; *opérateurs d'annihilation*	a, b, α, β	
creation operators; *opérateurs de création*	a^+, $b\dagger$, $\alpha\dagger$ $\beta\dagger$	

7.9 *Atomic and nuclear physics.*

mass number, nucleon number; *nombre de masse, nombre de nucléons*	A	
atomic number, proton number; *nombre atomique, nombre de protons*	Z	
neutron number; *nombre de neutrons:* $N = A - Z$	N	
elementary charge (of position); *charge élémentaire* (*du position*)	e	
electron mass; *masse de l'électron*	m, m_e	
proton mass; *masse du proton*	m_p	
neutron mass; *masse du neutron*	m_n	
meson mass; *masse du méson*	m_π,	
nuclear mass; *masse nucléaire* (of nucleus: ^{A}X)	m_N, m_N (^{A}X)	
atomic; mass; *masse atomique* (of nuclide: ^{A}X)	m_a, m_a (^{A}X)	
(unified) atomic mass constant; *constante* (*unifiée*) *de masse atomique:* $m_u = m_a(^{12}\mathrm{C})/12$	m_u	
relative atomic mass; *masse atomique relative:* m_a/m_u	A_r	
Planck constant; *constante de Planck* ($\hbar = h/2\pi$)	h	
principal quantum number; *nombre quantique principal*	n, n_i	
orbital angular momentum quantum number; *nombre quantique de moment angulaire orbital*	L, l_i	
spin quantum number; *nombre quantique de spin*	S, s_i	
angular momentum quantum number; *nombre quantique de moment angulaire* (including electron spin)	J, j_i	
magnetic quantum number; — — *magnétique*	M, m_i	
nuclear spin quantum number; — — *de spin nucléaire*	I, J	*)
hyperfine quantum number; — — *hyperfin*	F	
rotational quantum number; — — *de rotation*	J, K	
vibrational quantum number; — — *de vibration*	v	
quadrupole moment; *moment quadripolaire*	Q	
Rydberg constant; *constante de Rydberg***)	R_∞	
Bohr radius; *rayon de Bohr***)	a_0	
fine structure constant; *constante de structure fine***)	α	
mass excess; *excès de masse:* $m_a - Am_u$	Δ	
packing fraction; *packing fraction:* Δ/Am_u	f	
nuclear radius; *rayon nucléaire:* $R = r_0A^{1/3}$	R	
magnetic moment of particle: *moment magnétique d'une particule*	μ	
magnetic moment of proton; *moment magnétique du proton*	μ_p	
magnetic moment of neutron; *moment magnétique du neutron*	μ_n	
magnetic moment of electron; *moment magnétique électronique*	μ_e	
Bohr magneton; *magnéton de Bohr***)	μ_B	
nuclear magneton; *magnéton nucléaire*	μ_N	
g-factor; *facteur g:* e.g. $g = \mu/I\mu_N$	g	
gyromagnetic ratio; *rapport gyromagnétique***)	γ	
Larmor (angular) frequency; *fréquence* (*angulaire*) *de Larmor***)	ω_L	
level width; *largeur d'un niveau*	Γ	

*) I is used in atomic physics. J in nuclear physics.
**) See for definition: Appendix, section 2.

mean life; *vie moyenne*	τ
reaction energy; *énergie de réaction*	Q
cross section; *section efficace*	σ
macroscopic cross section; *section efficace macroscopique:* $\Sigma = n\sigma$	Σ
impact parameter; *paramètre de collision*	b
scattering angle; *angle de diffusion*	$\vartheta, \theta, \varphi$
internal conversion coefficient; *coefficient de conversion interne*	α
disintegration energy; *énergie de désintégration*	Q
half life; *demi-vie*	$T_{\frac{1}{2}}$
reduced half life; *demi-vie réduite*	$fT_{\frac{1}{2}}$
decay constant, disintegration constant; *constante de désintégration*	λ
activity; *activité*	A
Compton wavelength; *longueur d'onde de Compton:* $\lambda_C = h/mc$	λ_C
electron radius; *rayon de l'électron***)	r_e
linear attenuation coefficient; *coeff. d'atténuation linéique*	μ, μ_l
atomic attenuation coefficient; *coeff. d'atténuation atomique*	μ_a
mass attenuation coefficient; *coeff. d' atténuation massique*	μ_m
linear stopping power; *pouvoir d' arrêt linéaire*	S, S_l
atomic stopping power; *pouvoir d' arrêt atomique*	S_a
linear range; *distance de pénétration linéaire*	R, R_l
recombination coefficient; *coefficient de recombinaison*	α

7.10 Solid state physics.

fundamental translations for lattice; *translations fondamentales d'un réseau*	$\boldsymbol{a}, \boldsymbol{b}, \boldsymbol{c}$ $\boldsymbol{a}_1, \boldsymbol{a}_2, \boldsymbol{a}_3$
Miller indices; *indices de Miller*	h, k, l h_1, h_2, h_3
plane in lattice; *plan d'un réseau**)	(h, k, l) (h_1, h_2, h_3)
direction in lattice; *direction dans un réseau**)	$[h, k, l]$ $[h_1, h_2, h_3]$
fundamental translations in reciprocal lattice; *translations fondamentales de reseau réciproque*	$\boldsymbol{a}^*, \boldsymbol{b}^*, \boldsymbol{c}^*$ $\boldsymbol{b}_1, \boldsymbol{b}_2, \boldsymbol{b}_3$
vector in crystal lattice; *vecteur dans un réseau*	$\boldsymbol{r}$
distance between successive lattice planes; *distance de plans successifs d'un réseau*	d
Bragg angle; *angle de Bragg*	θ, ϑ
order of reflexion; *ordre du réflexion*	n
short range order parameter; *paramètre d'ordre (proche voisin)*	σ
long range order parameter; *paramètre d'ordre (longue distance)*	s
Burgers vector; *vecteur de Burgers*	$\boldsymbol{b}$
circular wave vector; propagation vector (of phonons); *vecteur d'onde circulaire, vecteur de propagation (de phonons)*	$\boldsymbol{q}$
circular wave vector, propagation vector (of particles); *vecteur d'onde circulaire, vecteur de propagation (de particules)*	$\boldsymbol{k}$
effective mass of electron; *masse effective d'une électron*	m^*, m_{eff}
Fermi energy; *énergie de Fermi*	E_F, ϵ_F
Fermi circular wave vector; *vecteur d'onde circulaire de Fermi*	$\boldsymbol{k}_F$
work function; *fonction de travail*	Φ
differential thermoelectric power; *force thermoélectrique différentielle*	$S, (\Sigma)$
Peltier coefficient; *coefficient de Peltier*	Π
Thomson coefficient; *coefficient de Thomson*	μ
piezoelectric coefficient; *coefficient piézoélectrique:* (polarization/stress)	d_{mn}
characteristic (Weiss) temperature; *température caractéristique (de Weiss)*	Θ, Θ_W
Curie temperature; *température de Curie*	T_C
Neel temperature; *température de Néel*	T_N
Hall coefficient; *coefficient de Hall*	R_H

7.11 Molecular spectroscopy.

qu.n. of component of electronic orbital angular momentum vector along symmetry axis; *n.qu. de la composante du moment angulaire orbital électronique suivant l'axe de symétrie*	Λ, λ_i
qu.n. of component of electronic spin along symmetry axis; *n.qu. de la composante du spin électronique suivant l'axe de symétrie*	Σ, σ_i
qu.n. of total electronic angular momentum vector along symmetry axis; *n.qu. du moment angulaire total électronique suivant l'axe de symétrie*	Ω, ω_i
qu.n. of electronic spin; *n.qu. du spin électronique*	S
qu.n. of nuclear spin; *n.qu. du spin nucléaire*	I
qu.n. of vibrational mode; *n.qu. d'une mode de vibration*	v
degeneracy of vibrational mode; *degré de dégénérescance d'une mode de vibration*	d
qu.n. of vibrational angular momentum; *n.qu. de moment angulairevibrationnel* (L.M.)	l
qu.n. of total angular momentum; *n.qu. de moment angulair total* (excluding nuclear spin)	J
qu.n. of component of $\boldsymbol{J}$ in direction of external field; *n.qu. de la composante de* $\boldsymbol{J}$ *dans la direction du champ extérieur*	M, M_J

*) Instead of the ordinary brackets ( ) and square brackets [ ], curly brackets { } and diamond brackets ⟨ ⟩ are also used respectively.

qu.n. of component of $\boldsymbol{S}$ in direction of external field; *n.qu. de la composante de* $\boldsymbol{S}$ *dans la direction du champ extérieur* M_S

qu.n. of total angular momentum; *n.qu. du moment angulaire total* (including nuclear spin) $\boldsymbol{F} = \boldsymbol{J} + \boldsymbol{I}$ F

qu.n. of component of $\boldsymbol{F}$ in direction of external field; *n.qu. de la composante de* $\boldsymbol{F}$ *dans la direction du champ extérieur* M_F

qu.n. of component of $\boldsymbol{I}$ in direction of external field; *n.qu. de la composante de* $\boldsymbol{I}$ *dans la direction du champ extérieur* M_I

qu.n. of component of angular momentum along axis; *n.qu. de la composante du moment angulaire suivant l'axe* (L.M. and S.T.M.; excluding electron- and nuclear spin; for L.M.: $(K = |\Lambda + l|)$ K

qu.n. of total angular momentum; *n.qu. du moment angulaire total* (L.M. and S.T.M.; excluding electron and nuclear spin: $\boldsymbol{J} = \boldsymbol{N} + \boldsymbol{S}$*) N

qu.n. of component of angular momentum along symmetry axis; *n.qu. de la composante du moment angulaire suivant l'axe de symétrie* (L.M. and S.T.M., excluding nuclear spin; for L.M.: $P = |K + \Sigma|$**) P

electronic term; *terme électronique:* $T_e = E_e/hc$†) T_e

vibrational term; *terme de vibration:* $G = E_{vibr}/hc$ G

coefficients in expression for vibrational term (for D.M.); *coefficients de l'expression d'un terme de vibration:*

$$G = \sigma_e(v + \tfrac{1}{2}) - x\sigma_e(v + \tfrac{1}{2})^2$$

σ_e, $x\sigma_e$

coefficients in expression for vibrational term (for P.M.) *coefficients de l'expression d'un terme de vibration:*

$$G = \Sigma\sigma_i(v_i + \tfrac{1}{2}d_i) + \tfrac{1}{2}\Sigma\Sigma x_{jk}(v_j + \tfrac{1}{2}d_j)(v_k + \tfrac{1}{2}d_k)$$

σ_i, x_{jk}

rotational term; *terme de rotation:* $F = E_{rot}/hc$ F

principal moments of inertia; *moments principaux d'inertie:* $I_A \leqslant I_B \leqslant I_C$††) I_A, I_B, I_C

rotational constants; *constants de rotation:* $A = h/8\pi^2 cI_A$, etc.††) A, B, C

total term; *terme total:* $T = T_e + G + F$ T

Remark: L.M. = linear molecules. *S.T.M.* = symmetric top molecules.
D.M. = diatomic molecules. P.M. = polyatomic molecules.
See for further details: Report on Notation for the Spectra of Polyatomic Molecules (Joint Commission for Spectroscopy of I.U.P.A.P. and I.A.U. 1954) J. Chem. Phys. **23** (1955) 1997.

7.12 Chemical physics.

amount of substance; *quantité de substance* ν, n *)

molar mass of substance B; *masse molaire de la substance* B M_B

molarity of subst. B; *molarité de la subst.* B c_B

mole fraction of subst. B; *fraction molaire de la subst.* B x_B

mass fraction of subst. B; *fraction massique de la subst.* B w_B

volume fraction of subst. B; *fraction volumique de la subst.* B φ_B

mole ratio of solution; *rapport molaire d'une solution* r

molality of solution; *molalité d'une solution* m

chemical potential of subst. B; *potentiel chimique de la subst.* B μ_B

absolute activity of subst. B (dimensionless); *activité absolue de la subst.* B (*sans dimension*) $\lambda_B = \exp(\mu_B/kT)$ λ_B

relative activity; *activité relative* a_B

activity coefficients; *coefficients d'activité* γ_B, f_B

osmotic pressure; *pression osmotique* Π

osmotic coefficient; *coefficient osmotique* g, φ

stoichiometric number of molec. B; *coeff. stœchiométrique des moléc.* B v_B

affinity; *affinité* A

extent of reaction; *état d'avancement d'une réaction* ξ

equilibrium constant; *constante d'équilibre* K

charge number of ion; *électrovalence d'un ion* z

Faraday constant; *constante de Faraday* F

ionic strength; *force ionique* I

activity of substance B; *activité de la substance* B: $z_B = (2\pi mkT/h^2)^{3/2}\lambda_B$ z_B

8. RECOMMENDED MATHEMATICAL SYMBOLS.

8.1 General symbols.

equal to; *égal à* $=$

not equal to; *différent de* $=|=$ $\neq$

identically equal to; *égal identiquement à* $\equiv$

corresponds to; *correspond à* $\geq$

*) case of loosely coupled electron spin. **) case of tightly coupled electron spin

†) All energies are taken here with respect to the ground state as reference level.

††) For diatomic molecules use I and $A = h/8\pi^2 cI$.

*) n is used in chemistry, but v may be used as an alternative to n, when n is used for number density of particles.

approximately equal to; *égal environ à*	$\approx$
asymptotically equal to; *asymptotiquement égal à*	$\simeq$
proportional to; *proportionnel à*	$\sim \quad \propto$
approaches; *tend vers*	$\rightarrow$
larger than; *supérieur á*	$>$
smaller than; *inférieur á*	$<$
much larger than; *très supérieur á*	$\gg$
much smaller than; *très inférieur á*	$\ll$
larger than or equal to; *supérieur ou égal á*	$\geqslant \quad \geqq \quad \geq$
smaller than or equal to; *inférieur ou égal á*	$\leqslant \quad \leqq \quad \leq$
plus; *plus*	$+$
minus; *moins*	$-$
plus or minus; *plus ou moins*	$\pm$
a multiplied by *b*; *a multiplié par b*	$ab,\ a.b,\ a \cdot b,\ a \times b$
a divided by *b*; *a divisé par b*	$a/b,\ \frac{a}{b},\ ab^{-1}$
a raised to the power *n*; *a puissance n*	a^n
magnitude of *a*; *valeur absolue de a*	$\lvert a \rvert$
square root of *a*; *racine carrée de a*	$\sqrt{\bar{a}},\ \sqrt{a},\ a^{\frac{1}{2}}$
mean value of *a*; *valeur moyenne de a*	$\bar{a}\ \langle a \rangle$
factorial *p*; *factorielle p*	$p!$
binomial coefficient; *coefficient binomial:* $n!/p!(n-p)!$	$\binom{n}{p}$
infinity; *infini*	∞

8.2 Letter symbols and letter expressions for *mathematical operations* should be written in *roman* (or *upright*) type

exponential of *x*; *exponentielle de x*	$\exp x,\ \mathrm{e}^x$	
base of natural logarithms; *base des logarithmes népériens*	e	
logarithm to the base *a* of *x*; *logarithme de base a de x*	$\log_a x$	
natural logarithm of *x*; *logarithme népérien de x*	$\ln x$	
common logarithm of *x*; *logarithme décimal de x*	$\lg x,\ \log x$	*)
binary logarithm of *x*; *logarithme binaire de x*	$\mathrm{lb}\ x \quad \log_2 x$	
summation; *somme*	Σ	
product; *produit*	Π	
finite increase of *x*; *accroissement fini de x*	Δx	**)
variation of *x*; *variation de x*	δx	
total differential of *x*; *différentielle totale de x*	$\mathrm{d}x$	
function of *x*; *fonction de x*	$f(x),\ \mathrm{f}(x)$	
limit of $f(x)$; *limite de* $f(x)$	$\lim f(x)$	
Dirac delta function; *fonction delta de Dirac* $\delta(\boldsymbol{r}) = \delta(\mathrm{x})\delta(\mathrm{y})\delta(\mathrm{z})$	$\delta(x),\ \delta(\boldsymbol{r})$	
Kronecker delta symbol; *symbole delta de Kronecker*	δ_{ij}	
Unit step function; *fonction unité*($\epsilon(x) = 1$ for $x > 0$, $\epsilon(x) = 0$ for $x < 0$)	$\epsilon(x)$	†)

8.3 Trigonometric functions.

sine of *x*, *sinus x*	$\sin x$
cosine of, *cosinus x*	$\cos x$
tangent of *x*, *tangente x*	$\tan x,\ \mathrm{tg}\ x$
cotangent of *x*, *cotangente x*	$\cot x,\ \mathrm{ctg}\ x$
secant of *x*, *sécante x*	$\sec x$
cosecant of *x*, *cosécante x*	$\mathrm{cosec}\ x$

Remarks:

a. It is recommended to use for the *inverse circular functions* the symbolic expressions for the corresponding circular function preceded by the letters: arc.
Examples: arcsin *x*, arccos *x*, arctan *x* or arctg *x*, etc. Sometimes the notation $\sin^{-1} x$, $\tan^{-1} x$, etc. is used.

b. It is recommended to use for the *hyperbolic functions* the symbolic expressions for the corresponding circular function, followed by the letter: h.
Examples: sinh *x*, cosh *x*, tanh *x* or tgh *x*, etc.

c. It is recommended to use for the *inverse hyperbolic functions* the symbolic expression for the corresponding hyperbolic function preceded by the letters: ar.

Example: arsinh *x*, arcosh *x*, etc.

8.4 Complex quantities.

imaginary unit; *unité imaginaire* ($\mathrm{i}^2 = -1$)	i, j
real part of *z*; *partie réelle de z*	Re *z*, *z*
imaginary part of *z*; *partie imaginaire de z*	Im *z*, z''
modulus of *z*; *module de z*	$\lvert z \rvert$
argument of *z*; *argument de z* : $z = \lvert z \rvert \exp \mathrm{i}\varphi$	arg *z*, φ
complex conjugate of *z*, conjugate of *z*; *complexe conjugué de z, conjugué de z*	z^*

Remark: Sometimes the notation $\bar{z}$ is used for the complex conjugate of *z*.

*) In case of ambiguity $\log_{10} x$. **) Greek capital delta, not triangle.
†) $\Theta(t)$ is used for unit step function of time.

8.5 Symbols for special values of periodic quantities.

	Normal case		Exceptional case	
instantaneous value	x	x	x	x
r.m.s. value	x_{eff}	$\tilde{x}$	X	X
maximum value	x_{m}	$\hat{x}$	x_{m}, X_{m}	$\hat{x}$, $\hat{X}$
average value	x_{av}	$\bar{x}$, $\langle x \rangle$	x_{av}	$\bar{x}$, $\langle x \rangle$

Remarks:

a. The "normal" case is the case in which only a small or only a capital letter may be used for the quantity. The "exceptional" case is the case in which both small and capital letters may be used for the same quantity.

b. The minimum value of x may be indicated by $x_{\min}$ or $\check{x}$.

c. The r.m.s.value is defined as $\tilde{x}^2 = T^{-1} \int_0^T x(t)^2 \mathrm{d}t$.

d. The average value is defined as $\bar{x} = T^{-1} \int_0^T x(t) \mathrm{d}t$.

8.6 Vector calculus. (see also *1.2.3*)

absolute value; *valeur absolue*	$\|\boldsymbol{A}\|$, A
differential vector operator; *vecteur opérateur différentiel*	$\partial/\partial \boldsymbol{r}$, ∇
gradient; *gradient*	grad φ, $\nabla\varphi$
divergence; *divergence*	div $\boldsymbol{A}$, $\nabla \cdot \boldsymbol{A}$
curl; *rotätionnel*	curl $\boldsymbol{A}$, rot $\boldsymbol{A}$, $\nabla \times \boldsymbol{A}$
Laplacian; *Laplacien*	$\triangle\varphi$, $\nabla^2\varphi$
Dalembertian; *Dalembertien*	$\Box\varphi$

8.7 Matrix calculus.

transpose of matrix A; *matrice transposée de A*: $A_{ij} = A_{ji}$	$\tilde{A}$
complex conjugate of A; *matrice complexe conjugueé de A*: $A^*_{ij} = (A_{ij})^*$	A^*
Hermitian conjugate of A; *matrice conjugueé Hermitienne de A*: $A\dagger_{ji} = A_{ji}^*$	$A\dagger$
Pauli matrices; *matrices de Pauli:*	σ, σ_x, σ_y, σ_z

$$\sigma_x = \begin{vmatrix} 0 & 1 \\ 1 & 0 \end{vmatrix} \qquad \sigma_y = \begin{vmatrix} 0 & -\mathrm{i} \\ \mathrm{i} & 0 \end{vmatrix} \qquad \sigma_z = \begin{vmatrix} 1 & 0 \\ 0 & -1 \end{vmatrix} \qquad \sigma_1, \sigma_2, \sigma_3$$

unit matrix; *matrice unité:* $I = \begin{vmatrix} 1 & 0 \\ 0 & 1 \end{vmatrix}$ — I

Dirac (4 × 4) matrices; (4 × 4) *matrices de Dirac* *)

$$\alpha_x = \begin{vmatrix} 0 & \sigma_x \\ \sigma_x & 0 \end{vmatrix} \qquad x_y = \begin{vmatrix} 0 & \sigma_y \\ \sigma_y & 0 \end{vmatrix} \qquad \alpha_z = \begin{vmatrix} 0 & \sigma_z \\ \sigma_z & 0 \end{vmatrix} \qquad \alpha, \alpha_x, \alpha_y, \alpha_z$$

$$\beta = \begin{vmatrix} I & 0 \\ 0 & -I \end{vmatrix} \qquad \beta$$

9. INTERNATIONAL SYMBOLS FOR UNITS.

9.1 Unit systems.

1. A *coherent system of units* is a system based on a certain set of "basic units" from which all "derived units" are obtained by multiplication or division without introducing numerical factors. In addition there are "dimensionless units," in particular the radian, symbol: rad, for plane angle and the steradian, symbol: sr, for solid angle.

2. The ***CGS system*** or cm-g-s system is a coherent system of units based on *three basic units* for the three basic quantities length, mass and time respectively:

centimetre	cm
gramme	g
second	s

*) Sometimes a different representation is used.

In the field of *mechanics* the following units of this system have special names and symbols, which have been approved by the Conférence Générale des Poids et Mesures:

l, b, h	centimetre; *centimètre*	cm
t	second; *seconde*	s
m	gramme; *gramme*	g
f, ν	hertz; *hertz* (= s^{-1})	Hz
F	dyne; *dyne* (= $g.cm/s^2$)	dyn
E, U, W, A	erg; *erg* (= $g.cm^2/s^2$)	erg
p	microbar; *microbar* (= dyn/cm^2)	μbar
η	poise; *poise* (= dyn. s/cm^2)	P

In the field of *electricity and magnetism* several variants of the CGS unit system have been developed, in particular the *electrostatic CGS system* and *electromagnetic CGS system*. Special names and symbols for some of the units of the second system are:*)

*H**	oersted; *oersted* (= $cm^{-}.g^{\frac{1}{2}}.s^{-1}$)	Oe
*B**	gauss; *gauss* (= $cm^{-\frac{1}{2}}.g^{\frac{1}{2}}.s^{-1}$)	G
Φ*	maxwell; *maxwell* (= $cm^{\frac{3}{2}}.g^{\frac{1}{2}}s^{-1}$)	Mx

For further information about the units and unit systems in electricity and magnetism see appendix, section 2.

3. The ***MKSA system*** or m-kg-s-A system is a coherent system of units for mechanics, electricity and magnetism, based on *four basic units* for the four basic quantities length, mass, time and electric current intensity:

metre	m
kilogramme	kg
second	s
ampere	A

Remark: The system based on these four units has been given the name *Giorgi system* by the International Electrotechenical Committee in 1958. The subsystem for mechanics, which is based on the first three units only, has the name *MKS system.*

The following units of the MKSA system have special names and symbols, which have been approved by the Conférence Générale des Poids et Measures:

l. b. h	metre; *mètre*	m
t	second; *seconde*	s
m	kilogramme; *kilogramme*	kg
ν, f	hertz; *hertz* (= s^{-1})	Hz
F	newton; *newton* (= $kg.m/s^2$)	N
E	joule; *joule* (= $kg.m^2/s^2$)	J
P	watt; *watt* (= J/s)	W

I	ampere; *ampère*	A
Q	coulomb; *coulomb* (= A.s)	C
V	volt; *volt* (= W/A)	V
C	farad; *farad* (= C/V)	F
R	ohm; *ohm* (= V/A)	Ω
L	henry; *henry* (= V.s/A)	H
Φ	weber; *weber* (= V.s)	Wb
B	tesla; *tesla* (= Wb/m^2)	T

4. The ***degree Kelvin.*** In the field of *thermodynamics* one introduces an additional basic unit, corresponding to the basic quantity:

   thermodynamic temperature, the unit being the degree Kelvin, symbol: °K
   When the *customary temperature* is used, defined by $t = T - T_0$, where $T_0 = 273.15$°K, this is usually expressed in degree Celsius symbol: °C. For *temperature interval* the name degree, symbol: deg is often used, the indications "Kelvin" or "Celsius," indicating the zeropoint of the temperature scale used, being irrelevant in this case.

5. The ***candela.***

   In the field of *photometry* one introduces an additional basic unit, corresponding to the basic quantity: *luminous intensity*, this unit being the candela, symbol: cd. Special names and symbols for units in this field are:

I	candela; *candéla*	cd
Φ	lumen; *lumen*	lm
E	lux; *lux* (= lm/m^2)	lx

*) See also Appendix.

6. The ***International System of Units.*** For the coherent system based on the six basic units:

metre	m	ampere	A
kilogramme	kg	degree Kelvin	°K
second	s	candela	cd

the name *International System of Units* has been recommended by the Conférence Générale des Poids et Mesures in 1960. The units of this system are called: *SI-units*

7. The ***mole.***

In the field of *chemical and molecular physics,* in addition to the basic quantities defined above having units defined by the Conférence Générale des Poids et Mesures, *amount of substance* is also treated as a basic quantity. The recommended basic unit is the mole, symbol: mol. The mole is defined as the amount of substance of a system, which contains the same number of molecules (or ions, or atoms, or electrons, as the case may be), as there are atoms in exactly 12 gramme of the pure carbon nuclide ^{12}C.

9.2 Other units.

l	angström; *angström*	Å
σ	barn; *barn* (= 10^{-24} cm²)	b
V	litre; *litre*	l
$t, \tau, T_{\frac{1}{2}}$	minute; *minute*	min
$t, \tau, T_{\frac{1}{2}}$	hour; *heure*	h
$t, \tau, T_{\frac{1}{2}}$	day; *jour*	d
$t, \tau, T_{\frac{1}{2}}$	year; *année*	a
p	atmosphere; *atmosphère*	atm
E	kilowatt-hour; *kilowatt-heure*	kWh
Q	calorie; *calorie*	cal
E, Q	electronvolt; *électronvolt*	eV
m	tonne; *tonne* (= 1000 kg)	t
m_a	(unified) atomic mass unit; *unité de masse atomique (unifiée)*	u
p	bar; *bar* (= 10^6 dyn/cm²)	bar
A	curie, *curie*	Ci

Remark: The (unified) atomic mass unit is defined as 1/12th of the mass of an atom of the ^{12}C nuclide.

APPENDIX*). SYSTEMS OF QUANTITIES AND UNITS IN ELECTRICITY AND MAGNETISM.

The CGS unit system with three basic units and the MKSA unit system with four basic units correspond respectively to two different sets of equations in the field of electricity and magnetism, which are developed starting from three and from four basic quantities respectively. These systems are denoted as three and four "dimensional" systems of equations respectively.

1. Systems of equations with 3 basic quantities.

Three distinct sets of equations with three basic quantities**) have been developed in the field of electricity and magnetism. These are:

(1.a) *The "electrostatic system" of equations,* defining the electric charge on the basis of Coulomb laws for the force between two electric charges, by taking the permittivity in vacuo equal to a dimensionless quantity, the number unity.

(1.b) *The "electromagnetic system" of equations,* defining the electric current on the basis of the interaction law for the force between two electric current elements, by taking the permeability in vacuo equal to a dimensionless quantity, the number unity.

(1.c) *The "symmetrical system" or Gaussian system of equations,* using the electric quantities from system (1.a) and the magnetic quantities from system (1.b). As a result of combining the two sets of quantities the velocity of light in vacuo appears explicitly in some of the equations interrelating electric and magnetic quantities.

The equations in these three systems are usually written in the *"non-rationalized"* form, which is called "non-rationalized," because in these equations often factors 2π or 4π appear in situations not involving circular or spherical symmetry respectively. These equations are sometimes written in a *"rationalized"* form, in which these factors appear only in those equations, where they could be expected from the geometry of the system. When three dimensional equations are used (e.g. in theoretical physics), one commonly uses those of the *non-rationalized symmetrical system.*

2. Systems of equations with 4 basic quantities.

In the equations with four basic quantities at least one quantity of electric or magnetic nature is included in the basic set. In such a system the permittivity and the permeability in vacuo appear explicitly as physical quantities with dimension in the relevant equations.

* The S.U.N. commission, after reproducing in the previous chapters all the recommendations on symbols, units and nomenclature approved by the I.U.P.A.P., gives in this Appendix some factual information about existing systems of quantities and units in the field of electricity and magnetism.

** Often length, time and mass are chosen as basic quantities, but also other choices, e.g. length, time and energy or length, time and force have been used.

Two different sets of equations are in use:

(2.a) The *"non-rationalized system" of equations*, in which the factors 4π and 2π often appear at unexpected places.

(2.b) The *"rationalized system" of equations*, in which these factors only appear in those equations, where they could be expected from the geometry.

When four-dimensional equations are used, one commonly writes these equations in the rationalized form (2.b).

Some characteristic expressions of the non rationalized three-dimensional "symmetrical system" (1.c) and the corresponding four-dimensional equations in the non-rationalized form (2.a) and the rationalized form (2.b) are given in the table. The quantities of the three-dimensional "symmetrical system" of equations (1.c) have been indicated with an asterisk (*), those of the non-rationalized four-dimensional equations (2.a) with a prime ('), as far as they are different from those of the system (2.b).

Non-rationalized symmetric system with 3 basic quantities	Non-rational. system with 4 basic quantities	Rationalized system with 4 basic quantities
$c \text{ rot } \boldsymbol{E}^* = -\partial \boldsymbol{B}^*/\partial t$	$\text{rot } \boldsymbol{E} = -\partial \boldsymbol{B}/\partial t$	$\text{rot } \boldsymbol{E} = -\partial \boldsymbol{B}/\partial t$
$\text{div } \boldsymbol{D}^* = 4\pi\rho$	$\text{div } \boldsymbol{D}' = 4\pi\rho$	$\text{div } \boldsymbol{D} = \rho$
$\text{div } \boldsymbol{B}^* = 0$	$\text{div } \boldsymbol{B} = 0$	$\text{div } \boldsymbol{B} = 0$
$c \text{ rot } \boldsymbol{H}^* = 4\pi \boldsymbol{j}^* + \partial \boldsymbol{D}^*/\partial t$	$\text{rot } \boldsymbol{H}' = 4\pi \boldsymbol{j} + \partial \boldsymbol{D}'/\partial t$	$\text{rot } \boldsymbol{H} = \boldsymbol{j} + \partial \boldsymbol{D}/\partial t$
$\boldsymbol{F} = Q^*\boldsymbol{E}^* + Q^*\boldsymbol{v} \times \boldsymbol{B}^*/c$	$\boldsymbol{F} = Q\boldsymbol{E} + Q\boldsymbol{v} \times \boldsymbol{B}$	$\boldsymbol{F} = Q\boldsymbol{E} + Q\boldsymbol{v} \times \boldsymbol{B}.$
$w = (\boldsymbol{E}^*.\boldsymbol{D}^* + \boldsymbol{B}^*.\boldsymbol{H}^*)/8\pi$	$w = (\boldsymbol{E}.\boldsymbol{D}' + \boldsymbol{B}.\boldsymbol{H}')/8\pi$	$w = (\boldsymbol{E}.\boldsymbol{D} + \boldsymbol{B}.\boldsymbol{H})/2$
$\boldsymbol{S} = c(\boldsymbol{E}^* \times \boldsymbol{H}^*)/4\pi$	$\boldsymbol{S} = (\boldsymbol{E} \times \boldsymbol{H}')/4\pi$	$\boldsymbol{S} = \boldsymbol{E} \times \boldsymbol{H}$
$\boldsymbol{E}^* = -\text{grad } V^* - (1/c)\partial \boldsymbol{A}^*/\partial t$	$\boldsymbol{E} = -\text{ grad } V - \partial \boldsymbol{A}/\partial t$	$\boldsymbol{E} = -\text{grad } V - \partial \boldsymbol{A}/\partial t$
$\boldsymbol{B}^* = \text{ rot } \boldsymbol{A}^*$	$\boldsymbol{B} = \text{rot } \boldsymbol{A}$	$\boldsymbol{B} = \text{ rot } \boldsymbol{A}$
$\epsilon_r \boldsymbol{E}^* = \boldsymbol{D}^*$	$\epsilon_0'\epsilon_r \boldsymbol{E} = \epsilon' \boldsymbol{E} = \boldsymbol{D}'$	$\epsilon_0\epsilon_r \boldsymbol{E} = \epsilon \boldsymbol{E} = \boldsymbol{D}$
$\boldsymbol{E}^* = \boldsymbol{D}^* - 4\pi \boldsymbol{P}^*$	$\epsilon_0' \boldsymbol{E} = \boldsymbol{D}' - 4\pi \boldsymbol{P}$	$\epsilon_0 \boldsymbol{E} = \boldsymbol{D} - \boldsymbol{P}$
$\boldsymbol{B}^* = \mu_r \boldsymbol{H}^*$	$\boldsymbol{B} = \mu' \boldsymbol{H}' = \mu_0'\mu_r \boldsymbol{H}'$	$\boldsymbol{B} = \mu \boldsymbol{H} = \mu_0\mu_r \boldsymbol{H}$
$\boldsymbol{B}^* = \boldsymbol{H}^* + 4\pi \boldsymbol{M}^*$	$\boldsymbol{B} = \mu_0'(\boldsymbol{H}' + 4\pi \boldsymbol{M})$	$\boldsymbol{B} = \mu_0(\boldsymbol{H} + \boldsymbol{M})$
$a_0 = \hbar^2/me^{*2}$	$a_0 = \hbar^2\epsilon_0'/me^2$	$a_0 = \hbar^2 4\pi\epsilon_0/me^2$
$hcR_\infty = e^{*2}/2a_0$	$hcR_\infty = e^2/2\epsilon_0'a_0$	$hcR_\infty = e^2/8\pi\epsilon_0'a_0$
$\mu_B{}^* = e^*\hbar/2mc$	$\mu_B = e\hbar/2m$	$\mu_B = e\hbar/2m$
$\gamma^* = g(e^*/2mc)$	$\gamma = g(e/2m)$	$\gamma = g(\epsilon/2m)$
$\omega_L = (e^*/2mc)B^* = \mu_B{}^*B^*/\hbar$	$\omega_L = (e/2m)B = \mu_B B/\hbar$	$\omega_L = (e/2m)B = \mu_B B/\hbar$
$\alpha = e^{*2}/\hbar c$	$\alpha = e^2/\epsilon_0'\hbar c$	$\alpha = e^2/4\pi\epsilon_0\hbar c$
$r_e = e^{*2}/mc^2$	$r_e = e^2/\epsilon_0'mc^2$	$r_e = e^2/4\pi\epsilon_0 mc^2$

Remark on rationalization:

The basis of these considerations in which the rationalization is connected with the writing of the equations between physical quantities is in agreement with the resolution accepted by the I.U.P.A.P. in 1951 in Copenhagen:

"The General Assembly of the Union of Physics considers that, in the case that the equations are rationalized, the rationalization should be effected by the introduction of new quantities."

("L'Assemblée Générale de l'Union de Physique considère que, dans le cas où les équations sont rationalisées, la rationalisation doit étre obtenue par l'introduction de grandeurs nouvelles.")

3. *CGS system of units.*

(3.a) The *electrostatic CGS system of units* forms a coherent system of units in combination with the three-dimensional "electrostatic system" of equations (1.a).

(3.b) The *electromagnetic CGS system of units* forms a coherent system of units in combination with the three-dimensional "electromagnetic system" of equations (1.b).

(3.c) The *mixed CGS system of units,* called *Gaussian units,* consisting of the set of electric units of the electrostatic CGS system on the one hand and the set of magnetic units of the electromagnetic CGS system on the other hand, form together also a coherent system of units, when used in combination with the three-dimensional "symmetrical system" of equations (1.c).

4. *MKSA system of units.*

The MKSA system of units forms a coherent system of units in either of the four-dimensional systems of equations mentioned under 2.

The MKSA system is, however, most commonly used together with the rationalized (four-dimensional) equations (2.b).

5. *centimetre-gramme-second-franklin system and centimetre-gramme-second-biot system*

Several investigators have pointed out over many years the advantages, which result from the use of the four-dimensional equations, i.e. equations with four basic quantities from which at least one is of electrical nature. These advantages are partly of a didactical nature, but many investigators consider the usage of four basic quantities also important for developing a clear representation of the field of electricity and magnetism.

It has often been considered as a disadvantage that the transition from the three- to the four-dimensional system of equations should be tied to the transition from the CGS system of units to the MKSA system of units as explained under 3, and 4, which as a consequence leads to a change in the numerical value of many well known physical quantities.

For that reason several investigators have advocated the introduction of a four-dimensional system of units, which is a generalization of the CGS system. This unit-system is chosen so that to any unit of the CGS system (with three basic units) there corresponds one particular unit of this "generalized CGS system" (with four basic units), such that the *numerical* values of the physical quantities in the field of electricity are invariant.

The introduction of such a "generalized CGS system" has the advantage that the relation between the units of this "generalized CGS system" and the units of the MKSA system can be expressed by ordinary conversion equations.

The use of the four-dimensional system of equations (with four basic quantities) accompanied by the usage of such a "generalized CGS system" of units (with four basic units) is therefore of advantage as an intermediate representation in this period of coexistence of the CGS system and the MKSA system, where transitions from one system to the other often have to be made.

As a result of all these considerations the General Assembly of the IUPAP, Copenhagen, 1951, approved by its resolution 5 the introduction of the following two "generalized CGS systems" based on four basic units:

(5.a) The system with the *centimetre, gramme, second* and the charge corresponding to one *electrostatic unit,* as basic units, being a 4-dimensional generalization of the electrostatic CGS system (3.a)

(5.b) The system with the *centimetre, gramme second* and the current corresponding to one *electromagnetic unit* or deca-ampere, as basic units, being a 4-dimensional generalization of the electromagnetic CGS system (3.b)

In practical applications of these systems it is of advantage to indicate the *basic electrical unit* in each of these two systems with a name and symbol.

The name franklin was proposed in 1941 for the electrostatic unit of electric charge considered as a basic unit of the system (5.a). The franklin, symbol: Fr, is thus defined as:

The *franklin* is that charge, which exerts on an equal charge at a distance of 1 centimetre in vacuo a force of 1 dyne.

According to this definition 1 franklin $= (10/\zeta)$ coulomb, where $\zeta = 2.997925 \times 10^{10}$ is the numerical value of the velocity of light in vacuum, measured in cm/s.

For the electromagnetic unit of electric current, considered as a basic unit of the system (5.b), the name biot has been used. The biot, symbol: Bi, is thus defined as:

The *biot* is that constant current intensity, which, when maintained in two parallel infinitely long rectilinear conductors of infinite length and of negligible circular section, placed at a mutual distance of 1 centimetre apart in vacuo, would produce between these conductors a force of 2 dyne per centimetre length.

According to this definition 1 biot = 10 ampere*).

These unit systems (5.a) and (5.b) are referred to as cm-g-s-Fr system and cm-g-s-Bi system respectively.

The definitions of the franklin in the system (5.a) and that of the biot in the system (5.b) which closely correspond to the electrostatic CGS unit of charge and the electromagnetic CGS unit of current respectively, ensure that the quantities of the *non-rationalized* four-dimensional system of equations (second column, table p. 27), when expressed in these units, have the same numerical value as the corresponding quantities of the non-rationalized three-dimensional system of equations (first column, table p. 27), when expressed in the corresponding CGS units.

Some of the units of the three-dimensional CGS system and the corresponding units of the two four-dimensional systems are given in the table:

	three-dimensional unit of mixed CGS system		corresponding four-dimensional unit of generalized CGS system
Q^*	$\text{erg}^{\frac{1}{2}} \cdot \text{cm}^{\frac{1}{2}}$	Q	Fr
I^*	$\text{erg}^{\frac{1}{2}} \cdot \text{cm}^{\frac{1}{2}}\ \text{s}^{-1\frac{1}{2}}$	I	Fr/s
V^*	$\text{erg}^{\frac{1}{2}} \cdot \text{cm}^{-\frac{1}{2}}$	V	erg/Fr
$\boldsymbol{E}^*$	$\text{erg}^{\frac{1}{2}} \cdot \text{cm}^{-\frac{3}{2}}$	$\boldsymbol{E}$	dyn/Fr
$\boldsymbol{D}^*$	$\text{erg}^{\frac{1}{2}} \cdot \text{cm}^{-\frac{3}{2}}$	$\boldsymbol{D}'$	Fr/cm^2
C^*	cm	C	Fr^2/erg
ϵ_r	1	ϵ'	$\text{Fr}^2/\text{erg} \cdot \text{cm}$
$\boldsymbol{B}^*$	G	$\boldsymbol{B}$	dyn/Bi · cm
$\boldsymbol{H}^*$	Oe	$\boldsymbol{H}'$	Bi/cm
μ_r	1	μ'	dyn/Bi^2

* The wording of this definition is, except for the replacement of 1 metre by 1 centimetre and of 2 newton per metre by 2 dyne per centimetre, the exact translation of the original french text defining the ampere (C. R. 9me Conf. Gén. Poids et Mesures, 1948, p. 49).

ALPHABET TABLE

Greek letter	Greek name	English equivalent	RUSSIAN letter	English equivalent
Α α	Alpha	(ä)	А а	(ä)
Β β	Beta	(b)	Б б	(b)
Γ γ	Gamma	(g)	В в	(v)
Δ δ	Delta	(d)	Г г	(g)
Ε ϵ	Epsilon	(e)	Д д	(d)
Ζ ζ	Zeta	(z)	Е е	(ye)
Η η	Eta	(ā)	Ж ж	(zh)
Θ θ	Theta	(th)	З з	(z)
Ι ι	Iota	(ē)	И и	(i, ē)
Κ κ	Kappa	(k)	Й й	(ē) 7
Λ λ	Lambda	(l)	К к	(k)
Μ μ	Mu	(m)	Л л	(l)
Ν ν	Nu	(n)	М м	(m)
Ξ ξ	Xi	(ks)	Н н	(n)
Ο ο	Omicron	(ɑ)	О о	(ô, o)
Π π	Pi	(p)	П п	(p)
Ρ ρ	Rho	(r)	Р р	(r)
Σ σ ς	Sigma	(s)	С с	(s)
Τ τ	Tau	(t)	Т т	(t)
Υ υ	Upsilon	(ü, o͞o)	У у	(o͞o)
Φ φ	Phi	(f)	Ф ф	(f)
Χ χ	Chi	(H)	Х х	(kh)
Ψ ψ	Psi	(ps)	Ц ц	(ts)
Ω ω	Omega	(ō)	Ч ч	(ch)
			Ш ш	(sh)
			Щ щ	(shch)
			Ъ ъ	8
			Ы ы	(ē)
			Ь ь	9
			Э э	(e)
			Ю ю	(ū)
			Я я	(yä)

AMERICAN STANDARD ABBREVIATIONS FOR SCIENTIFIC AND ENGINEERING TERMS

Reproduced by permission of The American Society of Mechanical Engineers

Introductory Notes

Scope and Purpose

1. The Executive Committee of the Sectional Committee on Scientific and Engineering Symbols and Abbreviations has made the following distinction between symbols and abbreviations: Letter symbols are letters used to represent magnitudes of physical quantities in equations and mathematical formulas. Abbreviations are shortened forms of names or expressions employed in texts and tabulations, and should not be used in equations.

Fundamental Rules

2. Abbreviations should be used sparingly in text and with due regard to the context and to the training of the reader. Terms denoting units of measurement should be abbreviated in the text only when preceded by the amounts indicated in numerals; thus "several inches," "one inch," "12 in." In tabular matter, specifications, maps, drawings, and texts for special purposes, the use of abbreviations should be governed only by the desirability of conserving space.

3. Short words such as ton, day, and mile should be spelled out.

4. Abbreviations should not be used where the meaning will not be clear. In case of doubt, spell out.

5. The same abbreviation is used for both singular and plural, as "bbl" for barrel and barrels.

6. The use of conventional signs for abbreviations in text is not recommended; thus "per," not /; "lb," not #; "in," not ″. Such signs may be used sparingly in tables and similar places for conserving space.

7. The period should be omitted except in cases where the omission would result in confusion.

8. The letters of such abbreviations as ASA should not be spaced (not A S A).

9. The use in text of exponents for the abbreviations of square and cube and of the negative exponents for terms involving "per" is not recommended. The superior figures are usually not available on the keyboards of typesetting and linotype machines and conposition is therefore delayed. There is also the liklihood of confusion with footnote reference numbers. These shorter forms are permissible in tables and are sometimes difficult to avoid in text.

10. A sentence should not begin with a numeral followed by an abbreviation. Abbreviations for names of units are to be used only after numerical values. such as 25 ft or 110 v.

Abbreviations*

absolute .... abs
acre .... spell out
acre-foot .... acre-ft
air horsepower .... air hp
alternating-current (as adjective) .... a-c
ampere .... amp
ampere-hour .... amp-hr
amplitude, an elliptic function .... am.
Angstrom unit .... A
antilogarithm .... antilog
atmosphere .... atm
atomic weight .... at. wt
average .... avg
avoirdupois .... avdp
azimuth .... az or α

barometer .... bar.
barrel .... bbl
Baumé .... Bé
board feet (feet board measure) .... fbm
boiler pressure .... spell out
boiling point .... bp
brake horsepower .... bhp
brake horsepower-hour .... bhp-hr
Brinell hardness number .... Bhn
British thermal unit[1] .... Btu or B
bushel .... bu

calorie .... cal
candle c
candle-hour .... c-hr
candlepower .... cp
cent .... c or ¢
center to center .... c to c
centigram .... cg
centiliter cl
centimeter .... cm
centimeter-gram-second (system) .... cgs
chemical .... chem
chemically pure .... cp
circular .... cir
circular mils .... cir mils
coefficient .... coef
cologarithm .... colog
concentrate .... conc
conductivity .... cond
constant .... const
continental horsepower .... cont hp
cord .... cd
cosecant .... csc
cosine .... cos
cosine of the amplitude, an elliptic function .... cn
cost, insurance, and freight .... cif
cotangent .... cot
coulomb .... spell out
counter electromotive force .... cemf
cubic .... cu
cubic centimeter .... cu cm, cm^3 (liquid, meaning milliliter. ml)
cubic foot .... cu ft
cubic feet per minute .... cfm
cubic feet per second .... cfs
cubic inch .... cu in.
cubic meter .... cu m or m^3
cubic micron .... cu μ or cu mu or μ^3
cubic millimeter .... cu mm or mm^3
cubic yard .... cu yd
current density .... spell out
cycles per second .... spell out or c
cylinder .... cyl

day .... spell out
decibel .... db
degree[1] .... deg or °
degree centigrade C
degree Fahrenheit .... F
degree Kelvin .... K
degree Réaumur .... R
delta amplitude, an elliptic function .... dn
diameter .... diam
direct-current (as adjective) .... d-c
dollar .... $
dozen .... doz
dram .... dr

* These forms are recommended for readers whose familiarity with the terms used makes possible a maximum of abbreviations. For other classes of readers editors may wish to use less contracted combinations made up from this list. For example, the list gives the abbreviation of the term "feet per second" as "fps." To some readers ft per sec will be more easily understood.

[1] Abbreviation recommended by the A.S.M.E. Power Test Codes Committee. B = 1 Btu, kB = 1000 Btu, mB = 1,000,000 Btu. The A.S.H. &V.E. recommends the use of Mb = 1000 Btu and Mbh = 1000 Btu per hr.

efficiency....eff
electric....elec
electromotive force....emf
elevation....el
equation....eq
external....ext
farad....spell out or f
feet board measure (board feet)....fbm
feet per minute....fpm
feet per second....fps
fluid....fl
foot....ft
foot-candle....ft-c
foot-Lambert....ft-L
foot-pound....ft-lb
foot-pound-second (system)....fps
foot-second (see cubic feet per second)
franc....fr
free aboard ship....spell out
free alongside ship....spell out
free on board....fob
freezing point....fp
frequency....spell out
fusion point....fnp

gallon....gal
gallons per minute....gpm
gallons per second....gps
grain....spell out
gram....g
gram-calorie....g-cal
greatest common divisor....gcd

haversine....hav
hectare....ha
henry....h
high-pressure (adjective)....h-p
hogshead....hhd
horsepower....hp
horsepower-hour....hp-hr
hour....hr
hour (in astronomical tables)....h
hundred....C
hundredweight (112 lb)....cwt
hyperbolic cosine....cosh
hyperbolic sine....sinh
hyperbolic tangent....tanh

inch....in.
inch-pound....in-lb
inches per second....ips
indicated horsepower....ihp
indicated horsepower-hour....ihp-hr
inside diameter....ID
intermediate-pressure (adjective)....i-p
internal....int

joule....j
kilocalorie....kcal
kilocycles per second....kc
kilogram....kg
kilogram-calorie....kg-cal
kilogram-meter....kg-m
kilograms per cubic meter....kg per cu m or kg/m³
kilograms per second....kgps
kiloliter....kl
kilometer....km
kilometers per second....kmps
kilovolt....kv
kilovolt-ampere....kva
kilowatt....kw
kilowatthour....kwhr

lambert....L
latitude....lat or ϕ
least common multiple....lcm
linear foot....lin ft
liquid....liq
lira....spell out
liter....l
logarithm (common)....log
logarithm (natural)....$\log_e$ or ln
longitude....long. or λ
low-pressure (as adjective)....l-p
lumen....l*
lumen-hour....l-hr*
lumens per watt....lpw

mass....spell out
mathematics (ical)....math
maximum....max
mean effective pressure....mep
mean horizontal candlepower....mhcp
megacycle....spell out
megohm....spell out
melting point....mp
meter....m
meter-kilogram....m-kg
mho....spell out
microampere....μa or mu a
microfarad....μf
microinch....μin.
micromicrofarad....$\mu\mu$f
micromicron....$\mu\mu$ or mu mu
micron....μ or mu
microvolt....μv
microwatt....μw or mu w
mile....spell out
miles per hour....mph
miles per hour per second....mphps
milliampere....ma
milligram....mg
millihenry....mh
millilambert....mL
milliliter....ml
millimeter....mm
millimicron....mμ or m mu
million....spell out
million gallons per day....mgd
millivolt....mv
minimum....min
minute....min
minute (angular measure)....′
minute (time)(in astronomical tables)....m
mole....spell out
molecular weight....mol. wt
month....spell out
National Electrical Code....NEC
ohm....spell out or Ω
ohm-centimeter....ohm-cm
ounce....oz
ounce-foot....oz-ft

[1] There are circumstances under which one or the other of these forms is preferred. In general the sign ° is used where space conditions make it necessary, as in tabular matter, and when abbreviations are cumbersome, as in some angular measurements, i.e., 59° 23′ 42″. In the interest of simplicity and clarity the Committee has recommended that the abbreviation for the temperature scale, F, C, K, etc., always be included in expressions for numerical temperatures, but, wherever feasible, the abbreviation for "degree" be omitted; as 69 F.

*** The International Commission on Illumination has changed the symbol for lumen to lm, and the symbol for lumen-hour to lm-hr. This nomenclature is used in American Standard for Illuminating Engineering Nomenclature and Photometric Standards (ASA Z7.1–1942).**

AMERICAN STANDARD ABBREVIATIONS (Continued)

ounce-inch .... oz-in.
outside diameter .... OD

parts per million .... ppm
peck .... pk
penny (pence) .... d
pennyweight .... dwt
per .... (See Fundamental Rules)
peso .... spell out
pint .... pt
potential .... spell out
potential difference .... spell out
pound .... lb
pound-foot .... lb-ft
pound-inch .... lb-in.
pound sterling .... £
pounds per brake horsepower-hour .... lb per bhp-hr
pounds per cubic foot .... lb per cu ft
pounds per square foot .... psf
pounds per square inch .... psi
pounds per square inch absolute .... psia
power factor .... spell out or pf

quart .... qt

radian .... spell out
reactive kilovolt-ampere .... kvar
reactive volt-ampere .... var
revolutions per minute .... rpm
revolutions per second .... rps
rod .... spell out
root mean square .... rms

secant .... sec
second .... sec
second (angular measure) .... ″
second-foot (see cubic feet per second)
second (time) (in astronomical tables) .... s
shaft horsepower .... shp
shilling .... s
sine .... sin
sine of the amplitude, an elliptic function .... sn
specific gravity .... sp gr
specific heat .... sp ht
spherical candle power .... scp
square .... sq
square centimeter .... sq cm or cm^2
square foot .... sq ft
square inch .... sq in.
square kilometer .... sq km or km^2
square meter .... sq m or m^2
square micron .... sq μ or sq mu or μ^2
square millimeter .... sq mm or mm^2
square root of mean square .... rms
standard .... std
stere .... s

tangent .... tan
temperature .... temp
tensile strength .... ts
thousand .... M
thousand foot-pounds .... kip-ft
thousand pound .... kip
ton .... spell out
ton-mile .... spell out

versed sine .... vers
volt v
volt-ampere .... va
volt-coulomb .... spell out

watt .... w
watthour .... whr
watts per candle .... wpc
week .... spell out
weight .... wt

yard .... yd
year .... yr

ABBREVIATIONS OF COMMON UNITS OF WEIGHT AND MEASURE

From NBS Miscellaneous Publication No. 233.

Spelling and Abbreviations of Units

The spelling of the names of units as adopted by the National Bureau of Standards is that given in the list below. The spelling of the metric units is in accordance with that given in the law of July 28, 1866, legalizing the metric system in the United States.

Following the name of each unit in the list below is given the abbreviation which the Bureau has adopted. Attention is particularly called to the following principles:

1. The period is omitted after all abbreviations of units, except where the abbreviation forms an English word.
2. The exponents "^{2}" and "^{3}" are used to signify "square" and "cubic," respectively, instead of the abbreviations "sq" or "cu," which are, however, frequently used in technical literature for the United States customary units. In conformity with this principle the abbreviation for cubic centimeter is "cm^3" (instead of "cc" or "c cm"). The term "cubic centimeter," as used in chemical work, is, in fact, a misnomer, since the unit actually used is the "milliliter" of which "ml" is the correct abbreviation.
3. The use of the same abbreviation for both singular and plural is recommended. This practice is already established in expressing metric units and is in accordance with the spirit and chief purpose of abbreviations.
4. It is also suggested that, unless all the text is printed in capital letters, only small letters be used for abbreviations, except in such case as, A for angstrom, etc., where the use of capital letters is general.

LIST OF THE MOST COMMON UNITS OF WEIGHT AND MEASURE AND THEIR ABBREVIATIONS

Unit	Abbreviation	Unit	Abbreviation
acre	acre	kiloliter	kl
angstrom	A	kilometer	km
are	a	link	li
avoirdupois	avdp	liquid	liq
barrel	bbl	liter	liter
board foot	fbm	meter	m
bushel	bu	metric ton	t
carat	c	microgram*	μg
centare	ca	microinch	$\mu in.$
centigram	cg	microliter*	μl
centiliter	cl	micron	μ
centimeter	cm	mile	mi
chain	ch	milligram	mg
cubic centimeter	cm^3	milliliter	ml
cubic decimeter	dm^3	millimeter	mm
cubic dekameter	dkm^3	millimicron	$m\mu$
cubic foot	ft^3	minim	min *or* ♏
cubic hectometer	hm^3	ounce	oz
cubic inch	$in.^3$	ounce, apothecaries	oz ap *or* ℥
cubic kilometer	km^3	ounce, avoirdupois	oz avdp
cubic meter	m^3	ounce, fluid	fl oz
cubic mile	mi^3	ounce, troy	oz t
cubic millimeter	mm^3	peck	pk
cubic yard	yd^3	pennyweight	dwt
decigram	dg	pint	pt
deciliter	dl	pound	lb
decimeter	dm	pound, apothecaries	lb ap
decistere	ds	pound, avoirdupois	lb avdp
dekagram	dkg	pound, troy	lb t
dekaliter	dkl	quart	qt
dekameter	dkm	rod	rd
dekastere	dks	scruple, apothecaries	s ap *or* ℈
dram	dr	square centimeter	cm^2
dram, apothecaries	dr ap *or* ʒ	square chain	ch^2
dram, avoirdupois	dr avdp	square decimeter	dm^2
dram, fluid	fl dr	square dekameter	dkm^2
fathom	fath	square foot	ft^2
foot	ft	square hectometer	hm^2
furlong	fur.	square inch	$in.^2$
gallon	gal	square kilometer	km^2
grain	grain	square link	li^2
gram	g	square meter	m^2
hectare	ha	square mile	mi^2
hectogram	hg	square millimeter	mm^2
hectoliter	hl	square rod	rd^2
hectometer	hm	square yard	yd^2
hogshead	hhd	stere	s
hundredweight	cwt	ton	ton
inch	in.	ton, metric	t
kilogram	kg	troy	t
		yard	yd

* The abbreviations γ and λ for microgram and microliter, respectively, have been advocated by some authorities.

CONVERSION FACTORS

(From various U. S. Government and IUPAC publications and from calculations based on values given in these publications)

To convert from	To	Multiply by
Abamperes	Amperes	10
"	E.M. cgs. units of current	1
"	E.S. cgs. units	2.997930×10^{10}
"	Faradays (chem.)/sec	1.036377×10^{-4}
"	Faradays (phys.)/sec	1.036086×10^{-4}
"	Statamperes	2.997930×10^{10}
Abamperes/cm	E.M. cgs. units of surface charge density	1
"	E.S. cgs. units	2.997930×10^{10}
Abamperes/sq. cm	Amperes/circ. mil	5.0670748×10^{-6}
"	Amperes/sq. cm	10
"	Amperes/sq. inch	64.516
Abampere-turns	Ampere-turns	10
Abampere-turns/cm	Ampere-turns/cm	10
Abcoulombs	Ampere-hours	0.0027777
"	Coulombs	10
"	Electronic charges	6.24196×10^{19}
"	E.M. cgs. units of charge	1
"	E.S. cgs. units	2.997930×10^{10}
"	Faradays (chem.)	1.036377×10^{-4}
"	Faradays (phys.)	1.036086×10^{-4}
"	Statcoulombs	2.997930×10^{10}
Abfarads	E.M. cgs. units of capacitance	1
"	E.S. cgs. units	8.987584×10^{20}
"	Farads	1×10^{9}
"	Microfarads	1×10^{15}
"	Statfarads	8.987584×10^{20}
Abhenries	E.M. cgs. units of induction	1
"	E.S. cgs. units	1.112646×10^{-21}
"	Henries	1×10^{-9}
Abmhos	E.M. cgs. units of conductance	1
"	E.S. cgs. units	8.987584×10^{20}
"	Megamhos	1000
"	Mhos	1×10^{9}
"	Statmhos	8.987584×10^{20}
Abohms	E.M.cgs. units of resistance	1
"	Megohms	1×10^{-15}
"	Microhms	0.001
"	Ohms	1×10^{-9}
"	Statohms	1.112646×10^{-21}
Abohm-cm	Circ. mil-ohms/ft	0.0060153049
"	E.M. cgs. units of resistivity	1
"	Microhm-inches	0.00039370079
"	Ohm-cm	1×10^{-9}
Abvolts	Microvolts	0.01
"	Millivolts	1×10^{-5}
"	Volts	1×10^{-8}
"	Volts (Int.)	9.99670×10^{-9}
Abvolts/cm	E.M. cgs. units of electric field intensity	1
"	E.S. cgs. units	3.335635×10^{-11}
"	Volts/cm	1×10^{-8}
"	Volts/inch	2.54×10^{-8}
"	Volts/meter	1×10^{-6}
Acres	Sq. cm	40468564
"	Sq. ft.	43560
"	Sq. ft. (U.S. Survey)	43559.826
"	Sq. inches	6272640
"	Sq. kilometers	0.0040468564
"	Sq. links (Gunter's)	1×10^{5}
"	Sq. meters	4046.8564
"	Sq. miles (statute)	0.0015625
"	Sq. perches	160
"	Sq. rods	160
"	Sq. yards	4840
Acre-feet	Cu. feet	43560
"	Cu. meters	1233.4818
"	Cu. yards	1613.333
Acre-inches	Cu. feet	3630
"	Cu. meters	102.79033
Acre-inches	Gallons (U.S.)	27154.286
Amperes	Abamperes	0.1
"	Amperes (Int.)	1.000165
"	Cgs. units of current	1
"	Mks. units of current	1
"	Coulombs/sec	1
"	Coulombs (Int.)/sec	1.000165
"	Faradays (chem.)/sec	1.036377×10^{-5}
"	Faradays (phys.)/sec	1.036086×10^{-5}
"	Statamperes	2.997930×10^{9}
Amperes (Int.)	Amperes	0.999835
"	Coulombs/sec	0.999835
"	Coulombs (Int.)/sec	1
"	Faradays (chem.)/sec*	1.03623×10^{-5}
"	Faradays (phys.)/sec*	1.03592×10^{-5}
Amperes/meter	Cgs. units of surface current density	0.01
"	E.M. cgs. units	0.001
"	E.S. cgs. units	2.997930×10^{7}
"	Mks. units	1
Amperes/sq. meter	Cgs. units of volume current density	0.0001
"	E.M. cgs. units	1×10^{-5}
"	E.S. cgs. units	299793.0
"	Mks. units	1
Amperes/sq. mil	Abamperes/sq. cm	15500.031
"	Amperes/sq. cm	1.5500031×10^{5}
Ampere-hours	Abcoulombs	360
"	Coulombs	3600
"	Faradays (chem.)*	0.373096
"	Faradays (phys.)*	0.372991
Ampere-turns	Cgs. units of magnetomotive force	1.2566371
"	E.M. cgs. units	1.2566371
"	E.S. cgs. units	3.767310×10^{10}
"	Gilberts	1.2566371
Ampere-turns/weber	Cgs. units of reluctance	1.256637×10^{-8}
"	E.M. cgs. units	1.256637×10^{-8}
"	E.S. cgs. units	1.129413×10^{13}
"	Gilberts/maxwell	1.256637×10^{-8}
Ångström units	Centimeters	1×10^{-8}
"	Inches	3.9370079×10^{-9}
"	Microns	0.0001
"	Millimicrons	0.1
"	Wave length of orange-red line of krypton 86	0.000165076373
"	Wave length of red line of cadmium	0.000155316413
Ares	Acres	0.024710538
"	Sq. dekameters	1
"	Sq. feet	1076.3910
"	Sq. ft. (U.S. Survey)	1076.3867
"	Sq. meters	100
"	Sq. miles	3.8610216×10^{-5}
Atmospheres	Bars	1.01325
"	Cm. of Hg (0°C.)	76
"	Cm. of H_2O (4°C.)	1033.26
"	Dynes/sq. cm	1.01325×10^{6}
"	Ft. of H_2O (39.2°F.)	33.8995
"	Grams/sq. cm	1033.23
"	In. of Hg (32°F.)	29.9213
"	Kg./sq. cm	1.03323
"	Mm. of Hg (0°C.)	760
"	Pounds/sq. inch	14.6960
"	Tons (short)/sq. ft.	1.05811
"	Torrs	760
Atomic mass units (chem.)*	Electron volts	9.31395×10^{8}
"	Grams*	1.66024×10^{-24}
Atomic mass units (phys.)*	Electron volts	9.31141×10^{8}
"	Grams*	1.65979×10^{-24}

* For factors for C = 12 scale or those which can be derived from same see table on Value for General Physical Constants.

CONVERSION FACTORS (Continued)

To convert from	To	Multiply by
Bags (Brit.)	Bushels (Brit.)	3
Barns	Sq. cm	1×10^{-24}
Barrels (Brit.)*	Bags (Brit.)	1.5
"	Barrels (U.S., dry)	1.415404
"	Barrels (U.S., liq.)	1.372513
"	Bushels (Brit.)	4.5
"	Bushels (U.S.)	4.644253
"	Cu. feet	5.779568
"	Cu. meters	0.1636591
"	Gallons (Brit.)	36
"	Liters	163.6546
Barrels (petroleum, U.S.)	Cu. feet	5.614583
"	Gallons (U.S.)	42
"	Liters	158.98284
Barrels (U.S., dry)	Barrels (U.S. liq.)	0.969696
"	Bushels (U.S.)	3.2812195
"	Cu. feet	4.083333
"	Cu. inches	7056
"	Cu. meters	0.11562712
"	Quarts (U.S., dry)	105
Barrels (U.S., liq.)	Barrels (U.S., dry)	1.03125
"	Barrels (wine)	1
"	Cu. feet	4.2109375
"	Cu. inches	7276.5
"	Cu. meters	0.11924047
"	Gallons (Brit.)	26.22925
"	Gallons (U.S., liq.)	31.5
"	Liters	119.23713
Bars	Atmospheres	0.986923
"	Baryes	1×10^{6}
"	Cm. of Hg (0°C.)	75.0062
"	Dynes/sq. cm	1×10^{6}
"	Ft. of H_2O (60°F.)	33.4883
"	Grams/sq. cm	1019.716
"	In. of Hg (32°F.)	29.5300
"	Kg./sq. cm	1.019716
"	Millibars	1000
"	Pounds/sq. inch	14.5038
Baryes	Atmospheres	9.86923×10^{-7}
"	Bars	1×10^{-6}
"	Dynes/sq. cm	1
"	Grams/sq. cm	0.001019716
"	Millibars	0.001
Bels	Decibels	10
Board feet	Cu. cm	2359.7372
"	Cu. feet	0.833333
"	Cu. inches	144
Bolts of cloth	Linear feet	120
"	Meters	36.576
Bougie decimales	Candles (Int.)	1.00
B.t.u	B.t.u. (IST.)**	0.999346
"	B.t.u. (mean)	0.998563
"	B.t.u. (39°F.)	0.994982
"	B.t.u. (60°F.)	0.999689
"	Cal. *gm*	251.99576
"	Cal., *gm.* (IST.)	251.831
"	Cal., *gm.* (mean)	251.634
"	Cal., *gm.* (20°C.)	252.122
"	Cu. cm.-atm	10405.6
"	Ergs	1.05435×10^{10}
"	Foot-poundals	25020.1
"	Foot-pounds	777.649
"	Gram-cm	1.07514×10^{7}
"	Hp.-hours	0.000392752
"	Hp.-years	4.48347×10^{-8}
"	Joules	1054.35
"	Joules (Int.)	1054.18
"	Kg.-meters	107.514
"	Kw.-hours	0.000292875

To convert from	To	Multiply by
B.t.u	Kw.-hours (Int.)	0.000292827
"	Liter-atm	10.4053
"	Tons of refrig. (U.S. std.)	3.46995×10^{-6}
"	Watt-seconds	1054.35
"	Watt-seconds (Int.)	1054.18
B.t.u. (IST.)	B.t.u	1.00065
B.t.u. (mean)	B.t.u	1.00144
"	B.t.u. (IST.)	1.00078
"	B.t.u. (39°F.)	0.996415
"	B.t.u. (60°F.)	1.00113
"	Hp.-hours	0.000393317
"	Joules	1055.87
"	Kg.-meters	107.669
"	Kw.-hours	0.000293297
"	Kw.-hours (Int.)	0.000293248
"	Liter-atm	10.4203
"	Watt-hours	0.293297
"	Watt-hours (Int.)	0.293248
B.t.u. (39°F.)	B.t.u	1.00504
"	B.t.u. (IST.)	1.00439
"	B.t.u. (mean)	1.00360
"	B.t.u. (60°F)	1.00473
"	Joules	1059.67
B.t.u. (60°F.)	B.t.u	1.00031
"	B.t.u. (IST.)	0.999657
"	B.t.u. (mean)	0.998873
"	B.t.u. (39°F.)	0.995291
B.t.u./hr	Cal., *kg.*/hr	0.251996
"	Ergs/sec	2.928751×10^{6}
"	Foot-pounds/hr	777.649
"	Horsepower	0.000392752
"	Horsepower (boiler)	2.98563×10^{-5}
"	Horsepower (electric)	0.000392594
"	Horsepower (metric)	0.000398199
"	Kilowatts	0.000292875
"	Lb. ice melted/hr	0.0069714
"	Tons of refrig. (U.S. comm.)	8.32789×10^{-5}
"	Watts	0.292875
B.t.u./min	Cal., *kg.*/min	0.251996
"	Ergs/sec	1.75725×10^{8}
"	Foot-pounds/min	777.649
"	Horsepower	0.0235651
"	Horsepower (boiler)	0.00179138
"	Horsepower (electric)	0.0235556
"	Horsepower (metric)	0.0238920
"	Joules/sec	17.5725
"	Kg.-meters/min	107.514
"	Kilowatts	0.0175725
"	Lb. ice melted/hr	0.41828
"	Tons of refrig. (U.S. comm.)	0.00499673
"	Watts	17.5725
B.t.u. (mean)/min	B.t.u. (mean)/hr	60
"	Cal., *kg.* (mean)/hr	15.1197
"	Cal., *kg.* (mean)/min	0.251996
"	Ergs/sec	1.75978×10^{8}
"	Foot-pounds/min	778.768
"	Horsepower	0.0235990
"	Horsepower (boiler)	0.00179396
"	Horsepower (electric)	0.0235895
"	Horsepower (metric)	0.0239264
"	Joules/sec	17.5978
"	Kg.-meters/min	107.669
"	Kilowatts	0.0175978
"	Lb. ice-melted/hr	0.41888
B.t.u./lb	Cal., *gm.*/gram	0.555555
"	Cu. cm.-atm./gram	22.9405
"	Cu. ft.-atm./lb	0.367471
"	Cu. ft.-(lb./sq. in.)/lb	5.40034
"	Foot-pounds/lb	777.649
"	Hp.-hr./lb	0.000392752

* Barrel (Brit., liq.) = Barrel (Brit., dry)
** International Steam Table.

1 BTU = .367 cu ft-atm

To convert from	To	Multiply by
B.t.u./lb	Joules/gram	2.32444
B.t.u. (mean)/lb	Cal., *gm.* (mean)/gram	0.555555
"	Cu. cm.-atm./gram	22.9735
"	Foot-pounds/lb	778.768
"	Hp.-hr./lb	0.000393317
"	Joules/gram	2.32779
B.t.u./sec	B.t.u./hr	3600
"	B.t.u./min	60
"	Cal., *kg.*/hr	907.185
"	Cal., *kg.*/min	15.1197
"	Cheval-vapeur	1.43352
"	Ergs/sec	1.05435×10^{10}
"	Foot-pounds/sec	777.649
"	Horsepower	1.41391
"	Horsepower (boiler)	0.107483
"	Horsepower (electric)	1.41334
"	Horsepower (metric)	1.43352
"	Kg.-meters/sec	107.514
"	Kilowatts	1.05435
"	Kilowatts (Int.)	1.05418
"	Watts	1054.35
"	Watts (Int.)	1054.18
B.t.u. (mean)/sec	Ergs/sec	1.05587×10^{10}
"	Foot-pounds/sec	778.768
"	Horsepower	1.41594
"	Horsepower (boiler)	0.107637
"	Horsepower (electric)	1.41537
"	Horsepower (metric)	1.43558
"	Watts	1055.87
B.t.u./sq. ft.	Cal., *gm.*/sq. cm	0.271246
B.t.u./sq.ft. × min.)	Hp./sq. ft	0.0235651
"	Kw./sq. ft	0.0175725
"	Watts/sq. in	0.122031
Buckets (Brit.)	Cu. cm	18184.35
"	Gallons (Brit.)	4
Bushels (Brit.)	Bags (Brit.)	0.333333
"	Bushels (U.S.)	1.032056
"	Cu. cm	36368.70
"	Cu. feet	1.284348
"	Cu. inches	2219.354
"	Dekaliters	3.636768
"	Gallons (Brit.)	8
"	Hectoliters	0.3636768
"	Liters	36.36768
Bushels (U.S.)*	Barrels (U.S.), dry	0.3047647
"	Bushels (Brit.)	0.9689395
"	Cu. cm	35239.07
"	Cu. feet	1.244456
"	Cu. inches	2150.42
"	Cu. meters	0.03523907
"	Cu. yards	0.04609096
"	Gallons (U.S., dry)	8
"	Gallons (U.S., liq.)	9.309177
"	Liters	35.23808
"	Ounces (U.S., fluid)	1191.575
"	Pecks (U.S.)	4
"	Pints (U.S., dry)	64
"	Quarts (U.S., dry)	32
"	Quarts (U.S., liq.)	37.23671
Butts (Brit.)	Bushels (U.S.)	13.53503
"	Cu. feet	16.84375
"	Cu. meters	0.4769619
"	Gallons (U.S.)	126
Cable lengths	Fathoms	120
"	Feet	720
"	Meters	219.456
Calories, *gm.***	B.t.u	0.0039683207
"	B.t.u. (IST.)	0.00396573
"	B.t.u. (mean)	0.00396262
"	B.t.u. (39°F.)	0.00394841
"	B.t.u. (60°F.)	0.00396709
"	Cal., *gm.* (IST.)	0.999346
"	Cal., *gm.* (mean)	0.998563
"	Cal., *gm.* (15°C.)	0.999570
"	Cal., *gm.* (20°C.)	1.00050
"	Cal., *kg.*	0.001
"	Cal., *kg.* (IST.)	0.000999346
"	Cal., *kg.* (mean)	0.000998563
"	Cal., *kg.* (15°C.)	0.000999570
"	Cal., *kg.* (20°C.)	0.00100050
"	Cu. cm.-atm	41.2929
"	Cu. ft.-atm	0.00145824
"	Ergs	4.184×10^{7}
"	Foot-poundals	99.2878
"	Foot-pounds	3.08596
"	Gram-cm	42664.9
"	Hp.-hours	1.55857×10^{-6}
"	Joules	4.184
"	Joules (Int.)	4.18331
"	Kg.-meters	0.426649
"	Kw.-hours	1.162222×10^{-6}
"	Liter-atm	0.0412917
"	Watt-hours	0.001162222
"	Watt-hours (Int.)	0.00116203
"	Watt-seconds	4.184
Calories, *gm.* (mean)	B.t.u	0.00397403
"	Cal., *gm*	1.00144
"	Cal., *gm.* (IST.)	1.00078
"	Cal., *gm.* (20°C.)	1.00194
"	Cal., *kg.* (mean)	0.001
"	Cu. cm.-atm	41.3523
"	Cu. ft.-atm	0.00146034
"	Ergs	4.19002×10^{7}
"	Foot-poundals	99.4308
"	Foot-pounds	3.09040
"	Hp.-hours	1.56081×10^{-6}
"	Joules	4.19002
"	Joules (Int.)	4.18933
"	Kg.-meters	0.427263
"	Kw.-hours	1.16390×10^{-6}
"	Liter-atm	0.0413511
"	Watt-seconds	4.19002
Calories, *gm.* (15°C.)	B.t.u	0.00397003
"	Cal., *gm*	1.00043
"	Cal., *gm.* (IST.)	0.999776
"	Cal., *gm.* (mean)	0.998992
"	Cal., *gm.* (20°C.)	1.00093
"	Joules	4.18580
"	Joules (Int.)	4.18511
Calories, *gm.* (20°C.)	B.t.u	0.00396633
"	Cal., *gm*	0.999498
"	Cal., *gm.* (IST.)	0.998845
"	Cal., *gm.* (mean)	0.998061
"	Cal., *gm.* (15°C.)	0.999068
"	Joules	4.18190
"	Joules (Int.)	4.18121
Calories, *kg*	B.t.u	3.9683207
"	B.t.u. (IST.)	3.96573
"	B.t.u. (mean)	3.96262
"	B.t.u. (60°F.)	3.96709
"	Cal., *gm*	1000
"	Cal., *kg.* (mean)	0.998563
"	Cal., *kg.* (15°C.)	0.999570
"	Cal., *kg.* (20°C.)	1.00050
"	Cu. cm.-atm	41292.86

* Stricken or struck bushel. A heaped bushel for apples of 2747.715 cu. inches was established by the U.S. Court of Customs Appeals on Feb. 15, 1912. A heaped bushel equal to $1\frac{1}{4}$ stricken bushels is also known.

** This is the calorie as defined by the U.S. National Bureau of Standards and is equal to 4.18400 joules.

To convert from	To	Multiply by
Calories, *kg*.	Ergs	4.184×10^{10}
"	Foot-poundals	99287.8
"	Foot-pounds	3085.96
"	Gram-cm	4.26649×10^{7}
"	Hp.-hours	0.00155857
"	Joules	4184
"	Kw.-hours	0.001162222
"	Liter-atm	41.2917
"	Watt-hours	1.162222
Calories, *kg*. (mean)	B.t.u	3.97403
"	B.t.u. (IST.)	3.97144
"	B.t.u. (mean)	3.9683207
"	B.t.u. (60°F.)	3.97280
"	Cal., *gm*	1001.44
"	Cal., *gm*. (IST.)	1000.78
"	Cal., *gm*. (mean)	1000
"	Cal., *gm*. (15°C.)	1000.10
"	Cal., *gm*. (20°C.)	1001.94
"	Ergs	4.19002×10^{10}
"	Foot-poundals	99430.8
"	Foot-pounds	3090.40
"	Gram-cm	4.27263×10^{7}
"	Hp.-hours	0.00156081
"	Joules	4190.02
"	Kg.-meters	427.263
"	Kw.-hours (Int.)	0.00116370
"	Liter-atm	41.3511
"	Watt-hours	1.16390
Cal., *gm*./°C	B.t.u./°F	0.00220462
"	Joules/°F	2.324444
"	Joules (Int.)/°F	2.32406
Cal., *gm*./gram	B.t.u./lb	1.8
"	Foot-pounds/lb	1399.77
"	Joules/gram	4.184
"	Watt-hours/gram	0.001162222
Cal., *gm*./(gram × °C)	B.t.u./(lb. × °C.)	1.8
"	B.t.u./(lb. × °F.)	1
"	Cal., *kg*./(kg. × °C.)	1
"	Joules/(gram × °C.)	4.184
"	Joules/(lb. × °F.)	1054.35
Cal., *gm*./hr	B.t.u./hr	0.0039683207
"	Ergs/sec	11622.222
"	Watts	0.001162222
Cal., *gm*. (mean)/hr	B.t.u. (mean)/hr	0.0039683207
"	Ergs/sec	11639.0
"	Watts	0.00116390
Cal., *kg*./hr	Watts	1.162222
Cal., *gm*./min	B.t.u./min	0.0039683207
"	Ergs/sec	697333.3
"	Watts	0.069733
Cal., *gm*. (mean)/min	B.t.u. (mean)/min	0.0039683207
"	Ergs/sec	698337
"	Joules/sec	0.0698337
"	Watts	0.0698337
Cal., *kg*./min	Kg. ice melted/min	0.012548
"	Lb. ice melted/min	0.027665
"	Watts	69.7333
Cal., *gm*./sec	B.t.u./sec	0.0039683207
"	Ergs/sec	4.184×10^{7}
"	Foot-pounds/sec	3.08596
"	Horsepower	0.00561084
"	Watts	4.184
Cal., *gm*. (mean)/sec	Ergs/sec	4.19002×10^{7}
"	Watts	4.19002
Cal., *gm*./(sec. × sq. cm.)	B.t.u./(hr. × sq. ft.)	13272.1
"	Cal., *gm*./(hr. × sq. cm.)	3600
"	Watts/sq. cm	4.184
Cal., *gm*./(sec. × sq. cm. × °C.)	B.t.u./(hr. × sq. ft. × °F.)	7373.38
Cal., *gm*./sq. cm	B.t.u./sq. ft	3.68669

To convert from	To	Multiply by
Cal., *gm*.-cm./(hr. × sq. cm. × °C.)	B.t.u.-ft./(hr. × sq. ft. × °F.)	0.0671969
"	B.t.u.-inch/(hr. × sq. ft. × °F.)	0.806363
Cal., *gm*.-cm./sq. cm	B.t.u.-inch/sq. ft	1.4514530
Cal., *gm*.-sec	Planck's constant	6.31531×10^{33}
Cal., *gm*.-sec./Avog. No. (chem.)*	Planck's constant	1.04849×10^{10}
Cal., *gm*.-sec./Avog. No. (phys.)*	Planck's constant	1.04821×10^{10}
Candles (English)	Candles (Int.)	1.04
"	Hefner units	1.16
Candles (German)	Candles (English)	1.01
"	Candles (Int.)	1.05
"	Hefner units	1.17
Candles (Int.)	Candles (English)	0.96
"	Candles (German)	0.95
"	Candles (pentane)	1.00
"	Hefner units	1.11
"	Lumens (Int.)/steradian	1
Candles (pentane)	Candles (Int.)	1.00
Candles/sq. cm	Candles/sq. inch	6.4516
"	Candles/sq. meter	10000
"	Foot-lamberts	2918.6351
"	Lamberts	3.1415927
Candles/sq. ft	Candles/sq. inch	0.0069444
"	Candles/sq. meter	10.763910
"	Foot-lamberts	3.1415927
"	Lamberts	0.0033815822
Candles/sq. inch	Candles/sq. cm	0.15500031
"	Candles/sq. foot	144
"	Foot-lamberts	452.38934
"	Lamberts	0.48694784
Candle power (spher.)	Lumens	12.566370
Carats (parts of gold per 24 of mixture)	Milligrams/gram	41.6666
Carats (1877)	Grains	3.168
"	Milligrams	205.3
Carats (metric)	Grains	3.08647
"	Grams	0.2
"	Milligrams	200
Carcel units	Candles (Int.)	9.61
Centals	Kilograms	45.359237
"	Pounds	100
Centares	Ares	0.01
"	Sq. feet	10.763910
"	Sq. inches	1550.0031
"	Sq. meters	1
"	Sq. yards	1.1959900
Centigrams	Grains	0.15432358
"	Grams	0.01
Centiliters	Cu. cm	10.00028
"	Cu. inches	0.6102545
"	Liters	0.01
"	Ounces (U.S., fluid)	0.3381497
Centimeters	Ångström units	1×10^{8}
"	Feet	0.032808399
"	Feet (U.S. Survey)	0.032808333
"	Hands	0.098425197
"	Inches	0.39370079
"	Links (Gunter's)	0.049709695
"	Links (Ramden's)	0.032808399
"	Meters	0.01
"	Microns	10000
"	Miles (naut., Int.)	5.3995680×10^{-6}
"	Miles (statute)	6.2137119×10^{-6}
"	Millimeters	10
"	Millimicrons	1×10^{7}
"	Mils	393.70079
"	Picas (printer's)	2.3710630
"	Points (printer's)	28.452756

* For factors for C = 12 scale or those which can be derived from same table on Values for General Physical Constants.

CONVERSION FACTORS (Continued)

To convert from	To	Multiply by
Centimeters	Rods	0.0019883878
"	Wave length of orange-red line of krypton 86	16507.6373
"	Wave length of red line of cadmium	15531.6413
"	Yards	0.010936133
Cm. of Hg (0°C.)	Atmospheres	0.013157895
"	Bars	0.0133322
"	Dynes/sq. cm	13332.2
"	Ft. of H_2O (4°C.)	0.446050
"	Ft. of H_2O (60°F.)	0.446474
"	In. of Hg (0°C.)	0.39370079
"	Kg./sq. meter	135.951
"	Pounds/sq. ft	27.8450
"	Pounds/sq. inch	0.193368
"	Torrs	10
Cm. of H_2O (4°C.)	Atmospheres	0.000967814
"	Dynes/sq. cm	980.638
"	Pounds/sq. inch	0.0142229
Centimeters/sec	Feet/min	1.9685039
"	Feet/sec	0.032808399
"	Kilometers/hr	0.036
"	Kilometers/min	0.0006
"	Knots (Int.)	0.019438445
"	Meters/min	0.6
"	Miles/hr	0.022369363
"	Miles/min	0.00037282272
Cm./(sec. × sec.)	Kilometers/(hr. × sec.)	0.036
"	Miles/(hr. × sec.)	0.022369363
Centimeters/year	Inches/year	0.39370079
Centipoises*	Grams/(cm. × sec.)	0.01
"	Poises	0.01
"	Pound/(ft. × hr.)	2.4190883
"	Pounds/(ft. × sec.)	0.00067196898
Centistokes*	Stokes	0.01
Chains (Gunter's)	Centimeters	2011.68
"	Chains (Ramden's)	0.66
"	Feet	66
"	Feet (U.S. Survey)	65.999868
"	Furlongs	0.1
"	Inches	792
"	Links (Gunter's)	100
"	Links (Ramden's)	66
"	Meters	20.1168
"	Miles (statute)	0.0125
"	Rods	4
"	Yards	22
Chains (Ramden's)	Centimeters	3048
"	Chains (Gunter's)	1.515151
"	Feet	100
"	Feet (U.S. Survey)	99.999800
Cheval-vapeur	Horsepower (metric)	1
Cheval-vapeur-heures	Joules	2647795
Circles	Degrees	360
"	Grades	400
"	Minutes	21600
"	Radians	6.2831853
"	Signs	12
Circular inches	Circular mm	645.16
"	Sq. cm	5.0670748
"	Sq. inches	0.78539816
Circular mm	Sq. cm	0.0078539816
"	Sq. inches	0.0012173696
"	Sq. mm	0.78539816
Circular mils	Circular inches	1×10^{-6}
"	Sq. cm	5.0670748×10^{-6}
"	Sq. inches	$7.8539816 + 10^{-7}$
"	Sq. mm	0.00050670748
"	Sq. mils	0.78539816
Circumferences	Degrees	360
"	Grades	400
Circumferences	Minutes	21600
"	Radians	6.2831853
"	Seconds	1296000
Cords	Cord-feet	8
"	Cu. feet	128
"	Cu. meters	3.6245734
Cord-feet	Cords	0.125
"	Cu. feet	16
Coulombs	Abcoulombs	0.1
"	Ampere-hours	0.0002777
"	Ampere-seconds	1
"	Coulombs (Int.)	1.000165
"	Electronic charge	6.24196×10^{18}
"	E.M. cgs. units of electric charge	0.1
"	E.S. cgs. units of electric charge	2.997930×10^{9}
"	Faradays (chem.)	1.036377×10^{-5}
"	Faradays (phys.)	1.036086×10^{-5}
"	Mks. units of electric charge	1
"	Statcoulombs	2.997930×10^{9}
Coulombs/cu. meter	E.M. cgs. units of volume charge density	1×10^{-7}
"	E.S. cgs. units	2997.930
Coulombs/sq. cm	Abcoulombs/sq. cm	0.1
"	Cgs. units of polarization, *and* surface charge density	1
Cubic centimeters	Board feet	0.00042377600
"	Bushels (Brit.)	2.749617×10^{-5}
"	Bushels (U.S.)	2.837759×10^{-5}
"	Cu. feet	3.5314667×10^{-5}
"	Cu. inches	0.061023744
"	Cu. meters	1×10^{-6}
"	Cu. yards	1.3079506×10^{-6}
"	Drachms (Brit., fluid)	0.28156080
"	Drams (U.S., fluid)	0.27051218
"	Gallons (Brit.)	0.0002199694
"	Gallons (U.S., dry)	0.00022702075
"	Gallons (U.S., liq.)	0.00026417205
"	Gills (Brit.)	0.007039020
"	Gills (U.S.)	0.0084535058
"	Liters	0.000999972
"	Ounces (Brit., fluid)	0.03519510
"	Ounces (U.S., fluid)	0.033814023
"	Pints (U.S., dry)	0.0018161660
"	Pints (U.S., liq.)	0.0021133764
"	Quarts (Brit.)	0.0008798775
"	Quarts (U.S., dry)	0.00090808298
"	Quarts (U.S., liq.)	0.0010566882
Cu. cm./gram	Cu. ft./lb	0.016018463
Cu. cm./sec	Cu. ft./min	0.0021188800
"	Gal. (U.S.)/min	0.015850323
"	Gal. (U.S.)/sec	0.00026417205
Cu. cm.-atm	B.t.u	9.61019×10^{-5}
"	B.t.u. (mean)	9.59637×10^{-5}
"	Cal., *gm*	0.0242173
"	Cal., *gm.* (mean)	0.0241824
"	Cu. ft.-atm	3.5314667×10^{-5}
"	Joules	0.101325
"	Watt-hours	2.81458×10^{-5}
Cu. cm.-atm./gram	B.t.u./lb	0.0435911
"	Cal., *gm.*/gram	0.0242173
"	Cu. ft.-(lb./sq. in.)/lb	0.235406
"	Ft.-lb./lb	33.8985
"	Joules/gram	0.101325
"	Kg.-meters/gram	0.0103323
"	Kw.-hr./gram	2.81458×10^{-8}
Cubic decimeters	Cu. cm	1000
"	Cu. feet	0.035316667
"	Cu. inches	61.023744
"	Cu. meters	0.001

* See also special table on viscosity.

To convert from	To	Multiply by
Cubic decimeters	Cu. yards	0.0013079506
"	Liters	0.999972
Cubic dekameters	Cu. decimeters	1×10^6
"	Cu. feet	35314.667
"	Cu. inches	6.1023744×10^7
"	Cu. meters	1000
"	Liters	999972
Cubic feet	Acre-feet	2.2956841×10^{-5}
"	Board feet	12
"	Bushels (Brit.)	0.7786049
"	Bushels (U.S.)	0.80356395
"	Cords (wood)	0.0078125
"	Cord-feet	0.0625
"	Cu. centimeters	28316.847
"	Cu. meters	0.028316847
"	Gallons (U.S., dry)	6.4285116
"	Gallons (U.S., liq.)	7.4805195
"	Liters	28.31605
"	Ounces (Brit., fluid)	996.6143
"	Ounces (U.S., fluid)	957.50649
"	Pints (U.S., liq.)	59.844156
"	Quarts (U.S., dry)	25.714047
"	Quarts (U.S., liq.)	29.922078
Cu. ft. of H_2O (39.2°F.)	Pounds of H_2O	62.4262
Cu. ft. ot H_2O (60°F.)	Pounds of H_2O	63.3663
Cu. ft./hr	Acre-feet/hr	2.2956841×10^{-5}
"	Cu. cm./sec	7.8657907
"	Cu. ft./day	24
"	Gal. (U.S.)/hr	7.4805195
"	Liters/hr	28.31605
Cu. ft./min	Acre-feet/hr	0.0013774105
"	Acre-feet/min	2.2956841×10^{-5}
"	Cu. cm./sec	471.94744
"	Cu. ft./hr	60
"	Gal. (U.S.)/min	7.4805195
"	Liters/sec	0.4719342
Cu. ft./lb	Cu. cm./gram	62.427961
"	Millimeters/gram	62.42621
Cu. ft./sec	Acre-inches/hr	0.99173553
"	Cu. cm./sec	28316.847
"	Cu. yards/min	2.222222
"	Gal. (U.S.)/min	448.83117
"	Liters/min	1698.963
"	Liters/sec	28.31605
Cu. ft. of H_2O (60°F.)/sec	Lb. of H_2O/min	3741.98
Cu. ft.-atm	B.t.u	2.72130
"	Cal., *gm*	685.756
"	Cu. cm.-atm	28316.847
"	Cu. ft.-(lb/sq. in.)	14.6960
"	Foot-pounds	2116.22
"	Hp.-hours	0.00106880
"	Joules	2869.20
"	Kg.-meters	292.577
"	Kw.-hours	0.000797001
Cubic inches	Barrels (Brit.)	0.0001001292
"	Barrels (U.S., dry)	0.00014172336
"	Board feet	0.0069444
"	Bushels (Brit.)	0.0004505815
"	Bushels (U.S.)	0.00046502544
"	Cu. cm	16.387064
"	Cu. feet	0.00057870370
"	Cu. meters	1.6387064×10^{-5}
"	Cu. yards	2.1433470×10^{-5}
"	Drams (U.S., fluid)	4.4329004
"	Gallons (Brit.)	0.003604652
"	Gallons (U.S., dry)	0.0037202035
"	Gallons (U.S., liq.)	0.0043290043
"	Liters	0.01638661
"	Milliliters	16.38661
"	Ounces (Brit., fluid)	0.5767444

To convert from	To	Multiply by
Cubic inches	Ounces (U.S., fluid)	0.55411255
"	Pecks (U.S.)	0.0018601017
"	Pints (U.S., dry)	0.029761628
"	Pints (U.S., liq.)	0.034632035
"	Quarts (U.S., dry)	0.014880814
"	Quarts (U.S., liq.)	0.017316017
Cu. in. of H_2O (4°C.)	Pounds of H_2O	0.0361263
Cu. in. of H_2O (60°F.)	Pounds of H_2O	0.0360916
Cubic meters	Acre-feet	0.00081071319
"	Barrels (Brit.)	6.110261
"	Barrels (U.S., dry)	8.648490
"	Barrels (U.S., liq.)	8.3864145
"	Bushels (Brit.)	27.49617
"	Bushels (U.S.)	28.377593
"	Cu. cm	1×10^6
"	Cu. feet	35.314667
"	Cu. inches	61023.74
"	Cu. yards	1.3079506
"	Gallons (Brit.)	219.9694
"	Gallons (U.S., liq.)	264.17205
"	Hogshead	4.1932072
"	Liters	999.972
"	Pints (U.S., liq.)	2113.3764
"	Quarts (U.S., liq.)	1056.6882
"	Steres	1
Cu. meters/min	Gal. (Brit.)/min	219.9694
"	Gal. (U.S.)/min	264.1721
"	Liters/min	999.972
Cu. millimeters	Cu. cm	0.001
"	Cu. inches	6.1023744×10^{-5}
"	Cu. meters	1×10^{-9}
"	Minims (Brit.)	0.01689365
"	Minims (U.S.)	0.016230731
Cu. yards	Bushels (Brit.)	21.02233
"	Bushels (U.S.)	21.696227
"	Cu. cm	764554.86
"	Cu. feet	27
"	Cu. inches	46.656
"	Cu. meters	0.76455486
"	Gallons (Brit.)	168.1787
"	Gallons (U.S., dry)	173.56981
"	Gallons (U.S., liq.)	201.97403
"	Liters	764.5335
"	Quarts (Brit.)	672.7146
"	Quarts (U.S., dry)	694.27926
"	Quarts (U.S., liq.)	807.89610
Cu. yd./min	Cu. ft./sec	0.45
"	Gal. (U.S.)/sec	3.3662338
"	Liters/sec	12.74222
Cubits	Centimeters	45.72
"	Feet	1.5
"	Inches	18
Daltons (chem.)	Grams	1.66024×10^{-24}
Daltons (phys.)	Grams	1.65979×10^{-24}
Days (mean solar)	Days (sidereal)	1.00273791
"	Hours (mean solar)	24
"	Hours (sidereal)	24.065710
"	Years (calendar)	0.0027397260
"	Years (sidereal)	0.0027378031
"	Years (tropical)	0.0027379093
Days (sidereal)	Days (mean solar)	0.99726957
"	Hours (mean solar)	23.934470
"	Hours (sidereal)	24
"	Minutes (mean solar)	1436.0682
"	Minute (sidereal)	1440
"	Second (sidereal)	86400
"	Years (calendar)	0.0027322454
"	Years (sidereal)	0.0027303277
"	Years (tropical)	0.0027304336
Decibels	Bels	0.1
Decimeters	Centimeters	10

To convert from	To	Multiply by
Decimeters	Feet	0.32808399
"	Feet (U.S. Survey)	0.328083333
"	Inches	3.9370079
"	Meters	0.1
Decisteres	Cu. meters	0.1
Degrees	Circles	0.0027777
"	Minutes	60
"	Quadrants	0.0111111
"	Radians	0.017453293
"	Seconds	3600
Degrees/cm	Radians/cm	0.017453293
Degrees/foot	Radians/cm	0.00057261458
Degrees/inch	Radian/cm	0.0068713750
Degrees/min	Degrees/sec	0.0166666
"	Radians/sec	0.00029088821
"	Revolutions/sec	4.629629×10^{-5}
Degrees/sec	Radians/sec	0.017453293
"	Revolutions/min	0.166666
"	Revolutions/sec	0.0027777
Dekaliters	Pecks (U.S.)	1.135136
"	Pints (U.S., dry)	18.16217
Dekameters	Centimeters	1000
"	Feet	32.808399
"	Feet (U.S. Survey)	32.808333
"	Inches	393.70079
"	Kilometers	0.01
"	Meters	10
"	Yards	10.93613
Demals	Gram-equiv./cu. decimeter	1
Drachms (Brit., fluid)	Cu. cm	3.551631
"	Cu. inches	0.2167338
"	Drams (U.S., fluid)	0.9607594
"	Milliliters	3.551531
Drams (apoth. *or* troy)	Drams (avdp.)	2.1942857
"	Grains	60
"	Grams	3887.9346
"	Ounces (apoth. *or* troy)	0.125
"	Ounces (avdp.)	0.13714286
"	Scruples (apoth.)	3
Drams (avdp.)	Drams (apoth. *or* troy)	0.455729166
"	Grains	27.34375
"	Grams	1.7718452
"	Ounces (apoth. *or* troy)	0.056966146
"	Ounces (avdp.)	0.0625
"	Pennyweights	1.1393229
"	Pounds (apoth. *or* troy)	0.0047471788
"	Pounds (avdp.)	0.00390625
"	Scruples (apoth.)	1.3671875
Drams (U.S., fluid)	Cu. cm	3.6967162
"	Cu. inches	0.22558594
"	Drachms (Brit., fluid)	1.040843
"	Gills (U.S.)	0.03125
"	Milliliters	3.696588
"	Minims (U.S.)	60
"	Ounces (U.S., fluid)	0.125
"	Pints (U.S., liq.)	0.0078125
Dynes	Grains	0.01573663
"	Grams	0.001019716
"	Newtons	0.00001
"	Poundals	7.2330138×10^{-5}
"	Pounds	2.248089×10^{-6}
Dynes/cm	Ergs/sq. cm	1
"	Ergs/sq. mm	0.01
"	Grams/cm	0.001019716
"	Poundals/inch	0.00018371855
Dynes/cu. cm	Grams/cu. cm	0.001019716
"	Poundals/cu. inch	0.0011852786
Dynes/sq. cm	Atmospheres	9.86923×10^{-7}
"	Bars	1×10^{-6}
"	Baryes	1
"	Cm. of Hg (0°C.)	7.50062×10^{-5}

To convert from	To	Multiply by
Dynes/sq. cm	Cm. of H_2O (4°C.)	0.001019745
"	Grams/sq. cm	0.001019716
"	In. of Hg (32°F.)	2.95300×10^{-5}
"	In. of H_2O (4°C.)	0.000401474
"	Kg./sq. meter	0.01019716
"	Poundals/sq. in	0.00046664510
"	Pounds/sq. in	1.450377×10^{-5}
Dyne-centimeters	Ergs	1
"	Foot-poundals	2.3730360×10^{-6}
"	Foot-pounds	7.37562×10^{-8}
"	Gram-cm	0.001019716
"	Inch-pounds	8.85075×10^{-7}
"	Kg.-meters	1.019716×10^{-8}
"	Newton-meters	1×10^{-7}
Electron volts	Ergs	1.60209×10^{-12}
"	Grams	1.78253×10^{-33}
Electronic charges	Abcoulombs	1.60209×10^{-20}
"	Coulombs	1.60209×10^{-19}
"	Statcoulombs	4.80296×10^{-10}
Electronic charges/kg	Statcoulombs/dyne	4.89766×10^{-16}
E.S. cgs. units of induction flux	E.M. cgs. units	2.997930×10^{10}
E.S. cgs. units of magnetic charge	E.M. cgs. units	2.997930×10^{10}
E.S. cgs. units of magnetic field intensity	E.M. cgs. units	3.335635×10^{-11}
Ells	Centimeters	114.3
"	Inches	45
Ergs	B.t.u	9.48451×10^{-11}
"	Cal., *gm*	2.39006×10^{-8}
"	Cal., *kg*	2.39006×10^{-11}
"	Cal., *kg.* (20°C.)	2.39126×10^{-11}
"	Cu. cm.-atm	9.86923×10^{-7}
"	Cu. ft.-atm	3.48529×10^{-11}
"	Cu. ft.-(lb./sq. in.)	5.12196×10^{-10}
"	Dyne-cm	1
"	Electron volts	6.24196×10^{11}
"	Foot-poundals	2.3730360×10^{-6}
"	Foot-pounds	7.37562×10^{-8}
"	Gram-cm	0.001019716
"	Joules	1×10^{-7}
"	Joules (Int.)	9.99835×10^{-8}
"	Kw.-hours	2.777777×10^{-14}
"	Kg.-meters	1.019716×10^{-8}
"	Liter-atm	9.86895×10^{-10}
"	Watt-sec	1×10^{-7}
Ergs/(gram-mol. × °C.)	Foot-pounds/(lb.-mol. × °F.)	1.85863×10^{-5}
Ergs/sec	B.t.u./min	5.69071×10^{-9}
"	Cal., *gm.*/min	1.43403×10^{-6}
"	Dyne-cm./sec	1
"	Foot-pounds/min	4.42537×10^{-6}
"	Gram-cm./sec	0.001019716
"	Horsepower	1.34102×10^{-10}
"	Joules/sec	1×10^{-7}
"	Kilowatts	1×10^{-10}
"	Watts	1×10^{-7}
Ergs/sq. cm	Dynes/cm	1
"	Ergs/sq. mm	0.01
Ergs/sq. mm	Dynes/cm	100
"	Ergs/sq. cm	100
Erg-sec	Planck's constant	1.50932×10^{26}
Farads	Abfarads	1×10^{-9}
"	E.M. cgs. units	1×10^{-9}
"	E.S. cgs. units	8.987584×10^{11}
"	Farads (Int.)	1.000495
"	Microfarads	1×10^{6}
"	Statfarads	8.98758×10^{11}
Farads (Int.)	Farads	0.999505
Fathoms	Centimeters	182.88

CONVERSION FACTORS (Continued)

To convert from	To	Multiply by
Fathoms	Feet	6
"	Inches	72
"	Meters	1.8288
"	Miles (naut., Int.)	0.00098747300
"	Miles (statute)	0.001136363
"	Yards	2
Feet	Centimeters	30.48
"	Chains (Gunter's)	0.01515151
"	Fathoms	0.166666
"	Feet (U.S. Survey)	0.99999800
"	Furlongs	0.00151515
"	Inches	12
"	Meters	0.3048
"	Microns	304800
"	Miles (naut., Int.)	0.00016457883
"	Miles (statute)	0.000189393
"	Rods	0.060606
"	Ropes (Brit.)	0.05
"	Yards	0.333333
Feet (U.S. Survey)	Centimeters	30.480061
"	Chains (Gunter's)	0.015151545
"	Chains (Ramden's)	0.010000020
"	Feet	1.0000020
"	Inches	12.000024
"	Links (Gunter's)	1.5151545
"	Links (Ramden's)	1.0000020
"	Meters	0.30480061
"	Miles (statute)	0.00018939432
"	Rods	0.060606182
"	Yards	0.33333400
Feet of air (1 atm., 60°F.)	Atmospheres	3.6083×10^{-5}
"	Ft. of Hg (32°F.)	0.00089970
"	Ft. of H_2O (60°F.)	0.0012244
"	In. of Hg (32°F.)	0.0010796
"	Pounds/sq. inch	0.00053027
Feet of Hg (32°F.)	Cm. of Hg (0°C.)	30.48
"	Ft. of H_2O (60°F.)	13.6085
"	In. of H_2O (60°F.)	163.302
"	Ounces/sq. inch	94.3016
"	Pounds/sq. inch	5.89385
Feet of H_2O (4°C.)	Atmospheres	0.0294990
"	Cm. of Hg (0°C.)	2.24192
"	Dynes/sq. cm	29889.8
"	Grams/sq. cm	30.4791
"	In. of Hg (32°F.)	0.882646
"	Kg./sq. meter	304.791
"	Pounds/sq. inch	0.433515
Feet/hour	Cm./hr	30.48
"	Cm./min	0.508
"	Cm./sec	0.0084666
"	Feet/min	0.0166666
"	Inches/hr	12
"	Kilometers/hr	0.0003048
"	Kilometers/min	5.08×10^{-6}
"	Knots (Int.)	0.0001645788
"	Miles/hr	0.000189393
"	Miles/min	3.156565×10^{-6}
"	Miles/sec	5.2609428×10^{-8}
Feet/minute	Cm./sec	0.508
"	Feet/sec	0.0166666
"	Kilometers/hr	0.018288
"	Meters/min	0.3048
"	Meters/sec	0.00508
"	Miles/hr	0.01136363
Feet/second	Cm./sec	30.48
"	Kilometers/hr	1.09728
"	Kilometers/min	0.018288
"	Meters/min	18.288
"	Miles/hr	0.68181818
"	Miles/min	0.01136363
Feet/(sec. × sec.)	Kilometers/(hr. × sec.)	1.09728

To convert from	To	Multiply by
Feet/(sec. × sec.)	Meters/(sec. × sec.)	0.3048
"	Miles/(hr. × sec.)	0.68181818
Firkins (Brit.)	Bushels (Brit.)	1.125
"	Cu. cm	40914.79
"	Cu. feet	1.444892
"	Firkins (U.S.)	1.200949
"	Gallons (Brit.)	9
"	Liters	40.91364
"	Pints (Brit.)	72
Firkins (U.S.)	Barrels (U.S., dry)	0.29464286
"	Barrels (U.S., liq.)	0.28571429
"	Bushels (U.S.)	0.96678788
"	Cu. feet	1.203125
"	Firkins (Brit.)	0.8326747
"	Liters	34.06775
"	Pints (U.S., liq.)	72
Foot-candles	Lumens/sq. ft	1
"	Lumens/sq. meter	10.763910
"	Lux	10.763910
"	Milliphots	1.0763910
Foot-lamberts	Candles/sq. cm	0.00034262591
"	Candles/sq. ft	0.31830989
"	Millilamberts	1.0763910
"	Lamberts	0.0010763910
"	Lumens/sq. ft	1
Foot-poundals	B.t.u	3.99678×10^{-5}
"	B.t.u. (IST.)	3.99417×10^{-5}
"	B.t.u. (mean)	3.99104×10^{-5}
"	Cal., *gm*	0.0100717
"	Cal., *gm.* (IST.)	0.0100651
"	Cal., *gm.* (mean)	0.0100573
"	Cu. cm.-atm	0.415890
"	Cu. ft.-atm	1.46870×10^{-5}
"	Dyne-cm	4.2140110×10^{5}
"	Ergs	4.2140110×10^{5}
"	Foot-pounds	0.0310810
"	Hp.-hours	1.56974×10^{-8}
"	Joules	0.042140110
"	Joules (Int.)	0.0421332
"	Kg.-meters	0.00429710
"	Kw.-hours	1.17056×10^{-8}
"	Liter-atm	0.000415879
"	B.t.u	0.00128593
Foot-pounds	B.t.u. (IST.)	0.00128509
"	B.t.u. (mean)	0.00128408
"	Cal., *gm*	0.324048
"	Cal., *gm.* (IST.)	0.323836
"	Cal., *gm.* (mean)	0.323582
"	Cal., *gm.* (20°C.)	0.324211
"	Cal., *kg*	0.000324048
"	Cal., *kg.* (IST.)	0.000323836
"	Cal., *kg.* (mean)	0.000323582
"	Cu. ft.-atm	0.000472541
"	Dyne-cm	1.35582×10^{7}
"	Ergs	1.35582×10^{7}
"	Foot-poundals	32.1740
"	Gram-cm	13825.5
"	Hp.-hours	5.05050×10^{-7}
"	Joules	1.35582
"	Kg.-meters	0.138255
"	Kw.-hours	3.76616×10^{-7}
"	Kw.-hours (Int.)	3.76554×10^{-7}
"	Liter-atm	0.0133805
"	Newton-meters	1.3558180
"	Lb. H_2O evap. from and at 212°F.	1.3245×10^{-6}
"	Watt-hours	0.000376616
Foot-pounds/hr	B.t.u./min	2.14321×10^{-5}
"	B.t.u. (mean)/min	2.14013×10^{-5}
"	Cal., gm./min	0.00540080
"	Cal., *gm.* (mean)/min	0.00539304
"	Ergs/min	2.25970×10^{5}

To convert from	To	Multiply by
Foot-pounds/hr	Foot-pounds/min	0.0166666
"	Horsepower	5.050505×10^{-7}
"	Horsepower (metric)	5.12055×10^{-7}
"	Kilowatts	3.76616×10^{-7}
"	Watts	0.000376616
"	Watts (Int.)	0.000376554
Foot-pounds/min	B.t.u./sec	2.14321×10^{-5}
"	B.t.u. (mean)/sec	2.14013×10^{-5}
"	Cal., *gm.*/sec	0.00540080
"	Cal., *gm.* (mean)/sec	0.00539304
"	Ergs/sec	2.25970×10^{5}
"	Foot-pounds/sec	0.0166666
"	Horsepower	3.030303×10^{-5}
"	Horsepower (metric)	3.07233×10^{-5}
"	Joules/sec	0.0225970
"	Joules (Int.)/sec	0.0225932
"	Kilowatts	2.25970×10^{-5}
"	Watts	0.0225970
Foot-pounds/lb	B.t.u./lb	0.00128593
"	B.t.u. (IST.)/lb	0.00128509
"	B.t.u. (mean)/lb	0.00128408
"	Cal., *gm.*/gm	0.000714404
"	Cal., *gm.* (IST.)/gram	0.000713937
"	Cal., *gm.* (mean)/gram	0.000713377
"	Hp.-hr./lb	5.05050×10^{-7}
"	Joules/gram	0.00298907
"	Kg.-meters/gram	0.000304800
"	Kw.-hr./gram	8.30296×10^{-10}
Foot-pounds/sec	B.t.u./min	0.0771556
"	B.t.u. (mean)/min	0.0770447
"	B.t.u./sec	0.00128593
"	B.t.u. (mean)/sec	0.00128408
"	Cal., *gm.*/sec	0.324048
"	Cal., *gm.* (mean)/sec	0.323582
"	Ergs/sec	1.35582×10^{7}
"	Gram-cm./sec	13825.5
"	Horsepower	0.00181818
"	Joules/sec	1.35582
"	Kilowatts	0.00135582
"	Watts	1.35582
"	Watts (Int.)	1.35559
Furlongs	Centimeters	20116.8
"	Chains (Gunter's)	10
"	Chains (Ramden's)	6.6
"	Feet	660
"	Inches	7920
"	Meters	201.168
"	Miles (naut., Int.)	0.10862203
"	Miles (statute)	0.125
"	Rods	40
"	Yards	220
Gallons (Brit.)	Barrels (Brit.)	0.027777
"	Bushels (Brit.)	0.125
"	Cu. centimeters	4546.087
"	Cu. feet	0.1605436
"	Cu. inches	277.4193
"	Drachms (Brit. fluid)	1280
"	Firkins (Brit.)	0.111111
"	Gallons (U.S., liq.)	1.200949
"	Gills (Brit.)	32
"	Liters	4.545960
"	Minims (Brit.)	76800
"	Ounces (Brit., fluid)	160
"	Ounces (U.S., fluid)	153.7215
"	Pecks (Brit.)	0.5
"	Lb. of H_2O (62°F.)	10
Gallons (U.S., dry)	Barrels (U.S., dry)	0.038095592
"	Barrels (U.S., liq.)	0.036941181
"	Bushels (U.S.)	0.125
"	Cu. centimeters	4404.8828
"	Cu. feet	0.15555700

To convert from	To	Multiply by
Gallons (U.S., dry)	Cu. inches	268.8025
"	Gallons (U.S., liq.)	1.16364719
"	Liters	4.404760
Gallons (U.S., liq.)	Acre-feet	3.0688833×10^{-6}
"	Barrels (U.S., liq.)	0.031746032
"	Barrels (petroleum, U.S.)	0.023809524
"	Bushels (U.S.)	0.10742088
"	Cu. centimeters	3785.4118
"	Cu. feet	0.133680555
"	Cu. inches	231
"	Cu. meters	0.0037854118
"	Cu. yards	0.0049511317
"	Gallons (Brit.)	0.8326747
"	Gallons (U.S., dry)	0.85936701
"	Gallons (wine)	1
"	Gills (U.S.)	32
"	Liters	3.785306
"	Minims (U.S.)	61440
"	Ounces (U.S., fluid)	128
"	Pints (U.S., liq.)	8
"	Quarts (U.S., liq.)	4
Gallons (U.S.) of H_2O (4°C.)	Lb. of H_2O	8.34517
Gallons (U.S.) of H_2O (60°F.)	Lb. of H_2O	8.33717
Gallons (U.S.)/day	Cu. ft./hr	0.0055700231
Gallons (Brit.)/hr	Cu. meters/min	7.576812×10^{-5}
Gallons (U.S.)/hr	Acre-feet/hr	3.0688833×10^{-6}
"	Cu. ft./hr	0.1336805
"	Cu. meters/min	6.3090197×10^{-5}
"	Cu. yd./min	8.2518861×10^{-5}
"	Liters/hr	3.785306
Gal. (Brit.)/sec	Cu. cm./sec	4546.087
Gal. (U.S.)/sec	Cu. cm./sec	3785.4118
"	Cu. ft./min	8.020833
"	Cu. yd./min	0.29706790
"	Liters/min	227.1183
Gammas	Grams	1×10^{-6}
"	Micrograms	1
Gausses	E.M. cgs. units of magnetic flux density	1
"	E.S. cgs. units	3.335635×10^{-11}
"	Gausses (Int.)	0.999670
"	Maxwells/sq. cm	1
"	Lines/sq. cm	1
"	Lines/sq. inch	6.4516
Gausses (Int.)	Gausses	1.000330
Gausses/oersted	E.M. cgs. units of permeability	1
"	E.S. cgs. units	1.112646×10^{-21}
Geepounds	Slugs	1
"	Kilograms	14.5939
Gigameters	Meters	1×10^{9}
Gilberts	Abampere-turns	0.079577472
"	Ampere-turns	0.79577472
"	E.M. cgs. units of mmf., *or* magnetic potential	1
"	E.S. cgs. units	2.997930×10^{10}
"	Gilberts (Int.)	1.000165
Gilberts (Int.)	Gilberts	0.999835
Gilberts/cm	Ampere-turns/cm	0.79577472
"	Ampere-turns/in	2.0212678
"	Oersteds	1
Gilberts/maxwell	Ampere-turns/weber	7.957747×10^{7}
"	E.M. cgs. units of reluctance	1
"	E.S. cgs. units	8.987584×10^{20}
Gills (Brit.)	Cu. cm	142.0652
"	Gallons (Brit.)	0.03125
"	Gills (U.S.)	1.200949
"	Liters	0.1420613
"	Ounces (Brit., fluid)	5

CONVERSION FACTORS (Continued)

To convert from	To	Multiply by
Gills (Brit.)	Ounces (U.S., fluid)	4.803764
"	Pints (Brit.)	0.25
Gills (U.S.)	Cu. cm	118.29412
"	Cu. inches	7.21875
"	Drams (U.S., fluid)	32
"	Gallons (U.S., liq.)	0.03125
"	Gills (Brit.)	0.8326747
"	Liters	0.1182908
"	Minims (U.S.)	1920
"	Ounces (U.S., fluid)	4
"	Pints (U.S., liq.)	0.25
"	Quarts (U.S., liq.)	0.125
Grades	Circles	0.0025
"	Circumferences	0.0025
"	Degrees	0.9
"	Minutes	54
"	Radians	0.015707963
"	Revolutions	0.0025
"	Seconds	3240
Grains	Carats (metric)	0.32399455
"	Drams (apoth. *or* troy)	0.016666
"	Drams (avdp.)	0.036571429
"	Dynes	63.5460
"	Grams	0.06479891
"	Milligrams	64.79891
"	Ounces (apoth. *or* troy)	0.0020833
"	Ounces (avdp.)	0.0022857143
"	Pennyweights	0.041666
"	Pounds (apoth. *or* troy)	0.000173611
"	Pounds (avdp.)	0.00014285714
"	Scruples (apoth.)	0.05
"	Tons (metric)	6.479891×10^{-8}
Grains/cu. ft	Grams/cu. meter	2.2883519
Grains/gal. (U.S.)	Parts/million*	17.11854
"	Pounds/million gal	142.8571
Grams	Carats (metric)	5
"	Decigrams	10
"	Dekagrams	0.1
"	Drams (apoth. *or* troy)	0.25720597
"	Drams (avdp.)	0.56438339
"	Dynes	980.665
"	Grains	15.432358
"	Kilograms	0.001
"	Micrograms	1×10^{6}
"	Myriagrams	0.0001
"	Ounces (apoth. *or* troy)	0.032150737
"	Ounces (avdp.)	0.035273962
"	Pennyweights	0.64301493
"	Poundals	0.0709316
"	Pounds (apoth. *or* troy)	0.0026792289
"	Pounds (avdp.)	0.0022046226
"	Scruples (apoth.)	0.77161792
"	Tons (metric)	1×10^{-6}
Grams/cm	Dynes/cm	980.665
"	Grams/inch	2.54
"	Kg./km	100
"	Kg./meter	0.1
"	Poundals/inch	0.180166
"	Pounds/ft	0.067196898
"	Pounds/inch	0.0055997415
"	Tons (metric)/km	0.1
Grams/(cm. × sec.)	Poises	1
"	Lb./(ft. × sec.)	0.06719690
Grams/cu. cm	Dynes/cu. cm	980.665
"	Grains/milliliter	15.43279
"	Grams/milliliter	1.000028
"	Poundals/cu. inch	1.16236
"	Pounds/circ. mil-ft	3.4049170×10^{-7}
"	Pounds/cu. ft	62.427961
"	Pounds/cu. inch	0.036127292
"	Pounds/gal. (Brit.)	10.02241
Grams/cu. cm	Pounds/gal. (U.S., dry)	9.7111064
"	Pounds/gal. (U.S., liq.)	8.3454044
Grams/cu. meter	Grains/cu. ft	0.43699572
Grams/liter	Parts/million*	1000
"	Lb./cu. ft.	0.06242621
"	Lb./gal. (U.S.)	8.345171×10^{-3}
Grams/milliliter	Grams/cu. cm	0.999972
"	Pounds/cu. ft	62.42621
"	Pounds/gallon (U.S.)	8.345171×10^{-3}
Grams/sq. cm	Atmospheres	0.000967841
"	Bars	0.000980665
"	Cm. of Hg. (0°C.)	0.0735559
"	Dynes/sq. cm	980.665
"	In. of Hg (32°F.)	0.0289590
"	Kg./sq. meter	10
"	Mm. of Hg (0°C.)	0.735559
"	Poundals/sq. inch	0.457623
"	Pounds/sq. inch	0.014223343
Grams/ton (long)	Milligrams/kg	0.98420653
Grams/ton (short)	Milligrams/kg	1.1023113
Grams-cm	B.t.u.	9.30113×10^{-8}
"	B.t.u. (IST.)	9.29505×10^{-8}
"	B.t.u. (mean)	9.28776×10^{-8}
"	Cal., *gm*	2.34385×10^{-5}
"	Cal., *gm.* (IST.)	2.34231×10^{-5}
"	Cal., *gm.* (mean)	2.34048×10^{-5}
"	Cal., *gm.* (15°C.)	2.34284×10^{-5}
"	Cal., *gm*, (20°C.)	2.34502×10^{-5}
"	Cal., *kg*	2.34385×10^{-8}
"	Cal., *kg.* (IST.)	2.34231×10^{-8}
"	Cal., *kg.* (mean)	2.34048×10^{-8}
"	Dyne-cm	980.665
"	Ergs	980.665
"	Foot-poundals	0.00232715
"	Foot-pounds	$7.2330138 \times 10^{-}$
"	Hp.-hours	3.65303×10^{-11}
"	Joules	9.80665×10^{-5}
"	Kw.-hours	2.72407×10^{-11}
"	Kw.-hours (Int.)	2.72362×10^{-11}
"	Newton-meters	9.80665×10^{-5}
"	Watt-hours	2.72407×10^{-8}
Gram-cm./sec	B.t.u./sec	9.30113×10^{-8}
"	Cal., *gm.*/sec	2.34385×10^{-5}
"	Ergs-sec	980.665
"	Foot-pounds/sec	7.2330138×10^{-5}
"	Horsepower	1.31509×10^{-7}
"	Joules/sec	9.80665×10^{-5}
"	Kilowatts	9.80665×10^{-8}
"	Kilowatts (Int.)	9.80503×10^{-8}
"	Watts	9.80665×10^{-5}
Gram/sq. cm	Pounds/sq. inch	0.000341717
Gram wt.-sec,/sq. cm.	Poises	980.665
Gravitational constants	Cm./(sec. × sec.)	980.621
"	Ft./(sec. × sec.)	32.1725
Hands	Centimeters	10.16
"	Inches	4
Hectares	Acres	2.4710538
"	Ares	100
"	Sq. cm	1×10^{8}
"	Sq. feet	107639.10
"	Sq. meters	10000
"	Sq. miles	0.0038610216
"	Sq. rods	395.36861
Hectograms	Grams	100
"	Poundals	7.09316
"	Pounds (apoth *or* troy)	0.26792289
"	Pounds (avdp.)	0.22046226
Hectoliters	Bushels (Brit.)	2.749694
"	Bushels (U.S.)	2.837839

* Based on density of 1 gram/ml.

To convert from	To	Multiply by
Hectoliters	Cu. cm	1.00028×10^5
"	Cu. feet	3.531566
"	Gallons (U.S., liq.)	26.41794
"	Liters	100
"	Ounces (U.S.) fluid	3381.497
"	Pecks (U.S.)	11.35136
Hectometers	Centimeters	10000
"	Decimeters	1000
"	Dekameters	10
"	Feet	328.08399
"	Meters	100
"	Rods	19.883878
"	Yards	109.3613
Hectowatts	Watts	100
Hefner units	Candles (English)	0.86
"	Candles (German)	0.85
"	Candles (Int.)	0.90
"	10-cp. pentane candles	0.090
Henries	Abhenries	1×10^9
"	E.M. cgs. units	1×10^9
"	E.S. cgs. units	1.112646×10^{-12}
"	Henries (Int.)	0.999505
"	Millihenries	1000
"	Mks. (r *or* nr) units	1
"	Stathenries	1.112646×10^{-12}
Henries (Int.)	Henries	1.000495
Henries/meter	Cgs. units of permeability	795774.72
"	E.M. cgs. units	795774.72
"	E.S. cgs. units	8.854156×10^{-16}
"	Gausses/oersted	795774.72
"	Mks. (nr) units	0.079577472
"	Mks. (r) units	1
Hogsheads	Butts (Brit.)	0.5
"	Cu. feet	8.421875
"	Cu. inches	14553
"	Cu. meters	0.23848094
"	Gallons (Brit.)	52.458505
"	Gallons (U.S.)	63
"	Gallons (wine)	63
"	Liters	238.47427
Horsepower*	B.t.u. (mean)/hr	2542.48
"	B.t.u./min	42.4356
"	B.t.u. (mean)/sec	0.706243
"	Cal., *gm.*/hr	6.41616×10^5
"	Cal., *gm.* (IST.)/hr	6.41196×10^5
"	Cal., *gm.* (mean)/hr	6.40693×10^5
"	Cal., *gm.*/min	10693.6
"	Cal., *gm.* (IST.)/min	10686.6
"	Cal., *gm.* (mean)/min	10678.2
"	Ergs/sec	7.45700×10^9
"	Foot-pounds/hr	1980000
"	Foot-pounds/min	33000
"	Foot-pounds/sec	550
"	Horsepower (boiler)	0.0760181
"	Horsepower (electric)	0.999598
"	Horsepower (metric)	1.01387
"	Joules/sec	745.700
"	Kilowatts	0.745700
"	Kilowatts (Int.)	0.745577
"	Tons of refrig. (U.S., comm.)	0.21204
"	Watts	745.700
Horsepower (boiler)	B.t.u. (mean)/hr	33445.7
"	Cal., *gm.*/min	140671.6
"	Cal., *gm.* (mean)/min	140469.4
"	Cal., *gm.* (15°C.)/min	140611.1
"	Cal., *gm.*, (20°C.)/min	140742.2
"	Ergs/sec	9.80950×10^{10}
"	Foot-pounds/min	434107
"	Horsepower	13.1548
"	Horsepower (electric)	13.1495
Horsepower (boiler)	Horsepower (metric)	13.3372
"	Horsepower (water)	13.1487
"	Joules/sec	9809.50
"	Kilowatts	9.80950
"	Lb. H_2O evap. per hr. from and at 212°F.	34.5
Horsepower (electric)	B.t.u./hr	2547.16
"	B.t.u. (IST.)/hr	2545.50
"	B.t.u. (mean)/hr	2543.50
"	Cal., *gm.*/sec	178.298
"	Cal., *kg.*/hr	641.874
"	Ergs/sec	7.46×10^9
"	Foot-pounds/min	33013.3
"	Foot-pounds/sec	550.221
"	Horsepower	1.00040
"	Horsepower (boiler)	0.0760487
"	Horsepower (metric)	1.0142777
"	Horsepower (water)	0.999942
"	Joules/sec	746
"	Kilowatts	0.746
"	Watts	746
Horsepower (metric)	B.t.u./hr	2511.31
"	B.t.u. (IST.)/hr	2509.66
"	B.t.u. (mean)/hr	2507.70
"	Cal., *gm.*/hr	6.32838×10^5
"	Cal., *gm.* (IST.)/hr	6.32425×10^5
"	Cal., *gm.* (mean)/hr	6.31929×10^5
"	Ergs/sec	7.35499×10^9
"	Foot-pounds/min	32548.6
"	Foot-pounds/sec	542.476
"	Horsepower	0.986320
"	Horsepower (boiler)	0.0749782
"	Horsepower (electric)	0.985923
"	Horsepower (water)	0.985866
"	Kg.-meters/sec	75
"	Kilowatts	0.735499
"	Watts	735.499
Horsepower (water)	Foot-pounds/min	33015.2
"	Horsepower	1.00046
"	Horsepower (boiler)	0.0760531
"	Horsepower (electric)	1.00006
"	Horsepower (metric)	1.01434
"	Kilowatts	0.746043
Horsepower-hours	B.t.u	2546.14
"	B.t.u. (IST.)	2544.47
"	B.t.u. (mean)	2542.48
"	Cal., *gm*	641616
"	Cal., *gm.* (IST.)	641196
"	Cal., *gm.* (mean)	640693
"	Foot-pounds	1.98×10^6
"	Joules	2.68452×10^6
"	Kg.-meters	273745
"	Kw.-hours	0.745700
"	Watt-hours	745.700
Hp.-hr./lb	B.t.u./lb	2546.14
"	Cal., *gm.*/gram	1414.52
"	Cu. ft.-(lb./sq. in.)/lb	13750
"	Foot-pounds/lb	1980000
"	Joules/gram	5918.35
Hours (mean solar)	Days (mean solar)	0.0416666
"	Days (sidereal)	0.041780746
"	Hours (sidereal)	1.00273791
"	Minutes (mean solar)	60
"	Minutes (sidereal)	60.164275
"	Seconds (mean solar)	3600
"	Seconds (sidereal)	3609.8565
"	Weeks (mean calendar)	0.0059523809
Hours (sidereal)	Days (mean solar)	0.41552899
"	Days (sidereal)	0.0416666
"	Hours (mean solar)	0.99726957
"	Minutes (mean solar)	59.836174

* Mechanical horsepower, equal to 550 ft.-lb./sec.

CONVERSION FACTORS (Continued)

To convert from	To	Multiply by
Hours (sidereal)	Minutes (sidereal)	60
Hundredweights (long)	Kilograms	50.802345
"	Pounds	112
"	Quarters (Brit., long)	4
"	Quarters (U.S., long)	0.2
"	Tons (long)	0.05
Hundredweights (short)	Kilograms	45.359237
"	Pounds (advp.)	100
"	Quarters (Brit., short)	4
"	Quarters (U.S., short)	0.2
"	Tons (long)	0.044642857
"	Tons (metric)	0.045359237
"	Tons (short)	0.05
Inches	Ångström units	2.54×10^8
"	Centimeters	2.54
"	Chains (Gunter's)	0.00126262
"	Cubits	0.055555
"	Fathoms	0.013888
"	Feet	0.083333
"	Feet (U.S. Survey)	0.083333167
"	Links (Gunter's)	0.126262
"	Links (Ramden's)	0.083333
"	Meters	0.0254
"	Mils	1000
"	Picas (printer's)	6.0225
"	Points (printer's)	72.27000
"	Wave length of orange-red line of krypton 86	41929.399
"	Wave length of the red line of cadmium	39450.369
"	Yards	0.027777
Inches of Hg (32°F.)	Atmospheres	0.0334211
"	Bars	0.0338639
"	Dynes/sq. cm	33863.9
"	Ft. of air (1 atm., 60°F.)	926.24
"	Ft. of H_2O (39.2°F.)	1.132957
"	Grams/sq. cm	34.5316
"	Kg./sq. meter	345.316
"	Mm. of Hg (60°C.)	25.4
"	Ounces/sq. inch	7.85847
Inches of Hg (32°F.)	Pounds/sq. ft	70.7262
Inches of Hg (60°F.)	Atmospheres	0.0333269
"	Dynes/sq. cm	39768.5
"	Grams/sq. cm	34.4343
"	Mm. of Hg (60°F.)	25.4
"	Ounces/sq. inch	7.83633
"	Pounds/sq. ft	70.5269
Inches of H_2O(4°C.)	Atmospheres	0.0024582
"	Dynes/sq. cm	2490.82
"	In. of Hg (32°F.)	0.0735539
"	Kg./sq. meter	25.3993
"	Ounces/sq. ft	83.2350
"	Ounces/sq. inch	0.578020
"	Pouncs/sq. ft	5.20218
"	Pounds/sq. inch	0.03612628
Inches/hr	Cm./hr	2.54
"	Feet/hr	0.0833333
"	Miles/hr	1.578282×10^{-5}
Inches/min	Cm./hr	152.4
"	Feet/hr	5
"	Miles/hr	0.000946969
Joules (abs.)	B.t.u	0.000948451
" "	B.t.u. (IST.)	0.000947831
" "	B.t.u. (mean)	0.000947088
" "	Cal., *gm*	0.239006
" "	Cal., *gm.* (IST.)	0.238849
" "	Cal., *gm.* (mean)	0.238662
" "	Cal., *gm.* (15°C.)	0.238903
" "	Cal., *gm.* (20°C.)	0.239126
Joules (abs)	Cal., *kg.* (mean)	0.000238662
" "	Cu. ft.-atm	0.000348529
" "	Ergs	1×10^7
" "	Foot-poundals	23.730360
" "	Foot-pounds	0.737562
" "	Gram-cm	10197.16
" "	Hp.-hours	3.72506×10^{-7}
" "	Joules (Int.)	0.999835
" "	Kg.-meters	0.1019716
" "	Kw.-hours	2.7777×10^{-7}
" "	Liter-atm	0.00986895
" "	Volt-coulombs (Int.)	0.999835
" "	Watt-hours (abs.)	0.0002777777
" "	Watt-hours (Int.)	0.000277732
" "	Watt-sec	1
" "	Watt-sec. (Int.)	0.999835
Joules (Int.)	B.t.u	0.000948608
"	B.t.u. (IST.)	0.000947988
"	B.t.u. (mean)	0.000947244
"	Cal. *gm*	0.239045
"	Cal., *gm.* (IST.)	0.238888
"	Cal., *gm.* (mean)	0.238702
"	C.h.u	0.000527004
"	C.h.u. (IST.)	0.000526660
"	C.h.u. (mean)	0.000526247
"	Cu. cm.-atm	9.87086
"	Cu. ft.-atm	0.000348586
"	Dyne-cm	1.000165×10^7
"	Ergs	1.000165×10^7
"	Foot-poundals	23.73428
"	Foot-pounds	0.737684
"	Gram-cm	10198.8
"	Joules (abs.)	1.000165
"	Kw.-hours	2.77824×10^{-7}
"	Liter-atm	0.00987058
"	Volt-coulombs	1.000165
"	Volt-coulombs (Int.)	1
"	Watt-sec	1.000165
"	Watt-sec. (Int.)	1
Joules/(abcoulomb) × °F.)	Joules/(coulomb × °C.)	0.18
Joules/amp.-hr	Joules/abcoulomb	0.002777
"	Joules/statcoulomb	9.265653×10^{-14}
Joules/coulomb	Joules/abcoulomb	10
"	Volts	1
Joules/(coulomb × °F.)	Joules/(coulomb × °C.)	1.8
Joules/°C	B.t.u./°F	0.000526917
"	Cal., *gm.*/°C	0.239006
"	Cal., *gm.* (mean)/°C	0.238662
Joules/electronic charge	Joules/abcoulomb	6.24196×10^{19}
Joules/(electronic charge × °C.)	Joules/(coulomb × °C.)	6.24196×10^{18}
Joules/(gram × °C.)	B.t.u./(lb. × °F.)	0.239006
"	Cal., *gm.*/(gram × °C.)	0.239006
Joules (Int.)/(gram °C.)	B.t.u./(lb. × °F.)	0.239045
"	Cal., *gm.* (mean)/(gram × °C.)	0.238702
Joules/sec. (abs.)	B.t.u./min	0.0569071
"	Cal., *gm.*/min	14.3403
"	Cal., *kg.*/min	0.0143403
"	Cal., *kg.* (mean)/min	0.0143197
"	Dyne-cm./sec	1×10^7
"	Ergs/sec	1×10^7
"	Foot-pounds/sec	0.737562
"	Gram-cm./sec	10197.16
"	Horsepower	0.00134102
"	Watts	1
"	Watts (Int.)	0.999835
Joules (Int.)/sec	B.t.u./min	0.0569165

To convert from	To	Multiply by
Joules (Int.)/sec	B.t.u. (mean)/min	0.0568347
"	Cal., *gm.*/min	14.3427
"	Cal., *kg.*/min	0.0143427
"	Dyne-cm./sec	1.000165×10^7
"	Ergs/sec	1.000165×10^7
"	Foot-pounds/min	44.2610
"	Foot-pounds/sec	0.737684
"	Gram-cm./sec	10198.8
"	Horsepower	0.00134124
"	Watts	1.000165
"	Watts (Int.)	1
Kilderkins (Brit.)	Cu. cm	81829.57
"	Cu. feet	2.889784
"	Cu. inches	4993.55
"	Cu. meters	0.08182957
"	Gallons (Brit.)	18
Kilograms	Drams (apoth. *or* troy)	257.20597
"	Drams (avdp.)	564.38339
"	Dynes	980665
"	Grains	15432.358
"	Hundredweights (long)	0.019684131
"	Hundredweights (short)	0.022046226
"	Ounces (apoth. *or* troy)	32.150737
"	Ounces (avdp.)	35.273962
"	Pennyweights	643.01493
"	Poundals	70.931635
"	Pounds (apoth. *or* troy)	2.6792289
"	Pounds (avdp.)	2.2046226
"	Quarters (Brit., long)	0.078736522
"	Quarters (U.S. long)	0.0039368261
"	Scruples (apoth.)	771.61792
"	Slugs	0.06852177
"	Tons (long)	0.00098420653
"	Tons (metric)	0.001
"	Tons (short)	0.0011023113
Kilograms/cu. meter	Grams/cu. cm	0.001
"	Lb./cu. ft	0.062427961
"	Lb./cu. inch	3.6127292×10^{-5}
Kg. of ice melted/hr	Tons of refrig. (U.S., comm.)	0.026336
Kilograms/sq. cm	Atmospheres	0.967841
"	Bars	0.980665
"	Cm. of Hg (0°C.)	73.5559
"	Dynes/sq. cm	980665
"	Ft. of H_2O (39.2°F.)	32.8093
"	In. of Hg (32°F.)	28.9590
"	Pounds/sq. inch	14.223343
Kilograms/sq. meter	Atmospheres	9.67841×10^{-5}
"	Bars	9.80665×10^{-5}
"	Dynes/sq. cm	98.0665
"	Ft. of H_2O (39.2°F.)	0.00328093
"	Grams/sq. cm	0.1
"	In. of Hg. (32°F.)	0.00289590
"	Mm. of Hg (0°C.)	0.0735559
"	Pounds/sq. ft	0.20481614
"	Pounds/sq. in	0.0014223343
Kilograms/sq. mm	Pounds/sq. ft	204816.14
"	Pounds/sq. in	1422.3343
"	Tons (short)/sq. in	0.71116716
Kilogram sq. cm	Pounds sq. ft	0.0023730360
"	Pounds sq. in	0.34171719
Kilogram-meters	B.t.u. (mean)	0.00928776
"	Cal., *gm.* (mean)	2.34048
"	Cal., *kg.* (mean)	0.00234048
"	Cu. ft.-atm	0.00341790
"	Dynes-cm	9.80665×10^7
"	Ergs	9.80665×10^7
"	Foot-poundals	232.715
"	Foot-pounds	7.23301
"	Gram-cm	100000
Kilogram-meters	Hp.-hours	3.65304×10^{-6}
"	Joules	9.80665
"	Joules (Int.)	9.80503
"	Kw.-hours	2.72407×10^{-6}
"	Liter-atm	0.0967814
"	Newton-meters	9.80665
"	Watt-hours	0.00272407
"	Watt-hours (Int.)	0.00272362
Kilogram-meters/sec	Watts	9.80665
Kilolines	Maxwells	1000
"	Webers	1×10^{-5}
Kiloliters	Cu. centimeters	1.000028×10^6
"	Cu. feet	35.31566
"	Cu. inches	61025.45
"	Cu. meters	1.000028
"	Cu. yards	1.307987
"	Gallons (Brit.)	219.9755
"	Gallons (U.S., dry)	227.0271
"	Gallons (U.S., liq.)	264.1794
"	Liters	1000
Kilometers	Astronomical units	6.68878×10^{-9}
"	Centimeters	100000
"	Feet	3280.8399
"	Feet (U.S. Survey)	3280.833
"	Light years	1.05702×10^{-13}
"	Meters	1000
"	Miles (naut., Int.)	0.53995680
"	Miles (statute)	0.62137119
"	Myriameters	0.1
"	Rods	198.83878
"	Yards	1093.6133
Kilometers/hr	Cm./sec	27.7777
"	Feet/hr	3280.8399
"	Feet/min	54.680665
"	Knots (Int.)	0.53995680
"	Meters/sec	0.277777
"	Miles (statute)/hr	0.62137119
Kilometers/(hr. × sec.)	Cm./(sec. × sec.)	27.7777
"	Ft./(sec. × sec.)	0.91134442
"	Meters/(sec. × sec.)	0.277777
Kilometers/min	Cm./sec	1666.666
"	Feet/min	3280.8399
"	Kilometers/hr	60
"	Knots (Int.)	32.397408
"	Miles/hr	37.282272
"	Miles/min	0.62137119
Kilovolts/cm	Abvolts/cm	1×10^{11}
"	Microvolts/meter	1×10^{11}
"	Millivolts/meter	1×10^8
"	Statvolts/cm	3.335635
"	Volts/inch	2540
Kilowatts	B.t.u./hr	3414.43
"	B.t.u. (IST.)/hr	3412.19
"	B.t.u. (mean)/hr	3409.52
"	B.t.u. (mean)/min	56.8253
"	B.t.u. (mean)/sec	0.947088
"	Cal., *gm.* (mean)/hr	859184
"	Cal., *gm.* (mean)/min	14319.7
"	Cal., *gm.* (mean)/sec	238.662
"	Cal., *kg.* (mean)/hr	859.184
"	Cal., *kg.* (mean)/min	14.3197
"	Cal., *kg.* (mean)/sec	0.238662
"	Cu. ft.-atm./hr	1254.70
"	Ergs/sec	1×10^{10}
"	Foot-poundals/min	1.42382×10^6
"	Foot-pounds/hr	2.65522×10^6
"	Foot-pounds/min	44253.7
"	Foot-pounds/sec	737.562
"	Gram-cm./sec	1.019716×10^7
"	Horsepower	1.34102

CONVERSION FACTORS (Continued)

To convert from	To	Multiply by
Kilowatts	Horsepower (boiler)	0.101942
"	Horsepower (electric)	1.34048
"	Horsepower (metric)	1.35962
"	Joules/hr	3.6×10^6
"	Joules (IST.)/hr	3.59941×10^6
"	Joules/sec	1000
"	Kg.-meters/hr	3.67098×10^5
"	Kilowatts (Int.)	0.999835
"	Watts (Int.)	999.835
Kilowatts (Int.)	B.t.u./hr	3414.99
"	B.t.u. (IST.)/hr	3412.76
"	B.t.u. (mean)/hr	3410.08
"	B.t.u. (mean)/min	56.8347
"	B.t.u. (mean)/sec	0.947244
"	Cal., *gm.* (mean)/hr	859326
"	Cal., *gm.* (mean)/min	14322.1
"	Cal., *kg.*/hr	860.563
"	Cal., *kg.* (IST.)/hr	860
"	Cal., *kg.* (mean)/hr	859.326
"	Cu. cm.-atm./hr	3.55351×10^7
"	Cu. ft.-atm./hr	1254.91
"	Ergs/sec	1.000165×10^{10}
"	Foot-poundals/min	1.42406×10^6
"	Foot-pounds/min	44261.0
"	Foot-pounds/sec	737.684
"	Gram-cm./sec	1.01988×10^7
"	Horsepower	1.34124
"	Horsepower (boiler)	0.101959
"	Horsepower (electric)	1.34070
"	Horsepower (metric)	1.35985
"	Joules/hr	3.60059×10^6
"	Joules (Int.)/hr	3.6×10^6
"	Kg.-meters/hr	367158
"	Kilowatts	1.000165
Kilowatt-hours	B.t.u. (mean)	3409.52
"	Cal., *gm.* (mean)	859184
"	Foot-pounds	2.65522×10^6
"	Hp.-hours	1.34102
"	Joules	3.6×10^6
"	Kg.-meters	367098
"	Lb. H_2O evap. from and at 212°F	3.5168
"	Watt-hours	1000
"	Watt-hours (Int.)	999.835
Kilowatt-hours (Int.)	B.t.u. (mean)	3410.08
"	Cal., *gm.* (IST.)	860000
"	Cal., *gm.* (mean)	859326
"	Cu. cm.-atm	3.55351×10^7
"	Cu. ft.-atm	1254.91
"	Foot-pounds	2.65566×10^6
"	Hp.-hours	1.34124
"	Joules	3.60059×10^6
"	Joules (Int.)	3.6×10^6
"	Kg.-meters	367158
Kw.-hr./gram	B.t.u./(lb	1.54876×10^6
"	B.t.u. (IST.)/lb	1.54774×10^6
"	B.t.u. (mean)/lb	1.54653×10^6
"	Cal., *gm.*/gram	860421
"	Cal., *gm.* (mean)/gram	859184
"	Cu. cm.-atm./gram	3.55292×10^7
"	Cu. ft.-atm./lb	569124
"	Hp.-hr./lb	608.277
"	Joules/gram	3.6×10^6
Knots (Int.)	Cm./sec	51.4444
"	Feet/hr	6076.1155
"	Feet/min	101.26859
"	Feet/sec	1.6878099
"	Kilometers/hr	1.852
"	Meters/min	30.8666
"	Meters/sec	0.514444
"	Miles (naut., Int.)/hr	1
"	Miles (statute)/hr	1.1507794
Lamberts	Candles/sq. cm	0.31830989
"	Candles/sq. ft	295.71956
"	Candles/sq. inch	2.0536081
"	Foot-lamberts	929.0304
"	Lumens/sq. cm	1
Lasts (Brit.)	Liters	2909.414
Leagues (naut., Brit.)	Feet	18240
"	Kilometers	5.559552
"	Leagues (naut., Int.)	1.0006393
"	Leagues (statute)	1.151515
"	Miles (statute)	3.454545
Leagues (naut., Int.)	Fathoms	3038.0577
"	Feet	18228.346
"	Kilometers	5.556
"	Leagues (statute)	1.1507794
"	Miles (statute)	3.4523383
Leagues (statute)	Fathoms	2640
"	Feet	15840
"	Kilometers	4.828032
"	Leagues (naut., Int.)	0.86897625
"	Miles (naut., Int.)	2.6069287
"	Miles (statute)	3
Light years	Astronomical units	63279.5
"	Kilometers	9.46055×10^{12}
"	Miles (statute)	5.87851×10^{12}
Lines	Maxwells	1
Lines (Brit.)	Centimeters	0.211666
"	Inches	0.083333
Lines/sq. cm	Gausses	1
Lines/sq. inch	Gausses	0.15500031
"	Webers/sq. inch	1×10^{-8}
Links (Gunter's)	Chains (Gunter's)	0.01
"	Feet	0.66
"	Feet (U.S. Survey)	0.65999868
"	Inches	7.92
"	Meters	0.201168
"	Miles (statute)	0.000125
"	Rods	0.04
Links (Ramden's)	Centimeters	30.48
"	Chains (Ramdens)	0.01
"	Feet	1
"	Inches	12
Liters	Bushels (Brit.)	0.02749694
"	Bushels (U.S.)	0.02837839
"	Cu. centimeters	1000.028
"	Cu. feet	0.03531566
"	Cu. inches	61.02545
"	Cu. meters	0.001000028
"	Cu. yards	0.001307987
"	Drams (U.S., fluid)	270.5198
"	Gallons (Brit.)	0.2199755
"	Gallons (U.S., dry)	0.2270271
"	Gallons (U.S., liq.)	0.2641794
"	Gills (Brit.)	7.039217
"	Gills (U.S.)	8.453742
"	Hogsheads	0.004193325
"	Minims (U.S.)	16231.19
"	Ounces (Brit., fluid)	35.19609
"	Ounces (U.S., fluid)	33.81497
"	Pecks (Brit.)	0.1099878
"	Pecks (U.S.)	0.1135136
"	Pints (Brit.)	1.759804
"	Pints (U.S., dry)	1.816217
"	Pints (U.S., liq.)	2.113436
"	Quarts. (Brit.)	0.8799021
"	Quarts (U.S., dry)	0.9081084
"	Quarts (U.S., liq.)	1.056718
Liters/min	Cu. ft./min	0.03531566
"	Cu. ft./sec	0.0005885943
"	Gal. (U.S., liq.)/min	0.2641794
Liters/sec	Cu. ft./min	2.118939
"	Cu. ft./sec	0.03531566

CONVERSION FACTORS (Continued)

To convert from	To	Multiply by
Liters/sec	Cu. yards/min	0.07847923
"	Gal. (U.S., liq.)/min	15.85077
"	Gal. (U.S., liq.)/sec	0.2641794
Liter-atm	B.t.u.	0.0961045
"	B.t.u. (IST.)	0.0960417
"	B.t.u. (mean)	0.0959664
"	Cal., *gm*	24.2179
"	Cal., *gm.* (IST.)	24.2021
"	Cal., *gm.* (mean)	24.1831
"	Cu. ft.-atm	0.0353157
"	Foot-poundals	2404.55
"	Foot-pounds	74.7356
"	Hp.-hours	3.77452×10^{-5}
"	Joules	101.328
"	Joules (Int.)	101.311
"	Kg.-meters	10.3326
"	Kw.-hours	2.81466×10^{-5}
Liter-atm. (lat. 45°)	Joules	101.323
Lumens	Candle power (spher.)	0.079577472
Lumens (at 5550 Å)	Watts	0.0014705882
Lumens/sq. cm	Lamberts	1
"	Phots	1
Lumens/(sq. cm. × steradian)	Lamberts	3.1415927
Lumens/sq. ft	Foot-candles	1
"	Foot-lamberts	1
"	Lumens/sq. meter	10.763910
Lumens/(sq. ft. × steradian)	Millilamberts	3.3815822
Lumens/sq. meter	Foot-candles	0.09290304
"	Lumens/sq. ft	0.09290304
"	Phots	0.0001
Lux	Foot-candles	0.09290304
"	Lumens/sq. meter	1
"	Phots	0.0001
Maxwells	E.M. cgs. units of induction flux	1
"	E.S. cgs. units	3.335635×10^{-11}
"	Gauss-sq. cm	1
"	Lines	1
"	Maxwells (Int.)	0.999670
"	Volt-seconds	1×10^{-8}
"	Webers	1×10^{-8}
Maxwells (Int.)	Maxwells	1.000330
Maxwells/sq. cm	Maxwells/sq. in	6.4516
"	Maxwells (Int.)/sq. cm	0.999670
Maxwells (Int.)/sq. cm	Maxwells/sq. cm	1.000330
Maxwells/sq. inch	Maxwells/sq. cm	0.15500031
Megalines	Maxwells	1×10^{6}
Megmhos/cm	Abmhos/cm	0.001
"	Megmhos/inch cube	2.54
"	$(\text{Microhm-cm.})^{-1}$	1
Megmhos/inch	Megmhos/cm	0.39370079
"	$(\text{Microhm-inches})^{-1}$	1
Megohms	Microhms	1×10^{12}
"	Ohms	1×10^{6}
"	Statohms	1.112646×10^{-6}
Megohms^{-1}	Micromhos	1
Meters	Ångström units	1×10^{10}
"	Centimeters	100
"	Chains (Gunter's)	0.049709695
"	Chains (Ramden's)	0.032808399
"	Fathoms	0.54680665
"	Feet	3.2808399
"	Feet (U.S. Survey)	3.280833
"	Furlongs	0.0049709695
"	Inches	39.370079
"	Kilometers	0.001
"	Links (Gunter's)	4.9709695

To convert from	To	Multiply by
Meters	Links (Ramden's)	3.2808399
"	Megameters	1×10^{-6}
"	Miles (naut., Brit.)	0.00053961182
"	Miles (naut., Int.)	0.00053995680
"	Miles (statute)	0.00062137119
"	Millimeters	1000
"	Millimicrons	1×10^{9}
"	Mils	39370.079
"	Rods	0.19883878
"	Yards	1.0936133
Meters of Hg (0°C.)	Atmospheres	1.3157895
"	Ft. of H_2O (60°F.)	44.6474
"	In. of Hg (32°F.)	39.370079
"	Kg./sq. cm	1.35951
"	Pounds/sq. inch	19.3368
Meters/hr	Feet/hr	3.2808399
"	Feet/min	0.054680665
"	Knots (Int.)	0.00053995680
"	Miles (statute)/hr	0.00062137119
Meters/min	Cm./sec	1.666666
"	Feet/min	3.2808399
"	Feet/sec	0.054680665
"	Kilometers/hr	0.06
"	Knots (Int.)	0.032397408
"	Miles (statute)/hr	0.037282272
Meters/sec	Feet/min	196.85039
"	Feet/sec	3.2808399
"	Kilometers/hr	3.6
"	Kilometers/min	0.06
"	Miles (statute)/hr	2.2369363
Meters/(sec. × sec.)	Kilometers/(hr. × sec.)	3.6
"	Miles/(hr. × sec.)	2.2369363
Meter-candles	Lumens/sq. meter	1
Mhos	Abmhos	1×10^{-9}
"	Cgs. units of conductance	1
"	E.M. cgs. units	1×10^{-9}
"	E.S. cgs. units	8.987584×10^{11}
"	Mhos (Int.)	1.000495
"	Mks. (r *or* nr) units	1
"	Ohms^{-1}	1
"	Siemen's units	1
"	Statmhos	8.987584×10^{11}
Mhos (Int.)	Abmhos	9.99505×10^{-10}
"	Mhos	0.999505
Mhos/meter	Abmhos/cm	1×10^{-11}
"	Mhos (Int.)/meter	1.000495
Mho-ft./circ. mil	Mhos/cm	6.0153049×10^{6}
Microfarads	Abfarads	1×10^{-15}
"	Farads	1×10^{-6}
"	Statfarads	8.987584×10^{5}
Micrograms	Grams	1×10^{-6}
"	Milligrams	0.001
Microhenries	Henries	1×10^{-6}
"	Stathenries	1.112646×10^{-18}
Microhms	Abohms	1000
"	Megohms	1×10^{-12}
"	Ohms	1×10^{-6}
"	Statohms	1.112646×10^{-18}
Microhm-cm	Abohm-cm	1000
"	Circ. mil-ohms/ft	6.0153049
"	Microhm-inches	0.39370079
"	Ohm-cm	1×10^{-6}
Microhm-inches	Circ. mil-ohms/ft	15.278875
"	Michrom-cm	2.54
Micromicrofarads	Farads	1×10^{-12}
Micromicrons	Ångström units	0.01
"	Centimeters	1×10^{-10}
"	Inches	$3.9370079 \times 10^{-11}$
"	Meters	1×10^{-12}
"	Microns	1×10^{-6}
Microns	Ångström units	10000

CONVERSION FACTORS (Continued)

To convert from	To	Multiply by
Microns	Centimeters	0.0001
"	Feet	3.2808399×10^{-6}
"	Inches	3.9370079×10^{-5}
"	Meters	1×10^{-6}
"	Millimeters	0.001
"	Millimicrons	1000
Miles (naut., Brit.)	Cable lengths (Brit.)	8.4444
"	Fathoms	1013.333
"	Feet	6080
"	Meters	1853.184
"	Miles (Adm., Brit.)	1
"	Miles (naut., Int.)	1.0006393
"	Miles (statute)	1.151515
Miles (naut., Int.)	Cable lengths	8.4390493
"	Fathoms	1012.6859
"	Feet	6076.1155
"	Feet (U.S. Survey)	6076.1033
"	Kilometers	1.852
"	Leagues (naut., Int.)	0.333333
"	Meters	1852
"	Miles (geographical)	1
"	Miles (naut. Brit.)	0.99936110
"	Miles (statute)	1.1507794
Miles (statute)	Centimeters	160934.4
"	Chains (Gunter's)	80
"	Chains (Ramden's)	52.8
"	Feet	5280
"	Feet (U.S. Survey)	5279.9894
"	Furlongs	8
"	Inches	63360
"	Kilometers	1.609344
"	Light years	1.70111×10^{-13}
"	Links (Gunter's)	8000
"	Meters	1609.344
"	Miles (naut., Brit.)	0.86842105
"	Miles (naut., Int.)	0.86897624
"	Myriameters	0.1609344
"	Rods	320
"	Yards	1760
Miles/hr	Cm./sec	44.704
"	Feet/hr	5280
"	Feet/min	88
"	Feet/sec	1.466666
"	Kilometers/hr	1.609344
"	Knots (Int.)	0.86897624
"	Meters/min	26.8224
"	Miles/min	0.0166666
Miles/(hr. × min.)	Cm./(sec. × sec.)	0.7450666
Miles/(hr. × sec.)	Cm./(sec. × sec.)	44.704
"	Ft./(sec. × sec.)	1.466666
"	Kilometers/(hr. × sec.)	1.609344
"	Meters/(sec. × sec.)	0.44704
Miles/min	Cm./sec	2682.24
"	Feet/hr	316800
"	Feet/sec	88
"	Kilometers/min	1.609344
"	Knots (Int.)	52.138574
"	Meters/min	1609.344
"	Miles/hr	60
Millibars	Atmospheres	0.000986923
"	Bars	0.001
"	Baryes	1000
"	Dynes/sq. cm	1000
"	Grams/sq. cm	1.019716
"	In. of Hg (32°F.)	0.0295300
"	Pounds/sq. ft	2.088543
"	Pounds/sq. inch	0.0145038
Milligrams	Carats (1877)	0.004871
"	Carats (metric)	0.005
"	Drams (apoth. *or* troy)	0.00025720597
"	Drams (advp.)	0.00056438339
Milligrams	Grains	0.015432358
"	Grams	0.001
"	Ounces (apoth. *or* troy)	3.2150737×10^{-5}
"	Ounces (avdp.)	3.5273962×10^{-5}
"	Pennyweights	0.00064301493
"	Pounds (apoth. *or* troy)	2.6792289×10^{-6}
"	Pounds (avdp.)	2.2046226×10^{-6}
"	Scruples (apoth.)	0.00077161792
Milligrams/assay ton	Milligrams/kg	34.285714
"	Ounces (troy)/ton (avdp.)	1
Milligrams/gm	Dynes/cm	0.980665
"	Pounds/inch	5.5997415×10^{-6}
Milligrams/gram	Carats (parts gold per 24 of mixture)	0.024
"	Grams/ton (short)	907.18474
"	Milligrams/assay ton	29.166666
"	Ounces (avdp.)/ton (long)	35.84
"	Ounces (avdp.)/ton (short)	32
"	Ounces (troy)/ton (long)	32.6666
"	Ounces (troy)/ton (short)	29.1666
Milligrams/inch	Dynes/cm	0.386089
"	Dynes/inch	0.980665
"	Grams/cm	0.00039370079
"	Grams/inch	0.0001
Milligrams/kg	Pounds (avdp.)/ton (short)	0.002
Milligrams/liter	Grains/gal. (U.S.)	0.05841620
"	Grams/liter	0.001
"	Parts/million*	1
"	Lb./cu. ft	6.242621×10^{-5}
Milligrams/mm	Dynes/cm	9.80665
Millihenries	Abhenries	1×10^{6}
"	Henries	0.001
"	Stathenries	1.112646×10^{-15}
Millilamberts	Candles/sq. cm	0.00031830989
"	Candles/sq. inch	0.0020536081
"	Foot-lamberts	0.9290304
"	Lamberts	0.001
"	Lumens/sq. cm	0.001
"	Lumens/sq. ft	0.9290304
Milliliters	Cu. cm	1.000028
"	Cu. inches	0.06102545
"	Drams (U.S., fluid)	0.2705198
"	Gills (U.S.)	0.008453742
"	Liters	0.001
"	Minims (U.S.)	16.23119
"	Ounces (Brit., fluid)	0.03519609
"	Ounces (U.S., fluid)	0.03381497
"	Pints (Brit.)	0.001759804
"	Pints (U.S., liq.)	0.002113436
Millimeters	Ångström units	1×10^{7}
"	Centimeters	0.1
"	Decimeters	0.01
"	Dekameters	0.0001
"	Feet	0.0032808399
"	Inches	0.039370079
"	Meters	0.001
"	Microns	1000
"	Mils	39.370079
"	Wave length of orange-red line of krypton 86	1650.76373
"	Wave length of red line of cadmium	1553.16413
Millimeters of Hg (0°C.)	Atmospheres	0.0013157895
"	Bars	0.00133322
"	Dynes/sq. cm	1333.224
"	Grams/sq. cm	1.35951
"	Kg./sq. meter	13,5951
"	Pounds/sq. ft	2.78450
"	Pounds/sq. inch	0.0193368
"	Torrs	1

* Density of 1 gram per milliliter of solvent.

CONVERSION FACTORS (Continued)

To convert from	To	Multiply by
Millimicrons	Ångström units	10
"	Centimeters	1×10^{-7}
"	Inches	3.9370079×10^{-8}
"	Microns	0.001
"	Millimeters	1×10^{-6}
Milliphots	Foot-candles	0.9290304
"	Lumens/sq. ft	0.9290304
"	Lumens/sq. meter	10
"	Lux	10
"	Phots	0.001
Millivolts	Statvolts	3.335635×10^{-6}
"	Volts	0.001
Minims (Brit.)	Cu. cm	0.05919385
"	Cu. inches	0.003612230
"	Milliliters	0.05919219
"	Ounces (Brit., fluid)	0.0020833333
"	Scruples (Brit., fluid)	0.05
Minims (U.S.)	Cu. cm	0.061611520
"	Cu. inches	0.0037597656
"	Drams (U.S., fluid)	0.0166666
"	Gallons (U.S., liq.)	1.6276042×10^{-5}
"	Gills (U.S.)	0.0005208333
"	Liters	6.160979×10^{-5}
"	Milliliters	0.06160979
"	Ounces (U.S., fluid)	0.002083333
"	Pints (U.S., liq.)	0.0001302083
Minutes (angular)	Degrees	0.0166666
"	Quadrants	0.000185185
"	Radians	0.00029088821
"	Seconds (angular)	60
Minutes (mean solar)	Days (mean solar)	0.0006944444
"	Days (sidereal)	0.00069634577
"	Hours (mean solar)	0.0166666
"	Hours (sidereal)	0.016712298
"	Minutes (sidereal)	1.00273791
Minutes (sidereal)	Days (mean solar)	0.00069254831
"	Minutes (mean solar)	0.99726957
"	Months (mean calendar)	2.2768712×10^{-5}
"	Seconds (sidereal)	60
Minutes/cm	Radians/cm	0.00029088821
Months (lunar)	Days (mean solar)	29.530588
"	Hours (mean solar)	708.73411
"	Minutes (mean solar)	42524.047
"	Second (mean solar)	2.5514428×10^{6}
"	Weeks (mean calendar)	4.2186554
Months (mean calendar)	Days (mean solar)	30.416666
"	Hours (mean solar)	730
"	Months (lunar)	1.0300055
"	Weeks (mean calendar)	4.3452381
"	Years (calendar)	0.08333333
"	Years (sidereal)	0.083274845
"	Years (tropical)	0.083278075
Myriagrams	Grams	10000
"	Kilograms	10
"	Pounds (avdp.)	22.046226
Newtons	Dynes	100000
"	Pounds	0.22480894
Newton-meters	Dyne-cm	1×10^{7}
"	Gram-cm	10197.162
"	Kg.-meters	0.10197162
"	Pound-feet	0.73756215
Noggins (Brit.)	Cu. cm	142.0652
"	Gallons (Brit.)	0.03125
"	Gills (Brit.)	1
Oersteds	Ampere-turns/inch	2.0212678
"	Ampere-turns/meter	79.577472
"	E.M. cgs. units of magnetic field intensity	1
"	E.S. cgs. units	2.997930×10^{10}

To convert from	To	Multiply by
Oersteds	Gilberts/cm	1
"	Oersteds (Int.)	1.000165
Oersteds (Int.)	Oersteds	0.999835
Ohms	Abohms	1×10^{9}
"	Cgs. units of resistance	1
"	Megohms	1×10^{-6}
"	Microhms	1×10^{6}
"	Ohms (Int.)	0.999505
"	Statohms	1.112646×10^{-12}
Ohms (Int.)	Ohms	1.000495
Ohms (mil, foot)	Circ. mil-ohms/ft	1
"	Ohm-cm	1.6624261×10^{-7}
Ohm-cm	Circ. mil-ohms/ft	6.0153049×10^{6}
"	Microhm-cm	1×10^{6}
"	Ohm-inches	0.39370079
Ohm-inches	Ohm-cm	2.54
Ohm-meters	Abohm-cm	1×10^{11}
"	E.M. cgs. units	1×10^{11}
"	E.S. cgs. units	1.112646×10^{-10}
"	Mks. units	1
"	Statohm-cm	1.112646×10^{-10}
Ounces (apoth. *or* troy)	Dekagrams	1.7554286
"	Drams (apoth. *or* troy)	8
"	Drams (avdp.)	17.554286
"	Grains	480
"	Grams	31.103486
"	Milligrams	31103.486
"	Ounces (avdp.)	1.0971429
"	Pennyweights	20
"	Pounds (apoth. *or* troy)	0.0833333
"	Pounds (avdp.)	0.068571429
"	Scruples (apoth.)	24
"	Tons (short)	3.4285714×10^{-5}
Ounces (avdp.)	Drams (apoth. *or* troy)	7.291666
"	Drams (avdp.)	16
"	Grains	437.5
"	Grams	28.349523
"	Hundredweights (long)	0.00055803571
"	Hundredweights (short)	0.000625
"	Ounces (apoth. *or* troy)	0.9114583
"	Pennyweights	18.229166
"	Pounds (apoth. *or* troy)	0.075954861
"	Pounds (avdp.)	0.0625
"	Scruples (apoth.)	21.875
"	Tons (long)	2.7901786×10^{-5}
"	Tons (metric)	2.8349527×10^{-5}
"	Tons (short)	3.125×10^{-5}
Ounces (Brit., fluid)	Cu. cm	28.41305
"	Cu. inches	1.733870
"	Drachms (Brit., fluid)	8
"	Drams (U.S., fluid)	7.686075
"	Gallons (Brit.)	0.00625
"	Milliliters	28.41225
"	Minims (Brit.)	480
"	Ounces (U.S., fluid)	0.9607594
Ounces (U.S., fluid)	Cu. cm	29.573730
"	Cu. inches	1.8046875
"	Cu. meters	2.9573730×10^{-5}
"	Drams (U.S., fluid)	8
"	Gallons (U.S., dry)	0.0067138047
"	Gallons (U.S., liq.)	0.0078125
"	Gills (U.S.)	0.25
"	Liters	0.029572702
"	Minims (U.S.)	480
"	Ounces (Brit., fluid)	1.040843
"	Pints (U.S., liq.)	0.625
"	Quarts (U.S., liq.)	0.03125
Ounces/sq. inch	Dynes/sq. cm	4309.22
"	Grams/sq. cm	4.3941849
"	In. of H_2O (39.2°F.)	1.73004
"	In. of H_2O (60°F.)	1.73166

CONVERSION FACTORS (Continued)

To convert from	To	Multiply by
Ounces/sq. inch	Pounds/sq. ft	9
"	Pounds/sq. inch	0.0625
Ounces (avdp.)/ton (long)	Milligrams/kg	27.901786
Ounces (avdp.)/ton (short)	Milligrams/kg	31.25
Paces	Centimeters	76.2
"	Chains (Gunter's)	0.0378788
"	Chains (Ramden's)	0.025
"	Feet	2.5
"	Hands	7.5
"	Inches	30
"	Ropes (Brit.)	0.125
Palms	Centimeters	7.62
"	Chains (Ramden's)	0.0025
"	Cubits	0.1666666
"	Feet	0.25
"	Hands	0.75
"	Inches	3
Parsecs	Kilometers	3.08374×10^{13}
"	Miles (statute	1.91615×10^{13}
Parts/million*	Grains/gal. (Brit.)	0.07015488
"	Grains/gal. (U.S.)	0.05841620
"	Grams/liter	0.001
"	Milligrams/liter	1
Pecks (Brit.)	Bushels (Brit.)	0.25
"	Coombs (Brit.)	0.0625
"	Cu. cm	9092.175
"	Cu. inches	554.8385
"	Gallons (Brit.)	2
"	Gills (Brit.)	64
"	Hogsheads	0.03812537
"	Kilderkins (Brit.)	0.111111
"	Liters	9.091920
"	Pints (Brit.)	16
"	Quarterns (Brit., dry)	4
"	Quarters (Brit., dry)	0.03125
"	Quarts (Brit.)	8
"	Quarts (U.S., dry)	8.256449
Pecks (U.S.)	Barrels (U.S., dry)	0.076191185
"	Bushels (U.S.)	0.25
"	Cu. cm	8809.7675
"	Cu. feet	0.311114005
"	Cu. inches	537.605
"	Gallons (U.S., dry)	2
"	Gallons (U.S., liq.)	2.3272944
"	Liters	8.809521
"	Pints (U.S., dry)	16
"	Quarts (U.S., dry)	8
Pennyweights	Drams (apoth. *or* troy)	0.4
"	Drams (avdp.)	0.87771429
"	Grains	24
"	Grams	1.55517384
"	Ounces (apoth. *or* troy)	0.05
"	Ounces (avdp.)	0.054857143
"	Pounds (apoth. *or* troy)	0.0041666
"	Pounds (avdp.)	0.0034285714
Perches (masonry)	Cu. feet	24.75
Phots	Foot-candles	929.0304
"	Lumens/sq. cm	1
"	Lumens/sq. meter	10000
"	Lux	10000
Picas (printer's)	Centimeters	0.42175176
"	Inches	0.166044
Pints (Brit.)	Cu. cm	568.26092
"	Gallons (Brit.)	0.125
"	Gills (Brit.)	4
"	Gills (U.S.)	4.803797
"	Liters	0.5682450

To convert from	To	Multiply by
Pints (Brit.)	Minims (Brit.)	9600
"	Ounces (Brit., fluid)	20
"	Pints (U.S., dry)	1.032056
"	Pints (U.S., liq.)	1.200949
"	Quarts (Brit.)	0.5
"	Scruples (Brit., fluid)	480
Pints (U.S., dry)	Bushels (U.S.)	0.015625
"	Cu. cm	550.61047
"	Cu. inches	33.6003125
"	Gallons, (U.S., dry)	0.125
"	Gallons (U.S., liq.)	0.14545590
"	Liters	0.5505951
"	Pecks (U.S.)	0.0625
"	Quarts (U.S., dry)	0.5
Pints (U.S., liq.)	Cu. cm	473.17647
"	Cu. feet	0.016710069
"	Cu. inches	28.875
"	Cu. yards	0.00061889146
"	Drams (U.S., fluid)	128
"	Gallons (U.S., liq.)	0.125
"	Gills (U.S.)	4
"	Liters	0.4731632
"	Milliliters	473.1632
"	Minims (U.S.)	7680
"	Ounces (U.S., fluid)	16
"	Pints (Brit.)	0.8326747
"	Quarts (U.S., liq.)	0.5
Planck's constant	Erg-seconds	6.6255×10^{-27}
"	Joule-seconds	6.6255×10^{-34}
"	Joule-sec./Avog. No. (chem.)	3.9905×10^{-10}
Points (printer's)	Centimeters	0.03514598
"	Inches	0.013837
"	Picas	0.0833333
Poises**	Cgs. units of absolute viscosity	1
"	Grams/(cm. × sec.)	1
Poise-cu. cm./gram	Sq. cm./sec	1
Poise-cu. ft./lb	Sq. cm./sec	62.427960
Poise-cu. in./gram	Sq. cm./sec	16.387064
Poles/sq. cm	E.M. cgs. units of magnetization	1
Pottles (Brit.)	Gallons (Brit.)	0.5
"	Liters	2.272980
Poundals	Dynes	13825.50
"	Grams	14.09808
"	Pounds (avdp.)	0.0310810
Pounds (apoth. *or* troy)	Drams (apoth. *or* troy)	96
"	Drams (avdp.)	210.65143
"	Grains	5760
"	Grams	373.24172
"	Kilograms	0.37324172
"	Ounces (apoth. *or* troy)	12
"	Ounces (avdp.)	13.165714
"	Pennyweights	240
"	Pounds (avdp.)	0.8228571
"	Scruples (apoth.)	288
"	Tons (long)	0.00036734694
"	Tons (metric)	0.00037324172
"	Tons (short)	0.00041142857
Pounds (avdp.)	Drams (apoth. *or* troy)	116.6666
"	Drams (avdp.)	256
"	Grains	7000
"	Grams	453.59237
"	Hundredweights (long)	0.00892857
"	Hundredweights (short)	0.01
"	Kilograms	0.45359237
"	Ounces (apoth. *or* troy)	14.583333
"	Ounces (avdp.)	16

* Based on density of 1 gram/ml. for the solvent.
** See separate table of "viscosity conversion factors".

CONVERSION FACTORS (Continued)

To convert from	To	Multiply by
Pounds (avdp.)	Pennyweights	291.6666
“	Poundals	32.1740
“	Pounds (apoth. *or* troy)	1.215277
“	Scruples (apoth.)	350
“	Slugs	0.0310810
“	Tons (long)	0.00044642857
“	Tons (metric)	0.00045359237
“	Tons (short)	0.0005
Pounds of H_2O evap. from and at 212°F	B.t.u.	970.9
“	B.t.u. (IST.)	970.2
“	B.t.u. (mean)	969.4
“	Joules	1.0237×10^6
“	Joules (Int.)	1.0234×10^6
Pounds/cu. ft.	Grams/cu. cm	0.016018463
“	Kg./cu. meter	16.018463
Pounds/cu. inch	Grams/cu. cm	27.679905
“	Grams/liter	27.68068
“	Kg./cu. meter	27679.905
Pounds/gal. (Brit.)	Pounds/cu. ft.	6.228839
Pounds/gal. (U.S., liq.)	Grams/cu. cm	0.11982643
“	Pounds/cu. ft.	7.4805195
Pounds/inch	Grams/cm	178.57967
“	Grams/ft.	5443.1084
“	Grams/inch	453.59237
“	Ounces/cm	6.2992
“	Ounces/inch	16
“	Pounds/meter	39.370079
Pounds/minute	Kilograms/hr	27.2155422
“	Kilograms/min	0.45359237
Pounds of H_2O (39.2°F.)/min	Cu. ft./min	0.01601891
“	Gal. (U.S.)/min	0.1198298
“	Liters/min	0.45359237
Pounds/sq. ft.	Atmospheres	0.000472541
“	Bars	0.000478803
“	Cm. of Hg (0°C.)	0.0359131
“	Dynes/sq. cm	478.803
“	Ft. of air (1 atm., 60°F.)	13.096
“	Grams/sq. cm	0.48824276
“	In. of Hg (32°F.)	0.0141390
“	In. of H_2O (39.2°F.)	0.192227
“	Kg./sq. meter	4.8824276
“	Mm. of Hg (0°C.)	0.359131
Pounds/sq. inch	Atmospheres	0.0680460
“	Bars	0.0689476
“	Cm. of Hg (0°C.)	5.17149
“	Cm. of H_2O (4°C.)	70.3089
“	Dynes/sq. cm	68947.6
“	Grams/sq. cm	70.306958
“	In. of Hg (32°F.)	2.03602
“	In. of H_2O (39.2°F.)	27.6807
“	Kg./sq. cm	0.070306958
“	Mm. of Hg (0°C.)	51.7149
Pound wt.-sec./sq. ft.	Poises	478.803
Pound wt.-sec./sq. in.	Poises	68947.6
Puncheons (Brit.)	Cu. meters	0.31797510
“	Gallons (Brit.)	69.94467
“	Gallons (U.S.)	84
Quadrants	Minutes	5400
“	Radians	1.5707963
Quarterns (Brit., dry)	Buckets (Brit.)	0.125
“	Bushels (Brit.)	0.0625
“	Cu. cm	2273.044
“	Gallons (Brit.)	0.5
“	Liters	2.272980
“	Pecks (Brit.)	0.25
Quarterns (Brit., liq.)	Cu. cm	142.0652
Quarterns (Brit., liq.)	Gallons (Brit.)	0.03125
“	Liters	0.1420613
Quarters (U.S., long)	Kilograms	254.0117272
“	Pounds (avdp.)	560
Quarters (U.S., short)	Kilograms	226.796185
“	Pounds	500
Quarts (Brit.)	Cu. cm	1136.522
“	Cu. inches	69.35482
“	Gallons (Brit.)	0.25
“	Gallons (U.S., liq.)	0.3002373
“	Liters	1.136490
“	Quarts (U.S., dry)	1.032056
“	Quarts (U.S., liq.)	1.200949
Quarts (U.S., dry)	Bushels (U.S.)	0.03125
“	Cu. cm	1101.2209
“	Cu. feet	0.038889251
“	Cu. inches	67.200625
“	Gallons (U.S., dry)	0.25
“	Gallons (U.S., liq.)	0.29091180
“	Liters	1.1011901
“	Pecks (U.S.)	0.125
“	Pints (U.S., dry)	2
Quarts (U.S., liq.)	Cu. cm	946.35295
“	Cu. feet	0.033420136
“	Cu. inches	57.75
“	Drams (U.S., fluid)	256
“	Gallons (U.S., dry)	0.21484175
“	Gallons (U.S., liq.)	0.25
“	Gills (U.S.)	8
“	Liters	0.9463264
“	Ounces (U.S., fluid)	32
“	Pints (U.S., liq.)	2
“	Quarts (Brit.)	0.8326747
“	Quarts (U.S., dry)	0.8593670
Quintals (metric)	Grams	100000
“	Hundredweights (long)	1.9684131
“	Kilograms	100
“	Pounds (avdp.)	220.46226
Radians	Circumferences	0.15915494
“	Degrees	57.295779
“	Minutes	3437.7468
“	Quadrants	0.63661977
“	Revolutions	0.15915494
“	Seconds	206264.81
Radians/cm	Degrees/cm	57.295779
“	Degrees/ft	1746.3754
“	Degrees/inch	145.53128
“	Minutes/cm	3437.7468
Radians/sec	Degrees/sec	57.295779
“	Revolutions/min	9.5492966
“	Revolutions/sec	0.15915494
Radians/(sec. × sec.)	Revolutions/(min. × min.)	572.95779
“	Revolutions/(min. × sec.)	9.5492966
“	Revolutions/(sec. × sec.)	0.15915494
Register tons	Cu. feet	100
“	Cu. meters	2.8316847
Revolutions	Degrees	360
“	Grades	400
“	Quadrants	4
“	Radians	6.2831853
Reyns*	Centipoises	6.89476×10^6
Rhes	$Poises^{-1}$	1
Rods	Centimeters	502.92
“	Chains (Gunter's)	0.25
“	Chains (Ramden's)	0.165
“	Feet	16.5
“	Feet (U.S. Survey)	16.499967
“	Furlongs	0.025
“	Inches	198
“	Links (Gunter's)	25

* See also separate table of "viscosity conversion factors".

CONVERSION FACTORS (Continued)

To convert from	To	Multiply by
Rods	Links (Ramden's)	16.5
"	Meters	5.0292
"	Miles (statute)	0.003125
"	Perches	1
"	Yards	5.5
Rods (Brit., volume)	Cu. feet	1000
"	Cu. meters	28.316847
Roods (Brit.)	Acres	0.25
"	Ares	10.117141
"	Sq. perches	40
"	Sq. yards	1210
Ropes (Brit.)	Feet	20
"	Meters	6.096
"	Yards	6.6666666
Scruples (apoth.)	Drams (apoth. *or* troy)	0.333333
"	Drams (avdp.)	0.73142857
"	Grains	20
"	Grams	1.2959782
"	Ounces (apoth. *or* troy)	0.041666
"	Ounces (avdp.)	0.045714286
"	Pennyweights	0.833333
"	Pounds (apoth. *or* troy)	0.003472222
"	Pounds (avdp.)	0.0028571429
Scruples (Brit., fluid)	Minims (Brit.)	20
Seams (Brit.)	Bushels (Brit.)	8
"	Cu. feet	10.27479
"	Liters	290.9414
Seconds (angular)	Degrees	0.000277777
"	Minutes	0.0166666
"	Radians	4.8481368×10^{-6}
Seconds (mean solar)	Days (mean solar)	1.1574074×10^{-5}
"	Days (sidereal)	1.1605763×10^{-5}
"	Hours (mean solar)	0.0002777777
"	Hours (sidereal)	0.00027853831
"	Minutes (mean solar)	0.0166666
"	Minutes (sidereal)	0.016712298
"	Seconds (sidereal)	1.00273791
Seconds (sidereal)	Days (mean solar)	1.1542472×10^{-5}
"	Days (sidereal)	1.1574074×10^{-5}
"	Hours (mean solar)	0.00027701932
"	Hours (sidereal)	0.000277777
"	Minutes (mean solar)	0.016621159
"	Minutes (sidereal)	0.0166666
"	Seconds (mean solar)	0.99726957
Siemen's units	*Same as* Mhos	
Skeins	Feet	360
"	Meters	109.728
Slugs	Geepounds	1
"	Kilograms	14.5939
"	Pounds (avdp.)	32.1740
Slugs/cu. ft.	Grams/cu. cm.	0.515379
Space (entire)	Hemispheres	2
"	Steradians	12.566371
Spans	Centimeters	22.86
"	Fathoms	0.125
"	Feet	0.75
"	Inches	9
"	Quarters (Brit. linear)	1
Spherical right angles	Hemispheres	0.25
"	Spheres	0.125
"	Steradians	1.5707963
Sq. centimeters	Ares	1×10^{-6}
"	Circ. mm	127.32395
"	Circ. mils	197352.52
"	Sq. chains (Gunter's)	2.4710538×10^{-7}
"	Sq. chains (Ramden's)	1.0763910×10^{-7}
"	Sq. decimeters	0.01
"	Sq. feet	0.0010763910
"	Sq. ft. (U.S. Survey)	0.0010763867
"	Sq. inches	0.15500031
"	Sq. meters	0.0001

To convert from	To	Multiply by
Sq. centimeters	Sq. mm	100
"	Sq. mils	155000.31
"	Sq. rods	3.9536861×10^{-6}
"	Sq. yards	0.00011959900
Sq. chains (Gunter's)	Acres	0.1
"	Sq. feet	4356
"	Sq. ft. (U.S. Survey)	4355.9826
"	Sq. inches	627264
"	Sq. links (Gunter's)	10000
"	Sq. meters	404.68564
"	Sq. miles	0.00015625
"	Sq. rods	16
"	Sq. yards	484
Sq. chains (Ramden's)	Acres	0.22956841
"	Sq. feet	10000
"	Sq. ft. (U.S. Survey)	9999.9600
"	Sq. inches	1.44×10^{6}
"	Sq. links (Ramden's)	10000
"	Sq. meters	929.0304
"	Sq. miles	0.00035870064
"	Sq. rods	36.730946
"	Sq. yards	1111.111
Sq. decimeters	Sq. cm	100
"	Sq. inches	15.500031
Square degrees	Steradians	0.00030461742
Sq. dekameters	Acres	0.024710538
"	Ares	1
"	Sq. meters	100
"	Sq. yards	119.59900
Sq. feet	Acres	2.295684×10^{-5}
"	Ares	0.0009290304
"	Sq. cm	929.0304
"	Sq. chains (Gunter's)	0.00022956841
"	Sq. ft. (U.S. Survey)	0.99999600
"	Sq. inches	144
"	Sq. links (Gunter's)	2.2956841
"	Sq. meters	0.09290304
"	Sq. miles	3.5870064×10^{-8}
"	Sq. rods	0.0036730946
"	Sq. yards	0.111111
Sq. feet (U.S. Survey)	Acres	$2.29569330 \times 10^{-5}$
"	Sq. centimeters	929.03412
"	Sq. chains (Ramden's)	0.00010000040
"	Sq. feet	1.0000040
Sq. hectometers	Sq. meters	10000
Sq. inches	Circ. mils	1273239.5
"	Sq. cm	6.4516
"	Sq. chains (Gunter's)	1.5942251×10^{-6}
"	Sq. decimeters	0.064516
"	Sq. feet	0.0069444
"	Sq. ft. (U.S. Survey)	0.0069444167
"	Sq. links (Gunter's)	0.01594225
"	Sq. meters	0.00064516
"	Sq. miles	$2.4909767 \times 10^{-10}$
"	Sq. mm	645.16
"	Sq. mils	1×10^{6}
Sq. inches/sec	Sq. cm./hr	23225.76
"	Sq. cm./sec	6.4516
"	Sq. ft./min	0.416666
Sq. kilometers	Acres	247.10538
"	Sq. feet	1.0763910×10^{7}
"	Sq. ft. (U.S. Survey)	1.0763867×10^{7}
"	Sq. inches	1.5500031×10^{9}
"	Sq. meters	1×10^{6}
"	Sq. miles	0.38610216
"	Sq. yards	1.1959900×10^{6}
Sq. links (Gunter's)	Acres	1×10^{-5}
"	Sq. cm	404.68564
"	Sq. chains (Gunter's)	0.0001
"	Sq. feet	0.4356
"	Sq. ft. (U.S. Survey)	0.43559826
"	Sq. Inches	62.7264

CONVERSION FACTORS (Continued)

To convert from	To	Multiply by
Sq. links (Ramden's)	Acres	2.2956841×10^{-5}
"	Sq. feet	1
Sq. meters	Acres	0.00024710538
"	Ares	0.01
"	Hectares	0.0001
"	Sq. cm	10000
"	Sq. feet	10.763910
"	Sq. inches	1550.0031
"	Sq. kilometers	1×10^{-6}
"	Sq. links (Gunter's)	24.710538
"	Sq. links (Ramden's)	10.763910
"	Sq. miles	3.8610216×10^{-7}
"	Sq. mm	1×10^{6}
"	Sq. rods	0.039536861
"	Sq. yards	1.1959900
Sq. miles	Acres	640
"	Hectares	258.99881
"	Sq. chains (Gunter's)	6400
"	Sq. feet	2.7878288×10^{7}
"	Sq. ft. (U.S. Survey)	2.78288×10^{7}
"	Sq. kilometers	2.5899881
"	Sq. meters	2589988.1
"	Sq. rods	102400
"	Sq. yards	3.0976×10^{6}
Sq. millimeters	Circ. mm	1.2732395
"	Circ. mils	1973.5252
"	Sq. cm	0.01
"	Sq. inches	0.0015500031
"	Sq. meters	1×10^{-6}
Sq. mils	Circ. mils	1.2732395
"	Sq. cm	6.4516×10^{-6}
"	Sq. inches	1×10^{-6}
"	Sq. mm	0.00064516
Sq. rods	Acres	0.00625
"	Ares	0.2529285264
"	Hectares	0.002529285264
"	Sq. cm	252928.5264
"	Sq. feet	272.25
"	Sq. ft. (U.S. Survey)	272.24891
"	Sq. inches	39204
"	Sq. links (Gunter's)	625
"	Sq. links (Ramden's)	272.25
"	Sq. meters	25.29285264
"	Sq. miles	9.765625×10^{-6}
"	Sq. yards	30.25
Sq. yards	Acres	0.00020661157
"	Ares	0.0083612736
"	Hectares	8.3612736×10^{-5}
"	Sq. cm	8361.2736
"	Sq. chains (Gunter's)	0.0020661157
"	Sq. chains (Ramden's)	0.0009
"	Sq. feet	9
"	Sq. ft. (U.S. Survey)	8.9999640
"	Sq. inches	1296
"	Sq. links (Gunter's)	20.661157
"	Sq. links (Ramden's)	9
"	Sq. meters	0.83612736
"	Sq. miles	$3.228305785 \times 10^{-7}$
"	Sq. perches (Brit.)	0.033057851
"	Sq. rods	0.033057851
Statamperes	Abamperes	3.335635×10^{-11}
"	Amperes	3.335635×10^{-10}
"	E.M. cgs. units of current	3.335635×10^{-11}
"	E.S. cgs. units	1
Statcoulombs	Ampere-hours	9.265653×10^{-14}
"	Coulombs	3.335635×10^{-10}
"	Electronic charges	2.082093×10^{9}
"	E.M. cgs. units of electric charge	3.335635×10^{-11}
Statfarads	E.M. cgs. units of capacitance	1.112646×10^{-21}
"	E.S. cgs. units	1

To convert from	To	Multiply by
Statfarads	Farads	1.112646×10^{-12}
"	Microfarads	1.112646×10^{-6}
Stathenries	Abhenries	8.987584×10^{20}
"	E.M. cgs. units of inductance	8.987584×10^{20}
"	E.S. cgs. units	1
"	Henries	8.987584×10^{11}
"	Millihenries	8.987584×10^{14}
Statohms	Abohms	8.987584×10^{20}
"	E.S. cgs. units	1
"	Ohms	8.987584×10^{11}
Statvolts	Abvolts	2.997930×10^{10}
"	Volts	299.7930
Statvolts/cm	Volts/cm	299.7930
"	Volts/inch	761.4742
Statvolts/inch	Volts/cm	118.0287
Steradians	Hemispheres	0.15915494
"	Solid angles	0.079577472
"	Spheres	0.079577472
"	Spher. right angles	0.63661977
"	Square degrees	3282.8063
Steres	Cubic meters	1
"	Decisteres	10
"	Dekasteres	0.1
"	Liters	999.972
Stilbs	Candles/sq. cm	1
"	Candles/sq. inch	6.4516
"	Lamberts	3.1415927
Stokes*	Cgs. units of kinematic viscosity	1
"	Sq. cm./sec	1
"	Sq. inches/sec	0.15500031
"	Poise cu. cm./gram	1
Stones (Brit., legal)	Centals (Brit.)	0.14
Tons (long)	Dynes	9.96402×10^{8}
"	Hundredweights (long)	20
"	Hundredweights (short)	22.4
"	Kilograms	1016.0469
"	Ounces (avdp.)	35840
"	Pounds (apoth. *or* troy)	2722.22
"	Pounds (avdp.)	2240
"	Tons (metric)	1.1060469
"	Tons (short)	1.12
Tons (metric)	Dynes	9.80665×10^{8}
"	Grams	1×10^{6}
"	Hundredweights (short)	22.046226
"	Kilograms	1000
"	Ounces (avdp.)	35273.962
"	Pounds (apoth. *or* troy)	2679.2289
"	Pounds (avdp.)	2204.6226
"	Tons (long)	0.98420653
"	Tons (short)	1.1023113
Tons (short)	Dynes	8.89644×10^{8}
"	Hundredweights (short)	20
"	Kilograms	907.18474
"	Ounces (avdp.)	32000
"	Pounds (apoth. *or* troy)	2430.555
"	Pounds (avdp.)	2000
"	Tons (long)	0.89285714
"	Tons (metric)	0.90718474
Tons of refrig. (U.S., comm.)	B.t.u. (IST.)/hr	12000
"	B.t.u. (IST.)/min	200
"	Cal., *kg.* (IST.)/hr	3023.949
"	Horsepower	4.71611
"	Kg. of ice melted/hr	37.971
"	Lb. of ice melted/hr	83.711
Tons of refrig. (U.S., std.)	B.t.u. (IST.)	288000
"	B.t.u. (mean)	287774
"	Cal., *kg.* (IST.)	72574.8

* See also separate table of "viscosity conversion factors".

CONVERSION FACTORS (Continued)

To convert from	To	Multiply by
Tons of refrig. (U.S., std.)	Cal., *kg.* (mean)	72517.9
"	Lb. of ice melted	2009.1
Tons (long)/sq. ft	Atmospheres	1.05849
"	Dynes/sq. cm	1.07252×10^{6}
"	Grams/sq. cm	1093.6638
"	Pounds/sq. ft	2240
Tons (short)/sq. ft	Atmospheres	0.945082
"	Dynes/sq. cm	957.605
"	Grams/sq. cm	976.486
"	Pounds/sq. inch	13.8888
Tons (long)/sq. in	Atmospheres	152.423
"	Dynes/sq. cm	1.54443×10^{8}
"	Grams/sq. cm	157487.59
Tons (short)/sq. in	Dynes/sq. cm	1.37895×10^{8}
"	Kg./sq. mm	1406.139
"	Pounds/sq. inch	2000
Torrs (*or* Tors)	Millimeters of Hg (0°C.)	1
Townships (U.S.)	Acres	23040
"	Sections	36
"	Sq. miles	36
Tuns	Gallons (U.S.)	252
"	Hogsheads	4
Volts	Abvolts	1×10^{8}
"	Mks. (r *or* nr) units	1
"	Statvolts	0.003335635
"	Volts (Int.)	0.999670
Volts (Int.)	Volts	1.000330
Volts/°C	Joules/(coulomb × °C.)	1
Volt-coulombs	Joules (Int.)	0.999835
Volt-coulombs (Int.)	Joules	1.000165
Volt-electronic charge-seconds	Planck's constant*	2.41814×10^{14}
Volt-faraday (chem.)-seconds	Planck's constant*	1.45650×10^{38}
Volt-faraday (phys.)-seconds	Planck's constant	1.45690×10^{38}
Volt-seconds	Maxwells	1×10^{8}
Watts	B.t.u./hr	3.41443
"	B.t.u. (mean)/hr	3.40952
"	B.t.u. (mean)/min	0.0568253
"	B.t.u./sec	0.000948451
"	B.t.u. (mean)/sec	0.000947088
"	Cal., *gm.*/hr	860.421
"	Cal., *gm.* (mean)/hr	859.184
"	Cal., *gm.* (20°C.)/hr	860.853
"	Cal., *gm.*/min	14.3403
"	Cal., *gm.* (IST.)/min	14.3310
"	Cal., *gm.* (mean)/min	14.3197
"	Cal., *kh.*/min	0.0143403
"	Cal., *kg.* (IST.)/min	0.0143310
"	Cal., *kg.* (mean)/min	0.0143197
"	Ergs/sec	1×10^{7}
"	Foot-pounds/min	44.2537
"	Horsepower	0.00134102
"	Horsepower (boiler)	0.000101942
"	Horsepower (elec.)	0.00134048
"	Horsepower (metric)	0.00135962
"	Joules/sec	1
"	Kilowatts	0.001
"	Liter-atm./hr	35.5282
Watts (Int.)	B.t.u./hr	3.41499
"	B.t.u. (mean)/hr	3.41008
"	B.t.u./min	0.569165
"	B.t.u. (mean)/min	0.0568347
"	Cal., *gm.*/hr	860.563
"	Cal., *gm.* (mean)/hr	859.326
"	Cal., *kg.*/min	0.0143427
"	Cal., *kg.* (IST.)/min	0.0143333
"	Cal., *kg.* (mean)/min	0.0143221

To convert from	To	Multiply by
Watts (Int.)	Ergs/sec	1.000165×10^{7}
"	Joules (Int.)/sec	1
"	Watts	1.000165
Watts/sq. cm	B.t.u./(hr. × sq. ft.)	3172.10
"	Cal., *gm.*/(hr. × sq. cm.)	860.421
"	Ft.-lb./(min. × sq. ft.)	41113.1
Watts/sq. in	B.t.u./(hr. × sq. ft.)	491.677
"	Cal., *gm.*/(hr. × sq. cm.)	133.365
"	Ft.-lb./(min. × sq. ft.)	6372.54
Watt-hours	B.t.u.	3.41443
"	B.t.u. (mean)	3.40952
"	Cal., *gm*	860.421
Watt-hours	Cal., *kg.* (mean)	0.859184
"	Cal., *gm.* (mean)	859.184
"	Foot-pounds	2655.22
"	Hp.-hours	0.00134102
"	Joules	3600
"	Joules (Int.)	3599.41
"	Kg.-meters	367.098
"	Kw.-hours	0.001
"	Watt-hours (Int.)	0.999835
Watt-sec	Foot-pounds	0.737562
"	Gram-cm	10197.16
"	Joules	1
"	Liter-atm	0.00986895
"	Volt-coulombs	1
Wave length of orange-red line of krypton 86	Ångström units	6057.80211
"	Millimeters	0.000605780211
Wave length of red line of cadmium	Ångström units	6438.4696
"	Millimeters	0.00064384696
Webers	Cgs. units of induction flux	1×10^{8}
"	E.M. cgs. units of induction flux	1×10^{8}
"	Lines	1×10^{8}
"	Maxwells	1×10^{8}
"	Mks. units of induction flux	1
"	Mks. units of magnetic charge	0.079577472
"	Mks. units of magnetic charge	1
"	Volt-seconds	1
Webers/sq. cm	Gausses	1×10^{8}
"	Lines	1×10^{8}
"	Lines/sq. inch	6.4516×10^{8}
Webers/sq. in	Gausses	1.5500031×10^{7}
Weeks (mean calendar)	Days (mean solar)	7
"	Days (sidereal)	7.0191654
"	Hours (mean solar)	168
"	Hours (sidereal)	168.45997
"	Minutes (mean solar)	10080
"	Minutes (sidereal)	10107.598
"	Months (lunar)	0.23704235
"	Months (mean calendar)	0.23013699
"	Years (calendar)	0.019178082
"	Years (sidereal)	0.019164622
"	Years (tropical)	0.019165365
Weys (Brit., mass.)	Pounds (avdp.)	252
Yards	Centimeters	91.44
"	Chains (Gunter's)	0.4545454
"	Chains (Ramden's)	0.03
"	Cubits	2
"	Fathoms	0.5
"	Feet	3
"	Feet (U.S. Survey)	2.9999940
"	Furlongs	0.00454545
"	Inches	36
"	Meters	0.9144
"	Poles (Brit.)	0.181818

* For factors for C = 12 scale or those which can be derived from same see table on Values for General Physical Constants.

CONVERSION FACTORS (Continued)

To convert from	To	Multiply by
Yards	Quarters (Brit., linear)	4
"	Rods	0.181818
"	Spans	4
Years (calendar)	Days (mean solar)	365
"	Hours (mean solar)	8760
"	Minutes (mean solar)	525600
"	Months (lunar)	12.360065
"	Months (mean calendar)	12
"	Seconds (mean solar)	3.1536×10^7
"	Weeks (mean calendar)	52.142857
"	Years (sidereal)	0.99929814
"	Years (tropical)	0.99933690
Years (leap)	Days (mean solar)	366
Years (sidereal)	Days (mean solar)	365.25636
Years (sidereal)	Days (sidereal)	366.25640
"	Years (calendar)	1.0007024
"	Years (tropical)	1.0000388
Years (tropical)	Days (mean solar)	365.24219
"	Days (sidereal)	366.24219
"	Hours (mean solar)	8765.8126
"	Hours (sidereal)	8789.8126
"	Months (mean calendar)	12.007963
"	Seconds (mean solar)	3.1556926×10^7
"	Seconds (sidereal)	3.1643326×10^7
"	Weeks (mean calendar)	52.177456
"	Years (calendar)	1.0006635
"	Years (sidereal)	0.99996121

DEFINED VALUES AND EQUIVALENTS

Meter	**(m)**	1 650 763.73 wave lengths in vacuo of the unperturbed transition $2p_{10} - 5d_5$ in ^{86}Kr
Kilogram	**(kg)**	mass of the international kilogram at Sèvres, France
Second	**(s)**	1/31 556 925.974 7 of the tropical year at 12^h ET, 0 January 1900
Degree Kelvin	**(°K)**	defined in the thermodynamic scale by assigning 273.16 °K to the triple point of water (freezing point, 273.15 °K = 0 °C)
Unified atomic mass unit	(u)	1/12 the mass of an atom of the ^{12}C nuclide
Mole	**(mol)**	amount of substance containing the same number of atoms as 12 g of pure ^{12}C
Standard acceleration of free fall	**(g_n)**	9.806 65 m s^{-2}, 980.665 cm s^{-2}
Normal atmospheric pressure	**(atm)**	101 325 N m^{-2}, 1 013 250 dyn cm^{-2}
Thermochemical calorie	**(cal_{th})**	4.1840 J, 4.1840×10^7 erg
International Steam Table calorie	**(cal_{IT})**	4.1868 J, 4.1868×10^7 erg
Liter	**(l)**	0.001 m^3, 1000 cm^3 (recommended by GCWM, 1964)
Inch	**(in)**	0.0254 m. 2.54 cm
Pound (avdp)	**(lb)**	0.453 592 37 kg, 453.592 37 g

FACTORS FOR THE CONVERSION OF ($LOG_{10}X$) TO ($RT\ LOG_eX$)

Units are in Calories

t°C	0	1	2	3	4	5	6	7	8	9
0	1249.4	1254.0	1258.6	1263.2	1267.7	1272.3	1276.9	1281.45	1286.0	1290.6
10	1295.2	1299.8	1304.3	1308.9	1313.5	1318.0	1322.6	1327.2	1331.8	1336.3
20	1340.9	1345.5	1350.1	1354.6	1359.2	1363.8	1368.4	1372.9	1377.5	1382.1
30	1386.7	1391.2	1395.8	1400.4	1405.0	1409.5	1414.1	1418.7	1423.2	1427.8
40	1432.4	1437.0	1441.5	1446.1	1450.7	1455.3	1459.8	1464.4	1469.0	1473.6
50	1478.1	1482.7	1487.3	1491.9	1496.4	1501.0	1505.6	1510.2	1514.7	1519.3
60	1523.9	1528.5	1533.0	1537.6	1542.2	1546.7	1551.3	1555.9	1560.5	1565.0
70	1569.6	1574.2	1578.8	1583.3	1587.9	1592.5	1597.1	1601.6	1606.2	1610.8
80	1615.4	1619.9	1624.5	1629.1	1633.7	1638.2	1642.8	1647.4	1651.9	1656.5
90	1661.1	1665.7	1670.2	1674.8	1679.4	1684.0	1688.5	1693.1	1697.7	1702.3
100	1706.8	1711.4	1716.0	1720.6	1725.1	1729.7	1734.3	1738.9	1743.4	1748.0
110	1752.6	1757.2	1761.7	1766.3	1770.9	1775.4	1780.0	1784.6	1789.2	1793.7
120	1798.3	1802.9	1807.5	1812.0	1816.6	1821.2	1825.8	1830.3	1834.9	1839.5
130	1844.1	1848.6	1853.2	1857.8	1862.4	1866.9	1871.5	1876.1	1880.6	1885.2
140	1889.8	1894.4	1898.9	1903.5	1908.1	1912.7	1917.2	1921.8	1926.4	1931.0
150	1935.5	1940.1	1944.7	1949.3	1953.8	1958.4	1963.0	1967.6	1972.1	1976.7
160	1981.3	1985.9	1990.4	1995.0	1999.6	2004.1	2008.7	2013.3	2017.9	2022.4
170	2027.0	2031.6	2036.2	2040.7	2045.3	2049.9	2054.5	2059.0	2063.6	2068.2
180	2072.8	2077.3	2081.9	2086.5	2091.1	2095.6	2100.2	2104.8	2109.4	2113.9
190	2118.5	2123.1	2127.6	2132.2	2136.8	2141.4	2145.9	2150.5	2155.1	2159.7
200	2164.2	2168.8	2173.4	2178.0	2182.5	2187.1	2191.7	2196.3	2200.8	2205.4

Differences

	Tenths	Units	Tens
1	.5	4.6	45.7
2	.9	9.1	91.5
3	1.4	13.7	137.2
4	1.8	18.3	183.0
5	2.3	22.9	228.7
6	2.7	27.4	274.4
7	3.2	32.0	320.2
8	3.7	36.6	365.9
9	4.1	41.2	411.7

RECIPROCALS OF NUMBERS

For given values of N the following table contains the corresponding values of 1000/N, N + 273.15 and 1000/N + 273.15. For convenience of converting from the Kelvin temperature scale to the Celsius temperature scale the table begins at −273.15. Thus for N one may read °C, and for N + 273.15 one may read °K.

N	1000/N	N + 273.15	1000/N + 273.15
−273.15	−3.66099	0.00	
.14	6113	.01	100000.000
.13	6126	.02	50000.000
.12	6139	.03	33333.333
.11	6153	.04	25000.000
.10	6166	.05	20000.000
.09	6180	.06	16666.667
.08	6193	.07	14285.714
.07	6206	.08	12500.000
.06	6220	.09	11111.111
.05	6233	.10	10000.000
.04	6247	.11	9090.909
.03	6260	.12	8333.333
.02	6274	.13	7692.308
.01	6287	.14	7142.857
−273.00	6300	.15	6666.667
−272.99	6314	.16	6250.000
.98	6327	.17	5882.353
.97	6341	.18	5555.556
.96	6354	.19	5263.158
.95	6367	.20	5000.000
.94	6381	.21	4761.905
.93	6394	.22	4545.455
.92	6408	.23	4347.826
.91	6421	.24	4166.667
.90	6435	.25	4000.000
.89	6448	.26	3846.154
.88	6461	.27	3703.704
.87	6475	.28	3571.429
.86	6488	.29	3448.276
.85	6502	.30	3333.333
.84	6515	.31	3225.806
.83	6529	.32	3125.000
.82	6542	.33	3030.303
.81	6555	.34	2941.176
.80	6569	.35	2857.143
.79	6582	.36	2777.778
.78	6596	.37	2702.703
.77	6609	.38	2631.579
.76	6623	.39	2564.103
.75	6636	.40	2500.000
.74	6650	.41	2439.024
.73	6663	.42	2380.952
.72	6676	.43	2325.581
.71	6690	.44	2272.727
.70	6703	.45	2222.222
.69	6717	.46	2173.913
.68	6730	.47	2127.660
.67	6744	.48	2083.333
.66	6757	.49	2040.816
.65	6771	.50	2000.000
.64	6784	.51	1960.784
.63	6797	.52	1923.077
.62	6811	.53	1886.792
.61	6824	.54	1851.852
.60	6838	.55	1818.182
.59	6851	.56	1785.714
.58	6865	.57	1754.386
.57	6878	.58	1724.138
.56	6892	.59	1694.915
.55	6905	.60	1666.667
.54	6919	.61	1639.344
.53	6932	.62	1612.903
.52	6946	.63	1587.302
.51	6959	.64	1562.500
.50	6972	.65	1538.462
.49	6986	.66	1515.152
.48	6999	.67	1492.537
.47	7013	.68	1470.588
.46	7026	.69	1449.275
.45	7040	.70	1428.571

N	1000/N	N + 273.15	1000/N + 273.15
−272.44	−3.67053	0.71	1408.451
.43	7067	.72	1388.889
.42	7080	.73	1369.863
.41	7094	.74	1351.351
.40	7107	.75	1333.333
.39	7121	.76	1315.789
.38	7134	.77	1298.701
.37	7148	.78	1282.051
.36	7161	.79	1265.823
.35	7175	.80	1250.000
.34	7188	.81	1234.568
.33	7202	.82	1219.512
.32	7215	.83	1204.819
.31	7229	.84	1190.476
.30	7242	.85	1176.471
.29	7255	.86	1162.791
.28	7269	.87	1149.425
.27	7282	.88	1136.364
.26	7296	.89	1123.596
.25	7309	.90	1111.111
.24	7323	.91	1098.901
.23	7336	.92	1086.957
.22	7350	.93	1075.269
.21	7363	.94	1063.830
.20	7377	.95	1052.632
.19	7390	.96	1041.667
.18	7404	.97	1030.928
.17	7417	.98	1020.408
.16	7431	.99	1010.101
.15	7444	1.00	1000.000
.14	7458	1.01	990.099
.13	7471	.02	980.392
.12	7485	.03	970.874
.11	7498	.04	961.538
.10	7512	.05	952.381
.09	7525	.06	943.396
.08	7539	.07	934.579
.07	7552	.08	925.926
.06	7566	.09	917.431
.05	7579	.10	909.091
.04	7593	.11	900.901
.03	7607	.12	892.857
.02	7620	.13	884.956
.01	7634	.14	877.193
−272.00	7647	.15	869.565
−271.99	7661	.16	862.069
.98	7674	.17	854.701
.97	7688	.18	847.458
.96	7701	.19	840.336
.95	7715	.20	833.333
.94	7728	.21	826.446
.93	7742	.22	819.672
.92	7755	.23	813.008
.91	7769	.24	806.452
.90	7782	.25	800.000
.89	7796	.26	793.651
.88	7809	.27	787.402
.87	7823	.28	781.250
.86	7836	.29	775.194
.85	7850	.30	769.231
.84	7863	.31	763.359
.83	7877	.32	757.576
.82	7891	.33	751.880
.81	7904	.34	746.269
.80	7918	.35	740.741
.79	7931	.36	735.294
.78	7945	.37	729.927
.77	7958	.38	724.638
.76	7972	.39	719.424
.75	7985	.40	714.286
.74	7999	.41	709.220

RECIPROCALS OF NUMBERS (Continued)

N	1000/N	N + 273.15	1000/N + 273.15
−271.73	−3.68012	1.42	704.225
.72	8026	.43	699.301
.71	8039	.44	694.444
.70	8053	.45	689.655
.69	8067	.46	684.932
.68	8080	.47	680.272
.67	8094	.48	675.676
.66	8107	.49	671.141
.65	8121	.50	666.667
.64	8134	.51	662.252
.63	8148	.52	657.895
.62	8161	.53	653.595
.61	8175	.54	649.351
.60	8188	.55	645.161
.59	8202	.56	641.026
.58	8216	.57	636.943
.57	8229	.58	632.911
.56	8243	.59	628.931
.55	8256	.60	625.000
.54	8270	.61	621.118
.53	8283	.62	617.284
.52	8297	.63	613.497
.51	8311	.64	609.756
.50	8324	.65	606.061
.49	8338	.66	602.410
.48	8351	.67	598.802
.47	8365	.68	595.238
.46	8378	.69	591.716
.45	8392	.70	588.235
.44	8406	.71	584.795
.43	8419	.72	581.395
.42	8433	.73	578.035
.41	8446	.74	574.713
.40	8460	.75	571.429
.39	8473	.76	568.182
.38	8487	.77	564.972
.37	8501	.78	561.798
.36	8514	.79	558.659
.35	8528	.80	555.556
.34	8541	.81	552.486
.33	8555	.82	549.451
.32	8568	.83	546.448
.31	8582	.84	543.478
.30	8596	.85	540.541
.29	8609	.86	537.634
.28	8623	.87	534.759
.27	8636	.88	531.915
.26	8650	.89	529.101
.25	8664	.90	526.316
.24	8677	.91	523.560
.23	8691	.92	520.833
.22	8704	.93	518.135
.21	8718	.94	515.464
.20	8732	.95	512.821
.19	8745	.96	510.204
.18	8759	.97	507.614
.17	8772	.98	505.051
.16	8786	1.99	502.513
.15	8800	2.00	500.000
.14	8813	.01	497.512
.13	8827	.02	495.050
.12	8840	.03	492.611
.11	8854	.04	490.196
.10	8868	.05	487.805
.09	8881	.06	485.437
.08	8895	.07	483.092
.07	8908	.08	480.769
.06	8922	.09	478.469
.05	8936	.10	476.190
.04	8949	.11	473.934
.03	8963	.12	471.698
.02	8976	.13	469.484
.01	8990	.14	467.290
−271.00	9004	.15	465.116

N	1000/N	N + 273.15	1000/N + 273.15
−270.99	9017	.16	462.963
.98	9031	.17	460.829
.97	9045	.18	458.716
.96	9058	.19	456.621
.95	9072	.20	454.545
.94	9085	.21	452.489
.93	9099	.22	450.450
.92	9113	.23	448.430
.91	9126	.24	446.429
.90	9140	.25	444.444
.89	9154	.26	442.478
.88	9167	.27	440.529
.87	9181	.28	438.596
.86	9194	.29	436.681
.85	9208	.30	434.783
.84	9222	.31	432.900
.83	9235	.32	431.034
.82	9249	.33	428.185
.81	9263	.34	427.350
.80	9276	.35	425.532
.79	9290	.36	423.729
.78	9303	.37	421.941
.77	9317	.38	420.168
.76	9331	.39	418.410
.75	9344	.40	416.667
.74	9358	.41	414.938
.73	9372	.42	413.223
.72	9385	.43	411.523
.71	9399	.44	409.836
.70	9413	.45	408.163
.69	9426	.46	406.504
.68	9440	.47	404.858
.67	9454	.48	403.226
.66	9467	.49	401.606
.65	9481	.50	400.000
.64	9495	.51	398.406
.63	9508	.52	396.825
.62	9522	.53	395.257
.61	9535	.54	393.701
.60	9549	.55	392.157
.59	9563	.56	390.625
.58	9576	.57	389.105
.57	9590	.58	387.597
.56	9604	.59	386.100
.55	9617	.60	384.615
.54	9631	.61	383.142
.53	9645	.62	381.679
.52	9658	.63	380.228
.51	9672	.64	378.788
.50	9686	.65	377.358
.49	9699	.66	375.940
.48	9713	.67	374.532
.47	9727	.68	373.134
.46	9740	.69	371.747
.45	9754	.70	370.370
.44	9768	.71	369.004
.43	9781	.72	367.647
.42	9795	.73	366.300
.41	9809	.74	364.964
.40	9822	.75	363.636
.39	9836	.76	362.319
.38	9850	.77	361.011
.37	9864	.78	359.712
.36	9877	.79	358.423
.35	9891	.80	357.143
.34	9905	.81	355.872
.33	9918	.82	354.610
.32	9932	.83	353.357
.31	9946	.84	352.113
.30	9959	.85	350.877
.29	9973	.86	349.650
.28	9987	.87	348.432
.27	−3.70000	.88	347.222
.26	0014	.89	346.021

RECIPROCALS OF NUMBERS (Continued)

N	1000/N	N + 273.15	1000/N + 273.15
−270.25	−3.70028	2.90	344.828
.24	0041	.91	343.643
.23	0055	.92	342.466
.22	0069	.93	341.297
.21	0083	.94	340.136
.20	0096	.95	338.983
.19	0110	.96	337.838
.18	0124	.97	336.700
.17	0137	.98	335.570
.16	0151	2.99	334.448
.15	0165	3.00	333.333
.14	0178	3.01	332.226
.13	0192	.02	331.126
.12	0206	.03	330.033
.11	0220	.04	328.947
.10	0233	.05	327.869
.09	0257	.06	326.797
.08	0261	.07	325.733
.07	0274	.08	324.675
.06	0288	.09	323.625
.05	0302	.10	322.581
.04	0316	.11	321.543
.03	0329	.12	320.513
.02	0343	.13	319.489
.01	0357	.14	318.471
−270.00	0370	.15	317.460
−269.99	0384	.16	316.456
.98	0398	.17	315.457
.97	0412	.18	314.465
.96	0425	.19	313.480
.95	0439	.20	312.500
.94	0453	.21	311.526
.93	0466	.22	310.559
.92	0480	.23	309.598
.91	0494	.24	308.642
.90	0508	.25	307.692
.89	0521	.26	306.748
.88	0535	.27	305.810
.87	0549	.28	304.878
.86	0563	.29	303.951
.85	0576	.30	303.030
.84	0590	.31	302.115
.83	0604	.32	301.205
.82	0617	.33	300.300
.81	0631	.34	299.401
.80	0645	.35	298.507
.79	0659	.36	297.619
.78	0672	.37	296.736
.77	0686	.38	295.858
.76	0700	.39	294.985
.75	0714	.40	294.118
.74	0727	.41	293.255
.73	0741	.42	292.398
.72	0755	.43	291.545
.71	0769	.44	290.698
.70	0782	.45	289.855
.69	0796	.46	289.017
.68	0810	.47	288.184
.67	0824	.48	287.356
.66	0837	.49	286.533
.65	0851	.50	285.714
.64	0865	.51	284.900
.63	0879	.52	284.091
.62	0892	.53	283.286
.61	0906	.54	282.486
.60	0920	.55	281.690
.59	0934	.56	280.899
.58	0947	.57	280.112
.57	0961	.58	279.330
.56	0975	.59	278.552
.55	0989	.60	277.778
.54	1002	.61	277.008
.53	1016	.62	276.243
.52	1030	.63	275.482

N	1000/N	N + 273.15	1000/N + 273.15
−269.51	−3.71044	3.64	274.725
.50	1058	.65	273.973
.49	1071	.66	273.224
.48	1085	.67	272.480
.47	1099	.68	271.739
.46	1112	.69	271.003
.45	1126	.70	270.270
.44	1140	.71	269.542
.43	1154	.72	268.817
.42	1168	.73	268.097
.41	1181	.74	267.380
.40	1195	.75	266.667
.39	1209	.76	265.957
.38	1223	.77	265.252
.37	1237	.78	264.550
.36	1250	.79	263.852
.35	1264	.80	263.158
.34	1278	.81	262.467
.33	1292	.82	261.780
.32	1306	.83	261.097
.31	1319	.84	260.417
.30	1330	.85	259.740
.29	1347	.86	259.067
.28	1361	.87	258.398
.27	1374	.88	257.732
.26	1388	.89	257.069
.25	1402	.90	256.410
.24	1416	.91	255.754
.23	1430	.92	255.102
.22	1443	.93	254.453
.21	1457	.94	253.807
.20	1471	.95	253.165
.19	1485	.96	252.525
.18	1499	.97	251.889
.17	1512	.98	251.256
.16	1526	3.99	250.627
.15	1540	4.00	250.000
.14	1554	4.01	249.377
.13	1568	.02	248.756
.12	1581	.03	248.139
.11	1595	.04	247.525
.10	1609	.05	246.914
.09	1623	.06	246.305
.08	1637	.07	245.700
.07	1650	.08	245.098
.06	1664	.09	244.499
.05	1678	.10	243.902
.04	1692	.11	243.309
.03	1706	.12	242.718
.02	1720	.13	242.131
.01	1733	.14	241.546
−269.00	1747	.15	240.964
−268.99	1761	.16	240.385
.98	1775	.17	239.808
.97	1789	.18	239.234
.96	1802	.19	238.663
.95	1816	.20	238.095
.94	1830	.21	237.530
.93	1844	.22	236.967
.92	1858	.23	236.407
.91	1872	.24	235.849
.90	1885	.25	235.294
.89	1899	.26	234.742
.88	1913	.27	234.192
.87	1927	.28	233.645
.86	1941	.29	233.100
.85	1955	.30	232.558
.84	1968	.31	232.019
.83	1982	.32	231.481
.82	1996	.33	230.947
.81	2010	.34	230.415
.80	2024	.35	229.885
.79	2038	.36	229.358
.78	2051	.37	228.833

RECIPROCALS OF NUMBERS (Continued)

N	1000/N	N + 273.15	1000/N + 273.15
−268.77	−3.72065	4.38	228.311
.76	2079	.39	227.790
.75	2093	.40	227.273
.74	2107	.41	226.757
.73	2121	.42	226.244
.72	2135	.43	225.734
.71	2148	.44	225.225
.70	2162	.45	224.719
.69	2176	.46	224.215
.68	2190	.47	223.714
.67	2204	.48	223.214
.66	2218	.49	222.717
.65	2232	.50	222.222
.64	2245	.51	221.729
.63	2259	.52	221.239
.62	2273	.53	220.751
.61	2287	.54	220.264
.60	2301	.55	219.780
.59	2315	.56	219.298
.58	2329	.57	218.818
.57	2342	.58	218.341
.56	2356	.59	217.865
.55	2370	.60	217.391
.54	2384	.61	216.920
.53	2398	.62	216.450
.52	2412	.63	215.983
.51	2426	.64	215.517
.50	2439	.65	215.054
.49	2453	.66	214.592
.48	2467	.67	214.133
.47	2481	.68	213.675
.46	2495	.69	213.220
.45	2509	.70	212.766
.44	2523	.71	212.314
.43	2537	.72	211.864
.42	2550	.73	211.416
.41	2564	.74	210.970
.40	2578	.75	210.526
.39	2592	.76	210.084
.38	2606	.77	209.644
.37	2620	.78	209.205
.36	2634	.79	208.768
.35	2648	.80	208.333
.34	2662	.81	207.900
.33	2675	.82	207.469
.32	2689	.83	207.039
.31	2703	.84	206.612
.30	2717	.85	206.186
.29	2731	.86	205.761
.28	2745	.87	205.339
.27	2759	.88	204.918
.26	2773	.89	204.499
.25	2787	.90	204.082
.24	2800	.91	203.666
.23	2814	.92	203.252
.22	2828	.93	202.840
.21	2842	.94	202.429
.20	2856	.95	202.020
.19	2870	.96	201.613
.18	2884	.97	201.207
.17	2898	.98	200.803
.16	2912	4.99	200.401
.15	2926	5.00	200.000
.14	2940	5.01	199.601
.13	2953	.02	199.203
.12	2967	.03	198.807
.11	2981	.04	198.413
.10	2995	.05	198.020
.09	3009	.06	197.628
.08	3023	.07	197.239
.07	3037	.08	196.850
.06	3051	.09	196.464
.05	3065	.10	196.078
.04	3079	.11	195.695

N	1000/N	N + 273.15	1000/N + 273.15
−268.03	−3.73093	5.12	195.313
.02	3106	.13	194.932
.01	3120	.14	194.553
−268.00	3134	.15	194.175
−267.99	3148	.16	193.798
.98	3162	.17	193.424
.97	3176	.18	193.050
.96	3190	.19	192.678
.95	3204	.20	192.308
.94	3218	.21	191.939
.93	3232	.22	191.571
.92	3246	.23	191.205
.91	3260	.24	190.840
.90	3274	.25	190.476
.89	3288	.26	190.114
.88	3301	.27	189.753
.87	3315	.28	189.394
.86	3329	.29	189.036
.85	3343	.30	188.679
.84	3357	.31	188.324
.83	3371	.32	187.970
.82	3385	.33	187.617
.81	3399	.34	187.266
.80	3413	.35	186.916
.79	3427	.36	186.567
.78	3441	.37	186.220
.77	3455	.38	185.874
.76	3469	.39	185.529
.75	3483	.40	185.185
.74	3497	.41	184.843
.73	3511	.42	184.502
.72	3525	.43	184.162
.71	3539	.44	183.824
.70	3552	.45	183.486
.69	3566	.46	183.150
.68	3580	.47	182.815
.67	3594	.48	182.482
.66	3608	.49	182.149
.65	3622	.50	181.818
.64	3636	.51	181.488
.63	3650	.52	181.159
.62	3664	.53	180.832
.61	3678	.54	180.505
.60	3692	.55	180.180
.59	3706	.56	179.856
.58	3720	.57	179.533
.57	3734	.58	179.211
.56	3748	.59	178.891
.55	3762	.60	178.571
.54	3776	.61	178.253
.53	3790	.62	177.936
.52	3804	.63	177.620
.51	3818	.64	177.305
.50	3832	.65	176.991
.49	3846	.66	176.678
.48	3860	.67	176.367
.47	3874	.68	176.056
.46	3888	.69	175.747
.45	3902	.70	175.439
.44	3916	.71	175.131
.43	3930	.72	174.825
.42	3944	.73	174.520
.41	3958	.74	174.216
.40	3972	.75	173.913
.39	3986	.76	173.611
.38	4000	.77	173.310
.37	4014	.78	173.010
.36	4028	.79	172.712
.35	4042	.80	172.414
.34	4056	.81	172.117
.33	4070	.82	171.821
.32	4083	.83	171.527
.31	4097	.84	171.233
.30	4111	.85	170.940

RECIPROCALS OF NUMBERS (Continued)

N	1000/N	N + 273.15	1000/N + 273.15
−267.29	−3.74125	5.86	170.648
.28	4139	.87	170.358
.27	4153	.88	170.068
.26	4167	.89	169.779
.25	4181	.90	169.492
.24	4195	.91	169.205
.23	4209	.92	168.919
.22	4223	.93	168.634
.21	4237	.94	168.350
.20	4251	.95	168.067
.19	4266	.96	167.785
.18	4280	.97	167.504
.17	4294	.98	167.224
.16	4308	5.99	166.945
.15	4322	6.00	166.667
.14	4336	.01	166.389
.13	4350	.02	166.113
.12	4364	.03	165.837
.11	4378	.04	165.563
.10	4392	.05	165.289
.09	4406	.06	165.017
.08	4420	.07	164.745
.07	4434	.08	164.474
.06	4448	.09	164.204
.05	4462	.10	163.934
.04	4476	.11	163.666
.03	4490	.12	163.399
.02	4504	.13	163.132
.01	4518	.14	162.866
−267.00	4532	.15	162.602
−266.99	4546	.16	162.338
.98	4560	.17	162.075
.97	4574	.18	161.812
.96	4588	.19	161.551
.95	4602	.20	161.290
.94	4616	.21	161.031
.93	4630	.22	160.772
.92	4644	.23	160.514
.91	4658	.24	160.256
.90	4672	.25	160.000
.89	4686	.26	159.744
.88	4700	.27	159.490
.87	4714	.28	159.236
.86	4728	.29	158.983
.85	4742	.30	158.730
.84	4756	.31	158.479
.83	4770	.32	158.228
.82	4784	.33	157.978
.81	4799	.34	157.729
.80	4813	.35	157.480
.79	4827	.36	157.233
.78	4841	.37	156.986
.77	4855	.38	156.740
.76	4869	.39	156.495
.75	4883	.40	156.250
.74	4897	.41	156.006
.73	4911	.42	155.763
.72	4925	.43	155.521
.71	4939	.44	155.280
.70	4953	.45	155.039
.69	4967	.46	154.799
.68	4981	.47	154.560
.67	4995	.48	154.321
.66	5009	.49	154.083
.65	5023	.50	153.846
.64	5038	.51	153.610
.63	5052	.52	153.374
.62	5066	.53	153.139
.61	5080	.54	152.905
.60	5094	.55	152.672
.59	5108	.56	152.439
.58	5122	.57	152.207
.57	5136	.58	151.976
.56	5150	.59	151.745

N	1000/N	N + 273.15	1000/N + 273.15
−266.55	−3.75164	6.60	151.515
.54	5178	.61	151.286
.53	5192	.62	151.057
.52	5206	.63	150.830
.51	5220	.64	150.602
.50	5235	.65	150.376
.49	5249	.66	150.150
.48	5263	.67	149.925
.47	5277	.68	149.701
.46	5291	.69	149.477
.45	5305	.70	149.254
.44	5319	.71	149.031
.43	5333	.72	148.810
.42	5347	.73	148.588
.41	5361	.74	148.368
.40	5375	.75	148.148
.39	5389	.76	147.929
.38	5404	.77	147.710
.37	5418	.78	147.493
.36	5432	.79	147.275
.35	5446	.80	147.059
.34	5460	.81	146.843
.33	5474	.82	146.628
.32	5488	.83	146.413
.31	5502	.84	146.199
.30	5516	.85	145.985
.29	5530	.86	145.773
.28	5545	.87	145.560
.27	5559	.88	145.349
.26	5573	.89	145.138
.25	5587	.90	144.928
.24	5601	.91	144.718
.23	5615	.92	144.509
.22	5629	.93	144.300
.21	5643	.94	144.092
.20	5657	.95	143.885
.19	5672	.96	143.678
.18	5686	.97	143.472
.17	5700	.98	143.266
.16	5714	6.99	143.062
.15	5728	7.00	142.857
.14	5742	.01	142.653
.13	5756	.02	142.450
.12	5770	.03	142.248
.11	5784	.04	142.045
.10	5799	.05	141.844
.09	5813	.06	141.643
.08	5827	.07	141.443
.07	5841	.08	141.243
.06	5855	.09	141.044
.05	5869	.10	140.845
.04	5883	.11	140.647
.03	5897	.12	140.449
.02	5912	.13	140.252
.01	5926	.14	140.056
−266.00	5940	.15	139.860
−265.99	5954	.16	139.665
.98	5968	.17	139.470
.97	5982	.18	139.276
.96	5996	.19	139.082
.95	6011	.20	138.889
.94	6025	.21	138.696
.93	6039	.22	138.504
.92	6053	.23	138.313
.91	6067	.24	138.122
.90	6081	.25	137.931
.89	6095	.26	137.741
.88	6110	.27	137.552
.87	6124	.28	137.363
.86	6138	.29	137.174
.85	6152	.30	136.986
.84	6166	.31	136.799
.83	6180	.32	136.612
.82	6194	.33	136.426

RECIPROCALS OF NUMBERS (Continued)

N	1000/N	N + 273.15	1000/N + 273.15
−265.81	−3.76209	7.34	136.240
.80	6223	.35	136.054
.79	6237	.36	135.870
.78	6251	.37	135.685
.77	6265	.38	135.501
.76	6279	.39	135.318
.75	6294	.40	135.135
.74	6308	.41	134.953
.73	6322	.42	134.771
.72	6336	.43	134.590
.71	6350	.44	134.409
.70	6364	.45	134.228
.69	6378	.46	134.048
.68	6393	.47	133.869
.67	6407	.48	133.690
.66	6421	.49	133.511
.65	6435	.50	133.333
.64	6449	.51	133.156
.63	6463	.52	132.979
.62	6478	.53	132.802
.61	6492	.54	132.626
.60	6506	.55	132.450
.59	6520	.56	132.275
.58	6534	.57	132.100
.57	6549	.58	131.926
.56	6563	.59	131.752
.55	6577	.60	131.579
.54	6591	.61	131.406
.53	6605	.62	131.234
.52	6619	.63	131.062
.51	6634	.64	130.890
.50	6648	.65	130.719
.49	6662	.66	130.548
.48	6676	.67	130.378
.47	6690	.68	130.208
.46	6705	.69	130.039
.45	6719	.70	129.870
.44	6733	.71	129.702
.43	6747	.72	129.534
.42	6761	.73	129.366
.41	6776	.74	129.199
.40	6790	.75	129.032
.39	6804	.76	128.866
.38	6818	.77	128.700
.37	6832	.78	128.535
.36	6847	.79	128.370
.35	6861	.80	128.205
.34	6875	.81	128.041
.33	6890	.82	127.877
.32	6903	.83	127.714
.31	6918	.84	127.551
.30	6932	.85	127.389
.29	6946	.86	127.226
.28	6960	.87	127.065
.27	6974	.88	126.904
.26	6989	.89	126.743
.25	7003	.90	126.582
.24	7017	.91	126.422
.23	7031	.92	126.263
.22	7045	.93	126.103
.21	7060	.94	125.945
.20	7074	.95	125.786
.19	7088	.96	125.628
.18	7102	.97	125.471
.17	7117	.98	125.313
.16	7131	7.99	125.156
.15	7145	8.00	125.000
.14	7159	.01	124.844
.13	7173	.02	124.688
.12	7188	.03	124.533
.11	7202	.04	124.378
.10	7216	.05	124.224
.09	7223	.06	124.069
.08	7245	.07	123.916
−265.07	−3.77259	8.08	123.762
.06	7273	.09	123.609
.05	7287	.10	123.457
.04	7302	.11	123.305
.03	7316	.12	123.153
.02	7330	.13	123.001
.01	7344	.14	122.850
−265.00	7358	.15	122.699
−264.99	7373	.16	122.549
.98	7387	.17	122.399
.97	7401	.18	122.249
.96	7415	.19	122.100
.95	7430	.20	121.951
.94	7444	.21	121.803
.93	7458	.22	121.655
.92	7472	.23	121.507
.91	7487	.24	121.359
.90	7501	.25	121.212
.89	7515	.26	121.065
.88	7529	.27	120.919
.87	7544	.28	120.773
.86	7558	.29	120.627
.85	7572	.30	120.482
.84	7586	.31	120.337
.83	7601	.32	120.192
.82	7615	.33	120.048
.81	7629	.34	119.904
.80	7644	.35	119.760
.79	7658	.36	119.617
.78	7672	.37	119.474
.77	7686	.38	119.332
.76	7701	.39	119.190
.75	7715	.40	119.048
.74	7729	.41	118.906
.73	7743	.42	118.765
.72	7758	.43	118.624
.71	7772	.44	118.483
.70	7786	.45	118.343
.69	7800	.46	118.203
.68	7815	.47	118.064
.67	7829	.48	117.925
.66	7843	.49	117.786
.65	7858	.50	117.647
.64	7872	.51	117.509
.63	7886	.52	117.371
.62	7900	.53	117.233
.61	7915	.54	117.096
.60	7929	.55	116.959
.59	7943	.56	116.822
.58	7958	.57	116.686
.57	7972	.58	116.550
.56	7986	.59	116.414
.55	8000	.60	116.279
.54	8015	.61	116.144
.53	8029	.62	116.009
.52	8043	.63	115.875
.51	8058	.64	115.741
.50	8072	.65	115.607
.49	8086	.66	115.473
.48	8100	.67	115.340
.47	8115	.68	115.207
.46	8129	.69	115.075
.45	8143	.70	114.943
.44	8158	.71	114.811
.43	8172	.72	114.679
.42	8186	.73	114.548
.41	8201	.74	114.416
.40	8215	.75	114.286
.39	8229	.76	114.155
.38	8243	.77	114.025
.37	8258	.78	113.895
.36	8272	.79	113.766
.35	8286	.80	113.636
.34	8301	.81	113.507

RECIPROCALS OF NUMBERS (Continued)

N	1000/N	N + 273.15	1000/N + 273.15
−264.33	−3.78315	8.82	113.379
.32	8329	.83	113.250
.31	8344	.84	113.122
.30	8358	.85	112.994
.29	8372	.86	112.867
.28	8387	.87	112.740
.27	8401	.88	112.613
.26	8415	.89	112.486
.25	8430	.90	112.360
.24	8444	.91	112.233
.23	8458	.92	112.108
.22	8472	.93	111.982
.21	8487	.94	111.857
.20	8501	.95	111.732
.19	8515	.96	111.607
.18	8530	.97	111.483
.17	8544	.98	111.359
.16	8558	8.99	111.235
.15	8573	9.00	111.111
.14	8587	.01	110.988
.13	8601	.02	110.865
.12	8616	.03	110.742
.11	8630	.04	110.619
.10	8644	.05	110.497
.09	8659	.06	110.375
.08	8673	.07	110.254
.07	8687	.08	110.132
.06	8702	.09	110.011
.05	8716	.10	109.890
.04	8730	.11	109.769
.03	8745	.12	109.649
.02	8759	.13	109.529
.01	8774	.14	109.409
−264.00	8788	.15	109.290
−263.99	8802	.16	109.170
.98	8817	.17	109.051
.97	8831	.18	108.932
.96	8845	.19	108.814
.95	8860	.20	108.696
.94	8874	.21	108.578
.93	8888	.22	108.460
.92	8903	.23	108.342
.91	8917	.24	108.225
.90	8931	.25	108.108
.89	8946	.26	107.991
.88	8960	.27	107.875
.87	8974	.28	107.759
.86	8989	.29	107.643
.85	9003	.30	107.527
.84	9018	.31	107.411
.83	9032	.32	107.296
.82	9046	.33	107.181
.81	9061	.34	107.066
.80	9075	.35	106.952
.79	9089	.36	106.838
.78	9104	.37	106.724
.77	9118	.38	106.610
.76	9133	.39	106.496
.75	9147	.40	106.383
.74	9161	.41	106.270
.73	9176	.42	106.157
.72	9190	.43	106.045
.71	9204	.44	105.932
.70	9219	.45	105.820
.69	9233	.46	105.708
.68	9248	.47	105.597
.67	9262	.48	105.485
.66	9276	.49	105.374
.65	9291	.50	105.263
.64	9305	.51	105.152
.63	9320	.52	105.042
.62	9334	.53	104.932
.61	9348	.54	104.822
.60	9363	.55	104.712

N	1000/N	N + 273.15	1000/N + 273.15
−263.59	−3.79377	9.56	104.603
.58	9391	.57	104.493
.57	9406	.58	104.384
.56	9420	.59	104.275
.55	9435	.60	104.167
.54	9449	.61	104.058
.53	9463	.62	103.950
.52	9478	.63	103.842
.51	9492	.64	103.734
.50	9507	.65	103.627
.49	9521	.66	103.520
.48	9535	.67	103.413
.47	9550	.68	103.306
.46	9564	.69	103.199
.45	9579	.70	103.093
.44	9593	.71	102.987
.43	9607	.72	102.881
.42	9622	.73	102.775
.41	9636	.74	102.669
.40	9651	.75	102.564
.39	9665	.76	102.459
.38	9680	.77	102.354
.37	9694	.78	102.249
.36	9708	.79	102.145
.35	9723	.80	102.041
.34	9737	.81	101.937
.33	9752	.82	101.833
.32	9766	.83	101.729
.31	9780	.84	101.626
.30	9795	.85	101.523
.29	9809	.86	101.420
.28	9824	.87	101.317
.27	9838	.88	101.215
.26	9853	.89	101.112
.25	9867	.90	101.010
.24	9881	.91	100.908
.23	9896	.92	100.806
.22	9910	.93	100.705
.21	9925	.94	100.604
.20	9939	.95	100.503
.19	9954	.96	100.402
.18	9968	.97	100.301
.17	9983	.98	100.200
.16	9997	.99	100.100
.15	−3.80011	10.00	100.000
.14	0026	.01	99.9001
.13	0040	.02	99.8004
.12	0055	.03	99.7009
.11	0069	.04	99.6016
.10	0084	.05	99.5025
.09	0098	.06	99.4036
.08	0113	.07	99.3049
.07	0127	.08	99.2063
.06	0141	.09	99.1080
.05	0156	.10	99.0099
.04	0170	.11	98.9120
.03	0185	.12	98.8142
.02	0199	.13	98.7167
.01	0214	.14	98.6193
263	3.80228	10.15	98.52216
262	3.81679	11.15	89.68610
261	3.83142	12.15	82.30453
260	3.84615	13.15	76.04563
259	3.86100	14.15	70.67138
258	3.87597	15.15	66.00660
257	3.89105	16.15	61.91950
256	3.90625	17.15	58.30904
255	3.92157	18.15	55.09642
254	3.93701	19.15	52.21932
253	3.95257	20.15	49.62779
252	3.96825	21.15	47.28132
251	3.98406	22.15	45.14673
250	4.00000	23.15	43.19654
249	4.01606	24.15	41.40787

RECIPROCALS OF NUMBERS (Continued)

N	1000/N	N + 273.15	1000/N + 273.15
+48	+20.83333	321.15	3.11381
49	20.40816	322.15	3.10414
50	20.00000	323.15	3.09454
51	19.60784	324.15	3.08499
52	19.23077	325.15	3.07550
53	18.86792	326.15	3.06607
54	18.51852	327.15	3.05670
55	18.18182	328.15	3.04739
56	17.85714	329.15	3.03813
57	17.54386	330.15	3.02893
58	17.24138	331.15	3.01978
59	16.94915	332.15	3.01069
60	16.66667	333.15	3.00165
61	16.39344	334.15	2.99267
62	16.12903	335.15	2.98374
63	15.87302	336.15	2.97486
64	15.62500	337.15	2.96604
65	15.38462	338.15	2.95727
66	15.15152	339.15	2.94855
67	14.92537	340.15	2.93988
68	14.70588	341.15	2.93126
69	14.49275	342.15	2.92269
70	14.28571	343.15	2.91418
71	14.08451	344.15	2.90571
72	13.88889	345.15	2.89729
73	13.69863	346.15	2.88892
74	13.51351	347.15	2.88060
75	13.33333	348.15	2.87233
76	13.15789	349.15	2.86410
77	12.98701	350.15	2.85592
78	12.82051	351.15	2.84779
79	12.65823	352.15	2.83970
80	12.50000	353.15	2.83166
81	12.34568	354.15	2.82366
82	12.19512	355.15	2.81571
83	12.04819	356.15	2.80781
84	11.90476	357.15	2.79994
85	11.76471	358.15	2.79213
86	11.62791	359.15	2.78435
87	11.49425	360.15	2.77662
88	11.36364	361.15	2.76893
89	11.23596	362.15	2.76129
90	11.11111	363.15	2.75368
91	10.98901	364.15	2.74612
92	10.86957	365.15	2.73860
93	10.75269	366.15	2.73112
94	10.63830	367.15	2.72368
95	10.52632	368.15	2.71628
96	10.41667	369.15	2.70893
97	10.30928	370.15	2.70161
98	10.20408	371.15	2.69433
99	10.10101	372.15	2.68709
100	10.00000	373.15	2.67989
101	9.90099	374.15	2.67272
102	9.80392	375.15	2.66560
103	9.70874	376.15	2.65851
104	9.61538	377.15	2.65146
105	9.52381	378.15	2.64445
106	9.43396	379.15	2.63748
107	9.34579	380.15	2.63054
108	9.25926	381.15	2.62364
109	9.17431	382.15	2.61677
110	9.09091	383.15	2.60994
111	9.00901	384.15	2.60315
112	8.92857	385.15	2.59639
113	8.84956	386.15	2.58967
114	8.77193	387.15	2.58298
115	8.69565	388.15	2.57632
116	8.62069	389.15	2.56970
117	8.54701	390.15	2.56312
118	8.47458	391.15	2.55656
119	8.40336	392.15	2.55004
120	8.33333	393.15	2.54356
121	8.26446	394.15	2.53711
+122	+8.19672	395.15	2.53068
123	8.13008	396.15	2.52430
124	8.06452	397.15	2.51794
125	8.00000	398.15	2.51162
126	7.93651	399.15	2.50532
127	7.87402	400.15	2.49906
128	7.81250	401.15	2.49283
129	7.75194	402.15	2.48663
130	7.69231	403.15	2.48047
131	7.63359	404.15	2.47433
132	7.57576	405.15	2.46822
133	7.51880	406.15	2.46214
134	7.46269	407.15	2.45610
135	7.40741	408.15	2.45008
136	7.35294	409.15	2.44409
137	7.29927	410.15	2.43813
138	7.24638	411.15	2.43220
139	7.19424	412.15	2.42630
140	7.14286	413.15	2.42043
141	7.09220	414.15	2.41458
142	7.04225	415.15	2.40877
143	6.99301	416.15	2.40298
144	6.94444	417.15	2.39722
145	6.89655	418.15	2.39149
146	6.84932	419.15	2.38578
147	6.80272	420.15	2.38010
148	6.75676	421.15	2.37445
149	6.71141	422.15	2.36883
150	6.66667	423.15	2.36323
151	6.62252	424.15	2.35766
152	6.57895	425.15	2.35211
153	6.53595	426.15	2.34659
154	6.49351	427.15	2.34110
155	6.45161	428.15	2.33563
156	6.41026	429.15	2.33019
157	6.36943	430.15	2.32477
158	6.32911	431.15	2.31938
159	6.28931	432.15	2.31401
160	6.25000	433.15	2.30867
161	6.21118	434.15	2.30335
162	6.17284	435.15	2.29806
163	6.13497	436.15	2.29279
164	6.09756	437.15	2.28754
165	6.06061	438.15	2.28232
166	6.02410	439.15	2.27713
167	5.98802	440.15	2.27195
168	5.95238	441.15	2.26680
169	5.91716	442.15	2.26168
170	5.88235	443.15	2.25657
171	5.84795	444.15	2.25149
172	5.81395	445.15	2.24643
173	5.78035	446.15	2.24140
174	5.74713	447.15	2.23639
175	5.71429	448.15	2.23140
176	5.68182	449.15	2.22643
177	5.64972	450.15	2.22148
178	5.61798	451.15	2.21656
179	5.58659	452.15	2.21166
180	5.55556	453.15	2.20677
181	5.52486	454.15	2.20192
182	5.49451	455.15	2.19708
183	5.46448	456.15	2.19226
184	5.43478	457.15	2.18747
185	5.40541	458.15	2.18269
186	5.37634	459.15	2.17794
187	5.34759	460.15	2.17320
188	5.31915	461.15	2.16849
189	5.29101	462.15	2.16380
190	5.26316	463.15	2.15913
191	5.23560	464.15	2.15448
192	5.20833	465.15	2.14984
193	5.18135	466.15	2.14523
194	5.15464	467.15	2.14064
195	5.12821	468.15	2.13607

RECIPROCALS OF NUMBERS (Continued)

N	1000/N	N + 273.15	1000/N + 273.15
+196	5.10204	469.15	2.13151
197	5.07614	470.15	2.12698
198	5.05051	471.15	2.12247
199	5.02513	472.15	2.11797
200	5.00000	473.15	2.11349
201	4.97512	474.15	2.10904
202	4.95050	475.15	2.10460
203	4.92611	476.15	2.10018
204	4.90196	477.15	2.09578
205	4.87805	478.15	2.09139
206	4.85437	479.15	2.08703
207	4.83092	480.15	2.08268
208	4.80769	481.15	2.07835
209	4.78469	482.15	2.07404
210	4.76190	483.15	2.06975
211	4.73934	484.15	2.06548
212	4.71698	485.15	2.06122
213	4.69484	486.15	2.05698
214	4.67290	487.15	2.05276
215	4.65116	488.15	2.04855
216	4.62963	489.15	2.04436
217	4.60829	490.15	2.04019
218	4.58716	491.15	2.03604
219	4.56621	492.15	2.03190
220	4.54545	493.15	2.02778
221	4.52489	494.15	2.02368
222	4.50450	495.15	2.01959
223	4.48430	496.15	2.01552
224	4.46429	497.15	2.01147
225	4.44444	498.15	2.00743
226	4.42478	499.15	2.00341
227	4.40529	500.15	1.99940
228	4.38596	501.15	1.99541
229	4.36681	502.15	1.99144
230	4.34783	503.15	1.98748
231	4.32900	504.15	1.98354
232	4.31034	505.15	1.97961
233	4.29185	506.15	1.97570
234	4.27350	507.15	1.97180
235	4.25532	508.15	1.96792
236	4.23729	509.15	1.96406
237	4.21941	510.15	1.96021
238	4.20168	511.15	1.95637
239	4.18410	512.15	1.95255
240	4.16667	513.15	1.94875
241	4.14938	514.15	1.94496
242	4.13223	515.15	1.94118
243	4.11523	516.15	1.93742
244	4.09836	517.15	1.93367
245	4.08163	518.15	1.92994
246	4.06504	519.15	1.92623
247	4.04858	520.15	1.92252
248	4.03226	521.15	1.91883
249	4.01606	522.15	1.91516
250	4.00000	523.15	1.91150
251	3.98406	524.15	1.90785
252	3.96825	525.15	1.90422
253	3.95257	526.15	1.90060
254	3.93701	527.15	1.89699
255	3.92157	528.15	1.89340
256	3.90625	529.15	1.88982
257	3.89105	530.15	1.88626
258	3.87597	531.15	1.88271
259	3.86100	532.15	1.87917
260	3.84615	533.15	1.87564
261	3.83142	534.15	1.87213
262	3.81679	535.15	1.86863
263	3.80228	536.15	1.86515
264	3.78788	537.15	1.86168
265	3.77358	538.15	1.85822
266	3.75940	539.15	1.85477
267	3.74532	540.15	1.85134
268	3.73134	541.15	1.84792
269	3.71747	542.15	1.84451

N	1000/N	N + 273.15	1000/N + 273.15
+270	3.70370	543.15	1.84111
271	3.69004	544.15	1.83773
272	3.67647	545.15	1.83436
273	3.66300	546.15	1.83100
274	3.64964	547.15	1.82765
275	3.63636	548.15	1.82432
276	3.62319	549.15	1.82100
277	3.61011	550.15	1.81769
278	3.59712	551.15	1.81439
279	3.58423	552.15	1.81110
280	3.57143	553.15	1.80783
281	3.55872	554.15	1.80457
282	3.54610	555.15	1.80131
283	3.53357	556.15	1.79808
284	3.52113	557.15	1.79485
285	3.50877	558.15	1.79163
286	3.49650	559.15	1.78843
287	3.48432	560.15	1.78524
288	3.47222	561.15	1.78205
289	3.46021	562.15	1.77888
290	3.44828	563.15	1.77573
291	3.43643	564.15	1.77258
292	3.42466	565.15	1.76944
293	3.41297	566.15	1.76632
294	3.40136	567.15	1.76320
295	3.38983	568.15	1.76010
296	3.37838	569.15	1.75701
297	3.36700	570.15	1.75392
298	3.35570	571.15	1.75085
299	3.34448	572.15	1.74779
300	3.33333	573.15	1.74474
301	3.32226	574.15	1.74171
302	3.31126	575.15	1.73868
303	3.30033	576.15	1.73566
304	3.28947	577.15	1.73265
305	3.27869	578.15	1.72965
306	3.26797	579.15	1.72667
307	3.25733	580.15	1.72369
308	3.24675	581.15	1.72073
309	3.23625	582.15	1.71777
310	3.22581	583.15	1.71482
311	3.21543	584.15	1.71189
312	3.20513	585.15	1.70896
313	3.19489	586.15	1.70605
314	3.18471	587.15	1.70314
315	3.17460	588.15	1.70025
316	3.16456	589.15	1.69736
317	3.15457	590.15	1.69448
318	3.14465	591.15	1.69162
319	3.13480	592.15	1.68876
320	3.12500	593.15	1.68591
321	3.11526	594.15	1.68308
322	3.10559	595.15	1.68025
323	3.09598	596.15	1.67743
324	3.08642	597.15	1.67462
325	3.07692	598.15	1.67182
326	3.06748	599.15	1.66903
327	3.05810	600.15	1.66625
328	3.04878	601.15	1.66348
329	3.03951	602.15	1.66072
330	3.03030	603.15	1.65796
331	3.02115	604.15	1.65522
332	3.01205	605.15	1.65248
333	3.00300	606.15	1.64976
334	2.99401	607.15	1.64704
335	2.98507	608.15	1.64433
336	2.97619	609.15	1.64163
337	2.96736	610.15	1.63894
338	2.95858	611.15	1.63626
339	2.94985	612.15	1.63359
340	2.94118	613.15	1.63092
341	2.93255	614.15	1.62827
342	2.92398	615.15	1.62562
343	2.91545	616.15	1.62298

RECIPROCALS OF NUMBERS (Continued)

N	1000/N	N + 273.15	1000/N + 273.15
+344	2.90698	617.15	1.62035
345	2.89855	618.15	1.61773
346	2.89017	619.15	1.61512
347	2.88184	620.15	1.61251
348	2.87356	621.15	1.60992
349	2.86533	622.15	1.60733
350	2.85714	623.15	1.60475
351	2.84900	624.15	1.60218
352	2.84091	625.15	1.59962
353	2.83286	626.15	1.59706
354	2.82486	627.15	1.59451
355	2.81690	628.15	1.59198
356	2.80899	629.15	1.58945
357	2.80112	630.15	1.58692
358	2.79330	631.15	1.58441
359	2.78552	632.15	1.58190
360	2.77778	633.15	1.57940
361	2.77008	634.15	1.57691
362	2.76243	635.15	1.57443
363	2.75482	636.15	1.57196
364	2.74725	637.15	1.56949
365	2.73973	638.15	1.56703
366	2.73224	639.15	1.56458
367	2.72480	640.15	1.56213
368	2.71739	641.15	1.55970
369	2.71003	642.15	1.55727
370	2.70270	643.15	1.55485
371	2.69542	644.15	1.55243
372	2.68817	645.15	1.55003
373	2.68097	646.15	1.54763
374	2.67380	647.15	1.54524
375	2.66667	648.15	1.54285
376	2.65957	649.15	1.54048
377	2.65252	650.15	1.53811
378	2.64550	651.15	1.53574
379	2.63852	652.15	1.53339
380	2.63158	653.15	1.53104
381	2.62467	654.15	1.52870
382	2.61780	655.15	1.52637
383	2.61097	656.15	1.52404
384	2.60417	657.15	1.52172
385	2.59740	658.15	1.51941
386	2.59067	659.15	1.51711
387	2.58398	660.15	1.51481
388	2.57732	661.15	1.51252
389	2.57069	662.15	1.51023
390	2.56410	663.15	1.50795
391	2.55754	664.15	1.50568
392	2.55102	665.15	1.50342
393	2.54453	666.15	1.50116
394	2.53807	667.15	1.49891
395	2.53165	668.15	1.49667
396	2.52525	669.15	1.49443
397	2.51889	670.15	1.49220
398	2.51256	671.15	1.48998
399	2.50627	672.15	1.48776
400	2.50000	673.15	1.48555
401	2.49377	674.15	1.48335
402	2.48756	675.15	1.48115
403	2.48139	676.15	1.47896
404	2.47525	677.15	1.47678
405	2.46914	678.15	1.47460
406	2.46305	679.15	1.47243
407	2.45700	680.15	1.47026
408	2.45098	681.15	1.46811
409	2.44499	682.15	1.46595
410	2.43902	683.15	1.46381
411	2.43309	684.15	1.46167
412	2.42718	685.15	1.45953
413	2.42131	686.15	1.45741
414	2.41546	687.15	1.45529
415	2.40964	688.15	1.45317
416	2.40385	689.15	1.45106
417	2.39808	690.15	1.44896

N	1000/N	N + 273.15	1000/N + 273.15
+418	2.39234	691.15	1.44686
419	2.38663	692.15	1.44477
420	2.38095	693.15	1.44269
421	2.37530	694.15	1.44061
422	2.36967	695.15	1.43854
423	2.36407	696.15	1.43647
424	2.35849	697.15	1.43441
425	2.35294	698.15	1.43236
426	2.34742	699.15	1.43031
427	2.34192	700.15	1.42827
428	2.33645	701.15	1.42623
429	2.33100	702.15	1.42420
430	2.32558	703.15	1.42217
431	2.32019	704.15	1.42015
432	2.31481	705.15	1.41814
433	2.30947	706.15	1.41613
434	2.30415	707.15	1.41413
435	2.29885	708.15	1.41213
436	2.29358	709.15	1.41014
437	2.28833	710.15	1.40815
438	2.28311	711.15	1.40617
439	2.27790	712.15	1.40420
440	2.27273	713.15	1.40223
441	2.26757	714.15	1.40027
442	2.26244	715.15	1.39831
443	2.25734	716.15	1.39636
444	2.25225	717.15	1.39441
445	2.24719	718.15	1.39247
446	2.24215	719.15	1.39053
447	2.23714	720.15	1.38860
448	2.23214	721.15	1.38667
449	2.22717	722.15	1.38475
450	2.22222	723.15	1.38284
451	2.21729	724.15	1.38093
452	2.21239	725.15	1.37903
453	2.20751	726.15	1.37713
454	2.20264	727.15	1.37523
455	2.19780	728.15	1.37334
456	2.19298	729.15	1.37146
457	2.18818	730.15	1.36958
458	2.18341	731.15	1.36771
459	2.17865	732.15	1.36584
460	2.17391	733.15	1.36398
461	2.16920	734.15	1.36212
462	2.16450	735.15	1.36027
463	2.15983	736.15	1.35842
464	2.15517	737.15	1.35658
465	2.15054	738.15	1.35474
466	2.14592	739.15	1.35291
467	2.14133	740.15	1.35108
468	2.13675	741.15	1.34925
469	2.13220	742.15	1.34744
470	2.12766	743.15	1.34562
471	2.12314	744.15	1.34382
472	2.11864	745.15	1.34201
473	2.11416	746.15	1.34021
474	2.10970	747.15	1.33842
475	2.10526	748.15	1.33663
476	2.10084	749.15	1.33485
477	2.09644	750.15	1.33307
478	2.09205	751.15	1.33129
479	2.08768	752.15	1.32952
480	2.08333	753.15	1.32776
481	2.07900	754.15	1.32600
482	2.07469	755.15	1.32424
483	2.07039	756.15	1.32249
484	2.06612	757.15	1.32074
485	2.06186	758.15	1.31900
486	2.05761	759.15	1.31726
487	2.05339	760.15	1.31553
488	2.04918	761.15	1.31380
489	2.04499	762.15	1.31208
490	2.04082	763.15	1.31036
491	2.03666	764.15	1.30864

RECIPROCALS OF NUMBERS (Continued)

N	1000/N	N + 273.15	1000/N + 273.15
+492	2.03252	765.15	1.30693
493	2.02840	766.15	1.30523
494	2.02429	767.15	1.30353
495	2.02020	768.15	1.30183
496	2.01613	769.15	1.30014
497	2.01207	770.15	1.29845
498	2.00803	771.15	1.29676
499	2.00401	772.15	1.29509
500	2.00000	773.15	1.29341
501	1.99601	774.15	1.29174
502	1.99203	775.15	1.29007
503	1.98807	776.15	1.28841
504	1.98413	777.15	1.28675
505	1.98020	778.15	1.28510
506	1.97627	779.15	1.28345
507	1.97239	780.15	1.28180
508	1.96850	781.15	1.28016
509	1.96464	782.15	1.27853
510	1.96078	783.15	1.27689
511	1.95695	784.15	1.27527
512	1.95313	785.15	1.27364
513	1.94932	786.15	1.27202
514	1.94553	787.15	1.27041
515	1.94175	788.15	1.26879
516	1.93798	789.15	1.26719
517	1.93424	790.15	1.26558
518	1.93050	791.15	1.26398
519	1.92678	792.15	1.26239
520	1.92308	793.15	1.26080
521	1.91939	794.15	1.25921
522	1.91571	795.15	1.25762
523	1.91205	796.15	1.25604
524	1.90840	797.15	1.25447
525	1.90476	798.15	1.25290
526	1.90114	799.15	1.25133
527	1.89753	800.15	1.24977
528	1.89394	801.15	1.24821
529	1.89036	802.15	1.24665
530	1.88679	803.15	1.24510
531	1.88324	804.15	1.24355
532	1.87970	805.15	1.24200
533	1.87617	806.15	1.24046
534	1.87266	807.15	1.23893
535	1.86916	808.15	1.23739
536	1.86567	809.15	1.23586
537	1.86220	810.15	1.23434
538	1.85874	811.15	1.23282
539	1.85529	812.15	1.23130
540	1.85185	813.15	1.22979
541	1.84843	814.15	1.22827
542	1.84502	815.15	1.22677
543	1.84162	816.15	1.22526
544	1.83824	817.15	1.22377
545	1.83486	818.15	1.22227
546	1.83150	819.15	1.22078
547	1.82815	820.15	1.21929
548	1.82482	821.15	1.21780
549	1.82149	822.15	1.21632
550	1.81818	823.15	1.21485
551	1.81488	824.15	1.21337
552	1.81159	825.15	1.21190
553	1.80832	826.15	1.21043
554	1.80505	827.15	1.20897
555	1.80180	828.15	1.20751
556	1.79856	829.15	1.20605
557	1.79533	830.15	1.20460
558	1.79211	831.15	1.20315
559	1.78891	832.15	1.20171
560	1.78571	833.15	1.20026
561	1.78253	834.15	1.19883
562	1.77936	835.15	1.19739
563	1.77620	836.15	1.19596
564	1.77305	837.15	1.19453
565	1.76991	838.15	1.19310
+566	1.76678	839.15	1.19168
567	1.76367	840.15	1.19026
568	1.76056	841.15	1.18885
569	1.75747	842.15	1.18744
570	1.75439	843.15	1.18603
571	1.75131	844.15	1.18462
572	1.74825	845.15	1.18322
573	1.74520	846.15	1.18182
574	1.74216	847.15	1.18043
575	1.73913	848.15	1.17904
576	1.73611	849.15	1.17765
577	1.73310	850.15	1.17626
578	1.73010	851.15	1.17488
579	1.72712	852.15	1.17350
580	1.72414	853.15	1.17213
581	1.72117	854.15	1.17075
582	1.71821	855.15	1.16939
583	1.71527	856.15	1.16802
584	1.71233	857.15	1.16666
585	1.70940	858.15	1.16530
586	1.70648	859.15	1.16394
587	1.70358	860.15	1.16259
588	1.70068	861.15	1.16124
589	1.69779	862.15	1.15989
590	1.69492	863.15	1.15855
591	1.69205	864.15	1.15721
592	1.68919	865.15	1.15587
593	1.68634	866.15	1.15453
594	1.68350	867.15	1.15320
595	1.68067	868.15	1.15187
596	1.67785	869.15	1.15055
597	1.67504	870.15	1.14923
598	1.67224	871.15	1.14791
599	1.66945	872.15	1.14659
600	1.66667	873.15	1.14528
601	1.66389	874.15	1.14397
602	1.66113	875.15	1.14266
603	1.65837	876.15	1.14136
604	1.65563	877.15	1.14006
605	1.65289	878.15	1.13876
606	1.65017	879.15	1.13746
607	1.64745	880.15	1.13617
608	1.64474	881.15	1.13488
609	1.64204	882.15	1.13359
610	1.63934	883.15	1.13231
611	1.63666	884.15	1.13103
612	1.63399	885.15	1.12975
613	1.63132	886.15	1.12848
614	1.62866	887.15	1.12721
615	1.62602	888.15	1.12594
616	1.62338	889.15	1.12467
617	1.62075	890.15	1.12341
618	1.61812	891.15	1.12215
619	1.61551	892.15	1.12089
620	1.61290	893.15	1.11963
621	1.61031	894.15	1.11838
622	1.60772	895.15	1.11713
623	1.60514	896.15	1.11588
624	1.60256	897.15	1.11464
625	1.60000	898.15	1.11340
626	1.59744	899.15	1.11216
627	1.59490	900.15	1.11093
628	1.59236	901.15	1.10969
629	1.58983	902.15	1.10846
630	1.58730	903.15	1.10724
631	1.58479	904.15	1.10601
632	1.58228	905.15	1.10479
633	1.57978	906.15	1.10357
634	1.57729	907.15	1.10235
635	1.57480	908.15	1.10114
636	1.57233	909.15	1.09993
637	1.56986	910.15	1.09872
638	1.56740	911.15	1.09751
639	1.56495	912.15	1.09631

RECIPROCALS OF NUMBERS (Continued)

N	1000/N	N + 273.15	1000/N + 273.15
+640	1.56250	913.15	1.09511
641	1.56006	914.15	1.09391
642	1.55763	915.15	1.09272
643	1.55521	916.15	1.09152
644	1.55280	917.15	1.09033
645	1.55039	918.15	1.08915
646	1.54799	919.15	1.08796
647	1.54560	920.15	1.08678
648	1.54321	921.15	1.08560
649	1.54083	922.15	1.08442
650	1.53846	923.15	1.08325
651	1.53610	924.15	1.08208
652	1.53374	925.15	1.08091
653	1.53139	926.15	1.07974
654	1.52905	927.15	1.07857
655	1.52672	928.15	1.07741
656	1.52439	929.15	1.07625
657	1.52207	930.15	1.07510
658	1.51976	931.15	1.07394
659	1.51745	932.15	1.07279
660	1.51515	933.15	1.07164
661	1.51286	934.15	1.07049
662	1.51057	935.15	1.06935
663	1.50830	936.15	1.06820
664	1.50602	937.15	1.06707
665	1.50376	938.15	1.06593
666	1.50150	939.15	1.06479
667	1.49925	940.15	1.06366
668	1.49701	941.15	1.06253
669	1.49477	942.15	1.06140
670	1.49254	943.15	1.06028
671	1.49031	944.15	1.05915
672	1.48810	945.15	1.05803
673	1.48588	946.15	1.05691
674	1.48368	947.15	1.05580
675	1.48148	948.15	1.05469
676	1.47929	949.15	1.05357
677	1.47710	950.15	1.05247
678	1.47493	951.15	1.05136
679	1.47275	952.15	1.05025
680	1.47059	953.15	1.04915
681	1.46843	954.15	1.04805
682	1.46628	955.15	1.04696
683	1.46413	956.15	1.04586
684	1.46199	957.15	1.04477
685	1.45985	958.15	1.04368
686	1.45773	959.15	1.04259
687	1.45560	960.15	1.04150
688	1.45349	961.15	1.04042
689	1.45138	962.15	1.03934
690	1.44928	963.15	1.03826
691	1.44718	964.15	1.03718
692	1.44509	965.15	1.03611
693	1.44300	966.15	1.03504
694	1.44092	967.15	1.03397
695	1.43885	968.15	1.03290
696	1.43678	969.15	1.03183
697	1.43472	970.15	1.03077
698	1.43266	971.15	1.02971
699	1.43062	972.15	1.02865
700	1.42857	973.15	1.02759
701	1.42653	974.15	1.02654
702	1.42450	975.15	1.02548
703	1.42248	976.15	1.02443
704	1.42045	977.15	1.02338
705	1.41844	978.15	1.02234
706	1.41643	979.15	1.02129
707	1.41443	980.15	1.02025
708	1.41243	981.15	1.01921
709	1.41044	982.15	1.01817
710	1.40845	983.15	1.01714
711	1.40647	984.15	1.01611
712	1.40449	985.15	1.01507
713	1.40252	986.15	1.01404

N	1000/N	N + 273.15	1000/N + 273.15
+714	1.40056	987.15	1.01302
715	1.39860	988.15	1.01199
716	1.39665	989.15	1.01097
717	1.39470	990.15	1.00995
718	1.39276	991.15	1.00893
719	1.39082	992.15	1.00791
720	1.38889	993.15	1.00690
721	1.38696	994.15	1.00588
722	1.38504	995.15	1.00487
723	1.38313	996.15	1.00386
724	1.38122	997.15	1.00286
725	1.37931	998.15	1.00185
726	1.37741	999.15	1.00085
727	1.37552	1000.15	.99985
728	1.37363	1001.15	.99885
729	1.37174	1002.15	.99785
730	1.36986	1003.15	.99686
731	1.36799	1004.15	.99587
732	1.36612	1005.15	.99488
733	1.36426	1006.15	.99389
734	1.36240	1007.15	.99290
735	1.36054	1008.15	.99192
736	1.35870	1009.15	.99093
737	1.35685	1010.15	.98995
738	1.35501	1011.15	.98897
739	1.35318	1012.15	.98800
740	1.35135	1013.15	.98702
741	1.34953	1014.15	.98605
742	1.34771	1015.15	.98508
743	1.34590	1016.15	.98411
744	1.34409	1017.15	.98314
745	1.34228	1018.15	.98217
746	1.34048	1019.15	.98121
747	1.33869	1020.15	.98025
748	1.33690	1021.15	.97929
749	1.33511	1022.15	.97833
750	1.33333	1023.15	.97737
751	1.33156	1024.15	.97642
752	1.32979	1025.15	.97547
753	1.32802	1026.15	.97452
754	1.32626	1027.15	.97357
755	1.32450	1028.15	.97262
756	1.32275	1029.15	.97168
757	1.32100	1030.15	.97073
758	1.31926	1031.15	.96979
759	1.31752	1032.15	.96885
760	1.31579	1033.15	.96791
761	1.31406	1034.15	.96698
762	1.31234	1035.15	.96604
763	1.31062	1036.15	.96511
764	1.30890	1037.15	.96418
765	1.30719	1038.15	.96325
766	1.30548	1039.15	.96232
767	1.30378	1040.15	.96140
768	1.30208	1041.15	.96048
769	1.30039	1042.15	.95955
770	1.29870	1043.15	.95863
771	1.29702	1044.15	.95772
772	1.29534	1045.15	.95680
773	1.29366	1046.15	.95589
774	1.29199	1047.15	.95497
775	1.29032	1048.15	.95406
776	1.28866	1049.15	.95315
777	1.28700	1050.15	.95224
778	1.28535	1051.15	.95134
779	1.28370	1052.15	.95043
780	1.28205	1053.15	.94953
781	1.28041	1054.15	.94863
782	1.27877	1055.15	.94773
783	1.27714	1056.15	.94684
784	1.27551	1057.15	.94594
785	1.27389	1058.15	.94505
786	1.27226	1059.15	.94415
787	1.27065	1060.15	.94326

RECIPROCALS OF NUMBERS (Continued)

N	1000/N	N + 273.15	1000/N + 273.15
+788	1.26904	1061.15	.94237
789	1.26743	1062.15	.94149
790	1.26582	1063.15	.94060
791	1.26422	1064.15	.93972
792	1.26263	1065.15	.93883
793	1.26103	1066.15	.93795
794	1.25945	1067.15	.93708
795	1.25786	1068.15	.93620
796	1.25628	1069.15	.93532
797	1.25471	1070.15	.93445
798	1.25313	1071.15	.93358
799	1.25156	1072.15	.93271
800	1.25000	1073.15	.93184
801	1.24844	1074.15	.93097
802	1.24688	1075.15	.93010
803	1.24533	1076.15	.92924
804	1.24378	1077.15	.92838
805	1.24224	1078.15	.92751
806	1.24069	1079.15	.92666
807	1.23916	1080.15	.92580
808	1.23762	1081.15	.92494
809	1.23609	1082.15	.92409
810	1.23457	1083.15	.92323
811	1.23305	1084.15	.92238
812	1.23153	1085.15	.92153
813	1.23001	1086.15	.92068
814	1.22850	1087.15	.91984
815	1.22699	1088.15	.91899
816	1.22549	1089.15	.91815
817	1.22399	1090.15	.91730
818	1.22249	1091.15	.91646
819	1.22100	1092.15	.91563
820	1.21951	1093.15	.91479
821	1.21803	1094.15	.91395
822	1.21655	1095.15	.91312
823	1.21507	1096.15	.91228
824	1.21359	1097.15	.91145
825	1.21212	1098.15	.91062
826	1.21065	1099.15	.90979
827	1.20919	1100.15	.90897
828	1.20773	1101.15	.90814
829	1.20627	1102.15	.90732
830	1.20482	1103.15	.90650
831	1.20337	1104.15	.90567
832	1.20192	1105.15	.90485
833	1.20048	1106.15	.90404
834	1.19904	1107.15	.90322
835	1.19760	1108.15	.90240
836	1.19617	1109.15	.90159
837	1.19474	1110.15	.90078
838	1.19332	1111.15	.89997
839	1.19190	1112.15	.89916
840	1.19048	1113.15	.89835
841	1.18906	1114.15	.89755
842	1.18765	1115.15	.89674
843	1.18624	1116.15	.89594
844	1.18483	1117.15	.89513
845	1.18343	1118.15	.89433
846	1.18203	1119.15	.89354
847	1.18064	1120.15	.89274
848	1.17925	1121.15	.89194
849	1.17786	1122.15	.89115
850	1.17647	1123.15	.89035
851	1.17509	1124.15	.88956
852	1.17371	1125.15	.88877
853	1.17233	1126.15	.88798
854	1.17096	1127.15	.88719
855	1.16959	1128.15	.88641
856	1.16822	1129.15	.88562
857	1.16686	1130.15	.88484
858	1.16550	1131.15	.88406
859	1.16414	1132.15	.88328
860	1.16279	1133.15	.88250
861	1.16144	1134.15	.88172
+862	1.16009	1135.15	.88094
863	1.15875	1136.15	.88017
864	1.15741	1137.15	.87939
865	1.15607	1138.15	.87862
866	1.15473	1139.15	.87785
867	1.15340	1140.15	.87708
868	1.15207	1141.15	.87631
869	1.15075	1142.15	.87554
870	1.14943	1143.15	.87478
871	1.14811	1144.15	.87401
872	1.14679	1145.15	.87325
873	1.14548	1146.15	.87249
874	1.14416	1147.15	.87173
875	1.14286	1148.15	.87097
876	1.14155	1149.15	.87021
877	1.14025	1150.15	.86945
878	1.13895	1151.15	.86870
879	1.13766	1152.15	.86794
880	1.13636	1153.15	.86719
881	1.13507	1154.15	.86644
882	1.13379	1155.15	.86569
883	1.13250	1156.15	.86494
884	1.13122	1157.15	.86419
885	1.12994	1158.15	.86345
886	1.12867	1159.15	.86270
887	1.12740	1160.15	.86196
888	1.12613	1161.15	.86122
889	1.12486	1162.15	.86047
890	1.12360	1163.15	.85973
891	1.12233	1164.15	.85900
892	1.12108	1165.15	.85826
893	1.11982	1166.15	.85752
894	1.11857	1167.15	.85679
895	1.11732	1168.15	.85605
896	1.11607	1169.15	.85532
897	1.11483	1170.15	.85459
898	1.11359	1171.15	.85386
899	1.11235	1172.15	.85313
900	1.11111	1173.15	.85241
901	1.10988	1174.15	.85168
902	1.10865	1175.15	.85096
903	1.10742	1176.15	.85023
904	1.10619	1177.15	.84951
905	1.10497	1178.15	.84879
906	1.10375	1179.15	.84807
907	1.10254	1180.15	.84735
908	1.10132	1181.15	.84663
909	1.10011	1182.15	.84592
910	1.09890	1183.15	.84520
911	1.09769	1184.15	.84449
912	1.09649	1185.15	.84378
913	1.09529	1186.15	.84306
914	1.09409	1187.15	.84235
915	1.09290	1188.15	.84164
916	1.09170	1189.15	.84094
917	1.09051	1190.15	.84023
918	1.08932	1191.15	.83952
919	1.08814	1192.15	.83882
920	1.08696	1193.15	.83812
921	1.08578	1194.15	.83742
922	1.08460	1195.15	.83672
923	1.08342	1196.15	.83602
924	1.08225	1197.15	.83532
925	1.08108	1198.15	.83462
926	1.07991	1199.15	.83392
927	1.07875	1200.15	.83323
928	1.07759	1201.15	.83254
929	1.07643	1202.15	.83184
930	1.07527	1203.15	.83115
931	1.07411	1204.15	.83046
932	1.07296	1205.15	.82977
933	1.07181	1206.15	.82908
934	1.07066	1207.15	.82840
935	1.06952	1208.15	.82771

RECIPROCALS OF NUMBERS (Continued)

N	1000/N	N + 273.15	1000/N + 273.15
+936	1.06838	1209.15	.82703
937	1.06724	1210.15	.82634
938	1.06610	1211.15	.82566
939	1.06496	1212.15	.82498
940	1.06383	1213.15	.82430
941	1.06270	1214.15	.82362
942	1.06157	1215.15	.82294
943	1.06045	1216.15	.82227
944	1.05932	1217.15	.82159
945	1.05820	1218.15	.82092
946	1.05708	1219.15	.82024
947	1.05597	1220.15	.81957
948	1.05485	1221.15	.81890
949	1.05374	1222.15	.81823
950	1.05263	1223.15	.81756
951	1.05152	1224.15	.81689
952	1.05042	1225.15	.81623
953	1.04932	1226.15	.81556
954	1.04822	1227.15	.81490
955	1.04712	1228.15	.81423
956	1.04603	1229.15	.81357
957	1.04493	1230.15	.81291
958	1.04384	1231.15	.81225
959	1.04275	1232.15	.81159
960	1.04167	1233.15	.81093
961	1.04058	1234.15	.81027
962	1.03950	1235.15	.80962
963	1.03842	1236.15	.80896
964	1.03734	1237.15	.80831
965	1.03627	1238.15	.80766
966	1.03520	1239.15	.80700
967	1.03413	1240.15	.80635
968	1.03306	1241.15	.80570
969	1.03199	1242.15	.80506
970	1.03093	1243.15	.80441
971	1.02987	1244.15	.80376
972	1.02881	1245.15	.80312
973	1.02775	1246.15	.80247
974	1.02669	1247.15	.80183
975	1.02564	1248.15	.80119
976	1.02459	1249.15	.80054
977	1.02354	1250.15	.79990
978	1.02249	1251.15	.79926
979	1.02145	1252.15	.79863
980	1.02041	1253.15	.79799
981	1.01937	1254.15	.79735
982	1.01833	1255.15	.79672
983	1.01729	1256.15	.79608
984	1.01626	1257.15	.79545
985	1.01523	1258.15	.79482
986	1.01420	1259.15	.79419
987	1.01317	1260.15	.79356
988	1.01215	1261.15	.79293
989	1.01112	1262.15	.79230
990	1.01010	1263.15	.79167
991	1.00908	1264.15	.79105
992	1.00806	1265.15	.79042
993	1.00705	1266.15	.78980
994	1.00604	1267.15	.78917
995	1.00503	1268.15	.78855
996	1.00402	1269.15	.78793
997	1.00301	1270.15	.78731
998	1.00200	1271.15	.78669
999	1.00100	1272.15	.78607
1000	1.00000	1273.15	.78545
1005	.99502	1278.15	.78238
1010	.99010	1283.15	.77933
1015	.98522	1288.15	.77631
1020	.98039	1293.15	.77331
1025	.97561	1298.15	.77033
1030	.97087	1303.15	.76737
1035	.96618	1308.15	.76444
1040	.96154	1313.15	.76153
1045	.95694	1318.15	.75864

N	1000/N	N + 273.15	1000/N + 273.15
+1050	.95238	1323.15	.75577
1055	.94787	1328.15	.75293
1060	.94340	1333.15	.75010
1065	.93897	1338.15	.74730
1070	.93458	1343.15	.74452
1075	.93023	1348.15	.74176
1080	.92593	1353.15	.73902
1085	.92166	1358.15	.73630
1090	.91743	1363.15	.73359
1095	.91324	1368.15	.73091
1100	.90909	1373.15	.72825
1105	.90498	1378.15	.72561
1110	.90090	1383.15	.72299
1115	.89686	1388.15	.72038
1120	.89286	1393.15	.71780
1125	.88889	1398.15	.71523
1130	.88496	1403.15	.71268
1135	.88106	1408.15	.71015
1140	.87719	1413.15	.70764
1145	.87336	1418.15	.70514
1150	.86957	1423.15	.70267
1155	.86580	1428.15	.70021
1160	.86207	1433.15	.69776
1165	.85837	1438.15	.69534
1170	.85470	1443.15	.69293
1175	.85106	1448.15	.69054
1180	.84746	1453.15	.68816
1185	.84388	1458.15	.68580
1190	.84034	1463.15	.68346
1195	.83682	1468.15	.68113
1200	.83333	1473.15	.67882
1205	.82988	1478.15	.67652
1210	.82645	1483.15	.67424
1215	.82305	1488.15	.67198
1220	.81967	1493.15	.66973
1225	.81633	1498.15	.66749
1230	.81301	1503.15	.66527
1235	.80972	1508.15	.66306
1240	.80645	1513.15	.66087
1245	.80321	1518.15	.65870
1250	.80000	1523.15	.65653
1255	.79681	1528.15	.65439
1260	.79365	1533.15	.65225
1265	.79051	1538.15	.65013
1270	.78740	1543.15	.64803
1275	.78431	1548.15	.64593
1280	.78125	1553.15	.64385
1285	.77821	1558.15	.64179
1290	.77519	1563.15	.63973
1295	.77220	1568.15	.63769
1300	.76923	1573.15	.63567
1305	.76628	1578.15	.63365
1310	.76336	1583.15	.63165
1315	.76046	1588.15	.62966
1320	.75758	1593.15	.62769
1325	.75472	1598.15	.62572
1330	.75188	1603.15	.62377
1335	.74906	1608.15	.62183
1340	.74627	1613.15	.61991
1345	.74349	1618.15	.61799
1350	.74074	1623.15	.61609
1355	.73801	1628.15	.61419
1360	.73529	1633.15	.61231
1365	.73260	1638.15	.61044
1370	.72993	1643.15	.60859
1375	.72727	1648.15	.60674
1380	.72464	1653.15	.60491
1385	.72202	1658.15	.60308
1390	.71942	1663.15	.60127
1395	.71685	1668.15	.59947
1400	.71429	1673.15	.59768
1405	.71174	1678.15	.59589
1410	.70922	1683.15	.59412
1415	.70671	1688.15	.59236

RECIPROCALS OF NUMBERS (Continued)

N	1000/N	N + 273.15	1000/N + 273.15
+1420	.70423	1693.15	.59062
1425	.70175	1698.15	.58888
1430	.69930	1703.15	.58715
1435	.69686	1708.15	.58543
1440	.69444	1713.15	.58372
1445	.69204	1718.15	.58202
1450	.68966	1723.15	.58033
1455	.68729	1728.15	.57865
1460	.68493	1733.15	.57698
1465	.68259	1738.15	.57532
1470	.68027	1743.15	.57367
1475	.67797	1748.15	.57203
1480	.67568	1753.15	.57040
1485	.67340	1758.15	.56878
1490	.67114	1763.15	.56717
1495	.66890	1768.15	.56556
1500	.66667	1773.15	.56397
1505	.66445	1778.15	.56238
1510	.66225	1783.15	.56081
1515	.66007	1788.15	.55924
1520	.65789	1793.15	.55768
1525	.65574	1798.15	.55613
1530	.65359	1803.15	.55459
1535	.65147	1808.15	.55305
1540	.64935	1813.15	.55153
1545	.64725	1818.15	.55001
1550	.64516	1823.15	.54850
1555	.64309	1828.15	.54700
1560	.64103	1833.15	.54551
1565	.63898	1838.15	.54403
1570	.63694	1843.15	.54255
1575	.63492	1848.15	.54108
1580	.63291	1853.15	.53962
1585	.63091	1858.15	.53817
1590	.62893	1863.15	.53673
1595	.62696	1868.15	.53529
1600	.62500	1873.15	.53386
1605	.62305	1878.15	.53244
1610	.62112	1883.15	.53103
1615	.61920	1888.15	.52962
1620	.61728	1893.15	.52822
1625	.61538	1898.15	.52683
1630	.61350	1903.15	.52544
1635	.61162	1908.15	.52407
1640	.60976	1913.15	.52270
1645	.60790	1918.15	.52134
1650	.60606	1923.15	.51998
1655	.60423	1928.15	.51863
1660	.60241	1933.15	.51729
1665	.60060	1938.15	.51596
1670	.59880	1943.15	.51463
1675	.59701	1948.15	.51331
1680	.59524	1953.15	.51199
1685	.59347	1958.15	.51069
1690	.59172	1963.15	.50939
1695	.58997	1968.15	.50809
1700	.58824	1973.15	.50680
1705	.58651	1978.15	.50552
1710	.58480	1983.15	.50425
1715	.58309	1988.15	.50298
1720	.58140	1993.15	.50172
1725	.57971	1998.15	.50046
1730	.57803	2003.15	.49921
1735	.57637	2008.15	.49797
1740	.57471	2013.15	.49673
1745	.57307	2018.15	.49550
1750	.57143	2023.15	.49428
1755	.56980	2028.15	.49306
1760	.56818	2033.15	.49185
1765	.56657	2038.15	.49064
1770	.56497	2043.15	.48944
1775	.56338	2048.15	.48825
1780	.56180	2053.15	.48706
1785	.56022	2058.15	.48587

N	1000/N	N + 273.15	1000/N + 273.15
+1790	.55866	2063.15	.48470
1795	.55710	2068.15	.48352
1800	.55556	2073.15	.48236
1805	.55402	2078.15	.48120
1810	.55249	2083.15	.48004
1815	.55096	2088.15	.47889
1820	.54945	2093.15	.47775
1825	.54795	2098.15	.47661
1830	.54645	2103.15	.47548
1835	.54496	2108.15	.47435
1840	.54348	2113.15	.47323
1845	.54201	2118.15	.47211
1850	.54054	2123.15	.47100
1855	.53908	2128.15	.46989
1860	.53763	2133.15	.46879
1865	.53619	2138.15	.46769
1870	.53476	2143.15	.46660
1875	.53333	2148.15	.46552
1880	.53191	2153.15	.46444
1885	.53050	2158.15	.46336
1890	.52910	2163.15	.46229
1895	.52770	2168.15	.46122
1900	.52632	2173.15	.46016
1905	.52493	2178.15	.45911
1910	.52356	2183.15	.45805
1915	.52219	2188.15	.45701
1920	.52083	2193.15	.45597
1925	.51948	2198.15	.45493
1930	.51813	2203.15	.45390
1935	.51680	2208.15	.45287
1940	.51546	2213.15	.45184
1945	.51414	2218.15	.45083
1950	.51282	2223.15	.44981
1955	.51151	2228.15	.44880
1960	.51020	2233.15	.44780
1965	.50891	2238.15	.44680
1970	.50761	2243.15	.44580
1975	.50633	2248.15	.44481
1980	.50505	2253.15	.44382
1985	.50378	2258.15	.44284
1990	.50251	2263.15	.44186
1995	.50125	2268.15	.44089
2000	.50000	2273.15	.43992
2005	.49875	2278.15	.43895
2010	.49751	2283.15	.43799
2015	.49628	2288.15	.43703
2020	.49505	2293.15	.43608
2025	.49383	2298.15	.43513
2030	.49261	2303.15	.43419
2035	.49140	2308.15	.43325
2040	.49020	2313.15	.43231
2045	.48900	2318.15	.43138
2050	.48780	2323.15	.43045
2055	.48662	2328.15	.42953
2060	.48544	2333.15	.42861
2065	.48426	2338.15	.42769
2070	.48309	2343.15	.42678
2075	.48193	2348.15	.42587
2080	.48077	2353.15	.42496
2085	.47962	2358.15	.42406
2090	.47847	2363.15	.42316
2095	.47733	2368.15	.42227
2100	.47619	2373.15	.42138
2105	.47506	2378.15	.42049
2110	.47393	2383.15	.41961
2115	.47281	2388.15	.41873
2120	.47170	2393.15	.41786
2125	.47059	2398.15	.41699
2130	.46948	2403.15	.41612
2135	.46838	2408.15	.41526
2140	.46729	2413.15	.41440
2145	.46620	2418.15	.41354
2150	.46512	2423.15	.41269
2155	.46404	2428.15	.41184

RECIPROCALS OF NUMBERS (Continued)

N	1000/N	N + 273.15	1000/N + 273.15
+2160	.46296	2433.15	.41099
2165	.46189	2438.15	.41015
2170	.46083	2443.15	.40931
2175	.45977	2448.15	.40847
2180	.45872	2453.15	.40764
2185	.45767	2458.15	.40681
2190	.45662	2463.15	.40598
2195	.45558	2468.15	.40516
2200	.45455	2473.15	.40434
2205	.45351	2478.15	.40353
2210	.45249	2483.15	.40271
2215	.45147	2488.15	.40191
2220	.45045	2493.15	.40110
2225	.44944	2498.15	.40030
2230	.44843	2503.15	.39950
2235	.44743	2508.15	.39870
2240	.44643	2513.15	.39791
2245	.44543	2518.15	.39712
2250	.44444	2523.15	.39633
2255	.44346	2528.15	.39555
2260	.44248	2533.15	.39477
2265	.44150	2538.15	.39399
2270	.44053	2543.15	.39321
2275	.43956	2548.15	.39244
2280	.43860	2553.15	.39167
2285	.43764	2558.15	.39091
2290	.43668	2563.15	.39014
2295	.43573	2568.15	.38939
2300	.43478	2573.15	.38863
2305	.43384	2578.15	.38788
2310	.43290	2583.15	.38712
2315	.43197	2588.15	.38638
2320	.43103	2593.15	.38563
2325	.43011	2598.15	.38489
2330	.42918	2603.15	.38415
2335	.42827	2608.15	.38341
2340	.42735	2613.15	.38268
2345	.42644	2618.15	.38195
2350	.42553	2623.15	.38122
2355	.42463	2628.15	.38050
2360	.42373	2633.15	.37977
2365	.42283	2638.15	.37905
2370	.42194	2643.15	.37834
2375	.42105	2648.15	.37762
2380	.42017	2653.15	.37691
2385	.41929	2658.15	.37620
2390	.41841	2663.15	.37550
2395	.41754	2668.15	.37479
2400	.41667	2673.15	.37409
2405	.41580	2678.15	.37339
2410	.41494	2683.15	.37270
2415	.41408	2688.15	.37200
2420	.41322	2693.15	.37131
2425	.41237	2698.15	.37062
2430	.41152	2703.15	.36994
2435	.41068	2708.15	.36926
2440	.40984	2713.15	.36858
2445	.40900	2718.15	.36790
2450	.40816	2723.15	.36722
2455	.40733	2728.15	.36655
2460	.40650	2733.15	.36588
2465	.40568	2738.15	.36521
2470	.40486	2743.15	.36454
2475	.40404	2748.15	.36388
2480	.40323	2753.15	.36322
2485	.40241	2758.15	.36256
2490	.40161	2763.15	.36191
2495	.40080	2768.15	.36125
2500	.40000	2773.15	.36060
2505	.39920	2778.15	.35995
2510	.39841	2783.15	.35931
2515	.39761	2788.15	.35866
2520	.39683	2793.15	.35802
2525	.39604	2798.15	.35738
+2530	.39526	2803.15	.35674
2535	.39448	2808.15	.35611
2540	.39370	2813.15	.35547
2545	.39293	2818.15	.35484
2550	.39216	2823.15	.35421
2555	.39139	2828.15	.35359
2560	.39063	2833.15	.35296
2565	.38986	2838.15	.35234
2570	.38911	2843.15	.35172
2575	.38835	2848.15	.35111
2580	.38760	2853.15	.35049
2585	.38685	2858.15	.34988
2590	.38610	2863.15	.34927
2595	.38536	2868.15	.34866
2600	.38462	2873.15	.34805
2605	.38388	2878.15	.34745
2610	.38314	2883.15	.34684
2615	.38241	2888.15	.34624
2620	.38168	2893.15	.34564
2625	.38095	2898.15	.34505
2630	.38023	2903.15	.34445
2635	.37951	2908.15	.34386
2640	.37879	2913.15	.34327
2645	.37807	2918.15	.34268
2650	.37736	2923.15	.34210
2655	.37665	2928.15	.34151
2660	.37594	2933.15	.34093
2665	.37523	2938.15	.34035
2670	.37453	2943.15	.33977
2675	.37383	2948.15	.33920
2680	.37313	2953.15	.33862
2685	.37244	2958.15	.33805
2690	.37175	2963.15	.33748
2695	.37106	2968.15	.33691
2700	.37037	2973.15	.33634
2705	.36969	2978.15	.33578
2710	.36900	2983.15	.33522
2715	.36832	2988.15	.33466
2720	.36765	2993.15	.33410
2725	.36697	2998.15	.33354
2730	.36630	3003.15	.33298
2735	.36563	3008.15	.33243
2740	.36496	3013.15	.33188
2745	.36430	3018.15	.33133
2750	.36364	3023.15	.33078
2755	.36298	3028.15	.33023
2760	.36232	3033.15	.32969
2765	.36166	3038.15	.32915
2770	.36101	3043.15	.32861
2775	.36036	3048.15	.32807
2780	.35971	3053.15	.32753
2785	.35907	3058.15	.32700
2790	.35842	3063.15	.32646
2795	.35778	3068.15	.32593
2800	.35714	3073.15	.32540
2805	.35651	3078.15	.32487
2810	.35587	3083.15	.32434
2815	.35524	3088.15	.32382
2820	.35461	3093.15	.32330
2825	.35398	3098.15	.32277
2830	.35336	3103.15	.32225
2835	.35273	3108.15	.32173
2840	.35211	3113.15	.32122
2845	.35149	3118.15	.32070
2850	.35088	3123.15	.32019
2855	.35026	3128.15	.31968
2860	.34965	3133.15	.31917
2865	.34904	3138.15	.31866
2870	.34843	3143.15	.31815
2875	.34783	3148.15	.31765
2880	.34722	3153.15	.31714
2885	.34662	3158.15	.31664
2890	.34602	3163.15	.31614
2895	.34542	3168.15	.31564

RECIPROCALS OF NUMBERS (Continued)

N	1000/N	N + 273.15	1000/N + 273.15
+2900	.34483	3173.15	.31514
2905	.34423	3178.15	.31465
2910	.34364	3183.15	.31415
2915	.34305	3188.15	.31366
2920	.34247	3193.15	.31317
2925	.34188	3198.15	.31268
2930	.34130	3203.15	.31219
2935	.34072	3208.15	.31171
2940	.34014	3213.15	.31122
2945	.33956	3218.15	.31074
2950	.33898	3223.15	.31026
2955	.33841	3228.15	.30977
2960	.33784	3233.15	.30930
2965	.33727	3238.15	.30882
2970	.33670	3243.15	.30834
2975	.33613	3248.15	.30787
2980	.33557	3253.15	.30739
2985	.33501	3258.15	.30692
2990	.33445	3263.15	.30645
2995	.33389	3268.15	.30598
3000	.33333	3273.15	.30552
3005	.33278	3278.15	.30505
3010	.33223	3283.15	.30459
3015	.33167	3288.15	.30412
3020	.33113	3293.15	.30366
3025	.33058	3298.15	.30320
3030	.33003	3303.15	.30274
3035	.32949	3308.15	.30228
3040	.32895	3313.15	.30183
3045	.32841	3318.15	.30137
3050	.32787	3323.15	.30092
3055	.32733	3328.15	.30047
3060	.32680	3333.15	.30002
3065	.32626	3338.15	.29957
3070	.32573	3343.15	.29912
3075	.32520	3348.15	.29867
3080	.32468	3353.15	.29823
3085	.32415	3358.15	.29778
3090	.32362	3363.15	.29734
3095	.32310	3368.15	.29690
3100	.32258	3373.15	.29646
3105	.32206	3378.15	.29602
3110	.32154	3383.15	.29558
3115	.32103	3388.15	.29515
3120	.32051	3393.15	.29471
3125	.32000	3398.15	.29428
3130	.31949	3403.15	.29385
3135	.31898	3408.15	.29341
3140	.31847	3413.15	.29298
3145	.31797	3418.15	.29256
3150	.31746	3423.15	.29213
3155	.31696	3428.15	.29170
3160	.31646	3433.15	.29128
3165	.31596	3438.15	.29085
3170	.31546	3443.15	.29043
3175	.31496	3448.15	.29001
3180	.31447	3453.15	.28959
3185	.31397	3458.15	.28917
3190	.31348	3463.15	.28875
3195	.31299	3468.15	.28834
3200	.31250	3473.15	.28792
3205	.31201	3478.15	.28751
3210	.31153	3483.15	.28710
3215	.31104	3488.15	.28668
3220	.31056	3493.15	.28627
3225	.31008	3498.15	.28587
3230	.30960	3503.15	.28546
3235	.30912	3508.15	.28505
3240	.30864	3513.15	.28464
3245	.30817	3518.15	.28424
3250	.30769	3523.15	.28384
3255	.30722	3528.15	.28343
3260	.30675	3533.15	.28303
3265	.30628	3538.15	.28263

N	1000/N	N + 273.15	1000/N + 273.15
+3270	.30581	3543.15	.28223
3275	.30534	3548.15	.28184
3280	.30488	3553.15	.28144
3285	.30441	3558.15	.28104
3290	.30395	3563.15	.28065
3295	.30349	3568.15	.28026
3300	.30303	3573.15	.27987
3305	.30257	3578.15	.27947
3310	.30211	3583.15	.27908
3315	.30166	3588.15	.27870
3320	.30120	3593.15	.27831
3325	.30075	3598.15	.27792
3330	.30030	3603.15	.27753
3335	.29985	3608.15	.27715
3340	.29940	3613.15	.27677
3345	.29895	3618.15	.27638
3350	.29851	3623.15	.27600
3355	.29806	3628.15	.27562
3360	.29762	3633.15	.27524
3365	.29718	3638.15	.27486
3370	.29674	3643.15	.27449
3375	.29630	3648.15	.27411
3380	.29586	3653.15	.27374
3385	.29542	3658.15	.27336
3390	.29499	3663.15	.27299
3395	.29455	3668.15	.27262
3400	.29412	3673.15	.27225
3405	.29369	3678.15	.27188
3410	.29326	3683.15	.27151
3415	.29283	3688.15	.27114
3420	.29240	3693.15	.27077
3425	.29197	3698.15	.27041
3430	.29155	3703.15	.27004
3435	.29112	3708.15	.26968
3440	.29070	3713.15	.26931
3445	.29028	3718.15	.26895
3450	.28986	3723.15	.26859
3455	.28944	3728.15	.26823
3460	.28902	3733.15	.26787
3465	.28860	3738.15	.26751
3470	.28818	3743.15	.26715
3475	.28777	3748.15	.26680
3480	.28736	3753.15	.26644
3485	.28694	3758.15	.26609
3490	.28653	3763.15	.26573
3495	.28612	3768.15	.26538
3500	.28571	3773.15	.26503
3505	.28531	3778.15	.26468
3510	.28490	3783.15	.26433
3515	.28450	3788.15	.26398
3520	.28409	3793.15	.26363
3525	.28369	3798.15	.26329
3530	.28329	3803.15	.26294
3535	.28289	3808.15	.26259
3540	.28249	3813.15	.26225
3545	.28209	3818.15	.26191
3550	.28169	3823.15	.26156
3555	.28129	3828.15	.26122
3560	.28090	3833.15	.26088
3565	.28050	3838.15	.26054
3570	.28011	3843.15	.26020
3575	.27972	3848.15	.25987
3580	.27933	3853.15	.25953
3585	.27894	3858.15	.25919
3590	.27855	3863.15	.25886
3595	.27816	3868.15	.25852
3600	.27778	3873.15	.25819
3605	.27739	3878.15	.25785
3610	.27701	3883.15	.25752
3615	.27663	3888.15	.25719
3620	.27624	3893.15	.25686
3625	.27586	3898.15	.25653
3630	.27548	3903.15	.25620
3635	.27510	3908.15	.25588

RECIPROCALS OF NUMBERS (Continued)

N	1000/N	N + 273.15	1000/N + 273.15
+3640	.27473	3913.15	.25555
3645	.27435	3918.15	.25522
3650	.27397	3923.15	.25490
3655	.27360	3928.15	.25457
3660	.27322	3933.15	.25425
3665	.27285	3938.15	.25393
3670	.27248	3943.15	.25360
3675	.27211	3948.15	.25328
3680	.27174	3953.15	.25296
3685	.27137	3958.15	.25264
3690	.27100	3963.15	.25232
3695	.27064	3968.15	.25201
3700	.27027	3973.15	.25169
3705	.26991	3978.15	.25137
3710	.26954	3983.15	.25106
3715	.26918	3988.15	.25074
3720	.26882	3993.15	.25043
3725	.26846	3998.15	.25012
3730	.26810	4003.15	.24980
3735	.26774	4008.15	.24949
3740	.26738	4013.15	.24918
3745	.26702	4018.15	.24887
3750	.26667	4023.15	.24856
3755	.26631	4028.15	.24825
3760	.26596	4033.15	.24795
3765	.26560	4038.15	.24764
3770	.26525	4043.15	.24733
3775	.26490	4048.15	.24703
3780	.26455	4053.15	.24672
3785	.26420	4058.15	.24642
3790	.26385	4063.15	.24611
3795	.26350	4068.15	.24581
3800	.26316	4073.15	.24551
3805	.26281	4078.15	.24521
3810	.26247	4083.15	.24491
3815	.26212	4088.15	.24461
3820	.26178	4093.15	.24431
3825	.26144	4098.15	.24401
3830	.26110	4103.15	.24372
3835	.26076	4108.15	.24342
3840	.26042	4113.15	.24312
3845	.26008	4118.15	.24283
3850	.25974	4123.15	.24253
3855	.25940	4128.15	.24224
3860	.25907	4133.15	.24195
3865	.25873	4138.15	.24165
3870	.25840	4143.15	.24136
3875	.25806	4148.15	.24107
3880	.25773	4153.15	.24078
3885	.25740	4158.15	.24049
3890	.25707	4163.15	.24020
3895	.25674	4168.15	.23991
3900	.25641	4173.15	.23963
3905	.25608	4178.15	.23934
3910	.25575	4183.15	.23905
3915	.25543	4188.15	.23877
3920	.25510	4193.15	.23848
3925	.25478	4198.15	.23820
3930	.25445	4203.15	.23792
3935	.25413	4208.15	.23763
3940	.25381	4213.15	.23735
3945	.25349	4218.15	.23707
3950	.25316	4223.15	.23679
3955	.25284	4228.15	.23651
3960	.25253	4233.15	.23623
3965	.25221	4238.15	.23595
3970	.25189	4243.15	.23567
3975	.25157	4248.15	.23540
3980	.25126	4253.15	.23512
3985	.25094	4258.15	.23484
3990	.25063	4263.15	.23457
3995	.25031	4268.15	.23429
4000	.25000	4273.15	.23402
4005	.24969	4278.15	.23375

N	1000/N	N + 273.15	1000/N + 273.15
+4010	.24938	4283.15	.23347
4015	.24907	4288.15	.23320
4020	.24876	4293.15	.23293
4025	.24845	4298.15	.23266
4030	.24814	4303.15	.23239
4035	.24783	4308.15	.23212
4040	.24752	4313.15	.23185
4045	.24722	4318.15	.23158
4050	.24691	4323.15	.23131
4055	.24661	4328.15	.23105
4060	.24631	4333.15	.23078
4065	.24600	4338.15	.23051
4070	.24570	4343.15	.23025
4075	.24540	4348.15	.22998
4080	.24510	4353.15	.22972
4085	.24480	4358.15	.22946
4090	.24450	4363.15	.22919
4095	.24420	4368.15	.22893
4100	.24390	4373.15	.22867
4105	.24361	4378.15	.22841
4110	.24331	4383.15	.22815
4115	.24301	4388.15	.22789
4120	.24272	4393.15	.22763
4125	.24242	4398.15	.22737
4130	.24213	4403.15	.22711
4135	.24184	4408.15	.22685
4140	.24155	4413.15	.22660
4145	.24125	4418.15	.22634
4150	.24096	4423.15	.22608
4155	.24067	4428.15	.22583
4160	.24038	4433.15	.22557
4165	.24010	4438.15	.22532
4170	.23981	4443.15	.22507
4175	.23952	4448.15	.22481
4180	.23923	4453.15	.22456
4185	.23895	4458.15	.22431
4190	.23866	4463.15	.22406
4195	.23838	4468.15	.22381
4200	.23810	4473.15	.22356
4205	.23781	4478.15	.22331
4210	.23753	4483.15	.22306
4215	.23725	4488.15	.22281
4220	.23697	4493.15	.22256
4225	.23669	4498.15	.22231
4230	.23641	4503.15	.22207
4235	.23613	4508.15	.22182
4240	.23585	4513.15	.22157
4245	.23557	4518.15	.22133
4250	.23529	4523.15	.22108
4255	.23502	4528.15	.22084
4260	.23474	4533.15	.22060
4265	.23447	4538.15	.22035
4270	.23419	4543.15	.22011
4275	.23392	4548.15	.21987
4280	.23364	4553.15	.21963
4285	.23337	4558.15	.21939
4290	.23310	4563.15	.21915
4295	.23283	4568.15	.21891
4300	.23256	4573.15	.21867
4305	.23229	4578.15	.21843
4310	.23202	4583.15	.21819
4315	.23175	4588.15	.21795
4320	.23148	4593.15	.21772
4325	.23121	4598.15	.21748
4330	.23095	4603.15	.21724
4335	.23068	4608.15	.21701
4340	.23041	4613.15	.21677
4345	.23015	4618.15	.21654
4350	.22989	4623.15	.21630
4355	.22962	4628.15	.21607
4360	.22936	4633.15	.21584
4365	.22910	4638.15	.21560
4370	.22883	4643.15	.21537
4375	.22857	4648.15	.21514

RECIPROCALS OF NUMBERS (Continued)

N	1000/N	N + 273.15	1000/N + 273.15
+4380	.22831	4653.15	.21491
4385	.22805	4658.15	.21468
4390	.22779	4663.15	.21445
4395	.22753	4668.15	.21422
4400	.22727	4673.15	.21399
4405	.22701	4678.15	.21376
4410	.22676	4683.15	.21353
4415	.22650	4688.15	.21330
4420	.22624	4693.15	.21308
4425	.22599	4698.15	.21285
4430	.22573	4703.15	.21262
4435	.22548	4708.15	.21240
4440	.22523	4713.15	.21217
4445	.22497	4718.15	.21195
4450	.22472	4723.15	.21172
4455	.22447	4728.15	.21150
4460	.22422	4733.15	.21128
4465	.22396	4738.15	.21105
4470	.22371	4743.15	.21083
4475	.22346	4748.15	.21061
4480	.22321	4753.15	.21039
4485	.22297	4758.15	.21017
4490	.22272	4763.15	.20995
4495	.22247	4768.15	.20972
4500	.22222	4773.15	.20951
4505	.22198	4778.15	.20929
4510	.22173	4783.15	.20907
4515	.22148	4788.15	.20885
4520	.22124	4793.15	.20863
4525	.22099	4798.15	.20841
4530	.22075	4803.15	.20820
4535	.22051	4808.15	.20798
4540	.22026	4813.15	.20776
4545	.22002	4818.15	.20755
4550	.21978	4823.15	.20733
4555	.21954	4828.15	.20712
4560	.21930	4833.15	.20690
4565	.21906	4838.15	.20669
4570	.21882	4843.15	.20648
4575	.21858	4848.15	.20626
4580	.21834	4853.15	.20605
4585	.21810	4858.15	.20584
4590	.21786	4863.15	.20563
4595	.21763	4868.15	.20542
4600	.21739	4873.15	.20521
4605	.21716	4878.15	.20500
4610	.21692	4883.15	.20479
4615	.21668	4888.15	.20458
4620	.21645	4893.15	.20437
4625	.21622	4898.15	.20416
4630	.21598	4903.15	.20395
4635	.21575	4908.15	.20374
4640	.21552	4913.15	.20354
4645	.21529	4918.15	.20333
4650	.21505	4923.15	.20312
4655	.21482	4928.15	.20292
4660	.21459	4933.15	.20271
4665	.21436	4938.15	.20250
4670	.21413	4943.15	.20230
4675	.21390	4948.15	.20210
4680	.21368	4953.15	.20189
4685	.21345	4958.15	.20169
4690	.21322	4963.15	.20148
4695	.21299	4968.15	.20128
4700	.21277	4973.15	.20108
4705	.21254	4978.15	.20088
4710	.21231	4983.15	.20068
4715	.21209	4988.15	.20048
4720	.21186	4993.15	.20027
4725	.21164	4998.15	.20007
4730	.21142	5003.15	.19987
4735	.21119	5008.15	.19967
4740	.21097	5013.15	.19948
4745	.21075	5018.15	.19928

N	1000/N	N + 273.15	1000/N + 273.15
+4750	.21053	5023.15	.19908
4755	.21030	5028.15	.19888
4760	.21008	5033.15	.19868
4765	.20986	5038.15	.19849
4770	.20964	5043.15	.19829
4775	.20942	5048.15	.19809
4780	.20921	5053.15	.19790
4785	.20899	5058.15	.19770
4790	.20877	5063.15	.19751
4795	.20855	5068.15	.19731
4800	.20833	5073.15	.19712
4805	.20812	5078.15	.19692
4810	.20790	5083.15	.19673
4815	.20768	5088.15	.19654
4820	.20747	5093.15	.19634
4825	.20725	5098.15	.19615
4830	.20704	5103.15	.19596
4835	.20683	5108.15	.19577
4840	.20661	5113.15	.19557
4845	.20640	5118.15	.19538
4850	.20619	5123.15	.19519
4855	.20597	5128.15	.19500
4860	.20576	5133.15	.19481
4865	.20555	5138.15	.19462
4870	.20534	5143.15	.19443
4875	.20513	5148.15	.19424
4880	.20492	5153.15	.19406
4885	.20471	5158.15	.19387
4890	.20450	5163.15	.19368
4895	.20429	5168.15	.19349
4900	.20408	5173.15	.19331
4905	.20387	5178.15	.19312
4910	.20367	5183.15	.19293
4915	.20346	5188.15	.19275
4920	.20325	5193.15	.19256
4925	.20305	5198.15	.19238
4930	.20284	5203.15	.19219
4935	.20263	5208.15	.19201
4940	.20243	5213.15	.19182
4945	.20222	5218.15	.19164
4950	.20202	5223.15	.19146
4955	.20182	5228.15	.19127
4960	.20161	5233.15	.19109
4965	.20141	5238.15	.19091
4970	.20121	5243.15	.19073
4975	.20101	5248.15	.19054
4980	.20080	5253.15	.19036
4985	.20060	5258.15	.19018
4990	.20040	5263.15	.19000
4995	.20020	5268.15	.18982
5000	.20000	5273.15	.18964
5005	.19980	5278.15	.18946
5010	.19960	5283.15	.18928
5015	.19940	5288.15	.18910
5020	.19920	5293.15	.18892
5025	.19900	5298.15	.18875
5030	.19881	5303.15	.18857
5035	.19861	5308.15	.18839
5040	.19841	5313.15	.18821
5045	.19822	5318.15	.18804
5050	.19802	5323.15	.18786
5055	.19782	5328.15	.18768
5060	.19763	5333.15	.18751
5065	.19743	5338.15	.18733
5070	.19724	5343.15	.18716
5075	.19704	5348.15	.18698
5080	.19685	5353.15	.18681
5085	.19666	5358.15	.18663
5090	.19646	5363.15	.18646
5095	.19627	5368.15	.18628
5100	.19608	5373.15	.18611
5105	.19589	5378.15	.18594
5110	.19569	5383.15	.18576
5115	.19550	5388.15	.18559

RECIPROCALS OF NUMBERS (Continued)

N	1000/N	N + 273.15	1000/N + 273.15
+5120	.19531	5393.15	.18542
5125	.19512	5398.15	.18525
5130	.19493	5403.15	.18508
5135	.19474	5408.15	.18491
5140	.19455	5413.15	.18474
5145	.19436	5418.15	.18456
5150	.19417	5423.15	.18439
5155	.19399	5428.15	.18422
5160	.19380	5433.15	.18406
5165	.19361	5438.15	.18389
5170	.19342	5443.15	.18372
5175	.19324	5448.15	.18355
5180	.19305	5453.15	.18338
5185	.19286	5458.15	.18321
5190	.19268	5463.15	.18304
5195	.19249	5468.15	.18288
5200	.19231	5473.15	.18271
5205	.19212	5478.15	.18254
5210	.19194	5483.15	.18238
5215	.19175	5488.15	.18221
5220	.19157	5493.15	.18204
5225	.19139	5498.15	.18188
5230	.19120	5503.15	.18171
5235	.19102	5508.15	.18155
5240	.19084	5513.15	.18138
5245	.19066	5518.15	.18122
5250	.19048	5523.15	.18106
5255	.19029	5528.15	.18089
5260	.19011	5533.15	.18073
5265	.18993	5538.15	.18057
5270	.18975	5543.15	.18040
5275	.18957	5548.15	.18024
5280	.18939	5553.15	.18008
5285	.18921	5558.15	.17992
5290	.18904	5563.15	.17975
5295	.18886	5568.15	.17959
5300	.18868	5573.15	.17943
5305	.18850	5578.15	.17927
5310	.18832	5583.15	.17911
5315	.18815	5588.15	.17895
5320	.18797	5593.15	.17879
5325	.18779	5598.15	.17863
5330	.18762	5603.15	.17847
5335	.18744	5608.15	.17831
5340	.18727	5613.15	.17815
5345	.18709	5618.15	.17799
5350	.18692	5623.15	.17784
5355	.18674	5628.15	.17768
5360	.18657	5633.15	.17752
5365	.18639	5638.15	.17736
5370	.18622	5643.15	.17721
5375	.18605	5648.15	.17705
5380	.18587	5653.15	.17689
5385	.18570	5658.15	.17674
5390	.18553	5663.15	.17658
5395	.18536	5668.15	.17642
5400	.18519	5673.15	.17627
5405	.18501	5678.15	.17611
5410	.18484	5683.15	.17596
5415	.18467	5688.15	.17580
5420	.18450	5693.15	.17565
5425	.18433	5698.15	.17550
5430	.18416	5703.15	.17534
5435	.18399	5708.15	.17519
5440	.18382	5713.15	.17503
5445	.18365	5718.15	.17488
5450	.18349	5723.15	.17473
5455	.18332	5728.15	.17458
5460	.18315	5733.15	.17442
5465	.18298	5738.15	.17427
5470	.18282	5743.15	.17412
5475	.18265	5748.15	.17397
5480	.18248	5753.15	.17382
5485	.18232	5758.15	.17367

N	1000/N	N + 273.15	1000/N + 273.15
+5490	.18215	5763.15	.17352
5495	.18198	5768.15	.17337
5500	.18182	5773.15	.17322
5505	.18165	5778.15	.17307
5510	.18149	5783.15	.17292
5515	.18132	5788.15	.17277
5520	.18116	5793.15	.17262
5525	.18100	5798.15	.17247
5530	.18083	5803.15	.17232
5535	.18067	5808.15	.17217
5540	.18051	5813.15	.17202
5545	.18034	5818.15	.17188
5550	.18018	5823.15	.17173
5555	.18002	5828.15	.17158
5560	.17986	5833.15	.17143
5565	.17969	5838.15	.17129
5570	.17953	5843.15	.17114
5575	.17937	5848.15	.17099
5580	.17921	5853.15	.17085
5585	.17905	5858.15	.17070
5590	.17889	5863.15	.17056
5595	.17873	5868.15	.17041
5600	.17857	5873.15	.17027
5605	.17841	5878.15	.17012
5610	.17825	5883.15	.16998
5615	.17809	5888.15	.16983
5620	.17794	5893.15	.16969
5625	.17778	5898.15	.16954
5630	.17762	5903.15	.16940
5635	.17746	5908.15	.16926
5640	.17730	5913.15	.16911
5645	.17715	5918.15	.16897
5650	.17699	5923.15	.16883
5655	.17683	5928.15	.16869
5660	.17668	5933.15	.16854
5665	.17652	5938.15	.16840
5670	.17637	5943.15	.16826
5675	.17621	5948.15	.16812
5680	.17606	5953.15	.16798
5685	.17590	5958.15	.16784
5690	.17575	5963.15	.16770
5695	.17559	5968.15	.16756
5700	.17544	5973.15	.16742
5705	.17528	5978.15	.16728
5710	.17513	5983.15	.16714
5715	.17498	5988.15	.16700
5720	.17483	5993.15	.16686
5725	.17467	5998.15	.16672
5730	.17452	6003.15	.16658
5735	.17437	6008.15	.16644
5740	.17422	6013.15	.16630
5745	.17406	6018.15	.16616
5750	.17391	6023.15	.16603
5755	.17376	6028.15	.16589
5760	.17361	6033.15	.16575
5765	.17346	6038.15	.16561
5770	.17331	6043.15	.16548
5775	.17316	6048.15	.16534
5780	.17301	6053.15	.16520
5785	.17286	6058.15	.16507
5790	.17271	6063.15	.16493
5795	.17256	6068.15	.16479
5800	.17241	6073.15	.16466
5805	.17227	6078.15	.16452
5810	.17212	6083.15	.16439
5815	.17197	6088.15	.16425
5820	.17182	6093.15	.16412
5825	.17167	6098.15	.16398
5830	.17153	6103.15	.16385
5835	.17138	6108.15	.16372
5840	.17123	6113.15	.16358
5845	.17109	6118.15	.16345
5850	.17094	6123.15	.16331
5855	.17079	6128.15	.16318

RECIPROCALS OF NUMBERS (Continued)

N	1000/N	N + 273.15	1000/N + 273.15
+5860	.17065	6133.15	.16305
5865	.17050	6138.15	.16292
5870	.17036	6143.15	.16278
5875	.17021	6148.15	.16265
5880	.17007	6153.15	.16252
5885	.16992	6158.15	.16239
5890	.16978	6163.15	.16225
5895	.16964	6168.15	.16212
5900	.16949	6173.15	.16199
5905	.16935	6178.15	.16186
5910	.16920	6183.15	.16173
5915	.16906	6188.15	.16160
5920	.16892	6193.15	.16147
5925	.16878	6198.15	.16134
5930	.16863	6203.15	.16121
5935	.16849	6208.15	.16108
5940	.16835	6213.15	.16095
5945	.16821	6218.15	.16082
5950	.16807	6223.15	.16069
5955	.16793	6228.15	.16056
5960	.16779	6233.15	.16043
5965	.16764	6238.15	.16030
5970	.16750	6243.15	.16018
5975	.16736	6248.15	.16005
5980	.16722	6253.15	.15992
5985	.16708	6258.15	.15979
5990	.16694	6263.15	.15966
5995	.16681	6268.15	.15954
6000	.16667	6273.15	.15941
6005	.16653	6278.15	.15928
6010	.16639	6283.15	.15916
6015	.16625	6288.15	.15903
6020	.16611	6293.15	.15890
6025	.16598	6298.15	.15878
6030	.16584	6303.15	.15865
6035	.16570	6308.15	.15853
6040	.16556	6313.15	.15840
6045	.16543	6318.15	.15827
6050	.16529	6323.15	.15815
6055	.16515	6328.15	.15802
6060	.16502	6333.15	.15790
6065	.16488	6338.15	.15777
6070	.16474	6343.15	.15765
6075	.16461	6348.15	.15753
6080	.16447	6353.15	.15740
6085	.16434	6358.15	.15728
6090	.16420	6363.15	.15715
6095	.16407	6368.15	.15703
6100	.16393	6373.15	.15691
6105	.16380	6378.15	.15679
6110	.16367	6383.15	.15666
6115	.16353	6388.15	.15654
6120	.16340	6393.15	.15642
6125	.16327	6398.15	.15630
6130	.16313	6403.15	.15617
6135	.16300	6408.15	.15605
6140	.16287	6413.15	.15593
6145	.16273	6418.15	.15581
6150	.16260	6423.15	.15569
6155	.16247	6428.15	.15557
6160	.16234	6433.15	.15544
6165	.16221	6438.15	.15532
6170	.16207	6443.15	.15520
6175	.16194	6448.15	.15508
6180	.16181	6453.15	.15496
6185	.16168	6458.15	.15484
6190	.16155	6463.15	.15472
6195	.16142	6468.15	.15460
6200	.16129	6473.15	.15448
6205	.16116	6478.15	.15437
6210	.16103	6483.15	.15425
6215	.16090	6488.15	.15413
6220	.16077	6493.15	.15401
6225	.16064	6498.15	.15389

N	1000/N	N + 273.15	1000/N + 273.15
+6230	.16051	6503.15	.15377
6235	.16038	6508.15	.15365
6240	.16026	6513.15	.15354
6245	.16013	6518.15	.15342
6250	.16000	6523.15	.15330
6255	.15987	6528.15	.15318
6260	.15974	6533.15	.15307
6265	.15962	6538.15	.15295
6270	.15949	6543.15	.15283
6275	.15936	6548.15	.15271
6280	.15924	6553.15	.15260
6285	.15911	6558.15	.15248
6290	.15898	6563.15	.15237
6295	.15886	6568.15	.15225
6300	.15873	6573.15	.15213
6305	.15860	6578.15	.15202
6310	.15848	6583.15	.15190
6315	.15835	6588.15	.15179
6320	.15823	6593.15	.15167
6325	.15810	6598.15	.15156
6330	.15798	6603.15	.15144
6335	.15785	6608.15	.15133
6340	.15773	6613.15	.15121
6345	.15760	6618.15	.15110
6350	.15748	6623.15	.15099
6355	.15736	6628.15	.15087
6360	.15723	6633.15	.15076
6365	.15711	6638.15	.15064
6370	.15699	6643.15	.15053
6375	.15686	6648.15	.15042
6380	.15674	6653.15	.15030
6385	.15662	6658.15	.15019
6390	.15649	6663.15	.15008
6395	.15637	6668.15	.14997
6400	.15625	6673.15	.14985
6405	.15613	6678.15	.14974
6410	.15601	6683.15	.14963
6415	.15588	6688.15	.14952
6420	.15576	6693.15	.14941
6425	.15564	6698.15	.14929
6430	.15552	6703.15	.14918
6435	.15540	6708.15	.14907
6440	.15528	6713.15	.14896
6445	.15516	6718.15	.14885
6450	.15504	6723.15	.14874
6455	.15492	6728.15	.14863
6460	.15480	6733.15	.14852
6465	.15468	6738.15	.14841
6470	.15456	6743.15	.14830
6475	.15444	6748.15	.14819
6480	.15432	6753.15	.14808
6485	.15420	6758.15	.14797
6490	.15408	6763.15	.14786
6495	.15396	6768.15	.14775
6500	.15385	6773.15	.14764
6505	.15373	6778.15	.14753
6510	.15361	6783.15	.14742
6515	.15349	6788.15	.14732
6520	.15337	6793.15	.14721
6525	.15326	6798.15	.14710
6530	.15314	6803.15	.14699
6535	.15302	6808.15	.14688
6540	.15291	6813.15	.14677
6545	.15279	6818.15	.14667
6550	.15267	6823.15	.14656
6555	.15256	6828.15	.14645
6560	.15244	6833.15	.14635
6565	.15232	6838.15	.14624
6570	.15221	6843.15	.14613
6575	.15209	6848.15	.14602
6580	.15198	6853.15	.14592
6585	.15186	6858.15	.14581
6590	.15175	6863.15	.14571
6595	.15163	6868.15	.14560

RECIPROCALS OF NUMBERS (Continued)

N	1000/N	N + 273.15	1000/N + 273.15
+6600	.15152	6873.15	.14549
6605	.15140	6878.15	.14539
6610	.15129	6883.15	.14528
6615	.15117	6888.15	.14518
6620	.15106	6893.15	.14507
6625	.15094	6898.15	.14497
6630	.15083	6903.15	.14486
6635	.15072	6908.15	.14476
6640	.15060	6913.15	.14465
6645	.15049	6918.15	.14455
6650	.15038	6923.15	.14444
6655	.15026	6928.15	.14434
6660	.15015	6933.15	.14423
6665	.15004	6938.15	.14413
6670	.14993	6943.15	.14403
6675	.14981	6948.15	.14392
6680	.14970	6953.15	.14382
6685	.14959	6958.15	.14372
6690	.14948	6963.15	.14361
6695	.14937	6968.15	.14351
6700	.14925	6973.15	.14341
6705	.14914	6978.15	.14330
6710	.14903	6983.15	.14320
6715	.14892	6988.15	.14310
6720	.14881	6993.15	.14300
6725	.14870	6998.15	.14289
6730	.14859	7003.15	.14279
6735	.14848	7008.15	.14269
6740	.14837	7013.15	.14259
6745	.14826	7018.15	.14249
6750	.14815	7023.15	.14239
6755	.14804	7028.15	.14228
6760	.14793	7033.15	.14218
6765	.14782	7038.15	.14208
6770	.14771	7043.15	.14198
6775	.14760	7048.15	.14188
6780	.14749	7053.15	.14178
6785	.14738	7058.15	.14168
6790	.14728	7063.15	.14158
6795	.14717	7068.15	.14148
6800	.14706	7073.15	.14138
6805	.14695	7078.15	.14128
6810	.14684	7083.15	.14118
6815	.14674	7088.15	.14108
6820	.14663	7093.15	.14098
6825	.14652	7098.15	.14088
6830	.14641	7103.15	.14078
6835	.14631	7108.15	.14068
6840	.14620	7113.15	.14058
6845	.14609	7118.15	.14049
6850	.14599	7123.15	.14039
6855	.14588	7128.15	.14029
6860	.14577	7133.15	.14019
6865	.14567	7138.15	.14009
6870	.14556	7143.15	.13999
6875	.14545	7148.15	.13990
6880	.14535	7153.15	.13980
6885	.14524	7158.15	.13970
6890	.14514	7163.15	.13960
6895	.14503	7168.15	.13951
6900	.14493	7173.15	.13941
6905	.14482	7178.15	.13931
6910	.14472	7183.15	.13921
6915	.14461	7188.15	.13912
6920	.14451	7193.15	.13902
6925	.14440	7198.15	.13892
6930	.14430	7203.15	.13883
6935	.14420	7208.15	.13873
6940	.14409	7213.15	.13864
6945	.14399	7218.15	.13854
6950	.14388	7223.15	.13844
6955	.14378	7228.15	.13835
6960	.14368	7233.15	.13825
6965	.14358	7238.15	.13816
6970	.14347	7243.15	.13806
6975	.14337	7248.15	.13797
6980	.14327	7253.15	.13787
6985	.14316	7258.15	.13778
6990	.14306	7263.15	.13768
6995	.14296	7268.15	.13759
7000	.14286	7273.15	.13749

DIMENSIONLESS GROUPS

John P. Catchpole and George Fulford

Reprinted from Industrial & Engineering Chemistry, Vol. 58, No. 3, pp. 47 to 60. Copyright 1966 by the American Chemical Society and reprinted by permission of the copyright owners and the authors.

Dimensionless groups are frequently generated in the analysis of a complex engineering problem. The more common groups thus generated are easily recognized, while the less common ones are not. Unless the less common existing groups are recognized, an already named group could unknowingly be renamed. Table A provides a tool that may be used to avoid this occurrence, by listing the groups by the variables of which they consist. These variables—i.e., length, density, diffusivity, viscosity, etc.—are further subdivided into their exponents to which they are raised in the groups in question. Thus, Reynolds number is listed under the exponent +1 for the variables, length, fluid velocity, and density, and the exponent −1 for viscosity.

To illustrate the use of the tables in the analysis of a problem, the group $(kE/\eta\sigma T^3)$ might be generated in the solution of a complex heat transfer problem. From Table A the groups containing the constituent variables are checked and the groups are listed:

Thermal conductivity	(k^{+1})	**F11, L7, R1**
Modulus of elasticity	(E^{+1})	**C1, E13, R1**
Stefan-Boltzmann coefficient	(η^{-1})	**R1, T6**
Surface tension	(σ^{-1})	**B9, C3, E8, L6, R1, W1**
Temperature	(T^{-3})	**R1, T6**

It is immediately apparent that the only group common to all the categories listed is the Radiation number, **R1**, which is equivalent to the previously unidentified group.

The symbol assigned to a dimensionless group is usually the first two letters of its names. Several groups, however, have nonstandard symbols, particularly in the groups which are named after persons. These symbols are listed in the nomenclature.

NOMENCLATURE

a = annulus or clearance width, L
A = area, L^2
A^* = cooling area/unit volume, $1/L$
b = bearing breadth, L
B = groups **B6, B11**
c = specific heat, L^2/θ^2T
c_A = concentration, M/L^3
c_b = specific vapor capacity (mass/unit mass/unit pressure change), $L\theta^2/M$
c_d, c_D = group **D13**
c_f = group **R9**
c_H = group **H4**
c_m = mass capacity, L^3/M
c_p, c_v = specific heats at constant pressure and volume, L^2/θ^2T
c_q = heat capacity, L^2/θ^2T
c_Q = group **F9**
cS = group **S11**
C = group **C10,** dimensional concentration, M/L^3
C_a = groups **C11, C4**
d = diameter, L
d_e = equivalent diameter (of particles, etc.), L
d_h = hydraulic diameter, L
D = diffusivity (molecular, unless noted otherwise), L^2/θ
D_{AB} = binary bulk diffusion coefficient, L^2/θ
D_{KA} = Knudsen diffusion coefficient, L^2/θ
e = voidage; porosity (⁻)
e^* = surface emissivity (⁻)
E = modulus of elasticity, $M/L\theta^2$
E_a = activation energy, L^2/θ^2
E_b = bulk modulus, $M/L\theta^2$
f = frequency, $1/\theta$, or Group **F1**
$f(M)$ = group **F11**
F = force, ML/θ^2
F_b = force per unit length of bearing, M/θ^2
$F(M)$ = group **F6**
F_R = resistance force in flow, ML/θ^2
g = acceleration due to gravity, L/θ^2
G = mass velocity (mass flux density; mass transfer coefficient), $M/\theta L^2$
h = heat transfer coefficient, $M/T\theta^3$
h_c = convective heat transfer coefficient, $M/T\theta^3$
H = energy change per unit mass (= $g \times$ head), L^2/θ^2
H' = fluid head, L
H_e = field strength, $Q/L\theta$
H_0 = homochronicity number
$I(M)$ = group **F8**
j = heat liberated per unit volume per unit time, $M/L\theta^3$
j_H, j_M = groups **J2, J3**
J = average free path/average velocity, θ, or group **L6**
k = thermal conductivity, $ML/T\theta^3$
k_c = mass transfer coefficient, L/θ
K = groups **K2, K10, N5**
K_1 = group **A4**
$\bar{K}_E$ = group **E4**
K_F = group **C2**
K_P = group **P13**
K_Q = group **H5**
K_r = group **E11**
$\bar{K}_g$ = group **A1**
$\bar{K}_E$ = group **E4**
$\bar{K}_r$ = group **E12**
$\bar{K}_{rE}$ = group **E13**
$\bar{K}_s$ = group **R1**
$\bar{K}_\sigma$ = group **C1**
$\bar{K}_{\sigma g}$ = group **C2**
L = characteristic dimension (except as noted), L
L_m = distance from midpoint to surface, L
m_T = group **T3**
M = group **M10**
M_H = group **H2**
n = concentration, wt./wt. (⁻)
n^* = specific mass content, mass/mass (⁻)
n_m = moisture content, wt./wt. bone dry gas (⁻)
N = rate of rotation, $1/\theta$, and groups **M3, N4**
N_{Bo} = groups **B7, B8**
N_c = group **C5**
N_{cv} = group **C10**
N_D = groups **D7, D14**
N_E = group **E1**
N_F = group **F4**
N_H = groups **H1, H9**
N_K = group **K1**
N_{KnA} = Knudsen number for diffusion (see Addendum)
N_1 = group **N2**
N_P = group **P7**
N_{rf} = group **R7**
N_{s1} = group **S8**
N_{s2} = group **S9**
N_T = group **N9**
P = pressure, $M/L\theta^2$
P = plasticity number (see Addendum)
p_b = bearing pressure, $M/L\theta^2$
p_s = static pressure, $M/L\theta^2$
p_v = vapor pressure, $M/L\theta^2$
p_σ = capillary pressure, $M/L\theta^2$
Δp_F = frictional pressure drop, $M/L\theta^2$
q = heat flux (heat flow/unit time), ML^2/θ^3
q^* = heat flux density (heat flux/unit area), M/θ^3
Q = heat liberated/unit mass, L^2/θ^2
r = latent heat of phase change, L^2/θ^2
r_v = heat of vaporization, L^2/θ^2
R = radius, L
R_H = hydraulic radius, L
R_2' = group **R5**
R_M = group **M7**
R_V = group **V3**
$\mathcal{R}$ = gas constant, L^2/θ^2T
s = humid heat, L^2/θ^2T
S = particle area/particle volume, L^2/L^3, and group **M6**
$\overline{St}$ = group **S14**
t = temperature, T
T = absolute temperature, T
Δt, ΔT = temperature difference, T
U^+ = group **P10**

DIMENSIONLESS GROUPS (Continued)

U = reaction rate, $M/L^3\theta$
v_s = velocity of surface (solid), L/θ
V = fluid velocity, L/θ, and group **V1**
V_A = velocity of Alfven magnetic waves, L/θ
V_f = volumetric flow rate, L^3/θ
V_l = velocity of light, L/θ
V_m = mass flow rate, M/θ
V_s = velocity of sound, L/θ
w = circumferential velocity, L/θ
W = volume of system, L^3
W^* = gross volume, L^3
x = entry length; distance from entrance, L
y^+ = group **P11**
Z = group **O2**
α = thermal diffusivity (temperature conductivity), L^2/θ
β = coefficient of bulk expansion, $1/T$, and group **D12**
β^* = Dufour coefficient, T
γ = specific gravity ($^-$) and group **R3**
$\dot{\gamma}$ = rate of shear, $1/\theta$
Γ = rate of change of temperature of medium, T/θ
δ = Soret or thermogradient coefficient, $1/T$, and group **D11**
Δ = difference in quantity
ϵ = height of roughness, L and group **A3**
ϵ_D = eddy mass diffusivity, L^2/θ
ζ = diffusion tortuosity ($^-$)
η = radiation coefficient (Stefan-Boltzmann coefficient), $M/T^4\theta^3$
θ = time, θ
θ_r = relaxation time, θ
λ = mean free path, L
μ = dynamic viscosity, $M/L\theta$
μ_e = magnetic permeability, ML/Q^2
μ_p = rigidity coefficient, $M/L\theta$
ξ = permeability, L^2
π = 3.1416. . . .
Π = power to agitator or impeller, ML^2/θ^3
ρ = density, M/L^3
ρ_n = group **P3**
σ = surface tension, M/θ^2 and group **S12**
σ_c = group **C6**
σ_e = electrical conductivity, $Q^2\theta/L^3M$
σ_t = group **T4**
τ = group **T8**
τ_w = wall shear stress, $M/L\theta^2$
τ_y = yield stress, $M/L\theta^2$
φ = group **D9**
ψ = groups **N8, P14, R10**
ω = angular velocity (of fluid, unless noted otherwise), $1/\theta$
Ω = mass transfer potential (concn.), M/L^3
— (bar over) = mean value
N.B.: $(F) = \left(\frac{ML}{\theta^2}\right)$; $(H) = \left(\frac{ML^2}{\theta^2}\right)$

TABLES FOR IDENTIFYING DIMENSIONLESS GROUPS
TABLE A

PHYSICAL PROPERTIES
General Physical Properties

Parameter	Symbol	Dimensions	Exponent	Group
Coefficient of bulk expansion	β	$1/T$	−1	**E4, 12, G2**
			+1	**G5, R5, K9, 6**
Density	ρ	M/L^3	−2	**M1**
			−1	**A1, B1, 10, C1, 2, 6, D3, 13, E3, 9, 10, 13, F1, 11, H5, J1, K4, K10, L7, M3, 6, N2, 3, 4, P7, 9, R10, 13, S4, 13, 15**
			$-\frac{2}{3}$	**C9, J3**
			$-\frac{1}{2}$	**E2, L11, O2, P13**
			$+\frac{1}{3}$	**J3, K5, 53**
			$+\frac{1}{2}$	**D10, G3, P10, 11, R4, W2**
			$+\frac{2}{3}$	**F2, N8, S2**
			+1	**A5, B1, B6, 9, C5, 7, D5, 7, 13, E4, 6, 7, 8, H6, 11, J1, 3, 4, K1, L9, P1, R5, 11, 13, S17, T1, T6, V1, W1, W3**
			+2	**C10, G1, 5, K9, R5, 6, T2**
Density gradient	$d\rho/dL$	M/L^4	+1	**R13**
Diffusivity (molecular unless noted otherwise)	D, a_m, ϵ_D	L^2/θ	−1	**B5, B7**(4)**, D2, K4**(1)**, K7**(2)**, L2, 7, 10, N7, P2, P9, S4, 13**
			$-\frac{2}{3}$	**J3**
			$-\frac{1}{2}$	**T3**
			+1	**D12, F12**(1)**, K7**(3)**, L2**(4)**, L9**(1)
Diffusivity (surface)	D_s	ML/θ^2	+1	**S18**
Diffusion tortuosity	ζ	....	−1	**K7**
Molecular weight	M	....	−1	**D14, K9, S6**
			+1	**D14**
Permeability (packed bed)	k	L^2	$+\frac{1}{2}$	**L4**
Porosity (voidage)	e	....	−1	**B4**
			$-\frac{1}{2}$	**L4**
			+1	**K7**

PHYSICAL PROPERTIES
General Physical Properties

Parameter	Symbol	Dimensions	Exponent	Group
Specific weight	γ	....	$\pm\frac{1}{3}$	**F2**
Surface tension	σ	M/θ^2	−3	**C2**
			−2	**C1**
			−1	**B9, C3, 13, E8, L6, R1, W1, 3**
			$-\frac{1}{2}$	**D10, G3, O2, P13, R4, W2**
			+1	**M9, S17, 18**

(1) Coefficient of potential diffusion in mass transfer.
(2) Knudsen diffusion coefficient.
(3) Binary bulk diffusion coefficient.
(4) Effective diffusivity ($D + \epsilon_D$) (molecular + eddy transfer).

Electrical and Magnetic Properties

Parameter	Symbol	Dimensions	Exponent	Group
Current density	I	$Q/L^2\theta$	+1	**K10**
Electrical conductivity	σ_e	$Q^2\theta/L^3M$	−1	**E6**
			$+\frac{1}{2}$	**H2**
			+1	**L11, M3, 7**
Field strength	H_e	$Q/L\theta$	−2	**J4**
			+1	**H2, L11**
			+2	**K9, M3, 6, S6**
Magnetic permeability	μ_e	ML/Q^2	−1	**E5, J4**
			+1	**H2, M6, 7**
			$+\frac{3}{2}$	**L11**
			+2	**M3**
Voltage	E	$ML/Q\theta^2$	+1	**K10**

Thermal Properties

Parameter	Symbol	Dimensions	Exponent	Group
Humid heat	S	L^2/θ^2T	−1	**P15**
Latent heats of phase change	λ, r	L^2/θ^2	−1	**A3, E11, 12, 13, J1, K10, M1**
			+1	**B13, C9, K8, 11, N5**
Ratio of specific heats	γ	....	$\pm\frac{1}{2}$	**L5**
Specific heat	C, c	L^2/θ^2T	−1	**B2, 13, D3, 15, E1, F11, J2, K8, 11, L7, M10, N5, R3, S13**
			$-\frac{1}{3}$	**J2**
			$+\frac{1}{2}$	**F6, 7, 8**

TABLE A (Continued)

Thermal Properties

Parameter	Symbol	Dimensions	Exponent	Group
Specific heat (*con't*)			$+\frac{2}{3}$	**J2**
			$+1$	**A3, E4, 12, G4, J1, 4, L9, P1, 8, R3, 5, 6, 7**
Surface emissivity	e^*	....	-1	**T6**
Temperature conductivity (thermal diffusivity)	α	L^2/θ	-1	**L9, P1, 12, R5**
			$+1$	**C13, L7, 10**
Thermal conductivity	k or λ	$ML/T\theta^3$	-3	**M1**
			-2	**R6**
			-1	**B4, 12, C7, 9, 10, D4, G4, K3, L9, N6, P1, 5, 8, R5, S14**

Rheological and Elastic Behavior

Parameter	Symbol	Dimensions	Exponent	Group
Modulus of elasticity	E	$M/L\theta^2$	-1	**C5, E4, H11**
			$+1$	**C1, E13, R1**
			$+3$	**A1**
Rate of shear	$\dot{\gamma}$	$1/\theta$	$+1$	**T8**
Rigidity coefficient	μ_p	$M/L\theta$	-2	**H6**
			-1	**B3**
			-1	**E5**
Shear stress	τ	$M/L\theta^2$	$-\frac{1}{2}$	**P10**
			$+\frac{1}{2}$	**P11**
Viscosity (in all cases kinematic viscosity has been written as μ/ρ)	μ	$M/L\theta$	-2	**A1, 5, G1, 5, K1, 9, S17, T2**
			-1	**B6, C10, D5, 7, E6, 7, H8, L1, 3, 9, O1, P4, 11, R5, 6, 11, S9, 19, T1, V1**
			$-\frac{2}{3}$	**F2, K5, N8**
			$-\frac{1}{2}$	**H2**
			$-\frac{1}{3}$	**S2, 3**
			$+\frac{1}{2}$	**E2**
			$+\frac{2}{3}$	**C9, J2, 3**
			$+1$	**B12, C3, 13, E3, 5, M1, O2, P8, 9, S4, 8, 15, T8**
			$+2$	**C1**
			$+4$	**C2**
Viscosity (surface)	μ_s	M/θ	$+1$	**S19**
Yield stress	τ_y	$M/L\theta^2$	$+1$	**B3, H6**

LENGTHS, AREAS AND VOLUMES

Characteristic Linear Dimensions

(In all cases kinematic viscosity has been written as μ/ρ)

Parameter	Symbol	Dimensions	Exponent	Group
General characteristic linear dimension	Various	L	-5	**P7**
			-3	**F9, L1**
			-2	**D9, 12, E3, F11, 12, H4, 5, N2, 3, 4, O1, R9, S8, 9, 15**
			-1	**B1, 10, E2, 5, 10, F1, 13, G4, H10, K6, L4, R6, 10, 15, 16, S6, 19, T5, W4**
			$-\frac{1}{2}$	**B11, D7, F14, O2**
			$+\frac{1}{2}$	**R4, T1, W2**
			$+1$	**B3, 4, 5, 7, 10, C7, D1, 3, 5, 11, 13, E7, 10, F1, 2, 5, G3, H2, K3, 4, 6, 10, L3, 4, 9, 11, M1, 3, 7, N6, 7, 8, O1, P1, 2, 11, R8, 10, 11, 16, S2, 7, 14, 16, 17, 18, T3, W1**
			$+\frac{3}{2}$	**D7, H7, T1**
			$+2$	**B9, D2, 4, E8, H6, K9, O1, P4, 5, 12, S8, 9, V1**
			$+3$	**A5, C10, G1, 5, K1, R5**
			$+4$	**T2**
			$+5$	**R6**
Dimension of agitator, impeller, etc.	Various	L	-5	**P7**
			-3	**F9**
			-2	**D7, H4**
			$+1$	**D11, F15**
			$+3$	**W3**
Film thickness	L_f	L	$+1$	**N8**
Furnace half-width	L	L	-1	**T7**
Larmor radius	L_L	L	$+1$	**L4**
Mean free path	λ	L	$+1$	**K6**
Particle dimension	d_e	L	-2	**S15**
			$+1$	**F2**
			$+3$	**A5, G1**
Pore or nozzle radius	R	L	$+1$	**M12, T7**
Reactor length	L	L	$+1$	**B7**
			-1	**C3, L9**
Thickness of liquid layer	L	L	$+1$	**H2, L11**
			$+2$	**L9**
			$+3$	**R5**
			$+4$	**T2**

Areas

Parameter	Symbol	Dimensions	Exponent	Group
Area	A	L^2	-1	**D9, F6, 7, 8**
			$+1$	**B2**
Area/unit volume	S, A^*	$1/L$	-1	**B6**
			$+1$	**M11, 12**

Volumes

Parameter	Symbol	Dimensions	Exponent	Group
Volume	....	L^3	$+$	**A5, H9, M11**

TIMES AND FREQUENCIES

Parameter	Symbol	Dimensions	Exponent	Group
Time	θ	θ	-1	**D8**
			$+1$	**D8, 12, E3, F11, 12, H1, 10, S15, T5**
Frequency	f'	$1/\theta$	$+1$	**H1, 10, S16, V1**

TEMPERATURES AND CONCENTRATIONS (DRIVING FORCES)

Concentrations and Related Quantities

Parameter	Symbol	Dimensions	Exponent	Group
Dimensional concentration	C, c_A	M/L^3	-1	**D1, 2, R2**
			$-\frac{1}{2}$	**T3**
			$+1$	**B10**
Dimensionless concentration—e.g., wt./wt. inert material, etc.	n	....	-1	**P6**
			$+1$	**A4, K8**
Mass capacity	C_m	L^3/M	-1	**R2**
Mole fraction	Y, Z	....	± 1	**D14**
Specific mass content, mass/unit mass	n^*	....	-1	**K4**
Surface concentration	Γ'	M/L^2	± 1	**S18**
Vapor capacity (porous body)	C_b	$L\theta^2/M$	$+1$	**B13, R2**

Temperatures, Temperature Differences

Parameter	Symbol	Dimensions	Exponent	Group
Temperature, temperature difference	T, ΔT	T	-3	**R1*, T6***
			-1	**A4, 6*, B12, 13, C4*, 7, 10, D3*, 4*, 15, E1, G2, 6*, K3, 8, 9, 11, L9, N5, P5*, 12*, R14*, S6**
			$-\frac{1}{2}$	**L5**
			$+\frac{1}{2}$	**F6*, 7*, 8***
			$+1$	**C4, G5, 6, J1, 4, L9, M1, P6, R5, 7, 14***
			$+3$	**S14***
Rate of temperature change	....	T/θ	$+1$	**P12**

* Absolute temperature; others—temperature differences.

TABLE A (Continued)

VELOCITIES, RATES, FLUXES, TRANSFER COEFFICIENTS

Velocities

Parameter	Symbol	Dimensions	Exponent	Group
Angular velocity (rate of rotation)	N	$1/\theta$	-3	**P7**
			-2	**H4, L1**
			-1	**F9, R15**
			$-\frac{1}{2}$	**E2**
			$+1$	**S8, 11, 12, T1**
			$+2$	**F15, T2, W3**
Fluid velocity	V	L/θ	-3	**H5**
			-2	**C6, 11, D13, E9, 10, F1, M6, N2, 3, 4, R7, 8, 9, 10**
			-1	**A2, B3, C12, 14, D1, 3, F10, H7, 8, K10, L3, M3, P4, S13, 16**
			$+1$	**A2, B6, 7, 11, C3, 12, 14, D5, 7, E5, 7, F10, 14, H10, K2, L5, 8, M2, 4, 7, P1, 2, 3, 10 R4, 11, 15, S1, 2, 3, T5, 6, V2, W2, 4**
			$+2$	**B1, 12, C5, 11, D15, E1, 11, F13, H11, W1**
			$+3$	**C7**
Impeller or agitator circumference	U_s	L/θ	-2	**P14**
			-1	**D9**
Light	V_l	L/θ	-1	**L8**
Sound	V_s	L/θ	-1	**M2, N1, S1**
Waves	V_w	L/θ	-1	**V1**
			-1	**P3**
Velocity gradient	dV/dL	$1/\theta$	-2	**R12**
Velocity of Alfven waves	V_A	L/θ	-1	**A2, K2, M4**
			$+1$	**A2, N1**
			$+2$	**C11**
Velocity of bearing surface	v_s	L/θ	$+1$	**H8, O1, S9**

Flow Rates (Mass Fluxes)

Parameter	Symbol	Dimensions	Exponent	Group
Mass flow rate (mass flux)	V_m	M/θ	-1	**B2, M11, T7**
			$+1$	**F6, 7, 8, G4, T7**
Mass flux density (mass flux/unit area)	G	$M/L^2\theta$	-1	**J2, 3, P15, S13**
			$+1$	**K4, M11**
Mass flux/unit volume (reaction rate)	U	$M/L^3\theta$	$+\frac{1}{2}$	**T3**
			$+1$	**D1, 2, 3, 4**
Reaction rate constant	K	L/θ	-1	**S5**
Volumetric flow rate	V_f	L^3/θ	-1	**H9**
			$-\frac{1}{2}$	**D11**
			$+\frac{1}{2}$	**S11, 12**
			$+1$	**D9, F9**

Heat Fluxes

Parameter	Symbol	Dimensions	Exponent	Group
Heat flux (heat flow/unit time)	q	ML^2/θ^3	$+1$	**H5**
Heat flux/unit area	q^*	M/θ^3	$+1$	**K3, R6**
Heat liberated/unit mass	Q	L^2/θ^2 (H/M)	$+1$	**D3, 4**
Rate of heat liberation/unit volume (heat source power)	j	$M/L\theta^3$	$+1$	**P5**

Transfer Coefficients

Parameter	Symbol	Dimensions	Exponent	Group
Heat transfer coefficient	h	$M/T\theta^3$	$+1$	**B2, 4, C9, J2, N6, P15, S13**
		$(H/L^2T\theta)$	$+4$	**M1**
Mass transfer coefficient	k_c	L/θ	$+1$	**B5, J3, L9, N7, S5, 7**

FORCE, HEAD, POWER, PRESSURE

Forces

Parameter	Symbol	Dimensions	Exponent	Group
Force (resistance)	F, F_R	ML/θ^2	$-\frac{1}{3}$	**S2**
			$+\frac{1}{3}$	**K5**
			$+1$	**N2, 3, 4, R9**
Force/unit length	F_b	M/θ^2	$+1$	**H8, O1, S9**

Heads, Power

Parameter	Symbol	Dimensions	Exponent	Group
Fluid head	H'	L	$-\frac{3}{4}$	**S11**
			$+1$	**H4**
Head (energy per unit mass of fluid = gH')	H	L^2/θ^2	-1	**P3, T4**
			$-\frac{3}{4}$	**S12**
			$+\frac{1}{4}$	**D11**
			$+1$	**A6, P14, T4**
Power	π	ML^2/θ^3 (LF/θ)	$+1$	**L1, P7**

Pressures

Parameter	Symbol	Dimensions	Exponent	Group
Pressure	P	$M/L\theta^2$	-1	**F6, 7, 8, H9, S8, T8**
			$+1$	**B13, C6, L6, P13, R2**
Pressure drop	ΔP, dP	$M/L\theta$	$+1$	**E9, 10, F1, H9, L3, R10**
Pressure gradient	$\Delta P/L$	$M/L^2\theta^2$	$+1$	**E10, F1, K1, P4, R10**

CONSTANTS AND MISCELLANEOUS QUANTITIES

Gravity Acceleration

Parameter	Symbol	Dimensions	Exponent	Group
Gravity acceleration	....	L/θ^2	-2	**A1**
			-1	**B1, F13, 15, M1, S6**
			$-\frac{3}{4}$	**S11**
			$-\frac{1}{2}$	**B11, F14, P13**
			$-\frac{1}{3}$	**C9, S3**
			$+\frac{1}{3}$	**F2, N8**
			$+1$	**A5, B9, C2, 10, D13, E8, G1, 5, H4, R5, 6, 8, 13**

Other Quantities

Parameter	Symbol	Dimensions	Exponent	Group
Avogadro's number	N	$1/M$	$+1$	**K9, S6**
Boltzmann's constant	k	....	-1	**K9, S6**
Dufour coefficient	β	T	$+1$	**A3, 4, F3**
Energy of activation	E_a	L^2/θ^2	$+1$	**A6**
Gas constant	$\mathcal{R}$	L^2/θ^2T	-1	**A6**
			$-\frac{1}{2}$	**L5**
Shape factor	ξ^*	....	$+1$	**V2**
Soret coefficient	δ	$1/T$	$+1$	**F3, P6**
Stefan constant	η	$M/T^4\theta^3$	-1	**R1, T6**
			$+1$	**S14**

TABLE B
ALPHABETICAL LIST OF NAMED GROUPS

Serial No.	Name	Symbol	Definition	Significance	Field of Use	Reference
A1	Acceleration number	$\overline{K}_g$	$E^3/\rho g^2\mu^2 = (N_{Re}N_{Fr1})^2/(H_o)^3$	Group dependent only on physical properties	Accelerated flow	(25)
A2	Alfven number	N_{Al}	V_A/V (or V/V_A) [*cf.* Cowling No. Kármán No. (2), magnetic mach number]	Ratio of Alfven wave velocity/fluid velocity	Magneto-fluid dynamics	(5, 6)
A3	Anonymous group (1)	ϵ	$\beta^* c/r$ [see also Fedorov No. (2)]		Transfer processes	(39)
A4	Anonymous group (2)	K_1	$\beta^* \Delta n/\Delta t$; Δt = temp. diff. [T]; Δn = conc. diff. [$^-$]		Transfer processes	(67)
A5	Archimedes number	N_{Ar}	$\frac{gL^3\rho}{\mu^2}(\rho - \rho_o)$; ρ_o = fluid density; ρ = particle density (*cf.* N_{Ga1})	N_{Re}, gravitational force/viscous force	Fluidization, motion of liquids due to density differences	(6, 13)
A6	Arrhenius group	....	$E_a/\Re T$	Activation energy/potential energy of fluid	Reaction rates	(5)
B1	Bagnold number	B	$3c\alpha\rho_g V^2/4d\rho_p g$; ρ_g = gas density; ρ_p = particle density	Drag force/gravitational force	Saltation studies	(12)
B2	Bansen number	N_{Ba}	$h_r A_w/V_{mc}$; h_r = radiant heat transfer coefficient; A_w = wall area of channel; (*cf.* N_{St})	Heat transferred by radiation/thermal capacity of fluid	Radiation	(1)
B3	Bingham number	N_{Bm}	$\tau_y L/\mu_P V$ (L = channel width)	Ratio of yield stress/viscous stress	Flow of Bingham plastics	(5)
B4	Biot number (heat transfer	N_{Bih}	hL_m/k (in French literature, "Biot No." = N_{Nu})	Midplane thermal internal resistance/surface film resistance	Unsteady state heat transfer	(5, 6, 13)
B5	Biot number (mass transfer)	N_{Bim}	$k_c L/D_{int}$; L = thickness of layer, D_{int} = diffusivity at interface	Mass transfer rate at interface/mass transfer rate in interior of solid wall thickness L	Mass transfer between fluid and solid	(13)
B6	Blake number	B	$V\rho/[\mu(1 - e)S]$	Inertial force/viscous force	Beds of particles	(6)
B7	Bodenstein number	N_{Bo}	$VL/D_a = N_{Pe_m}$; L = reactor length, D_a = axial diffusivity (effective) (L^2/θ)		Diffusion in reactors	(6)
B8	Boltzmann number	N_{Bo}	≡ Thring radiation group			(1)
B9	Bond number	N_{Bo}	$(\rho - \rho')L^2 g/\sigma = N_{We1}/N_{Fr1}$ if $\rho - \rho' \cong \rho$ (gas in liq.); ρ = drop or bubble density; ρ' = medium density	Gravitational force/surface tension force	Atomization, motion of bubbles and drops	(5)
B10	Bouguer number	N_{Bu}, B	$3C_D\lambda_r/4\rho_D\overline{R}$; C_D = wt. dust/unit bed volume (M/L^3), λ_r = mean path for radiation (L), ρ_D = dust density, $\overline{R}$ = mean particle radius. Also $N_{Bu} = kL$; L = characteristic dimension, k = absorption coefficient of medium		Radiant heat transfer to dust—gas streams	(66)
B11	Boussinesq number	B	$V/(2gR_H)^{1/2}$ (*cf.* N_{Fr2})	(Inertia force/gravitational force)$^{1/2}$	Wave behavior in open channels	(6)
B12	Brinkman number	N_{Br}	$\mu V^2/k\Delta t$; Δt = temp. diff.	Heat generation/heat transferred	Viscous flow	(6)
B13	Bulygin number	N_{Bu}	$r \cdot \frac{C_b}{C_q} \cdot \frac{P}{t_m - t_o}$ t_m = temp. of medium, t_o = init. temp. of body	Heat for vaporization/heat to bring liquid to boiling point	Heat transfer during evaporation	(6, 13, 14)
C1	Capillary number	$\overline{K}_\sigma$	$\mu^2 E/\rho\sigma^2 = (N_{We_1})^2/H_o.(N_{Re})^2$	Depends only on physical properties	Action of surface tension in flowing media	(25)
C2	Capillarity-buoyancy number (physical properties group) (film No.)[a]	$\overline{K}_{\sigma g}$, K_F	$g\mu^4/\rho\sigma^3. = \sqrt{\overline{K}\sigma/Kg} = (N_{We_1})^3/(N_{Fr_1})(N_{Re})^4$	Depends only on physical properties and g	Effects of surface tension and acceleration in flowing media (two-phase flow)	(21, 25)
C3	Capillary number	Ca	$\mu V/\sigma = N_{We}/N_{Re}$	Viscous force/surface-tension force	Atomization, two-phase flow	(6)
C4	Carnot number	Ca, N_{Ca}	$(T_2 - T_1)/(T_2)$; T_1, T_2 = abs. temp. of two heat sources or sinks	Theoretical efficiency of Carnot cycle operating between T_1 and T_2		(21)
C5	Cauchy number	N_c	$\rho V^2/E_b = (N_{Ma})^2$ = Hooke No.	Inertia force/compressibility force	Compressible flow	(5, 25)
C6	Cavitation number	σ_c	$[(p - P_v)/\rho]/(V^2/2)p$ = local static pressure (abs.); P_v = vapor pressure	Excess of local static head over vapor pressure head/velocity head	Cavitation	(5)

[a] Very similar to Hu and Kintner's pH factor for drops and bubbles [*A.I.Ch.E. J.* **1**, 42 (1955)].

TABLE B (Continued)

Serial No.	Name	Symbol	Definition	Significance	Field of Use	Reference
C7	Clausius number	Cl, N_{Cl}	$V^3L\rho/k\Delta T$; ΔT = temp. diff.		Heat conduction in forced flows	*(25)*
C8	Colburn number	....	Same as Schmidt number			*(5)*
C9	Condensation number (1)	N_{Co}	$(h/k)(\mu^2/\rho^2 g)^{1/3}$	$N_{Co} = N_{Nu}\left[\frac{\text{(viscous force)}}{\text{(gravity force)}} \times \frac{1}{Re}\right]^{1/3}$	Condensation	*(5)*
C10	Condensation number (2)	N_{Cv}	$L^3\rho^2 gr/k\mu\Delta t$; r = latent heat of condensation		Condensation on vertical walls	*(5)*
C11	Cowling number	C	$(V_A/V)^2 \equiv$ (Alfven number)^2		Magneto-fluid dynamics	*(6)*
C12	Craya-Curtet number	C_t	$V_k/(V_d^2 - V_k^2/2)^{1/2}$; V_k = kinematic mean velocity, V_d = dynamic mean velocity		Radiant heat transfer	*(3, 27)*
C13	Crispation group	N_{Cr}	$\mu\alpha/\sigma^* L$; σ^* = undisturbed surface tension; L = layer thickness		Convection currents	*(55)*
C14	Crocco number	N_{Cr}	$V/V_{\max} = \left[1 + \frac{2}{(\gamma - 1)(N_{Ma})^2}\right]^{-1/2}$ $V_{\max}$ = maximum velocity of gas expanding adiabatically	Velocity/maximum velocity	Compressible flow	*(5)*
D1	Damköhler group I	DaI	UL/V_{cA}	Chemical reaction rate/bulk mass flow rate	Chemical reaction, momentum, and heat transfer	*(5)*
D2	Damköhler group II	DaII	UL^2/D_{cA}	Chemical reaction rate/molecular diffusion rate	Chemical reaction, momentum, and heat transfer	*(5)*
D3	Damköhler group III	DaIII	$QUL/C_p\rho Vt$	Heat liberated/bulk transport of heat	Chemical reaction, momentum, and heat transfer	*(5)*
D4	Damköhler group IV	DaIV	QUL^2/kt	Heat liberated/conductive heat transfer	Chemical reaction, momentum, and heat transfer	*(5)*
D5	Damköhler group V	DaV	$= (N_{R_e})$			*(5)*
D6	Darcy number		$4f$; see Fanning friction factor			*(5)*
D7	Dean number	N_D	$(VL\rho/\mu)(L/2R)^{1/2}$; L = pipe diam.; R = radius of curvature of bend	N_{Re} (centrifugal force/inertial force)	Flow in curved channels	*(5, 6)*
D8	Deborah number	D	θ_r/θ_o; θ_o = observation time	Relaxation time/observation time	Rheology	*(49)*
D9	Delivery number	ϕ	V_f/Aw; A = impeller area = $\pi d^2/4 \equiv$ [Diameter No.]$^{-3}$ [Speed No.]$^{-1}$		Flow machines	*(25)*
D10	Deryagin number	De	$L(\rho g/2\sigma)^{1/2}$; L = film thickness	Film thickness/capillary length	Coating	*(58)*
D11	Diameter group	δ	$(\pi/4)^{1/2}(2H)^{1/4}\, d/(V_f)^{1/2}$, d = impeller diam. = [pressure No.]$^{1/4}$ $\times$ [delivery No.]$^{-1/2}$		Flow machines	*(25)*
D12	Diffusion group	β	$D\theta/L_m^2$; D = diffusivity of solute through stationary solution contained in solid; *cf.* N_{Fom}		Mass transfer	*(37)*
D13	Drag coefficient	$C_d = C_D$	$(\rho - \rho')L_g/\rho V^2$; ρ = density of object; ρ' = density of medium; *cf.* f, ψ, N_e	Gravity force/inertial force	Free settling velocities, etc.	*(5)*
D14	Drew number	N_D	$\frac{Z_A(M_A - M_B) + M_B}{(Z_A - Y_{AW})(M_B - M_A)} \ln\frac{M_V}{M_W}$: M_A, M_B = mol. wt. of components A and B; M_V, M_W = mol. wt. of mixture in vapor and at wall; Y_{AW} = mole fraction of A at wall; Z_A = mole fraction of A in diffusing stream		Boundary layer mass transfer rates; velocity profile distortion; drag coefficients for binary system	*(22)*
D15	Dulong number	Du, N_{Du}	$V^2/C_p\Delta T$ = Eckert No.			*(25)*
E1	Eckert number	N_E	$V_\infty^2/C_p\Delta T$, V_∞ = velocity of fluid far from body (= 2/recovery factor, *q.v.* $\equiv$ Dulong No.)		Compressible flow	*(5, 24)*
E2	Ekman number		$(\mu/2\rho\omega L^2)^{1/2} = (N_{Ro}/N_{Re})^{1/2}$	(Viscous force/Coriolis force)$^{1/2}$	Magneto-fluid dynamics	*(6)*

TABLE B (Continued)

Serial No.	Name	Symbol	Definition	Significance	Field of Use	Reference
E3	Elasticity number (1)	N_{El_1}	$\theta_r\mu/\rho L^2$, L = pipe radius	Elastic force/inertial force	Viscoelastic flow	*(5)*
E4	Elasticity number (2)	$\bar{K}_E$	$\rho C_p/\beta E$ ≡ [Gay Lussac No.] × [Hooke No.] ÷ [Dulong No.]	Depends on physical properties only	Effect of elasticity in flow processes	*(25)*
E5	Ellis number	N_{El}	$\mu_o V/2\tau_{1/2}R$; μ_o = zero shear viscosity, $\tau_{1/2}$ = shear stress when $\mu = \mu_o/2[M/L\theta^2]$; R = tube radius		Flow of non-Newtonian liquids	*(38)*
E6	Elsasser number	N_{El}	$\rho/\mu\sigma_e\mu_e \equiv N_{Re}$/[magnetic Reynolds No.]		Magneto-fluid dynamics	*(6)*
E7	Entry Reynolds Number	K_E	$\chi/d_h \,.\, N_{Re} = \dfrac{\chi V\rho}{\mu}$; χ = entry length	As N_{Re}	Entry or inlet processes	*(25)*
E8	Eötvös number	N_{Eo}	$(\rho - \rho')L^2g/\sigma$ ≡ Bond No., *q.v.*			*(6)*
E9	Euler number (1)	N_{Eu_1}	$\Delta P_F/\rho V^2$; ΔP_F = pressure drop due to friction	Friction head/2 × velocity head	Fluid friction in conduits	*(5, 13, 25)*
E10	Euler number (2)	N_{Eu_2}	$d(-dp/dL)\rho V^2$; d = pipe diam., dp/dL = pressure gradient ≡ 2 × Fanning friction factor		Fluid friction in conduits	
E11	Evaporation number	K_r	V^2/r [r = heat of vaporization (L^2/θ^2)]		Evaporation processes	*(25)*
E12	Evaporation number (2)	$\bar{K}_r$	$C_p/r\beta$ (r as in **E11**) ≡ (Gay Lussac No.) × (**E11**)/(Dulong No.)		Evaporation processes	*(25)*
E13	Evaporation-elasticity number	$\bar{K}_{rE}$	$E/r\rho = K_r$/Hooke number		Evaporation processes	*(25)*
F1	Fanning friction factor	f	$d\Delta p_F/2\rho V^2L$, d = dimension of cross section; L = length (*cf.* resistance coeff., Ne)	Shear stress at wall expressed as number of velocity heads	Fluid friction in conduits	*(5)*
F2	Fedorov number (1)	F_{e_1}, N_{Fe_1}	$d_e \sqrt[3]{\dfrac{4g\rho^2}{3\mu^2}\left(\dfrac{\gamma^{M-1}}{\gamma g}\right)}$; d_e = equiv. particle diam.; γM = sp. gr. of particles; γg = sp. gr. of gas (*cf.* N_{Ar})		Fluidized beds	*(6, 13)*
F3	Fedorov number (2)	F_{e_2} N_{Fe_2}	$\delta\beta^* = K_1Pn = \epsilon \times K_oPn$	Mass transfer analogy of Posnov number	Transport processes	*(39, 67)*
F4	Fenske number	N_F		Number of stages in separation process		*(6)*
F5	Fineness coefficient	ψ	$L/W_D^{1/3}$; W_D = volume displacement $[L^3]$		Ship modeling	*(26)*, *(43)*
F6 / **F7** / **F8**	Fliegner numbers	$F(M_a)$ / $f(M_a)$ / $I(M_a)$	Functions of ratio of specific heats and mach number; $V_m(cT)^{1/2}/A(p_s + \rho V^2) = \gamma/(\gamma - 1)^{1/2}\ Ma/1 + \gamma Ma^2 \times [1 + (\gamma - 1)/2\ Ma^2]^{1/2}$ = impulse Fliegner number; γ = ratio of specific heats, Ma = mach number, A = flow area			*(43)*
F9	Flow coefficient	C_Q	V_f/Nd, d = impeller diam.		Power required by fans, etc.	*(32)*
F10	Fluidization number		V/V_{init}, V_{init} = velocity for initial fluidization	Fluid velocity in fluidized bed/that at start of fluidization	Fluidization	*(59)*
F11	Fourier number (heat transfer)	N_{Foh}	$k\theta/\rho c_pL_m$		Unsteady state heat transfer	*(5, 13, 14, 25)*
F12	Fourier number (mass transfer)	N_{Fom}	$D\theta/L^2 = k_c\theta/L$ (*cf.* **D9**)		Unsteady state mass transfer	*(13)*
F13	Froude number (1)	N_{Fr_1}	$V^2/gL \equiv (N_{Fr_2})^2$ (*cf.* Reech No., Boussinesq No., Vedernikov No.)	Inertial force/gravitational force	Wave and surface behavior	*(5, 13, 25)*
F14	Froude number (2)	N_{Fr_2}	$V/\sqrt{gL} \equiv (N_{Fr_1})^{1/2}$ (*cf.* Boussinesq No.)	Velocity of open channel flow/speed of very small gravity wave	Open channel flow; free surfaces	*(7, 50)*
F15	Froude numbers (rotating)	Fr	DN^2/g; D = impeller diam.		Agitation	*(64)*
G1	Galileo number	N_{Ga_1}	$L^3g\rho^2/\mu^2$ (*cf.* N_{Ar}, Nusselt thickness group)	$N_{Ga_1} = N_{Re}$ × gravity force/viscous force	Circulation of viscous liquid, thermal expansion	*(5, 13)*
G2	Gay Lussac number	Ga, N_{Ga_2}	$1/\beta\Delta T$		Thermal expansion processes	*(25)*
G3	Goucher number	N_{Go}	$R(\rho g/2\sigma)^{1/2}$; R = wall or wire radius	Gravitational force/surface tension force$^{1/2}$	Coating	*(24)*

TABLE B (Continued)

Serial No.	Name	Symbol	Definition	Significance	Field of Use	Reference
G4	Graetz (Grätz) number	N_{Gz}	$V_m c_p/kL$	Thermal capacity fluid/convective heat transfer	Streamline flow	(*5, 25*)
G5	Grashof number	N_{Gr}	$L^3\rho^2 g\beta\Delta t/\mu^2 \equiv N_{Ga_1}/N_{Ga_2} \equiv (N_{Re})^2/(N_{Ga_2})(N_{Fr_1})$	$N_{Gr} = N_{Re}$ (buoyancy force/viscous force)	Free convection	(*5, 13, 25*)
G6	Gukhman number	Gu, N_{Gu}	$(t_o - t_m)/T_o$; t_o, T_o = temp. (°C., °K.) of hot gas stream, t_m = temp. of moist surface (wet bulb temp.)	Thermodynamic criterion of evaporation under isobaric adiabatic conditions	Convective heat transfer in evaporation	(*6, 23, 17, 53*)
G7	Guldberg-Waage group	N_{Gw}	Given by equation relating volumes of reacting gases and reaction products		Chemical reaction in blast furnaces	(*5*)
H1	Hall coefficient	N_H	$f_c J$ (f_c = cyclotron frequency, J = av. free path/av. veloc.)		Magneto-fluid dynamics	(*7*)
H2	Hartmann number	M_H	$(\mu_e^2 H_e^2 \sigma_e L^2/\mu)^{1/2} \equiv (SR_M N_{Re})^{1/2} \equiv (N_{Re} N)^{1/2}$	Magnetically induced stress/hydrodynamic shear stress (magnetic body force/viscous force)$^{1/2}$	Magneto-fluid dynamics	(*5, 6*)
H3	Hatta number	β	$\gamma/\tanh\gamma$; $\gamma = (rCD)^{1/2}/k_c$, r = reaction rate constant $[L^3/M\theta]$ [a modified Hatta number has also been defined (*35*)]		Gas absorption with chemical reaction	(*41*)
H4	Head coefficient	C_H	gH'/Nd^2 (d = impeller diam.)		Flow in pumps and fans	(*32*)
H5	Heat transfer number	K_Q	$q/V^3L^2\rho$		Heat transfer in stream	(*25*)
H6	Hedstrom number	N_{He}	$\tau_y L^2\rho/\mu_p{}^2 \equiv (N_{Re}) \times (N_{Bm})$		Flow of Bingham plastics	(*5*)
H7	Helmholtz resonator group		$(d^3/W)^{1/2}/Ma$	Proportional to frequency × residence time	Pulsating combustion	(*28*)
H8	Hersey numbers		$F_b/\mu v_s$ (*cf.* truncation number)	Load force/viscous force	Lubrication	(*6*)
H9	Hodgson number	N_H	$wf\Delta p_F/\bar{V}_f.\bar{p}_s$	Time constant of system/period of pulsation	Pulsating gas flow	(*5*)
H10	Homochronous number	H_{o_1}	$V\theta/L$ (θ = time for liquid to move characteristic distance L)	Duration of process/time for liquid to move through L	Choice of time scales	(*13*)
H11	Hooke number	H_{o_2}	$\rho V^2/E$ = Cauchy No., *q.v.*		Elasticity of flowing media	(*25*)
J1	Jakob modulus	Ja	$C_P\rho_L\Delta t/r\rho_v$ (ρ_L, ρ_v = densities of liquid and vapor; Δt = liquid superheat temperature diff.)	Maximum bubble radius/thickness of superheated film	Boiling	(*6*)
J2	*J*-factor (heat transfer)	j_H	$(h/c_pG)(c_p\mu/k)^{2/3} \equiv (N_{Nu})/(N_{Re})(N_{Pr})^{1/3}$		Heat, mass and momentum transfer theory	(*5*)
J3	*J*-factor (mass transfer)	j_M	$(k_c\rho/G)(\mu/\rho D)^{2/3} \equiv (k_c\rho/G)(N_{Sc})^{2/3}$			(*5*)
J4	Joule number	J	$2\rho C_p\Delta t/\mu_e H_e^2 \equiv 2(N_{Re})(R_M)/(M_H)^2(N_E)$	Joule heating energy/magnetic field energy	Magneto-fluid dynamics	(*6*)
K1	Kármán number (1)	N_K	$\rho d^3(-dp/dL)/\mu^2$ (d = pipe diam., dp/dL = pressure gradient) $\equiv 2(N_{Re}\sqrt{f})^2$		Fluid friction in conduits	(*5*)
K2	Kármán number (2)	K	V/V_A (see Alfven No.)		Magneto-fluid dynamics	(*6*)
K3	Kirpichev number for heat transfer	Ki_q, N_{Ki_q}	$q^*L/k\Delta t$ (*cf.* N_{Bih}, N_{Nu})	Intensity external heat transfer/internal heat transfer intensity	Heat transfer	(*6, 14*)
K4	Kirpichev number for mass transfer	Ki_m	$GL/D\rho n^*$ (*cf.* N_{Pem}, N_{Bim})	Intensity external mass transfer/internal mass transfer intensity	Mass transfer	(*13, 14*)
K5	Kirpitcheff number		$(\rho F_K/\mu^2)^{1/3} = 3\sqrt{(N_{Re})^2 cf}$		Flow around obstacles	(*34*)
K6	Knudsen number (1)	N_{Kn}	λ/L	Length of mean free path/characteristic dimension	Low pressure gas flow	(*5*)
K7	Knudsen number (2)	N_{Kn_A}	$\rho D/_{AB}D_{KA}\zeta$	Bulk diffusion/Knudsen diffusion	Gaseous diffusion in packed beds	(*6*)
K8	Kossovich number	K_o, N_{Ko}	$r_v\Delta n_m/c\Delta t$	Heat used for evaporation/heat used in raising temperature of body	Convective heat transfer during evaporation	(*6, 14*)

TABLE B (Continued)

Serial No.	Name	Symbol	Definition	Significance	Field of Use	Reference
K9	Kronig number	Kr	$4L^2\beta\rho^2\Delta tE_S^2N[\alpha + 2/3(p_o^2/k\Gamma)]/u^2M$ E_S = electric field at surface, N = Avogadro's Number, α = polarization coefficient, p_o = molecular dipole moment, k = Boltzmann's constant, M = molecular weight	(N_{Re}) (electrostatic force/viscous force	Convective heat transfer	(6, 44)
K10	Kutateladze number (1)	Ku	$IEL/\rho Vu'$; I = current density $[Q/L^2\theta]$, E = voltage $[ML/Q\theta^2]$, u' = enthalpy $[L^2/\theta^2]$		Electric arcs in gas streams	(65)
K11	Kutateladze number (2)	K	$r_v/c_p(t_o - t_w)$, (t_o, t_w = stream, wall temp.)		Combined heat and mass transfer in evaporation	(65)
L1	Lagrange group (1)	La_1	$\pi/\mu L^3N^2$; L = characteristic dimension of agitator = $N_{Re}.N_p$		Agitation	(6)
L2	Lagrange number (2)	La_2	$(D + E_D)/D$	Combined molecular and eddy mass transfer rate/molecular mass transfer rate	Mass transfer in turbulent systems	(31)
L3	Lagrange number (3)	La_3	$\Delta P\ R/\mu V$		Magneto-fluid dynamics	(6)
L4	Larmor number	R_{La}	L_L/L; (L_L = Larmor radius)		Magneto-fluid dynamics	(23)
L5	Laval number	La	$V/[^2\gamma/(\gamma + 1)RT]^{1/2}$; γ = ratio of specific heats	Linear velocity/critical velocity of sound	Compressible flow	(56)
L6	Leverett function	J	$(\xi/e)^{1/2}(P\sigma/\sigma)$	Characteristic dimension of surface curvature/characteristic dimension of pores	Two-phase flow in porous media	(6)
L7	Lewis	N_{Le}	$k/\rho c_pD = \alpha/D \equiv N_{Sc}/N_{Pr}$ (N.B.: Lewis number is sometimes defined as reciprocal of this quantity)		Combined heat and mass transfer	(5, 25)
L8	Lorentz number	N_{Lo}	V/V_1; (V_1 = velocity of light)	Fluid velocity/velocity of light	Magneto-fluid dynamics	(6)
L9	Luikov (Lŷkov) number	Lu	$k_cL/\alpha = k_cL\rho C_p/k$	Mass diffusivity/thermal diffusivity; rate of extension of mass transfer field/rate of extension of heat transfer field	Combined heat and transfer	(6, 13)
L10	Lukomskii number	Lu	α/a_m; a_m = potential conductivity of mass transfer $[L^2/\theta]$		Combined heat and mass transfer	(36)
L11	Lundquist number	N_{Lu}	$\sigma_eH_e\mu_e^{3/2}L/\rho^{1/2} = M_H(R_M/N_{Re})^{1/2}$ (L = thickness of fluid layer)		Magneto-fluid dynamics	(6)
L12	Lyashchenko number	Ly	$\equiv N^3{}_{Re}/N_{Ar}$		Fluidization	(19)
L13	Lykoudis number	N_{Ly}	$(\mu_eH_e)^2\dfrac{\sigma_e}{\rho}\left[\dfrac{L}{g\beta\Delta t}\right]^{1/2} \equiv (M_H)^2/(N_{Gr})^{1/2}$		Magneto-fluid dynamics	(6)
M1	McAdams group		$h^4L\mu\Delta t/k^3\rho^2gr$	Constant for given surface orientation	Condensation	(5)
M2	Mach number	N_{Ma}, Ma	V/V_s; (V_s = velocity of sound in fluid) $\equiv v/\sqrt{E_b/\rho}$; (E_b = bulk modulus of fluid) (*cf.* Sarrau number)	Linear velocity/sonic velocity	Compressible flow	(5–7, 25, 50)
M3	Magnetic force parameter	N	$\mu_e^2H_e^2\sigma_eL/\rho V$	Magnetic body force/inertia force; resistance time of fluid in field/relaxation time of lines of force	Magneto-fluid dynamics	(6)
M4	Magnetic mach number	M_{Ma}	V/V_a (see Alfven number)		Magneto-fluid dynamics	(6)
M5	Magnetic Oseen number	k	$\frac{1}{2}(1 - N_{Al}^2)R_M$	Magnetic force/inertia force	Magneto-fluid dynamics	(2)
M6	Magnetic pressure number	S	$\mu_eH_e^2/\rho V^2$	Magnetic pressure/2 × dynamic pressure	Magneto-fluid dynamics	(5)
M7	Magnetic Reynolds number	R_M	$\sigma_e\mu_eLV$ (*cf.* velocity number)	Mass transport diffusivity/magnetic diffusivity	Magneto-fluid dynamics	(5)
M8	Maievskii number		$\equiv Ma$		Compressible flow	(12)
M9	Marangoni number	N_{Ma}	$\dfrac{\Delta\sigma}{\Delta t}\dfrac{\Delta t}{\Delta L}L^2/\mu\alpha$; L = layer thickness		Cellular convection	(46)

TABLE B (Continued)

Serial No.	Name	Symbol	Definition	Significance	Field of Use	Reference
M10	Margoulis number	M	$\equiv N_{St}$		Forced convection	*(6, 7)*
M11	Merkel number	N_{Me}	$GA^*W^*/(V_m)$ gas	Mass of water transferred in cooling per unit humidity difference/mass of dry gas	Cooling towers, liquid-gas contact	*(5)*
M12	Miniovich number	Mn	SR/e; R = pore radius		Drying	*(36)*
M13	Mondt number	N_{Mo}		Convective/conductive heat transfer	Heat transfer	*(6)*
N1	Naze number	N_a	$V_A/V_s \equiv (N_{Ma}.N_{Al})$	Velocity Alfven wave/velocity of sound	Magneto-fluid dynamics	*(6)*
N2	Newton inertial force group	N_l	$F/\rho V^2L^2$	Imposed force/inertial force	Agitation	*(5)*
N3	Newton number	N_e	$F_R/\rho V^2L^2$; (*cf.* f, ψ)	Resistance force/inertia force	Friction in fluid flow	*(25)*
N4	Number of velocity heads	N	$(F/\rho L^2)/(V^2/2)$	Imposed head/velocity head	Friction in conduits	*(6)*
N5	Number for similarity of phys. and chem. changes	K	$\gamma/C_p\Delta t$	Heat flow for phase change/superheat (supercooling) of one of the phases	Changes of phase	*(14)*
N6	Nusselt number	N_{Nu}	$hL/k \equiv (N_{Re}N_{St})$ (*cf.* N_{Bi_h}, Ki_q)	Total heat transfer/conductive heat transfer	Forced convection	*(5, 13, 25)*
N7	Nusselt number for mass transfer	Nu_m, N_{Nu_m}	$k_cL/D = N_{Sh}$	Intensity of mass flux at interface/specific flux by pure molecular diffusion in layer of thickness, L	Mass transfer	*(13, 20)*
N8	Nusselt film thickness group	ψ, N_T	$L_f(\rho^2g/\mu^2)^{1/3} \equiv (N_{Ga})^{1/3}$; ($L_f$ = film thickness)	$= (N_{Re})^{1/3}$ (gravitational force/viscous force)$^{1/3}$	Falling films	*(6)*
O1	Ocvirk number		$(F_b/\mu V_s)(a/R)^2(D/b)^2$; ($v_s$ = shaft surface velocity; R = shaft radius; D = shaft diam.) (*cf.* N_S)	Load force/viscous force	Lubrication	*(6)*
O2	Ohnesorge number	Z	$\mu/(\rho L\sigma)^{1/2} = (N_{We_1})^{1/2}/(N_{Re})$	Viscous force/(inertia force × surface tension force)$^{1/2}$	Atomization	*(5)*
P1	Péclet number (heat)	Pe, N_{Pe_h}	$LV\rho C_p/k = LV/\alpha = (N_{Re}.N_{Pr})$	Bulk heat transfer/conductive heat transfer	Forced convection	*(5, 13, 25)*
P2	Péclet number (mass)	N_{Pe_m}	$LV/D = (N_{Re}.N_{Sc})$	Bulk mass transfer/diffusive mass transfer	Mass transfer	*(5)*
P3	Pipeline parameter	ρn	$V_WV_o/2\ H_s'$; (V_W = velocity water-hammer wave, V_o = initial velocity, H_s' = static head × $g[L^2/\theta^2]$	Maximum pressure rise in water hammer/2 × static pressure	Water hammer	*(5)*
P4	Poiseuille number		$D^2(-dp/dL)/\mu V$ (D = pipe diam., dp/dL = pressure gradient)	= 32 for laminar flow in round pipe	Laminar fluid friction	*(5)*
P5	Pomerantsev number	P_o	$jL^2/k(t_m - t_o)$ (t_m, t_o = temp. of medium, initial temp. of body) (*cf.* Damköhler Group IV)		Heat transfer with heat sources in medium	*(6, 14, 60)*
P6	Posnov number	Pn	$\delta\Delta t/(\Delta n_m)$ (*cf.* Fe_2)		Combined heat and mass transfer	*(13, 14, 67)*
P7	Power number	N_P	$\pi/L^5\rho N^3$	Drag on (agitator impeller) or inertial force	Power consumption by agitators, fans, pumps, etc.	*(5, 32)*
P8	Prandtl number	N_{Pr}	$C_p\mu/k$ = Da IV/Da III × Da V	Momentum diffusivity/thermal diffusivity	Forced and free convection	*(5, 13, 25)*
P9	Prandtl number (mass transfer)	Pr_m	$\mu/\rho D = N_{Sc}$, v (used in Russian, German literature)	See Schmidt number		*(13)*
P10	Prandtl velocity ratio	u^+	$V/(\tau w/\rho)^{1/2}$ (V = local fluid velocity)	Inertial force/wall shear force$^{1/2}$	Turbulence studies	*(5)*
P11	Prandtl dimensionless distance	y^+	$L(\rho\tau w)^{1/2}/\mu$ (L = distance from wall, etc.)		Turbulence studies	
P12	Predvoditelev number	Pd	$\Gamma L^2/\alpha t_o = \left(\frac{dt^*}{\alpha(N_{Fo})}\right)_{max}$ where t_o = init. temp. of body, t^* = temp. of medium relative to its initial temp.	Rate of change of temp. of medium/rate of change of temp. of body	Heat transfer	*(13, 14, 60)*

TABLE B (Continued)

Serial No.	Name	Symbol	Definition	Significance	Field of Use	Reference
P13	Pressure number (1)	K_P	$P/\{g\sigma(\rho' - \rho'')\}^{1/2}$ (ρ', ρ'' = density of liquid gas)	Absolute pressure in system (pressure jump on interface)		*(14)*
P14	Pressure number (2)		$H/\frac{1}{2}/_2U_s^2$ (U_s = circumferential velocity) ≡ [diameter No.]$^{-2}$ × [Speed No.]$^{-2}$		Flow machines (turbines, pumps, etc.)	*(25)*
P15	Psychrometric ratio		h_c/Gs	Heat transfer by convection/heat transfer by mass transfer	Wet and dry bulb thermometry	*(5)*
R1	Radiation number	$\bar{K}_s$	$kE/\eta\sigma T^3 = (N_{We_1})$/(Hooke No.) × (Stefan No.)		Radiant transfer	*(25)*
R2	Ramzin number	Ra	$\dfrac{C_bP}{C_m(\Delta\Omega)} = \dfrac{\text{(Bulygin No.)}}{\text{(Kosovich No.)}}$		Molar mass transfer	*(40)*
R3	Ratio of specific heats	γ	C_p/C_v (specific heats at constant pressure, volume)		Compressible flow	*(5)*
R4	Rayleigh number (1)	N_{Ra_1}	$V(\rho L/\sigma)^{1/2} = N_{We_2}$ (*q.v.*)	See N_{We}	Breakup of liquid jets	*(5)*
R5	Rayleigh number (2)	R_2'	$L^3\rho^2g\beta c_p\Delta t/\mu k = L^3\rho g\beta\Delta t/\mu\alpha = (N_{Gr})\cdot(N_{Pr})$		Free convection	*(5, 25)*
R6	Rayleigh number (3)	Ra_3	$q^*L^5\rho^2g\beta C_p/\mu k^2x = (N_{Gr})(N_{Pr})(N_{Nu})(L/x)$; ($L$ = pipe diam.)		Combined free and forced convection in vertical tubes	*(6)*
R7	Recovery factor	N_{rf}	$C_p(t_{aw} - t_m)/V^2$; t_{aw} = attained adiabatic wall temp. t_m = temp. of moving medium. (*cf.* Eckert No.)	Actual temp. recovery/theoretical temp. recovery	Convective heat transfer in compressible flow	*(5)*
R8	Reech number		$= 1/(N_{Fr_1})$ *q.v.*		Wave and surface behavior	*(6)*
R9	Resistance coefficient (1)	C_f	$F_R/\frac{1}{2}\rho V^2L^2$ (*cf.* drag coeff., Newton number, Fanning factor)		Flow resistance	*(25)*
R10	Resistance coefficient (2)	ψ	$\Delta p. D_H/\frac{1}{2}\rho V^2L$ (Δp = pressure drop over length, L) (*cf.* **R9**)		Fluid friction in conduits	*(25)*
R11	Reynolds number	N_{Re}	$LV\rho/\mu$	Inertia force/viscous force	Dynamic similarity	*(5, 25)*
R12	Reynolds number (rotating)	Re	$L^2N\rho/\mu$; L = impeller diam.		Agitation	*(45)*
R13	Richardson number	N_{Ri}	$-(g/\rho)(d\rho/dL)/(dV/dL)w^2$ [L = height of liquid layer, $(dV/dL)_w$ = velocity gradient at wall]	Gravity force/inertial force	Stratified flow of multilayer systems	*(54)*
R14	Romankov number	R_o'	T_D/T_{PROD}	Dry bulb temperature (abs.)/product temperature (abs.)	Drying	*(6, 52)*
R15	Rossby number	N_{Ro}	$V/2\omega_eL \sin\Lambda$ (ω_e = angular velocity of earth's rotation [1/θ]; Λ = angle between axis of earth's rotation and direction of fluid motion [−])	Inertia force/Coriolis force	Effect of earth's rotation on flow in pipes	*(5)*
R16	Roughness factor		ϵ/L		Fluid friction	*(5)*
S1	Sarrau number		≡ mach number, *q.v.*		Compressible flow	*(6)*
S2	Schiller number (1)		$LV(\rho^2/\mu F_R)^{1/3}$		Flow around obstacles	*(34)*
S3	Schiller number (2)	Sch	$V\left[\dfrac{3}{4}\cdot\dfrac{\rho\gamma_m}{g\mu(\gamma_M - \gamma_m)}\right]^{1/3}$; V = velocity in fluidized bed; γ_m, γ_M = specific gravity of medium and material in bed		Fluidization	*(66)*
S4	Schmidt number	N_{Sc}	$\mu/\rho D$ (*cf.* N_{Pr_m})(= DaII/DaI DaV)	Kinetic viscosity/molecular diffusivity	Diffusion in flowing	*(5, 25)*
S5	Semenov number	Sm	k_c/K; K = reaction rate constant [L/θ]		Reaction kinetics	*(11)*
S6	Senftleben number	S_e	$NE_s^2[\alpha + 2/3(\rho_0^2/kT)]\cdot[1/4LM_g]$ Kronig number, *q.v.*		Convective heat transfer	*(6)*
S7	Sherwood number	N_{Sh}	$k_cL/D = Nu_m$ (also termed Taylor number)	Mass diffusivity/molecular diffusivity	Mass transfer	*(5)*
S8	Sommerfeld number (1)	N_{S_1}	$(\mu N/P_b)(D/a)^2$ (D = shaft diam., (*cf.*) Ocvirk number)	Viscous force/load force	Lubrication	*(5)*
S9	Sommerfeld number (2)	N_{S_2}	$(F_b/\mu V_s)(a/R)^2$ (V_s = veloc. of shaft surface; R = shaft radius) ($N_{S_2} = 4/\pi N_{S_1}$)	Viscous force/load force	Lubrication	*(6)*

TABLE B (Continued)

Serial No.	Name	Symbol	Definition	Significance	Field of Use	Reference
S10	Spalding function	Sp	$-\left(\frac{\partial\theta}{\partial u^+}\right)_{u^+=0}$; $\theta = (T - T_\infty) \times (T_w - T_\infty)$, T_w = wall temperature, T_∞ = free stream temp., u^+ = Prandtl velocity ratio	Dimensionless temp. gradient at wall	Convection	*(18, 33)*
S11	Specific speed	C_s	$N(V_f)^{1/2}/(gH')^{3/4}$ (H' = head of liquid produced by one stage) (*cf.* speed number)		Pumps and compressors	*(8, 32)*
S12	Speed number	σ	$(4\eta)^{1/2}(V_f)^{1/2}\ N/(2H)^{3/4}$ = (delivery number)$^{1/2}$ × pressure number)$^{-3/4}$ (*cf.* specific speed)		Flow machines	*(25)*
S13	Stanton number	N_{St}	$h/C_p\rho V = h/C_pG = (N_{Nu})/(N_{Re})(N_{Pr})$	Heat transferred/thermal capacity of fluid	Forced convection	*(5, 6, 13, 25)*
S14	Stefan number	$\overline{St}$	$\eta LT^3/k$		Heat radiation	*(25)*
S15	Stokes number	St	$\mu\theta_v/\rho L^2$ (θ_v = vibration time) ≡ $(N_{S_1})^{-1}(N_{Re})^{-1}$		Particle dynamics	*(6)*
S16	Strouhal number	N_{S_1}, Sr	fL/V (*cf.* N_{Th})		Vortex streets; unsteady-state flow	*(6, 25)*
S17	Suratman number	Su	$\rho L\sigma/\mu^2 = (N_{Re})^2/(N_{We_1}) = (Z)^{-2}$		Particle dynamics	*(6)*
S18	Surface elasticity number	N_{El}	$-\frac{\Gamma'}{Ds}L\frac{(\partial\sigma)}{(\partial\Gamma)}$; Γ' = surface concentration of surfactant in undisturbed state, Ds = surface diffusivity, L = film thickness		Convection cells	*(10)*
S19	Surface viscosity number	N_{Vi}	$\mu s/\mu L$; μs = surface viscosity, $[M/\theta]$, L = film thickness		Convection cells	*(10)*
T1	Taylor number (1)	N_{Ta_1}	$\omega_c(R_a)^{1/2}a^{3/2}\rho/\mu$; ($\omega_c$ = angular velocity of cylinder; R_a = mean radius of annulus)		Stability of flow pattern in annulus with rotating cylinder	*(5)*
T2	Taylor number (2)	N_{Ta_2}	$(2\ \omega L^2\rho/\mu)^2[\omega$ = rate of spin $(1/\theta)$; L = height of fluid layer]	α(Coriolis force/viscous force)^2	Effect of rotation on free convection	*(6)*
T3	Thiele modulus	m_T	$Q^{1/2}U^{1/2}L/k^{1/2}\xi^{1/2}$ = (DaII)$^{1/2}$		Diffusion in porous catalysts	*(5)*
T4	Thoma number	σ_T	$(H_a - H_s - H_v)/H$ (H = total head; H_a = atm. pressure head; H_s = suction head; H_v = vapor pressure head)	Net positive suction head/total head	Cavitation in pumps	*(5)*
T5	Thomson number	N_{Th}	$\theta V/L$; θ = characteristic time (*cf.* N_{S_1})		Fluid flow	*(6)*
T6	Thring radiation group		$\rho C_pV/e^*\eta T^3$ (*cf.* Boltzmann number)	Bulk heat transport/heat transport by radiation	Radiation	*(5)*
T7	Thring-Newby criterion	θ	$[(V_{m_1} + V_{m0})/V_{m0}](R/L)$; V_{m_0}, V_{u_1} = mass flow rates of nozzle fluid and surrounding fluid $[M/\theta]$; R = equivalent nozzle radius; L = furnace half width		Combustion of fuels	*(4)*
T8	Truncation number	r	$\mu\dot{\gamma}/P$ (*cf.* Hersey number)	Shear stress/normal stress	Viscous flow	*(6)*
V1	Valensi number	V	$\omega L^2\rho/\mu$; ω = circular oscillation frequency when $\mu = 0$ $[1/\theta]$		Oscillations of drops and bubbles	*(57)*
V2	Vedernikov number	V	$\zeta^*\xi^*\bar{V}/(V_w - \bar{V}) \equiv \zeta^*\xi^*(N_{Fr_2})$ (ζ^* = exponent of hydraulic radius in formula [–]; ζ^* = shape factor of channel section; V_w = absolute velocity of disturbance wave)	Generalized Froude number	Instability of open-channel flow	*(6, 61)*
V3	Velocity number	R_v	≡ Magnetic Reynolds number, *q.v.*			*(5, 6)*
W1	Weber number (1)	N_{We_1}	$V^2\rho L/\sigma = (N_{We_2})^2$	Inertia force/surface tension force	Bubble formation, etc.	*(2, 25)*
W2	Weber number (2)	N_{We_2}	$V(\rho L/\sigma)^{1/2} = (N_{We_1})^{1/2}$			
W3	Weber number (rotating)	W_e	$L^3N^2\rho/\sigma$; L = impeller diameter		Agitation	*(64)*
W4	Weissenberg number	N_{We}	ω_3V/ω_1L; $\omega_3 \doteq \int_0^\infty sG(s)ds, \omega_1 = \int_0^\infty G(s)ds$, G = relaxation modulus of linear viscoelasticity, s = recoverable elastic strain	Viscoelastic force/viscous force	Viscoelastic flow	*(63)*

REFERENCES

(1) Adrianov, V. N., Shorin, S. N., *AIAA J.* **1**, 1729 (1963).
(2) Ahlstrom, H. G., *J. Fluid Mech.* **15**, 205 (1963).
(3) Becker, H. A., Hottel, H. C., Williams, G. C., "Ninth Symposium (International) on Combustion," p. 7, Academic Press, New York, 1963.
(4) Beer, J. M., Chigier, N. A., Lee, K. B., *Ibid.*, p. 892.
(5) Boucher, D. F., Alves, G. E., *Chem. Eng. Progr.* **55** (9), 55 (1959).
(6) *Ibid.*, **59** (8), 75 (1963).
(7) British Standard 1991, "Recommendations for Letter Symbols, Signs and Abbreviations. Part 2. Chemical Engineering, Nuclear Science, and Applied Chemistry," British Standards Institution, London, 1961.
(8) Brown, G. G., *et al.*, "Unit Operations," Wiley, New York, 1950.
(9) Buckingham, E., *Phys. Rev.* **4**, 345 (1914).
(10) Berg, J. C., Acrivos, A., *Chem. Eng. Sci.* **20**, 737 (1965).
(11) Chukhanov, Z. F., *Intern. J. Heat Mass Transfer* **6**, 691 (1963).
(12) Dallavalle, J. M., "Micromeritics," 2nd ed., Pitman, New York, 1948.
(13) El'perin, I. T., *Inzh. Fiz. Zh. Akad. Nauk Belorussk. SSR* **4** (1), 131 (1963).
(14) El'perin, I. T., *Intern. J. Heat Mass Transfer* **5**, 349 (1962).
(15) Engel, F. V. A., *Z.V.D.I.* **107**, 671, 793 (1965).
(16) Faller, A. J., *J. Fluid Mech.* **15**, 560 (1963).
(17) Fedorov, B. I., *Inzh. Fiz. Zh. Akad Nauk Belorussk. SSR* **7** (1), 21 (1964).
(18) Gardner, G. O., Kestin, J., *Intern. J. Heat Mass Transfer* **6**, 289 (1963).
(19) Gel'perin, I. T., Aĭnshteĭn, V. G., Goĭkhman, I. D., *Inzh. Fiz. Zh. Akad. Nauk Belorussk. SSR* **7** (7), 15 (1964).
(20) Grassmann, P., *Chem. Ing.-Tech.* **31**, 148 (1959).
(21) Grassman, P., Lemaire, L. H., *Ibid.*, **30**, 450 (1958).
(22) Greene, D. F., Ph.D. Thesis, Columbia Univ., 1961 [*Dissertation Abstr.* **24** (8), 3248 (1964)].
(23) Gukhman, A. A., "Introduction to the Theory of Similarity," Academic Press, New York, 1965.
(24) Gutfinger, C., Tallmadge, J. A., *A.I.Ch.E. J.* **10**, 774 (1965).
(25) Hahnemann, H. W., "Die Umstellung auf das internationale Einheitensystem in Mechanik und Wärmetechnik," VDI-Verlag, Düsseldorf, 1959.
(26) Holt, M., "Dimensional Analysis" in "Handbook of Fluid Dynamics," V. L. Streeter, ed., McGraw-Hill, New York, 1961.
(27) Hottel, H. C., Sarofim, A. F., *Intern. J. Heat Mass Transfer* **8**, 1153 (1965).
(28) Hottel, H. C., Williams, G. C., Jensen, W. P., Tobey, A. C., Burrage, P. M. R., p. 923 in "Ninth Symposium (International) on Combustion," Academic Press, New York, 1963.
(29) Huntley, H. E., "Dimensional Analysis," MacDonald & Co., London, 1952.
(30) Johnson, S. P., "Survey of Flow Calculation Methods," p. 98, Preprinted Papers & Program, Aeronautic & Hydraulic Divisions, A.S.M.E. Summer Meeting, June 19–21, Univ. of Calif. and Stanford Univ., 1934.
(31) Kafarov, V. V., *Zh. Prikl. Khim.* **29**, 40 (1956).
(32) Kay, J. M., "An Introduction to Fluid Mechanics & Heat Transfer," Cambridge Univ. Press, 1957.
(33) Kestin, J., Persen, L. N., *Intern. J. Heat Mass Transfer* **5**, 143 (1962).
(34) Klinkenberg, A., Mooy, H. H., *Chem. Eng. Progr.* **44**, 17 (1948).
(35) Koide, K., Kubota, H., Shindo, M., *Chem. Eng.* (Japan), **28** (8), 657 (1964).
(36) Lўkov, A. V., Mikhaĭlov, Yu. A., "Theory of Energy & Mass Transfer," Prentice-Hall, Englewood Cliffs, N.J., 1961.
(37) McCabe, W. L., Smith, J. C., "Unit Operations of Chemical Engineering," McGraw-Hill, New York, 1956.
(38) Matsuhisa, S., Bird, R. B., *A.I.Ch.E. J.* **11**, 588 (1965).
(39) Mikhaĭlov, Yu. A., Bornikova, R. M., *Inzh. Fiz. Zh. Akad. Nauk Belorussk. SSR* **6** (10), 45 (1963).
(40) Mikhaĭlov, Yu. A., Romanina, I. V., *Ibid.*, **7** (1), 49 (1964).
(41) Miyauchi, T., Nakano, K., Obata, K., Kimura, S., *Chem. Eng.* (Japan) **26** (9), 999 (1962).
(42) Mkhitaryan, A. M., "Hydraulics & Fundamentals of Gas Dynamics," Israel Program for Scientific Translations, Jerusalem, 1964.
(43) Mordell, D. L., Wu, J. H. T., *Can. Aeronaut. Space J.* **9** (4), 117 (1963).
(44) Motulevich, V. P., Eroshenko, V. M., Petrov, Yu. P., in "Physics of Heat Exchange & Gas Dynamics," A. S. Predvoditelev, ed., Consultants Bureau, New York, 1963.
(45) Nagata, S., *Chem. Eng.* (Japan) **27** (8), 592 (1962).
(46) Nield, D. A., *J. Fluid Mech.* **19**, 341 (1964).
(47) Potter, J. M. F., B.Sc. Thesis, Dept. of Chem. Engrg., Univ. of Birmingham, England, 1959.
(48) Rayleigh, Lord, *Phil. Mag.* **48**, 321 (1899).
(49) Reiner, M., *Phys. Today* **17** (1), 62 (1964).
(50) Rouse, H. (ed.), "Engineering Hydraulics," Wiley, New York, 1950.
(51) Rouse, H., Ince, S., "History of Hydraulics," Iowa Institute of Hydraulic Research, State University of Iowa, 1957.
(52) Sazhin, B. S., *Inzh. Fiz. Zh. Akad. Nauk Belorussk. SSR* **5** (6), 13 (1962).
(53) Sazhin, B. S., Miklin, Yu. A., *Ibid.*, **6** (10), 57 (1963).
(54) Schlichting, H., "Boundary Layer Theory," 4th ed., McGraw-Hill, New York, 1960.
(55) Scriven, L. E., Sternling, C. V., *J. Fluid Mech.* **19**, 321 (1964).
(56) Sillem, H., *Z.V.D.I.* **106**, 398 (1964).
(57) Szebehely, V. G., p. 771 in "Proc. 2nd U. S. Nat. Congress of Appl. Mech.," Ann Arbor, Mich., June 1954; A.S.M.E., New York, 1955.
(58) Tallmadge, J. A., Labine, R. A., Wood, B. H., IND. ENG. CHEM. FUNDAMENTALS **4**, 400 (1965).
(59) Tamarin, A. I., *Inzh. Fiz. Zh. Akad. Nauk Belorussk. SSR* **6** (7), 19 (1963).
(60) Tartakovskii, D. F., *Ibid.*, **7** (1), 71 (1964).
(61) Vedernikov, V. V., *Compt. Rend. Acad. Sci. U.R.S.S.* **48**, 239 (1945); 52, 207 (1946).
(62) Weber, M., *Jahrb. Schiffbautechn. Ges.* **20**, 355 (1919).
(63) White, J. L., *J. Appl. Polymer. Sci.* **8**, 2339 (1964).
(64) Yamaguchi, I., Yabuta, S., Nagata, S., *Chem. Eng.* (Japan) **27** (8), 576 (1963).
(65) Yas'ko, O. I., *Inzh.-Fiz. Zh. Akad. Nauk Belorussk. SSR* **7** (12), 112 (1964).
(66) Zabrodskiĭ, S. S., "Flow & Heat Transfer in Fluidized Beds," to be published shortly by MIT Press.
(67) Zhuravleva, V. P., *Inzh.-Fiz. Zh. Akad. Nauk Belorussk. SSR* **6** (9), 73 (1963).

NUMERICAL DATA PROJECTS

SOURCES OF CRITICAL DATA

The following is abstracted from *Continuing Numerical Data Projects, A Survey and Analysis*. This publication was prepared under the auspices of the Office of Critical Tables, National Academy of Sciences and National Research Council. Persons requesting further information and details regarding the projects listed may obtain the publication from Printing and Publishing Office, National Academy of Sciences, 2101 Constitution Avenue, Washington, D. C., 20418. Price $5.00.

THERMODYNAMIC PROJECTS

Selected Values of Chemical Thermodynamic Properties, NBS Circular 500, F. D. Rossini, D. D. Wagman, W. H. Evans, S. Levine, and I. Jaffe, Government Printing Office, Washington, D.C. 20402, 1952, iv + 1268 pp, $7.25 (out of print). Paperback reprint issued in two parts: I. Tables; and II. References; Government Printing Office, Washington, D.C. 20402, 1961 (out of print). Revisions of C500 are being issued provisionally as NBS Technical Notes, listed below.

Selected Values of Chemical Thermodynamic Properties

Part 1: Tables for the First Twenty-three Elements in the Standard Order of Arrangement, NBS Technical Note 270-1, D. D. Wagman, W. H. Evans, I. Halow, V. B. Parker, S. M. Bailey, and R. H. Schumm, U.S. Government Printing Office, Washington, D.C. 20402, October 1965, iv + 124 pp, $0.65.

Part 2: Tables for the Elements Twenty-four through Thirty-two in the Standard Order of Arrangement, NBS Technical Note 270–2, D. D. Wagman, W. H. Evans, I. Halow, V. B. Parker, S. M. Bailey, and R. H. Schumm, U.S. Government Printing Office, Washington, D.C. 20402, May 1966, iv + 62 pp, $0.40.

Selected Values of Properties of Hydrocarbons and Related Compounds, American Petroleum Institute Research Project 44, J. B. Zwolinski, dir., Texas A&M Research Foundation, College Station, Texas. As of June 1966 there were 2391 valid data sheets, complete set 6 volumes at $0.30 per sheet.

Selected Values of Properties of Chemical Compounds, Thermodynamics Research Center, B. J. Zwolinski, dir., Texas A&M Research Foundation, College Station, Texas. As of June 1966 there were 754 valid data sheets, complete set 2 volumes at $0.30 per sheet.

JANAF (Joint Army-Navy-Air Force) Thermochemical Tables, PB 168 370, D. R. Stull, dir., Clearinghouse for Federal Scientific and Technical Information, Springfield, Virg., 1965, 945 pp, $10.00.

First Addendum, PB 168 370–1, D. R. Stull, dir., Clearinghouse for Federal Scientific and Technical Information, Springfield, Virg., 1966, vii + 197 pp, $4.00 (Supplements 18, 19, 20, and 21).

Information on further supplements is available from Dr. D. R. Stull, dir., Thermal Research Laboratory, Dow Chemical Company, Midland, Michigan.

Contributions to the Data on Theoretical Metallurgy, I–XV, published as Bureau of Mines bulletins between 1932 and 1962. Many bulletins were superceded; six are revisions and Bulletin 601 is a reprinting of four previous bulletins and out of print for some years, but still in demand as essential in present-day research. The ones currently available are listed below.

Contributions to the Data on Theoretical Metallurgy

XIII High-Temperature Heat-Content, Heat Capacity, and Entropy Data for the Elements and Inorganic Compounds, K. K. Kelley, Bulletin 584 (Bureau of Mines), U.S. Government Printing Office, Washington, D.C. 20402, 1960, 232 pp, $1.25.

XIV Entropies of Elements and Inorganic Compounds, K. K. Kelley and E. G. King, Bulletin 592 (Bureau of Mines), U.S. Government Printing Office, Washington, D.C. 20402, 1961, 149 pp, $0.75.

XV A Reprint of Bulletins 383, 384, 393, and 406 (III, IV, V, and VII), Bulletin 601, Bureau of Mines (Publications Distribution Section, 4800 Forbes Avenue), Pittsburgh, Penna., 1962, 525 pp, single copies free.

383: The Free Energies of Vaporization and Vapor Pressures of Inorganic Substances, K. K. Kelley, 1935.

384: Metal Carbonates, Correlation and Applications of Thermodynamic Data, K. K. Kelley and C. T. Anderson, 1935.

393: Heats of Fusion of Inorganic Substances, K. K. Kelley, 1936.

406: The Thermodynamic Properties of Sulphur and Its Inorganic Compounds, K. K. Kelley, 1937.

Thermodynamic Properties of Chemical Substances, V. P. Glushko et al, eds, Vol. I, 1164 pp, and Vol. II, 916 pp, USSR Academy of Sciences, Moscow, 1962, approx $16.00.

Thermodynamic Constants of Substances, Handbook in 10 parts, V. P. Glushko, ed, USSR Academy of Sciences, All-Union Institute of Scientific and Technological Information, Moscow: Part I, 1965, 146 pp, 72 kopecks ($1.75); Part II, 1966, 96 pp, 52 kopecks ($1.40). Available in the United States from Victor Kamkin, Inc., Bookstore, 1410 Columbia Rd. N.W., Washington, D.C. 20009.

Chemical Thermodynamics in Non-Ferrous Metallurgy, J. I. Gerassimov, A. N. Krestovnikov, and A. S. Shakhov, Metallurgical Publishing House, Moscow.

Vol. I: Theoretical Introduction, Thermodynamic Properties of Important Gases, Thermodynamics of Zinc and Its Important Compounds, 1960, 231 pp.

THERMODYNAMIC PROJECTS (Continued)

Vol. II: Thermodynamics of Copper, Lead, Tin, Silver, and Their Important Compounds, 1960, 231 pp.

Vol. III: Thermodynamics of Tungsten, Molybdenum, Titanium, Zirconium, Niobium, Tantalum, and Their Important Compounds, 1963, 283 pp. English translation available as NASA-TT-F-285 or CFSTI-TT-65-50111, $6.00, from the Clearinghouse for Federal Scientific and Technical Information, Springfield, Virg. 22151.

Vol. IV: Thermodynamics of Aluminum, Antimony, Magnesium, Nickel, Bismuth, Cadmium, and Their Important Compounds, 1966, 428 pp.

Selected Values for the Thermodynamic Properties of Metals and Alloys, Ralph Hultgren, Raymond L. Orr, Philip D. Anderson, and Kenneth K. Kelley, John Wiley & Sons, Inc., New York, N.Y. 10016, 1963, xi + 963 pp, $12.50.

Thermochemistry for Steelmaking, Vol. I, J. F. Elliott and M. Gleiser, Addison-Wesley Publishing Co., Reading, Mass. 01867, 1960, viii + 296 pp, $17.50.

Thermochemistry for Steelmaking, Vol. II: Thermodynamics and Transport Properties, J. F. Elliott, M. Gleiser, and V. Ramakrishna, Addison-Wesley Publishing Co., Reading, Mass. 01867, 1963, xvi + 550 pp (pp 297–846), $25.00.

Thermodynamic Functions of Gases, F. Din, ed, Butterworth & Co. (Publishers) Ltd., 88 Kingsway, London, W.C.2, England, $12.50 per volume.

Vol. 1: Ammonia, Carbon Dioxide, Carbon Monoxide, 1956, viii + 175 pp.

Vol. 2: Air, Acetylene, Ethylene, Propane and Argon, 1956, vi + 201 pp.

Vol. 3: Methane, Nitrogen, Ethane, 1961, vi + 218 pp.

THERMOPHYSICAL PROJECTS

A Compendium of the Properties of Materials at Low Temperatures, V. J. Johnson, ed. Available from Clearinghouse for Federal Scientific and Technical Information, Springfield, Virg. 22151.

Phase I, Part I, Properties of Fluids, July 1960, 489 pp, WADD Technical Report 60-56, Part I (PB-171-618), $6.00.

Phase I, Part II, Properties of Solids, Oct. 1960, 330 pp, WADD Technical Report 60-56, Part II (PB-171-619), $4.00.

Phase I, Part III, Bibliography of References (cross-indexed), Oct. 1960, 161 pp, WADD Technical Report 60-56, Part III (PB-171-620), $3.00.

Phase II, R. B. Stewart and V. J. Johnson, eds, Dec. 1961, 501 pp, WADD Technical Report 60-56, Part IV (AD-272-769), $8.10.

Thermophysical Properties Research Center Data Book. Issued as loose-leaf sheets; sold on subscription basis at $0.10 per sheet, Thermophysical Properties Research Center, Purdue University, Research Park, 2595 Yeager Rd., West Lafayette, Ind. 47906.

Vol. I: Metallic Elements and Their Alloys.

Vol. II: Nonmetallic Elements, Compounds and Mixtures (In Liquid and Gaseous States at Normal Temperature and Pressure).

Vol. III: Nonmetallic Elements, Compounds and Mixtures (In the Solid State at Normal Temperature and Pressure).

Handbook of Thermophysical Properties of Solid Materials, A. Goldsmith, T. E. Waterman, and H. J. Hirschorn, revised edition, 1961 (8½″ × 11¼″), The Macmillan Co., 60 Fifth Ave., New York, N.Y. 10011, $90.00 per set.

Vol. I: Elements, vi + 752 pp.

Vol. II: Alloys, vi + 1,270 pp.

Vol. III: Ceramics, vi + 1,162 pp.

Vol. IV: Cermets, Intermetallics, Polymerics and Composites, vi + 798 pp.

Vol. V: Appendix (includes materials, author indexes, and a list of references), 286 pp.

PHYSICOCHEMICAL PROJECTS

Tables of Chemical Kinetics, Homogeneous Reactions, NBS Circular 510, U.S. Government Printing Office, Washington, D.C. 20402, 1951, xxiv + 732 pp (out of print).

Tables of Chemical Kinetics, Homogeneous Reactions, NBS Circular 510, Supplement 1, U.S. Government Printing Office, Washington, D.C. 20402, 1956, xiv + 422 pp (out of print).

Alphabetical Index to Tables of Chemical Kinetics, Homogeneous Reactions, NBS Circular 510, Supplement 2, U.S. Government Printing Office, Washington, D.C. 20402, 1960, iv + 37 pp, $0.35.

Tables of Chemical Kinetics, Homogeneous Reactions, Supplementary Tables, To Accompany Circular 510 and Supplements 1 and 2, NBS Monograph 34, U.S. Government Printing Office, Washington, D.C. 20402, 1961, $2.75.

PHYSICOCHEMICAL PROJECTS (Continued)

Tables of Chemical Kinetics, Homogeneous Reactions, Supplementary Tables, NBS Monograph 34, Vol. 2, U.S. Government Printing Office, Washington, D.C. 20402, 1964, $2.00.

Phase Diagrams for Ceramists, E. M. Levin, H. F. McMurdie, and F. P. Hall, The American Ceramic Society, 4055 N. High St., Columbus, Ohio 43214, 1956, 286 pp (811 phase diagrams).

Phase Diagrams for Ceramists, Part II, E. M. Levin and H. F. McMurdie, The American Ceramic Society, 4055 N. High St., Columbus, Ohio 43214, 1959, 153 pp (462 phase diagrams).

Phase Diagrams for Ceramists, E. M. Levin, C. R. Robbins, and H. F. McMurdie (7th compilation), The American Ceramic Society, 4055 N. High St., Columbus, Ohio 43214, 1964, 601 pp (2064 phase diagrams), $18.00 (discount to members).

Phase Equilibrium Diagrams of Oxide Systems, revised and redrawn by E. F. Osborn and A. Muan, The American Ceramic Society, 4055 N. High St., Columbus, Ohio 43214, (Ten 19″ × 23″ plates for three-oxide systems containing SiO_2, four of them for oxide phases in equilibrium with metallic iron. Reproductions of these plates appear in the 1964 compilation).

Constitution of Binary Alloys, M. Hansen and K. Anderko, McGraw-Hill Book Company, New York, N.Y. 10036, 1958, xix + 1305 pp, $39.50.

Constitution of Binary Alloys, First Supplement, R. P. Elliott, McGraw-Hill Book Company, New York, N.Y. 10036, 1965, xxxii + 877 pp, $35.00.

Stability Constants of Metal-Ion Complexes, Section I: Inorganic Ligands, compiled by Lars Gunnar Sillen; Section II: Organic Ligands, compiled by Arthur E. Martell; Special Publication No. 17, The Chemical Society, London, 1964, xviii + 754 pp, $23.00.

Solubilities of Inorganic and Metal Organic Compounds, A compilation of Solubility Data from the Periodical Literature, A. Seidell, 4th ed., W. F. Linke, American Chemical Society, 1155 16th St. N.W., Washington, D.C. 20036.

Vol. I, 1958, iv + 1287 pp, $32.50.
Vol. II, 1965, iv + 1914 pp, $32.50.

INDEXES

Consolidated Index of Selected Property Values: Physical Chemistry and Thermodynamics, Prepared by the Office of Critical Tables. A key to the contents of six compilations that present critically evaluated numerical property values. Publication 976, National Academy of Sciences—National Research Council Printing and Publishing Office, 2101 Constitution Ave., N.W., Washington, D.C. 20418, 1962, xxiii + 274 pp, $6.00.

CRYSTALLOGRAPHIC PROJECTS

Crystal Data, Classification of Substances by Space Groups and their Identification from Cell Dimensions, J. D. H. Donnay and Werner Nowacki, Memoir 60 of the Geological Society of America, 1954, ix + 719 pp, Geological Society of America, 419 W. 117 St., New York, N.Y. 10027 (out of print).

Crystal Data, Determinative Tables, second edition, J. D. H. Donnay, General Editor; Gabrielle Donnay, Assistant Editor; E. G. Cox, Inorganic Compounds; Olga Kennard, Organic Compounds; Murray Vernon King, Proteins; Monograph 5 of the American Crystallographic Association, 1963, x + 1,302 pp, $20.00. Polycrystal Book Service, P.O. Box 11567, Pittsburgh, Pa. 15238.

Crystal Structures, first edition, Ralph W. C. Wyckoff, Interscience Publishers, John Wiley & Sons, New York.

Section I: Chapters I to VIII, 1948, 378 pp, $13.50.
Section II: Chapters VIII–X, XIII, 1951, 509 pp, $17.00.
Section III: Chapters XIV, XV, Organic Index, 1953, 465 pp, $25.00.
Section IV: Chapters XI and XII, 1957, 261 pp, $8.00.
Supplement I: Additions to Chapters II to VII, 1951, 143 pp, $5.00.
Supplement II: Additions to Chapter XIII, 1953, 85 pp, $8.50.
Supplement III: Additions to Chapters II to VIII, 1958, 311 pp, $20.00.
Supplement IV: Additions to Chapters IX, X, XIII–XV, 1959, 509 pp, $22.00.
Supplement V: Additions to Chapters II to XV, Indexes, 1960, 513 pp, $26.50.

Crystal Structures, 2nd ed., Ralph W. C. Wyckoff, Interscience Publishers, John Wiley & Sons, New York.

Vol. I: Chapters I–IV, 1963, 467 pp, $17.50.
Vol. II: Chapters V–VII, 1964, 588 pp, $24.00.
Vol. III: Chapters VIII–X, 1965, 981 pp, $27.50.

CRYSTALLOGRAPHIC PROJECTS (Continued)

Powder Diffraction Data File (PD-1), 16 Sets, American Society for Testing and Materials, 1916 Race St., Philadelphia, Pa. 19103.

Sets	Indexes	Cards		
		Plain ($)	Keysort ($)	IBM ($)
1–16	Organic	715	1,125	300
	Inorganic	1,280	1,855	
16 only	Organic	60	95	25
	Inorganic	140	220	
Book Form				
1–5	Organic	25		
	Inorganic	45		
	Both	60		

Structure Reports, Vols. 8–16 and Vol. 18, A. J. C. Wilson, General Editor; Vol. 17 and Vols. 19–23, W. B. Pearson, General Editor; published for the International Union of Crystallography by N. V. A. Oosthoek's Uitgevers, Mij, Utrecht, the Netherlands. Agent in the U.S.—Polycrystal Book Service, Box 11567, Pittsburgh, Pa. 15238.

The years covered, year published, and price of each volume are as follows:

Volume	Years Covered	Year Published	Price	
			Dutch florin	$
8	1940–1941	1956	80	22.50
9	1942–1944	1955	70	19.50
10	1945–1946	1953	55	15.50
11	1947–1948	1951	100	28.00
12	1949	1952	70	19.50
13	1950	1954	100	28.00
14[a]	1940–1950	1959	35	10.00
15	1951	1957	110	31.00
16	1952	1959	120	33.50
17	1953	1963	125	35.00
18	1954	1961	120	33.50
19	1955	1963	100	28.00
20	1956	1963	100	28.00
21	1957	1964	100	28.00
22	1958	In preparation	...	.....
23	1959	1965	120	33.50

[a]Index.

A Handbook of Lattice Spacings and Structures of Metals and Alloys, No. 4 of Pergamon Press series of monographs on Metal Physics and Physical Metallurgy. W. B. Pearson, Pergamon Press, Ltd., Headington Hill Hall, Oxford, England, also London, Edinburgh, New York, Paris, and Frankfurt, 1958, x + 1044 pp, $38.00.

International Tables for X-Ray Crystallography, Kathleen Lonsdale, general editor, Published for the International Union of Crystallography by the Kynoch Press, Witton, Birmingham 6, England. Also available from Polycrystal Book Service, P.O. Box 11567, Pittsburgh, Pa. 15238.

Vol. I: Symmetry Groups, Norman F. M. Henry and Kathleen Lonsdale, eds, 1952, xii + 558 pp, £5.5.0. ($14.70).

Vol. II: Mathematical Tables, John S. Kasper and Kathleen Lonsdale, eds, 1959, xviii + 444 pp, £5.15.0. ($16.50).

Vol. III: Physical and Chemical Tables, Caroline H. MacGillavry and Gerard D. Rieck, eds, 1962, xvi + 362 pp, £5.15.0. ($16.10).

The Barker Index of Crystals, A Method for the Identification of Crystalline Substances, Published for the Barker Index Committee by W. Heffer & Sons, Ltd., Cambridge, England.

Vol. I, Crystals of the Tetragonal, Hexagonal, Trigonal and Orthorhombic Systems, M. W. Porter and R. C. Spiller, 1951.
Part 1: Introduction and Tables, ix + 350 pp, £1.10.0. net ($4.20).
Part 2: Crystal Descriptions, x + 1,086 pp, £4.10.0. net ($12.60).
The two parts—£6.0.0. net ($16.80).

Vol. II, Crystals of the Monoclinic System, M. W. Porter and R. C. Spiller, 1956.
Part 1: Introduction and Tables, v + 383 pp.
Part 2: Crystal Descriptions, M.1 to M.1800, viii + 760 pp.
Part 3: Crystal Descriptions, M.1801 to M.3572, viii + 688 pp.
The three parts—£10.0.0. net ($28.00).

CRYSTALLOGRAPHIC PROJECTS (Continued)

Vol. III, Crystals of the Anorthic System, M. W. Porter and L. W. Codd, 1964.

Part 1: Introduction and Tables, vi + 50 pp., and unpaginated tables.

Part 2: Crystal Descriptions, A.1 to A.831 and Atlas of Configurations, vii pp + unpaginated descriptions.

The two parts—£12.0.0. net ($33.60).

MINERALOGICAL PROJECTS

The System of Mineralogy of James Dwight Dana and Edward Salibury Dana, Yale University, 1837–1892, 7th edition, John Wiley & Sons, Inc., 605 Third Ave., New York, N.Y. 10016; and Glen House, Stag Place, London, S.W.1., England.

Vol. I: Elements, Sulfides, Sulfosalts, Oxides; Charles Palache, Harry Berman, and Clifford Frondel, 1944, xiii + 834 pp, $14.00.

Vol. II: Halides, Nitrates, Borates, Carbonates, Sulfates, Phosphates, Arsenates, Tungstates, Molybdates, etc.; C. Palache, H. Berman, and C. Frondel, 1951 xi + 1124 pp, $16.00

Vol. III: Silica Minerals, C. Frondel, 1962, xii + 334 pp, $7.95.

Rock-Forming Minerals, W. A. Deer, J. Zussman, and R. A. Howie, Longmans, Green and Co., Ltd., 48 Grosvenor St., London, W.1, England. Rights in USA, Philippines, and Central America—John Wiley & Sons, Inc., 605 Third Ave., New York, N.Y. 10016.

Vol. I: 1962, 333 pp, $16.50.

Vol. II: 1963, 379 pp, $19.50.

Vol. III: 1962, 270 pp, $16.50.

Vol. IV: 1963, 435 pp, $16.50.

Vol. V: 1962, 371 pp, $16.50.

INDEXES

Indexes to Crystallographic Data Compilations. The American Society for Testing and Materials publishes the following indexes to the Powder Diffraction File (see p. 278). All ASTM publications are available from the American Society for Testing and Materials, 1916 Race St., Philadelphia, Pa. 19103.

a. Index to the Powder Diffraction File; Organic (Hanawalt), includes Section 1–16, PDIS–15o, bound, $15.00.

b. Index to the Powder Diffraction File; Inorganic (Hanawalt), includes Section 1–15, PDIS–16i, bound, $20.00.

c. Fink Inorganic Index to the Powder Diffraction File, PDIS–16f, bound, $15.00.

d. Matthews Coordinate Index to the Powder Diffraction File, Inorganic: Termatrex Cards, $675.00; Negative-Line Cards, $185.00.

GENERAL NUCLEAR PROPERTIES PROJECTS

Nuclear Data Sheets, Jan. 1958—. Vols. 5, and 6, 1958, 1959, 1960, 1961, published by the National Academy of Sciences—National Research Council, 2101 Constitution Avenue, Washington, D.C. 20418. Section B of the new journal "Nuclear Data" will be devoted to Nuclear Data Sheets. Vol. 1, No. 1, containing sheets for A-chains from A = 182 through A = 185, was issued in Feb. 1966.

1959 Nuclear Data Tables, U.S. Government Printing Office, Washington, D.C. 20402, April 1959, viii + 151 pp, $1.00.

1960 Nuclear Data Tables, U.S. Government Printing Office, Washington, D.C. 20402.

Part 1: Consistent Set of Q Values, $A \leq 66$, 1961, 214 pp, $1.50.

Part 2: Consistent Set of Q Values, $67 \leq A \leq 199$, 1961, 456 pp, $2.75.

Part 3: Nuclear Reaction Graphs, 1960 (out of print).

Part 4: Short Tables, 249 pp, 1961, $1.50.

Energy Levels of Light Nuclei, May 1962, T. Lauritsen and F. Ajzenberg, issued as Sets 5 and 6 of the 1961 Nuclear Data Sheets, essentially an addendum to a review article in **Nucl. Phys., 11,** 1–340 (1959).

Members of the Nuclear Data Project contributed the section **Energy Levels of Nuclei,** A = 21 to A = 212 of Group I, Vol. 1 of the New Series of Landolt-Börnstein, **Numerical Data and Functional Relationships in Science and Technology.**

Table of Isotopes, J. J. Livingood and G. T. Seaborg, **Rev. Mod. Phys., 12,** 30–47 (1940).

Table of Isotopes, G. T. Seaborg and I. Perlman, **Rev. Mod. Phys., 16,** 1–32 (1944).

GENERAL NUCLEAR PROPERTIES PROJECTS (Continued)

Table of Isotopes, G. T. Seaborg and I. Perlman, **Rev. Mod. Phys., 20,** 585–666 (1948).

Table of Isotopes, J. M. Hollander, I. Perlman, and G. T. Seaborg, **Rev. Mod. Phys., 25,** 469–650 (1953).

Table of Isotopes, D. Strominger, J. M. Hollander, and G. T. Seaborg, **Rev. Mod. Phys., 30,** No. 2, Part 2, 585–904 (1958). Also available from the American Institute of Physics, 335 E. 45th St., New York, N.Y. 10017.

Tabellen der Atomkerne (Nuclear Tables), W. Kunz and J. Schintlmeister. Teil (Part) I, Eigenschaften der Atomkerne (Nuclear Properties); Band (Volume) 1, Die Elemente Neutron bis Zinn (The Elements Neutron to Tin), 1958, xliv + 465 pp, Akademie-Verlag, G.m.b.H., Berlin W.8, Mohrenstrasse 39, Germany, price, 105 DM ($26.25). Teil I, Band 2, Eigenschaften der Atomkerne, Die Elemente Antimon bis Nobelium (Antimony to Nobelium), 1959, xliv + 641 pp (466–1107), price, 130 DM ($32.50). Akademie-Verlag, Berlin. (Also available from Pergamon Press, Headington Hill Hall, Oxford, England; 4/5 Fitzroy Square, London, W.1, England; 122 E. 55th St., New York, N.Y. 10022, £20.0.0 per set.)

Part II, Nuclear Reactions, Vol. 1, The Elements from Neutron to Magnesium, Text Volume, xliv + 700 pp. Table Volume, 36 tables (five of which require 2 or 3 sheets), 1965. Price £25.0.0. ($75.00) for the two volumes.

CROSS-SECTION PROJECTS

Neutron Cross Sections, BNL (Brookhaven National Laboratory) 325, 2nd ed, U.S. Government Printing Office, Washington, D.C. 20402, July 1958, v + 373 pp, $4.50.

Supplement 1, U.S. Government Printing Office, Washington, D.C. 20402, January 1960, iv + 129 pp, $2.00.

Supplement 2, Vol. I, Z = 1 to 20, U.S. Government Printing Office, Washington, D.C. 20402, May 1964, Sectional pagination, $2.50.

Supplement 2, Vol. IIA, Z = 21 to 40, Clearinghouse for Federal Scientific and Technical Information, Springfield, Virg. 22151, February 1966, sectional pagination, $4.00.

Supplement 2, Vol. III, Z = 88 to 98, Clearinghouse for Federal Scientific and Technical Information, Springfield, Virg. 22151, February 1965, sectional pagination, $3.00.

Neutron Cross Sections—Angular Distributions, BNL 400, Clearinghouse for Federal Scientific and Technical Information, Virg. 22151, June 1956, 102 pp.
2nd ed, Angular Distributions in Neutron-Induced Reactions, Vol. I, Z = 1 to 22; Vol. II, Z = 23 to 94, Clearinghouse for Federal Scientific and Technical Information, Springfield, Virg. 22151, October 1962, 804 pp, sectional pagination, $8.50.

Part I, **Tabulated Neutron Cross Sections,** 0.001–14.5 MeV., R. J. Howerton, Vol. I, ${}_1$H–${}_{22}$Ti; Vol. II, ${}_{23}$V–${}_{50}$Sn; Vol. III, ${}_{51}$Sb–${}_{95}$Am; UCRL-5226, first edition, May 1958. UCRL-5226, revised, 1959; Vol. I, $5.00, Vol. II, $5.00, Vol. III, $4.00, Clearinghouse for Federal Scientific and Technical Information, Springfield, Virg. 22151.

Part II, **Semi-Empirical Neutron Cross Sections,** 0.5–15 MeV., ${}_1$H–${}_{94}$Pu, R. J. Howerton, Vol. I, Nov. 1958, UCRL-5351. (Photostat copy, $37.80, Microfilm copy, $11.10, available from the Library of Congress.)

Part III, **Tabulated Differential Neutron Cross Sections,** ${}_1$H–${}_{94}$Pu, R. J. Howerton, Vol. I, Jan. 1961, UCRL-5573, Clearinghouse for Federal Scientific and Technical Information, Springfield, Virg. 22151, $5.00.

Charged Particle Cross Sections, LA-2014, Nelson Jarmie and John D. Seagrave, eds, Clearinghouse for Federal Scientific and Technical Information, Springfield, Virg. 22151, Feb. 1957, 234 pp, $1.25.

Charged Particle Cross Sections, LA-2424, Neon to Chromium, Darryl B. Smith, Compiler and ed, Nelson Jarmie and John D. Seagrave, associate eds, Clearinghouse for Scientific and Technical Information, Springfield, Virg. 22151, Jan. 1961, iii + 137 pp, $2.50.

Nuclear Cross Sections for Charged-Particle Induced Reactions, ORNL-CPX-1, Mn, Fe, Co, F. K. McGowan, W. T. Milner, and H. J. Kim, compilers, Oak Ridge National Laboratory, P.O. Box X, Oak Ridge, Tenn. 37831, July 1964, 443 pp.

Nuclear Cross Sections for Charged-Particle Induced Reactions, ORNL-CPX-2, Ni, Cu, F. K. McGowan, W. T. Milner, and H. J. Kim, compilers, Oak Ridge National Laboratory, P.O. Box X, Oak Ridge, Tenn. 37831, Sept. 1964, iii + 511 pp.

ENERGY LEVEL PROJECTS

Energy Levels of Light Nuclei

III: W. F. Hornyak, Thomas Lauritsen, P. Morrison, and H. A. Fowler, **Rev. Mod. Phys., 22,** 291–372, (1950).

IV: F. Ajzenberg and T. Lauritsen, **Rev. Mod. Phys., 24,** 321–402 (1952).

ENERGY LEVEL PROJECTS (Continued)

V: F. Ajzenberg and T. Lauritsen, **Rev. Mod. Phys. 27,** 77–166 (1955).

VI: F. Ajzenberg-Selove and T. Lauritsen, **Nucl. Phys., 11,** 1–340 (1959). This report may be obtained for $10.00 from North Holland Publishing Company, P.O. Box 103, Amsterdam, the Netherlands.

Energy Levels of Light Nuclei, Z = 11 to Z = 20

I: P. M. Endt and J. C. Kluyver, **Rev. Mod. Phys., 26,** 95–166, (1954).

II: P. M. Endt and C. M. Braams, **Rev. Mod. Phys., 29,** 683–756 (1957).

III: P. M. Endt and C. van der Leun, **Nucl. Phys., 34,** 1–324 (1962). Reprints of this paper are obtainable for $10.00 from North Holland Publishing Company, P.O. Box 103, Amsterdam, the Netherlands.

Decay Schemes of Radioactive Nuclei, B. S. Dzhelepov and L. K. Peker, 1958, viii + 787 pp, USSR Academy of Sciences Press, Moscow and Leningrad. English translation of above, 1961, vi + 786 pp, Pergamon Press, New York, Oxford, London, Paris.

Decay Schemes of Radioactive Nuclei, A ≥ 100, B. S. Dzhelepov, L. K. Peker, and V. O. Sergejev, 1963, 1060 pp, USSR Academy of Sciences Press, Moscow and Leningrad.

ATOMIC SPECTRA

Atomic Energy Levels As Derived from the Analyses of Optical Spectra, Charlotte E. Moore, NBS Circular 467, U.S. Government Printing Office, Washington, D.C. 20402.

Vol. I: Hydrogen to Vanadium (Z = 1–23), 1949, 309 pp, $5.50.

Vol. II: Chromium to Niobium (Z = 24–41), 1952, 227 pp, $4.00.

Vol. III: Molybdenum to Lanthanum (Z = 42–57) and Hafnium to Actinium (Z = 72–89), 1958, 245 pp, $3.00.

Vol. IV: Lanthanide and Actinide groups; in preparation.

An Ultraviolet Multiplet Table, Charlotte E. Moore, NBS Circular 488, U.S. Government Printing Office, Washington, D.C. 20402.

Section 1, 1950, 85 pp, reprinted 1956.

Section 2, 1952, 120 pp, reprinted 1956.
Combined Sections 1 and 2, reprinted 1963, $1.25;

Section 3, 1962, 98 pp, 60 cents.
The first three sections include spectra of the elements in the corresponding volumes of Circular 467.

Section 4, 1962, 70 pp, 45 cents.
A Finding List for spectra of elements in Sections 1 and 2.

Section 5, 1962, 34 pp, 30 cents.
A Finding List for spectra of elements in Section 3.

Selected Tables of Atomic Spectra, Atomic Energy Levels and Multiplet Tables Si ii, Si iii, Si iv. NSRDS-NBS 3, Section 1, Charlotte E. Moore, U.S. Government Printing Office, Washington, D.C. 20402, June 1965, 40 pp, $0.35.

Atomic Transition Probabilities, A Critical Data Compilation; Vol. I, Hydrogen Through Neon, W. L. Wiese, M. W. Smith, and B. M. Glennon (NSRDS-NBS 4, Vol. I), U.S. Government Printing Office, Washington, D.C. 20402, May 1966, xi + 154 pp, $2.50.

INFRARED AND MICROWAVE SPECTRA

Selected Infrared Spectral Data, American Petroleum Institute Research Project 44, current loose-leaf sheets. As of June 30, 1966, there were 2813 valid sheets; sold as complete set 7 volumes at $0.30 per sheet, Texas A&M Research Foundation, College Station, Texas.

Selected Infrared Spectral Data, Thermodynamics Research Center, current loose-leaf sheets. As of June 30, 1966, there were 342 valid sheets. Sold as complete sets at $0.30 per sheet, Texas A&M Research Foundation, College Station, Texas.

Coblentz Society Spectra, set of 4,000 spectra, 1964, $115 per 1000. Indexes: Alphabetical Index, Molecular Formula Index, Chemical Classes Index, Numerical Index, $5.00 each, set of four, $10.00. Available from Sadtler Research Laboratories, Inc., 1517 Vine St., Philadelphia, Pa. 19102, to whom orders may be sent.

DMS (Documentation of Molecular Spectroscopy) Spectral Cards. The DMS card service is obtained by subscription from Butterworth & Co. (Publishers) Ltd., London, W.C.2, England, from Verlag Chemie, G.m.b.H., Weinheim an der Bergstrasse, West Germany. Volumes 1 through 6 contain 2000 cards, about 80 percent spectral and 20 percent literature. The spectral cards are supplied slotted or unslotted, the unslotted being marked according to the DMS code. The subscription includes a current literature list in a ring binder, the Junior Index, codes for the literature list, instruction and coding manual for the spectral cards, a formula list for each issue and the newsletters. Literature lists for nuclear magnetic resonance were begun in 1965. The prices for the DMS Spectral and Index cards are as follows:

Cards	Price		
	U.K. (£)	U.S. ($)	W. Germany (DM)
Spectral[a] (volume, each)			
1–4 (unslotted)	67. 0.0.	187.50	750.00
5–8 (unslotted)	75. 0.0.	210.00	840.00
1–4 (slotted)	80. 5.0.	225.00	900.00
5–8 (slotted)	88. 0.0.	247.00	988.00
9– (unslotted)	112.10.0.	315.00	1,260.00
9– (slotted)	130.10.0.	365.00	1,460.00
Index (sets)			
S-1 (1–4999) single	22. 5.0.	62.50	250.00
S-1 (1–4999) multiple	18. 0.0.	50.00	200.00
S-2 (5000–9999) single	26. 5.0.	73.50	295.00
S-2 (5000–9999) multiple	21. 0.0.	59.00	236.00

[a]Each volume consists of four issues, each issue containing about 400 spectral cards (pink) and 100 literature cards (yellow). From Issue 17 (beginning of Vol. 5), spectral cards for inorganic compounds (blue) are included and from Issue 25, the literature cards are replaced by literature lists.

The Infrared Data Committee of Japan Infrared Data Cards, published by Nankodo Co., Ltd., Harukicho, Bunkyo-ku, Tokyo, Japan. They are sold on a subscription basis at $238.00 for 1200 cards (postage included). Asian subscribers receive a special price of $150.00. Export agency is Sanyo Shuppan Bocki Co., Inc., P.O. Box 1705, Tokyo, Central Japan. The agent in the United States is the Preston Technical Abstracts Co., 909 Pitner Ave., Evanston, Ill. 60202. The agent in Europe is Heyden & Son Ltd., Spectrum House, Alderton Crescent, Hendon, London, N.W.4, England.

Standard Infrared Spectra (pure substances), Sadtler Research Laboratories, Inc., 3316 Spring Garden St., Philadelphia, Pa. 19104.

27 volumes of 1000 spectra each (complete through 1965), include alphabetical name, molecular formula, chemical classes, numerical, and commercial alphabetical indexes, $3,500.00 (or each volume $125.00, or indexes, set of five, $50.00).

1966 spectra by subscription, includes 2 volumes of 1000 spectra each, Nos. 27,001—29,000, five indexes and Spec-Finder, $350.00.

Standard spectra on 16-mm microfilm, prices same as above.

Sadtler Standard Grating Spectra, Sadtler Research Laboratories, Inc., 3316 Spring Garden St., Philadelphia, Pa. 19104.

Near Infrared Spectra
Completed set of 2000 spectra in 9 volumes, prices available upon request.

Far Infrared Spectra
Completed set of 500 spectra in 2 volumes, prices available upon request.

Commercial Spectra
Groups priced separately, prices available upon request.
Commercial Spec-Finder, $25.00.

The Red System ($A^2\pi - X^2\Sigma$) of the CN Molecule, Sumner P. Davis and John G. Phillips, University of California Press, Berkeley and Los Angeles, 1963, x + 214 pp, $9.50.

Spectral Data and Physical Constants of Alkaloids, J. Holubek and O. Štrouf, eds. Published by Heyden & Sons, Ltd., London, in cooperation with The Publishing House of the Czechoslovakia Academy of Sciences. Available in Western Europe and the Western Hemisphere from Heyden & Sons, Ltd., Spectrum House, Alderton Crescent, Hendon, London, N.W.4, England.

Vol. I, 1965, spectral cards 1–30 issued in two binders to take Vols. I and II, £23.0.0. ($69.00).

Vol. II, 1966, spectral cards 301–400, punched cards to be inserted in second binder, £7.0.0. ($21.00).

Microwave Spectral Tables, NBS Monograph 70, U.S. Government Printing Office, Washington, D.C. 20402.

Vol. I: Diatomic Molecules, Paul F. Wacker, Masataka Mizushima, Jean D. Peterson, and Joe R. Ballard, Dec. 1, 1964, xviii + 146 pp, 8″ × 10½″, $2.00.

Vol. II: Line Strengths of Asymmetric Rotors, Paul F. Wacker and Marlene R. Pratto, Dec. 15, 1964, xii + 340 pp, 8″ × 10½″, $3.00.

ELECTRONIC SPECTRA (ULTRAVIOLET AND VISIBLE)

Selected Ultraviolet Spectral Data, American Petroleum Institute Research Project 44, current loose-leaf sheets. As of June 30, 1966, there were 1076 valid sheets. Sold in complete sets, 3 volumes, at $0.30 per sheet. Texas A&M Research Foundation, College Station, Texas.

Selected Ultraviolet Spectral Data, Thermodynamics Research Center, current loose-leaf sheets. As of June 30, 1966, there were 126 valid sheets. Sold in complete sets at $0.30 per sheet. Texas A&M Research Foundation, College Station, Texas.

Organic Electronic Spectral Data, Interscience Publishers, a division of John Wiley & Sons, Inc., New York, London, and Sydney.

Vol. I, 1946–1952, M. J. Kamlet, ed, 1961, pp xiii + 1244, $25.00.

Vol. II, 1953–1955, H. E. Ungnade, ed, 1960, pp x + 919, $15.00.

Vol. III, 1956–1957, O. H. Wheeler and L. A. Kaplan, eds, 1966, xii + 1210 pp, $25.00.

Vol. IV, 1958–1959, J. P. Phillips and F. C. Nachod, eds, 1963, viii + 1179 pp, $20.00.

The "Sadtler Standard Ultraviolet Spectra" including alphabetical and numerical indexes are available on a subscription basis—$560.00 for the 1965 subscription (5000 spectra). The 11000 bound spectra previously issued sell for $1,265.00 including indexes. Individual spectra are priced at $0.15 each. The spectra are also available on 16-mm microfilm at the same price.

Orders and inquiries should be sent to Sadtler Research Laboratories, Inc., 3314–3320 Spring Garden St., Philadelphia, Pa. 19104.

Absorption Spectra in the Ultraviolet and Visible Region (a Theoretical and Technical Introduction), elaborated and edited by Dr. L. Láng; published in German and English in 1959, 2nd edition, 1961, 3rd edition, 1963, 80 pp, Co-production of the Academic Press, New York and London, and the Publishing House of the Hungarian Academy of Sciences, Budapest. Printed in English in Hungary. Available in both continents of America from Academic Press, Publishers, 111 Fifth Ave., New York, N.Y. 10003.

Absorption Spectra in the Ultraviolet and Visible Region, L. Láng, ed, Co-production of the Academic Press, New York and London, and the Publishing House of the Hungarian Academy of Sciences, Budapest Printed in English in Hungary. Available in both continents of America from Academic Press, Publishers, 111 Fifth Ave., New York, N.Y. 10003.

Vol. 1: Substances, 170, 1959, 414 pp + Index pamphlet, 24, 2nd edition, 1961, 3rd edition, 1963. Combined introduction and Vol. 1, $18.00.

Vol. 2: Substances, 179, 1961, 408 pp + Index pamphlet, 31, 2nd edition, 1964, $18.00.

Vol. 3: Substances, 172, 1962, 424 pp + Index pamphlet, 24, $20.00.

Vol. 4: Substances, 185, 1963, 414 pp + Index pamphlet, 24, $20.00.

Vol. 5: Substances, 192, 1965, 416 pp + Index pamphlet, 28. Cumulative Index, Vols. 1–5, 1965, 112 pp, $23.00.

Vol. 6: Substances, 197, 1966, 412 pp + Index pamphlet, 30, $23.00.

RAMAN SPECTRA

Selected Raman Spectral Data, American Petroleum Institute Research Project 44, current loose-leaf sheets. As of June 30, 1966, there were 501 valid data sheets. Sold in complete sets, 2 volumes, at $0.30 per sheet, Texas A&M Research Foundation, College Station, Texas.

Selected Raman Spectral Data, Thermodynamics Research Center, current loose-leaf sheets. As of June 30, 1966, there were 40 valid sheets. Sold as complete sets at $0.30 per sheet. Texas A&M Research Foundation, College Station, Texas.

MASS SPECTRA

Selected Mass Spectral Data, American Petroleum Institute Research Project 44, current loose-leaf sheets. As of June 30, 1966, there were 2342 valid sheets in the catalog. Sold as complete sets, 6 volumes, at $0.30 per sheet. Texas A&M Research Foundation, College Station, Texas.

Selected Mass Spectral Data, Thermodynamics Research Center, current loose-leaf sheets. As of June 30, 1966, there were 168 valid sheets. Sold as complete sets at $0.30 per sheet. Texas A&M Research Foundation, College Station, Texas.

Compilation of Mass Spectral Data (Index de Spectres de Masse), A. Cornu and R. Massot, 1966, xv + 617 pp, 8½″ × 11″. Published by Heyden & Sons Ltd., in cooperation with Presses Universitaires de France. Distributed by Heyden & Sons Ltd., Spectrum House, Alderton Crescent, London, N.W.4, England; $42.00.

NUCLEAR MAGNETIC RESONANCE SPECTRA

Selected Nuclear Magnetic Resonance Spectral Data, American Petroleum Institute Research Project 44, current loose-leaf sheets. As of June 30, 1966, there were 649 valid sheets. Sold as complete sets, 2 volumes, at \$0.30 per sheet. Texas A&M Research Foundation, College Station, Texas.

Selected Nuclear Magnetic Resonance Spectral Data, Thermodynamics Research Center, current loose-leaf sheets. As of June 30, 1966, there were 551 valid sheets. Sold as complete sets, 2 volumes, at \$0.30 per sheet. Texas A&M Research Foundation, College Station, Texas.

High Resolution NMR Spectra Catalog, Vol. I, 1962, N. S. Bhacca, L. F. Johnson, and J. N. Shoolery, compilers, Varian Associates, Instrument Division, Palo Alto, Calif. 94303, 43 pp + 368 spectra, \$6.00; 10 or more copies, \$4.50 each; student price, \$5.00.

High Resolution NMR Spectra Catalog, Vol. II, 1963, N. S. Bhacca, D. P. Hollis, L. F. Johnson, and E. A. Pier, compilers, Varian Associates, Instrument Division, Palo Alto, Calif. 94303, x + 62 pp + 332 spectra; price same as Vol. I; combined edition, Vols. I and II, hard cover, \$20.00.

Tables of Interatomic Distances and Configuration in Molecules and Ions, L. E. Sutton, scientific ed, Special Publication 11, The Chemical Society, Burlington House, London, W.1, England, 1958, 390 pp, £2.2.0. (\$6.00). Also available from Polycrystal Book Service, P.O. Box 11567, Pittsburgh, Pa. 15238.

Tables of Interatomic Distances and Configuration in Molecules and Ions, Supplement, 1956–1959, L. E. Sutton, scientific ed, Special Publication 18, The Chemical Society, Burlington House, London, W.1, England, 1965, 296 pp, £4.4.0. (\$12.00). Also available from Polycrystal Book Service, P.O. Box 11567, Pittsburgh, Pa. 15238.

LANDOLT-BÖRNSTEIN

Landolt-Börnstein: Zahlenwerte und Funktionen äus Physik, Chemie, Astronomie, Geophysik und Technik, sixth edition.

Volumes Published

Vol. I: Atom—und Molekularphysik
- Part 1, Atome und Ionen, 1950, xii + 441 pp., DM 126 (\$31.50).
- Part 2, Molekeln I: Kerngerüst, 1951, viii + 571 pp., DM 168 (\$42.00).
- Part 3, Molekeln II: Elektronenhülle, 1951, xi + 724 pp., DM 218 (\$54.50).
- Part 4, Kristalle, 1955, xi + 1,007 pp., DM 318 (\$79.50).
- Part 5, Atomkerne und Elementarteilchen, 1952, viii + 470 pp., DM 148 (\$37.00).

Vol. II: Eigenschaften der Materie in Ihren Aggregatzuständen
- Part 2, Gleichgewichte ausser Schmelzgleichgewichten.
  - 2a, Gleichgewichte Dampf-Kondensat und Osmotische Phänomene, 1960, xi + 974 pp., DM 448 (\$112.00).
  - 2b, Lösungsgleichgewichte I, 1962, x + 983 pp., DM 510 (\$127.50).
  - 2c, Lösungsgleichgewichte II, 1964, viii + 731 pp., DM 403.5 (\$100.88).
- Part 3, Schmelzgleichgewichte und Grenzflächenerscheinungen, 1956, xi + 535 pp., DM 198 (\$49.50).
- Part 4, Kalorische Zustandsgrössen, 1961, xii + 863 pp., DM 438 (\$109.50).
- Part 6, Elektrische Eigenschaften I, 1959, xv + 1,018 pp., DM 448 (\$112.00).
- Part 7, Elektrische Eigenschaften II, 1960, xii + 959 pp., DM 478 (\$119.50).
- Part 8, Optische Konstanten, 1962, xv + 901 pp., DM 476 (\$119.00).
- Part 9, Magnetische Eigenschaften I, 1962, xxv + 1,934 pp., DM 496 (\$124.00).

Vol. III: Astronomie und Geophysik, 1952, xviii + 795 pp., DM 248 (\$62.00).

Vol. IV: Technik
- Part 1, Stoffwerte und mechanisches Verhalten von Nichtmetallen, 1955, xvi + 881 pp., DM 288 (\$72.00).
- Part 2, Stoffwerte und Verhalten von metallischen Werkstoffen.
  - 2a, Grundlagen, Prüfverfahren, Eisenwerkstoffe, 1963, xii + 888 pp., DM 468 (\$117.00).
  - 2b, Sinterwerkstoffe, Schwermetalle (ohne Sonderwerkstoffe), 1964, xx + 1,000 pp., DM 530 (\$132.50).
  - 2c, Leichtmetalle, Sonderwerkstoffe, Halbleiter, Korrosion, 1965, xx + 976 pp., DM 518 (\$129.50).
- Part 3, Elektrotechnik, Lichttechnik, Röntgentechnik, 1957, xv + 1,076 pp., DM 396 (\$99.00).

Volumes in Preparation

Vol. II: Eigenschaften der Materie in Ihren Aggregatzuständen
- Part 1, Mechanisch-thermische Zustandgrössen.
- Part 5, Physikalische und chemische Kinetik.
- Part 10, Magnetische Eigenschaften II.

Vol. IV: Technik
- Part 4, Wärmetechnik.

SOURCES OF CRITICAL DATA (Continued)

LANDOLT-BÖRNSTEIN

Landolt-Börnstein: Zahlenwerte und Functionen aus Naturwissenschaften und Technik, Neue Serie—Numerical Data and Functional Relationships in Science and Technology, New Series, K. H. Hellwege, editor-in-chief.

Volumes Published

Group I: Kernphysik und Kerntechnik—Nuclear Physics and Technology

Vol. I, Energy Levels of Nuclei: A = 5 to A = 257, A. M. Hellwege and K. H. Hellwege, eds., 1961, xii + 813 pp., DM 212 ($53.00).

Group II: Atom und Molekularphysik—Atomic and Molecular Physics

Vol. I, Magnetische Eigenschaften Freier Radikale—Magnetic Properties of Free Radicals, K. H. Hellwege and A. M. Hellwege, eds., 1965, x + 154 pp., DM 68 ($17.00).

Group VI: Astronomie, Astrophysik und Weltraum-Forschung—Astronomy, Astrophysics and Space Research

Vol. I, Astronomie und Astrophysik—Astronomy and Astrophysics, H. H. Voigt, ed., 1965, xxxix + 711 pp., DM 314 ($78.50).

Other groups planned and volumes expected to be published in 1966 or 1967:

Group I: Nuclear Physics and Technology

Hilfstabellen für α-, β-, γ- Spektroskopie—Auxiliary Tables for α-, β-, γ- Spectroscopy.
Ladungs-und Dichteverteilung im Kern—Charge and Density Distribution in the Nucleus.

Group II: Atomic and Molecular Physics

Vol. I, Lumineszenz organischer Molekeln—Luminescence of Organic Molecules.
Mikrowellenspektroskopie von Molekeln—Microwave Spectroscopy of Molecules.
Molekularakustik—Molecular Acoustics.

Vol. II, Magnetische Eigenschaften der Koordinations und metallorganischen Verbindungen der Übergangselemente—Magnetic Properties of Co-ordination and Organometallic Compounds of Transition Elements.

Group III: Crystal and Solid State Physics

Elastische, elastooptische, piezoelektrische und andere Eigenschaften von Einkristallen—Elastic, Elasto-optic, Piezoelectric and Other Properties of Single Crystals.

Group IV: Macroscopic and Technical Properties of Matter

Phosphoreszenz anorganischer Substanzen—Phosphorescence (Luminescence) of Inorganic Substances.

Group V: Geophysics and Space Research

Aeronomy; Cosmic Radiation.

Published by Springer-Verlag, Berlin, Heidelberg, and New York. Also available from Walter J. Johnson, Inc., 111 Fifth Ave., New York, N.Y. 10003, and Stechert-Hafner, Inc., 31 E. 10th St., New York, N.Y. 10003.

TABLES DE CONSTANTES SELECTIONNEES—TABLES OF SELECTED CONSTANTS

Tables de Constantes Sélectionnées—Tables of Selected Constants

Vol. 1. Longueurs d'Onde d'Emissions X et des Ciscontinuités d'Absorption X-Wavelengths of Emission and Discontinuities in Absorption of X-Rays, Y. Cauchois and H. Hulubei, 1947, 199 pp. (out of print).

Vol. 2. Physique Nucléaire—Nuclear Physics, R. Grégoire, F. Joliot-Curie, and I. Joliot-Curie, 1948, 131 pp. (out of print).

Vol. 3. Pouvoir Rotatoire Magnétique (Effet Faraday)—Magnetic Rotatory Power (Faraday Effect); R. de Mallemann, Effet Magnéto-Optique de Kerr—Magneto-Optic Effect (Kerr), F. Suhner, 1951, 137 pp., 15 F, £1.10.0. ($4.50).

Vol. 4. Données Spectroscopiques Concernant les Molécules Diatomiques—Spectroscopic Data for Diatomic Molecules, B. Rosen, R. F. Barrow, A. D. Caunt, A. R. Downie, R. Herman, E. Huldt, A. MacKellar, E. Miescher, and K. Wieland, 1951, 361 pp., 48 F, £5.0.0. ($15.00).

Vol. 5. Atlas des Longueurs d'Onde Caractéristiques des Bandes d'Emission et d'Absorption des Molécules Diatomiques—Atlas of Characteristic Wavelengths for Emission and Absorption Bands of Diatomic Molecules (a continuation of Vol. 4 by the same authors), 1952, 389 pp., 56 F, £5.15.0 ($17.50).

Vol. 6. Pouvoir Rotatoire Naturel I-Steroïdes—Optical Rotatory Powers I-Steroids, J.-P. Mathieu and A. Petit, 1956, 507 pp. (out of print).

Vol. 7. Diamagnétisme et Paramagnétisme—Diamagnetism and Paramagnetism, G. Foëx; Relaxation Paramagnétique—Paramagnetic Relaxation, C. J. Gorter and L. J. Smits, 1957, 317 pp., 97 F, £9.15.9. ($29.00).

Vol. 8. Potentiels d'Oxydo-Réduction—Oxidation-Reduction Potentials, G. Charlot, D. Bézier, and J. Courtot, 1958, 41 pp., 21.60 F, £1.10.0 ($5.00).

Vol. 9. Pouvoir Rotatoire Naturel II-Triterpénoïdes—Optical Rotatory Power II-Triterpenoids, J.-P. Mathieu and G. Ourisson, 1958, 302 pp., 93.60F, £7.0.0. ($21.00)

Vol. 10. Pouvoir Rotatoire Naturel III-Amino-acides—Optical Rotatory Power III-Amino Acids, J.-P. Mathieu, J. Roche, and P. Desnuelle, 1959, 61 pp., 28 F, £2.0.0. ($6.50).

TABLES DE CONSTANTES SELECTIONNEES— TABLES OF SELECTED CONSTANTS (Continued)

Vol. 11. Pouvoir Rotatoire Naturel IV-Alcaloïdes—Optical Rotatory Power IV-Alkaloids, J.-P. Mathieu and M. M. Janot, 1959, 211 pp., 110.40 F, £8.0.0. ($24.00).

Vol. 12. Semi-Conducteurs—Semi-Conductors, P. Aigrain and J. Balkanski, 1961, 78 pp., 27 F, £2.0.0. ($6.50).

Vol. 13. Rendements Radiolytiques—Radiolytic Yields, M. Haïssinsky and M. Magat, 1963, 230 pp., 114 F, £8.10.0. ($25.50).

Vol. 14. Pouvoir Rotatoire Naturel Ia-Stéroïdes—Optical Rotatory Power Ia-Steroids, J. Jacques, H. Kagan, G. Ourisson, and S. Allard, 1965, 1,046 pp., 258 F.

Vol. 15. Données Relatives aux Sesquiterpenoïdes—Data Relative to Sesquiter-penoids, G. Ourisson, S. Munavalli, and C. Éhret, 1966, 70 pp.

In Preparation

Metals of High Purity, 1967

Refractory Compounds

There have been several publishers for the present series. Volume 8 and later volumes have been published by Pergamon Press from whom the earlier volumes as well as the titles, availability, and prices of the 40 installments of Tables Annuelles, covering the period 1931–1936, may be obtained.

Published by Pergamon Press, 4 and 5 Fitzroy Square, London, W.1, England; 122 E. 55th St., New York, N.Y., U.S.; and 24 rue des Ecoles, Paris V[e], France.

INTERNATIONAL CRITICAL TABLES

International Critical Tables of Numerical Data, Physics, Chemistry and Technology, published for the National Academy of Sciences—National Research Council, by McGraw-Hill Book Co., 330 W. 42nd St., New York, N. Y. 10036.

Vol. I, 1926, xx + 415 pp.
Vol. II, 1927, xviii + 616 pp.
Vol. III, 1928, xiv + 444 pp.
Vol. IV, 1928, viii + 481 pp.
Vol. V, 1929, ix + 465 pp.
Vol. VI, 1929, x + 471 pp.
Vol. VII, 1930, ix + 507 pp.
Index, 1933, vii + 321 pp.

Seven volumes and Index, $250.00; Vols. I-VII and Index, each $35.00.

Index

A

INDEX

INDEX

C

INDEX

INDEX

INDEX

INDEX

INDEX

INDEX

N

INDEX

INDEX

Q

R

INDEX

S

INDEX

INDEX

T

INDEX

ATOMIC WEIGHTS

For the sake of completeness all known elements are included in the list. Several of those more recently discovered are represented only by the unstable isotopes. The value in parenthesis in the atomic weight column is, in each case, the mass number of the most stable isotope.**

Name	Symbol	At. No.	International atomic weight 1966	International atomic weight 1959	Valence
Actinium	Ac	89	....	(227)	....
Aluminum	Al	13	26.9815	26.98	3
Americium	Am	95	....	(243)	3, 4, 5, 6
Antimony, stibium	Sb	51	121.75	121.76	3, 5
Argon	Ar	18	39.948	39.944	0
Arsenic	As	33	74.9216	74.92	3, 5
Astatine	At	85	....	(210)	1, 3, 5, 7
Barium	Ba	56	137.34	137.36	2
Berkelium	Bk	97	....	(247)	3, 4
Beryllium	Be	4	9.0122	9.013	2
Bismuth	Bi	83	208.980	208.99	3, 5
Boron	B	5	10.811	10.82	3
Bromine	Br	35	79.904[1]	79.916	1, 3, 5, 7
Cadmium	Cd	48	112.40	112.41	2
Calcium	Ca	20	40.08	40.08	2
Californium	Cf	98	....	(251)	....
Carbon	C	6	12.01115	12.011	2, 4
Cerium	Ce	58	140.12	140.13	3, 4
Cesium	Cs	55	132.905	132.91	1
Chlorine	Cl	17	35.453	35.457	1, 3, 5, 7
Chromium	Cr	24	51.996	52.01	2, 3, 6
Cobalt	Co	27	58.9332	58.94	2, 3
Columbium, see *Niobium*					
Copper	Cu	29	63.546[2]	63.54	1, 2
Curium	Cm	96	....	(247)	3
Dysprosium	Dy	66	162.50	162.51	3
Einsteinium	Es	99	....	(254)	....
Erbium	Er	68	167.26	167.27	3
Europium	Eu	63	151.96	152.0	2, 3
Fermium	Fm	100	....	(257)	....
Fluorine	F	9	18.9984	19.00	1
Francium	Fr	87	....	(223)	1
Gadolinium	Gd	64	157.25	157.26	3
Gallium	Ga	31	69.72	69.72	2, 3
Germanium	Ge	32	72.59	72.60	4
Gold, aurum	Au	79	196.967	197.0	1, 3
Hafnium	Hf	72	178.49	178.50	4
Helium	He	2	4.0026	4.003	0
Holmium	Ho	67	164.930	164.94	3
Hydrogen	H	1	1.00797	1.0080	1
Indium	In	49	114.82	114.82	3
Iodine	I	53	126.9044	126.91	1, 3, 5, 7
Iridium	Ir	77	192.2	192.2	3, 4
Iron, ferrum	Fe	26	55.847	55.85	2, 3
Krypton	Kr	36	83.80	83.80	0
Lanthanum	La	57	138.91	138.92	3
Lawrencium	Lr	103	(257)		
Lead, plumbum	Pb	82	207.19	207.21	2, 4
Lithium	Li	3	6.939	6.940	1
Lutetium	Lu	71	174.97	174.99	3
Magnesium	Mg	12	24.312	24.32	2
Manganese	Mn	25	54.9380	54.94	2, 3, 4, 6, 7
Mendelevium	Md	101	....	(256)	....
Mercury, hydrargyrum	Hg	80	200.59	200.61	1, 2
Molybdenum	Mo	42	95.94	95.95	3. 4. 6
Neodymium	Nd	60	144.24	144.27	3
Neon	Ne	10	20.183	20.183	0
Neptunium	Np	93	....	(237)	4, 5, 6
Nickel	Ni	28	58.71	58.71	2, 3
Niobium (columbium)	Nb	41	92.906	92.91	3, 5
Nitrogen	N	7	14.0067	14.008	3, 5
Nobelium	No	102	....	(254)	....
Osmium	Os	76	190.2	190.2	2, 3, 4, 8
Oxygen	O	8	15.9994	16.000	2
Palladium	Pd	46	106.4	106.4	2, 4, 6
Phosphorus	P	15	30.9738	30.975	3, 5
Platinum	Pt	78	195.09	195.09	2, 4
Plutonium	Pu	94	....	(244)	3, 4, 5, 6
Polonium	Po	84	....	(209)	....
Potassium, kalium	K	19	39.102	39.100	1
Praseodymium	Pr	59	140.907	140.92	3
Promethium	Pm	61	....	(145)	3
Protactinium	Pa	91	....	(231)	....
Radium	Ra	88	....	(226)	2
Radon	Rn	86	....	(222)	0
Rhenium	Re	75	186.2	186.22	....
Rhodium	Rh	45	102.905	102.91	3
Rubidium	Rb	37	85.47	85.48	1
Ruthenium	Ru	44	101.07	101.1	3, 4, 6, 8
Samarium	Sm	62	150.35	150.35	2, 3
Scandium	Sc	21	44.956	44.96	3
Selenium	Se	34	78.96	78.96	2, 4, 6
Silicon	Si	14	28.086	28.09	4
Silver, argentum	Ag	47	107.868[3]	107.873	1
Sodium, natrium	Na	11	22.9898	22.991	1
Strontium	Sr	38	87.62	87.63	2
Sulfur	S	16	32.064	32.066*	2, 4, 6
Tantalum	Ta	73	180.948	180.95	5
Technetium	Tc	43	....	(97)	6, 7
Tellurium	Te	52	127.60	127.61	2, 4, 6
Terbium	Tb	65	158.924	158.93	3
Thallium	Tl	81	204.37	204.39	1, 3
Thorium	Th	90	232.038	(232)	4
Thulium	Tm	69	168.934	168.94	3
Tin, stannum	Sn	50	118.69	118.70	2, 4
Titanium	Ti	22	47.90	47.90	3, 4
Tungsten (wolfram)	W	74	183.85	183.86	6
Uranium	U	92	238.03	238.07	4, 6
Vanadium	V	23	50.942	50.95	3, 5
Xenon	Xe	54	131.30	131.30	0
Ytterbium	Yb	70	173.04	173.04	2, 3
Yttrium	Y	39	88.905	88.91	3
Zinc	Zn	30	65.37	65.38	2
Zirconium	Zr	40	91.22	91.22	4

1. ±0.002
2. ±0.001
3. ±0.001

* Because of natural variations in the relative abundances of the isotopes of sulfur the atomic weight of this element has a range of ±0.003.

** The 1959 atomic weights are based on O = 16.000 whereas those of 1966 are based on the isotope C^{12}.